GLOSSARY OF GEOLOGY—Fifth Edition

This *Glossary* is dedicated to Robert L. Bates who set its standard for excellence through his work as science editor of the second and third editions.

Fifth Edition

GLOSSARY OF GEOLOGY

Editors:
Klaus K.E. Neuendorf
James P. Mehl, Jr.
Julia A. Jackson

American Geological Institute
Alexandria, Virginia
2005

Artwork at alphabet section openings courtesy of the U.S. Geological Survey: **A.** Asbestos, Gila County, Arizona; highly magnified, scanning electron microscope view of one of the three common types of commercially mined asbestos (U.S. Geological Survey, 1980). **B.** Boudinage, Rio Extorax, Queretaro, Mexico (K. Segerstrom, 1956). **C.** Columnar basalt, San Miguel Regla, Hidalgo, Mexico ©. Fries). **D.** Dike, Mount Clark, Lincoln County, Maine; pegmatite vein in diorite (T. N. Dale). **E.** Earthquake, Anchorage, Alaska, Cook Inlet region; collapse of 4th Avenue district near C Street, due to a landslide caused by the earthquake (1964). **F.** Fault, Minas Gerais, Brazil; steep reverse fault in dolomitic marble ®. M. Wallace, 1957). **G.** Gneiss, Gunnison area, Colorado; contorted gneiss, north rim of the Black Canyon near Colorado State Highway 92 (W. R. Hansen). **H.** Hogback, Niobrara Limestone, El Paso County, Colorado—west of Colorado Springs (N. H. Darton, 1926). **I.** Iceberg, Sohlstenholme Fjord, northeast of Thule Air Base, Nunatarssnaq region, Greenland ®. B. Colton, 1953). **J.** Joints, Las Animas canyon, San Juan County, Colorado; joints in granite cut by veins of quartz, feldspar, and biotite ©. W. Cross, 1900). **K.** Karst topography, along Li River, China (W. H. Monroe, 1983). **L.** Leaves, Metasequoia occidentalis (Newberry), Chaney, Stevens County, Washington; fossils from tuffaceous sandstones of the gerome andesite, Northwest Uranium Mine (G. E. Becraft, 1957). **M.** Moraines, Eastern Chugach Mountains, Chitina district, Copper River region, Alaska; view of Ross Green Lake and moraines (E. E. Brabb, 1959). **N.** Nunatak encircled by moraine, Alaska Gulf region, Alaska (D. J. Miller, 1958). **O.** Oxbow lake, Yukon region, Alaska; oblique aerial view of an oxbow lake and meanders of Hodzana River (J. R. Williams, 1948). **P.** Pahoehoe, Hawaii Volcanoes National Park, Hawaii; near view of ropy crust of pahoehoe lava on crater floor, Kilauea Volcano (H.R. Joesting, 1950). **Q.** Quartzite, Riverside Mountains, Riverside County, California; isoclinally folded quartzite (W.B. Hamilton, 1967). **R.** Rockfall, El Rancho rockfall on Interstate 70, Jefferson, County, Colorado (W.R. Hansen, 1973). **S.** Stalactites, Carlsbad Caverns National Park, Eddy County, New Mexico; pendulum stalactites in the "Rookery" (W.T. Lee, 1923). **T.** Till, Yosemite National Park, Mariposa County, California; glacial till, an unsorted mixture of boulders, sand, and clay exposed along Tioga Road at Siesta Lake (N.K. Huber). **U.** Unconformity, San Lorenzo Arroyo, Arizona (R.H. Chapman). **V.** V-shaped valley, Grand Canyon of the Yellowstone, Yellowstone National Park, Wyoming (W.T.Lee, 1921). **W.** Wind erosion, Illano de Caldera, Atacama Province, Chile; granite outcrop that has been etched and pitted by the abrasive action of windblown sand (K. Segerstrom). **X.** Xenolith, Moerakai, New Zealand; schist xenoliths (white) in layered gabbro (W.B. Hamilton, 1965). **Y.** Yosemite Valley from vicinity of Artist Point, Yosemite National Park, California (F.E. Matthes, 1923). **Z.** Zircon, Eureka County, Nevada; scanning electron microscope micrograph shows detrital zircon crystal in a matrix of porous authigenic clay minerals from Carlin Gold Mine (A.K. Armstrong, 1984).

Composed in Palatino Linotype using Corel Ventura 10
Printed offset on Woodlawn, by United Book Press
Bound in Arrestox

Copyright 2005 by American Geological Institute, 4220 King St., Alexandria, Va. 22302. Printed and bound in the United States of America. All rights reserved.

Cataloging-in-Publication Data applied for

A catalog record for this book is available from the Library of Congress.

Preface—Fifth Edition

The revising and updating of earth science terminology and of the concepts behind and communicated through it—these are continuous processes for which this fifth edition of the AGI *Glossary of Geology* provides just another frozen moment of panoramic overview. Naturally, the number of terms, definitions, and acronyms has increased again over the previous edition—from the previous more than 37,200 to this edition's nearly 39,300 entries. The main purpose of the *Glossary* is to assist readers and writers of contemporary earth science; yet antiquated and obsolete terms have been removed only sparingly.

Through the generous assistance of the Mineralogical Society of America and the late Dr. Richard A. Bideaux, mineral formulas from the *Handbook of Mineralogy* (Anthony et al., 1990-2003) are included in this edition of the Glossary. The formulas are expressed in a way that conveys information about a mineral's crystal chemistry and structure as well as its composition. The number of mineral names has been increased by about 1,000, to a new total of more than 5,300. As in the previous edition, for more information about mineral species, *Glossary* users are referred to the *Mineral Reference Manual* by Nickel and Nichols (1991).

The authority of this *Glossary* rests on the expertise of more than a hundred eminent geoscientists who participated as volunteers in the arduous task of a two-level review and whose names are listed in the "Acknowledgments" on pages xi and xii. Their contributions have been incorporated, as far as possible, in each reviewer's individual manner, so that some variance of form will be seen between different entries.

In view of the constant changes in terminology, we invite all our users to suggest additions and corrections for future versions of the *Glossary*. Such suggestions may be sent to the American Geological Institute or to <glossary@agiweb.org>.

Even this fifth edition of the *Glossary* owes much to Julia A. Jackson who, as collaborator with Robert L. Bates on the second and third editions and as sole editor of the fourth edition, has guided the progress of the *Glossary* most capably and with deep dedication through a quarter of a century. She has provided much of the impetus and the groundwork for the present edition. I feel honored by her participation and grateful for her contributions.

Jim Mehl deserves particular credit for coordinating the many divergent portions of the publication project at the AGI headquarters and for overseeing the updating of the database for the printed *Glossary*.

August 2005 — Klaus K.E. Neuendorf

Preface—Fourth Edition

The practice, technology, and terminology of the earth sciences have changed a lot in the 10 years, since the American Geological Institute published the third edition of the *Glossary of Geology*. Geoscientists use geographic information systems (GIS) and global positioning systems (GPS) as well as new tools and techniques for analysis, modeling, exploration, and communication. This fourth edition reflects both advances in scientific thought and changes in usage. Approximately 3,400 of the 37,000-plus entries are new and nearly 9,000 definitions have been updated. Although the *Glossary* has retained many obsolete terms for their historical value, more than 3,000 terms that have rarely been used have been dropped.

The *Glossary* has expanded coverage particularly in such active fields as carbonate sedimentology, environmental geology and geophysics, GIS, GPS, hydrology and hydraulics, marine and coastal geology, organic geochemistry, paleoecology, seismology, sequence stratigraphy, speleology and karst, and structural geology and tectonics. More than 10 per cent of the terms in the *Glossary* are from paleontology, and more than 100 conodont morphology terms have been added. Stratigraphic terms and definitions have been updated in accordance with the *International Stratigraphic Guide* (1994). Igneous rock terms have been updated following the classification of igneous rocks established by the International Union of Geological Sciences Subcommission on the Systematics of Igneous Rocks (Le Maitre et al., 1973; Woolley et al., 1996).

The 4,000-plus mineral names in the *Glossary* constitute its largest single group of terms, Since 1987, so much has been learned through the analysis of crystal structure that many of the mineral definitions required updating. Revised formulas are expressed in a form to emphasize crystal chemistry and structure. The fundamental building block of a mineral— the region of most highly charged (highest bond valence) units— is given within brackets. This fundamental building block along with the hydrogen bonding tells much about a mineral's properties. In expressing the structural formula of a mineral, hydroxyls and water ligands follow their bonded cations parenthetically; hydrate water follows at the end after a centered dot. Because hydrogen bonding occurs in at least half of the mineral species, the role of the hydrogen atom is extremely important in interpreting crystal structure and mineral paragenesis. The chemical formula for laueite, a mineral named in honor of Max von Laue, one of the co-discoverers of X-ray diffraction by crystals, illustrates the many roles of hydrogen played out in the crystal structure:

Laueite: $Mn^{2+}(H_2O)_4[Fe^{3+}(H_2O)(OH)(PO_4)]2.2H_2O$

Expressing formulas for minerals which are organic molecules or salts of organic acids or bases can be problematic. About 20 natural oxalates are known, and the oxalic acid radical is commonly presented as (C_2O_4). The oxalic acid radical appears int eh *Glossary* as $(OOC\text{-}COO)^{2-}$ to reflect the bonding and to avoid confusion. Glossary users seeking more information about mineral species are referred to the *Mineral Reference Manual* (1991) by Nickel and Nichols.

To produce this edition, the American Geological Institute invited eminent geoscientists to review and update sets of terms and definitions from the third edition by discipline. They also proposed terms and definitions to be added. As with earlier editions the accuracy and authority of the *Glossary of Geology* result from the scholarship and expertise of the reviewers. Nearly 120 experts worked on this revision, and their names appear on pages xi-xii. It speaks well for the earth sciences that so many outstanding geoscientists willingly make time as volunteers to work on each new edition. They, like former *Glossary* editor Robert L. Bates, science editors of the second and third editions. His meticulous editorial work, insistence on clarity, and breadth of knowledge set the *Glossary's* standard for quality. Bob and I began planning for this edition in 1994— just a few months before his death. He made a significant contribution by classifying terms *Glossary* users had submitted since 1987 and by reviewing lists of potential new terms generated by searching AGI's GeoRef database. Bob's philosophy and work permeate the *Glossary of Geology*, and I think he would approve of the fourth edition. It meets his standard for quality thanks to the unstinting help from reviewers, constructive criticism from users, and active support of the revision by AGI management and staff.

April 1997 Julia A. Jackson

Preface—Third Edition

As earth scientists continue to generate new facts, concepts, and interpretations, the geological language necessarily expands and evolves. This third edition of the *Glossary of Geology* incorporates more than 1,000 new terms and definitions, and expands or brings up to date some 650 definitions from the previous edition. In addition, it includes for the first time the division of terms into syllables, with accents to aid in pronunciation. Approximately 150 references have been added to the 2,000 in the second edition.

New entries are especially numerous in the fields of carbonate sedimentology, hydrogeology, marine geology, mineralogy, ore deposits, plate tectonics, and snow and ice. Terms and definitions recommended in the North American Stratigraphic Code (1983) are included.

The authority of this edition, like that of its predecessor, rests on the expertise of geoscientists from many specialties. their contributions make the *Glossary* an essential reference work for all in the geoscience community.

April 1987 Julia A. Jackson
Robert L. Bates

Preface—Second Edition

"If it were not for the occasional appearance of an authoritative glossary," wrote Ian Campbell in his preface to the first edition of this work, "our language. . . would rapidly degenerate into babel." He added that as the science advances and knowledge expands, we modify and improve our concepts, and with this must come modification and enlargement of our vocabulary.

The American Geological Institute had recognized the need for a glossary, and in 1957, in cooperation with the National Academy of Sciences, it published a 14,000-term supplement, appeared in 1960. These gave rise, in 1972, to AGI's one-volume *Glossary of Geology*, containing nearly 33,000 terms.

The present volume, the 36,000-term second edition, incorporates the modifications and growth of the geoscience vocabulary since 1972. Coverage has been expanded and updated, particularly in such active fields as biostratigraphy, caves and karst, igneous petrology, paleomagnetism, remote sensing, plate tectonics, and seismic stratigraphy.

Some 450 new mineral names join the 4,000 in the previous edition; more than 100 abbreviations commonly used by geoscientists make their first appearance; and the list of references includes about 400 additional entries.

We acknowledge the pioneering work of J.V. Howell, who set the course and standards of excellence for the *Glossary of Geology and Related Sciences*, and the editorial advice and counsel of Frank C. Calkins, who helped materially in the preparation of the first edition of the *Glossary of Geology*.

We hope this *Glossary* will meet the needs of the geoscience community, and will prove to be a bulwark against the babelisation of the geological language.

December 1979

Julia A. Jackson
Robert L. Bates

Introduction

All terms listed in this *Glossary* have appeared in English-language publications or other writings in English and generally reflect North-American usage, unless otherwise noted. Many obsolete terms have been retained, as they remain valuable to readers using the older literature. Besides giving the current or preferred meaning of a term, some definitions include information on original usage or historical development.

Alphabetization generally follows a letter-by-letter system, ignoring spaces or nonalphabet characters. Thus, e.g., *A horizon* is listed between *ahlfeldite* and *Ahren's prism*. Numbers are listed before letters.

Syllabication is usually given in parentheses following the first appearance of a term. As an aid to pronunciation, a stress mark (′) is inserted following the primary stressed syllable; in some cases, primary and secondary stresses are differentiated by heavy (′) and light (′) stress marks. In exceptional cases, pronunciation is suggested by "Pron: . . ."—usually in brackets within the syllabication parentheses.

Italicized words refer the reader to related terms or definitions that can be found elsewhere in the *Glossary*, except in cases where italics are used for Latin species names. This practice extends to some variants, such as plurals of listed terms.

Citations to the literature are given in many definitions, mainly as examples of usage. The works referred to are listed in the section "References Cited" at the end of the *Glossary*.

Multiple definitions for a term are separated by (a), (b), (c), and so on.

Bracketed identifiers are added to terms that have meanings in more than subject area. They are treated as separate entries. Such identifiers are either spelled out or abbreviated according to the following list of abbreviations:

arch—archaeology
astron—astronomy
biol—biology
bot—botany
cart—cartography
chem—chemistry
clim—climatology
cryst—crystallography
drill—drilling
eco geol—economic geology
env geol—environmental geology
evol—evolution
exp petrol—experimental petrology
exp struc geol—experimental structural geology
geochem—geochemistry
geochron—geochronology
geog—geography
geol—geology
geomorph—geomorphology
geophys—geophysics
glac geol—glacial geology
glaciol—glaciology
grd wat—ground water
hist—history and philosophy of geology
hydraul—hydraulics
hydrogr—hydrography
hydrol—hydrology
ign—igneous
intrus rocks—intrusive rocks
magnet—magnetism
marine geol—marine geology
mass move—mass movements
math—mathematics
meta—metamorphism
metal—metallurgy
meteorol—meteorology
mineral—mineralogy
oceanog—oceanography
ore dep—ore deposits
paleont—paleontology
palyn—playnology
part size—particle size
pat grd—patterned ground
philos—philosophy
photo—photography
phylogen—phylogenetics
phys—physics
planet—planetology
pyroclast—pyroclastics
rock mech—rock mechanics
sed—sedimentology
sed struc—sedimentary structures
seis—seismology
speleo—speleology
stat—statistics
stratig—stratigraphy
struc geol—structural geology
struc petrol—structural petrology
surv—surveying
taxon—taxonomy
tect—tectonics
topog—topography
volc—volcanism
weath—weathering

Abbreviations of certain common terms are used in the definitions (generally without regard to upper- or lower-case spelling) as follows:

abbrev:—abbreviation/abbreviated
adj:—adjective
ant:—antonym
cf:—compare/see also (Latin: "confer")
CGS (-System)—Centimeter-Gram-Second System
cgs (unit)—centimeter-gram-second unit(s)
e.g.—for example (Latin: "exempli gratia")
esp.—especially
et al.—and others (Latin: "et alii")
etc.—and so forth (Latin: "et cetera")
etymol: etymology
i.e.—that is (Latin: "id est")
n.—noun
pl:—plural
pron:—pronounced
q.v.—which see (Latin: "quod vide")
SI—International System of Units (French: "Système International d'Unités")
sing:—singular
specif.—specific(ally)
syn:—synonym(s)
v.—verb
var.—variant

Acknowledgments

Specialists in many fields of geoscience have volunteered their help in bringing the widely used fourth edition of the *Glossary* up to date, by reviewing definitions, adding new terms, recommending corrections, and citing references. The geoscience community owes special gratitude to the following persons for taking the responsibility of acting as group leaders and reviewing and editing fields with many subcategories and reviewers or those fields with massive amounts of data: Gail M. Ashley, Wolfgang H. Berger, Arthur L. Bloom, Robert W. Kay, Patricia H. Kelley, Ernest H. Nickel, Amos Salvador, Roy W. Schlische, Robert E. Sheriff, Al Traverse.

The fields covered in the *Glossary* that have been reviewed for the fifth edition, with their reviewers, are given below.

Archaeology George R. Rapp, Jr.

Astrogeology Richard A. F. Grieve, Ralph P. Harvey, James R. Zimbelman

Cartography/surveying/map projections John B. Conroy, Michael P. Finn, Jane A. Messenger, John P. Murphey, Leland A. Perry, Robert E. Rinehart

Climatology Reid Bryson, Edward J. Hopkins, Joseph M. Moran

Coal geology James C. Hower, Maria Mastalerz, John Popp

Economic geology Mark D. Barton, Keith R. Long

Energy sources Edward A. Beaumont, J. B. Thomas

Engineering geology Jeffrey R. Keaton, Roy J. Shlemon, Jim Slosson,

Environmental geology Alan Kehew, Edward A. Keller

Extraterrestrial Geology Al Harris, Kenneth L. Tanaka, John T. Wasson, John A. Wood

Forensic geology Sarah Andrews

General geology William R. Brice, James H. Shea

Geochronology Fred W. McDowell, Amos Salvador

Geomorphology/surficial geology Arthur L. Bloom, , Don J. Easterbrook, William L. Graf, Joseph Kelley, Daniel A. Livingstone, Henry J. Melosh, John Mylroie, Arthur N. Palmer, Jonathan D. Phillips, William Renwick, Ellen E. Wohl,

Geophysics, applied Serguei A. Goussev, Larry D. Lines, L. J. Patrick Muffler, Daniel R. Roman, Robert E. Sheriff, George A. Thompson, Ralph R. B. von Frese

Geophysics, general Alex Becker, Serguei A. Goussev, Robert E. Sheriff

Geophysics, solid earth Gunter Faure, John W. Geissman, Willian J. Hinze, Michael E. Wysession,

Geophysics, seismology Larry D. Lines, Robert E. Sheriff, Donald W. Steeples

History and philosophy Robert H. Dott, Jr., Sally Newcomb

Igneous petrology Susan M. DeBari, Marc J. Defant, Grant Heiken, Robert W. Kay, Wesley E. LeMasurier, Malcolm J. Rutherford, Gene Yogodzinski

Information systems/informatics G. Randy Keller

Marine geology/oceanography Daniel C. Abel, Wolfgang H. Berger, Alison Duxbury, Alyn Duxbury, Robert N. Ginsburg, Philip A. Meyers, Kenneth G. Miller, Robert H. Stewart

Metamorphic petrology Theodore Labotka

Mineralogy Carl A. Francis, Dana T. Griffin, Ernest H. Nickel, Caroline Nelms

Paleobotany/palynology David Dilcher, Lucy Edwards, Jan Jansonius, Edith Taylor, Al Traverse, Gordon Wood

Paleontology, general Loren Babcock, Rodney Feldmann, Roger L. Kaesler, Bruce S. Lieberman, Donald R. Prothero

Paleontology, morphology William I. Ausich, Joseph G. Carter, Pamela Hallock-Muller, Patricia H. Kelley, Jason A. Lillegraven

Quaternary geology Arthur L. Bloom, Roger Hooke, Joan M. Ramage

Remote sensing Jeffrey Moersch

Sedimentary petrology Gail M. Ashley, Henry S. Chafetz, Richard J. Cheel, Michael S. Fenster, Richard L. Hay, Charlotte Schreiber, Frederick L. Schwab, John Southard

Stratigraphy Amos Salvador, Mary L. Droser, Steven M. Holland, James G. Ogg

Structural geology Nancye Dawers, Terry Engelder, Eric A. Erslev, Steven Marshak, Gautam Mitra, Roy W. Schlische, Carol Simpson, Jan A. Tullis, Martha Withjack

To produce the fifth edition Inez Morgart and Gayatri Tetali made data corrections and lightened the load; Lawrence Berg refined the Glossary data base and wrote verification and print programs; Kay Yost and Karin Mills efficiently typeset and formatted the book; John Rasanen, Sharon Tahirkheli, and Marcus Milling provided resources within AGI that kept the revision on track.

GLOSSARY OF GEOLOGY—Fifth Edition

A *mass number.*

aa (a'-a [ah'-ah]) A Hawaiian term for lava flows typified by a rough, jagged, spinose, clinkery surface. Cf: *pahoehoe; block lava.* Etymol: Hawaiian. An expletive of pain when walking barefoot on such lava. Obs. syn: *aphrolith.*

Aalenian (Aa-le'-ni-an) A stage of the Standard Global Chronostratigraphic Scale: lowermost Middle Jurassic (above Toarcian, below Bajocian). The basal GSSP is at Fuentelsalz in central Spain (Cresta et al., 2001).

AAR *accumulation-area ratio.*

AAS *atomic absorption spectroscopy.*

a* axis That axis of a reciprocal crystal lattice which is perpendicular to (100). Cf: *b* axis; c* axis.*

***a* axis [cryst]** One of the crystallographic axes used as reference in crystal description. It is the axis that is oriented horizontally, front-to-back. In an orthorhombic or triclinic crystal, it is usually the *brachy-axis.* In monoclinic crystals, it is the *clinoaxis.* The letter *a* usually appears in italics. Cf: *b axis [cryst]; c axis [cryst].*

***a*-axis [ice]** In a hexagonal crystal are three *a*-axes. They are perpendicular to the c-axis and make angles of 60° to one another. Cf: *c-axis [ice].*

***a* axis [struc petrol]** One of three orthogonal reference axes, *a, b,* and *c,* that are used in two distinct ways. (a) To help describe the geometry of a fabric possessing orthorhombic or monoclinic symmetry. The unique symmetry plane is labelled the *a-c* plane, and *a* lies at the intersection of this plane with a prominent fabric surface. (b) In a kinematic sense, to describe a *deformation plan* that possesses orthorhombic or monoclinic symmetry, such as progressive pure or simple shear. In this case the *a* axis again lies in the unique plane of symmetry but parallel to the movement plane. It is the direction of maximum displacement and is commonly referred to as the direction of tectonic transport. In a progressive simple shear the *a* axis is the direction of shear. See also: *b axis [struc petrol]; c axis [struc petrol].* Syn: *a direction.*

abactinal (ab-ac'-ti-nal) A syn. of *aboral.* Ant: *actinal.*

abandoned channel (a-ban'-doned) (a) A drainage channel along which runoff no longer occurs, as on an alluvial fan. (b) *oxbow.*

abandoned cliff A sea cliff that is no longer undergoing wave attack, as a result of a relative drop of sea level or progradation at the cliff base.

abandoned meander *cutoff meander.*

abapertural (ab-ap-er'-tur-al) Away from the *aperture* of a gastropod, nautiloid, or tentaculite shell. Ant: *adapertural.*

abapical (ab-ap'-i-cal) Away from the *apex* of a gastropod shell and toward the base, along the axis of spiral or slightly oblique to it.

abathochroal eye (ab-ath'-o-chro'-al) A trilobite eye superficially similar to a *schizochroal eye* but possessing no scleral projections.

abaxial (ab-ax'-i-al) Directed or facing away from, or situated on the outside of, the axis or center of the axis, as of an organ, plant, or invertebrate; *dorsal* or *anterior.* Also, said of the abaxial side. Ant: *adaxial.*

Abbe refractometer An instrument used for determining the refractive index of liquids, minerals, and gemstones. Its operation is based on measurement of the *critical angle.*

abbreviation (ab-bre'-vi-a'-tion) Loss of the final ontogenetic stages during the course of evolution.

ABC method A seismic refraction method, often used for correcting seismic data for the effect of irregular thickness of the surficial *low-velocity layer* (Sheriff and Geldart, 1995, p.433).

ABC soil A soil with a distinctly developed profile, including A, B, and C horizons.

abdomen (ab'-do-men) (a) The posterior and often elongated region of the body of an arthropod, behind the *thorax* or *cephalothorax;* e.g. the *tagma* following the thorax of a crustacean, including the *telson,* and consisting of seven or fewer segments; or the complete, usually unsegmented part of the body of an arachnid or merostome, following the cephalothorax. Cf: *opisthosoma; pygidium.* (b) The third joint of the shell of a nasselline radiolarian. Pl: abdomens or abdomina. Adj: abdominal.

abelsonite (a'-bel-son-ite') A semi-metallic purple triclinic mineral: $C_{31}H_{32}N_4Ni$ (nickel porphyrin).

abenakiite-(Ce) A pale brown rhombohedral mineral: $Na_{26}(Ce,Nd,La)_6(SiO_3)_6(PO_4)_6(CO_3)_6(SO_2)O$.

aber The mouth of a river, or the confluence of two rivers. Etymol: Celtic.

abernathyite (ab-er-nath'-y-ite) A yellow tetragonal mineral: $K(UO_2)(AsO_4)\cdot 3H_2O$.

aberration (ab-er-ra'-tion) Any disturbance of the rays of a beam of light such that they cannot be brought to a sharp focus or form a clear image.

abhurite (ab-hur'-ite) A colorless rhombohedral mineral: $Sn^{2+}_{21}O_6Cl_{16}(OH)_{14}$.

abime (a) Abyss. (b) Wide, deep shaft in limestone, the walls of which are vertical or overhanging. Etymol: French "abîme", "chasm".

AB interray Right anterior interray in echinoderms situated between A ray and B ray and clockwise of A ray when the echinoderm is viewed from the *adoral* side; equal to interambulacrum 3 of the *Lovenian system.*

abiogenesis (a-bi'-o-gen'-e-sis) The development of living organisms from lifeless matter. Cf: *biogenesis.*

ablation [geomorph] (ab-la'-tion) Separation and removal of rock material, and formation of residual deposits, esp. by wind action or the washing away of loose and soluble materials. Most writers prefer to restrict the term to wasting of glaciers by melting and evaporation.

ablation [glaciol] (a) All processes by which snow and ice are lost from a glacier, from floating ice, or from a snow cover. These processes include melting, evaporation (sublimation), wind erosion, and calving. Sometimes ablation is restricted to surface phenomena, thus excluding *calving.* Cf: *accumulation.* (b) The amount of snow or ice removed by the process of ablation. Syn: *wastage [glaciol].*

ablation [meteorite] Removal of surface layers of meteorites and tektites resulting from the impacts of air molecules during flight through the atmosphere.

ablation area The part of a glacier or snowfield in which, over a year's time, ablation exceeds accumulation; the region below the *equilibrium line.* Cf: *accumulation area.* Syn: *zone of ablation.*

ablation breccia *solution breccia.*

ablation cave A type of *glacier cave,* a few meters in height and width, formed near a glacier terminus by circulating warm air, generally where a meltwater stream flows from beneath the ice.

ablation debris Rock material of all sizes, from blocks to clay, as isolated fragments or discontinuous piles on the glacier surface; not continuous enough to form *ablation moraine.* The term is also used for that material comprising ablation moraine.

ablation form A feature formed on a surface of snow, firn, or ice by melting or evaporation; e.g. *nieve penitente* or *ice pyramid.* Rarely used.

ablation funnel A closed depression, similar to a solution channel, formed by solution processes or by removal of particulate material by circulating groundwater.

ablation gradient A term apparently first introduced by Haefeli (1962) to refer to the variation in specific *net balance* with altitude up to the *equilibrium line* of a glacier, but redefined by Schytt (1967) as the change in specific *summer balance* with altitude over the entire glacier. It is usually expressed as millimeters of water equivalent per meter of altitude. Cf: *budget gradient.*

ablation moraine An uneven pile or continuous layer of *ablation till* or *ablation debris,* either overlying ice in the ablation area or resting on ground moraine derived from the same glacier.

ablation rate The amount of ice or snow loss per unit time from a glacier, from floating ice, or from snow cover. Usually expressed in millimeters of water equivalent per hour or day or year.

ablation season In glaciology, that period of a year when the *balance* of a glacier decreases from a maximum value to a minimum value for the year. This is a period when, on the average, ablation exceeds accumulation. Syn: *summer season.*

ablation spherule Glassy, spherical to semi-spherical particles produced by the erosional removal of melt droplets from the exterior of meteorites during passage through Earth's atmosphere. See: *cosmic spherule; meteoritic dust.*

ablation till Loosely consolidated rock debris, formerly in or on a glacier, that accumulated in place as the surface ice was removed by ablation.

ablatograph (ab-lat′-o-graph) An instrument that measures the loss of snow, firn, or ice from a surface by ablation during a given period.

ablykite (ab′-lyk-ite) A clay-mineral material consisting of an aluminosilicate of magnesium, calcium, and potassium. It resembles halloysite in its dehydration characteristics but differs from it in its thermal and X-ray diffraction properties. Also spelled: ablikite.

Abney level A rectangular telescope tube to one side of which is fastened a vertical, graduated arc with vernier and carrying a rotatable level. The instrument is small and light enough to be held in the hand. When used as a leveling instrument, the vernier is set at zero on the graduated arc. When used to determine vertical angles, the object is sighted through the tube and the vernier rotated until the bubble is centered in the vial. The angle is then read off the graduated arc. Named after William de Wiveleslie Abney (1843-1920), English physicist.

abnormal (ab-nor′-mal) Said of an anticlinorium in which the axial surfaces of the subsidiary folds converge upwards; said of a synclinorium in which the axial surfaces of the subsidiary folds converge downwards. Rarely used. Cf: *normal [struc geol].*

abnormal pore pressure Fluid pressure in the earth's crust that deviates substantially from hydrostatic pressure.

abnormal subaerial exposure In sequence stratigraphy, a type of basinward shift in facies marked by subaerial exposure of subtidal facies.

abnormal vertical association of facies *basinward shift in facies.*

aboral (ab-o′-ral) (a) Located opposite to or directed away from the mouth of an invertebrate; esp. applied to the surface (or to the structures on it) opposite that bearing the mouth and proximal ends of the ambulacral grooves of an echinoderm; or to the part of an echinoderm *theca* or plate directed away from the mouth (directed downward in an edrioasteroid). Cf: *adoral.* (b) Toward the underside of a conodont element; e.g. "aboral edge", "aboral groove", or "aboral attachment scar". Ant: *oral.*

aboral cup The bowl- to cone-shaped part of a crinoid *calyx* including the three major plate circlets, radials, basals, and infrabasals (if present).

aboral margin The trace of the aboral side of a conodont element in lateral view. The term has also been used for the aboral side itself.

aboral pole (a) The end of a flask-shaped chitinozoan that includes the chamber of the body and the base. Cf: *oral pole.* (b) The point of intersection of the oral-aboral axis with the aboral surface of the theca of echinoderms; it marks the center of the aboral surface.

aboral side The underside of a conodont element, to which the basal plate is attached or onto which the basal cavity or attachment scar opens. Cf: *oral side.*

aboral surface Morphologically related unit of an edrioasteroid *theca* distal to the oral surface plates; commonly forms only part of the lower side of the individual.

aboriginal (ab-o-rig′-i-nal) Said of the original race, fauna, or flora of a particular area, as distinguished from later immigrant or imported forms. n. aborigine.

abraded snow Snow crystals that are rounded mechanically by interaction with other particles in the saltation layer.

abrasion [geomorph] The mechanical wearing, grinding, scraping, or rubbing away (or down) of rock surfaces by friction and impact, in which the solid rock particles transported by wind, ice, waves, running water, or gravity are the tools of abrasion. The term *corrasion* is essentially synonymous. Also, an abraded place or the effect of abrading, such as the abrasion left by glacial action. v. abrade. Cf: *attrition.*

abrasion [streams] The physical erosion caused by sediment in transport, grinding against and colliding with the channel boundaries and other sediment within the channel.

abrasion pH A term proposed by Stevens and Carron (1948) to designate the characteristic pH achieved by a suspension of a pulverized mineral in water, resulting from a complex of hydrolysis and dissolution reactions.

abrasion platform An extensive, gently seaward-sloping intertidal surface produced by long-continued wave erosion. Term introduced by D.W. Johnson (1916, p.444); see also Trenhaile (1987). Cf: *erosion platform; wave-cut platform; plain of marine erosion.*

abrasion shoreline *retrograding shoreline.*

abrasion tableland A broad, elevated region in which the outcrops of various rocks have been reduced to nearly the same level by denuding agents (Stamp, 1961, p.2).

abrasive [geomorph] (ab-ra′-sive) n. A rock fragment, mineral particle, or sand grain used by natural agents in abrading rock material or land surfaces. adj. Possessing the characteristics of a tool for grinding or mechanical wear.

abrasive [materials] Any natural or artificial substance suitable for grinding, polishing, cutting, or scouring. Natural abrasives include diamond, emery, garnet, silica sand, diatomite, and pumice; manufactured abrasives include esp. silicon carbide, fused alumina, and boron nitride.

abrolho A term used for a mushroom-shaped barrier reef spreading widely near the surface. Etymol: Portuguese, "thorn; pointed rock". "Abrolho Islands" are found off the coasts of northwestern Brazil and southwestern Australia.

abrupt climate change Changes of hemispheric average temperature of 1-5 degrees Celsius over a period of decades to a few centuries associated with changes in the ocean's circulation, especially the circulation in the North Atlantic.

abrupt twist hackle The en echelon cracks produced upon reinitiation of rupture if there has been a stress field rotation following the arrest of a parent crack (Younes and Engelder, 1999). These structures may also form in more isotropic rocks such as granite. Cf: *twist hackle; gradual twist hackle.*

absarokite (ab-sa′-ro-kite) A *trachyandesite*, composed of phenocrysts of olivine and clinopyroxene in a groundmass of labradorite with alkali feldspar rims, olivine, and some leucite. Absarokite grades into *shoshonite* with a decrease in the olivine content and with the presence of some dark-colored glass, and into *banakite* with a decrease in the olivine and augite. It was named by Iddings in 1895 from the Absaroka Range, Wyoming.

abscission (ab-scis′-sion) Separation of plant parts, e.g. of a leaf from a stem, usually by cell-wall dissolution along a certain layer (*abscission layer*).

abscission layer The zone of cells, e.g. at the base of a petiole, along which separation of plant parts occurs. Syn: *separation layer.*

absite A thorian variety of *brannerite.*

absolute abundance (ab′-so-lute) The exact number of individuals of a taxon in a certain area or volume. See also: *abundance [ecol]; relative abundance.*

absolute age *numerical age.* The term is now in disfavor as it implies a certainty or exactness that may not be possible by present dating methods; e.g., two "absolute" ages for the same pluton may disagree by hundreds of millions of years.

absolute age determination *numerical age determination.*

absolute chronology *numerical chronology.*

absolute date *numerical date.*

absolute datum A geodetic datum in which the reference ellipsoid is Earth-centered with its minor axis corresponding to the Earth's axis of rotation. See also: *geocentric geodetic datum.*

absolute gravity An absolute value of the *gravitational acceleration* due to the attraction of the Earth's mass, as opposed to its relative values such as measured by conventional gravimeters. It varies from about 9.78 m/sec^2 (978,000 mGal) at the equator to about 9.83 m/sec^2 (983,000 mGal) at the poles. Measurements are usually made at regional base stations of the world-wide network using the same type instruments. Portable absolute gravimeters are now in use. See also: *International Gravity Standardization Net 1971; absolute-gravity instrument.*

absolute-gravity instrument A device for measuring the actual value of the Earth's gravitational acceleration at a point with all of the physical influences evaluated with extreme accuracy. Measurements are accomplished by various forms of reversible pendulums or by timing the motion of a body in free fall. Cf: *relative-gravity instrument.*

absolute humidity The content of water vapor in air, expressed as the mass of water per unit volume of the humid air. Cf: *relative humidity; specific humidity.*

absolute permeability The ability of a rock to conduct a fluid, e.g. gas, at 100% saturation with that fluid. See also: *effective permeability; relative permeability.*

absolute pollen frequency An estimate of the actual amount of pollen deposited per unit area in a given length of time, achieved by correcting the amount of pollen per gram of sediment by factors based on rate of sedimentation. Abbrev: APF.

absolute time *numerical time.* Jeletzky (1956, p.681) proposed that the term be abandoned, because its usage, based on criteria peculiar to the Earth and having the present part of geologic history as its starting point, is "incorrect and highly misleading".

absolute viscosity *viscosity coefficient.*

absolute vorticity The sum of *relative* and *planetary vorticity.*

absorbed water (ab-sorbed′) (a) Water retained mechanically within a soil mass and having properties similar to those of ordinary water at the same temperature and pressure. (b) Water entering the lithosphere by any means. Cf: *adsorbed water.*

absorptance (ab-sorp′-tance) The ratio of the energy absorbed by a material to that incident upon it. Syn: *absorption coefficient.*

absorption (ab-sorp′-tion) Taking up, assimilation, or incorporation; e.g., of liquids in solids or of gases in liquids. Cf: *adsorption.* Syn: *occlusion.*

absorption [grd wat] The entrance of surface water into the lithosphere by any method. Verb: to absorb. Cf: *adsorption.*

absorption [optics] The reduction of light intensity in transmission through an absorbing substance or in reflection from a surface. In crystals, the absorption may vary with the wavelength or vibration direction of the transmitted light.

absorption [phys] Any mechanism by which energy, e.g., electromagnetic or seismic, is converted into heat.

absorption band The wavelength interval at which electromagnetic radiation is absorbed by the atmosphere or by other media, e.g. an atmospheric absorption band at 5 to 8 μm, caused by water vapor that absorbs thermal infrared radiation of those wavelengths. Cf: *absorption spectrum; absorption line.*

absorption coefficient The rate of exponential decrease in amplitude. If the amplitude A is expressed as $A=A_0e^{-\alpha x}$, where x is the distance, α is the absorption coefficient. See: *Q [seis].*

absorption edge The wavelength at which there is an abrupt change in the intensity of an *absorption spectrum.* The term is usually applied to X-ray spectra.
absorption line Any of the dark lines in the *absorption spectrum* of a substance due to certain wavelengths in the spectrum being selectively absorbed on passing through a medium. Cf: *absorption band.*
absorption loss Water lost through *absorption* by rock and soil during the initial filling of a reservoir.
absorption spectroscopy The group of analytical techniques that involves determination and measurement of atomic and molecular energy levels (*spectrometry*), chemical identification, and molecular structure based on how the atoms and molecules absorb electromagnetic radiation. See also: *absorption spectrum; spectroscopy.*
absorption spectrum The characteristic spectrum that a substance capable of absorbing electromagnetic radiation produces when it is observed with a *spectroscope.* Black lines or bands appear where energy has been removed from the *continuous spectrum* by an absorbing medium. See also: *absorption spectroscopy; emission spectrum; spectroscopy; line spectrum; band spectrum.*
absorptivity (ab-sorp-tiv'-i-ty) The ability of a material to absorb energy incident upon it.
abstraction [streams] (ab-strac'-tion) The merging of two or more subparallel streams into a single stream course, as a result of competition between adjacent consequent gullies and ravines, as by the deepening and widening of one channel so that it absorbs a shallower and smaller one nearby; the simplest type of capture. It usually occurs at the upper end of a drainage line.
abstraction [water] That part of precipitation that becomes part of groundwater and is withdrawn from a well. Cf: *precipitation excess; rainfall excess.*
abswurmbachite A black tetragonal mineral: $Cu^{2+}Mn^{3+}_6O_8(SiO_4)$.
abtragung The part of degradation not resulting directly from stream erosion, i.e. preparation and reduction of rock debris by weathering and transportation of waste (Engeln, 1942, p.265). Etymol: German "Abtragung", "degradation; denudation".
abukumalite (ab-u-ku'-ma-lite) *britholite-(Y).*
Abukuma-type facies series (Ab-u-ku'-ma) Rocks produced in a type of dynamothermal regional metamorphism named after the Central Abukuma plateau of Japan, and characterized by the index minerals (in order of increasing metamorphic grade) biotite-andalusite-cordierite-sillimanite, representing the greenschist and amphibolite or hornblende-hornfels facies. Pressures are low, approaching those in contact metamorphism, i.e. 250-350 MPa (Hietanen, 1967, p.192). Cf: *Buchan-type facies series.*
abundance [ecol] (a-bun'-dance) In ecology, the number of individuals of a particular taxon in a certain area or volume of sediment. See also: *absolute abundance; relative abundance.*
abundance [geochem] The mean concentration of an element in a geochemical reservoir, e.g. the abundance of Ni in meteorites, or the crustal abundance of oxygen. Also used for relative average content, e.g. the order of abundance of elements in the Earth's crust is O, Si, Al, Fe, Ca, etc.; the estimated cosmic abundance of Li in atoms per 10,000 atoms of Si is 1.0 (Suess and Urey, 1956).
abundance, elemental [meteorite] Abundance, *in the strict sense,* means a doubly normalized atomic ratio (e.g., to the element Si and CI chondrites). Commonly also used as a synonym of weight concentration.
abundance of isotopes of an element Determined by mass spectrometry.
abundance-reserve relationship A method for estimating recoverable reserves of metals from crustal abundance data proposed by McKelvey (1960). McKelvey plotted recoverable reserves in the United States for several metals against their crustal abundance and proposed a general relationship $R = A \times 10^k$, where R is the tonnage of recoverable metal, A is crustal abundance in percent, and k is a number from 9 to 10 (Harris, 1984).
abundance zone (a) A *biozone* characterized by quantitatively distinctive maxima of relative abundance of one or more taxa (NACSN, 1983, Art. 52). Syn: *acme zone.* (b) A *stratum* or body of strata in which the abundance of a particular taxon or specified group of taxa is significantly greater than is usual in the adjacent parts of the section, regardless of either association or range (ISSC, 1994, p.63-64).
abundant (a-bun'-dant) In the description of coal constituents, 30-60% of a particular constituent occurring in the coal (ICCP, 1963). Cf: *rare; common; very common; dominant [coal].*
abyss [geomorph] (a-byss') *chasm.*
abyss [oceanog] Ocean environment at great depth.
abyssal [intrus rocks] (a-bys'-sal) (a) Pertaining to an igneous intrusion, of the resulting rock, formed at considerable depth; *plutonic.* (b) The same from the bottom of the ocean. Cf: *hypabyssal.*
abyssal [lake] Pertaining to great depths in a lake. Hutchinson (1967, p.241) questioned the utility of the term.
abyssal [oceanog] Pertaining to the ocean environment or depth zone of between 3,500 and 6,000 m; also, pertaining to the organisms of that environment.
abyssal benthic Pertaining to the benthos of the abyssal zone of the ocean. Syn: *abyssobenthic.*
abyssal cone A type of *submarine fan.*
abyssal fan Submarine fan at great depth.
abyssal gap A passage that connects two abyssal plains of different levels, through which clastic sediments are transported. Syn: *gap [marine geol].*
abyssal hill A common low-relief feature of the ocean floor, usually found in basins isolated by ridges, rises, or trenches. Abyssal hills are defined as less than 1 km in relief (>1 km is termed a seamount) and several kilometers in diameter. About 85% of the Pacific Ocean floor and 50% of the Atlantic Ocean floor are covered by abyssal hills.
abyssal pegmatite Feldspar-quartz pegmatite formed under conditions of granulite metamorphism (~4 to 9 kilobars pressure), and sometimes associated with migmatitic granite.
abyssal pelagic Pertaining to the open-ocean or pelagic environment at abyssal depths. Syn: *abyssopelagic.*
abyssal peridotite Serpentinized and altered peridotite of oceanic ridges. "Abyssal peridotites are widely accepted as residues of pressure-release melting accompanying upward convection of the mantle beneath mid-ocean ridges" (Dick, 1989).
abyssal plain A flat region of the deep ocean floor, whose slope is less than 1:1,000. It is formed by the deposition of gravity-current and pelagic sediments that obscure the preexisting topography.
abyssal theory A theory of mineral-deposit formation involving the separation and sinking of minerals below a silicate shell during the cooling of the Earth from a liquid stage, followed by their transport to and deposition in the crust as it was fractured (Shand, 1947, p.204). Modern thought has completely negated such theories.
abyssal tholeiite *mid-ocean-ridge basalt.*
abyssobenthic (a-byss'-o-ben'-thic) *abyssal benthic.*
abyssolith (a-byss'-o-lith) Obsolete syn. of *batholith.*
abyssopelagic (a-byss'-o-pe-lag'-ic) *abyssal pelagic.*
acadialite (a-ca'-di-a-lite') A flesh-red variety of chabazite, found in Nova Scotia.
Acadian (A-ca'-di-an) North American provincial series: Middle Cambrian (above Georgian, below Potsdamian). Obsolete syn. of *Albertan.*
Acadian Orogeny A middle Paleozoic deformation event, especially in the northern Appalachians; it is named for Acadia, the old French name for the Canadian Maritime Provinces. The climax of the orogeny can be dated stratigraphically as early in the Late Devonian, but deformational, plutonic, and metamorphic events occurred over a more extended period; the last two have been dated radiometrically as between 330 and 360 m.y. ago. The Acadian is probably a consequence of convergent-margin tectonism and collision of exotic terranes with the eastern margin of North America. For example, the Avalon platform probably docked during the Acadian. Cf: *Antler Orogeny.*
acalymmate (a-ca-lym'-mate) In obligate *tetrads* or *polyads,* not having the *ectexine/sexine* form a single continuous envelope around the unit. Cf: *calymma.* Ant: *calymmate.*
acantharian (a-can-tha'-ri-an) Any radiolarian belonging to the suborder Acantharina, characterized by a centrogenous skeleton composed of strontium sulfate and a central capsule enclosed by a thin simple membrane.
acanthine septum (a-can'-thine) A *corallite septum* composed of a vertical or steeply inclined series of trabeculae and commonly marked by spinose projections along the axially directed margin of the septum.
acanthite (a-can'-thite) A metallic black monoclinic mineral: Ag_2S. It is dimorphous with argentite and constitutes an ore of silver.
Acanthodii (Ac-an-tho'-di-i) A subclass of the Osteichthyes characterized by fixed paired fins supported anteriorly by spines; more than two pairs are usually present. It includes the oldest recorded gnathostomes (Upper Silurian). Range, Upper Silurian to Lower Permian.
acanthomamilla (a-can-tho-ma-mil'-la) In palynomorphs, a sculptural element consisting of a hemispheroidal base surmounted by a sharply contracted spine; e.g., in dibolisporites.
acanthomorph (a-can'-tho-morph) An *acritarch* having a spherical or ellipsoidal *vesicle* with prominent simple or branching *processes* (Mendelson, 1993; Williams et al., 2000).
acanthopore (a-can'-tho-pore) A small rodlike skeletal structure, originally believed to be hollow, consisting of a solid core surrounded by a sheath of cone-in-cone laminae lying within zooecial walls or extrazooidal skeleton in stenolaemate bryozoans. Acanthopores form spinelike projections at the colony surface.
acanthostyle (a-can'-tho-style) A monaxonic sponge *spicule* (style) covered with short or tiny spines over most of its surface.
acanthus (a-can'-thus) A secondary deposit in the chamber floor of certain foraminifers (such as *Endothyra*), sharply pointed but not curved toward the anterior (TIP, 1964, pt. C, p.58). Pl: acanthi.
acapulcoite A group of meteorites with chondritic mineral compositions in roughly chondritic proportions, thought to have been formed by a low degree of partial melting and segregation of Fe. See: *primitive achondrite.*
acarid (ac'-a-rid) Any arachnid belonging to the order Acarida, characterized by the absence of abdominal segmentation but with subdivision of the body into a proterosoma and hysterosoma. Range, Devonian to present.
acaustobiolith (a-caust'-o-bi'-o-lith) A noncombustible organic rock, or a rock formed by the organic accumulation of purely mineral matter (Grabau, 1924, p.280). Rarely used. Cf: *caustobiolith.*

acavate (a-ca′-vate) In *palynomorphs,* lacking a cavity between wall layers (Williams et al., 2000, p.2). Ant: *cavate.*

accelerated development (ac-cel′-er-at′-ed) The hypothetical change in a landscape where the rate of uplift is more rapid than the rate of downward erosion or where valley deepening exceeds valley widening, characterized by an increase of the relative relief and the formation of convex slopes. Cf: *declining development; uniform development.* Syn: *waxing development; ascending development.*

accelerated erosion Erosion occurring in a given region at a greater rate than *normal erosion,* usually brought about by the influence of human activities in disturbing or destroying the natural cover, thus sharply reducing resistance of the land surface and rate of infiltration. It may result from deforestation, improper cultivation of soil, dry-farming, overgrazing of rangelands, burning and clearance of natural vegetation, excavation for buildings and highways, urbanization of drainage areas, strip mining, or copper smelting; and by nonhuman influences, such as lightning-ignited wildfire or rodent invasion.

acceleration (ac-cel′-er-a′-tion) (a) During evolution, the appearance of modifications earlier and earlier in the life cycle of successive generations; adult characters of the ancestor appear earlier in immature stages of the descendants (*tachygenesis*), sometimes to the point that certain steps are omitted (*brachygenesis*). (b) In Paleozoic corals, the addition of more secondary septa in one pair of quadrants than in the other pair.

acceleration due to gravity The acceleration of a freely falling body in a vacuum as a result of gravitational attraction. Although its value varies with altitude, latitude, and the nature of the underlying rocks, the standard value of 980.665 cm/sec^2 has been adopted by the International Committee on Weights and Measures.

accelerator mass spectrometry A type of *mass spectrometry* that relies on high-voltage accelerators and provides ultra-high dynamic range. It is capable of measuring isotopic ratios as small as 10^{-16}. Abbrev: AMS.

accelerometer (ac-cel′-er-om′-e-ter) (a) A *seismometer* whose response is linearly proportional to the acceleration of earth materials with which it is in contact. (b) A *transducer* whose output is proportional to the acceleration. Accelerometers are used in shipboard gravimeter systems, in inertial navigation, and in some airborne gravity measuring systems (Fairhead and Odegard, 2002). The measured acceleration is relative to the rotation of the Earth; it is nonequilibrium acceleration, not simply acceleration of gravity. Cf: *aerogravity.*

accented contour (ac′-cent-ed) *index contour.*

accessible geothermal resource base That part of the geothermal resource base at depths shallow enough to be tapped by drilling in the foreseeable future (Muffler and Cataldi, 1978; Muffler, 1985). Includes identified and undiscovered components.

accessory [mineral] (ac-ces′-so-ry) *accessory mineral.*

accessory [paleont] adj. Said of a secondary or minor element of an ammonoid suture; e.g. "accessory lobe" or "accessory saddle". Also said of a secondary shelly addition to a bivalved molluscan shell. Cf: *auxiliary.* n. Such a lobe or saddle.

accessory [pyroclast] Said of pyroclastic materials that are formed from fragments of the volcanic cone or earlier lavas; it is part of a classification of volcanic ejecta based on mode of origin, and is equivalent to *resurgent* ejecta. Cf: *essential; accidental.* See also: *cognate [pyroclast].*

accessory aperture An opening in the test of a planktonic foraminifer that does not lead directly into a primary chamber but extends beneath or through accessory structures (such as bullae and tegilla); e.g. a *labial aperture,* an *infralaminal accessory aperture,* and an *intralaminal accessory aperture.*

accessory archeopyle suture An *archeopyle suture* that consists of a short cleft in the wall adjacent to the principal suture, or that may be more fully developed on the operculum of the dinoflagellate cyst, dividing that structure into two or more separate pieces.

accessory calcification [paleont] Any calcification by the molluscan mantle epithelium that is not an integral component of the normal shell plates, ligament, or periostracum, e.g., separate shell plates adjacent to the normal shell margins in corbulids and pholads, callum and pallet calcification in pholads, encrustations on the exterior of the periostracum in lithogaginids, tubes secreted by burrowing species, and burrow linings secreted by boring species in the Pholadacea and Gastrochaenacea (Carter, 1980a, p.77). The term is partly synonymous with the "Kalktapete" of Bandel and Keupp (1985).

accessory comb The line of large cilia within the preoral cavity in a tintinnid.

accessory element *trace element.*

accessory lobe Nodose projection of anterior part of platform of some pectiniform conodont elements, situated between posterior end of blade and that part of platform crossed by transverse ridges (TIP, 1981, pt. W, supp.2).

accessory mineral A mineral whose presence in a rock is not essential to the proper classification of the rock. Accessory minerals generally occur in minor amounts; in sedimentary rocks, they are mostly *heavy minerals.* Cf: *essential mineral [petrology].* Syn: *accessory [mineral].*

accessory muscle (a) A convenient noncommittal term for any muscle of a bivalve mollusk (other than an *adductor muscle* or a muscle withdrawing marginal parts of the mantle) of uncertain origin and having a scar of attachment to the shell. (b) One of a pair of diductor muscles branching posteriorly and ventrally from the main diductor muscles of a brachiopod and inserted in the *pedicle valve* posterior to the adductor bases (TIP, 1965, pt. H, p.139).

accessory spore A spore present in a rock only in very small quantities. Accessory spores may contain types with a restricted range and they have been used for correlation and for zoning (as of coal measures).

accident (ac′-ci-dent) (a) A departure from the normal cycle of erosion, caused by events that occur "arbitrarily as to place and time", such as climatic changes and volcanic eruptions (Davis, 1894). See also: *climatic accident; volcanic accident.* Cf: *interruption.* (b) An event, such as drowning, rejuvenation, ponding, or capture, that interferes with, or entirely puts an end to, the normal development of a river system (Scott, 1922, p.188). (c) An irregular feature in, or an undulation of, a land surface.

accidental (ac-ci-den′-tal) Said of pyroclastic materials that are formed from fragments of nonvolcanic rocks or from volcanic rocks not related to the erupting volcano; it is part of a classification of volcanic ejecta based on mode of origin, and is equivalent to *allothigenous* ejecta. Cf: *cognate [pyroclast]; accessory [pyroclast]; essential.* Syn: *noncognate.*

accidental error An unpredictable error that occurs without regard to any known mathematical or physical law or pattern and whose occurrence is due to chance only; e.g. an error ascribed to uncontrollable changes of external conditions. Syn: *random error.*

accidental inclusion *xenolith.*

accidented relief Rugged and irregular relief; probably a literal translation of the common French term "relief accidenté" (Stamp, 1961, p.4).

acclimation (ac-cli-ma′-tion) *acclimatization.*

acclimatization (ac-cli′-ma-ti-za′-tion) Physiologic adjustment by an organism to a change in its immediate environment. Syn: *acclimation.*

acclivity (ac-cliv′-i-ty) A slope that ascends from a point of reference. Ant: *declivity.*

accommodation The space made available for potential sediment accumulation. Accomodation is a function of sea-level rise, subsidence, or a combination of these two processes (Jervey, 1988). Accommodation refers to all the space available for sediment to fill, including old space (leftover from an earlier time) plus new space added; whereas new space added refers only to space contemporaneously being made available.

accommodation zone (a) A zone where local faulting and/or folding develops to accommodate room problems during the development of a larger structure, i.e. in the hinge zone of a fold. (b) A portion of a rift zone where the axis of extension and/or the regional dip of faults changes abruptly. (c) The area between two subparallel, non-collinear, overlapping faults along which displacement or strain is transferred from one fault to another. Rosendahl et al. (1986) specifically apply the term to the zone between *half grabens* whose *boundary faults* dip in opposite directions. Cf: *transfer zone.*

accordance of summit levels (ac-cord′-ance) *summit concordance.*

accordant (ac-cord′-ant) Said of topographic features that have the same or nearly the same elevation; e.g. an accordant valley whose stream enters the main stream at the same elevation as that of the main stream. Ant: *discordant [geomorph].*

accordant drainage Drainage that has developed in a systematic relationship with, and consequent upon, the present geologic structure. Ant: *discordant drainage.* Syn: *concordant drainage.*

accordant fold One of several folds having similar orientation.

accordant junction The joining of two streams or two valleys whose surfaces are at the same level at the place of junction. See also: *Playfair's law.* Ant: *discordant junction.* Syn: *concordant junction.*

accordant summit level A reconstructed level or gently sloping surface that regionally intersects hilltops or mountain summits. Accordant summit levels in a region of high topographic relief suggest that the summits are remnants of an erosion plain formed in a previous erosion cycle. See also: *summit concordance; even-crested ridge.* Syn: *concordant summit level.*

accordant summits Hilltops or mountain peaks that regionally reach the same hypothetical level or gently sloping surface. The term cannot be used in the singular.

accordion fold (ac-cor′-di-on) An old term, formerly used with genetic significance; now sometimes used as a syn. of *kink fold.* See also: *zigzag fold; chevron fold.* Syn: *angular fold; concertina fold.*

accreting plate boundary (ac-cret′-ing) A boundary between two plates that are moving apart, with new oceanic-type lithosphere being created at the seam (Dennis and Atwater, 1974, p.1033). See also: *mid-ocean ridge; spreading center; divergent plate boundary.* This older term is now avoided because it can be confused with the more recent concept of accretionary plate orogeny.

accretion [planet] (ac-cre′-tion) The process whereby small particles and gases in the solar nebula came together to form larger bodies, eventually of planetary size.

accretion [sed] (a) The gradual or imperceptible increase or extension of land by natural forces acting over a long period of time, as on a beach by the washing-up of sand from the sea or on a flood plain by the accumulation of sediment deposited by a stream. Legally, the added land belongs to the owner of the land to which it is added. Cf: *avulsion; reliction.* See also: *lateral* *accretion; vertical accretion.* Syn: *aggradation [geomorph]; alluvion.* (b) The land so added or resulting from accretion. (c) *continental accretion.*

accretion [sed struc] (a) The process by which an inorganic body increases in size by the external addition of fresh particles, as by adhesion. (b) A *concretion;* specif. one that grows from the center outward in a regular manner by successive additions of material (Todd, 1903). (c) Deposition of eolian sand on a continuous sand surface because of a decrease in wind intensity or an increase in surface roughness (Bagnold, 1941, p.127).

accretion [streams] The filling-up of a stream bed, due to such factors as silting or wave action. Cf: *degradation [streams].*

accretion [struc geol] The addition of island-arc or continental material to a continent by convergent and transform motion, i.e. by collision and welding or suturing. Locally, very large volumes of material, containing slivers of ophiolite, are added by accretion. The tectonic incorporation of terranes into a continental framework (Howell, 1995). Cf: *continental accretion; amalgamation.* Syn: *tectonic accretion.*

accretionary (ac-cre'-tion-a'-ry) Tending to increase by external addition or accumulation; esp. said of a secondary sedimentary structure produced by overgrowth upon a preexisting nucleus, such as a rounded form that originated through rolling, or said of a limestone formed in place by slow accumulation of organic remains.

accretionary lapilli Spheroidal pellets, mostly between 1 mm and 1 cm in diameter, of consolidated or cemented ash. Formed by accretion of particles around wet nuclei, e.g. raindrops falling through a cloud of ash (Macdonald, 1972, p.133) or accretion in a wet surge cloud. Syn. for individual spheroids: *pisolite [volc]; tuff ball.*

accretionary lava ball A rounded mass, ranging in diameter from a few centimeters to several meters, formed on the surface of a lava flow such as *aa,* or on cinder-cone slopes, by the molding of viscous lava around a core of already solidified lava.

accretionary orogeny A period of deformation and continental growth involving suturing of exotic terrane to a continental margin during collision and convergence.

accretionary prism A generally wedge-shaped mass of tectonically deformed sediment at a *convergent plate boundary* formed when pelagic sediment, oceanic-floor basalt, and trench-fill turbidite are scraped off the downgoing plate during the process of subduction.

accretionary terrane An allochthonous mass of continental or oceanic material added to the margin of a continent by collision and welding.

accretion ridge A beach ridge located inland from the modern beach, representing an ancient beach deposit and showing that the coast has been built out seaward (Fisk, 1959, p.111). It is often accentuated by the development of dunes. Cf: *chenier plain.*

accretion ripple An asymmetric ripple having a gentle and curved *lee* slope, with a maximum angle of dip less than the *angle of repose,* and composed of cross-strata without conspicuous sorting of particles (Imbrie and Buchanan, 1965, p.151, 153).

accretion ripple mark *accretion ripple.*

accretion till *basal till.*

accretion topography A landscape built by accumulation of sediment or lava.

accretion vein *sheeted vein.*

accumulated discrepancy (ac-cu'-mu-lat'-ed) The sum of the separate discrepancies that occur in the various steps of making a survey or of computing the results of a survey.

accumulation (ac-cu'-mu-la'-tion) (a) All processes that add snow or ice to a glacier, to floating ice, or to the snow cover, including snowfall, condensation, avalanching, drifting, and freezing of liquid water. Syn: *alimentation; nourishment [glaciol].* Cf: *ablation [glaciol]* (b) The amount of snow and other solid precipitation added to a glacier or snowfield by the processes of accumulation.

accumulation area The part of a glacier or snowfield in which, over a year's time, accumulation exceeds ablation; the region above the *equilibrium line.* Cf: *ablation area; névé.* Syn: *firn field; accumulation zone; zone of accumulation [snow].*

accumulation-area ratio The ratio of accumulation area to total area of a glacier for any given year, used as a rough guide to the balance between accumulation and ablation. Abbrev: AAR.

accumulation body A solid mass within the cell of a dinoflagellate that is assumed to be metabolic waste material (Williams et al., 2000, p.3).

accumulation mountain *mountain of accumulation.*

accumulation rate The amount of ice or snow gain per unit time to a glacier, floating ice, or snow cover. Usually expressed in millimeters of water equivalent per hour or day or year.

accumulation season In glaciology, that period of a year when the balance of a glacier increases to the maximum for the year. This is the part of the year when, on the average, accumulation exceeds ablation. Syn: *winter season.*

accumulation zone (a) *accumulation area.* (b) The area in which the bulk of the snow contributing to an avalanche was originally deposited. Syn: *zone of accumulation [snow].*

accumulative rock (ac-cu'-mu-la'-tive) *cumulate.*

accumulator plant (ac-cu'-mu-la'-tor) In geobotanical prospecting, a tree or plant that preferentially concentrates an element.

accuracy (ac'-cu-ra-cy) The degree of conformity with a standard, or the degree of perfection attained in a measurement. Accuracy relates to the quality of a result, and is distinguished from *precision,* which relates to the quality of the operation by which the result is obtained.

ACD *aragonite compensation depth.*

AC demagnetization *alternating-field demagnetization.*

A-centered lattice A *base-centered lattice* that is centered in the pair of (100) faces. Cf: *B-centered lattice; C-centered lattice.*

acequia (a-ce-qui'-a [ah-se-kee'-ah]) A Spanish word, of Arabic origin, for an irrigation ditch or canal.

acervuline (a-cer'-vu-line) Heaped, or resembling little heaps; e.g. said of some foraminifers (such as *Acervulina*) having chambers in irregular clusters.

acetamide (ac-et-am'-ide, ac-et'-am-ide) A colorless to gray trigonal organic mineral: CH_3CONH_2.

acetolysis (ac-e-tol'-y-sis) Any chemical reaction in which acetic anhydride plays a role similar to that of water in hydrolysis; e.g. a reaction used in maceration in which organic material such as peat is heated in a mixture of nine parts acetic anhydride and one part concentrated sulfuric acid. It breaks down cellulose especially vigorously.

ACF diagram A triangular diagram showing the simplified compositional character of metamorphic rocks and minerals by plotting the molecular quantities of the three components: $A = Al_2O_3 + Fe_2O_3 - (Na_2O + K_2O)$; $C=CaO - 3.3P_2O_5$; and $F=FeO + MgO + MnO$. A+C+F (in mols) are recalculated to 100%; the presence of excess SiO_2 is assumed. Cf: *AFM diagram; A'KF diagram.*

achavalite (a-cha-val'-ite) A metallic hexagonal mineral: FeSe.

achene (a-chene') A dry one-seeded indehiscent fruit developed from a simple ovary with unfused seed coat and fruit wall. Also spelled: *akene.*

achlamydate (ach-lam'-y-date) Said of a gastropod without a mantle.

achnelith (ach'-ne-lith) A term used by Walker and Croasdale (1972) to describe droplets from Strombolian lava fountaining. Etymol: Greek "achne," "spray, froth."

achoanitic (a-cho'-a-nit'-ic) Said of the condition in a nautiloid in which septal necks are vestigial or absent. Syn: *aneuchoanitic.*

achondrite (a-chon'-drite) A differentiated *stony meteorite;* as implied by the name, it lacks chondrules. Achondrites are commonly more coarsely crystallized than chondrites, and nickel-iron is almost completely lacking in most of them; most are brecciated. Adj: achondritic. Cf: *chondrite [meteorite]; iron meteorite.* Syn: *differentiated meteorite.*

achroite (a-chro'-ite) A colorless variety of tourmaline, used as a gemstone.

acicular [cryst] (a-cic'-u-lar) Said of a crystal that is needlelike in form. See also: *fascicular; sagenite.*

acicular [sed] Said of a sedimentary particle whose length is more than three times its width (Krynine, 1948, p.142). Cf: *platy.*

acicular ice Freshwater ice consisting of numerous long crystals and hollow tubes having variable form, layered arrangement, and a content of air bubbles; it forms at the bottom of an ice layer near its contact with water. Syn: *fibrous ice; satin ice.*

aciculate (a-cic'-u-late) Needle-shaped, or having a needlelike point; esp. said of a slender gastropod shell that tapers to a sharp point.

acid adj. (a) *silicic.* (b) *acidic.* (c) Said of a plagioclase that is sodic.

acid clay A clay that yields hydrogen ions in a water suspension; e.g. "Japanese acid clay", a variety of fuller's earth occurring in Kambara, Japan.

acidic (a-cid'-ic) (a) A descriptive term applied to those igneous rocks that contain more than 60% SiO_2, as contrasted with *intermediate* and *basic.* Sometimes loosely and incorrectly used as equivalent to *felsic* and to *oversaturated,* but these terms include rock types (e.g., nepheline syenite, quartz basalt) that are not generally considered acidic. This is not the chemist's usage; the term is deprecated by some because of its confusing nature. (b) Applied loosely to any igneous rock composed predominantly of light-colored minerals having a relatively low specific gravity. Cf: *felsic.* Syn: *acid; silicic.*

acidity coefficient (a-cid'-i-ty) *oxygen ratio.*

acidity quotient *oxygen ratio.*

acidization (ac'-id-i-za'-tion) The process of forcing acid down a well into a limestone or dolomite, in order to increase permeability and porosity by dissolving a part of the rock constituents. It is also used to remove mud injected during drilling. The general objective of acidization is to increase oil productivity. Syn: *acid treatment.*

acid job A method of well stimulation for an oil or gas well that may produce from a carbonate (e.g. limestone) reservoir or one cemented with carbonate minerals. Commonly used acids are hydrochloric, citric or acetic acids. The process is intended to dissolve the carbonate to increase productivity of that reservoir.

acid mine drainage Drainage with a pH of 2.0 to 4.5 from mines and mine wastes or from constructed excavations, such as road cuts or tunnels. It results from the oxidation of sulfides exposed during excavation, which produces sulfuric acid and sulfate salts.

acid neutralizing capacity The amount of acid that can be neutralized by the cations dissolved in lake water (Wetzel, 2001, p.191). Cf: *alkalinity [lake].*

acid plagioclase A variety of plagioclase having a relatively high content of SiO_2; e.g. an Ab-rich member such as albite or oligoclase.

acid rain Precipitation with a higher concentration of acids than the typical value of natural rain water (5.4). The high acidity is due to the disolution of naturally-occurring and human-generated oxides of sulfur and nitrogen in the atmosphere.

acid soil A soil with a pH of less than 7.0.

acid-sulfate type A type of *epithermal* gold mineralization characterized by advanced argillic to argillic alteration with abundant hypogene alunite, minor to absent chlorite or adularia, and a vein mineralogy of enargite + pyrite, e.g. Summitville, Colorado, and Goldfield, Nevada. Syn: *high-sulfidation epithermal; high-sulfur epithermal; kaolinite-alunite epithermal; quartz-alunite epithermal.*

acid treatment *acidization.*

ac-joint A joint in folded sedimentary rock that cuts fold axis at a high angle. Cf: *dip joint; cross-fold joint.*

aclinal (a-cli'-nal) A little-used term said of strata that have no inclination; horizontal.

acline [paleont] A syn. of *orthocline* as used to describe the hinge teeth or shell body of a bivalve mollusk, when the midumbonal line intersects the dorsal margin at 90°. Syn: *erect shell*. Cf: *prosocline*.

acline-A twin law *Manebach-Ala twin law.*

acline-B twin law A twin law for parallel twins in feldspar with twin axis *b* and composition plane (100). Cf: *x-Carlsbad twin law.*

aclinic line (a-clin'-ic) *magnetic equator.*

acme (ac'-me) That point in the phylogeny of a species, genus, or other taxon at which greatest abundance and/or variety occurs. See also: *paracme.*

acme zone *abundance zone.*

acmite (ac'-mite) *aegirine.*

acmite-augite A syn. of *aegirine-augite.*

acolpate (a-col'-pate) Said of pollen grains without *colpi.* In practice, such pollen grains are sometimes difficult to distinguish from *alete* spores. Cf: *inaperturate.*

acoustic adj. Relating to compressional waves, e.g., sound. Contrast with *elastic,* which includes S-waves also.

acoustic basement Generally the deepest more or less continuous seismic reflector, below which seismic energy returns are extremely poor to absent.

acoustic Doppler current meter An instrument that uses the Doppler shift of sound reflected from small particles in seawater within a narrow beam extending outward from the instrument to determine the radial component of the current velocity within the beam. Signals from several beams are combined to get the current velocity near the meter. Because the instrument has no moving mechanical parts, e.g., propellers of rotors, it is less susceptible to fouling, and it is replacing older mechanical current meters.

Acoustic Doppler Current Profiler A brand name for a current meter, on a ship, that uses the Doppler shift of sound reflected from small particles suspended in seawater to determine the radial component of the current velocity within a narrow beam, out to distances of a few hundred meters from the ship. Measurements from several beams are combined to get current velocity as a function of depth below the ship. Abbrev: ADCP.

acoustic impedance The product of seismic velocity and density. Syn: *impedance [seis].*

acoustic-impedance log A log of the product of seismic velocity and density. Reflectivity at normal incidence depends on changes in acoustic impedance. See also: *synthetic acoustic impedance log.*

acoustic log Generic term for a *well log* that displays any of several measurements of acoustic waves in rocks exposed in a borehole, e.g. compressional-wave transit time over an interval (*sonic log*) or relative amplitude (*variable density log; cement-bond log*).

acoustic positioning (a) Determining location using sonar waves, as by Dopplar-sonar or locating with respect to fixed sonar transponders. (b) Determining the distance between elements of a towed array by measuring traveltimes bewteen the elements. (c) Determining the positions of ocean-bottom seismometers by timing arrivals from first arrivals in production surveying or from specific positioning shots.

acoustics (a-cous'-tics) The study of sound, including its production, transmission, reception, and utilization, especially in fluid media such as air or water. With reference to Earth sciences, it is especially relevant to oceanography. The term includes seismic P-waves.

acoustic tomography The technique for measuring properties, such as temperature and currents, within an oceanic area hundreds to thousands of kilometers on a side, using measurements of the time it takes for sound to travel along many different rays through the area. The sound is transmitted in the *sound channel* using transmitters and receivers at many stations surrounding the area. Each pair of stations measures travel time of many different rays that sample water at different depths along a line between the stations, from the surface to the bottom, centered on the *sound channel.*

acoustic transparency A medium with constant acoustic impedance so that there are no reflections from within it.

acoustic wave A P-wave. It is often restricted to fluids such as air or water, but it also includes P-waves in the solid Earth. It excludes shear motions. Cf: *elastic wave.* Syn: *sound wave; sonic wave.*

acquired character (ac-quired') A character not inherited but acquired by an individual organism during its lifetime as a result of use or disuse according to its mode of life or the conditions under which it lived.

acquisition footprint A pattern in data, caused by the acquisition method or changes in the methods, that generally does not have geologic significance.

acre A unit of area in the English system of measure, defined as 10 square chains (1 chain is 4 rods or 66 feet). It is exactly equal to 43,560 ft or 4,840 sq yd and is approximately equal to 4046.8564224 sq m. There are 640 acres in one square mile. Although acre may be defined otherwise in some localities, it is a local definition and the word is usually qualified as builder's acre or Block Island acre. The English system of measures contains no legally defined unit of area. The size of the acre therefore depends on the length of the English yard, which is legally defined. By an ordinance of Edward I in 1303, the acre was defined as the area contained in a rectangle 40 rods long and 4 rods wide. With the rod defined as 5.5 ulnae (yards), as defined by Edward I's iron standard for the ulna, the acre is 4840 square yards.

acre-foot A measure of the volume of water in an area of one acre (43,560 square feet) with a height or depth of one foot. It is the unit commonly used in measuring volumes of water or reservoir storage space, and in measuring the volume of reservoir rock in an oil or gas field. See also: *acre-inch.*

acre-inch The volume of water required to cover and area of one acre to a depth of one inch. See also: *acre-foot.*

acrepid (a-cre'-pid) Said of a *desma* (of a sponge) that lacks an axial canal, implying that it was not formed about a *crepis.*

acre-yield The average amount of oil, gas, or water recovered from 1 acre of a reservoir.

acritarch (ac'-ri-tarch) A unicellular, or apparently unicellular, resistant-walled microscopic organic body of unknown or uncertain biologic relationship and characterized by varied sculpture, some being spiny and others smooth. Many if not most acritarchs are of algal affinity, but the group is artificial. They range from Precambrian to Holocene, but are esp. abundant in Precambrian and early Paleozoic. The term was proposed by Evitt (1963, p.300-301) as "an informal, utilitarian, 'catch-all' category without status as a class, order, or other suprageneric unit" consisting of "small microfossils of unknown and probably varied biological affinities consisting of a central cavity enclosed by a wall of single or multiple layers and of chiefly organic composition". See also: *hystrichosphaerid; dinoflagellate.*

acrodont (ac'-ro-dont) adj. Pertaining to vertebrate teeth fused to the occlusal margins of upper and lower jaws.

acrolamella (ac'-ro-la-mel'-la) A leaflike extension of the laesurae of megaspores. Cf: *gula; apical prominence.* Pl: acrolamellae.

acrolobe (ac'-ro-lobe) The central portion of the *cephalon* or *pygidium* of agnostid trilobites in which the axial furrows are effaced.

acron The anteriormost part of the *cephalon* of a crustacean, carrying the eyes and antennules.

acropetal (ac'-ro-pet-al) In palynomorphs, having *conidia* produced in succession toward the *apex.*

acrotelm The surface, active, and more water-permeable upper layer of peat in a peat deposit or mire (Ingram, 1978). Cf: *catotelm.*

acrotretacean (ac'-ro-tre-ta'-ce-an) Any inarticulate brachiopod belonging to the superfamily Acrotretacea, characterized by a conical to subconical, rarely convex, pedicle valve.

acrozone (ac'-ro-zone) *range zone.*

actinal (ac'-ti-nal) A syn. of *oral*. Ant: *abactinal.*

actine (ac'-tine) (a) One of the individual branches of the triaene or *triode* in the *ebridian* skeleton. (b) A star-shaped *spicule,* as of a sponge.

actinium series (ac-tin'-i-um) The radioactive series beginning with uranium-235.

actinodont (ac-tin'-o-dont) Said of the dentition of certain bivalve mollusks of early origin having hinge teeth radiating from the *beak* (the outer teeth being more or less elongate).

actinolite (ac-tin'-o-lite) A bright-green or grayish-green monoclinic mineral of the *amphibole* group: $Ca_2(Mg,Fe)_5(OH)_2[Si_8O_{22}]$. It may contain manganese. It sometimes occurs in the form of *asbestos,* and also in fibrous, radiated, or columnar forms in metamorphic rocks (such as schists) and in altered igneous rocks. Cf: *tremolite.* Syn: *kidney stone [mineral].*

actinomorphic (ac'-ti-no-mor'-phic) Said of an organism or organ that is radially symmetrical or capable of division into essentially symmetrical halves by any longitudinal plane passing through the axis. Cf: *zygomorphic.*

actinopod (ac-tin'-o-pod) Any protozoan belonging to the class Actinopoda and characterized by protoplasmic extensions radiating from the spheroidal main body. Cf: *rhizopod.*

Actinopterygii (Ac'-ti-nop'-te-ryg'-i-i) A subclass of the Osteichthyes characterized by movable paired fins supported by bony rays; ray-finned fish. Range, Lower Devonian to present.

actinosiphonate (ac'-ti-no-si'-phon-ate) Said of endosiphuncular structures of a nautiloid, consisting of radially arranged longitudinal lamellae.

actinostele (ac-tin'-o-stele) A type of *stele* consisting of alternating or radial groups of xylem and phloem within a pericycle and having a star shape in cross section.

activation [clay] (ac-ti-va'-tion) The act or process of treating clay (such as bentonite) with acid so as to improve its adsorptive properties or to enhance its bleaching action, as for use in removing colors from oils.

activation [radioactivity] The process of making a substance radioactive by bombarding it with nuclear particles. The radioactivity so produced is called *induced radioactivity.*

activation analysis A method of identifying stable isotopes of elements in a sample by irradiating the sample with neutrons, charged particles, or gamma rays to render the elements radioactive, after which the elements are identified by their characteristic radiations. Cf: *neutron activation.* Syn: *radioactivation analysis.*

activation energy The extra amount of energy which any particle or group of particles must have in order to go from one energy state into another, such as changes in phase and movement of particles in diffusion. The greater the amount of energy involved, the higher the resistance to the change, or the *potential barrier.*

active (ac′-tive) Said of a *karst* feature that contains moving water, or that is still being developed by the action of water.

active cave *live cave.*

active channel (a) A channel presently containing flow on a plain or valley bottom with multiple generations of channels. (b) The portion of a channel in which flow is present at the time of measurement, as opposed to the high-flow channel.

active diapirism Postdepositional diapir growth through prekinematic overburden. As the diapir increases in relief by growing upward, its base remains at a constant depth below the sedimentary surface, while its crest rises toward the sedimentary surface. Active diapirism involves forceful intrusion through a roof that is pushed upward and sideways. That is only possible in relatively tall diapirs overlain by relatively thin overburden, unless the overburden is being extended or is unusually weak or fluid (Jackson et al., 1988; Nelson, 1989). Syn: *upbuilding (b); upthrusting.* Cf: *downbuilding.*

active earth pressure The minimum value of lateral *earth pressure* exerted by soil on a structure, occurring when the soil is allowed to yield sufficiently to cause its internal shearing resistance along a potential failure surface to be completely mobilized. Cf: *passive earth pressure.*

active fault A fault on which slip has occurred recently and is likely to occur in the future. Active faults are the locus of contemporary elastic strain accumulation, seismicity, or fault creep.

active glacier (a) A glacier that has an *accumulation area,* and in which the ice is flowing. Ant: *dead glacier.* (b) A glacier that moves at a comparatively rapid rate, generally in a maritime environment at a low latitude where accumulation and ablation are both large. Ant: *passive glacier.*

active ice A glacier or part of a glacier that is moving at a geomorphically significant rate, e.g., meters per year.

active layer [eng] The surficial deposit that undergoes seasonal changes of volume, swelling when frozen or wet and shrinking when thawing and drying.

active layer [permafrost] A surface layer of ground, above the permafrost, that is alternately frozen each winter and thawed each summer; it represents *seasonally frozen ground* on permafrost. Its thickness ranges from several centimeters to a few meters. Syn: *annually thawed layer; mollisol.*

active margin A continental margin at which there is significant seismicity and deformation. An active margin occurs where the continental margin is a plate boundary. Cf: *passive margin.* Partial syn: *convergent plate boundary.*

active method (a) A seismic method that involves monitoring artificially induced signals. (b) A construction method in permafrost areas by which the frozen ground is thawed and removed or kept unfrozen at and near the structure. Cf: *passive method.*

active patterned ground Patterned ground that is still developing. Ant: *fossil patterned ground.*

active permafrost Permafrost that is able to revert to a perennially frozen state under present climatic conditions after having been thawed by natural or artificial means. Ant: *fossil permafrost.*

active remote sensing Remote-sensing methods that provide their own source of electromagnetic radiation. *Radar* is an example. Cf: *passive remote sensing.*

active speleothem *live speleothem.*

active volcano A volcano that is erupting; also, one that has erupted in historic time. There is no precise distinction between an active and a *dormant volcano.* Cf: *extinct volcano; inactive volcano.*

active water Water with corrosive capabilities.

activity [chem] (ac-tiv′-i-ty) (a) The ratio between the *fugacity* of a substance in some given state and its standard-state fugacity, f^o. The standard-state fugacity is unity for a gas or vapor; for solids and liquids, it is evaluated at each temperature as the fugacity of the pure substance at 1 atmosphere of pressure. Activity arises in consideration of chemical equilibrium problems involving solids or liquids. (b) The tendency of a substance to react spontaneously and energetically with other substances.

activity [radioactivity] The rate of decay of radionuclides measured as the number of disintegrations per unit time. The SI unit is the reciprocal second (s^{-1}), which is defined as the *bequerel* (Bq). See also: *specific activity.*

activity coefficient The ratio of chemical activity to concentration.

activity index *budget gradient.*

activity ratio In a sediment, the ratio of the *plasticity index* to the percentage of clay-sized minerals.

actual age (ac′-tu-al) The real age of a geologic feature or event as opposed to an age measured as an isotopic system that has been disturbed.

actual horizon A great circle on the celestial sphere whose plane is perpendicular to the direction of the plumb line (or the direction of gravity) at the point of observation. It is usually referred to as the horizon. Syn: *rational horizon.*

actualism (ac′-tu-al-ism′) The concept that the same forces presently in evidence acting over time are sufficient to explain the causes of all geologic records, relationships, and configurations. Possibly first invoked by J.G. Lehmann (1719-1767), but used especially in nineteenth-century France as "actualisme", appealing only to "causes actuelles". Leading early exponents in 1825 were Constant Prévost in France and George Poulett Scrope in England. Although commonly equated with Charles Lyell's *uniformitarianism,* continental Europeans (and Scrope) did not constrain the intensity of past processes as strictly as did Lyell (Hooykaas, 1963).

actualistic paleontology (ac′-tu-al-ist-ic pa′-le-on-tol′-o-gy) The paleontologic study of a present-day area, with the intent of discovering how it will appear later, after eventual burial and fossilization.

actun A Mayan word for *cave [speleo];* used in Mexico and Central America.

actuopalynology (ac′-tu-o-pal-y-nol′-o-gy) The study of extant palynomorphs and their distribution in atmosphere, hydrosphere, and lithosphere. Cf: *pollen analysis.*

acuminite (ac′-u-min-ite) A colorless monoclinic mineral: $SrAlF_4(OH){\cdot}H_2O$.

acute bisectrix (a-cute′) In optically biaxial minerals, the direction bisecting the acute angle between optic axes. Cf: *obtuse bisectrix.*

acyclic (a-cy′-clic) In plant morphology, pertaining to attachment of three or more parts, e.g. leaves, in a spiral or helical pattern. Cf: *whorled.*

A/D Analog to digital, as in conversion of a continuous flow of signal into one that is sampled periodically, at intervals and with a degree of precision determined by electronic circuits and possibly computer software.

adakite Andesitic to dacitic magma erupted or intruded in convergent margin settings, characterized by very fractionated REE (high La/Yb), high Sr content, and no Eu anomaly. Characteristics are consistent with partial melting of a mafic source at a high enough pressure (>about 50 km) that considerable garnet is stable in the residue (restite). Adakites appear to have diverse origins (e.g., Kay and Kay, 2002): (1) Melting of young (hot) subducted oceanic crust with MORB-like isotopic characteristics, at about 100 km depth (Defant and Drummond, 1990), with reaction with the mantle imparting the high Mg/Fe ratio and Ni and Cr content. These characteristics are those of first documented occurrence, on Adak Island, Aleutian Islands (Kay, 1978). This is the occurrence for which the rock type was named by Defant and Drummond (1990). (2) Melting of mafic, basal, thick continental crust of subduction-eroded crust. Cf: *boninite; sanukite.*

adamantine luster (ad-a-man′-tine) A brilliant mineral luster, characteristic of minerals with a high *index of refraction,* e.g. diamond and cerussite.

adamantine spar Silky brown *corundum.*

adambulacral (ad′-am-bu-lac′-ral) Said of the plate adjacent to the *ambulacral* plate in the arm of an asteroid.

adamellite (ad-a-mel′-lite) In English-speaking countries and the former USSR, a syn. of *quartz monzonite* of U.S. usage (i.e. quartz as 10-50% of the felsic minerals, and a plagioclase/total feldspar ratio of 35-65%). The term was first used in this sense by Brögger in 1895; it was originally used, however, by Cathrein in 1890 for an occurrence at Monte Adamello, Italy.

adamite (ad′-am-ite) A colorless, white, greenish or yellow orthorhombic mineral: $Zn_2(AsO_4)(OH)$. It is dimorphous with paradamite. Originally spelled: adamine.

adamsite-(Y) A colorless or white triclinic mineral: $NaY(CO){\cdot}6H_2O$.

adapertural (ad-ap-er′-tur-al) Toward the *aperture* of a gastropod or cephalopod shell. Ant: *abapertural.* Syn: *adoral.*

adapical (ad-ap′-i-cal) (a) Toward the *apex* of a gastropod or cephalopod shell. (b) Toward the *apical system* of an echinoid.

adaptation (ad-ap-ta′-tion) Modification, as the result of natural selection, of an organism or of its parts so that it becomes better fitted to exist under the conditions of its environment.

adaptive grid (a-dap′-tive) A changing series of ecologic zones having time as one dimension.

adaptive norm That part of an organic population that can survive and reproduce in the environment usually occupied by the species; the remainder may carry hereditary defects and diseases.

adaptive processing Data processing where the parameters are varied with arrival time or location as data statistics change.

adaptive radiation Increase in the number of taxa in diversified groups within a short geologic-time interval (or more or less simultaneously) as a result of evolution; often shown by the occupation of equivalent niches in comparable habitats by ecologically similar but taxonomically distinct organisms. Syn: *radiation [evol].* Cf: *convergence [evol]; divergence [evol].* See also: *evolutionary radiation.*

adaptive zone A unit of environment defined in terms of its occupation by a single kind of organism.

adaxial (ad-ax′-i-al) Facing, directed toward, or situated on the same side as, the axis or center of the axis, as of an organ, plant, or invertebrate; ventral or posterior. Also, said of the adaxial side. Ant: *abaxial.*

adcarinal groove More or less elongate depression or trough on either side of fixed blade or carina of some planate and scaphate pectiniform conodont elements (TIP, 1981, pt. W, supp.2).

ADCP *Acoustic Doppler Current Profiler.*

adcumulate (ad-cu'-mu-late) A *cumulate* formed by *adcumulus growth,* with intercumulus material comprising less than five percent of the rock. Cf: *mesocumulate.*

adcumulate texture Igneous texture consisting of early-formed accumulated crystals, commonly plagioclase, enlarged by overgrowths of the same mineral (and same composition) such that interstices are almost totally filled. Cf: *mesocumulate texture; orthocumulate texture.*

adcumulus growth (ad-cu'-mu-lus) Continued growth of cumulus crystals from liquid of the same composition so that the crystals are unzoned. This process reduces the intercumulus liquid by forcing it out of the intercumulus. See also: *adcumulate.*

addition solid solution (ad-di'-tion) The addition of small atoms or ions, at random, in some interstices between atoms of a crystal structure. It may result in an *interstitial defect.* Syn: *interstitial solid solution.*

additive color process In cartography, the mixing of three additive primary colors (red, green, blue) in various proportions to create an arbitrary color (Krzanowski et al., 1993).

additive primary colors The colors blue, green, and red. Filters of these colors transmit the primary color of the filter and absorb the other two. See also: *primary colors; subtractive primary colors.*

adductor (ad-duc'-tor) *adductor muscle.*

adductor muscle (a) A muscle, or one of a pair of muscles, that contracts and thereby closes and/or tends to draw or hold together the valves of a bivalved shell (as in ostracodes, brachiopods, and pelecypods). Two adductor muscles, each dividing dorsally, are commonly present in rhynchonelliform (formerly articulate) brachiopods; two pairs of adductor muscles (anterior and posterior), passing almost directly from the dorsal to the ventral side between the valves, are commonly present in inarticulate brachiopods. Two adductor muscles, each passing from left to right, one anterior and one posterior, or a single posterior adductor muscle, present in bivalved molluscs. (b) Any transverse muscle (esp. that of the maxillary segment) for closure of the aperture in a cirripede crustacean. (c) In vertebrates, a muscle that moves an appendage or other body part toward a point of reference, usually one on the midsagittal plane. (d) Among gnathostomes, a group of muscles that brings the jaws together. Cf: *diductor muscle.* Syn: *adductor.*

adductor-muscle scar A *muscle scar* showing the final site where an adductor muscle was attached. In rhynchenelliform (formerly articulate) brachiopods, a single pair of such scars is located between diductor impressions in the *pedicle valve* and two pairs (anterior and posterior) in the *brachial valve* (TIP, 1965, pt. H, p.139). In bivalved molluscs, an anterior and posterior adductor-muscle scar, or just a posterior adductor-muscle scar, is present. See also: *cicatrix.* Syn: adductor scar.

adductor pit A depression that develops on the interior of a *scutum* for attachment of an *adductor muscle* of a cirripede crustacean.

adductor ridge A linear elevation that develops (in association with an *adductor pit*) on the interior of a *scutum* of certain cirripede crustaceans (such as Balanomorpha).

adductor testis attachment The place of attachment of the adductor testis muscle, near the lower end of the *postcervical groove* in some decapod crustaceans; the position of this point is not clearly marked on the *carapace* (Holthuis, 1974, p.735).

adelite (a-del'-ite) A colorless to yellow orthorhombic mineral: $CaMg(AsO_4)(OH)$. It sometimes has appreciable fluorine.

adeoniform (ad-e-on'-i-form) Said of a rigid bilamellar lobate erect colony in cheilostome bryozoans, firmly attached by a calcified base.

ader wax *ozocerite.*

adfreezing (ad-freez'-ing) The process by which two objects adhere to one another due to the binding action of ice; applied in permafrost studies.

adherent (ad-her'-ent) *adnate.*

adhesion (ad-he'-sion) The molecular attraction between contiguous surfaces. Cf: *cohesion.*

adhesion ripple One of a series of irregularly parallel crests of sand, formed transverse to a wind blowing dry sand over a smooth moist surface. In cross-section the crests are asymmetrical, with the stoss side steeper than the lee side (i.e., the opposite of normal current ripples). The length (chord) of the ripple is generally less than 2 cm and the height less than 2 mm. Originally described by Van Straaten (1953a). Syn: *antiripplet; antiripple.*

adhesion wart A small, irregular wartlike sand accumulation made by wind that changes direction rapidly while blowing over a moist surface of sand (Glenn, 1979, p.399).

adhesive water (ad-he'-sive) *pellicular water.*

adiabat A temperature-pressure (or temperature-depth) curve where heat is neither gained nor lost.

adiabatic (ad'-i-a-bat'-ic) In thermodynamics, pertaining to the relationship of pressure and volume when a gas or fluid is compressed or expanded without either giving or receiving heat. In an adiabatic process, compression causes a rise in temperature, and expansion a drop in temperature (U.S. Naval Oceanographic Office, 1966). Within the *atmosphere,* expansional cooling and compressional warming of unsaturated air parcels are essentially adiabatic processes, i.e., no heat is exchanged between the air parcels and the surrounding (ambient) air. See also: *potential density; potential temperature [oceanog].*

adiabatic decompression Depressurization of a mineral-bearing fluid without heat loss, generally causing mineral precipitation.

adiagnostic (ad'-i-ag-nos'-tic) Nonrecommended synonym of *cryptocrystalline.* The term was originally used by Zirkel in German as "adiagnostisch". Obsolete. Ant: *eudiagnostic.*

adinole (ad'-i-nole) An argillaceous sediment that has undergone albitization as a result of contact metamorphism along the margins of a sodium-rich mafic intrusion. Rarely used. Cf: *desmosite; schalstein; spilosite; spotted slate.*

adipocere (ad'-i-po-cere') A waxy or unctuous, brownish or light-colored natural substance consisting mainly of free fatty acids, hydroxy acids, and their calcium and magnesium salts, formed on decay of human, animal, or fish remains in damp areas or in fresh or salt water. Its consistency may range from soft and pliable for recent material to hard and brittle for older material. It often replaces and takes the form of the decomposing body.

adipocerite (ad'-i-po-cer'-ite) *hatchettine.*

adipocire (ad'-i-po-cire') A syn. of *hatchettine.* Not to be confused with *adipocere.*

***a* direction** *a axis [struc petrol].*

Adirondack iron ore A term applied collectively to all types of magnetite ores found in the Adirondack region of New York and in adjacent states. These include titaniferous magnetite deposits in mafic rocks as well as low-titanium magnetite deposits in metamorphosed sediments.

adit A horizontal passage from the surface into a mine. Cf: *tunnel [mining].*

adjacency analysis Analysis performed to determine if a set of areas has a common boundary (Krzanowski et al., 1993).

adjusted position An adjusted value to the coordinates giving the location of a point on the Earth. In the adjustment of a horizontal control survey, discrepancies arising from errors in the observational data are removed to give adjusted coordinates in an adopted coordinate system (geodetic datum or plane coordinate system). The coordinates obtained by the adjustment are called adjusted positions; when used as control for other work they are referred to as fixed positions. In the adjustment of data from a vertical control survey, the values obtained are called adjusted elevations. When used to control other surveys, they are called fixed elevations.

adjusted stream (ad-just'-ed) A stream that carves its valley parallel to the strike of the least resistant rocks over which it flows.

adjustment (ad-just'-ment) The determination and application of corrections to a series of survey observations for the purpose of reducing errors, removing internal inconsistencies, and coordinating and correlating the derived results within the survey itself or with previously existing basic data; e.g. the determination and application of orthometric corrections in order to make the elevations of all bench marks consistent and independent of the circuit closures, or the positioning of public-land lines on a topographic map to indicate their true, theoretical, or approximate location relative to the adjacent terrain and culture. The term may refer either to mathematical procedures or to corrections applied to instruments used in making observations.

adjustment of cross section The tendency in glaciers and rivers to change the size of every cross section of the channel to accomodate the volume of ice or water that must pass through it.

adjustor muscle (ad-jus'-tor) One of a pair of two pairs of muscles in many rhynchonelliform (formerly articulate) brachiopods, branching from the *pedicle,* and responsible for moving the position of the shell on the pedicle. A ventral pair is attached posteriorly and laterally from the diductor muscles, and a dorsal pair is on hinge plates or floor of *brachial valve* behind posterior adductor muscles (TIP, 1965, pt. H, p.139).

admissable cross section A cross section that depicts structures whose style conforms with structures seen in outcrops, cliff faces, or seismic sections.

admission (ad-mis'-sion) *admittance [chem].*

admittance [chem] (ad-mit'-tance) In a crystal structure, substitution of a trace element for a major element of higher valence, e.g. Li^+ for Mg^{++}. Admitted trace elements generally have a lower concentration relative to the major element in the mineral than in the fluid from which the mineral crystallized. Cf: *capture [chem]; camouflage.* Syn: *admission.*

admittance [elect] The reciprocal of *impedance [elect],* or the ratio of complex current to voltage in a linear circuit.

admixture (a) A term applied by Udden (1914) to one of the lesser or subordinate of several particle-size grades of a sediment. See also: *coarse admixture; fine admixture; distant admixture; proximate admixture.* (b) A material that is added to another to produce a desired modification; e.g. a substance (other than aggregate, cement, or water) added during the mixing of concrete.

admontite (ad'-mont-ite) A colorless monoclinic mineral: $MgB_6O_{10}{\cdot}7H_2O$.

adnate (ad'-nate) In plant morphology, having unlike parts fused along their entire length, e.g., stamens fused to petals in a flower. Cf: *coherent [bot]; connate [bot].* Syn: *adherent.*

adobe (a-do'-be) (a) A fine-grained, usually calcareous, hard-baked clayey deposit mixed with silt, usually forming as sheets in the central or lower parts of desert basins, as in the playas of SW U.S. and in the arid parts of Mexico and Spanish America. It is probably a windblown deposit, although it is often reworked and redeposited by running water. The term was applied originally to a sun-dried brick made of adobe mixed with straw, and later to the clayey material constituting the brick. (b) A heavy-textured clay soil derived from adobe deposits. Etymol: Spanish. Cf: *loess.*

adobe flat A generally narrow plain formed by sheetflood deposition of fine sandy clay or adobe brought down by an ephemeral stream, and having a smooth, hard surface (when dry) usually unmarked by stream channels.

adolescence (ad-o-les'-cence) A hypothetical stage following youth and preceding maturity in a developmental sequence such as the cycle of erosion. It is sometimes considered "early maturity". Adolescence is characterized by incipient development of mature features, together with residual features of youth. It may be applied wherever the terms youth and maturity are appropriate, e.g. in the regional erosion cycle, the karst cycle, or development of a valley. Syn: *topographic adolescence.*

adolescent (ad-o-les'-cent) Pertaining to the stage of *adolescence* of the cycle of erosion; esp. said of a valley having a smoothly graded stream channel that may reach base level at its mouth, or of a coast marked by low but nearly continuous sea cliffs.

adont Said of a class of ostracode hinges having no teeth, with a ridge or bar in one valve fitting into a groove in the opposed valve.

adoral (ad-o'-ral) Located or directed toward or near the mouth of an invertebrate; e.g. an "adoral surface" of an echinoderm *theca* bearing the mouth or ambulacral grooves, or the "adoral direction" toward the *aperture* of an ammonoid shell, or an "adoral fiber" of cytoplasm tissue leading from neuromotorium to the edge of peristome in a ciliate protozoan. Cf: *aboral.*

adradial (ad-ra'-di-al) adj. (a) Said of the position corresponding to the boundary between *ambulacral* and *interambulacral* areas of an echinoid or edrioasteroid. (b) Directed toward the axis of the ray of an echinoderm. (c) Pertaining to a radius of the third order in a coelenterate. n. (a) One of the small plates lining ambulacra in certain edrioasteroids. (b) One of a series of ossicles on the *aboral* surface of an asterozoan ray.

adradial suture The zone of contact between the oral-ambulacral and interambulacral plates in echinoids and edrioasteroids (Bell, 1976).

adret A mountain slope so oriented as to receive the maximum available amount of light and warmth from the Sun; esp. a southward-facing slope of the Alps. Etymol: French dialect, "good side". Cf: *ubac.*

ADS *adularia-sericite epithermal.*

adsorbed water (ad-sorbed') Water adhering in ionic or molecular layers to the surfaces of soil or mineral particles. See also: *adsorption.* Cf: *absorbed water.*

adsorption (ad-sorp'-tion) Adherence of gas molecules, or of ions or molecules in solution, to the surface of solids with which they are in contact. Cf: *absorption.*

adularescence (ad'-u-la-res'-cence) A floating, billowy, white or bluish light, seen in certain directions as a gemstone (usually adularia) is turned; it is caused by diffused reflection of light from parallel intergrowths of another feldspar of slightly different refractive index from the main mass of adularia. It is often called *schiller.*

adularia (ad-u-lar'-i-a) A moderate to low-temperature mineral of the alkali feldspar group: $K[AlSi_3O_8]$. It is weakly triclinic (formerly regarded as apparently monoclinic) and typically occurs in well-developed, usually transparent, and colorless to milky-white (and often opalescent) pseudo-orthorhombic crystals in fissures in crystalline schists, esp. in the region of the Swiss Alps. Adularia displays pearly internal reflections and a fascinating variety of optical behavior between crossed nicols. It typically has a relatively high content of barium.

adularia-sericite epithermal *adularia-sericite type* deposit. Abbrev: ADS.

adularia-sericite type A type of *epithermal* gold mineralization characterized by vein adularia and sericite, an absence of hypogene alunite, presence of chlorite, and the absence of enargite + pyrite, e.g. Creede, Colorado and Tonopah and Round Mountain, Nevada.

advance [coast] (ad-vance') A continuing seaward movement of a shoreline as a result of progradation or emergence. Also, a net seaward movement of the shoreline during a specified period of time. Ant: *recession [coast].* Syn: *progression.*

advance [glaciol] (a) The forward movement of the terminus of a glacier, generally resulting from a period of positive *net balance,* or an excess of accumulation over ablation. Its rate is usually measured in meters per year. (b) A time interval marked by an advance or general expansion of a glacier. Cf: *readvance.* Syn: *glacial advance; glacier advance.*

advance-cut meander A meander in which the outer bank of the channel is eroded so rapidly that deposition on the inner bank fails to keep pace, thereby widening the channel (Melton, 1936, p.598-599). Cf: *forced-cut meander.* Syn: *induced meander.*

advanced argillic alteration Feldspar-destructive hydrothermal alteration in which kaolinite-quartz dominate at low temperatures (below 300°C) and pyrophyllite/andalusite-quartz dominate at high temperatures (above 300°C). Implies very low pH and low alkali contents of the hydrothermal fluids.

Advanced Very High Resolution Radiometer An instrument package flown on a series of polar orbit satellites used by the National Oceanic and Atmospheric Administration for weather observations. The sensors cover the visible, near- and mid-infrared, and far-infrared (thermal) bands. The sensor scans a swath 2,400 km wide with a ground resolution of 1.1 km per pixel. Data from these sensors can be used for analysis of land cover, vegetation, or mapping. In cloud-free areas, it measures sea-surface temperature globally with an accuracy of around 1.5 degrees Celsius, over an area four kilometers on a side. This is the most common source of sea-surface temperature data. Abbrev: AVHRR.

advanced wastewater treatment The final stage of wastewater treatment to remove remaining traces of such things as *BOD*, phosphate, odor, and taste (Studdard, 1973). Cf: *primary wastewater treatment; secondary wastewater treatment.*

advection [geotherm] *forced convection.*

advection [hydraul] The hydraulic mechanism by which dissolved constituents (i.e., solutes) and/or heat are transported in the direction of groundwater flow; the bulk mass of flowing fluid. Cf: *diffusion [grd wat].*

advection [meteorol] (ad-vec'-tion) Horizontal transport of air or of an atmospheric property within the Earth's atmosphere. Cf: *convection [meteorol].*

advection [oceanog] The horizontal or vertical flow of seawater as a current.

advection [ore dep] Unidirectional mass transport of solutes by bodily movement of hydrothermal fluids through permeable rocks, commonly across boundaries between volumes with contrasting chemical properties. It is contrasted with diffusion. Advection is measured in terms of fluid velocity and concentrations of solutes.

advection [tect] (a) Lateral mass movements of mantle material. (b) Transfer of heat by a heat-carrying material moving through a solid. For example, flow of magma or hot water through cracks in solid rock heats the rock by advection.

advection fog A type of *fog* that results when warm humid air moves across a cold or snow-covered surface. As the air loses heat to the ground, its temperature may drop below the dewpoint and the airborne water vapor condenses. Not to be confused with *steam fog.* Cf: *radiation fog.*

adventitious (ad-ven-ti'-tious) A plant part that arises from an unusual place, e.g. a root that arises from a leaf or stem rather than from a primary root.

adventitious avicularium An *avicularium* of a cheilostome bryozoan occupying a position on a wall of another *zooid.*

adventitious lobe A secondary lobe of an ammonoid *suture,* formed by subdivision of the first lateral saddle; also, all later lobes that develop from the first adventitious lobe. Syn: adventive lobe.

adventive (ad-ven'-tive) Obsolete syn. of *parasitic [volc].*

adventurine (ad-ven'-tu-rine) *aventurine.*

advolute (ad'-vo-lute) Said of a gastropod shell in which the whorls barely touch one another and are not distinctly embracing or overlapping, or said of a coiled cephalopod shell in which the outer whorl touches but does not cover any part of the adjacent inner whorls. Cf: *evolute; involute; convolute.*

adyr (a) A term used in Turkmenia for a part of a desert plain devoid of sands and having soft ground. Cf: *kyr.* (b) A term used in Kazakhstan for a small flat top of relict high ground or of a mesalike hill. (c) A term loosely applied in central Asia to a low mountain, a small hill, an eroded ridge with gentle slopes, etc. (Murzaevs and Murzeavs, 1959, p.20). Etymol: Turkish.

AE Abbrev. of *aeon,* referring to *eon (b),* or 1 billion years. Sometimes written as a digraph Æ.

Aegean (Ae-ge'-an) A European substage: lower substage of Ladinian Stage; Middle Triassic (above Spathian substage of Olenekian, below Bithynian).

aegirine (ae'-gir-ine) A brown or green monoclinic mineral of the clinopyroxene group: $NaFe^{3+}[Si_2O_6]$. It occurs in certain alkali-rich igneous rocks. Syn: *acmite.*

aegirine-augite A mineral intermediate between augite and aegirine. Syn: *acmite-augite.*

aegirite (ae'-gir-ite) *aegirine.*

Aeneolithic (Ae'-ne-o-lith'-ic) *Eneolithic.*

aenigmatite (ae-nig'-ma-tite) A vitreous black triclinic mineral: $Na_2Fe^{2+}_5TiSi_6O_{20}$. Also spelled: *enigmatite.*

aeolian (ae-o'-li-an) *eolian.*

aeolianite (ae-o'-li-an-ite) *eolianite.*

aeolotropy A syn. of *anisotropy.* Also spelled: eolotropy.

aeon *eon.*

aeration (aer-a'-tion) The process by which air in the soil is replaced by air from the atmosphere. In a well-aerated soil, the soil air is very similar in composition to the atmosphere above the soil. Poorly aerated soils usually contain a much higher content of CO_2 and a lower content of O_2 than the atmosphere above the soil. The rate of aeration depends largely on the volume and continuity of air-filled pores within the soil.

aeration porosity The volume of interstices that do not hold water at a specified low moisture tension (Jacks et al., 1960). Cf: *noncapillary porosity.*

aerial (aer'-i-al) Pertaining to the air; related to, located in, or consisting of, the Earth's atmosphere. Not to be confused with areal.

aerial arch An anticline, the crest of which has been eroded. Rarely used.

aerial magnetometer *airborne magnetometer.*

aerial mapping The taking of aerial photographs for making maps and for geologic interpretation.
aerial mosaic *mosaic [photo].*
aerial photograph Any photograph taken from the air, such as a photograph of a part of the Earth's surface taken by a camera mounted in an aircraft. Usually taken in strips of overlapping prints for mapping purposes. Syn: *air photograph.*
aerial survey (a) A survey using aerial photographs as part of the surveying operation. (b) The taking of aerial photographs for surveying purposes.
aërinite (aër'-in-ite) A blue monoclinic mineral: $(Ca,Na)_4Mg_3(Fe^{3+}Fe^{2+}Al)_3(Si,Al)_{18}O_{42}(OH)_6{\cdot}nH_2O$.
aerobic (aer-o'-bic) (a) Said of an organism (esp. a bacterium) that can live only in the presence of free oxygen; also, said of its activities. Noun: aerobe. (b) Said of conditions that can exist only in the presence of free oxygen. Cf: *anaerobic.*
aerobic decay Decomposition of organic substances, primarily by microorganisms, in the presence of free oxygen; the ultimate decay products are carbon dioxide and water.
aerobiology The study of organisms and parts of organisms in the atmosphere and their significance for human allergy. *Pollen* and *spores* are primary subjects for this study, which then is occasionally called aeropalynology.
aerogravity Measurements of the variations in the *Earth's gravity field* in a moving airplane or helicopter. It requires precise measurement of the aircraft's X, Y, Z position and velocity vector to calculate necessary corrections. Data are collected along a grid pattern of flight lines under stable air conditions. Free-air gravity is obtained after applying corrections for: (a) aircraft vertical and horizontal accelerations; (b) gravimeter platform velocity (*Eötvös correction*); (c) aircraft elevation above sea level (*free-air correction*); (d) variation of the Earth's radius and centrifugal force (*latitude correction*); (e) gyrostabilized platform off-leveling errors. Low-pass filtering is applied to the corrected data to remove residual high-frequency noise. Achievable accuracy is about 0.5-2.0 mGal at 2.0-10.0 km spatial resolution and is highly dependent on the speed of the aircraft.
aeroides (aer-oi'-des) Pale sky-blue aquamarine.
aerolite (aer'-o-lite) An obsolete syn. of *stony meteorite.*
aeromagnetics Measurements of the *Earth's magnetic field* in a moving aircraft (airplane or helicopter). Requires the precise measurement of the aircraft's X, Y, Z position during the flight. Data are collected along a grid pattern of flight lines under stable air and solar activity conditions. See also *high-resolution aeromagnetics*.
aeromagnetic survey A *magnetic survey* made with an *airborne magnetometer.*
Aeronian A stage of the Standard Global Chronostratigraphic Scale: middle stage of Llandoverian Series of Lower Silurian (above the Rhuddanian, below the Telychian). The basal GSSP is in Wales (Holland, 1985).
aerosiderolite (aer-o-sid'-er-o-lite') An obsolete syn. of *stony-iron meteorite.*
aerosol (aer'-o-sol) A *sol* in which the dispersion medium is a gas (usually air) and the dispersed or colloidal phase consists of solid particles or liquid droplets; e.g. mist, haze, most smoke, and some fog.
aerosol sinter Speleothemic material deposited from tiny (micrometer-size) particles in cave air.
aerospace (aer-o-space') A mnemonic term derived from aeronautics + space to denote both the Earth's atmosphere and the space beyond as a single unit.
aerugite (aer'-ug-ite) A blue-green rhombohedral mineral: $Ni_{17}As_6O_{32}$.
aeschynite (ae'-schyn-ite) A group name for orthorhombic minerals with the general chemical formula X $(Nb,Ta,Ti)_2(O,OH)_6$. Members include *aeschynite-(Ce), aeschynite-(Nd), aeschynite-(Y), niobo-aeschynite-(Ce), niobo-aeschynite-(Nd), tantalaeschynite-(Y)* and *rynersonite.* Also spelled: *eschynite.*
aeschynite-(Ce) (ae'-schyn-ite) An adamantine dark-brown orthorhombic mineral: $(Ce,Ca,Fe,Th)(Ti,Nb)_2(O,OH)_6$. A member of the *aeschynite* group.
aeschynite-(Nd) A metamict brown mineral: $(Nd,Ce)(Ti,Nb)_2(O,OH)_6$. A member of the *aeschynite* group.
aeschynite-(Y) (ae'-schyn-ite) A sub-metallic black orthorhombic mineral: $(Y,Ca,Fe,Th)(Ti,Nb)_2(O,OH)_6$. A member of the *aeschynite* group.
aethoballism (ae-tho-ball'-ism) A term proposed by Grabau (1904) for local metamorphism resulting from contact with a meteorite. The term is an obsolete syn. of *shock metamorphism.*
afanansyevaite A product of a burning coal heap: $Ca_8(Si_2O_7)_2{\cdot}Cl_2O$. Not approved by the International Mineralogical Association.
aff. Abbrev. of *affinity.* It implies less certain taxonomic similarity than does *cf.*
affine (af'-fine) adj. Said of a homogeneous deformation, i.e. one in which initially straight lines remain straight after deformation. Affine transformation is a mathematical transformation in which the coordinates of the deformed state are related to the coordinates of the undeformed state in a linear manner. Such a transformation describes a homogeneous deformation.
affinity (af-fin'-i-ty) In biology, the state of being akin to; used to indicate relationship without specifying identity. Abbrev: aff.
affluent (af'-flu-ent) adj. Said of a stream flowing toward or into a larger stream or into a lake. n. An affluent stream; esp. an *influent* flowing into a lake. The term, originally introduced by Jackson (1834, p.77-78) as distinct from a *confluent,* is obsolescent as the synonym *tributary* is more commonly used.
afflux (af'-flux) (a) The upstream rise of water level above the normal surface of the water in a channel, caused by contraction or obstruction of the normal waterway. (b) The difference between high flood levels upstream and downstream of a weir.
afghanite (af-ghan'-ite) A colorless to bluish hexagonal mineral: $(Na,Ca,K)_8(Si,Al)_{12}O_{24}(SO_4,Cl,CO_3,H_2O)_3$.
AFMAG method The use of natural audio-frequency electromagnetic noise to study lateral changes in Earth resistivity. Etymol: An acronym for *a*udio-*f*requency *mag*netic.
AFM diagram A triangular diagram showing the simplified compositional character of a metamorphosed pelitic rock by plotting molecular quantities of the three components: $A=Al_2O_3$; F=FeO; and M=MgO. Cf: *ACF diagram; A'KF diagram; AFM projection.*
AFM projection A triangular diagram showing molecular proportions, constructed by projecting the compositions of metamorphic pelitic minerals in the tetrahedron Al_2O_3 (=A), FeO(=F), MgO(=M), K_2O from the composition of muscovite or potassium feldspar onto the plane AFM. Devised by J. B. Thompson, Jr. (1957) and now widely used in the study of pelitic schists and gneisses. Syn: *Thompson diagram.*
A-form The megalospheric form of a foraminifer. Cf: *B-form.*
afterdamp (af'-ter-damp) The gas remaining in a coal mine after an explosion of *firedamp* or after a fire. It includes carbon monoxide and carbon dioxide. Cf: *whitedamp; blackdamp.*
aftershock (af'-ter-shock) An earthquake that follows a larger earthquake or *main shock* and originates at or near the focus of the larger earthquake. Generally, major earthquakes are followed by many aftershocks, which decrease in frequency and magnitude with time. Such a series of aftershocks may last many days for small earthquakes or many months for large ones. Cf: *foreshock.*
afterworking (af'-ter-work'-ing) *creep recovery.*
Aftonian (Af-ton'-i-an) Pertaining to the classical first interglacial stage of the Pleistocene Epoch in North America, following the Nebraskan and preceding the Kansan glacial stages. Etymol: Afton, town in Iowa. See also: *Günz-Mindel.* Obsolete.
afwillite (af-wil'-lite) A colorless monoclinic mineral: $Ca_3(SiO_3OH)_2{\cdot}2H_2O$.
agalite (ag'-a-lite) A fine fibrous variety of talc, pseudomorphous after enstatite. Syn: *asbestine.*
agalmatolite (ag-al-mat'-o-lite) A soft, waxy stone, such as pinite, pyrophyllite, or steatite, of a gray, green, yellow, or brown shade, used by the Chinese to simulate jade for carving small images, miniature pagodas, and similar objects. Syn: *pagodite; lardite; lard stone; pagoda stone.*
agardite-(Ce) A yellowish green hexagonal mineral: $CeCu_6(OH)_6(AsO_4)_3{\cdot}3H_2O$. A member of the *mixite* group.
agardite-(La) A colorless, green, or yellowish-green hexagonal mineral: $LaCu_6(OH)_6(AsO_4)_3{\cdot}3H_2O$. A member of the *mixite* group.
agardite-(Nd) A green hexagonal mineral: $NdCu_6(OH)_6(AsO_4)_3{\cdot}3H_2O$. A member of the *mixite* group. Not approved by the International Mineralogical Association.
agardite-(Y) (ag'-ard-ite) A blue-green hexagonal mineral: $YCu_6(OH)_6(AsO_4)_3{\cdot}3H_2O$. A member of the *mixite* group.
agate (ag'-ate) A translucent cryptocrystalline variety of quartz, being a variegated *chalcedony* frequently mixed or alternating with opal, and characterized by colors arranged in alternating stripes or bands, in irregular clouds, or in mosslike forms. Agate is found in virtually all colors, usually of low intensity; it commonly occupies vugs in volcanic rocks and cavities in some other rocks. Cf: *onyx.* See also: *banded agate; clouded agate; moss agate.*
agate jasper An impure variety of agate consisting of jasper with veins of chalcedony. Syn: *jaspagate.*
agate opal Opalized agate. Cf: *opal-agate.*
agathocopalite (ag'-a-tho-co'-pal-ite) *kauri.*
agatized wood (ag'-at-ized) *silicified wood.*
AGC Automatic gain control; see *gain control.*
age [geochron] (a) The formal *geochronologic unit* of lowest rank, below *epoch [geochron],* during which the rocks of the corresponding stage or substage (chronostratigraphic units) were formed. See also: *subage.* (b) A term used informally to designate a length of geologic time during which the rocks of any stratigraphic unit were formed. (c) A division of time of unspecified duration in the history of the Earth, characterized by a dominant or important type of life form; e.g. the "age of mammals". (d) The time during which a particular geologic event or series of events occurred or was marked by special physical conditions; e.g. the "Ice Age". (e) The position of anything in the geologic time scale; e.g. "the rocks of Miocene age". It is often expressed in years. Cf: *geologic age.*
age [ice] The stage of development of sea ice; the term usually refers to the length of time since the formation of the ice and to its thickness.
age determination The evaluation of geologic age by faunal or stratigraphic means, or by physical methods involving determination of the relative abundance of a radioactive parent and radiogenic daughter isotope. Some

methods of relative age determination are based on the extent of chemical change, like the hydration of obsidian or the fluorine uptake of bone.
age equation *general age equation.*
Agenian (A-gen'-ian) European mammal age: late Oligocene to earliest Miocene (before Orleanian).
age of amphibians An informal designation of the late *Paleozoic* Era, i.e., the *Carboniferous* and the *Permian.*
age of coal An informal designation of the *Carboniferous.* Syn: *coal age.*
age of cycads An informal designation of the *Jurassic.*
age of ferns An informal designation of the *Pennsylvanian.*
age of fishes An informal designation of the *Silurian* and the *Devonian.*
age of gymnosperms An informal designation of the *Mesozoic* Era.
age of mammals An informal designation of the *Cenozoic* Era.
age of man An informal designation of the *Quaternary.*
age of marine invertebrates An informal designation of the *Cambrian* and the *Ordovician.*
age of reptiles An informal designation of the *Mesozoic* Era.
age of the Earth The length of time that the Earth has existed essentially as it is now. High-precision Pb data give an age of 4.55 ± 0.01 Ga, with the oldest terrestrial material yielding an age of 4.4 Ga (Dalrymple, 2004).
age of the Moon The time elapsed in days since the last new Moon.
age of the Universe The term usually refers to the Hubble age of the Universe, which, given by the reciprocal of the *Hubble constant,* is currently 14.1 Ga. Other uses include the age of 13-15 Ga, the time when nucleosynthesis is thought to have begun; and 4.5-4.6 Ga., the age of meteorites (Dalrymple, 2004).
age of water The length of time since a water mass was last in contact with the atmosphere at the surface of the ocean.
age ratio The ratio of daughter to parent isotope upon which the *general age equation* is based.
age spectrum In isotopic dating methods where the radiogenic daughter product is a gas (argon-40/argon-39 dating method, xenon-xenon age spectrum method), it is possible to extract the gas from the sample by incremental heating, and to compute an apparent age for each fraction of gas released. A plot of apparent age versus temperature or versus cumulative percent of the gas released is termed the age spectrum.
agglomerate (ag-glom'-er-ate) A term originally used by Lyell in 1831 for a chaotic assemblage of coarse angular pyroclastic materials. The term has been variously defined since then, and should be defined in context to avoid confusion. Cf: *volcanic breccia.*
agglomerate ice Ice that has formed by congealing a mixture of floating ice fragments, as in a lake.
agglomerating (ag-glom'-er-at'-ing) Said of bituminous coal that softens when heated. See also: *caking coal; coke.*
agglomerating value A measure of the binding qualities of a coal, determined by fusing tests in which no inert material is heated with the sample. Cf: *agglutinating value.*
agglutinate [lunar] (ag-glu'-ti-nate) A term for certain particles in the lunar regolith that are held together by and largely composed of glass, probably spatter and melted ejecta from small hypervelocity impacts on the lunar surface, together with mineral grains and small rock fragments.
agglutinate [volc] A welded pyroclastic deposit characterized by vitric material binding the pyroclasts, or "sintered" vitric pyroclasts. Commonly used for deposits of bombs fused while hot and viscous, and said to be agglutinated.
agglutinate cone A small cone made of agglutinated fragments. Syn: *spatter cone.*
agglutinated (ag-glu'-ti-nat'-ed) Said of the tests of certain protists (some foraminifers, thecamoebians, and tintinnids) composed of foreign particles (sand grains, sponge spicules, mica flakes, etc.) bound together by cement.
agglutinating value (ag-glu'-ti-nat'-ing) A measure of the binding qualities of a coal and an indication of its caking or coking characteristics, determined by the ability of the coal, when fused, to combine with an inert material, such as sand. Cf: *agglomerating value.* Syn: *caking index.*
agglutination (ag-glu'-ti-na'-tion) A syn. of sedimentary cementation, esp. in regard to fine-grained pelletal muds and more rarely coarse-grained rocks, such as breccia or conglomerate.
aggradation [geomorph] (ag-gra-da'-tion) (a) The building-up of the Earth's surface by deposition; specif. the *upbuilding* performed by a stream in order to establish or maintain uniformity of grade or slope. Cf: *degradation [geomorph]; gradation [geomorph].* Syn: *upgrading.* (b) A syn. of vertical accretion, as in the development of a beach.
aggradation [permafrost] The spread or growth of permafrost, under present climatic conditions, due to natural or artificial causes. Ant: *degradation [permafrost].*
aggradation [stratig] The vertical component of coastal *onlap* (Mitchum, 1977, p.208). See also: *encroachment [stratig].*
aggradational ice (ag-gra-da'-tion-al) Ice incorporated in the upper part of permafrost during aggradation, specif. by a rise of the permafrost table, over a period of years. Commonly ice lenses are formed seasonally in the base of the active layer, and these can be incorporated in the permafrost if the depth of thaw over a period of years does not reach them (Mackay, 1972, p.10).
aggradational parasequence set In sequence stratigraphy, a parasequence set in which successively younger parasequences are deposited above one another with no significant lateral shifts; overall, the rate of accommodation approximates the rate of deposition (Van Wagoner et al., 1990).
aggradational plain A broad, fanlike plain with a gently sloping, concave-upward, longitudinal profile, formed by stream deposition.
aggradation recrystallization Recrystallization resulting in the enlargement of crystals. Ant: *degradation recrystallization.*
aggraded valley plain (ag-grad'-ed) An *alluvial plain,* or a stream-built flood plain; a plain formed by a stream aggrading its valley, the infilling with alluvium on the valley floor attaining a thickness greater than that of the depth of scour of the stream (Cotton, 1958, p.193).
aggrading neomorphism (ag-grad'-ing) A kind of *neomorphism* in which the crystal size increases (Folk, 1965, p.23); e.g. *porphyroid neomorphism* and *coalescive neomorphism.* Ant: *degrading neomorphism.*
aggrading stream (a) A stream that is actively building up its channel or flood plain by being supplied with more load than it is capable of transporting. (b) A stream that is upbuilding approximately at grade.
aggregate (ag'-gre-gate) (a) A mass or body of rock particles, mineral grains, or a mixture of both. (b) Any of several hard, inert materials, such as sand, gravel, slag, or crushed stone, used for mixing with a cementing or bituminous material to form concrete, mortar, or plaster; or used alone, as in railroad ballast or graded fill. The term sometimes includes rock material used as chemical or metallurgical fluxstone, or filtration medium. See also: *coarse aggregate; fine aggregate; lightweight aggregate.*
aggregated (ag'-gre-gat'-ed) Said of a massive sulfide or other ore deposit in which the sulfide or other valuable constituent makes up 20% or more of the total volume. The term is little used.
aggregate polarization In crystal optics, a pattern seen between crossed nicols in a fine-grained crystal aggregate, composed of the different interference colors of variously oriented grains. The presence of aggregate polarization may define a material as crystalline that in ordinary light appears amorphous.
aggregate structure A mass of separate crystals, scales, or grains that extinguish under the polarizing microscope at different intervals during the rotation of the stage.
aggregation [data] A process of grouping distinct data by spatial or nonspatial attributes (Krzanowski et al., 1993).
aggregation [soil] The process whereby primary soil particles (sand, silt, clay) are bound together, usually by natural forces and substances derived from root exudates and microbial activity.
aggressive water Water having the ability to dissolve rocks. In the context of limestone and dolomite, this term refers especially to water containing dissolved carbon dioxide (carbonic acid), or, rarely, other acids.
aging In lakes, formerly viewed as the process that characterizes a *dying lake.*
aglet A tiny plate pierced by a single pore in a radiolarian.
***a*-glide** A *glide plane* involving reflection across (010) or (001) combined with translation of one-half of a lattice translation along *a.* Cf: *b-glide; c-glide; d-glide; n-glide.*
aglime *agricultural lime.*
agmatite (ag'-ma-tite) A *migmatite* with appearance of breccia (Dietrich and Mehnert, 1961).
Agnatha (Ag-na'-tha) A class of vertebrates, the jawless fishes. A modern example is the lamprey. Range, Ordovician to the present.
agnostid (ag-nos'-tid) Any trilobite belonging to the order Agnostida, which includes small forms with subequal cephalons and pygidia and only two segments in the thorax. Range, Lower Cambrian to Upper Ordovician.
agonic line (a-gon'-ic) An *isogon* that connects points of zero magnetic declination. Its position changes according to the secular variation of the Earth's magnetic field.
agouni A term used in Morocco for a broad, generally dry gully carved by a torrent (Termier and Termier, 1963, p.399). Etymol: Berber. Cf: *asif.*
agpaite (ag'-pa-ite) A group of feldspathoid-bearing igneous rocks, first described from Ilimaussaq, Greenland, by Ussing in 1912, that includes sodalite-bearing nepheline syenite, *naujaite, lujavrite, kakortokite,* etc., and is distinguished by having (Na+K)>Al on a molecular or atomic basis. Obsolete.
agpaitic (ag-pa-it'-ic) (a) Said of crystallization in the presence of an excess of alkali (esp. sodium), so that the amount of aluminum oxide is insufficient for the formation of aluminum silicates (Thrush, 1968, p.18). (b) Said of rocks with an *agpaitic coefficient* > 1.
agpaitic coefficient The ratio Na + K/Al, where Na, K and Al are amounts of sodium, potassium, and aluminum atoms, respectively, in a rock, esp. an alkalic igneous rock.
Agrell dip In meteoritic metal, the decrease in kamacite Ni concentration near the interface to taenite.
agrellite (a-grel'-lite) A white or grayish triclinic mineral originally found in regionally metamorphosed agpaitic alkalic rocks: $NaCa_2Si_4O_{10}F$.
agric horizon (ag'-ric) A mineral soil horizon in which clay, silt and humus derived from an overlying cultivated and fertilized layer have accumulated. The wormholes and illuvial clay, silt and humus, occupy at least 5%

of the horizon by volume. The illuvial clay and humus occur as horizontal lamellae or fibers, or as coatings on ped surfaces or in wormholes.

agricolite (ag-ric'-o-lite) *eulytine.*

agricultural lime Pulverized limestone or dolomite added to crop land to correct acidity and promote plant growth. Syn: *aglime.*

agricultural pollution Pollution resulting from farming practices and agricultural wastes. Sources include pesticides and fertilizers; animal manure and carcasses; crop residues; sediment from erosion; and dust from plowing.

agrinierite (ag-rin-ier'-ite) An orange orthorhombic mineral: $(K_2,Ca,Sr)U_3O_{10}·4H_2O$.

agronomy The theory and practice of crop production and soil management.

aguada In the *karst* region of Yucatan, Mexico, a shallow depression, generally covering several hectares, that collects water in the rainy season; usually a solution sinkhole with a natural or artifically enhanced clay floor. Etymol: Spanish, "watering place".

aguilarite (a-gui-lar'-ite) A metallic black orthorhombic mineral: Ag_4SeS.

ahermatypic Not hermatypic, i.e., not reef-building like, e.g., a *hermatypic coral.*.

ahermatypic coral (a'-her-ma-typ'-ic) A coral lacking symbiotic algae; a coral capable of living in cold, deep, dark water. Ant: *hermatypic coral.*

aheylite (a-hey'-lite) A pale blue-green triclinic mineral: $(Fe^{2+},Zn)Al_6(PO_4)_4(OH)_8·4H_2O$. It is the ferrous analogue of *turquoise.*

ahlfeldite (ahl'-feld-ite) A brownish-pink monoclinic mineral: $NiSe^{4+}O_3·2H_2O$. It is isostructural with clinochalcomenite and cobaltomenite.

A horizons Mineral horizons that formed at the surface or below an O horizon that exhibit obliteration of all or much of the original rock structure and (1) are characterized by an accumulation of humified organic matter intimately mixed with the mineral fraction and not dominated by properties characteristic of E or B horizons; or (2) have properties resulting from cultivation, pasturing, or similar kinds of disturbance.

Ahren's prism A type of polarizing prism having three cut and recemented segments; the ordinary rays of the light are reflected to the sides, and the extraordinary ray forms the polarized light.

AI *artificial intelligence.*

aiguille (ai-guille' [ai-gweel]) A prominent sharp rock peak or pinnacle, of the type commonly found in intensely glaciated mountain regions (as near Mont Blanc in the French Alps); it is a remnant of a septum between two cirques, the rest of which has been largely or wholly removed by erosion. Cf: *gendarme.* Etymol: French, "needle". Syn: *needle [geol].*

aikinite (ai'-kin-ite) A metallic orthorhombic mineral: $PbCuBiS_3$. Syn: *needle ore.*

aillikite In the *IUGS classification* (Woolley et al., 1996) this dike rock is reclassified as a carbonatite rather than an ultramafic carbonate-rich *lamprophyre.* Obsolete.

ailsyte (ail'-syte) Fine-grained *quartz syenite* containing sodic amphibole, named for Ailsa Craig, Scotland, by Heddle in 1897. Obsolete.

aimless drainage (aim'-less) Drainage without a well-developed system, such as that in an area of glacial drift or karst topography. Cf: *deranged drainage pattern.*

ainalite (ai'-na-lite) A mixture of cassiterite and columbite.

air *atmosphere.*

air base An imaginary line connecting the points in space at which successive photos in a flight strip were taken; specif., the length of such a line (ASP, 1975, p.2064). Cf: *photobase.*

airborne magnetometer (air'-borne) An *magnetometer* for measuring the Earth's magnetic field from an aircraft. Syn: *aerial magnetometer.*

airborne profile recorder An electronic instrument that emits a pulsed-type radar signal from an aircraft to measure vertical distances between the aircraft and the Earth's surface. Abbrev: APR. Syn: *terrain profile recorder.*

air drilling *Rotary drilling* using high-velocity air pumped down the drillstring instead of conventional *drilling mud.* It is faster than drilling with mud but is unsuitable where significant volumes of water may be encountered or where natural gas may create an explosive mixture downhole.

air dry The condition of a substance whose moisture content has come into approximate equilibrium with the humidity of the surrounding air.

air entrainment The incorporation of air into moving water as a result of turbulence or the breaking of small waves.

air-entry pressure The *capillary pressure* at which air or gas begins to enter the pores in a porous medium and begins to displace water. Liquid saturation is 100% above the air-entry pressure and less than 100% otherwise. Syn: bubble point.

air-fall deposition Showerlike fall of pyroclastic fragments from an *eruption cloud.* Usage not recommended because term suggests falling air. See also: *ash fall; pumice fall.*

air gap *wind gap.*

air gun An energy source extensively used in marine seismic surveys. Air under high pressure is abruptly released into the water to generate a seismic wave. Modified air guns are also used in boreholes, and on the land surface.

air heave The process of deformation of plastic sediments by the enlargement of a pocket of air trapped in them, such as by the accretion of small air bubbles rising through sand exposed at low tide on a beach or tidal flat. See also: *gas heave.*

air-heave structure A crumpled sedimentary structure believed to have been produced by air heave, measuring several centimeters across, and characterized by an abrupt upward doming of laminae with a core of unlaminated sandstone (Stewart, 1956, p.159). Also called: gas-escape structure. See also: *gas-heave structure.*

air mass A large, widespread volume of air having relatively uniform horizontal temperature and moisture content that is characteristic of its *source region* and are modified as it moves away from its source; for example, a *polar air mass.*

air photograph *aerial photograph.*

air pocket An enclosed air space between the water surface and the roof of a cave.

air pressure *atmospheric pressure.*

air sac [palyn] Less preferred syn. of *saccus (a).*

air-sea gravity meter *surface-ship gravimeter.*

air shooting In seismic prospecting, a technique of applying a seismic pulse to the Earth by detonating explosive charges in the air. Also called: *Poulter method.*

air shrinkage The volume decrease that a clay undergoes in drying at room temperature.

air solubility The amount of any given gas which will dissolve in a unit mass of seawater of known salinity and temperature when the pressure of the gas is equal to its partial pressure in the atmosphere, assuming the relative humidity of the atmosphere is 100% and the total atmospheric pressure is 760 mm or 1 bar. Used as a standard when determining the percent saturation of dissolved gases in seawater or the extent of the production or consumption of dissolved gases by internal reactions in seawater.

air-space ratio In a soil, the ratio of the volume of water that can be drained by gravity from a saturated soil to the total volume of the soil's voids. Cf: *air-void ratio.*

air sparging In environmental context, a remediation technique for aquifers in which air is driven through an injection well into the contaminant-saturated zone. Volatile compounds are captured as the injected air rises through the aquifer. this procedure is frequently combined with *soil vapor extraction* to capture and remove contaminants that have risen to the *unsaturated zone.*

air-void ratio In a soil, the ratio of the volume of air-filled voids to the total volume of voids in the soil. Cf: *air-space ratio.*

air volcano A *mud volcano* characterized more by the gas it emits than by the mud and rocks thrown out.

air wave An acoustic wave in the air. Cf: *earthquake sound.*

air well A tower of loose rock, used in some desert countries to collect water, by condensation of moisture from the warm atmosphere on the relatively cooler rock surfaces within the tower, whose temperature fluctuates about the daily and annual mean in a smaller range than does that of the air.

Airy hypothesis A concept of the mechanism of *isostasy,* proposed by George Bedell Airy, that postulates an equilibrium of crustal blocks of the same density but of different thickness; thus the topographically higher mountains would be of the same density as other crustal blocks but would have greater mass and deeper roots. Cf: *Pratt hypothesis.*

Airy phase The amplitude build-up in a dispersive wavetrain corresponding to a maximum or minimum group velocity.

Aïstopoda (A-ïs'-to-po'-da) An order of the amphibian subclass Lepospondyli, characterized by wormlike or snakelike habitus and lacking appendicular skeletons. Range, Lower Mississippian to Lower Permian.

ait A British term for a small island in a lake or river; an islet. Syn: *eyot.*

Aitoff projection (Ai'-toff) (a) A map projection derived from the azimuthal equidistant projection by doubling the horizontal distances from the central meridian until the entire spherical surface is within an ellipse whose major axis (equator) is twice the length of its minor axis (central meridian). It resembles the Mollweide projection, but the parallels (except the equator) and meridians (except the central meridian) are represented by curved lines and there is less distortion at the margins. Named after David Aitoff (1854-1933), Russian geographer, who introduced the projection in 1889. (b) A name commonly, but incorrectly, given to the *Hammer-Aitoff projection.*

ajkaite (aj'-ka-ite) A pale-yellow to dark reddish-brown, sulfur-bearing fossil resin found in brown coal. Also spelled: ajkite.

ajoite (a-jo'-ite) A bluish-green triclinic mineral: $(K,Na)_3Cu_{20}Al_3Si_{29}O_{76}(OH)_{16}·8H_2O$.

akaganéite (a-ka-ga'-né-ite) A brown monoclinic mineral: beta-$Fe^{3+}O(OH,Cl)$.

akatoreite (a-ka-to'-re-ite) An orange-brown triclinic mineral: $Mn^{2+}_9Al_2Si_8O_{24}(OH)_8$.

Akchagylian (Ak-sha-gil'-ian) Ponto-Caspian stage: middle Pliocene (above Cimmerian, below Apsheronian). Syn: Aktschagylian.

akdalaite (ak-dal'-a-ite) A porcelaneous white hexagonal mineral: $4Al_2O_3·H_2O$.

akene (a-kene') *achene.*

akenobeite (ak-e-no'-be-ite) A granodioritic *aplite* composed of an aggregate of alkali feldspar and oligoclase, the former in excess of the latter, with an aggregate of fine quartz grains in the interstices and with little ferromagnesian material present (Johannsen, 1939, p.238). Its name is derived from Akenobe district, Japan. Obsolete.

akerite (a'-ker-ite) A *syenite* containing biotite, sodic clinopyroxene, and quartz. Its name, given by Brögger in 1895, is derived from Aker, Norway. Obsolete.

åkermanite (ak'-er-man-ite') A tetragonal greenish-brown mineral of the *melilite* group: $Ca_2MgSi_2O_7$. It is isomorphous with gehlenite.

A'KF diagram A triangular diagram showing the simplified compositional character of a metamorphic rock by plotting molecular quantities of the three components: $A'=Al_2O_3+Fe_2O_3-(Na_2O+K_2O+CaO)$; $K=K_2O$; and $F=FeO+MgO+MnO$. $A'+K+F$ (in mols) are recalculated to 100%; the diagram is used in addition to the *ACF diagram* when K minerals require representation. Cf: *AFM diagram.*

akhtenskite (akh-tensk'-ite) A dark to black hexagonal mineral: MnO_2.

akimotoite A black rhombohedral mineral found in meteorites: $(Mg,Fe)SiO_3$. A polymorph of enstatite.

akinetic surface (a-ki-net'-ic) The surface in a sedimentary rock layer that was the outer surface of the lithosphere at the place and time that oscillation of base level at that point passed through its maximum (Goldman, 1921, p.8-20).

aklé A pattern of sand dunes found in the western Sahara. It is a network of sinuous ridges made up of crescentic sections alternately facing into and away from the wind. It seems to require relatively unidirectional winds and a considerable quantity of sand (Cooke and Warren, 1973, p.285-288).

akmolith (ak'-mo-lith) An igneous intrusion along a zone of décollement, with or without tonguelike extensions into the overlying rock. Also spelled: acmolith. Obsolete.

akrochordite (ak-ro-chor'-dite) A pink to reddish-brown warty monoclinic mineral: $(Mn^{2+},Mg)_5(AsO_4)_2(OH)_4{\cdot}4H_2O$.

aksaite (ak'-sa-ite) A colorless to light orthorhombic mineral: $MgB_6O_7(OH)_6{\cdot}2H_2O$.

aktashite (ak-tash'-ite) A metallic gray rhombohedral mineral: $Cu_6Hg_3As_4S_{12}$. It is isotypic with gruzdevite and nowackiite.

aktology (ak-tol'-o-gy) Study of nearshore and shallow-water areas, conditions, sediments, life, and environments.

akyriosome (a-ky'-ri-o-some) A term used by Niggli (1954, p.191) for a subsidiary mass (such as a vein, nodule, band, lens, or block) of a complex rock; a minor part of a *migmatite.* The English translation of Niggli (1954) erroneously shows the spelling *"akyrosome"*. Rarely used. Cf: *kyriosome.*

akyrosome *akyriosome.*

ala [paleont] (a'-la) (a) A wing or winglike process or part of an organism; e.g. a winglike flange in the diatom *Surirella,* or the winglike extension of the ear of a brachiopod shell or of the ventral and lateral part of the *carapace* of an ostracode, or the triangular lateral part of a compartmental plate of a cirripede crustacean. (b) A smooth, commonly semicircular area flanking the posterior portion of the *glabella* in harpetid trilobites. (c) A winglike bone (alisphenoid) or bony extension (ala temporalis). Cf: *baccula.* Pl: alae.

ala [palyn] An equatorial flange or membranous expansion found in the *phycoma* stage of some modern *prasinophytes* and fossil *pteromorph acritarchs* (Mendelson, 1993).

Ala-A twin law A twin law in plagioclase, in which the twinning axis is the *a* axis and the composition plane is (001). An Ala-A twin is usually multiple and parallel, and often occurs with the Manebach twin law. Cf: *Manebach-Ala twin law; Ala-B twin law.*

alabandite (al-a-ban'-dite) A submetallic black cubic mineral: $Mn^{2+}S$. It usually occurs in iron-black massive or granular form. It possesses the rock salt structure. Syn: *manganblende.*

alabaster (al'-a-bas'-ter) (a) A firm, very fine-grained, massive or compact variety of gypsum, usually snow-white and translucent but sometimes delicately shaded or tinted with light-colored tones of yellow, brown, red, orange, or gray. It is used as an interior decorative stone (esp. for carved ornamental vases and figures) and for statuary. (b) *onyx marble.*

Ala-B twin law A twin law in plagioclase, in which the twinning axis is the *a* axis and the composition plane is (010). It often occurs with the albite twin law. Cf: *Ala-A twin law; albite-Ala twin law.*

alacranite (al'-ac-ran-ite) An adamantine brown or orange monoclinic fumarolic mineral: As_8S_9.

aladzha Impure ozokerite containing an admixture of country rocks, found in the region of the Caspian Sea (Tomkeieff, 1954, p.21). Etymol: Tatar.

alaite (al'-a-ite) A dubious mineral: $V_2O_5{\cdot}H_2O$.

alalite (al'-a-lite) A light-green variety of diopside.

alamandine (al-a-man'-dine) *almandine.*

alamosite (al-a-mos'-ite) A colorless, white, cream or light gray monoclinic mineral: $PbSiO_3$. It has a 12-repeat pyroxene-like chain.

alar fossula (a'-lar) A *fossula* developed in the position of an *alar septum* in a rugose coral or adjoining it on the side toward the *counter septum.*

alar projection A winglike extension of a foraminiferal test.

alar septum One of two protosepta of a rugose corallite located about midway between the *cardinal septum* and the *counter septum,* distinguished from other protosepta by pinnate insertion of newly formed metasepta on the side facing the *counter septum* (TIP, 1956, pt. F, p.245). Symbol: A.

alarsite A white, colorless, or creamy trigonal mineral: $AlAsO_4$. A member of the *berlinite* group.

alas A thermokarst depression with steep sides and a flat grass-covered floor (Washburn, 1973, p.237). Etymol: Yakutian. Anglicized plural: alases. Also spelled: alass.

alaskaite (a-las'-ka-ite) A mixture of sulfosalt minerals of lead, silver, copper, and bismuth. Not to be confused with *alaskite.*

Alaskan-type PGE deposit *zoned ultramafic PGE deposit.*

alaskite (a-las'-kite) Alkali feldspar granite with a low percentage of mafic minerals. In the U.S., a commonly used term for a granitic rock containing only a few percent of dark minerals. The term was introduced by Spurr in 1900 for a holocrystalline-granular plutonic rock characterized by essential alkali feldspar and quartz, and little or no dark component. Johannsen (1919) proposed two subdivisions of alaskite: kalialaskite, which lacks albite, and alaskite proper, which contains albite. In the recommendations of the Terminological Commission of the Petrographic Committee of the former USSR (1969), the term alaskite is used to designate granitoid rocks in which quartz constitutes 20-60% of the felsic minerals and in which the ratio of alkali feldspar to total feldspar is greater than 90%; i.e. the equivalent of alkali granite, or kaligranite. Cf: *aplogranite.* Obsolete syn. *tarantulite.*

alaskite-quartz Obsolete syn. of *tarantulite.*

alate (a'-late) Having wings or a winged form; e.g. said of an outward lateral extension in the ventral half of an ostracode valve, usually increasing in width backward and terminating abruptly, and tending to have a triangular shape (TIP, 1961, pt.Q, p.47); or said of the form of a brachiopod shell in which the valves are drawn out at the ends of the hinge line to form winglike extensions; or said of a bivalve-mollusk shell having auricles.

alate element A bilaterally symmetrical ramiform conodont element that lacks an anterior process, but has a posterior process and a lateral process on each side of the cusp (TIP, 1981, pt. W, supp.2).

Alaunian (A-laun'-i-an) A European substage: middle substage of Norian Stage, Upper Triassic (above Lacian, below Sevatian).

A layer The seismic region of the Earth equivalent to the crust, extending from the surface to the Mohorovicic discontinuity. It is part of a classification of the Earth's interior made up of layers A to G. Rarely used.

alb A flat or gently inclined narrow shelf separating the nearly vertical side of an alpine glacial trough from the mountain slope above. Cf: *alp.*

albedo (al-be'-do) The ratio of the amount of electromagnetic energy reflected by a surface to the amount of energy incident upon it. Usually albedo refers to the reflectivity in the solar (or short wave) region of the electromagnetic spectrum.

Albers equal-area conic projection *Albers projection.*

Albers projection (Al'-bers) An equal-area projection of the conical type, on which the meridians are straight lines that meet in a common point beyond the limits of the map and the parallels are concentric circles whose center is at the point of intersection of the meridians. Meridians and parallels intersect at right angles and the arcs of longitude along any given parallel are of equal length. The parallels are spaced to retain the condition of equal area. Along two selected parallels, called standard parallels, the scale is held exact; along the other parallels the scale varies with latitude, but is constant along any given parallel. Between the standard parallels the meridional scale is too large; beyond them, too small. When used for the map of the U.S., the projection normally has a maximum scale error of 1.25 percent along the northern and southern borders (Snyder, 1987, p.27). Named after Heinrich C. Albers (1773-1833), German cartographer, who devised the projection in 1805. Syn: *Albers equal-area conic projection.*

Albertan (Al-bert'-an) North American series or stage: Middle Cambrian (above Waucoban, below Croixian).

Albian (Al'-bi-an) Stage of the Standard Global Chronostratigraphic Scale: uppermost Lower Cretaceous, (above Aptian, below Cenomanian).

albic horizon (al'-bic) A mineral soil horizon from which clay and free iron oxides have been removed or in which the oxides have been segregated to the extent that the color of the horizon is determined primarily by the color of the primary sand and silt particles rather than by coatings on these particles.

albid element A conodont element containing white matter; contrasts with hyaline element (TIP, 1981, pt. W, supp.2).

albite (al'-bite) (a) A colorless or milky-white triclinic mineral of the feldspar group: $Na_{1.0\text{-}0.9}Ca_{0.0\text{-}0.1}Al_{1.0\text{-}1.1}Si_{3.0\text{-}2.9}O_8$. It is a variety of plagioclase with composition ranging from $Ab_{100}An_0$ to $Ab_{90}An_{10}$; it is also an alkali feldspar, representing the triclinic modification of sodium feldspar. Albite occurs in all groups of rocks, forming a common constituent of granite and of various acid-to-intermediate igneous rocks; it is widely distributed in low-temperature metamorphic rocks (greenschist facies), and is regularly deposited from hydrothermal solutions in cavities and veins. Albite crystals frequently exhibit polysynthetic twinning, predominantly after the *albite twin law.* Cf: *analbite.* Syn: *sodium feldspar; sodaclase; white feldspar; white schorl.* (b) The pure sodium-feldspar end member in the *plagioclase* series.

albite-Ala twin law A complex twin law in feldspar, in which the twin axis is perpendicular to [100] and the composition plane is (010). Cf: *Ala-A twin law; Ala-B twin law.*

albite-Carlsbad twin law A complex twin law in feldspar, in which the twin axis is perpendicular to [001] and the composition plane is (010).

albite-epidote-amphibolite facies *epidote-amphibolite facies.*

albite-epidote-hornfels facies The facies (set of metamorphic mineral assemblages) in which mafic rocks are represented by albite+

epidote+chlorite+actinolite. It is typical of the outermost parts of contact aureoles. It is not clearly distinct from the *greenschist facies,* and in fact is not given the status of a metamorphic facies by many workers (e.g., Miyashiro, 1973). Cf: *hornblende-hornfels facies.*

albite twin law A twin law in triclinic feldspars, in which the twin plane and composition plane are (010). An albite twin is usually multiple and lamellar, and shows fine striations on the (001) cleavage plane.

albitite (al'-bi-tite) A porphyritic igneous rock, defined by Turner in 1896, containing phenocrysts of albite in a groundmass chiefly consisting of albite. Muscovite, garnet, apatite, quartz, and opaque oxides are common accessory minerals. Obsolete.

albitization (al'-bit-i-za'-tion) Introduction of, or replacement by, albite, usually replacing a more calcic plagioclase.

Albolls Mollisols that have an albic horizon immediately below the mollic epipedon. These soils have an argillic or natric horizon and mottles, iron-manganese concretions, or both, within the albic, argillic or natric horizon. A suborder in the U.S. system of soil taxonomy.

alboranite (al-bo'-ran-ite) An olivine-free orthopyroxene-bearing basalt named by Becke in 1899 after Alboran Island of Spain. A *subalkaline basalt* or *tholeiitic basalt.* Obsolete.

albrechtschraufite (al'-brecht-schrauf'-ite) A yellow-green triclinic mineral: $Ca_4Mg(UO_2)_2(CO_3)_6F_2{\cdot}[17]H_2O$.

albrittonite (al'-brit-ton-ite') A former mineral name, now discredited.

alcove [geomorph] A large, deep niche formed in a precipitous face of rock; specif. a *spring alcove.*

alcove [speleo] A relatively small recess in the wall of a lava tube. See also: *cupola [karst].* Syn: *balcony.*

alcove lands An angular landscape characterized by terraced slopes consisting of resistant beds interbedded with deeply undercut softer rocks (Powell, 1875, p.149-150).

alcyonarian (al'-cy-o-nar'-i-an) Any coral of the subclass Alcyonaria, colonial forms with eight pinnate tentacles; an endoskeleton, and eight complete septa. Range, Permian to Holocene (Pennak, 1964). See also: *octocoral.*

aldanite (al'-dan-ite) A variety of thorianite containing lead and uranium.

aldermanite (al'-der-man-ite') A colorless orthorhombic mineral occurring in sedimentary phosphate deposits: $Mg_5Al_{12}(PO_4)_8(OH)_{22}{\cdot}32H_2O$.

aldzhanite (ald-zhan'-ite) A discredited borate mineral.

aleksite (a-lek'-site) A metallic hexagonal mineral: $PbBi_2Te_2S_2$.

alete (a-lete') Said of a spore without a laesura. In practice, such spores are sometimes difficult to distinguish from *acolpate* pollen. Cf: *inaperturate.*

aleurite (a-leu'-rite) An unconsolidated sedimentary deposit intermediate in texture between sand and clay, consisting of particles having diameters in the range of 0.01 to 0.1 mm. The term is common in Russian literature, and is frequently translated as "silt". Etymol: Greek "aleuron", "flour".

aleurolite (a-leu'-ro-lite) A consolidated *aleurite,* intermediate in texture between sandstone and shale; esp. siltstone.

aleurospore (a-leu'-ro-spore) A simple, lateral, conidioid fungal spore or sporelike body. Also spelled: aleurispore, aleuriospore.

Alexandrian (Al-ex-an'-dri-an) North American provincial series: Lowermost Silurian (above Cincinnatian of Ordovician, below Niagaran).

alexandrine sapphire (al-ex-an'-drine) An alexandrite-like sapphire that is blue in daylight, and violet, purple, or reddish under most artificial light.

alexandrite (al-ex-an'-drite) A transparent variety of chrysoberyl that has a grass-green or emerald-green color in daylight and wine-red to brownish-red color by transmitted or incandescent artificial light.

alexandrite effect Said of a mineral or gemstone whose color changes in hue when it is viewed under two different light sources.

alfisols Mineral soils that have umbric or ochric epipedons, argillic horizons, and that hold water at <1.5 MPa tension furing at least 90 days when the soil is warm enough for plants to grow outdoors. Alfisols have a mean annual soil temperature of <8° C or a base saturation in the lower part of the argillic horizon of 35% or more when measured at pH 8.2. An order in the U.S. system of soil taxonomy.

Alford rotation Rotating the information from orthogonal horizontal geophones into the natural coordinate system. Used in S-wave studies, especially birefringence studies.

alforsite (al'-fors-ite) A colorless hexagonal mineral of the *apatite* group: $Ba_5(PO_4)_3Cl$.

algae (al'-gae) A polyphyletic grouping of eukaryotic, photosynthetic, and almost entirely aquatic organisms, ranging from unicellular to complex, multicellular organisms (such as giant kelps). Groups are differentiated by photosynthetic pigments, biochemistry, and life cycles. Sing: alga. Cf: *brown algae; charophytes; green algae; red algae; yellow-green algae.*

algaenans Resistant biopolymers from fossil algae.

algal (al'-gal) Of, pertaining to, or composed of algae.

algal coal *boghead coal.*

algal mat *microbial mat.*

algal mound A local thickening of limestone attributed chiefly to the presence of a distinctive suite of rock types (such as massive calcilutite) containing algae.

algal reef An *organic reef* in which algae are or were the principal organisms secreting calcium carbonate. The reefs may be up to 10 m high and more than 15 m across.

algal ridge A low ridge or elevated margin at the seaward (outer) edge of a reef flat, composed of the calcium-carbonate secretions of actively growing calcareous algae. See also: *lithothamnion ridge.* Cf: *algal rim; coralgal ridge.*

algal rim A rim or ridge of crustose coralline algae developed on the margins of reefs, especially sea-facing or as separate ridges, and varying greatly in width (up to 500 m).

algarite (al'-gar-ite) A bitumen derived from algae.

algarvite (al-gar'-vite) A *melteigite* having more biotite and less nepheline than the original melteigite. Obsolete.

algebraic reconstruction technique A computational method for solving large systems of traveltime equations in tomography by using one traveltime row at a time. Abbrev: ART. Cf: *tomography.*

Algerian onyx (Al-ge'-ri-an) A distinctly banded, stalagmitic form of *onyx marble.*

algerite (al'-ge-rite) A pinitic pseudomorph after scapolite.

alginite (al'-gin-ite) A maceral of coal within the *liptinite* group, consisting of algal matter and characteristic of boghead coals. Cf: *algite; cutinite; sporinite; resinite.*

algite (al'-gite) A microlithotype of coal within the *liptite* group, consisting of at least 95% alginite. Cf: *boghead coal.*

algodonite (al-go-do'-nite) A metallic gray to white hexagonal mineral: Cu_6As.

ALGOL Algorithmic Language.

Algoman Orogeny (Al-go'-man) Orogeny and accompanying granitic emplacement that affected Precambrian rocks of northern Minnesota and adjacent Ontario about 2400 m.y. ago; it is synonymous with the *Kenoran Orogeny* of the Canadian Shield and was the final event of the Archean according to the current Canadian classification.

Algoma-type iron formation (Al-go'-ma-type) Chemically precipitated *iron formation* composed of thinly banded chert and iron minerals, with associated volcanic rocks and graywackes. It is found along volcanic arcs, chiefly Archean in age (Gross, 1980). Cf: *Lake Superior-type iron formation; sedex-type ore deposit.*

algon (al'-gon) The viscous, organic binding material of *vase,* consisting of finely divided remains of algae (or of land vegetation, as in the upper parts of the estuary) and iron principally in the form of FeS (Bourcart, 1939).

Algonkian (Al-gon'-ki-an) *Proterozoic.*

algorithm (al'-go-rithm) A rule or code for performing a calculation, usually in the form of an explicit step-by-step procedure, esp. by a computer.

alias (a'-li-as) An ambiguity in the frequency represented by sampled cyclic data. Where there are fewer than two samples per cycle, an input signal at one frequency yields the same sample values as, and hence appears to be, another frequency at the output of the system. Half of the frequency of sampling is called the "folding" or "Nyquist" frequency, f_N; a frequency larger than this, $f_N + Y$, appears to be the smaller frequency, $f_N - Y$. The two frequencies are "aliases" of each other. Aliasing can occur in the time or space domains.

alias filter A filter used before sampling to remove undesired frequencies which the sampling process would otherwide *alias.* Also called: *anti-alias filter.*

aliasing The "staircase" appearance of the edges or lines on a raster display (Krzanowski et al., 1993).

alidade (al'-i-dade) That part of a surveying instrument which consists of a sighting device, with index, and accessories for reading or recording data. The alidade or a theodolite or surveyor's transit is the upper part of the instrument, comprising the telescope, the microscopes or verniers, the vertical circle, and accessories, all mounted on what is called the upper motion of the instrument. It is used in measuring a direction or angle on a graduated circle mounted on the lower motion. The alidade used in some topographic surveying consists of a straightedge carrying a telescope or other sighting device, and used in recording a direction on the sheet fastened to the plane table. The moveable arm of a sextant is an alidade. See also: *peepsight alidade; telescopic alidade.*

aliettite (a-li-et'-tite) A regularly ordered, 1:1 mixed-layer talc/saponite mineral.

aligned current structure (a-ligned') *directional structure.*

alignment (a-lign'-ment) Locating points with respect to a straight line or to a set of straight lines. The term should be used, in surveying, only for operations associated with straight lines. Syn: *alinement.*

alimentation (al'-i-men-ta'-tion) *accumulation.*

alimentation facies Facies characteristics that provide evidence of the source of sediments, as revealed mainly by rock composition (such as "sandstone", "clay", and "chert") (Sonder, 1956). Cf: *precipitation facies.*

alinement (a-line'-ment) *alignment.*

alio (a'-li-o) A French term for an impervious, ferruginous crust formed by the precipitation of iron salts from subsurface water.

aliphatic hydrocarbon (al-i-phat'-ic) A straight or branched open-chain hydrocarbon having the empirical formula C_nH_{2n+2}, such as methane or phytane.

alive In ore deposits, a syn. of *quick [ore dep].*

alivincular (al-i-vin'-cu-lar) Said of a type of ligament of bivalve mollusks (e.g. *Ostrea*) in which the longer axis is transverse to the hinge axis, or that is not elongated in the longitudinal direction or necessarily situated entirely posterior to the beaks, but is located between the cardinal areas (where present) of the respective valves, with the lamellar component both anterior and posterior to the fibrous component.

alkali [chem] (al'-ka-li) n. Any strongly basic substance, such as a hydroxide or carbonate of an alkali metal (e.g. sodium, potassium). Plural: alkalies. Adj: *alkaline [chem]; alkalic [chem].*

alkali [mineral] Said of silicate minerals that contain alkali metals but little calcium, e.g. the alkali feldspars.

alkali [petrology] *alkalic [petrology].*

alkali basalt In the *IUGS classification*, a nepheline-normative basalt. Silica-undersaturated basalt, containing normative nepheline, diopside, and plagioclase with no normative hypersthene. The term was defined by Yoder and Tilley (1962). Cf: *alkali-olivine basalt; alkaline basalt; basalt [petrology]; olivine basalt.*

alkalic [chem] (al'-ka-lic) An adj. of *alkali [chem].*

alkalic [petrology] (a) Said of an igneous rock that contains more alkali metals than is considered average for the group of rocks to which it belongs. (b) Said of an igneous-rock series that contains more sodium and/or potassium than is required to form feldspar with the available silica. (c) Said of an igneous-rock series having an *alkali-lime index* below 51. Cf: *calcic [geol]; alkali-calcic; calc-alkalic.* (d) Said of an igneous rock belonging to the *Atlantic suite.* Syn: *alkali [petrology]; alkaline [petrology].*

alkali-calcic Said of a series of igneous rocks in which the weight percentage of silica is between 51 and 56 when the weight percentages of CaO and of $K_2O + Na_2O$ are equal. See also: *alkali-lime index.*

alkali-calc index *alkali-lime index.*

alkalic epithermal gold-silver deposit *alkalic intrusive-associated gold-silver deposit.*

alkali charnockite According to Tobi (1971), a rock of the *charnockite series* in which quartz constitutes 20-60% of the felsic constituents and in which the ratio of alkali feldspar to total feldspar is greater than 90%. He uses the term as a replacement for *enderbite.*

alkali-chloride deposit *adularia-sericite type* deposit.

alkalic intrusive-associated gold-silver deposit A *mineral-deposit type* characterized by gold telluride- and base-metal sulfide-bearing veins and stockworks hosted by alkalic intrusions, coeval volcanics, and surrounding sediments. Syn: *alkalic epithermal gold-silver deposit; Au-Ag-Te vein.*

alkalic porphyry copper deposit *Porphyry copper* deposit associated with alkalic intrusions and characterized by copper-gold rather than copper-molybdenum mineralization.

alkali feldspar (a) A group of feldspars composed of mixtures, or mixed crystals, of potassium feldspar ($KAlSi_3O_8$) and sodium feldspar ($NaAlSi_3O_8$) in any ratio; a group of feldspars containing alkali metals but little calcium. (b) A mineral of the alkali feldspar group, such as microcline, orthoclase, sanidine, adularia, albite, anorthoclase, and plagioclase in which the proportion of the An molecule is less than 20%. Cf: *plagioclase.* Syn: alkalic feldspar.

alkali-feldspar charnockite In the *IUGS classification*, the orthopyroxene-bearing equivalent of *alkali-feldspar granite*. Syn: *alkali charnockite*.

alkali-feldspar granite In the *IUGS classification*, a plutonic rock with Q between 20 and 60, and P/(A+P) less than 10.

alkali-feldspar quartz syenite In the *IUGS classification*, a plutonic rock with Q between 5 and 20, and P/(A+P) less than 10.

alkali-feldspar rhyolite In the *IUGS classification*, a *rhyolite* in which P/(P+A)<10%.

alkali-feldspar syenite In the *IUGS classification*, a plutonic rock with Q between 0 and 5, and P/(A+P) less than 10.

alkali-feldspar trachyte In the *IUGS classification*, a *trachyte* in which P/(P+A)<10%.

alkali flat A level area or plain in an arid or semiarid region, encrusted with alkali salts that became concentrated by evaporation and poor drainage; a *salt flat.* See also: *playa [geomorph].*

alkali gabbro The plutonic equivalent of *alkali basalt.* As defined by Wilkinson (1968), alkali gabbro contains plagioclase at least as calcic as An_{50}, clinopyroxene, and usually olivine; nepheline and/or analcime are present but make up less than 10 per cent of the rock, and alkali feldspar makes up less than 10 per cent of the total feldspar.

alkali-iron effect The formation of sodium-iron-oxygen-complex ions in a silicate melt which increases the iron content of the melt and acts as a flux.

alkali lake A *salt lake,* commonly found in an arid region, whose waters contain in solution large amounts of sodium carbonate, potassium carbonate, sodium chlorite and/or other alkaline compounds. Example: Lake Magadi in the Eastern Rift Valley of Kenya. See also: *potash lake; soda lake.*

alkali-lime index A means of classifying igneous rocks introduced by Peacock (1931) based on the weight percentage of silica present when the weight percentages of CaO and of $K_2O + Na_2O$ are equal. Four chemical classes of igneous rocks based on this index are recognized: *alkalic* (when the silica percentage is less than 51), *alkali-calcic* (when it is between 51 and 56), *calc-alkalic* (when it is between 56 and 61), calcic (when it is over 61). Syn: *alkali-calc index.*

alkali metal Any of the monovalent elements, lithium, sodium, potassium, rubidium, and cesium.

alkaline [chem] (al'-ka-line) An adj. of *alkali [chem].*

alkaline [petrology] *alkalic [petrology].*

alkaline basalt As proposed by Chayes (1964), a replacement for the terms *alkali basalt* and *alkali-olivine basalt.* Basalts with nepheline and/or acmite in the CIPW norm fall in this category.

alkaline magma series Magmas with high Na_2O plus K_2O, as discriminated from subalkaline magmas by (a) a discriminant line on total alkali vs. SiO_2 (TAS) plot; (b) presence of normative nepheline, not hypersthene; (c) modal presence of diagnostic silica-poor or alkali-rich minerals (e.g., sodic pyroxene, leucite). In specific occurrences, as in volcanoes or plutons, individual magmas in the series are often taken to be genetically related (e.g., by crystal fractionation, mixing, etc.). Not confined to any tectonic environment or to oceans or continents. There are several alkaline series (e.g., sodic and potassic), with many names given to actual members. For example, on Hawaii the series is sodic and comprises alkali basalt, hawaiite, mugearite, and trachyte.

alkaline rock As defined by Shand in 1922, any igneous rock in which the molecular ratio $[(Na_2O + K_2O):Al_2O_3:SiO_2]$ differs from 1:1:6 by deficiency in either Al_2O or SiO_2. The term therefore covers peralkaline silica-oversaturated rocks as well as all silica-undersaturated rocks.

alkaline series *alkaline magma series.*

alkaline soil A soil whose pH value is greater than 7.0.

alkaline-sulfide hypothesis A theory of ore-deposit formation that postulates complex sulfide ions in hydrothermal solutions as a means of ore transportation. The theory accounts satisfactorily for transport of only a few of the common metals (Krauskopf, 1967, p.501-502).

alkalinity [lake] (al-ka-lin'-i-ty) Refers to the excess of base cations over strong acid anions dissolved in lake water, which is balanced by other anions (primarily bicarbonate and hydroxide but also borate, silicate, phosphate, and the anions of fulvic and other organic acids) and which controls pH, i.e., the buffering capacity of the lake water. The alkalinity provides an index to the nature of the rocks within the drainage basin and the degree to which they are weathered (Wetzel, 2001, p.191).

alkalinity [oceanog] A measure of the capacity of seawater to resist changes in pH. Equal to the millimoles (or milliequivalents) of hydrogen ion from a strong acid that will be consumed (or neutralized) by the basic ions in a kilogram of seawater.

alkali-olivine basalt A term proposed by Tilley in 1950 to replace "olivine basalt" as used by Kennedy in 1933. The term was changed to *alkali basalt* by Yoder and Tilley (1962) and to *alkaline basalt* by Chayes (1964).

alkalitrophy (al-ka-lit'-ro-phy) The quality or state of an arid-region lake exhibiting alkaline characteristics (Wetzel, 2001, p.274). Adj: alkalitrophic.

alkenones Long-chained hydrocarbons with ketone groups; used for calculating sea surface temperature.

allabogdanite A light yellow orthorhombic mineral found in meteorites: $(Fe,Ni)_2P$. A dimorph of *barringerite.*

allactite (al-lac'-tite) A purplish-red pleochroic monoclinic mineral: $Mn_7(AsO_4)_2(OH)_8$. Syn: chlorarsenian.

allalinite (al-la'-li-nite) An altered *gabbro* containing saussurite as euhedral pseudomorphs after the original minerals, thus maintaining the original ophitic texture. Its name was derived from Allalin, Switzerland, by Rosenbusch in 1896. Obsolete.

allanite-(Ce) (al'-lan-ite) A sub-metallic pitchy brownish-black monoclinic mineral of the *epidote* group: $CeCa(Al,Fe)_3O(O,OH)(SiO_4)[Si_2O_7]$. It is typically an accessory mineral in igneous rocks (granite, syenite, diorite, pegmatite) and in their metamorphic equivalents. Syn: *orthite; cerine; bucklandite; treanorite.*

allanite-(La) A submetallic pitchy brownish-black monoclinic mineral of the *epidote* group: $LaCa(Al,Fe)_3O(O,OH)(SiO_4)[Si_2O_7]$.

allanite-(Y) A submetallic pitchy brownish-black monoclinic mineral of the *epidote* group: $YCa(Al,Fe)_3O(O,OH)(SiO_4)[Si_2O_7]$.

allargentum (al-lar-gen'-tum) A metallic silvery white hexagonal mineral: Ag_6Sb.

allcharite (all-cha'-rite) A discredited synonym of *goethite.*

alleghanyite (al-le-gha'-ny-ite) A bright pink or reddish-brown monoclinic mineral: $Mn^{2+}_5(SiO_4)_2(OH)_2$. A member of the *chondrodite* group.

Alleghany Orogeny (Al-le-gha'-ny) A Late Paleozoic (Carboniferous-Permain) mountain-building event that occurred as a consequence of the collision between Africa and North America. It is responsible for the fold-thrust belt deformation of the Appalachian Valley and Ridge province. It is also responsible for penetrative deformation, metamorphism, and plutonism in the internal portions of the Appalachians. The term was introduced by H.P. Woodward (1957, 1958), who intentionally used the spelling *Alleghany.*

Alleghenian (Al-le-ghe'-ni-an) North American provincial stage: middle Pennsylvanian (above Pottsvillian, below Conemaughian).

Allegheny Orogeny (Al-le-ghe'-ny) *Alleghany Orogeny.*

allemontite (al-le-mont'-ite) A mineral substance: a mixture of stibarsen (SbAs) and As or Sb. Syn: *arsenical antimony; stibarsen.*

allenite (al'-len-ite) *pentahydrite.*

Allen's rule In zoology, the statement that warm-blooded animals tend to have longer protruding body parts (i.e. legs, arms, tails) in warmer parts of the environment than in cooler. Named after Joel A. Allen (d.1921), American zoologist. Cf: *Bergmann's rule.*

Allerød n. A term used primarily in Europe for an interval of late-glacial time (centered about 13,500 calendar years ago) following the Older Dryas and preceding the Younger Dryas, during which the climate as inferred from stratigraphic and pollen data (Iversen, 1954) ameliorated, favoring the growth of birch, pine, and willow vegetation in Scandinavia. adj. Pertaining to the late-glacial Allerød interval and to its climate, deposits, biota, and events.

allevardite (al-le-vard′-ite) A discredited synonym of *rectorite.*

alliaceous (al-li-a′-ceous) Said of minerals that have an odor of garlic when rubbed, scratched, or heated; e.g. arsenical minerals.

Alling grade scale (Al′-ling) A metric *grade scale* designed by Harold L. Alling (1888-1960), U.S. geologist, for two-dimensional measurements (such as with thin sections or polished blocks) of sedimentary rocks. It has a constant geometric ratio of 10 for the major divisions (colloid, clay, silt, sand, gravel, cobble, boulder) and one of the fourth root of 10 for the four-fold subdivisions of each major unit (Alling, 1943).

allingite (al′-ling-ite) A fossil resin (retinite) containing no succinic acid but considerable sulfur, found at Allinges in Haute-Savoie, France.

allite (al′-lite) A rock name for *allitic* material; e.g. bauxite and laterite.

allitic (al-lit′-ic) adj. Pertaining to a rock or soil from which silica has been largely removed and which contains a high proportion of aluminum and iron compounds in the clay fraction.

allivalite (al′-li-va-lite′) A *troctolite* (containing anorthite and olivine), with pyroxene rare or absent and accessory apatite and opaque iron oxides. Its name was derived from Allival, Isle of Rhum, Scotland by Harker in 1908. Obsolete.

allochem (al′-lo-chem) A collective term introduced by Folk (1959, p.4) for one of several varieties of discrete and organized carbonate aggregates that serve as the coarser framework grains in most mechanically deposited limestones, as distinguished from sparry calcite (usually cement) and carbonate-mud matrix (micrite). Important allochems include: silt-, sand-, and gravel-size intraclasts; ooids; pellets; lumps; and fossils or fossil fragments (carbonate skeletons, shells, etc.). Adj: allochemical. Cf: *pseudoallochem; orthochem.*

allochemical metamorphism (al-lo-chem′-i-cal) Metamorphism that is accompanied by addition or removal of material so that the bulk chemical composition of the rock is changed (Mason, 1958). Cf: *metasomatism.*

allochetite (al-lo-chet′-ite) A porphyritic hypabyssal igneous rock composed of phenocrysts of labradorite, alkali feldspar, titanian augite, nepheline, magnetite, and apatite, in a fine-grained, felty groundmass of augite, biotite, magnetite, hornblende, nepheline, and alkali feldspar. It was named from the Allochet Valley, Italy, by Ippen in 1903. Obsolete.

allochromatic (al′-lo-chro-mat′-ic) Said of a mineral that is colorless in its pure state, but that has color due to submicroscopic inclusions, or to the presence of a *chromophore* that is not an essential chemical constant of the mineral. Cf: *idiochromatic.*

allochthon [sed] (al-loch′-thon) A mass of redeposited rock originating from distant sources.

allochthon [tect] A mass of rock or fault block that has been moved from its place of origin by tectonic processes, as in a *thrust sheet* or *nappe.* Allochthons commonly are underlain by *décollements.* Many allochthonous rocks have been moved so far from their original sites that they differ greatly in facies and structure from those on which they now lie. Ant: *autochthon.* Adj: *allochthonous.* Also spelled: *allochthone.*

allochthone (al-loch′-thone) *allochthon [tect].*

allochthonous (al-loch′-tho-nous) Formed or produced elsewhere than in its present place; of foreign origin, or introduced. The term is widely applied, e.g. to coal or peat that originated from plant material transported from its place of growth, to an allochthon on a low-angle thrust fault, to a sedimentary breccia (including klippes) of exotic origin, or to a *fallback breccia* at a meteorite impact crater. The term was first used by Naumann (1858, p.657) to designate rocks of distant origin; it is similar in meaning to *allogenic,* which refers to constituents rather than whole formations. Ant: *autochthonous.* See also: *parautochthonous; exotic [struc geol]; heterochthonous.*

allochthonous assemblage A fossil assemblage composed of specimens transported out of their life habitats and occurring in a foreign substratum (Kidwell et al., 1986). Cf: *thanatocoenosis.*

allochthonous mantle Loose debris of rock fragments or soil transported from elsewhere to its present site; not formed in situ. Cf: *autochthonous mantle; block field.*

allochthonous salt A sheetlike salt body emplaced at stratigraphic levels above the autochthonous source layer. Allochthonous salt overlies stratigraphically younger strata. The term can be applied even if the salt sheet remains attached to its source layer (Bally, 1981).

allochthony (al-loch′-tho-ny) Accumulation of plant materials elsewhere than at their site of growth. Ant: *autochthony.* See also: *primary allochthony; secondary allochthony; drift theory [coal]; hypautochthony.*

alloclasite (al-lo-clas′-ite) A metallic white monoclinic mineral: (Co,Fe)AsS. A polymorph of glaucodot.

alloclastic breccia (al-lo-clas′-tic) A breccia that is formed by disruption of nonvolcanic rocks by volcanic processes beneath the Earth's surface; a type of *volcanic breccia.*

allo-correlation Time correlation between different units in different localities (Henningsmoen, 1961, p.65-66).

allocyclicity (al′-lo-cy-clic′-i-ty) The state of cyclothemic deposition that results from changes in the supply of energy or material input into a sedimentary system (Beerbower, 1964, p.32). It involves such mechanisms as uplift, subsidence, climatic variation, eustatic change in sea level, and other changes external to the sedimentary unit. Cf: *autocyclicity.*

allodapic limestone (al-lo-dap′-ic) A limestone deposited by turbidity currents, e.g. off reef fronts (Meischner, 1964, p.156).

allofacial (al-lo-fa′-cial) Pertaining to rocks belonging to different metamorphic facies. It is inferred that the present juxtaposition of allofacial rocks has been brought about by tectonic transport. Cf: *isofacial [petrology].*

alloformation (al′-lo-for-ma′-tion) The fundamental *allostratigraphic unit.* It may be completely or partly divided into allomembers, or it may have no allomembers (NACSN, 1983, Art. 59). See also: *allogroup.*

allogene (al′-lo-gene) An *allogenic* mineral or rock constituent; e.g. a xenolith in an igneous rock, a pebble in a conglomerate, or a detrital mineral in a placer deposit. Ant: *authigene.* Syn: *allothigene.*

allogenic (al-lo-gen′-ic) (a) Formed or generated elsewhere, usually at a distant place; specif. said of rock constituents and minerals that were derived from pre-existing rocks and transported to their present depositional site, or that came into existence before the rock of which they now constitute a part and at some place other than where now found. Ant: *authigenic.* Cf: *chthonic; allochthonous.* Syn: *allothogenic; allothigenic; allothigenous; allothigenetic.* (b) Pertaining to a stream that derives much of its water from a distant terrain or from beyond its surface draining area, such as one originating in a humid or glacial region and flowing across an arid or desert region; e.g., River Nile. (c) Said of an ecologic succession that resulted from factors that arise from outside the natural community and alter its habitat, such as an allogenic drought of prolonged duration. Cf: *autogenic [ecol].*

allogenic recharge Water that collects on nonsoluble rocks or sediments and feeds *sinking streams* in adjacent *karst* (Ford and Williams, 1989).

allogenic succession A temporal succession that occurs at a location as a result of changing external influences (e.g., changing geophysicochemical forces) (Begon et al., 1996). Cf: *autogenic* succession.

allogroup (al′-lo-group) The *allostratigraphic unit* next in rank above an *alloformation.* An allogroup is established only if a unit of that rank is essential to elucidation of geologic history. An allogroup may consist entirely of named alloformations, or, alternatively, may contain one or more alloformations which jointly do not comprise the entire allogroup (NACSN, 1983, Art. 59).

allokite (al-lok′-ite) A clay mineral intermediate in structure between kaolinite and allophane.

allolistostrome (al-lo-list′-o-strome) A *mélange* of sedimentary origin; an *olistostrome* containing *exotic blocks* (Raymond, 1978). Cf: *endolistostrome.*

allomember (al′-lo-mem′-ber) The formal *allostratigraphic unit* next in rank below an *alloformation* (NACSN, 1983, Art. 59).

allomeric (al-lo-mer′-ic) *isomorphous.*

allomerism (al-lom′-er-ism) *isomorphism [cryst].*

allometry [biol] (al-lom′-e-try) (a) The differential growth of a part of an organism in relation to its entirety. Cf: *isometry.* (b) Measurement and study of the growth of a part of an organism in relation to its entirety.

allometry [geomorph] In general, the study of relationships between size and shape or form. It is sometimes applied literally, e.g., in relationships between the surface area or volume and the shape of alluvial fans. Allometry is also used to refer to the ratio between relative rates of change of variables within geomorphic systems.

allomorph (al′-lo-morph) (a) A polymorph or a *dimorph.* (b) A *pseudomorph.*

allomorphic (al-lo-mor′-phic) *polymorphic.*

allomorphism [cryst] (al-lo-mor′-phism) *paramorphism.*

allomorphism [paleont] A term used erroneously for *xenomorphism* in reference to bivalve mollusks.

allomorphite (al-lo-mor′-phite) A mineral substance consisting of barite pseudomorphous after anhydrite.

allomorphosis (al-lo-mor′-pho-sis) Biologic evolution characterized by a rapid increase in specialization; evolutionary allometry. Cf: *aromorphosis.*

allomorphous (al-lo-mor′-phous) *polymorphic.*

allopalladium (al′-lo-pal-la′-di-um) A discredited synonym of *stibiopalladinite.*

allopatric (al-lo-pat′-ric) Said of organisms or biologic events occurring in different geographic areas; e.g. the development of a distinct species from an isolated population. Noun: allopatry. Cf: *sympatric.*

allophane (al′-lo-phane) A semiordered hydrous aluminosilicate $r.Al_2O_3/s.SiO_2/t.H_2O$ (r, s, and t are rational numbers). Closely related to imogolite because both seem to be incipient 1:1 minerals both structurally and chemically. The ratio of SiO_2 to Al_2O_3 is usually 1.3 to 2.0 for allophane, but has been reported as low as 0.83, a greater variation than in imogolite. It appears that up to half of the Al in allophane is in fourfold coordination and half in sixfold coordination (Wada, 1989). Allophane has various colors (snow white, blue, green, brown, yellow, or colorless) and often occurs in incrustations, thin seams, or rarely stalactitic masses; occurs also as exceedingly small spherical particles especially in soils formed from volcanic ash. Syn: allophanite.

allophaneton (al-lo-phan'-e-ton) *kaolinton.*
allophanoid (al'-lo-phan'-oid) A group name for the clay minerals allophane, halloysite, and montmorillonite.
allostratigraphic unit (al'-lo-strat'-i-graph'-ic) A mappable stratiform body of sedimentary rock that is defined and identified on the basis of its bounding discontinuities. Allostratigraphic units, in order of decreasing rank, are *allogroup, alloformation,* and *allomember.* The fundamental unit is the alloformation (NACSN, 1983, Art. 58).
Allotheria (Al-lo-ther'-i-a) A subclass of Mammalia, characterized by elongate multituberculate molar teeth and peculiarly sectorial lower premolars. It includes a single order of rodentlike habitus, the longest-lived of all mammalian orders. Range, Upper Jurassic to Upper Eocene. Syn: *Multituberculata.*
allothigene (al-loth'-i-gene) *allogene.*
allothigenetic (al-lo'-thi-ge-net'-ic) *allogenic.*
allothigenic (al-lo-thi-gen'-ic) *allogenic.*
allothigenous (al-lo-thig'-e-nous) (a) *allogenic.* (b) In the classification of pyroclastics, the equivalent of *accidental.* Cf: *authigenous.*
allothimorph (al-loth'-i-morph) A constituent of a metamorphic rock which, in the new rock, has not changed its original crystal outlines (Johannsen, 1939).
allothogenic (al'-lo-tho-gen'-ic) *allogenic.*
allothrausmatic (al'-lo-thraus-mat'-ic) A descriptive term applied to igneous rocks with an orbicular texture in which the nuclei of the orbicules are xenoliths differing in composition from the groundmass (Eskola, 1938, p.476). Obsolete. Cf: *isothrausmatic; crystallothrausmatic; homeothrausmatic; heterothrausmatic.*
allotrio- (al-lot'-ri-o-) A prefix meaning "alien", "foreign".
allotrioblast (al-lot'-ri-o-blast') *xenoblast.*
allotriomorphic (al-lot'-ri-o-mor'-phic) Obsolete syn. of *xenomorphic.*
allotriomorphic-granular Obsolete syn. of *xenomorphic-granular.*
allotrope (al'-lo-trope) A crystal form of a substance that displays *allotropy.*
allotrophic (al-lo-troph'-ic) *heterotrophic.*
allotropy (al-lot'-ro-py) Polymorphism in an element, e.g. sulfur as both orthorhombic and monoclinic. See also: *allotrope.*
allowable (al-low'-a-ble) The amount of oil or gas that a well or leasehold is permitted to produce under *proration* by a regulatory body.
allowable bearing capacity In engineering geology, the load per unit area that the ground can support; it is reduced from the *ultimate bearing capacity* by a *factor of safety.*
alluaivite (al-lu-au'-dite) A colorless or pink rhombohedral mineral: $Na_{19}(Ca,Mn^{2+})_6(Ti,Nb)_3Si_{26}O_{74}Cl{\cdot}2H_2O$. A member of the *eudialyte* group.
alluaudite (al-lu-au'-dite) (a) A dirty brownish-green monoclinic mineral: $(Na,Ca)Mn^{2+}(Fe^{3+},Mn^{2+},Fe^{2+},Mg)_2(PO_4)_3$. It is a common sodium metasomatic reaction product of triphylite-heterosite. (b) A group name for monoclinic phosphate minerals of composition $ABCD(XO_4)_3$, where A = Na, Ca or Pb, B = Ca or Mg, C = Fe or Mn, D = Fe^{2+}, Fe^{3+} or Mn^{2+}, and X = P or As.
Alluvial (Al-lu'-vi-al) A name, now obsolete, applied by Jameson (1808) from the teachings of A.G. Werner in the 1790's to the group or series of rocks consisting of unconsolidated or poorly consolidated gravels, sands, clays, and peat that were believed to have been formed after the withdrawal of the ocean from the continents. It constituted the fourth (following the *Floetz*) of the divisions in which Werner placed the rocks of the geologic column. Syn: *Tertiary.*
alluvial [ore dep] (al-lu'-vi-al) Said of a *placer* formed by the action of running water, as in a stream channel or alluvial fan; also, said of the valuable mineral, e.g. gold or diamond, associated with an alluvial placer.
alluvial [sed] adj. Pertaining to or composed of *alluvium,* or deposited by a stream or running water; e.g. an "alluvial clay" or an "alluvial divide". Syn: alluvian; alluvious. n. *alluvium.*
alluvial apron A mass of sediment spread in front of another feature; often a *bajada.*
alluvial bench A *terrace* composed of alluvium. The term was used by Hobbs (1912, p.214) for the feature now known as a *bajada.*
alluvial channel A channel whose bed and banks are composed of alluvium.
alluvial cone An *alluvial fan* with very steep slopes; it is generally higher and narrower than a fan, and is composed of coarser and thicker material believed to have been deposited by larger streams. The term is sometimes used synonymously with alluvial fan. Syn: *cone of dejection; cone of detritus; hemicone; debris cone [geomorph]; cone delta; dry delta; wash [sed].*
alluvial dam A deposit of alluvium that is built by an overloaded stream and that obstructs its channel, thereby impounding water behind the deposit; esp. such a dam in a distributary on an alluvial fan.
alluvial deposit *alluvium.*
alluvial fan A low, outspread, relatively flat to gently sloping mass of loose rock material, shaped like an open fan or a segment of a cone, deposited by a stream (esp. in a semiarid region) at the place where it issues from a narrow mountain valley upon a plain or broad valley, or where a tributary stream is near or at its junction with the main stream, or wherever a constriction in a valley abruptly ceases or the gradient of the stream suddenly decreases; it is steepest near the mouth of the valley where its apex points upstream, and it slopes gently and convexly outward with gradually decreasing gradient. Cf: *alluvial cone; bajada.* Syn: *detrital fan; talus fan; dry delta.*
alluvial-fan shoreline A prograding shoreline formed where an alluvial fan is built out into a lake or sea. Syn: *fan delta.*
alluvial fill A deposit of alluvium occupying a stream valley, conspicuously thicker than the depth of the stream. Commonly consists of several deposition cycles.
alluvial flat A small *alluvial plain* bordering a river, on which alluvium is deposited during floods. Syn: *river flat.*
alluvial meander An extremely sinuous bend in an *alluvial river.*
alluvial plain A level or gently sloping tract or a slightly undulating land surface produced by extensive deposition of alluvium, usually adjacent to a river that periodically overflows its banks; it may be situated on a floodplain, a delta, or an alluvial fan. Cf: *alluvial flat.* Syn: *wash plain; waste plain; river plain; aggraded valley plain; bajada.*
alluvial-plain shoreline A prograding shoreline formed where the broad alluvial slope at the base of a mountain range is built out into a lake or sea.
alluvial river A river that occupies a broad floodplain belt over which the depth of alluvium deposited by the river equals or exceeds the depth to which scour takes place in time of flood.
alluvial slope A surface underlain by alluvium, which slopes down and away from the sides of mountains and merges with a plain or a broad valley floor (Bryan, 1923a, p.86); an alluvial surface that lacks the distinctive form of an alluvial fan or a bajada. Cf: *bajada.*
alluvial-slope spring *boundary spring.*
alluvial soil (a) A soil developing from recently deposited alluvium and exhibiting essentially no horizon development or modification of the recently deposited materials. (b) When capitalized, the term refers to a great soil group of the azonal order consisting of soils with little or no modification of the recent sediment in which they are forming. Not used in current U.S. system of soil taxonomy.
alluvial stone A rough gemstone that has been transported and deposited by a stream or river some distance from its primary source.
alluvial talus Accumulation of rock fragments of any size or shape, derived during and after heavy rains by rainwash, during storm flow, and by melting snow, passing through a gully, chute, or couloir in a cliff face or from a steep rocky slope above. Larger blocks collect at the top, but fine sizes reach the bottom as miniature debris flows (White, 1967, p.237). Cf: *scree; talus cone.*
alluvial terrace A *stream terrace* composed of unconsolidated alluvium, produced by renewed downcutting of the floodplain or valley floor by a rejuvenated stream or by the later covering of a terrace with alluvium. Cf: *rock terrace.* Syn: *stream-built terrace; fill terrace; drift terrace.*
alluvial tin *stream tin.*
alluviated spring A spring (usually a *karst spring*) whose outlet has been partly occluded by alluvial sediment (Powell, 1963).
alluviation (al-lu'-vi-a'-tion) (a) The subaerial deposition or formation of alluvium or alluvial features (such as cones or fans) at places where stream velocity is decreased or streamflow is checked; the process of aggradation or of building-up of sediments by a stream along its course, or of covering or filling a surface with alluvium. (b) A hydraulic effect on solids suspended in a current of water, whereby the coarsest and heaviest particles are the first to settle out, and the finest muds the last, as gradient or velocity of a stream is decreased (Pryor, 1963). Cf: *graded bedding.*
alluvion The formation of new land by the gradual or imperceptible action of flowing water or of waves and currents; accretion. Also, the land so added. Cf: *diluvion.*
alluvium (al-lu'-vi-um) (a) A general term for clay, silt, sand, gravel, or similar unconsolidated detrital material, deposited during comparatively recent geologic time by a stream or other body of running water, as a sorted or semisorted sediment in the bed of the stream or on its floodplain or delta, as a cone or fan at the base of a mountain slope; esp. such a deposit of fine-grained texture (silt or silty clay) deposited during time of flood. The term does not apply to subaqueous deposits in seas, estuaries, lakes, and ponds. Syn: *alluvial [sed]; alluvial deposit; alluvion.* (b) *alluvial soil.* Etymol: From Latin "alluere", "to wash against". Pl: alluvia; alluviums. Cf: *eluvium [sed]; diluvium.*
Almaden-type mercury deposit A *mineral-deposit type* characterized by stratabound disseminated cinnabar and native mercury in volcanoclastic rocks. Named for the Almadén deposit in Spain.
almandine (al'-man-dine) (a) The iron-aluminum end member of the *garnet* group, characterized by a deep-red to purplish color: $Fe_3^{2+}Al_2(SiO_4)_3$. It occurs in mica schists and other regionally metamorphosed rocks, and is used as a gemstone. Syn: *almandite; alamandine; almond stone.* (b) A violet or mauve variety of ruby spinel; a reddish-purple to purplish-red spinel. (c) A reddish-purple sapphire.
almandine-amphibolite facies A subdivision of Eskola's *amphibolite facies,* suggested by Fyfe et al. (1958) to distinguish high-pressure assemblages from those of low pressure, which these authors termed *hornblende-hornfels facies.* The term was later abandoned (Fyfe and Turner, 1966) in favor of *amphibolite facies,* although *hornblende-hornfels facies* was retained.
almandite (al'-man-dite) *almandine.*

almarudite A yellow or orange hexagonal mineral: $(K,Na)_3(Mn,Fe,Mg)_2(Be,Al)_3Si_{12}O_{30}$. A member of the *milarite* group.
almashite (al-mash'-ite) A green or black variety of amber that has a low content of oxygen (2.5-3%).
almeraite (al-mer'-a-ite) A dubious reddish mineral: $KNaMgCl_4 \cdot H_2O$. It may be a mixture of carnallite and halite.
almeriite (al-me-ri'-ite) A discredited synonym of natrolunite.
almond-shaped bomb (al'-mond) *spindle-shaped bomb.*
almond stone *almandine.*
almost-atoll An atoll with one or more noncoral islands, generally of volcanic origin, in the central lagoon. The classic example is Truk Atoll in the Caroline Islands, 50 km across, with 20 volcanic islets, its irregular lagoon floor suggesting a drowned Pleistocene karst (Guilcher, 1988, p.128).
alnoite (al'-no-ite) In the *IUGS classification,* a lamprophyre melilite with a groundmass (Woolley et al., 1996). Perovskite, apatite, nepheline, and garnet may be present. Its name (Rosenbusch, 1887) is derived from Alnö, Sweden. Also spelled: allnöite; alnöite.
alongshore (a-long'-shore) Along the shore or coast, such as an "alongshore drift" or "alongshore current"; *longshore.*
alp (a) A high, rugged, steep-sided mountain, esp. one that is snow-covered, resembling topographically those in the European Alps. (b) A high pasture or meadowland on a mountain side, between timberline and snowline, like those in the Swiss Alps; also a high *shoulder [glac geol]* or gentle slope, esp. in the Swiss Alps, commonly above a glaciated valley at a marked change of slope. Cf: *alb.*
alpestrine (al'-pes-trine) *montane.*
alpha [cryst] (al'-pha) (a) In a biaxial crystal, the smallest *index of refraction.* (b) The interaxial angle between the *b* and *c* crystallographic axes. Cf: *beta [cryst]; gamma [cryst].*
alpha [mineral] adj. Of or relating to one of two or more closely related minerals and specifying a particular physical structure (esp. a polymorphous modification); specif. said of a mineral that is stable at a temperature lower than those of its *beta* and *gamma* polymorphs (e.g. "alpha cristobalite" or "α-cristobalite", the low-temperature tetragonal phase of cristobalite). Some mineralogists reverse this convention, using α for the high-temperature phase (e.g. "alpha carnegieite", the isometric phase of carnegieite stable above 690°C).
alpha* angle The angle of the reciprocal lattice between the *b* axis* and the *c* axis,* which is equal to the dihedral angle between (010) and (001). Cf: *beta* angle; gamma* angle.*
alpha chalcocite *digenite.*
alpha decay Radioactive decay of an atomic nucleus by the emission of an *alpha particle.* Cf: *beta decay.*
Alpha diversity The number of taxa in a local region, community, or habitat (Whittaker, 1972; Sepkoski, 1988).
alpha level *significance.*
alpha particle (a) A particle, emitted from an atomic nucleus during one type of radioactive decay, which is positively charged and has two protons and two neutrons. It is physically identical with the nucleus of a ^4He atom. Cf: *beta particle; gamma radiation.* (b) By extension, the nucleus of a ^4He atom. Less-preferred syn: *alpha ray.*
alpha-particle recoil tracks The paths of *radiation damage* in a solid substance, caused by the recoil nucleus accompanying the alpha-particle decay of uranium and thorium impurities. The tracks are similar to *fission tracks,* but are much smaller and more numerous. An age determination can be made on the basis of the track density, usually examined with an electron microscope, plus determinations of the thorium and uranium contents of the sample (Huang and Walker, 1967, p.1103-1106). Syn: *alpha-recoil tracks.*
alpha quartz The trigonal polymorph of quartz that is stable below 573°C, that has a vertical axis of three-fold symmetry and three horizontal axes of two-fold symmetry, and that has a higher refractive index and birefringence than those of *beta quartz.* It occurs commonly as a constituent of igneous, metamorphic, and sedimentary rocks, and in veins, geodes, and large pegmatites. Also spelled: *α-quartz.* Syn: *low quartz.*
alpha ray A less-preferred syn. of *alpha particle.*
alpha-recoil tracks *alpha-particle recoil tracks.*
alpha scintillometry Measurement of the radon content of the atmosphere by observing the alpha particle activity.
alpha-vredenburgite A homogeneous, metastable mineral: $(Mn,Fe)_3O_4$. It has the same composition as that of *beta-vredenburgite,* and is regarded as an iron-rich hausmannite. Syn: *vredenburgite.*
Alpides (Al'-pi-des) A name used by Suess for the great orogenic belt or system of young folded mountains, including the Alps, that extends eastward from Spain to southeast Asia. Syn: *Alpine-Himalayan belt.*
alpine [ecol] (al'-pine) Characteristic or descriptive of the mountainous regions lying between timberline and snowline; said of the climate, flora, relief, ecology, etc. Less strictly, pertaining to high elevations and cold climates. Cf: *montane.*
alpine [geomorph] Pertaining to, characteristic of, or resembling the European Alps or any lofty mountain or mountain system, esp. one so modified by intense glacial erosion as to contain aiguilles, cirques, horns, etc.; e.g. an alpine lake resulting from glacial erosion and situated in or along the border of a high mountain region. Spelled Alpine when referring specif. to the European Alps.
alpine [struc geol] A general term for topographical and structural features that resemble in grandeur and complexity those of the European Alps, regardless of the age or location of the mountains and features so described.
alpine glacier Any *glacier* in a mountain range except an *ice cap* or *ice sheet.* It usually originates in a cirque and may flow down into a valley previously carved by a stream. Syn: *mountain glacier; valley glacier.*
Alpine-Himalayan belt *Alpides.*
alpine karst Karst formed at high altitude. Cf: *glaciokarst; nival karst.*
Alpine Meadow soils A great soil group of the intrazonal order, comprised of dark soils of grassy meadows at altitudes above the timberline. Not used in current U.S. system of soil taxonomy.
Alpine Orogeny A name for the relatively young orogenic events of southern Europe and Asia, by which the rocks of the Alps were strongly deformed and uplifted. Most geologists restrict the era to the Tertiary, with many episodes of varying strength from place to place, ending during the Miocene or Pliocene.
alpine peridotite A tectonic, not petrologic, term for ultramafic rock, commonly serpentinized and bounded by fault surfaces, marking convergent plate boundaries.
Alpine-type facies series Rocks produced in the highest-pressure type of dynamothermal regional metamorphism at no more than moderate temperature (150° to 400°C), characterized by the presence of the pumpellyite and glaucophane schist facies. It may also involve the zeolite facies in the low-temperature, high-pressure range and the deep-seated eclogite facies at the highest pressures with moderate temperatures (Hietanen, 1967, p.203).
Alpine-type lead-zinc deposit *Bleiberg-type lead-zinc deposit.*
alpinotype tectonics (al-pin'-o-type) The tectonics of orogenic belts, in which the internal parts of the orogen by are affected by metamorphism, deep-seated plastic folding and plutonism, and in the external parts by lateral thrusting, which has produced nappes, thrust sheets, and fault-related folds. Cf: *germanotype tectonics.* Syn: *orthotectonics.*
alpland (alp'-land) An area whose topography resembles that of the Alps.
alquifou (al'-qui-fou) A coarse-grained galena, used by potters in preparing a green glaze.
alsakharovite-Zn A colorless or pale brown monoclinic mineral: $NaSrKZn(Ti,Nb)_4(Si_4O_{12})_2(O,OH)_4 \cdot 7H_2O$. A member of the *vuoriyarvite* group.
alsbachite (als'-bach-ite) A porphyritic *granodiorite* chiefly composed of sodic plagioclase, quartz, and alkali feldspar, often with accessory garnet and mica. The quartz and alkali feldspar commonly form the phenocrysts in a granular groundmass. Named for Alsbach, Germany by Chelius in 1892. Obsolete.
ALSEP An acronym for Apollo Lunar Surface Experiments Package, a collection of geophysical and other instruments, with auxilary devices for data collection and transmission, powered by radioisotopic thermoelectric generators, and emplaced on the Moon during Apollo 12, 14, 15, 16, and 17 missions.
alstonite (al'-ston-ite) A variously colored triclinic mineral: $BaCa(CO_3)_2$. It is the pseudo-orthorhombic polymorph of barytocalcite and paralstonite. Syn: *bromlite.*
alta A miner's term for the black, shaly, highly sheared capping of quicksilver orebodies. Syn: *black alta.*
Altaides (Al-ta-i'-des) A late Paleozoic orogenic belt bordering the western and southern margins of the Siberian craton. It is named for the Altai Mountains of central Asia, where there was late Paleozoic deformation. The Altaides belt is very broad, and may have formed by convergent-margin tectonism and strike-slip duplexing.
altaite (al'-ta-ite) A tin-white cubic mineral: PbTe. It is a member of the *galena* group.
alteration (al-ter-a'-tion) (a) Any change in the mineralogic composition of a rock brought about by physical or chemical means, esp. by the action of hydrothermal solutions; also, a secondary, i.e. supergene, change in a rock or mineral. (b) Changes in the chemical or mineralogical composition of a rock produced by weathering.
alteration pipe A pipelike zone of alteration beneath a *massive sulfide deposit* that forms where mineralizing solutions rise to a point of isssue, generally along a fault.
altered rock (al'-tered) A rock that has undergone changes in its chemical and mineralogic composition through alteration (as opposed to metamorphic) processes since its original formation.
alterite (al'-ter-ite) A general term for altered, unrecognizable grains of heavy minerals.
alternate (al'-ter-nate) In plant morphology, pertaining to the attachment of parts, e.g. leaves, singly at each node; also, said of plant parts in regular occurrence between other organs.
alternate bar Referring to streams, one of several bars that form alternately along the left and right banks of a river, proceeding downstream.
alternate folding Deflection of the surface of a brachiopod shell in which the *fold* of one valve is opposed by the *sulcus* of the other valve.
alternate terrace One of several meander-scar terraces.

alternating-field demagnetization The process of progressive demagnetization by exposing rock samples to alternating-current (AC) fields that are continuously decreased in the absence of a direct-current (DC) field. This brings about randomization of magnetic remanence carried by material whose magnetization can be reversed by fields less than or equal to the maximum peak AC field. Cf: *thermal demagnetization; chemical demagnetization.* Syn: *AC demagnetization.*

alternation of generations (al-ter-na'-tion) The occurrence in the life cycle of a plant or animal of two or more forms having different types of development, usually involving the regular alternation of sexual and asexual generations. Syn: *heterogony; metagenesis [evol].*

alternative hypothesis In statistical testing, a hypothesis that is the complement of the *null hypothesis.* Together, the null and alternative hypotheses must encompass all possible outcomes of the test.

alternative metastable behavior A *polymorphic* transformation in which the final phase is not the most stable phase (that is, the one with the lowest *Gibbs free energy*) under the conditions of the transformation, but instead is *metastable.*

althausite (alt'-haus-ite) A vitreous colorless orthorhombic mineral: $Mg_2(PO_4)(OH,F,O)$.

althupite (al'-thu-pite) A yellow triclinic mineral: $ThAl(UO_2)_7O_2(PO_4)_4(OH)_5{\cdot}15H_2O$.

altimeter (al-tim'-e-ter) An instrument, usually an aneroid barometer, for determining height above ground or above mean sea level, based on the fall of atmospheric pressure accompanying an increase in altitude.

altimeter satellite A satellite that carries an instrument that measures the height of the satellite above the sea surface, an *altimeter,* and another instrument that measures the height of the satellite above Earth's center of mass. The difference of the two measurements gives the height of the sea surface. Satellites such as Topex/Poseidon and Jason can make both measurements with an accuracy of 1-3 cm. Sea-level measurements are used to calculate tides, *Kelvin waves, Rossby waves,* and surface *geostrophic currents.*

altimetric frequency curve (al-ti-met'-ric) A curve showing the altitudinal distribution of the highest points in a series of small squares that divide the map of a given region.

altiplanation (al'-ti-pla-na'-tion) A group of erosion processes, involving solifluction and related mass movement, which tend to produce flat or terracelike surfaces. Such processes are especially active at high elevations and latitudes where periglacial processes predominate. Cf: *equiplanation; cryoplanation.*

altiplanation terrace A hillside or summit bench that is cut in bedrock, lacks predominant structural control, and is confined to cold climates. These terraces have a veneer of solifluction debris, which may be imprinted with patterned ground. They appear to range in width from about 10 m to 2-3 km and to exceed 10 km in length. Their gradient is commonly 1° to 12 ° (Washburn, 1973, p.205). Syn: *equiplanation terrace.*

altiplano (al-ti-pla'-no) A high-lying plateau or tableland; specif. the high plateau of western Bolivia, consisting of a string of intermontane basins. Etymol: Spanish. Also called: altiplanicie.

altisite A vitreous colorless monoclinic mineral: $Na_3K_6Ti_2Al_2Si_8O_{26}Cl_3$.

Altithermal (Al-ti-ther'-mal) n. A term proposed by Antevs (1948, p.176) for a dry postglacial interval (from about 8,000 to 4,000 calendar years ago) following the Anathermal and preceding the Medithermal, during which temperatures were higher than at present. It corresponds to the *Atlantic* interval or the middle part of the *Hypsithermal.* See also: *thermal maximum; Megathermal; Xerothermic.* adj. Pertaining to the postglacial Altithermal interval and to its climate, deposits, biota, and events.

altithermal [climat] adj. (a) Pertaining or belonging to a climate characterized by rising or high temperatures; e.g. "altithermal soil" of postglacial time. (b) Pertaining to the postglacial Altithermal interval and to its climate, deposits, biota, and events.

altitude (al'-ti-tude) (a) The vertical distance of a level, a point, or an object considered as a point, above or below the surface of the Earth, measured from a given datum, usually mean sea level. Altitude is positive if the point or object is above the given datum, and negative if it is below it. Cf: *elevation [surv].* (b) The vertical angle between the horizontal plane of the observer and any higher point (such as the summit of a peak).

altitude matrix A rectangular grid containing elevation data. An altitude matrix can be obtained from the stereoscopic study of overlapping aerial photographs on analytical stereoplotters, or from interpolation of irregularly distributed elevation data points (Krzanowski et al., 1993). See also: *digital terrain model; digital elevation model.*

alto (al'-to) A term used in the SW U.S. for a bluff, height, or hill. Etymol: Spanish, "high ground".

Altonian (Al-ton'-ian) South Pacific stage; upper lower Miocene (above Otaian, below Clifdenian).

alum (al'-um) A group of hydrous aluminum sulfate minerals, including potassium alum, soda alum, tschermigite and lonecreekite.

alum coal A pyritic, argillaceous brown coal containing alum that formed as a weathering product. Cf: *alum shale.*

alum earth *alum shale.*

alumina Oxide of aluminum; Al_2O_3. A material in used the production of aluminum metal, refractories, abrasives, and chemicals.

aluminite (a-lu'-mi-nite) An earthy-white monoclinic mineral: $Al_2(SO_4)(OH)_4{\cdot}7H_2O$. Syn: *websterite.*

aluminium A metallic white cubic mineral: Al. Occurs in highly reducing environment.

aluminobarroisite A green monoclinic member of the amphibole mineral family; the aluminum-rich member of the *barroisite* series: $NaCa(Mg_3Al_2)(Si_7Al)O_{22}(OH)_2$.

aluminoceladonite A dioctahedral mica mineral with Al dominant: $KAl(Mg,Fe^{2+})Si_4O_{10}(OH)_2$.

aluminocopiapite (a-lu'-mi-no-co'-pi-a-pite') An earthy-yellow triclinic mineral of the *copiapite* group: $Al_{2/3}Fe^{3+}{}_4(SO_4)_6(OH)_2{\cdot}20H_2O$.

alumino-ferrobarroisite A hypothetical member of the amphibole mineral family; a barroisite with Al and Fe^{2+} dominant: $NaCa(Fe^{2+}{}_3Al_2)(Si_7Al)O_{22}(OH)_2$.

alumino-ferro-hornblende A green monoclinic member of the *amphibole* mineral family: $Ca_2(Fe^{2+}{}_4Al)(Si_7Al)O_{22}(OH)_2$. Probably an unnecessary name for *ferrohornblende.*

alumino-ferrotschermakite A hypothetical member of the *amphibole* mineral family; a highly aluminous tschermakite with Fe^{2+} dominant: $Ca_2(Fe^{2+}{}_3Al_2)(Si_6Al_2)O_{22}(OH)_2$.

alumino-katophorite A hypothetical member of the *amphibole* mineral family; probably an unnecessary name for *katophorite:* $Na_2Ca(Fe^{2+}{}_4Al)(Si^7Al)O_{22}(OH)_2$.

alumino-magnesio-hornblende A green monoclinic member of the *amphibole* mineral family; probably an unnecessary name for *magnesiohornblende:* $Ca_2(Mg_4Al)(Si_7Al)O_{22}(OH)_2$.

alumino-magnesiohulsite A strongly pleochroic monoclinic mineral; the aluminum-magnesium analogue of hulsite: $Mg_2(Al,Mg,Sn)O_2(BO_3)$.

alumino-magnesiosadanagaite A hypothetical member of the *amphibole* mineral family: $NaCa_2(Mg_3Al_2)(Si_5Al_3)O_{22}(OH)_2$. Not approved by the International Mineralogical Association.

alumino-magnesiotaramite A hypothetical member of the *amphibole* mineral family; the magnesium-aluminum end-member of the *taramite* series: $Na_2Ca(Mg_3Al_2)(Si_6Al_2)O_{22}(OH)_2$.

aluminosilicate (a-lu'-mi-no-sil'-i-cate) A silicate in which aluminum substitutes for some of the silicon in the SiO_4 tetrahedra.

aluminotaramite (a-lu'-mi-no-tar'-am-ite) A hypothetical monoclinic amphibole, the aluminum analogue of *taramite:* $Na_2Ca(Fe^{2+}{}_3Al_2)(Si_6Al_2)O_{22}(OH)_2$.

aluminotschermakite A green monoclinic member of the *amphibole* mineral family; a highly aluminous member of the *tschermakite* series: $Ca_2(Mg_3Al_2)(Si_6Al_2)O_{22}(OH)_2$.

alumino-winchite A hypothetical member of the *amphibole* mineral family; probably an unnecessary name for *winchite:* $NaCa(Mg_4Al)Si_8O_{22}(OH)_2$.

alumite (al'-um-ite) *alunite.*

alumocalcite (a-lu'-mo-cal'-cite) A variety of opal containing alumina and lime as impurities.

alumogel (a-lu'-mo-gel) An amorphous aluminum hydroxide that is a constituent of bauxite; *cliachite.*

alumohydrocalcite (a-lu'-mo-hy'-dro-cal'-cite) A triclinic mineral: $CaAl_2(CO_3)_2(OH)_4{\cdot}3H_2O$. It occurs as chalky white or pale-colored masses consisting of radially fibrous spherulites. Cf: *para-alumohydrocalcite.*

alumoklyuchevskite A dark green monoclinic mineral, the aluminum analogue of *klyuchevskite:* $K_3Cu_3AlO_2(SO_4)_4$.

alumopharmacosiderite (a-lu'-mo-phar'-ma-co-sid'-er-ite) A white cubic mineral, the Al-analogue of pharmacosiderite: $KAl_4(AsO_4)_3(OH)_4{\cdot}6.5H_2O$.

alumotantite (a-lu'-mo-tan'-tite) An adamantine colorless orthorhombic mineral: $AlTaO_4$.

alumotungstite (a-lu'-mo-tungs'-tite) A yellow cubic mineral of uncertain composition, possibly: $(W,Al)(O,OH)_3$.

alum rock (al'-um) *alunite.*

alum schist *alum shale.*

alum shale An argillaceous, often carbonaceous, rock impregnated with *alum,* originally containing iron sulfide (pyrite, marcasite) which, when decomposed, formed sulfuric acid that reacted with the aluminous and potassic materials of the rock to produce aluminum sulfates. Syn: *alum earth; alum schist; alum slate.*

alum slate *alum shale.*

alumstone (al'-um-stone) *alunite.*

alunite (al'-u-nite) (a) A vitreous white, yellowish, etc. rhombohedral mineral: $KAl_3(SO_4)_2(OH)_6$. It is isomorphous with natroalunite, sometimes contains appreciable sodium, generally occurs as a hydrothermal-alteration product in feldspathic igneous rocks, and is used in the manufacture of alum. Syn: *alumstone; alum rock; alumite.* (b) A group of minerals containing hydrous sulfates, including *alunite, natroalunite, jarosite, natrojarosite, ammoniojarosite, argentojarosite,* and *plumbojarosite.*

alunogen (a-lu'-no-gen) A colorless triclinic mineral: $Al_2(SO_4)_3{\cdot}17H_2O$. It occurs as a white, fibrous incrustation or efflorescence formed by volcanic action or by decomposition of pyrite in alum shales. Syn: *feather alum; hair salt.*

alurgite (a-lur'-gite) A manganiferous muscovite.

alushtite (a-lush'-tite) A bluish or greenish mineral, probably *tosudite,* a mixed-layer chorite-smectite.

alvanite (al′-van-ite) A vitreous bluish-green or bluish-black monoclinic mineral: $(Zn,Ni)Al_4(VO_3)_2(OH)_{12}\cdot 2H_2O$.

alveolar (al-ve-o′-lar) (a) In invertebrates, having small cavities or pits. (b) In vertebrates, pertaining to a tooth socket.

alveolar fabric Common structure in pedogenic carbonates occurring within tubular vugs, fractures, and former rootlets where thin threads of micrite form septa across the cavities.

alveolar weathering *honeycomb weathering.*

alveolate (al-ve-o′-late) A specimen or characteristic feature of a recently recognized eukaryote lineage that includes ciliates, dinoflagellates, and disease-causing parasites called apicomplexans. Common feature is a system of sacs (alveoli) beneath the cell membranes (Gajadhar et al., 1991; Speer, 1998).

alveole (al′-ve-ole) A space or cavity, such as a vacuole in a foraminiferal test wall; informal equivalent of *alveolus.*

alveolinid (al′-ve-o-lin′-id) Any foraminifer belonging to the family Alveolinidae, characterized by an imperforate, porcelaneous, axially elongated test that may be subcylindrical, fusiform, ellipsoidal, or spherical. Range, Lower Cretaceous to Holocene. Although this group resembles the fusulinids in shape, the two groups are not genetically related.

alveolitoid (al-ve-o′-lit-oid) Said of a type of reclined *corallite* having a vaulted upper wall and a nearly plane lower wall parallel to the surface of adherence of the colony (as in the genus *Alveolites*) (TIP, 1956, pt. F, p.245).

alveolization The pitting of a rock surface, as it is produced by wind loaded with sand, by water charged with carbon dioxide, or by plant roots. Etymol: Latin "alveolatus", "hollowed out".

alveolus (al-ve-o′-lus) (a) In invertebrates, a small cavity or pit; e.g., a minute blind cavity in the *keriotheca* of some fusulinids, the conical cavity in the anterior end of the *guard* of a belemnite, or a pit in the skeleton of a bryozoan colony. (b) In vertebrates, the socket of a tooth. Pl: alveoli. Adj: alveolar. Syn: *alveole.*

alveozone (al′-ve-o-zone) A trough just beneath the *selenizone* or periphery in certain gastropods (Batten, 1966, p.21).

alvikite (al′-vik-ite) A hypabyssal rock, the fine-grained equivalent of *sövite* (calcite carbonatite). The name, given by Eckermann in 1928, is for Alvik on Alnö Island, Sweden. Obsolete.

amagmatic (a-mag-mat′-ic) Said of a structure, region, or process that does not involve magmatic activity.

amakinite (a-mak′-i-nite) A yellow-green or pale green trigonal mineral: $(Fe^{2+},Mg)(OH)_2$.

amalgam (a-mal′-gam) (a) A naturally occurring alloy of silver with mercury; mercurian silver. It is found in the oxidation zone of silver deposits and as scattered grains in cinnabar ores. See also: *gold amalgam; moschellandsbergite.* Syn: *silver amalgam; argental mercury.* (b) A general term for alloys of mercury with one or more of the well-known metals (except iron and platinum); esp. an alloy of mercury with gold, containing 40-60% gold, and obtained from the plates in a mill treating gold ore.

amalgamate (a-mal′-gam-ate) Said of a type of wall structure in trepostome bryozoans in which the zooecial boundaries are not visible in tangential section. Cf: *integrate.*

amalgamation The joining together of two *terranes* in an oceanic setting away from a continental margin (Howell, 1995). Cf: *accretion [struc geol].*

amalgamation [metal] A method of gold and silver extraction from pulverized ores by alloying the precious metals with mercury. The resulting *amalgam* is then fire-refined to recover the gold and silver.

amalgamation surface The surface separating two *beds* or *layers* of similar *lithology* in vertical succession. Amalgamation surfaces are erosional surfaces that develop just prior to deposition of the overlying bed; such surfaces are commonly recognizable by a lag of coarse material immediately on the surface or by the truncation of structures in the underlying unit, immediately beneath the amalgamation surface.

amang A term used in Malaysia for the heavy iron and tungsten minerals (and associated minerals) found with placer cassiterite deposits.

amarantite (am-a-ran′-tite) A dark reddish-purple triclinic mineral: $Fe^{3+}{}_2O(SO_4)_2\cdot 7H_2O$.

amararhysis (am-a-ra′-rhy-sis) A *skeletal canal* in dictyonine hexactinellid sponges running longitudinally within the body wall, opening at intervals to the *cloaca* by slitlike apertures, and opening to the exterior by radial canals terminating in oscula. It is part of the *exhalant system.* Pl: amararhyses.

amargosite (a-mar-go′-site) A syn. of *bentonite.* Etymol: *Amargosite,* a trade name for a bentonitic clay (montmorillonite) from the Amargosa River, Calif. Rarely used.

amarillite (am-a-ril′-lite) A pale greenish-yellow monoclinic mineral: $NaFe^{3+}(SO_4)_2\cdot 6H_2O$.

amatrice (a-ma′-trice) A green gem cut from variscite and its surrounding matrix of gray, reddish, or brownish crystalline quartz or chalcedony.

Amazonian The youngest system of rocks and the period they represent in the geologic stratigraphy of the planet Mars. Named after Amazonis Planitia, a broad lowland area in the northern hemisphere of Mars which contains relatively smooth, moderately cratered plains that are among the youngest material units on Mars (Tanaka, 1986). See also: *Hesperian; Noachian.*

amazonite (am′-a-zon-ite′) An apple-green, bright green or blue-green variety of microcline, sometimes used as a gemstone. The color is caused by lead. Syn: *amazonstone.*

amazonstone (am′-a-zon-stone′) The earlier name for *amazonite.* Also spelled: Amazon stone.

amb The contour or outline of a pollen grain (less commonly of a spore) as viewed from directly above one of the poles. See also: *polar view.* Less preferred syn: *equatorial limb.*

ambatoarinite (am′-ba-to-ar′-i-nite) An incompletely described mineral: $Sr(Ce,La,Nd)_2O(CO_3)_3$ (?).

amber (am′-ber) (a) A hard, brittle fossil resin, usually yellowish to brownish and translucent or transparent, that is derived from coniferous trees, frequently encloses insects and other organisms, takes a fine polish, and is found in alluvial soils, clays, and recent sediments, in beds of lignite, and on some seashores (as of the Baltic Sea). It is used chiefly in making pipe mouthpieces, beads, and other small ornamental objects. Syn: *succinite; bernstein; electrum.* (b) A term applied to a group of fossil resins containing considerable succinic acid and having highly variable C:H:O ratios; e.g. almashite, simetite, delatynite, and ambrosine.

amberat (am′-ber-at) A blackish, acrid-scented, tarlike organic cave deposit derived from the urine of pack rats (genus *Neotoma*).

amberite (am′-ber-ite) *ambrite.*

amber mica *phlogopite.*

amberoid (am′-ber-oid) A gem material consisting of small fragments of genuine amber that have been artificially united or reconstructed by heat and pressure. It may be characterized by an obvious flow structure or by a dull spot left by a drop of ether. Also spelled: *ambroid.* Syn: *pressed amber.*

ambient surface temperature The average yearly temperature at the Earth's surface for a particular region or area.

ambilateral class (am-bi-lat′-er-al) A topological classification of channel networks of given magnitude: two topologically distinct channel networks belong to the same ambilateral class if one can be converted into the other by transposing the right-left arrangement of the tributary links at any of the junctions (Smart, 1969, p.1761).

ambitus (am′-bi-tus) The exterior edge or periphery; e.g. the greatest horizontal circumference of an echinoid *test* or edrioasteroid *theca,* or the thecal outline of a dinoflagellate as viewed from the dorsal or ventral side. Pl: ambitus.

amblygonite (am-blyg′-o-nite) (a) A vitreous greasy-white or pale-colored triclinic mineral: $LiAl(PO_4)(F,OH)$. It commonly occurs in cleavable masses in pegmatites, and is mined as an ore of lithium. Syn: *hebronite.* (b) A group name for minerals with the general chemical formula: $(Li,Na)(AlFe^{3+})PO_4(F,OH)$.

amblyproct (am′-bly-proct) Said of a sponge in which the exhalant surface is in the form of an open cup.

amblystegite (am-bly-steg′-ite) A dark brownish-green to black variety of hypersthene.

ambonite (am′-bo-nite) A group of porphyritic cordierite-bearing hornblende-biotite andesites and dacites originally described by Verbeck in 1905 from the Indonesian island of Ambon (or Amboina); also, any rock in that group, the three main subdivisions being orthopyroxene-bearing, mica-bearing, and hornblende-bearing andesites and dacites (Johannsen, 1939, p.239). Obsolete.

ambrite (am′-brite) A yellowish-gray, subtransparent variety of retinite resembling amber, occurring in large masses in the coalfields of New Zealand. Its approximate formula: $C_{40}H_{66}O_5$. Syn: *amberite.*

ambroid (am′-broid) *amberoid.*

ambrosine (am′-bro-sine) A yellowish or clove-brown variety of amber containing considerable succinic acid and occurring as rounded masses in phosphate beds.

ambulacral (am-bu-lac′-ral) adj. (a) Pertaining to an *ambulacrum* or to ambulacra collectively; e.g. a small needlelike "ambulacral spine" attached to the part of a side plate directed toward the main food groove of the ambulacrum of a blastoid. (b) Corresponding in position to an ambulacrum, or referring to the zone in which an ambulacrum is present; e.g. an "ambulacral ray" representing an area defined by direction of an ambulacrum radiating from the mouth of a crinoid. Cf: *interambulacral.* n. A small calcareous plate that covers part of an ambulacrum of an echinoderm; e.g. a side plate or the broader of the floor plates in a cystoid. Pl: ambulacralia.

ambulacral bifurcation plate One of two large, unpaired plates in edrioasteroids, one lying at the junction of each lateral pair of ambulacra (I-II and IV-V); the perradial tip of each bifurcation plate lies at the junction of the two ambulacral perradial lines with each other and with one end of the transverse oral midline (Bell, 1976).

ambulacral floor plate *floor plate.*

ambulacral groove A passageway, furrow, or depression along the course of an *ambulacrum* of an echinoderm, commonly covered by ambulacral plates, through which food particles are believed to have been conveyed to the mouth by means of ciliary currents. Syn: *food groove.*

ambulacral pore An opening, in or between ambulacral plates of an echinoderm, for the passage of tube feet or of podia or for the connection of a *podium* to an *ampulla.*

ambulacral radius Thecal radius in an edrioasteroid, defined by the midline of an *ambulacrum;* in curved ambulacra the radius is defined by the midline of the proximal end of the ambulacrum, which is extended directly toward the edge of the theca (Bell, 1976).

ambulacral tunnel The space enclosed between the ambulacral floor plates and the overlying cover plates of an edrioasteroid; the ambulacral food groove extends along the floor of this space (Bell, 1976).

ambulacrum (am-bu-lac'-rum) (a) One of the narrow, usually elongate areas extending radially from the mouth of an echinoderm, along which run the principal nerves, the blood vessels, the radial canals of the water-vascular system, and the *ambulacral groove.* Commonly it bears numerous tube feet. Most echinoderms have five such areas near the mouth that may branch or remain unbranched, e.g., the blastoids and edrioasteroids. See also: *ray [paleont].* (b) A trough of the *coenosteum* separating *collines* on the surface of some meandroid *coralla* of a scleractinian coral. Pl: ambulacra.

ameghinite (a-me-ghi'-nite) A colorless monoclinic mineral: $NaB_3O_3(OH)_4$.

amensalism (a-men'-sal-ism) *antibiosis.*

ament (am'-ent) A spike inflorescence of closely spaced, often intricate, apetalous flowers. It may be conelike and is usually deciduous. Syn: *catkin.*

American brilliant cut *Tolkowsky theoretical brilliant cut.*

American jade (a) Nephrite from Wyoming. (b) A syn. of *californite.*

American ruby A red garnet (pyrope) found in Arizona and New Mexico.

amerikanite (a-mer'-i-kan-ite) A natural glass from South America (Colombia and Peru), once classed as a tektite but now believed to be of volcanic origin.

amerospore (a-mer'-o-spore) A one-celled, asexual fungal spore of the *Saccardoan spore groups.*

amesite (ames'-ite) A trioctahedral serpentine-type mineral of ideal composition $Mg_2Al(OH)_4(AlSi)O_5$. It is found as well-crystallized hexagonal prisms in metamorphosed high-alumina rocks. Triclinic, hexagonal and rhombohedral polytypes have been reported. Cf: *kellyite.*

amethyst (am'-e-thyst) (a) A transparent to translucent, purple to pale violet variety of crystalline quartz, much used as a semiprecious gemstone. The color is due to iron compounds. Syn: *bishop's stone.* (b) A term applied to a deep-purple variety of corundum and to a pale reddish-violet beryl.

Amgaian European stage: Middle Cambrian above Toyonian and below Mayaian (Cowie and Bassett, 1989).

amherstite (am'-herst-ite) A *quartz monzodiorite* in which the feldspar is andesine antiperthite. It was described by Watson and Taber in 1913, from Amherst County, Virginia. Obsolete.

amianthus (am-i-an'-thus) A syn. of *asbestos,* applied esp. to a fine silky variety such as chrysotile. Syn: amiantus.

amicite (am-ic'-ite) A colorless monoclinic zeolite mineral: $K_2Na_2Al_4Si_4O_{16}{\cdot}5H_2O$.

amictic (a-mic'-tic) Said of a lake that is perennially covered with ice and does not undergo mixing of the water column (Wetzel, 2001, p.82). Cf: *dimictic; monomictic [lake]; polymictic [lake].*

amino acid (a-mi'-no) One of the group of organic compounds, containing both amine ($-NH_2$) and carboxyl ($-COOH$) groups, which are the building blocks of proteins and therefore essential to life processes. All but one (glycine) are optically active and most occur in nature in the L-form.

amino-acid racemization age method *racemization age method.*

aminochronology Using amino acid enantiomeric ratios to determine the age of a fossil.

aminoffite (am-i-noff'-ite) A vitreous colorless or yellowish tetragonal mineral: $Ca_3Be_2Si_3O_{10}(OH)_2$.

aminostratigraphy Using amino acid enantiomeric ratios to determine the stratigraphy of a geologic deposit.

Ammeburg-type deposit *Broken Hill-type deposit.*

ammersooite (am-mer-soo'-ite) A clay mineral (illite?) occurring in soil.

ammite (am'-mite) *ammonite [sed].*

ammonia alum (am-mo'-ni-a) *tschermigite.*

ammonioalunite (am-mo'-ni-o-al'-un-ite) A grayish-white rhombohedral mineral: $(NH_4)Al_3(SO_4)_2(OH)_6$.

ammonioborite (am-mo'-ni-o-bor'-ite) A monoclinic white mineral: $(NH_4)_3B_{15}O_{20}(OH)_8{\cdot}4H_2O$. It is found as aggregates of minute plates in fumarolic deposits.

ammoniojarosite (am-mo'-ni-o-ja'-ro-site) A pale-yellow earthy mineral of the *alunite* group: $(NH_4)Fe^{3+}{}_3(SO_4)_2(OH)_6$.

ammonioleucite A white tetragonal mineral: $(NH_4,K)(Si_2Al)O_6$. The ammonium-dominant analogue of *leucite.*

ammonite [paleont] (am'-mo-nite) Any ammonoid belonging to the suborder Ammonitina, characterized by a thick, strongly ornamented shell with sutures having finely divided lobes and saddles. Range, Jurassic to Cretaceous.

ammonite [sed] An obsolete term, applied in the 17th and 18th centuries to a sedimentary rock now known as oolite. Syn: *ammite.*

ammonitella (am'-mo-ni-tel'-la) The *protoconch* and first whorl of the ammonoid shell, terminating at the primary *varix (b).*

ammonitic suture (am-mo-nit'-ic) A type of *suture* in ammonoids characterized by complex fluting in which all of the smaller secondary and tertiary lobes and saddles (developed on a larger primary set) are denticulate or frilled; esp. a suture in ammonites. Cf: *goniatitic suture; ceratitic suture.*

ammonium alum (am-mo'-ni-um) *tschermigite.*

ammonoid (am'-mo-noid) Any extinct cephalopod belonging to the order Ammonoidea, characterized by an external shell that is symmetrical and coiled in a plane and has a bulbous protoconch, septa that form angular sutural flexures, and a small marginal siphuncle. Range, Lower Devonian to Upper Cretaceous.

amniote (am'-ni-ote) adj. Pertaining to a vertebrate egg characterized primitively by a large yolk and covered by a shell that is lined with extra-embryonic membranes (including the amnion) which are shed at hatching or birth and function to conserve water and for the exchange of gases. n. Any vertebrate reproducing by means of such an egg (or modifications therefrom, as in advanced mammals which have lost yolk and shell membranes); includes all tetrapod classes except Amphibia.

amoebocyte (a-moe'-bo-cyte) A sponge cell of amoeboid (irregular, changing) form. It includes such cells as archaeocytes, sclerocytes, trophocytes, and collencytes. Also called: amoeboid cell.

amoeboid (a-moe'-boid) Said of a fold that has no specific shape and a very shallow dip, e.g. a *placanticline.* Rarely used.

amorphous (a-mor'-phous) (a) Said of a mineral or other substance that lacks crystalline structure, or whose internal arrangement is so irregular that there is no characteristic external form. The term does not preclude the existence of any degree of order. Ant: *crystalline [cryst].* (b) A term formerly used to describe a body of rock occurring in a continuous mass, without division into parts. Cf: *massive [mineral].*

amorphous graphite Very fine-grained, generally sooty graphite from metamorphosed coal beds. The word "amorphous" is a misnomer because all graphite is crystalline. The term has also been applied to very fine particles of *flake graphite* that can be sold only for low-value uses (such as foundry facings), and to fine-grained varieties of Ceylon lump graphite.

amorphous magnesite *serpentine-hosted magnesite.*

amorphous material Noncrystalline constituents that either do not fit the definition of allophane or it is not certain if the constituent meets allophane criteria.

amorphous organic matter The unstructured kerogen component of sedimentary rocks that produces brownish, fluffy-appearing organic matter in palynological preparations; an important constituent of *palynodebris.* Abbrev: AOM.

amorphous peat Peat in which degradation of cellulose matter has destroyed the original plant structures. Cf: *pseudofibrous peat; fibrous peat.*

amosite (am'-o-site) A commercial term for an iron-rich, asbestiform variety of amphibole occurring in long fibers. It may consist of an orthorhombic amphibole (anthophyllite or gedrite) or of a monoclinic amphibole (cummingtonite or grunerite).

Amosnuria (Am-os-nur'-i-a) *Lemuria.*

ampangabeite (am-pan-ga'-be-ite) *samarskite-(Y).*

ampasimenite (am-pa'-si-me-nite') An igneous rock, probably equivalent to *ijolite* or *nephelinite,* characterized by the presence of nepheline, titanian augite, hornblende, and magnetite in a fine-grained or glassy groundmass. Named by Lacroix in 1922 for Ampasimena, Malagasy. Obsolete.

ampelite (am'-pe-lite) An obsolete term for a black carbonaceous or bituminous shale.

amphiaster (am'-phi-as'-ter) A siliceous sponge *spicule* (microsclere) composed of a straight rod with a group of radiating spines at each end.

amphibian (am-phib'-i-an) Any vertebrate of the class Amphibia: ectothermic tetrapods that breathe by means of gills in the early stages of life and by means of lungs in the later stages.

amphiblastula (am-phi-blas'-tu-la) A hollow, ovoid, free-swimming sponge larva, composed of small flagellated and large nonflagellated types of cells, one type grouped anteriorly and the other posteriorly. Pl: amphiblastulae.

amphibole (am'-phi-bole) (a) A group of dark rock-forming ferromagnesian silicate minerals, closely related in crystal form and composition and having the general formula: $A_{2\text{-}3}B_5(Si,Al)_8O_{22}(OH)_2$, where A = Mg, Fe^{2+}, Ca, or Na, and B = Mg, Fe^{2+}, Fe^{3+}, Li, Mn, or Al. It is characterized by a cross-linked double chain of tetrahedra with a silicon:oxygen ratio of 4:11, by columnar or fibrous prismatic crystals, and by good prismatic cleavage in two directions parallel to the crystal faces and intersecting at angles of about 56° and 124°; colors range from white to black. Most amphiboles crystallize in the monoclinic system, some in the orthorhombic. They constitute an abundant and widely distributed constituent in igneous and metamorphic rocks (some are wholly metamorphic), and they are analogous in chemical composition to the *pyroxenes.* (b) A mineral of the amphibole group, such as hornblende, anthophyllite, cummingtonite, tremolite, actinolite, riebeckite, glaucophane, arfvedsonite, etc. (c) A term sometimes used as a syn. of *hornblende.* Etymol: Greek "amphibolos", "ambiguous, doubtful", in reference to its many varieties.

amphibolite (am-phib'-o-lite) A crystalloblastic rock consisting mainly of amphibole and plagioclase. As the content of quartz increases, the rock grades into hornblende plagioclase gneiss. The term was originated by Brongniart. Cf: *feather amphibolite.*

amphibolite facies The facies (set of metamorphic mineral assemblages) in which mafic rocks are represented by hornblende + plagioclase, the plagioclase being oligoclase-andesine or some more calcic variety (Eskola, 1939). Epidote and almandine are common in amphibolites. Pelitic assemblages contain micas associated with almandine, staurolite, kyanite, or sillimanite, but not andalusite or cordierite (Turner, 1968). The facies is typical of regional dynamothermal metamorphism under moderate to high pressures (in excess of 300 MPa) with temperatures in the range 450°-700°C. Cf: *hornblende-hornfels facies.*

amphichoanitic (am'-phi-cho'-a-nit'-ic) Said of septal necks that are directed both forward and backward, as in early ontogenetic stages of ammonites.

amphidetic (am-phi-det'-ic) Said of a *ligament* or ligamental area of certain bivalve mollusks (e.g. *Arca*) extending on both the anterior and posterior sides of the beaks. Cf: *opisthodetic.*

amphidisc (am'-phi-disc) A siliceous sponge *spicule* (microsclere) consisting of a central shaft at each end of which is a transverse, stellate disk or an umbrella-like structure (umbel) containing multiple recurved teeth. Also spelled: amphidisk.

amphidont (am'-phi-dont) Said of a class of ostracode hinges consisting of three elements, of which the terminal elements are teeth (or sockets in the opposed valve), and a median element is subdivided into an anterior socket and a bar (or a tooth and a groove in the opposed valve).

amphidrome (am'-phi-drome) A point in the sea where there is zero tidal amplitude due to canceling of tidal waves. *Cotidal lines* radiate away from the amphidrome and *corange lines* surround it. Syn: *nodal point.*

amphidromic system A system of tidal action in which the tide wave rotates around a point or center of little or no tidal movement. Cf: *amphidrome.*

amphiesma (am-phi-es'-ma) The dinoflagellate cell covering, consisting of a series of membranes, vesicles, and related structures (Edwards, 1993; Williams et al., 2000, p.6-7).

amphigene (am'-phi-gene) *leucite.*

Amphineura (Am-phi-neu'-ra) A subphylum of marine mollusks with radula, gut with mouth and anus widely separated at opposite ends of body, and prominent primitive spicular skeleton. Veliger larval stage lacking. The subphylum includes both epifaunal and infaunal representatives. Two small classes are included that have other, more detailed anatomic similarities. Range, Late Cambrian to Holocene.

amphineuran (am-phi-neu'-ran) A marine mollusk belonging to the subphylum Amphineura. See also: *polyplacophoran; aplacophoran.*

amphioxea (am-phi-ox'-e-a) A slightly curved oxea (sponge *spicule*).

amphiphloic (am-phi-phlo'-ic) Pertaining to the siphonostele of certain vascular plants, having phloem both internal and external to the xylem. Cf: *ectophloic.*

amphipod (am'-phi-pod) Any crustacean belonging to the order Amphipoda, the members of which resemble the isopods by the absence of a carapace and by the presence of unstalked sessile eyes but differ from them in having laterally, rather than dorsoventrally, compressed bodies. Range, Upper Eocene to Holocene.

amphi-sapropel (am'-phi-sap'-ro-pel) Sapropel containing coarse plant debris (Tomkeieff, 1954, p.23).

amphistylic (am-phi-sty'-lic) Pertaining to a jaw suspension in which the cartilaginous or cartilage-replacement skeleton of the upper jaw (quadrate or palatoquadrate) is directly articulated to the ear region of the neurocranium and, in addition, is articulated via the hyoid element of the visceral skeleton; both sets of articulation are movable.

amphitheater (am-phi-the'-a-ter) A concave landform, generally oval or circular in outline, nearly surrounded by steep slopes, and having a relatively flat floor; e.g. a *cirque [glac geol].* Cf: *theater-headed valley.*

amphitropous (am-phit'-ro-pous) Characterizing a plant ovule or seed that is recurved, so that the *micropyle* is close to the base of the *funiculus* or stalk. Cf: *anatropous; campylotropous; orthotropous.*

amphoteric (am-pho-ter'-ic) Having both basic and acidic properties.

amphoterite (am-pho-ter'-ite) An obsolete term for *LL chondrite.*

amplexoid septum (am-plex'-oid) A rugose corallite *septum* characterized by extreme shortness except where it is extended axially on the distal side of a *tabula,* as in *Amplexus.*

amplitude [ecol] (am'-pli-tude) The degree of adaptability exhibited by a particular kind or group of organisms to its surroundings.

amplitude [struc geol] For a symmetrical, periodic fold system it is analogous to the amplitude of a wave form, i.e. half the orthogonal distance between the antiformal and synformal enveloping surfaces. For asymmetrical and nonperiodic systems, various definitions have been proposed (see Fleuty, 1964a).

amplitude [waves] (a) Half the vertical distance between the crest of a wave and the adjacent trough. (b) The maximum departure of a wave from zero, or sometimes its average value.

amplitude equalization A procedure by which the gain of each trace is adjusted to produce the same average amplitude as other traces.

amplitude extraction Determining and displaying relative amplitude values over some surface (usually a *horizon slice*), often by color coding. Cf: *amplitude map.*

amplitude map A display of amplitude measurements over a horizon slice.

amplitude spectrum Amplitude-versus-frequency relationship such as computed in a *Fourier analysis.*

amplitude variation with angle The variation in the amplitude of a seismic reflection with angle of incidence. Often the *amplitude variation with offset* (source-geophone distance) is used instead. Abbrev: AVA.

amplitude variation with offset The variation in the amplitude of a seismic reflection with source-geophone distance. Depends on the velocity, density, and Poisson's ratio contrasts. Often used as a hydrocarbon indicator for gas because a large change in Poisson's ratio (as may occur when the pore fluid is a gas) tends to produces an increase in amplitude with offset. With respect to AVO, reservoir sands in a clastic sequence are sometimes classified according to their normal-incidence impedance contrast with surrounding rocks: class 1 are high-impedance sands, class 2 are near-zero impedance contrast, and class 3 are low-impedance sands (Rutherford and Williams, 1989). Abbrev: AVO.

ampulla (am-pul'-la) One of the muscular vesicles of the *water-vascular system* of an echinoderm (such as an asteroid), being a contractile bulb or saclike structure of a tube foot, either seated externally in a small cup-shaped depression on the surface of an ambulacral plate or internally, and connecting with the *podium* by a branch canal through the podial pore or passageway. Pl: ampullae.

AMS *accelerator mass spectrometry.*

amstallite (am'-stall-ite) A colorless monoclinic mineral: $CaAl(Si,Al)_4O_8(OH)_4{\cdot}(H_2O,Cl)$.

Amstelian (Am-stel'-ian) North Sea stage: upper Pliocene, in some usage also lower Pleistocene (above lower Scaldisian, and below upper Icenian); based on cold-climate mollusk faunas.

AMT *audio-magneto-telluric method.*

amygdale (a-myg'-dale) *amygdule.*

amygdaloid (a-myg'-da-loid) n. An extrusive or intrusive rock containing numerous amygdules. adj. Said of a rock having numerous *amygdules.* Syn: *amygdaloidal.*

amygdaloidal (a-myg'-da-loid'-al) Said of the textures of rocks containing amygdules.

amygdule (a-myg'-dule) A gas cavity or vesicle in an igneous rock, which is filled with secondary minerals. The term *amygdale* is preferred in British usage. Adj: *amygdaloidal.*

ana- A prefix meaning "toward" or "up to".

ana- [palyn] A prefix indicating the position of features (such as apertures) to be on the distal face.

Anabar block (An'-a-bar) A syn. of *Angaraland,* used by Shatsky and Bogdanoff (1957).

anabatic wind (an-a-bat'-ic) A local wind that moves upslope as a result of buoyancy differences resulting from local surface heating; e.g. a *valley wind.* Ant: *katabatic wind.*

anabohitsite (an'-a-bo-hit'-site) An olivine-bearing *pyroxenite* containing hornblende and orthopyroxene and a high proportion of magnetite and/or ilmenite, approximately 30 percent according to Johannsen (1939, p.240). Its name was given by Lacroix in 1914 from Anabohitsy, Malagasy. Obsolete.

anaboly (a-nab'-o-ly) Acceleration of the ontogeny. Syn: *hypermorphosis.*

anabranch (an'-a-branch) (a) A diverging branch flowing out of a main stream and later rejoining it farther downstream; esp. one of the several branches composing a braided stream. The term was coined by Jackson (1834, p.79). Etymol: *ana*stomosing + branch. Cf: *braid.* Syn:*valley braid; anastomosing branch* (b) A branch of a stream that loses itself in sandy soil.

anabranching A generic term characterizing any multi-channel river.

anacline (an'-a-cline) Said of the dorsal and posterior *inclination [paleont]* of the *cardinal area* in either valve of a brachiopod, lying in the top left or first quadrant moving clockwise from the *orthocline* position (TIP, 1965, pt. H, p.60, fig. 61).

anadiagenesis (an'-a-di'-a-gen'-e-sis) Postdepositional alteration of a sedimentary rock during deep burial, by physical or chemical processes, under temperatures and pressures that approach but do not achieve low-grade metamorphism. See also: *epidiagenesis; syndiagenesis.* Adj: anadiagenetic. Syn: *middle diagenesis.*

anaerobic (an-aer-o'-bic) adj. (a) Said of an organism (esp. a bacterium) that can live in the absence of free oxygen; also, said of its activities. n. anaerobe. (b) Said of conditions that exist only in the absence of free oxygen. Cf: *aerobic.*

anaerobic decay Decomposition of organic substances in the absence or near absence of oxygen; the ultimate decay products are enriched in carbon.

anaerobic sediment A highly organic sediment characteristic of some fjords and basins where restricted circulation of the water results in the absence or near absence of oxygen at the sediment surface, and bottom water is rich in hydrogen sulfide.

anagenesis (an-a-gen'-e-sis) The process by which a genetic or phenotypic character changes within a species, whether that change in random or nonrandom, or slow or rapid; progressive evolution between branches in an evolving lineage (Wiley, 1981, p.8). Cf: *cladogenesis, reticulate speciation.*

anaglyph (an'-a-glyph) A stereoscopic image with the right component portrayed in a red color superimposed on the left component portrayed in another color (usually bright green). When viewed through the correspondingly colored filters, a 3-dimensional image is rendered (Krzanowski et al., 1993).

anal (a'-nal) adj. Pertaining to or situated near the anus of an animal; e.g. an "anal opening" representing a large thecal orifice marking the position of the anus in the CD interray of a blastoid or crinoid. n. *anal plate.*

analbite (a-nal'-bite) The triclinic polymorph of albite, having a disordered Al-Si distribution, and obtained by heating albite.

analcime (a-nal'-cime) A glassy colorless to pink mineral: $NaAlSi_2O_6 \cdot H_2O$. It is a zeolite, occurring as cubic, tetragonal, orthorhombic and monoclinic crystallographic variants; commonly found in diabase and in alkali-rich basalts. Syn: *analcite.*

analcime basanite (a-nal'-cime bas'-a-nite) In the *IUGS classification,* a variety of basanite in which analcime is the most abundant feldspathoid.

analcime diorite (a-nal'-cime di'-o-rite) In the *IUGS classification,* a variety of *foid diorite,* in which analcime is the most abundant feldspathoid.

analcime gabbro (a-nal'-cime gab'-bro) In the *IUGS classification,* a variety of *foid gabbro* in which analcime is the most abundant feldspathoid. Syn: *teschenite.*

analcime monzosyenite (a-nal'-cime mon'-zo-sy'-e-nite) In the *IUGS classification,* a variety of *foid monzosyenite* in which analcime is the most abundant feldspathoid. Syn: *analcime plagisyenite.*

analcime phonolite (a-nal'-cime pho'-no-lite) In the *IUGS classification,* a variety of phonolite in which analcime is the most abundant *feldspathoid.*

analcime plagisyenite (a-nal'-cime plag'-i-sy'-e-nite) *analcime monzosyenite.*

analcimite (a-nal'-ci-mite) In the *IUGS classification,* a volcanic rock with F/(F+A+P)>90%, and in which the most abundant feldspathoid is analcime. An extrusive or hypabyssal igneous rock consisting mainly of analcime and pyroxene (usually titanian augite). Feldspathoids, plagioclase, and/or olivine may be present. Apatite, titanite, and opaque oxides may be present as accessories.

analcite (a-nal'-cite) *analcime.*

anal cover plate One of the plates covering the anal opening in the *theca* of echinoderms, e.g. one of many small polygonal plates that may extend over and conceal the anal opening of a blastoid and that are bordered by *anal deltoids* or *deltoid plates.*

anal deltoid An undivided interradial plate on the posterior (CD) part of a blastoid *theca* below the circlet of oral plates or the mouth opening.

anal fasciole (a) A *fasciole* generated on gastropod-shell whorls by the indentation of the outer lip (such as a sinus) situated close to the adapical suture. (b) A *fasciole* adoral and lateral to the *periproct* of an echinoid.

analog (an'-a-log) Said of the representation of a range of numbers by directly measurable quantities such as voltage or rotation, as in an *analog computer* or analog recording. Cf: *digital.*

analog computer A *computer* that operates with numbers represented by directly measurable quantities (such as length, voltage, or resistance) in a one-to-one correspondence; a measuring device that operates on continuous variables represented by physical or mathematical analogies between the computer variables and the variables of a given problem to be solved. Cf: *digital computer; hybrid computer.*

analogous [paleont] The adj. of analogy. Cf: *homologous [paleont].*

analogous pole (a-nal'-o-gous) In crystallography, that pole of a crystal which becomes electrically positive when the crystal is heated or expanded by decompression. Cf: *antilogous pole.*

analogue model A term commonly used to refer to a *scaled physical model.*

analogy (a-nal'-o-gy) In evolution, likeness in form or function but not in origin. Cf: *homology.*

anal plate Any plate covering the anus or anal opening of an echinoderm; a plate of the CD interray of a crinoid, mostly confined to the aboral cup but excluding fixed pinnulars incorporated in the calyx. Syn: *anal.*

anal pyramid A cone-shaped structure composed of several elongate, more or less triangular plates, commonly imbricate, serving to close the *anus* in echinoderms such as cystoids, blastoids, and edrioasteroids.

anal sac A variously shaped, generally inflated, and highly elevated part of *theca* encompassing the anus, as developed among cladid, disparid, and flexible crinoids.

anal tube A conical to cylindrical structure on a crinoid *tegmen,* possibly of considerable height, bearing the anal opening at its summit. It is typically developed in many camerate crinoids.

anal X A special anal plate in cladid and flexible crinoids, typically located between posterior (CD) radial plates, adjacent or next adjacent to *tegmen.*

analysis of variance (a-nal'-y-sis) A statistical technique for simultaneously partitioning the total variance of a set of data into components that can be attributed to different sources of variation and used to test for differences among several samples. Acronym: *ANOVA.*

analytical model A model in which the relationships among parameters can be solved using classical mathematical tools, such as differential equations (National Academy, 1994). Cf: *numerical model.*

analytical paleontology The branch of paleontology emphasizing the use of quantitative techniques to analyze patterns of change in taxonomic diversity and morphology.

analytical triangulation Triangulation accomplished by computational routines using measured coordinates and appropriate formulas; e.g. aerotriangulation by computing positions and/or elevations of ground stations from measurements made on aerial photographs and known locations of control points.

analytic group (an-a-lyt'-ic) A rock-stratigraphic unit formerly classed as a formation but now called a *group* because subdivisions of the unit are considered to be formations (Weller, 1960, p.434). Cf: *synthetic group.*

analyzer (an'-a-lyz'-er) That part of a polariscope that receives the light after polarization and exhibits its properties. In a petrographic microscope, it is the polarizing mechanism (*Nicol prism, Polaroid,* etc.) that intersects the light after it has passed through the object under study. See also: *polarizer.*

anamigmatization (an'-a-mig'-ma-ti-za'-tion) High-temperature, high-pressure remelting of preexisting rock to form *migma.* Cf: *anatexis; metatexis.* Rarely used.

anamniote (an-am'-ni-ote) adj. Pertaining to a vertebrate egg that lacks protective extra-embryonic membranes (including the amnion) which are shed at hatching or birth. Any vertebrate reproducing by means of such an egg; includes all vertebrate classes except Reptilia, Aves, and Mammalia.

anamorph (an'-a-morph) A mitotic, asexual morph of non-lichen-forming ascomycetous and basidiomycetous fungi (including Ustilaginales). Cf: *holomorph; teleomorph.*

anamorphic zone (an-a-mor'-phic) The zone deep in the Earth's crust in which *anamorphism* takes place. The term, originated in 1898 by Van Hise, is now little used. Cf: *katamorphic zone.*

anamorphism (an-a-mor'-phism) Intense metamorphism in the *anamorphic zone* in which rock flowage takes place and simple minerals of low density are changed into more complex ones of greater density by silication, decarbonation, dehydration, and deoxidation. The term was originated by Van Hise in 1904 and is little used. Cf: *katamorphism.*

anamorphosis (an'-a-mor'-pho-sis) In the evolution of a group of animals or plants, the gradual change from one form to another; e.g., certain arthropods acquire an additional body segment after hatching.

anandite (an-and'-ite) A black lustrous monoclinic or orthorhombic trioctahedral brittle mica of ideal composition $BaFe_3^{2+}(OH)S[Fe^{3+}Si_3O_{10}]$.

anapaite (an-a-pa'-ite) A pale green or greenish white triclinic mineral: $Ca_2Fe^{2+}(PO_4)_2 \cdot 4H_2O$.

anapeirean (an-a-pei'-re-an) Said of rocks of the *Pacific suite.*

anaphoresis (an-a-phor'-e-sis) *Electrophoresis* in which the movement of suspended negative particles in a fluid is toward the anode. Cf: *cataphoresis.*

anaprotaspis (an'-a-pro-tas'-pis) A small *protaspis* trilobite in which the protopygidium is poorly defined. Cf: *protaspis; metaprotaspis.* Pl: anaprotaspides.

Anapsida (An-ap'-si-da) A reptilian subclass that includes turtles and tortoises in addition to diverse forms; of late Paleozoic and early Mesozoic age.

anaptychus (an-ap'-tych-us) A single plate with rounded edges, made of calcite, originally lodged in the body chamber of some ammonoid conchs; it probably formed part of the lower jaw of the animal. It was formerly interpreted as an opercular structure. See also: *aptychus.*

anarakite (an-a-rak'-ite) A zincian *paratacamite*(?).

anascan (an-as'-can) adj. Pertaining to the cheilostome bryozoan suborder Anasca, characterized by parietal muscles attached to the frontal membrane, which is generally exposed. n. An anascan cheilostome (bryozoan). Cf: *ascophoran.*

anaseism (an'-a-seism) Earth movement away from the focus of an earthquake. Cf: *kataseism.*

Anaspida (An-as'-pi-da) An order of one-nostril jawless vertebrates characterized by fusiform habitus and unexpanded head shield. Range, Middle Silurian to Upper Devonian.

anastable (an'-a-sta'-ble) adj. Stable, with a tendency towards uplift. Cf: *catastable.*

anastomose v. To merge and diverge so as to give the impression of being braided, as in "The cleavage domains anastomose."

anastomosing (a-nas'-to-mos'-ing) (a) Pertaining to a network of branching and rejoining fault surfaces or surface traces. (b) Said of the channel pattern of a *braided stream.* (c) Said of a leaf whose veins form a netlike pattern; pertaining to an interveined leaf. Sometimes the vein branches meet only at the margin.

anastomosing branch *anabranch.*

anastomosing stream Branching and reconnecting characterized by multiple channels separated by semi-permanent islands. Distinct from braiding in that braided channels are separated by ephemeral mid-channel bars and that anastomosing rivers are the low-energy, fine-sediment end of the range of multi-channeled rivers. Cf: *anabranch; braided stream.*

anastomosis (a) braiding. (b) A product of braiding; esp. an interlacing network of branching and reuniting stream channels.

anastomosis tube One of many small, irregular, repeatedly interconnected solution tubes commonly found along bedding planes. Collectively referred to as an anastomosis.

anastomotic cave (a-nas'-to-mot'-ic) A cave consisting of braided interconnected tubular passages. See also: *network cave; spongework cave.*

anatase (an'-a-tase) An adamantine brown, dark-blue, or black tetragonal mineral: TiO_2. It is trimorphous with rutile (which has different facial angles) and brookite, and occurs as an alteration product of other titanium minerals. Syn: *octahedrite.*

anatectic magma (an-a-tec′-tic) Magma formed as a result of *anatexis.*
anatectite (an-a-tec′-tite) *anatexite.*
anatexis (an-a-tex′-is) The partial melting of preexisting rock. Anatexis implies in situ partial melting. This term is commonly modified by terms such as intergranular, partial, differential, selective, crustal, or complete (Dietrich and Mehnert, 1961). Cf: *metatexis; diatexis; palingenesis [petrology]; syntexis; anamigmatization.*
anatexite (an-a-tex′-ite) Rock formed by *anatexis.* Var: *anatectite.* Cf: *arterite; syntectite.*
anathermal (an-a-ther′-mal) n. A period of time during which temperatures are rising. The term was used by Emiliani (1955, p.547) for a part of a cycle as displayed in a deep-sea sediment core. Ant: *catathermal.* adj. Pertaining to the postglacial Anathermal interval and to its climate, deposits, biota, and events.
Anathermal [paleoclim] (An-a-ther′-mal) n. A term proposed by Antevs (1948, p.176) for a postglacial interval (from about 10,000 to 8,000 calendar years ago) preceding the Altithermal and representing the period of generally rising temperatures following the last major advance of the continental glaciers. It is equivalent to the Preboreal and Boreal. adj. Pertaining to the postglacial Anathermal interval and to its climate, deposits, biota, and events.
anatriaene (an-a-tri′-aene) A sponge triaene in which the cladi are bent back toward the rhabdome.
anatropous (a-nat′-ro-pous) A plant ovule that is reversed, i.e. one whose opening (micropyle) is close to the point of funiculous attachment (Lawrence, 1951, p.738). Cf: *amphitropous; campylotropous; orthotropous.*
anauxite (an-aux′-ite) A clay consisting of a mixture of kaolinite and amorphous silica. Syn: *ionite.*
anaxial (an-ax′-i-al) Lacking a distinct axis or axes; e.g. said of an arm or branch of a sponge spicule that has no axial filament or canal.
ancestral river (an-ces′-tral) A term applied to a major, ancient river system, esp. in glaciated regions, e.g., the ancestral Ohio or Missouri Rivers.
ancestral species *stem species.*
ancestral taxon A species that gave rise to at least one new daughter species during speciation, either through cladogenesis or reticulate evolution (Wiley et al., 1991, p.3).
ancestrula (an-ces′-tru-la) The primary or first-formed *zooid* of a bryozoan colony, derived by metamorphosis of a free-swimming larva, from which secondary individuals are formed by budding. Pl: ancestrulae.
anchi- (an′-chi-) In petrologic terminology, a prefix signifying "almost".
anchieutectic (an′-chi-eu-tec′-tic) Said of a rock whose minerals are in almost completely eutectic proportions.
anchimetamorphism (an′-chi-met′-a-mor′-phism) A term introduced by Harrassowitz (1927) to indicate changes in mineral content of rocks under temperature and pressure conditions prevailing in the region between the Earth's surface and the zone of true metamorphism; i.e., approximately in the zones of weathering and groundwater circulation. "Anchimetamorphism or anchizone is recognized only on the basis of illite 'crystallinity' data, and this method is used as the main indicator of metamorphic grade for very low grade metaclastites" (Frey, 1987).
anchimonomineralic (an′-chi-mon′-o-min′-er-al′-ic) Said of an igneous rock (such as anorthosite or dunite) that consists essentially of a single mineral. Term originated by Vogt (1905). Cf: *monomineralic.*
anchizone The region or zone of *anchimetamorphism.*
anchor (an′-chor) A holothurian *sclerite* in the shape of an anchor, consisting of a shank, two or more flukes, and usually a stock.
anchorate (an′-chor-ate) n. (a) A sponge *spicule* (hexactin) with one long ray and two (coplanar) or four recurved, short rays at one end. (b) A strongly dentate *chela* of a sponge. adj. Said of a sponge spicule having one or more processes like the fluke of an anchor.
anchor branch A curved hooklet in the radiolarian suborder Phaeodarina.
anchored dune (an′-chored) A sand dune whose movement is arrested or whose form is protected from further wind action, as by a growth of vegetation or by cementation of the sand. Cf: *attached dune; wandering dune.* Syn: *fixed dune; established dune; stabilized dune.*
anchor ice Spongy underwater ice formed on a submerged object or structure, or attached to the bottom of a body of water (as a stream, lake, or very shallow sea) which itself is not frozen; usually forms in cold, clear, and still water. Syn: *bottom ice; ground ice; underwater ice.*
anchor-ice dam An accumulation of anchor ice raising the water level of a river or stream.
anchylosis (an-chy-lo′-sis) *ankylosis.*
ancillary data In remote sensing, the data used to enhance the analysis of the primary spectral data (Krzanowski et al., 1993).
ancylite (an′-cy-lite) A group name for rare-earth carbonate minerals with the general chemical formula $X(CO_3)(OH,H_2O)$, where X = rare-earth elements or Sr.
ancylite-(Ce) (an′-cy-lite) A vitreous yellowish-brown orthorhombic mineral: $(Ce,Sr)CO_3(OH,H_2O)$. A member of the *ancylite* group.
ancylite-(La) A light yellow or yellowish-brown orthorhombic mineral: $(La,Sr)CO_3(OH,H_2O)$. A member of the *ancylite* group.
andalusite (an-da-lu′-site) A brown, yellow, green, red, or gray orthorhombic mineral: Al_2SiO_5. It is trimorphous with kyanite and sillimanite. Andalusite occurs in thick, nearly square prisms in schists, gneisses, and hornfelses; it forms at medium temperatures and pressures of a regionally metamorphosed sequence and is characteristic of contact-metamorphosed argillaceous rocks. In transparent gem quality, andalusite has a very strong pleochroism: brownish green in one direction and brownish red at 90°. See also: *chiastolite.*
Andepts Previous to 1994 this term was used to indicate Inceptisols that have formed either in vitric pyroclastic materials, or have low bulk density and large amounts of amorphous materials, or both. The term was dropped as a suborder in the 1994 revision of the U.S. Department of Agriculture, Soil Taxonomy.
andersonite (an′-der-son-ite′) A bright yellow-green rhombohedral secondary mineral: $Na_2Ca(UO_2)(CO_3)_3{\cdot}6H_2O$.
andesine (an′-de-sine) A mineral of the plagioclase *feldspar* group with composition ranging from $Ab_{70}An_{30}$ to $Ab_{50}An_{50}$. It occurs as a primary constituent of intermediate igneous rocks, such as andesites and diorites.
andesinite (an′-de-si-nite′) A coarse-grained igneous rock almost entirely composed of andesine. It was named by Turner in 1900. Cf: *anorthosite.* Obsolete.
andesite (an′-de-site) (a) In the *IUGS classification,* a volcanic rock defined modally by Q/(Q+A+P) between 0 and 20% or F/(F+A+P) between 0 and 10%, P/(A+P)>65%, and M<35%. Because modes are difficult to estimate for these rocks, andesite is also defined in the TAS diagram as rock falling in the area bounded by points with the SiO_2 and total alkali coordinates: 57, 0; 57, 5.9; 63, 0; 63, 7. (b) A dark-colored, fine-grained extrusive rock that, when porphyritic, contains phenocrysts composed primarily of zoned sodic plagioclase (esp. andesine) and one or more of the mafic minerals (e.g. biotite, hornblende, pyroxene), with a groundmass composed generally of the same minerals as the phenocrysts, although the plagioclase may be more sodic and quartz is generally present; the extrusive equivalent of *diorite.* It was named by Buch in 1826 from the Andes Mountains, South America.
andesite copper deposit A *mineral-deposit type* formed in calc-alkaline volcanic rocks characterized by *stratabound* and *stratiform* zones of disseminated, open-space filling, and vein-hosted copper minerals.
andesite line Obsolete term for the geographic-petrographic boundary between basalts of the *Atlantic suite* and the mainly andesitic rocks of the *Pacific suite.* The boundary on the west is generally drawn from Alaska to the east of New Zealand and Chatham Island, by way of Japan, the Marianas, Palau Islands, Bismarck Archipelago, and the Fiji and Tonga groups. The boundary on the east is less clearly defined but probably runs along the coasts of North and South America; it has not been traced in the South Pacific. Syn: *Marshall line.*
andesitoid In the *IUGS classification,* a general term in the provisional "field" classification for rocks tentatively identified as *andesite.*
andic Soil material containing volcanic glass, having a low bulk density and less than 25% organic carbon.
Andisols Mineral soils that are dominated by andic soil properties in 60 percent or more of their thickness. An order in the U.S. system of soil taxonomy.
andorite (an-dor′-ite) A metallic dark gray or black sulfosalt mineral: Two compositional and crystallographic variants have been reported: orthorhombic $Pb_{10}Ag_{12}CuSb_{37}S_{72}$, called *senandorite* or andorite VI, and monoclinic $Pb_9Ag_7Sb_{23}S_{48}$, called quatrandorite or andorite IV.
andradite (an′-dra-dite) The calcium-iron end member of the *garnet* group: $Ca_3Fe_2^{3+}(SiO_4)_3$. It has a variety of colors, ranging from yellow, red, and green to brown and black; it often occurs in contact-metamorphosed limestones. Varieties include topazolite, demantoid, melanite, aplome, and bredbergite.
andremeyerite (an-dre-mey′-er-ite) A pale-green monoclinic mineral: $BaFe^{2+}{}_2Si_2O_7$.
andrewsite (an′-drews-ite) A mixture of *hentschelite, rockbridgeite* and *chalcosiderite.*
androsite-(La) A vitreous brown-red monoclinic mineral, a La- and Mn-bearing member of the *epidote* group: $(Mn^{2+},Ca)(La,Ce,Ca)(Al,Mn^{3+},Mn^{2+})_3(SiO_4)(Si_2O_7)O(OH)$.
andryobertsite A vitreous blue monoclinic mineral: $K(Cd,Ca)Cu_5(AsO_4)_4[As(OH)_2O_2]{\cdot}2H_2O$.
anduoite (an′-du-o-ite) A metallic lead-gray orthorhombic mineral: $(Ru,Os)As_2$. A member of the *löllingite* group. Not approved by the International Mineralogical Association.
anelasticity [exp struc geol] (an′-e-las-tic′-i-ty) The inelastic relaxation in time of very small deformations, usually studied as forced vibrations.
anelasticity [seis] A deviation from a linear proportionality between stress and strain. Often the effect of absorption of seismic energy. See also: *absorption; absorptance; Q [seis].*
anemochore (a-nem′-o-chore) A plant whose seeds or spores are distributed by the wind. Cf: *anemophily.*
anemoclast (a-nem′-o-clast) A rock fragment that was broken off and more or less rounded by wind action (Grabau, 1904).
anemoclastic rock (a-nem′-o-clas′-tic) A rock consisting primarily of anemoclasts.
anemolite (a-nem′-o-lite) An obsolete term for *helictite.*

anemometer (an-e-mom'-e-ter) Any type of instrument that measures the speed of wind.
anemophily (an-e-moph'-i-ly) Pollination by wind. Adj: anemophilous. Cf: *entomophily.*
anemosilicarenite (a-nem-o-sil-i-car'-e-nite) An eolian sand of siliceous composition (Grabau, 1904; 1913, p.293).
anemousite (an-e-mous'-ite) A silica-deficient variety of albite.
aneroid barometer (an'-er-oid) A type of *barometer* that measures atmospheric pressure and change by its effect on the thin sides of a partially evacuated short hollow cylinder. The *altimeter,* which is commonly used to measure altitude, is a barometer of this type. Cf: *mercury barometer.*
aneuchoanitic (an'-eu-cho'-a-nit'-ic) *achoanitic.*
Angaraland (An-gar'-a-land) A name used by Suess for a small shield exposing ancient Precambrian rocks in north-central Siberia, supposed to have been the *vertex* or high point around which all other structures of Asia were built. Modern Russian geologists ascribe less significance to the feature. Syn: *Anabar block; Angara Shield.*
angaralite (an-gar'-a-lite) An obsolete name for a mineral of the *chlorite* group, occurring in thin black plates.
Angara Shield (An-gar'-a) *Angaraland.*
angelellite (an-gel-el'-lite) An adamantine blackish-brown triclinic mineral: $Fe^{3+}_4O_3(AsO_4)_2$.
angiosperm (an'-gi-o-sperm') A plant with true flowers, in which the seeds, resulting from double fertilization, are enclosed in an ovary, comprising the fruit. Angiosperms range from Early Cretaceous (megafossils) or possibly Late Jurassic (pollen) to the present. Examples include grasses, orchids, elms, and roses. Syn: *Anthophyta; flowering plant.* Cf: *gymnosperm.*
angle (an'-gle) The difference in direction between two convergent lines or surfaces; a measure of the amount of rotation required to make either of two intersecting lines coincide with or become parallel to the other, the rotation being in the plane of the lines and about the point of intersection.
angle of departure An obsolete term used for the acute angle between a structural plane, e.g. a bed or a fault, and the vertical plane of a geologic cross section, as measured in an imaginary plane that is perpendicular to the trace of the structural plane on the cross section (Knutson, 1958). Cf: *angle of penetration.*
angle of dip *dip [struc geol].*
angle of emergence An angle formed between a seismic wavefront and the horizontal. Cf: *apparent dip [seis].* Syn: *emergence angle.*
angle of incidence (a) The angle that a ray of energy electromagnetic or acoustic makes with the normal to a boundary surface. It is the complement of the *angle of emergence.* See also: *critical angle.* (b) In SLAR, the angle between the vertical and a line connecting the antenna and a target. Syn: *aspect angle [remote sensing].*
angle of internal friction *soil friction angle.*
angle of penetration An obsolete term used for the minimum angle between a structural plane and the plane of a geologic cross section, comparable to the angle of plunge on a geologic map (Knutson, 1958). Cf: *angle of departure.*
angle of reflection (a) In optics, the angle between the reflected ray and the normal to the reflecting surface. (b) *Bragg angle.*
angle of refraction In optics, the angle of a refracted ray of light, measured from a perpendicular to the surface from which the ray is refracted. Syn: *refraction angle.*
angle of repose The maximum angle of slope (measured from a horizontal plane) at which loose, cohesionless material will come to rest on a pile of similar material. This angle is somewhat less than the slope angle at which sliding will be initiated (angle of slide) and is generally 5° to 10° less than the angle of internal friction of the same material. The angle of repose commonly ranges between 33° and 37° on natural slopes, and is rarely less than 30° or more than 39°. The angle depends on the frictional properties of the material and increases slightly as the size and angularity of the fragments increase. Cf: *angle of slide.* Syn: *angle of rest.*
angle of rest *angle of repose.*
angle of slide The angle (usually measured from a horizontal plane) of minimum slope at which any loose material (such as earth or talus) will start to slide; it is slightly greater than the *angle of repose; critical slope angle.*
angle of ultimate stability *critical slope angle.*
anglesite (an'-gle-site) A white or slightly colored orthorhombic mineral: $PbSO_4$. It is a common secondary mineral formed by the oxidation of galena and is a valuable ore of lead. Syn: *lead vitriol; lead spar.* A member of the *barite* group.
angrite (an'-grite) A rare group of basaltic achondrites characterized by very old (4.6 Ga) crystallization ages, fassaitic pyroxenes rich in Ti and Al, and the occasional presence of kirschsteinite.
angstrom A unit of length equal to 10^{-10} meter, traditionally used in expressing X-ray wavelengths, interatomic distances in crystal structures, ionic sizes, and so forth. It is equal to 0.1 nanometer (nm), which is the preferred international standard unit in some professional publications. Symbol: Å.
anguclast (an'-gu-clast) An angular *phenoclast,* such as a large fragment within a breccia. Cf: *spheroclast.*
angulaperturate (an-gul-ap-er'-tur-ate) Having equatorial apertures positioned at the radial angles of the outline of a pollen grain in polar view. Ant: *planaperturate.*
angular Having sharp angles or borders; specif. said of a sedimentary particle showing very little or no evidence of abrasion, with all of its edges and corners sharp, such as blocks with numerous (15-30) secondary corners and a roundness value between zero and 0.15 (midpoint at 0.125) (Pettijohn, 1957, p.58-59). Powers (1953) gives values between 0.17 and 0.25 (midpoint at 0.21). Also, said of the *roundness class* containing angular particles. Cf: *angularity.*
angular cross-bedding Cross-bedding in which the inclined beds appear in section as nearly straight lines meeting the underlying surface at high, sharp, or discordant angles; it often implies deposition by water, as in *torrential cross-bedding.* Cf: *planar cross-bedding; tangential cross-bedding.*
angular discordance *angular unconformity.*
angular distance The angle, measured at the Earth's center, that subtends the great-circle path between the earthquake's epicenter and the receiver. Cf: *epicentral distance.*
angular distortion The change in shape of an area on a globe when it is represented on a map projection. See also: *distortion [cart].*
angular drift "Rock debris formed by intensive frost action, derived from underlying or adjacent bedrock" (ADTIC, 1955, p.4).
angular field of view The angle subtended by lines from a remote sensing system to the outer margins of the strip of terrain that is viewed by the system. Cf: *instantaneous field of view.*
angular fold *accordion fold.*
angular frequency Repetition rate measured in radians/s.
angularity A term often used for the property of a sedimentary particle now commonly known as *roundness,* but used by Lamar (1928, p.148-151) for the property now referred to as *sphericity.* Cf: *angular.*
angular shear A measure of the change in angle between two lines, normally determined by measuring the change in angle between two lines that were originally orthogonal.
angular spreading The lateral extension of ocean waves as they move out of the generating area as swell.
angular unconformity An *unconformity* between two groups of rocks whose bedding planes are not parallel or in which the older, underlying rocks dip at a different angle (usually steeper) than the younger, overlying strata; specif. an unconformity in which younger sediments rest upon the eroded surface of tilted or folded older rocks. It is sometimes regarded as a type of *nonconformity.* Cf: *discordance.* Syn: *angular discordance; clinounconformity; structural unconformity.*
angulate drainage pattern (an'-gu-late) A modified *rectangular drainage pattern* developed where streams follow joints or faults that join each other at acute or obtuse angles, rather than at right angles (Zernitz, 1932, p.517). Examples are found in the Timiskaming and Nipissing areas of Ontario.
angulate element An arched pectiniform conodont element with two primary processes, which are anterior and posterior (TIP, 1981, pt. W, supp.2).
anguliplanate element An angulate conodont element with lateral platform extensions and the attachment surface characteristic of planate pectiniform elements. See also: *angulate element; planate element.*
anguliscaphate element An angulate conodont element with lateral platform extensions and the attachment surface characteristic of scaphate pectiniform elements. See also: *angulate element; planate element; scaphate element.*
anhedral (an-he'-dral) A grain lacking well-developed crystal faces. Syn: *xenomorphic* and *allotriomorphic* (as these were originally defined). Cf: *subhedral; euhedral.*
anhydrite (an-hy'-drite) A mineral consisting of anhydrous calcium sulfate: $CaSO_4$. It represents *gypsum* without its water of crystallization, and it alters readily to gypsum, from which it differs in crystal form (anhydrite is orthorhombic) and in being harder and slightly less soluble. Anhydrite usually occurs in white or slightly colored, granular to compact masses, forming large beds or seams in sedimentary rocks or associated with gypsum and halite in evaporites. Syn: *cube spar.*
anhydrous (an-hy'-drous) Said of a substance, e.g. magma or a mineral, that is completely or essentially without water. An anhydrous mineral contains no water in chemical combination.
anhysteretic remanent magnetization (an-hys'-ter-et'-ic) Remanent magnetization produced by simultaneous application of a constant weak magnetic field and a considerably larger alternating magnetic field whose amplitude decreases continuously to zero.
anideltoid (an-i-del'-toid) Externally visible *anal deltoid,* which is unaccompanied by any others and lies on the aboral side of the anal opening or the *anispiracle* (TIP, 1967, pt. S, p.345).
anidiomorphic (an-id'-i-o-mor'-phic) Obsolete syn. of *xenomorphic.*
anilite (an'-i-lite) A metallic bluish-gray orthorhombic mineral: Cu_7S_4.
Animikean (A-nim'-i-ke'-an) (a) An obsolete geochronometric subdivision of the *Precambrian,* covering the time span from 2,200 to 1,650 Ma (Harland et al., 1990), therefore approximately equivalent to the middle through late *Paleoproterozoic* Era (early Proterozoic) of the Standard Global Geochronometric Scale. (b) *Animikie.*

Animikie (A-nim'-i-kie) A provincial series of the *Proterozoic* of the Canadian Shield. Also called: *Animikean.*
animikite (a-nim'-i-kite) A silver ore consisting of a mixture of sulfides, arsenides, and antimonides showing striking intergrowth relations and occurring in white or gray granular masses. It contains nickel and lead. Cf: *macfarlanite.*
anion An atom that has acquired a negative charge by virtue of having gained one or more electrons.
anion exchange (an'-i-on) The displacement of an anion bound to a site on the surface of a solid by an anion in solution. See also: *ion exchange.*
anion exclusion The exclusion or repulsion of anions from the vicinity of negatively charged soil particle surfaces.
Anisian (A-ni'-sian) A stage of the Standard Global Chronostratigraphic Scale: lower Middle Triassic (above Olenekian, below Ladinian).
anisochela (an-i'-so-che'-la) A sponge *chela* having unequal or dissimilar ends. Cf: *isochela.*
anisodesmic (an-i'-so-des'-mic) Said of a crystal or compound in which the ionic bonds are of unequal strength. Cf: *isodesmic.*
anisomerism (an-i-som'-er-ism) (a) Repetition of parts that differ more or less importantly among themselves. (b) Reduction in number and differentiation of similar parts in organisms.
anisomyarian (an-i'-so-my-ar'-i-an) adj. Said of a mollusk with anterior adductor muscles much reduced or absent. n. A mollusk with such muscles. Cf: *heteromyarian.*
anisotropic [cryst] (an-i'-so-trop'-ic) Having some physical property that varies with direction. All crystals are anisotropic relative to some properties, e.g. propagation of sound waves. Unless otherwise stated, however, the term refers to optical properties. In this sense, all crystals except those of the isometric system are anisotopic. Ant: *isotropic [cryst].* Syn: aeolotropic.
anisotropic [grd wat] A condition in which a property, e.g., *hydraulic conductivity,* varies with the direction of measurement at a point in a geologic formation (Freeze and Cherry, 1979, p.32). Ant: *isotropic [grd wat].* See also: *anisotropy.*
anisotropic temperature factor *temperature factor.*
anisotropy (an-i-sot'-ro-py) The condition of having different properties in different directions, as in geologic strata that transmit seismic waves with different velocities in the vertical and horizontal directions. Adj: *anisotropic.*
anisotropy [seis] Variation of seismic velocity depending on be either the direction of travel (for P or S waves) or the direction of polarization (for S waves). Velocity anisotropy (or *coefficient of anisotropy*) is sometimes taken as the fractional difference between the maximum and minimum velocities in different directions, $(V_{max} - V_{min})/V_{max}$, often expressed as a percentage. P-wave anisotropy is usually meant unless S-wave anisotropy is specified, although anisotropy of P waves usually implies anisotropy for S waves. The most common anisotropy observed is polar anisotropy (also called *transverse isotropy* or azimuthal isotropy); it involves properties that are the same in any direction perpendicular to a symmetry axis. Layering and parallel fracturing tend to produce this situation. The next most common anisotropy is orthorhombic asymmetry (the symmetry of a brick). An S wave of arbitrary polarization entering a polar anisotropic region in a direction other than along the symmetry axis splits into two S waves; this is called *shear-wave splitting* or *birefringence.*
anispiracle (an-i-spi'-ra-cle) An enlarged opening in the summit part of the posterior interray of a blastoid, formed by the union of anal opening and posterior *spiracle* or spiracles.
anitaxis (an-i-tax'-is) A linear succession of crinoid anal plates. Pl: anitaxes.
ankangite An adamantine black tetragonal mineral: $Ba(Ti,V^{3+},Cr)_8O_{16}$. A member of the *hollandite* group.
ankaramite (an-ka'-ra-mite) An olivine-bearing basanite containing numerous pyroxene (usually clinopyroxene) phenocrysts, the former being more abundant than the latter, in a fine-grained groundmass composed of clinopyroxene microlites and calcic plagioclase. It was named by Lacroix in 1916 from Ankaramy, Malagasy.
ankaratrite (an-ka'-ra-trite) Olivine *nephelinite* containing biotite. Named by Lacroix in 1916 for Ankaratra, Malagasy. Obsolete.
ankerite (an'-ker-ite) A white, red, or grayish iron-rich rhombohedral mineral: $Ca(Fe^{2+},Mg,Mn^{2+})(CO_3)_2$. It is associated with iron ores and commonly forms thin veins of secondary matter in some coal seams. Syn: *ferroan dolomite; cleat spar.* A member of the *dolomite* group.
ankylosis (an-ky-lo'-sis) (a) Fusion of columnals or other skeletal elements of an echinoderm, commonly obscuring the sutures. (b) In vertebrates, comparable fusion between adjacent bones, or between base of tooth and supporting bone. Also spelled: *anchylosis.*
annabergite (an'-na-berg-ite') An apple-green monoclinic mineral: $Ni_3(AsO_4)_2{\cdot}8H_2O$. It is isomorphous with erythrite, and usually occurs in incrustations as an alteration product of nickel arsenides. A member of the *vivianite* group. Syn: *nickel bloom; nickel ocher.*
annealing (an-neal'-ing) A high-temperature adjustment to mineral grains including *static recrystallization* and growth of grains.
annealing recrystallization (an-neal'-ing) The formation of new grains in a rock after solid-state deformation, while the temperature is still high. The process commonly leads to grain enlargement and more polygonal shape; common in marble and dunite. See also: *dynamic recrystallization [ign].*
annelid (an'-ne-lid) Any wormlike invertebrate belonging to the phylum Annelida, characterized by a segmented body with a distinct head and appendages. Because the annelids lack skeletal structures (except for chitinous jaws, called scolecodonts), they are usually known as fossils only from their burrows and trails.
annerödite (an-ne-röd'-ite) A black mineral mixture consisting of samarskite with parallel overgrowths of columbite. Also spelled: annerodite.
annihilation Spontaneous mutual destruction of a positron (positive beta particle) and an extranuclear electron to form two gamma rays which are emitted in opposite directions having energies of 0.51 MeV each.
annite (an'-nite) A vitreous black monoclinic trioctahedral mica of ideal composition $KFe_3^{2+}(OH,F)_2[AlSi_3O_{10}]$. Cf: *ferri-annite.*
annivite An ill-defined sulfosalt mineral with a composition approximating (Cu,Zn,Ag)12(Bi,As,Sb)4S13. Probably a member of the tetrahedrite-*tennantite* series.
annual balance (an'-nu-al) The change in mass of a glacier from the beginning to the end of a hydrologic year (usually October 1 to September 30), or other measurement year defined by fixed calendar dates, determined at a point, as an average for an area, or as a total mass change for the glacier. Millimeters, meters, or cubic meters are the units normally used. Cf: *mass balance; net balance.*
annual flood (a) The highest discharge of a stream in a given water year. (b) That flood in a given water year which has been equalled or exceeded on the average of once a year. (ASCE, 1962).
annual growth ring *growth ring [geochron].*
annual layer (a) A sedimentary layer deposited or presumed to have been deposited during the course of a year; e.g. a glacial varve. (b) More general: laminae consisting of alternating light and dark bands caused by compositional differences, e.g., dark coloring caused by iron sulfide in strongly reduced environments (Müller, 1966) or calcite-rich, light-colored layers in the silty, clayey deposits of Lake Constance (Müller and Gees, 1970). The bedding is a case of seasonal rhythmites (annual varves).
annually thawed layer (an'-nu-al-ly) *active layer [permafrost].*
annual wave Yearly cyclical heating and cooling of the upper 3 to 5 meters of the Earth, in response to the annual solar cycle. Below this level the temperature is constant. Cf: *diurnal wave.*
annular Cyclical or ringlike.
annular drainage pattern (an'-nu-lar) A drainage pattern in which subsequent streams follow a roughly circular or concentric path along a belt of weak rocks, resembling in plan a ring-like pattern. It is best displayed by streams draining a maturely dissected structural dome or basin where erosion has exposed rimming sedimentary strata of greatly varying degrees of hardness, as in the Red Valley which nearly encircles the domal structure of the Black Hills, South Dakota (Howard, 1967, Fig. 1).
annular lobe A small secondary dorsal lobe in the center of the main *internal lobe* of a suture of some coiled nautiloid conchs.
annular tracheid In plants, commonly the first *tracheid* to mature, characterized by the deposition of secondary wall material in ringlike form. Cf: *helical tracheid; protoxylem.*
annulation (an-nu-la'-tion) A ringlike structure; e.g. a ringlike expansion of an ammonoid conch, either transverse or slightly oblique to the longitudinal axis of the conch.
annulus [bot] (an'-nu-lus) (a) In the ferns, a specialized ring of cells on the sporangium that is involved in sporangial dehiscence. (b) Variously used for other annulate structures in mosses, horsetails, mushrooms, and diatoms.
annulus [drill] The space between the casing in a well and the wall of the hole, or between two concentric strings of casing, or between casing and tubing.
annulus [paleont] (a) A thin, ring-shaped endosiphuncular deposit, semicircular in cross section, on the inner side of a *septal neck* of a nautiloid (TIP, 1964, pt. K, p.54). (b) The *periphract* of a nautiloid. (c) A ring-shaped plate taking part in construction of the wall of archaeocyathids.
annulus [palyn] A ring bordering a pore of a pollen grain, in which the ektexine is modified (usually thickened). Cf: *margo.* See also: *endannulus.*
anomalous dispersion (a-nom'-a-lous) The extraordinary scattering effects that an atom displays when the wavelength of impinging X-rays is close to a natural absorption edge of the atom. Normal atomic scattering factors may be corrected for anomalous dispersion using correction factors listed in International Tables for Crystallography, vol. C, p.219-222. Syn: *X-ray dispersion; anomalous scattering.*
anomalous iron an *ungrouped iron* meteorite.
anomalous lead Lead with isotopic ratios that indicate a model age either older or younger than the accepted age of the lead-bearing mineral or whole-rock sample. Cf: *lead-lead age; J-type lead; B-type lead.*
anomalous scattering *anomalous dispersion.*
anomaly (a-nom'-a-ly) (a) A departure from the expected or normal. (b) The difference between an observed value and the corresponding computed value. (c) A geological feature, esp. in the subsurface, which is different from the general surroundings and possibly of potential economic value;

e.g., a magnetic anomaly. It may be distinguished by geological, geophysical, or geochemical means.

anomite (an'-o-mite) A variety of biotite differing in optic orientation. It needs further investigation.

anomoclone (an'-o-mo-clone') A sponge *desma* consisting of a short arm (brachyome) and several longer arms directed at various angles away from the short arm.

Anomodontia (An'-o-mo-dont'-i-a) A suborder of therapsid synapsid reptiles that includes the infraorder Dicynodontia, highly specialized for herbivorous habit, and a variety of forms, mostly large, of generally primitive aspect. Range, Upper Permian to Upper Triassic.

anomphalous (a-nom'-pha-lous) Said of a gastropod shell lacking an *umbilicus.* Cf: *phaneromphalous.*

anorogenic (an'-or-o-gen'-ic) Not orogenic, lacking in or unrelated to orogenic disturbance; e.g. an anorogenic area, time, or granite.

anorogenic pluton Intrusive igneous body emplaced in a tectonic setting which is not undergoing mountain building. Cf: *postorogenic pluton.*

anorthic (an-or'-thic) Said of crystals having unequal oblique axes; i.e., triclinic crystals.

anorthite (an-or'-thite) (a) A white or grayish triclinic mineral of the plagioclase *feldspar* group: $Na_{0.1-0.0}Ca_{0.9-1.0}Al_{1.9-2.0}Si_{2.1-2.0}O_8$. It is the most basic member of the plagioclases, its composition ranging from $Ab_{10}An_{90}$ to Ab_0An_{100}. Anorthite occurs in basic and ultrabasic igneous rocks (gabbro, norite, anorthosite), rarely as a well-developed druse mineral, sometimes in tuffs, and very rarely in metamorphic rocks (skarns). Syn: *calcium feldspar; calciclase.* (b) The pure calcium-feldspar end-member in the *plagioclase* series.

anorthoclase (an-or'-tho-clase) A triclinic mineral of the alkali *feldspar* group: $(Na,K)AlSi_3O_8$. It is a sodium-rich feldspar ($Or_{40}Ab_{60}$ to $Or_{10}Ab_{90}$) that shows deviations from monoclinic symmetry and that contains very fine-grained intergrowths; it is widespread as a groundmass constituent of slightly alkalic lavas. The term is usually applied to a mixture of several phases that may not even have a stability field of their own at any temperature. Cf: *orthoclase.* Syn: *anorthose; soda microcline.*

anorthominasragrite A vitreous pale blue triclinic mineral: $V^{4+}O(SO_4)\cdot 5H_2O$. A dimorph of *minasragrite.*

anorthose (an-or'-those) *anorthoclase.*

anorthosite (an-or'-tho-site) (a) In the *IUGS classification,* a plutonic rock with Q between 0 and 5, P/(A+P) greater than 90, and M less than 10. (b) A group of essentially monomineralic plutonic igneous rocks composed almost entirely of plagioclase feldspar, which is usually labradorite but may be as calcic as bytownite or as sodic as andesine or oligoclase, and little or no dark-colored minerals; also, any rock in that group. Anorthosites occur as large nonstratiform plutonic bodies and as stratiform intrusions; they are the main rock type of the lunar highlands. Syn: *plagioclase rock.*

ANOVA *analysis of variance.*

ansermetite An adamantine red triclinic mineral: $MnV_2O_6\cdot 4H_2O$.

ANT An acronym for a suite of rock types abundant in the lunar highlands, including anorthosite, norite, and troctolite, as well as gradational mixtures of these (Taylor, 1975).

antagonism (an-tag'-o-nism) In ecology, the relationship that exists between two organisms in which one or both are harmed, usually as a result of their trying to occupy the same ecologic niche.

antapex (ant-a'-pex) (a) The posterior end of a *dinoflagellate* when oriented in the direction of swimming; the corresponding end of a dinoflagellate cyst. (b) The most posterior point of the hypocyst of an *acritarch* (Williams et al., 2000, p.9). Adj: *antapical.*

antapical (ant-ap'-i-cal) Relating to the *antapex.*

antapical series (ant-ap'-i-cal) The series of plates forming the terminal group behind the postcingular series in a dinoflagellate theca. Cf: *apical series.*

antarctic (ant-arc'-tic) n. The area within the Antarctic Circle; the region of the South Pole. adj. Pertaining to features, climate, vegetation, and animals characteristic of the antarctic region.

Antarctic Circle (Ant-arc'-tic) The parallel of latitude falling at approx. 66°32′ S; it marks the latitude where the Sun's rays are tangent to Earth's surface at the June solstice. Cf: *Arctic Circle.*

Antarctic convergence A natural and distinct oceanographic boundary around the continent of Antarctica, more or less equivalent to the 10°C isotherm for the warmest month. The colder, denser Antarctic waters sink sharply below the warmer, lighter sub-Antarctic waters, with little mixing. The oceanographic aspect of the boundary is reflected in water and air temperatures and in the distribution of marine organisms.

antarcticite (ant-arc'-tic-ite) A white trigonal mineral: $CaCl_2\cdot 6H_2O$. Highly hygroscopic.

Antarctic meteorite Meteorite recovered from Antarctica, generally from the blue ice fields.

Antarctic ozone hole A large area over the Antarctic continent, of significant stratospheric ozone depletion, that typically develops annually between late August and early October and generally ends in mid-November. Ozone thinning is attributed to the action of chlorine (Cl) liberated from a group of chemicals known as chlorofluorocarbons (CFCs).

antecedent (an-te-ced'-ent) Said of a stream, valley, or drainage system that maintains its original course or direction despite subsequent deformation or uplift. The term was first applied by Powell (1875, p.163) to the valley thus formed.

antecedent moisture The amount of moisture present in a soil mass at the beginning of a runoff period; often expressed in terms of an *antecedent precipitation index.*

antecedent-platform theory A theory of coral-atoll and barrier-reef formation according to which reefs are built upward to the water surface from an extensive submarine platform (perhaps consisting of volcanic debris rapidly leveled by wave erosion), situated 50 m or more below sea level, and formed before its colonization by corals, without the intervention of relative changes in sea level (Hoffmeister and Ladd, 1944). Cf: *glacial-control theory; subsidence theory.*

antecedent precipitation index The amount of moisture in a drainage basin before a storm. Abbrev: API. See also: *antecedent moisture.*

antecedent stream A stream that was established before local uplift or diastrophic movement was developed across it and that maintained its original course after and in spite of the deformation by incising its channel at approximately the same rate as the land was rising; a stream that existed prior to the present topography.

anteclise (an'-te-clise) A rarely used term for a positive or uplifted structure of the continental platform; it is of broad, regional extent (tens to hundreds of thousands of square kilometers) and is produced by slow crustal upwarp during the course of several geologic periods. The term is used mainly in the Russian literature; e.g. the Belorussian anteclises of the Volga-Urals. Also spelled: *anticlise.* Ant: *syneclise.*

antediluvian (an'-te-di-lu'-vi-an) Pertaining to or produced before the flood described in the Bible; antedating the *Noachian flood.* Syn: *prediluvian.*

antenna (an-ten'-na) (a) One of a pair of anterior sensory appendages of the *cephalon* of a crustacean, preceded by *antennule* and followed by *mandible.* (b) One of a pair of slender multijointed sensory appendages attached to the ventral surface of the *cephalon* in front of the mouth of a trilobite. (c) An obsolete term for a *chelicera* of an arachnid, used in the past to emphasize its homology with an antenna of a crustacean or of an insect. Pl: antennae or antennas.

antennal carina (an-ten'-nal) A ridge extending backward from the antennal spine on some decapod crustaceans (Holthuis, 1974, p.735).

antennal groove In decapod crustaceans, a groove that extends forward from the lower end of the *cervical groove,* and curves around the lower border of the elevation of the *carapace* behind the antennal spine. Anteriorly it may split into two branches, often indistinct, one going straight forward, the other curving up towards the antennal spine (Holthuis, 1974, p.732-733).

antennal region The anterior marginal part of the *carapace* of some decapod crustaceans, bordering the orbital region laterally and also touching the hepatic, pterygostomial, and, in some forms, the frontal regions (TIP, 1969, pt. R, p.92).

antennule (an-ten'-nule) A small antenna; specif. one of a pair of the most anterior appendages of the *cephalon* of a crustacean, followed by antennae. Adj: antennular. Syn: *first antenna.*

anter (an'-ter) Part of the orifice in ascophoran cheilostome bryozoans that is distal to the condyles or the sinus. Cf: *poster.*

anterior (an-te'-ri-or) adj. (a) Situated toward the front of an animal, or near or toward the head or head region, as opposed to *posterior;* e.g. in a direction (in the plane of symmetry or parallel to it) away from the *pedicle* and toward the mantle cavity of a brachiopod, or in a direction (in the plane of bilateral symmetry) parallel to the *cardinal axis* of a bivalve mollusk and approximating the direction in which the mouth faces, or in a direction toward the *aperture* of a foraminifer. n. The forward-moving or head region of an animal. (b) Toward front end of a conodont element in conventional orientation. In coniform elements, convex side of cusp or side of element opposite that toward which cusp apex points. In ramiform and pectiniform elements with discernible cusp, convex side of cusp or side of element opposite that toward which denticles incline. In platformed pectiniform elements with blade, side toward which distal end of blade is directed. In some elements, convex side of basal cavity defines anterior side (TIP, 1981, pt. W, supp.2).

anterior arch Archlike conodont structure formed by junction of lateral processes at anterior end of alate ramiform element (TIP, 1981, pt. W, supp.2).

anterior face Front side of a conodont element or feature of an element in conventional orientation. For determination, see *anterior.* (b) (TIP, 1981, pt. W, supp.2).

anterior lateral muscle One of a pair of retractor muscles in some lingulid brachiopods, originating on the *pedicle valve* posteriorly and laterally to the central muscles, and converging dorsally to their insertions anteriorly on the *brachial valve* (TIP, 1965, pt. H, p.139).

anterior margin Trace of anterior side of a conodont element or feature of an element in lateral view (TIP, 1981, pt. W, supp.2).

anterior oral midline The perradial line between opposing oral plates in edrioasteroids; it extends anteriorly from the oral pole (Bell, 1976).

anterior side The front end of a conodont; e.g. the convex side of a cusp (the side facing in the direction opposite that toward which the tip of the cusp points) in simple conodont elements, the convex side of the cusp and denticles in compound conodont elements, or the distal end of the free blade in platelike conodont elements. Ant: *posterior side.*

anterior trough margin Portion of platform in planate conodont elements of some *Polygnathus* species in front of the geniculation point (TIP, 1981, pt. W, supp.2).

anterior tubercle A polygenetic swelling or small protuberance in the anterior region of the *carapace* of a phyllocarid crustacean. It includes the "optic tubercle" of some authors.

anterolateral region (an'-ter-o-lat'-er-al) The lateral part of the *carapace* of some decapod crustaceans, bordering the subhepatic or hepatic regions (TIP, 1969, pt. R, p.92).

antetheca (an-te-the'-ca) The final septal face in a fusulinid; e.g., the front wall of the last-formed volution of the test of *Triticites.*

anteturma One of two groupings in which turmae are classified in the turmal system: Sporites (spores) and Pollenites (pollen). See also: *turma.*

anther (an'-ther) The pollen-bearing part of a stamen.

antheridia Pl. of *antheridium.*

antheridium (an-ther-id'-i-um) (a) In cryptogamous plants, the male reproductive organ, within which the male sexual cells are organized. (b) In primitive seed plants, a minute structure of only a few cells developed within the pollen grain. (c) The male *gametangium* of fungi. Pl: antheridia. Cf: *oogonium.*

anthill (ant'-hill) A common term for *termitarium.*

anthoblast (an'-tho-blast) The basal portion of the zooid in certain solitary corals, from which the *anthocyathus* is pinched off to form a new *zooid;* e.g. the stage of ontogeny of *Acrosmilia* derived by transverse division from a solitary *Fungia* individual by extratentacular budding. Cf: *anthocaulus.*

anthocaulus (an-tho-caul'-us) The stalklike basal portion of the *zooid* in certain solitary corals, from which the *anthocyathus* is pinched off to form a new zooid; e.g., the stage of ontogeny of *Acrosmilia* derived by transverse division from a solitary *Fungia* individual by sexual generation. Pl: anthocauli. Cf: *anthoblast.*

anthocyathus (an'-tho-cy-a'-thus) The oral disk that is pinched off from the basal portion in some solitary corals and enlarges to become a new *zooid;* e.g. the neanic stage of *Fungia* after separation from an *anthocaulus* or *anthoblast.* Pl: anthocyathi.

anthodite (an'-tho-dite) A pencil-like *speleothem,* fed through a small central canal, usually composed of aragonite, typically in clusters radiating out from a common base (Henderson, 1949).

anthoinite (an-thoin'-ite) A chalky white triclinic mineral of uncertain composition, possibly $WAlO_3(OH)_3$.

anthonyite (an'-tho-ny-ite') A lavender-colored mineral of undetermined crystallography: $Cu(OH,Cl)_2 \cdot 3H_2O$.

anthophyllite (an-tho'-phyll-ite) A clove-brown to colorless orthorhombic mineral of the *amphibole* group: $(Mg,Fe^{2+})_2(Mg,Fe^{2+})_5Si_8O_{22}(OH)_2$. It is dimorphous with cummingtonite; with increase in aluminum it grades into gedrite. Anthophyllite occurs in metamorphosed ultrabasic rocks, typically with olivine or talc or in monomineralic aggregates of parallel or radiating asbestiform fibers. It has been mined for asbestos. Syn: *bidalotite.*

Anthophyta (An-tho-phy'-ta) *angiosperm.*

anthozoan (an-tho-zo'-an) Any coelenterate belonging to the class Anthozoa, which includes marine, polypoid, solitary or colonial, mostly sedentary forms and is characterized by the presence of a stomodaeum. Range, Ordovician to Holocene.

anthracite (an'-thra-cite) Coal of the highest metamorphic rank, in which fixed-carbon content is between 92% and 98% (on a dry, mineral-matter-free basis). It is hard and black, and has a semimetallic luster and semiconchoidal fracture. Anthracite ignites with difficulty and burns with a short blue flame, without smoke. Syn: *hard coal; stone coal; kilkenny coal.*

anthracitic (an-thra-cit'-ic) Pertaining to *anthracite.*

anthracitization (an'-thra-cit-i-za'-tion) The metamorphic transformation of bituminous coal into anthracite.

anthracology (an-thra-col'-o-gy) Coal petrography; the analysis of coals by type.

anthraconite (an-thrac'-o-nite) A black *bituminous limestone* (or marble) that usually emits a fetid smell on being struck or rubbed; a *stinkstone.* Obsolete syn: *swinestone; lacullan.*

Anthracosauria (An'-thra-co-saur'-i-a) An order of labyrinthodont amphibians characterized by vertebrae with unreduced intercentrum, distinctive pattern of dermal bones of skull, and reptilian phalangeal formula. It includes the most probable candidates for reptilian ancestry. Range, Upper Mississippian to Upper Permian.

anthracoxene (an-thrac'-o-xene) A brownish resin which, when treated with ether, dissolves into an insoluble portion, *anthracoxenite,* and a soluble portion, *schlanite.*

anthracoxenite (an'-thra-co-xen'-ite) The insoluble resin remaining when *anthracoxene* is treated with ether. See also: *schlanite.*

anthraxolite (an-thrax'-o-lite) A hard, black asphaltite with a high fixed-carbon content; it occurs in veins and masses in sedimentary rocks, esp. in association with oil shales.

anthraxylon (an-thra-xy'-lon [an-thra-zy'-lon]) A composite term for the vitreous coal components derived from woody tissues of plants and forming lustrous bands interlayered with dull *attritus* in banded coal. Etymol: Greek "anthrax", "coal", and "xylon", "wood".

anthraxylous-attrital coal (an-thra-xy'-lous-at-tri'-tal) A bright coal in which the ratio of anthraxylon to attritus ranges from 3:1 to 1:1. Cf: *attrital-anthraxylous coal; anthraxylous coal; attrital coal.*

anthraxylous coal A bright coal in which the ratio of anthraxylon to attritus is greater than 3:1. Cf: *attrital coal; attrital-anthraxylous coal; anthraxylous-attrital coal.*

anthrinoid (an'-thri-noid) Vitrinite that occurs in noncaking anthracites and that has a reflectance higher than 2.0% (ASTM, 1970, p.19). Cf: *xylinoid; vitrinoid.*

anthropic epipedon (an-throp'-ic) A surface layer of mineral soil that has the same requirements as the mollic epipedon with respect to color, thickness, organic carbon content, consistence and base saturation, but that has > 110 mg Pkg^{-1} soluble in 0.05 M citric acid, or is dry > 300 days (cumulative) during the period when not irrigated. The anthropic epipedon forms under long continued cultivation and fertilization.

Anthropogene (An-thro'-po-gene) An informal term for the latest interval of the *Cenozoic,* formerly widely used in the former USSR in place of *Quaternary.* The term refers to the anthropological characterization of the "ice age" interval, but was later defined by a level in the Ukraine which is closely equivalent to the international GSSP base of the *Pleistocene* at about 1.8 Ma. Cf: *Pleistogene.*

anthropozoic (an'-thro-po-zo'-ic) An informal descriptor of geologic time since the appearance of man; also, of the rocks formed during that time.

anthrosol (an'-thro-sol) A soil that has been materially affected in its physical or chemical properties by human activities. Cf: *anthropic epipedon.*

anti-alias filter *alias filter.*

antibiosis (an'-ti-bi-o'-sis) Passive action by one organism which is harmful to another (Ager, 1963, p.313). Syn: *amensalism.*

anticenter (an'-ti-cen'-ter) That point on the Earth's surface that is diametrically opposite the *epicenter* of an earthquake.

anticlinal [bot] (an'-ti-cli'-nal) adj. At right angles to the surface or circumference of a plant organ. Cf: *periclinal [bot].*

anticlinal [struc geol] n. An obsolete form of *anticline.* adj. Pertaining to an anticline.

anticlinal nose *nose [struc geol].*

anticlinal theory The theory that oil and gas tend to accumulate in anticlinal structures. It was well set forth by I. C. White in 1885.

anticlinal valley A valley that follows the axis of an anticline.

anticline (an'-ti-cline) A fold, generally convex upward, whose core contains the stratigraphically older rocks. Ant: *syncline.* See also *antiform; synformal anticline.*

anticlinoria Pl. of *anticlinorium.*

anticlinorium (an'-ti-cli-no'-ri-um) A composite anticlinal structure of regional extent composed of lesser folds. Cf: *synclinorium.* Pl: *anticlinoria.*

anticlise (an'-ti-clise) *anteclise.*

anticonsequent stream (an-ti-con'-se-quent) *obsequent stream.*

anticrack The crack-tip structure (usually stress-solution cleavage) that resembles a tail crack. This structure is found in the compressional region of the crack tip when the tip is subject to mode II crack-tip displacement (Fletcher and Pollard, 1981).

anticusp (an'-ti-cusp) The downward projection of the base of a conodont cusp. It may or may not bear denticles.

anticyclone (an-ti-cy'-clone) An atmospheric high-pressure system with closed isobars, the pressure gradient being directed away from the center so that the near surface wind spirals outward in a clockwise direction in the Northern Hemisphere, counterclockwise in the Southern. It was named by Francis Galton in 1861. See also: *cyclone.*

antidip stream (an'-ti-dip) A stream flowing in a direction opposite to that of the general dip of the strata. It is frequently but not necessarily an *obsequent stream.*

antidune (an'-ti-dune) (a) A term used by Gilbert (1914, p.31) for an ephemeral or transient *bed form* formed on a stream bed (rarely preserved in sediments), similar to a dune but traveling upstream as the individual sand particles move downcurrent, and characterized by erosion on the downstream slope and deposition on the upstream slope. Its profile is more symmetrical than that of a subaqueous dune; it is indicated on the water surface by a regular undulating wave. Syn: *regressive sand wave.* (b) Any bed form (whether it moves upstream, downstream, or not at all) that is produced by unidirectional flow and is in phase with surface water waves (Kennedy, 1963). Syn: *in-phase wave; sinusbed.* (c) A term used by Lamont (1957) for *flame structure [sed].* This usage is not recommended.

antidune phase The part of stream *traction* (transitional to the *smooth phase*) whereby a mass of sediment travels as a bed form, having an eroded downcurrent slope and a depositional upcurrent slope (Gilbert, 1914, p.30-34); it develops when the bed load is large or in rapid flow conditions. Named "antidune" by Gilbert (1914) because the bed form moves upstream as the individual particles move downstream. Cf: *dune phase.*

anti-estuarine circulation The pattern of water circulation in an inverse estuary, in which ocean water flows landward into the estuary along the surface and exits seaward along the bottom. The flow pattern is created by a negative water balance in which evaporation increases the salinity and hence the density of the seawater. The prime example for large-scale anti-estuarine circulation is the Mediterranean Sea with its low nutrient content.

antiferromagnetism (an'-ti-fer'-ro-mag'-ne-tism) A type of *magnetic order* in which the moments of neighboring arrays of magnetic ions are aligned antiparallel, so that there is no macroscopic spontaneous magnetization. Cf: *ferromagnetism; ferrimagnetism.* See also: *weak ferromagnetism.*

antiform (an'-ti-form) Any convex-upward, concave downward fold. The term is usually used when the folded layers do not possess a stratigraphic order, when the stratigraphic order of the folded layers is not known, or when the fold core also contains the stratigraphically younger rock.

antiformal stack A *duplex* in which the trailing branch lines coincide and the leading branch lines either nearly coincide or lie vertically above each other. This results in a stack of horses and leads to a local culmination in thrust sheet (Boyer and Elliott, 1982). Cf: *foreland-dipping duplex; hinterland-dipping duplex.*

antiformal syncline (an-ti-form'-al) A *syncline* the limbs of which close upward as in an *antiform* (Turner and Weiss, 1963, p.106).

antigorite (an-tig'-o-rite) A macroscopically lamellar brown to green monoclinic serpentine mineral, which consists structurally of alternating wave forms in which the 1:1 T-O layer reverses sides and direction of curvature at each wave null point. In most specimens the repeat distance of the wave pattern measures between 25.5 and 51.0 Å: $(Mg,Fe^{2+})_3Si_2O_5(OH)_4$.

antigravitational gradation (an'-ti-grav'-i-ta'-tion-al) A term introduced by Keyes (1913) for wind erosion and deposition operating mainly from a lower to a higher elevation, as on broad intermont valleys of arid regions where "the wind is able to blow sands erodingly and extensively up-hill". See also: *planorasion.*

antilogous pole (an-til'-o-gous) In crystallography, that pole of a crystal which becomes electrically negative when the crystal is heated or is expanded by decompression. Cf: *analogous pole.*

antimeridian (an'-ti-me-rid'-i-an) The meridian that is 180 degrees of longitude from a given meridian. A meridian and its antimeridian constitute a great circle.

antimonate (an'-ti-mon-ate') A mineral compound characterized by the presence of antimony and oxygen in the radical. An example is stibiconite, $Sb^{3+}Sb^{5+}{}_2O_6(OH)$.

antimonite (an'-ti-mon-ite') *stibnite.*

antimonpearceite (an'-ti-mon-pearce'-ite) A black submetallic monoclinic mineral: $(Ag,Cu)_{16}(Sb,As)_2S_{11}$. The antimony analogue of *pearceite.* Cf: *arsenpolybasite.*

antimonselite A metallic black orthorhombic mineral: Sb_2Se_3. The selenium analogue of *stibnite.*

antimony (an'-ti-mo'-ny) A metallic white rhombohedral mineral, the native metallic element Sb. It is brittle and commonly occurs in silvery or tin-white granular, lamellar, or shapeless masses.

antimony blende *kermesite.*

antimony bloom *valentinite.*

antimony glance *stibnite.*

antimony ocher Any of several native antimony oxides such as stibiconite or cervantite.

antimorph (an'-ti-morph) A smoothly arched surface on an undulatory thrust fault. It does not reflect warping or deformation of a once-planar fault surface. Antimorphs are usually exposed in fensters (Kopf, 1982). Cf: *synmorph.*

antinode (an'-ti-node) That point on a standing wave where the amplitude is greatest and the horizontal velocity is least. Ant: *node [waves].*

antipathetic (an'-ti-pa-thet'-ic) Said of two or more minerals that are far apart from each other in a crystallization sequence and thus will not be commonly found in association. See also: *antipathies of minerals.*

antipathies of minerals (an-tip'-a-thies) An aspect of the theory of fractional crystallization, which states that minerals that are far apart in a crystallization sequence will not be found in association to any great extent. Such minerals are said to be *antipathetic.*

antiperthite (an-ti-perth'-ite) A variety of alkali feldspar consisting of parallel or subparallel intergrowths in which the sodium-rich phase (albite, oligoclase, or andesine) appears to be the host from which the potassium-rich phase (usually orthoclase) exsolved. Cf: *perthite.*

antiphase boundary The boundary between *antiphase domains.* Abbrev: APB.

antiphase domains Adjoining volumes of a crystal displaying atomic *ordering,* such that the ordering sequence on one side of the boundary between these volumes is out of phase (discontinuous, or out of step) with the ordering sequence on the other side. Antiphase domains usually develop during a *substitutional transformation* when various regions of a crystal order independently in equivalent but physically different ways.

antipodal bulge (an-tip'-o-dal) The *tidal bulge* not facing the Moon.

antipodal point *antipode.*

antipode (an'-ti-pode) The opposite point with respect to any given point; specif. one of two diametrically opposite parts of the Earth. The term is usually used in the plural and is often extended to include the whole region at the opposite end of a diameter of the Earth, such as Australia and New Zealand which lie roughly opposite to the British Isles. Syn: *antipodal point.*

antiprism (an'-ti-prism) An *n*-gonal form with faces consisting of two parallel regular *n*-gons connected by 2*n* isosceles triangles.

antiripple (an'-ti-rip'-ple) *adhesion ripple.*

antiripplet (an'-ti-rip'-plet) *adhesion ripple.*

antiroot (an'-ti-root) According to the *Pratt hypothesis,* crustal material under the oceans of higher density as isostatic compensation for its lesser mass and lower topographic elevation. Cf: *root [tect].*

antistress mineral (an'-ti-stress) A term suggested by Harker (1918) for minerals such as cordierite, the feldspars, the pyroxenes, forsterite, andalusite, etc. whose formation in metamorphosed rocks is believed to be favored by conditions that are not controlled by shearing stress but by thermal action and by hydrostatic pressure that is probably no more than moderate. Obsolete. Cf: *stress mineral.*

antitaenite A low-spin paramagnetic Ni-Fe alloy associated with tetrataenite. Not approved by the International Mineralogical Association.

antitaxial vein (an-ti-tax'-i-al) A vein that fills by growth of minerals from the center of the vein toward the walls. Vein minerals tend to be different from those found in the wall, and commonly have a fibrous habit. Cf: *syntaxial vein.* Etymol: Greek, "taxis," arrangement, order, orientation.

antithetic fault (an-ti-thet'-ic) A fault that is subsidiary to a larger fault and whose dip is opposite to the dip of the larger fault. Term introduced by H. Cloos to describe faults that dip in a direction opposite to the dip of the rocks displaced, and that rotate fault-bounded blocks so that the net slip on each fault is greater than it would be without rotation (Dennis, 1967). Many authors now use the term to describe a fault that is subsidiary to a major fault, formed in the same stress regime, is oriented at a high angle to the major fault, and (for strike-slip faults) has a sense of displacement opposite that of the major fault or (for normal faults) dips in the opposite direction (Biddle and Christie-Blick, 1985). Cf: *synthetic fault.*

antitrades (an'-ti-trades) A layer of westerly winds in the troposphere, above the *trade winds* of the tropics.

antler helictite (ant'-ler) A type of helictite having thick branches which commonly bifurcate in near vertical or horizontal directions.

antlerite (ant'-ler-ite) An emerald-green to blackish-green orthorhombic mineral: $Cu_3(SO_4)(OH)_4$. It occurs in interlaced aggregates of needle-like crystals and rarely constitutes an ore of copper. Syn: *vernadskite.*

Antler Orogeny (Ant'-ler) An *orogeny* that extensively deformed Paleozoic rocks of the Great Basin in Nevada during late Devonian and early Mississippian time; named by R. J. Roberts (1951) for relations in the Antler Peak Quadrangle near Battle Mountain, Nevada. Equivalent tectonism can be traced further north, along the length of the North American Cordillera, into Canada. Its main expression is the emplacement of eugeosynclinal western rocks over miogeosynclinal eastern rocks along the Roberts Mountains Allochthon. Minor orogenic pulses followed the main event, extending into the Permian. The Antler Orogeny marks the cessation of the passive-margin phase of the Cordillera and the beginning of the convergent-margin phase. Cf: *Acadian Orogeny.*

antofagastite (an'-to-fa-gas'-tite) *eriochalcite.*

antozonite (an-to-zon'-ite) A dark-violet to black semiopaque variety of fluorite that emits a strong odor when crushed, perhaps due to free fluorine. It is produced by alpha bombardment, as in the inner bands of halos surrounding uraninite and thorite inclusions.

Anura (An-u'-ra) Order of tailless lissamphibians that includes frogs and toads. Range, Middle Jurassic to Holocene.

anus (a'-nus) The posterior or terminal opening of the alimentary canal or digestive tract of an animal. In vertebrates, the orifice of the digestive tube that is separated from the orifice(s) for excretory and reproductive egress.

anvil cell A device consisting of two opposed anvils (often the culets of matched, brilliant-cut diamonds, in which case it is called a *diamond anvil cell*) and an intervening gasket containing a hole into which a single crystal or other small sample may be placed in a fluid. By driving the anvils together against the gasket and fluid, the sample is subjected to elevated hydrostatic pressures.

anvil-type fibrous prismatic structure [paleont] A fibrous *prismatic structure* in which the fibers show anvil-like cross sections, i.e., combining convex and concave lateral faces.

anyuiite (an'-yu-i-ite) A metallic gray tetragonal mineral: $Au(Pb,Sb)_2$.

AOM *amorphous organic matter.*

Aouelloul glass Glass associated with Aouelloul crater, Mauritania, thought to have been produced by melting of local target rocks during the impact event. See also: *impactite.*

AP *arboreal pollen.*

Apache tear (A-pach'-e) (a) An obsidian nodule weathered from a lava flow. (b) A widely used syn. of marekanite.

apachite [mineral] (a-pach'-ite) A monoclinic blue mineral: $Cu_9Si_{10}O_{29}{\cdot}11H_2O$.

apachite [rock] A peralkaline *phonolite,* named by Osann in 1896 for the Apache (now Davis) Mountains, Texas. Obsolete.

apalhraun (ap'-al-hraun) An Icelandic term for both *block lava* and *aa.* Cf: *helluhraun.*

apatite (ap'-a-tite) (a) A group of variously colored hexagonal minerals consisting of calcium phosphate together with fluorine, chlorine, hydroxyl, or carbonate in varying amounts and having the general formula:

$Ca_5(F,OH,Cl)(PO_4,CO_3)_3$. Also, any mineral of the *apatite* group, such as fluorapatite, chlorapatite, hydroxylapatite, carbonate-apatite, and francolite; when not specified, the term usually refers to *fluorapatite.* The apatite minerals occur as accessory minerals in almost all igneous rocks, in metamorphic rocks, and in veins and other ore deposits; and most commonly as fine-grained and often impure masses as the chief constituent of phosphate rock and of most or all bones and teeth. Syn: *calcium phosphate.* (b) A group of hexagonal minerals having the general formula: $A_5(F,OH,Cl)(RO_4)_3$, where A = Ca, Sr, or Pb, and R = P, As, V, or less commonly Si. Examples include svabite, hedyphane, mimetite, pyromorphite, and vanadinite.

apatite iron ore A variety of *iron oxide copper-gold deposit* consisting of magnetite-apatite with actinolite or pyroxene mineralization. Syn: *Kiruna-type iron.*

apertural bar (ap-er'-tur-al) A fused pair of costae immediately proximal to the *orifice* in cribrimorph cheilostome bryozoans.

aperture [paleont] (ap'-er-ture) (a) The opening of a univalve shell; e.g. the opening at the last-formed margin of a gastropod shell, through which the head-foot mass is extended or withdrawn. (b) Any of the major openings through the *theca* or *calyx* of echinoderms, such as the mouth and anus, and sometimes including hydropores and gonopores. (c) The terminal skeletal openings of zooids in stenolaemate bryozoans. Also used as a syn. of *orifice* in cheilostome bryozoans. (d) An opening in the *test* or shell of a foraminifer, such as a relatively large opening to the exterior in the last-formed chamber. Also, the large main opening of a radiolarian shell. (e) The opening into the mantle cavity of a cirripede crustacean.

aperture [palyn] Any of various structures (usually thinning) in the exine of palynomorphs that facilitate the exit of the living contents. An aperture developed in only one wall layer is called simple; one that affects more than one wall layer is called compound. See also: *germinal aperture.*

aperture [seis] (a) An opening, gate, or window that limits the information affecting a measuring device. (b) In processing, the spatial range of the data considered in a calculation or the time range of the data considered (e.g., in *deconvolution*).

aperture membrane The *exine* layer that forms the floor of an *ectoaperture* (pore or colpus) of a pollen grain.

apex [geomorph] The tip, summit, or highest point of a landform, as of a mountain; specif. the highest point on an alluvial fan, usually the point where the stream that formed the fan emerged from the mountain or from confining canyon walls.

apex [mining] In mining, the highest point of a vein relative to the surface, whether it crops out or not. The concept is used in mining law. See also: *apex law.*

apex [paleont] (a) The first-formed, generally pointed end of an elongate or conical form of an organism, such as the small end of the shell or spire of a gastropod. (b) The first-formed part of a brachiopod valve, around which the shell has grown subsequently. The term is "usually restricted to valves having this point placed centrally or subcentrally" (TIP, 1965, pt. H, p.140). (c) The tip of the basal cavity or of a denticle of a conodont. Also, the juncture of bars, blades, or other processes of conodont elements. (d) The upper (umbonal) angle of a valve or plate of a cirripede crustacean. (e) The pointed beginning of a straight or slightly curved cephalopod shell.

apex [palyn] (a'-pex) (a) The anterior end of a *dinoflagellate* when oriented in the direction of swimming; the corresponding and of a dinoflagellate cyst. (b) The most anterior point of the epicyst of an *acritarch* (Williams et al., 2000, p.13).

apex [struc geol] (a'-pex) *culmination.*

apex law In U.S. mining law, the individual whose claim contains the apex of a vein may follow and exploit the vein indefinitely along its dip, even if it passes downdip under adjoining surface property lines. Syn: *law of extralateral rights.*

APF *absolute pollen frequency.*

aphanic (a-phan'-ic) Said of the texture of a carbonate sedimentary rock characterized by individual crystals or clastic grains whose diameters are less than 0.01 mm (Bissell and Chilingar, 1967, p.150) or 0.005 mm (Chilingar et al., 1967, p.311). Obsolete.

aphanite (aph'-a-nite) Any fine-grained igneous rock whose components are not distinguishable with the unaided eye; a rock having *aphanitic* texture. The obsolescent syn. *felsite* has been sometimes restricted to the light-colored rocks with this texture and *aphanite* to the dark-colored (Johanssen, 1939, p.201). Adj: *aphanitic [ign].* Cf: *felsite; phanerite.*

aphanitic [ign] (aph-a-nit'-ic) Said of the texture of an *aphanite.* Cf: *phaneritic; felsitic.* Syn: *fine-grained (a).*

aphanitic [sed] A term formerly loosely applied to a sedimentary carbonate-rock texture.

aphanophyric (a'-phan-o-phyr'-ic) *microcrystalline.*

Aphebian (A-pheb'-i-an) In a three-part division of the *Proterozoic* of Canada, the earliest division, before the *Helikian.* Cf: *Hadrynian,* from 2,390 to 1,640 Ma.

aphelion (a-phe'-li-on) The point at which the Earth (or any planetary body) is most distant from the sun in its elliptical orbit. Cf: *perihelion.*

aphlebia (aph-le'-bi-a) Anomalous pinnae, usually located at the base of the rachis of certain fossil ferns and pteridosperms (Swartz, 1971, p.36).

aphodus (a-phod'-us) A short canal of uniform diameter in a sponge, leading to an exhalant canal from an *apopyle* of approximately the same cross-sectional area. Pl: aphodi. Cf: *prosodus.*

aphotic (a-pho'-tic) Literally, "without light". It refers to a depth in the sea below approximately 600-1,000 m. Photosynthesis is negligible below about 200-m depth.

aphotic zone [lake] The zone in a lake below the *compensation depth.* Cf: *euphotic zone [lake]; tropholytic.*

aphotic zone [oceanog] That part of the ocean in which there isn't enough penetration of light for photosynthesis. Cf: *disphotic zone; euphotic [oceanog].*

aphrite (aph'-rite) A foliated, lamellar, scaly, or chalky variety of calcite having a white pearly luster. Syn: *earth foam; foaming earth.*

aphrizite (aph'-ri-zite) A black variety of tourmaline containing iron.

aphrodite (aph'-ro-dite) *stevensite.*

aphroid (aph'-roid) Said of a massive *corallum* similar to *astreoid* type but with septa shortened peripherally and adjacent *corallites* united by a dissepimental zone.

aphrolith (aph'-ro-lith) An obsolete syn. of *aa.* Cf: *dermolith.*

aphrosiderite (aph-ro-sid'-er-ite) *ripidolite.*

aphthitalite (aph-thit'-a-lite) A white or variously colored trigonal mineral: $(K,Na)_3Na(SO_4)_2$, easily confounded with *glaserite.*

aphylactic projection (a-phy-lac'-tic) A map projection that does not possess any one of the three special properties of equivalence, conformality, or equidistance; e.g. a gnomonic projection.

aphyllous (a-phyl'-lous) Said of a leafless plant.

aphyric (a-phyr'-ic) Said of the texture of a fine-grained or aphanitic igneous rock that lacks phenocrysts; also a rock exhibiting such texture.

aphytal (a-phy'-tal) Said of those parts of a body of water that are not shallow enough for the growth of rooted green plants. Cf: *phytal.*

Aphytic (A-phyt'-ic) A paleobotanic division of geologic time, signifying the time that preceded the first occurrence of plant life. Cf: *Archeophytic; Eophytic; Paleophytic; Mesophytic; Cenophytic.*

API (a) The American Petroleum Institute. (b) *antecedent precipitation index.*

apical (ap'-i-cal) (a) Situated at or in the direction or vicinity of the *apex* of a shell; e.g. "apical horn", a spine at the apex of the shell of a nasselline radiolarian. (b) Located away from the mouth of an echinoderm; *aboral.*

apical angle The angle between the axis of a cone and a side of the cone. On a stereonet, it is used to describe the opening angle of a small circle.

apical archeopyle An *archeopyle* formed in a dinoflagellate cyst by the loss of the entire apical series of plates.

apical area On an embryophytic spore, the proximal area associated with the *laesura.*

apical axis The lengthwise axis of a pennate diatom. Cf: *pervalvar axis.*

apical denticle *cusp [paleont].*

apical papilla A dotlike thickening of the interradial area of a spore. When present, there is generally one apical papilla per interradial area, hence three per spore.

apical prominence In megaspores, mostly Paleozoic, a variously constructed proximal projection at the intersection of the contact areas. Cf: *acrolamella; gula.*

apical series The series of plates grouped about an open apical pore or forming an apical cluster in the epitheca of a dinoflagellate. Cf: *antapical series.*

apical system A system of primordial plates at the aboral terminus of ambulacra and interambulacra of echinoids, sometimes including an outer circlet of *ocular plates* surrounding an inner circlet of *genital plates,* and one or more supplementary plates. See also: *oculogenital ring.*

apiculate (a-pic'-u-late) (a) Pertaining to surface ornamentation, as in palynomorphs, consisting of short, sharp spines. (b) Said of pollen grains with slightly protuberant poles.

apiculus (a-pic'-u-lus) An open-end process extending from the valve surface in a diatom. Plural: apiculi.

API gravity A standard adopted by the American Petroleum Institute for expressing the specific weight of oils. API gravity = (141.5/specific gravity at 60°F) - 131.5. This arbitrary scale simplifies the construction of hydrometers because it enables the stems to be calibrated linearly. The lower the specific gravity, the higher the API gravity. Cf: *Baumé gravity.*

API unit (a) A counting unit for gamma-ray logs based on the API calibration pit at the University of Houston. (b) A counting unit for neutron logs. The reading in the Indiana limestone portion of the API neutron log calibration pit (which has 19 percent porosity and is saturated with fresh water) is defined as 1,000 API units.

API well number A unique number assigned by the American Petroleum Institute to each well drilled in the United States.

apjohnite (ap'-john-ite) A silky-white, faintly green, or yellow monoclinic mineral: $Mn^{2+}Al_2(SO_4)_4{\cdot}22H_2O$. It occurs in crusts, fibrous masses, or efflorescences. Syn: *manganese alum.*

aplacophoran (a-pla-coph'-o-ran) A marine mollusk belonging to the class Aplacophora and known only from living forms. *Amphineura* with worm-like body form, and no head or shell; body covered with cuticle in which aragonitic spicules are embedded. Range, Holocene; about 70 genera. See also: *polyplacophoran.*

aplanospore (a-pla′-no-spore) In Fungi, a nonmotile spore formed within a sporangium.

aplite (ap′-lite) A light-colored hypabyssal igneous rock characterized by a fine-grained allotriomorphic-granular (i.e. aplitic) texture. Aplites may range in composition from granitic to gabbroic, but the term "aplite" with no modifier is generally understood to mean granitic aplite, consisting essentially of quartz, potassium feldspar, and sodic plagioclase. The term, from a Greek word meaning "simple", was in use before 1823.

aplitic (ap-lit′-ic) Fine- to medium-grained *saccharoidal* or xenomorphic-granular texture characteristic of some granitic rocks (aplites) which lack dark minerals and micas.

aplogranite (ap-lo-gran′-ite) A light-colored plutonic rock, named by Bailey in 1916, having granitic texture and essentially composed of alkali feldspar and quartz, with lesser amounts of biotite and with or without muscovite. Cf: *alaskite; two-mica granite.* Obsolete.

aplome (ap′-lome) A dark-brown, yellowish-green, or brownish-green variety of andradite garnet containing manganese. Syn: *haplome.*

aplowite (ap′-low-ite) A pink monoclinic mineral: $(Co,Mn^{2+},Ni)SO_4{\cdot}4H_2O$.

apo- In petrologic terminology, a prefix signifying metasomatic change without destruction of original texture.

apocarpous (ap-o-car′-pous) Said of a plant ovary whose carpels are separate rather than united; also, said of a gynoecium of separate pistils (Lawrence, 1951, p.739). Cf: *syncarpous.*

apochete (ap′-o-chete) An *exhalant canal* of a sponge.

apocolpia Pl. of *apocolpium.*

apocolpial field A region at the pole of a *parasyncolpate* pollen grain delineated by the margins of anastomosing colpi.

apocolpium (a-po-col′-pi-um) A region at the pole of a *parasyncolpate* pollen grain delimited by lines drawn between the apices of the colpi. Pl: *apocolpia.*

apocolpium index *polar-area index.*

apodeme (ap′-o-deme) One of the ingrowths from the *exoskeleton* of many arthropods that provide points of muscle attachment, such as an invagination of the body wall of an arachnid, an inward deflection of a *sclerite* of a merostome, a downward projection from the dorsal interior of a thoracic segment of a trilobite, or an infold of the exoskeleton of a crustacean.

apogean tide (ap-o-ge′-an) A tide of decreased range when the Moon is farthest from Earth in its orbit. Ant: *perigean tide.*

apogee (ap′-o-gee) That point on the orbit of an Earth satellite, natural or man-made, which is farthest from the Earth. Cf: *perigee.*

apogranite (ap′-o-gran′-ite) Albitized and greisenized granite, located at the peripheral and apical parts of certain intrusives, and commonly mineralized in rare elements (Nb, Ta, Li, Rb, Be, Sn, W, Mo, etc.). The term was originated in 1962 by A. A. Beus and co-workers in the Soviet Union. Obsolete.

apogrit (ap′-o-grit) *graywacke.*

apolar (a-po′-lar) In palynomorphs, lacking a clear polarity.

apomorphic character Descendant character in an ancestor-descendant relationship that is described as preexisting-novel (Wiley et al., 1991, p.9). Cf: *homologue; plesiomorphic character.*

apomorphy (ap′-o-mor′-phy) A character derived from its preexisting homologue, and having relatively derived or special traits; an evolutionary novelty (Brooks et al., 1984, p.79). Adj: apomorphic. Cf: *plesiomorphy.*

aponeurotic band (ap′-o-neu-rot′-ic) An anterior or posterior area of attachment of ligaments (mantle and visceral) on the inside of the body chamber of a nautiloid (TIP, 1964, pt. K, p.54).

apophyllite (a-poph′-yl-lite) A mineral group including *fluorapophyllite, hydroxapophyllite,* and *natroapophyllite.* These platy, colorless, tetragonal secondary minerals are related to and occur with zeolites in geodes in decomposed basalts and other igneous rocks. Syn: *fish-eye stone*, ichthyophthalmite.

apophysis [intrus rocks] An offshoot from a larger, intrusive body. Syn: *tongue [intrus rocks].* Cf: *epiphysis [intrus rocks].*

apophysis [paleont] (a) An internal projection from interambulacral basicoronal plates of an echinoid for the attachment of muscles supporting *Aristotle's lantern.* (b) A lateral transverse process of a radial spine in acantharian radiolarians. (c) In vertebrates, a projection on bone or cartilage, generally for tendinous attachment by muscles. Pl: apophyses.

apopore (ap′-o-pore) The exit opening of an *exhalant canal* of a sponge, located either within the sponge (as on the lining of a larger exhalant canal or the *spongocoel*) or on its surface, in which case it is equivalent to an *osculum.* Cf: *prosopore.*

apoporium (ap-o-po′-ri-um) Polar area of a *zonoporate* pollen grain delimited by the tangential lines connecting the borders of the pores.

apopyle (ap′-o-pyle) Any opening through which water passes out of a *flagellated chamber* of a sponge. Cf: *prosopyle.*

aporhyolite (ap-o-rhy′-o-lite) A *rhyolite* in which the once glassy groundmass has become devitrified. Not recommended usage.

aporhysis (a-po-rhy′-sis) A *skeletal canal* in dictyonine hexactinellid sponges running radially through the body wall, opened at the *spongocoel* end but closed over at the outer end. Pl: aporhyses.

apotaphral (ap-o-taph′-ral) Descriptive of a type of *tectonics* involving lateral outward spreading (under gravity) of an orogenic zone, away from the axis of a geosyncline (Carey, 1963, p. A6). Obsolete. Cf: *syntaphral; diataphral.*

Appalachian relief (Ap-pa-la′-chi-an) Structurally controlled topography developed on folded strata of dissimilar erosional resistance, characterized by homoclinal ridges and subsequent valleys that are adapted to the structure and differential resistance of the rocks. Type examples: the relief of the Ridge and Valley region of the Appalachian Mountains in North America; the Zagros Mountains in Iran. Cf: *Jurassian relief; ridge-and-valley topography.*

Appalachian Revolution An outdated concept that Paleozoic time was closed by a profound crustal disturbance, which especially deformed the rocks in the central and southern Appalachians. *Alleghany Orogeny* is now the preferred term.

Appalachian zinc A variety of *Mississippi Valley-type deposit* characterized by stratabound deposits of sphalerite with little or no galena filling primary and secondary porosity in platform dolomite and limestone.

apparatus The entire complement of elements composing the endoskeleton of a conodont (TIP, 1981, pt. W, supp.2). Syn: *conodont apparatus.*

apparent ablation (ap-par′-ent) *summer balance.*

apparent accumulation *winter balance.*

apparent crater An explosion or impact crater as it appears after modification of the original shape by late-stage formational processes such as slumping and deposition of material ejected during crater formation; a crater that is visible on the surface. Measurements of the "apparent diameter" and "apparent depth" are based on the inner diameter at the pre-impact surface level and the depth of the crater floor below this level. Cf: *true crater.*

apparent dip [seis] In seismology, the angle that an emerging wavefront makes with the horizontal, which relates to the dip of the reflector associated with it. The apparent dip is the angle whose tangent is equal to the ratio of the horizontal and vertical components of the displacement produced by the wavefront. Cf: *dip [seis]; moveout.*

apparent dip [struc geol] The angle that a structural surface, e.g. a bedding or fault plane, makes with the horizontal, measured in any random, vertical section rather than perpendicular to the strike. It varies from nearly zero to nearly the *true dip,* depending on whether the random section is close to the direction of the strike or of the dip. Syn: *false dip.*

apparent horizon The visible line of demarcation between land or sea and sky. Strictly, it is the circle that bounds the part of the Earth's surface which would be visible from a given point if no irregularities or obstructions were present. The apparent horizon is extended slightly downward because of atmospheric refraction. In popular usage, the term horizon usually signifies the "apparent horizon". Cf: *true horizon.* Syn: *visible horizon; local horizon; sensible horizon; geographic horizon; topocentric horizon; natural horizon.*

apparent movement The apparent sense of slip observed in any chance section across a fault. It is a function of several variables: the attitude of the fault, of the disrupted strata, and of the section on which the fault is observed, as well as the *net slip* of the fault. Cf: *separation.*

apparent onlap *Onlap* observed in a randomly oriented vertical geologic section, which may not be parallel to the depositional dip and thus may give only one component of the true onlap direction (Mitchum, 1977, p.208).

apparent optic angle The *optic angle* as it appears under the conoscope, after being refracted upon emergence from the crystal. It is larger than the actual optic angle within the crystal. Symbol: 2E.

apparent plunge An obsolete term used for the angle assumed by a normal projection of a geologic structure in the plane of a vertical cross section (Knutson, 1958).

apparent polar wandering A time-sequence distribution of paleomagnetic poles for a tectonic plate, where each pole represents a known time interval that is short but sufficient to average paleosecular variation of the geomagnetic field. Abbrev: APWP. See also: *true polar wandering.*

apparent resistivity The resistivity of homogeneous, isotropic ground that would give the same voltage-current relationship as measured.

apparent thermal inertia An approximation of thermal inertia calculated as one minus albedo divided by the difference between daytime and nighttime radiant temperatures. Abbrev: ATI.

apparent thickness The *thickness [geol]* of a stratigraphic unit or other tabular body, measured at right angles to the surface of the land. Cf: *true thickness.*

apparent velocity The velocity with which the phase of a seismic wavetrain appears to travel along the surface of the Earth. It exceeds the actual velocity if the wavetrain is not traveling parallel to the surface and appears to be infinite for waves traveling perpendicular to the surface. Filtering is sometimes done using apparent velocity as the criterion. Cf: *f-k space.*

apparent-velocity filtering Attenuating events based on their *apparent velocity.* See: *velocity filter.*

apparent water table *perched water table.*

apparent wavelength The distance between correlative points on a wave train as seen by observations along a line. Differs from actual wavelength if the wave direction makes an angle with the line. Cf: *dominant wavelength.*

appendicular skeleton (ap-pen-dic′-u-lar) In the vertebrates, the skeleton of the shoulder and hip girdles and of the fore and hind limbs.

appinite (ap′-pin-ite) A group of dark-colored hornblende-rich plutonic rocks, such as certain syenites, monzonites, and diorites, in which the hornblende occurs as large prismatic phenocrysts and also in the finer-grained groundmass. It was named by Bailey in 1916 for Appin, Loch Linnhe, Scotland. Obsolete.

applanation (ap-pla-na'-tion) All processes that tend to reduce the relief of an area, causing it to become more and more plainlike. These include lowering of the high parts by erosion and raising of the low parts by addition of material; the latter is usually more effective.

applied geology (ap-plied') The application of various fields of geology to economic, engineering, water-supply, or environmental problems; geology related to human activity.

applied geophysics *geophysical exploration.*

applied seismology The use of artificially generated seismic waves in the search for economic deposits such as coal, oil, and gas, or in engineering studies such as determining depth to bedrock or the presence of potentially active faults. Syn: *seismic exploration; prospecting seismology.*

apposition beach (ap-po-si'-tion) One of a series of roughly parallel beaches successively formed on the seaward side of an older beach. Cf: *beach ridge.*

appraisal well A well drilled following a discovery to determine the physical extent or producibility of the resource.

appressed [paleont] (ap-pressed') Said of very closely set conodont denticles, each partly or entirely fused to those adjoining. Cf: *discrete [paleont].*

appressed [struc geol] Said of a fold whose limbs are almost parallel. Rarely used. Cf: *isoclinal.*

approach (ap-proach') The area of the sea extending indefinitely seaward from the shoreline at mean low-water spring tide (Wiegel, 1953, p.2).

approximate contour *indefinite contour.*

apron [geomorph] (a'-pron) An extensive blanketlike deposit of alluvial, glacial, eolian, marine, or other unconsolidated material derived from an identifiable source, and deposited at the base of a mountain, in front of a glacier, etc.; e.g. a *bajada* or an *outwash plain.* Syn: *frontal apron.*

apron [glaciol] *ice apron.*

apron [ice] *ram.*

apron [speleo] (a) A smooth bulging mass of flowstone covering sloping projections from walls of caves or limestone cliffs. (b) A top surface that slopes down inward from a lava tube wall.

apsacline (ap'-sa-cline) Said of the ventral and posterior inclination of the *cardinal area* in either valve of a brachiopod, lying in the bottom left or first quadrant moving counterclockwise from the *orthocline* position (TIP, 1965, pt. H, p.60, fig. 61).

Apsheronian (Ap-she-ron'-ian) Ponto-Caspian stage: uppermost Pliocene-lower Pleistocene (above Akchagylian, below Bakuan).

aptation (ap-ta'-tion) The general, static phenomenon of being fit, or appropriate to the survival of the organism. Cf: *adaptation; exaptation.*

Aptian (Apt'-i-an) Stage of the Standard Global Chronostratigraphic Scale: Lower Cretaceous, (above Barremian, below Albian).

aptychus (ap'-tych-us) A pair of symmetrical calcareous or horny plates, joined along straight edges, originally lodged in the body chamber of some ammonoid conchs; it probably formed part of the lower jaw of the animal. It was formerly interpreted as an opercular structure. See also: *anaptychus.*

apuanite (ap'-uan-ite) A metallic black tetragonal mineral: $Fe^{2+}Fe^{3+}_4Sb^{3+}_4O_{12}S$.

aquafact (aq'-ua-fact) An isolated boulder or cobble, commonly on a sandy beach, that has been worn smooth on its seaward face by wave action, so that a sharp ridge parallel to the shore is developed along the exposed surface of the boulder or cobble; a *water-faceted stone* produced by *wet blasting* (Kuenen, 1947).

aquagene tuff (aq'-ua-gene) *hyaloclastite.*

Aqualfs Alfisols that are saturated with water for periods long enough to limit their use for most crops other than pasture or woodland unless they are artificially drained. Aqualfs have mottles, iron-manganese concretions or gray colors immediately below the A1 or Ap horizons and gray colors in the argillic horizon. A suborder in the U.S. system of soil taxonomy.

aquamarine (aq'-ua-ma-rine') (a) A transparent and pale-blue, greenish-blue, or bluish-green gem variety of beryl. (b) A pale-blue, greenish, light greenish-blue, or bluish-green color designation applied to mineral names; e.g. "aquamarine sapphire" (a pale-blue sapphire), "aquamarine topaz" (a greenish topaz), and "aquamarine tourmaline" (a pale-blue or pale greenish-blue tourmaline).

Aquands Andisols that are saturated with water for periods long enough to limit their use for most crops other than pasture unless they are artificially drained. Aquands have low chromas in redox depletions or on ped faces. A suborder in the U.S. system of soil taxonomy.

aquathermal pressure *thermal pressure.*

aquatic (a-quat'-ic) (a) Living entirely or primarily in or on water. (b) Growing in or on water. (c) Living near or frequenting water.

aquatolysis (aq-ua-tol'-y-sis) A term proposed by Müller (1967, p.130) for the chemical and physicochemical processes that occur in a freshwater environment during transportation, weathering, and preburial diagenesis of sediments. Rarely used. Cf: *halmyrolysis.*

Aquents Entisols that are saturated with water for periods long enough to limit their use for most crops other than pasture unless they are artificially drained. Aquents have low chromas or distinct mottles within 50 cm of the surface, or are saturated with water at all times. A suborder in the U.S. system of soil taxonomy.

aqueoglacial (aq'-ue-o-gla'-cial) *glacioaqueous.*

aqueo-residual sand (aq'-ue-o-re-sid'-u-al) A term used by Sherzer (1910, p.627) for a sand in which the particles, produced by various residual agents, were subsequently modified by the action of water. It includes "all water-transported sand, for residual agencies have been present to some extent in the derivation of all sand from the parent material" (Allen, 1936, p.12). Cf: *residuo-aqueous sand.*

aqueous (aq'-ue-ous) (a) Of, or pertaining to, water. (b) Made from, with, or by means of water; e.g. aqueous solutions. (c) Produced by the action of water; e.g. aqueous sediments.

aqueous fusion Melting in the presence of water, as a magma (Thrush, 1968, p.48).

aqueous ripple A ripple (3-D form) made by waves or currents of water, as opposed to one made by air currents (wind).

aqueous ripple mark *aqueous ripple.*

Aquepts Inceptisols that are saturated with water for periods long enough to limit their use for most crops other than pasture or woodland unless they are artificially drained. Aquepts have either a histic or umbric epipedon and gray colors within 50 cm of the surface, or an ochric epipedon underlain by a cambic horizon with gray colors, or have sodium saturation of 15% or more. A suborder in the U.S. system of soil taxonomy.

aquibed (aq'-ui-bed) The smallest formal hydrostratigraphic unit next in rank below aquimember. Designation should be limited to certain distinctive beds whose recognition is particularly useful and of more than local economic significance (Seaber, 1996). See also: *aquimember.*

aquiclude (aq'-ui-clude) A saturated geologic unit that is incapable of transmitting significant quantities of water under ordinary hydraulic gradients (Freeze and Cherry, 1979). *Confining bed* has replaced the term aquiclude. Syn: *aquitard.*

aquic moisture regime (aq'-uic) A reducing soil-moisture regime nearly free of dissolved oxygen owing to groundwater saturation, when soil temperature at a depth of 50 cm is more than 5°C (USDA, 1975).

aquifer (aq'-ui-fer) A body of rock that contains sufficient saturated permeable material to conduct groundwater and to yield significant quantities of water to wells and springs (Lohman, 1972, p.2). Syn: *water-bearing formation; groundwater reservoir.*

aquiferous system (aq-uif'-er-ous) The entire water-conducting system between the ostia and the oscula of a sponge, including the *inhalant system* and the *exhalant system.* Syn: *canal system.*

aquifer system A heterogeneous body of intercalated permeable and less permeable material that acts as a water-yielding hydraulic unit of regional extent.

aquifer test A test involving the withdrawal of measured quantities of water from, or addition of water to, a well and the measurement of resulting changes in *head [hydraul]* in the aquifer both during and after the period of discharge or addition. The results are used to estimate hydraulic properties of the aquifer. Cf: *pumping test; step drawdown test.*

aquiformation The fundamental formal unit of hydrostratigraphic classification. It may by an aquifer, aquitard, or aquifuge. It may be completely or partly divided into aquimembers if some useful purpose is served, or it may have no aquimembers (Seaber, 1996). See also: *hydrostratigraphic unit.*

aquifuge (aq'-ui-fuge) An obsolete term for a rock that contains no interconnected openings or interstices and therefore neither store nor transmits water (Fetter, 1994). See: *confining bed.*

aquigroup (aq'-ui-group) The formal hydrostratigraphic unit next in rank above aquiformation. It may consist of any combination of aquifers, aquitards, or aquifuges and need not be composed entirely of named aquiformations. It may be established to express the natural relationships of associated aquiformations if it is essential to elucidation of the hydrogeology of a large regional groundwater body (Seaber, 1996). See also: *aquiformation.*

aquimember (aq'-ui-mem-ber) The formal hydrostratigraphic unit next in rank below aquiformation. It may be an aquifer, aquitard, or aquifuge (Seaber, 1996). See also: *aquiformation.*

Aquitanian (Aq-ui-ta'-ni-an) A stage of the Standard Global Chronostratigraphic Scale: lowermost Miocene (above Chattian Stage of Oligocene, below Burdigalian) The basal GSSP is north of Genoa, Italy (Steininger et al., 1997).

aquitard (aq'-ui-tard) *aquiclude; confining bed.*

Aquods Spodosols that are saturated with water for periods long enough to limit their use for most crops other than pasture or woodland unless they are artificially drained. Aquods may have a histic epipedon, an albic horizon that is mottled or contains a duripan, or mottling or gray colors within or immediately below the spodic horizon. A suborder in the U.S. system of soil taxonomy.

Aquolls Mollisols that are saturated with water for periods long enough to limit their use for most crops other than pasture unless they are artificially drained. Aquolls may have a histic epipedon, a sodium saturation in the upper part of the mollic epipedon of >15% that decreases with depth or mottles or gray colors within or immediately below the mollic epipedon. A suborder in the U.S. system of soil taxonomy.

Aquox (Aq'-uox) Oxisols that have continuous plinthite near the surface, or that are saturated with water sometime during the year if not artificially drained. Aquox have either a histic epipedon, or mottles or colors indicative of poor drainage within the oxic horizon or both. A suborder in the U.S. system of soil taxonomy.

Aquults Ultisols that are saturated with water for periods long enough to limit their use for most crops other than pasture or woodland unless they are artificially drained. Aquults have mottles, iron-manganese concretions or gray colors immediately below the A1 or Ap horizons and gray colors in the argillic horizon. A suborder in the U.S. system of soil taxonomy.

arabesquitic (ar'-a-bes-quit'-ic) Said of the texture of certain porphyritic rocks in which an apparently homogeneous groundmass breaks up, under crossed nicols, into irregular patches, supposedly resembling arabesques (Johannsen, 1939, p.202). Obsolete.

arable land Land suited to the economical and practical production of cultivated crops.

arachnid (a-rach'-nid) Any terrestrial *chelicerate* belonging to the class Arachnida, characterized by the presence of one pair of preoral appendages with two to three joints. Cf: *merostome.*

arachnoid (a-rach'-noid) On Venus, a magmatic structure characterized by a combination of radial fractures and ridges extending outward for several radii and concentric fractures and ridges. In many cases, flows and small shield volcanoes occur in the central region (Crumbler and Aubele, 2000).

Aragonian (Ar-a-gon'-ian) European continental stage: middle Miocene (above Ramblian, below Vallesian). Obs. syn: *Vindobonian.*

aragonite (a-rag'-o-nite) (a) A white, yellowish, or gray orthorhombic mineral: $CaCO_3$. It is trimorphous with calcite and vaterite. Aragonite has a greater density and hardness, and a less distinct cleavage, than calcite, and is also less stable and less common. It occurs in fibrous aggregates in beds of gypsum and iron ore; as a deposit from hot springs; and as a major constituent of shallow marine muds and the upper parts of coral reefs. Aragonite is also an important constituent of the pearl, and of some shells. Syn: *Aragon spar.* (b) A group of orthorhombic carbonate minerals, including *aragonite, alstonite, witherite, strontianite,* and *cerussite.*

aragonite compensation depth A depth in the ocean below which the dissolution rate of *aragonite* particles equals or exceeds their delivery rate. It is significantly shallower than the *calcite compensation depth.* Abbrev: ACD.

Aragon spar (Ar'-a-gon) *aragonite.*

arakawaite (ar-a-ka'-wa-ite) *veszelyite.*

arakiite A resinous red-brown to orange-brown monoclinic mineral: $(Zn,Mn)(Mn^{2+},Mg)_{12}(Fe,Al)_2AsO_3(AsO_4)_2(OH)_{23}$.

aramayoite (ar-a-may'-o-ite) An iron-black triclinic mineral: $Ag_3Sb_2(Sb,Bi)S_6$.

araneid (ar-a-ne'-id) Any *arachnid* belonging to the order Araneida, characterized by the presence of maxillary lobes and glands and by the similarity of the first pair of legs to the other legs. Range, Carboniferous (possibly Devonian) to Holocene.

arapahite (a-rap'-a-hite) A dark-colored, porous, fine-grained basalt that is microscopically holocrystalline, poikilitic, and composed of magnetite (about 50 percent), bytownite, and augite. It was named by Washington and Larsen in 1913 for the Arapaho Indians, Colorado. Obsolete.

aravaipaite (a-ra-vai'-pa-ite) A colorless monoclinic mineral: $Pb_3AlF_9{\cdot}H_2O$.

A ray Anterior ray in echinoderms, located opposite the CD interray; equal to ambulacrum III of the *Lovenian system.*

arbitrary cutoff (ar'-bi-trar-y) An arbitrary boundary separating two laterally intergrading stratigraphic units that differ from each other in some arbitrarily defined way (Wheeler and Mallory, 1953, p.2412). See also: *cutoff [stratig].*

arbitrary line A vertical seismic section along a line across a 3-D data volume that is not necessarily straight or in in-line or cross-line directions. Also called: *random line; user line.*

arbitrary line [surv] *random line.*

arboreal pollen (ar-bo'-re-al) Pollen of trees. Abbrev: TP (sometimes AP). Syn: *arborescent pollen.*

arborescent (ar-bo-res'-cent) *dendritic.*

arborescent lava A form of pahoehoe lava characterized by a surface pattern reminiscent of tiny horizontal conifers or chicken feet. More common in lava tubes than on the surface (Stearns and Clark, 1930).

arborescent pollen *arboreal pollen.*

Arbuckle Orogeny (Ar'-buck-le) A name used by Van der Gracht and van Waterschoot (1931) for the last major deformation in the Wichita orogenic belt of southern Oklahoma (Wichita and Arbuckle mountains, and subsurface). It is placed in the Late Pennsylvanian (Virgil) by its relations to limiting fossiliferous strata. The nearby Ouachita Mountains were not supposed to have been materially affected by this orogeny, but to have been deformed later.

arc (a) A portion of the perimeter of a two-dimensional closed figure lying between two nodes at which two or more arcs intersect. An arc may represent a continuous boundary between two adjoining mapping units. (b) A line, representing a linear feature, that connects any two consecutive nodes. (c) An ordered set of points having a beginning and an end, that represents the location of linear features like faults or unconformities, or the boundary of an area (polygon) on a map. The start and end points of an arc are called nodes; while intermediate points are called vertices (Krzanowski et al., 1993). (d) A locus of points that forms a curve that is defined by mathematical expression (FGDC, 1998a, p.61). Cf: *chain [cart].*

arcanite (ar'-ca-nite) An orthorhombic colorless mineral: K_2SO_4.

arch [geomorph] *natural arch.*

arch [struc geol] A broad, open anticlinal fold on a regional scale; it is usually a basement doming, e.g. the Cincinnati Arch. Cf: *dome [struc geol].* Less-preferred syn: *swell [struc geol].*

Archaea (Ar-chae'-a) The members of the *domain* Archaea, a group of mostly anaerobic, prokaryotic microorganisms that generally but not always live in extreme environments, e.g., around hot springs and hydrothermal vents. The group includes thermophiles and methanogens. Sing: archaeum.

archaeoclimatology A term coined by R. A. Bryson in 2003 to describe a systematic estimation of past climates at specific locations and time intervals for archeological investigations, using an economical model predicated on orbital forcing, variations in atmospheric transparency, and synoptic climatology.

archaeocyathid (ar'-chae-o-cy-a'-thid) Any marine organism belonging to the phylum Archaeocyatha and characterized chiefly by a cone-, goblet-, or vase-shaped skeleton composed of calcium carbonate. The archaeocyathids have been variously classified as corals, sponges, protozoans, and calcareous algae. Range, Lower to Middle Cambrian; worldwide in distribution. Syn: *pleosponge; cyathosponge.*

archaeocyathid reef Early Middle Cambrian reefs in which archaeocyathids comprise the primary framework builders.

archaeocyte (arch'-ae-o-cyte') An *amoebocyte* of a sponge that has a large nucleus and cytoplasm rich in ribonucleic acid, is capable of ingesting particulate material and serving as the origin of any other type of cell, and is believed to be a persistent undifferentiated embryonic cell. Also spelled: *archeocyte.*

archaeology (ar-chae-ol'-o-gy) The study of human cultures through the recovery and analysis of their material relics. Also spelled: *archeology.* Cf: *geoarchaeology.*

archaeomagnetism (ar'-chae-o-mag'-net-ism) The study of natural remanent magnetism of baked clays, hearths, and other human-affected materials to determine intensity and direction of the Earth's magnetic field in the archaeologic past.

archaeometry (ar-chae-om'-e-try) The application of methods from the natural sciences to archaeological measurements; e.g., thermoluminescent dating of pottery sherds, or geophysical prospecting for archaeological sites. Also spelled: *archeometry.*

archaeomineralogy (ar-chae-o-min-er-al'-o-gy) A subdiscipline of *geoarchaeology;* the study of the rock and mineral makeup and the alteration of ancient artifacts, including stone tools, decorations, ceramics, and building materials. It can include the study of age, source, in-situ alteration, or general characterization of such artifacts (Rapp, 2000). Also spelled: archeomineralogy.

Archaeopteryx (Ar-chae-op'-ter-yx) A genus of birds characterized by teeth, elongate caudal skeleton, and very dinosaurlike details of trunk and limbs. Range, Upper Jurassic.

Archaeozoic (Ar'-chae-o-zo'-ic) *Archeozoic.*

Archaic (Ar-cha'-ic) In Old World archaeology, the latest *Neolithic* stage. In Egyptian archaeology, the very early period in the Egyptian chronology. In Mediterranean archaeology, the period between 700 and 480 B.C. and the corresponding style of art. In New World archaeology, a prehistoric cultural stage that follows the Paleo-Indian/Lithic and is characterized by a foraging pattern of existence and numerous types of stone implements. It is followed by the Formative and/or Woodland. Correlation of relative cultural levels with actual age (and, therefore, with the time-stratigraphic units of geology) varies from region to region.

arch dam A dam built in the form of a horizontal arch that abuts against the side walls of a gorge and that has its convex side upstream.

Archean (Ar-che'-an) The lowest eon of the Standard Global Geochronometric Scale, below the *Proterozoic* Eon. The lower boundary has not been defined; the upper boundary has been established geochronometrically at 2,500 Ma.

Archean lode gold (a) *low-sulfide gold-quartz vein;* (b) *arsenical sulfide-silicate iron formation.*

arched iceberg An iceberg eroded in such a manner that a large opening at the water line extends through the ice, forming an arch.

archegonium (ar-che-go'-ni-um) In plants, the multicellular female gametangium, commonly flask-shaped, produced by gametophytes of all higher cryptogams and gymnosperms; it contains the egg (Swartz, 1971, p.41).

archeocyte (ar'-che-o-cyte) *archaeocyte.*

archeology (ar-che-ol'-o-gy) *archaeology.*

archeometry (ar-che-om'-e-try) *archaeometry.*

Archeophytic (Ar'-che-o-phyt'-ic) A paleobotanic division of geologic time, signifying the time of initial plant evolution, specif. algae. Cf: *Aphytic; Cenophytic; Eophytic; Mesophytic; Paleophytic.*

archeopyle (arch'-e-o-pyle) An opening in the wall of a dinoflagellate cyst by means of which the cell contents can emerge from the cyst. It is usually more or less polygonal in shape, and operculate. See also: *apical archeopyle; cingular archeopyle; precingular archeopyle; combination archeopyle.*

archeopyle suture A line of dehiscence on the dinoflagellate cyst that more or less completely separates a part of the cyst wall to form an operculum that covers the archeopyle. See also: *accessory archeopyle suture.*

Archeozoic (Ar'-che-o-zo'-ic) The initial era of geologic history preceeding the *Paleozoic,* therefore a syn. of *Precambrian.* Also spelled: *Archaeozoic.*

archerite (arch'-er-ite) A colorless or buff tetragonal mineral occurring in cave deposits: $H_2(K,NH_4)PO_4$.

archetype (ar'-che-type) *prototype.*

archetypical mollusk (ar-che-typ'-i-cal) *hypothetical ancestral mollusk.*

arch-gravity dam A solid-masonry arch dam that has sufficient mass and breadth of base so that its own weight combines with the arch design to provide stability.

archibenthic (ar-chi-ben'-thic) Pertaining to the benthos of the continental slope. Cf: *bathybenthic.*

archibole (ar'-chi-bole) An obsolete syn. of *positive element.*

Archie equations (Ar'-chie) Three empirical relationships between *well log* resistivity measurements and the properties of porous, nonshaly sandstone, formulated by G. E. Archie (1942). Equation 1: $R_o = FR_w$, where R_o is the resistivity of fully water-saturated rock, F is the *formation factor,* and R_w is the resistivity of the interstitial water. Equation 2: $F = \phi^{-m}$, where ϕ is the porosity of Gulf Coast and other sandstone core samples and m is the "cementation component". From equation 1, equation 2 may be restated: $R_o = R_w\phi^{-m}$. Equation 3: $S_w = (R_o/R_t)^{1/n}$, where S_w is the percent formation-water saturation of a porous sandstone, R_t is the true resistivity of the sandstone underground, and n is the "saturation exponent". The Archie equations are the foundations of quantitative well-log analysis in hydrocarbon reservoirs. They have been elaborated and their constants refined for specific reservoirs, rock types, and subsurface conditions, e.g. by Winsauer et al. (1952).

Archimedes' principle (Ar-chi-me'-des) The statement in fluid mechanics that a fluid buoys up a completely immersed solid so that the apparent weight of the solid is reduced by an amount equal to the weight of the fluid that it displaces.

arching (arch'-ing) The transfer of stress from a yielding part of a soil or rock mass to adjoining less yielding or restrained parts of the mass (ASCE, 1958, term 22).

archipelagic apron (ar'-chi-pe-lag'-ic) A smooth, fanlike slope or broad cone surrounding a seamount or an island. It is comparable to a continental rise in its topography and sedimentary processes.

archipelago (ar-chi-pel'-a-go) A sea or area in a sea that contains numerous islands; also, the island group itself.

architype (ar'-chi-type) The type of a genus or species named in a publication prior to the time of the establishment of the current interpretation of types.

Archosauria (Ar-cho-saur'-i-a) A subclass of reptiles in which the skull is diapsid and the teeth are set in sockets. It includes the dinosaurs, crocodilians, pterosaurs, and a variety of Triassic forms including the ancestors of these groups.

arc measurement The measurement that follows a given meridian to determine the shape and size of the Earth along that line.

arcose (ar'-cose) *arkose.*

arc second A minute or 1/60 of arc, or 1/3600 of a degree (Maune, 2001, p.472).

arc shooting *fan shooting.*

arc spectrum The spectrum of light emitted by a substance at the temperature of an electric arc when the substance is placed in an arc or applied to one of the poles of the arc as a coating. The spectrum is representative of non-ionized atoms owing to the low potential difference of the arc. Cf: *spark spectrum.*

arctic (arc'-tic) n. The area within the Arctic Circle; the area of the North Pole. adj. (a) Pertaining to cold, frigid temperatures. (b) Pertaining to features, climate, vegetation, and animals characteristic of the arctic region. (c) Said of a climate in which the mean temperature of the coldest month is less than 0° C, and the mean temperature for the warmest month is below 10° C. See also: *polar climate.*

Arctic [paleoclim] The oldest subunit of the *Blytt-Sernander sequence* (Post, 1924), preceding the Preboreal, characterized primarily by tundra vegetation in Scandinavia, and recording the cold climate of full and late glacial time (prior to about 10,000 years B.P.). It is subdivided into Oldest Dryas, Bølling, Older Dryas, Allerød, and Younger Dryas. The term is rarely used.

Arctic Circle The parallel of latitude falling at approx. 66°32′ N; it marks the latitude where the Sun's rays are tangent to Earth's surface at the December solstice. Cf: *Antarctic Circle.*

arctic desert *polar desert.*

arctic karst Karst formed in polar regions.

arctic pack *polar ice.*

Arctic suite Obsolete name for a group of basaltic and associated igneous rocks intermediate in composition between rocks of the *Atlantic suite* and the *Pacific suite.*

arctite (arc'-tite) A colorless rhombohedral mineral: $Na_5Ca_7Ba(PO_4)_6F_3$.

arc-trench gap The area between a *volcanic arc* and an *oceanic trench* at a convergent plate boundary.

arc triangulation Triangulation designed to progress in a single, general direction and to produce a network of limited width but of considerable length. Arc triangulation is done to connect independent and widely separated survey networks, to coordinate and correlate local networks along the arc, to furnish data for the determination of a geodetic datum, and to provide a network of control points for a country-wide survey. Cf: *area triangulation.*

arcuate (ar'-cu-ate) Said of a fold whose axial trace is curved or bent.

arcuate delta A curved or bowed delta with its convex outer margin facing the sea or lake, such as that formed at the mouth of the Nile River. Cf: *cuspate delta; wave-dominated delta.* Syn: *lobate delta.*

arcuate fault A fault that has a curved trace on any given transecting surface. Cf: *peripheral fault.* See also: *listric fault.*

arcubisite (ar-cu'-bis-ite) A metallic gray mineral occurring in the cryolite deposit at Ivigtut, Greenland: Ag_6CuBiS_4.

arculite (ar'-cu-lite) A textural term for a bow-shaped aggregate of crystallites.

arcus (ar'-cus) A bandlike thickening in the exine of a pollen grain (as in *Alnus*), running from one pore apparatus to another.

ardaite (ar'-da-ite) A metallic monoclinic mineral: $Pb_{19}Sb_{13}S_{35}Cl_7$.

ardealite (ar'-de-al-ite') An earthy light yellow or brown monoclinic mineral: $Ca_2(HPO_4)(SO_4){\cdot}4H_2O$.

Ardennian Orogeny (Ar-den'-ni-an) One of the 30 or more short-lived orogenies during Phanerozoic time identified by Stille, in this case late in the Silurian, within the Ludlovian Stage.

ardennite (ar-den'-nite) A yellow to yellowish-brown orthorhombic mineral: $Mn^{2+}_4(Al,Mg)_6(SiO_4)_2(Si_3O_{10})[(As,V)O_4](OH)_6$.

arduinite (ar-du'-i-nite) *mordenite.*

are A metric unit of area equal to 100 square meters, 0.01 hectare, or 119.60 square yards. Abbrev: a.

area (a) A bounded, continuous two-dimensional object that may or may not include its boundary. (b) The two-dimensional extent of such an object expressed in units of length squared. (c) A term for *polygon* used by some geographic information systems.

area-altitude analysis *hypsometric analysis.*

area cladogram A *dendrogram* that illustrates an array of hypothesized relationships among organisms and is congruent with the geographic areas that the species occupied (Wiley, 1981, p.292). Cf: *vicariance biogeography.*

area contagionis *contact area.*

area curve In hydraulics, a curve that expresses the relation between area and some other variable, as between cross-sectional area of a stream and water-surface elevation, or between surface area of a reservoir and water-surface elevation (ASCE, 1962).

areal aperture An aperture in the face of the final chamber of a foraminiferal test.

areal bulla A blisterlike structure covering multiple *areal apertures* (e.g., *Globigerinatella*). Pl: bullae.

areal geology (ar'-e-al) The geology of an area, esp. the spatial distribution and position of stratigraphic units, structural features, and surface forms. Cf: *regional geology.*

areal map A geologic map showing the horizontal extent and distribution of rock units exposed at the surface.

area of influence The area surrounding a pumping or recharging well within which the potentiometric surface is affected by the withdrawal or injection of water at the well. Beyond it, no groundwater flows toward or from the well.

area slope The generalization of slope conditions within a given area.

area source The origin of air or water *nonpoint-source pollution* that is diffused, such as runoff from an urban area or automobile exhaust.

area triangulation Triangulation designed to extend in every direction from a control point and to cover the region surrounding it. Cf: *arc triangulation.*

areg Arabic plural of *erg.*

arena (a-re'-na) A term used in Uganda for a large, undulating, relatively low-lying central area more or less completely surrounded by a hilly rim of resistant rock, and representing a dome of softer rock that has been worn away (Wayland, 1920, p.36-37).

arenaceous (ar-e-na'-ceous) (a) Said of a sediment or sedimentary rock consisting wholly or in part of sand-size fragments, or having a sandy texture or the appearance of sand; pertaining to sand or arenite. Also said of the texture of such a sediment or rock. The term implies no special composition, and should not be used as a syn. of "siliceous". Syn: *psammitic; sandy; sabulous; arenarious.* (b) Said of organisms growing in sandy places.

arenarious (ar-e-nar'-i-ous) Composed of sand; *arenaceous.*

arenated (ar'-e-nat'-ed) Said of a substance that is mixed with sand or that has been reduced to sand.

arendalite (ar-en'-dal-ite) A dark-green variety of *epidote* from Arendal, southern Norway.

arenicolite (ar-e-nic'-o-lite) A sand-filled, U-shaped hole in a sedimentary rock (generally a sandstone), interpreted as a burrow of a marine worm and resembling the U-shaped burrow or trail of the modern worm *Arenicola.* It has also been regarded as the trail of a mollusk or crustacean.

Arenigian (Ar-e-nig'-i-an) A series or stage of the Standard Global Chronostratigraphic Scale: Lower Ordovician (above Tremadocian, below Llanvirnian).

arenilitic (a-re'-ni-lit'-ic) Pertaining to, having the quality of, or resembling sandstone.

arenite (ar'-e-nite) (a) A general name used for consolidated sedimentary rocks composed of sand-sized fragments (irrespective of composition) with a pure or nearly pure chemical cement and little or no interstitial matrix material; e.g., sandstone, graywacke, arkose, and calcarenite. The term is equivalent to the Greek-derived term *psammite* and was introduced as *arenyte* by Grabau (1904, p.242) who used it with appropriate prefixes in classifying medium-grained rocks (e.g., "autoarenyte," "autocalcarenyte," "hydrarenite," and "hydrosilicarenyte"). See also: *lutite; rudite.* (b) A "clean" sandstone that is well sorted, contains little or no matrix material, and has a relatively simple mineralogic composition; specif. a pure or nearly pure, chemically cemented sandstone containing <10% argillaceous matrix (Williams et al., 1954, p.290). The term is used for a major category of sandstone, as distinguished from *wacke.* Etymol: Latin "arena" = sand. Adj: arenitic.

arenose (ar'-e-nose) Full of grit or sand. Cf: *gritty.*

Arents Entisols that contain recognizable fragments of pedogenic horizons that have been mixed by mechanical disturbance. Arents are not saturated with water for periods long enough to limit their use for most crops. A suborder in the U.S. system of soil taxonomy.

arenyte (ar'-e-nyte) Var. of *arenite.*

areola (ar-e-o'-la) (a) One of the thinner chamberlike or boxlike structures arranged in characteristic pattern within the shell wall of diatoms, being larger and more complex than a *puncta,* lying perpendicular to the valve surface, and permitting the diffusion of gases and nutrients. An areola may be subcircular, elliptical, or hexagonal, and wholly or partly closed on either inner or outer surface. Syn: *areole.* (b) A marginal opening in the *frontal shield* in some ascophoran cheilostome bryozoans containing an intrazooidal *septula.* (c) A generally smooth, featureless area of a crinoid columnal *articulum,* situated between *lumen* and inner margin of *crenularium.* It may be granulose or marked by fine vermicular furrows and ridges. Pl: areolae.

areolate Divided into small spaces on a larger surface.

areole (ar'-e-ole) (a) A *scrobicule* or depression around a *boss* of an echinoid for the attachment of muscles controlling the movement of spines. (b) An *areola* of a diatom.

arête (a-rête') A narrow serrate mountain crest or rocky sharp-edged ridge or spur, commonly present above the snowline in rugged mountains (as in the Swiss Alps) sculptured by glaciers, and resulting from the continued backward growth of the walls of adjoining cirques. Etymol: French, "fish bone". See also: *horn [glac geol]; comb ridge; grat.* Syn: *crib [glac geol]; arris.*

arete and pinnacle karst *pinnacle karst.*

aretic (a-ret'-ic) *arheic.*

arfvedsonite (arf-ved'-son-ite) (a) A greenish black to black monoclinic mineral of the *amphibole* group, approximately: $Na_{2-3}(Fe,Mg,Al)_5Si_8O_{22}$. It may contain some calcium, and it occurs in strongly pleochroic prisms in certain sodium-rich igneous rocks. Syn: *soda hornblende.* (b) An end-member of the amphibole group: $Na_3(Fe_4^{2+}Fe^{3+})Si_8O_{22}(OH)_2$.

argental mercury (ar-gen'-tal) Naturally occurring *amalgam.*

argentian (ar-gen'-tian) *argentiferous.*

argentiferous (ar-gen-tif'-er-ous) Said of a substance that contains or yields silver, e.g. "argentiferous galena". Syn: *argentian.*

argentine (ar'-gen-tine) n. A pearly-white variety of calcite with undulating lamellae. adj. Pertaining to, containing, or resembling silver; silvery.

argentite (ar'-gen-tite) A dark lead-gray cubic dimorph of acanthite: Ag_2S. Cubic above 179°C, it inverts to acanthite below this temperature. Unstable at ambient temperatures. Syn: *silver glance; vitreous silver; argyrite.*

argentojarosite (ar'-gen-to-ja'-ro-site) A yellow or brownish rhombohedral mineral of the *alunite* group: $AgFe^{3+}_3(SO_4)_2(OH)_6$.

argentopentlandite (ar'-gen-to-pent'-land-ite) A metallic bronze-brown cubic mineral: $Ag(Fe,Ni)_8S_8$. A silver-bearing member of the *pentlandite* group.

argentopyrite (ar'-gen-to-py'-rite) A metallic grayish-white orthorhombic mineral: $AgFe_2S_3$.

argentotennantite (ar'-gen-to-ten'-nant-ite) A submetallic grayish black cubic mineral: $(Ag,Cu)_{10}(Zn,Fe)_2(As,Sb)_4S_{13}$. The silver-bearing analogue of *tennantite.*

argic water (ar'-gic) A syn. of *intermediate vadose water* proposed by Meinzer (1939) as one of the three classes of kremastic water.

Argids Aridisols that have an argillic or a natric horizon. A suborder in the U.S. system of soil taxonomy.

argillaceous (ar-gil-la'-ceous) (a) Pertaining to, largely composed of, or containing clay-size particles or clay minerals, such as an "argillaceous ore" in which the gangue is mainly clay; esp. said of a sediment (such as marl) or a sedimentary rock (such as shale) containing an appreciable amount of clay. Cf: *shaly; lutaceous; pelitic; argillic.* Syn: *clayey; pelolithic* (rarely used); argillous. (b) Said of the peculiar odor emitted by an argillaceous rock when breathed upon. (c) Pertaining to argillite.

argillaceous hematite A brown to deep-red variety of natural ferric oxide containing an appreciable portion of clay (or sand).

argillaceous limestone A limestone containing an appreciable amount (but less than 50%) of clay.

argillaceous sandstone (a) A term applied loosely to an impure sandstone containing an indefinite amount of fine silt and clay. (b) A relatively weak sandstone, not suitable for building purposes, containing a considerable amount of clay that serves as the cementing material. Cf: *clayey sandstone.*

argillan (ar'-gil-lan) A *cutan* composed dominantly of clay minerals (Brewer, 1964, p.212); e.g. a clay skin. Syn: argitan.

argillation (ar-gil-la'-tion) A term used by Keller (1958, p.233) for the development of kaolinite and other clay minerals by weathering of primary aluminum-silicate minerals.

argille scagliose (ar-gil'-le scag-lio'-se [ar-jee-leh skehl-yo'seh]) A thick sheet of chaotic, allochthonous material consisting of highly plastic, churned, and slickensided clays that have been displaced many kilometers by lateral or vertical stresses aided by gravity sliding or by diapiric movement; specif. a stratigraphic unit of Jurassic to Oligocene age exposed along parts of the Apennines, on which great slabs of rock have slid. Etymol: Italian, "scaly shale".

argillic (ar-gil'-lic) Pertaining to clay or clay minerals; e.g. "argillic alteration" in which certain minerals of a rock are converted to minerals of the clay group. Cf: *argillaceous.*

argillic alteration Alteration characterized by conversion of rock minerals to clay minerals.

argillic horizon A diagnostic subsurface soil horizon characterized by an accumulation of clay. Its thickness and clay content depend on the thickness and clay content of the overlying eluvial horizon. It usually has clay coatings on the surface of peds or pores, or as bridges between sand grains (USDA, 1975).

argillic zone The zone in porphyry ore deposits characterized by a quartz-kaolin-montmorillonite-chlorite-biotite alteration assemblage (Lowell and Guilbert, 1970). Cf: *phyllic zone; potassic zone; propylitic zone.*

argillite (ar'-gil-lite) (a) A compact rock, derived either from mudstone (claystone or siltstone) or shale, that has undergone a somewhat higher degree of induration than mudstone or shale but is less clearly laminated than shale and without its fissility, and that lacks the cleavage distinctive of slate. Flawn (1953, p.563-564) regards argillite as a weakly metamorphosed argillaceous rock, intermediate in character between a claystone and a *meta-argillite,* in which less than half of the constituent material (clay minerals and micaceous paste) has been reconstituted to combinations of sericite, chlorite, epidote, or green biotite, the particle size of the reconstituted material ranging from 0.01 to 0.05 mm. Cf: *clay slate.* (b) A term that has been applied to an argillaceous rock cemented by silica (Holmes, 1928, p.35) and to a claystone composed entirely of clay minerals. Also spelled: *argillyte.*

argillith (ar'-gil-lith) A term suggested by Grabau (1924, p.298) for claystone. Syn: *argillyte.*

argillization (ar'-gil-li-za'-tion) Replacement or alteration of feldspars to form clay minerals, esp. that occurring in wall rocks adjacent to mineral veins.

argillutite (ar-gil-lu'-tite) A pure lutite. Term introduced as "argillutyte" by Grabau (1904, p.243).

argillyte (ar'-gil-lyte) (a) Var. of *argillite.* (b) *argillith.* (c) Obsolete syn. of *orthoclase.*

argon-40/argon-39 age method (ar'-gon) A variation of the *potassium-argon age method* in which the sample to be dated is first irradiated with neutrons, converting some potassium-39 to argon-39. Argon is then extracted from the sample (either in one step or incrementally), and its isotopic composition analyzed. The amount of argon-39 is a measure of potassium content, and the ratio of radiogenic argon-40 to argon-39 is a function of age. It is sometimes possible to detect extraneous argon, and to determine whether or not the dated material has been disturbed by later thermal or chemical events (Miller, 1972).

Argovian (Ar-go'-vi-an) Substage in Great Britain: Upper Jurassic (lower Lusitanian; above Oxfordian Stage, below Rauracian Substage).

argutite (ar'-gu-tite) A colorless tetragonal mineral occurring as prismatic crystals in sphalerite: GeO_2. A member of the *rutile* group.

argyrite (ar'-gyr-ite) *argentite.*

argyrodite (ar-gyr'-o-dite) A steel-gray orthorhombic mineral: Ag_8GeS_6. It is isomorphous with canfieldite.

arhbarite (arh'-bar'-ite) A vitreous blue triclinic mineral: $Cu_2MgAsO_4(OH)_3$.

arheic (a-rhe'-ic) Said of a drainage basin or region characterized by arheism; without flow. Syn: *aretic.*

arheism (a'-rhe-ism) The condition of a region (such as a desert) in which runoff is nil or surface drainage is almost completely lacking, or where rainfall is so infrequent that the water sinks into the ground or evaporates.

arich (ar'-ich) A term used in Algeria for a rocky, sand-cloaked butte on which an isolated dune is based (Capot-Rey, 1945, p.399).

arid (ar'-id) Said of a climate characterized by dryness, variously defined as rainfall insufficient for plant life or for crops without irrigation; less than 25 cm of annual rainfall; or a higher evaporation rate than precipitation rate. Syn: *dry.*

arid cycle The cycle of erosion in an arid region. Cf: *normal cycle.*

aridic moisture regime (a-rid'-ic) A soil moisture regime characteristic of arid climates. There is little or no leaching, and soluble salts may accumulate. There is no plant-available water for more than half the time the soil is warm enough for plant growth (above 5°C), and the soil never has water continuously available for as long as 90 consecutive days when the soil temperature is above 8°C (USDA, 1975).

Aridisols Mineral soils that have an aridic moisture regime, an ochric epipedon, and other pedogenic horizons but no oxic horizon. An order in the U.S. system of soil taxonomy.

ariegite (ar-i-eg'-ite) A group of pyroxenites chiefly composed of clinopyroxene, orthopyroxene, and spinel, with pyrope and/or hornblende as possible accessories and lacking primary feldspar; a spinel *pyroxenite.* The term is more commonly used by European (esp. French) petrologists. The rock was named by Lacroix in 1901 for Ariège in the Pyrenees. Obsolete.

Arikareean (A-rik'-a-ree'-an) North American land mammal age. Late Oligocene and Early Miocene (after Whitneyan, before Hemingfordian).

aristarainite (a-ris'-ta-rain'-ite) A colorless monoclinic mineral: $Na_2MgB_{12}O_{16}(OH)_8{\cdot}4H_2O$.

Aristotle's lantern (Ar'-is-tot'-le's) A complex masticatory system of as many as forty calcareous skeletal elements that surround the mouth of an echinoid and function as jaws. Etymol: from a belief of Aristotle that the shape of a sea urchin resembles the frame of a lantern. Syn: *lantern.*

arithmetic mean (a-rith-met'-ic) In *geostatistics,* the sum of the values of n numbers divided by n. This statistic is usually referred to simply as the "mean" and denoted by a bar, as in x. Syn: *average.*

arithmetic mean diameter An expression of the average particle size of a sediment or rock, obtained by summing the products of the size-grade midpoints and the frequency of particles in each class, and dividing by the total frequency. Syn: *equivalent grade.*

Arizona ruby (Ar-i-zo'-na) A deep-red or ruby-colored pyrope garnet of igneous origin from the SW U.S.

arizonite [mineral] (ar-i-zon'-ite) (a) A doubtful mineral: $Fe_2Ti_3O_9$. It is found in irregular metallic steel-gray masses in pegmatite veins near Hackberry, Ariz. Cf: *kalkowskite; pseudorutile.* (b) A mixture of hematite, rutile, ilmenite, and anatase. (c) A type of ore, discovered in Yavapai County, Ariz., whose principal vein material consists of micaceous iron oxide, silver iodide, gold, iron sulfides, and antimony sulfide.

arizonite [rock] A light-colored hypabyssal rock composed chiefly of quartz (80 percent) and alkali feldspar (18 percent), with mica and apatite as possible accessories. It was named by Spurr and Washington in 1917 for Arizona, where the rock was first found. Obsolete.

Arkansas stone (Ar'-kan-sas) A common name for *novaculite* found in the Ouachita Mountains of western Arkansas. Also, a whetstone made of Arkansas stone.

arkansite (ar'-kan-site) A brillant iron-black variety of brookite from Magnet Cove, Arkansas.

arkelite Microscopic cubic crystals of ZrO_2 seen in some zircons. Inadequately characterized.

arkite (ar'-kite) A porphyritic plutonic foidite with the same general composition as *fergusite* but distinguished from it by the presence of andradite. Leucite usually forms the phenocrysts. It was named by Washington in 1901 from Magnet Cove, Arkansas.

arkose (ar'-kose) A feldspar-rich sandstone, commonly coarse-grained and pink or reddish, that is typically composed of angular to subangular grains that may be either poorly or moderately well sorted, is usually derived from the rapid disintegration of granite or granitic rocks, and often closely resembles granite; e.g. the Triassic arkoses of the eastern United States. Quartz is usually the dominant mineral, with feldspars constituting at least 25%. Cement (silica or calcite) is commonly rare, and matrix material (usually less than 15%) includes clay minerals (esp. kaolinite), mica, and iron oxide; fine-grained rock fragments are often present. Modern definitions of arkose include those by Krynine (1940); Folk (1954); Williams et al. (1954, p.294-295); Pettijohn (1957; 1975, p.214); McBride (1963, p.667); and Folk (1968, p.124). The term "arkose" was introduced by Brongniart (1823, p.497-498) in an attempt to limit use of "grés" (sandstone) and was defined by him as a rock of granular texture formed principally by mechanical aggregation and composed essentially of large grains of feldspar and glassy quartz mixed together unequally, with mica and clay as fortuitous constituents (see Oriel, 1949, p.825). Roberts (1839, p.11) attributes the term to Bonnard. Etymol: French, probably from Greek "archaios", "ancient, primitive" (Oriel, 1949, p.826). Adj: arkosic. Cf: *graywacke; feldspathic sandstone; subarkose.* Also spelled: *arcose.*

arkose-quartzite *arkosite.*

arkosic arenite (ar-ko'-sic) A sandstone containing abundant quartz, chert, or quartzite, less than 10% argillaceous matrix, and more than 25% feldspar (chiefly unaltered sodic and potassic varieties), and characterized by an abundance of unstable materials in which the feldspar grains exceed the rock fragments (Williams et al., 1954, p.294). It is more feldspathic and less mature than *feldspathic arenite.* See also: *arkosic sandstone.*

arkosic conglomerate A poorly sorted, lithologically homogeneous *orthoconglomerate* consisting of immature gravels derived from granites; an arkose with scattered pebbles of granite or with lenses of granitic gravel. The sand and silt matrix has the composition of arkose, consisting of quartz and feldspar particles with some finer kaolinitic material. Syn: *granite-pebble conglomerate.*

arkosic graywacke A graywacke characterized by abundant unstable materials; specif. a sandstone containing more feldspar grains than fine-grained rock fragments, the feldspar content exceeding 25% (Williams et al., 1954, p.294). It is more feldspathic than *feldspathic graywacke.*

arkosic sandstone A sandstone with considerable feldspar, such as one containing minerals derived from coarse-grained quartzo-feldspathic rocks (granite, granodiorite, gneiss) or from highly feldspathic sedimentary rocks; specif. a sandstone containing more than 25% feldspar and less than 20% matrix material of clay, sericite, and chlorite (Pettijohn, 1949, p.227). It is more feldspathic than *feldspathic sandstone.* The term is used also as a general term to include *arkosic arenite, arkosic wacke,* and *arkose* (Williams et al., 1954, p.310), or arkose and *subarkose* (Pettijohn, 1954, p.364). See also: *arkosite.*

arkosic wacke A sandstone containing abundant quartz, chert, or quartzite, more than 10% argillaceous matrix, and more than 25% feldspar (chiefly sodic and potassic varieties), and characterized by an abundance of unstable materials in which the feldspar grains exceed the fine-grained rock fragments (Williams et al., 1954, p.291-292). It is more feldspathic and less mature than *feldspathic wacke.*

arkosite (ark'-o-site) A quartzite with a notable amount of feldspar; e.g. a well-indurated *arkosic sandstone* (Pettijohn, 1949, p.227) or a well-cemented arkose (Tieje, 1921, p.655). Syn: *arkose-quartzite; quartzitic arkose.*

arkositite (ar-ko'-si-tite) A term used by Tieje (1921, p.655) for an arkose so well-cemented that the particles are interlocking.

arm [coast] A long, narrow inlet of water extending inland from another body of water, such as an "arm of the sea". It is usually longer and narrower than a bay, often a fjord.

arm [geomorph] (a) A ridge or spur extending from a mountain. (b) The trailing outer extension of a parabolic dune; the forward-projecting extension of a *barchan.*

arm [paleont] (a) One of several radially disposed appendages bearing an extension of an *ambulacrum* and mounted on the oral surface of an echinoderm, such as one of the five radial extensions of the body of an asteroid; also, a free, usually pinnule-bearing extension of a crinoid. It is a major element in the food-gathering structure of many echinoderms, and may or may not be distinct from the disc. Cf: *brachiole.* (b) A raylike structure of a sponge *spicule,* whether or not it is a true ray or a pseudoactin. (c) A *brachium* of a brachiopod. (d) A flat extension from the central region of a radiolarian shell.

arm [streams] A tributary or branch of a stream.

armalcolite (ar-mal'-co-lite) An opaque orthorhombic mineral of the pseudobrookite group found in Apollo 11 lunar samples: $(Mg,Fe^{2+})Ti_2O_5$. Originally called ferropseudobrookite. Syn: *karooite.*

armangite (ar'-man-gite) A brown to black trigonal mineral: $Mn^{2+}{}_{26}As^{3+}{}_{18}O_{50}(CO_3)(OH)_4$.

armenite (ar'-me-nite) A colorless hexagonal mineral: $BaCa_2Al_6Si_9O_{30}{\cdot}2H_2O$.

armor In streams, (a) a coarse surface layer on the bed of a stream that is mobilized only during high discharges; (b) a coarse *lag* layer developed at waning flows that is regularly disrupted. Cf: *censored layer, pavement [streams].*

armored (ar'-mored) Said of dinoflagellates possessing a rigid, tabulate layer below the outer membrane of the theca (Edwards, 1993). Ant: *unarmored.*

armored clay ball *armored mud ball.*

armored mud ball A large subspherical mass of silt or clay, which becomes coated or studded with coarse sand and fine gravel as it rolls along downstream; it is generally 5-10 cm in diameter, although the size ranges between 1 cm and 50 cm. See also: *clay ball; till ball.* Syn: *armored clay ball; mud ball.*

armored relict An *unstable relict* that is prevented from further reaction by a rim of reaction products.

armored snowball A fragment of rock or ice falling from the cliffs above Greely Fiord, Ellesmere Island, "gathering snow by accretion during their descent of the scree and acquiring a final coating or armor of shingle when crossing the beach before rolling out onto the ice of the fiord" (Ferguson, 1970). Armored snowballs are believed to be the precursors of moated ice mounds.

Armorican Orogeny (Ar-mor'-i-can) A name used by Suess for late Paleozoic deformation in western Europe; it is based on relations in Brittany (Armorica). The term is now little used; modern geologists prefer the names *Hercynian Orogeny* or *Variscan Orogeny.*

armoring (ar'-mor-ing) Formation of a reaction rim resulting from, e.g., a loss of equilibrium in a discontinuous reaction series. Cf: *zoning [cryst].*

armstrongite (arm'-strong-ite) A brown monoclinic mineral: $CaZrSi_6O_{15}{\cdot}2.5H_2O$.

arnhemite A pearly white hexagonal mineral found in a cave: $(K,Na)_4Mg_2(P_2O_7)_2{\cdot}5H_2O$. Formed by the combustion of guano. Not approved by the International Mineralogical Association.

arnimite (ar'-nim-ite) An ill-defined hydrous copper-sulfate mineral, perhaps identical with antlerite.

aromatic hydrocarbon (ar-o-mat'-ic) A monocyclic or polycyclic relatively stable hydrocarbon having the empirical formula C_nH_{2n-6}, of which the simplest example is benzene. The compounds of higher molecular weight in this series are solid and may fluoresce or be slightly colored. See also: *benzene series.*

aromatite (a-ro′-ma-tite) A bituminous stone resembling a fragrant gum resin in color and odor. It was a precious stone in ancient Arabia and Egypt.

aromorphosis (a-ro-mor′-pho-sis) Biologic evolution characterized by an increase in the degree of organization without marked specialization. Cf: *allomorphosis.*

arpent (ar′-pent) (a) An old French unit of land area, still used in certain French sections of Canada and U.S., equal to about 0.85 acre depending on local custom and usage; e.g. a unit of area in Missouri and Arkansas equal to 0.8507 acre, and in Louisiana, Mississippi, Alabama, and NW Florida equal to 0.84725 acre. The arpent was used, in the U.S., in surveys of land granted by the French crown. (b) An old French unit of distance, taken as the length of a side of a square region one arpent in area, 192.50 ft (58.67 m) in Missouri and Arkansas, and 191.99 ft (58.52 m) in Louisiana, Mississippi, Alabama, and NW Florida. In Canada, the arpent is exactly 180 French ft, which is about 191.85 English ft.

arquerite (ar′-que-rite) A soft, malleable, silver-rich variety of native amalgam, containing about 87% silver and 13% mercury.

arrastre A primitive form of grinding mill, consisting of a circular rock-lined pit in which broken ore is pulverized by stones hung from horizontal poles attached to a central post and dragged around the pit.

array [geophys] (ar-ray′) An ordered arrangement of geophysical instruments such as electrodes, magnetometers, seismic sources, or geophones, the data from which feed into a central point or receiver.

array [stat] An orderly arrangement of numbers, characters, symbols, etc., such that the relative position of an element of the array has some bearing on the operation that will be performed on the element. A *matrix* is an array.

arris (ar′-ris) A term used in the English Lake District for *arête.*

arrival (ar-ri′-val) The appearance of a seismic event on a seismic record, the buildup of amplitude and the coherent lineup of energy signifying the passage of a wavefront. See also: *first arrival.* Syn: *break [seis]; kick [seis].*

arrival time In seismology, the time at which a particular wave phase arrives at a detector, usually measured from the time of generation of the wave.

arrojadite (ar-ro-ja′-dite) (a) A dark-green monoclinic mineral: $KNa_4Ca(Fe^{2+},Mn^{2+})_{14}Al(PO_4)_{12}(OH)_2$. It is isostructural with dickinsonite and sigismundite, and occasionally occurs as enormous crystalline masses in pegmatites. (b) A group of monoclinic phosphate minerals of the general formula $ANa_{3-4}\,CaB_{14}Al(PO_4)_{12}(OH)_2$, where A = K, Ba or Pb, and B = Fe, Mn or Mg.

arrow (ar′-row) *pin [surv].*

arroyo (ar-roy′-o) A term applied in the arid and semiarid regions of the SW U.S. to the small, deep, flat-floored channel or gully of an ephemeral stream or of an intermittent stream, usually with vertical or steeply cut banks of unconsolidated material at least 60 cm high; it is usually dry, but may be transformed into a temporary watercourse or short-lived torrent after heavy rainfall. Cf: *dry wash.* (b) The small intermittent stream or rivulet that occupies such a channel. Etymol: Spanish, "stream, brook; gutter, watercourse of a street". Cf: *wadi; nullah.*

arroyo-running A term applied in the SW U.S. to the phase of local flooding characterized by a temporary mountain torrent debouching from a canyon and spreading out over a great fan (Keyes, 1910, p.572). Obsolete.

arsenate (ar′-se-nate) A mineral compound characterized by pentavalent arsenic and oxygen in the anion. An example is mimetite, $Pb_5(AsO_4)_3Cl$. Cf: *phosphate; vanadate.*

arsenate-belovite *talmessite.*

arsenbrackebuschite (ar-sen-brack′-e-busch-ite′) An adamantine yellow monoclinic mineral: $Pb_2(Fe^{3+},Zn)(AsO_4)_2(OH,H_2O)$.

arsendescloizite (ar′-sen-des-cloi′-zite) An adamantine pale yellow orthorhombic mineral: $PbZn(AsO_4)(OH)$. It is the arsenate analogue of descloizite and a member of the *adelite* group.

arsenic (ar′-se-nic) A rhombohedral mineral, the native metallic element As. It is brittle and commonly occurs in steel-gray and granular or kidney-shaped masses.

arsenical antimony (ar-sen′-i-cal) *allemontite.*

arsenical nickel *nickeline.*

arsenical pyrites *arsenopyrite.*

arsenical sulfide-silicate iron formation A *mineral-deposit type* characterized by *stratabound* to *stratiform* gold-bearing arsenical sulfide mineralization in silicate iron formation. Syn: *Homestake-type deposit; iron-formation-hosted gold.*

arsenic bloom (a) *arsenolite.* (b) *pharmacolite.*

arseniopleite (ar-se′-ni-o-ple′-ite) A brownish-red monoclinic mineral: $NaCaMn^{2+}(Mn^{2+},Mg)_2(AsO_4)_3$. A member of the *alluaudite* group.

arseniosiderite (ar-se′-ni-o-sid′-er-ite) A yellowish-brown to black monoclinic mineral: $Ca_2Fe^{3+}{}_3O_2(AsO_4)_3{\cdot}3H_2O$. See also: *mitridatite; robertsite.*

arsenite (ar′-se-nite) A mineral compound characterized by trivalent arsenic and oxygen in the anion. An example is trigonite, $Pb_3MnH(AsO_3)_3$.

arsenobismite (ar′-se-no-bis′-mite) A mixture of preisingerite and other minerals.

arsenoclasite A red orthorhombic mineral: $Mn^{2+}{}_5(AsO_4)_2(OH)_4$. Isomorphous with *gatehouseite.*

arsenocrandallite (ar′-se-no-cran′-dal-lite) A bluish-green rhombohedral mineral: $(Ca,Sr)Al_3[(As,P)O_4]_2(OH)_5{\cdot}H_2O$. A member of the *crandallite* group.

arsenoflorencite-(Ce) (ar′-se-no-flor′-enc-ite) A colorless to light brown rhombohedral mineral: $(Ce,La)Al_3(AsO_4,PO_4)_2(OH)_6$. The arsenic analogue of florencite-(Ce), and a member of the *crandallite* group.

arsenoflorencite-(La) The lanthanum and arsenic analogue of florencite-(Ce), and a member of the *crandallite* group. Inadequately characterized, and not approved by the International Mineralogical Association.

arsenoflorencite-(Nd) The neodymium and arsenic analogue of florencite-(Ce), and a member of the *crandallite* group. Inadequately characterized, and not approved by the International Mineralogical Association.

arsenogorceixite A vitreous white rhombohedral mineral: $BaAl_3(AsO_3OH)(AsO_4,PO_4)(OH)_6$. The arsenic analogue of gorceixite, and a member of the *crandallite* group.

arsenogoyazite (ar′-se-no-goy′-az-ite) A vitreous white rhombohedral mineral: $(Sr,Ca,Ba)Al_3(AsO_4,PO_4)_2(OH,F)_5{\cdot}H_2O$. The arsenic analogue of goyazite, and a member of the *crandallite* group.

arsenohauchecornite (ar′-se-no-hau′-che-cor′-nite) A metallic bronze tetragonal mineral, the arsenic analogue of hauchecornite: $Ni_{18}Bi_3AsS_{16}$, and a member of the *hauchecornite* group.

arsenoklasite (ar′-se-no-kla′-site) *arsenoclasite.*

arsenolamprite (ar′-se-no-lam′-prite) A lead-gray orthorhombic polymorph of native arsenic: As. It was formerly regarded as a mixture of arsenic and arsenolite.

arsenolite (ar-sen′-o-lite) A white to colored cubic mineral: As_2O_3. It usually occurs as a white bloom or crust, and is dimorphous with claudetite. Syn: *arsenic bloom.*

arsenopalladinite (ar-se′-no-pal-la′-di-nite) A metallic white triclinic mineral: $Pd_8(As,Sb)_3$.

arsenopyrite (ar′-se-no-py′-rite) (a) A tin-white or silver-white to steel-gray orthorhombic mineral: FeAsS. It occurs chiefly in crystalline rocks and esp. in lead and silver veins, and it constitutes the principal ore of arsenic. Syn: *arsenical pyrites; mispickel; white pyrites; white mundic.* (b) a group of minerals with the general formula $RX(As,Sb)S$, where R = Fe, Co or platinum-broup metals.

arsenosulvanite (ar′-se-no-sul′-van-ite) A metallic bronze cubic mineral: $Cu_{12}VAs_3S_{16}$. It is the arsenic analogue of *sulvanite.* Syn: *lazarevicite.*

arsenowaylandite An ill-defined arsenic analogue of *waylandite:* $(Bi^{3+},Sr)(Al,Fe)_3(AsO_4,PO_4)_2(OH)_6$. A member of the *crandallite* group. Not approved by the International Mineralogical Association.

arsenpolybasite (ar′-sen-po′-ly-ba′-site) A submetallic black monoclinic mineral: $(Ag,Cu)_{16}(As,Sb)_2S_{11}$. The arsenic analogue of polybasite. Cf: *antimonpearceite.* Syn: arsenopolybasite.

arsentsumebite (ar′-sen-tsum′-eb-ite) A green monoclinic mineral: $Pb_2Cu(AsO_4SO_4)(OH)$. The arsenic analogue of tsumebite.

arsenuranospathite (ar′-sen-u-ran′-o-spath-ite) A yellow pseudo-tetragonal mineral: $HAl(UO_2)_4(AsO_4)_4{\cdot}[40]H_2O$. The arsenic analogue of phosphuranylite.

arsenuranylite (ar′-sen-u-ran′-y-lite) An orange-yellow orthorhombic mineral: $Ca(UO_2)_4(AsO_4)_2(OH)_4{\cdot}6H_2O$(?). The arsenic analogue of phosphuranylite.

arsoite (ar′-so-ite) A *trachyte* containing phenocrysts of alkali feldspar, andesine, clinopyroxene, and olivine in a groundmass of alkali feldspar, oligoclase, clinopyroxene, magnetite, and sodalite. Its name is derived from the Arso lava flow of 1302 on Ischia, Italy. Obsolete.

ART Abbrev. for *algebraic reconstruction technique.* See also: *tomography.*

artefact A variant spelling of *artifact.*

arterite (ar′-ter-ite) A *migmatite,* the more mobile portion of which was injected magma (Dietrich and Mehnert, 1961). Originally proposed by Sederholm, along with the term *venite* (Holmquist), to replace *veined gneiss* with terms of genetic connotation (Mehnert, 1968, p.17). Rarely used. Cf: *anatexite; composite gneiss; diadysite; phlebite; venite.*

artesian (ar-te′-sian) An adjective referring to groundwater confined under hydrostatic pressure. Etymol: French "artésien", "of Artois", a region in northern France.

artesian aquifer *confined aquifer.*

artesian basin A terrane, often but not necessarily basin-shaped, including an artesian aquifer whose potentiometric surface typically is above the land surface in the topographically lower portion of the terrane. Examples range in size from areas a few hundred feet across to several hundred miles across. Cf: *groundwater basin.*

artesian discharge Discharge of water from a well, spring, or aquifer under artesian pressure.

artesian head The *hydrostatic head* of an artesian aquifer or of the water in the aquifer.

artesian leakage Slow *percolation* from a *confined aquifer* into confining beds.

artesian pressure Hydrostatic pressure of artesian water, often expressed in terms of pounds per square inch at the land surface; or height, in feet above the land surface, of a column of water that would be supported by the pressure.

artesian-pressure surface A potentiometric surface that is above the zone of saturation. Cf: *normal-pressure surface; subnormal-pressure surface.*

artesian province A region within which structure, stratigraphy, topography, and climate combine to produce conditions favorable to the existence of one or more artesian aquifers; e.g. the Atlantic and Gulf Coastal Plain.

artesian spring A spring from which the water flows under artesian pressure, usually through a fissure or other opening in the confining bed above the aquifer.

artesian system (a) A structure permitting water confined in a body of rock to rise within a well or along a fissure. (b) Any system incorporating a water source, a body of permeable rock bounded by bodies of distinctly less permeable rock, and a structure enabling water to percolate into and become confined in the permeable rock under pressure distinctly greater than atmospheric.

artesian water *confined groundwater.*

artesian weathering Chemical weathering produced by artesian water moving down an aquifer, as in a sandstone overlying an unconformity or in the permeable zone along the unconformable surface.

artesian well A well tapping *confined groundwater.* Water in the well rises above the level of the top of the aquifer under artesian pressure, but does not necessarily reach the land surface. Sometimes restricted to mean only a *flowing artesian well.* Cf: *water-table well; nonflowing artesian well.*

arthrodire (ar′-thro-dire) One of a group of extinct Paleozoic fishes abundant in the Devonian. They had heavily armored heads movably jointed to similar armor covering the anterior parts of the body. Arthrodires grew to lengths of as much as 30 feet.

arthrolite (ar′-thro-lite) A cylindrical concretion with transverse joints, sometimes found in clays or shales.

arthrophycus (ar-thro-phy′-cus) A sand-filled rounded furrow, curving and branching, with faint but regularly spaced transverse ridges commonly bearing a median depression, probably representing a feeding burrow but also variously regarded as an inorganic structure or a trail produced by a worm, mollusk, or arthropod crawling over a soft-mud surface. The "branches" of the trace fossil may reach 60 cm in length. It was originally described as a plant fossil (seaweed) and assigned to the genus *Arthrophycus.*

arthropod (ar′-thro-pod) Any one of a group of solitary marine, freshwater, and aerial invertebrates belonging to the phylum Arthropoda, characterized chiefly by jointed appendages and segmented bodies. Among the typical arthropods are trilobites, crustaceans, chelicerates, and myriapods. Range, Lower Cambrian to Holocene.

arthurite (ar′-thu-rite) An apple-green to yellowish gray monoclinic mineral: $CuFe^{3+}_2(AsO_4)_2(OH)_2{\cdot}4H_2O$. A member of the whitmoreite group.

article (ar′-ti-cle) An articulated *segment* of an appendage in an arthropod.

articular (ar-tic′-u-lar) adj. Pertaining to a joint or joints, e.g. "articular furrow" and "articular groove" occurring on the margins of the *scutum* and *tergum* of a cirripede crustacean and together forming the articulation between these valves. n. In some vertebrates, e.g. amphibians and reptiles, a cartilage-replacement (from the first visceral arch) bone of the lower jaw that provides the joint for articulation between it and the skull (via the quadrate bone); in mammals, the ancestral articular bone of the lower jaw is transformed into the malleus of the middle ear.

articular facet *articulum.*

articulate [bot] (ar-tic′-u-late) Said of a plant that has nodes or joints, or places where separation may occur.

articulate [paleont] (a) n. Any brachiopod belonging to the formerly recognized class Articulata, characterized by calcareous valves united by hinge teeth and dental sockets. adj. Said of a brachiopod possessing such valves, or of the valves themselves. Cf: *inarticulate.* (b) n. Any crinoid belonging to the subclass Articulata, characterized by highly differentiated brachial articulations. adj. Said of a crinoid having highly differentiated brachial articulation.

articulated brachiopod Any brachiopod with a distinct hinge and usually with a shell of low-magnesium calcite. Articulated brachiopods were formerly regarded as forming a distinct class Articulata. Cf: *inarticulated brachiopod.*

articulation (ar-tic′-u-la′-tion) (a) The action or manner of jointing, or the state of being jointed; e.g. the interlocking of two brachiopod valves by two ventral teeth fitting into sockets of the brachial valve, or the union of adjoined cirrals and columnals of a crinoid stem effected by ligaments attached to articula. (b) Any movable joint between the rigid parts of an invertebrate, as between the segments of an insect appendage, or of a vertebrate, as between the bones of limbs or jaws.

articulite (ar-tic′-u-lite) *itacolumite.*

articulum (ar-tic′-u-lum) A smooth or sculptured surface of a crinoid *columnal* or *cirral* serving for articulation with the contiguous stem. Pl: articula. Syn: *articular facet.*

articulus (ar-tic′-u-lus) The hinge in bivalve mollusks, including the hinge plate, hinge teeth, and ligament. Pl: articuli.

artifact (ar′-ti-fact) (a) An object made or used by humans. Var. spelling: *artefact.* (b) An incidental, unintended effect produced in data acquisition or processing. Sometimes called "footprint."

artificial avalanche (ar-ti-fi′-cial) An avalanche triggered by active control measures; field or laboratory simulation of an avalanche.

artificial brine (ar-ti-fi′-cial) Brine produced from an underground deposit of salt or other soluble rock material in the process of *solution mining.* Cf: *brine [geol].*

artificial classification A biological classification system containing one or more artificial groups; i.e., paraphyletic or polyphyletic groups (Wiley, 1981, p.8). Ant: *natural classification.*

artificial horizon A device for indicating the horizontal, as a bubble, gyroscope, pendulum, or the flat surface of a liquid. It is sometimes simply called a horizon. Syn: *false horizon.*

artificial illumination A surface attribute display of a relief map showing shadows as if it was illuminated from a particular direction at a low sun angle. The shadows emphasize subtle relief such as involved in faulting.

artificial intelligence The concept that computers can be programmed to assume capabilities normally thought to be like human intelligence, such as learning, adaptation, and self-correction. Includes *expert systems.* Abbrev: *AI.*

artificial levee Dike along the side of a river channel erected to prevent overflow during floods, ordinarily running more or less along the channel direction and near the natural levee crests of alluviating streams. Term is also used for structures designed to prevent flooding by seas, lakes, etc.

artificial magnetic anomaly Non-geologic magnetic anomaly caused by a man-made object such as a pipleine, railway, well casing, power transmission line, tank, etc. Syn: cultural magnetic anomaly.

artificial neural network A hardware or software system that simulates certain functions and topology of a central nervous system. Artificial neural networks, because of their capability to generalize and make abstractions, are often used as pattern-recognition tools in remote sensing and spatial-information systems (Krzanowski et al., 1993).

artificial radioactivity The radioactivity of radionuclides formed by nuclear reactions under human control such as in a nuclear reactor. Cf: *induced radioactivity.*

artificial recharge Artificial replenishment of an aquifer; recharge at a rate greater than natural, resulting from human activities, e.g., spreading water, recharge wells, or ditches.

artificial taxon A taxon of any rank that does not correspond to a unit involved in the evolutionary process or to a unit of evolutionary history (Wiley et al., 1991, p.3). Ant: *natural taxon.*

artinite (ar′-ti-nite) A snow-white monoclinic mineral: $Mg_2(CO_3)(OH)_2{\cdot}3H_2O$.

Artinskian (Ar-tinsk′-i-an) A stage of the Standard Global Chronostratigraphic Scale: middle Cisuralian Series (Lower Permian; above Sakmarian, below Kungurian).

artroeite A vitreous colorless triclinic mineral: $PbAlF_3(OH)_2$.

artsmithite A vitreous colorless monoclinic mineral: $Hg^{1+}_4Al(PO_4)_{1.8}(OH)_{1.7}$.

arupite An earthy blue monoclinic meteorite mineral: $Ni_3(PO_4)_2{\cdot}8H_2O$. A member of the *vivianite* group.

arzakite (ar′-zak-ite) An adamantine brown monoclinic mineral: $Hg_3S_2(Br,Cl)_2$. The bromine-dominant analogue of lavrentievite. Not approved by the International Mineralogical Association.

arzrunite (arz-ru′-nite) A bluish-green mineral: $Pb_2Cu_4(OH)_4(SO_4)Cl_6{\cdot}2H_2O$. Inadequately characterized.

ås (Pron: auss) A Swedish term for *esker.* Pl: *åsar.* Cf: *os [glac geol].*

åsar Plural of *ås.*

asbecasite (as-be′-ca-site) A yellow trigonal mineral: $Ca_3(Ti,Sn^{4+})As^{3+}_6Si_2Be_2O_{20}$.

asbestiform (as-bes′-ti-form) Said of a mineral that is composed of separable fibers, e.g. *chrysotile.*

asbestine (as-bes′-tine) adj. Pertaining to or having the characteristics of asbestos. n. A variety of talc; specif. *agalite.*

asbestos (as-bes′-tos) (a) A commercial term applied to a group of silicate minerals that readily separate into thin, strong fibers that are flexible, heat resistant, and chemically inert, and therefore are suitable for uses (as in yarn, cloth, paper, paint, brake linings, tiles, insulation, cement, fillers, and filters) where incombustible, nonconducting, or chemically resistant material is required. (b) A mineral of the asbestos group, principally chrysotile (best adapted for spinning) and certain fibrous varieties of amphibole (esp. amosite, anthophyllite, and crocidolite). (c) A term strictly applied to the fibrous variety of actinolite. Certain varieties are deleterious to health. Syn: asbestus; *amianthus; earth flax; mountain flax.*

asbolane (as′-bo-lane) A soft, black, earthy mineral aggregate, often classed as a variety of wad, containing hydrated oxides of manganese and cobalt, the content of cobalt sometimes reaching as much as 32% (or 40% cobalt oxide). Syn: *asbolane; earthy cobalt; black cobalt; cobalt ocher.*

asbolite (as′-bo-lite) *asbolane.*

ascending branch (as-cend′-ing) Either of two ventral elements of a brachiopod *loop,* continuous anteriorly with ventrally recurved descending branches and joined posteriorly by a transverse band.

ascending development *accelerated development.*

ascension theory (as-cen′-sion) A theory of hypogene mineral-deposit formation in which mineralizing solutions rise through fissures from magmatic sources in the Earth's interior. Cf: *descension theory.*

aschamalmite (asch′-am-alm′-ite) A metallic gray monoclinic mineral: $Pb_6Bi_2S_9$. It is closely related to heyrovskyite and other phases in the system Bi_2S_3-PbS. It occurs with exsolution lamellae of galena and cosalite.
ascharite (a-schar′-ite) *szaibelyite.*
aschistic (a-schist′-ic) Said of the rock of a minor intrusion which has a composition equivalent to that of the parent magma, i.e. in which there has been no significant differentiation. Cf: *diaschistic.* Obsolete.
asci Pl. of *ascus.*
ASCII American Standard Code for Information Interchange.
ascocarp (as′-co-carp) The *ascospore*-bearing structure of *Ascomycetes.*
ascoceroid conch (as-co′-cer-oid) The mature *conch* of the nautiloid group Ascocerida, typically consisting of an expanded exogastric *brevicone* having an inflated posterior part with dorsal *phragmocone,* an anterior cylindric neck, and an apical end formed by a transverse partition of conch comprising specialized thick septum (septum of truncation) (TIP, 1964, pt. K, p.54, 263). Cf: *ascocone.*
ascocone (as′-co-cone) A cephalopod shell (like that of *Ascoceras*) in which the early portion is slender, curved, and deciduous; its later portion is the *ascoceroid conch.*
Ascomycetes (As-co-my-ce′-tes) The largest class of Fungi, characterized by the presence of a distinctive, saclike cell, called the ascus, that typically contains eight free ascospores (which may be one-celled or multi-celled). Generally, Ascomycetes have two reproductive phases: the (sexual) *ascus* stage and the (asexual) conidial stage. Cf: *Fungi imperfecti.*
ascon (as′-con) A thin-walled sponge or sponge larva with a single *flagellated chamber* that is also the *spongocoel* and lacking either inhalant or exhalant canals. Cf: *leucon; sycon.* Adj: *asconoid.*
asconoid Adj. of *ascon.*
ascophoran (as-coph′-o-ran) adj. Pertaining to the cheilostome bryozoan suborder Ascophora, characterized by parietal muscles attached to the floor of the ascus. n. An ascophoran cheilostome (bryozoan). Cf: *anascan.*
ascopore (as′-co-pore) A median frontal pore that serves as the inlet of the *ascus* in some ascophoran cheilostomes (bryozoans).
ascospore (as′-co-spore) A sexual fungal spore produced as a free cell by *meiosis* and meiotic processes in an *ascus.*
ascus (as′-cus) (a) A flexible floored sac beneath the *frontal shield* in ascophoran cheilostomes (bryozoans), variously developed to serve in protruding the *lophophore.* (b) An enlarged saclike cell of Ascomycetes fungi, containing a specific number of ascophores, commonly four but typically eight. Pl: *asci.* Syn: *compensation sac; compensatrix.*
ascus [palyn] An enlarged, saclike cell of Ascomycetes fungi, containing a specific number of ascospores, often four but typically eight.
aseismic (a-seis′-mic) Said of an area that is not subject to earthquakes.
aseismic ridge A submarine ridge that is a fragment of continental crust or an anomalously shallow oceanic crust, usually associated with a hot spot; it is so named to distinguish it from a seismically active mid-ocean ridge. Cf: *microcontinent.*
aseismic slip Gradual movement on a fault without generating any significant seismic waves.
asexual reproduction (a-sex′-u-al) Reproduction that does not involve or follow the union of individuals or of germ cells of two different sexes.
ash [coal] The inorganic residue after burning. Ignition generally alters both the weight and the composition of the inorganic matter. See also: *extraneous ash; inherent ash.*
ash [volc] Fine pyroclastic material (under 2.0 mm diameter; under 0.063 mm diameter for fine ash). The term usually refers to the unconsolidated material but is sometimes also used for its consolidated counterpart, *tuff.* Syn: *dust [volc]; pumicite; volcanic ash;* volcanic dust.
ashanite (ash′-a-nite) *ixiolite.*
ashburtonite A vitreous blue tetragonal mineral: $HCu_4Pb_4Si_4O_{12}(HCO_3)_4(OH)_4Cl$.
ash cloud *eruption cloud.*
ash-cloud surge Dilute *pyroclastic flow.* A type of *pyroclastic surge.*
ashcroftine-(Y) (ash′-croft-ine) A pink tetragonal mineral: $K_5Na_5(Y,Ca)_{12}(OH)_2[Si_{28}O_{70}](CO_3)_8 \cdot 8H_2O$.
ash fall Airborne ash that falls from an eruption cloud, and the resulting deposit. Cf: *pumice fall.* Syn: *ash shower.*
ash field A more or less well-defined area that is covered by volcanic ash. Cf: *lava field; volcanic field.*
ash flow A density current, generally a hot mixture of volcanic gases and tephra that travels across the ground surface; produced by the explosive disintegration of viscous lava in a volcanic crater, or from a fissure or group of fissures. The solid materials contained in a typical ash flow are generally unsorted and ordinarily include volcanic dust, pumice, scoria, and blocks in addition to ash. Cf: *pumice flow; pyroclastic density current.* Syn: *pyroclastic flow; incandescent tuff flow; glowing avalanche.* Obsolete syn: sand flow.
ash-flow tuff A tuff deposited by an ash flow; a type of *ignimbrite.* It is a consolidated but not necessarily welded deposit.
Ashgillian (Ash-gil′-li-an) A series of the Standard Global Chronostratigraphic Scale: uppermost Ordovician (above Caradocian, below Llandoverian of the Silurian).
ashlar (ash′-lar) Rectangular pieces of stone of nonuniform size that are set randomly in a wall.
ashoverite (ash′-ov-er-ite) A colorless tetragonal mineral: $Zn(OH)_2$. A polymorph of sweetite.
ash shower *ash fall.*
ashstone (ash′-stone) An indurated deposit of fine volcanic ash. Obsolete.
ashtonite (ash′-ton-ite) *mordenite.*
ash tuff Pyroclastic rock made up chiefly of consolidated ash.
ashy grit (a) A deposit of sand-size and smaller pyroclastic materials. (b) A deposit of sand and volcanic ash.
ash yield The percentage of incombustible material in coal and peat, determined by burning a sample and measuring the ash.
asif A term used in Morocco for a large, generally dry valley in a mountainous region. Etymol: Berber. Cf: *agouni.*
asisite (a-sis′-ite) An adamantine yellow or greenish-yellow tetragonal mineral: $Pb_7SiO_8Cl_2$.
askeletal (a-skel′-e-tal) Without a skeleton; said esp. of sponges.
asparagus stone (as-par′-a-gus) A yellow-green variety of apatite. Syn: asparagolite.
aspect [cart] Conceptual placement of a projection system in relation to the Earth's axis direct, normal, polar, equatorial, oblique, and so on (Snyder and Voxland, 1989, p.2).
aspect [slopes] (as′-pect) The direction toward which a slope faces with respect to the compass or to the rays of the Sun. Cf: *exposure [geol].*
aspect [stratig] The general appearance of a particular geologic entity or fossil assemblage, considered more or less apart from relations in time and space; e.g. the gross or overall lithologic and/or paleontologic character of a stratigraphic unit as displayed at any single geographic point of observation (as in a borehole or outcrop section), and representing the summation or "flavor" of a *facies.* The aspect of a facies generally has environmental significance.
aspect angle [remote sensing] *angle of incidence.*
aspect angle [slopes] The angle between the aspect of a slope and geographic south (if measured in the Northern Hemisphere), usually reckoned positive eastward and negative westward.
aspect ratio A ratio of one dimension to another, for example, the ratio of length to width of the rupture zone of a fault.
aspect ratio [volc] The ratio of the mean thickness of a lava flow (or pyroclastic deposit) to the diameter of a circle with an area equal to the area covered by the flow (Walker, 1973; Peterson and Tilling, 2000).
asperity (a) A protrusion above the surface of a fault, which inhibits slip. In order for sliding to begin, asperities must either break off or plow a groove into the opposite wall of the fault. (b) An irregularity on a fracture surface, typically a projection that reduces the aperture size. (c) A point where the walls of a round-walled natural fracture touch each other.
asphalt (as′-phalt) A dark brown to black viscous liquid or low-melting solid bitumen that consists almost entirely of carbon and hydrogen and is soluble in carbon disulfide. Natural asphalt, formed in oil-bearing rocks by the evaporation of the volatiles, occurs in Trinidad, near the Dead Sea, and in the Uinta Basin of Utah. Asphalt can be prepared by the pyrolysis of coals or shales. Var: *asphaltum.* Syn: *pitch [petroleum].* Obsolete syn: *mineral pitch.* See also: *tabbyite.*
asphalt-base crude Crude oil containing a high percentage of naphthenic and asphaltic hydrocarbons. Cf: *paraffin-base crude; mixed-base crude.* Syn: *naphthene-base crude.*
asphaltene (as′-phal-tene) Any of the components of a *bitumen* that are soluble in carbon disulfide but not in paraffin naphtha; high-molecular-weight hydrocarbons that constitute the solid dispersed particles of the bitumen.
asphaltic (as-phal′-tic) Pertaining to or containing *asphalt;* e.g. "asphaltic limestone" or "asphaltic sandstone" impregnated with asphalt, or "asphaltic sand" representing a natural mixture of asphalt with varying proportions of loose sand grains.
asphaltic pyrobitumen A bitumen, usually black and structureless, similar in composition to *asphaltite,* but infusible and insoluble in carbon disulfide; it generally contains less than 5% of oxygen. Examples are albertite, elaterite, impsonite, and wurtzilite.
asphaltic sand A natural mixture of asphalt and sand.
asphaltite (as′-phal-tite) Any one of the naturally occurring black solid bitumens that are soluble in carbon disulfide and fuse above 230°F. Examples are uintahite, glance pitch, and grahamite.
asphalt rock A porous rock, such as a sandstone or limestone, that is impregnated naturally with asphalt. Syn: *asphalt stone; rock asphalt; tar sand.*
asphalt seal The clogging of porosity where an oil-bearing rock crops out, as a result of loss of the lighter fractions of the oil and accumulation of the heavier asphaltic residues.
asphalt stone *asphalt rock.*
asphaltum (as-phal′-tum) *asphalt.*
aspidolite An olive-green trioctahedral mica: $NaMg_3(Si_3Al)O_{10}(OH)_2$. Syn: *sodium phlogopite; wonesite.*
aspis (as′-pis) A prominently protruding thickening of pollen *exine* bordering a pore. Pl: aspides. Adj: aspidate.
aspondyle (a-spon′-dyle) In dasycladacean algae, an irregular arrangement of branches (laterals) on the central axis.
assay (as′-say) v. In economic geology, to analyze the proportions of metals in an ore; to test an ore or mineral for composition, purity, weight, or

other properties of commercial interest. n. The test or analysis itself; its results.

assay foot In determining the *assay value* of an orebody, the multiplication of its *assay grade* by the number of feet along which the sample was taken. Cf: *assay inch.*

assay grade The percentage of valuable constituents in an ore, determined from assay. Cf: *assay value; value.*

assay inch In determining the *assay value* of an orebody, the multiplication of its *assay grade* by the number of inches along which the sample was taken. Cf: *assay foot.*

assay limit The limit of an ore body as determined by assay, rather than by structural, stratigraphic, or other geologic controls. Cf: *cutoff grade.* Syn: *cutoff limit; economic limit.*

assay ton A weight of 29.1667 grams, used in assaying to represent proportionately the *assay value* of an ore. Since it bears the same ratio to one milligram that a ton of 2,000 pounds bears to the troy ounce, the weight in milligrams of precious metal obtained from an assay ton of ore equals the number of ounces to the ton. Abbrev: *AT.*

assay value (a) The quantity of an ore's valuable constituents, determined by multiplying its *assay grade* or percentage of valuable constituents by its dimensions. Cf: *assay inch; assay foot.* The figure for precious metals is generally given in troy ounces per ton of ore, or *assay ton.* See also: *value.* (b) The monetary value of an orebody, calculated by multiplying the quantity of its valuable constituents by the market price (von Bernewitz, 1931).

asselbornite (as'-sel-born-ite) A brown or yellow cubic mineral: $(Pb,Ba)(UO_2)_6(BiO)_4(AsO_4)_2(OH)_{12}{\cdot}3H_2O$.

Asselian A series of the Standard Global Chronostratigraphic Scale: lowermost stage of the Cisuralian (Lower Permian; above Carboniferous System, below Sakmarian Stage). The basal GSSP (base of Permian) is in northern Kazakhstan (Davydov et al., 1998).

assemblage [arch] (as-sem'-blage) Collection of archaeological artifacts found together that helps to reconstruct past culture, environment, or practices.

assemblage [ecol] (a) A group of fossils that occur at the same stratigraphic level; often with a connotation also of localized geographic extent. For its accurate interpretation, an assemblage should be relatively homogeneous or uniformly heterogeneous. Cf: *congregation; association [ecol]; community; faunule.* Syn: *fossil assemblage.*; (b) Physical association, on a bedding plane, of several types of discrete conodont elements interpreted as skeletal parts of one animal (TIP, 1981, pt. W, supp.2). Syn: *natural assemblage.*

assemblage [petrology] (a) *mineral assemblage.* (b) *metamorphic assemblage.*

assemblage zone (a) A *biozone* characterized by the association of three or more taxa. It may be based on all kinds of fossils present, or restricted to only certain kinds of fossils (NACSN, 1983, Art. 51). It may consist of a geographically or stratigraphically restricted assemblage or may incorporate two or more contemporaneous assemblages with shared characteristic taxa. (b) A *stratum* or body of strata characterized by a distinctive assemblage or association of three or more fossil taxa that, taken together, distinguish it in biostratigraphic character from adjoining strata (ISSC, 1994, p.62-63). The assemblage zone is named for one or more of the fossils particularly representative of the assemblage. Cf: *faunizone; florizone; range zone; abundance zone; Oppel zone.* Syn: *cenozone.*

assembled stone (as-sem'-bled) Any gemstone constructed of two or more parts, whether natural, synthetic, imitation, or a combination thereof; e.g. a doublet or triplet. Syn: *composite stone.* Cf: *imitation.*

assimilated (as-sim'-i-lat'-ed) Said of an ore-forming fluid or mineralizer that is derived from crustal, palingenic magmas (Smirnov, 1968). Cf: *juvenile [ore dep]; filtrational.*

assimilation (as-sim'-i-la'-tion) The incorporation and digestion of xenoliths and their chemical constituents into a body of magma. Such a magma, or the rock it produces, may be called hybrid or *contaminated.* See also: *hybridization; contamination [ign].* Cf: *differentiation [intrus rocks].* Syn: *magmatic dissolution.*

assimilation-fractional crystallization Simultaneous wall-rock assimilation and fractional crystallization. Assimilation draws heat from the magma which necessarily leads to magma cooling, crystal nucleation, and fractional crystallization. DePaolo (1981) gave equations by which the trace-element and isotopic effects of these coupled processes can be calculated.

assise (as-sise' [a-sees']) (a) A term approved by the 2nd International Geological Congress in Bologna in 1881 for a stratigraphic unit next in rank below stage and equivalent to a substage. (b) A term suggested by P. F. Moore (1958) to replace *format.* Etymol: French, "course, layer, base", from the architectural usage for a continuous row or layer of stones or brick placed side by side between layers of mortar.

associated gas (as-so'-ci-at'-ed) Natural gas that occurs in association with oil in a reservoir, either as free gas or in solution. Cf: *nonassociated gas.*

association [ecol] (as-so'-ci-a'-tion) A group of organisms (living or fossil) occurring together because they have similar environmental requirements or tolerances, and usually having one or more dominant species. Cf: *assemblage [ecol]; community.* Syn: *fossil association.*

association [petrology] *rock association.*

associes (as-so'-cies [a-so'-sees]) A *seral* community in which there are two or more dominant forms.

assorted *poorly sorted.*

assortment The inverse of sorting or of the measure of dispersion of particle sizes within a frequency distribution.

assured mineral (as-sured') *developed reserves.*

Assyntian Orogeny (As-syn'-ti-an) A name proposed by Stille and widely used in western Europe for orogenies and disturbances at the end of the Precambrian. It is named for the district of Assynt in the northwest highlands of Scotland, and based on the angular truncation of Torridonian strata by Lower Cambrian strata. The name is unfortunate, because the age of the Torridonian is undetermined; it may be very much older than the Cambrian.

assyntite (as-syn'-tite) A *nepheline syenite* rich in titanite, containing augite, and with biotite and apatite among the accessories. The name is from Assynt, Scotland. Not recommended usage.

Astaracian (As-tar-a'-cian) European mammal age: medial Miocene (after Orleanian, before Vallesian); equiv. upper Aragonian stage. Obs. syn: *Maremmian;* upper *Vindobonian; Tortonian* (of Vienna Basin).

astatic (a-stat'-ic) Said of a geophysical instrument that has a negative restoring force which aids a deflecting force, making the instrument more sensitive and/or less stable.

astatic gravimeter An instrument with a mechanical system designed to produce relatively large motions for small changes in gravity. Syn: *unstable gravimeter.*

astatic magnetometer An old instrument designed to measure the *remanent magnetization* of rock samples. The measuring system consists of two magnets of equal magnetic moments and opposed polarity suspended by a torsion fiber. Angular deflections of magnets over a variously oriented rock sample are measured (Telford et al., 1984). See also *spinner magnetometer.*

astatic pair A pair of magnets of identical strength and opposed polarity, geometrically positioned so that they experience zero net torque in a homogeneous magnetic field.

aster (as'-ter) A sponge *spicule* (microsclere) that has a starlike appearance, with a relatively large number of rays or pseudoactins radiating from a relatively restricted central area.

asteria (a-ste'-ri-a) Any gemstone that, when cut *en cabochon* in the correct crystallographic direction, displays a rayed figure (a star) by reflected and/or transmitted light; e.g. a *star sapphire.* Syn: *star stone.*

asteriated (a-ste'-ri-at'-ed) Said of a cabachon-cut natural or synthetic gemstone that exhibits asterism; e.g. "asteriated beryl". Syn: *star.*

asterism [cryst] (as'-ter-ism) Elongation of Laue X-ray diffraction spots produced by stationary single crystals as a result of internal crystalline deformation. The size of the Laue spot is determined by the solid angle formed by the normals to any set of diffracting lattice planes; this angle increases with increasing crystal deformation, producing progressively by elongated ("asterated") spots. Measurements of asterism are used as indicators of deformation in crystals subjected to slow stress or to shock waves.

asterism [gem] The optical phenomenon of a rayed or star-shaped figure of light displayed by some crystals when viewed in reflected light, as in a star sapphire or a gemstone cut *en cabochon,* or in transmitted light, as in some quartz or mica. It is caused by minute oriented acicular inclusions. See also: *star.*

asteroid [astron] (as'-ter-oid) One of the many small rocky celestial bodies in orbit around the Sun. Most asteroid orbits are between those of Mars and Jupiter. The largest asteroid is Ceres, nearly 1,000 km in diameter. There are likely 10 million or more 1 km in diameter or larger, of which nearly 100,000 have currently been discovered. Syn: *planetoid; minor planet.*

asteroid [paleont] Any asterozoan belonging to the subclass Asteroidea, characterized by relatively broad arms usually not separable from the central disc; e.g. a starfish. Range, Lower Ordovician to Holocene.

asterolith (as'-ter-o-lith') A star- or rosette-shaped coccolith with a concave face, formed of a single crystal with the *c*-axis perpendicular to the plane of the disk. See also: *discoaster.*

asterozoan (as'-ter-o-zo'-an) Any free-living echinoderm belonging to the subphylum Asterozoa, having a characteristic depressed star-shaped body composed of a central disc and symmetrical radiating arms. Range, Cambrian to Holocene.

asthenolith (as-then'-o-lith) A body of magma that was formed by melting in response to heat generated by radioactive disintegration (Willis, 1938). See also: *asthenolith hypothesis.* Obsolete.

asthenolith hypothesis An outdated theory of magmatic activity, both intrusive and extrusive, that postulates local asthenoliths, or areas of melting by radioactive heat, that have a repetitive cycle of melting, growth, migration, cooling, solidification, and remelting. Asthenolithic activity is postulated as the cause of uplift and subsidence, orogeny, earthquakes, and metamorphism (Willis, 1938, p.603). Cf: *blister hypothesis.*

asthenosphere (as-then'-o-sphere) The layer or shell of the Earth below the lithosphere, which is weak and in which isostatic adjustments take place, magmas may be generated, and seismic waves are strongly attenuated. It is a part of the *upper mantle.* Syn: *zone of mobility.*

Astian (As'-ti-an) European stage: Upper Pliocene (above Plaisancian, below Calabrian). Equivalent to Redonian.

astite (as'-tite) A variety of *hornfels* in which mica and andalusite dominate (Holmes, 1928, p.37). Rarely used. Type locality: Cima d'Asta, Italian Alps. Cf: *aviolite; edolite.*

astogenetic stage (as'-to-ge-net'-ic) A stage in the development of a colony, as applied to bryozoa by Cumings in 1904.

astogeny (as-tog'-e-ny) The course of development of a sequence of asexual generations of zooids and extrazooidal parts, which together form a colony of attached members, as in bryozoans and graptolites.

astomate (a-sto'-mate) In fungal fructifications, without an *ostiole.*

astrakhanite (as-tra'-khan-ite) A syn. of *blödite.* Also spelled: astrakanite.

astreoid (as'-tre-oid) Said of a massive rugose *corallum* in which the septa of each *corallite* are fully developed but walls between corallites are lacking, and characterized by septa of adjacent corallites generally in alternating position. Cf: *aphroid.*

astringent (as-trin'-gent) (a) Said of a mineral (such as alum) having a taste that tends to pucker the tissues of the mouth. (b) Said of a clay containing an astringent salt.

astrobleme (as'-tro-bleme) An ancient erosional scar on the Earth's surface, produced by the impact of a cosmic body, and usually characterized by a circular outline and highly disturbed rocks showing evidence of intense shock (Dietz, 1961, p.53); an eroded remnant of an impact crater. Term introduced by Dietz (1960); now obsolete. Etymol: Greek astron, "star", + blema, "wound from a thrown object such as a javelin or stone". Cf: *geobleme.* Syn: *fossil meteorite crater; impact structure.*

astrobleme-associated nickel-copper deposit A *mineral-deposit type* characterized by disseminated to massive nickel-copper sulfide ores in rocks formed by shock metamorphism of the Earth's crust during a meteorite impact. Syn: *Sudbury-type nickel-copper deposit.*

astrocyanite-(Ce) A blue or blue-green hexagonal mineral: $Cu_2(Ce,Nd,La)_2(UO_2)(CO_3)_5(OH)_2{\cdot}1.5H_2O$.

astrogeodetic measurement (as'-tro-ge'-o-det'-ic) The direct measurement of the Earth to determine the deflection of the vertical, and hence the separation of the geoid and the ellipsoid. The method contrasts with gravimetric or indirect measurement.

astrogeology (as'-tro-ge-ol'-o-gy) *planetary geology.*

astrolabe (as'-tro-labe) A compact optical instrument designed for measuring the altitudes of celestial bodies; e.g. a "prismatic astrolabe" consisting of a telescope in a horizontal position, with a prism (generally of 45 or 60 degrees) in front of the object glass, immediately underneath which is an artificial horizon (pool of mercury). It was formerly used to fix latitude precisely by observing the apparent transit of the Sun across the meridian at midday, but has been superseded by the *sextant.*

astronomical position (as-tro-nom'-i-cal) The latitude and longitude of a point on the Earth as determined from astronomical measurements based on the position of the stars.

astronomical unit A unit of planetary distance equal to the length of the *semimajor axis* of the Earth's orbit, or the mean distance of the Earth from the Sun: 1.496×10^8 km, or approximately 93 million statute miles. Abbrev: AU.

astronomic azimuth (as-tro-nom'-ic) The angle between the astronomic meridian plane of the observer and the plane containing the observed point and the true normal (vertical) of the observer, measured in the plane of the horizon, preferably clockwise from north. Cf: *azimuth [surv]; bearing.*

astronomic equator The line on the Earth's surface whose astronomic latitude at every point is zero degrees. When corrected for station error, it becomes the *geodetic equator.*

astronomic horizon A great circle on the celestial sphere formed by the intersection of the celestial sphere and a plane passing through any point (such as the eye of an observer) and perpendicular to the zenith-nadir line; the plane that passes through the observer's eye and is perpendicular to the zenith at that point. It is the projection of a horizontal plane in every direction from the point of orientation. Cf: *celestial horizon.* Syn: *sensible horizon.*

astronomic latitude The *latitude* or angle between the plane of the celestial equator and the plumb line (direction of gravity) at a given point on the Earth's surface; the angle between the plane of the horizon and the Earth's axis of rotation. It represents the latitude resulting directly from observations on celestial bodies; when corrected for station error, it becomes the *geodetic latitude.* Symbol: Φ.

astronomic longitude The *longitude* or angle between the plane of the celestial meridian and the plane of an arbitrarily chosen prime meridian (generally the Greenwich meridian). It represents the longitude resulting directly from observations on celestial bodies; when corrected for station error, it becomes the *geodetic longitude.* Symbol: Λ.

astronomic measurement The determination of latitude, longitude, or azimuth from observations on stars, referred to a plumb line or gravity vector at the point of observation. The sun is sometimes used in lower-order surveying. Cf: *deflection of the vertical.*

astronomic meridian A line on the Earth's surface having the same astronomic longitude at every point. It is an irregular line, not lying in a single plane. Syn: *terrestrial meridian.*

astronomic parallel A line or circle on the Earth's surface having the same astronomic latitude at every point. It is an irregular line, not lying in a single plane.

astronomic tide *equilibrium tide.*

astronomy (as-tron'-o-my) The study of celestial bodies: their positions, sizes, movements, relative distances, compositions and physical conditions, interrelationships, and history.

astrophyllite (as-tro-phyl'-lite) (a) A bronze-yellow to brown platy triclinic mineral: $(K,Na)_3(Fe^{2+},Mn)_7Ti_2Si_8O_{24}(O,OH)_7$. (b) The name for a group of silicates with the general formula $A_{3\text{-}4}B_7C_2Si_8O_{24}(O,OH,F)_7$, where A = K, Na or Cs; B = Fe^{2+}, Mn^{2+} or Mg; and C = Ti or Nb.

astrophysics (as-tro-phys'-ics) That aspect of astronomy which is concerned with the physics and chemistry of celestial bodies, and with their origins.

astropyle (as'-tro-pyle) A nipplelike projection from the central capsule of a radiolarian of the suborder Phaeodarina.

astrorhiza (as-tro-rhi'-za) A system of regularly spaced, stellate, generally sinuous surficial impressions and internal canals of some living sponges and fossil stromatoporoids. They typically lack walls, may be crossed by thin tabulae, and decrease in diameter and branch distally from their point of divergence, which is marked by one or more vertical small tubes. Pl: astrorhizae.

astrum (as'-trum) A term established by the International Astronomical Union for radial-patterned features on Venus (USGS, 2003). Pl: astra.

Asturian Orogeny (As-tu'-ri-an) One of the 30 or more short-lived orogenies during Phanerozoic time identified by Stille, in this case late in the Carboniferous, between the Westphalian and Stephanian Stages.

asylum (a-sy'-lum) *refugium.*

asymmetric (a-sym-met'-ric) Said of crystals of the hemihedral class of the triclinic system, which have no symmetry elements; also, said of any irregular crystal.

asymmetric bedding Bedding characterized by lithologic types or facies that succeed each other vertically in a cyclical arrangement illustrated by the sequence 1-2-3-1-2-3-1-2-3. Cf: *symmetric bedding.*

asymmetric fold (a) A fold whose axial surface is not perpendicular to the *enveloping surface.* (b) A fold whose limbs have different angles of dip relative to the *axial surface.* Ant: *symmetrical fold.*

asymmetric prismatic structure [paleont] A simple *prismatic structure* in which the prisms are flattened in alternating directions in adjacent prisms, thereby producing a crisscross (herringbonelike) pattern in horizontal sections (Batten and Dumont, 1976).

asymmetric ripple A *ripple* having an asymmetric profile in cross section, characterized by a short steep slope facing downcurrent and a long gentle slope facing upcurrent; specif. *current ripple.* In plan view, the crest may be relatively straight or markedly curved. Ant: *symmetric ripple.*

asymmetric ripple mark *asymmetric ripple.*

asymmetric unit The smallest volume of simply connected space in a *unit cell* that, through repetition by each symmetry element present, forms the entire *unit cell.* The symmetry elements of the *space group* form vertices, edges, and boundary planes of the asymmetric unit. (b) Those atoms of a *crystal structure* that are within that volume.

asymmetric valley A valley with one side steeper than the other.

AT *assay ton.*

atacamite (at-a-cam'-ite) A green orthorhombic mineral: $Cu_2Cl(OH)_3$. It is trimorphous with partacamite and botallackite, and is formed by weathering of copper lodes, esp. under desert conditions. Syn: *remolinite.*

atavism (at'-a-vism) *reversion.*

ataxic (a-tax'-ic) Said of an unstratified mineral deposit. Cf: *eutaxic.*

ataxite [meteorite] (a-tax'-ite) An *iron meteorite* having no structure visible to the naked eye. Detailed microscopy shows that ataxites with Ni contents >~12 wt% generally have fine plessitic structures, whereas those with lower Ni contents generally show "quench" structures, often associated with impact heating.

ataxite [volc] A *taxite* whose components have mixed in a breccialike manner. Obsolete. Cf: *eutaxite.*

Atdabanian Russian stage: Lower Cambrian (above Tommotian, below Lenian) (Harland et al., 1990) or below Botomian (Cowie and Bassett, 1989).

atectate (a-tec'-tate) In pollen *exine,* having little or no internal structure. See also: *intectate; tectum [palyn].*

atelestite (at-e-les'-tite) A yellow or yellowish-green monoclinic mineral: $Bi_2O(AsO_4)(OH)$. Isomorphous with smrkovecite and hechtsbergite.

atexite (a-tex'-ite) A basic material that is unchanged during *anatexis* (Dietrich and Mehnert, 1961). Var: atectite. Rarely used.

at grade [eng] At the same level or degree of rise; at design level or slope. The term is applied to highways, walks, culverts, etc., or combinations of these, at the point where they intersect.

at grade [geomorph] *graded [geomorph].*

athabascaite (ath-a-bas'-ca-ite) A metallic grayish-white orthorhombic mineral: Cu_5Se_4.

atheneite (a-the'-ne-ite) A metallic white hexagonal mineral: $(Pd,Hg)_3As$.

Atlantic (At-lan'-tic) n. A term used primarily in Europe for an interval of Holocene time (from about 8,000 to 5,000 years ago) following the Boreal and preceding the Subboreal, during which the inferred climate was warmer than at present and generally wet (i.e., influenced by more maritime-Atlantic-air masses); a subunit of the *Blytt-Sernander sequence,* characterized by oak, elm, linden, and ivy vegetation in Scandinavia. It corresponds to most of the *Altithermal* and the middle part of the Hypsithermal. adj. Pertaining to the postglacial Atlantic interval and to its climate, deposits, biota, and events.

Atlantic flysch succession Compositionally mature passive margin turbidites rich in quartz and metamorphic rock fragments derived from a continental source (Friedman and Sanders, 1978 p.458).

Atlantic suite Obsolete name for one of two large groups of igneous rocks, characterized by alkalic and alkali-calcic rocks. Harker (1909) divided all Tertiary and Holocene igneous rocks of the world into two main groups, the Atlantic suite and the Pacific suite, the former being so named because of the predominance of alkalic and alkali-calcic rocks in the nonorogenic areas of crustal instability around the Atlantic Ocean. Because there is such a wide variety of tectonic environments and associated rock types in the areas of Harker's Atlantic and Pacific suites, the terms are now seldom used to indicate kindred rock types; e.g. Atlantic-type rocks are widespread in the mid-Pacific volcanic islands. Cf: *Mediterranean suite.* See also: *andesite line.*

Atlantic-type coastline A *discordant coastline,* esp. as developed in many areas around the Atlantic Ocean; e.g. the SW coastline of Ireland and the NW coastlines of France and Spain. Ant: *Pacific-type coastline.*

atlas (at'-las) A collection of maps usually bound into a volume. The use of the term is derived from the figure of Atlas (a Titan of Greek mythology, often represented as supporting the heavens) used as a frontispiece to certain early collections of maps, first appearing on the general title page of Mercator's *Atlas* (1595).

atlasovite (at'-las-ov-ite) A dark brown tetragonal mineral: $KCu_6Fe^{3+}Bi^{3+}O_4(SO_4)_5Cl$.

atmidometer (at-mid-om'-e-ter) *atmometer.*

atmoclast (at'-mo-clast) A rock fragment broken off in place by atmospheric weathering, either chemically or mechanically.

atmoclastic rock (at-mo-clas'-tic) A clastic rock consisting of atmoclasts that have been recemented without rearrangement by wind or water (Grabau, 1924, p.292).

atmodialeima (at'-mo-di'-a-lei'-ma) A term proposed by Sanders (1957, p.295) for an unconformity caused by subaerial processes.

atmometer (at-mom'-e-ter) Any device used to measure the rate of evaporation in the atmosphere. It may be a large evaporation tank, a small evaporation pan, a porous porcelain body, or a porous paper-wick device. Syn: *atmidometer; evaporimeter; evapograph.*

atmophile (at'-mo-phile) (a) Said of those elements that are most typical in the Earth's atmosphere: H, C, N, O, I, Hg, and inert gases (Rankama and Sahama, 1950, p.88). (b) Said of those elements that occur in the uncombined state, or that "will concentrate in the gaseous primordial atmosphere" (Goldschmidt, 1954, p.26).

atmosphere (at'-mo-sphere) (a) The mixture of gases that surrounds the Earth, being held thereto by gravity. At altitudes below 80 km, dry atmosphere (air) consists of a relatively uniform concentration by volume of 78% nitrogen, 21% oxygen, 0.9% argon, 0.03% carbon dioxide, and minute quantities of helium, krypton, neon, and xenon. Water vapor can represent up to 4% by volume. The atmosphere is so compressed by its own weight that half of its mass is below 5.5 km from the Earth's surface. (b) A unit of normal or standard pressure at sea level. Syn: *air.*

atmospheric argon (at-mo-spher'-ic) Argon in the atmosphere and argon absorbed on the surfaces of rocks and minerals that have been exposed to the atmosphere. It has a specific isotopic composition. Cf: *excess argon; inherited argon; radiogenic argon.*

atmospheric pressure The pressure, or force per unit area, exerted by the atmosphere on any surface beneath or within it. Normal or standard pressure at sea level is 1013.25 (millibars) hectopascal or 101325 Newton per square meter, where 1 Pascal = 1 $N/m^2$and 1 hectopascal = 100 N/m^2 . Equivalent common measures are 14.66 lb per in^2; 76.0 cm, or 29.92 in., of mercury; and 1033.3 cm, or 33.9 ft, of water. Syn: *air pressure; barometric pressure.*

atmospheric radiation The infrared radiation emitted by constituents in the atmosphere in two directions: upward into space and downward toward the Earth. The latter is known as *counterradiation.*

atmospheric scattering Multiple interactions between light rays and the gases and particles in the atmosphere.

atmospheric tide The rhythmic vertical oscillation of the atmosphere, produced primarily by thermal effects from absorption of radiation at different levels and secondarily by gravitational attraction by the sun and moon. Syn: *tide.*

atmospheric water Water in the atmosphere, in gaseous, liquid, or solid state.

atmospheric weathering Weathering occurring at the surface of the Earth.

atmospheric window A wavelength interval at which the atmosphere transmits most (atmospheric gases absorb little or no) *electromagnetic radiation..* Syn: *transmission window.*

Atokan (A-to'-kan) North American provincial series: lower middle Pennsylvanian (above Morrowan, below Desmoinesian).

atokite (a-to'-kite) A metallic cubic mineral: $(Pd,Pt)_3Sn$.

atoll (at'-oll) A coral reef appearing in plan view as a ring or horseshoe-shaped reef, rising from deep water of (1) the open sea (such as central Pacific), (2) a submarine bank or plateau of intermediate depth (e.g. Queensland Plateau), or (3) a continental shelf (e.g. Great Barrier Reef or Sahul Shelf). The three categories are identified respectively as "classic" or "open-sea" type, *bank atolls,* and *shelf atolls.* Atolls vary in size from about 1 km to more than 100 km. Smaller atolls built on the rim of a larger one are *faros.* Miniature growths of biological reef builders in the range of 1 to 10 m are called *microatolls.* Atoll reefs are commonly crowned by a discontinuous ring of islets built of coral sands, breccia or a pre-exisiting (emerged) reef. Cf: *motu.* Etymol: native name in the Maldive Islands (Indian Ocean) which are typical examples of this structure. Syn: *lagoon island; ring reef; reef ring.*

atollon (at'-ol-lon) A term used in the Maldive Islands of the Indian Ocean for a large atoll consisting of many smaller ones; the term "atoll" was derived from this name.

atoll structure In a metamorphic rock, porphyroblasts with hollow centers resembling atolls. The ring may be almost complete or consist of a chain of granules (Joplin, 1968).

atoll texture In mineral deposits, the surrounding of one mineral by a ring of one or more other minerals. It commonly results from replacement of pyrite by another mineral, with the outermost pyrite unaffected and constituting the "atoll". Syn: *core texture.* Cf: *tubercle texture.*

atomic absorption spectrometer Instrument used to conduct atomic absorption spectrometry.

atomic absorption spectrometry (a-tom'-ic) The recording and measuring components that are a normal component of *atomic absorption spectroscopy.* The term is rarely used.

atomic absorption spectrophotometer *atomic absorption spectrometer.*

atomic absorption spectroscope *atomic absorption spectrometer.*

atomic absorption spectroscopy An analytical technique whereby a sample is vaporized and its nonexcited atoms are identified and quantified by the *electromagnetic radiation* they absorb at characteristic wavelengths. Abbrev: AAS. See also: *absorption spectrum.*

atomic absorption spectrum A type of *absorption spectrum* based on how atoms absorb electromagnetic radiation.

atomic clock *radioactive clock.*

atomic coordinates *fractional coordinates.*

atomic emission spectroscopy An analytical technique whereby a sample is vaporized and heated or bombarded by ions, electrons, or photons. Its atoms are then identified by the *electromagnetic radiation* they emit at characteristic wavelengths. See also: *emission spectrum.*

atomic mass unit The unit of mass of atoms defined as $^1/_{12}$ of the mass of $^{12}_{6}C$.

atomic number The number of protons in the nucleus of an atom. Abbrev: Z.

atomic parameters A set of numbers that includes the *fractional coordinates* of an atom and an expression of the extent of its thermal vibration.

atomic plane In a crystal, any plane that contains a regular array of atomic units (atoms, ions, or molecules); it is a potential crystal face or cleavage plane. Cf: *lattice plane.*

atomic resolution microscopy All types of microscopy used to produce images of the atoms in a sample and their relative dimensions, distribution, and topography. See also: *scanning force microscopy; scanning tunneling microscopy.*

atomic time scale A *geologic time scale* calibrated on the basis of radioactive decay in rocks. Measurements are made in years. Cf: *relative time scale.*

atomous (a'-tom-ous) Said of a crinoid arm that does not branch.

atom percent (at'-om) The percentage of an atomic species in a substance, calculated with reference to number of atoms rather than weight, number of molecules, or other criteria.

atopite (at'-o-pite) A yellow or brown variety of *roméite* containing fluorine.

at rest In engineering geology, said of lateral *earth pressure* when the soil is neither compressed nor allowed to expand and the structure (such as a wall) does not move. Cf: *neutral pressure.*

atria Pl. of *atrium.*

atrium [paleont] (a'-tri-um) The *spongocoel* of a sponge. Pl: *atria.*

atrium [palyn] In a pollen grain with a complex pore structure, the space contained between the external opening (*exopore*) and a much larger internal opening (*endopore*) in the endexine. Pl: *atria.* Cf: *vestibulum.*

atrium oris A preoral cavity in a crustacean, bounded ventrally by the posteriorly directed *labrum,* dorsally by the ventral surface of the *cephalon* just behind the mouth, and laterally by *metastoma* and mandibles.

atrypoid (a'-try-poid) Any articulated brachiopod belonging to the family Atrypidae, characterized by costate or plicate shells that are unequally biconvex or convexo-plane, the brachial valve being the more convex. Range, Middle Ordovician to Upper Devonian.

attached dune (at-tached') A dune that accumulates around a rock or other obstacle in the path of windblown sand, occurring on either the windward or the leeward sides of the obstacle, or on both sides, and varying widely in size and form. Cf: *anchored dune.*

attached groundwater Ground water retained on the walls of interstices in the *unsaturated zone.* It is considered equal in amount to the *pellicular water* and is measured by specific retention. Cf: *pendular water.*

attached operculum The part of a dinoflagellate cyst that is surrounded by archeopyle sutures and hence remains joined to the main part of the theca where the suture is not developed. Cf: *free operculum.*

attachment scar (at-tach'-ment) (a) An expanded *basal cavity* of a conodont, or one that is larger than a small-sized pit. (b) The part of the aboral side of a conodont element to which the basal plate was attached. Syn: *attachment surface.*

attachment surface *attachment scar.*

attakolite (at-tak'-o-lite) A white to pale red orthorhombic mineral: $(Ca,Sr)Mn^{2+}(Al,Fe^{3+})_4(HSiO_4)(PO_4)_3(OH)_4$. It is structurally related to carminite. Also spelled: attacolite.

attapulgite (at-ta-pul'-gite) *palygorskite.*

attenuation [grd wat] The process by which a compound is reduced in concentration over time, through absorption, adsorption, degradation, dilution, and/or transformation (U.S. Environmental Protection Agency, 1994).

attenuation [seis] (at-ten'-u-a'-tion) (a) A reduction in the amplitude or energy of a signal, such as might be produced by passage through a filter. (b) A reduction in the amplitude of seismic waves, as produced by *divergence [seis],* reflection and scattering, and *absorption.* (c) That portion of the decrease in seismic or sonar signal strength with distance that is not dependent on geometrical divergence, but on the physical characteristics of the transmitting medium.

attenuation coefficient If the loss of amplitude A with distance x is expressed as an exponential, $A/A_o = e^{-\alpha x}$, then α is the attenuation coefficient. Cf: *Q [seis].*

attenuation coefficient [oceanog] The relative rate at which the energy in a sound or light signal decreases as the signal travels through a fluid. The decrease is due to absorption and scatter.

attenuation constant *Q [seis].*

Atterberg grade scale (At'-ter-berg) A geometric and decimal *grade scale* devised by Albert Atterberg (1846-1916), Swedish soil scientist; it is based on the unit value 2 mm and involves a fixed ratio of 10 for each successive grade, yielding the diameter limits of 200, 20, 2.0, 0.2, 0.02, and 0.002 (Atterberg, 1905). Subdivisions are the geometric means of the grade limits. The scale has been widely used in Europe and was adopted in 1927 by the International Commission on Soil Science (but not by the U.S. Bureau of Soils).

Atterberg limits In a sediment, the water-content boundaries between the semiliquid and plastic states (known as the *liquid limit*) and between the plastic and semisolid states (known as the *plastic limit*). Syn: *consistency limits.*

attic (at'-tic) The very small, uppermost (abaxial) chamberlet in superposed chamberlets in a volution of a fusiform foraminiferal shell (as in *Flosculinella* and *Alveolinella*).

Attic Orogeny (At'-tic) One of the 30 or more short-lived orogenies during Phanerozoic time, identified by Stille; in this case in the Miocene, between the Sarmatian and Pontian Stages.

attitude [photo] (at'-ti-tude) (a) The angular orientation of a camera, or of the photograph taken with that camera, with respect to some external reference system. It is usually expressed as "tilt", "swing", and "azimuth", or as "roll", "pitch", and "yaw". (b) The angular orientation of an aerial or space vehicle with respect to a reference system (ASP, 1975, p.2066).

attitude [struc geol] The orientation of a *structural element* (plane or line) relative to a horizontal plane. Attitudes of planes are expressed by *strike* and *dip* (or dip direction and dip), whereas attitudes of lines are expressed by *bearing* (or trend, azimuth) and *plunge.*

attribute [geol] (at'-tri-bute) Any detail that seems to qualify, identify, classify, quantify, or express the state of a quantity. A qualitative characteristic of an object, such as the presence or absence of a specific feature or membership in a specific class or group. The presence or absence of spines may be an attribute of certain organisms. A feature code used to indicate the color or style of a line in automated cartography is an attribute.

attribute [seis] Any property or combination of properties that can be measured on seismic data. Trace attributes include envelope amplitude, instantaneous phase, instantaneous frequency, apparent polarity, and velocity determined by inversion; surface attributes include dip magnitude, dip azimuth, artificial illumination, and coherence (or incoherence).

attribute tagging An assignment of attributes to objects, e.g. points, lines, and polygons, in spatial information systems (Krzanowski et al., 1993).

attrital-anthraxylous coal (at-tri'-tal-an'-thra-xy'-lous) A bright coal in which the ratio of anthraxylon to attritus varies from 1:1 to 1:3. Cf: *anthraxylous coal; anthraxylous-attrital coal; attrital coal.*

attrital coal (at-tri'-tal) (a) A coal in which the ratio of anthraxylon to attritus varies from 1:1 to 1:3. Cf: *anthraxylous coal; anthraxylous-attrital coal; attrital-anthraxylous coal.* (b) The groundmass or matrix of *banded coal,* in which vitrain and commonly fusain are embedded.

attrition (at-tri'-tion) The act or process of wearing down by friction; specif. the mutual wear and tear that loose rock fragments or particles, moved about by wind, waves, running water, or ice, undergo by rubbing, grinding, knocking, scraping, and bumping against one another, with resulting reduction in size and increase in roundness. Although the term has been used synonymously with *abrasion* and *corrasion,* strictly it is the wearing away and reduction in size undergone by rock fragments serving as tools of abrasion or corrasion.

attritus (at-tri'-tus) A composite term for dull gray to nearly black coal components of varying maceral content, unsorted and with fine granular texture, that forms the bulk of some coals or is interlayered with bright bands of *anthraxylon* in others. It is formed of a tightly compacted mixture of altered vegetal materials, especially those that were relatively resistant to complete degradation. Cf: *attrital coal.* Syn: *durain.*

A-type earthquake Event with clear P waves and S waves occurring under volcanoes at depths of 1-10 km (McNutt, 2000).

A-type granitoid A granitic rock thought to be emplaced in *anorogenic* settings. The term, proposed by Loiselle and Wones, is thoroughly discussed by Clarke (1992).

AU *astronomical unit.*

Au-Ag-Te vein *alkalic intrusive-associated gold-silver deposit.*

aubertite (au'-bert-ite) A blue triclinic mineral: $CuAl(SO_4)_2Cl \cdot 14H_2O$. Isomorphous with magnesioaubertite and zvyazhinite.

aubrite (au'-brite) A rare group of achondritic stony meteorites consisting almost entirely of enstatite. Chemical and petrologic studies suggest that aubrites are the product of igneous processing of an enstatite chondrite precursor. See also: *enstatite chondrite.*

audio-magneto-telluric method (au'-di-o-mag-ne'-to-tel-lu'-ric) A magnetic-telluric survey using frequencies in the natural electromagnetic field above 20 Hz. Abbrev: AMT.

auerlite (au'-er-lite) A variety of thorite containing phosphorus, with a PO_4/SiO_4 ratio of about 0.8:1.

aufeis (auf'-eis) A syn. of *icing.* Etymol: German.

aufwuchs (auf'-wuchs) Aquatic organisms that are attached to or coat but do not penetrate the substrate; e.g. crustaceans. Cf: *periphyton.* Etymol: German, "growth".

auganite (au'-ga-nite) An augite-bearing, olivine-free *andesite;* according to Johannsen (1939, p.242). Obsolete.

augelite (au'-ge-lite) A colorless, white, or pale-red monoclinic mineral: $Al_2(PO_4)(OH)_3$.

augen (au'-gen [Pron: ow-gn]) In foliate metamorphic rocks such as schists and gneisses, large lenticular mineral grains or mineral aggregates having the shape of an eye in cross section, in contrast to the shapes of other minerals in the rock. Cf: *augen structure.* Etymol: German, "eyes".

augen-blast An *augen*-forming porphyroblast in dynamically metamorphosed rocks (Bayly, 1968). Little used. Cf: *augen-clast.*

augen-clast In dynamically metamorphosed rock, *augen* consisting of clastic fragments in a clastic matrix (Bayly, 1968). Cf: *augen-blast.*

augen gneiss A general term for a gneissic rock containing *augen.*

augen schist A metamorphic rock characterized by the presence of recrystallized minerals as *augen* or lenticles parallel to and alternating with schistose streaks.

augen structure In some gneissic and schistose metamorphic rocks, a structure consisting of minerals like feldspar, quartz, or garnet that have been squeezed into elliptical or lens-shaped forms resembling eyes (*augen*), which are commonly enveloped by essentially parallel layers of contrasting constituents such as mica or chlorite. Cf: *flaser structure [meta].* Syn: *eyed structure; phacoidal structure.*

auger (au'-ger) (a) A screwlike boring tool resembling a carpenter's auger bit but much larger, usually motor-driven, designed for use in clay, soil, and other relatively unconsolidated near-surface materials. (b) A rotary drilling device for making seismic shotholes or geophone holes by which the cuttings are mechanically and continuously removed from the bottom of the borehole during the drilling operation without the use of fluids.

Auger effect The emission of an electron from an atom without the emission of an X-ray or gamma ray *photon,* resulting from the de-excitation of an excited electron of that atom. The effect was discovered by Pierre Auger in 1925.

Auger electron (Au-ger' [o-jay]) An electron emitted from an atom by an X-ray produced during the electron-capture decay of its parent.

Auger spectrum (Au'-ger) A series of discrete lines representing the energies of emitted electrons arising from the *Auger effect.*

augite (au'-gite) (a) A common mineral of the *clinopyroxene* group: $(Ca,Na)(Mg,Fe,Al,Ti)(Si,Al)_2O_6$. It may contain titanium and ferric iron. Augite is usually black, greenish black, or dark green, and occurs as an essential constituent in many basic igneous rocks and in certain metamorphic rocks. Dana (1892) confined the name "augite" to clinopyroxenes containing appreciable $(Al,Fe)_2O_3$, but petrologists have applied it to members of the system $(Mg,Fe,Ca)SiO_3$. Cf: *pigeonite.* (b) A term often used as a syn. of *pyroxene.* Syn: *basaltine; violaite.*

augitite (au'-gi-tite) A *tephrite* containing abundant phenocrysts of clinopyroxene with lesser amounts of amphibole, magnetite or ilmenite, apatite, and sometimes nepheline, hauyne, or feldspar in a dark-colored glassy groundmass probably of the composition of analcime. Obsolete.

aulacocerid (au-la-co'-cer-id) Said of cephalopods belonging to the order Aulacocerida of the subclass Coleoidea.

aulacogen (au-lac'-o-gen) (a) A sediment-filled continental rift that trends at a high angle to the adjacent continental margin or orogen. Some aulacogens initiate as the failed arm of a three-armed spreading system, whereas others are abandoned rifts that were later truncated by rifting at a substantially different orientation. (b) Term introduced by Shatski (1946a, b) to describe a narrow, elongate basin that extends into the craton either from a passive-margin basin or from a mountain belt that formed from a passive-margin basin (Biddle and Christie-Blick, 1985). It is a tectonic trough on a craton, bounded by convergent normal faults. Aulacogens have a radial orientation relative to cratons and are open outward. Also spelled: *avlakogene.* Cf: *rift [tect].* Etymol: Greek "furrow".

aulocalycoid (au-lo-cal'-y-coid) Said of the skeleton of a hexactinellid sponge in which presumed *dictyonal strands* are diagonally interwoven and connected by synapticulae.

aulos (au'-los) An *axial structure* in a rugose coral, consisting of a tube commonly formed by abrupt sideward deflection of the inner edges of septa and their junction with neighbors.

aurantimonate An olive-gray mineral: $AuSbO_3$. Inadequately characterized, and not approved by the International Mineralogical Association.

aureole (au'-re-ole) A zone surrounding an igneous intrusion in which the country rock shows the effects of *contact metamorphism.* Syn: *contact aureole; contact zone; exomorphic zone; metamorphic aureole; thermal aureole.*

auric (au'-ric) Said of a substance that contains gold, esp. gold in its trivalent state. Cf: *auriferous.*

aurichalcite (au-ri-chal'-cite) A pale-green or pale-blue orthorhombic mineral: $(Zn,Cu)_5(CO_3)_2(OH)_6$. Syn: *brass ore.*

auricle (au'-ri-cle) (a) An earlike extension of the dorsal region of the shell in certain bivalve mollusks, commonly separated from the body of the shell by a notch or sinus. (b) An internal process arising from basicoronal ambulacral plates of an echinoid and serving for attachment of muscles supporting *Aristotle's lantern.* (c) In the vertebrates, a blind pocket projecting from a receiving or venous chamber (atrium) of the heart; or this chamber itself.

auricula (au-ric'-u-la) One of the thickened, equatorial, radially projecting "ears" of certain *zonate* spores with triangular *amb.* Pl: auriculae. Cf: *valva; zona.*

auricular sulcus (au-ric'-u-lar) An external furrow at the junction of an *auricle* with the body of the shell of a bivalve mollusk.

auriculate (au-ric'-u-late) Having auriculae.

auricupride (au-ri-cu'-pride) A metallic yellow cubic mineral: Cu_3Au. Syn: *cuproauride.*

auriferous (au-rif'-er-ous) Said of a substance that contains gold, esp. said of gold-bearing mineral deposits. Cf: *auric.*

auriferous pyritic quartz-pebble conglomerate A *mineral-deposit type* characetrized by pyrite and native gold in Precambrian quartz-pebble conglomerate. A variety of *paleoplacer* deposit.

auriform (au'-ri-form) Said of a mollusk shell (such as that of a gastropod) shaped like the human ear.

aurivilliusite A metallic gray-black monoclinic mineral: $Hg^{1+}Hg^{2+}OI$.

auroantimonate An ill-defined olive-gray mineral of composition $AuSbO_3$.

aurora (au-ro'-ra) (a) The time interval from the first appearance of a new phylogenetic stage (subspecies) until the first appearance of the next subspecies (Tan Sin Hok, 1931, p.643). (b) A division of geologic time from the appearance of a mutant in a bioseries to the appearance of the following mutant (Glaessner, 1945, p.225). Pl: aurorae.

aurorite (au-ror'-ite) A sub-metallic black rhombohedral mineral: $(Mn^{2+},Ag,Ca)Mn^{4+}{}_3O_7{\cdot}3H_2O$. A member of the *chalcophanite* group.

aurostibite (au-ro-stib'-ite) A metallic gray cubic mineral: $AuSb_2$. It has the pyrite structure.

austausch (aus'-tausch) A measure of turbulent mixing, equal to the product of mass and transverse distance traveled in a unit of time by a fluid in turbulent motion as it passes through a unit area parallel to the general direction of flow (Twenhofel, 1939, p.187). Syn: *eddy conductivity; mixing coefficient; austausch coefficient; eddy coefficient; exchange coefficient.*

austausch coefficient *austausch.*

Austinian (Aus-tin'-i-an) North American (Gulf Coast) stage: Upper Cretaceous (above Eaglefordian, below Tayloran).

austinite (aus'-tin-ite) A colorless or yellowish orthorhombic mineral: $CaZn(AsO_4)(OH)$. A member of the *adelite* group.

austral (aus'-tral) Pertaining to the south, or located in southern regions; southern.

Australasian tektite A tektite from the largest and youngest (ca. 0.8 Ma) of the four main tektite fields.

australite (aus-tral'-ite) A tektite from Australia. Some have flanges leading to button-like shapes. The flanges are produced by frictional melting during atmospheric passage and thus require solidification prior to atmospheric entry.

Austrian Orogeny (Aus'-tri-an) One of the 30 or more short-lived orogenies during Phanerozoic time identified by Stille, in this case at the end of the Early Cretaceous.

autapomorphy (au'-ta-po-mor'-phy) In *cladism,* a *derived character* that is unique to a single taxon, usually a single species. Adj: autapomorphic; autapomorphous.

autecology (aut-e-col'-o-gy) The study of the relationships between individual organisms or species (or a particular taxon) and their environment. Also spelled: *autoecology.* Cf: *synecology.*

authalic projection (au-thal'-ic) *equal-area projection.*

authigene (au'-thi-gene) An *authigenic* mineral or rock constituent, e.g. a mineral of an igneous rock; the cement of a sedimentary rock if deposited directly from solution; or a mineral resulting from metamorphism. The term was introduced by Kalkowsky (1880, p.4). Ant: *allogene.*

authigenesis (au-thi-gen'-e-sis) The process by which new minerals form in place within an enclosing sediment or sedimentary rock during or after deposition, as by replacement or recrystallization, or by secondary enlargement such as quartz overgrowths. Cf: *neogenesis.*

authigenetic (au'-thi-ge-net'-ic) *authigenic.*

authigenic (au-thi-gen'-ic) Formed or generated in place; specif. said of rock constituents and minerals that have not been transported or that crystallized locally at the spot where they are now found, and of minerals that came into existence at the same time as, or subsequently to, the formation of the rock of which they constitute a part. The term, as used, often refers to a mineral (such as quartz or feldspar) formed during *diagenesis* after deposition of the original sediment. Ant: *allogenic.* Cf: *halmeic; autochthonous.* Syn: *authigenous; authigenetic.*

authigenic recharge *Karst* groundwater derived from infiltration through the overlying surface. Cf: *allogenic recharge.*

authigenous (au-thig'-e-nous) (a) *authigenic.* (b) In the classification of pyroclastics, the equivalent of *essential.* Cf: *allothigenous.*

authimorph (au'-thi-morph) A constituent of a metamorphic rock which, in the formation of the new rock, had its outlines or boundaries altered. The term is obsolete.

autoarenite (au-to-ar'-e-nite) A sand produced by crushing due to tectonic movements; the sand-size equivalent of an autoclastic breccia.

autoassociation (au'-to-as-so'-ci-a'-tion) A measure of similarity between a nonnumeric or qualitative sequence and itself, offset by a specified *lag.* Cf: *autocorrelation [stat].*

auto-biotope The total of all environments which one and the same taxonomic unit characteristically inhabits (van Morkhoven, 1966, p.306).

autobreccia (au-to-brec'-cia) A breccia formed by some process that is contemporaneous with the formation or consolidation of the rock unit from which the fragments are derived; specif. a *flow breccia.*

autobrecciation (au'-to-brec'-ci-a'-tion) Formation of an autobreccia; e.g. the fragmentation process whereby portions of the first consolidated crust of a lava mass are incorporated into the still-fluid portion.

autochthon [sed] (au-toch'-thon) A residual deposit produced in place by decomposition.

autochthon [tect] A body of rocks in the footwall of a fault that has not moved substantially from its site of origin. Although not moved from their original site, autochthonous rocks may be mildly to considerably deformed. Adj: *autochthonous.* Ant: *allochthon [tect].* Cf: *parautochthon.* Also spelled: *autochthone.*

autochthone (au-toch'-thone) *autochthon [tect].*

autochthonous (au-toch'-tho-nous) Formed or produced in the place where now found. The term is widely applied, e.g. to a coal or peat that originated at the place where its constituent plants grew and decayed, to rocks that have not been displaced by overthrust faulting, or to a breccia at an explosion crater that remains in its original position, with only minor rotation or translation of the fragments. The term was first used by Naumann (1858, p.657) to designate rock units remaining at the site of their formation; it is similar in meaning to *authigenic,* which refers to constituents rather than whole formations. Ant: *allochthonous.* See also: *parautochthonous.*

autochthonous assemblage A fossil assemblage composed of specimens derived from the local community and preserved in life positions (Kidwell et al., 1986). Cf: *biocoenosis.*

autochthonous mantle Loose debris of rock fragments or soil derived from the underlying bedrock and formed in situ. Cf: *allochthonous mantle; block field.*

autochthony (au-toch'-tho-ny) Accumulation of plant remains in their original environment or in the place of their growth. Ant: *allochthony.* See also: *euautochthony; hypautochthony; in-situ theory.*

autoclast (au'-to-clast) A rock fragment in an *autoclastic rock.*

autoclastic (au'-to-clas'-tic) Applied to fragments formed by mechanical friction during movement of lava, or gravity crumbling of spines and domes.

autoclastic rock (au-to-clas'-tic) A term originated by Smyth (1891) for a rock having a broken or brecciated structure, formed in the place where it is found as a result of crushing, shattering, orogenic forces, or other mechanical processes; e.g. a *fault breccia,* or a brecciated dolomite produced by diagenetic shrinkage followed by recementation. Cf: *cataclastic rock; epiclastic rock.*

autocoel (au'-to-coel) In a *dinoflagellate cyst,* the cavity in the capsule formed by the *autophragm.* Cf: *endocoel; pericoel.*

autoconsequent (au-to-con'-se-quent) Said of a stream whose course is guided by the slopes of material it has deposited; also said of the topographic features (such as waterfalls) developed by such a stream.

autocorrelation [seis] (au'-to-cor'-re-la'-tion) A measure of the dependence of a later portion of a waveform on an earlier portion, or the extent to which future values can be predicted from past values; the *correlation [seis]* of a signal with itself. Cf: *crosscorrelation [seis].*

autocorrelation [stat] A measure of the correlation between adjacent values in a sequence of quantitative observations that are ordered in time or space. The autocorrelation usually is calculated for successively greater lags and shown as a correlogram, a plot of autocorrelation versus lag number. Cf: *autoassociation.*

autocorrelation [stratig] Time-correlation of the same unit in different localities (Henningsmoen, 1961, p.65-66).

autocovariance A measure of the *covariance* (joint variation) between adjacent quantitative values in a sequence of observations that are ordered in time or space. If the values are standardized, the autocovariance is equal to the autocorrelation.

autocyclicity (au'-to-cy-clic'-i-ty) The state of cyclothemic deposition that requires no change in the total energy and material input into a sedimentary system but involves simply the redistribution of these elements within the system (Beerbower, 1964, p.32). It involves such mechanisms of deposition as channel migration, channel diversion, and bar migration. Cf: *allocyclicity.*

autocyst (au'-to-cyst) In dinoflagellate *cysts,* the body formed by the *autophragm* (Williams et al., 2000, p.23).

autoecology (au'-to-e-col'-o-gy) *autecology.*

autogenetic (au'-to-ge-net'-ic) (a) Said of landforms that have developed or evolved under strictly local conditions, without interference by orogenic movements; esp. a topography resulting from the action of falling rains and flowing streams upon land surfaces having free drainage to the sea. (b) Said of a type of drainage (and of its constituent streams) that is determined entirely by the conditions of the land surface over which the streams flow, as a drainage system developed solely by headwater erosion. Syn: *autogenous; autogenic [geomorph].*

autogenic [ecol] (au-to-gen'-ic) Said of an ecologic succession that resulted from factors originating within the natural community and altering its habitat. Cf: *allogenic (c).*

autogenic [geomorph] *autogenetic.*

autogenous (au-tog'-e-nous) *autogenetic.*

autogenous grinding Grinding of ore without the use of media such as steel balls or rods.

autogeosyncline (au'-to-ge'-o-syn'-cline) A *parageosyncline* without an adjoining uplifted area, containing mostly carbonate sediments (Kay, 1947, p.1289-1293); in modern parlance, it is a basin formed in a continental interior, i.e., an *intracratonic basin.* Obsolete. Syn: *residual geosyncline.*

autointrusion [ign] (au'-to-in-tru'-sion) A process wherein the residual liquid of a differentiating magma is injected into rifts formed in the crystallized fraction at a late stage by deformation of unspecified origin.

autointrusion [sed] Sedimentary intrusion of rock material from one part of a bed or set of beds in process of deposition into another part.

autolith (au'-to-lith) (a) An inclusion in an igneous rock to which it is genetically related. Cf: *xenolith.* Syn: *cognate inclusion; cognate xenolith; endogenous inclusion.* (b) In a granitoid rock, an accumulation of Fe-Mg minerals of uncertain origin (Balk, 1937, p.10-12). It may appear as a round, oval, or elongate segregation or clot.

autolysis (au-tol'-y-sis) (a) The process of "self-digestion", as in the albitization of plagioclase in a lava by sodium from the lava itself rather than by newly introduced sodium. (b) Return of a substance to solution, as phosphate removed from seawater by plankton and returned when these organisms die and decay.

automated cartography The process of drawing maps with the aid of computer driven display devices such as plotters and graphics screens. No information processing is implied. Syn: *digital cartography.*

automatic gain control *gain control.*

automatic name placement The automated lettering of objects on a map. Simple automatic name placement systems locate the text in certain predefined positions, relative to graphic objects. Complex automatic name placement systems resolve label conflicts and overlaps to improve the readability of textual information and the aesthetic quality of maps (Krzanowski et al., 1993).

autometamorphism (au'-to-met'-a-mor'-phism) (a) A process of recrystallization of an igneous mineral assemblage under conditions of falling temperature, attributed to the action of its own volatiles, e.g., serpentinization of peridotite or spilitization of basalt. (b) The alteration of an igneous rock by its own residual fluid (Tyrrell, 1926). This process should rather be called *deuteric* because it is not considered to be metamorphic.

autometasomatism (au'-to-met'-a-so'-ma-tism) Alteration of a recently crystallized igneous rock by its own last water-rich liquid fraction, trapped within the rock, generally by an impermeable chilled border. Cf: *autopneumatolysis; autometamorphism.*

automolite (au-tom'-o-lite) A dark-green to nearly black variety of *gahnite.*

automorphic (au-to-mor'-phic) Said of a grain bounded by perfect crystal faces; well-formed. Obsolete. Syn: *euhedral; idiomorphic.*

automorphic-granular A phaneritic igneous texture dominated by euhedral equant grains. Obsolete. Syn: *idiomorphic-granular.* Cf: *hypidiomorphic-granular; xenomorphic-granular.*

Autonomous Underwater Vehicle An unmanned undersea vehicle that is untethered and free to perform underwater tasks in response to pre-programmed directions and can be reprogrammed in response to what it observes. Abbrev: AUV.

autophragm (au'-to-phragm) The wall present in dinoflagellate cysts with a single-layer wall; also the inner wall layer in a two-layer cyst with an *ectophragm* (Williams et al., 2000, p.23-24).

autopiracy (au-to-pi'-ra-cy) Capture of an upper part of a stream by its lower part, as by the cutting-off of a meander, generally resulting in a shortening of its own course.

autopneumatolysis (au'-to-pneu'-ma-tol'-y-sis) Obsolete term for *autometamorphism* involving the crystallization of minerals or the alteration of a rock by gaseous emanations originating in the magma or rock itself.

autopore (au'-to-pore) Tubular *autozooecium* in Paleozoic bryozoans. Obsolete.

autopotamic (au'-to-po-tam'-ic) Said of an aquatic organism adapted to living in flowing fresh water. Cf: *eupotamic; tychopotamic.*

autoradiograph (au-to-ra'-di-o-graph') An image on photographic film of an object that contains radionuclides that emit gamma rays, X-rays, or high-energy beta particles. Less-prefered syn: *radiograph.*

autoregressive series A time series generated from another time series where previous values of the output enter into the determination of a current value.

autoskeleton (au-to-skel'-e-ton) The *endoskeleton* of sponges, consisting of spicules or spongin secreted by the cells. Cf: *pseudoskeleton.*

autospore (au'-to-spore) Nonmotile *spore* occurring within the mother cell.

auto-thanatotope The total area is which the dead specimens of one particular taxon are normally deposited (van Morkhoven, 1966, p.309).

autotheca (au-to-the'-ca) The largest tube of three regularly produced at each budding in the development of a graptolithine colony. It may have contained a female *zooid.* Cf: *bitheca; stolotheca; metatheca.*

autotrophic [ecol] Said of an organism that nourishes itself by using inorganic material to synthesize living matter, an autotroph. Includes both photoautotrophic organisms (e.g., green plants, some protozoans, some microbes) and chemoautotrophic organisms (e.g., some microbes). Cf: *heterotrophic.*

autotrophic [lake] Said of a lake that derives its organic matter mainly from its primary production (Kalff, 2002, p.22, 351-352; Wetzel, 2001, p.137).

autozooecium (au'-to-zo-oe'-ci-um) The skeleton of a bryozoan *autozooid.*

autozooid (au-to-zo'-oid) (a) A fully formed octocorallian *polyp* with eight well developed tentacles and septs. It is the only type of polyp in monomorphic species and it is the major type in dimorphic species. Cf: *siphonozooid.* (b) A feeding bryozoan *zooid.*

autumn ice (au'-tumn) Sea ice in an early stage of formation and not yet affected by lateral pressure; it is relatively salty and is crystalline in appearance.

Autunian (Au-tun'-i-an) European stage: Lower Permian (above Stephanian of Carboniferous, below Saxonian).

autunite (au'-tun-ite) (a) A lemon-yellow or sulfur-yellow radioactive tetragonal mineral: $Ca(UO_2)_2(PO_4)_2 \cdot 10\text{-}12H_2O$. It is isomorphous with torbernite and a number of other hydrous uranyl phosphates and arsenates. Autunite is commonly a secondary mineral and occurs as tabular plates or in micalike scales. Syn: *lime uranite; calcouranite.* (b) A group of isomorphous tetragonal minerals of general formula: $R^{2+}(UO_2)_2(XO_4)_2 \cdot nH_2O$, where R =Ca, Cu, Mg, Ba, Na_2, or other elements, and X = P or As. The group includes minerals such as autunite, torbernite, uranocircite, saléeite, sodium autunite, zeunerite, uranospinite, novacekite, and kahlerite.

AUV *Autonomous Underwater Vehicle.*

auwai (au'-wai) A Hawaiian term for a watercourse or channel, esp. one used for irrigation.

auxiliary (aux-il'-ia-ry) adj. Said of an inflection (any lateral lobe or lateral saddle) of the ammonoid *suture* added later than the first two or three pairs; e.g. "auxiliary lobe" springing from the umbilical lobe and occurring between the second lateral saddle and the *umbilicus* (TIP, 1959, pt. L, p.18). Cf: *accessory [paleont].* n. An auxiliary lobe or an auxiliary saddle.

auxiliary contour *supplementary contour.*

auxiliary fault A minor fault abutting against or branching from a major one. Syn: *branch fault.*

auxiliary mineral In the Johannsen classification of igneous rocks, a light-colored, relatively less abundant mineral such as apatite, muscovite, corundum, fluorite, or topaz.

auxiliary plane The plane in an earthquake focal-mechanism plot that is perpendicular to the actual fault plane, as determined from seismic data for earthquakes.

auxiliary reference section Equivalent to hypostratotype (ISSC, 1994, p.28).

auxotrophic (aux-o-troph'-ic) Said of a microscopic organism that requires certain specific nutrients.

ava Polynesian term for a narrow channel traversing the rim of an atoll or barrier reef that is navigable for boats or shipping (Guilcher, 1988, p.55); they are generally located on the leeward side and are interpreted as sites of karst drainage channels formed during Pleistocene low sea-level stands.

AVA [seis] *amplitude variation with angle.*

available moisture (a-vail'-a-ble) *available water.*

available relief The total relief available for stream dissection in a given area, equal to the vertical distance between the height of the remnants of an original upland surface and the level at which grade is first attained by adjacent streams (Glock, 1932). Cf: *local relief.*

available water Water available to plants; the difference between *field capacity* and *wilting point*. Syn: *available moisture.*

avalanche (av'-a-lanche) A large mass of snow, ice, soil, or rock, or mixtures of these materials, falling, sliding, or flowing very rapidly under the force of gravity. Velocities may sometimes exceed 500 km/hr.

avalanche air blast A very destructive *avalanche wind* preceding a *powder avalanche* or occuring when an avalanche is stopped abruptly, as when it falls vertically onto a valley floor, or displaces a large volume of air by flowing horizontally for a long distance.

avalanche bedding Steeply inclined bedding in barchan and related dune forms, produced by an avalanche of sand down the slip face of the dune.

avalanche boulder tongue An *avalanche talus* of a narrow, railroad-embankment shape, controlled by winter snowbanks on both sides of the avalanche track. The snowbanks confine the avalanche, and when they melt the talus stands above the surrounding slopes with steep edges (Rapp, 1959). Cf: *avalanche talus; avalanche track.*

avalanche chute The track formed by an avalanche. Cf: *avalanche track.*

avalanche class A method of quantifying avalanche size. The USFS classification classes are class 1, sluff, any slide running less than 50 meter slope distance regardless of other dimensions; class 2, small, relative to path size; class 3, medium, relative to path size; class 4, large, relative to path size; class 5, major or maximum, relative to path size.

avalanche cone The mass of material deposited where an avalanche has fallen, consisting of snow, ice, rock, and all other material torn away and carried along by the avalanche.

avalanche cycle A distinct period of time during which a sequence of natural avalanches occur.

avalanche hazard An estimate of how recognized avalanche potential will affect people or structures in their path.

avalanche path The entire area down which avalanches occur; made up of the starting zone, track, and runout zone.

avalanche talus An accumulation of rock fragments of any size or shape, derived from snow and ice mixed with soil and rock debris avalanched from a cliff or rocky slope above. It usually occurs on the downwind side of a ridgetop or arête, in mountains where drifting snow builds cornices that collapse and bring down snow, ice, and rocks. Angle of slope is much less than 30° and the slope profile is concave upward (White 1967, p.237). Cf: *avalanche boulder tongue; scree; avalanche track.*

avalanche track The central channel-like corridor along which an avalanche has moved; it may take the form of an open path in a forest, with bent and broken trees, or an eroded surface marked by pits, scratches, and grooves. Cf: *avalanche chute; avalanche talus; avalanche boulder tongue; devil's slide.*

avalanche wind A high wind or rush of air produced in front of a large landslide or of a fast-moving dry-snow avalanche, and sometimes causing destruction at a considerable distance from the avalanche itself.

Avalonian Orogeny (Av-a-lon'-i-an) An orogenic event affecting rocks of the Avalon platform. The Avalon platform is a microcontinent that was sutured to eastern North America during the Devonian-age Acadian orogeny. Indications of the event, represented by radiometric dates ranging from slightly earlier to later, occur in accreted terranes along the southeastern edge of the Appalachian orogenic belt, and in rocks beneath the sediments of the Atlantic Coastal Plain as far southwest as Florida. It was named by Lilly (1966) and Rodgers (1967) for the Avalon Peninsula, southeastern Newfoundland.

avanturine (a-van'-tu-rine) Var. (error) of *aventurine.*

avatar A population of organisms that interacts with the environment to obtain food and resources.

avelinoite (a-ve-li'-no-ite) *cyrilovite.*

aven (a'-ven) (a) A blind cupola or fissure in the ceiling of a cave. (b) A *vertical shaft,* as viewed from below, open to the surface. Partial syn: *vertical cave.* Etymol: French.

aventurescence (a-ven'-tu-res'-cence) In certain translucent minerals, a display of bright or strongly colored reflections from included crystals. Examples are aventurine quartz and aventurine feldspar (sunstone).

aventurine (a-ven'-tu-rine) (a) A translucent quartz spangled throughout with tiny inclusions of another mineral; a grayish, greenish, brown, or yellowish quartzite that exhibits aventurescence from minute crystals, platelets, flakes, or scales of minerals such as green mica, ilmenite, hematite, and limonite. Syn: *aventurine quartz.* (b) *aventurine feldspar.* The term is also used as an adjective in referring to the brilliant, spangled appearance of a glass or mineral containing gold-colored or shiny inclusions. Misspelled: *avanturine.* Syn: *adventurine.*

aventurine feldspar A variety of feldspar (oligoclase, albite, andesine, or adularia) characterized by a reddish luster produced by fiery, golden reflections or firelike flashes of color from thin disseminated mineral particles (such as flakes of hematite) oriented parallel to structurally defined planes and probably formed by exsolution; specif. *sunstone.* Syn: *aventurine.*

aventurine glass *Goldstone,* a laboratory-grown product.

aventurine quartz *aventurine.*

average (av'-er-age) *arithmetic mean.*

average deviation *mean deviation.*

average discharge As used by the U.S. Geological Survey, the arithmetic average of all complete water years of record of discharge whether consecutive or not.

average igneous rock A theoretical rock whose chemical composition is believed to be similar to the average composition of the outermost layer of the Earth.

average linear velocity An estimate of the mean rate that groundwater molecules flow; the *specific discharge* of groundwater divided by the *effective porosity.*

average seawater Seawater which has the average salinity of the oceans (35). Cf: *salinity; standard seawater.*

average velocity [hydraul] (a) For a stream, discharge divided by the area of a cross section normal to the flow. (b) For groundwater, the volume of groundwater passing through a given cross-sectional area, divided by the porosity of the material through which it moves. Syn: *mean velocity.*

average velocity [seis] The ratio of the distance traversed along a ray path by a seismic pulse to the time required for that traverse. Cf: *seismic velocity; interval velocity.*

averievite A resinous black trigonal mineral: $Cu_5O_2(VO_4)_2 \cdot n(Cu,Cs)Cl$.

avezacite (a-ve-zac'-ite) A dike rock intermediate in composition between *pyroxenite* and *hornblendite,* with amphibole in excess of pyroxene and with ilmenite constituting approximately 20 percent of the rock. Named by Lacroix in 1901 for Avezac in the Pyrenees. Obsolete.

AVHRR *Advanced Very High Resolution Radiometer.*

avicennite (av-i-cen'-nite) A black cubic mineral: Tl_2O_3. Isomorphous with bixbyite.

avicularium (a-vic'-u-lar'-i-um) A polymorph in cheilostome bryozoans, resembling a bird's head. It is commonly a *heterozooid,* having the equivalent of the operculum (the *mandible*) and associated muscles relatively larger than those of an *autozooid.* Pl: avicularia.

aviolite (av-i-o'-lite) A type of *hornfels* whose main constituents are mica and cordierite. Rarely used. Type locality: Monte Aviolo, Italian Alps. Cf: *astite; edolite.*

avlakogene (av-la'-ko-gene) *aulacogen.*

AVO *amplitude variation with offset.*

avogadrite (a-vo-ga'-drite) A translucent colorless orthorhombic mineral: $(K,Cs)BF_4$. Isomorphous with barberiite.

avogadro One avogadro number of items (Emiliani, 1991, p.31-33). Symbol: Av.

Avogadro number A unit of numerical quantity, equal to the number of atomic mass unit in one gram. It is equal to $6.00221367(\pm 36) \times 10^{23}$ (Emiliani, 1991, p.31-33).

avon (a'-von) A river. Etymol: Celtic.

avulsion (a-vul'-sion) (a) A sudden cutting off or separation of land by a flood or by an abrupt change in the course of a stream, as by a stream breaking through a meander or by a sudden change in current whereby the stream deserts its old channel for a new one. Legally, the part thus cut off or separated belongs to the original owner. Cf: *accretion [sed].* (b) Rapid erosion of the shore by waves during a storm (Wiegel, 1953, p.4).

awaruite (a-wa-ru'-ite) A metallic white cubic mineral: Ni_3Fe. *nickel-iron.*

axial angle (ax'-i-al) *optic angle.*

axial canal (a) A longitudinal passageway penetrating columnals of an echinoderm and connecting with the body cavity, and generally but not invariably located centrally. (b) An intraspicular cavity left by decay of an axial filament in a sponge.

axial compression In experimental work with cylinders, a compression applied parallel with the cylinder axis.

axial cross The orthogonal crossing of the six axial filaments at the center of a hexactinellid sponge spicule. It is used esp. where some of the axial filaments are reduced to the area about the center, as in spicules with fewer than six rays.

axial diameter The anterior-posterior thecal diameter in edrioasteroids. Cf: *transverse diameter.*

axial dipole field A hypothetical magnetic field, consisting of an ideal *dipole field* at the Earth's center, with its axis along the Earth's rotational axis. While the actual geomagnetic field does not have this ideal form, it is hypothesized that it would after averaging thousands of years of *secular variation.*

axial elements In crystallography, the ratio of unit distances along crystallographic axes and the angles between these axes.

axial figure An *interference figure* in which one optic axis is centered in the figure. Syn: *centered optic axis figure.*

axial filament An organic fiber about which the mineral substance of a sponge-spicule ray is deposited.

axial filling A deposit of dense calcite developed in the axial region of some fusulinacean foraminifers and formed probably at the same time as excavation of tunnel or foramina and formation of chomata and parachomata (TIP, 1964, pt. C, p.58).

axial furrow (a) One of the two longitudinal grooves bounding the axis of a trilobite. Syn: *dorsal furrow.* (b) A longitudinal groove separating the median lobe or axis of a *merostome* from the pleural area.

axial glide plane An *a-glide, b-glide,* or *c-glide.*

axial increase A type of *increase* (offset formation of corallites) in *coralla* characterized by the appearance of dividing walls between newly formed corallites approximately in position of the axis of the parent corallite.

axial jet A flow pattern characteristic of *hypopycnal inflow,* in which the inflowing water spreads as a cone with an apical angle of about 20° (Moore, 1966, p.88). Cf: *plane jet.*

axial lobe The *axis* of a trilobite.

axial plane [cryst] (a) The plane of the optic axes of an optically biaxial crystal. (b) A crystallographic plane that includes two of the crystallographic axes.

axial plane [struc geol] An *axial surface* that is planar.

axial-plane cleavage Cleavage that is, in the hinge of a fold, parallel to the hinge surface of the fold and whose intersection lineation with bedding is parallel to the fold hinge. Axial-plane cleavage may be everywhere parallel to the fold hinge surface, but normally cleavage fans or changes its orientation systematically with position about the fold. Cf: *bedding-plane cleavage; fan cleavage; transecting cleavage.*

axial-plane separation The distance between axial surfaces of adjacent antiforms and synforms where the folds occur in the same layer or surface.

axial ratio The ratio of the lengths of the crystallographic axes of a crystal, stated in terms of one axis as unity.

axial section A slice bisecting a foraminiferal test in a plane coinciding with the axis of coiling, and intersecting the proloculus.

axial septulum A secondary or tertiary septum located between primary septa of a foraminifer, its plane approximately parallel to the axis of coiling and thus observable in sagittal, parallel, and tangential sections. See also: *primary axial septulum; secondary axial septulum.*

axial shortening The joint-driving mechanism in which the energy release for driving the joint is derived from an applied compression parallel to the direction of crack growth (Engelder and Fischer, 1996). Natural joints produced by this mechanism are sheet fractures.

axial skeleton In the vertebrates, the vertebral column, ribs, braincase, and median fins when present. In common usage the axial skeleton also includes the entire skull and visceral skeleton.

axial splitting crack A crack generated by crack-parallel compression rather than crack-normal tension. This term arises from laboratory experiments on cylinders which were observed to split along the axis when compressed uniaxially or at low confining pressure (Griggs, 1936). Sheet fractures in the near surface are natural examples.

axial stream (a) The main stream of an intermontane valley, flowing in the deepest part of the valley and parallel to its longest dimension. (b) A stream that follows the axis of a syncline or anticline.

axial structure A collective term for various longitudinal structures in the axial region of a *corallite,* whether a solid or spongy rodlike columella or an *axial vortex.* See also: *clisiophylloid; aulos.*

axial surface A surface that connects the hinge lines of the strata in a *fold.* Syn: *hinge surface.*

axial symmetry In structural petrology, a fabric having a unique axis of symmetry, an infinite number of mirror planes passing through that axis, and a single mirror plane normal to it. Syn: *spheroidal symmetry.*

axial trace The intersection of the *axial surface* of a fold with the surface of the Earth or other given surface. Syn: *surface axis.*

axial vortex A longitudinal structure in the axial region of a *corallite,* formed by the twisting together of the inner edges of major septa associated commonly with the transverse skeletal elements.

axil (ax'-il) The distal angle formed between two plant parts, specif. a stalk or petiole and the stem (axis) from which it grows (Swartz, 1971, p.54).

axillary (ax'-il-lar-y) A *brachial plate* supporting two crinoid arm branches.

axinellid (ax-i-nel'-lid) Said of a sponge skeleton built of spiculofibers in which the component spicules are all directed obliquely outward from the axes of the fibers.

axinite (ax'-i-nite) A group name for brown, violet, blue, green, or gray triclinic borosilicate minerals: $(Ca,Mn,Fe,Mg)_3Al_2(OH)[BSi_4O_{15}]$. See: *ferro-axinite; magnesioaxinite; manganaxinite; tinzenite.* Syn: *glass schorl.*

axiolite (ax'-i-o-lite') A spherulitic aggregate in which the needles radiate from a central line or axis, rather than from a point; e.g. a spherulite in a rhyolite composed of welded glass fragments to whose outlines minute acicular crystals or fibers of feldspar are approximately perpendicular and radiate inward; or a subspherical oolith or pisolith in a carbonate sediment, around which acicular needles are axially grouped. Term proposed by Zirkel (1876, p.167) is now obsolete.

axiolitic (ax'-i-o-lit'-ic) Volcanic igneous texture characterized by tiny crystal fibers oriented normal to a central axis. Obsolete.

axiometer (ax-i-om'-e-ter) A device that permits accurate location and measurement of pebble and cobble axes by means of a clamp and a track-mounted caliper (Schmoll and Bennett, 1961).

axis Midline of a conodont element, a process, a cusp, or a denticle (TIP, 1981, pt. W, supp.2).

axis [cryst] (ax'-is) *crystal axis.*

axis [geomorph] (a) The central or dominant region of a mountain chain. (b) A line that follows the trend of large landforms, such as one following the crest of a ridge or mountain range, or of the bottom or trough of a depression.

axis [paleont] (a) The median lobe of a trilobite, consisting of the longitudinal raised portion of the exoskeleton lying between the pleural regions, particularly of the *thorax* and the *pygidium.* Syn: *axial lobe.* (b) The central supporting structure of certain octocorals, such as a spicular structure in a gorgonian, or a horny structure in the order Pennatulacea. (c) A straight line with respect to which an invertebrate is radially or bilaterally symmetrical; e.g. the oral-aboral axis of a corallite, or the axis formed by ambulacral plates in the sheath of radial water vessel in an asterozoan ray (TIP, 1966, pt. U, p.28). (d) An imaginary line through the apex of a gastropod shell, about which the whorls of conispiral and discoid shells are coiled. (e) An imaginary line around which a spiral or cyclical shell of a protist is coiled, transverse to the plane of coiling. (f) In certain algae, the central portion of the thallus, from which branches may originate. (g) In the tetrapods, the second cervical vertebra. Pl: axes.

axis [struc geol] (a) The line which, moved parallel to itself, generates the form of a *cylindrical fold.* (b) The line which, moved parallel to itself, approximates the shape of any fold. (c) The trace of the axial surface of a fold on the fold profile plane. This definition is now rarely used.

axis culmination *culmination.*

axis of divergence The generally vertical or oblique line in a coral septum from which *trabeculae* incline inward and outward. See also: *fan system.*

axis of symmetry *symmetry axis.*

axis of tilt The line along which a tilted photo intersects the plane of an imaginary vertical photo taken with the same camera from the same point, and along which the tilted photo has the same scale as would the vertical photo (ASP, 1975, p.2067).

axoblast (ax'-o-blast) An individual *scleroblast* that produces the axis of certain octocorals, e.g. *Holaxonia.*

axopodium (ax-o-po'-di-um) A semipermanent *pseudopodium,* typically present in radiolarian and heliozoan cells, consisting of an axial rod surrounded by a protoplasmic envelope. Pl: axopodia. Informal syn: axopod.

azimuth [seis] (az'-i-muth) A horizontal angle, measured clockwise, between the north meridian and the arc of the great circle connecting a seismic source, such as the epicenter of an earthquake, and the receiver. When measured at the epicenter, it is called the azimuth from epicenter to receiver; when measured at the receiver, the *back azimuth.*

azimuth [surv] (a) Direction of a horizontal line as measured on an imaginary horizontal circle; the horizontal *direction* reckoned clockwise from the meridian plane of the observer, expressed as the angular distance between the vertical plane passing through the point of observation and the poles of the Earth and the vertical plane passing through the observer and the object under observation. In the basic control surveys of U.S., azimuths are measured clockwise from south, a practice not followed in all countries. Cf: *bearing.* See also: *true azimuth; magnetic azimuth.* (b) *astronomic azimuth.*

azimuthal equal-area projection (az-i-muth'-al) *Lambert azimuthal equal-area projection.*

azimuthal equidistant projection A map projection (neither equal-area nor conformal) in which all points are placed at their true distances and true directions from the central point of the projection. Any point on the globe may be placed at the center, and a straight line radiating from this point to any other point represents the shortest distance (a great circle in its true azimuth from the center), whose length can be measured to scale. Syn: zenithal equidistant projection.

azimuthal projection (a) A map projection in which a portion of the sphere is projected upon a plane tangent to it at the pole or any other point (which becomes the center of the map) and on which the azimuths (directions) of all lines radiating from the central point to all other points are the same as the azimuths of the corresponding lines on the sphere. Distortion at the central point is zero and scale distortions are generated radially from the central point. All great circles through the central point are straight lines intersecting at true angles. (b) A similar projection used in structural petrology. Syn: *zenithal projection.*

azimuthal survey A resistivity or induced-polarization survey in which an area is traversed by a voltage-measuring electrode pair along azimuths away from a fixed current electrode which may be in a drill hole or in contact with a metallic ore. The second current electrode is placed so far away that its location does not affect the measurements. Syn: *radial array.*

azimuthal vertical seismic profiling A vertical seismic profile where the sources are offset in different directions.

azimuth angle (a) The horizontal angle, less than 180 degrees, between the plane of the celestial meridian and the vertical plane containing the observation point and the observed object (celestial body), reckoned from the direction of the elevated pole. In the astronomic triangle (composed of the pole, the zenith, and the star), it is the spherical angle at the zenith. (b) An angle in triangulation or in a traverse, through which the computation of azimuth is carried.

azimuth circle An instrument for measuring azimuth, consisting of a horizontal graduated circle divided into 360 major divisions; e.g. one attached to a compass to show magnetic azimuths.

azimuth compass A magnetic compass, supplied with vertical sights, for measuring the angle that a line on the Earth's surface, or the vertical circle through a heavenly body, makes with the magnetic meridian; a compass used for taking the magnetic azimuth of a celestial body.

azimuth line A term used in radial triangulation for a radial line from the principal point, isocenter, or nadir point of a photograph, representing the direction to a similar point on an adjacent photograph in the same flight line.

azimuth map A surface attribute display of the azimuth of dip of a horizon. Cf: *dip magnitude map.*

azimuth mark A mark set at a significant distance from a triangulation or traverse station to mark the end of a line for which the azimuth has been determined and to serve as a starting or reference azimuth for later use.

Azoic (A-zo'-ic) (a) That earlier part of *Precambrian* time, represented by rocks in which there is no trace of life. (b) The entire Precambrian.

azonal peat (a-zon'-al) *local peat.*

azonal soil In the 1938 soil-classification system, one of the three soil orders. Azonal soils lack distinct genetic horizons and resemble the parent material (USDA, 1938). Not used in current U.S. system of soil taxonomy. Cf: *intrazonal soil; zonal soil; Entisols.* Syn: *immature soil.*

azonate (a-zon'-ate) Said of spores without a zone or a similar (usually equatorial) extension.

azoproite (a-zo-pro'-ite) A black orthorhombic mineral of the *ludwigite* group: $(Mg,Fe^{2+})_2(Fe^{3+},Ti,Mg)BO_5$.

AZRAN Azimuth and range; abbrev. for the components of a polar coordinate system (Krzanowski et al., 1993).

azulite (az'-u-lite) A translucent pale-blue variety of *smithsonite* often found in large masses (as in Arizona and Greece).

azurchalcedony (az'-ur-chal-ced'-o-ny) *azurlite.*

azure quartz (az'-ure) A blue variety of quartz; specif. *sapphire quartz.*

azure stone A term applied to lapis lazuli and to blue minerals such as lazulite and azurite.

azurite (az'-ur-ite) (a) A deep-blue to violet-blue monoclinic mineral: $Cu_3(CO_3)_2(OH)_2$. It is an ore of copper and is a common secondary mineral associated with malachite in the upper (oxidized) zones of copper veins. Syn: *chessylite; blue copper ore; blue malachite.* (b) A semiprecious stone derived from compact azurite and used chiefly for ornamental objects. (c) A trade name for a sky-blue gem variety of *smithsonite.*

azurlite (az'-ur-lite) A variety of chalcedony colored blue by chrysocolla and used as a gemstone. Syn: *azurchalcedony.*

azurmalachite (az-ur-mal'-a-chite) An intimate mixture or intergrowth of azurite and malachite, usually occurring massive and concentrically banded, and used as an ornamental stone.

azygous basal plate (az-y'-gous) The smallest of the three plates of the basal circlet of a blastoid or the basal or infrabasal circlet of crinoids, normally located in the AB interray but sometimes in the DE interray. Cf: *zygous basal plate.*

azygous node A special kind of cusp located directly above the basal cavity of certain conodonts (such as *Palmatolepis* and *Panderodella*); forms part of carina. Syn: *central node.*

B

babefphite (ba-bef'-phite) A white triclinic mineral: $BaBe(PO_4)F$.

Babel quartz (Ba'-bel) A variety of crystalline quartz so named for its fancied resemblance to the successive tiers of the Tower of Babel. Syn: Babylonian quartz.

babingtonite (bab'-ing-ton-ite') A greenish-black triclinic mineral: $Ca_2(Fe^{2+},Mn)Fe^{3+}Si_5O_{14}(OH)$.

babkinite A silvery gray metallic trigonal mineral: $Pb_2Bi_2(S,Se)_3$. A member of the *tetradymite* group.

bacalite (ba'-ca-lite) A variety of amber from Baja California, Mexico.

baccula (bac'-cu-la) A swollen area, generally sagittally elongate, within the *axial furrow* and alongside the posterior of the *glabella* in some trilobites, especially the Agnostidae, Raphiophoridae, Trinucleidae and Telephinidae. It has been commonly referred to as an *ala,* but this term should be restricted to a different structure, in harpetid trilobites. Pl: bacculae.

bache A term used in England for the valley of a small stream.

bacillite (ba-cil'-lite) A rodlike crystallite composed of a group of parallel longulites. Obsolete.

back (a) The ceiling or roof of a mine passageway or stope. (b) That part of a vein or lode between a mine working and the surface or the next mine level above.

back arc The region adjacent to a subduction-related volcanic arc, on the side of the arc opposite the trench and subducting plate. If back-arc tectonic stress is tensional, a back-arc basin opens. If it is compressional, a back-arc fold-thrust belt develops. Partial syn: *remnant arc.*

back-arc basin A basin floored by oceanic crust formed by sea-floor spreading on the opposite side of a volcanic arc from an oceanic trench.

back azimuth The horizontal direction from receiver to epicenter of an earthquake. See also: *azimuth [seis].*

back-barrier marsh A relatively flat, low-lying, intermittently water-covered area with generally halophytic grasses and existing landward of a barrier island.

back bay A small, shallow bay into which coastal streams drain and which is connected to the sea through a pass between barrier islands, as along the coast of Texas. Cf: *front bay; lagoon [coast].*

backbeach *backshore.*

back bearing (a) A *bearing* along the reverse direction of a line; the reverse or reciprocal of a given bearing. If the bearing of line AB is N 58° W, the back bearing (bearing of line BA) is S 58° E. Syn: *reverse bearing; reciprocal bearing.* (b) A term used by the U.S. Public Land Survey system for the reverse direction of a line as corrected for the curvature of the line from the forward bearing at the preceding station.

backbone A ridge serving as the main axis of a mountain; the principal mountain ridge, range, or system of an area.

backdeep An obsolete syn. of *epieugeosyncline,* so named because of its relative position, away from the craton (Aubouin, 1965, p.34).

backfill (a) Earth or other material used to replace material removed temporarily during construction or permanently during mining, such as stones and gravel used to fill pipeline trenches or placed behind structures such as bridge abutments, or waste rock used to support the roof after removal of ore from a stope. Also, material such as sand or dirt placed between an old structure and a new lining, as in a shaft or tunnel. Backfill in excavations may or may not be the material originally removed. (b) The process of refilling an excavation, a mine opening, or the space around a foundation.

background [geochem] The abundance of an element, or any chemical property of a naturally occurring material, in an area in which the concentration is not anomalous (Hawkes, 1958, p.336).

background [radioactivity] *background radiation.*

background radiation The radiation of the environment, e.g., from cosmic rays and from the Earth's naturally radioactive elements. Also, any radiation that is not part of a controlled experiment. Syn: *background [radioactivity].*

backing A counterclockwise change in wind direction, e.g. east to northeast to north. Cf: *veering.*

backland [geomorph] The lowland along either side of a river, behind the natural levee; the part of a floodplain extending from the base of a valley slope and separated from the river by the natural levee.

backland [tect] An obsolete term for *hinterland.*

back lead A *lead [eco geol]* or deposit of coastline sands above the high-water mark. Pron: back leed.

backlimb The less steep of the two limbs of an asymmetrical, anticlinal fold. Cf: *forelimb.*

back-limb thrust fault A thrust fault developed on the more gently dipping "back limb" of an asymmetric anticline, in which the direction of tectonic transport is uplimb and the fault dips in the same direction as the limb but at a steeper angle (Douglas, 1950, p.88-89). See also: *front-limb thrust fault; contraction fault.* Also spelled: backlimb thrust fault.

back marsh *backswamp.*

back radiation *counterradiation.*

back reef The landward side of a reef. The term is often used adjectivally to refer to deposits within the restricted lagoon behind a barrier reef, such as the "back-reef facies" of lagoonal deposits. In some places, as on a platform-edge reef tract, "back reef" refers to the side of the reef away from the open sea, even though no land may be nearby. Cf: *fore reef.* Also spelled: backreef.

back-reef moat *boat channel.*

backrush *backwash.*

backscatter The scattering of radiant energy into the hemisphere bounded by a plane normal to the direction of the incident radiation and lying on the same side as the incident ray; the opposite of *forward scatter.* In radar usage, backscatter refers to the microwave radiation scattered back toward the antenna.

backset bed A cross-bed that dips against the direction of flow of a depositing current; e.g. an inclined layer of sand deposited on the gentle windward slope of a transverse dune, often trapped by tufts of sparse vegetation; or a glacial deposit that formed, as the ice retreated, on the rear slope of an apron or sand plain, consequently dipping toward the retreating ice.

backset eddy A small current revolving in the direction opposite to that of the main ocean current.

backshore (a) The upper or inner, usually dry, zone of the shore or beach, lying between the high-water line of mean spring tides and the upper limit of shore-zone processes; it is acted upon by waves or covered by water only during exceptionally severe storms or unusually high tides. It is essentially horizontal or slopes gently landward, and is divided from the *foreshore* by the crest of the most seaward berm. (b) The area lying immediately at the base of a sea cliff. (c) berm. Syn: *backbeach.*

backshore terrace A wave-built terrace on the backshore of a beach; a berm.

backsight (a) A sight or bearing on a previously established survey point (other than a closing or check point), taken in a backward direction. (b) A reading taken on a level rod held in its unchanged position on a survey point of previously determined elevation when the leveling instrument has been moved to a new position. It is used to determine the height of instrument prior to making a foresight. Syn: *plus sight.* Abbrev: BS. Ant: *foresight.*

back slope (a) A syn. of *dip slope;* the term is used where the angle of dip of the underlying rocks is somewhat divergent from the angle of the land surface. (b) The slope at the back of a scarp; e.g. the gentler slope of a cuesta or of a fault block. It may be unrelated to the dip of the underlying rocks. Also spelled: backslope. Cf: *scarp slope.*

backsteinbau (back'-stein-bau) Nacreous tablets in the mollusks arranged in a brick-and-mortar pattern (Wise, 1970). Etymol: German, "brick structure."

backstripping A technique that progressively removes the effects of sediment loading (including the effects of compaction), eustasy, and paleoenvironment from basin subsidence to obtain tectonic subsidence (Miller et al., 1998).

backswamp A swampy or marshy depressed area developed on a flood plain, with poor drainage due to the natural levees of the river. Syn: *back marsh.*

backswamp deposits Sediment deposited in the flood basin behind the natural levees of a river; typically laminated silt and clay.
backswamp depression A low, usually swampy area adjacent to a leveed river. Syn: *levee-flank depression.*
back thrust A thrust fault in a fold-thrust belt that dips opposite to the general dip of faults and results in displacement back toward the hinterland of the fold-thrust belt.
backwall *headwall.*
backward obliquity *opisthocline.*
backwash The seaward return of water running down the foreshore of a beach following a wave *uprush* or surge; also, the seaward-flowing mass of water so moved. Syn: *backrush.*
backwash mark A term used by Johnson (1919, p.517) for a "criss-cross ridge" developed on a beach slope by the return flow of the uprush. Probable syn: *rhomboid ripple.*
backwash ripple mark A term used by Kuenen (1950, p.292) for a broad, flat ripple between narrow, shallow troughs, formed on a beach by backrush above the level of maximum wave retreat; its crest is parallel to the shoreline.
backwasting (a) Wasting that causes a slope to retreat without changing its declivity. (b) The recession of the front of a glacier. Cf: *downwasting.*
backwater (a) Water that is retarded, backed up, or turned back in its course by an obstruction (such as a bridge or dam), an opposing current, or the movement of the tide; e.g. the water in a reservoir or the water obtained at high tide to be discharged at low tide. Also, the resulting *backwater effect.* (b) A body of currentless or relatively stagnant water, parallel to a river and usually fed from it through a single channel at the lower end by the back flow of the river. Loosely, any tranquil body of water joined to a main stream but little affected by its current, such as the water collected in side channels or flood-plain depressions after it overflowed the lowland. (c) A creek, arm of the sea, or series of connected lagoons, usually parallel to the coast, separated from the sea by a narrow strip of land but communicating with it through barred outlets. (d) A backward current of water. Also, the motion of water that is turned back; a *backwash.*
backwater curve (a) The form of the water surface along a longitudinal profile, assumed by a stream above the point where depth is made to exceed the normal depth by a constriction or obstruction in the channel (ASCE, 1962). Cf: *drop-down curve.* (b) A generic term for all surface profiles of water; esp. *flow profile.*
backwater effect The upstream increase in height of the water surface of a stream, produced when flow is retarded above a temporary obstruction (such as a bridge or dam) or when the main stream overflows low-lying land and backs up water in its tributaries. The effect is also characterized by an expansion in width of the body of water and by a slackening in the current. Syn: *backwater.*
backwearing Erosion that causes the parallel retreat of an escarpment or of the slope of a hill or mountain, or the sideways recession of a slope without changing its declivity; a process contributing to the development of a pediment or pediplain. Cf: *downwearing.*
backweathering Weathering that contributes to slope retreat.
bacon [speleo] A color-banded variety of *drapery* speleothem in caves.
bacon-rind drapery A type of *dripstone* that projects from the cave walls and ceiling in thin translucent sheets and is characterized by parallel colored bands. Syn: *bacon [speleo].* Partial syn: *drapery.*
bacon stone *speckstone.*
bacteria (bac-te'-ri-a) The members of the *domain* Bacteria, which includes all non-*Archaea* prokaryotic microorganisms, ranging from Archean to the present.They represent the most widespread and abundant organisms on Earth, being decomposers, symbionts, and pathogens. They are important in mineral formation and mediate many reactions in the geosphere. Sing: *bacterium.*
bacterial shrub *shrub [sed].*
bacteriogenic (bac-te'-ri-o-gen'-ic) Said of ore deposits formed by the action of anaerobic bacteria, by the reduction of sulfur or the oxidation of metals (Park and MacDiarmid, 1970, p.105-107). See also: *iron bacteria; sulfur bacteria.*
bacterium (bac-te'-ri-um) Sing. of *bacteria.*
bactritoid (bac'-tri-toid) Small cephalopods having thin straight to curved external shells. Sutures simple usually having a midventral lobe. Protoconch calcareous small, commonly globular and constricted at attachment to conch. Range, Silurian to Upper Triassic.
bacula Pl. of *baculum.*
baculate (bac'-u-late) Said of sculpture of palynomorphs consisting of bacula. Cf: *baculum.*
baculite (bac'-u-lite) A crystallite that appears as a dark rod. Obsolete.
baculum (bac'-u-lum) One of the tiny rods, varying widely in size and either isolated or clustered, that make up the ektexine sculpture of pollen or spores. Pl: *bacula.*
baddeleyite (bad'-de-ley-ite') A colorless, yellow, brown, or black monoclinic mineral: ZrO_2. It may contain some hafnium, titanium, iron, and thorium.
Badenian (Ba-den'-ian) Central European stage: lower Middle Miocene, planktonic zone N.9/12 (above Karpatian, below Sarmatian). Obs. syn: *Tortonian* (Vienna Basin usage); *Vindobonian* (upper part).
badenite (bad'-e-nite) A mixture of bismuth, safflorite, and modderite.
badlands (bad'-lands) Intricately stream-dissected topography, characterized by a very fine drainage network with high drainage densities (50-500 km/km^2) and short steep slopes with narrow interfluves. Badlands develop on surfaces with little or no vegetative cover, overlying unconsolidated or poorly cemented clays or silts, sometimes with soluble minerals such as gypsum or halite (Fairbridge, 1968, p.43). They may also be induced in humid areas by removal of the vegetative cover through overgrazing, or by air pollution from sulfide smelting, as at Ducktown, Tennessee (Strahler, 1956, p.630). The term was first applied to an area in western South Dakota, which was called "mauvaises terres" by the early French fur traders.
Baer's law *von Baer's law.*
bafertisite (ba-fer'-ti-site) An yellowish-red monoclinic mineral: $Ba(Fe^{2+},Mn^{2+})_2TiOSi_2O_7(OH,F)_2$.
bafflestone (baf'-fle-stone) A type of *reef rock* or *boundstone* composed of upright, branching colonies, closely to widely spaced but not in contact, separated by fine sediment or coarse skeletal debris. In life, the colonies functioned as a sediment trap or baffle (Embry and Klovan, 1971; Cuffey, 1985). Cf: *branchstone.*
baghdadite (bagh'-dad-ite) A colorless or grayish monoclinic mineral: $Ca_3ZrSi_2O_9$.
Bagnold effect (Bag'-nold) A shear stress between two layers in a fluid, caused by the impact between cohesionless particles that are free to collide with each other during current flow in the absence of applied body force. The stress increases as the diameter squared and the large particles, subjected to highest stress, are forced to the bed surface where the stress is zero. The term was proposed by Sanders (1963, p.174) and named for Ralph A. Bagnold (b. 1896), British geographer, who quantified the influence of collective mutual collisions on current flow (Bagnold, 1956). Syn: *dispersive stress.*
baguette (ba-guette') A *step cut* used for small, narrow rectangular or tapered gemstones, principally diamonds. Also, the gem so cut. Etymol: French, "rod".
bahada (ba-ha'-da) Anglicized var. of *bajada.*
bahamite (ba-ham'-ite) A term proposed by Beales (1958, p.1851-1852) for a shallow marine deposit that consists of limestone grains closely resembling the predominant deposits (described by Illing, 1954) now accumulating in the interior of the Bahama Banks. Obsolete. See also: *grapestone.*
bahianite (ba-hi'-a-nite) A tan, cream, or brown dense adamantine monoclinic mineral: $Al_5Sb^{5+}{}_3O_{14}(OH)_2$.
bahr A body of water as found in the Saharan region; esp. a deep natural spring, often in the form of a small, crater-shaped lake of great depth, as in some oases of eastern Algeria. Etymol: Arabic, "sea". Pl: bahar; bahrs.
Baikalian Orogeny (Bai-kal'-i-an) A name widely used throughout Russia for orogeny that occurred about the time of the Precambrian-Cambrian transition and was completed in the Middle Cambrian. Initial movements are known from the lower Riphean. Different phases are defined by relations within the many late Precambrian stratified sequences of Russia and adjoining counties. Named after Lake Baikal in Siberia.
baikalite (bai'-kal-ite) A dark-green variety of diopside containing iron and found near Lake Baikal, Russia.
baikerinite (bai-ker'-in-ite) A thick, tarry hydrocarbon that makes up about one third of *baikerite* and from which it may be separated by alcohol.
baikerite (bai'-ker-ite) A variety of *ozocerite.* See also: *baikerinite.*
bail-down test A type of *slug test* in which a known volume of water is instantaneously withdrawn from a well and the water level in the well is measured periodically as it recovers.
bailer (bail'-er) (a) A cylindrical steel container with a valve at the bottom. It is attached to a sand wire and used in *cable-tool drilling* for recovering and removing water, cuttings, and mud from the bottom of a well. (b) A plastic or stainless steel cylinder with a foot valve at the base used to develop or sample monitor wells.
baileychlore (bai'-ley-chlore) A yellow-green triclinic chlorite mineral containing essential Zn: $(Zn,Fe^{2+},Al,Mg)_6(Si,Al)_4O_{10}(OH)_8$.
baiyuneboite-(Ce) (bai-yun'-e-bo-ite) An unnecessary name for *cordylite-(Ce)*: $NaBaCe_2F(CO_3)_4$.
bajada (ba-ja'-da [ba-hah'-da]) A broad, continuous *alluvial slope* or gently inclined detrital surface extending from the base of mountain ranges out into and around an inland basin, formed by the lateral coalescence of a series of separate but confluent alluvial fans, and having an undulating character due to the convexities of the component fans; it occurs most commonly in semiarid and desert regions, as in the SW U.S. A bajada is a surface of deposition, as contrasted with a *pediment* (a surface of erosion that resembles a bajada in surface form), and its top often merges with a pediment. Originally, the term was used in New Mexico for the gentler of the two slopes of a cuesta. Etymol: Spanish, "descent, slope". Syn: *bahada; apron [geomorph]; alluvial apron; mountain apron; fan apron; debris apron; alluvial plain; compound alluvial fan; piedmont alluvial plain; piedmont plain; waste plain; piedmont slope; gravel piedmont; alluvial bench.*
bajada breccia A term used by Norton (1917, p.167) for a wedge-shaped, imperfectly stratified accumulation of coarse, angular, poorly sorted rock fragments mixed with mud, formed, for example, in an arid region. Cf: *fanglomerate.*

bajir A term applied in the deserts of central Asia to a lake occupying a flat-bottomed basin separating sand hills or dunes.

Bajocian (Ba-joc'-i-an) A stage of the Standard Global Chronostratigraphic Scale: Middle Jurassic (above Aalenian, below Bathonian). The basal GSSP is at Cabo Mondego in Portugal (Pavia and Enay, 1997).

bakelite A resinoid of phenol and formaldehyde used as an amber substitute.

bakerite (ba'-ker-ite) A monoclinic mineral: $Ca_4B_4(BO_4)(SiO_4)_3(OH)_3{\cdot}H_2O$. It occurs in white or gray compact nodules resembling unglazed porcelain. A member of the *datolite* group.

bakhchisaraitsevite A vitreous colorless, yellowish, or greenish monoclinic mineral: $Na_2Mg_5(PO_4)_4{\cdot}7H_2O$.

baking (bak'-ing) The hardening of rock material by *recrystallization* heat from magmatic intrusions or lava flows. Prolonged baking leads to contact-metamorphic effects. Cf: *caustic metamorphism.*

baksanite A metallic steel gray triclinic mineral: $Bi_6Te_2S_3$. A member of the *tetradymite* group.

Bakuan Ponto-Caspian stage: lower middle Pleistocene (above Apsheronian, below Chaudan). Syn: Bakunian, Bakinian.

balaghat (ba'-la-ghat) A term used in India for a tableland situated above mountain passes.

balance *mass balance.*

balanced cross section A cross section that can be restored, i.e. for which one can construct an undeformed equivalent that comprises a volume of rock equal to that in the deformed-state cross section and where the lengths and thicknesses of strata restore to a coherent undeformed section.

balanced rock (bal'-anced) (a) A large rock resting more or less precariously on its base, formed by weathering and erosion in place. See also: *pedestal rock; rocking stone.* (b) *perched block.*

balance rate The rate of change of mass of a glacier at any time, equal to the difference between *accumulation rate* and *ablation rate.*

balance year The period from the time of minimum mass in one year to the time of minimum mass in the succeeding year for a glacier; the period of time between the formation of one *summer surface* and the next. See also: *net balance.* Syn: *budget year.*

balancing (bal'-anc-ing) The process of systematically distributing corrections through any traverse to eliminate the error of closure and to obtain an adjusted position for each traverse station. Also known as "balancing a survey".

balangeroite (ba'-lan-ger'-o-ite) A brown fibrous monoclinic mineral: $(Mg,Fe^{2+},Fe^{3+},Mn^{2+})_{21}Si_8O_{27}(OH)_{20}$. See: *gageite.*

balas ruby (bal'-as) A pale rose-red or orange gem variety of spinel, found in Badakhshan (or Balascia) province in northern Afghanistan. See also: *ruby spinel.* Syn: *ballas.*

balcony Any projection on the wall of a cave large enough to support at least one person.

bald n. A local term, esp. in the southern U.S., for an elevated, grassy area, as a mountain top or high meadow, that is devoid of trees.

bald-headed anticline Rarely used petroleum geology term for an anticline whose crest was eroded prior to deposition of an unconformably overlying sedimentary unit.

baldite (bald'-ite) The hypabyssal equivalent of analcime-bearing basalt, composed of pyroxene phenocrysts in a groundmass of analcime, augite, and iron oxide. The name (Johannsen, 1938) is derived from Big Baldy Mountain, in the Little Belt Mountains, Montana. Obsolete.

balipholite (ba-liph'-o-lite) A yellowish-white orthorhombic mineral: $BaMg_2LiAl_3(Si_2O_6)_2(OH,F)_8$. A member of the *carpholite* group.

balk A low ridge of earth that marks a boundary line (ASCE, 1954, p.20).

balkanite (bal'-kan-ite) A metallic gray orthorhombic mineral: $Cu_9Ag_5HgS_8$.

ball [sed] A primary sedimentary structure consisting of a spheroidal mass of material; e.g. an armored mud ball, a slump ball, a lake ball.

ball-and-pillow structure A primary sedimentary structure found in sandstones and some limestones, characterized by hemispherical or kidney-shaped masses resembling balls and pillows, and commonly attributed to foundering; the ball- and pillow-form structures of Smith (1916); e.g., a *flow roll* or a *pseudonodule.* See also: *pillow structure [sed]; ball structure.*

ball-and-socket joint In basalt columns, cross-joint surfaces that are concave either upward or downward. Syn: *cup-and-ball joint.*

ballas (bal'-las) (a) A dense, globular aggregate of minute diamond crystals, having a confused radial or granular structure, whose lack of through-going cleavage planes imparts a toughness that makes it useful as an *industrial diamond.* Cf: *bort; carbonado.* (b) A term incorrectly applied to a rounded, single-crystal form of diamond. (c) *balas ruby.*

ballast (bal'-last) (a) Broken stone, gravel, or other heavy material used to provide weight in a ship and therefore improve its stability or control its draft. Jettisoned ballast may be found in samples of marine sediments. Also, similar material used to supply weight in equipment for use in lunar-gravity studies. (b) Gravel, broken stone, expanded slag or similar material used as a foundation for roads, esp. that laid in the roadbed of a railroad to provide a firm bed for the ties, distribute the load, and hold the track in line, as well as to facilitate drainage.

ball clay A plastic, kaolinitic claystone with minor to abundant organic matter producing high green strength and high-fired shrinkage in ceramic bodies and usually firing to white. The term comes from the old English practice of rolling this clay into ~25-cm balls weighing 13-22 kg (Burst and Hughes, 1994).

ball coal Coal occurring in spheroidal masses that are probably formed by jointing. Not to be confused with *coal ball.* Syn: *pebble coal.*

balled-up structure A term used by O.T. Jones (1937) for a knot, several centimeters to a few meters in diameter, of highly contorted silty material lying isolated in mud and produced by subaqueous slump. See also: *ball structure.*

ball ice Sea ice, either *frazil ice* or *pancake ice,* consisting of numerous soft, spongy, floating spheres (2.5-5 cm in diameter) shaped by waves, and usually occurring in belts.

ball ironstone A sedimentary rock containing large argillaceous nodules of ironstone.

ballistic magnetometer (bal-lis'-tic) A type of magnetometer that uses the transient voltage induced in a coil when the magnetized specimen is moved relative to the coil or vice versa.

ballistic sedimentation The process by which the *ejecta blanket* of *impact craters* on airless bodies is deposited. Since these ejecta were originally thought to be emplaced by *base surges,* it was unclear how deposition from density currents could occur in a vacuum. Oberbeck (1975) demonstrated that a purely ballistic mechanism could account for ejecta emplacement on the Moon.

ball jasper (a) Jasper showing a concentric banding of red and yellow. (b) Jasper occurring in spherical masses.

ballon (bal-lon') A rounded, dome-shaped hill, formed either by erosion or by uplift. Etymol: French, "balloon".

ballstone A nodule or large rounded lump of rock in a stratified unit; specif. an ironstone nodule in a coal measure. Syn: *ball [sed].*

ball structure (a) A ball-shaped primary sedimentary structure. See also: *ball-and-pillow structure; balled-up structure.* (b) The structure of ball coal.

bally (bal'-ly) A term used in northern California for a mountain. It is thought to be a corruption of "buli", an American Indian term for mountain. Syn: *bolly.*

balm A concave cliff or precipice, forming a shelter beneath the overhanging rock; a cave. Etymol: Celtic.

balneology (bal-ne-ol'-o-gy) The science of the healing qualities of baths, esp. with natural mineral waters.

baltimorite (bal-ti-mor'-ite) An unnecessary name for a grayish-green and silky, fibrous, or splintery serpentine mineral found near Baltimore, Md.; *antigorite.*

balyakinite (bal-ya'-ki-nite) A bluish-green orthorhombic mineral: $CuTeO_3$.

bambollaite (bam-bol'-la-ite) A metallic brownish-gray tetragonal mineral: $Cu(Se,Te)_2$. Also spelled: *bombollaite.*

bamfordite A yellowish green triclinic mineral: $Fe^{3+}Mo_2O_6(OH)_3{\cdot}H_2O$. An oxidation product of molybdenite.

banakite (ba'-nak-ite) A *trachyandesite* rock composed of olivine and clinopyroxene phenocrysts in a groundmass of labradorite with alkali feldspar rims, olivine, clinopyroxene, and some leucite, or possibly quartz. Banakite grades into *shoshonite* with an increase in the olivine and clinopyroxene and with less alkali feldspar, and into *absarokite* with more olivine and clinopyroxene. It was named by Iddings in 1895 from the Bannock (or Robber) Indians.

banalsite (ba-nal'-site) A white orthorhombic mineral related to the *feldspar* group: $BaNa_2Al_4Si_4O_{16}$.

banana hole A term used in the Bahamas for a sinkhole produced by collapse of a shallow phreatic cave formed at the top of a past higher fresh-water lens. The hole is used to cultivate specialty crops such as bananas (Harris, et al., 1995). Cf: *collapse sinkhole; flank-margin cave.*

banatite (ban'-a-tite) A *quartz diorite* containing alkali feldspar. The term has been variously defined since its original usage. Obsolete. (Johannsen, 1939, p.242).

banco (ban'-co) A term applied in Texas to the part of a stream channel or floodplain cut off and left dry by a change in the stream course; an *oxbow lake.* Etymol: Spanish, "sandbank, shoal, bench".

band [coal] *dirt band [coal].*

band [geophys] (a) A range of frequencies such as those passed or rejected by a filter. Measurements are usually made between points where the amplitude is down by 3 dB (or down 30%) from the peak value; see also *bandwidth.* (b) The names given to the frequency ranges of the electromagnetic spectrum. (c) Landsat data channel.

band [glaciol] *foliation [glaciol].*

band [phys] A frequency or wavelength interval, e.g. the infrared band, comprising electromagnetic radiation ranging in wavelength from 0.7 μm to 1.0 mm.

band [stratig] (a) A thin stratum with a distinctive lithology or color. (b) A widespread thin stratum that is useful in correlating strata. (c) A deprecated term for any bed or stratum of rock (BSI, 1964, p.5).

bandaite (ban'-da-ite) An *andesite* containing labradorite or bytownite. It was named by Iddings in 1913 for Bandai San, Japan. Obsolete.

banded (band'-ed) Said of a vein, sediment, or other deposit having alternating layers that differ in color or texture and that may or may not differ in mineral composition, e.g. *banded iron formation.* Cf: *ribbon [ore dep].*

banded agate An *agate* whose various colors (principally different tones of gray, but also white, pale and dark brown, bluish, and other shades) are arranged in delicate parallel alternating bands or stripes of varying thicknesses. The bands are sometimes straight but more often wavy or zigzag and occasionally concentric; they may be sharply demarcated or grade imperceptibly into one another. Banded agate is formed by deposits of silica (from solutions intermittently supplied) in irregular cavities in rocks, and it derives its concentric pattern from the Liesegang phenomenon. Cf: *onyx.*

banded coal Heterogeneous coal, containing bands of varying luster. Banded coal is usually bituminous although banding occurs in all ranks of coal. Cf: *banded ingredients.* See also: *bright-banded coal; dull-banded coal; intermediate coal.*

banded constituents *banded ingredients.*

banded gneiss A regularly layered metamorphic or composite rock with alternating layers of different composition or texture. The thickness of individual layers is usually not more than a few meters (Dietrich, 1960, p.36).

banded hematite quartzite A term used in India and Australia for *iron formation.* See also: *banded quartz-hematite.*

banded ingredients Vitrain, clarain, fusain, and durain as they appear as macroscopically visible and separable bands of varying luster in *banded coal.* Cf: *lithotype.* Syn: *banded constituents; primary-type coal; rock type [coal].*

banded iron formation *Iron formation* that shows marked banding, generally of iron-rich minerals and chert or fine-grained quartz. Abbrev: *bif.*

banded ironstone A term used in South Africa for *iron formation* consisting essentially of iron oxides and chert occurring in prominent layers or bands of brown or red and black. This usage of the term *ironstone* is at variance with that applied in the U.S. and elsewhere.

banded peat Peat that consists of alternating bands of plant debris and sapropelic matter. Cf: *mixed peat; marsh peat.*

banded quartz-hematite A syn. of *itabirite.* See also: *banded hematite quartzite.*

banded structure An outcrop feature developed in igneous and metamorphic rocks as a result of alternation of layers, stripes, flat lenses, or streaks differing conspicuously in mineral composition and/or texture. See also: *banding [ign]; banding [meta].*

banding [cart] A technique of progressively bringing together contours into a single *carrying contour* to indicate vertical escarpments, such as cliffs and vertical embankments. It is the favoured technique when the spacing between the top and bottom contours of the escarpment would be less than 0.005 inches (10 feet at 1:24,000 scale; USGS, 1996, p.7A-22). Cf: *feathering [cart].*

banding [glaciol] *foliation [glaciol].*

banding [ign] Intersection of igneous or metamorphic layering with an outcrop surface. It may be produced by such processes as flow of heterogeneous material (e.g. *flow layering* of rhyolites) or successive deposition of layers of different materials (cf: *phase layering*). Although the term strictly describes the appearance of a two-dimensional feature on a rock exposure, it is commonly used for any "roughly planar heterogeneity in igneous rocks, whatever its origin" (Wager and Brown, 1967, p.5).

banding [meta] A *banded structure* of metamorphic rocks consisting of nearly parallel bands having different textures or minerals or both. It may be produced by incomplete segregation of constituents during recrystallization or may be inherited from bedding in sediments or from layering in igneous rocks. Cf: *ribbon [petrology].*

banding [sed] Thin bedding produced by deposition of different materials in alternating layers, and conspicuous in a cross-sectional appearance of laminated sedimentary rocks; e.g. *ribbon banding.* Also, color banding formed during diagenesis, commonly crosscuts the primary structures. Cf: *Liesegang rings.*

band interleaving format A method of organizing a raster data base such that the data for different layers are stored together in specific patterns such as by row or by pixel (Krzanowski et al., 1993). Abbrev: BIL.

band sequential format A method of organizing a raster data base such that each complete layer (often all the data for one spectral band) is stored in contiguous sequence (Krzanowski et al., 1993). Abbrev: BSQ.

band spectrum A spectrum that appears to be a number of bands because the array of intensity values occurs over broad ranges of wavelengths of the ordering variable. The term is also applied to lines in a *line spectrum* that are grouped closely and cannot be resolved by available instruments, so that the groups of lines appear as bands. An optical band spectrum arises mainly in molecular transitions.

bandwidth (band'-width) (a) A range of wavelengths or frequencies, such as that to which a detector responds. (b) The frequencies at which the power drops to half the peak power (3 dB). (c) The rate at which a computer resource can carry (accept or deliver) data, usually expressed in bytes per second or bits per second.

bandylite (ban'-dy-lite) A dark-blue tetragonal mineral: $CuB(OH)_4Cl$.

bangar (ban'-gar) *bhangar.*

bank [coast] (a) An embankment; a *sandbank.* (b) A *barrier island* along the Atlantic coast of the U.S., composed of sand, and forming a barrier between the inland lagoon or sound and the ocean. (c) A *shoal,* e.g. Georges Bank. (d) An obsolete term for a *seacoast.* (e) The rising ground bordering a sea.

bank [geomorph] (a) A steep slope or face, as on a hillside, usually of sand, gravel, or other unconsolidated material; rarely of bedrock. (b) A term used in northern England and in Scotland for a hill or a hillside. (c) A term used in South Africa for a moderately high scarp (up to 500 m) consisting of resistant rock layers and forming a high hill or a low mountain, often occurring in groups of two or more with broad longitudinal valleys in between. The term is most often used in the plural: *banke.* Etymol: Afrikaans.

bank [lake] (a) The sharply rising ground, or abrupt slope, bordering a lake. (b) The scarp of a littoral shelf of a lake. (c) Shoal bottom of a lake.

bank [mining] A coal deposit; the surface or face of a coal deposit that is being worked.

bank [oceanog] A mound-like or ridge-like submerged elevated area on the sea floor of modest to substantial extent, usually close to sea level (e.g., Bahama banks). See also: *mudbank; shell bank; lithoherm.* Syn: *organic bank.*

bank [sed] A moundlike or ridgelike limestone deposit consisting of skeletal matter, largely unbroken, formed in place by organisms such as crinoids or brachiopods that lack the ecologic potential to erect a rigid, wave-resistant structure (Nelson et al., 1962, p.242). Banks therefore lack the structural framework of organic reefs, and may develop in cold or deep waters far from true reefs. See also: *mudbank; shell bank; lithoherm.* Syn: *organic bank.*

bank [streams] The sloping margin of, or the ground bordering, a stream, and serving to confine the water to the natural channel during the normal course of flow. It is best marked where a distinct channel has been eroded in the valley floor, or where there is a cessation of land vegetation. A bank is designated as right or left as it would appear to an observer facing downstream.

bankable Said of a feasibility study for a proposed mine that is sufficiently detailed and documented to warrant consideration by a bank or other source of funding.

bank atoll A ring-shaped coral reef or reef complex rising from an oceanic plateau (e.g., Queensland Plateau) or submarine bank (e.g., Maldive and Laccadive Islands, site of the original term *atollon.*)

bank-barrier reef (a) A *coral reef* consisting of a coral-algal frame capping a carbonate rubble-and-sand bar that previously accumulated on a shallow limestone platform. (b) An organic reef developed on the margins of "banks" (Bahamas) and carbonate platform margins (Florida).

bank caving The slumping or sliding into a stream channel of masses of sand, gravel, silt, or clay, caused by a highly turbulent current undercutting or undermining the channel wall on the outside of a stream bend. See also: *caving [geomorph].*

banke Plural of *bank [geomorph].*

banker (bank'-er) An Australian term for a stream flowing full to the top of its banks.

banket A general term for a compact, siliceous conglomerate of vein-quartz pebbles of about the size of a pigeon's egg, embedded in a quartzitic matrix. The term was originally applied in the Witwatersrand area of South Africa to the mildly metamorphosed gold-bearing conglomerates containing rounded, oblong quartz pebbles and resembling an almond cake made by the Boers. Etymol: Afrikaans, "a kind of confectionery".

bankfull discharge (a) A discharge that fills a channel to *bankfull stage.* (b) A discharge that recurs, on average, every 1-2 years.

bankfull stage The elevation of the water surface of a stream flowing at *channel capacity.* Discharge at this stage is called *bankfull discharge.*

bank gravel Gravel found in natural deposits, usually more or less intermixed with sand, silt or clay. Syn: *pit run.*

bank-inset reef A coral reef situated on a submarine flat (such as the shelf of a continent or island, or an offshore bank), well within its locally unrimmed outer margin (Kuenen, 1950, p.426). Cf: *bank reef.*

bank reef Any large reef growth, generally irregular in shape, developed over submerged highs (whether of tectonic or other origin), and more or less completely surrounded by water too deep to support the growth of reef-forming organisms (Henson, 1950, p.227). See also: *bank-inset reef; shelf-edge reef.*

bank stability The resistance of a streambank to changes in slope and contour. It can be attained by benching, growth of vegetation, and artificial protections such as retaining walls, drainage systems, and fences. Cf: *slope stability.*

bank storage Water absorbed and retained in permeable material adjacent to a stream during periods of high water and returned as seepage or flow during periods of low water. Syn: *lateral storage.*

bank subsidence *caving [geomorph] (a).*

bannermanite (ban'-ner-man-ite') A sub-metallic black monoclinic mineral occurring in fumaroles: $(Na,K)_xV^{4+}{}_xV^{5+}{}_{6-x}O_{15}(x=0.7)$. Classed as a "vanadium bronze".

bannisterite (ban'-nis-ter-ite') A dark brown platy monoclinic mineral: $KCaMn_{21}(OH)_{16}[(Si,Al)_{32}O_{76}]{\cdot}12H_2O$.

banquette (ban-quette') An embankment at the toe of the land side of a levee, constructed to protect the levee from sliding when saturated with water.

baotite (ba'-o-tite) A light-brown to black tetragonal mineral: $Ba_4(Ti,Nb)_8Si_4O_{28}Cl$.

bar [coast] A generic term for any of various elongate offshore ridges, banks, or mounds of sand, gravel, or other unconsolidated material, submerged at least at high tide, and built up by the action of waves or currents on the water bottom, esp. at the mouth of a river or estuary, or at a slight distance from the beach. A bar commonly forms an obstruction to water navigation. Cf: *barrier [coast].*

bar [eco geol] (a) Any band of hard rock, such as a vein or dike, crossing a lode. (b) A hard band of barren rock crossing a stream bed. (c) A mass of inferior rock in a workable deposit of granite. (d) A fault across a coal seam or orebody. (e) A banded ferruginous rock; specif. *jasper bar; jaspilite.*

bar [geophys] A unit of pressure, 10^5 pascals or 10^5 N/m^2; approximately one atmosphere or 14.5 psi.

bar [paleont] (a) The slender shaft of a compound conodont element, commonly bearing denticles, and having an anterior, lateral, or posterior position. Also, any conodont with discrete denticles and with a single large denticle near one end. Cf: *blade [paleont].* (b) An elongate, thin skeletal element in archaeocyathids, rectangular in section.

bar [streams] A ridgelike accumulation of sand, gravel, or other alluvial material formed in the channel, along the banks, or at the mouth, of a stream where a decrease in velocity induces deposition; e.g. a *channel bar* or a *meander bar.* See also: *river bar.*

barachois (ba-ra-chois') A French term used esp. in the NW of Canada for a *lagoon* or *beach pond* separated from the sea by a sandbar or gravel beach. Var. spelling: *barrachois.*

bararite (ba'-rar-ite) A white trigonal low-temperature mineral: $(NH_4)_2SiF_6$. Cf: *cryptohalite.*

baratovite (ba'-ra-tov'-ite) A white monoclinic, pseudo-hexagonal mineral: $KLi_3Ca_7(Ti,Zr)_2Si_{12}O_{36}F_2$.

barbed drainage pattern A drainage pattern produced by tributaries that join the main stream in sharp boathook bends that point upstream; it is usually the result of stream piracy that has reversed the of flow of the main stream.

barbed tributary A stream that joins the main stream in an upstream direction, forming a sharp bend that points upstream and an acute angle that points downstream at the point of junction.

barberiite A vitreous colorless orthorhombic mineral, formed as a fumarolic encrustation: $(NH_4)BF_4$. The ammonium analogue of avogadroite.

barbertonite (bar'-ber-ton-ite') A rose-pink to violet hexagonal mineral: $Mg_6Cr_2(CO_3)(OH)_{16}{\cdot}4H_2O$. It is dimorphous with stichtite, and a member of the *manasseite* group.

barbierite (bar-bier'-ite) A name formerly applied to a hypothetical high-temperature monoclinic form of albite, and later changed to *monalbite.* The name was originally applied to a mineral later shown to be finely twinned microcline with about 20% of unmixed albite.

barbosalite (bar-bo'-sa-lite) A dark blue-green to black monoclinic mineral: $Fe^{2+}Fe^{3+}_2(PO_4)_2(OH)_2$. It is also known as "ferrous ferric lazulite". A member of the *lazulite* group.

bar-built estuary An *estuary* formed by the growth of a *spit* or migration of a *barrier island* across its seaward side; e.g., Pamlico Sound, Galveston Bay.

barcan (bar'-can) *barchan.*

barcenite (bar'-cen-ite) A mixture of *roméite* and *metacinnabar.*

barchan (bar'-chan [bar'-kan]) An isolated, crescent-shaped sand dune lying transverse to the direction of the prevailing wind, with a gently sloping convex side facing the wind, wings or horns of the crescent pointing downwind, and a steep concave leeward slope inside the horns. It can grow to heights of >30 m and widths up to 350 m from horn to horn. A barchan forms on a flat, hard surface where the sand supply is limited and the wind is constant with only moderate velocity. It is among the commonest of the dune types, characteristic of very dry desert regions the world over. Etymol: Russian *barkhan,* from Kirghiz; originally a Turkish word meaning "sand hill" in central Asia. Cf: *parabolic dune; snow barchan.* Syn: *barchan dune; barcan; barchane; barkan; barkhan; horseshoe dune; crescentic dune.*

barchan dune *barchan.*

barchane (bar-chane') A French variant of *barchan.*

bar diggings A term applied in the western U.S. to diggings for gold or other precious minerals located on a bar or in the shallows of a stream, and worked when the water is low.

bare ice Ice without a snow cover.

bare karst *naked karst.*

barentsite (bar'ents-ite) A colorless triclinic mineral: $H_2Na_7Al(CO_3)_4F_4$.

bar finger An elongated, lenticular body of sand underlying, but several times wider than, a distributary channel in a bird-foot delta; it is produced and lengthened by the seaward advance of the *lunate bar* at the distributary mouth. Examples in the Mississippi River delta are as much as 30 km long, 8 km wide, and 80 m thick. Syn: *bar-finger sand.*

bar-finger sand (a) A deposit of sand in the form of a bar finger. (b) *bar finger.*

bariandite (ba-ri-and'-ite) A dark greenish-brown monoclinic mineral: $Al_3(V^{5+},V^{4+})_{40}O_{100}{\cdot}45H_2O$. Classed as a "vanadium bronze".

baricite (bar'-i-cite) A colorless to pale blue monoclinic mineral: $(Mg,Fe^{2+})_3(PO_4)_2{\cdot}8H_2O$. A member of the *vivianite* group.

baric type (ba'-ric) The inferred geothermal gradient corresponding to the different *metamorphic facies series.* Miyashiro (1973) recognizes three baric types: low-, medium-, and high-pressure.

barinophytalean *barinophyte.*

barinophyte (bar-i'-no-phyte) A member of a group of heterosporous Devonian *lycophytes* with micro- and megaspores in the same sporangium.

bariomicrolite (bar'-i-o-mic'-rol-ite) A variously colored cubic mineral: $Ba(Ta,Nb)_2(O,OH)_7$. A member of the *pyrochlore* group; originally called *rijkeboerite.*

bario-orthojoaquinite (bar'-i-o-or-tho-joa-quin'-ite) A yellow-brown orthorhombic mineral: $Ba_4Fe_2Ti_2O_2[Si_4O_{12}]_2{\cdot}H_2O$. The barium analogue of orthojoaquinite, and a member of the joaquinite group.

bariopyrochlore (bar'-i-o-py'-ro-chlore) A yellowish-gray cubic mineral: $(Ba,Sr)_2(Nb,Ti)_2(O,OH)_7$. A member of the *pyrochlore* group. Originally called *pandaite.*

bariosincosite A pale green tetragonal mineral: $BaVO_2(PO_4)_2{\cdot}4H_2O$. The barium analogue of *sincosite.*

barite (bar'-ite) (a) A white, yellow, or colorless orthorhombic mineral: $BaSO_4$. Strontium and calcium are often present. Barite occurs in tabular crystals, in granular form, or in compact masses resembling marble, and it has a specific gravity of 4.5. It is used in paint, drilling mud, and as a filler for paper and textiles, and is the principal ore of barium. Syn: *barytes; heavy spar; cawk.* (b) A group of sulfates with the general formula *R*(S,Se)O4, where *R* = Ba, Pb or Sr.

barite dollar A term used esp. in Texas and Oklahoma for a small disk-shaped mass of barite formed in a sandstone or sandy shale.

barite rosette A rosette consisting of a cluster or aggregate of tabular sand-filled crystals of barite, usually forming in sandstone. Syn: *petrified rose.*

barium-pharmacosiderite A yellowish brown to reddish brown tetragonal mineral: $Ba_{0.5}Fe^{3+}_4(AsO_4)_3(OH)_4{\cdot}5H_2O$. The barium analogue of pharmacosiderite, and a member of the *pharmacosiderite* group.

barium-zinc alumopharmocosiderite A vitreous colorless or yellow tetragonal mineral: $(Ba,K)_{0.5}(Zn,Cu)_{0.5}(Al,Fe)_4(AsO_4)_3{\cdot}5H_2O$. The Ba-Zn analogue of pharmacosiderite, and a member of the *pharmacosiderite* group. Not approved by the International Mineralogical Association.

bark Outermost tissues of woody stems, external to the vascular cambium.

barkan (bar'-kan) *barchan.*

barkevikite (bar-ke-vik'-ite) A brownish or velvet-black monoclinic mineral of the amphibole group, near arfvedsonite in composition and appearance. An unnecessary name for an iron-bearing or ferropargasitic hornblende.

barkhan (bar'-khan) Var. of *barchan.* Etymol: Russian.

bar lake *beach pond.*

barn Unit of neutron-capture cross-section; 1 barn = 10^{-24} cm^2. Usually designated by σ and derived from the apparent size of the nucleus (i.e., barn door).

barnesite (barnes'-ite) A dark red monoclinic mineral: $(Na,Ca)_2V^{5+}_6O_{16}{\cdot}3H_2O$. It is the sodium analogue of hewettite and is classed as a "vanadium bronze".

baroclinic (bar-o-clin'-ic) Property of a fluid whereby the surfaces of constant pressure are inclined to surfaces of constant density. Such a fluid must be in motion and the velocity varies with depth.

barograph (bar'-o-graph) A *barometer* that makes a continuous record of changes in atmospheric pressure. It usually consists of a special type of *aneroid barometer.*

barometer (ba-rom'-e-ter) An instrument that is used to measure atmospheric pressure. It may be either a *mercury barometer* or an *aneroid barometer.* See also: *barograph.*

barometric altimeter (bar-o-met'-ric) An instrument that indicates elevation or height above sea level, or above some other reference level, by measuring the weight of air (atmospheric pressure) on the instrument. Also called *aneroid barometer.* Syn: *pressure altimeter.*

barometric efficiency The ratio of the fluctuation of water level in a well to the change in atmospheric pressure causing the fluctuation, expressed in the same units such as feet of water. Symbol: B. Cf: *tidal efficiency.*

barometric elevation An elevation above mean sea level estimated by the use of a barometer to measure the difference in air pressure between the point in question and a reference base of known value.

barometric leveling A type of indirect leveling in which differences of elevation are determined from differences of atmospheric pressure observed with altimeters or barometers. Cf: *thermometric leveling.*

barometric pressure *atmospheric pressure.*

barometric tendency *pressure tendency.*

barophilic (bar-o-phil'-ic) Referring to marine organisms that live under conditions of high pressure. Etymol: Greek, "weight" and "to love."

baroque (ba-roque') adj. Said of a pearl, or of a tumble-polished gem material, that is irregular in shape. n. A baroque pearl.

baroque dolomite *saddle dolomite.*

baroseismic storm (bar-o-seis'-mic) Strong microseisms caused by barometric changes.

barotropic (bar-o-trop'-ic) Property of a fluid whereby surfaces of constant pressure are parallel to surfaces of constant density. If the surfaces are level, then the fluid is at rest. If they are not level, the fluid is in uniform motion, the velocity is independent of depth.

bar plain A term introduced by Melton (1936, p.594, 596) for a relatively smooth flood plain with neither a low-water channel nor an alluvial cover, and characterized by a network of elongate and irregularly sized "bars" built from "the tractional and suspended load in the declining stages of the last flood". Cf: *meander plain; covered plain.*

barquillite A metallic gray tetragonal mineral: Cu_2CdGeS_4. A member of the luzonite group.

barrachois (bar-ra-chois' [bar-ra-shwa]) Var. spelling of *barachois.*

barrage (bar-rage') A syn. of *dam,* archaic in the U.S. but used elsewhere, particularly in Europe, North Africa, and parts of Asia. Etymol: French, "a barring".

barranca (bar-ran'-ca) (a) Var. of barranco. (b) A large rift in a *piedmont glacier* or in an *ice shelf.* Cf: *donga [glaciol].*

barranco [geomorph] (bar-ran'-co) (a) A term used in the SW U.S. for a deep, steep-sided, usually rock-walled ravine, gorge, or small canyon, and for a deep cleft, gully, or arroyo made by a heavy rain. (b) A term used in the SW U.S. for a steep bank or a precipice; in New Mexico it is equivalent to cliff. Etymol: Spanish. Cf: *quebrada.* Syn: *barranca.*

barranco [volc] A deep, steep-sided drainage valley on the slope of a volcanic cone, formed by erosion and coalescence of smaller channels. Barrancos form a radiating pattern around a volcanic cone. Etymol: Spanish "barranca", "ravine".

barrandite (bar-ran'-dite) An unnecessary name for a mineral intermediate in composition between *strengite* and *variscite.*

barred basin *restricted basin.*

barred basin theory A theory advanced by Ochsenius in 1877 to account for thick deposits of evaporites. It assumes a lagoon separated from the ocean by a bar, in an arid or semiarid climate. As water is lost by evaporation, additional water of normal salinity flows in from the ocean. Because some water in the lagoon is evaporating, the salinity there constantly increases, and it finally reaches a point where gypsum, salt, and other evaporites are deposited.

barrel (bar'-rel) As used in the petroleum industry, a volumetric unit of measurement equivalent to 42 U.S. gallons (158.76 liters).

barrel copper Pieces of native copper occurring in sizes large enough to be extracted from the gangue, and of sufficient purity to be smelted without mechanical concentration. Syn: *barrel work.*

barrels of oil equivalent A unit of energy equivalent to 5,604 ft^3 natural gas; 5.8 x 10^6 BTU; 1,700 kWh; 0.22 ton bituminous coal. Abbrev: BOE.

barrel work A syn. of *barrel copper,* used in the Lake Superior mining region.

Barremian (Bar-re'-mi-an) Stage of the Standard Global Chronostratigraphic Scale: Lower Cretaceous (above Hauterivian, below Aptian).

barren (bar'-ren) A word, usually used in the plural, for rugged and unproductive land that is devoid of significant vegetation compared to adjacent areas because of environmental factors such as adverse climate, poor soil, or wind.

barren interval A stratigraphic interval without fossils or devoid of all diagnostic fossils or representatives of the taxonomic categories on which the remainder of the stratigraphic section is subdivided (ISSC, 1994, p.56). Syn: *barren zone.*

barren interzone An interval lacking in fossils between successive *biozones.* It is referred to informally by reference to the adjacent biozones, e.g. Exus parvus to Exus magnus barren interzone (ISG, 1976, p.49); term discarded in ISSC, 1994. Cf: *barren intrazone.*

barren intrazone A barren interval of substantial thickness within a *biozone,* e.g. the barren intrazone near the top of the Exus albus assemblage zone (ISG, 1976, p.49); term discarded in ISSC, 1994. Cf: *barren interzone.*

barren zone *barren interval.*

barrerite (bar'-rer-ite) A colorless or pale pink orthorhombic zeolite mineral: $(Na,K,Ca)_2Al_2Si_7O_{18}·7H_2O$.

barrier [coast] (bar'-ri-er) An elongate, offshore ridge or mass, usually of sand, rising above the high-tide level, generally extending parallel to, and at some distance from, the shore, and separated from it by some type of salt marsh, coastal lagoon, or bay. It is built up by the action of waves and currents. Examples include *barrier beach, barrier island,* and *barrier spit.* Cf: *bar [coast].*

barrier [ecol] A condition, such as a topographic feature or physical quality, that tends to prevent the free movement and mixing of populations or individuals.

barrier [glaciol] *ice barrier.*

barrier [grd wat] *groundwater barrier.*

barrier beach A narrow, elongate sand or gravel ridge rising slightly above the high-tide level and extending generally parallel with the shore, but separated from it by a lagoon (Shepard, 1952, p.1904), estuary, or marsh; it is extended by longshore transport and is rarely more than several kilometers long. Cf: *barrier island.* This feature was termed an *offshore bar* by Johnson (1919, p.259, 350). Syn: *offshore barrier; offshore beach; spit.*

barrier chain A series of barrier islands, barrier spits, and barrier beaches extending along a coast for a considerable distance (Shepard, 1952, p.1908).

barrier flat A relatively flat area, often occupied by pools of water, separating the exposed or landward edge of a barrier from the lagoon behind it.

barrier ice (a) A term used by Robert F. Scott in 1902 for the ice constituting the Antarctic ice shelf (which was then called an "ice barrier"). The syn. *shelf ice* seems to be more commonly used for the actual ice itself. (b) A term sometimes used improperly as a syn. of *ice shelf.*

barrier island (a) A long, narrow, sandy coastal island, representing a broadened barrier beach that is above high tide and parallel to the shore, and that commonly has dunes and marshy terranes extending landward from the beach. Examples include Long Beach, N.J., and the Lido in Venice. This feature was termed an *offshore bar* by Johnson (1919). (b) A detached portion of a *barrier beach* between two inlets (Wiegel, 1953, p.5).

barrier-island marsh A salt or brackish marsh on the low landward margin of a barrier island.

barrier lagoon (a) A *lagoon* that is roughly parallel to the coast and is separated from the open ocean by a strip of land or by a barrier reef. Tidal influence is typically restricted and the lagoon is commonly hypersaline. (b) A lagoon encircled by coral islands or coral reefs, esp. one enclosed within an atoll. Syn: *back bay; sound.*

barrier reef A long, narrow coral reef roughly parallel to the shore and separated from it by a lagoon of considerable depth and width. It may enclose a volcanic island (either wholly or in part), or it may lie a great distance from a continental coast (such as the Great Barrier Reef off the coast of Queensland, Australia). Generally, barrier reefs follow the coasts for long distances, often with short interruptions, termed passes or channels. Cf: *fringing reef.*

barrier spit A *barrier beach* that is connected at one end to the mainland. Cf: *spit.*

barrier spring A spring resulting from the diversion of a flow of groundwater over or underneath an impermeable barrier in the floor of a valley (Schieferdecker, 1959, term 0308). Cf: *contact spring.*

barrier well (a) A pumping well used to intercept a plume of contaminated groundwater. (b) A recharge well that delivers water to, or in the vicinity of, a zone of contamination under sufficient head to prevent the further spreading of the contaminant.

barringerite (bar'-ring-er-ite') A metallic white hexagonal meteorite mineral: $(Fe,Ni)_2P$.

barringtonite (bar'-ring-ton-ite') A transparent colorless triclinic carbonate mineral of uncertain composition: $MgCO_3·2H_2O$ (?). Not approved by the International Mineralogical Association.

barroisite (bar-rois'-ite) A dark green to black monoclinic mineral of the *amphibole* group: $NaCa(Mg_3Al_2)(Si_7Al)O_{22}(OH)_2$.

Barrovian metamorphic zone (Bar-ro'-vi-an) In complexes of regional dynamothermal metamorphism, one of the belts of progressively increasing metamorphic grade, based on the first appearance of the index minerals chlorite, biotite, almandine, staurolite, kyanite, and sillimanite (Tilley, 1925). This now classical sequence, representing the commonest type of regional metamorphism, is named after George Barrow who in 1893 first described and mapped zones of progressive regional metamorphism in the Grampian Highlands of Scotland.

Barrovian-type facies series Rocks produced in the most common type of dynamothermal regional metamorphism, characterized by the appearance of metamorphic zones of the greenschist and amphibolite facies. Typical index minerals (in order of increasing metamorphic grade) are chlorite-biotite-garnet (almandine)-staurolite-kyanite-sillimanite (but not andalusite). Cf: *Barrovian metamorphic zone; Idahoan-type facies series; Saxonian-type facies series.*

barrow pit (bar'-row) A term used locally in the U.S. for a *borrow pit;* more frequently used elsewhere for a ditch or excavation near a road or other construction site that serves as a source of fill.

barsanovite (bar-san'-o-vite) *eudialyte.*

barshawite (bar'-shaw-ite) An analcime nepheline monzosyenite containing alkali feldspar and andesine; named by Johannsen (1931) for Barshaw, Renfrewshire, Scotland. Obsolete.

Barstovian (Bar-sto'-vi-an) North American land mammal age. Late early Miocene (after Hemingfordian, before Clarendonian).

barstowite An adamantine colorless monoclinic mineral: $Pb_4CO_3Cl_6·H_2O$.

bar switching A type of morphological process within the surf zone in which longshore sandbars become discontinuous and landward bars on one side of the discontinuity realign and join with the seaward bars on the other side (Shand and Baily, 1999).

bartelkeite (bar-tel'-ke-ite) A colorless to pale green monoclinic mineral: $PbFe^{2+}Ge_3O_8$.

barthite (barth'-ite) A variety of *austinite* containing copper. It was formerly thought to be a variety of veszelyite.

Bartonian (Bar-to'-ni-an) A stage of the Standard Global Chronostratigraphic Scale: upper middle Eocene (above Lutetian, below Priabonian).

bartonite (bar'-ton-ite) A submetallic blackish-brown tetragonal mineral: $K_6Fe_{20}S_{26}(Cl,S)$. Related to djerfisherite.

barycenter (bar'-y-cen'-ter) The location of the center-of-mass of a collection of bodies. That of the Earth-Moon pair remains below the Earth's surface but about $^3/_4$ of the Earth's radius from the center of the Earth; that of the

Solar System varies from inside the Sun to a distance about 1.5×10^6 km from the heliocenter over a cycle that averages 19.9 years in length, due mainly to the relative positions of Jupiter and Saturn, the two most massive planets.

barylite (bar'-y-lite) A colorless to pale blue orthorhombic mineral: $BaBe_2Si_2O_7$.

barysilite (bar'-y-sil'-ite) A white to pink platy rhombohedral mineral: $Pb_8Mn(Si_2O_7)_3$.

barysphere (bar'-y-sphere) An archaic term for the interior of the Earth beneath the lithosphere, including both the mantle and the core. However, it is sometimes used to refer only to the core or only to the mantle. Cf: *pyrosphere*.

barytes (ba-ry'-tes) A syn. of *barite.* Also spelled: baryte; barytine; barytite.

barytocalcite (ba-ry'-to-cal'-cite) (a) A variously colored monoclinic mineral: $BaCa(CO_3)_2$. It is the monoclinic dimorph of alstonite. (b) A mixture of calcite and barite.

barytolamprophyllite (ba-ry'-to-lam'-pro-phyl'-lite) A dark brown platy monoclinic mineral: $Na_3Ba_2(Ti,Fe^{3+})_2O_2(Si_2O_7)_2(O,OH,F)_2$. The barium analogue of lamprophyllite.

basal (ba'-sal) adj. Pertaining to, situated at, or forming the base of an animal structure; e.g. referring to the *aboral* part of the *theca* of an echinoderm, or pertaining to the under or reverse side of an encrusting or freely growing bryozoan colony. n. A basal structure of an animal; esp. a *basal plate* of an echinoderm.

basal arkose An arkosic sandstone basal to a sedimentary sequence, resting unconformably on a granitic terrane; the arkosic equivalent of a granitic *basal conglomerate.* It may grade downward into sedentary or residual arkose (Pettijohn et al., 1972, p.163).

basal attachment scar *Attachment surface* on lower, or *under side* of a conodont element (TIP, 1981, pt. W, supp.2). See also: *attachment scar*.

basal canalules Minute spherical or tubular voids in basal portion of certain coniform conodont elements (TIP, 1981, pt. W, supp.2).

basal cavity A pit or concavity about which a conodont element was built through accretion of lamellae, opening onto the aboral side, and present on all true conodont elements. See also: *cup; attachment scar.* Improper syn: *pulp cavity; escutcheon.*

basal circlet (a) A circlet of basal plates in the *theca* of a blastoid, normally consisting of two large zygous plates and one small azygous plate. (b) In a crinoid, the circlet of plates immediately beneath the radials. (c) Prostalia (spicules) protruding from the base of a sponge and serving to stabilize or anchor it to the substrate.

basal cleavage Mineral cleavage parallel to the basal pinacoid; e.g. in molybdenite.

basal clinker The zone of *autobreccia* that forms the base of an *aa* flow.

basal cone *basal funnel.*

basal conglomerate A lithologically homogeneous conglomerate that forms the bottom stratigraphic unit of a sedimentary series and that rests on a surface of erosion, thereby marking an unconformity; esp. a coarse-grained beach deposit of an encroaching or transgressive sea. Cf: *marginal conglomerate.*

basal diaphragm In stenolaemate bryozoans, a partition that acts as the floor of the living chamber.

basal disk An expanded basal part by which certain stalked sessile organisms are attached to the substrate; specif. the aboral fleshy part of a scleractinian coral *polyp,* typically subcircular in outline, that closes off the lower end of the polyp; or the *proancestrula* in stenolaemate bryozoans. Cf: *oral disk.* Also spelled: basal disc.

basal filling In conodont morphology a general term for various types of basal material (TIP, 1981, pt. W, supp.2). See also: *basal cone; basal funnel; basal plate*.

basal funnel An excavated conelike basal plate, the tip of which fits into the basal cavity of a conodont.

basal glide In glaciology, translation or slip along the *basal planes* of an ice crystal subject to plastic deformation. Cf: *crystal gliding*.

basal granule A dotlike body forming part of the neuromotor system in tintinnids.

basal groove Groovelike extension of basal cavity or basal pit beneath a process or beneath some part of the carina of palmate pectiniform conodont elements (TIP, 1981, pt. W, supp.2).

basal groundwater A term that originated in Hawaii and refers to a major body of ground water floating on and in hydrodynamic equilibrium with salt water. Syn: *basal water.*

basal ice Ice at the bottom of a glacier or ice sheet.

basal lamina The generally encrusting, calcified colony wall of stenolaemate bryozoans, extending to the growing edge of the colony. See also: *median lamina.*

basal layer Varyingly thick, comparatively uniform deposit of shell substance in *adaxial* part of test *volution,* adherent to *abaxial* wall (roof) of preceding volution (e.g., Alveolinidae).

basal leaf cross Broad wings on radial spines of acantharian radiolarians.

basal lobe One of two lobes set off by furrows in the posterior and lateral parts of the *glabella* of a trilobite, just in advance of the *occipital ring.*

basal margin Lower edge of last-formed lamella on a conodont element; coincides with trace of *under side* of unit in lateral view if attachment surface lacks zone(s) of recessive basal margin, otherwise coincident with outer margin of attachment surface (TIP, 1981, pt. W, supp.2).

basal pinacoid In all crystals except those of the isometric system, the *{001}* pinacoid. Cf: *front pinacoid; side pinacoid.* Syn: *basal plane [cryst].*

basal pit An upwardly concave, basically conical invagination of the *under side* of a conodont element, which is bordered or surrounded by a zone, or zones, of recessive basal margin and, together with these forms an attachment surface. If entire attachment surface is upwardly concave and there are no contiguous zones of recessive basal margin, the term *basal cavity* is used (TIP, 1981, pt. W, supp.2).

basal plane [cryst] *basal pinacoid.*

basal plane [ice] The plane perpendicular to the *c*-axis in an *ice crystal.* Bonding parallel to the *c*-axis is much weaker than that normal to the *c*-axis, so deformation of ice crystals occurs much more easily on the basal plane than on other crystallographic planes. Cf: *basal glide.*

basal plate (a) One of a circlet of certain chiefly ventral skeletal plates of an echinoderm; e.g. a plate composing the aboral end of the *theca* of a blastoid and articulated to the stem aborally and to radial plates on oral borders, or an interambulacral plate just below the arm-bearing radial plates of a crinoid. Syn: *basal.* (b) A thin skeletal plate formed initially as a part of a *corallite* immediately below the basal disk of the *polyp* of a scleractinian coral, from which the septa and walls begin to extend upward and outward. (c) A laminated platelike structure of organic material attached to the aboral side of a conodont element along the attachment scar or basal cavity.

basal platform A term used by Linton (1955) as a syn. of *basal surface.*

basal pore One of the pores outlined by connector bars joining the *basal ring* of a radiolarian skeleton (as in the subfamily Trissocyclinae).

basal reflection A diffraction peak from layered materials originating from those atomic planes that compose the layers. For most phyllosilicates, basal reflections are of the Miller index type: 00R, where R is an integer. The d(001) value is usually an indicator of the structure unit and thus of the type of phyllosilicate, assuming that no interstratifications occur in the phase. For example, mica minerals have a d(001) value of about 10 Å, and chlorite minerals have a d(001) value of about 14 Å.

basal ring A ring at or below the base of the sagittal ring of a radiolarian skeleton, commonly with basal spines projecting from it (as in the subfamily Trissocyclinae).

basal sapping The undercutting, or breaking away of rock fragments, along the headwall of a cirque, due to frost action at the bottom of the *bergschrund.* Syn: *sapping [glac geol].*

basal shear stress *Shear stress* that acts at the base of a glacier or ice sheet.

basal sliding The slip of a glacier over its bed. It may occur at subfreezing temperatures (Shreve, 1984; Echelmeyer and Wang, 1987) but is most important when the bed is at the *pressure-melting temperature.* Bumps on the bed are accommodated by a combination of *regelation* and plastic flow. Syn: *basal slip.*

basal slip *basal sliding.*

basal slope *wash slope.*

basal surface The generalized boundary between weathered and unweathered rock, or the lower limit to active weathering (Ruxton and Berry, 1959). This contact, which may be regular or irregular, indicates a very rapid or sudden change upward into the base of the mass of weathering debris. Syn: *basal platform; weathering front.*

basal suture Boundary line between lower edge of the lamella of a conodont element and the upper edge of the simultaneously secreted lamella of its basal plate or funnel (TIP, 1981, pt. W, supp.2).

basalt [lunar] An igneous rock from the Moon that is composed chiefly of nearly equal amounts of augite, plagioclase, and ilmenite. The plagioclase is characteristically highly calcic (An_{80}-An_{90}). Lunar basalt contains more titanium dioxide, rare-earth elements, and zirconium, and less nickel than terrestrial basalt.

basalt [petrology] (ba-salt', ba'-salt) (a) In the *IUGS classification,* a volcanic rock defined modally by Q/(Q+A+P) between 0 and 20% or F/(F+A+P) between 0 and 10%, P/(A+P)>65%, and M>35%. Because modes are difficult to estimate for these rocks, basalt is also defined in the TAS diagram as rock falling in the area bounded by points with the SiO_2 and total alkali coordinates: 45, 0; 45, 5; 52, 0; 52, 5. (b) A general term for dark-colored mafic igneous rocks, commonly extrusive but locally intrusive (e.g. as dikes), composed chiefly of calcic plagioclase and clinopyroxene; the fine-grained equivalent of *gabbro.* Nepheline, olivine, orthopyroxene, and quartz may be present in the *CIPW norm,* but not all simultaneously: nepheline and olivine can occur together, as can olivine and orthopyroxene, and orthopyroxene and quartz, but nepheline does not occur with orthopyroxene or quartz, nor quartz with nepheline or olivine. These associations and incompatibilities are discussed by Yoder and Tilley (1962) and by Muir and Tilley (1961). Cf: *basaltic rocks.*

basalt-andesite copper *basaltic copper deposit.*

basal thrust plane *sole fault.*

basaltic (ba-sal'-tic) Pertaining to, composed of, containing, or resembling basalt; e.g. "basaltic lava".

basaltic achondrite An achondritic meteorite with a basaltic composition. The most common classes of basaltic achondrites are *eucrite, shergottite,* and *angrite.*

basaltic andesite (a) In the *IUGS classification*, a volcanic rock defined in the TAS diagram as rock falling in the area bounded by points with the SiO_2 and total alkali coordinates: 52, 0; 52, 5; 57, 0; 57, 5.9. (b) "Rocks of the calc-alkaline kindred that are intermediate, in one or more respects, between typical basalt and typical andesite" (Coats, 1968, p.689-690).

basaltic copper deposit A *mineral-deposit type* characterized by disseminated native copper and copper sulfide minerals in subaerial basalt and copper sulfide minerals in overlying sedimentary rocks. Syn: *copper manto; volcanic-hosted copper; volcanic redbed copper.*

basaltic dome An obsolete term for a basaltic *shield volcano*. Use not recommended. Cf: *lava dome; shield volcano.*

basaltic glass A volcanic glass that may be black, green, or brown because of abundant crystallite. It is formed from basaltic magma and is commonly found as chilled margins of dikes, sills, or flows. Syn: *basalt obsidian; sideromelane.*

basaltic hornblende A black or brown variety of hornblende rich in ferric iron (ferrous iron having been oxidized) and occurring in basalt and other iron-rich basic igneous (volcanic) rocks; a type of *brown hornblende* characterized optically by strong pleochroism and birefringence, high refractive indices, and a small extinction angle. Syn: *lamprobolite; oxyhornblende; basaltine.*

basaltic kimberlite A term introduced by Wagner in 1914 for *kimberlite* containing less than 5% mica phenocrysts. The term is considered inappropriate. Later called Type I *kimberlite.*

basaltic komatiite A lava, associated with *komatiite*, with MgO between 5 and 15%. It is approximately intermediate between *tholeiitic basalt* and *boninite.*

basaltic layer A syn. of sima, so named for its supposed petrologic composition. It is also called the *gabbroic layer*, and may be equivalent to the *Conrad layer*. Cf: *granitic layer*. A layer is sometimes termed "basaltic layer" if it possesses the appropriate seismic velocity (~6.5-7.0 km/s), although nothing may be known about its composition. In marine usage, equivalent to *Layer 2* of oceanic crust.

basaltic plateau *lava plateau.*

basaltic rocks A general term incorporating fine-grained compact dark-colored extrusive igneous rocks such as *basalt [petrology], diabase, dolerite*, and dark-colored *andesite.*

basaltic trachyandesite In the *IUGS classification*, a volcanic rock intermediate between *trachybasalt* and *trachyandesite*; defined in the TAS diagram as the field bounded by four corners with the SiO_2 and total alkali wt% coordinates 49.4, 7.3; 52, 5; 53, 9.3; and 57, 5.9.

basal till A firm fine-grained till containing many abraded stones dragged along beneath a moving glacier and deposited upon bedrock or other glacial deposits. Cf: *lodgment till*. Syn: *accretion till.*

basaltine (ba-sal'-tine) n. (a) *basaltic hornblende*. (b) *augite*. adj. *basaltic.*

basalt obsidian *basaltic glass.*

basaltoid In the *IUGS classification*, a general term in the provisional "field" classification for rocks tentatively identified as basalt.

basal tunnel A water-supply tunnel excavated along the basal water table in basaltic areas, esp. Hawaii. Cf: *Maui-type well.*

basaluminite (bas-a-lu'-mi-nite) A monoclinic white mineral: $Al_4(SO_4)(OH)_{10}{\cdot}5H_2O$. It occurs in veinlets lining crevices in ironstone. Cf: *felsöbanyite.*

basal water *basal groundwater.*

basal water table The water table of a body of basal groundwater.

basal wrinkles Coarse longitudinal ribs that form a narrow band, or zone, adjacent to the basal margin of conodont elements proper of *Panderodus, Neopanderodus, Belodina* and related genera (TIP, 1981, pt. W, supp.2).

basanite [ign] (bas'-a-nite) (a) In the *IUGS classification*, a volcanic rock with >10% modal olivine, defined in the QAPF diagram by F/(F+A+P) between 10 and 60%, P/(P+A)>90%; also defined in the TAS diagram as rock falling in the irregular field bounded by points with the SiO_2 and total alkali coordinates: 41, 3; 41, 7; 45, 9.4; 49.4, 7.3; 45, 5; and 45, 3. Cf: *limburgite; tephrite*. (b) A group of basaltic rocks characterized by calcic plagioclase, clinopyroxene, a feldspathoid (nepheline, leucite), and olivine; also, any rock in that group. Without the olivine, the rock would be called a *tephrite*. The term was coined by Brongniart in 1813.

basanite [sed] (a) A *touchstone* consisting of flinty jasper or finely crystalline quartzite. Syn: *Lydian stone*. (b) A black variety of jasper.

basanitic foidite In the *IUGS classification*, a collective term for volcanic rocks consisting of feldspathoids with some plagioclase and at least 10% modal olivine. Cf: *tephritic foidite.*

basanitoid (ba-san'-i-toid) n. A term proposed by Bücking in 1881 for a group of rocks intermediate in composition between basanite and basalt (Johannsen, 1939, p.243), i.e. having the chemical composition of basanite but without modal feldspathoids and with a glassy groundmass. adj. Said of a rock resembling basanite, having normative but no modal feldspathoids.

basculating fault (bas'-cu-lat'-ing) *wrench fault.*

base [eng] That part of an engineering structure resting on the subgrade, on supporting soil, or on solid rock; the *base course.*

base [gem] *pavilion.*

base [paleont] (a) The aboral end of an echinoderm *theca*. In cystoids the term is restricted by some to the columnar facet but extended by others to include thecal plates of the basal circlet or aboral circlets. (b) The aboral end of a crinoid *calyx*, or the part of a crinoid dorsal cup between radial plates and stem, normally composed of basal plates or of basal and infrabasal plates, but in articulate crinoids including the centrale. (c) Portion of a conodont element that includes the basal cavity of pit; commonly used to refer to the part of an element adjacent to the *under side*, but has also been used to refer to the under side itself (TIP, 1981, pt. W, supp.2).

base [petroleum] An informal term for the hydrocarbon series that is dominant in a given crude oil. Cf: *asphalt-base crude; mixed-base crude; paraffin-base crude.*

base [surv] *baseline.*

base apparatus Any apparatus (such as wood tubes, metal wires, iron bars, steel rods, or invar tapes) used in geodetic surveying and designed to measure with accuracy and precision the length of a base line in triangulation or the length of a line in a traverse.

base bullion Crude lead containing recoverable silver, with or without gold.

base-centered lattice A type of *centered lattice* that is centered in one pair of the (001), (010), or (100) faces. Syn: *end-centered lattice; one-face-centered lattice*. Cf: *A-centered lattice; B-centered lattice; C-centered lattice.*

base correction A correction or adjustment of geophysical measurements to express them relative to the values at a *base station.*

base course (a) A granular soil, usually well-graded (in engineering terms) or well-sorted (in geologic terms) that maximizes density when compacted and minimizes the potential for grains to migrate in response to percolating water. (b) A bottom layer of coarse gravel or crushed stone, generally of specified character and thickness, constructed on the subgrade or subbase of a roadway or structure, serving such functions as distributing load, providing drainage, and minimizing frost action. Syn: *base [eng].*

base discharge As used by the U.S. Geological Survey, that discharge above which peak discharge data are published.

base discordance A term used in seismic stratigraphy to refer to a lack of parallelism between a sequence of strata and its lower boundary, owing to either *onlap* or *downlap* (Mitchum, 1977, p.206). Cf: *top discordance.*

base exchange *cation exchange.*

base flow That part of stream discharge that is not attributable to direct runoff from precipitation or melting snow. It is sustained by groundwater discharge (Ground Water Subcommittee, 1989). Syn: *base runoff.*

base-height ratio The ratio between the *air base* and the *flight height* of a stereoscopic pair of aerial photographs.

baselap (base'-lap) A term used in seismic stratigraphy to describe the termination of a sequence of strata along its lower boundary where discrimination between *onlap* and *downlap* is difficult or impossible (Mitchum, 1977, p.205).

base level (a) The theoretical limit or lowest level toward which erosion of the Earth's surface constantly progresses but seldom, if ever, reaches; esp. the level below which a stream cannot erode its bed. The general or *ultimate base level* for the land surface is sea level, but temporary base levels may exist locally. The base level of eolian erosion may be above or below sea level; that of marine erosion is the lowest level to which marine agents can cut a bottom. Cf: *base level of erosion*. (b) A curved or planar surface extending inland from sea level, inclined gently upward from the sea and representing the theoretical limit of stream erosion. (c) The surface toward which external forces strive, at which neither erosion nor deposition takes place (Barrell, 1917); a surface of equilibrium.

base level of deposition The highest level to which a sedimentary deposit can be built (Twenhofel, 1939, p.8); if built of marine deposits, it coincides with the *base level of erosion.*

base level of erosion (a) A term introduced by Powell (1875, p.203) for an irregular "imaginary surface, inclining slightly in all its parts toward the lower end of the principal stream", below which the stream and its tributaries were supposed to be unable to erode. (b) A syn. of *base level.*

base-level peneplain *peneplain.*

base-level plain A flat surface, area, or lowland produced by the wearing-down of a region to or near its base level; a plain that cannot be materially reduced in elevation by erosion. Cf: *peneplain.*

baseline (a) A surveyed line established with more than usual care, which serves as a reference to which surveys are coordinated and correlated. Syn: *base [surv]*. (b) The initial measurement in triangulation, being an accurately measured distance constituting one side of one of a series of connected triangles, and used, together with measured angles, in computing the lengths of the other sides. (c) One of a pair of coordinate axes (the other being the *principal meridian*) used in the U.S. Public Land Survey system. It consists of a line extending east and west along the true parallel of latitude passing through the initial point, along which standard township, section, and quarter-section corners are established. (d) An aeromagnetic profile flown at least twice in opposite directions and at the same level, in order to establish a line of reference of magnetic intensities on which to base an aeromagnetic survey. (e) The center line of location of a railway or highway; the reference line for the construction of a bridge or other engineering structure.

base map (a) A map of any kind showing essential outlines necessary for adequate geographic reference, on which additional or specialized information is plotted for a particular purpose; esp. a topographic map on which geologic information is recorded. (b) *master map.* (c) Obsolete syn. of *outline map.*

basement (base'-ment) n. (a) The undifferentiated *complex* of rocks that underlies the rocks of interest in an area. Cf: *basement terrane.* (b) The crust of the Earth below sedimentary deposits, extending downward to the Mohorovicic discontinuity. In many places the rocks of the complex are igneous and metamorphic and of Precambrian age, but in some places they are Paleozoic, Mesozoic, or even Cenozoic. Syn: *basement rock; basement complex.* adj. Said of material, processes, or structures originating or occurring in the basement.

basement complex *basement.*

basement fold An obsolete term for a compressional structure (fold or thrust) that formed within a continent and affected the entire thickness of the continental crust. This obsolete term is the English translation of Argand's plis de fond.

basement-involved fault A fault that cuts the basement, which are those rocks that formed before the cover rocks accumulated. Cf: *detached fault.*

basement rock *basement.*

basement terrane The lowest mappable mass of rock, generally with complex structure, that underlies the other major rock sequences of a region, specif. of the California Coast Ranges (Berkland et al., 1972, p.2296). No unconformable upper contact is required. Cf: *basement.*

base metal (a) Any of the more common and more chemically active metals; e.g., lead, copper. (b) The principal metal of an alloy; e.g., the copper in brass. Cf: *noble metal.*

base net A small net of triangles and quadrilaterals, starting from a measured *baseline* and connecting with a line of the main scheme of a *triangulation net;* e.g. a triangle formed by sighting a point from both ends of a base line, or two adjacent triangles with the base line common to both. It is the initial figure in a triangulation system.

base of drift A seismic-velocity discontinuity often encountered between glacial drift material and the competent formation beneath.

base of weathering In seismic work, the boundary between a low-velocity surface layer and an underlying, comparatively high-velocity layer. The velocity contrast is often about 1:3. It often corresponds to the water table. Important in deriving time corrections for seismic records.

base runoff Sustained or fair-weather *runoff [water]* (Langbein and Iseri, 1960). It is primarily composed of effluent groundwater, but also of runoff delayed by slow passage through lakes or swamps. The term refers to the natural flow of a stream, unaffected by the works of man. Cf: *base flow; direct runoff.* Syn: *fair-weather runoff; sustained runoff.*

base-saturation percentage The extent to which the adsorption complex of a soil is saturated with exchangeable cations other than hydrogen. It is expressed as a percentage of the total cation-exchange capacity.

base station An observation point used in geophysical surveys as a reference, to which measurements at additional points can be compared. See also: *base correction.*

base surge A ring-shaped cloud of gas and suspended solid debris that moves radially outward at high velocity as a density flow from the base of a vertical explosion column. It accompanies a volcanic eruption or crater formation by a hyrovolcanic explosion or hypervelocity impact.

base temperature The temperature in a region of uniform temperature normally occurring in the lower part of a convecting hydrothermal system.

Bashkirian (Bash-kir'-i-an) A stage of the Standard Global Chronostratigraphic Scale: lowest stage of the Pennsylvanian Subsystem (Upper Carboniferous; above Serpukhovian, below Moscovian).

basic (ba'-sic) (a) Said of an igneous rock having a relatively low silica content, sometimes delimited arbitrarily as 44 to 51% to 45 to 52%; e.g. gabbro, basalt. Basic rocks are relatively rich in iron, magnesium, and/or calcium, and thus include most mafic rocks as well as other rocks. "Basic" is one of four subdivisions of a widely used system for classifying igneous rocks based on their silica content: *acidic, intermediate, basic,* and *ultrabasic.* Cf: *femic.* (b) Said loosely of dark-colored minerals. Cf: *silicic; mafic.* (c) Said of a plagioclase that is calcic.

basic behind In granitization, a zone in which residual mafic components are concentrated.

basic border The marginal area of an igneous intrusion, characterized by a more basic composition than its interior. Cf: *chill zone.* Syn: *mafic margin.*

basic front In granitization, an advancing zone enriched in calcium, magnesium, and iron, which is said to represent those elements in the rock being granitized that are in excess of those required to form granite. During granitization, these elements are believed to be displaced and moved through the rock ahead of the granitization front, to form a zone enriched in minerals such as hornblende and pyroxene. Cf: *basic behind.* Syn: *mafic front; magnesium front.*

basic hydrologic data Data including inventories of land and water features that vary from place to place (e.g. topographic and geologic maps), and records of processes that vary with time and from place to place (e.g. precipitation, evapotranspiration, streamflow, and groundwater levels) (Langbein and Iseri, 1960). Cf: *basic hydrologic information.*

basic hydrologic information A broader term than *basic hydrologic data,* including surveys and appraisals of the water resources of an area and a study of its physical, chemical, and biological characteristics and related economic processes, interrelations, and mechanisms (Langbein and Iseri, 1960).

basicoronal (ba'-si-cor-o'-nal) Pertaining to the *corona* of an echinoid at the edge of the peristome.

basic plagioclase A variety of plagioclase having relatively low content of SiO_2; e.g. an An-rich member such as bytownite or anorthite.

basic rock In the *IUGS classification,* any igneous rock with wt% SiO_2 between 45 and 52.

basic structural unit [paleont] (= "Grundelement") One of the elemental structural units, commonly about 0.2 microns in diameter, that comprise many aragonitic shell microstructures (Bandel, 1977a,b, 1979a,b, 1981a,b).

basic wash A driller's term for material eroded from outcrops of basic igneous rocks (gabbro, basalt) and redeposited to form a rock having approximately the same major mineral constituents as the original rock (Taylor and Reno, 1948, p.164). Cf: *granite wash.*

basidia Pl. of *basidium.*

Basidiomycetes (Ba-sid-i-o-my-ce'-tes) A class of Fungi characterized by the production of haploid sexual spores (basidiospores) externally, on specialized (diploid) structures (*basidia*).

basidiospore (ba-sid'-i-o-spore') A thin-walled, often *apiculate* and asymmetrical *fungal spore* produced by the *basidium* of a *basidiomycete.* Such spores rarely are preserved as microfossils.

basidium (ba-sid'-i-um) A structure on the fruiting body of a basidiomycete, where fungal spores are formed. Pl: *basidia.*

basification (ba'-si-fi-ca'-tion) Enrichment of a rock in elements such as calcium, magnesium, iron, and manganese.

basin (ba'-sin) As a geologic term, it is used in four major contexts: In *topography,* basins (e.g., the Great Basin of the western U.S.) are low areas where surface drainage is confined to the basin and streams do not leave the basin. Relating to *structure,* basins (e.g., the Michigan Basin or the Williston Basin) are areas where the dip of strata is toward the center of the basin. *Drainage* basins (e.g., the Mississippi River drainage basin or the Clear Creek drainage basin) are areas that are drained by a single stream of any size. Relating to *sedimentation,* basins (e.g., the Appalachian Basin or the Anadarko Basin) are areas where thick sequences of sediments have accumulated.

basin-and-range Said of a topography, landscape, or physiographic province characterized by a series of tilted fault blocks forming longitudinal, asymmetric ridges or mountains and broad, intervening basins. Type example: Basin and Range physiographic province in SW U.S. Cf: *basin-and-range structure; basin range.*

basin-and-range structure Regional geologic structure dominated by subparallel fault-block mountains separated by sediment-filled basins; e.g., basin-range faulting characterized by movement on normal faults, as in the *Basin-and-Range* province of SW United States.

basin area For a given stream order (u), the total area, projected upon a horizontal plane, of a drainage basin bounded by the basin perimeter and contributing overland flow to the stream segment of order (u), including all tributaries of lower order (Strahler, 1964, p.4-48). Cf: *watershed area.* Cf: *law of basin areas; stream order.*

basin-area ratio The ratio of mean basin area of a given order to the mean basin area of the next lower order within a specified larger drainage basin (Schumm, 1956, p.606).

basin-circularity ratio Ratio of the area of a drainage basin to the area of a circle with the same perimeter as the basin. Syn: *circularity ratio.*

basin-elongation ratio Ratio of the diameter of a circle having the same area as a drainage basin to the maximum length of that basin.

basin facies Sediments deposited beyond the outer limits of a land-bordering submarine shelf. See also: *facies.*

basin-floor fan In sequence stratigraphy, a portion of the lowstand systems tract characterized by deposition of submarine fans on the lower slope or basin floor. Fan formation is associated with the erosion of canyons into the slope and the incision of fluvial valleys into the shelf. Siliciclastic sediment bypasses the shelf and slope through the valleys and canyons to feed the basin-floor fan. The basin-floor fan may be deposited at the mouth of a canyon, although it may occur widely separated from the canyon mouth, or a canyon may not be evident. It has no age-equivalent rocks on the slope or shelf. The base of the basin-floor fan (coincident with the base of the lowstand systems tract) is the type-1 sequence boundary; the top of the fan is a downlap surface (Van Wagoner et al., 1987).

basining (ba'-sin-ing) The bending down or settling of part of the Earth's crust in the form of a basin, as by tectonism, erosion, or by solution and transportation of underground deposits of salt or gypsum.

basin length Horizontal distance of a straight line from the mouth of a stream to the farthest point on the drainage divide of its basin, parallel to the principal drainage line (Schumm, 1956, p.612).

basin-margin fault *border fault.*

basin modeling In geophysical exploration, a procedure integrating inferred or historical sediment influx and compaction, erosion and deformation, heat transfer, and water flow in order to work out the history of hydrocarbon generation, expulsion, and migration. The resulting pressure-temperature history is important in determining hydrocarbon generation and hence a basin's prospectivity.

basin order The number assigned to an entire drainage basin contributing to the stream segment of a given order and bearing an identical integer designation; e.g., a first-order basin contains all of the drainage area of a first-order stream. See also: *stream order.*
basin peat *local peat.*
basin perimeter The length of the line enclosing the area of a drainage basin.
basin-plain deposit A turbidite sequence formed far from its turbidity-current sources and characterized by sequences of graded beds that are thin, fine-grained, and nonsystematic with respect to upward fining or upward coarsening of successive beds, because more than one submarine-fan system contributes to the sequence.
basin range A mountain range that owes its elevation and structural form mainly to faulting and tilting of strata and that is surrounded by alluvium-filled basins or valleys. Etymol: from the Great Basin, a region in SW U.S. characterized by fault-block mountains. See also: *basin-and-range.*
basin relief The difference in elevation between the mouth of a stream and the highest point within, or on the perimeter of, its drainage basin (Strahler, 1952b, p.1119); the maximum relief in the basin.
basin valley A broad, shallow valley with gently sloping sides.
basinward shift in facies In sequence stratigraphy, the vertical juxtaposition of facies such that shallow-marine to nonmarine strata deposited above a sequence boundary, lie directly on strata deposited in a much deeper depositional environment, with no intervening rocks deposited in intermediate depositional environments. A basinward shift in facies is commonly marked by abnormal subaerial exposure (Van Wagoner et al., 1990).
basiophthalmite (ba'-si-oph-thal'-mite) The proximal segment (lowest joint) of the eyestalk of a decapod crustacean, articulating with the *podophthalmite.*
basipetal (ba-si-pet'-al) In *conidia,* said of *conidiospores* produced in succession from the proximal base.
basipinacoderm (ba'-si-pi-nac'-o-derm) The pinacoderm delimiting a sponge at the surface of fixation.
basipod (ba'-si-pod) The *basis* of a crustacean limb.
basis The limb segment of a crustacean just distal to the *coxa,* commonly bearing the *exopod* and *endopod;* a membranous or calcareous structure (in nonpedunculate cirripedes) that contacts the substratum. Pl: bases. Syn: *basipod.*
basis rami The calcified structure from which a node emanates to start a new branch in a jointed colony of cheilostome bryozoans. It commonly consists of the calcified proximal ends of one or more zooids, the more distal parts of which lie in the node and following internode.
basis vectors The set of three vectors that define the orientation and lattice parameters of a crystal lattice.
basket-of-eggs topography (bas'-ket) A landscape characterized by swarms of closely spaced drumlins, distributed more or less en echelon, and commonly separated by small marshy tracts. Syn: *drumlin field.*
bass *batt [coal].*
bassanite (bas'-sa-nite) A white monoclinic or trigonal mineral: $2CaSO_4 \cdot H_2O$. Syn: *vibertite.*
basset (bas'-set) An obsolete term for the noun "outcrop" and the verb "to crop out".
basset edge Area of obliquely truncated outcrop, as in cross-bedded strata exposed transverse to their depositional planes.
bassetite (bas'-set-ite) A yellow monoclinic mineral: $Fe^{2+}(UO_2)_2(PO_4)_2 \cdot 8H_2O$.
bastard (bas'-tard) adj. (a) Said of an inferior or impure rock or mineral, or of an ore deposit that contains a high proportion of noncommercial material. (b) Said of any metal or ore that gives misleading assays or values. (c) Said of a vein or other deposit close to and more or less parallel to a main vein or deposit, but thinner, less extensive, or of a lower grade.
bastard coal Thin partings of impure coal occurring in the lower part of shale strata immediately overlying a coal seam; any coal with a high ash content. Syn: *batt [coal].*
bastard ganister A silica rock having the superficial appearance of a true *ganister* but characterized by more interstitial matter, a greater variability of texture, and often an incomplete secondary silicification.
bastard quartz (a) A syn. of *bull quartz.* (b) A round or spherical boulder of quartz embedded in soft or decomposed rock.
bastard shale *cannel shale.*
bastinite (bas'-tin-ite) *huréaulite.*
bastion (bas'-tion) A knob or mass of bedrock projecting into a main glacial valley at the junction with a hanging valley, at or below the level of the hanging valley floor.
bastite (bas'-tite) An olive-green, blackish-green, or brownish variety of serpentine mineral resulting from the alteration of orthorhombic pyroxene (esp. enstatite), occurring as foliated masses in igneous rocks, and characterized by a schiller (metallic or pearly luster) on the chief cleavage face of the pyroxene. Syn: *schiller spar.*
bastnäsite A group of rare-earth carbonate minerals with the general formula $ACO_3(F,OH)$, where A = rare-earth elements, Y or Th. Also spelled: bastnaesite.
bastnäsite-(Ce) (bast'-nas-ite-Ce) A greasy, wax-yellow to reddish-brown mineral: $(Ce,La)CO_3(F,OH)$ crystallized in various hexagonal and rhombohedral polytypes. It occurs in alkaline igneous rocks, esp. carbonatite, as at Mountain Pass, Calif. Bastnäsite is the chief U.S. source of rare-earth elements.
bastnäsite-(La) A lanthanum-dominant member of the *bastnäsite* group: $(La,Ce)CO_3(F,OH)$.
bastnäsite-(Y) A brick-red hexagonal mineral: $(Y,Ce)(CO_3)F$. A member of the *bastnäsite* group.
bat *batt [coal].*
batch melting A process involving continuous equilibrium melting of rock until accumulation of a sufficient volume of liquid such that the liquid generated then physically separates from the remaining rock.
batholith (bath'-o-lith) A large, generally discordant plutonic body having an aerial extent of 40 mi^2 (100 km^2) or more and no known floor. Obsolete syn: *abyssolith.*
Bathonian (Bath-o'-ni-an) Stage of the Standard Global Chronostratigraphic Scale: Middle Jurassic (above Bajocian, below Callovian).
Båth's law A generalization in seismology that the largest aftershock occurring within a few days of a main shock has a magnitude of 1.2 units lower than that of the main shock (Richter, 1958, p.69).
bathtub-ring manganese *sedimentary manganese.*
bathvillite (bath'-vil-lite) An amorphous, opaque, very brittle woody resin occurring as fawn-brown porous lumps in *torbanite* at Bathville, Scotland.
bathy- A prefix meaning "deep".
bathyal (bath'-y-al) Pertaining to the ocean environment or depth zone between 200 and 3500 meters; also, pertaining to the organisms of that environment.
bathybenthic (bath-y-ben'-thic) Pertaining to the benthos of the bathyal zone of the ocean. Cf: *archibenthic.*
bathydermal (bath-y-der'-mal) An obsolete term describing deformation or gliding of the lower part of the *sialic* crust. Cf: *dermal [tect]; epidermal.*
bathygenesis (bath-y-gen'-e-sis) An obsolete term for negative or subsident tectonic movement; tectonic lowering of marine basins. It is analogous to *epeirogeny* or positive tectonic movement associated with the continents. Adj: *bathygenic.*
bathygenic (bath-y-gen'-ic) Adj. of *bathygenesis.*
bathylimnion (bath-y-lim'-ni-on) The deeper part of a *hypolimnion* characterized by constant rates of heat absorption at different depths. Cf: *clinolimnion.*
bathymetric chart (bath-y-met'-ric) A topographic map of the bottom of a body of water (such as the sea floor), with depths indicated by contours (isobaths) drawn at regular intervals.
bathymetric contour *isobath [oceanog].*
bathymetric tint A distinctive shading or coloring of the area between bathymetric contour lines to emphasize the distribution of high and low areas and the range of water depths. Cf: *hypsometric tint.*
bathymetry (ba-thym'-e-try) The measurement of ocean depths and the charting of the topography of the ocean floor. Commonly (mis-)used to mean bottom topography.
bathyorographical (bath'-y-or'-o-graph'-i-cal) (a) Pertaining to ocean depths and mountain heights considered together; said of a map that shows both the relief of the land and the depths of the ocean. (b) Pertaining to the description of the relief features on the ocean floor.
bathypelagic (bath'-y-pe-lag'-ic) Pertaining to the open water of bathyal zone.
bathyphreatic A cave passage traversing from insurgence to resurgence as a single, deep loop beneath a stable water table (Ford and Williams, 1989, p.261). See also: *phreatic passage.*
bathyscaph (bath'-y-scaph) A manned, submersible vehicle for deep-sea exploration; it is somewhat navigable, in contrast to a *bathysphere.*
bathyseism (bath'-y-seism) A deep-focus earthquake that is instrumentally detected worldwide. The term is little used.
bathysphere (bath'-y-sphere) A manned submersible sphere that is lowered into the deep ocean by cable for observations; unlike the *bathyscaph,* it is not navigable.
bathythermograph (bath-y-ther'-mo-graph) An oceanographic instrument that measuress temperature as a function of depth within a few hundred meters of the sea surface. Obsolete, except as *expendable bathythermograph.*
batiferrite A submetallic black hexagonal mineral: $BaTi_2Fe_{10}O_{19}$. A member of the *magnetoplumbite* group.
batisite (ba'-tis-ite) A dark brown orthorhombic mineral: $(Na,K)_2BaTi_2(Si_2O_7)_2$.
Batoidea (Ba-toi'-de-a) An order, largely marine, of elasmobranch fishes, characterized by flattened body form, expanded pectoral fins, and crushing dentitions; it includes the skates and rays. Range, Upper Jurassic to Holocene.
batt [coal] (a) An English term for a compact black carbonaceous shale, which splits into fine laminae and is often interstratified with thin layers of coal or ironstone. Syn: *bass.* (b) *bastard coal.* Also spelled: *bat.*
battered pile In engineering geology, a *pile* that is driven into the ground at a nonvertical angle to resist lateral loads.

battery ore (bat′-ter-y) A type of manganese ore, generally a pure crystalline manganese dioxide (pyrolusite or nsutite), that is suitable for use in dry cells.

batture (bat′-ture) An elevated part of a riverbed, formed by gradual accumulation of alluvium; esp. the land between low-water stage and a levee along the banks of the lower Mississippi River. Etymol: Louisiana French, from French "battre", "to strike upon or against".

batukite (ba-tuk′-ite) A dark-colored *leucitite* composed of phenocrysts of clinopyroxene and minor olivine in a groundmass of clinopyroxene, magnetite, and leucite. It was named by Iddings and Morley in 1917 for Batuku, Celebes. Obsolete.

Baumé gravity (Bo-may′) The specific weight of a liquid, measured on a scale based on the weight of water; it is used in the petroleum industry for denoting the specific weight of oils. For liquids lighter than water, degrees Baumé=140/(specific gravity of the liquid at 60°F)-130. Cf: *API gravity.*

baumhauerite (baum′-hau-er-ite′) A lead- to steel-gray mineral of variable composition approximating $Pb_3As_4S_9$, sometimes containing appreciable silver. It crystallizes in orthorhombic, monoclinic and triclinic forms.

baumite (baum′-ite) An unnecessary name for *lizardite;* approximate composition: $(Mg,Mn,Fe,Zn)_3(OH)_4Si_2O_5$.

baum pot (a) A calcareous concretion in the roof of a coal seam; a *bullion.* (b) A cavity left in the roof of a coal seam due to the dropping downward of a cast of a fossil tree stump after removal of the coal. Cf: *coal pipe; saddle [coal].*

baumstarkite A metallic black triclinic mineral: $Ag_3Sb_2(Bi,Sb)S_6$.

bauranoite (ba-u-ran′-o-ite) A reddish-brown mineral: $BaU_2O_7{\cdot}4\text{-}5H_2O$.

bauxite (baux′-ite) An off-white, grayish, brown, yellow, or reddish-brown rock composed of a mixture of various amorphous or crystalline hydrous aluminum oxides and aluminum hydroxides (principally gibbsite, some boehmite and diaspore), along with free silica, silt, iron hydroxides, and esp. clay minerals; a highly aluminous *laterite.* It is a common residual or transported constituent of clay deposits in tropical and subtropical regions, and occurs in concretionary, compact, earthy, pisolitic, or oolitic forms. Bauxite is the principal commercial source of aluminum; the term is also used collectively for lateritic aluminous ores. Bauxite was formerly regarded as an amorphous clay mineral consisting essentially of hydrated alumina, $Al_2O_3{\cdot}2H_2O$. Named after Les Baux de Provence, a locality near Arles in southern France.

bauxitic (baux-it′-ic) Containing much bauxite; e.g. a "bauxitic clay" containing 47% to 65% alumina on a calcined basis, or a "bauxitic shale" abnormally high in alumina and notably low in silica.

bauxitization (baux′-it-i-za′-tion) Development of bauxite from primary aluminum silicates (such as feldspars) or from secondary clay minerals under aggressive tropical or subtropical weathering conditions of good surface drainage, such as the dissolving (usually above the water table) of silica, iron compounds, and other constituents from alumina-containing material.

bavenite (ba′-ve-nite) A white fibrous orthorhombic mineral: $Ca_4Be_2Al_2Si_9O_{26}(OH)_2$. Syn: *duplexite.*

Baveno twin law (Ba-ve′-no) An uncommon twin law in feldspar, in which the twin plane and composition surface are (021). A Baveno twin usually consists of two individuals.

Baventian (Ba-ven′-ti-an) North Sea stage: upper Pliocene (above Ludhamian, below Beestonian); based on shallow-marine and continental beds of Tiglian warm-climate phase. Syn: *Norwich Crag* (upper).

b* axis That axis of a reciprocal crystal lattice which is perpendicular to (010). Cf: *a* axis; c* axis.*

***b* axis [cryst]** One of the crystallographic axes used as reference in crystal description. It is the axis that is oriented horizontally, right-to-left. In an orthorhombic or triclinic crystal, it is usually the *macro-axis.* In a monoclinic crystal, it is the *orthoaxis.* The letter *b* usually appears in italics. Cf: *a axis [cryst]; c axis [cryst].*

***b* axis [struc petrol]** One of three orthogonal reference axes, *a, b,* and *c,* that are used in two distinct ways. (a) To help describe the geometry of a fabric possessing orthorhombic or monoclinic symmetry. The *b* axis is that axis at right angles to the unique plane of symmetry. In many instances this direction is parallel to a fold axis, but in complicated fabrics this need not necessarily be so. (b) In a kinematic sense, to help describe a *deformation plan* that possesses orthorhombic or monoclinic symmetry, such as progressive pure or simple shear. In this case the *b* axis is again the line at right angles to the unique symmetry plane. In a progressive simple shear the *b* axis lies in the shear plane at right angles to the direction of shear. See also: *a axis [struc petrol]; c axis [struc petrol].* Syn: *b direction.*

bay [coast] (a) A wide, curving open indentation, recess, or arm of a sea or lake into the land or between two capes or headlands, larger than a cove, and usually smaller than, but of the same general character as, a gulf. (b) A large tract of water that penetrates into the land and around which the land forms a broad curve. By international agreement (for purposes of delimiting territorial waters), a bay is a water body having a baymouth less than 24 nautical miles wide and an area that is equal to or greater than the area of a semicircle whose diameter is equal to the width of the baymouth. See also: *bight [coast]; embayment [coast].*

bay [geog] (a) Any terrestrial formation resembling a bay of the sea, as a recess or extension of lowland along a river valley or within a curve in a range of hills, or an arm of a prairie extending into and partly surrounded by a forest. Also, a piece of low marshy ground producing many bay trees (such as laurel). (b) A *Carolina bay.* (c) A term used in southern Georgia and in Florida for an arm of a swamp extending into the upland as a baylike indentation (Veatch and Humphrys, 1966, p.23).

bay [ice] (a) *bight [ice].* (b) A part of the sea partly enclosed by ice.

bay [magnet] A transient magnetic disturbance, lasting typically an hour. On a magnetic record it has the appearance of a V or of a bay of the sea.

bayankhanite (bay-ank′-han-ite) A metallic black mineral of variable composition: $Cu_{3\text{-}8}HgS_{3\text{-}5}$. Not approved by the International Mineralogical Association.

bay bar An obsolete syn. of *baymouth bar* and *bay barrier.* The term is "confusing" because it "fails to indicate whether or not the sand ridge is submerged or stands above the water level" (Shepard, 1952, p.1908).

bay barrier A term proposed by Shepard (1952, p.1908) to replace *bay bar,* signifying a spit that has grown "entirely across the mouth of a bay so that the bay is no longer connected to the main body of water". Cf: *baymouth bar.* Obsolete.

bay cable *bottom cable.*

bay delta A delta formed at the mouth of a stream entering, and filling or partially filling, a bay or drowned valley. Cf: *bayhead delta.* Obsolete.

bayerite (bay′-er-ite) A white monoclinic mineral: $Al(OH)_3$. It is a polymorph of gibbsite. Not to be confused with *beyerite.*

Bayesian statistics Statistical methods that utilize *Bayes theorem* and information, about a prior conditional event to estimate the probability of a later, or posterior, conditional event. The term is sometimes applied to all statistical methods that use subjective estimates of probabilities.

Bayes theorem A statement of the relationship between conditional, joint, and marginal probabilities of two or more nonindependent events. Bayes theorem is usually expressed as $p(A\backslash B) = p(A,B)/p\ (B)$ where $p(A\backslash B)$ is the conditional probability that *A* will occur, given that *B* has occurred; $p\ (A,B)$ is the joint probability that *A* and *B* both occur; and $p(B)$ is the marginal probability that *B* occurs. Bayes theorem is widely used in petroleum risk assessment to estimate the probability of a discovery conditional upon specific geologic conditions. Named for the Rev. Thomas Bayes, (1701 1761), an English minister.

bayhead (bay′-head) (a) The part of a bay that lies farthest inland from the larger body of water with which the bay is in contact. (b) A local term in southern U.S. for a swamp at the bayhead.

bayhead bar A feature similar to a *bayhead barrier,* but smaller and generally submerged.

bayhead barrier A barrier formed a short distance from the shore and across a bay near its head. It commonly has a narrow inlet. Cf: *bayhead bar.*

bayhead beach A small, crescentic beach formed at the head of a bay by materials eroded from adjacent headlands. Syn: *pocket beach; cove beach.*

bayhead delta A delta at the head of a bay or estuary into which a river discharges. See also: *bay delta.*

bay ice (a) Relatively smooth sea ice of more than one winter's growth. (b) A term sometimes used in Antarctica for thick ice floes recently broken away from an ice shelf. (c) A term used in Labrador for one-year ice that forms in bays and inlets. (d) An obsolete term for young sea ice sufficiently thick to impede navigation.

baykovite A brown triclinic mineral of uncertain composition, possibly: $Ca_2(Fe,Mg,Ti)_6(Si,Al)_6O_{20}$. Not approved by the International Mineralogical Association.

bayldonite (bayl′-don-ite) A grass-green to yellowish green monoclinic mineral: $PbCu_3O(AsO_3OH)_2(OH)_2$.

bayleyite (bay′-ley-ite) A yellow monoclinic mineral: $Mg_2(UO_2)(CO_3)_3{\cdot}18H_2O$.

baylissite (bay′-lis-site) A colorless monoclinic mineral: $K_2Mg(CO_3)_2{\cdot}4H_2O$.

baymouth (bay′-mouth) The entrance to a bay; the part of a bay that is in contact, and serves as a connection, with the main body of water.

baymouth bar A long, narrow accumulation of sand or gravel, deposited entirely or partly by waves across the mouth or entrance of a bay, so that the bay is no longer connected or is connected only by a narrow inlet with the main body of water; it usually connects two headlands, thus straightening the coast. It can be produced by the convergent growth of two spits from opposite directions, by a single spit extending in one direction, or by a longshore bar being driven shoreward. Cf: *bay barrier; bay bar.*

bayou (bay′-ou) (a) A term variously applied to many local water features in the lower Mississippi River basin and in the Gulf Coast region of the U.S., esp. Louisiana. Its general meaning is a creek or secondary watercourse that is tributary to another body of water; esp. a sluggish and stagnant stream that follows a winding course through alluvial lowlands, coastal swamps, or river deltas. (b) An effluent branch, esp. sluggish or stagnant, of a main river; e.g., a distributary flowing through a delta. Also, the distributary channel that carries floodwater or affords a passage for tidal water through swamps or marshlands. (c) A *bayou lake* or an *oxbow lake.* (d) A *slough* in a salt marsh. (e) An estuarine creek (generally tidal), or an inlet, bay, or open cove on the Gulf Coast. (f) A term used in northern Arkansas and southern Missouri for a clear brook or rivulet that rises in the hills. Etymol: American French "boyau", "gut"; from Choctaw "bayuk", "small stream".

bayou lake A lake or pool in an abandoned and partly closed channel of a stream, as on the Mississippi River delta. Syn: *bayou.*

bay ravinement A disconformity that marks a break in sedimentation in a bay environment; e.g., a surface formed as a result of erosion of underlying beds during relative sea-level rise; a "tidal ravinement unconformity" (Allen and Posamentier, 1993).

bayside beach A beach formed along the side of a bay, generally by materials eroded from adjacent headlands.

bazhenovite (ba'-zhen-ov'-ite) An yellow to orange monoclinic mineral: $[CaS_5]\cdot[CaS_2O_3]\cdot[6Ca(OH)_2]\cdot[20H_2O]$. A product of a burning coal dump.

bazirite (ba-zir'-ite) A colorless hexagonal mineral: $BaZrSi_3O_9$. A member of the *benitoite* group.

bazzite (baz'-zite) An azure-blue hexagonal mineral: $Be_3(Sc,Fe)_2Si_6O_{18}\cdot n(Na,H_2O)$. It is the scandium analogue of beryl, and a member of the *beryl* group.

B-centered lattice A *base-centered lattice* that is centered in the pair of (010) faces. Cf: *A-centered lattice; C-centered lattice.*

BC interray Right posterior interray in echinoderms, situated between B ray and C ray and clockwise of B ray when the echinoderm is viewed from the adoral side; equal to interambulacrum 4 in the *Lovenian system.*

bc-joint A joint in folded sedimentary rock that is parallel to the fold axis. Cf: *longitudinal joint; strike joint.*

BCL *branching crossed lamellar structure.*

BC soil A soil having only B and C horizons.

***b* direction** *b axis [struc petrol].*

beach (a) The unconsolidated material at the shoreline that covers a gently sloping zone, typically with a concave profile, extending landward from the low-water line to the place where there is a definite change in material or physiographic form (such as a cliff), or to the line of permanent vegetation (usually the effective limit of the highest storm waves); at the shore of a body of water, formed and washed by waves or tides, usually covered by sand or gravel, and lacking a bare rocky surface. See also: *strand [coast].* (b) The relatively thick and temporary accumulation of loose water-borne material (usually well-sorted sand and pebbles, accompanied by mud, cobbles, boulders, and smoothed rock and shell fragments) that is in active transit along, or deposited on, the shore zone between the limits of low water and high water. The term was originally used to designate the loose wave-worn shingle or pebbles found on English shores, and is so used in this sense in some parts of England (Johnson, 1919, p.163). (c) A term used locally for a low sand island along a coast. (d) A term commonly used for a seashore or lake-shore area, esp. that part of the shore used for recreation.

beach ball A diagram showing the *first motion* of fault displacement involved in earthquakes.

beach berm *berm [beach].*

beach breccia A breccia formed on a beach where wave action is inefficient and angular blocks are supplied from cliffs, and produced under conditions of rapid submergence (Norton, 1917, p.181).

beach concentrate A natural accumulation in beach sand of heavy minerals selectively concentrated (by wave, current, or surf action) from the ordinary beach sands in which they were originally present as accessory minerals; esp. a *beach placer.*

beach crest A temporary ridge or berm marking the landward limit of normal wave activity (Veatch and Humphrys, 1966, p.24). Cf: *berm crest.*

beach cusp A low seaward projection of sand, pebbles, gravel, or boulders, formed on the foreshore of a beach by wave action; specif. a relatively small cusp along a straight beach. Distance between beach cusps is commonly 10-60 m; it generally increases with increase in wave height.

beach cycle The periodic retreat and outbuilding of a beach under the influence of tides and waves: cutting back occurs during periods of spring tides and of high waves produced by winter storms; building out occurs during periods of neap tides and of low waves characteristic of summer.

beach erosion The destruction and/or removal of beach materials by wave action, tidal currents, littoral currents, or wind.

beach face The section of the beach normally exposed to the action of wave uprush; the *foreshore* of a beach. Not to be confused with *shoreface.*

beach firmness The ability of beach sand to resist pressure; the "strength" of the sand. It is controlled by the degree of packing and sorting, by moisture content, by the quantity of trapped air, and by sand-particle size: the damper and finer-grained the sand, the firmer the beach.

beach-head map A map used by the military that indicates the character of beaches, sediments, and other factors which affect beach mobility as interpreted from topographic and geologic maps, aerial photography, and ground reconnaissance (Rose, 1994).

beach mining The extraction and concentration of beach placer ore, usually by *dredging.*

beach ore *beach placer.*

beach pad A discrete mass of shoreline sand, with the form of an asymmetrical triangle in plan view. The base lies along the beach, the gentle slope faces updrift, and the short steep slope faces downdrift. Pads have a quasi-regular spacing of tens or hundreds of meters and move slowly along the shoreline (Tanner, 1975, p.175).

beach placer A placer deposit of heavy minerals, e.g. zircon, ilmenite, or rutile, on a contemporary or ancient beach or along a coastline; a *beach concentrate* containing valuable minerals. Syn: *beach ore; marine placer; mineral sands.*

beach plain A continuous and level or undulating area formed by closely spaced successive embankments of wave-deposited beach material added more or less uniformly to a prograding shoreline, such as to a growing compound spit or to a cuspate foreland (Johnson, 1919, p.297, 319); a *wave-built terrace.*

beach platform *wave-cut bench.*

beach pond A coastal lake ponded by a *sandbar.* Syn: *bar lake; barrachois.*

beach pool (a) A small, usually temporary, body of water between two beaches or two beach ridges, or a lagoon behind a beach ridge. (b) A pool adjoining a lake and resulting from wave action.

beach profile The topographic trace of a beach surface perpendicular to the shoreline. The profile is variable, depending on position within the *beach cycle.* An accreting beach has a slightly concave-upward foreshore and a gently landward-sloping backshore; an erosional profile is commonly all foreshore, with the backshore removed. Cf: *Emery method.*

beach rhythm The sinuous shape commonly displayed on sandy shorelines where the wavelength is typically in tens of meters and the amplitude is up to 5 meters. Syn: *rhythmic topography; beach cusp.*

beach ridge A low, essentially continuous mound of beach or beach-and-dune material (sand, gravel, shingle) on the backshore of a beach beyond the present limit of storm waves or the reach of ordinary tides, and occurring singly or as one of a series of approximately parallel deposits. The ridges are roughly parallel to the shoreline and represent successive positions of an advancing shoreline. Syn: *full.* Cf: *apposition beach.*

beachrock (beach'-rock) A friable to well-cemented sedimentary rock, formed in the intertidal zone in a tropical or subtropical region, consisting of sand or gravel (detrital and/or skeletal) cemented with calcium carbonate, i.e., beach sediment cemented in situ. Also spelled: beach rock. Cf: *calcrete.*

beach scarp An almost vertical slope on a beach, caused by wave erosion. It may range in height from several centimeters to a few meters, depending on the character of the wave action and the nature and composition of the beach; common on recently nourished beaches.

beach width The horizontal dimension of the beach as measured normal to the shoreline.

bead In blowpipe analysis of minerals, a drop of a fused material, such as a *borax bead* used as a solvent in color testing for various metals. The addition of a metallic compound to the bead will cause the bead to assume the color that is characteristic of the metal. See also: *blowpiping.*

beaded drainage (bead'-ed) A series of small pools connected by short watercourses. The pools result from the thawing of ground ice that commonly represents the intersections of ice wedges. Pools may be 1-4 m deep and 100 m across (Washburn, 1973, p.235). Syn: *beaded stream.*

beaded esker An esker with numerous bulges or swellings (commonly representing fans or deltas) along its length. It may have been formed in lakes or other water bodies during pauses in the retreat of the glacier that nurtured the esker-forming stream.

beaded helictite A variety of *helictite* containing tiny beads of aragonite or calcite, arranged like beads on a string.

beaded lake One of a string of lakes, as a *paternoster lake* or a long narrow lake between sand dunes.

beaded stream A stream consisting of a series of small pools or lakes connected by short stream segments; e.g. a stream commonly found in a region of paternoster lakes or an area underlain by permafrost. Syn: *beaded drainage.*

beak [bot] A long, prominent point of a plant part, e.g. of a fruit or pistil.

beak [paleont] (a) The generally pointed extremity of a bivalve shell, marking the point of the initial growth of the shell; specif. the projection of the dorsal part of a pelecypod valve, located along or above the hinge line, and typically showing strong curvature and pointing anteriorly, or the tip of the *umbo* of a brachiopod, located adjacent or posterior to the hinge line and in the midline of the valve. Cf: *umbo.* (b) The prolongation of certain univalve shells, containing the canal (as in a gastropod). The term in this sense is not generally used by paleontologists. (c) The generally calcified tips of the otherwise horny jaws of *Nautilus* and some coleoid cephalopods; also found in fossil cephalopods. Cf: *rhyncholite.* (d) The skeletal rim or *rostrum* around the palate of an avicularium in cheilostome bryozoans. (e) The toothless jaws (of bone sheathed with horn) of a bird, turtle, dicynodont, dinosaur, or pterosaur.

beaked apex The upper angle of *tergum* of a cirripede crustacean, produced into a long narrow point (TIP, 1969, pt. R, p.91).

beak ridge A more or less angular linear elevation of a brachiopod shell, extending from each side of the *umbo* so as to delimit all or most of the *cardinal* area (TIP, 1965, pt. H, p.140).

Beaman stadia arc (Bea'-man) A specially graduated arc attached to the vertical circle of an alidade or transit to simplify the computation of elevation differences for inclined stadia sights (without the use of vertical angles). The arc is so graduated that each division on the arc is equal to 100(0.5 sin 2*A*), where *A* is the vertical angle. Named after William M. Beaman (1867-1937), U.S. topographic engineer, who designed it in 1904.

beam balance *Westphal balance.*

beam steering A method for emphasizing seismic energy coming from a particular direction by delaying successive channels so that events of a certain dip moveout, or apparent velocity, occur at the same time, and then summing them. This procedure can be repeated for a succession of different dip moveouts to determine direction of approach. See also: *steer.*
beam width In *SLAR,* the angle subtended in the horizontal plane by the radar beam.
bean ore A loose, coarse-grained pisolitic iron ore; limonite occurring in lenticular aggregations. See also: *pea ore.*
beard A fibrous mineral growth in a *pressure shadow* or on one or both ends of a mineral grain.
bearing (bear'-ing) (a) The angular *direction* of any place or object at one fixed point in relation to another fixed point; esp. the horizontal direction of a line on the Earth's surface with reference to the cardinal points of the compass, usually expressed as an angle of less than 90 degrees east or west of a reference meridian adjacent to the quadrant in which the line lies and referred to either the north or south point (e.g. a line in the NE quadrant making an angle of 50 degrees with the meridian will have a bearing of N 50° E). Cf: *azimuth [surv].* See also: *true bearing; magnetic bearing; compass bearing; back bearing.* (b) The horizontal direction of one terrestrial point from another (such as an observer on a ship), usually measured clockwise from a reference direction and expressed in degrees from zero to 360; specif. *astronomic azimuth.* (c) Relative position or direction, as in reference to the compass or to surrounding landmarks. Usually used in the plural.
bearing capacity The maximum load per unit of area that the ground can safely support without failing in shear or undergoing unacceptable settlement. Cf: *allowable bearing capacity; ultimate bearing capacity; California bearing ratio.*
bearing tree A tree forming a *corner accessory,* its distance and direction from the corner being recorded. It is identified by prescribed marks cut into its trunk. Syn: *witness tree.*
bearsite (be-ar'-site) A white monoclinic mineral: $Be_2(AsO_4)(OH){\cdot}4H_2O$.
bearthite A white or yellowish monoclinic mineral: $(Ca,Sr)_2Al(PO_4)_2OH$. A member of the *brackebuschite* group.
Beaufort wind scale (Beau'-fort) A wind scale commonly used at sea. It is based on the visible effects of wind on the sea surface or on fixed objects. Code numbers and descriptive terms are assigned to various ranges of wind speed, e.g. a wind speed of 8-10 mph (or 7-10 knots) is Beaufort Code Number 3, and is called a 'gentle breeze'. The scale is a modernized version of that devised by Admiral Beaufort of the British Navy early in the nineteenth century.
beaverite (bea'-ver-ite) A canary-yellow rhombohedral mineral: $Pb(Fe^{3+},Cu,Al)_3(SO_4)_2(OH)_6$. A member of the *alunite* group.
beaver meadow (bea'-ver) An area of soft, moist ground resulting from the construction of a beaver dam; a beaver pond that has been changed into a marsh of grass or sedge upon abandonment of the dam by the beavers.
bebedourite (be-be-dour'-ite) A *pyroxenite* that contains biotite, with accessory perovskite, apatite, and titanomagnetite. Obsolete.
bechererite A vitreous light green trigonal mineral: $(Zn,Cu)_6Zn_2(OH)_{13}[(S,Si)(O,OH)_4]_2$.
beck A British term for a small stream or brook, often with a stony bed, a rugged or winding course, and a rapid flow. Etymol: Old Norse.
Becke line In optics, a concentration of light, usually taking the form of a thin, bright line about a fragment submerged in an *immersion liquid.* If the stage of a petrographic microscope is moved downward, the Becke line moves into the medium of higher *index of refraction,* and thus may be used to determine the relative indices of refraction of the crystal and the immersion liquid.
beckelite-(Ce) (beck'-e-lite) A wax-yellow to brown isotropic (metamict?) mineral: $Ca_3(Ce,La,Y)_4(Si,Zr)_3O_{15}$. It may be britholite.
beckerite (beck'-er-ite) A brown variety of *retinite* having a very high oxygen content (20-23%).
Becke test In optical mineralogy, a test under the microscope, at moderate or high magnification, for comparing the indices of refraction of two contiguous minerals, or of a mineral and a mounting medium or immersion liquid, in a thin section or other mount. If these substances differ materially in refractive index, they are separated by a bright line (the *Becke line*), which moves toward the more refractive substance when the stage of the microscope is lowered, and away from that substance when the stage is raised.
beckite (beck'-ite) Original, but incorrect, spelling of *beekite.*
becquerelite (bec-que-rel'-ite) An amber to yellow secondary mineral: $Ca(UO_2)_6O_4(OH)_6{\cdot}8H_2O$. It occurs in small orthorhombic crystals and crusts on pitchblende.
Becquerel ray (Bec-que-rel') A term used, before the terms alpha, beta, and gamma rays were introduced, for the particles emitted during radioactive decay. Obsolete.
bed [geomorph] (a) The ground upon which any body of water rests, or the land covered by the waters of a stream, lake, or ocean; the *bottom* of a watercourse or of a stream channel. (b) The land surface marking the site of a former body of water, or representing land recently exposed by recession or by drainage. Syn: *floor [geomorph]; bottom [geomorph].*
bed [stratig] (a) A layer of sediments or sedimentary rocks bounded above and below by more or less well-defined bedding surfaces. A bed (or beds) is the smallest formal lithostratigraphic unit of sedimentary rocks. The designation of a bed or a unit of beds as a formally named lithostratigraphic unit generally should be limited to certain distinctive beds whose recognition is particularly useful. Coal beds, oil sands, and other beds of economic importance commonly are named, but such units and their names usually are not a part of formal stratigraphic nomenclature (NACSN, 1983, Art. 26; ISSC, 1994, p.34-35). (b) A relatively conformable succession of genetically related laminae or lamina-sets bounded by surfaces (called bedding surfaces) of erosion, non-deposition or their correlative conformities (Campbell, 1967). (c) *key bed.*
bed configuration A group of bed forms, produced by flow and making up a particular bed geometry. Cf: *bed phase.*
bedded [ore dep] (bed'-ded) (a) Said of a vein or other mineral deposit that follows the bedding plane in a sedimentary rock. (b) Said of a layered replacement deposit. Cf: *stratiform [ore dep]; strata-bound.*
bedded [stratig] Formed, arranged, or deposited in layers or beds, or made up of or occurring in the form of beds; esp. said of a layered sedimentary rock, deposit, or formation. The term has also been applied to nonsedimentary material that exhibits depositional layering, such as the "bedded deposits" of volcanic tuff alternating with lava in the mantle of a stratovolcano. See also: *stratified; well-bedded.*
bedded barite A *mineral-deposit type* characterized by stratiform lenses of barite interbedded with cherty and calcareous sedimentary rocks.
bedded chert Brittle, closely jointed, rhythmically layered *chert* occurring in areally extensive deposits, with thicknesses measured in tens of meters, and consisting of distinct, usually even-bedded layers (3-5 cm thick), separated by partings of dark siliceous shale or by layers of siderite. Most bedded cherts are believed to be the result of crystallization of biogenic opaline skeletons such as those of diatoms and radiolaria. Examples include the Monterey (Miocene) and the Franciscan (Jurassic?) cherts of California, the Mesozoic radiolarian cherts of the Alps and Apennines, and precipitation of hydrous sodium silicate and conversion to bedded chert in volcanic terrains such as Lake Magadi, Kenya. See also: *novaculite.*
beddedness index (bed'-ded-ness) *stratification index.*
bedded vein A *vein* or *mineral deposit* oriented parallel to the stratification of the host rock. Cf: *blanket deposit [ore dep]; blanket vein.*
bedded volcano A less preferred syn. of *stratovolcano.*
bedding [mining] (bed'-ding) (a) A quarrying term applied to a structure occurring in granite and other crystalline rocks that tend to split in well-defined planes more or less horizontally or parallel to the land surface. (b) The storing and mixing of different ores in thin layers in order to blend them more uniformly in reclamation.
bedding [stratig] (a) The arrangement of sedimentary rocks in beds or layers of varying thickness and character; the general physical and structural character or pattern of the beds and their contacts within a rock mass, such as *cross-bedding* and *graded bedding;* a collective term denoting the existence of beds. Also, the structure so produced. The term may be applied to the layered arrangement and structure of an igneous or metamorphic rock. See also: *stratification [sed].* Syn: *layering [stratig].* (b) *bedding plane.*
bedding cleavage Little-used term for cleavage that parallels bedding in a rock body. Syn: *bedding-plane cleavage.*
bedding fault A fault whose surface is parallel to the bedding plane of the constituent rocks. Cf: *bedding glide.* Syn: *bedding-plane fault.*
bedding fissility The property possessed by a sedimentary rock (esp. shale) of tending to split more or less parallel to the bedding; *fissility* along bedding planes. It is a primary foliation that forms in a sedimentary rock while the sediment is being deposited and compacted, and is a result of the parallelism of the platy minerals to the bedding plane. Fissility is particularly well developed in organic-rich black shales and its presence implies absence of bioturbation.
bedding glide A little-used term for a nearly horizontal overthrust fault produced by bedding-plane slip (Nelson, 1965). Cf: *bedding fault.* Syn: *bedding thrust.*
bedding index *stratification index.*
bedding joint In sedimentary rock, a joint that is parallel to the bedding plane; a joint that follows a bedding plane. Syn: *bed joint.*
bedding plane (a) A planar or nearly planar *bedding surface* that visibly separates each successive layer of stratified rock (of the same or different lithology) from the preceding or following layer; a plane of deposition. It often marks a change in the circumstances of deposition, and may show a parting, a color difference, or both. (b) A term commonly applied to any bedding surface, even when conspicuously bent or deformed by folding. (c) A term commonly applied to a plane of discontinuity (usually the bedding plane) along which a rock tends to split or break readily. Syn: *bedding [stratig]; bed plane; stratification plane; plane of stratification.*
bedding-plane cave (a) A cave with *passages,* generally much wider than high, that have developed along a bedding plane, usually by dissolution of one of the beds. (b) A passage formed along a bedding plane, especially when there is a difference in susceptibility to corrosion in the two beds.
bedding-plane cleavage *bedding cleavage.*
bedding-plane fault *bedding fault.*
bedding-plane parting A parting or surface of separation between adjacent beds or along a bedding plane.

bedding-plane sag Depressed and disturbed strata or laminae of tuff or other deposit into which a volcanic block or bomb has fallen. Cf: *secondary crater.* Syn: *bomb sag.*

bedding-plane slide A submarine slide in which the mass moves more or less intact, along a planar slip surface parallel to the bedding, without backward rotation of the mass (Saxov, 1982).

bedding-plane slip The slipping of sedimentary strata along bedding planes during folding. It produces *disharmonic folding* and, in extreme form, *décollement.* Cf: *flexural slip.*

bedding surface A surface, usually conspicuous, within a mass of stratified rocks, representing an original surface of deposition; the surface of separation or interface between two adjacent beds of sedimentary rock. If the surface is more or less regular or nearly planar, it is called a *bedding plane.*

bedding thrust *bedding glide.*

bedding void An open space between successive lava flows, formed where an overlying flow does not completely conform to the solidified crust of lava beneath it.

bederite A vitreous dark brown to black orthorhombic mineral: $Ca_2Mn^{2+}{}_4Fe^{3+}{}_2(PO_4)_6{\cdot}2H_2O$. A member of the *wicksite* group.

bed form (a) A bedding surface feature that is an individual element of the morphology of a mobile granular or cohesive bed that develops due to local deposition and/or erosion in response to the interaction of a flowing current of air or water. Bed forms range from flat, almost featureless surfaces (e.g., *upper plane bed*) to complex forms spanning a wide range of sizes that are characterized by topographic highs and lows of varying morphology. The nature of the currents that produce bed forms range from simple, *unidirectional currents* (e.g., in a flowing stream) to complex, multidirectional currents. Examples of bed forms include *current ripple, oscillation ripple, dune, current crescent, draa,* and *flute.* See also: *bed configuration.* (b) Any deviation from a flat bed, generated by the flow on the bed of an alluvial channel (Middleton, 1965, p.247).

bediasite (be-dias'-ite) A tektite from east-central Texas named after the town of Bedias.

bed joint *bedding joint.*

bed load The part of the total *stream load* that is moved on or immediately above the stream bed, such as the larger or heavier particles (boulders, pebbles, gravel) transported by traction or saltation along the bottom; the part of the load that is not continuously in suspension or solution. See also: *bed-material load; contact load; saltation load.* Also spelled: bedload. Syn: *bottom load; traction load.*

bed-load function The rate at which various streamflows for a given channel will transport the different particle sizes of the *bed-material load.*

bed material The material of which the bed of a stream is composed; it may originally have been the material of *suspended load* or of *bed load,* or may in some cases be partly residual.

bed-material load The part of the total *sediment load* (of a stream) composed of all particle sizes found in appreciable quantities in the bed material; it is the coarser part of the load, or the part that is most difficult to move by flowing water. See also: *bed load.* Cf: *wash load.*

bed moisture *inherent moisture.*

Bedoulian (Be-dou'-li-an) A European substage (France, Switzerland): Lower Cretaceous (lower Aptian; below Gargasian Substage).

bed phase "The aggregate of all bed states that involve a particular kind of bed form or combination of bed forms" (Southard, 1971, p.904). Thus a bed phase is a general class of bed configurations produced over a particular range of hydraulic conditions, by flow acting on a bed composed of a particular range of bed materials. For example, the ripple bed phase can be defined, for quasi-equilibrium unidirectional flows, by a certain range of conditions of depth, velocity, and grain size. See also: *bed form; bed state.*

bed plane *bedding plane.*

bedrock A general term for the rock, usually solid, that underlies soil or other unconsolidated, superficial material. A British syn. of the adjectival form is *solid,* as in *solid geology.*

bedrock channel A channel with bedrock exposed along at least half of the bed and banks or with a sediment veneer that is largely mobilized during high flows, so that the underlying bedrock geometry strongly influences channel processes.

bedrock valley A valley eroded in bedrock.

bed roughness The degree to which beds under currents depart from being planes, from gross to microscopic scale. Sediment cover of differing sizes increases bed roughness and hence promotes ease of entrainment; it is more difficult for currents of wind or water to entrain sand from a floor of sand than from a floor of sand mixed with gravel. Bed roughness is increased by irregularities of any kind, e.g. the presence of ripple marks or irregular masses of bedrock.

beds (a) An informal term for strata that are incompletely known, constitute a lithologically similar succession, or are of economic significance. (b) A formal stratigraphic unit consisting of several continuous strata of similar lithology (ISSC, 1994, p.34-35).

bed separation In mining, the parting of strata along bedding planes caused by differential subsidence above a mine roof. It is an important factor in the engineering structure of mines.

bedset A relatively conformable succession of genetically related beds bounded by surfaces (called bedset surfaces) of erosion, non-deposition, or their correlative conformities (Campbell, 1967).

bed state The average of all particular bed configurations that can be formed by a given flow over a bed (Southard, 1971, p.904). See also: *bed form; bed configuration; bed phase.*

beef A quarrying term, used originally in Purbeck, southern England, for thin, flat-lying veins or layers of fibrous calcite, anhydrite, gypsum, halite, or silica, occurring along bedding planes of shale, giving a resemblance to beef. It appears to be due to rapid crystallization in lenticular cavities.

beegerite (bee'-ger-ite) A mixture of *matildite* and *schirmerite.*

beekite (beek'-ite) (a) White, opaque silica occurring in the form of subspherical, discoid, rosettelike, doughnut-shaped, or botryoidal accretions, commonly found as bands or layers on silicified fossils or along joint surfaces as a replacement of organic matter; e.g. chalcedony pseudomorphous after coral, shell, or other fossils. See also: *ooloid.* (b) Concretionary calcite commonly occurring in small rings on the surface of a fossil shell that has weathered out of its matrix. Named after Dr. Beek, dean of Bristol. Originally spelled: *beckite.*

beerbachite (beer'-bach-ite) A *hornfels,* originally described as a hypabyssal dike rock, that is typically light-colored, similar to aplite in appearance but chiefly composed of fine-grained labradorite, orthopyroxene, clinopyroxene, and magnetite.

Beestonian (Bee-ston'-ian) North Sea stage: basal Pleistocene (above Baventian, below Pastonian); based on cold-climate deposits of Eburonian glacial phase.

beetle stone (bee'-tle) (a) A *septarium* of coprolitic ironstone, the enclosed coprolite resembling the body and limbs of a beetle. (b) An old name for *turtle stone.*

beforsite (be-for'-site) A hypabyssal dolomitic carbonatite, named by Eckermann for Bergeforseu, Alnö, Sweden. Cf: *rauhaugite.* Obsolete.

beheaded stream (be-head'-ed) The diminished lower part of a stream whose headwaters have been captured by another stream.

beheading (be-head'-ing) (a) The cutting-off of the upper part of a stream and the diversion of its headwaters into another drainage system by capture. (b) The removal of the upper part of a stream's drainage area by wave erosion. Cf: *betrunking.*

behierite (be-hier'-ite) An adamantine pinkish-gray tetragonal mineral: $(Ta,Nb)BO_4$.

behoite (be-ho'-ite) A colorless orthorhombic mineral: $Be(OH)_2$. A dimorph of clinobehoite.

beidellite (bei'-del-lite) A member of the dioctahedral *smectite* group of minerals in which the layer charge is the result of substitution primarily in the tetrahedral sheet. An ideal formula is $R_{0.33}{+}Al_2(Si_{3.67}Al_{0.33})O_{10}(OH)_2$. Cf: *smectite.*

bekinkinite (be-kin'-kin-ite) An igneous rock composed chiefly of sodic amphibole, clinopyroxene, nepheline, and olivine, along with plagioclase, biotite, and analcime; essentially a *theralite.* Cf: *fasinite.* Named by Rosenbusch in 1907 for Bekinkina, Malagasy. Obsolete.

bel A term applied in India and Pakistan to "sandy islands in the beds of rivers" (Stamp, 1961, p.60). Etymol: Panjabi. Also spelled: *bhel.*

Belanger's critical velocity (Bel'-an-ger's) *critical velocity* (c).

belemnite (bel'-em-nite) Any member of an order of coleoid cephalopods characterized by a well-developed internal shell consisting of a guard, phragmocone, and forward-projecting daggerlike or spadelike proostracum. The body has a ten-armed crown, each arm equipped with a double row of arm hooks. Fossil phragmocones are cigar-shaped. Range, Mississippian to Eocene.

belemnoid (bel'-em-noid) A broad term applied to any one of the belemnitelike coleoids including, besides the belemnites, those forms having a tripartite proostracum and those with a living chamber and tentacles without arm hooks.

belendorffite A metallic blackish brown rhombohedral mineral: Cu_7Hg_6.

belite (be'-lite) A calcium orthosilicate found as a constituent of portland-cement clinkers; specif. *larnite.* Syn: *felite.*

belkovite An adamantine brown hexagonal mineral: $Ba_3(Nb,Ti)_6(Si_2O_7)_2O_{12}$.

bell A cone-shaped nodule or concretion in the roof of a coal seam, which may fall without warning. Cf: *caldron bottom; pot bottom; camel back.* See also: *bell hole [coal]; coal pipe.*

bellbergite A vitreous colorless hexagonal zeolite mineral: $(K,Ba,Sr)_2Sr_2Ca_2(Ca,Na)_4(Si,Al)_{36}O_{72}{\cdot}30H_2O$.

bell canopy An overhanging, lateral extension of flowstone that is shaped like a bell or mushroom.

bell hole [coal] A cavity in the roof of a coal seam, produced by the falling of a *bell.*

bell hole [speleo] A vertical dissolutional cylinder developed in a cave roof; mainly reported from the humid tropics.

bellidoite (bel-li'-do-ite) A metallic tetragonal mineral: Cu_2Se. A dimorph of berzelianite.

bellingerite (bel'-lin-ger-ite') A light-green or bluish-green triclinic mineral: $Cu_3(IO_3)_6{\cdot}2H_2O$.

bellite (bell'-ite) A red-orange hexagonal mineral: $Pb_5Cl[(Cr,As,Si)O_4]_3$. May be a chromium-bearing mimetite.
bell-jar intrusion An igneous intrusion similar to a *bysmalith* but differing in that the adjacent strata have become domed and severely fractured.
bell-metal ore A syn. of *stannite,* esp. the bronze-colored variety.
belloite A vitreous yellowish- to olive-green monoclinic mineral: $Cu(OH)Cl$.
bell-shaped distribution A *frequency distribution* whose plot has the shape of a bell. A normal or *Gaussian distribution* is bell-shaped.
belly (bel'-ly) *pocket [eco geol].*
beloeilite (bel-oei'-lite) A granular plutonic rock composed of sodalite, lesser amounts of potassium feldspar, and a small amount of mafic minerals. Nepheline may or may not be present. Cf: *tawite.* The rock was named by Johannsen (1931) for Mont Beloeil (now Mont St. Hilaire), Quebec. Obsolete.
belonosphaerite (bel'-o-no-sphae'-rite) An obsolete term for a spherulite whose minerals are in radial arrangement.
belovite *talmessite.*
belovite-(Ce) (be'-lov-ite) A yellow trigonal mineral: $NaSr_3Ce(OH)(PO_4)_3$. Originally called strontiapatite. A member of the *apatite* group.
belovite-(La) A yellow or greenish yellow trigonal mineral: $NaSr_3La(PO_4)_3(F,OH)$. A member of the *apatite* group.
belt A long area of pack ice, measuring 1 km to more than 100 km in width. Cf: *strip [ice].*
belted coastal plain A broad, maturely dissected coastal plain on which there are a series of roughly parallel cuestas alternating with subsequent lowlands or vales; e.g. the Gulf Coastal Plain through Alabama and Mississippi.
belted metamorphics A geomorphic term proposed by Strahler (1946) to describe mountain areas of folded and faulted metamorphosed sedimentary and igneous rocks that have been differentially eroded in distinct elongate subparallel ridges and valleys.
belteroporic fabric (bel'-ter-o-por'-ic) A rock fabric in which the preferred orientation of its mineral constituents was determined solely by the direction of easiest growth. Obsolete.
Beltian orogeny A name proposed by Eardley (1962) for a supposed orogeny at the end of the Precambrian in western North America, based on unconformable relations between the Belt Series and the Cambrian in northwestern Montana. The existence of an orogeny at the time proposed is dubious; strata overlying the Belt Series are Middle Cambrian or younger, and the Belt itself has now been dated radiometrically as much older than 900 m.y.; moreover, it is overlain unconformably farther west by thick sequences of younger Precambrian strata.
belt of no erosion A zone adjacent to a drainage divide where no erosion by overland flow occurs because of insufficient depth and velocity of flow, and a slope that is too gentle to overcome the initial resistance of the soil surface to sheet erosion (Horton, 1945, p.320); its width is equal to *critical length.*
belt of soil moisture *belt of soil water.*
belt of soil water The upper subdivision of the *unsaturated zone,* limited above by the land surface and below by the intermediate belt. This zone contains plant roots and water available for plant growth. Syn: *belt of soil moisture; discrete-film zone; zone of soil water; soil-water zone; soil-water belt.*
belt of wandering The whole breadth of the valley floor that may be worn down or aggraded by a stream.
belyankinite (bel-yan'-kin-ite) A white, yellowish-brown, or black isotropic (amorphous?) mineral of uncertain composition, possibly $Ca_{1-2}(Ti,Zr,Nb)_5O_{12}·9H_2O(?)$.
bementite (be-ment'-ite) A lamellar grayish-yellow or grayish-brown monoclinic mineral: $Mn^{2+}{}_7Si_6O_{15}(OH)_8$. It may contain small amounts of zinc, magnesium, and iron.
ben A Scottish term for a high hill or mountain, or a mountain peak; it is used only in proper names, esp. in those of the higher summits of mountains, as *Ben* Nevis or *Ben* Lomond. Etymol: Gaelic "beann" or "beinn", "peak".
benauite A yellow or brown rhombohedral mineral: $SrFe^{3+}{}_3(PO_4)_2(OH,H_2O)_6$. Originallly called lusungite. A member of the *crandallite* group.
benavidesite (be-na-vi'-de-site) A metallic gray monoclinic mineral: $Pb_4(Mn,Fe)Sb_6S_{14}$. It is the manganese analogue of jamesonite.
bench [coal] A layer of coal; either a coal seam separated from nearby seams by an intervening noncoaly bed, or one of several layers within a coal seam that may be mined separately from the others.
bench [coast] (a) *wave-cut bench.* (b) A nearly horizontal area at about the level of maximum high water on the ocean side of an artificial dike (CERC, 1966, p. A3).
bench [geomorph] A long, narrow, relatively level or gently inclined strip or platform of land, earth, or rock, bounded by steeper slopes above and below, and formed by differential erosion of rocks of varying resistance or by a change of base-level erosion; a small terrace or steplike ledge breaking the continuity of a slope; an eroded bedrock surface between valley walls. The term sometimes denotes a form cut in solid rock as distinguished from one (as a terrace) cut in unconsolidated material. See also: *berm [geomorph]; mesa.*
bench [mining] Step-like zones for access and mining along the walls of an openpit mine.
bench [speleo] A bank lining the side of a *lava tube,* joining both floor and wall (Larson, 1990, p.22). Cf: *curb.*
bench gravel A term applied in North America to gravel beds, on the side of a valley above the present stream bottom, which represent part of the stream bed when it was at a higher level. See also: *bench placer.*
benchland (bench'-land) (a) A bench, esp. one along a river. Also, the land situated in, or forming, a bench. (b) A land surface composed largely of benches; e.g. a *piedmont benchland.*
bench mark (a) A relatively permanent metal tablet or other mark firmly embedded in a fixed and enduring natural or artificial object, indicating a precisely determined elevation above or below a standard datum (usually sea level) and bearing identifying information, and used as a reference in topographic surveys and tidal observations; e.g. an embossed and stamped disk of bronze or aluminum alloy, about 3.75 in. in diameter, with an attached shank about 3 in. in length, which may be cemented in natural bedrock, in a massive concrete post set flush with the ground, or in the masonry of a substantial building. Abbrev: BM. See also: *permanent bench mark; temporary bench mark.* (b) A well-defined, permanently fixed point in space, used as a reference from which measurements of any sort (such as of elevations) may be made.
bench-mark soil A soil that is representative of many similar soils and has been selected for detailed characterization because of its historical significance, wide geographic extent, or importance to soil classification.
bench placer A *bench gravel* that is mined as a placer. Syn: *river-bar placer; terrace placer.*
bench reef A *coral reef* consisting of a coral-algal frame arising directly from the margin of a pre-existing terrace or shelf generally less than 10 m deep.
bench terrace A shelflike embankment of earth with a flat or gently inclined top and often a steep or vertical downhill side, constructed along the contour of sloping land to control runoff and erosion, or to improve stability, as in a road cut. It is also used in series to convert mountainous slopes to arable land or to building lots.
bend [geomorph] (a) A curve or turn in the course, bed, or channel of a stream, not yet developed into a meander. Also, the land area partly enclosed by a bend or meander. (b) A curved part of a lake, inlet, or coastline.
bend [paleont] A rather sharp angulation in the ventral part of an ostracode valve, usually parallel to the free edge and commonly in the position of a carina, in which case it is called a *carinal* bend (Kesling, 1951, p.120).
bend gliding Slip on planes that are being bent about an axis. Usually applied to slip on an atomic scale, but may also refer to slip on bedding or other planes during flexural slip folding.
Bendigo-type vein *turbidite-hosted gold vein.*
bending Folding that occurs in response to forces applied oblique to layering. Cf: *buckle folding.*
bendway (bend'-way) A term applied along the Mississippi River to one of the deep-water pools occurring at the bends of meanders on alternate sides of the river. See also: *crossing.*
beneficiation (ben'-e-fi'-ci-a'-tion) Improvement of the grade of ore by milling, flotation, sintering, gravity concentration, or other processes. The resulting product is a *concentrate.*
Benioff zone *Wadati-Benioff zone.*
benitoite (be-ni'-to-ite) A blue to colorless, transparent, hexagonal mineral: $BaTiSi_3O_9$. It is strongly dichroic, resembles sapphire in appearance, and is sometimes used as a gem. A group name for hexagonal minerals with the general formula $BaRSi_3O_9$, where *R* is Ti, Sn or Zr.
benjaminite (ben'-ja-min-ite') A metallic gray monoclinic mineral: $(Ag,Cu)_3(Bi,Pb)_7S_{12}$.
benleonardite (ben-leon'-ard-ite) A metallic tetragonal mineral: $Ag_8(Sb,As)Te_2S_3$.
benmoreite (ben-more'-ite) (a) In the *IUGS classification,* a *trachyandesite* with Na_2O-$2.0 \geq K_2O$ in wt%. (b) A silica-saturated to undersaturated igneous rock intermediate between *mugearite* and *trachyte,* with a differentiation index between 65 and 75 and K_2O:Na_2O less than 1:2. It is named for Ben More, Mull, Scotland (Tilley and Muir, 1964).
Bennettitales (Ben-net-ti-ta'-les) A group of Mesozoic *gymnosperms* (Cycadeoidales) with pinnate leaves and lateral cones. Vegetatively, they appeared similar to the true *cycads.* Plants were either short and squat (Cycadeoidaceae) or slender and branched (Williamsoniaceae), with the stem covered by an armor of persistent leaf bases. Range from Triassic to Cretaceous. Cf: *cycadophyte, cycad.* Obsolete syn: *cycadeoids; Cycadeoidales.*
Bennittitales (Ben-net-ti-ta'-les) A group of Mesozoic *gymnosperms* (Cycadeoidales) with pinnate leaves and lateral cones. Vegetatively, they appeared similar to the true *cycads.* Plants were either short and squat (Cycadeoidaceae) or slender and branched (Williamsoniaceae), with the stem covered by an armor of persistent leaf bases. Range from Triassic to Cretaceous. Cf: *cycadophyte, cycad.* Obsolete syn: *cycadeoids; Cycadeoidales.*
benstonite (ben'-ston-ite) A white rhombohedral mineral: $(Ba,Sr)_6Ca_6Mg(CO_3)_{13}$.
benthic (ben'-thic) Pertaining to the ocean bottom or organisms living on or in the substrate; also, referring to that environment. Syn: *benthonic; demersal.*
benthonic (ben-thon'-ic) *benthic.*

benthos (ben'-thos) Those forms of marine life that are bottom-dwelling. Certain fish that are closely associated with the benthos may be included. Adj: *benthic.*

bentonite (ben'-ton-ite) (a) Soft clay or greasy claystone composed largely of smectite formed by the chemical alteration of glassy volcanic ash in contact with water. It often contains accessory crystal grains that were originally phenocrysts in the parent rock. The rock commonly has the ability to absorb large quantities of water accompanied by a large increase in volume that can result in a thixotropic gel. The term "taylorite" was used by Knight (1898), after the owner of a quarry near Rock Springs in the Wyoming territory (see Taylor, 1897). It was later renamed "bentonite" after the Benton Formation (formerly Fort Benton Formation) in eastern Wyoming. Syn: *volcanic clay; soap clay; mineral soap; amargosite.* Cf: *fuller's earth.* (b) A commercial term applied to clay deposits (especially bentonite) containing smectite as the essential mineral. This clay presents a very large total surface area, swells in water, and is used chiefly to thicken oil-well drilling mud. (c) Any clay composed dominantly of a smectite clay mineral whose physical properties are dictated by this mineral (Grim and Güven, 1978, p.1).

bentonite debris flow A *debris flow* associated with the seasonal freezing and thawing and extreme cold of the arctic region, formed where easily hydrated bentonite-rich sediments are exposed to surface water (in moderate quantities for at least several weeks) on slopes of 5-30°, and developed in a smooth-sided, fluted, leveed, and U-shaped mudflow channel. Term proposed by Anderson et al. (1969, p.173) for such features near Umiat, Alaska.

bentonitic arkose (ben-ton-it'-ic) A term used by Ross and Shannon (1926, p.79) for a sandy volcanic ash containing less than 25% bentonitic clay minerals.

bentorite (ben'-tor-ite) A violet hexagonal mineral: $Ca_6(Cr,Al)_2(SO_4)_3(OH)_{12}\cdot 26H_2O$.

benyacarite A vitreous greenish yellow orthorhombic mineral: $(K,Na)TiMn^{2+}Fe^{3+}_2(PO_4)(O,F)_2\cdot 15H_2O$. A member of the *mantiennéite* group.

benzene (ben'-zene) A colorless, volatile, highly inflammable toxic liquid that is the simplest member (formula C_6H_6) of the *aromatic hydrocarbon* series. It is usually produced from coal tar or coke-oven gas or synthesized from open-chain hydrocarbons, and is used chiefly as a solvent, as a motor fuel, as a material in the manufacture of dyes, and in organic synthesis. See also: *benzol.*

benzene series The *aromatic hydrocarbon* series of liquids and solids, empirical formula C_nH_{2n-6}, containing the benzene ring; i.e. it consists of benzene, the simplest member, and the homologues of benzene.

benzene, toluene, ethyl benzene, and xylene Volatile, monocyclic aromatic compounds present in coal tar, petroleum products, and various organic chemical product formulations (Cohen and Mercer, 1993). These are the most soluble of the major gasoline components and, therefore, are common indicators of gasoline contamination in groundwater. Abbrev: BTEX.

benzol (ben'-zol) A commercial form of *benzene,* which is at least 80 percent benzene but also contains its homologues toluene and xylene.

bequerel The SI unit of activity equal to one disintegration per second. Abbrev: Bq.

beraunite (be-raun'-ite) A deep green to dark-red or brown monoclinic mineral: $Fe^{2+}Fe^{3+}_5(PO_4)_4(OH)_5\cdot 6H_2O$.

berborite (ber-bor'-ite) A colorless trigonal or hexagonal mineral: $Be_2(BO_3)(OH)\cdot H_2O$.

berdesinskiite (ber-de-sin'-ski-ite) A black monoclinic mineral: $V^{3+}_2TiO_5$.

Berek compensator (Ber'-ek) In a microscope used for optical analysis of minerals, a type of *compensator* used for the measurement of the path difference produced by a crystal plate; it is a calcite plate that is cut at right angles to the optic axis and mounted on a rotating axis in the tube slit above the objective. The angle through which it is rotated to reach compensation is the measurement of the path difference.

berezanskite A vitreous white hexagonal member of the *milarite* group: $KLi_3Ti_2Si_{12}O_{30}$.

berezovite (be-re-zov'-ite) A syn. of *phoenicochroite.* Also spelled: beresovite.

berezovskite (be-re-zov'-skite) A variety of chromite with Fe:Mg from 3 to 1. Also spelled: beresofskite.

berg [geomorph] (a) A term used in the Hudson River valley, N.Y., for a mountain or hill. Etymol: Dutch "bergh", akin to German "Berg", "mountain". (b) A term used in South Africa for a mountain or mountain range. Etymol: Afrikaans.

berg [glaciol] Shortened form of *iceberg.*

bergalite (ber'-ga-lite) A *lamprophyre* containing phenocrysts of melilite, hauyne, biotite, and rare clinopyroxene in a fine-grained groundmass of the same minerals and also nepheline, magnetite, perovskite, apatite, calcite, and glass. Named by Soellner in 1913. Obsolete.

bergenite (ber'-gen-ite) A yellow monoclinic secondary mineral: $Ca_2Ba_4(UO_2)_9O_6(PO_4)_6\cdot 16H_2O$.

Bergeron-Findeisen process *Precipitation* formation in cold clouds whereby ice crystals grow at the expense of supercooled water droplets in response to differences in vapor pressure relative to water and ice surfaces. Also known as the ice-crystal process.

Bergmann's rule (Berg'-mann's) In zoology, the statement that warm-blooded animals tend to be larger in colder parts of the environment than in warmer. Named after Carl Bergmann (d. 1865), German biologist. Cf: *Allen's rule.*

bergschrund (berg'-schrund) A deep and often wide gap or *crevasse* in ice or firn at or near the head of an *alpine glacier* or snowfield. The bergschrund separates moving ice and snow from the relatively immobile ice and snow (*ice apron*) adhering to the confining headwall of a cirque. It may be covered by or filled with snow during the winter but visible and reopened in the summer. Etymol: German "Bergschrund", "mountain crack". Cf: *randkluft.* Syn: *schrund.*

bergschrund action Enlargement of a *cirque* occupied by a glacier through such processes as frost action and abrasion along the *bergschrund.*

bergslagite (bergs'-la-gite) A colorless monoclinic mineral: $CaBe(AsO_4)(OH)$. A member of the *herderite* group.

berg till (a) A glacial *till* deposited intact by grounded icebergs in fresh or saline water bordering an ice sheet. (b) A lacustrine or marine clay containing boulders and stones dropped into it by melting icebergs. Cf: *glacionatant.* Syn: *floe till; subaqueous till.*

bergy bit (berg'-y) A piece of floating ice, generally less than 5 m above sea level and not more than about 10 m across, larger than a *growler.* It is generally glacier ice but may be a massive piece of sea ice or disrupted hummocked ice.

beringite (ber'-ing-ite) A dark-colored sodic-amphibole-bearing *trachyte* containing albite and a smaller amount of potassium feldspar. Its name was derived from Bering Island, Russia. Obsolete.

berkeyite (ber'-key-ite) A transparent gem lazulite from Brazil.

berlinite (ber-lin'-ite) (a) A colorless to rose-red trigonal mineral: $AlPO_4$. It possesses the ordered quartz structure. (b) A group name for trigonal phosphates with the general formula RXO_4, where R = Al,Fe^{3+} and X = P, As.

berm [beach] A low, impermanent, nearly horizontal or landward-sloping bench, shelf, ledge, or narrow terrace on the backshore of a beach, formed of material thrown up and deposited by storm waves. Some beaches have no berms, others have one or several. See also: *storm berm.* Syn: *beach berm; backshore; backshore terrace.*

berm [eng] (a) A relatively narrow, horizontal man-made shelf, ledge, or bench built along an embankment, situated part way up and breaking the continuity of a slope. (b) The bank of a canal opposite the towing path. (c) The margin or shoulder of a road, adjacent to and outside the paved portion. Etymol: Dutch, "strip of ground along a dike".

berm [geomorph] (a) A term introduced by Bascom (1931) for a terracelike or benchlike remnant of a surface developed to middle or late maturity in a former erosion cycle which has since been interrupted by renewed downward cutting following uplift; e.g. the undissected remnant of an earlier valley floor of a rejuvenated stream, or the remnant of an uplifted abrasion platform that underwent wave erosion along the coast. Bascom intended that the term replace strath, although "berm" sometimes includes the *valley shoulder* of a new valley together with the remnant of the old valley floor (Engeln, 1942, p.221). See also: *bench [geomorph]; strath terrace.* (b) A horizontal ledge of land bordering either bank of the Nile River and inundated when the river overflows.

Berman balance In mineralogy, a type of torsion balance used to measure the specific gravity of specimens in the mass range of 10-100 milligrams.

bermanite (ber'-man-ite) A reddish-brown monoclinic mineral: $Mn^{2+}Mn^{3+}_2(PO_4)_2(OH)_2\cdot 4H_2O$.

berm crest The seaward or outer limit or edge, and generally the highest part, of a berm on a beach; a line representing the intersection of two berms or of a berm and the foreshore. The crest of the most seaward berm separates the foreshore from the backshore. Cf: *beach crest.* Syn: *berm edge; crest.*

berm edge *berm crest.*

bermudite (ber-mu'-dite) A lamprophyric extrusive rock composed of small biotite phenocrysts in a groundmass presumably composed of analcime, nepheline, and alkali feldspar; a biotite *nephelinite.* Named by Pirsson in 1914 for the island of Bermuda. Obsolete.

bernalite A vitreous to adamantine dark green orthorhombic mineral: $Fe(OH)_3\cdot nH_2O$.

bernardite (ber-nard'-ite) A metallic black monoclinic mineral: $Tl(As,Sb)_5S_8$.

berndtite (berndt'-ite) A yellow-brown mineral: SnS_2. It occurs as trigonal and hexagonal polytypes.

Bernoulli effect (Ber-noul'-li) The observation that, in a stream of fluid, pressure is reduced as velocity of flow increases. Cf: *Bernoulli's theorem.*

Bernoulli's theorem The statement in hydraulics that under conditions of uniform steady flow of water in a conduit or stream channel, the sum of the velocity head, the pressure head, and the head due to elevation at any given point is equal to the sum of these heads at any other point plus or minus the losses in head between the two points due to friction or other causes (plus if the latter point is upstream and minus if downstream). It was developed by the Swiss engineer Daniel Bernoulli in 1738 (ASCE, 1962). Cf: *Bernoulli effect.*

bernstein (bern'-stein) A syn. of *amber.* Etymol: German "Bernstein", orig. "burn (burnable/burning)" + "stone, gem".

berondrite (be-rond'-rite) A *theralite* similar to *luscladite* but characterized by the presence of amphibole laths and titanian augite as mafic phases, and modal nepheline. Cf: *fasinite; mafraite.* It was named by Lacroix in 1920 for Berondra, Malagasy. Obsolete.

Berriasian (Ber-ri-a'-sian) A stage of the Standard Global Chronostratigraphic Scale: lowermost Cretaceous (above Tithonian of Jurassic, below Valanginian).

berryite (ber'-ry-ite) A metallic bluish gray monoclinic mineral: $(Ag,Cu)_3Pb_2Bi_5S_{11}$.

berthierine (ber'-thi-er-ine') A 1:1 layer silicate mineral of the *serpentine* group $(Fe^{+2},Fe^{+3},Mg)_{2-3}(Si,Al)_2O_5(OH)_4$, found in many of the same geological circumstances as chamosite but is not chamosite. It occurs as monoclinic and trigonal polytypes. Cf: *chamosite.*

berthierite (ber'-thi-er-ite') (a) A dark steel-gray orthorhombic mineral: $FeSb_2S_4$. (b) A group name for orthorhombic minerals with the composition ABS_4, where A = Fe or Mn, and B = Sb or Bi.

berthonite (ber'-thon-ite) *bournonite.*

Bertillon pattern (Ber'-til-lon) A pattern of fine raised terrace lines on the dorsal surface of a trilobite exoskeleton, arranged subconcentrically and resembling a fingerprint.

bertossaite (ber-tos'-sa-ite) A pale pink orthorhombic mineral: $(Li,Na)_2CaAl_4(PO_4)_4(OH,F)_4$. Isotypic with palermoite.

bertrandite (ber'-trand-ite) A colorless to pale yellow orthorhombic mineral: $Be_4Si_2O_7(OH)_2$.

Bertrand lens (Ber'-trand) A removable lens in the tube of a petrographic microscope that is used in conjunction with convergent light to form an interference figure.

beryl (ber'-yl) (a) A mineral: $Be_3Al_2Si_6O_{18}$. It usually occurs in green or bluish-green, sometimes yellow or pink, or rarely white, hexagonal prisms in metamorphic rocks and granitic pegmatites and as an accessory mineral in acid igneous rocks. Transparent and colored gem varieties include emerald, aquamarine, heliodor, golden beryl, and vorobievite. Beryl is the principal ore of beryllium. (b) A group name for analogous silicates, including *stoppaniite, bazzite* and *indialite.*

beryllite (ber'-yl-lite) A white orthorhombic mineral: $Be_3SiO_4(OH)_2{\cdot}H_2O$. Inadequately characterized.

beryllium-10 age method A *radiometric* method of age determination based on measurement of the activity of beryllium-10 (half life = 1.5×10^6 years), a nuclide formed in the upper atmosphere. It has been used in dating deep-sea sediments, and in determining sedimentation rates (Amin et al., 1975, p.1187-1191).

beryllium detector (be-ryl'-li-um) An instrument, commonly portable, that uses the principles of gamma-ray *activation analysis* to detect and analyze for beryllium. An enclosed gamma-ray source, generally ^{124}Sb, transforms ^9Be to ^8Be (2^{α}) plus a neutron (n). Measurement of the neutron production rate allows for quantitative evaluation of beryllium. A popular term for the instrument is *berylometer.*

beryllonite (ber-yl'-lon-ite) A colorless or yellow monoclinic mineral: $NaBePO_4$. It occurs in transparent, topaz-like pseudo-orthorhombic crystals.

beryllosodalite (ber'-yl-lo-so'-da-lite) *tugtupite.*

berylometer (ber-yl-om'-e-ter) *beryllium detector.*

berzelianite (ber-ze'-li-an-ite') A metallic silver-white cubic mineral: $Cu_{1.9}Se$.

berzeliite (ber-ze'-li-ite) (a) A bright-yellow cubic mineral possessing the garnet structure type: $NaCa_2(Mg,Mn^{2+})_2(AsO_4)_3$. It is isomorphous with manganberzeliite. Syn: berzelite, pyrrhoarsenite. (b) A group name for analogous arsenates and vanadates, including *manganberzeliite, palenzonaite* and *schäferite.*

Bessarabian (Bess-ar-ab'-ian) Ponto-Caspian substage: lower upper Miocene, middle Sarmatian (above Volhynian, below Khersonian).

Bessemer ore (Bes'-se-mer) An iron ore that contains very little phosphorus (generally less than 0.045%). It is so named because it was suitable for use in the Bessemer process of steelmaking, a process no longer in use.

Besshi-type massive sulfide deposit A variety of *massive sulfide deposit* associated with submarine calcalkaline basalt and andesite tuffs and flows. Named for the Besshi deposit in Japan.

bessmertnovite (bess-mert'-nov-ite) *bezsmertnovite.*

beta [cryst] (a) In a biaxial crystal, the intermediate *index of refraction.* (b) The interaxial angle between the *a* and *c* crystallographic axes. Cf: *alpha [cryst]; gamma [cryst].*

beta [mineral] adj. Of or relating to one of two or more closely related minerals and specifying a particular physical structure (esp. a polymorphous modification); specif. said of a mineral that is stable at a temperature intermediate between those of its *alpha* and *gamma* polymorphs (e.g. "beta cristobalite" or "β-cristobalite", the high-temperature isometric phase of cristobalite). Some mineralogists reverse this convention, using β for the low-temperature phase (e.g. "beta carnegieite", the triclinic phase of carnegieite produced from alpha carnegieite at temperatures below 690°C).

beta* angle The angle of the reciprocal lattice between the *a* axis* and the *c* axis,* which is equal to the dihedral angle between (100) and (001). Cf: *alpha* angle; gamma* angle.*

beta axis An estimate of the axis of a cylindrical fold, found by determining on a stereographic projection the intersection of two great circles representing two distinct orientations of the folded layers. Also written: β-axis.

beta chalcocite *chalcocite.*

beta decay Radioactive decay of an atomic nucleus involving the emission of a positively or negatively charged *beta particle* or *electron capture.* Cf: *alpha decay.*

beta diagram A diagram used to fix the orientation of the axis of a cylindrical fold, by finding, on a stereographic projection, the intersections of great circles representing distinct orientations of the folded layers. Also written: β-diagram.

beta diversity The change in taxon diversity from place to place or habitat to habitat, along an environmental gradient and arising from differences in habitats. Typically calculated as the ratio of gamma and alpha diversity (Whittaker, 1972). The term is sometimes used to indicate any change in diversity among sites or communities (Sepkoski, 1988).

beta-fergusonite (be-ta-fer'-gu-son-ite') A group name for monoclinic $RNbO_4$ minerals where R is a rare-earth element or Y.

beta-fergusonite-(Ce) A red to reddish brown monoclinic mineral: $(Ce,La)NbO_4$. Syn: *brocenite.*

beta-fergusonite-(Nd) A yellowish brown to brownish red monoclinic mineral: $(Nd,Ce)NbO_4$. A member of the beta-fergusonite group.

beta-fergusonite-(Y) A monoclinic mineral: $YNbO_4$. This represents the original beta-fergusonite.

betafite (bet'-a-fite) A yellow, brown, greenish, or black cubic mineral of the *pyrochlore* group: $(Ca,U)_2(Ti,Nb,Ta)_2O_6(OH)$. It is a uranium-rich member of the group originally found in granitic pegmatites near Betafo, Madagascar. Betafite probably forms a continuous series with pyrochlore; the name is assigned to members of the series with uranium greater than 15%. Syn: *ellsworthite; hatchettolite; blomstrandite; scheteligite.*

beta-iridisite (be'-ta-ir-id'-is-ite) A metallic steel-gray cubic mineral: $Ir_{0.75}S_2$. Not approved by the International Mineralogical Association.

beta-moissanite (be'-ta-mois-san'-ite) A metallic green cubic mineral: SiC.

beta particle A particle, emitted from an atomic nucleus during radioactive decay, which is physically identical with either the electron or the positron. Cf: *alpha particle; gamma radiation.* Less-preferred syn: *beta ray.*

beta quartz The polymorph of quartz that is stable from 573°C to 870°C, that has a vertical axis of six-fold symmetry and six horizontal axes of two-fold symmetry, and that has a lower refractive index and birefringence than those of *alpha quartz.* It occurs as phenocrysts in quartz porphyries, graphic granite, and granite pegmatites. Also spelled: β-quartz. Syn: *high quartz.*

beta ray A less-preferred syn. of *beta particle.*

beta-roselite (be-ta-ro'-se-lite) A rose-red triclinic mineral: $Ca_2Co(H_2O)_2(AsO_4)_2$. It is dimorphous with roselite, and a member of the *fairfieldite* group.

beta-sulfur The yellow monoclinic polymorph of native sulfur. Originally called sulfurite.

beta-uranophane (be'-ta-u-ran'-o-phane) A yellow, yellowish green, or brownish yellow monoclinic secondary mineral: $Ca(UO_2)_2[Si_2O_7]{\cdot}6H_2O$. It is a dimorph of uranophane. Syn: *beta-uranotile.*

beta-uranotile (be'-ta-u-ran'-o-tile) *beta-uranophane.*

beta-vredenburgite (be'-ta-vred'-en-burg-ite) An exsolution mixture or oriented intergrowth of jacobsite and hausmannite. Cf: *alpha-vredenburgite.* Syn: *vredenburgite.*

betekhtinite (be-tekh'-tin-ite) A metallic black orthorhombic mineral: $(Cu,Fe)_{21}Pb_2S_{15}$.

Betfian (Bet-fi'-an) European mammal age: earliest medial Pleistocene (after Villanyan, before Biharian); Syn: early *Biharian.*

bet lands A term used in India and Pakistan for a flood plain. Etymol: anglicization of Panjabi "bét".

betpakdalite (bet-pak'-dal-ite) A lemon-yellow or greenish yellow monoclinic mineral: $Ca_2MgFe^{3+}{}_3As^{5+}{}_2Mo^{6+}{}_8O_{36}(OH){\cdot}23H_2O$.

betrunked river (be-trunked') A river that is shorn of its lower course by *betrunking.* Cf: *dismembered stream.*

betrunking (be-trunk'-ing) The removal of the lower part of a stream course by submergence of a valley or by recession of the coast, leaving the several upper branches of the drainage system to enter the sea as independent streams. Cf: *dismembering; beheading.*

betwixt mountains (be-twixt') An obsolete synonym for *Zwischengebirge.*

beudantite (beu'-dant-ite) (a) A green to black rhombohedral mineral: $PbFe^{3+}{}_3(AsO_4)(SO_4)(OH)_6$. (b) A group name for rhombohedral minerals with the general formula $AB^{3+}{}_3(XO4)(SO4)(OH)6$ where A = Ca, Sr, Ba or Pb, B = Al, Fe^{3+} or Ga, and X = P or As.

beusite (be'-us-ite) A reddish-brown monoclinic mineral: $(Mn^{2+},Fe^{2+},Ca,Mg)_3(PO_4)_2$. It is the manganese analogue of graftonite. Syn: *magniophilite.*

bevel cut Any style of cutting for a gemstone having a very large table joined to the girdle by one or two bevel facets, and a pavilion that may be step cut, brilliant cut, or any other style; it is used predominantly for less valuable gems. See also: *table cut.*

beveled (bev'-eled) Said of a geologic structure or landform that is truncated or cut across by an erosion surface.

beveling (bev'-el-ing) An act or instance of cutting across a geologic structure or landform; e.g. the planation of an anticline or of the outcropping edges of strata on a mountain summit. Cf: *truncation [geomorph].*

beyerite (bey'-er-ite) A yellow, white, or gray tetragonal mineral: $(Ca,Pb)Bi_2O_2(CO_3)_2$. Not to be confused with *bayerite.*

bezel (bez′-el) (a) Inclined surface on the crown of a round brilliant-cut gem between the table and the girdle. (b) Polisher's term for the four facets between the top corner facets. (c) The setting edge, that part of the crown just above the girdle which is gripped by prongs or a bezel mounting. (d) Thin strip of metal encircling a gem stone to secure it in a mounting; used frequently for cabochons, occasionally for faceted diamonds (Liddicoat et al., 1993).

bezel facet One of the eight large four-sided kite-shaped facets on the crown of a round brilliant-cut gem, the upper points joining the table and the lower points joining the girdle. Cf: *star facet.*

bezsmertnovite (bez-smert′-nov-ite) A metallic orthorhombic mineral: $Au_4Cu(Te,Pb)$.

BF *bright field.*

B-form The microspheric form of a foraminifer. Cf: *A-form.*

***b*-glide** A *glide plane* involving reflection across (100) or (001) combined with translation of one-half of a lattice translation along *b*. Cf: *a-glide; c-glide; d-glide; n-glide.*

bhabar (bha-bar′) A great piedmont composed of gravel and fringing the outer margin of the Siwalik Range in northern India. Etymol: Urdu-Hindi "bhabar", "porous". Also spelled: bhabbar.

bhangar (bhan′-gar) A term used in India for a high area (as a terrace, scarp, or hill) consisting of an old alluvial plain so situated within a river valley that it is beyond the reach of river floods. Etymol: Urdu-Hindi "bangar". Cf: *khadar.* Also spelled: bangar.

bheel *bhil.*

bhel *bel.*

bhil A term applied in the Ganges delta region of India to a brackish and stagnant body of water (such as an oxbow lake or marsh) occupying an *interdistributary bay* that is often below sea level. Etymol: Bengali. Cf: *jhil.* Also spelled: bil; bheel.

bhit A term used in West Pakistan for a sand hill or sand ridge. Etymol: Sindhi.

B horizons Horizons that formed below an A, E, or O horizon and are dominated by obliteration of all or much of the original rock structure and show one or more of the following: (1) illuvial concentration of silicate clay, iron, aluminum, humus, carbonates, gypsum, or silica, alone or in combination; (2) evidence of removal of carbonates; (3) residual concentration of sesquioxides; (4) coatings of sesquioxides that make the horizon conspicuously lower in value, higher in chroma, or redder in hue than overlying and underlying horizons without apparent illuviation of iron; (5) alteration that forms silicate clay or liberates oxides or both and that forms granular, blocky, or prismatic structure if volume changes accompany changes in moisture content; or (6) brittleness.

BHP *bottomhole pressure.*

bhur A term used in India and Pakistan for a hill or patch of windblown sandy soil, frequently capping the high bank of a river. Etymol: Urdu-Hindi.

bialite *wavellite.*

bianchite (bi-an′-chite) A monoclinic mineral occurring as white or yellowish crystalline crusts: $(Zn,Fe^{2+})(SO_4)\cdot 6H_2O$. A member of the *hexahydrite* group.

Biarritzian (Bi-ar-ritz′-i-an) European stage: upper part of the Middle Eocene, essentially equivalent to Bartonian stage.

bias (bi′-as) (a) A purposeful or accidental distortion of observations, data, or calculations in a systematic or nonrandom manner. (b) The amount by which the average of a set of values departs from a reference value. (c) Superposing an additional magnetic field upon the magnetic field associated with the signal during magnetic tape recording, a procedure used to obtain a linear relationship between the amplitude of the signal and the remanent flux density in the recording medium. (d) A voltage that is maintained at a point in a circuit so that the device will operate with desired characteristics.

biaxial (bi-ax′-i-al) Said of a crystal having two optic axes and three principal indices of refraction, e.g. of an orthorhombic, monoclinic, or triclinic crystal. Cf: *uniaxial.*

biaxial figure An *interference figure* that may display both optic axes or no optic axes.

biaxial stress A stress system in which only one of the three principal stresses is zero.

bicchulite (bic-chu′-lite) A colorless cubic mineral: $Ca_2Al_2SiO_6(OH)_2$. It is dimorphous with kamaishilite.

biconvex (bi′-con-vex′) Convex on both sides; e.g. said of a bivalved shell having both valves convex.

bidalotite (bi-dal′-o-tite) *anthophyllite.*

bideauxite (bi-deaux′-ite) A colorless cubic mineral: $Pb_2AgCl_3(F,OH)_2$.

bidirectional current A current which causes a back-and-forth motion of the fluid and sediment in transport by changing direction by 180° (or nearly so) over periods ranging from seconds or tens of seconds (e.g., the motion on the bed produced by water surface waves) to several hours (e.g., the currents produced by tides). Cf: *unidirectional current.*

bieberite (bie′-ber-ite) A flesh-red to rose-red monoclinic mineral occurring esp. in crusts and stalactites: $CoSO_4\cdot 7H_2O$. Syn: *red vitriol; cobalt vitriol.* A member of the *melanterite* group.

biehlite A fibrous white monoclinic mineral: $(Sb^{3+},As^{3+})_2MoO_6$.

bif Banded iron formation. Also written: BIF.

bifacies (bi′-fa-cies) A term used by Bailey and Childers (1977) in uranium exploration to refer to "varicolored formations characterized by both oxidized and reduced facies." Cf: *monofacies.*

bifoliate (bi-fo′-li-ate) In stenolaemate Bryozoa, said of an erect colony in which two layers of zooids bud back to back from an interior median wall. Cf: *bilamellar (a).*

biforaminate (bi-fo-ram′-i-nate) Said of a foraminifer (such as *Discorbis*) having both *protoforamen* and *deuteroforamen.*

biform (bi′-form) Having forms of two distinct kinds; e.g. said of the *rhabdosome* of a graptoloid (esp. of a monograptid) with thecae of two conspicuously different shapes, or said of foraminiferal shells having a growth plan that changes during ontogeny.

biform processes In a spore, exine projections with broad bases that abruptly terminate in sharply pointed tips. The term is most commonly applied to Paleozoic spores.

bifurcating link (bi′-fur-cat′-ing) A link of magnitude μ that is formed at its upstream fork by the confluence of two links, each of a magnitude of ½ μ, and that flows at its downstream fork into a link of a magnitude of less than 2μ (Mock, 1971, p.1559). Cf: *link [geomorph]; magnitude [geomorph].*

bifurcation (bi-fur-ca′-tion) (a) The separation or branching of a stream into two parts. (b) A stream branch, common on deltas, produced by bifurcation.

bifurcation ratio The ratio of the number of streams of a given order to the number of streams of the next higher order. According to the law of stream numbers, the ratio tends to be constant for all orders of streams in the basin. It is a measure of the degree of branching within a drainage network.

"big bang" hypothesis The hypothesis that the currently observed expansion of the Universe may be extrapolated back to a primeval cosmic fireball. Depending on the ratio of the initial expansion velocity to the mass of the Universe, which is relatable to currently observable parameters (the deceleration parameter), the Universe may or may not reach a maximum distension and collapse in on itself. Current observational evidence favors an "open" universe that will not collapse in on itself. Indeed, there is evidence that the rate of expansion may actually be accelerating due to "dark energy." Syn: *fireball hypothesis; primeval-fireball hypothesis.* Cf: *steady-state theory.*

bigcreekite A vitreous to pearly white orthorhombic mineral: $BaSi_2O_5\cdot 4H_2O$.

bight [coast] (a) A long gradual bend or gentle curve, or a slight crescent-shaped indentation, in the shoreline of an open coast or of a bay; it may be larger than a bay, or it may be a segment of or a smaller feature than a bay. (b) A tract of water or a large bay formed by a bight; an *open bay.* Example: the Great Australian Bight. (c) A term sometimes, although rarely, applied to a bend or curve in a river, or in a mountain chain.

bight [ice] An extensive crescent-shaped indentation in the ice edge, formed by either wind or current. Syn: *bay [ice]; ice bay.*

bigwoodite (big′-wood-ite) A medium-grained plutonic rock consisting chiefly of microcline, microcline-microperthite, sodic plagioclase (albite), and sodic amphibole, with sodic pyroxene or biotite sometimes substituting for the amphibole; an alkalic *syenite.* It was named by Quirke in 1936 for Bigwood, Ontario, Canada. Obsolete.

BIH *Bureau International de l'Heure.*

Biharian (Bi-har′-i-an) European stage: Middle Quaternary (above Villafranchian, below Olderburgian).

bijvoetite-(Y) (bij′-voet-ite) A yellow orthorhombic mineral: $Y_8(UO_2)_{16}O_8(CO_3)_{16}(OH)_8\cdot 39H_2O$.

bikitaite (bi-ki′-ta-ite) A white mineral of composition $LiAlSi_2O_6\cdot H_2O$. It occurs in monoclinic and triclinic froms.

bil *bhil.*

bilamellar (bi-la-mel′-lar) (a) Said of a cheilostome bryozoan colony consisting of two layers of zooids growing back to back with separate but touching basal walls. Cf: *bifoliate.* Also called: bilaminar. (b) Said of the walls of each chamber (in hyaline calcareous foraminifers) consisting of two primary formed layers.

bilateral symmetry (bi-lat′-er-al) The condition, property, or state of having the individual parts of an organism arranged symmetrically along the two sides of an elongate axis or having a median plane dividing the organism or part into equivalent right and left halves so that they are counterparts one of the other. Cf: *radial symmetry.* Also called: bilateralism.

bilibinskite (bi-li-binsk′-ite) A semi-metallic brown cubic mineral with a crystal structure related to that of gold, and previously referred to as *rickardite*: $Au_3Cu_2Pb\cdot nTeO_2$.

bilinite (bi-lin′-ite) A white to yellowish monoclinic mineral: $Fe^{2+}Fe^{3+}_2(SO_4)_4\cdot 22H_2O$. It occurs in radially fibrous masses.

bill A long, narrow promontory or headland, or a small peninsula, resembling a beak or ending in a prominent spur; e.g. Portland Bill in Dorset, England.

billabong (bil′-la-bong) (a) A term applied in Australia to a blind channel leading out from a river and to a usually dry stream bed that may be filled seasonally. (b) An Australian term for an elongated, stagnant backwater or pool produced by a temporary overflow from a stream, or for an *oxbow lake* that may not be permanently filled with water. Etymol: aboriginal term meaning "dead river".

billietite (bil′-lie-tite) An amber-yellow secondary mineral: $Ba(UO_2)_6O_4(OH)_6 \cdot 4H_2O$. It occurs in orthorhombic plates and is closely related to becquerelite.

billingsleyite (bil′-lings-ley-ite′) A metallic gray cubic mineral: $Ag_7(As,Sb)S_6$.

billitonite (bil′-li-ton-ite′) An Indonesian tektite from Belitung (Billiton) Island, near Sumatra; a tektite from the East Indies.

bilobite (bi′-lo-bite) A *trace fossil* consisting of a two-lobed (bilobate) trail; esp. a shallow pocketlike pit, passage, or burrow shoveled or scratched by a trilobite, or a coffee-bean form with a median groove and transverse wrinkles representing a resting trail made by a trilobite.

biloculine (bi-loc′-u-line) Having two chambers; specif. pertaining to or shaped like *Pyrgo* ("Biloculina"), a genus of calcareous imperforate foraminifers having a two-chambered exterior part of the *test* and found abundantly in the North Sea where their remains form much of the ooze covering the bottom.

bimaceral (bi-mac′-er-al) Said of a coal microlithotype consisting of two macerals. Cf: *monomaceral; trimaceral.*

bimagmatic (bi-mag-mat′-ic) Obsolete generic term for porphyritic rocks in which the minerals occur in two generations. A translation of the German term "bimagmatisch" (Johannsen, 1939, p.203).

bimembrate apparatus A conodont skeletal apparatus consisting of two morphologically distinct element types, regardless of the total number of elements that may have composed the apparatus (TIP, 1981, pt. W, supp.2).

bimodal distribution (bi-mod′-al) A *frequency distribution* with two modes, or intervals having higher frequencies of occurrence than other adjacent intervals. Cf: *polymodal distribution.*

bimodal sediment A sediment whose particle-size distribution shows two maxima (modes).

bin One of a set of discrete rectangular areas into which a survey area is divided. Three-dimensional seismic data are sorted among bins (often 10-25 m in size) according to midpoint locations for unmigrated data or according to reflecting points for migrated data. The data elements within each bin are subsequently summed (stacked) and divided by the number of elements to obtain the output trace for the particular bin.

binary Composed of only two elements.

binary sediment A sediment consisting of a mixture of two components or end members; e.g. a sediment with one clastic component (such as quartz) and one chemical component (such as calcite); or an aggregate containing sand and gravel.

binary system A chemical system containing two components, e.g. the MgO SiO_2 system.

bind A British coal miner's term for any fine-grained, well-laminated rock (such as shale, clay, or mudstone, but not sandstone) associated with coal. See also: *blaes.*

binder (bind′-er) (a) The material that produces or promotes consolidation in loosely aggregated sediments; e.g. a mineral cement that is precipitated in the pore spaces between grains and that holds them together, or a primary clay matrix that fills the interstices between grains. (b) soil binder. (c) A term used in Ireland for a bed of sand in shale, slate, or clay. (d) A coal miner's term used in Pembrokeshire, England, for shale.

bindheimite (bind′-heim-ite) An earthy yellow, brown, white, gray, or greenish cubic secondary mineral: $Pb_2Sb_2O_6(O,OH)$. A member of the *stibiconite* group.

binding coal (bind′-ing) *caking coal.*

bindstone (bind′-stone) A type of *reef rock* or *boundstone* composed of sheetlike colonies encrusting large fossil fragments or finer sediment, forming a layered mass that is partly in-place skeletal and partly bioclastic (Embry and Klovan, 1971; Cuffey, 1985). Cf: *cruststone.*

Bingham substance (Bing′-ham) An idealized material showing linear-viscous behavior above a yield stress. Below the yield stress the material is presumed to be rigid.

binning Assigning a seismic trace to a *bin.*

binnite (bin′-nite) A variety of *tennantite* containing silver.

binocular microscope (bin-oc′-u-lar) A microscope adapted to the simultaneous use of both eyes.

binomen (bi-no′-men) Two Latin or Latinized words which, taken together, are the name of a species. The first word is its *generic* name and the second its *specific* name.

binomial distribution A discrete *frequency distribution* governing the occurrence of independent events that have only two possible outcomes, such as zero and one, or success and failure, when the probability of an outcome does not change from event to event.

binominal nomenclature A system of naming plants, animals, and other organisms in which the name of each species consists of two words (i.e., a *binomen*), the first designating the genus and the second the particular species; e.g., *Phacops rana.* Nonrecommended syn: binomial nomenclature.

biocalcarenite (bi′-o-cal-car′-e-nite) A calcarenite containing abundant fossils or fossil fragments; e.g. a crinoidal limestone. Rarely used.

biocalcilutite (bi′-o-cal′-ci-lut′-ite) A *calcilutite* containing abundant fossils or fossil fragments. Rarely used.

biocalcilyte (bi-o-cal′-ci-lyte) A term used by Grabau (1924, p.297) for a calcareous biogenic clastic rock, such as coral rock, shell rock, or calcareous ooze. Obsolete.

biocalcirudite (bi′-o-cal′-ci-rud′-ite) A *calcirudite* containing abundant fossils or fossil fragments. Rarely used.

biocalcisiltite (bi′-o-cal′-ci-silt′-ite) A *calcisiltite* containing abundant fossils or fossil fragments. Rarely used.

biocementstone (bi′-o-ce-ment′-stone) A type of *reef rock, boundstone,* or *framestone* characterized by evidence of soft-bodied strands, mostly microbial, extending through the rock mass, which is predominantly micstone (Tsien, 1981; Cuffey, 1985).

biocenology (bi′-o-ce-nol′-o-gy) The branch of ecology concerned with all aspects of natural communities and the relationships between the members of those communities. Also spelled: *biocoenology.* Cf: *biosociology.*

biocenosis (bi′-o-ce-no′-sis) *biocoenosis.*

biochemical oxygen demand (bi-o-chem′-i-cal) The amount of oxygen, measured in parts per million, that is removed from aquatic environments rich in organic material by the metabolic requirements of aerobic microorganisms. Abbrev: BOD. Cf: *chemical oxygen demand.* Syn: *biological oxygen demand.*

biochemical rock A sedimentary rock characterized by, or resulting directly or indirectly from, the chemical processes and activities of living organisms; e.g. bacterial iron ores and certain limestones.

biochore (bi′-o-chore) (a) A region with a distinctive fauna and/or flora; specif. one or more similar biotopes. (b) The part of the Earth's surface having a life-sustaining climate, characterized by a major type of vegetation. It consists largely of the *dendrochore.*

biochron (bi′-o-chron) First defined by Williams (1901, p.579-580) as the time of duration of a fauna or flora, the term today signifies the "total time represented by a biozone" (ISG, 1976, p.48); a term discarded in ISSC, 1994.

biochronologic unit (bi′-o-chron′-o-log′-ic) (a) A division of time distinguished on the basis of biostratigraphic or objective paleontologic data; a *geologic-time unit* during which deposition of a biostratigraphic unit took place (Teichert, 1958a, p.117). Examples include a moment corresponding to a biostratigraphic zone, and a biochron equivalent to a time-stratigraphic biozone. (b) A term used by Jeletzky (1956, p.700) to replace *biostratigraphic unit* and *time-stratigraphic unit,* being a material rock unit defined in its type locality "by agreement among specialists, elsewhere by criteria of time-correlation found in the contained rocks, which in practice means geochronologically valuable fossils"; a time-stratigraphic unit considered as a biostratigraphic unit.

biochronology (bi′-o-chron-ol′-o-gy) *Geochronology* based on the relative dating of geologic events by biostratigraphic or paleontologic methods or evidence; i.e. the study of the relationship between geologic time and organic evolution (Teichert, 1958a, p.103). Cf: *orthochronology; parachronology.*

biochronostratic unit (bi′-o-chron′-o-strat′-ic) A chronostratigraphic unit based on a biostratic unit (Henningsmoen, 1961, p.66).

biochronozone (bi-o-chron′-o-zone) A *chronozone* based on a biostratigraphic unit, e.g. *Cardioceras cordatum* biochronozone (NACSN, 1983, Art. 75). Cf: *lithochronozone; polarity chronozone.*

bioclast (bi′-o-clast) (a) A single fossil fragment (Carozzi and Textoris, 1967, p.3). (b) Material derived from "the supporting or protective structures of animals or plants, whether whole or fragmentary" (Sander, 1967, p.327).

bioclastic rock (bi-o-clas′-tic) (a) A rock consisting primarily of fragments that are broken from pre-existing rocks, or are pulverized or arranged, by the action of living organisms, such as plant roots or earthworms (Grabau, 1904). The rock need not consist of biogenic material. The term includes "rocks" (such as concrete) that owe their existence to human activities. (b) A sedimentary rock consisting of fragmental or broken remains of organisms, such as a limestone composed of shell fragments or bedded phosphate deposits composed largely of vertebrate skeletal fragments, coprolites, and similar remains (Boggs, 1987). Cf: *biogenic rock.*

biocoenology (bi′-o-coe-nol′-o-gy) *biocenology.*

biocoenosis (bi′-o-coe-no′-sis) (a) A set of fossil remains found in the same place where the organisms lived (ISSC, 1994, p.54). Cf: *thanatocoenosis.* Syn: *life assemblage.* (b) A group of organisms that live closely together and form a natural ecologic unit. The term was first defined and introduced by the German zoologist Moebius in 1877. Cf: *community.* Var: *biocenosis.* Plural: biocoenoses. Etymol: Greek "bios", "mode of life" + "koinos", "general, common".

biocorrelation (bi′-o-cor′-re-la′-tion) Biostratigraphic correlation. Demonstration of similarity of fossil content and biostratigraphic position between geologic units (NACSN, 1983, p.851). Cf: *lithologic correlation; chronocorrelation.*

biocycle (bi′-o-cy′-cle) A group of related biochores that comprise one of the major divisions of the biosphere; salt water, fresh water, and dry land are biocycles.

biodegradable (bi′-o-de-grad′-a-ble) Said of materials that can be readily decomposed by organisms.

biodegradation Breakdown of organic matter by living organisms.

bioecology (bi′-o-e-col′-o-gy) The branch of ecology concerned with the relationships between plants and animals in their common environment.

bioecozone A body of strata characterized by a certain paleoenvironment that is inferred from its fossil content and that can be distinguished from the paleoenvironment of adjacent strata (Poag and Valentine, 1976, p.188).

bioerosion (bi'-o-e-ro'-sion) Removal of consolidated mineral or lithic substrate by the direct action of organisms (Neumann, 1966).

biofacial zone A biostratigraphic zone that recurs in a single rock succession due to recurrence of the environment in which the organisms represented by the fossil assemblage that characterize the biozone lived (Woodford, 1965, p.164-169).

biofacies (bi-o-fa'-cies) (a) A subdivision of a stratigraphic unit, distinguished from adjacent subdivisions on the basis of its fossils, without respect to nonbiologic features; esp. such a body of sediment or rock recognized by characters that do not affect lithology, such as the taxonomic identity or environmental implications of fossils (Weller, 1958, p.634). (b) The biologic aspect or fossil character of a facies of some definite stratigraphic unit, esp. considered as an expression of local biologic conditions; "the total biological characteristics of a sedimentary deposit" (Moore, 1949, p.17). (c) A distinctive assemblage of organisms formed under one set of environmental conditions, as compared with another assemblage formed at the same time but under different conditions; an ecological association of fossils, or the fossil record of a biocoenosis. See also: *facies.* Syn: *biologic facies.*

biofacies map A *facies map* based on paleontologic attributes, showing areal variation in the overall paleontologic character of a given stratigraphic unit. It may be based on proportions or population discriminants of the fossil organisms present, or on ratios among them.

biogas (bi'-o-gas) Any gas of organic origin that is primarily methane, such as natural gas, marsh gas, or gas from organic matter in landfill deposits.

biogenesis (bi-o-gen'-e-sis) (a) Formation by the action of organisms; e.g. coral reefs. (b) The doctrine that all life has been derived from previously living organisms. Cf: *abiogenesis.*

biogenetic law (bi'-o-ge-net'-ic) The so-called "law" of recapitulation: ontogeny recapitulates phylogeny.

biogenetic rock *biogenic rock.*

biogenic (bi-o-gen'-ic) Referring to minerals or sediments that were produced by organisms. Syn: *biogenous.*

biogenic rock (bi-o-gen'-ic) An *organic rock* produced directly by the physiological activities of organisms, either plant or animal (Grabau, 1924, p.280); e.g., coral reefs, shelly limestone, pelagic ooze, coal, and peat. Cf: *bioclastic rock; biolith.* See also: *phytogenic rock; zoogenic rock.* Syn: *biogenous rock; biogenetic rock.*

biogenous (bi-og'-e-nous) *biogenic.*

biogenous rock (bi-og'-e-nous) *biogenic rock.*

biogeochemical cycling (bi'-o-ge'-o-chem'-i-cal) The cycling of chemical constituents through a biological system.

biogeochemical prospecting *Geochemical exploration* based on the chemical analysis of systematically sampled plants in a region, in order to detect biological concentrations of elements that might reflect hidden ore bodies. The trace-element content of one or more plant organs is most often measured. Cf: *geobotanical prospecting.*

biogeochemistry (bi'-o-ge'-o-chem'-is-try) A branch of geochemistry that deals with the effects of life processes on the distribution and fixation of chemical elements in the biosphere.

biogeography (bi'-o-ge-og'-ra-phy) The science that deals with the geographic distribution of all living organisms. See also: *zoogeography; phytogeography.* Syn: *chorology.*

biogeology (bi'-o-ge-ol'-o-gy) The biological aspects of geology, e.g. systematic paleontology, the study of organically influenced sedimentation, or the identification of a concealed rock unit by the type of plant growing at the surface above it. See also: *geobiology; paleobiology; paleontology.*

bioglyph (bi'-o-glyph) A *hieroglyph* produced by an organism or of biologic origin (Vassoevich, 1953, p.38). Ant: abioglyph. Cf: *trace fossil.* Syn: *organic hieroglyph.*

bioherm (bi'-o-herm) A moundlike, domelike, lenslike, or reeflike mass of rock built up by sedentary organisms (such as corals, algae, foraminifers, mollusks, gastropods, and stromatoporoids), composed almost exclusively of their calcareous remains, and enclosed or surrounded by rock of different lithology; e.g. an *organic reef* or a nonreef limestone mound. Term proposed by Cumings and Shrock (1928, p.599), and defined by Cumings (1930, p.207), as a structural term, although as applied it often stresses calcareous composition. Cf: *biostrome.* Syn: *organic mound.*

biohermal (bi-o-her'-mal) Pertaining to a bioherm, such as a "biohermal limestone" of restricted extent.

biohorizon (bi'-o-ho-ri'-zon) Biostratigraphic horizon. A surface of biostratigraphic change or of distinctive biostratigraphic character, esp. valuable for correlation; it is commonly a *biozone* boundary, although biohorizons are sometimes recognized within biozones. In theory, a biohorizon is strictly a surface or interface; in practice, it may be a thin biostratigraphically distinctive bed. Features on which biohorizons are commonly based include "first appearance" of a given fossil form, "last occurrence", change in frequency and abundance, evolutionary change, and change in the character of individual taxons, e.g. in direction of coiling in foraminifers, or in number of septa in corals (ISG, 1976, p.49). Cf: *chronohorizon; lithohorizon.*

biohydrology (bi'-o-hy-drol'-o-gy) The study of the interactions of water, plants, and animals, including both the effects of water on biota and the physical and chemical changes in water or its environment caused by biota. Cf: *hydrobiology.*

biointerval zone (bi'-o-in'-ter-val) *interval zone.*

biokarst *phytokarst.*

biokinematic (bi'-o-kin'-e-mat'-ic) Said of sedimentary operations in which "the largest displacement vectors occur between a living organism and the unmodified deposit surrounding the structure produced" (Elliott, 1965, p.196); e.g. the activities shown by trace fossils. Also, said of the sedimentary structures produced by biokinematic operations.

bioleaching In *extractive metallurgy,* the use of organisms such as bacteria to facilitate the *leaching* of an ore.

biolite [mineral] (bi'-o-lite) A group name for minerals formed by biologic action (Hey, 1963, p.92).

biolite [sed] (a) *biolith.* (b) An old term for a concretion formed through the action of living organisms.

biolith (bi'-o-lith) A rock of organic origin or composed of organic remains; specif. *biogenic rock.* See also: *phytolith [sed]; zoolith [sed].* Syn: *biolite [sed].*

biolithite (bi-o-lith'-ite) A limestone constructed by organisms (faunal or floral) that grew and remained in place, characterized by a rigid framework of carbonate material that binds allochem grains and skeletal elements. It is typical of reef cores. The major organism should be specified when using the term; e.g. "coral biolithite", or "rudist biolithite". See also: *boundstone.*

biological oceanography (bi-o-log'-i-cal) The study of marine organisms and their interactions with each other and their physical, chemical, and geological environment.

biological oxygen demand *biochemical oxygen demand.*

biologic artifact (bi-o-log'-ic) An organic compound whose chemical structure demonstrates its derivation from living matter.

biologic facies A syn. of *biofacies* as that term is used in stratigraphy and in ecology; e.g. coral reefs and shell banks are "biologic facies" characterized by the organisms themselves.

biologic time scale An uncalibrated *geologic time scale,* based on organic evolution, giving the relative order for a succession of events. Cf: *relative time scale.*

biologic weathering *organic weathering.*

biology (bi-ol'-o-gy) The study of all organisms, esp. living ones; includes *neontology* and *paleontology,* but most often is used to imply neontology alone.

biolysis (bi-ol'-y-sis) Death and subsequent disintegration of the body.

biomagnification A term describing the process whereby chemicals accumulate in the food chain at higher and higher concentrations. Also referred to as biological concentration (Keller, 2002).

biomarker Organic compound with a specific structure that can be related to a particular source organism.

biomass (bi'-o-mass) The amount of living material in a particular area, stated in terms of the weight or volume of organisms per unit area or of the volume of the environment. Syn: *standing crop.*

biome (bi'-ome) A climax community that characterizes a particular natural region; esp. a particular type of vegetation, climatically bounded, which dominates a large geographic area. Partial syn: *biotic formation.*

biomechanical (bi'-o-me-chan'-i-cal) Said of a rock or deposit formed by detrital accumulation of organic material.

biomere (bi'-o-mere) A regional biostratigraphic unit bounded by abrupt nonevolutionary changes in the dominant elements of a single phylum. The changes are not necessarily related to physical discontinuities in the sedimentary record, and they may be diachronous (Palmer, 1965, p.149-150).

biometrics (bi-o-met'-rics) Statistics as applied to biologic observations and phenomena.

biomicrite (bi-o-mic'-rite) A limestone consisting of a variable proportion of skeletal debris and carbonate mud (micrite); specif. a limestone containing less than 25% intraclasts and less than 25% ooids, with a volume ratio of fossils and fossil fragments to pellets greater than 3 to 1, and the carbonate-mud matrix more abundant than the sparry-calcite cement (Folk, 1959, p.14). It is characteristic of environments of relatively low physical energy. The major organism should be specified when using the term; e.g. "crinoid biomicrite" or "brachiopod biomicrite". See also: *sparse biomicrite.*

biomicrosparite (bi'-o-mic'-ro-spar'-ite) A biomicrite in which the carbonate-mud matrix has recrystallized to microspar (Folk, 1959, p.32); a *microsparite* containing fossils or fossil fragments.

biomineral (bi-o-min'-er-al) A mineral substance "of obviously organic origin", e.g. francolite in a *Lingula* shell or the apatitic mineral of teeth and bones (McConnell, 1973, p.425-426).

biomineralization The process by which organisms fix minerals in their skeletons or internal organs.

biomineralogy (bi'-o-min'-er-al'-o-gy) The systematic study of biominerals (McConnell, 1973).

biomorphic (bi-o-mor'-phic) Pertaining to or incorporating the forms of organisms; e.g. "biomorphic sediments" containing fossil forms.

bionomics (bi-o-nom'-ics) *ecology.*
biopelite (bi-o-pel'-ite) An organic pelite; specif. a *black shale.*
biopelmicrite (bi'-o-pel-mic'-rite) A limestone intermediate in content between biomicrite and pelmicrite. Rarely used.
biopelsparite (bi'-o-pel-spar'-ite) A limestone intermediate in content between biosparite and pelsparite. Rarely used.
biophile (bi'-o-phile) (a) Said of those elements that are the most typical in organisms and organic material (Rankama & Sahama, 1950, p.88). (b) Said of those elements that are concentrated in and by living plants and animals (Goldschmidt, 1954, p.26).
biopolymer Complex organic structure with repeating subunits that has been synthesized in a living organism.
biopyribole (bi-o-py'-ri-bole) A mineral whose structural architecture includes single and double silicate chains (characteristic of pyroxenes and amphiboles) and silicate sheets (characteristic of biotite), e.g., jimthompsonite (Veblen et al., 1977). Etymol: *bio*tite+*pyr*oxene+amph*ibole*.
bioremediation With respect to soil pollution, the technology that uses natural or enhanced microbial action in the soil to degrade organic contaminants in situ, not requiring excavation of the soil (Keller, 2002).
biorhexistasy (bi'-o-rhex-is'-ta-sy) The name given to a theory of sediment production related to variations in the vegetational cover of the land surface, and characterized by long-term stable subtropical weathering conditions resulting in lateritic soils accompanied by removal of calcium, silica, alkalies, and alkaline earths (Erhart, 1956). See also: *rhexistasy; biostasy.*
bioseries (bi-o-se'-ries) "A morphogenetic sequence marked by stages in the progressive structural development of morphological characters of index fossils" (Glaessner, 1945, p.225).
biosiliceous ooze (bi'-o-si-li'-ceous) A deep-sea ooze containing at least 30% biosiliceous skeletal remains, such as diatoms, radiolarians, and silicoflagellates.
biosociology (bi'-o-so'-ci-ol'-o-gy) The branch of ecology concerned with the social behavior of organisms in communities. Cf: *biocenology.*
biosome (bi'-o-some) (a) An ecologically controlled biostratigraphic unit that is mutually intertongued with one or more biostratigraphic units of differing character; the biostratigraphic equivalent of *lithosome* (Wheeler, 1958a, p.647-648). Cf: *holosome.* (b) A "body of sediment deposited under uniform biological conditions" (Sloss, in Weller, 1958, p.625); the record of a uniform biologic environment or of a biotope; a three-dimensional rock mass of uniform paleontologic content. Not to be confused with *biostrome.* Cf: *biotope [stratig].*
biospace (bi'-o-space) As used by Valentine (1969, p.686), that part of the *environmental hyperspace lattice* that actually represents conditions existing on the Earth. Syn: *realized ecological hyperspace.*
biosparite (bi-o-spar'-ite) A limestone consisting of a variable proportion of skeletal debris and calcite cement (spar); specif. a limestone containing less than 25% intraclasts and less than 25% ooids, with a volume ratio of fossils and fossil fragments to pellets greater than 3 to 1, and the sparry-calcite cement more abundant than the carbonate-mud matrix (micrite) (Folk, 1959, p.14). It is generally characteristic of high-energy carbonate environments, with the spar being normally a pore-filling cement. According to Folk (1962), further textural subdivision may be made into "unsorted biosparite", "sorted biosparite", and "rounded biosparite". The major organism should be specified when using the term; e.g. "trilobite biosparite" or "pelecypod biosparite".
biosparrudite (bi'-o-spar-rud'-ite) A *biosparite* containing fossils or fossil fragments that are more than one millimeter in diameter.
biospecies (bi-o-spe'-cies) A species defined on the basis of observed interbreeding capability and potential.
biospeleology (bi'-o-spe'-le-ol'-o-gy) The scientific study of the organisms that live in caves.
biosphere (bi'-o-sphere) (a) All the area occupied or favorable for occupation by living organisms. It includes parts of the lithosphere, hydrosphere, and atmosphere. Cf: *ecosphere.* (b) All living organisms of the Earth and its atmosphere.
biostasy (bi-os'-ta-sy) Maximum development of organisms at a time of tectonic repose when residual soils form extensively on land and deposition of calcium carbonate is widespread in the sea. See also: *rhexistasy; biorhexistasy.*
biostratic unit (bi-o-strat'-ic) A stratigraphic unit based on paleontologic data; an abbreviation of the term *biostratigraphic unit* (Henningsmoen, 1961, p.63, 68).
biostratigraphic classification (bi'-o-strat'-i-graph'-ic) The systematic subdivision and organization of strata into named units based on their fossil content (ISSC, 1994, p.55).
biostratigraphic correlation *biocorrelation; biohorizon.*
biostratigraphic horizon *biohorizon.*
biostratigraphic unit A body of rock defined or characterized by its fossil content. Its boundaries may or may not coincide with those of a lithostratigraphic unit, but they bear no inherent relation to them. The basic unit is the *biozone* (NACSN, 1983, Art. 48).
biostratigraphic zone A general term for any kind of biostratigraphic unit. The fundamental biostratigraphic unit (ISSC, 1994, p.55). See also: *biozone.*
biostratigraphy (bi'-o-stra-tig'-ra-phy) The element of stratigraphy that deals with the distribution of fossils in the stratigraphicc record and the organization of strata into units of the basis of their contained fossils. The term was apparently proposed by Louis Dollo, Belgian paleontologist, in 1904 in a wider sense for the entire research field in which paleontology exercises a significant influence upon historical geology. Cf: *stratigraphic paleontology.*
biostratinomy (bi'-o-stra-tin'-o-my) The branch of paleoecology concerned with all processes occurring after the death of an organism until its final burial (Deevey, 1942). Cf: *taphonomy; fossil diagenesis.*
biostratonomy (bi'-o-stra-ton'-o-my) *biostratinomy.*
biostromal (bi-o-strom'-al) Pertaining to a biostrome, e.g. "biostromal limestone".
biostrome (bi'-o-strome) A distinctly bedded and widely extensive or broadly lenticular, blanketlike mass of rock built by and composed mainly of the remains of sedentary organisms, and not swelling into a moundlike or lenslike form; an "organic layer", such as a bed of shells, crinoids, or corals, or a modern reef in the course of formation, or even a coal seam. Term proposed by Cumings (1932, p.334). Cf: *bioherm.* Not to be confused with *biosome.*
biota (bi-o'-ta) All living organisms of an area; the flora and fauna considered as a unit.
Biot-Fresnel law (Bi'-ot-Fres'-nel) A statement in optics that the directions of extinction in any section of a biaxial crystal are parallel to the traces, on that section, of the planes bisecting the angles between the two planes containing the normal to the section and the optic axes.
biotic (bi-ot'-ic) Of or pertaining to life or the mode of living of plants and animals collectively.
biotic community *community.*
biotic factor A factor of a biological nature, such as availability of food, competition between species, and predator-prey relationships, that affects the distribution and abundance of species.
biotic formation (a) *biome.* (b) In botany, a broad natural unit consisting of distinctive plants in a climax community.
biotic province A geographic area that supports one or more ecologic associations which are distinct from those of adjacent provinces.
biotic succession An ambiguous term improperly used for either *seral* succession or *faunal succession;* its use is not recommended.
biotite (bi'-o-tite) (a) A widely distributed and important rock-forming mineral of the mica group: $K(Mg,Fe^{2+})_3(Al,Fe^{3+})Si_3O_{10}(OH,F)_2$. It is generally black, dark brown, or dark green, and occurs in various monoclinic polytypes. It forms a constituent of crystalline rocks (either as an original crystal in igneous rocks of all kinds or a product of metamorphic origin in gneisses and schists) or a detrital constituent of sandstones and other sedimentary rocks. Biotite is useful in the potassium-argon method of age determination. (b) A general term to designate all ferromagnesian micas. Syn: *black mica; iron mica; magnesia mica.*
biotope [ecol] (bi'-o-tope) (a) An area of uniform ecology and organic adaptation (Hesse et al., 1937, p.135); the habitat, or physical basis, of a uniform community of animals and plants adapted to its environment; a limited region characterized by certain environmental conditions under which the existence of a given biocoenosis is possible. It is more or less ephemeral and at any moment it is circumscribed by a boundary that is subject to expansion, contraction, or other shift in position. Cf: *biochore.* (b) An association of organisms characteristic of a particular geographic area. See also: *paleobiotope.*
biotope [stratig] The environment under which an assemblage of plants or animals lives or lived (Wells, 1944, p.284; 1947, p.119). It is an ecologic term, which should be used in stratigraphy in a sense of paleontologic environment. However, it has been applied to a biostratigraphic surface or area (Wheeler, 1958a, p.653-654), a faunal or floral unit (Sloss et al., 1949, p.95-96), and numerous other entities (see Weller, 1958, p.614-615, 636). Cf: *lithotope; biosome.*
bioturbation (bi'-o-tur-ba'-tion) The reworking of a sediment by organisms.
biotype (bi'-o-type) Any one of a group of organisms having the same genetic constitution.
biozone (bi'-o-zone) (a) The range of organisms in time as indicated by their entombment in the strata (Buckman, 1902). (b) An abbreviation of *biostratigraphic zone.* (c) The basic unit in biostratigraphic classification. Three principal kinds are recognized: *interval zone, assemblage zone,* and *abundance zone.* A biozone may be subdivided into subzones, which may or may not have formal status (NACSN, 1983, Art. 49). See also: *zone [stratig].*
bipartite oolith (bi-par'-tite) An oolith whose central part is divided into two more or less distinct fractions that differ in texture and/or grain size, so that it displays an asymmetric appearance (Choquette, 1955, p.338).
bipennate element A ramiform conodont element with an anterior and a posterior process; posterior process commonly longer than anterior process, which may also be curved or deflected toward one side and may even be adenticulate (TIP, 1981, pt. W, supp.2).
biphosphammite (bi-phos'-pham-mite) A white or buff tetragonal mineral found in guano: $(NH_4,K)H_2PO_4$.
bipolarity (bi-po-lar'-i-ty) The similarity or identity of groups of organisms occurring north and south of, but not in, the equatorial zone.

bipolar oblique conic conformal projection A conformal conic map projection on which the scale is true along two lines that do not lie along any parallel. It is mathematically based on two cones whose apexes are 104° apart and which conceptually are obliquely secant to the globe along lines following the trend of North and South America. The graticule spacing increases away from the lines of true scale but retains the property of conformality where the two conic projections join. The projection has been used to show one or both of the American continents, but, since the advent of computer plotting, other projections are preferable and practical for regional maps (Snyder and Voxland, 1989, p.99). Presented by Osborn Maitland Miller (1897-1979) and William A. Briesemeister of the American Geographical Society in 1941.

bipole A *dipole* electrode arrangement in which the electrodes of the dipole are an appreciable distance apart when compared to source-receiver separation.

bipole-dipole array (bi'-pole-di'-pole) A direct-current resistivity array in which the Earth is energized using a fixed bipole source, and the resultant electric field is mapped at numerous locations using orthogonal pairs of dipoles (closely spaced pairs of electrodes) to determine the magnitude and direction of the electric field.

bipyramid (bi-pyr'-a-mid) *dipyramid.*

biquartz plate (bi-quartz') A type of *compensator* in a polarizing microscope, one half of which is *right-handed [cryst]* and the other half *left-handed [cryst],* with superimposed wedges. It is used to detect accurately the position of extinction. Syn: *Wright biquartz wedge.*

biramous (bi-ra'-mous) Two-branched; said of a crustacean limb in which the *basis* bears both *exopod* and *endopod,* or said of a trilobite appendage consisting of an outer and an inner branch.

birbirite (bir'-bir-ite) A brownish quartzite produced by the silicification of dunite. It occurs in Ethiopia and was first described by Duparc and Borloz in 1927.

bird A geophysical measuring device such as a magnetometer, plus the housing in which it is towed behind an aircraft or ship.

bird-foot delta A delta formed by many levee-bordered distributaries extending seaward and resembling in plan the outstretched claws of a bird; e.g. the Mississippi River delta. Syn: *digitate delta; bird's-foot delta.* Also called: *fluvially-dominated delta.*

bird's-eye A spot, bleb, tube, or irregular patch of sparry calcite commonly found in limestones (such as dismicrites) and some dolomites as a precipitate that infills cavities resulting from localized disturbances, such as microbial or burrowing activity, escaping gas bubbles, shrinkage cracking, or plant roots. Also applied to the porosity created by the presence of bird's-eyes in a rock. Also spelled: birdseye. Syn: *calcite eye.*

bird's-eye coal (a) Anthracite with numerous small fractures that display its semiconchoidal fracture pattern. (b) Cannel coal with semiconchoidal fracture pattern.

bird's-eye limestone A syn. of *dismicrite.* The term was applied in a titular sense in early New York State reports to the Lowville Limestone, a very fine-textured limestone containing spots or tubes of crystalline calcite or having light-colored specks due in part to a characteristic fossil supposed to be a form of coral and now known as *Tetradium cellulosum* (Wilmarth, 1938, p.192). Cf: *loferite.*

bird's-eye ore A miner's term used in Arkansas for a variety of pisolitic bauxite characteristic of residual deposits. Also spelled: birdseye ore.

bird's-foot delta *bird-foot delta.*

bird track A term applied in the mid-19th century to a dinosaur track before its true character was recognized, and now used to denote footprints or tracks made by ancient birds.

bireflectance (bi-re-flec'-tance) The ability of a mineral to change color in reflected polarized light with change in crystal orientation. Cf: *pleochroism.* Syn: *reflection pleochroism.*

birefracting (bi-re-frac'-ting) *birefringent.*

birefraction (bi-re-frac'-tion) *birefringence [cryst].*

birefractive (bi-re-frac'-tive) *birefringent.*

birefringence [cryst] (bi-re-frin'-gence) The ability of crystals other than those of the *isometric system* to split a beam of ordinary light into two beams of unequal velocities; the difference between the greatest and the least indices of refraction of a crystal. Cf: *single refraction.* See also: *positive birefringence; refraction [cryst].* Adj: *birefringent.* Syn: *double refraction; birefraction.*

birefringence [seis] Splitting of an incident wave into two waves with different polarizations. When entering a region of *transverse isotropy* in a direction other than along, or perpendicular to, the symmetry axis, an S-wave splits into two S-waves with orthogonal polarizations. Also called *shear-wave splitting.* The mode with the faster velocity is called S_1, that with the slower velocity S_2. Involved in examining fractured reservoirs and plate motion.

birefringent (bi-re-frin'-gent) Said of a crystal that displays *birefringence;* such a crystal has more than one *index of refraction.* Syn: *birefractive; birefracting.*

biringuccite (bi-rin-guc'-cite) A colorless monoclinic mineral: $Na_2B_5O_8(OH){\cdot}H_2O$.

birnessite (bir'-nes-site) A black hexagonal mineral: $(Na,Ca)_{0.5}(Mn^{4+},Mn^{3+})_2O_4{\cdot}1.5H_2O$.

birthstone (birth'-stone) A stone that has been chosen as appropriate to the time (month) of one's birth. The commonly used list specifies: January (garnet); February (amethyst); March (bloodstone or aquamarine); April (diamond); May (emerald); June (pearl, moonstone, or alexandrite); July (ruby); August (sardonyx or peridot); September (sapphire); October (opal or tourmaline); November (topaz or citrine); December (turquoise or zircon).

birunite A white mineral of approximate composition: $Ca_{18}(SiO_3)_{8.5}(CO_3)_{8.5}(SO_4){\cdot}15H_2O$. Inadequately characterized.

bisaccate (bi-sac'-cate) Pertaining to pollen, having two *sacci.* Bisaccate pollen are usually the pollen of conifers but can also occur in other gymnosperms, e.g., Caytoniales and other seed ferns. Syn: *disaccate.* Less preferred syn: *bivesiculate.*

bisbeeite (bis'-bee-ite) *chrysocolla.*

bischofite (bisch'-of-ite) A white to colorless monoclinic mineral: $MgCl_2{\cdot}6H_2O$.

biscuit-board topography A glacial landscape characterized by a rolling upland on the sides of which are cirques that resemble the bites made by a biscuit-cutter in the edge of a slab of dough; e.g. the Wind River Mountains in Wyoming. It may represent an early or partial stage in glaciation.

bisectrix (bi-sec'-trix) In a biaxial crystal, a line that bisects either of the complementary angles between the two optic axes of a biaxial crystal. See also: *acute bisectrix; obtuse bisectrix.*

biserial (bi-se'-ri-al) Arranged in, characterized by, or consisting of two rows or series; e.g. a "biserial arm" of a crinoid composed of brachial plates arranged in a double row, a "biserial brachiole" of a cystoid consisting of plates arranged in two rows, a "biserial test" of a foraminifer with chambers in a two-row series, or a "biserial rhabdosome" of a *scandent* graptoloid composed of two rows of thecae in contact either back-to-back (dipleural) or side-by-side (monopleural). Cf: *uniserial.*

bishop's stone (bish'-op's) *amethyst.*

bismite (bis'-mite) A sub-adamantine or earthy, variously colored monoclinic mineral: Bi_2O_3. Syn: *bismuth ocher.*

bismoclite (bis'-mo-clite) A pale-grayish or creamy-white tetragonal mineral: BiOCl. It is isomorphous with daubréeite, and a member of the *matlockite* group.

bismuth (bis'-muth) A rhombohedral mineral, the native metallic element Bi. It is brittle and heavy and commonly occurs in silvery-white or grayish-white (with a pinkish or reddish tinge) and arborescent, foliated, or granular forms.

bismuth blende *eulytine.*

bismuth glance *bismuthinite.*

bismuth gold *maldonite.*

bismuthide (bis'-muth-ide) A mineral compound that is a combination of bismuth with a more positive element.

bismuthine (bis'-muth-ine) *bismuthinite.*

bismuthinite (bis-muth'-in-ite) A lead-gray to tin-white orthorhombic mineral: Bi_2S_3. It has a metallic luster and an iridescent tarnish, and it usually occurs in foliated, fibrous, or shapeless masses associated with copper, lead, and other ore minerals. Syn: *bismuth glance; bismuthine.* A member of the *stibnite* group.

bismuth ocher A group name for earthy oxides and carbonates of bismuth; specif. *bismite.*

bismuth spar *bismutite.*

bismutite (bis'-mut-ite) A tetragonal mineral: $Bi_2O_2(CO_3)$. It is earthy and usually dull white, yellowish, or gray. Syn: *bismuth spar.*

bismutocolumbite A submetallic black orthorhombic mineral: $Bi(Nb,Ta)O_4$.

bismutoferrite (bis'-mut-o-fer'-rite) A yellow or green monoclinic mineral: $BiFe^{3+}{}_2(SiO_4)_2(OH)$. Isotypic with chapmanite.

bismutohauchecornite (bis'-mut-o-hau'-che-cor'-nite) A metallic tetragonal mineral, and a member of the *hauchecornite* group: $Ni_9Bi_2S_8$.

bismutomicrolite A variously colored cubic mineral, a member of the *pyrochlore* group: $(Bi,Ca,[])_2(Ta,Nb)_2(O,OH)_7$. Syn: *westgrenite.*

bismutopyrochlore A brown or dark greenish brown metamict mineral, a member of the *pyrochlore* group: $(Bi,U,Ca,Pb)^{1+x}(Nb,Ta)_2O_6(OH){\cdot}nH_2O$.

bismutostibiconite (bis'-mu-to-stib'-i-con-ite) A yellow, yellowish brown, or greenish cubic mineral, a member of the *stibiconite* group: $Bi(Sb^{5+},Fe^{3+})_2O_7$.

bismutotantalite (bis'-mut-o-tan'-ta-lite) (a) A black or dark brown orthorhombic mineral: $Bi(Ta,Nb)O_4$. (b) A group name for *ABO4* minerals where *A* = Sb or Bi, and *B* = Nb,Ta or Sb.

bisphenoid (bi-sphe'-noid) *disphenoid.*

bistatic radar (bi-stat'-ic) A *radar* with its transmitter and receiver spatially separated.

bit A general term for drill bit or *core bit.*

biteplapallidite (bi-te-pla-pal'-lid-ite) *merenskyite.*

biteplatinite (bi-te-plat'-i-nite) *moncheite.*

bitheca (bi-the'-ca) The smallest of three tubes regularly produced at each budding in the development of a graptolithine colony. It may have contained a male *zooid.* Cf: *autotheca; stolotheca.*

Bithynian (Bi-thyn'-i-an) A European substage: middle of Ladinian Stage, Middle Triassic (above Aegean, below Pelsonian).

bitter lake (bit'-ter) A *salt lake* whose waters contain in solution a high content of sodium sulfate and lesser amounts of the carbonates and chlorides ordinarily found in salt lakes; a lake whose water has a bitter taste. Examples include Carson Lake in Nevada, and the Great Bitter Lake in Egypt.

bittern (bit'-tern) (a) The bitter liquid remaining after seawater has been concentrated by evaporation until most of the sodium chloride has crystallized out. See also: *bittern salt.* (b) A natural solution, in an evaporite basin, that resembles a saltworks liquor, esp. in its high magnesium content.

bittern salt Any of the salts that may be extracted from the *bittern* of a saltworks or from a comparable natural solution; e.g., magnesium chloride, magnesium sulfate, bromides, iodides, and calcium chloride.

bitter salts *epsomite [mineral].*

bitter spar *dolomite [mineral].*

bitumen (bi'-tu-men) A generic term applied to natural inflammable substances of variable color, hardness, and volatility, composed principally of a mixture of hydrocarbons substantially free from oxygenated bodies. Bitumens are sometimes associated with mineral matter, the nonmineral constituents being fusible and largely soluble in carbon disulfide, yielding water-insoluble sulfonation products. Petroleums, asphalts, natural mineral waxes, and asphaltites are all considered bitumens.

bitumenite (bi-tu'-men-ite) *torbanite.*

bitumenization (bi-tu'-men-i-za'-tion) (a) *coalification.* (b) Enrichment in hydrocarbons.

bitumicarb (bi-tu'-mi-carb) Low-rank bituminous matter in coal that is derived from resins, waxes, spores, exines, etc. (Tomkeieff, 1954).

bituminous (bi-tu'-mi-nous) Pertaining to *bituminous coal.*

bituminous [mineral] Said of a mineral having an odor like that of bitumen.

bituminous [sed] (a) Said of a sedimentary rock that is naturally impregnated with, contains, or constitutes the source of *bitumen.* (b) Loosely, said of a substance containing much organic or carbonaceous matter; e.g. "bituminous ore" (iron ore whose gangue consists principally of coaly matter).

bituminous brown coal *pitch coal.*

bituminous coal Coal that ranks between subbituminous coal and anthracite and that contains more than 14% volatile matter (on a dry, ash-free basis) and has a calorific value of more than 11,500 BTU/lb (26.75 MJ/kg) (moist, mineral-matter-free) or more than 10,500 BTU/lb (24.42 MJ/kg) if agglomerating (ASTM, 1970). It is dark brown to black in color and burns with a smoky flame. Bituminous coal is the most abundant rank of coal; much is Carboniferous in age. Cf: *high volatile bituminous coal; medium volatile bituminous coal; low volatile bituminous coal.* Syn: *soft coal.*

bituminous fermentation Fermentation of vegetable matter under conditions of no air and abundant moisture. Volatiles are retained, resulting in the formation of bitumens, i.e. peat, coal.

bituminous lignite *pitch coal.*

bituminous limestone A dark, dense limestone containing abundant organic matter, believed to have accumulated under stagnant conditions and emitting a fetid odor when freshly broken or vigorously rubbed; e.g. the Bone Spring Limestone of Permian age in west Texas. See also: *stinkstone; anthraconite.*

bituminous wood *woody lignite.*

bityite (bit'-y-ite) A colorless monoclinic trioctahedral brittle mica, of ideal composition $Ca(Li,Al)_3(OH,F)_2(Si,Al,Be)_4O_{10}$.

biumbilicate (bi-um-bil'-i-cate) Having a central depression (umbilicus) on each side of a foraminiferal test (as in planispiral forms).

biumbonate (bi-um'-bo-nate) Said of a foraminifer having two raised umbonal bosses (as in *Lenticulina*).

bivalve (bi'-valve) adj. Having a shell composed of two distinct and usually movable valves, equal or subequal, that open and shut. Cf: *univalve.* Syn: bivalved. n. A bivalve animal, such as a rostroconch, a brachiopod, or an ostracode; specif. a mollusk of the class Bivalvia (Pelecypoda), including the clams, oysters, scallops, and mussels, generally sessile or burrowing into soft sediment, having no distinct head, and possessing a hatchet-shaped foot and a sheetlike or lamelliform gill on each side of a bilaterally symmetrical body. The class was formerly named class Pelecypoda or class Lamellibranchia. Range, Cambrian (limited), Ordovician to Holocene. See also: *pelecypod.*

bivariate (bi-var'-i-ate) Pertaining to or involving two mathematical variables; e.g. "bivariate distribution".

bivesiculate (bi-ve-sic'-u-late) Less preferred syn. of *bisaccate.*

bivium (biv'-i-um) (a) The two posterior ambulacra of a crinoid, echinoid, or other echinoderm. (b) The part of an asterozoan containing the *madreporite* and the rays on each side of it. This usage is not recommended (TIP, 1966, pt. U, p.29). Pl: bivia. Cf: *trivium.*

bixbyite (bix'-by-ite) A black cubic mineral: $(Mn^{3+},Fe^{3+})_2O_3$. Syn: *partridgeite; sitaparite.*

bizardite (bi-zard'-ite) An *alnoite* that contains nepheline as an essential phase. The name is for the Ile Bizard, Quebec. Obsolete.

bjarebyite (bja'-re-by'-ite) A vitreous green monoclinic mineral: $(Ba,Sr)(Mn^{2+},Fe^{2+},Mg)_2Al_2(PO_4)_3(OH)_3$.

bjerezite (bjer'-e-zite) A porphyritic igneous rock in which the phenocrysts of nepheline, pyroxene with aegirine rims, elongated andesine laths, and alkali feldspar are contained in a fine-grained groundmass of pyroxene, brown mica, andesine, potassium feldspar, nepheline, analcime, and indeterminate zeolites. The rock was named by Erdmannsdoerfer in 1928 for Bjerez, Russia. Obsolete.

black alkali An old term for an alkali soil whose sodium tends to disperse organic matter and give a black color.

black alta *alta.*

black amber (a) *jet [coal].* (b) *stantienite.*

black-and-white iceberg An iceberg made up of sharply defined alternating layers of dark, opaque ("black") ice containing dirt and stones, and cleaner, light-colored ("white") ice.

blackband (black'-band) (a) A dark, earthy variety of the mineral siderite, occurring mixed with clay, sand, and considerable carbonaceous matter, and frequently associated with coal. Syn: *blackband ore.* (b) A thin layer (up to 10 cm in thickness) of blackband interbedded with clays or shales in blackband ironstone. (c) *blackband ironstone.*

blackband ironstone A dark variety of *clay ironstone* containing sufficient carbonaceous matter (10-20%) to make it self-calcining (without the addition of extra fuel). Syn: *blackband; blackband ore.*

blackband ore (a) *blackband.* (b) *blackband ironstone.*

Blackbird-type cobalt-copper deposit *Sediment-hosted copper-cobalt deposit.* Named for the Blackbird mine in Idaho, United States.

blackbody (black'-bod-y) An ideal emitter that radiates energy at the maximum possible rate per unit area at each wavelength for any given temperature. A blackbody also absorbs all the radiant energy incident upon it. No actual substance behaves in this way, although some materials, such as lampblack, approach it.

black chalcedony The correct designation for most so-called *black onyx.*

black chalk A bluish-black carbonaceous clay, shale, or slate, used as a pigment or crayon.

black chert (a) Carbonaceous chert, such as that occurring in South Africa. (b) A term used in England for flint. Cf: *white chert.*

black chlorite alteration A zone of dark, finely-crystalline, massive chlorite alteration around conduits that feed massive sulfide deposits.

black cobalt *asbolite.*

black copper *tenorite.*

black cotton soil *Regur.*

blackdamp (black'-damp) A coal mine gas that is nonexplosive and consists of about 15% carbon dioxide and about 85% nitrogen. Cf: *whitedamp; afterdamp; firedamp.* Syn: *chokedamp.*

black diamond [coal] A syn. of *coal.*

black diamond [mineral] (a) *carbonado.* (b) A black gem diamond. (c) Dense black hematite that takes a polish like metal.

black drift *forest bed.*

black earth Brown coal that is finely ground and used as a pigment. Syn: *Cologne earth; Cologne umber; Cassel brown; Cassel earth; Vandyke brown.*

Black Earth A term used by some as synonymous with "Chernozem;" by others (in Australia) to describe self-mulching black clays. Not used in current U.S. system of soil taxonomy.

black gold (a) A fanciful name for *crude oil.* (b) *maldonite.* (c) Placer gold coated with a black or dark-brown substance (such as a film of manganese oxide) so that the yellow color is not visible until the coating is removed.

black granite A *commercial granite* that when polished is dark gray to black. It may be a diabase, diorite, or gabbro.

black hematite A syn. of *romanèchite.* The term is a misnomer because romanèchite contains no iron. Cf: *red hematite; brown hematite.*

black ice (a) A clear ice layer, usually <10 cm thick, formed on the sea, in rivers or lakes, or on land, that appears dark because of its transparency. (b) Dark glacier ice formed by freezing of silt-laden water. Cf: *blue ice; white ice.* (c) A thin (mm-scale) sheet or glaze of dark ice formed when a light rain or drizzle falls on a surface whose temperature is below freezing. Cf: *verglas.*

blackjack [coal] (black'-jack) (a) A thin stratum of coal interbedded with layers of *slate;* a slaty coal with a high ash content. (b) A British term for a variety of cannel coal. Also spelled: black jack.

blackjack [mineral] A syn. of *sphalerite.* esp. a dark variety. The term was originated by miners who regarded sphalerite as an impish intrusion ("jack") of worthless material in lead ores. Also spelled: black jack.

blackjack [sed] A term used in Arkansas for a soft, black carbonaceous clay or earth associated with coal.

blackland (black'-land) A term used in Texas for a Vertisol.

black lead *graphite.*

black-lead ore An old name for the black variety of cerussite.

black light (a) A prospector's and miner's term for ultraviolet light, used in exploration and evaluation to detect mineral fluorescence. (b) An instrument, usually portable, that produces ultraviolet light for this purpose.

black lignite *lignite A.*

black manganese A term applied to dark-colored manganese minerals, such as pyrolusite, hausmannite, and psilomelane.

black metal A *black shale* associated with coal measures.

black mica *biotite.*

blackmorite (black'-mor-ite) A yellow variety of opal from Mount Blackmore, Mont.

black mud A type of *mud [marine geol]* whose dark color is due to fine iron sulfide minerals from hydrogen sulfide, developed under anaerobic conditions. Syn: *hydrogen-sulfide mud; reduced mud.*

black ocher *wad [mineral].*

black oil Oil with a low gas/oil ratio (GOR < 2,000 ft^3/barrel), which is easy to deal with in fluid-flow simulation because of its simple thermodynamic properties. Other categories of hydrocarbons include volatile oils, retrograde gas, wet gas, and dry gas.

black onyx The popular name for *black chalcedony,* usually artificially colored. Although the word "onyx" is not quite accurate (except for banded material), it has come to be accepted as the usual term for solid-color chalcedony.

black opal A form of precious opal whose *play of color* (usually red or green) is displayed against a dark gray (rarely black) body color; e.g. the fine Australian blue opal with flame-colored flashes. Cf: *white opal.*

black prairie A *prairie* with rich, dark soil.

Blackriverian (Black-ri-ver′-i-an) A North American stage: Upper Ordovician.

blacks (a) Highly carbonaceous black shale; impure cannel coal. (b) A British term for dark coaly shale, clay, or mudstone.

black sand (a) An alluvial or beach sand consisting predominantly of grains of heavy, dark minerals or rocks (e.g. magnetite, ilmenite, chromite, rutile, garnet, or basaltic glass), concentrated chiefly by wave, current, or surf action. It may yield valuable minerals. (b) An asphaltic sand.

black-sand beach A beach containing a large quantity of *black sand,* predominantly grains of heavy minerals or volcanic rock fragments (as in Hawaii) concentrated by the action of waves and currents.

black shale (a) A dark, thinly laminated carbonaceous shale, exceptionally rich in organic matter (5% or more carbon content) and sulfide (esp. iron sulfide, usually pyrite), and commonly containing unusual concentrations of certain trace elements (U, V, Cu, Ni); also composed of manganese oxides (Friedman and Sanders, 1978, p.556). Fossil organisms (principally planktonic and nektonic forms) are commonly preserved as a graphitic or carbonaceous film or as pyrite replacements. Syn: *biopelite.* (b) A finely laminated, sometimes canneloid, carbonaceous shale often found as a roof to a coal seam (Tomkeieff, 1954, p.29). Syn: *black metal.*

black silver *stephanite.*

black smoker (a) A submarine hot spring that occurs on the deep sea floor near spreading centers. It ejects hot water, hydrogen sulfide, and mineral particles under great pressure and provides nutrients for local for sulfur bacteria. (b) A hydrothermal vent at the crest of an oceanic ridge, e.g. the East Pacific Rise at the mouth of the Gulf of California. Waters blackened by sulfide precipitates jet out at 1 to 5 m/s at temperatures of 350°C and more (Macdonald et al., 1980). Typically, these vents consist of chimneys of sulfate deposits encrusted with base metal sulfides. When active, rapid cooling of saturated thermal brines results in precipitation of pyrite and other sulfide phases making the vent appear to be exhaling black smoke. The term refers to uprushing black turbulent suspension. Cf: *white smoker.*

black tellurium *nagyágite.*

black tin *cassiterite.*

blackwall (black′-wall) Originally a quarrymen's term for a tabular body of black or dark-colored mica- or chlorite-rich rock, it is now used in metamorphic petrology for black or dark-colored almost monomineralic biotite, chlorite, and amphibole rocks, formed by reaction between ultramafic rock and a rock of contrasting bulk composition.

bladder (blad′-der) *saccus (a).*

blade (a) A carminate or angulate pectinform conodont element. (b) In palmate or scaphate pectiniform conodont elements, that part of the longitudinal axis anterior to the basal cavity or pit. That part of the blade bordered by platform extensions is commonly termed a fixed blade; any portion of the blade that protrudes anteriorly beyond platformed portions of the element is a free blade (TIP, 1981, pt. W, supp.2).

blade [bot] The widened portion of a leaf or of a plant structure that resembles a leaf.

blade [mineral] A flattened, elongate mineral crystal.

blade [paleont] A laterally compressed structure of a conodont; e.g., a denticle-bearing posterior or anterior process (based on position with reference to the basal cavity) in a compound conodont, or a generally compressed and denticulate part of the axis anterior to the basal cavity in a platelike conodont. Cf: *bar [paleont].*

blade [sed] A bladed or triaxial shape of a sedimentary particle, defined in *Zingg's classification* as having width/length and thickness/width ratios less than ⅔.

blade [speleo] A thin, sharp projection jutting out from any surface of a cave.

bladed (blad′-ed) Said of a mineral in the form of aggregates of flattened blades or elongate crystals; e.g. "bladed calcite" having elongate crystals somewhat wider than those of fibrous calcite, elongation 1 ½:1 to 6:1.

blade parapet High narrow inner-platform conodont structure, to which the blade of platformed *P* elements of *Cavusgnathus* is joined (TIP, 1981, pt. W, supp.2).

blaes (a) A Scottish term for a gray-blue carbonaceous shale that weathers to a crumbly mass and eventually to a soft clay. See also: *bind.* (b) A Scottish term for a hard, joint-free sandstone. Syn: *blaize.*

blairmorite (blair′-mor-ite) A porphyritic *analcimite* rock consisting predominantly of analcime megacrysts in a groundmass of analcime, alkali feldspar, and sodic clinopyroxene, with accessory titanite, andradite, and nepheline. The name, given by Knight in 1904, is for Blairmore, Alberta. Obsolete.

blaize *blaes.*

blakeite (blake′-ite) (a) A reddish-brown mineral consisting of a ferric tellurite, found sparingly as crusts from Goldfield, Nev. (b) *zirconolite.*

Blancan (Blanc′-an) North American land mammal age: Pliocene (after Hemphillian, before Irvingtonian).

blanchardite (blan′-chard-ite) *brochantite.*

blanket (blan′-ket) A thin, widespread sedimentary body whose width/thickness ratio is greater than 1,000:1 and may be as great as 50,000:1 (Krynine, 1948, p.146). Cf: *tabular.* Syn: *sheet [sed].*

blanket bog (a) A bog covering a large, fairly horizontal area and depending on high rainfall or high humidity, rather than local water sources, for its supply of moisture. See also: *highmoor bog.* (b) An extensive peatland type over undulating terrain, a "blanket" of *peat,* not confined to depressions and usually formed in response to a very humid or maritime climate and rarely exceeding 2 m in thickness. Syn: *blanket peat; climatic bog.*

blanket deposit [ore dep] A flat mineral deposit, the length and breadth of which are relatively great compared with the thickness. The term has no genetic connotation.

blanket deposit [sed] A sedimentary deposit of great lateral or areal extent and of relatively uniform thickness; esp. a *blanket sand* and associated blanket limestones.

blanket moss An accumulation of dead algae, often forming peat. See also: *blanket peat.*

blanket peat *blanket bog.*

blanket sand A blanket deposit of sand or sandstone of unusually wide distribution, typically a sheetlike sandstone deposited by a transgressive sea advancing for a considerable distance over a stable shelf area; e.g. the St. Peter Sandstone of the east-central U.S. Syn: *sheet sand.*

blanket vein A horizontal *vein* or *mineral deposit.* Cf: *blanket deposit [ore dep].*

blanking In geophysical exploration, an observed area of relatively low amplitudes, such as a gas-hydrate zone seen in deep water.

blast (a) A suffix signifying a texture formed entirely by metamorphism. (b) A prefix signifying a relict texture.

blastetrix (blas′-te-trix) In an anisotropic medium, any surface to which a direction of greatest ease of growth is perpendicular (Turner, 1948, p.223). Rarely used.

blastic deformation (blas′-tic) One of the processes of dynamothermal metamorphism that operates by recrystallization according to *Riecke's principle,* in such a way that previously existing minerals are elongated perpendicular to the direction of greatest pressure, and new minerals grow in the same plane. Cf: *clastic deformation; plastic deformation.*

blasting (blast′-ing) Abrasion or attrition effected by the impact of fine particles moved by wind or water against or past an exposed, stationary surface; esp. *sandblasting.*

blastogeny (blas-tog′-e-ny) In corals, "the phenomena of development of the offset (or asexually developed corallite in the colony) from the parent corallite" (Fedorowski and Jull, 1976, p.39). Cf: *astogeny; hystero-ontogeny.*

blastogranitic (blas′-to-gra-nit′-ic) (a) A relict texture in a metamorphic rock in which remnants of the original granitic texture remain. (b) A less preferred syn. of *blastogranular.* Obsolete.

blastogranular (blas-to-gran′-u-lar) A *heterogranular* metamorphic texture characterized by volumetrically significant amounts of large paleoblasts and smaller neoblasts, and by a weak foliation. The term is regarded as a syn. of *porphyroclastic* or is applied to a textural subgroup for small strains. Less preferred syn: *blastogranitic.*

blastoid (blas′-toid) Any crinozoan belonging to the class Blastoidea, characterized chiefly by highly developed quinqueradiate symmetry, a dominant meridional growth pattern, uniform arrangement of thecal plates in four cycles, specialized recumbent ambulacral areas, and the presence of hydrospires. Range, Ordovician to Permian.

blastolaminar (blas-to-lam′-i-nar) A *heterogranular* metamorphic texture characterized by volumetrically significant amounts of large paleoblasts and smaller neoblasts, and by a strong foliation. The term is regarded as a textural subgroup of *porphyroclastic* for large strains.

blastomylonite (blas-to-my′-lo-nite) A mylonitic rock in which some recrystallization or neomineralization has taken place. Cf: *mylonite.*

blastopelitic (blas′-to-pe-lit′-ic) Said of a texture of a metamorphosed argillaceous rock in which there are relicts of the parent rock. Rarely used.

blastophitic (blas-to-phit′-ic) Said of a relict texture in a metamorphic rock in which traces of an original ophitic texture remain. Rarely used.

blastoporphyritic (blas′-to-por′-phy-rit′-ic) Said of a relict texture in a metamorphic rock in which traces of an original porphyritic texture remain. Rarely used.

blastopsammitic (blas′-to-psam-mit′-ic) Said of a texture of a metamorphosed sandstone that contains relicts of the parent rock. Rarely used.

blastopsephitic (blas′-to-pse-phit′-ic) Said of a texture of a metamorphosed conglomerate or breccia that contains relicts of the parent rock. Rarely used.

blast wave A sharply defined wave of increased atmospheric pressure rapidly propagated through a surrounding medium from a center of detonation or similar disturbance. See also: *shock wave.*

blatonite A hexagonal mineral occurring as silky yellow fibres: $UO_2CO_3{\cdot}H_2O$.

blatterite (blatt'-er-ite) A sub-metallic black orthorhombic mineral: $(Mn^{2+},Mg)_{35}(Mn^{3+},Fe^{3+})_9Sb^{5+}{}_3(BO_3)_{16}O_{32}$. A member of the *orthopinakiolite* group.

blattfuss (blatt'-fuss) A platelike mesosomal appendage associated with gill structures of eurypterids (Waterston, 1975, p.243). Etymol: German, "leaf (or sheet) foot."

B layer The seismic region of the Earth from the Mohorovicic discontinuity to 410 km. It is part of a classification of the Earth's interior made up of layers A to G. Cf: *upper mantle.* Rarely used.

blaze An man-made mark made on a tree trunk, usually at about breast height, in which a piece of the bark and a very small amount of the live wood tissue is removed leaving a flat scar that permanently marks the tree. It is made for the purpose of guiding the course of a survey or of a trail in wooded country.

bleached Said of sand and silt that has become pale because of leaching, with consequent loss of iron-oxide coatings.

bleaching clay (bleach'-ing) A clay or earth that, either in its natural state or after chemical activation, has the capacity for adsorbing or removing coloring matter or grease from liquids (esp. oils). Syn: *bleaching earth.* Cf: *fuller's earth.*

bleaching earth *bleaching clay.*

bleach spot A greenish or yellowish area in a red rock, developed by the reduction of ferric oxide around an organic particle (Tyrrell, 1926). Syn: deoxidation sphere.

bleasdaleite A resinous dark brown monoclinic mineral: $(Ca,Fe^{2+})_2Cu_5(Bi,Cu)(PO_4)_4(H_2O,OH,Cl)_{13}$. The phosphate analogue of richelsdorfite.

bleb In petrology, a small, usually rounded inclusion, e.g. olivine that is poikilitically enclosed in pyroxene.

bleeding (bleed'-ing) n. (a) The process of giving off oil or gas from pore spaces or fractures; it can be observed in drill cores. (b) The exudation of small amounts of water from coal or a stratum of some other rock.

Bleiberg-type lead (Blei'-berg) *B-type lead.*

Bleiberg-type lead-zinc deposit A variety of *Mississippi Valley-type deposit* occurring in the eastern Alps of Austria, Italy, and Slovenia. Syn: *Alpine-type lead-zinc deposit.*

bleicherde The light-colored, leached A2 (E) horizon of Podzol soils.

blende (a) *sphalerite.* (b) Any of several minerals (chiefly metallic sulfides) with bright or resinous but nonmetallic luster, such as zinc blende (sphalerite), antimony blende (kermesite), bismuth blende (eulytite), cadmium blende (greenockite), pitchblende, and hornblende. Etymol: German "Blende" (translation of Greek "sphaleros"), "deceiver/deceptive".

blended unconformity (blend'-ed) An unconformity having no distinct surface of separation or sharp contact, such as at an erosion surface that was originally covered by a thick residual soil, which graded downward into the underlying rocks and was partly incorporated in the overlying rocks; e.g. a nonconformity between granite and overlying basal arkosic sediments derived as a product of its disintegration. Syn: *graded unconformity.*

blind [eco geol] Said of a mineral deposit that does not crop out. The term is more appropriate for a deposit that terminates below the surface than for one that is simply hidden by unconsolidated surficial debris.

blind [struc geol] A term applied to a fault that has a *tip line* entirely within the subsurface.

blind apex The near-surface end of a mineral deposit, e.g. the upper end of a seam or vein that is truncated by an unconformity. Syn: *suboutcrop.*

blind coal (a) Anthracite or other coal that burns without a flame. (b) *Natural coke* that resembles anthracite.

blind creek A creek that is dry except during a rainfall. The term is "obsolete or obsolescent and better avoided because of confusion with blind valley" (Stamp, 1961, p.66).

blind estuary A term used in Australia and South Africa for an *estuarine lagoon.*

blind fault A fault that does not appear at the Earth's surface as an outcrop.

blind joint In apparently massive rock that is being quarried, a plane of potential fracture along which the rock may break during excavation.

blind lead A long narrow passage in pack ice with only one outlet. Syn: *cul-de-sac [ice].*

blind shaft In speleology, a *vertical shaft* with no passage leading from its bottom.

blind thrust A thrust fault that does not reach the surface and decreases in upward.

blind valley (a) A valley in karst that ends abruptly downstream at the point at which its stream disappears underground as a *sinking stream.* (b) In older usage, a syn. of *pocket valley.*

blind zone (a) A layer that cannot be detected by refraction seismic methods, also called "hidden layer". It may have a velocity lower than that of shallower refractors, which will lead to an overestimate of the depth of deeper refractors; or it may have a velocity intermediate between those of layers above and below but with insufficient velocity difference or thickness to produce first arrivals, thus tending to cause an underestimate of the depth of deeper refractors. (b) A zone from which reflections do not occur; a shadow zone. (c) A portion of a volume that is not illuminated by seismic energy with given source/receiver geometry.

blink (a) A brightening of the sky near the horizon or the underside of a cloud layer, caused by reflection of light from a snow- or ice-covered surface. See also: *iceblink [meteorol]; snowblink; landblink.* (b) A dark appearance of the sky near the horizon or the underside of a cloud layer, caused by the relative absence of reflected light from a water or land surface. See also: *water sky; land sky.*

blister [coal] (blis'-ter) In a coal seam, a downward protrusion of roof rock into the seam, probably formed as the filling of a pothole in a stream bed.

blister [volc] A surficial swelling of the crust of a lava flow formed by the puffing-up of gas or vapor beneath the flow. A blister is usually about 1 m in diameter, and is hollow. Syn: *lava blister.* Cf: *shelly pahoehoe; tumulus [volc].*

blister hypothesis A hypothesis of the cause of *orogeny,* which proposes that, in a zone not more than 80 km deep in the crust, heat from radioactive disintegration creates a large convex-upward lens of heated and expanded rock, which produces doming of the overlying crust and in turn deforms near-surface rocks. The "melting spot" or *asthenolith hypothesis* of B. Willis is similar. Geophysical evidence indicates that the existence of such blisters is unlikely, and the hypothesis is obsolete. Cf: *undation theory.*

blixite (blix'-ite) A pale yellow orthorhombic mineral: $Pb_2Cl(O,OH)_2$.

blizzard A severe winter weather event that is characterized by low temperatures and strong winds (at least 35 miles per hour) that cause blowing snow which reduces visibility to less than 0.25 miles (400 m).

block [exploration] (a) v. To subdivide a structure so that the different parts can be mapped separately, the subdivision often being along faults. (b) v. To average data over different regions, as to block sonic-log values into larger units. (c) n. An exploration concession area. Blocks 3 mi on a side on the United States continental shelf are numbered to designate locations.

block [ice] A fragment of floating sea ice ranging in size from 2 m to 10 m across; the term is being replaced by *ice cake.*

block [part size] (a) A large, angular rock fragment, showing little or no modification by transporting agents, its surfaces resulting from breaking of the parent mass, and having a diameter greater than 256 mm (10 in.); it may be nearly in place or transported by gravity, ice, or other agents. Cf: *boulder.* (b) A term used by Woodford (1925) for a nearly equidimensional, angular rock fragment of any diameter greater than 4 mm. (c) A rock or mineral particle in the soil, having a diameter range of 200-2,000 mm (Atterberg, 1905). (d) A layer of sedimentary rock, from 60 cm to 120 cm (2-4 ft) thick, produced by splitting (McKee and Weir, 1953, p.383).

block [tect] *fault block.*

block [volc] A pyroclast ejected in a solid state; having a diameter greater than 64 mm. It may be essential, accessory, or accidental. Cf: *lapilli; volcanic gravel; cinder.*

block-and-ash flow deposit A small-volume pyroclastic flow deposit characterized by a large fraction of dense to moderately vesicular juvenile blocks in a medium to coarse ash matrix of the same composition.

block caving A large-production low-cost method of mining, in which the greater part of the bottom area of a block of ore is undercut, the supporting pillars are blasted away, and the ore settles as it caves downward and is removed. As the block caves and settles, the cover follows. Syn: *caving [eco geol].* See also: *sublevel caving.*

block diagram (a) A plane figure representing an imaginary rectangular block of the Earth's crust (depicting geologic and topographic features) in what appears to be a three-dimensional perspective, showing a surface area on top and including one or more (generally two) vertical cross sections. The top of the block gives a bird's-eye view of the ground surface, and its sides give the underlying geologic structure (Lobeck, 1924). (b) A sketch of a relief model; a representation of a landscape in perspective projection.

block disintegration *joint-block separation.*

blocked-out ore *developed reserves.*

block faulting An imprecise term typically used in reference to high-angle faulting (generally normal faulting) in which the crust is broken into separate blocks that move relative to one another. Typically, surfaces of adjacent blocks end up with different elevations and/or tilts.

block field A thin accumulation of usually angular blocks, with no fine sizes in the upper part, over solid or weathered bedrock, colluvium, or alluvium, without a cliff or ledge above as an apparent source. Block fields occur on high mountain slopes above treeline, and in polar regions; they are most extensive along slopes parallel to the contour; and they exist on slopes of less than 5°. Blocks may be subround to subangular suggesting abrasion during transport or in-situ derivation (White, 1976, p.89). Cf: *allochthonous mantle; autochthonous mantle; block slope; block stream; scree.* Syn: *block sea; felsenmeer; blockmeer; mountain-top detritus; stone field.*

block glide A translational landslide in which the slide mass remains essentially intact, moving outward and downward as a unit, most often along a pre-existing plane of weakness, such as bedding, foliation, joints, or faults. In contrast to rotational slides, the various points within a displaced block-glide slide have predominantly maintained the same mutual difference in elevation in relation to points outside the slide mass.

blocking out (block′-ing) In economic geology, delimitation of an orebody on three sides in order to develop it, i.e., to make estimates of its tonnage and quality. The part so prepared is an *ore block.* See also: *developed reserves.*
blockite (block′-ite) *penroseite.*
block kriging In *geostatistics*, estimating the average value of a property (such as the concentration of an ore mineral) within a two- or three-dimensional block, usually from observations having smaller volumes (such as drill cores).
block lava Lava having a surface of angular blocks; it is similar to *aa* but the fragments are more regular in shape, somewhat smoother, and less vesicular.
blockmeer (block′-meer) A syn. of *block field.* Etymol: German "Blockmeer", "sea of blocks".
block mountain A linear mountain range that is formed by normal *block faulting.* Syn: *fault-block mountain.*
block movement [glaciol] A syn. of *Block-Schollen movement* generally used in English-language publications.
block movement [mining] In mining, a general failure of the hanging wall.
Block-Schollen movement (Block-Schol′-len) A type of glacier flow in which the greater portion of the glacier moves as a solid mass with a nearly uniform velocity; blocks of ice are produced by movement over irregularities in the glacier bed (Finsterwalder, 1950). Etymol: German. Syn: *block movement [glaciol].*
block sea *block field.*
block slope A thin accumulation of usually angular blocks, on high mountains and in polar regions, on slopes of 5° to 25° (White, 1976, p.89). Cf: *block field.*
block stream An accumulation of boulders or angular blocks, with no fine sizes in the upper part, over solid or weathered bedrock, colluvium, or alluvium. Block streams usually occur at the heads of ravines, as narrow bodies more extensive downslope than along the slope. They may extend into forests or fill a valley floor; and they may exist on any slope angle, but ordinarily not steeper than 40° (White, 1976, p.91). Sometimes incorrectly referred to as *block field* or *rock glacier.* Cf: *block field.* Syn: *boulder field; boulder stream; rock stream; stone stream.*
block stripe A short, broad *sorted stripe* containing material that is coarser, and of less uniform size, than that in a *stone stripe.*
block talc A general term for any massive talc or soapstone that can be worked by machines.
blocky iceberg (block′-y) An iceberg with steep, precipitous sides and a horizontal or nearly horizontal upper surface.
blocky prismatic structure [paleont] A simple *prismatic structure* in which each first-order prism has a low length/width ratio and consists of irregularly stacked, more or less equidimensional structural units (Bandel, 1977a).
blödite (a) A white or colorless monoclinic mineral: $Na_2Mg(SO_4)_2{\cdot}4H_2O$. Also spelled: bloedite; blodite. Syn: *astrakhanite.* (b) A group name for minerals of composition $Na_2B(SO_4)_2{\cdot}4H_2O$, where B = Mg, Ni, or Zn.
blomstrandine (blom′-strand-ine) *priorite.*
blomstrandite (blom′-strand-ite) *betafite.*
Blondeau method (Blon-deau′) A seismic method of determining vertical time to a predetermined depth based on first-break data and the assumption that the instantaneous velocity is proportional to a power of the depth.
blood agate (a) Flesh-red, pink, or salmon-colored agate from Utah. (b) *hemachate.*
blood rain Rain with a reddish color caused by dustlike material, containing iron oxide, picked up from the air by raindrops during descent, often leaving a red stain on the ground; e.g. the blood rain of Italy, containing red dust carried north by great storms from the Saharan desert region. Syn: *dust fall.*
bloodstone (blood′-stone) (a) A semitranslucent leek-green or dark-green variety of chalcedony speckled with red or brownish-red spots of jasper resembling drops of blood. Cf: *plasma [mineral].* Syn: *heliotrope [mineral]; oriental jasper.* (b) *hematite.*
bloom [oceanog] *plankton bloom.*
blossite (bloss′-ite) A metallic black orthorhombic mineral: $Cu_2V^{5+}{}_2O_7$.
blow *blowhole [coast].*
blowhole [coast] (blow′-hole) A nearly vertical hole, fissure, or natural chimney in coastal rocks, leading from the inner end of the roof of a sea cave to the ground surface above, through which incoming waves and the rising tide forcibly compress the air to rush upward or spray water to spout intermittently, often with a noise resembling a geyser outburst. It is usually formed by wave erosion concentrated along planes of weakness, as in a well-jointed rock. Syn: *puffing hole; blow; boiler; buller; spouter.*
blowhole [glaciol] An opening that passes through a snowbridge into a crevasse, or system of crevasses, that is otherwise sealed by snowbridges (Armstrong et al., 1966, p.11). It is commonly characterized by a current of moving air.
blowhole [karst] (a) A hole on land near the shore through which air and water are forced by incoming waves. (b) In Australia, a small hole in the surface of the Nullarbor Plain through which air blows in and out with observable force, sometimes audibly.
blowhole [volc] A minute gas vent on the surface of a lava flow.
blowing cave A cave that has a current of air moving out of or into its entrance. See also: *breathing cave; cave breathing.*
blowing well A water well that has a movement of air through its entrance. Syn: *breathing well.*
blown-out land In soil surveying, a map-unit which is a type of miscellaneous area from which most of the soil has been removed by wind erosion. The areas are generally shallow depressions with flat, irregular floors, which in some instances have a layer of pebbles or cobbles.
blown sand Sand that has been transported by the wind; sand consisting of wind-borne particles; eolian sand. See also: *dune sand.*
blowoff (blow′-off) The removal of humus and loose topsoil by wind action. Also, the material so moved.
blowout [drilling] An uncontrolled flow of gas, oil, and other fluids from a *borehole,* resulting from excessive *formation pressure.*
blowout [geomorph] (blow′-out) (a) A general term for a small saucer-, cup-, or trough-shaped hollow or depression formed by wind erosion on a preexisting dune or other sand deposit, esp. in an area of shifting sand or loose soil, or where protective vegetation is disturbed or destroyed; the adjoining accumulation of sand derived from the depression, where recognizable, is commonly included. Some blowouts may be many kilometers in diameter. (b) A *butte,* the top of which has been blown out by the wind until it resembles a volcanic crater. (c) A shallow basin formed where vegetation has been destroyed by fire or by overgrazing. Cf: *deflation basin.* Syn: *deflation hollow.*
blowout [grd wat] *sand boil.*
blowout [ore dep] (a) A prospector's term for a weathered exposure considered to be indicative of a mineral deposit. (b) A large mineral-deposit outcrop beneath which the deposit is smaller.
blowout dune A dune consisting of a large accumulation of sand, with or without trailing arms, derived from the formation of a *blowout.* An "elongate blowout dune" is characterized by a slight migration of the blowout and its crescent-shaped rim in the direction of the prevailing wind (Stone, 1967, p.226).
blowout pond A form of *deflation lake.*
blowover (blow′-o-ver) (a) Sand blown by onshore winds across a barrier and deposited on its landward side or as a veneer in the lagoon, e.g. along the Gulf Coast of Texas. Cf: *washover.* (b) The process of forming a blowover.
blowpipe (blow′-pipe) A plain brass tube that produces an intense heat by combining a flame from a bunsen burner or other heat source with a stream of air; it is used in simple qualitative analysis of minerals. See also: *blowpiping.*
blowpipe reaction The indicative changes of a mineral specimen as it undergoes *blowpiping;* e.g. color of the flame, odor, nature of the encrustation.
blowpiping (blow′-pip-ing) In mineralogy, a qualitative test of a mineral made by heating a specimen in the flame of a *blowpipe* and observing its *blowpipe reaction,* such as color of the flame or color of the encrustation, to determine what elements may be present. See also: *bead.*
blue amber A variety of osseous amber with a bluish tinge that is probably due to the presence of calcium carbonate.
blue asbestos *crocidolite.*
blue band [glaciol] (a) A sharply bounded lens or layer of relatively bubble-free glacier ice; a bluish band marking the appearance of such a lens or layer on the surface of a glacier. The bluish tint is due to the low content of air in the ice. Cf: *white band.* (b) The dark-ribbon effect produced on the surface of a glacier by the exposure of blue bands.
blue band [sed] The thin but persistent bed of bluish clay found throughout the Illinois-Indiana coal basin.
blue-black ore *corvusite.*
blue chalcocite *digenite.*
blue copper ore *azurite.*
blue earth *blue ground.*
blue elvan A Cornish term for greenstone occurring in dikes.
blue-green algae Obsolete syn. of *cyanobacteria.*
blue ground Unoxidized slate-blue or blue-green *kimberlite,* usually a breccia (as in the diamond pipes of South Africa) that is found below the surficial oxidized zone of *yellow ground.* Cf: *hardebank.* Syn: *blue earth.*
blue hole (a) A subsurface void developed in carbonate banks and islands, also offshore. It is open to the Earth's surface; contains tidally influenced waters of fresh, marine, or mixed chemistry; extends below sea level for the majority of its depth; and may provide access to cave *passages* (Mylroie et al., 1995). Cf: *inland blue hole; ocean hole.* (b) An obsolete term for a *resurgence.*
blue ice (a) Nonbubbly, unweathered, coarse-grained glacier ice, often occurring as blue bands; it is distinguished by a slightly bluish or greenish color. Cf: *black ice; white ice.* (b) An ablation area created by wind erosion on the Antarctic Ice Sheet, characterized by bare glacier ice at the surface.
blue iron earth Pale-blue powdery *vivianite.*
blue ironstone A bluish iron-bearing mineral; specif. crocidolite and vivianite.
blue john A massive, fibrous, or columnar and blue or purple variety of fluorite found in Derbyshire, England. It is frequently banded, and is used esp. for the manufacture of vases. Syn: *derbystone.*
blue lead [mineral] A syn. of *galena,* esp. a compact variety with a bluish-gray color. Syn: blue lead ore.

blue lead [ore dep] A bluish, gold-bearing *lead [eco geol]* or gravel deposit found in Tertiary river channels of the Sierra Nevada, California. Pron: blue leed.
blue malachite A misnomer for *azurite.*
blue metal A term used in England for a hard bluish-gray shale or mudstone lying at the base of a coal bed and often containing pyrite.
blue mud A hemipelagic type of *mud [marine geol]* whose bluish-gray color is due to iron compounds in various reduced states.
blue ocher *vivianite.*
blue ore Residually enriched cherty oxide iron formation composed of fine-grained, blue to dark gray-black hematite with lesser amounts of red hematite, martite, and brown goethite found in the Knob Lake iron range of Quebec and Labrador, Canada. Cf: *yellow to brown ore; red ore; rubble ore; treat rock.*
blue quartz (a) A faintly blue or lavender variety of crystalline quartz, containing needlelike inclusions of rutile. It occurs as grains in metamorphic and igneous rocks. (b) *sapphire quartz.*
blue-rock phosphate A term used for the Ordovician bedded phosphate rock of Tennessee.
blueschist (blue'-schist) A schistose metamorphic rock with a blue color owing to the presence of sodic amphibole, e.g. glaucophane or crossite, and commonly mottled bluish-gray lawsonite. Cf: *glaucophane schist.*
blueschist facies *glaucophane-schist facies.*
blue spar *lazulite.*
bluestone [mineral] (blue'-stone) *chalcanthite.*
bluestone [rock] (a) A commercial name for a building or paving stone of bluish-gray color; specif. a dense, tough, fine-grained, dark blue-gray or slate-gray feldspathic sandstone that splits easily into thin, smooth slabs and that has been extensively quarried in New York State for use as flagstone. The color is due to the presence of fine black and dark-green minerals, chiefly hornblende and chlorite. The term is applied locally to other rocks, such as dark-blue shale and blue limestone. (b) A highly argillaceous sandstone of even texture and bedding, formed in a lagoon or lake near the mouth of a stream (Grabau, 1920a, p.579). (c) A local term used in Great Britain for a hard shale or clay (as in south Wales), and for a basalt.
blue vitriol *chalcanthite.*
blue-white Traditionally used to describe colorless diamonds. Due to flagrant misuse, Federal Trade Commission rulings limit its application in the U.S. to stones which are blue, or bluish, in color; similar restrictions apply in some other countries as well (Liddicoat et al., 1993).
bluff (a) A high bank or bold headland with a broad, precipitous, sometimes rounded cliff face overlooking a plain or a body of water; esp. on the outside of a stream meander; a *river bluff.* (b) Any cliff with a steep broad face.
bluff line The side of a valley formed by a river or by ice cutting away the heads of *interlocking spurs* (Swayne, 1956, p.25).
bluntness A measure of the relative curvature of a folded layer near its hinge and along its limbs, determined by comparing the radius of curvature at the fold closure versus the radius of the circle tangent to the folded layer at its inflection points (Twiss, 1988).
blythite (blyth'-ite) A hypothetical member of the *garnet* group: $Mn_3^{+2}Mn_2^{+3}(SiO_4)_3$. Not approved by the International Mineralogical Association.
Blytt-Sernander sequence (Blytt'-Ser-nan'-der) A classification of late-glacial and Holocene climate inferred originally from bog stratigraphy and plant fossils from Norway and Sweden, and later refined by Post (1924) from pollen evidence. It is the classic system for subdividing the climate of postglacial time. It includes six subunits: Arctic, Preboreal, Boreal, Atlantic, Subboreal, and Subatlantic. Named after Axel Gudbrand Blytt (1843-1898), Norwegian botanist, and Johan Rutger Sernander (1866-1944), Swedish botanist.
BM *bench mark.*
board coal *woody lignite.*
boar's back A *horseback [glac geol]* or *esker* in northern New England, esp. Maine.
boart Var. of *bort.*
boat channel A channel, on or behind a reef flat, separating a *fringing reef* from the shore to which the channel is parallel. It is generally only a few meters in depth and width. Cf: *shipping channel; moat [reef].* Syn: *back-reef moat.*
boathook bend (boat'-hook) The sharp curvature of a tributary where it joins the main stream in an upstream direction in a *barbed drainage pattern,* resembling in plan a boathook.
bobfergusonite (bob-fer'-gu-son-ite) A green-brown monoclinic mineral: $Na_2Mn^{2+}_5Fe^{3+}Al(PO_4)_6$. A member of the *wyllieite* group.
bobierrite (bo'-bi-err-ite') A colorless monoclinic mineral: $Mg_3(PO_4)_2 \cdot 8H_2O$. It occurs massive or in crystals in guano.
bobjonesite A blue-green monoclinic mineral occurring as crusts and efflorescences: $VO(SO_4) \cdot 3H_2O$.
bobkingite A vitreous pale blue monoclinic mineral occurring as thin plates with other secondary copper minerals: $Cu_5Cl_2(OH)_8 \cdot 2H_2O$.
bobtraillite A vitreous gray, pale green, or brown hexagonal mineral: $Na_{15}Sr_{12}Zr_{14}Si_{42}B_6O_{138}(OH)_6 \cdot 12H_2O$.
boca (bo'-ca) The mouth of a stream, esp. the point where a stream or its channel emerges from a canyon, gorge, or other precipitous valley and flows onto or enters a plain. Etymol: Spanish, "mouth".
bocanne (bo-kahn') A naturally burning shale (Crickmay, 1967, p.626). Etymol: French-Canadian, "smoke".
bocca (boc'-ca) An aperture on any part of a volcano from which magma or gas escapes. Etymol: Italian, "mouth". Pl: bocche.
BOD *biochemical oxygen demand.*
bodden (bod'-den) A broad shallow irregularly shaped inlet or bay along the southern Baltic coast, typically produced by partial submergence of an uneven lowland surface and characterized by seaward islands. Pl: bodden. Etymol: German "Bodden", "shallow bay". Cf: *förde.*
bodily tide (bod'-i-ly) *Earth tide.*
body (bod'-y) The fatty, inflammable property that makes a coal combustible; e.g. bituminous coal has more body than anthracite.
body cavity A cavity or major space within an animal body, such as a *coelom;* e.g. the principal part of the coelomic space in a brachiopod, situated posteriorly, bounded by the body wall, and containing the alimentary tract, internal organs, etc.
body-centered lattice (bod'-y-cen'-tered) A type of *centered lattice* in which the unit cell contains two lattice points; the point at the intersection of the four body diagonals is identical with those at the corners. Syn: *I-centered lattice.*
body chamber (a) The undivided anterior space in a cephalopod shell occupied by the living body of the animal, bounded at the back by a *septum* and open at the front through the *aperture.* Syn: *living chamber; chamber [paleont].* (b) The interior of the shell containing the soft parts of a cirripede crustacean.
body force Any force acting on a material proportional to the mass of the substance, e.g. gravity, centrifugal force, magnetic force.
body plan The assemblage of morphological features found among members of a higher taxon, generally either a phylum or a class, which define the basic architecture of the taxon.
body wall The external surface of the body in animals, enclosing the body cavity; e.g. the part of a sponge between the exterior and a central *spongocoel,* or the *perisome* of an echinoderm.
body wave A *seismic wave* that travels through the interior of the Earth, with a propagation mode that does not depend on any boundary surface. A body wave may be either longitudinal (a *P wave*) or transverse (an *S wave*).
body-wave magnitude An *earthquake magnitude* determined at teleseismic distances, using the logarithm of the ratio of amplitude to period for body waves. Abbrev: m_b.
body whorl The outer, last-formed, and typically largest *whorl* of a univalve shell; e.g. the last complete loop in the spiral of a gastropod shell, terminating in the *aperture.*
BOE *barrels of oil equivalent.*
boehmite *böhmite.*
Boehm lamellae Planar structural features produced in deformed mineral grains. They resemble deformation lamellae except that they are decorated by trains of fluid inclusions. Cf: *Tuttle lamellae.* Also spelled: *Böhm lamellae.*
bog (a) Waterlogged, spongy ground, consisting primarily of mosses, containing acidic, decaying vegetation that may develop into peat. (b) The vegetation characteristic of this environment, esp. sphagnum, sedges, and heaths. Cf: *fen; marsh; swamp.*
bogan (bo'-gan) *pokelogan.*
bogaz (bo'-gaz) A Slavic term for *solution corridor.*
bog burst The bursting of a bog under the pressure of its swelling, due to water retention by a marginal dam of growing vegetation. The escaping water produces muddy peat flows over the surrounding area.
bog butter A substance, found preserved in Irish peat bogs, that was formerly believed to be a native hydrocarbon but is now known to be "fossil" butter that had been buried for storage and found at a much later date (Tomkeieff, 1954, p.30). Syn: *butyrellite; butyrite.*
bog coal An earthy type of brown coal.
bogdanovite (bog-dan'-o-vite) A submetallic brown cubic mineral with a structure related to that of gold: $(Au,Te,Pb)_3(Cu,Fe)$.
bog expanse The inner part of a raised bog within the *rand.* Cf: *peat plateau.*
bog flow The outflow from a *bog burst.* Cf: *peat flow.*
Bøggild intergrowth Plagioclase feldspar consisting of alternating exsolution lamellae of An_{-48} and An_{-58}, where An_x represents percent anorthite composition ($CaAl_2Si_2O_8$) and the implied remainder is albite composition ($NaAlSi_3O_8$). Bøggild intergrowths produce the *play of color* known as *labradorescence.* Cf: *Huttenlocher intergrowth; Peristerite intergrowth.*
bøggildite (bøg'-gild-ite) A flesh-red monoclinic mineral: $Sr_2Na_2Al_2(PO_4)F_9$.
boggsite (boggs'-ite) A colorless orthorhombic zeolite mineral: $Ca_8Na_3(Si,Al)_{96}O_{192} \cdot 70H_2O$.
boghead coal (bog'-head) A *sapropelic coal* resembling *cannel coal* in its physical properties but consisting dominantly of algal matter rather than spores. Name derived from Boghead, Scotland. Cf: *torbanite; algite.* Syn: *algal coal; gélosic coal; sapromyxite; tomite.*
boghedite (bog'-hed-ite) An old syn. of *torbanite.*
bog iron ore (a) A general term for a soft, spongy, and porous deposit of impure hydrous iron oxides formed in bogs, marshes, swamps, peat mosses, and shallow lakes by precipitation from iron-bearing waters and by the oxi-

dizing action of algae, iron bacteria, or the atmosphere; a *bog ore* composed principally of limonite that is often impregnated with plant debris, clay, and clastic material. It is a poor-quality iron ore, found in tubular, pisolitic, nodular, concretionary, or thinly layered forms, or in irregular aggregates, in level sandy soils, and is esp. abundant in the glaciated northern regions of North America and Europe (Scandinavia). See also: *murram.* Syn: *limnite; morass ore; meadow ore; marsh ore; lake ore; swamp ore.* (b) A term commonly applied to a loose, porous, earthy form of *limonite* occurring in wet ground.

bog lake A lake surrounded or nearly surrounded by bog and characterized by low pH, low nutrient content, brown-stained water, and low productivity. Cf: *sphagnum bog.*

bog lime An obsolete syn. for the *marl* of freshwater lakes.

bog manganese A *bog ore* of variable composition, but consisting chiefly of hydrous manganese oxide; specif. wad formed in bogs or marshes by the action of minute plants.

bog-mine ore A syn. of *bog ore.* Also called: bog mine.

bog moat *lagg.*

bog ore A poorly stratified accumulation of earthy metallic-mineral substances, mainly oxides, that are formed in bogs, marshes, swamps, and other low-lying moist places, usually by direct chemical precipitation from surface or near-surface percolating waters; specif. *bog iron ore* and *bog manganese.* Cf: *lake ore.* Syn: *bog-mine ore.*

bog peat *highmoor peat.*

Bog soil A great soil group in the 1938 classification system, an intrazonal, hydromorphic group of soils having a mucky or peaty surface horizon and an underlying peat horizon. These soils developed in swamps or marshes in humid or subhumid climates (USDA, 1938). Bog soils are now classified as *Histosols.* Cf: *Half-Bog soil.* Not used in current U.S. system of taxonomy.

bogue A term used in Alabama and Mississippi for the mouth or outlet of a stream, or for the stream itself, or for a bayou. Etymol: American French, from Choctaw bouk, "stream, creek".

bogusite (bo'-gus-ite) An intrusive rock of the same general composition as *teschenite* but of lighter color. Named by Johannsen (1931) for Boguschowitz, Poland. Obsolete.

bøgvadite A vitreous colorless orthorhombic mineral: $Na_2Ba_2SrAl_4F_{20}$. It occurs in the Ivigtut cryolite deposit in Greenland.

bohdanowiczite (boh-dan'-o-wicz-ite') A metallic rhombohedral mineral: $AgBiSe_2$. The selenium analogue of matildite, and a member of the *matildite* group.

Bohemian garnet (Bo-he'-mi-an) A yellowish-red to dark, intense-red gem variety of *pyrope* obtained from Bohemia.

Bohemian ruby A red variety of crystalline quartz; specif. *rose quartz* cut as a gem.

Bohemian topaz *citrine.*

böhmite (boehm'-ite) A grayish, brownish, or reddish orthorhombic mineral: $AlO(OH)$. It is a major constituent of some bauxites and it represents the gamma phase dimorphous with diaspore. Also spelled: *boehmite.*

Böhm lamellae *Boehm lamellae.*

boil In streams, a zone of flow separation characterized by strongly vertical, upward flow.

boiler (boil'-er) (a) A small submerged coral reef, esp. one occurring where the sea breaks frequently (that is, "boils"). Syn: *breaker [reef]; cup reef.* (b) A *blowhole* along the coast.

boiling spring [grd wat] (a) A spring, the water from which is agitated by the action of heat. (b) A spring that flows so rapidly that strong vertical eddies develop.

boiling spring [karst] Any fountaining *resurgence.* Cf: *vauclusian spring; blue hole.*

bojite (boj'-ite) A *gabbro* in which primary hornblende substitutes for most of the pyroxene, although some augite and biotite may be present; a *hornblende gabbro.* The name, given by Weinschenk in 1899, is for the Boii, "a Celtic tribe that settled in Germany" (Johannsen, 1937, p.227). Obsolete.

bokite (bok'-ite) A submetallic dull black mineral: $(Al,Fe^{3+})_{1.3}(V^{5+},V^{4+},Fe^{3+})_8O_{20}\cdot 7.4H_2O$. Classed as a "vanadium bronze".

bold coast A prominent landmass, such as a cliff or promontory, rising or sloping steeply from a body of water, esp. along the seacoast.

boldyrevite (bol-dy-rev'-ite) A yellow amorphous mineral: $NaCaMgAl_3F_{14}\cdot 4H_2O$.

bole Any of several varieties of fine, compact, friable, and earthy or unctuous clay (impure halloysite), usually colored red, yellow, or brown due to the presence of iron oxide, and consisting essentially of hydrous silicates of aluminum or less often of magnesium; a waxy decomposition product of basaltic rocks, having the variable composition of lateritic clays. Rarely used. Adj: bolar. Syn: *terra miraculosa.*

boleite (bo'-lé-ite) An indigo-blue mineral: $KPb_{26}Ag_9Cu_{24}Cl_{62}(OH)_{48}$. Also spelled: boléite.

bolide (bo'-lide) (a) An exploding or exploded meteor or meteorite; a detonating *fireball.* (b) An impacting projectile large enough to produce a hypervelocity crater or similarly catastrophic event.

bolivarite (bo'-li-var-ite) A non-crystalline pale greenish yellow mineral: $Al_2(OH)_3(PO_4)\cdot 4\text{-}5H_2O$.

Bolivian polymetallic vein A *mineral-deposit type* characterized by polymetallic sulfide with cassiterite veins hosted by subvolcanic intrusives and associated extrusive rocks.

Bølling (Bøll'-ing) n. A term used primarily in Europe for an interval of late-glacial time (centered about 14,000 calendar years ago) following the Oldest Dryas and preceding the Older Dryas, during which the climate ameliorated favoring birch and park-tundra vegetation in Scandinavia. Also spelled: Bölling. adj. Pertaining to the late-glacial Bølling interval and to its climate, deposits, biota, and events.

bolly (bol'-ly) *bally.*

Bologna stone (Bo-lo'-gna) A nodular, concretionary, or rounded form of barite, composed of radiating fibers; phosphorescent when calcined with charcoal. Also called: Bolognan stone; Bologna spar.

bolometer (bo-lom'-e-ter) A detector used to measure the radiant temperature or flux by measuring the change in electrical resistance of a metal (e.g., platinum) or of a semiconductor (e.g., a thermistor).

bolson (bol'-son, bol-son') (a) A term applied in the desert regions of the SW U.S. to an extensive, flat, alluvium-floored basin or depression, into which drainage from the surrounding mountains flows centripetally, with gentle gradients toward a playa or central depression; an *interior basin,* or a basin with internal drainage. See also: *semibolson.* Syn: *playa basin.* (b) A temporary lake, usually saline, formed in a bolson. Etymol: Spanish "bolsón", "large purse".

bolson plain A broad, intermontane plain in the central part of a bolson or semibolson, composed of deep alluvial accumulations washed into the basin from the surrounding mountains.

boltonite (bol'-ton-ite) A greenish or yellowish granular variety of *forsterite* from Bolton, Mass.

boltwoodite (bolt'-wood-ite) A yellow monoclinic mineral: $HK(UO_2)(SiO_4)\cdot 1.5H_2O$.

bolus alba A syn. of *kaolin.* Etymol: Latin, "white clay".

bomb [geochem] A vessel in which experiments can be conducted at high temperature and pressure. It is used in geochemistry and in experimental petrology. Syn: *pressure vessel.*

bomb [geophys] (a) An explosive charge used as a seismic energy source. (b) A device to collect a sample of fluid from a borehole.

bomb [pyroclast] A pyroclast ejected while viscous and shaped while in flight. It is larger than 64 mm in diameter, and may be vesicular to hollow inside. Actual shape or form varies greatly, and is used in descriptive classification, e.g. rotational bomb; spindle bomb.

bombiccite (bom-bic'-cite) *hartite.*

bombite (bom'-bite) A blackish-gray aluminosilicate of ferric iron and calcium from Bombay, India. It resembles Lydian stone and is probably a glassy rock.

bombollaite (bom-bol'-la-ite) *bambollaite.*

bomb sag *bedding-plane sag.*

bomby (bom'-by) An Australian term for a large, submerged pinnacle-like reef development, generally a few to several tens of square meters in area, found in a back-reef lagoon and constituting a hazard for navigation and fishing (Maxwell, 1968, p.133). Syn: *pinnacle [reefs].* See also: *bommy.*

bommy (bom'-my) A large *coral head.* See also: *bomby.*

bonaccordite (bon-ac-cord'-ite) A reddish-brown orthorhombic mineral: $Ni_2Fe^{3+}BO_5$. A member of the *ludwigite* group.

bonamite (bon-am'-ite) A trade name for an apple-green gem variety of *smithsonite,* resembling the color of chrysoprase.

bonanza (bo-nan'-za) A miner's term for a rich body of ore or a rich part of a deposit; a mine is "in bonanza" when it is operating profitably. Also, discontinuous locally rich ore deposits, esp. epithermal ones. Spanish, "prosperity, success". Cf: *borasca.*

bonanza gold-silver deposit *adularia-sericite type* deposit.

bonattite (bo-nat'-tite) A pale blue monoclinic mineral: $CuSO_4\cdot 3H_2O$.

bonchevite (bon-chev'-ite) A metallic gray orthorhombic mineral: $(Pb,Cu)_3Bi_{11}S_{18}$. Probably a mixture of galenobismutite and pekoite.

bond clay A clay which, because of its plasticity, serves to bond relatively nonplastic materials in the fabrication of ceramic or other molded products ("green bond"). Also, a clay which, on firing to furnace or vitrification temperature, bonds adjacent ceramic materials that vitrify at a still higher temperature ("fired bond").

bonding clay A clay or a mixture of clay minerals added to sand to form molding sands for use in casting metals (Grim, 1962; Grimshaw, 1971, p.304) Cf: *bond clay.*

bone A tough, very fine-grained, gray, white, or reddish quartz.

bone amber *osseous amber.*

bone bed Any sedimentary stratum (usually a thin bed of sandstone, limestone, or gravel) in which fossil bones or bone fragments are abundant, and often containing other organic remains, such as scales, teeth, and coprolites.

bone breccia An accumulation of bones or bone fragments, often mixed with earth and sand, and cemented with calcium carbonate; esp. such a deposit formed in limestone caves or other animal retreats. Syn: *osseous breccia.*

bone cave A cave that has served as a trap for fossil vertebrates.

bone chert A weathered, residual chert that appears chalky and somewhat porous, and that is usually white but may be stained with red or other colors. When found in insoluble residues, it is an indicator of an unconformity.

bone coal (a) Coal that has a high ash content. It is hard, compact, and nonbanded. Syn: *bony coal.* (b) Argillaceous partings in coal, sometimes called *slate.*

bone magnesite *serpentine-hosted magnesite.*

bone phosphate of lime Tricalcium phosphate, $Ca_3(PO_4)_2$. The phosphate content of phosphorite may be expressed as percentage of bone phosphate of lime. Abbrev: BPL.

bone turquoise *odontolite.*

boneyard A three-dimensional network of cave voids that is not fracture controlled. It may be large enough for humans to traverse. Preferred syn: *spongework.*

boninite (bo-nin'-ite) (a) In the *IUGS classification*, a volcanic rock defined in the TAS diagram with SiO_2>52, Na_2O+K_2O<3, MgO>8, and TiO_2<0.5 wt%. A "high-Mg" *basaltic andesite* or *andesite.* (b) A glassy olivine-orthopyroxene *andesite* that contains little or no modal feldspar. Named by Petersen in 1891 for the Bonin Islands, Japan. Cf: *sanukite.*

Bonne projection An equal-area, modified-conic map projection having one standard parallel intersecting the central meridian (a straight line along which the scale is exact) near the center of the map. All parallels are represented by equally spaced arcs of concentric circles (divided to exact scale) and all meridians (except the central meridian) are curved lines connecting corresponding points on the parallels. The projection is commonly used for mapping compactly shaped areas in middle latitudes (such as France) and for mapping continents such as North America and Eurasia. Named after Rigobert Bonne (1727-1795), French cartographer, who used the projection beginning in 1752 (though it was used by others in the 16th century). See also: *sinusoidal projection.*

bonshtedtite (bon'-shtedt-ite) A colorless monoclinic mineral: $Na_3(Fe^{2+},Mg)(PO_4)(CO_3)$. It is the ferrous analogue of bradleyite and sidorenkite, and a member of the *bradleyite* group.

bony coal (bon'-y) *bone coal.*

bony fish *Osteichthyes.*

book *mica book.*

book clay Clay deposited in thin, leaflike laminae. Syn: *leaf clay.*

bookhouse structure A term introduced by Sloane and Kell (1966, p.295) for a fabric found in compacted kaolin clays, consisting of parallel and random arrangements of packets of oriented clay particles (flakes). Cf: *cardhouse structure.*

books Petrographic description of appearance of kaolinite crystals.

bookshelf faulting A geometric model of crustal extension in which movement on a series of parallel normal faults is accompanied by rigid rotation of the fault-bounded blocks. Syn: *domino-style faulting.* Cf: *step fault.*

book structure In ore deposits, the alternation of ore with gangue, usually quartz, in parallel sheets. Cf: *ribbon [ore dep].*

boolgoonyakh (bool-goo'-nyakh) A syn. of *pingo.* Etymol: Yakutian. Also spelled: boolyunyakh; bulgunniakh.

boomer (boom'-er) (a) A marine seismic-energy source in which a high-voltage discharge causes two metal plates to separate abruptly in a body of water. (b) A very strong, usually low-frequency event on a seismic record.

booming dune A term used by Criswell et al. (1975) for a dune of *booming sand.*

booming sand A *sounding sand,* found on a desert, that emits a low-pitched note of considerable magnitude and duration as it slides (either spontaneously or when induced) down the slip face of a dune or drift (Humphries, 1966, p.135). See also: *roaring sand; booming dune.*

boort *bort.*

boothite (booth'-ite) A blue monoclinic mineral: $CuSO_4{\cdot}7H_2O$. Its blue color is lighter than that of chalcanthite. It dehydrates to chalcanthite in air.

bootstrap A computer-implemented iterative procedure for estimating standard errors and confidence intervals for statistics calculated from a sample whose population has an unknown form. Random subsets of the sample are repeatedly assembled and statistics calculated from the subsets. The statistics will form a distribution which reflects the population distribution. The procedure has been used to estimate confidence intervals around surfaces estimated by *kriging.* Cf: *jackknife.*

bora (bo'-ra) A *katabatic wind* of the northern Adriatic coast.

boracite (bo'-ra-cite) A white, yellow, greenish, or bluish orthorhombic mineral: $Mg_3B_7O_{13}Cl$. It is strongly pyroelectric, becomes cubic at high temperatures, and occurs in evaporites and saline deposits. See also: *stassfurtite.*

Boralfs Alfisols that have formed in cool places. Boralfs have frigid or cryic but not pergelic temperature regimes, and have udic moisture regimes. Boralfs are not saturated with water for periods long enough to limit their use for most crops. A suborder in the U.S. system of soil taxonomy.

boralsilite A colorless or white monoclinic mineral: $Al_{16}B_6O_{30}(Si_2O_7)$.

borasca (bo-ras'-ca) A miner's term for an unproductive area of a mine or orebody; a mine is "in borasca" when it is exhausted. Etymol: Mexican Spanish borrasca, "exhaustion of a mine". Cf: *bonanza.*

borate (bo'-rate) A mineral characterized by the presence of BO_3^{-3} and/or BO_4^{-5} oxyanions in its chemical formula. An example of a borate is boracite, $Mg_3B_7O_{13}Cl$. Cf: *carbonate [mineral]; nitrate.*

borax (bor'-ax) A white, yellowish, blue, green, or gray mineral: $Na_2B_4O_5(OH)_4{\cdot}8H_2O$. It is an ore of boron and occurs as a surface efflorescence or in large monoclinic crystals embedded in muds of alkaline lakes. Borax is used chiefly in glass, ceramics, agricultural chemicals, and pharmaceuticals, and as a flux, cleansing agent, water softener, preservative, and fire retardant. Syn: *tincal.*

borax bead The type of *bead* commonly used in blowpipe analysis of metallic compounds.

borax lake (a) A lake whose shores are encrusted with deposits rich in borax (sodium borate). (b) A dry, borax-rich bed of a lake.

borcarite (bor-car'-ite) A bluish-green monoclinic mineral: $Ca_4MgB_4O_6(CO_3)_2(OH)_6$.

border belt (bor'-der) A term used by Chamberlin (1893, p.263) for superficial glacial deposits now known as a *boulder belt.*

bordered pit (bor'-dered) A *pit* in a *tracheid* in which the margin projects over a chamber separated from the pit-closing membrane, as in tracheids of coniferous wood. Cf: *simple pit.*

border facies The marginal portion of an igneous intrusion, which differs in texture and composition from the main body of the intrusion, possibly due to more rapid cooling or to assimilation of material from the country rock.

border fault (a) *boundary fault.* (b) *peripheral fault.* (c) normal fault bounding a graben or half-graben basin. Syn: *basin-margin fault.*

borderland (bor'-der-land) According to a concept widely held in the first part of the 20th Century, and championed by Schuchert (1923), a crystalline landmass on the seaward borders of the Phanerozoic orogenic belts near the edges of the North American continent. These borderlands served as a source for syntectonic sediment, were said to be tectonically much more active than the Canadian Shield, and were subsequently lost by foundering into the oceans. The concept is now discredited. So-called borderlands are actually crustal blocks that converged and collided with the continental margin. These blocks later rifted off the continent and drifted away as sea-floor spreading occurred. Cf: *hinterland; tectonic land.* See also: *Cascadia; Llanoria.*

bore [tides] A tidal wave which propagates as a solitary wave with a steep leading edge up certain rivers. Formation is favored in wedge-shaped shoaling estuaries at times of spring tides. Syn: *tidal bore.*

bore [volc] The outlet of a geyser at the Earth's surface.

boreal (bo'-re-al) (a) Pertaining to the north, or located in northern regions; northern. (b) Pertaining to the northern biotic area (or Boreal region) characterized by *tundra* and *taiga* and by dominant coniferous forests. (c) Pertaining to the *Boreal* postglacial period, characterized by a cool climate like that of the present Boreal region. Also, said of the climate of such a period. (d) Pertaining to a Boreal climatic zone, or to the climate of such a zone. One having a definite winter with snow and a short summer that is generally hot, and characterized by a large annual range of temperature. It includes large parts of North America, central Europe, and Asia, generally between latitudes 60°N and 40°N. Etymol: Boreas, Greek god of north wind.

Boreal [paleoclim] n. A term used primarily in Europe for an interval of Holocene time (from about 10,000 to 8,000 calendar years ago) following the Preboreal and preceding the Atlantic, during which the inferred climate was relatively warm and dry; a subunit of the *Blytt-Sernander sequence,* characterized by pine and hazel vegetation in Scandinavia. adj. Pertaining to the postglacial Boreal interval and to its climate, deposits, biota, and events.

bored well Generally, a shallow water well, 3 to 30 m deep and 20 to 90 cm in diameter, constructed by hand-operated or power-driven augers.

borehole (bore'-hole) A circular hole made by drilling; esp. a deep hole of small diameter, such as an oil well or a water well. Syn: *hole [drill]; wellbore.*

borehole breakout The spalling at the edge of a borehole as a consequence of the concentration of the maximum horizontal stress (Plumb and Hickman, 1985). The stress concentration is so large that induced differential stress causes shear fractures within the rock next to the borehole wall. Spalling releases the fractured rock to cause the elongation of the borehole in the direction of the least horizontal stress.

borehole geophone A geophone that can withstand the pressure and temperature conditions in a borehole. It often consists of three orthogonal elements so that it can read all three components of wave motion and usually includes provision for clamping to the borehole wall to ensure good coupling. Borehole hydrophones are also used but they are much more subject to tube waves.

borehole gravimeter A remote reading *gravimeter* that is lowered through a borehole as a logging tool. The difference between the gravity reading at two different depths gives the apparent density ρ in g/cm³: $\rho = 3.686 - 128.5\ \Delta g/\Delta h$, where Δg is the gravity reading difference in mGal and Δh is the depth difference in meters.

borehole gravity meter A *gravity meter* adapted to the borehole environment that measures the gravitational force while hanging stationary at selected depths within a well. Differences in readings at different depths derive from a large volume of the intervening rock; the effects of washed-out zones, mud, filtrate invasion, or casing are not significant. Borehole gravity data may be solved for bulk formation density. Abbrev: BHGM. Syn: *borehole gravimeter.*

borehole storage effect Immediately following the start of pumping water from a well, part of the water pumped comes from dead storage in the well, and this condition affects water-level changes in the well.

borehole survey A *directional survey.*

borehole-to-borehole measurements *crosshole method.*

borengite A *trachyte* dike rock dominated by K-feldspar with accessory fluorite. Named by Von Eckermann in 1960 at Alnö, Sweden. Obsolete.

borickyite *delvauxite.*

boring (bor′-ing) A *trace fossil* consisting of an etching, groove, or hollow, produced by plants (fungi, algae) or animals (sponges, worms, bryozoans, barnacles) in shells, bones, or other hard parts of invertebrates and vertebrates. Cf: *burrow.*

boring porosity Porosity resulting from organisms boring into lithified substrate.

borishanskiite (bo-ri-shan′-ski-ite) A metallic steel-gray orthorhombic mineral: $Pd(As,Pb)_2$. The arsenic analogue of polarite.

Bormidian (Bor-mid′-ian) Italian stage: lower Miocene (below Langhian). Equiv. to Aquitanian and Burdigalian of Bordeaux basin.

bornemanite (bor′-ne-man-ite′) A pale yellow monoclinic mineral: $Na_3Ba(Na,Ti,Mn)_4(Ti,Nb)_2O_2(Si_4O_{14})(PO_4)(F,OH)_2$.

bornhardt (born′-hardt) A residual peak having the characteristics of an *inselberg;* specif. a large granite-gneiss inselberg associated with the second cycle of erosion in a rejuvenated desert region (King, 1948). Named in honor of F. Wilhelm C.E. Bornhardt (1864-1946), German explorer of Tanganyika, who first described the feature.

bornhardtite (born′-hardt-ite) A metallic rose cubic mineral: Co_3Se_4. A member of the *linnaeite* group.

bornite (born′-ite) A brittle, metallic-looking orthorhombic mineral: Cu_5FeS_4. It has a reddish-brown or coppery-red color on fresh fracture, but tarnishes rapidly to iridescent purple or blue. Bornite is a valuable ore of copper. Syn: *erubescite; variegated copper ore; peacock ore; horseflesh ore; purple copper ore.*

borocookeite A light pinkish gray monoclinic mineral: $LiAl_4(Si_3B)O_{10}(OH)_8$. The boron analogue of cookeite, a chlorite-like mineral.

borodaevite A metallic monoclinic mineral: $Ag_5(Pb,Fe)Bi_7(Sb,Bi)_2S_{17}$. A member of the *lillianite* homologous series.

borolanite (bo-ro-lan′-ite) A plutonic rock composed chiefly of alkali feldspar and andradite, with lesser amounts of nepheline, biotite, and pyroxene; an andradite *nepheline syenite.* The feldspar and nepheline commonly form aggregates that resemble phenocrysts of leucite. The term was originated by Horne and Teall in 1892, for Loch Borolan, Scotland. Obsolete.

Borolls Mollisols with a mean annual soil temperature of $<8°$ C that are never dry for 60 consecutive days or more within the 90 days following the summer solstice. Borolls do not contain material that has a $CaCO_3$ equivalent > 400 $g kg^{-1}$ unless they have a calcic horizon, and they are not saturated with water for periods long enough to limit their use for most crops. A suborder in the U.S. system of soil taxonomy.

boromuscovite A porcelaneous buff or pale cream-colored mineral: $KAl_2(Si_3B)O_{10}(OH,F)_2$. The boron analogue of muscovite, and a member of the *mica* group.

boronatrocalcite (bo′-ro-na′-tro-cal′-cite) *ulexite.*

borovskite (bo-rov′-skite) A metallic dark gray cubic mineral: Pd_3SbTe_4.

borrow (bor′-row) Earth material (sand, gravel, etc.) taken from one location (such as a borrow pit) to be used for fill at another location; e.g., embankment material obtained from a pit when sufficient excavated material is not available nearby to form the embankment. The borrow material usually has suitable or desirable physical properties for its intended purpose.

borrow pit An excavated area where *borrow* has been obtained. Cf: *barrow pit.*

bort (a) A granular to very finely crystalline aggregate consisting of imperfectly crystallized diamonds or of fragments produced in cutting diamonds. It often occurs as spherical forms, with no distinct cleavage, and having a radial fibrous structure. (b) A diamond of the lowest quality, so flawed, imperfectly crystallized, or off-color that it is suitable only for crushing into abrasive powders for industrial purposes (as for saws and drill bits); an *industrial diamond.* Originally, any crystalline diamond (and later, any diamond) not usable as a gem. (c) A term formerly used as a syn. of *carbonado.* Cf: *ballas.* Syn: *boart; boort; bortz; bowr.*

bortz *bort.*

böschung (bösch′-ung) A syn. of *gravity slope.* Etymol: German "Böschung", term used by Penck (1924) for a rock slope maintaining constant gradient as it retreats.

Bosphorian Eastern European brackish-marine substage: upper Miocene, upper Pontian (below Getian substage of Dacian stage).

boss [geomorph] A smooth and rounded mound, hillock, or other mass of resistant bedrock, usually bare of soil or vegetation.

boss [ign] A plutonic body having an areal extent of less than 40 sq mi (100 km^2) and a roughly circular outline in map view. Cf: *stock [intrus rocks].*

boss [mining] A large, round or irregular bulge in the mine roof, usually composed of limestone (Nelson, 1983).

boss [paleont] (a) A rounded and raised knoblike ornamental structure in foraminifera. (b) The part of an echinoid *tubercle,* below the *mamelon,* shaped like a truncated cone and supporting the spheroidal summit of the tubercle. (c) A coarse, short nodule occurring on the spire of a gastropod.

boss structures Small raised features distributed over part of the upper surfaces of conodont basal plates or basal funnels; most have an oblong outline and, at high magnification, a rough surface texture (TIP, 1981, pt. W, supp.2).

bostonite (bos′-ton-ite) A light-colored hypabyssal rock composed chiefly of alkali feldspar; a fine-grained *trachyte* with few or no mafic components. The name, given by Hunter and Rosenbusch in 1890, is derived from Boston, Massachusetts, for no clear reason. Obsolete.

bostonitic (bos-to-nit′-ic) The texture of *bostonite* in which feldspar microlites occur as crudely radial clusters associated with patches in which the microlites are aligned. Obsolete.

bostwickite (bost′-wick-ite) A dark brownish red mineral: $CaMn^{3+}{}_6Si_3O_{16}\cdot 7H_2O$.

botallackite (bo-tal′-lack-ite) A bluish-green monoclinic mineral: $Cu_2Cl(OH)_3$. A polymorph of atacamite and paratacamite.

botanical anomaly (bo-tan′-i-cal) A local increase above the normal variation in the chemical composition, distribution, ecological assemblage, or morphology of plants, indicating the possible presence of an ore deposit. See also: *geobotanical prospecting.*

botn A Norwegian and Swedish term for the "bottom" of a glacial lake or of a fjord, but used as an equivalent of *cirque [glac geol].*

Botomian Russian stage: Lower Cambrian (above Atdabanian, below Toyonian) (Cowie and Bassett, 1989).

botryococcoid alga (bot-ry-o-coc′-coid) A colonial alga of fresh to brackish water (*Botryococcus* - green algae), with a distinctive molecular geochemical signature called botryococcane (Batten and Grenfell, 1996).

botryogen (bot′-ry-o-gen′) A deep red or deep orange monoclinic mineral, usually botryoidal: $MgFe^{3+}(SO_4)_2(OH)\cdot 7H_2O$.

botryoid (bot′-ry-oid) In speleology, a variety of subaerial *coralloid* of *botryoidal* shape. Syn: *cave coralloid.*

botryoidal (bot-ry-oi′-dal) Having the form of a bunch of grapes. Said of mineral deposits, e.g. hematite, having a surface of spherical shapes; also said of a crystalline aggregate in which the spherical shapes are composed of radiating crystals. Cf: *colloform; mammillary [mineral]; reniform.*

botryolite (bot′-ry-o-lite′) A radiated, columnar variety of *datolite* with a botryoidal surface.

bottinoite A vitreous light blue-green triclinic mineral: $NiSb^{5+}{}_2(OH)_{12}\cdot 6H_2O$.

bottleneck bay (bot′-tle-neck) A bay with a narrow entrance which is guarded from the waves by features other than barrier islands.

bottom [geog] (bot′-tom) Low-lying, level land, usually highly fertile, esp. in the Mississippi Valley region and farther west where the term signifies a grassy lowland formed by deposition of alluvium along the margin of a watercourse; an alluvial plain or a flood plain; the floor of a valley. The term is usually used in the plural. Syn: *bottomland; flat [geog]; lowland.*

bottom [geomorph] (a) *bed [geomorph].* (b) *valley floor.* (c) A term used in England for the former head of a lake in a U-shaped valley, now covered with sediment deposited by inflowing streams.

bottom [ore dep] (a) Syn. of *gutter [ore dep].* (b) The lower limit of an orebody, either structurally or by economic grade. Syn: *root [ore dep].* See also: *bottoming.*

bottom cable A marine seismic cable that is laid in place on the water bottom, as opposed to a drag cable or *streamer,* which is towed into place. Also called: *bay cable.*

bottomhole pressure The pressure recorded in a wellbore on the bottom of the well. It may be measured as a "flowing bottom-hole pressure" or as *shut-in pressure* to record the rate of pressure build-up during the survey period. Cf: *formation pressure.* Abbrev: BHP.

bottomhole temperature The temperature of the fluid measured near or at the bottom of a borehole using a maximum-reading thermometer attached to a wireline cable. Those temperatures usually are obtained commercially in hydrocarbon exploration wells to obtain an estimate of temperature for calculation of fluid resistivity. Temperatures are significantly lower than the formation temperature if the measurement was performed shortly after drilling has ceased or borehole fluids have been circulated recently. Abbrev: BHT. Cf: *drill-stem test.*

bottom ice *anchor ice.*

bottoming (bot′-tom-ing) The downward pinching-out or termination of an orebody, either structurally or by economic grade. See also: *bottom [ore dep].*

bottomland A syn. of *bottom.*

bottom load *bed load.*

bottom lock A situation where doppler-sonar measurements are based on reflections from the sea bottom. It is the normal operational mode, as opposed to the "water-scatter mode" that occurs in deeper water.

bottom moraine *ground moraine.*

bottom peat Peat that is associated with lakes or streams and is derived mainly from mosses such as Hypnum.

bottomset (bot′-tom-set) *bottomset bed.*

bottomset bed One of the horizontal or gently inclined layers of sediment deposited in front of the advancing foreset beds of a delta or bed form. Syn: *bottomset.*

bottom-simulating reflection A reflection that roughly parallels the sea-floor reflection, caused by the contrast between an overlying clathrate (gas hydrate) and underlying gas-saturated sediments. Occurs in deep water where subseafloor sediments at low temperature are under substantial pressure.

bottom terrace A depositional landform produced by streams having moderate or small bed loads of coarse sand and gravel. It is characterized by a broad surface (a few meters wide), gently sloping in the direction of flow, and a steep escarpment (about 1 m high) facing downstream and generally trending at right angles to the flow (Russell, 1898b, p.166-167).

bottom topography The depth pattern of the ocean. Commonly, the landscape of the sea floor. Cf: *bathymetry.*

bottom water [oceanog] The deepest and densest seawater of the oceans, formed by surface cooling and/or sea ice formation in high latitudes. Since much of this water has a narrow characteristic temperature and salinity range, it is often called a water type and sometimes a water mass. Cf: *deep water; intermediate water; surface water.*

bottom water [petroleum] The water immediately underlying the oil or gas in an *oil pool* or a *gas pool.* Cf: *edge water.*

Bottrycoccus braunei A bloom-forming, unicellular green alga that can produce a certain type of kerogen.

boudin (bou-din'[boo-dan']) One of a series of elongate, sausage-shaped segments occurring in *boudinage,* structure, either separate or joined by pinched connections, and having barrel-shaped cross sections. Boudins form where a relatively rigid layer, embedded between two relatively ductile layers, undergoes extension. Cf: *tectonic lens.* Etymol: French, "bag; blood sausage".

boudinage (bou-din-age') A structure common in strongly deformed sedimentary and metamorphic rocks, in which an original continuous competent layer or bed between less competent layers has been stretched, thinned, and broken at regular intervals into bodies resembling boudins or sausages, elongated parallel to the fold axes. Syn: *sausage structure.*

Bouguer anomaly (Bou-guer' [boo-gay']) The gravity field values after the *latitude correction, free-air correction, Bouguer correction, Eötvös correction* (for marine and airborne measurements) and, usually, *terrain correction* have been applied to the observed gravity field. If the terrain correction has not been applied, the anomalous gravity field is sometimes called simple Bouguer anomaly. Syn: Bouguer gravity. Named for Pierre Bouguer (1698-1758), French mathematician.

Bouguer correction A correction applied to the gravity data to compensate for the attraction of rock masses between the station (point of measurement) and the datum elevation (commonly, sea level), or, if stations are below the datum elevation, for the rock missing between the station and datum. The Bouguer correction is 0.04192 ρh, where ρ is the density of rock and h is the difference between the station and datum elevations in meters. This correction does not "reduce the data to the datum (i.e., the value that would have been obtained had the station been located on the datum)." Syn: *slab correction.* See also: *double Bouguer correction.*

Bouguer plate An imaginary layer of constant density, having infinite length and a thickness equal to the height of the observation point above the reference surface, which is usually sea level.

boulangerite (bou-lan'-ger-ite) A bluish-gray or lead-gray metallic-looking orthorhombic mineral: $Pb_5Sb_4S_{11}$. It occurs in plumose masses.

boulder (a) A detached rock mass larger than a cobble, having a diameter greater than 256 mm (10 in., or -8 phi units, or about the size of a volleyball), being somewhat rounded or otherwise distinctively shaped by abrasion in the course of transport; the largest rock-fragment size recognized by sedimentologists. In Great Britain, the limiting size of 200 mm (8 in.) has been used. Cf: *block [part size].* See also: *small boulder; medium boulder; large boulder; very large boulder.* (b) *glacial boulder.* (c) *boulder of weathering.* (d) *boulder stone.* (e) A general term for any rock that is too heavy to be lifted readily by hand. Also spelled: *bowlder.*

boulder barricade An accumulation of many large boulders visible along a coast (such as that of Labrador) between low tide and half tide (Daly, 1902, p.260).

boulder barrier A shore ridge created by great pressure from floating ice under the influence of strong winds and gentle shore slopes, and measuring over 6 m in height and 800 m in length (Hamelin and Cook, 1967, p.97).

boulder beach A beach consisting mostly of boulders.

boulder bed (a) A boulder-bearing conglomerate. (b) A glacial deposit, such as a till or tillite, containing a wide range of particle sizes; e.g. the Talchir boulder beds of India.

boulder belt A long, narrow accumulation of glacial boulders derived from distant sources, lying transverse to the direction of movement of the glacier by which it was deposited; also, a zone of such boulders. Cf: *boulder train.* Syn: *border belt.*

boulder clay A term used in Great Britain as an equivalent of *till,* but applied esp. to glacial deposits consisting of striated, subangular boulders of various sizes embedded in stiff, hard, pulverized clay or rock flour. The term "till" is preferable as a general term, applicable not only to material of the character described above but to glacial deposits that contain no boulders or that may be so sandy as to have very little clay. Syn: *drift clay.*

boulder cluster A group of interlocking clasts formed around exceptionally large streambed particles and standing slightly above an otherwise planar gravel bed.

boulder conglomerate A consolidated rock consisting mainly of boulders. Syn: *boulderstone.*

boulder depression A type of *block field* situated in a shallow depression, displaying a flat surface of pure boulder material that gradually decreases in size downward, and found mainly below the timberline. Diameter: a few meters to hundreds of meters.

boulderet (boul-der-et') A term suggested by Chamberlin (1883, p.324) for a rounded, coarse fragment of glacial drift, having a diameter range of 6-15 in. (15-38 cm).

boulder facet One of the small plane surfaces on a *faceted boulder.*

boulder fan A fan-shaped assemblage of clasts (mainly boulders) diverging from their bedrock source in the direction of movement of the glacier by which they were transported and deposited.

boulder field *block stream.*

boulder flat A level tract covered with boulders.

boulder gravel An unconsolidated deposit consisting mainly of boulders.

boulder of decomposition A *boulder of weathering* produced by chemical weathering; e.g. a joint block of basalt, modified and rounded by *spheroidal weathering,* leaving a relatively fresh spherical core surrounded by shells of decayed rock.

boulder of disintegration A *boulder of weathering* produced by mechanical weathering; e.g. a boulder fashioned by *exfoliation.*

boulder of weathering A large, detached rock mass whose corners and edges have been rounded in place, at or somewhat below the surface of the ground, by chemical or mechanical weathering; e.g. *boulder of decomposition* and *boulder of disintegration.* Cf: *boulder.* Syn: *residual boulder; weathering boulder.*

boulder opal A miner's term applied in Queensland, Australia, to siliceous ironstone nodules of concretionary origin, containing precious opal and occurring in sandstone or clay.

boulder pavement [geomorph] (a) An accumulation of boulders produced on a terrace by the eroding action of waves or river currents in removing finer material from littoral or fluvial deposits. (b) A slightly inclined surface composed of randomly spaced, flat-surfaced, usually frost-shattered blocks resulting from solifluction or other mass movement. (c) A *desert pavement* consisting of boulders.

boulder pavement [glac geol] (a) An accumulation of glacial boulders once contained in a moraine and remaining nearly in their original positions when the finer material has been removed by waves and currents. (b) A relatively smooth surface strewn with striated and polished boulders, abraded to flatness by the movement of an overriding glacier. Cf: *glacial pavement.*

boulder prospecting The use of boulders and boulder trains from outcrops of mineral deposits as a guide to ore.

boulder quarry A quarry in which weathering has produced so much jointing in the stone that it is not possible to mine large blocks from it.

boulder rampart A rampart or narrow ridge of boulders built along the seaward margin of a reef flat, esp. on the side from which the prevailing winds blow. The rampart, which seldom exceeds 1 or 2 m. in height, occurs close behind the *lithothamnion ridge* where present. Syn: *boulder ridge.*

boulder ridge (a) A *beach ridge* composed of boulders. (b) *boulder rampart.*

boulder saprolite Blocks of fresh parent rock surrounded by material weathered to *saprolite.*

boulder size A term used in sedimentology for a particle volume greater than that of a sphere with a diameter of 256 mm (10 in.).

boulderstone (boul'-der-stone) A consolidated sedimentary rock consisting of boulder-size particles (Alling, 1943, p.265). Syn: *boulder conglomerate.*

boulder stone An obsolete term for any large rock mass lying on the surface of the ground or embedded in the soil, differing from the country rock of the region, such as an erratic. Syn: *boulder.*

boulder stream *block stream.*

boulder train A line or series of glacial boulders and smaller clasts extending from the same bedrock source, often for many kilometers, in the direction of movement of the glacier by which they were transported and deposited. Cf: *boulder belt; boulder fan; indicator fan.*

boulder wall A boulder-built glacial *moraine.*

bouldery (boul'-der-y) Characterized by boulders; e.g. a "bouldery soil" containing stones having diameters greater than 60 cm (24 in.) (SSSA, 1965, p.333).

boule A pear-shaped or carrot-shaped mass, as of sapphire, ruby, spinel, or rutile, that forms during the production of synthetic gem material by the Verneuil flame-fusion process. Etymol: French, "ball".

Bouma cycle (Bou'-ma) A fixed, characteristic succession, of five intervals, that makes up a complete sequence of a turbidite (Bouma, 1962). One or more of the intervals may be missing. The five intervals, from the top: (e) pelitic; (d) upper parallel laminations; (c) current ripple laminations; (b) lower parallel laminations; and (a) graded. Named after Arnold H. Bouma, Dutch sedimentologist.

bounce cast The cast of a *bounce mark*, consisting of a short ridge that fades out gradually at both ends.

bounce mark A shallow *tool mark* oriented parallel to the current and produced by an object that struck or grazed against the bottom, rebounded, and was carried upward. The longitudinal profile is symmetrical. The mark is widest and deepest in the middle and fades out gradually in both directions. The term was proposed by Wood and Smith (1958, p.168). Cf: *prod mark [sed struc].* See also: *brush mark.*

boundary current (bound'-a-ry) A deep-ocean geostrophic current running parallel to a coast. Boundary currents are fast along western boundaries of the ocean and slow along eastern boundaries. The Gulf Stream is a western boundary current. The California Current is an eastern boundary current.

boundary curve *boundary line [geochem].*

boundary fault A descriptive term used in coal-mining geology for a fault along which there has been sufficient displacement to truncate the coal-bearing strata and thus bound the coalfield. Syn: *marginal fault.* Partial syn: *border fault.*

boundary layer [hydraul] In a fluid, a region of concentrated velocity variation and shear stress close to a solid that is moving relatively to the fluid. It is thin and its flow may be either turbulent or laminar (Middleton, 1965, p.247).

boundary layer [petrology] Two types of boundary layers, thermal and compositional, may occur in a magma body. A thermal boundary layer is a thin region of cool magma adjacent to a magma-wallrock contact or in front of an advancing *solidification front.* The thickness of the layer is controlled by the temperature contrast between the magma and the wall-rock or solidification front, the rate of thermal diffusion through both the magma and wall-rock or solidification front, and time. Liquid density increases within the boundary layer and may lead to *thermal convection.* A compositional boundary layer is a thin region of fractionated magma in front of an advancing solidification front. The thickness of the layer is controlled by the nature and rate of crystallization in the solidification front and the rate of chemical diffusion through the magma. Liquid density may either increase or decrease within the boundary layer (see *fractionation density*) and may lead to *compositional convection.*

boundary layer fractionation Fractional crystallization brought about by in-situ crystallization in a magmatic boundary layer coupled with buoyancy driven separation of the fractionated interstitial liquid. Continuous mixing of the fractionated interstitial liquid with the main body of magma may lead to a liquid path that is different from that generated through *homogeneous fractionation.*

boundary line [geochem] In a binary system, the line along which any two phase areas adjoin; in a ternary system, the line along which any two liquidus surfaces intersect. In a condensed ternary system, the boundary line represents equilibrium, typically with two solid phases and one liquid phase. See also: *reaction line.* Syn: *boundary curve; phase boundary.*

boundary line [surv] A line along which two areas meet; a line of demarcation between contiguous political or geographic entities.

boundary map A map prepared specifically for delineating a boundary line and the adjacent territory.

boundary migration recrystallization Recrystallization resulting from the migration of high-angle grain boundaries separating highly strained crystals from less strained or unstrained crystals of the same material. Syn: *migration recrystallization.*

boundary monument A monument placed on or near a boundary line for the purpose of preserving and identifying its location on the ground.

boundary spring A type of *gravity spring* whose water issues from the lower slope of an alluvial cone. Syn: *alluvial-slope spring.*

boundary stratotype A specific sequence of strata in which a specific point is designated that serves as the standard for the definition and recognition of a stratigraphic boundary (ISSC, 1994, p.26); the upper or lower limit of a unit stratotype. Cf: *stratotype.*

boundary survey A survey made to establish or re-establish a boundary line on the ground or to obtain data for constructing a map showing a boundary line; esp. such a survey of boundary lines between political territories. Cf: *land survey; cadastral survey.*

boundary-value component *perfectly mobile component.*

boundary-value problem One of three problems of potential theory that have great significance in geodesy. See: *Dirichlet's problem; Neumann's problem.* The third (unnamed) problem is to determine a function that is harmonic outside of a given surface and is such that a certain linear combination of it and its normal derivative assumes prescribed boundary values on the surface.

boundary vista A lane cleared along a boundary line passing through a wooded area.

boundary wave A seismic wave propagated along a free surface or an interface between layers.

bound gravel A hard, lenticular, cemented mass of sand and gravel occurring in the region of the water table; it is often mistaken for bedrock.

bounding surface (bound'-ing) An erosional truncation surface that separates groups of cross beds on various scales (Brookfield, 1977).

boundstone (bound'-stone) A term used by Dunham (1962) for a sedimentary carbonate rock whose original components were bound together during deposition and remained substantially in the position of growth (as shown by such features as intergrown skeletal matter and lamination contrary to gravity); e.g. most reef rocks and some biohermal and biostromal rocks. It is composed of bioclasts, over 2 mm in diameter and forming more than 10 percent of the rock, that are organically attached or cemented to each other; these may form a structural framework or be interspersed among coarse skeletal debris (Cuffey, 1985). Cf: *biolithite.*

bound water Water present in such materials as animal and plant cells and soils, which cannot be removed without changing the structure or composition of the material and cannot react as does *free water* in such ways as dissolving sugar and forming ice crystals.

bourne A small stream or brook; specif. an intermittent stream that flows on the chalk downs and limestone heights of southern England after a heavy rainfall. Syn: *gypsey; chalk stream.* Cf: *burn; winterbourne; woebourne.* Also spelled: bourn.

bournonite (bour'-non-ite) (a) A steel-gray to iron-black orthorhombic mineral: $PbCuSbS_3$. It commonly occurs in wheel-shaped twin crystals associated with other copper ores. Syn: *wheel ore; cogwheel ore; endellionite; berthonite.* (b) A group name for minerals with composition $PbCuXS_3$, where X = As, Sb or Bi.

bourrelet (bour'-re-let) (a) An externally inflated or elevated part of an interambulacral area of an echinoid, situated adjacent to the *peristome.* Cf: *phyllode [paleont].* (b) Either of two parts of the ligamental area of a bivalve flanking the *resilifer* on its anterior and posterior sides. Each bourrelet comprises a growth track and a seat of the lamellar ligament.

Boussinesq approximation An important simplification of the equations of motion for flow in the ocean whereby the compressibility of seawater is neglected. The resulting equations are much simpler than equations for compressible flow.

boussingaultite (bous'-sin-gaul-tite') A colorless or yellowish pink monoclinic mineral: $(NH_4)_2Mg(SO_4)_2 \cdot 6H_2O$. A member of the *picromerite* group.

bowenite (bow'-en-ite) A hard, compact, greenish-white to yellowish-green mineral of the *serpentine* group, representing a translucent, massive, fine-grained variety of antigorite resembling nephrite jade in appearance and composed of a dense feltlike aggregate of colorless fibers, with occasional patches of magnesite, flakes of talc, and grains of chromite. The term has also been applied to a serpentine rock in New Zealand. Syn: *tangiwai.*

Bowen ratio (Bow'-en) For a moist surface, the ratio of heat energy used for sensible heating (conduction and convection) to the heat energy used for latent heating (evaporation of water or sublimation of snow). The Bowen ratio varies from one locatity to another depending on the amount of surface moisture, ranging from 0.1 for the ocean surface to more than 5.0 for deserts; negative values are also possible.

Bowen's reaction series *reaction series.*

Bowie effect (Bow'-ie) The indirect effect on gravity due to a warping of the geoid resulting from the application of gravity corrections. Syn: *indirect effect.*

bowieite (bow'-ie-ite) A metallic gray orthorhombic mineral: $(Rh,Ir,Pt)_2S_3$.

bowlder (bowl'-der) *boulder.*

bowlingite (bow'-ling-ite) *saponite.*

bowr *bort.*

bowralite (bow'-ra-lite) A syenitic *pegmatite* composed chiefly of tabular euhedral alkali feldspar crystals with lesser amounts of sodic amphibole and aegirine, and with quartz, perovskite, zircon, and ilmenite as possible accessories. It was named by Mawson in 1906 for Bowral, New South Wales. Obsolete.

bow tie The appearance of a buried focus on a seismic record section, which can be collapsed by seismic migration.

box (a) A hollow limonitic concretion. (b) *box-stone.*

box canyon (a) A narrow gorge or canyon containing a stream following a zigzag course, characterized by high, steep rock walls and typically closed upstream with a similar wall, giving the impression as viewed from its bottom of being surrounded or "boxed in" by almost-vertical walls. (b) A steep-walled canyon heading against a cliff; a dead-end canyon, a *blind valley.* Cf: *cajon.*

boxcar [geophys] A rectangular (square-wave) window function that selects the data within the window without attenuation and completely rejects the data outside.

box corer A type of *corer* that retrieves relatively undisturbed sediment samples from the seafloor in a block rather than in a cylinder.

box fold A fold with the approximate profile form of three sides of a rectangle.

box-in-box structure A sedimentary feature, 60 cm to 2 m across, consisting of concentric layers of quartzitic sandstone resembling nested boxes as seen in plan view on the top surface of the bed. It is found at one locality in the Dhandraul Quartzite of India; origin is uncertain (Mathur, 1981b).

box level *circular level.*

box-stone A British term applied to a ferruginous concretion (found in Jurassic and Tertiary sands), often of rounded and rectangular or boxlike form, having a hollow interior in which white, powdery sand is sometimes present (P.G.H. Boswell in Wentworth, 1935, 241). Syn: *box.*

box the compass To name or repeat the 32 points of the compass in their contact order, clockwise from north.

boxwork [eco geol] In mineral deposits, a network of intersecting blades or plates of limonite or other iron oxide, deposited in cavities and along fracture planes, which forms one or more layers of adjacent boxy cavities. Typically assumed to have formed by dissolution of sulfides by processes associated with the oxidation and leaching of sulfide ores, esp. porphyry copper deposits.

boxwork [speleo] In dissolution caves, a network of resistant intersecting fins, usually of calcite (also quartz, gypsum, anhydrite, limonite, silica, or other minerals) that project outward from the intervening, more deeply weathered bedrock. Common in present or former sulfate or sulfide zones in karst (Palmer and Palmer, 1995).

boyleite (boyle'-ite) A white monoclinic mineral: $(Zn,Mg)SO_4{\cdot}4H_2O$. It dehydrates to gunningite. A member of the *starkeyite* group.

BP Before present (with "present" fixed as the year 1950).

BPL *bone phosphate of lime.*

Bq *bequerel.*

brabantite (bra-ban'-tite) A reddish-brown monoclinic mineral, a member of the monazite group: $CaTh(PO_4)_2$. Syn: *lingaitukuang.*

braccianite (brac'-cian-ite) A melilite-free tephritic leucitite, named by Lacroix in 1917 for Bracciano, Italy. Obsolete.

bracewellite (brace'-well-ite) A deep red orthorhombic mineral: $CrO(OH)$. It is a member of the diaspore group, and is polymorphic with guyanaite and grimaldiite.

brach (a) *brachial plate.* (b) *brachiopod.*

brachia (bra'-chi-a) (a) Plural of *brachium.* (b) A term sometimes used as a syn. of *lophophore.*

brachial (bra'-chi-al) adj. Pertaining to an arm or armlike structure of an animal (such as to the rays of a starfish or the *brachium* of a brachiopod). n. A brachial part; esp. *brachial plate.*

brachial plate One of the plates that form the arms of a crinoid; any crinoid-ray plate above the radial plates (exclusive of *pinnulars*). Syn: *brachial; brach.*

brachial process An anteriorly directed bladelike or rodlike projection from the *cardinalia* of pentameracean brachiopods.

brachial ridge A narrow elevation of the secondary shell of some rhynchonelliform (formerly articulate) brachiopods, extending laterally or anteriorly as an open loop from the dorsal adductor-muscle field. The brachial ridges are thought to be the region of attachment of the lophophore.

brachial valve The valve of a brachiopod that invariably contains any skeletal support (*brachidium*) for the *lophophore* and never wholly accommodates the *pedicle,* that is commonly smaller than the *pedicle valve,* and that has a distinctive muscle-scar pattern (TIP, 1965, pt. H, p.141). It typically has a small or indistinguishable beak. Syn: *dorsal valve.*

brachidium (bra-chid'-i-um) The looplike internal calcareous skeletal support structure of the *lophophore* of certain brachiopods. Pl: brachidia.

brachinite A group of olivine-rich *primitive achondrites* whose mineralogy, bulk chemistry and O-isotopic compositions are similar to those of the Brachina meteorite.

brachiolar facet (bra-chi-o'-lar) An elliptical or subcircular facet, indentation, or scarlike area where a brachiole was attached, as in a cystoid or blastoid. Also spelled: brachiole facet.

brachiolar plate One of the biserially arranged plates of a brachiole of a blastoid or other primitive stalked echinoderm, semielliptical in cross section and subquadrangular in side view, with a basal pair attached at the *brachiolar facet* (TIP, 1967, pt. S, p.346).

brachiole (bra'-chi-ole) A biserial, nonpinnulate exothecal appendage of primitive stalked echinoderm, springing independently from its surface and containing no extension of the body systems; esp. an erect, food-gathering structure arising from a rhombiferan thecal plate at the end or along the side of an ambulacrum, and bearing an extension of the *ambulacral groove.* Cf: *arm [paleont].*

brachiophore (bra'-chi-o-phore') One of the short, typically stout, bladelike processes of secondary shell projecting from either side of the *notothyrium* and forming anterior and median boundaries of sockets in the brachial valves of certain brachiopods.

brachiophore base The basal (dorsal) part of a *brachiophore* that joins the floor of a brachiopod valve (TIP, 1965, pt. H, p.141).

brachiophore process A distal rodlike extension of a *brachiophore* that possibly supported the *lophophore* in some brachiopods.

brachiopod (bra'-chi-o-pod') Any solitary marine invertebrate belonging to the phylum Brachiopoda, characterized by a lophophore and by two bilaterally symmetrical valves that may be calcareous or composed of chitinophosphate and that are commonly attached to a substratum but may also be free. Range, Lower Cambrian to Holocene. Syn: *brach; lamp shell.*

brachistochrone (bra-chis'-to-chrone) *minimum-time path.*

brachitaxis (bra-chi-tax'-is) A series of brachial plates in the crinoids extending from the radial plate or an axillary to and including the next axillary. Pl: brachitaxes.

brachium (bra'-chi-um) (a) Either of the two armlike, coiled, muscular projections from the mouth segment of the *lophophore* of a brachiopod, variably disposed but symmetrically placed about the mouth. (b) Any process of an invertebrate similar to an arm, such as a tentacle of a cephalopod. (c) In the tetrapods, the humerus and its surrounding soft tissue. Pl: *brachia.*

brachy- (brach'-y-) A prefix meaning "short".

brachyanticline (brach-y-an'-ti-cline) A short, broad anticline. Cf: *brachysyncline.* Rarely used.

brachy-axis The shorter lateral axis of an orthorhombic or triclinic crystal; it is usually the *a* axis [cryst]. Cf: *macro-axis.* Also spelled: brachyaxis.

brachydome (brach'-y-dome) An obsolete term for a *first-order prism* in the orthorhombic system; it is rhombic, with four faces parallel to the brachy-axis. Its indices are {*0kl*}. Cf: *clinodome.*

brachygenesis (brach-y-gen'-e-sis) The phenomenon in evolution in which part of a presumed recapitulated sequence has evolved outward and no longer appears in the course of development. Cf: *acceleration.*

brachygeosyncline (brach'-y-ge'-o-syn'-cline) A deep, oval depression formed during the later stages of geosynclinal deformation; a type of secondary geosyncline (Peyve and Sinitzyn, 1950). Obsolete.

brachylinear (brach-y-lin'-e-ar) "Any lineation....ranging in length from less than two kilometers to the lower limit of visibility of the unaided eye" (El-Etr, 1976, p.485).

brachyome (brach'-y-ome) The short arm of an anomoclone or ennomoclone of a sponge, or the different fourth ray of a *trider* of a sponge.

brachypinacoid (brach-y-pin'-a-coid) *side pinacoid.*

brachysyncline (brach-y-syn'-cline) A short, broad syncline. Cf: *brachyanticline.* Rarely used.

brachyuran (brach-y-ur'-an) Any *decapod* belonging to the infraorder Brachyura, characterized by a carapace that becomes progressively shortened and widened, developing a lateral margin; e.g. a crab. Range, Lower Jurassic to Holocene.

brackebuschite (brack'-e-busch-ite') (a) A submetallic dark brown to black monoclinic mineral: $Pb_2(Mn^{3+},Fe^{3+})(VO_4)_2(OH)$. (b) A group name for minerals of composition $A_2BXO_4(OH)$, where A = Ca, Sr, Ba or Pb; B = Al, Fe^{3+} or Mn^{3+}; and X = P, V or As.

brackish water (brack'-ish) An indefinite term for water with a salinity intermediate between that of average seawater (35) and that of fresh water (0).

bract (a) A modified leaf associated with a flower or inflorescence, e.g. bearing a flower on its axis or being borne on a floral axis (subtending the flower or inflorescence). (b) A scooplike or spoonlike extension from the lower half of the rim of pores in the wall of an archaeocyathid.

bracteate (brac'-te-ate) Said of a plant, or plant structure, having *bracts.*

bradaczekite An adamantine dark blue monoclinic mineral: $NaCu_4(AsO_4)_3$.

Bradfordian (Brad-ford'-i-an) *Conewangoan.*

bradleyite (brad'-ley-ite) (a) A colorless to gray monoclinic mineral: $Na_3Mg(PO_4)(CO_3)$. (b) A group name for minerals of composition $Na_3B(PO_4)CO_3$, where B = Mg, Fe, Mn or Sr.

Bradydonti (Brad-y-don'-ti) A taxon of marine cartilaginous fishes of uncertain affinity and poorly known general structure, characterized by a dental battery that consists of a few heavy crushing teeth. The teeth are distinctive. Lower taxa are of limited duration but wide distribution and are useful in upper Paleozoic biostratigraphy. Range, Upper Devonian to Lower Permian.

bradygenesis (brad-y-gen'-e-sis) *bradytely.*

bradytely (brad'-y-tel'-y) Retardation in the development of a group of organisms that may gradually cause certain individuals to fall behind the normal rate of evolution or progress in some or all of their characteristics. Cf: *horotely; tachytely; lipogenesis.* Syn: *bradygenesis.*

Bragg angle In the *Bragg equation,* the angle between the diffracted beam of X-rays and the diffracting crystal planes. It is symbolized by θ (theta). Syn: *reflection angle; angle of reflection; diffraction angle.*

Bragg equation A statement in crystallography that the X-ray diffractions from a three-dimensional lattice may be thought of as reflecting from the lattice planes: $n\lambda=2d \sin\theta$, in which n is any integer, λ is the wavelength of the X-ray, d is the crystal plane separation, also known as *d-spacing,* and θ is the angle between the crystal plane and the diffracted beam, also known as the *Bragg angle.* Syn: *Bragg's law.*

braggite (bragg'-ite) A steel-gray tetragonal mineral: $(Pt,Pd,Ni)S$.

Bragg reflection A diffracted beam of X-rays by a crystal plane according to the Bragg equation.

Bragg's law *Bragg equation.*

braid v. To branch and rejoin repeatedly to form an intricate pattern or network of small interlacing stream channels. n. A reach of a braided stream, characterized by relatively stable branch islands and hence two or more separate channels. Cf: *anabranch.*

braid bar Any exposed sand or gravel bar that divides flow and causes a braided pattern in a stream (Rust, 1972, p.232). Cf: *channel bar; unit bar; meander bar; river bar; point bar.*

braided channel A stream channel that frequently branches and rejoins after separation by visible bars or islands with lenticular shapes.

braided drainage pattern (braid'-ed) A drainage pattern consisting of braided streams. Syn: *interlacing drainage pattern.*

braided stream A stream that divides into or follows an interlacing or tangled network of several small branching and reuniting shallow channels separated from each other by ephemeral branch islands or channel bars, resembling in plan the strands of a complex braid. Such a stream is generally believed to indicate an inability to carry all of its load, such as an overloaded and aggrading stream flowing in a wide channel on a floodplain. Cf: *anastomosing stream; split stream.*

braiding index A measure of the number of channels within a multi-channel river.

braincase (brain'-case) *neurocranium.*

brait A rough diamond.

braitschite-(Ce) (braitsch'-ite) A colorless, white, or pink hexagonal mineral of uncertain composition, approximately $(Ca,Na_2)_7(Ce,La)_2B_{22}O_{43}{\cdot}7H_2O$.

brammallite (bram'-mall-ite) A white or pale green monoclinic micaceous clay mineral, representing the sodium analogue of illite. Syn: *sodium illite; hydroparagonite.*

branch [seis] (a) A set of values for a multivalued function. (b) One of two or more reflecting events that may be observed at a given location from a reflector that is concave upward or discontinuous. See: *buried focus.* (c) A refraction event that may be observed at a given point because of the configuration of the refractor.

branch [streams] (a) A small stream that flows into another, usually larger, stream; a tributary. (b) A term used in the southern U.S. for a creek, or a stream normally smaller than and often tributary to a river. (c) A stream flowing out of the main channel of another stream and not rejoining it, as on a delta or alluvial fan; a distributary. (d) A stream flowing out of another stream and rejoining it, such as an anabranch; a by-channel. (e) A fork of a tidal river; e.g. the fork of the Severn River, Md.

branch fault *auxiliary fault.*

branch gap An interruption in the vascular tissue of a stem at the point at which a *branch trace* occurs. It is most evident in cross section, at the point of branch-trace departure.

branchia (bran'-chi-a) (a) A thin-walled, fingerlike or leaflike structure extending outward from a crustacean limb or secondarily from a side of the body, typically occurring in pairs, and functioning for respiration. Syn: *gill [paleont].* (b) A slender, hollow, fingerlike extension of the body wall of an asteroid. Pl: branchiae.

branchial carina (bran'-chi-al) A longitudinal ridge extending over the *branchial region* behind the *postcervical groove* on some decapod crustaceans (Holthuis, 1974, p.735).

branchial chamber The space between the body and the wall of *carapace* enclosing the branchiae of a crustacean. Syn: *gill chamber.*

branchial region The lateral part of the carapace of some decapod crustaceans, behind the pterygostomial region and overlying the branchiae; it is divided by some authors into epibranchial, mesobranchial, and metabranchial subregions (TIP, 1969, pt. R, p.92).

branchial slit A *gill slit* of an echinoid.

branching bay (branch'-ing) A bay having a dendritic pattern, produced by drowning or flooding of a river valley by the sea. See also: *estuary.*

branching crossed lamellar structure A *crossed lamellar structure* in which the first-order lamellae are elongate, as in LCL structure, but are more highly branched (Carter, 1976). Abbrev: BCL.

branching fault A fault that splits into two or more parts or branches.

branching ratio The ratio of the decay constants for each of two competing modes of *radioactive decay.*

branchiocardiac groove (bran'-chi-o-car'-di-ac) The groove in some decapod crustaceans that separates the *branchial region* from the *cardiac region* of the carapace. In its posterior part it runs longitudinally, parallel with the median dorsal line of the carapace, and anteriorly it curves down to meet the *postcervical groove* (Holthuis, 1974, p.732).

branchiopod (bran'-chi-o-pod') Any crustacean belonging to the class Branchiopoda, characterized by the morphologic similarity of their numerous somites and limbs and by their filter-feeding mode of nourishment. Range, Lower Devonian to Holocene.

branchiostegite (branch-i-os'-te-gite) The extended portion of *carapace* covering the *branchial chamber* of a decapod crustacean.

branch island An island formed by the braiding of the branches of a stream; an island formed between a tributary and the main stream. Term was introduced by Jackson (1834, p.79).

branchite (branch'-ite) *hartite.*

branch line A line along which two fault surfaces join.

branch line map A map showing the locations and orientations of the branch lines for a system of faults (Diegel, 1986).

branch point The point on a map or a cross section where two faults separate or join. Normally, the branch point in a map or cross section is only one point along the line of intersection of two faults.

branchstone (branch'-stone) A type of *reef rock, boundstone,* or *framestone* composed of upright, branching colonies in contact with one another so as to form a lattice-like framework. The interstices between the branches may be open, or filled with fine sediment or crystalline cement (Cuffey, 1985). Cf: *bafflestone.*

branch trace Vascular tissue extending from a stem into a branch. Cf: *leaf trace.* See also: *branch gap.*

branch water Water from a small stream or branch.

branchwork cave (branch'-work) A cave in which the passages intersect as tributaries. See also: *network cave; maze cave.*

brandbergite (brand'-berg-ite) A hypabyssal granite having aplitic texture and being composed of potassium feldspar (as whitish Carlsbad twins), quartz grains, arfvedsonite, and aggregates of biotite in a micrographic groundmass. The name, given by Chudoba in 1930, is for Brandberg, Namibia. Obsolete.

brandholzite A vitreous colorless trigonal mineral: $MgSb_2(OH)_{12}{\cdot}6H_2O$.

brandtite (brandt'-ite) A colorless monoclinic mineral: $Ca_2(Mn^{2+},Mg)(AsO_4)_2{\cdot}2H_2O$. It is a member of the *roselite* group.

brannerite (bran'-ner-ite) A black, brownish olive green, or yellowish brown monoclinic mineral: $(U,Ca,Y,Ce)(Ti,Fe)_2O_6$. The uranium analogue of thorutite.

brannockite (bran'-nock-ite) A colorless hexagonal mineral of the milarite group: $KLi_3Sn_2Si_{12}O_{30}$.

brash *brash ice.*

brash ice An accumulation of floating fragments not more than 2 m across, and representing the wreckage of other forms of ice; occurs esp. near an ice pack or floe. Syn: *brash; debris ice; mush.*

brass (a) An English term for yellowish iron pyrites (pyrite and marcasite) found in coal or coal seams. Syn: brasses. (b) A metallic yellow cubic mineral: β-(Cu,Zn).

brassil (bras'-sil) *brazil.*

brassite (brass'-ite) A white orthorhombic mineral: $Mg(AsO_3OH){\cdot}4H_2O$.

brass ore (a) *aurichalcite.* (b) A mixture of *sphalerite* and *chalcopyrite.*

braunite (braun'-ite) A brittle brownish-black or steel-gray tetragonal mineral: $Mn^{2+}Mn^{3+}{}_6SiO_{12}$. It sometimes has appreciable ferric iron, and is an occasional ore for manganese.

braunite-II A submetallic black tetragonal mineral: $CaMn_{14}{}^{3+}O_{20}(SiO_4)$.

bravaisite (bra'-vais-ite) A name proposed for a micaceous clay mineral having about half the potassium of muscovite, and later used as a synonym to replace *illite.* Material from the type locality has been shown to be a mixture of montmorillonite and illite, with illite predominating; therefore, bravaisite is not a specific mineral and has no standing as a distinct mineral species (Grim, 1953, p.36). Cf: *sarospatakite.*

Bravais lattice (Bra'-vais) A syn. of *crystal lattice;* it is named for the nineteenth-century French physicist, Auguste Bravais, who demonstrated that there are only 14 possible unique kinds of crystal lattices.

bravoite (bra'-vo-ite) A varietal name for nickel-bearing pyrite. It is a yellow cubic mineral: $(Fe,Ni)S_2$.

B ray Right anterior ray in echinoderms situated clockwise of A ray when the echinoderm is viewed from the adoral side; equal to ambulacrum IV in the *Lovenian system.*

brazil (bra-zil') An English dialectal term for iron pyrite, esp. associated with coal. Also, by extension, a term applied to a coal seam containing much pyrite. Adj: brazilly. Also spelled: brazzil; brazzle; *brassil.*

Brazilian emerald (Bra-zil'-i-an) A transparent green variety of tourmaline occurring in Brazil and used as a gemstone.

brazilianite (bra-zil'-i-an-ite') A yellowish-green to greenish-yellow monoclinic mineral: $NaAl_3(PO_4)_2(OH)_4$.

Brazilian ruby (Bra-zil'-i-an) A reddish mineral resembling ruby in appearance and occurring in Brazil; e.g. a light rose-red spinel, or a pink to rose-red or deep-red topaz (either natural or artificially heated), or a reddish tourmaline.

Brazilian sapphire A transparent blue variety of tourmaline occurrring in Brazil and used as a gemstone.

Brazilian topaz Topaz mined in Brazil and ranging in color from pure white to blue; esp. yellowish topaz.

brazilite (bra-zil'-ite) A mixture of baddeleyite, zircon, and altered zircon. The term has also been applied to an oil shale, to a fibrous variety of baddeleyite, and as a syn. of *baddeleyite.*

Brazil test [rock mech] A test where a disk is split by a line load across the diameter of the disk. This is one technique for determining the tensile strength of rock.

Brazil twin law (Bra-zil') A type of twin law in quartz in which the twin plane is perpendicular to one of the *a* crystallographic axes; an example of *optical twinning.* Cf: *Dauphiné twin law.*

bre The Norwegian term for *glacier.* Etymol: Norwegian "breen" [Pron: bre'-en], "glacier". Example: Nigardsbreen, the Nigards glacier.

brea A rarely used term for a viscous asphalt formed by the evaporation of volatile components from oil in seepages. In Trinidad, it is used as the name for maltha. Etymol: Spanish, "pitch".

breach v. To cut a deep opening in a landform, esp. by erosion.

breached Said of a volcanic cone or crater, the rim of which has been broken through by the outpouring lava.

breached anticline An anticline whose crest has been eroded, so that it is flanked by inward-facing erosional cliffs. Cf: *bald-headed anticline.* Syn: *unroofed anticline; scalped anticline.*

breachway A connecting channel.

bread-crust bomb A type of volcanic bomb characterized by a network of opened cracks on its surface, due to continued expansion of the interior after solidification of the crust. See also: *explosive bomb.*

break [drill] A change in the penetration rate of a drill; usually said of an increase, i.e. a "fast break". The penetration rate is generally measured in minutes per length drilled (e.g., 6 minutes per foot). Cf: *shale break.*

break [geomorph] A marked variation of topography, or a tract of land distinct from adjacent land, or an irregular and rough piece of ground; e.g. a deep valley, esp. a ravine or gorge cutting through a ridge or mountain. See also: *breaks.*
break [mining] A general term used in mining geology for any discontinuity in the rock, such as a fault, a fracture, or a small cavity.
break [seis] *arrival.*
break [slopes] A marked or abrupt change or inflection in a slope or profile; a *knickpoint.* Term is used in the expressions "break of slope" and "break of profile".
break [stratig] (a) An abrupt change at a definite horizon in a sequence of sedimentary rocks, usually indicative of an unconformity (esp. a disconformity) or hiatus; esp. a marked change in lithology, such as one separating a channel sand from an underlying shale. See also: *faunal break.* (b) An interruption of a normal geologic sequence, esp. of stratigraphic continuity; a *discontinuity [stratig].* Syn: *gap [stratig]; stratigraphic break.*
breakaway (break'-a-way) The steep surface along which the allochthonous rocks above a detachment surface have separated from the rocks that remain in place.
break-back sequence *hanging-wall imbrication.*
breakdown *cave breakdown.*
breaker [reef] (break'-er) A string of *boilers.*
breaker [waves] A sea-surface wave that has become so steep (wave steepness of $^1/_7$) that the crest outraces the body of the wave and collapses into a turbulent mass on shore or over a reef or rock. Breaking usually occurs when the water depth is less than 1.3 times the wave height. See also: *plunging breaker; spilling breaker; surging breaker; surf.* Syn: *breaking wave.*
breaker depth The still-water depth at the point where a wave breaks. Syn: *breaking depth.*
breaker line The axis along which a wave breaks as it approaches the shore.
breaker zone *surf zone.*
break-forward sequence *footwall imbrication.*
break frequency *corner frequency.*
breaking depth (break'-ing) *breaker depth.*
breaking strength *fracture strength.*
breaking wave *breaker [waves].*
break line A string of x,y coordinates used to model a linear change in the slope or continuity of a terrain surface. Break lines may be modeled using two-dimensional or three-dimensional line modeling techniques. Two-dimensional break lines are used to represent breaks in the terrain along lines of constant elevation such as water body shorelines and footprints of buildings on level terrain. They cannot be used to model linear terrain breaks that trend up or down slopes. Three-dimensional break lines allow the z-value to vary for each x,y coordinate and are typically used to model gentle to abrupt surface breaks that trend up or down slope. However, they can also be used to model terrain breaks of constant elevation. Three-dimensional break lines may be used to model features such as stream banks, water body shorelines, dams, sharp or gentle ridges, and footprints of buildings on sloping terrain (Geo-One-Stop, 2003, p.2).
break of slope *break [slopes].*
breakover (break'-o-ver) A rounded crest that is both structurally and topographically high.
breakpoint bar (break'-point) A *longshore bar* formed at the breakpoint of waves, where there is a sudden decrease of sand moving landward outside the breakpoint but where sand is moving seaward to this point (King and Williams, 1949, p.80).
breaks (a) A term used in the western U.S. for a tract of rough or broken land dissected by ravines and gullies, as in a badlands region. (b) Any sudden change in topography, as from a plain to hilly country, or a line of irregular cliffs at the edge of a mesa or at the head of a river; e.g. Cedar Breaks, Utah. See also: *break [geomorph].*
breakthrough (break'-through) (a) The erosive action of water in wearing or cutting a passage. (b) The channel made by such a breakthrough.
break thrust A fault that breaks through the forelimb of an asymmetric anticline during development of the fold, typically after the forelimb strata attained steep to overturned dips.
breakup (break'-up) (a) The melting, loosening, fracturing, or destruction of snow or floating ice during the spring; specif. the destruction of the ice cover on a river during the spring thaw. (b) The period during the spring thaw when a breakup occurs.
breakwater (break'-wa-ter) An offshore, generally shore-parallel structure (such as a mole) that, by breaking the force of the waves, protects a harbor, anchorage, beach, or shore area (Bush et al., 1996). Syn: *water-break.*
breast A miner's term for the face of a mine working.
breast wall A wall designed to withstand the force of a natural bank of earth, as of timber used to support the face of a tunnel.
breathing cave A *cave [speleo]* that has an alternating movement of air through its passages. Cf: *blowing cave.*
breathing well A well, generally a water well, that, in response to changes in atmospheric pressure, alternately takes in and emits a strong current of air, often with an alternating sucking and blowing sound. It penetrates, but is uncased in at least part of, a thick *unsaturated zone* that is porous and permeable enough to exchange air freely with the well but otherwise is poorly connected with the atmosphere because of the presence of tight soil or other low-permeability material above the unsaturated material. Syn: *blowing well.*
breccia [geol] (brec'-cia [bret'-shia]) A coarse-grained clastic rock, composed of angular broken rock fragments held together by a mineral cement or in a fine-grained matrix; it differs from *conglomerate* in that the fragments have sharp edges and unworn corners. Breccia may originate as a result of sedimentary processes such as talus accumulation (*sedimentary breccia*); igneous processes, esp. explosive (*igneous breccia, volcanic breccia*); disturbance during sedimentation (intraclastic breccia); collapse of rock material (*solution breccia, collapse breccia*); or tectonic processes (*fault breccia*). Etymol: Italian, "broken stones, rubble". Syn: *rubblerock.* Cf: *conglomerate.*
breccia [meteorite] A meteorite texture consisting of fragments, commonly angular, crushed, mixed, compacted and lithified during impact processes.
breccia [planet] A common rock on planetary surface, containing clasts of angular rock fragments and fine-grained matrix produced by meteoroid impact. As impact energy serves to lithify unconsolidated debris, a single breccia specimen may record several successive breaking and annealing events.
breccia-conglomerate (brec'-cia-con-glom'-er-ate) A sedimentary rock consisting of both angular and rounded particles (Norton, 1917, p.181); a sedimentary rock that is not clearly referable to either breccia or conglomerate. Syn: *breccio-conglomerate.*
breccia dike A sedimentary dike composed of breccia injected into the country rock.
breccial (brec'-cial) Pertaining to breccia.
breccia lens In *impacts,* a lens-shaped accumulation of broken rock, with or without melt, lying beneath the floor of a *simple crater.* This accumulation forms as the *transient crater* collapses and the steep crater walls collapse inward into the crater. The breccia contains rocks from all horizons intersected by the crater. Exotic rocks are rare or absent. Cf: *mixed breccia.*
breccia marble Any marble composed of angular fragments. The term was in use before the separate use of "breccia" in geology.
breccia pipe A cylindrical chimney filled with breccia fragments of the country rock. Often found in mineralized epithermal systems and associated with porphyry-type mineral deposits. Syn: *pipe [volc].* Cf: *diatreme.*
breccia porosity *Interparticle porosity* in a breccia (Choquette and Pray, 1970, p.244).
breccia-sandstone (brec'-cia-sand'-stone) A sandstone containing "some small angular fragments" (King, 1899, p.105).
brecciated [geol] (brec'-ci-at-ed) Converted into, characterized by, or resembling a breccia; esp. said of a rock structure marked by an accumulation of angular fragments, or of an ore texture showing mineral fragments without notable rounding.
brecciated [meteorite] A term incorrectly applied to a meteorite of intermediate type (between iron and stony-iron), in which the main mass is iron with octahedral or hexahedral structure but there are also relatively large silicate inclusions of rounded or angular form. Also, said of such a texture occurring in a meteorite, including stony meteorites.
brecciation (brec-ci-a'-tion) Formation of a breccia, as by crushing or breaking a rock into angular fragments.
breccia vein A fissure containing numerous wall-rock fragments, with mineral deposits in the interstices.
brecciform (brec'-ci-form) In the form or shape of a breccia, or resembling a breccia.
breccio-conglomerate (brec'-cio-con-glom'-er-ate) *breccia-conglomerate.*
breccioid (brec'-ci-oid) Having the appearance of a breccia.
brecciola (brec-ci-o'-la) A well-graded, intraformational breccia consisting of small, angular limestone fragments in well-defined beds separated by dark shale, such as the breccia occurring in the northern Apennines. Etymol: Italian, diminutive of breccia.
bredbergite (bred'-berg-ite) A variety of andradite garnet containing magnesium.
Breddin graph A graph showing variations of angular shear strain with orientation of the principal extension axis (Breddin, 1956). The graph is used to determine finite strain from deformed markers where angular shear strain can be measured, e.g. fossils originally possessing bilateral symmetry.
bredigite (bre'-dig-ite) A white mineral: $(Ca,Ba)Ca_{13}Mg_2(SiO_4)_8$. It is a metastable orthorhombic phase of calcium orthosilicate (but not isomorphous with olivine), stable from about 800° to 1447°C on heating and from 1,447° to 670°C on cooling. Cf: *calcio-olivine; larnite.*
breithauptite (breit'-haupt-ite) A copper-red hexagonal mineral: NiSb. A member of the *nickeline* group.
Breithaupt twin law (Breit'-haupt) A rare type of normal twin law in feldspar, in which the twin plane is $(\bar{1}11)$.
brendelite A vitreous to adamantine dark brown to black monoclinic mineral: $(Bi,Pb)_2(Fe^{3+},Fe^{2+})O_2(OH)PO_4$.
brenkite (brenk'-ite) A colorless orthorhombic mineral: $Ca_2(CO_3)F_2$.
Bretonian Orogeny (Bre-to'-ni-an) One of the 30 or more orogenies during Phanerozoic time identified by Stille. It consisted of several phases, from the Late Devonian to the end of the Devonian; it is considered to be the earliest part of the Variscan orogenic era, which continued to the end of the Paleozoic.

breunnerite (breun′-ner-ite) A variety of magnesite containing 5-30% iron carbonate.

Brevaxones (Brev′-ax-ones) A group of mid-Cretaceous and younger angiosperm pollen in which the polar axis is shorter than the equatorial diameter, representing an evolutionary advance over *Longaxones,* and including such forms as Normapolles.

breviaxal (brev′-i-ax-al) Of *pollen grains,* where a polar axis is shorter than the equatorial diameter.

brevicone (brev′-i-cone) A straight or slightly curved shell characteristic of certain Paleozoic cephalopods, having a short, blunt form; it expands rapidly from the apex to the base of the body chamber, or to a point a short distance before or behind that base. Cf: *longicone.*

Brewster angle (Brew′-ster) *polarizing angle.*

brewsterite-Ba Ba-dominant brewsterite, a colorless or pale yellow zeolite mineral: $(Ba,Sr,Ca)Al_2Si_6O_{16}{\cdot}5H_2O$.

brewsterite-Sr (brew′-ster-ite) Sr-dominant brewsterite, a colorless monoclinic zeolite mineral: $(Sr,Ba,Ca)Al_2Si_6O_{16}{\cdot}5H_2O$. It usually contains some calcium.

brewsterlinite (brew-ster′-lin-ite) Liquid CO_2, found as inclusions in cavities in minerals such as quartz, topaz, and chrysoberyl. It will expand so as to fill cavities under the warmth of the hand.

Brewster's law A statement in optics that when unpolarized light is incident on a surface, it acquires maximum plane polarization at a particular *angle of incidence* whose tangent equals the refractive index of the substance. This angle is called the *polarizing angle,* or the *Brewster angle.*

brezinaite (bre-zin′-a-ite) (a) A metallic brownish-gray monoclinic meteorite mineral: Cr_3S_4. (b) A group name for A_3S_4 minerals where A = Cr, Fe, Ti or Ni.

Brezina lamellae (Bre-zi′-na) Lamellae of schreibersite oriented parallel to dodecahedral planes in parent taenite of iron meteorites. Named after M. Aristides S.F. Brezina (1848-1909), Austrian mineralogist.

brianite (bri′-an-ite) A colorless monoclinic meteorite mineral: $Na_2CaMg(PO_4)_2$. It is isostructural with merwinite.

brianroulstonite A vitreous colorless or white monoclinic mineral: $Ca_3B_5O_6(OH)_7Cl_2{\cdot}8H_2O$.

brianyoungite A vitreous white orthorhombic mineral: $Zn_3(CO_3,SO_4)(OH)_4$.

briartite (bri′-art-ite) A metallic gray tetragonal mineral: $Cu_2(Fe,Zn)GeS_4$. A member of the *luzonite* group.

brick clay (a) Any clay suitable for the manufacture of bricks or coarse pottery; a *brick earth.* (b) An impure clay containing iron, calcium, magnesium, and other ingredients.

brick earth Earth, clay, or loam suitable for making bricks; specif. a fine-grained brownish deposit consisting of quartz and flint sand mixed with ferruginous clay and found on river terraces as a result of reworking by water of windblown material, such as that overlying the gravels on certain terraces of the River Thames in England. See also: *brick clay.*

bridal-veil fall (brid′-al-veil) A cataract of great height and such small volume that the falling water is largely dissipated in spray before reaching the lower stream bed, and having a form that suggests a bridal veil. Type example: Bridalveil Fall in Yosemite Valley, Calif.

bridge [drill] n. A rock fragment, *cavings,* or other obstruction that lodges (either accidentally or intentionally) part way down in a drill hole v. To form a bridge in a drill hole.

bridge [geomorph] *natural bridge.*

bridge [speleo] In a cave, a solutional remnant of rock that spans a passage from wall to wall. See also: *partition [speleo].*

bridge plug A mechanical device or a volume of cement deliberately set in a wellbore at a selected depth in order to seal off the hole below.

Bridgerian (Brid-ger′-i-an) North American land mammal age: Medial Eocene (after Wasatchian, before Uintan).

bridging factor (bridg′-ing) A term used by Gruner (1950) for a number that expresses the manner by which SiO_4 tetrahedra are tied together in a mineral. It is equal to 0.8 plus twenty percent of the quotient of the sum of the valence bonds of a silicate divided by the number of cations. A bridging factor of 1.00 is assigned to quartz, whose SiO_4 tetrahedra are all directly tied to other tetrahedra, resulting in the highest number of bridges possible; all other structures have smaller factors. See also: *energy index [mineral].*

bridging oxygen In a *crystal structure* containing cations in *tetrahedral coordination,* an oxygen atom that is shared between two such cations (i.e., it "bridges" between two tetrahedra). Cf: *nonbridging oxygen.*

bridle (bri′-dle) *parallel shot.*

brigg An English term for a headland formed by "a scarp of hard rock cropping out at or near tide marks" (Stamp, 1961, p.78).

bright-banded coal (bright′-band-ed) *Banded coal* consisting mainly of vitrain and clarain, with some durain and minor fusain. Cf: *dull-banded coal; semisplint coal.*

bright coal A type of banded coal defined microscopically as consisting of more than 5% of anthraxylon and less than 20% of opaque matter; banded coal in which translucent matter predominates. Bright coal corresponds to the microlithotypes vitrite and clarite and in part to duroclarite and vitrinerite (ICCP, 1963). Cf: *dull coal; semibright coal; semidull coal; intermediate coal.* Syn: *brights.*

bright field An image obtained in the transmission electron microscope by deliberately excluding the diffracted beams. This is accomplished by placing an aperture in the back focal plane of the objective lens, permitting only the transmitted beam to form the image. Cf: *dark field.*

brightness temperature (bright′-ness) (a) The temperature of a blackbody radiating the same amount of energy per unit area at the wavelengths under consideration as the observed body. Cf: *color temperature.* (b) The apparent temperature of a nonblackbody determined by measurement with an optical pyrometer or radiometer.

brights *bright coal.*

bright spot A local increase of amplitude on a seismic section. An increase of amplitude is sometimes caused by hydrocarbon accumulation. See: *hydrocarbon indicator.*

brilliance Intensity of the internal and external reflections of white light from the crown of a polished diamond or other gemstone. Hardness, refractive index, reflectivity, polish, luster, and proportions all affect a gemstone's brilliance (Liddicoat et al., 1993). See also: *critical angle; dispersion [gem]; refraction [cryst]; scintillation [gem].*

brilliant (bril′-liant) n. A *brilliant cut* diamond. The term is less correctly applied to any brilliant-cut gemstone.

brilliant cut A gemstone cut with a facet arrangement that radiates from the center of the stone towards the girdle, and triangular or kite-shaped facets. The most common brilliant cut is the standard round brilliant; modifications include the marquise, half moon, pear shape, heart shape, oval, cushion, and antique cushion brilliant cuts (Liddicoat et al., 1993). Cf: *Tolkowsky theoretical brilliant cut; step cut; single cut; mixed cut; pear-shaped cut; pendeloque.*

brim The flared or recurved portion of a *cyrtochoanitic septal neck* of a nautiloid, measured transverse to the longitudinal axis of the siphuncle.

brimstone (brim′-stone) A common or commercial name for *sulfur,* esp. native sulfur or fine sulfur fused into rolls, sticks, or blocks.

brindleyite (brind′-ley-ite) A green monoclinic serpentine mineral, the Ni-analogue of *berthierine,* with Ni in place of octahedral Fe^{2+}: $(Ni,Mg,Fe^{2+})_2Al(SiAl)O_5(OH)_4$.

brine [geol] A term used for pore fluids in deep sedimentary basins, stratified hot fluids in restricted basins such as the Red Sea, oil-field waters, and geothermal mineralizing fluids. It denotes warm to hot highly saline waters containing Ca, Na, K, Cl, and minor amounts of other elements. Syn: *natural brine.* Cf: *artificial brine.*

brine [oceanog] (a) Seawater that, owing to evaporation or freezing, has considerably more than the average concentration of dissolved ions, i.e., a salinity of 40 or higher, 40 being the highest salinity commonly found. (b) Seawater beneath the seabed in sediments or rock fractures with a high concentration of dissolved ions due to water-rock reactions within the ocean floor.

brine cell A small inclusion, usually in the shape of an elongated tube about 0.05 mm in diameter, containing residual liquid more saline than seawater, formed in sea ice as it develops. Syn: *brine pocket.*

brine content Relative volume of ice composed of brine, expressed as an absolute ratio or in parts per thousand.

brine lake *salt lake.*

Brinell hardness test (Bri-nell′) A test of resistance to deformation of metals or minerals in which a sphere of hardened steel is forced into a polished surface of the specimen to be tested under a specific static load. The result is measured by the size of the resulting indentation. Cf: *Vickers hardness test; Knoop hardness test.*

brine pit A *salt well,* or an opening at the mouth of a salt spring, from which water is taken to be evaporated for making salt. Archaic.

brine pocket *brine cell.*

brine slush "A mixture of ice crystals and salt water, which retards or prevents complete freezing, often found between young sea ice and a cover of newly fallen snow" (ADTIC, 1955, p.14).

brine spring *salt spring.*

Bringewoodian (Brin-ge-wood′-i-an) European stage: Upper Silurian (above Eltonian, below Leintwardinian).

brink (a) A bank, edge, or border of a body of water, esp. of a stream. (b) The top of the slip face of a dune. It need not be the same as the crest.

brinkpoint (brink′-point) The point on a cross section of a ripple or dune that separates the steeply inclined *lee* side from the gently inclined *stoss* side or crestal platform. Term introduced by Allen (1968). Cf: *summitpoint.*

brinrobertsite A dull gray to yellowish gray mixed-layer mineral comprising smectite and pyrophyllite layers: $(Na,K,Ca)_{0.3}(Al,Fe,Mg)_4(Si,Al)_8O_{20}(OH)_4{\cdot}3.5H_2O$.

briolette (bri-o-lette′) A pear-shaped or drop-shaped gemstone with a circular cross section, having its entire surface cut in triangular, or sometimes rectangular, facets. Cf: *pendeloque.*

britholite (brith′-o-lite) A group name for apatite-type minerals with SiO_4 partially or completely substituting for PO_4.

britholite-(Ce) The Ce-dominant member of the *britholite* group: $(Ce,Ca)_5(OH,F)(SiO_4,PO_4)_3$. It is monoclinic and variously colored.

britholite-(Y) A reddish brown monoclinic mineral of the *britholite* group: $(Y,Ca)_5(SiO_4,PO_4)_3(OH,F)$. Syn: *abukumalite.*

brittle (brit'-tle) Said of a rock that fractures at less than 3-5% deformation or strain. Cf: *ductile.*

brittle behavior The response of a rock to stress during which the rock loses continuity; it reflects the occurrence of *brittle-deformation mechanisms.*

brittle-deformation mechanism The mechanism by which brittle deformation occurs, including the formation and growth of *fractures* and *frictional sliding.*

brittle-ductile transition The depth or zone at which rocks change from deforming by primarily brittle processes, to deforming by primarily ductile processes. Typically a couple of tens of km deep, depending upon temperature and composition of rocks.

brittle mica (a) A group of minerals resembling the true micas in crystallographic characters, but having the cleavage flakes less elastic and containing calcium (instead of potassium) as an essential constituent. Their structures are closest-packed. Syn: *clintonite.* (b) A mineral of the brittle-mica group, such as clintonite, margarite, and ephesite. (c) A micaceous mineral occurring in brittle folia; e.g. chloritoid.

brittle silver ore *stephanite.*

brizziite A colorless rhombohedral mineral formed by the weathering of slag residues: $NaSbO_3$. A member of the *corundum* group.

broad A British term for a lake or wide sheet of shallow, reed-fringed fresh water, forming a broadened part of, or joined to, a sluggish river near its estuary; often used in the plural. The feature is typically found in East Anglia (Norfolk and Suffolk), and is believed to have been produced artificially by the cutting and removal of peat in the Middle Ages.

broadband Including a substantial range of frequencies.

brocenite (bro'-cen-ite) *beta-fergusonite-(Ce).*

brochantite (bro-chant'-ite) An emerald-green to dark-green monoclinic mineral: $Cu_4(SO_4)(OH)_6$. It is common in the oxidation zone of copper-sulfide deposits. Syn: *blanchardite; kamarezite.*

brock An English term for a *brook.*

brockite (brock'-ite) A reddish brown hexagonal mineral: $(Ca,Th,Ce)(PO_4)\cdot H_2O$. A member of the *rhabdophane* group.

Brodelboden (Bro'-del-bo'-den) *involution [sed].*

brodtkorbite A metallic dark gray monoclinic mineral: Cu_2HgSe_2.

bröggerite (brög'-ger-ite) A variety of thorian *uraninite,* $(U,Th)O_2$.

broken belt (bro'-ken) The transition zone between open water and consolidated pack ice.

broken formation A rock formation broken by faults but retaining substantial continuity of contacts and internal stratigraphic units (Raymond, 1984). No distinction is made between units deformed by gravity-induced fracturing and those with tectonically created fractures. Cf: *dismembered formation.*

brokenhillite (bro'-ken-hill'-ite) A brown hexagonal mineral: $Mn_8(OH,Cl)_{10}[Si_6O_{15}]$. It is related to the *friedelite* group. Not approved by the International Mineralogical Association.

Broken Hill-type deposit A *mineral-deposit type* characterized by layers or stacked lenses of massive to semimassive galena, sphalerite, pyrrhotite, and pyrite and/or magnetite in sedimentary rocks. Known deposits are highly metamorphosed, obscuring their origin. Named for the Broken Hill deposit, New South Wales, Australia. Syn: *Ammeburg-type deposit; Shushwap-type deposit; Jervois-type deposit.*

broken round A term used by Bretz (1929, p.507) for a roundstone (such as a pebble or cobble) that has undergone breakage and whose spalled corners are believed to indicate exceptionally high-velocity currents.

broken sand A sandstone containing a mixed sequence of deposits (such as shaly layers).

broken-stick model A model comparing the relative abundances (or niche sizes) of several species competing within a habitat to the lengths of segments of a straight stick broken randomly into as many pieces as there are species (MacArthur, 1957). Its principal implication that species diversity and abundance result from nonoverlapping, randomly sized ecologic niches has stimulated much controversy among population ecologists.

broken stream A stream that repeatedly disappears and reappears, as in an arid region.

broken water Water whose surface is covered with ripples and eddies.

bromargyrite (brom-ar'-gy-rite) A variously colored cubic mineral: AgBr. Syn: *bromyrite.* The bromine analogue of chlorargyrite.

bromellite (bro-mell'-ite) A white or creamy white hexagonal mineral: BeO. The Be analogue of zincite.

bromlite (brom'-lite) *alstonite.*

bromoform (bro'-mo-form) Tribromethane: $CHBr_3$. It is used as a *heavy liquid;* its specific gravity is 2.9. Cf: *methylene iodide; Clerici solution; Sonstadt solution; Klein solution.*

bromyrite (bro'-my-rite) *bromargyrite.*

brongniardite (brong-ni-ard'-ite) A metallic monoclinic mineral of uncertain composition, possibly $Ag_2PbSb_2S_5$.

Bronze Age In archaeology, a cultural level that was originally the middle division of the *three-age system,* and is characterized by the technology of bronze in the Old World. Correlation of relative cultural levels with actual age (and, therefore, with the time-stratigraphic units of geology) varies from region to region. The term is used mainly in Europe and the Near East for the post-Stone Age period until about 1100 B.C.; but in East Asian archaeology it coincides with written history.

bronzite (bronz'-ite) (a) A brown or green variety of *enstatite* containing iron and often having a bronzelike or pearly metallic luster; an orthopyroxene intermediate in composition between enstatite and hypersthene. (b) A synonym for *clintonite.*

bronzitite (bronz'-it-ite) A *pyroxenite* composed almost entirely of orthopyroxene with bronzite composition.

brood chamber A space within which embryos develop in bryozoans, partly or entirely enclosed by body walls of one or more polymorphs or extrazooidal parts of a colony. It may be part of the body cavity or outside the body cavity of the colony. Cf: *brood pouch.*

brood pouch A sac or cavity of the body of an animal where the eggs or embryos are received and undergo a part of their development; e.g. the gently to strongly swollen part of the heteromorphous (presumed female) *carapace* of an ostracode, thought to be used for containing the not yet independent young. Partial syn: *brood chamber.*

brook (a) A small stream or rivulet, commonly swiftly flowing in rugged terrain, of lesser length and volume than a creek; esp. a stream that issues directly from the ground, as from a spring or seep, or that is produced by heavy rainfall or melting snow. Also, one of the smallest branches or ultimate ramifications of a drainage system. (b) A term used in England and New England for any tributary to a small river or to a larger stream. Syn: *brock; bruik.* (c) A general literary term for a creek.

brookite (brook'-ite) A brown, reddish, or sometimes black orthorhombic mineral: TiO_2. It is trimorphous with rutile and anatase, and occurs in druses and cavities. Syn: *pyromelane.*

brooklet (brook'-let) A small brook; a rill.

brookside (brook'-side) The land adjacent to or bordering on a brook. Syn: *burnside.*

broomstick stalagmite A variety of tall, thin stalagmite.

brow (a) The projecting upper part or margin of a steep slope just below the crest; the edge of the top of a hill or mountain, or the place at which a gentle slope becomes abrupt. (b) An English term for a steep slope.

brown algae A group of complex, almost entirely marine, noncalcareous, multicellular *algae* belonging to the *phylum* Phaeophyta and ranging from the Ordovician to the present.Their brown to olive-green color stems from carotenoid pigments, especially xanthophylls. Modern forms are mostly temperate and include the giant kelps and other seaweeds of the Northern Hemisphere. An exception is the tropical genus *Sargassum.* Sing: alga. Cf: *cyanobacteria; green algae; red algae; yellow-green algae.*

brown body A colored mass formed in many bryozoan zooids by the aggregation of the residue of a degenerated *polypide.*

brown clay *red clay.*

brown coal A brown to brownish black coal, intermediate in rank between peat and lignite A, in which original plant structures may usually be seen. The term is generally used in Europe, Australia, and Great Britain. Cf: *lignite.* Partial syn: *lignite B.* Old syn: *fulvurite.*

Brown Earths Soils with a mull horizon but having no horizon of accumulation of clay or sesquioxides. Not used in current U.S. system of soil taxonomy.

Brown Forest soils A great soil group of the intrazonal order and calcimorphic suborder, formed on calcium-rich parent materials under deciduous forest, and possessing a high base status but lacking a pronounced illuvial horizon. Not used in current U.S. system of soil taxonomy.

brown hematite A syn. of *limonite.* The term is a misnomer, because true hematite (unlike limonite) is anhydrous. Cf: *red hematite; black hematite.*

brown hornblende A brown variety of hornblende rich in iron; specif. *basaltic hornblende.*

brown iron ore *limonite.*

brown ironstone *Minette-type ironstone.*

brown lignite *lignite B.*

brown matter *Humic degradation matter;* cell-wall degradation matter that is brown and translucent in thin section.

brown mica *phlogopite.*

brownmillerite (brown'-mil-ler-ite') A reddish-brown orthorhombic mineral: $Ca_2(Al,Fe^{3+})_2O_5$. It is a constituent of portland cement. Syn: *celite.*

brown ocher A *limonite* that is used as a pigment.

brown ore A brown-colored ore; specif. the limonite group of iron ores.

Brown Podzolic soil A great soil group in the 1938 classification, a group of zonal soils that is similar to a Podzol but lacks the leached, light-colored A2 horizon. It is considered by some to be a type of Podzol rather than a separate soil group (USDA, 1938). Most of these soils are now classified as *Andepts* and *Orthods.*

brown rock A term used in Tennessee for dark brown to black phosphorite resulting from the weathering of phosphatic limestone. Cf: *white-bedded phosphate.*

Brown soils A great soil group of the temperate to cool arid regions, composed of soils with a brown surface and a lightcolored transitional subsurface horizon over calcium carbonate accumulation. Not used in current U.S. system of soil taxonomy.

brown spar Any light-colored crystalline carbonate mineral that is colored brown by the presence of iron; e.g. ankerite, dolomite, magnesite, or siderite.

brownstone A brown or reddish-brown sandstone whose grains are generally coated with iron oxide; specif. a dark reddish-brown, ferruginous quartz sandstone, principally of Triassic age, once extensively quarried in the Connecticut River valley for use as building stone.

browser (brows'-er) (a) A terrestrial vertebrate that eats the leaves and shoots from branches well above the ground surface. (b) A marine invertebrate that eats tiny algae scraped off the solid substrate.

brucite (bru'-cite) (a) A variously colored platy trigonal mineral: $Mg(OH)_2$. It commonly occurs in thin pearly folia and in fibrous form, as in serpentine and impure limestone. (b) A group name for trigonal minerals of composition $A(OH)_2$, where A = Mg, Mn^{2+}, Fe^{2+}, Ni or Ca.

brucite-like A term that describes the trioctahedral sheet in 1:1 and 2:1 clay minerals. This sheet is analogous to brucite which consists of two planes of closest packed oxygens with all of the octahedral sites between the two planes occupied by divalent cations. In a brucite-like sheet, some of the oxygens are replaced by hydroxyls.

Brückner cycle (Brück'-ner) A climatic cycle of 33-35 years, first noted by Sir Francis Bacon in 1625 and restated by E. Brückner in 1890. Applies to cold, damp episodes alternating with warm dry conditions in northwestern Europe. It is not seen in many records, but is suggested by tree-ring analyses and solar-cycle indices.

brüggenite (brüg'-gen-ite) A colorless to bright yellow monoclinic mineral: $Ca(IO_3)_2{\cdot}H_2O$.

brugnatellite (brug-na-tel'-lite) A flesh-pink, yellowish or brownish white platy brucite-like hexagonal mineral: $Mg_6Fe^{3+}(CO_3)(OH)_{13}{\cdot}4H_2O$. A member of the *manasseite* group.

bruik A Scottish var. of *brook.*

brunckite (brunck'-ite) A colloidal variety of *sphalerite.*

Brunhes Chron An interval of geologic time from about 780,000 years ago to the present, delimited by a geomagnetic reversal, and having the normal magnetic field of the present. Named for Bernard Brunhes, French geophysicist. Symbol: C1n.

brunogeierite (bru'-no-gei'-er-ite) A black cubic mineral of the *spinel* group: $(Ge^{2+},Fe^{2+})Fe^{3+}{}_2O_4$.

Brunssumian European paleofloral stage: upper Miocene (below Reuverian).

Brunton compass (Brun'-ton) A compact pocket instrument that consists of an ordinary compass, folding open sights, a mirror, and a rectangular spirit-level clinometer, which can be used in the hand or on a staff or light rod for reading horizontal and vertical angles, for leveling, and for reading the magnetic bearing of a line. It is used in sketching mine workings, and in preliminary topographic and geologic surveys on the surface, e.g. in determining elevations, stratigraphic thickness, and strike and dip. Named after its inventor, David W. Brunton (1849-1927), U.S. mining engineer. Usually called a "Brunton". Syn: *pocket transit.*

brush (a) Numerous fine cytoplasmic strands radiating from the distal end of the *caecum* and connected to the *periostracum* of the *punctate* shells of rhynchonelliform (formerly articulate) brachiopods (TIP, 1965, pt. H, p.141). (b) A bunch of fine terminal branches in phaeodarian radiolarians.

brush cast The cast of a *brush mark,* characterized by a crescentic depression around the downcurrent end. Originally defined by Dzulynski and Slaczka in 1959.

brush hook A short, stout, heavy hooked blade with a sharpened iron edge, attached to an axe handle, and used by surveyors for cutting brush.

brushite (brush'-ite) A colorless or pale yellow monoclinic mineral: $Ca(HPO_4){\cdot}2H_2O$. A member of the *churchite* group.

brush mark A *bounce mark* whose downcurrent end has a small crescentic ridge of mud pushed up by and in front of the impinging object.

brute-force radar *real-aperture radar.*

brute stack A preliminary *common-midpoint stack* where final static and normal-moveout corrections and processing have not yet been applied.

Bruxellian (Brux-el'-li-an) European stage: lower Middle Eocene (above Ypresian, below Auversian). It includes Cuisian and Lutetian.

bryalgal (bry-al'-gal) Said of a rigid, wave-resistant limestone composed largely of materials constructed in place by frame-building bryozoans and calcareous algae that commonly encrust one another. The material so formed is intimately associated with reefs. Term proposed by Bissell (1964, p.586). Cf: *coralgal.*

bryochore (bry'-o-chore) A climatic term for the part of the Earth's surface represented by tundras.

Bryophyta (Bry'-o-phy'-ta) A phylum of nonvascular plants that may have differentiated stems and leaves but have no true roots. It includes liverworts and mosses.

bryophyte (bry'-o-phyte) A small, nonvascular plant that may have stemlike, leaflike, and rootlike structures. Some plants form flattened, green, ribbonlike organisms on the soil surface. Liverworts, hornworts, and mosses are bryophytes. Cf: *thallophyte; pteridophyte.*

bryozoan (bry'-o-zo'-an) Any invertebrate belonging to the phylum Bryozoa and characterized chiefly by colonial growth, a calcareous skeleton, or, less commonly, a chitinous membrane, and a U-shaped alimentary canal, with mouth and anus. Range, Ordovician to Holocene, with a possible downward extension into the Upper Cambrian. Syn: *sea mat; moss animal; moss coral; moss polyp; polyzoan.* See also: *ectoproct; entoproct.*

BS *backsight.*

B-tectonite A tectonite whose fabric is dominated by linear elements. Not in common use, having been largely replaced by the term L-tectonite. Cf: *L-tectonite; S-tectonite.*

BTEX *benzene, toluene, ethyl benzene, and xylene.*

B-type earthquake Event with weak P waves and no S waves, a low-frequency content, and occurring under volcanoes at depths of <1 km (McNutt, 2000).

B-type lead *Anomalous lead* that gives model ages older than the age of the enclosing rock. Cf: *J-type lead.* Syn: *Bleiberg-type lead.*

bubble (bub'-ble) (a) A small air-filled cavity, or the globule of air or gas, in the glass tube of a spirit level. When the level is adjusted to the horizontal, the center of the bubble comes to rest under a fixed mark or etched line at the highest point possible in the tube. (b) A term sometimes applied to a bubble tube and its contents.

bubble impression A small, shallow depression (2.5 cm in diameter) formed on a beach or sedimentary surface by a bubble of gas after it has been dislodged into the air or water above. It has a smooth surface, is not margined by a raised rim, and may pass downward into a tube. Syn: *bubble mark.*

bubble mark *bubble impression.*

bubble noise Unwanted seismic energy generated by the oscillation of a bubble of high-pressure gas in water. The waste gases from an explosion, or from other seismic-energy sources such as air guns, can give rise to a sequence of bubble pulses.

bubble orientation A preferred orientation of air bubbles in glacier ice when the bubbles become elongate due to shear.

bubble-point pressure The pressure at which gas, held in solution in crude oil, breaks out of solution as free gas. Syn: *saturation pressure.*

bubble pulse An implosion attributable to the collapse of a gas bubble produced by a seismic charge fired in deep water. The bubble may pulsate several times with a period proportional to the cube root of the charge, each oscillation producing an unwanted seismic effect.

bubble tube The circular or slightly curved glass tube containing the liquid and bubble in a spirit level and mounted with the bend convex upward.

bubble-wall texture Texture shown by phenocrysts that have a thin coating of vesicular glass.

Bubnoff unit (Bub'-noff) A standard measure of geologic time-distance rates (as for geologic movements and increments), proposed by Fischer (1969) and defined as 1 micrometer/year (1 mm/thousand years, or 1 m/million years). Named in honor of Serge von Bubnoff (1888-1957), Russian-born German geologist.

bucaramangite (bu'-ca-ra-man'-gite) A pale-yellow variety of retinite that is insoluble in alcohol, found at Bucaramanga, Colombia.

buccal aperture A tunnel opening between the chambers of fusulinids. See also: *foramen.*

buccal cavity (buc'-cal) In crustaceans, a hollow space on the ventral side of the body, containing the mouth parts. In Malacostraca, it is bounded by the *epistome* in front and the free edges of the *carapace* on the sides. (b) In Ostreostraci, the hollow space containing the gill bars on the ventral side of the head shield. (c) In vertebrates generally, that part of the alimentary canal, usually separated from nasal structures, between the mouth and esophagus. Not to be confused with the buccal space, which in jawed vertebrates is the potential space between the teeth and margin of the mouth (cheek and lips, when present).

buccal frame A structure enclosing the mouth parts of brachyuran decapod crustaceans, bounded laterally by free anterior and lateral edges of the *carapace* and in front by the *epistome* and commonly by closed maxillipeds.

buccal mass The mouth parts of mollusks, except bivalves, and the muscles with which they are attached.

buccal plate (a) One of ten large primordial plates of an echinoid, located on the tissue between peristomial margin and mouth, and containing pores for passage of tube feet. (b) *buccal shield.*

buccal shield A large, more or less triangular *ossicle* in the interradial position, adjoining the mouth in an ophiuroid (TIP, 1966, pt. U, p.29). Syn: *buccal plate.*

Buchan-type facies series (Buch'-an-type) Rocks produced in a type of dynamothermal regional metamorphism rather similar to that of the *Abukuma-type facies series* but in a somewhat higher pressure environment, of 300-400 MPa (Hietanen, 1967, p.192).

buchite (buch'-ite) A vitrified hornfels produced by fusion of an argillaceous rock by intense local thermal metamorphism. Cf: *hyalomylonite.*

buchonite (buch'-o-nite) A dark-colored extrusive rock containing hornblende and biotite in addition to plagioclase, nepheline, and augite; an alkali-feldspar-bearing *tephrite.* Named by Sandberger in 1872 for Buchonia, Germany. Obsolete.

buchwaldite (buch'-wald-ite) A white orthorhombic meteorite mineral: $NaCaPO_4$.

bucket auger A tool used to drill a large-diameter borehole, typically 18-30 in. (460-760 mm) in diameter, in which cuttings accumulate in a hollow, cylindrical bucket and are brought to the ground surface by lifting the drill tools out of the hole. Bucket-auger borings are large enough to permit geologists to enter the borehole, on a plate or in a cage suspended from a cable operated from the drill, for direct observation of subsurface conditions exposed in the borehole wall.

bucket temperature Temperature of water at the surface of the sea measured by putting a thermometer in a bucket, lowering the bucket into the water from a ship, letting the bucket sit for a few minutes, then bringing it on board and recording the temperature measured by the thermometer before the water in the bucket has time to change temperature.

buckhornite A metallic black orthorhombic mineral: $AuPb_2BiTe_2S_3$.

bucking (buck'-ing) Pulverization of a representative sample of rock, as for assay or quantitative analysis.

bucking electrodes In drill logging, extra electrodes (as on a logging sonde) for directing the current flow into the *formation.*

Buckingham-Darcy equation The equation of motion for flow under unsaturated conditions. Similar to *Darcy's law,* except that *hydraulic conductivity* is a function of pressure head of water.

bucklandite (buck'-land-ite) (a) A black variety of epidote containing iron and having nearly symmetric crystals. (b) *allanite.*

buckle folding (buck'-le) Folding that occurs in response to forces parallel to competent layers. Cf: *bending.*

buck quartz *bull quartz.*

buckshot (buck'-shot) *shot [soil].*

bud [bot] The undeveloped or meristematic state of a branch or flower cluster, with or without covering scales; contains an apical *meristem* and young, developing leaves. Buds may be apical, axillary, or adventitious in location on the stem (Swartz, 1971, p.75).

bud [paleont] (a) An asexual reproductive body (including various types of cells) that is eventually isolated from a parent sponge or *polyp.* (b) A newly developing, asexually produced *zooid* in bryozoans.

budding (bud'-ding) *vegetative reproduction.*

buddingtonite (bud'-ding-ton-ite') A colorless monoclinic mineral: $(NH_4)AlSi_3O_8$. It is isostructural with orthoclase, and a member of the *feldspar* group.

budget gradient The rate of change with altitude of the *net balance* of a glacier, measured in the vicinity of the *equilibrium line.* High values indicate vigorous transfer of excess *accumulation* to lower altitudes (e.g., temperate maritime glaciers); low values indicate minimal or sluggish transfer (e.g., polar continental glaciers). It is usually measured in millimeters per meter. Cf: *ablation gradient.* Syn: *activity index; energy of glacierization.*

budget year *balance year.*

bud scale A protective modified leaf, covering or enclosing a bud.

bud scar A scar left on a twig by the falling-away of a bud or a group of bud scales (Fuller and Tippo, 1954, p.952).

buergerite (buer'-ger-ite) A brown rhombohedral mineral of the *tourmaline* group: $NaFe^{3+}_3Al_6(BO_3)_3Si_6O_{18}(O,F,OH)_4$.

Buerger precession method (Buer'-ger) The recording on film of a single level of the reciprocal lattice of an individual crystal by means of X-ray diffractions, for the purpose of determining unit-cell dimensions and space groups. See also: *precession camera.* Syn: *precession method.*

buetschliite (buetsch'-li-ite) *bütschliite.*

Buffalo gun An impulsive energy source, which involves detonation of a blank shotgun shell in a shallow hole.

buffalo wallow (buf'-fa-lo) (a) One of the small undrained shallow depressions that were once common on the Great Plains of the western U.S., usually containing water after a rain (and often remaining as a stagnant water hole for most of the year). It is generally believed to have been deepened or modified, and perhaps initially formed, by the trampling and wallowing of buffalo herds in mud and dust. The diameter ranges from about 1 m to as much as 20 m and the depth from several centimeters to a few meters. (b) A term improperly applied to one of the large natural depressions widely distributed throughout the Great Plains of the western U.S. (esp. on the High Plains), often containing an intermittent pond or temporary lake (Veatch and Humphrys, 1966, p.49).

buffer (a) An intermediate storage device that accomodates differences between the rate at which information is fed into a computer and the rate at which the computer can receive it, or which performs the same function for information output by the computer or between component parts of a computer. (b) An isolating circuit to prevent a reaction of the circuit which is driven from affecting the circuit which is doing the driving. A buffer amplifier may follow a critical stage to isolate subsequent stages from load impedance variations. (c) A solution that minimizes the changes in hydrogen-ion concentration that would otherwise result from a chemical reaction.

buffered Somewhat isolated from effects that would otherwise occur.

buffered reaction (buf'-fered) An equilibrium among metamorphic minerals and pore fluid in which an intensive parameter, such as temperature or the partial pressure of a fluid component, is determined or fixed by the equilibrium constant for the reaction.

buffer-zone (buff'-er-zone) A chrono-stratigraphic boundary, placed as precisely as possible in a continuous type section but still not truly isochronous in other continuous sections. "It would perhaps be more realistic to think of isochronous 'buffer-zones' instead of isochronous surfaces" (Hornibrook, 1965, p.1199). Cf: *chronohorizon.*

bug hole A miner's term for vug.

bugor A hill or succession of small hills separating creeks or ravines, as on the shore of the Black Sea. Etymol: Russian "bugori", "hillock".

building stone (build'-ing) A general, nongeneric term for any rock suitable for use in construction. Whether igneous, metamorphic, or sedimentary, a building stone is chosen for its properties of durability, attractiveness, and economy. See also: *dimension stone; facing stone.*

buildup (build'-up) (a) A development of limestone which has some topographic relief above the sea floor. Buildups commonly contain fossils in their growth position; many are reefs, with a framework of carbonate skeletons. Others are accumulations of skeletal debris and lime mud where there is no obvious framework. (b) A nongeneric term used by Merriam (1962, p.73) for "any extra, stray, or super" limestone bed or beds, in addition to the "normal" sequence, as exemplified in the rhythmic (cyclic) deposits of the northern midcontinent region of U.S.; e.g. a marine bank, a bioherm, and an organic reef.

bukovite (bu'-kov-ite) A metallic grayish-brown tetragonal mineral: $Tl(Cu,Fe)_2S_2$.

bukovskyite (bu-kov'-sky-ite) A yellowish green to grayish green triclinic mineral: $Fe^{3+}_2(AsO_4)(SO_4)(OH){\cdot}7H_2O$.

bulachite (bu'-lach-ite) A silky white orthorhombic mineral: $Al_2(AsO_4)(OH)_3{\cdot}3H_2O$.

bulawan (bu'-la-wan) A Philippine term for gold or gold ore.

bulb Swollen, crestal portion of a *salt diapir.* Extremely broad bulbs grade into *salt sheets* (Jackson and Talbot, 1991).

bulbar scar (bul'-bar) In archaeology, a bulb-shaped scar on the *bulb of percussion* on a *flake* from a worked stone.

bulb glacier *expanded-foot glacier.*

bulb of percussion In archaeology, the arc-shaped area emanating from the *point of percussion* on a *flake,* showing the effect of shock waves from an impact.

bulbous dome (bul'-bous) Obsolete syn. of *volcanic dome.*

bulge (a) A tumescence of lava. (b) A landmass projecting beyond the general outline of the body of which it is a part; e.g. the "bulge" of Brazil. (c) A diapiric structure with a clay core beneath more competent overlying strata. (d) An asymmetric uplift of the seaward side of a rifted continental margin, e.g. as originally applied to southern Africa. Cf: *camber.*

Bulitian (Bu-lit'-i-an) North American stage: Paleocene (above Ynezian, below Penutian).

bulk density The weight of an object or material divided by its volume, including pore spaces and in situ moisture. It is identified as dry density for the weight per unit volume of a soil mass that has been oven-dried to a constant weight at 105°C.

bulk formula Equation relating transfer of heat and momentum through the sea surface, using variables that can be measured easily by observers on a ship; e.g., latent heat flux is very difficult to measure, so it is calculated from an equation that uses relative humidity, air temperature, and wind speed, all of which can be measured from a ship.

bulkhead (bulk'-head) A stone, steel, wood, or concrete wall-like structure primarily designed to resist earth or water pressure, as a retaining wall holding back the ground from sliding into a channel, or a partition preventing water from entering a working area in a mine.

bulking The erosion and incorporation of secondary, exotic debris by lahars as they move downstream (Vallance, 2000).

bulk modulus A *modulus of elasticity* which relates a change in volume to a change in hydrostatic state of stress. It is the reciprocal of *compressibility.* Symbol: k. Syn: *volume elasticity; incompressibility modulus; modulus of incompressibility.*

bulk partition coefficient A *partition coefficient* "weighted according to the relative abundances of coexisting phases in a particular assemblage" (Beattie et al., 1993).

bulk silicate Earth *primitive mantle.*

bulla (bul'-la) (a) One of the blisterlike structures that partly or completely cover the apertures in planktonic foraminifers, that are not closely related to primary chambers, and that may be umbilical, sutural, or areal in position and may have one or more marginal accessory apertures. (b) A radially elongated *tubercle* of an ammonoid. (c) In many mammals, the auditory or tympanic bulla, a cup- or shell-shaped bone that floors the tissues of the air-filled middle ear and supports the ear drum (tympanic membrane). Pl: bullae. Adj: bullate.

Bullard discontinuity (Bul'-lard) The seismic-velocity interface between the *outer core* and the *inner core.*

Bullard's method Computation of the effect of topography for the Hayford zones, by first calculating the effect of the *spherical cap* of height equal to the station height, then of the topographic deviations of this cap (Schieferdecker, 1959, term 3489).

bulldust An Australian term for fine floury dust.

buller (bull'-er) *blowhole [coast].*

bullet crystal (bul'-let) A snow crystal in the shape of a short hexagonal prism with one pointed end, characteristically formed at very cold temperatures.

bullette (bul-lette′) A siphonal deposit of a nautiloid, similar to an *annulus* but flatter and more elongated in cross section, in which it appears knoblike or bosslike (TIP, 1964, pt. K, p.54).

bullion (bul′-lion) (a) A concretion found in some types of coal. It is composed of carbonate or silica, stained brown by humic derivatives, and may be several centimeters to a meter or more in diameter. Well-preserved plant structures often form the nucleus. Cf: *coal ball.* (b) A nodule of clay, shale, ironstone, or pyrite that generally encloses a fossil.

bull mica Large clusters of diversely oriented and partially intergrown crystals of muscovite with a little interstitial albite and quartz (Skow, 1962, p.169).

bull quartz A miner's or prospector's term for white massive quartz, essentially free of accessory minerals and valueless as ore. Syn: *bastard quartz; buck quartz.*

bull's-eye level *circular level.*

bultfonteinite (bult-fon′-tein-ite) A colorless to pale pink triclinic mineral: $Ca_2SiO_2(OH,F)_4{\cdot}H_2O$.

bummock (bum′-mock) A downward projection from the underside of sea ice; the submariner's counterpart of a *hummock [ice].*

Bumstead head (Bum′-stead) A lightweight tripod, adapted to use on foot traverses. It carries a 15-inch-square plane-table board, which can be revolved about its axis, but provides no means of leveling. It is used with a peep-sight alidade. Named after Albert H. Bumstead (1875-1940), U.S. cartographer.

bund Any artificial embankment used to control the flow of water in a river or on irrigated land. The term is applied extensively in India to large low dams and dikes and also to the small ridges between rice fields. Also, an embanked causeway or thoroughfare along a river or the sea.

bundle scar (bun′-dle) In a *leaf scar,* a scar indicating the position of a vascular bundle that had connected the stem and the stalk.

bunsenite (bun′-sen-ite) A pistachio-green cubic mineral: NiO. A member of the *periclase* group.

Bunter (Bun′-ter) A term used esp. in Germany to refer to what is now approximately the Lower Triassic (above Permian, below Muschelkalk); essentially a lithostratigraphic term, a colloquial abbreviation of "Buntsandstein".

buoyancy frequency *stability frequency.*

burangaite (bu-ran′-ga-ite) A blue to bluish green monoclinic mineral: $(Na,Ca)(Fe^{2+},Mg)Al_5(PO_4)_4(OH,O)_6{\cdot}2H_2O$. It is the Al-analogue of dufrénite, and a member of the *dufrénite* group.

burbankite (bur′-bank-ite) (a) A variously colored hexagonal mineral: $(Na,Ca)_3(Sr,Ba,Ce)_3(CO_3)_5$. (b) A group name for hexagonal minerals with the general formula $A_3B_3(CO_3)_5$, where A = Na or Ca; and B = Ca, Sr, Ba or rare-earth elements.

burckhardtite (burck′-hardt-ite) A violet-red or pale pink monoclinic mineral: $Pb_2(Fe^{3+},Mn^{3+})Te^{4+}(AlSi_3)O_{12}(OH)_2{\cdot}H_2O$.

Burdigalian (Bur-di-gal′-i-an) A stage of the Standard Global Chronostratigraphic Scale: upper Lower Miocene (above Aquitanian, below Langhian).

Bureau International de l'Heure The organization responsible, by international agreement, for coordinating the measurements of time by national observatories and for providing an internationally acceptable and common time. It is also responsible for maintaining the international atomic second. As part of its function, it calculates the position of the Earth's axis of rotation and changes in the Earth's rate of rotation. Abbrev: BIH.

Burgers circuit (Burg′-ers) A loop that encloses a dislocation line in a crystal and that fails to close by an amount known as the *Burgers vector.*

Burgers vector The vector required to complete a *Burgers circuit.* It is the same as the vector of, and independent of the position of, the crystal dislocation.

Burgess Shale fauna A Middle Cambrian fossil assemblage in a rock unit found in the Rocky Mountains of British Columbia, famous for its preservation of soft-bodied fossils of unusual Cambrian animals that are not normally preserved.

burial (bur′-i-al) Covering up or concealing geologic features by the process of *sedimentation.*

burial diagenesis The processes of diagenesis operating as a result of burial; includes pressure dissolution, compaction, dissolution, and cementation.

burial joints A term for those joints forming during a phase of sedimentation, downwarping, or diagenesis. They may develop as vertical, planar joints with regular spacing (Bahat, 1991). Syn: *neptunian joint; syneresis joint.*

burial metamorphism A type of low-grade regional metamorphism affecting sediments and interlayered volcanic rocks in a geosyncline without any influence of orogenesis or magmatic intrusion. Original rock fabrics are largely preserved but mineralogical compositions are generally changed (Coombs, 1961). Cf: *dynamothermal metamorphism.*

buried channel (bur′-ied) An old channel concealed by surficial deposits; esp. a preglacial channel filled with glacial drift.

buried-channel placer A *paleoplacer* deposit within a *buried channel.*

buried erosion surface An *erosion surface,* such as a peneplain, that has been covered by younger sediments; it may represent a surface of unconformity at depth. Cf: *fossil erosion surface.*

buried focus A situation where the concave-upward curvature of a seismic reflector is large enough that the energy focuses before it reaches the recording plane. Several branches (usually three) of a reflection may be observable from the same surface location. Focusing can also be produced by certain velocity distributions that act as lenses. See also: *reverse branch.*

buried ice Any relatively distinct ice mass buried in the ground, esp. surface ice, as that of sea, lake, river, or glacier origin, that has been buried syngenetically by sediments, esp. in a permafrost region.

buried karst Karst covered by recent sediments such as loess or moraine deposits. Cf: *paleokarst.*

buried river A riverbed that has been concealed beneath alluvium, lava, pyroclastic rocks, or till.

buried soil Soil covered by an alluvial, loessal, or other surface mantle of more recent depositional material, usually to a depth greater than 50 cm.

buried valley A depression in an ancient land surface or in bedrock, now covered by younger deposits; esp. a preglacial valley filled with glacial drift.

burkeite (burke′-ite) A white, buff, or grayish mineral: $Na_6(SO_4)_2(CO_3)$.

burl An nodule in fireclay. It may have a high content of alumina, iron oxide, or siderite.

burley clay (bur′-ley) A clay containing burls; specif. a diaspore-bearing clay in Missouri, usually averaging 45 to 65% alumina. See also: *diaspore clay.*

burmite (bur′-mite) A dark brown, pale yellow, or reddish variety of *retinite,* found in Burma, that resembles amber but is tougher and harder. It has also been regarded as a variety of amber low in succinic acid.

burn A term used in Scotland and northern England for a small stream or brook, such as a *bourne.*

burned Said of shale that adheres tightly to and is difficult to remove from the coal with which it is associated. Syn: *frozen [coal].*

burnie (burn′-ie) A Scottish term for a little stream or brook.

burnside (burn′-side) *brookside.*

burnsite A vitreous dark red hexagonal mineral: $KCdCu_7O_2(SeO_3)_2Cl_9$.

burnt stone A gemstone whose color has been altered by heating; e.g. amethyst, which changes from purple to near-colorless or yellow (citrine) sapphire which changes from colorless or light blue to dark blue, or ruby which changes from brownish or dark red to red. Cf: *heated stone; stained stone.*

burpalite A colorless to pale yellow monoclinic mineral: $Na_2CaZrSi_2O_7F_2$. A member of the *låvenite* group.

burr (a) A term used in England for a rough or hard stone, such as a compact siliceous sandstone especially hard to drill. (b) A knob, boss, nodule, or other hard mass of siliceous rock in a softer rock; a hard lump of ore in a softer vein. (c) *whetstone.*

burr ball *lake ball.*

burrow (bur′-row) A tubular or cylindrical hole or opening, made in originally soft or loose sediment, by a mud-eating worm, a mollusk, or other invertebrate, extending along a bedding plane or penetrating a rock, and often later filled with clay or sand and preserved as a filling; it may be straight or sinuous, and vertical, horizontal, or inclined. Cf: *boring.*

burrow porosity Porosity resulting from burrowing organisms. Although this type of porosity is uncommon in ancient carbonate rocks due to infilling of burrows, other types such as interparticle porosity can develop within the burrow-filling material, esp. if its permeability is greater than that of the host sediment (Choquette and Pray, 1970, p.244). Syn: *boring porosity.*

burr rock An aggregate of muscovite books and quartz (Skow, 1962).

bursa (bur′-sa) An internal gill pouch in ophiuroids, entered by the gill slit. In vertebrates, a pouch or saclike encapsulating tissue (such as at bony joints, in which the capsule contains a viscous lubricating fluid). Pl: bursae.

bursaite (bur′-sa-ite) A metallic gray orthorhombic mineral: $Pb_5Bi_4S_{11}$.

burtite (burt′-ite) A colorless cubic or rhombohedral mineral: $CaSn(OH)_6$.

buryatite An earthy light gray trigonal mineral: $Ca_3(Si,Fe^{3+},Al)SO_4B(OH)_4(OH,O)_6{\cdot}12H_2O$. A member of the *ettringite* group.

Burzyan An obsolete geochronometric subdivision of the *Precambrian,* covering the time span from 1,650 to 1,350 Ma (Harland et al., 1982; 1990), therefore approximately equivalent to the *Calymmian* Period (early *Mesoproterozoic* Era) of the Standard Global Geochronometric Scale.

buserite A black orthorhombic mineral with a formula approximating $Na_4Mn_{14}O_{27}{\cdot}21H_2O$. Not approved by the International Mineralogical Association.

bushmakinite A vitreous yellow monoclinic mineral: $Pb_2(Al,Cu)(PO_4)(V,Cr,P)O_4(OH)$. A member of the *brackebuschite* group.

bushveld (bush′-veld) A large, flat grassy area with scattered trees, found in tropical or subtropical regions, esp. in Africa. Etymol: Afrikaans, "bosveld". Cf: *savanna.*

Bushveld-type chromite deposit *stratiform chromite deposit.*

Bushveld-type Fe-Ti-V deposit *stratiform iron-titanium-vanadium deposit.*

bussenite A yellowish brown triclinic mineral: $Na_2Ba_2Fe^{2+}TiSi_2O_7(CO_3)(OH)_3F$.

bustamite (bus′-tam-ite) A pink or brownish red triclinic pyroxenoid mineral: $CaMnSi_2O_6$.

butane (bu′-tane) A gaseous inflammable paraffin hydrocarbon, formula C_4H_{10}, occurring in either of two isomeric forms: n-butane, $CH_3CH_2CH_2CH_3$, or isobutane, $CH_3CH(CH_3)_2$. The butanes occur in petroleum and natural gas.

butlerite (but'-ler-ite) An orange monoclinic mineral: $Fe^{3+}(SO_4)(OH)\cdot 2H_2O$. Cf: *parabutlerite.*

bütschliite (bütsch'-li-ite) A white rhombohedral mineral: $K_2Ca(CO_3)_2$. It is dimorphous with fairchildite. Also spelled: *buetschliite.*

butt cleat The minor cleat system, or jointing, in a coal seam, usually at right angles to the *face cleat.* Syn: *end cleat.* See also: *cleat; face cleat.*

butte [Pron. bewt] (a) A conspicuous, usually isolated, generally flat-topped hill or small mountain with relatively steep slopes or precipitous cliffs, often capped with a resistant layer of rock and bordered by talus, and representing an erosion remnant carved from flat-lying rocks. The summit is smaller in extent than that of a *mesa,* and many buttes in the arid and semiarid regions of the western U.S. result from the wastage of mesas. (b) An isolated hill having steep sides and a craggy, rounded, pointed, or otherwise irregular summit; e.g., a volcanic cone (as Mount Shasta, Calif., formerly known as Shasta Butte), or a *volcanic butte.* Etymol: French.

butter rock *halotrichite.*

Butterworth filter A frequency filter characterized by a very flat passband, often used as an alias filter. Sometimes called a maximally flat filter or trapezoidal filter (Sheriff and Geldart, 1995).

butte témoin A flat-topped hill representing the former extension of an escarpment edge or plateau, now detached by stream erosion, its surface in broadly the same plane as the main mass; an outlier. In most cases, a *butte* is a "butte témoin". Etymol: French, "witness hill". Syn: *witness butte; zeugenberg.*

buttgenbachite (butt'-gen-bach-ite') A sky-blue hexagonal mineral: $Cu_{37}(NO_3)_4Cl_8(OH)_{62}\cdot 6H_2O$. It is isomorphous with connellite.

buttress [geomorph] (but'-tress) A protruding rock mass on, or a projecting part of, a mountain or hill, resembling the buttress of a building; a spur running down from a steep slope. Example: a prominent salient produced in the wall of a gorge by differential weathering.

buttress [paleont] (a) An internal ridgelike projection from the shell wall of a bivalve mollusk which supports the *hinge plate* or *adductor muscle.* Syn: *clavicle.* (b) A ridge of skeletal material extending adapically from an echinoid *auricle* on the inner surface of the test (TIP, 1966, pt. U, p.253). (c) An aboral tongue-shaped extension of the apical denticle of a conodont, generally on the inner and outer sides.

buttress unconformity A surface on which onlapping strata abut against a steep topographic scarp of regional extent. A buttress unconformity may be produced on a submarine escarpment along which older rocks crop out, between a deep basin and a platform (Enos, 1974, p.807).

butyrellite (bu-tyr-el'-lite) *bog butter.*

butyrite (bu-tyr'-ite) *bog butter.*

Buys Ballot's law The statement in meteorology that describes the relationship between horizontal wind direction and barometric pressure: if an observer stands with his or her back to the wind in the Northern Hemisphere, the pressure is lower to the left than to the right; the reverse relationship holds true for an observer in the Southern Hemisphere. The law is named after the Dutch meteorologist who formulated it in 1857.

by-channel A stream or branch along one side of the main stream.

byelorussite-(Ce) A yellow to yellow-brown orthorhombic mineral: $NaBa_2Ce_2Mn^{2+}Ti_2Si_8O_{26}(F,OH)\cdot H_2O$. A member of the *joaquinite* group.

byerite (by'-er-ite) A bituminous coal that resembles albertite.

Byerlee's law A nearly linear relationship between shear stress and normal (perpendicular) stress during faulting that holds for most types of shallow crustal rocks. Named for James D. Byerlee, U.S. geophysicist.

bypassing (by'-pass-ing) (a) A term applied by Eaton (1929) to sedimentary transport across areas of nondeposition, as in the case where one particle size passes another that is being simultaneously transported or continues in motion after the other has come to rest; e.g., the normal decrease in average particle size of sediments away from a source area. The term is also applied to transport of coarser particles farther than finer particles ("reverse" bypassing); e.g., gravel bypassing sand along the edge of the continental shelf, probably the result of *density current* deposition. Cf: *total passing.* Also spelled: by-passing. (b) Flow of reservoir-flushing fluids around hydrocarbon pockets because of inhomogeneities or faults in a reservoir.

bysmalith (bys'-ma-lith) A discordant, roughly vertical cylindrical igneous intrusion. Cf: *bell-jar intrusion.* Obsolete.

by-spine A small accessory spine additional to a radial spine in acantharian and spumellarian radiolarians.

byssal gape (bys'-sal) An opening between the margins of a bivalve-mollusk shell for the passage of the *byssus.*

byssal notch The indentation, below the anterior *auricle* of the right valve in many pectinacean bivalve mollusks, serving for the passage of the *byssus* or for the protrusion of the foot.

byssiferous (bys-sif'-er-ous) Possessing a *byssus.*

byssolite (bys'-so-lite) An olive-green, fibrous variety of amphibole. The term is used in the gem trade for a variety of quartz containing greenish, fibrous inclusions of actinolite or asbestos.

byssus (bys'-sus) A tuft or bundle of long, tough hairlike strands or filaments, secreted by a gland in a groove of the foot of certain bivalve mollusks and issuing from between the valves, by which a temporary attachment of the bivalve can be made to rocks or other solid objects. Pl: byssi or byssuses.

by-stream Said of the part of a flood plain consisting of a narrow belt of levee deposits immediately adjacent to the stream channel, and composed usually of sandy alluvium.

bystrite A vitreous yellow trigonal mineral: $(Na,K,Ca)_8(Si_6A_6)O_{24}S_{1.5}\cdot H_2O$. A member of the *cancrinite* group.

byströmite (by'-ström-ite) (a) A pale blue-gray or yellowish tetragonal mineral: $MgSb^{5+}_2(O,OH)_6$. (b) A monoclinic polymorph of pyrrhotite. Also spelled: bystromite.

by-terrace Said of the part of a flood plain consisting of a narrow belt of clayey deposits adjacent to a bounding terrace relatively distant from the stream channel.

bytownite (by'-town-ite) A bluish to dark-gray triclinic mineral of the plagioclase *feldspar* group with composition ranging from $Ab_{30}An_{70}$ to $Ab_{10}An_{90}$. It occurs in basic and ultrabasic igneous rocks.

by-water A yellow-tinted diamond.

C

c *carat.*

caballing (ca-bal'-ling) The mixing of two water masses to produce a blend that sinks because it is denser than either of its original components. This occurs when the two water masses that have the same density but different temperatures and salinities are mixed.

cabalzarite A vitreous orange-brown mineral: $Ca(Mg,Al,Fe^{2+})_2(AsO_4)_2(H_2O,OH)_2$.

cabbage-leaf mark (cab'-bage-leaf') *frondescent cast.*

cable (ca'-ble) (a) A heavy multiple-strand steel rope used in *cable-tool drilling* as the line between the tools and the *walking beam.* (b) A term used to signify a *wireline* able to carry information or energy. (c) A *wireline.*

cable bolt An element of rock-slope stabilization or reinforcing that consists f a wire-rope tendon of which the end is grouted into a rock mass below a detached block. Tension is applied to the free end of the cable bolt; the rock surface is used for reaction; and the assembly is held in place by a clamping device.

cable break An event on a well geophone record caused by elastic energy being transmitted down the supporting cable to the well geophone.

cable tool Any component of a set of bottomhole tools used in *cable-tool drilling,* e.g. a *drill bit,* a *drill stem,* a set of *jars,* or a *bailer.*

cable-tool drilling A method of drilling, now almost entirely replaced by *rotary drilling,* in which the rock at the bottom of the hole is broken up by a steel bit with a blunt, chisel-shaped cutting edge. The bit is at the bottom of a heavy string of steel tools suspended on a cable connected to a *walking beam.* The bit is raised and lowered to chip the rock by regularly repeated blows. A bailer is used at regular intervals to remove the well cuttings. The method is adapted to drilling water wells and relatively shallow oil wells in mixed geologic units such as talus debris flow deposits or alluvial fan deposits. Syn: *churn drilling.*

cabochon (cab-o-shon') (a) An unfaceted cut gemstone of domed or convex form. The top is smoothly polished; the back, or base, is usually flat or slightly convex, sometimes concave, and often unpolished. The girdle outline may be round, oval, square, or of any other shape. (b) The style of cutting such a gem. Etymol: French. See also: *en cabochon.*

cabocle (cab'-o-cle) A compact rolled pebble resembling red jasper, supposed to be a hydrated phosphate of calcium and aluminum, found in the diamond-producing sands of Bahia, Brazil.

cabriite (ca-bri'-ite) A metallic white orthorhombic mineral occurring in massive Cu-Ni sulfide ores: Pd_2SnCu.

cacholong (cach'-o-long) An opaque or feebly translucent and bluish-white, pale-yellowish, or reddish variety of common opal containing a little alumina. Syn: *pearl opal.*

cacoxenite (ca-cox'-e-nite, ca-co-xen'-ite) A variously colored hexagonal acicular mineral: $Fe^{3+}_{24}AlO_6(PO_4)_{17}(OH)_{12}{\cdot}75H_2O$.

CAD Computer-aided design.

cadastral (ca-das'-tral) Delineating or recording property boundaries, and sometimes subdivision lines, buildings, and other details. Etymol: French cadastre, an official register of the real property of a political subdivision with details of area, ownership, and value, and used in apportioning taxes.

cadastral map A large-scale map showing the boundaries of subdivisions of land, usually with the directions and lengths thereof and the areas of individual tracts, compiled for the purpose of describing and recording ownership. It may also show culture, drainage, and other features relating to use of the land. Syn: *plat [cart] b.*

cadastral survey A survey relating to land boundaries and subdivisions, made to create units suitable for transfer or to define the limitations of title; esp. a survey of the public lands of the U.S., such as one made to identify or restore property lines. Cf: *land survey; boundary survey.*

cadicone (cad'-i-cone) An evolute, coiled cephalopod shell with a strongly depressed whorl section, wide venter, and a deep *umbilicus* (as in the ammonoid *Cadoceras*).

cadmia (cad'-mi-a) (a) *calamine.* (b) A chemical compound: CdO. (c) An impure zinc oxide that forms on the walls of furnaces in the smelting of ores containing zinc.

cadmium The metallic element Cd. It is metallic white in appearance, and has the hexagonal zinc structure.

cadmium blende (cad'-mi-um) *greenockite.*

cadmium ocher *greenockite.*

cadmoselite (cad-mo-sel'-ite) A black hexagonal mineral: CdSe. A member of the *wurtzite* group.

caducous (ca-duc'-ous) Said of the calyx of a flower that is shed before the flower expands (Swartz, 1971, p.78).

cadwaladerite (cad-wal'-a-der-ite') A yellow amorphous mineral: $AlCl(OH)_2{\cdot}4H_2O$. Its species status is disputed.

Caecilia (Cae-cil'-i-a) An order of burrowing caudate lissamphibians in which limbs have been lost and eyes reduced greatly. Alone among extant amphibians, caecilians retain rudimentary scales and reproduce by means of a large-yolked egg. Syn: *Gymnophiona.*

caecum (cae'-cum) (a) The sac-shaped apical end of the *siphuncle* of a nautiloid or ammonoid. Also, a cavity associated with the digestive system in the living *Nautilus.* (b) The evagination of the outer epithelium projecting into the *endopuncta* of a brachiopod shell. Pl: caeca. Adj: caecal.

Caerfaian (Caer-fai'-an) European series or stage: Lower Cambrian (above Precambrian, below Solvan). Syn: Caerfai.

cafarsite (ca-far'-site) A dark brown cubic mineral: $Ca_6(Ti,Fe^{2+},Fe^{3+},Mn)_8(As^{3+}O_3)_{12}{\cdot}4H_2O$.

cafetite (caf'-e-tite) An adamantine colorless to pale yellow monoclinic mineral: $CaTi_2O_5{\cdot}H_2O$.

cage A void in a crystal structure that is large enough to trap one or more atoms (such as argon or xenon) foreign to the structure.

cahemolith (ca-he'-mo-lith) *humic coal.*

cahnite (cahn'-ite) A colorless tetragonal mineral occurring in white sphenoidal crystals: $Ca_2B(AsO_4)(OH)_4$.

CAI Computer-aided (or -assisted) instruction.

CAI [meteorite] *calcium aluminum-rich inclusions.*

caichengyunite A white to grayish monoclinic mineral: $Fe^{2+}_3Al_2(SO_4)_6{\cdot}30H_2O$. The name has not been approved by the International Mineralogical Association.

caino- Equivalent to (and older spelling than) ceno-.

Cainophyticum (Cai-no-phyt'-i-cum) A paleobotanic division of geologic time, corresponding approximately to, and characterized by the plant life of, the Cenozoic. Cf: *Palaeophyticum; Mesophyticum.*

Cainozoic (Cai'-no-zo'-ic) *Cenozoic.*

cairn An artificial mound of rocks, stones, or masonry, usually conical or pyramidal, used in surveying to aid in the identification of a point or boundary.

cairngorm (cairn'-gorm) A type of *smoky quartz* from Cairngorm, a mountain southwest of Banff in Scotland. Syn: *Scotch topaz.* Also called: cairngorm stone.

caisson (cais'-son) A lined excavation extending below the groundwater table to facilitate dewatering. The term is sometimes applied to a relatively large-diameter drilled shaft foundation. Cf: *shaft [eng].*

cajon (ca-jon'[ca-hone']) (a) *box canyon.* (b) A defile leading up to a mountain pass; also, the pass itself. Etymol: Spanish "cajón", "large box". The term is used in the SW U.S.

cake [drill] *mud cake [drill].*

cake [ice] *ice cake.*

cake ice An accumulation of ice cakes.

caking coal (cak'-ing) Coal that softens and agglomerates when heated, and on quenching produces a hard gray cellular coke. Not all caking coals are good coking coals. Syn: *binding coal.*

caking index *agglutinating value.*

cal A term used in Cornwall, England, for iron *tungstate* (wolframite).

cala [coast] (ca'-la) A short, narrow *ria* formed in a limestone coast; a small semicircular shallow bay along a *cala coast,* as along the coast of Majorca. Etymol: Spanish, "cove, small bay, inlet". See also: *caleta [coast].*

cala [streams] A term applied in SW U.S. to a creek corresponding to a lateral stream of a main drainage. Etymol: Spanish, "creek". See also: *caleta [streams].*

Calabrian (Ca-lab′-ri-an) A European stage (Mediterranean): lowermost Pleistocene (above Gelasian, below Ionian). The base is at isotope stage 64. Syn: *Selinuntian.*

cala coast A coast formed by the submergence of many small valleys having steep slopes so that *calas,* separated by narrow peninsulas, are formed under the influence of breakers. Examples occur along several coasts of the Mediterranean Sea.

calaite (ca′-la-ite) *turquoise.*

calamine (cal′-a-mine) (a) A name used in the U.S. for *hemimorphite.* This name is disapproved by the Commission on New Minerals and Mineral Names of the International Mineralogical Association (Fleischer, 1966, p.1263). (b) A name frequently used in Great Britain for *smithsonite.* (c) *hemimorphite.* (d) A commercial, mining, and metallurgical term for the oxidized ores of zinc (including silicates and carbonates), as distinguished from the sulfide ores of zinc. Syn: *cadmia.*

calamistrum (cal-a-mis′-trum) A spinose comb or row of special bristles on the fourth (hind) metatarsi of certain spiders, used for drawing out a band of special silk from the *cribellum.* Pl: calamistra.

calanque (ca-lanque′) (a) A French term for a cove or small inlet. (b) And/or formed by collapse of the roof of a cave valley excavated in limestone and later submerged by a rise of sea level. Examples occur along the Mediterranean coast of France.

calaverite (cal-a-ver′-ite) A metallic yellowish white monoclinic mineral: $AuTe_2$. It often contains silver, and is an important source of gold.

calc-alkalic (calc-al′-ka-lic) (a) Said of a series of igneous rocks in which the weight percentage of silica is between 56 and 61 when the weight percentages of CaO and of $K_2O + Na_2O$ are equal. See also: *alkali-lime index.* (b) Said of an igneous rock containing plagioclase feldspar.

calc-alkaline magma series (calc-al′-ka-line) (a) Subalkaline basalts, andesites, dacites, and rhyolites that frequently (and perhaps exclusively) occur at convergent margins and show a trend of low iron enrichment with increasing silica content, as commonly discriminated from the tholeiitic magma series (also subalkalic) by Miyashiro's (1977) discriminant line on a plot of FeO/MgO vs. SiO_2. In specific occurrences, as in volcanoes or plutons, individual magmas in the series are often taken to be genetically related (e.g., by crystal fractionation, mixing, etc.). (b) A series as in (a) discriminated from calcic and alkaline magma series by relative amounts of CaO and alkalis, following Peacock (1931). This classification line is little used at present. Usage is contentious at present, as, in addition to the above, some authors use modal mineralogy (e.g., presence of hornblende) or equate hypersthene series (vs. pigeonite series) with calk-alkaline series. Cf: *tholeiitic magma series.*

calcarenaceous (cal-car′-e-na′-ceous) Said of a sandstone containing abundant sand-sized chemical and biochemical clasts such as intraclasts, pellets, and ooids. The calcareous components constitute up to 50% of the total clastic particles (Pettijohn, 1957, p.404-405).

calcarenite (cal-car′-e-nite) A limestone consisting predominantly (more than 50%) of sand-size carbonate grains; a consolidated calcareous sand. The term was introduced by Grabau (1903). Cf: *calcareous sandstone.*

calcareous (cal-car′-e-ous) Said of a substance that contains calcium carbonate. When applied to a rock name it implies that as much as 50% of the rock is calcium carbonate (Stokes and Varnes, 1955).

calcareous algae A group of algae that remove calcium carbonate from the shallow water in which they live and secrete or deposit it around the thallus as a more or less solid calcareous structure. See also: *coralline algae; dasycladacean.*

calcareous clay A clay containing a significant amount of calcium carbonate; specif. a *marl.*

calcareous crust An indurated soil horizon cemented with calcium carbonate; caliche. Syn: *calc-crust.*

calcareous dinoflagellate (calc-car′-e-ous) A *dinocyst* that has calcareous cell walls (Sarjeant, 1970).

calcareous dolomite A term used by Leighton and Pendexter (1962, p.54) for a carbonate rock containing 50-90% dolomite. Cf: *calcitic dolomite.*

calcareous nannoplankton Any of the chromatophore-bearing protists that normally produce coccoliths (calcium carbonate plates) during some phase in their life cycle; also, in a broader sense, the morphologically diverse group of minute calcareous skeletal elements produced by coccolithophores. Cf: *nannoplankton.*

calcareous ooze A deep-sea pelagic sediment containing at least 30% calcareous skeletal remains, such as pteropods, cocolithophorids, and foraminifera. Cf: *siliceous ooze.*

calcareous peat *eutrophic peat.*

calcareous rock A sedimentary rock containing an appreciable amount of calcium carbonate, such as limestone, chalk, travertine, or shelly sandstone. See also: *carbonate rock.*

calcareous sandstone (a) A sandstone cemented with calcite. (b) A sandstone containing abundant calcium carbonate, but in which clastic quartz is present in excess of 50% (Pettijohn, 1957, p.381). Cf: *calcarenite.*

calcareous shale A shale containing at least 20% calcium carbonate in the form of finely precipitated materials or small organically-fixed particles (Pettijohn, 1957, p.368-369).

calcareous sinter *travertine.*

calcareous soil Soil containing sufficient free $CaCO_3$ and other carbonates to effervesce visibly or audibly when treated with cold 0.1M HCl. These soils usually contain from 10 to almost 1000g kh^{-1} $CaCO_3$ equivalent.

calcareous spar *calcspar.*

calcareous tufa *tufa.*

calc-crust *calcareous crust.*

calcdolomite (calc-do′-lo-mite) *calcitic dolomite.*

calcedony (cal-ced′-o-ny) *chalcedony.*

calceoloid (cal′-ce-o-loid′) Said of a solitary *corallite* shaped like the tip of a pointed slipper (as in *Calceola*), with angulated edges between flattened and rounded sides.

calc-flinta (calc-flin′-ta) A fine-grained calc-silicate rock of flinty appearance formed by thermal metamorphism of a calcareous mudstone, possibly with some accompanying pneumatolytic action. Rarely used.

calcian dolomite (cal′-ci-an) A dolomite mineral that contains at least 8% calcium in excess of the ideal composition (Ca:Mg = 1:1 molar) of dolomite.

calciborite (cal-ci-bor′-ite) A white orthorhombic mineral: CaB_2O_4.

calcibreccia (cal-ci-brec′-cia) A limestone breccia, or a consolidated calcareous rubble; a *calcirudite* whose constituent particles are angular (Carozzi and Textoris, 1967, p.3).

calcic [geochem] (cal′-cic) Said of minerals and igneous rocks containing a relatively high proportion of calcium; the proportion required to warrant use of the term depends on circumstances.

calcic [geol] Obsolete term for a series of igneous rocks in which the weight percentage of silica is greater than 61 when the weight percentages of CaO and of $K_2O + Na_2O$ are equal. See also: *alkali-lime index.* Cf: *alkalic [petrology].*

calcic horizon A mineral soil horizon of secondary carbonate enrichment that is >15 cm thick, has a $CaCO_3$ equivalent of >150 gkg^{-1}, and has at least 50 gkg^{-1} more calcium carbonate equivalent than the underlying C horizon.

calciclase (cal′-ci-clase) *anorthite.*

calciclastic (cal-ci-clas′-tic) Pertaining to a clastic carbonate rock (Braunstein, 1961).

calcicole (cal′-ci-cole) A plant requiring a lime-rich, i.e. alkaline, soil. Cf: *calcifuge.* Syn: *calciphile.*

calcicrete (cal-ci-crete′) An obsolete syn. of *calcrete.*

Calcids Aridisols that have a calcic or petrocalcic horizon that has its upper boundary within 100 cm of the soil surface. A suborder in the U.S. system of soil taxonomy.

calciferous (cal-cif′-er-ous) In a stratigraphic sense, pertaining to a series of strata containing limestone, e.g. the Calciferous Sandstone series of Scotland (Challinor, 1978, p.39).

calcification [paleont] (cal′-ci-fi-ca′-tion) (a) Deposition of calcium salts in living tissue. (b) Replacement of organic material, generally original hard parts, by calcium carbonate ($CaCO_3$) in fossilization.

calcification [soil] A general term used for those processes of soil formation in which the surface soil is kept sufficiently supplied with calcium to keep the soil colloids nearly saturated with exchangeable calcium and thus render them relatively immobile and nearly neutral in reaction. The process is best expressed in *Borolls, Ustolls,* and other soils with calcic horizons.

calcifuge (cal′-ci-fuge) A plant surviving, but not thriving, on a lime-rich soil; it grows better on an acid soil. Cf: *calcicole.*

calcigravel (cal-ci-grav′-el) The unconsolidated equivalent of *calcirudite.* Obsolete.

calcilith (cal′-ci-lith) (a) A term suggested by Grabau (1924, p.298) for a limestone. (b) A sedimentary rock composed principally of the calcareous remains of organisms (Pettijohn, 1957, p.429). Obsolete.

calcilutite (cal-ci-lut′-ite) A limestone consisting predominantly (more than 50%) of detrital calcite particles of silt and/or clay size; a consolidated carbonate mud. The term was introduced by Grabau (1903). See also: *micritic limestone.* Cf: *calcisiltite.*

calcimixtite (cal-ci-mix′-tite) A term proposed by Schermerhorn (1966, p.835) for a *mixtite* that is dominantly calcareous.

calcimorphic (cal-ci-mor′-phic) Said of an intrazonal soil whose characteristics reflect the influence of the calcification process on its development. Examples of calcimorphic soils are Brown Forest and Rendzina soils.

calcination (cal-ci-na′-tion) The heating of a substance to its temperature of dissociation, e.g. of limestone to CaO and CO_2 or of gypsum to lose its water of crystallization.

calcioancylite-(Ce) (cal′-ci-o-an′-cy-lite) A variously colored orthorhombic mineral: $(Ca,Sr)(Ce,La)_3(CO_3)_4(OH)_3{\cdot}H_2O$. A member of the *ancylite* group.

calcioancylite-(Nd) A vitreous pale pink monoclinic mineral: $(Nd,Ce,Ca)_3(CO_3)_4(OH)_3{\cdot}H_2O$. A member of the *ancylite* group.

calcioandyrobertsite A vitreous blue monoclinic or orthorhombic mineral: $K(Ca,Cd)Cu_5(AsO_4)_4[As(OH)_2O_2]{\cdot}2H_2O$. The calcium analogue of andyrobertsite.

calcioaravaipaite A vitreous colorless or white triclinic mineral: $PbCa_2Al(F,OH)_9$. In spite of the name, the mineral is not an exact calcium analogue of aravaipaite, although the structures are similar.

calciobetafite (cal′-ci-o-bet′-a-fite) A reddish-brown cubic mineral: $Ca_2(Nb,Ti)_2(O,OH)_7$. A member of the *pyrochlore* group.

calcioburbankite A vitreous deep orange hexagonal mineral: $Na_3(Ca,Ce,Sr,La)_3(CO_3)_5$. A member of the *burbankite* group.

calciocarbonatite A chemically defined carbonatite in which wt% $CaO/(CaO+MgO+FeO+Fe_2O_3+MnO)>0.8$ (Woolley and Kempe, 1989).
calciocarnotite (cal'-ci-o-car'-no-tite) *tyuyamunite.*
calciocopiapite (cal'-ci-o-co-pi-a'-pite) A grayish or brownish yellow triclinic mineral of the *copiapite* group: $CaFe^{3+}_4(SO_4)_6(OH)_2{\cdot}20H_2O$.
calcioferrite (cal'-ci-o-fer'-rite) A variously colored monoclinic mineral: $Ca_4Mg(Fe^{3+},Al)_4(PO_4)_6(OH)_4{\cdot}12H_2O$. It is the iron analogue of *montgomeryite* and a member of the *montgomeryite* group.
calciohilairite (cal'-ci-o-hil-air'-ite) A white rhombohedral mineral: $CaZrSi_3O_9{\cdot}3H_2O$. A member of the *hilairite* group.
calcio-olivine (cal'-ci-o-ol'-i-vine) An orthorhombic phase of calcium orthosilicate: γ-Ca_2SiO_4. It is stable below 780° to 830°C, and is isomorphous with olivine. The term has also been applied to a highly calciferous variety of olivine, and to any of the polymorphs of Ca_2SiO_4. Cf: *larnite; bredigite.* Syn: *lime olivine.*
calciosamarskite A vitreous black metamict mineral: $(Ca,Fe,Y)(Nb,Ta,Ti)O_4$. A member of the *samarskite* group.
calciotantite (cal'-ci-o-tan'-tite) An adamantine colorless hexagonal mineral: $CaTa_4O_{11}$. A member of the *natrotantite* group. Syn: *ungursaite.*
calciouranoite (cal'-ci-o-u-ran'-o-ite) A yellow, orange, or brown metamict mineral: $(Ca,Ba,Pb)U_2O_7{\cdot}5H_2O$.
calciovolborthite (cal'-ci-o-vol'-borth-ite) A green, yellow, or gray orthorhombic mineral: $CaCu(OH)(VO_4)$. A dubious species; it may be *tangeite.*
calciphile (cal'-ci-phile) *calcicole.*
calciphyre (cal'-ci-phyre) A marble containing conspicuous crystals of calcium silicate or magnesium silicate or both. Rarely used.
calcirudite (cal-ci-rud'-ite) A limestone consisting predominantly (more than 50%) of detrital calcite particles larger than sand size (larger than 2 mm in diameter), and often also cemented with calcareous material; a consolidated calcareous gravel or rubble, or a limestone conglomerate or breccia. Some authors (e.g. Folk, 1968, p.162) use one millimeter as the lower limit. The term was introduced by Grabau (1903). Cf: *calcigravel; calcibreccia.*
calcisiltite (cal-ci-sil'-tite) A limestone consisting predominantly of detrital calcite particles of silt size; a consolidated calcareous silt. Cf: *calcilutite.*
calcisponge (cal-ci-sponge') Any sponge belonging to the class Calcispongea (=Calcarea) and characterized mainly by a skeleton composed of spicules of calcium carbonate. Range, Cambrian to Holocene.
calcitan A cutan composed of calcite.
calcite (cal'-cite) (a) A common rock-forming mineral: $CaCO_3$. It is trimorphous with aragonite and vaterite. Calcite is usually white, colorless, or pale shades of gray, yellow, and blue; it has perfect rhombohedral cleavage, a vitreous luster, and a hardness of 3 on the Mohs scale, and it readily effervesces in cold dilute hydrochloric acid. It is the principal constituent of limestone; calcite also occurs crystalline in marble, loose and earthy in chalk, spongy in tufa, and stalactitic in cave deposits. It is commonly found as a gangue mineral in many ore deposits and as a cementing medium in clastic sedimentary rocks; it is also a minor constituent of many igneous rocks and the chief constituent of some carbonatites. Calcite crystallizes in a variety of forms, such as nailhead spar, dogtooth spar, and Iceland spar. Symbol: Cc. Cf: *dolomite [mineral].* Syn: *calcspar.* (b) A group name for minerals with the general formula ACO_3, where A = Ca, Mg, Fe, Mn, Co, Ni, Zn or Cd.
calcite bubble A nonattached hollow sphere, usually of calcite, that has formed around a gas bubble on the surface of a cave pool. Syn: *cave bubble.*
calcite carbonatite A mineralogically defined carbonatite in which calcite is at least 90% of the modal carbonates.
calcite compensation depth (a) A depth in the ocean below which calcium carbonate particles are present in very low abundance in sediments even though such particles are formed and/or sink out of the overlying water column. (b) A depth in the ocean below which calcium carbonate dissolution rates equal or exceed calcium carbonate deposition rates. Varies from ocean to ocean, but is usually hundreds to thousands of meters below the sea surface. Abbrev: CCD. Cf: *aragonite compensation depth.* Syn: *calcium-carbonate compensation depth.*
calcite eye (a) One of the rounded bodies of clear calcite occurring sporadically in the radial zone and central area of foraminifers of the family Orbitolinidae. The term is usually used in the plural. (b) *bird's-eye.*
calcite raft A thin mineral film, usually of calcite, formed on the surface of a cave pool. Syn: *cave raft.*
calcitic dolomite (cal-cit'-ic) A dolomite rock in which calcite is conspicuous, but the mineral dolomite is more abundant; specif. a dolomite rock containing 10-50% calcite and 50-90% dolomite and having an approximate magnesium-carbonate equivalent of 22.7-41.0% (Pettijohn, 1957, p.418), or a dolomite rock whose Ca/Mg ratio ranges from 2.0 to 3.5 (Chilingar, 1957). Cf: *dolomitic limestone; calcareous dolomite.* Syn: *calcdolomite.*
calcitic limestone A limestone that consists essentially of calcite; specif. a limestone whose Ca/Mg ratio exceeds 105 (Chilingar, 1957).
calcitite (cal'-cit-ite) A term used by Kay (1951) for a rock composed of calcite; e.g., a limestone. Obsolete.
calcitization (cal'-cit-i-za'-tion) (a) The act or process of forming calcite, as by alteration of aragonite. (b) The alteration of existing rocks to limestone, due to the replacement of mineral particles by calcite; e.g., of dolomite in dolomite rocks or of feldspar and quartz in sandstones.
calcitostracum (cal-cit-os'-tra-cum) An internal layer of various mollusk shells, consisting chiefly of calcite. Cf: *nacre.*
calcium aluminum-rich inclusions Refractory inclusions in (especially carbonaceous) chondritic meteorites, so named because Al and Ca are the two most abundant cations. Typical phases are spinel, Ti-rich pyroxene, melilite, hibonite, and anorthite. Syn: *refractory inclusions.* Abbrev: CAI.
calcium bentonite A *bentonite* with a high calcium to magnesium ratio, imparting low- or non-swelling properties.
calcium-carbonate compensation depth (cal'-ci-um-car'-bon-ate) *calcite compensation depth; aragonite compensation depth.*
calcium-catapleiite (cal'-ci-um-cat'-a-plei'-ite) A pale yellow hexagonal mineral: $CaZrSi_3O_9{\cdot}H_2O$. Also called: calciocatapleiite. See also: *catapleiite.*
calcium feldspar A plagioclase feldspar rich in the An molecule $(Ca[Al_2Si_2O_8])$; specif. *anorthite.* See also: *lime feldspar.* Syn: *Ca-spar.*
calcium-larsenite (cal'-ci-um-lar'-sen-ite) *esperite.*
calcium mica *margarite [mineral].*
calcium-pharmacosiderite (cal'-ci-um-phar-ma-co-sid'-er-ite) A yellowish cubic mineral: $CaFe_4^{3+}(OH)_5(AsO_4)_3{\cdot}5H_2O$.
calcium phosphate A mineral name applied to naturally occurring *apatite.*
calcjarlite (calc-jar'-lite) A white monoclinic mineral: $Na(Ca,Sr)_3Al_3(F,OH)_{16}$. A member of the *jarlite* group.
calclacite (cal-cla'-cite) A white monoclinic mineral: $Ca(C_2H_3O_2)Cl{\cdot}5H_2O$. It is calcium chloride acetate found as a fibrous efflorescence on calcareous museum specimens.
calclithite (calc-lith'-ite) (a) A litharenite in which carbonates constitute the most abundant rock fragments (Folk, 1968, p.124); e.g. the Oakville Sandstone (Miocene) in S. Texas, derived largely from erosion of Cretaceous limestones. Not to be confused with a limestone consisting of intraclasts. Syn: *carbonate-arenite.* (b) A term originally defined by Folk (1959, p.36) as a limestone that was derived mainly from erosion of older, lithified limestones and that contains more than 50% carbonate rock fragments (extraclasts); e.g. a terrigenous rock formed as an alluvial fan in an area of intense tectonism or very dry climate.
calcolistolith (cal-co-lis'-to-lith) A limestone olistolith.
calcouranite (cal'-co-u-ran'-ite) *autunite.*
calcrete (cal-crete') (a) A term for a pedogenic calcareous soil, e.g., limestone consisting of surficial sand and gravel cemented into a hard mass by calcium carbonate precipitated from solution and redeposited through the agency of infiltrating waters, or deposited by the escape of carbon dioxide from vadose water. (b) A calcareous *duricrust;* caliche. Etymol: *cal*careous + con*crete.* Cf: *silcrete; ferricrete.* The term caliche has been widely used in the SW U.S. for calcrete. Obsolete syn: *calcicrete.*
calcretization The production of calcrete.
calc-sapropel (calc-sap'-ro-pel) Sapropel containing calcareous algae.
calc-schist A metamorphosed argillaceous limestone with a schistose structure produced by parallelism of platy minerals (Holmes, 1928, p.52).
calc-silicate marble A *marble* in which calcium- or magnesium-silicate minerals are conspicuous.
calc-silicate rock A metamorphic rock consisting mainly of calcium-bearing silicates such as diopside and wollastonite, and formed by metamorphism of impure limestone or dolomite. Syn: *lime-silicate rock.*
calc-sinter (calc-sin'-ter) *travertine.*
calcspar (calc'-spar) Crystalline *calcite.* Also spelled: calc-spar. Syn: *calcareous spar.*
calcsparite (calc-spar'-ite) A sparry calcite crystal, as distinguished from *dolosparite.* The term is synonymous with *sparite* when the latter is understood to mean the calcareous variety.
calc-tufa (calc-tu'-fa) *tufa.*
calculated vertical gradient *first vertical derivative.*
calcurmolite (calc-ur'-mo-lite) An ill-defined honey-yellow secondary mineral of approximate composition $Ca(UO_2)_3(MoO_4)_3(OH)_2{\cdot}11H_2O$.
calcybeborosilite-(Y) (cal'-cy-be-bor'-o-sil-ite-y) A colorless to greenish-gray monoclinic mineral: $(Ca,Y)(B,Be)(OH)(SiO_4)$. A member of the *datolite* group.
caldera (cal-der'-a) A large, basin-shaped volcanic depression, more or less circular or cirquelike in form, the diameter of which is many times greater than that of the included vent or vents, no matter what the steepness of the walls or form of the floor (Williams, 1941). It is formed by collapse during an eruption. Etymol: Spanish, "kettle." See also: *collapse caldera; erosion caldera; explosion caldera; cauldron [volc].*
caldera collapse breccia Broken rock derived from slumping of oversteepened walls during subsidence of a caldera. It includes *megabreccia (c)* and *mesobreccia,* which may be gradational (Lipman, 1984).
caldera complex The diverse rock assemblage underlying a caldera, comprising dikes, sills, stocks, and vent breccias; crater fills of lava; talus beds of tuff, cinder, and agglomerate; fault gouge and fault breccias; talus fans along fault escarpments; cinder cones; and other products formed in a *caldera.*
caldera lake A term that may be used for a *crater lake* in a caldera.
calderite (cal'-der-ite) The reddish cubic manganese-iron end member of the *garnet* group: $(Mn^{2+},Ca)_3(Fe^{3+},Al)_2(SiO_4)_3$.

calderonite A vitreous red-orange to red-brown monoclinic mineral: $Pb_2Fe^{3+}(VO_4)_2(OH)$. A member of the *brackebuschite* group.
caldron [coal] (cal'-dron) *caldron bottom.*
caldron [marine geol] A small steep-sided pot-shaped depression in the ocean floor. Also spelled: cauldron.
caldron bottom The mud cast of a fossil root or trunk of a tree or fern extending upward into the roof of a coal seam, and resembling the bottom of a caldron or pot; it is commonly surrounded by a film of coal, and may fall without warning. Cf: *pot bottom; bell; coal pipe.* Syn: *kettle bottom; caldron [coal].*
Caledonian Orogeny (Cal-e-do'-ni-an) A name commonly used for the early Paleozoic deformation in western Europe that created an orogenic belt, the *Caledonides,* extending from Ireland and Scotland northeastward through Scandinavia. The classical Caledonian Orogeny has been dated as Late Silurian and Early Devonian. It roughly correlates with the *Acadian Orogeny* of North America.
Caledonides (Cal'-e-do-nides') The orogenic belt, named by Suess, extending from Ireland and Scotland northeastward through Scandinavia, formed by the *Caledonian Orogeny.*
caledonite (cal'-e-do-nite') A green or bluish green orthorhombic mineral: $Pb_5Cu_2(SO_4)_3(CO_3)(OH)_6$. Not to be confused with *celadonite.*
caleta [coast] (ca-le'-ta) A small *cala;* a *cove* or *inlet.*
caleta [streams] The ultimate and smallest headwater ramification of a cala; a draw, drain, or coulee. Etymol: Spanish, "creek".
calf A small mass of *calved ice;* specif. a piece of ice that has risen to the surface after breaking loose from the submerged part of an *iceberg.*
calice (ca'-lice) The oral (upper or distal), generally bowl-shaped surface of a *corallite,* on which the basal disk of a *polyp* rests. Pl: calices. See also: *calyx.*
caliche [eco geol] (ca-li'-che) (a) Gravel, rock, soil, or alluvium cemented with soluble salts of sodium in the nitrate deposits of the Atacama Desert of northern Chile and Peru; it contains sodium nitrate (14-25%), potassium nitrate (2-3%), sodium iodate (up to 1%), sodium chloride, sodium sulfate, and sodium borate, mixed with brecciated clayey and sandy material in beds up to 2 m thick. It may form by leaching of bird guano, by bacterial fixation of nitrogen, by leaching from volcanic tuffs, or by drying up of former shallow lakes. (b) A term used in various geographic areas for: a thin layer of clayey soil capping a gold vein (Peru); whitish clay in the selvage of veins (Chile); feldspar, white clay, or a compact transition limestone (Mexico); a mineral vein recently discovered, or a bank composed of clay, sand, and gravel in placer mining (Colombia). The term has been extended by some authors to quartzite and kaolinite. Etymol: Spanish, flake of limestone.
caliche [soil] A syn. of *calcrete.* The term caliche has been applied broadly in SW U.S. (esp. Arizona) to a reddish-brown to buff or white calcareous material of secondary accumulation, commonly found in layers on or near the surface of stony soils of arid and semiarid regions, but also occurring as a subsoil deposit in subhumid climates. It is composed largely of crusts of soluble calcium salts in addition to such materials as gravel, sand, silt, and clay. It may occur as a thin porous friable horizon within the soil, but more commonly it is several centimeters to a meter or more in thickness, impermeable, and strongly indurated; the cementing material is essentially calcium carbonate, but it may include magnesium carbonate, silica, or gypsum. The term has also been used for the calcium-carbonate cement itself. Caliche appears to form by a variety of processes, e.g. capillary action, in which soil solutions rise to the surface and on evaporation deposit their salt content on or in the surface materials. It is called *hardpan,* calcareous *duricrust,* or *calcrete* in some localities, and *kankar* in parts of India. Syn: *soil caliche; calcareous crust; nari; sabach; tepetate.* Etymol: American Spanish, from a Spanish word for almost any porous material (such as gravel) cemented by calcium carbonate. "The Spanish word originally was used for a small stone or pebble accidentally burned with the clay mass when brick or tile was made, and it also was used for a crust of lime or similar material flaking from a wall" (Cottingham, 1951, p.162).
calichification (ca-li'-chi-fi-ca'-tion) The production process that produces caliche.
calico rock (cal'-i-co) (a) A term used in South Africa for *iron formation.* (b) A local term used in eastern Pennsylvania for a quarry rock of the Helderberg Limestone.
California bearing ratio A measure of the relative resistance of a soil to penetration under controlled conditions of density and moisture content. It is the ratio of the force per unit area required to penetrate a subgrade soil to that required for corresponding penetration of a standard material (crushed-rock base material) whose resistance under standardized conditions is well established. Abbrev: CBR. Cf: *bearing capacity.*
California onyx A dark, amber-colored or brown variety of *aragonite* used in ornamentation.
californite (cal-i-for'-nite) (a) A compact, massive, translucent to opaque variety of vesuvianite, typically dark-green, olive-green, or grass-green, usually mottled with white or gray, closely resembling jade, and used as an ornamental stone. Principal sources are in Fresno, Siskiyou, and Tulare counties in California. Syn: *American jade.* (b) A white variety of grossular garnet from Fresno County, Calif.
caliper log (cal'-i-per) A *well log* that shows the variations with depth in the diameter of an uncased borehole. It is produced by spring-activated arms that measure the varying widths of the hole as the device is drawn upward. Syn: *section-gage log.*
calk (a) *cawk.* (b) cauk. Also spelled: *caulk.*
calkinsite-(Ce) (cal'-kins-ite) A pale yellow orthorhombic mineral: $(Ce,La)_2(CO_3)_3{\cdot}4H_2O$.
callaghanite (cal'-la-ghan-ite') A blue monoclinic mineral: $Cu_2Mg_2(CO_3)(OH)_6{\cdot}2H_2O$.
callainite (cal-lai'-nite) A massive, waxlike, translucent, apple-green to emerald-green hydrated aluminum phosphate, possibly a mixture of wavellite and turquoise.
callais (cal'-la-is) An ancient name used by Pliny for a precious green or greenish-blue stone, probably turquoise. The name is still sometimes used for turquoise. Pl: callaides.
callose (cal'-lose) n. A carbohydrate component of cell walls in certain plants; e.g. the amorphous cell-wall substance that envelops the pollen mother cell during pollen-grain development and acts as a barrier between mother cells but that disappears as the ektexine structure is completed and impregnated with sporopollenin. adj. Having protuberant hardened spots; e.g. "callose leaves".
Callovian (Cal-lo'-vi-an) Stage of the Standard Global Chronostratigraphic Scale: uppermost Middle Jurassic (above Bathonian, below Oxfordian).
calluna peat (cal-lu'-na) Peat that is derived mainly from the common heather *Calluna vulgaris.* Syn: *heath peat.*
callus (cal'-lus) (a) The thickened inductura on the parietal region of a gastropod shell, or a growth of shelly material extending from the inner lip over the base of the shell and perhaps into the *umbilicus* in a gastropod. (b) Any excessive thickening of secondary shell (fibrous inner layer) located on the valve floor of a brachiopod and covering interior structures.
calm A Scottish term for a light-colored shale or mudstone, such as a baked shale or clay slate used for making pencils, or a shale that is easily cut with a knife.
calomel (cal'-o-mel) A tetragonal mineral: HgCl. It is a colorless, white, or faintly tinted tasteless salt, and is used as a cathartic, fungicide, and insecticide; when fused, it has a horny appearance. Syn: *horn quicksilver; horn mercury.*
calorific value (cal-o-rif'-ic) For solid fuels and liquid fuels of low volatility, the amount of heat produced by combustion of a specified quantity under specified conditions. See also: *net calorific value; gross calorific value.*
calotte (ca-lotte') The ring of surrounding rock in a tunnel which has become weakened or pressure-relieved by the excavation (Nelson and Nelson, 1967, p.57).
calthrops (cal'-throps) A tetraxonic sponge spicule in which the rays are equal or nearly equal in length. Pl: calthrops. See also: *candelabrum; cricocalthrops.*
caltonite (cal'-to-nite) A dark-colored analcime-bearing basanite that contains microphenocrysts of olivine and clinopyroxene in a trachytic groundmass composed of feldspar laths, clinopyroxene, iron oxides, and analcime. It was named by Johannsen (1931) for Calton Hill, Derbyshire, England. Obsolete.
calumetite (cal-u-met'-ite) An azure-blue mineral: $Cu(OH,Cl)_2{\cdot}2H_2O$.
calved ice A fragment or fragments of ice, floating in a body of water after *calving* from a larger ice mass. Cf: *calf.*
calving (calv'-ing) The breaking away of a mass of ice from a near-vertical *ice face* (as from an *ice shelf* or *iceberg*); the process of iceberg formation. Cf: *calved ice.*
calymma (ca-lym'-ma) A frothy layer of cytoplasm in radiolarians.
calymmate (ca-lym'-mate) In *obligate tetrads* or *polyads,* having the *ectexine/sexine* form a single continuous envelope around the unit. Cf: *calymma.* Ant: *acalymmate.*
Calymmian A period of the Standard Global Geochronometric Scale: oldest period of the *Mesoproterozoic* Era; characterized by expansion of *platform* covers ("calymma", "cover"). It is defined as the time span between 1,600 Ma, the end of the *Statherian* Period of the *Paleoproterozoic* Era, and 1,400 Ma, the beginning of the *Ectasian* Period (Plumb, 1991).
calyptra [palyn] (ca-lyp'-tra) A coating or covering of adherent, flocculent organic material surrounding a *cyst* (Williams et al., 2000, p.33, 34, 117).
calyptrolith (ca-lyp'-tro-lith) A basket-shaped *coccolith,* opening proximally.
calyx (ca'-lyx) (a) The skeletal cover of an echinoderm including the body and internal structures, but excluding the stem and appendages such as free arms or *brachioles;* e.g. the plated skeletal structure surrounding the viscera of a crinoid and comprising the aboral cup and *tegmen.* (b) A small cup-shaped structure or living cavity in which a coral *polyp* sits. See also: *calice.*
calzirtite (cal-zir'-tite) An adamantine light to dark brown tetragonal or orthorhombic mineral: $Ca_2Zr_5Ti_2O_{16}$.
cam A term used in the English Lake District for the crest of a mountain. Etymol: German "Kamm", "crest", ridge".
camarophorium (ca'-ma-ro-pho'-ri-um) A spoon-shaped, adductor-bearing platform in the *brachial valve* of brachiopods of the superfamily Stenoscismatacea, supported by the median septum, and derived independently of the *cardinalia* (TIP, 1965, pt. H, p.141).

camarostome (ca-ma′-ro-stome) A concave space found in certain arachnids, formed by a depression in the common wall of fused pedipalpal coxae and the convex rostrum fitting into it, and serving as a filter of liquefied food before it reaches the mouth (TIP, 1955, pt. P, p.61).

camber (cam′-ber) A superficial structure resembling an arch or ridge, caused by gravitational sagging toward topographically lower areas. Cf: *bulge.* Rarely used.

cambic horizon (cam′-bic) A mineral soil horizon that has a texture of loamy very fine sand or finer, has soil structure rather than rock structure, contains some weatherable minerals, and is characterized by the alteration or removal of mineral material as indicated by mottling or gray colors, stronger chromas or redder hues than in underlying horizons, or the removal of carbonates. The cambic horizon lacks cementation or induration and has too few evidences of illuviation to meet the requirements of the argillic or spodic horizon.

Cambids Aridisols that are not in cryic temperature regimes and do not have the following diagnostic subsurface horizons or features: argillic, salic, duripan, gypsic, petrogypsic, calcic, petrocalcic. A suborder in the U.S. system of soil taxonomy.

cambium (cam′-bi-um) In woody plants, a layer of persistent meristematic tissue beneath the bark, from which new bark and wood originate; it gives rise to secondary xylem, secondary phloem, and parenchyma (Swartz, 1971, p.82). See also: *cork cambium; vascular cambium.*

Cambrian (Cam′-bri-an) The lowest system of the *Paleozoic* Erathem of the Standard Global Geochronometric Scale. Also the time during which these rocks were formed, the Cambrian Period, covering the time span between 540 and 500 Ma. Named after "Cambria", the Roman name for Wales, where rocks of this age were first studied.

Cambrian evolutionary fauna The classes of marine invertebrates which account for the bulk of marine diversity during the Cambrian Period, includes trilobites, inarticulate brachiopods, and monoplacophoran molluscs.

camel back (cam′-el) A coal miner's term for a bell, potbottom, kettle bottom, or other rock mass that tends to fall easily from the roof of a coal seam. Syn: *tortoise.*

cameo (cam′-e-o) A carved gem that is actually a miniature bas-relief sculpture, commonly cut from materials of differently colored layers (esp. chalcedony onyx or a gastropod shell), the upper layer being used for the figure and the lower layer serving as the background. Cf: *intaglio.*

cameo mountain A mountain composed of elevated horizontal strata left behind by two or more subparallel streams carving out deep channels that eventually unite. Used rarely.

camera (cam′-er-a) The space enclosed within a cephalopod shell between two adjacent septa but excluding the *siphuncle.* It represents a portion of an earlier living space now closed off by a *septum.* Pl: camerae. Syn: *chamber.*

camera axis A line perpendicular to the focal plane of the camera and passing through the interior perspective center or emergent (rear) nodal point of the lens system.

cameral aperture The opening of a chamber. Cf: *camera.*

cameral deposit (cam′-er-al) A calcareous deposit secreted against the septa and/or the original walls of camerae during the life of a cephalopod; e.g. *mural deposit.*

camera obscura A simple monocular instrument for hand-copying or tracing a map or diagram onto a sheet of paper. It consists of a half-silvered mirror (or a prism or the optical equivalent) attached to the eyepiece of a microscope, thereby causing a virtual image of an external object to appear as if projected upon a plane. Also called: camera lucida. Etymol: Latin, "light chamber".

camera station The point in the air or on the ground occupied by the camera lens at the moment of exposure.

camerate [paleont] (cam′-er-ate) n. Any crinoid belonging to the subclass Camerata, characterized by rigidly united calyx plates forming a chamber. adj. Divided into chambers or forming a chamber.

camerate [palyn] adj. (a) In *acritarchs,* possessing a cavity between inner and outer wall layers. (b) Pertaining to a *dinoflagellate plate,* having five sides of which one side is parallel to the *cingulum* and two sides form a roof-shaped point away from the cingulum (Lister, 1970; Eisenack et al., 1973; Below, 1987; Williams et al., 2000, p.34).

cameronite (ca′-mer-on-ite) A metallic gray tetragonal mineral: $AgCu_7Te_{10}$.

camgasite A vitreous colorless monoclinic mineral: $CaMgAsO_4(OH)\cdot 5H_2O$.

caminite (ca′-min-ite) A colorless tetragonal mineral: $Mg_7(SO_4)_5(OH)_4\cdot H_2O$.

camouflage (cam′-ou-flage) In crystal structure, substitution of a trace element for a major element of the same valence, e.g. Ga for Al, and Hf for Zr. The trace element is then said to be camouflaged by the common element. Cf: *admittance [chem]; capture [chem].*

campagiform (cam-pag′-i-form) Said of the *loop,* or of the growth stage in the development of the loop, of a dallinid brachiopod, marked by a proportionally large hood without lateral lacunae, with the position of attachment of the descending branches to the *septum* and *hood* varying in different genera (TIP, 1965, pt. H, p.141). It precedes the *frenuliniform* stage.

campagna (cam-pa′-gna [cam-pahn′-ya]) An Italian term for a nearly level, open plain; esp. the undulating, uncultivated plain surrounding Rome.

Campanian (Cam-pa′-ni-an) A stage of the Standard Global Chronostratigraphic Scale: Upper Cretaceous (above Santonian, below Maastrichtian).

campanite (cam′-pa-nite) An extrusive rock originally described as a *tephrite* containing large leucite crystals, and later defined as a pseudoleucite-bearing extrusive equivalent of *nepheline syenite* (Johannsen, 1939, p.245); Tröger considered it the extrusive equivalent of essexite-foidite (Streckeisen, 1967, p.208). Obsolete.

Campbell's law (Camp′-bell's) The general law of migration of drainage divides, which states that the divide tends to migrate toward an axis of uplift or away from an axis of subsidence (Campbell, 1896, p.580-581). "Where two streams that head opposite to each other are affected by an uneven lengthwise tilting movement, that one whose declivity is increased cuts down vigorously and grows in length headward at the expense of the other" (Cotton, 1948, p.342). Named in honor of its originator, Marius R. Campbell (1858-1940), American geologist.

campigliaite (cam-pig′-li-a-ite) A blue monoclinic mineral: $Cu_4Mn^{2+}(SO_4)_2(OH)_6\cdot 4H_2O$.

camptonite (camp′-ton-ite) (a) In the *IUGS classification,* a *lamprophyre* in which plagioclase is more abundant than alkali feldspar, and total feldspar exceeds feldspathoids. (b) A *lamprophyre,* similar in composition to *nepheline diorite,* being composed essentially of plagioclase (usually labradorite) and brown hornblende (usually barkevikite). Its name, given by Rosenbusch in 1887, is derived from Campton, New Hampshire.

camptospessartite (camp-to-spes′-sart-ite) A dark-colored spessartite in which the pyroxene is titanian augite. Obsolete.

campylite (cam′-py-lite) A yellow or brown variety of mimetite sometimes crystallizing in barrel-shaped forms.

campylotropous (cam-py-lot′-ro-pous) Said of a plant ovule curved by uneven growth so that its axis is approximately at right angles to its funiculus. Cf: *amphitropous; anatropous; orthotropous.*

cañada (ca-ña′-da) (a) A term used in the western U.S. for a ravine, glen, or narrow valley, smaller and less steep-sided than a canyon, such as the V-shaped valley of a dry riverbed; a dale or open valley between mountains. (b) A term used in the western U.S. for a small stream; a creek. Etymol: Spanish "caña", "cane, reed".

canada balsam (can′-a-da) A natural cement used in mounting specimens for microscopic analysis; it is exuded as a viscous, yellow-green oleoresin by the balsam fir tree. Almost totally replaced by epoxy resins in current petrologic work. Syn: *canada turpentine.*

canada turpentine *canada balsam.*

Canadian (Ca-na′-di-an) (a) North American series: Lowermost Ordovician (above Croixian of Cambrian), below Whiterockian or Champlainian.

canadite (can′-a-dite) A *nepheline syenite* containing albite as the main feldspar, along with calcian and aluminian mafic minerals, so that the normative plagioclase is much more anorthite-rich than the modal plagioclase. Named by Quensel in 1913. Obsolete.

canal [coast] A long, narrow channel or arm of the sea connecting two larger stretches of water, usually extending far inland (sometimes between islands or between an island and the mainland), and approximately uniform in width; e.g., Lynn Canal in Alaska.

canal [paleont] A hollow vessel, tube, passage, channel, or groove of an invertebrate; e.g., a *ring canal* and *stone canal* of an echinoderm, a gutterlike extension of the lower end of a gastropod shell that carries the siphon, a tube running lengthwise along the walls of a foraminifer test, or a tube leading from an external pore of a sponge to the *spongocoel* and serving for water flow.

canal [speleo] A passage in a cave that is partly filled with water.

canal [streams] An artificial watercourse of relatively uniform dimensions, cut through an inland area, and designed for navigation, drainage, or irrigation by connecting two or more bodies of water; it is larger than a ditch.

canalarium (ca-na-lar′-i-um) A specialized sponge *spicule* lining a canal. Pl: canalaria.

canali (ca-na′-li) Channels on Venus generally a few kilometers wide and hundreds to thousands of kilometers in length. Canali are hypothesized to have been formed by large discharges of low-viscosity lava over prolonged periods (Bougher et al., 1997). Etymol: Italian. Pl. of canale.

canaliculate (ca-na-lic′-u-late) Grooved or channeled longitudinally; e.g. said of the sculpture of pollen and spores consisting of more or less parallel grooves, or said of a foraminifer possessing a series of fine tubular cavities or of the system of tiny, interconnected tubules characteristic of *Haversian* units in bone.

canalicules Extremely tiny channels extending from the central canal of a *helictite* to its outer walls.

canal system A system of passages connecting various cavities of an invertebrate body; e.g. the *aquiferous system* of a sponge.

canaphite (ca′-na-ph-ite) A colorless monoclinic mineral: $CaNa_2(P_2O_7)\cdot 4H_2O$.

canary (ca-na′-ry) A pale-yellow diamond.

canary stone A yellow variety of carnelian.

canasite (can′-a-site) A greenish yellow or grayish green monoclinic mineral: $Na_3K_3Ca_5Si_{12}O_{30}(OH,F)_4$.

Canastotan (Can-a-sto'-tan) Stage in New York State: Upper Silurian (lower Cayugan; below Murderian).
canavesite (ca-nav'-e-site) A white monoclinic mineral: $Mg_2(HBO_3)(CO_3)\cdot 5H_2O$.
cancalite A variety of *lamproite* named by Fuster et al. in 1967 for occurrences at Cancarix and Calasparra, Spain. Mitchell and Bergman (1991) suggest that this name should be replaced by orthopyroxene-sanidine-phlogopite *lamproite*. Obsolete.
cancellate (can-cel'-late) Having a honeycomblike structure, divided into small spaces by laminae, or marked with numerous crossing plates, bars, lines, threads, etc.; e.g. said of the shell surface of a mollusk marked by subequal, intersecting concentric and radial markings. Cancellous (cancellate) bone is the irregular network of struts characteristic of spongy (as opposed to compact) parts of the skeleton. Syn: *cancellous.* Etymol: Latin, "cancellus/cancelli", grid, lattice.
cancellous *cancellate.*
cancellus (can'-cel-lus) A cylindrical, interzooidal tubelike cavity developed in some cyclostome bryozoans (as in the family Lichenoporidae). Pl: cancelli. Syn: *alveolus.*
cancrinite (can'-cri-nite) (a) A variuosly colored hexagonal feldspathoid mineral: $Na_6Ca_2Al_6Si_6O_{24}(CO_3)_2\cdot 2H_2O$. (b) A group name for hexagonal or trigonal minerals with the above general formula, but with Na replaced by Ca or K; CO_3 replaced by SO_4, OH, Cl or S; and containing variable amounts of H_2O.
cancrinite monzosyenite In the *IUGS classification,* a variety of *foid monzosyenite* in which cancrinite is the most abundant feldspathoid.
cancrisilite A vitreous lilac hexagonal mineral: $Na_7(Si,Al)_{12}O_{24}(CO_3)\cdot 3H_2O$. Syn: carbonate-vishnevite. A member of the *cancrinite* group.
cand A term used in Cornwall, England, for fluorite occurring in a vein.
candela (can-de'-la) A term used in the SW U.S. for a candlelike rocky pinnacle. Etymol: Spanish, "candle".
candelabrum (can-de-la'-brum) A *calthrops* (sponge spicule) with multiply branched rays, the branches of one ray often differing from those of the others. Pl: candelabra.
candite (can'-dite) A blue spinel; *ceylonite.*
candle coal (can'-dle) A syn. of *cannel coal,* so named because it burns with a steady flame.
candle ice Disintegrating sea or lake ice consisting of ice prisms oriented perpendicular to the ice surface; a form of *rotten ice.* Syn: *needle ice.*
caneolith (can'-e-o-lith') A *heterococcolith* having a central area with laths, a simple or complex wall, and petaloid upper and lower rims. See also: *complete caneolith; incomplete caneolith.*
canfieldite (can'-field-ite) A black orthorhombic mineral: Ag_8SnS_6. It is isomorphous with argyrodite.
canga (can'-ga) A Brazilian term for a tough, well-consolidated, unstratified, iron-rich rock composed of varying amounts of fragments derived from *itabirite,* high-grade hematite, or other ferruginous material, and cemented by limonite (which may range from about 5% to more than 95%). It occurs as a near-surface or surficial deposit and blankets older rocks on or near the present or an ancient erosion surface, and is very resistant to erosion and chemical weathering. Park (1959, p.580) restricts the term to a rock formed by the cementation by hematite of *rubble ore* into "a hard ironstone conglomerate". Some writers use the term to refer to a ferruginous laterite developed from any iron-bearing rock, commonly basalt or gabbro; e.g. as used in Sierra Leone, "canga" is equivalent to "lateritic iron ore".
cannel (can'-nel) *cannel coal.*
cannel bass An English term for an inferior or impure carbonaceous shale approaching the character of an oil shale.
cannel coal A compact, tough *sapropelic coal* that consists dominantly of spores and pollen and is characterized by dull to waxy luster, conchoidal fracture, and massiveness. It is attrital and high in volatiles; by American standards it must contain less than 5% anthraxylon. Cf: *boghead coal; torbanite.* Syn: *candle coal; kennel coal; cannel; cannelite; parrot coal; curly cannel.* Adj: *canneloid.*
cannelite (can'-nel-ite) *cannel coal.*
canneloid (can'-nel-oid) Of or pertaining to *cannel coal.*
cannel shale A black shale or oil shale formed by the accumulation of sapropelic sediments, accompanied by an approximately equal amount of silt and clay. Syn: *bastard shale.*
cannilloite A hypothetical hyper-calcic end-member of the *amphibole* group: $CaCa_2(Mg,Al)_3(Si,Al)_8O_{22}(OH)_2$.
cannizzarite (can-niz-zar'-ite) A metallic silver gray monoclinic mineral: $Pb_{46}Bi_{54}S_{127}$.
cannonball (can'-non-ball) A large, dark concretion, as much as 3 m in diameter, resembling a cannonball, as in the Cannonball Member (Paleocene) of the Fort Union Formation in the Dakotas.
cannonite An adamantine colorless monoclinic mineral: $Bi_2O(SO_4)(OH)_2$.
canoe fold (ca-noe) A closely folded syncline, the surface expression of which is elongate. Rarely used.
cañon (ca'-ñon) An obsolete spelling of *canyon.*
canonical correlation A measure of the strength of the relationship between two sets of random variables in a multivariate extension of *regression analysis* called canonical analysis. Using methods related to factor analysis and principal components, linear combinations of one set of variables are estimated from linear combinations of another set of variables.
canopy (can'-o-py) In a cave, a *flowstone* deposit left hanging after clastic fill has been washed away from beneath it.
cant An inclination from a horizontal, vertical, or other given line; a slope or tilt.
cantalite (can'-ta-lite) An obsolete name for a glassy sodic *rhyolite.* The name is from Cantal, France.
cantonite (can'-ton-ite) A variety of covellite occurring in cubes and probably pseudomorphous after chalcopyrite that had replaced galena.
canyon [geomorph] (can'-yon) (a) A long, deep, relatively narrow steep-sided valley confined between lofty and precipitous walls in a plateau or mountainous area, often with a stream at the bottom; similar to, but larger than, a *gorge.* It is characteristic of an arid or semiarid area (such as western U.S.) where stream downcutting greatly exceeds weathering; e.g. Grand Canyon. (b) Any valley in a region where canyons abound. Etymol: Anglicized form of Spanish "cañón". Cf: *cañada.*
canyon [speleo] (a) In a cave, a passage that is much higher than wide, generally meandering, formed by downward incision of a vadose stream. (b) A chasm that has been formed by a cave stream. (c) A valley formed by collapse of the roof of a long, fairly straight cave; a karst valley.
canyon bench One of a series of relatively narrow, flat landforms occurring along a canyon wall and caused by differential erosion of alternating strong and weak horizontal strata. See also: *step [geomorph].*
canyon dune A dune formed in a *box canyon* (Stone, 1967, p.219).
canyon fill Unconsolidated material filling a canyon, consisting of sediment in transport and sediment permanently or temporarily deposited.
caoxite A vitreous colorless triclinic oxalate mineral, thought to be of hydrothermal origin: $CaC_2O_4\cdot 3H_2O$.
cap [geophys] A small explosive designed to be detonated by an electrical current, which in turn detonates another explosive. Seismic caps are designed to detonate with little uncertainty in their time of detonation.
cap [glaciol] *ice cap.*
cap [palyn] *cappa.*
capable fault (ca'-pa-ble) A fault defined by the Nuclear Regulatory Commission as one that is "capable" of "near future" movement; in general, a fault on which there has been movement within the last 35,000 years. The definition was developed for use in the siting of nuclear power plants. Cf: *active fault.*
capacitance (ca-pac'-i-tance) That property of an electrically nonconducting material which permits energy storage, due to electron displacement when opposite surfaces of the nonconductor are maintained at a difference of potential.
capacitive coupling (ca-pac'-i-tive) The capacity between two adjacent circuit elements. In induced-polarization surveys, capacitive coupling between a current wire and a potential wire, or between either wire and the ground, can lead to fictitious anomalies.
capacity (ca-pac'-i-ty) (a) The ability of a soil to hold water. (b) The yield of a pump, well, or reservoir. (c) The ability of a current of water or wind to transport detritus, as shown by the amount measured at a given point per unit of time. Capacity may vary according to the detrital grain size. Cf: *competence [hydraul].* See also: *efficiency.*
capacity curve In hydraulics, a graphic presentation of the relationship between the water-surface elevation of a reservoir and the volume of water below; also, a graphic presentation of the rate of discharge in a pipe or conduit or through porous material (ASCE, 1962). Syn: *storage curve.*
cape [coast] An extensive, somewhat rounded irregularity of land jutting out from the coast into a large body of water, either as a peninsula (e.g. Cape Cod, Mass.) or as a projecting *point* (e.g., Cape Hatteras, N.C.); a *promontory* or *headland,* generally more prominent than a point. Also, the part of the projection extending farthest into the water.
cape [mineral] *Cape diamond.*
Cape blue Crocidolite asbestos from South Africa.
Cape diamond A diamond having a yellowish tinge. Etymol: Cape of Good Hope. See also: *silver Cape.* Syn: *cape [mineral].*
cap effect *caprock effect.*
Cape ruby A misleading name applied to a ruby-colored garnet; specif. gem-quality *pyrope* that is perfectly transparent and ruby red in color, such as that obtained in the diamond mines of the Kimberley district of South Africa. Syn: *South African ruby.*
capgaronnite A submetallic black orthorhombic mineral: AgHg(Cl,Br,I)S.
capilla (cap-il'-la) A very fine radial ridge on the external surface of a brachiopod shell; there are usually more than 25 capillae in a width of 10 mm. Cf: *costa [paleont]; costella.*
capillarity (cap-il-lar'-i-ty) (a) The action by which a fluid, such as water, is drawn up (or depressed) in small interstices or tubes as a result of surface tension. Syn: *capillary action.* (b) The state of being capillary.
capillary [mineral] (cap'-il-lar'-y) Said of a mineral that forms hairlike or threadlike crystals, e.g. millerite. Syn: *filiform; moss [eco geol]; wire;* wiry.
capillary [water] adj. Said of tubes or interstices with such small openings that they can retain fluids by *capillarity.*
capillary action *capillarity.*

capillary attraction The adhesive force between a liquid and a solid in capillarity.

capillary barrier A contact in the *unsaturated zone* between an overlying geologic unit containing relatively small-diameter openings and an underlying unit containing relatively large-diameter openings and across which water does not flow (Montazer and Wilson, 1984, p.26-28).

capillary condensation The formation of rings of *pendular water* around point contacts of grains, and, when the rings around adjacent contacts become large enough to touch, of *funicular water* filling clusters of interstices surrounded by a single closed meniscus.

capillary conductivity The ability of an unsaturated soil or rock to transmit water or another liquid. As the larger interstices are partly occupied by air or other gas, the liquid must move through bodies surrounding point contacts of rock or soil particles. For water, the conductivity increases with the moisture content, from zero in a perfectly dry material to a maximum equal to the hydraulic conductivity, or *permeability coefficient*.

capillary ejecta *Pele's hair.*

capillary flow *capillary migration.*

capillary fringe The lower subdivision of the *unsaturated zone*, immediately above the water table, in which the interstices are filled with water under pressure less than that of the atmosphere, being continuous with the water below the water table but held above it by surface tension. Its upper boundary with the intermediate belt is indistinct, but is sometimes defined arbitrarily as the level at which 50 percent of the interstices are filled with water. Syn: *zone of capillarity; capillary-moisture zone.*

capillary head The *capillary potential* expressed as head of water.

capillary interstice An *interstice* small enough to hold water by surface tension at an appreciable height above a free water surface, yet large enough to prevent molecular attraction from extending across the entire opening. There are no definite size limitations (Meinzer, 1923, p.18). Cf: *subcapillary interstice; supercapillary interstice.*

capillary migration The movement of water by *capillarity*. Syn: *capillary flow; capillary movement.*

capillary-moisture zone (cap'-il-lar'-y-mois'-ture) *capillary fringe.*

capillary movement (a) The rise of water in the subsoil above the water table by capillarity (Nelson, 1965, p.66). (b) *capillary migration.*

capillary percolation *imbibition [water].*

capillary porosity The volume of interstices in a soil mass that hold water by capillarity (Jacks et al., 1960).

capillary potential A number representing the work required to move a unit mass of water from the soil to an arbitrary reference location and energy state (SSSA, 1965, p.348). Symbol: M. Cf: *capillary head.*

capillary pressure The difference in pressure across the interface between two immiscible fluid phases jointly occupying the interstices of a rock. It is due to the tension of the interfacial surface, and its value depends on the curvature of that surface.

capillary pyrites *millerite.*

capillary ripple *capillary wave.*

capillary rise The height above the free water level to which water will rise as a result of *capillarity*.

capillary water *fringe water.*

capillary wave A wave whose wavelength is shorter than 1.7 cm, and whose propagation velocity is controlled mainly by the surface tension of the liquid in which the wave is traveling. Cf: *gravity wave.* Syn: *capillary ripple; ripple.*

capillary yield The amount of *capillary water*, in mm per day or liters per second per hectare, that rises through a plane parallel to the *water table*, at a given distance below the land surface (Schieferdecker, 1959, term 0301).

Capitanian (Ca-pi-ta'-ni-an) A stage of the Standard Global Chronostratigraphic Scale: upper stage of the Guadalupian Series (middle Permian; above Wordian, below Wuchiapingian). The basal GSSP is in the Guadalupe Mountains, Texas, USA.

capitulum (ca-pit'-u-lum) (a) A part of the *carapace* of a cirripede crustacean, enclosing trophic structures and commonly armored by calcareous plates. (b) An obsolete syn. of *gnathosoma.* (c) The head of a vertebrate rib, i.e., the articular surface in contact with the vertebral centrum. Pl: capitula.

cappa (cap'-pa) The thick-walled proximal side of the body of a *saccate* pollen grain.

capped column A snow crystal in the shape of a hexagonal column with thin hexagonal plates or stars at each end.

capped deflection A sharp bend in the trend of a mountain range, in which the arcs meet approximately at right angles and the junction consists of a *cap range* of very high mountains (Wilson, 1950, p.155).

capped quartz A variety of quartz containing thin layers of clay.

cappelenite-Y (cap-pe'-len-ite-y) A pale yellowish to greenishbrown trigonal mineral: $BaY_6F_2[B_6Si_3O_{24}]$.

capping (cap'-ping) (a) A syn. of *overburden [eco geol]*, usually used for consolidated material. (b) A syn. of *gossan.*

cappula (cap'-pu-la) The thick-walled distal side of the body of a *saccate* pollen grain.

cap range A secondary mountain arc that curves around the junction of two primary arcs, as in a *capped deflection.*

capricorn (cap'-ri-corn) An ammonoid shell resembling a goat's horn, encircled by widely spaced blunt ribs and subequal rounded areas between adjacent ribs.

cap rock A shale or other impervious rock that acts as a barrier to the movement of oil or gas; it overlies a reservoir to form a trap. Also called: *roof rock.*

cap rock [coast] Estuarine sandstones along the Yorkshire coast of England (Nelson and Nelson, 1967, p.58).

cap rock [eco geol] (a) A syn. of *overburden [eco geol]*, usually used for consolidated material. (b) A hard rock layer, usually sandstone, overlying the shale above a coal bed. (c) A comparatively impermeable stratum immediately overlying an oil- or gas-bearing rock. Also spelled: caprock.

cap rock [tect] In a *salt dome*, an impervious body of anhydrite and gypsum, with minor calcite and sometimes with sulfur, that overlies the salt body, or plug. It probably results from accumulation of the less soluble minerals of the salt body during leaching in the course of its ascent.

caprock effect (cap'-rock) A sharp positive *gravity anomaly* superimposed on a broader negative, indicative of a salt dome. It is commonly produced by the dense caprock of the dome, but very shallow salt is denser than the surrounding sediments, so caprock is not essential in producing this effect. Syn: *cap effect.*

cap-rock waterfall A *waterfall* descending over a lip of strong and resistant rock.

captation The trapping of irregular sand masses against cliffs or mountains to form climbing or hanging dunes (Glenn, 1979, p.400). Very common in arid regions with relief and strong winds.

captor stream (cap'-tor) *capturing stream.*

capture [chem] (cap'-ture) In a crystal structure, the substitution of a trace element for a major element of lower valence, e.g. Ba^{++} for K^+. Captured trace elements generally have a higher concentration relative to the major element in the mineral than in the fluid from which it crystallized. Cf: *admittance [chem]; camouflage.*

capture [streams] The natural diversion of the headwaters of one stream into the channel of another stream having greater erosional activity and flowing at a lower level; esp. diversion effected by a stream eroding headward at a rapid rate so as to tap and lead off the waters of another stream. Cf: *abstraction [streams]; beheading; intercision.* Syn: river capture; *piracy; stream piracy; robbery; stream robbery.*

capture cross-section The atomic capture cross-section for neutrons is the effective area within which a neutron has to pass in order to be captured by an atomic nucleus. It is a probabilistic value dependent on the nature and energy of the particle as well as the nature of the capturing nucleus. Atomic capture cross-section is often measured in barns (1 barn = 10^{-28} m^2). The macroscopic capture cross-section is the effective cross-sectional area per unit volume for capture of neutrons; hence it depends on the number of atoms present as well as their atomic capture cross-sections. The unit of measure is $cm^2/cm^3 = 10^3$ cu, where cu = capture unit.

captured stream (cap'-tured) A stream whose former upper course has been diverted into the channel of another stream by capture. Syn: *pirated stream.*

capture theory The theory that holds that the Moon originated as an independent planet whose orbit around the Sun lay so close to the Earth that it strayed into the gravitational field of the Earth and was captured by it.

capture zone The area within an aquifer in which groundwater flows to a well under the influence of pumping conditions. The procedure is used in the capture and removal of aqueous-phase contaminants from an aquifer.

capturing stream (cap'-tur-ing) A stream into which the headwaters of another stream have been diverted by capture. Syn: *pirate; pirate stream; captor stream.*

caracolite (car-a-co'-lite) A colorless, gray or greenish hexagonal or monoclinic mineral: $Na_3Pb_2(SO_4)_3Cl$. It possesses the apatite structure type.

Caradocian (Ca-ra-do'-ci-an) Series or stage of the Standard Global Chronostratigraphic Scale: Upper Ordovician (above Llandeilian, below Ashgillian).

carapace (car'-a-pace) (a) A bony or chitinous case or shield covering the whole or part of the back of certain animals, such as the dorsal covering of the *cephalothorax* in arachnids or the dorsal bony shell of diverse vertebrates such as ankylosaurs, turtles, and glyptodonts. (b) The fossilized remains of an ostracode whose calcified cephalothorax covering has become divided laterally into two subsymmetrical parts (valves) joined at the dorsum. The term has also been applied to trilobites, eurypterids, and crustaceans other than ostracodes, but for these organisms the term *dorsal exoskeleton* is preferred.

carapace carina A narrow ridge variously located on the *carapace* of a decapod crustacean.

carapace groove A furrow, generally dorsal, on the *carapace* of a decapod crustacean.

carapace horn The anteriorly dorsal termination of the valves of the *carapace* of phyllocarid crustaceans.

carapace spine A sharp projection of the *carapace* of a decapod crustacean.

carat (car'-at) A unit of weight for diamonds and other gemstones. It formerly varied somewhat in different countries, but the *metric carat* or international carat equal to 0.2 gram, or 200 mg, was adopted in the U.S. in 1913 and is now standard in the principal countries of the world. Abbrev: ct. Cf: *point [gem]; grain [gem].* Not to be confused with *karat.*

caratiite (car'-a-tite) *piypite.*

carbankerite (carb-an'-ker-ite) Any coal microlithotype containing 20-60% by volume of carbonate minerals (calcite, siderite, dolomite, and ankerite) (ICCP, 1963).

carbapatite (carb-ap'-a-tite) *carbonate-apatite.*

carbargillite (carb-ar'-gil-lite) Any coal microlithotype containing 20-60% by volume of clay minerals and mica, and lesser proportions of quartz, all with particle size averaging 1 to 3 micrometers (ICCP, 1963). Partial syn: *carbominerite.*

carbene (car'-bene) An *asphaltene* that is insoluble in carbon tetrachloride, but soluble in carbon disulfide, benzene, or chloroform.

carbide (car'-bide) A mineral compound that is a combination of carbon with a metal. An example is cohenite, $(Fe,Ni,Co)_3C$.

carbite (car'-bite) A general term, now obsolete, applied to diamond and graphite.

carboborite (car-bo-bor'-ite) A colorless monoclinic mineral: $Ca_2Mg[B(OH)_4]_2(CO_3)_2{\cdot}4H_2O$.

carbocer (car'-bo-cer) A pitchy, ocherous, and carbonaceous mineral substance containing rare-earth elements.

carbocernaite (car-bo-cer'-na-ite) A variously colored orthorhombic mineral: $(Ca,Na)(Sr,Ce,Ba)(CO_3)_2$.

carbohumin (car-bo-hu'-min) *ulmin.*

carbohydrate (car-bo-hy'-drate) A polyhydroxy aldehyde or ketone, or a compound that can be hydrolyzed to such a form. Carbohydrates, of which sugars, starches, and cellulose are examples, are produced by all green plants and form an important animal food.

carboirite (car-boir'-ite) A paleo to dark green triclinic mineral: $Fe^{2+}Al_2(Ge,Si)O_5(OH)_2$. It is found enclosed in sphalerite and associated with Ge-bearing quartz. There is limited solid solution with chloritoid to which it is structurally related.

carbokentbrooksite A vitreous yellow rhombohedral mineral of the *eudialyte* group: $(Na,[])_{12}(Na,Ce)_3Ca_6Mn_3Zr_3NbSi_{25}O_{73}(OH)_3(CO_3)DH_2O$.

carbominerite (car-bo-min'-e-rite) A microlithotype of coal containing different minerals in association with macerals (ICCP, 1971). Cf: *carbopyrite; carbankerite.* Partial syn: *carbargillite.*

carbon (car'-bon) (a) A nonmetallic, chiefly tetravalent chemical element (atomic number 6; atomic weight 12.01115) occurring native in the crystalline form (as diamond and as graphite) or amorphous, and forming a constituent of coal, petroleum, and asphalt, of limestone and other carbonates, and of all organic compounds. Symbol: C. (b) *carbonado.*

carbon-14 A heavy radioactive isotope of carbon having a mass number of 14 and a half-life of 5,730±40 years (Godwin, 1962). (The figure 5,568±30 is also used.) It is produced in nature by the reaction of atmospheric nitrogen with neutrons produced by cosmic-ray collisions and artificially by atmospheric nuclear explosions. Carbon-14 is useful in dating and tracer studies of materials directly or indirectly involved with the Earth's carbon cycle during the last 50,000 years. Symbol: ^{14}C. Partial syn: radiocarbon. Popularly referred to as *carbon clock.*

carbon-14 age An *isotopic age* expressed in years and calculated from the quantitative determination of the amount of carbon-14 remaining in an organic material. Syn: *radiocarbon age.*

carbon-14 dating A method of determining an age in years by measuring the concentration of carbon-14 remaining in an organic material, usually formerly living matter, but also dissolved bicarbonate, etc. The method, worked out by Willard F. Libby, U.S. chemist, in 1946-1951, is based on the assumption that assimilation of carbon-14 ceases abruptly upon removal of the material from the Earth's carbon cycle (i.e., on the death of an organism) and that it thereafter remains a closed system. Most carbon-14 ages are calculated using a half-life of 5,730 ± 40 years or 5,568 ± 30 years. Thus the method is useful in determining ages in the range of 500 to 30,000 or 40,000 years, although it may be extended to 70,000 years by using special techniques involving controlled enrichment of the sample in carbon-14. Syn: *radiocarbon dating; carbon dating.*

carbonaceous (car-bo-na'-ceous) (a) Said of a rock or sediment that is rich in carbon; coaly. (b) Said of a sediment containing organic matter.

carbonaceous chondrite A *chondrite* characterized by high refractory-lithophile-element abundances and low-to-moderate contents of carbonaceous matter. Many carbonaceous chondrites show evidence of aqueous alteration such as the presence of phyllosilicates. Carbonaceous chondrites are commonly subdivided into CI, CK, CM, CO, CR, and CV groups based on mineralogy, elemental abundances, relative sizes and abundances of chondrules, and oxygen isotopic signatures.

carbonaceous coal Coal that is intermediate in composition between metabituminous coal and anthracite (Tomkeieff, 1954, p.35).

carbonaceous rock A sedimentary rock that consists of, or contains an appreciable amount of, original or subsequently introduced organic material, including plant and animal residues and organic derivatives greatly altered (carbonized or bituminized) from the original remains; e.g. the coal series, black shale, asphaltic sediments, sapropel, certain clays, various solid substances derived from altered plant remains, and esp. *carbonaceous shale.* Syn: *carbonolite.*

carbonaceous shale A dark-gray or black shale with abundant carbon in the form of small disseminated particles or flakes; it is commonly associated with coal seams.

carbonado (car-bo-na'-do) An impure opaque dark aggregate composed of minute diamond particles, forming a usually rounded mass with a granular to compact structure, and displaying a superior toughness as a result of its cryptocrystalline character and lack of cleavage planes. It is used as an *industrial diamond.* Cf: *bort; ballas.* Syn: *black diamond [mineral]; carbon diamond; carbon.*

carbonate [mineral] (car'-bon-ate) A mineral compound characterized by a fundamental anionic structure of CO_3^{-2}. Calcite and aragonite, $CaCO_3$, are examples of carbonates. Cf: *borate; nitrate.*

carbonate [sed] A sediment formed by the biotic or abiotic precipitation from aqueous solution of carbonates of calcium, magnesium, or iron; e.g. limestone and dolomite. See also: *carbonate rock.*

carbonate-apatite (car'-bon-ate-ap'-a-tite) An apatite mineral containing a considerable amount of carbonate. Cf: *collophane; dahllite; francolite.* Syn: *carbapatite; podolite; tavistockite.*

carbonate-arenite (car'-bon-ate-ar'-e-nite) *calclithite.*

carbonate buildup A carbonate rock mass that is thicker than, and different from, laterally equivalent strata, and probably stood above the surrounding sea floor during some or all of its depositional history. Cf: *reef; bank [sed]; submarine bank.* Syn: *limestone buildup.*

carbonate-cyanotrichite (car'-bon-ate-cy'-a-no-trich'-ite) A pale-blue mineral: $Cu_4(OH)_{12}Al_2(CO_3,SO_4){\cdot}2H_2O$. A carbonate-dominant analogue of cyanotrichite.

carbonate cycle The biogeochemical pathways of carbonate, involving transformation from or to CO_2 and HCO_3^-, and its solution, deposition in minerals, and metabolism and regeneration in biological fixation.

carbonated spring (car'-bon-at'-ed) A spring whose water contains carbon dioxide gas. This type of spring is especially common in volcanic areas (Comstock, 1878, p.34).

carbonate-facies iron formation (car'-bon-ate-fa'-cies) An *iron formation* characterized by alternating laminae of chert and iron-rich carbonate minerals (James, 1954, p.251-256).

carbonate-fluorapatite (car'-bon-ate-flu'-or-ap'-a-tite) A colorless mineral of the apatite group: $Ca_5(F,OH)(PO_4,CO_3)_3$. Syn: *francolite; kurskite; staffelite.* Discredited syn: *dehrnite; lewistonite.*

carbonate hardness Hardness of water, expressed as $CaCO_3$, that is equivalent to the carbonate and bicarbonate alkalinity. When the total alkalinity expressed as $CaCO_3$ equals or exceeds the total hardness, all the hardness is carbonate. It can be removed by boiling and hence is sometimes called *temporary hardness,* although this synonym is becoming obsolete. Cf: *noncarbonate hardness; hardness [water].*

carbonate-hydroxylapatite A variously colored hexagonal mineral of the *apatite* group: $Ca_5(PO_4,CO_3)_3(OH,F,O)$. Syn: *dahllite.*

carbonate platform A general term for the large-scale depositional environment of shallow-water limestones. The common types are shelves, ramps, and epeiric platforms.

carbonate ramp A gently sloping surface (a few meters/km) from shallow to deeper water, underlain largely by redeposited shallow-water carbonate sediments (*cascajo, grainstone.*) Carbonate ramps occur along coasts in most tectonic settings (J.F. Read, 1981).

carbonate rock A rock consisting chiefly of carbonate minerals, such as limestone, dolomite, or carbonatite; specif. a sedimentary rock composed of more than 50% by weight of carbonate minerals. See also: *calcareous rock; carbonate [sed].*

carbonate shelf A type of carbonate platform where many limestones were deposited, bordered by a shelf margin and relatively steep slope into deep water.

carbonate thermometer The temperature-dependent oxygen-18/oxygen-16 isotope ratio in the carbonate shells of fossil marine animals, as used to indicate the water temperature that existed at the time the shell was deposited (i.e. the paleotemperature). Accurate determination depends on the isotopic composition of the shell being in equilibrium with the surrounding water at the time of deposition, on knowledge of the isotopic composition of the water, and on the material being preserved without further isotopic fractionation or substitution. See also: *oxygen-isotope fractionation.*

carbonatite (car-bon'-a-tite) An igneous rock composed of at least 50% carbonate minerals (Bell, 1989).

carbonatization (car-bon'-a-ti-za'-tion) Introduction of, or replacement by, carbonates.

carbon clock *radiocarbon clock.*

carbon cycle The continued exchange and reactions of carbon in the biosphere, atmosphere, and hydrosphere (Pettijohn, 1949, p.363).

carbon dating *carbon-14 dating.*

carbon diamond *carbonado.*

carbon fixation The process by which organisms (usually photosynthetic) convert components of the carbonate system, e.g. carbon dioxide or bicarbonate, to organic matter. Occurs most frequently in the upper sunlit portion of the ocean (euphotic zone) and is accompanied by the consumption of dissolved inorganic nutrients such as nitrate, ammonia, phosphate, and silicate and by the production of dissolved oxygen. Dissolved trace metals such as zinc, cadmium, and iron may be consumed along with inorganic nutrients.

carbon-hydrogen ratio (car'-bon-hy'-dro-gen) The ratio of carbon to hydrogen in coal and peat. It is the basis for a method of coal classification. Abbrev: C/H ratio.

Carboniferous (Car-bon-if'-er-ous) A system of the late *Paleozoic* Erathem of the Standard Global Chronostratigraphic Scale, above the *Devonian* and below the *Permian.* Also the time during which these rocks were formed, the Carboniferous Period, covering the time span between 355 and 295 Ma. Named after its coal-rich deposits in Europe and North America. The system is subdivided into two subsystems: the Mississippian (older) and the Pennsylvanian (younger).

carbonification (car-bon'-if-i-ca'-tion) A syn. of *coalification,* suggested for standard use by the International Committee for Coal Petrology (ICCP, 1963).

carbonite *natural coke.*

carbonization (car'-bon-i-za'-tion) (a) In the process of *coalification,* the accumulation of residual carbon by the changes in organic matter and decomposition products. (b) The accumulation of carbon by the slow, underwater decay of organic matter. (c) The conversion into carbon of a carbonaceous substance such as coal by driving off the other components, either by heat under laboratory conditions or by natural processes.

carbonization [paleont] Process of fossilization wherein the original organic material of an organism has been reduced to a film of carbon compounds.

carbonolite (car-bon'-o-lite) *carbonaceous rock.*

carbonolith (car-bon'-o-lith) A term suggested by Grabau (1924, p.298) for a carbonaceous sedimentary rock.

carbon ratio [coal] (a) The percentage of fixed carbon in a coal or peat. (b) The ratio of fixed carbon in a coal or peat to the fixed carbon plus the volatile hydrocarbons. Syn: *fixed-carbon ratio.* Cf: *fuel ratio.*

carbon ratio [geochron] The ratio of the most common carbon isotope, carbon-12, which is nonradioactive, to either of the less common isotopes, carbon-13 (nonradioactive) or carbon-14 (radioactive), or the reciprocal of one of these ratios. If unspecified, the term generally refers to the ratio (carbon-13/carbon-12).

carbon-ratio theory (car'-bon-ra'-tio) The hypothesis that in any region, the specific gravity of oil varies inversely with the carbon ratio of the associated coals. As the percentage of fixed carbon in the coal increases as a result of metamorphism, the oil becomes lighter, i.e., higher in volatile hydrocarbons. The theory was most effectively stated by David White in 1915.

carbon spot (a) A misnomer referring to any apparently black inclusion or imperfection in a diamond. Under dark-field illumination, most "carbon spots" are found to be neither black nor graphite. (b) A black flecklike or flakelike graphite inclusion in the body of a diamond crystal (Liddicoat et al., 1993).

carbon trash Carbon remains of plant life found in sedimentary rocks and often associated with uranium and *red-bed copper* mineralization. See also: *tree ore.*

carbopyrite (car-bo-py'-rite) Any coal microlithotype containing 5-20% by volume of iron disulfide (pyrite and marcasite) (ICCP, 1963).

Carborundum (Car-bo-run'-dum) Trade name for a synthetic substance (silicon carbide) used as an abrasive and as a refractory material. It corresponds to the mineral moissanite.

carbuncle (car'-bun-cle) (a) A cabochon-cut red garnet. (b) An old name, now obsolete, for any of several precious stones of a fiery red color, such as ruby or spinel.

carbunculus (car-bun'-cu-lus) A term applied to ruby, ruby spinel, almandine garnet, and pyrope.

carburan (car'-bu-ran) A pitchy hydrocarbon containing uranium.

cardella (car-del'-la) A *condyle* in the orifice of an autozooid in ascophoran cheilostome bryozoans. Pl: cardellae. Syn: cardelle.

cardhouse structure A structure found in certain marine sediments, consisting of platy aggregates of clay minerals arranged in edge-to-face fashion and resembling a "house" of playing cards (Lambe, 1953, p.38). Cf: *bookhouse structure.*

cardiac lobe (car'-di-ac) The median lobe of the *prosoma* and the *opisthosoma* in merostomes.

cardiac region The median part of the *carapace* of some decapod crustaceans, behind the *cervical groove* or *suture,* between the urogastric and intestinal areas (TIP, 1969, pt. R, p.92).

cardinal (car'-di-nal) adj. Pertaining to the hinge of a bivalve shell; e.g., "cardinal extremity" (termination of posterior hinge margin of a brachiopod shell). n. A cardinal part; e.g., a cardinal tooth.

cardinal angle The angle formed at each of the extremities of the hinge of a bivalved shell; e.g. the angle formed between the hinge line and the anterior or posterior free margin of an ostracode valve, or between the hinge line and the posterolateral margins of a brachiopod shell.

cardinal area (a) A flat or slightly concave, commonly triangular surface extending between the beak and the hinge margin in many bivalve mollusks, and partly or wholly occupied by ligament. It is set off from the remainder of the shell by a sharp angle. (b) The flattened, posterior sector of a valve of a rhynchonelliform (formerly articulate) brachiopod, exclusive of the *delthyrium* or the *notothyrium.* It may be the *interarea,* a *planarea,* or the *palintrope.* See also: *hinge area.*

cardinal axis The *hinge axis* in a bivalve mollusk.

cardinal fossula A *fossula* developed in the position of the *cardinal septum* in a rugose coral. It is most commonly due to abortion of the cardinal septum.

cardinalia (car-di-nal'-i-a) A collective term for the varied internal outgrowths and structures of secondary shell located in the posterior and median region of the *brachial valve* near the *beak* of a brachiopod, and associated with articulation, support of the *lophophore,* and muscle attachment. It may include, for example, the *cardinal processes, socket ridges, crural bases,* and *hinge plates.*

cardinal margin The curved posterior margin of a brachiopod shell along which the valves are hinged, homologous with the hinge line of *strophic* shells but not parallel to the *hinge axis* (TIP, 1965, pt. H, p.141). See also: *hinge [paleont].*

cardinal muscle scar A posteriorly and laterally placed muscle scar in certain brachiopods (as in the superfamilies Acrotretacea and Obolellacea).

cardinal plate A plate extending across the posterior end of the *brachial valve* of a brachiopod, consisting laterally of outer hinge plates and medially of either conjunct inner hinge plates or a single plate, and commonly perforated posteriorly (TIP, 1965, pt. H, p.141).

cardinal platform The *hinge plate* in a bivalve mollusk.

cardinal point One of the four principal points of the compass (viz: the north, south, east, or west points) that lie in the direction of the Earth's two poles and of sunrise and sunset and that indicate the four principal astronomic directions on the Earth's surface, spaced at 90-degree intervals.

cardinal process A blade or variably shaped boss, ridge, or projection of secondary shell of a brachiopod, situated medially in the posterior end of the *brachial valve,* and serving for separation or attachment of *diductor muscles.*

cardinal septum The *protoseptum* lying in the plane of bilateral symmetry of a rugose *corallite,* distinguished from other protosepta by pinnate insertion of newly formed metasepta adjacent to it on both sides. Symbol: C. Cf: *alar septum; counter septum.*

cardinal tooth A *hinge tooth,* often relatively large, situated close to and directly beneath the *beak* of a bivalve mollusk. Its long axis is perpendicular or oblique to the hinge line. Cf: *lateral tooth.*

cardiophthalmic region (car'-di-oph-thal'-mic) The space between the ophthalmic ridges in merostomes.

Cardium clay (Car'-di-um) A Pleistocene glacial clay of northern Europe, characterized by fossil shells of the genus Cardium, a marine bivalve mollusk of the family Cardiidae.

caresite-3T A vitreous green-brown trigonal mineral: $Fe_4Al_2(OH)_{12}CO_3{\cdot}3H_2O$. A member of the *quintinite* group.

carex peat (car'-ex) Peat that is derived mainly from Carex, a genus of sedges of the family Cyperaceae. Cf: *eriophorum peat.* Syn: *sedge peat.*

cargneule (car-gneule') A French term for a porous or cavernous carbonate sedimentary rock (esp. a cellular dolomite), its cavities filled with soft, friable, evaporitic material that easily dissolves or falls out, leaving a rough, corroded surface. Typically associated with thrust faults. Syn: *cornieule.*

Cariboo Orogeny (Car'-i-boo) A name proposed by W.H. White (1959) for an orogeny that is believed to have occurred during early Paleozoic time in the Cordillera of British Columbia, especially in the Selkirk and Omineca mountains, where Permian strata overlie deformed and metamorphosed Proterozoic and lower Paleozoic rocks.

caridoid (car'-i-doid) adj. Said of a decapod crustacean of the infraorder Caridea containing most shrimps, prawns, and related forms; e.g. "caridoid facies", an aspect of primitive Eumalacostraca distinguished by enclosure of thorax by carapace, movably stalked eyes, biramous antennules, scaphocerite-bearing antennae, thoracopods with natatory exopods, ventrally flexed and powerfully muscled elongate abdomen, and caudal fan. n. A caridoid crustacean.

caries texture (car'-ies) In ore microscopy, a replacement pattern in which the younger mineral forms a series of scallop-shaped incursions into the host mineral, which resemble filled dental cavities. Syn: *cusp-and-caries texture.*

carina (ca-ri'-na) (a) An unpaired compartmental plate adjacent to the terga of a cirripede crustacean. (b) A flangelike elevation on the side of a septum of a rugose coral, formed by thickened trabecula. See also: *yardarm carina; zigzag carina.* (c) A keel-shaped structure or a flange going around the edge of some foraminiferal tests. (d) A prominent keel-like ridge on the exterior shell of a mollusk; e.g. an extended, somewhat angular linear elevation on the exterior of a whorl at the edge of a gastropod shell. (e) The central and denticulated, nodose, or smooth ridge extending down the middle of the platform or blade of a conodont. (f) A major angular elevation on the surface of a brachiopod valve, externally convex in transverse profile and radial from the umbo. Cf: *fold [paleont].* (g) A median ridge or keel-like structure on the frontal side of a bryozoan branch, chiefly in Cryptostomata. Syn: *keel.* (h) A frill-like or ridgelike structure more or less parallel to the free edge in an ostracode valve, usually situated above the velar structure; present in many genera of the order Palaeocopida. (i) The ventrally projecting, keel-like ridge on the midline of the breast bone (sternum) of carinate birds (as opposed to ratite birds which lack the carina). Adj: carinate. Etymol: Latin, keel. Pl: carinae.

carinal (ca-ri′-nal) adj. Pertaining to a carina; e.g. a "carinal band" representing an imperforate marginal area between the carinae of a foraminiferal test. n. One of a series of ossicles along the midline of the aboral surface of an asterozoan ray.

carinate [paleont] Having a *carina* or shaped like a carina (esp. related to carina(i)).

carinate [struc geol] (ca′-ri-nate) (a) Said of a fold that is almost isoclinal. (b) Said of an *antiform* or *synform* that is confined to incompetent strata, so that adjacent strata are undisturbed. Rarely used.

carinolateral (ca′-ri-no-lat′-er-al) A compartmental plate in certain cirripede crustaceans, located on either side of the carina. Syn: carinal latus.

Carixian (Ca-rix′-i-an) A European substage: lower substage of the Pliensbachian Stage, Lower Jurassic (above Lotharingian substage of Sinemurian, below Domerian).

carletonite (carle′-ton-ite) A pink to blue tetragonal mineral: $KNa_4Ca_4Si_8O_{18}(CO_3)_4(F,OH){\cdot}H_2O$.

carlfriesite (carl-fries′-ite) A yellow monoclinic mineral: $CaTe^{4+}{}_2Te^{6+}O_8$.

carlhintzeite (carl-hint′-ze-ite) A colorless triclinic mineral: $Ca_2AlF_7{\cdot}H_2O$.

carlinite (car′-lin-ite) A very soft gray rhombohedral mineral: Tl_2S.

Carlin-type gold (Car′-lin) Gold occurring as microscopic particles (up to 30 micrometers) that must be identified by chemical analysis as it is not recoverable by panning. The term is taken from its occurrence at Carlin, Nevada. Syn: *invisible gold.*

carloruizite A colorless to pale yellow trigonal mineral: $K_3(Na,K)_2Na_3Mg_5(IO_3)_6(SeO_4,SO_4,CrO_4)_6{\cdot}6H_2O$. The selenium analogue of *fuenzalidaite.*

carlosturanite (car′-lo-stur′-an-ite) A light-brown monoclinic mineral: $(Mg,Fe^{2+},Ti)_{21}(Si,Al)_{12}O_{28}(OH)_{34}{\cdot}H_2O$.

Carlsbad B twin law (Carls′-bad) A twin law that is now equated with the *x-pericline twin law.*

Carlsbad twin law A twin law in feldspar, especially orthoclase, that defines a penetration twin in which the twin axis is the *c* crystallographic axis and the composition surface is irregular. Also spelled: *Karlsbad twin law.*

carlsbergite (carls′-berg-ite) A cubic meteorite mineral: CrN. The chromium analogue of osbornite.

Carman-Kozeny equation An empirical relationship for instrinsic permeability as functions of *porosity* and mean grain size, primarily.

carmeloite (car-mel′-o-ite) A basalt or an *andesite,* depending on whether the plagioclase is andesine or labradorite, that contains iddingsite as an alteration product of olivine phenocrysts. Its name, given by Lawson in 1893, is derived from Carmel Bay, California. Obsolte.

carmichaelite A metallic black monoclinic mineral: $(Ti,Cr,Fe)(O,OH)_2$.

carminate element A pectiniform conodont element with two primary processes, which are anterior and posterior, and a longitudinal axis that is straight, or essentially so, in lateral view (TIP, 1981, pt. W, supp.2).

carminiplanate element A carminate conodont element with lateral platform extensions and the attachment surface characteristic of planate pectiniform elements (TIP, 1981, pt. W, supp.2). See also: *carminate element; planate element.*

carminiscaphate element A carminate conodont element with lateral platform extensions and the attachment surface characteristic of scaphate pectiniform elements (TIP, 1981, pt. W, supp.2). See also: *carminate element; scaphate element.*

carminite (car′-min-ite) A carmine to tile-red orthorhombic mineral: $PbFe^{3+}{}_2(AsO_4)_2(OH)_2$.

carnallite (car′-nall-ite) A variously colored orthorhombic mineral: $KMgCl_3{\cdot}6H_2O$. It occurs as a saline residue and is a raw material of fertilizer manufacture in some European districts.

carnegieite (car-ne′-gie-ite) A synthetic compound: $Na[AlSiO_4]$. It is the high-temperature equivalent of nepheline. It is triclinic at low temperatures, cubic at high temperatures.

carnelian (car-nel′-i-an) A translucent red or orange-red variety of chalcedony, pale to deep in shade, containing iron impurities. It is used for seals and signet rings. Cf: *sard.* Also spelled: *cornelian.* Syn: *carneol.*

carneol (car′-ne-ol) *carnelian.*

Carnian (Car′-ni-an) A stage of the Standard Global Chronostratigraphic Scale: Upper Triassic (above Ladinian, below Norian). Also spelled: *Karnian.*

carnivore (car′-ni-vore) A heterotrophic organism that nourishes itself mainly by feeding on other animals, living or dead. Adj: carnivorous. Cf: *herbivore.*

carnotite (car′-no-tite) A strongly radioactive, canary-yellow to greenish-yellow monoclinic secondary mineral: $K_2(UO_2)_2(V_2O_8){\cdot}3H_2O$. An ore of uranium and vanadium, and a source of radium, it occurs as a powdery incrustation or in loosely coherent masses, chiefly in sandstone (as in the western U.S.).

carobbiite (ca-rob′-bi-ite) A colorless cubic mineral: KF. A member of the *halite* group.

Carolina bay (Car-o-li′-na) Any of various shallow, often oval or elliptical, generally marshy, closed depressions in the Atlantic coastal plain (from southern New Jersey to NE Florida, esp. developed in the Carolinas). They range from about 100 m to many kilometers in length, are rich in humus, and contain trees and shrubs different from those of the surrounding areas. They are characterized by the approximate parallel orientation of their long axes. Their origin is much debated and has been attributed to meteorites, upwelling springs, eddy currents, and solution. Syn: *bay [geog].*

carotenoid Long-chained pigment used in photosynthesis.

Carpathian *Karpatian.*

carpathite (car-path′-ite) A pale yellow to red-violet organic mineral: $C_{24}H_{12}$. Syn: *pendletonite.* Also spelled: *karpatite.*

carpel (car′-pel) A modified leaf in the center of a flower that encloses one or more ovules. A carpel consists of *ovary, style,* and *stigma;* all the carpels together form the *pistil.* A pistil with only one carpel is a simple pistil; with two or more carpels fused together it is a compound pistil. Either sort of pistil is synonymously called a *gynoecium.*

carpholite (car′-pho-lite) (a) A straw-yellow orthorhombic mineral: $Mn^{2+}Al_2Si_2O_6(OH)_4$. (b) A group name for minerals of composition $AAl_2Si_2O_6(OH)_4$, where A = Mg, Fe^{2+}, Mn^{2+}, Ba or Li.

carphosiderite (car-pho-sid′-er-ite) *hydronium jarosite.*

carpoid (car′-poid) Any homalozoan echinoderm having an ambulacral groove and a stereome composed of crystalline calcite with reticular microstructure.

carpolite (car′-po-lite) (a) A fossil fruit, nut, or seed. (b) An ellipsoidal concretion or similar diagenetic structure, 1-2 cm in diameter, originally believed to be a fossil seed and assigned the generic name *Carpolites.*

carpopod (car′-po-pod) The *carpus* of a malacostracan crustacean. Syn: carpopodite.

carpus (car′-pus) (a) The fifth pereiopodal segment from the body of a malacostracan crustacean, located distal to the *merus* and proximal to the *propodus.* It comprises the third segment of a typical *endopod.* (b) The wrist of a tetrapod. Pl: carpi. Syn: *carpopod; wrist.*

carr An isolated mass of rock found off the coast, esp. in the British Isles. Syn: *carrig.*

carraraite A vitreous white hexagonal mineral: $Ca_3Ge(SO_4)(CO_3)(OH)_6{\cdot}12H_2O$.

carrboydite (carr-boyd′-ite) A yellowish-green hexagonal mineral: $(Ni,Al)_9(SO_4)_2(OH)_{18}{\cdot}10H_2O(?)$. Related to woodwardite.

carrig (car′-rig) *carr.*

carrollite (car′-roll-ite) A light steel-gray mineral of the *linnaeite* group: $Cu(Co,Ni)_2S_4$. Syn: *sychnodymite.*

carrying capacity (car′-ry-ing) The biomass of fish or other aquatic life that a lake can support.

carrying contour A single contour line representing two or more contours, used to show a vertical or near-vertical topographic feature such as a cliff, cut, or fill (USGS, 1965, p.26).

carse A Scottish term for low, level, fertile land; esp. a tract of alluvial land or river bottom bordering an estuary or near a river mouth, representing a marine terrace or a raised beach. Example: Carse of Gowrie along the coast of Scotland.

carstone (car′-stone) A British term for a hard, firmly cemented ferruginous sandstone, esp. one of Cretaceous age used as a building stone.

Cartesian coordinate system A coordinate system in which the location of a point on a plane is expressed by two coordinates (x,y) that measure the point's distance from two intersecting, usually perpendicular, straight-line axes, or three (x,y,z) coordinates that locate a point in space by its distance from three fixed planes that intersect one another at right angles.

cartilage-replacement bone (car′-ti-lage-re-place′-ment) Bone that replaces a cartilaginous precursor; includes the axial skeleton and most of the appendicular skeleton, but only limited parts of the skull. Syn: *endochondral bone.*

cartogram (car′-to-gram) An abstracted, simplified map with areas or distances systematically distorted in order to portray in a diagrammatic way such concepts as travel time or population size. Syn: *diagrammatic map.*

cartographic content The representation of only those content-worthy features that can be symbolized, taking into account the space and legibility constraints of the map (USGS, 2002b, p.1A-2). Cf: *geographic content.*

cartographic generalization A variety of methods of limiting the amount of information symbolized on a map, taking into consideration space and legibility issues. Methods include simplification, exaggeration, and selection/elimination (USGS, 2002b, p.1A-2). Cf: *cartographic license.*

cartographic license The extent to which a cartographer can change the appearance, layout, and content of a map without making it less accurate (Kennedy et al., 2001). Cf: *Cartographic generalization.*

cartographic position The position of a feature after it has been displaced or modified to permit legible symbolization on a map (USGS, 2002b, p.1A-2). Cf: *geographic position [cart], source position, true position.*

cartographic unit (car-to-graph′-ic) A rock or group of rocks that is shown on a geologic map by a single color or pattern. The standard cartographic unit is a *formation.*

cartography (car-tog′-ra-phy) (a) The science and art of making maps and charts. Syn: *chartology.* (b) The study of maps as scientific documents and works of art.

cartology (car-tol′-o-gy) A graphic method of coal-seam correlation, involving the mapping and drawing of both vertical and horizontal sections. Cf: *composite map.*

cartouche (car-touche') A decorative frame or scroll-shaped embellishment on a map or chart, enclosing the title, scale, legend, and other descriptive matter.

caryinite (car'-y-in-ite') A nut-brown or yellowish brown monoclinic mineral: $(Na,Pb)(Ca,Na)(Ca,Mn^{2+})(Mn^{2+},Mg)_2(AsO_4)_3$. It is isostructural with hagendorfite. Also spelled: karyinite.

caryopilite (car'-y-o-pil'-ite) A tan to reddish brown lamellar monoclinic mineral of the *serpentine* group: $(Mn^{2+},Mg)_3Si_2O_5(OH)_4$. Also spelled karyopilite.

caryopsis (car-y-op'-sis) A small dry one-seeded indehiscent fruit that has a completely united seed coat and pericarp and forms a single grain; e.g., in grasses.

cascade [glaciol] (cas-cade') *glacial stairway.*

cascade [streams] (a) A *waterfall,* esp. a small fall or one of a series of small falls descending over steeply slanting rocks; a shortened rapids. Also, a stepped series of small, closely spaced waterfalls or very steep rapids. Cf: *cataract.* (b) A short, rocky declivity in a stream bed over which water flows with greater rapidity and a higher fall than through a *rapids.*

cascade decay A little-used term for radioactive decay from a parent isotope through several daughter isotopes to a stable isotope.

cascade fold One of a series of folds that is formed by gravity collapse along the limb of larger folds.

cascade stairway *glacial stairway.*

Cascadia (Cas-ca'-di-a) One of the *borderlands* proposed by Schuchert (1923), in this case along the western margin of North America, partly at sea, partly inland. Most of the evidence adduced for the existence of Cascadia can now be otherwise interpreted.

Cascadian Revolution (Cas-ca'-di-an) A name used by Schuchert and others for a supposed profound crustal disturbance in western North America which brought an end to the Tertiary. The concept is now known to be untenable. The type area (the Cascade Range) is unfortunate, as no notable crustal events seem to have occurred there at this time. The term should be abandoned.

cascading glacier (cas-cad'-ing) A glacier passing over a steep, irregular bed, and therefore crossed by numerous crevasses and suggestive in appearance of a cascading stream. Cf: *icefall.*

cascadite (cas-ca'-dite) A sodic *minette* containing biotite, olivine, and augite phenocrysts in a groundmass composed almost entirely of alkali feldspar. It was named by Pirsson in 1905 for Cascade Creek, Highwood Mountains, Montana. Obsolete.

cascajo (cas-ca-jo[kahs-kah'-ho]) Reef-derived material composed of coral detritus and other sediment, occurring in old deposits. Etymol: Spanish, "gravel".

cascalho (cas-ca'-lho) A term used in Brazil for alluvial material, including gravel and ferruginous sand, in which diamonds are found. Etymol: Portuguese, "pebbles, small stones, coarse sand, gravel, grit".

cascandite (cas'-cand-ite) A pale pink triclinic mineral: $CaScSi_3O_8(OH)$.

cased hole A well in which casing has been run and cemented into it. Cf: *open hole.*

case hardening The process by which the surface of a porous rock (esp. tuff and certain sandstones) is coated with a cement or desert varnish formed by evaporation of mineral-bearing solutions. Adj: case-hardened. Also spelled: casehardening.

casing (cas'-ing) Heavy, large-diameter pipe of metal, plastic, or fiber; it is lowered into a *borehole* during or after drilling and cemented to the sides of the well. It prevents the sides of the hole from caving, prevents loss of drilling mud or other fluids into porous formations, and prevents unwanted fluids from entering the hole. The metal, plastic, or stainless steel casing consists of sections that are fastened together to form a casing string. A well may contain several strings of casing, the inner and smaller-diameter strings extending progressively deeper. All casing strings run up to the surface. Cf: *liner; surface pipe.* Syn: *well casing.*

casing-collar log (cas'-ing-col'-lar) A *well log* of relative magnetic intensity used in cased holes to identify the threaded junctions between consecutive lengths of well casing. It is used with a *gamma-ray log* run simultaneously to correlate between the geologic section and the sequence of collars, for depth control of perforating or other operations. Abbrev: CCL. Syn: *collar log.*

casinghead A steel fitting attached to the top of the *casing* set in an oil or gas well. A *Christmas tree* can be bolted to it. It may carry a valve to control pressure in the annulus between casing and *tubing.*

casing-head gas Unprocessed natural gas produced from a reservoir containing oil. Such gas contains gasoline vapors and is so called because it is usually produced under low pressure through the casing head of an oil well.

casing point The depth in a drill hole to which a given string of *casing* extends.

Ca-spar *calcium feldspar.*

Cassadagan (Cas-sa-dag'-an) North American stage: Upper Devonian (above Chemungian, below Conewangoan).

cassedanneite (cass'-e-danne-ite) An orange-red monoclinic mineral: $Pb_5(VO_4)_2(CrO_4)_2{\cdot}H_2O$.

Cassel brown (Cas'-sel) Black earth. Etymol: source near Cassel, Germany.

Cassel earth Black earth. Etymol: source near Cassel, Germany.

Cassiar Orogeny (Cas-si-ar') A name proposed by W.H. White (1959) for an orogeny that is believed to have occurred near the end of Paleozoic time in the Cordillera of British Columbia, especially in the Omineca and Cassiar districts. It was characterized by uplift, folding, and ultramafic intrusion. Obsolete.

cassidyite (cas'-sid-y-ite') A green triclinic meteorite mineral: $Ca_2(Ni,Mg)(PO_4)_2{\cdot}2H_2O$.

Cassinian curve (Cas-sin'-i-an) An *isochromatic curve* in the interference figure of a biaxial crystal.

Cassini projection (Cas-si'-ni) A map projection constructed by computing the lengths of arcs along a central meridian and along each great circle perpendicular to that meridian, and plotting these as rectangular coordinates on a plane. The scale is thereby preserved along these lines, but there is distortion in all other directions. It was formerly used as the base for much topographic and cadastral mapping. Named after C.F. Cassini de Thury (1714-1784), French astronomer, who introduced the projection in 1745.

cassinite (cas'-sin-ite) (a) A bluish variety of orthoclase containing barium and occurring in Delaware County, Penna. (b) Perthitic intergrowths of hyalophane and plagioclase.

cassiterite (cas-sit'-er-ite) A variously colored tetragonal mineral: SnO_2. It is the principal ore of tin. Cassiterite occurs in prismatic crystals of adamantine luster, and also in massive forms, either compact with concentric fibrous structure (*wood tin*) or in rolled or pebbly fragments (*stream tin*). Syn: *tinstone; tin ore; black tin.*

cast [paleont] Secondary rock or mineral material that fills a *natural mold;* specif. a replica or reproduction of the external details (size, shape, surface features) of a fossil shell, skeleton, or other organic structure, produced by the filling of a cavity formed by the decay or dissolution of some or all of the original hard parts of which the organism consisted. Cf: *mold [paleont].*

cast [sed] A sedimentary structure representing the infilling of an original mark or depression made on top of a soft bed and preserved as a solid form on the underside of the overlying and more durable stratum; e.g. a *flute cast* or a *load cast.* Cf: *mold [sed].* Syn: *counterpart.*

castellated (cas'-tel-la'-ted) Said of a physiographic feature, such as a cliff, peak, or iceberg, that displays a towering or battlementlike structure.

castile A conical hill of relatively resistant limestone within a gypsum terrain, formed by the local reduction of gypsum. The term is widely used in the Delaware basin of Texas-New Mexico (Kirkland and Evans, 1980).

casting (cast'-ing) Something that is cast out or off, esp. a worm casting or a *fecal pellet.*

castle (cas'-tle) A natural rock formation bearing a fancied resemblance to a castle.

Castlecliffian (Cas-tle-cliff'-ian) South Pacific stage: Pleistocene (above Pliocene Nukumaruan, below Recent Haweran).

castle kopje *castle koppie.*

castle koppie A pointed *koppie,* or a small *bornhardt,* with a castellated profile, often occurring as a jumbled pile of joint-bounded granite blocks. Also spelled: *castle kopje.*

castor (cas'-tor) *castorite.*

castorite (cas'-tor-ite) A transparent variety of *petalite.* Syn: *castor.*

casuzone (cas'-u-zone) A *biostratigraphic unit* defined as "a body of rock with upper and lower boundaries marked by reversible faunal changes that are essentially parallel to time planes" (Vella, 1964, p.622).

caswellite (cas'-well-ite) A mineral mixture consisting of mica and manganese-bearing andradite.

caswellsilverite (cas'-well-sil'-ver-ite) A metallic gray rhombohedral mineral found in a meteorite: $NaCrS_2$.

cat *cat claw.*

cataclasis [meta] Rock deformation accomplished by fracture and rotation of mineral grains or aggregates without chemical reconstitution. Cf: *cataclastic metamorphism; cataclastic rock.*

cataclasis [struc geol] A process typically occurring along fault planes or in fault zones involving brittle fragmentation of minerals, with rotation of grain fragments, grain-size reduction, grain-boundary sliding, and volume increase (Sibson, 1977).

cataclasite (cat-a-clas'-ite) A fine-grained, cohesive *cataclastic rock,* normally lacking a penetrative foliation or microfabric. A cataclasite forms by pervasive fracturing, milling, crushing, and grinding by brittle deformation, typically under conditions of elevated pressure, sufficient to keep the material from losing coherence and becoming a *gouge* (Sibson, 1977; Wise et al., 1984).

cataclasm (cat'-a-clasm) A breaking down or rending asunder; a violent disruption.

cataclast (cat'-a-clast) A *megacryst* that is a remnant of incomplete cataclasis.

cataclastic (cat-a-clas'-tic) (a) Pertaining to the structure produced in a rock by the action of severe mechanical stress during dynamic metamorphism; characteristic features include the bending, breaking, and granulation of the minerals. Also, said of the rocks exhibiting such structures. Cf: *mortar structure.* (b) Pertaining to clastic rocks, the fragments of which have been produced by the fracture of preexisting rocks by Earth stresses; e.g., crush breccia (Teall, 1887).

cataclastic breccia *crush breccia.*

cataclastic conglomerate *crush conglomerate.*

cataclastic flow *Flow [exp struc geol]* involving intergranular fracturing and movement, i.e. mechanical displacement of particles relative to each other; a brittle flow mechanism.

cataclastic metamorphism A type of local metamorphism confined to the vicinity of faults and overthrusts, involving purely mechanical forces causing crushing and granulation of the rock fabric (*cataclasis*). Cf: *dislocation metamorphism; kinetic metamorphism.*

cataclastic rock A rock, such as a tectonic breccia, containing angular fragments that have been produced by the crushing and fracturing of preexisting rocks as a result of mechanical forces in the crust; a metamorphic rock produced by cataclasis. Its fabric is a structureless rock powder. Pettijohn (1957, p.281) would include glacial till as a "cataclastic deposit" as it is "an extensive gouge caused by the grinding along the base of an overthrust ice sheet". See also: *cataclasis [struc geol]; autoclastic rock; mylonite.* Syn: *cataclasite.*

cataclastic structure *mortar structure.*

cataclastic texture A texture in a dynamically metamorphosed rock produced by severe mechanical crushing and differential movement of the component grains and characterized by granular, fragmentary, deformed, or strained mineral crystals, commonly flattened in a direction at right angles to the mechanical stress. Rarely used syn: *pressure texture.*

cataclinal (cat-a-cli'-nal) Said of a *dip stream* or of a valley that descends in the same direction as that of the general dip of the underlying strata it traverses. Term introduced by Powell (1875, p.160).

catacline (cat'-a-cline) Said of the inclination of the *cardinal area* in either valve of a brachiopod lying at right angles to the *orthocline* position.

cataclysm (cat'-a-clysm) (a) Any geologic event that produces sudden and extensive changes in the Earth's surface; e.g. an exceptionally violent earthquake. Adj: cataclysmic; cataclysmal. Cf: *catastrophe.* (b) Any violent, overwhelming flood that spreads over the land; a *deluge.*

catagenesis [evol] (cat-a-gen'-e-sis) Evolution leading to decadence and decreased vigor. Also spelled: *katagenesis.*

catagenesis [geochem] Breakdown of geopolymers to make oil and gas.

catagenesis [sed] The changes occurring in an already formed sedimentary rock buried by a distinct (though sometimes thin) covering layer, characterized by pressure-temperature conditions that are much different from those of deposition (Fersman, 1922); specif. the breakdown of rocks. The term is more or less equivalent to epigenesis as applied by Russian geologists.

catagraphite (cat-a-graph'-ite) A complex structure made up of traces of canals and cavities thought to be the result of the activity of algae and bacteria in the late Precambrian and Early Cambrian. Cf: *oncolite.*

catalanoite A colorless orthorhombic mineral: $Na_2HPO_4{\cdot}8H_2O$.

catamorphism (cat-a-mor'-phism) *katamorphism.*

catanorm (cat'-a-norm) The theoretical calculation of minerals in a metamorphic rock of the katazone, as indicated by chemical analyses. Cf: *mesonorm; epinorm; Niggli molecular norm.*

cataphoresis (cat-a-pho'-re-sis) *Electrophoresis* in which the movement of suspended positive particles in a fluid is toward the cathode. Cf: *anaphoresis.*

cataphorite (cat-a-phor'-ite) *kataphorite.*

catapleiite (cat-a-plei'-ite) A colorless, gray, or blue monoclinic pseudo-hexagonal mineral: $Na_2ZrSi_3O_9{\cdot}2H_2O$. See also: *calcium-catapleiite.* Also spelled: catapleite, katapleiite.

cataract (cat'-a-ract) (a) A *waterfall,* esp. one of great volume in which the vertical descent has been concentrated in one sheer drop over a precipice. Cf: *cascade [streams].* (b) A series of steep *rapids* in a large river; e.g., the Nile. (c) An overwhelming rush of water; a flood.

cataract lake A lake occupying the plunge basin of an extinct cataract whose stream has been diverted above the fall.

cataspire (cat'-a-spire) A foldlike respiratory structure with thin calcite walls extending beneath the *deltoid plate* of a parablastoid, from its aboral margin to the edge of an adjacent *ambulacrum.* Seawater apparently flowed adorally through each cataspire (Sprinkle, 1973, p.32).

catastable (cat'-a-sta'-ble) adj. Stable, with a tendency towards sinking. Cf: *anastable.*

catastrophe (ca-tas'-tro-phe) A sudden, violent disturbance of nature, ascribed to exceptional or supernatural causes, affecting the physical conditions and the inhabitants of the Earth's surface; e.g. the Noachian flood, or an extinction of an entire fauna. Cf: *cataclysm; paroxysm.* Syn: *convulsion.*

catastrophe theory A mathematical theory in which continuous changes in one variable lead to abrupt discontinuities in another variable. It has been used to model the transition from laminar to turbulent flow in fluids and other phenomena that are characterized by a threshold.

catastrophic advance (cat-a-stroph'-ic) *surge [glaciol].*

catastrophism (ca-tas'-tro-phism) (a) The doctrine that sudden violent, short-lived, more or less worldwide events outside our present experience or knowledge of nature have greatly modified the Earth's crust; that "...not one of [Nature's] agents now at work would have sufficed to have effected her ancient works" (Cuvier). (b) The doctrine that the present configuration of the Earth's crust, as well as the distribution of living beings, is mainly the result of "a great and sudden revolution" (Cuvier) of 5,000 or 6,000 years ago and, by extension, that geologic processes of the past were of substantially greater intensity and number than those of the present, as phrased by William Whewell in 1831. (c) The doctrine that changes in the Earth's fauna and flora are explained by recurring catastrophes followed by the creation of different organisms. Cf: *uniformitarianism.* Syn: *convulsionism.*

catastrophist (ca-tas'-tro-phist) A proponent of *catastrophism.*

catathermal (cat-a-ther'-mal) n. A period of time during which temperatures are declining. The term was used by Emiliani (1955, p.547) for a part of a cycle as displayed in a deep-sea sediment core. Ant: *anathermal.*

catawberite (ca-taw'-ber-ite) A metamorphic rock of South Carolina that consists mainly of talc and magnetite. Rarely used.

catazonal pluton An intrusive body that formed at medium to low crustal depth, typically concordant, with gradational contacts, and associated with high-grade metamorphic rocks and migmatites. Cf: *epizonal pluton; mesozonal pluton.*

catazone (cat'-a-zone) *katazone.*

catch basin (a) A reservoir or basin into which surface water may drain and be collected. (b) A basin to collect and retain material from a street gutter that would not readily pass through the sewer system.

catchment (catch'-ment) (a) A term used in Great Britain for an area that collects and drains rainwater; a *drainage basin.* (b) A depression that collects rainwater; a reservoir. (c) The act of catching water; also, the amount of water that is caught.

catchment area [grd wat] (a) The *recharge area* and all areas that contribute water to it. (b) An area paved or otherwise waterproofed to provide a water supply for a storage reservoir. Syn: *collecting area.*

catchment area [streams] *drainage basin.*

catchment basin *drainage basin.*

catchment glacier *drift glacier.*

catchwater drain (catch'-wa-ter) A ditch or surface drain designed to intercept and collect surface water and to divert it for protection of adjacent property or for use in irrigation or for other purposes.

cat claw A miner's term used in Peoria County, Ill., for an irregular protuberance (2.5-7.5 cm in height and width) in the lower surface of a bed of marcasite overlying a coal seam (Cady, 1921, p.164). Syn: *cat.*

cat clay Poorly drained, clayey soils, commonly formed in an estuarine environment, that become very acidic when drained due to oxidation of ferrous sulfide.

cat coal Coal that contains pyrite.

catena [planet] (ca-te'-na) A term established by the International Astronomical Union for a chain of craters on planetary surfaces (USGS, 2003). Most are thought to be of volcanic origin, tectonically controlled (Mutch et al., 1976, p.57). Etymol: Latin, "catena", a chain. Pl: catenae.

catena [soil] A sequence of soils of about the same age, derived from similar parent material and occurring under similar climatic conditions, but having different characteristics due to variation in relief and drainage. See also: *toposequence; soil association.*

catenary (cat'-e-nar'-y) The curve assumed by a perfectly flexible, inextensible cord of uniform density and cross section when suspended freely from two fixed points both at the same level; e.g. such a curve as formed by a surveyor's tape hanging between adjacent supports. See also: *sag correction.*

catenary ripple "A ripple is described as catenary if the trace of its crestline has the pattern of a chain of catenary waves, such that the more pointed segments of the crestline face downcurrent" (Allen, 1968, p.62-63). It differs from a lunate ripple in having lesser curvature of the crestline. Catenary ripples may be in or out of phase with the ripples immediately upstream or downstream.

catenary ripple mark *catenary ripple.*

catenate (ca-te'-nate) (a) Of fungal spores, being produced in chains. (b) Of dinoflagellates or acritarchs, forming chains of cells.

catenicelliform (ca-ten-i-cel'-li-form) Said of a jointed, delicately branching, erect colony in cheilostome bryozoans, the internodes of which consist of few zooids. It is attached by rootlets (Lagaaij and Gautier, 1965, p.51).

cateniform (ca-ten'-i-form) Said of a tabulate corallum with the corallites united laterally as palisades that appear chainlike in cross section and commonly form a network.

catenulate colony (ca-ten'-u-late) Colonial growth of archaeocyathids in which the cups are contiguous and in a chainlike row and the outer wall does not develop between neighboring cups (TIP, 1972, pt. E, p.6).

cat face A miner's term for glistening balls or nodules, or small discontinuous veinlets, of pyrite in the mining face of coal; also, lenticular deposits of pyrite associated with coal. Also spelled: catface.

Cathaysia (Ca-thay'-sia) An early Paleozoic landmass in the area now occupied by southeastern China, Indochina, and the southeast Asian archipelago. It lay to the east of *Gondwana* (C.R. Jones, 1973, p.27, 55).

cathodoluminescence (cath-o'-do-lu'-mi-nes'-cence) Characteristic visible luminescence produced when high-energy electrons bombard a material in a vacuum.

cathole (cat'-hole) A local term used in southern Michigan for a shallow boggy depression less than an acre in extent, esp. one formed by a glacier in a till plain (Veatch and Humphrys, 1966, p.59).

cation An atom that has acquired a positive charge by virtue of having lost one or more electrons.

cation exchange (cat′-i-on) The displacement of a cation bound to a site on the surface of a solid, as in silica-alumina clay-mineral packets, by a cation in solution. Syn: *base exchange.* See also: *ion exchange.*

cation exchange capacity [cryst] The capacity of a mineral (generally a clay or zeolite) to exchange cations in their crystal structures for different cations in a surrounding solution. The surface charge of, for example, a clay mineral that reflects an internal imbalance of charges that is negative. Therefore there is a cation excess and an anion deficit at the surface. Cations may be exchanged by exposing the material holding the excess cations to a solution containing a concentration of a replacing cation. Measurements of the exchanged cations indicate the amount of charge on the exchanging material. Changing cations changes some of the characteristics of the exchanging material (Laudelout, 1987). Abbrev: CEC.

cation exchange capacity [soil] The sum of exchangeable bases plus total soil acidity at a specific pH, values, usually 7.0 or 8.0. When acidity is expressed as salt extractable acidity, the cation exchange capacity is called the effective cation exchange capacity (ECEC) because this is considered to be the CEC of the exchanger at the native pH value. It is usually expressed in centimoles of charge per kilogram of exchanger ($cmol_c\ kg^{-1}$) or millimoles of charge per kilogram of exchanger. Abbrev: CEC.

catkin (cat′-kin) *ament.*

catlinite (cat′-lin-ite) A red, siliceous, indurated clay or argillite from the upper Missouri River valley region (SW Minnesota), formerly used by the Dakota Indians for making tobacco pipes; a *pipestone.* Named after George Catlin (1796-1872), American painter of Indians. See Berg, 1938.

catoctin (ca-toc′-tin) A residual knob, hill, or ridge of resistant material rising above a peneplain and preserving on its summit a remnant of an older peneplain. Named after Catoctin Mountain, Maryland and Virginia. Cf: *monadnock.*

catophorite (cat-o-phor′-ite) *katophorite.*

catoptrite (ca-top′-trite) *katoptrite.*

catotelm The subsurface fossilized layers of a peat in a peat deposit or mire (Ingram, 1978). Cf: *acrotelm.*

Ca-Tschermak molecule A synthetic pyroxene, $CaAl(AlSi)O_6$; a hypothetical component of natural pyroxenes. Not to be confused with *tschermakite.* Syn: *Tschermak molecule.*

cat's-eye [gem] Any gemstone that, when cut *en cabochon,* exhibits under a single strong point-source of light a narrow, well-defined chatoyant band or streak that moves across the summit of the gemstone, shifts from side to side as it is turned, and resembles a slit pupil of the eye of a cat. Internal reflection of light from parallel inclusions of tiny fibrous crystals or from long parallel cavities or tubes causes the "cat's-eye". Some of the gemstones that can exhibit this phenomenon include chrysoberyl, quartz, sillimanite, scapolite, cordierite, orthoclase, albite, beryl, and tourmaline. See also: *chatoyancy.*

cat's-eye [mineral] (a) A greenish gem variety of chrysoberyl that exhibits *chatoyancy.* Syn: *cymophane; oriental cat's-eye.* (b) A variety of minutely fibrous, grayish-green quartz (chalcedony) that exhibits an opalescent play of light. Syn: *occidental cat's-eye.* (c) A yellowish-brown silicified form of crocidolite. Cf: *tiger's-eye.* The term, when used alone, is properly applied only to chrysoberyl.

catstep (cat′-step) A terracette, esp. one produced by slumping of deep loess deposits as in western Iowa. Also spelled: cat step.

cattierite (cat′-tier-ite) A metallic pinkish cubic mineral: CoS_2. A member of the *pyrite* group.

cattiite A vitreous colorless triclinic mineral: $Mg_3(PO_4)_2{\cdot}22H_2O$.

Caucasian (Cau-cas′-ian) Ponto-Caspian stage: upper Oligocene (below Sakaraulian).

cauda (cau′-da) The slender proximal portion of an *autozooecium* in some cheilostome bryozoans, comprising part of the *gymnocyst* and adjacent parts of lateral and basal walls (Thomas and Larwood, 1956, p.370). Pl: caudae.

caudal fan (cau′-dal) A powerful swimming structure in malacostracan crustaceans, consisting of a combination of laterally expanded *uropods* and *telson,* and constituting a means of steering and balancing. Syn: *tail fan.*

caudal furca A crustacean *furca* consisting of a pair of caudal rami.

caudal process In certain ostracodes, a prolongation (usually pointed) of the posterior end of the carapace, commonly situated at or near mid-height.

caudal ramus One of a pair of appendages of the *telson* of a crustacean. It is usually rodlike or bladelike, but may be filamentous and multiarticular. Syn: caudal appendage; *cercus.*

Caudata (Cau-dat′-a) A superorder of the lissamphibians, characterized by a tail and nonhopping gait. It includes the salamanders, newts, and caecilians.

caudex (cau′-dex) A basal part of the axis of an erect plant where it is neither clearly stem, rhizome, or root; the persistent base of an otherwise annual herbaceous stem (Fernald, 1950, p.1571).

cauldron [marine geol] (caul′-dron) *caldron [marine geol].*

cauldron [volc] As used by Smith and Bailey (1968), an inclusive term for all volcanic subsidence structures regardless of shape or size, depth of erosion, or connection with the surface. The term thus includes cauldron subsidences, in the classical sense, and collapse calderas. See also: *caldera.*

cauldron subsidence (a) A structure resulting from the lowering along a steep ring fracture of a more or less cylindrical block into a magma chamber; usually associated with ring dikes. The ring fracture may or may not reach the surface of the Earth. (b) The process of forming such a structure. Magmatic stoping involving the sinking, into a magma body of a single, large block of country rock (1 km diameter). Cf: *piecemeal stoping.* See also: *ring-fracture stoping; surface cauldron subsidence; underground cauldron subsidence.*

caulescent (cau-les′-cent) In plants, producing or having a stem above ground (Lawrence, 1951, p.743). Etymol: Latin "caulis," stem, stalk.

cauline (cau-line′) In a plant leaf, flower, or fruit, being attached directly to a stem or major axis without *petiole* or *peduncle.*

caulk *cawk.*

caunter lode (caun′-ter) *cross vein.*

causal stratigraphy An approach to stratigraphic problems based on ecosystem analysis of interrelations between geologic events and organic evolution (Krassilov, 1974, p.170).

causse A limestone plateau in the southeastern part of France characterized by *closed depressions,* caves, and avens. A number of such plateaus in and around the basin of the river Tarn constitute Les Grandes Causses. Etymol: French dialect for "chaux", "lime". Cf: *karst.*

caustic (caus′-tic) n. (a) A curve representing a locus of points where rays (as of light or sound) are brought exceptionally close together owing to the reflective or refractive properties of the medium through which they travel; specif. the curve or surface to which adjacent orthogonals to waves are tangents. (b) The envelope of the system of such rays.

caustic metamorphism The indurating, baking, burning, and fritting effects of lava flows and small dikes on the rocks with which they come in contact. The term was originated by Milch in 1922. Cf: *baking.* Rarely used.

caustobiolith (caus-to-bi′-o-lith) A combustible organic rock (Grabau, 1924, p.280). It is usually of plant origin. Cf: *acaustobiolith.*

caustolith (caus′-to-lith) A rock that has the property of combustibility (Grabau, 1924, p.280). It is usually of organic origin (e.g. coal and peat), but inorganic deposits (e.g. sulfur, asphalt, graphite) also occur.

caustophytolith (caus-to-phy′-to-lith) A *caustobiolith* formed by the direct accumulation of vegetal matter; e.g. peat, lignite, and coal.

caustozoolith (caus-to-zo′-o-lith) A rare *caustobiolith* formed by the direct accumulation of animal matter (Grabau, 1924, p.280); e.g. some oils.

cavaedium (ca-vae′-di-um) An irregular space within a sponge, communicating directly with the exterior, but morphologically outside the sponge in that it is lined by *exopinacoderm.* Pl: cavaedia.

cavansite (ca-van′-site) A sky-blue to greenish blue orthorhombic mineral: $Ca(V^{4+}O)Si_4O_{10}{\cdot}4H_2O$. It is dimorphous with pentagonite.

cavate (ca′-vate) (a) Descriptive of spores whose exine layers are separated by a cavity, including a rather slight separation as well as a more extensive separation producing a bladderlike protuberance (pseudosaccus). Syn: *camerate [palyn].* (b) Said of a dinoflagellate cyst with space or spaces of notable size between periphragm and endophragm (as in *Deflandrea phosphoritica*).

cave [coast] *sea cave.*

cave [speleo] (a) A natural underground open space; it generally has a connection to the surface, is large enough for a person to enter, and extends into darkness. The most common type of cave is formed in limestone by dissolution. Cf: *cavern.* (b) A similar feature that was formed artificially. (c) In informal use, any natural rock shelter, e.g., a cliff overhang. (d) The term is also used in statistical studies for subterranean cavities of any size, with caves large enough for a person to enter being called *proper caves.* Cf: *karst feature.*

cavea (ca′-vea) A cavity between two layers of pollen *exine,* extending to the *colpus* margin where the two layers meet. Pl: caveae. Adj: caveate.

cave balloon A hollow bubble in cave, usually of hydromagnesite, which has been inflated by a gas (air or biogenic carbon dioxide).

cave beard A variety of *cave cotton* where parallel, fibrous strands of gypsum or other sulfate minerals are free-hanging, usually from a cave ceiling.

cave blister A partly or completely hollow hemispherical to nearly spherical *speleothem* attached to a cave wall. May be composed of calcite, gypsum, chalcedony or opal.

cave breakdown The collapse of the ceiling or walls of a cave; also, the accumulation of debris thus formed. See also: *cave breccia.* Syn: *breakdown.*

cave breathing The back-and-forth movement of air in the constricted passages of caves, usually with a cycle of a few minutes. See also: *blowing cave.*

cave breccia Angular fragments (*cave breakdown*) of limestone that have fallen to the floor from the roof and sides of a cave and that are cemented with calcium carbonate or occur in a matrix of cave earth. See also: *collapse breccia; solution breccia.*

cave bubble *calcite bubble.*

cave coralloid A rough, knobby *speleothem,* usually of calcite, that resembles *coral.* Syn: *botryoid; cave popcorn.*

cave cotton Masses of intergrown, fibrous crystals in cotton-like mounds; composed of sulfate and other minerals.

cave cup A hollow, circular to pyramidal, calcite *speleothem* that projects outward from a cave wall or upward from a shallow cave pool. Etymol: French "coupelle".

cave earth Fine-grained, generally unconsolidated detrital material partly filling a cave; also, similar material regardless of grain size. See also: *fill [speleo].* Syn: *cave soil.*

cave flower A mineral that occurs as a curved elongate deposit extending from a cave wall; it is commonly gypsum or epsomite but can be many other minerals, as well as ice. Growth of the structure occurs at the attached end.
cave formation A *mineral* formation in caves. An obsolete syn. for *speleothem.*
cave guano Accumulations of dung in caves, generally from bats; in some places partially mineralized.
cave-in (a) The partial or complete collapse of earth material into a large underground opening, such as an excavation or a mine. (b) The sudden slumping of wall material into a pit. (c) A place where material has collapsed or fallen in or down.
cave-in lake *thermokarst lake.*
cave marble *cave onyx.*
cave onyx A compact banded deposit of calcite or aragonite found in caves, capable of taking a high polish, and resembling true onyx in appearance. See also: *dripstone; flowstone; onyx marble; speleothem; travertine.* Syn: *cave marble.*
cave pearl A nonattached rounded cave deposit, usually of calcite, formed by precipitation of concentric layers around a nucleus, and characterized by radial crystal structure (Hill, 1992).
cave popcorn A variety of *cave coralloid.*
caver (cav′-er) One who engages in cave exploration, or *caving [speleo].* Cf: *spelunker; potholer; speleologist.*
cave raft *calcite raft.*
cave rim A rim of carbonate or sulfate material which occurs at a constriction in a cave passage. Cave rims are formed by aerial transport of dissolved minerals from a source area to a deposition site.
cavern (cav′-ern) A *cave [speleo],* generally understood to be of large size; a system or series of caves or cave chambers. Commonly used to denote a *show cave.*
cavern flow Movement, often turbulent, of groundwater through caves, coarse sorted gravel, or large open conduits, either by gravity or under pressure.
cavernous [speleo] (cav′-ern-ous) Said of an area or geologic formation, e.g. limestone, that contains caverns, or caves.
cavernous [volc] Said of the texture of a volcanic rock that is coarsely *porous* or *cellular.*
cavernous rock Any rock that has many cavities, cells, or large interstices; e.g. a cliff face pitted with shallow holes resulting from cavernous weathering.
cavernous weathering Chemical and mechanical weathering on a cliff face, in which grains and flakes of rock are loosened so as to enlarge hollows and recesses "opened through a chemically hardened shell" on the surface of the cliff face (Cotton, 1958, p.15). It produces the tafoni in seaside cliffs. Cf: *honeycomb weathering.* See also: *tafone.*
cavern porosity A pore system having large, cavernous openings. The lower size limit, for field analysis, is practically set at "about the smallest opening an adult person can enter" (Choquette and Pray, 1970, p.244).
cavern system *cave system.*
cave rope A sulfate *speleothem* with parallel fibers aligned in a ropelike manner.
cave soil *cave earth.*
cave spring A spring rising in a cave.
cave system (a) A group of caves that are connected or hydrologically related. (b) A complex cave. Syn: *cavern system.*
cave velvet Small scalenohedra of calcite on the surface of a *speleothem* which impart a velvety texture and luster.
caving [eco geol] (cav′-ing) *block caving; sublevel caving.*
caving [geomorph] (a) Detachment and movement of bank materials into a channel or onto a *batture,* commonly as slump blocks. Normally a result of removal of support for banks above scour pools, particularly during high stages of flow, and when the stage after flood crest drops rapidly. Called *bank subsidence* by many engineers. See also: *bank caving.* (b) A falling in; the action of caving in, as of the walls of an excavation or bore hole.
caving [speleo] The exploration of caves; sometimes used to characterize the activity more as a hobby or sport than as a scientific study. Cf: *caver; speleology.* Syn: *spelunking; potholing.*
cavings (cav′-ings) Rock fragments that fall from the walls of a borehole. They can contaminate the *drill cuttings* or block the hole.
cavitation [seis] The collapse of bubbles in a fluid, caused when the static pressure is less than the fluid vapor pressure.
cavitation [streams] A form of physical erosion occurring in rivers where fluctuations in velocity and pressure cause vapor bubbles to alternately form and implode. Implosion of the bubbles generates shock waves that weaken adjacent sediment and bedrock.
cavity (cav′-i-ty) (a) A solutional hollow in a limestone cave. (b) A small hollow in cavernous lava.
cavity dweller *Coelobite.* An organism that lives in a recessed cavity or void within reef-rock. Cf: *cryptic; endolithic; sciaphilic.*
cavoite A colorless to olive green-brown orthorhombic mineral: CaV_3O_7.
cavus (ca′-vus) A term established by the International Astronomical Union for irregular steep-sided depressions on planetary surfaces (USGS, 2003). Pl: cavi.
cawk A syn. of *barite;* esp. a white, massive, opaque variety of barite found in Derbyshire, England. Syn: cauk; *caulk; calk.*
c* axis That axis of a reciprocal crystal lattice which is perpendicular to (001). Cf: *a* axis; b* axis.*
***c* axis [cryst]** One of the crystallographic axes used for reference in crystal description; it is oriented vertically. In tetragonal and hexagonal crystals, it is the unique symmetry axis. It is usually the principal axis. The letter *c* usually appears in italics. Cf: *a axis [cryst]; b axis [cryst].*
***c*-axis [ice]** Ice is a hexagonal mineral, thus single crystals grown slowly under controlled conditions will be hexagonal rods. The long axis of such a rod is the *c*-axis. Cf: *a*-axis.
***c* axis [struc petrol]** One of three orthogonal reference axes, *a, b,* and *c,* that are used in two distinct ways. (a) To help describe the geometry of a fabric that possesses orthorhombic or monoclinic symmetry. The *c* axis lies in the unique symmetry plane at right angles to a prominent fabric plane; thus in many tectonites the *c* axis is normal to the schistosity. (b) In a kinematic sense, to describe a *deformation plan* that possesses orthorhombic or monoclinic symmetry, such as a progressive pure or simple shear. Here the *c* axis lies in the unique symmetry plane and normal to the movement plane. In a progressive simple shear the *c* axis lies normal to the shear plane. See also: *a axis [struc petrol]; b axis [struc petrol].* Syn: *c direction.*
cay A small, low, coastal island or emergent reef of sand or coral; a flat mound of sand and admixed coral fragments, built up on a reef flat at or just above high-tide level. Term is used esp. in the Caribbean where it is pronounced "key" and sometimes spelled "kay". Etymol: Spanish "cayo", "shoal or reef". Cf: *key [coast]; quay.*
cayeuxite (ca-yeux′-ite) A nodular variety of pyrite containing silicon, arsenic, antimony, and germanium.
cay sandstone A friable to firmly cemented coral sand formed near the base of a coral-reef *cay* and reaching above high-tide level; it is cemented by calcium carbonate deposited from fresh water.
caysichite-(Y) (cay′-sich-ite-y) A white, pale yellow, or greenish orthorhombic mineral: $Ca_3(Yb,Er)Y_4Si_8O_{20}(CO_3)_6(OH){\cdot}7H_2O$.
Cayugan (Ca-yu′-gan) North American series: Upper Silurian (above Niagaran, below Helderbergian of Devonian).
Cazenovian (Caz-e-no′-vi-an) A North American stage: Middle Devonian (above Oncequethwan, below Tiouchniogan).
CBR *California bearing ratio.*
CCD *calcite compensation depth.*
C-centered lattice A *base-centered lattice* that is centered in the pair of (001) faces. Cf: *A-centered lattice; B-centered lattice.*
CCF *complex cross foliated structure.*
CCL *complex crossed lamellar structure.*
CDD *coherent diffracting domain.*
CD interray Posterior interray in echinoderms situated between C ray and D ray and clockwise of C ray when viewed from the oral side. It differs frequently in shape from the other interrays and contains the anal opening; equal to ambulacrum IV of the *Lovenian system.*
***c* direction** *c axis [struc petrol].*
cebaite-(Ce) (ce′-ba-ite-ce) A yellow to orange-yellow monoclinic mineral: $Ba_3Ce_2(CO_3)_5F_2$.
cebaite-(Nd) A vitreous beige Nd analogue of cebaite-(Ce). Not approved by the International Mineralogical Association.
cebollite (ce′-bol-lite) A variously colored fibrous mineral: $Ca_5Al_2Si_3O_{12}(OH)_4$.
CEC *cation exchange capacity [cryst]; cation exchange capacity [soil].*
cechite (cech′-ite) A resinous black orthorhombic mineral of the descloizite-pyrobelonite group: $Pb(Fe^{2+},Mn)(VO_4)(OH)$.
cecilite (ce′-cil-ite) A basaltic rock with few phenocrysts, composed of leucite, augite, melilite, nepheline, olivine, anorthite, magnetite, and apatite. Leucite comprises about 50 percent of the total rock, followed by the mafic minerals, with melilite comprising about one-eighth, nepheline about 8%, and anorthite 4% (Johannsen, 1939, p.246). The rock is essentially a *leucitite.* It was named by Cordier in 1868 for the tomb of Cecilia Metella, Capo di Bove, Italy. Obsolete.
cedarite (ce′-dar-ite) *chemawinite.*
cedar-tree structure A term applied to a laccolith or volcanic neck in which sill-like intrusive layers taper away from a central intrusive mass, the whole structure resembling the outline of a cedar tree in cross section. Cf: *compound laccolith.* Obsolete.
cedricite (ced′-ric-ite) A *lamproite* that contains leucite and clinopyroxene phenocrysts in a very fine-grained groundmass containing phlogopite. Named by Wade and Prider in 1940 after Mount Cedric, Western Australia. Mitchell and Bergman (1991) suggest that this name should be replaced by diopside-leucite *lamproite.* Obsolete.
ceiling block A roughly cubical joint-bounded large block, which has fallen from the ceiling of a cave. See also: *cave breakdown.*
ceiling cavity (ceil′-ing) A solutional hollow in the ceiling of a cave. Some are clearly joint controlled. Cf: *joint cavity; pocket [speleo].*
ceiling channel A solutional groove on the ceiling of a cave that presumably was filled with water when it formed, most commonly formed by water flowing over a surface of sediment that once nearly filled the cave. Syn: *ceiling meander.*

ceiling meander *ceiling channel.*

ceiling slab A thin but extensive piece of rock that has fallen from the ceiling of a cave. Cf: *cave breakdown; ceiling block.*

ceiling tube A *half-tube* remaining in the ceiling of a cave.

ceja (ce'-ja[say'-ha]) A term used in the SW U.S. for the jutting edge along the top of a mesa or upland plain, and also for the cliff at this edge; an escarpment, esp. the steeper of the two slopes of a cuesta (if the slope is a cliff) or part of this slope that is a cliff. Etymol: Spanish, "eyebrow; mountain summit".

cejkaite A pale yellow to beige triclinic mineral: $Na_4UO_2(CO_3)_3$.

celadonite (cel'-a-don-ite) A soft, green or gray-green, earthy, dioctahedral mineral of the mica group, of ideal composition $K(Mg,Fe)^{2+}(Fe,Al)^{3+}(OH)_2[Si_4O_{10}]$, generally occurring in cavities in basaltic rocks. It has a structure very similar to that of glauconite. Not to be confused with *caledonite.* Syn: *svitalskite.*

celerity (ce-ler'-i-ty) *phase velocity.*

celestial coordinate (ce-les'-tial) Any member of any system of coordinates used to locate a point on the celestial sphere; e.g. altitude, azimuth, declination, right ascension.

celestial equator The great circle on the celestial sphere whose plane is perpendicular to the Earth's axis of rotation. It is formed by the intersection of the celestial sphere and the extension of the Earth's equatorial plane. Often simply called the equator. Syn: *equinoctial circle.*

celestial horizon That circle of the celestial sphere formed by the intersection of the celestial sphere and a plane through the center of the Earth and perpendicular to the zenith-nadir line. Syn: *astronomic horizon; rational horizon.*

celestialite (ce-les'-tial-ite) A variety of *ozocerite* found in some iron meteorites.

celestial latitude Angular distance north or south of the ecliptic; the arc of a circle of latitude between the ecliptic and a point on the celestial sphere, measured northward or southward from the ecliptic through 90°, and labeled N or S to indicate the direction of measurement. Cf: *latitude.* Syn: *ecliptic latitude.*

celestial longitude Angular distance east of the vernal equinox, along the ecliptic, between the circle of latitude of the vernal equinox and the circle of latitude of a point on the celestial sphere; measured eastward from the circle of latitude of the vernal equinox, through 360°. Cf: *longitude.* Syn: *ecliptic longitude.*

celestial mechanics The study of the motions of celestial bodies primarily under the influence of gravity but including the effects of other forces as well.

celestial meridian One half of a great circle of the celestial sphere passing through the zenith of a given place and terminating at the celestial poles; the hour circle that contains the zenith, or the vertical circle that contains the celestial pole. Usually called simply the *meridian.*

celestial pole Either of the two points of intersection of the celestial sphere and the extended axis of rotation of the Earth, around which the diurnal rotation of the stars appears to take place; specif. north pole and south pole. The altitude of an observer's celestial pole is equal to his geographic latitude.

celestial sphere An imaginary sphere of infinite radius, described around an assumed center (usually the center of the Earth), and upon whose "inner surface" all the heavenly bodies (except the Earth) appear to be projected along radii passing through them. The apparent dome of the visible sky forms half of the celestial sphere. Since the radius of the Earth is negligible in comparison to the distances to most celestial bodies, the center of the celestial sphere is assumed for most surveying purposes to coincide with the point of observation on the Earth's surface.

celestine (cel-es'-tine) A variously colored orthorhombic mineral of the *barite* group: $SrSO_4$. It often occurs in residual clays and in deposits of salt, gypsum, and associated dolomite and shale. Celestine is the principal ore of strontium. Syn: *celestite; coelestine.*

celestite (cel-es'-tite) *celestine.*

celite (cel'-ite) *brownmillerite.*

cellariiform (cel-lar'-ii-form) Said of a jointed, delicately or stoutly branching, erect colony in cheilostome bryozoans, the internodes of which consist of many zooids. It is attached by radicles on an encrusting base (Lagaaij and Gautier, 1965, p.51).

cell dimension *lattice constant.*

celleporiform (cel-le-por'-i-form) Said of a multilamellar, commonly massive colony in cheilostome bryozoans, formed by repeated self-overgrowth or budding in a frontal direction, from an encrusting basal layer of zooids (Lagaaij and Gautier, 1965, p.51).

cell texture In mineral deposits, a network pattern formed by solution or replacement of organic structures, e.g. cell walls.

cellular (cel'-lu-lar) Said of the texture of a rock (e.g. a cellular dolostone) characterized by openings or cavities, which may or may not be connected. Although there are no specific size limitations, the term is usually applied to cavities larger than pores and smaller than caverns. Obsolete. The syn. *vesicular* is preferred when describing igneous rocks. Cf: *porous; cavernous [volc].*

cellular automata (au-tom'-a-ta) The term is used in geomorphology to explore properties of emergence, *self-organization,* and complex system dynamics. It refers to discrete dynamic systems whose behavior is completely specified in terms of a local relation. Space is represented by a uniform grid, with each cell containing data (e.g., elevation). Time advances in discrete steps, and laws are applied at each step in each cell to determine its new state from that of its neighbors. Thus, the system's laws are local and uniform (Phipps, 1989; Smith, 1991). Usually used in the plural; sing: automaton.

cellular porosity A term applied originally by Howard and David (1936, p.1406) to equidimensional openings formed by dissolution. The term has since been applied to intraparticle porosity formed biotically within fossils. Choquette and Pray (1970, p.245) suggest that the term be abandoned because of its infrequent use and diverse application.

cellular soil *polygonal ground.*

cellule (cel'-lule) A subdivision of a marginal chamberlet in the outer part of the marginal zone of a foraminifer (as in Orbitolinidae), formed by primary and secondary partitions.

cellulose (cel'-lu-lose) A polymeric carbohydrate composed of glucose units, formula $(C_6H_{10}O_5)_x$, of which the permanent cell membranes of plants are formed, making it the most abundant carbohydrate.

cell wall A rigid wall outside the cytoplasmic membrane of the cells in most plants; commonly composed of cellulose but sometimes lignified.

cell-wall degradation matter *humic degradation matter.*

celsian (cel'-si-an) A colorless, white, or yellow monoclinic mineral of the *feldspar* group: $BaAl_2Si_2O_8$. It is the barium analogue of anorthite and is dimorphous with paracelsian.

cement [materials] (ce-ment') A manufactured gray powder which when mixed with water makes a plastic mass that will "set" or harden. It is combined with an *aggregate* to make *concrete.* Nearly all of today's production is *portland cement.* See also: *cement rock.*

cement [ore dep] Ore minerals, e.g. gold, that are part of or have replaced cement in the sedimentary use of the word.

cement [sed] Mineral material, usually chemically precipitated, that occurs in the spaces among the individual grains of a consolidated sedimentary rock, thereby binding the grains together as a rigid, coherent mass; it may be derived from the sediment or its entrapped waters, or it may be brought in by solution from outside sources. The most common cements are silica (quartz, opal, chalcedony), carbonates (calcite, dolomite, siderite), and various iron oxides. Others include clay minerals, barite, gypsum, anhydrite, and pyrite. Detrital clay minerals and other fine clastic particles may also serve as cements.

cementation [eng] (ce-men-ta'-tion) *grouting.*

cementation [metal] A process for separating gold and silver in which natural alloys are heated with cement.

cementation [sed] The diagenetic process by which coarse clastic sediments become lithified or consolidated into hard, compact rocks, usually through deposition or precipitation of minerals in the spaces among the individual grains of the sediment. It may occur simultaneously with sedimentation, or at a later time. Cementation may occur by secondary enlargement. Syn: *agglutination.*

cementation [soil] The binding-together of the particles of a soil by such cementing agents as colloidal clay, hydrates of iron, or carbonates. Three degrees of cementation are recognized: weakly cemented, strongly cemented, and indurated.

cement-bond log An *acoustic log* of compressional wave amplitude run shortly after cementing well casing to evaluate the casing-to-cement and cement-to-formation bonds. Abbrev: CBL.

cement clay A clay with a variable amount of calcium carbonate, used in the manufacture of cement.

cement deposits In the Black Hills of the U.S., gold-bearing Cambrian conglomerates believed to be ancient beach or stream-channel deposits.

cement gravel Gravel that is consolidated by some binding material such as clay, silica, or calcite.

cementing The operation whereby cement slurry is pumped into a *drill hole* and forced up behind the *casing,* to seal the casing to the walls of the hole and to prevent unwanted leakage of fluids into the hole or migration of oil, gas, or water between formations.

cement rock Any rock that is capable of furnishing cement when properly treated, with little or no addition of other material; specif. a massive, sparsely fossiliferous, clayey limestone that contains the ingredients (alumina, silica, lime) for cement in approximately the required proportions. Example: the Blackjack Creek Limestone Member of the Fort Scott Limestone in Kansas. Syn: *cement stone.*

cement stone A type of boundstone containing a high percentage of marine cement. Also spelled: cementstone.

cemetery mound (cem'-e-ter-y) A small hillock caused by the melting of surrounding ice wedges in areas of patterned ground. Cf: *thermokarst mound.*

cenology (ce-nol'-o-gy) An old term for the geology of surficial deposits.

Cenomanian (Ce-no-man'-i-an) A stage of the Standard Global Chronostratigraphic Scale: lowermost Upper Cretaceous (above Albian, below Turonian). The basal GSSP is in southeast France.

Cenophytic (Ce-no-phyt'-ic) A paleobotanic division of geologic time, signifying the time since the development of the angiosperms in the middle or late *Cretaceous*. Cf: *Aphytic; Archeophytic; Eophytic; Paleophytic; Mesophytic.* Syn: *Neophytic.*

cenosis (ce-no'-sis) *coenosis.*

cenosite (cen'-o-site) *kainosite-(Y).*

cenosphere (ce'-no-sphere) An accumulation of ash particles resulting from incomplete high-temperature combustion of pitch, fuel oil, and coal that can be common in palynological preparations (Miller and Jansonius, 1996).

cenote (ce-no'-te) Steep-walled *natural well* open to the surface and usually floored with water; generally caused by collapse of a cave roof. The term is used mainly for features in Yucatán, Mexico. Partial syn: *karst window; sótano.* Etymol: Spanish form of a Mayan word.

cenotypal (ce'-no-typ'-al) Said of a fine-grained porphyritic igneous rock having the appearance of fresh or nearly fresh extrusive rocks, such as those of Tertiary and Holocene age. This term and the term *paleotypal* were introduced to distinguish *neovolcanic* and *paleovolcanic* fine-grained igneous rocks; both are little used.

Cenozoic (Ce-no-zo'-ic) The upper erathem of the *Phanerozoic* Eonothem of the Standard Global Chronostratigraphic Scale, Also the time during which these rocks were formed, the Cenozoic Era, covering the time span between 65 Ma and the present. It is characterized paleontologically by the evolution and abundance of mammals and angiosperm plants. Also spelled: *Cainozoic; Kainozoic.*

cenozone (ce'-no-zone) *assemblage zone.*

cenozonite An alternative term for *assemblage zone* (Henningsmoen, 1961, p.68).

cenozoology (ce'-no-zo-ol'-o-gy) Zoology of existing animals without regard to those that are extinct. Cf: *neontology; paleobiology.*

censored layer A coarse surface layer on the *bed* of a river. It forms as matrix material is removed from around the surface-framework particles as *stage [hydraul]* increases. Cf: *armor; pavement [streams].*

centare (cen'-tare) A metric unit of area equal to one square meter, 0.01 are, or 10.76 square feet. Abbrev: ca.

Centaur A minor planet or asteroid in an orbit that crosses the orbit(s) of one or more of the outer (giant) planets, Jupiter, Saturn, Uranus, or Neptune. These objects are in very unstable, short-lived orbits and are probably derived from the *Kuiper Belt* beyond the orbit of Neptune.

centered lattice (cen'-tered) A crystal lattice in which the axes have been chosen according to the rules for the crystal system, and in which there are lattice points at the centers of certain planes as well as at their corners; thus a centered lattice has two, three, or four lattice points per unit instead of one, as in a *primitive lattice.* Syn: multiply centered lattice. See also: *base-centered lattice; one-face-centered lattice; face-centered lattice; body-centered lattice.*

centered optic axis figure *axial figure.*

center line (cen'-ter) A straight or curved line that continuously bisects a feature or figure (such as a stream, a strip of land, or the bubble tube in a spirit level); specif. the line connecting opposite corresponding corners of a quarter section or quarter-quarter section, or the line extending from the true center point of overlapping aerial photographs through each of the transposed center points.

center of gravity That point in a body or system of bodies through which the resultant attraction of gravity acts when the body or system is in any position; that point from which the body can be suspended or poised in equilibrium in any position.

center-of-gravity map (cen'-ter-of-grav'-i-ty) A *vertical-variability map,* or *moment map,* that shows the relative, weighted mean position of a lithologic type in terms of its distance from the top of a given stratigraphic unit, expressed as a percentage of total thickness of the unit. Cf: *standard-deviation map.*

center of instrument The point on the vertical axis of rotation (of a surveying instrument) that is at the same elevation as that of the collimation axis when that axis is in a horizontal position. In a transit or theodolite, it is close to or at the intersection of the horizontal and vertical axes of the instrument.

center of symmetry A point in a crystal structure through which every aspect of an array is repeated by *inversion [cryst];* it is a symmetry element. Syn: *inversion center; symmetry center.*

center point (a) The *principal point* of a photograph. (b) The central point from which a map projection is geometrically based.

center-to-center methods A group of methods for the determination of strain that use the spatial relationships between the centers of markers.

central axis (cen'-tral) In dasycladacean and cyclocrinitid algae and in receptaculitids, a vesicular, generally uncalcified central supporting structure on which branches or meroms are borne in tightly packed whorls or spirals.

central body The main part of a pollen grain or spore; e.g. the main part of a vesiculate pollen grain, as distinct from the vesicles, or the compact central part of a dinoflagellate cyst from which the projecting structures extend.

central capsule The mucoid or chitinous perforated internal skeleton or sac of a radiolarian, enclosing the nucleus and intracapsular cytoplasm.

central cavity The opening or space enclosed by the inner wall along the axis of archaeocyathid cups (TIP, 1972, pt. E, p.40). Cf: *internal cavity.*

central complex The core or central zone in which foraminiferal chamber passages bifurcate and anastomose in a reticulate pattern (as in Orbitolinidae).

central cylinder *stele.*

centrale (cen-trale') (a) A prominent plate at the center of the aboral surface of the disc in many asterozoans. It is the central plate of the primary circlet. Syn: *centrodorsal.* (b) A noncirriferous thecal plate typically occurring inside the infrabasal circlet of some articulate crinoids (such as *Marsupites* and *Uintacrinus*). (c) Any of up to four bones present in the wrist and ankle of primitive tetrapods, set between the proximal and distal rows of carpal or tarsal elements, respectively. Pl: centralia.

central eruption Ejection of debris and lava flows from a central point, forming a more or less symmetrical *volcano.*

Central European tektite A tektite from one of the four main tektite fields, formed about 15 Ma ago.

central illumination method A method for determining relative index of refraction using the *Becke test.*

central induction loop method An inductive electromagnetic method in which the transmitting and receiving coils are concentric.

centrallasite (cen-tral'-la-site) *gyrolite.*

central lumen The opening in an edrioasteroid *theca* that extends from the proximal ends of the ambulacral tunnels down through the oral frame into the underlying thecal cavity (Bell, 1976).

central meridian The line of longitude at the center of a map projection; the meridian about which the geometric properties of a map projection are symmetric and which is a straight line on the map. It is used to determine the directions of axes of plane coordinates. See also: *principal meridian.*

central muscle One of an anteriorly or medially placed pair of muscles in lingulid brachiopods, originating on the *pedicle valve,* and passing anteriorly and dorsally to the *brachial valve* (TIP, 1965, pt. H, p.142).

central node *azygous node.*

central peak A central area in a large crater, topographically higher than the surrounding crater floor. One or more peaks are commonly observed in the approximate centers of large lunar craters (such as Copernicus and Tycho); their average height is about 2% of the crater diameter. In terrestrial craters produced by impact or explosion, the central peaks are formed by uplift of material originally below the crater floor. Also called: central mound. Cf: *central uplift.*

central pit A variant of the *central peak* crater, in which the peak is replaced by a pit of comparable dimensions. It is common on the icy satellites and Mars.

central plate A small thin plate inside the marginal rim of some flattened early echinoderms, such as stylophorans, ctenocystoids, and some eocrinoids (TIP, 1968, pt. S, p.537).

central projection *gnomonic projection.*

central tendency The tendency for quantitative data to cluster around some central value. A measure of central tendency is any statistic that indicates the location of the central value; e.g., *median, mode,* and *mean.*

central uplift A central high area produced in an *impact crater* or an explosion crater by inward and upward movement of material below the crater floor. It is formed at a relatively late stage during the crater-forming event and is not due to long-term slow adjustment. Central uplifts are characteristic of *complex impact craters.* The total stratigraphic uplift is typically 10% of the rim-to-rim diameter. Cf: *central peak.*

central valley *rift valley [oceanog].*

central vent The opening at the Earth's surface of a volcanic conduit of cylindrical or pipelike form. Cf: *central eruption.*

centric (cen'-tric) Said of the texture of a rock in which the components are arranged about a center, either radially (e.g. *spherulitic*) or concentrically (e.g. *orbicular*). Obsolete.

centric diatom A diatom having basically radial symmetry; a member of the diatom order Centrales. Cf: *pennate diatom.*

centrifugal drainage pattern (cen-trif'-u-gal) *radial drainage pattern.*

centrifugal replacement Mineral replacement in which the host mineral is replaced from its center outward. Cf: *centripetal replacement.*

centrifugal selection Natural selection resulting in increasing variation.

centrifuge moisture equivalent (cen'-tri-fuge) *moisture equivalent.*

centripetal drainage pattern (cen-trip'-e-tal) A drainage pattern in which the streams converge inward toward a central depression; it may be indicative of a volcanic crater or caldera, a structural basin, a breached dome, a sinkhole, or a bolson. Cf: *radial drainage pattern.* See also: *internal drainage.*

centripetal replacement Mineral replacement in which the host mineral is replaced from its periphery inward. Cf: *centrifugal replacement.*

centripetal selection Natural selection resulting in decreasing variation.

centroclinal (cen-tro-cli'-nal) adj. Said of strata and structures that dip towards a common center. Ant: *quaquaversal.* Cf: *periclinal [geol].*

centroclinal fold *centrocline.*

centrocline (cen'-tro-cline) An equidimensional basin characteristic of cratonic areas, in which the strata dip towards a central low point. The term is little used in the U.S. Cf: *pericline [struc geol].* Adj: centroclinal. Ant: *quaquaversal.* Syn: *centroclinal fold.*

centrodorsal (cen-tro-dor'-sal) n. (a) A commonly cirriferous crinoid *columnal,* or semifused to fused columnals attached to the *theca* of comatulid articulate crinoids. (b) The *centrale* in many asterozoans. adj. (a) Central and dorsal. (b) Pertaining to a centrodorsal; e.g. "centrodorsal cavity" consisting of a depression on the ventral surface of a crinoid centrodorsal and containing a chambered organ and accessory structures.

centrogenous skeleton (cen-trog'-e-nous) The supporting rods generated at the cell center in acantharian radiolarians.

centroid (a) In physics, the center of mass, area, or volume. By extension, the multivariate *mean,* which is calculated in a manner analogous to calculation of the center of mass. (b) The geometric center of a polygon. The location of the centroid can be calculated as the average location of vertices defining polygon boundaries. May in some cases lie outside the polygon (Krzanowski et al., 1993). (c) In spatial information systems, the centroid is a point in a polygon to which attribute information about that specific area is linked (Krzanowski et al., 1993).

centronelliform (cen-tro-nell'-i-form) Said of a simple spear-shaped *loop* of a brachiopod (as in the subfamily Centronellinae), suspended free of the valve floor, and commonly bearing a median vertical plate in addition to the *echmidium* (TIP, 1965, pt. H, p.142).

centrosphere (cen'-tro-sphere) *barysphere.*

centrosymmetric (cen'-tro-sym-met'-ric) Said of a crystal having a center of symmetry.

centrum [paleont] (cen'-trum) (a) A differentiated central part of a sponge spicule. (b) The substance of a stem plate (columnal or cirral) of a crinoid, including luminal flanges. (c) The cylindrical part of a vertebra that chondrifies or ossifies around the notochord. Pl: centra.

centrum [seis] *focus [seis].*

CEP *circular error probable.*

cephalic (ce-phal'-ic) Pertaining to the head; esp. directed toward or situated on, in, or near the *cephalon,* such as the "cephalic shield" of a crustacean exoskeleton covering the head region and formed of fused tergites, or a "cephalic spine" carried by the cephalon of a trilobite.

cephalis (ceph'-a-lis) The first or apical chamber of a nasselline radiolarian.

cephalon (ceph'-a-lon) (a) The anterior or head region of an exoskeleton of a trilobite, consisting of several fused segments and bearing the eyes and mouth. (b) The most anterior *tagma* of a crustacean, bearing eyes, mouth, two pairs of antennae, mandibles, and two pairs of maxillae. See also: *head [paleont].* Pl: cephala.

cephalopod (ceph'-a-lo-pod') Any marine mollusk belonging to the class Cephalopoda, characterized by a definite head, with the mouth surrounded by part of the foot that is modified into lobelike processes with tentacles or armlike processes with hooklets or suckers or both. The external shell, if present, as in nautiloids, is univalve and resembles a hollow cone, which may be straight, curved, or coiled and is divided into chambers connected by a siphuncle; the shell is internal in nearly all present-day cephalopods and their fossil ancestors, such as the belemnites. The living *Nautilus* has an external shell. Most nautiloids and ammonoids are extinct cephalopods, generally valuable as index fossils; octopuses, squids, and cuttlefishes are common living cephalopods. Range, Upper Cambrian to Holocene.

cephalothorax (ceph'-a-lo-tho'-rax) The fused head and *thorax* of certain arthropods; e.g. the anterior part of the body of a crustacean, composed of united cephalic and thoracic somites and covered by the carapace, or the fore part of the body of a merostome in front of the *opisthosoma,* or the anterior part of the body of an arachnid, bearing six pairs of appendages. Cf: *prosoma; gnathothorax.*

cepstrum The Fourier-transform of a log frequency distribution (Sheriff and Geldart, 1995, p.298, 554).

cerargyrite (ce-rar'-gy-rite) *chlorargyrite.*

ceratite (cer'-a-tite) Any ammonoid belonging to the order Ceratitida, characterized by a shell having sutures with serrate lobes and, in some groups, by an ornamented shell. Range, Permian to Triassic.

ceratitic suture (ce-ra-tit'-ic) A type of *suture* in ammonoids characterized by small, rounded, unbroken saddles and finely denticulate lobes developed on a major set; specif. a suture in ceratites. Cf: *ammonitic suture; goniatitic suture; pseudoceratitic suture.*

ceratoid (cer'-a-toid) Said of a very slenderly conical, horn-shaped *corallite* of a solitary coral.

ceratolith (ce-rat'-o-lith) A horseshoe-shaped skeletal element of the coccolithophorid *Ceratolithus,* acting optically as a single unit of calcite.

cerchiaraite A vitreous deep green tetragonal mineral: $Ba_4(Mn,Fe,Al)_4O_3(OH)_3(Si_4O_{12})[Si_2O_3(OH)_4]Cl$.

cercopod (cer'-co-pod) *cercus.*

cercus (cer'-cus) Either of a pair of simple or segmented appendages situated at the posterior end of certain arthropods; e.g. a *caudal ramus* of a crustacean. Pl: cerci. Syn: *cercopod.*

ceresine (cer'-e-sine) A white wax that results from the bleaching of *ozocerite.*

cerianite-(Ce) (ce'-ri-an-ite'-ce) An amber, yellow, or dark greenish yellow cubic mineral: $(Ce^{4+},Th)O_2$. It usually contains some thorium. A member of the uraninite group.

cerine (ce'-rine) (a) *allanite.* (b) *cerite-(Ce).*

cerioid (cer'-i-oid) Said of a massive *corallum* in which the walls of adjacent polygonal corallites are closely united.

ceriopyrochlore-(Ce) A resinous brown cubic mineral: $(Ce,Ca)_2(Nb,Ta)_2O_6(O,OH)$. A member of the *pyrochlore* group.

cerite-(Ce) (ce'-rite-ce) A resinous reddish brown rhombohedral mineral: $(Ce,La,Ca)_9(Mg,Fe^{3+})(SiO_4)_6(SiO_3OH)(OH)_3$. Syn: *cerine.*

cerite-(La) A vitreous light yellow or pinkish brown rhombohedral mineral: $(La,Ce,Ca)_9(Fe,Ca,Mg)(SiO_4)(SiO_3OH)_4(OH)_3$.

cerium The native element cerium, found as a mineral in a lunar sample. Not approved by the International Mineralogical Association.

cernyite (cer'-ny-ite) A metallic gray tetragonal mineral occurring in pegmatites, intergrown with kesterite: Cu_2CdSnS_4.

cerolite (ce'-ro-lite) A yellow or greenish waxlike mixture of serpentine and stevensite.

cerotungstite-(Ce) (ce-ro-tung'-stite-ce) *yttrotungstite-(Ce).*

cerrito (cer-ri'-to) A small *cerro.* Also called: cerrillo.

cerro (cer'-ro) A term used in the SW U.S. for a hill, esp. a craggy or rocky eminence of moderate height. Etymol: Spanish.

ceruleite (ce-ru'-le-ite) A turquoise-blue triclinic mineral: $Cu_2Al_7(AsO_4)_4(OH)_{13}·12H_2O$. Also spelled: ceruléite.

cerulene (ce'-ru-lene) (a) A trade name for a form of calcite colored blue or green by azurite or malachite and used as a gemstone. (b) A term used less correctly for a blue variety of satin spar.

cerussite (ce-rus'-site) A colorless, white, yellowish, or grayish orthorhombic mineral of the *aragonite* group: $PbCO_3$. It is a common alteration product of galena and is a valuable ore of lead. Syn: *white lead ore; lead spar.*

cervandonite-(Ce) (cer'-van-don-ite-ce) An adamantine black monoclinic mineral: $(Ce,Nd,La)(Fe^{3+},Ti,Fe^{2+},Al)_3(Si,As)_3O_{13}$.

cervantite (cer-van'-tite) A white or yellow orthorhombic mineral: $Sb^{3+}Sb^{5+}O_4$. It was formerly regarded as identical with stibiconite.

cervelleite (cer-velle'-ite) A metallic white cubic mineral: Ag_4TeS.

cervical groove (cer'-vi-cal) In decapod crustaceans, a transverse groove somewhat parallel to the *postcervical groove* and placed before it. It extends upward from the confluence of the *hepatic* and *antennal grooves* (Holthuis, 1974, p.733).

cervical sinus An indentation at the front of the *carapace* of a cladoceran crustacean, exposing the rear part of the head.

cesanite (ce'-sa-nite) A white or colorless hexagonal mineral: $Na_7Ca_3(SO_4)_6(OH)·H_2O$. It is isotypic with apatite.

cesarolite (ce-sa'-ro-lite) A submetallic steel-gray mineral: $PbMn^{4+}{}_3O_6(OH)_2$. It occurs in spongy masses.

cesbronite (ces-bron'-ite) A green orthorhombic mineral: $Cu_5(Te^{4+}O_3)_2(OH)_6·2H_2O$.

cesium kupletskite (ce'-si-um ku-plet'-skite) A platy brown triclinic mineral of the *astrophyllite* group: $Cs_2Na(Mn,Fe)_7{}^{2+}(Ti,Nb)_2O_2(Si_4O_{12})_2(OH,F)_5$. It forms a series with kupletskite.

cesium-vapor magnetometer (ce'-si-um-va'-por) A type of *optically pumped magnetometer* that measures the magnitude of the absolute total magnetic intensity with extreme sensitivity by determining the Larmor frequency of cesium atoms. Cf: *rubidium-vapor magnetometer.*

cesplumbtantite (ces'-plumb-tant-ite) An adamantine colorless tetragonal mineral: $(Cs,Na)_2(Pb,Sb)_3(Ta_4O_{12})_2$.

cesstibtantite (ces-stib-tant'-ite) An adamantine colorless cubic mineral isostructural with microlite, and a member of the *pyrochlore* group: $Cs(Sb^{3+},Na)Ta_2(O,OH,F)_7$. Cf: *natrobistantite.*

cetineite (ce-tin'-e-ite) An orange-red hexagonal mineral formed by the weathering of slag and mined material: $K_7(Sb_2O_3)_6(SbS_3)_2(OH)·4H_2O$.

ceylonite (cey-lon'-ite) A dark green, brown, or black variety of spinel containing iron. Syn: *pleonaste; candite.* Also spelled: ceylanite; *zeylanite.*

cf. (a) Used in paleontology to indicate that a specimen is very closely comparable to, but not certainly the same as, those of a named species; it implies more certain similarity than does *aff.* (b) Used in this glossary and other reference works to mean "compare". Etymol: Latin "conferre", "to compare".

***c*-glide** A *glide plane* involving reflection across a plane containing the *c axis* combined with translation of one-half of a lattice translation along *c.* Cf: *a-glide; b-glide; d-glide; n-glide.*

chabazite (chab'-a-zite) A group name for zeolites of composition $A_{1-2}(Si,Al)_6O_{12}·H_2O$, where A = Ca, Na, K or Sr. Also spelled: chabasite.

chabazite-Ca A variously colored triclinic pseudo-rhombohedral zeolite mineral of the *chabazite* group: $CaAl_2Si_4O_{12}·6H_2O$.

chabazite-K A rhombohedral zeolite of the *chabazite* group with dominant K: $(K,Na,Ca)_2(Si,Al)_6O_{12}·6H_2O$.

chabazite-Na A white rhombohedral zeolite of the *chabazite* group with dominant Na: $(Na,K,Ca)_2(Si,Al)_6O_{12}·6H_2O$.

chabazite-Sr A colorless rhombohedral zeolite of the *chabazite* group with dominant Sr: $Sr(Si_4Al_2)O_{12}·6H_2O$.

chabourneite (cha'-bourne-ite) A submetallic black triclinic mineral: $(Tl,Pb)_{23}(Sb,As)_{91}S_{147}$.

chadacryst (chad'-a-cryst) (a) The enclosed crystal in a poikilitic texture. (b) A syn. of *xenocryst.*

Chadronian (Chad-ron'-i-an) North American land mammal age: early Oligocene (after Duchesnean, before Orellan).

chadwickite A dull yellow tetragonal mineral: UO_2HAsO_3.

chaemolith (chae'-mo-lith) *humic coal.*

chaetetid (chae-tet'-id) Any organism characterized by massive coralla composed of very slender aseptate corallites with imperforate walls and complete tabulae. The chaetetids are currently regarded as being related to the sponges (phylum Porifera) but have been variously classified as hydrozoans, anthozoans, and bryozoans. Range, Ordovician to Permian.

chaff peat Peat that is derived from fragments of plants.

chagrenate (cha'-gre-nate) Said of a smooth and translucent sculpture of palynomorphs. Its use is considered confused and is less preferred. Also spelled: *shagrinate;* shagreen(ate); chagranate.

chaidamuite A vitreous brown to yellowish brown triclinic mineral: $ZnFe^{3+}(SO_4)_2(OH){\cdot}4H_2O$.

chain [cart] A directed nonbranching sequence of nonintersecting line segments and (or) arcs bounded by nodes, not necessarily distinct, at each end. Area chain, complete chain, and network chain are special cases of *chain,* and share all characteristics of the general case as defined (FGDC, 1998a, p.61). Cf: *arc.*

chain [geomorph] A general term for any series or sequence of related natural features arranged more or less longitudinally, such as a chain of lakes, islands, seamounts, or volcanoes; esp. a *mountain chain* or other extended group of more or less parallel features of high relief.

chain [ore dep] adj. In mineral deposits, e.g. chromite, said of a crystal texture or structure in which a series of connected crystals resembles a linked or chainlike pattern.

chain [surv] (a) A measuring device used in land surveying, consisting of 100 links joined together by rings; specif. *Gunter's chain.* The term is commonly used interchangeably with *tape* although strictly a chain is a series of links and a tape is a continuous strip. (b) A unit of length prescribed by law for the survey of U.S. public lands and equal to 66 feet or 4 rods. It is a convenient length for land measurement because 10 square chains equals one acre.

chain codes A method of coordinate compaction used to represent lines on a regular grid. Each line segment in a chain code is represented by its length (in grid units) and a unique directional code (Krzanowski et al., 1993).

chain coral Any coral (esp. one belonging to the family Halysitidae) characterized, in plan view, by cylindrical, oval, or subpolygonal corallites joined together on two or three sides to form a branching, chainlike network.

chain crater One of several small aligned depressions on the surface of a planetary body, believed to be formed by either volcanic activity or by secondary impacts; more commonly applied to those of probable volcanic origin. See also: *crater chain.*

chain gage A type of *gage* used in determining water-surface elevation, consisting of a tagged or indexed chain, tape, or other form of line. It is used in situations in which the water surface is difficult to reach. Cf: *staff gage.*

chaining (chain'-ing) A term that was applied originally to measuring distances on the ground by means of a surveyor's chain, but later to the use of either a chain or a surveyor's tape. The term was formerly synonymous with *taping,* but "chaining" is now preferred (for historical and legal reasons) for surveys of the U.S. public-lands system and "taping" for all other surveys.

chainperson A surveyor's assistant who measures distances, marks measuring points, and performs related duties; specif. one who marks the tape ends in chaining or who measures distances with a tape. See also: *rod person.* Syn: *tape person.*

chain silicate *inosilicate.*

chain termination The termination of a *chain-width error* in a *biopyribole* by adjustment of chain widths in the *a* crystallographic direction.

chain width The number of single tetrahedral chains that are laterally linked to form the basic structural feature of inosilicates. Pyroxenes (chain width = 1) and micas (chain width = ∞) represent the extreme chain widths; amphiboles have chain width = 2, and nonclassical biopyriboles have finite chain widths > 2.

chain-width error A "mistake" in the crystal structure of a *biopyribole,* ranging from the occasional presence of a chain of anomalous width to a completely disordered sequence of chain widths.

chalcanthite (chal-can'-thite) A blue triclinic mineral: $CuSO_4{\cdot}5H_2O$. It is a minor ore of copper. Syn: *blue vitriol; copper vitriol; bluestone [mineral]; cyanosite.*

chalcedonic chert (chal-ce-don'-ic) A transparent, translucent, vitreous, milky, smoky, waxy, or greasy variety of *smooth chert,* generally buff or blue-gray, sometimes mottled (Ireland et al., 1947, p.1484); it breaks into splintery fragments with smooth conchoidal surfaces. In thin section, it is seen to consist of microcrystalline quartz crystals with a fibrous crystal habit. Cf: *novaculitic chert.*

chalcedonite (chal-ced'-o-nite) *chalcedony.*

chalcedony (chal-ced'-o-ny) (a) A cryptocrystalline variety of quartz. It is commonly microscopically fibrous, may be translucent or semitransparent, and has a nearly waxlike luster, a uniform tint, and a white, pale-blue, gray, brown, or black color; it has a lower density and lower indices of refraction than ordinary quartz. Chalcedony is the material of much chert, and often occurs as an aqueous deposit filling or lining cavities in rocks. In the gem trade, the name refers specif. to the light blue-gray or "common" variety of chalcedony. Varieties include carnelian, sard, chrysoprase, prase, plasma, bloodstone, onyx, and sardonyx. See also: *agate.* Var:*calcedony.* Syn: *chalcedonite.* (b) A general name for crystalline silica that forms concretionary masses with radial-fibrous and concentric structure and that is optically negative (unlike true quartz). (c) A trade name for a natural blue onyx.

chalcedony patch A milklike, semitransparent blemish in a ruby.

chalcedonyx (chal-ce-don'-yx) An onyx with alternating stripes or bands of gray and white. It is valued as a semiprecious stone.

chalchihuitl (chal'-chi-huitl) A Mexican term for any green stone that has been carved into a decorative or useful object, and sometimes any stone, regardless of color, that has been carved. It refers esp. to jadeite or chalchuite (turquoise), but sometimes to porphyry, serpentine, or smithsonite.

chalchuite (chal'-chu-ite) A blue or green variety of turquoise.

chalco- (chal'-co-) A prefix meaning "copper".

chalcoalumite (chal-co-al'-u-mite) A turquoise-green to pale-blue monoclinic mineral: $CuAl_4(SO_4)(OH)_{12}{\cdot}3H_2O$. Its Ni and NO_3 analogue is mbobomkulite.

chalcocite (chal'-co-cite) A black or dark lead-gray mineral: Cu_2S. It has a metallic luster, occurs in tetragonal or monoclinic crystals or as anhedral masses, and is an important ore of copper. Syn: *copper glance; chalcosine; redruthite; beta chalcocite; vitreous copper.*

chalcocyanite (chal-co-cy'-a-nite) A variously colored orthorhombic mineral: $CuSO_4$. Syn: *hydrocyanite.*

chalcodite (chal'-co-dite) *stilpnomelane.*

chalcolite (chal'-co-lite) *torbernite.*

Chalcolithic (Chal-co-lith'-ic) In archaeology, a transitional period between the *Neolithic* and *Bronze* ages; formerly sometimes called the Copper Age. Syn: *Eneolithic.*

chalcomenite (chal-co'-menite) A blue orthorhombic mineral: $CuSe^{4+}O_3{\cdot}2H_2O$.

chalconatronite (chal-co-nat'-ro-nite) A greenish-blue monoclinic mineral: $Na_2Cu(CO_3)_2{\cdot}3H_2O$. It occurs as incrustations on ancient bronze objects from Egypt.

chalcopentlandite (chal-co-pent'-land-ite) A hypothetical high-temperature sulfide of copper, nickel, and iron, now represented by mixtures of pentlandite and chalcopyrite.

chalcophanite (chal-co-phan'-ite) A purplish black or iron-black rhombohedral mineral: $(Zn,Fe^{2+},Mn^{2+})Mn^{4+}{}_3O_7{\cdot}3H_2O$.

chalcophile (chal'-co-phile) (a) Said of an element concentrated in the sulfide rather than in the metallic and silicate phases of meteorites. Cf: *lithophile; siderophile.* (b) Said of an element tending to concentrate in sulfide minerals and ores. Such elements have intermediate electrode potentials and are soluble in iron monosulfide. (Goldschmidt, 1954, p.24; Krauskopf, 1967, p.580). Examples are: S, Se, As, Fe, Pb, Zn, Cd, Cu, Ag.

chalcophyllite (chal-co-phyl'-lite) A green or bluish green rhombohedral mineral: $Cu_{18}Al_2(AsO_4)_4(SO_4)_3(OH)_{24}{\cdot}[36]H_2O$. Syn: *copper mica.*

chalcopyrite (chal-co-py'-rite) (a) A bright brass-yellow tetragonal mineral: $CuFeS_2$. It is generally found massive and constitutes the most important ore of copper. Syn: *copper pyrites; yellow copper ore; yellow pyrites; fool's gold.* (b) A group name for minerals with the formula $ABX2$, where A = Cu or Ag, B = Fe, Ga or In, and X = S or Se.

chalcosiderite (chal-co-sid'-er-ite) A green triclinic mineral: $CuFe^{3+}{}_6(PO_4)_4(OH)_8{\cdot}4H_2O$. It is a member of the *turquoise* group.

chalcosine (chal'-co-sine) *chalcocite.*

chalcosphere (chal'-co-sphere) An archaic term for that zone or layer of the Earth containing heavy-metal oxides and sulfides; it is the equivalent of *stereosphere.*

chalcostibite (chal-co-stib'-ite) A lead-gray orthorhombic mineral: $CuSbS_2$. It is the Sb analogue of emplectite. Syn: *wolfsbergite.*

chalcothallite (chal-co-thal'-lite) A metallic gray tetragonal mineral: $Tl_2(Cu,Fe)_6SbS_4$.

chalcotrichite (chal-co-trich'-ite) A capillary variety of cuprite occurring in fine slender interlacing fibrous crystals. Syn: *plush copper ore; hair copper.*

chalk (a) A soft, pure, earthy, fine-textured, usually white to light gray or buff limestone of marine origin, consisting almost wholly (90-99%) of calcite, formed mainly by calcareous tests of floating microorganisms (chiefly foraminifers) and of comminuted remains of calcareous algae (such as coccoliths and rhabdoliths), set in a structureless matrix of very finely crystalline calcite. The rock is porous, somewhat friable, and only slightly coherent. It may include the remains of bottom-dwelling forms (e.g. echinoderms and bivalves), and nodules of chert and pyrite. The best known and most widespread chalks are of Cretaceous age, such as those exposed in cliffs on both sides of the English Channel. Syn: *creta.* (b) A white, pure (or nearly pure), natural calcium carbonate, breaking into crumbly or powdery pieces. Etymol: Old English "cealc", from Latin "calx", "lime".

Chalk A stratigraphic term used in NW Europe for Upper Cretaceous. In Great Britain, it is divided into Lower Chalk (Cenomanian), Middle Chalk (Turonian), and Upper Chalk (Senonian).

chalkland (chalk'-land) A region underlain by chalk deposits, characterized by rolling hills, undulating plateaus, open expanses of pastureland, and dry valleys; e.g. SE England.

chalk rock (a) A soft, milky-colored rock resembling white chalk, such as talc, calcareous tufa, diatomaceous shale, volcanic tuff, or a bed of white limestone. (b) A chalky rock; specif. the Chalk Rock, a bed of hard nodular chalk, in places containing glauconite or phosphatic nodules, occurring at or near the base of the Upper Chalk in England (Himus, 1954, p.24).

chalk stream A stream flowing across or among the strata of a chalk deposit; a *bourne.*

chalky (chalk'-y) (a) Said of a soil or rock consisting of, rich in, or characterized by chalk. (b) Said of a limestone having the appearance of chalk. (c) Said of the porosity of such finely textured rocks as chalk and marl.

chalky chert A commonly dull or earthy, soft to hard, sometimes finely porous chert of essentially uniform composition, having an uneven or rough fracture surface, and resembling chalk (Ireland et al., 1947, p.1487). It is common in insoluble residues. Cf: *smooth chert; granular chert.* Syn: *dead chert; cotton chert.*

chalky marl A grayish marly rock rich in chalk and containing up to 30% clayey material; specif. the Chalk Marl near the base of the English Chalk.

challantite (chal'-lan-tite) *ferricopiapite.*

chalmersite (chal'-mers-ite) *cubanite.*

chalybeate (cha-lyb'-e-ate) An adj. applied to water strongly flavored with iron salts or to a spring yielding such water. Etymol: Greek, an ancient tribe of ironworkers in Asia Minor.

chalybite (chal'-y-bite) British syn. of siderite (ferrous-carbonate mineral).

chamber [paleont] (cham'-ber) (a) The fundamental unit of a foraminiferal *test,* consisting of a cavity and its surrounding walls. It is a variously shaped enclosure that invariably is connected by pores, intercameral foramina, or other passages leading to similar enclosures or to the exterior. (b) One of the regular, juxtaposed, hollow structures formed by the skeleton of sphinctozoan sponges. Also, a term that is often used as an abbreviated form of *flagellated chamber;* this usage is not recommended. (c) A *camera* of a cephalopod. Also, the *body chamber* of a cephalopod.

chamber [speleo] A *room* in a cave.

chambered (cham'-bered) Said of a vein or lode of brecciated, irregular texture, e.g. a *stockwork.*

chambered level A spirit-level tube with a partition near one end which cuts off a small air reservoir, so arranged that the length of the bubble can be regulated.

chamberlet (cham'-ber-let) A small chamber in a foraminifer, created by subdivision of chambers by axial or transverse secondary septula.

chamber passage One of the radial corridors consisting of centrally directed extensions of marginal chamberlets of foraminifers (such as Orbitolinidae).

chambersite (cham'-bers-ite) A colorless to deep purple orthorhombic mineral: $Mn_3B_7O_{13}Cl$. A member of the boracite group.

chaméanite (cha-mé-an'-ite) A metallic gray cubic mineral occurring in intergrowths with giraudite: $(Cu,Fe)_4As(Se,S)_4$.

chameleon-type diamond Diamond which changes color, generally from grayish green to bright yellow, when heated slightly. Such stones generally become yellower when left in darkness and have a strong yellow phosphorescence to ultraviolet light.

chameolith (cham'-e-o-lith') *humic coal.*

chamosite (cham'-o-site) An Fe-rich member of the *chlorite* group: $(Fe^{2+},Mg,Al,Fe^{3+})_6(Si,Al)_4O_{10}(OH,O)_8$. It is an important constituent of many oolitic and other bedded iron ores and frequently the chlorite of grain-coating chlorites in sandstones. It is not *berthierine.*

Champlainian (Cham-plain'-i-an) North American provincial series: Middle Ordovician (above Canadian, below Cincinnatian).

chance packing A random combination of systematically packed grains surrounded by or alternating with haphazardly packed grains (Graton and Fraser, 1935). The average porosity of a chance-packed aggregate of uniform solid spheres is slightly less than 40%.

chandelier Irregularly twisting, coarsely-crystalline *stalactite;* commonly branching; normally composed of gypsum.

Chandler motion (Chand'-ler) *polar wandering.*

Chandler wobble An aspect of the Earth's rigid body motion departing from simple or pure spin, because its angular-momentum vector is not precisely colinear with a principal axis of inertia. It is the free *nutation* of the Earth's spin, with a 428-day cycle, and causes a variation of instantaneous astronomical latitude (as defined in accordance with an instantaneous angular-velocity vector or axis of rotation) of amplitude about 0.4 sec-arc.

changbaiite (chang-bai'-ite) An adamantine rhombohedral colorless mineral: $PbNb_2O_6$.

change of color (a) An optical phenomenon consisting of a difference in color when a mineral or gemstone is moved about; specif. *labradorescence.* (b) An optical phenomenon consisting of a difference in color from daylight to artificial light, caused by selective absorption; e.g. that shown by alexandrite. Cf: *play of color.*

Changhsingian A stage of the Standard Global Chronostratigraphic Scale: uppermost stage of the Permian (uppermost stage of Lopingian Series; above Wuchiapingian, below Induan Stage of the Triassic). Some authors spell the term "Changxingian".

channel [coast] (chan'-nel) (a) A relatively narrow sea or stretch of water between two close landmasses and connecting two larger bodies of water (usually seas); e.g. the English Channel between England and France. (b) The deeper part of a moving body of water (as a bay, estuary, or strait) through which the main current flows or which affords the best passage through an area otherwise too shallow for navigation; it is often deepened by dredging. Also, a navigable waterway between islands or other obstructions, as on a lake.

channel [drill] A cavity or passage in a faulty cementing job behind the casing in a borehole. Syn: *holiday.*

channel [geophys] (a) A system of interconnected devices through which data may flow from source to recorder; for example, geophone, cable, amplifier, recorder. (b) A layer whose velocity is such that seismic energy may be trapped; see *channel wave.* (c) A paleo-river or submarine channel whose stratigraphy differs from the rock into which it is cut. The different stratigraphy may create velocity anomalies in earlier sediments. (d) A time or energy gate during which measurements are made.

channel [ice] *lead [ice].*

channel [ore dep] *channelway [ore dep].*

channel [paleont] (a) A groove of an invertebrate, such as one that winds down the *columella* near its base in some gastropod shells and terminates in the siphonal notch or in the canal. (b) An area lying between septal pinnacles and the peripheral ends of septa in the zone between parent corallite and offset during the early stages of increase in corals (Fedorowski and Jull, 1976, p.41).

channel [sed struc] (a) A linear current mark, larger than a groove, produced on a sedimentary surface, parallel to the current, and often preserved as a *channel cast.* It is 0.5-2 m wide, 20-50 cm deep, and up to 30 m long and is best developed in a *turbidite* sequence. (b) An erosional feature "that may be meandering and branching and is part of an integrated transport system" (Pettijohn and Potter, 1964, p.288).

channel [streams] (a) The bed where a natural body of surface water flows or may flow; a natural passageway or depression of perceptible extent containing continuously or periodically flowing water, or forming a connecting link between two bodies of water; a watercourse. Syn: *channelway [streams].* (b) The deepest or central part of the bed of a stream, containing the main current, and occupied more or less continuously by water; the thalweg. (c) A term used in quantitative geomorphology for a line or pattern of lines, without regard to width or depth, in the analysis of streams. Syn: *stream.* (d) An abandoned or buried watercourse represented by stream deposits of gravel and sand. (e) An artificial waterway, such as an open conduit, an irrigation ditch or canal, or a floodway. (f) An obsolete term for a stream or small river.

channel [volc] A narrow, sinuous flow channel, commonly formed in lava flows.

channel bar An elongate deposit of sand and gravel located in the course of a stream, esp. of a braided stream. Cf: *point bar.*

channel basin An obsolete term for a long, narrow proglacial valley, trench, or channel.

channel capacity The maximum flow that a given channel can transmit without overflowing its banks. See also: *bankfull stage.*

channel cast The cast of a channel that is generally cut in shale and filled with sand. Cf: *washout [sed struc].* Syn: *channel fill; gouge channel.*

channeled scabland (chan'-neled) *Scabland* deeply eroded. On the Columbia Plateau of eastern Washington it represents intense scouring by glacial meltwater. Cf: *coulee [geomorph].*

channeled upland *grooved upland.*

channel erosion Erosion in which material is removed by water flowing in well-defined courses; erosion caused by channel flow. Cf: *sheet erosion; rill erosion; gully erosion.* Syn: *concentrated wash.*

channel fill (a) An alluvial deposit in a stream channel, esp. one in an abandoned cutoff channel or where the transporting capacity of the stream is insufficient to remove material supplied to it. Several types of channel fills are also found in vertical facies sequences associated with ancient deep clastic seas such as slope-apron deposits and deep-water fan deposits (with olistostromes) (Reading, 1986). (b) *channel cast.*

channel flow Movement of surface runoff in a long narrow troughlike depression bounded by banks or valley walls that slope toward the channel; specif. *streamflow.* Cf: *overland flow.* Syn: *concentrated flow.*

channel frequency *stream frequency.*

channel geometry The description of the shape (form) of a given cross section within a limited reach of a river channel. See also: *river morphology.*

channel-gradient ratio *stream-gradient ratio.*

channeling The flow of fluid along preferred paths in the plane of a fracture.

channelization (chan'-nel-i-za'-tion) The straightening and deepening of a stream channel, to permit the water to move faster or to drain marshy acreage for farming.

channel length *stream length.*

channel line The line of the fastest current or the strongest flow of a stream; it generally coincides with (and is sometimes known as) the thalweg. Cf: *thread.*

channel maintenance constant *constant of channel maintenance.*

channel-maintenance flow The discharge necessary to maintain specified components of river form, such as pools and riffles.

channel-margin linear bars *Sandbars* that border the outer edges of an ebb channel on an *ebb delta.*

channel morphology *river morphology.*

channel-mouth bar A bar built where a stream enters a body of standing water; it results from a decrease in the stream's velocity.
channel net The pattern of all stream channels within a drainage basin. Syn: *channel network.*
channel network *channel net.*
channel order *stream order.*
channel pattern The configuration in plan view of a limited reach of a river channel as seen from an airplane (Leopold and Wolman, 1957, p.39-40). Recognized patterns include meandering, braided, sinuous, and relatively straight. See also: *river morphology.* Syn: *river pattern.*
channel porosity A system of pores in which the openings are markedly elongate and have developed independently of the textural or fabric elements of the rock (Choquette and Pray, 1970, p.245).
channel precipitation Part of direct runoff; precipitation that falls directly onto the surfaces of lakes and streams. It is usually considered to be a part of surface runoff (Chow, 1964, p.14-2).
channel sample A composite rock sample, generally taken across the face of a formation or vein to give an average value.
channel sand A sand or sandstone deposited in a stream bed or other channel eroded into the underlying strata. If exposed, such sands may contain gold or other valuable minerals; if buried, they may be oil or gas reservoirs. See also: *shoestring sand.*
channel segment *stream segment.*
channel splay *floodplain splay.*
channel spring A type of *depression spring* issuing from the bank of a stream that has cut its channel below the water table.
channel storage In a stream channel, or over its flood plains, the volume of water at a given moment.
channel unit In streams, a bed feature that is one or more channel widths in length, e.g., a *riffle* or *pool.*
channel wave A type of *guided wave* that is propagated in a low-velocity layer within the Earth, or in the ocean or atmosphere. Syn: *normal mode propagation; trapped wave.* See also: *seam wave; wave guide.* Cf: *Stoneley wave; sofar.*
channelway [ore dep] (chan'-nel-way) An opening or passage in a rock through which mineral-bearing solutions or gases may move. Syn: *channel [ore dep]; feeder [eco geol]; feeding channel.*
channelway [streams] *channel [streams].*
channel width The distance across a channel or a stream, measured from bank to bank near bankfull stage. Symbol: w.
channery (chan'-ner-y) (a) Thin, flat coarse fragments of limestone, sandstone, or schist, having diameters as large as 150 mm (6 in.). (b) A term used in Scotland and Ireland for gravel.
chantalite (chan'-ta-lite) A colorless tetragonal mineral: $CaAl_2SiO_4(OH)_4$.
chaoite (cha'-o-ite) A mineral: a hexagonal form of elemental carbon. Polymorphous with diamond, graphite, and lonsdaleite, and of meteoritic origin.
chaos [geol] (a) The disorganized state of primordial matter and infinite space before the ordered universe was created. Ant: *cosmos.* (b) A branch of mathematical physics involving applications of non-linear mathematics to random, disorderly or irregular phenomena such as discontinuities, critical points, fracture, explosion, etc. characteristic of geologic materials, processes and records.
chaos [planet] A term established by the International Astronomical Union used to describe a distinctive area of broken terrain on planetary surfaces (USGS, 2003).
chaos [struc geol] (cha'-os) A structural term proposed by Noble (1941, p.963-977) for a gigantic breccia associated with thrusting, consisting of a mass of large and small blocks of irregular shape with very little fine-grained material, in a state of semidisorder. Type example: the Amargosa chaos, a widespread deposit in the Death Valley area of California, consisting of an extremely complex mosaic of enormous, tightly packed, often unshattered but internally coherent, random blocks and masses of formations of different ages occupying a definite zone above a major thrust fault. The blocks range in size from pods a meter in diameter to blocks more than 800 m in length. Cf: *mélange [sed]; megabreccia.*
chaos theory (ka'-os) Theory dealing with the unpredictability that is intrinsic to non-linear systems. The effects of a small (perhaps unmeasurable) perturbation may grow progressively larger and thus prevent long-term predictability. Unstable behavior in dynamic systems that is neither random nor deterministic. Cf: *fractal.*
chaotic [geomorph] (cha-ot'-ic) Said of a surface or land area consisting of short, jumbled ridges and valleys.
chaotic [petrology] Said of a massive, unstratified tuff consisting of a mixture of fine and coarse material distributed within the mass without recognizable physical pattern or *fabric [sed].*
chaotic crystal texture Randomly oriented crystals of saline minerals within a muddy or sandy matrix.
chaotic terrain Regions on Mars, first seen in 1969 on Mariner 6 images, that are topographically low and consist of irregular ridges, apparently formed at the expense of higher cratered terrain (Leighton et al., 1969). It has been interpreted to have formed by removal, perhaps catastrophically, of large amounts of subsurface materials.
chapeau de fer (cha-peau'-de-fer) *gossan.* Etymol: French, iron hat or iron cap.
chapeiro (cha-pei'-ro) An isolated coral reef with a mushroomlike morphology. Etymol: Portuguese "chapeirão", "broad-brimmed hat".
chapmanite (chap'-man-ite) An olive green or deep yellow monoclinic mineral: $Sb^{3+}Fe^{3+}_2(SiO_4)_2(OH)$. It is the Sb analogue of bismutoferrite.
char [coal] The solid carbonaceous residue that results from incomplete combustion of organic material. It can be burned for heat, or, if pure, processed for production of activated carbon for use as a filtering medium. See also: *charcoal; coke.*
char [streams] A term applied in India to a newly formed alluvial tract or flood-plain island formed of silt and sand deposited in the bed of a deltaic river, such as a sandbank left dry on the subsidence of a river after the flood season. Etymol: Hindu. Syn: *chur; diara.*
character [paleont] (char'-ac-ter) Any specifiable, definable, or recognizable attribute (often a morphologic feature) of an organism or taxon. A character may appear in several possible expressions or aspects (each termed a *character state*) in different organisms or taxa.
character [seis] A recognizable aspect of a sequence of seismic events (or waveforms) that distinguishes it from others. It is usually a frequency or phasing effect, and is often not defined precisely and hence is dependent on subjective judgment.
character argumentation *character polarization.*
characteristic fossil (char'-ac-ter-is'-tic) A fossil species or genus that is characteristic of a stratigraphic unit (formation, zone series, etc.) or time unit. It is either confined to the unit or is particularly abundant in it. Inappropriate syn: *index fossil.* Syn: *diagnostic fossil.*
characteristic value *eigenvalue.*
characteristic vector *eigenvector.*
character polarization In cladistic analysis, the process of determining which of two homologous characters is plesiomorphic or apomorphic by determination of the transformation of an evolutionarily novel character state from the preexisting or plesiomorphic character state (Brooks et al., 1984, p.8; Wiley et al., 1991, p.9).
character state Specific expression of *character* exhibited by an individual taxon (Brooks et al., 1984, p.79).
charco (char'-co) (a) A term applied in SW U.S. to a small natural depression in which water collects, as in a desert alluvial plain; a *tank* or a water hole. Syn: *represo.* (b) A natural or artificial pool of water occupying a charco and supplied by desert floods. Also, a pool in a stream bed or a puddle in a playa. Etymol: Spanish, "pond, small lake".
charcoal (char'-coal) An impure carbon residue of the burning of wood or other organic material in the absence of air. It is black, often porous, and able to absorb gases. Like *coke,* it can be used as a fuel. See also: *char [coal].*
charge (a) The sediment that is carried into a channel in unit time. (b) In seismic work, the explosive combination employed for generating seismic energy, specified by the quantity and type of explosive used.
chargeability (charge-a-bil'-i-ty) The primary unit of measurement in time-domain induced-polarization surveys. It is the area under the decay curve between two delay times after cessation of the transmitted current. Usually expressed in millivolt-seconds per volt.
charlesite (charles'-ite) A colorless trigonal mineral: $Ca_6(Al,Si)_2(SO_4)_2B(OH)_4(OH,O)_{12}\cdot 26H_2O$. A member of the *ettringite* group.
charmarite-2H A variously colored hexagonal mineral of the *quintinite* group: $Mn_4Al_2(OH)_{12}CO_3\cdot 3H_2O$.
charmarite-3T A variously colored trigonal mineral of the *quintinite* group: $Mn_4Al_2(OH)_{12}CO_3\cdot 3H_2O$.
Charmouthian (Char-mouth'-i-an) Stage in Great Britain: Lower Jurassic (above Sinemurian, below Domerian).
Charnian Orogeny (Char'-ni-an) An orogeny that supposedly occurred late in Precambrian time in the English Midlands; the dating is questionable, however, and the term has only local significance. The Charnian folds trend NW-SE and seem to have had a posthumous influence on structures in the surrounding Paleozoic. It is named for Charnwood Forest, Leicestershire, where small inliers of Precambrian sedimentary rocks emerge.
charnockite (char'-nock-ite) An orthopyroxene-bearing granite. Most classifications require that quartz constitute at least 20% of the felsic constituents and that the ratio of alkali feldspar to total feldspar fall between 40% and 90%. Tobi (1971) places the quartz content at 10-60% and the ratio of alkali feldspar to total feldspar at 35-90% to correspond with Streckeisen's (1967) definition of granite. Although its origin (igneous or metamorphic) is controversial, charnockite is commonly found only in granulite-facies terranes, and high temperature and pressure are generally thought to be essential to its formation. The name is derived from that of Job Charnock (d.1693), the founder of Calcutta, India, from whose tombstone the rock was first described by Holland in 1893.
charnockite series A series of plutonic rocks compositionally similar to the granitic rock series but characterized by the presence of orthopyroxene.
charoite (cha'-ro-ite) A violet monoclinic mineral: $(K,Na)_5(CA,Ba,Sr)_8Si_8O_{46}(OH,F)\cdot nH_2O$. It occurs in Siberia near Lake Baikal and is used as an ornamental stone.
Charophyta (Cha'-ro-phy-ta) *charophyte.*

charophyte (char'-o-phyte) A member of a group of complex, mainly freshwater *green algae* of the phylum Charophyta (stoneworts). Their range is Late Silurian to the present, and they are the probable ancestors of the land plants. The egg-containing structure or oogonium of these algae is commonly fossilized and termed a *gyrogonite.*

chart (a) A special-purpose map; esp. one designed for purposes of navigation, such as a *hydrographic chart* or a *bathymetric chart.* (b) A base map conveying information about something other than the purely geographic. (c) *weather map.* (d) Obsolete syn. of *map.*

chart datum The *datum* to which levels on a nautical chart and tidal predictions are referred. In the United States, the National Ocean Service's nautical charts use a datum calculated from tidal observations averaged over a 19-year period called the *National Tidal Datum Epoch.* The chart datum is a legally defined term used to draw official charts used for navigation and to delineate the extent of the public and private lands in the littoral zone. The definitions vary from country to country, from state to state within the United States, and even within some states.

chart of the nuclides A plot of the atomic numbers and mass numbers of all known isotopes of all elements.

chartology (char-tol'-o-gy) *cartography.*

chartometer (char-tom'-e-ter) An instrument for measuring distances on charts or maps, such as the length of a stream in a drainage basin on a topographic map. See also: *opisometer.* Syn: *map measurer.*

chasm (a) A deep breach, cleft, or opening in the Earth's surface, such as a yawning fissure or narrow gorge; e.g. the Ausable Chasm near Keeseville, N.Y. (b) A deep, wide elongated gap in the floor of a cave. Syn: *abyss [geomorph].*

chasma (chas'-ma) A term established by the International Astronomical Union for a large canyon on planetary surfaces (USGS, 2003). Most are thought to be of structural as opposed to fluvial origin. Generally used as part of a formal name for a landform, such as Coprates Chasma on Mars (Mutch et al., 1976, p.57). Etymol: Greek *chasma,* gulf, open mouth. Pl: chasmata.

chasmic archeopyle (chas'-mic arch'-e-o-pyle) An *excystment* opening in *dinocysts* and *acritarchs* formed by a linear to slightly curved rupture, usually in a medial position (Williams et al., 2000, p.39).

chasmophyte (chas'-mo-phyte) A plant growing in the crevices of a rock; a *saxifragous* plant.

chassignite (chas'-si-gnite) A group of olivine-rich cumulate achondrites similar to the Chassigny Meteorite, whose young age and petrologic, chemical, and isotopic composition suggest origin on the same body as the shergottites and the nakhlites, probably the planet Mars. See: *SNC meteorite.*

chathamite (chat'-ham-ite) A variety of nickel-skutterudite containing much iron.

chatkalite (chat'-ka-lite) A metallic tetragonal mineral isotypic with mawsonite: $Cu_6FeSn_2S_8$.

chatoyancy (cha-toy'-an-cy) An optical phenomenon, possessed by certain minerals in reflected light, in which a movable wavy or silky sheen is concentrated in a narrow band of light that changes its position as the mineral is turned. It results from the reflection of light from minute, parallel fibers, cavities or tubes, or needlelike inclusions within the mineral. The effect may be seen on a cabochon-cut gemstone, either distinct and well-defined (as the narrow, light-colored streak in a fine chrysoberyl cat's-eye) or less distinct (as in the usual tourmaline or beryl cat's-eye). Var: chatoyance.

chatoyant (cha-toy'-ant) adj. Said of a mineral or gemstone possessing *chatoyancy* or having a changeable luster or color marked by a narrow band of light. n. A chatoyant gem.

chattermark [beach] (chat'-ter-mark) A crescent-shaped mark on a wave-worn pebble, such as flint, caused by "hammering" of a beach by wave action.

chattermark [glac geol] One of a series of small, closely spaced, short curved scars or cracks (smaller than a crescentic fracture) made by vibratory chipping of a firm but brittle bedrock surface by rock fragments carried in the base of a glacier. Each mark is roughly transverse to the direction of ice movement (although a succession of such marks is parallel to that direction), and usually convex toward the direction from which the ice moved (its "horns" point in the direction of ice movement). The term has been applied loosely to any glacial crescentic mark.

chattermark [struc geol] Any mark, pit, or scratch made on a rock surface by frictional sliding or indentation. Chattermarks can be caused by the material embedded in the bottom of a glacier, or they can occur on a fault surface. Cf: *slip-mark; vibration mark.*

Chattian (Chat'-ti-an) A stage of the Standard Global Chronostratigraphic Scale: upper Oligocene (above Rupelian, below Aquitanian Stage of the Miocene).

Chautauquan (Chau-tau'-quan) North American provincial series: Uppermost Devonian (above Senecan, below Mississippian).

chayesite (chayes'-ite) A blue hexagonal mineral: $K(Mg,Fe^{2+})_4Fe^{3+}Si_{12}O_{30}$. A member of the *milarite* group.

Chayes point counter An instrument used for petrographic modal analysis. A pattern of regularly spaced traverses, along which are regularly spaced points, is placed over a thin section; at each point the mineral is identified and then mechanically tabulated. Cf: *Hunt-Wentworth recording micrometer.*

Chazyan (Cha-zy'-an) North American stage: Middle Ordovician (below Blackriverian, above Whiterockian).

CH chondrite (or CH meteorite) Meteorite closely related to the carbonaceous chondrites, but with some properties (e.g., high metal, low FeS and volatile abundances) that seem inconsistent with formation in the solar nebula.

check dam A dam designed to retard the flow of water in a channel, used esp. for controlling soil erosion. Syn: *sabo dam.*

checkerboard topography (check'-er-board) A landscape characterized by a repeating pattern in the relief, such as the "diaper pattern of rectangles" due to a fracture system as seen on the Elizabethtown topographic quadrangle, Adirondack Mountains, N.Y. (Hobbs, 1901, p.150; 1911b, p.131 and plate 9).

checker coal (check'-er) Rectangular grains of anthracite.

check shot A shot into a well seismometer to determine the time-depth relationship. Used to check the results of integrating a continuous velocity or sonic log. See also: *well survey.*

cheek One of the two lateral or pleural parts of the *cephalon* of a trilobite, anterior to and typically much lower and flatter than the *glabella.* See also: *fixed cheek; free cheek.* Syn: *gena.*

cheilostome (chei'-lo-stome) Any ectoproct bryozoan belonging to the order Cheilostomata and characterized by the presence of a movable operculum over the orifice of the zooecia. Adj: cheilostomatous.

chekhovichite An adamantine gray monoclinic mineral: $Bi_2Te^{4+}{}_4O_{11}$.

chela (che'-la) (a) The pincer-like claw or organ borne by certain of the limbs of arthropods; e.g. the distal part of a crustacean limb consisting of opposed movable and immovable fingers and usually involving *dactylus* and *propodus,* or the pincer of an arachnid appendage formed by a rigid process of the penultimate joint and a movable last joint. (b) A siliceous, monaxonic sponge spicule (microsclere) consisting of an arcuate shaft at each end of which is a recurved, cup-like expansion, either lobed or toothed. See also: *isochela; anisochela.* Pl: chelae.

chelation (che-la'-tion) The reaction between a metallic ion and a complexing agent, generally organic, with the formation of a ring structure and the effective removal of the metallic ion from the system. It is significant in chemical weathering.

chelicera (che-lic'-er-a) One of the pre-oral appendages of all Chelicerata (subphylum of Arthropoda), corresponding to the second antennae of crustaceans, but modified for piercing or biting, and composed of two or three segments (as in arachnids) or of three or four(?) joints with the distal ones forming a *chela* (as in merostomes). Pl: chelicerae. Cf: *antenna.*

chelicerate (che-lic'-er-ate) Any terrestrial (*arachnid*) or aquatic (*merostome*) arthropod belonging to the subphylum Chelicerata, characterized chiefly by paired preoral appendages. Range, Cambrian to Holocene. Cf: *pycnogonid.*

cheliped (che'-li-ped) Any thoracopod-bearing chelae; e.g. one of the pair of legs that bears the large chelae in decapod crustaceans.

chelkarite (chel-kar'-ite) A colorless orthorhombic mineral: $CaMgB_2O_4Cl_2{\cdot}7H_2O$(?).

chelogenic (che-lo-gen'-ic) Said of a cycle of continental evolution; shield-forming. The term is rarely used.

Chelonia (Che-lo'-ni-a) An order of the reptilian subclass Anapsida, characterized by a carapace formed from modified ribs and dermal plates; it includes the turtles and tortoises. Range, Upper Triassic to Holocene.

cheluviation (che-lu'-vi-a'-tion) *Eluviation* under the influence of chelating agents.

chelyabinskite A white orthorhombic mineral, product of a burning coal dump: $(Ca,Mg)_3Si(OH)_6(SO_4,CO_3)_2{\cdot}9H_2O$. Not approved by the International Mineralogical Association.

chemawinite (che-ma'-wi-nite) A pale-yellow to dark-brown variety of retinite found in decayed wood at Cedar Lake in Manitoba. Syn: *cedarite.*

chemical activity (chem'-i-cal) *activity [chem].*

chemical composition [mineral] The weight percent of the elements (generally expressed as oxides) composing a mineral.

chemical composition [petrology] The weight percent of the elements (generally expressed as certain oxide molecules) composing a rock. Syn: *composition [petrology].*

chemical demagnetization A technique of partial *demagnetization* involving treatment by acid or other reagents to selectively remove one magnetically ordered mineral, usually fine-grained hematite, while leaving the remanent magnetization of another unaffected. Cf: *alternating-field demagnetization; thermal demagnetization.*

chemical equilibrium The state of a chemical system under which there are no net spontaneous changes occurring in the physical or chemical properties or in the quantities of any of the substances making up the system.

chemical erosion *corrosion [geomorph].*

chemical exfoliation A type of *exfoliation* caused by a volume increase induced by changes in the bulk chemical composition of the rock.

chemical fossil A chemical trace of an organism that has been mostly destroyed by diagenetic processes. Cf: *molecular paleontology.*

chemical gaging A type of *stream gaging* in which velocity of flow is measured by introducing a chemical of known saturation into the stream and then measuring the amount of dilution.

chemical grout Any grouting material characterized by being a pure solution with no particles in suspension (ASCE, 1980, p.805).

chemical limestone A limestone formed by direct chemical precipitation.

chemical magnetization *chemical remanent magnetization.*

chemical mining The extraction of valuable constituents of an orebody in place, by chemical methods such as leaching or dissolution. See also: *solution mining; in-situ mining.*

chemical oceanography The study of the composition and chemical changes of ocean water, its dissolved and suspended material, and the geographic distribution and temporal variation of its chemical features.

chemical oxygen demand The amount of oxygen required for the oxidation of the organic matter in a water sample or a water body. Abbrev: COD. Cf: *biochemical oxygen demand.* Syn: *oxygen demand.*

chemical potential (a) An intensive variable, defined as the change in Gibbs free energy of a component of the system or a single phase with the change in moles of that component, while the number of moles of all other components, the pressure, and the temperature remain constant. (b) Partial molar *Gibbs free energy,* usually symbolized as μ: i.e., the increase in free energy of a system or phase (due to addition or removal of an infinitesimal amount of a component in the absence of change in P or T), divided by the amount of the component added or removed.

chemical remanence *chemical remanent magnetization.*

chemical remanent magnetization A geologically stable remanent magnetization caused by the volume growth of magnetically ordered mineral grains in the presence of a magnetic field, e.g., during such processes as oxidation, reduction, exsolution, and precipitation. Syn: *chemical remanence; chemical magnetization; crystallization remanent magnetization; crystallization magnetization.* Abbrev: CRM.

chemical residue A *residue* formed by chemical weathering in place; e.g. a deposit of sand resulting from the removal by solution of nitrates from a Chilean niter bed.

chemical rock (a) A sedimentary rock composed primarily of material formed directly by precipitation from solution or colloidal suspension (as by evaporation) or by the deposition of insoluble precipitates (as by mixing solutions of two soluble salts); e.g. gypsum, rock salt, chert, or tufa. It generally has a crystalline texture. (b) A sedimentary rock having less than 50% detrital material (Krynine, 1948, p.134). Cf: *detrital rock.*

chemical unconformity An unconformity or stratigraphic boundary determined by chemical analysis, such as in the case of a limestone formation whose basal part has a higher concentration of impurities (silica, magnesia, sulfur) due to the presence of inwashed fine clastic and organic material (Landes, 1957).

chemical water *water of hydration.*

chemical weathering The process of weathering by which chemical reactions (hydrolysis, hydration, oxidation, carbonation, ion exchange, and solution) transform rocks and minerals into new chemical combinations that are stable under conditions prevailing at or near the Earth's surface; e.g. the alteration of orthoclase to kaolinite. Cf: *mechanical weathering.* Syn: *decomposition; decay [weath].*

chemoautotrophic (chem'-o-au'-to-troph'-ic) Said of an organism that obtains nourishment from chemical reactions of inorganic substances. Syn: *chemotrophic.*

chemocline (chem'-o-cline) The zone or layer of a lake or marine basin in which the chemical properties (concentration of major ions) change rapidly with water depth and which corresponds to the boundary between well-mixed and poorly mixed parts of the water column, such as the *mixolimnion* and the *monimolimnion* of a *meromictic lake* (Wetzel, 2001, p.83). Cf: *halocline; pycnocline.*

chemofacies (chem-o-fa'-cies) A term used by Keith and Degens (1959, p.40) to designate "all the chemical elements that are collected, precipitated, or adsorbed from the aqueous environment or fixed by chemical reactions within the bottom muds" and intended "only as a convenience in discussing chemical differences among environmental groups of sediments" (as in differentiating between marine and freshwater sediments).

chemogenic (chem-o-gen'-ic) Said of a rock or mineral that was deposited directly from solution without biological mediation, e.g. travertine, in contrast to clastic, bioclastic, or organogenic limestones.

chemography (chem-og'-ra-phy) The graphical representation of the compositions of minerals in terms of their components. Minerals in a binary system are depicted as points along a straight line, in a ternary system as points on or within a triangle, etc.

chemolithotrophic (chem'-o-lith'-o-troph'-ic) Said of an organism that obtains its nourishment by oxidation of inorganic compounds. Cf: *chemoorganotrophic.*

chemoorganotrophic (chem'-o-or-ga'-no-troph'-ic) Said of an organism that obtains its nourishment by the oxidation of organic compounds. Cf: *chemolithotrophic.*

chemostratigraphy The element of stratigraphy that deals with chemical signals, such as carbonate cycles and isotope cycles, as recorded in the stratigraphic record.

chemosynthesis The process of producing basic nutrients by the breakdown of chemicals such as methane or hydrogen sulfide, rather than by photosynthesis involving light and chlorophyll.

chemotaxis (chem-o-tax'-is) *Taxis [ecol]* resulting from chemical stimuli. Cf: *chemotropism.*

chemotrophic (chem-o-troph'-ic) *chemoautotrophic.*

chemotropism (chem-ot'-ro-pism) *Tropism* resulting from chemical stimuli. Cf: *chemotaxis.*

Chemungian (Che-mung'-i-an) North American stage: Upper Devonian (above Fingerlakesian, below Cassadagan).

chenevixite (chen-e-vix'-ite) A dark-green to greenish-yellow monoclinic mineral: $CuFe^{3+}(AsO_4)(OH)_2$. It is the iron analogue of *luetheite.*

chengbolite (cheng'-bol-ite) *moncheite.*

chengdeite A metallic steel-black cubic mineral: Ir_3Fe. The Ir analogue of isoferroplatinum.

chenier (che'-nier) A long, narrow, wooded *beach ridge* or sandy or shelly hummock, 1-6 m high, forming roughly parallel to a prograding shoreline seaward of marsh and mud-flat deposits (as along the coast of SW Louisiana), enclosed on the seaward side by fine-grained sediments, and resting on foreshore or mudflat deposits. It is well drained and fertile, often supporting large evergreen oaks or pines on higher areas; its width ranges from 45 to 450 m, and its length may be several tens of kilometers. Etymol: French "chêne", "oak". Cf: *marsh bar.*

chenier plain A strand plain, occupied by cheniers and intervening mud flats with marsh and swamp vegetation (Otvos and Price, 1979). Bight-coast chenier plains exist in SW Louisiana and in Guiana with a maximum length of 700 km. Chenier plains develop when (1) substantial quantities of river-supplied mud become available for nearshore marine transport and coastal mudflat deposition; (2) a balance exists between longshore sand transport, deposition, and erosional sand-winnowing; and (3) these two conditions alternate.

chenite (chen'-ite) A blue triclinic mineral: $Pb_4Cu(SO_4)_2(OH)_6$.

Chenj'iang fauna A Lower Cambrian assemblage of marine invertebrates from Yunnan Province, China, largely dominated by arthropods, in which the soft parts of the organisms are well-preserved in the fossils. This fauna includes many of the same species as the *Burgess Shale fauna.*

chenxianite A metallic brownish-black rhombohedral mineral: $AlMn_{11}O_{16}(OH)_9$. Not approved by the International Mineralogical Association.

cheralite (cher'-a-lite) A green monoclinic mineral: $(Ca,Ce,Th)(P,Si)O_4$. It is a member of the *monazite* group, and is essentially an intermediate member of a solid-solution series apparently extending between $CePO_4$ (monazite) and $CaTh(PO_4)_2$ (an artificial compound).

cheremnykhite A greenish yellow to dark green orthorhombic mineral: $Pb_3Zn_3TeO_6(VO_4)_2$.

cherepanovite (cher-e-pan'-ov-ite) A metallic black orthorhombic mineral: RhAs. A member of the *westerveldite* group.

chernikovite (cher'-ni-kov-ite) A pale yellow to greenish yellow tetragonal mineral: $(H_3O)(UO_2)(PO_4)\cdot 3H_2O$. A member of the *natroautunite* group.

chernovite-(Y) (cher'-nov-ite-y) A colorless, pale yellow, or greenish yellow tetragonal mineral: $YAsO_4$. A member of the *xenotime* group.

Chernozem (Cher'-no-zem) A great soil group of the 1938 classification system, a group of zonal soils whose surface horizon is dark and highly organic, below which is a lighter-colored horizon and an accumulation of lime. It is developed under conditions of temperate to cool subhumid climate (USDA, 1938). Most Chernozems are now classified as *Borolls* and *Ustolls.* Etymol: Russian tschernosem, "black earth". Also spelled: Chernozyom; Tchornozem; Tschernosiom; Tschernosem. Cf: *Chestnut soil.* Partial syn: *black earth.* Not used in current U.S. system of soil taxonomy.

chernykhite (cher'-ny-khite) A dioctahedral olive to dark green monoclinic vanadium-bearing mica of ideal composition $(Ba,Na,NH_4)(V^{3+},Al)_2(OH)_2[(Si,Al)_4O_{10}]$.

cherokite (cher'-o-kite) The dense brown residual sand constituting the cement of the chert breccias in the zinc-mining district of Joplin, Missouri.

cherry coal (cher'-ry) A soft, black, noncaking bituminous coal with a resinous luster that ignites and burns readily.

chert A hard, extremely dense or compact, dull to semivitreous, microcrystalline or cryptocrystalline sedimentary rock, consisting dominantly of interlocking crystals of quartz less than about 30 µm in diameter; it may contain amorphous silica (opal). It sometimes contains impurities such as calcite, iron oxide, and the remains of siliceous and other organisms. It has a tough, splintery to conchoidal fracture, and may be white or variously colored gray, green, blue, pink, red, yellow, brown, and black. Chert occurs principally as nodular or concretionary segregations (chert nodules) in limestones and dolomites, and less commonly as areally extensive layered deposits (*bedded chert*); it may be an original organic or inorganic precipitate or a replacement product. The term flint is essentially synonymous, although it has been used for the dark variety of chert (Tarr, 1938). Cf: *jasper; black chert.* Syn: *hornstone [rock]; white chert; silexite [sed].*

chert-arenite (chert-ar'-e-nite) (a) A term used by McBride (1963, p.668) for a *quartzarenite* containing more than 25% chert. (b) A term used by Folk (1968, p.124) for a *litharenite* in which the main rock fragment is chert.

chertification (chert'-i-fi-ca'-tion) A type of silicification in which fine-grained quartz or chalcedony is introduced into limestones, as in the Tri-State mining district of the Mississippi Valley (Fowler and Lyden, 1932).

chert nodule A dense, irregular, usually structureless, sometimes fossiliferous diagenetic segregation of *chert,* ranging from regular disks up to 5 cm in diameter to large, highly irregular, tuberous bodies up to 30 cm in length, frequently occurring distributed through calcareous strata. The larger nodules, of rounded contour, are marked by warty or knobby extensions. Examples include the cherts in the Mississippian limestones of the upper Mississippi Valley region, and the flint nodules of the Cretaceous chalk of England and France. See also: *nodular chert.*

cherty (chert'-y) Containing chert; e.g. a "cherty limestone" so siliceous as to be worthless for the limekiln, or a "cherty iron carbonate" consisting of siderite intimately interbedded with chert.

chervetite (cher-ve'-tite) A colorless or yellow monoclinic mineral: $Pb_2V^{5+}{}_2O_7$.

chesofite A product of a burning coal heap: $Ca_9(Si_2O_7)_3{\cdot}CaCl_2$. Not approved by the International Mineralogical Association.

chessexite (ches'-sex-ite) A silky white orthorhombic mineral: $Na_4Ca_2(Mg,Zn)_3Al_8(SiO_4)_2(SO_4)_{10}(OH)_{10}{\cdot}40H_2O$.

chessman spicule (chess'-man) *discorhabd.*

chessylite (ches'-sy-lite) A term commonly used in France and elsewhere for *azurite.*

Chesterian (Ches-ter'-i-an) A North American stage: uppermost Mississippian (above Meramecian, below Morrowan of Pennsylvanian).

chesterite (ches'-ter-ite) A colorless to light pinkish-brown orthorhombic biopyribole mineral: $(Mg,Fe^{2+})_{17}Si_{20}O_{54}(OH)_6$.

chesterlite (ches'-ter-lite) Microcline feldspar from Chester County, Penna.

chestermanite (ches'-ter-man-ite) A grayish green to black orthorhombic mineral: $Mg_2(Fe^{3+},Mg,Al,Sb^{5+})O_2BO_3$. A member of the *orthopinakiolite* group.

Chestnut soil (Chest'-nut) A zonal great soil group consisting of soils with a moderately thick, dark-brown A horizon over a lighter colored horizon that is above a zone of calcium carbonate accumulation. Not used in current U.S. system of soil taxonomy.

chevee (che-vee') A flat carved cameo gemstone with a smooth, concave depression. Cf: *cuvette [gem].*

chevkinite (chev'-kin-ite) A group name for monoclinic minerals with the general formula $A_4B_5O_8(Si_2O_7)_2$, where A = Ce, La, Ca or Sr, and B = Mg, Fe, Al, Ti or Zr.

chevkinite-(Ce) A reddish brown to black monoclinic mineral of the *chevkinite* group: $(Ce,La)_4(Ti,Fe)_5O_8(Si_2O_7)_2$.

chevron cast (chev'-ron) The cast of a *chevron mark.*

chevron cross-bedding Cross-bedding that dips in different or opposite directions in alternating or superimposed beds, forming a chevron or herringbone pattern. Syn: *herringbone cross-bedding; zigzag cross-bedding.*

chevron dune A V-shaped dune formed in a vegetated area where strong winds blow in a constant direction.

chevron fold A *kink fold,* the limbs of which are of equal length. Cf: *zigzag fold.*

chevron groove A V-shaped furrow on the *cardinal area* for the insertion of *ligament* in certain bivalve mollusks (as in some of the superfamily Arcacea and in early forms of the superfamilies Pteriacea and Pectinacea). Cf: *duplivincular.*

chevron halite Halite crystals, usually in continuous layers, with liquid, vapor, and/or solid inclusions arranged in a chevron (or rarely triradiate) pattern. Inclusions form along growing crystal faces during very rapid crystal growth, usually in saturated or supersaturated solutions. Such crystal crusts are commonly observed growing along the floor or sides of shallow brine water bodies. Crystals growing more slowly have few inclusions and are clear.

chevron mark A *tool mark* consisting of chevron-like depressions arranged in a row, the points of the chevrons generally but not always pointing downstream. Originally proposed by Dunbar and Rodgers (1957, p.195). Cf: *reversed chevron mark; vibration mark.* Syn: *herringbone mark.*

Chézy equation (Che'-zy) An equation used to compute the velocity of uniform flow in an open channel: mean velocity of flow (V) equals the Chézy coefficient (C) times the square root of the product of hydraulic radius in feet (R) times the slope of the channel (S). Cf: *Manning equation.* See also: *Kutter's formula.*

chiastoclone (chi-as'-to-clone) A desma (of a sponge) in which several subequal, zygome-bearing arms radiate from a very short central shaft, giving the spicule an X-shaped profile.

chiastolite (chi-as'-to-lite) An opaque variety of andalusite containing black carbonaceous impurities arranged in a regular manner so that a section normal to the longer axis of the crystal shows a black Maltese cross formed as a result of the pushing aside of the impurities into definite areas as the crystal grew in metamorphosed shales. It has long been used for amulets, charms, and other inexpensive novelty jewelry. Syn: *cross-stone; crucite; macle [mineral].*

chiastolite slate A rock formed by contact metamorphism of carbonaceous shale, characterized by prominent cleavage or schistosity and the presence of conspicuous chiastolite (andalusite variety) crystals in a fine-grained groundmass.

chiavennite (chia-venn'-ite) An orange-yellow orthorhombic mineral: $CaMn^{2+}Be_2Si_5O_{13}(OH)_2{\cdot}2H_2O$.

chibinite (chib'-i-nite) A eudialyte-bearing *nepheline syenite* distinguished from *lujavrite* by its smaller amount of mafic components, which are in compact aggregates of thick rather than acicular crystals, and by having the eudialyte in patches in the interstices rather than as euhedral crystals. Also spelled: khibinite. The rock was named by Ramsay in 1898 for Khibina (Uruptek), Kola Peninsula, Russia. Obsolete.

Chickasawhay (Chick-as'-a-whay) North American (Gulf Coast) stage: Oligocene (above Vicksburgian). Syn: Chickawhayan.

chickenwire anhydrite (chick'-en-wire) An evaporite texture in which irregularly polygonal nodules of anhydrite (or gypsum), 1-5 cm in diameter, are separated by thin, darker stringers of other minerals, generally carbonates or clays. It is believed by some to be diagnostic of sabkha deposition; it also may be the result of porphyroblastic recrystallization (Dean et al., 1975). See also: *nodular texture.*

Chideruan (Chi-der'-u-an) European stage: Upper Permian (above Kazanian, below Triassic). Syn: Tatarian.

childrenite (chil'-dren-ite) A vitreous yellowish-brown brittle bladed monoclinic (pseudo-orthorhombic) mineral: $Fe^{2+}Al(PO_4)(OH)_2{\cdot}H_2O$. The manganese analogue is *eosphorite.*

Chile-loeweite (Chil'-e-loe'-we-ite) *humberstonite.*

Chile saltpeter Naturally occurring sodium nitrate; *soda niter* occurring in caliche in northern Chile. Cf: *saltpeter.*

chilidial plate (chi-lid'-i-al) One of a pair of posterior platelike extensions of the walls of the *notothyrium* of certain brachiopods, commonly forming lateral boundaries of the *cardinal process.*

chilidium (chi-lid'-i-um) The triangular plate covering the apex of the *notothyrium* of certain brachiopods, commonly convex externally and extending for a variable distance ventrally over the proximal end of the *cardinal process.*

chillagite (chil'-la-gite) A variety of wulfenite containing tungsten.

chilled margin The border or outer zone of an igneous intrusion that is finer grained than the interior due to more rapid cooling. Syn: *chill zone.*

chill zone *chilled margin.*

chiluite An adamantine yellow hexagonal mineral: $Bi_3Te^{6+}Mo^{6+}O_{10.5}$.

Chimaeriformes (Chi'-mae-ri-form'-es) The sole order of the chondrichthyan subclass *Holocephali,* including the living ratfishes or chimaerae and three extinct groups of uncertain relations. All are characterized by holostylic jaw suspension and teeth reduced to a few crushing plates.

chimney [coast] (chim'-ney) (a) An angular, columnar mass of rock, smaller than a *stack,* isolated on a *wave-cut platform* by differential wave erosion of a *sea cliff.* (b) A *blowhole;* a *spouting horn.*

chimney [ore dep] *pipe [ore dep].*

chimney [speleo] In a cave, a rounded vertical passage or opening. Cf: *dome; domepit; vertical shaft.*

chimney [volc] A conduit through which magma reaches the Earth's surface. Cf: *vent; pipe [volc].* Syn: *feeder [volc].*

chimney rock A chimney-shaped column of rock rising above its surroundings or isolated on the face of a steep slope; a small, weathered outlier shaped like a sharp pinnacle; a stack formed by wave erosion. Syn: *pulpit rock.*

china clay A commercial term for low iron *kaolin* obtained from china-clay rock after washing, and suitable for use in the manufacture of chinaware. Sometimes spelled: *China clay.* Cf: *porcelain clay.*

china-clay rock Kaolinized granite composed chiefly of quartz and kaolin, with muscovite and tourmaline as possible accessories. The rock crumbles easily in the fingers. Also spelled: *China-clay rock.* Cf: *china stone [ign].*

china stone [clay] A rock composed mainly of sericite and quartz or sericite, kaolinite, and quartz for use in the manufacture of porcelain (Yanyi, 1987). Syn: *porcelain stone; petuntse.*

china stone [ign] Partially kaolinized granite containing quartz, kaolin, and sometimes mica and fluorite. It is harder than *china-clay rock* and is used to make a glaze in the manufacture of china. Syn: *petuntse; porcelain stone.* Cf: *Cornish stone.*

china stone [sed] A fine-grained, compact Carboniferous mudstone or limestone found in England and Wales.

chine (a) A term used in England (esp. in Hampshire and in the Isle of Wight) for a narrow, deep ravine, gorge, or cleft, cut in a soft, earthy cliff by a stream descending steeply to the sea. (b) A ridge or crest of rocks.

chinglusite A black amorphous mineral of doubtful validity: $(Na,K)_2(Mn,Ca)_5(Ti,Zr)_3Si_{14}O_{41}{\cdot}9H_2O$.

chinook (chi-nook') A warm dry downslope wind. It descends the leeward or east slopes of the Rocky Mountains and warms by adiabatic compression. See also: *foehn.*

chiolite (chi'-o-lite) A snow-white tetragonal mineral: $Na_5Al_3F_{14}$.

chip (a) A shallow break on a diamond or colored gemstone that extends from a facet junction or girdle edge. (b) A small fragment from a crystal; specif. a *diamond chip.*

chipping (chip'-ping) Abrasion of a rock fragment resulting in the flaking-off of its corners.

chip sample A series of chips of ore or rock taken at regular intervals across an exposure.
chip yard *forest bed.*
chiral twinning (chi'-ral) *optical twinning.*
chi-square test (chi'-square) A statistical test that employs the sum of values given by the quotients of the squared difference between observed and expected (theoretical) frequencies divided by the expected frequency. It enables assessment of *goodness-of-fit,* association, or commonality in a population, and is used to determine equivalency of observed sample and expected population.
chitin (chi'-tin) A resistant organic compound with the same basic carbohydrate structure as cellulose, but nitrogenous because some hydroxyl groups are replaced by acetamide groups (i.e. it is a repeating unit of N-acetylglucosamine instead of glucose). It is a common constituent of various invertebrate skeletons such as insect exoskeletons and foraminiferal inner tests, and also occurs in hyphae and spores of fungi.
chitinous (chi'-tin-ous) Consisting of *chitin.*
chitinozoan (chi'-tin-o-zo'-an) A marine fossil of the group Chitinozoa (ca. 25-1,000 µm), of unknown animal affinity, Late Cambrian to latest Devonian; cylindrical to flask-shaped to subspherical with a wide variety of sculpture. Chitinozoans occur as individuals, in chains, or in clusters and are of unknown, robust organic-chemical composition. Named by Eisenack (1931) who noted the resemblance of their walls to chitin.
chiton (chi'-ton) An invertebrate marine molluscan animal, class Amphineura, the shell of which consists of eight overlapping calcareous valves or plates. It is popularly called the coat-of-mail shell. Syn: *polyplacophoran.*
chkalovite (chka'-lov-ite) A vitreous white orthorhombic mineral: $Na_2BeSi_2O_6$.
chladniite A colorless rhombohedral meteorite mineral: $Na_2CaMg_7(PO_4)_6$. A member of the *fillowite* group.
chladnite [meteorite] (chlad'-nite) An obsolete designation of achondritic stony meteorites (aubrites and diogenites) composed essentially of enstatitic pyroxene.
chladnite [mineral] Pure meteoritic *enstatite.*
chlamydospore (chla-myd'-o-spore) An acid-resistant, thick-walled, nondeciduous spore rich in stored reserves, such as a unicellular *resting spore* developed by *hyphae* of certain fungi, usually intercalary but occasionally terminal.
chloanthite (chlo-an'-thite) *nickel-skutterudite.*
chloraluminite (chlor-a-lu'-mi-nite) A white to yellow rhombohedral mineral: $AlCl_3{\cdot}6H_2O$.
chlorapatite (chlor-ap'-a-tite) (a) A variously colored hexagonal or monoclinic mineral of the *apatite* group: $Ca_5(PO_4)_3Cl$. (b) An apatite mineral in which chlorine predominates over fluorine and hydroxyl.
chlorargyrite (chlor-ar'-gyr-ite) A white, pale yellow, greenish, or gray cubic waxlike mineral that darkens on exposure to light: AgCl. It occurs in the weathering zones of silver-sulfide deposits and it represents an important ore of silver. Syn: *cerargyrite; horn silver.*
chlorartinite A white rhombohedral mineral of the *artinite* group: $Mg_2CO_3Cl(OH){\cdot}3H_2O$.
chlorastrolite (chlor-as'-tro-lite) A mottled, green variety of *pumpellyite* used as a semiprecious stone, occurring as grains or small nodules of a radial, fibrous structure in geodes in basic igneous rocks. It resembles prehnite, and is found in the Lake Superior region (esp. on Isle Royale).
chlorbartonite A submetallic brown-black tetragonal mineral: $K_6Fe_{24}S_{26}(Cl,S)$. The Cl analogue of bartonite.
chlorellestadite (chlor-ell'-e-stad-ite) A yellowish green or pale rose hexagonal mineral with the apatite structure: $Ca_5(SiO_4,SO_4,PO_4)_3(Cl,F)$. A member of the *britholite* group.
chloride equivalent The mass of dissolved chloride ion that would be present in a given mass of seawater if each bromide ion and iodide ion were replaced by a chloride ion. Usual units are grams/kilogram or ppt.
chlorides (chlo'-rides) A miner's or prospector's term for ores containing silver chloride.
chlorine log A *radioactivity log* designed to indicate chlorine content and hence salinity of pore water. Now largely replaced by the *pulsed-neutron-capture log.* Syn: *salinity log.*
chlorinity (chlo-rin'-i-ty) The chloride content of seawater, measured by mass, or grams per kilogram of seawater, and including the chloride equivalent of all the halides.
chlorite (chlor'-ite) A group of platy, monoclinic, usually greenish minerals with the general formula $(R_{2+},R_{3+})_{4\text{-}6}(Si,Al)_4O_{10}(OH,O)_8$. There are four subgroups of this 2:1 layer clay mineral (the interlayer hydroxyl sheet is to be treated like other interlayer material), (1) trioctahedral chlorite (the most common chlorites) i.e., both the octahedral sheet sandwiched between tetrahedral sheets and the interlayer one are trioctahedral; (2) dioctahedral chlorite with both octahedral sheets dioctahedral, e.g. donbassite; (3) di,trioctahedral chlorite with the octahedral sheet in the 2:1 layer dioctahedral, but with the hydroxyl sheet trioctahedral, e.g. cookeite or sudoite; and (4) tri,dioctahedral with the 2:1 layer trioctahedral, but with the hydroxyl sheet dioctahedral. No examples of this mineral have yet been found. The most common chlorites, the trioctahedral ones, are named according to the dominant cation: Fe-rich is chamosite, Mg-rich is clinochlore, Ni-rich is nimite, and Mn-rich is pennantite (Bailey et al., 1979). Cf: *structural terms.*
chlorite schist A schist in which the main constituent, chlorite, imparts a schistosity by parallel arrangement of its flakes. Quartz, epidote, magnetite, and garnet may be present, the last two often as conspicuous porphyroblasts.
chlorite/smectite A mixed-layered clay mineral with chlorite and smectite components. Some workers prefer to think of part of this series as mixed-layered chlorite/corrensite. Abbrev: C/S. Cf: *chlorite; corrensite; smectite.*
chloritic shale (chlo-rit'-ic) A poorly laminated shale containing a variety of angular to subrounded mineral particles of silt size, including unstable types, characterized by feldspar sometimes exceeding quartz in abundance, and by chlorite often abundant in the finer matrix. It is commonly associated with graywacke (high-rank or feldspathic graywacke).
chloritoid (chlor'-i-toid) Micaceous brittle monoclinic or triclinic minerals: $(Fe^{2+},Mg,Mn)Al_2SiO_5(OH)_2$. They occur in dull-green or dark-green to gray or grayish-black masses of brittle folia in metamorphosed argillaceous sedimentary rocks, and are related to the brittle micas. Their structures are based on local closest-packing.
chlormagaluminite (chlor'-mag-a-lu'-mi-nite) A colorless or yellow-brown hexagonal mineral: $(Mg,Fe^{2+})_4Al_2(OH)_{12}(Cl_2,CO_3){\cdot}2H_2O$. A member of the *manasseite* group.
chlormanasseite A colorless to greenish brown mineral of the *manasseite* group: $(Mg,Fe^{2+})_5Al_3(OH)_{16}(Cl,OH)_3{\cdot}3H_2O$.
chlormanganokalite (chlor-man'-ga-no-ka'-lite) A yellow rhombohedral mineral: K_4MnCl_6. The manganese analogue of rinneite.
chlorocalcite (chlor-o-cal'-cite) A white orthorhombic mineral: $KCaCl_3$.
Chlorococcales (Chlo-ro-coc-ca'-les) An order of the Division Chlorophyta (green algae), containing unicellular and colonial species. Colonial chlorococcales include *Botryococcus* (Precambrian to Holocene), *Pediastrum* (Early Cretaceous to Holocene), *Scenedesmus* (Early Cretaceous to Holocene), and *Tetrastrum* (Triassic?; Early Cretaceous to Holocene).
chloromagnesite (chlor-o-mag'-ne-site) A colorless rhombohedral mineral: $MgCl_2$.
chloromelanite (chlor-o-mel'-a-nite) (a) A dark-green to nearly black variety of jadeite. (b) A solid solution of roughly equal amounts of diopside, jadeite, and acmite.
chloromenite A vitreous green monoclinic mineral: $Cu_9O_2(Se^{4+}O_3)_4Cl_6$.
chloropal (chlor-o'-pal) (a) A name originally applied to a deep-green, opal-like mineral that was later shown to be a crystalline clay mineral and renamed *nontronite.* (b) A greenish variety of common opal from Silesia.
chlorophaeite (chlor-o-phae'-ite) A mineraloid closely related to chlorite in composition (hydrous silicate of magnesium, iron, and calcium) and found in the groundmass of tholeiitic basalts, where it occupies spaces between feldspar laths, forms pseudomorphs after olivine, or occurs in veinlets and amygdules. It is pale green when fresh, but may be dark green, brown, or red in weathered rocks.
chlorophane (chlor'-o-phane) A variety of fluorite that emits a bright-green light when heated.
chlorophoenicite (chlo-ro-phoe'-ni-cite) A gray-green to purplish pleochroic acicular monoclinic mineral: $(Mn,Mg)_3Zn_2(AsO_4)(OH,O)_6$. It is isostructural with magnesium-chlorophoenicite.
chlorophyll (chlo'-ro-phyll) Generally a mixture of two waxy pigments: chlorophyll a, $C_{55}H_{72}O_5N_4Mg$, blue-black, and chlorophyll b, $C_{55}H_{70}O_6N_4Mg$, yellow-green, which occurs in plasmic bodies (chloroplasts) of plants and serves as a catalyst in photosynthesis. Other forms of chlorophyll occur in diatoms, algae, etc.
chlorophyll *a* A pigment in phytoplankton that can be used to measure the abundance of phytoplankton.
chlorophyll coal A variety of dysodile which contains chlorophyll that can be extracted by alcohol.
Chlorophyta (Chlo'-ro-phy-ta) *green algae.*
chlorospinel (chlor'-o-spi-nel') A grass-green variety of spinel containing some copper.
chlorothionite (chlor-o-thi'-o-nite) A bright-blue orthorhombic secondary mineral: $K_2Cu(SO_4)Cl_2$.
chlorotile (chlor'-o-tile) *agardite-(Y).*
chloroxiphite (chlor-o-xiph'-ite) A dull-olive or pistachio-green monoclinic mineral: $Pb_3CuO_2Cl_2(OH)_2$.
cho A rainy-season torrent carrying sand from the Himalayan foothills onto a plain below. Etymol: Panjabi, connoting "a bed of loose boulders, gravel, and sand, indicating rapid erosion" (Stamp, 1961, p.103). Also spelled: choh.
Choanichthyes (Cho-a-nich'-thy-es) *Sarcopterygii.*
choanocyte (cho-an'-o-cyte) An endoderm cell of a sponge, bearing a distinct tubular collarlike contractile protoplasmic rim, surrounding the base of a flagellum. Choanocytes line the inner surfaces of canals and/or spongocoel. Adj: choanocytal. Syn: *collar cell.*
choanoderm (cho-an'-o-derm) A single layer of choanocytes in a sponge; a choanocytal membrane.

choanosome (cho-an′-o-some) The inner layer of a sponge containing choanocyte-lined cavities (flagellated chambers).

chockstone A rock wedged between the walls of a cave passage.

chocolate tablet structure A type of *boudinage* where a relatively rigid layer is stretched in all directions, resulting in crossing boudin necks. The surface of the layer resembles that of a sheet of chocolate tablets.

choke [drill] An orifice or constriction. It can be used in the *tubing* of a well and is produced to measure or control the flow rate and producing pressure, especially of oil or gas.

choke [speleo] An area in a cave that is blocked by debris.

chokedamp (choke′-damp) *blackdamp.*

Chokrakian (Chok-rak′-ian) Ponto-Caspian stage: lower middle Miocene (above Tarkhanian, below Konkian). Equiv. Langhian.

choloalite (cho-lo′-a-lite) A green cubic mineral: $(Pb,Ca)_3(Cu,Sb)_3Te_6O_{18}Cl$.

choma (cho′-ma) A ridgelike deposit of dense shell substance delimiting a tunnel in a fusulinid. Pl: chomata. Cf: *parachoma.*

Chondrichthyes (Chon-drich′-thy-es) A class of vertebrates including fish with skeletons of cartilage rather than bone; esp. the sharks.

chondrite [meteorite] (chon′-drite) A *meteorite* containing chondrules and a primitive (solar-like) bulk composition. Chondrites are sediments consisting of particles formed in the solar nebula, but often including secondary alteration features resulting from impact, heat and interactions with aqueous fluids. Adj: chondritic. Cf: *achondrite.* See also: *ordinary chondrite; carbonaceous chondrite; enstatite chondrite.*

chondrite [paleont] A common *trace fossil* of the "genus" *Chondrites,* consisting of plantlike, regularly ramifying tunnel structures that neither cross each other nor anastomose but radiate around a central vertical tube. It is interpreted as a dwelling or feeding burrow, probably made by a marine worm. It is commonly called a *fucoid.*

chondrodite (chon′-dro-dite) A dark-red, orange-red, or yellow monoclinic mineral of the humite group: $(Mg,Fe^{2+})_5(SiO_4)_2(F,OH)_2$. It commonly occurs in contact-metamorphosed dolomites. Also spelled: *condrodite.*

chondrophore (chon′-dro-phore) A relatively prominent process with a hollowed-out surface for holding or attaching the internal ligament (*resilium*) of a bivalve mollusk. See also: *resilifer.*

Chondrostei (Chon-dro′-ste-i) An infraclass of ray-finned bony fish which includes the living sturgeons and a variety of late Paleozoic forms near the stem of all ray-fins.

chondrule (chon′-drule) A ferromagnesian silicate spheroidal grain in meteorites, the shape produced by surface-tension effects in melt. Most chondrules are porphyritic, with olivine and orthopyroxene the main phenocryst phases. Mean diameters vary from group to group and range from 0.2 to 1.3 mm.

chone An inhalant canal penetrating a cortex in a sponge, often leading from a vestibule to a *subcortical crypt.*

chonetid (cho-ne′-tid) Any articulated brachiopod belonging to the suborder Chonetidina, characterized chiefly by a functional foramen located outside the delthyrium. Range, Lower Silurian (possibly Upper Ordovician) to Lower Jurassic.

chop hill A term used in Nebraska for a *sand hill.*

chorate cyst (chor′-ate) A *dinoflagellate* cyst that bears processes or septa with a length of >30% of the diameter of the inner body (Fensome et al., 1993; Williams, 2000, p.39). See also: *proximate cyst; proximochorate cyst.*

chord The horizontal distance between ripple crests, measured normal to the crestlines. The term was introduced by Allen (1968, p.61) to replace the term "wavelength," which is still commonly used. Syn: *ripple wavelength; ripple-mark wavelength; ripple spacing.*

Chordata (Chor-da′-ta) A phylum of animals characterized by a notochord at some stage in the life cycle. In adults of most taxa the notochord is enclosed or replaced by a bony or cartilaginous spinal column. The phylum may or may not be considered to include the *Protochordata.*

chorismite (cho-ris′-mite) Megascopically composite rock that consists of two or more petrogenetically dissimilar materials of uncertain or doubtful origin (Dietrich and Mehnert, 1961). The term was first introduced without genetic connotation, as a replacement for *migmatite.* Five types of chorismite were outlined. Rarely used.

choristid (cho-ris′-tid) adj. Said of a sponge having a skeleton containing tetraxonic *megascleres* and lacking *desmas.* n. A choristid sponge; specif. a sponge of the order Choristida, class Demospongiae.

choristopore (cho-ris′-to-pore) The most advanced position of reproductive organs in dasycladacean algae, in which cysts are in gametangia located in the second- or third-order branches (laterals). Cf: *cladospore.*

C horizons Horizons or layers, excluding hard bedrock, that are little affected by pedogenic processes and lack properties of O, A, E, or B horizons. The material of C horizons may be either like or unlike that from which the solum presumably formed. The C horizon may have been modified even if there is no evidence of pedogenesis. Syn: C layers.

chorochromatic map (cho′-ro-chro-mat′-ic) A British term for a map in which broad distributions or variations are shown qualitatively over an area by means of different colors, tints, or shadings. Syn: *color-patch map.*

chorogram (chor′-o-gram) A generic term suggested by Wright (1944, p.653) for "any and all quantitative areal symbols" on a map.

chorographic (cho-ro-graph′-ic) Pertaining to *chorography;* specif. relating to an area of regional or continental extent, or said of a map representing a large region on a small scale (such as one between 1:500,000 and 1:5,000,000).

chorography (cho-rog′-ra-phy) (a) The art or practice of describing or mapping a particular region or district, esp. one larger than that considered by *topography* but smaller than that by geography. The term was widely used in the 17th and 18th centuries. (b) A broad account, description, map, or chart of a region considered by chorography. Also, the physical conformation or configuration, and the features, of such a region. Etymol: Greek "choros", "place," + graphein, "to write".

chorology (cho-rol′-o-gy) *biogeography.*

choropleth map A map on which distributions are classified quantitatively and shown by means of colors, shading, or patterns within specific areal units. A map portraying properties of a surface using area symbols. Area symbols on a choropleth map usually represent discrete categories, such as rock units on a geological map (Krzanowski et al., 1993).

C/H ratio *carbon-hydrogen ratio.*

chrisstanleyite A metallic silvery monoclinic mineral: $Ag_2Pd_3Se_4$.

christelite A vitreous greenish blue triclinic mineral: $Zn_3Cu_2(SO_4)_2(OH)_6{\cdot}4H_2O$.

christensenite (chris′-ten-sen-ite′) *tridymite.*

Christiansen effect (Chris′-tian-sen) In optical mineralogy, a dispersion phenomenon in which the boundary of a mineral grain that is immersed in a liquid of the same refractive index appears blue on one side and red to orange on the other.

christite (christ′-ite) An adamantine red monoclinic mineral: $TlHgAsS_3$.

Christmas tree (christ′-mas) The assemblage of valves, pipes, gages, and fittings at the top of the wellhead of a flowing oil or gas well, used to control the flow of fluids from the well and to prevent *blowouts.*

christophite (chris′-to-phite) *marmatite.*

chroma Color dimension that refers to the strength or purity of the dominant color or, the degree of saturation of the dominant color (Krzanowski et al., 1993).

chromate (chro′-mate) A mineral containing the chromate ion CrO_4^{-2}. An example is crocoite, $PbCrO_4$. Cf: *sulfate.*

chromatic aberration (chro-mat′-ic) In crystal optics, the production of color fringes due to the failure of rays of different wavelengths to converge at the same point.

chromatite (chro′-ma-tite) A citron-yellow tetragonal mineral: $CaCrO_4$.

chromatography (chro-ma-tog′-ra-phy) A general name for several processes of separating components of a sample by moving the sample in a mixture or solution over or through a medium using adsorption, partition, ion exchange, or other property in such a way that the different components have different mobilities and thus become separated. One of the earliest applications was in the separation of components of dye mixtures, giving rise to bands of different colors and hence to the name chromatography. See also: *column chromatography; electrochromatography; gas chromatography; liquid chromatography; paper chromatography; thin-layer chromatography.*

chrombismite An adamantine orange or yellowish brown tetragonal mineral: $Bi_{16}CrO_{27}$.

chromceladonite A green monoclinic mineral of the *mica* group: $KMgCrSi_4O_{10}(OH)_2$.

chromdravite (chrom-dra′-vite) A dark green rhombohedral mineral, a member of the *tourmaline* group: $NaMg_3(Cr^{3+},Fe^{3+})_6(BO_3)_3Si_6O_{18}(OH)_4$.

chrome A term commonly used to indicate ore of chromium, consisting esp. of the mineral chromite, or chromium-bearing minerals such as chrome mica or chrome diopside.

chrome diopside A bright-green variety of diopside containing a small amount of Cr_2O_3.

chrome iron ore Var: chrome iron; *chromic iron.* Syn: *chromite.*

chrome mica *fuchsite.*

chrome ocher A chromiferous clay; specif. a bright-green clay material containing 2-10.5% Cr_2O_3.

chrome spinel *picotite.*

chromferide (chrom-fer′-ide) A metallic light gray cubic mineral: $Fe_{7.5}Cr$.

chromic iron (chro′-mic) *chrome iron ore.*

chromite (chro′-mite) A brownish-black to iron-black cubic mineral of the spinel group: $Fe^{2+}Cr_2O_4$. It occurs in octahedral crystals as a primary accessory mineral in basic and ultrabasic igneous rocks; it also occurs massive, and it forms detrital deposits. Chromite is isomorphous with magnesiochromite, and is the most important ore of chromium. Syn: *chrome iron ore.*

chromium (chro′-mi-um) (a) A hard, corrosion-resistant metallic element, symbol Cr. (b) A metallic white cubic mineral that occurs in contact zones between ultrabasic rock and marble.

chromomphacite An ill-defined mineral, possibly a Cr-bearing omphacite. Not approved by the International Mineralogical Association.

chromophore A chemical element that causes color in a mineral (e.g., Fe^{2+} in olivine).

chromphyllite An emerald green mineral of the *mica* group: $KCr_2AlSi_3O_{10}(OH,F)_2$.

chromrutile (chrom-ru-tile′) *redledgeite.*

chron (a) The time span of a *chronozone.* The term was introduced by Williams (1901, p.583-584) for an indefinite division of geologic time and used by Wheeler et al. (1950, p.2362) as a general geologic-time unit. Cf: *polarity chron.* (b) The interval of geologic time (geochronologic unit) during which the rocks of a chronozone (chronostratigraphic unit) were formed (ISSC, 1994, p.83).

chronocline (chron'-o-cline) A gradational series of changes in the members of a natural group of organisms in successive stratigraphic units.

chronocorrelation (chron'-o-cor'-re-la'-tion) Chronostratigraphic correlation. Demonstration of correspondence in age and chronostratigraphic position between geologic units (NACSN, 1983, p.851). Cf: *biocorrelation; lithologic correlation.* Syn: *time-correlation; chronostratigraphic correlation.*

chronofauna (chron-o-fau'-na) A geographically restricted natural assemblage of interacting animal populations that maintained its basic structure over a geologically significant period of time.

chronogenesis (chron-o-gen'-e-sis) The time sequence of appearance of organisms in stratified rocks.

chronohorizon (chron'-o-ho-ri'-zon) Chronostratigraphic horizon. A stratigraphic surface or interface that is everywhere of the same age. Although theoretically without thickness, it is commonly a very thin and distinctive interval that is essentially isochronous over its whole geographic extent and thus constitutes an excellent time-reference or time-correlation horizon (ISG, 1976, p.67-68). Examples of horizons that may have strong chronostratigraphic significance include many biohorizons, bentonite beds, horizons of *magnetic reversal,* and coal beds. Cf: *buffer-zone; datum level [stratig].* See also: *moment; instant.* Syn: *chronostratigraphic horizon.*

chronolith (chron'-o-lith) *chronostratigraphic unit.*

chronolithologic unit (chron'-o-lith'-o-log'-ic) *chronostratigraphic unit.*

chronology (chro-nol'-o-gy) The arrangement of events in their proper sequence in time; also, the subject of considering or measuring time in discrete units. Cf: *geochronology.*

chronomere (chron'-o-mere) Any interval of geologic time, irrespective of duration (Geological Society of London, 1967, p.82) Cf: *stratomere.*

chronomeric standard terms The terms of the formal geochronologic hierarchy (Geological Society of London, 1967, p.82).

chronometric unit *geochronometric unit.*

chronosome The body of rock enclosed between two markers that may approximate isochronous surfaces and so define the upper and lower limits of a single period of deposition within a basin. It is a special type of chronozone recognized primarily in subsurface practice (Schultz, 1982, p.29).

chronostratic unit (chron-o-strat'-ic) Chronostratigraphic unit. A stratigraphic unit with time boundaries (Henningsmoen, 1961, p.66, 75-79).

chronostratigraphic classification (chron'-o-strat'-i-graph'-ic) The organization of rock strata into units on the basis of their age or time of origin (ISG, 1976, p.66).

chronostratigraphic correlation *chronocorrelation; time-correlation.*

chronostratigraphic horizon *chronohorizon.*

chronostratigraphic unit A body of rocks established to serve as the material reference for all rocks formed during the same span of time. Each of its boundaries is synchronous. The body also serves as the basis for defining the specific interval of time, or *geochronologic unit,* represented by the referent (NACSN, 1983, Art. 66). Chronostratigraphic units in order of decreasing rank: *eonothem, erathem, system, series, stage.* Syn: *chronostratic unit; chronolithologic unit; time-stratigraphic unit; time-rock unit; chronolith.* See also: *chronozone.*

chronostratigraphic zone *chronozone.*

chronostratigraphy (chron'-o-stra-tig'-ra-phy) The branch of stratigraphy that deals with the age of strata and their time relations (ISG, 1976, p.66). Syn: *time-stratigraphy.*

chronotaxial (chron-o-tax'-i-al) Pertaining to, characterized by, or exhibiting chronotaxy.

chronotaxis (chron-o-tax'-is) Similarity in sequence with respect to age (Henbest, 1952, p.310). Proposed as a complementary term for *homotaxis.* See: *chronotaxy.*

chronotaxy (chron'-o-tax-y) Equivalent to chronotaxis as originally proposed (Henbest 1952, p.310). Cf: *homotaxy.*

chronozone (chron'-o-zone) (a) A non-hierarchical, commonly small, *chronostratigraphic unit;* its boundaries may be independent of those of ranked units. Although a chronozone is an isochronous unit, it may be based on a biostratigraphic unit, e.g. *Cardioceras cordatum* biochronozone on a lithostratigraphic unit, e.g. Woodbend Lithochronozone; or on a magnetopolarity unit, e.g. Gilbert Reversed-Polarity Chronozone (NACSN, 1983, Art. 75). (b) A body of rocks formed anywhere during the time span of some designated stratigraphic unit or geologic feature. It is a formal chronostratigraphic unit of unspecified rank, not part of the hierarchy of chronostratigraphic units (ISSC, 1994, p.83-85). The temporal equivalent of a chronozone is a *chron.* Syn: *chronostratigraphic zone.*

chrysoberyl (chrys-o-ber'-yl) (a) A hard orthorhombic mineral: $BeAl_2O_4$. It is usually yellow, pale green, or brown, contains a small amount of iron, and is used as a gem. Principal varieties are cat's-eye and alexandrite. It possesses the olivine structure type. Syn: *chrysopal; gold beryl; cymophane.* (b) An obsolete syn. of *heliodor.*

chrysocolla (chrys-o-col'-la) (a) A blue, blue-green, or emerald-green amorphous mineral: $Cu_2(OH)_4[Si_2O_3(OH)_2]$. It occurs as incrustations and thin seams in the zone of weathering of copper ores. Its chemical composition was formerly given as: $CuSiO_3{\cdot}2H_2O$. (b) An old name given to a mineral or minerals (such as chrysocolla, borax, and malachite) used for soldering gold (Hey, 1962, p.384).

chrysocolla chalcedony Translucent to semitranslucent, vivid-blue to greenish-blue chalcedony. Minutely distributed chrysocolla causes the color.

chrysolite [gem] (chrys'-o-lite) A misleading term applied in gemology to several yellow to yellowish-green gems. Correctly used it refers to the pale-yellow to yellowish-green gem variety of olivine (=peridot).

chrysolite [mineral] (a) A yellowish-green, reddish, or brownish variety of olivine in which the ratio of magnesium to total magnesium plus iron is between 0.90 and 0.70 or in which the Fe_2SiO_4 content is 10-30 mole percent. The name has at times carried a wider meaning, as a syn. of *olivine.* (b) A name that has been applied at various times to topaz, prehnite, and apatite. This usage is obsolete. Not to be confused with *chrysotile.*

chrysomonad (chrys-o-mon'-ad) Any of a group of microscopic organisms, either protozoans or algae, usually with flagella at some stage of their life history; many are autotrophic. *Holozoic* forms are assigned to the protozoan order Chrysomonadina.

chrysopal (chrys-o'-pal) (a) A translucent variety of common opal, colored apple-green by the presence of nickel. (b) *chrysoberyl.* (c) A gemstone-trade name for opalescent chrysolite (olivine).

chrysoprase (chrys'-o-prase) (a) An apple-green or pale yellowish-green variety of chalcedony containing finely dispersed nickel minerals, and valued as a gem. Syn: *green chalcedony.* (b) A misleading name used in the gem trade for green-dyed chalcedony having a much darker color than natural chrysoprase.

chrysoquartz (chrys-o-quartz') Green *aventurine.*

chrysotile (chrys-o-tile') A white, gray, or greenish orthorhombic or monoclinic mineral of the *serpentine* group: $Mg_3(OH)_4Si_2O_5$. It is a highly fibrous, silky variety of serpentine, and constitutes the most important type of *asbestos.* Not to be confused with *chrysolite.* Syn: *serpentine asbestos; clinochrysotile.*

chrystocrene (chryst'-o-crene) A term introduced as *crystocrene* by Tyrrell (1904, p.234) for a surface mass of ice formed each winter by the overflow of springs; also, the ice formed in the interstices of a mass of loose rock fragments (such as talus) by the freezing of a subjacent spring. The term is not synonymous with *rock glacier* (Tyrrell, 1910). Cf: *crystosphene.*

chthonic (chtho'-nic) Said of deep-sea sediments and clastic debris derived from preexisting rocks. Ant: *halmeic.* Cf: *allogenic.*

chuco (chu'-co) A term used in Chile for the upper part of a caliche deposit, composed mainly of sodium sulfate.

chudobaite (chu-do'-ba-ite) A pink triclinic mineral: $(Mg,Zn)_5(AsO_4)_2(AsO_3OH)_2{\cdot}10H_2O$.

chukhrovite-(Ce) A colorless to pale brown cubic mineral: $Ca_3(Ce,Y)Al_2(SO_4)F_{13}{\cdot}10H_2O$.

chukhrovite-(Y) (chukh'-rov-ite-y) A colorless cubic mineral: $Ca_3(Y,Ce)Al_2(SO_4)F_{13}{\cdot}10H_2O$.

chur *char [streams].*

Churchill-type vein A *mineral-deposit type* characterized by copper veins associated with mafic intrusive rocks. Named for the Churchill copper deposit in British Columbia, Canada.

churchite-(Nd) A yellowish monoclinic mineral: $NdPO_4{\cdot}2H_2O$.

churchite-(Y) (church'-ite-y) A colorless, yellowish, gray, or reddish monoclinic mineral: $YPO_4{\cdot}2H_2O$. Syn: *weinschenkite.*

churn drilling *cable-tool drilling.*

churn hole *pothole [streams].*

chursinite (chur'-sin-ite) An adamantine yellow or brown monoclinic mineral: $Hg^{1+}Hg^{2+}(AsO_4)$.

chute [geomorph] A term used in the Isle of Wight for a steep cutting affording a passage from the surface above a cliff to the lower undercliff ground (Stamp, 1961, p.104).

chute [hydraul] An inclined water course, either natural or artificial.

chute [ore dep] Var. of shoot, as in *ore shoot.*

chute [speleo] An inclined channel or passage in a cave.

chute [streams] (a) A fall of water; a rapid or quick descent in a river; a steep channel by which water falls from a higher to a lower level; a rapids. See also: *shoot [streams].* (b) A narrow channel through which water flows rapidly, esp. along an overflow river (such as the lower Mississippi River); specif. a *chute cutoff.*

chute cutoff A narrow "short cut" across a meander bend, formed at time of flood when the main flow of a stream is diverted to the inside of the bend, along or through a trough between adjacent parts of a point bar. Cf: *neck cutoff; chute [streams].*

chvaleticeite (chval'-e-tice-ite) A white, pinkish, or yellowish green monoclinic mineral: $(Mn^{2+},Mg)SO_4{\cdot}6H_2O$. A member of the *hexahydrite* group.

chvilevaite (chvil'-ev-a-ite) A metallic bronze trigonal mineral: $Na(Cu,Fe,Zn)_2S_2$.

chymogenic (chy'-mo-gen'-ic) Said of that portion of a composite rock formed by crystallization from an ionic- or molecular-dispersed phase or from a fluid, whether a gas, hydrothermal solution, or magma (Mehnert, 1968, p.353). Rarely used. Var: chymogenetic. Cf: *metatect; mobilizate; neosome; stereosome.*

CI (a) *crystallization index.* (b) *contour interval.*

cianciulliite A dark reddish brown monoclinic mineral: $Mg,Mn^{2+}{}_2Mn^{2+}Zn_2(OH)_{10}{\cdot}2\text{-}4H_2O$.

cicatricose (ci-cat'-ri-cose) Marked with scars; esp. said of sculpture of palynomorphs consisting of more or less parallel ridges.

cicatrisation (cic'-a-tri-sa'-tion) Obsolete term for reconstruction of a broken or corroded crystal as a result of a secondary deposit of the same mineral in optical continuity.

cicatrix (cic'-a-trix) (a) A scar in an echinoderm; esp. the scar indicating the former position of the column in some echinoderms (such as cystoids) autotomizing. (b) A small groove or scar on the apex of some nautiloid conchs. (c) The impression on the inside of a bivalve shell caused by the insertion of the adductor muscle. See also: *adductor-muscle scar.* Pl: cicatrices. Syn: *scar [paleont].*

CI chondrite A member of the Ivuna group of *carbonaceous chondrites.*

cienaga (ci-e'-na-ga) A marshy area where the ground is wet due to the presence of seepage or springs, often with standing water and abundant vegetation. The term is commonly applied in arid regions such as the southwestern U.S. Etymol: Spanish ciénaga, "marsh, bog, miry place". Also spelled: cienega.

cigar-shaped mountain (ci-gar') An anticlinal ridge plunging at each end.

cilia (cil'-i-a) Plural of *cilium.*

ciliate (cil'-i-ate) n. Any protozoan belonging to the class Ciliata and characterized by the presence of cilia throughout its life cycle. Range, Upper Jurassic to Holocene. adj. Possessing cilia.

cilifer (cil'-i-fer) Said of a variant of radulifer type of brachiopod crura, flattened in the plane of commissure, forming direct prolongations of horizontal hinge plates, then turning parallel to the plane of symmetry as slightly crescentic blades.

cilium (cil'-i-um) One of numerous short hairlike processes found on the surface of cells, capable of rhythmic vibratory or lashing movement, and serving as organs of locomotion in free-swimming unicellular organisms and in some small multicellular forms and as producers of currents of water in higher animals. Pl: *cilia.* Cf: *flagellum.*

cima (ci'-ma [chee'-ma]) A mountain peak or dome. Etymol: Italian.

ciminite (cim'-i-nite) A trachydolerite composed of olivine, clinopyroxene, and labradorite with alkali feldspar rims in a trachytic groundmass; named by Washington in 1896 for Monte Cimino, Italy. Obsolete.

Cimmerian (Kim-mer'-ian) A Ponto-Caspian stage: uppermost Miocene-lower Pliocene (above Pontian, below Akchagylian).

Cimmerian Orogeny (Cim-me'-ri-an) One of the 30 or more short-lived orogenies during Phanerozoic time identified by Stille; in this case, two orogenies are included, the early Cimmerian late in the Triassic, between the Norian and Rhaetian stages, and the late Cimmerian at the end of the Jurassic. Also spelled: *Kimmerian Orogeny.*

cimolite (cim'-o-lite) A white, grayish, or reddish hydrous aluminum silicate mineral occurring in soft, claylike masses. Probably not a valid mineral species.

Cincinnatian (Cin-cin-nat'-i-an) North American provincial series: Upper Ordovician (above Champlainian, below Alexandrian of Silurian).

cinder (cin'-der) A juvenile vitric vesicular pyroclastic fragment that falls to the ground in an essentially solid condition (Macdonald, 1972, p.128). Cf: *block [volc]; scoria [volc]; lapilli; volcanic gravel.* Syn: *scoria [volc]* is preferred in more recent papers.

cinder coal *natural coke.*

cinder cone A conical hill formed by the accumulation of cinders and other pyroclasts, normally of basaltic or andesitic composition. Steepness of the slopes depends on coarseness of the ejecta, height of eruption, wind velocity, and other factors, but is normally greater than 10 degrees.

cinerite (cin'-er-ite) A deposit of volcanic cinders. Obsolete.

cingula Pl. of *cingulum.*

cingular archeopyle (cin'-gu-lar) (a) An *archeopyle* formed in a dinoflagellate cyst by the loss of one or more cingular *plates,* e.g., in the genus *Nannoceratopsis.* (b) In obsolete usage, also one formed by breakage along and within the *cingulum* (Downie and Sarjeant, 1966; Evitt, 1985, p.175; Piel and Evitt, 1980).

cingular series The series of plates along the girdle in a dinoflagellate possessing a theca.

cingulate (cin'-gu-late) Having a girdle; esp. said of a spore possessing a cingulum.

cingulum (cin'-gu-lum) (a) Either of the two connecting bands forming the sides of the two valves of a diatom; a *girdle [paleont].* (b) An annular, more or less equatorial extension of a spore in which the wall is thicker than that of the main body of the spore. (c) In a living dinoflagellate, the more or less equatorially located furrow that houses the *transverse flagellum;* also the corresponding feature on a dinoflagellate cyst. Cf: *crassitude; zone [palyn].* (d) Zooecial lining that results from secondary skeletal thickening in some *stenolaemate* bryozoans. Pl: *cingula.*

cinnabar (cin'-na-bar) A rhombohedral mineral: HgS. It is dimorphous with metacinnabar and represents the most important ore of mercury. Cinnabar occurs in brilliant red acicular crystals or in red, brownish, or gray masses in veins and alluvial deposits. Syn: *vermilion.*

cinnamon stone (cin'-na-mon) Yellow-brown to reddish-brown *essonite.*

cipolin (ci'-po-lin) (a) *cipolino.* (b) A term used in France for any crystalline limestone.

cipolino (ci-po-li'-no) A siliceous marble containing micaceous layers. Partial syn: *cipolin.*

ciprianiite A vitreous pale brown monoclinic mineral: $Ca_4(Th,U,REE)_2Al_2Si_4B_4O_{22}(OH,F)_2$.

CIPW classification A system for classifying and naming igneous rocks based on the *CIPW norm.* The initials represent the initial letters of the names of the men who devised the system in 1902, Cross, Iddings, Pirsson, and Washington. Syn: *quantitative system; norm system.*

CIPW norm A norm in which the reported content of a mineral represents its weight percentage and in which the minerals are all anhydrous minerals of simplified composition (Cross et al., 1902). Cf: *norm.*

circadian (cir-ca'-di-an) Said of a time period approximately 24 hours in length, or of an event occurring at roughly 24-hour or daily intervals; e.g. "circadian rhythms". Cf: *circannian.*

circannian (cir-can'-ni-an) Said of a time period approximately one year in length, or of an event that occurs annually; e.g. "circannian rhythms". Cf: *circadian.*

circinate (cir'-ci-nate) (a) Pertaining to the unrolling of a developing fern frond. (b) Pertaining to a protist that is curled downward from the apex.

circinate suture (cir'-ci-nate) An *acritarch excystment* consisting of a *suture* with a coiled *levorotatory* configuration. It produces an *operculum* that remains attached, e.g., in *Circinatisphaera* (Miller, 1987).

circle [pat grd] (cir'-cle) A form of patterned ground whose horizontal mesh is dominantly circular. See: *sorted circle; nonsorted circle.*

circle [surv] The graduated disk of a surveying instrument, perpendicular to and centered about an axis of rotation, and calibrated to read the amount of rotation; e.g. a horizontal circle or a vertical circle of a theodolite or transit.

circle of declination *hour circle.*

circle of latitude A meridian of the terrestrial sphere, along which latitude is measured. Cf: *parallel of latitude.*

circle of longitude A circle, parallel to the plane of the Earth's equator; along which longitude is measured.

circle of right ascension *hour circle.*

circle shooting A marine technique for acquiring three-dimensional seismic data wherein the boat track forms a spiral or a series of overlapping circles. Occasionally used to survey the flanks of a salt dome or similar structure. This technique avoids time lost in changing from one line to another.

circlet (cir'-clet) A series of plates that forms a ring entirely or partially around the *theca* of an echinoderm.

circuit (cir'-cuit) (a) A continuous series of connected survey lines that form a closed loop. (b) A line or series of lines connecting two fixed survey points.

circuit closure The *error of closure* of a level circuit, being the algebraic sum of all the junction closures in a circuit, usually reckoned counterclockwise around the circuit; hence, the accumulated error (before adjustment) of measured differences of elevation around the circuit, or the amount by which the last computed elevation fails to equal the initial elevation.

circular coal (cir'-cu-lar) *eye coal.*

circular error probable The radius of a circle centered at the most probable point, such that half the measurements fall within the circle. Abbrev: CEP.

circularity ratio (cir-cu-lar'-i-ty) *basin-circularity ratio.*

circular level A *spirit level* having the inside surface of its upper part ground to a spherical shape, the outline of the bubble formed being circular, and the graduations being concentric circles. It is used where a high degree of precision is not required, as in setting an instrument in approximate position. Syn: *bull's-eye level; box level.*

circular normal distribution In *geostatistics,* a theoretical *distribution* of a variable measured in a polar coordinate system that is analogous to a *normal distribution* or a *Gaussian distribution.* Also called: *von Mises distribution.*

circular polarization In optics, circularly polarized light consisting of upward-spiraling vibration vectors that define a surface similar to the thread of a screw. It is caused by the interaction of mutually perpendicular wave motions whose path differences differ in phase by $(2n+1)/4\lambda$ on emergence from a crystal. Cf: *elliptical polarization.*

circular section [cryst] In a *uniaxial* crystal, an equatorial section perpendicular to the *optic axis;* in a *biaxial* crystal, one of two sections containing the Y axis of the biaxial indicatrix and perpendicular to one optic axis (Wahlstrom, 1948).

circular section [exp struc geol] One of the two circular cross sections through a strain ellipsoid.

circular slide A landslide whose slip surface follows the arc of a circle.

circulation [drill] (cir-cu-la'-tion) In *rotary drilling,* the continuous cycling of drilling mud (less commonly air or foam) down the drillstring, out through the drill bit, and up to the surface through the *annulus* between the drill pipe and the casing or walls of the hole.

circulation [lake] The general mixing of the water in a lake, which commonly occurs when the water column is of nearly the same density from top to bottom (Wetzel, 2001, p.75). Cf: *overturn.*

circulation [oceanog] In oceanography, a general term for the flow of water in a large area, usually in a closed pattern or gyre. Circulation along horizontal surfaces is due to the force of the wind acting on the sea surface. Circulation in the vertical is fed by sinking of cold dense water at high latitudes and driven by mixing of deep water back to the surface by turbulence created by winds and tides.

circulus (cir'-cu-lus) A *cameral deposit* on the concave surface of a *cyrtochoanitic septal neck* of a nautiloid (TIP, 1964, pt. K, p.54).

circumdenudation (cir'-cum-den'-u-da'-tion) The denudation or erosion of a landmass such that a part of the ground is left isolated and upstanding; e.g. denudation around a resistant rock mass. Syn: *circumerosion.*

circumdenudation mountain *mountain of circumdenudation.*

circumerosion (cir'-cum-e-ro'-sion) *circumdenudation.*

circumferential wave (cir-cum'-fer-en'-tial) An obsolete syn. of *surface wave [seis]..*

circumferentor (cir-cum'-fer-en'-tor) A type of *surveyor's compass* having vertical slit sights on projecting arms.

circumgranular cracking Development of spar-filled cracks surrounding pebbles, sand grains, etc. Typical of calcrete.

circummural budding (cir-cum-mu'-ral) A type of *polystomodaeal budding* in which indirectly linked stomodaea are arranged around discontinuous *collines* or *monticules* of *corallum.* Cf: *intramural budding.*

circumoral budding (cir-cum-o'-ral) A type of *polystomodaeal budding* in which directly linked stomodaea are arranged concentrically around the central parent *stomodaeum.*

circum-Pacific belt (cir'-cum-Pa-cif'-ic) The belt of tectonically active regions bordering the Pacific Ocean. Its presence is the result of interaction along *plate boundaries.*

circumpolar vortex The planetary-scale atmospheric circulation regime that surrounds the cold pool of air in the polar regions.

Circumpolles (Cir-cum-pol'-les) A group of spherical gymnospermous pollen commonly occurring as tetrads and characterized by a *rimula,* a distal pseudopore, and a proximal triangular area (e.g., in *Classopollis*). Late Triassic to Paleogene.

circumvallation (cir'-cum-val-la'-tion) The process whereby mountains are formed by streams incising a featureless plain (Hobbs, 1912, p.442).

cirque [geomorph] A semicircular, topographic *hollow* eroded by alpine glaciers. Also rarely used for a amphitheaterlike or armchair-shaped hollow of nonglacial origin but resembling a glacial cirque; e.g., a *doline* in a limestone region, a blowout in an arid region, or a depression formed by *landslide sapping.* Cf: *pseudocirque.*

cirque [glac geol] A deep steep-walled half-bowl-like recess or hollow, variously described as horseshoe- or crescent-shaped or semicircular in plan, situated high on the side of a mountain and commonly at the head of a glacial valley, and produced by the erosive activity of a mountain glacier. It often contains a small round lake, and it may or may not be occupied by ice or snow. Etymol: French, from Latin "circus", "ring". Syn: *corrie; cwm; coire; kar; botn; amphitheater; combe; oule; van.*

cirque floor The nearly flat surface at the bottom of a *cirque.* Cf: *cirque niveau.*

cirque glacier A small glacier occupying a cirque, or resting against the headwall of a cirque. It is the most common type of glacier in the mountains of the western U.S. Cf: *glacieret; niche glacier.* Syn: *corrie glacier.*

cirque lake A small, deep, commonly circular glacial lake occupying a cirque; it has no prominent inlet, being fed by runoff from the surrounding slopes and dammed by a lip of bedrock or by a small moraine. Syn: *tarn.*

cirque mountain *horn [glac geol].*

cirque niveau The level of a *cirque floor* representing the surface of a terrace developed by preglacial erosion (Swayne, 1956, p.34); it is the approximate altitude at which most cirques in a region have excavated their floors. Etymol: French.

cirque platform A relatively level surface formed by the coalescence of several cirques.

cirque stairway A succession of cirques situated in a row at different levels in the same glacial valley. Cf: *glacial stairway.*

cirral (cir'-ral) adj. Pertaining to a *cirrus.* n. A single segment or plate of a crinoid cirrus.

cirri (cir'-ri) Plural of *cirrus.*

cirriped (cir'-ri-ped) Any marine crustacean belonging to the class Cirripedia, characterized chiefly by the permanent attachment of the adult stage to some substrate; e.g. a barnacle. Range, Upper Silurian to Holocene. Also spelled: cirripede.

cirrolite (cir'-rol-ite) An ill-defined yellow mineral: $Ca_3Al_2(OH)_3(PO_4)_3$. Probably not a valid species.

cirrus (cir'-rus) (a) Any of the flexible rootlike jointed appendages attached to the side of the stem (and sometimes to the aboral surface) of a crinoid, exclusive of the *radix.* It is composed of small articulated plates, or cirrals. (b) A multiarticulate food-gathering thoracic appendage in a cirripede crustacean. Pl: *cirri.*

cis link A link in a trunk stream channel bounded by tributaries that enter from the same side (James and Krumbein, 1969). The trunk channel is traced upstream by following the link of greater magnitude at each fork. Cf: *cis-trans link; trans link.*

cislunar (cis-lu'-nar) Pertaining to phenomena, or to the space, between the Earth and the Moon or the Moon's orbit. Cf: *translunar.*

cistern (cis'-tern) (a) An artificial reservoir or tank for storing water. (b) A natural reservoir; a hollow containing water.

cis-trans link A link of magnitude μ that is formed at its upstream fork by the confluence of two links of unequal magnitude and that flows at its downstream fork into a link of a magnitude of less than 2μ (Mock, 1971, p.1559). Cf: *cis link; trans link.*

Cisuralian (Cis-u-ra'-lian) A series of the Standard Global Chronostratigraphic Scale: Lower Permian (below Guadalupian Series, above Carboniferous System).

cis-vacant Said of a layered mineral with a vacant (M)2 site. Most dioctahedral phyllosilicates are monoclinic. The mirror present in the C2/m minerals, and the position of this mirror for other phyllosilicates that are not C2/m, is perpendicular to the platy or 001 surface of these minerals. This mirror bisects one of the three octahedral sites in the octahedral sheet. This is the (M)1 site and the other two are (M)2 sites. When either of the (M)2 sites is vacant, the mineral is referred to a cis-vacant. Cf: *trans-vacant.*

citrine (cit-rine') A transparent yellow to orange-brown variety of crystalline quartz closely resembling topaz in color. It can be produced by heating amethyst or dark smoky quartz. Syn: *topaz quartz; false topaz; Bohemian topaz; quartz topaz; yellow quartz.*

civil engineering (civ'-il) A branch of engineering concerned primarily with the investigation, design, construction, operation, and maintenance of civil-works projects (public and private) such as highways, bridges, tunnels, waterways, harbors, dams, water supply, irrigation, railways, airports, buildings, sewage disposal, and drainage. Civil engineering originally referred to all engineering that was not intended for military use. It is now subdivided into specialties such as structural, mechanical, electrical, chemical, geological, and geotechnical engineering.

CK chondrite A member of the Karoonda group of *carbonaceous chondrites.*

CL *crossed lamellar structure.*

clade [evol] (a) A *monophyletic* higher taxon, one consisting of an ancestral species and all its known descendant species. (b) In *cladism,* all taxa distal to a given *node [evol]* in a *cladogram.*

clade [paleont] A branch at the extremity of an actine in an ebridian skeleton that may connect adjacent actines. See also: *proclade; opisthoclade; mesoclade.*

clade displacement The displacement of one group of organisms by a second, ecologically similar group of organisms due to ecological competition between the two groups. Clade displacement is a subset of clade replacement and requires some ecological interaction between the two groups.

clade replacement The replacement of one group by a second, ecologically similar group of organisms. This replacement may be due to clade displacement, or it may be due to the disappearance of the first group by extinction.

cladi (cla'-di) Plural of *cladus.*

cladism (clad'-ism) That method in *systematics* wherein ancestor-descendant (phylogenetic or evolutionary) relationships among taxa are analyzed strictly on the basis of the distribution of shared derived characters, and consequently where genealogy is the sole criterion for the definition of taxa. Syn: *cladistics; phylogenetic systematics.*

cladistic analysis A method of evolutionary analysis that attempts to recover genealogical relationships among groups of organisms, and attempts to produce trees that reflect these relationships (Brooks et al., 1984, p.79). Syn: *phylogenetics.*

cladistics (cla-dis'-tics) *cladism.*

cladoceran (cla-doc'-er-an) Any crustacean belonging to the order Cladocera, characterized by a univalve carapace that is bent along the back giving a bivalve appearance. Cladocerans are commonly found in fresh- to brackish-water postglacial deposits. Range, Oligocene to Holocene.

cladogenesis (clad-o-gen'-e-sis) (a) Phylogenetic splitting or branching. Cf: *speciation.* (b) Progressive evolutionary specialization. (c) Evolutionary process resulting in a branching pattern of divergence; two or more species, of which at least one species is new, result from a single preexisting species (Wiley, 1981, p.8). Cf: *anagenesis, reticulate speciation.*

cladogram (clad'-o-gram) (a) A *dendrogram* expressing the classificatory relationships among a group of organisms, as based on their inferred phylogenetic or evolutionary relationships. Syn: *phylogenetic tree.* (b) In *cladism,* a dendrogram showing phylogenetic relationships among taxa in terms of recency of common ancestry. Identities of nodes (ancestors) are not specified, connecting lines usually represent shared derived characters, and all taxa are terminal in position. (c) A type of phylogentic tree that has specific connotations of implied ancestry and a relative time axis; a common ancestry phylogentic tree (Wiley et al., 1991, p.6). Cf: *phylogenetic tree.*

cladome (clad'-ome) The group of similar rays of a diaene, triaene, or tetraene sponge spicule.

Cladoselachii (Clad'-o-se-la'-chi-i) An order of marine elasmobranch fishes with fusiform body, broadly based fins, and terminal mouth; the most primitive true sharks. Stratigraphic range, Middle Devonian to Mississippian.

cladospore (clad′-o-spore) The position of reproductive organs in dasycladacean algae, in which cysts are located in primary branches. Cf: *choristopore.*

cladus (cla′-dus) One of the rays of a cladome; a branch of a ramose spicule. The term is usually used in the plural: *cladi.*

Claibornian (Clai-born′-i-an) North American (Gulf Coast) stage: Eocene (above Sabinian, below Jacksonian).

claim In mining law, a portion of public land on which an individual may have mining rights; a *mining claim.* Size and other legal restrictions vary from country to country.

Clairaut's theorem (Clai-raut's′) An expression for the variation of normal gravity on the Earth that is the basis for standard gravity formulas such as the International Gravity Formula. It establishes the relationship between normal gravity and the flattening of the Earth, from which it becomes possible to compute the flattening from surface gravity observations.

clairite (clair′-ite) An orange-yellow triclinic mineral occurring with lonecreekite and sabieite: $(NH_4)_2Fe^{3+}{}_3(SO_4)_4(OH)_3 \cdot 3H_2O$.

clam A popular term for a bivalve mollusk, sometimes applied to equivalved edible ones that live partially or completely buried in sand or mud.

CLAMP *Climate Leaf Analysis Multivariate Program.*

clamp connection A hook-shaped short hyphal branch that arises at the time of cell division, as characteristically found in *Basidiomycetes* fungi.

clamshell snapper (clam′-shell) A term for various large or small weight- or spring-loaded devices lowered by wire and designed to snap shut collecting a sample of sea floor sediment. A *grab sampler.*

clan [ecol] (a) A category in the hierarchy of classification used by some zoologists; it ranks below the *subfamily* and above the *genus.* (b) A small ecologic community, usually a *climax community,* that occupies only a few square meters of space and has only one dominant species.

Clansayesian (Clan-say-e′-sian) A European substage (France, Switzerland): uppermost Aptian Stage of Lower Cretaceous (above Gargasian Substage, below Albian Stage).

Clapeyron equation (Cla-pey′-ron) A statement in chemistry, developed by Clausius in 1850, that relates the slope of a reaction in P-T space (dP/dT) to the thermodynamics of the reaction in a closed system. The equation states that the change of pressure with temperature is equal to the heat of reaction (ΔH) divided by the product of the temperature (in K) and the volume change (ΔV). Syn: *Clausius-Clapeyron equation.* See also: *Ehrenfest relation; Poynting's law.*

clarain (clar′-ain) A coal *lithotype* characterized macroscopically by semibright, silky luster and sheetlike, irregular fracture. It is distinguished from *vitrain* by containing fine intercalations of a duller lithotype, *durain.* Its characteristic microlithotype is *clarite.* Cf: *fusain.*

claraite (clar′-a-ite) A blue triclinic mineral: $(Cu,Zn)_3(CO_3)(OH)_4 \cdot 4H_2O$.

Clarendonian (Clar-en-don′-i-an) North American continental stage: Middle-upper Miocene (above Barstovian, below Hemphillian).

claringbullite (cla-ring-bull′-ite) A blue hexagonal mineral: $Cu_4Cl(OH)_7$. Its cell parameters suggest a relation to connellite and buttgenbachite.

clarinite (clar′-i-nite) The major maceral of clarain, according to the Stopes classification; the term is no longer in general use.

clarite (clar′-ite) A coal microlithotype that contains a combination of vitrinite and liptinite totalling at least 95%, and containing more of each than of inertinite. Cf: *clarain.*

clarity A gemstone's relative freedom from inclusions and blemishes (Liddicoat et al., 1993).

clarity grade Professional assessment of the clarity characteristics observable in or on a diamond or other gemstone, with reference to a systematic set of standards. Grades are usually based on how visible the clarity characteristics are, and how they affect the stone's durability; they are typically represented by a standardized nomenclature of terms (or their abbreviations), letters, numbers, or a combination thereof, which describe the extent to which a diamond varies from flawless (Liddicoat et al., 1993).

clarity grading Process of examining a diamond's clarity characteristics, determining their effect on its overall quality, and assigning a grade with reference to a systematic set of standards described in terms of a standardized nomenclature (Liddicoat et al., 1993).

Clark degree A British unit for measuring hardness of water, equal to one grain per British gallon or 14.3 ppm as $CaCO_3$. Cf: *grain [water]; degree [hydrol].*

clarke The average abundance of an element in the crust of the Earth. It is named in honor of F.W. Clarke. Cf: *clarke of concentration.* Syn: *crustal abundance.*

Clarke-Bumpus plankton sampler One of many plankton nets with an opening–closing mechanism and a flow meter to record the volume of water filtered.

Clarke ellipsoid of 1866 The ellipsoid of reference for geodetic surveys in North and Central America, the Hawaiian Islands, and the Philippines. It was the basis of the North American datum of 1927. Cf: *ellipsoid.*

clarkeite (clarke′-ite) A dark brown or reddish brown rhombohedral mineral: $(Na,Ca,Pb)(UO_2)O(OH) \cdot nH_2O$.

clarke of concentration The concentration of an element in a mineral or rock relative to its crustal abundance. The term is applied to specific as well as average occurrences. Cf: *clarke.*

Clarkforkian (Clark-fork′-i-an) North American land mammal age: Lower Eocene (after Tiffanian, before Wasatchian).

clarocollain (clar-o-col′-lain) A transitional lithotype of coal characterized by the presence of collinite with lesser amounts of other macerals. Cf: *colloclarain.* Syn: *clarocollite.*

clarocollite (clar-o-col′-lite) *clarocollain.*

clarodurain (clar-o-du′-rain) A transitional lithotype of coal characterized by vitrinite, but more of other macerals such as micrinite and liptinite than of vitrinite; it corresponds to *semisplint coal.* Cf: *duroclarain.*

clarodurite (clar-o-du′-rite) A coal microlithotype containing at least 5% each of vitrinite, liptinite, and inertinite, with more inertinite than vitrinite and liptinite. It is a variety of *trimacerite,* intermediate in composition between clarite and durite, but closer to durite. Cf: *duroclarite.*

clarofusain (clar-o-fu′-sain) A transitional lithotype of coal characterized by the presence of fusinite and vitrinite, with other macerals; fusinite is more abundant than it is in *fusoclarain.* Syn: *clarofusite.*

clarofusite (clar-o-fu′-site) *clarofusain.*

clarotelain (clar-o-te′-lain) A transitional lithotype of coal characterized by the presence of telinite, with lesser amounts of other macerals. Cf: *teloclarain.* Syn: *clarotelite.*

clarotelite (clar-o-te′-lite) *clarotelain.*

clarovitrain (clar-o-vit′-rain) A transitional lithotype of coal characterized by the presence of vitrinite, with lesser amounts of other macerals. Cf: *vitroclarain.* Syn: *clarovitrite.*

clarovitrite (clar-o-vit′-rite) *clarovitrain.*

clasmoschist (clas′-mo-schist) A term suggested by W.D. Conybeare to replace "graywacke" (an arenaceous rock in the lower part of the Secondary strata) (Roberts, 1839, p.72).

clasper (clasp′-er) (a) An appendage of a crustacean, modified for attachment in copulation or for fixation of parasites. (b) A part of the pelvic fin of many sharks, modified for copulation.

class [cryst] One of thirty-two possible crystallographic combinations of the non-translational elements of symmetry. Crystal classes are divided among the six crystal systems, and deal with outward symmetry. Syn: *crystallographic point group.*

class [petrology] In the CIPW classification of igneous rocks, a subdivision based on the relative proportions of salic and femic standard minerals. The classes correspond approximately to the color-based divisions leucocratic, melanocratic, and mesocratic. The basic unit of the class is the *order [petrology].*

class [stat] A subdivision of the observed range of a variable, having stated limits.

class [taxon] A category in the hierarchy of classification of animals and plants, intermediate between *phylum* (or division) and order. Super- and sub-categories may be introduced as needed, provided they do not produce confusion (Blackwelder, 1967, p.57). Cf: *subclass.*

class 1,2,3 reservoirs A classification of reservoir sands according to their normal-incidence impedance contrasts with surrounding rocks (and their AVO responses). Class 1 is a reservoir sand that has an impedance larger than that of surrounding rocks, Class 2 has only a very small impedance contrast, and Class 3 has an impedance smaller than surrounding rocks. Cf: *amplitude variation with angle/offset*(AVA/AVO). (Rutherford and Williams, 1989).

Classic (Clas′-sic) In New World archaeology, a cultural stage that follows the Formative and is characterized by the rise of civilizations such as the Mayan. It is followed by the Post-Classic. Cf: *Classical.*

Classical In Mediterranean archaeology, the period between 480 and 330 B.C. and the corresponding style of art. Cf: *Classic.*

classical equilibrium constant (clas′-si-cal) An *equilibrium constant* that is defined by concentrations rather than by activities.

classification [cart] Grouping cells (often by spectral characteristics) from or a set of coregistered raster objects to isolate and identify image features such as wetlands, forest, crop varieties, or other surface cover.

classification [taxon] (clas′-si-fi-ca′-tion) That part of *systematics* that deals chiefly with the grouping of like things within a system (Blackwelder, 1967, p.4).

clast (a) An individual constituent, grain, or fragment of a sediment or rock, produced by the mechanical or chemical disintegration of a larger rock mass; e.g. a *phenoclast.* (b) *pyroclast.* (c) *bioclast.*

clast [meteorite] An angular fragment in a meteoritic impact breccia.

clastation (clas-ta′-tion) (a) The breaking-up of rock masses in situ by physical or chemical means (Grabau, 1924, p.17); *weathering.* (b) The disrupting of rocks to form clastic sediments (Galloway, 1922). Obsolete.

clastic (clas′-tic) adj. (a) Pertaining to a rock or sediment composed principally of broken fragments that are derived from preexisting rocks or minerals and that have been transported some distance from their places of origin; also said of the texture of such a rock. The term has been used to indicate a source both within and outside the depositional basin. (b) *pyroclastic.* (c)

Said of a bioclastic rock. (d) Pertaining to the fragments (clasts) composing a clastic rock. n. A clastic rock. Term is usually used in the plural; e.g. the commonest "clastics" are sandstone and shale.

clastic breccia A breccia formed by transportation and deposition (McKinstry, 1948, p.634).

clastic deformation One of the processes of dynamothermal metamorphism, which involves the fracture, rupture, and rolling-out of mineral and rock particles. In the extreme case, the rock may be thoroughly pulverized (Tyrrell, 1926). Cf: *blastic deformation; plastic deformation.*

clastic dike A *sedimentary dike* consisting of a variety of clastic materials derived from underlying or overlying beds; esp. a *sandstone dike* or a *pebble dike.*

clasticity (clas-tic'-i-ty) (a) The quality, state, or degree of being clastic. (b) The maximum apparent particle size in a sediment or sedimentary rock (Carozzi, 1957).

clastic pipe A cylindrical body of clastic material, having an irregular columnar or pillarlike shape, standing approximately vertical through enclosing formations (usually limestone), and measuring a few centimeters to 50 m in diameter and a meter to 60 m in height; esp. a *sandstone pipe.* Syn: *cylindrical structure.*

clastic ratio A term introduced by Sloss et al. (1949, p.100) for the ratio of the thickness or percentage of clastic material (conglomerate, sandstone, shale) to that of nonclastic material (limestone, dolomite, evaporites) in a stratigraphic section; e.g. a ratio of 5 indicates that the section contains an average of 5 m of clastics per meter of nonclastics. Cf: *sand-shale ratio.* Syn: *detrital ratio.*

clastic rock (a) A consolidated sedimentary rock composed principally of broken fragments that are derived from preexisting rocks (of any origin) or from the solid products formed during chemical weathering of such rocks, and that have been transported mechanically to their places of deposition; e.g. a sandstone, conglomerate, or shale, or a limestone consisting of particles derived from a preexisting limestone. See also: *epiclastic rock.* Syn: *fragmental rock.* (b) pyroclastic rock. (c) *bioclastic rock.*

clastic sediment A sediment formed by the accumulation of fragments derived from preexisting rocks or minerals and transported as separate particles to their places of deposition by mechanical agents (such as water, wind, ice, and gravity); e.g. gravel, sand, mud, clay. Cf: *detrital sediment.* Syn: *mechanical sediment.*

clastic wedge A wedge-shaped deposit of sediments shed toward the foreland region of an active thrust belt. Clastic wedges typically fill foreland basins, regions of crust that have subsided in response to loading by thrust sheets. Formerly defined as the sediments of an *exogeosyncline,* derived from the tectonic land masses of the adjoining orthogeosynclinal belt (King, 1959, p.59). Cf: *geosynclinal prism.*

clastogenic lava A lava flow formed by the rheomorphic flow of coalesced and agglutinated hot pyroclastics, typically fed by a lava fountain. Lava may contain obvious spatter fragments (Wolff and Sumner, 2000). Cf: *spatter [pyroclast].*

clast support A sedimentary texture in which the larger particles are in contact. Cf: *matrix support.*

clathrate (clath'-rate) The texture commonly found in leucite-bearing rocks in which leucite crystals are surrounded by tangential augite crystals giving the appearance of a net or sponge, the augite representing the threads or walls and the leucite the holes (Johannsen, 1939, p.205). Obsolete.

clathrate wall The outer wall in archaeocyathids, in which vertical or longitudinal laths are applied to oblique annuli which are in turn attached to the outer edges of septa (TIP, 1972, pt. E, p.11).

claudetite (clau'-de-tite) A colorless monoclinic mineral: As_2O_3. It is dimorphous with arsenolite.

Clausius-Clapeyron equation (Clau'-si-us-Cla-pey'-ron) *Clapeyron equation.*

clausthalite (claus'-thal-ite) A metallic gray cubic mineral: PbSe. It is a member of the *galena* group, and resembles galena in appearance.

clavate (cla'-vate) adj. (a) Club-shaped, being slender at one end and gradually thickening near the other end, like a baseball bat; e.g. said of spores and pollen having sculpture consisting of processes that widen to a knob at the end. Cf: *pilate.* (b) Pertaining to a *clavus* of an ammonoid. n. A club-shaped thecal form in edrioasteroids, in which the oral surface plates form an upper gibbous "head", a lower constricted pedunculate zone, and an outward flaring peripheral rim (Bell, 1976).

clavicle (clav'-i-cle) (a) A shelly buttress or heavy internal ridge supporting the *hinge plate* or *adductor muscle* in some bivalve mollusks. Syn: *buttress [paleont].* (b) A paired, platelike bone of the pectoral girdle of Osteichthyes and the more rodlike collar bone of tetrapods (united at the midline as the *furcula* in birds).

clavidisc (clav'-i-disc) A sponge spicule (microsclere) in the form of an ovate disk with a central perforation.

clavula (clav'-u-la) A small ciliated spine in a *fasciole* of an echinoid. Pl: clavulae.

clavule (clav'-ule) A *sceptrule* in which the end bearing the axial cross is swollen or bears a ring of recurved teeth.

clavus (cla'-vus) An ammonoid *tubercle* elongated longitudinally in the direction of coiling. Pl: clavi.

clay [clay sci] A naturally occurring material composed primarily of fine-grained minerals. It is generally plastic at appropriate water contents and will harden when dried or fired (Guggenheim and Martin, 1995).

clay [eng] A term used to describe cohesive particles having diameters less than 0.074 mm, passing U.S. standard sieve no. 200, in accordance with test method D422 of the American Society for Testing and Materials (ASTM). Cf: *silt [eng].*

clay [geol] (a) As a size term, different disciplines use different boundaries, but most place the boundary between silt and clay at either 2 or 4 micrometers. (b) Sediments or rocks with >67% clay-size or colloidal particles. If the rock is unindurated it is clay; it the rock is indurated it is claystone; if the rock is indurated and laminated it is shale (Folk, 1974). (c) *clay mineral.*

clay [soil] A soil containing a high percentage of fine particles and colloidal substances, becoming sticky and plastic when wet and forming hard lumps or clods when dry; specif. a soil containing 40% or more of clay and not more than 45% of sand or 40% of silt. The term has also been used for a soil containing 30% or more of clay.

clay ball A chunk of clay released by erosion of a clayey bank and rounded during transport by currents; esp. an *armored mud ball.* Also spelled: clayball.

clay band A light-colored, *argillaceous* layer in clay ironstone. Also spelled: clayband.

clay-band ironstone A variety of *clay ironstone* characterized by abundant clay bands.

clay boil A *mud circle* that suggests a welling-up or heaving of the central core.

clay colloid (a) A clay particle having a diameter less than 1 micrometer (0.001 mm) (Jacks et al., 1960, p.24). (b) A colloidal substance consisting of clay-size particles.

claycrete (clay'-crete) Weathered argillaceous material forming a layer immediately overlying bedrock.

clay dike A clay-filled vertical or inclined fracture in coal and adjacent rocks. Usually very irregular; can be a meter or more wide (Nelson, 1981).

clay-dike fault A nontectonic normal fault associated with clay dikes and believed to have been formed by the same process as clay dikes (Nelson, 1981).

clay dune A dune composed of clay fragments heaped up by the wind, as in the lower Rio Grande Valley, Tex. (Coffey, 1909). Cf: *lunette.*

C layer The seismic region of the Earth between 410 km and 660 km, equivalent to the *transition zone* of the upper mantle. It is a part of a classification of the Earth's interior made up of layers A to G. Rarely used.

clayey (clay'-ey) Abounding in, consisting of, characterized by, or resembling clay; *argillaceous.*

clayey breccia A term used by Woodford (1925, p.183) for a breccia containing at least 80% rubble and 10% clay.

clayey sand (a) An unconsolidated sediment containing 50-90% sand and having a ratio of silt to clay less than 1:2 (Folk, 1954, p.349). (b) An unconsolidated sand containing 40-75% sand, 12.5-50% clay, and 0-20% silt (Shepard, 1954).

clayey sandstone (a) A consolidated *clayey sand.* (b) A sandstone containing more than 20% clay (Krynine, 1948, p.141). Cf: *argillaceous sandstone.*

clayey silt (a) An unconsolidated sediment containing 40-75% silt, 12.5-50% clay, and 0-20% sand (Shepard, 1954). (b) An unconsolidated sediment containing more particles of silt size than of clay size, more than 10% clay, and less than 10% of all other coarser sizes (Wentworth, 1922).

clay fill Dry or wet clay that fills a cave passage.

clay films Coatings of oriented clay on the surfaces of peds and mineral grains and lining pores. Also called clay skins, clay flows, illuviation cutans, or argillans.

clay gall (a) A markedly flattened and somewhat rounded pellet or curled fragment, chip, or flake of clay, generally embedded in a sandy matrix and esp. abundant at the base of some sandy beds. It may arise from the drying and cracking of a thin layer of coherent mud, the fragment commonly being rolled or blown into sand and buried, and forming a lenticular bleb upon wetting. (b) An ocherous, sometimes hollow inclusion of clay or mudstone, occurring esp. in oolitic limestones (Woodward, 1894, p.340). Syn: *gall [sed].*

clay gouge (a) A clayey deposit in a fault zone; *fault gouge.* (b) A thin seam of clay separating masses of ore, or separating ore from country rock. See also: *gouge [ore dep].*

clay gravel Gravel containing fine-grained silica and clay.

clay ironstone (a) A compact, hard, dark gray or brown fine-grained sedimentary rock, consisting of a mixture of argillaceous material (up to 30%) and iron carbonate (siderite), occurring in layers of nodules or concretions or as relatively continuous irregular thin beds, and usually associated with carbonaceous strata, esp. overlying a coal seam in the coal measures of the U.S. or Great Britain; a clayey iron carbonate, or an impure siderite ore occurring admixed with clays. The term has also been applied to an argillaceous rock containing iron oxide (such as hematite or limonite). See also: *blackband ironstone; clay-band ironstone.* (b) A sideritic concretion or nodule occurring in clay ironstone and other argillaceous rocks, often displaying septarian structure. (c) A sheet-like deposit of clay ironstone. Syn: *ironstone.*

clay loam A soil containing 27-40% clay, 20-45% sand, and the remainder silt. See also: *silty clay loam; sandy clay loam.*
clay marl A whitish, smooth, chalky clay; a marl in which clay predominates.
clay mineral Any mineral occurring in the clay-sized fraction of a soil, a suspension, or a rock, with the understanding that size imposes physical and chemical characteristics, e.g., colloidal and catalytic properties. However, minerals so designated are mostly 1:1 or 2:1 hydrous aluminosilicates or phyllosilicates with some important exceptions, such as hydrated oxides of iron and aluminum, zeolites, and a few others. The most common clay minerals belong to the chlorite, illite, kaolin, or smectite groups, or are mixed-layered clay minerals formed from these components. Syn: *sialite.*
clay pan *dry mudflat.*
claypan [geomorph] (clay′-pan) A term used in Australia for a shallow depression containing clayey and silty sediment and having a hard, sun-baked surface; a *playa* formed by deflation of alluvial topsoils in a desert, in which water collects after a rain.
claypan [soil] A dense, heavy, relatively impervious subsurface soil layer that owes its hardness to a relatively higher clay content than that of the overlying material, from which it is separated by a sharply defined boundary. It is usually hard when dry and plastic when wet, and is presumably formed by the concentration of clay by percolating waters or by in-situ synthesis. Cf: *hardpan; iron pan.* Also spelled: *clay pan.*
clay parting (a) Clayey material between a vein and its wall. Syn: *parting [ore dep].* (b) A seam of hardened carbonaceous clay between or in beds of coal, or a thin layer of clay between relatively thick beds of some other rock (such as sandstone).
clay plug A mass of silt, clay, and organic muck, deposited in and eventually filling an oxbow lake.
clay pocket A clay-filled cavity in rock; a mass of clay in rock or gravel.
clay rock An indurated clay, composed of argillaceous detrital material and sufficiently hardened to be incapable of being worked without grinding, but not chemically altered or metamorphosed; a claystone.
clay shale (a) A consolidated sediment consisting of no more than 10% sand and having a silt/clay ratio of less than 1:2 (Folk, 1954, p.350); a fissile claystone. (b) A shale that consists chiefly of clayey material and that becomes clay on weathering.
clay-silicate nickel laterite A *nickel laterite* composed of nickel-rich clay minerals such as nontronite.
clay size A term used in sedimentology for a particle having an equivalent spherical diameter equal to or less than $^1/_{256}$ mm (0.00016 in.) See also: *dust size.*
clay-size fraction The portion of a soil or disaggregated rock consisting of particles that are less than 2µm in equivalent spherical diameter (Moore and Reynolds, 1997).
clay slate (a) A low-grade, essentially unreconstituted slate, as distinguished from the more micaceous varieties that border on phyllite; specif. an *argillite,* less than 50% reconstituted, with a parting, slaty cleavage or incipient foliation; or a weakly metamorphosed rock intermediate in character between a shale and a slate (Flawn, 1953, p.564). (b) A slate derived from an argillaceous rock, such as shale, rather than from volcanic ash; a metamorphosed clay, with cleavage developed by shearing or pressure, as distinguished from "mica slate". (c) An English term much used in the early 19th century for true slate.
claystone [ign] (clay′-stone) (a) A dull, altered, feldspathic igneous rock in which the groundmass or the whole rock has been reduced to a compact mass of earthy or clayey alteration products (Holmes, 1928, p.61-62). (b) A term used in Australia for a soft earthy feldspathic rock occurring in veins and having the appearance of indurated clay. Also spelled: clay stone.
claystone [sed] (a) An indurated rock with more than 67 per cent clay-sized minerals. See also: *clay [geol]; clay mineral; shale.* (b) An indurated clay having the texture and composition of shale but lacking its fine lamination or fissility; a massive *mudstone* in which clay predominates over silt; a nonfissile *clay shale.* Flawn (1953, p.562-563) regards claystone as a weakly indurated sedimentary rock whose constituent particles have diameters less than 0.01 mm. Shrock (1948a) describes it as a somewhat unctuous, conchoidally fracturing sedimentary rock composed largely of clay material. Syn: *clay rock.* (c) A concretionary clay found in alluvial deposits in the form of flat rounded disks that are variously united to give rise to curious shapes (Fay, 1918, p.160). Also, a calcareous concretion frequently found in a bed of clay. (d) An old English term for an argillaceous limestone (Arkell and Tomkeieff, 1953, p.24). Also spelled: clay stone.
claystone porphyry An old and indefinite name for a *porphyry* whose fine groundmass is more or less kaolinized "so as to be soft and earthy, suggesting hardened clay" (Kemp, 1934, p.203). Obsolete.
clay vein A body of clay, usually roughly tabular in form like a dike or vein, that fills a crevice in a coal seam. It is believed to originate where clay from the roof or floor has been forced into a small fissure, often altering or enlarging it. Cf: *horseback [coal]; spar [mining]; clastic dike.* Syn: *dirt slip.*
clay-with-flints (a) A term used in southern England (as on the North Downs and in Dorset) for a residual deposit of reddish-brown, tenacious clay containing mechanically unworn flint fragments, lying unevenly and directly on the surface of chalk or occurring in funnel-shaped pipes penetrating to considerable depths. It represents in part the insoluble residue of chalk subjected to prolonged subaerial weathering and in part admixed waste material derived from formerly overlying Tertiary rocks. There is much variation in the relative proportions of flints and clay. (b) A term applied loosely to any clay-flint drift deposit that rests on chalk.
clean (a) Said of a diamond or other gemstone that is free from noticeable internal flaws. (b) Said of a mineral that is virtually free of undesirable non-ore or waste-rock material.
clean sandstone A relatively pure or well-washed sandstone containing little matrix; specif. an arenite with less than 10% argillaceous matrix (Williams et al., 1954) or an orthoquartzite with less than 15% detrital clay matrix (Pettijohn, 1954). The particles are held together by a mineral cement. Cf: *dirty sandstone.*
clearcreekite A pale greenish yellow monoclinic mineral: $Hg^{1+}_3CO_3(OH)\cdot 2H_2O$.
clearing (clear′-ing) *polynya.*
cleat In a coal seam, a joint or system of joints along which the coal fractures. There are usually two cleat systems developed perpendicular to each other. See also: *face cleat; end cleat; butt cleat.* Also spelled: *cleet.*
cleat spar Crystalline mineral matter occurring in the *cleat* planes of a coal seam; specif. *ankerite.*
cleavage [mineral] (cleav′-age) The breaking of a mineral along its crystallographic planes owing to either paucity of, or systematically weak, atomic bonds, thus reflecting crystal structure. The types of cleavage are named according to the morphology, e.g. *prismatic cleavage.* Cf: *fracture [mineral]; parting [cryst].*
cleavage [struc geol] A locally planar fabric in an unmetamorphosed or weakly metamorphosed, fine-grained rock defined by either the tendency of a rock to split in a particular direction, a preferred orientation of crystal planes in mineral grains, or a preferred orientation of inequant grain shapes; it is a product of deformation or metamorphism. Cf: *fissility.* See also: *foliation [struc geol]; schistosity.* Obsolete syn: *secondary cleavage.*
cleavage banding A compositional banding that is parallel to the cleavage rather than to the bedding. It results from the mechanical movement of incompetent material, such as argillaceous rocks, into the cleavage planes in a more competent rock, such as sandstone. The argillaceous bands are commonly only a few millimeters thick (Billings, 1954). Cf: *segregation banding.*
cleavage domain Thin, typically phyllosilicate-rich laminae in cleaved rock that bound *microlithons.* In addition to phyllosilicates, domains commonly contain insoluble materials such as carbonaceous matter and opaque minerals.
cleavage face In a crystal, a smooth surface produced by cleavage. It may be essentially planar, e.g. in mica.
cleavage fan *fan cleavage.*
cleavage fold A *shear fold* in which the shear occurs along cleavage planes of secondary foliation. This folding-mechanism concept is now discredited.
cleavage fragment A fragment of a crystal that is bounded by cleavage faces.
cleavage mullion An obsolete term for a *mullion* formed by the intersection of cleavage planes with bedding.
cleavage plane One of the surfaces along which a rock tends to split because of cleavage. Cleavage planes may be planar or wavy. If planar, cleavage planes parallel one another. If wavy, cleavage planes *anastomose.* Cf: *cleavage domain.*
cleavelandite (cleave′-land-ite) A white, lamellar or leaflike variety of albite, having an almost pure Ab content and often forming fan-shaped aggregates of tabular crystals that show mosaic development and look as though bent. It is formed as a late-stage mineral in pegmatites, replacing other minerals. Also spelled: *clevelandite.*
cleavings (cleav′-ings) The partings in a coal seam that separate it into beds.
cleet A less-preferred spelling of *cleat.*
cleft [geomorph] An abrupt chasm, cut, breach, or other sharp opening, such as a craggy fissure in a rock, a wave-cut gully in a cliff, a trench on the ocean bottom, a notch in the rim of a volcanic crater, or a narrow recess in a cave floor. Obsolete syn: *clift [geomorph].*
cleft [paleont] A tension fracture formed during growth on either side of the shell, anterior and/or posterior to the beak, of some rostroconch mollusks (Pojeta and Runnegar, 1976).
cleft deposit A pocket, specifically a fissure filling, in alpine regions.
cleft girdle On a fabric diagram, an annular maximum occupying a small circle of the net (Turner and Weiss, 1963, p.58). Cf: *girdle [struc petrol]; maximum [struc petrol].* Syn: *small-circle girdle.*
cleft lava *lava-inflation cleft.*
cleftstone (cleft′-stone) *flagstone.*
Clerici solution (Cle-ri′-ci) A solution of thallium malonate and thallium formate in water that is used as a *heavy liquid;* its specific gravity is 4.15. Cf: *Sonstadt solution; Klein solution; bromoform; methylene iodide.*
clerite A metallic black orthorhombic mineral: $MnSb_2S_4$. A member of the *berthierite* group.
cleve (a) An English syn. of *cliff [geomorph].* (b) An English term for *brow* or steeply sloping ground.
cleveite (cleve′-ite) A variety of uraninite containing rare earths (cerium).
clevelandite (cleve′-land-ite) *cleavelandite.*

cliachite (cli′-ach-ite) (a) A ferruginous bauxite. (b) A group name for colloidal aluminum hydroxides constituting most bauxite. Also spelled: *kliachite.* Syn: *alumogel.*

Clifdenian (Clif-den′-ian) South Pacific stage: lower middle Miocene (above Altonian, below Lillburnian).

cliff [geomorph] (a) *sea cliff.* (b) Any high, very steep to perpendicular or overhanging face of rock; a precipice. A cliff is usually produced by erosion, less commonly by faulting. (c) A British term for a steep slope or declivity (>45°), or a hill. English syn: *cleve.* Dialectical var: *clift [geomorph].*

cliff [sed] (a) *clift [sed].* (b) An old term used in SW England for rock lying directly above or between coal seams. Pl: *clives.*

cliffed headland A headland characterized by a cliff, such as one formed by erosion during the early development of an embayed coast.

cliff erosion *sapping [geomorph].*

cliff-foot cave A *solution cave* at the foot of a *karst tower* or cliff. Syn: *foot cave.*

cliff glacier A short glacier that occupies a niche or hollow on a steep slope and does not reach a valley, such as a glacier perched on a ledge or bench on the face of a cliff. Cf: *hanging glacier.* Less preferred syn: *cornice glacier.*

cliffline (cliff′-line) The *coastline* on a steep coast, represented by an imaginary line along the base of the cliffs.

cliff of displacement *fault scarp.*

cliffordite (clif′-ford-ite) An adamantine yellow cubic mineral: $(UO_2)Te^{4+}{}_3O_7$.

cliff overhang A rock mass jutting out from a slope; esp. the upper part or edge of an eroded cliff projecting out over the lower, undercut part, as above a wave-cut notch.

clift [geomorph] (a) Obsolete var. of *cleft [geomorph].* (b) Dialectal var. of *cliff [geomorph].*

clift [sed] A term used in southern Wales for various kinds of shale, esp. a strong, usually silty, mudstone. Syn: *cliff [sed].*

cliftonite (clif′-ton-ite) A black polycrystalline aggregate of graphite, with cubic morphology, representing a minutely crystalline form of carbon occurring in meteorites. It is considered by some to be a pseudomorph after diamond.

climate (cli′-mate) The characteristic weather of a region, particularly as regards temperature and precipitation, averaged over some significant interval of time and including extremes. See also: *climate classification; climatic province.*

climate classification An arrangement or description of the various climate types by particular descriptive factors, such as temperature, rainfall, vegetation, or position relative to land and sea; e.g. *Thornthwaite's classification; Köppen's classification.* See also: *climatic zone; temperature zone.*

Climate Leaf Analysis Multivariate Program A method of leaf margin analysis for fossil leaf floras, utilizing multivariate techniques for estimating paleotemperature and paleoprecipitation. Abbrev: CLAMP.

climate-stratigraphic unit A term used by the ACSN (1959, p.669) for a time interval now known as a *geologic-climate unit.* It is not strictly a stratigraphic unit.

climatic (cli-mat′-ic) Said of ecologic formations resulting from or influenced by differences in climate. Cf: *edaphic.*

climatic accident A departure from the normal cycle of erosion, caused by marked changes in the climate, such as those effected by glaciation or by a change to aridity, independent of the normal climatic change due to loss of relief from youth to old age. "The term has become virtually obsolete since the idea of humid temperate climate being 'normal' no longer holds credence" (Monkhouse, 1965, p.3). See also: *accident; volcanic accident.*

climatic amelioration A qualitative term designating a change to a warmer climate, commonly applied to climatic trends of late glacial and Holocene time; an "improvement" of climate as seen by someone in polar latitudes. Cf: *climatic deterioration.*

climatic bog *blanket bog.*

climatic deterioration A term designating a change to a colder climate, applied specif. to the primary and secondary climatic trends of late glacial and Holocene time such as occurred during the Little Ice Age; a "degeneration" or "worsening" of climate as seen by someone in polar latitudes. Cf: *climatic amelioration.*

Climatic Optimum An informal term designating the postglacial interval of most equable climate, with elevated summer temperatures and abundant rainfall in some regions. The concept is derived from the mid-Holocene warm interval between 8,000 and 4,500 calendar years ago. As it is believed to have begun earlier in low latitudes, it should not be used as a precise time interval (Fairbridge, 1972). See also: *thermal maximum; Hypsithermal.*

climatic peat Peat that characteristically occurs in a certain climatic zone.

climatic province A region characterized by a particular *climate.*

climatic snowline (a) The average line or altitude above which horizontal surfaces have more than 50 percent snow cover, averaged over a time period of climatic significance (e.g., 10 years). Cf: *snowline.* (b) The same line as observed in late summer so that it approximately coincides with the firn line or *equilibrium line* on glaciers. Cf: *regional snowline.*

climatic terrace A stream terrace whose formation is controlled by climatic changes that may induce aggradation or degradation of a valley.

climatic zone A general term for a latitudinal region characterized by a relatively homogeneous climate, e.g. any of the zones delimited by the tropics of Cancer and Capricorn and the Arctic and Antarctic circles, or a zone or province in some climatic classification. Cf: *temperature zone.*

climax (cli′-max) In ecology, the final stable or equilibrium stage of development that a sere, community, species, flora, or fauna attains in a given environment. The major world climaxes correspond to formations and *biomes.* Cf: *pioneer.*

climax avalanche A snow avalanche of maximum size, containing a large portion of old snow, and arising from conditions that developed over a period of time longer than one year. The term is part of an obsolete classification of avalanches.

climax community An assemblage of species that represents the permanent or usual long-term inhabitants of a region during ecologic or seral succession; it is normally preceded by one or more shorter-lived assemblages whose life activities prepare the habitat. Cf: *climax; pioneer.*

Climax-type Referring to a sub-group of porphyry molybdenum deposits characterized by: (1) an association with highly evolved rhyolitic porphyries, (2) relatively high grades (typically 0.35-0.45% MoS_2), and (3) high concentrations of large ion lithophile, *LIL,* elements, especially fluorine in fluorite and topaz. Named for classic locality at Climax, Colorado.

climbing bog (climb′-ing) An elevated boggy area on the margin of a swamp, usually in a region characterized by a short summer and considerable rainfall, caused by the upward growth of sphagnum from the original level of the swamp to higher ground.

climbing dune A dune formed by the piling-up of sand by wind against a cliff or mountain slope. Syn: *rising dune.*

climbing ripple One of a series of cross-laminae produced by superimposed ripples migrating over an aggrading bed in which the crests of vertically succeeding laminae appear to be advancing upslope. See also: *ripple drift; cross-stratification; type-A climbing ripple; type-B climbing ripple; type-C climbing ripple; sinusoidal climbing ripple.*

climograph (cli′-mo-graph) A graphic representation of climatic data, containing at least two climate elements, esp. mean monthly values of precipitation and temperature. Also spelled: climagraph.

cline (a) A gradational series of variant forms (e.g. morphologic or physiologic variations) within a group of closely related (usually conspecific) organisms, generally developing as a result of environmental, geographic, chronological, or stratigraphic transition. (b) The formal, hierarchical *diachronic unit* of lowest rank and shortest scope (NACSN, 1983, Art. 93). Cf: *episode; phase [stratig]; span [stratig].*

clinker [coal] (clink′-er) (a) Coal that has been altered by igneous intrusion. Cf: *natural coke.* Syn: *scoria [coal].* (b) Masses of coal ash that are a by-product of combustion. Cf: *coke.*

clinker [volc] A rough, jagged pyroclastic or autobrecciated fragment, such as *aa* that resembles the clinker or slag of a furnace. Adj: *clinkery.*

clinkertill (clink′-er-till) Glacial *till* baked by the burning of lignite beds.

clinkery (clink′-er-y) Adj. of *clinker [volc];* it is used to describe the surface of a lava flow.

clinkstone (clink′-stone) An obsolete syn. of *phonolite* (in its broadest sense).

clino (cli′-no) adj. A term applied by Rich (1951, p.2) to the environment of sedimentation that lies on the sloping part of the floor of a water body, extending from wave base down to the more or less level deeper parts. It may be used alone or as a combining form. See also: *clinoform; clinothem.* Cf: *unda; fondo.*

clinoamphibole (cli-no-am′-phi-bole) (a) A group name for amphiboles crystallizing in the monoclinic system. (b) Any monoclinic mineral of the amphibole group, such as hornblende, cummingtonite, grunerite, tremolite, actinolite, riebeckite, glaucophane, and arfvedsonite. Cf: *orthoamphibole.*

clinoatacamite A green to greenish black monoclinic mineral: $Cu_2(OH)_3Cl$. A polymorph of atacamite.

clinoaugite (cli-no-au′-gite) *clinopyroxene.*

clinoaxis (cli′-no-ax′-is) In a monoclinic crystal, the lateral axis that is oblique to the vertical; it is the *a* axis [cryst]. Cf: *orthoaxis.*

clinobarrandite (cli′-no-bar-ran′-dite) A variety of the mineral phosphosiderite containing aluminum.

clinobarylite A vitreous colorless monoclinic mineral: $BaBe_2Si_2O_7$. A dimorph of barylite.

clinobehoite (cli′-no-be′-ho-ite) A colorless monoclinic mineral: $Be(OH)_2$. A polymorph of behoite.

clinobisvanite (cli-no-bis′-van-ite) A yellow or orange monoclinic mineral: $BiVO_4$. It is dimorphous with pucherite.

clinobronzite (cli-no-bronz′-ite) A variety of clinoenstatite containing iron; a clinopyroxene intermediate in composition between clinoenstatite and clinohypersthene, having less than 20 mole percent of $FeSiO_3$.

clinocervantite A vitreous colorless monoclinic mineral: $Sb^{3+}Sb^{5+}O_4$. A dimorph of cervantite.

clinochalcomenite (cli′-no-chal-co′-me-nite) A bluish-green monoclinic mineral: $CuSe^{4+}O_3 \cdot 2H_2O$. It is a dimorph of chalcomenite, and a member of the *cobaltomenite* group.

clinochlore (cli′-no-chlore) A trioctahedral member of the chlorite group of minerals, in which Mg is the dominant divalent octahedral cation: $(Mg,Fe^{2+},Al)_6(Si,Al)_4O_{10}(OH)_8$. It occurs as monoclinic and triclinic polytypes. Cf: *ripidolite.* Syn: *penninite; sheridanite.*

clinochrysotile (cli-no-chrys'-o-tile) A white, pale green to dark green monoclinic mineral of the serpentine group, originally called chrysotile: $Mg_3Si_2O_5(OH)_4$. Cf: *orthochrysotile.*

clinoclase (cli'-no-clase) A dark green monoclinic mineral: $Cu_3(AsO_4)(OH)_3$. A dimorph of gilmarite.

clinodome (cli'-no-dome) An obsolete term for a *first-order prism* in the monoclinic system. Its indices are {0kl} and its symmetry is 2/m. Cf: *brachydome.*

clinoenstatite (cli-no-en'-sta-tite) A colorless, yellow, brown, or green monoclinic mineral of the *pyroxene* group: $Mg_2Si_2O_6$.

clinoferroholmquistite A theoretical member of the *amphibole* group, the Fe^{2+}-Al member of the *clinoholmquistite* series: $Li_2(Fe^{2+},Mg,Al)_5(Si,Al)_8O_{22}(OH,F)_2$.

clinoferrosilite (cli'-no-fer'-ro-sil'-ite) A colorless or amber monoclinic mineral of the pyroxene group: $(Fe^{2+},Mg)_2Si_2O_6$. See also: *ferrosilite.* Cf: *orthoferrosilite.*

clinoform (cli'-no-form) The subaqueous landform analogous to the continental slope of the oceans or to the foreset beds of a delta (Rich, 1951, p.2). It is the site of the *clino* environment of deposition. Cf: *undaform; fondoform.*

clinoform rollover A point on a cross section and a line on a three-dimensional clinoform where there is a sharp increase in downslope gradient (Miller et al., 1998).

clinoform surface A sloping depositional surface, commonly associated with strata prograding into deep water (Mitchum, 1977, p.205).

clinograde (cli'-no-grade) Pertaining to the decreasing concentration of oxygen or other chemicals in the *hypolimnion* of a lake. Cf: *orthograde.*

clinographic (cli-no-graph'-ic) Pertaining to a representation of a crystal in which no crystal face is projected as a line.

clinographic curve A curve representing the slope or slopes of an area of the Earth's surface as it varies with altitude; in practice it is designed to show the actual variation of the average slope within each contour interval.

clinographic projection An oblique projection used for representing crystals in such a manner that no crystal face will be projected as a line.

clinohedral class (cli-no-hed'-ral) *domatic class.*

clinohedrite (cli-no-hed'-rite) A colorless, white, or purplish monoclinic mineral: $CaZnSiO_4 \cdot H_2O$.

clinoholmquistite (cli-no-holm'-quist-ite) A vitreous blue monoclinic mineral of the amphibole group: $Li_2(Mg_3Al_2)Si_8O_{22}(OH)_2$.

clinohumite (cli-no-hum'-ite) A white, orange, or brown monoclinic mineral of the *humite* group: $(Mg,Fe^{2+})_9(SiO_4)_4(F,OH)_2$. A dimorph of humite.

clinohypersthene (cli-no-hy'-per-sthene) An unnecessary name for a monoclinic mineral of the *pyroxene* group: $(Mg,Fe)SiO_3$. Syn: *clinoferrosilite, clinoenstatite.*

clinojimthompsonite (cli'-no-jim-thomp'-son-ite) A colorless to pale pinkish brown monoclinic pyribole mineral: $(Mg,Fe^{2+})_5Si_6O_{16}(OH)_2$.

clinokurchatovite (cli'-no-kur'-cha-to-vite) A colorless to light gray monoclinic mineral: $Ca(Mg,Fe^{2+},Mn^{2+})B_2O_5$. It is a polymorph of kurchatovite.

clinolimnion (cli-no-lim'-ni-on) The upper part of a *hypolimnion,* where the rate of heat absorption falls off almost exponentially with depth. Cf: *bathylimnion.*

clinometer (cli-nom'-e-ter) Any of various instruments used for measuring angles of slope, elevation, or inclination (esp. the dip of a geologic stratum or the slope of an embankment); e.g. a simple hand-held device consisting of a tube with cross hair, with a graduated vertical arc and an attached spirit level so mounted that the inclination of the line of sight can be read on the circular scale by centering the level bubble at the instant of observation. A clinometer is usually combined with a compass (e.g. the *Brunton compass*). See also: *inclinometer [drill].*

clinomimetite A yellowish-white monoclinic mineral of the *apatite* group: $Pb_5(AsO_4)_3Cl$. A dimorph of mimetite.

clinophosinaite (cli'-no-pho-sin'-a-ite) A pale lilac monoclinic mineral: $Na_3Ca(SiO_3)(PO_4)$. A dimorph of phosinaite.

clinopinacoid (cli-no-pin'-a-coid) In a monoclinic crystal, a pinacoid that is parallel to the mirror plane of symmetry and perpendicular to the axis of symmetry.

clinoplain (cli'-no-plain) An inclined plain projecting from the mountains and forming a low bluff on the side of a flood plain, as in the Rio Grande valley (Herrick, 1904, p.379).

clinoptilolite (cli-no-ptil'-o-lite) A group name for a monoclinic zeolite mineral with the general formula $A_{2\text{-}3}(Si,Al)_{18}O_{36} \cdot 11H_2O$, where A = Na, K, or Ca.

clinoptilolite-(Ca) A colorless monoclinic zeolite mineral of the *clinoptilolite* group: $(Ca,Na,K)_{2\text{-}3}(Si,Al)_{18}O_{36} \cdot 11H_2O$.

clinoptilolite-(K) A colorless monoclinic zeolite mineral of the *clinoptilolite* group: $(K,Na,Ca)_{2\text{-}3}(Si,Al)_{18}O_{36} \cdot 11H_2O$.

clinoptilolite-(Na) (cli-no-ptil'-o-lite) A colorless monoclinic zeolite mineral: $(Na,K,Ca)_{2\text{-}3}(Si,Al)_{18}O_{36} \cdot 11H_2O$. A member of the *clinoptilolite* group.

clinopyroxene (cli-no-py'-rox-ene) (a) A group name for pyroxenes crystallizing in the monoclinic system and sometimes containing considerable calcium with or without aluminum and the alkalies. (b) Any monoclinic mineral of the *pyroxene* group, such as diopside, hedenbergite, clinoenstatite, clinohypersthene, clinoferrosilite, augite, acmite, pigeonite, spodumene, jadeite, and omphacite. Cf: *orthopyroxene.* Syn: *monopyroxene; clinoaugite.*

clinopyroxene norite In the *IUGS classification,* a *gabbro* consisting mostly of calcic plagioclase and orthopyroxene, with clinopyroxene exceeding 5%.

clinopyroxenite (cli'-no-py-rox'-e-nite) In the *IUGS classification,* a plutonic rock with M ≥ 90, and cpx/(ol+opx+cpx) > 90.

clinosafflorite (cli-no-saf'-flor-ite) A white metallic monoclinic mineral: $(Co,Fe,Ni)As_2$. It is dimorphous with safflorite.

clinostrengite (cli-no-streng'-ite) *phosphosiderite.*

clinothem (cli'-no-them) Rock units formed in the *clino* environment of deposition (Rich, 1951, p.2). Cf: *undathem; fondothem.*

clinotobermorite A colorless or white mineral existing as monoclinic and triclinic polytypes: $Ca_5Si_6(O,OH)_{18} \cdot 5H_2O$. A dimorph of tobermorite.

clinotyrolite (cli-no-ty-rol'-ite) A green monoclinic mineral: $Ca_2Cu_9(AsO_4,SO_4)_4(OH)_{10} \cdot 10H_2O$. A dimorph of tyrolite; not approved by the International Mineralogical Association.

clinounconformity (cli'-no-un'-con-form'-i-ty) An obsolete syn. of *angular unconformity.* Term proposed by Crosby (1912, p.296). Also spelled: *clinunconformity.*

clinoungemachite (cli'-no-un'-ge-mach-ite') A pale yellow monoclinic mineral of doubtful validity, possible a variety of ungemachite: $K_3Na_8Fe^{3+}(SO_4)_6(NO_3)_2 \cdot 6H_2O$ (?). It may be a variety of ungemachite.

clinozoisite (cli-no-zo'-is-ite) A grayish-white, pink, or green mineral of the *epidote* group: $Ca_2Al_3(SiO_4)(Si_2O_7)O(OH)$. It is the monoclinic dimorph of *zoisite* and grades into, but is lighter in color than, epidote.

clint [geomorph] A Scottish term used in a general sense for any hard or flinty rock, such as a ledge projecting from a hillside or in a stream bed; also, a rocky cliff.

clint [karst] A term used in England for slabs of limestone, parallel to the bedding, forming a *pavement.* Widened joints, or *grikes,* isolate individual clints. Cf: *crevice karst.*

Clintonian (Clin-ton'-i-an) A North American stage: Lower Silurian (above Alexandrian, below Cliftonian).

clintonite (clin'-ton-ite) (a) A trioctahedral brittle mica, of ideal composition $Ca(MgAl)_3(OH,F)_2[(SiAl_3)O_{10}]$. It has a reddish-brown, copper-red, or yellowish color, and occurs in monoclinic crystals and foliated masses. Its crystal structure is based on cubic closest-packing of the anions. Syn: *seybertite; xanthophyllite.* (b) A group name for the *brittle micas.*

Clinton ore (Clin'-ton) A red, fossiliferous sedimentary iron ore; e.g., the Clinton Formation (Middle Silurian) or correlative rocks of the east-central U.S., containing lenticular or oolitic grains of hematite. Cf: *fossil ore; flaxseed ore.*

Clinton-type ironstone An *ironstone* composed principally of hematite. Syn: *Clinton ore.*

clinunconformity (clin'-un-con-form'-i-ty) *clinounconformity.*

clipping The process of extracting a subset of spatial data from a larger data set by selecting only those data located inside (or outside) a selected boundary. The area enclosed by this boundary may be referred to as a *clipping window* (Krzanowski et al., 1993). In GIS, it is also referred to as selecting a "region of interest" (ROI).

clipping window A polygon used to perform *clipping* (Krzanowski et al., 1993).

clisere (cli'-sere) A *sere* that develops as a result of great physiographic changes.

clisiophylloid (cli'-si-o-phyl'-loid) Said of an *axial structure* in a rugose coral (such as in *Clisiophyllum*) characterized by a short medial plate joining the *cardinal* and *counter septa* and resembling a spider web in transverse section.

clives Plural of *cliff [sed],* so called because of the "easy cleavage" of the rocks overlying the coal.

cloaca (clo-a'-ca) (a) An exhalant chamber in an invertebrate; esp. the large central cavity of a sponge into which pores and/or canals empty and which communicates through the *osculum* directly and externally with the surface of the sponge. Syn: *atrium; spongocoel; paragaster.* (b) The common chamber into which the excretory, reproductive, and digestive tracts discharge in all vertebrates except most mammals. Pl: cloacae.

clockwise inclination (clock'-wise) The inclination to the right of a heterococcolith suture as it proceeds to the periphery. Ant: *counterclockwise inclination.*

clod [mining] A miner's term applied to a soft, weak, or loosely consolidated shale (or to a hard, earthy clay), esp. one found in close association with coal or immediately overlying a coal seam. It is so called because it falls away in lumps when worked.

clod [soil] A compact, coherent mass of soil varying in size, usually produced by plowing, digging, etc., especially when these operations are performed on soils that are either too wet or too dry and usually formed by compression, or breaking off from a larger unit, as opposed to a building-up action as in aggregation.

clone A group term for genetically identical individuals that are descended from a single founding parent by any form of asexual reproduction.

Cloosian dome (Cloos'-i-an) An elliptical uplift or upwarping, usually with a central graben, described by Cloos (1939) and exemplified in the East African and Rhine rift valleys. The term is no longer used.

closed basin An enclosed area having no drainage outlet, from which water escapes only by evaporation, as in an arid region. Cf: *interior basin.*
closed-cavity ice (closed-cav'-i-ty) Ice, commonly as large crystals of hoarfrost, formed in underground cavities within permafrost.
closed contour A contour line that forms a closed loop and does not intersect the edge of the map area on which it is drawn; e.g. a *depression contour* indicating a closed depression, or a normal contour indicating a hilltop.
closed depression An area of lower ground indicated on a topographic map by a hachured *depression contour* line forming a closed loop; e.g. a fault sag, or a hollow below the general land surface, with no surface outlet. Syn: *topographic depression.*
closed drainage *internal drainage.*
closed fold An old, rarely used syn. for *isocline.*
closed form A crystal form whose faces enclose space; e.g., a dipyramid. Cf: *open form.*
closed-in pressure *shut-in pressure.*
closed lake A lake that does not lose water by a surface outlet or by seepage but only by evaporation (Wetzel, 2001, p.46-47). Cf: *enclosed lake.* Ant: *open lake.*
closed ridge A circular, elliptical, or irregularly shaped ridge of glacial material surrounding a central depression (or sometimes a mound of glacial material or a moraine plateau), and resulting from the melting of a block of stagnant ice (Gravenor and Kupsch, 1959, p.52-53).
closed structure A structure which, when represented on a map by contour lines, is enclosed by one or more closed contours; e.g. a closed anticline or a closed syncline. Ant: *open structure.*
closed system [chem] A chemical system in which, during a process under consideration, thermal and mechanical energy may be transferred into or out of the system, but no transfer of mass occurs. Cf: *open system [chem].*
closed system [permafrost] A condition of freezing of thawed ground within permafrost in which no additional groundwater is available (Muller, 1947, p.214), exemplified by the pingos of the Mackenzie Delta, Canada. Ant: *open system [permafrost].*
closed traverse A surveying traverse that starts and terminates upon the same station or upon stations whose relative locations have been determined by other surveys. Cf: *open traverse.*
close fold A fold with an inter-limb angle between 30° and 70° (Fleuty, 1964, p.470).
close-grained Said of a rock, and of its texture, characterized by fine, tightly packed particles.
close-joints cleavage An obsolete term for an array of closely spaced joints (tensile cracks).
close-packed structure A type of crystal structure that provides the tightest possible packing: a first layer of atoms in which each atom has six similar atoms touching it, a second layer of atoms fitting into the indentations of the first, and a third layer either as a repetition of the first or in a third position. If the third layer repeats the first, the result is *hexagonal close packing* (hcp); if it is in a third position the result is *cubic close packing* (ccp). Cf: *open-packed structure.*
close pack ice Pack ice in which the ice cover or concentration is 7/10 to 9/10 and composed of floes mostly in contact. Syn: *packed ice.*
close packing The manner of arrangement of uniform solid spheres packed as closely as possible so that the porosity is at a minimum; e.g. the packing of a face-centered cubic lattice or of a close-packed hexagonal lattice. See also: *rhombohedral packing.* Ant: *open packing.*
closing error (clos'-ing) *error of closure.*
closing the horizon Measuring the last angle of a series of adjacent horizontal angles at a station, required to make the series complete around the horizon. See also: *horizon closure.*
closterite (clos'-ter-ite) A dense, laminated, brownish-red canneloid material from the Irkutsk River basin of Siberia. The organic matter is largely *Pila bibractensis* colonies (Twenhofel, 1950, p.475).
closure [paleont] Calcification of the frontal membrane of an anascan cheilostome (bryozoan) *autozooid,* with loss of its *lophophore* and feeding function. Also, the calcified frontal structure so formed (Cook, 1965, p.159).
closure [struc geol] In a subsurface fold, dome, or other *structural trap,* the vertical distance between the structure's highest point and its lowest closed *structure contour.* It is used in the estimation of petroleum reserves. Syn: *structural closure.*
closure [surv] A cumulative measure of the various individual errors in survey measurements; the amount by which a series of survey measurements fails to yield a theoretical or previously determined value for a survey quantity. See also: *error of closure; discrepancy.*
closure temperature Temperature at which chemical exchange between minerals and their surroundings cease.
clot A group of ferromagnesian minerals in igneous rock, from a few inches to a foot or more in size, commonly drawn out longitudinally, that may be a segregation or an altered xenolith (Balk, 1937).
clotted (clot'-ted) Formed of clustered, aggregated, or flocculated grains. It is characterized by patches of dark, dense, fine-grained areas that are ultimately surrounded by sparry calcite. Commonly formed by the squashing together of soft micrite pellets.
cloud [meteorol] A visible aggregate of minute water droplets or ice crystals above the Earth's surface. Water droplets form by condensation around nuclei such as salt crystals, dust particles, or pollen; ice crystallizes on dust particles or on other ice crystals. Clouds are classified according to altitude and appearance.
cloud [speleo] In a cave, a subaqueous carbonate coating over a bedrock projection. Cf: *mammillary [speleo].*
cloudburst flood (cloud'-burst) An ephemeral flood commonly occurring during an abrupt summer rain of high intensity, usually in an arid or semiarid region.
cloud condensation nuclei Tiny solid and liquid particles that promote the condensation of water at temperatures both above and below the freezing point of water; may include *hygroscopic nuclei.*
clouded agate (cloud'-ed) A transparent or semitransparent light-gray *agate* with irregular, indistinct, or more or less rounded patches of darker gray resembling dark clouds. Syn: cloud agate; cloudy agate.
clouding (cloud'-ing) The effect produced in crystals (as of plagioclase) by the presence of numerous minute dark particles (microlites, dustlike specks, short rods, thin hairlike growths, needles) distributed throughout. The particles consist of one or more minerals recognized with difficulty and seldom with certainty.
clough [Pron. kluf or klow, as in "now"] (a) A British term for a cleft in a hill; esp. a ravine, gorge, or glen with precipitous and rocky sides. (b) The cliff or precipitous face of a clough. Also spelled: cleugh; cleuch; claugh.
clove A term used in the Catskill Mountains, N.Y., for a narrow, deep valley; esp. a ravine or gorge. Etymol: Dutch "kloof", "cleft".
Clovelly (Clo-vel'-ly) North American (Gulf Coast) stage: Miocene (above Duck Lakean, below Foleyan).
club moss One of a group of vascular *cryptograms* (chiefly the genus *Lycopodium*) that have small, simple leaves with a single midvein and sporangia borne in the leaf axils. Sometimes the *sporophylls* are modified and grouped into terminal cones. See also: *lycopod.*
cluse [Pron. klooz] A narrow deep gorge, trench, or water gap, cutting transversely through an otherwise continuous ridge; esp. an antecedent or superposed valley crossing an anticlinal limestone ridge in the Jura Mountains of the European Alps. Etymol: French. Cf: *combe; val.*
cluster (clus'-ter) A group of criss-crossing dikes demonstrably related to an exposed pluton. Cf: *dike swarm; dike set.*
cluster analysis A procedure for arranging individual objects into homogeneous groups based on their mutual similarities. Hierarchical clustering algorithms are interactive procedures in which objects are combined into increasingly larger groups, resulting in a *dendrogram* or tree-like diagram of similarities. Partitioning cluster algorithms iteratively divide the variable space into relatively homogenous compartments.
cluster particles Extremely friable and porous micrometeorites composed of an aggregate of small mafic silicates, phyllosilicates, carbonaceous material, metal, and sulfide particles. See also: *interplanetary dust.*
cluster reef Reef in which essentially in-place skeletons are adjacent but not in contact (Riding, 2002).
clymenid (cly'-me-nid) Any ammonoid belonging to the order Clymeniida, characterized by the dorsal, rather than ventral, position of the marginal siphuncle. Range, Upper Devonian.
clypeus (clyp'-e-us) (a) The labrum-carrying part of the *cephalon* of a crustacean. (b) The part of the *carapace* of an arachnid between its anterior edge and the eyes. Pl: clypei.
Clysmic (Clys'-mic) A term proposed by W.F. Hume in 1921 to refer to the rift faults paralleling and extending far beyond the Gulf of Suez (Robson, 1971). Clysma was the name of the Roman town that stood at the head of the gulf.
CMB *core-mantle boundary.*
CMBR *core-mantle boundary region.* Syn: *core-mantle boundary.*
CM chondrite A member of the Murchison group of *carbonaceous chondrites.*
CM diagram *CM pattern.*
CMP *common midpoint.*
CM pattern A two-dimensional sample-point plot designed to distinguish different depositional environments of sediments and to define, compare, and correlate clastic sediments (Passega, 1957, p.1952). The plot ordinate is the maximum or one-percentile particle size (C) and the abscissa is median particle size (M). Syn: *CM diagram.*
cnidarian (cni-da'-ri-an) Any coelenterate belonging to the phylum Cnidaria, characterized mainly by the presence of nematocysts and simple muscles. All known fossil coelenterates belong to this subphylum.
cnidoblast (cni'-do-blast) A cell that produces a coral nematocyst or that develops into a *nematocyst.*
Coahuilan (Co-a-hui'-lan) North American provincial series: Lower Cretaceous (above Upper Jurassic, below Comanchean).
coak An obsolete var. of *coke.*
coal A readily combustible rock containing more than 50% by weight and more than 70% by volume of carbonaceous material including inherent moisture, formed from compaction and induration of variously altered plant remains similar to those in peat. Differences in the kinds of plant materials (type), in degree of metamorphism (rank), and in the range of impu-

rity (grade) are characteristic of coal and are used in classification (ASTM, 1970, p.70). Syn: *black diamond [coal].*

coal age *age of coal.*

coal ball A concretion of permineralized plant debris, occurring in a coal seam or in adjacent rocks. Not to be confused with *ball coal.* Cf: *sulfur ball [coal].*

coal basin A coal field with a basinal structure, e.g. the Carboniferous Coal Measures of England.

coal bed A *coal seam.* Also spelled: coalbed.

coal blende *coal brass.*

coal blossom *coal smut.*

coal brass Iron pyrites (pyrite) found in coal or coal seams. Syn: *coal blende.*

coal breccia Naturally fragmented coal in a seam. The fragments often show polished or slickensided surfaces (Stutzer and Noé, 1940, p.248).

coal classification (a) The analysis or grouping of coals according to a particular property, such as degree of metamorphism (rank), constituent plant materials (type), or degree of impurity (grade). (b) The analysis or grouping of coals according to the percentage of volatile matter, caking properties, and coking properties.

coal clay *underclay.*

coal equivalent The heat energy of fuels other than coal, expressed in terms of comparable heat energy of coal.

coalescing fan (co-a-les'-cing) One of a series of confluent alluvial fans that form a *bajada.*

coalescing pediment One of a series of expanding pediments that join to produce a continuous pediment surrounding a mountain range or that merge over a broad region to ultimately reduce a desert mountain mass to an approximately continuous level; one of a number of pediments that make up a *pediplain.*

coalescive neomorphism (co-a-les'-cive) A term introduced by Folk (1965, p.22) for *aggrading neomorphism* in which small crystals are converted to large ones by gradual enlargement maintaining a uniform crystal size at all times (all crystals are consuming or being consumed); e.g. the process that forms most microspar calcite. Cf: *porphyroid neomorphism.*

coalfield (coal'-field) A region in which coal deposits of known or possible economic value occur.

coal gas The fuel gas produced from *gas coal;* its average composition, by volume, is 50% hydrogen, 30% methane, 8% carbon monoxide, 4% other hydrocarbons, and 8% carbon dioxide, nitrogen, and oxygen (Nelson, 1965, p.89).

coal gravel A secondary coal deposit consisting of transported and redeposited fragments. Cf: *float coal.*

coalification (coal'-i-fi-ca'-tion) The alteration or metamorphism of plant material into coal; the biochemical processes of diagenesis and the geochemical process of metamorphism in the formation of coal. See also: *carbonization; incorporation; vitrinization; fusinization; peat-to-anthracite theory; coalification break.* Syn: *carbonification; incarbonization; incoalation; bitumenization.*

coalification break That point in the process of *coalification* at which the liptinites change from dark to pale gray in reflected light, owing to loss of methane. This change coincides with a significant change in the slope of curves of various coalification parameters, and corresponds to a volatile-matter content in vitrain of about 29% (on a dry, ash-free basis). Syn: *coalification jump.*

coalification jump *coalification break.*

coalingite (co-a-ling'-ite) A brown or reddish-brown rhombohedral mineral: $Mg_{10}Fe^{3+}_2(CO_3)(OH)_{24}\cdot 2H_2O$.

coal land An area containing coal beds which falls within the public domain.

coal measures (a) A succession of sedimentary rocks (or *measures*) ranging in thickness from a meter or so to a few thousand meters, and consisting of claystones, shales, siltstones, sandstones, conglomerates, and limestones, with interstratified beds of coal. (b) A group of coal seams.

Coal Measures A stratigraphic term used in Europe (esp. in Great Britain) for Upper Carboniferous, or for the sequence of rocks (typically, but not necessarily, coal-bearing) occurring in the upper part of the Carboniferous System. It is broadly synchronous with the Pennsylvanian of North America.

coal-measures unit A sequence (from oldest to youngest) of coal, shale, and sandstone, occurring in coal measures.

coal pipe A cylindrical extension from a coal seam into the overlying rock, representing a tree stump that was rapidly buried. Cf: *baum pot; bell; caldron bottom.*

coal plant A fossil plant found in association with, or contributing by its substance to the formation of, coal deposits, e.g., in the coal *measures.*

coal seam A stratum or bed of coal. Syn: *coal bed.*

coal seat *seat earth.*

coal smut An earthy coal stratum near the surface; the weathered outcrop of a *coal seam.* Syn: *coal blossom.*

coal split *split [coal].*

coaly (coal'-y) Covered with coal, or containing or resembling coal; e.g. "coaly rashings", small pieces of soft, dark shale containing much carbonaceous material.

coarse Composed of or constituting relatively large particles; e.g. "coarse sandy loam". Ant: *fine.*

coarse admixture A term applied by Udden (1914) to an *admixture* (in a sediment of several size grades) whose particles are coarser than those of the dominant or maximum grade; material coarser than that found in the maximum histogram class.

coarse aggregate The portion of an *aggregate* consisting of particles with diameters greater than approximately ¼ inch or 4.76 mm. Cf: *fine aggregate.*

coarse clay A geologic term for a clay particle having a diameter in the range of 1/512 to 1/256mm (2-4 micrometers, or 9 to 8 phi units). Also, a loose aggregate of clay consisting of coarse clay particles.

coarse fragment A rock or mineral particle in the soil with an equivalent diameter greater than 2 mm; it may be gravelly, cobbly, stony, flaggy, cherty, slaty, or shaly. See also: *channery.* Cf: *soil separate.*

coarse-grained (a) Said of a crystalline rock, and of its texture, in which the individual minerals are relatively large; specif. said of an igneous rock whose particles have an average diameter greater than 5 mm (0.2 in.). Johannsen (1931, p.31) earlier used a minimum diameter of 1 cm, and referred to igneous rocks having walnut-size to coconut-size grains as "very coarse-grained". Syn: *phaneritic.* (b) Said of a sediment or sedimentary rock, and of its texture, in which the individual constituents are easily seen with the unaided eye; specif. said of a sediment or rock whose particles have an average diameter greater than 2 mm (0.08 in., or granule size and larger). The term is used in a relative sense, and various size limits have been suggested and used. Cf: *fine-grained; medium-grained.* (c) Said of a soil in which gravel and/or sand predominates. In the U.S., the minimum average diameter of the constituent particles is 0.05 mm (0.002 in.), or, as used by engineers, 0.074 mm (retained on U.S. standard sieve no.200); the International Society of Soil Science recognizes a diameter limit of 0.02 mm. Cf: *fine-grained.*

coarse-grained ripple A type of *oscillation ripple* that forms in coarse sand to gravel size material under the action of water surface waves. Ripple spacing ranges from decimeters to a few meters and amplitude ranges from several to a few tens of centimeters. The term was introduced by Leckie (1988). Syn: *gravel wave.*

coarse gravel An engineering term for *gravel* whose particles have a diameter in the range 19-76 mm (¾ to 3 in.).

coarse pebble A geologic term for a pebble having a diameter in the range 16-32 mm (0.6-1.3 in., or -4 to -5 phi units).

coarse sand (a) A geologic term for a sand particle having a diameter in the range 0.5-1 mm (500-1000 micrometers, or 1 to zero phi units). Also, a loose aggregate of sand consisting of coarse sand particles. (b) An engineering term for a sand particle having a diameter in the range of 2 mm (retained on U.S. standard sieve no.10) to 4.76 mm (passing U.S. standard sieve no. 4) (c) A soil term used in the U.S. for a sand particle having a diameter in the range of 0.5-1 mm. The diameter range recognized by the International Society of Soil Science is 0.2-2 mm. (d) Soil material containing 85% or more of sand-size particles (percentage of silt plus 1.5 times the percentage of clay not exceeding 15), 25% or more of very coarse sand and coarse sand, and less than 50% of any other one grade of sand (SSSA, 1965, p.347).

coarse silt A geologic term for a silt particle having a diameter in the range of 1/32 to 1/16mm (31-62 micrometers, or 5 to 4 phi units). In Great Britain, the range 1/20 to 1/10 mm has been used. Also, a loose aggregate of silt consisting of coarse silt particles.

coarse-tail grading In a sedimentary bed, a progressive upward shift toward the finer grain sizes in the coarse tail (1-5%) of the distribution only (Middleton, 1967, p.487). Cf: *distribution grading.*

coarse topography A topography with coarse *topographic texture,* characterized by low drainage density and widely spaced streams. It is common in regions of resistant rocks where the surface is incompletely dissected or the erosional features are on a large scale.

coast (a) A strip of land of indefinite width (may be many kilometers), bordering the ocean, that extends from the low-tide line inland to the first major change in landform features. Note that lakes and rivers have *shores,* while coasts are generally understood as bordering only oceans. (b) The part of a country regarded as near the coast, often including the whole of the coastal plain; a littoral district having some specific feature, such as the Gold Coast. Adj: *coastal.*

coastal (coast'-al) Pertaining to a coast; bordering a coast, or located on or near a coast, as *coastal* waters, *coastal* zone management, or *coastal* shipping routes.

coastal aggradation The growth of the shoreline seaward.

coastal area The areas of land and sea bordering the shoreline and extending seaward through the breaker zone (CERC, 1966, p. A6).

coastal current Drift of water roughly parallel to the shore and outside the surf zone, whatever the cause.

coastal desert Generally, any desert area bordering an ocean. See also: *west-coast desert.*

coastal dune A sand dune on low-lying land recently abandoned or built up by the sea; the dune may ascend a cliff and travel inland.

coastal energy The total energy, including that of wind, waves, tides, and currents, available for work along the coast. Most well-known coasts are characterized as dominated by *wave energy.*

coastal lake A lake produced by shoreline processes, as by the formation of a bar across a bay or by the joining of an offshore island to the mainland by a *double tombolo.*

coastal marsh A marsh bordering a seacoast, generally formed under the protection of a barrier beach, or enclosed in the sheltered part of an estuary. Cf: *open-coast marsh; salt marsh.*

coastal onlap *onlap.*

coastal plain (a) A low, generally broad plain that has its margin on an oceanic shore and its strata either horizontal or very gently sloping toward the water, and that generally represents a strip of recently prograded or emerged sea floor; e.g., the coastal plain of SE U.S. extending for 3,000 km from New Jersey to Texas. (b) In less restricted usage any lowland area bordering a sea or ocean, extending inland to the nearest elevated land, and sloping very gently seaward; it may result from the accumulation of material, as along the Adriatic coast of northern Italy.

coastal ramp An area of no significant bathymetric variation that slopes gently away from the coast, starting at the base of the foreshore.

Coastal Zone Color Scanner Multispectral scanner on NOAA polar-orbiting satellite designed to measure chlorophyll concentrations in the oceans. Abbrev: CZCS.

coastland (coast'-land) Land along a coast; esp. a section of seacoast.

coastline (coast'-line) (a) The line that forms the boundary between land and salt water. (b) A general term to describe the appearance or configuration of the land along a coast, esp. as viewed from the sea; it includes bays, but crosses narrow inlets and river mouths. (c) A broad zone of land and water extending indefinitely both landward and seaward from a shoreline. Cf: *shoreline.*

Coast Range Orogeny (a) A name proposed by W.H. White (1959) for major deformation, metamorphism, and plutonism during Jurassic and Early Cretaceous time in the Coast Mountains of the Cordillera of British Columbia. It is broadly equivalent to the *Nevadan Orogeny* of the United States. (b) A term sometimes used for the late Cenozoic orogenic events in southern California called *Pasadenan Orogeny.*

coated grain (coat'-ed) A sedimentary particle possessing concentric or enclosing layers (cortices) of calcium carbonate; e.g., an ooid, a pisoid, or oncolite (Wolf, 1960, p.1414). The cortices may be either biotic (oncolite) or abiotic (pisoid).

coated stone (a) A gemstone partly or entirely covered by some transparent substance to heighten or reduce the effects of an existing color, improve phenomenal effects, conceal defects, improve durability, or improve appearance. (b) A diamond crystal with a natural opaque diamond overgrowth containing a high percentage of impurities and inclusions; usually found in alluvial deposits. The stone's surface has the granular appearance; the coating (which may be a secondary growth) may be several millimeters thick and require the polishing of a "window" to determine the natural color and/or clarity of the stone. Coated crystal colors range from light greenish gray through light yellow green, dark green, gray or black. Coated crystals are common in Zaire and west Africa and are sometimes known as "Congo coated".

coaxial (co-ax'-i-al) Said of a progressive deformation in which the principal directions of total and incremental strain are parallel at all times (Hsu, 1966, p.217). Ant: *noncoaxial.*

coaxial progressive deformation Deformation in which the principal strain directions of the incremental and finite strain ellipsoids are parallel during all stages of deformation. Cf: *noncoaxial progressive deformation.*

cob v. To carry on the process of *cobbing.*

coba Uncemented rock or gravel underlying the nitrate (caliche) deposits of Chile. Cf: *congela.*

cobaltarthurite A yellow to dark brown monoclinic mineral: $CoFe^{3+}_2(AsO_4)_2(OH)_2{\cdot}4H_2O$. The cobalt analogue of arthurite, and a member of the *whitmoreite* group.

cobaltaustinite (co'-balt-aus'-tin-ite) A dull green orthorhombic mineral: $Ca(Co,Cu)(AsO_4)(OH)$. The cobalt analogue of austenite, and a member of the *adelite* group.

cobalt bloom (co'-balt) *erythrite.*

cobalt glance *cobaltite.*

cobaltite (co'-bal-tite) A grayish to silver-white orthorhombic (pseudo-cubic) mineral with a reddish tinge: CoAsS. It usually occurs massive and in association with smaltite, and represents an important ore of cobalt. Syn: *cobalt glance; white cobalt; gray cobalt.*

cobaltkieserite A pink monoclinic mineral, a member of the *kieserite* group: $CoSO_4{\cdot}H_2O$.

cobaltkoritnigite (co'-balt-ko-rit'-nig-ite) A purple triclinic mineral isostructural with koritnigite: $(Co,Zn)(AsO_3OH){\cdot}H_2O$.

cobaltlotharmeyerite A vitreous brown monoclinic mineral: $Ca(Co,Fe,Ni)_2(AsO_4)_2(OH,H_2O)_2$. The cobalt analogue of lotharmeyerite, and a member of the *tsumcorite* group.

cobaltmalanite A metallic cubic mineral of doubtful validity: $CuCoPtS_4$. It was not approved by the International Mineralogical Association, and probably is dayingite.

cobaltneustädtelite An adamantine brownish triclinic mineral: $Bi_2Fe^{3+}(Co,Fe^{3+})(O,OH)_4(AsO_4)_2$. The cobalt analogue of neustädtelite, and a member of the *medenbachite* group.

cobaltocalcite (co'-bal-to-cal'-cite) (a) A red variety of calcite containing cobalt. (b) *spherocobaltite.*

cobalt ocher (a) *erythrite.* (b) *asbolite.*

cobaltomenite (co-bal-to'-men-ite) (a) A red or pink monoclinic mineral: $CoSe^{4+}O_3{\cdot}2H_2O$. It is isostructural with ahlfeldite and clinochalcomenite. (b) A group name for monoclinic minerals of the general formula $ASeO_3{\cdot}2H_2O$, where A = Co, Cu or Ni.

cobalt pentlandite A metallic yellow cubic mineral: $(Co,Ni,Fe)_9S_8$. It is the cobalt analogue of pentlandite, and a member of the *pentlandite* group.

cobalt pyrite A variety of pyrite containing cobalt.

cobalt pyrites *linnaeite.*

cobalttsumcorite A brown to red-brown monoclinic mineral: $Pb(Co,Fe)_2(AsO_4)_2(OH,H_2O)_2$. The cobalt analogue of tsumcorite.

cobalt-type vein A *mineral-deposit type* characterized by native silver, nickel-cobalt-iron arsenides, nickel-cobalt-iron-antimony sulfarsenides, and bismuth minerals in carbonate veins.

cobalt vitriol *bieberite.*

cobalt-zippeite (co'-balt-zip'-pe-ite) An orange-yellow monoclinic mineral: $Co_2(UO_2)_4O_4(SO_4)_2{\cdot}7H_2O$. A member of the *zippeite* group.

cobb *promontory.*

cobbing (cob'-bing) The separation, generally with a hand-held hammer, of worthless minerals from desired minerals in a mining operation, e.g. quartz from feldspar.

cobble [part size] (a) A rock fragment larger than a pebble and smaller than a boulder, having a diameter in the range of 64-256 mm (2.5-10 in., or -6 to -8 phi units) being somewhat rounded or otherwise modified by abrasion in the course of transport; in Great Britain, the range of 60-200 mm has been used. Also, a similar rock fragment rounded in place by weathering at or somewhat below the surface of the ground; e.g. a "cobble of exfoliation" or a "cobble of spheroidal weathering". See also: *large cobble; small cobble.* (b) A rock or mineral fragment in the soil, having a diameter in the range of 20-200 mm (Atterberg, 1905). In the U.S., the term is used for a soil particle having a diameter in the range of 75-250 mm (3-10 in.) (SSSA, 1965, p.333). Syn: *cobblestone.* (c) An engineering term for a particle having a diameter greater than 76 mm. (d) *cobblestone.*

cobble beach *shingle beach.*

cobble conglomerate A consolidated rock consisting mainly of rounded cobbles.

cobble size A term used in sedimentology for a particle size greater than that of a sphere with a diameter of 64 mm (2.5 in.) and less than that of a sphere with a diameter of 256 mm (10 in.).

cobblestone (a) A naturally rounded, usually waterworn stone suitable for use in paving a street or in other construction. Syn: *cobble; roundstone.* (b) A cobble in the soil. (c) A consolidated sedimentary rock consisting of cobble-size particles (Alling, 1943, p.265).

cobbly (cob'-bly) Characterized by cobbles; e.g. a "cobbly soil" or a "cobbly land" containing an appreciable quantity of cobbles (SSSA, 1965, p.333).

Coble creep *grain-boundary diffusion creep.*

Coblenzian (Co-blenz'-i-an) European stage: upper Lower Devonian (above Gedinnian, below Eifelian). It includes Siegenian and Emsian. Obsolete. Also spelled: Coblencian.

cocarde ore (co-carde') *cockade ore.*

coccinite (co-ccin'-ite) An orange-red tetragonal mineral: HgI_2.

coccocylinder (coc-co-cyl'-in-der) A cylindrical test of a coccolithophore in which the coccoliths are arranged in an ordered and repetitive fashion (Covington, 1985). Cf: *coccosphere.*

coccolite (coc'-co-lite) A granular variety of diopside of various colors.

coccolith (coc'-co-lith) (a) A general term applied to various microscopic calcareous structural elements or buttonlike plates having many different shapes and averaging about 3 micrometers in diameter (though some have diameters as large as 35 micrometers), constructed of minute calcite or aragonite crystals, and constituting the outer skeletal remains of a *coccolithophore.* Coccoliths are found in chalk and in deep-sea oozes of the temperate and tropical oceans, and were probably not common before the Jurassic. See also: *rhabdolith.* (b) Two shields connected by a central tube in a coccolithophore. (c) A term loosely applied to a coccolithophore.

coccolithophore (coc-co-lith'-o-phore) Any of numerous minute, mostly marine, planktonic biflagellate algae, class Chrysophyceae, having brown pigment-bearing cells; at some phase of their life cycle they are encased in a sheath enclosing calcareous platelets, called coccoliths, which form a complex calcareous shell. Coccolithophores are autotrophic, but they have been variously classified as algae and protozoans. Var: coccolithophorid.

coccosphere (coc'-co-sphere) (a) The entire spherical or spheroidal test or skeleton of a coccolithophore, composed of an aggregate of interlocking coccoliths that are external to or embedded within an outer gelatinous layer of the cell. (b) A coccolithophore.

CO chondrite A member of the Ornans group of *carbonaceous chondrites.*

cochromite (co-chro'-mite) A metallic black cubic mineral, a member of the *spinel* group: $(Co,Ni,Fe^{2+})(Cr,Al)_2O_4$.

cocinerite (co-cin'-er-ite) A mixture of chalcocite and silver.

cocite (coc'-ite) A *lamprophyre* containing olivine, biotite, and clinopyroxene phenocrysts in a groundmass of leucite, alkali feldspar, biotite, and magnetite. Named by Lacroix in 1933 for Coc-Pia, Vietnam. Obsolete.

cockade ore (cock-ade') An open-space vein filling in which the ore and gangue minerals are deposited in successive comblike crusts around rock fragments, e.g. around vein breccia fragments. Syn: *cocarde ore; ring ore; sphere ore.*

cockpit (cock'-pit) (a) In Jamaica, any *closed depression* having steep sides. (b) A star-shaped depression having a conical or slightly concave floor; the surrounding hill slopes are steep and convex. Cockpits are the common types of closed depressions in a *kegelkarst.*

cockpit karst Typical karst topography containing many *closed depressions* surrounded by conical hills. It is divided by French and German geographers into several types depending on the shape of the hills. Cf: *cone karst; halbkugelkarst; kegelkarst; tower karst; polygonal karst.*

cockscomb barite (cocks'-comb) A comblike variety of barite displaying tabular crystals disposed roughly parallel to one another.

cockscomb pyrites A crestlike form of marcasite in twin crystals. Cf: *spear pyrites.*

cockscomb ridge A topographic ridge of bedrock composed of resistant, inclined sedimentary beds cut by V-shaped notches, giving the appearance of a cock's comb. Also, a term used in South Africa for a wind-chiseled ridge similar to a *yardang.*

cockscomb structure *hacksaw structure.*

coconinoite (co-co-nin'-o-ite) A light creamy-yellow to olivine-green orthorhombic secondary mineral: $Fe^{3+}{}_2Al_2(UO_2)_2(PO_4)_4(SO_4)(OH)_2{\cdot}20H_2O$.

coconut-meat calcite (co'-co-nut-meat') Calcite forming crusts, a few millimeters to a few centimeters thick, made of white, very thin parallel fibers, perpendicular to the encrusted surface. It commonly has growth bands running parallel to the substrate. Fibers commonly are optically length-slow. This calcite is found in modern and ancient cave deposits (Folk and Assereto, 1976, p.486-496).

COCORP *Consortium for Continental Reflection Profiling.*

COD *chemical oxygen demand.*

coda (co'-da) The concluding part of a seismogram following the early, identifiable waves. Long trains of waves may last for hours following an earthquake, especially if long oceanic paths are involved.

coefficient of acidity (co-ef-fi'-cient) *oxygen ratio.*

coefficient of anisotropy The square root of the ratio of the true transverse resistivity to true longitudinal resistivity in an anisotropic material.

coefficient of earth stress The principal-stress ratio at a given point in a soil.

coefficient of fineness The ratio of suspended solids to turbidity; a measure of the size of particles causing turbidity, the particle size increasing with coefficient of fineness (ASCE, 1962).

coefficient of kinematic viscosity *kinematic viscosity.*

coefficient of permeability *permeability coefficient.*

coefficient of storage *storage coefficient.*

coefficient of thermal expansion The relative increase of the volume of a system with increasing temperature in an isobaric process.

coefficient of transmissibility A less preferred term for *transmissivity.* Syn: *transmissibility coefficient.*

coefficient of variation A descriptive statistic expressing the variability of data relative to its magnitude; the *standard deviation* divided by the *mean.*

coefficient of viscosity The ratio of the shear stress in a substance to the rate of shear strain; *viscosity.*

coefficient of volume change *Compressibility* in which deformation is restricted to the vertical direction, in conformity with Terzaghi's theory of one-dimensional consolidation.

coefficient of volume compressibility The ratio of the volume of a rock or soil mass to its original volume after it has been subjected to a unit increase in uniaxial pressure.

coelacanth (coe'-la-canth) A member of a suborder of crossopterygian fish that entered marine waters during the Mesozoic; it includes the sole living representative of the order *Crossopterygii,* the genus *Latimeria.* Range, Upper Devonian to Holocene.

coelentera Pl. of *coelenteron.*

coelenterate (coe-len'-ter-ate) Any multicelled invertebrate animal, solitary or colonial, belonging to the phylum Coelenterata, characterized by a body wall composed of two layers of cells connected by a structureless mesogloea, by a single body cavity with a single opening for ingestion and egestion, and by radial or biradial symmetry. Range, Precambrian to Holocene. Cf: *cnidarian.*

coelenteron (coe-len'-ter-on) The spacious internal cavity enclosed by the body wall of a *coelenterate* and opening externally through the mouth. Pl: *coelentera.*

coelestine (coel'-es-tine) *celestine.*

coelobite A *cavity dweller.*

coeloconoid (coe-lo-co'-noid) Said of a gastropod shell that approaches a conical shape but has concave sides. Cf: *cyrtoconoid.*

Coelolepida (Coe-lo-lep'-i-da) Order of diplorhinate jawless fishes characterized by flattened body form and armor consisting of a loosely articulated coat of small scales. Range, Upper Silurian to Lower Devonian.

coelom (coe'-lom) The body cavity of triploblastic animals, lined by the *mesoderm.* Where well developed, the coelom forms a fluid-filled potential space between the alimentary viscera and the body walls. Adj: coelomic. Also spelled: *coelome.*

coelome (coe'-lome) *coelom.*

coenenchymal increase (coe-nen-chy'-mal) In some Paleozoic corals, esp. heliolitids, "offsets arise from the coenosteum that unites individual corallites in coenenchymal massif coralla. The offset cannot be related to any one parent" (Oliver, 1968, p.20).

coenenchyme (coe'-nen-chyme) The complex *mesogloea* uniting the polyps of a compound coral; a collective term for both the *coenosteum* and the *coenosarc.* Adj: coenenchymal.

coenobia (coe-no'-bi-a) Pl. of *coenobium.*

coenobium (coe-no'-bi-um) A colony, usually of algae, with a fixed arrangement and number of cells, commonly in sheet form, i.e., all cells in one plane, such as in the genus *Pediastrum,* of the green algal family Hydrodictyaceae. Pl: coenobia.

coenocorelation (coe-no-cor-re-la'-tion) The stratigraphic correlation of horizons among sections, based on their fossil communities; it is typically produced through *gradient analysis* of position along an environmental gradient (Cisne and Rabe, 1978).

coenocyte (coe'-no-cyte) An organism such as certain filamentous green algae that consists of continuous, multinucleate protoplasm, lacking walls to separate protoplasts.

coenoecium A tubular exoskeleton of colonies or associations of Pterobranchia.

coenosarc (coe'-no-sarc) Common soft tissue connecting coral polyps in a colony (TIP, 1956, pt. F, p.246).

coenosis (coe-no'-sis) A population that is held together by ecologic factors in a state of unstable equilibrium (Stamp, 1966, p.115). Also spelled: *cenosis.* Plural: coenoses.

coenosteum (coe-nos'-te-um) (a) Calcareous skeletal deposits formed between the individual corallites of a colonial coral. (b) Vesicular or dense calcareous skeletal material between zooecia of some stenolaemate bryozoans, esp. in the exozone. Partial syn: *extrazooidal skeleton.* (c) The calcareous skeleton secreted by a millepore colony or a stromatoporoid. Pl: coenostea.

coercive force (co-er'-cive) The opposing applied magnetic field H required to reduce the saturation magnetization of a substance to zero in that applied field. See also: *hysteresis.* Syn: *coercivity.*

coercivity (co-er-civ'-i-ty) *coercive force.*

coeruleolactite (coe-ru'-le-o-lac'-tite) A triclinic mineral: $(Ca,Cu)Al_6(PO_4)_4(OH)_8{\cdot}4\text{-}5H_2O$. It has a milk-white to sky-blue color, and is a member of the *turquoise* group.

coesite (coes'-ite) A monoclinic mineral, a dense (2.93 g/cm^3) polymorph of SiO_2, which is stable at room temperature only at pressures above 20 kilobars. The silicon is coordinated to 4 oxygens. Coesite is found in impact craters, or in rocks (such as suevite) associated with such structures. It also occurs in xenoliths in kimberlite, and as inclusions in garnet and pyroxene in unusual eclogite-facies rocks. Cf: *stishovite.*

coetaneous (co-e-ta'-ne-ous) A suggested replacement for the term *isochronous,* in the sense of being equal in duration or uniform in time. Rare.

Coeur d'Alene-type vein A *mineral-deposit type* characterized by veins rich in lead, silver, copper, and zinc sulfides and sulfosalts in carbonate and/or quartz gangue hosted by metasedimentary clastic rocks.

coevolution (co'-ev-o-lu'-tion) A pattern of evolution in which two unrelated lineages profoundly influence each other's evolutionary directions and rates, so that the two evolve together as an integrated complex; e.g. the relation between flowering plants and pollinating insects through time.

cofferdam (cof'-fer-dam) In engineering geology, a watertight enclosure used in construction from which water is pumped to expose the bottom of a body of water.

coffinite (cof'-fin-ite) A black tetragonal, although often metamict, mineral: $U(SiO_4)_{1-x}(OH)_{4x}$. An important ore of uranium, it occurs in many sandstone deposits and in hydrothermal veins.

COG *cutoff grade.*

cogeoid (co-ge'-oid) *compensated geoid.*

cognate [pyroclast] (cog'-nate) Said of pyroclastics that are *essential* or accessory. Cf: *accidental.*

cognate [struc geol] Said of shear fractures in a system that have the same time and deformational mechanism of origin. Cf: *conjugate [struc geol]; complementary [struc geol].*

cognate inclusion *autolith.*

cognate xenolith *autolith.*

cogwheel ore (cog'-wheel) The mineral *bournonite* esp. when occurring in wheel-shaped twin crystals.

cohenite (co'-hen-ite) A metallic orthorhombic mineral: $(Fe,Ni,Co)_3C$. It occurs as an accessory mineral in iron meteorites. Syn: cementite.

coherence (co-her'-ence) The property of wave trains being in phase. The coherence of adjacent seismic traces, along with an increase in amplitude, is the principal evidence for an event. Coherence is used qualitatively in picking events, but it can also be quantified. See: *semblance.*

coherence map A display of discontinuities in coherence. It is used to delineate subtle faults and stratigraphic features. Also called: coherence cube.

coherent [bot] (co-her'-ent) *connate [bot].*

coherent [geochem] Said of a group of elements which, owing to similarity in radius and valence, occur intimately associated in nature, such as entering into the same minerals at about the same stage of fractional crystallization; e.g. zirconium and hafnium form a "coherent pair" (Goldschmidt, 1937, p.662).

coherent [geol] Said of a rock or deposit that is consolidated, or that is not easily shattered.

coherent diffracting domain The volume of crystal that is without defects and yet sufficiently large to serve as a diffraction grating, producing a coherent X-ray beam (Klug and Alexander, 1974, chap.9). Sometimes called "coherent scattering domain." Abbrev: CDD.

coherent structure In streams, a zone of persistent flow separation, such as an *eddy*.

cohesion (co-he'-sion) *Shear strength* of a rock not related to interparticle friction. Cf: *adhesion*.

cohesionless (co-he'-sion-less) Said of a soil that has relatively low shear strength when air-dried and low cohesion when wet, e.g. a sandy soil. Cf: *cohesive*. Syn: *noncohesive; frictional*.

cohesive (co-he'-sive) Said of a soil that has relatively high shear strength when air-dried, and high cohesion when wet, e.g. a clay-bearing soil. Cf: *cohesionless*.

cohesiveness (co-he'-sive-ness) A mass property of unconsolidated, fine-grained sediments by which like or unlike particles (having diameters less than 0.01 mm) cohere or stick together by surface forces.

cohesive strength Inherent strength of a material when normal stress across the prospective surface of failure is zero.

cohesive zone The region at a crack tip where crack opening is resisted by molecular forces of cohesion, resulting in partial, but not complete crack opening (Barenblatt, 1962).

coignimbrite (co-ig-nim'-brite) Refers to fallout tephra deposited from a pyroclastic flow. Near-vent breccias composed of large lithic clasts that dropped from pyroclastic flows and fine-grained ash elutriated from the top of a pyroclastic flow by the turbulent rise of hot gases.

coiled tubing A specialized type of flexible pipe or tubing often used in oil or gas drilling or production.

coiling direction (coil'-ing) The direction (dextral or sinistral) in which a gastropod shell or a foraminiferal test is coiled. Changes in the coiling directions of planktonic foraminiferal tests are applied in stratigraphy to interpret paleoclimates or to determine correlations.

coincident-range zone *concurrent-range zone*.

coire A var. of *corrie*. Etymol: Gaelic, "large kettle".

coke A combustible material derived from *agglomerating* coal, consisting of mineral matter and fixed carbon fused together. It is produced by driving off by heat the coal's volatile matter, i.e., by carbonization. Coke is gray, hard, and porous, and as a fuel it is practically smokeless. It occurs in nature, but most is manufactured. Also spelled: *coak* (obsolete). Cf: *clinker [coal]; natural coke; charcoal*.

coke coal *natural coke*.

cokeite (coke'-ite) *natural coke*.

coking coal (cok'-ing) A *caking coal* suitable for the production of coke for metallurgical use.

cokriging A geostatistical technique for interpolating where two types of data are available, the one exact but sparse, e.g. well data, the other fuzzy but dense data, e.g. seismic data. Exactly honors the exact data and uses the fuzzy data as a guide where exact data are not available. Cf: *kriging*.

col (a) A high, narrow, sharp-edged pass or depression in a mountain range, generally across a ridge or through a divide, or between two adjacent peaks; esp. a deep pass formed by the headward erosion and intersection of two cirques, as in the French Alps. Also, the highest point on a divide between two valleys. (b) A marked, saddle-like depression in the crest of a mountain ridge; the lowest point on a ridge. Syn: *saddle [geomorph]*. (c) A short ridge or elevated neck of land connecting two larger and higher masses. Etymol: French, from Latin collum, "neck". Cf: *gap [geomorph]; notch [geomorph]*. Syn: *joch*.

cold avalanche A snow avalanche involving the movement of dry snow and occurring during the time of greatest cold, usually coinciding with a drop in temperature; e.g. a *dry-snow avalanche*. The term is part of an obsolete classification of avalanches.

cold desert A desert in a high latitude or at a high altitude, whose low temperature restricts or prohibits plant and animal life. The term is often used for *tundra* areas. Cf: *polar desert*.

cold front The steeply sloping boundary surface between an advancing cold air mass and a warmer air mass beneath which it pushes like a wedge. Its passage is usually accompanied by a rise of air pressure, a fall in temperature, a veer of wind, and a heavy shower or thunderstorm. Cf: *warm front*.

cold fumarole A fumarole whose steam is less than 100°C in temperature.

cold glacier *polar glacier*.

cold ice Ice below the *pressure-melting temperature*.

cold lahar A flow of cooled volcanic materials down the slope of a volcano, produced by heavy rains, or by collapse of an unstable section of the volcano. Cf: *hot lahar*. Syn: *cold mudflow*.

cold loess Periglacial loess derived from glacial outwash and formed in fringes about the Pleistocene ice sheets, as in northern Europe and in north-central U.S. Cf: *warm loess*.

cold mudflow *cold lahar*.

cold region An area where the temperature is sufficiently low to affect engineering design, construction, and operation.

cold spring A spring whose water has a temperature appreciably below the mean annual atmospheric temperature in the area; also, a nonpreferred usage for any *nonthermal spring* in an area having thermal springs (Meinzer, 1923, p.55).

cold working Plastic deformation primarily by dislocation glide with significant work hardening. Characteristic of low-temperature deformation.

colemanite (cole'-man-ite) A colorless, yellowish, or gray monoclinic mineral: $CaB_3O_4(OH)_3 \cdot H_2O$. It is an important source of boron, occurring in crystalline masses or as nodules in clay.

coleoid (co'-le-oid) Any member of the subclass Coleoidea (=Dibranchiata) of the cephalopods, having a muscular mantle, internal shell, fins, ink bag, chromatophores, suction cups, closed funnel, and cameralike eyes (Jeletsky, 1966). See also: *decapod*. Syn: *dibranchiate; endocochlian*. Range, Lower Carboniferous to Holocene.

colina (co-li'-na) A term used in the SW U.S. for a hillock or other small *eminence*. Etymol: Spanish. Syn: *collado*.

collabral (col-lab'-ral) Conforming to the shape of the outer lip of a gastropod shell, as indicated by growth lines.

collado (col-la'-do) (a) *colina*. (b) A term used in the SW U.S. for a saddle, gap, or pass. Etymol: Spanish.

collage A patchwork map pattern of accreted terranes (Howell, 1995). A geotectonic assemblage of crustal blocks separated by major faults; e.g., the Philippine Archipelago (based on a French term for a work of art consisting of a selection of fragments glued together).

collagen Dominant structural protein found in bone and other mineralized tissues; survives in the fossil record.

collain (col'-lain) A kind of *euvitrain* that consists of ulmin compounds that are redeposited by precipitation from solution. Cf: *ulmain*.

collapse breccia (col-lapse') A breccia formed by the collapse of rock overlying an opening, as by foundering of the roof of a cave or of the roof of country rock above an intrusion; e.g. a *solution breccia*. Syn: *founder breccia*.

collapse caldera A type of *caldera* produced by collapse of the roof of a magma chamber due to removal of magma by voluminous pyroclastic or lava eruptions or by subterranean withdrawal of magma. Most calderas are of this type. Cf: *erosion caldera; explosion caldera*. Syn: *subsidence caldera*.

collapse crater A large crater on planetary surfaces believed to have formed by roof subsidence of lava-filled cavities.

collapse depression An elliptical to elongate depression in the surface of a lava flow, resulting from partial or complete collapse of the roof of a long lava tunnel or of several short ones. Blocks from the roof may rest on a collapse-depression floor, and water and/or alluvium cover the blocks (Nichols, 1946, p.1064). Not to be confused with a *kipuka*.

collapse fault A normal, gravity fault on the margin of a salt-withdrawal basin in the Gulf Coast (Seglund, 1974, p.2389). Sediments overlying the salt collapse periodically as salt is withdrawn into clusters of domes or large intrusions.

collapse sink A round to elongate depression with broken, rocky walls due to structural failure of the roof of a lava tube, subsidence cave, or shelly pahoehoe tube, with a length slightly greater than its width.

collapse sinkhole A type of *sinkhole* that is formed by collapse of an underlying cave. See also: *solution sinkhole*.

collapse structure Any rock structure resulting from removal of support and consequent collapse by the force of gravity, e.g., gravitational sliding on fold limbs, salt solution causing collapse of overlying rocks in salt basins, sink-hole collapse, or collapse into mine workings.

collapsible soil (col-laps'-ible) A deposit of unconsolidated sediment with low unit weight and low water content that is susceptible to substantial reduction in volume upon soaking with water. Collapsible soil commonly is deposited as loess or, on certain alluvial fans, as a thin, sheetlike accumulation of silt- or clay-rich sediment with significant amounts of entrained air, which dries before it is buried by a subsequent similar deposit. Also called *hydrocollapsible soil* and soil subject to hydroconsolidation or hydrocompaction.

collar [drill] (col'-lar) *drill collar*.

collar [paleont] The smooth tapering part of an echinoid spine located above the milled ring.

collar cell *choanocyte*.

collar log *casing-collar log*.

collar pore A tiny aperture that occurs in a horizontal plate at the base of the cephalis in some nasselline radiolarians.

collecting area (col-lect'-ing) *catchment area [grd wat]*.

collective group (col-lec'-tive) A zoological term for a collection of animals that can be divided into identifiable species but whose generic position is uncertain. No *type species* is required but the collective name is treated as having generic rank (ICZN, 1964, p.43, 148; Cowan, 1968, p.23).

collective species A botanical term for a collection of species that are individually recognizable but closely similar. The term is informal and has no nomenclatural status (Cowan, 1968, p.23).

collector well (col-lec'-tor) A large-diameter well consisting of a concrete cylinder, sealed at the bottom, with perforated pipes extending radially into an aquifer. "The radial pipes are jacked hydraulically into the formation. Fine material around the pipes is removed by washing during construction. Collector wells are most often constructed in alluvial formations adjoining rivers. The radial pipes extend toward and under the river, thereby inducing movement of water downward through the stream bed to the pipes" (Chow, 1964, p.13-31). This well was developed by the engineer Leo Ranney and is also known as the *Ranney collector.*

collenchyma (col-len-chy'-ma) A strengthening tissue in a plant, composed of cells with walls usually thickened at the angles of the walls (Fuller and Tippo, 1954, p.954).

collenia (col-len'-i-a) A markedly convex, slightly arched or turbinate stromatolite, about 10 cm in diameter and less than 3 cm in height, produced by late Precambrian blue-green algae of the genus *Collenia* by the addition of external calcareous layers of varying thickness. It is associated with beds of flat-pebble conglomerate.

colles (col'-les) A term established by the International Astronomical Union for small hills or knobs on planetary surfaces (USGS, 2003). Etymol: Latin, pl. of collis.

collimate (col'-li-mate) (a) To make refracted or reflected rays of light parallel to a certain line or direction, such as by means of a lens or concave mirror. (b) To adjust the line of sight of a surveying instrument or the lens axis of an optical instrument so that it is in its proper position relative to other parts of the instrument, such as by means of a collimator. (c) To adjust the fiducial marks of a surveying camera so that they define the principal point.

collimating mark (col'-li-mat'-ing) *fiducial mark.*

collimation axis (col-li-ma'-tion) The straight line passing through the rear nodal point of the objective lens, perpendicular to the axis of rotation of the telescope of a surveying instrument. It is perpendicular to the horizontal axis of the telescope in a transit or theodolite, and perpendicular to the vertical axis in a leveling instrument.

collimation error The angle by which the line of sight of an optical instrument differs from its *collimation axis.* Syn: *error of collimation.*

collimation line *line of collimation.*

collimation plane The plane described by the collimation axis of the telescope of a transit when the telescope is rotated about its horizontal axis.

collimator (col'-li-ma'-tor) An optical device for producing a beam of parallel rays of light or for artificially creating an infinitely distant target that can be viewed without parallax, usually consisting of a tube having an objective converging lens with an arrangement of cross hairs placed in the plane of its principal focus. It is used in testing and adjusting certain optical surveying instruments. See also: *vertical collimator.*

colline (col'-line) A protuberant ridge of *corallum* surface between corallites of a scleractinian coral. Cf: *monticule [paleont].*

collinite (col'-lin-ite) (a) A maceral of coal within the *vitrinite* group, consisting of homogeneous jellified and precipitated plant material, lacking cell structure and of middle-range reflectance under normal reflected-light microscopy. Cf: *ulminite.* (b) A preferred syn. of *euvitrinite.*

collinsite (col'-lins-ite) A colorless to light brown triclinic mineral: $Ca_2(Mg,Fe^{2+})(PO_4)_2 \cdot 2H_2O$. A member of the *fairfieldite* group.

collision-coalescence process Growth of cloud droplets into raindrops within a warm cloud; droplets merge upon impact. This process takes place in a cloud made up of droplets of different sizes; larger droplets with higher terminal velocity overtake, then collide and coalesce with smaller droplets in their paths.

collite (col'-lite) A microlithotype of coal; a variety of *vitrite.*

collobrierite (col-lo-bri'-er-ite) A metamorphic rock composed of fayalite, garnet (almandine-spessartine), grunerite, magnetite, and some feldspar. The term was originated by Lacroix in 1917. Cf: *gondite.*

colloclarain (col-lo-clar'-ain) A transitional lithotype of coal characterized by the presence of collinite, but in lesser amounts than other macerals. Cf: *clarocollain.* Syn: *colloclarite.*

colloclarite (col-lo-clar'-ite) *colloclarain.*

collocryst (col'-lo-cryst) A crystal formed by recrystallization of aggregated colloidal parent material, as in a mobilized sediment.

collodetrinite The compact vitrinitic groundmass of *clarite, trimacerite,* and *vitrinertite;* a maceral of the *detrovitrinite* group (ICCP, 1998).

colloform (col'-lo-form) Said of the rounded, finely banded kidneylike mineral texture formed by ultra-fine-grained rhythmic precipitation once thought to denote deposition of colloids. Cf: *botryoidal; reniform.*

colloid (col'-loid) (a) A particle-size range of less than 0.00024 mm, i.e. smaller than clay size (U.S. Naval Oceanographic Office, 1966). (b) Originally, any finely divided substance that does not occur in crystalline form; in a more modern sense, any fine-grained material in suspension, or any such material that can be easily suspended (Krauskopf, 1967).

colloidal complex (col-loid'-al) In a soil, a mixture of humus and clay.

colloidal dispersion (a) A suspension of particles of colloidal size in a medium, usually liquid; a *sol.* (b) An *aerosol.*

colloid plucking A mechanical-weathering process in which small fragments are pulled off or loosened from rock surfaces by soil colloids in contact with them (Reiche, 1945, p.14).

colloidstone (col'-loid-stone') A consolidated sedimentary rock consisting of colloid-size particles (Alling, 1943, p.265).

collophane (col'-lo-phane) Any of the massive cryptocrystalline varieties of apatite, often opaline, horny, dull, colorless, or snow-white in appearance, that constitute the bulk of phosphate rock and fossil bone and that are used as a source of phosphate for fertilizers; esp. *carbonate-apatite* or a hydroxylapatite containing carbonate, and sometimes francolite. The chemical formula $Ca_3P_2O_8 \cdot 2H_2O$ is sometimes given for collophane. It is not a true mineral, but more properly a mixture. Syn: *collophanite.*

collophanite (col'-lo-pha-nite', col-loph'-a-nite) *collophane.*

collotelinite A maceral of the *telovitrinite* group with homogenized vitrinite which may exhibit poorly defined structure (ICCP, 1998).

colluvial (col-lu'-vi-al) Pertaining to colluvium; e.g. "colluvial deposits".

colluviation (col-lu'-vi-a'-tion) The formation of colluvium.

colluvium (col-lu'-vi-um) (a) A general term applied to any loose, heterogeneous, and incoherent mass of soil material and/or rock fragments deposited by rainwash, sheetwash, or slow continuous downslope creep, usually collecting at the base of gentle slopes or hillsides. (b) Alluvium deposited by unconcentrated surface runoff or sheet erosion, usually at the base of a slope. Cf: *slope wash.* Etymol: Latin colluvies, "collection of washings, dregs".

colmatage (col'-ma-tage) A term used in New Zealand for the artificial impounding of silt-laden water in order to build up the banks in the lower part of a river by deposition of alluvium; originally the term referred to the natural process of bank growth. Cf: *warping [sed].*

Cologne earth (Co-logne') Black earth. Etymol: source near Cologne, Germany.

Cologne umber Black earth. Etymol: source near Cologne, Germany.

cololite (co'-lo-lite) A *trace fossil* now assigned to the "genus" *Lumbricaria,* consisting of a cylindrical, stringlike, and tortuous or convoluted body approximately 3 mm wide, probably representing a fossil cast of a worm, but formerly regarded as the petrified intestines of a fish or the contents of such intestines. It occurs esp. in *lithographic limestone* such as the Solenhofen stone.

colonial (co-lo'-ni-al) Said of an animal that lives in close association, sometimes with morphologically differing castes, with others of the same species and that usually cannot exist as a separate individual; esp. a "colonial coral" in which the individuals are attached as a unit. Cf: *solitary coral.*

colonization (co'-lo-ni-za'-tion) A natural phenomenon wherein a species invades an area previously unoccupied by it and becomes established there.

colonnade (col-on-nade') Thick, well-formed columnar jointing normal to the base of a lava flow (Cas and Wright, 1987). Cf: *entablature.*

colony (col'-o-ny) (a) A morphological and functional aggregate of individual organisms that interacts with the environment as a single organism. All zooids and other parts are assumed to be genetically uniform, as in bryozoans, graptolites, and anthozoan corals. (b) A group of living or fossil organisms found in an area or rock unit other than that of which they are characteristic, or that migrate into and become established in a barren area.

colophonite (col'-o-phon-ite') (a) A coarse, cloudy, yellow-brown variety of andradite garnet. (b) A nongem variety of vesuvianite.

color (col'-or) In mining, a trace of gold found in *panning.* Syn: *colors.*

coloradoite (col-o-rad'-o-ite) A metallic black cubic mineral: HgTe. A member of the *sphalerite* group.

color center In crystal optics, a defect in the atomic structure that selectively absorbs a component of visible light. See also: *F center.*

color composite In satellite imagery, a color negative, transparency, or print produced by allowing the reflectance recorded for each band of a spectral image to be represented by a proportionate intensity of one of the primary colors.

color-composite image (col'-or-com-pos'-ite) A color image prepared by combining three individual images in blue, green, and red.

colored stone (col'-ored) A gemstone of any species other than diamond. This usage illogically classifies all varieties of such species as colored stones, including colorless varieties, but does not include colored diamonds. However, it has proved a practicable and satisfactory classification.

color grade (a) The grade or classification into which a gem is placed by examination of its color in comparison to the color of other gems of the same variety (Shipley, 1974). (b) The relative position of a diamond's bodycolor on a colorless-to-light-yellow scale, denoted by standard nomenclature such as letters, numbers, words, or a combination thereof. Color grades are established by comparing a diamond to a set of standard master diamonds under controlled conditions. Such grades are normally assigned only to colorless, near colorless, or light yellow, light brown, and light gray diamonds; other colors are considered fancy diamond colors and are described differently.

color grading Process of evaluating the body-color of colorless to light yellow, light brown, or light gray diamonds by comparing them to a set of "master" diamonds under controlled conditions, and assigning a grade. Color grading is usually done only on natural, untreated faceted diamonds. Special nomenclature is used for fancy color diamonds.

colorimeter (col-or-im′-e-ter) An instrument for measuring and comparing the intensity of color of a compound for quantitative chemical analysis, usually based on the relationship between concentration of a chemical solution and the amount of absorption of certain characteristic colors of light. See also: *Dubosq colorimeter; spectrocolorimeter.*

colorimetric (col′-or-i-met′-ric) Pertaining to *colorimetry.*

colorimetric analysis Quantitative chemical analysis performed by adding a certain amount of a substance to both an unknown and a standard solution and then comparing color intensities.

colorimetry (col-or-im′-et-ry) The art or process of measuring and/or comparing colors, usually with a *colorimeter,* for quantitative chemical analysis.

color index In petrology, esp. in the classification of igneous rocks, a number that represents the percent, by volume, of dark-colored (i.e. mafic) minerals in a rock. According to this index, rocks may be divided into "leucocratic" (color index, 0-30), "mesocratic" (color index, 30-60), and "melanocratic" (color index, 60-100).

color-patch map *chorochromatic map.*

color ratio image Color composite image prepared by combining three ratio images.

colors (col′-ors) *color.*

color table A table of values that a computer uses to look up and assign display colors to features in a digital file. For any particular color, the intensities for the mix of red, green, and blue are given (Krzanowski et al., 1993).

color temperature (a) An estimate of the temperature of an incandescent body, determined by observing the wavelength at which it is emitting with peak intensity (its color) and applying *Wien's displacement law.* For an ideal blackbody, the temperature so estimated would be its true temperature and would agree with its *brightness temperature;* but for actual bodies, the color temperature is generally only an approximate value. (b) The temperature to which a blackbody radiator must be raised in order that the light it emits may match a given light source in color. It is usually expressed in degrees Kelvin.

colpa (col′-pa) A nonrecommended syn. of *colpus.* Pl: colpae.

colpate (col′-pate) Said of pollen grains having more or less elongated, longitudinal furrows (colpi) in the exine.

colpi (col′-pi) Plural of *colpus.*

colporate (col-por′-ate) Said of pollen grains having colpi in which there is a pore or some other modification of the exine, such as a transverse furrow, usually at the equator.

colpus (col′-pus) A longitudinal furrowlike or groovelike modification in the exine of pollen grains, associated with germination. It either encloses a germ pore or serves directly as the place of emergence of the pollen tube, often with harmomegathic swelling. It may be distal (as in monocolpate pollen), meridional (as in tricolpate pollen), or otherwise disposed. The membrane of the colpus consists of exine in which ektexine and/or endexine are thinned or absent. Pl: *colpi.* Cf: *pore [palyn]; pseudocolpus.* Syn: *germinal furrow; sulcus [palyn].*

colpus equatorialis A ring-shaped continuous *endoaperture* lying in the equatorial plane of a pollen grain.

colpus transversalis *transverse furrow.*

colquiriite (col-qui′-ri-ite) A white trigonal mineral: $LiCaAlF_6$.

columbite (co-lum′-bite) A group name for members of the columbite-tantalite series: $(Fe,Mn,Mg)(Nb,Ta)_2O_6$. It occurs in granites and pegmatites, and is an ore of niobium as well as a source of tantalum. Syn: *niobite; dianite; greenlandite.*

columbotantalite (co-lum′-bo-tan′-ta-lite) A noncommittal term for minerals of the columbite-tantalite series.

columbretite (co-lum′-bre-tite) A leucite *phonolite* composed of sanidine and altered hornblende laths in a dense groundmass of corroded oligoclase microlites with interstitial sanidine enclosing rounded leucite grains. The embayments of the microlites are filled with analcime, augite, and magnetite (Johannsen, 1939, p.247). The name is from the Columbrete Islands, Spain. Obsolete.

columella [paleont] (col-u-mel′-la) (a) A pillarlike calcareous axial structure of a corallite, formed by various modifications of the inner edges of septa. It commonly projects into the central part of the calice of many corals in the form of a sharp-pointed protuberance. See also: *trabecular columella; lamellar columella; fascicular columella; styliform columella.* (b) The medial pillar surrounding the axis of a spiral gastropod shell, formed by the coalescence of the inner (adaxial) walls of the whorls. (c) A vertical rod between two horizontal rings, or within the shell cavity, in certain radiolarians. (d) The *stapes* (columella auris) or middle-ear bone of lower tetrapods and birds, esp. when slender and rodlike as in lizards; also, the rodlike epipterygoid (columella cranii) of lower tetrapods. Pl: columellae.

columella [palyn] (a) One of the rodlets of ektexine that may branch and fuse distally to produce a tectum on pollen grains with complex exine structure. (b) A similar supporting column on a dinoflagellate cyst, partially or completely fused with others distally (Below, 1987; Williams, 2000). Pl: columellae.

columellar fold (col-u-mel′-lar) A *fold [paleont]* or spirally wound ridge on the columella of a gastropod, projecting into the shell interior.

columellar lip The adaxial part of the *inner lip* of a gastropod shell.

columellate (col-u-mel′-late) Possessing or forming a columella or columellae; e.g. said of pollen grains with a complex ektexine structure consisting of columellae.

column [paleont] (col′-umn) (a) A cylindrical structure consisting of a series of disklike plates mounted one on top of the other and attached to the aboral end of the *calyx* in crinoids and the *theca* of blastoids and most cystoids, and presumably used for anchoring or as a means of support. The distal end is known to be variously modified in some species. Syn: *stem.* (b) The smooth cylindrical body wall of a scleractinian coral *polyp* between the basal and oral disks.

column [speleo] A columnar deposit formed by the union of a *stalactite [speleo]* with its complementary *stalagmite [speleo].* Cf: *pillar [speleo].*

column [stratig] *geologic column.*

columnal (col-um′-nal) One of the numerous individual vertical segments (ossicles or plates) that make up the *column* or *stem* of an echinoderm. Columnals are circular or polygonal, discoid, or barrel-shaped.

columnar (col-um′-nar) Said of a crystal habit that is a subparallel arrangement of columnar individuals.

columnar coal Coal that has developed a columnar fracture, usually due to metamorphism by an igneous intrusion.

columnar facet A normally circular indentation in the basal plates of a cystoid theca to accommodate the proximal end of the column.

columnar ice Ice that has been built by columnar ice crystals, mostly broader in the lower part than in the upper part of the ice cover.

columnar ice crystal A vertical ice column. Massed together, columnar ice crystals form *columnar ice* and disintegrate to form *candle ice.*

columnar joints Parallel, prismatic columns, polygonal in cross section, in basaltic flows and sometimes in other extrusive and intrusive rocks. They form as the result of contraction during cooling. Syn: *columnar structure; prismatic joints; prismatic structure; cooling crack.*

columnar section A *vertical section,* or graphic representation on a vertical strip, of the sequence of rock units that occurs in an area or at a specific locality. Thicknesses are drawn to scale, and lithology is indicated by standard or conventional symbols, usually supplemented by brief descriptive notes indicating age, rock classification, fossil contents, etc. See also: *geologic column.*

columnar seminacreous structure A generally, if not exclusively, calcitic *seminacreous structure* in which the tablets or groups of tablets show vertical stacking in all vertical sections through the shell layer (Carter, 1990).

columnar structure [mineral] A columnar, subparallel arrangement shown by aggregates of slender, elongate mineral crystals.

columnar structure [sed] A primary sedimentary structure found in some calcareous shales or argillaceous limestones, consisting of columns (9-14 cm in diameter, and 1-1.4 m in length) perpendicular to bedding and oval to polygonal in section (Hardy & Williams, 1959).

columnar structure [struc geol] *columnar joints.*

column chromatography A chromatographic technique for separating components of a sample by moving it in a mixture or solution through tubular structures, packed with appropriate substrates, in such a way that the different components have different mobilities and thus become separated. See also: *chromatography.*

column crystal A snow crystal in the shape of a hexagonal prism, either solid or hollow.

colusite (co-lu′-site) A bronze-colored cubic mineral: $Cu_{26}V_2(As,Sn,Sb)_6S_{32}$.

comagmatic (co-mag-mat′-ic) Said of igneous rocks that have a common set of chemical and mineralogic features, and thus are regarded as having been derived from a common parent magma. Also, said of the region in which such rocks occur. See also: *consanguinity.* Obsolete syn: *consanguineous [petrology].*

comagmatic region *petrographic province.*

Comanchean (Co-man′-che-an) (a) North American provincial series: Lower and Upper Cretaceous (above Coahuilan, below Gulfian). (b) Obsolete name applied to a period (or system of rocks) between the Jurassic below and the Cretaceous above. Also spelled: Comanchian.

comancheite (co-man′-che-ite) An yellow, orange-red, or red orthorhombic mineral: $Hg_{13}(Cl,Br)_8O_9$.

comb [geomorph] (a) The crest of a mountain or hill; a mountain ridge. Syn: *combe.* (b) A var. of *combe,* a valley. (c) A var. of *cwm.*

comb [ore dep] n. A vein filling in which subparallel crystals, generally of quartz, have grown perpendicular to the vein walls and thus resemble the teeth of a comb. adj. Said of such a crystal texture or structure in a vein.

comb [paleont] (a) An arachnid structure resembling a comb; e.g. a row of serrated bristles on the fourth tarsi of an araneid, found only in the family Theridiidae, or a pair of abdominal appendages situated on the sternite following upon the genital opercula, present in all scorpions but not in any other arachnids. (b) A radial series of knobs or projections in acantharian radiolarians.

comb [seis] An infinite sequence of regularly-spaced impulses.

combe (a) A British term, also used in France, for a small, deep valley running down to the sea. Also, a bowl-shaped, generally unwatered valley or hollow on the flank of a hill, esp. a dry, closed-in valley on the chalk downs of southern England. Etymol: Celtic. Syn: *comb [geomorph]; coom; coomb; coombe.* (b) A large longitudinal depression or valley along the crest or side of an anticline in the folded Jura Mountains of the European Alps, formed by downfaulting or more generally by differential erosion, often occurring along the line of junction of a hard crystalline rock with one that is soft. Cf: *cluse.* See also: *val; crêt.* (c) The amphitheaterlike steep bank of an incised meandering stream. (d) Var. of *comb [geomorph],* a mountain crest. (e) A term used in England and southern Scotland for a glacial valley and for a cirque. Also, a var. of *cwm.*

combeite (combe'-ite) A colorless trigonal mineral: $Na_2Ca_2Si_3O_9$.

comber (comb'-er) (a) A long, curling, deep-water ocean wave whose high, breaking crest (much larger than a whitecap) is pushed forward by a strong wind. (b) A long-period *spilling breaker.* Cf: *roller [waves].*

combination (com-bi-na'-tion) In *geostatistics,* any set that can be made by using all or part of a given collection of objects without regard to sequence. Cf: *permutation.*

combination archeopyle An *archeopyle* formed by the release of a part of the dinoflagellate-cyst wall that corresponds to plates of more than one thecal plate series (such as combining the plates of the apical series and the precingular series).

combination trap A trap for oil or gas that has both structural and stratigraphic elements.

combined flow The resultant pattern of flow over a *bed [geomorph]* or in the water column that is created by the superposition of two or more currents (e.g., a combination of a *unidirectional current* and a *bidirectional current*). Cf: *bidirectional current; unidirectional current.*

combined flow ripple A *ripple [sed struc]* produced on a bed under a combined flow. Most commonly the combined flow is produced by superposition of a *bidirectional current* on a *unidirectional current.* Cf: *compound ripple; cross ripple.*

combined twinning (com-bined') A rare type of twinning in quartz in which there appears to be a 180° rotation around *c* with reflection over *{1120}* or over *{0001}.* The crystal axes are parallel but the electrical polarity of the *a* axes is not reversed in the twinned parts.

combined water "Water of solid solution and water of hydration which does not freeze even at the temperature of -78°C" (Muller, 1947, p.214).

comblainite (com-blain'-ite) A turquoise-blue rhombohedral mineral of the *pyroaurite* group: $Ni_6Co^{3+}{}_2(CO_3)(OH)_{16}{\cdot}4H_2O$.

comb layering *Willow Lake layering.*

comb ridge A jagged, sharp-edged, steep-sided mountain ridge whose crest resembles a cockscomb because it bears pinnacles alternating with notches; an *arête* marked by a series of aiguilles. It commonly separates adjacent cirques in glaciated mountain regions.

comb texture A row of mineral grains oriented with their long axis perpendicular to a surface.

combustible shale (com-bus'-ti-ble) *tasmanite [coal].*

comendite (co-mend'-ite) (a) In the *IUGS classification,* a peralkaline *rhyolite* in which Al_2O_3>1.33 (total iron as FeO)+4.4. Cf: *pantellerite.* (b) A peralkaline *rhyolite* or *quartz trachyte,* less mafic than *pantellerite.* Macdonald and Bailey (1973) provide criteria for distinguishing comendite from pantellerite. The rock was named by Bertolio in 1895 from Le Comende, San Pietro Island, Sardinia. The spelling "commendite" (Irvine and Baragar, 1971) is in error.

comet A frozen mass of rock, dust and ice, usually orbiting the sun well beyond the orbits of the planets. Occasionally comets are perturbed into orbits that pass through the inner solar system where heat from the sun causes gas and dust to be ejected from the comet, forming a long tail that points away from the sun. Two sources of comets are recognized, the *Oort Cloud* and the *Kuiper Belt.*

commarginal On a shell surface, oriented so as to correspond to former position of shell margin.

commensal (com-men'-sal) Said of organisms living in a state of *commensalism.*

commensalism (com-men'-sal-ism) The relationship that exists between two organisms in which the first benefits from the second, the second being neither benefited nor harmed. Adj: *commensal.* Cf: *inquilinism; mutualism; symbiosis.*

commercial granite A general term for a decorative building stone that is hard and crystalline. It may be a granite, gneiss, syenite, monzonite, granodiorite, anorthosite, or larvikite. See also: *black granite.*

comminution (com-mi-nu'-tion) (a) The gradual diminution of a substance to a fine powder or dust by crushing, grinding, or rubbing; specif. the reduction of a rock to progressively smaller particles by weathering, erosion, or tectonic movements. (b) The breaking, crushing, or grinding by mechanical means of stone, coal, or ore, for direct use or further processing. Syn: *pulverization; trituration.*

comminution till Compact subglacial *till* manufactured more or less in place, composed of rock debris created by crushing and shearing, and made dense by water from melting of basal ice.

commissural plane (com-mis-su'-ral) (a) The plane containing the *cardinal margin* of a brachiopod and either the commissure of a *rectimarginate* shell or points on the anterior commissure midway between crests of folds in both valves (TIP, 1965, pt. H, p.142). (b) The plane of junction along which the valves of pelecypods can be separated from each other.

commissure (com'-mis-sure) (a) The line of junction between the edges or margins of valves in a brachiopod or bivalve mollusk. (b) In a plant, the surface of joining of two mericarps, or of appressed stigmas or style branches. (c) The groove of the laesura along which an embryophytic spore germinates. It is essentially equivalent to suture.

common (com'-mon) In the description of coal constituents, 5-10% of a particular constituent occurring in the coal. (ICCP, 1963). Cf: *rare; very common; abundant; dominant [coal].*

common bud One of the confluent coelomic spaces and its protective outer membranous or membranous and skeletal wall, into which a new *zooid* is budded in stenolaemate bryozoans.

common depth point The reflecting point that is common to a subset of data. The use of this term as a synonym for common-midpoint is discouraged because the reflecting point is not common if reflectors dip. Abbrev: CDP.

common factor In *factor analysis,* any factor that appears on more than one variate is a common factor. Any common factor that appears on all variates is a general factor. Ant: specific factor.

common feldspar *orthoclase.*

common lead Any lead (Pb) from a *phase* with a low value of U/Pb and/or Th/Pb, such that no significant *radiogenic lead* has been generated in situ since the phase formed. Such phases include galena and other sulfides such as pyrite; feldspars, in particular K-feldspar; micas; and most abundant rock types of Cenozoic age. Common-lead isotopic data are used in determining ages and, more importantly, in the solution of genetic problems (Doe, 1970, p.35). Syn: *ordinary lead.*

common-lead age method The determination of an age in years for a lead by isotopically determining the lead ratios $^{206}Pb/^{204}Pb$, $^{207}Pb/^{204}Pb$, and $^{208}Pb/^{204}Pb$, which can be plotted and compared to proposed growth curves.

common mica *muscovite.*

common midpoint A set of data for which the point midway between the source and receiver is common. Abbrev: CMP. Obsolete syn: *common reflection point; common depth point.*

common-midpoint shooting A type of seismic field layout designed for multiple subsurface coverage to yield reflections associated with common midpoints.

common-midpoint stack A sum of seismic traces that have the same *common midpoint.* The summing is done after appropriate statics and normal-moveout corrections have been applied to each trace. The objective is to attenuate noise and multiple reflections while accentuating reflection events.

common-offset gather A side-by-side display of traces which have the same source-to-geophone distance (offset).

common opal A variety of opal that never exhibits play of color, is found in a wide variety of colors and patterns, sometimes occurs as an earthy form, and generally is not suitable for gem use. Cf: *precious opal.*

common pyrites *pyrite.*

common reflection point *common midpoint.*

common salt A colorless or white crystalline cubic compound consisting of sodium chloride (NaCl), occurring abundantly in nature as a solid mineral (halite), or in solution (constituting about 2.6% of seawater), or as a sedimentary deposit (such as in salt domes or as a crust around the margin of a *salt lake*).

common strontium Strontium-87 in a rock or mineral which was present at the time the rock or mineral formed, and is not the result of in-situ decay of rubidium-87 after the formation of the rock or mineral. Cf: *radiogenic strontium.*

communality In *factor analysis,* the portion of the variance that is attributable to the factors.

communication pore (com-mu'-ni-ca'-tion) An opening in the wall or pore plate of a bryozoan *zooid* or extrazooidal structure.

community (com-mu'-ni-ty) A group of organisms (living or fossil) occurring together because they possess an integrated system (*food chain* or *food web*) of energy transfer operating through several different feeding or *trophic* levels. Not all ecologists and paleoecologists accept this integrated view of a community; instead, some consider community and assemblage to be synonymous. Cf: *assemblage [ecol]; association [ecol]; biocoenosis.* Syn: *biotic community.*

community evolution Change in the composition and/or structure of a *community* of organisms through geologic time; it may appear as the development of new roles for the organisms to fill within the community, or as the development of new organisms to fill existing community roles.

compact (com-pact') (a) Said of any rock or soil that has a firm, solid, or dense texture, with particles closely packed. (b) Said of a *close-grained* rock in which no component particles or crystals can be recognized by the unaided eye. (c) Said of a finely textured rock with low matrix porosity.

compactability (com-pact'-a-bil'-i-ty) A property of a sedimentary material that permits it to decrease in volume or thickness under pressure; it is a function of the size, shape, hardness, onset of cementation, and nature of the constituent particles.

compaction [sed] (com-pac'-tion) (a) Reduction in bulk volume or thickness of, or the pore space within, a body of fine-grained sediments in response to the increasing weight of overlying material that is continually being deposited or to the pressures resulting from earth movements within the crust. It is expressed as a decrease in porosity brought about by a tighter *packing* of the sediment particles. Cf: *differential compaction.* (b) The process whereby fine-grained sediment is converted to consolidated rock, such as a clay lithified to a shale.(c) Can result in expulsion of interstitial water with a decrease in the volume of sediment. See: Rieke and Chilingarian, 1974; Chilingarian and Wolf, 1975, 1976.

compaction [soil] (a) Any process, such as burial or desiccation, by which a soil mass loses pore space and becomes denser. (b) The densification of a soil by mechanical means, accomplished by rolling, tamping, or vibrating, usually at controlled water content. Cf: *consolidation [eng].*

compaction curve A laboratory curve that shows the relationship between dry density and the water content of a soil for a given compactive effort. "Water content" refers to gravimetric water content (the weight of water divided by the weight of solids), as opposed to volumetric water content (the volume of water divided by the total volume). Syn: *moisture-density curve.*

compaction fold A type of *supratenuous fold* developed by differential compaction of sedimentary material over more resistant rock, over a subsurface structure such as a buried hill, or over an active fault or fold.

compaction grout *Grout* injected with <1 in. (25 mm) slump. It is normally a soil cement with sufficient silt sizes to provide plasticity together with sufficient sand sizes to develop internal friction. The grout generally does not enter soil pores but remains in a homogeneous mass that gives controlled displacement to compact loose soils or lift structures or both (ASCE, 1980, p.805). Syn: *limited-mobility displacement grout.*

compaction grouting *grouting.*

compaction test A laboratory compacting procedure to determine the optimum water content at which a soil can be compacted so as to yield the maximum dry density. The method involves placing (in a specified manner) a soil sample at a known water content in a mold of given dimensions, subjecting it to a compactive effort of controlled magnitude, and determining the resulting unit weight (ASCE, 1958, term 74). The procedure is repeated for various water contents sufficient to establish a relation between water content and unit weight. The maximum dry density for a given compactive effort will usually produce a sample whose saturated strength is near maximum. The laboratory procedures are usually based on standard compaction tests of the American Society for Testing and Materials and the American Association of State Highway and Transportation Officials (ASTM D698; AASHO T99) or on modified compaction tests (ASTM D1557; AASHO T188). Syn: *moisture-density test.*

comparative planetology The investigation of Solar System bodies, comparing their structures, history of development, and processes that have resulted in their current surface conditions (ICA, 2003).

compartmental plate (com-part-men'-tal) A rigid articulated skeletal element forming part of the shell wall in certain cirripede crustaceans; e.g. a lateral and a *carinolateral.* Syn: *mural plate.*

compass [paleont] (com'-pass) A slender, arched radial rod in ambulacral position at the top of *Aristotle's lantern* in an echinoid.

compass [surv] (a) An instrument or device for indicating horizontal reference directions relative to the Earth by means of a magnetic needle or group of needles; specif. *magnetic compass.* Also, a nonmagnetic device that serves the same purpose; e.g. a *gyrocompass.* (b) A simple instrument for describing circles, transferring measurements, or subdividing distances, usually consisting of two pointed, hinged legs (one of which generally having a pen or pencil point) joined at the top by a pivot.

compass bearing A *bearing* expressed as a horizontal angle measured clockwise from north as indicated by a magnetic compass.

compass error The amount by which a compass direction differs from the true direction, usually expressed as the number of degrees east or west of true azimuth north and marked plus or minus according to whether the compass direction is less or greater than true azimuth. It combines the effects, or is the algebraic sum, of the deviation and variation of the compass.

compass rose A graduated circle, usually marked in 360 degrees, printed or inscribed on a nautical chart for reference, and indicating directions. It may be oriented with respect to true north, to magnetic north, or to both. Compass roses are used to give directions of prevailing winds. On geologic maps or diagrams, they may be adapted to show alignment of structures such as faults and joints. Syn: *rose [surv].*

compass traverse A surveying traverse in which a number of straight lines are measured by tape or pace and their bearings are taken by a magnetic or prismatic compass. It is executed where lines are not shown or not clearly defined on a map, or when a base map of an area is not available.

compatible element (a) An element for which the *partition coefficient* for a mineral or assemblage of minerals relative to a liquid is much greater than unity; i.e., the element is concentrated in the solid phases. (b) A chemical element that enters into early crystals when a magma crystalizes. Cf: *incompatible element.*

compensated geoid (com'-pen-sat'-ed) One of various surfaces approximating the *geoid* and obtained from *Stokes' formula* or similar equations using gravity anomalies in the calculations. Syn: *cogeoid.* A different cogeoid is obtained for each system of reduction: free-air cogeoid, isostatic cogeoid, etc.

compensation depth [lake] The depth in a lake at which phytoplankton *photosynthesis* just balances *respiration* on a daily basis. Cf: *euphotic.*

compensation depth [oceanog] (com-pen-sa'-tion) Depth in the upper ocean above which sufficient light is present that photosynthetic carbon fixation can equal or exceed respiration. Usually found a few tens to 200 meters below the sea surface, depending on location. The depth of the bottom of the *euphotic* zone. In situations of sluggish vertical mixing, may be the same depth as the *critical depth.* Cf: *carbon fixation; respiration; calcite compensation depth.* Syn: depth of compensation.

compensation depth [tect] *depth of compensation.*

compensation level [oceanog] The depth of the ocean at which the consumption and production of oxygen are equal. It is the deepest level at which phytoplankton, which produce oxygen, can persist; i.e., the bottom of the *euphotic* zone.

compensation level [tect] *depth of compensation.*

compensation point In crystal optics, the point at which an interference color is compensated by introduction of a quartz wedge or by rotation of a Berek compensator.

compensation sac The *ascus* of an ascophoran cheilostome. Syn: *compensatrix.*

compensator An apparatus in a polarizing microscope that measures the phase difference between two components of polarized light, e.g. a *Berek compensator* or a *biquartz plate.*

compensatrix (com-pen-sa'-trix) *compensation sac.*

competence [hydraul] The ability of a current of water or wind to transport detritus, in terms of particle size rather than amount, measured as the diameter of the largest particle transported. It depends on velocity: a small but swift stream, for example, may have greater competence than a larger but slower-moving stream. Cf: *capacity.* Adj: *competent [hydraul].*

competence [struc geol] A term used to compare the resistance to deformation among different rocks. Adj: competent. Examples of sedimentary rocks, in increasing order of competence, are rock salt, shale, limestone, graywacke, sandstone, and dolomite; similarly, examples of igneous and metamorphic rocks are schist, marble, quartzite, gneiss, granite, and basalt. More competent rock layers tend to have more widely spaced cleavage planes.

competent [hydraul] (com'-pe-tent) Pertaining to the *competence* of a stream or of a current of air.

competent rock A volume of rock which under a specific set of conditions is able to support a tectonic force. Such a volume may be competent or incompetent a number of times in its deformational history depending upon the environmental conditions, degree and time of fracturing, etc. Cf: *incompetent rock.*

competition In ecologic context, a direct or indirect negative interaction between two or more organisms or species seeking the same resource. It results in slower growth, fewer progeny, or increased death risk for the inferior competitor (Begon et al., 1996).

compilation (com-pi-la'-tion) The selection and assembly of map detail from various source materials (such as existing maps, aerial photographs, surveys, and new data), and the preparation and production of a new or improved map (or a part of a map) based on this detail. See also: *delineation.*

compiled map (com-piled') A map (esp. a small-scale map of a large area) incorporating information collected from various source materials and not developed from original survey data for the map in question; a map prepared by *compilation.*

complementary [petrology] (com-ple-men'-ta-ry) Said of different rocks or groups of rocks differentiated from the same magma, whose total composition is that of the parent magma.

complementary [struc geol] Said of sets of fractures that are considered to be conjugate although their origin is unknown. Cf: *cognate [struc geol].*

complete caneolith (com-plete') A *caneolith* having upper and lower rim elements and a wall. Cf: *incomplete caneolith.*

complete flower A flower having all four types of floral appendages: sepals, petals, stamens, and carpels. Cf: *incomplete flower.*

complete overstep A term proposed by Swain (1949, p.634) for an *overstep* in which an unconformity (partly angular, partly parallel) is universal throughout a basin of deposition or in which the older rocks of the basin are entirely blanketed unconformably by younger rocks.

complete tabula A coral *tabula* consisting of a single platform. Cf: *incomplete tabula.*

complex (com'-plex) n. (a) A lithostratigraphic unit comprising diverse types of any age or any class or classes or rocks (sedimentary, igneous, metamorphic) and characterized by irregularly mixed lithology or by highly complicated structural relations to the extent that the original sequence of the component rocks may be obscured and the individual rocks or rock sequences cannot be readily mapped (ISG, 2nd ed., 1994, p.36, 42). See also: *igneous complex; injection complex; metamorphic complex; basement complex.* (b) A *lithodemic unit* that consists of a mixture of rocks of two or more genetic classes, i.e. igneous, sedimentary, or metamorphic, with or without highly complicated structure; example: Franciscan Complex. Complex is comparable in rank to suite or *supersuite* and is named in the same manner (NACSN, 1983, Art. 37).

complex crossed lamellar structure A *crossed structure* with three or more nonvertical dip directions of the elongate structural units, excluding decussate lamellar structures (Carter, 1980b). Abbrev: CCL.

complex cross foliated structure A form of *complex crossed lamellar structure* in which the elongate subunits are calcitic blades or laths (Carter, 1980b). Abbrev: CCF.

complex cuspate foreland A *cuspate foreland* in which erosion on one side of the cusp has truncated beach ridges and swales but which is later prograded so that beach ridges, swales, and other symmetrical lines of growth are parallel to the new shoreline (Johnson, 1919, p.325); e.g. Cape Canaveral, Fla. Cf: *simple cuspate foreland.*

complex drainage pattern A drainage pattern that shows variations among component parts, such as one in an area of complicated geologic structure and geomorphic history (Thornbury, 1954, p.123).

complex dune A dune formed by multidirectional winds, resulting in the intersection of two or more dunes.

complex fold *superposed fold.*

complex impact crater An *impact crater* with a relatively low depth/diameter ratio; a central uplift; an annular trough; and a down-faulted, terraced rim structure. Central uplift can be expressed topographically as a peak and/or ring. Cf: *central peak; peak-ring crater; rim terrace.*

complex mountain A mountain that includes a combination of structures and a variety of landforms; a mountain whose structures defy simple classification.

complex ore (a) An ore that yields several metals. Cf: *simple ore.* (b) An ore that is difficult to utilize because it contains more than one metal or because of the presence of unusual metals.

complex prism A prism composed of fibrils radiating from an axis, found in many mollusks (MacClintock, 1967, p.15).

complex resistivity Representation of apparent resistivity as having real and imaginary parts, to accommodate observed variations in resistivity with frequency or time in induced-polarization surveys.

complex response In streams, a process whereby a single external change causes successive responses along a river over a period of time. These responses may be out of phase along the river. For example, a drop in *base level* initiates a *headcut* that erodes progressively upstream. As the headcut proceeds upstream, portions of the channel that are downstream begin to aggrade in response to increased sediment supply from the headcut erosion. Once the initial headcut erosion ceases, the resulting decrease in sediment supply causes another headcut to form and move upstream.

complex ripple A syn. of *cross ripple.* Term used by Kelling (1958, p.121) for an "interference ripple-pattern" of any kind.

complex ripple mark *complex ripple.*

complex spit A large *recurved spit* with minor or secondary spits developed at the end of the main spit (as by minor currents). The lines of growth of the minor and the main spits do not merge or curve into each other, but intersect at distinct angles. Example: Sandy Hook, N.J. Cf: *compound spit.*

complex stream A stream that has entered a second or later cycle of erosion (Davis, 1889b, p.218); e.g. the headwaters of a compound stream.

complex tombolo The system of islands and beaches that results when several islands are united with each other and with the mainland by a complicated series of tombolos. Syn: *tombolo cluster.*

complex twin A twin that is the result of both normal twinning and parallel twinning. Cf: *normal twin; parallel twin.*

complex valley A valley of which part may be parallel and part transverse to the general structure of the underlying strata. Term introduced by Powell (1874, p.50). Cf: *simple valley; compound valley.*

component (com-po'-nent) One of a set of arbitrarily chosen chemical entities that taken together completely describe all chemical variation within a system. No component may be a linear algebraic combination of two or more other components in the same system. In a phase-rule sense, a minimum number of components must be chosen for a system, and numbers of components must not exceed phases. Components may be algebraic entities that are not physically realizable and which may have negative concentrations. Cf: *exchange operator.*

componental movement (com-po-nen'-tal) In the deformation of a rock, the relative movements of component particles.

components analysis *principal components analysis.*

component-stratotype (com-po'-nent-strat'-o-type) One of several specified intervals of strata making a *composite-stratotype* (ISG, 1976, p.24).

composite (com-pos'-ite) n. A term occasionally used in ecology to denote an association of algae and fungi in an aqueous environment.

composite assemblage zone A *biozone* that incorporates two or more contemporaneous assemblages with shared characterizing taxa (NACSN, 1983, Art. 51; Kauffman, 1969).

composite coast A term used by Cotton (1958, p.441) for an initial coast resulting from deformation (either upwarping or subsidence) occurring along lines transverse to the coast, characterized by salients and embayments developed on a very large scale; a "coast of transverse deformation" consisting of alternate zones of submergence and emergence, as along the coast of New Zealand near Wellington.

composite cone *stratovolcano.*

composite fault scarp A fault scarp whose height results from the combined effects of differential erosion and fault movement.

composite fold *compound fold.*

composite gneiss Gneiss that is constituted of materials of at least two different phases (Dietrich and Mehnert, 1961). Cf: *arterite; phlebite; venite; veined gneiss.* Syn: *mixed gneiss.*

composite grain A sedimentary particle formed by aggregation of two or more discrete particles; esp. a carbonate particle resulting from clustering of lumps, pellets, coated grains, or detrital, skeletal, or microbial particles (Bissell and Chilingar, 1967, p.153).

composite intrusion An igneous intrusion composed of two or more kinds of rock with different chemical and mineralogical composition and formed by separate injections. Cf: *multiple intrusion.*

composite map In mining, a map that shows several levels of a mine on one sheet; a map that vertically projects data from different elevations in the mine to one level. Cf: *cartology.*

composite prismatic structure A *prismatic structure* in which each first-order prism consists of elongate second-order prisms radiating toward the depositional surface from a longitudinal prism axis rather than from a single spherulite or single point of origin (Carter, 1980b).

composite profile A plot consisting of the highest points of a series of profiles drawn along several regularly spaced and parallel lines on a map (Monkhouse and Wilkinson, 1952); it represents the surface of any relief area as viewed in the horizontal plane of the summit levels from an infinite distance. Cf: *superimposed profile; projected profile.* Syn: *zonal profile.*

composite seam A coal seam composed of two or more distinct coal beds that come into contact where dirt bands or other intervening strata wedge out.

composite section A single inclined or vertical *section [geol],* prepared by projecting data from various, more or less parallel, sections.

composite sequence (a) A term used by Duff and Walton (1962) for a sequence of beds that comprises all lithologic types (in a succession displaying cyclic sedimentation) in the order in which they tend to occur. It is constructed from statistical data based on actual rock successions. (b) In sequence stratigraphy, a succession of genetically related sequences in which the individual sequences stack into lowstand, transgressive, and highstand sequence sets. Composite sequences form in 1-3 million years (Mitchum and Van Wagoner, 1991).

composite set A term proposed by McKee and Weir (1953, p.384) for a large sedimentary unit of similar or gradational lithology, "compounded from both stratified and cross-stratified units", and consisting of horizontal strata together with cosets of cross-strata.

composite stone *assembled stone.*

composite-stratotype (com-pos'-ite-strat'-o-type) A *stratotype* formed by the combination of several specified type intervals of strata or type areas known as component-stratotypes (ISG, 1976, p.24).

composite stream A stream whose drainage basin receives waters from areas of different geomorphologic structures (Davis, 1889b).

composite terrane Two or more distinct terranes that became amalgamated and subsequently shared a common geologic history before their accretion to a continental margin (D.L. Jones et al., 1983).

composite topography A landscape whose topographic features have developed in two or more *cycles of erosion.*

composite unconformity An unconformity representing more than one episode of nondeposition and possible erosion.

composite volcano *stratovolcano.*

composite wedge An infilling of sediment from different sources in the space formerly occupied by an *ice wedge.*

compositing (com-pos'-it-ing) *mixing.*

composition [mineral] (com-po-si'-tion) *chemical composition [mineral].*

composition [petrology] (a) *chemical composition [petrology].* (b) The make-up of a rock in terms of the species and number of minerals present; mineralogic composition. Cf: *mode [petrology].*

compositional convection Density driven convection within a magma chamber wherein density contrasts are generated by local differences in magma composition. Most commonly envisioned as occurring along the margins of a crystallizing magma body. See also: *liquid fractionation .*

compositional layering (com-po-si'-tion-al) Alternating layers of different minerals in igneous and metamorphic rocks. See: *phase layering.* Syn: *primary layering.*

compositional maturity A type of sedimentary maturity in which a clastic sediment approaches the compositional end product to which it is driven by the formative processes that operate upon it, specif. a sediment containing no original minerals and/or recognizable rock fragments as a results of chemical weathering or abrasion. It may be expressed as a ratio between chemical compounds (e.g. alumina/soda) or between mineral components (e.g. quartz/feldspar) (Pettijohn, 1957, p.286, 508). Cf: *mineralogic maturity; textural maturity.*

composition face *composition surface.*

composition plane In a crystal twin, a *composition surface* that is planar.

composition point In a plot of phase equilibria, that point whose coordinates represent the chemical composition of a phase or mixture.

composition surface The surface along which the individuals of a crystal twin are joined. It may or may not be a plane, and it is usually identical with the *twin plane.* See also: *composition plane.* Syn: *composition face.*

compound alluvial fan (com'-pound) *bajada.*

compound aperture A pollen aperture with two or more components that are situated in more than one wall layer.

compound atoll A coral reef complex comprising one or more smaller atolls superimposed on the rim of a pre-existing giant or chain of atolls, as in Maldive Islands, site of administrative term *atollon,* with its subdivisions, *faros.*

compound composite prismatic structure A *composite prismatic structure* in which each first-order prism consists of many second-order nondenticular composite prisms which diverge in a fanlike arrangement as a consequence of their deposition on a curved, denticulated shell margin (Carter, 1980b).

compound conodont element A bladelike or barlike *conodont element,* commonly bearing denticles.

compound cooling unit An ash-flow unit or series of such units that record breaks in the history of cooling, compaction, and welding (Smith, 1960, p.157).

compound coral The skeleton of a colonial coral.

compound cross-stratification A complex type of cross-stratification that combines forms differing substantially in scale and orientation (Harms et al., 1975, p.51). Cf: *compound foreset bedding.*

compound cuspate bar A bar formed by a *compound spit* that unites with the shore at its distal end (Johnson, 1919, p.319).

compound elements A general term for ramiform and pectiniform conodont elements; contrasts with simple elements (TIP, 1981, pt. W, supp.2).

compound eye An eye of an arthropod, consisting essentially of a great number of minute eyes crowded together; e.g. an array of contiguous visual units (ommatidia) of a crustacean having a common optic-nerve trunk, or the lateral eye of a merostome composed of many facets, or a *holochroal eye* of a trilobite. See also: *ocellus [paleont].*

compound fault A little-used term for *fault zone.*

compound fold A fold upon which minor folds with similar axes have developed. Cf: *simple fold.* Syn: *composite fold.*

compound foreset bedding Cross-bedding characterized by a concave base and by several foreset beds dipping in more than one direction, as where tangential foresets recently deposited may have their upper ends truncated and may then be overlain by new beds of similar nature dipping in the same or in a different direction (Lahee, 1923, p.80). It may develop by changes in the stream or by fluctuations of water level of a lake, such as interference among adjacent lobes on a delta front. Cf: *compound cross-stratification.*

compound laccolith A laccolith having several parts that are separated from each other by thick layers of country rock but that were formed by a single intrusion. Cf: *cedar-tree structure.*

compound leaf A leaf composed of several distinct parts or leaflets.

compound operculum A dinoflagellate operculum that is divided into two or more pieces that are separable from one another. Cf: *simple operculum.*

compound pellet A pellet of silt, sand, or granule size or larger, originating from pelleted limestone with micritic or sparry cement, and sometimes containing matrix or interstitial material (Bissell and Chilingar, 1967, p.153).

compound plate An ambulacral-plate unit of an echinoid, composed of two or more individual *ambulacral* plates, each with a pore for a tube foot, bound together by a single large *tubercle* articulating with the *primary spine* (TIP, 1966, pt. U, p.253).

compound ripple (a) A *cross ripple* resulting from the simultaneous interference of wave oscillation with current action, and characterized by a systematic breaking or offsetting of the crests of the *current ripples* (Bucher, 1919, p.195). (b) A *ripple* produced by modification of a pre-existing set of ripples by a later set of either wave or current origin (Kelling, 1958, p.122); e.g., *wave-current ripple mark.* Cf: *combined flow ripple.*

compound ripple mark *compound ripple.*

compound shoreline A shoreline showing a very marked development of the features characteristic of at least two of the following: a *shoreline of emergence,* a *shoreline of submergence,* a *neutral shoreline* (Johnson, 1919, p.172); e.g., where a formerly submerged shoreline is elevated slightly but not enough to destroy the effect of submergence, or where a dissected shoreline of emergence undergoes a slight submergence so that the coastal-plain valleys are drowned. The concept is obsolete and generally regarded as erroneous.

compound skeletal wall As seen in sections of stenolaemate bryozoans, a wall calcified by epidermis located on growing edges and both sides, generally producing a bilaterally symmetrical microstructural pattern (Boardman and Cheetham, 1969, p.211). Cf: *simple skeletal wall.*

compound spit A *recurved spit* whose inner side shows a series of intermittent landward-deflected points representing successive recurved termini; e.g,. Cape Cod. Cf: *complex spit.*

compound stream A stream that is of different ages in its different parts (Davis, 1889b); e.g., a stream with old headwaters rising in the mountains and a young lower course traversing a coastal plain.

compound structure A term used by Cotton (1948) for an arrangement of rocks characterized by a simple cover mass resting unconformably on more complex undermass.

compound trabecula A *trabecula* of a scleractinian coral, composed of bundles of sclerodermites. Cf: *simple trabecula.*

compound valley A valley whose main course may be a *simple valley* or a *complex valley,* but whose tributary valleys are of a different kind. Term introduced by Powell (1874, p.50).

compound valley glacier A glacier composed of two or more individual valley glaciers emanating from tributary valleys.

compound volcano A volcano that consists of a complex of two or more vents, or a volcano that has an associated volcanic dome, either in its crater or on its flanks. Examples are Vesuvius and Mont Pelée.

compreignacite (com-preig'-nac-ite) A yellow orthorhombic secondary mineral: $K_2(UO_2)_6O_4(OH)_6{\cdot}7H_2O$.

compressed (com-pressed') Said of the shape of the whorl section in a chambered cephalopod *conch* in which the dorsoventral diameter is greater than its lateral diameter. Ant: *depressed.*

compressed crossed lamellar structure A *crossed lamellar structure* in which the first-order lamellae appear compressed, with their length axes more or less perpendicular to the dip directions of their second-order lamellae, on the depositional surface and in horizontal sections. The first-order lamellae appear to be radially oriented, but the dip directions of their second-order lamellae are more or less parallel or slightly oblique to the shell margins.

compressibility (com-pres'-si-bil'-i-ty) The reciprocal of *bulk modulus;* ratio of the change in volume for a change in hydrostatic component of stress equals 1/k where k is the bulk modulus. Its symbol is β. Syn: *modulus of compression.*

compressing flow (com-press'-ing) A flow pattern on glaciers in which the velocity decreases with distance downstream; thus the longitudinal strain rate (velocity gradient) is compressive. This condition requires a transverse or vertical expansion, because ice is almost incompressible (Nye, 1952). Ant: *extending flow.*

compression [exp struc geol] (com-pres'-sion) A system of forces or stresses that tends to decrease the volume of, or shorten, a substance.

compression [paleont] The remains of a fossil plant that has been flattened by the vertical pressure of overlying rocks.

compressional wave (com-pres'-sion-al) *P wave.*

compression fault A general term for a fault that has been produced by lateral crustal compression, e.g., a reverse fault. Little used and obsolete. Cf: *tension fault.*

compression test *triaxial compression test.*

compressive strength (com-pres'-sive) The maximum *compressive stress* that can be applied to a material, under given conditions, before failure occurs.

compressive stress A *normal stress* that tends to push together material on opposite sides of a real or imaginary plane. See also: *compressive strength.* Cf: *tensile stress.*

compromise boundary (com'-pro-mise) A surface of contact, not corresponding to a crystal face, between two mutually growing but differently oriented crystals.

Compton continuum That part of gamma energy spectrum formed by photons that have lost part of their original energies through Compton scattering (IAEA, 2003).

Compton correction coefficient (stripping correction) The correction applied to an elemental window count rate to eliminate ineterference from gamma rays due to other elements in that window (IAEA, 2003).

comptonite (comp'-ton-ite) An opaque variety of thomsonite from the Lake Superior region.

Compton scattering The interaction of a photon with an orbit electron of an atom, in which the photon loses part of its energy and changes its direction (IAEA, 2003).

computer (com-put'-er) A programmable device capable of accepting data, applying prescribed processes to them, and presenting the results of these processes. Without modifiers, the term applies almost exclusively to digital electronic computers. See also: *analog computer; digital computer; hybrid computer.*

computer contouring Construction of a contour map of spatially distributed data using a mathematical model of a surface generated by computer. Contours or isolines showing the form of the surface are displayed on the monitor of a personal computer or terminal, or printed using a digital plotter.

computer graphics Maps, graphs, and diagrams plotted directly from data stored in computer banks and suitable for publication with minimal cartographic effort.

Comstock epithermal vein An *adularia-sericite type* mineral deposit hosted by felsic to intermediate volcanic rocks overlying clastic sedimentary rocks or their metamorphic equivalents. Comstock epithermal vein deposits, compared with Creede epithermal veins, have very low copper, lead, and zinc grades, much higher gold and silver grades, and are smaller in size. Compared with Sado epithermal veins, Comstock epithermal veins are larger in size and have higher gold and silver grades. Named for the Comstock Mining District in Nevada. Cf: *Creede epithermal vein; Sado epithermal vein.*

conate (co'-nate) Said of sculpture of palynomorphs consisting of coni. Cf: *conus.*

concave bank (con-cave') The outer bank of a curved stream, with the center of the curve toward the channel; e.g. an *undercut* slope. Ant: *convex bank.*

concave cross-bedding Cross-bedding with concave, foreset beds formed by the deposition of cross beds into a scour hole by migrating *bed form.* This type of cross-bedding is very common and is used as a criterion for distin-

guishing top from bottom in sedimentary rocks. Cf: *convex cross-bedding; festoon cross-bedding.*
concave slope *waning slope.*
concavo-convex (con-ca'-vo-con-vex') Concave on one side and convex on the other; e.g. said of a brachiopod shell having a concave *brachial valve* and a convex *pedicle* valve. Cf: *convexo-concave.*
concealed coalfield (con-cealed') Deposits of coal that do not crop out. Cf: *exposed coalfield.*
concealed pediment A pediment that is buried by a thin layer of alluvium from an encroaching bajada; it is generally caused by a local rise of base level. Cf: *suballuvial bench; fan-topped pediment.*
concentrate Enriched ore material collected after a removal of waste in a mill or concentrator. The rejected waste material is known as tailings.
concentrated flow (con'-cen-trat'-ed) *channel flow.*
concentrated wash *channel erosion.*
concentration [ice] (con-cen-tra'-tion) The ratio of the area actually covered by sea ice to the total area of the sea surface, both covered by ice and ice free, at a specific location or over a defined area. Expressed in eighths or tenths. Cf: *ice cover.*
concentration [sed] *sediment concentration.*
concentration boundary "A line approximating the transition between two areas of pack ice with distinctly different concentrations" (U.S. Naval Oceanographic Office, 1968, p. B32).
concentration curve The rising limb on a hydrograph.
concentration point The point at which all runoff from a given area passes.
concentration time (a) The time required for water to flow from the farthest point in a watershed to a gaging station. (b) That time at which the rate of runoff equals the rate of precipitation of a storm. Syn: *time of concentration.*
concentric [paleont] In invertebrates, oriented to coincide with growth lines (by no means concentric in literal and geometrical sense).
concentric fold (con-cen'-tric) *parallel fold.*
concentric shearing surface *surface of concentric shearing.*
concentric weathering *spheroidal weathering.*
conceptual model (con-cep'-tu-al) *model.*
concertina fold (con-cer-ti'-na) *kink fold.*
conch (a) The portion of a cephalopod shell developed after the embryonic shell; e.g. the complete shell of an ammonoid exclusive of the *protoconch,* or all the hard calcareous parts of a nautiloid (including the external shell, septa, and *siphuncle*) exclusive of *cameral deposits* and any structures within the siphuncle. (b) Any of various large spiral-shelled marine gastropods, often of the genera *Strombus* or *Cassis.* Also, the shell of such a conch, often used for making cameos and formerly made into horns. (c) Any of various marine shells of invertebrates, including bivalve mollusks and brachiopods. Pl: conchs.
conchal furrow (con'-chal) A shallow, midventral groove on the inside wall of a nautiloid conch.
conchiform (conch'-i-form) Shell-shaped; esp. shaped like one half of a bivalve shell.
conchilite (conch'-i-lite) A bowl-shaped body of limonite or goethite growing in an inverted position on mineralized bedrock and resembling the shell of an oyster or clam coated with a rusty deposit (Tanton, 1944). It is roughly oval or circular in plan, with a smooth or irregular and scalloped outline; it ranges from 2.5 cm to 1 m in diameter and from 2 cm to 7.5 cm in height.
conchiolin (con-chi-o'-lin) A fibrous protein, $C_3H_{48}N_9O_{11}$, that constitutes the organic basis of most mollusk shells; e.g. the material of which the *periostracum* and organic matrix of the calcareous parts of a mollusk shell are composed. Syn: conchyolin.
conchoidal (con-choi'-dal) Smoothly curving fracture surface of certain minerals and rock in three-dimensional form, either concave or convex, and is in contrast to the planar and angular cleavage of most samples. It is a characteristic habit of quartz, opal, quartzite, and of obsidian. Etymol: like the curve of a conch (seashell). The detail of the fracture surface provides diagnostic information for archaeological study of "flint" tools.
conchology (con-chol'-o-gy) The study of shells of both fossil and existing animals. Cf: *malacology.*
conchorhynch (conch'-o-rhynch) The calcified hard tip of the horny lower jaw, probably only in nautiloids.
conchostracan (con-chos'-tra-can) Any branchiopod belonging to the order Conchostraca, characterized by a translucent bivalve shell with a clawlike furca developed at the posterior end. Range, Lower Devonian to Holocene.
concordance (con-cor'-dance) Parallelism of strata with sequence boundaries, with no terminations of strata against the boundary surfaces (Mitchum, 1977, p.206). Cf: *discordance.*
concordance of summit levels *summit concordance.*
concordant [geochron] (con-cor'-dant) (a) Said of radiometric ages, determined by more than one method, that are in agreement within the analytical precision for the determining methods. (b) Said of radiometric ages given by coexisting minerals, determined by the same method, that are in agreement. (c) In a more restricted sense, the term has been used to indicate agreement of $^{238}U/^{206}Pb$, $^{235}U/^{207}Pb$, $^{207}Pb/^{206}Pb$, and $^{232}Th/^{208}Pb$ ages determined, within experimental error, for the same mineral sample. Ant: *discordant [geochron].*
concordant [hydraul] Said of flows, e.g. floods, which, at different points along a channel, have the same frequency of occurrence (Langbein and Iseri, 1960, p.6).
concordant [intrus rocks] Said of an igneous intrusion possessing contacts which are parallel to foliation or bedding in the country rock. Cf: *discordant [intrus rocks]; conformable [intrus rocks].*
concordant [stratig] Structurally conformable; said of strata displaying parallelism of bedding or structure. The term may be used where a hiatus cannot be recognized but cannot be dismissed. Ant: *discordant [stratig].*
concordant bedding A sedimentary structure marked by beds that are parallel and without angular junctions. Ant: *discordant bedding.* Syn: *parallel bedding.*
concordant coastline A coastline that is broadly parallel to the main trend of the land structure (such as mountain ranges or fold belts) forming the margin of the ocean basin; it is generally linear and regular. Cf: *Dalmatian coastline.* Ant: *discordant coastline.* Syn: *Pacific-type coastline; longitudinal coastline.*
concordant drainage *accordant drainage.*
concordant junction *accordant junction.*
concordant summit level *accordant summit level.*
concordia (con-cor'-di-a) The graphed curve formed when the $^{206}Pb/^{238}U$ ratio is plotted against the $^{207}Pb/^{235}U$ ratio as both increase in value due to nuclear decay of uranium to lead with passage of time, assuming a closed U-Pb system. The curve (concordia) is the locus of all concordant U-Pb ages, thus it is a time curve. Syn: *concordia plot; concordia diagram; concordia curve.*
concordia curve *concordia.*
concordia diagram *concordia.*
concordia intercept The intersections of the *concordia curve* (time curve) and a straight line (chord) which depicts the plot (locus) of *discordant* U-Pb ages, normally for analyses of minerals (mostly zircon) from a single sample. The intercepts may indicate significant events in the history of the mineral under study, the older intercept representing the start of the U-Pb clock, and the younger age a time when the clock was disturbed.
concordia plot *concordia.*
concrescence (con-cres'-cence) A growing-together or coalescence of originally separate parts; e.g. union of radial spines in a radiolarian skeleton.
concrete (con-crete') A mixture of *cement [materials],* an *aggregate,* and water, which will "set" or harden to a rocklike consistency.
concretion (con-cre'-tion) (a) A hard, compact mass or aggregate of mineral matter, normally subspherical but commonly oblate, disk-shaped, or irregular with odd or fantastic outlines; formed by precipitation from aqueous solution about a nucleus or center, such as a leaf, shell, bone, or fossil, in the pores of a sedimentary or fragmental volcanic rock, and usually of a composition widely different from that of the rock in which it is found and from which it is rather sharply separated. It represents a concentration of some minor constituent of the enclosing rock or of cementing material, such as silica (chert), calcite, dolomite, iron oxide, pyrite, or gypsum, and it ranges in size from a small pellet-like object to a great spheroidal body as much as 3 m in diameter. Most concretions were formed during diagenesis, and many (especially in limestone and shale) shortly after sediment deposition. Cf: *nodule [sed]; secretion [sed struc].* See also: *accretion [sed struc]; incretion; intercretion; excretion.* (b) A collective term applied loosely to various primary and secondary mineral segregations of diverse origin, including irregular nodules, spherulites, crystalline aggregates, geodes, septaria, and related bodies.
concretionary (con-cre'-tion-a'-ry) Characterized by, consisting of, or producing concretions; e.g. a "concretionary ironstone" composed of iron carbonate with clay and calcite, or a zonal "concretionary texture" (of an ore) characterized by concentric shells of slightly varying properties due to variation during growth.
concurrent-range zone (con-cur'-rent) (a) The *interval zone* included between the documented lowest occurrence of one taxon and the documented highest occurrence of another taxon, resulting in stratigraphic overlap of the two taxa (NACSN, 1983, Art. 50). Cf: *taxon-range zone; partial range zone; lineage zone; Oppel zone.* (b) A body of strata including the concurrent, coincident, or overlapping parts of the *range zones* of two specified taxa selected from among the total forms contained in a *sequence* of strata (ISSC, 1994, p.58).
concussion fracture (con-cus'-sion) One of a system of fractures in individual grains of a shock-metamorphosed rock which are generally radial to the grain surface and related to the contacts with adjacent grains. They are apparently formed by violent grain-to-grain contacts in the initial stages of passage of a shock wave and are formed by the resulting tensile stresses parallel to the surfaces of the impacting grains (Kieffer, 1971, p.5456, 5468).
concyclothem (con-cy'-clo-them) A cyclic sequence of strata resulting from the local coalescence of two or more cyclothems (Gray, 1955).
condensate (con-den'-sate) Liquid hydrocarbons, generally clear or pale straw-colored and of high API gravity (above 60°), that are produced with *wet gas.* Syn: *distillate; natural gasoline.*
condensation [phys] (con-den-sa'-tion) The process by which a vapor becomes a liquid or solid; the opposite of *evaporation.*
condensation [stratig] A stratigraphic process in which the thinning of a sedimentary deposit or succession takes place contemporaneously with de-

position, as by strong hydrostatic pressures resulting in solution along grain boundaries.

condensation corrosion Process by which water contained in the air and charged with a high level of carbon dioxide condenses on bedrock or *speleothem* surfaces and corrodes them. Gas weathering. A significant process in the enlargement of some cave passages (Hill, 1987).

condensation room In a cave, an area in which water vapor condenses on the ceiling and walls because of a lower temperature of the wall rock than of the air that enters the room.

condensation zone A concentration of several biozones in a single bed (Heim, 1934).

condensed deposit (con-densed') A sedimentary deposit in a *condensed succession.*

condensed section [sed] A thin interval of sediment or sedimentary rock that had a slow rate of sedimentation; bioturbation, glauconite, pyrite, or phosphate are common in such sections.

condensed section [stratig] In sequence stratigraphy, a facies consisting of thin hemipelagic or pelagic sediments deposited as the shelf is starved of terrigenous sediment, during a rise in sea level. The greatest diversity and abundance of fauna within a sequence are found in this terrigenous-starved interval. Deposition within the condensed section is continuous although the section commonly is thin, accumulates at very slow rates, and encompasses a great deal of time (Loutit et al., 1988).

condensed succession A relatively thin but uninterrupted stratigraphic succession of considerable time duration, in which the deposits accumulated very slowly; it is generally represented by a time-equivalent thick succession elsewhere in the same sedimentary basin or region. Ant: *extended succession.*

condensed system (a) A chemical system in which the vapor pressure is negligible, and can thus be ignored. (b) A chemical system in which the pressure maintained on the system is greater than the vapor pressure of any portion.

conditional probability The nonindependent probability that event A will occur, given that event B has occurred; it is different from the independent probability that event A will occur regardless of the state of event B. Conditional probability implies that the two events A and B are dependent in some way.

conditional resources (con-di'-tion-al) *identified subeconomic resources.*

conditional simulation A geostatistical technique for interpolating which procudes a set of estimates that honor data exactly and reproduce the spatial characteristics of the *variogram.* A Monte Carlo process.

condrodite (con'-dro-dite) *chondrodite.*

conductance (con-duct'-ance) The inverse of electrical *resistance,* measured in mhos.

conduction The transport of heat in static groundwater, controlled by the thermal conductivity of the geologic formation and the contained groundwater and described by a linear law relating heat flux to temperature gradient. Cf: *convection [grd wat].*

conductivity (con-duc-tiv'-i-ty) (a) *electrical conductivity.* (b) *thermal conductivity.*

conductivity log *induction log.*

conductivity ratio The principal parameter used to indicate the total concentration of dissolved ions, i.e., salinity, in a sample of seawater. Equal to the conductivity of the sample divided by the conductivity of a standard ionic solution, which is usually standard seawater but can be an ultra-pure KCl solution. Cf: *salinity; standard seawater.*

conductivity-thickness product (con-duc-tiv'-i-ty-thick'-ness) The product of average conductivity and thickness of a rock layer; it is used to characterize the response of an electromagnetic or direct-current method to the presence of a highly conductive layer. Measured in siemens.

conduit (con'-duit) In a cave, a subsurface bedrock channel. The term is sometimes used synonymously with *cave [speleo]* and *passage [speleo]* but is generally understood to refer to a space either too small for human entry or of explorable size but inaccessible.

conduit flow An underground water flow within conduits; it is generally turbulent but may also be a *laminar flow.* Cf: *conduit.*

condyle [marine geol] (con'-dyle) One of a series of knucklelike humps of gravity and bathymetry with a wavelength of about 300 km that are distributed along the lengths of the outer swells of oceanic trenches and seamount troughs (G.W. Moore, 1985).

condyle [paleont] (a) In some cheilostome bryozoans, one of a pair of oppositely placed teeth or protuberances on which the operculum of an *autozooid,* the mandible of an *avicularium,* or the seta of a *vibraculum* pivots or is slung. Cf: *cardella.* (b) A swollen knob on the shell surface of an acantharian radiolarian. (c) A rounded articular prominence on a bone that normally is covered with cartilage during life. See also: *occipital condyle.*

cone [bot] A fertile branch that bears an imbricated group of seed- or spore-bearing scales or sporophylls, as in pine or club moss. Cf: *strobilus.*

cone [geomorph] A mountain, hill, or other landform shaped like a cone, having relatively steep slopes and a pointed top; specif. an *alluvial cone.*

cone [glac geol] In glacial geology, a steep-sided pile of sand, gravel, and sometimes boulders, with a fanlike outwash base, deposited against the front of a glacier by meltwater streams.

cone [marine geol] A type of *submarine fan.* The term is frequently used to describe a deep-sea fan associated with a major active *delta;* e.g. Mississippi, Nile, Ganges (Kelling and Stanley, 1976, p.387).

cone [volc] *volcanic cone.*

cone complex crossed foliated structure A calcitic *cone complex crossed lamellar structure.* This structure is similar to aragonitic cone complex crossed lamellar structure, but the first-order lamellae (vertical columns) are generally much broader, and the second-order lamellae (laminae comprising the cones) show a much lower dip angle relative to the depositional surface (Carter, 1980b). This includes both cone CCF structure in the strict sense and spiral cone CCF structure.

cone complex crossed lamellar structure A *complex crossed lamellar structure* in which the first-order lamellae constitute cone-in-cone or spiral-conical structures, with their longitudinal axes generally perpendicular to the depositional surface and with strongly interdigitating lateral boundaries. The conical or spiral-conical second-order lamellae consist of elongate, flattened third-order lamellae radiating from the cone apices toward the depositional surface (Carter and Tevesz, 1978; Carter, 1980b).

cone cup The inside of an enclosing cone of a sedimentary cone-in-cone structure.

cone delta *alluvial cone.*

cone dike *cone sheet.*

cone-in-cone coal Coal exhibiting cone-in-cone structure. Syn: *crystallized coal.*

cone-in-cone structure [sed] (a) A minor sedimentary structure in thin, generally calcareous layers of some shales and in the outer parts of some large concretions, esp. septaria; it resembles a set of concentric, right circular cones fitting one into another in inverted positions (base upward, apex downward), commonly separated by clay films, and consisting usually of fibrous calcite and rarely of siderite or gypsum. The apical angles are between 30 and 60 degrees and the cone axes are normal to the bedding; the height of the cones usually ranges from 10 mm to 10 cm, and their sides are often ribbed, fluted, or grooved, and marked by annular depressions and ridges that are more pronounced near the bases and finer and more obscure near the apices. The structure appears to be due to pressure aided by crystallization and weathering (solution) along intersecting conical shear zones. (b) A similar structure developed in coal and consisting of a set of interpenetrating cones packed closely together.

cone-in-cone structure [volc] A volcanic structure in which a younger cone or cones have developed within the original one; also, a similar pattern of crater development. Adj: *nested.*

cone karst A type of karst topography, common in the tropics. The term emphasizes the cone-shaped hills between the star-shaped depressions (cockpits). A variety of *kegelkarst.* Cf: *cockpit karst.*

conella (co-nel'-la) A tiny prism or cone of spherulitic aragonite or secondary calcite, found occasionally on the surface of steinkerns of ammonoid conchs, less often on those of nautiloid conchs, produced by partial solution of the inner prismatic shell layer. Pl: conellae.

Conemaughian (Co-ne-maugh'-i-an) North American provincial stage: lower Upper Pennsylvanian (above Allegheny, below Monongahelan).

cone of dejection An *alluvial cone* consisting of coarse material, formed where a mountain torrent emerges from a narrow valley upon a plain or plunges over a valley-side bench. Syn: *dejection cone.*

cone of depression A depression in the *potentiometric surface* of a body of groundwater that has the shape of an inverted cone and develops around a well from which water is being withdrawn. It defines the *area of influence* of a well. The shape of the depression is due to the fact that the water must flow through progressively smaller cross sections as it nears the well, and hence the hydraulic gradient must be steeper. Cf: *drawdown.*

cone of detritus *alluvial cone.*

cone of pressure relief A *cone of depression* in the potentiometric surface of a body of confined groundwater. As defined by Tolman (1937, p.562), "an imaginary surface indicating pressure-relief conditions in a confined aquifer due to pumping".

cone penetration test A soil *penetration test* in which a steel cone of standard shape and size is pushed into the soil, and recordings are made of the force required to advance the cone either at a predetermined, usually slow and constant, rate or for a specified distance; or, in some designs, the penetration resulting from various loads is recorded. It is usually referred to as a "CPT" where separate measurements are made of the penetration resistance at the cone's tip and along its sleeve. The tip resistance and the ratio of sleeve and tip resistance are often used to classify soil type and physical properties. Interpretation of laterally continuous CPTs (ASTM D3441) may indicate the relative thickness and characteristics of a particular stratigraphic marker useful for site or regional correlation.

cone sheet A dike that is arcuate in plan and dips at 30°-45° toward the center of the arc. Cone sheets occur in concentric sets, which presumably converge at a magmatic center. They are associated with ring dikes to form a *ring complex.* Syn: *cone dike.*

Conewangoan (Co-ne-wan'-go-an) North American provincial stage: uppermost Devonian (above Cassadagan, below Mississippian).

conferva peat (con-fer'-va) Peat that is derived mainly from filamentous algae of the genus Tribonema.

confidence interval (con′-fi-dence) The interval between *confidence limits.*
confidence limit Either the upper or the lower value of the range within which an observation parameter is expected to fall with a stated probability.
configuration (con-fig′-u-ra′-tion) (a) The form or shape of a part of the Earth's surface with regard to its horizontal outline, its elevation, and its relative disposition or arrangement with other parts of the surface. (b) The topography of a region as shown by the typical contour map, where the contour interval is well suited to express the slope of the terrain of the area (ASCE, 1954, p.39-40).
confined aquifer (con-fined′) An aquifer bounded above and below by confining beds; an aquifer containing *confined groundwater*. Syn: *artesian aquifer.*
confined groundwater Groundwater under pressure significantly greater than that of the atmosphere. Its upper surface is the bottom of a confining bed. Ant: *unconfined groundwater.* Syn: *artesian water; confined water.*
confined water *confined groundwater.*
confining bed (con-fin′-ing) A body of relatively impermeable or distinctly less permeable material stratigraphically adjacent to one or more aquifers. Cf: *aquitard; aquifuge; aquiclude.* Syn: *confining unit.*
confining pressure An equal, all-sided pressure around a specific volume of material; e.g., *geostatic pressure* or *hydrostatic pressure.* It is the pressure applied to fluid surrounding a sample in a *triaxial test* apparatus.
confining unit *confining bed.*
conflation A process by which digital map files may be matched and merged into one by selecting the best features from each individual map to be added to the composite new map (Krzanowski et al., 1993).
confluence [glaciol] (con′-flu-ence) (a) A flowing-together of two or more glaciers. (b) A junction or place where confluence occurs. Ant: *diffluence.*
confluence [streams] (a) A place of meeting of two or more streams; the point where a tributary joins the main stream; a fork. (b) A flowing together of two or more streams. (c) The stream or other body of water produced by confluence; a combined flood. See also: *junction [streams].*
confluence plain A plain formed by the merging of the valley floors of two or more streams.
confluence step A *rock step* that rises upstream toward the heads of two glacial valleys at their place of confluence. It is probably caused by the strengthening of glacial action downvalley from that point. Ant: *diffluence step.*
confluent (con′-flu-ent) adj. Said of a stream, glacier, vein, or other geologic feature that combines or meets with another like feature to form one stream, glacier, vein, etc. Ant: *diffluent.* n. A confluent stream, usually a stream uniting with another of nearly equal size; a fork or branch of a river. The term is sometimes loosely applied to an *affluent.*
conformability (con-form′-a-bil′-i-ty) The quality, state, or condition of being conformable, such as the relationship of conformable strata; *conformity.*
conformable [intrus rocks] (con-form′-a-ble) An intrusive igneous contact which has the same attitude as that of the intrusion's internal structures. Cf: *disconformable [intrus rocks]; concordant [intrus rocks].*
conformable [stratig] Said of strata or stratification characterized by an unbroken sequence in which the layers are formed one above the other in parallel order by regular, uninterrupted deposition under the same general conditions; also said of the contacts (abrupt, gradational, or intercalated) between such strata. The term is often applied to a later formation having bedding planes that are parallel with those of an earlier formation and showing an arrangement in which disturbance or erosion did not take place at the locality during deposition. Dennis (1967, p.26) regards the term as descriptive, primarily referring to succession without disturbance and not necessarily implying parallelism of succeeding beds or continuity of deposition; e.g. cross-bedding without intervening erosion is conformable, but not parallel. Cf: *unconformable; concordant [stratig].*
conformable transgressive surface In sequence stratigraphy, a surface that separates underlying, upward-shallowing, regressive strata from overlying, upward-deepening, transgressive strata. There is no significant depositional hiatus associated with the surface (Embry and Johannessen, 1992).
conformably superimposed stream (con-form′-a-bly) A term suggested by Kümmel (1893, p.380) for a stream superimposed from a conformable cover.
conformality (con-for-mal′-i-ty) The unique property of a conformal map projection in which all small or elementary figures on the surface of the Earth retain their original shapes on the map. Cf: *equivalence [cart].* Syn: *orthomorphism.*
conformal projection (con-for′-mal) A map projection on which the shape of any very small area of the surface mapped is preserved unchanged on the map and the scale at any point is the same in every direction (as along the meridian and the parallel at that point), although the scale may vary from point to point. It always shows a right-angle intersection of any parallel with any meridian. Examples include: stereographic projection, Mercator projection, and Lambert conformal conic projection. Cf: *equal-area projection.* Syn: *orthomorphic projection.*
conformity (con-form′-i-ty) (a) The mutual and undisturbed relationship between adjacent sedimentary strata that have been deposited in orderly sequence with little or no evidence of time lapses; true stratigraphic continuity in the sequence of beds without evidence that the lower beds were folded, tilted, or eroded before the higher beds were deposited. Syn: *conformability.* (b) An uninterrupted sequence of strata displaying conformity. (c) A surface that separates younger strata from older ones, along which there is no physical evidence of erosion or nondeposition, and no significant hiatus (Mitchum, 1977, p.206). Cf: *unconformity.* (d) A surface across which the process of deposition was not interrupted by erosion, but nondeposition may be possible (Crosby, 1912, p.289-290).
confound In a statistical experiment, two effects are said to be confounded if it is impossible to differentiate between them. Hence, in general, a factor is confounded with another if the variance attributable to each cannot be estimated.
confused sea (con-fused′) A highly disturbed sea surface with indeterminate direction and period of wave travel. It is caused by the superimposition of various wave trains.
congela (con-ge′-la) A term used in Chile for *coba* with a high salt content.
congelation (con-ge-la′-tion) The change from a fluid to a solid state; freezing. Also, the product of such a change. Syn: *gelation [ice].*
congelation crystallization Magmatic crystallization along the walls and floor of a magma chamber which concentrates residual elements in the remaining magma at the center and/or top of the chamber. Obsolete.
congelifluction (con-gel′-i-fluc′-tion) *gelifluction.*
congelifract (con-gel′-i-fract) An angular rock fragment split off by frost action (*congelifraction*), ranging from very large blocks to finely comminuted material. Syn: gelifract.
congelifractate (con-gel′-i-fract′-ate) A mass of rock fragments (congelifracts) of different sizes produced by frost action.
congelifraction (con-gel′-i-frac′-tion) *gelifraction.*
congeliturbate (con-gel′-i-tur′-bate) *cryoturbate.*
congeliturbation (con-gel′-i-tur-ba′-tion) *cryoturbation.*
congeneric (con-ge-ner′-ic) Belonging to the same genus.
conglomerate (con-glom′-er-ate) A coarse-grained clastic sedimentary rock, composed of rounded to subangular fragments larger than 2 mm in diameter (granules, pebbles, cobbles, boulders) typically containing fine-grained particles (sand, silt, clay) in the interstices, and commonly cemented by calcium carbonate, iron oxide, silica, or hardened clay; the consolidated equivalent of *gravel* both in size range and in the essential roundness and sorting of its constituent particles. The rock or mineral fragments may be of varied composition and range widely in size. Conglomerates may be classified according to nature or composition of fragments, proportion of matrix, degree of size sorting, type of cement, and agent or environment of formation. Etymol: Latin "conglomeratus", "heaped, rolled, or pressed together". Cf: *breccia [geol].* Syn: *puddingstone.*
conglomerated ice (con-glom′-er-at′-ed) Several forms of floating ice compacted into one mass. The term refers to the contents of an ice mass, not to the concentration.
conglomeratic (con-glom′-er-at′-ic) Pertaining to a conglomerate; composed or having the properties of conglomerate.
conglomeratic mudstone A mudstone with a sparse to liberal sprinkling of pebbles or cobbles; e.g. a consolidated gravelly mud containing 5-30% gravel and having a ratio of sand to mud less than 1:1 (Folk, 1954, p.346), such as a *pebbly mudstone* or a *tilloid.* Pettijohn (1957, p.261) considered the term synonymous with *paraconglomerate.* Cf: *mudstone conglomerate.*
conglomeratic sandstone (a) A sandstone containing 5-30% gravel and having a ratio of sand to mud (silt + clay) greater than 9:1 (Folk, 1954, p.347); a consolidated gravelly sand. (b) A sandstone containing more than 20% pebbles (Krynine, 1948, p.141). Cf: *pebbly sandstone.*
conglomerite (con-glom′-er-ite) A term suggested by Willard (1930) for a conglomerate that has reached the same stage of induration as that of a quartzite, characterized by the welding together of matrix and clasts as evidenced by fractures passing through both. Examples include the partially metamorphosed or "stretched" Carboniferous conglomerates of Rhode Island.
Congo copal (Con′-go) A hard, yellowish to colorless *copal* derived from certain trees of the genus Copaifera, found as a fossil resin in the Congo, and used in making varnish.
congolite (con′-go-lite) A pale red rhombohedral mineral: $(Fe^{2+},Mg)_3B_7O_{13}Cl$. It is dimorphous with ericaite.
congregation (con-gre-ga′-tion) Those fossil species occurring together in and characterizing the rocks of a particular *zone [stratig].* A congregation may include all or part of an *assemblage [ecol],* or more than one assemblage.
congressite (con′-gres-site) A coarsely granular *urtite* consisting chiefly of nepheline, with minor amounts of sodalite, plagioclase, mica, and calcite. Cf: *craigmontite.* Named by Adams and Barlow in 1913 for Congress Bluff, Craigmont Hill, Ontario. Obsolete.
congruent (con-gru′-ent, con′-gru-ent) Said of a parasitic fold, the axis and axial surface of which are parallel to the axis and axial surface of the main fold to which it is related. Ant: *incongruous.*
congruent melting Melting of a substance directly to a liquid that is of the same composition as the solid. Cf: *incongruent melting.*
coni (co′-ni) Plural of *conus.*
Coniacian (Co-ni-a′-ci-an) Stage of the Standard Global Chronostratigraphic Scale: Upper Cretaceous (above Turonian, below Santonian).

coniatolite (co-ni-at'-o-lite) A hard, sheetlike crust of aragonite found in supratidal saline environments in the Arabian Gulf area (Purser and Loreau, 1973, p.375).

conical fold (con'-i-cal) A fold model that can be described geometrically by the rotation of a line about one of its ends, which is fixed. Cf: *cylindrical fold.*

conical wave *head wave.*

conichalcite (con-i-chal'-cite) A pistachio-green to emerald-green orthorhombic mineral: $CaCu(AsO_4)(OH)$. It often contains phosphorus. A member of the *adelite* group. Syn: *higginsite.*

conic projection (con'-ic) One of a group of map projections produced by projecting the geographic meridians and parallels onto the surface of a cone that is tangent to, or intersects, the surface of the sphere, and then developing (unrolling and laying flat) the cone as a plane. True distances are measured along the lines of tangency or secancy; everywhere else on the map the scale is too large or too small. Examples include: Lambert conformal conic projection, and Albers projection. See also: *Bonne projection; polyconic projection; simple conic projection.* Syn: conical projection.

conidia Pl. of *conidium.*

conidiophore (co-nid'-i-o-phore') A structure that bears conidia; specif. a specialized, typically erect *hypha* that produces successive conidia in certain fungi.

conidiospore (co-nid'-i-o-spore') A *conidium;* a *fungal spore* that may have chitinous walls and therefore occur as a microfossil in palynologic preparations.

conidium (co-nid'-i-um) A nonmotile, exogenous, asexual fungal spore, produced from the tip or side of a *conidiophore;* broadly, any asexual fungal spore not borne within an enclosing structure (i.e., not produced in a sporangium). Pl: *conidia.* Syn: *conidiospore.*

conifer (co'-ni-fer) A gymnosperm, a member of the class Coniferae, having needlelike or scalelike leaves and naked seeds borne in cones. Conifers include pines, firs, and spruces.

coniferous (co-nif'-er-ous) Bearing cones, as in conifers.

coniform element Basically conical conodont elements without greatly produced or denticulated margins (TIP, 1981, pt. W, supp.2). Obsolete syn: *simple cone.*

coning (con'-ing) (a) The cone-shaped rise of saltwater beneath fresh water in an aquifer as fresh water is produced from a well. Syn: *upconing.* (b) The cone-shaped rise of water underlying oil or gas in a reservoir as the oil or gas is withdrawn from a well.

conispiral (co-ni-spi'-ral) adj. Said of a gastropod shell with a cone-shaped spire; said of a cephalopod shell characterized by a spiral coiled on the surface of a cone, the whorls not in a single plane. n. A conispiral shell.

conjugate [struc geol] (con'-ju-gate) (a) Said of a shear fracture system, the sets of which are related in deformational origin, usually compression. Also, said of the mineral deposits which may form in such joints. Cf: *cognate [struc geol]; complementary [struc geol].* (b) Said of two faults formed under the same stress field and intersecting with an acute dihedral angle bisected by the maximum principal stress (Anderson, 1951).

conjugate fold system Two sets of minor folds, the axial surfaces of which are inclined towards each other. Cf: *kink band.*

conjugate margin A margin that matches another on the opposite side of an ocean or basin and to which it was joined before rifting.

conjugate solutions Two solutions coexisting in equilibrium whose compositions are separated by a miscibility gap in a potentially continuous compositional field. The possibility of a critical point at which the two phases would become identical is implicit.

conjunct (con'-junct) In a rhombiferan echinois, said of a pore rhomb in which externally visible slits are continuous across the suture between plates bearing the rhomb. Cf: *disjunct.*

connate [bot] (con'-nate) In plant morphology, having like parts fused, e.g., petals fused together into a floral tube. Cf: *adnate.*

connate [petrology] Said of fluids derived from the same magma.

connate [sed] Originating at the same time as adjacent material; esp. pertaining to waters and volatile materials (such as carbon dioxide) entrapped in sediments at the time the deposits were laid down and out of contact with the atmosphere for an appreciable part of a geologic period.

connate water Water entrapped in the interstices of a sedimentary rock at the time of its deposition; White (1957, p.1661) has recommended that it be defined as "water that has been out of contact with the atmosphere for at least an appreciable part of a geologic period". Commonly misused by reservoir engineers and well-log analysts to mean *interstitial water* or *formation water.* Syn: *fossil water; fossilized brine; native water.*

connecting band (con-nect'-ing) (a) One of the two hooplike bands at the edge of the flange or valve mantle in a diatom frustule. See also: *girdle [paleont]; cingulum.* (b) A part of the *loop* of terebratellacean brachiopods that joins descending branches to the median septum or that joins ascending branches and descending branches posterior to their anterior curvature (TIP, 1965, pt. H, p.142).

connecting bar A bar that is connected at both ends to a landmass; esp. a *tombolo.*

connecting lobe A rounded linear elevation of the valve surface of an ostracode, confluent with two or more subvertically trending lobes (TIP, 1961, pt. Q, p.49).

connecting ring The partly calcareous and partly conchiolinous delicate tubular membrane formed between septa; the wall of a cephalopod *siphuncle,* such as one that connects the *septal neck* of an *ellipochoanitic* nautiloid conch with the septum immediately behind it. Vestiges of the connecting ring are also found in *holochoanitic* forms.

connecting tubule A subhorizontal tubular connection between neighboring corallites in a *fasciculate* corallum.

connective suture (con-nec'-tive) One of a pair of trilobite cephalic sutures, generally entirely on the ventral side, bounding the sides of the *rostral plate.* Cf: *median suture.*

connectivity analysis An analysis which is performed in a spatial information system to determine which features in a set are connected to each other (Krzanowski et al., 1993).

connector bar (con-nec'-tor) A bar that joins the sagittal ring of a radiolarian skeleton to the lattice shell (as in the subfamily Trissocyclinae).

connellite (con'-nel-lite) A deep-blue hexagonal mineral: $Cu_{37}(SO_4)_2Cl_8(OH)_{62}{\cdot}6H_2O$. It is isomorphous with buttgenbachite. Syn: *footeite.*

conode (co'-node) *tie line [chem].*

conodont (co'-no-dont) One of a large number of small, disjunct fossil elements assigned to the order Conodontophorida, phosphatic in composition, and commonly toothlike in form but not necessarily in function; produced in bilaterally paired, serial arrangement by small vagile marine Chordata. See: Hass (1962). Range, Cambrian (possibly Late Precambrian) to Upper Triassic; commonly abundant, widespread, and useful biostratigraphically.

conodont apparatus *apparatus.*

conodont color alteration index The experimentally calibrated color sequence caused by thermal maturation and breakdown of the organic matter in euconodont elements. The sequence is progressive and irreversible, ranging from pale yellow (CAI=1) through brown and black (5) to crystal clear (8). Abbrev: CAI.

conodont element A unit or complete specimen of a conodont, e.g. *compound conodont element; platelike conodont element; lamellar conodont element; fibrous conodont element.*

conoplain (co'-no-plain) A rarely used syn. of *pediment.* The term was used by Ogilvie (1905, p.28) for an erosion surface peripheral to and sloping radially away in all directions from a laccolithic mountain mass.

conoscope (co'-no-scope) A polarizing microscope using convergent light with the Bertrand lens inserted, used to test the interference figures of crystals. Cf: *orthoscope.*

conotheca (co-no-the'-ca) (a) Conoidal *theca* with a small circular aperture located at the terminus of a short neck developed irregularly in colonies of tuboid graptolithines (Whittington and Rickards, 1968). (b) The outer shell of ectocochleate cephalopods and of the phragmocone of endocochleate cephalopods.

Conrad discontinuity (Con'-rad) The seismic-velocity discontinuity within the Earth's crust that is equivalent to the boundary between the *sial* and the sima, at which velocities increase from ~6.1 km/sec to 6.4-6.7 km/sec. It is not always discernible; i.e. there may be a general increase in velocity with depth in the crust, with no evidence of layer boundaries. See also: *Conrad layer.* A possible equivalent is the *Riel discontinuity.*

Conrad layer The seismic region of the Earth between the *Conrad discontinuity* and the *Mohorovicic discontinuity.* Also called the *intermediate layer* or the *lower crustal layer.*

consanguineous [petrology] (con-san-guin'-e-ous) The adj. of *consanguinity* and obsolete syn. of *comagmatic.*

consanguineous [sed] Said of a natural group of sediments or sedimentary rocks related to one another by origin; e.g. a "consanguineous association" (such as flysch, molasse, or paralic sediments) interrelated by common ancestry, environment, and evolution, representing a "facies" in a broad sense but "not equivalent to a single lithologic type" (Fairbridge, 1958, p.319).

consanguinity (con-san-guin'-i-ty) Obsolete term for the genetic relationship that exists between igneous rocks that are presumably derived from the same parent magma. Such rocks are closely associated in space and time and commonly have similar geologic occurrence and chemical and mineralogic characteristics. Adj: consanguineous. See also: *comagmatic.*

consecutive-range zone (con-sec'-u-tive-range') (a) "A body of strata within the range of a fossil group such that it forms the first part of the range of that fossil group before the first appearance of its immediate evolutionary descendant" (Geological Society of London, 1967, p.85). (b) A biozone characterized by one taxon of a lineage, in such a way that the precursor taxon ceases at the lower limit of the biozone and the successor taxon begins at the upper limit (Holland et al., 1978, p.13).

consensus tree In cladistic analysis, a phylogenetic tree that combines the information about topologically different but equally parsimonious trees into a single tree. Alternatively, it is a phylogenetic tree that combines the information from a set of topologically different trees derived from different data sets into a single tree (Wiley et al., 1991, p.80).

consequent [geomorph] (con'-se-quent) Said of a geologic or topographic feature that originated as a result of and in harmony with preexisting conditions or features; e.g. a consequent ridge (such as an anticlinal arch that retained its axial eminence), or a consequent island in a lake basin (such as an elevation that remained above the water level at the time of the formation of the lake), or a consequent waterfall (such as one resulting from the irregularities of the surface over which the stream originally flowed). Cf: *subsequent [geomorph]; resequent [geomorph]; obsequent [geomorph].*

consequent [streams] adj. Said of a stream, valley, or drainage system whose course or direction is dependent on or controlled by the general form and slope of an existing land surface. The term was first applied by Powell (1875, p.163-166) to the valley thus formed. n. *consequent stream.*

consequent divide A divide between two consequent streams.

consequent fault scarp A *fault scarp* whose face approximates the initial formation of the scarp.

consequent lake A lake occupying a depression that represents an original inequality in any new land surface; e.g. a lake existing in a depression in glacial deposits or among sand dunes or in an irregularity of a recently uplifted sea floor.

consequent stream A stream that originates on a newly exposed or recently formed surface and that flows along a course determined entirely by the initial slope and configuration of that surface; a stream whose direction of flow is directly related to and a necessary consequence of the original dip-slope surface of the land and the geologic structure of the area. Syn: *consequent [streams]; original stream.*

consertal (con-ser'-tal) A syn. of *sutured,* preferred in European usage but obsolete in American usage.

conservative elements (con-serv'-a-tive) In seawater, those elements whose total amounts, and average concentrations, tend to be constant over time scales of millions of years. They constitute the vast majority of the dissolved ions in the ocean. Their abundance ratios relative to each other and to salinity are the same throughout the ocean, and the concentration of any one of them could in principle be used to calculate the total salinity or chlorinity of a seawater sample. Dominant among them are the major cations (sodium, potassium, magnesium, and calcium) and major anions (chloride and sulfate) in the sea. Salinity, being the total of their abundances, and chlorinity are also conservative. Cf: *constancy of relative proportions.* Cf: *nonconservative elements; major elements; salinity; chlorinity.* Syn: *conservative solutes; conservative ions.*

conservative ions *conservative elements.*

conservative solutes *conservative elements.*

consistence The attributes of soil material as expressed in degree of cohesion and adhesion or in resistance to deformation or rupture.

consistency (con-sist'-en-cy) The relative ease with which a soil can be deformed. The term expresses the degree of firmness or cohesion of soil particles and their resistance to rupture or deformation, usually by hand pressure.

consistency index The ratio of the difference between the *liquid limit* and the natural water content of a soil to the *plasticity index.* Syn: *relative consistency.*

consistency limits *Atterberg limits.*

consociation (con-so'-ci-a'-tion) An ecologic community, within an association, that has one dominant species, usually one of several common or abundant species.

consolidated pack ice (con-sol'-i-dat'-ed) *Pack ice* in which the *concentration* is $^8/_8$ (or $^{10}/_{10}$) and the floes are frozen together. Obsolete syn: *field ice.*

consolidation [eng] (con-sol-i-da'-tion) The gradual reduction in volume and increase in density of a soil mass in response to increased load or effective compressive stress; e.g., the squeezing of fluids from pore spaces. Cf: *compaction [soil].*

consolidation [geol] (con-sol-i-da'-tion) (a) Any process whereby loosely aggregated, soft, or liquid earth materials become firm and coherent rock; specif. the *solidification* of a magma to form an igneous rock, or the lithification of loose sediments to form a sedimentary rock.

consolidation [tect] A term sometimes used for diastrophic processes by which mobile belts are converted into rigid parts of the craton. The more specific terms *orogeny* and *orogenesis* are preferable.

consolidation grouting *grouting.*

consolute (con'-sol-ute) Said of liquids that are miscible in all proportions.

consolute point A point that represents the composition and maximum or minimum temperature or pressure of a miscibility gap. Syn: *critical solution point.*

consortium (con-sor'-ti-um) An intimately associated group of individuals of different kinds of organisms, generally belonging to different phyla, that live together.

Consortium for Continental Reflection Profiling (Con-sor'-ti-um) A program for obtaining seismic reflection profiles of the Earth's crust, begun about 1973. It is funded by the National Science Foundation, and Cornell University is the operating institution. Abbrev: COCORP.

conspecific (con-spe-cif'-ic) Belonging to the same species.

constancy of composition (con'-stan-cy) *constancy of relative proportions.*

constancy of interfacial angles *law of constancy of interfacial angles.*

constancy of relative proportions The principle that the ratios of the abundances of *conservative elements* (dissolved solids) in seawater are virtually constant, regardless of the absolute total concentration of dissolved solids; i.e., salinity. Syn: *law of constancy of relative proportions; constancy of composition.*

constant error (con'-stant) A *systematic error* that is the same in both magnitude and sign throughout a given series of observations (the observational conditions remaining unchanged) and that tends to have the same effect upon all the observations of the series or part thereof under consideration; e.g. the *index error* of a precision instrument.

constant-head permeameter A *permeameter* in which fluid moves under steady-state flow conditions.

constant of channel maintenance The ratio of the area of a drainage basin to the total stream lengths of all the stream orders within the basin (Schumm, 1956, p.607); approximately the reciprocal of the *drainage density* (Shreve, 1969, p.412). It expresses the minimum limiting area required for the development of a unit length of a drainage channel.

constant slope The straight part of a hillside surface, lying below the *free face* and determined by the angle of repose of the material eroded from it (Wood, 1942); it merges downslope into the *waning slope.* Cf: *gravity slope.* Cf: *debris slope.*

constant-sum data Multivariate data in which all variables measured on each object in a set sum to a constant, usually 1 or 100%. Examples include geochemical data in which oxide percentages sum to 100%, and grain-size distributions in which the sum of the grain-size classes sum to 100%. Such data are said to be closed, and there are inherent negative relationships between variables that result from the fact that one variable can increase only if other variables decrease in value. Conventional statistical tests are not valid for constant-sum data.

constant sum effect An attribute of wt% oxide plots wherein the dependent and independent variables sum to 100%, e.g. MgO Variation Diagram, Harker (SiO_2) Variation Diagram (Chayes, 1966). As the independent variable approaches 100% the value of any dependent variable necessarily approaches 0%. This effect may result in apparent oxide correlations that are actually due to the plotting procedure itself.

constant velocity stack In *velocity analysis,* one of several trial stacks assuming several trial velocities that are constant in time and space, in order to determine the stacking velocities that produce the best section. Abbrev: CVS.

constitutional ash (con-sti-tu'-tion-al) *inherent ash.*

constitutive equation An equation relating either stress and strain or stress and strain rate. It may be empirically derived from laboratory data or based on theoretical relations, e.g. Hooke's Law.

constrictional strain Three-dimensional strain described by a prolate ellipsoid.

constructional (con-struc'-tion-al) (a) Said of a landform that owes its origin or general character to the processes of upbuilding, such as accumulation by deposition (e.g. a plain or an alluvial terrace), volcanic eruption (e.g. a volcanic cone), and diastrophism or orogenic activity (e.g. a mountain, a fault block, or a tectonic valley); also said of a surface whose form was not acted upon by erosion. Ant: *destructional.* Cf: *initial landform.* (b) Said of a stream or drainage pattern that is formed by runoff from a constructional landform or surface (Davis, 1894, p.74).

constructional void porosity Primary porosity in a carbonate framework (Murray, 1960, p.61). Cf: *growth-framework porosity.*

constructive metamorphism (con-struc'-tive) An archaic term for the development of *kinetic growth form (faceted form)* in snow crystals.

constructive waterfall In a cave, a waterfall over a *rimstone dam* or a *canopy.*

constructive wave A wave that builds up a beach by moving material landward, as a gentle wave with a more powerful uprush than backwash. On sandy beaches, a wave with a steepness of less than 0.25 is considered a constructive wave. Ant: *destructive wave.*

consumer (con-sum'-er) An organism that is unable to manufacture its food from nonliving matter but is dependent on the energy stored in other living things. Cf: *producer [ecol].*

consuming plate margin (con-sum'-ing) *convergent plate boundary.*

consumptive use (con-sump'-tive) Removal of water from the water environment by evaporation, transpiration, incorporation into products or crops, or consumption by man or livestock (Solley et al., 1983).

consumptive waste Water returned to the atmosphere without having benefited man (Langbein and Iseri, 1960).

contact (con'-tact) n. (a) A plane or irregular surface between two types or ages of rock; examples are faults, intrusive borders, bedding planes separating distinct strata, and unconformities. (b) The surface between two fluids in a reservoir, i.e. oil and gas, oil and water, or gas and water. Syn: *interface.* adj. Said of a mineral deposit that occurs at the contact of two unlike rock types.

contact-and-compression stage The first of three stages in the *impact* cratering process. During this stage, a meteoritic projectile transfers its kinetic energy to the target. The stage is the shortest of the three stages and ends when the projectile has embedded itself into the target by a distance equal to its own diameter. Cf: *excavation stage; modification stage.*

contact area One of the areas of the proximal side of a spore or pollen grain that are in contact with the other members of the tetrad. Contact areas are seldom visible in mature pollen grains but are frequently apparent in spores. Trilete spores have three contact areas; monolete spores have two. Syn: *area contagionis.*

contact aureole *aureole.*

contact breccia Breccia at the contact of an igneous intrusion, caused by country-rock fragmentation, and consisting of both intrusive material and country rock; *intrusion breccia.* Cf: *agmatite.*

contact erosion valley A valley eroded along a line of weakness at the contact between two different types of rock.

contact goniometer A *goniometer* that measures the solid angle between two crystal planes by contact with the surface. It is a 180° protractor, divided in degrees and accurate to half a degree, with a straightedge pivoted at the center. Cf: *two-circle goniometer; reflection goniometer.*

contactite (con-tac′-tite) A general term applied to rocks formed by contact metamorphism. Little used.

contact load The part of the *bed load* that is in substantially continuous contact with the stream bed.

contact log A *well log* produced by closely spaced electrodes or source-to-detector components of the *sonde* placed in direct contact with the walls of the borehole. Contact resistivity devices produce the various microresistivity logs, dipmeter, and electrical imaging logs. Logs of contact design include the *density log* and certain neutron logs. *Caliper logs* require wall contact. Syn: *wall-contact log.*

contact-metamorphic (con′-tact-met′-a-mor′-phic) Said of a rock or mineral that has originated through the process of *contact metamorphism.* In older literature, used as a syn. of *contact-metasomatic* and *pyrometasomatic.*

contact metamorphism One of the principal local processes of thermal metamorphism, genetically related to the intrusion and extrusion of magmas and taking place in rocks at or near their contact with a body of igneous rock. Metamorphic changes are effected by the heat and fluids emanating from the magma and by some deformation connected with the emplacement of the igneous mass (Holmes, 1920). Adj: *contact-metamorphic.* Cf: *endomorphism; exomorphism; thermal metamorphism.*

contact-metasomatic (con′-tact-met′-a-so-mat′-ic) Said of a rock or mineral that has originated through the process of *contact metasomatism.*

contact metasomatism A change in the composition of rocks in contact with an intruding magma, from which fluid constituents combine with the country-rock constituents to form a new suite of minerals. The term was originated by Barrell (1907). Cf: *regional metasomatism.*

contact mineral A mineral formed by contact metamorphism.

contact resistance The resistance observed between a grounded electrode and the ground, or between an electrode and a rock specimen.

contact spring A type of *gravity spring* whose water flows to the land surface from permeable strata over less permeable or impermeable strata that prevent or retard the downward percolation of the water (Meinzer, 1923, p.51). Syn: *hillside spring; outcrop spring.* Cf: *barrier spring.*

contact twin A twinned crystal, the two individuals of which are symmetrically arranged about a twin plane. Syn: *juxtaposition twin.*

contact zone *aureole.*

contaminant (a) Any physical, chemical, biological, or radiological substance or matter that has an adverse effect on air, water, or rocks (U.S. Environmental Protection Agency, 1994), a pollutant. (b) An undesirable substance in water, air, or rocks that is either not normally present or is an unusually high concentration of a naturally occurring substance.

contaminated (con-tam′-i-nat′-ed) *hybrid [petrology].*

contamination [ign] (con-tam′-i-na′-tion) The process whereby the chemical composition of a magma is altered as a result of the *assimilation* of inclusions or country rock. Cf: *hybridization.*

contamination [water] The addition to water of any substance or property that changes the physical and/or chemical characteristics of the water and prevents the use or reduces its usability for ordinary purposes such as drinking, preparing food, bathing, washing, recreation, and cooling. Sometimes arbitrarily defined differently from *pollution,* but generally considered synonymous.

contemporaneous (con-tem′-po-ra′-ne-ous) Formed or existing at the same time. Said of lava flows interbedded in a single time-stratigraphic unit, and generally of any feature or facies that develops during the formation of the enclosing rocks. Cf: *penecontemporaneous.*

contemporaneous erosion Local erosion that goes on while elsewhere deposition is occurring generally or continuously.

contemporaneous fault *growth fault.*

contemporary carbon *modern carbon.*

content graded bedding (con′-tent) A type of graded bedding due to variation in the relative amounts ("contents") of different grains rather than the size: "although the mean size decreases upwards as in normal graded bedding, there is generally no gradation of the maximum size" (McBride, 1962, p.50). The opposite of *coarse-tail grading,* in which the gradation is restricted to the maximum size.

contiguity analysis *adjacency analysis.*

continent (con′-ti-nent) (a) One of the Earth's major land masses, including both dry land and continental shelves. (b) An obsolete syn. of *terra.*

continental (con-ti-nen′-tal) Formed on land rather than in the sea. Continental deposits may be of lake, swamp, wind, stream, or volcanic origin.

continental accretion A theory proposed by J.D. Dana in the 19th century, that continents have grown at the expense of the ocean basins by the gradual addition of new continental material around an original nucleus. The new material was believed to have accumulated in concentric belts, each in turn consolidated by orogeny and succeeded by a new belt farther out. More recently, the term has been used in reference to the process by which exotic terranes are added to the margins of a continent during convergent tectonics. Cf: *accretion [struc geol].*

continental apron *continental rise.*

continental basin A region, in the interior of a continent, that comprises one or more closed basins.

continental borderland An area of the *continental margin* between the shoreline and the continental slope that is topographically more complex than the *continental shelf.* It is characterized by ridges and basins, some of which are below the depth of the continental shelf. An example is the southern California continental borderland; many geologists feel that the term can refer only to the southern California example which is unique.

continental climate The climate of the interior of a continent, characterized by seasonal temperature extremes and by the occurrence of maximum and minimum temperature about one month after summer and winter solstice, respectively. Cf: *marine climate.*

continental crust That type of the Earth's crust which underlies the continents and the continental shelves; it is equivalent to the *sial* and continental *sima* and ranges in thickness from about 25 km to more than 70 km under mountain ranges, averaging ~40 km. The density of the continental crust averages ~2.8 g/cm³ and is ~2.7 g.cm³ in the upper layer. The velocities of compressional seismic waves through it average ~6.5 km/s and are less than ~7.0 km/sec. Cf: *oceanic crust.*

continental deposit A sedimentary deposit laid down on land (whether a true continent or only an island) or in bodies of water (whether fresh or saline) not directly connected with the ocean, as opposed to a marine deposit; a glacial, fluvial, lacustrine, or eolian deposit formed in a nonmarine environment. See also: *terrestrial deposit.*

continental divide A drainage divide that separates streams flowing toward opposite sides of a continent, often into different oceans; e.g. in North America, the divide separating the watersheds of the Pacific Ocean from those of the Atlantic Ocean, and extending from the Yukon Territory, along the British Columbia-Alberta boundary, through western Montana, Wyoming, Colorado, western New Mexico, and into Mexico.

continental drift A general term, which can be used for many aspects of a theory originally propounded by Wegener (1912). It states that continents move relative to one another. Wegener could not provide an adequate explanation for the process. It is now clear that drift is a consequence of plate tectonics. The continents move away from each other due to *sea-floor spreading* along a mid-ocean ridge, producing new oceanic areas between the continents. They move toward each other when ocean lithosphere between them is subducted and they move laterally past each other along transform faults. Syn: *drift [tect]; epeirophoresis; continental migration.* See also: *expanding Earth.*

continental edge The seward boundary of the continent, shelf edge.

continental flexure A hinge-line structure along the contact of continent and sea floor, in which warping steepens the angle of slope of the continental shelf, causing relative uplift of the continent and eventual formation of coastal ranges.

continental flood basalt *flood basalt.*

continental glacier (a) A glacier of considerable thickness completely covering a large part of a continent or an area of at least 50,000 sq km, obscuring the relief of the underlying surface, such as the ice sheets covering Antarctica and Greenland. Cf: *inland ice.* Syn: *continental ice sheet; ice sheet.* (b) Any glacier in a continental, as opposed to a maritime, climatic environment. This usage is less preferred.

continental ice sheet *continental glacier.*

continentality A condition of climate or paleoclimate that is dominated by persistent dry descending air. Its dominance over certain regions in the geologic past has given rise to distinctive lithic facies, including *loess* in glacial times, large dune tracts, and characteristic geomorphic features.

continental margin The ocean floor that is between the shoreline and the abyssal ocean floor, including various provinces: the *continental shelf, continental borderland, continental slope,* and the *continental rise.*

continental migration *continental drift.*

continental nucleus *shield [tect].*

continental period The interval of time when a specific area was above sea level, forming part of a continent.

continental platform *continental shelf.*

continental polar air mass Cool, dry air derived from land areas at high latitudes. Abbrev: cP.

continental rise That part of a *continental margin* that is between the continental slope and the abyssal plain. It is best developed on trailing edges. It is a gentle incline with slopes of 1:40 to 1:2,000, and generally smooth topography, although it may bear submarine canyons. Syn: *continental apron.*

continental river A river with no outlet to the sea, its water disappearing by percolation or evaporation.

continental sea *epicontinental sea.*

continental shelf That part of the *continental margin* that is between the shoreline and the continental slope (or, when there is no noticeable continental slope, a depth of 200 m). It is characterized by its very gentle slope of 0.1°. Cf: *insular shelf; marginal plateau; continental borderland.* Syn: *continental platform; shelf [marine geol].*

continental shield *shield [tect].*

continental slope That part of the *continental margin* that is between the continental shelf and the continental rise if there is one. It is characterized by its relatively steep slope of 1.5° to 6°. Cf: *insular slope.*

continental terrace The sediment and rock mass underlying the coastal plain, the continental shelf, and the continental slope (Curray, 1966, p.207).

continental time A term used by Kobayashi (1944b, p.477) for *fossil time* as indicated by nonmarine organisms. Cf: *marine time.*

continental transgression Any enlargement of the area of continental deposition (such as of piedmont deposition on a bajada) in which areas previously subjected to erosion, or areas of equilibrium between erosion and deposition, are covered with sediments. Cf: *transgression.*

continental tropical air mass Warm, dry air derived from land areas at low latitudes. Abbrev: cT.

contingency table A two-way table in which rows represent states or categories of one variable and columns represent states or categories of another variable. Data consist of the relative number of observations that fall into each row/column category or cell. If the two variables are independent, the expected number in each cell can be calculated from the product of the row and column totals.

continuation (con-tin'-u-a'-tion) A mathematical data processing technique based on the potential field continuity, i.e., the observed gravity or magnetic field can be re-computed at the elevation higher (*upward continuation*) or lower (*downward continuation*) than that of the original measurements.

continuity equation (con-ti-nu'-i-ty) An axiom stating that the rate of flow past one section of a conduit is equal to the rate of flow past another section of the same conduit plus or minus any additions or subtractions between the two sections (ASCE, 1962).

continuous Said of a deformation in which the displacements of material points vary smoothly from one location in a body of rock to another; this circumstance leads to the formation of fabric elements whose geometry and character vary gradually with position. See also: *continuous deformation.*

continuous-creation hypothesis (con-tin'-u-ous-cre-a'-tion) *steady-state theory.*

continuous deformation Deformation that occurs without the loss of continuity at the scale of observation. Cf: *discontinuous deformation.*

continuous mining The excavation and removal of ore and rock in an underground or open-pit mine by a system of machinery that advances into the mining face.

continuous permafrost A zone of permafrost that, for the most part, is uninterrupted by pockets or patches of unfrozen ground. Cf: *discontinuous permafrost; sporadic permafrost.*

continuous profiling A seismic method in which geophone groups are placed uniformly along the length of the line and so spread that a uniformly spaced set of points in the subsurface is sampled once. Cf: *correlation shooting.*

continuous reaction A metamorphic reaction that is continuously at equilibrium over a range of temperatures at constant pressure (or vice versa), owing to compositional variability among the minerals, e.g., in Fe/Mg ratio. A paragenetic diagram, e.g., the *AFM projection*, shows continuous variation in tie line orientations but no topological change. Cf: *discontinuous reaction.*

continuous reaction series A *reaction series* in which reaction of early-formed crystals with later liquids takes place without abrupt phase changes; e.g. the plagioclase feldspars form a continuous reaction series. Cf: *discontinuous reaction series.*

continuous sampling tube system A thin-wall sampling tube attached in advance of the cutting head of a hollow-stem *auger* that allows undisturbed samples to be taken continuously while the auger is rotated.

continuous spectrum A complete sequence of electromagnetic frequencies observed over a wide range of the electromagnetic spectrum. See also: *absorption spectrum; emission spectrum.*

continuous stream A stream that does not have interruptions in space; it may be perennial or ephemeral, but it does not have wet and dry reaches. Ant: *interrupted stream.*

continuous transformation A *phase transformation* that takes place simultaneously throughout a crystal, usually taking the form of small deviations from the parent structure, which become progressively more pronounced with time or temperature. Cf: *discontinuous transformation; modulated structure; incommensurate phase.*

continuous variable A quantity that can assume any value within a specified range. Most measurements of real properties are continuous, because a measurement always can be refined to greater numbers of decimal places by the use of a measuring instrument of higher precision. Cf: *discrete variable.*

continuous-velocity log *sonic log.*

contorted bedding (con-tort'-ed) *convolute lamination.*

contorted drift A glacial deposit that exhibits folding, thrusting, and other irregularities resulting from pressure during ice movement. The contortions are usually aligned in the direction of ice movement.

contortion (con-tor'-tion) (a) The intricate folding, bending, or twisting-together of laminated sediments on a considerable scale, the laminae being drawn out or compressed in such a manner as to suggest kneading more than simple folding; esp. *intraformational contortion.* Also, the state of being contorted. It occurs on a larger scale than *corrugation.* (b) A structure produced by contortion.

contour [cart] n. An imaginary line, or a line on a map or chart, that connects points of equal elevation of the land surface above or below some reference value or datum plane, generally sea level. Cf: *form line; isohypse; structure contour.* Syn: *contour line; topographic contour.* v. To provide a map with contour lines; to draw a contour line.

contour [geog] The outline or configuration of a surface feature seen two-dimensionally; e.g., the contour of a mountain pass or a coastline.

contour diagram An equal-area projection of structural data in which the poles have been contoured according to their density per unit area on the projection.

contour horizon *datum horizon.*

contouring In the strictest sense, contouring is the construction of lines (contours) connecting points of equal elevation on a map representing topography. The term has been generalized to the presentation of any spatial property on a map by *isolines* that connect points of equal value, in order to represent the form of a real or imaginary surface.

contour interval The difference in value between two adjacent contours; specif. the vertical distance between the elevations represented by two successive contour lines on a topographic map. It is generally a regular unit chosen according to the amount of vertical distance involved and the scale of the map, but it need not be constant over the entire map (a variable contour interval may be used for optimum portrayal of relief features). Syn: *vertical interval.*

contour line *contour [cart].*

contour map A map that portrays surface configuration by means of contour lines; esp. a topographic map that shows surface relief by means of contour lines drawn at regular intervals above mean sea level, or a *structure-contour map* that shows the configuration of a specified rock surface underground and the inferred configuration of that surface where it has been removed by erosion.

contour mining A technique of *opencut mining* in which coal seams are mined in relatively level benches along a hillside.

contour sketching Freehand delineation, on a map, of the surface relief as seen in perspective view and controlled by map locations corresponding to salient points on the ground. See also: *field sketching.*

contour value A numerical value placed upon a contour line, such as a figure denoting elevation relative to mean sea level.

contracting Earth (con-tract'-ing) A theory widely believed in the 19th and early 20th centuries that mountain belts of the Earth were produced by compression of the crust during its gradual contraction on the surface of a cooling but originally molten globe (a familiar textbook analogy of the time was a dried apple). The theory is now discredited, as the Earth is not cooling or contracting in the manner once believed. Cf: *expanding Earth; wedge theory.*

contractional bend A spatial variation in the orientation of a fault plane that causes local contraction of the wall rocks as they are displaced around the bend (Peacock et al., 2000). Cf: *convergent bend; restraining bend.*

contractional overstep An *overstep* that has to be accommodated by contraction, e.g., a right-step between left-lateral faults. This type of overstep occurs in cross section along normal faults, with the lower fault overstepping or projecting into the hanging wall of the upper fault (Peacock et al., 2000). Syn: *convergent overstep; restraining overstep.* Cf: *extensional overstep.*

contraction crack (con-trac'-tion) *frost crack.*

contraction fault A fault in sedimentary rocks along which there has been bed-parallel shortening (Norris, 1958, 1964), giving rise to tectonic thickening.

contraction fissure A fissure that is formed as a result of cooling, drying, or syneresis and consequent contraction of the rock. Cf: *cooling crack; mud crack; syneresis joint.*

contragradation (con'-tra-gra-da'-tion) Stream aggradation caused by an obstruction (Shaw, 1911). Syn: *dam gradation.*

contrast (con'-trast) In photography, the difference in *density [optics]* between the shadows and the highlights (darkest parts) of a photographic negative, or the ratio of reflecting power between the shadows and the highlights (lightest parts) of a photographic print.

contrasted differentiation (con-trast'-ed) Differentiation of magma into basic and acidic magmas. Reactions between these contrasted magmas may produce intermediate rock types which resemble intermediate types usually considered as the product of progressive fractionation (Nockolds, 1934). Obsolete.

contrast stretching An image enhancement procedure in which the original range of gray scale presented on an image is expanded to the full range of the display device or recording media (Krzanowski et al., 1993).

contratingent (con-tra-tin'-gent) Said of a *minor septum* (of a rugose coral) that leans against the adjoining *major septum* on the side toward the *counter septum.*

contributing area The portion of a *drainage basin* that is actually contributing runoff to a channel during a storm.

contributory (con-trib'-u-to'-ry) An obsolete syn. of *tributary.* Also spelled: contributary.

control [hydraul] (con-trol') (a) A section or reach of an open channel in which natural or artificial conditions make the water level above it a stable index of discharge. It may be either complete (i.e., water-surface elevation above the control is completely independent of downstream water-level fluctuations) or partial; it may also shift. (b) That waterway cross section that is the bottleneck for a given flow and determines the energy head required to produce the flow. In an open channel, it is the point at which flow is at critical depth; in a closed conduit, it is the point at which hydrostatic pressure and cross-sectional area of flow are definitely fixed, except where the flow is limited at some other point by a hydrostatic pressure equal to the greatest vacuum that can be maintained unbroken at that point. (ASCE, 1962).

control [surv] (a) The coordinated and correlated dimensional data that are used to establish the position, elevations, scale, and orientation of the detail of a map and that are responsible for the interpretations placed on the map. (b) The assemblage of accurately located points that determines the accuracy of a map and with which local secondary surveys may be tied in to insure their essential accuracy; a system of relatively precise field measurements of points, marks, or objects on the ground, whose horizontal and/or vertical positions have been (or will be) more or less accurately determined by surveying instruments. A map that includes many such points is said to have "good control".

control base A surface on which a map projection and ground control are plotted and on which templets have been assembled or aerotriangulation has been accomplished and the control points thus determined have been marked.

controlled blasting (con-trolled') Blasting that is done in such a way as to minimize damage to the final rock slope. Typical controlled blasting techniques are cushion blasting and presplitting.

controlled mosaic (con-trolled') A *mosaic [photo]* in which the photographs or images have been adjusted, oriented, and scaled to horizontal ground control in order to provide an accurate representation with respect to distances and distortions. It is usually assembled from photographs that have been corrected for tilt and for variations in flight altitude.

control network Geodetic control together with the measured or adjusted values of the distances, angles, directions, or heights used in determining the coordinates of the control.

control point Any station in a horizontal and/or vertical control system that is identified on a photograph and used for correlating the data shown on that photograph (ASP, 1975, p.2074). See also: *pass point.*

control station An accurately located point, mark, or object on the ground, whose horizontal and/or vertical position is used as a base for a dependent survey; any surveyed point used for horizontal and/or vertical control.

control survey A survey that provides horizontal- and/or vertical-position data for the support or control of subordinate surveys or for mapping; e.g. a survey that provides the geographic positions and/or plane coordinates of triangulation and traverse stations and the elevations of bench marks. Control surveys are classified according to their precision and accuracy: the highest prescribed order is designated first order, the next lower is second order, and so on.

conulariid (con-u-lar'-i-id) Any marine fossil belonging to the order Conulariida, characterized by a tetramerous cone-shaped to elongate, pyramidal or subcylindrical, chitinophosphatic periderm, which may be smooth or have longitudinal markings. Range, Middle Cambrian to Lower Triassic.

conulite (con'-u-lite) A hollow cylindrical or conical *speleothem* with the apex downward, formed as the lining of a drip-drilled hole in cave sediment.

conus (co'-nus) A cone-shaped, pointed projection in a palynomorph sculptural type that is a variant of *echinate.*

convection [eco geol] (con-vec'-tion) In hydrothermal systems, the flow of waters around and through heated zones adjacent to plutons in response to thermal gradients and controlled by porosity-permeability, salinity, fluid viscosity, and allied factors. The flow is generally down along the periphery, toward the system at depth, and upward along and through its central portions, possibly completing more than one loop.

convection [grd wat] The transport of heat by flowing groundwater. Cf: *free convection; forced convection; conduction.*

convection [meteorol] (a) Transfer of heat by vertical movements within the Earth's atmosphere owing to density differences caused by heating from below. (b) The mixing and transport of properties other than heat. Cf: *advection [meteorol].*

convection [oceanog] A general term for mass fluxes of water that transport and mix water in the vertical plane. It may be free gravitational convection of fluids driven by density differences or forced convection caused by wind. It is also a means of energy transfer.

convection [tect] A process of heat transfer involving movement of mass. During convection, warmer, less dense material rises while cooler, denser material sinks. Convection takes place when buoyancy forces exceed the strength of the material, and heat flow is too great to be accommodated by conduction. Convection occurs in the Earth's mantle beneath the lithosphere; it is not clear whether whole-mantle convection occurs, or whether upper-mantle convection is independent of lower-mantle convection. Convective flow may exert shear stress to the base of plates and thus may affect plate velocity, but shear due to convection is probably not the main cause of plate motion. Cf: *advection [tect].*

convection cell A pattern of movement of convecting material in which the central area is upwelling and the outer area is downwelling, due to heat variations. See also: *convection [tect].*

convection fractionation Crystal fractionation brought about by the buoyancy driven convection of fractionated liquid away from stationary crystals that are growing "in place" as opposed to removal of crystals from a stationary liquid, e.g. crystal setting. Typically associated with boundary layer crystallization along the floor and sidewalls of an active magma body. See also: *compositional convection.*

convective fractionation Magmatic crystallization in which convective flow, driven by heat within the magma chamber, feeds crystallization along the floor, walls, and roof of the chamber.

conventional potassium-argon age A *potassium-argon age* for which argon analysis is performed on one aliquot of sample and potassium analysis on a different aliquot of sample.

conventional unit cell The smallest *unit cell* that is defined by a right-handed coordinate system, best shows the symmetry displayed by the *space lattice*, and is consistent with the metric requirements of the *crystal system.* Such a unit cell may be either primitive or centered, and it is not unique if the symmetry of the lattice does not constrain the choice of all lattice parameters. Cf: *reduced unit cell; primitive unit cell.*

convergence [currents] (con-ver'-gence) A zone where ocean currents or water types or masses meet and mix. The denser water mass will sink below the other, entrining some of the mixture. Also, the line or area in which convergence occurs. See also: *polar convergence.* Cf: *divergence [currents].*

convergence [evol] (a) The acquisition or possession of similar characteristics by animals or plants of different groups as a result of similarity in habits or environment. Cf: *adaptive radiation; parallelism; radiation [evol].* See also: *convergent evolution.* (b) In *cladism,* a syn. of *parallelism.*

convergence [meteorol] A wind pattern that brings about a net inflow of air toward some point, accompanied by vertical motion. Ant: *divergence [meteorol].*

convergence [petrology] The production, generally during metamorphism, of petrographically similar rocks from different original rocks. Obsolete.

convergence [phylogen] Two apparently similar characters that developed from different preexisting characters; recognizable on a cladogram as a character occuring in two taxa separated by at least two nodes (Brooks et al., 1984, p.80). A type of homoplasy. Cf: *homoplasy, parallelism.*

convergence [stratig] The gradual decrease in the vertical distance or interval between two specified rock units or geologic horizons as a result of the thinning of intervening strata; e.g. the reduction in thickness of sedimentary beds (as measured in a given direction and at right angles to the bedding planes), caused by variable rates of deposition or by unconformable relationship.

convergence [surv] *convergence of meridians.*

convergence map *isochore map.*

convergence of meridians (a) The angular drawing-together of the geographic meridians in passing from the equator to the poles. (b) The difference between the two angles formed by the intersection of a great circle with two meridians. Also, the relative difference of direction of meridians at specific points on the meridians. Syn: *convergence [surv].*

convergent bend (con-ver'-gent) A bend in a strike-slip fault that results in over-all crustal shortening in the vicinity of the bend (Biddle and Christie-Blick, 1985). Cf: *contractional bend; divergent bend; restraining bend.*

convergent configuration A pattern in which a seismic unit thins laterally by nonsystematic reflection terminations, due to thinning of individual strata beyond the resolvable limit.

convergent evolution The development of similar-appearing forms in genetically unrelated lineages: convergence. Cf: *parallel evolution; homeomorphy.*

convergent fan A type of *fan cleavage* where cleavage orientations converge in the core of the fold.

convergent fault A *strike-slip fault* or a *wrench fault* along which strike-slip deformation is accompanied by a component of shortening transverse to the fault (Wilcox et al., 1973). Cf: *divergent fault; transpressional fault.*

convergent-isogon fold A fold in which the dip isogons, when traced along a traverse from the outer arc of the fold to the inner arc of the fold, converge.

convergent margin A continental edge that is also a plate edge and is being deformed by collision with another plate; *active margin.*

convergent overstep *contractional overstep.*

convergent plate boundary A boundary between two plates that are moving toward each other. It is essentially synonymous with *subduction zone,* but is used in different contexts (Dennis and Atwater, 1974, p.1034). Syn: *consuming plate margin* Partial syn: *active margin.*

conversion point In seismic exploration, the reflecting point where P waves convert to S waves or vice-versa. The set of data that have the same conversion point constitute a common-conversion-point (CCP) gather; after the proper NMO correction, they can be stacked together to display converted waves. The conversion point is moved when source and receiver are interchanged.

converted wave (con-vert'-ed) A seismic wave that has been converted from a P-wave to an S-wave or vice versa by reflection or refraction at an interface. Such waves are sometimes designated PS or SP. Cf: *reflected wave.*

converter plant (con-vert'-er) A plant that incorporates into its structure an insoluble element from the soil, and later, when the plant decays, returns that element to the soil in a soluble form.

convex bank (con-vex') The inner bank of a curved stream, with the center of the curve away from the channel; e.g. a *slip-off slope.* Ant: *concave bank.*

convex cross-bedding Cross-bedding with convex (upward-arching) foreset beds. Cf: *concave cross-bedding.*

convexo-concave (con-vex'-o-con-cave') Convex on one side and concave on the other; e.g. said of a resupinate brachiopod shell having a convex *brachial valve* and a concave *pedicle valve.* Cf: *concavo-convex.*

convexo-plane Convex on one side and flat on the other; e.g., said of a brachiopod shell having a convex *brachial valve* and a flat *pedicle valve.* Cf: *plano-convex.*

convex slope *waxing slope.*

conveyor-belt model A three-dimensional depiction of a mature *extratropical cyclone* and its corresponding *fronts* in terms of three interacting airstreams, often referred to as conveyor belts. This model developed out of a better understanding of upper-air circulation in the years since the *Norwegian cyclone model.*

convolute (con'-vo-lute) Coiled or wound together one part upon another; e.g. said of a coiled foraminiferal *test* in which the inner part of the last *whorl* extends to the center of the spiral and covers the inner whorls, or said of a coiled gastropod shell whose inner or earlier whorls are entirely concealed or embraced by the outer or later whorls. Cf: *involute; evolute; advolute.*

convolute bedding *convolute lamination.*

convoluted organ (con-vo-lut'-ed) A loose, calcareous, spicular, coiled structure within the calyx in many camerate crinoids.

convolute lamination A descriptive term used by Kuenen (1953) for the wavy, extremely disorganized, and markedly and intricately crumpled, twisted, or folded laminae that are confined within a single, relatively thin, well-defined, undeformed layer, that die out both upward and downward, and that are overlain and underlain by parallel undisturbed layers. The structure appears to result from deformation during deposition of sediments that become partially liquefied but still retain some cohesion. Convolutions may be truncated by erosion surfaces that are themselves convoluted, demonstrating the penecontemporaneous nature of the deformation. Axes of the convolutions generally have a preferred orientation normal to the paleocurrent. See also: *slip bedding; slump bedding; intraformational contortion.* Syn: *convolute bedding; convolution [sed struc]; contorted bedding; crinkled bedding; curly bedding; hassock structure; intrastratal flow structure.*

convolution [sed struc] (con-vo-lu'-tion) (a) The process of producing convolute bedding; the state of being convoluted. (b) A structure produced by convolution, such as a small-scale but intricate fold. (c) *convolute bedding.*

convolution [seis] (a) An operation in which two time series are combined to create a third time series; convolution is used in time-series analysis, including geophysical signal processing. Each element in one of the time series is replaced with the other time series scaled according to the amplitude of the element. (b) A change in wave shape as a result of passing through a linear *filter [seis];* the mathematical operation called "linear superposition". The Earth acts as a filter during the passage of seismic waves, and the movement that a seismic detector senses may be thought of as the result of the convolution of an Earth filter with the input seismic wave shape. Cf: *filtering.*

convolutional ball (con-vo-lu'-tion-al) A "nearly closed, more or less elliptical, concentrically laminated structure" (Ten Haaf, 1956, p.191) within a convolutely laminated bed. If the *convolute lamination* is not noticed, these features are easily mistaken for concretions or slump structures. Syn: *roll-up structure.*

convolutional model The model in which a seismic trace is given by the convolution of an embedded (equivalent) wavelet with a reflectivity function, often with random noise added. The convolutional model is implied in most seismic processing and interpretation.

convulsion (con-vul'-sion) *catastrophe.*

convulsionism (con-vul'-sion-ism) *catastrophism.*

convulsive event (con-vul'-sive) A violent disturbance of regional or even global extent. Because convulsive events affect a broad area, their sedimentologic consequences provide excellent chronologic markers in the stratigraphic record. "The term 'convulsive' is used to avoid the connotation of disaster and any doctrinal implications associated with the term 'Rcatastrophic'" (Clifton, 1985). Not to be confused with *convulsion* or *convulsionism.*

cookeite (cooke'-ite) A member of the *chlorite* group of minerals, in which the octahedral sheet within the 2:1 T-O-T layer is dioctahedral and the interlayer sheet is trioctahedral. It exists in monoclinic and triclinic polytypes. Cookeite is Li-rich relative to the species *sudoite*: $LiAl_4(Si_3Al)O_{10}(OH)_8$.

coolgardite (cool'-gard-ite) A mixture of coloradoite, sylvanite, and calaverite, found at Kalgoorlie in Western Australia.

cooling crack (cool'-ing) A joint that is formed as a result of cooling of an igneous rock. These can also be *microcracks* in which case a set of such microcracks constitutes the *rift* of a granite. Syn: *columnar joints.* Cf: *contraction fissure; rift [struc geol].*

cooling unit An ash-flow unit or series of such units that were deposited rapidly enough to share a common history of cooling, compaction, and welding (Smith, 1960, p.157).

coom (a) Var. of *cwm.* (b) An English term for a hollow in the side of a hill or mountain; a cove; a *combe.*

coomb (a) A var. of *combe,* a valley. (b) A var. of *cwm.*

coombe A var. of *combe,* a valley.

coombe rock An irregular mass of unstratified rock debris of any type accumulating as a result of solifluction; esp. the angular mass of unrolled and unweathered flints mixed with chalk rubble and other earthy matrix material, partly filling a dry valley (coombe), and spreading out onto the coastal plain, as in SW England. See also: *head [mass move]; rubble drift.* Syn: *elephant rock.*

coombsite A vitreous brown rhombohedral mineral: $K(Mn^{2+},Fe^{2+},Mg)_{13}(Si,Al)_{18}O_{42}(OH)_{14}$. It is an isotype of zussmannite.

coontail ore (coon'-tail) A term used in the Cave-in-Rock district of southern Illinois for a light- and dark-banded ore of sphalerite and fluorite.

cooperite (coop'-er-ite) A steel-gray tetragonal mineral: (Pt,Pd)S. It occurs in minute irregular grains in igneous rocks. A dimorph of braggite.

coordinate (co-or'-di-nate) n. Any one of a set of numbers designating linear and/or angular quantities that specify the position of a point on a line, in space, or on a given plane or other surface in relation to a given reference system; e.g. latitude and longitude are coordinates of a point on the Earth's surface. The term is usually used in the plural, esp. to designate the particular kind of reference system (such as "spherical coordinates", "plane coordinates", and "polar coordinates").

coordinated stasis A pattern described from the fossil record in which groups of coexisting species lineages exhibit joint morphologic and ecologic stability over prolonged intervals of geologic time separated by much shorter intervals of relatively abrupt changes in morphology and ecologic structure (Brett et al., 1996).

coordinated universal time (a) The time kept by a clock whose rate is controlled by atomic clocks so as to be as uniform as possible for one year, but with the rate chosen by the Bureau International de l'Heure and subject to change at the beginning of a calendar year. This time was used for many broadcast signals from 1960 through 1971. (b) The time kept by a clock controlled by atomic clocks and running at the correct rate (zero offset in frequency) but changed by the infrequent addition or deletion of 1 second to keep the time within $0.^s7$ of universal time 1 (UT1); adopted by the International Radio Consultive Committee in Geneva in February 1971 and became effective January 1, 1972. Abbrev: UTC. See also: *WWV.*

coordinate geometry A set of procedures for encoding and manipulating bearings, distances and angles of survey data into a graphic representation. COGO is commonly a subsystem of *GIS.* Abbrev: COGO.

coordinate system A reference system for defining points in space or on a particular surface by means of distances and/or angles with relation to designated axes, planes, or surfaces.

coordination number (co-or'-di-na'-tion) In crystallography, the number of nearest neighbor ions that surround a given ion in the crystal structure, e.g. four, six, or eight.

coorongite (coo'-rong-ite) A soft, brown, rubbery variety of *elaterite,* originating from deposits of Elaeophyton algae in salt water bodies near the Coorong in Australia; it may represent a stage in the formation of boghead coal. See also: *n'hangellite.*

cop A term used in England for the steep-sided top or summit of a hill; also, a small hill with a rounded top.

copal (co'-pal) An inclusive term for a wide variety of hard, brittle, semitransparent, yellowish to red *fossil resins* from various tropical trees (such as *Copaifera* and *Agathis*), being nearly insoluble in the usual solvents, and resembling amber in appearance; e.g. *Congo copal* and *kauri.* Copal also occurs as modern resinous exudations. Syn: *gum copal.*

copalite (co'-pal-ite) A clear, pale-yellow, dirty-gray, or dirty-brown fossil resin resembling copal in hardness, color, transparency, and difficult solubility in alcohol, containing succinic acid, and being much poorer in oxygen than most amber; e.g. the "Highgate resin" found in the blue Tertiary clay of Highgate Hill in London. Syn: *fossil copal.*

coparsite A metallic black orthorhombic mineral: $Cu_4O_2(As,V)O_4Cl$.

Copenhagen water (Co'-pen-ha'-gen) *standard seawater.*

copepod (co'-pe-pod) Any crustacean belonging to the class Copepoda, characterized by the absence of both a carapace and compound eyes. The only known pre-Pleistocene fossil copepods have been found in Miocene lake deposits.

Copernican (Co-per'-ni-can) (a) Pertaining to the youngest lunar topographic features and lithologic map units constituting a system of rocks mostly made up of scattered crater deposits and formed during the period of formation of ray craters (such as Copernicus) less than about 1.1 Ga (Wilhelms, 1987). (b) Said of the stratigraphic period during which the Copernican System was developed.

Copernican system A concept of planetary motion according to which the Earth rotates on its axis once a day and revolves in orbit about the Sun once a year. In this system all other planets also have Sun-centered orbits. It is named after the Polish astronomer Copernicus (d.1543).

cophenetic correlation A measure of the distortion introduced in hierarchical *cluster analysis* by comparison of the apparent similarities in the *dendrogram* to the original similarities between the objects.

copiapite (co'-pi-a'-pite) (a) A yellow, orange, or greenish-yellow triclinic mineral: $Fe^{2+}Fe^{3+}_4(SO_4)_6(OH)_2{\cdot}20H_2O$. Syn: *ferrocopiapite; yellow copperas; ihleite; knoxvillite.* (b) A group of minerals containing hydrous ferric sulfates, with the general formula $A_{0.7\text{-}1.0}Fe^{3+}_4(SO_4)_6(OH)_2{\cdot}20H_2O$, where A = Mg, Fe^{2+}, Cu^{2+}, Zn, Al, Fe^{3+} or Ca.

coppaelite (cop-pae'-lite) An olivine-free extrusive melilitite containing small clinopyroxene phenocrysts in a holocrystalline groundmass of melilite, clinopyroxene and phlogopite. The name, given by Sabatini in 1903, is for Coppaeli di Sotto, Umbria, Italy. Woolley et al. (1996) replace this name with kalsilite-phlogopite melilitite. Obsolete.

copper (cop'-per) A reddish or salmon-pink cubic mineral, the native metallic element Cu. It is ductile and malleable, a good conductor of heat and electricity, usually dull and tarnished, and formerly an important ore. Copper is the only metal that occurs native abundantly in large masses; it frequently occurs in dendritic clusters or mossy aggregates, in sheets, or in plates filling narrow cracks or fissures. It has many uses, notably as an electric conductor and as the base metal in brass, bronze, and other alloys.

copperas (cop'-per-as) (a) *melanterite.* (b) A name sometimes applied to other sulfate minerals, such as copiapite ("yellow copperas") and goslarite ("white copperas").

copper glance *chalcocite.*

copper manto *basaltic copper deposit.*

copper mica *chalcophyllite.*

copper nickel *nickeline.*

copper pyrites *chalcopyrite.*

copper shale *sediment-hosted copper deposit.*

copper uranite *torbernite.*

copper vitriol *chalcanthite.*

coppice (cop'-pice) (a) A thicket of small trees or shrubs. (b) *coral coppice.*

coppice mound *shrub-coppice dune.*

coprocoenosis (cop'-ro-coe-no'-sis) An accumulation of microvertebrate fossil remains that first passed through the digestive tracts of carnivores (mainly mammalian) and were deposited as fecal droppings and subsequently preserved. (Mellett, 1974). Etymol: Greek "copros", "dung", + "koinos", "common".

coprogenic (cop-ro-gen'-ic) Said of a deposit formed of animal excrement. Etymol: Greek "kopros", "dung".

coprolite (cop'-ro-lite) (a) The fossilized excrement of vertebrates such as fishes, reptiles, and mammals; larger than a *fecal pellet,* measuring up to 20 cm in length, characterized by an ovoid to elongate form, a surface marked by annular convolutions, and a brown or black color; commonly composed largely of calcium phosphate; petrified excrement. The term is incorrectly used to refer to desiccated or fresh fecal remains. Term introduced by Buckland (1829a). (b) An English term applied commercially and popularly to any *phosphatic nodule* mined for fertilizer.

copropalynology (cop-ro-pal-y-nol'-o-gy) The study of palynomorphs contained in coprolites.

copropel (cop'-ro-pel) A term introduced by Swain and Prokopovich (1954, p.1184) to replace the vaguely applied term *gyttja,* and signifying a "dark-brown or gray coprogenic ooze, containing chitinous exoskeletons of benthonic arthropods in addition to reworked organic matter".

copula (cop'-u-la) A short, hollow, funnel-like tube in the position of the mucron at the base of the chamber of chain-forming chitinozoans. Copulas connect to prosomes of preceding chambers to form chains.

coquandite A colorless or white triclinic mineral: $Sb^{3+}_6O_8(SO_4){\cdot}H_2O$.

coquimbite (co-quim'-bite) A pale violet, yellowish, or greenish trigonal mineral: $Fe^{3+}_2(SO_4)_3{\cdot}9H_2O$. It sometimes contains appreciable aluminum, and it is dimorphous with paracoquimbite. Syn: *white copperas.*

coquina (co-qui'-na [ko-kee'-na]) A detrital limestone composed wholly or chiefly of mechanically sorted fossil debris that experienced abrasion and transport before reaching the depositional site. Wentworth (1935, p.244) recommended a particle size greater than 2 mm. Etymol: Spanish, "cockle, shellfish". Cf: *coquinoid limestone; micrococquina; coquinite.*

coquinite (co-qui'-nite) Compact, well-indurated, and firmly cemented equivalent of coquina.

coquinoid limestone (co-qui'-noid) A limestone consisting of coarse, unsorted, and often unbroken shelly materials that have accumulated in place without subsequent transportation or agitation, and generally having a fine-grained matrix. It is autochthonous, unlike the allochthonous *coquina;* under certain conditions it can develop into a biostrome.

coracite (cor'-a-cite) *uraninite.*

coral (cor'-al) (a) A general name for any of a large group of bottom-dwelling, sessile, marine invertebrate organisms (polyps) that belong to the class Anthozoa (phylum Cnidaria (formerly Coelenterata)), are common in warm intertropical modern seas and abundant in the fossil record in all periods later than the Cambrian, produce external skeletons of calcium carbonate, and exist as solitary individuals or grow in colonies. (b) A hard calcareous substance consisting of the continuous skeleton secreted by coral polyps for their support and habitation, and found in single specimens growing plantlike on the sea bottom or in extensive solidified accumulations (coral reefs). Also, any marine deposit like coral resulting from vital activities of various organisms (such as certain algae, or bryozoans and worms). (c) A piece of coral; e.g. "precious coral", a semitranslucent to opaque mass usually red to orange red, but sometimes white, cream, brown, blue, or black.

coral cap A thick deposit of coral-reef material overlying material of noncoral origin. Cf: *reef cap.*

coral coppice A *coral thicket* within which broken coral branches have accumulated in significant quantities and thereby thickly carpet the sea floor around the bases of the colonies.

coralgal (cor-al'-gal) Said of a firm carbonate rock formed by an intergrowth of frame-building corals and algae (esp. coralline algae). The material so formed is an excellent sediment binder in a coral reef. Cf: *bryalgal.*

coralgal ridge A low ridge or elevated margin at the seaward (outer) edge of a reef flat, composed of the calcium-carbonate secretions of actively growing calcareous algae thickly encrusting and thus binding coral and shell rubble into a limestone mass. Cf: *algal ridge.*

coral head (a) A single massive, rounded coral colony. (b) A rounded, massive, often knobby or mushroom-shaped protuberance or growth of coral, usually forming on the submerged part of a coral reef, and frequently large enough to be dangerous to navigation; a small *reef patch* of coralline material. Syn: *coral knob; coral knoll; bommy; hardhead.*

coral horse An elongate remnant of a former reef tract, characterized by a flat top (or by a flat side if it has been undercut and tilted); a *coral head* formed by solution and dismemberment of a former reef platform.

coral island (a) A coral reef that appears above sea level, situated far from any other kind of land. (b) An oceanic island formed from coral accumulations lying atop volcanic peaks.

coral key A subaerial mound of coral sand and associated skeletal debris resting on a flat coral reef. Variant spellings: coral cay; coral kay.

coral knob *coral head.*

coral knoll *coral head.*

coralla (co-ral'-la) Plural of *corallum.*

Corallian (Co-ral'-li-an) An obsolete European stage: Upper Jurassic (above Callovian, below Kimmeridgian). Equivalent to *Oxfordian.*

coralline (cor'-al-line) n. Any organism that resembles a coral in forming a massive calcareous skeleton or base, such as certain algae or stromatoporoids. adj. Pertaining to, composed of, or having the structure of corals, as coralline limestone.

coralline algae *Calcareous algae* that form encrustations resembling coral.

corallite (cor'-al-lite) The calcareous exoskeleton formed by an individual coral *polyp,* consisting of walls, septa, and accessory structures such as tabulae and dissepiments. It is embedded in the general structure of the *corallum.*

corallith (cor'-al-lith) A subspheroidal, unattached coral colony that can be readily moved or rolled about (Glynn, 1974, p.184). Cf: *rhodolith; oncolite.*

coralloid (cor'-al-loid) A term describing a variety of nodular, globular, botryoidal, or coral-like *speleothems* with concentric crystal growth (Hill and Forti, 1997, p.257). Cf: *popcorn.*

corallum (co-ral'-lum) The calcareous *exoskeleton* of a coral colony, or the *corallite* of a solitary coral; the entire skeleton of a *coral.* Pl: *coralla.*

coral pavement A hard ground of limestone bearing scattered corals or coral colonies.

coral pinnacle *pinnacle [reefs].*

coral pipe A *speleothem* which forms as a lining over silt or mud pillars.

coral reef A coral-algal or coral-dominated *organic reef;* a mound or ridge of in-place coral colonies and accumulated skeletal fragments, carbonate sand, and limestone resulting from organic secretion of calcium carbonate that lithifies colonies and sands. A coral reef is built up around a potentially wave- and surf-resistant framework, especially of coral colonies but often including many algae; the framework may constitute less than half of the reef volume. Coral reefs occur today throughout the tropics, wherever the temperature is suitable (generally above about 18°C, a winter minimum).

coral-reef coast A coast formed by deposits of coral and algae, partly exposed at low tide, and characterized by reefs built upward from a submarine floor or outward from the margin of a land area.

coral thicket An aggregation of openly branching coral colonies, closely spaced and covering much of the sea bottom; esp. an aggregation in which the sea floor between the corals' bases is still largely exposed and not yet covered by coral debris. Such thickets range from shallow to deep, and from tropical to sub-polar. Cf: *thicket reef; coral coppice.*

corange line (co-range') A line on a chart joining points of equal tide range.

corbiculoid (cor-bic'-u-loid) Said of heterodont dentition of a bivalve mollusk with three cardinal teeth in each valve, the middle tooth of the right valve occupying a median position below the beaks. Cf: *lucinoid.* Obsolete syn: *cyrenoid.*

cordate (cor'-date) Said of structures, e.g. certain leaves, that are heart-shaped. Cf: *obcordate.*

corded pahoehoe (cord'-ed) The typical kind of *pahoehoe,* having a surface resembling coils of rope.

corderoite (cor-de′-ro-ite) A light orange-pink cubic mineral: $Hg_3S_2(Cl,Br)_2$. The color darkens to light gray or black on exposure to light.

cordierite (cor′-di-er-ite′) A variously colored orthorhombic mineral: $(Mg,Fe^{2+})_2Al_4Si_5O_{18}$. It exhibits strong pleochroism, is easily altered by exposure, and is an accessory mineral in granites and a common constituent in metamorphic rocks formed under low pressure. Syn: *iolite; dichroite.*

cordierite-amphibolite facies (cor′-di-er-ite′-am-phib′-o-lite) A subdivision of the *amphibolite facies* in which pelitic rocks characteristically contain andalusite, cordierite, or sillimanite, but not kyanite or almandine (Winkler, 1967). It represents the low-pressure part of the amphibolite facies of Eskola. Rarely used. Cf: *hornblende-hornfels facies.*

cordillera (cor-dil-le′-ra, cor-dil′-le-ra) (a) A comprehensive term for an extensive series or broad assemblage of more or less parallel ranges, systems, and chains of mountains (together with their associated valleys, basins, plains, plateaus, rivers, and lakes), the component parts having various trends but the mass itself having one general direction; esp. the main mountain axis of a continent, as the great mountain region of western North America from the eastern face of the Rocky Mountains to the Pacific Ocean, or the parallel chains of the Andes in South America; a mountain province. (b) An individual *mountain chain* with closely connected, distinct summits resembling the strands of a rope or the links of a chain; e.g. one of the parallel chains of the Rocky Mountains. (c) A term also used in South America for an individual *mountain range.* Etymol: Spanish, "chain or range of mountains", from Latin "chorda", "cord".

Cordilleran vein-type deposit (Cor-dil-le′-ran) One of a group of hydrothermal deposits of base and precious metals, chiefly Cu-W-Pb-Zn-Ag-Au, which are strongly vein-controlled and lack association with porphyry base-metal deposits. They may involve open-space filling, replacement or both; are typically mesothermal; and occur in Cordilleran continental margins. Examples: Mayflower, Utah; Magma, Arizona; Casapalca, Peru.

cordon littoral French term for beach, including both its upper and lower parts.

cordylite-(Ce) (cor′-dyl-ite) A colorless to wax-yellow hexagonal mineral: $(Ce,La)_2Ba(CO_3)_3F_2$.

core [arch] In archaeology, the interior of a pebble from which *flakes* have been removed to make a stone tool.

core [drill] n. A full-diameter core is a cylindrical section of rock, usually 5-10 cm in diameter and up to several meters in length, taken as a sample of the interval penetrated by a *core bit,* and brought to the surface for geologic examination and/or laboratory analysis. A sidewall core is only 2.5 cm in diameter and is taken from the side of a well. Cf: *core [oceanog].* v. To obtain a core in drilling.

core [eng] A wall or structure of impervious material forming the central part of an embankment, dike, or dam, the outer parts of which are pervious. Syn: *core wall.*

core [interior Earth] The central zone or nucleus of the Earth's interior, below the *Gutenberg discontinuity* at a depth of 2,890 km. It is divided into an *inner core* and an *outer core,* and is equivalent to the E, F, and G layers. Since only compressional waves propagate in the outer core, it is a fluid. The inner core, having a radius of approximately 1,200 km, is solid, as shear waves have been observed to propagate through it. The magnetic field originates within the core.

core [oceanog] A relatively undisturbed, cylindrical sample of ocean-bottom sediment collected by an oceanographic *corer.* See also: *box corer; piston corer; gravity corer.*

core [struc geol] The inner or central part of a fold, especially of a folded structure that includes some sort of structural break. Cf: *envelope.*

core and mantle structure The microstructure consisting of a central core of original grain grading outward into subgrains and/or newly recrystallized grains. It is indicative of *subgrain rotation recrystallization.*

core-and-shell structure A term used by McKee (1954, p.65) for a sedimentary structure resembling a concretion, developed in massive silty mudstone, and characterized by a rounded lump (core), ovoid to elongate and 3 to 30 cm in diameter, surrounded by a series of concentric layers (shells) each with a thickness ranging from 3 to 12 mm and appearing to be due to shrinkage in the uniform structureless mud. It is characteristic of subaerial conditions following flooding in a region of considerable aridity.

core barrel (a) Two concentric tubes above the bit of a *core drill,* the outer rotating with the bit, the inner receiving and preserving a continuous section or core of the material penetrated. (b) The tubular section of a *corer,* in which ocean-bottom sediments are collected either directly in the tube or in a plastic liner placed inside it.

core bit A hollow, cylindrical *drill bit* for cutting a full-diameter core of rock in a drill hole; the cutting end of a *core drill.*

core box The wooden, metal, or cardboard box divided into narrow parallel sections, used to store the cores at the surface as they are extracted from a *core barrel.*

core catcher A device used in rotary drilling, commonly consisting of upward-directed metal leaves on a ring in a *core barrel,* designed to prevent cored rock or sediment from slipping from the core barrel when it is raised to the surface.

core complex *metamorphic core complex [meta].*

cored bomb A type of *bomb [pyroclast]* that has a core of nonvolcanic rock or already solidified lava, around which the lava has molded itself. Obsolete syn: *perilith.*

core drill (a) A rotary *drilling rig* that cuts and brings to the surface a core from the drill hole. It is equipped with a *core bit* and a *core barrel.* (b) A lightweight, usually mobile drill that uses *tubing* instead of *drill pipe* and that can core down from the grass roots.

core hole Any hole drilled for the purpose of obtaining cores; loosely, a well, generally shallow, drilled for geological information only. Syn: *core test.*

core-mantle boundary Boundary marking the separation between the Earth's core (radius ~3,480 km) and the mantle at ~2,890 km depth. Abbrev: CMB. Syn: *core-mantle boundary region.* Cf: *Gutenberg discontinuity.*

core-mantle boundary region *core-mantle boundary.* Abbrev: CMBR.

core method A method for tracing a water mass or type from its origin across an area through which it is spreading, by tracing the path of minimum change in a characteristic parameter such as temperature, salinity or oxygen content.

corer (cor′-er) An ocean-bottom sampler that is a metal or plastic cylinder or box, lowered by wire and driven into the ocean floor by impact of attached weights. See also: *piston corer; gravity corer; box corer.* Cf: *dredge [oceanog]; grab sampler.*

core record A record showing the depth, character, lithology, porosity, permeability, and fluid content of cores.

core recovery The amount of the drilled rock withdrawn as core in core drilling, generally expressed as a percentage of the total length of the interval cored.

core sample One or several pieces of whole or split parts of a core, selected for analysis; a sample obtained by coring.

core-stone An ellipsoidal joint block of granite formed by subsurface weathering (Linton, 1955). Cf: *elephant rock; tor.*

core test *core hole.*

core texture *atoll texture.*

core wall *core [eng].*

Corey shape factor A measure of the sphericity of a sediment particle equal to $D_S/(D_LD_I)^{1/2}$, where D_S = shortest diameter, D_L = longest diameter, and D_I = intermediate diameter (Corey, 1949). Cf: maximum projection sphericity.

corindon (co-rin′-don) *corundum.*

Coriolis acceleration (Co-ri-o′-lis) The acceleration of a body in motion with respect to the Earth resulting from the rotation of the Earth, as seen by an observer on the Earth. A moving gravimeter has a Coriolis acceleration. The vertical component of the Coriolis acceleration is the *Eötvös effect.*

Coriolis force The apparent deflective component of the centrifugal force produced by the rotation of the Earth. It is named for Gustave Gaspard Coriolis (1792-1843), French mathematician who studied its effects. Cf: *pole-fleeing force.* See also: *Ferrel's law.* Also spelled: *coriolis force.* Syn: *geostrophic force.*

Coriolis parameter Two times the sine of the latitude times earth's rotation rate in radians per second. Cf: *planetary vorticity.*

cork A protective layer of dead suberized tissue on the outside of older stems and roots, which is formed by a *cork cambium.* See also: *periderm [bot].* Cf: *phelloderm.*

cork cambium In woody plants, a *cambium* or lateral *meristem* in the outer bark that produces cork (phellem) to the outside and phelloderm toward the inside. Syn: *phellogen.* Cf: *periderm [bot]; vascular cambium.*

corkite (cork′-ite) A pale yellow to green rhombohedral mineral: $PbFe^{3+}_3(PO_4)(SO_4)(OH)_6$. A member of the *beudantite* group.

corkscrew flute cast (cork′-screw) A corkscrew-shaped flute cast, with a "twisted" beak at the upcurrent end.

cornelian (cor-nel′-i-an) *carnelian.*

corner (cor′-ner) (a) A point of intersection of two boundary lines of a tract of land; esp. a point on the Earth's surface, determined by surveying, that marks an extremity of a boundary of a subdivision of the public lands, usually at the intersection of two or more surveyed lines. Corners are described in terms of the points they represent; e.g. "township corner" located at the extremity of a township boundary. See also: *witness corner.* (b) A term that is often incorrectly used to denote the physical station, or monument, erected to mark the corner.

corner accessory A physical object adjacent to a corner, to which such a corner is referred for its future identification or restoration; e.g. a *bearing tree.* Corner accessories include mounds, pits, ledges, rocks and other natural features to which distances or directions, or both, from the corner or monument are recorded.

corner frequency (a) The frequency at which a graph of the logarithm of the amplitude of the frequency response has an abrupt change of slope. Syn: *break frequency.* (b) For a seismic wave generated from an earthquake, the frequency at which the spectral field begins to decrease. It is related theoretically to the dimensions or other properties of the source.

corner mark *register mark.*

corner prism *corner reflector.*

corner reflector A special optical reflector used with electro-optical distance-measuring instruments to return a light beam to its source. A corner is cut from a highly polished cube of glass in which the three intersecting planes are precisely perpendicular. Light enters the cube through the cut surface and is reflected to the source by the polished cube faces. Syn: *corner prism; retrodirective prism.*

corner tick *register mark.*

cornet halite Flat-topped, upward-widening cornet- or cone-shaped growth habit of halite produced by crystallization at the floor of a brine body with growth on upward-facing crystal faces. Crystals are skeletal in form and marked by copious fluid inclusions, and the hollow centers are commonly filled by clear halite.

cornetite (cor'-ne-tite) A peacock-blue orthorhombic mineral: $Cu_3(PO_4)(OH)_3$.

cornice (cor'-nice) An overhanging mass of snow on the edge of a steep ridge or cliff face. Syn: *snow cornice.*

cornice glacier *cliff glacier.*

cornieule (cor-nieule') *cargneule.*

Cornish diamond (Cor'-nish) A rock crystal (clear quartz) from Cornwall, England.

Cornish stone A variety of china stone composed of feldspar, mica, and quartz and used as a bond in the manufacture of pottery. Syn: *Cornwall stone.*

corn snow *Spring snow* consisting of large, wet crystals.

cornstone (corn'-stone) An obsolete term for *calcrete*. See also: *caliche [soil].*

cornubianite (cor-nu'-bi-a-nite') A *hornfels* formed by contact metamorphism, and consisting of micas, quartz, and feldspar (Holmes, 1928, p.69). Rarely used. Cf: *leptynolite.*

cornubite (cor-nu'-bite) A light to dark green triclinic mineral: $Cu_5(AsO_4)_2(OH)_4$. A dimorph of cornwallite.

cornulitid (cor-nu-lit'-id) An invertebrate animal known only by a tapering, flexuous tube composed of calcium carbonate, with circular cross section, and with transverse rings developing in later growth stages. Cornulitids belong to the family Cornulitidae and are taxonomically unassigned in the TIP (1962, pt. W, p.137), but were originally described as being closely related to certain annelids. Range, Middle Ordovician to Mississippian.

cornuspirine (cor-nu-spi'-rine) Having a tubelike planispirally coiled test; specif. pertaining to the foraminifer *Cyclogyra* ("*Cornuspira*").

cornutate (cor'-nu-tate) Said of a diatom valve that has hornlike extensions.

cornwallite (corn'-wall-ite) An emerald-green to blackish-green monoclinic mineral: $Cu_5(AsO_4)_2(OH)_4$. A dimorph of cornubite.

Cornwall stone (Corn'-wall) *Cornish stone.*

corona [bot] (co-ro'-na) In charophyte algae, the outer layer of the nucule, formed by cells cut off from the sheath cells of the female sex organ.

corona [paleont] The principal skeletal structure or main part of the calcareous *test* of an echinoid, including all ambulacra and interambulacra, but excluding the apical system of plates, the periproctal and peristomial systems, Aristotle's lantern, and appendages.

corona [palyn] A more or less equatorial extension of a spore, similar in disposition to a *zone [palyn]* but divided in fringelike fashion.

corona [petrology] A zone of minerals, usually with radial arrangement, around another mineral. It is a general term that has been applied to reaction rims, corrosion rims, and originally crystallized minerals. Cf: *kelyphytic rim.*

corona [struc geol] A term established by the International Astronomical Union for a circular to elliptical structural feature, often displaying an elevated central plain surrounded by a shallow topographic moat on planetary surfaces (USGS, 2003). Interpreted to be related to surface manifestations of the impingement of mantle plumes on a thin lithospheric crust. Originally applied to features on Venus, but also applicable to other planetary surfaces. Syn: *ovoid.* Pl: coronae.

coronadite (co-ro-na'-dite) A submetallic dark gray to black monoclinic mineral: $Pb(Mn^{4+},Mn^{2+})_8O_{16}$. It is a member of the *hollandite* group.

coronal (co-ro'-nal) (a) Pertaining to a *corona.* (b) Pertaining to certain openings of organisms, such as fringing an *osculum* of a sponge; e.g. "coronal pores" or tiny openings at the periphery of shields in acantharian radiolarians.

coronate (cor'-o-nate) Having or resembling a crown; esp. said of a cephalopod whorl section resembling a crown as viewed from the side, or said of a gastropod shell when the spire is surrounded by a row of spines or tubercles.

corona texture A concentric ring of one or several minerals enclosing a core mineral grain. Syn: *reaction rim.* Cf: *kelyphitic.*

coronite (cor'-o-nite) A rock containing mineral grains surrounded by coronas.

coronula (co-ron'-u-la) In charophyte oogonia, one or two tiers of small cells resting on the apical ends of the enveloping cells to form a more or less erect, elevated ring around the summit.

corpogelinite A *gelovitrinite* maceral comprising structureless bodies of humic cell fillings, in situ or isolated (ICCP, 1998).

corrasion (cor-ra'-sion) (a) A process of erosion whereby rocks and soil are mechanically removed or worn away by the abrasive action of solid materials moved along by wind, waves, running water, glaciers, or gravity; e.g., the wearing-away of the bed and banks of a stream by the cutting, scraping, scratching, and scouring effects of a sediment load carried by the stream, or the sawing and grinding action of sand, gravel, and boulders hurled by waves and currents against a shore. The term has also been used for the loosening of rock material by the impact of rushing water itself, and was used by Penck (1953, p.112) for the "freeing of loosened rock fragments from their place of origin". The term "abrasion" is essentially synonymous. Syn: *mechanical erosion.* (b) A term sometimes used as a syn. of *attrition.* (c) A term formerly used as a syn. of *corrosion,* or as including the work of corrosion. The term was first used by Powell (1875, p.205) for channel cutting or the deepening of any valley floor and was extended by Gilbert (1877, p.101) to the work of all running water (including lateral corrasion). Verb: corrade.

corrasion valley An elongated hollow or furrow excavated by the corrading action of a moving mass of material (Penck, 1953, p.112).

correction (cor-rec'-tion) In the analysis of physical measurements, a quantity that is applied to a measured quantity in order to negate the effects of a known interference, or to reduce the measurement to some arbitrary standard.

correction line A *standard parallel* in the U.S. Public Land Survey system.

correlate (cor'-re-late) v. To show correspondence in character and stratigraphic position between such geologic phenomena as formations or fossil faunas of two or more separated areas. adj. Belonging to the same stratigraphic position or level.

correlation [geol] (cor-re-la'-tion) (a) Demonstration of correspondence in character and in stratigraphic position between geographically separated stratigraphic sections or rock bodies (ISSC, 1994, p.15). "There are different kinds of correlation depending on the feature to be emphasized. Lithologic correlation demonstrates correspondence in lithologic character and lithostratigraphic position; a correlation of two fossil-bearing beds demonstrates correspondence in their fossil content and in their biostratigraphic position; and chronocorrelation demonstrates correspondence in age and in chronostratigraphic position" (ISG, 1976, p.14). See also: *stratigraphic correlation.* (b) The condition or fact of being correlated, such as the correspondence between two or more geologic phenomena, features, or events.

correlation [geomorph] The concept applied by Penck (1953, p.419) which states that strata formed from the products of denudation are related to the period during which the denudation occurred.

correlation [seis] (a) Identifying a phase of a seismic record as representing the same phase on another record, thus relating reflections from the same stratigraphic sequence or refractions from the same marker horizon. (b) A measurement of the degree of linear relationship between a pair of traces, or of the extent to which one can be considered as a linear function of the other. Cf: *autocorrelation [seis]; crosscorrelation [seis].*

correlation [stat] The intensity of association or interdependence between two or more mathematical variables; e.g. *linear correlation.*

correlation coefficient A measure of the linear relationship between two variables. It is defined as the ratio of the covariance to the square root of the product of the two variances and ranges from -1 to +1, with 0 indicating no relationship between the two variables. The correlation coefficient is estimated by the *Pearson product-moment correlation* and other statistics.

correlation log The generic term applied to any *well log* curve used in identifying equivalent subsurface geologic sections or individual lithologic units in wells. The *spontaneous-potential curve* and the *gamma-ray log* are often used because they can usually be compared directly to distinguish sandstone or limestone beds from shale intervals.

correlation shooting A seismic surveying method in which isolated profiles are shot and correlated to obtain relative structural positions of the horizons mapped. Cf: *continuous profiling.*

correlative conformity In sequence stratigraphy, a surface separating younger from older strata along which there is no evidence of erosion (neither subaerial nor submarine) or nondeposition, and along which no significant hiatus is indicated. It includes surfaces onto which there is very slow deposition or low rates of sediment accumulation, with long periods of geologic time being represented by very thin deposits (Mitchum et al., 1977; Van Wagoner et al., 1987, 1988, 1990). The correlative conformity is the basinward extension of the unconformity. Together the unconformity and its correlative conformity compose the sequence boundary. Cf: *correlative surface.*

correlative estimate (cor-rel'-a-tive) In hydraulics, a likely discharge value estimated by correlation for a particular span of time.

correlative surface In sequence stratigraphy, the landward extension of the marine-flooding surface defining the parasequence boundary. The correlative surface in the coastal plain is not marked by significant subaerial erosion, stream rejuvenation, downward shift in coastal onlap, or onlap of overlying strata; it may be marked by evidence of subaerial exposure such as soil horizons, coals, or root horizons. This surface can be traced over a significant regional extent. The correlative surface on the shelf is a conformable surface with no significant hiatus indicated and is marked by thin pelagic or hemipelagic deposits. These deposits include thin carbonates, organic-rich mudstones, glauconites, and volcanic ashes indicating terrigenous-sediment starvation. Strata across correlative surfaces may not indicate a change in water depth. In some cases the correlative surfaces in the coastal plain or on the shelf can be identified only by correlating updip or downdip from a marine-flooding surface (Van Wagoner et al., 1990). Cf: *correlative conformity.*

corrensite (cor′-rens-ite) A yellowish green mixed-layer mineral consisting of interstratified smectite and chlorite layers: $(Mg,Fe,Al)_9(Si,Al)_8O_{20}(OH)_{10}\cdot nH_2O$.
correspondence analysis (cor-re-spond′-ence) A method of factor analysis primarily designed for frequency data. Data are standardized across rows and columns before factoring.
corridor [karst] (cor′-ri-dor) *solution corridor.*
corridor [remote sensing] A buffer along a line feature (Krzanowski et al., 1993). See also: *proximity analysis.*
corridor [speleo] In a cave, a traversable *passage [speleo]* that is generally large, long, tunnel-like, narrow and straight. See also: *gallery [speleo].*
corridor stack A summation of the traces in an upgoing vertical seismic profile (VSP) that has been processed to retain only primary reflection events and that has been time-shifted to their two-way arrival times at the surface. Summation is over a data window (a corridor) beginning at the first-break time and usually ending 200-300 ms later. This eliminates most multiples (Sheriff, 2002).
corrie (cor′-rie) A term used in Scotland as a syn. of *cirque [glac geol]*, esp. a *hanging cirque.* Etymol: Gaelic *coire,* "kettle". Also spelled: *coire; corry.*
corrie glacier *cirque glacier.*
corrosion [geomorph] (cor-ro′-sion) (a) A process of erosion whereby rocks and soil are removed or worn away by natural chemical processes, esp. by the solvent action of running water, but also by other reactions such as hydrolysis, hydration, carbonation, and oxidation. Syn: *chemical erosion.* (b) A term formerly used interchangeably with *corrasion* for the erosion ("gnawing away") of land or rock, including both mechanical and chemical processes. The mechanical part is now properly restricted to "corrasion" and the chemical to "corrosion". Verb: corrode.
corrosion [petrology] The partial resorption, dissolution, fusion, modification, or eating-away of the outer parts of early-formed crystals (such as quartz phenocrysts), or of xenoliths, by the solvent action of the residual magma in which they are contained. It sometimes results in the formation of corrosion borders. See also: *embayment [petrology].*
corrosion rim One of a series of borders of one or more secondary minerals around an original crystal, representing the modification of a phenocryst due to the corrosive action of its magma. Cf: *reaction rim.* Syn: *corrosion zone; resorption rim.*
corrosion surface A pitted, irregular bedding surface found only in certain carbonate sediments, characterized by a black manganiferous stain, and presumed to result from cessation of lime deposition and from submarine solution or resorption of some of the previously deposited materials. Syn: *corrosion zone.*
corrosion zone (a) *corrosion surface.* (b) *corrosion border.*
corrugate (cor′-ru-gate) A descriptive term for a *sculpture* of ribs, ridges, or folds, the surfaces of which are more or less irregular due to superimposed thickenings, bumps, or verrucae (wartlike projections). The term is partly synonymous with *rugulate* and less regular than *cicatricose.*
corrugated lamination (cor′-ru-gat′-ed) A form of convolute bedding that differs from *convolute lamination* in that the lamination is intricately contorted and does not show a regular subparallel orientation of anticlines and synclines.
corrugated ripple A *longitudinal ripple* with a sigmoidal profile, an equally rounded and usually symmetric crest and trough, and a ridge that rarely branches (Kelling, 1958, p.124). Cf: *mud-ridge ripple.*
corrugated ripple mark *corrugated ripple.*
corrugation (cor-ru-ga′-tion) (a) The process of deforming or crumpling sedimentary beds into small-scale folds, wrinkles, or furrows; esp. *intraformational corrugation.* Also, the state of being corrugated. It occurs on a smaller scale than *contortion.* (b) A structure produced by corrugation.
corry (cor′-ry) *corrie.*
corsite (cors′-ite) An orbicular *diorite* or *gabbro.* Cf: *esboite.* The name is from Corsica. Obsolete.
cortex (cor′-tex) (a) A layer of the *ectosome* of a sponge, consolidated by a distinctive skeleton, either organic or mineral, or both. Also, a layer of specialized sponge spicules or modified structure at the outer surface of the skeleton. (b) An outer coenenchymal layer of certain octocorals (esp. the Gorgonacea); the outer horny layer of the axis in the Holaxonia, as opposed to its medulla. (c) The dense, differentiated outer layer of an echinoid spine, usually bearing ornamentation. It is nonliving material on a mature spine. (d) The tissue, composed mostly of parenchyma cells, between the central vascular cylinder and the epidermis of a stem or root (Cronquist, 1961, p.873). (e) The outer layer of an organ or structure (as in the brain).
cortical (cor′-ti-cal) Pertaining to, located in or on, or consisting of a *cortex* or outer part of an invertebrate; e.g. "cortical shell" (outermost of the concentric shells of a spumellinid radiolarian).
cortical fabric One of three major types of materials recognized in electron-microscope study of graptolithine periderm as a fundamental structural element in the *periderm.* Cortical fabric is layered, each layer being formed of relatively long fibers arranged in parallel; the fibers in one layer are commonly oriented at an angle to fibers in the subjacent and superjacent layers (Urbanek and Towe, 1974, p.4-5). Cf: *fusellar fabric; sheet fabric.*
cortical tissue The outer part of the *periderm* in graptolithines; it is formed of layers of *cortical fabric, sheet fabric,* and, rarely, *fusellar fabric* (Urbanek and Towe, 1974, p.5).
cortlandtite (cort′-landt-ite) A *peridotite* that contains hornblende poikilitically enclosing olivine. Syn: *hudsonite.* The name, given by Williams in 1886, is from the Cortlandt complex near Peekskill, New York. Obsolete.
corundophilite (co-run′-do-phi′-lite) A ferroan variety of clinochlore.
corundum (co-run′-dum) A mineral: Al_2O_3. It occurs as shapeless grains and masses, or as variously colored rhombohedral crystals (such as prisms or tapering hexagonal pyramids), including the gem varieties such as ruby and sapphire. Corundum is extremely tough, has a hardness of 9 on the Mohs scale, and is used industrially as an abrasive. See also: *emery.* Syn: *adamantine spar; diamond spar; corindon.*
corvusite (cor′-vus-ite) A submetallic blue-black, brown, or purplish mineral: $(Na,Ca,K)(V^{5+},V^{4+},Fe^{2+})O_{20}\cdot 4H_2O$. It is classed as a "vanadium bronze", and is an ore of vanadium. Syn: *blue-black ore.*
Coryell-Masuda diagram A logarithmic plot of rare-earth-element concentrations normalized to a standard, versus atomic number. Named for two researchers who independently developed this format.
corynite (cor′-y-nite) A variety of gersdorffite containing antimony.
corystosperm (co-rys′-to-sperm) A member of a group of Mesozoic *seed ferns* widespread in Gondwana during the Triassic, characterized by pinnate fronds and seeds borne in uniovulate cupules. The most common foliage genus is *Dicroidium,* which has been used as an index fossil for the Triassic of Gondwana.
cosalite (co′-sa-lite) A lead-gray or steel-gray orthorhombic mineral: $Pb_2Bi_2S_5$. It often contains copper.
cosedimentation (co-sed′-i-men-ta′-tion) Contemporaneous sedimentation, such as the precipitation of iron-bearing minerals at the same time as deposition of fine argillaceous sediment to produce a shale or mudstone with a high iron content (Pettijohn, 1957, p.368).
coseismal line (co-seis′-mal) A line connecting points on the Earth's surface at which an earthquake wave has arrived simultaneously. Cf: *isoseismal line.*
coset (co′-set) A term proposed by McKee and Weir (1953, p.384) for a sedimentary unit composed of two or more sets, either of strata or of cross-strata, "separated from other strata or cross-strata by original flat surfaces of erosion, nondeposition, or abrupt change in character". See also: *composite set.*
cosine theta coefficient Properties measured on objects can be considered as coordinates of vectors in multidimensional space. The vectors of two similar objects will have similar coordinates so the vectors will be close together and form a narrow angle at their origin; dissimilar objects will differ in their coordinates and their vectors will form a large angle, so the cosine of the angle can be used as a measure of similarity between the objects represented by the vectors. The cosine theta coefficient is used widely in *numerical taxonomy* as a measure of similarity between objects.
coskrenite-(Ce) A pale pink, blue, or cream triclinic oxalate mineral: $(Ce,Nd,La)_2(SO_4)_2(C_2O_4)\cdot 8H_2O$. It exhibits a weak alexandrite-like effect.
cosmic (cos′-mic) Of, from, or relating to the cosmos; esp. pertaining to phenomena or features that occur, exist, or originate beyond the Earth's atmosphere or in the universe in contrast to the Earth alone, such as "cosmic sediment" found in the oceans and containing particles of extraterrestrial origin.
cosmic dust (a) Very finely divided particles of solid matter in interplanetary space; (b) The smallest particles that enter the Earth's atmosphere from interplanetary or interstellar space. Because of their high surface/volume ratios, many are decelerated without melting. These reach the Earth's surface (and can be collected from the sea floor or from polar ice) in an essentially unaltered state. Cf: *meteoritic dust, zodiacal dust.*
cosmic radiation Natural radiation derived from high-energy particles and photons coming from outer space, which bombard the Earth's atmosphere. Primary cosmic rays (atomic nuclei) are almost all absorbed in the upper atmosphere; secondary cosmic rays, which have less energy, provide a part of natural background radiation.
cosmic spherule A small particle formed by the detachment of a molten droplet from the surface of a meteorite during atmospheric passage; first collected from deep-sea sediments. See also: *magnetic spherule.*
cosmic water Water that is accreted to the Earth with asteroidal and (especially) cometary materials.
cosmochemistry (cos-mo-chem′-is-try) The study of the origin, distribution, and abundance of elements in the universe.
cosmochlore (cos′-mo-chlore) *ureyite.*
cosmoclastic rock (cos-mo-clas′-tic) One of the original rocks of the Earth (Fairchild, 1904).
cosmogenic nuclide (cos-mo-gen′-ic) A *nuclide,* radiogenic or stable, that has been produced by the action of energetic nuclear particles in cosmic rays.
cosmogony (cos-mog′-o-ny) A scientific theory or a cultural mythology regarding the origin of the Universe, either as a whole or in more limited scope, such as the solar system of the Earth. Cf: *cosmology.*
cosmological principle (cos-mo-log′-i-cal) The principle that the universe presents essentially the same picture throughout all space and has done so throughout all time. Cf: *actualism.*
cosmology (cos-mol′-o-gy) The study, both theoretical and observational, of the space-time structure of the Universe as a whole. Cf: *cosmogony.*

cosmopolitan (cos-mo-pol'-i-tan) Said of a kind of organism or a species that is widely distributed, typically occurring in several biogeographic or paleobiologic provinces. n. *cosmopolite.*

cosmopolite (cos-mop'-o-lite) A *cosmopolitan* organism.

cosmos (cos'-mos) The universe considered as an orderly and harmonious system. Ant: *chaos.*

COSPEC An ultraviolet correlation spectrometer used to detect the concentration and emission rate of SO2 from degassing volcanoes (William-Jones and Rymer, 2000).

cossyrite (cos'-sy-rite) Sodium-rich variety of aenigmatite, occurring in minute black crystals in lava.

costa [bot] (cos'-ta) (a) In a simple leaf, the midrib. (b) In a pinnately compound leaf, the rachis. (c) In a diatom, a wall mark formed by two well defined ridges and containing fine pores.

costa [paleont] (a) One of the usually paired spines united medially and commonly laterally with neighboring spines to form a *frontal shield* overarching the *frontal membrane* in cribrimorph cheilostomes (bryozoans). Syn: costula; costule. (b) A round-topped elevation of moderate width and prominence disposed collabrally on the surface of a gastropod shell. (c) One of several moderately broad and prominent elevations of the surface of a bivalve-mollusk shell, directed radially or otherwise from the beak. Syn: *rib [paleont].* (d) A radial ridge on the external surface of a brachiopod shell, originating at the margin of the protegular node. Also, any coarse rib of a brachiopod, without reference to origin (usually fewer than 15 costae in a width of 10 mm). Cf: *costella; capilla.* (e) A ridge on the external surface of a foraminiferal *test.* It may run along a suture or be transverse to it. (f) The prolongation of a *septum* on the outer side of a corallite wall. (g) A long narrow raised area or ridge of a conodont. (h) A transverse ridge, straight or slightly curved, on the surface of whorls of coiled cephalopod conchs. (i) Anything that might be riblike in shape, including vertebrate ribs. Pl: costae.

costa [palyn] One of the riblike thickenings in the endexine of pollen, associated with colpi or pores. Costae are most often meridional and border colpi, but they may be transverse in association with transverse furrows.

costean (cos-tean') In prospecting, to dig shallow pits or trenches designed to expose bedrock. Etymol: Cornish.

costella (cos-tel'-la) A radial ridge on the external surface of a brachiopod shell, not extending to the margin of the *protegulal node* but arising by bifurcation of the existing costae or costellae or by intercalation between them. Also, a fine rib of a brachiopod or pelecypod, without reference to origin (usually 15-25 costellae in a width of 10 mm). Pl: costellae. Cf: *costa [paleont]; capilla.*

costibite (co-stib'-ite) A metallic orthorhombic mineral: CoSbS.

costula (cos'-tu-la) (a) A small ridge of an invertebrate; e.g. a marking that makes up the sculpture of a mollusk shell. (b) A *costa* of a cribrimorph cheilostome (bryozoan). Pl: costulae. Also called: costule.

coteau (co-teau') A French word used in parts of the U.S. for a variety of features: a range or sharp ridge of hills; a high plateau; a hilly upland, including the divide between two valleys; a morainal hill; an elevated pitted plain of rough surface (as in Missouri); a low, dry ridge within a swampy area (as in Louisiana); a side of a valley, esp. a prominent and dissected escarpment forming the edge of a plateau (as in north-central U.S.). Etymol: Canadian French, "slope of a hill; hillock".

cotectic (co-tec'-tic) Said of conditions of temperature, pressure, and composition under which two or more solid phases crystallize simultaneously and without resorption from a single liquid over a finite range of falling temperature; also, said of the geometric form (e.g. line or surface) representing the corresponding phase boundary on the liquidus of a phase diagram.

coticule (cot'-i-cule) A fine-grained metasedimentary rock consisting mainly of quartz and manganiferous garnet; a spessartine quartzite. The term was first used in Europe in the early 1800s (Kramm, 1976).

cotidal line (co-tid'-al) A line on a tidal chart connecting places where the tide has the same phase; for example, where high water occurs at the same time. Usually drawn for a particular tidal constituent.

cotterite (cot'-ter-ite) A variety of quartz having a peculiar metallic pearly luster.

cotton ball (cot'-ton) *ulexite.*

cotton chert *chalky chert.*

cotton rock (a) A term used in Missouri for a soft, fine-grained, siliceous, white to slightly gray or buff magnesian limestone having a chalky or porous appearance suggestive of cotton. (b) The white or light-colored decomposed exterior surrounding the dense black interior of a chert nodule.

cotton stone *mesolite.*

cotunnite (co-tun'-nite) A soft white, yellowish, or greenish orthorhombic mineral: $PbCl_2$.

cotyledon (cot-y-le'-don) The first leaf of the embryo of a seed plant, developed within the seed. It commonly serves as a storage organ. Syn: *seed leaf.*

Cotylosauria (Cot'-y-lo-saur'-i-a) An order of anapsid reptiles of generalized structure and lizardlike or turtlelike habitus, found mostly in the upper Paleozoic. Range, Lower Pennsylvanian to Upper Triassic.

cotype (co'-type) A term originally used for either a *syntype* or a *paratype* but recommended for rejection by the International Commission on Zoological Nomenclature because of its dual meaning (ICZN, 1964, p.79, 148).

coulee [geomorph] (cou'-lee) (a) A term applied in the western U.S. to a small stream, often intermittent. Also, the bed of such a stream when dry. (b) A term applied in the NW U.S. to a dry or intermittent stream valley, gulch, or wash of considerable extent; esp. a long, steep-walled, trenchlike gorge or valley representing an abandoned overflow channel that temporarily carried meltwater from an ice sheet; e.g., the Grand Coulee (formerly occupied by the Columbia River) in Washington State. (c) A small valley or a low-lying area. Etymol: French "coulée", "flow or rush of a torrent".

coulee [mass move] A tongue-like mass of debris moved by solifluction (Monkhouse, 1965, p.81).

coulee [volc] A flow of viscous lava that has a blocky, steep-fronted form. Also spelled: coulée. Etymol: French, "outflowing".

coulee lake A lake produced by the damming of a water course by lava.

coulisse (cou-lisse' [koo-lease']) A term introduced by Scrivenor (1921, p.354) for one of the prominent features, formed by the erosion of folded stratified rocks and of igneous intrusions, that are arranged en echelon on the Earth's surface "like the wings of the stage in a theatre". Etymol: French, "a side scene or wing on a theater stage".

couloir (cou-loir' [kool-wahr']) (a) A deep, narrow valley; esp. a gorge or gully on a mountain side in the Alps. (b) A French term for a passage in a cave, or a vertical cleft in a cliff. Etymol: French, "passage".

Coulomb-Mohr envelope *Mohr envelope.*

Coulomb's equation (Cou-lomb's') An equation describing the Coulomb failure criterion, or the failure of a material in shear fracture: critical shear stress for failure equals cohesion plus the coefficient of internal friction times normal stress across potential failure surface.

Coulomb's failure criterion A criterion of brittle shear failure based on the concept that shear failure will occur along a surface when the shear stress acting in that plane is large enough to overcome the cohesive strength of the material plus the resistance to movement. Cohesive strength is equal to inherent shear strength when the stress normal to the shear surface is zero; resistance to movement is equal to stress normal to the shear surface multiplied by the coefficient of internal friction of the material. See also: *Coulomb's equation.*

Coulomb's modulus *modulus of rigidity.*

coulometry (cou-lom'-e-try) A method of quantitative chemical analysis utilizing the number of coulombs necessary to release a substance during electrolysis to determine the amount of substance released. One faraday (96,500 coulombs) will release or deposit 1 gram-equivalent weight of an ion.

coulsonite (coul'-son-ite) A metallic bluish-gray cubic mineral of the *spinel* group: $Fe^{2+}V^{3+}{}_2O_4$. Syn: *vanado-magnetite.*

counter (count'-er) In structural petrology, an instrument used to contour the density per unit area of a distribution of poles on an equal-area projection. It consists of a piece of plastic or similar material with a circular hole whose area usually equals 1.0% of that of the diagram. Cf: *peripheral counter.*

counterclockwise inclination (coun-ter-clock'-wise) The inclination to the left of a heterococcolith suture as it proceeds to the periphery. Ant: *clockwise inclination.*

countercurrent (coun-ter-cur'-rent) A current flowing upwind. For example, the North Equatorial Countercurrent.

counter fossula A *fossula* developed in the position of the *counter septum* in a rugose coral.

counter-lateral septum (coun-ter-lat'-er-al) One of two protosepta of a rugose *corallite* that adjoin the *counter septum* on either side. Symbol: KL.

counterlode (coun'-ter-lode) *cross vein.*

counterpart (coun'-ter-part) *cast [sed].*

counterradiation (coun'-ter-ra'-di-a'-tion) The downward flux of infrared radiation emitted by the atmosphere after prior absorption from the Earth's surface. Syn: *back radiation.*

counterscarp (coun'-ter-scarp) In landslides underlain by noncircular shear surfaces, a scarp that is parallel to the *crown scarp* and that occurs on the downslope side. Together, the scarps bound the structural trough or graben that is formed by such a landslide.

counter septum The *protoseptum* lying directly opposite the *cardinal septum* in the plane of bilateral symmetry of a rugose corallite. Symbol: K. Cf: *alar septum.*

countervein (coun'-ter-vein) *cross vein.*

country rock [eco geol] The rock enclosing or transversed by a mineral deposit. Originally a miner's term, it is somewhat less specific than *host* rock.

country rock [intrus rocks] The rock surrounding an igneous intrusion.

counts Response of a radiometric detector due to either particles or photons of energy (IAEA, 2003).

coupled wave (cou'-pled) A type of surface wave that is continuously generated by another wave which has the same phase velocity. Syn: *C wave.*

couplet (cou'-plet) Genetically related paired sedimentary laminae, generally occurring in repeating series, as varves, but applied to laminated nonglacial shales, evaporites, and other sediments as well.

coupole The French term for *cupola [karst].*

course [stratig] (a) An old term used in Great Britain and applied to a stratum or outcrop, and to stratification. (b) A British term for a coal seam.

course [streams] The path followed by water, or the channel through which it flows; a *watercourse.*

course [surv] (a) A term used in surveying with several meanings: the bearing of a line; the length of a line; and the bearing (or azimuth) and length of a line, considered together. Also, the line connecting two successive stations in a traverse. (b) A term used in navigation for the azimuth or bearing of a line along which a ship or aircraft is to travel, without change of direction; the line drawn on a chart or map as the intended track.

cousinite (cous'-in-ite) An ill-defined black mineral: $MgU_2Mo_2O_{13}·6H_2O$(?).

coutinhoite A flaky yellow orthorhombic mineral: $Ba(UO_2)_2Si_5O_{13}·H_2O$.

Couvinian (Cou-vin'-i-an) Equivalent to *Eifelian.*

covariance (co-var'-i-ance) The arithmetic *mean* or expected value of the product of the deviations of two variables from their respective means. A measure of the joint variation of two variables around their *bivariate* mean.

cove [coast] (a) A small narrow sheltered bay, inlet, creek, or recess in a coast, often inside a larger embayment; it usually affords anchorage to small craft. (b) A small, often circular, wave-cut indentation in a cliff; it usually has a restricted or narrow entrance. (c) A fairly broad, looped embayment in a lake shoreline (Veatch and Humphrys, 1966, p.73). (d) A shallow tidal river, or the backwater near the mouth of a tidal river.

cove [geomorph] (a) Any precipitously walled and rounded or cirquelike opening, as at the head of a small steep valley; specif. a deep recess, hollow, or nook in a cliff or steep mountainside, or a small, straight valley extending into a mountain or down a mountainside. Examples: among the foothills of the Blue Ridge in Virginia, and in the English Lake District. Cf: *rincon.* (b) A term used in the southern Appalachian Mountains for a relatively open area sheltered by hills or mountains; e.g. Cades Cove in eastern Tennessee. Also, in the Southern Appalachians, a narrow steep-sided karst valley flanking limestone plateaus. (c) A gap or pass between hills. (d) A basin or hollow where the land surface has undergone differential weathering or subsidence, as from solution of underlying rock. (e) A Scottish term for a hollow or recess in rock, as a cave or cavern.

cove beach A *bayhead beach* formed in a cove.

covelline (co'-vel-line) *covellite.*

covellite (co'-vel-lite) An indigo-blue hexagonal mineral: CuS. It is a common secondary mineral and represents an ore of copper. Syn: *covelline; indigo copper.*

coverage (a) The representation of a map composed of graphic and attribute files in a digital mapping system. (b) In remote sensing, often used to describe the extent of the Earth's surface represented on an image or a set of images, as in areal coverage. (Krzanowski et al., 1993) See also: *scene.*

covered flagellar field (cov'-ered) The part of a flagellate coccolithophore having a complete cover of coccoliths in the flagellar region. Cf: *naked flagellar field.*

covered karst A terrain of karst features, usually subdued, resulting from the development of solution features in limestone covered by soil. It is contrasted with *naked karst* which is soil free. Cf: *interstratal karst; paleokarst.*

covered plain A term introduced by Melton (1936, p.594) for an alluvial flood plain through which a low-water channel does not actively meander; it usually has a natural levee and a thick alluvial cover derived from the suspended load. Cf: *meander plain; bar plain.*

cover head A thick accumulation of debris, consisting of talus cones and alluvial fans, resting on an elevated marine terrace; the material is deposited during and after the emergence of the terrace.

cover plate Any of small polygonal biserially arranged plates forming the walls and roof of the *ambulacral groove* of an echinoderm, such as a blastoid. Also spelled: coverplate.

cover-plate passageway A tubular canal in edrioasteroids that extends along lateral sutural faces of contiguous ambulacral and oral cover plates. These passageways connect the exterior to the interior of the theca (not in direct communication with the *ambulacral tunnel*) (Bell, 1976).

cover rock The thickness of rock between mine workings and the surface.

cover sand An eolian deposit of fine to very fine sand, usually containing more than 90% quartz, and believed to have been deposited by storms during the glacial epoch.

covite (cov'-ite) A *nepheline syenite* in which the mafic minerals are sodic clinopyroxene and amphibole. The term is more frequently used by Russian and European petrographers. The name, given by Washington in 1901, is for Magnet Cove, Arkansas. Obsolete.

cow-dung bomb A type of volcanic bomb whose flattened shape is due to its impact while still viscous. Its surface is somewhat scoriaceous. Syn: *cowpat bomb.*

cowlesite (cowles'-ite) A colorless orthorhombic mineral of the zeolite group: $CaAl_2Si_3O_{10}·6H_2O$.

cowpat bomb *cow-dung bomb.*

coxa The first segment of the leg of an arthropod, by which the leg articulates with the body; e.g. the proximal (basal) segment of the limb of a crustacean (except rarely where a *precoxa* is distinguishable), or the basal segment of all cephalothoracic appendages of an arachnid (the name is rarely used in the case of the chelicerae), or the proximal (basal) joint of the thoracic appendage of a merostome, directly attached to the body. Pl: coxae. Adj: coxal. See also: *coxopod; maxilla.*

coxopod (cox'-o-pod) The *coxa* of a crustacean limb.

coyoteite (coy'-ote-ite) A metallic black triclinic mineral: $NaFe_3S_5·2H_2O$.

crab The condition indicated on a vertical photograph by the lateral edges not being parallel to the air-base lines, caused by failure to orient the camera with respect to the track of the aircraft.

crack [ice] Any fracture in sea ice, not sufficiently wide to be described as a *lead,* and usually narrow enough to jump across. Cf: *tide crack.*

crack [struc geol] A parting with crack-normal motion. Cf: *joint [struc geol].*

cracking (crack'-ing) The breaking-up of more complex chemical compounds into simpler ones, usually by heating; i.e. the subjecting of compounds to pyrolysis.

crackle breccia (a) An incipient breccia having fragments parted by planes of rupture but showing little or no displacement (Norton, 1917, p.161). (b) *shatter breccia.*

crack mouth opening displacement The distance between crack walls when the crack is subject to a net tensile stress. In this context the crack may be a joint, dike, or vein.

crack propagation A rupturing process involving the breakage of bonds and an increase in surface area due to a high stress concentration located near the tip of a crack or flaw. In this context the crack may be a joint, dike, or vein.

crack-seal A mechanism of vein filling by repetitions of microcracking followed by cementation (Ramsay and Huber, 1983).

crack-seal vein A vein formed by multiple cracking and precipitation events along the same plane (Ramsay, 1980).

crag [geomorph] (a) A steep precipitous point or eminence of rock, esp. one projecting from the side of a mountain. Syn: *craig.* (b) An obsolete term for a sharp, rough, detached or projecting fragment of rock.

crag [sed] A shelly sandstone or a compacted fossiliferous sandy marl of marine origin and of Pliocene and Pleistocene age, found in eastern England and used as a fertilizer; e.g. the Coralline Crag.

crag and tail An elongate hill or ridge resulting from glaciation, having at the *stoss* end a steep, often precipitous, face or knob of ice-smoothed, resistant bedrock (the "crag") obstructing the movement of the glacier, and at the *lee* end a tapering, streamlined, gentle slope (the "tail") of intact weaker rock and/or drift protected by the crag. Cf: *knob and trail.*

craig A Scottish term for crag; e.g. Ailsa Craig. Etymol: Celtic, "rock".

craigmontite (craig'-mont-ite) A light-colored nepheline *diorite* composed of nepheline, oligoclase, and muscovite with minor amounts of calcite, magnetite, corundum, and biotite. Craigmontite has more nepheline and less plagioclase and corundum than *congressite* or *raglanite.* It was named by Adams and Barlow in 1913 for Craigmont Hill, Ontario. Obsolete.

crandallite (cran'-dal-lite) (a) A white, yellow, or gray rhombohedral mineral: $CaAl_3(PO_4)(PO_3OH)(OH)_6$. It sometimes contains appreciable strontium, barium, iron, or rare earths. Syn: *pseudowavellite.* (b) A group name for minerals with the analogous formula, but with Ca replaced by Sr, Ba, Pb, REE, Bi, or Th; and P replaced by As or Si.

craniacean (cra-ni-a'-ce-an) Any inarticulated brachiopod belonging to the superfamily Craniacea, characterized generally by a strongly punctate, calcareous shell.

cranidium (cra-nid'-i-um) The central part of the *cephalon* of a trilobite, consisting of the *glabella* and its two fixed *cheeks,* and bounded by the facial sutures. Pl: cranidia. Adj: cranidial.

crassitude (cras'-si-tude) A local exinous thickening, e.g., in the equatorial region of *spores.* The term is not commonly used. Cf: *cingulum; zona.*

crater [geophys] (cra'-ter) A typically bowl-shaped or saucer-shaped pit or depression, generally of considerable size and with steep inner slopes, formed on a surface or in the ground by the explosive release of gas, chemical, nuclear, or kinetic energy; e.g. an *impact crater* or an *explosion crater.*

crater [grd wat] *geyser crater.*

crater [impact] An approximately circular or polygonal depression in the surface of a planetary object, having a diameter that may range from a few centimeters to hundreds of kilometers and a depth that is small relative to its diameter. Large lunar craters often have lofty, rugged rim crests, terraced walls, and prominent central peaks. Impact craters are believed to have formed in several ways, including primary meteoritic and cometary impacts (e.g. Kepler and Tycho on the Moon) and secondary debris impacts. ways, including meteoritic or cometary impact (e.g. Kepler and Tycho), secondary debris impact, volanic activity, and subsidence.

crater [volc] A basinlike, rimmed structure that is usually at the summit of a volcanic cone. It commonly forms over a vent during an eruption. During hydrovolcanic eruptions, a crater may form by gradual accumulation of pyroclastic material into a surrounding rim. Cf: *caldera.*

crater chain A distinctive, chainlike, linear group of small craters first observed in most regions of the Moon's surface (such as east of the crater Copernicus). Lunar chains may be as much as 325 km long. Crater chains have also been found on many other solid surfaces throughout the solar system. See also: *chain crater.*

crater depth (a) In an artificial crater, the maximum depth measured from the deepest point to the original ground surface (Flanders and Sauer, 1960, p.5). (b) In a natural crater, in which the original ground level may be uncertain, the depth measured from the highest point on the rim crest to the deepest part of the observable depression.

crater fill Solidified lava at the bottom of a volcanic crater, with associated cinders and weathering debris.

cratering (cra′-ter-ing) (a) The dynamic process or mechanism of formation of an individual crater. (b) The process of modification of a surface by repeated crater formation.
crater lake A lake, usually of fresh water, that has formed in a volcanic crater or caldera by the accumulation of rain and groundwater. See also: *caldera lake.*
crater radius (a) In an artificial crater, the average radius, measured at a level corresponding to the original ground surface (Flanders and Sauer, 1960, p.6). (b) In a natural crater, in which the original ground level may be uncertain, the crater radius is measured to the rim crest, if present. In eroded craters, the center-to-rim radius and apparent crater radius may be impossible to differentiate. Cf: *apparent crater.*
crater rays Bright streaks of impact crater *ejecta* that extend radially many tens of crater diameters from a fresh *impact crater.* Their exact nature is still not known, but they form prominent features on the Moon, where some of the most spectacular examples occur around the crater Tycho.
crater rim Circular elevated terrain surrounding an impact crater, with internal slopes in fresh craters that are much steeper than the external slopes (ICA, 2003).
crater ring A low-relief rim of fragmental material surrounding a *maar.*
cratogene (cra′-to-gene) A rarely used term for *craton.*
craton (cra′-ton) A part of the Earth's crust that has attained stability and has been little deformed for a prolonged period. The term is now restricted to continental areas that have not been pervasively metamorphosed and deformed for at least about one billion years. Cratons include shield areas, where Precambrian rocks are exposed, and platform areas, where Precambrian rocks are overlain by a thin layer of Phanerozoic strata. Also spelled: *kraton.* See also: *thalassocraton.* Obsolete syn: *kratogen.*
cratonic basin (cra-ton′-ic) *intracratonic basin.*
crawfordite A vitreous colorless or white monoclinic mineral: $Na_3Sr(PO_4)(CO_3)$.
crawlway (crawl′-way) A cave passage that is traversable only by crawling.
C ray Right posterior ray in echinoderms situated clockwise of B ray when the echinoderm is viewed from the adoral side; equal to ambulacrum V in the *Lovenian system.*
craze plane A planar void in a soil material, having a highly complex conformation of the walls due to the interconnection of numerous short, flat and/or curved planes (Brewer, 1964, p.198).
CR chondrite A member of the Renazzo group of *carbonaceous chondrites.*
crease A term used by Woodworth (1901, p.665) for an *overflow channel* that formerly contained meltwater.
creaseyite (crea′-sey-ite) A pale green or yellowish orthorhombic mineral: $Pb_2Cu_2Fe^{3+}{}_2Si_5O_{17}{\cdot}6H_2O$.
crednerite (cred′-ner-ite) A steel-gray to iron-black monoclinic mineral: $CuMnO_2$.
Creede epithermal vein An *adularia-sericite type* mineral deposit hosted by felsic to intermediate volcanic rocks overlying older evaporite rocks and/or basinal brine-bearing sedimentary rocks. Creede epithermal veins are larger in size, have lower gold and silver grades, and much higher copper, lead, and zinc grades than Comstock and Sado epithermal veins. Named for the deposits at Creede, Colorado. Cf: *Comstock epithermal vein; Sado epithermal vein.*
creedite (creed′-ite) A white, colorless, or purple monoclinic mineral: $Ca_3Al_2(SO_4)(F,OH)_{10}{\cdot}2H_2O$.
creek [coast] (a) A British term for a small inlet, narrow bay, or arm of the sea, longer than it is wide, and narrower and extending farther into the land than a cove. The term is used in the U.S. (as in Maryland and Virginia) in names given during the earliest period of English colonization for a narrow recess in the shore of the sea, a river, or a lake, and often offering port or anchorage facilities for vessels. (b) A small, narrow tidal inlet or estuary, esp. on a low-lying coast or on the lower reaches of a wide river; e.g. Napa Creek, Calif. Syn: *tidal creek.*
creek [streams] (a) A term generally applied over most of the U.S. (except New England) and in Canada and Australia to any natural stream of water, normally larger than a *brook* but smaller than a river; a branch or tributary of a main river; a lowland watercourse of medium size; a flowing rivulet. Also, a wide or short arm of a river, such as one filling a short ravine that joins the river. (b) A term used in the SW U.S. and in Australia for a long, shallow stream of intermittent flow; an *arroyo.*
creekology (creek-ol′-o-gy) A slang term in oil exploration for the study of stream drainage patterns and gradients to obtain information on the dip and structure of surface and subsurface formations.
creek placer A placer gold deposit occurring in a modern stream.
creep [mass move] The slow, more or less continuous downslope movement of mineral, rock, and soil particles under gravitational stresses. Many types of creep have been described, on the basis of material properties, stress level, stage and rate of deformation, fundamental mechanics of failure, geometric patterns, and cause of deformation. However, the term should not be limited by a presumption of mechanism, depth, velocity profile, thickness of creep zone, or lateral extent. Syn: *rock drift.*
creep [mining] Gradual strain failure of rock, as in mine pillars and roofs, owing to the pressure of superincumbent load. It generally involves lateral mass transfer in response to vertical stress. Cf: *flash [mining]; crown-in; inbreak; heave [mining].*
creep [struc geol] Continuously increasing, usually slow, deformation (strain) of solid rock resulting from a small, constant stress acting over a long period of time.
creeping (creep′-ing) The slow shifting of a *divide* from one position to another. It occurs when a stream, because of its greater steepness and volume or the weaker nature of the rocks over which it flows, cuts down more rapidly than another stream on the opposite side of the divide (Cotton, 1958, p.69-70). Cf: *leaping.*
creep limit The maximum differential stress that a material can withstand without observable creep.
creep recovery The gradual recovery of anelastic strain when stress is released. Syn: *elastic aftereffect; elastic afterworking; afterworking; transient strain.*
creep strength The load per unit area leading to a specified steady creep strain rate at a given temperature.
creep wrinkle One of a series of small-scale corrugations of a bedding-plane surface, oriented at right angles to the direction of movement (slumping or creep) (McIver, 1961, p.227). See also: *crinkle mark; pseudo ripple mark.*
crenate (cre′-nate) Having the edge, margin, or crest cut into rounded scallops or shallow rounded notches; e.g. "crenate costae" of bivalve mollusks.
crenella (cre-nel′-la) (a) A narrow furrow between culmina of a crinoid columnal *articulum.* (b) A small radially disposed groove on the stem impression at the base of a blastoid *theca* and on the distal and proximal surfaces of columnals (TIP, 1967, pt. S, p.346). Pl: crenellae.
crenula (cren′-u-la) A ridge (*culmen*) combined with the adjacent furrow (*crenella*) of a crinoid columnal *articulum.* Pl: crenulae.
crenularium (cren-u-lar′-i-um) The entire area of a crinoid columnal *articulum* bearing crenulae. Pl: crenularia.
crenulate shoreline (cren′-u-late) A minutely irregular shoreline, characterized by crenulate lines and sharp headlands, developed during a youthful stage of submergence by differential wave erosion acting upon less resistant rocks; e.g. the shoreline of SW Ireland.
crenulation (cren-u-la′-tion) Small-scale folding (wavelength up to a few millimeters). Crenulations may occur along the cleavage planes of a deformed rock. Cf: *plication [struc geol].*
crenulation cleavage A *foliation* that results from a preferred orientation of rock layers or mineral grains due to microscale folding. The cleavage is defined by microlithons of reoriented or crenulated preexisting foliation. In a zonal crenulation cleavage, the aligned components that define the cleavage may be either the limbs or the axial regions of the small folds, and they may curve continuously into each other. In a discordant crenulation cleavage, sharply discordant zones of aligned minerals cut across small folds of layering. See also: *slip cleavage; spaced cleavage.*
crenulation lineation A lineation defined by the hinges of small folds (crenulations) associated with *crenulation cleavage.*
creolite (cre′-o-lite) (a) Red-and-white banded jasper from Shasta and San Bernardino counties, Calif. (b) A silicified rhyolite from Baja California.
crepis (cre′-pis) The initial fiber about which a desma of a sponge is secreted. Pl: crepides.
crerarite A metallic cubic mineral of the *galena* group: $(Pt,Pb)Bi_3(S,Se)_{4-x}$ (x = 0.4-0.8).
crescent beach (cres′-cent) A curving beach, concave toward the sea, formed along a hilly or mountainous coast at a bayhead or at the mouth of a stream entering a bay.
crescent cast *current crescent.*
crescentic crack (cres-cen′-tic) *crescentic fracture.*
crescentic dune *barchan.*
crescentic fracture A crescentic mark in the form of a hyperbolic crack, of larger size (up to 10-12 cm long) than a chattermark; it is convex toward the direction from which the ice moved (its "horns" point in the direction of ice movement) and consists of a single fracture without removal of any rock. Cf: *lunate fracture; crescentic gouge.* Syn: *crescentic crack.*
crescentic gouge A *crescentic mark* in the form of a groove or channel with a somewhat rounded bottom, formed by glacial plucking on a bedrock surface; it is concave toward the direction from which the ice moved (its "horns" point away from the direction of ice movement) and it consists of two fractures from between which rock has been removed. Cf: *crescentic fracture.* Syn: *gouge [glac geol]; gouge mark; lunoid furrow.*
crescentic lake A lake occupying a crescent-shaped depression; e.g. an *oxbow lake.*
crescentic levee lake A lake confined between old natural levees inside a meander and resulting from the enlargement of the meander.
crescentic mark [glac geol] Any curved or lunate mark produced by a glacier moving over a bedrock surface; e.g. a *crescentic fracture* and a *crescentic gouge.* Syn: *lunate mark.*
crescentic mark [sed] A crescent-shaped scour mark; specif. *current crescent.*
crescent scour *current crescent.*
crescent-type cross-bedding *trough cross-bedding.*
crescumulate (cres-cu′-mu-late) n. A general term for any plutonic rock formed by crystal accumulation and exhibiting *crescumulate texture.*

crescumulate texture Igneous texture in which large, elongated crystals are oriented roughly normal to cumulate layering in the rock. The term *harrisitic* was originally applied to this texture, which was thought to be limited to olivine crystals, as observed in the igneous rock harrisite; "crescumulate" has been suggested as a more general term without mineralogic restrictions (Wager, 1968, p.579).

crest [beach] (a) *berm crest.* (b) *beach crest.*

crest [geomorph] The highest point or line of a landform, from which the surface slopes downward in opposite directions; esp. the highest point of a mountain or hill, or the highest line or culminating ridge of a range of mountains or hills. See also: *summit.*

crest [palyn] In referring to dinoflagellates, used as a syn. of both *list* and *septum.*

crest [sed struc] The part of a *ripple* with an elevation of greater than half its total height (Blatt et al., 1980).

crest [struc geol] The highest point of a given stratum in any vertical section through a fold. Cf: *crest surface; crest line.*

crest line The line joining the crest points in a given stratum. Cf: *crest [struc geol].*

crest plane Planar *crest surface.*

crest-stage recorder A streamflow measuring device that records the maximum stage of flow.

crest surface A surface that connects the crest lines of the beds of an anticline. Cf: *crest plane.*

crêt In the French Jura, an in-facing cliff or escarpment; the wall of a *combe.*

creta (cre′-ta) *chalk.* Etymol: Latin, "chalk".

Cretaceous (Cre-ta′-ceous) The upper system of the *Mesozoic* Erathem of the Standard Global Chronostratigraphic Scale, above the *Jurassic* and below the *Tertiary* System of the Cenozoic Erathem. Also the time during which these rocks were formed, the Cretaceous Period, covering the time span between about 145 and 65 Ma. It is named after the Latin word for chalk ("creta") because of the English chalk beds of this age.

crevasse [geomorph] (cre-vasse′) (a) A wide breach or crack in the bank of a river or canal; esp. one in a natural levee or an artificial bank of the lower Mississippi River. Etymol: American French. (b) A wide, deep break or fissure in the Earth after an earthquake. (c) A fissure in the surface of a glacier or icefall.

crevasse [glaciol] A deep, nearly vertical crack in a glacier or other mass of land ice, or in a snowfield, caused by tensile stresses resulting from differential movement over an uneven surface. Crevasses may be meters in width and tens of meters in depth. They may be concealed beneath *snow bridges.* Etymol: French. Cf: *transverse crevasse; longitudinal crevasse; marginal crevasse; splaying crevasse.* Syn: *fissure [glaciol].*

crevasse channel A stream branch created at time of bank overflow, ordinarily lasting for a brief interval and commonly located on a concave bank. If persisting for some years it may become a more or less permanent *distributary* or even a site where *diversion* occurs.

crevasse filling A short, straight ridge of stratified sand and gravel believed to have been deposited in a crevasse of a wasting glacier and left standing after the ice melted; a variety of kame (Flint, 1928, p.415). May also occur as long sinuous ridges and linear complexes of till or drift. Cf: *ice-channel filling; till crevasse filling.*

crevasse hoar A type of frost consisting of large leaf-, plate-, or cup-shaped ice crystals that form and grow below the surface of a snowfield or glacier in a crevasse or other large open space where water vapor can condense under calm, still conditions. Cf: *depth hoar.*

crevasse ridge A mass of fluvial material originally deposited in a crevasse and now forming a more or less straight ridge that stands above the general land surface and extends parallel to the direction of ice movement (Leighton, 1959, p.340). May also occur in long sinuous ridges and linear complexes of till or drift in various directions as related to direction of ice movement.

crevasse splay A low energy delta, oriented perpendicular to a main channel, formed by a break in a natural levee during flood stage. Found in deltas and subsea fans.

crevasse topography Modification of normal natural-levee topography as a result of one or more crevasses, characterized by small branching channels along sides of which are low natural levees leading down the original main natural-levee backslope. The deposits are likely to be useful for agriculture because they are well drained and have loose texture. Best developed near natural-levee crests, crevasse topography ordinarily fans out and becomes less evident toward the adjacent basin.

crevice (crev′-ice) (a) A narrow opening or recess, as in a wave-eroded cliff or glacier. (b) A colloquial syn. of *crevasse.*

crevice karst An intricate irregular crevice system that has formed by solution widening of closely spaced joints. Cf: *clint [geomorph]; grike.*

crib [eng] (a) A bin-type retaining wall consisting of interlocking members of steel, concrete, or wood, used to stabilize slopes and protect road cuts. (b) An engineering structure enclosing a water intake and filter offshore in a lake.

crib [glac geol] A Welsh term for *arête.*

crib [mining] In mining, a structure, usually made of interlocking timbers, that forms a brace between the roof and floor. Also, short lengths of wood or other material placed in tunnels to transfer loads to steel load sets or other braces.

cribellum (cri-bel′-lum) A single or paired perforated plate in a small group of spiders, corresponding to the anterior and medial spinnerets of Liphistiina and serving as an outlet for special silk glands (TIP, 1955, pt. P, p.61). Pl: cribella.

cribrate (crib′-rate) (a) Like a sieve; e.g. said of a foraminifer perforated with round holes. (b) Said of an erect frondescent bryozoan colony with broad flattened anastomosing branches separated by large fenestrules. (c) Sievelike, as in the cribriform plate of the ethmoid or mesethmoid bone, through which pass many branches of the olfactory nerve.

cribrilith (crib′-ri-lith) A *discolith* with numerous central perforations and a lamellar rim.

cribrimorph (crib′-ri-morph) adj. Pertaining to the cheilostome bryozoan Cribrimorpha, characterized by a *frontal shield* of costae. n. A cribrimorph cheilostome bryozoan.

crichtonite (crich′-ton-ite) (a) A black rhombohedral mineral: $(Sr,La,Ce,Y)(Ti,Fe^{3+},Mn)_{21}O_{38}$. It was long supposed to be identical with ilmenite. (b) A group name for minerals with the general formula $AB_{21}O_{38}$, with A = K, Ca, Sr, Ba, Pb, REE or Y; and B = Ti, Fe^{3+}, Cr, Mn, Zn or Zr.

cricocalthrops (cri-co-cal′-throps) A *calthrops* (sponge spicule) bearing a series of annular ridges on each ray.

cricolith (cri′-co-lith) A *heterococcolith* having units arranged in a simple ring; any elliptical ring coccolith.

criddleite (crid′-dle-ite) A metallic monoclinic mineral: $TlAg_2Au_3Sb_{10}S_{10}$.

crimp A marginal band on the aboral side of a plate in a platelike conodont element, representing the area covered by the last lamella accreted to the element.

crinanite (cri′-nan-ite) An olivine-analcime *teschenite* in which the ophitic texture is well developed. It was named by Flett in 1909 for Loch Crinan, Scotland. Obsolete.

crinkled bedding (crink′-led) (a) *convolute bedding.* (b) Bedding that displays minute wrinkles; in carbonate rocks, it is believed to be related to algal mats.

crinkle mark (crink′-le) One of a series of subparallel corrugations of a bedding-plane surface, related to very small-scale and crumpled internal laminae, and produced by subaqueous solifluction (Williams and Prentice, 1957, p.289). See also: *creep wrinkle; pseudo ripple mark.*

crinoid (cri′-noid) Any pelmatozoan echinoderm belonging to the class Crinoidea, characterized by quinqueradiate symmetry, by a disk-shaped or globular body enclosed by calcareous plates from which appendages, commonly branched, extend radially, and usually by the presence of a stem, or column, more common in fossil than in living forms. Syn: *encrinite [paleont].* Range, Ordovician to Holocene.

crinoid-type pressure shadow A *pressure shadow* containing crystal fibers growing from a rigid object towards the wall (Ramsay and Huber, 1983). Fibers generally are of the same mineral as in the rigid object and are growing syntaxially on grains along the wall of the rigid object. Cf: *pyrite-type pressure shadow.*

crinozoan (cri-no-zo′-an) Any attached echinoderm belonging to the subphylum Crinozoa, characterized by a partial meridional growth pattern tending to produce an aboral cup-shaped or globoid plated theca and a partial radially divergent growth pattern forming appendages. Among the major groups included in the subphylum are blastoids, crinoids, and cystoids.

cripple (crip′-ple) A swampy area in the Pine Barrens, N.J., supporting a growth of Atlantic white cedar. It is sometimes defined as always having flowing water. Cf: *spong.*

criquina (cri-qui′-na) A coquina composed of crinoid fragments.

criquinite (cri-qui′-nite) Compact, well-indurated, and firmly cemented equivalent of *criquina.* See also: *encrinite [sed].*

crisscross-bedding (criss′-cross-bed′-ding) A kind of *cross-bedding* characteristic of eolian deposits, in which the layers dip in opposite directions.

crisscross crossed lamellar structure A *crossed lamellar structure* in which the first-order lamellae show a crisscross pattern on the depositional surface and in horizontal sections. The two dip directions of the third-order lamellae are generally oblique rather than parallel to the shell margin (Carter, 1976). Abbrev: XCL.

crista (cris′-ta) One of the crestlike elements making up the *sculpture* of certain *palynomorphs* with a sharp, toothy upper edge. The long, curved bases may meander and fuse with neighboring cristae. Pl: cristae. Adj: *cristate.*

cristate (cris′-tate) Crested or having a crest or consisting of *cristae,* esp. said of the sculpture of palynomorphs.

cristobalite (cris′-to-bal-ite′) A mineral: SiO_2. It is a high-temperature polymorph of quartz and tridymite, and occurs as white octahedrons in the cavities and fine-grained groundmasses of acidic volcanic rocks. Cristobalite is stable only above 1470°C; it has a tetragonal structure (alpha-cristobalite) at low temperatures and a cubic structure (beta-cristobalite) at higher temperatures. Cf: *tridymite.*

cristula *rostral ridge.*

critical angle (crit′-i-cal) The least *angle of incidence* at which there is total reflection when an electromagnetic or acoustic wave strikes the interface between one medium and another medium that is less refractive. The angle of

incidence qc such that sin $\theta_c=V_1/V_2$, where V_1 and V_2 are the velocities above and below the interface.
critical damping *Damping* such that the displaced mass just returns to its original position without oscillation.
critical density [chem] The density of a substance at its critical temperature and under its critical pressure.
critical density [exp struc geol] That density of a saturated, granular material below which, under rapid deformation, it will lose strength and above which it will gain strength.
critical depth [hydraul] In a channel of water, the depth at which flow is at a *critical velocity*, e.g. the depth at which a flow is at its minimum energy with respect to the channel bottom. Cf: *critical flow.*
critical depth [oceanog] The depth in the ocean below which there is no net photosynthetic *carbon fixation*, and *respiration* prevails.
critical dip [seis] (a) A *dip* in the direction opposite to the regional attitude, possibly indicating a closure. (b) Dip in the direction which is most critical in establishing closure.
critical distance The source-receiver distance for reflection at the *critical angle*.
critical end point A point at which two of three or more phases participating in a univariant equilibrium become identical in composition and physical properties, thus terminating the line representing the equilibrium.
critical flow Fluid flow at a *critical velocity*, e.g. flow at the point at which it changes from laminar to turbulent; when the *Froude number* = 1. Cf: *critical depth [hydraul]; streaming flow [hydraul]; subcritical flow; supercritical flow.*
critical gradient *critical slope.*
critical height In engineering geology, the maximum height at which a vertical or sloped bank of soil will usually stand unsupported under a given set of conditions (ASCE, 1958, term 98).
critical hydraulic gradient In a cohesionless soil, that *hydraulic gradient* at which intergranular pressure is reduced to zero by the upward flow of water (ASCE, 1962).
critical length The maximum horizontal distance over which sheet erosion does not occur, measured in the direction of overland flow from the drainage divide to a point at which the eroding stress becomes equal to the resistance of the soil to erosion (Horton, 1945, p.320); it determines the width of the *belt of no erosion.* Cf: *length of overland flow.*
critically climbing translatent strata A term proposed by Hunter (1977) for *climbing ripple* with an angle of climb that is equal to the angle of the *stoss* slope of the ripple *bed form.* Cf: *subcritically climbing translatent strata; supercritically climbing translatent strata.*
critically undersaturated (crit'-i-cal-ly) Said of a rock having feldspathoids and olivine, but no hypersthene, in its norm. Cf: *undersaturated.*
critical mineral [eco geol] A mineral essential to national defense but which, in all likelihood, can be obtained domestically or in adequate quantities from secure foreign suppliers in times of national emergency.
critical mineral [meta] A mineral that is stable only under the conditions of one metamorphic facies or zone. Cf: *typomorphic mineral; index mineral.*
critical moisture In a soil, that degree of moisture below which, under constant load increase, deformation will increase, and above which it will decrease.
critical point A point representing a set of conditions (pressure, temperature, composition) at which two phases become physically indistinguishable; in a system of one component, the temperature and pressure at which a liquid and its vapor become identical in all properties.
critical points (a) Points on the line that remain stable under orthographic projection. Critical points include curvature minima and maxima, end points, and points of intersections. Critical points are the bases for efficient line generalization algorithms. (b) Frequently used to refer to those points on a line that capture its basic character (Krzanowski et al., 1993).
critical pressure The pressure required to condense a gas at the critical temperature, above which, regardless of pressure, the gas cannot be liquefied.
critical reflection [seis] A reflection at the critical angle. Amplitude may be exceptionally large in this vicinity. Reflection at angles in this vicinity is called wide-angle reflection.
critical resolved shear stress Shear stress on a crystallographic slip plane, acting in the slip direction, which is required for the initiation of slip.
critical shear stress The *threshold* shear stress (generally of fluid flows) necessary to entrain a given particle or to erode a given surface.
critical slope The slope or grade of a channel that is exactly equal to the loss of head per foot resulting from flow at a depth that will give uniform flow at critical depth; the slope of a conduit which will produce critical flow (ASCE, 1962). Syn: *critical gradient.*
critical slope angle The local maximum slope inclination which the soil and rock materials underlying the slope can support without failure under existing climate, vegetation, and land use. Syn: *angle of ultimate stability.* Cf: *angle of repose.*
critical solution point *consolute point.*
critical strain energy release rate The strain energy release per unit crack length associated with unstable joint propagation. Equivalent to *fracture toughness* multiplied by *elastic constants.*
critical-taper theory A theory for the development of a fold-thrust belt in which the belt is a wedge that maintains a characteristic taper angle, measured between its topographic surface and its basal detachment. Tapered wedges develop in response to a push from the rear as well as stresses arising from their inclined topographic surfaces. If the taper angle is less than the critical taper, the wedge thickens to attain critical taper. If the taper angle exceeds the critical taper, the wedge adds new material to its toe by new fault propagation. If the taper angle equals the critical taper, the wedge slides without internal deformation. Tapered wedges have been modeled as materials that follow a Coulomb failure criterion, as linearly or non-linearly viscous liquids, and as plastic solids.
critical temperature The temperature of a system at its critical point; for a one-component system, that temperature above which a substance can exist only in the gaseous state, no matter what pressure is exerted.
critical tractive force The minimum *tractive force* required to set sediment particles of a stream bed moving.
critical velocity (a) That velocity of fluid flow at which the flow changes from laminar to turbulent. (b) That velocity of fluid flow at which the flow changes from laminar to turbulent, and at which friction becomes proportional to a power of the velocity higher than the first power. Syn: *Reynolds critical velocity.* (c) That velocity of fluid flow at which the fluid's minimum energy value is attained. Syn: *Belanger's critical velocity.* (d) In an open channel, that velocity of fluid flow at which silt is neither picked up nor deposited. Syn: *Kennedy's critical velocity.* (e) In an open channel, that velocity of fluid flow at which the velocity head equals one half the mean depth, and at which the energy head is at a minimum. Syn: *Unwin's critical velocity.* See also: *critical flow; critical depth [hydraul].*
CRM *chemical remanent magnetization.*
crocidolite (cro-cid'-o-lite) A lavender-blue, indigo-blue, or leek-green asbestiform variety of riebeckite, occurring in silky fibers and in massive and earthy forms. Also spelled: *krokidolite.* Syn: *blue asbestos; Cape blue.*
Crocodilia (Croc-o-dil'-i-a) An order of the reptilian subclass Archosauria, characterized by persistently aquatic habit and conservative body form. Range, Upper Triassic to Holocene.
crocoite (croc'-o-ite) A bright-red, yellowish-red, or orange monoclinic mineral: $PbCrO_4$. Syn: *red lead ore.*
crocydite (cro'-cyd-ite) A *Migmatite* with a flakelike or flufflike light-colored part (Dietrich and Mehnert, 1961). Rarely used. Var: *krokydite.*
Croixian (Croix'-ian) A North American series or stage: uppermost Cambrian (above Albertan, below Canadian of Ordovician).
cromaltite (cro'-malt-ite) A *pyroxenite* in which sodic clinopyroxene is the predominant mineral, with andradite as a characteristic phase, and smaller amounts of biotite, perovskite, and oxide minerals. The name, given by Shand in 1906, is for the Cromalt Hills, Scotland. Obsolete.
Cromerian (Cro-mer'-i-an) (a) North European warm-climate paleofloral stage: middle Pleistocene (above Menapian, below Elsterian). (b) British continental stage: lower Pleistocene (above Beestonian, below Anglian).
cromfordite (crom'-ford-ite) *phosgenite.*
Cromwell current (Crom'-well) *equatorial undercurrent.*
cronstedtite (cron'-stedt-ite) A trioctahedral Fe-rich phyllosilicate based on a 1:1 T-O layer approximately 7 Å thick. Its composition is $(Fe^{2+},Fe^{3+})_3OH_4(Si,Fe^{3+})_2O_5$. Cronstedtite occurs as jet-black crystals in a wide variety of layer stacking sequences, primarily in low-temperature hydrothermal sulfide veins. A member of the *serpentine* group.
cronusite A submetallic black rhombohedral meteorite mineral: $Ca_{0.2}CrS_2{\cdot}2H_2O$.
crooked hole (crook'-ed) A borehole that has deviated beyond the allowable limit from the vertical or from its intended course. Cf: *directional well.*
crookesite (crookes'-ite) A metallic lead-gray tetragonal mineral: $Cu_7(Tl,Ag)Se_4$.
crop n. Deprecated syn. of *outcrop.* v. To appear at the surface of the ground; to *outcrop.*
crop coal (a) That part of a coal seam that is near the surface. (b) A coal deposit that crops out, as in an *exposed coalfield.* (c) Inferior, weathered coal occurring near the surface. Cf: *deep coal.*
crop out v. To appear exposed and visible at the Earth's surface.
cropping (crop'-ping) n. Deprecated syn. of *outcrop.*
cross association A method for pairwise comparison of two strings or sequences of nonnumeric data. Cf: *cross correlation.*
crossbar (cross'-bar) The complete skeletal rim on which the fixed edge of the *mandible* of the *avicularium* pivots in some cheilostome bryozoans.
cross bar (a) A short, bifurcating "ridge" of a transverse ripple mark. (b) A low ridge trending across a *blind valley.*
cross-bed A single bed, inclined at an angle to the main planes of stratification. The term is restricted by McKee and Weir (1953, p.382) to a bed that is more than 1 cm in thickness. See also: *cross-stratum; cross-lamina; cross-stratification.* Also spelled: crossbed.
cross-bedding (cross-bed'-ding) (a) *Cross-stratification* in which the cross-beds are more than 1 cm in thickness (McKee and Weir, 1953, p.382). (b) A cross-bedded structure; a cross-bed. See also: *current bedding; inclined bedding; discordant bedding; crisscross-bedding.* Syn: *false bedding; diagonal bedding; oblique bedding; foreset bedding.* Also spelled: crossbedding.
cross channel A transverse drainageway cutting across an interstream area or connecting two successive low areas.

cross correlation A measure of the correlation between two sequences of observations, on the same or different variables, that are ordered in time or space. The cross correlation usually is calculated for successively greater lags or offsets between the two sequences and shown in a cross-correlogram, a plot of cross correlation versus lag number. Cf: *autoassociation.*

crosscorrelation [seis] (cross'-cor-re-la'-tion) (a) A measure of the similarity of two waveforms, the degree of linear relationship between them, or the extent to which one is a linear function of the other. Cf: *autocorrelation [seis].* (b) A method for comparison of two strings or sequences of numerical data. Cf: *correlation [seis].*

cross-coupling effect The effect produced by acceleration relative to the Earth of a moving gravity meter.

cross course *cross vein.*

crosscut (cross'-cut) (a) A small passageway that may be driven at an angle to the main entry of a mine, to connect it with a parallel entry or an air course. (b) A level driven across the course of a vein or across the general direction of the workings; thus, a mine opening that intersects a vein or ore-bearing structure at an angle.

crossed acicular structure A *crossed structure* consisting of acicular crystallites (generally aragonitic) which show two predominant dip directions; the crystallites are not aggregated into first- or second-order lamellae, but they may locally form groups of a few mutually parallel crystallites (Erben, 1971, p.56; Flajs, 1972, p.59; Bandel, 1977b, p.74; Carter, 1980b).

crossed bladed structure A *microstructure* of sheets of laths or blades in which the blades or laths form parallel aggregations oriented in more or less random directions in adjacent sheets and in portions of the same sheet. This is common in strophomenid brachiopods, and it occurs in an irregular form in tentaculitids, bryozoans, and locally in the outer layer of limacean bivalves (Armstrong, 1969, p.319).

crossed composite prismatic structure A *composite prismatic structure* in which adjacent groups of minute, mutually parallel nondenticular composite prisms dip predominantly in two directions relative to the depositional surface, as seen in horizontal sections through the shell layer. This structure commonly grades toward the exterior and interior parts of the shell layer into regular *nondenticular composite prismatic structure* (Carter, 1990, Fig. 3F).

crossed foliated structure A calcitic *crossed lamellar structure* in which each second-order lamella consists of *regularly foliated structure* (Carter, 1980b).

crossed-lamellar Said of the type of mollusk shell structure composed of primary and secondary lamellae, the latter inclined in alternate directions in successive primary lamellae.

crossed lamellar structure A *microstructure* of mutually parallel rods, laths, or blades aggregated into variably shaped but commonly lensatic to branching first-order lamellae. Each first-order lamella is oriented with its height axis more or less perpendicular to the depositional surface and with its length axis usually parallel or perpendicular to the shell margin, depending on the variety. Adjacent second-order lamellae show two predominant dip directions, uniformly alternating and opposite or alternating and oblique in adjacent first-order lamellae. Unlike in *crossed acicular structure,* each first-order lamella consists of a large number of structural subunits. Calcitic crossed lamellar structures are called crossed foliated or crossed semifoliated. Other varieties include simple, rod-type, diffuse, linear, branching, irregular, crisscross, compressed, and triangular. As traditionally used, the unmodified term "crossed lamellar" excludes *complex crossed lamellar structure* (Boggild, 1930, p.251; Cox, 1969, p. N104; Carter, 1980a,b). Abbrev: CL.

crossed nicols In a polarizing microscope, two Nicol prisms or Polaroid plates that are oriented so that the transmission planes of polarized light are at right angles; light that is transmitted from one will be intercepted by the other, unless there is an intervening substance. Syn: *crossed polars.*

crossed polars *crossed nicols.*

crossed semifoliated structure A calcitic *crossed lamellar structure* in which each second-order lamella consists of *semifoliated structure.* The dip angle of the opposing sets of second-order lamellae is typically low, i.e., commonly 10°-20° relative to the depositional surface (Carter, 1990, Fig. 45).

crossed structure A *microstructure* that shows two or more clearly nonhorizontal dip directions of its elongate structural units relative to the depositional surface (Carter, 1980a,b).

crossed twinning *cross-hatched twinning.*

cross-equalizing [geophys exploration] (a) Filtering one channel to match the frequency spectrum of adjacent channels. The matching involves a phase shift as well as adjustment of the amplitude of frequency components. Tends to align coherent events better but may increase short-period reverberations. (b) Filtering one data set to match the frequency spectrum and arrival times of another data set, as may be done with time-lapse sets in order to match regions that have not been changed.

cross fault (a) A fault whose strike crosses at a high angle the strike of the constituent strata or the general trend of the regional structure. (b) A minor fault that intersects a major fault.

crossfeed (cross'-feed) Interference resulting from the unintentional pickup by one channel of information or noise on another channel. Syn: *crosstalk.*

cross fiber Veins of fibrous minerals, esp. asbestos, in which the fibers are at right angles to the walls of the vein. Cf: *slip fiber.*

cross fold A fold that intersects a preexisting fold of different orientation; the resulting structure is a *superposed fold.* Syn: *subsequent fold.*

cross-fold joint A systematic, bedding-normal joint set oriented perpendicular to strike; typically develops in foreland fold-thrust belts. Syn: *dip joint.*

cross fracture A small-scale joint structure developed between fringe joints.

cross-grading The process of slope dissection by rills and gullies in which an original slope (alongside, and parallel with, a stream) is replaced by a new slope deflected toward the stream (Horton, 1945, p.335). See also: *micropiracy.*

cross-hatched twinning Repeated twinning after two laws, e.g. microcline twin law, with intersecting composition planes. Syn: *gridiron twinning; crossed twinning.*

cross-hatching (a) The process of drawing or shading (on a map) with a pattern consisting of two sets of parallel lines crossing each other at a predetermined angle (obliquely or at right angles) so that the space between the lines in one set is identical to the space between the lines in the other set. (b) The effect produced by cross-hatching, such as a pattern indicating abrupt gradients. See also: *hatching.*

crosshole method Technique for investigating the region between two or more boreholes by measurements involving sources in one borehole and detectors in others. Seismic measurement of transit times and/or amplitudes often is resolved by *tomography.*

cross-hole method [geophys exploration] (a) A technique for investigating the region between two or more boreholes by measuring the transit times and/or amplitudes of P and/or S waves from a source located in one borehole to geophones in other boreholes. Usually implies *crosshole method* and *tomography.* Three-component geophones may be used. Also called cross-well method. (b) A technique for resistivity or electromagnetic measurements between boreholes, used for fracture and cavity detection, reservoir studies, and orebody delineation.

crossing (cross'-ing) A term applied to the shallow part of a river channel, separating deeper pools of water (or bendways) at the bends of meanders.

crossing canal A proximal (prothecal) part of a graptoloid *theca,* growing across the sicula to develop on that side of the *sicula* opposite from the side from which it originated.

crossite (cross'-ite) A blue monoclinic mineral of the *amphibole* group, intermediate in composition between glaucophane and riebeckite; a variety of glaucophane rich in iron.

cross joint A joint that is perpendicular to the major lineation of the rock or a nonsystematic joint roughly orthogonal to and cutting between two systematic joints (Gross, 1993). Syn: *transverse joint.* Cf: *Q-joint; ac-joint.*

cross-joint fan In igneous rock, a fanlike pattern of cross joints that follow the arching of the flow lineation.

cross-lamina A *cross-bed.* The term is restricted by McKee and Weir (1953, p.382) to a *cross-stratum* that is less than 1 cm in thickness. Syn: *cross-lamination.*

cross-lamination (a) *Cross-stratification* characterized by cross-beds that are <1 cm in thickness (McKee and Weir, 1953, p.382). (b) A cross-laminated structure; a *cross-lamina.* See also: *flow-and-plunge structure.* Syn: *oblique lamination; diagonal lamination.*

crossline [geophys exploration] (a) The direction at right angles to the direction in which the data were acquired. (b) Generally the direction parallel to that of the source lines in 3-D surveys; where receiver lines are perpendicular to source lines, the direction of the receiver lines. (c) A vertical section extracted from a 3-D data volume in the crossline direction. (d) A *cross spread.*

cross lode *cross vein.*

crossopodium (cros-so-po'-di-um) A *trace fossil* of the "genus" *Crossopodia,* consisting of a sinuous or meandering marking about 1 cm wide with a median furrow, and believed to be a trail left by a creeping marine animal. Pl: crossopodia.

Crossopterygii (Cros'-sop-te-ryg'-i-i) An order of lobefinned bony fish characterized by a hinged braincase, amphistylic jaw suspension, and sharp, conical teeth with labyrinthodontine infolding of enamel. It includes the coelacanths.

crossover [geophys exploration] (a) The reversal of the dip direction of the electromagnetic field over the apex of a conductor. The undisturbed electromagnetic field of stations in the plane of a vertical source loop is horizontal. In the presence of a subsurface conductor, the field will be horizontal over the apex but will have vertical components in opposite directions on either side of the conductor. (b) The intersection of two curves. (c) The intersection of two lines.

crossover distance (cross'-o-ver) That distance at which a refracted wave becomes the *first arrival.* Cf: *critical distance.*

crossplot [geophys exploration] A graph used to determine the relationship between two different measurements, e.g., a crossplot of porosity measured from one type of log against porosity from another type of log is used to show lithology or secondary porosity, which affects the two logs differently.

cross profile A plot of elevation drawn against distance along a line at right angles to the long direction of a valley, stream, or ridge or orthogonal to the general trend of a set of contour lines. Cf: *longitudinal profile.* Syn: *transverse profile.*

cross ripple A *ripple* resulting from the interference of at least two sets of ripples, one set forming after the completion of, or simultaneously with, the other; e.g. current cross ripple and *oscillation cross ripple.* Term introduced by Bucher (1919, p.190). See also: *compound ripple; tadpole nest.* Syn: *interference ripple; dimpled current mark; complex ripple.*

cross ripple mark *cross ripple.*

cross sea A confused, choppy state of the ocean, occurring where waves from two or more different directions meet. Wave direction may appear to be the same as one of the original directions, or it may be a new direction.

cross section [geol] (a) A diagram or drawing that shows features transected by a given plane; specif. a vertical section drawn at right angles to the longer axis of a geologic feature, such as the trend of an orebody, the mean direction of flow of a stream, or the axis of a fossil. Cf: *longitudinal section.* Syn: *transverse section.* (b) An actual exposure or cut that shows transected geologic features. Adj: cross-sectional. Also spelled: cross-section.

cross section [radioactivity] The probability that a particular nuclear reaction will occur.

cross-shore current Nearshore current moving in a direction roughly perpendicular to the shoreline.

cross-shore transport Nearshore sediment transport in a direction at a right angle to the shoreline. Wave motion can produce onshore transport because of differences in wave speed and duration during the onshore and offshore parts of the wave cycle. Rip currents can produce offshore transport.

cross spread (a) A seismic spread that makes a large angle to the line of traverse; it is used to determine the component of dip perpendicular to that line. (b) A seismic spread that is laid out in the pattern of a cross.

cross-stone (a) *chiastolite.* (b) *staurolite.* (c) *harmotome.*

cross-stratification Arrangement of strata inclined at an angle to the main stratification. In modern usage, following McKee and Weir (1953, p.382), this is considered to be the general term, and to have two subdivisions: *cross-bedding,* in which the cross-strata are thicker than 1 cm, and *cross-lamination,* in which they are thinner than 1 cm. A single group of related cross-strata is a set and a group of similar, related sets is a *coset.* There are many types of cross-stratification; Allen (1963) recognized fifteen, to which he attached Greek-letter designations. Probably most cross-stratification is produced by the migration of bed forms, particularly ripples (which form small-scale cross-lamination) and dunes (which form medium- to large-scale cross-lamination or cross-bedding). Syn: *false stratification; diagonal stratification.*

cross-stratum A *cross-bed.* McKee and Weir (1953, p.382) consider it a general term that includes cross-bed and *cross-lamina.*

crosstalk (cross'-talk) *crossfeed.*

cross validation Methods for assessing uncertainty in which observations are removed one at a time from a sample, and the remaining observations are used to compute a statistic. The variance in the statistic from the censored data is used as an estimate of the standard error.

cross valley *transverse valley.*

cross vein (a) A vein or lode that intersects a larger or more important one. Syn: *cross course; cross lode; countervein; counterlode; caunter lode.* (b) A vein that crosses the bedding planes in a sedimentary sequence.

croton pattern A pollen *sculpture* consisting of rings of four to seven raised, more or less triangular, sexinous elements arranged around circular areas. It is generally formed by capitate columellae and is characteristic of some members of the family Euphorbiaceae.

crotovina (cro-to-vi'-na) *krotovina.*

crowfoot (crow'-foot) Obsolete syn. of *stylolite.*

crown [gem] The portion of any faceted gemstone above the girdle. Cf: *pavilion.* Syn: *bezel; top [gem].*

crown [geomorph] The top or highest part of a mountain or an igneous intrusion; the summit.

crown [mass move] The practically undisturbed material still in place and adjacent to the highest parts of the scarp along which a landslide or avalanche moved.

crown [paleont] (a) The whole of a crinoid exclusive of the *pelma;* the part of a crinoid skeleton above the *column,* including the *dorsal cup, tegmen,* and arms. (b) A conodont element minus its basal funnel or plate (TIP, 1981, pt. W, supp.2). (c) The exposed part of a tooth, distal to the root, that usually functions in piercing or shearing food. Syn: *element proper.*

crown group A group of organisms represented on a cladogram that includes the crown species. Cf: *crown species.*

crown-in In mining, a falling of the mine roof or a heave of the mine floor due to the pressure of overlying strata. Cf: *flash [mining]; inbreak; creep [mining].*

crown scarp The outward-facing scarp, bordering the upper portion of a landslide. It is almost always concave in a downslope direction. The scarp surface may be slickensided, indicating the downward displacement of the sublandslide shear surface; or it may have a rough, sugary surface, indicating a tension or cleavage fracture formed by downslope displacement of the slide mass (Varnes, 1958).

crown species In a monophyletic group, a derived species identified as a named entity on a cladogram. Cf: *crown group, stem species.*

crucite (cru'-cite) (a) *chiastolite.* (b) Pseudomorph of hematite or limonite after arsenopyrite.

crude Said of a mineral material in its natural, unrefined state, e.g. crude oil, crude ore.

crude oil *Petroleum* in its natural state as it emerges from a well, or after passing through a gas-oil separator but before refining or distillation.

crumble coal (crum'-ble) An incoherent brown coal that lacks cementing material. Syn: *formkohle.*

crumble peat Peat that is friable and earthy.

crumb structure A type of soil structure in which the peds are spheroids or polyhedrons that have little or no accommodation to surrounding peds, are porous, and range in size from less than 1.0 mm to 5 mm. Cf: *granular structure.* Not currently used in the U.S.

crumina (cru'-mi-na) A saclike semienclosed space developed in the ventral part of the carapace of female ostracodes belonging to some Paleozoic species. Pl: cruminae.

crump Ground movement, perhaps violent, caused by failure under stress of the ground surrounding an underground working (usually in coal), and so named because of the sound produced.

crumpled (crum'-pled) *plicate.*

crumpled ball A highly irregular, crumpled-up mass of laminated sandstone, measuring 5-25 cm across, and flattened parallel to the bedding (Kuenen, 1948, p.371). Cf: *slump ball.*

crumpled mud-crack cast A *mud-crack cast* that displays tortuous and contorted crumpling produced by the adjustment of the sand filling to the compaction of the enclosing mud matrix (Bradley, 1930).

crura (cru'-ra) Plural of *crus.*

crural base (cru'-ral) A part of the *crus* of a brachiopod, united to a *hinge plate,* and separating the inner and outer hinge plates when present.

cruralium (cru-ral'-i-um) A spoon-shaped structure of the *brachial valve* of a pentameracean brachiopod, formed by the dorsal union of outer plates (or homologues) and bearing adductor muscles.

crural plate A plate extending from the inner edge of an outer hinge plate or crural base to the floor of the *brachial valve* of a brachiopod (TIP, 1965, pt. H, p.143). See also: *septalial plate.*

crural process The pointed part of the *crus* of a brachiopod directed obliquely inward and ventrally.

crus (a) Either of a pair of short, curved, calcareous basal processes that extend from the *cardinalia* or *septum* of a brachiopod to give support to the posterior end of the *lophophore.* The distal end may also be prolonged into primary lamella of *spiralium* or descending branch of *loop* (TIP, 1965, pt. H, p.142). (b) In tetrapods, the shank of the hind limb, i.e. tibia and fibula and surrounding soft structures. Pl: *crura.*

crush belt A belt of crushed rock characterized by intense cataclasis and mylonitization.

crush border A microscopic, granular metamorphic structure sometimes characterizing adjacent feldspar particles in granite due to their having been crushed together during or subsequent to crystallization (Dale, 1923).

crush breccia A breccia formed in place, or nearly in place, by mechanical fragmentation of rocks during crustal movements; a *tectonic breccia* associated with planes of movement and formed as a result of folding or faulting. Syn: *cataclastic breccia.*

crush conglomerate A rock formed essentially in place by deformation (folding or faulting) of brittle, closely jointed rocks, containing lozenge-shaped fragments produced by granulation of rotated joint blocks and rounded by attrition, and closely simulating a normal (sedimentary) conglomerate; a rock similar to a *crush breccia* but having fragments that are more rounded. It is characterized by similarity in composition between fragments and matrix (generally one rock type). Term was introduced by Lamplugh (1895, p.563). Syn: *tectonic conglomerate; cataclastic conglomerate.*

crushing strength (crush'-ing) Unconfined compressive strength of a material.

crushing test *unconfined compression test.*

crush zone An area of fault breccia or fault gouge.

crust [ice] *ice rind.*

crust [interior Earth] The outermost layer or shell of the Earth, defined according to various criteria, including seismic velocity, density and composition; that part of the Earth above the Mohorovicic discontinuity, made up of the *sial* and the *sima.* It represents less than 0.1% of the Earth's total volume. See also: *continental crust; oceanic crust.* Cf: *tectonosphere.*

crust [sed] A laminated, commonly crinkled deposit of algal dust, filamentous or bladed algae, or clots (from slightly arched forms to bulbous cabbagelike heads) of algae, formed on rocks, fossils, or other particulate matter by accretion, aggregation, or flocculation.

crust [snow] *snow crust.*

crust [speleo] (a) A thin mineral layer overlying bedrock or clastic material. Similar to a coating but, in general, rough, brittle, and flaky. (b) The outer dessicated layers of a *speleothem.*

crust [volc] The outer part of a lava flow that has solidified after losing heat to the exterior (Kilburn, 2000).

crustacean (crus-ta'-ce-an) Any arthropod belonging to the subphylum Crustacea, characterized chiefly by the presence of two pairs of antennae on the head. Most forms occur in marine environments. Crustaceans are second only to insects in numbers of individuals. Range, Cambrian to present.

crustaceous (crus-ta'-ce-ous) (a) Having, suggesting, or of the nature of a crust or shell. (b) Belonging to the Crustacea; crustacean.

crustal abundance (crust'-al) *clarke.*

crustal dichotomy Strong hemispheric contrast in the physical nature of a planet's crust. On Mars, the division of the planet into heavily cratered highlands and younger lowland plains (Kieffer et al., 1992).

crustal plateau Large, steep-sided elevated regions on Venus hundreds to a few thousand kilometers across (Bougher et al., 1997).

crustified (crust'-i-fied) Said of a vein in which the mineral filling is deposited in layers on the wall rock. Syn: *healed.*

crust-mound A small low-standing compact *bioherm* or *patch reef* consisting of a cruststone, bindstone, or lettucestone core, sometimes surrounded by coarse rudstone flanking beds (Cuffey, 1985). Cf: *frame-thicket; mud-mound.*

cruststone (crust'-stone) A type of *reef rock, boundstone,* or *framestone* composed of sheetlike colonies encrusting one another, forming a layered in-place skeletal mass (Cuffey, 1985). Cf: *bindstone.*

Cryands Andisols that have a cryic or pergelic soil temperature regime. A suborder in the U.S. system of soil taxonomy.

cryergic (cry-er'-gic) A term recommended by Baulig (1956, paragraph 77) as a syn. of the broad meaning of *periglacial* as when the latter word is commonly but "rather incorrectly" applied to processes and deposits "in some regions not actually peripheral to glaciated regions". The term may also be used "to denote phenomena due to cold conditions" (Stamp, 1966, p.141).

Cryerts Vertisols that have a cryic soil temperature regime. A suborder in the U.S. system of soil taxonomy.

cryic temperature regime (cry'-ic) A soil temperature regime in which the mean annual temperature (measured at 50cm depth) is more than 0°C but less than 8°C, with cold summer temperatures and a summer-winter variation of more than 5°C (USDA, 1975). Cf: *frigid temperature regime.*

Cryids Aridisols that have a cryic soil temperature regime. A suborder in the U.S. system of soil taxonomy.

cryochore (cry'-o-chore) An old climatic term for the part of the Earth's surface covered with perpetual snow and ice. Syn: *cryosphere [permafrost].*

cryoconite [glaciol] (cry-oc'-o-nite) A combination of fine organic and inorganic particles, once thought to be of cosmic origin, transported by the wind and deposited on a snow or ice surface (e.g., on the Greenland Ice Cap). Photosynthesis by algae and cyanobacteria provides nutrients for a complex community of bacteria, diatoms, fungi, etc. (Christner et al., 2003). It is found mainly in *cryoconite holes,* but may form long stripes or an almost continuous cover. Var. spelling: *kryokonite.*

cryoconite [mineral] A mixture of garnet, sillimanite, zircon, pyroxene, quartz, and other minerals.

cryoconite hole A small hollow or pit on the surface of glacier ice containing particles of *cryoconite;* a *dust well,* commonly centimeters in depth and decimeters in plan dimension. The particles of cryoconite absorb solar radiation and may generate heat biologically, causing increased ablation of glacier ice around and below them.

Cryods Spodosols that have a cryic or pergelic soil temperature regime. A suborder in the U.S. system of soil taxonomy.

Cryogenian (Cry-o-ge'-nian) A period of the Standard Global Geochronometric Scale: middle period of the *Neoproterozoic* Era, characterized by global glaciation ("cryos", "ice", and "genesis", "birth"). It is defined as the time span between 850 Ma, the end of the *Tonian* Period, and the base of the *Ediacaran* Period, which is estimated as ~600 to ~560 Ma (Plumb, 1991).

cryogenic fabric The distinct soil micromorphology resulting from the effects of freezing and thawing processes.

cryogenic lake (cry-o-gen'-ic) *thermokarst lake.*

Cryogenic period (Cry-o-gen'-ic) An informal designation for a period in geologic history "when large bodies of ice formed at or near the poles and the climate was generally suitable for the growth of continental glaciers" (ADTIC, 1955, p.22).

cryogenics (cry-o-gen'-ics) The branch of physics pertaining to the production and effects of very low temperatures; the science of extreme cold. Adj: cryogenic. Obsolete syn: cryogeny.

cryogenic soil Soil that has formed under the influence of cold soil temperatures.

cryokarst (cry'-o-karst) The European equivalent of *thermokarst.*

cryolaccolith (cry'-o-lac'-co-lith) *frost blister.*

cryolite (cry'-o-lite) A variously colored monoclinic mineral: Na_3AlF_6. It may contain iron and has been found chiefly in a pegmatite at Ivigtut, Greenland, in cleavable masses of waxy luster. Natural and synthetic cryolite is used in the manufacture of aluminum. Syn: *Greenland spar; ice stone.*

cryolithionite (cry'-o-lith'-i-o-nite') A colorless cubic mineral: $Na_3Li_3Al_2F_{12}$. It belongs to the garnet structure type.

cryolithology (cry'-o-lith-ol'-o-gy) The study of the development, nature, and structure of underground ice, esp. ice in permafrost regions; a branch of *geocryology.*

cryolithosphere The top part of the crust of Mars in which the temperature is below the freezing point of water (Kieffer et al., 1992). Syn: *cryosphere [planet].*

cryology (cry-ol'-o-gy) (a) A general term for the study of all aspects of the *cryosphere,* including snow, ice, and frozen ground. (b) The science of refrigeration.

cryomorphology (cry'-o-morph-ol'-o-gy) The part of geomorphology "pertaining to the various processes and products of cold climates" (Black, 1966, p.332). See also: *periglacial geomorphology.*

cryonival (cry-o-ni'-val) (a) Pertaining to the combined action of frost and snow. (b) *periglacial* (Hamelin, 1961, p.200).

cryopedology (cry'-o-pe-dol'-o-gy) The study of the processes of intensive frost action and the occurrence of frozen ground, esp. permafrost, including the civil-engineering methods used to overcome or minimize the difficulties involved; term introduced by Bryan (1946).

cryophilic (cry-o-phil'-ic) Said of an organism that prefers low temperatures, esp. below 10°C. Syn: *psychrophilic.*

cryophyllite (cry-o-phyl'-lite) A variety of zinnwaldite with some deficiency in the (Li,Fe,Al) group and containing some ferric iron.

cryoplanation (cry'-o-pla-na'-tion) The reduction and modification of a land surface by processes associated with intensive frost action, such as solifluction, supplemented by the erosive and transport actions of running water, moving ice, and other agents (Bryan, 1946, p.640). Cf: *altiplanation.*

cryoseism (cry'-o-seism) A nontectonic earthquake caused by freezing action in ice, ice-soil, and ice-rock materials. It can occur under perennial or seasonal frost conditions. Cryoseisms of intensity IV MM or V MM are characterized by ground rupture and are felt over a localized area (Lacroix, 1980).

cryosol Soil formed in either mineral or organic materials having permafrost either within 1 m below the surface or, if the soil is strongly cryoturbated, within 2 m below the surface, and having a mean annual ground temperature below 0° C. Cryosols are divided into turbic cryosols developed on mineral soils and strongly cryoturbated; static cryosols developed on mineral soils but with little or no cryoturbation; and organic cryosols developed on organic (peat) materials (Canada Soil Survey Committee, 1978).

cryosphere [permafrost] The part of the Earth's surface that is perennially frozen; the zone of the Earth where ice and frozen ground are formed. The component of Earth that is covered by permanent ice. The cryosphere interacts with Earth's other components, the atmosphere, hydrosphere, and lithosphere.

cryosphere [planet] *cryolithosphere.*

cryostatic (cry-o-stat'-ic) Descriptive of frost-induced hydrostatic phenomena (Washburn, 1956, p.842), e.g. the movement of water-saturated material confined between downward-advancing seasonal frost and an impermeable surface such as the permafrost table or bedrock. Such progressive downward freezing may generate large hydrostatic pressures.

cryostructure The structural characteristics of frozen, fine-grained earth materials. The cryostructure is determined by the amount and distribution of pore ice (or ice cement) and lenses of segregated ice. The type and arrangement of ice in the frozen material will depend largely on the initial total water content of the material and the extent of moisture migration during subsequent freezing (Kudryavtsev, 1978).

cryotectonic (cry'-o-tec-ton'-ic) Said of complicated and deranged features and deposits found at glacier borders, and consisting of material that has been overturned, inverted, folded, and transported by the shoving action of glaciers. Syn: *glaciotectonic.*

cryotexture The textural characteristics of frozen, fine-grained organic and mineral earth materials cemented together with ice.

cryotic ground Ground (soil or rock) having a temperature below 0° C.

cryoturbate (cry-o-tur'-bate) A mass of soil or other unconsolidated earth material moved or disturbed by frost action, and usually coarser than the underlying material; esp. a rubbly deposit formed by solifluction. See also: *head [mass move]; pocket [soil]; rubble drift; coombe rock.* Syn: *frost soil; warp [mass move]; trail [mass move]; congeliturbate.*

cryoturbation (cry'-o-tur-ba'-tion) A collective term to describe the stirring, churning, modification, and all other disturbances of soil, resulting from frost action; it involves frost heaving, solifluction, and differential and mass movements, and it produces patterned ground. Syn: *frost stirring; frost churning; geliturbation; congeliturbation.* Also spelled: *kryoturbation.*

cryovolcanism Icy-cold volcanism, usually involving aqueous solutions with or without admixture with solids and gases (Kargel, 1995). See also: *volcanism [planet].*

cryovolcanism [extraterrestrial] Eruption of liquid or vapor phases (with or without entrained solids) of water or other volatiles that would be frozen solid at the normal temperature of an icy satellite's surface (Geissler, 2000).

cryptacanthiiform (crypt-a-canth'-ii-form) Said of a brachiopod *loop* composed of descending branches fused distally to form an *echmidium,* which bears a *hood* on the ventral anterior end (TIP, 1965, pt. H, p.143). With continued growth, the echmidium becomes deeply cleft anteriorly but still connected with descending branches.

cryptalgal (crypt-al'-gal) Said of rocks or rock structures formed "through the sediment-binding and/or carbonate-precipitating activities of nonskeletal algae" (Aitken, 1967, p.1163). The influence of these organisms is more commonly inferred than observed, hence the etymol: Greek "kryptos", "hidden, secret", + *algal.*

cryptalgalaminate (crypt′-al-ga-lam′-i-nate) Said of carbonate rocks "displaying a distinctive form of discontinuous, more or less planar lamination believed to have resulted from the activities upon and within the sediments of successive mats or films of blue-green and green algae" (Aitken, 1967, p.1164). See also: *cryptalgal; stromatolite.*

cryptarch (cryp′-tarch) A group of organic-walled microfossils of uncertain or ambiguous affinity, common in the Precambrian and consisting of hollow vesicles or cellular bodies that occur as individuals, pairs, clusters, septate or aseptate filaments, spheroids, or ellipsoids. They do not have processes but may have surface texture (Diver and Peat, 1973; Mendelson, 1993). Cf: *acritarch; leiosphere; sphaeromorph.*

crypthydrous (crypt-hy′-drous) Said of vegetal matter deposited on a wet substratum. Cf: *phenhydrous.*

cryptic (cryp′-tic) Said of reef organisms, esp. invertebrates, that live inconspicuously under corals, shells, sponges, and rocks. Cf: *coelobite; cryptofaunal; sciaphilic.*

cryptic layering Invisible layering, within and commonly parallel to the sides and bottom of igneous intrusions, identified by a regular vertical change in chemical composition of the minerals; so named because it is less obvious than *rhythmic layering.* Typical element ratios defining cryptic layering are Fe/Mg in mafic minerals and Ca/Na in plagioclase.

cryptic zoning Zoning in minerals that is not visible to the unaided eye. It may involve variation in major elements or in trace-element populations. The term is used with respect to igneous and hydrothermal systems.

cryptoaperture (cryp-to-ap′-er-ture) A pollen *endoaperture* lacking a distinct overlying *ectoaperture.*

cryptochron One of small-scale magnetic anomalies in the ocean-floor magnetic-anomaly profiles, affectionately referred to as "tiny wiggles." They may reflect short polarity subchrons or intensity fluctuations of the geomagnetic field.

cryptoclastic rock (cryp-to-clas′-tic) (a) A clastic rock whose extremely fine constituents can be seen only under the microscope. Ant: *macroclastic rock.* Cf: *microclastic rock.* (b) A carbonate sedimentary rock having an aphanic clastic texture and discrete particles whose diameters are less than 0.001 mm, and displaying little or no crystallinity under high-power magnification (Bissell and Chilingar, 1967, p.154); e.g. extremely finely comminuted carbonate dust. Cf: *cryptograined.*

cryptocrystalline (cryp′-to-crys′-tal-line) (a) Said of the texture of a rock consisting of crystals that are too small to be recognized and separately distinguished even under the ordinary microscope (although crystallinity may be shown by use of the electron microscope); indistinctly crystalline, as evidenced by a confused aggregate effect under polarized light. Also, said of a rock with such a texture. Cf: *microcrystalline; dubiocrystalline.* Syn: *microaphanitic ; microcryptocrystalline; microfelsitic; felsophyric.* (b) Said of the texture of a crystalline rock in which the crystals are too small to be recognized megascopically. This usage is not recommended "since it cannot be known that an aphanitic rock is cryptocrystalline until the microscope has shown that it is actually microscopically crystalline" (Johannsen, 1939, p.206). (c) Descriptive of a crystalline texture of a carbonate sedimentary rock having discrete crystals whose diameters are less than 0.001 mm (Bissell and Chilingar, 1967, p.103) or less than 0.01 mm (Pettijohn, 1957, p.93). Some petrographers use an upper limit of 0.004 mm.

cryptocrystalline magnesite *serpentine-hosted magnesite.*

cryptocyst (cryp′-to-cyst) A *frontal shield* in cheilostome bryozoans, formed by calcification of an inner wall grown into the zooidal cavity subparallel to and beneath the frontal wall.

cryptodeltoid (cryp-to-del′-toid) One of two plates on either side of the anal opening of a blastoid, generally overlapped aborally by the *hypodeltoid* and adjacent radial limbs, abutting against the *superdeltoid* plate adorally, bordering the *lancet plate* laterally, and infolded into *hydrospire* folds on the inner side. By adoral extension and fusion above the anal opening, cryptodeltoids may form a horseshoe-shaped *subdeltoid* developed in some genera (TIP, 1967, pt. S, p.346).

cryptodepression (cryp′-to-de-pres′-sion) A lake basin whose bottom lies below sea level (although the water surface may be above sea level).

cryptodome (cryp′-to-dome) A shallow intrusion of magma in a volcanic cone, as of dacite at Mount St. Helens, Washington, U.S.A. (Moore and Albee, 1981).

cryptoexplosion structure (cryp′-to-ex-plo′-sion) A nongenetic, descriptive term suggested by Dietz (1959, p.496-497) to designate a roughly circular structure formed by the sudden, explosive release of energy and exhibiting intense, often localized rock deformation with no obvious relation to volcanic or tectonic activity. Such structures typically show some or all of the following: wide variation in diameter (less than 1.5 km to more than 50 km); a central dome-shaped uplift with intense structural deformation, often surrounded by a *ring depression;* complex faulting and subordinate folding; widespread brecciation and shearing; and occurrence of shatter cones. Cryptoexplosion structures that contain shock-metamorphic effects, including shatter cones, are the result of hypervelocity impact by extraterrestrial bodies; others may have been produced by obscure volcanic activity. The term largely replaces the earlier term *cryptovolcanic structure.* Also called: cryptoexplosive structure. See also: *astrobleme.*

cryptofaunal (cryp-to-fau′-nal) Pertaining to small cryptic animals.

cryptogam (cryp′-to-gam) A plant that reproduces by spores rather than seeds. Examples include thallophytes, bryophytes, and pteridophytes. Cf: *phanerogam.*

cryptogene (cryp′-to-gene) adj. Said of a rock whose origin cannot be determined. Obsolete.

cryptograined (cryp′-to-grained) Said of the texture of a carbonate sedimentary rock having discrete clastic (or precipitated or flocculated) particles whose diameters are less than 0.001 mm (Bissell and Chilingar, 1967, p.103; DeFord, 1946) or less than 0.01 mm (Thomas, 1962). Some petrographers use an upper limit of 0.004 mm. Cf: *cryptoclastic rock.*

cryptohalite (cryp-to-hal′-ite) A white to gray, cubic, high-temperature mineral: $(NH_4)_2SiF_6$. The ammonium analogue of hieratite. Cf: *bararite.*

cryptolite (cryp′-to-lite) *monazite.*

cryptolithic (cryp-to-lith′-ic) Pertaining to organism living within rocks, including shells and coral.

cryptomelane (crypt′-o-mel′-ane) A black mineral: $K_{1-1.5}(Mn^{4+},Mn^{3+})_8O_{16}$. It is isostructural with hollandite and coronadite. It is tetragonal below 710°C, and monoclinic above this temperature.

cryptomphalus (cryp-tom′-pha-lus) An *umbilicus* filled with callus in certain gastropods.

cryptonelliform (cryp′-to-nel′-li-form) Said of a long brachiopod *loop* (as in the superfamily Cryptonellacea) unsupported in adults by a median septum and having a narrow transverse band.

cryptoolitic (cryp′-to-o-lit′-ic) Pertaining to an oolitic texture of such fine grain that it can be recognized only under the microscope. Also, said of a rock with such a texture.

cryptoperthite (cryp-to-perth′-ite) An extremely fine-grained variety of perthite in which the lamellae are of submicroscopic dimensions (1-5 µm wide) and are detectable only by X-rays or with the aid of the electron microscope. The potassium-rich host can be sanidine, orthoclase, or microcline; the sodium-rich lamellae can be analbite or albite. Cryptoperthite frequently displays a bluish to whitish milky luster. Cf: *microperthite.*

cryptopylome (cryp-to-py′-lome) A circular or subcircular *operculate exsystment* type in *acritarchs.* Neither the operculum nor the opening is evident prior to exsystment (e.g., *Cymbosphaeridium pilar, Gracilisphaeridium encantador*).

cryptorheic (cryp-to-rhe′-ic) Said of drainage by subterranean streams. Also spelled: cryptoreic.

cryptorhomb (cryp′-to-rhomb) In the rhombiferan echinoderms, a specialized type of pore rhomb that has external openings of simple or compound pores. Cryptorhombs are found in the rhombiferan superfamily *Hemicosmitida* (Paul, 1968, p.705).

cryptospore (cryp′-to-spore) A sporelike body often occurring in tetrads or dyads but lacking trilete or monolete haptotypic marks, ranging from the Cambrian to the Early Devonian.

cryptostome (cryp′-to-stome) Any ectoproct bryozoan belonging to the order Cryptostomata and resembling a trepostome but having a short endozone and an aperture at the bottom of the vestibule. Adj: cryptostomatous.

cryptotabulate (cryp-to-tab′-u-late) Of a dinoflagellate *cyst,* having *tabulation* expressed only by the outline of the *archeopyle,* although there may be some suggestion of a *cingulum* or *sulcus.*

cryptovolcanic structure (cryp-to-vol-can′-ic) (a) A term introduced by Branco and Fraas (1905) and originally applied to a highly deformed, strongly brecciated, generally circular structure believed to have been produced by volcanic explosions, but lacking any direct evidence of volcanic activity; e.g. volcanic rocks, hydrothermal alteration, contact metamorphism, or mineralization; type example: Steinheim Basin, Germany. Many of these structures, originally termed cryptovolcanic, are now known to have been formed by meteorite impact, and the nongenetic term *cryptoexplosion structure* is preferred. Also, the term cryptovolcanic structure "has tended to become a 'wastebasket' term and now includes many structures which are unquestionably of volcanic origin" (Dietz, 1959, p.496). (b) A circular structure lacking evidence of shock metamorphism or of meteorite impact and therefore presumed to be of igneous origin, but lacking exposed igneous rocks or obvious volcanic features; a rock structure produced by concealed volcanic activity.

Cryptozoic (Cryp-to-zo′-ic) An obsolete term for that part of geologic time represented by rocks in which evidence of life is only slight and of primitive forms (Chadwick, 1931).

cryptozoon (cryp-to-zo′-on) (a) A structure in Precambrian rocks, believed to be the remains of primitive life. (b) A hemispherical or cabbagelike algal structure of variable size, spreading somewhat above its base, composed of irregular and concentric laminae of calcite of very unequal thicknesses traversed by minute canals that branch irregularly, produced by the problematical Cambrian and Ordovician reef-forming calcareous alga of the genus *Cryptozoon.* Pl: cryptozoa.

crystal (crys′-tal) A homogeneous, solid body of a chemical element, compound, or isomorphous mixture, having a regularly repeating atomic arrangement that may be outwardly expressed by plane faces.

crystal accumulation In a magma, the development of layering by the process of *crystal settling.*

crystal axial indices *indices of lattice row.*

crystal axis (a) *crystallographic axis.* (b) One of the three edges of the chosen unit cell in a crystal lattice. (c) Any *lattice row;* it can be considered a zone axis. Syn: *axis [cryst].*

crystal-body playa (crys'-tal-bod'-y) A playa with one or more thick salt bodies at or near the surface, formed by the evaporation of a lake that once occupied the area (Stone, 1967, p.220); e.g. Searles Lake in California. Archaic.

crystal cast The filling of a crystal mold; e.g. *ice-crystal cast, salt-crystal cast.*

crystal cave A cave in which much of the surface of the roof, walls, and floor is covered with well-formed mineral crystals.

crystal chemistry The study of the relations among chemical composition, internal structure, and the physical properties of crystalline matter.

crystal class *class [cryst].*

crystal defect An imperfection in the ideal crystal structure. See also: *line defect; point defect; plane defect.* Syn: *lattice defect.*

crystal dislocation *line defect.*

crystal face (a) A planar face bounding a crystal. A crystal face may be symbolized by *Miller indices* enclosed in parentheses, e.g. (*hkl*). Syn: *face [cryst].* (b) A *rational face.* The terms are equivalent for igneous crystals, but the faces of crystals developed by recrystallization may not be rational.

crystal field An electrostatic field produced by anions located at the vertices of a polyhedron about a central cation.

crystal field splitting The energy difference between d-orbitals in a transition metal ion, produced by the influence of a *crystal field.*

crystal field stabilization energy The amount by which the total energy of the d-electrons in a transition-metal *cation* is lowered by virtue of the arrangement of surrounding anions, relative to the energy inherent in some other arrangement of anions.

crystal field theory A theory of chemical bonding in which anions surrounding a transition-metal atom are treated as point negative charges that establish a *crystal field.* The interaction between the crystal field and the d-orbitals of the cation result in modifications to the orbital energy levels that can be related to various physical and chemical effects.

crystal flotation In petrology, the floating of lighter-weight crystals in a body of magma. Cf: *crystal settling.* Syn: *flotation.*

crystal form (a) The geometric shape of a crystal. (b) An assemblage of symmetrically equivalent crystal faces. A crystal may be bounded by one or more forms, each consistent with the internal symmetry of the crystal. Crystal form may be symbolized by *Miller indices* enclosed in braces, e.g., {*hkl*}.

crystal fractionation Magmatic differentiation resulting from the separation of liquid from crystals during crystallization. Cf: *fractional crystallization [petrology].* Syn: *gravitational differentiation.*

crystal gliding Deformation of crystalline material by orderly displacement of atoms such that good crystal structure remains after the process is finished. It often produces crystal twins. See also: *twin gliding.* Syn: *gliding [cryst]; translation gliding.*

crystal habit The general shape of crystals, e.g. cubic, prismatic, fibrous. For a given type of crystal, the habit may vary from locality to locality depending on environment of growth.

crystal indices *Miller indices.*

crystallaria (crys-tal-la'-ri-a) A general term proposed by Brewer (1964, p.284) for a group of soil features consisting of single crystals or arrangements of crystals of relatively pure fractions of the soil plasma that do not enclose the matrix of the soil material but form cohesive masses whose shape and internal fabric are "consistent with their formation and present occurrence in original voids in the enclosing soil material"; e.g. spherulites, rosettes, crystal tubes, and intercalary crystals embedded in a dense soil matrix. See also: *crystal tube.*

crystal lattice The three-dimensional regularly repeating set of points that represent the translational periodicity of a crystal structure. Each lattice point has identical surroundings. There are fourteen possible lattice patterns. Lattice is the abstract pattern used to describe the internal geometric structure of crystals. Lattice and structure are not synonymous as structure refers to the concrete, real mineral material. Syn: *Bravais lattice; space lattice; direct lattice; lattice [cryst]; translation lattice.*

crystalline [cryst] (crys'-tal-line) (a) Pertaining to or having the nature of a crystal, or formed by crystallization; specif. having a crystal structure or a regular arrangement of atoms in a space lattice. Ant: *amorphous.* (b) Said of a mineral particle of any size, having the internal structure of a crystal but lacking well-developed crystal faces or an external form that reflects the internal structure. (c) Resembling a crystal; clear, transparent, pure.

crystalline [petrology] Igneous, metamorphic, or sedimentary rock texture consisting of interlocking, crystalline grains; also, a rock with such texture. Obsolete.

crystalline carbonate A carbonate sedimentary rock in which the depositional texture is not recognizable, owing to recrystallization and replacement; e.g. dolomite rock and dolomitic limestone (Dunham, 1962).

crystalline chert Obsolescent syn. of *granular chert* (Ireland et al., 1947, p.1486).

crystalline chondrite An obsolete term for a hard, metamorphically recrystallized chondrite probably to petrologic type 5 or 6.

crystalline flake *flake graphite.*

crystalline-granular (crys'-tal-line-gran'-u-lar) *granular [geol].*

crystalline-granular texture (a) A primary granular texture of a sedimentary rock, produced by crystallization from an aqueous medium; it may be exhibited by rock salt, gypsum, or anhydrite. (b) A texture of a carbonate sedimentary rock produced by dolomitization of a limestone containing packed granules (Thomas, 1962, p.197).

crystalline limestone (a) A metamorphosed limestone; a *marble* formed by recrystallization of limestone as a result of metamorphism. Syn: *cipolin.* (b) A calcarenite with crystalline calcite cement formed in optical continuity with crystalline fossil fragments by diagenesis (Pettijohn, 1957, p.407-408). Syn: *sedimentary marble.*

crystalline magnesite *serpentine-hosted magnesite.*

crystalline rock (a) An inexact but convenient term designating an igneous or metamorphic rock, as opposed to a sedimentary rock. (b) A rock consisting wholly of relatively large mineral grains, e.g. a plutonic rock, an igneous rock lacking glassy material, or a metamorphic rock. (c) The term has also been applied to sedimentary rocks, e.g. some limestones, that are composed of coarsely crystalline grains or exhibit a texture formed by partial or complete recrystallization.

crystalline structure *crystal structure.*

crystallinity [cryst] The state of being crystalline, i.e., of consisting of a regular, periodic array of atoms.

crystallinity [ign] (crys-tal-lin'-i-ty) The degree to which an igneous rock is composed of crystalline material in relation to glass.

crystallinoclastic rock (crys-tal'-li-no-clas'-tic) A clastic rock containing abundant crystalline material, such as one having a crystalline cement.

crystallite (crys'-tal-lite) (a) A broad term applied to a minute body whose small size precludes optical characterization. Crystallites represent the initial stage of crystallization of a magma or of a glass. Adj: crystallitic. Cf: *microlite [cryst]; crystalloid.* (b) An obsolete syn. of *stylolite.*

crystal-lithic tuff (crys'-tal-lith'-ic) A tuff that is intermediate between *crystal tuff* and *lithic tuff* or that is predominantly the latter. Cf: *lithic-crystal tuff.*

crystallization (crys'-tal-li-za'-tion) The process by which matter becomes crystalline, from a gaseous, fluid, or dispersed state.

crystallization differentiation The progressive change in composition of the liquid fraction of a magma as a result of the crystallization of mineral phases that differ in composition from the magma. The process may be *equilibrium crystallization* or fractional crystallization or some combination of the two.

crystallization fabric A term used by Friedman (1965, p.643) for the size and mutual relations of mineral crystals in sedimentary rocks such as evaporites, chemically deposited cements, and recrystallized limestones and dolomites. Cf: *crystallization texture.*

crystallization index In igneous petrology, specif. igneous differentiation, the number that is calculated from the system anorthite-diopside-forsterite and that represents the sum (in weight percent) of normative anorthite, magnesian diopside, normative forsterite, normative enstatite converted to forsterite, and magnesian spinel calculated from normative corundum in ultramafic rocks (Poldervaart and Parker, 1964, p.281). Abbrev: CI. Cf: *petrogeny's primitive system.*

crystallization interval (a) The interval of temperature (or of pressure) between the formation of the first crystal and the disappearance of the last drop of a magma upon cooling or during magma ascent, usually excluding late-stage aqueous fluids. (b) More specifically, with reference to a given mineral, the temperature range or ranges over which that particular phase is in equilibrium with liquid. In the case of equilibria along reaction lines or reaction surfaces, crystallization intervals defined in this way include temperature ranges in which certain solid phases are actually decreasing in amount with temperature decrease. Syn: *freezing interval.*

crystallization layering Obsolete syn. of *phase layering.*

crystallization magnetization *chemical remanent magnetization.*

crystallization remanent magnetization *chemical remanent magnetization.*

crystallization texture A term used by Friedman (1965, p.643) for the shape of mineral crystals in sedimentary rocks such as evaporites, chemically deposited cements, and recrystallized limestones and dolomites. Cf: *crystallization fabric.*

crystallized coal (crys'-tal-lized) *cone-in-cone coal.*

crystallizing force (crys'-tal-liz'-ing) The expansive force of a crystal that is forming within a solid medium. The force varies according to crystallographic direction. Syn: *force of crystallization.*

crystalloblast (crys'-tal-lo-blast') A crystal of a mineral produced entirely by metamorphic processes. Cf: *idioblast; holoblast; hypidioblast; xenoblast.*

crystalloblastesis (crys'-tal-lo-blas'-te-sis) Deformation accomplished by metamorphic recrystallization (Knopf and Ingerson, 1938).

crystalloblastic (crys'-tal-lo-blas'-tic) (a) Pertaining to a *crystalloblast.* (b) Said of a crystalline texture produced by metamorphic recrystallization under conditions of high viscosity and directed pressure, in contrast to igneous rock textures that are the result of successive crystallization of minerals under conditions of relatively low viscosity and nearly uniform pressure (Becke, 1903). Cf: *heteroblastic; homeoblastic.*

crystalloblastic order *crystalloblastic series.*

crystalloblastic series An arrangement of metamorphic minerals in order of decreasing tendency to form crystal faces, so that minerals high on the list tend to assume *idioblastic* outlines at surfaces of contact with simultaneously developed minerals lower in the series (Becke, 1913). Syn: *crystalloblastic order; idioblastic series.*

crystalloblastic strength *form energy.*

crystallogeny (crys-tal-log'-e-ny) That branch of crystallography which deals with crystal growth.

crystallographic (crys'-tal-lo-graph'-ic) Pertaining to crystallography or to the properties of a crystal.

crystallographic axis One of three imaginary lines in a crystal (four in a hexagonal crystal) that pass through its center; it is used as a reference in describing crystal structure and symmetry. One or all of the crystallographic axes may coincide with axes of symmetry. Syn: *crystal axis.*

crystallographic orientation The relation of the axes or planes of a given crystal to some other established directions in space, e.g. geographic or geologic lines or planes.

crystallographic plane Any plane, crystal face, cleavage or lattice plane that can be described mathematically in terms of the lengths and directions of the crystallographic axes.

crystallographic point group *class [cryst].*

crystallographic preferred orientation Alignment of crystallographic axes of crystals in a deformed rock body, e.g. a quartzite in which the *c*-axis of most grains parallel a particular direction or limited range of directions.

crystallographic texture A texture of mineral deposits formed by replacement or exsolution, in which the distribution and form of the inclusions are controlled by the crystallography of the host mineral.

crystallography (crys-tal-log'-ra-phy) The study of crystals, including their growth, structure, physical properties, and classification by form.

crystalloid (crys'-tal-loid) n. A microscopic crystal which, when examined under a microscope, polarizes light but has no crystal outline or readily determinable optical properties. Cf: *crystallite; microlite [cryst].*

crystallolith (crys-tal'-lo-lith) A crystalline *coccolith;* e.g. a disciform holococcolith.

crystallothrausmatic (crys-tal'-lo-thraus-mat'-ic) Igneous rocks with an orbicular texture in which early phenocrysts form the nuclei of the orbicules (Eskola, 1938, p.476). Obsolete. Cf: *allothrausmatic; isothrausmatic; heterothrausmatic; homeothrausmatic.*

crystal mold A cavity left by solution or sublimation of a crystal (as of salt, ice, or pyrite) embedded in soft, fine-grained sediment.

crystal mush [petrology] Partially crystallized magma; "an aggregate of solid crystals lubricated by compressed water vapor" (Krauskopf, 1967, p.419).

crystal mush [salt] A water-saturated loose aggregate of saline mineral crystals with little or no muddy matrix.

crystal optics The study of the transmission of light in crystals; the concern of *optical crystallography.*

crystal plastic deformation A deformation caused by one or more of the mechanisms of dislocation glide, dislocation creep, or diffusional creep on the scale of the crystal lattice.

crystal plasticity The grain-scale deformation mechanism that involves the breaking, rearrangement, and formation of chemical bonds without the loss of continuity.

crystal pool In a cave, standing water lined with crystals of calcite.

crystal raft Thin, usually millimetric, platelike agregate of crystals formed at the air-water interface during the evaporation of saline water, e.g., halite, mirabilite, epsomite, and other minerals. Commonly rafts float, attached to the meniscus surface of the water, sink with minor wind action, and may form chiplike, imbricate breccias.

crystal sandstone (a) A sandstone in which the quartz grains have been enlarged by deposition of silica so that the grains show regenerated crystal facets and sometimes nearly perfect quartz euhedra. Crystal sandstones of this nature sparkle in bright sunlight. (b) A sandstone in which calcite has been deposited in the pores in large patches or units having a single crystallographic orientation, resulting in a "poikiloblastic" or "luster-mottling" effect. In some rare sandstones with incomplete cementation, the carbonate occurs as sand-filled scalenohedra of calcite sand crystals (Pettijohn et al., 1972, p.164).

crystal sedimentation *crystal settling.*

crystal seeding The use of a *seed crystal* or foreign particle in a solution to initiate crystallization of the solute.

crystal settling In a magma, the sinking of crystals due to their greater density, sometimes aided by magmatic convection. It results in *crystal accumulation,* which develops layering. Cf: *crystal flotation.* Syn: *crystal sedimentation.*

crystal shrub *Shrub* form, commonly composed of calcite, that exhibits crystal habits and regular repeating morphologies. Crystal shrubs commonly have attriburtes of noncrystallographic as well as crystallographic dendrites. They are the products of both bacterially induced as well as abiotic mineral precipitation (Chafetz and Guidry, 1999).

crystal sorting The separation, by any process, of crystals from a magma, or of one crystal phase from another during crystallization of the magma.

crystal structure The regular, orderly, and repeated arrangement of atoms in a crystal, the translational properties of which are described by the crystal lattice or space lattice. Syn: *crystalline structure.*

crystal-structure refinement A procedure for determining, from diffraction data, the *atomic parameters* in a *crystal structure* with high precision. It is usually carried out by a least-squares procedure applied to data obtained by X-ray or neutron diffraction from a single crystal; adjustable parameters are modified until the observed and calculated structure amplitudes are close enough to produce a small *R-factor.* Also, the results of such a procedure. Syn: *structure refinement.* Cf: *Rietveld refinement.*

crystal-structure solution The determination of the approximate *fractional coordinates* for the atoms of a *crystal structure* by any of several methods, usually involving diffraction data. Syn: *structure solution.*

crystal system One of six groups or classifications of crystals according to the symmetry of their crystal faces, and having characteristic dimensional equivalences in the lattices or axes of reference. The systems are: *isometric system, hexagonal system, tetragonal system, orthorhombic system, monoclinic system,* and *triclinic system.* Within the six systems there is a total of 32 crystal classes. Syn: *system [cryst].*

crystal tube A type of *crystallaria* consisting of masses of crystals filling or partly filling relatively large tube-shaped or acicular voids in soil material. It is usually formed by crystallization from the walls inward.

crystal tuff A tuff that consists predominantly of crystals and fragments of crystals. Cf: *crystal-vitric tuff; crystal-lithic tuff.*

crystal-vitric tuff (crys'-tal-vit'-ric) A tuff that consists of fragments of crystals and volcanic glass. Cf: *crystal tuff; vitric tuff.*

crystal zone (a) Three or more nonparallel crystal faces, the edges of intersection of which are parallel to a common line or lattice row called the *zone axis.* (b) A result of *zoning [cryst].*

crystocrene (crys'-to-crene) *chrystocrene.*

crystosphene (crys'-to-sphene) A buried mass or sheet of clear ice developed by a wedging growth between beds of other material (Tyrrell, 1904, p.234), such as the freezing of springwater rising and spreading laterally beneath alluvial deposits or under swamps in a tundra region. Cf: *chrystocrene.*

CSAMT Controlled-source audio-magnetotelluric technique. A technique similar to magnetotellurics in the range 1 Hz to 20 kHz using a source transmitter at least five skin-depths distant from the receiver.

csiklovaite (csik-lo'-va-ite) A mixture of tetradymite, galenobismutite and bismuthinite.

ct *carat.*

CTD An instrument composed of sensors for conductivity, temperature, and pressure and used to measure conductivity and temperature (from which salinity can be calculated) as a function of depth in the ocean.

ctenoid cast (cte'-noid) A very rare, toothlike *sole mark* having the form of an obliquely cut, longitudinally ribbed cylinder and probably representing a *bounce cast* made in mud by plant stems of *Equisetites* that intermittently touched bottom as they were carried along by a current of water. The term was introduced as "ctenoid marking" by Beasley (1914), who likened the structure to the large tortoise-shell comb worn in women's hair during the early Victorian period. Etymol: Greek "ktenos", "comb".

ctenolium (cte-no'-li-um) A comblike row of small teeth on the lower side of the *byssal notch* in some pectinacean mollusks.

ctenostome (cte'-no-stome) Any ectoproct bryozoan belonging to the order Ctenostomata and characterized by the presence of comblike protrusions at the mouth.

cualstibite (cu-al'-stib-ite) A blue-green trigonal mineral: $Cu_6Al_3Sb^{5+}{}_3O_{12}{\cdot}12H_2O$.

cubanite (cu'-ban-ite) A bronze-yellow orthorhombic mineral: $CuFe_2S_3$. Syn: *chalmersite.*

cube A crystal form of the *isometric system,* consisting of six symmetrically equivalent (ideally square) and mutually perpendicular faces, with indices of *{100}.*

cube ore *pharmacosiderite.*

cube spar *anhydrite.*

cubic cleavage (cu'-bic) Mineral cleavage parallel to the faces of a cube; e.g., in galena.

cubic close packing In a crystal, close packing of spheres by stacking close-packed layers in the sequence ABCABC etc. Cf: *hexagonal close packing.* See also: *close-packed structure.*

cubic coordination An atomic structure or arrangement in which an ion is surrounded by eight ions of opposite sign, whose centers form the vertices of a cube. An example is found in the structure of cesium chloride. Partial syn: *hexahedral coordination.*

cubic law An empirical observation that the volumetric flux of water through a fracture bounded by smooth parallel walls is proportional to the cube of its aperture. Originally derived by Boussinesq in 1868.

cubic packing The "loosest" manner of systematic arrangement of uniform solid spheres in a clastic sediment or crystal lattice, characterized by a unit cell that is a cube whose eight corners are the centers of the spheres involved (Graton and Fraser, 1935). An aggregate with cubic packing has the maximum porosity (47.64%). Cf: *rhombohedral packing; open packing.*

cubic plane In a crystal of the cubic system, any plane at right angles to a crystallographic axis.

cubic system *isometric system.*

cuboargyrite A metallic grayish black cubic mineral of the *galena* group: $AgSbS_2$.

cuboctahedron (cub'-oc-ta-he'-dron) A cubic crystal shape bounded by both the six equal squares of the cube and the eight equal triangles of the octahedron, the latter cutting off the corners of the former.

cubo-dodecahedron (cub'-o-do'-dec-a-he'-dron) A crystal in the cubic system that is bounded by cube and dodecahedron forms.

cuchilla (cu-chil'-la) A term used in the SW U.S. for a sharply edged crest of a *sierra.* Etymol: Spanish, "large knife".

cuesta (cues'-ta) (a) A hill or ridge with a gentle slope on one side and a steep slope on the other; specif. an asymmetric ridge (as in the SW U.S.) with one face (dip slope) long and gentle and conforming with the dip of the resistant bed or beds that form it, and the opposite face (scarp slope) steep or even clifflike and formed by the outcrop of the resistant rocks, the formation of the ridge being controlled by the differential erosion of the gently inclined strata. Originally, the term applied to the steep slope or scarp that terminates a gently sloping plain at its upper end. (b) A ridge or belt of low hills formed between lowlands in a region of gently dipping sedimentary rocks (as on a coastal plain), having a gentle slope conforming with the dip of the rocks and a relatively steep slope descending abruptly from its crest. Etymol: Spanish, "flank or slope of a hill, hill, mount, sloping ground". Cf: *hogback [geomorph].* Syn: *wold; scarped ridge; escarpment.*

cueva A Spanish word for cave, especially one that is horizontal, or nearly so.

cuirasse A ferruginous *duricrust.*

Cuisian (Cuis'-i-an) Obs. European stage: Lower Eocene (above Ypresian, below Lutetian). See also: *Bruxellian.*

cul-de-sac [ice] (cul'-de-sac) *blind lead.*

cul-de-sac [karst] A closed abandoned *sinking stream* that has been partially filled.

cul-de-sac [speleo] A passage in a cave that has only one entrance.

culet (cu-let') The small facet that is polished parallel to the girdle plane across what would otherwise be the sharp point or ridge that terminates the pavilion of a diamond or other gemstone. Its principal function is to reduce the possibility of damage to the gem.

culm (a) *kolm.* (b) The anthracite contained in the series of shales and sandstones of North Devon, England, known as the Culm measures. (c) Fine particles of anthracite. (d) Coal dust or fine-grained waste from anthracite mines.

culmen (cul'-men) A narrow ridge between adjoining *crenellae* on the articular surface of a *columnal* of a crinoid or blastoid. Pl: culmina.

culmination (cul-mi-na'-tion) The highest point of a structural feature, e.g. of a dome, anticlinal crest, synclinal trough, or nappe. The axis of an anticline may have several culminations that are separated by saddles. See also: *crest [struc geol].* Syn: *axis culmination; apex [struc geol].*

cultural features Items in a spatial data base or shown on a map representing human-made objects or features (Krzanowski et al., 1993).

culture (cul'-ture) The details of a map, representing human works (such as roads, railroads, buildings, canals, trails, towns, and bridges), as distinguished from natural features; they are usually printed in black on a topographic map. The term also includes political boundary lines and place names.

culvert (cul'-vert) Any covered structure, not classified as a bridge, that constitutes a transverse drain, waterway, or other opening under a road, railroad, canal, or similar structure.

cumacean (cu-ma'-ce-an) Any malacostracan belonging to the order Cumacea, characterized chiefly by the long, slim, subcylindrical pleon which is usually strongly differentiated from the broad, commonly inflated pereion and cephalon (TIP, 1961, pt.R, p.368).

cumberlandite (cum'-ber-land-ite') A coarse-grained ultramafic rock with olivine crystals (approximately 50 percent of the total rock) in a groundmass of ilmenite and magnetite (together, 40 percent), labradorite, and accessory spinel. The state rock of Rhode Island; the name, given by Wadsworth in 1884, is for the town of Cumberland in that state. Obsolete.

cumbraite (cum'-bra-ite) A porphyritic extrusive rock composed of phenocrysts of calcic plagioclase (bytownite, anorthite) in a groundmass of labradorite, orthopyroxene, clinopyroxene, and abundant glass; its chemical composition is andesitic rather than basaltic. See also: *inninmorite.* Named by Tyrrell in 1917 for Great Cumbrae, Scotland. Obsolete.

cumengéite A deep-blue or light indigo-blue tetragonal mineral: $Pb_{21}Cu_{20}(OH)_{40}Cl_{42}$. Also spelled: cumengeite; cumengite.

cummingtonite (cum'-ming-ton-ite') A dark green, brown, gray, or beige monoclinic mineral of the *amphibole* group: $(Mg,Fe^{2+})_2(Mg,Fe^{2+})_5Si_8O_{22}(OH)_2$. It is dimorphous with anthophyllite, and typically contains calcium and manganese. Cummingtonite occurs in metamorphosed ironstone, mafic and ultrabasic rocks, some dacites and rhyolites, and as a component of uralite. Its iron-rich variety is *grunerite.*

cumulate (cu'-mu-late) n. An igneous rock formed by the accumulation of crystals that settle out from a magma by the action of gravity. Syn: *accumulative rock; cumulus.*

cumulate crystals Saline mineral crystals formed by precipitation at the air-water interface that accumulate when they sink to the bottom of a saline body of water.

cumulate texture Aggregate of touching, medium to large, subhedral to euhedral mineral grains (cumulus grains) and interstitial patches of other minerals (post cumulus grains) formed by accumulation of crystals in a magma chamber.

cumulative curve (cu'-mu-la'-tive) *cumulative frequency distribution.*

cumulative frequency distribution For grouped data, a step function in which each group is added to the preceding one until the total number of observations is included. For a continuous random variable, a function whose values are the probabilities that the variable is equal to or less than the specified value. Both functions usually sum to 100%. Syn: *cumulative curve.*

cumulative-size distribution For all unconsolidated aggregate of sediment, soil, or rock particles, a curve that shows the percentage of the material that is finer than or coarser than a particular grain size, for the entire range of particle sizes represented in the sample. Cf: *particle-size distribution.*

cumulite (cu'-mu-lite) A cloudy aggregate of globulites commonly found in glassy igneous rocks. Obsolete.

cumulo-dome (cu'-mu-lo-dome') Obsolete syn. of *volcanic dome.*

cumulophyre (cu'-mu-lo-phyre') An igneous rock characterized by *cumulophyric* texture. Obsolete.

cumulophyric (cu'-mu-lo-phyr'-ic) Porphyritic igneous texture in which the phenocrysts, not necessarily of the same mineral, are clustered in irregular groups. Obsolete. Cf: *synneusis; gregaritic.* Syn: *glomeroporphyritic; glomerophyric.*

cumulo-volcano (cu'-mu-lo-vol-ca'-no) Obsolete syn. of *volcanic dome.*

cumulus (cu'-mu-lus) *cumulate.*

cumulus crystal A unit of the *cumulate.*

cuneate (cu'-ne-ate) Said of structures, e.g. certain leaves, that are wedge-shaped or triangular, with the narrower end at the base.

cunette (cu-nette') A small channel dug in the bottom of a larger channel or conduit for the purpose of concentrating the flow at low-water stages.

cuniculus (cu-nic'-u-lus) A continuous tunnel-like cavity formed in foraminifers (such as Verbeekinidae) by strong septal fluting, the opposed folds of adjacent septa meeting to form continuous spiral sutures with vaulted arches between, and serving to connect adjoining chambers from one foramen to the next. Pl: cuniculi.

Cunnersdorf twin law (Cun'-ners-dorf) A rare type of normal twin law in feldspar, in which the twin plane is $(\bar{2}01)$.

cup (a) A greatly expanded *basal cavity* beneath the anterior or posterior half of some conodont elements. (b) The calyx of a crinoid. (c) A skeleton in archaeocyathids, generally a double-walled inverted cone with an open central cavity and the space between the walls divided by septa, rods, tabulae or other structures (TIP, 1972, pt. E, p.3).

cupalite (cup'-al-ite) A metallic yellowish gray orthorhombic mineral: (Cu,Zn)Al.

cup-and-ball joint *ball-and-socket joint.*

cup coral *solitary coral.*

cup crystal An ice crystal with stepped surfaces in the form of a hollow hexagonal cup, one side of which may be undeveloped and appear to be rolled up. It is a common form of *depth hoar.*

cupel (cu'-pel) A ceramic cup form in which *cupellation* is performed; it is similar in function to a crucible but more of a hearth than a vessel (Ramage and Craddock, 2000).

cupellation (cu-pel-la'-tion) A process of refining precious metals in which the raw materials are mixed with lead, heated, and exposed to air, allowing the metals to oxidize, form lead oxide, and separate from the precious metals. It has been practiced since ancient times and was described as early as the Renaissance (Ramage and Craddock, 2000).

cupid's dart (cu'-pid's) *flèche d'amour.*

cupola [eco geol] The dome-shaped apex of a steep-sided stock or plug with diameters from 200 m to 1000 m which typically host post-magmatic hydrothermal mineralization. Especially important ore controls in *Climax-type* molybdenum deposits.

cupola [intrus rocks] (cu'-po-la) A large upward projection of the roof of an igneous intrusion into the country rock above. The cone-shaped apex of steep-sided stocks or plugs with diameters from 200-1000 m which typically host post-magmatic hydrothermal mineralization. Especially important ore controls in *Climax-type* molybdenum deposits. Cf: *roof pendant.*

cupola [karst] A hemispheric hill of limestone. Etymol: French cupole.

cupola [paleont] A large vaulted dome in nassellarian and spumellarian radiolarians.

cupolate (cu'-po-late) Said of a button-shaped scleractinian *corallite* with a flat base and a highly convex oral surface.

cupped pebble A pebble whose upper side has been subject to solution, often being so corroded that it becomes a mere shell (Scott, 1947). Cf: *pitted pebble.*

cup reef A small vase-shaped reef; in Bermuda, these reefs range up to 10 m high by 30 m across, and crustose coralline algae area a principal component. Cf: *boiler; breaker [reef].*

cuprite (cu'-prite) A red (crimson, scarlet, vermilion, brownish-red) cubic mineral: Cu^{1+}_2O. It is an important ore of copper, and occurs as a secondary mineral in the zone of weathering of copper lodes. Syn: *red copper ore; red oxide of copper; ruby copper; octahedral copper ore.*

cuproadamite (cu'-pro-ad'-am-ite) A pale green mineral of doubtful validity: $(Cu,Zn)_2(OH)AsO_4$. It may be olivenite or copper-bearing adamite.

cuproauride (cu-pro-aur′-ide) A metallic mineral of doubtful validity, possibly Cu_3Au_2.
cuprobismutite (cu-pro-bis′-mut-ite) A metallic bluish-gray monoclinic mineral: $Cu_2Bi_7S_{12}$. It is dimorphous with emplectite, and was formerly regarded as a mixture of bismuthinite and emplectite.
cuprocopiapite (cu′-pro-co-pi-a′-pite) An yellow, orange, greenish-yellow triclinic mineral of the *copiapite* group: $CuFe^{3+}_4(SO_4)_6(OH)_2{\cdot}20H_2O$.
cuprodescloizite (cu′-pro-des-cloi′-zite) *mottramite.*
cuproiridsite (cu′-pro-ir-id′-site) A metallic gray cubic mineral: $CuIr_2S_4$. A member of the *linnaeite* group.
cupromakovickyite A metallic gray monoclinic mineral: $Cu_4AgPb_2Bi_9S_{18}$.
cupropavonite (cu′-pro-pav′-on-ite) A metallic gray monoclinic mineral occurring as an exsolution product with pavonite: $AgPbCu_2Bi_5S_{10}$.
cuprorhodsite (cu′-pro-rhod′-site) A metallic gray cubic mineral: $CuRh_2S_4$. A member of the *linnaeite* group.
cuprorivaite (cu′-pro-ri′-va-ite) A blue tetragonal mineral: $CaCuSi_4O_{10}$. A member of the *gillespite* group.
cuproscheelite (cu-pro-schee′-lite) A mixture of scheelite & cuprotungstite.
cuprosklodowskite (cu′-pro-sklo-dow′-skite) A strongly radioactive, greenish-yellow or grass-green, triclinic secondary mineral: $(H_3O)_2Cu(UO_2)_2(SiO_3OH)_2{\cdot}6H_2O$.
cuprospinel (cu′-pro-spi-nel′) A metallic black cubic mineral of the *spinel* group: $(Cu,Mg)Fe^{3+}_2O_4$.
cuprostibite (cu′-pro-stib′-ite) A metallic steel-gray tetragonal mineral with a violet-red tint: $Cu_2(Sb,Tl)$.
cuprotungstite (cu′-pro-tung′-stite) A green tetragonal mineral: $Cu_2WO_4(OH)_2$.
cuprouranite (cu′-pro-u′-ran-ite) *torbernite.*
cupule (cu′-pule) (a) A cuplike *involucre,* esp. of the acorn, whose bracts are adherent at the base and may or may not be free upwards. (b) A free sheathing structure from the peduncle, investing one or more seeds.
curb In a *lava tube,* a low, narrow *bench* (Larson, 1990, p.24).
curetonite (cur′-e-ton-ite) A yellow-green monoclinic mineral: $Ba(Al,Ti)(PO_4)(OH,O)F$.
curie (cu′-rie) A unit of measurement of radioactivity, defined as the amount of a *radionuclide* in which the decay rate is 37 billion disintegrations per second, which is approximately equal to the decay rate of one gram of pure radium.
Curie balance (Cu′-rie) A *magnetic balance* used to determine *saturation magnetization* as a function of temperature, hence to determine *Curie point.*
curienite (cu′-rie-nite) A yellow orthorhombic mineral: $Pb(UO_2)_2(VO_4)_2{\cdot}5H_2O$. It forms a series with francevillite.
Curie point The temperature above which thermal agitation prevents spontaneous magnetic ordering. Specifically, the temperature at which the phenomenon of ferromagnetism (or ferrimagnetism) disappears and the substance becomes simply paramagnetic. Cf: *Néel point.* Syn: *Curie temperature.*
Curie's law The statement that magnetic susceptibility is inversely proportional to absolute temperature. It is applicable to substances that do not show spontaneous magnetic order at low temperatures. Cf: *Curie-Weiss Law.*
Curie temperature *Curie point.*
Curie-Weiss law The statement that the magnetic susceptibility of a ferromagnetic material above its Curie point is inversely proportional to the difference between actual temperature and the Curie point. Cf: *Curie's law.*
curite (cu′-rite) An orange-red, orthorhombic radioactive mineral: $Pb_3(UO_2)_8O_8(OH)_6{\cdot}2H_2O$. It is an alteration product of uraninite.
curl (a) A term used in southern U.S. for a bend in a stream. (b) An eddy in a stream.
curly bedding (curl′-y) *convolute bedding.*
curly cannel A syn. of *cannel coal,* so named for its conchoidal fracture.
current (cur′-rent) (a) The part of a fluid body, esp. as air or water, that is moving continuously in a definite direction, often with a velocity much swifter than the average, or in which the progress of the fluid is principally concentrated. (b) A horizontal movement or continuous flow of water in a given direction with a more or less uniform velocity, producing a perceptible mass transport, set in motion by winds, waves, gravity, or differences in temperature and density, and of a permanent or seasonal nature; esp. an *ocean current.* (c) The velocity of flow of a fluid in a stream.
current bedding Any bedding or bedding structure produced by current action; specif. cross-bedding resulting from water or air currents of variable direction. The term is used (esp. in Great Britain) as a syn. of *cross-bedding,* but such usage is not recommended (Middleton, 1965, p.248). See also: *ripple bedding.* Syn: *false bedding.*
current crescent (a) A small, semicircular or U-shaped rounded ridge, convex upcurrent, commonly with a pit in the center, and developed on a muddy surface by current action (Peabody, 1947, p.73). (b) A flute cast of a horseshoe-shaped moat eroded on the upcurrent side of a pebble, shell, or other obstacle. Syn: *horseshoe flute cast; crescent cast; crescent scour; crescentic mark [sed].*
current density The current per unit area perpendicular to the direction of current flow.
current direction *set [currents].*
current electrode A metal contact with the ground used to facilitate current flow through the ground.
current lineation (a) A form of *parting lineation* that is characterized by low (a few grain diameters high) linear ridges spaced millimeters to centimeters apart. The long axes of the ridges extend parallel to the direction of flow. Current lineation is a common structure on surfaces of *upper plane bed* deposits but has also been reported on hummocky surfaces associated with *hummocky cross-stratification.* Cf: *parting lineation; parting-step lineation; primary current lineation.* (b) A term used for *parting lineation* by Stokes (1947).
current mark (a) Any structure formed by the action of a current of water, either directly or indirectly, on a sedimentary surface; e.g. a *scour mark* made by the current itself, or a *tool mark* formed by solid objects swept along by the current (Dzulynski and Sanders, 1962). See also: *flow mark.* Syn: current marking. (b) An irregular structure made by a tidal current in the beach zone, consisting of a small depression extending toward the shore from the lee side of an obstruction. (c) A *current ripple.* Kindle (1917, p.36) used the term to designate the linguoid variety of current ripple.
current meter Any one of numerous instruments for measuring the speed alone, or both speed and direction, of flowing water, as in a stream or the ocean; it is usually activated by a wheel equipped with a set of revolving vanes or cups whose rate of turning is proportional to the velocity of the current.
current rip A *rip* consisting of small waves formed on the sea surface by the meeting of opposite currents.
current ripple An *asymmetric ripple* with a sharp or rounded crest between rounded troughs, formed by currents of air or water moving more or less constantly in a uniform direction over a sandy surface (such as a stream bar, tidal flat, beach, or sand dune), the ripple slowly migrating downcurrent much like a miniature sand dune. See also: *ripple; linguoid ripple; rhomboid ripple; normal ripple.* Cf: *oscillation ripple.* Syn: *current mark; parallel ripple.*
current-ripple cast (cur′-rent-rip′-ple) A term used by Kuenen (1957, fig.6) for a sedimentary structure now known as a *transverse scour mark.*
current ripple mark *current ripple.*
current rose A graphic representation of current direction for a given ocean area over a period of time; it indicates by means of proportional radiating arrows the direction toward which the prevailing current flows and the percentage frequency of any given direction of flow.
current scour A term introduced by Dzulynski and Walton (1965, p.40) to refer to scour marks produced by a current, without the presence of an obstacle being essential to their formation. Cf: *scour mark; current mark; obstacle scour.*
curtain fold Cylindrical fold having radial axial trace and steeply plunging hinge within a diapir, possibly incorporating sheath folds originally formed in the source layer (Jackson and Talbot, 1991).
curtain of fire A row of coalescing lava fountains along a fissure; a typical feature of a Hawaiian-type eruption (Macdonald, 1972, p.214).
curtisite (cur′-tis-ite) *idrialite.*
curvatura (cur-va-tu′-ra) A visible line of some mid-Paleozoic trilete spores that connects the extremities of the ends of the laesura and outlines the contact areas; e.g. a "curvatura perfecta" having three lines complete all around the spore's proximal face, or a "curvatura imperfecta" having forklike projections from the radial ends of the laesura but not joining with their neighbors. Pl: curvaturae.
curvature (cur′-va-ture) (a) *terminal curvature.* (b) *outcrop curvature.* (c) *earth curvature.*
curvature correction An adjustment applied to an observation or computation (e.g. of difference in elevation) to allow for *earth curvature.* In geodetic spirit leveling, the effects of curvature and of atmospheric refraction are considered together, and tables have been prepared from which combined corrections can be taken.
curvature of gravity A vector calculated from torsion-balance data indicating the shape of the gravity equipotential surface. It points in the direction of the longer radius of curvature.
curved fracture cleavage A pattern of cleavage surfaces in graded beds which cut more directly across the lower, coarser parts and curve to a more diagonal direction in the upper, finer parts, thus being convex outward from an anticlinal axis (Muller, 1965).
curved path The curvature of a seismic *raypath* because of variation in velocity. The curvature is usually concave upward because velocity usually increases with depth.
curvilinear regression Statistical models in which one variable is expressed as a *polynomial* function of another variable. Parameters of the model equation usually are estimated by *least squares* methods. The equations are referred to as "curvilinear" because polynomials are linear functions, and higher order polynomials define curved lines. Cf: *regression analysis.*
curviplanar (cur-vi-pla′-nar) Pertaining to a surface or form derived from the curving of a plane about one or more axes. Cf: *conchoidal.*
cushion (cush′-ion) An artificial pool designed to absorb the kinetic energy of falling water and so prevent erosion and reduce vibration.
cusp [coast] One of a series of sharp, seaward-projecting points of beach material, built by wave action and separated from its neighbors by smoothly curved shallow re-entrants. Cusps are spaced at more or less regular intervals along the shoreline, and generally at right angles to it; distance between

them ranges from less than a meter to many kilometers. Cf: *beach cusp; storm cusp; giant cusp; cuspate spit; cuspate foreland.* The term "beach cusp" is frequently and loosely used as a synonym.

cusp [geomorph] A landform characterized by a projection with indentations of crescent shape on either side; e.g. a *meander cusp.*

cusp [paleont] A spinelike, fanglike, or cone-shaped structure (i.e., a large denticle) located above the basal cavity of conodont elements. It comprises the entire element in simple conodont elements. See also: *anticusp.* Syn: *apical denticle.*

cusp-and-caries texture *caries texture.*

cuspate bar (cusp'-ate) A seaward-pointing, doubly crescentic bar uniting with the shore at each end. It may be formed by a single spit growing from the shore and then turning back to meet it, or by two spits growing obliquely from the shore and converging to form a bar of sharply cuspate form. Cf: *V-bar; looped bar.*

cuspate delta A tooth-shaped delta in which a single dominant river builds the delta forward into a lake or sea while vigorous wave action spreads the deltaic deposits uniformly on either side of the river mouth to form two curving beaches, each concave toward the water; e.g., the delta of the Tiber River on the Mediterranean Sea or the Sao Francisco in Brazil. Cf: *arcuate delta; wave-dominated delta.*

cuspate foreland The largest cusp, occurring as a cape or as a broadly triangular point of sand or shingle, with the apex pointing seaward, along an open coast. On some coasts, cuspate forelands measure many kilometers from apex to apex and extend seaward for several kilometers. They are formed by long-continued shore drifting of sediment, as by the convergence of separate spits or beach ridges from opposed directions, or by the progradation of cuspate bars or barriers. Term originated by Gulliver (1896, p.401). Examples: Cape Canaveral, Fla., and Cape Hatteras, N.C.

cuspate-foreland bar (cusp'-ate-fore'-land) (a) A transition form between a *compound cuspate bar* and a *cuspate foreland.* (b) A *bar* produced where a cuspate bar enclosing a triangular lagoon or marsh is prograded by the addition of successive beach ridges on its seaward side (Johnson, 1919, p.324-325). Cf: *barrier [coast].*

cuspate-lobate folds Folds in which adjacent hinges are alternately rounded and pointed; thought to be formed by buckling layers with low competence relative to their matrix.

cuspate reef A *wall reef* whose ends curve leeward and border the passages between adjacent reefs (Maxwell, 1968, p.99, 101).

cuspate ripple One of a series of linguoid ripples arranged in phase, in rows parallel to the flow (Allen, 1968, p.66).

cuspate ripple mark *cuspate ripple.*

cuspate spit A prominent point commonly extending from a barrier island into a bay or lagoon (Shepard, 1952, p.1911); the distance between the crescentic tips is 2 km or more.

cusp cast A term introduced by Spotts and Weser (1964, p.199) for a crescentic, asymmetric sole mark that lacks a deeper upstream end and that is not elongated in the downstream direction. The original depression responsible for the cast may represent current scour.

cuspidine (cus'-pi-dine) A colorless, white, green, or pink monoclinic mineral: $Ca_4Si_2O_7(F,OH)_2$. Syn: *custerite.*

cusplet (cusp'-let) A minor *beach cusp,* measuring 1.5 m or less between the tips of the crescent, occurring in the swash zone; it has a short life-span, appearing and disappearing with the turn of the tide.

cusp-ripple A term used by McKee (1954, p.60) for a crescent-shaped current ripple resulting from an "irregular and fluctuating" stream. See also: *linguoid ripple.*

custerite (cus'-ter-ite) *cuspidine.*

cut [gem] (a) The shape and style in which a gem has been fashioned; e.g. *brilliant cut* or *step cut.* (b) The proportions and finish of a diamond.

cut [geol] v. To excavate or hollow out a depression, channel, or furrow by erosion. n. (a) A notch, depression, channel, inlet, or other incision produced by erosion or natural excavation, as by water or waves. (b) A passage or space from which material has been excavated, such as a road cut. Also, the material excavated.

cutan (cu'-tan) A modification of the texture, structure, or fabric of a soil material (such as a soil aggregate, ped, or skeleton grain) along a natural surface within it, caused by a concentration of a particular soil constituent. It can be composed of any of the component substances of the soil material. Examples: *argillan; mangan.* Etymol: Latin cutis, "a coating, surface, or skin".

cut and fill [eng] The procedure of excavating material in one place and depositing it, usually compacted, as fill in an adjacent place; e.g., during construction of a road, canal, or embankment or of residential, commercial and industrial sites; also used in stope mining.

cut and fill [geomorph] A process of leveling whereby material eroded from one place by waves, currents, streams, or winds is deposited nearby until the surfaces of erosion and deposition are continuous and uniformly graded; esp. lateral erosion on the concave banks of a meandering stream accompanied by deposition within its loops. Cf: *scour and fill [geomorph].*

cut and fill [sed struc] A sedimentary structure consisting of a small erosional channel that is subsequently filled; a small-scale *washout.*

cutbank (cut'-bank) A local term in the western U.S. for a steep bare slope formed by lateral erosion of a stream. Also spelled: cut bank.

cut grading Process of evaluating and describing the proportions and finish of a polished diamond, principally with regard to their overall effect on brilliance and dispersion and the balance between them.

cuticle (cu'-ti-cle) The superficial, noncellular layer secreted by the *epidermis* and consisting of a mixture of oxidation and condensation products of fatty acids. It protects the underlying tissues from undue desiccation. It shows the imprint of underlying cellular tissues, including stomata and lenticels, which often can be used to identify botanical affinity.

cuticle [paleont] (cu'-ti-cle) A secretion of epidermal cells comprising the noncalcified, acellular parts of the body wall of a *bryozoan.*

cutin (cu'-tin) The waxy material of the *cuticle* covering external cell surfaces of vascular plants and some mosses (Scagel et al., 1965, p.614).

cutinite (cu'-tin-ite) A maceral of coal within the *liptinite* group, consisting of plant cuticles. Cf: *sporinite; resinite; alginite.*

cutoff [eng] (cut'-off) An impermeable wall, collar, or other structure placed beneath the base or within the abutments of a dam to prevent or reduce losses by seepage along a construction interface or through porous or fractured strata. A cutoff also reduces the velocity or pressure of water moving in permeable material under a dam by increasing the distance the water flows before emerging at the toe of the dam. Cutoffs may be made of concrete, compacted clay, interlocking *sheet piles,* or *grout* injected along a line of holes.

cutoff [stratig] A boundary, oriented normal to bedding planes, that marks the areal limit of a specific stratigraphic unit where the unit is not defined by erosion, pinch-out, faulting, or other obvious means. Cutoffs are applicable to map, cross-sectional, and three-dimensional views, and are in effect specialized facies boundaries. See also: *arbitrary cutoff.* Syn: *stratigraphic cutoff.*

cutoff [streams] (a) The new and relatively short channel formed when a stream cuts through a narrow strip of land and thereby shortens the length of its channel. See also: *neck cutoff; chute cutoff.* Syn: *meander cutoff; cutoff channel.* (b) A channel constructed to straighten a stream or to bypass large bends, thereby relieving an area normally subjected to flooding or channel erosion. See also: *pilot channel.* (c) The crescent-shaped body of water separated from the main stream by a cutoff. (d) The formation of a cutoff.

cutoff [struc geol] The intersection between a once-continuous planar or linear geological feature and a fault surface.

cutoff channel A meander cutoff.

cutoff grade In economic geology, the lowest grade of mineralized material that qualifies as ore in a given deposit; ore of the lowest assay value that is included in an ore estimate. Cf: *assay limit.* Abbrev: COG.

cutoff lake *oxbow lake.*

cutoff limit *assay limit.*

cutoff line The hanging wall or footwall cutoff of a particular stratum.

cutoff line map A map showing the relative positions of hanging wall and footwall cutoffs for several different strata cut by a single fault or a system of faults (Woodward and Beets, 1988).

cutoff meander A meander that has been abandoned by its stream after the formation of a neck cutoff. See also: *oxbow.* Syn: *abandoned meander.*

cutoff spur The remnant of a *meander spur,* formed when a vigorously downcutting stream breaks through a narrow strip of land between adjacent curves in the stream course; it usually stands as an isolated hill. Syn: *meander core.*

cutout (cut'-out) A mass of shale, siltstone, or sandstone filling an erosional channel cut into a coal seam. Cf: *roll [coal]; washout [mining].* Syn: *horseback [coal]; want.*

cutover (cut'-over) Said of a forest area where most of the saleable timber has been cut.

cut plain A *stratum plain* on any hard rock that has been much dissected by erosion; the original surface is approximately represented by the summits of the least-eroded parts (Hill, 1900, p.7).

cutter [eco geol] (cut'-ter) A *solution fissure,* esp. as used in Tennessee for a solution crevice in limestone underlying a residual phosphate deposit.

cutter [karst] A solution-widened fracture. Depths range up to tens of meters. It may or may not be soil covered. Cf: *karren; grike.*

cuttings (cut'-tings) *drill cuttings.*

cuvette [gem] (cu-vette') A gemstone with a raised cameolike figure in a polished depression. Cf: *chevee.*

cuvette [sed] A large-scale *basin* in which sedimentation has occurred or is taking place, as distinguished from a tectonic basin due to folding of preexisting rocks; e.g. the Anglo-Parisian cuvette of SE England and NE France, in which Cenozoic rocks accumulated and were later folded into several distinct but smaller basins, such as the Paris Basin and the London Basin. Etymol: French, "small tub or vat".

Cuvier's principle (Cu'-vi-er's) The theory that certain very different characteristics are commonly associated, e.g. kinds of feet and teeth among the vertebrates. Named after the French naturalist Georges Cuvier (1769-1832).

cuyamite (cu-yam'-ite) A *teschenite* composed of labradorite, analcime, hauyne, hornblende, augite, and magnetite. Its name (Johannsen, 1939) is derived from Cuyamas Valley, California. Obsolete.

cuzticite (cuz'-ti-cite) A yellow hexagonal mineral: $Fe^{3+}_2Te^{6+}O_6 \cdot 3H_2O$.

CV chondrite A member of the Vigarano group of *carbonaceous chondrites.*

CVS *constant velocity stack.*
C wave (a) *converted wave;* (b) *coupled wave.*
cwm (Pron: koom) Welsh term for *cirque [glac geol].* The term is occasionally used in Wales for a narrow, deep valley of nonglacial origin in a mountain region (Stamp, 1961, p.142). Pl: cwms. Var. spellings: *coom; coomb; combe;* comb.
cyanidation The use of a solution of sodium or potassium cyanide to recover gold from ore. Cf: *leaching.*
cyanite (cy'-a-nite) *kyanite.*
cyanobacteria (cy-a'-no-bac-te'-ri-a) A group of aerobic, photoautotrophic *bacteria* that contain chlorophyll and other pigments. These bacteria were formerly called *blue-green algae.* Cyanobacteria include unicellular and colonial forms and are involved in *stromatolite* formation (along with green algae). Their range is from the Archean to the present. Sing: cyanobacterium. Cf: *brown algae; green algae; red algae; yellow-green algae.*
cyanochroite (cy'-a-no-chro'-ite) A greenish blue monoclinic mineral of the *picromerite* group: $K_2Cu(SO_4)_2{\cdot}6H_2O$.
cyanolith An oncoid, or *microbial biscuit,* formed by calcified cyanophytes (blue-green algae) or cyanobacteria. Also called cyanoids (Riding, 1983).
cyanophyllite (cy'-a-no-phyl'-lite) A greenish-blue orthorhombic mineral: $Cu_5Al_2(OH)(Sb^{3+}O_4)_3{\cdot}12H_2O$.
cyanosite (cy-an'-o-site) A syn. of *chalcanthite.* Also spelled: cyanose.
cyanotrichite (cy'-a-no-trich'-ite) A bright-blue or sky-blue orthorhombic mineral: $Cu_4Al_2(SO_4)(OH)_{12}{\cdot}2H_2O$. Syn: *lettsomite; velvet copper ore.*
cyatholith (cy-ath'-o-lith) *placolith.*
cyathosponge (cy-ath'-o-sponge) *archaeocyathid.*
cyberinfrastructure Information technology infrastructure that is needed to manage, preserve, and efficiently access large amounts of data and create an environment in which data-computational resources are shared. The goal: to foster integrated scientific analysis, to accelerate the pace of scientific discovery, and to facilitate innovation.
cycad (cy'-cad) A member of a group of gymnospermous plants (Cycadales) with pinnately compound leaves, poorly developed wood, and ovules borne on loosely or compactly arranged *sporophylls* in terminal cones. Cycads range from the Late Carboniferous(?) to the present.
Cycadeoidales (Cy-ca-de-oi-da'-les) *Bennettitales.*
cycadeoids (cy-ca'-de-oids) *Bennettitales.*
cycadophyte (cy-cad'-o-phyte) A gymnosperm having compound leaves, and naked seeds borne separately on sporophylls or in simple cones. It is an informal term that includes both *cycads* and *Bennettitales.*
cycle [geol] (cy'-cle) (a) A succession of events that are normally recurrent and return to a starting point, that are repeated in the same order at more or less regular intervals, and that end under conditions that are the same as they were at the beginning; e.g., the cycle of the seasons, a geochemical cycle, or a cycle of sedimentation. (b) An interval of time during which one sequence of a regularly recurring succession of events or phenomena runs to completion, with the last stage or event being quite different from the first; e.g., a cycle of erosion, or an *orogenic cycle.* (c) A group of rock units that occur in a certain order, with one unit being repeated frequently throughout the succession (Duff and Walton, 1962, p.239); esp. a *cyclothem.* Adj: *cyclic.*
cycle [paleont] (a) A ring of segments in heterococcoliths. (b) A set of septa or tentacles of like age in a coral.
cycle [stratig] In sequence stratigraphy, a period of time during which sea level falls from a highstand position, through a lowstand, and returns to a highstand. Cycle is not used as a rock term (Vail et al., 1977; Mitchum, 1977).
cycle of denudation *cycle of erosion.*
cycle of erosion (a) The complete, progressive, and systematic sequence of natural changes or stages in a landscape from the start of its erosion on a newly uplifted or exposed surface through its dissection into mountains and valleys until it has been reduced in the final stage to a low, featureless plain or to a base level (such as sea level) that limits the activity of the agents involved; according to some authors, a complete cycle is from base level back to base level. The cycle, usually subdivided into youthful, mature, and old-age stages, is largely hypothetical because it is normally interrupted before it runs to completion; the landforms produced and destroyed during the sequence are a function of climate, geography, and geologic structure. The concept was first developed and formalized by Davis (1889b). See also: *normal cycle.* (b) The interval of time during which the cycle of erosion is completed; the time involved in the reduction of a newly uplifted land area to a base level. Syn: *geomorphic cycle; geographic cycle; erosion cycle; cycle of denudation; physiographic cycle.*
cycle of fluctuation *phreatic cycle.*
cycle of relative change of sea level The interval of time during which a relative rise and fall of sea level takes place. Regional and global cycles are recognized (Mitchum, 1977, p.206). See also: *paracycle of relative change of sea level; supercycle.*
cycle of sedimentation (a) A sequence of related processes and conditions, repeated in the same order, that is recorded in a sedimentary deposit; e.g. the processes and conditions that determine the ordered sequence of small-scale features such as varves and large-scale *cyclothems.* Examples: rhythmically bedded turbidites; laminated evaporites; limestone-shale rhythmic sequences; orthoquartzite, graywacke, and arkose. (b) The deposition of sediments in a basin between the beginnings of two successive marine transgressions, comprising the deposits formed initially on dry land, followed by shallow-water and then deep-water deposits that in turn gradually change to shallow-water and then dry-land type during a marine regression. (c) A *cyclothem.* Syn: *sedimentary cycle.* Cf: *rhythmic sedimentation.*
cycleology (cy-cle-ol'-o-gy) A term used by Elias (1965, p.339) to designate the detection and study of cycles in paleontologic and geologic phenomena. The basic unit in cycleology is the phase.
cycle skip An occurrence in acoustic or sonic logging, when the first arrival is strong enough to trigger the receiver closest to the transmitter but not the farthest receiver, which may then be triggered by a later cycle, resulting in an erroneously large transit time.
cyclic (cy'-clic) Adj. of cycle; recurrent rather than *secular.*
cyclic cover-plate series A serially repeated set of cover plates in edrioasteroids, along each side of the ambulacral perradial line. Each set is a cycle; cycles on one side of the *ambulacrum* are a mirror image of those on the other side; opposing cycles alternate, offset by half a cycle (Bell, 1976).
cyclic crystallization A process of recurring crystallization of some mineral phases during magmatic settling that produces *rhythmic layering.* See also: *rhythmic crystallization.*
cyclic evolution Evolution, supposed by some to have occurred in many lineages, involving successively initial rapid and vigorous expansion, a long stable or slowly changing phase, and a final short episode in which overspecialized, degenerate, or inadaptive forms became extinct.
cyclic salt Salt that is lifted from the sea as spray, is blown inland, and returns to the sea via drainage.
cyclic sedimentation A syn. of *rhythmic sedimentation.* The term is used esp. for sedimentation involving a circuitous sequence of conditions, such as found in a megacyclothem exhibiting asymmetric bedding.
cyclic terrace One of several stream terraces representing former valley floors formed during periods when downcutting had essentially stopped for a time and lateral erosion had become dominant; e.g. a *valley-plain terrace.* Terraces on opposite sides of the valley are paired or correspond in altitude along any given section of the valley. Cf: *noncyclic terrace.*
cyclic twinning *Repeated twinning* of three or more individual crystals according to the same twin law but with the twin axes or twin planes not parallel. Cyclic twinning often results in threefold, fourfold, fivefold, sixfold, or eightfold twins, which, if equally developed, display geometrical symmetry not formed in single crystals. Cf: *polysynthetic twinning.*
cyclocrinitid (cy-clo-cri'-ni-tid) Any of a group of Ordovician and Silurian spherical radially symmetrical calcareous algae of the order Cyclocrinales, characterized by a single erect central axis and tightly packed meroms (branches) arranged in whorls or circlets. Cyclocrinitid meroms are without stellate structures and terminate with lids.
cyclocystoid (cy-clo-cyst'-oid) Any small, discoid echinozoan belonging to the class Cyclocystoidea, characterized by a theca composed of calcareous plates separable into central oral and aboral discs, submarginal ring, and marginal ring, by a flat aboral surface, and by a multiple branching ambulacral system. Range, Middle Ordovician to Middle Devonian.
cyclodont (cy'-clo-dont) Said of the dentition of bivalve mollusks characterized by arched hinge teeth curving out from below the hinge margin and by a small or absent *hinge plate.*
cyclographic projection (cy-clo-graph'-ic) A term used in structural geology for the representation of planes on a stereogram by means of great circles (Dennis, 1967, p.140).
cyclohexane (cy-clo-hex'-ane) A colorless liquid, a saturated homocyclic hydrocarbon of the cycloparaffin series, formula C_6H_{12}. It has a pungent odor and is found in petroleum. Syn: *hexamethylene.*
cyclolith (cy'-clo-lith) Any elliptical or circular ring *coccolith.* The term should be restricted to circular forms.
cyclone (cy'-clone) An atmospheric low-pressure system with a closed, roughly circular wind motion that is counterclockwise in the Northern Hemisphere and clockwise in the Southern. Ant: *anticyclone.* See also: *extratropical cyclone; trough [meteorol]; tropical cyclone.*
cycloparaffin (cy-clo-par'-af-fin) A saturated homocyclic hydrocarbon having the empirical formula C_nH_{2n}. Examples are cyclopentane and cyclohexane, both of which are found in petroleum. Syn: *naphthene.*
cyclopean texture (cy-clo'-pe-an) *mosaic texture [meta].*
cyclosilicate (cy-clo-sil'-i-cate) A class or structural type of *silicate* characterized by the linkage of the SiO_4 tetrahedra in rings, with a ratio of Si:O=1:3. An example of a cyclosilicate is benitoite, $BaTiSi_3O_9$. Cf: *nesosilicate; sorosilicate; inosilicate; phyllosilicate; tectosilicate.* Syn: *ring silicate.*
Cyclostomata (Cy'-clo-sto-ma'-ta) An order of monorhinate jawless fishes characterized by rasping or sucking mouth parts, predaceous or carrion-feeding habit, and lack of ossification. Lampreys and hagfishes are the sole surviving members. Range, Pennsylvanian to Holocene.
cyclostome (cy'-clo-stome) Any ectoproct bryozoan belonging to the order Cyclostomata and characterized by calcareous tubular zooecia with circular apertures and no operculum. Adj: cyclostomatous.
cyclosystem (cy'-clo-sys'-tem) A *gastrozooid* of a hydrozoan, surrounded by a circular row of five to seven individual dactylozooids.
cyclothem (cy'-clo-them) (a) A term proposed by Weller (in Wanless and Weller, 1932, p.1003) for a series of beds deposited during a single sedimentary cycle of the type that prevailed during the Pennsylvanian Period. It is

an informal lithostratigraphic unit equivalent to "formation". Because of extremely variable development, a cyclothem cannot be defined rigidly in terms of the members actually present at any locality (Weller, 1956, p.27-28). Cyclothems are typically associated with unstable shelf or interior basin conditions in which alternate marine transgressions and regressions occur; nonmarine sediments usually occur in the lower half of a cyclothem, marine sediments in the upper half. The term has also been applied to rocks of different ages and of different lithologies from the Pennsylvanian cyclothems. The cyclothem concept was developed by Weller (1930). Cf: *ideal cyclothem; megacyclothem; rhythmite.* (b) A cycle applied to sedimentary rocks (Duff and Walton, 1962).

cyclowollastonite *pseudowollastonite.*

cylindrical divergence (cy-lin'-dri-cal) The decrease in amplitude of seismic surface waves with distance from the source; often characteristic of surface waves. The energy density varies inversely as the distance and the amplitude inversely as the square root of the distance. Cf: *spherical divergence.*

cylindrical fault A fault whose plane is curved and whose displacement is rotational about an axis parallel with the fault plane (Dennis, 1967). Little used.

cylindrical fold A fold that can be generated or reproduced by movement of a line with fixed orientation through space. Cf: *conical fold; cylindroidal fold.*

cylindrical projection A projection on the surface of a cylinder; esp. any of numerous map projections of the Earth, produced by projecting the geographic meridians and parallels onto the surface of a cylinder that is tangent to, or intersects, the surface of the sphere, and then developing (unrolling and laying flat) the cylinder as a plane. The principal scale is preserved along the lines of tangency or secancy. Examples: *Mercator projection; Miller cylindrical projection; Gall projection.*

cylindrical structure A vertical sedimentary structure with an irregular columnar or pillar-like shape; e.g. a *clastic pipe.*

cylindrite (cyl'-in-drite) A blackish lead-gray mineral: $Pb_3Sn_4FeSb_2S_{14}$. It occurs in cylindrical forms that separate under pressure into distinct shells or folia.

cylindroidal fold (cyl-in-droi'-dal) A fold with a curved hinge line that closely approximates a cylindrical fold. Thus, segments of the fold are closely approximated by movement of a straight line with fixed orientation through space. Cf: *conical fold; cylindrical fold.*

cymatogeny (cy-ma-tog'-e-ny) Undulating movement or warping of the Earth's crust to produce regional, linear arching or doming, but with minimal deformation. The concept was introduced by L.C. King (1959, p.117) and has since been discounted. Cf: *orogeny; epeirogeny.*

cymoid (cy'-moid) adj. In economic geology, said of a vein that in cross section forms a reverse curve; a vein that swerves from its course, then returns. A pair of veins in such a pattern forms a cymoid loop.

cymophane (cy'-mo-phane) A syn. of *chrysoberyl.* The name is applied esp. to chrysoberyl exhibiting a girasol or chatoyant effect, and more specif. to cat's-eye only.

cymrite (cym'-rite) A variously colored monoclinic mineral: $Ba(Si,Al)_4(O,OH)_8{\cdot}H_2O$. It is related to the feldspars.

cyphonautes (cy-pho-nau'-tes) Pelagic bivalved larva of some bryozoans (such as *Membranipora*).

cyprine (cyp-rine') A light-blue variety of vesuvianite containing a trace of copper.

Cyprus-type deposit (Cy'-prus) A pyritic copper deposit associated with underlying serpentinite and pillow basalt and with overlying cherts and ferruginous sediments. It is thought to form on oceanic crust of the sea floor.

cyrenoid (cy'-re-noid) Obsolete syn. of *corbiculoid.*

cyrilovite (cy'-ril-o-vite) An orange, orange-brown, or yellow tetragonal mineral: $NaFe^{3+}_3(PO_4)_2(OH)_4{\cdot}2H_2O$. It is a member of the *wardite* group. Syn: *avelinoite.*

cyrtochoanitic (cyr'-to-cho'-a-nit'-ic) Said of a comparatively short, *retrochoanitic septal neck* of a nautiloid, curved so as to be concave outward (TIP, 1964, pt. K, p.55). See also: *ellipochoanitic; orthochoanitic.*

cyrtocone (cyr'-to-cone) A curved, slender cephalopod *conch* (like that of *Cyrthoceratites*) that completes less than one whorl. Also called: cyrtoceracone.

cyrtoconoid (cyr-to-con'-oid) (a) Said of a gastropod shell in which the spire approaches a conical shape but has convex sides. Cf: *coeloconoid.* (b) Resembling a *cyrtocone.*

cyrtolite (cyr'-to-lite) An altered variety of zircon containing uranium, beryllium, and rare earths.

cyrtolith (cyr'-to-lith) A basket- or calotte-shaped *heterococcolith* with laths and a projecting central structure.

Cyrtosoma A subphylum of mollusks generally with conical univalved shell, spiral in many; single shell opening or aperture usually small. Anus usually close to mouth; gut usually bent or twisted into U-shape; radula in all classes. Body laterally compressed in few species. Range, Lower Cambrian to Holocene.

cyrtosome (cyr'-to-some) Any mollusk of the subphylum Cyrtosoma. It includes the classes Monoplacophora, Gastropoda, and Cephalopoda. Obsolete.

cyst [paleont] A sac or capsule that is secreted by many protozoans and other minute animals as a prelude to a resting or a specialized reproductive phase and that envelops protoplasm and protects it from adverse environmental conditions.

cyst [palyn] A microscopic *resting spore* with a resistant wall, formed in dinoflagellates, presumably in acritarchs, in desmids and other algae, and in cyanobacteria, by the breaking-up of parts of the filaments or by the enclosing of a cell group and their investment by a sheath or envelope. Dinoflagellate cysts exist abundantly as fossils (Cramer and Diez Rodriguez, 1979; Williams, 2000, p.55). See also: *statospore.*

cystid (cys'-tid) (a) The combined cellular and skeletal layers of the body wall of a bryozoan *zooid.* (b) A *cystoid.*

cystiphragm (cys'-ti-phragm) A skeletal partition in stenolaemate bryozoans, extending from zooecial wall into zooecial cavity and recurved inwardly to form a compartment that is generally closed by abutting zooecial wall or *diaphragm,* or by another cystiphragm. Cystiphragms encircle the living chamber with a hollow collar or ring.

cystocarp (cys'-to-carp) In the red algae, the fertile structure and surrounding pericarp.

cystoid (cyst'-oid) Any crinozoan belonging to the class Cystoidea, characterized by diplopores, pore rhombs, and brachioles. Range, Lower Ordovician to Upper Devonian. Var: *cystid.*

cystoporate (cys-to-por'-ate) An ectoproct bryozoan belonging to the order Cystoporata and characterized by calcareous tubular zooecia separated by vesicular tissue made up of small chambers (cystopores).

Czochralski A crystal-growth technique where the crystal solidifies as it is slowly withdrawn vertically (or pulled) from a melt of the same chemical composition. Used to produce gem materials such as synthetic ruby, sapphire, spinel, and alexandrite.

D

d As in d(001) is the repeat distance between planes in a diffracting domain. This is the d in the *Bragg equation.* To use the expression d-spacing is redundant.

D" The lowest part of the mantle, roughly 250 km thick. D" was originally determined as a thermal boundary layer, analogous to the lithosphere. D" also contains chemical heterogeneities, structural discontinuities, large lateral variations, and a high degree of anisotropy. See also: *D layer.*

D/A Digital to analog.

dachiardite-Ca (da-chi-ar′-dite) A variously colored monoclinic zeolite mineral: $(Ca,Na,K)_4(Si,Al)_{24}O_{48}\cdot 18H_2O$.

dachiardite-Na A red-orange monoclinic zeolite mineral: $(Na,K,Ca)_4(Si,Al)_{24}O_{48}\cdot 18H_2O$.

Dacian (Da′-ci-an) Eastern European stage: lower Pliocene (above Pontian, below Romanian). Approx. equiv. Akchagylian.

dacite (da′-cite) (a) In the *IUGS classification,* a volcanic rock defined in the QAPF diagram as having Q/(Q+A+P) between 20 and 60 and P/(P+A)>65, and in the TAS diagram as falling in the field bounded by points with the SiO_2, total alkali coordinates: 63, 0; 63, 7; 69, 8; and 77, 0. (b) A fine-grained extrusive rock with the same general composition as *andesite* but having a less calcic plagioclase and more quartz; according to many, it is the extrusive equivalent of *granodiorite.* The name, given by Stache in 1863, is from the ancient Roman province of Dacia (now part of Romania).

dactylethra (dac-ty-leth′-ra) A polymorph in the stenolaemate bryozoan genus *Terebellaria* that is apparently the *zooecium* of a feeding zooid covered by a terminal *diaphragm* (Boardman and Cheetham, 1973, p.155).

dactylitic (dac-ty-lit′-ic) A rock texture produced by a *symplectic* intergrowth, in which one mineral is penetrated by fingerlike projections from another mineral. Obsolete.

dactylopod (dac-tyl′-o-pod) The *dactylus* of a malacostracan crustacean. Syn: dactylopodite.

dactylopore (dac-tyl′-o-pore) A relatively small or narrow tubular cavity of certain hydrozoans, occupied by a *dactylozooid.* Cf: *gastropore.*

dactylous (dac′-ty-lous) (a) Pertaining to a *dactylus* of a crustacean. (b) Said of a *pedicellaria* of an echinoid having spoon-shaped jaws mounted on individual stalks. (c) In tetrapods, a suffix signifying "toes"; e.g., pentadactylous, five-toed.

dactylozooid (dac′-ty-lo-zo′-oid) An elongate, slender, mouthless polyp housed in the *dactylopore* of a hydrozoan, equipped with numerous stinging cells, and performing protective, food-capturing, and tactile functions for the colony. Cf: *gastrozooid.*

dactylus (dac′-ty-lus) The seventh and most distal segment of the *pereiopod* of a malacostracan crustacean. Pl: dactyli. Syn: *dactylopod.*

dadsonite (dad′-son-ite) A metallic gray triclinic or monoclinic mineral: $Pb_{10+x}Sb_{14-x}S_{31-x}Cl_x$.

dagala (da-ga′-la) A syn. of *steptoe,* used in the region of Mt. Etna, Italy.

dahllite (dahl′-lite) A resinous, yellowish-white carbonate-apatite mineral, sometimes occurring as concretionary spherulites.

daily variation (dai′-ly) Daily, short-period fluctuation of the Earth's *external magnetic field* caused mainly by solar activity. Stationary magnetometer records are used to remove variations from airborne or ground magnetic measurements. Syn: *diurnal variation.*

daily wave *diurnal wave.*

dakeite *schröckingerite.*

dal A Scandinavian term for a valley.

dale (a) A term used in northern England and southern Scotland for a broad, open river valley. (b) A poetic var. of valley; a *vale* or a small valley. Cf: *dell.* (c) A district whose slope lines run to the same pit (Warntz, 1975, p.213).

dalles (a) The rapids in a deep, narrow stream confined between the rock walls of a canyon or gorge; e.g. The Dalles of the Columbia River where it flows over columnar basalt. (b) A steep-sided part of a stream channel, near the dalles proper, marked by clefts, ravines, or gorges; e.g. along the Wisconsin River, Wisc. Etymol: French plural of dalle, "gutter". Syn: *dells.*

dallinid (dal-lin′-id) Any articulated brachiopod belonging to the family Dallinidae, characterized by loop development passing through precampagiform, campagiform, frenuliform, terebratelliiform, and dalliniform, or variations of these.

dalliniform (dal-lin′-i-form) Said of the *loop,* or of the growth stage in the development of the loop, of a dallinid brachiopod (as in *Dallina*), consisting of long descending branches recurved into ascending branches that meet in a transverse band, all free of the valve floor (TIP, 1965, pt. H, p.143). The dalliniform loop is morphologically similar to the *magellaniform* and *cryptonelliform* loops.

dallol (dal′-lol) A term applied in Nigeria to a flat-bottomed steep-sided dry valley, usually many kilometers wide; specif. a part of an old drainage system on the left bank of the Niger River.

Dalmatian coastline (Dal-ma′-tian) A *concordant coastline* developed where drowning or a rise of sea level has produced lines of narrow islands (representing the outer mountain ranges) separated by long inlets or straits (representing parallel valleys) lying peripheral, and trending roughly parallel, to the coast. Type region: the eastern coastline of the Adriatic Sea in the region of Dalmatia.

Daly Gap A compositional hiatus between basalt and trachyte on some oceanic islands, named for R. A. Daly who described the gap on Ascension Island. Its cause, and even its existence, have been topics of debate. Current opinion is that the gap is real, but is expressed by some parameters (such as SiO_2, *differentiation index,* and CaO) that are nonlinear measures of continuous magmatic fractionation, and is not expressed by the more sensitive changes recorded by incompatible elements.

dalyite (da′-ly-ite) A variously colored triclinic mineral: $K_2ZrSi_6O_{15}$.

dam [eng] n. (a) An artificial barrier or wall constructed across a watercourse or valley for one or more of the following purposes: creating a pond or lake for storage of water; diverting water from a watercourse into a conduit or channel; creating a hydraulic head that can be used to generate power; improving river navigability; controlling floods; or retention of debris. It may be constructed of wood, earth materials, rocks, or solid masonry. Archaic syn: *barrage.* (b) A body of standing water confined or held by a dam. v. To provide a dam; to obstruct or confine a flow of water.

dam [geomorph] An obstruction formed by a natural agent (such as a glacier, a lava flow, or the work of beavers) across a stream so as to produce a lake or pond; e.g. an *ice dam.*

damage zone The area of fracturing around and related to a fault, typically caused by accumulation of displacement on a fault as well as fault-propagation and linkage (Peacock et al., 2000).

damaraite A colorless orthorhombic mineral: $Pb_3O_2(OH)Cl$.

dambo (dam′-bo) A term used in central Africa (esp. Zambia) for a small, ill-defined flood plain or channelless drainageway that is extremely flat with broad, grassy clearings, swampy during the wet season but dry for the greater part of the year. Etymol: Bantu.

dam gradation *contragradation.*

damiaoite A metallic white cubic mineral: In_2Pt.

damkjernite (dam-kjern′-ite) A hypabyssal rock composed of phenocrysts of biotite and titanian augite in a fine-grained groundmass of pyroxene, biotite, perovskite, and magnetite, with interstitial nepheline, microcline, and calcite. The name, given by Brögger in 1921, is for the locality Damkjern (or Damtjern), Fen complex, Norway. Also spelled: damtjernite. Obsolete.

damourite (da-mour′-ite) A variety of muscovite, which gives off water more readily and has less elastic folia and a more pearly or silky luster. It is unctuous or talclike to the touch. Syn: *talcite.*

damp A general term for coal-mine gases.

damping (damp′-ing) The loss of amplitude of an oscillation, owing to energy converting to some other form, usually heat. See also: *critical damping; damping factor.*

damping capacity The ability of a material to dissipate strain within itself.

damping factor The ratio of the observed damping to that required for *critical damping.*

danaite (da′-na-ite) A variety of arsenopyrite containing cobalt.

danalite (da′-na-lite) A variously colored cubic mineral: $Fe^{2+}_4Be_3(SiO_4)_3S$. It is the iron end-member isomorphous with helvite and genthelvite.

danbaite (dan′-ba-ite) A metallic cubic mineral, the gamma form of brass: $CuZn_2$.

danburite (dan′-bur-ite) (a) An orange-yellow, yellowish-brown, grayish, or colorless orthorhombic mineral: $CaB_2(SiO_4)_2$. It is structurally related to the feldspars, resembles topaz in crystal habit, physical properties, and appearance, and is used as a gemstone. (b) A group name for minerals of the general formula $AB_2Si_2O_8$, where A = Ca, Sr or Ba.

dancalite (dan′-cal-ite) An extrusive rock containing phenocrysts of oligoclase, clinopyroxene and rare amphibole in a trachytic groundmass composed of plagioclase laths with interstitial analcime; an analcime *trachyandesite* or feldspar-bearing analcime *tephrite.* The name, given by De Angelis in 1925, is for Dancala, Ethiopia. Obsolete.

dangle A topological error that occurs when an arc has a node with a valency of 1 (Krzanowski et al., 1993).

dangling node A node of valency 1. A dangling node is considered to be a topological error in graphic files containing area features (Krzanowski et al., 1993).

Danian (Dan′-i-an) A stage of the Standard Global Chronostratigraphic Scale: lowermost Paleocene and base of Cenozoic (above Maastrichtian Stage of Cretaceous, below Selandian). The basal GSSP in Tunisia coincides with global iridium anomaly.

danielsite (dan′-iels-ite) A metallic gray orthorhombic mineral: $(Cu,Ag)_{14}HgS_8$.

dannemorite (dan-ne-mor′-ite) *manganogrunerite.*

dans A broad, shallow valley in South Africa. Etymol: Afrikaans.

d'ansite (d'ans′-ite) A colorless cubic mineral: $Na_{21}Mg(SO_4)_{10}Cl_3$.

daomanite (da-o′-man-ite) A metallic pale yellowish-green orthorhombic mineral: $CuPtAsS_2$. Not approved by the International Mineralogical Association.

daphnite (daph′-nite) A variety of chlorite.

daqingshanite-(Ce) (da-qing-shan′-ite) A pale yellow rhombohedral mineral with a calcite-like crystal structure: $(Sr,Ca,Ba)_3(RE)Ce(PO_4)(CO_3)_3$.

darapiosite (da-rap′-i-o-site′) A variously colored hexagonal mineral of the *milarite* group: $KNa_2(Li,Zn,Fe)_3(Mn,Zr,Y)_2Si_{12}O_{30}$.

darapskite (da-rap′-skite) A colorless monoclinic mineral: $Na_3(NO_3)(SO_4)\cdot H_2O$.

darcian velocity *specific discharge.*

darcy (dar′-cy) A standard unit of permeability, equivalent to the passage of one cubic centimeter of fluid of one centipoise viscosity flowing in one second under a pressure differential of one atmosphere through a porous medium having an area of cross-section of one square centimeter and a length of one centimeter. Cf: *millidarcy.*

Darcy's law (Dar′-cy's) An empirical law that describes the flow of fluids through porous media. In groundwater flow, it states that the velocity of flow through a unit area of a permeable media is directly proportional to the hydraulic gradient normal to that area, assuming that the flow is laminar and inertia can be neglected (Department of Interior, 1989). In gas flow, it states that the velocity of the flow is proportional to the pressure gradient multiplied by the ratio of permeability times density, divided by the viscosity of the gas. Named for Henry Darcy, a French engineer, who developed the relationships.

Darcy-Weisbach formula (Dar′-cy-Weis′-bach) A formula for determining the loss of head in flowing water: loss of head is equal to a coefficient that varies with the surface roughness of the conduit and the Reynolds number, times the length of the conduit, divided by its diameter, times the velocity head of the flowing fluid. In the case of a noncircular conduit or a circular conduit not flowing full, four times the hydraulic radius is substituted for the diameter (ASCE, 1962).

dark-colored Said of a rock-forming mineral having a dark color as viewed megascopically, but being transparent in thin section; also, said of the rock that such minerals form. By convention, dark-colored aphanites include those that are dark gray, dark green, black, and brownish black. Cf: *light-colored; mafic.* Syn: *melanocratic.*

dark field An image obtained in the transmission electron microscope by deliberately excluding all beams except the particular diffracted beam of interest. This is accomplished by placing an aperture in the back focal plane of the objective lens, permitting only a diffracted beam to form the image. Cf: *bright field.*

darkfield illumination A system for lighting a diamond from the side against a black background, to reduce surface reflections and make inclusions in a gemstone stand out in sharp relief.

dark halo crater A small crater surrounded by material with a lower albedo than that of the adjacent terrain.

dark mineral Any one of a group of rock-forming minerals that are dark-colored in thin section, e.g. biotite, hornblende, augite.

dark red silver ore *pyrargyrite.*

dark ruby silver *pyrargyrite.*

Darriwilian (Dar-ri-wil′-i-an) A stage of the Standard Global Chronostratigraphic Scale: upper Middle Ordovician. The basal GSSP is in Zhejiang Province, China, and was ratified 1997 (Mitchell et al., 1997).

Darwin glass (Dar′-win) A highly siliceous vesicular frothy glass, found in the Mt. Darwin range in western Tasmania (near Queenstown) in the form of blobs, drops, and twisted shreds. It is probably a locally derived impact glass, although believed by some to be a kind of tektite.

Darwinism (Dar′-win-ism) The theory that evolution resulted from variation and the survival of favored individuals through *natural selection.* Named after the English naturalist Charles Darwin (1809-1882).

dashkesanite (dash-ke′-san-ite) *potassic-chlorohastingsite.*

dashkovite A colorless monoclinic formate mineral: $Mg(HCOO)_2\cdot 2H_2O$.

dasycladacean (da′-sy-cla-da′-ce-an) Pertaining to a type of green algae of the family Dasycladaceae, whose filaments are whorled about a central axis and often encrusted with calcium carbonate.

data assimilation The incorporation of data into a model that are of the same type as, or directly related to, the variable calculated by the model. It helps define the initial conditions needed by the model, and it reduces the growth of errors in the calculation of the variable; e.g., the introduction of satellite measurements of wave height into numerical models that forecast wave height.

data dictionary A database that serves as a catalogue containing information about map features or attributes. The catalogue may define data file and element names, sources, accuracy, date of entry or update, or other characteristics of the data and its sources.

data layer *layer [cart].*

date n. An instant of time in the past specified by the number of years that have passed. v. To assign a specific or approximate position on the geologic time scale to a past geologic event.

date line A longitudinal line on the Earth's surface that closely follows the 180° meridian, and is taken as the line along which the calendar day begins. Syn: *international date line.*

dating (dat′-ing) Age determination of naturally occurring substances or relicts by any of a variety of methods mostly based on the amount of change, happening at a constant and measurable rate, in a component. The changes may be chemical, or induced or spontaneous nuclear.

datolite (dat′-o-lite) (a) A variously colored monoclinic mineral: $CaBSiO_4(OH)$. It commonly occurs in cracks and cavities in diabase or basalt; it is sometimes used as a minor gem. Syn: *humboldtite; dystome spar.* (b) A group name for minerals with an analogous formula, but with Ca replaced by rare earth elements or Y, B replaced by Be, and sometimes containing Fe^{2+} or Mg.

datum [geodesy] (da′-tum) The astronomic latitude and longitude of an initial point, the astronomic azimuth of a line from this point, the *deflection of the vertical,* the *geoidal separation* (frequently these quantities are assumed to be zero for lack of more complete information), and the two constants necessary to define the reference ellipsoid on which horizontal control surveys are to be computed. In simplest terms, a (geodetic) datum consists of a reference ellipsoid that is fixed in some manner with respect to the physical Earth. More modern (geocentric) datums consist of an ellipsoid approximating the shape of the global geoid and having its center at the Earth's center of gravity (the geocenter). See also: *geodetic coordinates.* Syn: *geodetic datum.*

datum [geol] (a) The top or bottom of a bed of rock, or any other surface, on which structure contours are drawn, e.g. a *datum horizon.* (b) *sea-level datum.* On topographic maps, datum is normally mean sea level (MSL), known in U.K. as "Ordnance Datum" (OD), but on marine charts datum is commonly mean low spring tide. Pl: datums.

datum [oceanog] The fixed starting point on a map from which vertical and horizontal distances are calculated. The *chart datum* is the starting point for calculating vertical elevations on nautical charts.

datum [photo] A direction, level, or position from which angles, heights, depths, speeds, or distances are conventionally measured; e.g. in photographic mapping, the assumed horizontal plane on which the map is constructed.

datum [surv] Any numerical or geometric quantity or value that serves as a base or reference for other quantities or values; any fixed or assumed position or element (such as a point, line, or surface) in relation to which others are determined, such as a level surface to which depths or heights are referred in leveling. Pl: *datums;* the plural "data" is used for a group of statistical or inclusive references, such as "geographic data" for a list of latitudes and longitudes. See also: *datum plane [surv]; mean sea level.*

datum gravity The arbitrary reference surface to which corrections are applied.

datum horizon A geologic horizon used as a reference plane for the position of rock strata or for the comparative measurement of the thickness of strata; the *key horizon* or bed on which elevations are taken or to which all elevations are finally referred in making a structure-contour map. Syn: *structural datum; contour horizon.*

datum level [stratig] The base or top of a range of fossils that can be correlated in sections over a wide area. "The first evolutionary appearance of a fossil species immediately above a known ancestor may be accorded first order of reliability in defining a datum level" (Bolli, 1969, p.200). Cf: *chronohorizon.* Syn: *datum plane [stratig].*

datum level [surv] Any level surface, such as mean sea level, used as a reference from which elevations are reckoned; a datum plane.

datum line *reference line.*

datum plane [seis] An arbitrary reference surface, used to minimize or eliminate local topographic and near-surface effects. Seismic times and velocity determinations are adjusted to the datum plane to approximate as if source and seismometers had been located on this plane and as if no low-velocity layer existed.

datum plane [stratig] *datum level [stratig].*

datum plane [surv] A permanently established horizontal plane, surface, or level to which soundings, ground elevations, water-surface elevations, and tidal data are referred; e.g. *mean sea level* is a common datum plane used in topographic mapping. See also: *tidal datum; chart datum.* Syn: *datum level [surv]; reference level; reference plane.*

datum point An assumed or fixed point used as a reference from which calculations or measurements may be taken.

daubrée A unit of intensity of wear of a sedimentary particle, equivalent to the removal of 0.1 gram from a 100-gram sphere of quartz (Wentworth, 1931, p.25). Named in honor of Gabriel Auguste Daubrée (1814-1896), French mining engineer and mineralogist, who showed experimentally (Daubrée, 1879) that quartz grains lost only one part in 10,000 per kilometer of travel. Rarely used.

daubreeite (dau-bree'-ite) A creamy white, grayish, or yellowish brown tetragonal mineral: BiO(OH,Cl). It is isomorphous with bismoclite and a member of the *matlockite* group. Also spelled: daubréeite; daubreite.

daubreelite (dau-bree'-lite) A metallic black cubic meteorite mineral with the spinel structure type and a member of the *linnaeite* group: $Fe^{2+}Cr_2S_4$. Also spelled: daubréelite.

daugh A Scottish term for soft coaly fireclay.

daughter (daugh'-ter) A *nuclide* formed by disintegration of a radioactive precursor (*parent*). See also: *radioactive series; end product.*

daughter mineral A crystalline mineral in a *fluid inclusion* of a phase different from that of the walls. The most common is halite, but carbonates and sulfates are also found (Roedder, 1979, p.700).

daunialite (dau'-ni-a-lite') A sedimentary rock consisting of clay rich in silica and smectite, as distinct from bentonite of volcanic origin. It contains 25% organic silica (opal, chalcedony, quartz) and small amounts of sericite, chlorite, and kaolinite. Rarely used.

Dauphiné twin law (Dau-phi-ne') A twin law in quartz in which two right-handed or two left-handed individuals form an interpenetration twin by a 180° rotation about the *c* crystal axis. The result is termed *electrical twinning.* Cf: *Brazil twin law.*

davanite (dav'-a-nite) A colorless triclinic mineral: $K_2TiSi_6O_{15}$.

davidite (da'-vid-ite) A subgroup name for dark-brown to brownish-black rhombohedral iron-titanate minerals of the *crichtonite* group, with the general formula: $(REE)YM_{20}O_{38}$ Davidite is a primary mineral in high-temperature hydrothermal lodes; it occurs in pegmatites and basic igneous rocks, and in all stages of intergrowth and exsolution with ilmenite and hematite. Syn: *ferutite.*

davidite-(Ce) A dark brown to black rhombohedral mineral: $(Ce,La)(Y,U)(Ti,Fe^{3+})_{20}O_{38}$. A member of the *crichtonite* group.

davidite-(La) A dark brown to black rhombohedral mineral: $(La,Ce)(Y,U)(Ti,Fe^{3+})_{20}O_{38}$. A member of the *crichtonite* group. Syn: *ferutite; ufertite.*

davidite-(Y) A vitreous black rhombohedral mineral of the *crichtonite* group: $Y(Ti,Fe)_{21}O_{38}$.

davidsonite (da'-vid-son-ite') A greenish or greenish-yellow variety of beryl.

Davisian (Da-vis'-i-an) Pertaining to the "American" school of geomorphology based on the teachings and writings of William Morris Davis (1850-1934), Harvard geologist and geographer; esp. said of the concepts of *peneplanation* and the *cycle of erosion,* and of the genetic method of landform description whereby differences are largely explainable in terms of geologic structure, geomorphic process, and stage of development.

davisonite (da'-vi-son-ite) A mixture of apatite and crandallite. Syn: *dennisonite.*

davreuxite (da-vreux'-ite) A creamy white or pale rose monoclinic mineral: $Mn^{2+}Al_6Si_4O_{17}(OH)_2$.

davyne (da-vyne') A chlorine-bearing colorless hexagonal feldspathoid mineral of the *cancrinite* group: $(Na,Ca,K)_8Al_6Si_6O_{24}(Cl,SO_4,CO_3)_{2-3}$. Syn: *pharaonite.*

dawn stone *eolith.*

dawsonite (daw'-son-ite) A white bladed orthorhombic mineral: $NaAl(CO_3)(OH)_2$.

daylighting (day'-light-ing) In engineering geology, the exposure of a planar feature, such as bedding or a fault, by an open cut whose angle is steeper than that of the exposed feature. Such exposure may increase the likelihood of landsliding by removal of buttressing strata, but it may also reduce sliding tendencies by promoting drainage.

DCL *diffuse crossed lamellar structure.*

D-coal Microscopic coal particles that are predominantly durain, as found in miners' lungs. Cf: *F-coal; V-coal.*

dead (a) In economic geology, said of an economically valueless area, in contrast to a quick area or ore; barren ground. (b) In mining, said of an area of subsidence that is thought to be completely settled and will not move again; dead ground.

dead-burned dolomite A refractory product, CaO.MgO, produced by *calcination* of dolomite or dolomitic limestone.

dead cave A cave in which there is no longer any moisture or any growth of speleothems associated with the presence of moisture. See also: *live cave.* Syn: *dry cave.*

dead chert *chalky chert.*

dead fault A fault along which movement has ceased. Cf: *active fault.*

dead glacier A glacier that is without an accumulation area or is no longer receiving material from one. It may continue to spread or creep downhill due to its bulk and topographic setting. Ant: *active glacier.* Cf: *dead ice.* Syn: *stagnant glacier.*

dead ground Rock in a mine that must be removed in order to get at productive ground.

dead ice (a) Ice that is not flowing significantly and is not receiving material from an accumulation area; the ice of a *dead glacier.* (b) Detached blocks of ice left behind by a retreating glacier, commonly buried in moraine and melting slowly. Cf: *fossil ice.* Syn: *stagnant ice.*

dead lake Part of the terminology in an obsolete concept of a lake's life history. Cf: *dying lake.* Partial syn: *extinct lake.*

dead line The level above which a batholith is metalliferous and below which it is economically barren. It is exposed during the epibatholithic stage of erosion (Emmons, 1933).

dead oil Crude oil containing very little dissolved gas.

dead reckoning [surv] Determining position by extrapolation of the track and direction from a previously known point. Inertial navigation and Doppler sonar are sophisticated versions of dead reckoning. Cf: *doppler positioning; inertial surveying system.*

dead sea A body of water devoid of normal aquatic organisms, from which evaporites have been or are being precipitated. Seasonally it may be very rich in cyanobacterial and algal life forms which leave little fossil trace. Type locality: Dead Sea in the Near East. Archaic.

dead trace A trace on a seismic record that shows no appreciable deflection, usually because of an instrument failure.

dead valley *dry valley.*

dead water In a stream, water that is or appears to be standing or still.

deanesmithite An adamantine orange-red triclinic mineral: $(Hg^{1+})_2(Hg^{2+})_3S_2O(CrO_4)$.

death assemblage *thanatocoenosis.*

debacle (de-ba'-cle) (a) A *breakup* on a river, esp. on the great rivers of the former U.S.S.R. and of North America. (b) The rush of water, broken ice, and debris in a stream immediately following a breakup. Syn: *ice gang; ice run.* (c) Any sudden, violent, destructive flood, deluge, or rush of water that breaks down opposing barriers and sweeps before it debris of all kinds. Etymol: French "débâcle", "unblocking".

debitumenization (de'-bi-tu'-me-ni-za'-tion) *devolatilization.*

debouchment (de-bouch'-ment) (a) The issuing forth of a stream, as the *debouchment* of a tributary into the main stream. (b) The mouth of a river or channel. Syn: *debouchure.* Etymol: French "débouchement".

debouchure (de-bouch-ure') (a) *debouchment.* (b) The place where an underground stream reaches the surface; the opening from which a spring issues. See also: *resurgence.* (c) The point in a cave where a tubular passage connects with a larger passage or chamber.

debris [geol] Any surficial accumulation of loose material detached from rock masses by chemical and mechanical means, as by decay and disintegration. It consists of rock fragments, soil material, and sometimes organic matter. The term is often used synonymously with *detritus,* although "debris" has a broader connotation. Etymol: French "débris". Pl: debris. Syn: *rock waste.*

debris [glaciol] The rocks, earth, and other material lying on the surface or incorporated in the body of a glacier, or pushed ahead of the glacier front. Syn: *glacial debris.*

debris apron *bajada.*

debris avalanche The very rapid and usually sudden sliding and flowage of incoherent, unsorted mixtures of soil and weathered bedrock.

debris cone [geomorph] (a) *alluvial cone.* (b) A conical mound of tightly packed, fine-grained debris piled at the angle of repose atop certain boulders moved by a landslide.

debris cone [glaciol] A cone or mound of ice or snow on a glacier, covered with a veneer of debris thick enough to protect the underlying material from the ablation that has lowered the surrounding surface. Cf: *dirt cone; sand cone [glaciol].* Rarely used syn: *glacier cone.*

debris dam A mass of coarse alluvium deposited at the mouth of a tributary stream, commonly during a flash flood, and forming an obstruction in the main valley.

debris fall The relatively free collapse of predominantly unconsolidated or weathered mineral and rock material from a steep slope or cliff; it is esp. common along the undercut banks of streams. Cf: *soilfall.*

debris flood A disastrous flood, intermediate between the turbid flood of a mountain stream and a true mudflow, of the type that has occurred in southern California (Strahler, 1963, p.465-466).

debris flow A moving mass of rock fragments, soil, and mud, more than half of the particles being larger than sand size. Slow debris flows may move less that 1 m per year; rapid ones reach 160 km per hour, as in the 1977

Huascaran flow in the Peruvian Andes. Cf: *mudflow [mass move]; sturzstrom; bentonite debris flow; rockfall avalanche; lahar.*

debris flow [marine geol] (a) A jumbled mass of sediment, unsorted, traveling downslope on the sea floor in the manner of a slide but with less cohesion. (b) A jumbled mass of sedimentary materials resulting from such a movement.

debris ice (a) Sea ice containing soil, mud, stones, shells, and other material. (b) *brash ice.*

debris island A *sorted circle* having a diameter of about 1 m and consisting of an isolated patch of fine-textured, compact material surrounded by frost-shattered boulders; term introduced by Washburn (1956, p.827). Syn: *earth island; rubble island.*

debris line A *swash mark* composed of debris washed up on a beach by storm waves. Cf: *trash line.*

debris plain A plain covered with rock waste.

debris slide A slide involving a slow to rapid downslope movement of comparatively dry and predominantly unconsolidated and incoherent earth, soil, and rock debris in which the mass does not show backward rotation (as in a slump) but slides or rolls forward, forming an irregular hummocky deposit resembling a moraine (Sharpe, 1938, p.74). It is often called an "earth slide", but this is incorrect because the moving mass of a debris slide is greatly deformed or consists of many small units.

debris slope A *constant slope* with debris on it from the free face above. Cf: *talus slope.*

debris stream *debris flow.*

debris torrent A sediment-laden flood. The term is sometimes used synonymously with *debris flow* or *hyperconcentrated flow.* Cf: *torrent.*

Debye-Scherrer method (De-bye'-Scherr'-er) A method for recording the X-ray diffraction pattern of a crystalline material on film by using a specimen in the form of a powder. Cf: *powder diffraction.*

decalcified (de-cal'-ci-fied) Said of a soil that has been leached of its calcium carbonate.

decapod (dec'-a-pod) (a) Any *eumalacostracan* belonging to the order Decapoda, characterized by the presence of five pairs of uniramous limbs behind the maxillipeds. Range, Permian to present. (b) An early name for a *coleoid,* now discontinued.

decapod horizontal laminate structure A *microstructure* of more or less mutually parallel, calcified, chitin-protein fibers arranged into horizontal laminae which give rise to, and are interconnected by, numerous mutually parallel, curved (reclined to vertical to inclined) laminae. These laminae themselves consist of more or less mutually parallel, calcified, chitin-protein fibers and are penetrated by vertical pore canals (Mutvei, 1974, 1981).

decarbonation A form of *metasomatism* whereby silicate and oxide minerals are produced by the removal of CO_2 from carbonate rocks.

decay [radioactivity] (de-cay') *radioactive decay.*

decay [waves] The attenuation or loss of energy from wind-generated ocean waves after they leave the generating area and pass into a region of lighter winds; it is accompanied by a gradual increase in wavelength and a gradual decrease in wave height.

decay [weath] The general weathering or wasting away of rock; specif. *chemical weathering.*

decay constant [elect] The time τ for an exponentially decaying quantity to decrease to 1/e of its initial value (e is the base of the Napierian logarithm).

decay constant [radioactivity] The proportionality constant that relates the rate of decay of a radionuclide to the number of atoms of that radionuclide remaining: $dN/dt = -\lambda N$, where λ is the decay constant, N is the number of radionuclides remaining, and t is the time. See also: *branching ratio; mean life.* Syn: *disintegration constant.*

decay curve A plot in linear coordinates of the number of radioactive atoms remaining as a function of time expressed in terms of multiples of half-lives.

decementation (de'-ce-men-ta'-tion) The dissolving-out or leaching of the cement of a sedimentary rock, as in a sandstone whose void-filling fluids and solid grains do not form a closed system, thereby permitting fluids to move (or ions to diffuse) in and out (Pettijohn, 1957, p.659); e.g. the removal of carbonates from a calcareous sandstone.

deciduous (de-cid'-u-ous) (a) Said of plants that shed their leaves annually. (b) Said of plant parts that are shed regularly within a year of their production.

decimal degree A *degree* of latitude or longitude expressed in decimal form, rather than in degrees, minutes, and seconds (Kennedy et al., 2003).

decimation A process in which a data set is reduced in size by resampling; e.g., selecting every other data value.

decke (deck'-e) The German equivalent of *nappe [struc geol],* rarely used in the English-language literature.

deckenkarren (deck'-en-kar'-ren) In speleology, a German term for solutional *pendant* features on the ceiling of a cave.

declination (dec-li-na'-tion) A deviation of the geomagnetic field vector from true north, i.e., the angle between geographic north and magnetic north. May also be defined as an azimuth of the *magnetic meridian.* It is one of the *magnetic elements.* Syn: *magnetic variation.*

declination diagram A diagram, usually found in the marginalia of large-scale maps, illustrating the angular differences between true, grid, and magnetic north, expressed in degrees.

decline Passage or adit driven on a decline from the surface to provide access to a mine.

decline curve (de-cline') A graph of the decline in production of an oil or gas well or group of wells. Production rate (ordinate) is plotted against time (abscissa). It is used to predict *ultimate recovery.*

declined (de-clined') Said of a graptoloid *rhabdosome* with stipes hanging below the *sicula* and enclosing an angle of <180° between their ventral sides. Cf: *deflexed; reclined; reflexed.*

declining development (de-clin'-ing) The hypothetical production of a landscape where the rate of downward erosion is more rapid than the rate of uplift or where valley widening exceeds valley deepening, characterized by a decrease of the relative relief and the formation of concave slopes. Cf: *accelerated development; uniform development.* Syn: *waning development; descending development.*

declinometer (dec-li-nom'-e-ter) An instrument that measures magnetic declination.

declivity (de-cliv'-i-ty) (a) A slope that descends from a point of reference; esp. a steep or overhanging slope, as of a cliff. Ant: *acclivity.* (b) A gradient of a surface; a deviation downward from the horizontal; an inclination.

décollement (de-colle'-ment) A large-displacement (kilometers to tens of kilometers) shallowly dipping to subhorizontal fault or shear zone. Typically, faults and folds in rocks above the décollement do not extend across it. Syn: *detachment fault; décollement fault; glide horizon.* Obsolete syn: *strip thrust.*

décollement fault *décollement.*

décollement fold *detachment fold.*

décollement zone (a) An interval in a stratigraphic succession in which a décollement occurs, typically in a mechanically weak rock like shale or an evaporite. (b) A zone of intensely sheared rock (*mylonite* if formed by plastic deformation, or *breccia* if formed by brittle deformation) bordering a décollement. Syn: *glide plane [struc geol].*

decomposers (de-com-pos'-ers) Organisms, usually microscopic, that break down organic matter and thus aid in recycling nutrients.

decomposition (de'-com-po-si'-tion) *chemical weathering.*

decomposition [geophys exploration] (a) Separating effects of different kinds or attributable to different causes. (b) Separating a potential (e.g., gravity) map into regional and residual, *residualizing.* (c) Removing the effects of filtering from a filtered waveform, *deconvolution.* (d) Deriving resistivity stratification from a kernel function.

decompression Reduction in total pressure during ascent of a rock or magma body through the mantle or crust. Rock decompression is a common cause of partial melting (decompression melting). Magma decompression can exert important controls on mineral compositions and stabilities and volatile solution (or exsolution) and *degassing.*

decontamination A variety of processes used to clean equipment that has contacted formation material, groundwater, or surface water that is known to be or suspected of being contaminated.

deconvolution (de'-con-vo-lu'-tion) A process designed to restore a wave shape to the form it is assumed to have had before it underwent a filtering action or convolution. It is a data-processing technique applied to seismic-reflection and other data for the purpose of improving the visibility and resolution of reflected events.

decorated planar deformation features Annealed *planar deformation features* consisting of discontinuously aligned vugs and inclusions formed during recrystallization of the originally amorphous lamellae; typically in shocked tectosilicates on the floor of impact craters. Also known from shocked olivine in thermally annealed chondrites, where the decorations consist of ultra-fine grained troilite and metal droplets.

decoration An obsolete term for *speleothem.*

decorative stone (dec'-o-ra'-tive) A stone used for architectural decoration, as in mantels, columns, and store fronts.

decoupling [geophys exploration] With respect to vibrators, when the pad loses contact with the ground. Generally, peak ground force is set so as not to exceed 90% of the hold-down weight.

decrement (dec'-re-ment) *groundwater discharge.*

decrepitation (de'-crep-i-ta'-tion) The breaking up of a mineral, usually violently and noisily, when it is heated.

decrespignyite-(Y) A blue monoclinic mineral: $(Y,REE)_4Cu(CO_3)_4Cl(OH)_5 \cdot 2H_2O$.

decurrent (de-cur'-rent) Said of parts that extend downward below the point of insertion, as in decurrent leaves adnate to a stem.

decussate (de-cus'-sate) Said of plants with appendages opposite one another on an axis, alternating appendages in sequence being inserted at right angles (Swartz, 1971, p.142). Cf: *distichous.*

decussate tetrad (de-cus'-sate) A *tetrad* of pollen or spores consisting of two pairs of grains that cross each other, more or less at right angles, in two different planes.

decussate texture In metamorphosed rocks, a microtexture in which axes of contiguous crystals lie in diverse, criss-cross directions that are not random but rather are part of a definite mechanical expedient for minimizing internal stress. It is most noticeable in rocks composed largely of minerals with a flaky or columnar habit (Harker, 1939).

dedolomitization (de-do'-lo-mit'-i-za'-tion) A process resulting from metamorphism, wherein part or all of the magnesium in a dolomite or dolomitic limestone is used for the formation of magnesium oxides, hydroxides, and silicates (e.g. brucite, forsterite) and resulting in an enrichment in calcite (Teall, 1903). The term was originally used by Morlot (1847) for the replacement of dolomite by calcite during diagenesis or chemical weathering.

deduction (de-duc'-tion) Reasoning from the general to the particular; inferring consequences from evidence; deriving applications from general principles. "It is a mental process not always given its proper priority in considering geological questions. An explanation or hypothesis in accord with deduction must be preferred to any other" (Challinor, 1978, p.79). Ant: *induction [philos].*

deep n. A clearly discernible depression of the ocean floor.

deep coal Coal that is far enough below the surface to require underground mining. Cf: *crop coal.*

deep earthquake *deep-focus earthquake.*

deeper-pool test (deep'-er-pool) A well located within the known limits of an oil or gas pool and drilled with the object of searching for new producing zones below the producing zone of the pool (Lahee, 1962, p.134). Cf: *shallower-pool test.*

deep-focus earthquake An earthquake whose focus is at a depth of 300-700 km. Cf: *shallow-focus earthquake; intermediate-focus earthquake.* Syn: *deep earthquake.*

deep lead A *lead [eco geol]* or alluvial placer that is buried under soil or rock. Pron: deep leed.

deep percolation Precipitation moving downward, below the plant-root zone, toward storage in subsurface strata. Cf: *shallow percolation.*

deep phreatic Said of cave formation at considerable depth below the top of the water-saturated zone. See also: *shallow phreatic.*

deep scattering layer A stratified aggregation of marine organisms (*zooplankton* and *nekton*) in the open ocean that scatters sound waves from an echo sounder and that is involved in *diurnal vertical migration.* Cf: *shallow scattering layer; surface scattering layer.* Syn: *false bottom [oceanog]; phantom bottom.*

deep-sea channel A trough-shaped, low-relief valley on the deep-sea floor beyond the continental rise. It has few tributaries, and may be either parallel or at an angle to the continental margin. Syn: *mid-ocean canyon.*

deep-sea deposit Sediment found on the deep-sea floor. See: *turbidite; eupelagic deposit; hemipelagic deposit; pelagic deposit.*

Deep Sea Drilling Project Multi-institutional program to drill into the deep-sea floor at numerous sites using the ship "Glomar Challenger". Initiated in 1965, and managed by the Scripps Institution of Oceanography with advice and guidance by *JOIDES.* Abbrev: DSDP. Between 1968 and 1983, DSDP recovered cores from 624 sites, curated over 70 miles of core, and published 96 volumes of "Initial Reports of the Deep Sea Drilling Project." DSDP was terminated in April 1987, and succeeded by the *Ocean Drilling Program.*

deep-sea fan *submarine fan.*

deep-seated Said of geologic features and processes that originate or are situated at depths of one kilometer or more below the Earth's surface; *plutonic.*

deep seismic sounding A long seismic profile (usually refraction) which has the objective of studying the crust, Moho, and upper mantle. Abbrev: DSS.

deep tow A submersible designed to be towed in deep water a short distance above the sea floor, and to take continuous oceanographic, geophysical and bathymetric measurements.

deep water A dense *water mass* overlying bottom water. It is formed by cooling, then sinks and spreads at great depth. Cf: *intermediate water; bottom water [oceanog]; surface water.*

deep-water wave A wave on the surface of a body of water, the wavelength of which is less than twice the depth of the water, and for which the water depth is not an influence on the velocity or on the shape of the orbital. Cf: *shallow-water wave; transitional-water wave.* Syn: *short wave; surface wave [oceanog].*

deep well A water well, generally drilled, extending to a depth greater than that typical of shallow wells in the vicinity. The term may be applied to a well 20 m deep in an area where shallow wells average 7 or 8 m deep, or to a much deeper well in an area where the shallowest aquifer supplies wells 100 m deep or more. Cf: *shallow well.*

deep-well disposal Disposal of liquid waste by injection into wells, usually constructed especially for the purpose, that penetrate deep, porous and permeable formations that are confined vertically by relatively impermeable beds. The method is used for disposal of saline water brought to the surface in oil wells, and for disposal of a variety of liquid wastes from industrial processes. Also called: deep-well injection.

deerite (deer'-ite) A black monoclinic mineral with a chain silicate structure: $(Fe^{2+},Mn^{2+})_6(Fe^{3+},Al)_3Si_6O_{20}(OH)_5$.

Deerparkian (Deer-park'-i-an) North American stage: Lower Devonian (above Helderbergian, below Esopusian).

defeated stream (de-feat'-ed) A stream that, owing to uplift or other cause, is unable to degrade as fast as the land rises and thereby fails to maintain its original course, becomes ponded and diverted into a new course, and resumes as a consequent stream.

defect lattice (de'-fect) A crystal lattice in which the expected systemic repetition is interrupted by an omission, an inclusion of an extra item, or the substitution of an unexpected item. Cf: *Schottky defect.*

defect-lattice solid solution (de'-fect-lat'-tice) *omission solid solution.*

defernite (de-fern'-ite) A colorless, red, or rose-brown orthorhombic mineral occurring in skarn: $Ca_6(CO_3,SiO_4)_2(OH)_5$.

deferred junction (de-ferred') The place on a flood plain where the main stream is joined by a tributary whose course is prolonged downstream for a considerable distance by a barrier along the main stream; esp. the junction of a *yazoo stream* with the main stream, as along the convex side of a major meander. Syn: *yazoo.*

deferred tributary *yazoo stream.*

defilade (def'-i-lade) A military plan or construction for protection from direct or reverse fire; e.g., a fortification. See also: *enfilade.*

defile (de'-file) A long, narrow pass or passage through hills or mountains, often forming the approach to a larger pass; esp. a passage enclosed between high, precipitous walls, as a gorge.

definition (def-i-ni'-tion) The degree of clarity and sharpness of an image in a photograph; also, the ability of a lens to record fine detail.

deflation (de-fla'-tion) The sorting out, lifting, and removal of loose dry fine-grained particles (clay and silt sizes) by the turbulent eddy action of the wind, as along a sand-dune coast or in a desert; a form of *wind erosion.*

deflation armor A *desert armor* whose surface layer of coarse particles is concentrated chiefly by deflation.

deflation basin A topographic basin excavated and maintained by wind erosion which removes unconsolidated material. Cf: *blowout [geomorph].* Syn: *wind-scoured basin.*

deflation hollow *blowout [geomorph].*

deflation lake A lake occupying a basin formed mainly by wind erosion, esp. in an arid or semiarid region; it is usually very shallow and may contain water only during certain seasons. Cf: *deflation basin; dune lake.*

deflation residue *desert pavement.*

deflation ripple *granule ripple.*

deflection [drill] (de-flec'-tion) *deviation [drill].*

deflection [geodesy] *deflection of the vertical.*

deflection [geomorph] A sharp change in the trend of a mountain chain. The term was introduced by Bucher (1933) as a translation of Staub's term Beugung. It differs from an *orocline* by not necessarily being a strain imposed on the completed orogen. See also: *capped deflection; fractured deflection.* Cf: *linkage [geomorph].*

deflection [streams] A relatively spontaneous diversion of a stream, as by warping, alluviation, glaciation, lateral corrasion, volcanic action, or shoreline changes.

deflection angle [photo] A vertical angle, measured in the vertical plane containing the flight line, by which the datum of any model in a stereotriangulated strip departs from the datum of the preceding model.

deflection angle [surv] A horizontal angle measured from the forward prolongation of the preceding line to the following line; the angle between one survey line and the extension of another survey line that meets it. A deflection angle to the right is positive; one to the left is negative.

deflection basin A basin hollowed out by the erosive action of ice in front of a barrier obstructing the path of a glacier (Geikie, 1898, p.297).

deflection of the vertical The angle at a given point on the Earth between the vertical, defined by gravity, and the direction of the normal to the reference ellipsoid through that point. It is sometimes referred to as *deviation of the vertical* or deflection of the plumb line. Cf: *datum [geodesy]; astronomic measurement.* Syn: *station error; deflection [geodesy].*

deflection pool A pool occupying a depression scooped out by a stream in its obstructing bed at a curve in its course (Miller, 1883, p.275).

deflexed (de-flexed') Said of a graptoloid *rhabdosome* with initial part of stipes hanging below the *sicula* and enclosing an angle of <180° between their ventral sides and distal parts of stipes tending to the horizontal. Cf: *declined; reflexed; reclined.*

defluent (de'-flu-ent) A stream that flows from a lake or glacier.

deformation [exp struc petrol] A change in position (rigid-body translation), orientation (rigid-body rotation), shape (distortion), and/or size (dilation). A change in shape and/or size is also known as *strain.*

deformation [tect] (a) A general term for the process of folding, faulting, shearing, or fabric development of the rocks as a result of Earth stresses. (b) The change in the geometry of a body of rock that occurs as a consequence of stress, e.g. translation, rigid body rotation about an axis, and *strain* or distortion.

deformation band (a) A planar zone of distributed deformation in which the strain is approximately simple shear, typically involving micro-fracturing and cataclasis (Peacock et al., 2000). (b) A brittle shear zone that develops in highly porous rocks (e.g., eolian sandstone) and is characterized by a reduction of porosity. In consequence of strain hardening, the deformation band widens as shear displacement accumulates (Aydin and Johnson, 1978). Cf: *kink band; shear zone.*

deformation ellipsoid *strain ellipsoid.*

deformation fabric A rock fabric that has resulted from deformation; the fabric of a *secondary tectonite.* Typical fabrics are composed of lineations, schistosities, cleavages, fold axes, and crystallographic preferred orientations. Cf: *depositional fabric.* Syn: *tectonic fabric.*

deformation front The boundary between deformed and undeformed regions. It can occur at the foreland edge of an orogen or on the trench side of an *accretionary prism.*

deformation lamella One of a series of narrow planar features developed by deformation within single mineral grains; they have a slightly different refractive index than the host grain. Pl: lamellae.

deformation path The path taken by a deforming object in going from one strain state to another. The path can be plotted on a Flinn or other deformation diagram.

deformation plan The array of displacement vectors that connect points in the deformed state to those same points in the undeformed state. Syn: *movement picture; movement plan.*

deformation plane In structural petrology, a term used in a kinematic sense to describe a deformation that has monoclinic symmetry. It is the unique symmetry plane and is parallel to the *a* axis and normal to the *b* axis. During progressive simple shear the deformation plane is parallel to the direction of shear and normal to the planes of shear. It is also called the a-c plane.

deformation twin A crystal twin that is produced by gliding, i.e. deformation within a preexisting crystal. Syn: *glide twin; mechanical twin.*

deformation twinning In a crystal, twinning that is produced by gliding. Syn: *secondary twinning.*

deformed cross-bedding (de-formed') Cross-bedding in which the foreset beds are "overturned or buckled in the downcurrent direction usually prior to deposition of the overlying bed" (Pettijohn and Potter, 1964, p.299). The foreset dip angle may also be altered by subsequent tectonic folding.

deformed ice A term used by the U.S. Naval Oceanographic Office (1968, p. B32) for *pressure ice.*

degassing The process of volatile exsolution from a magma. Exsolution may be induced by decompression or the concentration of volatiles in the liquid through crystallization of anhydrous minerals.

degaussing *demagnetization.*

De Geer moraine One of a series of recessional moraines produced individually during a single year as determined by varve chronology in Sweden.

degenerative recrystallization (de-gen'-er-a-tive') *degradation recrystallization.*

deghosting *deconvolution.*

Deglacial (De-gla'-cial) n. A term used by Antevs (1953) for a time unit that covers in North America the time from the greatest extension of Wisconsin glaciation to the beginning of recession from the Cochrane moraines.

deglaciation (de'-gla-ci-a'-tion) The uncovering of a land area from beneath a glacier or ice sheet by the withdrawal of ice due to shrinkage by melting or calving of icebergs. As used in Great Britain, the term is restricted to a process that occurred in the past, in contrast to *deglacierization.* Also, the result of deglaciation.

deglacierization (de-gla'-cier-i-za'-tion) A term used in Great Britain for the gradual withdrawal, going on at the present time, of a glacier or an ice sheet from a land area. Cf: *deglaciation.*

degradation [geomorph] (deg-ra-da'-tion) (a) The progressive wearing down or away, and the general lowering or reduction, of the Earth's surface by the natural processes of weathering and erosion; e.g. the deepening by a stream of its channel. The term sometimes includes the process of transportation; and sometimes it is used synonymously with *denudation,* or used to signify the results of denudation. Cf: *gradation [geomorph].* (b) Less broadly, the vertical erosion or *downcutting* performed by a stream in order to establish or maintain uniformity of grade or slope. Cf: *aggradation [geomorph].*

degradation [grd wat] The breakdown of substances by biological action (Rogers, 1981, p.94).

degradation [permafrost] The shrinkage or disappearance of permafrost due to natural or artificial causes. Ant: *aggradation [permafrost].* Cf: *depergelation.*

degradation [soil] A decrease in exchangeable bases in a soil, and destruction of layer-silicate clay, as a result of leaching (Jacks et al., 1960, p.162) in publications. Term no longer used in publications of the Soil Science Society of America.

degradation [streams] The lowering of a stream bed, due to such factors as increased scouring. Cf: *accretion [streams].*

degradation recrystallization Recrystallization resulting in a relative decrease in the size of crystals. Ant: *aggradation recrystallization.* Syn: *degenerative recrystallization.*

degradation vacuity The space-time value of the degradationally removed part of a transgressive-regressive depositional succession; e.g. the part of a *lacuna [stratig]* resulting from degradation of formerly existing rocks at an unconformity. The term was used by Wheeler (1964, p.602) to replace *erosional vacuity.* Cf: *hiatus.* Syn: *vacuity.*

Degraded Chernozem (De-grad'-ed) A zonal great soil group consisting of soils with a very dark brown or black A1 (A) horizon underlain by a dark gray, weakly expressed A2 (E) horizon and a brown B (?) horizon; formed in the forest-prairie transition of cool climates. Not used in current U.S. system of soil taxonomy.

degraded illite (de-grad'-ed) Illite that has had much of its potassium removed from the interlayer position as a result of prolonged leaching. Syn: *stripped illite.*

degrading neomorphism (de-grad'-ing) A kind of *neomorphism* in which the crystal size decreases (Folk, 1965, p.23). Ant: *aggrading neomorphism.*

degrading stream (a) A stream that is actively cutting down its channel or valley and that is capable of transporting more load than it is supplied with. (b) A stream that is downcutting approximately at grade.

degree [cart] A unit of angular measure equal to 1/360 of a complete revolution and represented by the symbol ° (Kennedy et al., 2001; McGraw-Hill, 1984).

degree [hydrol] (de-gree') In hydrologic terminology, a unit for expressing the hardness of water. Cf: *Clark degree.*

degree-day unit A measure of the difference between the mean daily temperature and an arbitrary temperature such as 18.3°C (65°F). It is used most frequently by heating engineers in terms of heating degree units, calculated when the mean daily temperature is below the base temperature. Others include cooling degree day units and growing degree day units. It is normally applied to mean temperatures that are below the standard.

degree of curve A measure of the sharpness of curvature; e.g. the angle at the center of a circle subtended by a chord 100 ft long (as for U.S. railroads) or by an arc 100 ft long (as in highway surveying).

degree of freedom The capability of variation in a chemical system. The number of degrees of freedom in a system may be defined as the number of independent, intensive variables (e.g. temperature, pressure, and concentration in the various phases) necessary to define the system completely, or as the number of variables that may be changed independently without causing a change in phase. Syn: *variance [exp petrol].* Cf: *phase rule.*

degree of meraspid period A developmental stage of a trilobite during the *meraspid* period, when the number of thoracic segments are increased. The degree is defined by the number of thoracic segments present and ranges from zero to the holaspid number of thoracic segments minus one.

degree of slope The angular measurement (expressed in degrees) of slope from a horizontal plane (Van Riper, 1962, p.603).

degree of sorting A measure of the spread or range of variation of the particle-size distribution in a sediment. It is defined statistically as the extent to which the particles are dispersed on either side of the average: the wider the spread, the poorer is the sorting. It may be expressed by *sigma phi.* Cf: *sorting.*

degrees of freedom The number of independent items of information in a collection of observations. Usually, the number of observations minus the number of statistics calculated from the observations.

dehiscent (de-his'-cent) Said of certain fruits and sporangia that split open along definite seams at maturity.

dehrnite (dehrn'-ite) A discredited syn. of *carbonate-fluorapatite.*

dehydration reaction (de-hy-dra'-tion) A metamorphic reaction that results in the loss of H_2O from a mineral to the fluid phase. Cf: *hydration reaction.*

DE interray Left posterior interray in echinoderms situated between D ray and E ray and clockwise of D ray when the echinoderm is viewed from the adoral side; equal to interambulacrum 1 in the *Lovenian system.*

dejection cone (de-jec'-tion) *cone of dejection.*

delafossite (de-la-foss'-ite) A metallic black rhombohedral mineral: $Cu^{1+}Fe^{3+}O_2$.

delatorreite (de-la-tor'-re-ite) *todorokite.*

delatynite (de-la-tyn'-ite) A variety of amber rich in carbon, low in succinic acid, and lacking sulfur, found at Delatyn in the Carpathian Mountains of Galicia.

delawarite (de'-la-war-ite') Pearly orthoclase from Delaware County, Penna.

delayed runoff (de-layed') Water from precipitation that sinks into the ground and discharges later into streams through seeps and springs (Tarr and Von Engeln, 1926, p.70). As defined above, delayed runoff is a syn. of *groundwater runoff,* but it could also be defined as runoff delayed by any means, such as temporary storage of precipitation in the form of snow and ice.

delay time (de-lay') In seismic refraction work, the additional time required to traverse any raypath over the time which would be required to traverse the horizontal component at the refractor velocity. Source delay time refers to the source end of the trajectory; receiver delay time refers to the receiver end of the trajectory. See also: *intercept time.*

deldoradite (del-do-rad'-ite) A light-colored cancrinite *syenite,* named by Johannsen (1939) for Deldorado Creek, Colorado. Obsolete.

delessite (de-less'-ite) A ferroan variety of clinochlore; it occurs in cavities and seams in basic igneous rocks.

delhayelite (del-hay'-e-lite) A colorless orthorhombic mineral: $(K,Na)_{10}Ca_5Al_2Si_{14}O_{38}F_4Cl_2$.

deliensite A pale yellow or grayish white orthorhombic mineral: $Fe^{2+}(UO_2)_2(SO_4)_2(OH)_2{\cdot}3H_2O$.

delindeite (de-linde'-ite) A pinkish-gray monoclinic mineral: $(Na,K)_3Ba_2Ti_3Si_4O_{14}(O,OH,H_2O)_6$.

delineation (de-lin'-e-a'-tion) A step in map *compilation* in which mapworthy features are distinguished and outlined on various possible source materials or are visually selected (as when operating a stereoscopic plotting instrument).

deliniation tolerance [cart] An allowable threshold of *source position* accuracy for previously mapped existing features. The positions of previously mapped existing features that do not exceed the *delineation tolerance* are not modified or adjusted during revision mapping (USGS, 2002b, p.1A-2).

dell (a) A small, secluded wooded valley or natural hollow. The term is often used in a literary sense with no definite meaning. Cf: *dale.* (b) A depression upvalley from the source of a stream (Penck, 1953, p.421). Etymol: German "Delle", "dent".

dellaite (del′-la-ite) A white triclinic mineral: $Ca_6Si_3O_{11}(OH)_2$.

dellenite (del′-len-ite) A volcanic rock intermediate between rhyolite and dacite. The term, proposed by Brögger in 1896 from Dellen Lake, Sweden, is roughly synonymous with *rhyodacite,* but neither name is incorporated in the *IUGS classification.* Obsolete.

dells A corruption of *dalles,* as applied along the Wisconsin River, Wisc.

Delmontian (Del-mont′-i-an) A Pacific Coast provincial stage: middle Miocene (lateral facies of lower Mohnian).

deloneite-(Ce) A bright yellow trigonal mineral of the *apatite* group: $Na(Ca,Sr)_3Ce(PO_4)_3(F,OH)$.

delorenzite (de-lor-en′-zite) *tanteuxenite-(Y).*

deloryite A dark green to black monoclinic mineral: $Cu_4(UO_2)Mo_2O_8(OH)_6$.

delphinite (del′-phi-nite) A yellowish-green epidote variety from France.

delrioite (del-ri′-o-ite) A pale yellow-green monoclinic mineral: $CaSrV^{5+}{}_2O_6(OH)_2 \cdot 3H_2O$. A dimorph of metadelrioite.

delta (del′-ta) The low, nearly flat, alluvial tract of land at or near the mouth of a river, commonly forming a triangular or fan-shaped plain of considerable area, crossed by many distributaries of the main river, perhaps extending beyond the general trend of the coast, and resulting from the accumulation of sediment supplied by the river in such quantities that it is not removed by tides, waves, and currents. Most deltas are partly subaerial and partly below water. The term was introduced by Herodotus in the 5th century B.C. for the tract of land, at the mouth of the Nile River, whose outline broadly resembled the Greek capital letter "delta", Δ, with the apex pointing upstream. Syn: *delta lobe.*

delta bar A "bar" formed by a tributary stream that is building a delta into the channel of the main stream.

delta bedding The bedding characteristic of a delta, consisting of comparatively flat topset beds and bottomset beds, between which are steeper foreset beds leading from close to the delta surface to the bottom of surrounding water; the inclined bedding "presumed to originate as foresets of small deltas" (Pettijohn and Potter, 1964, p.299).

delta complex A low, nearly flat, alluvial tract of land formed as a fan-shaped plain and composed of two or more delta lobes or subdeltas.

delta cycle A term used by Barrell (1912, p.397) for a two-phase tectonic cycle dependent on stream erosion and changing sea level, and involving deposition with a stationary crust followed by vertical movement (normally subsidence) of the bottom. For a deposit that is essentially a deltaic accumulation, the cycle is one of increasing coarseness from the base upward, of an increasing volume of clastic material, and of eventual overtake of subsidence by sedimentation.

delta diversity The change in taxon diversity from geographic region to geographic region, arising from biotic provinciality. It is typically calculated as the ratio of epsilon to gamma diversity (Whittaker, 1972).

delta fan A deposit formed by the merging of an alluvial fan with a delta. Syn: *fan delta.*

delta front A narrow zone where deposition in deltas is most active, consisting of a continuous sheet of sand, and occurring within the effective depth of wave erosion (10 m or less). It is the zone separating the *prodelta* from the *delta plain,* and it may or may not be steep.

delta-front platform A zone of shallows, up to about 5 km wide, in front of the advancing distributaries of a delta.

delta-front trough A trough-shaped submarine valley formed off a large river delta on the continental shelf and slope. It has straight walls of soft rock, few if any tributaries, and a flat, seaward-sloping floor.

deltaic (del-ta′-ic) Pertaining to or characterized by a delta; e.g. "deltaic sedimentation". Also, constituting a delta; e.g. a "deltaic coast".

deltaic coastal plain A coastal plain composed of a series of coalescing deltas; it consists initially of natural levee ridges separated by basins.

deltaic deposit A sedimentary deposit laid down in a delta, characterized by well-developed local cross-bedding and by a mixture of sand, clay, and the remains of brackish-water organisms and of organic matter. Cf: *estuarine deposit.*

deltaic tract An extension of the *plain tract* of a stream, characterized by the formation of a delta and the deposition of fine sediments.

deltaite (del′-ta-ite) A mixture of crandallite and hydroxylapatite.

delta kame A flat-topped, steep-sided hill of well-sorted sand and gravel deposited by a meltwater stream flowing into a proglacial or other ice-marginal lake; the proximal margin of the delta was built in contact with glacier ice. Cf: *esker delta.* Syn: *kame delta; ice-contact delta; sand plateau.*

delta lake A lake formed along the margin of or within a delta, as by the building of barriers across a shallow embayment or by the enclosure of part of the sea by the growth of deltaic deposits.

delta levee lake A lake on an advancing delta, formed between sandbars or natural levees deposited at the mouths of distributaries. Example: Lake Pontchartrain on the Mississippi River delta.

delta lobe *delta.*

delta-mooreite (del′-ta-moore′-ite) *torreyite.*

delta plain The level or nearly level surface composing the landward part of a large or compound delta; strictly, an alluvial plain characterized by repeated channel bifurcation and divergence, multiple distributary channels, and interdistributary flood basins. Syn: *intradelta.* Cf: *marine delta plain.*

delta plateau A raised or abandoned delta plain.

delta porphyroclast An asymmetric porphyroclast with "wings" or "tails" of fine grained recrystallized material that extend tangentially away from the edge. The tails cross a line drawn through the center of the porphyroclast parallel to the external foliaton. The asymmetry of the porphyroclast can be used to determine the shear sense. Cf: *sigma porphyroclast.*

Delta seventeen-O ($\Delta^{17}O$) In studies of meteorites, the measure, on a $\delta^{17}O$-$\delta^{18}O$ diagram, of the vertival separation between a datum and the terrestrial fractionation (reference) line. It is expressed as $(\Delta^{17}O) = \delta^{17}O - 0.52 \cdot \delta^{18}O$.

delta shoreline A prograding shoreline produced by the advance of a delta into a lake or the sea.

delta structure The sedimentary structure produced by the three sets of beds in a delta: bottomset, foreset, and topset beds.

delta-switching The fundamental process of river course diversion that results in the formation of a new delta as water and sediment are systematically diverted from a previous river course and its delta.

delta terrace A fan-shaped terrace composed of a delta that remained after the disappearance of the stream that produced it (Chamberlin, 1883, p.304).

delta t (Δt) [seis] (a) The time difference between the arrivals of the same phase of a seismic event at two locations. Cf: *dip moveout; dip calculation.* Delta t ordinarily does not imply normal moveout unless specifically stated. (b) Interval transit time, as used with the sonic log.

delta unit *delta value.*

delta value The difference between the *isotope ratio* in a sample and that in a standard, divided by the ratio in the standard, and expressed as parts per thousand per mil. Syn: *delta unit.*

delthyrial chamber (del-thy′-ri-al) The cavity beneath the *umbo* of the *pedicle valve* of a brachiopod, bounded by dental plates or by posterior and lateral shell walls if dental plates are absent. It corresponds to the *notothyrial chamber* of the brachial valve.

delthyrial plate A plate within the *delthyrial chamber* of some spiriferoid brachiopods, extending a variable distance from the apex between dental plates. It is probably homologous with the pedicle collar (TIP, 1965, pt. H, p.143).

delthyrium (del-thy′-ri-um) The median triangular or subtriangular opening beneath the beak of the *pedicle valve* of a brachiopod, bisecting the ventral *cardinal area* or *pseudointerarea,* and commonly serving as a *pedicle* opening. Pl: delthyria. Cf: *notothyrium.*

deltidial plate (del-tid′-i-al) One of two plates growing medially (inward) from the margins of the *delthyrium* of a brachiopod and partly or completely closing it.

deltidium (del-tid′-i-um) The cover of the delthyrium of a brachiopod, formed by conjunct *deltidial plates* (in contact anteriorly and dorsally of the *pedicle*) whose line of junction is visible. Pl: deltidia. Cf: *pseudodeltidium.*

deltohedron (del-to-hed′-ron) A *tristetrahedron* whose faces are quadrilateral, rather than triangular, as in the *trigonal tristetrahedron.* Syn: *deltoidal dodecahedron; tetragonal tristetrahedron.*

deltoid [paleont] *deltoid plate.*

deltoidal cast (del-toi′-dal) A term used by Birkenmajer (1958, p.143) for *frondescent cast.*

deltoidal dodecahedron *deltohedron.*

deltoid branch A branch of a stream, enclosing a whole delta (Jackson, 1834, p.79).

deltoid island A *branch island* formed on a delta (Jackson, 1834, p.79).

deltoid plate One of a circlet of *interradial,* subtriangular plates of a blastoid, situated near the summit (oral end) of the *theca* but aboral to the oral plates, between adjacent *ambulacra,* and above radial plates. Syn: deltoid.

deluge (del′-uge) A great inundation or overflowing of the land by water; specif. The Deluge (the Noachian flood).

deluvium (de-lu′-vi-um) *diluvium.*

delvauxite (del-vaux′-ite) A yellowish brown, brownish black, or reddish amorphous mineral, perhaps: $CaFe^{3+}{}_4(PO_4,SO_4)_2(OH)_8 \cdot 4\text{-}5H_2O$(?).

delve A surface depression or hollow.

DEM *digital elevation model.*

demagnetization (de-mag′-net-i-za′-tion) A reduction in the magnetic field intensity inside a body because of the magnetization of the body (Sheriff, 2002). Progressive reduction of remanent magnetization utilizes the fact that different geologically important magnetizations in rocks are carried by assemblages of magnetized grains of different grain size, shape, and/or composition, hence coercivity and/or unblocking temperatures. Techniques employed include *alternating-field demagnetization, thermal demagnetization,* and *chemical demagnetization.* See also: *magnetic cleaning.*

demagnetization diagram The graphical representation of the response, in magnetization direction and intensity, of a rock specimen to progressive demagnetization. An orthogonal demagnetization diagram is constructed by simultaneously projecting the endpoint of the magnetization vector measured onto a horizontal plane (using the N-S, E-W coordinates), where declination can be directly measured, and a vertical plane (using the V and either N-S or E-W coordinates).

demantoid (de-man'-toid) A bright-green to yellowish-green transparent gem variety of andradite garnet, characterized by a brilliant luster, a dispersion stronger than that of diamond, and a hardness less than that of other garnets. Syn: *Uralian emerald.*

demarcation line (de-mar-ca'-tion) An imaginary line on the surface of a valve of a bivalve mollusk, originating at the *beak* and marking the locus of points on successive positions of the margin where the transverse-growth component has had maximum effect. It forms a dorsal/ventral profile when the valve is viewed from one end.

demersal (de-mer'-sal) *Benthic,* usually with reference to fish.

demesmaekerite (de-mes-mae'-ker-ite) A bottle-green to clear olive-green triclinic mineral: $Pb_2Cu_5(UO_2)_2(Se^{4+}O_3)_6(OH)_6{\cdot}2H_2O$.

demic (dem'-ic) Said of a randomly interbreeding population smaller than the species population. The term is usually preceded by a prefix indicating the nature or cause of the separation of such a group; e.g., "topodemic," applied to a population restricted to a particular geographic area.

demicolpus (dem-i-col'-pus) A *meridional colpus* of which both margins are raised and connected in the equatorial region, in this manner forming a ridge that divides the colpus into two parts.

demigration [geophys exploration] (a) Creating a zero-offset synthetic seismic section from a structural model. (b) Inverting the seismic migration process, e.g., in order to use a different velocity model.

demiplate (dem'-i-plate) An ambulacral plate of an echinoid, touching an adradial suture but not touching a perradial suture.

demipyramid (dem-i-pyr'-a-mid) One of the ten elements that support the teeth in *Aristotle's lantern* of an echinoid. See also: *pyramid [paleont].*

demkha (dem'-kha) A term used in Algeria for an almond-shaped *dune massif* (Capot-Rey, 1945, p.395).

demodulation The process of retrieving an original signal from a modulated signal.

demoiselle (dem-oi-selle' [duh-mwa-zel]) A term used in the French Alps for an *earth pillar* capped by a large boulder, esp. one formed by weathering of volcanic breccia or of glacial till. Etymol: French, "young lady".

demoiselle hill A symmetrical beehive-shaped elevation with a grassy surface, bordered by a cauldronlike depression, occurring on the Magdalen Islands in the Gulf of St. Lawrence (Clarke, 1911, p.145); it may be a knob, knoll, or hill 175 m high.

demonstrated resources A term for the sum of measured resources plus indicated resources.

demorphism (de-mor'-phism) An obsolete syn. of *weathering.*

demosponge (dem'-o-sponge) Any sponge belonging to the class Demospongea and characterized chiefly by the presence of spongin as all or part of its skeleton. "Most demosponges contain siliceous spicules, with or without spongin. A few contain only spongin, and still fewer produce neither" (TIP, 1955, pt.E, p.34).

dendrite (den'-drite) A mineral, e.g. a surficial deposit of an oxide of manganese, or an inclusion, that has crystallized in a branching pattern. Adj: *dendritic.* Syn: *dendrolite.* See also: *spotty dendrite.*

dendritic (den-drit'-ic) Said of a mineral that has crystallized in a branching pattern; pertaining to a *dendrite.* Syn: *arborescent.*

dendritic drainage pattern A drainage pattern in which the streams branch randomly in all directions and at almost any angle, resembling in plan the branching habit of certain trees; it is produced where a consequent stream receives several tributaries which in turn are fed by smaller tributaries. It is indicative of insequent streams flowing across horizontal and homogeneous strata or complex crystalline rocks offering uniform resistance to erosion. Cf: *pinnate drainage pattern.*

dendritic glacier A *trunk glacier* joined by many tributary glaciers to form a pattern that resembles a branching tree.

dendritic ridge A type of *longitudinal ridge* that has a dendritic pattern. Bifurcation of most ridges is in the upcurrent direction (Dzulynski and Walton, 1963; 1965, p.69).

dendritic tufa *dendroid tufa.*

dendrochore (den'-dro-chore) The part of the Earth's surface having a tree-sustaining climate. It constitutes the bulk of the *biochore.*

dendrochronology (den'-dro-chro-nol'-o-gy) The study of annual growth rings of trees for dating of the recent past. Cf: *dendroclimatology.* Syn: *tree-ring chronology.*

dendroclimatology (den'-dro-cli'-ma-tol'-o-gy) The study of the patterns and relative sizes of annual growth rings of trees for paleoclimatologic data of the recent past. Cf: *dendrochronology.*

dendroclone (den'-dro-clone) A desma (of a sponge) having a smooth, straight shaft with a tree-like, branching process at each end.

dendrodate (den'-dro-date) A date calculated by *dendrochronology.*

dendrogram (den'-dro-gram) A treelike diagram depicting relationships between objects, especially the degree of similarity between objects that have been clustered by a hierarchical clustering algorithm. Dendrograms were originally introduced in *numerical taxonomy,* where they were interpreted as analogous to *phylogenetic trees.*

dendrograph (den'-dro-graph) A *dendrogram* in which the spacing between branches reflects the degree of dissimilarity between groups. Cf: *dendrogram.*

dendroid (den'-droid) adj. (a) Said of certain invertebrates that form many-branched colonies or that have a treelike habit of growth, such as the irregularly branched bushy colonies formed by graptolites. (b) Said of an irregularly branching type of fasciculate *corallum.* (c) A syn. of *ramose* in stenolaemate bryozoans. n. A sessile graptolite of the order Dendroidea ranging from Middle Cambrian to Carboniferous, characterized by a typically erect, dendroid *rhabdosome* having many stipes which are composed of autothecae, bithecae, and stolothecae arranged in regularly alternating triads along each branch.

dendroid colony An archaeocyathid colony in which each individual cup is isolated from others, except at its origin. It contrasts with massive or catenulate colonies (TIP, 1972, pt. E, p.6).

dendroid tufa Gray tufa occurring as spheroidal, mushroom, or dome-shaped masses with concentric macrostructure and less pronounced internal dendritic structure; e.g. the tufa along the shore of the extinct Lake Lahontan in Nevada, where it constitutes the major part of the dome-like mass. Cf: *thinolitic tufa; lithoid tufa.* Syn: *dendritic tufa.*

dendrolite (den'-dro-lite) *dendrite.*

denisovite (den-i-sov'-ite) A greenish-gray monoclinic mineral occurring in alkaline rocks of the Kola Peninsula: $(K,Na)Ca_2Si_3O_8(F,OH)$.

denitrification (de-ni-tri-fi-ca'-tion) In lakes, the bacterial reduction of dissolved nitrate to a gas, commonly molecular nitrogen (Last and Smol, 2001, p.453).

denivellation (de'-ni-vel-la'-tion) A variation in the level of a body of water, esp. of a lake; e.g. "wind denivellation" is a rise of water level due to wind drift, as along the windward shore of a lake. Etymol: French "dénivellation", "unleveling".

denningite (den'-ning-ite) A colorless to pale-green, brownish, or purplish tetragonal mineral: $(Ca,Mn^{2+})(Mn^{2+},Zn)(Te^{4+}{}_2O_5)_2$.

dennisonite (den'-ni-son-ite') *davisonite.*

dense [geol] (a) Said of a rock or mineral possessing a relatively high specific gravity. (b) Said of a fine-grained, aphanitic igneous rock whose particles generally average less than 0.05 to 0.1 mm in diameter, or whose texture is so fine that the individual particles cannot be recognized by the unaided eye. (c) Said of a rock whose constituent grains are crowded close together. The rock may be fine- or coarse-grained.

dense [optics] Said of a substance that is highly refractive.

dense-rock equivalent Volume of igneous rock erupted after all pore space caused by vesiculation, fracturing, and intergranular porosity has been subtracted. The calculation is used to determine the volume of magma that reached the surface during an eruption. Abbrev: DRE.

denser-than-water nonaqueous-phase liquid A polluting liquid with a density greater that water that sinks to the base of the aquifer, e.g. creosote and trichloro-ethylene (Stanger, 1994). Abbrev: DNAPL.

density [oceanog] (den'-si-ty) The mass of water per unit volume, in grams per cubic centimeter or kg per m^3. Density of water ranges from about 1.0 for fresh water to 1.05 for water in the deep ocean (5 km). Density may be written as *sigma-t* or σ_t. Cf: *in-situ density; neutral density; potential density.*

density [optics] A measure of the degree of opacity of any translucent medium, defined strictly as the common logarithm of the opacity; e.g. the degree of blackening of an exposed photographic film, plate, or paper after development, or of the direct image. Cf: *contrast.*

density [phys] Mass per unit volume.

density [stat] The relative number of observations, points or occurrences per unit interval of a variable, or unit interval of time, area, or volume.

density current A gravity-induced flow, owing to density differences with surrounding fluids. Factors affecting density differences include temperature, salinity, and concentration of suspended particles. See also: *turbidity current; nuée ardente.*

density log The *well log* curve of induced radioactivity showing the bulk density of rocks and their contained fluids. It is a *porosity log* of the wall-contact type, indicating formation density by recording the back scatter of gamma rays. Syn: *gamma-gamma log.*

density profile A line of gravity readings taken over a topographic feature with appreciable relief that is not associated with variations in density or structure. A set of several density profiles is constructed for different values of rock density between the datum (usually, sea level) and station elevation in order to determine the optimal density value for the *Bouguer correction.* The most appropriate density is the one that minimizes the correlation of gravity readings with topography. Also called: Nettleton profile.

density slicing The classification of digital multispectral, remotely-sensed images or elevation data in which pixels with values in designated bands are assigned to different categories. This technique has been used to produce "elevation maps" from satellite imagery.

density stratification The stratification of a lake produced as a result of density differences, the lightest layer occurring near the top and the heaviest layer at the bottom. It is usually brought about by temperature changes, but may also be caused by differences at different depths in the amount of suspended or dissolved material; e.g., where a surface layer of freshwater overlies salt water (Wetzel, 2001, p.73). Cf: *thermal stratification.*

densofacies (den-so-fa'-cies) A term used by Vassoevich (1948) for *metamorphic facies.* Little used.

densospore (den'-so-spore) A trilete spore, chiefly Paleozoic, with a pronounced cingulum that has a tendency to be "doubled", i.e. a thicker part toward the center of the spore and a thinner, more external part; e.g. the genus Densosporites and similar genera.

dental plate (den'-tal) (a) One of variably disposed plates of secondary shell underlying and supporting the hinge teeth of a brachiopod and extending to the floor of the *pedicle valve.* (b) In gnathostomes, a large crushing element formed by the expansion of a single tooth or the aggregation of several teeth.

dental socket A shallow excavation in the posterior margin of the *brachial valve* of a brachiopod for the reception of a *hinge tooth* of the *pedicle valve.* Syn: *socket.*

dentate (den'-tate) Toothed, or having small conical or toothlike projections; e.g. "dentate chela" of a sponge with toothed terminal expansions, or bivalved molluscs with hinge teeth.

denticle (den'-ti-cle) (a) A small spinelike, needlelike, or sawtoothlike structure of compound and platelike conodont elements, similar to a cusp but commonly smaller. (b) One of the small sharp protruding ridges that alternate with complementary dental sockets located along the cardinal margin or the hinge line of both valves of a brachiopod. (c) A primary or secondary toothlet on the sutural edge of the radius of a compartmental plate of a cirripede crustacean, serving to strengthen the articulation of the plates; a small, delicate, spinelike or toothlike projection on the carapace of an ostracode. (d) In vertebrates, a small toothlike scale; also, a small toothlike structure attached to the gill bars.

denticular composite prismatic structure A *composite prismatic structure* in which divergence of the second-order prisms is a consequence of their deposition on a curved, denticulated shell margin. The second-order prisms may be fibrous prismatic or simple prismatic but not composite prismatic, i.e., unlike *compound composite prismatic structure* (Carter, 1980b).

denticulate (den-tic'-u-late) (a) Having small teeth, or bearing a series of small spinelike or toothlike projections; e.g. said of a shell covered with such projections. (b) Minutely or finely dentate, as of a leaf.

denticulate sculpture [palyn] (den-tic'-u-late) Dinoflagellate or acritarch *sculpture* of toothlike spines (Eisenack et al., 1973).

denticulation (den-tic'-u-la'-tion) The state of being denticulate; also, a series of small toothlike structures, such as the denticles on the anterior and posterior margins of the shell of a cytherid ostracode.

dentition (den-ti'-tion) The number, kind, and arrangement of teeth or toothlike structures in invertebrates (e.g., hinge teeth of bivalved molluscs), and of teeth in vertebrates.

denudation (den-u-da'-tion) (a) The sum of the processes that result in the wearing away or the progressive lowering of the Earth's surface by various natural agencies, which include weathering, erosion, mass wasting, and transportation; also the combined destructive effects of such processes. The term is wider in its scope than *erosion,* although it is commonly used as a syn. of that term. It is also used as a syn. of *degradation,* although some authorities regard "denudation" as the proper term for the actual processes and "degradation" for the results produced. Davis (1909, p.408) distinguished between "denudation" as the active processes operating early in the cycle of landform development, and "degradation" as the more leisurely processes operating later. Obsolete. (b) Strictly, the laying bare, uncovering, or exposure of bedrock or a designated rock formation through the removal of overlying material by erosion. This is the original, and etymologically correct, meaning of the term, which was often applied in a catastrophic sense; e.g., the "Great Denudation" resulting from the universal flood.

denudation chronology The study of the timing and sequence of events leading to the formation and evolution of an existing landscape.

departure (de-par'-ture) The projection of a line onto an east-west axis of reference. The departure of a line is the difference of the meridian distances or longitudes of the ends of the line. It is east or positive (sometimes termed the *easting*) for a line whose azimuth or bearing is in the northeast or southeast quadrant; it is west or negative (sometimes termed the *westing*) for a line whose azimuth or bearing is in the southwest or northwest quadrant. Syn: *longitude difference.*

departure curve In well logging, a graph of correction factors to be applied to recorded (apparent) log measurements of particular sonde design to estimate "true" measurements under specific geometric and physical conditions. The term in the plural is usually applied to the collections of charts used in *resistivity log* analysis.

depauperate fauna (de-pau'-per-ate) (a) An *assemblage [ecol]* exhibiting substantially reduced *diversity.* Syn: *impoverished fauna.* (b) A *dwarf fauna.*

dependable yield (de-pend'-a-ble) The minimum water supply to a given area that is available on demand and which may decrease on the average of once every *n* number of years (Langbein and Iseri, 1960).

dependent variable (de-pend'-ent) In *regression analysis* and similar procedures where a variable y is treated as a function of other variables x_i plus a random error, the y variable is the dependent variable and the x_i variables are the *ndependent variables.*

depergelation (de'-per-gel-a'-tion) The act or process of thawing permafrost (Bryan, 1946, p.640). Cf: *degradation [permafrost].*

dephasing Converting to zero phase.

depleted mantle A part of the Earth's mantle in which the concentrations of incompatible elements have been decreased, usually by partial melting.

depleted MORB-source mantle An end-member source for mantle-derived magmas, postulated by Zindler and Hart (1986). Abbrev: DMM.

depletion [eco geol] (de-ple'-tion) The exhaustion of a natural resource, i.e. ore or oil, by commercial exploitation, measured quantitatively in monetary terms.

depletion [water] Loss of water from surface water or groundwater reservoirs at a rate greater than that of recharge.

depletion allowance A proportion of income derived from mining or oil production that is considered to be a return of capital not subject to income tax. It is a way of recognizing that mining or petroleum production ultimately exhausts the reserve.

depletion curve A *hydrograph* showing the loss of water from groundwater storage, by seepage or flowage into streams, or from any storage or channel. See also: *recession curve.*

depletion drive *solution gas drive.*

depocenter (de'-po-cen-ter) An area or site of maximum deposition; the thickest part of any specified stratigraphic unit in a depositional *basin.*

depopod (de'-po-pod) A sediment-filled shallow basin on or between salt uplifts on the continental slope offshore Louisiana (Spindler, 1977, p.180). Depopods contain shallow-water deltaic or shelf sediments. They develop later than depotroughs.

deposit (de-pos'-it) n. (a) Earth material of any type, either consolidated or unconsolidated, that has accumulated by some natural process or agent. The term originally applied to material left by water, but it has been broadened to include matter accumulated by wind, ice, volcanoes, and other agents. Cf: *sediment.* (b) An informal term for an accumulation of ore or other valuable earth material of any origin. v. To lay down or let drop by a natural process; to become precipitated.

deposit feeder An animal that obtains its food from detrital material and associated microorganisms on the sea floor; e.g., sea cucumber, scaphopod. Syn: *detritovore; sediment feeder.*

deposition [meteorol] The process in which a substance (usually assumed to be water) in a vapor state forms a solid phase without going through a liquid phase. In meteorology, this process is the opposite of sublimation.

deposition [sed] (de-po-si'-tion) (a) The laying, placing, or accumulation of any material; specif. the constructive process of accumulation into beds, veins, or irregular masses of any kind of loose rock material by any natural agent, such as the mechanical settling of sediment from suspension in water, the chemical precipitation of mineral matter by evaporation from solution, or the accumulation of organic material on the death of plants and animals. See also: *sedimentation.* (b) Material that is deposited; a deposit or sediment.

depositional (de-po-si'-tion-al) (a) Pertaining to the process of deposition; e.g. a "depositional basin" or a "depositional surface". (b) Formed by the process of deposition; e.g. a "depositional topography".

depositional dip *primary dip.* See also: *regional dip.*

depositional fabric Rock fabric or that fabric element or elements that result from deposition during the rock's formation, e.g. sedimentary grains in an unmetamorphosed, current-laid sediment or crystals deposited by crystal settling in a magma chamber; the fabric of a *primary tectonite.* Cf: *deformation fabric.*

depositional fault *growth fault.*

depositional interface The interface between the water or other medium and the bottom where sediments are deposited (Plumley et al., 1962, p.86).

depositional magnetization *depositional remanent magnetization.*

depositional mark An irregularity formed on the bedding plane of a sediment during deposition; e.g. a scour mark or a tool mark.

depositional remanent magnetization Remanent magnetization resulting from mechanical alignment of magnetic moments in ferrimagnetic mineral grains along the ambient field during sedimentation. Its inclination, due to compaction as well as other phenomena, is often shallower than that of the ambient field. Abbrev: DRM. Syn: *depositional magnetization; detrital remanent magnetization.*

depositional sequence *sequence.*

depositional-shoreline break In sequence stratigraphy, the position on the shelf landward of which the depositional surface is at or near base level, usually sea level, and seaward of which the depositional surface is below sea level (Van Wagoner, 1991). The depositional-shoreline break is the surface on the shelf to which the fluvial systems are graded. This is usually some small distance seaward of the shoreline, e.g. less than a half mile. Generally, the depositional-shoreline break and the shoreline can be considered synonymous.

depositional systems Three-dimensional assemblages of lithofacies, genetically linked by active (modern) processes or inferred (ancient) processes and environments (Fisher and McGowan, 1967).

depositional topography Topography formed as a result of sediments being dropped from a moving medium, e.g. coastal bars and barriers, kame terraces, or sand dunes.
deposition zone The area where the bulk of snow carried down by an avalanche comes to rest or the runout zone of an avalanche path.
depotrough (de'-po-trough) A narrow anastomosing depression between flat-topped salt massifs on the continental slope offshore Louisiana (Spindler, 1977, p.180). Depotroughs contain significant thicknesses of slope-derived sediment. Cf: *depopod.*
depressed (de-pressed') Said of the shape of a whorl section in a cephalopod *conch* in which the lateral diameter is greater than the dorsoventral diameter. Ant: *compressed.*
depressed flute cast A flat or weakly developed flute cast.
depressed moraine An irregular moraine "developed along the ice front in line with the normal morainal ridge, but failing to rise above the adjacent outwash", due more to "the nonaccumulation of marginal material than to an excess of outwash" (Fuller, 1914, p.33-34); e.g. those on Long Island, New York.
depression [geomorph] (de-pres'-sion) Any relatively sunken part of the Earth's surface; esp. a low-lying area surrounded by higher ground and having no natural outlet for surface drainage, as an interior basin or a karstic sinkhole.
depression [meteorol] An atmospheric region of relatively low pressure. It may refer to the closed low-pressure area of cyclone type or to the open V-shaped trough of low pressure. Syn: *low [meteorol].*
depression [surv] (a) The angular distance of an object beneath the horizontal plane that passes through the observer. (b) The angular distance of a celestial object below the horizon.
depression [tect] (a) A structurally low area in the crust, produced by negative movements that sink or downthrust the rocks. The term includes basins and furrows. (b) An along-strike saddle in a fold belt toward which fold hinges plunge. Cf: *uplift [tect].*
depression angle In *SLAR,* the angle between the horizontal plane passing through the antenna and the line connecting antenna and target.
depression contour A *closed contour,* inside of which the ground or geologic structure is at a lower elevation than that outside, and distinguished on a map from other contour lines by hachures marked on the downslope or downdip side. See also: *closed depression.*
depression spring A type of *gravity spring,* with its water flowing onto the land surface from permeable material as a result of the land surface sloping down to the water table (Meinzer, 1923, p.51).
depression storage Accumulation of water from precipitation in depressions in the land surface (Langbein and Iseri, 1960); accumulation of rainwater or snowmelt in depressions when the soil has reached its infiltration capacity (Chow, 1964, p.20-5). Cf: *detention.*
depressor muscle crest (de-pres'-sor) The elevated denticles, on the interior of a *tergum* of a cirripede crustacean, for attachment of muscles that depress or draw down.
depth The vertical distance from a specified datum to the bottom of a body of water. Syn: *water column.*
depth contour *isobath [oceanog].*
depth controller [geophys exploration] A device with moveable wings that fastens to a marine streamer to maintain it at a predetermined depth. Often includes a magnetic compass. Sometimes called a *bird.*
depth gather (a) A collection of the seismic traces that all reflect from the same point, often corrected for normal moveout (velocity). (b) The same sort of collection for converted-wave data except for a common conversion point.
depth hoar Ice crystals in the form of prisms, pyramids, or hollow hexagonal prisms, formed at the base of the seasonal snowpack in dry *snow covers.* Such crystals form by upward transfer of water molecules in low-density snow, when temperature gradients are high in autumn; they may also form in summer by radiative heating of near-surface snow (Alley et al., 1990). Crystals may have striations and scrolls, are typically 2-5 mm in diameter and may be larger, but are poorly bonded. They develop more complex structure and greater size at higher growth rates (Colbeck, 1986). Cf: *crevasse hoar; cup crystal; faceted form; kinetic growth form; surface hoar.*
depth migration Any migration which accommodates lateral velocity variations as well as vertical variations. The objective of depth migration is to output an accurate image of the subsurface, usually in depth, occassionally in vertical 2-way traveltime.
depth of compensation According to the concept of *isostasy,* a horizontal plane within the Earth below which other horizontal planes are isobaric. In general, the depth of compensation lies in the asthenosphere. See also: *isostatic compensation.* Syn: *compensation level [tect]; compensation depth [tect]; level of compensation.*
depth of exploration The maximum depth of the cause of a geophysical anomaly that is just recognizable above *noise.*
depth of focus The distance from the focus of an earthquake to the epicenter, i.e, the depth below sea level of the first rupture in the earthquake. Syn: *focal depth.*
depth of investigation Lateral distance from the logging tool to some point in the formation that represents the range of predominant measurement by the tool.
depth of penetration (a) The depth below the surface of a material by which the incident radiation has been attenuated to l/e, or to 37 percent. This is sometimes called *skin depth,* or attenuation distance. (b) The depth at which the integrated temperature differentials of adjacent spatial resolution cells are equivalent to the temperature resolution of the sensor. (c) The depth from which radiation may still reach the receiver. Such radiation is integrated over the distance downwards from the surface.
depth point In seismic work, a position at which a depth determination of a mapped horizon has been calculated. See also: *common midpoint.*
depth probe [geophys exploration] A group of profiles for which the measuring system dimensions are successively increased, designed to obtain information on the layering pattern in an area. Syn: *sounding.* Also called: expander, depth profile. A refraction depth probe determines approximate depths and velocities of refraction markers. An electrical surveying technique of exploring vertically down into the earth employs an orderly horizontal expansion of the interelectrode interval or changes the frequencies. Also called: vertical electrical sounding (VES). Cf: *expanding spread.*
depth rule A rule relating the depth of a source body to some measured features of the observed anomaly shape or profile. Depth rules are used in graphic methods of estimating depth to the top of a source of an assumed isolated anomaly. See also: *Peters' length.*
depth section A seismic section plotted with its vertical scale linear in depth units rather than time units. The section is usually (but not always) migrated, and the migration may be either time or depth migration, and the operation may be performed on post- or prestacked data. Cf: *depth migration.*
depth slice A display of the seismic measurements (usually amplitude) corresponding to a particular depth through a volume of depth-converted 3-D data that may, or may not, have been migrated. Cf: *time slice; horizon slice.*
depth sounder *echo sounder.*
depth-velocity curve *vertical-velocity curve.*
depth zone [meta] One of the characteristic physicochemical environments at various depths in the Earth that give rise to different metamorphic phenomena (Grubenmann, 1904): *epizone; mesozone; katazone.*
depth zone [oceanog] One of five oceanic environments, or ranges of oceanic depths: the littoral zone, between high and low tides; the *neritic* zone, between low-tide level and 200 m above the continental shelf; the *bathyal* zone, between 200 and 3,500 m; the *abyssal* zone, between 3,500 and 6,000 m; and the *hadal* zone 6,000 m and deeper.
deranged drainage pattern (de-ranged') A distinctively disordered drainage pattern in a recently glaciated area whose former surface and preglacial drainage have been remodeled and effaced, and in which the new drainage system shows a complete lack of underlying structural and bedrock control. It is characterized by irregular streams that flow into and out of lakes, by only a few short tributaries, and by swampy interstream areas. Cf: *aimless drainage.*
derangement (de-range'-ment) The process by which changes in a stream course are effected by agents other than streams, such as by glaciation, wind deposition, or diastrophism. Cf: *diversion.*
derbylite (der'-by-lite) A submetallic black or brown monoclinic mineral: $Fe^{3+}_4Ti^{4+}_3Sb^{3+}O_{13}(OH)$.
Derbyshire spar (Der'-by-shire) A popular name for fluorite from Derbyshire, England. Also called: Derby spar.
derbystone (der'-by-stone) *blue john.*
derecho A family of straight-line *downburst* winds in excess of 95 km (58 mi) per hour that impacts a path up to hundreds of kilometers long; may be produced by a cluster of thunderstorm cells (e.g., a squall line).
derelict (der'-e-lict) (a) A tract of dry land formed by *dereliction.* (b) Any property abandoned at sea, often of sufficient size as to be a menace to navigation.
dereliction (der-e-lic'-tion) A recession of water from the sea or other body of water so that land is left dry. Cf: *reliction.*
derivative map [cart] (de-riv'-a-tive) Any map that uses information from another base map, i.e., a map derived from geologic data on several maps and based on interpretation of geologic structure, lithology, topography, rainfall, vegetation, groundwater levels, or other features.
derivative map [geophys] A map of one of the derivatives of the potential field, such as the Earth's gravity or magnetic field. This map emphasizes mid- and short-wavelength anomalies. See *first-derivative map,* and *second-derivative map.*
derivative structure A crystal structure having a multiple unit cell and/or a suppression of some symmetry elements, formed by the substitution of a simple fraction of one atom by another; e.g. the structure of chalcopyrite as derivative after that of sphalerite; or by distortion of the atomic arrangement, e.g. the structure of alpha quartz as derivative after that of beta quartz.
derived (de-rived') Said of geologic materials that are not native or that have been displaced or brought from elsewhere; esp. said of a fossil washed out from its original site and redeposited in a later formation at a different lo-

cality. Also said of geologic materials that are not primary or original. Cf: *reworked; remanié.*

derived character In *cladism,* a *character [paleont]* possessed by a descendant taxon and modified from a homologous ancestral *primitive character.* Syn: *apomorphy.*

derived till A till-like deposit "formed from the erosion of substantially older tillites, not necessarily with any glacial transport involved in the second formation"; e.g. the deposit of the so-called "Cretaceous glaciation" of South Australia (Harland et al., 1966, p.232).

dermal [paleont] (der'-mal) (a) Pertaining to the *exopinacoderm,* a *cortex,* or the *ectosome* of a sponge; e.g. "dermal skeleton" (differentiated skeleton at the outer surface), "dermal membrane" (roofing over a vestibule), or "dermal pores" (minute openings, or prosopores, in the surface). (b) Pertaining to mesodermal (and neural-crest) derivatives of deeper layers of the integument or skin of vertebrates; also, originating there, as "dermal bone".

dermal [tect] Said of deformation or gliding in the upper part of the sialic crust. Obsolete. Cf: *epidermal; bathydermal.*

dermal bone Bone deposited in a more or less superficial layer of connective tissue, without an intervening phase of preformation in cartilage. It includes primarily the mesodermally or mesenchymally (of neural-crest origin) derived superficial flat bones of the skull, and the clavicle and interclavicle of the appendicular skeleton. In armored forms, dermal bone includes scales, ossicles, plates, and other elements of the superficial skeleton.

dermolith (der'-mo-lith) An obsolete syn. of *pahoehoe.* Cf: *aphrolith.*

dermoskeleton (der-mo-skel'-e-ton) *exoskeleton.*

derrick (der'-rick) A framework tower of steel (on a drill floor), erected over a deep drill hole (such as an oil well). It is used to support the various tools and tackle employed in hoisting and lowering the equipment used in drilling and completing the well. Cf: *drilling rig.*

derriksite (der'-rik-site) A green orthorhombic mineral: $Cu_4(UO_2)(Se^{4+}O_3)_2(OH)_6$.

dervillite (der-vill'-ite) A metallic white monoclinic mineral: Ag_2AsS_2.

desalination (de'-sal-i-na'-tion) The removal of dissolved ions from seawater in order to make fresh water that is suitable for drinking or other uses. Common methods are reverse osmosis, distillation, and freezing.

desaulesite (de-saules'-ite) *pimelite.*

desautelsite (des'-au-tels'-ite) An orange rhombohedral mineral of the *hydrotalcite* group: $Mg_6Mn^{3+}{}_2(CO_3)(OH)_{16}·4H_2O$.

descendant (de-scend'-ant) A topographic feature carved from the mass beneath an older topographic form that has been removed (Willis, 1903, p.74).

descending branch (de-scend'-ing) Either of two dorsal elements of a brachiopod *loop,* extending distally from crura and recurved ventrally at anterior ends. Cf: *ascending branch.*

descending development *declining development.*

descension theory (de-scen'-sion) A theory of formation of supergene mineral deposits involving the descent from above of mineral-bearing solutions. The theory originated with the Neptunian school of thought of the 18th Century, which postulated an aqueous origin for all rocks. Cf: *ascension theory.*

descloizite (des-cloi'-zite) (a) A variously colored orthorhombic mineral: $PbZn(VO_4)(OH)$. It is isomorphous with mottramite. Syn: *eusynchite.* (b) A group name for minerals with an analogous formula, but with Zn replaced by Cu, Fe^{2+} or Mn^{2+}.

descrespignyite-(Y) A blue monoclinic mineral: $(Y,REE)_4Cu(CO_3)_4Cl(OH)_5·2H_2O$.

description (a) A written description of the boundaries of a piece of land, presumably sufficient and accurate enough that the boundaries can be found and identified from the description. Four types of description are common in the US: description by metes and bounds, verbal description by reference to a map, description by reference to natural objects and adjoiners, and the United States Public Land System. If the map to which reference is made is a plat, the description is said to be by platting. In particular, a document listing the metes and bounds of a property, usually prepared by a surveyor and included in a conveyance by the lawyer preparing the conveyance. (b) The formal, published data on each triangulation station, bench mark, etc. The data include sufficient information of the location and type of mark that anyone can to the immediate locality and identify the mark with certainty.

desert (des'-ert) A region with a mean annual precipitation of 10 inches or less, and so devoid of vegetation as to be incapable of supporting any considerable population. Four kinds of deserts may be distinguished: (1) polar or high-latitude deserts, marked by perpetual snow cover and intense cold; (2) middle-latitude deserts, in the basinlike interiors of the continents, such as the Gobi, characterized by scant rainfall and high summer temperatures; (3) trade-wind deserts, notably the Sahara, with negligible precipitation and a large daily temperature range; and (4) coastal deserts, as in Peru, where there is a cold current on the western coast of a large land mass. Adj: desertic; *eremic.*

desert arch An elongate *desert dome.*

desert armor A *desert pavement* whose surface of stony fragments protects the underlying finer-grained material from further wind erosion; a common feature of stony deserts. Cf: *pebble armor; deflation armor.*

desert crust (a) A hard layer, containing algae, calcium carbonate, gypsum, or other binding matter, exposed at the surface in a desert region. (b) *rock varnish.* (c) *desert pavement.*

desert dome A convex rock surface with uniform smooth slopes, representing the result of prolonged exposure of a mountain mass to desert erosion; e.g. Cima Dome in the Mojave Desert, Calif. See also: *desert arch; granite dome.* Syn: *pediment dome.*

desertification Conversion of arable land to desert due to some combination of climate change and human mismanagement of the land.

desert lacquer *rock varnish.*

desert mosaic A *desert pavement* characterized by tightly interlocking and evenly set fragments, covering the surface in the manner of a mosaic; e.g. a *pebble mosaic.*

desert patina *rock varnish.*

desert pavement A natural residual concentration of wind-polished, closely packed pebbles, boulders, and other rock fragments, mantling a desert surface (such as an area of *reg*) where wind action and sheetwash have removed all smaller particles, and usually protecting the underlying finer-grained material from further deflation. The fragments commonly are cemented by mineral matter. See also: *desert armor; desert mosaic; lag gravel; boulder pavement [geomorph]; reg.* Syn: *desert crust; deflation residue.*

desert peneplain A syn. of *pediplain,* esp. in reference to a wind-erosion surface in southern Africa. The term is inappropriate because such a surface is produced under different conditions and by different processes than those of a humid-land peneplain.

desert plain (a) A general term used by Blackwelder (1931) for any plain commonly found in a desert; e.g. a flood plain, structural plain, playa, bajada, or pediment. (b) *pediplain.*

desert polish (a) A smooth, shiny or glossy surface imparted to rocks of desert regions by windblown sand and dust. Syn: *wind polish.* (b) A term sometimes used as a syn. of desert varnish.

desert rind *rock varnish.*

desert ripple One of a system of slightly arcuate ridges produced by the wind, arranged roughly en echelon about 15 m apart with crests supporting vegetation and troughs plated with caliche; it may be as high as 1 m and as long as 150 m.

desert rose A radially symmetric group of crystals with a fancied resemblance to a rose, formed in sand, soft sandstone, or clay. The crystals are usually of calcite, less commonly of barite, gypsum, or celestite.

Desert soil A zonal great soil group consisting of soils with a very thin, light-colored surface horizon, which may be vesicular and is ordinarily underlain by calcareous material; formed in arid regions under sparse shrub vegetation. Not used in current U.S. system of soil taxonomy.

desert varnish *rock varnish.*

desiccation (des-ic-ca'-tion) A complete or nearly complete drying-out or drying-up, or a deprivation of moisture or of water not chemically combined; e.g. the loss of water from pore spaces of soils or sediments as a result of compaction, or the formation of evaporites as a result of direct evaporation from bodies of water in an arid region, or the progressive increase in aridity of an area as a result of a climatic change (such as decreasing rainfall) or of accelerated erosion (such as deforestation). Cf: *exsiccation.*

desiccation breccia A breccia formed where dried-out and cracked mud polygons have broken into angular fragments that have then been deposited with other sediments. Syn: *mud breccia.*

desiccation conglomerate A term used by Shrock (1948, p.208) for a conglomerate consisting of fragments eroded from a mud-cracked layer of sediment and rounded by transportation. Syn: *mudstone conglomerate.*

desiccation crack A crack in sediment, produced by drying; esp. a *mud crack.* Syn: *drying crack; desiccation mark.*

desiccation mark A *desiccation crack;* esp. a *mud crack.*

desiccation polygon A small *nonsorted polygon* formed by drying of moist, clayey soil or sediment, thus producing contortion resulting in cracking. The polygon normally has three to five sides which may measure 2-30 cm in length. See also: *giant desiccation polygon.* Syn: *mud-crack polygon; mud-flat polygon; drought polygon; shrinkage polygon.*

design flood (de-sign') A flood against which protective measures are taken.

design of experiments A branch of applied statistics concerned with partitioning variance among different sources by means of planned experiments.

desilication [petrology] (de'-sil-i-ca'-tion) The removal of silica from a rock or magma by the breakdown of silicates and the resultant freeing of silica, or by reaction between a body of magma and the surrounding wall rock. A form of, but not synonymous with, silicification. Ant: *silication.*

desilication [soil] The removal of silica from soils in a warm, humid climate by the percolation of large amounts of rainwater, resulting in a soil relatively rich in hydroxides of iron, aluminum, and manganese, i.e. an Oxisol.

desilting basin (de-silt-ing) A *settling basin* consisting of an enlargement in a stream where silt carried in suspension may be deposited. Cf: *stilling basin.*

desma (des'-ma) An irregularly branched, siliceous sponge spicule that bears knotty outgrowths (zygomes) which interlock with adjacent spicules. Pl: desmas or desmata.

desmid (des'-mid) One of a family of unicellular, microscopic *green algae,* commonly composed of semicells that are mirror images of each other.

desmine (des'-mine) *stilbite-Ca.*

desmite (des'-mite) *Residuum [coal]* that is transparent; it is characteristic of higher coal grades.

desmodont (des'-mo-dont) adj. Said of the dentition of a bivalve mollusk characterized by the prominence of a large *chondrophore* inside the hinge line. n. Any bivalve mollusk belonging to the order Desmodonta, characterized by the presence of two equal muscle scars and a *pallial sinus* and by the absence of hinge teeth or with irregular hinge teeth intimately connected with the *chondrophore.*

desmoid (des'-moid) A siliceous sponge spicule bearing outgrowths like those of a desma but not interlocking with adjacent spicules.

Desmoinesian (Des-moines'-i-an) A North American stage: upper Middle Pennsylvanian (above Atokan, below Missourian).

desmosite (des'-mo-site) A banded *adinole.* Rarely used.

despujolsite (des-pu-jol'-site) A yellow hexagonal mineral of the *fleischerite* group: $Ca_3Mn^{4+}(SO_4)_2(OH)_6{\cdot}3H_2O$.

desquamation (des-qua-ma'-tion) An obsolescent syn. of *exfoliation* characterized by the peeling-off or detachment of scaly rock fragments.

dess A term used in Morocco for silt deposited by a new stream flowing in an arid region (Termier and Termier, 1963, p.403).

dessauite-(Y) A metallic black rhombohedral mineral of the *crichtonite* group: $(Sr,Pb)(Y,U)(Ti,Fe^{3+})_{20}O_{38}$.

destinezite (des-ti-nez'-ite) *diadochite.*

destructional (de-struc'-tion-al) Said of a landform that owes its origin or general character to the removal of material by erosion and weathering; e.g. a mesa, canyon, cliff, or plain resulting from the wearing down or away of the land surface. Ant: *constructional.* Cf: *sequential landform.*

destructive metamorphism (de-struc'-tive) An archaic term for the development of *equilibrium form* (*rounded form*) in snow crystals.

destructiveness index The logarithm of the area covered by lava, pyroclastic flows, and surges or buried under more than 100 kg/m² of tephra during an eruption (Pyle, 2000).

destructive wave A wave that erodes a beach by moving material seaward, as a storm wave with a more powerful backrush than uprush; on a sandy beach, a wave with a steepness greater than 0.25. Ant: *constructive wave.*

desulfidation Removal of sulfur by means of a chemical reaction between rock and fluid.

detached core (de-tached') The inner bed or beds of a fold that become separated or pinched off from their source due to extreme folding and compression.

detached fault A fault that does not cut the basement, which are those rocks formed before the cover rocks accumulated. Cf: *basement-involved fault.*

detachment [soil] Separation of transportable particles from a soil layer, usually by running water, raindrop impact, or wind. Cf: *dispersion [soil].*

detachment [struc geol] (de-tach'-ment) *décollement.*

detachment fault Commonly used as a synonym for *décollement,* it is also widely used for a regionally extensive, gently dipping normal fault that is commonly associated with extension in a metamorphic core complex (Peacock et al., 2000).

detachment fold A fold developed by the shortening of strata above a subhorizontal *detachment.* Syn: *décollement fold; Jura-type fold; superficial fold.*

detachment horizon A mechanically weak stratigraphic horizon, such as a shale or an evaporite, in which a décollement forms or is likely to form.

detachment thrust *sole fault.*

detail log (de'-tail) Any well log plotted on a scale larger than the conventional scale (2 inches per 100 ft of depth) in order to portray minor variations in the formations penetrated by the borehole; specif. an electric log plotted on a scale of 5, 10, or 25 inches per 100 ft of depth.

detectability (de-tect'-a-bil'-i-ty) A measure of the smallest object that can be discerned on an image.

detectable limit [seis] The minimum thickness for a bed to give a reflection that stands out above the background. Often of the order of $^1/_{25}$ of the dominant wavelength. Also called: limit of visibility.

detection limit The smallest concentration of a constituent in water that can be detected. Three types of detection limit have been defined: IDL (instrument detection limit) is the constituent concentration that produces a signal greater than five times the signal to noise ratio of the instrument; MDL (method detection limit) is the constituent concentration that when processed through a complete method, produces a signal with a 99 percent probability that it is different from a blank; PQL (practical quantification limit) is the lowest constituent concentration achievable among laboratories within specified limits during routine laboratory operations. The ratios of these limits are approximately: IDL:MDL:PQL=1:4:20.

detection monitoring A program of monitoring for the express purpose of determining whether or not a contaminant release to groundwater has occurred.

detectivity (de-tec-tiv'-i-ty) In infrared-detector terminology, the reciprocal of noise-equivalent power.

detector (de-tec'-tor) (a) The component of a remote-sensing system that converts electromagnetic radiation into a signal that can be recorded. Syn: *radiation detector.* (b) *seismic detector.*

detector [radiation] Sensor having the capacity to register ionizing radiation directly or to transform the energy of incident nuclear particles to electrical quantities (IAEA, 2003).

detention (de-ten'-tion) The amount of water from precipitation existing as overland flow. Sheet flow of water is overland flow before channel is reached. *Depression storage* is not considered part of the detention (Rechard and McQuisten, 1968). Syn: *detention storage; surface detention.*

detention storage *detention.*

deterministic process (de-ter'-min-is'-tic) A process in which there are exact mathematical relationships between variables and there is no random or stochastic component in the system. Ant: *stochastic process.*

detrital (de-tri'-tal) adj. Pertaining to or formed from *detritus;* said esp. of rocks, minerals, and sediments. The term may indicate a source outside the depositional basin (Krynine, 1948, p.133) or a source within it.

detrital fan *alluvial fan.*

detrital mineral Any mineral grain resulting from mechanical disintegration of parent rock; esp. a heavy mineral found in a sediment or weathered and transported from a vein or lode and found in a placer or alluvial deposit.

detrital ratio *clastic ratio.*

detrital remanent magnetization *depositional remanent magnetization.*

detrital rock A rock composed primarily of particles or fragments detached from pre-existing rocks either by erosion or by weathering; specif. a sedimentary rock having more than 50% detrital material (Krynine, 1948, p.134). Cf: *chemical rock.*

detrital sediment A sediment formed by the accumulation of *detritus,* esp. that derived from pre-existing rocks and transported to the place of deposition. Cf: *clastic sediment.*

detrition (de-tri'-tion) A general term for the processes involved in producing detritus; a wearing-away by breaking or rubbing of rock masses.

detritovore (de-tri'-to-vore) *deposit feeder.*

detritus (de-tri'-tus) A collective term for loose rock and mineral material that is worn off or removed by mechanical means, as by disintegration or abrasion; esp. fragmental material, such as sand, silt, and clay, derived from older rocks and moved from its place of origin. Cf: *debris [geol].* Pl: *detritus.* See also: *reef detritus.*

detrohuminite A subgroup of *huminite* comprising particulate huminitic substances (Pickel, 2004).

detrovitrinite A subgroup of *vitrinite* comprising particulate vitrinitic substances (Pickel, 2004).

deuteric (deu-ter'-ic) Referring to reactions between primary magmatic minerals and the water-rich solutions that separate from the same body of magma at a late stage in its cooling history. Obsolete syn: *paulopost; epimagmatic.* Cf: *multopost.* See also: *synantectic; synantexis; autometamorphism.*

deuteroconch (deu'-ter-o-conch') The chamber in foraminifers immediately adjoining the *proloculus* and formed next after it.

deuteroforamen (deu-ter'-o-fo-ra'-men) A secondary aperture independent of the tooth plate in some enrolled foraminifers. Cf: *protoforamen.*

deuterogene (deu'-ter-o-gene') An old term for a secondary rock; a rock formed from a pre-existing rock. Adj: *deuterogenic.* Cf: *protogene.*

deuterogenic (deu'-ter-o-gen'-ic) Adj. of *deuterogene.*

deuteroglacial (deu'-ter-o-gla'-cial) Pertaining to the last great glaciation following the *proteroglacial* period (Hansen, 1894, p.128).

deuterolophe (deu'-ter-o-lophe') A spirally coiled part of a brachiopod *lophophore,* bearing a double brachial fold and a double row of paired filamentary appendages. It is homologous with the side arms of *plectolophe.* Cf: *spirolophe.*

deuteromorphic (deu'-ter-o-mor'-phic) A general term applied to crystals whose shapes have been acquired or modified by mechanical or chemical processes acting on the original forms (Loewinson-Lessing, 1899). Depending on the nature of the secondary agent, they may be described as clastomorphic, lytomorphic, schizomorphic, tectomorphic, or neomorphic. These terms are now obsolete.

Deuteromycetes A syn. of *Fungi imperfecti.*

deuteropore (deu'-ter-o-pore') One of a group of protopores fusing into a single larger pore cavity in the outer wall of foraminifers.

deuterostome Coelomate animals in which the embryonic blastopore becomes the anus.

deutonymph (deu'-to-nymph) The second developmental stage in the arachnid order Acarida.

deval A minor *transform fault* at which the ends of the offset sea-floor-spreading axis curve inward somewhat like interlocking numerals 6 and 9. Acronym for: deviation from axial lineality.

developable surface A surface that can be flattened without tearing or stretching. The cone and the cylinder, after cutting along a generating element, are examples of developable surfaces. The two principal geodetic applications of developable surfaces are (a) for use as intermediaries in the mapping of a spheroid onto the plane, and (b) for sectioning the Earth's surface and so creating a profile of that surface.

developed ore (de-vel'-oped) *developed reserves.*

developed reserves Ore that has been exposed on three sides and for which tonnage and quality estimates have been made; ore essentially ready for mining. Cf: *positive ore; proved reserves.* Syn: *developed ore; measured ore; ore in sight; blocked-out ore; assured mineral.*

development [eco geol] (de-vel'-op-ment) (a) The preparation of a mining property for mining, generally by removing overburden, constructing surface and underground access, and building mineral processing facilities and other surface plants. Development is an intermediate stage between *exploration* and *mining.* (b) In older usage, underground work necessary to provide access to an ore deposit.

development [grd wat] (a) In the construction of a water well, the removal of fine-grained material adjacent to a drill hole, enabling water to enter the hole more freely. (b) Exploitation of groundwater.

development well A well drilled within the known or proved productive area of an oil field, with the expectation of obtaining oil or gas from the producing formation or formations in that field. Cf: *exploratory well; offset well.*

deviation [drill] (de-vi-a'-tion) (a) The departure of a drilled hole from its straight pathway. The hole may be either vertical or inclined and the departure may be in any direction. Deviation may be intentional, as in *directional drilling,* or undesirable as in crooked hole. Syn: *deflection [drill].* (b) In more general use, the angle of departure of a wellbore from the vertical, without reference to direction. Cf: *inclination [drill].*

deviation [stat] An informal expression indicating the difference between individual observations and the mean of the sample in which they are included. Cf: *mean deviation; standard deviation.*

deviation log A log that records the amount and direction that a borehole deviates from the vertical. If the deviation exceeds 2% or 3%, this information is required in order that depth data from the hold can be used to construct maps or cross sections showing the true thicknesses of the strata penetrated by the borehole (Friedman et al., 1992, p.11).

deviation of the vertical *deflection of the vertical.*

deviatoric stress [exp struc geol] A stress tensor of nine components where mean stress is subtracted from each of the three normal stress components. Incorrectly used term for differential stress.

deviatoric stress [glaciol] The stress obtained by subtracting the *hydrostatic stress* from the total stress. Most solids deform minimally under a pure hydrostatic stress. Deviatoric stresses are responsible for large-scale deformation, as in a *glacier.* Also called: stress deviator; nonhydrostatic stress.

devilline (de-vil'-line) A green to bluish green monoclinic mineral: $CaCu_4(SO_4)_2(OH)_6{\cdot}3H_2O$.

devil's slide (dev'-il's) (a) An *avalanche track* down a steep slope. (b) A long narrow mass of talus material descending a steep mountain.

devitrification (de-vit'-ri-fi-ca'-tion) Conversion of glass to crystalline material.

devolatilization (de-vol'-a-til'-i-za'-tion) A process of metamorphism resulting in the loss of volatile constituents and a consequent proportional increase in nonvolatile constituents, e.g. the increase in carbon content in *coalification* or the loss of H_2O and CO_2 in the metamorphism of shales and carbonate sediments.

Devonian (De-vo'-ni-an) A system of the *Paleozoic* Erathem of the Standard Global Chronostratigraphic Scale, above the *Silurian* and below the *Carboniferous.* Also the time during which these rocks were formed, the Devonian Period, covering the time span between 410 and 355 Ma. It is named after Devonshire, England, where rocks of this age were first studied.

dew Liquid water droplets caused by condensation of atmospheric water vapor on a surface whose temperature is below the dewpoint but above the freezing point. Cf: *frost.*

dewatering The expulsion during diagenesis or metamorphism of water from sediments. The water may have been present in the form of interstitial pore waters or water bound to hydrous minerals, such as certain clays or gypsum (Friedman et al., 1992, p.11).

deweylite (dew'-ey-lite) A mixture of 2:1 and 1:1 trioctahedral phyllosilicates. Syn: *gymnite.*

dewindtite (de-windt'-ite) A canary-yellow orthorhombic secondary mineral:$H_2Pb_3(UO_2)_6O_4(PO_4)_4{\cdot}12H_2O$. The Pb analogue of phosphuranylite.

dewpoint The temperature to which air must be cooled, at constant pressure and constant water-vapor content, in order for *saturation* to occur; the temperature at which the saturation pressure is the same as the existing vapor pressure.

dextral (dex'-tral) Pertaining, inclined, or spiraled to the right; specif. pertaining to the normal or clockwise direction of coiling of gastropod shells. A dextral gastropod shell in apical view (apex toward the observer) has the whorls apparently turning from the left toward the right; when the shell is held so that the axis of coiling is vertical and the apex or spire is up (as in orthostrophic shells) or down (as in *hyperstrophic* shells), the aperture is open toward the observer to the right of the axis. Actually, the definition depends on features of soft anatomy: with genitalia on the right side of the head-foot mass, the soft parts and shell are arranged such that the aperture is on the right when viewed with the apex (of orthostrophic shells) uppermost (TIP, 1960, pt. I, p.130). Ant: *sinistral.* Syn: *right-handed [paleont].*

dextral element Component of a conodont skeletal apparatus on the right-hand side of the plane of bilateral symmetry (TIP, 1981, pt. W, supp.2).

dextral fault *right-lateral fault.*

dextral fold An asymmetric fold with the asymmetry of an S as opposed to that of a Z when seen in profile. The long limb is apparently offset to the right. Cf: *sinistral fold.*

dextral imbrication The condition in a heterococcolith in which each segment overlaps the one to the right when viewed from the center of the cycle. Ant: *sinistral imbrication.*

dextrorotatory (dex'-tro-ro'-ta-to'-ry) *right-handed [cryst].*

DF *dark field.*

DFA *discriminant function analysis.*

d-glide A *glide plane* in which reflection across a mirror plane is combined with a translation (parallel to the mirror plane) of one-fourth of the vector sum or difference of either two or three of the *basis vectors.* This type of glide plane may occur only in space groups based on orthorhombic face-centered lattices, tetragonal body-centered lattices, and isometric face-centered or body-centered lattices. Cf: *a-glide; b-glide; c-glide; n-glide.*

DGPS *differential global positioning system.*

dhand A term used in Sind (region of West Pakistan) for a salt lake, esp. an alkali lake. Etymol: Sindhi.

DI [ign] *differentiation index.*

DI [sed] *durability index.*

diabantite (di-a-ban'-tite) A variety of the mineral chlorite. It occurs in cavities in basic igneous rocks.

diabase (di'-a-base) In the U.S., an intrusive rock whose main components are labradorite and pyroxene and which is characterized by ophitic texture. As originally applied by Brongniart in 1807, the term corresponded to what is now recognized as *diorite.* "The word has come to mean a pre-Tertiary basalt in Germany, a decomposed basalt in England, and a dike-rock with ophitic texture in the United States and Canada" (Johannsen, 1939, p.248). Cf: *trap [ign].* Syn: *dolerite.*

diabasic (di-a-bas'-ic) Igneous texture characterized by dominant lath-shaped plagioclase grains and interstitial, anhedral pyroxene. Syn: *doleritic; intergranular.* See: *ophitic; subophitic.*

diablastic (di-a-blas'-tic) Pertaining to a texture in metamorphic rock that consists of intricately intergrown and interpenetrating constituents, usually with rodlike shapes (Becke, 1903).

diaboleite (di-a-bo'-le-ite) A sky-blue tetragonal mineral: $Pb_2CuCl_2(OH)_4$.

diabrochite (di-a-broch'-ite) A metamorphic rock that owes its mineralogical composition to intensive penetration by ascending fluids or to partial fusion, but without injection of visible granitic material. Rarely used.

diachron (di'-a-chron) The fundamental and non-hierarchical *diachronic unit* (NACSN, 1983, Art. 93). Diachrons may differ greatly in magnitude because they are the spans of time represented by individual or grouped lithostratigraphic, allostratigraphic, biostratigraphic, or pedostratigraphic units.

diachronic (di-a-chron'-ic) Pertaining to, or during, the time of the Earth's existence. The term is used in considering events or changes as they happen or develop over time. Ant: *prochronic.* Cf: *synchronous.*

diachronic unit A unit comprising the unequal spans of time represented either by a specific lithostratigraphic, allostratigraphic, biostratigraphic, or pedostratigraphic unit, or by an assemblage of such units (NACSN, 1983, Art. 91). One or both of the boundaries of a diachronic unit are time-transgressive. The fundamental and non-hierarchical unit is the *diachron;* if a hierarchy is needed, the terms episode, phase, span, and *cline,* in order of decreasing rank, are recommended.

diachronism (di-ach'-ro-nism) A term introduced by Wright (1926) for the transgression, across time planes or biozones, by a rock unit whose age differs from place to place; the state or condition of being diachronous.

diachronous (di-ach'-ro-nous) Said of a rock unit that is of varying age in different areas or that cuts across time planes or biozones; e.g. said of a sedimentary formation related to a narrow depositional environment, such as a marine sand that was formed during an advance or recession of a shoreline and becomes younger in the direction in which the sea was moving. Syn: *time-transgressive.*

diachyte (di'-a-chyte) A rock product of marked mechanical or chemical contamination of *anatectic magma* by cognate mafic material (Dietrich and Mehnert, 1961). Rarely used.

diaclinal (di-a-cli'-nal) Said of a stream or valley that passes through or across a fold, with a direction at right angles to the strike of the underlying strata. Also said of a region having diaclinal streams. Term introduced by Powell (1874, p.50). Ant: *paraclinal.*

diacromorph In *acritarchs,* having a spherical to ellipsoidal *vesicle* characterized by ornamented polar areas and an equatorial zone free of *sculpture;* common in the Ordivician, e.g., *Acanthodiacrodian* (Downie et al., 1963).

diacrystallic (di'-a-crys-tal'-lic) Pertaining to the texture of a diagenetically recrystallized and essentially monomineralic rock in which contiguous crystals interpenetrate in a complicated manner (Phemister, 1956, p.72).

diactine (di-ac'-tine) A sponge spicule having two rays, usually monaxonic. See also: *rhabdodiactine.*

diad (di'-ad) Noun or adjective for a symmetry axis that requires a rotation of 180° to repeat an object's appearance. Also spelled: *dyad.* Cf: *triad; tetrad [cryst]; hexad.* Syn: *digonal; twofold axis.*

diadactic (di-a-dac'-tic) *diatactic.*

diadochite (di-a-doch'-ite) A variously colored triclinic mineral: $Fe^{3+}_2(PO_4)(SO_4)(OH)\cdot 6H_2O$. It is isomorphous with sarmientite. Syn: *destinezite.*

diadochy (di-ad'-o-chy) *ionic substitution.*

diadysite (di-ad'-y-site) A *migmatite* consisting of granitic veins and metamorphic parent rock (Jung and Roques, 1936, 1952). Rarely used. Cf: *arterite; phlebite; venite.*

diaene (di'-aene) A sponge spicule with two rays of equal length and one of a different length, usually longer; a triaene with one ray reduced or absent.

diagenesis [mineral] (di-a-gen'-e-sis) The sum of all chemical and physical changes in minerals during and after their initial accumulation, a process limited on the high temperatures, high pressure side by the lowest grade of metamorphism. Diagenesis involves addition and removal of material, transformation by dissolution and recrystallization or replacement, or both, and by phase changes. This definition includes pedogenesis, the changes and formation of minerals in the soil environment, within diagenesis. Authigenic refers to minerals formed in place. It is also applied to minerals that are clearly the result of new crystal growth on older crystals of the same kind, e.g. K-feldspar overgrowths are referred to as authigenic overgrowths. Authigenesis is a subprocess of diagenesis, and in most cases diagenesis is the better term. Weaver and Broeckstra (1984) proposed a boundary between diagenesis and metamorphism: that point at which all illite-1M has been converted to $2M_1$. This definition for diagenesis does not adequately cover the changes in hydrothermal, geothermal, or simple thermal situations.

diagenesis [sed] All the chemical, physical, and biologic changes undergone by a sediment after its initial deposition, and during and after its lithification, exclusive of surficial alteration (weathering) and metamorphism. This is the definition as applied by most geologists in the U.S. (Twenhofel, 1939, p.254-255) and Germany (Correns, 1950). It embraces those processes (such as compaction, cementation, reworking, authigenesis, replacement, crystallization, leaching, hydration, bacterial action, and formation of concretions) that occur under conditions of pressure (up to 1 kb) and temperature (maximum range of 100°C to 300°C) that are normal to the surficial or outer part of the Earth's crust; and it may include changes occurring after lithification under the same conditions of temperature and pressure. The father of this concept was Walther (1893-1894, p.693-711), although the term "Diagenese" was first used by Gümbel (1868, p.838) for a postsedimentary transformation of sediments into individual crystalline minerals, leading to the creation of metamorphic rocks such as gneiss and schist. Russian (and some U.S.) geologists restrict the term to the initial phase of postsedimentary changes, occurring in the zone where the sediment is still unconsolidated, the process being complete when the sediment has been converted to a more or less compact sedimentary rock (Fersman, 1922); in this usage, the term is equivalent to *early diagenesis* as used in the U.S. There is no universally accepted definition of the term, and no delimitation (such as the boundary with metamorphism). For a historical discussion and review, see Larsen and Chilingar (1967) and Dunoyer de Segonzac (1968). Cf: *epigenesis [sed].*

diagenetic (di'-a-ge-net'-ic) Pertaining to or caused by *diagenesis;* e.g. a "diagenetic change" resulting from compaction, a "diagenetic structure" (such as a stylolite) formed after deposition, a "diagenetic deposit" (such as dolomitized limestone or one consisting of manganese nodules), or a "diagenetic environment" of rock consolidation.

diagenetic differentiation The redistribution of material within a sediment by solution and diffusion toward centers or nuclei where reprecipitation occurs, leading to segregation of minor constituents into diverse forms and structures, such as chert nodules in limestone or concretions in shale (Pettijohn, 1957, p.672).

diagenetic facies A *facies* that includes all rocks or sedimentary materials that have, by a process of diagenesis, developed "mineral assemblages that are the result of adjustment to a particular diagenetic environment" (Packham and Crook, 1960, p.400). A "low-rank" facies corresponds to an early stage of alteration, a "high-rank" facies to a late stage. Cf: *parfacies.*

diagenodont (di-a-gen'-o-dont) Said of the dentition of a bivalve mollusk (e.g., *Astarte*) having differentiated cardinal teeth and lateral teeth located on the hinge plate, with the lateral teeth not exceeding two and the cardinal teeth not exceeding three in either valve.

diaglyph (di'-a-glyph) A *hieroglyph* formed during diagenesis (Vassoevich, 1953, p.33).

diagnosis (di-ag-no'-sis) A statement of the characteristics of a *taxon* which in the opinion of its author distinguish it from others (McVaugh et al., 1968, p.12).

diagnostic fossil (di-ag-nos'-tic) *characteristic fossil.*

diagnostic horizons As used in the U.S. system of soil taxonomy. Combinations of specific soil characteristics that are indicative of certain classes of soils. Those which occur at the soil surface are called epipedons, those below the surface, diagnostic subsurface horizons.

diagnostic mineral (a) A mineral, such as olivine or quartz, whose presence in an igneous rock indicates whether the rock is undersaturated or oversaturated. (b) A mineral, such as garnet in granite, whose presence indicates a sedimentary protolith.

diagonal bedding (di-ag'-o-nal) An archaic syn. of *inclined bedding,* or bedding diagonal to the principal surface of deposition; specif. *cross-bedding.*

diagonal glide plane *n-glide.*

diagonal joint A joint whose strike is oblique to the strike of the sedimentary strata, or to the cleavage plane of the metamorphic rocks, in which it occurs. Syn: *oblique joint.* Partial syn: *(hkO) joint.*

diagonal lamination *cross-lamination.*

diagonal scour mark One of a series of scour marks arranged diagonally to the main direction of flow and formed by concentration of smaller scour marks (usually longitudinal flutes) into distinct rows that alternate with areas where scour marks are absent or less abundant.

diagonal-slip fault *oblique-slip fault.*

diagonal stratification *cross-stratification.*

diagrammatic map (di'-a-gram-mat'-ic) *cartogram.*

dial (di'-al) n. A compass used for surface and underground surveying, fitted with sights, spirit levels, and *vernier,* and mounted on a tripod. v. To survey or measure with a dial and chain.

diallage (di'-al-lage) (a) A dark-green or grass-green, brown, gray, or bronze-colored clinopyroxene (usually a variety of augite or of aluminum-bearing diopside) occurring in lamellae or in foliated masses and often having a metallic or brassy luster. It is characterized by a conspicuous parting parallel to the front pinacoid, and is typically found in basic igneous rocks such as gabbro. (b) A term applied to various poorly defined alteration products of pyroxene.

diallagite (di'-al-lag-ite') A *pyroxenite* composed almost entirely of altered diopside. Other pyroxenes, hornblende, spinel, and garnet may be present as accessories. Obsolete.

dialogite (di-al'-o-gite) A syn. of *rhodochrosite.*

dialysis (di-al'-y-sis) A method of separating compounds in solution or suspension by their differing rates of diffusion through a semipermeable membrane, some colloidal particles not moving through at all, some moving slowly, and others diffusing quite readily. Cf: *osmosis.* See also: *electrodialysis.*

diamagnetic (di'-a-mag-net'-ic) Said of a mineral with a negative magnetic susceptibility. All materials that do not show paramagnetism or permanent magnetic order are diamagnetic. Typical diamagnetic minerals are quartz and feldspar. Cf: *paramagnetic.*

diamantaire (a) French term, used throughout the diamond industry for a manufacturer or dealer. (b) Any knowledgeable or experienced person in the diamond industry.

diamantiferous (di'-a-man-tif'-er-ous) *diamondiferous.*

diametral spine (di-am'-e-tral) A basally fused spine opposite radial spines and passing through the diameter of the central capsule of an acantharian radiolarian.

diamict (di'-a-mict) A general term proposed by Harland et al. (1966, p.229) to include *diamictite* and *diamicton.*

diamictite (di-a-mic'-tite) A comprehensive, nongenetic term proposed by Flint et al. (1960b) for a nonsorted or poorly sorted, noncalcareous, terrigenous sedimentary rock that contains a wide range of particle sizes, such as a rock with sand and/or larger particles in a muddy matrix; e.g. a tillite or a pebbly mudstone. Cf: *diamicton.* Syn: *mixtite.* Originally termed symmictite by Flint et al. (1960a).

diamicton (di-a-mic'-ton) A general term proposed by Flint et al. (1960b) for the nonlithified equivalent of a *diamictite;* e.g. a till. Originally termed *symmicton* by Flint et al. (1960a). Syn: *mixton* (Edwards, 1986, p.445).

diamond (di'-a-mond) (a) A cubic mineral, representing a naturally occurring crystalline form of carbon dimorphous with graphite and being the hardest natural substance known (hardness of 10 on the Mohs scale). It often occurs in octahedrons with rounded edges or curved faces. Diamonds form under extreme temperatures and pressures and are found in ultrabasic breccias, pipes in igneous rocks, and alluvial deposits. Pure diamond is colorless or nearly so, color is imparted by impurities. When transparent and more or less free from flaws, it is the most cherished and among the most highly valued gemstones; its high refractive index and dispersive powers result in remarkable brilliance and play of prismatic color when faceted. Off-color or flawed diamonds are used for industrial purposes (such as in rock drills, abrasive powder, and cutting tools). (b) Artificially produced crystallized carbon similar to the native form. (c) A crystalline mineral that resembles diamond in brilliance, such as "Alençon diamond" (a smoky quartz sometimes valued as a jewel); esp. any of various kinds of rock crystal such as "Bristol diamond", "Herkimer diamond", "Lake George diamond", and "Arkansas diamond".

diamond anvil cell *anvil cell.*

diamond bit A solid metal rotary-drilling bit studded with diamonds (usually bort). It is used for drilling and coring in extremely hard rock.

diamond chip A thin, tabular *chip* or broken piece of an uncut diamond crystal, weighing less than 1.8 carat.

diamond drilling A variety of rotary drilling in which diamond bits are used as the rock-cutting tool. It is a common method of prospecting for mineral deposits, esp. in development work where core samples are desired and in deviation drilling with a turbine motor.

diamond dust [ice] Minute ice crystals, usually columnar or bullet-shaped, precipitated out of clear air at very low temperatures as on the polar plateau of Antarctica.
diamond dust [materials] Powdered, crushed, or finely fragmented diamond material used as a cutting, grinding, and polishing abrasive medium.
diamond glide plane *d-glide.*
diamondiferous (di'-a-mond-if'-er-ous) Said of any substance (esp. rock or alluvial material) containing or yielding diamonds. Syn: *diamantiferous.*
diamond simulant Any material which is not diamond or synthetic diamond, but which simulates a faceted diamond's appearance and is used in its place. Common diamond simulants include cubic zirconia (CZ), gadolinium gallium garnet (GGG), synthetic spinel, and yttrium aluminum garnet (YAG). Also called diamond substitute.
diamond spar *corundum.*
diamond structure A type of crystal structure in which each atom or ion is four-coordinated; minerals having this structure are characteristically rigid and have low electrical conductivity.
dianite (di'-a-nite) *columbite.*
diaoyudaoite (di'-ao-yu-dao'-ite) A colorless to light green hexagonal mineral found in sea-floor mud: $NaAl_{11}O_{17}$.
diaphaneity (di'-aph-a-ne'-i-ty) The light-transmitting quality of a mineral.
diaphanotheca (di-aph'-a-no-the'-ca) The relatively thick, light-colored to transparent, intermediate layer of the spirotheca next below the tectum in fusulinids.
diaphorite (di-aph'-o-rite) A metallic gray-black orthorhombic mineral: $Pb_2Ag_3Sb_3S_8$. Syn: *ultrabasite.*
diaphragm (di'-a-phragm) (a) Any of various more or less rigid partitions in the bodies or shells of invertebrates; e.g. a skeletal membranous partition extending across the zooidal chamber in stenolaemate bryozoans; a thin crescentic plate of secondary brachiopod shell developed around the visceral disc of the brachial valve; an imperforate partition crossing the *siphuncle* of a nautiloid; or a partial *septum* just below the *aperture* in a thecamoebian, perforated for protrusion of pseudopodia. (b) The transversly oriented muscular wall in mammals that separates the pleural cavity from the abdominal cavity and aids in breathing.
diaphthoresis (di-aph-thor'-e-sis) A little used syn. of *retrograde metamorphism.*
diaphthorite (di-aph-thor'-ite) A crystalline rock in which minerals characteristic of a lower metamorphic grade have developed by retrograde metamorphism at the expense of minerals peculiar to a higher metamorphic grade. The term was originated by Becke in 1909.
diapir [intrus rocks] A spherical, elliptical, or teardrop-shaped body of magma or flowing (ductile) solid rock which rises toward the surface due to its low density compared to that of the enclosing rocks.
diapir [struc geol] (di'-a-pir) A dome or anticlinal fold in which the overlying rocks have been ruptured by the squeezing-out of plastic core material. Diapirs in sedimentary strata usually contain cores of salt or shale; igneous intrusions may also show diapiric structure. See also: *diapirism.* Syn: *piercement dome; diapiric fold; piercing fold.*
diapir fall Subsidence of a diapir's crest or flanks that occurs when regional extension causes widening. The diapir subsides because salt cannot be imported rapidly enough from the adjoining source layer to fill the widening diapir.
diapiric Said of the process by which relatively low-density material rises by piercing and/or doming the overlying rocks.
diapiric fold (di-a-pir'-ic) *diapir [struc geol].*
diapirism (di'-a-pir-ism') The process of piercing or rupturing of domed or uplifted rocks by mobile core material, by tectonic stresses as in anticlinal folds, by the effect of geostatic load in sedimentary strata as in salt domes and shale diapirs, or by igneous intrusion, forming diapiric structures such as plugs. The concept was first applied to salt structures, which are the most common type of diapir. Obsolete syn: *tiphon.*
diaplectic (di-a-plec'-tic) Said of minerals in which the crystal structure has been affected by the passage of shock wave ; e.g. a "diaplectic mineral" whose disordered and deformed crystals have been modified by shock waves without melting and have characteristics such as planar deformation features, lowered refractive indices, and lowered birefringence, or a "diaplectic glass" (of quartz, feldspar, or other minerals) representing an amorphous phase produced by shock waves without melting. Diaplectic materials represent intermediate stages of structural order between the crystalline and the normal glassy phases. Term proposed by Wolf von Engelhardt in 1966 (Engelhardt and Stöffler, 1968, p.163). Etymol: Greek "diaplesso", "to destroy by beating or striking".
diaplectic glass Amorphous form of crystals, "solid state glass", resulting from shock wave compression and subsequent pressure release of single crystals or polycrystalline rocks; most commonly observed for tectosilicates.
diapositive (di-a-pos'-i-tive) A positive image on a transparent medium such as glass or film; a *transparency [photo].* The term originally was used primarily for a transparent positive on a glass plate used in a plotting instrument, a projector, or a comparator, but now is frequently used for any positive transparency (ASP, 1975, p.2077).
diapsid (di-ap'-sid) adj. Pertaining to a reptilian skull characterized by two temporal fenestrae, upper and lower; includes the *Archosauria.* n. Loosely, any reptile with such a skull.
diara (di-ar'-a) *char [streams].*
diaresis (di-ar'-e-sis) A transverse groove on the posterior part of the *exopod* (rarely also the *endopod*) of a uropod of a malacostracan crustacean.
diarhysis (di-a-rhy'-sis) A radial *skeletal canal* in *dictyonine* hexactinellid sponges, penetrating the body wall completely, open at each end, and containing a flagellated chamber. Pl: diarhyses.
diaschistic (di-a-schist'-ic) The rock of a minor intrusion that consists of a differentiate, i.e. its composition is not the same as that of the parent magma. Obsolete. Cf: *aschistic.*
Diasoma A subphylum of mollusks with pseudobivalved, bivalved, or univalved shell that often gapes at anterior and posterior ends. Gut relatively straight; mouth and anus widely separated from one another and opening at opposite ends of the shell so that the gut is not bent or twisted into U-shape. Head poorly developed or absent, and radula present or absent. Body primitively laterally compressed. Range, Lower Cambrian to Holocene.
diasome (di'-a-some) Any benthic mollusk of the subphylum Diasoma. It includes the classes Rostroconchia, Scaphopoda, and Bivalvia.
diaspore (di'-a-spore) (a) A variously colored orthorhombic mineral: AlO(OH). It represents the alpha base dimorphous with boehmite. Diaspore is found in bauxite and is associated with corundum and dolomite; it occurs in lamellar masses with pearly luster or in prismatic crystals. Syn: diasporite. (b) A group name for minerals of composition *A*O(OH), where *A* = Al, Fe^{3+}, Mn^{3+}, Cr or V.
diaspore clay A high-alumina refractory clay consisting essentially of the mineral diaspore. It has been interpreted as a desilication product of associated *flint clay* and other kaolinitic materials (Keller et al., 1954). Commercial diaspore of first-grade quality contains more than 68% alumina. See also: *burley clay.*
diastem (di'-a-stem) A relatively short interruption in sedimentation, involving only a brief interval of time, with little or no erosion before deposition is resumed; a depositional break of lesser magnitude than a *paraconformity,* or a paraconformity of very small time value. Diastems are not ordinarily susceptible of individual measurement, even qualitatively, because the lost intervals are too short; they are often deduced solely on paleontologic evidence. The term was introduced by Barrell (1917, p.794). The synonymous term *nonsequence* is preferred in Great Britain. Adj: diastemic. Etymol: Greek "diastema", "interval".
diastrophic (di-a-stroph'-ic) Adj. of *diastrophism.* Cf: *orographic [tect].*
diastrophic eustatism *tectono-eustasy.*
diastrophic plateau A plateau formed by the upheaval of a plain, and cut, broken, or divided into parts by rivers, faults, or flexures (Powell, 1895, p.39-40).
diastrophism (di-as'-tro-phism) A general term, rarely used anymore, for all movement of the crust produced by tectonic processes, including the formation of ocean basins, continents, plateaus, and mountain ranges. *Orogeny, epeirogeny,* and *taphrogeny* are major subdivisions. The use of this general term for small-scale features (e.g. diastrophic event, diastrophic ridge, diastrophic structure) is vague and undesirable; more specific terms should be substituted. Adj: *diastrophic.* Syn: *tectonism.*
diastrophy (di-as'-tro-phy) "An event, localized in time and space, that produces significant deformation of the lithosphere" (Stokes, 1983). Cf: *orogeny.*
diatactic (di-a-tac'-tic) Said of a sedimentary structure, like that shown by varves, characterized by the repetition of a pair of unlike laminae showing a gradation in grain size from coarse below to fine above. Also spelled: *diadactic.*
diataphral (di-a-taph'-ral) An obsolete descriptive term for a type of tectonics in which *syntaphral* folds and faults are refolded by upward diapiric regurgitation of the axial zone of a geosyncline (Carey, 1963, p. A6). Cf: *apotaphral.*
diatexis (di-a-tex'-is) Extensive *anatexis* (partial melting), involving rock components with high melting points; e.g., mafic minerals (Dietrich and Mehnert, 1961). Rarely used. Cf: *metatexis.*
diatexite (di-a-tex'-ite) A rock formed by *diatexis.* Var: diatectite.
diathermic (di-a-ther'-mic) Said of a substance or "wall" between two thermodynamic systems that is able to transmit heat. Two systems separated by a diathermic barrier will eventually reach thermal equilibrium.
diatom (di'-a-tom) A microscopic, single-celled *alga* of the class Bacillariophyceae (ochrophytes), which lives in freshwater or marine environments. Diatoms secrete walls of silica, called *frustules.* Although diatoms are known from the Jurassic, they first become abundant in the Cretaceous, ranging to the present.
diatomaceous (di'-a-to-ma'-ceous) Composed of or containing numerous diatoms or their siliceous remains.
diatomaceous chert A *diatomite* that has a well-developed siliceous cement or groundmass.
diatomaceous earth *diatomite.*
diatomaceous shale An impure diatomite with much clayey matter and with shaly partings.

diatomite (di-at′-o-mite) A light-colored soft friable siliceous sedimentary rock, consisting chiefly of opaline frustules of the diatom, typically white, but also pale yellow, gray, or tan. Its low density enables it to float on water. Some deposits are of lake origin but the largest are marine. Owing to its high surface area, high absorptive capacity, and relative chemical stability, diatomite has a number of uses, esp. as a filter aid and as an extender in paint, rubber, and plastics. The term is generally reserved for deposits of actual or potential commercial value. Syn: *diatomaceous earth; kieselguhr.* Obsolete syn: *tripoli-powder.* See also: *tripoli; tripolite.*

diatom ooze A deep-sea pelagic sediment containing at least 30% diatom frustules; it is a *siliceous ooze.*

diatreme (di′-a-treme) Funnel-shaped breccia pipe that reaches as much as 2,500 m depth. Diatremes are thought to form by hydrovolcanic fragmentation and wall-rock collapse. Diatremes may underlie maars and grade at depth into dikes (Vespermann and Schmincke, 2000).

diazo print (di-az′-o) A print on light-sensitized material, made directly by exposure to strong light from a positive transparency and developed dry in the presence of ammonia vapor (USGS, 1965, p.103). Generally scale-stable when made in a vacuum frame, but elongated in direction of movement through a circular-drum printer. Syn: *ozalid.*

dibranchiate (di-bran′-chi-ate) *coleoid.*

dicalycal (di-cal′-y-cal) Said of a graptoloid *theca* from which two others originate.

dice mineral A term used in Wisconsin for galena occurring in small cubes.

dicentric (di-cen′-tric) Said of a *corallite* formed by a polyp retaining a distomodaeal condition permanently.

dichotomous (di-chot′-o-mous) Said of a crinoid arm that is divided into two branches, which may be equal (isotomous) or unequal (heterotomous).

dichotomy (di-chot′-o-my) A repeated, twofold equal branching of the main axis of a plant, e.g. in liverworts, seaweeds, and many pteridophytes.

dichotriaene (di-cho-tri′-aene) A sponge *triaene* with dichotomously branched cladi.

dichroic (di-chro′-ic) Said of a mineral that displays *dichroism.*

dichroism (di-chro′-ism) *Pleochroism* of a crystal that is indicated by two different colors. A mineral showing dichroism is said to be *dichroic.* Cf: *trichroism.*

dichroite (di-chro′-ite) *cordierite.*

dichroscope (di′-chro-scope) An optical instrument that is used to analyze the colors of a pleochroic crystal; it consists of a calcite rhomb and a lens.

dickinsonite (dick′-in-son-ite′) A green platy monoclinic mineral of the *arrojadite* group: $KNa_4Ca(Mn^{2+},Fe^{2+})_{14}Al(PO_4)_{12}(OH)_2$.

dickite (dick′-ite) A well-crystallized white to variously colored monoclinic clay mineral of the kaolin group: $Al_2Si_2O_5(OH)_4$. It is polymorphous with kaolinite and nacrite. Dickite is structurally distinct from other members of the *kaolin* group, having a more complex order of stacking in the *c*-axis direction than kaolinite. It usually occurs in hydrothermal veins.

dickthomssenite A platy vitreous light golden brown monoclinic mineral: $MgV_2O_6{\cdot}7H_2O$.

dicolpate (di-col′-pate) Said of pollen grains having two colpi.

dicolporate (di-col′-por-ate) Said of pollen grains having two colpi, with at least one colpus provided with a pore or transverse furrow. Dicolporate pollen are rare.

dicot (di′-cot) *dicotyledon.*

dicotyledon (di′-cot-y-le′-don) An angiosperm whose seeds contain an embryo with two embryonic leaves. Such a plant has flower parts in fours or fives, reticulate leaf venation, and tricolpate or tricolpate-derived pollen. Examples include roses, thistles, and oaks. Dicotyledons range from the Early Cretaceous to the present. Cf: *monocotyledon.* Syn: *dicot.*

dicranoclone (di-cran′-o-clone) (a) A tuberculate monaxonic desma (of a sponge) of dipodal to polypodal form with rootlike terminal zygomes. (b) A desma (of a sponge) with arms that diverge from one side of a central point.

dictyonal framework (dic-ty-o′-nal) The rigid interior skeleton of *dictyonine* sponges between the dermalia and gastralia, built of fused dictyonal strands, commonly but not always forming a cubic lattice.

dictyonalia (dic′-ty-o-nal′-i-a) The sponge spicules of a dictyonal framework.

dictyonal strand A linear series of hexactines (sponge spicules) in parallel orientation and fused to form a continuous strand.

dictyonine (dic′-ty-o-nine′) adj. Said of a hexactinellid sponge whose parenchymalia (spicules) form a rigid framework composed of dictyonal strands; more loosely, said of a sponge whose skeleton is composed of subparallel hexactines rigidly fused so that the limits of individual spicules are not apparent. Ant: *lyssacine.* n. A dictyonine sponge.

dictyonite (dic′-ty-o-nite′) A *migmatite* with reticulated character, i.e. with a veinlet network (Dietrich and Mehnert, 1961). Rarely used. Var: diktyonite.

dictyospore (dic′-ty-o-spore′) A multicellular, asexual *fungal spore* with both traverse and longitudinal septa, producing a pattern reminiscent of a brick wall. Cf: *Saccardoan spore groups.*

dictyostele (dic′-ty-o-stele′) A *stele* consisting of separate vascular bundles, or of a network of bundles (Fuller and Tippo, 1954, p.955). See also: *meristele.*

dicyclic (di-cy′-clic) (a) Said of a crinoid having two circlets of plates proximal to the radial plates or (in some crinoids that lack radial plates) proximal to the oral plates. (b) Said of the apical system of an echinoid in which ocular and genital plates are arranged in two concentric circles, the genital plates alone in contact with the periproctal margin. Cf: *monocyclic.*

didodecahedron (di-do′-dec-a-he′-dron) *diploid.*

diductor muscle (di-duc′-tor) A muscle that opens the valves in articulate brachiopods; commonly, one of a pair of two pairs of muscles attached to the *brachial valve* immediately anterior to the beak, usually to the *cardinal process.* The principal pair is commonly inserted in the *pedicle valve* on either side of the adductor muscles and the accessory pair is inserted posterior to them. Syn: *divaricator.*

didymoclone (did′-y-mo-clone′) A desma (of a sponge) consisting of a short, straight shaft, from the enlarged ends of which several zygome-bearing arms project, predominantly on one side of the spicule.

didymolite (did′-y-mo-lite′) *anorthite.*

didymospore (did′-y-mo-spore) An asexual *fungal spore* with both transverse and longitudinal septa which produce a pattern reminiscent of a brick wall. Cf: *Saccardoan spore groups.*

diel (di′-el) Of or pertaining to a 24-hour day.

dielectric (di-e-lec′-tric) adj. Said of material in which displacement currents predominate over conduction currents, i.e. an insulator n. Such a material.

dielectric constant A complex electrical property of matter that influences radar returns. A measure of the displacement currents resulting from application of an electric field. Symbol: K_e.

dielectric polarization The response of a dielectric material to an electric field, producing an induced dipole-moment per unit volume. In an insulating dielectric material, no net electric charge need be transferred by the exciting field. By some definitions, induced polarization is a lossy type of dielectric polarization with a long time constant.

Dienerian (Die-ner′-i-an) A Canadian stage and suggested upper substage of the Induan Stage: Lower Triassic (above Griesbachian, below Smithian).

dienerite (die′-ner-ite) A metallic gray-white cubic mineral of doubtful validity, as the type specimen is lost, no other occurrences have been found, and it is not a stable phase in the synthetic Ni-As system. Reported formula: Ni_3As.

Diestian (Di-es′-ti-an) North Sea stage: uppermost Miocene (below Pliocene Scaldisian).

dietella (di-e-tel′-la) A small partly enclosed chamber near the base of the vertical walls of zooids in some cheilostome bryozoans. Its walls include one or more plates of interzooidal septulae. Syn: *pore chamber.*

dietrichite (die′-trich-ite) A silky white or brownish yellow monoclinic mineral of the *halotrichite* group: $(Zn,Fe^{2+},Mn^{2+})Al_2(SO_4)_4{\cdot}[22]H_2O$.

dietzeite (diet′-ze-ite) A dark golden-yellow monoclinic mineral: $Ca_2(IO_3)_2(CrO_4){\cdot}H_2O$.

difference maps Maps (sections) made by subtracting maps (sections) from surveys made at different times. See: *time-lapse seismology.*

differential compaction (dif-fer-en′-tial) Results from central differences in pore water (clays vs. sands) and produces dip of the more compactable sediment away from the less compactable.

differential curvature A measure of a curved shape of the gravity equipotential surface. It is defined as the difference between the curvature of a surface in the direction in which it curves the most and the curvature at right angles to this direction, multiplied by the *gravitational constant.*

differential entrapment The control of oil and gas migration and accumulation by selective trapping in interconnected reservoirs. A trap filled with oil is an effective gas trap but a trap filled with gas is not an effective oil trap. As a result, gas may be trapped downdip and oil updip (Gussow, 1954).

differential erosion Erosion that occurs at irregular or varying rates, caused by the differences in the resistance and hardness of surface materials: softer and weaker rocks are rapidly worn away, whereas harder and more resistant rocks remain to form ridges, hills, or mountains. Syn: *etching.*

differential fault *scissor fault.*

differential global positioning system Employing the signals received from a nearby fixed reference station to remove the effects of deviations in a satellite's location or in the travel path of signals through the atmosphere. Abbrev: DGPS.

differential leveling The process of measuring the difference of elevation between any two points by spirit leveling.

differential loading Creation of lateral pressure gradients caused by lateral variations in density, strength, or especially thickness of the overburden. Such variations may be sedimentary, erosional, or structural in origin. This mechanism effectively initiates *halokinesis,* especially near the surface (Harrison, 1927; Hollingworth et al., 1945).

differential melting Partial melting of a rock.

differential pressure The difference in pressure between the two sides of an orifice; the difference between reservoir and sand-face pressure; between pressure at the bottom of a well and at the wellhead; between flowing pressure at the wellhead and that in the gathering line; any difference in pressure between upstream and downstream where a restriction to flow exists.

differential settlement In engineering geology, uneven consolidation or compaction of foundation soils under a structure, often caused by variability of compacted fill or of the relief of underlying rock or sediments; and by the weight of building and fill itself.
differential solution *intrastratal solution.*
differential stress In experimental rock deformation, the maximum principal stress minus the minimum principal stress. Syn: *stress difference.*
differential thermal analysis *Thermal analysis* carried out by uniformly heating or cooling a sample that undergoes chemical and physical changes, while simultaneously heating or cooling in identical fashion a reference material that undergoes no changes. The temperature difference between the sample and the reference material is measured as a function of the temperature of the reference material. Abbrev: DTA.
differential weathering Weathering that occurs at different rates, as a result of variations in composition and resistance of a rock or differences in intensity of weathering, and usually resulting in an uneven surface where more resistant material stands higher or protrudes above softer or less resistant parts. Syn: *selective weathering.*
differentiate (dif-fer-en'-ti-ate) n. A rock formed as a result of magmatic differentiation.
differentiated (dif-fer-en'-ti-at'-ed) Said of an igneous intrusion composed of more than one rock type, owing to differentiation.
differentiated meteorite *achondrite.*
differentiation [intrus rocks] All processes by which multiple rock types form from a single magma. Cf: *assimilation.* Syn: *magmatic differentiation.*
differentiation [planet] (dif'-fer-en'-ti-a'-tion) The processes by which planets and satellites develop concentric layers or zones of different chemical and mineralogical composition. An undifferentiated body would have a more homogeneous composition than a differentiated one (Lowman, 1976).
differentiation [sed] (a) *sedimentary differentiation.* (b) *diagenetic differentiation.*
differentiation index In igneous petrology, the number that represents the sum of the weight percentages of normative quartz, orthoclase, albite, nepheline, leucite, and kalsilite (Thornton and Tuttle, 1960); a numerical expression of the extent of differentiation of a magma. Abbrev: DI. See also: *petrogeny's residua system.*
diffluence (dif'-flu-ence) (a) A lateral branching or flowing-apart of a glacier in its ablation area. This separation may result from the glacier's spilling over a preglacial divide or through a gap made by basal sapping of a cirque wall, or from downvalley blocking at the junction of a tributary glacier. (b) A place at which diffluence occurs. Ant: *confluence [glaciol].*
diffluence pass The lower part of a *trough end,* where a *distributary glacier* has left the main valley.
diffluence step A rock step that rises downstream away from the main glacial valley at the place of diffluence. It is probably caused by the weakening of glacial action at that point. Ant: *confluence step.*
diffluent (dif'-flu-ent) Said of a stream or glacier that flows away or splits into two or more branches. Ant: *confluent.*
diffraction [cryst] In crystallography, the coherent scattering of X-rays, neutrons, or electrons by the electrons, the nuclei, or the electrostatic potential fields, respectively, of atoms in crystals.
diffraction [phys] (dif-frac'-tion) (a) The departure of a ray from the path predicted by reflection and refraction. See also: *X-ray scattering.* (b) The process by which the direction of wave motion in any medium is modified by bending around the edges of an obstacle or scattering by a point discontinuity, and the resultant formation of an interference pattern within the geometric shadow of the obstacle. See also: *electron diffraction; X-ray diffraction.* (c) An event produced by the termination of a reflector or refractor.
diffraction [seis] (a) The generation and transmission in all directions of seismic wave energy in accordance with Huygens' principle. (b) An event observed on seismic data produced by diffracted energy. Diffraction events result at the termination of reflectors and are characterized on seismic records and sections by a distinctive curved alignment.
diffraction [waves] The bending of a water wave around an obstacle, e.g., the bending of waves into the wave shadow behind an island.
diffraction angle *Bragg angle.*
diffraction chart A chart showing the time-distance relationships that seismic diffractions should obey. It is used to identify diffractions.
diffraction grating *grating.*
diffraction indices A set of three symbols (integers or letters) used to express the orientation and location within a *unit cell* of any set of planes from which X-rays may be diffracted. Diffraction indices resemble *Miller indices* in form and function, except that they are written without parentheses or braces, to distinguish them from Miller indices for a *crystal face* and for a *crystal form,* respectively. Cf: *Miller indices.*
diffraction pattern In crystallography, the set of intensities and diffraction angles obtained by performing a X-ray-, neutron-, or electron-diffraction experiment on a crystalline substance. Each crystalline substance has a characteristic diffraction pattern, which, when found, is taken to be evidence that that substance is present.
diffraction spacing In a crystal lattice, interplanar spacings given by a diffraction pattern.
diffractogram (dif-frac'-to-gram) A record of diffraction of a crystalline sample, obtained by electronic detectors and recorded on a paper chart.
diffractometer (dif-frac-tom'-e-ter) In mineral and crystallographic analysis, an instrument that records either powder or single-crystal X-ray diffraction patterns.
diffuse crossed lamellar structure A *crossed lamellar structure* in which the first- and second-order lamellae show indistinct boundaries as viewed on the depositional surface and in horizontal sections. Fractures through this structure show poorly defined first-, second-, and third-order lamellae. This structure is intermediate between simple crossed lamellar and homogeneous. In certain studies of nuculanacean bivalves it has been called "homogeneous." However, not all "homogeneous" layers in the Nuculanacea are diffuse crossed lamellar (Carter, 1976). Abbrev: DCL.
diffuse flow The circulation of groundwater in *karst* aquifers (or other aquifers) under conditions in which all, or almost all, openings in the karstified rock intercommunicate and are full of water but have not been selectively enlarged in specific zones by dissolution, so that no concentration of groundwater occurs in restricted conduits (Lowe and Waltham, 1995). The groundwater flow is generally slow moving, may be laminar, and may have a uniform discharge and slow response to storms. The term is being replaced by the term "slow flow" because of significant confusion regarding its usage, especially when thought of in terms of porous-media flow (Field, 2002).
diffuse layer (dif-fuse') The outer, mobile layer of ions in an electrolyte, required to satisfy a charge unbalance within a solid with which the electrolyte is in contact. It constitutes part of the *double layer* of charge adjacent to the electrolyte-solid interface. Cf: *fixed layer.*
diffuse-porous wood (dif-fuse'-po'-rous) A type of wood in which the vessels are more or less uniform in size and distribution throughout each annual ring (Fuller and Tippo, 1954, p.956).
diffusion [grd wat] The process by which both ionic and molecular species dissolved in water move from areas of higher concentration to areas of lower concentration (Fetter, 1994). Movement is random and is proportional to the gradient of concentration. The process tends to distribute the particles more uniformly. Cf: *diffusion coefficient; advection [hydraul].*
diffusion [struc petrol] The movement of atoms or *vacancies* through a crystallographic lattice.
diffusional mass transfer *diffusion creep.*
diffusion coefficient The coefficient that relates solute flux due to *diffusion* to the concentration gradient in groundwater.
diffusion creep Grain-scale, ductile deformation accomplished by the motion of atoms through crystals, along grain boundaries, and through pore fluids. It results in viscous flow in solids. Syn: *diffusional mass transfer.* Cf: *solid flow; viscous creep; viscous flow.*
diffusion metasomatism A process of mass transfer in which chemical components move by diffusion along grain boundaries or through a stationary aqueous solution occupying pores in rocks. Cf: *infiltration metasomatism.*
diffusion porosity The volume of pores through which mass can be transferred only by *diffusion* divided by the total volume of porous medium.
diffusive convection An instability of the interface between a cooler fresher layer above a warmer saltier layer, due to more rapid diffusion of heat than salt across the interface. Cf: *salt fingering.*
diffusivity (dif-fu-siv'-i-ty) *thermal diffusivity.*
digenite (dig'-e-nite) A blue to black rhombohedral mineral: Cu_9S_5. It often occurs with chalcocite, and is stable below 73°C; it converts to a cubic phase above this temperature. Syn: *blue chalcocite; alpha chalcocite.*
digestion (di-ges'-tion) Partial or complete assimilation of wall rock into a magma.
digital (dig'-it-al) Said of the representation of measured quantities in discrete or quantized units. A digital computer is one in which information is stored and manipulated as a series of discrete numbers, as opposed to an *analog* computer.
digital cartography *automated cartography.*
digital computer A *computer* that operates with numbers and characters expressed directly as digits in a decimal or binary or other system and performs the basic arithmetic operations in a manner similar to a human mathematician. Cf: *analog computer; hybrid computer.*
digital elevation model An array of regularly spaced elevation values referenced horizontally either to a Universal Transverse Mercator (UTM) projection or to a geographic coordinate system. The grid cells are spaced at regular intervals along south to north profiles that are ordered from west to east. Abbrev: DEM. The U. S. Geological Survey produces five primary types of elevation data: 7.5-minute DEM, 30-minute DEM, 1-degree DEM, 7.5-minute Alaska DEM, and a 15-minute Alaska DEM (USGS, 2000). See also: *altitude matrix; digital terrain model.*
digital image A two-dimensional array of regularly spaced picture elements (pixels) constituting a picture (FGDC, 1998a, p.62).
digital image processing Computer manipulation of digital images.
digital line graph A vector file of containing coordinates of zero-, one-, or two-dimensional elements representing human-made and natural features. The features have attributes describing them. The file is topologically structured to provide information about the relationship of a feature to

other features in the file. The files are the equivalents of standard quadrangle limits for each scale digital line graph. Abbrev: DLG.

digital log A *well log* whose analog curves have been discretely sampled and recorded on a magnetic tape or disc for use in computer-processed interpretation and plotting.

digital orthophotoquad A raster image, corresponding to the area covered by a U.S. Geological Survey 7.5-minute quadrangle, of remotely sensed data in which displacement in the image due to sensor orientation and terrain relief have been removed. Abbrev: DOQ.

digital orthophoto quarter-quadrangle A *digital orthophotoquad* corresponding to one-quarter the area covered by a U. S. Geological Survey 7.5-minute quadrangle. Abbrev: DOQQ.

digital raster graphic A georeferenced, rectified image of a scanned U. S. Geological Survey topographic or planimetric map (USGS, 2001). Abbrev: DRG.

digital terrain analysis Automated *terrain analysis* utilizing quantitative data and methods.

digital terrain model The term used by U. S. Department of Defense and other organizations to describe digital elevation data. DTMs are not necessarily formatted to U. S. Geological Survey quadrangle sizes. Abbrev: DTM. See also: *altitude matrix, digital elevation model.*

digitate delta (dig'-it-ate) A *bird-foot delta* whose seaward-extending margin has a fingerlike outline in plan.

digitate process tip branching A feature in an acritarch or dinoflagellate in which the *process* tips are arranged like the fingers (digits) of a hand (Williams et al., 1973; Cramer and Diez Rodriguez, 1979).

digitization (dig'-i-ti-za'-tion) The process of converting analog data (such as an image on photographic film) into numerical format.

digitize (dig'-i-tize) To sample a continuous function at discrete time intervals and to record the values as a sequence of numbers.

digitized map (dig'-i-tized) (a) A map expressed or stored in digital form. (b) A map prepared from cartographic information that has been converted from analog to digital form for use in automatic plotters. See also: *GIS.*

digitizer A device for converting point locations on a graphic image to plane (x,y) coordinates for digital processing.

digitizing The process of converting an analog image, map, diagram, or text into a digital format usable by a computer.

digonal (di'-go-nal) *diad.*

digue A French term for *dike [eng],* generally used in British English to denote a structural enclosure in harbor areas or where land reclamation has taken place.

digyrate element Bilaterally asymmetric ramiform conodont element with short, adenticulate posterior process and longer, denticulate inner- and outer-lateral processes, the distal extremities of which commonly twist in opposite directions (TIP, 1981, pt. W, supp.2).

dihedral angle The angle between two planes measured in a third plane that is perpendicular to their intersection, or the angle between two planes measured by the angle between the normals to the two planes.

dihexagonal (di-hex-ag'-o-nal) Said of a symmetrical twelve-sided figure, the alternate angles of which are equal. Such a figure is common in crystals of the hexagonal system.

dihexagonal dipyramid A crystal form that is a dipyramid of 24 faces, in which any section perpendicular to the sixfold axis is dihexagonal. Its indices are {*hkl*} and its symmetry is 6/*m* 2/*m* 2/*m*.

dihexagonal dipyramidal class That crystal class of the hexagonal system having symmetry 6/*m* 2/*m* 2/*m*.

dihexagonal prism A crystal form of twelve faces parallel to the symmetry axis, in which any cross section perpendicular to the prism axis is dihexagonal. Its indices are {*hk*0} with symmetry 6/*m* 2/*m* 2/*m*.

dihexagonal pyramid A crystal form consisting of a pyramid of 12 faces, in which any cross section perpendicular to the sixfold axis is dihexagonal. Its indices are {*hkl*} or {$hk\bar{l}$} in symmetry 6*mm*.

dihexagonal-pyramidal class That crystal class in the hexagonal system having symmetry 6*mm*.

dihydrite (di-hy'-drite) *pseudomalachite.*

dikaka Dune sand covered by scrub or grass; may include the plant-root tubes (after Glenn, 1979, p.400).

dikaryon (di-kar'-y-on) In fungal reproduction, one cell with a pair of unfused nuclei, one from each parent.

dike [eng] An artificial wall, embankment, ridge, or mound, usually of earth or rock fill, built around a relatively flat, low-lying area to protect it from flooding; a levee. A dike may also be constructed on the shore or border of a lake to prevent inflow of undesirable water. Syn: *digue; dyke.*

dike [intrus rocks] A tabular igneous intrusion that cuts across the bedding or foliation of the country rock. Also spelled: *dyke.* Cf: *sill [intrus rocks]; sheet [intrus rocks].* See also: *dikelet.*

dike [sed] *sedimentary dike.*

dike [streams] (a) An artificial watercourse; esp. a deep drainage ditch. The term has also been applied to any channel, including those formed naturally. (b) A pool or small pond.

dikelet (dike'-let) A small dike. There is no agreement on specific size distinctions.

dike ridge (a) *dike wall.* (b) A small wall-like ridge produced by differential erosion (as, e.g., along a shore).

dike rock The intrusive rock comprising a dike.

dike set A group of linear or parallel dikes. Cf: *dike swarm; cluster.*

dike spring A spring issuing from the contact between a dike composed of an impermeable rock, such as basalt or dolerite, and a permeable rock into which the dike was intruded.

dike swarm A group of dikes in radial, parallel, or en echelon arrangement. Their relationship with the parent plutonic body may not be observable. Cf: *cluster; dike set.*

dike wall A ridge consisting of a *dike* that formed in a nearly vertical fracture and was left standing after the rocks on either side were removed by erosion. Syn: *dike ridge.*

diktytaxitic (dik'-ty-tax-it'-ic) Volcanic igneous texture characterized by numerous jagged, irregular vesicles bounded by crystals, some of which protrude into the cavities (Dickinson, 1965, p.101).

dilatancy (di-lat'-an-cy) An increase in the bulk volume during deformation, caused by a change from close-packed structure to open-packed structure, accompanied by an increase in the pore volume. The latter is accompanied by rotation of grains, microfracturing, and grain boundary slippage.

dilatated septum (di'-lat-at-ed) A partly or wholly thickened *septum* of a rugose coral.

dilatation [exp struc geol] (dil-a-ta'-tion) *dilation.*

dilatation [seis] Change of volume per unit volume. Syn: *rarefaction.*

dilatational transformation (dil-a-ta'-tion-al) In a crystal, usually rapid thermal dilation and a rearrangement of the anion from cubic coordination to octahedral coordination, due to heating, e.g. the *transformation [cryst]* of CsCl to the NaCl structure at 460°C.

dilatational wave *P wave.*

dilatation theory The theory that attributed glacier movement to infiltration and freezing of water in cracks and other openings. That process is no longer considered to contribute significantly to glacier movement.

dilation (di-la'-tion) Deformation by a change in volume but not shape. Also spelled: dilatation.

dilation vein A mineral deposit in a vein space formed by bulging of the walls, contrasted with veins formed by wall-rock replacement.

dilete (di-lete') In spores, having a *laesura* with two radial arms arranged like the letter V, a rare spore type. Cf: *trilete.*

dillnite (dill'-nite) A fluorine-rich variety of zunyite.

Diluvial (Di-lu'-vi-al) An archaic term for geologic time since the appearance of humans but before historical times. The term "diluvial" was originally assigned to "deposits from the Biblical Great Flood". Cf: *anthropozoic.*

diluvialist (di-lu'-vi-al-ist') An adherent of *diluvianism.* Cf: *fluvialist.*

diluvianism (di-lu'-vi-an-ism') The doctrine that the widespread surficial deposits now known to be glacial drift, and other geologic phenomena, can be explained by a former worldwide flood or deluge, esp. the *Noachian Flood.*

diluvion (di-lu'-vi-on) (a) *diluvium.* (b) A term used in India as an ant. of *alluvion;* "it appears to mean loss of land by river erosion after flooding" (G.T. Warwick in Stamp, 1961, p.157).

diluvium (di-lu'-vi-um) (a) An archaic term applied during the early 1800s to certain widespread surficial deposits that could not be explained by the normal action of rivers and seas but were believed to be produced by extraordinary floods of vast extent, esp. the Noachian Flood; these deposits are now known to be mostly *glacial drift.* (b) A general term used in continental Europe for the older Quaternary, or Pleistocene, glacial deposits, as distinguished from the younger *alluvium.* Syn: *drift [glac geol].* Var. spellings: *deluvium; diluvion.*

dimensional orientation (di-men'-sion-al) In structural petrology, a tendency for planar and/or linear fabric elements to have a preferred orientation.

dimension stone (di-men'-sion) Building stone that is quarried and prepared in regularly shaped blocks according to specifications.

dimictic (di-mic'-tic) Said of a lake that circulates, or *overturns,* twice a year, as dimictic temperate lakes do in the spring and fall (Wetzel, 2001, p.79). Cf: *monomictic [lake]; polymictic [lake].*

diminutive fauna (di-min'-u-tive) *dwarf fauna.*

dimorph (di'-morph) Either of two crystal forms displaying dimorphism. Syn: *allomorph.*

dimorphic [bot] (di-mor'-phic) Said of a plant or plant part that is normally produced in two forms, e.g. juvenile and adult types of foliage.

dimorphic [cryst] *dimorphous.*

dimorphism [biol] (di-mor'-phism) The characteristic of having two distinct forms in the same species, such as male and female, or megaspheric and microspheric stages.

dimorphism [cryst] That type of *polymorphism [cryst]* in which two crystalline species, known as dimorphs, occur. Adj: *dimorphous.* Cf: *trimorphism; tetramorphism.*

dimorphite (di-mor'-phite) An orange-yellow orthorhombic mineral: As_4S_3. It occurs as dimorphs, α and β, both orthorhombic.

dimorphous (di-mor'-phous) Adj. of *dimorphism [cryst].* Syn: *dimorphic [cryst].*

dimple crater (dim'-ple) A small, almost circular craterlike feature restricted to mare regions of the Moon's surface and attributed to volcanic activity (possibly to withdrawal of molten subsurface lava). It lacks the raised rim of most lunar impact craters.

dimpled current mark (dim'-pled) An obsolete syn. of *cross ripple mark.*

dim spot A local decrease of amplitude on a seismic section. A decrease of amplitude is sometimes caused by hydrocarbon accumulation. See: *hydrocarbon indicator.*

dimyarian (di-my-ar'-i-an) adj. Said of a bivalve mollusk or its shell with two adductor muscles, whether equal (*isomyarian*) or unequal (*anisomyarian*) in size. Cf: *monomyarian.* n. A dimyarian mollusk.

Dinantian (Di-nan'-tian) European stage: Lower Carboniferous. It includes Tournaisian and Viséan.

Dinarides (Di-nar'-i-des) The mountain belt that stretches from northeastern Italy to Greece, paralleling the eastern Adriatic coast. It constitutes the southwest-vergent branch of the Alpine mountain system.

dinite (din'-ite) A colorless or yellowish orthorhombic hydrocarbon mineral with a low melting point (30°C), found in lignite: $C_{20}H_{36}$.

dinocyst (di'-no-cyst) Dinoflagellate cyst; informal contraction of *dinoflagellate* and *cyst.*

dinoflagellate (din-o-flag'-el-late) A member of the group of primarily single-celled organisms that during some part of their life cycle have a free-swimming stage with two dissimilar flagella. The vast majority of these organisms have a special type of nucleus in which the chromosomes remain condensed even when the cell is not dividing. Some are plantlike, containing chloroplasts and producing organic compounds by photosynthesis, and some are animal-like, ingesting other organisms for food. Dinoflagellates are considered to represent their own division (phylum), the Dinoflagellata. Certain dinoflagellates have a theca or *test [paleont]* that is resistant to decay; it may be simple and smooth or variously sculptured and divided into characteristic plates and grooves. Others produce a resting stage or *cyst [palyn]* with a resistant organic wall that is often spiny and may differ markedly from the theca of the same species. Cysts exist abundantly as fossils, and have a range primarily Triassic to present. Dinoflagellates also have been reported from the Paleozoic, but are mainly important for correlating and dating Jurassic, Cretaceous, and Tertiary deposits. They inhabit all water types and are capable of extensive diurnal vertical migrations in response to light; they constitute a significant element in marine plankton, including certain brilliantly luminescent forms and those that cause *red tide.* See also: *hystrichosphaerid.*

dinokaryon (di-no-kar'-y-on) A special eukaryotic nucleus that is shared among most dinoflagellate cells and involves, along with other distinctive features, fibrillar chromosomes that remain condensed during the mitotic cycle.

dinosaur (di'-no-saur) Any reptile of the subclass Archosauria distinguished from other reptiles especially by features of the pelvic bones. Dinosaurs were carnivorous or herbivorous, bipedal or quadrupedal, land dwelling, and of moderate to very large size. Range, Triassic to Cretaceous. See also: *Ornithischia; Saurischia.*

dinosaur leather A local term applied by Chadwick (1948) to complex sole marks, probably including both flute casts and load casts. Cf: *squamiform cast.*

dinosporin (di-no-spo'-rin) A compound of carbon, hydrogen, and oxygen that forms the resistant wall of fossilizable dinoflagellate cysts and has a high molecular weight. It is similar to *sporopollenin* but differs in its response to various stains and in its fluorescence properties (Williams et al., 2000, p.64).

dioctahedral (di'-oc-ta-he'-dral) Said of a layered-mineral structure in which two thirds of the available octahedrally coordinated positions are occupied. Cf: *trioctahedral.*

dioecious (di-oe'-cious) Said of a taxonomic unit of plants with male and female reproductive organs entirely separated on different individuals.

diogenite (di-og'-e-nite) One of a group of coarse-grained achondritic meteorites consisting almost entirely of low-Ca pyroxene. Chemical, petrologic, isotopic studies indicate that they are the product of igneous processing on the same parent body from which the eucrite and howardite meteorites were derived. See also: *HED meteorite.*

diomignite (di-o-mig'-nite) A colorless tetragonal mineral: $Li_2B_4O_7$.

diopside (di-op'-side) A monoclinic mineral of the *pyroxene* group: $CaMgSi_2O_6$. It contains little or no aluminum and may contain some iron. It ranges in color from white to green; transparent varieties are used in jewelry. Diopside occurs in some metamorphic rocks, and is found esp. as a contact-metamorphic mineral in crystalline limestones. Symbol: Di. Syn: *malacolite.*

diopside-jadeite (di-op'-side-jade'-ite) *omphacite.*

dioptase (di-op'-tase) An emerald-green rhombohedral mineral: $CuSiO_2(OH)_2$. It occurs in the zone of weathering of copper lodes in Chile, Namibia, and Siberia. Syn: *emerald copper.*

diorite (di'-o-rite) (a) In the *IUGS classification,* a plutonic rock with Q between 0 and 5, P/(A+P) greater than 90, and plagioclase more sodic than An_{50}. (b) A group of plutonic rocks intermediate in composition between acidic and basic, characteristically composed of dark-colored amphibole (esp. hornblende), sodic plagioclase (oligoclase, andesine), pyroxene, and sometimes a small amount of quartz; also, any rock in that group; the approximate intrusive equivalent of *andesite.* Diorite grades into *monzonite* with an increase in the alkali feldspar content. In typical diorite, plagioclase contains less than 50% anorthite, hornblende predominates over pyroxene, and mafic minerals total less than 50% of the rock. Etymol: Greek "diorizein", "to distinguish", in reference to the fact that the characteristic mineral, hornblende, is usually identifiable megascopically. Cf: *dolerite; gabbro.* See also: *diabase.*

dip [geomorph] (a) A low place or marked depression in the land surface; e.g. a steep-sided hollow among hills or a gap in a ridge. (b) A pronounced depression in a highway at the point of intersection with a dry stream bed, once common in the western U.S.

dip [magnet] *inclination [magnet].*

dip [seis] The angle between an interface associated with a seismic event and the horizontal. Cf: *apparent dip [seis].*

dip [struc geol] n. The maximum angle that a structural surface, e.g. bedding or a fault plane, makes with the horizontal; measured perpendicular to the strike of the structure and in the vertical plane. Syn: *true dip; angle of dip.* Cf: *regional dip; primary dip.* v. To be tilted or inclined at an angle.

dip [surv] (a) The vertical angle, at the eye of an observer, between the plane of the horizon and the line of sight tangent to the apparent (visible or sensible) horizon; the angular distance of the apparent horizon below the horizontal plane through the observer's eye. See also: *dip angle.* Also called: "dip of horizon". (b) The apparent depression of the visible horizon due to the observer's elevation and to the convexity of the Earth's surface. (c) The first detectable decrease in the altitude of a celestial body after reaching its maximum altitude on or near the meridian transit.

dip angle The vertical angle, measured at an observation point in surveying or at an exposure station in photogrammetry, between the plane of the true horizon and a line of sight to the apparent horizon. See also: *dip [surv].*

dip-angle method [geophys] An electromagnetic surveying method in which the *dip angle* of the total magnetic field is measured. The secondary field resulting from eddy currents in a conducting body changes the direction of the total field (horizontal in the case of a vertical transmitter loop) that would be measured in the absence of the conductor (Keller and Frischknecht, 1966).

dip-azimuth map *azimuth map.*

dip calculation Calculation of the dip of a reflecting interface from observations of the variation of the arrival time of seismic events as the observing point is moved. See also: *moveout.*

dip circle An obsolete type of *inclinometer [magnet].*

dip equator *magnetic equator.*

dip filter *velocity filter.*

diphyletic (di-phy-let'-ic) Said of a higher taxon that artificially includes descendants of two separate evolutionary lineages.

dip isogon A line drawn on a cross section of a folded surface so as to connect points of equal dip on the top and bottom of the surface. A classification of folds by Ramsay (1967, p.363) is based on dip isogons.

dip joint A joint whose strike is approximately perpendicular to the strike of bedding or cleavage of the containing rock. Cf: *strike joint.*

dipleural (di-pleu'-ral) Said of the arrangement of the two rows of thecae (the stipes) in the biserial *rhabdosome* of a *scandent* graptoloid in which the rows are in contact back-to-back. Cf: *monopleural.*

diploblastic (dip-lo-blas'-tic) Said of the structure of lower invertebrates (sponges, coelenterates) having ectodermal and endodermal layers but lacking a true *mesoderm.*

diploconical (dip-lo-con'-i-cal) Said of a radiolarian shell formed by fusion of the bases of two cones opposite in one axis (TIP, 1954, pt. D, p.14).

dip log *dipmeter.*

diplogenesis (dip-lo-gen'-e-sis) Formation of a mineral deposit by partly syngenetic and partly epigenetic processes, such that some mineral components are original, some added. An example is fluorite replacing limestone by addition of fluoride ions (Guilbert and Park, 1986, p.303).

diplogenetic (dip'-lo-ge-net'-ic) Said of a mineral deposit that is in part syngenetic and in part epigenetic in origin (Lovering, 1963, p.315-316).

diplohedron (dip-lo-he'-dron) *diploid.*

diploid (dip'-loid) n. A crystal form of the isometric system having 24 similar quadrilateral faces in a paired arrangement. Each face intersects the crystallographic axes at unequal lengths. Its indices are *{hkl}* in symmetry $2/m\bar{3}$. Syn: *didodecahedron; diplohedron; dyakisdodecahedron.*

diploidal class (dip-loid'-al) That crystal class in the isometric system having symmetry $2/m\bar{3}$.

diplopore (dip'-lo-pore) Any of double pores piercing a thecal plate in diploporan echinoderms and mostly confined to that plate. It may be unbranched but it usually consists of a Y-shaped branching canal or tube that is oblique or perpendicular to the surface of the plate and that has two external openings. Cf: *pore rhomb; haplopore.*

Diplorhina (Di-plo-rhi'-na) A subclass of Agnatha characterized presumably by paired nostrils and assuredly by uniquely acellular bone.

diploxylonoid (dip-lo-xy'-lo-noid) Said of bisaccate pollen, in which the outline of the sacci in distal-proximal view is discontinuous with the body outline so that the grain appears to consist of three distinct, more or less oval figures. Cf: *haploxylonoid.*

dip magnitude map A horizon slice that displays the dip magnitude, often by color coding. Also called: *dip map*. Cf: *azimuth map*.
dip map *dip magnitude map*.
dipmeter (dip′-me-ter) A 3-4 or 6-pad *wall-contact log* whose finely detailed *microresistivity log* curves are correlated to measure depth offsets relative to each other. In conjunction with simultaneous measurements of the caliper, inclination, and direction of the borehole, such measurements can be solved for dip and strike of the strata. Both the borehole curves as measured and the subsequent graphic plot of computed dip-strike symbols are called dipmeters, the former a "continuous dipmeter" or dipmeter log, the latter a "computed dipmeter" or "tadpole plot".
dip moveout A change in the arrival time of a reflection because of dip of the reflector. See also: *DMO processing*.
dip needle An obsolete type of magnetometer used for mapping high-amplitude magnetic anomalies. It consists of a magnetized needle pivoted to rotate freely in a vertical plane, with an adjustable weight on the south side of the magnet.
Dipnoi (Dip′-noi) An order of lobefinned bony fish characterized by nonhinged braincase, holostylic jaw suspension, and teeth distinctively modified into durophagous crushing plates, one up and one down on each side. Syn: *lungfish*. Range, Lower Devonian to Holocene.
dipole (di′-pole) Two poles of opposite charge an infinitesimal distance apart.
dipole-dipole array An electrode array in which one dipole provides current to the ground and an adjacent dipole is used to measure a potential difference. The separation between dipoles is usually comparable to or greater than the spacing within each electrode pair constituting a dipole. The potential dipole lies entirely outside the current dipole.
dipole field (a) A mathematically simple magnetic field, having an axis of symmetry, with magnetic-field lines pointing outward along one half of the axis (*positive pole*) and inward along the other half (*negative pole*). Most magnetic fields that are sufficiently remote from their source resemble a dipole field. (b) The field of a dipole located at (or near) the center of the Earth that approximates the actual Earth's magnetic field. See also: *axial dipole field*.
dipole moment (di′-pol) For a pair of equal charges or poles of opposite signs that ideally are infinitesimally close together, the product of the charge and the separation distance. Dipole moment per unit volume is a measure of the intensity of polarization of a material. Units are ampere-meters per cubic meter.
dipole sonic log [geophys exploration] A sonic array tool in which sources are dipoles as opposed to radially symmetric monopoles, so that it generates both P and S waves in formations. Dipoles are directional and are sensitive to azimuthal velocity variations.
diporate (di-por′-ate) Said of pollen grains having two germinal pores.
dipping rod (dip′-ping) *divining rod*.
dip plain A *stratum plain* coincident in slope with the dip of the underlying resistant rock (Hill, 1891, p.522).
dip pole One of the locations on the Earth where the horizontal magnetic field is zero and the magnetic inclination is ±90°. Cf: *magnetic poles*.
dip separation The distance or *separation* of formerly adjacent beds on either side of a fault surface, measured in the dip direction of the fault. Cf: *dip slip; strike separation*. See also: *dip-separation fault*.
dip-separation fault A little-used term for a fault on which the displacement has been *dip separation*. Cf: *lateral fault*.
dip shooting A system of seismic surveying in which the primary concern was determining the dip and position of reflecting interfaces rather than in tracing such interfaces continuously. It usually involved the use of cross spreads.
dip slip In a fault, the component of the movement or slip that is parallel to the dip of the fault. Cf: *dip separation; strike slip; oblique slip*.
dip-slip fault A fault on which the movement is parallel to the dip of the fault. Cf: *strike-slip fault*.
dip slope A slope of the land surface, roughly determined by and approximately conforming with the direction and the angle of dip of the underlying rocks; specif. the long, gently inclined face of a cuesta. Cf: *scarp slope*. Syn: *back slope; outface*.
dip stream A consequent stream flowing in the general direction of dip of the strata it traverses; a *cataclinal* stream.
dip throw The component of the slip of a fault measured parallel with the dip of the strata.
dip valley A valley trending in the general direction of dip of the strata of a region; a valley at right angles to a subsequent stream.
dipyramid (di-pyr′-a-mid) A closed crystal form consisting of two *pyramids* that are arranged base to base so that they appear as mirror images across the plane of symmetry. Adj. *dipyramidal*. Syn: *bipyramid*.
dipyramidal (di-py-ram′-i-dal) Having the symmetry of a *dipyramid*.
dipyre (di′-pyre) A syn. of *mizzonite;* specif. a term applied to a variety of scapolite with the components marialite and meionite in a ratio of about 3:1 to 3:2. Syn: *dipyrite*.
dipyrite (di-py′-rite) (a) *dipyre*. (b) *pyrrhotite*.
direct-action avalanche (di-rect′-ac′-tion) A snow avalanche that occurs during or immediately after a storm and is the direct result of that storm. The term is part of an obsolete classification of avalanches.
direct angle (di-rect′) An angle measured directly between two lines; e.g. a horizontal angle measured clockwise from a preceding surveying line to a following one.
direct intake Recharge to the aquifer directly through the *saturated zone*.
direction (di-rec′-tion) (a) The position of one point relative to another without reference to the distance between them. It may be three-dimensional or two-dimensional. (b) The angle between a line or plane and an arbitrarily chosen reference line or plane; specif. the angle between a great circle passing through both the position of the observer and a given point on the Earth's surface and a true north-south line passing through the observer. When the reference line is north and the angle is designated east or west, the direction is called the *bearing;* when the reference line is south and the angle is reckoned clockwise, the direction is called the azimuth.
directional charge [geophys exploration] An explosive charge, charge array, or shaped charges in which the explosion front travels at approximately the velocity of seismic waves in the surrounding rock, so that energy traveling in a desired direction (usually vertically) adds constructively as opposed to that traveling in other directions. The charge must have a length that is a significant fraction of a wavelength to achieve appreciable directivity.
directional data Observations expressed as vectors in a circular or spherical coordinate system. Directional data may be measured in degrees, angles, strikes and dips, or equivalent systems.
directional drilling (di-rec′-tion-al) The intentional drilling of a well at controlled departures from the vertical and at controlled azimuths, often utilizing a *whipstock* or a special downhole assembly with a bent (angled) subassembly and a turbine motor located close to the drill bit. It is done to establish multiple wells from a single location such as an offshore platform, and for other purposes. Cf: *deviation [drill]; sidetracking*.
directional hardness The tendency for the hardness of a mineral to vary with crystallographic direction. Hardness depends on the number and strength of atomic bonds, which are usually greater within a crystal plane than between planes. Diamond's hardest polishing direction is parallel to octahedral planes; the easiest is on the dodecahedral face. A direction of intermediate hardness, which is also used in polishing, is parallel to a cubic face.
directional load cast A term originally applied to a structure interpreted as a flowage cast, but now regarded as a flute cast (Pettijohn and Potter, 1964, p.301).
directional log A *well log* that shows the inclination of a borehole, and the direction of the inclination. It is usually obtained with the *dipmeter* log.
directional structure Any sedimentary structure that indicates the direction of the current that produced it; e.g. *cross-bedding, current marks, and ripples*. Syn: *paleocurrent structure; aligned current structure; vector structure*.
directional survey (a) Determination of the direction and *deviation* from the vertical of a borehole by precise measurements at various points. Also, the record of the information thus obtained. Syn: *borehole survey*.
directional well A well that is not vertical and was drilled by *directional drilling*.
direction instrument theodolite A *theodolite* in which the graduated horizontal circle remains fixed during a series of observations, the telescope being pointed on a number of signals or objects in succession, and the direction of each read on the circle, usually by means of micrometer microscopes. Instrument theodolites are used almost exclusively in first- and second-order triangulation. Syn: *direction theodolite*.
direction of dip *line of dip*.
direction of the wind *wind direction*.
direction theodolite *direction instrument theodolite*.
directive couple (di-rec′-tive) A pair of mesenteries in the so-called dorsoventral plane (extending from the dorsal toward the ventral side) of a coral *polyp*, characterized by pleats on the opposite rather than the facing sides of the mesenteries.
directivity graph (di-rec-tiv′-i-ty) (a) A plot of relative intensity versus direction of an outgoing seismic wave, such as that resulting from a directional charge or from a source pattern. The directivity results from the interference of waves from the various components of the pattern. (b) A plot of the directivity response of a geophone pattern or of directivity resulting from mixing.
direct lattice A syn. of *crystal lattice*, used when comparison is made with the *reciprocal lattice*.
direct leveling A type of *leveling* in which differences of elevation are determined by means of a continuous series of short horizontal lines, the vertical distances from these lines to adjacent ground marks being determined by direct observations on graduated rods with a leveling instrument equipped with a spirit level. Cf: *indirect leveling*.
direct linkage A type of linkage in scleractinian corals with mesenterial strands connecting the adjacent stomodaea. See also: *lamellar linkage*. Cf: *indirect linkage*.
direct methods A class of techniques for obtaining an approximate *crystal-structure solution* in X-ray crystallography by making use of relationships among the phases of different reflections to assign trial phases to them.
direct position computation On an ellipsoid, given the geodetic coordinates of a point P_1 and the azimuth and the length of the geodesic from it to a

point P_2, the geodetic coordinates of P_2 and the azimuth from P_2 to P_1 can be computed.

direct-replacement nonsulfide zinc deposit A zinc deposit formed by replacement of primary sulfide or zinc silicate minerals by *hemimorphite* and *smithsonite* (Hitzman et al., 2003).

direct runoff The runoff reaching stream channels immediately after rainfall or snow melting (Langbein and Iseri, 1960). Cf: *base runoff.* Syn: *direct surface runoff; immediate runoff; stormflow; storm runoff; storm water.*

direct space A mathematical space defined by the coordinate system of the *direct lattice.* Cf: *reciprocal space.*

direct stratification *primary stratification.*

direct surface runoff *direct runoff.*

direct wave [seis] A wave which travels directly by the shortest path. Other waves traveling by longer routes may arrive earlier because they travel at higher velocity.

Dirichlet's problem (Di-rich'-let's) A classical problem in geodesy: to determine a function that is harmonic outside of a given surface and assumes prescribed boundary values on the surface. Cf: *boundary-value problem; Neumann's problem.*

dirt band [coal] A thin stratum of shale or other inorganic rock material in a coal seam. Syn: *band [coal]; dirt bed; dirt parting; stone band.*

dirt band [glaciol] (a) Any dark layer in a glacier, usually the trace of silt or debris along a *summer surface.* (b) A dark band below an *icefall* that may be related to dirt collected in the broken ice of the icefall or between the ridge of one *ogive* and another. Cf: *dust band.* (c) A term that was originally applied to a *Forbes band.* Cf: *dirt-band ogive.*

dirt-band ogive A curved band or *ogive* composed of debris-laden or dirt-laden ice that may be related to dirt collected in the broken ice of an icefall or between the ridge of one *wave ogive* and another. Cf: *dirt band [glaciol]; Forbes band.*

dirt bed *dirt band [coal].*

dirt cone A cone or mound of ice or snow on a glacier, covered with a veneer of silt thick enough to protect the underlying material from the ablation that has lowered the surrounding surface. Cf: *debris cone [glaciol].*

dirt parting *dirt band [coal].*

dirt slip *clay vein.*

dirty arkose *impure arkose.*

dirty sand A term used in electrical prospecting for a sandstone that contains abundant clay and hence can exhibit appreciable membrane polarization and abnormally high electrical conductivity because of surface conduction along the clay minerals.

dirty sandstone A sandstone containing much matrix; specif. a *wacke* with more than 10% argillaceous matrix or a *graywacke* with more than 15% detrital clay matrix (Pettijohn, 1954). The particles are held together by primary or secondary, fine-grained interstitial detritus of clay-like nature, or by authigenic derivatives of such material. Cf: *clean sandstone.*

disaccate (di-sac'-cate) *bisaccate.*

disaggregation (dis'-ag-gre-ga'-tion) Separation or reduction of an aggregate into its component parts; specif. *mechanical weathering.*

disappearing stream (dis-ap-pear'-ing) *sinking stream.*

disaster species A blossom of opportunistic species immediately after a mass extinction.

disc (a) The central part of the body of an echinoderm, especially in asteroids and ophiuroids, more or less distinctly separable from its arms. Sometimes spelled disk. (b) A discoidal, typically imperforate *sclerite* of a holothurian.

discharge [hydraul] (dis'-charge) The rate of flow of surface water or groundwater at a given moment, expressed as volume per unit of time. See also: *specific discharge.*

discharge [sed] *sediment discharge.*

discharge area An area in which subsurface water, including both groundwater and vadose water, is discharged to the land surface, to bodies of surface water, or to the atmosphere. Cf: *recharge area.*

discharge coefficient That coefficient by which a theoretical discharge must be multiplied to obtain the actual discharge. It is the product of the contraction coefficient and the velocity coefficient (ASCE, 1962).

discharge efficiency *drainage ratio.*

discharge-rating curve *stage-discharge curve.*

discharge velocity The rate of discharge of water through a porous medium, measured per unit of total area perpendicular to the direction of flow (ASCE, 1962).

discharge zone Region of a lake bottom where groundwater discharges into a lake.

disciform (dis'-ci-form) Of round or oval shape; e.g. "disciform holococcolith" having a discolithlike shape and a raised margin two or more cycles of microcrystals high.

discinacean (dis-ci-na'-cean) Any inarticulated brachiopod belonging to the superfamily Discinacea, characterized by holoperipheral growth of the brachial valve, with the beak marginal to central.

discoaster (dis'-co-as'-ter) One of the tiny star- or rosette-shaped calcareous plates, 10-35 micrometers in diameter, that are generally believed to be the remains of a planktonic organism and that may be isolated coccolithlike bodies of either a motile cell or a cyst. Discoasters are common in Tertiary deposits but are apparently absent in the Pleistocene; the level where they disappear has been suggested as a Pliocene-Pleistocene boundary. See also: *asterolith.*

discohexaster (dis-co-hex'-as-ter) A hexactinal sponge spicule (microsclere) in which the ray tips bear branches terminated by umbels.

discoid (dis'-coid) adj. Having the shape of a disk; e.g. a solitary coral-lite. Syn: *discoidal.* n. An object having such a shape.

discoidal (dis-coi'-dal) *discoid.*

discoidal system A biconvex, flattened crystal habit of gypsum precipitated in sediment that has been attributed to inhibition of growth perpendicular to the c-axis by organic matter (Cody, 1979).

discolith (dis'-co-lith) A discoidal *coccolith* with a single, apparently imperforate elliptical or circular shield and a thickened margin. Cf: *tremalith.*

disconformable [intrus rocks] (dis-con-form'-a-ble) Said of an intrusive igneous contact which does not have the same attitude as that of the intrusion's internal structures. Cf: *conformable [intrus rocks]; discordant [intrus rocks].*

disconformable [stratig] Pertaining to a *disconformity.* Term proposed by Grabau (1905, p.534) to refer to formations that exhibit parallel bedding but "comprise between them a time break of greater or less magnitude".

disconformity (dis-con-form'-i-ty) An *unconformity* in which the bedding planes above and below the break are essentially parallel, indicating a significant interruption in the orderly sequence of sedimentary rocks, generally by a considerable interval of erosion (or sometimes of nondeposition), and usually marked by a visible and irregular or uneven erosion surface of appreciable relief; e.g. an unconformity in which the older rocks remained essentially horizontal during erosion or during simple vertical rising and sinking of the crust (without tilting or faulting). The tendency is to apply the term to breaks represented elsewhere by rock units of at least formation rank (Stokes and Varnes, 1955, p.157). The term formerly included what is now known as *paraconformity.* Syn: *parallel unconformity; erosional unconformity; nonangular unconformity.*

discontinuity [seis] (dis'-con-ti-nu'-i-ty) A surface at which seismic-wave velocities abruptly change; a boundary between seismic layers of the Earth. Syn: *interface [seis]; seismic discontinuity; velocity discontinuity.*

discontinuity [stratig] Any interruption in sedimentation, whatever its cause or length, usually a manifestation of nondeposition and accompanying erosion; an unconformity. Syn: *break [stratig].*

discontinuity [struc geol] A surface separating two unrelated groups of rocks; e.g. a fault or an unconformity. See also: *discrete [struc geol].*

discontinuity (410-km discontinuity) A seismic discontinuity resulting from the transition of alpha olivine to beta olivine.

discontinuity (520-km discontinuity) A seismic discontinuity resulting from the transition of beta olivine to gamma spinel.

discontinuity (660-km discontinuity) A seismic discontinuity resulting from the transition of gamma spinel to perovskite.

discontinuity surface A minor break in the sedimentary column, chiefly intraformational but including interformational junctions that have not involved large-scale erosion. It is lower in rank than disconformity (Bromley, 1975). Cf: *omission surface.*

discontinuous deformation (dis-con-tin'-u-ous) (a) Deformation involving the loss of continuity at the scale of observation. Cf: *continuous deformation.* (b) Deformation with development of kinematic discontinuities, e.g., fractures or cleavage planes.

discontinuous gully A gully with a vertical headcut at the upstream end and a fan at the point where its floor intersects the more steeply sloping plane of the original valley floor (Leopold et al., 1964, p.448-449). The depth of its channel decreases downstream.

discontinuous permafrost A zone of permafrost containing patches of unfrozen ground, as beneath large rivers or lakes; it occurs in an intermediate zone between the northerly *continuous permafrost* and the southerly *sporadic permafrost.*

discontinuous reaction A metamorphic reaction that, despite compositional variability among the minerals, is at equilibrium, given a fixed pressure, at only one temperature. A paragenetic diagram, for example the *AFM projection,* shows a change in topology, such as a "tie line flip". Cf: *continuous reaction.*

discontinuous reaction series A *reaction series* in which reaction of early-formed crystals with later liquid represents an abrupt phase change; e.g., the minerals olivine, pyroxene, amphibole, and biotite form a discontinuous reaction series. Cf: *continuous reaction series.*

discontinuous transformation A *phase transformation* in which nucleation of a new phase occurs at discrete locations within the host, or parent, phase, and the new phase then grows at the expense of the host phase. Cf: *continuous transformation.*

discordance (dis-cord'-ance) (a) Lack of parallelism between adjacent strata. The term was used by Willis (1893, p.222) in cases where the cause is in doubt. Although the term has not been widely adopted, "it appears eminently suitable for descriptive use where there is insufficient evidence to decide between stratigraphic discordance (unconformity) and tectonic discordance (e.g. overthrust, slide, detachment fault)" (Dennis, 1967, p.36). Cf: *concordance.* (b) *angular unconformity.*

discordance index A numeric statistic used by Pearn (1964, p.401) to represent the amount of deviation of any actual rock sequence from the *ideal cyclothem*. It is defined as the minimum value of the number of missing lithologic units. Symbol: G.

discordant [geochron] (dis-cord'-ant) (a) Said of *isotopic ages,* determined by more than one method for the same sample or coexisting minerals, that are in disagreement beyond experimental error. (b) Said of ages given by coexisting minerals determined by the same method that are in disagreement. (c) In a more restricted sense, the term has been used to indicate disagreement of $^{238}U/^{206}Pb$, $^{235}U/^{207}Pb$, $^{207}Pb/^{206}Pb$, and $^{232}Th/^{208}Pb$ ages determined for the same mineral sample (mainly zircon). The term usually implies that one (or more) of the isotopic systems used for dating purposes has been disturbed by some geologic event (metamorphism, weathering) following the initial formation of the geologic material or by inadvertent laboratory procedures. Ant: *concordant [geochron].*

discordant [geomorph] Said of topographic features that do not have the same or nearly the same elevation; e.g. a discordant valley whose stream enters the main stream via a waterfall, or a discordant lip over which the floor of a hanging valley passes into the floor of the main valley. Ant: *accordant.*

discordant [intrus rocks] Said of an igneous intrusion possessing contacts which are not parallel to foliation or bedding in the country rock. Cf: *concordant [intrus rocks]; disconformable [intrus rocks].*

discordant [stratig] Structurally *unconformable;* said of strata lacking conformity or parallelism of bedding or structure. Ant: *concordant [stratig].*

discordant ages Ages given by various dating methods (esp. *uranium-lead age method*) for the same igneous or metamorphic body. They are the result of thermal and/or dynamic changes that affected the body at some time after its formation. Cf: *hybrid age; overprint [geochron].* Syn: *mixed ages.*

discordant bedding A sedimentary structure in which parallelism of beds is lacking or in which sedimentary layers are inclined to the major lines of deposition, such as bedding developed by rapid deposition of material from heavily laden currents of air or water; specif. *cross-bedding.* Ant: *concordant bedding.* See also: *inclined bedding.*

discordant coastline A coastline that develops where the general structural grain of the land (such as mountain chains or folded belts) is transverse to the margin of the ocean basin, and that represents rifting, faulting, subsidence, or other interruption of a formerly continuous and harmonious structure; it is generally irregular, with many inlets. Ant: *concordant coastline.* Syn: *Atlantic-type coastline; transverse coastline.*

discordant drainage Drainage that has not developed in a systematic relationship with, and is not consequent upon, the present geologic structure. Ant: *accordant drainage.*

discordant fold A fold whose axis is inclined to that of the *longitudinal fold* axes of the area.

discordant junction The joining of two streams or two valleys whose surfaces are at markedly different levels at the place of junction, as the abrupt entry of a tributary flowing at a high level into a main stream at a lower level. Ant: *accordant junction.*

discordant margin A margin of closed valves (of a bivalve mollusk) not in exact juxtaposition, but with one overlapping the other.

discordia (dis-cor'-di-a) A line on a plot of $^{206}Pb/^{238}U$ versus $^{207}Pb/^{235}U$, formed by data on *phases* that have lost lead or gained uranium during a period of time that is short compared with the age of the phase. Also a mixture of ages within a phase.

discordogenic fault (dis'-cord-o-gen'-ic) A fault in a tectonic belt that separates zones of uplift and subsidence, and that remains active during several geologic periods (Nikolaev, 1959). Obsolete and little used.

discorhabd (dis'-co-rhabd) A sponge spicule (streptaster) consisting of a straight shaft bearing whorls of spines or transverse discoidal flanges. Syn: *chessman spicule.*

discotriaene (dis-co-tri'-aene) A sponge *triaene* in which the cladome is represented by a transverse disk containing the axial canals of the three cladi.

discovery (dis-cov'-er-y) The actual finding of a valuable mineral, indicative of a deposit (lode, placer, or coal seam). Legally, a discovery is a prerequisite to making a mining claim on an area.

discovery claim A claim containing the original discovery of exploitable mineral deposits in a given locale, which may lead to claims being made on adjoining areas.

discovery process model A method of estimating undiscovered mineral resources that assumes the probability of discovering a mineral deposit of a given size is proportional to the ratio of that deposit size to the size of an area permissive for the discovery of deposits of that type.

discovery rate The rate of discovery of mineral deposits of a given type measured either as discoveries per unit of time or per unit of effort (e.g., money expended, footage drilled).

discovery well The first well to encounter gas or oil in a hitherto unproven area or at a hitherto unproductive depth; a successful *wildcat, outpost well, deeper-pool test,* or *shallower-pool test.*

discrepancy (dis-crep'-an-cy) In surveying, the difference in computed values of a quantity obtained by different processes using data from the same survey; also, the difference between results of duplicate or comparable measures of a quantity. See also: *accumulated discrepancy.*

discrepancy index *R-factor.*

discrete [paleont] (dis-crete') Said of conodont denticles that are not closely set, each denticle being separated from adjacent ones by open space. Ant: *fused.* Cf: *appressed [paleont].*

discrete [struc geol] Said of any body of rock which has a definite boundary with adjacent rocks in space. See also: *discontinuity [struc geol].*

discrete [weath] A term proposed by Gilbert (1898) to describe the surficial, weathered, and unconsolidated material composing the regolith.

discrete element A conodont element that has become dissociated from the skeletal apparatus of which it was once a part (TIP, 1981, pt. W, supp.2).

discrete-film zone *belt of soil water.*

discrete spectrum In *harmonic analysis* of time series, a plot of power or variance at discrete frequencies or harmonics.

discrete variable A discontinuous variable which can only assume discrete values.

discriminant analysis (dis-crim'-i-nant) A statistical procedure for classifying samples in categories previously defined and differentiated on the basis of samples from known populations. Syn: *discriminant function analysis.*

discriminant function analysis A multivariate statistical procedure for classifying observations into categories that have been previously defined and differentiated based on the characteristics of samples from known populations. Syn: *discriminant analysis; DFA.*

disembogue (dis-em-bogue') To discharge water through an outlet or into another body of water, such as a stream disemboguing into the ocean. Cf: *bogue.*

disequilibrium assemblage (dis'-e-qui-lib'-ri-um) An association of minerals not in thermodynamic equilibrium.

disequilibrium system A geomorphic system that tends toward equilibrium but has not attained a steady state (Renwick, 1992). Cf: *nonequilibrium system.*

disharmonic fold (dis-har-mon'-ic) A fold that varies noticeably in profile form and wavelength in the various layers through which it passes. Ant: *harmonic fold.*

disharmonic folding Folding in which there is an abrupt change in *fold profile* when passing from one folded surface or layer to another. It is characteristic of rock layers that have significant contrasts in viscosity (Whitten, 1966, p.606). An associated structure is *décollement.* Ant: *harmonic folding.*

dish structure A primary sedimentary structure, generally found in sandstone, consisting of small meniscus-shaped lenses (4-50 cm long and one to a few centimeters thick) that are oval in plan, oriented parallel to the bedding, and defined by slightly finer-grained, concave-up bottoms each of which truncates the underlying lenses. It is thought to form as a result of elutriation of clay by pore water escaping soon after deposition of the sand. Term introduced by Wentworth (1967); see also Lowe and LoPiccolo (1974).

disintegration [coal] (dis'-in-te-gra'-tion) The decomposition of vegetable matter by slow combustion, in which there is no formation of carbon compounds and in which only volatile substances (carbon dioxide and water) are produced. Cf: *peat formation; putrefaction.*

disintegration [glaciol] *ice disintegration.*

disintegration [radioactivity] *radioactive decay.*

disintegration [weath] A syn. of *mechanical weathering.* Sometimes the term includes chemical action, in which case it is practically synonymous with *weathering.*

disintegration constant *decay constant [radioactivity].*

disintegration rate The rate of decay of a quantity of radionuclides measured as the number of atoms decaying per unit time.

disjunct (dis'-junct) (a) Said of a rhombiferan echinoderm pore rhomb in which externally visible slits forming parts of the rhomb are separated by solid areas of plates. Cf: *conjunct.* (b) Said of the apical system of an echinoid whose anterior part is separated from its posterior part.

disjunctive cleavage Cleavage where the domainal nature may be seen with the unaided eye. Cf: *crenulation cleavage; slaty cleavage.*

disjunctive fold (dis-junc'-tive) A fold in which the more brittle strata have fractured and separated and the more plastic beds have flowed under the forces of deformation.

disk [paleont] (a) The flattened circumoral part of a cnidarian (such as a sea anemone). See also: *oral disk; basal disk.* (b) *disc.*

disk [sed] A notably discoidal (flat and circular), or oblate or tabular, shape of a sedimentary particle, defined in *Zingg's classification* as having a width/length ratio greater than 2/3 and a thickness/width ratio less than 2/3.

disk hardness gage A device for measuring the penetration resistance of snow. Metal disks of various sizes are pressed against a snow surface by a calibrated spring until collapse occurs.

disk-tension infiltrometer A *constant-head permeameter* designed to estimate hydraulic parameters of unsaturated soil near the land surface by allowing *infiltration* at a pressure head that is less than zero (i.e., under positive moisture suction). Syn: disk-tension permeameter.

dislocation [cryst] (dis-lo-ca'-tion) *line defect.*

dislocation [struc geol] A general term for a displacement caused by movement on a fault.

dislocation [struc petrol] A linear array of crystallographic lattice imperfections. A dislocation bounds an area of a crystallographic lattice that has moved, or slipped, relative to the rest of the lattice. The crystallographic lattice is distorted around dislocations, resulting in *undulose extinction.*

dislocation breccia A little-used term for *fault breccia.*

dislocation climb Movement of a dislocation to a higher or lower glide plane. Requires diffusion of atoms through the crystal lattice which is favored by deformation at high temperatures.

dislocation creep Grain-scale, ductile deformation accomplished by the motion of dislocation, including both glide and climb.

dislocation glide The movement of a *dislocation [struc petrol]* on a *glide plane [struc petrol].*

dislocation metamorphism A form of dynamic regional metamorphism concentrated along narrow belts of shearing or crushing without an appreciable rise in temperature. The term was originated by Lossen in 1883 and is considered to be equivalent to *dynamometamorphism.* Little used. Cf: *dynamic metamorphism; cataclastic metamorphism.*

dismal (dis'-mal) n. *pocosin.*

dismembered drainage (dis-mem'-bered) A complex drainage system that has been altered by *dismembering,* thus creating a series of independent streams that enter the sea by separate mouths.

dismembered formation A body of rock mappable at a scale of 1:24,000 or smaller that lacks continuity of internal contacts or strata and is characterized by the inclusion of native blocks and fragments of all sizes in a matrix of finer-grained material (Raymond, 1984). Cf: *endolistostrome.* See also: *mélange [sed]; broken formation.*

dismembered stream A tributary that is left as an independent stream after the lower part of the drainage system to which it formerly belonged was submerged by an invasion of the sea, e.g. tributaries to Chesapeake Bay. Cf: *betrunked river.*

dismembering (dis-mem'-ber-ing) The making of a tributary into an independent stream by a change of geologic conditions, esp. by the submergence of the lower part of a valley by the sea. Cf: *betrunking.*

dismicrite (dis-mic'-rite) A fine-textured limestone with less than 1% allochems, consisting mainly of lithified carbonate mud (micrite), and containing irregular patches or bird's-eyes of sparry calcite filling cavities generally formed by desiccation (Folk, 1959, p.28). Syn: *bird's-eye limestone.*

disomatic (di-so-mat'-ic) Obsolete term that used to be applied to a crystal now called a *xenocryst.*

disorder in minerals (dis-or'-der) In a substitutional solid solution, the random occupation of one atom site in a crystal by two or more different atoms of similar size and charge, or of similar size and different charge if there is a concomitant substitution to balance charges, as in plagioclase, in which (Na and Si) in albite is substituted by (Ca and Al) as the composition approaches anorthite. Cf: *order in minerals.*

dispersal [ecol] (dis-per'-sal) The spreading of a species by migration into new areas having conditions favorable for its existence. Cf: *dispersion [ecol].*

dispersal [glac geol] *glacial dispersal.*

dispersal center The place on a delta at which the first stream distributary branches off from the main channel (Moore, 1966, p.92).

dispersal map A stratigraphic map that shows the inferred source area and the direction or distance of transportation of clastic materials (Krumbein and Sloss, 1963, p.484).

dispersal shadow An accumulation of sediments formed downcurrent from a generating source (Pettijohn, 1957, p.574); e.g. a boulder train on the lee side of a resistant knob overridden by ice. Cf: *crag and tail; sedimentary petrologic province.*

dispersed element (dis-persed') An element that is generally too rare and unconcentrated to become an essential constituent of a mineral, and that therefore occurs principally as a substituent of the more abundant elements.

dispersed phase Colloidal material suspended in another phase, which in turn is known as the *dispersion medium.*

dispersion [ecol] (dis-per'-sion) The pattern of geographic or spatial distribution of individuals within a species; may be uniform, random or irregular, or aggregated or clumped. Cf: *dispersal [ecol].*

dispersion [gem] The property of a transparent gemstone to separate white light into the spectral colors and is measured as the difference in indices of refraction of red (Frauenhofer line B) and violet (Frauenhofer line G) light.

dispersion [grd wat] The spread of a solute in flowing groundwater along and perpendicular to the mean flow path by the combined processes of *diffusion* and physical mixing of the groundwater with water that has lower concentration of the solute. Cf: *longitudinal dispersion, transverse dispersion; dispersion length.*

dispersion [optics] The differences in the optical constants, e.g. *optic angle* and indices of refraction, of a given mineral for different wavelengths of the spectrum. See also: *dispersion curve [optics].*

dispersion [seis] (a) Distortion of the shape of a seismic-wave train because of variation of velocity with frequency. The peaks and troughs may advance toward or recede from the beginning of the wave as it travels. (b) Angular dispersion results from the change of velocity with direction because of anisotropy.

dispersion [soil] Breaking down or separation of soil aggregates into single grains. Cf: *detachment [soil].*

dispersion [stat] The range or scatter of values about a central tendency; a statistical spread or variability. Common measures of dispersion are *standard deviation* and sorting.

dispersion [tect] Tectonic disruption that results in large-scale horizontal separation of terranes (Howell, 1995).

dispersion [waves] Tendency of waves of different wavelength or period to have different *phase velocity* and *group velocity. Shallow-water waves* are non-dispersive, *deep-water waves* are dispersive.

dispersion coefficient The product of *dispersion length* and mean flow velocity of groundwater.

dispersion curve [optics] The plotting on a logarithmic scale of a crystal's dispersion.

dispersion curve [seis] A plot of seismic-wave velocity versus frequency or period. Often both phase velocity and group velocity are plotted.

dispersion flow Flow of granular sediment in which collisions between particles maintain the fluidity of the material.

dispersion length A geometric attribute of a porous medium that pertains to its ability to spread or disperse constituents dissolved or suspended in groundwater. The factor which, when multiplied by the average linear velocity, estimates the dispersivity coefficient. Syn: *dispersivity.*

dispersion medium That material (solid, liquid, or gas) in which colloidal material, known as the *dispersed phase,* is suspended.

dispersion pattern The pattern of distribution of chemical elements, especially trace elements, in the wall rocks of an orebody or in the surface materials surrounding it. Cf: *halo.*

dispersive power (dis-per'-sive) The refractive ability of a transparent substance, usually isotropic; it is symbolized by δ and equals n_F-n_C/n_{D-1}, in which n is the refractive index for Fraunhofer lines C, F, and D. (C = hydrogen discharge at wavelength 656.3, F = hydrogen discharge at wavelength 486.1, and D = sodium flame at wavelength 589.3, measured in nanometers.) Syn: *relative dispersion.*

dispersivity *dispersion length.*

disphenoid (di-sphe'-noid) A closed crystal form consisting of two sphenoids, in which the two faces of the upper sphenoid alternate with those of the lower. Adj: *disphenoidal.* Syn: *bisphenoid.*

disphenoidal (di-sphe-noi'-dal) Having the symmetry of a disphenoid, e.g. rhombic 222.

disphotic zone (dis-phot'-ic) The zone in the ocean between depths of about 80 and 200 meters, where penetration of light is so low as to severely limit the rate of photosynthesis. It is the zone of transition between the *euphotic* zone above and the *aphotic* zone below. Syn: *twilight zone.*

displacement [photo] (dis-place'-ment) Any shift in the position of an image on a photograph that does not alter the perspective characteristics of the photograph. It may be caused by the relief of the objects photographed, the tilt of the photograph, changes of scale, or atmospheric refraction. Cf: *distortion [photo].*

displacement [struc geol] (a) A general term for the relative movement of the two sides of a fault, measured in any chosen direction; also, the specific amount of such movement. Syn: *dislocation.* (b) *net slip; slip.* (c) The amount and direction of translation, going from the undeformed state to the deformed state.

displacement gradient A mathematical expression that describes how displacement vectors vary from point to point in a body of rock.

displacement meter A seismometer designed to respond to the displacement of earth particles.

displacement parameters A number or set of numbers expressing the extent of thermal vibrations of an atom in a *crystal structure,* or alternatively, the average static positional disorder of equivalent atoms in different unit cells. Isotropic displacement parameters are single numbers that express the average magnitude of the thermal displacement or static positional disorder, modeling as spherical the effective volume of the vibrating or disordered atom. Anisotropic displacement parameters model the vibrational volume as a triaxial ellipsoid and consist of six numbers, three of which relate to the magnitudes of vibration along the axes of the ellipsoid and three of which relate to the orientation of the ellipsoid. Displacement parameters are determined during a *crystal-structure refinement* and are the coefficients in the exponential expressions for the temperature factors. Syn: *temperature factor coefficients; vibrational parameters; thermal parameters.* Cf: *temperature factor.*

displacement pressure The minimum capillary pressure required to force a nonwetting fluid into capillary openings in a porous medium saturated with a wetting fluid; specif., to force oil or gas from one water-filled pore to the next. Syn: *entry pressure.*

displacement shear A fracture surface that often occurs in the marginal zones of earthflows. It develops on shearing, parallel or subparallel to the direction of relative movement.

displacement vector A line joining the initial position of a material point with its position after deformation.

displacive crystals A saline mineral occurrence wherein crystals push aside sediment as they grow. Rapid growth permits incorporation of matrix along advancing crystal faces, but slow growth pushes all sediment to the side.

displacive transformation (dis-plac′-ive) A high-low type of crystal transformation that involves no breaking of bonds, e.g. in high-low quartz at 573°C, involving rotation of SiO_4 tetrahedra. It is usually a rapid transformation. Cf: *dilatational transformation; reconstructive transformation; rotational transformation; substitutional transformation.*

disrupted (dis-rupt′-ed) Said of a metamorphic texture in which some phase or phases occur "in discontinuous stringers or groups of relatively small grains (partially separated by other minerals) that appear to have formed by disaggregation (with or without recrystallization) of initially larger grains. Although the disrupted grains may approach a lenticular shape, it is not intended to include this texture within the meaning of laminated" (Harte, 1977). In the past, the term has been used mainly with reference to the distribution and size of spinel or garnet in peridotites.

disrupted terrane A terrane characterized by blocks of heterogeneous lithology and age, usually set in a matrix of sheared shale, flysch, or serpentinite (D.L. Jones et al., 1983).

dissakisite-(Ce) A pale yellow-brown monoclinic mineral of the *epidote* group: $Ca(Ce,La)MgAl_2(Si_2O_7)(SiO_4)(O,OH)_2$.

dissected crossed prismatic structure A *crossed structure* with one or two predominant oblique structural trends which dissect vertical simple or fibrous prisms, subdividing them into parallelogram-shaped segments. A variety of this structure with more closely spaced oblique dissections of the vertical prisms was called "dissected crossed acicular" by Bandel (1979b, p.51), but Bandel considered this latter term to be synonymous with his "crossed prismatic," i.e., the present dissected crossed prismatic structure (Bandel, 1977a; Carter, 1990).

dissected pediment (dis-sect′-ed) An eroded *pediment.* It is generally regarded as a product of second-cycle erosion of an originally gently sloping pediment, although it may also be "born dissected".

dissected peneplain An ancient and uplifted peneplain that has become the initial surface upon which erosion begins to cut the forms of a new cycle; a partially destroyed peneplain represented in a maturely dissected region by only a few remnants, such as plateaus or occasional flat-topped mountains and ridges.

dissected plateau A plateau in which a large part of the original level surface has been deeply cut into by streams.

dissection (dis-sec′-tion) The process of erosion by which a relatively even topographic surface is gradually sculptured or destroyed by the formation of gullies, ravines, canyons, or other kinds of valleys; esp. the work of streams in cutting or dividing the land into hills and ridges, or into flat upland areas separated by fairly close networks of valleys. The process is applicable esp. to surfaces, such as plains, that have been uplifted. Adj: dissected.

disseminated (dis-sem′-i-nat′-ed) Said of a mineral deposit (esp. of metals) in which the desired minerals occur as scattered particles in the rock, but in sufficient quantity to make the deposit an ore. There is no genetic connotation. Cf: *impregnated.*

dissepiment (dis-sep′-i-ment) (a) A small domed calcareous plate forming a *vesicle* or cystlike enclosure typically occurring between radiating septa in the peripheral region of a *corallite.* Its convex surface faces inward and upward. (b) A skeletal crossbar connecting branches of a fenestrate bryozoan colony. (c) An imperforate sagging or bubblelike plate in the *intervallum* or central cavity of an archaeocyathid. (d) A crossbar or strand of periderm uniting adjacent branches (stipes) in a dendroid graptolite colony or *rhabdosome,* as in *Dictyonema.* (e) A thin, upwardly or obliquely convex blisterlike internal structure partially filling the gallery space in most stromatoporoid coenostea or composing the entire coenosteum of a few stromatoporoids. Syn: *cyst [paleont].* Cf: *stromatoporoid; coenosteum; gallery [paleont].* Pl: dissepiments.

dissepimentarium (dis-sep′-i-men-tar′-i-um) The peripheral zone of the interior of a *corallite,* occupied by dissepiments. See also: *regular dissepimentarium.*

dissociation constant (dis′-so-ci-a′-tion) The equilibrium constant for a dissociation reaction, defined as the product of activities of the products of dissociation divided by the activity of the original substance. When used for ionization reactions, it is called an *ionization constant;* when it refers to a very slightly soluble compound, it is called a *solubility product.*

dissociation point That temperature at which a compound breaks up reversibly to form two or more other substances, e.g. $CaCO_3$ becoming CaO plus CO_2. All variables should be stated in order to define the point precisely. The term dissociation refers to the breakup itself, and covers a wide variety of types, such as the breakup of molecular groupings in gases or liquids.

dissociation temperature A temperature point at which a given dissociation presumably occurs; in fact, it is usually a range of temperature owing to variations in composition or pressure, and may refer merely to the temperature at which the rate of a given dissociation becomes appreciable, under stated conditions.

dissoconch (dis′-so-conch) The postlarval shell of a bivalve mollusk.

dissolution (dis-so-lu′-tion) The process of dissolving into a homogenous solution, as when an acidic solution dissolves limestone. In karst, refers to the process of dissolving rock to produce landforms, in contrast to solution, the chemical product of dissolution.

dissolve The aggregation of polygons on the basis of selected common attributes, resulting in the generation of a simplified polygon coverage (Krzanowski et al., 1993).

dissolved-gas drive (dis-solved′) Energy within an oil pool, resulting from the expansion of gas liberated from solution in the oil. Cf: *gas-cap drive; water drive.*

dissolved load (dis-solved′) The part of the total *stream load* that is carried in solution, such as chemical ions yielded by erosion of the landmass during the return of rainwater to the ocean. Syn: *dissolved solids; solution load.*

dissolved organic carbon Carbon present in seawater in the form of dissolved organic molecules, i.e., the elemental carbon in *dissolved organic matter.* Abbrev: DOC.

dissolved organic matter Nonparticulate organic matter in ocean or lake water. It is operationally defined as the organic matter that passes through a 0.45-µm sieve, and includes truly dissolved material, *gelbstoff,* and colloidal material. Abbrev: DOM.

dissolved organic nitrogen Nitrogen present in seawater in the form of dissolved organic molecules. Abbrev: DON.

dissolved organic phosphorus Phosphorus present in seawater in the form of dissolved organic molecules. Abbrev: DOP.

dissolved oxygen The amount of oxygen, in parts per million by weight, dissolved in water, now generally expressed in mg/l. It is a critical factor for fish and other aquatic life, and for self-purification of a surface-water body after inflow of oxygen-consuming pollutants. Abbrev: DO.

dissolved solids (a) *dissolved load.* (b) A term that expresses the quantity of dissolved material in a sample of water, "either the residue on evaporation, dried at 180°C, or, for many waters that contain more than about 1000 parts per million, the sum of determined constituents" (USGS, 1958, p.50). Syn: *total dissolved solids.*

distal [eco geol] (dis′-tal) Said of an ore deposit formed at a considerable distance, typically tens of kilometers, from a volcanic source to which it is related and from which its constituents have been derived. Cf: *proximal [eco geol].*

distal [paleont] Remote or away from the point of attachment, plane of reference, or point of view. Examples in invertebrate morphology: "distal direction" away from a crinoid *theca* toward the holdfast or free lower extremity of the column; "distal portion" of the *rhabdosome* of a graptolite colony, farthest away from the point of origin; and "distal side" away from the *ancestrula* or origin of growth of a bryozoan colony. Ant: *proximal [paleont].*

distal [palyn] Said of the parts of pollen grains or spores away from the center of the original tetrad; e.g. said of the side of a monocolpate pollen grain upon which the colpus is borne, or said of the side of a spore opposite the laesura. For dinoflagellates and acritarchs, being that part of a process or sculptural element farthest from its area of origin on the body (Williams et al., 2000, p.65). Ant: *proximal [palyn].*

distal [sed] Said of a sedimentary deposit consisting of fine clastics and formed farthest from the source area; e.g. a "distal turbidite" consisting of thin silty varves, or the most remote foreland deposit derived from the borderland in a geosynclinal region. Cf: *proximal [sed].*

distal downlap *Downlap* in the direction away from the source of clastic supply (Mitchum, 1977, p.206).

distal onlap *Onlap* in the direction away from the source of clastic supply (Mitchum, 1977, p.208). Cf: *proximal onlap.*

distance coefficient A multivariate measure of similarity between two objects determined as the square root of the sum of their squared differences. The measure is analogous to the Pythagorean equation for distance between two points and increases in value with increasing dissimilarity. Two identical objects will have a distance of zero.

distance-function map (dis′-tance-func′-tion) A term used by Krumbein (1955) for a map now known as *facies-departure map.*

distance meter (dis′-tance) A device for measuring line-of-sight distances, generally by transmitting a laser pulse to a prism reflector at a station and timing the arrival of the reflected pulse.

distant admixture (dis′-tant) A term applied by Udden (1914) to an *admixture* (in a sediment of several size grades) whose particles are most different in size from those of the dominant or maximum grade; material in one of the two classes at the extreme ends of a histogram.

disthene (dis′-thene) *kyanite.*

disthenite (dis′-then-ite) A metamorphic rock composed almost entirely of kyanite (disthene) and some quartz, often associated with magnetiferous quartzite and amphibolite (LaCroix, 1922, p.497). Rarely used.

distichous (dis′-ti-chous) Said of a plant with two-ranked appendages, e.g. with leaves, leaflets, or flowers on opposite sides of the same point on a stem. Cf: *decussate.*

distillate (dis-til′-late) *condensate.*

distillation [paleont] (dis-til-la′-tion) A process of fossilization whereby the liquid and/or gaseous components of an organic substance are removed leaving a carbonaceous residue.

distillation [water] Conversion of liquid to vapor by the addition of heat, and returning the vapor to a liquid by cooling, as in the purification of water.

distinctive mineral (dis-tinc′-tive) *varietal mineral.*

dististele (dis'-ti-stele) (a) The distal region of a crinoid column. (b) The distal part of the stele of certain homalozoans. Cf: *proxistele.*

distomodaeal budding (di-sto-mo'-dae-al) A type of budding in scleractinian corals in which two stomodaea are developed within a common tentacular ring and two interstomodaeal couples of mesenteries are located between the original and each new *stomodaeum.*

distortion [cart] (dis-tor'-tion) The change in shape, distance, and size of a land area on a map due to the flattening of the curved earth surface to fit a plane. Distortion is inevitable and is controlled in the development of a projection to produce the characteristics of equal area, conformality, equidistance or a compromise among them. Cf: *angular distortion.*

distortion [photo] (a) Any shift in the position of an image on a photograph that alters the perspective characteristics of the photograph. It may be caused by lens aberration, differential shrinkage of film or paper, or motion of the film or camera. (b) Compression or expansion of the scale of the imagery in the azimuth direction, perhaps caused by incorrect film speed. (c) A change in scale from one part of the imagery to another. Cf: *displacement [photo].*

distortional wave (dis-tor'-tion-al) *S wave.*

distributary [marine geol] (dis-trib'-u-tar'-y) A channel in the depositional regime of a deep-sea fan, usually branching off a leveed fan valley.

distributary [streams] (a) A divergent stream flowing away from the main stream and not returning to it, as in a delta or on an alluvial plain. It may be produced by stream deposition choking the original channel. Ant: *tributary.* (b) One of the channels of a braided stream; a channel carrying the water of a stream distributary.

distributary glacier Any *ice stream* or *lobe* that flows away or forks off from the lower part of a glacier; a subsidiary terminus or outlet of a trunk glacier. Cf: *glacial lobe; outlet glacier.*

distributary mouth bar The coarse fluvial sediments deposited at the mouth of a delta distributary channel as fluvial discharge moves from confined to unconfined flow conditions.

distributed deformation Deformation that is homogeneous at the scale of observation (Rutter, 1986).

distributed fault (dis-trib'-ut-ed) A little-used term for *fault zone.*

distributed system [geophys exploration] (a) A seismic recording system where signals from one or more geophone groups are collected at remote units before they are transmitted to the recording truck. (b) A system where digitization occurs at the individual geophone rather than after transmission (via dedicated channels) from each geophone group to the recording system.

distribution [ecol] (dis-tri-bu'-tion) *range [ecol].*

distribution [stat] *frequency distribution.*

distribution coefficient *distribution ratio.*

distribution function The distribution function $F(x)$ of a variable x is the relative proportion of values that are equal to or less than x, for all values of x. The cumulative distribution.

distribution grading In a sedimentary bed, a progressive upward shift toward the finer grain sizes for almost all percentiles of the distribution (Middleton, 1967, p.487). Cf: *coarse-tail grading.*

distribution ratio The ratio of concentrations of a solute in two immiscible solvents. Syn: *distribution coefficient.*

distribution scatter *scatter diagram [stat].*

distributive fault (dis-trib'-u-tive) An obsolete term for a fault zone in which displacement is distributed among several subparallel fault surfaces or splays. Cf: *step fault.*

distributive province The environment embracing all rocks that contribute to the formation of a contemporaneous sedimentary deposit and the agents responsible for their distribution (Milner, 1922, p.366). Cf: *provenance.*

disturbance (dis-turb'-ance) A term used by some geologists for a minor orogeny. Schuchert (1924) used *revolution* for a major orogeny at the end of an era, and *disturbance* for an orogeny within an era; this usage is obsolete. Cf: *event [tect]; pulsation [tect].*

disturbed-neighborhood assemblage (dis-turbed'-neigh'-bor-hood) An *assemblage [ecol]* in which the specimens have been transported only a short distance from where they originally lived, and then deposited in essentially the same kind of sediment on which they had lived, in roughly the same proportions as when alive. Cf: *fossil community; winnowed community; transported assemblage; mixed assemblage.*

disturbing potential (dis-turb'-ing) The difference between the actual value of the potential at a point and the value predicted by theory or used for reference. In geopotential theory, it is the difference between the actual value of the gravity potential at a point (*geopotential*) and the value calculated from a standard formula for the gravity potential (*spheropotential*). Syn: potential disturbance; potential of random masses.

ditch A long, narrow excavation artificially dug in the ground; esp. an open and usually unpaved waterway, channel, or trench for conveying water for drainage or irrigation, and usually smaller than a canal. Some ditches may be natural watercourses.

ditch sample A sample of well cuttings brought to the surface by the drilling mud in rotary drilling for oil, caught in the ditch where the mud returns to the pit before being pumped back into the drill pipe.

ditetragonal (di-te-trag'-o-nal) Said of a crystal form having eight similar faces, the alternate interfacial angles of which are equal.

ditetragonal dipyramid A crystal form that is a dipyramid of 16 faces in which any section perpendicular to the fourfold axis is ditetragonal. Its indices are *{hkl}* in symmetry $4/m\ 2/m\ 2/m$.

ditetragonal-dipyramidal class (di-te-trag'-o-nal-di-py-ram'-i-dal) That crystal class in the tetragonal system having symmetry $4/m\ 2/m\ 2/m$.

ditetragonal prism A crystal form of eight faces parallel to the symmetry axis in which any section perpendicular to the prism axis is ditetragonal. Its indices are *{hk0}* with symmetry $4/m\ 2/m\ 2/m$.

ditetragonal pyramid A crystal form consisting of eight faces in a pyramid, in which any section perpendicular to the fourfold symmetry axis is ditetragonal. Its indices are *{hkl}* or $\{hk\bar{l}\}$ in symmetry $4mm$.

ditetragonal-pyramidal class (di-te-trag'-o-nal-py-ram'-i-dal) That crystal class in the tetragonal system having symmetry $4mm$.

ditrigonal (di-trig'-o-nal) Said of a symmetrical, six-sided figure, the alternate angles of which are equal. Such a figure is characteristic of certain crystal forms in the hexagonal system.

ditrigonal dipyramid A crystal form that is a dipyramid of twelve faces in which any section perpendicular to the threefold or sixfold symmetry axis is ditrigonal. Its indices are *{hkl}* or *{khl}* in symmetry $\bar{6}m2$.

ditrigonal-dipyramidal class (di-trig'-o-nal-di-py-ram'-i-dal) That class in the hexagonal system having symmetry $\bar{6}m2$.

ditrigonal prism A crystal form of six faces parallel to the symmetry axis, in which any section perpendicular to the axis is ditrigonal. Its indices are *{hk0}* or *{kh0}* in symmetry $\bar{6}m2$ or $3m$.

ditrigonal pyramid A crystal form consisting of six faces in a pyramid, in which any section perpendicular to the symmetry 3 axis is ditrigonal. Its indices are {hkl}, $\{hk\bar{l}\}$, *{khl}*, or $\{kh\bar{l}\}$ in symmetry 3.

ditrigonal-pyramidal class (di-trig'-o-nal-py-ram'-i-dal) That class in the rhombohedral division of the hexagonal system having symmetry $3m$.

ditrigonal-scalenohedral class (di-trig'-o-nal-sca-le'-no-he'-dral) *hexagonal-scalenohedral class.*

ditroite (dit'-ro-ite) A *nepheline syenite* containing sodalite, biotite, and cancrinite. The term was introduced by Zirkel in 1866, and later Brögger proposed applying it to nepheline syenites having granular texture (Johannsen, 1939, p.249). Its name is derived from Ditrau or Ditró, Romania. Cf: *foyaite.* Obsolete.

dittmarite (ditt'-mar-ite) A white orthorhombic mineral: $(NH_4)Mg(PO_4)\cdot H_2O$.

Dittonian (Dit-to'-ni-an) A series in the Old Red Sandstone of England: Lower Devonian (below Breconian, above Downtonian).

diurnal current (di-ur'-nal) A tidal current that has only one flood period and one ebb period during a tidal day.

diurnal inequality The difference between the heights and durations of the two successive high waters or of the two successive low waters of a tidal day.

diurnal tide A tide with only one high water and one low water occurring during a tidal day, as in the Gulf of Mexico. Cf: *diurnal current.*

diurnal variation *daily variation.*

diurnal vertical migration A pattern of daily movement of certain marine organisms that is upward at sunset and downward at sunrise. Cf: *deep scattering layer.*

diurnal wave The daily cyclical heating and cooling of the upper 30 to 50 cm of the Earth in response to the daily solar cycle. Below this depth, daily temperature is constant. Cf: *annual wave.* Syn: *daily wave.*

divariant (di-var'-i-ant) Said of a chemical system having two *degrees of freedom.*

divaricator (di-var'-i-ca'-tor) A muscle that causes divergence or separation of parts; specif. a *diductor muscle* of a brachiopod. The term is rarely used.

divergence [currents] (di-ver'-gence) The separation of ocean currents by horizontal flow of water in different directions from a common source. Usually requires upwelling of water from depth to supply water to the divergence zone. Also, the area in which divergence occurs. Cf: *convergence [currents].*

divergence [evol] An evolutionary trend in which a lineage evolves into two morphologically distinct lineages, or in which two lineages become less similar morphologically through time. Cf: *adaptive radiation.*

divergence [glac geol] Interruption of a *drainage pattern* by the advance of a glacier or ice sheet.

divergence [meteorol] The spreading of air away from a central region or zone. Vertically moving air replaces the air that moves outward horizontally. Ant: *convergence [meteorol].*

divergence [seis] The decrease in amplitude of a wavefront because of geometrical spreading. See also: *spherical divergence; cylindrical divergence.*

divergent bend (di-ver'-gent) A bend in a strike-slip fault that results in over-all crustal extension in the vicinity of the bend (Biddle and Christie-Blick, 1985). Cf: *convergent bend; extensional bend.* Syn: *releasing bend.* See also: *pull-apart basin.*

divergent fan A type of *fan cleavage* where cleavage orientations diverge from the core of the fold.

divergent fault A little-used term for a *strike-slip fault* or a *wrench fault* along which strike-slip deformation is accompanied by a component of extension transverse to the fault (Wilcox et al., 1973). Cf: *convergent fault; transtensional fault.*

divergent-isogon fold A fold in which the dip isogons, when traced along a traverse from the outer arc of the fold to the inner arc of the fold, diverge.

divergent margin The trailing edge of a continent; *passive margin.*

divergent plate boundary A boundary between two plates that are moving apart. Such boundaries are delineated by *mid-ocean ridges* at which *sea-floor spreading* occurs. Cf: *accreting plate boundary.*

diversilite-(Ce) A tabular yellowish orange rhombohedral mineral: $Na_2(Ba,K)_6Ce_2Fe^{2+}Ti_3Si_{12}O_{36}(OH,H_2O)_{12}$.

diversion (di-ver'-sion) (a) The process by which a stream actively effects changes in the drainage or course of another stream, as by aggradation or capture. Cf: *derangement.* (b) The artificial draining, pumping, siphoning, or other removal of water from a stream, lake, or other body of water, into a canal, pipe, or other conduit. (c) A channel designed to divert water from a body of water for purposes such as prevention of flooding, reduction of erosion, or promotion of infiltration.

diversity (di-ver'-si-ty) The number of different kinds of organisms in an assemblage. It can be measured in various ways, some of which separate diversity in the strict sense (the number of species within the assemblage) from species evenness or *equitability.*

diversity dependent Ecologic and evolutionary factors which affect individuals through depletion of required ecologic factors, such as oxygen or habitat area.

diversity independent Ecologic and evolutionary factors which affect all individuals equally without respect to the number of individuals, i.e. diversity independent extinction, diversity selection.

diversity stack A *vertical stack* in which amplitudes that exceed a preset threshhold are set to zero before stacking. It is used to prevent occasional large bursts of noise (such as traffic noise) from dominating the stacked record. See also: *mute.*

diverted stream (di-vert'-ed) A stream whose course or drainage has been affected by another stream; e.g. a captured stream.

diverter (di-vert'-er) *diverting stream.*

diverting stream (di-vert'-ing) A stream that effects diversion; e.g. a capturing stream. Syn: *diverter.*

divide [grd wat] (di-vide') A ridge in the *water table* or other *potentiometric surface* from which the groundwater represented by that surface moves away in both directions. Water in other aquifers above or below, and even in the lower part of the same aquifer, may have a potentiometric surface lacking the ridge, and so may flow past the divide. Syn: *water-table divide; groundwater divide; groundwater ridge.*

divide [streams] (a) The line of separation, or the ridge, summit, or narrow tract of high ground, marking the boundary between two adjacent drainage basins or dividing the surface waters that flow naturally in one direction from those that flow in the opposite direction; the line forming the rim of or enclosing a drainage basin; a line across which no water flows. An "anomalous" divide is one that does not follow the crest of the highest mountain range of a mountain chain. See also: *continental divide.* British syn: *watershed.* Syn: *drainage divide; height of land; topographic divide; watershed line.* (b) A tract of relatively high ground between two streams; a line that follows the summit of an interfluve. (c) The highest summit of a pass or gap. (d) *groundwater divide.*

dividing wall (di-vid'-ing) The wall dividing offset from parent *corallite,* jointly formed by both during increase in corals. It consists of two layers separated by a central epitheca (Fedorowski and Jull, 1976, p.41). Cf: *partition [paleont].*

diviner (di-vin'-er) *dowser.*

diving wave (div'-ing) A seismic wave that is refracted in a strong velocity-gradient zone and returned to the surface; it may be observed as a refraction arrival even though it has had no appreciable path through a distinctive refractor.

divining (di-vin'-ing) *dowsing.*

divining rod Traditionally, a forked wooden stick, cut from a willow or other water-loving plant, used in *dowsing.* Dowsers report that it dips downward sharply when held over a body of groundwater or a mineral deposit, thus revealing the presence of these substances. Syn: *witching stick; wiggle stick; dowsing rod; mineral rod; dipping rod; twig; dowser.* Cf: *water witch; water-finder.*

division [bot] (di-vi'-sion) A category in the hierarchy of botanical classification intermediate in rank between *kingdom* and *subdivision* (ICBN, 1972, p.17). It is generally regarded as equivalent to the rank of *phylum* in zoology.

division [sed] A portion of a *bed [geomorph]* that is distinguished from other parts of the bed by the presence of a specific type of *sedimentary structure* (Blatt et al., 1980).

division [stratig] (a) A term proposed by Stormer (1966, p.25) for a chronostratigraphic unit equivalent to stage (and possibly series), but having a regional or more limited geographic range. (b) Any, as yet informal, stratigraphic unit of any kind (Geol. Soc. London, 1967, p.79).

divisional plane (di-vi'-sion-al) A general term that includes joints, cleavage, faults, bedding planes, and other surfaces of separation.

dixenite (di'-xen-ite) A deep copper-red platy rhombohedral mineral $CuFeMn_{14}(OH)_6(AsO_3)_5(AsO_4)(SiO_4)_2$.

Dix equation A relationship that gives the velocity of an interval between two parallel seismic reflectors from measurements of their associated rms velocities.

djalmaite (djal'-ma-ite) *uranmicrolite.*

djebel (dje'-bel) *jebel.*

djerfisherite (djer-fish'-er-ite) A submetallic olive cubic meteorite mineral: $K_6(Na,Li)(Fe,Cu,Ni)_{24}S_{26}Cl$. It occurs in enstatite chondrites. Its thallium analogue is thalfenisite.

djurleite (djur'-le-ite) A metallic gray monoclinic mineral: $Cu_{31}S_{16}$. Its X-ray diffraction pattern is near to, but distinct from, that of chalcocite.

D layer The seismic region of the Earth between 660 km and 2,890 km, equivalent to the *lower mantle.* At a depth of 2,700 km, there is a change from chemical homogeneity to inhomogeneity; the upper division is the D' layer, and the lower is the D" layer. It is a part of a classification of the Earth's interior made up of layers A to G. Rarely used.

DLG *digital line graph.*

dmisteinbergite A vitreous colorless hexagonal mineral formed in a burning coal heap: $CaAl_2Si_2O_8$. A dimorph of anorthite.

DMO processing Dip moveout processing is a seismic processing operation to correct for the fact that the component traces of a common midpoint gather do not involve a common reflecting point when the reflectors dip.

DNA Deoxyribonucleic acid; concentrated mainly in the nuclear structures of organisms. See also: *gene.*

DNAPL *denser-than-water nonaqueous-phase liquid.*

dneprovskite (dne-provsk'-ite) *wood tin.*

DO *dissolved oxygen.*

doab [sed] An Irish term for a dark sandy clay or shale found in the vicinity of bogs (Power, 1895). Rarely used.

doab [streams] (a) A term applied in the Indo-Gangetic Plain of northern India to the tongue of low-lying, alluvial land between two confluent rivers; specif. the Doab, the tract between the Ganges and Jumna rivers. The term is commonly restricted to the alluvial-plains portion characterized by very little relief. Cf: *interfluve.* (b) The confluence of two rivers. Etymol: Persian, "two waters".

DOC *dissolved organic carbon.*

Docodonta (Doc-o-don'-ta) One of two orders of mammals (the other being Triconodonta) of primitive structure and uncertain subclass assignment, mostly of Triassic and Jurassic age.

docrystalline (do-crys'-tal-line) A term, now obsolete, suggested by Cross et al. (1906, p.694) for porphyritic rocks that are dominantly crystalline, the ratio of crystals to glass being less than 7 to 1 but greater than 5 to 3.

dodecahedral cleavage (do'-dec-a-he'-dral) Mineral cleavage parallel to the faces of the dodecahedron (110); e.g. in sphalerite.

dodecahedron (do'-dec-a-he'-dron) A crystal form with 12 faces that are either pentagonal or rhombic; if pentagonal, the faces are not regular. Each face is parallel to one crystallographic axis and intersects the other two. See also: *pyritohedron; rhombic dodecahedron.*

dodecant (do-dec'-ant) In the hexagonal crystal system, one of the 12 spatial divisions made by the four reference axes.

dodging (dodg'-ing) The process of holding back light from certain areas of sensitized paper in making a print, in order to avoid overexposure.

dofemic (do-fem'-ic) Obsolete term for one of five classes in the *CIPW classification* of igneous rocks, in which the ratio of salic minerals to femic is less than three to five but greater than one to seven. Cf: *perfemic.*

dogger (dog'-ger) (a) A large, irregular nodule, usually of clay ironstone, sometimes containing fossils, found in a sedimentary rock, as in the Jurassic rocks of Yorkshire, England. (b) An English term for any large, lumpy mass of sandstone longer than it is broad, with steep rounded sides.

Dogger (Dog'-ger) Middle part of the Jurassic; corresponds approximately with the standard Middle Jurassic. Above Liassic, or Lias, below Malm.

dogger stone A miner's term for a brown, compact, relatively pure, nonoolitic clay ironstone interbedded with oolitic ironstones in the British Middle Jurassic.

dog hole A prospector's informal term for a small prospect pit.

dogleg (dog'-leg) An abrupt angular change in course or direction, as of a borehole or in a survey traverse. Also, a deflected borehole, survey course, or anything with an abrupt change in direction resembling the hind leg of a dog.

dogtooth spar (dog'-tooth) A variety of calcite in sharply pointed crystals of acute scalenohedral form resembling the teeth of a dog. Syn: *hogtooth spar.*

dohyaline (do-hy'-a-line) In the *CIPW classification* of igneous rocks, those rocks in which the ratio of crystals to glassy material is greater than three to five but less than one to seven. Obsolete. Cf: *perhyaline; hyalocrystalline.*

dolabrate element A ramiform conodont element consisting of a cusp and a posterior process; commonly pick-shaped in lateral view (TIP, 1981, pt. W, supp.2).

dolarenaceous (dol'-ar-e-na'-ceous) Said of the texture of a dolarenite.

dolarenite (dol-ar'-e-nite) A dolomite rock consisting predominantly of detrital dolomite particles of sand size (Folk, 1959, p.16); a consolidated dolomitic sand.

dolerite (dol'-er-ite) (a) In the U.S., a syn. of *diabase.* (b) In British usage, the preferred term for what is called *diabase* in the U.S. Etymol: Greek "doleros", "deceitful", in reference to the fine-grained character of the rock which makes it difficult to identify megascopically. Cf: *diorite; trap [ign].*

doleritic (dol-er-it'-ic) The *ophitic* texture typical of dolerites or diabases.

dolerophanite (dol-er-o-phan'-ite) A brown to black monoclinic mineral: $Cu_2O(SO_4)$.

Dolgellian (Dol-gel'-li-an) A British stage: uppermost Cambrian (above Festiniogian, below Ordovician) (Harland et al., 1990).

dolimorphic (dol-i-mor'-phic) Said of an igneous rock in which released minerals are prominent; e.g. a lamprophyre composed chiefly of biotite and quartz, with a little hornblende. Obsolete.

dolina (do-li'-na) *doline.*

doline (do-line') A basin- or funnel-shaped *hollow* in limestone, ranging in width from a few meters to a kilometer and in depth from a few to several hundred meters. Some dolines are gentle, grassy hollows; others are rocky, cliff-bounded basins. A distinction may be made between dolines formed mainly by direct dissolution of the limestone surface zone (solution dolines) and those produced by collapse of the surface into underlying voids (collapse dolines); however, it is generally not possible to establish the origin of individual examples. *Closed depressions* receiving a stream are known as swallow holes. In America most dolines are referred to as *sinks* or *sinkholes.* Also spelled: *dolina.* Etymol: German transliteration of Slovene "dolina", "little valley".

doline lake A *karst lake* occupying a doline.

dollaseite-(Ce) A brown monoclinic mineral of the *epidote* group: $CeCaMg_2Al(OH)F[SiO_4][Si_2O_7]$.

Dollo's law (Dol'-lo's) A syn. of *irreversibility.* It is named after the Belgian paleontologist Louis Dollo (d.1931).

dolocast (do'-lo-cast) A cast or impression of a dolomite crystal, preserved in an insoluble residue. Rarely used. Adj: dolocastic. Cf: *dolomold.*

doloclast (do'-lo-clast) A lithoclast derived by erosion from an older dolomite rock; also, an intraclast disrupted from partly consolidated dolomitic mud. Rarely used.

dolocrete A type of *calcrete* composed of dolomite.

dololithite (do-lo-lith'-ite) A dolomite rock containing 50% or more of fragments of older dolomitic rocks that have been eroded and redeposited (Hatch and Rastall, 1965, p.223). Rarely used.

dololutite (do-lo-lut'-ite) A dolomitic rock consisting predominantly of detrital dolomite particles of silt and/or clay size (Folk, 1959, p.16); a consolidated dolomitic mud. It is commonly interlayered with dense primary dolomites in evaporitic sequences. Cf: *dolosiltite.* Rarely used.

dolomicrite (do-lo-mic'-rite) A sedimentary rock consisting of clay-sized dolomite crystals, interpreted as a lithified dolomite mud (analogous to calcite mud or micrite), and containing less than 1% allochems (Folk, 1959, p.14). See also: *primary dolomite.* Syn: *dolomite mudstone.*

dolomite [mineral] (do'-lo-mite) A common rock-forming rhombohedral mineral: $CaMg(CO_3)_2$. Part of the magnesium may be replaced by ferrous iron and less frequently by manganese. Dolomite is white, colorless, or tinged yellow, brown, pink, or gray; it has perfect rhombohedral cleavage and a pearly to vitreous luster, effervesces feebly in cold dilute hydrochloric acid, and forms curved, saddlelike crystals. Dolomite is found in extensive beds as dolomite rock; it is a common vein mineral, and is found in *serpentinite* and other magnesian rocks. Cf: *calcite.* Syn: *bitter spar; pearl spar; magnesian spar; rhomb spar.*

dolomite [sed] A carbonate sedimentary rock of which more than 50% by weight or by areal percentages under the microscope consists of the mineral dolomite, or a variety of limestone or marble rich in magnesium carbonate; specif. a carbonate sedimentary rock containing more than 90% dolomite and less than 10% calcite, or one having a Ca/Mg ratio in the range of 1.5-1.7 (Chilingar, 1957), or one having an approximate MgO equivalent of 19.5-21.6% or magnesium-carbonate equivalent of 41.0-45.4% (Pettijohn, 1957, p.418). Dolomite occurs with well-developed crystalline habits as well as anhedrally, is clearly associated and commonly interbedded with limestone, and usually represents a postdepositional replacement of limestone. Pure dolomite (unless finely pulverized) will effervesce very slowly in cold hydrochloric acid. Named after Déodat Guy de Dolomieu (1750-1801), French geologist, and first applied to certain carbonate rocks of the Tyrolean Alps. See also: *primary dolomite; magnesian limestone.* Syn: *dolostone; dolomite rock.*

dolomite carbonatite A mineralogically defined carbonatite in which dolomite is at least 90% of the modal carbonates (LeMaitre, 1989). Cf: *beforsite; rauhaugite.*

dolomite limestone *dolomitic limestone.*

dolomite mudstone *dolomicrite.*

dolomite rock *dolomite [sed].*

dolomitic (do-lo-mit'-ic) (a) Dolomite-bearing, or containing dolomite; esp. said of a rock that contains 5-50% of the mineral dolomite in the form of cement and/or grains or crystals. (b) Containing magnesium; e.g. "dolomitic lime" containing 30-50% magnesium.

dolomitic conglomerate A conglomerate consisting of limestone pebbles and dolomite cement (Nelson and Nelson, 1967, p.112).

dolomitic limestone (a) A limestone in which the mineral dolomite is conspicuous, but calcite is more abundant; specif. a limestone containing 10-50% dolomite and 50-90% calcite and having an approximate magnesium-carbonate equivalent of 4.4-22.7% (Pettijohn, 1957, p.418), or a limestone whose Ca/Mg ratio ranges from 4.74 to 60 (Chilingar, 1957). Cf: *calcitic dolomite; magnesian limestone.* Syn: *dolomite limestone.* (b) A limestone that has been incompletely dolomitized (Chilingar et al., 1967, p.314).

dolomitic marble A variety of marble composed largely of dolomite and formed by the metamorphism of dolomitic or magnesian limestone. Cf: *magnesian marble.*

dolomitic mottling A textural feature resulting from incipient or arrested dolomitization of limestones, characterized by preferential alteration that leaves patches, blotches, bird's-eyes, laminae, allochems and/or other structures unaffected. Also, a similar phenomenon resulting from arrested or incomplete *dedolomitization.*

dolomitization (do-lo'-mit-i-za'-tion) The process by which limestone is wholly or partly converted to dolomite rock or dolomitic limestone by the replacement of the original calcium carbonate (calcite) by magnesium carbonate (mineral dolomite), usually through the action of magnesium-bearing water (seawater or percolating meteoric water). It can occur penecontemporaneously or shortly after deposition of the limestone, or during lithification at a later period.

dolomold (do'-lo-mold) A rhombohedral opening in an insoluble residue, formed by the dissolution of a dolomite crystal. Adj: dolomoldic. Cf: *dolocast.*

dolon (do'-lon) In certain Paleozoic ostracodes, the strong curvature of the vela or frills such that the two meet outside the contact margin to form a false pouch; also the cavity formed by this curvature of the frills (TIP, 1961, pt. Q, p.49). Syn: *false pouch.*

doloresite (do-lo'-res-ite) A submetallic black monoclinic mineral: $V^{4+}{}_3(OH)_4$.

dolorudite (do-lo-rud'-ite) A dolomite rock consisting predominantly of detrital dolomite particles larger than sand size (Folk, 1959, p.15); a consolidated dolomitic gravel.

dolosiltite (do-lo-silt'-ite) A dolomite rock consisting predominantly of detrital dolomite particles of silt size; a consolidated dolomitic silt. Cf: *dololutite.*

dolosparite (do-lo-spar'-ite) A sparry dolomite crystal. Cf: *calcsparite.*

dolostone (do'-lo-stone) A term proposed by Shrock (1948a, p.126) for the sedimentary rock dolomite, in order to avoid confusion with the mineral of the same name. Not recommended; obsolete.

DOM *dissolved organic matter.*

domain [magnet] (do-main') A region within a grain of magnetically ordered mineral, within which the spontaneous magnetization has a constant orientation and moment characteristic of the mineral composition and temperature. Syn: *magnetic domain.*

domain [meta] (a) A chemically distinct part of a rock at any scale. (b) A macroscopically recognizable part of an altered rock, frequently of a mafic volcanic or plutonic igneous rock, that can be regarded as having a distinctive lithologic or bulk chemical composition (R.E. Smith, 1968).

domain [paleont] A level of classification above the level of kingdom. There are three domains: *Bacteria, Eucarya,* and *Archaea.*

domain [sed] The areal extent of a given lithology or environment; specif. the area in which a given set of physical controls combined to produce a distinctive sedimentary facies.

domain [struc petrol] *fabric domain.*

domal (dom'-al) Said of a thecal shape in edrioasteroids, in which the oral surface is convex upward and is confined to the upper side of the theca; the distal edge of the oral surface forms the thecal ambitus. The nonplated aboral surface forms the entire lower side of the theca (Bell, 1976).

domatic class (do-mat'-ic) That crystal class in the monoclinic system having symmetry m. Syn: *clinohedral class.*

dome [beach] A miniature elevation (2.5 cm or more high, and 5 to 30 cm in diameter), composed of sand, with a hollow center, formed on beaches by the rush of waves entrapping and confining air (Shepard, 1967, p.58). Syn: *sand dome.*

dome [cryst] An open crystal form composed of two nonparallel faces that are related by a symmetry plane. Cf: *sphenoid.*

dome [geomorph] (a) A general term for any smoothly rounded landform or rock mass, such as a rock-capped mountain summit, that roughly resembles the dome of a building; e.g., the rounded granite peaks of Yosemite, Calif. The term is also applied to broadly up-arched regions, e.g., the English Lake District or the Black Hills of South Dakota. (b) A rounded snow peak, esp. in the French Alps. (c) A high central area within an ice sheet or ice cap.

dome [marine geol] A general, nonrecommended term for such ocean-floor features as a *seamount* or a *knoll [marine geol].*

dome [petrology] A large magmatic or migmatitic intrusion whose surface is convex upward and whose sides slope away at low but gradually increasing angles. Intrusive igneous domes include laccoliths and batholiths which cannot be differentiated where evidence is lacking regarding their nature at depth.

dome [planet] A small, almost circular surface bulge, generally several kilometers wide and tens to a few hundred meters high, found in volcanic regions of planetary surfaces. Domes commonly have summit craters; they are generally believed to be formed by local extrusive igneous activity. Cf: *tholus.*

dome [speleo] (a) A high shaft in a cave *room* or *passage* formed by dissolution. (b) A large, hemispheroidal hollow in the roof of a cave formed mechanically by breakdown and/or mineral weathering.

dome [struc geol] n. An uplift or anticlinal structure, either circular or elliptical in outline, in which the rocks dip gently away in all directions. A dome may be small, e.g. a Gulf Coast salt dome, or many kilometers in diameter. Domes include diapirs, volcanic domes, and cratonic uplifts. Type structure: Nashville Dome, Tennessee. See also: *pericline [struc geol]; arch [struc geol].* Syn: *dome structure; structural dome; quaquaversal fold.* Less preferred syn: *swell [struc geol].* Ant: *basin.* v. To bend, push, or thrust up into a dome, e.g. underlying magma doming the surface by upward pressure.

dome [volc] (a) *volcanic dome.* (b) *lava dome.*

dome-and-keel structure A regional-scale arrangement of domal bodies composed of basement metamorphic and intrusive rocks surrounded by steep-sided troughs of supracrustal rocks (volcanics and sedimentary rocks). Domes typically range from 10 to 50 km across, and may have circular to elliptical shapes in plan. Dome-and-keel structure is characteristic of Archean granite-greenstone terranes. In such cases, the domes are crystalline rocks, whereas the keels consist of mafic and ultramafic volcanics and associated sediments. The structure also occurs in Paleoproterozoic orogens.

domed peat Peat whose upper surface is markedly convex and rises above the local topography. The water table in the peat is generally at or just below the surface and rises above the local water table (Cecil et al., 1985). Cf: *ombrogenous peat; highmoor peat.*

dome mountain A mountain produced where a region of flat-lying sedimentary rocks is warped or bowed upward to form a structural dome; a mountain resulting from dissection of a structural dome. Examples: the Black Hills in South Dakota, and the Weald uplift in SE England.

domepit (dome'-pit) An obsolete term for *vertical shaft.*

Domerian (Do-mer'-i-an) A European substage: upper substage of the Pliensbachian Stage, Lower Jurassic (above Carixian, below Toarcian Stage).

domestone Scleractinian reef growth framework dominated by domal and irregular massive colonies which have the same calcification potential in all free directions, and these growth forms constitute >60% of the skeletal volume (Insalaco, 1998).

dome structure *dome [struc geol].*

domeykite (do-mey'-kite) A tin-white or steel-gray cubic mineral: Cu_3As. It is dimorphous with metadomeykite.

domiciliar dimorphism (dom-i-cil'-i-ar) In certain ostracodes, a kind of dimorphism in which the adult form presumed to be the female has a much larger and more spacious posterior half of the carapace that the presumed adult male; also called "kloedenellid dimorphism", although not restricted to the family Kloedenellidae.

domicilium (dom-i-cil'-i-um) The main part of the carapace of an ostracode exclusive of alae or other accessory projecting structures.

dominant [coal] (dom'-i-nant) In the description of coal constituents, more than 60% of a particular constituent occurring in the coal (ICCP, 1963). Cf: *rare; common; very common; abundant.*

dominant [ecol] A species or group of species that is numerically very abundant, or largely controls the energy flow, or strongly affects the environment, within a community or association.

dominant discharge The discharge of a natural channel that determines the characteristics and principal dimensions of the channel. It depends on the sediment characteristics, the relationship between maximum and mean discharge, duration of flow, and flood frequency (ASCE, 1962). Syn: *effective discharge.*

domino-style faulting *bookshelf faulting.*

domite (dom'-ite) An altered porphyritic oligoclase-biotite *trachyte* that contains tridymite in the groundmass. It is named for Puy de Dome in the Auvergne district of France. Obsolete.

domoikic (do-moi'-kic) In the *CIPW classification* of igneous rocks, those rocks in which the ratio of oikocrysts to chadacrysts is less than seven to one but greater than five to three. Obsolete. Cf: *peroikic.*

DON *dissolved organic nitrogen.*

donathite (do'-nath-ite) Originally reported as a metallic black tetragonal mineral: $(Cr,Fe)_2(Fe,Mg)O_4$, but later discredited as a mixture of two spinels.

donbassite (don'-bass-ite) A white monoclinic dioctahedral member of the *chlorite* group of minerals, in which the octahedral sheet within the 2:1 T-O-T layer and the octahedral interlayer are both dioctahedral: $Al_2(Si_3Al)O_{10}(OH)_2{\cdot}Al_{2.33}(OH)_6$.

donga [geomorph] (a) A term used in South Africa for a small narrow steep-sided ravine or gorge formed by turbulent water flow; it is usually dry except in the rainy season. A donga is similar to a wadi or a mullah. (b) A term used in South Africa for a gully formed by soil erosion. Etymol: Afrikaans, from Bantu (Zulu).

donga [glaciol] (don'-ga) A small, steep-walled rift in a piedmont glacier or in an ice shelf. Cf: *barranca.*

donharrisite (don-har'-ris-ite) A metallic brown monoclinic mineral: $Ni_8Hg_3S_9$.

donnayite-(Y) (don-nay'-ite-y) A white to yellow triclinic mineral occurring in triclinic and rhombohedral polytypes: $NaCaSr_3Y(CO_3)_6{\cdot}3H_2O$. The triclinic form is isomorphous with weloganite.

donpeacorite (don-pea'-cor-ite) A buff orthorhombic pyroxene mineral: $(Mn^{2+},Mg)MgSi_2O_6$.

doodlebug (doo'-dle-bug) A popular term for any of various kinds of geophysical prospecting equipment.

DOP *dissolved organic phosphorus.*

dopatic (do-pat'-ic) In the *CIPW classification* of igneous rocks, those rocks in which the ratio of groundmass to phenocrysts is less than seven to one but greater than five to three. Obsolete. Cf: *perpatic;* sempatic.

Doppler effect (dop'-pler) A change in the observed frequency of electromagnetic or sound waves, caused by relative motion between the source and the observer. See also: *doppler signal.*

dopplerite (dop'-pler-ite) An amorphous brownish-black gelatinous calcium salt of a humic acid that is found at depth in marsh and bog deposits. It may represent an accumulation of *phytocollite* concentrated by groundwater (Swain, 1963, p.105).

Doppler navigation Positioning in which frequency shift because the Doppler effect is involved. Usually refers to Doppler radar (for aircraft) or Doppler sonar (for ships), occasionally to satellite navigation. Syn: *Doppler positioning.*

Doppler positioning *Doppler navigation.*

Doppler radar A weather radar system that has the capacity for determining the radial velocity of precipitation or aerosol targets as a result of the Doppler shift in returned power. These sensors also measure the reflectivity.

Doppler signal The difference in frequency of waves produced by the *Doppler effect.*

Doppler surveying Determining the coordinates of points on the Earth, or the angles and/or distances between these points, by measuring the Doppler shift in radio waves emitted by a satellite whose orbit is known.

DOQ *digital orthophotoquad.*

DOQQ *digital orthophoto quarter-quadrangle.*

dorallcharite A golden yellow rhombohedral mineral of the *alunite* group: $(Tl,K)(Fe^{3+})_3(SO_4)_2(OH)_6$.

dorbank (dor'-bank) A term used in southern Africa for a calcareous and siliceous concretion occurring beneath the surface layer of sandy loam. Etymol: Afrikaans, "dry layer".

doré A gold and silver bullion, the first saleable product of gold mining and metallurgical processing.

doreite (dore'-ite) An andesitic lava containing approximately equal amounts of potassium and sodium; the extrusive equivalent of *mangerite* (Streckeisen, 1967, p.209). The name (for Mont Dore, Auvergne, France) was proposed by Lacroix in 1923 for olivine-bearing *trachyandesite.* Obsolete.

dorfmanite (dorf'-man-ite) A white orthorhombic mineral: $Na_2(PO_3OH){\cdot}2H_2O$.

dormant volcano (dor'-mant) A volcano that is not now erupting but is considered likely to do so in the future. There is no precise distinction between a dormant and an *active volcano.* Cf: *extinct volcano; inactive volcano.*

dorr A glacial trough, open at both ends, across a ridge or mountain range, lying in a pass; it is formed through overdeepening by the crowding of ice through the pass. The term was introduced by Chadwick (1939, p.362) to describe a fjord-like trough that may or may not have been submerged, and is named in honor of George B. Dorr (b.1853), executive of the Hancock County (Me.) Trustees of Public Reservations. Example: Somes Sound on Mount Desert Island, Me.

dorrite (dorr'-ite) A submetallic dark brown triclinic mineral: $CaMgFe^{3+}{}_2Al_2SiO_{10}$. A member of the *aenigmatite* group.

dorsal (dor'-sal) (a) Pertaining to, or situated near or on, the back (not the hind end) or upper surface of an animal or of one of its parts; e.g. toward the brachial valve from the pedicle valve of a brachiopod, or pertaining to the hinge region of the shell of a bivalve mollusk, the side of the uniserial stipe opposite the thecal apertures of a graptoloid, or the spiral side of a trochoid foraminifer. (b) Referring to the direction or side of an echinoderm away from the mouth, normally downward and outward, e.g. toward the point of attachment of the column with blastoid theca or to the part of the crinoid or cystoid calyx located toward the column; aboral. (c) In vertebrates, toward the part of the animal that bears the spinal cord (rather than toward the belly). Ant: *ventral.*

dorsal area The part of the *whorl* of a coiled cephalopod *conch* in contact with the preceding whorl (TIP, 1964, pt. K, p.55).

dorsal cleft A triangular area on the dorsal side of a *septum* in which cameral deposits are absent (if such deposits are otherwise present); occurs in orthocerid cephalopods (Fischer and Teichert, 1969).

dorsal cup *aboral cup.*

dorsal exoskeleton (a) The resistant mineralized dorsal integument of a trilobite. The term is commonly used to include the reflexed border or *doublure* on the ventral side. (b) The commonly calcified part of the covering of a crustacean, including the cephalic shield and fold of integument arising from the posterior border of the maxillary somite and extending over the trunk, usually covering it laterally as well as dorsally and in many forms having a dorsal longitudinal hinge, and often fused to one or more thoracic somites. Less-preferred syn: *carapace.*

dorsal furrow (a) An *axial furrow* of a trilobite. (b) A *septal furrow* of a nautiloid.

dorsal lobe The median primary lobe of a suture on the *dorsum* of a cephalopod shell. See also: *internal lobe.* Cf: *ventral lobe.*

dorsal shield (a) One of a series of ossicles along the midline of the aboral surface of an arm in an ophiuroid. Cf: *ventral shield.* (b) The entire dorsal test of a trilobite, including the *cephalon, thorax,* and *pygidium.*

dorsal valve The *brachial valve* of a brachiopod.

dorsomyarian (dor'-so-my-ar'-i-an) Said of a nautiloid in which the retractor muscles of the head-foot mass are attached to the shell along the interior areas of the body chamber adjacent to, or coincident with, its dorsal midline (TIP, 1964, pt. K, p.55). Cf: *pleuromyarian; ventromyarian.*

dorsum [paleont] The back or dorsal surface of an animal; e.g. the dorsal side of a cephalopod *conch,* opposite the ventral side and equivalent to the impressed area in slightly involute shells but referring only to that part of the conch adjacent to the venter of the preceding whorl in deeply involute shells; or the more or less flattened area of the *carapace* surface of an ostracode, adjacent to the hinge line and set off from the lateral surface of the valves. The term is somewhat loosely applied in fossil nautiloids, to the concave side of a whorl in coiled forms and to the side farthest removed from the *siphuncle* in straight or curved forms with eccentric siphuncle (TIP, 1964, pt. K, p.55). Pl: dorsa. Cf: *venter [paleont].*

dorsum [planet] (dor'-sum) A term established by the International Astronomical Union for an elongated ridge on planetary surfaces (USGS, 2003), such as Gordii Dorsum on Mars (Mutch et al., 1976, p.57). Pl: dorsa.

dosalic (do-sal'-ic) Obsolete term for one of five classes in the *CIPW classification* of igneous rocks, in which the ratio of salic to femic minerals is less than seven to one but greater than five to three. Cf: *persalic.*

dosemic (do-sem'-ic) In the *CIPW classification* of igneous rocks, those rocks in which the ratio of groundmass to phenocrysts is less than three to five but greater than one to seven. Obsolete. Cf: *persemic.*

dot chart A graphic aid used in the correction of station gravity for terrain effect, or for computing gravity effects of irregular masses. It can also be used in magnetic interpretation. Now replaced by computerized techniques.

dot map A map utilizing dots (usually of uniform size), with each dot representing a specific number of the objects whose distribution is being mapped.

double biseries (dou'-ble) Two distinct sets of coverplates in edrioasteroids, the pairs of one set alternating with those of the other to form an integrated system (Bell, 1976).

double Bouguer correction A *Bouguer correction* to sea level for measurements made on the ocean floor. It involves a correction to remove the upward attraction of the sea water above the meter, and another correction to replace the sea water with the replacement density. Similar "double" corrections are required for measurements made in mines or boreholes.

double-chain silicate An *inosilicate* with Si:O ratio of 4:11.

double island silicate *sorosilicate.*

double layer A layer of ions in an electrolyte, required to satisfy a charge imbalance within a solid with which the electrolyte is in contact. See also: *diffuse layer; fixed layer.*

double-line stream A watercourse drawn to scale (on a map) by two lines representing the banks. Cf: *single-line stream.*

double-mass curve A graphical means of detecting a change in hydrologic or meteorologic responses through time. A dependent variable, e.g. runoff or precipitation, is accumulated in time sequence and plotted against cumulative values of several neighboring stations. A break in slope indicates change.

double porosity Two sets of *porosity* occuring within a hydrogeological system. An example is a porous rock with a fracture set; such a system has two characteristic porosities, one for the fractures and one for the porous matrix. Syn: *dual porosity.*

double reflection The phenomenon whereby an X-ray diffracted by a set of lattice planes fortuitously satisfies the *Bragg equation* for another set of lattice planes in the crystal, and thus undergoes a second diffraction. The phenomenon occurs in neutron diffraction and electron diffraction as well. Syn: *Renninger effect.*

double-refracting spar (dou'-ble-re-fract'-ing) *Iceland spar.*

double refraction *birefringence [cryst].*

double-serrate (dou'-ble-ser'-rate) Said of a leaf having coarse marginal serrations with additional smaller teeth on the coarser serration lobes.

doublet (dou'-blet) A gem substitute composed of two pieces of gem material, or one of gem material and a second of glass or synthetic, fused or cemented together; e.g. a glass imitation with a thin layer of natural garnet fused on the top. Cf: *triplet; assembled stone.*

double tide (a) A high tide consisting of two high-water maxima of nearly the same height separated by a slight lowering of water. (b) A low tide consisting of two low-water minima separated by a slight rise of water.

double tombolo Two separate bars or barriers connecting an island (usually of large extent and close to the shore) with the mainland. Cf: *tombolo.*

double valley A valley with a low divide on its floor, from which one stream flows in one direction and a second stream flows in another.

doublure (dou-blure') An infolded margin of the *exoskeleton* of an arthropod, such as the reflexed ventral margin of the carapace integument of a crustacean or the inwardly deflected marginal part of the dorsal exoskeleton of a merostome; esp. a generally narrow band extending around the border of the dorsal exoskeleton of a trilobite, turned or bent under to the ventral side. Etymol: French, "doubling, lining."

doubly plunging fold (dou'-bly) A fold, either an anticline or a syncline, that reverses its direction of plunge within the observed area (Billings, 1954, p.46).

doughnut (dough'-nut) (a) A small circular *closed ridge* of glacial origin (Gravenor and Kupsch, 1959, p.52). (b) *rock doughnut.*

douglasite (doug'-la-site) A light green monoclinic mineral: $K_2Fe^{2+}Cl_4{\cdot}2H_2O$.

dousing (dous'-ing) *dowsing.*

doverite (do'-ver-ite) *synchysite-(Y).*

down An upland in southeastern England, generally treeless, underlain by chalk. Commonly used in the plural, e.g. South Downs.

downbuckle (down'-buck-le) An obsolete term for compressional downfolding of sialic crust, associated with oceanic trenches. Syn: *tectogene (b).*

downbuilding (down'-build-ing) *passive diapirism.*

downburst A strong and potentially destructive *thunderstorm* downdraft that impacts a horizontal distance ranging from less than 1 to 10 km.

downcutting (down'-cutting) Stream erosion in which the cutting is directed in a downward direction (as opposed to lateral erosion). Cf: *degradation [geomorph].* Syn: *vertical erosion; incision.*

downdip (down-dip') A direction that is downwards and parallel to the dip of a structure or surface. Cf: *updip.*

downdip block The rocks on the *downthrown* side of a fault. Cf: *updip block.*

downeyite (dow'-ney-ite) An adamantine colorless tetragonal mineral: SeO_2. A coal-seam combustion product.

downfaulted (down'-fault-ed) Said of the rocks on the *downthrown* side of a fault, or the *downdip block.* Cf: *upfaulted.*

downgoing plate At a *convergent plate boundary,* the plate that is being subducted and is sinking down into the mantle. Downgoing plates must be composed of oceanic lithosphere. Syn: *slab.*

downhole (down'-hole) adj. Located or used in a borehole; e.g., "downhole equipment". adv. Deeper; e.g. "to perforate downhole". Cf: *uphole.*

downhole assembly The drill collars, subs and drill bit on the bottom of a drillstring used in rotary drilling.

downhole ground In exploratory drilling, a long electrode attached to the logging cable some distance (perhaps 100 ft) above an electrical logging sonde or hung just below the casing and used as the reference electrode instead of a reference electrode at the surface.

downhole method [seis] (a) Measuring near-surface velocity with geophones or hydrophones in a borehole and a source on or near the surface . (b) Engineering seismology measurement of P- and S-wave velocities utilizing a source at the surface and a clamped triaxial geophone in a borehole. (c) *Induced polarization* (IP) method that explores the region near a drill hole using a single potential or current electrode in the drill hole and other electrodes on the ground surface.

downlap (down'-lap) A base-discordant relation in which initially inclined strata terminate downdip against an initially horizontal or inclined surface (Mitchum, 1977, p.206). See also: *distal downlap.*

downslope (down'-slope) n. A slope that lies downward; downhill. adj. In a downward direction, or descending; e.g. a *downslope* ripple that migrated down a sloping surface.

downstream fining In streams, a commonly observed pattern in which the average size of sediment present on the bed of a river becomes progressively smaller downstream.

down-structure method (down'-struc-ture) The method of examining structures on a geologic map by orienting the map so as to look down "into" it along the direction of plunge of the structures. The structures then appear in much the same attitude as they would in a profile plane, since any plane, e.g. one parallel to the ground surface, that intersects plunging structures produces such a pattern, though with different proportions. The method is very useful for quick interpretation of complex folding and faulting (Mackin, 1950). Syn: down-plunge projection.

downthrow (down'-throw) n. (a) The downthrown side of a fault. (b) The amount of downward vertical displacement of a fault. Cf: *upthrow; heave [struc geol].*

downthrown (down'-thrown) Said of that side of a fault that appears to have moved downward, compared with the other side. Cf: *upthrown.*

down-to-basin fault A term used in petroleum geology for a fault whose downthrown side is toward the adjacent basin.

Downtonian (Down-ton'-i-an) A series in the Old Red Sandstone of England: Lower Devonian (lowermost Gedinnian; below Dittonian). It was originally assigned to the uppermost Silurian (upper Ludlowian).

downward continuation Re-computing the observed potential field, such as the Earth's gravity or nagnetic field, at the elevation (level) that is lower than the level of the actual measurements. The process emphasizes both short-wavelength anomalies and noise. It also demands that there are no sources between two levels. See also: *continuation; upward continuation.*

downwarping (down'-warp-ing) Subsidence of a regional area of the Earth's crust, as in an *orogenic belt* or a *centrocline.* Cf: *upwarping.*

downwash (down′-wash) Fine-grained surface material (such as soil) moved down a mountain slope or hillside by rain, esp. where there is little vegetation. Cf: *rainwash; sheetwash.*
downwasting (down′-wast-ing) The thinning of a glacier during ablation. Cf: *backwasting.*
downwearing (down′-wear-ing) Erosion that causes the flattening-out of a hill or mountain and the decline of its slope; a process contributing to the development of a peneplain. Cf: *backwearing.*
downwelling (down′-welling) *sinking.*
dowser (dows′-er) (a) One who practices *dowsing.* Syn: *diviner.* (b) *divining rod.*–Syn: *water witch; waterfinder.*
dowsing (dows′-ing) The practice of searching for groundwater, mineral deposits, or other objects by means of a *divining rod* or a pendulum. Some dowsers claim also to be able to diagnose diseases, determine the sex of unborn babies, find lost items, locate buried treasure, etc. Syn: *dousing; divining; water witching.* Cf: *rhabdomancy; pallomancy.*
dowsing rod *divining rod.*
doxenic (dox-en′-ic) In the *CIPW classification* of igneous rocks, those rocks in which the ratio of oikocrysts to phenocrysts is less than three to five but greater than one to seven. Obsolete. Cf: *perxenic.*
doyleite (doyle′-ite) A pearly white triclinic mineral: $Al(OH)_3$. A dimorph of nordstrandite.
dozyite A regularly ordered, 1:1 mixed-layer of amesite and clinochlore: $(Mg,Al,Fe^{2+})_9(Si,Al)_6O_{15}(OH)_{12}$.
draa The largest class of eolian sand bedforms. These forms have lengths on the order of kilometers and heights in tens or hundreds of meters (Collinson, 1986). Their windward and leeward faces commonly have superimposed sand dunes (Cooke and Warren, 1973, p.282).
draft A term used in eastern U.S. for a gully or gorge, and for a small stream or creek.
drag [eco geol] *drag ore.*
drag [hydraul] (a) The friction of moving air against a water surface which tends to pull the water-surface layer in the direction of the wind. (b) The force exerted by a flowing fluid on an object in or adjacent to the flow. Cf: *push [hydraul].*
drag [struc geol] The folding of strata on either side of a fault; also, the folds so formed.
drag cast The *cast* of a *drag mark;* a more appropriate term for *drag mark* as used by Kuenen (1957, p.243-245).
drag coefficient The ratio of the force per unit area exerted on a body by a flowing liquid to the pressure at the stagnation point (ASCE, 1962).
drag fold A minor fold, formed due to shear resulting from slip on a fault. Cf: *normal drag; reverse drag; parasitic fold; intrafolial fold.*
drag groove A *drag mark* consisting of a long, narrow, even groove.
drag line A short, feeble glacial striation formed on the lee side of an older glacial groove.
dragma (drag′-ma) A siliceous monaxonic sponge spicule (microsclere) occurring in bundles and produced within a single sclerocyte. Pl: dragmata.
drag mark (a) A long, even groove or *striation* made by a solid body dragged over a soft sedimentary surface, as by a stone or shell pulled along the mud bottom by attached algae; it tends to be narrower and deeper than a typical *slide mark.* See also: *drag groove; drag striation.* (b) A term used by Kuenen (1957, p.243-245) for the structure called *groove cast* by Shrock (1948, p.162-163), being a broad and rounded or flat-topped or sharp-crested ridge, commonly with longitudinal striations, formed on the underside of an overlying bed by the filling of a *drag groove* probably under turbidity-current conditions. A more appropriate term for this feature would be *drag cast.*
Dragonian (Dra-go′-ni-an) North American continental stage: Lower Paleocene (above Puercan, below Torrejonian).
dragonite (drag′-o-nite) A rounded quartz pebble, representing a quartz crystal that has lost its brilliancy and form; it was once once believed to be a fabulous stone obtained from the head of a flying dragon.
drag ore Crushed and broken fragments of rock or ore torn from an orebody and contained in and along a fault zone. See also: *trail of a fault.* Syn: *drag [eco geol].*
drag striation A *drag mark* consisting of a short, narrow striation, curved or straight. Dzulynski and Slaczka (1958, p.234) used "drag stria" for a feature that is essentially a *striation cast.*
drain n. (a) An opening for a complete emptying of a reservoir or container. (b) A small, narrow natural watercourse. (c) A channel, conduit, or waterway, either natural or artificial, for draining or carrying off excess water from an area, such as a ditch designed to lower the water table so that land may be farmed; a sewer or trench. v. To carry away the surface water or discharge of streams in a given direction or to an outlet.
drainage (drain′-age) (a) The manner in which the waters of an area pass or flow off by surface streams or subsurface conduits. (b) The processes of surface discharge of water from an area by streamflow and sheet flow, and the removal of excess water from soil by downward flow. Also, the natural and artificial means for effecting this discharge or removal, such as a system of surface and subsurface conduits. (c) A collective term for the streams, lakes, and other bodies of surface water by which a region is drained; a *drainage system.* (d) The water features of a map, such as seas, lakes, ponds, streams, and canals. (e) An area or district drained of water, as by a stream; a drainage area. (f) The act or an instance of removing water from a previously marshy land area.
drainage area [petroleum] That area from which one well can produce the hydrocarbons contained in the reservoir rock.
drainage area [streams] The horizontal projection of the area whose surface directs water toward a stream above a specified point on that stream; a *drainage basin.*
drainage basin A region or area bounded by a *drainage divide* and occupied by a *drainage system;* specif. the tract of country that gathers water originating as precipitation and contributes it to a particular stream channel or system of channels, or to a lake, reservoir, or other body of water. Cf: *river basin.* Syn: *watershed; drainage area [streams]; catchment; catchment area [streams]; catchment basin; gathering ground; hydrographic basin.*
drainage coefficient The amount of runoff (expressed in water-depth or other units) removed or drained from an area in 24 hours.
drainage composition Quantitative description of a drainage basin in terms of stream order, drainage density, bifurcation ratio, and stream-length ratio. Term introduced by Horton (1945, p.286) to imply "the numbers and lengths of streams and tributaries of different sizes or orders, regardless of their pattern".
drainage density The ratio of the total stream lengths of all the stream orders within a drainage basin to the area of that basin projected to the horizontal; approximately the reciprocal of the *constant of channel maintenance* (Shreve, 1969, p.412). It is an expression of topographic texture: high density values are favored in regions of weak or impermeable surface materials, sparse vegetation, mountainous relief, and high rainfall intensity. Term introduced by Horton (1932, p.357) to represent the average stream length within the basin per unit area.
drainage district A governmental corporation or other public body created by a state to control drainage in a specified area; it functions under legal regulations for financing, constructing, and operating a drainage system.
drainage divide The boundary between adjacent drainage basins; a divide.
drainage lake An *open lake* that loses water through a surface outlet or whose level is largely controlled by the discharge of its effluent (Wetzel, 2001, p.46). Cf: *seepage lake.*
drainage line The course or channel of a major stream in a drainage system.
drainage network *drainage pattern.*
drainage pattern The configuration or arrangement in plan view of the natural stream courses in an area. It is related to local geologic and geomorphologic features and history. Syn: *drainage network.*
drainage ratio The ratio between runoff and precipitation in a given area for a given period of time. Syn: *discharge efficiency.*
drainage system A surface stream, or a body of impounded surface water, together with all other such streams and water bodies that are tributary to it and by which a region is drained. An artificial drainage system includes also surface and subsurface conduits.
drainage varve An abnormally thick and sandy varve formed by the drainage of lakes ponded between the ice edge and higher land, or behind dams of glacial deposits.
drainageway (drain′-age-way) A channel or course along which water moves in draining an area.
drainage well A type of *inverted well* used to drain excess soil or surface water, where the aquifer penetrated is permeable enough, and has a head far enough below the land surface, to remove the water at a satisfactory rate. Drainage wells have been used to dispose of some untreated domestic and other wastes, but such uses are now largely prohibited. Cf: *relief well.*
drain tile Porous pipe used for collection of excess groundwater.
drakonite (drak′-o-nite) An extrusive rock composed of phenocrysts of alkali feldspar, plagioclase, and biotite and/or hornblende, in a trachytic groundmass of alkali feldspar microlites and interstitial alkali amphibole or pyroxene. The plagioclase ranges from oligoclase to labradorite. Apatite, titanite, magnetite, and zircon may be present as accessories. Obsolete.
drape fold (a) A *supratenuous fold* or *compaction fold.* (b) A fold produced in layered rocks by movement of an underlying brittle block at high angles to the layering (Friedman et al., 1976, p.1049); a type of *forced fold.* In this usage, the term excludes supratenuous or compaction folds.
drape lamination Sedimentary unit deposited from suspension blanketing underlying bed topography.
draper point (drap′-er) The temperature, 977°F (525°C), at which red light first becomes visible from a heated object in darkened surroundings; hence, the minimum temperature of incandescent lava (Draper, 1847; Siegel and Howell, 1968).
drapery (drap′-er-y) In a cave,a thin, translucent sheet of travertine formed when drops of water flow down an inclined cave ceiling and leave behind a sinuous trail of calcite. Syn: *drip curtain.* Cf: *bacon-rind drapery.*
draping [reef] (drap′-ing) The general structural concordance of warped strata, lying above a limestone reef or other hard core, to the upper surface of that reef or core.
draping [remote sensing] The display of selected two-dimensional data on a perspective view of terrain relief or any other spatially distributed variable. For example, a geological map may by draped over a perspective view of surface elevations (Krzanowski et al., 1993).

dravite (dra'-vite) A brown, magnesium-rich rhombohedral mineral of the *tourmaline* group: $NaMg_3Al_6(BO_3)_3Si_6O_{18}(OH)_4$.

draw [geomorph] (a) A small natural watercourse or gully, generally shallower or more open than a ravine or gorge; a shallow gulch; a valley or basin. (b) A usually dry stream bed; a coulee whose water results from periodic rainfall. (c) A sag or troughlike depression leading up from a valley to a gap between two hills.

draw [mining] The horizontal distance, measured on the surface ahead of an underground coal face, over which the rocks are influenced by subsidence.

drawdown (draw'-down) (a) The lowering of the water level in a well as a result of withdrawal. (b) The difference between the height of the static (undisturbed) water table and that of the water level in a pumped well. (c) Reduction of the pressure head as a result of the withdrawal of water from a well. (d) In a stream or conduit, the difference between the water-surface elevation at a constriction and what the elevation would be if there were no constriction (ASCE, 1962). (e) The amount that the level of a reservoir, water level in a well, or head in an aquifer is lowered by the withdrawal of water. Cf: *cone of depression.*

drawdown curve A graph showing the relation between drawdown and radial distance from a pumping well, often used in the analysis of a pumping test.

drawdown model A conceptual model by Maiklem (1971) to account for thick deposits of marine evaporites with shallow-water textures. It suggests that a deep-water basin is isolated from the ocean by a drop in sea level, allowing it to evaporate to near dryness. Additional water may be added by seepage through a barrier or by storms or spill over a sill.

draw slate In coal mining, shale that occurs above a coal seam and collapses during or shortly after removal of the coal.

drawworks The powered winch used in *rotary drilling* located on the drillfloor and used for lifting and lowering the drillstring and other equipment in the well.

D ray Left posterior ray in echinoderms situated clockwise of C ray when the echinoderm is viewed from the adoral side; equal to ambulacrum I in the *Lovenian system.*

DRE *dense-rock equivalent.*

dredge [eng] (a) A floating machine for sucking or scooping up or excavating earth material from the bottom of a body of water, raising it to the surface, and discharging it to the bank through a floating pipeline or conveyor, into a scow for removal, or, in the case of certain mining dredges, into the same body of water after removal of the ore mineral. A "hydraulic dredge" uses a centrifugal pump; other dredges use dippers, clamshells, bucket chains, and scrapers. They may or may not be self propelled. Dredges are used to excavate or deepen harbor channels; to raise the level of lowland areas; to dig ditches and improve drainage; and to obtain sand, gravel, placer gold, and other materials. (b) A ship designed as such a machine.

dredge [oceanog] An ocean-bottom sampler that scrapes material from the sea floor as the device is dragged behind a slowly moving ship. It may be a heavy metal frame with an attached chain mail bag for collecting rock or a heavy frame with a strong net bag protected by anti-chafing material for collecting biological samples. Cf: *corer; grab sampler.*

dredge peat *sedimentary peat.*

dredging A form of *openpit mining* in which the digging machinery and processing plant are situated on a floating barge or hull. Cf: *dredge [eng].*

dreikanter (drei'-kan-ter) (a) A doubly pointed ventifact or wind-worn stone, having three curved faces intersecting in three sharp edges, resembling a Brazil nut. Syn: *pyramid pebble.* (b) A term loosely applied as a syn. of *ventifact.* Etymol: German "Dreikanter", "one having three edges". Pl: dreikanters; dreikanter.

Dresbachian (Dres-bach'-i-an) North American stage: Middle-Upper Cambrian (above Mayaian, below Franconian) (Cowie and Bassett, 1989).

dresserite (dres'-ser-ite) A silky white orthorhombic mineral: $Ba_2Al_4(CO_3)_4(OH)_8{\cdot}3H_2O$.

drewite (drew'-ite) An obsolete term for white neritic impalpable calcareous mud or ooze, consisting chiefly of minute aragonite needles a few micrometers in length, believed to have been precipitated directly from seawater through the action of nitrate- and sulfate-reducing bacteria. Named after George Harold Drew (1881-1913), British scientist.

dreyerite (drey'-er-ite) An adamantine orange-yellow tetragonal mineral of the *xenotime* group: $BiVO_4$.

DRG *digital raster graphic.*

driblet (drib'-let) *spatter [pyroclast].*

driblet cone *hornito.*

dries n. "An area of a reef or other projection from the bottom of a body of water which periodically is covered and uncovered by the water" (Baker et al., 1966, p.51). Syn: *uncovers.*

drift [coast] Detrital material moved and deposited by waves and currents; e.g. *littoral drift.* Also, floating material (e.g. driftwood or seaweed; flotsam) that has been washed ashore by waves and left stranded on a beach.

drift [drill] (a) *inclination [drill].* (b) The deviation of a well from its intended direction. See also: *deviation [drill].*

drift [geophys] A gradual change in a reference reading that is supposed to remain constant. An instrument such as a gravimeter may show drift as a result of elastic aging, long-term creep, hysteresis, or other factors. Different from tare, which is a sharp sudden change of reference value. See also: *drift correction.*

drift [glac geol] A general term applied to all rock material (clay, silt, sand, gravel, boulders) transported by a glacier and deposited directly by or from the ice, or by running water emanating from a glacier. Drift includes unstratified material (*till*) that forms moraines, and stratified deposits that form *outwash plains, eskers, kames, varves, fluvioglacial* sediments, etc. The term is generally applied to Pleistocene glacial deposits in areas (as large parts of North America and Europe) that no longer contain glaciers. The term "drift" was introduced by Murchison (1839, v.1, p.509) for material, then called *diluvium,* that he regarded as having drifted in marine currents and accumulated under the sea in comparatively recent times; this material is now known to be a product of glacial activity. Cf: *glacial drift; fluvioglacial drift.*

drift [hydraul] "The effect of the velocity of fluid flow upon the velocity (relative to a fixed external point) of an object moving within the fluid" (Huschke, 1959, p.178).

drift [mining] A horizontal or nearly horizontal underground opening driven from a shaft to an orebody.

drift [oceanog] (a) One of the wide, slower movements of surface oceanic circulation under the influence of, and subject to diversion or reversal by, prevailing winds; e.g. the easterly drift of the North Pacific. Syn: *drift current; wind drift; wind-driven current.* Cf: *Ekman current.* (b) The slight motion of ice or vessels resulting from ocean currents and wind stress. (c) The speed of an ocean current or ice floe, usually given in nautical miles per day or in knots. (d) Sometimes used as a short form of *littoral drift.*

drift [photo] Apparent offset of aerial photographs with respect to the true flight line, caused by the displacement of the aircraft owing to cross winds, and by failure to orient the camera to compensate for the angle between the flight line and the direction of the aircraft's heading. The photograph edges remain parallel to the intended flight line, but the aircraft itself drifts farther and farther from that line.

drift [sed] n. (a) A general term, used esp. in Great Britain, for all surficial, unconsolidated rock debris transported from one place and deposited in another, and distinguished from solid bedrock; e.g. *river drift.* It includes loess, till, river deposits, etc., although the term is often used specif. for glacial deposits. (b) Any surface movement of loose incoherent material by the wind; also, an accumulation of such material, such as a *snowdrift* or a *sand drift.* v. To accumulate in a mass or be piled up in heaps by the action of wind or water.

drift [stat] Term used in *geostatistics* to indicate spatial nonstationarity or change in average value of a property with location. The drift can be modeled as an arbitrary deterministic function of the geographic coordinates of the observations. Cf: *trend [stat].*

drift [streams] (a) In South Africa, a ford in a river. The term is used in many parts of Africa to indicate a ford or a sudden dip in a road over which water may flow at times (Stamp, 1961, p.162). (b) As used by aquatic ecologists, the passive movement downstream by organisms such as invertebrates.

drift [tect] *continental drift.*

drift avalanche *dry-snow avalanche.*

drift-barrier lake A glacial lake formed upstream from a moraine that has blockaded a valley or a drainage course (Fairchild, 1913, p.153). Cf: *valley-moraine lake.*

drift bed A layer of drift "of sufficient uniformity to be distinguished from associated ones of similar origin" (Fay, 1918, p.231).

drift bottle A bottle containing a record of the date and place at which it was released into the sea and a card requesting return by the finder with the date and place of recovery. It is used in studying surface currents.

drift clay *boulder clay.*

drift coal Coal formed according to the drift theory; allochthonous coal.

drift copper Native copper transported from its source by a glacier.

drift correction Adjustment to remove the effects of geophysical drift, usually by repeated observations at a base station.

drift current *drift [oceanog].*

drift curve A graph of a series of gravity values read at the same station at different times and plotted in terms of instrument reading versus time; a procedure to determine the internal drift appropriate to the instrument used to correct gravity readings between drift readings.

drift dam A dam formed by the accumulation of glacial drift in a pre-existing stream valley.

drift epoch A syn. of *glacial epoch;* specif. the "Drift epoch", also known as the Pleistocene Epoch. See also: *Drift period.*

drifter (a) Free floating instrument having the same density as water at some specified depth in the ocean, which moves with the water at that depth, making a *Lagrangean* measurement of current at that depth; e.g., Profiling Autonomous Lagrangean Circulation Explorers (PALACE) are used to map circulation at a depth of two kilometers. (b) Free-floating instrument at the sea surface that moves with the water at the surface, making a *Lagrangean* measurements of the surface current. Commonly called a *drogue.*

drift glacier A small mass of flowing ice in a mountain area nourished primarily with windblown snow from adjacent snowfields, slopes, or ridges. Syn: *Ural-type glacier; snowdrift glacier; glacieret.* Less preferred syn: *catchment glacier.*

drift ice (a) Any ice that has been been broken apart and drifted from its place of origin by winds and currents, e.g., a fragment of a *floe* or a detached *iceberg;* loose, unattached pieces of *floating ice,* with the area of open water exceeding that of ice; navigable with ease. (b) A syn. of *pack ice* as that term is used in a broad sense.

drifting [mining] Driving a *drift [mining].*

drifting ice station (drift'-ing) An oceanographic research base established on ice in the Arctic Ocean.

drift lake A glacial lake occupying a depression left in the surface of glacial drift after the disappearance of the ice (White, 1870, p.70).

driftless area (drift'-less) A region that was surrounded, but presumably not covered, by continental ice sheets of the Pleistocene Epoch, and is supposedly devoid of glacial deposits; specif. the "Driftless Area" occupying SW Wisconsin and parts of Illinois, Iowa, and Minnesota.

drift line A line of drifted material washed ashore and left stranded. It marks the highest stage of water, such as of a flood.

drift map A 19th century British term for a geologic map representing a true picture of the visible ground, including all surficial deposits and only those rocks exposed at the surface (Nelson and Nelson, 1967, p.114). Cf: *solid map.*

driftmeter (drift'-me-ter) An instrument for determining the inclination of a drill pipe from the vertical and the depth of measurement.

drift mining (a) The extraction of *placer* ore by underground horizontal or inclined tunneling rather than by the use of water. Cf: *placer mining.* (b) The extraction of near-surface coal seams by underground inclined tunneling rather than by opencut mining or vertical-shaft methods.

drift peat Peat that occurs in association with glacial drift.

Drift period A term formerly used to designate the Pleistocene Epoch. See also: *drift epoch.*

drift plain A plain, e.g. a *till plain,* underlain by glacial drift.

drift scratch *glacial striation.*

drift sheet A widespread sheetlike body of glacial drift, deposited during a single glaciation (e.g. the Cary drift sheet) or during a series of closely related glaciations (e.g. the Wisconsin drift sheet).

drift terrace A term used in New England for an *alluvial terrace.*

drift theory [coal] The theory that coal originates from the accumulation of plant material that has been transported from its place of growth and deposited in another locality, where coalification occurs. Ant: *in-situ theory.* See also: *allochthony.*

drift theory [glac geol] A theory of the early 19th century which attributed the origin of widespread surficial deposits, including the erratic boulders, to the action of marine currents and floating ice. Cf: *glacier theory.*

driftwood (drift'-wood) Woody material, such as parts of trees, drifted or floated by water and cast ashore or lodged on beaches by storm waves.

drill n. A device used for making circular holes in rock or earth material by a succession of blows or by rotation of the cutting surface. Syn: *drill bit;* rock bit. v. To make a circular hole with a drill or other cutting tool.

drill bit *drill.*

drill collar A length of extra-heavy, thick-walled *drill pipe* in a rotary *drillstring* directly above either the bit or the core barrel, to concentrate weight and give rigidity so that the bit will cut properly. Syn: *collar [drill].*

drill cuttings *well cuttings.*

drilled well A well constructed by cable-tool or rotary drilling methods in the search for water, minerals, oil, or gas.

driller's log (drill'-er's) The brief, often vernacular notations, included as part of a driller's *tour report,* that describe the gross characteristics of the well cuttings noted by the drilling crew as a well is drilled. It is useful only if a detailed sample log is not available.

drill hole A circular hole made by drilling; to explore for valuable minerals, water or petroleum or to obtain geologic information. Cf: *borehole.* Syn: *hole [drill].*

drilling (drill'-ing) The act or process of making a circular hole with a drill or other cutting tool, for purposes such as blasting, exploration, prospecting, valuation, or obtaining oil, gas, or water.

drilling and spacing unit The unit allotted by a regulatory agency for the positioning of producing oil or gas wells or exploratory wells intended to be productive. Common spacings are 40 acres between oil wells and 360 acres between gas wells. Abbrev: DSU.

drilling in The process of completing, or drilling into, a potential oil-, water-, or gas-bearing formation by setting *casing* just above the zone of this formation. If the zone proves to be productive, it commonly will be cased off.

drilling mud A carefully formulated heavy suspension, usually in water but sometimes in oil, or an oil-in-water emulsion, used in rotary drilling. It commonly consists of bentonitic clays, chemical additives, and weighting materials such as barite. It is pumped continuously down the drillstring, out through openings in the drill bit, and back up in the annulus between the pipe and the walls of the hole to the mud tanks where it is screened and reintroduced through the *mud pump.* The mud is used to lubricate and cool the bit; to carry the cuttings up from the bottom; and to prevent blowouts and cave-ins by plastering friable or porous formations with mud cake, and maintaining a hydrostatic pressure in the borehole offsetting pressures of fluids that may exist in the formation. Cf: *oil-base mud.* Syn: *mud [drill].*

drilling rig (a) In the petroleum industry, a general term for all the equipment necessary in rotary or cable-tool drilling. (b) In environmental/geotechnical usage, a truck or trailer mounted, portable drilling system and accessories. Syn: *rig.*

drilling time (a) The time required for a rotary drill bit to penetrate a specified thickness (usually one foot or one meter) of rock. (b) The elapsed time required to drill a well, excluding periods when not actually drilling.

drilling-time log A *strip log* showing the times required to drill an increment of depth, e.g. minutes per foot, in a borehole. Rapid drilling (a fast break) often indicates a porous reservoir rock.

drill pipe The heavy steel pipe that turns the drill bit in *rotary drilling* by transmitting the motion from the rotary table of the drilling rig to the bit at the bottom of the hole, and that conducts the *drilling mud* from the surface to the bottom. It is normally formed of 30-ft sections called joints that are connected end to end. Cf: *drill collar.*

drill stem (a) A term used in *rotary drilling* for the *drillstring.* (b) A term used in *cable-tool drilling* for a solid shaft or cylindrical bar of steel or iron attached to the drill bit to give it weight. Syn: *stem [drill].*

drill-stem test A procedure for determining the potential productivity of an oil or gas reservoir by measuring reservoir pressures and flow capacities while the drill pipe is still in the hole, the well is still full of drilling mud, and usually the well is uncased. The tool consists of a packer to isolate the section to be tested and a chamber to collect a sample of fluid. It is similar to a temporary completion of the well. Abbrev: DST. Cf: *wireline test.*

drill stem test temperature That temperature taken during a drill stem test in a well. Usually closer to the formation temperature than a bottom-hole temperature if formation fluids have been produced with expansion into the borehole. Abbrev: DST temperature.

drillstring A term used in *rotary drilling* for the assemblage in a borehole of drill pipe, drill collars, subs, drill bit, and core barrel (if in use), connected to and rotated by the *drilling rig* at the surface. Syn: *drill stem.*

drip curtain *drapery.*

driphole (drip'-hole) (a) A small hole or niche in clay or rock beneath a point where water drips. (b) The center hole in a feature built up beneath dripping water.

dripstone (drip'-stone) A general term for a deposit calcite or other mineral formed in caves by dripping water, including *stalactites* and *stalagmites.* Cf: *cave onyx; speleothem; travertine.*

dripstone cement A cement occurring on the underside of grains indicating precipitation in the meteoric vadose zone, normally composed of calcite.

driven well (driv'-en) A shallow well, usually of small diameter (3-10 cm), constructed by driving a series of connected lengths of pipe into unconsolidated material to a water-bearing stratum, without the aid of any drilling, boring, or jetting device. Syn: *tube well.*

driving stress The sum of the remote tensile stress and internal fluid pressure acting on a crack or joint (Pollard and Segall, 1987). Compressive stress is a negative driving stress which inhibits crack or joint propagation.

drizzle A form of liquid *precipitation* consisting of water droplets having diameters between 0.2 and 0.5 mm (0.01 and 0.02 in.); most often falls from low stratus clouds.

DRM *depositional remanent magnetization.*

drogue *drifter.*

droogmansite (droog'-man-site) *kasolite.*

drop [coal] In the roof of a coal seam, a funnel-shaped downward intrusion of sedimentary rock, usually sandstone. Cf: *stone intrusion.*

drop [hydraul] The difference in water-surface elevations measured upstream and downstream from a constriction in the stream.

drop-down curve The form of the water surface along a longitudinal profile, assumed by a stream or open conduit upstream from a sudden fall. In a uniform channel, the curve is convex upward (ASCE, 1962). Cf: *backwater curve.*

dropline A process of graphic elimination of boundaries between polygons having common selected attributes. This results in the generation of simplified polygon maps (Krzanowski et al., 1993). See also: *dissolve.*

dropoff Localized and sharp increase in slope, as on the outer side of a tidal flat, over a ledge, between areas separated by a cliff, etc.

dropout (drop'-out) A loss of information in reading from or recording on magnetic tape or other recording media.

dropped block *dropstone.*

dropped coverage The portion of a seismic line that is not shot, perhaps because of permit problems, access difficulties, or danger of damage.

dropstone (drop'-stone) (a) An oversized stone in laminated sediment that depresses the underlying laminae and may be covered by "draped" laminae. Most dropstones originate through *ice-rafting;* other sources are floating tree roots and kelp holdfasts (Crowell, 1957; Ojakangas, 1985). Cf: *lonestone.* (b) *dripstone.*

drought In general, an extended period of dry weather, or a period of deficient rainfall that may extend over an indefinite number of days, without any quantitative standard by which to determine the degree of deficiency needed to constitute a drought. Qualitatively, it may be defined by its effects as a dry period sufficient in length and severity to cause a least partial crop failure or to impair the ability to meet a normal water demand.

drought index A computed value used as a measure of drought conditions in a locale. Typically the index refers to the Palmer Drought Severity Index designed by Wayne Palmer in 1965 that determines the deficit or excess of soil moisture compared with average climatic values based upon observed precipitation and on potential evaporation as determined from the average temperature.

drought polygon *desiccation polygon.*

drowned Said of a land surface or land feature that has undergone *drowning;* e.g. a *drowned* coast.

drowned atoll An atoll occurring at great depth, so that further coral-algal reef growth is prevented. See also: *drowned reef.*

drowned reef A reef situated at such great depth that reef growth is prevented or greatly hampered. See also: *drowned atoll.*

drowned river mouth The lower end of a river valley that is widened or submerged by seawater invading the coast; an *estuary.* Example: Chesapeake Bay.

drowned valley A valley that is partly submerged by the intrusion of a sea or lake. Syn: *submerged valley.*

drowning (drown'-ing) The submergence of a land surface or topography beneath water, either by a rise in the water level or by a sinking or subsidence of the land.

drugmanite (drug'-man-ite) An adamantine pale yellow monoclinic mineral: $HPb_2(Fe^{3+},Al)(PO_4)_2(OH)_2$.

druid stone (dru'-id) A *sarsen* used in the building of ancient stone circles at Stonehenge and elsewhere in Great Britain.

drum (a) A Scottish term for a long narrow ridge. (b) *drumlin.*

drumlin (drum'-lin) (a) A low, smoothly rounded, elongate oval hill, mound, or ridge of compact glacial till or, less commonly, other kinds of drift (sandy till, varved clay), built under the margin of the ice and shaped by its flow, or carved out of an older moraine by readvancing ice; its longer axis is parallel to the direction of movement of the ice. It usually has a blunt nose pointing in the direction from which the ice approached, and a gentler slope tapering in the other direction. Height is 8-60 m, average 30 m; length is 400-2000 m, average 1500 m. Syn: *drum.* (b) *rock drumlin.* Etymol: Irish and Gaelic, diminutive of "druim", "back, ridge".

drumlin field *basket-of-eggs topography.*

drumlinoid (drum'-lin-oid) A *rock drumlin* or a drift deposit whose form approaches that of a true drumlin but does not fully attain it even though it seemingly results from the work of moving ice.

drumloid (drum'-loid) An oval hill or ridge of glacial till whose shape resembles that of a *drumlin* but is less regular and symmetrical.

drumstick barrier island A *barrier island* on a *mesotidal* coast with a bulbous up-drift end and a narrow down-drift end. It results from wave refraction around the *ebb delta* and leads to downdrift offset (Hayes, 1979).

drupe A fruit with a fleshy pericarp and a stony pit, e.g. a peach. The pits are readily fossilized.

druse (a) An irregular cavity or opening in a vein or rock, having its interior surface or walls lined (encrusted) with small projecting crystals usually of the same minerals as those of the enclosing rock, and sometimes filled with water; e.g. a small solution cavity, a steam hole in lava, or a lithophysa in volcanic glass. Cf: *geode; vug.* (b) A mineral surface covered with small projecting crystals; specif. the crust or coating of crystals lining a druse in a rock, such as sparry calcite filling pore spaces in a limestone. Etymol: German. Adj: *drusy.*

drusy (dru'-sy) (a) Pertaining to a *druse,* or containing many druses. Cf: *miarolitic.* (b) Pertaining to an insoluble residue or encrustation, esp. of quartz crystals; e.g. a "drusy oolith" covered with subhedral quartz.

dry In climatology, *arid.*

dry assay Any type of assay procedure that does not involve liquid as a means of separation. Cf: *wet assay.*

dry avalanche *dry-snow avalanche.*

dry basin An *interior basin* (as in an arid region) containing no perennial lake because the drainage is "occasional only and not continuous" (Gilbert, 1890, p.2).

dry-bone ore An earthy, friable, honeycombed variety of smithsonite, usually found in veins or beds in stratified calcareous rocks, accompanying sulfides of zinc, iron, and lead. The term is sometimes applied to hemimorphite.

dry bulk density The density of a dry substance, e.g., a rock or sediment without water or other fluid in the porosity.

dry calving The breaking-away of a mass of ice from a near-vertical glacier face on dry land.

dry cave A cave without flowing water; dripping water may be present. Cf: *fossil cave.*

dry delta (a) *alluvial fan.* (b) *alluvial cone.*

drydock iceberg *valley iceberg.*

dry frozen ground Relatively loose and crumbly ground (or soil) that has a temperature below freezing but contains no ice.

dry gap A *gap [geomorph]* eroded but not occupied by a stream; specif. a *wind gap.*

dry gas Natural gas with a very low content of liquid hydrocarbons. Cf: *wet gas.*

dry hole The universal term in the petroleum industry for an unsuccessful well, i.e. one that does not produce oil or gas in commercial quantities.

dry ice (a) Ice at a temperature below the freezing point; specif. bare glacier ice on which there is no slush or standing water (ADTIC, 1955, p.26). (b) Solidified carbon dioxide.

drying crack (dry'-ing) *desiccation crack.*

dry lake (a) A *lake basin* that formerly contained a lake. (b) A *playa;* a tract of salt-encrusted land in an arid or semiarid region, occasionally covered by an intermittent lake.

dryline A boundary betweeen *continental tropical air mass* and *maritime tropical air mass* in the southeast sector of a mature *extratropical cyclone;* it is a likely site for *squall line* and severe *thunderstorm* development.

dry mudflat A subaerially exposed plain of fine-grained sediment with abundant features produced by desiccation and in which there is no significant accumulation of evaporites because the groundwater table is too deep. Syn: *dry playa; clay pan.*

dry peat Peat derived from humic matter and formed under drier conditions than those of a moor.

dry permafrost Loose and crumbly permafrost containing little or no ice or moisture.

dry placer A placer in arid regions that is mined by dry washing with air rather than water.

dry playa A playa that is normally hard, buff in color, and smooth as a floor (Thompson, 1929); the water table is at a considerable distance beneath the dense, sun-baked surface. Syn: *dry mudflat; clay pan.* Cf: *wet playa.*

dry quicksand A sand accumulation that offers no support to heavy loads due to entrapped air.

drysdallite (drys'-dall-ite) A metallic grayish black hexagonal mineral of the *molybdenite* group: $Mo(Se,S)_2$.

dry snow Deposited snow that has not been subject to melting or to infiltration of liquid water. Cf: *wet snow.*

dry-snow avalanche An avalanche composed of dry, loose or powdery snow that is set in motion by the wind and is sometimes drifted but not wind-packed; the driving-ahead of a column of compressed air creates a vacuum in its wake. It is the fastest-moving of the snow avalanches, capable of reaching a speed of 450 km/hr. Syn: *dry avalanche; drift avalanche; powder avalanche.*

dry-snow line The boundary on a glacier or ice sheet between the *dry-snow zone* and an area where surface melting occurs.

dry-snow zone The area on a glacier or *ice sheet* where no surface melting occurs even in summer, delimited by the *dry-snow line.*

dry steam Steam that has an enthalpy greater than that for equilibrium with water for the existing pressure.

dry-steam reservoir *vapor-dominated reservoir.*

dry surge Pyroclastic surge containing steam that is mostly superheated. Surge deposits show little textural evidence of the presence of moisture during emplacement.

dry unit weight The *unit weight* of soil solids per unit of total volume of soil mass. See also: *maximum unit weight.* Syn: *unit dry weight.*

dry valley A valley that is devoid or almost devoid of running water; a streamless valley. It may be the result of stream capture, climatic change, or a fall in the water table. Dry valleys are common in areas underlain by chalk and limestone; other examples include wind gaps and glacial overflow channels. Syn: *dead valley.*

dry wash A wash that carries water only at infrequent intervals and for short periods, as after a heavy rainfall. Cf: *arroyo.*

dry weathering Mechanical weathering of a rock without the action of water, as in an arid region.

DSDP *Deep Sea Drilling Project.*

d-spacing In diffraction of X-rays by a crystal, the distance or separation between the successive and identical parallel planes in the crystal lattice. It is expressed as d in the *Bragg equation.*

DST *drill-stem test.*

DSU *drilling and spacing unit.*

DTA *differential thermal analysis.*

DTM *digital terrain model.*

dual induction log An *induction log* consisting of two induction curves with different depths of investigation.

dual porosity *double porosity.*

dubiocrystalline (du'-bi-o-crys'-tal-line) Said of the texture of a rock whose crystallinity can be determined only with difficulty or uncertainty; e.g. said of the texture of a porphyry whose groundmass is too fine to be resolved into its constituents under a microscope, but shows faint anisotropism or polarizes light like an aggregate. Also, said of a rock with such a texture. Term introduced by Zirkel (1893, p.455) as "dubiokrystallinisch". Cf: *cryptocrystalline.*

dubiofossil (du'-bi-o-fos'-sil) A structure of undetermined or uncertain origin, possibly biogenic (Hofmann, 1972, p.27); a *problematic fossil.* Etymol: Latin "dubius", "doubtful", + fossil.

Dubosq colorimeter (Du-bosq') An instrument that compares visually the color intensity of a solution of unknown strength with that of a variable depth of standard solution. From the depth of standard solution required to obtain a visual match, the strength of the unknown can be determined. See also: *colorimeter.*

Duchesnian (Du-ches'-ni-an) North American continental stage: Upper Eocene (above Uintan, below Chadronian).

Duck Lake North American (Gulf Coast) stage: Miocene (above Napoleonville, below Clovelly).

ducktownite (duck'-town-ite) A term used in Tennessee for an intimate mixture of pyrite and chalcocite, or for the matrix of a blackish copper ore containing grains of pyrite.

ductile (duc'-tile) Said of a rock that is able to sustain, under a given set of conditions, 5-10% strain without losing strength. Cf: *brittle.*

ductile behavior The response of rock to stress in which the rock appears to have deformed without the loss of continuity at the scale of observation. Ductile behavior may occur as a result of the *brittle-deformation mechanism* of *cataclastic flow* or *plastic deformation mechanisms* (Rutter, 1986).

ductility (duc-til'-i-ty) A measure of the degree to which a rock exhibits *ductile* behavior under given conditions, commonly expressed by the strain at which fracture or faulting commences.

ductolith (duc'-to-lith) A more or less horizontal igneous intrusion that resembles a tear drop in cross section. Obsolete.

duff mull A forest humus type, transitional between mull and mor, characterized by an accumulation or organic matter on the soil surface in friable Oe horizons, reflecting the dominant zoogenous decomposers. They are similar to mors in that they generally feature an accumulation of partially to well-humified organic materials resting on the mineral soil. They are similar to mulls in that they are zoologically active. Duff mulls usually have four horizons: Oi(L), Oe(F), Oa(H), and A. Sometimes differentiated into the following groups: Mormoder, Leptomoder, Mullmoder, Lignomoder, Hydromoder, and Saprimoder.

dufrénite (du-fré'-nite) (a) A blackish green or olive-brown platy monoclinic mineral: $Ca_{0.5}Fe^{2+}Fe^{3+}_5(PO_4)_4(OH)_6{\cdot}2H_2O$. Syn: *kraurite; green iron ore.* (b) A group name for minerals with the analogous formula, with Ca replaced by Na, and Fe^{3+} by Al.

dufrénoysite (du-fré-noy'-site) A metallic gray monoclinic mineral: $Pb_2As_2S_5$.

duftite (duft'-ite) An olive-green orthorhombic mineral: $PbCu(AsO_4)(OH)$. It is sometimes called α-duftite; β-duftite may be a intergrowth of α-duftite and conichalcite.

dugganite (dug'-gan-ite) An adamantine colorless to green trigonal mineral of composition $Pb_3(Zn,Cu)_3Te^{6+}O_6(AsO_4)_2$.

dug well A shallow, generally large-diameter well constructed by excavating with hand tools or power machinery instead of by drilling or driving, such as a well for individual domestic water supplies.

duhamelite (du-ham'-el-ite) Originally described as a greenish-yellow orthorhombic mineral of composition $Cu_4(OH)_3Pb_2Bi(VO_4)_4{\cdot}8H_2O$, but later shown to be a Bi- and Ca-bearing mottramite.

Duhem's theorem (Du'-hem's) The statement in chemistry that the state of any closed system is completely defined by the values of any two independent variables, extensive or intensive, provided the initial masses of each component are given. The choice of variables, however, must not conflict with the phase rule.

dukeite A yellow to yellow-brown trigonal mineral: $Bi^{3+}_{24}Cr^{6+}_8O_{57}(OH)_6{\cdot}3H_2O$.

dull-banded coal (dull'-band-ed) *Banded coal* consisting mainly of vitrain and durain, some clarain, and minor fusain. Cf: *bright-banded coal.*

dull coal A type of banded coal defined microscopically as consisting mainly of clarodurain and durain and of 20% or less of bright materials such as vitrain, clarain, and fusain. Cf: *bright coal; semidull coal; semibright coal; intermediate coal.* Syn: *dulls.*

dull luster The luster of a mineral or rock surface that diffuses rather than reflects light, even though the surface may appear smooth. Earthy materials have a dull luster.

dulls *dull coal.*

Duluth-type copper-nickel-PGE deposit A *mineral-deposit type* characterized by disseminated to sporadically distributed massive nickel-copper and PGE-bearing sulfides in the basal portions of layered mafic-ultramafic intrusions associated with incipient continental rift zones. Named for deposits in the Duluth Ultramafic Complex, northern Minnesota.

dumalite (du'-ma-lite) A *trachyandesite* characterized by intersertal texture and a glassy mesostasis which possibly has the composition of nepheline. It was named by Loewinson-Lessing in 1905 for Dumala in the Caucasus. Obsolete.

dumbbell (dumb'-bell) Two land areas connected by a relatively narrow isthmus of sand which is never below the high-water mark in any part of its length, and whose highest points are higher above sea level than any part of the isthmus (Schofield, 1920).

dummy observation An artificial observation given the average value of the category or group in which it is inserted, used to achieve better symmetry or balance in a statistical design. In *computer contouring,* dummy observations may be placed in a map to control *edge effects* near the limits of the map, or to control instability of interpolations in areas of limited control, or to force desired trends.

dummy variable An artificial variable expressing a qualitative state such as the presence or absence of an attribute, or indicating membership in a group or class.

dumontite (du-mon'-tite) A yellow monoclinic mineral: $Pb_2(UO_2)_3O_2(PO_4)_2{\cdot}5H_2O$. It is the phosphate analogue of hügelite.

dumortierite (du-mor'-ti-er-ite') A bright-blue or greenish-blue hard orthorhombic pseudo-hexagonal mineral: $Al_{27}B_4Si_{12}O_{69}(OH)_3$. It may contain iron, and it occurs principally in schists and gneisses.

dumortierite-quartz A massive opaque blue variety of crystalline quartz, colored by intergrown crystals of dumortierite.

dumped deposit An unsorted sediment deposited directly below wave base or current base, or brought in at a rate too rapid for waves and currents to distribute it (Weeks, 1952, p.2107); e.g. a shaly sand or a sandy shale.

dumping Abnormally rapid precipitation of minerals from a hydrothermal fluid.

dump moraine An *end moraine* consisting of *englacial* and *superglacial* material dropped by a glacier at its front.

dumpy level (dump'-y) A leveling instrument in which the telescope is permanently attached (either rigidly or by a hinge) to the vertical spindle or leveling base and that is capable only of rotatory movement in a horizontal plane. The dumpy level takes its name from the dumpy appearance of the early type of this instrument, the telescope of which was short and had a large object glass. Cf: *wye level.*

dun An inconspicuous hill in the English Lake District. Etymol: Gaelic.

dunal A nonrecommended adj. meaning of or pertaining to dunes.

dundasite (dun'-das-ite) A silky white orthorhombic mineral: $PbAl_2(CO_3)_2(OH)_4{\cdot}H_2O$.

dune [geomorph] A low mound, ridge, bank, or hill of loose, windblown granular material (generally sand), either bare or covered with vegetation, capable of movement from place to place but retaining its characteristic shape. Etymol: French. Cf: *sand dune.*

dune [streams] A term used by Gilbert (1914, p.31) for a *sand wave* formed on a stream bed and usually transverse to the direction of flow, traveling downstream by the erosion of sand from the gentle upstream slope and its deposition on the steep downstream slope, and having an approximately triangular cross section in a vertical plane in the direction of flow; a large-scale mound or ridge of sand, similar to an eolian sand dune, but formed in a subaqueous environment. Most modern authors follow Simons and Richardson (1961) in considering that dunes are features greater than 5 cm in height (as compared with ripples which are less than 5 cm and spacing > 60 cm). Most authors recognize straight crested forms produced under lower energy conditions (sometimes called transverse bars and sandwaves) and sinuous crested forms produced under higher energy conditons. Similar aqueous dunes form in response to tidal currents. See: Ashley (1990). Cf: *antidune.* Syn: *subaqueous sand dune.*

dune complex An aggregate of moving and fixed sand dunes in a given area, together with sand plains and the ponds, lakes, and swamps produced by the blocking of streams by the sand.

dune field Extensive deposits on sand in an area where the supply is abundant. As a characteristic, individual dunes somewhat resemble barchans but are highly irregular in shape and crowded; *erg* areas of the Sahara are an example.

dune lake A lake that is ponded by dune sand (Wetzel, 2001, p.31-32). Cf: *bajir.*

dune massif A large irregular cone- or pyramid-shaped dune with curved slopes and steep sides consisting of small hollows and terraces (Stone, 1967, p.225).

dune movement In hydraulics, the movement of sediment along the bed of a stream in the form of a wave or dune which travels downstream. The upstream face of the wave is eroded and the eroded material is deposited on the downstream face of the wave. The water surface has only a slight undulation (ASCE, 1962).

dune phase The part of stream traction whereby a mass of sediment travels in the form of a small, dunelike body having a gentle upcurrent slope and a much steeper downcurrent slope (Gilbert, 1914, p.30-34); it develops when the bed load is small or the current is weak. The dune form moves downstream. Cf: *smooth phase; antidune phase.*

dune ridge A series of parallel dunes, whose movements are arrested by the growth of vegetation, along the shore of a retreating sea. See also: *foredune.*

dune rock An *eolianite* consisting of dune sand.

dune sand A type of *blown sand* that has been piled up by the wind into a sand dune, usually consisting of rounded mineral grains, commonly quartz, having diameters ranging from 0.1 to 1 mm.

dune slack A damp slack or depression between dunes or dune ridges on a shore; a *dune valley.*

dune valley A hollow, furrow, or depression between dunes or dune ridges. Syn: *dune slack.*

dungannonite (dun-gan'-non-ite) An alkalic corundum-bearing *diorite.* Its name, given by Adams and Barlow in 1910, is derived from Dungannon, Ontario. Obsolete.

dunite (dun'-ite) (a) In the *IUGS classification,* a plutonic rock with M equal to or greater than 90 and ol/(ol+opx+cpx+hbd) greater than 90. (b) *Peridotite* in which the mafic mineral is almost entirely olivine, with accessory chromite almost always present. Named by Hochstetter in 1864 from Dun Mountain, New Zealand.

Dunkardian (Dun-kard'-i-an) North American provincial series: uppermost Pennsylvanian-Lower Permian (above Monongahelan).

duns A term used in SW England for a shale or massive clay associated with coal.

dunstone (a) A term used for a hard granular yellowish or cream-colored magnesian limestone which occurs near Matlock, England. (b) A term used in Wales for a hard fireclay or underclay, and in England for a shale.

Duntroonian (Dun-troon'-ian) South Pacific stage: upper Oligocene (below Waitakian).

DUPAL anomaly A region in the Earth's mantle that has anomalous Sr and Pb isotopic ratios and lies between 0 and 60°S. Named for Dupre and Allegre by S. R. Hart in 1984.

duplex *duplex fault zone.*

duplex fault zone (du'-plex) A structural complex consisting of a *roof thrust* at the top and a *floor thrust* at the base, within which a suite of more steeply dipping imbricate thrust faults thicken and shorten the intervening panel of rock (Dahlstrom, 1970, p.418-421; Boyer, 1976). Syn: *duplex.*

duplexite (du'-plex-ite) *bavenite.*

duplicature (du'-pli-ca-ture') A doubling or a fold; e.g. the distal portion of an ostracode *carapace* that is folded toward the interior of the valve around the free margin to form a doubling of the lamella (if fused, the radial pore canals can be seen extending from the inner to the outer margin, otherwise a cavity or vestibule is present, esp. in the anterior and posterior regions). See also: *skeletal duplicature.*

duplivincular (du-pli-vin'-cu-lar) Said of a type of *ligament* of a bivalve mollusk (e.g. *Arca*) in which the lamellar component is repeated as a series of bands, each with its two edges inserted in narrow grooves in the cardinal areas of the respective valves. Cf: *chevron groove.*

durability index (du-ra-bil'-i-ty) The relative resistance to abrasion exhibited by a sedimentary particle in the course of transportation, represented by the ratio of the *reduction index* of a standard (such as quartz) to that of a given rock or mineral under the same conditions (Wentworth, 1931, p.26). Abbrev: DI.

durain (du'-rain) A coal lithotype characterized macroscopically by dull, matte luster, gray to brownish black color, and granular fracture. It occurs in bands up to many centimeters in thickness; its characteristic microlithotype is *durite.* Cf: *vitrain; clarain; fusain.* Syn: *attritus.*

durangite (du-ran'-gite) An orange-red monoclinic mineral: $NaAl(AsO_4)F$. A member of the *tilasite* group.

Durangoan (Du-ran'-go-an) North American (Gulf Coast) stage: Lower Cretaceous (above LaCasitan of Jurassic, below Nuevoleonian).

duranusite (du-ran'-us-ite) A metallic red orthorhombic mineral: As_4S.

duration-area curve (du-ra'-tion-ar'-e-a) A curve which shows the area beneath a *duration curve,* and any value of the flow, and is therefore the integral of duration with respect to stream flow. When the duration curve is plotted as a percentage of time, the resulting duration area shows the average flow available below a given discharge (ASCE, 1962).

duration curve A graphic illustration of how often a given quantity is equaled or exceeded during a given span of time, e.g. a *flow-duration curve.* It is used in hydraulics.

durbachite (dur'-bach-ite) A plutonic rock composed chiefly of alkali feldspar, biotite, and hornblende, with a smaller amount of plagioclase and accessory quartz, apatite, titanite, zircon, and opaque oxides; a dark-colored biotite-hornblende *syenite.* The orthoclase phenocrysts form Carlsbad twins in a groundmass that is essentially an aggregate of biotite flakes and orthoclase (Johannsen, 1939, p.249). Its name is derived from Durbach, in the Black Forest, Germany. Obsolete.

durdenite (dur'-den-ite) *emmonsite.*

duricrust (du'-ri-crust) A general term for a hard crust on the surface of, or layer in the upper horizons of, a soil in a semiarid climate. It is formed by the accumulation of soluble minerals deposited by mineral-bearing waters that move upward by capillary action and evaporate during the dry season. See also: *ferricrete; silcrete; calcrete; caliche [soil].* Etymol: Latin "durus", "hard", + crust. Cf: *hardpan.*

Durids Aridisols which have a duripan that has its upper boundary within 100 cm of the soil surface. A suborder in the U.S. system of soil taxonomy.

durinite (du'-ri-nite) The major maceral of durain, according to the Stopes classification; the term is no longer in general use.

durinodes Weakly cemented to indurated soil nodules cemented with SiO_2. Durinodes break down in concentrated KOH after treatment with HCl to remove carbonates, but do not break down on treatment with concentrated HCl alone.

duripan (du'-ri-pan) A subsurface soil horizon that is cemented by illuvial silica, usually opal or microcrystalline forms of silica, to the degree that less than 50 percent of the volume of air-dry fragments will slake in water or HCl.

durite (du'-rite) A coal microlithotype that contains a combination of inertinite and liptinite totalling at least 95%, and containing more of each than of vitrinite. Cf: *durain.*

duroclarain (du-ro-clar'-ain) A transitional lithotype of coal, characterized by the presence of vitrinite with lesser amounts of other macerals such as inertinite and liptinite; it corresponds to *semisplint coal.* Cf: *clarodurain.*

duroclarite (du-ro-clar'-ite) A coal microlithotype containing at least 5% each of vitrinite, liptinite, and inertinite, with more vitrinite than inertinite and liptinite. It is a variety of *trimacerite,* intermediate in composition between clarite and durite, but closer to clarite. Cf: *clarodurite.*

durofusain (du-ro-fu'-sain) A coal lithotype transitional between fusain and durain, but predominantly fusain. Cf: *fusodurain.*

durotelain (du-ro-te'-lain) A coal lithotype transitional between telain and durain, but predominantly telain. Cf: *telodurain.*

durovitrain (du-ro-vit'-rain) A coal lithotype transitional between vitrain and durain, but predominantly vitrain. Cf: *vitrodurain.*

dusmatovite A dark blue to violet-brown hexagonal mineral of the *milarite* group: $K(K,Na,[\,])_2(Mn,Zr,Y)_2(Zn,Li)_3Si_{12}O_{30}$.

dussertite (dus'-sert-ite) A green rhombohedral mineral of the *crandallite* group: $Ba(Fe^{3+},Sb^{5+})_3(AsO_4)_2(OH,H_2O)_6$.

dust [sed] (a) Dry, solid matter consisting of clay- and silt-size earthy particles (diameters less than 1/16 mm, or 62 micrometers) so finely divided or comminuted that they can be readily lifted and carried considerable distances by the wind, staying aloft a long time but eventually falling back to the Earth's surface. Terrestrial sources of atmospheric dust include volcanic eruptions; salt spray from the seas; mineral particles; pollen and bacteria; and smoke and fly ash. (b) Small extraterrestrial particles that invade the Earth's atmosphere, such as *cosmic dust* and *meteoritic dust.* (c) *gold dust.* (d) Diamond dust. (e) A syn. of *volcanic ash,* esp. the finer fractions of ash.

dust [volc] A syn. of volcanic ash, esp. the finer fractions of ash.

dust band Layers of ice with abundant dust-sized particles. Cf: *dirt band [glaciol].*

dust basin A large, shallow *dust well.*

dust cloud [mass move] A mixture of air and snow particles accompanying an avalanche.

dust cloud [volc] *eruption cloud.*

dust-cloud hypothesis A theory of the formation of the planets by the accretion of a cloud of small, cold bodies which are sometimes called "planetesimals." The hypothesis was advanced by T.C. Chamberlin and R.R. Moulton in 1900. Syn: *planetesimal hypothesis.*

dust devil A swirling mass of dust triggered by intense solar heating of a dry surface. The most common dust devil is less than 1 m (about 3 feet) in diameter, lasts less than a minute, and is too weak to cause serious property damage.

dust fall (a) *dusting.* (b) *blood rain.*

dust hole A small *cryoconite hole.*

dusting (dust'-ing) The process by which dust and dustlike particles are deposited from the atmosphere; e.g. the deposition of opal phytoliths in the ocean. Syn: *dust fall.*

dust ring A ring of tiny inclusions seen in thin section, marking the original surface of a detrital sand grain that has grown by secondary enlargement.

dust sand A term used by Searle (1923, p.1) for a material whose particles have diameters in the range of 0.025-0.04 mm. The particle sizes correspond to medium silt and coarse silt.

dust size A term used in sedimentology for a particle size less than that of a sphere with a diameter of 1/16 mm (0.0025 in.); it includes *silt size* and *clay size.*

dust tuff A tuff of very fine fraction; an indurated deposit of volcanic dust. Syn: *mud tuff.*

dust veil The stratospheric pall which results from ejection of volcanic dust into the stratosphere as an implicit consequence of paroxysmal eruptions (Lamb, 1970, p.425-533). See also: *volcanic aerosol.*

dust-veil index A standard for classifying the degree by which dust in the atmosphere, esp. volcanic dust, forms a barrier to incoming solar radiation.

dust well *cryoconite hole.*

duttonite (dut'-ton-ite) A pale brown monoclinic mineral: $V^{4+}O(OH)_2$.

duty of water (du'-ty) The quantity (or depth) of irrigation water required for a given area for the purpose of producing a particular crop; it is commonly expressed in acre-inches or acre-feet per acre, or simply as depth in inches or feet. Syn: duty.

duxite (dux'-ite) An opaque, dark-brown variety of retinite containing about 0.5% sulfur, found in lignite at Dux in Bohemia, Czechoslovakia.

dwarf fauna A fossil assemblage consisting of specimens of small size. Many dwarf faunas result from sedimentary sorting, others from pathologies or environmentally influenced growth patterns. Also spelled: dwarfed fauna. Syn: *diminutive fauna; stunted fauna.* Partial syn: *depauperate fauna; impoverished fauna.*

dwip The basic unit of deposition in a tidal channel, consisting of a circular bank with a horseshoe outline and a hollow center, the toe pointing upstream (Strickland, 1940). It is caused primarily by channel bifurcation and reversing tidal currents.

dwornikite (dwor'-nik-ite) A white monoclinic mineral: $(Ni,Fe^{2+})SO_4 \cdot H_2O$. It is a member of the *kieserite* group.

dy [Pron: dyh] A sediment deposited in *oligotrophic lakes* that are also generally dystrophic ones. It is composed largely of a blackish-brown colloidal precipitate that dissolves in KOH to form a dark-brown extract (Faegri and Iversen, 1975, p.74). Etymol: Swedish, "silt". Cf: *gyttja; sapropel.*

dyad (dy'-ad) An uncommon grouping, characteristically seen in Ordovician-Devonian *cryptospores,* in which mature spores or pollen grains are shed in fused pairs. Cf: *tetrad [palyn]; polyad.*

dyakisdodecahedron (dy-a'-kis-do'-dec-a-he'-dron) *diploid.*

dying lake The terminal stage in the hypothetical life history of a lake, as it becomes more *eutrophic* with time. this concept was based on the classic study of Linsley Pond in Connecticut by Deevey (1942). The idea that artificial eutrophication by phosphorus-rich sewage effluent accelerated the *aging* of lakes appealed to journalists so strongly that Lake Erie became famous in the popular mind as a dying lake, soon to be dead or even extinct. By that time, paleolimnologists, including Deevey himself, knew that there was no demonstrable general tendency for lakes to become more eutrophic with age. They all tend to disappear through *sedimentation* in their depths and *downcutting* at their outlets.

dyke *dike [intrus rocks].*

dynamic analysis The study of the stress field responsible for strain that has developed in a rock.

dynamic denudation A concept of landscape and regolith evolution combining elements of several basic geomorphic and pedologic theories. A key element is the idea that denudation occurs along three surfaces: the land surface itself, the bedrock weathering front, and an intermediate surface associated with textural contrasts which promote subsurface lateral mass fluxes (Johnson, 1993).

dynamic ellipticity A ratio expressed as the difference between the moments of inertia about the polar and equatorial axes, and the moment of inertia about the polar axis.

dynamic equilibrium A condition of a system in which there is a balanced inflow and outflow of materials. Cf: *stable equilibrium.*

dynamic geology A general term for the branch of geology that deals with the causes and processes of geologic phenomena; physical geology.

dynamic geomorphology (a) The quantitative analysis of geomorphic processes treated as "gravitational or molecular shear stresses acting upon elastic, plastic, or fluid earth materials to produce the characteristic varieties of strain, or failure, that constitute weathering, erosion, transportation, and deposition" (Strahler, 1952a, p.923); (b) The application of dynamic systems theory and methods to the study of geomorphic systems.

dynamic head That head of fluid which would produce statically the pressure of a moving fluid (ASCE, 1962).

dynamic height The distance above the geoid of points on the same equipotential surface, expressed in linear units measured along a plumb line at a specified latitude, usually 45°. Cf: *height [geodesy]; orthometric elevation; geopotential number.* Syn: *geopotential height.*

dynamic magnetization A syn. in Russian literature of *shock remanent magnetization.*

dynamic metamorphism The total of the processes and effects of orogenic movements and differential stresses in producing new rocks from old, with marked structural and mineralogical changes due to crushing and shearing at low temperatures and extensive recrystallization at higher temperatures. It may involve large areas of the Earth's crust, i.e., be regional in character. Cf: *dislocation metamorphism; dynamothermal metamorphism; regional metamorphism.* Syn: *dynamometamorphism.*

dynamic positioning Referring to staying on location, most commonly in regard to a drilling vessel, by correcting the drift of the vessel, using thrusters, as calculated from the changing arrivals of sound pulse from an acoustic transponder on the sea floor.

dynamic pressure The pressure of a flowing fluid against a surface. Reaction to dynamic pressure affects direction and velocity of flow (ASCE, 1962).

dynamic range In seismic recording, the ratio of the maximum to minimum signal amplitude that can be recorded faithfully. It is usually specified over a certain frequency-band width and measured in decibels.

dynamic recrystallization [ign] The formation of new grains in a rock during solid-state deformation. It is essentially a recovery process, in which simultaneous *nucleation* and *grain growth* compete, ultimately to yield a uniform grain size inversely proportional to the applied differential stress. See also: *annealing recrystallization.*

dynamic recrystallization [struc petrol] The formation of relatively low-strain or unstrained grains during applied *differential stress* by the processes of *boundary migration recrystallization* and *rotation recrystallization.*

dynamic rejuvenation Renewal of the effectiveness of erosion processes, caused by uplift of a landmass with accompanying tilting and warping.

dynamic similarity A condition in which all forces, stresses, and strengths in *scaled physical models* are reduced by the same amount relative to the forces, stresses, and strengths in the natural prototype (Hubbert, 1937).

dynamic topography *sea-surface topography.*

dynamic viscosity *viscosity coefficient.*

dynamofluidal (dy'-na-mo-flu'-id-al) Pertaining to a texture in dynamometamorphosed rocks showing parallel arrangement in one direction only. Rarely used.

dynamogranite (dy'-na-mo-gran'-ite) A seldom-used term for an *augen gneiss* containing much microcline and orthoclase (Krivenko and Lapchik, 1934).

dynamometamorphism (dy'-na-mo-met'-a-mor'-phism) The equivalent of *dislocation metamorphism* and a syn. of *dynamic metamorphism.*

dynamo theory (dy'-na-mo) The theory that the Earth's main magnetic field is sustained by self-exciting dynamo action in the fluid core. The conducting liquid is hypothesized to flow in such a pattern that the electric current induced by its motion through the magnetic field sustains that field.

dynamothermal metamorphism (dy'-na-mo-ther'-mal) A common type of metamorphism that involves the effects of directed pressures and shearing stress as well as a wide range of confining pressures and temperatures. It is related both geographically and genetically to large orogenic belts and hence is regional in character. Cf: *burial metamorphism; dynamic metamorphism; regional metamorphism.*

dypingite (dy'-ping-ite) A white mineral: $Mg_5(CO_3)_4(OH)_2{\cdot}5H_2O$.

dysaerobic facies Transitional zone of sediments and fauna between aerobic and anaerobic settings, characterized by low diversity, low abundance, and a small-sized fauna. Typical oxygen values range down to 0.1 ml/l dissolved oxygen (Thompson et al., 1985). Cf: *exaerobic facies.*

dysanalyte (dys-an'-a-lyte) A variety of perovskite containing niobium and tantalum.

dyscrasite (dys-cra'-site) A metallic silver-white orthorhombic mineral: Ag_3Sb.

dyscrystalline (dys-crys'-tal-line) *microcrystalline.*

dysgeogenous (dys-ge-og'-e-nous) Not easily weathered; said of a rock that produces by weathering only a small amount of detritus. Ant: *eugeogenous.*

dysluite (dys-lu'-ite) A brown variety of gahnite containing manganese and iron.

dystome spar (dys'-tome) *datolite.*

dystrophic lake (dys-troph'-ic) A *bog lake* (Wetzel, 2001, p.274). Juday and Birge rejected this term, and Hutchinson (1967, p.380) called it "unfortunate". Cf: *oligotrophic lake.*

dysyntribite (dy-syn'-tri-bite) A hydrated aluminosilicate of sodium and potassium, probably a variety of pinite or an impure muscovite.

dzhalindite (dzhal-ind'-ite) A yellow-brown cubic mineral: $In(OH)_3$. It is an alteration product of indite, and the indium analogue of söhngeite.

dzharkenite A metallic black cubic mineral of the *pyrite* group: $FeSe_2$.

dzhetymite A term proposed by Dzholdoshev (1964; see Schermerhorn, 1966, p.833) as a quantitative designation for a nonsorted rock composed of approximately equal proportions (25-35% each) of angular "gravel" (1-10 mm diameter in Russian literature), sand (0.1-1 mm), and mud (under 0.1 mm).

dzhezkazganite (dzhez-kaz'-gan-ite) A discredited mineral reported to be a lead rhenium sulfide.

E

Eaglefordian (Ea-gle-ford′-i-an) North American (Gulf Coast) stage: Upper Cretaceous (above Woodbinian, below Austinian).

eaglestone (ea′-gle-stone) A walnut-sized concretionary nodule (usually of clay ironstone or flint), often containing a loose stone in its hollow interior, and believed by the ancients to be taken by an eagle to her nest to facilitate egg-laying.

EA interray Left anterior interray in echinoderms, situated between E ray and A ray and clockwise of E ray when the echinoderm is viewed from the adoral side; equal to interambulacrum 2 in the *Lovenian system.*

eakerite (ea′-ker-ite) A colorless monoclinic mineral: $Ca_2Sn^{4+}Al_2Si_6O_{18}(OH)_2{\cdot}2H_2O$.

ear (a) The flattened or pointed extremity of a brachiopod or pelecypod shell, subtended between the hinge line and the lateral part of the commissure. (b) In vertebrates, used loosely to include any or all anatomical elements related to the functions of hearing, equilibrium, or the collection of air- or waterborne mechanical vibrations. The term may be applied to the outer, middle, or inner ear, or collectively to all three.

eardleyite *takovite.*

earlandite (ear′-land-ite) A pale-yellow to white monoclinic mineral consisting of a hydrous calcium citrate: $Ca_3(C_6H_5O_7)_2{\cdot}4H_2O$. It is found in ocean-bottom sediments from the Weddell Sea in the south Atlantic Ocean.

earlshannonite (earl-shan′-non-ite) A reddish-brown monoclinic mineral of the *whitmoreite* group: $(Mn^{2+},Fe^{2+})Fe^{3+}{}_2(PO_4)_2(OH)_2{\cdot}4H_2O$. It is the Mn-analogue of whitmoreite.

early (ear′-ly) Pertaining to or occurring near the beginning of a segment of time. The adjective is applied to the name of a *geochronologic unit* (era, period, epoch) to indicate relative time designation and corresponds to lower as applied to the name of the equivalent *chronostratigraphic unit;* e.g., rocks of a Lower Jurassic batholith were intruded in Early Jurassic time. The initial letter of the term is capitalized to indicate a formal subdivision (e.g., "Early Devonian") and is set in lower case to indicate an informal subdivision (e.g., "early Miocene"). The informal term may be used for eras and epochs and for periods where there is no formal subdivision. Cf: *middle [geochron]; late.*

early diagenesis Diagenesis occurring immediately after deposition or immediately after burial. It is equivalent to *diagenesis* as interpreted by Russian and some U.S. geologists. Syn: *syngenesis; syndiagenesis.*

early wood Xylem formed in initial stages of a growth zone, which is not so dense as that produced later. Cf: *late wood.* Syn: *springwood.*

Earth That planet of the solar system which is fifth in size of the 9 major planets, and the third (between Venus and Mars) in order of distance from the Sun (about 150 x 10^6 km, or 93 million mi). Major data for the Earth: equatorial radius: 6,378 km (3,963.5 mi); polar radius: 6,357 km (3,941 mi); mean radius: 6,371 km (3,950 mi); equatorial circumference: 40,075 km (24,902 mi); surface area: 5.101 x 10^8 km^2 (197 x 10^6 sq mi).

earth [chem] (a) A difficultly reducible metallic oxide (such as alumina), formerly regarded as an element. (b) One of the four elements of the alchemists (the others: air, water, fire). See also: *rare earths.*

earth [eng] Material that can be removed and handled economically with pick and shovel, or loosened and removed with a power shovel, scraper, or end loader.

earth [geog] A general term for the solid materials that make up the physical globe, as distinct from water and air. Also, the firm land or ground of the Earth's surface, uncovered by water.

earth [sed] (a) An organic deposit that has remained unconsolidated although it is no longer in the process of accumulation; e.g. *radiolarian earth* and *diatomaceous earth.* (b) An amorphous fine-grained material, such as a clay or a substance resembling clay; e.g. *fuller's earth.*

Earth-centered ellipsoid An ellipsoid respresenting the Earth, having its center at the Earth's center of mass, and having its minor axis coincident with Earth's axis of rotation.

earth circle *soil circle.*

earth current Static or alternating electric current flowing through the ground and arising either in natural or artificial electric or magnetic fields. Syn: *ground current; telluric current.*

earth curvature The divergence of the surface of the Earth (spheroid) from a horizontal plane tangent at the point of observation. See also: *curvature correction.*

earth dam A *dam* generally constructed of gravel; broken, weathered rock; sand; silt; or soil. It usually has a core of clay or other impervious material and a rock facing of *riprap* to protect against wave erosion; less commonly, it may be zoned to have a core, filters, and shell material. If made of durable rock fragments, it is generally called a rock-fill dam. Archaic syn: earthen dam. Preferably called: *embankment* dam.

earth dike An artificial *levee* or *embankment* built of earth fill.

earth finger A miniature "earth pillar" in a nearly horizontal attitude, produced by wind-driven rain falling upon clayey material (Cotton, 1958, p.31, 33).

earth fissure In engineering geology, a linear opening or crack in the ground. Earth fissures often are associated with ground subsidence. Causes of earth fissures can range from introduction of water at the ground surface into *collapsible soil* deposits; to withdrawal of groundwater, with a resulting increase in *effective stress* in compressible deposits; and locally to aseismic *creep.* Earth fissures often are located along mountain fronts above buried faults with substantial differences in unconsolidated sediment thickness. Compared to tension cracks caused by landslide movement, earth fissures can be similar in nature but larger in scale. Also called: *ground fissure.* Syn: *fissure [eng].*

earth flax A fine silky *asbestos.*

earthflow (earth′-flow) A mass-movement landform and process characterized by downslope translation of soil and weathered rock over a discrete basal shear surface (landslide) within well defined lateral boundaries. The basal shear surface is more or less parallel with the ground surface in the downslope portion of the flow, which terminates in lobelike forms. Overall, little or no rotation of the slide mass occurs during displacement, although, in the vicinity of the crown scarp, minor initial rotation is usually observed in a series of slump blocks. Earthflows grade into mudflows through a continuous range in morphology associated with increasing fluidity. Also spelled: earth flow.

earth foam Soft or earthy *aphrite.*

Earth history *geologic history.*

earth holography Recording of the wave pattern from a constant frequency source along with a reference wave. The resulting hologram can be viewed by light to allow one to "see" the structure which generated the wave pattern (Hoover, 1972). See also: *holography.*

earth hummock A low, dome-shaped *frost mound,* consisting of a fine-textured earthen core covered by a tight mass of vegetation, esp. mosses, but also humus, grasses, sedges, and scrubby plants, and produced by hydrostatic pressure of groundwater or by heaving from growth of ice lenses in arctic and alpine regions; the general height is 10-20 cm and the diameter ranges from 1/2 m to 1 m. Earth hummocks form in groups to produce a nonsorted patterned ground. Cf: *turf hummock.* Syn: *earth mound; thufa.*

earth inductor A type of *inclinometer [magnet],* based on the principle that a voltage is induced in a coil rotating in the Earth's field whenever the rotation axis does not coincide with the field direction.

earth island *debris island.*

earthlight (earth′-light) The faint illumination of the dark part of the Moon, produced by sunlight reflected onto the Moon from the Earth's surface and atmosphere. It is best seen during the Moon's crescent phases. Syn: *earthshine.*

earth mound *earth hummock.*

earth pillar A tall, conical column of unconsolidated to semiconsolidated earth materials (e.g. clay, till, or landslide debris), produced by differential erosion in a region of sporadic heavy rainfall (as in a badland or a high alpine valley), and usually capped by a flat, hard boulder that shielded the underlying softer material from erosion, it often measures 6-9 m in height, and its diameter is a function of the width of the protective boulder. Cf: *hoodoo.* Syn: *earth pyramid; demoiselle; fairy chimney; hoodoo column; penitent [geomorph].*

earth pressure The pressure exerted on a structure such as a retaining wall by earth materials such as soil or sediments (*active earth pressure*) or the resistance provided by soil to the movement of a wall or building footing (*passive earth pressure*); also, the pressure exerted by soil on any boundary. Cf: *at rest.*

earth pyramid A less steep-sided and more conical *earth pillar,* occurring esp. where rainwash is the chief agent of erosion.

earthquake (earth'-quake) A sudden motion or trembling in the Earth caused by the abrupt release of slowly accumulated strain. Syn: *shock; quake; seism; macroseism; temblor.*

earthquake engineering The study of the behavior of foundations and structures relative to seismic ground motion, and the attempt to mitigate the effect of earthquakes on structures.

earthquake first motion The direction of the first motion of the body waves from an earthquake. The areal distribution of these motions gives information about the fault movement associated with the earthquake. Syn: *first motion.*

earthquake intensity A measure of the effects of an earthquake at a particular place. Intensity depends not only on the *earthquake magnitude,* but also on the distance from earthquake to epicenter and on the local geology. See also: *intensity scale.*

earthquake lights Lights of varied appearance that have been reported during, but also before and after, earthquake shocks, generally in the epicentral region. There is no agreement on a mechanism to explain them. See: Derr (1977); Noszticzius (1979).

earthquake magnitude A measure of the strength of an earthquake, or the strain energy released by it, as determined by seismographic observations. C. F. Richter first defined *local magnitude* as the logarithm, to the base 10, of the amplitude in micrometers of the largest trace deflection that would be observed on a standard torsion seismograph 100 km from the epicenter. Magnitudes are called body-wave magnitude (m_b) and surface-wave magnitude (M_s) depending on the type of wave involved. The two magnitudes do not necessarily have the same numerical value. See also: *Richter scale.* Cf: *earthquake intensity, earthquake moment.* Syn: *magnitude [seis].*

earthquake moment A measure of the size of an earthquake relating to the force couples across the area of fault slip. See: *seismic moment.*

earthquake prediction That aspect of seismology which deals with the physical conditions or indications that precede an earthquake, in order to predict the size, time, and location of an impending shock.

earthquake record *seismogram.*

earthquake rent Popular syn. of *reverse scarplet.*

earthquake scarplet A visible vertical ground offset, low, nearly straight fault scarp or step; often continuous for many kilometers, formed simultaneously with an earthquake. Cf: *reverse scarplet.*

earthquake sound An *air wave* associated with an earthquake.

earthquake swarm A series of minor earthquakes occurring in the same area and time, none of which may be identified as the main shock or with the same fault. Syn: *swarm [seis].*

earthquake tremor *tremor.*

earthquake volume The volume of an earthquake's major potential energy content and equal to the total volume of aftershocks following each shock (Båth, 1966, p.79).

earthquake wave *seismic wave.*

earthquake zone An area of the Earth's crust in which fault movements and sometimes associated volcanism occur; a *seismic belt* or *seismic area.*

Earth radiation *terrestrial radiation.*

Earth rotation The turning of the Earth about its axis, described either as counterclockwise about the North Pole or as eastward according to the movement of the Equator. The Earth rotates once in 23h 56m relative to the stars (a sidereal or inertial day), or once in 24 hours relative to the position of the Sun (a solar day).

earth run A lobe of earth material that has flowed downhill beneath the vegetation cover, forming a sloping step whose front is often $\frac{1}{3}$-2 m high (Sharpe, 1938, p.42). Not preferred usage. Cf: *earthflow.*

earth science An all-embracing term for sciences related to the Earth (analogous, in educational parlance, to "life science"). It is occasionally used as a syn. for *geology* or geological sciences, but this usage is misleading because in its wider scope earth science may be considered to include such subjects as meteorology, physical oceanography, soil chemistry, and agronomy. The term is generally used in the singular.

Earth's gravity field A vector field g of the *gravitational acceleration* of the unit mass m due to the attraction of the Earth's mass M. In gravity exploration, it is measured by the gravitational force F on the unit mass in the gravimeter: $g = F/m = -\gamma M/R^2$, where $F = -\gamma mM/R^2$ (see *law of universal gravitation*), γ is the *gravitational constant,* R is the Earth's radius at the point of measurement. The absolute value of the Earth's gravity field varies from about 978,000 mGal at the Equator to 983,000 mGal at the poles. Theoretical value at any latitude is defined by the *International Gravity Formula.* The Earth's gravity field value, measured for exploration purposes, depends on six main factors: (1) latitude; (2) elevation above or below sea level; (3) topography of the surrounding area; (4) *Earth tide;* (5) velocity of the moving observation platform; (6) variations of rock density in the subsurface, i.e., geological structure. The objective of *gravity reduction* is to isolate the last factor. Cf: *absolute gravity; observed gravity.*

Earth shell Any one of the concentric shells that constitute the structure of the Earth.

earthshine (earth'-shine) *earthlight.*

Earth's magnetic field The field that is often represented by a dipole at the Earth's center. Intersections of the axis of this dipole with the Earth's surface are *geomagnetic poles.* Locations where the magnetic dip is 90° are called magnetic dip poles and do not coincide with the geomagnetic poles. Three basic components: (1) *main magnetic field* which is of the internal (core) origin and varies slowly over hundreds and thousands of years. Includes *epoch, reversal,* and *normal magnetic field;* (2) *external magnetic field* which originates outside the Earth and features rapid changes in time, partly cyclic and partly random. Includes *bay, daily variation, magnetic storm,* and *spheric* with variations from about 1-2 nT (small daily variations) up to 1,000 nT and more (magnetic storms); (3) *local magnetic field* that represents the result of interaction between above two components and local distribution of magnetic materials in the upper Earth's crust. Varies from 0.1 to 100 nT over magnetically quiet deep sedimentary basins up to several thousand nanoteslas over highly magnetized crystalline basement outcrops. Also called: *geomagnetic field.* See also: *nanotesla.*

Earth's orbit The path through space of the Earth in its annual journey around the Sun. This path is an ellipse, with a *semimajor axis* of about 1.496×10^8 km (92,700,000 miles), an eccentricity of 0.0167, and the Sun at one focus of the ellipse. See also: *ellipse.*

earth stripe *soil stripe.*

Earth systems science The study of the Earth as a system, with particular attention to linkages between atmosphere, lithosphere, hydrosphere, and biosphere (Keller, 2002).

Earth tide The response of the solid Earth to the forces that produce the tides of the sea; semidaily Earth tides have a fluctuation of between seven and fifteen centimeters (Strahler, 1963, p.110-111). Syn: *bodily tide.*

earth tremor *tremor.*

earth wax *ozocerite.*

earthwork (earth'-work) (a) The operations involved in constructing earth embankments, preparing foundations for structures, and placing and compacting earth materials; also called grading operations. (b) An embankment or other construction made of earth.

earthy [geol] (earth'-y) Composed of or resembling earth, or having the properties or nature of earth or soil; e.g. an "earthy limestone" containing argillaceous material and characterized by high porosity, loosely aggregated particles, and close association with chalk.

earthy [mineral] (a) Said of minerals with a *dull luster;* the surface may feel rough to the touch. (b) Said of a type of fracture similar to that of a hard clay.

earthy breccia A term used by Woodford (1925, p.183) for a breccia in which sand and mud each constitute more than 10% of the rock.

earthy calamine *hydrozincite.*

earthy cobalt *asbolite.*

earthy lignite Lignite that is structurally amorphous and is soft and friable. Cf: *woody lignite.*

earthy manganese *wad [mineral].*

easting (east'-ing) A *departure* (difference in longitude) measured to the east from the last preceding point of reckoning; e.g. a linear distance eastward from the north-south (vertical) grid line that passes through the origin of a grid system. Cf: *false easting.*

eastonite (eas'-ton-ite) A colorless to brown monoclinic mineral of the *mica* group: $KMg_2Al(Al_2Si_2)O_{10}(OH)_2$.

Eaton tabulation system A way of describing the number and arrangement of plates in *dinoflagellates.* It was developed by Eaton (1980), building on the system of Kofoid. Successive series are apical closing, apical, precingular, cingular, postcingular, antapical, and antapical closing.

ebb and flow spring A spring or flowing well exhibiting periodic variation in volume of flow; the periodicity, which is often irregular, is attributed to siphonic action or flushing of sediment plugs. Syn: *tidewell.*

ebb current The tidal current associated with the decrease in the height of a tide, generally moving seaward or down a tidal river or estuary. Cf: *flood current.* Erroneous syn: *ebb tide.*

ebb delta *ebb-tidal delta.*

ebbing-and-flowing spring (ebb'-ing-and-flow'-ing) *periodic spring.*

ebbing well A well in permeable rocks near the coast, in which the water level fluctuates with the tide.

ebb shield A high rim or margin around a *tidal delta* that protects parts of it from modification by ebb currents (Hayes, 1980).

ebb spit A spit on a flood-tidal delta formed by ebb-tidal currents (Hayes, 1980).

ebb-tidal delta A *tidal delta* formed on the seaward side of a *tidal inlet* (Davis and FitzGerald, 2004). Cf: *flood-tidal delta.*

ebb tide (a) *falling tide.* (b) An erroneous syn. of *ebb current.*

ebridian (eb-rid'-i-an) A marine protist characterized by the presence of flagella, a skeleton of solid silica rods, and the absence of chromatophores.

Eburonian (Eb-ur-on'-ian) European cold-climate paleofloral stage: basal Pleistocene (above Tiglian, below Waalian).

ecandrewsite (ec-an'-drewsite) A submetallic brown to black rhombohedral mineral: $(Zn,Fe^{2+},Mn^{2+})TiO_3$. It is the Zn analogue of ilmenite, and a member of the *ilmenite* group.

ecardinal (e-car'-di-nal) Said of certain brachiopods (or of their shell) without a hinge.

eccentric (ec-cen′-tric) *helictite.*

eccentricity (ec-cen-tric′-i-ty) (a) The condition, degree, amount, or instance of deviation from a center or of not having the same center; e.g. the horizontal displacement of a surveying instrument or signal from a triangulation station mark at the time an observation is made, or an effect seen in a surveyor's compass when the line of sight fails to pass through the vertical axis of the compass or when a straight line through the ends of the magnetic needle fails to pass through the center rotation of the needle. (b) The distance of the center of figure of a body from an axis about which it turns; e.g. the "eccentricity of alidade" represented by the distance between the center of figure of the index points on the alidade and the center of figure of the graduated circle. (c) The ratio of the distances from any point of a conic section to a focus and the corresponding directrix; e.g. the "eccentricity of ellipse" represented by the ratio of the distance between the center and a focus of an ellipse to the length of its semimajor axis, or $e^2 = (a^2-b^2)/a^2$, where e = eccentricity, a = semimajor axis, and b = semiminor axis. A measure of the ellipticity of a gravitationally bound orbit around a central mass. If the eccentricity if less than one, the orbit is an ellipse with the central mass at one foci. If the eccentricity is exactly one, the orbit is a parabola. If the eccentricity if greater than one, the orbit is a hyperbola.

eccentricity [planet] A measure of the deviation of an elliptical orbit from circularity. See *ellipse* for definitions of related terms.

eccentricity reduction The correction that must be applied to an observed direction made with an eccentric instrument or signal, or both, in order to reduce the observed value to what it would have been if there had been no eccentricity. Also known as "eccentric correction" or "reduction-to-center".

eccentric signal A survey signal (target) which is not in the same vertical line with the station which it represents, such as a signal placed in triangulation at some point other than directly over the triangulation station and not in line with the station and the instrument.

eccentric station A survey point over which an instrument is centered and observations are made, and which is not in the same vertical line with the station which it represents and to which the observations will be reduced before being combined with observations at other stations (Mitchell, 1948, p.26). It is established and occupied when it is not practicable to set up directly over the actual station center or when it becomes necessary in order to see points that are not visible from the station center.

ecdemite (ec′-de-mite) A yellow, greenish yellow, or green tetragonal mineral: $Pb_6As^{3+}{}_2O_7Cl_4$. Also spelled: *ekdemite.*

ecdysis (ec′-dy-sis) The process of molting of the *exoskeleton* in arthropods.

ecesis (e-ce′-sis) The establishment of a plant in a new location as a result of successful germination, growth, and reproduction. Syn: *establishment.*

echinate (ech′-i-nate) Spiny-surfaced, or densely covered with stiff, stout, or bluntish bristles, prickles, or spines; e.g. "echinate pollen" having a sculpture consisting of spines.

echinating (ech′-i-nat′-ing) Said of sponge spicules that protrude at an angle from the surface of a skeletal fiber.

echinoderm (e-chi′-no-derm) Any solitary marine benthic (rarely pelagic) invertebrate, belonging to the phylum Echinodermata, characterized by radial symmetry, an endoskeleton formed of plates or ossicles composed of crystalline calcite, and a water-vascular system. Echinozoans, asterozoans, crinozoans, and homalozoans are echinoderm subphyla.

echinoid (ech′-i-noid) Any echinozoan belonging to the class Echinoidea, characterized by a subspherical to modified spherical shape, interlocking calcareous plates, and movable appendages; e.g. a *sea urchin.*

echinozoan (ech′-i-no-zo′-an) Any free-living globoid, discoid, or cylindroid echinoderm belonging to the subphylum Echinozoa, characterized by the absence of arms, brachioles, and outspread rays and by the dominance of a meridional growth pattern over bilateral symmetry. Range, Lower Cambrian to Holocene.

echmidium (ech-mid′-i-um) A spear-shaped plate formed during ontogeny of a brachiopod *loop* by fusion of anterior ends of descending branches.

echo dune (ech′-o) A dune formed in front of a continuous impermeable obstacle, such as a wind-facing cliff, at a distance slightly less than the obstacle's height, because of the formation of a reverse eddy (Goldsmith, 1985, p.314-316).

echogram (ech′-o-gram) The graphic record made by an *echo sounder,* in the form of a continuous profile.

echo logging Ultrasonic surveying of a cavity such as a storage reservoir in a salt dome.

echo sounder In oceanography, an instrument, on a ship, that records the depth of the ocean below the ship by measuring the time required for sound to travel to and return from the sea floor. See also: *echogram, fathometer, precision depth recorder.* Syn: *depth sounder; sonic depth-finder.*

eckermannite (eck′-er-mann-ite′) A dark bluish-green to black monoclinic mineral of the *amphibole* group: $NaNa_2(Mg_4Al)Si_8O_{22}(OH)_2$.

Eckert projection (Eck′-ert) One of a series of six map projections of the entire Earth in each of which the geographic poles are represented by parallel straight lines that are one half the length of the equator. The parallels are rectilinear and the meridians may be rectilinear or curved. They are broadly similar in appearance to the Mollweide projection. Named after Max Eckert (1868-1938), German cartographer, who developed the projections in 1906.

eckrite (eck′-rite) *winchite.*

eclarite (e-clar′-ite) A metallic lead-gray orthorhombic mineral: $Pb_9(Cu,Fe)Bi_{12}S_{28}$.

eclipse (e-clipse′) One celestial body casting a shadow on another, as when the Moon obscures the Sun (solar eclipse), or the Earth shadows the moon (lunar eclipse).

ecliptic The mean plane of the Earth's orbit around the Sun, used as the reference plane from which the *inclinations* of other planetary orbits are defined. Most of the other planetary orbits lie within a few degrees of the ecliptic. See also: *inclination [astron].*

ecliptic latitude (e-clip′-tic) *celestial latitude.*

ecliptic longitude *celestial longitude.*

eclogite (ec′-lo-gite) A granular rock composed essentially of garnet (almandine-pyrope) and sodic pyroxene (omphacite). Rutile, kyanite, and quartz may also be present. Var. spelling: *eklogite.*

eclogite facies The set of metamorphic mineral assemblages (facies) in which basic rocks are represented by omphacitic pyroxene and almandine-pyrope garnet. Also common, although not essential, is the association pyrope+olivine+diopside+enstatite. Phase-equilibrium work has shown that these high-density mineral associations indicate high pressure of crystallization, although the range of geologic environments in which the facies has been encountered, and the variation in mineral composition (Coleman et al., 1965), point to a broad range of possible pressure-temperature conditions. Many workers have suggested that low H_2O pressures are required, and that pressures and temperatures overlap those of several other metamorphic facies.

eclogitic (ec-lo-git′-ic) (a) Pertaining to an *eclogite.* (b) Said of a rock having an association of garnet and clinopyroxene with a proportion of jadeite molecule in the pyroxene. In this case, no genetic connotation is implied or bulk composition considered (Church, 1968, p.757).

ecochronology (ec′-o-chro-nol′-o-gy) A term introduced as "Ökochronologie" by Schindewolf (1950, p.35) for *geochronology* based on the ecology of life forms.

ecocline (ec′-o-cline) A *cline* related to the gradation between two different niches.

ecography (e-cog′-ra-phy) The strictly descriptive part of *ecology.* The term is seldom used.

ecological hierarchy A nested set of ecologic entities including organisms, avatars, and communities that interact with the environment and are involved in energy and matter transfer.

ecologic evolutionary unit A long interval of Phanerozoic time during which marine communities maintained stable ecologic structures (Boucot, 1983).

ecologic facies (ec-o-log′-ic) *environmental facies.*

ecologic locking A term for the hypothesis that species lineages are bound by ecologic interactions within tightly integrated ecologic communities, with the effect that communities are granted relative stability until a critical threshold of disturbance is reached. The hypothesis is one of several proposed mechanisms of the pattern of *coordinated stasis* (Brett et al., 1996).

ecologic niche *niche [ecol].*

ecologic potential A term used by Lowenstam (1950) for the capability of an organism to directly control or modify its environment, such as that possessed by a reef-building organism that is able to erect a rigid and resistant framework in the zone of wave action.

ecologic reef A fossil *reef* recognizable as having been built and bound by organisms into a rigid, wave-resistant topographic high on the sea floor (Dunham, 1970, p.1931). Cf: *stratigraphic reef.*

ecologic succession *succession [ecol].*

ecology (e-col′-o-gy) The study of the relationships between organisms and their environment, including the study of communities, patterns of life, natural cycles, relationships of organisms to each other, biogeography, and population changes. See also: *paleoecology; ecography.* Adj: ecologic; ecological. Syn: *bionomics.*

economic geology (e-co-nom′-ic) The study and analysis of geologic bodies and materials that can be utilized profitably by man, including fuels, metals, nonmetallic minerals, and water; the application of geologic knowledge and theory to the search for and the understanding of mineral deposits.

economic limit *assay limit.*

economic yield The maximum estimated rate at which water may be withdrawn from an aquifer without creating a deficiency or affecting the quality of the supply. Cf: *safe yield.*

economy (e-con′-o-my) The input and consumption of energy within a system, and the changes that result; e.g. the *balance* of a glacier.

ecophenotype (ec-o-phe′-no-type) A variant of a species, produced by nongenetic modification of the *phenotype* by particular ecologic conditions.

ecoregion A region defined on the basis of ecologic characteristics. Syn: *natural region.*

ecospace (ec′-o-space) As used by Valentine (1969, p.687), that volume within the *environmental hyperspace lattice* corresponding to the environmental conditions under which a particular organism may live. See also: *realized ecospace; prospective ecospace.*

ecosphere (ec′-o-sphere) Portions of the universe favorable for the existence of living organisms; esp. the *biosphere.*

ecostratigraphic unit (ec′-o-strat′-i-graph′-ic) A stratigraphic unit based on the mode of origin or the environment of deposition of rocks that compose it (Hedberg, 1958, p.1893); e.g. a marine zone, a brackish-water zone, or a glacially deposited zone. The terminology of most ecostratigraphic units has not reached formal status. Syn: *ecozone.*

ecostratigraphy (ec′-o-stra-tig′-ra-phy) (a) A term introduced as "Oekostratigraphie" by Schindewolf (1950, p.35) for stratigraphy based on the ecology of life forms; the stratigraphic occurrence of local or regional faunal or floral assemblages that are valuable for ecologic considerations. (b) A term used by Hedberg (1958, p.1893) for the study and classification of stratified rocks according to their mode of origin or their environment of deposition. (c) "The evolutionary, ecologic, biogeographic, stratigraphic correlation, and basin analysis consequences of the basic fact of biofacies" (Boucot, 1984).

ecosystem (ec′-o-sys′-tem) A unit in ecology consisting of the environment with its living elements, plus the nonliving factors that exist in and affect it.

ecotone (ec′-o-tone) A transition zone that exists between two ecologic communities. Members of both communities may compete within this zone, thus yielding an apparent enrichment known as the *edge effect.* Syn: *tension zone.*

ecotope (ec′-o-tope) The habitat of a particular organism. See also: *ecotopic.*

ecotopic (ec-o-top′-ic) Having the tendency to adjust to the specific conditions of the *ecotope.*

ecotype (ec′-o-type) An ecologic variant of a species that has adapted to local environmental conditions.

écoulement (é-coule′-ment) A syn. of *gravitational sliding.* Etymol: French, "flowing".

ecozone (ec′-o-zone) (a) A stratigraphic unit defined on the basis of its fossil content and sedimentological characteristics as they indicate the environment of deposition of the unit. Syn: *ecostratigraphic unit.* (b) "A body of rock with upper and lower boundaries marked by reversible faunal changes that are essentially oblique to time planes. These faunal changes were probably caused by secular lateral shift of facies belts" (Vella, 1964, p.183-199; 1964, p.622-623).

Ectasian A period of the Standard Global Geochronometric Scale: the middle period of the *Mesoproterozoic* Era, characterized by continued expansion of *platform* covers ("ektasis", "extension"). It is defined as the time span between 1,400 Ma, the end of the *Calymmian* Period, and 1,200 Ma, the beginning of the *Stenian* Period (Plumb, 1991).

ectexine (ec-tex′-ine) *ektexine.*

ectexis (ec-tex′-is) Migmatization with in situ formation of the mobile part (Dietrich and Mehnert, 1961). Rarely used. Var. spelling: *ektexis.* Cf: *entexis.*

ectexite (ec-tex′-ite) Rock formed by *ectexis.* Rarely used. Var: ectectite.

ectinite (ec′-ti-nite) A rock formed as a result of essentially isochemical regional metamorphism, i.e., with no notable associated metasomatism (Dietrich and Mehnert, 1961). Rarely used.

ectoaperture (ec-to-ap′-er-ture) An aperture in the outer layer of the *sporoderm* of pollen.

ectocochleate (ec-to-coch′-le-ate) Said of cephalopods whose body is lodged within a shell; sometimes grouped as Ectocochleata (e.g., *Nautilus*).

ectocyst (ec′-to-cyst) (a) The outer, cuticular layer of the body wall of a bryozoan, with or without incorporated calcification. (b) The outer body of a *dinoflagellate cyst.* Cf: *endocyst.*

ectoderm [paleont] (ec′-to-derm) The outer body layer of a multicellular organism; e.g. the outer layer of the oral and basal disks, tentacles, and column wall of a coral polyp. In the vertebrates, the term refers exclusively to the originally most superficial of the embryonic primary germ layers that give rise in adults to epidermis, nervous system, and important parts of the organs of special sense. Cf: *endoderm; mesoderm.*

ectoderm [palyn] The outermost layer in acritarchs, with more than one wall layer (Williams, 1978; Cramer and Diez Rodrigues, 1979). Cf: *endoderm [palyn].*

ectoderre (ec′-to-derre) The principal layer in the external covering of a chitinozoan. Cf: *endoderre; periderre.*

ectoexine (ec-to-ex′-ine) *ektexine.*

ectooecium (ec′-to-oe′-ci-um) The outer, generally calcified part of the body-wall fold forming an *ovicell* in cheilostome bryozoans. Cf: *entooecium.*

ectophloic (ec-to-phlo′-ic) Pertaining to the siphonostele of certain vascular plants, having phloem external to the xylem. Cf: *amphiphloic.*

ectophragm (ec′-to-phragm) A thin outer membrane partially or completely connecting distal ends of processes or other structures on a dinoflagellate cyst. Cf: *endophragm; periphragm.*

ectoplasm The outer zone of protoplasm of a foraminifer, hyaline and homogeneous. Cf: *hyaline [paleont].*

ectoproct (ec′-to-proct) Any bryozoan belonging to the subphylum Ectoprocta and characterized by a circular or horseshoe-shaped lophophore around the mouth but not the anus. Range, Ordovician (or possibly Upper Cambrian) to Holocene. Cf: *entoproct.*

ectosiphuncle (ec-to-si′-phun-cle) The wall of the *siphuncle* of certain cephalopods, consisting generally of septal necks and connecting rings. Cf: *endosiphuncle.*

ectosolenian (ec′-to-so-le′-ni-an) Said of a foraminifer (e.g. *Lagena*) having an external tubelike neck. Cf: *entosolenian.*

ectosome (ec′-to-some) The peripheral region of a sponge beneath the inhalant surface and devoid of flagellated chambers; the cortical part of a sponge. Cf: *endosome.*

ectyonine (ec′-ty-o-nine′) Said of a sponge skeleton built of spiculofibers made up of both coring and echinating monaxons.

ecumeme (ec′-u-meme) That part of the Earth that is permanently inhabited.

edaphic (e-daph′-ic) Said of ecologic formations or effects resulting from or influenced by local conditions of the soil or substrate; also, an old term applied to any soil characteristic that affects plant growth, e.g. acidity, alkalinity. Cf: *climatic.*

edaphology The science that deals with the influence of soils on living things; particularly plants, including man's use of land for plant growth.

edaphon (ed′-a-phon) All the animals and plants living in the soil.

Edaphosauria (Ed′-a-pho-sau′-ri-a) A suborder of pelycosaurian synapsid reptiles characterized by small heads, a tendency toward large size, and dentitions that suggest herbivorous habit. Stratigraphic range, Upper Pennsylvanian to Lower Permian.

eddy (ed′-dy) Circular current in the ocean, most frequently around 50-200 km in diameter. Commonly found downstream of western boundary currents and in the Antarctic Circumpolar Current. Cf: *mesoscale eddy.*

eddy-admitting model A *numerical model* of the ocean circulation having grid points sufficiently close in the horizontal plane that the model can calculate *mesoscale eddies.*

eddy coefficient *austausch.*

eddy conductivity *austausch.*

eddy diffusion Mixing by turbulent flow. Syn: *turbulent diffusion.*

eddy diffusivity The coefficient of the proportionality of the rate of transfer of mass to the gradient of the average concentration. It depends on the nature of the turbulent motion (Fairbridge, 1966, p.230).

eddy diffusivity [oceanog] The rate at which momentum, salt, or thermal energy is spread through the ocean by turbulent mixing. The rate in the vertical is much smaller than the rate in the horizontal.

eddy flux The rate of transport or flux of fluid properties, e.g. momentum or suspended matter, by turbulent flow. Syn: *turbulent flux.*

eddy mark One of numerous superimposed or overlapping loops (0.3-1 m in diameter) forming a spiral impression on a sedimentary (sandstone) surface, believed to result from "dragging of a small limb of a larger floating log caught in a vortex or eddy current" of a stream, or from movement of a pebble or stick caught in circular winds of a "dust devil" after the sands had been exposed along the stream bank (Rigby, 1959).

eddy mill *pothole [streams].*

eddy spectrum Turbulent flow described in terms of the frequency distribution of eddy size, or of the partition of kinetic energy among eddies of various sizes. Syn: *turbulence spectrum.*

eddy viscosity The transfer coefficient for momentum, corresponding to *kinematic viscosity.*

edenharterite A metallic to submetallic brownish black or dark red orthorhombic mineral: $TlPbAs_3S_6$.

Edenian (E-den′-i-an) North American stage: Upper Ordovician (above Mohawkian, below Maysvillian). Lowest substage of Cincinnatian.

edenite (e′-den-ite) (a) A light-colored, iron-free variety of hornblende. (b) An end member in the *amphibole* mineral group: $NaCa_2(Mg,Fe^{2+})_5(Si_7Al)O_{22}(OH)_2$. Cf: *pargasite.*

edentulous (e-den′-tu-lous) Said of a bivalve mollusk lacking hinge teeth.

edgarbaileyite A vitreous to resinous yellow, olive-green, or dark green-brown monoclinic mineral: $Hg^{1+}_6Si_2O_7$. It is photo-sensitive.

edgarite A metallic dark gray to black hexagonal mineral: $FeNb_3S_6$.

edge (a) A sharply pointed ridge; also the crest of such a ridge. (b) The escarpment that terminates a plateau; the extreme margin of a cliff. (c) The highest part of an elevated tract of land of great extent; esp. a ridge or divide between two streams.

edge coal An English and Scottish term for a steeply inclined to vertical coal seam.

edge detection An image-processing technique to emphasize abrupt discontinuities such as caused by faulting.

edge dislocation [cryst] In a crystal, a row of atoms marking the edge of a crystallographic plane and extending only part way through the crystal; it is a type of *line defect.*

edge dislocation [struc petrol] A *dislocation [struc petrol]* involving the termination of an extra half-plane of atoms in the crystal lattice. The dislocation is perpendicular to the glide, or slip, direction.

edge effect In ecology, the apparent increase in number of species inhabiting an *ecotone* as compared with the smaller number occupying either adjacent community.

edge effects Unconstrained extrapolation of the slope of a contoured surface near the edges of an area being mapped, the result of estimating values of the surface beyond the limits of control. Edge effects are most pronounced outside the part of the map that contains observations, but also extend into the map for distances that depend upon the local density of control points.

edge enhancement Analytical image processing techniques which emphasize tonal transition between features in an image (Krzanowski et al., 1993).
edge joining *edge matching.*
edge line A heavy line on a relief map, depicting a sudden sharp change or break of slope.
edge matching The comparison and adjustment of features to obtain agreement along the edges of adjoining mapped regions. Edge matching may be done on both graphic and digital products. Also called: *edge joining.*
edge water The water around the margins of an *oil pool* or a *gas pool.* Cf: *bottom water [petroleum].* Also spelled: edgewater.
edge wave An ocean wave with crests perpendicular to the shoreline and with amplitude that decreases exponentially with distance from shore. Usually driven by groups of sea-surface waves reaching shore.
edgewise conglomerate (edge'-wise) A conglomerate exhibiting *edgewise structure;* e.g. an intraformational conglomerate containing elongated mudstone pebbles that are transverse to the bedding.
edgewise structure A primary sedimentary structure characterized by an arrangement of flat, tabular, or disc-shaped fragments whose long axes are set at varying steep angles to the bedding. It may be due to running water or to sliding or slumping soon after deposition. See also: *edgewise conglomerate.*
edge zone A fold of the body wall of a coral *polyp,* extending laterally and/or downward over the edge of the wall.
Ediacaran (E-di-a-ca'-ran) A period of the Standard Global Geochronometric Scale: the youngest period of the *Neoproterozoic* Era and of the *Precambrian,* characterized by the earliest, well-documented soft-bodied forms of metazoan life. The basal GSSP to the underlying *Cryogenian* Period coincides with the termination of the *Varangian* (or Marinoan) near-global glaciation at ~600 to 560 Ma. The upper boundary at the beginning of the *Cambrian* Period is dated as ~540 Ma. Syn: *Vendian.*
Ediacaran fauna The general term applied to soft-bodied fossils dating from the late Neoproterozoic interval. These fossils were first discovered in the Ediacaran Hills of South Australia but have since been described from rocks of the same age from many other regions.
edifice [volc] The constructional mass of a volcano (Davidson and De Silva, 2000).
edingtonite (ed'-ing-ton-ite') A white, grayish, or pink zeolite mineral: $BaAl_2Si_3O_{10}·4H_2O$. An orthorhombic and tetragonal dimorphous pair occurs. It sometimes contains appreciable calcium.
EDM *electronic distance-measuring instrument.*
edolite (ed'-o-lite) A type of *hornfels* consisting mainly of feldspar and mica. There are varieties that also contain cordierite (*aviolite*) or andalusite (*astite*) (Holmes, 1928, p.87). Type locality: Edolo, Italian Alps. Rarely used.
edoylerite A yellow to orange monoclinic mineral: $Hg^{2+}_3Cr^{6+}O_4S_2$. It is photosensitive, changing color to olive-green on exposure to light.
edrioasteroid (ed'-ri-o-as'-ter-oid) Any many-plated, attached echinozoan belonging to the class Edrioasteroidea, having a well-developed quinqueradial endothecal ambulacral system. Range, Lower Cambrian to Pennsylvanian.
eduction (e-duc'-tion) Lateral flow of mantle material beneath a continent and its emergence at the continental margin. It provides a mechanism for exhumation of formerly subducted blueschists (Dixon and Farrar, 1980).
Edwards tabulation system A way, developed by Edwards (1990), of describing the number and arrangement of plates in *dinoflagellates.* Successive series are D-series, M-series, E-series, C-series, H-series, and T-series.
eel A series of hydrophones in a tube, often up to about 50 m in length, used in seismic reflection profiling in shallow water.
EELS *electron energy loss spectroscopy.*
Eemian (Eem'-i-an) European warm-climate paleofloral stage: upper Pleistocene (above Warthe, below Weichselian). Equiv. Tyrrhenian marine stage.
EEZ *exclusive economic zone.*
effect hypothesis The pattern whereby species in one clade proliferate at a faster rate than the species within another clade that is driven by the process of natural selection of organisms.
effective diameter (a) The diameter of particles in an assumed rock or soil that would transmit water at the same rate as the rock or soil under consideration, and that is composed of spherical particles of equal size and arranged in a specified manner. (b) The approximate diameter of a rock or soil particle equal to the sieve size that allows 10% (by weight) of the material to pass through; the particle diameter of the 90 percent line of a cumulative curve, or the maximum diameter of the smallest 10% of the particles of sediment. Syn: *effective size.*
effective discharge *dominant discharge.*
effective drainage porosity *effective porosity.*
effective permeability The ability of a rock to conduct one fluid, e.g. gas, in the presence of other fluids, e.g. oil or water. See also: *absolute permeability; relative permeability.*
effective pore volume The pore space in rocks that is available for the free circulation of water. This excludes pore space taken up by air and pellicular water.
effective porosity The percent of the total volume of a given mass of soil or rock that consists of interconnected interstices. The use of this term as a syn. of *specific yield* is to be discouraged. Syn: *effective drainage porosity.* Cf: *porosity.*
effective precipitation (a) That part of precipitation producing runoff. (b) That part of precipitation falling on an irrigated area that meets the demands of consumptive use. Cf: *precipitation excess.*
effective pressure *effective stress.*
effective radiation *effective terrestrial radiation.*
effective size *effective diameter.*
effective stress The average normal force per unit area transmitted directly from particle to particle of a soil or rock mass. It is the stress that is effective in mobilizing internal friction. In a saturated soil in equilibrium, the effective stress is the difference between the total stress and the neutral stress of the water in the voids (pore water pressure). It attains a maximum value at complete consolidation and before shear failure. Syn: *effective pressure; intergranular pressure; intergranular stress.*
effective tension A stress state in which high pore pressures within flaws or joints result in local tensile stresses at the crack tip in otherwise fully compressive stress fields. Natural hydraulic fractures are driven when cracks are in a state of effective tension.
effective terrestrial radiation The difference between the outgoing infrared *terrestrial radiation* and the downward infrared *counterradiation* from the Earth's atmosphere. Syn: *effective radiation.*
effective unit weight The *unit weight* of a soil or rock mass that, when multiplied by the height of the overlying column of soil or rock, yields the effective stress caused by the weight of the overburden.
effective velocity The actual velocity of groundwater percolating through water-bearing material. "It is measured by the volume of groundwater passing through a unit cross-sectional area divided by effective porosity" (Tolman, 1937, p.593). The velocity is the average for water moving through the interstices.
effenbergerite A blue tetragonal mineral of the *gillespite* group: $BaCuSi_4O_{10}$.
efficiency (ef-fi'-cien-cy) The *capacity* of a stream per unit discharge and unit gradient, or the quotient of capacity by the product of discharge and gradient (Gilbert, 1914, p.36). It is a measure of the stream's potential work of transportation in relation to its potential energy. Symbol: E.
efflorescence (ef-flo-res'-cence) (a) A whitish fluffy or crystalline powder, produced as a surface encrustation on a rock or soil in an arid region by evaporation of water brought to the surface by capillary action or by loss of water of crystallization on exposure to the air. It may consist of one or several minerals, commonly soluble salts such as gypsum, thermonatsite, and halite. (b) The process by which an efflorescent salt or crust is formed. See also: *efflorescent crust; powdery efflorescence; puffy ground.*
efflorescent crust A surface pavement comprised of fine interlocking saline mineral crystals formed by efflorescence. Rigid crusts dominated by halite commonly display complex surface morphology of pinnacles and pits. Other mineralogies form crusts similar to those composed of halite to soft powdery surface coatings.
effluent (ef'-flu-ent) adj. Flowing forth or out; emanating. n. (a) A surface stream that flows out of a lake (e.g. an outlet), or a stream or branch that flows out of a larger stream (e.g. a distributary). Ant: *influent.* Cf: *effluent stream.* (b) A liquid discharged as waste, such as contaminated water from a factory or the outflow from a sewage works; water discharged from a storm sewer or from land after irrigation.
effluent cave *outflow cave.*
effluent flow Flow of water from the ground into a body of surface water; e.g., the flow of water to an *effluent stream.*
effluent lava flow A lava flow that is discharged from a volcano by way of a lateral fissure (Dana, 1890); an obsolete term. Cf: *interfluent lava flow; superfluent lava flow.*
effluent seepage Diffuse discharge of groundwater to the land surface; seepage of water from out of the ground.
effluent stream (a) A stream or reach of a stream that receives water from the zone of saturation and provides *base flow;* its channel lies below the water table. Syn: *gaining stream.* (b) *effluent.*
efflux (ef'-flux) *outflow.*
effusion (ef-fu'-sion) The emission of relatively fluid lava onto the Earth's surface; also, the rock so formed. Cf: *extrusion.*
effusive (ef-fu'-sive) *extrusive.*
efremovite (ef-rem-ov'-ite) A dull gray cubic mineral of the langbeinite group: $(NH_4)_2Mg_2(SO_4)_3$. It is the product of a burning coal dump.
egeran (eg'-er-an) A brown or yellowish-green variety of vesuvianite.
Egerian (E-ger'-ian) Central European stage: upper Oligocene-lower Miocene (below Eggenburgian).
egg *ovum.*
Eggenburgian (Eg-gen-burg'-ian) Central European stage: lower Miocene (above Egerian, below Ottnangian).
eggletonite (eg'-gle-ton-ite') A dark brown monoclinic mineral, the Na-analogue of ganophyllite: $(Na,K,Ca)_2(Mn,Fe)_8(Si,Al)_{12}O_{29}(OH)_7·11H_2O$.
eggstone (egg'-stone) *oolite.*
eglestonite (eg'-le-ston-ite') An adamantine brownish-yellow cubic mineral: $Hg^{1+}_6OCl_3(OH)$.

egueiite (e-guei′-ite) A yellowish-brown material consisting of a hydrous phosphate of ferric iron with a little calcium and aluminum and occurring in small nodules in clay. Probably an alteration product of vivianite.

Egyptian jasper (E-gyp′-tian) A brown or banded jasper occurring as pebbles or small boulders scattered over the desert surface between Cairo and the Red Sea.

Eh The potential of a half-cell, measured against the standard hydrogen half-cell. Syn: *oxidation potential.*

EH chondrite An *enstatite chondrite* having a high content of metallic Fe-Ni.

E horizons Mineral horizons in which the main feature is loss of silicate clay, iron, aluminum, or some combination of these, leaving a concentration of sand and silt particles of quartz or other resistant materials.

Eh-pH diagram A diagram in which equilibria and stability fields of minerals, complexes, and aqueous ions are represented as functions of oxidation potential on the ordinate versus pH on the abscissa.

Ehrenfest relation (Eh′-ren-fest) A modified *Clapeyron equation* that is used for second-order transitions.

ehrleite (ehrle′-ite) A colorless triclinic mineral: $Ca_2ZnBe(PO_4)_2(PO_3OH)·4H_2O$.

Eifelian (Ei-fe′-li-an) A stage of the Standard Global Chronostratigraphic Scale: Middle Devonian (above Emsian, below Givetian). The basal GSSP is in the Eifel plateau, western Germany (Ziegler and Klapper, 1985).

eifelite (ei′-fel-ite) A colorless to light green or yellow hexagonal mineral of the *milarite* group: $KNa_3Mg_4Si_{12}O_{30}$.

eigenmodes *normal modes.*

eigenvalue The product of a matrix T and a nonzero vector v is equal to the product of a scalar λ and the vector c, when λ is the root of the polynomial equation [T-λI] = 0. In this circumstance, λ is called an eigenvalue of T and v is an *eigenvector.* In multivariate statistics, T is the *covariance* matrix and each eigenvalue represents the variance of a composite variable that is a weighted combination of the original variables; the weights assigned each original variable is given by the terms in the associated eigenvector. Syn: *characteristic value.*

eigenvector In multivariate statistics, eigenvectors can be regarded as the axes of hyper ellipses defined by the rows of a *covariance* matrix. Syn: *characteristic vector.* Cf: *eigenvalue.*

eightling (eight′-ling) A crystal twin, either cyclic or interpenetrating, that consists of eight individuals. Cf: *twoling; trilling; fourling; fiveling.*

eikonal equation A form of the wave equation for harmonic waves in which the local velocity V is compared to a reference velocity V_R (analogous to comparing a velocity to the speed of light in vacuum). Valid only where the variation of properties is small within a wavelength, sometimes called the "high-frequency condition."

eilyma (ei′-ly-ma) The *acritarch vesicle* wall (Sarjeant and Stancliffe, 1994). Etymol: Greek, cover, clothing.

einkanter (ein′-kan-ter) A *ventifact* or wind-worn stone having only one face or a single sharp edge; it implies a steady, unchanging wind direction. Etymol: German "Einkanter", "one having one edge". Pl: einkanters; einkanter.

eiscir (eis′-cir) An Irish term for ridge; esp. *esker.*

eitelite (ei′-tel-ite) A colorless rhombohedral mineral: $Na_2Mg(CO_3)_2$.

ejecta [impact] (e-jec′-ta) Solid, liquid, or vaporized rock ejected ballistically from an explosion or impact crater during formation. Such material may be distributed around a crater in distinctive patterns, forming "ejecta rays" or "ejecta loops", as well as partially building the rim.

ejecta [pyroclast] Material explosively ejected from a volcano; *pyroclastics.* Obsolete syn: *ejectamenta.*

ejecta blanket A continuous *ejecta* deposit around an *impact crater.*

ejecta curtain The inverted-cone-shaped configuration of *ejecta* expelled from an *impact crater* during the *excavation stage.* The trajectories of individual ejected fragments form a large angle to the cone, which appears to expand rapidly as the crater grows in size. The *ejecta blanket* is deposited from the base of the cone, as the ejecta curtain sweeps across the surface. Cf: *ballistic sedimentation.*

ejectamenta (e-jec′-ta-men′-ta) *ejecta [pyroclast].*

ekanite (ek′-a-nite) A yellow, green, or brown tetragonal mineral: $Ca_2ThSi_8O_{20}$.

ekaterinite (e-kat′-e-rin′-ite) A white hexagonal mineral: $Ca_2B_4O_7(Cl,OH)_2·2H_2O$.

ekatite A brownish black hexagonal mineral: $(Fe^{3+},Fe^{2+},Zn)_{12}(AsO_3)_6(AsO_3,HOSiO_3)_2(OH)_6$.

ekdemite (ek′-de-mite) *ecdemite.*

ekerite (ek′-er-ite) A *syenite* or *quartz syenite* that contains arfvedsonite as an essential component along with aegirine, microperthite, and soda microcline, and with little quartz. The name was taken by Brögger in 1906 from Eker in the Oslo district, Norway. Obsolete.

eklogite (ek′-lo-gite) *eclogite.*

Ekman current (Ek′-man) A theoretical description of the current at the top of the ocean where friction, *wind stress* and *Coriolis forces* are important. The description assumes the wind blows constantly for several days over an area of many kilometers in extent. The current decreases exponentially with depth below the surface, and the direction of the current turns clockwise (counterclockwise) as depth increases in the northern (southern) hemisphere. The direction of the current at the surface is 45 degrees to the right (left) of the wind in the northern (southern) hemisphere. Speed at the surface is about 2% of the wind speed. A similar current exists just above the sea floor where bottom stress is important.

ekmanite (ek′-man-ite) A complex green, grayish white, or black platy hexagonal mineral possibly related to stilpnomelane: $(Fe,Mg)_3(OH)_2(Si,Al)_4O_{10}·2H_2O$.

Ekman layer (Ek′-man) The thin layer, approximately 50-100 m thick at the sea surface and above the sea floor, where friction, *wind stress* or bottom drag, and *Coriolis force* are all important. The average flow of water in the layer, the *mass transport,* is perpendicular to the *wind stress.*

Ekman pumping The vertical velocity at the base of the Ekman layer due to inhomogeneous winds blowing on the sea. It drives the deeper *geostrophic* circulation, which must have an equal and opposite velocity at the base of the Ekman layer; e.g., the downward flux of warm surface water at the center of the mid-latitude gyre in the North Atlantic due to northward Ekman transport in the trades and southward Ekman transport in the westerlies.

Ekman spiral A theoretical, graphic description of the way in which a wind, blowing uniformly and steadily over a homogeneous ocean of unlimited depth and extent and of constant viscosity, would cause currents in the surface layers to vary with depth, the water at the very surface drifting at an angle of 45° to the right of the wind direction in the Northern Hemisphere (and to the left in the Southern Hemisphere) and water at successive depths drifting in directions farther away from the wind direction (as a spiral) with a rapidly decreasing speed until it would move in the direction opposite to the wind; the net water transport (Ekman transport) is 90° to the right of the wind direction in the Northern Hemisphere. It is named for Vagn Walfrid Ekman, Swedish oceanographer, who in 1902 developed the theory of the spiral, which has also been applied to atmospheric motion.

eksedofacies (ek′-sed-o-fa′-cies) Facies of the weathering environment (Vassoevich, 1948).

ektexine (ek-tex′-ine) The outer of the two layers of the *exine* of palynomorphs, normally more densely or deeply staining than the *endexine,* and characterized by richly detailed external sculpture and often by complex internal structure of granules, columellae, and other elements. Also spelled: *ectexine.* Syn: *ectoexine; sexine.*

ektexis (ek-tex′-is) *ectexis.*

elaeolite (e-lae′-o-lite) *eleolite.*

élan vital (e-lan′ vi-tal′) In early evolutionary theories, the internal vital or driving force that supposedly stimulates the process of evolution. It was regarded as an inherent property of living matter. Etymol: French.

Elasmobranchii (E-las′-mo-branch′-i-i) A subclass of cartilaginous fishes characterized by hyostylic or amphistylic jaw suspension and numerous teeth; if modified for crushing, the teeth are not reduced to a few plates as in the Bradydonti and Holocephali. The subclass includes sharks, skates, and rays.

elastic (e-las′-tic) Said of a body in which strains are instantly and totally recoverable and in which deformation is independent of time. Cf: *plastic [struc geol].*

elastic aftereffect *creep recovery.*

elastic afterworking *creep recovery.*

elastic behavior The *elastic* response of rock to stress.

elastic bitumen *elaterite.*

elastic compliance The reciprocal of *Young's modulus;* or more generally, a coefficient that relates a component of strain to a component of stress in an elastic material.

elastic constant One of various coefficients that define the elastic properties of a body, including the *Lamé constants, Poisson's ratio,* or one of the *moduli of elasticity.*

elastic contraction The joint-driving mechanism in which the energy release for driving the joint is derived from thermoelastic strains generated in the rock during cooling and denudation (Engelder and Fischer, 1996).

elastic deformation Deformation of a substance, which disappears when the deforming forces are removed. Commonly, that type of deformation in which stress and strain are linearly related, according to *Hooke's law.* Cf: *plastic deformation.*

elastic deformation mechanism A mechanism by which *elastic deformation* occurs, namely the change in length of chemical bonds holding atoms or molecules together.

elastic discontinuity A boundary between strata of different elastic moduli and/or density at which seismic waves are reflected and refracted.

elastic energy The energy stored within a solid by elastic deformation.

elastic limit The greatest stress that can be developed in a material without permanent deformation remaining when the stress is released.

elastic material *Hookean substance.*

elastic medium A material that supports both P- and S-wave travel.

elastic modulus *modulus of elasticity.*

elasticoplastic (e-las′-ti-co-plas′-tic) Said of deformation that has a perfectly elastic phase and a perfectly plastic phase. It is demonstrated by the model of a *Saint Venant substance.*

elasticoviscous behavior The response of a rock to stress such that application of a stress causes an instantaneous strain (*elastic behavior*) followed by continuously developed permanent strain (*viscous behavior*), so long as the stress is maintained; after the stress is removed, only the elastic component of strain is recovered. Syn: Maxwellian behavior. See also: *Maxwell liquid.*

elastic rebound Elastic recovery from strain.

elastic-rebound theory (e-las'-tic-re'-bound) The statement that movement along a fault is the result of an abrupt release of a progressively increasing elastic strain between the rock masses on either side of the fault. Such a movement returns the rocks to a condition of little or no strain. The theory was proposed by Harry Fielding Reid in 1911. Syn: *Reid mechanism.*

elastic recoil detection analysis An analytical technique measuring the energies of forward elastic recoil of atoms out of a sample of interest. Similar to *Rutherford backscattering*, and useful for investigating surficial concentrations of elements. Especially useful for light elements such as H and He. Abbrev: ERDA.

elastic strain The strain developed during the elastic behavior of a material.

elastic wave *Seismic wave* including both compressional (P-wave) and shear (S-waves).

elater (el'-a-ter) A ribbonlike band (or bands), spirally twisted around certain *spores* (e.g., *Equisetum*), assisting in dispersal. Somewhat similar appendages may project from certain Late Cretaceous angiosperm pollen.

elaterite (e-lat'-er-ite) A brown asphaltic pyrobitumen, soft and elastic when fresh but hard and brittle on exposure to air. It is derived from the metamorphism of petroleum. See also: *coorongite.* Syn: *elastic bitumen; liverite; mineral caoutchouc.*

elatolite (e-lat'-o-lite) A supposedly high-temperature modification of calcite, but probably crystal casts of calcium carbonate after villiaumite (Hey, 1962, p.412).

E layer The seismic region of the Earth from 2,890 km to 4,710 km, equivalent to the *outer core.* It is a part of a classification of the Earth's interior made up of layers A to G. Rarely used.

elb A *transverse dune* in the desert of Algeria.

elbaite (el'-ba-ite) A variously colored rhombohedral mineral of the *tourmaline* group: $Na(Al,Li)_3Al_6(BO_3)_3Si_6O_{18}(OH)_4$. It is the principal gem tourmaline.

Elbe (El'-be) Term formerly applied in northern Europe to the first glacial stage of the Pleistocene Epoch, followed by the *Elster;* probably equivalent to the *Günz* and *Nebraskan.* Superseded by *Menap.*

elbow of capture (el'-bow) The point at which capture was effected along a stream course, characterized by an abrupt or sharp bend where the course turns from the captured part of its valley into the valley of the capturing stream.

elbow twin *geniculate twin.*

EL chondrite An *enstatite chondrite* having a low content of metallic Fe-Ni.

electrical conductivity (e-lec'-tri-cal) A measure of the ease with which a conduction current can be caused to flow through a material under the influence of an applied electric field. It is the reciprocal of resistivity and is measured in mhos per meter.

electrical imaging log A gray-level, high-resolution resistivity image of the borehole wall processed from multiple microresistivity logs as an extension of *dipmeter* technology.

electrical method A geophysical prospecting method that depends on the electrical, electromagnetic, or electrochemical properties of rocks. The resistivity, spontaneous-polarization, induced-polarization, and inductive-electromagnetic methods are the principal electrical methods.

electrical resistivity The electrical resistance per unit length of a unit cross-sectional area of a material.

electrical-resistivity sounding (e-lec'-tri-cal-re-sis-tiv'-i-ty) A procedure for determining depths to geological interfaces, wherein separations of electrodes in an array are increased by increments. Observed apparent resistivity versus electrode separation is inverted to give the depths to the interfaces and the resistivities of the strata. See also: *geometric sounding; parametric sounding; inversion [geophys].* Syn: *resistivity sounding.*

electrical twinning Twinning in quartz according to the *Dauphiné twin law.*

electric calamine (e-lec'-tric) *hemimorphite.*

electric-field intensity The strength of an electric field at any point. It is defined by the force exerted on a unit positive charge placed at that point. Syn: *voltage gradient.*

electric log The generic term for a *well log* that displays electrical measurements of induced current flow (*resistivity log, induction log*) or natural potentials (*spontaneous-potential curve*) in the rocks of an uncased borehole. An electric log typically consists of the spontaneous-potential (SP) curve and one or more resistivity or induction curves. The *Archie equations* form the basis for interpretation of electric logs. Abbrev: E-log. Informal syn: *resistivity log.*

electrochemical induration (e-lec'-tro-chem'-i-cal) A method of strengthening and consolidating saturated and unconsolidated soils or other granular earth materials by passing a direct current of electricity through probes placed in them (Titkov et al., 1965).

electrochromatography (e-lec'-tro-chro'-ma-tog'-ra-phy) *Chromatography* wherein an applied electric potential is used to produce differential electrical migration.

electrode array (e-lec'-trode) A configuration of electrodes on or in the ground for the purpose of making an electrical survey.

electrodeless discharge (e-lec'-trode-less) Emission of light from matter energized by induced electrical currents.

electrodiagenesis (e-lec'-tro-di'-a-gen'-e-sis) Diagenesis affected or stimulated by electric currents and potentials.

electrodialysis (e-lec'-tro-di-al'-y-sis) *Dialysis* assisted by the application of an electric potential across the semipermeable membrane. Two important uses of electrodialysis are in water desalination and in removing electrolytes from naturally occurring colloids such as proteins. Cf: *electro-osmosis.*

electrofiltration *electrostatic precipitation.*

electrofiltration potential (e-lec'-tro-fil-tra'-tion) An electrical potential that is caused by movement of fluids through porous formations. Syn: *streaming potential; electrokinetic potential.*

electrographic (e-lec'-tro-graph'-ic) Pertaining to a method for analyzing minerals and metals by transferring a small amount of the sample by electrical means to a prepared surface where the ions are identified.

electrokinetic potential (e-lec'-tro-ki-net'-ic) *electrofiltration potential.*

electrolysis (e-lec-trol'-y-sis) A method of breaking down a compound in its natural form or in solution by passing an electric current through it, the ions present moving to one electrode or the other where they may be released as new substances.

electromagnetic energy Energy that travels at the speed of light in a harmonic wave pattern. See also: *electromagnetic radiation.*

electromagnetic induction (e-lec'-tro-mag-net'-ic) The generation of electric field or current in an electric conductor when that body is in a changing magnetic field or is moving through a magnetic field; a phenomenon of *electromagnetism.*

electromagnetic method An electrical exploration method based on the measurement of alternating magnetic fields associated with currents artificially or naturally maintained in the subsurface. If these currents are induced by a primary alternating magnetic field, the name inductive electromagnetic method applies, whereas if they are conducted into the ground via electrodes, the name conductive electromagnetic method applies. See also: *inductive method.*

electromagnetic propagation log A log that records the propagation time and attenuation of electromagnetic energy microwaves through a formation. The radically differenct propagation time of water results in a log that is useful in the detection of hydrocarbons regardless of formation water salinity.

electromagnetic radiation Energy that travels in the form of electomagnetic waves and photons. Energy propogated in the form of an advancing interaction between electric and magnetic fields. Abbrev: EMR.

electromagnetic sounding An electromagnetic technique designed to determine variations in electrical conductivity with depth, usually assuming horizontal layering. Measurements are usually made at a number of frequencies (in frequency-domain sounding) or at several time intervals after a transient pulse (in time-domain or transient electromagnetic sounding) using a fixed source and receiver. Soundings can also be made at constant frequency by varying the spacing between source and receiver. See: *transient electromagnetic method.*

electromagnetic spectrum The entire distribution of wavelengths of *electromagnetic radiation* including gamma (short wavelengths), X-ray, ultraviolet, visible, infrared and radio waves (long wavelengths). See also: *spectrum [phys].*

electromagnetism (e-lec'-tro-mag'-net-ism) The totality of electric and magnetic phenomena, or their study; particularly those phenomena with both electric and magnetic fields, such as *electromagnetic induction.*

electrometallurgy The use of electrical current to chemically separate the components of a beneficiated ore. Cf: *hydrometallurgy.*

electromigration (e-lec'-tro-mi-gra'-tion) A method of separating isotopes or ions by their differing rates of movement during electrolysis.

electron capture (e-lec'-tron) A mode of radioactive decay in which an orbital electron is captured by the nucleus. The resulting nuclear transformation is identical with that in β^+ emission.

electron diffraction The diffraction of a beam of electrons, usually by the three-dimensional periodic array of atoms in a crystal that has periodic repeat distances (lattice dimensions) of the same order of magnitude as the wavelength of the electron beam. See also: *X-ray diffraction.*

electron diffraction analysis Analysis by *electron diffraction.*

electron diffraction pattern The interference pattern seen when a beam of electrons is sent through a substance, each pattern being characteristic for a particular substance. Electron diffraction patterns contain basic crystallographic information as well as information about orientation, defects, crystal size, and additional phases. See also: *X-ray diffraction pattern.*

electron energy loss spectroscopy A type of spectroscopy using the inelastic scattering of a primary electron beam transmitted through a thin specimen; used to determine compositions, especially in metals. Most useful for light elements such as H, He, and Li, but more generally used for elements lighter that Si. Abbrev: EELS.

electronic distance-measuring instrument (e-lec-tron'-ic) A device that measures the phase differences between transmitted and returned (i.e. reflected or retransmitted) electromagnetic waves, of known frequency and speed, or the round-trip transit time of a pulsed signal, from which distance is computed. Commercial trade names are Electrotape, Geodimeter, and Tellurometer. Abbrev: EDM.

electron magnetic resonance *electron spin resonance.*

electron microprobe An analytical instrument that uses a finely focused beam of electrons to excite X-ray emission from selected portions of a sample. From the emitted X-ray spectrum the composition of the sample at the point of excitation can be determined. Spots as small as 1 micrometer in diameter can be analyzed, with sensitivities around 50ppm or less for most metals. Syn: *electron probe; microanalyzer.*

electron microscope An electron-optical instrument in which a beam of electrons, focused by systems of electrical or magnetic lenses, is used to produce enlarged images of minute objects on a fluorescent screen or photographic plate in a manner similar to that in which a beam of light is used in a compound microscope. The electron microscope, because of the very short wavelength of the electrons, is capable of resolving much finer structures than the optical instrument, and is capable of magnifications on the order of 100,000X. See also: *scanning electron microscope.*

electron microscopy Determining and identifying the structure of substances by using the *electron microscope.*

electron paramagnetic resonance A syn. of *electron spin resonance.* Abbrev: EPR.

electron probe *electron microprobe.*

electron spin resonance *Resonance* occurring when electrons that are undergoing transitions between energy levels in a substance are irradiated with electromagnetic energy of a proper frequency to produce maximum absorption. Abbrev: ESR. Syn: *electron magnetic resonance; electron paramagnetic resonance; paramagnetic resonance.*

electro-osmosis (e-lec'-tro-os-mo'-sis) The motion of liquid through a membrane under the influence of an applied electric field. See also: *osmosis.*

electrophoresis (e-lec'-tro-pho-re-sis) The movement toward electrodes of suspended charged particles in a fluid by applying an electromotive force to the electrodes that are in contact with the suspension. See also: *cataphoresis; anaphoresis.*

electroprecipitation (e-lec'-tro-pre-cip'-i-ta'-tion) *electrostatic precipitation.*

electrostatic precipitation (e-lec'-tro-stat'-ic) A method for removing suspended solid or liquid particles from a gas by applying a strong electric field to the mixture that charges the particles and precipitates them. Syn: *electrofiltration; electroprecipitation.*

electrostriction (e-lec'-tro-stric'-tion) Physical deformation induced in materials on polarization by an applied electric field.

Electrotape (E-lec'-tro-tape) A trade name for a precise electronic surveying device that transmits a radio-frequency signal to a responder unit which in turn transmits the signal back to the interrogator unit. The time lapse between original transmission and receipt of return signal is measured and displayed in a direct digital readout for eventual reduction to a precise linear distance. It operates on the same principle as the *Tellurometer* and *Geodimeter.*

electroviscosity (e-lec'-tro-vis-cos'-i-ty) The viscosity of a fluid as influenced by electric properties, e.g. greater viscosity of a low-conductivity fluid than of a high-conductivity fluid flowing through narrow capillaries.

electrowinning The recovery of metals from a solution by electrolysis, commonly following *solvent extraction.*

electrum (e-lec'-trum) (a) A naturally occurring, deep-yellow to pale-yellow cubic alloy of gold with silver; argentiferous gold, containing more than 20% silver. Also spelled: "elektrum". (b) An ancient Greek name, "elektron", now obsolete, for *amber.*

electrum-tarnish method (e-lec'-trum-tar'-nish) A technique for direct determination of sulfur fugacity in experimental study of sulfide systems. The method consists of determining the temperature at which electrum of a given gold-silver ratio tarnishes when equilibrated with a vapor in equilibrium with the sulfide assemblage being measured (Barton and Toulmin, 1964b).

element An individual component of the skeletal apparatus of a conodont; consists of the commonly preserved "element proper" and the less commonly preserved basal funnel or plates (TIP, 1981, pt. W, supp.2). Syn: *holoconodont.*

element proper *crown [paleont] (c).*

elements of symmetry (el'-e-ments) *symmetry elements.*

eleolite (e-le'-o-lite) A syn. of *nepheline,* esp. of a translucent, massive or coarsely crystalline, and dark variety (grayish, bright-green, or brown to brownish-red), having a greasy luster and sometimes used as an ornamental stone. Also spelled: *elaeolite.*

elephant-head dune (el'-e-phant) A *sand shadow* or small dune resembling the head of an elephant, having a rounded windward face covered with vegetation and a long, tapering snout of bare sand on the leeward side; examples occur in the Coachella Valley of the Colorado Desert in California.

elephant-hide pahoehoe A type of *pahoehoe* having a wrinkled and draped surface.

elephant rock A term used in SE Missouri for a *rocking stone,* not necessarily delicately balanced, formed in place by the weathering and removal of surrounding material. Cf: *core-stone; tor.*

eleutheromorph (e-leu'-ther-o-morph') A new mineral in a metamorphic rock that has been freely developed and thus has gained its form independently. Little used. Cf: *pseudomorph.*

eleutherozoan (e-leu'-ther-o-zo'-an) n. Any echinoderm that does not live attached to a substrate. adj. Said of an echinoderm having a free mode of life. Var: eleutherozoic. Cf: *pelmatozoan.*

elevated shoreline (el'-e-vat'-ed) A shoreline whose development has been interrupted by a relatively sudden rise of the coast or by a rapid lowering of the water level; it is not a true shoreline because it is no longer being shaped by waves and currents. Examples: a broad marine terrace (common along the continental and insular coasts of the Pacific Ocean), or a narrow strandline. Not to be confused with *shoreline of elevation.*

elevation [geomorph] (el-e-va'-tion) A general term for a topographic feature of any size that rises above the adjacent land or the surrounding ocean bottom; a place or station that is elevated.

elevation [surv] The vertical distance from a datum (usually mean sea level) to a point or object on the Earth's surface; esp. the height of a ground point above the level of the sea. The term is used synonymously with *altitude* in referring to distance above sea level, but in modern surveying practice the term "elevation" is preferred to indicate heights on the Earth's surface whereas "altitude" is used to indicate the heights of points in space above the Earth's surface.

elevation correction The correction applied to time values observed in reflection or refraction seismic surveys due to differences of station elevations, in order to reduce the observations to an arbitrary reference datum.

elevation head The elevation of the point at which the hydrostatic pressure is measured, above or below an arbitrary horizontal datum. Syn: potential head. Cf: *static head.*

elevation meter A mechanical or electromechanical device on wheels that measures slope and distance and that automatically and continuously integrates their product into difference of elevation.

elevation-relief ratio It is the ratio: (Mean elevation - minimum elevation) ÷ (maximum elevation - minimum elevation) (Wood and Snell, 1960). Pike and Wilson (1971) have shown it to be identical to the *hypsometric integral.*

elevator tectonics (el'-e-va'-tor) A rarely used term introduced by Dietz and Holden (1966, p.353) for the rise and fall of blocks of sialic crust from surficial to abyssal levels or vice versa.

elkhornite (elk'-horn-ite) A hypabyssal labradorite-bearing augite *syenite.* It was named by Johannsen in 1937 for the Elkhorn district of Montana. Obsolete.

ellenbergerite (el'-len-ber'-ger-ite) A lilac hexagonal mineral: $Mg_6TiAl_6Si_8O_{28}(OH)_{10}$.

ellestadite (el'-le-stad-ite') A sub-group name for a hexagonal or monoclinic mineral of the *britholite* group (apatite structure) in which the PO_4 of apatite is largely replaced by SiO_4 and/or SO_4.

ellipochoanitic (el-lip'-o-cho'-a-nit'-ic) Said of a relatively short, *retrochoanitic septal neck* of a nautiloid that does not reach as far as the preceding septum. See also: *cyrtochoanitic; orthochoanitic.*

ellipse The closed mathematical curve that describes the path of a body in a gravitationally bound orbit about another body, e.g., a planet around the Sun. An elliptical orbit is characterized by its semimajor axis *a* (half the long axis of the ellipse), and eccentricity *e,* which measures the range of distances between the two bodies, from a minimum of $a(1 - e)$ to a maximum of $a(1 + e)$. Cf: *parabola [astron], hyperbola.* See also: *Kepler's laws of planetary motion.*

ellipsoid (el-lip'-soid) (a) A mathematical figure closely approaching the *geoid* in form and size. It is generally defined by its equatorial radius and by the reciprocal of the flattening, a/(a-b), where a and b are the equatorial and polar radii. A task of geodesists is the determination of more exact parameters of the ellipsoid. Cf: *spheroid [geodesy]; Clarke ellipsoid of 1866; reference ellipsoid.* (b) A surface whose plane sections (cross sections) are all ellipses or circles, or the solid enclosed by such a surface. In geology, ellipsoid and spheroid are used interchangeably.

ellipsoidal height The sum of the elevation of a point and the geodetic height of the geoid above the reference ellipsoid at the corresponding point on the geoid.

ellipsoidal lava (el-lip-soi'-dal) (a) *pillow lava.* (b) An inclusive term for any lava flow with an ellipsoidal pattern, i.e. pillow lava and the lava toes of pahoehoe (Macdonald, 1953).

ellipsoid of revolution The simple mathematical figure that would be produced by an ellipse revolving around its minor axis. It is often used as reference surface for the Earth. See also: *spheroid [geodesy].*

elliptical polarization (el-lip'-ti-cal) In optics, elliptically polarized light consisting of upward-spiraling vibration vectors, the surface of which is elliptical rather than circular, as in *circular polarization.* It is caused by the inconstant lengths of vibration vectors of mutually perpendicular plane-polarized waves whose path differences differ in phase by amounts other than $(n+1)/4\lambda$ on emergence from a crystal.

elliptical projection One of several map projections showing the Earth's surface upon the interior of an ellipse; e.g. *Mollweide projection; Aitoff projection.*

ellipticity (el-lip-tic'-i-ty) (a) The degree of flattening of the *reference ellipsoid* as expressed by the equation e=(a-b)/a, where a and b are the equatorial and polar radii. Cf: *equatorial bulge.* (b) The ratio of minor to major axes of an ellipse. This is one of the three "Milankovitch parameters" that contribute to long-term climatic variability, specifically for the ice ages; the principal period of this Earth-Moon relationship is 90,000 to 100,000 years. See also: *strain ratio.*

ellipticone (el-lip'-ti-cone) A coiled cephalopod shell having elliptic coiling of the last whorl or half whorl which breaks away from the spiral or slightly breaks the regularity of the spiral form.

ellisite (el'-lis-ite) A metallic gray rhombohedral mineral: Tl_3AsS_3.

ellsworthite (ells'-worth-ite) *uranpyrochlore.*

El Niño (Ni'-ño [neenyo]) An anomalous warming of the surface water of the eastern tropical Pacific with accompanying heavy precipitation over the normally dry coastal zones of Ecuador and Peru. This event often starts near Christmas, hence its name. It shuts down upwelling off Peru and its highly detrimental to fishing there. See also: *El Niño Southern Oscillation.* [Etymol: Spanish, The (Christ)child. English transliteration: El Ninyo, not El Nino.]

El Niño Southern Oscillation Oscillation of atmospheric pressure, wind, and sea-surface temperature in the coupled ocean-atmosphere system of the tropics. Originally based on the pronounced negative correlation between air pressure series in Darwin, Australia, and Tahiti in the South Pacific. The warm phase, known as El Niño, occurs roughly twice per decade. The cool phase occurs when the atmospheric pressure difference is reversed from the El Niño stage and is called (arbitrarily) "La Niña". Weather consequences of this oscillation are felt world wide and include changes in the atmospheric jet streams, droughts, floods, fish harvests, and hurricane frequency. Abbrev: ENSO.

elongation (e-lon-ga'-tion) *extension [exp struc geol].*

elongation sign *sign of elongation.*

elpasolite (el-pas'-o-lite) A colorless cubic mineral: K_2NaAlF_6.

elphidiid (el-phid'-i-id) Any foraminifer belonging to the family Elphidiidae, characterized by having a sutural canal system opening into rows of sutural pores. Range, Paleocene to Holocene.

elpidite (el'-pi-dite) A fibrous variously colored orthorhombic mineral: $Na_2ZrSi_6O_{15}{\cdot}3H_2O$.

Elsonian Orogeny (El-so'-ni-an) An orogenic event in the Canadian Shield about 1400 m.y. ago.

Elster (El'-ster) The term applied in northern Europe to the second glacial stage of the Pleistocene Epoch, following the Elbe and preceding the Saale glacial stage; equivalent to the *Mindel* and *Kansan.*

Elsterian (El-ste'-ri-an) European cold-climate paleofloral phase: upper middle Pleistocene (above Cromerian, below Holsteinian). The name may be applied to successive cold maxima in this interval. Equiv. Mindel glacial system(s).

Eltonian (El-to'-ni-an) European stage: Upper Silurian (above Wenlockian, below Bringewoodian).

elutriation (e-lu-tri-a'-tion) (a) A method of mechanical analysis of a sediment, in which the finer and/or lightweight particles are separated from the coarser and/or heavy particles by means of a slowly rising current of air or water of known and controlled velocity, carrying the lighter particles upward and allowing the heavier ones to sink. (b) Purification, or removal of material from a mixture or in suspension in water, by washing and decanting, leaving the heavier particles behind. (c) The washing away of the lighter-weight or finer particles in a soil by the splashing of raindrops.

eluvial [eco geol] (e-lu'-vi-al) Said of an incoherent ore deposit, such as a placer, resulting from the decomposition or disintegration of rock in place. The material may have slumped or washed downslope for a short distance but has not been transported by a stream.

eluvial [sed] Pertaining to or composed of wind-deposited *eluvium;* e.g. the "eluvial (or passive) phase" of a dune cycle, marked by sufficient vegetation to check deflation. Cf: *eolian.*

eluvial [weath] Pertaining to eluvium; *residual [weath].*

eluvial deposit A secondary mineral deposit resulting from the disintegration or decomposition of the original host rock, with minimal transportation of the material; thus eluvial deposits remain relatively close to the primary deposit from which they are derived.

eluvial horizon A soil horizon from which material has been removed by the process of *eluviation.* Cf: *illuvial horizon.*

eluviated (e-lu'-vi-at'-ed) Said of a soil horizon or of materials that have been subjected to the process of *eluviation.*

eluviation (e-lu'-vi-a'-tion) The removal of soil material in suspension (or in solution) from a layer or layers of a soil. Usually, the loss of material in solution is described by the term "leaching."

eluvium [sed] (e-lu'-vi-um) Fine soil or sand moved and deposited by the wind, as in a sand dune. Cf: *alluvium.*

eluvium [weath] An accumulation of rock debris produced in place by the decomposition or disintegration of rock; a weathering product; a *residue.*

elvan (el'-van) A Cornish term (from Celtic, "white rock") for hypabyssal rocks having the composition of granite, esp. a *quartz porphyry.* Tourmaline, fluorite, and topaz may be accessories. Syn: elvanite. Not recommended usage.

Elvis taxa Fossils that have converged in their morphology on some previously extinct taxon, giving the erroneous appearance that the earlier fossil's range has been extended.

elyite (e'-ly-ite) A silky violet monoclinic mineral: $Pb_4CuO_2(SO_4)(OH)_4{\cdot}H_2O$.

emanation (em-a-na'-tion) *exhalation.*

emanometry (e-ma-nom'-e-try) A method of geophysical prospecting for radioactive minerals based on a measurement of the radioactive daughter gas, radon. Adj. emanometric.

emarginate (e-mar'-gin-ate) Having a notched margin; e.g. said of a gastropod having a variously excavated margin of the outer lip, a bivalve mollusk whose margin is interrupted by a notch or sinus, the posteriorly deflected median segment of the anterior commissure of a brachiopod, or a leaf with a shallow notch at the apex.

embankment [coast] (em-bank'-ment) A narrow depositional feature, such as a spit, barrier, or bar, built out from the shore of a sea or lake by the action of waves and currents that deposit excess material at its deep end; it may be emerged or submerged. Syn: *bank [coast].*

embankment [eng] A linear structure, usually of earth or gravel, constructed so as to extend above the natural ground surface and designed to hold back water from overflowing a level tract of land, to retain water in a reservoir, tailings in a pond, or a stream in its channel, or to carry a roadway or railroad; e.g. a dike, seawall, or fill.

embayed coast (em-bayed') A coast with many projecting headlands, bays, and outlying islands, usually resulting from submergence.

embayed mountain A mountain that has been partly submerged, so that seawater enters the valleys; e.g. on the coast of SW Ireland.

embayment [coast] (em-bay'-ment) (a) The formation of a bay, as by the sea overflowing a depression of the land near the mouth of a river. (b) A bay, either the deep indentation or recess of a shoreline, or the large body of water (as an open bay) thus formed.

embayment [petrology] (a) Penetration of microcrystalline groundmass material into phenocrysts, making their "normal" euhedral boundaries incomplete. (b) An irregular corrosion or modification of the outline of a crystal by the magma from which it previously crystallized or in which it occurs as a foreign inclusion; esp. the deep corrosion into the sides of a phenocryst. (c) The penetration of a crystal by another, generally euhedral, crystal. Such a crystal is called an "embayed crystal".

embayment [struc geol] (a) A downwarped area containing stratified rocks, either sedimentary or volcanic or both, that extends into a terrain of other rocks, e.g. the Mississippi Embayment of the U.S. Gulf Coast. (b) *recess [struc geol].*

embedded (em-bed'-ded) Covered or enclosed by sediment in a matrix, such as gravel *embedded* in silt.

embedded wavelet The wavelet shape which would result from reflection of an actual wavetrain by a single sharp interface with positive reflection coefficient. Often called equivalent wavelet or basic wavelet. See: *convolutional model.*

embolite (em'-bo-lite) A yellow-green cubic mineral: $Ag(Cl,Br)$. It is intermediate in composition between chlorargyrite and bromargyrite.

embossed rock (em-bossed') A term introduced by Hitchcock (1843, p.180) as a syn. of *roche moutonnée.*

embouchure (em-bou-chure') (a) The mouth of a river, or that part where it enters the sea. (b) An expansion of a river valley into a plain. Etymol: French. Syn: embouchement.

embrechite (em'-brech-ite) A *migmatite* in which some textural components of the preexisting rocks are preserved (Dietrich and Mehnert, 1961); a migmatite with preserved parallel layering, often including feldspar phenoblasts or granitic layers and lenses (Mehnert, 1968, p.354). Rarely used.

embreyite (em'-brey-ite) An orange monoclinic mineral: $Pb_5(PO_4)_2(CrO_4)_2{\cdot}H_2O$.

embryo (em'-bry-o) A young sporophytic plant; the *germ* of a seed.

embryonic (em-bry-on'-ic) Said of the earliest growth stage in the life history of an animal; the stage preceding the *nepionic* stage.

embryonic apparatus A group of chambers at the center of some megalospheric tests of foraminifers, larger in size and different in shape and arrangement from other chambers. See also: *juvenarium.* Syn: *nucleoconch.*

embryonic volcano A breccia-filled volcanic pipe without surface expression and considered to be produced by phreatic explosions. Examples of Permian age occur in Scotland.

Embryophyta (Em'-bry-o-phy'-ta) The subkingdom of plants with embryos; includes all seed plants, pteridophytes, and bryophytes (Tippo, 1942, p.204).

embryophytic (em'-bry-o-phyt'-ic) Said of plants of the subkingdom *Embryophyta.*

emeleusite (em-el-e'-us-ite) A colorless to creamy pinkish orthorhombic mineral of the tuhualite group: $Na_2LiFe^{3+}Si_6O_{15}$.

emerald (em'-er-ald) (a) A brilliant green variety of beryl, highly prized as a gemstone. The color, which is caused by the presence of chromium or possibly vanadium, ranges from medium-light or medium-dark tones of slightly bluish-green to those of slightly yellowish-green. Syn: *smaragd.* (b) Any of various gemstones having a green color, such as "oriental emerald" (sapphire), "copper emerald" (dioptase), "Brazilian emerald" (tourmaline), and "Uralian emerald" (demantoid). (c) Said of a gemmy and richly

green-colored mineral, such as "emerald jade" (jadeite), "emerald spodumene" (hiddenite), and "emerald malachite" (dioptase).
emerald copper *dioptase.*
emerald cut A *step cut* in which the finished gem is square or rectangular and the rows (steps) of elongated facets on the crown and pavilion are parallel to the girdle with sets on each of the four sides and always at the corners. It is commonly used on diamonds to emphasize the absence of color and on emeralds and other colored stones to enhance the color. See also: *square emerald cut.*
emerald nickel *zaratite.*
emerged bog (e-merged') A bog which tends to grow vertically, i.e. increase in thickness, by drawing water up through the mass of plants to above the water table where the growth takes place. Cf: *immersed bog.*
emerged shoreline *shoreline of emergence.*
emergence [bot] (e-mer'-gence) Any outgrowth of cortical and epidermal plant tissues that lack a vascular supply. See also: *enation.*
emergence [coast] A change in the levels of water and land such that the land is relatively higher and areas formerly under water are exposed; it results either from an uplift of the land or from a fall of the water level. Ant: *submergence.*
emergence [karst] In a *karst* region, the point where an underground stream appears at the surface to become a surface stream. Cf: *resurgence; rise [karst]; rising.*
emergence angle (a) *angle of emergence.* (b) *apparent dip [seis].*
emergence velocity (a) The sum of the vertical component of glacier motion, measured at the surface, and the product of horizontal velocity times the tangent of ice slope. (b) The rate the surface would rise if there were no ablation.
emergent [bot] (e-mer'-gent) Said of a plant that rises above its substrate; e.g., an emergent aquatic plant. Syn: *emersed.*
emergent [struc geol] Term applied to a fault that intersects and displaces the ground surface during deformation.
emergent aquatic plant A rooted plant growing in shallow water, with part of its stem and leaves above the water surface; e.g., bulrush, cattail.
emergent evolution Evolution characterized by the appearance of completely new and unpredictable characteristics or qualities at different levels due to a rearrangement of pre-existing entities.
emersed (e-mersed') *emergent [bot].*
emery (em'-er-y) (a) A gray to black granular impure variety of corundum, which contains varying amounts of iron oxides (usually magnetite or hematite). It is used in granular form for polishing and grinding. It occurs as masses in marble and as segregations in igneous rocks. (b) A natural abrasive composed essentially of pulverized impure corundum. Also, the commercial product obtained by crushing emery rock. (c) *emery rock.*
Emery method A technique for two people to rapidly measure a topographic profile across a beach, or beach profile, with minimal equipment (Emery, 1961).
emery rock A granular rock that is composed essentially of an impure mixture of corundum, magnetite, and spinel, and that may be formed by magmatic segregation or by metamorphism of highly aluminous sediments. Syn: *emery.*
emigrant (em'-i-grant) In ecology, a migrant plant or animal.
emildine (em'-il-dine) A variety of spessartine garnet containing yttrium.
Emilian (E-mi'-li-an [E-meel-ian]) An Italian substage: middle substage of the regional Calabrian stage, lower Pleistocene (above Santernian, below Sicilian).
emilite A metallic tin-white orthorhombic mineral: $Cu_{10.7}Pb_{10.7}Bi_{21.3}S_{48}$. A derivative of the bismuthinite-aikine solid-solution series.
eminence (em'-i-nence) (a) An elevated area of any size, shape, or height; a mass of high land; a mountain or a hill. (b) The high point of an elevated feature.
emission spectroscopy (e-mis'-sion) The observation of an *emission spectrum* and all processes of recording and measuring that go with it.
emission spectrum The characteristic electromagnetic spectrum that a substance emits when observed with a *spectroscope* while being heated or bombarded with photons, ions, or electrons. Lines or bands appear where energy has been emitted at specific frequencies or groups of frequencies by an emitting medium. See also: *emission spectroscopy; line spectrum; band spectrum; spectroscopy.*
emittance (e-mit'-tance) (a) The ratio of the emitted *radiant flux* per unit area of a substance to that of a blackbody radiator at the same temperature. See also: *spectral emittance.* (b) An obsolete term for the radiant flux per unit area emitted by a body.
emmonsite (em'-mons-ite) A triclinic yellow-green mineral: $Fe^{3+}_2Te^{4+}_3O_9 \cdot 2H_2O$. Syn: *durdenite.*
empirical formula [mineral] The chemical formula written with the simplest ratio of atoms.
emplacement [intrus rocks] (em-place'-ment) *intrusion [ign].*
emplacement [ore dep] The localization of ore minerals, by whatever process; ore deposition.
emplectite (em-plec'-tite) A metallic grayish or white orthorhombic mineral: $CuBiS_2$. It is the Bi analogue of chalcostibite.
empolder (em-pol'-der) v. To reclaim land by the creation of polders; to make low-lying or periodically flooded land cultivable by adequate drainage and the erection of dikes to prevent or control inundation. See also: *polderization.* Syn: *impolder.* n. A tract of empoldered land; a polder.
empoldering (em-pol'-der-ing) *polderization.*
empressite (em'-press-ite) A pale-bronze orthorhombic mineral: AgTe.
EMR *electromagnetic radiation.*
Emscherian (Em-sche'-ri-an) An obsolete syn. of Coniacian-Santonian.
Emsian (Em'-si-an) A stage of the Standard Global Chronostratigraphic Scale: Lower Devonian (above Pragian, below Eifelian). The basal GSSP is near Smarkand, Uzbekistan (Yolkin et el., 1997).
emulsion A suspension of photosensitive silver halide grains in gelatin that constitutes the image-forming layer on photographic film.
emulsion stage (e-mul'-sion) Obsolete term for that stage in the crystallization of some magmas in which the concentration of water exceeds the solubility and a new, water-rich phase is formed, either as a gas or as liquid droplets (Shand, 1947).
emulsion texture An ore texture showing minute blebs or rounded inclusions of one mineral irregularly distributed in another.
enalite (en'-a-lite) A variety of thorite containing uranium.
enantiomer Chemical mirror image, used in reference to amino acids.
enantiomorph (en-an'-ti-o-morph') Either of two crystals that display *enantiomorphism.*
enantiomorphism (en-an'-ti-o-mor'-phism) The characteristic of two crystals to be mirror images of each other, e.g. right-handed and left-handed quartz. Such crystals are called enantiomorphs. Enantiomorphism is produced by the improper crystallographic operations of reflection across a plane and inversion through a point. Adj: *enantiomorphous.*
enantiomorphous (en-an'-ti-o-mor'-phous) Adj. of *enantiomorphism.*
enantiotropy (en-an'-ti-ot'-ro-py) The relationship between polymorphs that possess a stable transition point, and that therefore can be stably interconverted by changes of temperature and/or pressure. Although the term was originally applied only in systems with a vapor present, modern usage seems to give the term the more general meaning above. Cf: *monotropy.*
enargite (en-ar'-gite) A grayish-black or iron-black orthorhombic mineral: Cu_3AsS_4. It is dimorphous with luzonite. Enargite is an important ore of copper, occurring in veins in small crystals or granular masses, and often containing antimony (up to 6%) and sometimes iron and zinc.
enargite gold deposit *acid-sulfate type* deposit.
enation (e-na'-tion) An emergence or outgrowth from the surface of a plant. An enation does not have vascular tissue. Cf: microphyll.
en cabochon (en ca-bo-chon') adv. Cut in a style characterized by a smooth-domed, but unfaceted, surface; e.g. a ruby cut "en cabochon" in order to bring out the star. Etymol: French. See also: *cabochon.*
enclosed lake A lake that has neither surface influent nor effluent and that never overflows the rim of its basin; e.g. a *kettle lake* or a *crater lake* (Pennington, 1979). Cf: *closed lake.*
enclosed meander *inclosed meander.*
encrinal (en-cri'-nal) *encrinital.*
encrinital (en-cri'-ni-tal) Pertaining to, or made up of, encrinites; specif. said of a carbonate rock or sediment containing stem and/or plate fragments of crinoids. Syn: *encrinal.*
encrinite [paleont] (en'-cri-nite) A syn. of *crinoid,* esp. a fossil crinoid belonging to the genus *Encrinus.*
encrinite [sed] A crinoidal limestone; specif. a limestone in which crinoidal fragments constitute more than 50% of the bulk (Bissell and Chilingar, 1967, p.156).
encroachment [petroleum] (en-croach'-ment) The movement of bottom water or *edge water* into a petroleum reservoir as the oil and gas is removed.
encroachment [sed] Deposition of eolian sand from surface creep behind an obstruction of any sort, e.g. on the slip face of a dune (Bagnold, 1941, p.127). Cf: *accretion [sed struc].*
encroachment [stratig] The horizontal component of coastal *onlap* (Mitchum, 1977, p.208). See also: *aggradation [stratig].*
encrustation (en-crus-ta'-tion) (a) A crust or coating of minerals formed on a rock surface, e.g. calcite on cave objects or soluble salts on a playa or coastal splash zone. (b) A thin sheetlike organic growth, esp. a colonial invertebrate such as a bryozoan or coral, or a calcareous alga, closely adhering to the underlying solid substrate and mirroring its irregularities. (c) An *external mold* of a plant, usually in some incompressible rock such as sandstone or tufa (Walton, 1940); this usage is not recommended. (d) The process by which a crust or coating is formed. Also spelled: *incrustation.* Cf: *efflorescence.*
endannulus (en-dan'-nu-lus) An *annulus [palyn]* in the endexine of a pollen grain.
endarch (en'-darch) Position of protoxylem indicating centrifugal sequence of maturation in primary wood (Foster and Gifford, 1974, p.56).
end-centered lattice *base-centered lattice; one-face-centered lattice.*
end cleat *butt cleat.*
endellionite (en-del'-li-on-ite') *bournonite.*
endellite (en-dell'-ite) A name used in the the U.S. for a clay mineral: $Al_2Si_2O_5(OH)_4 \cdot 2H_2O$. It is the more hydrous form of halloysite, and is synon-

ymous with *halloysite* of European authors. Syn: *hydrated halloysite; hydrokaolin.* Cf: *hydrohalloysite.*

endemic (en-dem'-ic) Said of an organism or group of organisms that is restricted to a particular region or environment. Syn: *indigenous; native.*

enderbite (en'-der-bite) A plagioclase-rich member of the *charnockite* series containing quartz, plagioclase (commonly antiperthitic), orthopyroxene, and a small amount of magnetite. Most classification systems require that quartz constitute 10-65% of the felsic constituents and that the ratio of alkali feldspar to total feldspar be greater than 87.5%. Tobi (1971) has abandoned the term in favor of *alkali charnockite.* The name, proposed by Tilley in 1936, is for Enderby Land, Antarctica. Obsolete.

endexine (en-dex'-ine) The inner, usually homogeneous and smooth layer of the two layers of the *exine* of palynomorphs, normally less deeply staining than the *ektexine.* Syn: *intexine; nexine.*

endite (en'-dite) One of the appendages of the inner side of the limb of an arthropod, such as the medially directed lobe of *precoxa, coxa, basis,* or *ischium* of a crustacean, or the median or inner lobe or segment of the biramous appendage of a trilobite. Cf: *exite [paleont].*

end lap *overlap [photo].*

endlichite (end'-lich-ite) A variety of vanadinite in which the vanadium is partly replaced by arsenic; a mineral intermediate in composition between mimetite and vanadinite.

end member (a) One of the two or more simple compounds of which an isomorphous (solid-solution) series is composed. For example, the end members of the plagioclase feldspar series are albite, $NaAlSi_3O_8$, and anorthite, $CaAl_2Si_2O_8$. Syn: *minal.* (b) One of the two extremes of a series, e.g. types of sedimentary rock or of fossils.

end moraine A ridgelike accumulation that is being produced at the margin of an actively flowing glacier at any given time; a moraine that has been deposited at the lower or outer end of a glacier. Cf: *terminal moraine.* Syn: *frontal moraine.*

endo- (en'-do-) A prefix meaning "within".

endoaperture (en-do-ap'-er-ture) An aperture in the inner layer of the *sporoderm* of pollen; often the inner aperture of a *compound aperture.*

endobiontic (en'-do-bi-on'-tic) Said of an organism living in bottom sediments. Cf: *epibiontic.*

endoblastesis (en-do-blas'-te-sis) Obsolete term for late or epimagmatic crystallization in an igneous rock, from *residual liquid.* Not in common use. Adj: endoblastic.

endocast (en'-do-cast) *steinkern.*

endochondral bone (en-do-chon'-dral) *cartilage-replacement bone.*

endocochleate (en-do-coch'-le-ate) Said of cephalopods with a shell or other hard parts lodged inside the soft body; sometimes grouped as Endocochleata (e.g., Spirula).

endocochlian (en-do-coch'-li-an) *coleoid.*

endocoel (en'-do-coel) The central cavity in the capsule formed by the *endophragm* in a dinoflagellate cyst. Cf: *autocoel; pericoel.*

endocone (en'-do-cone) One of a series of concentric conical calcareous deposits or structures formed within the posterior or adapical part of the *siphuncle* of certain cephalopod conchs (e.g. *Endoceras*). The apices of the cones point toward the apex of the *conch,* and are usually perforated.

endocyclic (en-do-cy'-clic) Said of a *regular* echinoid in which the *periproct* is located within the *oculogenital ring.* Ant: *exocyclic.*

endocyst (en'-do-cyst) (a) The soft layer of the body wall of a bryozoan, lining the interior of a *zooecium* and enclosing the *polypide,* and giving rise to the *ectocyst* (TIP, 1953, pt. G, p.10). (b) The inner body of a *dinoflagellate cyst,* formed by the endophragm (Williams et al., 2000, p.73). Cf: *ectocyst.*

endoderm (en'-do-derm) The inner body layer of an organism; e.g. the inner layer of the outer body walls of a coral polyp, occurring as a double lamina in mesenteries. In the vertebrates the term refers exclusively to the originally innermost of the embryonic primary germ layers that give rise in adults principally to the gut and related outpocketings such as the liver tubules. Cf: *ectoderm [paleont]; mesoderm.*

endoderm [paleont] (en'-do-derm) The inner body layer of an organism; e.g. the inner layer of the outer body walls of a coral polyp, occurring as a double lamina in mesenteries. In the vertebrates the term refers exclusively to the originally innermost of the embryonic primary germ layers that give rise in adults principally to the gut and related outpocketings such as the liver tubules. Cf: *ectoderm; mesoderm.*

endoderm [palyn] The innermost layer in acritarchs, with more than one wall layer (Williams, 1978; Cramer and Diez Rodrigues, 1979). Cf: *ectoderm [palyn].*

endodermal epityche (en-do-der'-mal ep'-i-tyche) An *acritarch excystment* (e.g., *Beromia*) where the *epityche* appears on the endoderm, and a simple lateral split is present on the ectoderm (Wood, 1996).

endodermis (en-do-der'-mis) A layer of specialized cells in many roots and some stems, delimiting the inner margin of the cortex and exhibiting *suberin* thickenings on the cell walls (Cronquist, 1961, p.874). Cf: *epidermis [bot].*

endoderre (en'-do-derre) The wall of the prosome of a chitinozoan. It is sometimes regarded as a third layer of the wall. Cf: *ectoderre; periderre.*

endogastric (en-do-gas'-tric) (a) Said of a cephalopod shell that is curved or coiled, so that the venter is on or near the inner or concave side or area of whorls. (b) Said of a gastropod shell that is coiled so as to extend backward from the aperture over the extruded head-foot mass, as in most adult forms (TIP, 1960, pt. I, p.130). Cf: *exogastric.*

endogene effect (en'-do-gene) The contact-metamorphic effect of an igneous intrusion on the margin of the intrusive body itself (Bateman, 1959). Cf: *exogene effect.*

endogenetic (en'-do-ge-net'-ic) Derived from within; said of a geologic process, or of its resultant feature or rock, that originates within the Earth, e.g. volcanism, volcanoes, extrusive rocks. The term is also applied to chemical precipitates, e.g. evaporites, and to ore deposits that originate within the rocks that contain them. Cf: *exogenetic (a); hypogene.* Syn: *endogenic; endogenous.*

endogenic (en-do-gen'-ic) *endogenetic.*

endogenic explosion Forceful ejection of volcanic material originating from the interior of an advancing lava flow or dome (Fink and Anderson, 2000).

endogenous (en-dog'-e-nous) *endogenetic.*

endogenous dome A volcanic dome that has grown primarily by expansion from within and is characterized by a concentric arrangement of flow layers (Williams, 1932). Cf: *exogenous dome.*

endogenous inclusion *autolith.*

endokarst The part of a karst system that is beneath the surface, including all underground voids and dissolutional features present in the surrounding rock. Cf: *exokarst.*

endokinematic (en'-do-kin'-e-mat'-ic) Said of sedimentary operations in which "the largest displacement vectors occur between some matter within that part of the deposit destined to form the structure and the unmodified deposit" (Elliott, 1965, p.196); e.g. translational slumping, and horizontal or vertical transposition. Also, said of the sedimentary structures produced by endokinematic operations. Cf: *exokinematic.*

endokinetic (en'-do-ki-net'-ic) Said of a fissure in a rock that is the result of strain within the rock unit itself. Archaic. Cf: *exokinetic.*

endolistostrome (en-do-lis'-to-strome) A *broken formation* of sedimentary origin; an *olistostrome* without exotic blocks (Raymond, 1978). Cf: *allolistostrome.*

endolithic (en-do-lith'-ic) Pertaining to organisms, generally microscopic algae or fungi, that live in minute burrows in corals, shells, or reef rock. Syn: *petricolous.* Cf: *epilithic.*

endolithic breccia A breccia formed by forces acting within the Earth's crust, as by tectonic movements, by swelling or by hydration.

endometamorphism (en'-do-met'-a-mor'-phism) *endomorphism.*

endomorph (en'-do-morph) A crystal that is surrounded by a crystal of a different mineral species. Adj: endomorphic.

endomorphic metamorphism (en-do-mor'-phic) *endomorphism.*

endomorphism (en-do-mor'-phism) Changes within an igneous rock produced by the complete or partial assimilation of country-rock fragments or by reaction upon it by the country rock along the contact surfaces. It is a form of *contact metamorphism* with emphasis on changes produced within the igneous body rather than in the country rock. The term was originated by Fournet in 1867. Rarely used. Cf: *exomorphism.* Partial syn: *endogene effect.* Syn: *endometamorphism; endomorphic metamorphism.*

endopelos (en-do-pel'-os) Animals that lie on or burrow in soft mud.

endophragm (en'-do-phragm) (a) The complex internal skeletal structure of a crustacean, formed by the fusion of apodemes, and providing a framework for muscle attachment. Also called: endophragmal skeleton. (b) The inner-wall layer of a dinoflagellate cyst. Cf: *autophragm; ectophragm;* mesophragm; *periphragm.*

endopinacoderm (en'-do-pi-nac'-o-derm) The *pinacoderm* lining the inhalant and exhalant systems of a sponge. Cf: *exopinacoderm.*

endoplicae (en-do-pli'-cae) Folds in the endexine of palynomorphs. Sing: endoplica.

endopod (en'-do-pod) The medial or internal *ramus* of a limb of a crustacean, arising from the *basis.* Cf: *exopod.*

endopore [paleont] (en'-do-pore) A pore in the inner wall of the chamber in sphinctozoan sponges; an opening from the chamber into the central tube. Cf: *exopore [paleont].*

endopore [palyn] The internal opening in the endexine of a pollen grain with a complex porate structure. See also *vestibulum.* Syn: *os.* Cf: *exopore [palyn].*

endopsammon (en-do-psam'-mon) Animals that lie on or burrow in sand.

endopuncta (en-do-punc'-ta) A *puncta* of a brachiopod shell not extending to its external surface, occupied by a *caecum.* These pores are common over the whole inner surface of the shell but are not visible on the outer surface if the primary layer is intact. Cf: *exopuncta.* Pl: endopunctae.

endorheic (en-do-rhe'-ic) Said of a basin or region characterized by internal drainage; relating to endorheism.

endorheism (en-do-rhe'-ism) (a) *internal drainage.* (b) The condition of a region in which little or no surface drainage reaches the ocean. Ant: *exorheism.*

endosaturation The soil is saturated with water in all layers from the upper boundary of saturation to a depth of 200 cm or more from the mineral soil surface. See also: *episaturation.*

endosiphotube (en-do-si'-pho-tube) A fine canal near the center of the *siphuncle* of certain eurysiphonate cephalopods, esp. endoceroids, in which the rest of the siphuncle is filled with organic deposits, e.g. endocones or bullettes.

endosiphuncle (en-do-si'-phun-cle) The space within the *ectosiphuncle* of certain cephalopods, including all organic tissues and calcareous structures.

endoskarn (en'-do-skarn) That portion of *skarn* mineralization that involves metamorphic, metasomatic, and hybrid mixing effects on the igneous side of the contact between an intrusive body and the country rock. Cf: *endomorphism; exoskarn.*

endoskeleton (en-do-skel'-e-ton) An internal *skeleton* in an animal, serving as a supporting framework; e.g. any internal hard parts serving for the attachment of muscles in a crustacean, or the internal system of articulated bones in a vertebrate. Cf: *exoskeleton.*

endosome (en'-do-some) The inner part of the body of various sponges; e.g. the choanosome with few if any supporting spicules, or the part of a sponge internal to a cortex, or the part surrounding the *spongocoel* and devoid of flagellated chambers. Because of conflicting usage, the term is not recommended. Cf: *ectosome.*

endosperm (en'-do-sperm) The food-storage tissue in a seed. In the gymnosperms, it is a part of the female gametophyte and is haploid; in the angiosperms, it results from the fusion of a sperm with two polar nuclei and is triploid (Fuller and Tippo, 1954, p.957). Syn: albumen.

endosphere (en'-do-sphere) An archaic term for all that part of the Earth below the lithosphere.

endospore (en'-do-spore) (a) The innermost, nonresistant wall layer of a *spore,* in contradistinction to that of a pollen (which is termed intine). Cf: *exospore; perispore.* (b) An asexual spore developed within the cell, esp. in bacteria. (c) A thin-walled spore of cyanobacteria. (d) The most primitive position of reproductive organs in dasycladacean algae, in which cysts are located in the central axis. Cf: *exospore.* (e) A *fungal spore* developed endogenously and parthenogenetically, often resembling an *ascospore.*

endosternite (en-do-ster'-nite) A part of the endoskeleton of an arthropod; e.g. a tendinous endoskeletal plate in the *cephalon* of a crustacean.

endostratic (en-do-strat'-ic) Bedded within; e.g. said of the formation of bedding in clays as a result of "alternating desiccation and saturation by groundwater" (Becker, 1932, p.85), or said of a breccia bedded within a distinct stratum (Norton, 1917).

endotheca (en-do-the'-ca) A collective term for the dissepiments inside the wall of a scleractinian *corallite.* Cf: *exotheca.*

endothecal (en-do-the'-cal) Said of edrioasteroid ambulacra between thecal plates, in contact with both the interior and the exterior of the theca (Bell, 1976).

endothermic (en-do-ther'-mic) Pertaining to a chemical reaction that occurs with an absorption of heat. Cf: *exothermic.*

endotomous (en-dot'-o-mous) Characterized by bifurcation in two main crinoid arms that give off branches only on their inner sides. Ant: *exotomous.*

endowment That portion of the mineral *resource base* of a region that meets specified physical characteristics such as quality, size, and depth. Endowment is intermediate between *resource base* and *reserves* (Harris, 1984). Syn: *mineral endowment.*

endozone (en'-do-zone) The inner part of a bryozoan colony, usually characterized by thin vertical walls, relative scarcity of intrazooidal skeletal structures, and a combination of zooidal growth directions at low angles to the colony growth directions or surfaces. Cf: *exozone.* Syn: *immature region.*

end peneplain *endrumpf.*

end product As applied to radioactivity, the stable nuclide at the end of a *radioactive series.* Cf: *parent; daughter.*

endrumpf (end'-rumpf) A term proposed by W. Penck (1924) for the final landscape or plain that results from the erosion of a landmass that had high relief; it represents the end product of a period of degradation marked by waning uplift. Although Penck considered the Davisian *peneplain* as an equivalent term, *endrumpf* differs in that it does not imply a particular sequence of development leading up to leveling of the original relief. *Endrumpf* may be more likened to an extended use of *pediplain.* Etymol: German *Endrumpf,* "terminal torso". Cf: *primärrumpf.* Syn: *end peneplain.*

endurance limit (en-dur'-ance) That stress below which a material can withstand hundreds of millions of repetitions of stress without fracturing. It is lower than rupture strength. Syn: *fatigue limit.*

endurance ratio The ratio of the *endurance limit* of a material to its static, compressive strength. Syn: *fatigue ratio.*

en echelon (en ech'-e-lon) adj. Said of geologic features that are in an overlapping or staggered arrangement, e.g. faults. Each is relatively short but collectively they form a linear zone, in which the strike of the individual features is oblique to that of the zone as a whole. Etymol: French, "en échelon", "in steplike arrangement".

en echelon cracks A series of parallel joints or veins which are systematically offset in either a left-stepping or right-stepping manner. In some instances they may form an incipient fault zone (Hancock, 1972).

Eneolithic (E'-ne-o-lith'-ic) *Chalcolithic.* Also spelled: *Aeneolithic.*

energy grade line (en'-er-gy) *energy line.*

energy gradient The slope of the *energy line* of a body of flowing water, with reference to any plane. Syn: *energy slope.*

energy index [mineral] A term used by Gruner (1950) for a number that expresses the stability of a silicate. It is equal to the *bridging factor* multiplied by the electronegativity. Quartz has the highest energy index (1.80).

energy index [sed] The inferred degree of water agitation in the sedimentary environment of deposition. Roughly inferred from grain size and ratio of sand to mud, etc. Abbrev: EI.

energy level The kinetic energy (due to wave or current action) that existed or exists in the water of a sedimentary environment, either at the interface of deposition or a meter or two above it. See also: *high-energy environment; low-energy environment.*

energy line In hydraulics, a line joining the elevations of the energy heads of a stream when referred to the stream bed. It lies above the water surface at any cross section; the vertical distance is equal to the velocity head at that cross section (ASCE, 1962). See also: *energy gradient.* Syn: *energy grade line.*

energy loss The difference between energy input and output as a result of transfer of energy between two points. In the flow of water, the rate of energy loss is represented by the slope of the *hydraulic grade line.*

energy of glacierization *budget gradient.*

energy slope *energy gradient.*

enfilade (en'-fi-lade') (a) A fire that can rake a line of troops. (b) A position exposed to direct fire. See also: *defilade.*

engineering geology (en-gi-neer'-ing) Geology as applied to engineering practice, esp. mining and civil engineering. As defined by the Council of Professional Geological Organizations, Suggested Geologist Practice Act of 1993, engineering geology is a specialty of geology relevant to (1) the planning, design, operation, and maintenance of civil engineering works; (2) the development, protection, and remediation of ground- and surface-water resources; and (3) other human activities where geologic factors and conditions impact the public welfare and the safeguarding of life, health, property, and the environment. Cf: *geologic engineering.*

engineering geophysics Use of geophysical methods to get information for civil engineering. The aim is usually to describe not only the geometry of the subsurface but also its nature (e.g., its densities and elastic characteristics as determined by measurements of seismic velocities). Shallow seismic-reflection and refraction, gravity, magnetic, electric, and sampling methods are commonly employed to find the depth of bedrock and the strength of sediments for foundations, to determine the rippability of rock, to measure the degree of fracturing, to detect underground cavities and near-surface gas pockets, to locate buried pipeline hazards below the sea floor, buried barrels of pollutants, the integrity of landfills, etc. In water-covered areas high-powered fathometers, sparkers, gas guns, and other seismic-reflection methods employing high frequencies (up to 5 kHz) are used to obtain reflections from shallow interfaces so that bedrock and the nature of fill material can be diagnosed.

englacial (en-gla'-cial) Contained, embedded, or carried within the body of a glacier or ice sheet; said of meltwater streams, till, drift, moraine, etc. Syn: *intraglacial.*

englacial drift Rock material contained within a glacier or ice sheet.

englishite (eng'-lish-ite) A white platy monoclinic mineral: $K_3Na_2Ca_{10}Al_{15}(OH)_7(PO_4)_{21}·26H_2O$.

engrafted stream (en-graft'-ed) A stream composed of the waters of several previously independent streams that unite before reaching the sea; esp. a main stream consisting of several separate extended streams flowing from an oldland and merging with each other on an uplifted coastal plain. Also spelled: *ingrafted stream.*

enhanced oil recovery Thermal, miscible, and chemical methods for producing oil left behind in the ground after primary and *secondary recovery.* Thermal methods include steam stimulation ("huff and puff"), steam flooding, and in-situ combustion. Miscible recovery methods include injection of hydrocarbons, carbon dioxide, and nitrogen. Chemically-enhanced recovery includes flooding with polymers, surfactants, and alkaline fluids. Abbrev: EOR.

enhancement [gem] Any post-recovery process on rough or fashioned gem material which alters the appearance, especially the color or clarity, of a natural diamond or other gem material. Coating, fracture filling, irradiation, heating (annealing), and lasering are all forms of enhancement. Also called treatment.

enhancement [photo] (en-hance'-ment) The process of altering the appearance of an image so that the interpreter can extract more information from it. Enhancement may be done by digital or photographic methods.

enhydrite (en-hy'-drite) (a) A mineral or rock having cavities containing water. (b) *enhydros.*

enhydros (en-hy'-dros) A hollow nodule or geode of chalcedony containing water, sometimes in large amount. Syn: *enhydrite; water agate.*

enhydrous (en-hy'-drous) Said of certain crystalline minerals containing water or having drops of included fluid; e.g. "enhydrous chalcedony". Not to be confused with *anhydrous.*

enigmatite (e-nig'-ma-tite) *aenigmatite.*

enneri (en'-ne-ri) A term used in northern Africa (esp. Libya) for a wadi or dry river valley.

ennomoclone (en'-no-mo-clone') A *desma* (of a sponge) consisting of one short distal arm (*brachyome*) and three or six longer proximal arms directed symmetrically away from it; e.g. a *tricranoclone* or a *sphaeroclone.*

enriched mantle A part of the Earth's mantle in which the concentrations of incompatible elements have been increased, enhancing the capacity to yield magma by partial fusion.

enriched mantle 1 An end-member source for mantle-derived magmas, postulated by Zindler and Hart (1986). Abbrev: EM1.

enriched mantle 2 An end-member source for mantle-derived magmas, postulated by Zindler and Hart (1986). Abbrev: EM2.

enrichment (en-rich'-ment) *supergene enrichment.*

enrockment (en-rock'-ment) A mass of large stones placed in water to form a base, as for a pier.

ensialic geosyncline (en-si-al'-ic) An obsolete term for a sedimentary basin which contains clastics accumulating on sialic crust (Wells, 1949). Cf: *ensimatic geosyncline.* See also: *miogeosyncline.*

ensialic orogen An *orogen* that appears to be bounded on at least three sides by cratonic crust. Ensialic orogens probably represent the closure of a rift basin, or a narrow oceanic basin that tapers into a rift basin. Many ensialic orogens are Proterozoic.

ensimatic geosyncline (en-si-mat'-ic) An obsolete term for a geosyncline which contains effusive rocks accumulating on simatic crust (Wells, 1949, p.1927). Cf: *ensialic geosyncline.* See also: *eugeosyncline.*

ENSO *El Niño Southern Oscillation.*

enstatite (en'-sta-tite) A common rock-forming mineral of the *orthopyroxene* group: $MgSiO_3$. It is isomorphous with hypersthene, and may contain a little iron replacing the magnesium. Enstatite ranges from grayish white to yellowish, olive green, and brown. It is an important primary constituent of intermediate and basic igneous rocks. Symbol: En. Cf: *bronzite.* Syn: *chladnite [mineral].*

enstatite chondrite A type of highly reduced chondrite containing very little oxidized iron; the mafic silicate mineral assemblage is usually dominated by enstatite. The redox state is so low that about 5-10% of the Si is metallic, dissolved in Fe-Ni. Enstatite meteorites are subdivided into EL and EH classes according to the relative abundance of metallic Fe-Ni. Syn: *EL chondrite; EH chondrite.*

enstenite (en'-ste-nite) An unnecessary name for orthopyroxenes of the $MgSiO_3$-$FeSiO_3$ isomorphous series.

entablature (en-tab'-la-ture) A layer of relatively thin, irregular columnar jointing above the thicker, more regular columnar joints of the basal colonnade, in two- or three-tiered lava flows (Cas and Wright, 1987). Cf: *colonnade.*

enterolithic (en'-ter-o-lith'-ic) (a) Said of a sedimentary structure consisting of ribbons of intestinelike folds that resemble those produced by tectonic deformation but that originate through chemical changes involving an increase or decrease in the volume of the rock; e.g. said of a small fold or local crumpling formed in an evaporite by flowage or by the swelling of anhydrite during hydration. See also: *tepee structure.* (b) Said of the deformation or folding that produces enterolithic structures.

enterolithic anhydrite Folded single or multiple laminae of anhydrite giving the appearance of intestinal convolutions. Some spectacular examples from the Permian Castile Formation are illustrated by Kirkland and Anderson (1970), who called them "microfolds" and concluded that they were the result of tectonic compression. Microfolded anhydrite laminae are common features in ancient and modern anhydrite deposits. Many sedimentary and diagenetic origins have been proposed for this type of structure, but none has been generally accepted.

enteron (en'-ter-on) The digestive cavity or alimentary system of an animal, generally consisting of esophagus, stomach, and intestine. Cf: *coelom.*

entexis (en-tex'-is) *Migmatization* with introduction from without of the more mobile part (Dietrich and Mehnert, 1961). Rarely used. Cf: *ectexis.*

entexite (en-tex'-ite) Rock formed by *entexis.* Rarely used. Var: entectite.

enthalpy (en'-thal-py) A thermodynamic variable (typically symbolized as H) that is defined as the sum of the internal energy of a body plus the product of its volume multiplied by the pressure, measured in joules (J). Syn: *heat content.*

entire (en-tire') Said of a leaf with a continuous smooth margin, not lobed or dentate.

Entisols Mineral soils that have no distinct subsurface diagnostic horizons within 1 m of the soil surface. An order in the U.S. system of soil taxonomy.

entocoele (en'-to-coele) The space within a pair of mesenteries of a coral. Cf: *exocoele.*

entomodont (en-tom'-o-dont) Said of a class of ostracode hinges intermediate in form between the merodont hinge and the amphidont hinge, having denticulate terminal and median elements with a partial subdivision of the median element.

entomophily (en-to-moph'-i-ly) Pollination by insects. Adj: entomophilous. Cf: *anemophily.*

entomostracan (en-to-mos'-tra-can) An obsolete term originally applied to branchiopod, copepod, and cirriped crustaceans. Cf: *malacostracan.*

entooecium (en-to-oe'-ci-um) Calcified or uncalcified inner part of the body-wall fold forming an *ovicell* in cheilostome bryozoans. Cf: *ectooecium.*

entoolitic (en'-to-o-lit'-ic) Pertaining to oolitic structures or grains that have formed or grown inward by the filling of small cavities, as by the deposition of successive coats on the cavity walls. Ant: *extoolitic.*

entoproct (en'-to-proct) Any bryozoan belonging to the subphylum Entoprocta and lacking hard parts and a body cavity. These bryozoans are not found as fossils. Cf: *ectoproct.*

entoseptum (en-to-sep'-tum) A scleractinian-coral *septum* developed within an *entocoele.* Cf: *exoseptum.*

entosolenian (en'-to-so-le'-ni-an) Said of a foraminifer (e.g. *Oolina*) having an internal tubelike apertural extension (siphon). Cf: *ectosolenian.*

entotoichal (en-to-toi'-chal) Said of a cheilostome bryozoan *ovicell* that appears to be immersed in the distal zooid while opening independently to the exterior.

entozooecial (en'-to-zo-oe'-ci-al) *entozooidal.*

entozooidal (en'-to-zo-oi'-dal) Said of a cheilostome bryozoan *ovicell* that appears to be immersed in the distal zooid while opening below the operculum of the maternal zooid. Syn: *entozooecial.*

entrail pahoehoe (en'-trail) A type of *pahoehoe* that has a surface resembling an intertwined mass of entrails, formed on steep slopes as dribbles around and through cracks in the flow crust.

entrain Where flowing avalanche snow incorporates adjacent snow along its descending path.

entrainment (en-train'-ment) The process of picking up and carrying along, as the collecting and movement of sediment by currents, or the incorporation of air bubbles into a cement *slurry.*

entrapment burrow (en-trap'-ment) A term used by Kuenen (1957, p.253) for a burrow occupied by an animal buried below the sandy deposit of a passing turbidity current.

entrenched meander (en-trenched') (a) An *incised meander* carved downward into the surface of the valley in which the meander originally formed; it exhibits a symmetric cross profile. Such a deepened meander, which preserves its original pattern with little modification, suggests rejuvenation of a meandering stream, as when there has been a rapid vertical uplift or a lowering of base level. Cf: *ingrown meander.* Syn: *inherited meander.* (b) A generic term used as a syn. of *incised meander.* Syn: *intrenched meander.*

entrenched stream A stream, often meandering, that flows in a narrow trench or valley cut into a plain or relatively level upland; e.g. a stream that has inherited its course from a previous cycle of erosion and that cuts into bedrock with little modification of the original course. Also spelled: *intrenched stream.*

entrenchment (en-trench'-ment) The process whereby a stream erodes downward so as to form a trench or to develop an entrenched meander. Also, the results of such a process. Cf: *incision.*

entropy [phys] (en'-tro-py) A macroscopic, thermodynamic quantity which measures the unavailable energy, ultimately reflecting the degree of microscopic randomness or disorder. Higher entropy represents increased disorder. Entropy never decreases in a reaction, according to the second law of thermodynamics. Symbol: S.

entropy [stratig] A measure of the degree of "mixing" of the different kinds of rock components in a stratigraphic unit (Pelto, 1954). The entropy value of a given component is the product of its proportion in the unit and the natural logarithm of that proportion. A stratigraphic unit with equal parts of each component has an entropy value of 100; as the composition approaches that of a single component, the entropy value approaches zero.

entropy [streams] A quantity that is expressed by the probability of a given distribution of energy utilization within or along a stream from headwaters to a downstream point, the most probable condition existing when the stream is graded or the energy is as uniformly distributed as may be permitted by physical constraints (Leopold and Langbein, 1962).

entropy map A *facies map* that is based on the degree of "mixing" of three end members (rock components) of a given stratigraphic unit, but that does not distinguish the natures of these end members (see Forgotson, 1960; Krumbein and Sloss, 1963). Cf: *entropy-ratio map.*

entropy-ratio map (en'-tro-py-ra'-ti-o) A *facies map* that is based on the degree of "mixing" of three end members (rock components) of a given stratigraphic unit and that indicates by map pattern the nature of the lithologic "mixture" through which a given end member is approached (Forgotson, 1960, p.93). Cf: *entropy map.*

entropy unit A unit of measurement defined as one calorie per mole-degree. It is essentially equivalent to the *gibbs.*

entry (en'-try) (a) *portal.* (b) The mouth of a river.

entry pressure *displacement pressure.*

envelope (en'-ve-lope) The outer or covering part of a fold, especially of a folded structure that includes some sort of structural break. Cf: *core [struc geol].* Rarely used.

envelope amplitude The amplitude of the low-frequency curves encompassing or bounding deflections of higher frequency. The curves can be drawn by smoothly connecting adjacent peaks and adjacent troughs. Often calcuated by complex-trace analysis (Taner et al., 1979).

enveloping surface (en-vel'-op-ing) An imaginary surface tangent to antiformal and synformal hinges in a single folded surface.

environment [ecol] (en-vi'-ron-ment) All those external factors and conditions which may influence an organism or a community. Syn: *habitat.*

environment [sed] A geographically restricted complex where a sediment accumulates. Environments are described in terms of their geomorphic setting and the physical, chemical, and biological conditions that prevail there, e.g. beach or a foreshore and backshore, or outer shelf.

environmental assessment (en-vi'-ron-men'-tal) A detailed statement prepared by an organization for its own use to appraise the effect of a proposed project on the aggregate of social and physical conditions that influence a community or ecosystem. The assessment is often prepared to determine the need for a formal environmental impact statement.

environmental facies Facies that are concerned solely with environment or determined by the nature of environment; e.g. lithotopes, biotopes, and tectotopes (Weller, 1958, p.628). They are not material units or bodies of rock, but areas inferred from the results of a combination of mutually interacting influences and conditions as these are exhibited in the form of sedimentary types and organic communities. See also: *facies.* Syn: *ecologic facies.*

environmental geochemistry The effect on humans of the distribution and interrelations of the chemical elements and radioactivity among surficial rocks, water, air, and biota.

environmental geology A specialty of geology concerned with Earth processes, Earth resources, and engineering properties of Earth materials and relevant to (1) the protection of human health and natural ecosystems from adverse biochemical and/or geochemical reactions to naturally occurring chemicals or to chemicals and chemical compounds released into the environment by human activities and (2) the protection of life, safety, and well-being human from natural processes, such as floods, hurricanes, earthquakes and landslides, through land-use planning. Cf: *urban geology.*

environmental geology map As used in environmental geology, a map prepared for the general public that classifies the suitability of land for a particular use on the basis of geologic characteristics. Examples: general construction, sand and gravel development, land burial of waste, groundwater development.

environmental geophysics The use of geophysical measurements to evaluate or help remediate problems in the Earth's environment.

environmental hyperspace lattice In a geometric model of the environment and its ecologic units, a multidimensional space containing as many dimensions as there are possible environmental factors (Valentine, 1969, p.685). See also: *biospace; ecospace.*

environmental impact statement A document prepared by industry or a political entity on the environmental impact of its proposals for legislation and other major actions significantly affecting the quality of the human environment. Environmental impact statements are used as tools for decision-making and are required by the National Environmental Policy Act.

environmental resistance Factors in the environment that tend to restrict the development of an organism or group of organisms and to limit its numerical increase.

environmental-resource unit A portion of the *environment [ecol]* with a homogeneous set of physical and biological characteristics; a supposedly natural division characterized by specific patterns or assemblages of structural components, such as rocks, soils, and vegetation, and natural processes, such as erosion, runoff, and soil processes (Keller, 2002). Abbrev: ERU.

environmental science A multi-disciplinary field of study, often founded within geology, biology or cultural programs, that incorporates aspects of life, social, and earth science courses to deal with (1) social and political decision making and policy setting relative to the environment or environmental ethics, (2) the planning and management of natural and human resources with an emphasis on the social and political aspects, and (3) legal and regulatory aspects of environmental protection activities carried out by the government under its police powers.

environmental site assessment A set of procedures to identify environmental risks and potential cleanup liability specified by the American Society of Testing and Materials (ASTM). Phase I environmental site assessments include a site inspection and an environmental records review. They are commonly required for commercial, industrial, and multi-family mortgage transactions and for asset sales, mergers, and acquisitions, real estate developments, and public works projects. Abbrev: ESA.

Eoarchean (E-o-ar'-che-an) An era of the Standard Global Geochronometric Scale: the oldest formally defined era of the *Archean* Eon. The older limit is not formally defined; the younger limit to the *Paleoarchean* Era is assigned as 3,600 Ma (Plumb, 1991).

Eocene (E'-o-cene) The middle series of the *Paleogene* Subsystem of the *Tertiary* System of the Standard Global Chronostratigraphic Scale, above the *Paleocene* and below the *Oligocene.* Also the time during which these rocks were formed, the Eocene Epoch, covering the time span between 53 and 34 Ma.

eogenetic (e'-o-ge-net'-ic) Pertaining to the period of time between final deposition of a sediment and burial of that sediment below the depth to which surface or near-surface processes are effective. The upper limit of the eogenetic zone is the land surface; the lower boundary, less clearly defined due to the gradual diminishing of surface-related processes, is the *mesogenetic* zone. Also applied to the porosity that develops during the eogenetic stage. (Choquette and Pray, 1970, p.219-220). Cf: *telogenetic.*

eogenetic karst (eo-ge-net'-ic) The product of the solutional modification of young rocks undergoing meteoric diagenesis, soon after deposition (Vacher and Mylroie, 2002).

eohypse (e'-o-hypse) A contour line on a former land surface, reconstructed or "restored" on a map by plotting the contours of the surviving portions of the land and by extrapolation of those contours. Syn: eohyps; *eoisohypse.*

eoisohypse (e-o-i'-so-hypse) *eohypse.*

eolation (e-o-la'-tion) The gradational work performed by the wind in modifying the land surface, such as the transportation of sand and dust, the formation of dunes, the effects of sandblasting, and, indirectly, the action of water waves generated by wind currents.

eolian (e-o'-li-an) (a) Pertaining to the wind; esp. said of such deposits as loess and dune sand, of sedimentary structures such as wind-formed ripple marks, or of erosion and deposition accomplished by the wind. (b) Said of the active phase of a dune cycle, marked by diminished vegetal control and increased dune growth. Cf: *eluvial [sed].* Etymol: Aeolus, god of the winds. Syn: *aeolian; eolic.*

eolianite (e-o'-li-an-ite') A consolidated sedimentary rock consisting of clastic material deposited by the wind; e.g. dune sand cemented by calcite. Also spelled: *aeolianite.* Cf: *dune rock.*

eolic (e-o'-lic) *eolian.*

eolith (e'-o-lith) In archaeology, one of the most primitive human-made stone implements. Etymol: Greek "eos", "dawn" + "lithos", "stone". Syn: *dawn stone.*

Eolithic (E-o-lith'-ic) In archaeology, a cultural level that is pre-*Paleolithic* or at the very beginning of the Paleolithic. It is characterized by *eoliths.*

eolium (e-o'-li-um) A general term to include deposits of windblown silt (loess), and wind-drifted sand (dunes), granules, and pebbles (Colton, 1978). Little used.

eolomotion (e'-o-lo-mo'-tion) A relatively slow downwind or downhill movement of sand due to direct or indirect wind action on surface rock particles (Kerr and Nigra, 1952).

eometamorphism (e'-o-met'-a-mor'-phism) Early metamorphism or the very beginnings of metamorphism, esp. as affecting hydrocarbons which are highly vulnerable (Landes, 1967, p.832).

eon (e'-on) (a) The largest unit of geologic time; immediately above *era* in the geochronologic hierarchy; the time during which the rocks of an *eonothem* were formed (ISSC, 1994, p.83). The Phanerozoic Eon encompasses the Paleozoic, Mesozoic, and Cenozoic Eras. (b) One billion (10^9) years. Also spelled: *aeon.*

eonothem (e-on'-o-them) The formal *chronostratigraphic unit* of highest rank. The Phanerozoic Eonothem encompasses the Paleozoic, Mesozoic, and Cenozoic Erathems; all older rocks have been assigned to the Precambrian Eonothem, or to Archean or Proterozoic Eonothems (NACSN, 1983, Art. 70). The body of rocks formed during the time span of an eon.

Eophytic (E-o-phyt'-ic) A paleobotanic division of geologic time, signifying that time during which algae were abundant. Cf: *Aphytic; Archeophytic; Cenophytic; Mesophytic; Paleophytic.*

EOR *enhanced oil recovery.*

eosphorite (e-os'-pho-rite) A pink to rose-red orthorhombic mineral: $Mn^{2+}Al(PO_4)(OH)_2{\cdot}H_2O$. It is the manganese analogue of childrenite.

Eosuchia (E-o-su'-chi-a) An order of lepidosaurian reptiles, of generally lizardlike habitus and Late Paleozoic and Early Mesozoic age, believed to include the ancestors of living lizards and snakes. Range, Upper Permian to Eocene.

Eötvös correction (E-öt'-vös) A correction applied to the observed gravity data to compensate for the velocity of *gravimeter* during shipborne and airborne surveys: E.C. $= 7.503\ V \cos\phi \sin\alpha + 0.004154\ V^2$ (mGal), where V - velocity in knots; ϕ - latitude in degrees; α - azimuth angle in degrees from the North. The largest values are obtained when the observation platform (gravimeter) has a velocity component in the east-west direction. See also: *Eötvös effect.*

Eötvös effect A motion-related effect defined as a vertical component of the *Coriolis acceleration.* Observed when measuring the *Earth's gravity field* while in motion, as with marine (shipborne) and airborne surveys. The gravimeter's velocity over the Earth's surface adds vectorially to the velocity due to the Earth's rotation, varying the centrifugal acceleration and, hence, the apparent gravitational attraction of the Earth's mass measured by gravimeter. Compensated by applying the *Eötvös correction.*

Eötvös torsion balance *torsion balance gradiometer.*

Eötvös unit A unit of gravitational gradient or curvature; 10^{-6} mgal/cm.

eozoan (e-o-zo'-an) An obsolete term for a *protozoan,* or a one-celled organism of the subkingdom Eozoa. Syn: *eozoon.*

eozoon (e-o-zo'-on) (a) An inorganic banded structure of coarsely crystalline calcite and serpentine, occurring in the Grenville Series of Canada, originally interpreted as the remains of a gigantic foraminifer, *Eozoon canadense.* Pl: eozoons; eozoa. Adj: eozoonal. (b) *eozoan.*

epanticlinal fault (ep'-an-ti-cli'-nal) A longitudinal or transverse fault that is associated with a doubly plunging minor anticline and formed concurrently with the folding (Irwin, 1926). Also spelled: *epi-anticlinal fault.*

epaulet (ep-au-let') A modified five-sided step cut of a gem, resembling a shoulder ornament (epaulet) in outline.

epeiric sea (e-pei'-ric) *epicontinental sea.*

epeirocratic (e-pei'-ro-crat'-ic) Said of a period of low sea level in the geologic past. Cf: *thalassocratic.* Obsolete syn: *geocratic.*

epeirogenesis (e-pei'-ro-gen'-e-sis) *epeirogeny.*

epeirogenetic (e-pei'-ro-ge-net'-ic) Adj. of *epeirogenesis.*

epeirogenic (e-pei'-ro-gen'-ic) Adj. of *epeirogeny.* Cf: *orographic [tect].*

epeirogeny (ep-ei-rog'-e-ny) As defined by Gilbert (1890), a form of *diastrophism* that has produced broad regional uplifts and basins of the continents and oceans, in contrast to the more localized process of *orogeny,* which has produced mountain chains. Epeirogenic movements are primarily vertical, either upward or downward, and have affected large parts of the continents, not only in the cratons but also in stabilized former orogenic belts, where they have produced most of the present mountainous topography. Some epeirogenic and orogenic structures grade into each other in detail, but most of them contrast strongly. Adj: *epeirogenic.* Syn: *epeirogenesis.* Cf: *bathygenesis; cymatogeny.*

epeirophoresis *continental drift.*

epharmone (ep-har'-mone) An organism that has undergone morphologic change as a result of changes in the environment and therefore differs from the normal or usual form.

ephebic (e-phe'-bic) Said of the adult stage in the life history of an animal; i.e., the stage when the animal is normal in size and able to reproduce. The stage follows the *neanic* stage and precedes the *gerontic* stage.

ephemeral data The part of the data broadcast by a navigation satellite which varies with each broadcast. The ephemeral data include the time and corrections to the satellite location.

ephemeral lake (e-phem'-er-al) A short-lived lake. Cf: *evanescent lake.*

ephemeral stream A stream or reach of a stream that flows briefly only in direct response to precipitation in the immediate locality and whose channel is at all times above the water table. The term "may be arbitrarily restricted" to a stream that does "not flow continuously during periods of as much as one month" (Meinzer, 1923, p.58). Cf: *intermittent stream.*

ephemeris (e-phem'-er-is) A publication giving coordinates of celestial bodies at uniform time intervals, e.g., during one calendar year; or more generally, the computed positions vs. time of a moving celestial body such as a planet.

ephemeris second The fundamental invariable unit of time. It is defined as $^1/_{31,556,925.9747}$ of the tropical year for 1900 January 0^d12^h ephemeris time. The ephemeris day is 86,400 seconds. The former unit of time was the mean solar second, defined as $^1/_{86,400}$ of the mean solar day.

ephemeris time A uniform measure of time determined by relative changes in the positions of Earth, Moon, and stars. Cf: *ephemeris second.*

ephesite (eph'-e-site) A pink triclinic or monoclinic trioctahedral brittle mica of ideal composition $Na(Li_1Al_2)(OH,F)_2(Si_2Al_2)O_{10}$.

ephippium (e-phip'-pi-um) The dorsal brood pouch of various cladoceran crustaceans that is shed with the eggs and serves for protection until hatching. Pl: ephippia.

epi- (ep'-i-) A prefix signifying "on" or "upon".

epi-anticlinal fault (ep'-i-an'-ti-cli'-nal) *epanticlinal fault.*

epibiontic (ep'-i-bi-on'-tic) Said of an organism living on the surface of bottom sediments, rocks, or shells. Cf: *endobiontic.*

epibole (ep'-i-bole) (a) A syn. of *abundance zone* and *acme zone.* (b) The deposits accumulated during a *hemera* (Trueman, 1923, p.200).

epibolite (e-pib'-o-lite) A term introduced by Jung and Roques (1936, 1952) for a migmatite with granitic layers concordant with the gneissosity of its nongranitic parent rock. Rarely used.

epibystra (ep-i-bys'-tra) The main element of an *acritarch excystment* structure that involves one *process* whose distal extremity has disappeared, leaving a thickened basal plug, e.g., in the gennus *Onodagaella* (Playford, 1977).

epicenter (ep'-i-cen'-ter) The point on the Earth's surface that is directly above the focus of an earthquake. Cf: *anticenter.*

epicentral distance (ep'-i-cen'-tral) The distance from the epicenter of an earthquake to the receiver. It may be measured in angular units (*angular distance*) or in linear distance along a great-circle path.

epiclastic rock (ep-i-clas'-tic) A rock formed at the Earth's surface by consolidation of fragments of pre-existing rocks; a sedimentary rock whose fragments are derived by weathering or erosion. See also: *clastic rock.*

epicontinental (ep'-i-con'-ti-nen'-tal) Situated on the continental shelf or on the continental interior, as an *epicontinental sea.*

epicontinental sea A sea on the continental shelf or within a continent. Cf: *mediterranean sea.* Syn: *inland sea; continental sea; epeiric sea.*

epicotyl (ep-i-co'-tyl) The part of the embryo of the seed that lies above the cotyledon and becomes the growing point of the shoot. Syn: *plumule.*

epicycle (ep'-i-cy'-cle) A minor or secondary cycle within a major or primary cycle; specif. a subdivision of a cycle of erosion, initiated by a small change of base level, such as an episode of stillstand of sufficient duration to rank as a part of a cycle and recorded as a terrace.

epicyst (ep'-i-cyst) The part of a *dinoflagellate* cyst anterior to the *cingulum.* Also the part of an *acritarch* anterior to the equator and containing the excystment opening. Cf: *hypocyst.*

epideltoid (ep-i-del'-toid) The anal *deltoid* between mouth and anus in blastoids that have only one or two anal deltoids.

epidermal (ep-i-der'-mal) Said of shallow or surficial deformation of the sialic crust. Cf: *dermal [tect]; bathydermal.*

epidermis [bot] (ep-i-der'-mis) The characteristic outermost tissue of a plant, usually one cell thick, covering leaves, stems, and other parts (Cronquist, 1961, p.874). Cf: *endodermis.*

epidermis [paleont] Any of various animal integuments; e.g. the *periostracum* of a mollusk, the imperforate outer layer in foraminifera, or the external cellular layer in the body wall of a cnidarian. Cf: *hypodermis.*

epidesmine (ep-i-des'-mine) *stilbite-Ca.*

epidiabase (ep-i-di'-a-base) A name proposed as a replacement for *epidiorite.*

epidiagenesis (ep'-i-di'-a-gen'-e-sis) A term used by Fairbridge (1967) for the final emergent phase of diagenesis, in which sediments are lithified during and after uplift or emergence but before erosion. It is characterized by modification of connate solutions (by deeply penetrating and downward-migrating groundwaters) and by reintroduction of oxidizing conditions; near the Earth's surface, it passes into a zone where weathering processes become dominant. It is equivalent to *late diagenesis.* See also: *syndiagenesis; anadiagenesis.* Cf: *epigenesis [sed].* Adj: epidiagenetic.

epididymite (ep-i-did'-y-mite) A variously colored orthorhombic mineral: $NaBeSi_3O_7(OH)$. It is dimorphous with eudidymite.

epidiorite (ep-i-di'-o-rite) A metamorphosed gabbro or diabase in which generally fibrous amphibole (uralite) has replaced the original clinopyroxene (commonly augite). It is usually massive but may have some schistosity. Cf: *epidiabase.*

epidosite (ep-i-do'-site) A metamorphic rock consisting of epidote and quartz, and generally containing other secondary minerals such as uralite and chlorite.

epidote (ep'-i-dote) (a) A yellowish-green, pistachio-green, or blackish-green mineral: $Ca_2Al_2(Fe^{3+},Al)(SiO_4)(Si_2O_7)O(OH)$. It commonly occurs associated with albite and chlorite as formless grains or masses or as monoclinic crystals in low-grade metamorphic rocks (derived from limestones), or as a rare accessory constituent in igneous rocks, where it represents alteration products of ferromagnesian minerals. Syn: *pistacite; arendalite; delphinite; thallite.* (b) A name for a mineral group with an analogous formula, but with Ca partially replaced by rare-earth elements, and Fe^{3+} replaced by Al, Mn^{3+}, V^{3+}, Fe^{2+} or Mg.

epidote-amphibolite facies The facies (set of metamorphic mineral assemblages) in which mafic rocks are represented by hornblende+albite+epidote. Equivalent to Eskola's *epidote-amphibolite facies,* it is of uncertain status, transitional between the *greenschist facies* and the *amphibolite facies.* It is generally believed to be favored by the low temperature end of amphibolite facies of regional metamorphism. Syn: *albite-epidote-amphibolite facies.*

epidotization (ep'-i-do'-ti-za'-tion) The formation of epidote in rocks by hydrothermal alteration. Also the alteration of rocks in which plagioclase feldspar is albitized, whereby the anorthite molecule is freed for the formation of epidote and zoisite, often accompanied by chloritization.

epieugeosyncline (ep'-i-eu'-ge-o-syn'-cline) A deeply subsiding trough with limited volcanism, associated with rather narrow uplifts and overlying a deformed and intruded eugeosyncline (Kay, 1947, p.1289-1293); in modern terminology, a rift basin formed in an orogenic belt. Obsolete syn: *backdeep.* Cf: *secondary geosyncline.*

epifauna (ep'-i-fau'-na) (a) Fauna living upon rather than below the surface of the sea floor. Cf: *infauna.* (b) Fauna living attached to rocks, seaweed, pilings, or to other organisms in shallow water and along the shore.

epifaunal tiering Sessile filter-feeders of several different heights, feeding at different levels above the sediment-water interface.

epigene (ep'-i-gene) (a) Said of a geologic process, or of its resultant features, occurring at or near the Earth's surface. Cf: *hypogene.* Syn: *epigenic.* (b) Pertaining to a crystal that is not natural to its enclosing material, e.g. a *pseudomorph.*

epigenesis [meta] (ep-i-gen'-e-sis) The change in the mineral character of a rock as a result of external influences operating near the Earth's surface, e.g. mineral replacement during metamorphism.

epigenesis [sed] The changes, transformations, or processes, occurring at low temperatures and pressures, that affect sedimentary rocks subsequent to their compaction, exclusive of surficial alteration (weathering) and metamorphism; e.g. postdepositional dolomitization. The term is equivalent to *late diagenesis* (as used in the U.S.) and to *metharmosis,* but is considered by Russian geologists to include those changes occurring subsequent to *diagenesis* (as defined in a restricted sense). Cf: *epidiagenesis.* Syn: *metagenesis [sed]; metadiagenesis; catagenesis [sed].*

epigenetic [eco geol] (ep'-i-ge-net'-ic) Said of a mineral deposit of origin later than that of the enclosing rocks. Cf: *syngenetic [ore dep]; diplogenetic.* Syn: *xenogenous.*

epigenetic [sed] (a) Said of a sedimentary mineral, texture, or structure formed after the deposition of the sediment. (b) Pertaining to sedimentary epigenesis. Cf: *syngenetic [sed].*

epigenetic ice Ice commonly in lenses in material that predates the growth of the ice. The common form of ice produced during the aggradation of permafrost; a form of *segregated ice.*

epigenic (ep-i-gen'-ic) *epigene.*

epigenic cave A dissolution cave formed by water whose solutional capacity is derived from surface or near-surface sources, generally carbon dioxide (Palmer, 1991). Cf: *hypogenic cave.*

epigenite (ep'-i-gen'-ite) A mixture of tennantite, chalcopyrite, and pyrite.

epiglacial bench (ep-i-gla'-cial) A terrace cut by the lateral erosion of a supraglacial meltwater stream originating on a large glacier; also, the valley-side channel developed by such a stream (Stamp, 1961, p.179).

epiglacial epoch A period of time that closes a "great glacial series", representing a "constant phase" of glacial activity (Hansen, 1894, p.131). Not in current usage.

epigynous (e-pig'-y-nous) Said of a flower in which the sepals, petals, and stamens appear to arise from the top of the ovary. Such an ovary is called an *inferior ovary.* Cf: *hypogynous; perigynous.*

epiianthinite (ep-i-ian'-thi-nite) *schoepite.*

epikarst The uppermost zone of *dissolution* in a soil-covered or bare karst (Williams, 1983). Syn: *subcutaneous zone.*

epilimnetic (ep'-i-lim-net'-ic) Pertaining to an *epilimnion.* Syn: *epilimnial.*

epilimnial (ep-i-lim'-ni-al) *epilimnetic.*

epilimnion (ep-i-lim'-ni-on) The uppermost layer of water in a lake, characterized by an essentially uniform temperature that is generally warmer than elsewhere in the lake and by a relatively uniform mixing caused by wind and wave action; specif. the light (less dense), oxygen-rich layer of water that overlies the *metalimnion* in a thermally stratified lake (Wetzel, 2001, p.74). The oceanographic equivalent is *mixed layer.*

epilithic (ep-i-lith'-ic) Said of an organism that lives on or attached to rock or other stony matter. Syn: *petrophilous.* Cf: *endolithic.*

epilithon The organic layer covering the substrata (cobble and sediment) of a streambed, composed of algae, bacteria, and dutrital organic material. Syn: *periphyton.*

epimagma (ep-i-mag'-ma) Obsolete term for a vesicular magmatic residue that is relatively gas-free and of semisolid, pasty consistency, commonly formed by the cooling of lava in a lava lake. Cf: *hypomagma; pyromagma.*

epimagmatic (ep'-i-mag-mat'-ic) *deuteric.*

epimatrix (ep-i-ma'-trix) A term introduced by Dickinson (1970, p.702) for "inhomogeneous interstitial materials grown in originally open interstices during diagenesis" of graywackes and arkoses.

epimere (ep'-i-mere) A lateral downfold of a crustacean *tergite.* Syn: *pleurite; pleuron; tergal fold.*

epimerization (e-pim'-er-i-za'-tion) A process in which there is an alteration of the configuration at only one asymmetric center in an organic compound containing more than one asymmetric center (Cram and Hammond, 1959, p.131). Cf: *racemization age method.*

epinorm (ep'-i-norm) Theoretical calculation of minerals in a metamorphic rock of the *epizone,* as indicated by chemical analyses (Barth, 1959). Cf: *catanorm; mesonorm; Niggli molecular norm.*

epipedon (ep-i-pe'-don) *diagnostic horizons.*

epipelagic (ep'-i-pe-lag'-ic) Pertaining to the pelagic environment of the ocean to a depth of 200 meters; also, pertaining to the organisms inhabiting that zone. Cf: *mesopelagic.*

epipelic (ep-i-pel'-ic) Said of an organism growing on sediment, esp. soft mud.

epiphreas The zone in a cave system, directly above the *phreatic zone* affected morphologically and hydrologically by floods too large for the cave to absorb at once.

epiphysis [intrus rocks] (e-piph'-y-sis) An *apophysis* or tongue of a larger intrusion. Obsolete.

epiphysis [paleont] An interambulacral element at the top of *Aristotle's lantern* in echinoids.

epiphyte (ep'-i-phyte) A plant not growing from the soil but living attached to another plant or some inanimate object such as a pole or wire. Adj: epiphytic.

epiplankton (ep-i-plank'-ton) Organisms that are attached to floating vegetation or to mobile swimmers, for example, turtles and cetaceans. Syn: *pseudoplankton.*

epipsammon (ep-i-psam'-mon) Animals that live on a sandy surface.

epirhysis (ep-i-rhy'-sis) A *skeletal canal* in dictyonine hexactinellid sponges corresponding to an inhalant canal. Pl: epirhyses.

epirostrum (ep-i-ros'-trum) The adapical part of the *rostrum* of some belemnites, esp. those of the Jurassic, in which only the outer part shows concentric lamellar structure, the inner part being structureless (Spaeth et al., 1971).

episaturation The soil is saturated with water in one or more layers within 200 cm of the mineral soil surface and also has one or more unsaturated layers with an upper boundary above 200 cm depth, below the saturated layer(s) (a perched water table). See also: *endosaturation.*

episeptal deposit (ep-i-sep'-tal) A proximal *cameral deposit* on the concave (adoral) side of the septum of a nautiloid. Ant: *hyposeptal deposit.*

episkeletal (ep-i-skel'-e-tal) Above or outside the endoskeleton of an animal.

episode (ep'-i-sode) (a) A term used informally and without time implication for a distinctive and significant event or series of events in the geologic history of a region or feature; e.g. "glacial episode", "volcanic episode". (b) The formal, hierarchical *diachronic unit* of highest rank and greatest scope (NACSN, 1983, Art. 93). Cf: *phase [stratig]; span [stratig]; cline.*

episodic evolution (ep-i-sod'-ic) That evolutionary pattern in which the rate of change producing new taxa fluctuates or varies greatly as time proceeds, so that some periods or "episodes" exhibit much change while others show little. When applied to the species level, it may be synonymous with *punctuated gradualism.* Cf: *gradualistic evolution; punctuational evolution.*

episome (ep'-i-some) That part of a dinoflagellate theca or cyst anterior to the cingulum. This term encompasses the terms "epicyst," "epitheca," and "epitract" (Fensome et al., 1993; Williams et al., 2000, p.77 and 104). Cf: *hyposome.*

epispire (ep'-i-spire) A small round or elliptical sutural pore between the thecal plates of many primitive echinoderms such as eocrinoids, some edrioasteroids, and some stylophorans. Each epispire is thought to have housed an outward projecting soft-bodied *papula* used for respiration (Sprinkle, 1973, p.28).

epispore (ep'-i-spore) The fundamental layer of the fungal spore wall.

epistilbite (ep-i-stil'-bite) A colorless, pinkish, or yellowish platy monoclinic zeolite mineral: $Ca(Si_{18}Al_6)O_{48}{\cdot}16H_2O$.

epistolite (e-pis'-to-lite) A white, gray, or brownish triclinic mineral: $Na_5TiNb_2(Si_2O_7)_2FO_3{\cdot}5H_2O$.

epistoma (e-pi-sto'-ma) *epistome.*

epistome (ep'-i-stome) (a) The region between the antennae and the mouth of a crustacean. Also, a plate covering this region, such as that between the labrum and the bases of the antennae in brachyuran decapods. (b) The *rostral plate* of a trilobite. (c) A small labiate organ overlapping the mouth in bryozoans of the class Phylactolaemata. (d) The median plate of the prosomal *doublure* of eurypterids. Pl: epistomes. Syn: *epistoma.*

episyenite A term used in France for pervasively albitized and carbonatized syenite intrusions, usually associated with desilicification of a syenite intrusion.

episyenitization Albitization and desilicification of a syenite intrusion.

epitactic (ep-i-tac'-tic) Recommended adj. of *epitaxy.*

epitaxial (ep-i-tax'-i-al) Adj. of *epitaxy.*

epitaxic Adj. of *epitaxy.*

epitaxy (ep'-i-tax'-y) Orientation of one crystal with that of the crystalline substrate on which it grew. It is a type of overgrowth in which the two nets in contact share a common mesh. Adj: *epitactic; epitaxic; epitaxial.* Cf: *topotaxy; syntaxy.*

epitheca (ep-i-the'-ca) (a) An external calcareous layer or sheath of skeletal tissue surrounding a *corallite* and comprising an extension of the basal plate. (b) A dark secondary deposit in the inner wall of a fusulinid foraminifer; a *tectorium.* (c) The thin outermost calcareous layer of a thecal plate of a cystoid. It is thinner than the *stereotheca.* (d) The anterior part of a dinoflagellate theca, above the girdle. Cf: *hypotheca.* (e) *epivalve.*

epithelium (ep-i-the'-li-um) (a) In an animal, a cellular tissue that forms a covering or lining and that may serve various functions such as protection or secretion. (b) In a plant, "any cellular tissue covering a free surface or lining a tube or cavity" (Swartz, 1971, p.178); e.g. lining a resin duct and excreting resin.

epithermal (ep-i-ther'-mal) Said of a hydrothermal mineral deposit formed within about 1 kilometer of the Earth's surface and in the temperature range of 50°-200°C, occurring mainly as veins (Park and MacDiarmid, 1970, p.344). Also, said of that environment. Cf: *hypothermal; mesothermal [eco geol]; leptothermal; telethermal; xenothermal.*

epithermal-neutron log (ep-i-ther'-mal-neu'-tron) A *well log* of the *wall-contact log* type that measures radioactivity induced by neutron-neutron reactions. Fast neutrons of several Mev energy, emitted by a source in the *sonde,* may be slowed and backscattered as epithermal neutrons (0.4 to 100 ev) by collisions with hydrogen nuclei. Epithermal-neutron abundance is interpreted by assuming a constant lithology, usually limestone, that produces a continuous curve in porosity units. The epithermal-neutron log is largely insensitive to the differences in chlorinity of formation water that may hamper the interpretation of the *thermal-neutron log* and *neutron-gamma log.* Syn: *n-en log.*

epithet (ep'-i-thet) A term used in biological nomenclature to designate the single word that names a species, which is written following the generic name as part of a *binomen.*

epithyridid (ep-i-thy'-rid-id) Said of a brachiopod *pedicle opening* lying wholly within the ventral umbo and ventral from the beak ridges (TIP, 1965, pt. H, p.144).

epityche (ep'-i-tyche) An excystment aperture in the acritarch genus *Veryhachium.* It originates as an arched slit between two processes, in which rupture allows the folding-back of a relatively large flap.

epivalve (ep'-i-valve) The outer half of a diatom frustule. Cf: *hypovalve.* Syn: *epitheca.*

epixenolith (ep-i-xen'-o-lith) Obsolete term for a xenolith that is derived from the adjacent wall rock (Goodspeed, 1947, p.1251). Cf: *hypoxenolith.*

epizoic (ep-i-zo'-ic) Said of an organism growing on the body surface of an animal.

epizonal pluton An intrusive body that formed at shallow crystal depth (<6.5 km); typically composite, discordant, with chilled contacts and contact metamorphism. Cf: *mesozonal pluton; catazonal pluton.*

epizone (ep'-i-zone) According to Grubenmann's classification of metamorphic rocks (1904), the uppermost depth zone of metamorphism, characterized by low to moderate temperatures (less than 300°C) and hydrostatic pressures with low to high shearing stress. Typical rocks are slate, phyllite, and sericite and chlorite schist. The concept was modified by Grubenmann and Niggli (1924) to include effects of low-temperature contact metamorphism and metasomatism. Modern usage stresses pressure-temperature conditions (low metamorphic grade) rather than the likely depth of zone. Cf: *mesozone; katazone.*

epoch [geochron] (ep'-och) (a) The formal *geochronologic unit* longer than an *age [geochron]* and shorter than a *period [geochron],* during which the rocks of the corresponding series were formed. (b) A term used informally to designate a length (usually short) of geologic time; e.g. *glacial epoch.*

epoch [magnet] (a) A date to which measurements of a time-varying quantity are referred, e.g., "a chart of magnetic declination for epoch 1965.0". (b) *polarity chron.*

epontic (e-pon'-tic) Said of an organism that grows attached to some substratum.

EPR *electron paramagnetic resonance.*

epsilon cross-stratification Large-scale (most commonly), solitary sets of cross-strata discordantly overlying a commonly planar, erosional bounding surface. In vertical section the internal cross-strata may be straight, but they are commonly convex upward. The surfaces on internal cross-strata are *lateral accretion surfaces,* as the cross-strata commonly form by the lateral accretion of a channel bank into the main channel (Allen, 1963). Cf: *point bar.*

epsilon diversity The total taxon *richness* of the Earth (Whittaker, 1972).

epsomite [mineral] (ep'-som-ite) A mineral: $Mg(H_2O)_6(SO_4)\cdot H_2O$. It consists of native Epsom salts, and occurs in colorless, pink, or greenish orthorhombic prismatic crystals, botryoidal masses, incrustations in gypsum mines or limestone caves, or in solution in mineral waters. Syn: *Epsom salt; bitter salts; hair salt.*

epsomite [sed] Obsolete syn. of *stylolite.*

Epsom salt (Ep'-som) (a) A chemical: $Mg(H_2O)_6(SO_4)\cdot H_2O$. It is a bitter, colorless or white, crystalline salt with cathartic qualities. (b) *epsomite [mineral].*

equal-area projection (e'-qual-ar'-e-a) (a) A map projection on which a constant ratio of areas is preserved, so that any given part of the map has the same relation to the area on the sphere it represents as the whole map has to the entire area represented. Examples include: Bonne projection, Albers projection, and Mollweide projection. Cf: *conformal projection.* Syn: *authalic projection; equivalent projection; homolographic projection.* (b) *equiareal projection.*

equant (e'-quant) (a) Said of a crystal having the same or nearly the same diameter in all directions. Cf: *tabular; prismatic.* Syn: *equidimensional; isometric.* (b) Said of a sedimentary particle whose length is less than 1.5 times its width (Krynine, 1948, p.142). (c) Said of a rock in which the majority of grains are equant (Harte, 1977).

equant element A fabric element all of whose dimensions are approximately equal. Cf: *linear element; planar element.*

equation of state (e-qua'-tion) An equation interrelating the thermodynamic variables that define the state of a system. It is classically applied to simple gases and liquids in terms of pressure, volume, and temperature, but in modern geochemistry and petrology it is commonly extended to solids and to solutions, in which case the equations must contain terms describing the composition of the phase.

equation of state of seawater The equation relating density of seawater to in situ temperature, salinity, and pressure.

equator [astron] (e-qua'-tor) The great circle of a celestial sphere, having a plane that is perpendicular to the axis of the Earth.

equator [palyn] An imaginary line connecting points midway between the poles of a spore or pollen grain.

equator [surv] (a) The great circle formed on the surface of a sphere or spheroid by a plane drawn through its center and perpendicular to its polar axis, such as the great circle midway between the poles of rotation of a celestial body; specif. the Equator, which divides the Earth's surface into northern and southern hemispheres and is the line from which latitudes are reckoned, its own latitude being everywhere zero degrees. The Equator is the largest of the parallels of latitude, having a length on the Earth's surface of 40,075.76 km (24,901.92 miles). Syn: *terrestrial equator.* (b) *celestial equator.*

equatorial [clim] (e-qua-to'-ri-al) Said of a climate characterized by uniformly high temperature and humidity and heavy rainfall, and occurring in lowland areas within five to ten degrees of the equator. Cf: *tropical.*

equatorial [paleont] Pertaining to or located in the median plane normal to the axis of coiling or symmetry in a foraminifer; e.g. "equatorial section" representing a slice through a foraminiferal test passing through the proloculus. See also: *sagittal.*

equatorial aperture A symmetrical aperture in a planispiral foraminiferal test. It is commonly an *interiomarginal aperture,* but it may be areal or peripheral.

equatorial bulge An expression used to describe the *ellipticity* or flattening of the earth. Not recommended usage.

equatorial countercurrent A narrow and often variable ocean-surface current near the equator, flowing eastward between the westward-flowing equatorial currents to the north and south.

equatorial current Any of the broad ocean-surface currents in the tropical areas just north or just south of the equator, driven southwest or west in the Northern Hemisphere by northeast trade winds (North Equatorial Current), or northwest or west in the Southern Hemisphere by southeast trade winds (South Equatorial Current).

equatorial Kelvin wave An *internal wave* on the equatorial thermocline, influenced by gravity and *Coriolis force* on either side of the equator. It travels eastward, and generates a coastal *Kelvin wave* when it reaches the eastern boundary of the ocean. Cf: *Kelvin wave [oceanog].*

equatorial limb A term sometimes applied to the *amb* of a pollen grain or spore. It is undesirable because of possible confusion. Syn: *limb [palyn].*

equatorial opening *median split.* Also the part of an *acritarch* anterior to the equator and containing the excystment opening. Cf: *hypocyst.*

equatorial plane The plane of a planet's equator. The Earth's equatorial plane is inclined about 23.5° to the plane of the *ecliptic.*

equatorial projection One of a group of projections that have their center points on the equator and their polar axes vertical; e.g. an "equatorial cylindrical conformal map projection" (also known as the *Mercator projection*).

equatorial space A four-sided region resulting from formation of the *basal leaf cross* in an acantharian radiolarian.

equatorial spine A radial spine arising on the shell equator in an acantharian radiolarian.

equatorial undercurrent A narrow (400 km wide), thin (200 m thick), fast (0.5-1.5 m/s) current precisely on the equator flowing just below the surface from west to east within the westward flowing equatorial currents. Syn: *Cromwell current.*

equatorial view The lateral view of a spore or pollen grain from an aspect more or less midway between the poles. Ant: *polar view.*

equator system of coordinates A system of curvilinear celestial coordinates (usually declination and right ascension) based on the celestial equator as the primary great circle. Cf: *horizon system of coordinates.*

equiareal projection (e-qui-ar'-e-al) A term used in structural petrology for an *equal-area projection* developed from the center of a sphere through points on its surface to a plane that is tangent at the south pole of the sphere and so constructed that areas between meridians and parallels on the plane are equal to corresponding areas on the surface of the sphere.

equidimensional (e'-qui-di-men'-sion-al) *equant.*

equidistant conic projection *simple conic projection.*

equidistant projection (e-qui-dis'-tant) A map projection in which distances are represented true to scale and without length distortion in all directions from one or two given points.

equiform (e'-qui-form) Crystals that have the same (or nearly the same) shape. Obsolete.

equiglacial line (e-qui-gla'-cial) A line drawn on a map or chart to show coincidence of ice conditions, as in a lake or river, at a given time. Cf: *isopag; isopectic; isotac.*

equigranular (e-qui-gran'-u-lar) Texture consisting of grains roughly equal in size; used primarily for igneous and metamorphic rocks. Syn: *granoblastic* (metamorphic).

equilateral (e-qui-lat'-er-al) Bilaterally symmetrical; specif. said of a bivalve-mollusk shell whose parts anterior and posterior to the beaks are subequal or equal in length and nearly symmetrical. Cf: *equivalve.* Ant: *inequilateral.*

equilibrated chondrite A chondrite whose major mineral constituents (chondrules, matrix, and metal) have equilibrated during thermal metamorphism. These meteorites are of petrographic (petrologic) grade 4, 5, 6, or 7.

equilibrium (e-qui-lib'-ri-um) That state of a chemical system in which the phases do not undergo any spontaneous change in properties or proportions with time; a dynamic balance.

equilibrium constant A number representing equilibrium of a chemical reaction, defined as the result of multiplying the activities of the equation, each raised to a power indicated by its coefficient in the equation, and dividing by a similar product of the activities of the reactants. It may be referred to as the *thermodynamic equilibrium constant;* when concentrations instead of activities are used, it may be called the *classical equilibrium constant.* See also: *van't Hoff equation.*

equilibrium crystallization Crystallization in which crystals formed on cooling continually react and re-equilibrate with the liquid. Cf: *fractional crystallization [petrology]; crystallization differentiation.*

equilibrium diagram *phase diagram.*

equilibrium form The form of rounded ice crystals in a dry snow cover undergoing slow recrystallization due to a low temperature gradient (Colbeck, 1986). Syn: *rounded form.* Ant: *kinetic growth form.*

equilibrium fusion Fusion in which the liquid produced on heating continually reacts and re-equilibrates with the crystalline residue (Presnall, 1969). Cf: *fractional fusion.*

equilibrium limit *equilibrium line.*

equilibrium line A line on a glacier where the specific *net balance* is zero. The boundary between the *ablation area* and the *accumulation area.* At the end of the melt season, snow or superimposed ice from the previous winter persists above the equilibrium line. Below it, there may be bare glacier ice or either firn or superimposed ice, or both, from a preceding balance year. Cf: *climatic snowline; snowline; firn line.* Less preferred syn: *equilibrium limit.*

equilibrium moisture content The moisture content of a soil mass at a time when there is no moisture movement (Nelson and Nelson, 1967, p.127).

equilibrium path On a liquidus phase diagram, the crystallization sequence in which all crystals maintain contact with evolving liquid and react continuously and completely with this liquid, so that adjustment of crystal compositions throughout the crystallization interval is perfect. This path assumes constancy of chemical composition and ultimate intracrystalline homogeneity of crystalline products.
equilibrium profile *profile of equilibrium [coast].*
equilibrium shoreline A shoreline that has a local vertical profile of equilibrium and also an equilibrium shape in plan view; a *graded shoreline.*
equilibrium species A species that tends to be specialized in its adaptations, disperses and reproduces at moderate rates, but achieves large population size (close to the carrying capacity "K" in the logistic population-growth equation) because of the stable, predictable, or permanent nature of the environment occupied. Syn: *specialist species, K strategist.* Ant: *opportunistic species.*
equilibrium stage In hypsometric analysis of drainage basins, the stage in which a steady state is developed and maintained as relief slowly diminishes, and corresponding to maturity and old age in the geomorphic cycle (Strahler, 1952b, p.1130); the hypsometric integral is stable between 35% and 60%. Cf: *inequilibrium stage; monadnock phase.*
equilibrium tide A hypothetical global form of the sea-surface elevation which would be in equilibrium with the lunar and solar tide-generating forces in the absence of inertia and currents, and no continents. Syn: *gravitational tide; astronomic tide.*
equinoctial circle (e-qui-noc'-tial) *celestial equator.*
equinoctial tide A tide occurring when the Sun is near equinox, characterized by greater-than-average ranges of the spring tide.
equiplanation (e'-qui-pla-na'-tion) Those processes that operate at high latitudes and tend toward reduction of the land (usually resulting in the reduction of *local relief*) without reference to a base-level control and without involving any loss or gain of material (Cairnes, 1912, p.76). Cf: *altiplanation.*
equiplanation terrace A syn. of *altiplanation terrace* (G.T. Warwick in Stamp, 1961, p.21).
equipotential line (e'-qui-po-ten'-tial) A contour line on the potentiometric surface; a line along which the pressure head of groundwater in an aquifer is the same. Fluid flow is approximately normal to these lines in the direction of decreasing fluid potential. Syn: *isopotential line; potentiometric contour.*
equipotential-line method An early electrical-survey method wherein lines of equal potential, near the current electrodes, were searched for and mapped using a pair of potential electrodes, one of which was held stationary for each line mapping.
equipotential surface A surface on which the potential is everywhere constant for the attractive forces concerned. The gravity vector is everywhere normal to a gravity equipotential surface; the geoid is an "equipotential". Syn: *gravity equipotential surface; niveau surface; level surface.*
equirectangular projection (e'-qui-rec-tan'-gu-lar) *plate carrée projection.*
equisetophyte (e-qui-se'-to-phyte) *sphenopsid.*
equitability (e'-qui-ta-bil'-i-ty) Uniformity of abundance of taxa in a sample, location, or habitat (Hayek and Buzas, 1997). Syn: evenness.
equivalence [cart] (e-quiv'-a-lence) The unique property of an equal-area map projection in which the ratio between areas on the map is the same as that between the corresponding areas on the surface of the Earth. Cf: *conformality.* Syn: *orthembadism.*
equivalence [stratig] Geologic contemporaneity, esp. as indicated by identical fossil content. Syn: equivalency.
equivalency factor Used to convert rads to rems. For beta and gamma radiation: 1.0; thermal neutrons: 3.0; fast neutrons: 10.0; alpha particles: 20.0.
equivalent (e-quiv'-a-lent) adj. Corresponding in geologic age or stratigraphic position; esp. said of strata or formations (in regions far from each other) that are contemporaneous in time of formation or deposition or that contain the same fossil forms. n. A stratum that is contemporaneous or equivalent in time or character.
equivalent diameter Twice the *equivalent radius.*
equivalent grade A term used by Baker (1920, p.367), and synonymous with *arithmetic mean diameter.*
equivalent projection *equal-area projection.*
equivalent radius A measure of particle size, equal to the computed radius of a hypothetical sphere of specific gravity 2.65 (quartz) having the same settling velocity as those calculated or measured for a given sedimentary particle in the same fluid; one half of the *equivalent diameter.* Cf: *nominal diameter; sedimentation diameter.*
equivalent uranium, equivalent thorium The concentration of uranium/thorium estimated by gamma-ray spectrometry under the assumption that the U and Th decay series are in secular equilibrium (IAEA, 2003).
equivalent wavelet *embedded wavelet.*
equivalve (e'-qui-valve) Having valves equal in size and form; specif. said of a bivalve mollusk or its shell in which the right valve and the left valve are subequal or equal and symmetrical about the plane of commissure. Cf: *equilateral.* Ant: *inequivalve.*
equivoluminal wave (e'-qui-vo-lu'-mi-nal) *S wave.*
era (e'-ra) The formal *geochronologic unit* next in order of magnitude below an *eon,* during which the rocks of the corresponding *erathem* were formed; e.g. the Paleozoic Era, the Mesozoic Era, and the Cenozoic Era. Each of these includes two or more periods, during each of which a system of rocks was formed. Long-recognized Precambrian eras are the Archeozoic (older) and Proterozoic (younger).
erathem (e'-ra-them) The formal *chronostratigraphic unit* next lower than *eonothem* and higher than system. Also, the body of rocks formed during the time span of an era. An erathem consists of several adjacent systems (NACSN, 1983, Art. 71). See also: *supersystem.* Obsolete syn: *group; sequence.*
Eratosthenian (E'-ra-tos-the'-ni-an) (a) Pertaining to lunar topographic features and lithologic map units constituting a system of rocks including mare basalts and crater deposits deposited during the period of formation of large craters (such as Eratosthenes) whose rays are no longer visible (about 3.2 to 1.1 Ga). Eratosthenian rocks are older than those of the Copernican System but younger than those of the Imbrian System. (Wilhelms, 1987). (b) Said of the stratigraphic period during which the Eratosthenian System was developed.
E ray [cryst] In a uniaxial crystal, the ray of light that vibrates in a plane containing the optic axis and at an angle with the basal pinacoid, and whose velocity or refraction approaches that of the O ray as the angle approaches zero; the *extraordinary ray.* Cf: *O ray.*
E ray [paleont] Left anterior ray in echinoderms situated clockwise of the D ray when the echinoderm is viewed from the adoral side; equal to ambulacrum II in the *Lovenian system.*
ercitite A vitreous dark brown to black mineral: $NaMn^{3+}PO_4(OH){\cdot}2H_2O$.
ERDA *elastic recoil detection analysis.*
erdite (erd'-ite) A metallic copper-red monoclinic mineral: $NaFeS_2{\cdot}2H_2O$.
erect When used to describe the cusp or denticles of a conodont element, this term implies that the axis of the cusp or of the denticles is normal to that of the upper edge of the base or of a posterior process (TIP, 1981, pt. W, supp.2).
erect shell *acline [paleont].*
eremacausis (er'-e-ma-cau'-sis) The gradual transformation by oxidation of plant material into humus from exposure to air and moisture.
eremeyevite (er-e-me'-yev-ite) A syn. of *jeremejevite.* Also spelled: eremeevite.
eremic (er-e'-mic) Pertaining to a *desert* or deserts, or to sandy regions.
eremology (er-e-mol'-o-gy) The scientific study of deserts and their phenomena.
eremophyte (er-e'-mo-phyte) *xerophyte.*
erg A region in the Sahara, deeply covered with shifting sand and occupied by complex sand dunes; an extensive tract of *sandy desert;* a *sand sea.* Etymol: Hamitic. Pl: *areg;* ergs. See also: *koum; nefud.* Also spelled: *ergh.*
ergeron (erg'-e-ron) A French term for a very fine argillaceous sand, or variety of loess, containing a substantial amount of calcium carbonate, commonly occurring in northern France and the Belgian province of Hainaut.
ergh *erg.*
ergodic [stat] (er-god'-ic) Having the same statistical properties throughout the ensemble. An ergodic system will eventually return arbitrarily close to any prior state. The statistical properties measured over sufficiently long intervals anywhere throughout the ensemble will be the same. Etymol: Greek "work" and "path."
ergodicity A form of the law of large numbers, applied to stationary processes.
Erian (E'-ri-an) North American provincial series: Middle Devonian (above Ulsterian, below Senecan).
Erian Orogeny One of the 30 or more short-lived orogenies during Phanerozoic time identified by Stille, in this case at the end of the Silurian; the last part of the Caledonian orogenic era. Syn: *Hibernian Orogeny.*
ericaite (er'-i-ca-ite') A variously colored orthorhombic mineral of the *boracite* group: $(Fe^{2+},Mg)_3B_7O_{13}Cl$.
ericophyte (er'-i-co-phyte') A plant growing on a heath or moor.
ericssonite (er'-ics-son-ite') A reddish-black platy monoclinic mineral: $BaMn^{2+}{}_2Fe^{3+}OSi_2O_7(OH)$. It is dimorphous with orthoericssonite.
erikite (er'-ik-ite) A yellow-green, greenish-yellow, or brown mineral: $(La,Ce)_x(P,Si)O_4{\cdot}H_2O$. It is perhaps a silicate-rich rhabdophane. The type erikite from Greenland has been shown to be monazite.
eriochalcite (er'-i-o-chal'-cite) A bluish-green to greenish-blue orthorhombic mineral: $CuCl_2{\cdot}2H_2O$. Syn: *antofagastite.*
erionite-Ca (er'-i-o-nite') A white hexagonal zeolite mineral: $(Ca,K_2Na_2)_2Al_4Si_{14}O_{36}{\cdot}15H_2O$.
erionite-K (er'-i-o-nite') A white hexagonal zeolite mineral: $(K_2,Ca,Na_2)_2Al_4Si_{14}O_{36}{\cdot}15H_2O$.
erionite-Na (er'-i-o-nite') A white hexagonal zeolite mineral: $(Na_2K_2Ca)_2Al_4Si_{14}O_{36}{\cdot}15H_2O$.
eriophorum peat (er-i-oph'-o-rum) Peat formed mainly from *Eriophorum,* also known as cotton grass, a genus of sedges of the family Cyperaceae. Cf: *carex peat.*
erlianite (er'-li-an-ite) A silky black orthorhombic mineral: $(Fe^{2+},Mg)_4(Fe^{3+},V^{3+})_2Si_6O_{15}(OH,O)_8$.
erlichmanite (er'-lich-man-ite') A metallic gray cubic mineral of the *pyrite* group: OsS_2.
ernienickelite A vitreous black rhombohedral mineral of the *chalcophanite* group: $NiMn^{4+}{}_3O_7{\cdot}3H_2O$.
erniggliite A gray-black trigonal mineral: $Tl_2SnAs_2S_6$.

ernstite (ernst′-ite) A yellow-brown monoclinic mineral: $(Mn^{+2}Fe^{+3})Al(PO_4)(OH,O)_2$. It is an oxidation product of eosphorite.

erode (e-rode′) (a) To wear away the land, as by the action of streams, waves, wind, or glaciers. (b) To produce or modify a landform by the wearing-away of the land.

erodibility (e-rod′-i-bil′-i-ty) (a) The quality, degree, or capability of being eroded or yielding more or less readily to erosion. (b) The tendency of soil to be detached and carried away; the rate of soil erosion. In the universal soil-loss equation, the soil erodibility factor, K, indicates the inherent erodibility of a soil. K gives an indication of the soil loss from a unit plot 22 m (72 ft) long with a 9% slope and continuous fallow culture. The soil erodibility, or K factor, normally varies from near zero to about 0.6. Cf: *erosiveness*. Adj: erodible. Syn: *erosibility*. See also: *erosivity*.

eroding channel (e-rod′-ing) A channel in which the energy of a stream is greater than that required to move the sediment available for transport.

eroding stress The shear stress of overland flow available to dislodge or tear loose soil material per unit area. Originally defined by Horton (1945, p.319) as the "eroding force" exerted parallel with the soil surface per unit of slope length and width.

erosibility (e-ros′-i-bil′-i-ty) *erodibility*.

erosion (e-ro′-sion) (a) The general process or the group of processes whereby the materials of the Earth's crust are loosened, dissolved, or worn away, and simultaneously moved from one place to another, by natural agencies, which include weathering, solution, corrasion, and transportation, but usually exclude mass wasting; specif. the mechanical destruction of the land and the removal of material (such as soil) by running water (including rainfall), waves and currents, moving ice, or wind. The term is sometimes restricted by excluding transportation (in which case "denudation" is the more general term) or weathering (thus making erosion a dynamic or active process only). Cf: *denudation*. (b) An instance or product, or the combined effects, of erosion.

erosional (e-ro′-sion-al) Pertaining to or produced by the wearing-away of the land.

erosional floodplain A *floodplain* produced by the lateral erosion and gradual retreat of the valley walls.

erosional shoreface retreat The landward and upward movement of the shoreface in response to relative rising sea level and littoral wave and current processes.

erosional unconformity An unconformity made manifest by erosion, or a surface that separates older rocks that have been subjected to erosion from younger sediments that cover them; specif. *disconformity*.

erosional vacuity A term formerly used by Wheeler (1958, p.1057) later replaced by *degradation vacuity* (Wheeler, 1964, p.602).

erosion caldera A type of *caldera* that is developed by the erosion and resultant widening of a caldera or by erosion of a volcanic cone, resulting in a large, central cirquelike depression. It is considered by some not to be a true caldera type, since it is not formed by volcanic processes. Cf: *explosion caldera; collapse caldera*.

erosion crater *makhtésh*.

erosion cycle *cycle of erosion*.

erosion fault scarp *fault-line scarp*.

erosion groove A sedimentary structure formed by "closely spaced lines of straight-sided scour marks" (Dzulynski and Sanders, 1962, p.66). The scouring may be initially concentrated by a pre-existing groove.

erosion integral An expression of the relative volume of a landmass removed by erosion at a given contour; the inverse of the *hypsometric integral*.

erosion intensity The quantity (or depth) of solid material actually removed from the soil surface by sheet erosion per unit of time and area; originally termed "erosion rate" by Horton (1945, p.324).

erosionist (e-ro′-sion-ist) A believer in the obsolete theory that the irregularities of the Earth's surface are mainly the result of erosion.

erosion lake A lake occupying a basin excavated by erosion.

erosion pavement A surficial concentration of pebbles, gravel, and other rock fragments that develops after sheet erosion or rill erosion has removed the finer soil particles and that tends to protect the underlying soil from further erosion.

erosion plain A general term for any plain produced by erosion, such as a peneplain, a pediplain, a panplain, or a plain of marine erosion.

erosion platform (a) A relatively level surface of limited extent formed by shore-zone erosion. (b) A *wave-cut platform* along a coast. Cf: *abrasion platform*.

erosion ramp A sloping belt of reef rock immediately above the reef flat on an atoll islet, where marine erosion is active.

erosion remnant A topographic feature that remains or is left standing above the general land surface after erosion has reduced the surrounding area; e.g. a monadnock, butte, mesa, or stack. Syn: *residual [geomorph]; relic [geomorph]; remnant*.

erosion ripple A minor wavelike feature produced by the cutting action of the wind on a lower and somewhat more coherent layer of a sand dune.

erosion scarp A *scarp* produced by erosion; e.g. a *fault-line scarp* or a *beach scarp*.

erosion surface A land surface shaped and subdued by the action of erosion, esp. by running water. The term is generally applied to a level or nearly level surface. Syn: *planation surface*.

erosion terrace A *terrace* produced by erosion; e.g., a *rock terrace*.

erosiveness (e-ro′-sive-ness) The quality, degree, or capability of accomplishing erosion; the power or tendency to effect erosion.

erosive velocity (e-ro′-sive) That velocity of water in a channel above which erosion of the bed or banks will occur.

erosivity In the universal soil-loss equation, erosivity, the rainfall and runoff factor (R), measures the erosive force of rainfall and runoff (Brady, 1990).

erpoglyph (er′-po-glyph) *worm casting*.

erratic (er-rat′-ic) n. A rock fragment carried by glacial ice, or by floating ice, deposited at some distance from the outcrop from which it was derived, and generally though not necessarily resting on bedrock of different lithology. Size ranges from a pebble to a house-size block. See also: *perched block; exotic [struc geol]*. Syn: *erratic block; erratic boulder; glacial erratic*. adj. Transported by a glacier from its place of origin. Syn: *traveled*.

erratic block *erratic*.

erratic boulder *erratic*.

error (er′-ror) (a) The difference between an observed, calculated, or measured value of a quantity and the ideal or true value of that quantity or of some conventional or standard value determined by established procedure or authority and used in lieu of the true value. (b) An inaccuracy or variation in the measurement, calculation, or observation of a quantity due to mistakes, imperfections in equipment or techniques, human limitations, changes of surrounding conditions, or other uncontrollable factors. (c) The amount of deviation of a measurement from some standard, arbitrary, estimated, or other reference value. See also: *random error; systematic error; personal error*.

errorchron (er′-ror-chron) An *isochron* about which data are scattered, not only because of analytical error, but also because of departures of the geologic system investigated from an ideal model.

error of closure The amount by which a quantity obtained by a series of related measurements differs from the true or theoretical value or a fixed value obtained from previous determinations; esp. the amount by which the final value of a series of survey observations made around a closed loop differs from the initial value. The surveyed quantities may be angles, azimuths, elevations, or traverse-station positions. See also: *closure [surv]; circuit closure; triangle closure; horizon closure; mis-tie*. Syn: *misclosure; closing error*.

error of collimation *collimation error*.

ershovite A vitreous green triclinic mineral: $K_3Na_4(Fe,Mn,Ti)_2Si_8O_{20}(OH)_4{\cdot}4H_2O$.

ertixiite (er-tix′-i-ite) A colorless or white cubic mineral: $Na_2Si_4O_9$.

ERTS Earth Resource Technology Satellite. Now called: *Landsat*.

ERU *environmental-resource unit*.

erubescite (er-u′-bes-cite) *bornite*.

eruption (e-rup′-tion) The ejection of volcanic materials (lava, pyroclasts, and volcanic gases) onto the Earth's surface, either from a central vent or from a fissure or group of fissures. Cf: *central eruption; fissure eruption*.

eruption breccia *explosion breccia*.

eruption cloud A cloud of volcanic gases, with ash and other pyroclastic fragments (tephra), that forms by volcanic explosion; the ash may fall from it the eruption cloud to deposit *fallout tuff*. Syn: *explosion cloud; ash cloud; dust cloud [volc]; volcanic cloud*. See also: *eruption column*.

eruption column The initial form that an *eruption cloud* takes at the time of explosion; the lower portion of such a cloud. It results from initial ejection velocity rather than buoyant rise of the hot gases.

eruption cycle The sequence of events that occurs during a volcanic eruption; the regular change in the behavior of the eruptions in a period of activity.

eruption intensity The mass eruption rate, expressed as kg/sec. Syn: *intensity [volc]*.

eruption magnitude The size of an eruption, expressed as km^3 of magma or as kg mass.

eruption plume A convoluted, rolling mass of partly condensed water vapor, vapor, dust, and ash that often rises to great heights. Syn: *eruption cloud*.

eruption rain A rain following a volcanic eruption that results from condensation of the eruption's associated steam. Syn: *volcanic rain*.

eruption-time In stochastic treatment of volcanologic data based on renewal theory, the age at which a repose period is ended by an outbreak (Wickman, 1966, p.293).

eruptive (e-rup′-tive) Said of a rock formed by the solidification of magma; i.e. either an *extrusive* or an *intrusive* rock. Most writers restrict the term to its extrusive or volcanic sense.

eruptive evolution Evolution characterized by the relatively sudden appearance of varied new stocks from a common ancestral line.

erythraean (er-y-thrae′-an) Pertaining to the ancient sea that occupied the Arabian Sea, the Red Sea, and the Persian Gulf areas. Also spelled: *erythrean*.

erythrean (er-y-thre′-an) *erythraean*.

erythrine (er′-y-thrine) *erythrite*.

erythrite (er′-y-thrite) A red or pink mineral of the *vivianite* group: $Co_3(AsO_4)_2{\cdot}8H_2O$. It is isomorphous with annabergite, and may contain some nickel. Erythrite occurs in monoclinic crystals, in globular and reniform masses, or in earthy forms, as a weathering product of cobalt ores in the upper (oxidized) parts of veins. Syn: *erythrine; cobalt bloom; red cobalt; cobalt ocher; peachblossom ore.*

erythrosiderite (er'-y-thro-sid'-er-ite) A red or brownish red orthorhombic mineral: $K_2Fe^{3+}Cl_5{\cdot}H_2O$. It may contain some aluminum.

erythrozincite (er'-y-thro-zinc'-ite) A variety of wurtzite containing manganese.

erzbergite (erz'-berg-ite) Calcite and aragonite in alternate layers.

Erzgebirgian Orogeny (Erz-ge-birg'-i-an) One of the 30 or more short-lived orogenies during Phanerozoic time identified by Stille, in this case in the early Late Carboniferous (Namurian or Westphalian).

ESA *environmental site assessment.*

esboite (es'-bo-ite) An orbicular *diorite* in which andesine or oligoclase is the dominant plagioclase and forms the orbicules. Its name is derived from Esbo, Finland. Cf: *corsite.* Obsolete.

esboitic crystallization (es-bo-it'-ic) The process by which the orbicules of an *esboite* attain oligoclasitic composition (Eskola, 1938, p.449). Not recommended usage.

escalation Coevolutionary response between predator and prey.

escar (es'-car) *esker.*

escarpment (es-carp'-ment) (a) A long, more or less continuous cliff or relatively steep slope facing in one general direction, breaking the continuity of the land by separating two level or gently sloping surfaces, and produced by erosion or faulting. The term is often used synonymously with *scarp,* although *escarpment* is more often applied to a cliff formed by differential erosion. (b) A steep, abrupt face of rock, often presented by the highest strata in a line of cliffs, and generally marking the outcrop of a resistant layer occurring in a series of gently dipping softer strata; specif. the steep face of a *cuesta.* Cf: *scarp slope.* (c) A term used loosely in Great Britain as a syn. of *cuesta.* Etymol: French, "escarpement", "steep face or slope".

eschar (es'-char) *esker.*

escharan (es'-cha-ran) *eschariform.*

eschariform (es-char'-i-form) Said of a rigid, bilamellar, foliaceous, erect colony; in cheilostome bryozoans, attached firmly by a calcareous base or loosely by radicles. Syn: *escharan.*

eschwegeite (esch-wege'-ite) A syn. of *tanteuxenite.*

eschynite (es'-chy-nite) *aeschynite.*

escutcheon (es-cutch'-eon) (a) The flat or simply curved, typically lozenge-shaped, dorsal differentiated area extending posteriorly from the beaks of certain bivalve mollusks and sometimes bordered by a ridge in each valve. It corresponds to the posterior part of the cardinal area and is separated from the remainder of the shell surface by a sharp change in angle. (b) An improper term for the *basal cavity* of a conodont.

eskar (es'-kar) *esker.*

eskebornite (es-ke-born'-ite) A metallic brass yellow tetragonal mineral: $CuFeSe_2$. It is isomorphous with chalcopyrite, and a member of the *chalcopyrite* group.

esker (es'-ker) A long, narrow, sinuous, steep-sided ridge composed of irregularly stratified sand and gravel that was deposited by a subglacial or englacial stream flowing between ice walls or in an ice tunnel of a stagnant or retreating glacier, and was left behind when the ice melted. It may be branching and is often discontinuous, and its course can be at a high angle to the edge of the glacier. Eskers range in length from less than 100 m to more than 500 km (if gaps are included), and in height from 3 to more than 200 m. Etymol: Irish *eiscir,* "ridge". Cf: *ice-channel filling; kame.* Syn: *ås; os [glac geol]; serpent kame; Indian ridge; morriner.* Also spelled: *eskar; eschar; escar.*

esker delta A flat-topped deposit of sand and gravel formed by, and at the mouth of, a glacial stream as it issued from an ice tunnel and flowed into a lake or sea; associated with an esker or ice-channel filling made at the same time. Cf: *kame delta; delta kame.* Syn: *sand plain [glac geol]; sand plateau.*

esker fan A small plain of gravel and sand built at the mouth of a subglacial stream, and associated with an esker formed at the same time.

eskerine (es'-ker-ine) Characteristic of an esker; e.g. *eskerine* topography.

esker lake A lake enclosed or dammed by an esker (Dryer, 1901, p.129).

esker trough A term applied in Michigan to a shallow valley, cut in till, that contains an esker (Leverett, 1903, p.118).

eskimoite (es'-ki-mo-ite') A metallic gray monoclinic mineral: $Ag_7Pb_{10}Bi_{15}S_{36}$. It is a member of the lillianite homologous series.

eskolaite (es'-ko-la-ite) A dark green to black rhombohedral mineral of the *corundum* group: Cr_2O_3.

esmeraldite (es-me-ral'-dite) A coarse- to medium-grained rock having hypidiomorphic-granular texture, in which quartz and muscovite are the essential phases. Its name is derived from Esmeralda County, Nevada. Syn: *nordfieldite.* Obsolete.

Esopusian North American stage: Lower Devonian, above Deerparkian below Erian, uppermost part of Ulsterian.

espalier drainage pattern (es-pa'-lier) *trellis drainage pattern.*

esperanzaite A pale blue-green monoclinic mineral: $NaCa_2Al_2(AsO_4)_2F_4(OH){\cdot}2H_2O$. It is the arsenate analogue of morinite.

esperite (es'-per-ite) A white monoclinic mineral: $PbCa_3Zn_4(SiO_4)_4$. It produces intense lemon-yellow response to short-wave ultraviolet radiation. Syn: *calcium-larsenite.*

espichellite (es-pi-chel'-lite) A lamprophyric rock, similar to *camptonite,* in which hornblende, augite, olivine, magnetite, and pyrite phenocrysts are contained in a compact groundmass of magnetite, hornblende, augite, mica, and labradorite with orthoclase rims. Analcime may also occur in the groundmass. Espichellite resembles *teschenite* but has less analcime and is porphyritic rather than granular. It was named by Souza-Brandão in 1907 from Cape Espichel, Portugal. Obsolete.

esplanade (es-pla-nade') (a) A term used in the SW U.S. for a broad bench or terrace bordering a canyon, esp. in a plateau region. (b) A level stretch of open and grassy or paved ground, often designed for providing a vista.

ESR *electron spin resonance.*

esseneite (es-sene'-ite) A reddish-brown monoclinic pyroxene mineral: $CaFe^{3+}AlSiO_6$.

essential (es-sen'-tial) Said of pyroclastic materials that are formed from magma; it is part of a classification of ejecta based on mode of origin, and is equivalent to the terms *juvenile [volc]* and *authigenous.* Cf: *accessory [pyroclast]; accidental; cognate [pyroclast].*

essential element An element whose presence is necessary in order for an organism to carry out its life processes.

essential mineral [eco geol] A mineral essential to the national defense for which no difficulty of procurement during a national emergency is foreseen.

essential mineral [petrology] A mineral component of a rock that is necessary to its classification and nomenclature, but that is not necessarily present in large amounts. Cf: *accessory mineral.*

essexite (es'-sex-ite) (a) In the *IUGS classification,* a plutonic rock with F between 10 and 60, and P/(A+P) between 50 and 90. It is synonymous with both *foid monzodiorite* and *foid monzogabbro.* (b) An alkali gabbro primarily composed of plagioclase, hornblende, biotite, and titanian augite, with lesser amounts of alkali feldspar and nepheline. Essexite grades into theralite with a decrease in potassium feldspar and an increase in the feldspathoid minerals. Its name is derived from Essex County, Massachusetts, from where it was originally defined by Sears in 1891 (Johannsen, 1939, p.250). Cf: *glenmuirite.*

essonite (es'-son-ite) A yellow-brown or reddish-brown transparent gem variety of grossular garnet containing iron. Syn: *cinnamon stone; hyacinth (b);* jacinth. Also spelled: *hessonite.*

established dune (es-tab'-lished) *anchored dune.*

establishment (es-tab'-lish-ment) *ecesis.*

estavelle (es-ta-velle') A surface feature that either recharges or discharges groundwater, depending on the level of the water table or potentiometric surface (Ford and Williams, 1989). Cf: *insurgence; resurgence.*

esterellite (es-te-rel'-lite) A porphyritic *quartz diorite* also containing zoned andesine and hornblende. It was named by Michel-Lévy in 1897 from Esterel, France. Obsolete.

Estérel twin law (Es-te'-rel) A twin law for parallel twins in feldspar with twin axis *a* and composition plane (0*kl*) parallel to *a.*

estero (es-te'-ro) (a) An estuary or inlet, esp. when marshy. (b) Land adjoining an estuary inundated by the tide. Etymol: Spanish.

estuarine (es'-tu-a-rine') Pertaining to or formed or living in an estuary; esp. said of deposits and of the sedimentary or biological environment of an estuary.

estuarine circulation The pattern of water circulation in an *estuary,* in which ocean water flows landward into the estuary along the bottom and mixes with less dense river water within the estuary, and the resulting *brackish water* flows seaward on the surface. The pattern describes a form of upwelling and is associated with high productivity. A prime example is the Baltic Sea.

estuarine delta A delta that has filled, or is in the process of filling, an estuary.

estuarine deposit A sedimentary deposit laid down in the brackish water of an estuary, characterized by fine-grained sediments (chiefly clay and silt) of marine and fluvial origin mixed with a high proportion of decomposed terrestrial organic matter; it is finer-grained and of more uniform composition than a *deltaic deposit.*

estuarine lagoon A lagoon produced by the temporary sealing of a river estuary by a storm barrier. Such lagoons are usually seasonal and exist until the river breaches the barrier; they occur in regions of low or spasmodic rainfall. Syn: *blind estuary.*

estuarine salinity Salinity that varies according to tidal or seasonal conditions, as in an estuary.

estuary (es'-tu-a'-ry) A semi-enclosed coastal body of water that has a free connection with the open sea and within which seawater is measurably diluted with freshwater from land drainage (Pritchard, 1952). Estuaries are classified in terms of their mixing as stratified, partially stratified, well mixed, and reverse; or by their origins as bar-built, fjord, tectonic, fjard, and ria (Woodroffe, 2002). Cf: *ria; branching bay; liman; fjord.* Etymol: Latin aestus, "tide".

étang (e'-tang) A French term for a shallow pool, pond, or lake, esp. one lying among sand dunes, formed by the ponding of inland drainage by beach material thrown up by the sea, and gradually becoming filled with silt, like those along the Mediterranean coast of France (e.g. in Languedoc).

etch figure A marking, usually in the form of minute pits, produced by a solvent on a crystal surface; the form varies with the mineral species and the solvent, but conforms to the symmetry of the crystal.

etching (etch'-ing) (a) The reduction of the Earth's surface by the slow processes of differential weathering, mass wasting (esp. creep), sheetwash, and deflation, so that areas underlain by more resistant rocks are brought into relief as the less resistant rocks are lowered (Rich, 1951). (b) A general term for the formation of a landform by erosion or chiseling, as the *etching* of a canyon by a stream.

etchplain (etch'-plain) A relatively extensive erosion surface, believed to develop by the comparatively rapid but local differential lowering, during uplift, of a peneplain surface kept at or near base level by the removal of a deep overlying cover of weathered rock. The feature was originally described as an "etched plain" by Wayland (1934).

ethane (eth'-ane) A colorless, odorless, water-insoluble, gaseous paraffin hydrocarbon, formula C_2H_6, which occurs in natural gas or can be produced as a by-product in the cracking of petroleum.

ethmolith (eth'-mo-lith) A discordant, funnel-shaped pluton which tapers downward. Obsolete.

ethnostratigraphy The stratigraphic classification into units characterized by their content of archaeological artifacts; archaeological stratigraphy (Gasche and Tunca, 1983, p.332).

ethology (e-thol'-o-gy) The science concerned with animal behavior, some effects of which are preserved in the fossil record, esp. as trace fossils.

etindite (e-tind'-ite) A dark-colored extrusive rock intermediate in composition between *leucitite* and *nephelinite*, with phenocrysts of clinopyroxene in a dense groundmass of leucite, nepheline, and clinopyroxene. Named by Lacroix in 1923 for Etinde, Cameroon. Obsolete.

etnaite (et'-na-ite) A *trachyandesite* named by Rittmann in 1960. Obsolete.

ettringite (et'-tring-ite) A colorless, yellow, or light brown rhombohedral mineral: $Ca_6Al_2(SO_4)_3(OH)_{12}\cdot 26H_2O$. Syn: *woodfordite.*

eu- A prefix meaning "well" or "well developed".

euaster (eu'-as-ter) A sponge spicule (microsclere) having the form of a modified aster in which the rays arise from a common center. Cf: *streptaster.*

euautochthony (eu-au-toch'-tho-ny) Accumulation of plant remains (such as roots, stumps, tree trunks) that are now found in the exact place, and more or less in the correct relative positions, in which they grew. Cf: *hypautochthony.*

eu-bitumen (eu-bi'-tu-men) A collective name for those fluid, viscid, or solid bitumens that are easily soluble in organic solvents. Examples are petroleum, ozokerite, elaterite, and asphalt (Tomkeieff, 1954).

eucairite (eu'-cair-ite) A metallic creamy-white orthorhombic mineral: CuAgSe. Also spelled: *eukairite.*

Eucarya (Eu-car'-y-a) *eukaryote.*

euchlorine (eu-chlor'-ine) An emerald-green monoclinic mineral: $KNaCu_3O(SO_4)_3$.

euchroite (eu'-chro-ite) An emerald-green or leek-green orthorhombic mineral: $Cu_2(AsO_4)(OH)\cdot 3H_2O$.

euclase (eu'-clase) A brittle monoclinic mineral: $BeAlSiO_4(OH)$. It occurs in pale tones of blue, green, yellow, or violet, and is sometimes colorless; the blue variety is greatly esteemed by gem collectors.

eucolite (eu'-co-lite) A variety of eudialyte that is optically negative. Also spelled: *eukolite.*

euconodont A "true conodont". A conodont in which the individual elements consist of a lamellar apatitic crown and a basal body (or basal filling) which is less highly mineralized. Lamellae of the crown are continuous around the upper surface and are separated by thinner organic layers. Conodonts with euconodont elements range from Upper Cambrian through Triassic. Some workers consider only euconodonts as members of the major animal group Conodonta. See also: *conodont*. Cf: *protoconodont; paraconodont.*

eucrite [ign] (eu'-crite) A very basic *gabbro* composed chiefly of calcic plagioclase (bytownite, anorthite) and clinopyroxene, with accessory olivine. The name, given by Rose in 1864, is from Greek, "easily discerned," for a class of achondritic meteorite.

eucrite [meteorite] A group of augite- and pigeonite-rich basaltic achondrites whose chemical and petrological characteristics indicate that they are the product of igneous processing on the same parent body from which the diogenite and howardite meteorites were derived. See also: *HED meteorite.*

eucryptite (eu-cryp'-tite) A colorless or white rhombohedral mineral of the *phenakite* group: $LiAlSiO_4$.

eucrystalline (eu-crys'-tal-line) *macrocrystalline.*

eudiagnostic (eu'-di-ag-nos'-tic) Said of the texture of a rock (esp. an igneous rock) in which all mineral components are of such size and shape as to be identifiable. The term includes both macrocrystalline and microcrystalline textures. It was originally used by Zirkel in German as "eudiagnostisch". Ant: *adiagnostic.* Obsolete.

eudialyte (eu-di'-a-lyte) (a) A pale-pink, garnet red, to brownish-red rhombohedral zeolitic mineral: $Na_{15}Ca_6Fe_3Zr_3Si(Si_{25}O_{73})(O,OH,H_2O)_3(Cl,OH)_{22}$. It occasionally occurs in great quantity in certain nepheline syenites. Cf: *eucolite.* Syn: *barsanovite.* (b) A group name for minerals with the analogous formula, with Na replaced by rare-earth elements or H_2O; Ca replaced by Mn, Sr or Fe; Fe replaced by Mn; and Zr replaced by Ti or Nb.

eudicots (eu'-di-cots) An informal grouping of derived *angiosperms* that includes most of the *dicotyledons,* with the exception of the *magnoliid complex* and a few other taxa. Eudicots have *tricolpate* pollen or other forms directly derived from the tricolpate condition. The term includes, e.g., Rosaceae and Asteraceae and refers especially to *angiosperm* evolution.

eudidymite (eu-did'-y-mite) A white, glassy, monoclinic mineral: $Na_2Be_2Si_6O_{15}\cdot H_2O$. It is dimorphous with epididymite.

eudiometer (eu-di-om'-e-ter) An instrument such as a graduated glass tube for measuring the amounts of different gases in a gas mixture by exploding the gases one at a time by passing an electric spark through the mixture.

eugenesis (eu-gen'-e-sis) The period of development and death of the organic material found in coal-ball concretions (McCullough, 1977, p.133). It is followed in order by syngenesis, diagenesis, and epigenesis.

eugenite A metallic cubic mineral: $Ag_{11}Hg_2$. It is an amalgam of Ag and Hg.

eugeocline (eu-ge'-o-cline) A shortened version of the term *eugeosyncline.* "Tentatively used in the Great Basin to describe lower Paleozoic siliceous assemblage rocks that may be continental rise deposits" (Stewart and Poole, 1974, p.29). Obsolete. Cf: *miogeocline.*

eugeogenous (eu-ge-og'-e-nous) Easily weathered; said of a rock that produces by weathering a large amount of detritus. Ant: *dysgeogenous.*

eugeosyncline (eu'-ge-o-syn'-cline) (a) A geosyncline in which volcanism is associated with clastic sedimentation; the volcanic part of an *orthogeosyncline,* located away from the craton (Stille, 1940). (b) The deep-water, offshore portion of a passive-margin basin. Cf: *miogeosyncline.* Syn: *pliomagmatic zone.* See also: *ensimatic geosyncline.*

euglenoid (eu'-gle-noid) One of a group of unicellular flagellates with a gullet and with the cell usually bounded by a fairly firm but flexible membrane instead of a cell wall.

eugsterite (eug'-ster-ite) A colorless monoclinic mineral: $Na_4Ca(SO_4)_3\cdot 2H_2O$.

euhedral (eu-hed'-ral) A grain bounded by perfect crystal faces; well-formed. A syn. of *idiomorphic* and *automorphic* (as they were originally defined). Cf: *anhedral; subhedral.*

euic High level of bases in soil material, specified at family level of classification.

eukairite (eu-kai'-rite) *eucairite.*

eukaryote (eu-kar'-y-ote) One of a major group of organisms, characterized by a complex protoplasmic organization, a vesicular nucleus, and various sorts of membrane-bounded cytoplasmic organelles. Cf: *prokaryote.*

eukaryote [paleont] Any organism of the kingdoms *Protoctista,* Fungi, Plantae, or Animalia characterized by one or more nucleated cells, typically a larger size than typical *prokaryotes,* and organelles sometimes separated from the cytoplasm by membranes. Range, Proterozoic to Holocene.

eukolite (eu'-ko-lite) *eucolite.*

eukrite (eu'-krite) Obsolete spelling of *eucrite [meteorite].*

Euler deconvolution A method of inverting the profiles or gridded (map) magnetic or gravity data for the locations and depths of sources of particular model geometry defined by the *Euler structural index.* On profiles (2-D Euler deconvolution), each calculation is run for different window lengths to obtain solutions for different depths. On maps (3-D Euler deconvolution), the calculations are run over square windows that are moved by a grid unit for the next depth calculation. Values cluster together at a valid structural index. A fault might be indicated by a vertical alignment of depth solutions. Sometimes called *Euler method.* See also: *Werner deconvolution.*

eulerhabd (eu'-le-rhabd) A sinuous stout U-shaped oxea (sponge spicule); a more sharply curved variety of *ophirhabd.*

Eulerian (Eu-ler'-i-an) A way of describing fluid motion whereby the motion and properties of the fluid are calculated or measured at fixed points in space as the fluid moves past the points. This is the most common way of describing fluid dynamics. Named in honor of Leonhard Euler (1707-1783), Swiss mathematician. Cf: *Lagrangean.*

Euler method *Euler deconvolution.*

Euler number (Eu'-ler) Inertial force divided by pressure-gradient force.

Euler pole In plate tectonics, the imaginary pole of rotation around which one plate on a sphere moves with respect to another plate. Cf: *Euler's theorem.*

Euler's theorem A statement that any displacement of a spherical surface can be defined as a rotation about an axis that passes through the center of the sphere. It is much used in *plate tectonics* because any displacement of a rigid body such as a lithospheric plate on the Earth's surface may be considered as a rotation about a properly chosen axis through a point on the sphere.

Euler structural index The exponent in a power law expressing the falloff of the potential field strength versus distance from source. For magnetic data, the values of the Euler structural index "N" range from 0 to 3: for contact, N=0; for a thin dike or sill edge, N=1; for a narrow kimberlite pipe or thin-bed fault, N=2; for a sphere or compact body, N=3 (Reid, 2003). Observed anomalies are generated by sources approximated with various model geometries and they therefore require solutions for a range of the Euler structural inices. See also: *Euler deconvolution.*

eulytine (eu'-ly-tite) A colorless, yellow, gray, or brown cubic mineral: $Bi_4(SiO_4)_3$. Syn: *bismuth blende; agricolite.* Var. spelling: eulytite.

eumalacostracan (eu'-mal-a-cos'-tra-can) A shrimplike crustacean belonging to the subclass Eumalacostraca, differing mainly from other malacostracans by the nonbivalve nature of the carapace and the presence of biramous thoracic appendages with a single joint in the protopod. Range, Middle Devonian to Holocene.

eumorphism (eu-mor'-phism) A property of an equal-area map projection in which undue distortion of shapes is not shown, as in the arithmetic mean by S.W. Boggs (1889-1954) of the sinusoidal and Mollweide projections (BNCG, 1966, p.17, 47).

eumycete (eu-my'-cete) A plant of the subdivision Eumycetes, which comprises the true fungi. Cf: *myxomycete; schizomycete.*

euosmite (eu-os'-mite) A brownish-yellow resin, with a low oxygen content and a characteristic pleasant odor, found in brown coal.

eupelagic deposit (eu-pe-lag'-ic) Deep-sea sediment in which less than 25% of the fraction coarser than 5 μm is of terrigenous, volcanogenic, and neritic origin. Such deposits usually form far from the continents, beyond the continental margin and associated abyssal plain. They accumulate by vertical settling of particulate matter, are highly oxidized, and include pelagic clays and oozes.

euphotic [oceanog] A layer starting at the sea surface in which sufficient sunlight is present such that photosynthesis prevails over respiration, i.e., the oceanic region in which photosynthetic carbon fixation and the associated chemical or biochemical reactions occur. Usually occupies the upper 50 to 150 meters of the ocean, and the light intensity at its bottom boundary is 1% of the incident sunlight intensity at the sea surface.

euphotic zone [lake] (eu-phot'-ic) The zone in a lake above the *compensation depth.* In the euphotic zone there is enough light for *photosynthesis* to exceed *respiration* of the phytoplankton. Syn: *photic zone.* Cf: *aphotic zone [lake]; trophogenic.*

eupotamic (eu-po-tam'-ic) Said of an aquatic organism adapted to living in both flowing and quiet fresh water. Cf: *autopotamic; tychopotamic.*

Eurasian-Melanesian belt (Eur-a'-sian-Mel-a-ne'-sian) An old term for a belt of orogens that includes the Alps and Himalayas and extends from the Mediterranean to southern Asia (Strahler, 1963). Cf: *circum-Pacific belt; great-circle belt (b).*

euripus (eu'-ri-pus) A strait or narrow channel of water where the tide or a current flows and counterflows with turbulent force. Pl: euripi. Etymol: Greek euripos, "strait, channel".

europium anomaly Deviation of normalized europium concentration in a sample, relative to the expected concentration. The expected concentration is calculated by interpolation between normalized concentrations of the neighboring elements, Sm and Gd, on a *Coryell-Masuda diagram.*

Euryapsida (Eu-ry-ap'-si-da) A subclass of reptiles characterized by a single upper temporal fenestra; it includes large marine forms such as plesiosaurs, nothosaurs, and placodonts, as well as smaller and less specialized terrestrial forms of uncertain relations. Range, Lower Permian to Cretaceous.

eurybathic (eu-ry-bath'-ic) Referring to a marine organism that tolerates a wide range of depth. Cf: *stenobathic.*

eurybiontic (eu'-ry-bi-on'-tic) *eurytropic.*

eurygeographic (eu'-ry-ge'-o-graph'-ic) Said of plant or animal species with wide geographic ranges. Cf: *stenogeographic.*

euryhaline (eu-ry-ha'-line) Referring to a marine organism that tolerates a wide range of salinities. Inhabitants of estuaries are typically euryhaline. Cf: *stenohaline.*

eurypalynous (eu-ry-pal'-y-nous) Of plant taxa (families, etc.), having palynomorphs with a great diversity of morphology. Ant: *stenopalynous.*

euryplastic (eu-ry-plas'-tic) Having great capacity for modification and adaptation to a wide range of environmental conditions; capable of major evolutionary differentiation. Cf: *stenoplastic.*

euryproct (eu'-ry-proct) Said of a sponge in which the *cloaca* is conical, with the widest part forming the *osculum.*

eurypterid (eu-ryp'-ter-id) Any *merostome,* typically brackish or freshwater, belonging to the subclass Eurypterida, characterized by an elongate segmented lanceolate body that is only rarely trilobed, and a thin chitinous integument with tubercles or scalelike ornamentation. Range, Ordovician to Permian. Cf: *xiphosuran.*

eurysiphonate (eu-ry-si'-phon-ate) Said of nautiloids with relatively large siphuncles. Cf: *stenosiphonate.*

eurythermal (eu-ry-ther'-mal) Referring to a marine organism that tolerates a wide range of temperatures. Cf: *stenothermal.*

eurytopic (eu-ry-top'-ic) Said of an organism occurring in many different habitats. Cf: *eurytropic.*

eurytropic (eu-ry-trop'-ic) Said of an organism that can tolerate a wide range of a particular environmental factor. Cf: *eurytopic.* Syn: *eurybiontic.* Ant: *stenotropic.*

euspondyle (eu-spon'-dyle) In dasycladacean algae, a regular whorled arrangement of branches (laterals) on the central axis.

eusporangiate (eu-spo-ran'-gi-ate) In a strict sense, said of isosporous ferns of subclass Eusporangiatae in which the sporangium wall is two or more cell layers thick. Also applied as a comparative term to similar sporangial walls in higher plants (Melchior and Werdermann, 1954, p.287).

eustasy (eu'-sta-sy) (a) A condition of world-wide sea-level. Fluctuations in eustasy may be caused by change in the global hydrologic balance (glacier ice, groundwater, soil moisture), a volumetric change; or a change in the capacity of ocean basins, such as by plate tectonic activity (acceleration of heat flow at spreading centers or crustal deformation at subduction zones). Adj: *eustatic.* (b) Synchronous global changes in sea level. For example, a change in the level of the ocean surface measured by a geosynchronous satellite is a eustatic change. Cf: *glacio-eustasy; tectono-eustasy.*

eustatic (eu-stat'-ic) adj. Pertaining to worldwide changes of sea level that affect all the oceans. Eustatic changes may have various causes, but the changes dominant in the past few million years were caused by additions of water to, or removal of water from, the continental icecaps. n. *eustasy.*

eustatic rejuvenation A renewal of the effectiveness of erosion processes resulting from causes that produce worldwide lowering of sea level.

eustatism (eu'-stat-ism) *eustasy.*

eustratite (eu-strat'-ite) A dense lamprophyric rock having rare phenocrysts of olivine, corroded hornblende, augite, and possibly oligoclase, in a groundmass composed of idiomorphic augite and magnetite with interstitial feldspar, mica, and colorless glass. The name, given by Ktenas in 1928, is for Haghios Eustratios Island in the Aegean Sea. Obsolete.

eusynchite (eu-synch'-ite) *descloizite.*

eutaxic (eu-tax'-ic) Said of a stratified mineral deposit. Cf: *ataxic.*

eutaxite (eu-tax'-ite) A *taxite* whose components have aggregated into separate bands. Cf: *ataxite [volc].*

eutaxitic (eu-tax-it'-ic) Said of the banded structure in certain extrusive rocks, resulting from the parallel arrangement and alteration of layers of different textures, mineral composition, or color. Commonly applied to banded structures such as *fiamme* in *welded tuff* and defined by pumice fragments which have been compacted into parallel, disc-shaped lenses.

eutectic (eu-tec'-tic) Said of a system consisting of two or more solid phases and a liquid whose composition can be expressed in terms of positive quantities of the solid phases, all in equilibrium (coexisting) at an (isobaric) invariant point, which is the minimum melting temperature for the assemblage of solids. Addition or removal of heat causes an increase or decrease, respectively, of the proportion of liquid to solid phases, but does not change the temperature of the system or the composition of any phases. See also: *eutectoid.*

eutectic point An invariant point that represents the lowest temperature at which a eutectic mixture will melt at any given pressure, and the composition of the resulting eutectic liquid. Also the composition of the lowest-temperature liquid in a eutectic system.

eutectic ratio The ratio of solid phases forming from the eutectic liquid at the eutectic point; it is such as to yield a gross composition for the crystal mixture that is identical with that of the liquid. It is most frequently stated in terms of weight percent.

eutectic texture A pattern of intergrowth of two or more minerals, formed as they coprecipitate during crystallization, e.g. the quartz and feldspar of *graphic granite.* See also: *exsolution texture.* Syn: *eutectoid texture.*

eutectoid The syn. of *eutectic,* or its equivalent when applied to a system all of whose participating phases are crystalline. Also applied to the symplectic intergrowth texture typical of simultaneous crystallization of solid phases at a eutectic point.

eutectoid texture (eu-tec'-toid) *eutectic texture.*

eutectoperthite (eu-tec'-to-perth'-ite) *mesoperthite.*

Eutheria (Eu-the'-ri-a) A subclass of mammals characterized by a placenta. It includes most living forms, which comprise about 24 orders. Range, Middle Cretaceous to Holocene. See also: *placental; Marsupialia.*

eutrophic (eu-troph'-ic) Referring to a body of water characterized by a high level of nutrients (phosphate, nitrate), with correspondingly high primary productivity.

eutrophication (eu'-troph-i-ca'-tion) The process by which waters become more eutrophic; esp. the artificial or natural enrichment of a lake by an influx of nutrients required for the growth of aquatic plants such as algae that are vital for fish and animal life.

eutrophic lake A lake with an abundant supply of nutrients to support a high level of *productivity.* If it is also *dimictic,* the deep water is likely to be seasonally deficient in oxygen (Wetzel, 2001, p.273-274). Cf: *dystrophic lake; mesotrophic lake; oligotrophic lake.*

eutrophic layer A portion of the euphotic zone in which organic matter production via carbon fixation is so high that respiration processes utilizing the organic matter cause a marked decline in dissolved oxygen concentration. The process is called *eutrophication,* and is frequently the result of pollution by nutrients. Found most often in nearshore waters. Cf: *respiration; carbon fixation.*

eutrophic peat Peat containing abundant plant nutrients, such as nitrogen, potassium, phosphorus, and calcium. Cf: *mesotrophic peat; oligotrophic peat.* Syn: *calcareous peat.*

eutrophy (eu'-tro-phy) The quality or state of a *eutrophic lake.*

euvitrain (eu-vit'-rain) Structureless, amorphous *vitrain.* It is the more common type. Cf: *provitrain.* See also: *collain; ulmain.* Syn: *xylovitrain.*

euvitrinite (eu-vit'-ri-nite) A variety of the maceral *vitrinite* characteristic of euvitrain and including the varieties ulminite and collinite. Plant material has been completely jellified and shows no cell structure. The term *collinite* has been proposed as a preferable synonym. Cf: *provitrinite.*

euxenite-(Y) (eux'-e-nite-y) (a) A submetallic brownish-black orthorhombic mineral, commonly metamict: $(Y,Ca)(Nb,Ta)_2O_6$. It is isomorphous with polycrase and occurs in granite pegmatites. (b) A group name for minerals with an analogous formula, but with Y replaced by U and rare-earth elements, and Nb replaced by Ti.

euxinic (eux-in'-ic) (a) Pertaining to an environment of restricted circulation and stagnant or anaerobic conditions, such as a fjord or a nearly isolated or silled basin with toxic bottom waters. Also, pertaining to the material (such as black organic sediments and hydrogen-sulfide muds) deposited in such an environment or basin, and to the process of deposition of such material (as in the Black Sea). (b) Pertaining to a rock facies that includes black shales and graphitic sediments of various kinds. Etymol: From Latin, "Pontus Euxinus" = Black Sea; orig. Greek name "axenos" = inhospitable. Cf: *pontic.*

evaluation map (e-val'-u-a'-tion) A stratigraphic map that summarizes the results of stratigraphic analyses made for economic purposes (Krumbein and Sloss, 1963, p.484).

evanescent lake (ev-a-nes'-cent) A short-lived lake formed after a heavy rain. Cf: *ephemeral lake.*

evansite (ev'-ans-ite) A colorless or milky-white to brown or reddish-brown amorphous mineral: $Al_3(PO_4)(OH)_6 \cdot 6H_2O$(?). It has a bluish, greenish, or yellowish tinge. Evansite may contain small amounts of uranium and thorium.

evapograph (e-vap'-o-graph) *atmometer.*

evaporates (e-vap'-o-rates) Goldschmidt's name for the evaporite group of sediments; sedimentary salts precipitated from aqueous solutions and concentrated by evaporation. Archaic. The synonymous term *evaporite* is more commonly used. Cf: *reduzates; oxidates; resistates; hydrolyzates.*

evaporation (e-vap'-o-ra'-tion) The process, also called *vaporization,* by which a substance passes from the liquid or solid state to the vapor state. Limited by some to vaporization of a liquid, in contrast to sublimation, the direct vaporization of a solid. Also limited by some (e.g. hydrologists) to vaporization that takes place below the boiling point of the liquid. The opposite of condensation (Langbein and Iseri, 1960).

evaporation discharge The release of water from the *saturated zone* by evaporation from the soil (soil discharge) or by the *transpiration* of plants (vegetal discharge).

evaporative pumping A mechanism proposed by Hsu and Siegenthaler (1969) wherein evaporation of pore water near the sediment surface induces capillary upward movement of groundwater, causing a net increase of groundwater salinity lower in the sediment, as water is progressively removed.

evaporimeter (e-vap'-o-rim'-e-ter) *atmometer.*

evaporite (e-vap'-o-rite) A sedimentary rock composed primarily of minerals produced from a saline solution as a result of extensive or total evaporation of the solvent. Examples include gypsum, anhydrite, other diverse sulfates, halite (rock salt), primary dolomite, and various nitrates and borates. The term commonly includes rocks with saline minerals formed by other mechanisms, e.g., mixing of waters or temperature change. Mineral grains formed by these mechanisms may be reworked by wind or saline waters as clastic particles but are also considered evaporitic deposits. Syn: evaporate; *saline deposit.*

evaporite mineral A mineral precipitated as a result of evaporation, e.g., halite. The term is commonly used synonymously with saline mineral, despite knowledge of other mechanisms of formation. Satin spar veins (fibrous gypsum) formed along bedding planes by the oxidation of pyrite within black shales are mistakenly termed evaporite minerals.

evaporite-solution breccia (e-vap'-o-rite-so-lu'-tion) A term used by Sloss and Laird (1947, p.1422-1423) for a *solution breccia* formed where soluble evaporites (rock salt, anhydrite, gypsum, etc.) have been removed.

evapotranspiration (e-vap'-o-tran'-spi-ra'-tion) Loss of water from a land area through transpiration of plants and evaporation from the soil and surface-water bodies. Also, the volume of water lost through evapotranspiration.

eveite (eve'-ite) An apple-green orthorhombic mineral of the *libethenite* group: $Mn^{2+}_2(AsO_4)(OH)$.

even-crested ridge (e'-ven-crest'-ed) One of the high fold ridges, as in the Appalachian Mountains of Pennsylvania, whose tops all rise to an approximately uniform elevation. From even-crested ridges it can be inferred that a plain reconstructed by filling the valleys to the level of the ridgetops is an old *peneplain.* See also: *summit concordance; accordant summit level.*

even-grained Obsolete syn. of *equigranular; homogranular; granuloblastic; granoblastic.*

evening emerald (e'-ve-ning) Olivine (peridot or chrysolite) that loses some of its yellow tint in artificial light, appearing more greenish (like an emerald) and used as a gem. Syn: *night emerald.*

evenkite (e-venk'-ite) A colorless or yellowish monoclinic hydrocarbon mineral: $C_{24}H_{50}$. It was formerly regarded as a paraffin wax: $C_{21}H_{42}$.

evenness *equitability.*

event [magnet] (e-vent') *polarity subchron.*

event [seis] (a) The arrival of a seismic wave, an *arrival,* denoted by a definite phase change or amplitude buildup on a seismic record. (b) An *earthquake.* Syn: *seismic event.*

event [tect] A noncommittal term used for any incident of probable tectonic significance that is suggested by geologic, isotopic, or other evidence, but whose full implications are unknown. It is used especially for minor clusters of radiometric dates whose relations to geologic structures or processes have not been precisely evaluated. Cf: *pulsation [tect]; disturbance.*

event stratigraphy The used of inferred geological events rather than the intrinsic petrological characters of the rocks, or their contained fossils, for correlation (Ager, 1963, p.63). Numerous subsequent authors have defined event stratigraphy as the organization, classification, and correlation of the Earth's rocks on the basis of short-term phenomena–explosive volcanism, rapid tectonic movements, abrupt changes of sea level, climatic cycles, storms, distinctive sedimentologic and biologic events, and even extraterrestrial or other "rare events" at any scale.

event stratinomy A term proposed by Seilacher (1984) for "the level of research centered on the analysis of individual beds" that record rare events. Cf: *event stratigraphy.*

everglade (ev'-er-glade) A term used esp. in the southern U.S. for a large expanse of marshy land, covered mostly with tall grass, e.g. the Florida Everglades.

eveslogite A light brown to yellow monoclinic mineral: $(Ca,K,Na,Sr,Ba)_{24}(Ti,Nb,Fe,Mn)_6Si_{24}O_{72}(OH,F,Cl)_7$.

Evian water (E'-vi-an) Noneffervescent alkaline mineral water. The term is derived from Évian-les-Bains, a town in southeastern France where it is found.

evolute (e'-vo-lute) Loosely coiled or tending to uncoil; e.g. said of a foraminiferal test with nonembracing chambers, a gastropod shell whose whorls are not in contact, or a cephalopod conch with little or no overlapping of the whorls. Cf: *involute; advolute; convolute.*

evolution (ev-o-lu'-tion) (a) The change of a group of related organisms toward adaptation to the environmental conditions to which they have been exposed with the passage of time. (b) The theory that life on Earth has descended from a common ancestor. Syn: *organic evolution.* (c) The permanent change in the form and function of organisms of successive ancestor-descendant generations or populations, over geologic time, so that the latest members of the succession differ significantly from the earliest.

evolutionary fauna Groups of classes of marine animals with similar diversity histories Cambrian, Paleozoic and Modern (Sepkoski, 1981).

evolutionary guild A collection of species, genera, or other taxa that share a common mode of life, including feeding strategies and other ecological roles.

evolutionary momentum (ev-o-lu'-tion-a'-ry) The tendency of evolution to appear to continue along the same trend after the external stimulus provoking it has apparently diminished or disappeared. The term originated in the idea that organic function is activity, and activity is motion, which led to the notion that evolution may continue to inadaptive lengths and result in extinction.

evolutionary novelty An inherited change from one previously existing character; it is the homologue of the previously existing character in an ancestor-descendant relationship, and an apomorphy at the time that it originates (Wiley et al., 1991, p.9).

evolutionary plexus A complex *lineage* of organisms that consists of minor lines that repeatedly divide and reunite. Syn: *plexus [evol].*

evolutionary radiation Rapid increase in diversity or morphology by one or more clades.

evolutionary series In paleontology, a morphologic series that corresponds with time to a significant degree. Cf: *lineage.*

evolutionary systematics A traditional systematic approach which recognizes relationships based on overall similarity and shared homologous characters.

evolutionary zone *lineage zone.*

evorsion (e-vor'-sion) The formation of potholes in a streambed by the erosional action of vortices and eddies, whirling water that may carry sand and gravel. Etymol: Latin, "e" (="ex"), "from", + "vorsare/versare", "to turn often, spin". Syn: *pothole erosion.*

evorsion hollow *pothole [streams].*

ewaldite (e'-wald-ite) A bluish-green hexagonal mineral: $Ba(Ca,Y,Na,K)(CO_3)_2 \cdot 2.6H_2O$.

Ewald sphere *sphere of reflection.*

Ewing corer (Ew'-ing) A variety of *piston corer.* Named for Maurice Ewing.

exact interpolator If the location where an estimate is to be made coincides with the location of an observation, an exact interpolator will yield a value that is identical to the observation. In mapping, an exact interpolator is said to "honor the data points."

exaerobic facies (ex-aer-o'-bic) Zone of sediments and fauna with anomalous occurrences of shelly faunas in laminated strata deposited in very low oxygen conditions. Shelly organisms are interpreted to be chemosymbiotic (Savrda and Bottjer, 1987). Cf: *dysaerobic facies.*

exaptation A feature that now enhances fitness but was not shaped by natural selection for its present role. Cf: *preadaptation.*

exaration (ex-a-ra'-tion) The general process of *glacial erosion.* Grabau (1924, p.263-264) suggested that the term be restricted to "glacial denudation, i.e. the removal and transport of weathered material" by glaciers. Rarely used.
excavation (ex-ca-va'-tion) (a) The act or process of removing soil and/or rock materials from one location and transporting them to another. It includes digging, blasting, breaking, loading, and hauling, either at the surface or underground. (b) A pit, cavity, hole, or other uncovered cutting produced by excavation. (c) The material dug out in making a channel or cavity.
excavation depth The maximum depth from which material beneath an *impact crater* can be ejected. Often wrongly confused with the *crater depth,* the maximum depth of excavation is seldom more than 1/10 of the *transient-crater diameter.*
excavation stage The second of the three stages of the impact-cratering process, during which the *shock wave* expands away from the impact site, imparts a velocity to the target material, expels the *ejecta,* and opens the transient crater. Cf: *contact-and-compression stage; modification stage.*
excentric (ex-cen'-tric) Not centrally located; e.g. said of an ammonoid *umbilicus* characterized by an abrupt opening-up of the spiral described by the umbilical seam, or by a tendency toward closing of this spiral while the peripheral spiral is relatively unchanged (TIP, 1959, pt. L, p.4).
excess argon (ex'-cess) *Radiogenic argon* that is incorporated as a gas or solid into rocks and minerals by processes other than in-situ radioactive decay of potassium-40. The term does not include *atmospheric argon.* Cf: *radiogenic argon; inherited argon; extraneous argon.*
excess ice The volume of ice in the ground which exceeds the total pore volume that the ground would have under natural unfrozen conditions.
excess pore pressure Transient pore pressure at any point in an aquitard or aquiclude in excess of the pressure that would exist at that point if steady-flow conditions had been attained throughout the bed (Poland et al., 1972).
excess water *rainfall excess.*
exchange (ex-change') In glaciology, the sum of *accumulation* and *ablation,* both taken as positive numbers, averaged over a glacier for a balance year or a hydrologic year; a measure of the intensity of mass exchange with the atmosphere or hydrosphere. Usually given in millimeters or meters of water equivalent.
exchange capacity A quantitative measure of surface charge of a substance, reported in equivalents of exchangeable ions per unit weight of the solid. See also: *ion exchange.*
exchange coefficient *austausch.*
exchange component *exchange operator.*
exchange force A quantum-mechanical apparent interaction between electrons that is the cause of *magnetic order.*
exchange operator A *component* that changes, by addition, one compound or mineral into another or others, or that expresses ionic substitutions that can occur in a given mineral. Generally written as a formula containing negative coefficients, as $MgCa_{-1}$, and F_2O_{-1}, read "Mg, Ca minus one" and "F two, 0 minus one." Such operators can either be physically unrealizable (as in the above two examples) or may correspond to real substances, as $(OH)_2O_{-1}$ (equals water, H_2O). These components can be used to describe exchange processes, including diffusion, and reciprocal ternary chemical systems. Term originated by Burt (1974); concept originated earlier by J.B. Thompson, Jr. Syn: *exchange component* (Brady, 1975).
exchange reaction A balanced reaction among phases and components of phases in which progress of the reaction results in no change in the modal amounts of the phases. The reaction can be described in terms of an *exchange operator.* It is of value in geothermometry (Thompson, J.B., Jr., et al., 1982). Cf: *net-transfer reaction.*
excitation-at-the-mass *mise a la masse.*
excitation potential (ex-ci-ta'-tion) The characteristic minimum energy required to remove an electron from an atom.
exclusive economic zone A marine area where mineral exploration and exploitation activities are reserved to the adjacent country but which is generally open to shipping. Abbrev: EEZ.
excretion (ex-cre'-tion) A term proposed by Todd (1903) for a *concretion* that grows progressively inward from the exterior; e.g. a shell of sand cemented by iron oxide and generally filled by unconsolidated sand or containing other shells of cemented sand.
excurrent (ex-cur'-rent) (a) Said of a plant growth or structure that extends beyond the margin or tip, e.g. a midrib developing into a mucro or awn. (b) Said of the growth habit of plants having a continuous unbranched axis, e.g. the excurrent habit of spruces and firs (Lawrence, 1951, p.752).
excurrent canal *exhalant canal.*
excystment (ex-cyst'-ment) The process of cyst abandonment. In dinoflagellates and acritarchs, the opening through which excystment occurs is the *archeopyle* (Williams et al., 2000, p.81).
excystment opening Presumed escape opening found in some *acritarchs.* Excystment modes include *lateral rupture, median split, epityche, pylome, cryptopylome,* and *epibystra* (Mendelson, 1993; Williams et al., 2000). Cf: *trochospiral suture; circinate suture.*
exdysis [palyn] In some dinoflagellates, the shedding of the *theca* related to vegetative division, environmental stress, or cyst formation (Williams et al., 2000, p.69).
exfoliation (ex-fo'-li-a'-tion) The process by which concentric scales, plates, or shells of rock, from less than a centimeter to several meters in thickness, are successively spalled or stripped from the bare surface of a large rock mass. It is caused by physical or chemical forces producing differential stresses within the rock, as by expansion of minerals as a result of near-surface chemical weathering, or by the release of confining pressure of a once deeply buried rock as it is brought nearer to the surface by erosion (*pressure-release jointing*). It often results in a rounded rock mass or dome-shaped hill. Cf: *spheroidal weathering.* Syn: *spalling; scaling; desquamation;* sheet jointing.
exfoliation cave A cave formed by the partial destruction of a plate, sheet, or slab of rock produced by exfoliation, having a planar back wall and a continuation of the joint (of exfoliation) up into the roof of the cave (Bradley, 1963, p.525). Examples are found on the Colorado Plateau in SW U.S.
exfoliation dome A large dome-shaped form, developed in massive homogeneous rocks, esp. granite, by exfoliation; well-known examples occur in Yosemite Valley, Calif.
exhalant canal (ex-hal'-ant) (a) Any canal forming part of the *exhalant system* of a sponge. Syn: *excurrent canal; apochete.* (b) A channel formed by the outer and the parietal lips in certain gastropods.
exhalant system The part of the *aquiferous system* of a sponge between the apopyles and the oscula, characterized by water flowing outward toward the oscula. Cf: *inhalant system.*
exhalation (ex-ha-la'-tion) The streaming-forth of volcanic gases; also, the escape of gases from a magmatic fluid (Schieferdecker, 1959, terms 4462, 4826). Syn: *emanation.*
exhalite (ex-hal'-ite) A chemical sediment, usually containing oxide, carbonate, or sulfide as anions, and iron, manganese, base metals, and gold as cations, formed by the issuance of volcanically-derived fluids onto the sea floor or into the sea; thus, the product of *exhalation.* Adj: oxhalitine, e.g. exhalitive massive sulfide ore.
exhumation (ex-hu-ma'-tion) The uncovering or exposure by erosion of a pre-existing surface, landscape, or feature that had been buried. See also: *resurrected; relict [geomorph].*
exhumed (ex-humed') *resurrected.*
exhumed karst Karst features reexposed by erosion from beneath former covering strata.
exilazooecium (ex-i'-la-zo-oe'-ci-um) A polymorph in stenolaemate bryozoans, generally smaller than feeding zooecia in cross section, occurring in the exozone only, and containing few or no diaphragms so that an appreciable living chamber occurs (Boardman and Cheetham, 1973, p.154). Cf: *mesozooecium.*
exine (ex'-ine) The outer, very resistant layer of the two major layers forming the wall (sporoderm) of palynomorphs, consisting of sporopollenin, and situated immediately outside the *intine.* It is divided into two layers (*ektexine* and *endexine*) on the basis of staining characteristics. See also: *perisporium.* Syn: *extine; exospore.*
exinite (ex'-i-nite) Old syn. of *liptinite.*
exinoid (ex'-i-noid) A maceral group that includes the macerals in the liptinite series.
exinonigritite (ex'-i-no-nig'-ri-tite) A type of *nigritite* that is derived from spore exines. Cf: *humonigritite; polynigritite; keronigritite.*
exinous (ex'-i-nous) Consisting of exine.
exite [coal] (ex'-ite) *liptite.*
exite [paleont] A movable lobe on the exterior side of the limb of an arthropod, such as a lateral *ramus* (e.g. an exopod) of the protopodal limb segments of a crustacean, or the lateral or outer lobe or joint of the biramous appendage of a trilobite. Cf: *endite.*
exitèle *valentinite.*
exitèlite *valentinite.*
exitus [palyn] The site where a *pollen tube* exits through the *sporoderm.*
exo- (ex'-o-) A prefix meaning "outside" or "out of".
exoadaptation (ex'-o-ad'-ap-ta'-tion) Adaptation of organisms to their external environments.
exocast (ex'-o-cast) *external cast.*
exocoele (ex'-o-coele) The space between adjacent pairs of mesenteries of a coral. Cf: *entocoele.*
exocyathoid expansion (ex-o-cy'-a-thoid) Growth of an additional intervallum in an archaeocyathid cup, which appears to have had an adherent function (TIP, 1972, pt. E, p.7).
exocyclic (ex-o-cy'-clic) Said of an *irregular* echinoid in which the *periproct* is located outside of the *oculogenital ring.* Ant: *endocyclic.*
exodiagenesis (ex'-o-di'-a-gen'-e-sis) A term used by Shvetsov (1960) for diagenesis in subaerial environments and shallow stable seas. It is characterized by dehydration, coagulation of colloids, rapid growth of crystals (recrystallization), formation of concretions, and preservation of textural properties of sediments.
exogastric (ex-o-gas'-tric) (a) Said of a cephalopod shell that is coiled so that the venter is on or near the outer or convex side of the whorls. (b) Said of a gastropod shell that is coiled so as to extend forward from the aperture over

the front of the extruded head-foot mass, as in the earliest forms (TIP, 1960, pt. I, p.130). Cf: *endogastric.*

exogene effect (ex'-o-gene) The effect of an igneous mass on the rock that it invades (Bateman, 1959). Cf: *endogene effect; contact metamorphism.*

exogenetic (ex'-o-ge-net'-ic) (a) Said of processes originating at or near the surface of the earth, such as weathering and denudation, and of rocks, ore deposits, and landforms that owe their origin to such processes. Cf: *endogenetic.* Syn: *exogenic; exogenous.* (b) Said of energy sources and objects of extraterrestrial origin, as solar radiation, cosmic rays, meteorites, and cosmic dust.

exogenic (ex-o-gen'-ic) *exogenetic.*

exogenic explosion Forceful ejection of material from beneath a volcano through a vent on its surface (Fink and Anderson, 2000).

exogenite (ex-og'-e-nite) A little-used term for an epigenetic mineral deposit differing in composition from the enclosing rock.

exogenous (ex-og'-e-nous) *exogenetic.*

exogenous dome A volcanic dome that is built by surface effusion of viscous lava, usually from a central vent or crater (Williams, 1932). Cf: *endogenous dome.*

exogeosyncline (ex'-o-ge'-o-syn'-cline) A *parageosyncline* accumulating clastic sediments from the uplifted orthogeosynclinal belt adjacent to it but outside the craton (Kay, 1947, p.1289-1293); an obsolete term for a *foreland basin.*. Syn: *foredeep; transverse basin.* Cf: *secondary geosyncline.*

exokarst (ex'-o-karst) All features found on the surface of a karst landscape, ranging in size from tiny *karren* forms to extensive *poljes.* Cf: *endokarst.*

exokinematic (ex'-o-kin'-e-mat'-ic) Said of sedimentary operations in which "the largest displacement vectors occur between matter outside the deposit and the unmodified deposit surrounding the structure produced" (Elliott, 1965, p.196); e.g. types of streamflow. Also, said of the sedimentary structures produced by exokinematic operations. Cf: *endokinematic.*

exokinetic (ex'-o-ki-net'-ic) Said of a fissure in a rock that is the result of strain in an adjacent rock unit. Archaic. Cf: *endokinetic.*

exometamorphism (ex'-o-met'-a-mor'-phism) *exomorphism.*

exomorphic metamorphism (ex-o-mor'-phic) *exomorphism.*

exomorphic zone *aureole.*

exomorphism (ex-o-mor'-phism) Changes in country rock produced by the intense heat and other properties of magma or lava in contact with them; *contact metamorphism* in the usual sense. The term was originated by Fournet in 1867. Cf: *endomorphism.* Syn: *exometamorphism; exomorphic metamorphism.*

exoolitic (ex'-o-o-lit'-ic) *extoolitic.*

exopinacoderm (ex'-o-pi-nac'-o-derm) The *pinacoderm* covering the free surface of a sponge. Cf: *endopinacoderm.*

exopod (ex'-o-pod) The lateral or external *ramus* of a limb of a crustacean, arising from the *basis.* Cf: *endopod.*

exopore [paleont] A pore in the outer wall of a chamber in a sphinctozoan sponge. Cf: *endopore [paleont].*

exopore [palyn] The external opening in the ektexine of a pollen grain with a complex porate structure. See also: *vestibulum.* Cf: *endopore [palyn].*

exopuncta (ex-o-punc'-ta) A *puncta* of the external shell surface of a brachiopod, commonly restricted to the primary layer, and never penetrating to the internal surface. Cf: *endopuncta.* Pl: exopunctae.

exorheic (ex-o-rhe'-ic) Said of a basin or region characterized by external drainage; relating to *exorheism.*

exorheism (ex-o-rhe'-ism) (a) *external drainage.* (b) The condition of a region in which its water reaches the ocean directly or indirectly. Ant: *endorheism.*

exoseptum (ex-o-sep'-tum) A scleractinian-coral septum developed within an *exocoele.* Cf: *entoseptum.*

exoskarn (ex-o-skarn') That portion of *skarn* mineralization that involves metamorphic and metasomatic effects on the country-rock (nonintrusive) side of the contact between an intrusive body and the country rock. Cf: *endoskarn; exomorphism.*

exoskeleton (ex-o-skel'-e-ton) A supportive and protective framework that lies outside the body tissues of an animal, forming an external covering; e.g., a shell. Ant: *endoskeleton.* Occasionally applied in vertebrates to parts of the dermal skeleton that, in life, are covered principally by epidermal integument. Syn: *dermoskeleton.*

exosphere (ex'-o-sphere) The outermost portion of a planet's atmosphere, above which the air density is so low that an upward traveling molecule is unlikely to suffer a collision, and if traveling fast enough would escape from the planet. For a planet the size of the Earth, this leads to a slow preferential escape from the atmosphere of lighter, faster moving molecules such as hydrogen and helium. Its lower boundary for the Earth is estimated at 500 to 1,000 km above the surface. It is also called the "region of escape".

exospore (ex'-o-spore) (a) The outer layer of sporoderm of spores. The probably homologous layer of pollen grains is the *exine.* Cf: *endospore; perispore.* (b) One of the asexual spores formed by abstriction from a parent cell (as in certain fungi). (c) One of the spores formed from above downward and one at a time in certain cyanobacteria. Cf: *endospore.*

exostome (ex'-o-stome) The outer peristome of a protist.

exotheca (ex-o-the'-ca) A collective term for the dissepiments outside the wall of a scleractinian corallite. Cf: *endotheca.*

exothermic (ex-o-ther'-mic) Pertaining to a chemical reaction that occurs with a liberation of heat. Cf: *endothermic.*

exotic [ecol] (ex-ot'-ic) Said of an organism that has been introduced into a new area from an area where it grew naturally. Ant: *indigenous.*

exotic [struc geol] Applied to a boulder, block, or larger rock body unrelated to the rocks with which it is now associated, which has been moved from its place of origin by one of several processes. Exotic masses of tectonic origin are also *allochthonous;* those of glacial or ice-rafted origin are generally called erratics.

exotic block A mass of rock occurring in a lithologic association foreign to that in which the mass formed (Berkland et al., 1972, p.2296).

exotic copper A type of secondary copper deposit formed by the lateral migration of copper leached from a nearby primary copper deposit.

exotic limonite Limonite precipitated in rock that did not formerly contain any iron-bearing sulfide. Cf: *indigenous limonite.*

exotic marker An easily recognized spore (*Lycopodium*) or pollen (*Eucalyptus globosus*), of which a fixed number is added during the preparation of a known amount of sample, so that the relative abundance of palynomorphs obtained from that sample can be determined.

exotic stream A stream that derives much of its waters from a drainage system in another region; e.g. a stream that has its source in a humid or well-weathered area but that flows across a desert before reaching the sea. Example: the Nile. Cf: *indigenous stream.*

exotic terrane A terrane that has undergone significant motion with respect to the stable craton to which it is accreted (McWilliams and Howell, 1982).

exotomous (ex-ot'-o-mous) Characterized by bifurcation in two main crinoid arms that give off branches only on their outer sides. Ant: *endotomous.*

exozone (ex'-o-zone) The outer parts of a stenolaemate bryozoan colony, characterized by thick vertical walls, concentrations of intrazooidal skeletal structures, and zooidal growth directions at high angles to the colony growth directions or surfaces. Cf: *endozone.* Syn: *mature region.*

expandables Used generically for clay minerals that incorporate polar liquids in their interlayer space causing an increase in the repeat distance perpendicular to the 001 surface compared to this dimension when liquids are removed by drying. The distinction between smectite and vermiculite cannot always be made, therefore the term expandables can be used. Cf: *smectite; vermiculite.*

expanded foot (ex-pand'-ed) A broad, bulb-shaped lobelike or fanlike mass of ice formed where the lower part of a valley glacier leaves its confining walls and extends onto an adjacent lowland at the foot of a mountain slope. Syn: *piedmont bulb.* Less preferred syn: *glacier bulb; ice fan.*

expanded-foot glacier A small *piedmont glacier* consisting of an expanded foot. Syn: *bulb glacier; foot glacier.*

expanding Earth (ex-pand'-ing) A hypothesis favored by a small minority of geologists, that the diameter of the Earth has grown larger, perhaps by a third or more during recorded geologic time, as a result of changes in atomic and molecular structure in the core and lower mantle, without change in actual mass. The hypothesis has been invoked to explain sea-floor spreading and continental drift without requiring subduction. Cf: *contracting Earth.*

expanding spread (a) A *spread* moved to greater offsets for successive energy releases from the same location, to give the equivalent of recording more geophone groups from a single source. Used in refraction work, in noise analysis, and in velocity analysis. (b) *Depth probe [geophys exploration].*

expanding-Universe hypothesis (ex-pand'-ing-U'-ni-verse) Interpretation of the universally extragalactic red shift as a Doppler shift, suggesting that the Universe is expanding at a uniform rate.

expansion breccia (ex-pan'-sion) A breccia formed by increase of volume due to chemical change, as by recrystallization or by hydration (Norton, 1917, p.191).

expansion fissure In petrology, one of a system of fissures that radiate irregularly through feldspar and other minerals adjacent to olivine crystals that have been replaced by serpentine. The alteration of olivine to serpentine involves considerable increase in volume, and the stresses so produced are relieved by the fissuring of the surrounding minerals. This phenomenon is common in norite and gabbro (Tyrrell, 1950).

expansion wave Perturbation in fluid flow that propagates at the local sound speed while causing the flow to accelerate, expand in volume, and decrease in density and pressure.

expansive (ex-pan'-sive) In engineering geology, said of soil and rock materials that swell in volume upon application of moisture and shrink upon drying. Expansive soils and rocks typically contain smectite clay minerals that are able to incorporate water molecules into their crystal structure; some contain complex sulfate salts that expand and contract in response to temperature changes, similar to *frost heave* in certain silty soil materials.

expectation (ex-pec-ta'-tion) If a process leads to a random variable or function of a random variable, the expectation of the process is the average on N outcomes as N tends toward infinity. The average of all possible outcomes weighted by their probabilities of occurrence.

expected value The long-term mean of a very large number of outcomes.

expendable bathythermograph An instrument that measures temperature as a function of depth in the upper layers of the ocean down to depths of 1,800 m, using a thermistor in the nose of a streamlined weight falling through the water. The weight is linked to a recorder on a ship by a fine copper wire spooled out by the weight and the ship. Introduced in the 1960s. Abbrev: *XBT.*

experimental petrology (ex-per'-i-men'-tal) A branch of petrology dealing with the laboratory study of reactions designed to elucidate rock-forming processes. The term includes experiments relating to the physical properties or physical chemistry of minerals, rocks, rock melts, vapors, gases, or solutions coexisting with solid or molten rock materials.

experimental semivariogram A graph of *semivariance* versus *lag* or distance, calculated from observations.

experimental structural geology The study of the deformation of samples of rock under various conditions of temperature, pressure, strain rate, and pore-fluid pressure. Also the study of deformation using *scaled physical models* or the construction of dynamic models that illustrate structural processes.

expert system A computer system that can perform tasks at or near the level of a human expert. An expert system uses rule-based logic to simulate human intelligence. A part of *artificial intelligence.* Also called: knowledge-based system.

explanate (ex'-pla-nate) Said of the flattened growth habit of erect bryozoan colonies.

explanation (ex-pla-na'-tion) A term used by the U.S. Geological Survey in preference to *legend* or *key [cart]* (except on international maps).

exploding-bomb texture (ex-plod'-ing) In mineral deposits, a pattern of pyrite replacement by copper sulfides, usually chalcopyrite-bornite, in which scattered, residual pyrite fragments are surrounded by the copper minerals in a manner suggesting a time-lapse photograph of an exploding bomb. Var: exploded-bomb texture.

exploration (ex-plo-ra'-tion) (a) The search for deposits of useful minerals or fossil fuels; *prospecting.* It may include geologic reconnaissance, e.g. remote sensing, photogeology, geophysical and geochemical methods, and both surface and underground investigations. (b) Establishing the nature of a known mineral deposit, preparatory to development. In the sense that exploration goes beyond discovery, it is a broader term than prospecting.

exploration geologist A geologist who is involved in the identification and location of economic mineral deposits. Typically referring to an employee of a large corporation or company in contrast to a *prospector.*

exploration seismology *applied seismology.*

exploratory well (ex-plor'-a-tor'-y) A well drilled to an unexplored depth or in unproven territory, either in search of a new pool of oil or gas or with the expectation of greatly extending the known limits of a field already partly developed. Cf: *development well.* Syn: *test well [grd wat]; wildcat well.*

explorer's alidade (ex-plor'-er's) A lightweight, compact alidade with a low pillar and a reflecting prism through which the ocular may be viewed from above. Syn: *Gale alidade.*

explosion breccia (ex-plo'-sion) A type of volcanic breccia that is formed by a volcanic explosion. Syn: *eruption breccia; pyroclastic breccia.*

explosion caldera A type of *caldera* that is formed by explosive removal of the upper part of a volcanic cone. It is extremely rare, and is small in size. Cf: *collapse caldera; erosion caldera.*

explosion cloud *eruption cloud.*

explosion crater (a) A saucer-shaped to conical crater produced experimentally by detonation of a nuclear device or a chemical explosive. (b) A volcanic crater, e.g. a *maar.* Incorrectly used to describe some impact craters.

explosion tuff A tuff whose pyroclastic fragments are in the place in which they fell, rather than having been washed into place after they landed.

explosive bomb (ex-plo'-sive) A *bread-crust bomb* that throws off fragments of its crust due to continued expansion of its interior after solidification of its crust.

explosive eruption An eruption or eruption phase that is characterized by the energetic ejection of pyroclastic material. Cf: *lava eruption.*

explosive evolution (a) Within a group or lineage of organisms, morphologic or ecologic change at an extremely rapid rate compared to the usual or normal rate. (b) Sometimes used to denote *adaptive radiation.*

explosive radiation An *adaptive radiation* that appears to have occurred very rapidly.

exponential distribution (ex-po-nen'-tial) A *frequency distribution* of the form $F(\chi)=1-e^{-\chi/\beta}$, in which β is the mean. A graph of the exponential distribution has a high initial value that declines with increasing χ. Used to model the time between events or the distances between locations.

exponentiate (ex-po-nen'-ti-ate) To use an empirically derived exponential time-dependent gain in seismic playback.

exposed coalfield (ex-posed') Deposits of coal that crop out at the surface, as along the rim of a coal basin. Cf: *concealed coalfield; crop coal.*

exposure [geol] (ex-po'-sure) (a) An area of a rock formation or geologic structure that is visible ("hammerable"), either naturally or artificially, i.e. is unobscured by soil, vegetation, water, or the works of man; also, the condition of being exposed to view at the Earth's surface. Cf: *outcrop.* (b) The nature and degree of openness of a slope or place to wind, sunlight, weather, oceanic influences, etc. The term is sometimes regarded as a syn. of *aspect [slopes].*

exposure [photo] (a) The total quantity of light received per unit area on a sensitized plate or film, usually expressed as the product of the light intensity and the time during which the light-sensitive material is subjected to the action of light. (b) A loosely used term "generally understood to mean the length of time during which light is allowed to act on a sensitive surface" (Smith, 1968, p.496). (c) The act of exposing a light-sensitive material to a light source. (d) An individual picture of a strip of photographs.

exposure interval The time interval between the taking of successive photographs.

exsert (ex-sert') adj. Projecting beyond an enclosing part or organ, such as having the ocular plates of an echinoid not in contact with the periproctal margin. Ant: *insert.*

exsiccation (ex-sic-ca'-tion) The drying-up of an area due to a change that drives out, or decreases the amount of, available moisture without reducing appreciably the average rainfall; e.g. the draining of a swamp or marsh, or the migration of sand dunes across cultivated ground. Cf: *desiccation.*

exsolution (ex-so-lu'-tion) The process whereby an initially homogeneous solid solution separates into two (or possibly more) distinct crystalline phases without addition or removal of material, i.e., without change in the bulk composition. It generally, though not necessarily, occurs on cooling. Syn: *unmixing [chem].*

exsolutional (ex-so-lu'-tion-al) Pertaining to sediments or sedimentary rocks that solidified from solution either by precipitation or by secretion.

exsolution lamella A *lamella* produced by *exsolution,* such as a lamella of pigeonite in orthopyroxene or cummingtonite in hornblende.

exsolution texture In mineral deposits, a general term for the texture of any mineral aggregate or intergrowth formed by exsolution. It is generally fairly homogeneous, ranging from perthitic to geometrically regular. See also: *eutectic texture.*

exsolve To separate or unmix from a solid solution into two distinct phases by *exsolution.*

exsudation (ex-su-da'-tion) A kind of *salt weathering* by which rock surfaces are scaled off owing to growth of salines by capillary action (Thornbury, 1954, p.39).

exsurgence (ex-sur'-gence) The rising of a stream from a cave, the water having entered the cave by downward percolation through the overlying limestone. Cf: *karst spring; resurgence.*

extended consequent stream (ex-tend'-ed) A consequent stream that flows seaward across a newly emerged coastal plain and that forms an extension of an earlier, larger stream with headwaters in the older land behind the coastal plain.

Extended Core Barrel A special core barrel used in the Ocean Drilling Program to recover core samples from soft to moderately hard formation. Typically deployed when the formation becomes too stiff for piston coring but is not hard enough to permit efficient recovery with the Rotary Core Barrel. The XCB cutting shoe extends ahead of the main bit in soft sediments but retracts into the main bit as the weight on bit increases when firm lithologies are encountered.

extended reach well A well with a long horizontal distance from the surface location to the bottom of the well.

extended stream A stream lengthened by the extension of its course downstream across newly emerged land (such as a coastal plain, a delta, or a plain of glacial deposition).

extended succession A relatively thick and uninterrupted stratigraphic succession in which the deposits accumulated rapidly. Ant: *condensed succession.*

extending flow (ex-tend'-ing) A flow pattern on glaciers in which the velocity increases with distance downstream; thus the longitudinal strain rate (velocity gradient) is extending. This condition requires a transverse or vertical compression to maintain the continuity of (Nye, 1952). Ant: *compressing flow.*

extensiform (ex-ten'-si-form) Said of a graptoloid (such as *Didymograptus*) with two stipes that are horizontal.

extension [exp struc geol] (ex-ten'-sion) (a) A strain term signifying increase in length. Cf: *tension,* which is a stress term. (b) A measure of the change in length of a line, specif. the ratio of the change in length to the original length. Cf: *stretch [exp struc geol].* Syn: *elongation.*

extension [streams] The lengthening of a stream by headward erosion and the multiplication of tributaries, or by regression of the sea or uplift of the coastal area.

extensional bend A spatial variation in the orientation of the fault plane that causes local extension in the wall rocks as they are displaced around the bend (Peacock et al., 2000). Cf: *divergent bend; releasing bend.*

extensional collapse The phenomenon that occurs in an orogenic belt when crustal thickening leads to a condition where the mid-crustal region becomes warm and soft such that the upper crust collapses under its own weight and develops extensional strain. Syn: orogenic collapse.

extensional overstep An *overstep* that has to be accommodated by extension, e.g., a left-step between left-lateral faults. This type of overstep occurs in cross section along normal faults, with the lower fault overstepping or projecting into the footwall of the upper fault (Peacock et al., 2000). Cf: *contractional overstep.*

extension fault A little-used term for a fault in sedimentary rocks along which there has been bed-parallel elongation (Norris, 1958, 1964), giving rise to tectonic thinning.

extension fracture A fracture that develops perpendicular to the direction of greatest tensile stress and parallel to the direction of compression; a *tension fracture.* This term is often reserved for brittle fracture during laboratory experiments. Cf: *axial splitting crack.* See also: *extension joint; tension crack; joint [struc geol].*

extension joint A joint that forms parallel to the direction of compression during a laboratory experiment; a joint that is an *extension fracture.*

extension ore The *possible ore* ahead of or beyond an exposure in a mine. Cf: *probable ore.*

extension test *triaxial extension test.*

extension vein An *extension fracture* filled by secondary minerals.

extension well Any well located as an outpost well or as a wildcat well that extends the productive area of a pool. The term cannot logically be applied until after the fact is demonstrated (Lahee, 1962, p.134). Cf: *step-out.*

extensiveness (ex-ten'-sive-ness) In quantification of hydrothermal alteration, the degree to which susceptible minerals are converted to alteration phases; the amount of alteration in a host rock, regardless of mineralogy or distribution. It is analogous to extensive parameters in physical chemistry. Cf: *intensiveness; pervasiveness.*

extensive quantity (ex-ten'-sive) A thermodynamic variable such as enthalpy or mass that depends on the total quantity of matter in the system.

extensometer (ex-ten-som'-e-ter) (a) An instrument used for measuring small deformations, as in tests of stress. (b) A field instrument used for measuring the compaction of a soil column between the land surface and a prescribed depth.

extent A rectangle bounding a map, the size of which being determined by the minimum and maximum map coordinates (Krzanowski et al., 1993). Cf: *neat line.*

exterior link A link in a channel network emanating from a source; equivalent to a first-order stream segment of Strahler (Shreve, 1967). Cf: *link [geomorph]; stream order; interior link.*

exterior wall A body wall in bryozoans that extends the body of a *zooid* and of the colony. It includes an outermost cuticular layer.

extermination (ex-ter'-mi-na'-tion) The local or even regional disappearance of a species that still exists elsewhere, as a result of changing environmental conditions, disease, competitors, or other adverse conditions. Cf: *extinction [evol].*

external cast (ex-ter'-nal) An improper term sometimes used as a syn. of *external mold.* Syn: *exocast.*

external contact The planar or irregular surface between a pluton and the country rock (Compton, 1962, p.277). Obsolete. Cf: *internal contact.*

external drainage Drainage whereby the water reaches the ocean directly or indirectly. Ant: *internal drainage.* Syn: *exorheism.*

external drift Use of a regionalized variable (e.g., a seismic attribute) to control the shape of a map created by *kriging* or simulation. A geostatistical linear regression technique that uses a spatial model of covariance.

external furrow One of the shallow, linear axial depressions or grooves on the outer surface of a fusulinid test, formed at the point of union between successive chambers and corresponding in position to a septum, and dividing the outer surface into melonlike lobes. Syn: *septal furrow.*

external lobe The *ventral lobe* in normally coiled cephalopod conchs.

external magnetic field The component of the *Earth's magnetic field* originating from an interaction of the Earth's internal (core) magnetic field with *solar wind* coupled with the Earth's rotation, tidal forces, and thermal effects. It is the main cause of *daily variation* that can strongly affect the magnetic measurements.

external mold A mold or impression in the surrounding earth or rock, showing the surface form and markings of the outer hard parts of a fossil shell or other organic structure; also, the surrounding rock material whose surface receives the external mold. Cf: *internal mold; external cast.*

external rotation A change in orientation of structural features during deformation referred to coordinate axes external to the deformed body. Cf: *internal rotation.*

external suture The part of a suture of a coiled cephalopod *conch* exposed on the outside of whorls between the umbilical seams. Cf: *internal suture.*

externides (ex-ter'-ni-des) Kober's term for the outer part of an orogenic belt, nearest to the craton or foreland, commonly the site of a continental shelf during early stages and of a foredeep basin during late stages. It is usually subjected to marginal deformation (folding and lateral thrusting) during the orogenic phase. Cf: *internides; secondary arc.*

extinction [evol] (ex-tinc'-tion) The total disappearance of a species or higher taxon, so that it no longer exists anywhere. Cf: *extermination.*

extinction [lake] The disappearance of a lake, by drying up or by destruction of the lake basin. Cf: *dying lake.*

extinction [optics] The more or less complete darkness obtained in a *birefringent* mineral at two positions during a complete rotation of a section between *crossed nicols.* Also, the darkness that persists for a complete rotation if the line of sight is parallel to the optic axis. See also: *extinction angle; extinction direction; inclined extinction; parallel extinction; undulatory extinction.*

extinction angle The angle through which a section of a *birefringent* mineral must be rotated from a known crystallographic plane or direction to the position at which it gives maximum extinction or darkness under a polarizing microscope. The extinction angle can be diagnostic in the identification of a mineral.

extinction coefficient *attenuation coefficient [oceanog].*

extinction direction One of the two positions at which a section of a birefringent crystal shows extinction between *crossed nicols.*

extinct lake (ex-tinct') Part of the terminology in an obsolete concept of a lake's life history, a *dying lake.* Cf: *dead lake.*

extinct radionuclide A radionuclide that was present when a rock or meteorite was formed, but whose half-life is far smaller than the intervening period, with the result that no measurable amount of this nuclide is still present. Its former presence is indicated by an excess in the abundance of the daughter isotope. Cf: *isotopic anomaly.*

extinct volcano A volcano that is not now erupting and that is not considered likely to erupt in the future. Cf: *active volcano; dormant volcano; inactive volcano.*

extine (ex'-tine) Var. of *exine.* The term is not in good usage in palynology.

extoolitic (ex'-to-o-lit'-ic) Pertaining to oolitic structures or grains that have formed or grown outward by deposition of material around a core or center, as in the formation of a small concretion. Ant: *entoolitic.* Syn: *exoolitic.*

extraclast (ex'-tra-clast) A fragment of calcareous sedimentary material, produced by erosion of an older rock outside the area in which it accumulated; a component of calclithite (Wolf, 1965). Cf: *intraclast.*

extractive metallurgy The separation of a desired metal from its ore by such means as *electrometallurgy, hydrometallurgy,* and *pyrometallurgy.*

extragalactic nebula (ex'-tra-ga-lac'-tic) *galaxy.*

extraglacial (ex-tra-gla'-cial) Said of glacial deposits formed by meltwater beyond the farthest limit of the ice, or of glacial phenomena displayed in an area never covered by ice. Ant: *intraglacial.*

extramorainal (ex'-tra-mo-rain'-al) Said of deposits and phenomena occurring outside the area occupied by a glacier and its lateral and end moraines. Ant: *intramorainal.*

extraneous argon (ex-tra'-ne-ous) *Inherited argon* and *excess argon* taken collectively.

extraneous ash Ash in coal that is derived from inorganic material introduced during formation of the seam, such as sedimentary particles, or filling cracks in the coal. Cf: *inherent ash.* Syn: *secondary ash; sedimentary ash.*

extraordinary ray (ex-traor'-di-nar'-y) *E ray [cryst].*

extrapolation (ex-trap'-o-la'-tion) The estimation of the value of a function outside its sampling domain. If a variate y is considered a function of one or more variables X_i, extrapolation is the estimation of y beyond the range(s) of X_i. The extension of a curve beyond the data on a graph, or of a contoured surface beyond the outermost control points on a map (Krzanowski et al., 1993). Cf: *interpolation.*

extratentacular budding (ex'-tra-ten-tac'-u-lar) Formation of new scleractinian coral polyps by invagination of the *edge zone* or *coenosarc* outside of the ring of tentacles surrounding the mouth of the parent. Cf: *intratentacular budding.*

extraterrestrial (ex'-tra-ter-res'-tri-al) Existing, occurring, or originating beyond the Earth or its atmosphere; e.g. "extraterrestrial radiation" or solar radiation received "on top of" the Earth's atmosphere.

extra-terrestrial impact event The collision of a large (R>~1 km) extraterrestrial object, sometimes termed a bolide, with the Earth. The identification of the element iridium in a thin clay layer at the Cretaceous/Tertiary boundary led to suggestions that at least some mass extinctions were caused by large extra-terrestrial impacts.

extratropical cyclone A low-pressure system that forms in mid-latitudes, typically along a zone of horizontal temperature contrast which imparts a frontal structure to the system. These systems are typically migratory. Cf: *tropical cyclone.*

extraumbilical aperture (ex'-tra-um-bil'-i-cal) An aperture in the final chamber of a foraminiferal test not connecting with the umbilicus. It is commonly sutural midway between the umbilicus and the periphery. See also: *interiomarginal aperture.*

extraumbilical-umbilical aperture (ex'-tra-um-bil'-i-cal-um-bil'-i-cal) An aperture in the final chamber of a foraminiferal test that extends along its forward margin from the umbilicus toward the periphery, thus reaching a point outside the umbilicus (as in *Globorotalia*).

extravasation (ex'-tra-va-sa'-tion) The eruption of molten or liquid material, e.g. lava, or water from a geyser, onto the surface of the Earth.

extrazooidal skeleton (ex'-tra-zo-oid'-al) A protective or supportive colony structure in bryozoans which, once developed, remains outside zooidal boundaries throughout the life of the colony.

extreme desert An area without a regular season of rainfall and where 12 consecutive months without rainfall have been recorded.

extremely coarsely crystalline (ex-treme'-ly) Descriptive of an interlocking texture of a carbonate sedimentary rock having crystals whose diameters exceed 4 mm (Folk, 1959).

extremely finely crystalline *nannocrystalline.*

extrusion (ex-tru'-sion) The emission of lava onto the Earth's surface; also, the rock so formed. Cf: *effusion.*

extrusion flow A concept in a discredited hypothesis for a type of glacier flow in which the pressure of overlying ice was supposed to force the basal part of the glacier to flow faster than the upper part. *Gravity flow,* originally contrasted with extrusion flow, is now considered to account for all glacier flow.

extrusive (ex-tru'-sive) adj. Said of igneous rock that has been erupted onto the surface of the Earth. Extrusive rocks include lava flows and pyroclastic material such as volcanic ash. n. An extrusive rock. Cf: *intrusive.* Syn: *effusive; volcanic; eruptive.*

extrusive ice Ice formed subaerially from water that emerges from the ground or below other ice, as in an icing.

exudation basin (ex-u-da'-tion) A spoon-shaped depression on the ice surface at the head of an *outlet glacier.* Examples are found on the Greenland and Antarctic ice sheets.

exudation vein *segregated vein.*

exuvia (ex-u'-vi-a) All or part of the exoskeleton of an arthropod that has been shed, molted, or cast off. Pl: exuviae.

exuviation (ex-u'-vi-a'-tion) The removal of the theca of a dinoflagellate, either plate by plate or as small groups of plates.

eye [meteorol] The approximately circular area of relatively light wind and fair weather in the center of a *tropical cyclone.* Such an area may be 10 to 60 km in diameter.

eye [paleont] The ringlike part of a hook of a holothurian, sometimes partly closed by a bar. Also, a ringlike end of a rod of a holothurian.

eye agate Agate displaying concentric bands, usually of various colors, around a dark center, suggesting an eye. Also called: aleppo stone; eyed agated.

eyebrow scarp (eye'-brow) A fault scarp that crosses a piedmont alluvial fan near its apex and that seldom maintains the dip of the fault surface in the unconsolidated gravels of the fan (Davis, 1927, p.62).

eye coal Coal that contains structural disks in circular or elliptical shapes, either parallel or normal to the bedding, with concentric, bending rims and radiating striae. They reflect light in a mirrorlike way. Syn: *circular coal.* Etymol: German "Augenkohle", "eye coal".

eyed structure *augen structure.*

eyepiece (eye'-piece) The lens (or lenses) in a microscope or telescope through which the image formed by the *objective* is viewed. Syn: *ocular [optics].*

eye ridge A raised line or narrow band extending from the forward and inner part of a trilobite eye to the anterior part of the *glabella.*

eyestalk (eye'-stalk) One of the movable peduncles in a decapod crustacean, carrying the eye at its distal extremity. Syn: *ophthalmite [paleont].*

eye tubercle A polished, transparent, rounded protuberance in the anterior and dorsal region of an ostracode valve, and on the cheek regions of some trilobites, forming the lens of the eye. See also: *tubercle.*

eylettersite (ey-let'-ter-site) A white to creamy rhombohedral member of the *crandallite* group: $(Th,Pb)_{1-x}Al_3(PO_4,SiO_4)_2(OH)_6$(?).

eyot (ey'-ot) *ait.*

ezcurrite (ez-cur'-rite) A colorless triclinic mineral: $Na_2B_5O_7(OH)_3 \cdot 2H_2O$.

E-zone With respect to coastal erosion, the zone that is expected to erode within a particular time period (Keller, 2002).

eztlite (ez'-tl-ite) A brilliant red monoclinic mineral: $Pb_2Fe^{3+}_6(Te^{4+}O_3)_3(Te^{6+}O_6)(OH)_{10} \cdot 8H_2O$.

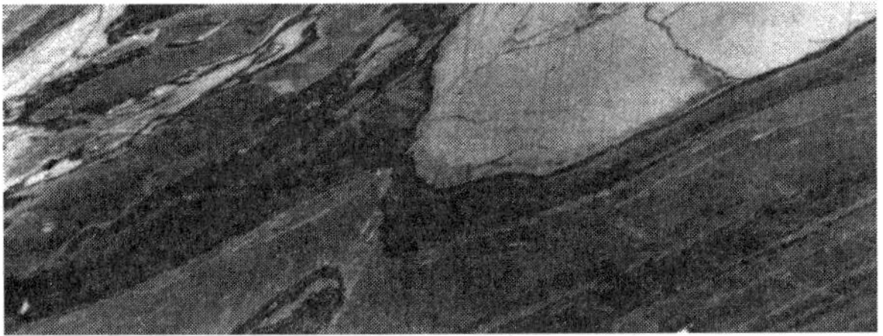

F

fabianite (fa'-bi-an-ite') A colorless monoclinic mineral: $CaB_3O_5(OH)$.
fabric [sed] (fab'-ric) The orientation (or lack of it) of discrete particles, crystals and cement of which a sedimentary rock is composed. Cf: *packing.* See also: *crystallization fabric.*
fabric [soil] The physical nature of a soil according to the spatial arrangement of its particles and voids.
fabric [struc geol] The complete spatial and geometrical configuration of all those components that make up a deformed rock. It covers such terms as texture, structure, and preferred orientation, and so is an all-encompassing term that describes the shapes and characters of individual parts of a rock mass and the manner in which these parts are distributed and oriented in space. The term was first defined by Sander (1930), who used the German word "Gefüge". Syn: *petrofabric.*
fabric analysis A term usually used as a synonym for *structural petrology.* In some instances, however, it may refer to the geometrical part of the much broader study of structural analysis.
fabric axis One of three orthogonal axes used in structural petrology as references in the orientation of fabric elements, and in the description of folding and of the movement symmetry of deformed rocks. Cf: *a* axis [struc petrol]; *b* axis [struc petrol]; *c* axis [struc petrol]. Syn: *reference axis; tectonic axis.*
fabric diagram In structural petrology, a stereographic or equal-area projection of fabric elements; an *orientation diagram.* See also: *point diagram; contour diagram.* Syn: *petrofabric diagram.*
fabric domain In a deformed rock, a volume defined by boundaries such as structural or compositional discontinuities, within which the fabric is more or less homogeneous. Syn: *domain [struc petrol].*
fabric element A component of a rock fabric, such as fold axes, foliation, lineation, or crystallographic axes. See also: *fabric [sed]; fabric [struc geol]; subfabric.*
face [cryst] *crystal face.*
face [geomorph] The principal side or surface of a landform; esp. *rock face.*
face [mining] Any surface on which mining operations are in progress. Syn: *breast; highwall.*
face [struc geol] n. A term used by Shrock (1948, p.17-18) for the original top or upper surface of a layer of rock, esp. if it has been raised to a vertical or steeply inclined position. v. To be directed toward or to present an aspect of. Sedimentary beds are said "to face" in the direction of the stratigraphic top of the succession. Folds are said "to face" in the direction of the stratigraphically younger rocks along their axial surfaces and normal to their axes (Shackleton, 1958, p.363). Cf: *vergence.*
face-centered lattice (face'-cen-tered) A type of *centered lattice* in which each unit cell has a lattice point at the center of each face as well as those at each corner, i.e. it has four lattice points per unit cell. Syn: *F-centered lattice.*
face cleat The major systematic joint in a coal seam. Cf: *end cleat.* See also: *face of coal.*
facellite (fa-cel'-lite) *kaliophilite.* Also spelled: *phacellite.*
face of coal The plane or surface of the coal in situ at the advancing end of the working face.
face pole In crystallography, a point on a projection surface that represents the intersection of a crystal pole with the crystal face.
facet [gem] (fac'-et) One of the plane polished surfaces on a finished gemstone.
facet [geog] Any part of a landscape defined as a unit for geographic study on the basis of homogeneous topography.
facet [geomorph] (a) A nearly plane surface produced on a rock fragment by abrasion, as by wind sandblasting, by the grinding action of a glacier, or by a stream that differentially removes material from the upstream side of a boulder or pebble; inclined at an angle of 50 degrees or less to the direction of the impinging current (Maxson, 1940, p.721). (b) Any plane surface produced by erosion or faulting, and intersecting a general slope of the land; e.g. a *triangular facet.*
facet [paleont] (a) A nearly flat surface on an echinoderm plate, serving for articulation with contiguous skeletal elements; e.g. *brachiolar facet.* (b) A small, circular or oval protuberance set within a channel of a cyclocystoid and connected with the *water-vascular system.* (c) A small, flat surface developed in trilobites on the anterolateral region of the *pygidium* and thoracic pleurae.
facet [speleo] *scallop [speleo].*
faceted boulder (fac'-et-ed) A boulder that has been ground flat on one or more sides by the action of natural agents, such as by glacier ice, streams, or wind. Cf: *faceted pebble.*
faceted form *kinetic growth form.*
faceted pebble A pebble on which facets have been developed by natural agents, such as by wave erosion on a beach or by the grinding action of a glacier; specif. a *windkanter.* Cf: *faceted boulder.*
faceted spur A spur or ridge with an inverted-V face that was produced by faulting or by the trimming, beveling, or truncating action of streams, waves, or glaciers. See also: *truncated spur.*
facial (fa'-cial) (a) Pertaining to facies. The term is used by some non-English European geologists. (b) Pertaining to an outer surface.
facial suture One of the two symmetrical sutures that open at the time of molting of a trilobite, extending from the anterior margin of the *cephalon* around the eye and outward or backward to the lateral or posterior margin. It separates the *free cheek* from the *fixed cheek.*
faciation (fa-ci-a'-tion) Part of an ecologic association, usually a large community, which is characterized by the dominance of two or more but not all of the more abundant organisms.
facies (fa'-cies) (a) The aspect, appearance, and characteristics of a rock unit, usually reflecting the conditions of its origin; esp. as differentiating the unit from adjacent or associated units. Cf: *stratigraphic facies; lithofacies; igneous facies.* (b) A mappable, areally restricted part of a lithostratigraphic body, differing in lithology or fossil content from other beds deposited at the same time and in lithologic continuity. Cf: *sedimentary facies.* (c) A distinctive rock type, broadly corresponding to a certain environment or mode of origin, e.g. "red-bed facies", "black-shale facies". Cf: *petrographic facies.* (d) A body of rock distinguished on the basis of its fossil content. Cf: *biofacies (a), (b).* (e) A local assemblage or association of living or fossil organisms, esp. one characteristic of some type of marine conditions. Cf: *biofacies (c).* (f) The environment or area in which a rock was formed, e.g. "sandy-bottom facies", "eolian facies", "volcanic facies". Cf: *environmental facies.* (g) Rocks broadly defined on a paleogeographic or paleotectonic basis. Cf: *geosynclinal facies; shelf facies.* (h) Rocks of any origin formed within certain pressure-temperature conditions. Cf: *mineral facies; metamorphic facies.* The concept of stratigraphic facies was first defined by Gressly (1836; 1838, p.10-12, 20-25) as the different lithologic and biologic aspects of rocks of essentially the same age. The different lithologic aspects may be termed lithofacies, the biologic aspects, biofacies (Steinker and Steinker, 1972, p.46). Usages and definitions have been reviewed by Moore (1949), Weller (1958), Teichert (1958), and Steinker and Steinker (1972). "The general term 'facies' has been greatly overworked. . If the term is used, it is desirable to make clear the specific kind of facies to which reference is made" (ISG, 1976, p.15). The term facies should not be used for bodies of rock (lithostratigraphic units) of any rank (Teichert, 1958; Weller, 1958). Etymol: Latin. Pron: faysheez or fayseez; or with short *a* as in "cat". Pl: *facies.*
facies change A lateral or vertical variation in the lithologic or paleontologic characteristics of contemporaneous sedimentary deposits. It is caused by, or reflects, a change in the depositional environment. Cf: *facies evolution.*
facies contour The trace (on a map) of a vertical surface that cuts a three-dimensional rock body into facies segments; a line indicating equivalence in lithofacies development.
facies-departure map A *facies map* based on the degree of similarity to some particular sedimentary rock composition (optimum facies or single-component end member). Term suggested by Forgotson (1960, p.94) to replace *distance-function map.*
facies evolution A gradual change of facies over a period of time, indicating gradually changing depositional conditions (Teichert, 1958, p.2723). Cf: *facies change.*

facies family A term used by Teichert (1958, p.2737) for several genetically interconnected facies tracts; e.g. coral-atoll deposits and desert deposits. See also: *facies suite.*

facies fauna A group of animals characteristic of a given stratigraphic facies or adapted to life in a restricted environment; e.g. the black-shale fauna of the Middle and Upper Devonian of the Appalachian region of U.S.

facies fossil A fossil, usually a single species or a genus, that is restricted to a defined stratigraphic facies or is adapted to life in a restricted environment. Although often applied only in extreme cases, the term could be applied to most organisms, given that all organisms are limited to some degree to particular environments.

facies map A broad term for a map showing the gross areal variation or distribution (in total or relative content) of observable attributes or aspects of different rock types occurring within a designated stratigraphic unit, without regard to the position or thickness of individual beds in the vertical succession; specif. a *lithofacies map.* Conventional facies maps are prepared by drawing lines of equal magnitude through a field of numbers representing the observed values of the measured rock attributes. Cf: *vertical-variability map.* See also: *biofacies map; isofacies map; isolith map; percentage map; ratio map; entropy map; entropy-ratio map; facies-departure map.*

facies plane A term used by Caster (1934, p.19, 24) for the boundary between *magnafacies* or *parvafacies,* although it is usually not sharp enough to be recognizable as a plane in any section. Cf: *plane of contemporaneity.*

facies sequence A term used by Teichert (1958, p.2723) for a vertical succession of different (heteropic) facies formed at different times.

facies strike The compass direction of a facies contour at a given point on a map.

facies suite (a) A term suggested by Oriel (in Teichert, 1958, p.2737) for several genetically interconnected facies families; e.g. all marine deposits or all continental deposits. (b) A collection or group of rocks that shows variations within a single rock mass.

facies tract A system of different but genetically interconnected sedimentary facies of the same age (Teichert, 1958, p.2723); e.g. the outer-slope deposits of a coral atoll, or dry-channel deposits. It includes the areas of erosion from which the sediments of these facies are derived, so that an erosional interval represents part of a facies tract. The concept was developed by Walther (1893-1894). See also: *facies family.* Syn: *macrofacies.*

facing (fac'-ing) (a) The direction toward which a rock unit or layer youngs. (b) The direction along the axial plane of a fold in which it passes through younger layers. Syn: *fold facing.*

facing stone Building stone prepared in relatively thin slabs and applied to buildings for decorative appearance and protective covering.

faciostratotype (fa'-ci-o-strat'-o-type) A supplemental stratotype designated for local reference or reference to different facies, and which distinguishes different ecologic conditions that existed during the time of the chronostratigraphic unit represented by the stratotype (Sigal, 1964).

facsimile crystallization (fac-sim'-i-le) *mimetic crystallization.*

factor analysis (fac'-tor) The application of techniques that attribute the variation and interrelations observed within a set of multivariate observations to correlations between the observed variables and a smaller set of hypothetical variables or underlying "factors". The factors usually are considered to be uncorrelated and sometimes interpreted as being casual influences, although there is no mathematical basis such an interpretation. Cf: *Q-mode factor analysis; R-mode factor analysis.*

factor of safety In engineering geology, the ratio of forces resisting movement to those causing or driving movement, usually in the consideration of slope stability. Resisting forces are usually limited to the shear strength of soil or rock along a potential slip surface, whereas driving forces relate to the weight of material in the slope above a potential slip surface and to the geometry of the slope and slip surface. External forces can resist or cause movement, such as reinforcing elements (e.g., *rock bolts*) or earthquake shaking. Building codes typically define a "stable" slope as having a minimum factor of safety of 1.5 under static loading conditions and 1.15-1.3 under earthquake loading conditions. Regulatory agencies in some seismically active areas have recognized that minor, but acceptable, deformations caused by earthquakes indicate that the factor of safety against sliding actually was <1.0; they may now consider "acceptable deformation" in lieu of "factor of safety" as the definition of a "stable" slope. Syn: *safety factor.*

facula (fac'-u-la) A term established by the International Astronomical Union to describe a bright spot on planetary surfaces (USGS, 2003). Pl: faculae.

facultative (fac'-ul-ta'-tive) Said of an organism capable of growth under a number of specific conditions; adaptable to alternate environments. Cf: *obligate.*

FAD *first appearance datum.*

fadama A term used in western Africa for a flood plain in a wide river valley, subject to annual inundation and characterized by savanna vegetation (J.C. Pugh in Stamp, 1961, p.187). Etymol: Hausa.

faecal pellet (fae'-cal) *fecal pellet.*

faheyite (fa'-hey-ite) A white, bluish white, or brownish white hexagonal mineral: $(Mn^{2+},Mg)Fe^{3+}_2Be_2(PO_4)_4 \cdot 6H_2O$.

fahlband (fahl'-band) A band of sulfide impregnation in metamorphic rocks. The sulfides are too abundant to be classed as accessory minerals, but too sparse to form an ore lens (Gammon, 1966, p.177). Fahlbands have a characteristic rusty-brown appearance on weathering. Etymol: German, "fahl," fallow (color) + "Band," ribbon, band.

fahleite A yellow, gray, or bright green orthorhombic mineral: $CaZn_5Fe^{3+}_2(AsO_4)_6 \cdot 14H_2O$.

fahlerz (fahl'-erz) A syn. of *fahlore.* Etymol: German "Fahlerz", "pale ore".

fahlore (fahl'-ore) Any gray-colored ore mineral consisting essentially of sulfantimonides or sulfarsenides of copper; specif. tetrahedrite and tennantite. Syn: *fahlerz.*

fahlunite (fah'-lu-nite) An altered form of cordierite.

faikes *fakes.*

failure (fail'-ure) Fracture or rupture of a rock or other material that has been stressed beyond its *ultimate strength.* Syn: *rock failure.*

failure stress *fracture strength.*

fairbankite (fair'-bank-ite) A colorless triclinic mineral: $PbTe^{4+}O_3$. It is a dimorph of plumbotellurite.

fairchildite (fair'-child-ite) A white hexagonal mineral: $K_2Ca(CO_3)_2$. It is found in fused wood ash in partly burned trees. Cf: *buetschliite.*

fairfieldite (fair'-field-ite) (a) A white, pale yellow or greenish-white platy triclinic mineral: $Ca_2(Mn^{2+},Fe^{2+})(PO_4)_2 \cdot 2H_2O$. It is isomorphous with collinsite. (b) A group name for minerals with the analogous formula, but with Mn and Fe replaced by Mg, Ni, Co or Zn; and PO4 replaced by AsO_4.

fairing Artificial irregularities used to induce turbulent flow. Drag on a cable pulled transversely through the water is less for turbulent than for nonturbulent flow. Irregularities (such as short strings fastened to the cable) create small vortices which reduce fluid friction. A pulled cable without fairing alternately tends to shed large vortices accompanied by spurts of motion, causing vibration and inducing noise.

fairway (fair'-way) The main navigable channel (usually buoyed) of a river or bay, through which vessels enter or leave a port or harbor; the part of a waterway that is kept open and unobstructed for navigation.

fairway [geophys] (a) The region within which effort is to be concentrated, such as (in velocity analysis) a band of possible velocities within which one searches for velocity picks from normal-moveout measurements. (b) A trend of hydrocarbon accumulations. (c) The region within which data values must lie to be considered valid.

fair-weather runoff (fair'-weath'-er) *base runoff.*

fairy chimney (fai'-ry) A translation of the French term "cheminée de fée", used in the French Alps for *earth pillar.*

fairy stone (a) A cruciform-twinned crystal of staurolite, used as a curio stone without fashioning for adornment. The term is also applied as a syn. of *staurolite,* and esp. to the variety occurring in the form of a twinned crystal. (b) Any of various odd or fantastically shaped calcareous or ferruginous concretions formed in alluvial and eolian clays. Cf: *loess doll.* (c) A fossil sea urchin. (d) A stone arrowhead.

fakes A British vernacular term for a platy rock, such as a fissile sandy shale, a well-laminated siltstone, or a flaggy sandstone or micaceous flagstone. Adj: fakey. Also spelled: *faikes.*

falaise (fa-laise') An old, low sea cliff, on an emergent coast, that reestablishes contact with the open sea; the type example is the coast of Normandy. Etymol: French, "cliff".

falcate (fal'-cate) Hooked or curved like a sickle; e.g. said of a sickle-shaped cephalopod rib. Syn: falciform.

falcifer (fal'-ci-fer) A brachiopod *crus* that arises on the dorsal side of the *hinge plate* and projects into the *brachial valve* as a broad bladelike process.

falcondoite (fal-con'-do-ite) A whitish green orthorhombic mineral of the *sepiolite* group: $(Ni,Mg)_4Si_6O_{15}(OH)_2 \cdot 6H_2O$.

falcon's-eye (fal'-con's) *hawk's-eye.*

falkmanite (falk'-man-ite) A metallic lead-gray monoclinic mineral: $Pb_5Sb_4S_{11}$. See: *boulangerite.*

fall [mass move] (a) A very rapid downward movement of a mass of rock or earth that travels mostly through the air by free fall, leaping, bounding, or rolling, with little or no interaction between one moving unit and another; e.g.: *rockfall; debris fall.* (b) The mass of material moved by a fall.

fall [meteorite] Meteorites that are picked up soon after they have reached the Earth's surface and for which information (place and time of fall) are archived. Cf: *find.*

fall [slopes] (a) The descent of land or of a hill; a slope or declivity. (b) The distance to which a stream or physiographic feature slopes.

fall [streams] A *waterfall* or other precipitous descent of water. The plural "falls" is commonly used in place names, esp. where applied to a series of waterfalls.

fallback (fall'-back) Fragmental material ejected from an impact or explosion crater during formation and redeposited within, and partly filling, the *transient impact crater* immediately after formation. It includes slide-block deposits, talus, and aerially transported material. Cf: *fallout [impact]; throwout; transient impact crater.*

fallback breccia An allochthonous breccia composed of *fallback* located within an impact or explosion crater.

fall diameter The diameter of a sphere that has a specific gravity of 2.65 and the same uniform settling velocity as that of a given particle having any specific gravity "when each is allowed to settle alone in quiescent distilled water of infinite extent and at a temperature of 24°C" (Simons et al., 1961, p. vii).

falling dune (fall′-ing) An accumulation of sand that is formed as sand is blown off a mesa top or over a cliff face or steep slope, forming a solid wall, sloping at the angle of repose of dry sand, or a fan extending downward from a re-entrant in the mesa wall.

falling-stage systems tract In sequence stratigraphy, a *systems tract* that lies above and basinward of the *highstand systems tract*, yet below the sequence boundary and the overlying *lowstand systems tract*. The sediment that is bypassed on the landward side of a regressing shoreline feeds an advancing and descending shoreface, the falling-stage systems tract. It is characterized by stratal *offlap*, unlike other systems tracts that are characterized by *onlap* (Plint and Nummedal, 2000).

falling star *shooting star.*

falling tide That part of a tide cycle between high water and the following low water, characterized by seaward or receding movement of water. Also, an outgoing tidal river. Ant: *rising tide.* Syn: *ebb tide.*

fall line An imaginary line or narrow zone connecting the waterfalls on several adjacent near-parallel rivers, marking the points where these rivers make a sudden descent from an upland to a lowland, as at the edge of a plateau; specif. the Fall Line marking the boundary between the ancient, resistant crystalline rocks of the Piedmont Plateau and the younger, more erodible sediments of the Atlantic Coastal Plain in the eastern U.S. Syn: *fall zone.*

fallout [impact] (fall′-out) Fragmental material ejected from an impact or explosion crater during formation and redeposited in and around the crater. It may have undergone considerable atmospheric sorting before deposition. Cf: *throwout; fallback.*

fallout [radioactivity] The descent of usually radioactive particles through the Earth's atmosphere, following a nuclear explosion; also, the particles themselves.

fallout breccia An allochthonous breccia composed of fallout from a crater. It is generally one of the last ejecta units to be deposited, and it characteristically contains small amounts of glass fragments and a limited range of fragment sizes.

fallout tuff Consolidated or cemented ash formed by ash-fall deposition. May be deposited in water or on land (Fisher and Schmincke, 1984, p.8).

fall overturn The seasonal *circulation* of a *dimictic* lake, as the water column cools to approach equal density from top to bottom (Wetzel, 2001, p.78). Also called: fall turnover.

fall velocity *settling velocity.*

fall zone *fall line.*

false Becke line A bright line, visible under the microscope, which moves in an opposite direction to that of the *Becke line* as the microscope stage is raised or lowered. It is especially conspicuous when the mineral fragment is thick and irregular or displays conspicuous cleavage; when the difference between the refractive indices of fragment and immersion medium is considerable; and when strongly converging light from the substage condensing lens passes through and near the fragment and then enters an objective lens of large angular aperture (Wahlstrom, 1969, p.108). See also: *Becke line; Becke test.*

false bedding An old term for *cross-bedding* and *current bedding*, or bedding affected by currents that were often erratic and frequently changed direction. The term is not recommended in this sense because it might refer to pseudobedding (Hills, 1963, p.10).

false body *thixotropic clay.*

false bottom [eco geol] An apparent bedrock underlying an alluvial deposit that conceals a lower alluvial deposit; e.g. a bed of clay or sand cemented by hydrous iron oxides, on which a gold placer deposit accumulates, and under which there is another alluvial deposit resting on bedrock.

false bottom [lake] The poorly defined bottom of a lake occurring where a firm bottom grades upward to a suspended or soupy mass of muck, colloidal sludge, soft marl, or organic matter, through which a weight easily sinks; e.g., in a *bog lake* (Wetzel, 2001, p.821).

false bottom [oceanog] *deep scattering layer.*

false cleavage A quarrying term for minor cleavage in a rock, e.g. *spaced cleavage* or *slip cleavage*, to distinguish it from the dominant or *true cleavage*. Geologically, the term is misleading and should be avoided.

false color Colors used to represent different frequency bands, or other measurable characteristics, where the colors are not those naturally characteristic of the frequencies. False color makes infrared differences evident and is useful in other applications.

false diamond A colorless mineral (such as zircon, white sapphire, white topaz, and quartz) that superficially resembles diamond when cut and polished.

false dip *apparent dip [struc geol].*

false drumlin *rock drumlin.*

false easting An adjustment constant added to coordinate values so that only positive values of *easting* are recorded.

false esker A feature resembling an esker but "composed of till instead of water-laid drift" (Woodworth, 1894b).

false floor In a cave, a more or less horizontal layer of flowstone that has open space beneath it.

false folding An imprecise and therefore less desirable syn. of *bending.*

false form *pseudomorph.*

false galena *sphalerite.*

false gossan A laterally or vertically displaced, rather than indigenous, iron-oxide zone. It may be confused with the iron oxide of a *gossan*, which is weathered from underlying sulfide deposits.

false horizon (a) *artificial horizon.* (b) A line resembling the apparent horizon, but situated above or below it.

false lapis (a) *lazulite.* (b) Agate or jasper artificially dyed blue.

false mud crack A sedimentary structure resembling a mud crack, such as a polygonal pattern formed in soil or a fucoidal network. See also: *pseudo mud crack.*

false northing A value assigned the east-west line through the origin on a map grid to avoid the inconvenience of negative northings at some points. Equivalently, a constant value added to the *northing* of points on a map grid with origin at (0,0).

false oolith *pseudo-oolith.*

false origin An arbitrary point to the south and west of the true origin of a grid system, from which grid distances are measured eastward and northward to insure that all points have positive coordinate values.

false pouch *dolon.*

false shoreline The line of contact between the open water of a lake and the front or edge of a floating mat of vegetation built out from the true shore (Veatch and Humphrys, 1966, p.111).

false stratification An old term for *cross-stratification*. It was used by Lyell (1838, p.38) for the diagonal arrangement of "minor layers placed obliquely to the general planes of stratification". The term is rarely used today.

false stream An accumulation of water in a hollow along the side of a flood plain that slopes away from the main stream toward the side of the valley (Swayne, 1956, p.58).

false synapomorphy *parallelism (b).*

false topaz (a) A yellow transparent variety of quartz resembling the color of topaz: specif. citrine. (b) A yellow variety of fluorite.

famatinite (fa-ma′-ti-nite) A metallic pinkish brown orthorhombic mineral of the luzonite group: Cu_3SbS_4.

Famennian (Fa-men′-ni-an) A stage of the Standard Global Chronostratigraphic Scale: uppermost Devonian (above Frasnian, below Tournaisian of the Carboniferous System). The basal GSSP is at Montagne Noir, southern France (Klapper et al., 1993).

family [soil] In soil classification one of the categories intermediate between the subgroup and the soil series. Families provide groupings of soils with ranges in texture, mineralogy, temperature, and thickness.

family [taxon] A category in the hierarchy of zoological and botanical classification intermediate between *order [taxon]* and *genus*. In zoology, the name of a family characteristically ends in -idae, e.g. Cytheridae; in botany the ending is -aceae, e.g. Rosaceae. Cf: *subfamily; superfamily.*

fan [geomorph] (a) A gently sloping, fan-shaped mass of detritus forming a section of a very low cone, commonly at a place where a notable decrease in gradient occurs; specif. an *alluvial fan.* (b) A fan-shaped mass of congealed lava that formed on a steep slope by the continually changing direction of flow.

fan [marine geol] *submarine fan.*

fan apron *bajada.*

fan bay The head of an alluvial fan that extends a considerable distance into a mountain valley. Cf: *fanhead.*

fan cleavage Cleavage in which the orientation changes systematically with position about a fold. Cleavage may fan convergently, where cleavage surfaces converge toward the inner arc of a folded layer, or fan divergently, where cleavage surfaces diverge away from the inner arc of a folded layer. Syn: *cleavage fan; fan cleavage; fanning cleavage.*

fan coral Any coral that forms flat, fanlike colonies.

fancy-color diamond Any diamond with a natural body color strong enough to be attractive rather than off-color. Red, pink, blue, and green are very rare; orange and violet, rare; strong yellow, yellowish-green, greenish yellow, brown, and black stones are more common.

fancy cut (fan′-cy) Any style of diamond cutting other than the round brilliant cut or single cut. It includes, among others, the marquise, emerald cut, pear-shape cut, baguette, heart-shape, oval, and fancy cuts.

fancy sapphire A *sapphire* of any hue other than blue.

fan delta (a) A gently sloping alluvial deposit produced where a mountain stream flows out onto a lowland. (b) *alluvial-fan shoreline; delta fan.* Cf: *arcuate delta.*

fan-filter *velocity filter.*

fan fold A fold with a broad hinge region and limbs that converge away from the hinge. Rarely used.

fangite A vitreous to metallic deep red to maroon orthorhombic mineral: Tl_3AsS_4.

fanglomerate (fan-glom′-er-ate) A sedimentary rock consisting of waterworn, fragments of various sizes, deposited in an *alluvial fan* and later cemented into a firm rock; it may be characterized by persistence parallel to depositional strike and by rapid thinning downdip. The term was proposed by Lawson (1913, p.329) for the coarser, consolidated rock material occurring in the upper part of an alluvial fan. Cf: *bajada breccia.*

fanhead (fan′-head) The area on an alluvial fan close to its apex. Cf: *fan bay.*

fanhead trench A linear depression formed by a drainage line that is incised considerably below the surface of an alluvial fan.
fan mesa An alluvial-fan remnant left standing after dissection of the fan.
fanning cleavage *fan cleavage.*
fan scarp A *piedmont scarp* formed by faulting, occurring entirely within alluvium and not observed to cross bedrock in any part of its course (Billings, 1954, p.156).
fan shooting Seismic shooting in which detectors are laid out along an arc so that each detector is in a different direction at roughly the same distance from a single source point. It was used in the 1920s and 1930s to detect the presence of shallow salt domes intruding low-velocity sediments. Syn: *arc shooting.*
fan system A fan-shaped pattern formed by diverging trabeculae in the plane of the septum of a scleractinian coral. See also: *axis of divergence.*
fantasy cut A faceting style of fashioning gemstones which involves a free-form shape with alternate curved and flat surfaces. Etymol: Dutch, "Fantasie".
fan-topped pediment A pediment with a thin (15-60 m) covering of alluvial fans built upon it in response to some minor change of climate or of other controlling conditions (Blackwelder, 1931, p.139). Cf: *concealed pediment.*
fan valley A valley in a submarine fan; a continuation of a submarine canyon. It is either V-shaped or trough-shaped, and commonly has natural levees and distributaries.
faradaic path (far'-a-da'-ic) One of the two available paths for transfer of energy across an electrolyte-metal interface. Energy is carried by conversion of atom to ion or vice versa, due to electrochemical reaction and ion diffusion. Cf: *nonfaradaic path.*
faratsihite (fa-rat'-si-hite) A pale-yellow clay substance: $(Al,Fe)_2(OH)_4Si_2O_5$. It has been regarded as an iron-bearing variety of kaolinite, a mixture of kaolinite and nontronite, or identical with nontronite.
farinaceous (far-i-na'-ceous) Pertaining to a texture or structure of a rock or sediment that is mealy, soft, and friable, e.g. a limestone or a pelagic ooze; also, said of a lava flow in which the particles seem to be in a state of mutual repulsion. Syn: *mealy.*
far infrared Pertaining to the longer wavelengths of the *infrared* region, from 15 µm to 1 mm. The atmosphere transmits very little radiation at these wavelengths, so terrestrial use of this spectral band is severely limited. Cf: *reflected infrared.*
faro (far'-o) A small, atoll-shaped or oblong *reef* with a lagoon up to 30 m deep, forming part of the rim of a fringing reef, a barrier reef or an atoll. Etymol: political subdivision of an *atollon* in the Maldive Islands of the Indian Ocean.
far range The portion of a SLAR image farthest from the aircraft flight path.
farringtonite (far'-ring-ton-ite') A colorless, wax-white, or yellow monoclinic meteorite mineral: $Mg_3(PO_4)_2$.
farrisite (far'-ris-ite) A fine-grained hypabyssal rock composed primarily of a melilite, making up approximately one third of the rock, and amphibole, with smaller amounts of biotite, olivine, and magnetite. Feldspars and nepheline are almost entirely replaced by zeolites. The name, given by Brögger in 1898, is from Lake Farris in the Oslo district, Norway. Obsolete.
farrum (far'-rum) A term established by the International Astronomical Union for a pancake-like structure, or a row of such structures on planetary surfaces (USGS, 2003). Pl: farra. Syn: *pancake dome.*
farrum [volc] (far'-rum) Pl. farra. On Venus, a volcano with diameter less than 100 km, with a steep perimeter slope and relatively low-angle flanks. Also called a pancake dome (Crumbler and Aubele, 2000).
farsundite (far'-sund-ite) An orthopyroxene- and hornblende-bearing granite of the charnockite series. Cf: *opdalite.* The name, given by Kalderup in 1904, is for Farsund, Norway. Obsolete.
fascicular (fas-cic'-u-lar) Said of an aggregate of acicular crystals.
fascicular columella A *columella* in scleractinian corals, formed by twisted vertical ribbons or rods resembling pali or paliform lobes.
fasciculate (fas-cic'-u-late) Arranged in fascicles; e.g. said of a *corallum* with cylindrical corallites that are somewhat separated from one another but may be joined by connecting tubules, or descriptive of ribbing in coiled ammonoid conchs having bunched or bundled ribs near the umbilical margin.
fasciole (fas'-ci-ole) (a) A heavily ciliated tract in an echinoderm; esp. a narrow band of small tubercles bearing densely ciliated spines (clavulae) on the denuded test of certain echinoids. The term is also applied to a narrow band of such spines in which cilia beat to create currents. (b) A band generated on a gastropod shell by a narrow sinus or notch in, or a lamellose projection of, successive growth lines (TIP, 1960, pt. I, p.131). See also: *anal fasciole.*
fashioning (fash'-ion-ing) A general term for the entire process of sawing, cleaving, rounding up, facet-grinding, cobbing, and polishing, and other operations employed in preparing rough gem material for use in jewelry.
fasibitikite (fas-i-bit'-i-kite) A medium-colored riebeckite-aegirine granite that also contains eudialyte and zircon. The name, given by Lacroix in 1915, is derived from the locality Ampasibitika, Malagasy. Obsolete.
fasinite (fas'-i-nite) A coarse-grained *melteigite* that contains titanian augite and nepheline as its main components along with alkali feldspar, olivine, and biotite. It has the same chemical composition as *berondrite* and differs from *bekinkinite* by the absence of hornblende and analcime. The name, given by Lacroix in 1916, is derived from Ampasindava, Malagasy. Obsolete.
fassaite (fas'-sa-ite) A pale-green to dark-green variety of monoclinic pyroxene containing considerable aluminum substituting for silicon: $(Ca,Mg,Fe^{3+},Al,Ti)_2(Si,Al)_2O_6$.
Fassanian (Fas-san'-i-an) A European substage: lower substage of the Ladinian Stage, Middle Triassic (above Illyrian substage of Anisian, below Longobardian).
fast ice Any sea ice that forms along and remains attached to the coast (e.g. *icefoot*), or that forms between grounded icebergs, or is attached to the bottom in shallow water (e.g. *anchor ice*). Fast ice may form in situ from seawater or by freezing of *pack ice* to the shore. It may extend a few meters to several hundred kilometers from the coast.
fastigate (fas'-ti-gate) Said of an ammonoid with a roof-shaped venter, the periphery of the shell being sharpened but not keeled.
fastland (fast'-land) A *mainland;* esp. one that is high and dry near water, such as an *upland.*
fast ray In crystal optics, that component of light in any birefringent crystal section that travels with the greater velocity and has the lower index of refraction. Cf: *slow ray.*
fat clay A cohesive and compressible clay of high plasticity, containing a high proportion of minerals that make it greasy to the feel. It is difficult to work when damp, but strong when dry. Rarely used. Ant: *lean clay.* Syn: *long clay.*
fathometer (fa-thom'-e-ter) A copyrighted name for a type of *echo sounder.*
fatigue (fa-tigue') Failure of a material after many repetitions of a stress that of itself is not strong enough to cause failure.
fatigue limit *endurance limit.*
fatigue ratio *endurance ratio.*
fatty acid (fat'-ty) Any one of a group of organic acids that occur in animal and vegetable oils and fats. Common examples are butyric acid ($C_4H_8O_2$); palmitic acid ($C_{16}H_{32}O_2$); stearic acid ($C_{18}H_{36}O_2$); and oleic acid ($C_{18}H_{34}O_2$).
fauces terrae (fau'-ces ter'-rae) A term used in international law for headlands and promontories that enclose territorial water that is not part of the high seas. Etymol: Latin, "gulf".
faujasite-Ca A vitreous colorless cubic zeolite mineral: $CaAl_2Si_4O_{12}{\cdot}8H_2O$.
faujasite-Mg A vitreous colorless cubic zeolite mineral: $MgA_{12}Si_4O_{12}{\cdot}8H_2O$.
faujasite-Na A vitreous colorless cubic zeolite mineral: $(Na_2,Ca)Al_2Si_4O_{12}{\cdot}8H_2O$. Cf: *gmelinite.*
fault [cryst] (a) A general term for a dislocation in a crystal. (b) A *stacking fault.*
fault [struc geol] A discrete surface or zone of discrete surfaces separating two rock masses across which one mass has slid past the other. Cf: *shear zone; fault zone.* Obsolete syn: *paraclase.*
fault apron A mass of rock waste deposited along the base of a fault scarp, formed by numerous coalescing alluvial cones.
fault array A set or multiple sets of faults with regular orientations and interrelated kinematic character.
fault basin A depression separated from the surrounding area by faults.
fault bench A small *fault terrace.*
fault-bend fold A fold formed by movement on a nonplanar fault surface.
fault block A block of the crust that has been displaced by movement on a fault. See also: *tilt block.* Syn: *block [tect].*
fault-block mountain *block mountain.*
fault breccia Angular rock fragments resulting from fracturing and frictional slip along a fault; may be cohesive or incohesive. Syn: *dislocation breccia.* Obsolete syn: *fault rubble.*
fault cave A cave that developed along a fault or fault zone.
fault cliff A cliff formed by faulting; esp. a *fault scarp.*
fault coast A coast formed directly by faulting (Cotton, 1916), as one along a fault scarp, or a narrow arm of the sea that floods a fault trough (e.g., the coast of the Red Sea). Cf: *fault-line coast.*
fault-dam spring *fault spring.*
fault-displacement fold A fold that results from variations in the amount of slip on a *fault surface.*
fault embayment A fault trough, graben, or other depressed region in a fault zone or between two faults, invaded by the sea; e.g. the Red Sea.
fault escarpment *fault scarp.*
fault fissure A fissure that is the result of faulting. It may or may not be filled with vein material.
fault gap A depression between the offset ends of a ridge, formed by a transverse fault that laterally displaces the ridge so that the two parts are no longer continuous (Lahee, 1961, p.356). Cf: *fault-line gap.*
fault gouge Soft, uncemented pulverized clayey or claylike material, commonly a mixture of minerals in finely divided form, found along some faults or between the walls of a fault, and filling or partly filling a fault zone; a slippery mud that coats the fault surface or cements the *fault breccia.* It is formed by frictional wear of rock material during slip; it may undergo subsequent decomposition and alteration by fluid circulation. Cf: *gouge; clay gouge; selvage.*

fault-graded beds Gradationally compacted marine muds that have been deformed by earthquake shock (Seilacher, 1969). A liquefied zone is underlain by a rubble zone and this by a step-faulted zone. See also: *seismite.*

fault growth Propagation and enlargement of a fault surface, and the associated increase in cumulative displacement. Fault growth processes include segment linkage and tip-line propagation, both of which are associated with increases in total displacement. Generally, intermittent, small-scale movement along a fault surface that, accumulated, results in considerable displacement.

faulting (fault'-ing) (a) The process of fracturing, frictional slip, and displacement accumulation that produces a fault. (b) The displacement of the crust on a fault or *fault array*, as in "Active faulting occurs in California".

fault-line adj. Said of a secondary or subsequent landform or feature created solely by the action of weathering and erosion upon faulted materials (Sharp, 1954, p.27).

fault line The trace of a fault plane on the ground surface or on a reference plane. Syn: *fault trace.*

fault-line coast A coast formed by the partial submergence of a *fault-line scarp* so that the waters of a sea or lake rest against the scarp. The fault-line scarp may be the result of erosion along an ancient fault line. Cf: *fault coast.*

fault-line gap A gap produced solely by erosion of a resistant ridge laterally offset by earlier faulting; e.g. such a gap located along the line of outcrop of a dip fault or of a diagonal fault that intersects the rock layer of the ridge (Lahee, 1961, p.367). Cf: *fault gap.*

fault-line saddle A saddle created by rapid erosion of a ridge crest where it is crossed by a fault (Sharp, 1954, p.27). Examples occur along the San Gabriel Fault within the San Gabriel Range in southern California. Cf: *fault saddle.*

fault-line scarp (a) A steep slope or cliff formed by differential erosion along a fault, as by the more rapid erosion of soft rock on one side of a fault as compared to that of more resistant rock on the other side; e.g., the east face of the Sierra Nevada in California. Cf: *obsequent fault-line scarp; resequent fault-line scarp.* Syn: *erosion fault scarp.* (b) A *fault scarp* that has been modified by erosion.

fault-line valley A valley that is formed along or follows a fault line; e.g. a subsequent valley developed by headward erosion in the soft, crushed, relatively weak material along a fault zone. Cf: *fault valley.*

fault-line-valley shoreline A shoreline formed by the partial submergence of a valley that has been eroded along the crushed zone of a fault or along a narrow strip of faulted weak rock, such as along the coast of northern Nova Scotia.

fault plane A *fault surface* that is more or less planar.

fault-plane solution Determination of the orientation of a fault plane and the direction of slip motion on it from an analysis of the first motion of the P waves and/or the amplitudes of the P waves, S waves, and surface waves. The solution also gives the orientation of the principal axes of compression and tension. See also: *slip-vector analysis.* Syn: *focal mechanism.*

fault population A collection of geometrically and kinematically related faults that have grown concurrently during a deformation episode. Faults in the population may or may not intersect, but the combined effect of slip on them is an approximately homogeneous change in the shape of a particular volume of rock.

fault-propagation fold A fold that forms just in front of the tip of an actively propagating fault.

fault-related fold A general term describing folds such as fault-bend folds, fault-displacement folds, fault-propagation folds, and detachment folds, that have a clear geometric relationship and probable genetic relationship with faults.

fault rock One of several rock types physically associated with fault surfaces and genetically linked to fault movement. See: *breccia [geol]; cataclasite; mylonite; pseudotachylyte.*

fault rubble An assemblage of detached, jumbled, and crushed or shattered angular fragments torn from the walls of a fault; an incohesive *fault breccia.*

fault saddle A particular type of *kerncol*, being a notch, col, or saddle in a ridge, created by actual displacement of the ridge crest by faulting (Sharp, 1954, p.25). Cf: *fault-line saddle.*

fault sag A small, enclosed depression along an active or recent fault. It is caused by differential movement between slices and blocks within the fault zone or by warping and tilting associated with differential displacement along the fault, and it forms the site of a *sag pond.* Term introduced by Lawson et al. (1908, p.33). Syn: *sag [struc geol].*

fault scarp (a) A steep slope or cliff formed directly by movement along a fault and representing the exposed surface of the fault before modification by erosion and weathering. It is an initial landform. Cf: *fault-line scarp.* Syn: *fault escarpment; fault cliff; cliff of displacement.* (b) A term used loosely in England for any scarp that is due to the presence of a fault, even though the relief may be erosional.

fault-scarp shoreline A shoreline produced by recent faulting.

fault segment An individual fault that is part of a set of subparallel faults that together form a *fault zone* (Peacock et al., 2000).

fault set A group of faults that are parallel or nearly so, and that are related to a particular deformational episode. Cf: *fault system.* See also: *fault population.*

fault shadow [geophys exploration] Distorted or relatively dead reflection in the footwall zone of a fault, resulting when reflection raypaths bend differently as they traverse a fault at locations where the velocity contrast differs (e.g., for the traces that make up a CMP gather).

fault slice Slicing through a three-dimensional data volume along a curved surface parallel to a fault plane. Fault slices are made for both sides of a fault to ascertain closure against the fault (Brown, Edwards, and Howard, 1987).

fault-slice ridge *slice ridge.*

fault splinter An obsolete term for a narrow, ramplike connection between the opposite ends of two parallel normal faults. The feature occurs in major fault zones (Strahler, 1963, p.596); *relay ramp* is now preferred. Cf: *fault step.*

fault spring A spring flowing onto the land surface from a fault that brings a permeable bed into contact with an impermeable bed. Cf: *fracture spring; fissure spring.* Syn: *fault-dam spring.*

fault step Along a normal fault expressed at the surface, one of a series of thin rock slices along which the fault's total displacement is dispersed (Strahler, 1963, p.596). Obsolete and little used.

fault strain The change of shape that results from the combined slip on numerous faults. In order to define the fault strain in a rock volume, the total displacements of different material points within a limited volume of rock must yield nearly constant displacement gradients within the rock volume. The amount of strain contributed by an individual fault is proportional to the fault surface area and total displacement.

fault strand An individual fault of a set of closely spaced parallel or subparallel faults of a fault system (Biddle and Christie-Blick, 1985).

fault surface The fracture surface along which displacement has occurred. Cf: *fault plane.*

fault system (a) An array of interconnected and kinematically related faults. (b) Two or more geometrically related fault sets, e.g. a conjugate fault system. See also: *fault array; fault population; fault set.*

fault terrace An irregular, terrace-like tract between two fault scarps, produced on a hillside by step faulting in which the downthrow is systematically on the same side of two approximately parallel faults. Cf: *fault bench.*

fault tip A point on a map or a cross section where a fault terminates.

fault trace *fault line.*

fault trap A trap for oil or gas in which the closure results from the presence of one or more faults.

fault-trellis drainage pattern (fault-trel'-lis) A *trellis drainage pattern* developed where a series of parallel faults have brought together alternating bands of hard and soft rocks (Thornbury, 1954, p.123).

fault trench A cleft or crack formed on the Earth's surface as a result of faulting. It is a smaller-scale feature than a fault trough (rift valley), and is not necessarily related to normal faulting.

fault-trough lake *sag pond.*

fault valley A linear depression produced by faulting; e.g. a small, narrow valley created within a major fault zone by relative depression of narrow slices, or a large graben situated between tilted block mountains, or a valley created by relative uplift on opposite sides of two parallel thrust faults. Cf: *fault trough; fault-line valley.*

fault wall *wall [struc geol].*

fault wedge A wedge-shaped rock mass bounded by at least two faults.

fault weld A fault surface or fault zone joining strata originally separated by *autochthonous* or *allochthonous salt;* equivalent to a *salt weld* along which there has been significant fault slip or shear (Jackson and Talbot, 1991).

fault zone A fault that is expressed as a zone of numerous small fractures, breccia, and/or fault gouge. A fault zone may be as wide as hundreds of meters. Cf: *step fault.* Syn: *distributed fault; distributive fault.* Obsolete syn: *shatter belt.*

fauna (fau'-na) (a) The entire animal population, living or fossil, of a given area, environment, formation, or time span. Cf: *flora.* (b) Sometimes incorrectly used to include both the animal and plant fossils of a particular rock unit, i.e. the *biota.* Adj: faunal.

faunal break (fau'-nal) An abrupt change or *break [stratig]* from one fossil assemblage to another at a definite horizon in a stratigraphic sequence, usually produced by an unconformity or hiatus or sometimes by a change in bottom ecology without interruption of deposition; e.g. a gap in the orderly evolution of a single organism through a vertical series of beds.

faunal diversity (a) *Diversity* of a fauna. (b) Occasionally used more precisely to denote the number of species, either estimated or counted, whose combined totals comprise 95% of the total population.

faunal dominance That percentage of a population constituted by the most common species.

faunal evolution Change in the composition of a fauna with time.

faunal province A geographic region characterized by a specific assemblage of animals more or less widely distributed within it.

faunal stage A chronostratigraphic unit (stage) based on a faunizone.

faunal succession The observed chronologic sequence of life forms (esp. animals) through geologic time. See also: *law of faunal succession.*

faunal zone *faunizone.*

faunichron (fau'-ni-chron) A term used by Dunbar and Rodgers (1957, p.300) for the geologic-time unit corresponding to faunizone of Buckman (1902).

faunizone (fau'-ni-zone) (a) A biostratigraphic unit or body of strata characterized by a particular assemblage of fossils (specif. fossil faunas), regardless of whether it is inferred to have chronological or only environmental significance. (b) A term sometimes used for the strata equivalent in age to a certain overlap of "biozones" and having dominantly chronostratigraphic significance. (c) A term that has been used in the sense of "zone" regarded as a time unit of biochronologic significance. The term was introduced by Buckman (1902) for "belts of strata, each of which is characterized by an assemblage of organic remains", and has been generally regarded as the animal-based variety of (biostratigraphic) zone of Oppel (1856-1858). The ACSN (1961, art.21d) states that the term is "not generally accepted" and that its correct definition is "in dispute". See also: *assemblage zone.* Cf: *florizone.* Syn: *faunal zone; faunizonite.*

faunizonite *faunizone.*

faunula (fau'-nu-la) (a) A set of animal species found in a relatively small and isolated region, and not peculiar to it. (b) *faunule.*

faunule (fau'-nule) (a) A collection of fossil animals obtained from a stratum over a very limited geographic area, esp. from only one outcrop. Syn: *local fauna.* (b) A term used by Fenton and Fenton (1928) for an assemblage of fossil animals associated in a single stratum or a few contiguous strata of limited thickness and dominated by the representatives of one community; the faunal assemblage of a *zonule.* Cf: *florule.* Syn: *faunula.*

faustite (faust'-ite) An apple-green triclinic mineral of the *turquoise* group: $(Zn,Cu)Al_6(PO_4)_4(OH)_8{\cdot}4H_2O$.

favositid (fav-o-sit'-id) Any tabulate coral belonging to the family Favositidae, characterized by massive colonies (usually without coenenchyme) of slender corallites with mural pores, short equal spinose septa, and complete tabulae. Range, Upper Ordovician to Permian, possibly Triassic.

***f* axis** A term used in crystal plasticity to denote a line in the crystal slip plane at right angles to the slip direction (*t* direction). It is commonly an axis of rotation of the crystal lattice during deformation.

fayalite (fa'-yal-ite) A greenish yellow, brown to black orthorhombic mineral of the *olivine* group: Fe_2SiO_4. It is isomorphous with forsterite, and occurs chiefly in igneous rocks and in marbles which host magnetite ores. Symbol: Fa. Syn: *iron olivine.*

FCCL *fine complex crossed lamellar structure.*

F center A type of *color center* in a crystal that is formed by a negative ion vacancy with two bound electrons. The F stands for "Farbe", the German word for "color".

F-centered lattice *face-centered lattice.*

FCI *fuel-coolant interaction.*

F-coal Microscopic coal particles that are predominantly fusain, as found in miners' lungs. Cf: *V-coal; D-coal.*

F distribution A statistical *distribution* that describes the ratio of two independent chi-square random variables, each divided by its associated degrees of freedom. The distribution is named in honor of R.A. Fisher (1890-1962), English statistician, and is widely used in analysis of *variance.* Cf: *F test.*

F-distribution test *F test.*

feasibility study [eco geol] A formal evaluation of the anticipated profits from mining a mineral deposit. These studies require a detailed estimate of extraction, mineral processing, marketing costs, and a forecast of future mineral prices.

feather (feath'-er) (a) A series of elongated and irregular liquid inclusions in a gemstone, grouped together in orderly proximity to each other so as to resemble the overall pattern of a bird's feather. (b) A flaw inside a gemstone, such as a jagged fracture that is whitish in appearance. (c) In diamonds, a cleavage, crack, or fracture that has a feathery appearance when viewed at right angles to the separation plane.

feather alum (a) *halotrichite.* (b) *alunogen.*

feather amphibolite A metamorphic rock in which porphyroblastic crystals of amphibole (usually hornblende) tend to form stellate or sheaflike groups on the planes of foliation or schistosity. Cf: *amphibolite.* Syn: *garbenschiefer.*

feather edge The thin edge of a bed of sedimentary rock where it disappears by thinning, pinching, or wedging out. Syn: *knife edge.*

featheredging (feath-er-edg'-ing) *feathering [cart].*

feather fracture A less-preferred syn. of *plume structure.* Although it was the original term, its use would lead to confusion with the term *feather joints.*

feather ice *pipkrake.*

feathering [cart] (feath'-er-ing) The technique of progressively dropping contour lines to avoid congestion on steep slopes. A former practice, of tapering the line weight near the end of the contour line to be dropped, was abandoned with the advent of the scribing technique. Syn: *featheredging.* Cf: *banding [cart].*

feathering [seis] En-echelon arrangement of successive spreads, such as produced in marine shooting when a cross current causes the cable to drift at an angle to the seismic line.

feather joints A set of en echelon joints formed in a fault zone by tension. The joints appear to the fault as the barbs of a feather to its shaft. The syn. *pinnate joints* is preferred.

feather ore A capillary, fibrous, or feathery form of an antimony sulfide mineral, such as stibnite or boulangerite; specif. jamesonite.

feather out To end irregularly. The term is applied to lenticular bodies of rock.

feather quartz Quartz in imperfect crystals, the bases of which meet at an angle along a crystal plane so that a cross section looks somewhat like a feather.

feather zeolite *hair zeolite.*

feature class A group of phenomena based on common properties or relationships (Gupthill et al., 1990, p.9).

feature code (a) A unique alphanumeric string representing a category of an object in a spatial data base (Krzanowski et al., 1993). (b) A number assigned to represent a category of features on a map, such as hydrography, boundaries, culture, etc.

fecal pellet (fe'-cal) An organic excrement, mainly of invertebrates, occurring esp. in modern marine deposits but also fossilized in some sedimentary rocks, usually of a simple ovoid form less than a millimeter in length, or more rarely rod-shaped with longitudinal or transverse sculpturing, devoid of internal structure, and smaller than a *coprolite.* Also spelled: *faecal pellet.* Cf: *casting.*

fedorite (fe-do'-rite) A colorless to pink triclinic mineral: $(K,Na)_{2.5}(Ca,Na)_7Si_{16}O_{38}(OH,F)_2{\cdot}3.5H_2O$.

fedorovskite (fed-o-rov'-skite) A brown orthorhombic mineral isostructural with roweite: $Ca_2(Mg,Mn^{2+})_2B_4O_7(OH)_6$.

Fedorov stage (Fed'-o-rov) *universal stage.*

fedotofite (fe-do-tof'-ite) A green monoclinic mineral isostructural with euchlorine: $K_2Cu_3O(SO_4)_3$.

feedback (feed'-back) Partial reversal of a certain process to its beginning or to a preceding stage as a means of reinforcement or modification, esp. in biologic, psychologic, and social systems.

feeder [eco geol] (feed'-er) *channelway [ore dep].*

feeder [intrus rocks] The conduit through which magma passes from the magma chamber to some localized intrusion, e.g. a feeder dike.

feeder [streams] *tributary.*

feeder [volc] *chimney [volc].*

feeder beach An artificially widened beach serving to nourish downdrift beaches by natural littoral currents.

feeder current The part of a rip current that flows parallel to the shore (inside the breakers) before converging with other feeder currents to form the neck of the rip current. Cf: *longshore current.*

feeding channel (feed'-ing) *channelway [ore dep].*

feeding esker A small esker joining a larger one.

feidj A term used in the Saharan region for a sand-covered interdune passage. Cf: *gassi.* Also spelled: feidsh; feij; fejj.

feinglosite A pale olive-green monoclinic mineral of the *brackebuschite* group: $Pb_2(Zn,Fe^{2+})(AsO_4,SO_4)_2(OH,H_2O)$.

feitknechtite (feit'-knecht-ite) A black or brownish black trigonal mineral: $\beta Mn^{3+}O(OH)$. Cf: *manganite.*

fei ts'ui An emerald- or bluish-green variety of jadeite from Burma, esp. one resembling the color of the brilliant blue-green back of the kingfisher. Etymol: Chinese, "kingfisher jade".

feklichevite A vitreous dark brown rhombohedral mineral of the *eudialyte* group: $Na_{11}Ca_9(Fe^{3+},Fe^{2+})_2Zr_3Nb(Si_{25}O_{73})(OH,H_2O,Cl,O)_5$.

felbertalite A metallic gray monoclinic mineral: $Cu_2Pb_6Bi_8S_{19}$.

felder (fel'-der) An obsolete term for crustal blocks of a polygonal, mosaic pattern that are produced by *taphrogeny.* Syn: *tesserae.*

feldmark (feld'-mark) *fell-field.*

feldspar (feld'-spar) (a) A group of abundant rock-forming minerals of general formula: $MAl(Si,Al)_3O_8$, where M = K, Na, Ca, Ba, Rb, Sr, or rarely Fe. Feldspars are the most widespread of any mineral group and constitute 60% of the Earth's crust; they occur as components of all kinds of rocks (crystalline schists, migmatites, gneisses, granites, most magmatic rocks) and as fissure minerals in clefts and druse minerals in cavities. Feldspars are usually white or nearly white and clear and translucent (they have no color of their own but are frequently colored by impurities), have a hardness of 6 on the Mohs scale, frequently display twinning, exhibit monoclinic or triclinic symmetry, and possess good cleavage in two directions (intersecting at 90° as in orthoclase and at about 86° as in plagioclase). On decomposition, feldspars yield a large part of the clay of soil and also the mineral kaolinite. (b) A mineral of the *feldspar* group, such as alkali feldspar (orthoclase, microcline), plagioclase (albite, anorthite), and celsian. Syn: *felspar; feldspath.*

feldspath (feld'-spath) *feldspar.*

feldspathic (feld-spath'-ic) Said of a rock or other mineral aggregate containing feldspar.

feldspathic arenite A sandstone containing abundant quartz, chert, and quartzite, less than 10% argillaceous matrix, and 10-25% feldspar, and characterized by an abundance of unstable materials in which the feldspar grains exceed the fine-grained rock fragments (Williams et al., 1954, p.294, 316). It is less feldspathic and more mature than *arkosic arenite.* The rock is roughly equivalent to *subarkose.* See also: *feldspathic sandstone.*

feldspathic graywacke (a) A graywacke characterized by abundant unstable materials; specif. a sandstone containing a heterogeneous mixture of rock fragments and angular grains of quartz and feldspar, and having feldspar grains in greater abundance than rock fragments (indicating a supracrustal provenance) (Pettijohn, 1954; 1957, p.303). Williams et al. (1954, p.294) give a feldspar content of 10-25% and an argillaceous matrix greater than 10%; the rock is less feldspathic than *arkosic graywacke*. It is equivalent to the *high-rank graywacke* of Krynine (1945). (b) A term used by Folk (1954, p.354) for a sandstone containing 25-90% micas and metamorphic rock fragments, 10-50% feldspars and igneous-rock fragments, and 0-65% quartz and chert. Cf: *impure arkose.* (c) A term used by Hubert (1960, p.176-177) for a sandstone containing 25-90% micas and micaceous metamorphic-rock fragments, 10-50% feldspars and feldspathic crystalline-rock fragments, and 0-65% quartz, chert, and metamorphic quartzite. Cf: *micaceous arkose.* Cf: *lithic graywacke.* Syn: *lithic arkosic wacke.*

feldspathic litharenite (a) A term used by McBride (1963, p.667) for a litharenite containing appreciable feldspar; specif. a sandstone containing 10-50% feldspar, 25-90% fine-grained rock fragments, and 0-65% quartz, quartzite, and chert. (b) A term used by Folk (1968, p.124) for a sandstone containing less than 75% quartz and metamorphic quartzite and having a "F/R ratio" between 1:1 and 1:3, where "F" signifies feldspars and fragments of gneiss and granite, and "R" signifies all other fine-grained rock fragments. Cf: *lithic arkose.*

feldspathic lithwacke Essentially a *lithic graywacke* (over 15 percent matrix) in which rock fragments exceed feldspar but the latter forms 10 percent or more of the sand fraction (Pettijohn et al., 1972, p.164).

feldspathic polylitharenite A *polylitharenite* containing more than 10% feldspar (Folk, 1968, p.135).

feldspathic quartzite (a) A term used by Hubert (1960, p.176-177) for a sandstone containing 70-95% quartz, chert, and metamorphic quartzite, 5-15% feldspars and feldspathic crystalline-rock fragments, and 0-15% micas and micaceous metamorphic-rock fragments. Cf: *micaceous quartzite.* (b) A term used by Pettijohn (1949, p.227) for a well-indurated *feldspathic sandstone,* and later (1954, p.364) as a syn. of *subarkose.*

feldspathic sandstone A feldspar-rich sandstone; specif. a sandstone intermediate in composition between an *arkosic sandstone* and a quartz sandstone, containing 10-25% feldspar and less than 20% matrix material of clay, sericite, and chlorite (Pettijohn, 1949, p.227). Pettijohn (1957, p.322) redefined the term and used it as a less-preferred syn. of *subarkose.* Krumbein and Sloss (1963, p.170) used the term for a quartzose sandstone with 10-25% feldspar, and Williams et al. (1954, p.316) used it as a general term to include *feldspathic arenite* and feldspathic wacke. See also: *feldspathic quartzite; arkose.*

feldspathic shale A shale characterized by a feldspar content greater than 10% in the silt size and by a finer matrix of kaolinitic clay minerals, commonly associated with arkose (Krumbein and Sloss, 1963, p.175). Syn: *kaolinitic shale.*

feldspathic subgraywacke A term used by Folk (1954, p.354) for a sandstone composed of subequal amounts of rock fragments of igneous and metamorphic derivation; specif. a sandstone containing 10-25% feldspars and igneous-rock fragments, 10-25% micas and metamorphic-rock fragments, and 50-80% quartz and chert.

feldspathic sublitharenite *lithic subarkose.*

feldspathic wacke A sandstone containing abundant quartz, chert, or quartzite, more than 10% argillaceous matrix, and 10-25% feldspar, and characterized by an abundance of unstable materials in which the feldspar grains exceed the fine-grained rock fragments (Williams et al., 1954, p.292, 316). It is less feldspathic and more mature than *arkosic wacke.* Syn: *subarkosic wacke.*

feldspathide (feld'-spath-ide) *feldspathoid.*

feldspathization (feld'-spath-i-za'-tion) The formation of feldspar in a rock, usually as a result of metamorphism. Material for the feldspar may come from the country rock or be introduced by magmatic or other solutions.

feldspathoid (feld'-spath-oid) (a) A group of comparatively rare rock-forming minerals consisting of aluminosilicates of sodium, potassium, or calcium and having too little silica to form feldspar. Feldspathoids are chemically related to the feldspars, but differ from them in crystal form and physical properties; they take the places of feldspars in igneous rocks that are undersaturated with respect to silica or that contain more alkalies and aluminum than can be accommodated in the feldspars. Feldspathoids may be found in the same rock with feldspars but never with quartz or in the presence of free magmatic silica. (b) A mineral of the *feldspathoid* group, including leucite, nepheline, sodalite, nosean, hauyne, lazurite, cancrinite, and melilite. Var. spellings: *felspathoid; feldspathide.*

felite (fel'-ite) *belite.*

fell (a) A term used in Scotland and northern England for a bare, uncultivated, open hillside or mountain. (b) A term used in Great Britain for an elevated tract of wasteland or a mountain moorland; a *fell-field.* Etymol: Scandinavian. See also: *fjeld; fjäll.*

fell-field An open, treeless, rock-strewn area that is above the timberline or in a high latitude and that has a sparse ground cover of low plants or grasses and sedges. Syn: *fell; feldmark; fjeldmark.*

fellside A hillside or mountainside. Rarely used.

feloid (fel'-oid) A group name for the feldspar and feldspathoid minerals.

fels A rarely used term applied to massive metamorphic rock lacking schistosity or foliation, e.g., calcsilicate fels (Winkler, 1967). Cf: *granofels.*

felsenmeer (fel'-sen-meer) *block field.* Etymol: German "Felsenmeer", "sea of rocks".

felsic (fel'-sic) A mnemonic adjective derived from *fe*ldspar + *l*enad (feldspathoid) + *si*lica + *c*, and applied to an igneous rock having abundant light-colored minerals in its mode; also, applied to those minerals (quartz, feldspars, feldspathoids, muscovite) as a group. It is the complement of *mafic.*

felsic index A chemical parameter of igneous rocks, equal to 100 × (Na_2O+K_2O)/(Na_2O+K_2O+CaO). It ranges from about 25 (basalt) to 100 (rhyolite). It is most commonly plotted as the abscissa on variation diagrams, on which the ordinate represents the mafic index. It reflects changes produced by fractional crystallization of the felsic minerals (Simpson, 1954). Cf: *mafic index.* Abbrev: FI.

felsiphyric (fel-si-phyr'-ic) A syn. of *cryptocrystalline* originally used by Cross et al. (1906, p.703). Obsolete.

felsite (fel'-site) A general term for any light-colored, fine-grained or aphanitic extrusive or hypabyssal rock, with or without phenocrysts and composed chiefly of quartz and feldspar; a rock characterized by *felsitic* texture. Cf: *aphanite; mafite.*

felsitic (fel-sit'-ic) A syn. of *aphanitic* applied to light-colored microcrystalline rocks, with aphanitic then being reserved for similar dark-colored rocks. The term is not recommended because in its original use it was applied to a mineral substance now known to be a mixture of quartz and feldspar.

felsöbanyite (fel-sö-ban'-y-ite) A snow-white to yellow monoclinic mineral: $Al_4(OH)_{10}(SO_4)\cdot 4H_2O$. It has a similar unit cell to that of basaluminite, but one less H_2O molecule in the formula.

felsophyric (fel-so-phyr'-ic) Volcanic igneous texture in which phenocrysts occur in a dense ground mass of intergrown quartz and feldspar. Obsolete.

felsosphaerite (fel-so-sphae'-rite) A spherulite composed of a felsitic substance. Obsolete.

felspar (fel'-spar) A chiefly British spelling of *feldspar.*

felspathoid (fel'-spath-oid) *feldspathoid.*

felty (felt'-y) *pilotaxitic.*

femag (fe'-mag) *mafic.*

femic (fem'-ic) Said of an igneous rock having one or more normative, dark-colored iron-, magnesium-, or calcium-rich minerals as the major components of the norm; also, said of such minerals. Etymol: a mnemonic term derived from *ferric* + *magnesium* + *ic.* Cf: *basic; salic; mafic; felsic.*

femur (fe'-mur) (a) The thigh-bone of a tetrapod. (b) The third segment of a leg of an arachnid, forming the "hip" articulation with the preceding segment (trochanter) and the "knee" articulation with the following segment (patella) (TIP, 1955, pt. P, p.61). (c) A joint belonging to the proximal part of a prosomal appendage of a merostome.

fen Waterlogged, spongy ground containing alkaline decaying vegetation, characterized by reeds, that may develop into peat. It sometimes occurs in the sinkholes of karst regions. Cf: *bog; marsh; swamp.*

fenaksite (fe-nak'-site) A pale-rose triclinic mineral: $KNaFe^{2+}Si_4O_{10}$. Not to be confused with *phenakite.*

fence diagram [geochem] A diagram of chemical factors, such as Eh and pH, that influence mineral stability, having discrete fields defined by boundaries between phases in an assemblage of minerals, rocks, or compounds. Cf: *geochemical fence.*

fence diagram [geol] A drawing in perspective of three or more geologic sections, showing their relationships to one another.

fence effect [geophys exploration] An IP, resistivity, or electromagnetic anomaly produced by the presence of a nearby grounded conductor, such as a metal fence.

fencooperite A jet-black to dirty gray-brown trigonal mineral: $Ba_6Fe^{3+}_3Si_8O_{23}(CO_3)_2Cl_3\cdot H_2O$.

fenestra [paleont] (fe-nes'-tra) (a) A small opening in an invertebrate; e.g. an open space in a reticulate or anastomosing bryozoan colony, or an open or closed window in the wall or lorica of a tintinnid. Pl: fenestrae. Syn: *fenestrule.* (b) A relatively large opening in a bone or between adjacent bones that ordinarily does not serve as a passage for blood vessels or nerves.

fenestra [sed] A term used by Tebbutt et al. (1965, p.4) for *shrinkage pore,* or "primary or penecontemporaneous gap in rock framework, larger than grain-supported interstices". It may be an open space in the rock, or be completely or partly filled with secondarily introduced sediment or cement. Also used to describe the porosity and fabric of rocks with fenestral features (Choquette and Pray, 1970, p.246). Etymol: Latin, "window". Pl: fenestrae.

fenestrate (fe-nes'-trate, fe'-nes-trate) Having openings or transparent areas; perforated or reticulated. The term has been applied esp. to bryozoans possessing small windowlike openings between branches, arranged in a reticulate or anastomosing pattern; to corals having regularly perforated septa; and to pollen exhibiting large geometrically arranged holes in the exine. Syn: *fenestrated.*

fenestrated (fen'-es-trat'-ed) *fenestrate.*

fenestrule (fe-nes'-trule) *fenestra [paleont].*

fenêtre (fe-nêtre') A rarely used syn. of window. Etymol: French, "window".

fengcong (Pron: fung-sung) The Chinese term for a cluster of karst towers rising from a common base, separated by large *sinkholes.* Cf: *tower karst.* Etymol: Chinese, "peak cluster".

fenglin (Pron: fung-lin) The Chinese term for *karst towers* rising above an intervening alluvial plane. Cf: *tower karst.* Etymol: Chinese, "peak forest".

fengluanglite (feng-lu-ang'-lite) A thorian britholite-(Ce).

fenite (fen'-ite) A quartzo-feldspathic rock that has been altered by alkali metasomatism at the contact of a carbonatite intrusive complex. The process is called *fenitization.* Fenite is mostly alkali feldspar, with some aegirine, subordinate alkali-hornblende, and accessory sphene and apatite. The name "fenite" for the altered rock was originated by Brögger (Turner and Verhoogen, 1960).

fenitization (fen'-it-i-za'-tion) As generally used today, widespread alkali metasomatism of quartzo-feldspathic country rocks in the environs of carbonatite complexes.

fen peat *lowmoor peat.*

fenster (fen'-ster) The German equivalent of *window;* used, e.g., in structural geology.

feral (fe'-ral) Said of an "unsubdued" landform or landscape in early maturity, when the crests of ridges and spurs are shaped by the intersection of the slopes of valley sides that are for the most part still steep, so that the ridges are sharp and serrate (Cotton, 1958, p.110). Cf: *subdued.*

ferberite (fer'-ber-ite) A submetallic black monoclinic mineral of the *wolframite* group: $Fe^{2+}WO_4$. It is isomorphous with hübnerite, and may contain up to 20% manganese tungstate.

ferchromide (fer-chrom'-ide) A metallic gray cubic mineral: $Cr_3Fe_{0.4}$. Isomorphous with chromferide.

ferdisilicite (fer-di-sil'-i-cite) A metallic steel-gray tetragonal mineral: $FeSi_2$.

ferghanite (fer'-gha-nite) A sulfur-yellow secondary substance, possibly: $(UO_2)_3(VO_4)_2 \cdot 6H_2O$. It is perhaps a leached or weathered product of tyuyamunite. Also spelled: ferganite.

fergusite (fer'-gus-ite) (a) In the *IUGS classification,* a plutonic rock in which F is between 60 and 100, M is between 30 and 50, and potassium exceeds sodium. Cf: *ijolite.* (b) A plutonic foidite containing leucite and 30 to 60 percent mafic minerals, such as olivine, apatite, and biotite, with accessory opaque oxides. Its name, given by Pirsson in 1905, is derived from Fergus County, Montana. Cf: *arkite; missourite; italite.*

fergusonite-beta-(Ce) A red to reddish brown monoclinic mineral; dimorphous with fergusonite-(Ce).

fergusonite-beta-(Nd) A brownish-black tetragonal mineral of the *scheelite* group: $(Nd,Ce)NbO_4$. Dimorphous with fergusonite-(Nd).

fergusonite-beta-(Y) A light yellow monoclinic mineral; dimorphous with fergusonite-(Y).

fergusonite-(Ce) A dark red to black tetragonal mineral of the *scheelite* group: $CeNbO_4$.

fergusonite-(Nd) An ill-defined dark brown mineral of the *scheelite* group: $Nd(Nb,Ti)O_4$.

fergusonite-(Y) (fer'-gu-son-ite') A brownish-black tetragonal mineral of the *scheelite* group: $Y(Nb,Ta)O_4$. It is isomorphous with formanite and dimorphous with beta-fergusonite. It may contain erbium, cerium, iron, titanium, and uranium.

Fermat's principle (Fer'-mat's) The statement that a seismic wave will follow the path between two points that takes less time than variations of this path. Such a path is called a *minimum-time path.* Occasionally involves a maximum-time path, as with reversed branches, or an inflection point.

fermorite (fer'-mor-ite) A pale pinkish-white monoclinic (pseudo-hexagonal) mineral of the *apatite* group: $(Ca,Sr)_5(PO_4,AsO_4)_3(F,OH)$.

fern A vascular, nonflowering plant of the class Filicineae. Many modern ferns have complex fronds growing from an underground rhizome, and sporangia grouped on the surface of the leaf.

fernandinite (fer-nan'-di-nite) A dull green monoclinic mineral: $(Ca,Na,K)_{0.9}(V^{5+},V^{4+},Fe^{2+},Ti)_8O_{20} \cdot 4H_2O$. Classed as a "vanadium bronze".

feroxyhyte (fer-ox'-y-hyte) A brown hexagonal mineral: $\delta Fe^{3+}O(OH)$. It is a polymorph of goethite, lepidocrocite, and akaganéite, and is a constituent of oceanic "ferromanganese" nodules.

ferran A cutan composed of iron oxides, hydroxides, or oxyhydroxides.

ferrarisite (fer-rar'-i-site) A colorless triclinic mineral: $Ca_5(AsO_4)_2(AsO_3OH)_2 \cdot 9H_2O$. It is a dimorph of guérinite.

ferrazite (fer'-ra-zite) *gorceixite.*

Ferrel's law (Fer'-rel's) The statement that the centrifugal force produced by the rotation of the Earth (Coriolis force) causes a *rotational deflection* of currents of water and air to the right in the Northern Hemisphere and to the left in the Southern Hemisphere.

ferriallanite-(Ce) A black monoclinic mineral of the *epidote* group, representing allanite-(Ce) with essential Fe3+: $CaCe(Fe^{3+},Fe^{2+},Al)_3(SiO_4)(Si_2O_7)O(OH)$.

ferrian (fer'-ri-an) Containing ferric iron. Cf: *ferroan.*

ferri-annite (fer'-ri-an'-nite) *tetra-ferri-annite.*

ferri-argillan A cutan consisting of a mixture of clay minerals and iron oxides, hydroxides, or oxyhydroxides.

ferribarroisite A dark green to black monoclinic mineral of the *amphibole* group, representing barroisite with essential Fe^{3+}: $NaCa(Mg_3Fe^{3+}_2)(Si_7Al)O_{22}(OH)_2$.

ferric-ferronyböite A theoretical member of the *amphibole* group, representing the Fe^{2+}-Fe^{3+} end-member of the *nyböite* series: $Na_3(Mg_3Fe^{3+}_2)(Si_7Al)O_{22}(OH)_2$.

ferri-clinoferroholmquistite A vitreous black monoclinic member of the *amphibole* group, representing clinoholmquistite with essential Fe^{2+} and Fe^{3+}: $Li_2(Fe^{2+}_3Fe^{3+}_2)Si_8O_{22}(OH)_2$.

ferri-clinoholmquistite A vitreous black monoclinic member of the *amphibole* group, representing clinoholmquistite with essential Fe^{3+}: $Li_2(Mg_3Fe^{3+}_2)Si_8O_{22}(OH)_2$.

ferric-nyböite A theoretical member of the *amphibole* group, representing the Fe^{3+} end-member of the *nyböite* series: $Na_3(Mg_3Fe^{3+}_2)(Si_7Al)O_{22}(OH)_2$.

ferricopiapite (fer-ri-co'-pi-a-pite') A yellow to orange-yellow mineral of the *copiapite* group: $Fe^{3+}_{2/3}Fe^{3+}_4(SO_4)_6(OH)_2 \cdot 20H_2O$.

ferricrete (fer'-ri-crete) (a) A term suggested by Lamplugh (1902) for a sandstone or *conglomerate* consisting of surficial sand and gravel cemented into a hard mass by iron oxide derived from the oxidation of percolating solutions of iron salts. (b) A ferruginous *duricrust.* Etymol: *fer*ruginous + concrete. Cf: *calcrete; silcrete; laterite.*

ferricrust (fer'-ri-crust) (a) A general term for an indurated soil horizon cemented with iron oxide, mainly hematite. (b) The hard crust of an iron concretion.

ferrierite-K A monoclinic zeolite mineral with K as the principal interchangeable cation: $(K,Na,Mg)_{4.4}(Si,Al)_{36}O_{72} \cdot 20H_2O$.

ferrierite-Mg A white monoclinic zeolite mineral with Mg as the principal interchangeable cation: $(Mg,K,Ca)_{4.4}(Si,Al)_{36}O_{72} \cdot 20H_2O$.

ferrierite-Na (fer'-ri-er-ite') A platy colorless to pink or red monoclinic zeolite mineral with Na as the principal interchangeable cation: $(Na,K)Mg_2Ca_{0.5}(Si,Al)_{36}O_{72} \cdot 20H_2O$.

ferrifayalite (fer-ri-fay'-a-lite) *laihunite.*

ferriferous (fer-rif'-er-ous) Iron-bearing; said esp. of a mineral containing iron, or of a sedimentary rock that is richer in iron than is usually the case, such as a shale whose iron-oxide content is greater than 15%. Cf: *ferruginous.*

ferri-ferrobarroisite A dark green to black monoclinic member of the *amphibole* group, representing barroisite with Fe^{2+} and Fe^{3+} dominant: $NaCa(Fe^{2+}_3Fe^{3+}_2)(Si_7Al)O_{22}(OH)_2$.

ferri-ferrotschermakite A theoretical monoclinic member of the *amphibole* group, representing tschermakite with Fe^{2+} and Fe^{3+} dominant: $Ca_2(Fe^{2+}_3Fe^{3+}_2)(Si_6Al_2)O_{22}(OH)_2$.

ferrihydrite (fer-ri-hy'-drite) A yellow-brown to dark brown hexagonal mineral: $5Fe_2O_3 \cdot 9H_2O$.

ferri-katophorite A theoretical monoclinic member of the *amphibole* group, representing katophorite with essential Fe^{3+}: $Na_2Ca(Fe^{2+}_4Fe^{3+})(Si_7Al)O_{22}(OH)_2$.

ferrilith (fer'-ri-lith) A term suggested by Grabau (1924, p.298) for an iron-rich sedimentary rock (ironstone). Syn: ferrilyte.

ferrilotharmeyerite A yellow to brownish yellow monoclinic mineral: $Ca(Zn,Cu)(Fe^{3+},Zn)(AsO_3OH)_2(OH,H_2O)_2$. It is the ferric analogue of lotharmeyerite, and a member of the *tsumcorite* group.

ferri-magnesiotaramite A theoretical monoclinic member of the *amphibole* group, representing the Mg-Fe^{3+} end-member of *taramite* series: $Na_2Ca(Mg_3Fe^{3+}_2)(Si_6Al_2)O_{22}(OH)_2$.

ferrimagnetism (fer-ri-mag'-net-ism) A type of *magnetic order,* macroscopically resembling *ferromagnetism.* Magnetic cations at different crystal sites are antiferromagnetically coupled, yet a net magnetization exists because of inequality in the number or magnitude of atomic magnetic moments at the opposing sites. This type of magnetic order occurs in magnetite. Cf: *antiferromagnetism.*

ferrimolybdite (fer'-ri-mo-lyb'-dite) A yellowish orthorhombic mineral: $Fe^{3+}_2(MoO_4)_3 \cdot 8H_2O$(?). It occurs as an earthy powder or incrustation, or as silky, fibrous, and radiating crystals, and is formed by the oxidation of molybdenite. Cf: *molybdite.* Syn: *molybdic ocher.*

ferrinatrite (fer-ri-nat'-rite) A grayish-white or whitish-green trigonal mineral: $Na_3Fe^{3+}(SO_4)_3 \cdot 3H_2O$.

ferri-ottoliniite A vitreous black monoclinic mineral of the *amphibole* group: $NaLi(Fe^{3+}_2Mg_3)Si_8O_{22}(OH)_2$.

ferripedrizite A vitreous black Li-bearing monoclinic member of the *amphibole* group, representing pedrizite with essential Fe^{3+}: $NaLi_2(Fe^{3+}_2Mg_2Li)Si_8O_{22}(OH)_2$.

ferripyrophyllite A brownish yellow monoclinic mineral of the *pyrophyllite* group: $Fe^{3+}Si_2O_5(OH)$.

ferrisicklerite (fer-ri-sick'-ler-ite) A yellowish brown to dark-brown orthorhombic mineral of the *triphylite* group: $(Li,[])(Fe^{3+},Mn^{2+})PO_4$. It is the Fe^{3+}-dominant analogue of sicklerite.

ferristrunzite (fer-ri-strunz'-ite) A light brownish yellow triclinic mineral of the *strunzite* group: $Fe^{3+}Fe^{3+}_2(PO_4)_2(OH)_3 \cdot 5H_2O$.

ferrisurite A light to dark green monoclinic mineral: $(Pb,Ca)_{2.4}(Fe^{3+},Al)_2Si_4O_{10}(CO_3)_{1.7}(OH,F)_3 \cdot H_2O$. It is the ferric analogue of surite.

ferrisymplesite (fer-ri-sym'-ples-ite) An poorly defined amorphous amber-brown mineral of the *vivianite* group: $Fe^{3+}{}_3(AsO_4)_2(OH)_3{\cdot}6H_2O$(?). Cf:*symplesite.*

ferritaramite A rose-red, dark red-brown, or bluish black monoclinic mineral of the *amphibole* group, representing taramanite with essential Fe^{3+}: $Na_2Ca(Fe^{2+}{}_3Fe^{3+}{}_2)(Si_6Al_2)O_{22}(OH)_2$.

ferrite [ign] (fer'-rite) An obsolete term applied to grains, scales, and threads of unidentifiable, more or less transparent or amorphous, red, brown, or yellow iron oxide in the groundmass of a porphyritic rock (Johannsen, 1939, p.177). Cf: *opacite; viridite.*

ferrite [sed] A term used by Tieje (1921, p.655) for a cemented iron-rich sediment whose particles do not interlock.

ferritschermakite A dark green monoclinic mineral of the *amphibole* group, representing tschermakite with essential Fe^{3+}: $Ca_2(Mg_3Fe^{3+}{}_2)(Si_6Al_2)O_{22}(OH)_2$.

ferritungstite (fer-ri-tung'-stite) An ocherous yellow, orange, or brownish yellow cubic mineral: $(W,Fe^{3+})_2(O,OH)_6{\cdot}nH_2O$. It occurs as a pale-yellowish to brownish-yellow earthy powder. Syn: *tungstic ocher.*

ferriturquoise (fer-ri-tur'-quoise) A variety of turquoise containing 5% Fe_2O_3.

ferriwhittakerite A bluish-black monoclinic mineral of the *amphibole* group: $Na(NaLi)(Mg_2Fe^{3+}{}_2Li)Si_8O_{22}(OH)_2$.

ferri-winchite A dark green to black monoclinic member of the *amphibole* group, representing winchite with essential Fe^{3+}: $NaCa(Mg_4Fe^{3+})Si_8O_{22}(OH)_2$.

ferro-actinolite (fer'-ro-ac-tin'-o-lite) A green-black monoclinic mineral component representing a theoretical end-member of the *amphibole* group: $Ca_2Fe^{2+}{}_5Si_8O_{22}(OH)_2$. Syn: ferrotremolite.

ferroalluaudite (fer-ro-all-u-aud'-ite) An olive-green to greenish black monoclinic mineral of the *alluaudite* group: $(Na,Ca)Fe^{2+}Fe^{3+}{}_2(PO_4)_3$.

ferro-alumino-barroisite A theoretical monoclinic member of the *amphibole* group, representing barroisite with essential Fe^{2+} and Al: $NaCa(Fe^{2+}{}_3Al_2)(Si_7Al)O_{22}(OH)_2$. Syn: *alumino-ferrobarroisite.*

ferro-aluminoceladonite An earthy blue-green monoclinic mineral of the *mica* group: $K(Fe^{2+},Mg)(Al,Fe^{3+})Si_4O_{10}(OH)_2$.

ferro-alumino-tschermakite A green monoclinic mineral of the *amphibole* group, representing tschermakite with essential Fe^{2+} and Al: $Ca_2(Fe^{2+}{}_3Al_2)(Si_6Al_2)O_{22}(OH)_2$.

ferro-alumino-winchite A theoretical monoclinic mineral of the *amphibole* group, representing winchite with essential Fe^{2+} and Al: $NaCa(Fe^{2+}{}_4Al)Si_8O_{22}(OH)_2$.

ferroan (fer'-ro-an) Containing ferrous iron. Cf: *ferrian.*

ferroan dolomite A mineral that is intermediate in composition between dolomite and ferrodolomite; specif. *ankerite.*

ferro-anthophyllite (fer-ro-an-tho'-phyll-ite) A green, grayish-green to dark brown orthorhombic mineral of the *amphibole* group, representing the Fe^{2+} analogue of anthophyllite: $Fe^{2+}{}_7Si_8O_{22}(OH)_2$.

ferro-axinite (fer-ro-ax'-i-nite) A lilac-brown triclinic mineral of the *axinite* group: $Ca_2(Fe,Mn)Al_2BSi_4O_{15}(OH)$.

ferrobarroisite A theoretical monoclinic mineral of the *amphibole* group, representing barroisite with essential Fe^{2+}: $NaCa(Fe^{2+}{}_3AlFe^{3+})(Si_7Al)O_{22}(OH)_2$.

ferrobasalt (fer'-ro-ba-salt') A lava marked by strong relative and absolute enrichment in iron. "Total iron normally exceeds 12 or 13%, and MgO is less than 6%. Silica, which shows little enrichment with respect to primitive tholeiites, ranges from about 48 to 50%" (McBirney and Williams, 1969, p.144).

ferrobustamite (fer-ro-bus'-tam-ite) A colorless, pink or brown triclinic mineral of the *wollastonite* group: $Ca(Fe^{2+},Ca,Mn^{2+})Si_2O_6$. It is isostructural with bustamite.

ferrocarbonatite A chemically defined carbonatite in which wt% $CaO/(CaO+MgO+FeO+Fe_2O_3+MnO)<0.8$ and $MgO<(FeO+Fe_2O_3+MnO)$ (Woolley and Kempe, 1989).

ferrocarpholite (fer-ro-car'-pho-lite) A dark green or grayish green orthorhombic mineral of the *carpholite* group: $(Fe^{2+},Mg)Al_2Si_2O_6(OH)_4$.

ferroceladonite A micaceous blue-green monoclinic mineral of the *mica* group: $K(Fe^{2+},Mg)(Fe^{3+},Al)Si_4O_{10}(OH)_2$.

ferrocolumbite (fer'-ro-co-lum'-bite) A black orthorhombic mineral of the *columbite* group: $(Fe^{2+},Mn^{2+})(Nb,Ta)_2O_6$.

ferrocopiapite (fer-ro-co'-pi-a-pite') *copiapite.*

ferrodiorite (fer-ro-di'-o-rite) A dioritic rock "in which the actual (not normative) plagioclase is less calcic than about An_{50}, and the ferromagnesian minerals iron-rich" (Wager and Brown, 1967, p.78).

ferrodolomite (fer-ro-do'-lo-mite) *ankerite.*

Ferrods Spodosols that have more than six times as much free iron (elemental) than organic carbon in the spodic horizon. Ferrods are rarely saturated with water or do not have characteristics associated with wetness. A suborder in the U.S. system of soil taxonomy.

ferro-eckermannite A theoretical monoclinic member of the *amphibole* group, representing eckermannite with essential Fe^{2+}: $Na_3(Fe^{2+}{}_4Al)Si_8O_{22}(OH)_2$.

ferro-edenite A dark green to black monoclinic member of the *amphibole* group, representing edenite with essential Fe^{2+}: $NaCa_2Fe^{2+}{}_5(Si_7Al)O_{22}(OH)_2$.

ferroelastic Said of a crystal that exhibits elastic hysteresis.

ferroelastic phase transformation A *phase transformation* that occurs at a well defined temperature below which a crystal is *ferroelastic*, and above which it is not. Ferroelastic crystals exhibit lattice distortions that can be reoriented by external stress.

ferroelectric Said of a crystal that exhibits hysteresis with respect to an applied electric field and an induced dipole field.

ferroelectric phase transformation A *phase transformation* that occurs at a well defined temperature below which a crystal is *ferroelectric*, and above which it is not.

ferro-ferri-tschermakite A theoretical monoclinic member of the *amphibole* group, representing tschermakite with essential Fe^{2+} and Fe^{3+}: $Ca_2(Fe^{2+}{}_3Fe^{3+}{}_2)(Si_6Al_2)O_{22}(OH)_2$.

ferro-ferri-winchite A green monoclinic member of the *amphibole* group, representing winchite with essential Fe^{2+} and Fe^{3+}: $NaCa(Fe^{2+}{}_4Fe^{3+})Si_8O_{22}(OH)_2$.

ferrogabbro (fer-ro-gab'-bro) A name given by Wager and Deer in 1939 to igneous rocks in the upper zone of the Skaergaard intrusion, East Greenland, containing iron-rich pyroxenes and olivine. The term was withdrawn by Wager and Brown (1967), in favor of *ferrodiorite.*

ferrogedrite A gray, yellowish brown to dark brown orthorhombic mineral of the *amphibole* group, representing the Fe^{2+} analogue of gedrite: $Fe^{2+}{}_5Al_2(Si_6Al_2)O_{22}(OH)_2$.

ferroglaucophane A blue to bluish gray monoclinic member of the *amphibole* group, representing glaucophane with essential Fe^{2+}: $Na_2(Fe^{2+}{}_3Al_2)Si_8O_{22}(OH)_2$.

ferrohagendorfite (fer-ro-hag'-en-dorf-ite) A hypothetical member of the *alluaudite* mineral group, representing the Fe^{2+} analogue of hagendorfite : $NaCaFe^{2+}Fe_2{}^{2+}(PO_4)_3$.

ferrohexahydrite (fer'-ro-hex'-a-hy'-drite) A colorless monoclinic mineral of the *hexahydrite* group: $Fe^{2+}SO_4{\cdot}6H_2O$.

ferrohögbomite-2N2S A chestnut-brown hexagonal mineral of the *högbomite* group: $(Fe,Mg,Zn,Al)_6Al_{14}(Ti,Fe)_2O_{30}(OH)_2$. The crystal structure comprises two nolanite and two spinel modules.

ferroholmquistite A theoretical orthorhombic member of the *amphibole* group, representing holmquistite with essential Fe^{2+}: $Li_2(Fe^{2+}{}_3Al_2)Si_8O_{22}(OH)_2$.

ferrohornblende A green, greenish brown, dark brown, or black monoclinic member of the *amphibole* group, representing hornblende with essential Fe^{2+}: $Ca_2(Fe^{2+}{}_4Al)(Si_7Al)O_{22}(OH)_2$.

ferrohortonolite (fer'-ro-hor-ton'-o-lite) An unnecessary name for a mineral of the forsterite-fayalite solid-solution series, containing 70 to 90% of the fayalite component.

ferrokaersutite A dark brown to black monoclinic member of the *amphibole* group, representing kaersutite with essential Fe^{2+}: $NaCa_2(Fe^{2+}{}_4Ti)(Si_6Al_2)O_{22}(OH)$.

ferrokentbrooksite A red to reddish brown rhombohedral mineral of the *eudialyte* group: $Na_{15}Ca_6Fe^{2+}{}_3Zr_3Nb(Si_{25}O_{73})(O,OH,H_2O)_3(F,Cl)_2$.

ferrokësterite (fer-ro-kes'-ter-ite) A steel gray tetragonal mineral: $Cu_2(Fe,Zn)SnS_4$. It is the Fe^{2+} analogue of kësterite.

ferrokinoshitalite A micaceous dark green monoclinic member of the *brittle mica* group, and the Fe^{2+} analogue of kinoshitalite: $(Ba,K)(Fe^{2+},Mg)_3(Si,Al)_4O_{10}(OH,F)_2$.

ferroleakeite A theoretical monoclinic member of the *amphibole* group, representing leakeite containing essential Fe^{2+}: $Na_3(Fe^{2+}{}_2Fe^{3+}{}_2Li)Si_8O_{22}(OH,F)_2$.

ferromagnesian (fer'-ro-mag-ne'-sian) Containing iron and magnesium; applied to *mafic* minerals.

ferromagnetic Said of a crystal that exhibits magnetic hysteresis.

ferromagnetic phase transformation A *phase transformation* that occurs at a well defined temperature below which a crystal is *ferromagnetic*, and above which it is not.

ferromagnetism (fer-ro-mag'-net-ism) A type of *magnetic order* in which all magnetic cations in a domain have moments aligned in the same direction; loosely, any type of magnetic order. Cf: *ferrimagnetism; antiferromagnetism.*

ferronickelplatinum (fer'-ro-nick'-el-plat'-i-num) A metallic silvery-white tetragonal mineral of the *tetraferroplatinum* group: (Ni,Fe)Pt.

ferronigerite-2N1S A dark brown, reddish brown, or red trigonal mineral of the *nigerite* group: $Al_{10.9}Sn_2Fe_{1.7}Zn_{0.7}O_{22}(OH)_2$. The crystal structure comprises two nolanite modules and one spinel module.

ferronigerite-6N6S A colorless rhombohedral mineral of the *nigerite* group: $(Fe,Zn)_4Sn_2(Al,Fe)_{15}O_{30}(OH)_2$. The crystal structure comprises six nolanite and six spinel modules.

ferronordite-(Ce) A colorless to light brown orthorhombic mineral of the *nordite* group: $Na_3Sr(Ce,La)(Fe^{2+},Mn^{2+},Zn)Si_6O_{17}$.

ferronordite-(La) A colorless to pale brown orthorhombic mineral of the *nordite* group: $Na_3Sr(La,Ce)Fe^{2+}Si_6O_{17}$.

ferronyböite A theoretical monoclinic mineral of the *amphibole* group, representing nyböite with essential Fe^{2+}: $Na_3(Fe^{2+}{}_3Al_2)(Si_7Al)O_{22}(OH)_2$.

ferropargasite (fer-ro-par'-gas-ite) A brown monoclinic mineral of the *amphibole* group, representing pargasite with essential Fe^{2+}: $NaCa_2(Fe^{2+}{}_4Al)(Si_6Al_2)O_{22}(OH)_2$.

ferropumpellyite *pumpellyite-(Fe²⁺).*

ferropyrosmalite A brown, gray, or green trigonal mineral of the *pyrosmalite* group: $(Fe^{2+},Mn^{2+})_8Si_6O_{15}(OH,Cl)_{10}$.

ferrorhodsite A metallic black cubic mineral of the *linnaeite* group: $(Fe,Cu)(Rh,Ir,Pt)_2S_4$.

ferrorichterite A brown or green monoclinic mineral of the *amphibole* group, representing richterite with essential Fe^{2+}: $Na_2CaFe^{2+}{}_5Si_8O_{22}(OH)_2$.

ferrosaponite A dark green or brownish green monoclinic mineral of the *montmorillonite* group, representing saponite with dominant Fe^{2+}: $Ca_{0.3}(Fe^{2+},Mg,Fe^{3+})_3(Si,Al)_4O_{10}\cdot 4H_2O$.

ferroselite (fer-ro-sel'-ite) A metallic white to rose-cream orthorhombic mineral of the *marcasite* group: $FeSe_2$.

ferrosilite (fer-ro-sil'-ite) A dark green, dark brown, or black orthorhombic mineral of the *pyroxene* group: $(Fe^{2+},Mg)_2Si_2O_6$. It is the iron analogue of enstatite.

ferrospinel (fer'-ro-spi-nel') (a) *hercynite.* (b) A synthetic magnetic substance of spinel structure, containing iron, and being a poor conductor of electricity.

ferrostrunzite (fer-ro-strunz'-ite) A light brown triclinic mineral of the *strunzite* group: $Fe^{2+}Fe^{3+}{}_2(PO_4)_2(OH)_2\cdot 6H_2O$.

ferrotaaffeite-6N'3S A vitreous green rhombohedral mineral of the *taaffeite* group: $Be(Fe,Zn,Mg)_2Al_6O_{12}$. The crystal structure comprises six modules of nolanite-like structure and 3 modules of spinel structure. Syn: *pehrmanite.*

ferrotantalite (fer-ro-tan'-ta-lite) An black or brownish black orthorhombic mineral of the *columbite* group: $(Fe^{2+},Mn^{2+})(Ta,Nb)_2O_6$. It is a dimorph of ferrotantalite.

ferrotapiolite (fer-ro-tap'-i-ol-ite) A black or brownish black orthorhombic mineral of the *tapiolite* group: $(Fe^{2+},Mn^{2+})(Ta,Nb)_2O_6$. It is a dimorph of ferrotantalite.

ferrotellurite A yellow mineral of doubtful validity: $FeTeO_4$ (?).

ferrotitanowodginite A dark brown to black monoclinic mineral of the *wodginite* group: $(Fe^{2+},Mn^{2+})(Ti,Sn^{4+},Ta,Fe^{3+})(Ta,Nb)_2O_8$.

ferrotschermakite A green monoclinic mineral of the *amphibole* group, representing tschermakite with essential Fe^{2+}: $Ca_2(Fe^{2+}{}_3AlFe^{3+})(Si_6Al_2)O_{22}(OH)_2$.

ferrotychite (fer-ro-ty'-chite) A colorless to pale yellow cubic mineral, the Fe^{2+} analogue of tychite, and a member of the *northupite* group: $Na_6(Fe^{2+},Mg,Mn^{2+})_2(SO_4)(CO_3)_4$.

ferrous metal Those metals [that are] typically mined for their alloying properties with iron in the manufacture of steel. Customarily, this includes iron, chromium, cobalt, manganese, molybdenum, nickel, tungsten, and vanadium.

ferrowinchite A theoretical monoclinic mineral of the *amphibole* group, representing winchite with essential Fe^{2+}: $NaCa(Fe^{2+}{}_4Al)Si_8O_{22}(OH)_2$.

ferrowodginite A dark brown to black monoclinic mineral of the *wodginite* group: $(Fe^{2+},Mn^{2+})(Sn^{4+},Ti,Ta,Fe^{3+})(Ta,Nb)_2O_8$.

ferrowyllieite (fer-ro-wyl'-lie-ite) A submetallic bluish green to grayish green monoclinic mineral of the *wyllieite* group: $(Na,Ca,Mn^{2+})_2(Fe^{2+},Mn^{2+})(Fe^{2+},Fe^{3+},Mg)Al(PO_4)_3$.

ferruccite (fer-ruc'-cite) A colorless or white orthorhombic mineral: $NaBF_4$.

ferruginate (fer-ru'-gi-nate) adj. A term restricted by Allen (1936, p.22) to designate the iron-bearing cement of a sedimentary rock. v. To stain a rock with an iron compound.

ferruginous (fer-ru'-gi-nous) (a) Pertaining to or containing iron, e.g. a sandstone that is cemented with iron oxide. Cf: *ferriferous; siderose.* (b) Said of a rock having a red or rusty color due to the presence of ferric oxide (the quantity of which may be very small).

fersilicite (fer-sil'-i-cite) A metallic white cubic mineral: FeSi.

fersmanite (fers'-man-ite) A dark brown or golden yellow monoclinic or triclinic mineral: $Ca_4(Na,Ca)_4(Ti,Nb)_4(Si_2O_7)_2O_8F_3$. Sometimes misspelled: fersmannite.

fersmite (fers'-mite) A black, brown, or yellow orthorhombic mineral of the *euxenite* group: $(Ca,Ce,Na)(Nb,Ta,Ti)_2(O,OH,F)_6$.

fertile mantle A part of the Earth's mantle which retains its original concentrations of incompatible elements, or has been enriched in them, so that it is capable of yielding magma by partial fusion.

ferutite (fer-u'-tite) *davidite-(La).*

feruvite (fer-uv'-ite) A brown-black rhombohedral mineral of the *tourmaline* group: $(Ca,Na)(Fe^{2+},Mg)_3(Al,Mg)_6(BO_3)_3Si_6O_{18}(OH)_4$.

fervanite (fer'-van-ite) A golden-brown monoclinic mineral: $Fe^{3+}{}_4V_4O_{16}\cdot 5H_2O$. It occurs with radioactive minerals but is not itself radioactive.

Festiniogian (Fes'-ti-ni-o'-gi-an) European stage: Upper Cambrian (above Maentwrogian, below Dolgellian).

festoon (fes-toon') The upfolded or pointed part of a layer in a congeliturbate. Ant: *pocket [soil].*

festoon cross-bedding A variety of *trough cross-bedding* described by Knight (1929); it consists of elongate, semi-ellipsoidal, eroded, plunging troughs or scooplike structures that are filled by sets of thin laminae conforming in general to the shapes of the troughs, and that crosscut each other so that only parts of each unit are preserved, resulting in a festoonlike (looped or curved) appearance in section. The cross-beds are deposited on concave surfaces so that both the lower bounding surfaces and the cross-beds are trough-shaped.

festooned pahoehoe (fes-tooned') A type of *pahoehoe,* the ropy surface of which has been dragged by flow of underlying molten lava into festoon patterns.

festoon flow On Venus, a large radar-bright lava-flow field characterized by organized patterns of internal ridges and flow bands (Bougher et al., 1997).

fetch (a) The area of the ocean surface over which a constant and uniform wind generates waves. (b) The distance over which the wind blows measured horizontally in the direction of the wind.

fetiasite A submetallic brown or black monoclinic mineral: $(Fe^{2+},Fe^{3+},Ti^{4+})_3O_2(As^{3+}{}_2O_5)$.

fettelite A submetallic dark violet or scarlet hexagonal mineral: $Ag_{24}HgAs_5S_{20}$.

feuerstein (feu'-er-stein) A syn. of *firestone.* Etymol: German *Feuerstein,* "flint".

FFT Acronym for Fast Fourier Transform. A computer algorithm that rapidly transforms a sequence of observations given in Cartesian coordinates (x and y) or as a *time series* to a set of orthogonal harmonics characterized by frequency and phase angle. Cf: *Fourier analysis.*

FI *felsic index.*

fiamme (fi-am'-me) Dark, vitric lenses in welded tuffs, averaging a few centimeters in length, perhaps formed by the collapse of fragments of pumice. The presence of fiamme may be called *flame structure [pyroclast].* Etymol: Pl. of Italian "fiamma", "flame". Cf: *piperno.*

fianelite A vitreous orange-red monoclinic mineral: $Mn^{2+}{}_2(V,As)_2O_7\cdot 2H_2O$.

fiard Anglicized variant of *fjard.*

fiasconite (fi-as'-co-nite) An anorthite-bearing leucitite-basanite that also contains augite, olivine, nepheline, and iron oxides. Its name (Johannsen, 1939) is derived from Montefiascone, Italy. Obsolete.

fiber (fi'-ber) A strengthening cell, usually elongated, tapering, and thick-walled, occurring in various parts of vascular plants.

fiber tracheid A type of *tracheid* that occurs in angiospermous secondary wood, with a thick secondary wall, pointed ends, and small bordered pits having lenticular to slitlike, usually extended, inner apertures (Record, 1934, p.30).

Fibrists Histosols that have a high content of undecomposed plant fibers and a bulk density less than about 0.1 g cm^{-3}. Fibrists are saturated with water for periods long enough to limit their use for most crops unless they are artificially drained. A suborder in the U.S. system of soil taxonomy.

fibroblastic (fi-bro-blas'-tic) Pertaining to a *homeoblastic* type of texture of a metamorphic rock due to the development during recrystallization of minerals with a fibrous habit. Cf: *nematoblastic.*

fibrocrystalline (fi-bro-crys'-tal-line) Characterized by the presence of fibrous crystals. Obsolete.

fibroferrite (fi-bro-fer'-rite) A silky yellowish rhombohedral mineral: $Fe^{3+}(SO_4)(OH)\cdot 5H_2O$.

fibrolite (fi'-bro-lite) *sillimanite.*

fibrous conodont element (fi'-brous) A *lamellar conodont element* in which the lamellae are thick and *white matter* is absent or reduced to a thin column along cusp and denticle axes, and whose broken edges are typically frayed. Elements that lack appreciable white matter but are not known to fracture with fibrous habitus are best described as hyaline.

fibrous habit The tendency of certain minerals, e.g. asbestos, to crystallize in needlelike grains or fibers.

fibrous ice *acicular ice.*

fibrous layer The *secondary layer* in many articulate brachiopods, secreted intracellularly as fibers bounded by cytoplasmic sheaths.

fibrous ligament The part of a *ligament* of a bivalve mollusk characterized by fibrous structure and in which *conchiolin* is commonly impregnated with calcium carbonate. It is secreted by the epithelium of the mantle and is elastic chiefly to compressional stresses. Cf: *lamellar ligament.*

fibrous peat Peat in which original plant structures are only slightly altered by degradation of cellulose matter. It is tough and nonplastic. Cf: *pseudofibrous peat; amorphous peat.* Syn: *woody peat.*

fibrous prismatic structure A *prismatic structure* in which each first-order prism is like a simple prism in showing a nonspherulitic prismatic and noncomposite prismatic substructure, but the prisms have much higher length/width ratios than typical simple prisms, appearing as long fibers (Kobayashi, 1964, 1969, 1971; Carter and Tevesz, 1978; Uozumi and Suzuki, 1981; Carter, 1980b).

fibrous texture In mineral deposits, a pattern of finely acicular, rodlike crystals, e.g. in chrysotile and amphibole asbestos. See also: *cross fiber.*

fichtelite (fich'-tel-ite) A white, translucent, monoclinic, non-aromatic, fully saturated hydrocarbon: $C_{19}H_{34}$. It is found in fossil wood.

fiducial mark (fi-du'-cial) An index or point used as a basis of reference; one of usually four index marks connected with a camera lens (as on the metal frame that encloses the negative) that form an image on the negative or print such that lines drawn between opposing points intersect at and thereby define the *principal point* of the photograph. Syn: *collimating mark.*

fiducial time A time marked on a record to correspond to some arbitrary time. Such marks may aid in synchronizing different records or may indicate a reference, such as a *datum plane [seis].*

fiedlerite (fied'-ler-ite) An adamantine colorless monoclinic or triclinic mineral: $Pb_3Cl_4F(OH)\cdot H_2O$.

field [eco geol] A region or area that possesses or is characterized by a particular mineral resource, e.g. gold field, coal field.

field [geol] A broad term for the area, away from the laboratory and esp. outdoors, in which a geologist makes firsthand observations and collects data, rock and mineral samples, and fossils.

field [geophys] That space in which an effect, e.g., gravitational or magnetic, exists, or the potential to exert a force on a measuring instrument that is placed within that space. It is characterized by continuity, i.e., there is a value associated with every location within the space. The field at a point is the gradient of the potential at that point; it is a vector quantity. See also: *field intensity; potential [geophys].*

field [ice] (a) *ice field [ice].* (b) A very large floe or other unbroken area of sea ice.

field capacity The quantity of water held by soil or rock against the pull of gravity. It is sometimes limited to a certain drainage period, thereby distinguishing it from *specific retention* which is not limited by time. Syn: *field-moisture capacity; normal moisture capacity.*

field classification A preliminary analysis of fossils or hand specimens of rocks or minerals in the field, usually with the aid of a hand lens.

field coefficient of permeability *field permeability coefficient.*

field completion Obtaining additional information in the field in order to edit a topographic map from a compiled manuscript or to fill in and confirm that part of a map manuscript prepared by stereocompilation.

field contouring Contouring of a topographic map by field methods accomplished by planetable surveys. It is usually done for terrain unsuitable for contouring by photogrammetric methods. Cf: *field sketching.*

field focus The total area or volume that is the source of an earthquake, inferred from the area of shaking as observed in the field. The concept is inexact and the term is not commonly used.

field geology Geology as practiced by direct observation in the field; original, primary reconnaissance; *field work.*

field ice (a) An obsolete term for *consolidated pack ice* consisting of very large, relatively flat floes many kilometers across. (b) A general term used for all types of *sea ice* except that newly formed.

field intensity The force of attraction exerted on a unit particle at a point by a field.

field map A preliminary or original map of the geology of an area, made in the field, on which a final map may be based.

field moisture Water present in the ground above the water table.

field-moisture capacity *field capacity.*

field-moisture deficiency The amount of water required to raise the moisture content of the soil to field capacity.

field-moisture equivalent The minimum water content of a soil mass, expressed as a percentage of its dry weight, at which a drop of water placed on a smoothed surface of the soil will not be absorbed but will spread out, giving a shiny appearance to the soil.

field of force The force that would be exerted on a unit particle at any location within the field.

field of view (a) *angular field of view.* (b) *instantaneous field of view.*

field permeability coefficient An obsolete term for *permeability coefficient,* defined for prevailing conditions rather than for a temperature of 60°F. Syn: coefficient of field permeability; *intrinsic permeability (b).*

field reversal *geomagnetic reversal.*

field sketching The art of drawing contours based on the elevations of selected features located on a planetable sheet. Cf: *field contouring.* See also: *contour sketching.*

fieldstone Rough stone collected as found in nature, with limited processing and shaping, for use in landscaping.

field well A well drilled for oil or gas within the area of a pool that has already been essentially proved for production (Lahee, 1962, p.132-133).

field work *field geology.*

fiery cloud (fi'-er-y) *nuée ardente.*

figure of the Earth (fig'-ure) The *geoid,* or surface of the Earth, as approximated by mean sea level over the oceans and the sea-level surface extended continuously through the continents. Irregularities or undulations of the geoid have been called humps and hollows, but the surface is nowhere concave.

filamented pahoehoe (fil'-a-ment'-ed) A type of *pahoehoe,* the surface of which displays threadlike strands that are formed by escaping gas bubbles and are recumbent and aligned with the direction of flow. It is a common type and is often found superimposed on other forms.

filiform (fil'-i-form) *capillary [mineral].*

filiform lapilli *Pele's hair.*

filipstadite (fil'-ip-stad-ite) A metallic black cubic mineral of the *spinel* group: $(Mn^{2+},Mg)_4Sb^{5+}Fe^{3+}O_8$.

fill [eng] (a) Man-made deposits of natural earth materials (e.g. rock, soil, gravel) and waste materials (e.g. tailings or spoil from dredging), used to fill an enclosed space such as an old stope or chamber in a mine, to extend shore land into a lake or bay, or in building dams. See also: *backfill; made land.* (b) Soil or loose rock used to raise the surface of low-lying land, such as an embankment to fill a hollow or ravine in railroad construction. Also, the place filled by such an embankment. (c) The depth to which material is to be placed to bring the surface to a predetermined grade.

fill [sed] Any sediment deposited by any agent so as to fill or partly fill a valley, sink, or other depression.

fill [speleo] Detrital material partly or completely filling a cave. Cf: *cave earth.*

filled-lake plain A swampy plain formed by the filling of a lake by sediments, aided by the growth of plants (Tarr, 1902, p.82).

filled valley A wide-basin valley, in an arid or semiarid region, that contains abundant alluvium in the form of fans, flood plains, and lake deposits.

filler [materials] (fill'-er) *mineral filler.*

filler [streams] A stream that empties into and fills a lake.

fill-in fill terrace A terrace left by a stream that, having incised its valley fill, partly fills up the new valley and incises anew (Schieferdecker, 1959, term 1514).

filling temperature (fill'-ing) The temperature at which fluid was originally trapped in a *fluid inclusion,* estimated by observing the homogenization temperature and applying a pressure correction (Roedder, 1979, p.724-725).

fillowite (fil'-low-ite) (a) A brown, reddish brown, or yellow rhombohedral mineral: $Na_2CaMn^{2+}_7(PO_4)_6$. (b) A group name for minerals with the analogous formula but with Ca replaced by Fe, and Mn replaced by Fe^{2+} or Mg.

fillstrath terrace (fill'-strath) A *fill terrace* whose surface has been eroded to a level below the original depositional surface (Howard, 1959, p.242); it consists of alluvial material as contrasted with a *strath terrace* formed in bedrock.

fill terrace (a) A term used by Bucher (1932, p.131) for a remnant, resulting from stream rejuvenation, of a flat valley bottom (or of an alluvial plain) that had been produced by stream aggradation; e.g. an *alluvial terrace* or a glacial terrace. (b) The part of a former alluvial valley floor built upward by deposition of valley-filling sediments (Howard, 1959, p.242); it includes *filltop terrace* and *fillstrath terrace.* Leopold et al. (1964, p.460) noted that the term is "confusing" and "should probably be abandoned".

filltop terrace (fill'-top) A *fill terrace* whose flat surface is the original depositional surface (Howard, 1959, p.242).

film separate *separate [cart].*

film water *pellicular water.*

filose Threadlike.

filter [math] A mathematical operator in the form of a vector or matrix which is convolved with a data sequence to emphasize some feature or component of the data. For example, filters are used for edge detection, the removal of unwanted frequencies, and to emphasize specific orientations.

filter [photo] (fil'-ter) Any transparent material that, by absorption or reflection, selectively modifies the radiation transmitted through an optical system; specif. a glass or gelatin plate placed in front of, in, or behind a camera lens to reduce or eliminate the effect of light of a certain color or colors on the film or plate.

filter [seis] A device or system that changes the waveform or amplitude of a signal. A filter discriminates most commonly on the basis of frequency, but dip, wavelength, amplitude, apparent velocity, and other bases are sometimes used. The device may be electrical or mechanical, natural, or it may be a computer. The Earth acts as a filter to seismic waves. Cf: *convolution [seis].*

filter bridge A narrow *land bridge* that permits the selective migration of some organisms.

filter cake *mud cake [drill].*

filter feeder An animal that obtains its food by removing particulate organic matter from water that passes through a part of its body used as a filter; e.g., bryozoans, barnacles, bivalves. Cf: *suspension feeder.*

filtering (fil'-ter-ing) (a) The attenuation of certain components and the enhancement of others. Usually refers to discriminating on basis of frequency, which may be done electrically, or numerically in a digital computer after the signal has been recorded, or naturally. Cf: *convolution [seis].* (b) As used in image analysis, the selective process of removing certain spectral or spatial frequencies to highlight or enhance features in an image (Krzanowski et al., 1993).

filter pressing A process of magmatic differentiation wherein a magma, having crystallized to a "mush" of interlocking crystals in liquid, is compressed by earth movements and the liquid moves toward regions of lower pressure, thus becoming separated from the crystals.

filtration (fil-tra'-tion) Removal of suspended and/or colloidal material from a liquid by passing it through a relatively fine porous medium.

filtrational (fil-tra'-tion-al) Said of an ore-forming fluid or mineralizer that is a nonmagmatic underground water (Smirnov, 1968). Cf: *juvenile [ore dep]; assimilated.*

filtration spring A spring whose water percolates from numerous small openings in permeable material. It may have either a small or a large discharge (Meinzer, 1923, p.50). Cf: *fracture spring; seepage spring.*

filum aquae (fi'-lum aq'-uae) The thread of a stream. Etymol: Latin, "thread of water". Pl: fila aquarum.

fimmenite (fim'-men-ite) A peat that is derived mainly from spores.

finandranite (fin-an'-dra-nite) A coarse-grained potassium-rich *syenite* composed of alkali feldspar, amphibole, and some biotite, ilmenite, and apatite; described from Malagasy in 1922 by Lacroix. Obsolete.

find A meteorite not seen to fall, but recognized as such by its composition and structure. Cf: *fall [meteorite].*

fine Composed of or constituting relatively small particles; e.g. "fine sandy loam". Ant: *coarse.*

fine admixture A term applied by Udden (1914) to an *admixture* (in a sediment of several size grades) whose particles are finer than those of the dominant or maximum grade; material finer than that found in the maximum histogram class.

fine aggregate The portion of an *aggregate* consisting of particles with diameters smaller than approximately 1/4 in. or 4.76 mm. Cf: *coarse aggregate.*

fine clay A geologic term for a clay particle having a diameter in the range of 1/2048 to 1/1024 mm (0.5-1 micrometer, or 11 to 10 phi units). Also, a loose aggregate of clay consisting of fine clay particles.

fine complex crossed lamellar structure A *complex crossed lamellar structure* in which the first-order lamellae consist of only a few parallel, elongate structural subunits (Taylor et al., 1973, p.265; Carter, 1980b). A similar structure occurs in the chitino-phosphatic "B zone" of the inarticulate brachiopods *Lingula unguis* and *Lingula reevei* (Iwata, 1981) and in the innermost enamel zone in shark fangs and cutting teeth (Preuschoft et al., 1974). Abbrev: FCCL.

fine-grained (a) Said of a crystalline or glassy rock, and of its texture, in which the individual minerals are relatively small; specif. said of an igneous rock whose particles have an average diameter less than 1 mm (0.04 in.). Syn: *aphanitic.* (b) Said of a sediment or sedimentary rock, and of its texture, in which the individual constituents are too small to distinguish with the unaided eye; specif. said of a sediment or rock whose particles have an average diameter less than 1/16 mm (62 micrometers, or silt size and smaller). The term is used in a relative sense, and various size limits have been suggested and used. Cf: *coarse-grained; medium-grained.* (c) Said of a soil in which silt and/or clay predominate. In the U.S., the maximum average diameter of the constituent particles is 0.05 mm (0.002 in.), or as used by engineers, 0.074 mm (passing U.S. standard sieve No. 200); the International Society of Soil Science recognizes a diameter limit of 0.02 mm. Cf: *coarse-grained.*

fine-granular *microgranular [ign].*

fine gravel (a) A soil term used in the U.S. for *gravel* whose particles have a diameter in the range of 2-12.5 mm (1/12 to 1/2 in.); it was formerly applied to soil particles (now called *very coarse sand*) having diameters of 1-2 mm. (b) An engineering term for *gravel* whose particles have a diameter in the range of 4.76 mm (retained on U.S. standard sieve no.4) to 19 mm (3/4 in.).

fineness [ore dep] The degree of purity of gold expressed in parts per thousand. Pure gold is 1000 fine; gold with 0.45% impurities is 955 fine.

fineness [part size] (fine'-ness) The state of subdivision of a substance; the size of the constituent particles of a substance. The term is applied in describing sedimentary texture.

fineness factor A measure of the average particle size of clay and ceramic material, computed by summing the products of the reciprocal of the size-grade midpoints and the weight percentage of material in each class (expressed as a decimal part of the total frequency) (Purdy, 1908). The measure is based on the assumption that the surface areas of two powders are inversely proportional to their average particle sizes. Syn: *surface factor.*

fineness modulus A means of evaluating sand and gravel deposits, consisting of passing samples through standardized sets of sieves, accumulating percentages retained, dividing by 100, and comparing the resultant fineness-modulus number to various specification requirements.

fine pebble A geologic term for a pebble having a diameter in the range of 4-8 mm (1/6 to 0.3 in., or -2 to -3 phi units) (AGI, 1958).

fines [mining] Finely crushed or powdered material, e.g. of coal, crushed rock, or ore, as contrasted with the coarser fragments; esp. material smaller than the minimum specified size or grade, such as coal with a maximum particle size less than 8 mm, or ores too pulverulent to be smelted in the ordinary way; or material passing through a given screen or sieve.

fines [sed] (a) Very small particles, esp. those smaller than the average in a mixture of particles of various sizes; e.g. the silt and clay fraction in glacial drift, or the fine-grained sediment that settles slowly to the bottom of a body of water. (b) An engineering term for the clay- and silt-sized soil particles (diameters less than 0.074 mm) passing U.S. standard sieve no. 200.

fine sand (a) A geologic term for a sand particle having a diameter in the range of 0.125-0.25 mm (125-250 micrometers, or 3 to 2 phi units). Also, a loose aggregate of sand consisting of fine sand particles. (b) An engineering term for a sand particle having a diameter in the range of 0.074 mm (retained on U.S. standard sieve no. 200) to 0.42 mm (passing U.S. standard sieve no. 40). (c) A soil term used in the U.S. for a sand particle having a diameter in the range of 0.10-0.25 mm. The diameter range recognized by the International Society of Soil Science is 0.02-0.2 mm. (d) Soil material containing 85% or more of sand-size particles (percentage of silt plus 1.5 times the percentage of clay not exceeding 15) and 50% or more of fine sand or less than 25% of very coarse sand, coarse sand, and medium sand together with less than 50% of very fine sand (SSSA, 1965, p.347).

fine silt A geologic term for a silt particle having a diameter in the range of 1/128 to 1/64 mm (8-16 micrometers, or 7 to 6 phi units). In Great Britain, the range 1/100 to 1/20 mm has been used. Also, a loose aggregate of silt consisting of fine silt particles.

fine texture (a) A broad group of textures consisting of or containing large quantities of the fine fractions, particularly of silt and clay (Includes all sandy clay, silty clay, and clay textural classes). (b) When used in reference to family particle-size classes in U.S. soil taxonomy, is specifically defines as having 35 to 60 percent clay. See also: *texture [soil].*

fine topography A topography with fine *topographic texture,* characterized by high drainage density and closely spaced streams, and common in regions of weak rocks. An "ultra-fine" topography is characterized by the extremely fine dissection of badlands topography.

finger (fin'-ger) One of two pincerlike blades of the distal end of a *cheliped* of a crustacean or of a *chela* of an arachnid. One finger is movable and the other is fixed.

finger coal *Natural coke* occurring as small, hexagonal columns associated with igneous intrusion.

finger gully One of a group of very small gullies that forms a fan-shaped extension at the head of a system of gullies.

fingerite (fing'-er-ite) A metallic black triclinic mineral: $Cu_{11}O_2(VO_4)_6$.

finger lake A long, relatively narrow lake, usually of glacial origin, which may occupy a rock basin in the floor of a glacial trough or be held in by a morainal dam across the lower end of the valley; esp. one of a group of such lakes disposed somewhat like the fingers of a hand, such as the Finger Lakes in central New York State.

Fingerlakesian (Fin-ger-lakes'-i-an) North American stage: lower Upper Devonian (above Erian, below Chemungian). Syn: Finger Lakes.

fingertip channel (fin'-ger-tip) One of the smaller, unbranched stream channels at the head of a drainage network.

finite-strain theory (fi'-nite-strain') A theory of material deformation which considers displacements and strains too large to be evaluated through *infinitesimal-strain theory.*

finnemanite (fin'-ne-man-ite) A subadamantine gray, olive-green, or black hexagonal mineral: $Pb_5(As^{3+}O_3)_3Cl$. It possess an apatite-like structure.

fiord Anglicized variant of *fjord.*

fiorite (fi'-o-rite) *siliceous sinter.*

fire Flashes of the different spectral colors seen in diamonds and other gemstones as the result of dispersion. Cf: *play of color.*

fire assay Any type of assay procedure that involves the heat of a furnace.

fireball (fire'-ball) A bright or brilliant meteor with luminosity that equals or exceeds that of the brightest planets. Cf: *bolide.*

fireball hypothesis *"big bang" hypothesis.*

fireblende (fire'-blende) A syn. of *pyrostilpnite.* Also spelled: fire blende.

fireclay (fire'-clay) A plastic, kaolinitic claystone with sufficient Al_2O_3 to be refractory. It usually occurs as an *underclay. Flint clay* should not be included in this category (Burst and Hughes, 1994). Fireclays commonly occur as underclays. However, not all underclays are fireclays and not all underclays are refractory.

fireclay mineral A disordered variety of kaolinite. See also: *mellorite.*

firedamp (fire'-damp) A coal-mine gas that is explosive and consists mainly of *methane.* Cf: *blackdamp; whitedamp; afterdamp.*

fire fountain Rise-and-fall eruption of incandescent lava, either from a central volcanic vent or along a fissure, as a "jet" of molten material that breaks into a spray of melt droplets and bombs.

fire opal A transparent to translucent and yellow, orange, red, or brown variety of opal that gives out fiery reflections in bright light and that may or may not have play of color. See also: *gold opal.* Syn: *sun opal; pyrophane.*

firestone (fire'-stone) (a) Any fine-grained siliceous stone formerly used for striking fire; specif. flint. Syn: *feuerstein.* (b) A nodule of pyrite formerly used for striking fire. (c) A fine-grained siliceous rock that can resist or endure high heat and that is used for lining furnaces and kilns, such as certain Cretaceous and Jurassic sandstones in southern England. (d) *fireclay.*

firmatopore (fir-mat'-o-pore) A polymorph that is slender and proximally directed on the reverse sides of the colonies of some stenolaemate bryozoans. Cf: *hematopore.*

firm ground A tunnelman's term for materials in which a tunnel can be advanced without the aid of initial roof support, and the permanent lining can be constructed before the crown or walls begin to deform. Firm ground may become unstable if construction methods are changed or tunnel size is increased. See also: *flowing ground; raveling ground; running ground; squeezing ground; swelling ground.*

firmoviscosity (fir'-mo-vis-cos'-i-ty) The *viscoelastic* state, as modeled by a Kelvin body. In response to a given stress, elastic strain is produced only over a finite period of time; unloading is time-dependent, although all of the strain is recoverable. At a constant strain, stress may be supported indefinitely (Turner and Weiss, 1963, p.279).

firn (a) Wetted snow that has survived one summer without being transformed into ice. Firn becomes glacier ice when interconnecting air passages between grains are sealed off (Paterson, 1994). The term has also been defined, although rarely, on the basis of certain physical properties, such as density. Syn: *névé; firn snow.* (b) A geographic term applied to the accumulation area or upper region of a glacier. This usage is being supplanted by *firn field,* or by *névé* (in Great Britain). Etymol: German, adjective meaning "old, of last year".

firn basin *firn field.*

firn edge The boundary on a glacier between glacier ice and firn during the ablation season.

firn field The *accumulation area* of a glacier; a broad expanse of glacier surface over which snow accumulates and firn is created; an area of *firn*. Syn: *firn basin; névé.*
firn ice *iced firn.*
firnification (firn'-i-fi-ca'-tion) The process whereby snow is transformed into firn and then into ice in a glacier.
firn limit *firn line.*
firn line Commonly used as a synonym for *equilibrium line*. This is not preferred, however, as firn from previous years may be exposed in the upper part of the ablation area. The lower boundary of the firn may then be significantly lower in elevation than the equilibrium line. The preferred syn. is *firn limit.*
firn snow (a) *firn.* (b) *old snow.*
firnspiegel (firn'-spie-gel) A thin sheet or film of clear ice on a snow surface, bridging hollows in the snow, formed when surface meltwater is immediately refrozen as a thin ice film, and the snow below continues to melt by radiation passing through the transparent ice sheet. Etymol: German "firn", "old, last year's", + "Spiegel", "mirroring surface".
first antenna *antennule.*
first appearance datum Commonly used in its acronym form, FAD; mostly applied to fossils in cored sedimentary rock sequences. Cf: *LAD.*
first arrival The first energy to arrive from a seismic source. First arrivals on reflection records are used for information about a surficial low-velocity or weathering layer; refraction studies are often based on first arrivals. Syn: *first break.*
first bottom The normal flood plain of a river. Cf: *second bottom.*
first break *first arrival.*
first-class ore An ore of sufficiently high grade to be acceptable for shipment to market without preliminary treatment. Cf: *second-class ore.* Syn: *shipping ore.*
first-derivative map A map of the *first vertical derivative* of the gravity or magnetic field, calculated after all proper corrections have been applied to the observed data. Being less resolving and much less enhancing noise than a *second-derivative map,* it enhances mid-to-short wavelength "residual" components and delineates areas of fast decay of high frequencies (short wavelengths) often associated with anomalies of relatively shallow origin.
first law of thermodynamics The statement describing the *internal energy* of a chemical system, sometimes referred to as the law of conservation of energy. It says that the change of internal energy of a system is equal to the amount of energy given to, or received from, external sources, in the form of heat exchanged by the system and mechanical work (P-V work) done on the system.
first maxilla *maxillule.*
first meridian *prime meridian.*
first motion *earthquake first motion.*
first-order leveling Leveling of high precision and accuracy in which, for a section of 1-2 km in length, the maximum allowable difference obtained by running the line first forward to the objective point and then backward to the starting point is 4.0 mm times the square root of the distance in kilometers separating the ends of the line (or 0.017 ft times the square root of the distance in miles). Cf: *second-order leveling; third-order leveling.*
first-order pinacoid In a triclinic crystal, the $\{0kl\}$ pinacoid and the $\{0\bar{k}l\}$ pinacoid. Cf: *second-order pinacoid; third-order pinacoid; fourth-order pinacoid.*
first-order prism A crystal form: in a tetragonal crystal, the $\{110\}$ prism; in a hexagonal crystal, the $\{10\bar{1}0\}$ prism; in an orthorhombic crystal, any $\{0kl\}$ prism; and in a monoclinic crystal, any $\{0kl\}$ prism. Cf: *second-order prism; third-order prism; fourth-order prism.* See also: *clinodome; brachydome.*
first-order transformation A polymorphic transformation in which the first derivatives (with respect to T and P) of the Gibbs free energy function are discontinuous at the transformation temperature. It is accompanied by finite changes in entropy (S) and volume (V), and consequently in enthalpy (H) (Putnis and McConnell, 1980), and is generally a quenchable transformation. Cf: *second-order transformation.*
first setting An orientation of a *crystal* or a *crystal structure* with monoclinic symmetry in which the *c axis* is the symmetrically unique axis (i.e., it is parallel to the *diad* or perpendicular to the *reflection plane*). This is no longer the standard orientation for the *monoclinic system*. Cf: *second setting.*
first vertical derivative A rate of change of the potential field, such as the Earth's magnetic or gravity field, in the vertical direction. Syn: *calculated vertical gradient.* See also: *first-derivative map.*
first water An obsolete term occasionally used, esp. in England, for the highest quality, color, or transparency of a diamond that is flawless, and colorless or almost blue-white. Cf: *second water; third water.*
first-year ice Sea ice, not more than one winter's growth, developing from young ice; it is subdivided on the basis of thickness: "thin" (30-70 cm; also known as *white ice*); "medium" (70-120 cm); and "thick" (120 cm to 2 m). See also: *one-year ice.*
firth A long, narrow arm of the sea; also, the opening of a river into the sea. Along the Scottish coast, it is usually the lower part of an *estuary* (e.g. Firth of Forth), but sometimes it is a fjord (e.g. Firth of Lorne) or a strait (e.g. Pentland Firth). Etymol: Scottish. Var. spelling: *frith.*
fischerite (fisch'-er-ite) A substance, consisting of a green hydrous aluminum phosphate, that is probably identical with *wavellite.*
Fischer's rule The arrangement in tetrads of tri-aperturate pollen grains whereby the apertures occur in pairs at six points in the tetrad (e.g., Ericaceae). Cf: *Garside's rule.*
fischesserite (fisch-es'-ser-ite) A metallic cubic mineral: Ag_3AuSe_2. It is the selenium analogue of petzite.
fish [drill] Broken or dropped equipment in a well bore, recoverable only by *fishing.* Syn: *junk.*
fish [oceanog] Any oceanographic sensing device that is towed behind a ship, unless it is a "string".
fish [seis] A sensor that is towed in the water, such as that used with side-scan sonar.
Fisher distribution A frequency distribution describing directional data in three dimensions. The distribution has three parameters; *a* and *b* specify the mean direction, and *k* is a dispersion parameter or concentration. The Fisher distribution is a spherical equivalent of the *normal distribution.*
fish-eye stone *apophyllite.*
fishhook dune (fish'-hook) A dune consisting of a long, sinuous, sigmoidal ridge forming the "shaft" and a well-defined crescent forming the "hook" (Stone, 1967, p.228). Syn: *hooked dune.*
fishing In drilling, an attempt to recover, by the use of specially prepared tools, a piece or pieces of equipment (such as sections of pipe, cables, or casing) that have become broken or been accidentally dropped into the hole.
fish kill Destruction of fish in lakes or ponds, due to a decrease in oxygen resulting from snow or from excessive amounts of suspended organic matter; to toxic pollutants; or to the total freezing of shallow lakes or ponds.
fishnet A method of perspective representation of surface by lines drawn in X and Y directions (Krzanowski et al., 1993).
fishtail structure (fish'-tail) The ragged lateral termination of a coal seam, produced where wedges of clastic sediment entered the parent peat deposit parallel to the bedding.
fissiculate (fis-sic'-u-late) Said of a blastoid having exposed or partly exposed *hydrospire* slits or spiracular slits.
fissile (fis'-sile) (a) Capable of being easily split along closely spaced planes; exhibiting *fissility.* (b) Said of bedding that consists of laminae less than 2 mm in thickness (Payne, 1942). Cf: *cleavage [struc geol].*
fissility (fis-sil'-i-ty) A general term for the property possessed by some rocks of splitting easily into thin layers along closely spaced, roughly planar, and approximately parallel surfaces, such as bedding planes in shale or cleavage planes in deformed rocks; its presence distinguishes shale from mudstone. The term includes such phenomena as *bedding fissility* and *fracture* or *spaced cleavage.* Etymol: Latin "fissilis", "that which can be cleft or split". Adj: *fissile.*
fission [chem] The spontaneous or induced splitting, by particle collision, of a heavy nucleus into a pair (only rarely more) of nearly equal fission fragments plus some neutrons. Fission is accompanied by the release of a large amount of energy. Cf: *fusion [chem].* See also: *spallation.* Syn: *nuclear fission.*
fission [evol] (fis'-sion) Asexual reproduction occurring when a single cell or polyp divides into two theoretically equal parts.
fissionable (fis'-sion-a-ble) Said of nuclei, such as uranium and plutonium, that are capable of fission.
fission-track dating A method of calculating an age in years by determining the ratio of the spontaneous fission-track density to induced fission tracks. The method, which has been used for ages from 20 years to 1.4×10^9 years, works best for zircon, apatite, and glass and is also useful for determining the amount and distribution of the uranium in the sample. Syn: *fission-track method; spontaneous fission-track dating.*
fission-track method *fission-track dating.*
fission tracks The paths of *radiation damage* made by nuclear particles in a mineral or glass by the spontaneous fission of uranium-238. They are similar in occurrence and formation to *alpha-particle recoil tracks* but are larger and less numerous. Fission-track density is established by etching and subsequent microscopic examination and counting.
fissura (fis-su'-ra) A sharp, straight split that appears during germination in certain inaperturate pollen grains (e.g., TCT pollen).
fissure [eng] *earth fissure.*
fissure [geol] (fis'-sure) A surface of fracture or a crack in rock along which there is a distinct separation. It is often filled with mineral-bearing material. Syn: *open joint.*
fissure [glaciol] *crevasse [glaciol].*
fissure eruption An eruption that takes place from an elongate fissure, rather than from a central vent. Cf: *central eruption; plateau eruption.* See also: *fissure flow; fissure vent.*
fissure flow A flow formed as a result of a *fissure eruption.*
fissure passage A high, linear cave passage formed along one or more joints.
fissure polygon A *nonsorted polygon* marked by intersecting grooves or fissures producing a gently convex polygonal surface pattern and by the absence of a well-defined stone border. The term is inadequate because some polygons with fissures have stone borders coincident with the fissures (Washburn, 1956, p.825). The feature is typical of broad areas of NW Canadian lowlands. See also: *mud polygon; ice-wedge polygon.*
fissure spring A spring issuing from a crack or joint. Several springs of this type may flow out along the same fissure line. Cf: *fault spring; fracture spring.*

fissure system A group of fissures of the same age and of more or less parallel strike and dip.
fissure theory An early theory, now partially discredited, that oil and gas migrate extensively through fissures resulting from the arching of beds into anticlines.
fissure vein A type of mineral deposit of veinlike shape with the implication of clearly defined walls rather than extensive host-rock replacement.
fissure vent The opening at the Earth's surface of a volcanic conduit having the form of a crack or fissure. Cf: *fissure eruption.*
fissure volcano One of a series of volcanic vents in a pattern of eruption along a fissure.
fistulose (fis'-tu-lose) Said of a foraminifer having tubular irregular growth in the apertural region.
fitting The process of estimating values of a model equation that best describes a set of observations. The parameters of a fitted model usually are estimated by least-squares or maximum likelihood.
fitzroyite (fitz-roy'-ite) A *lamproite* composed of phenocrysts of leucite and phlogopite in a very fine-grained groundmass. Named by Wade and Prider in 1940 for the Fitzroy Basin, Western Australia. Obsolete.
five-component system A chemical system having five main components, e.g., Na_2O-CaO-K_2O-$Al2O_3$-SiO_2. Syn: *quinary system.*
fiveling (five'-ling) A crystal twin consisting of five individuals. Cf: *twoling; trilling; fourling; eightling.*
fix (a) A position determined from terrestrial, electronic, or astronomical data. Also, the point thus established. (b) The act of determining a fix.
fixed ash *inherent ash.*
fixed blade Portion of longitudinal axis on upper side of platformed pectiniform conodont elements between free blade and carina; fixed blade is flanked by platforms (TIP, 1981, pt. W, supp.2).
fixed carbon In coal, peat, coke, and bituminous materials, the solid combustible matter remaining after removal of moisture, ash, and volatile matter. It is expressed as a percentage.
fixed-carbon ratio *carbon ratio [coal].*
fixed channel A stream whose course is determined by its incision into bedrock or other materials that effectively resist erosion. This is the normal condition of eroding channels in regions of hills, mountains, or flat territory where sides and beds of channels consist of solid rock. In alluvial regions channels are fixed in tenacious clay or where they have cut down locally into bedrock below the alluvium. Channel positions are also fixed along faults and other geologic structures.
fixed cheek The part of the *cheek* of a trilobite inside the facial suture, remaining attached to the *glabella* at the time of molting. Cf: *free cheek.* Syn: *fixigena.*
fixed dune *anchored dune.*
fixed elevation An elevation that has been adopted either as a result of tide observations or previous adjustment of spirit leveling and that is held at its accepted value in any subsequent adjustment (Mitchell, 1948, p.27).
fixed form A crystal form whose indices are fixed relative to length, e.g. cube *{100}*, or octahedron *{111}* of the cubic system. Syn: *singular crystal form.*
fixed grips In fracture toughness testing, the grips on a rock sample are configured to remain immobile during crack propagation. This configuration leads to a decreasing load on the grips as the crack propagates (Lawn and Wilshaw, 1975).
fixed groundwater Groundwater in material having interstices so small that the water is held permanently to the walls of the interstices, or moves so slowly that it is not available for withdrawal at useful rates. Outside the saturation zone, material with very small openings can hold water indefinitely against the pull of gravity, whereas within the saturation zone there is apparently always movement, even though at very low rates.
fixed layer The inner, relatively immobile layer of ions in an electrolyte, required to satisfy a charge unbalance within a solid with which the electrolyte is in contact. It constitutes part of the *double layer* of charge adjacent to the electrolyte-solid interface. Cf: *diffuse layer.*
fixed moisture Moisture retained in the soil in a quantity that is less than the *hygroscopic coefficient.*
fixigena (fix-i-ge'-na) *fixed cheek.*
fixity of species (fix'-i-ty) The theory that a species, once created, remained constant in its characteristics throughout its span of existence. Cf: *evolution.*
fizelyite (fi-ze'-ly-ite) A metallic lead-gray monoclinic mineral: $Pb_{14}Ag_5Sb_{21}S_{48}$(?). It is closely related to andorite and ramdohrite. Also spelled: fizélyite.
fjäll A Swedish word for a mountain rising above the timberline and having flat undissected areas (Stamp, 1961, p.193). See also: *fjeld; fell.*
fjard (Pron: fyard) A small, narrow, irregular inlet or bay, typically formed by submergence of a glacial valley excavated in a lowland along the margin of a flat rocky coast, such as that of SE Sweden; often accompanied by numerous fringing islands. A fjard is shorter, shallower, and broader in profile than a *fjord,* but deeper than a ria. Etymol: Swedish "fjärd", a large continuous area of water surrounded by *skerry-guard* islands (Stamp, 1961, p.193); usage in English therefore has taken a more specialized meaning not apparent in Sweden. Var. spelling: *fiard.*
fjeld A Norwegian word for "field" having a wide meaning, but when used in English refers to an elevated, rocky, almost barren plateau above the timberline, covered with snow during the winter, as in the Scandinavian upland. Cf: *fell; fjäll.*
fjeldbotn (fjeld'-botn) A Norwegian term for a cirque carved by an ice field (Termier and Termier, 1963, p.405).
fjeldmark (fjeld'-mark) *fell-field.*
fjord (Pron: fyord) (a) A long, narrow, winding, glacially-eroded inlet or arm of the sea, U-shaped and steep-walled, generally several hundred meters deep, between high rocky cliffs or slopes along a mountainous coast; typically with a shallow sill or threshold of solid rock or earth material submerged near its mouth, and becoming deeper inland. A fjord usually represents the seaward end of a deeply excavated glacial-trough valley that is partially submerged by drowning after the melting of the ice. Examples: along the glaciated coasts of Alaska, Greenland, and Norway. (b) Any embayment of the seacoast in a Scandinavian country regardless of the adjacent topography, as a *fjard* in the low flat Swedish coast or a *förde* in eastern Denmark. Etymol: Norwegian. Cf: *estuary; ria; sea loch.* Var. spelling: *fiord.*
fjord coast A deeply indented, glaciated coast characterized by a partial submergence of glacial troughs and by the presence of steep parallel walls, truncated spurs, and hanging valleys. Its development is favored by marine west-coast climates combined with strong relief, as in Norway, Alaska, and southern Chile.
fjord ice Ice formed during the winter in a fjord and melting in place during the summer. Cf: *sikussak.*
fjord lake A lake in a glacially excavated rock basin of a U-shaped valley near sea level.
fjord shoreline A shoreline of submergence characterized by the development of numerous fjords; e.g. along the west coast of Norway. Cf: *fjord coast.*
fjord strait A strait between two fjords opening in opposite directions; e.g. the Straits of Magellan.
f-k space A domain in which the independent variables are frequency (f) and wavenumber (k). Seismic data analysis sometimes involves f-k plots, in which energy density within a given time interval is plotted and contoured on a frequency-versus-wavenumber basis. Apparent velocity filtering can be done in f-k space. Also written: f,k.
flabellate (fla-bel'-late) Fanlike; e.g. said of a fan-shaped corallite, or of a meandroid corallum with a single continuous, laterally free, linear series of corallites.
flabelliform Said of a foraminifer that is shaped like a fan.
flabellum (fla-bel'-lum) A body part resembling a fan; esp. the thin, most distal exite of the limb of a branchiopod crustacean. Pl: flabella.
fladen (Fla'-den) (a) A distinctive, pancake-shaped body resembling a volcanic bomb and composed of glass and fragments of rocks and minerals, found in the suevites at the Ries crater in Germany. It exhibits flow structure and surface sculpturing apparently produced by aerodynamic forces. Cf: *impact bomb.* (b) Any similar, glass-rich, aerodynamically shaped body, formed by impact and found associated with other *impact structures.* Etymol: German "Fladen", "flat cake". Pl: fladen.
flag A syn. of *flagstone.* The term is often used in the plural, such as the "Lingula flags" of the European Upper Cambrian.
flagella Pl. of *flagellum.*
flagellar field (fla-gel'-lar) The area around the flagella of a coccolithophore; e.g. *covered flagellar field* and *naked flagellar field.*
flagellar pore The opening on the *dinoflagellate theca* where one or both of the flagella emerge; usually located near the junction of cingulum and sulcus.
flagellate (flag'-el-late) n. An organism, esp. a protozoan or alga, that bears flagella. adj. Possessing flagella.
flagellated chamber (flag'-el-lat'-ed) Any cavity, in a sponge, lined by choanocytes. See also: *chamber [paleont].*
flagellum (fla-gel'-lum) (a) Any of various threadlike appendages of animals, such as the multiarticulate distal part of the limb of a crustacean, or a long whiplike extension in a sponge. (b) A long, whiplike protoplasmic process that projects singly or in groups from a cell or microorganism, is possibly equivalent to a much enlarged *cilium,* and is the primary organ controlling the movement (through water) of a flagellated protozoan and of many algae, bacteria, and zoospores. Pl: flagella.
flagging Flags of cloth, plastic, or paper used by surveyors to mark certain locations.
flaggy (flag'-gy) (a) Splitting or tending to split into layers of suitable thickness for use as flagstones; specif. descriptive of a sedimentary rock that splits into layers from 1 cm to 5 cm in thickness (McKee and Weir, 1953, p.383). (b) Said of bedding that consists of layers from 1 cm to 10 cm in thickness (Payne, 1942). (c) Pertaining to a flag or flagstone. (d) Said of a soil full of flagstone fragments.
flagstaffite (flag'-staff-ite) A colorless, orthorhombic, organic mineral: $C_{10}H_{20}O_2 \cdot H_2O$.
flagstone (a) A hard sandstone, usually micaceous and fine-grained, that occurs in extensive thin beds with shale partings; it splits uniformly along bedding planes into thin slabs suitable for use in terrace floors, retaining walls, and the like. Cf: *bluestone [rock]; freestone [rock].* (b) A flat slab of flag-

stone used for paving; esp. a thin piece split from flagstone. Also, a surface of such stone. (c) A relatively thin flat fragment (of limestone, sandstone, shale, slate, or schist) occurring in the soil, having a length in the range of 15-38 cm (6-15 in.) (SSSA, 1965, p.336). Syn: *flag; slabstone; grayband; cleftstone.*

flajolotite (flaj'-o-lo-tite') *tripuhyite.*

flake In archaeology, an exterior fragment chipped off a pebble during work to create a stone tool.

flake graphite In economic geology, graphite disseminated in metamorphic rock as thin, visible flakes that are separable from the rock by mechanical means. Syn: *crystalline flake.*

flake mica Finely divided mica recovered from mica schist or sericite schist or obtained as a byproduct of beneficiation of feldspar or kaolin.

flakestone A type of limestone composed of flakes, interclasts composed of lime mud.

flake tectonics Term introduced by Oxburgh (1972), to refer to the movement of large sheetlike masses ("flakes") of less than lithospheric thickness in the eastern Alps. These are thought to be formed when two continents collide and flakes are "sheared from the top of one of them and driven over the other for more than 100 km." The term has also been applied in California (Yeats, 1981).

flame emission spectrometry *flame photometry.*

flame photometer The instrument used to conduct *flame spectroscopy* and involving the excitation of samples in a flame, usually compounds in a vaporized solution, to provide spectra for observation and quantitative analysis. See also: *photometer.*

flame photometry Measurement of the intensity of the lines in a spectrum produced by a *flame photometer*, and now considered a normal component of *flame spectroscopy*. Syn: *flame emission spectrometry.*

flame spectroscopy The observation of a *flame spectrum* and all aspects of comparison, recording, and measurement (flame spectrometry) involved with it.

flame spectrum The spectrum of light emitted by a substance by heating it in a flame.

flame structure [pyroclast] The presence of *fiamme* in a welded tuff; e.g., piperno.

flame structure [sed] A term introduced by Walton (1956, p.267) for a sedimentary structure consisting of wave- or flame-shaped plumes of mud that have been squeezed irregularly upward into an overlying layer. It is probably formed by load casting accompanied by horizontal slip or drag. See also: *load wave.*

flame test A qualitative analysis of a mineral made by intensely heating a sample in a flame and observing the flame's color, which will be indicative of the element involved, e.g. green from copper.

Flandrian (Flan'-dri-an) European climatostratigraphic stage: *Holocene* (post-10,000 ^{14}C years; above Devensian).

flange (a) A projecting rim or edge of an organism or a bone; e.g. a platelike marginal extension along foraminiferal chambers (as in Sphaeroidinella), the part of a coccolith that spreads out like a rim, a lateral projection from a brachiopod *crus*, or the outwardly bent rim of the aperture in some cephalopod conchs. (b) In conodont morphology, shelflike lateral projection of the side of a process; wider than a lateral ridge, but narrower than a platform (TIP, 1981, pt. W, supp.2).

flank [paleont] (a) The lateral side of a coiled cephalopod *conch* between the venter and the *umbilical seam;* the *whorl side.* (b) The sloping surface between the venter and the lateral margin of a brachiopod valve. (c) In a bivalve mollusk, the median part of the surface of the valve.

flank [struc geol] *limb [struc geol].*

flank eruption An eruption on the side of a volcano. Cf: *summit eruption.*

flanking moraine (flank'-ing) A moraine left by a glacial lobe or by a tonguelike projection of an ice sheet (Fairchild, 1932, p.629). Cf: *lateral moraine.*

flank-margin cave A cave developed by nonconduit flow, within a carbonate island or coastline, in the discharging margin of a freshwater lens (Mylroie and Carew, 1990). Cf: *banana hole.*

flaser (fla'-ser) The streaky layers of parallel, scaly aggregates surrounding the lenticular bodies of granular material in flaser structure. Etymol: German, "streak".

flaser gabbro Coarse-grained *blastomylonite* formed by dislocation metamorphism of a gabbro. Flakes of mica or chlorite sweep around augen of feldspar or quartz with much recrystallization and neomineralization (Joplin, 1968, p.21). Cf: *zobtenite.*

flaser structure [meta] A structure in dynamically metamorphosed rock in which lenses and layers of original or relatively unaltered granular minerals are surrounded by a matrix of highly sheared and crushed material, giving the appearance of a crude flow structure, e.g. flaser gabbro. Cf: *mylonitic structure; augen structure.* Syn: *phacoidal structure.*

flaser structure [sed] *Ripple cross-lamination* in which mud streaks or lenses are preserved in the troughs but incompletely or not at all on the crests. A detailed classification is given by Reineck and Wunderlich (1968). Cf: *wavy bedding.*

flash [mining] (a) A subsidence of the surface resulting from underground mining, esp. from the working of rock salt and pumping of brine. Cf: *inbreak; heave [mining]; crown-in.* (b) A small lake or shallow reach of water occupying a hollow produced by such a subsidence.

flash [water] (a) A sudden rise of water in a stream, as where water is held back by a dam. (b) A pool of water; a marsh or marshy place. Cf: *flush.*

flash figure In optic analysis of crystals under a *conoscope*, a vague and broad black cross that appears when the *optic axis* of a *uniaxial* crystal is parallel to the *privileged direction* of either *nicol*, or when the optic normal of a *biaxial* crystal is verticle and X and Z are parallel to the privileged directions of the nicols. Only slight rotation from this position causes the cross to separate into two hyperbolic segments and leave the field. Syn: *optic normal* figure (biaxial).

flash flood A local and sudden flood or torrent of relatively great volume and short duration, overflowing a stream channel in a usually dry valley (as in a semiarid area), carrying an immense load of mud and rock fragments, and generally resulting from a brief but heavy rainfall over a relatively small area having steep slopes. It may also be caused by ice jams and by dam failure. See also: *freshet.*

flashy stream (flash'-y) One prone to flash floods.

flat [eco geol] n. A horizontal orebody, regardless of genetic type.

flat [geog] (a) A tract of low-lying, level *wetland;* e.g. a marsh or swamp in a river valley. (b) A term used in northern and central U.S. as a syn. of *bottom,* or low-lying land along a watercourse.

flat [geomorph] adj. Having or marked by a continuous surface or stretch of land that is smooth, even, or horizontal, or nearly so, and that lacks any significant curvature, slope, elevations, or depressions. n. A general term for a level or nearly level surface or small area of land marked by little or no relief, as a plain; specif: *mud flat; valley flat.* Also, a nearly level region that visibly displays lower relief than its surroundings.

flat [lake] (a) The low-lying, exposed, flat land of a lake delta or of a lake bottom. (b) The flat bottom of a desiccated lake in the arid parts of western U.S. Commonly used in the plural.

flat [struc geol] That part of a step-shaped thrust fault or normal fault that is nearly parallel to the enveloping layering; thrust fault and (less ubiquitous) normal fault flats are also commonly nearly horizontal.

flatiron (flat'-i-ron) One of a series of short, triangular hogbacks forming a spur or ridge on the flank of a mountain, having a narrow apex and a broad base, resembling (when viewed from the side) a huge flatiron; it usually consists of a plate of steeply inclined resistant rock on the *dip slope.*

flat joint In igneous rock, joint dipping at an angle of 45° or less. Rarely applied to joints dipping more than 20°.

flatland (flat'-land) A region, or tract of land, characterized by predominant levelness or by no significant variation in elevation, as along a river or a coast.

flatness (flat'-ness) A term used by Wentworth (1922a) to express the shape of a pebble, defined as the ratio of the radius of curvature of the most convex portion of the flattest developed face to the mean radius of the pebble. Wentworth (1922b) also expressed flatness as the arithmetic mean of the long and intermediate diameters (length and width) of the pebble divided by the short diameter (thickness). Cf: *roundness.*

flat-on-flat geometry The situation where a flat in the hanging-wall template of a fault is juxtaposed against a flat in the footwall template of the fault.

flat-pebble conglomerate *intraformational conglomerate.*

flats and pitches (a) A phrase descriptive of the structure of the lead and zinc deposits in dolomite of the Upper Mississippi Valley region of the U.S., esp. in Winconsin. The "flats" are nearly horizontal solution openings; the "pitches" are the inclined, interconnecting joints. (b) A slump structure of both horizontal and steeply inclined cracks in sedimentary strata. Syn: *pitches and flats.*

flat spot A horizontal seismic reflection attributed to an interface between two fluids such as gas and water or gas and oil. See: *hydrocarbon indicator.*

flattened section A seismic record section in which a particular event has been made flat (or sometimes merely smooth) by introducing arbitrary time shifts. If flattened on an event which represents a bed that was deposited horizontally, the flattened section resembles a paleosection showing the attitude of deeper events at the time of deposition of the flattened bed.

flattening of ellipse The ratio f of difference between the lengths a and b of the semimajor and semiminor axes, respectively, of an ellipse to the length of the semimajor axis: $f = (a-b)/a$. Sometimes called *ellipticity.*

flattening of the Earth The flattening of the rotational ellipsoid taken to represent the Earth or the geoid.

flattening strain Three-dimensional strain defined by a strain tensor in which the magnitude of the maximum and intermediate strains are approximately equal and greatly exceed that of the minimum strain. Syn: *oblate strain.* Cf: *prolate strain; plane strain.*

flat-topped ripple A *ripple* with a flat, wide crest between narrow troughs; e.g. a shallow-water ripple whose crest was planed off during ebb tide or a fall in water level.

flat-topped ripple mark *flat-topped ripple.*

flaw [gem] A general trade term for any internal or external *imperfection* of a fashioned diamond or other gemstone. It includes cracks, inclusions, visibly imperfect crystallization, internal twinning, and cleavage, as well as surface abrasions.

flaw [ice] (a) A narrow separation zone (*fracture [ice]*) between pack ice and fast ice, formed when pack ice at this boundary undergoes shearing due to a strong wind or current, and characterized by pieces of ice in a chaotic state; it is not wide enough to permit passage of a navigable vessel. (b) Obsolete syn. of *flaw lead.*

flaw [struc geol] (a) Any hole, cavity, or discontinuity in a rock which may serve to concentrate the tensile stress necessary for joint propagation. Pores, cracks, fossils, bedding boundaries, and other inclusions may all serve as flaws in rock (Engelder and Lacazette, 1990). (b) A defect in a crystal, grain, or rock body that can cause a stress concentration and/or can localize deformation. Examples of flaws include microcracks, inclusions, and crystal defects. (c) An obsolete term for a strike-slip fault.

flaw lead A navigable passage between *pack ice* and *fast ice.* Cf: *shore lead.* Obsolete syn: *flaw [ice].*

flawless (flaw'-less) Said of a diamond or other gemstone that is free from flaws of any description when observed by a trained eye under efficient illumination with a fully corrected magnifier of not less than 10 power.

flaxseed ore (flax'-seed) An iron-bearing sedimentary deposit, e.g. the *Clinton ore,* composed of disk-shaped hematitic oolites that have been somewhat flattened parallel to the bedding plane. Cf: *fossil ore.*

F layer The seismic region of the Earth from 4,710 km to 5,150 km, at the bottom of the *outer core.* It is a part of a classification of the Earth's interior made up of layers A to G. Archaic.

flèche d'amour (fleche d'a-mour') An acicular, hairlike crystal of rutile embedded in sagenitic quartz. The term is used loosely as a syn. of "sagenitic quartz", and was used formerly for amethyst containing brown needles of goethite. Etymol: French, "arrow of love". Syn: *cupid's dart; love arrow.*

fleckschiefer (fleck'-schief-er) A type of *spotted slate* characterized by minute flecks or spots of indeterminate material. Etymol: German. Cf: *knotenschiefer; garbenschiefer; fruchtschiefer.*

fleet (a) A term used in England for a small, shallow inlet, estuary, creek, or arm of the sea. Also, a place where water flows; a small rapid stream. (b) A small, usually salty or brackish lagoon behind the coastline, separated from the open sea by a long bank of sand or shingle parallel to the coast (Monkhouse, 1965, p.125).

fleischerite (fleisch'-er-ite) (a) A silky white to pale rose hexagonal mineral: $Pb_3Ge^{4+}(SO_4)_2(OH)_6{\cdot}3H_2O$. (b) A group name for minerals with the analogous formula, but with Pb replaced by Ca, and Ge by Mn^{4+}.

fleshy sponge (flesh'-y) Any sponge that lacks a skeleton, i.e. one of the demosponges.

fletcherite (fletch'-er-ite) A metallic gray cubic mineral of the *linnaeite* group: $Cu(Ni,Co)_2S_4$.

fleur-de-lis (fleur-de-lis') A sedimentary structure consisting of frondescent spatulate elevations and resembling the appearance of an iris. Etymol: French "fleur de lis", "lily flower". Also spelled: fleur-de-lys.

flexible (flex'-i-ble) Said of a mineral which can be bent without breaking but will not return to its original form; e.g. talc.

flexible crinoid Any crinoid belonging to the subclass Flexibilia, characterized by the incorporation, but not firm attachment, of the lower brachials in the dorsal cup and by a flexible tegmen.

flexible sandstone A fine-grained, thin-layered variety of *itacolumite.*

flexible silver ore *sternbergite.*

flexostyle (flex'-o-style) In foraminifera, a tubular, enrolled *deuteroconch* (as in *Amphisorus*).

flexural fold (flex'-u-ral) A general term for *flexure-flow folds* and *flexure-slip folds.* Cf: *passive fold.*

flexural slip Slip between rock layers during flexural folding of a multilayered sequence. Slip occurs on discrete surfaces, normally under sub-metamorphic or low-grade metamorphic conditions. Cf: *bedding-plane slip.*

flexural-slip thrust fault A syn. of *uplimb thrust fault.* See: Price (1965).

flexure (flex'-ure) *hinge [struc geol].*

flexure-flow fold A flexural fold in which the mechanism of folding includes displacement parallel to layer boundaries and some flow within layers, resulting in thickening of hinge areas and thinning of limbs. Cf: *flexure-slip fold.*

flexure line A line, extending from the beak to the anterior border of both ventral propareas in some lingulacean brachiopods, marked by deflection of growth lines (TIP, 1965, pt. H, p.145).

flexure-slip fold A flexure fold in which the mechanism of folding is slip along bedding planes or along surfaces of foliation. There is no change in thickness of individual strata, and the resultant folds are parallel. Cf: *flexure-flow fold.*

flexus (flex'-us) A term established by the International Astronomical Union for a very low curvilinear ridge with a scalloped pattern on planetary surfaces (USGS, 2003). Pl: flexus.

flight *terrace flight.*

flight altitude The vertical distance above a given datum, usually mean sea level, of an aircraft in flight or during a specified portion of a flight (ASP, 1975, p.2082). See also: *flight height.*

flight auger A tool used to drill a borehole, in which cuttings accumulate on a spiral fin welded to the drill stem and are brought to the ground surface by the rotating action of the drill tools. A hollow-stem flight auger permits a soil-sampling device to be lowered to the bottom of the borehole and driven into the formation below the bit for conducting penetration tests and collecting relatively undisturbed samples for laboratory testing. A pilot bit on a solid steel rod inserted into the hollow-stem auger prevents soil material from entering the hollow stem, as the flight auger is advanced during drilling operations.

flight height A term used in aerial photography for the *flight altitude* when the datum is mean ground level of the area being photographed.

flight line A line drawn on a map or chart to represent the planned or actual track of an aircraft during the taking of aerial photographs.

flight map A map that shows the desired flight lines and/or air stations before the taking of aerial photographs, or a map on which are plotted, after photography, selected air stations and the tracks between them.

flight path The line on the ground directly beneath a remote sensing aircraft or satellite.

flight strip A succession of overlapping aerial photographs taken along a single course. Syn: *strip [photo].*

flimmer (flim'-mer) *mastigoneme.*

flinkite (flink'-ite) A greenish-brown or green orthorhombic mineral: $Mn^{2+}_2Mn^{3+}(AsO_4)(OH)_4$.

Flinn diagram A plot of maximum/intermediate stretches versus intermediate/minimum stretches used to represent three-dimensional strain (Flinn, 1962).

flint [mineral] (a) A term that has been considered as a mineral name for a massive, very hard, somewhat impure variety of chalcedony, usually black or of various shades of gray, breaking with a conchoidal fracture, and striking fire with steel. See also: *chert.* Syn: *firestone.* (b) Pulverized quartz of any kind; e.g. "potter's flint" in the form of powdered quartz, made by pulverizing flint pebbles.

flint [sed] A term that is widely used as a syn. of *chert* or for the homogeneous, dark-gray or black variety of chert. According to Tarr (1938), the term "flint" should either be discarded or be reserved for siliceous artifacts (such as the "flint arrowheads" used by primitive humans) because rocks described as flint are identical with chert in texture and composition, despite the fact that the term "flint" has been in use since about A.D. 700 for "anything hard" and since A.D. 1000 for "a variety of stone" and that it antedates "chert" by almost 1000 years. Flint has been described as having a denser texture, a more perfect (smooth) conchoidal or less splintery fracture, a smaller quartz content, and greater infusibility than chert, and as having thin translucent splinters or sharp cutting edges. The term is commonly used in southern England for one of the siliceous nodules occurring in the Cretaceous chalk beds, and elsewhere in England for any hard rock (such as in Shropshire for a fine-grained sandstone suitable for building). Syn: *black chert; silex; hornstone [rock].*

flint clay A hard, smooth, nonplastic, usually kaolinitic claystone that breaks with conchoidal fracture and does not disperse in water. "Burley flint clay" is a claystone in which aluminum-rich minerals such as diaspore raise the Al_2O_3 content above that of kaolinite (~40% Al_2O_3) (Burst and Hughes, 1994). Volcanic ash deposited in a coal swamp can form a flint clay (Bohor and Triplehorn, 1993). Flint clays generally have an aggregate particle size of >75 μm.

flint curtain A concentration of silica derived from flints, occurring along a vertical joint plane in the chalk beds of eastern Denmark; it results from dehydration of silica gel and flowage of silica, immediately subsequent to jointing.

flint meal Fine, flourlike material consisting primarily of minute fossils (such as sponge spicules) and occurring in an enclosed cavity of a flint nodule from the chalk beds of southern England.

flinty (flint'-y) (a) Composed of flint, or containing more than the normal percentage of silica; e.g. a "flinty slope". (b) Resembling flint in hardness or fracture; flintlike.

flinty crush rock *ultramylonite.*

flinty slate A *touchstone* consisting of siliceous slate.

float A general term for isolated, displaced fragments of a rock, esp. on a hillside below an outcropping ledge or vein. Cf: *floating reef; colluvium.* Syn: *floater.*

float coal Small isolated bodies of coal in sandstone or shale, probably deposited as pieces of peat that were eroded and transported from the original deposit. Cf: *coal gravel.* Syn: *raft [coal].*

float coccolith A modified coccolith serving as a suspension organ in nonmotile coccolithophores exhibiting dimorphism (such as *Thorosphaera*).

floater (float'-er) *float.*

floating bog *quaking bog.*

floating ice (float'-ing) Any form of ice floating in water, including ice that is stranded or grounded and ice formed on land but drifting in the sea. The term formerly excluded icebergs and other forms of land ice. See also: *drift ice.*

floating island A mass or mat of vegetation with little or no soil, floating freely in a lake or tropical sea, usually due to detachment from a marshy or boggy shore during a storm or a rise in the water level.

floating marsh *flotant.*

floating peat Peat that is derived from floating plants.

floating reef An isolated, displaced rock mass in alluvium. Cf: *float.*

floating sand grain An isolated grain of quartz sand that is not, or does not appear to be, in contact with neighboring sand grains scattered throughout the finer-grained matrix of a sedimentary rock, esp. of a limestone; e.g. a grain surrounded on all sides by coarse mosaic of calcite cement.

float ore Scattered fragments of vein material broken from outcrops and dispersed in soil; a type of *float.*

floatstone [mineral] (float'-stone) A lightweight, porous, friable variety of opal that floats on water and occurs in white or grayish, spongy, and concretionary or tuberous masses. Also spelled: float stone. Syn: *swimming stone.*

floatstone [sed] A carbonate rock containing a few bioclasts or other fragments more than 2 mm in diameter, widely spaced, embedded in sand- or mud-size carbonate sediment that forms over 90 percent of the rock (Embry and Klovan, 1971). Cf: *rudstone; boundstone; grainstone.*

floc (a) A loose, open-structured mass formed in a suspension by *flocculation;* e.g. a small aggregate of tiny sedimentary grains or colloidal clay particles. (b) A term used by Brewer (1964, p.367) for soil plasma that has a relatively low luster and a rough surface, giving it the appearance of clusters of silt-size grains in reflected light up to magnifications of approximately 20 times.

flocculation (floc-cu-la'-tion) The process by which a number of individual minute suspended particles are tightly held together in clotlike masses, or are loosely aggregated or precipitated into small lumps, clusters, or granules; e.g. the joining of soil colloids into a small group of soil particles, or the deposition or settling out of suspension of clay particles in salt water. See also: *floc.*

floe A piece of floating ice other than fast ice or glacier ice, larger than an *ice cake* and smaller than an *ice field.* Floes are subdivided according to horizontal extent and many varying size limits have been used; the U.S. Naval Oceanographic Office (1968, p. A27) gives the following dimensions: "giant" (over 10 km); "vast" (2-10 km); "big" (500 m to 2 km); "medium" (100-500 m); "small" (20-200 m). Syn: *ice floe.*

floeberg (floe'-berg) (a) A massive piece of sea ice composed of a *hummock [ice]* or a group of hummocks, frozen together and separated from any ice surroundings, and floating with its highest point up to 5 m above sea level (U.S. Naval Oceanographic Office, 1968, p. B33). It resembles a small iceberg. (b) In the older literature, a thick mass of well-hummocked sea ice originating from an ice floe, and sometimes projecting more than 15 m above sea level.

floe till *berg till.*

Floetz A name applied by A.G. Werner in the 1790s to the group or series of rocks that included most of the obviously stratified, comparatively flat, fossiliferous rocks (and certain associated trap rocks) and that were believed to represent the emergence of mountains from beneath the receding ocean, with products of the resulting erosion deposited on the mountain flanks. The rocks succeeded the *Transition* series and included the whole range of strata from the Devonian through the Tertiary. Etymol: German "Flötz" (now "Flöz"), "flat layer, stratum, seam, bed". Syn: *Secondary.*

floitite (flo'-i-tite) A rock consisting of *biotite* and those minerals that are typical of the *greenschist facies.* The term was originated by Becke in 1922. Rarely used.

flokite (flo'-kite) *mordenite.*

flood [sed] A term used by Milner (1940, p.457) to describe the occurrence, in a sedimentary rock, of a mineral species "so far in excess of all others as to constitute almost a pure concentrate".

flood [water] (a) A rising body of water (as in a stream, lake, or sea, or behind a dam) that overtops its natural or artificial confines and that covers land not normally under water; esp. any relatively high streamflow that overflows its banks in any reach of the stream, or that is measured by gage height or discharge quantity. (b) A flood of special severity or local interest; specif. the Flood, or the *Noachian flood.* (c) An archaic term for a large body of moving water, such as a river.

flood absorption A reduction in discharge resulting from the storage of flowing water in a reservoir, channel, or lake (ASCE, 1962).

flood basalt A term applied to those basaltic lavas that occur as vast accumulations of horizontal or subhorizontal flows, which were erupted from fissures in rapid succession over large areas, resulting in the buildup of regional-scale basaltic plateaus. Syn: *continental flood basalt; plateau basalt.* Cf: *shield basalt.*

flood basin (a) The tract of land covered by water during the highest known flood. (b) The broad, flat area between a sloping, low plain and the natural levee of a river, "occupied by heavy soils and commonly having either no vegetation or strictly swamp vegetation" (Bryan, 1923b, p.39). Syn: *tule land.*

flood-compatible land use Land uses, such as parks and open spaces or elevated structures, that can be occasionally flooded without property damage or loss.

flood control The prevention or reduction of damage caused by flooding, as by containing water in reservoirs removed from areas where it would do damage, improving channel capacity to convey water past or through critical areas with the least amount of damage, and diverting excess water into bypasses or floodways.

flood crest The highest stage of a flood. The term is nearly synonymous with *flood peak,* but does not refer to discharge since it connotes the top of the flood wave (Langbein and Iseri, 1960, p.10).

flood current The tidal current associated with the increase in the height of a tide, generally set toward the shore or up a tidal river or estuary. Cf: *ebb current.* Erroneous syn: *flood tide.*

flood dam A dam to store floodwaters temporarily, or to supply a surge of water as for clearing a channel. It is often specifically designed to withstand rapid changes in reservoir water level.

flood delta *flood-tidal delta.*

flood frequency The average occurrence of flooding of a given magnitude, over a period of years.

flood-frequency curve A graphic illustration of the number of times per year that a flood of a given magnitude is equaled or exceeded.

flood fringe *pondage land.*

floodgate (flood'-gate) (a) A gate for shutting out, admitting, releasing, or otherwise regulating a body of water, such as excess water in times of flood; specif. the lower gate of a lock. See also: *sluice.* (b) A stream stopped by or allowed to pass by a floodgate.

flood icing *icing.*

flooding (flood'-ing) (a) The covering or causing to be covered with a fluid, such as the covering of flat lands with a thin sheet of water; the filling or becoming full with water, esp. to excess. (b) A method of injecting water via select wells into the *pay zone* of an oil pool in order to force the oil toward producing wells; *water flooding.*

flooding ice *icing.*

flooding surface [coast] The surface that is flooded during rising sea level or water level.

flooding surface [stratig] In sequence stratigraphy, a surface separating younger from older strata across which there is evidence of an abrupt increase in water depth. This deepening is commonly accompanied by minor submarine erosion or nondeposition. This is not accompanied by subaerial erosion due to stream rejuvenation or a basinward shift in facies, including abnormal subaerial exposure, with a minor hiatus indicated. The flooding surface has a correlative surface in the coastal plain and a correlative surface on the shelf (Van Wagoner, et al., 1990). Syn: *marine-flooding surface.*

floodland (flood'-land) The land along a river that may be submerged by floodwaters; a *flood plain.*

flood lavas A term applied to the very large scale lava flows in flood basalt fields; also to extensive trachyte and phonolite flows described from the East African rift system (Cas and Wright, 1987). Cf: *flood basalt.*

flood peak The highest discharge or stage value of a flood. Cf: *flood crest.* Syn: *peak discharge.*

floodplain (a) The surface or strip of relatively smooth land adjacent to a river channel, constructed by the present river in its existing regimen and covered with water when the river overflows its banks. It is built of alluvium carried by the river during floods and deposited in the sluggish water beyond the influence of the swiftest current. A river has one floodplain and may have one or more terraces representing abandoned floodplains. Cf: *valley flat; erosional floodplain.* (b) Any flat or nearly flat lowland that borders a stream and that may be covered by its waters at flood stages; the land described by the perimeter of the maximum probable flood. Syn: *floodland.* (c) The part of a lake-basin plain between the shoreline and the shore cliff, subject to submergence during a high stage of the lake.

flood-plain deposit Sandy and muddy sediment deposited by river water that was spread out over a flood plain; a deposit beneath and forming a low, flat valley bottom being thickest near the river and thinning out toward the valley slopes. See also: *overbank deposit.* Syn: *vertical-accretion deposit.*

flood-plain icing *icing.*

floodplain lobe The part of a floodplain enclosed by a stream meander.

floodplain meander scar A crescentic mark indicating the former position of a river meander on a floodplain.

floodplain scroll One of a series of short, crescentic, slightly sinuous strips or patches of coarse alluvium formed along the inner bank of a stream meander and representing the beginnings of a floodplain. Syn: *flood scroll.*

floodplain splay A small alluvial fan or other outspread deposit formed where an overloaded stream breaks through a levee (artificial or natural) and deposits its material (often coarse-grained) on the floodplain. Syn: *overbank splay; sand splay; channel splay; splay [geomorph].*

flood plane (a) The position occupied by the water surface of a stream during a specific flood. (b) Loosely, the elevation of the water surface at any of various points along the stream during a specific flood.

flood probability The probability, determined statistically, that a flood of a given size will be equaled or exceeded in a given period, e.g. a 10% probability would be called a ten-year flood.

flood profile A graph of elevation of the water surface of a river in flood, plotted as ordinate, against distance, measured in the downstream direction, plotted as abscissa. A flood profile may be drawn to show elevation at a given time, crests during a particular flood, or to show stages of concordant flows (Langbein and Iseri, 1960, p.10).

flood-prone area Land on a flood plain that is subject to inundation during a flood of a designated frequency or magnitude; for example, a 100-year flood or a *standard project flood.*

floodproofing Any combination of structural and nonstructural additions, changes, or adjustments to structures which reduce or eliminate risk of flood damage to real estate or improved real property, waste and sanitation facilities, or structures with their contents.
flood routing Progressive determination of the timing and shape of a flood wave at successive points along a river (Langbein and Iseri, 1960, p.10).
flood scroll *floodplain scroll.*
flood series A listing of flood events for a given period of time, arranged in order of magnitude.
flood stage (a) The height of the gage at the lowest bank of the reach (other than an unusually low place or break). (b) The stage at which stream overflow begins to cause damage.
flood-tidal delta A *tidal delta* formed on the landward side of a *tidal inlet* (Davis and FitzGerald, 2004). Cf: *ebb-tidal delta.*
flood tide (a) *rising tide.* (b) A tide at its greatest height. (c) An erroneous syn. of *flood current.*
flood tuff A widespread sheet of *ignimbrite.* Obsolete.
floodwall (flood'-wall) A wall, often of reinforced concrete, built to prevent flooding. Cf: *levee [streams].*
floodwater (flood'-wa-ter) (a) Water that has overflowed its confines; the water of a flood. (b) The flooded area behind a dam; an impoundment.
flood wave A rise in the stage of a stream that culminates in a crest before receding.
floodway (flood'-way) (a) A large-capacity channel constructed to divert floodwaters or excess streamflow from populous or flood-prone areas, such as a bypass route bounded by levees. (b) The part of a flood plain kept clear of encumbrances and reserved for emergency diversion of floodwaters. (c) *flowage land.*
flood zone [stratig] *abundance zone; acme zone.*
flood zone [streams] (a) The land bordering a stream, subject to floods of about equal frequency; e.g. a strip of the flood plain, subject to flooding more often than once but not as frequently as twice in a century (Langbein and Iseri, 1960, p.11). (b) The land bordering a reservoir or stream impoundment, subject to inundation above the normal operating level.
floor [eco geol] The *footwall* of a horizontal orebody.
floor [geomorph] *bed [geomorph].*
floor [intrus rocks] The country rock bordering the lower surface of an igneous intrusion. Cf: *roof [intrus rocks].*
floor plate Any plate, in a double or single row, that forms an *ambulacral groove* in an echinoderm. Syn: *ambulacral floor plate.*
floor-plate passageway A tubular canal in edrioasteroids that extends along lateral sutural faces of contiguous ambulacral floor plates. Floor-plate passageways connect the *ambulacral tunnel* to the interior of the theca (Bell, 1976).
floor thrust The basal *detachment* of a *duplex.*
flora (flo'-ra) The entire plant population of a given area, environment, formation, or time span. Cf: *fauna.*
floral stage (flo'-ral) A chronostratigraphic unit (stage) based on a *florizone* or commonly on a floral assemblage; e.g. the Ravenian (upper Eocene) of Washington State.
floral zone *florizone.*
florencite-(Ce) A pale yellow rhombohedral mineral of the *crandallite* group: $CeAl_3(OH)_6(PO_4)_2$.
florencite-(La) A pale yellow rhombohedral mineral of the *crandallite* group: $LaAl_3(OH)_6(PO_4)_2$.
florencite-(Nd) A pale yellow rhombohedral mineral of the *crandallite* group: $NdAl_3(OH)_6(PO_4)_2$.
florenskyite A metallic creamy white orthorhombic meteorite mineral: FeTiP.
florensovite (flor-en'-sov-ite) A metallic black cubic mineral of the *linnaeite* group: $(Cu,Zn)(Cr,Sb)_2S_4$.
Florida earth (Flor'-i-da) A variety of fuller's earth containing palygorskite, from Florida (esp. from Quincy and Jamieson) or resembling that from Florida.
florizone (flor'-i-zone) A biostratigraphic unit or body of strata characterized by a particular assemblage of fossil floras, regardless of whether it is inferred to have chronological or only environmental significance. Although the term, like *faunizone,* has been given different meanings, it is close in concept to *assemblage zone* and has been generally regarded as the plant-based variety of (biostratigraphic) zone. ACSN (1961, art.21d) states that the term is "not generally accepted" and that its correct definition is "in dispute". Syn: *floral zone; florizonite.*
florizonite *florizone.*
florule (flo'-rule) (a) An assemblage of fossil plants obtained from a stratum over a very limited geographic area, esp. from only one exposure. (b) A term used by Fenton and Fenton (1928, p.15) for an assemblage of fossil plants associated in a single stratum or a few contiguous strata of limited thickness and dominated by the representatives of one community; the floral assemblage of a *zonule.* Cf: *faunule.* Syn: florula.
floscelle (flos-celle') A star-shaped area around the *peristome* of an echinoid, formed by phyllodes and bourrelets.
flos ferri (flos fer'-ri) An arborescent variety of aragonite occurring in delicate white coralloid masses that commonly encrust hematite, forming picturesque snow-white pendants and branches.
floss A British term for a stream.
flotant (flo'-tant) A coastal marsh formed along an abandoned channel or in a low basin between natural levees of active and inactive stream channels, as in Louisiana south of New Orleans; it is not as firm as marshland. Syn: *floating marsh.*
flotation (flo-ta'-tion) *crystal flotation.*
flotsam Debris that floats on the sea, particularly that derived from trash or wrecks of ships. Cf: *jetsam.*
flour A finely powdered rock or mineral mass, resulting from pulverization and grinding; e.g. *rock flour,* of glacial origin, or chalky-looking finely comminuted carbonate mud formed under intense wave or current action in shoal areas.
flour sand *very fine sand.*
flow [coast] A Scottish term for an arm of the sea; used chiefly in place names, e.g. Scapa Flow.
flow [exp struc geol] Any rock deformation that is not instantly recoverable. Various types of flow in which the mechanism is known include *cataclastic flow, gliding flow,* and *recrystallization flow.* Syn: *rock flowage.*
flow [glaciol] *glacier flow.*
flow [hydraul] The movement of water, and the moving water itself; also, the rate of movement.
flow [mass move] (a) A mass movement of unconsolidated material that exhibits a continuity of motion and a plastic or semifluid behavior resembling that of a viscous fluid; e.g. *creep; solifluction; earthflow; mudflow; debris flow; sturzstrom.* Water is usually required for most types of flow movement. (b) The mass of material moved by a flow.
flow [stratig] The smallest formal lithostratigraphic unit of volcanic flow rocks. A flow is a discrete, extrusive, volcanic body distinguishable by texture, composition, order of superposition, paleomagnetism, or other objective criteria. It is part of a *member* and thus is equivalent in rank to a bed or beds of sedimentary-rock classification. Many flows are informal units. Designation of flows as formal units should be limited to those that are distinctive and widespread (NACSN, 1983, Art. 27).
flow [volc] *lava flow.*
flowage (flow'-age) (a) An act of flowing or flooding, such as the overflowing of a stream onto adjacent land; the state of being flooded. (b) A body of water resulting from flowage; the floodwater of a stream. The term is used locally in Wisconsin for the backwater of an artificial lake. (c) The area affected by a previous flooding.
flowage cast A term used by Birkenmajer (1958, p.141) for a sole mark believed to result from the flowage of mobile, hydroplastic sand over the uneven bottom in the direction of slope; it may be transverse, longitudinal, or multidirectional. See also: *flow cast.*
flowage fold *flow fold.*
flowage land The part of a flood plain that will be covered by the water impounded by a proposed dam, exclusive of the river channel; the principal flow-carrying part of the natural cross section of a stream. Syn: *floodway.*
flow-and-plunge structure A variety of *cross-lamination,* consisting of short, obliquely laminated beds deposited irregularly at various angles of slope, resulting from tidal action accompanied by plunging waves.
flow bog A peat bog whose surface level fluctuates with rain and the tides.
flow breccia A breccia that is formed contemporaneously with the movement of a lava flow; the cooling crust becomes fragmented while the flow is still in motion. It is a type of *autobreccia.*
flow cast (a) A term introduced by Shrock (1948, p.156) for a *sole mark* consisting of a lobate ridge or other raised feature and representing the filling of a depression produced by the flowage or warping of the soft and hydroplastic underlying sediment. Kuenen (1953, p.1058) applied the term *load cast* to such a structure produced by vertical adjustments. Prentice (1956) revived "flow cast" for a sole mark resulting from a combination of load casting and current-oriented flow, such as a load cast modified by horizontal flowage of sediment during or after settling. See also: *flowage cast.* (b) *flute cast.*
flowchart (flow'-chart) A graphic representation or schematic diagram of steps in a sequence of operations that are represented by symbols, as for a computer program.
flow cleavage An obsolete term that was once used as a synonym for *slaty cleavage,* so called because of the largely incorrect assumption that formation of slaty cleavage involved flow of rock.
flow differentiation A differentiation process whereby crystals are separated from magma by their concentration in the high velocity center of a narrow conduit, dike, or sill by grain dispersive pressure developed during flow of the magma.
flow direction The axis parallel to the direction of relative displacement in both igneous and metamorphic rocks. In the former it is a *flow line [petrology].* In the latter, it is usually subparallel to lineations visible on hand specimens, and corresponds to the average orientation of the slip direction of individual crystals, even in conditions of *dynamic recrystallization.*
flow-duration curve A type of *duration curve* showing how often a particular stream discharge is equaled or exceeded.

flow earth *solifluction mantle.*
flower (flow′-er) The reproductive structure of an angiosperm. In a morphologic sense, it is considered to be a specialized branch system.
flowering plant (flow′-er-ing) An informal designation of an *angiosperm.*
flower structure An array of upward-diverging fault splays within a strike-slip zone (Biddle and Christie-Blick, 1985). See also: *positive flower structure; negative flower structure.* Syn: *palm-tree structure.*
flow failure A form of slope movement involving the transport of earth materials in a fluid manner over distances of at least several tens of feet.
flow field A collection of lava flows produced by the same effusion (Kilburn, 2000).
flow finger A vertical channel in snow with percolating water.
flow fold A fold composed of relatively plastic rocks that have flowed towards the synclinal trough. In this type of deformation, there are no apparent surfaces of slip. Syn: *flowage fold.* Cf: *reverse-flowage fold; rheid fold.*
flow-foot breccia Foreset-bedded hyaloclastite breccia that passes upward into, and is structurally continuous with, overlying subhorizontal sheet lavas; a principal component of lava-fed deltas, commonly found on the distal flanks of intraglacial volcanoes (Jones and Nelson, 1970). Cf: *intraglacial volcano; passage zone; table mountain [volc]; tindar; tuya.*
flow gneiss Gneiss whose structure was produced by flowage in an igneous mass before complete solidification.
flowing artesian well (flow′-ing) An *artesian well* that taps an artesian aquifer in which the head is sufficient to raise the water in the well above the land surface. Cf: *nonflowing artesian well; flowing well.* Syn: blow well.
flowing ground In engineering geology, a term for soil that flows into a tunnel from floor, walls, or roof, driven by water seepage. A typical example of flowing ground is the cohesionless soil below the water table. Cf: *firm ground; raveling ground; running ground; squeezing ground; swelling ground.*
flowing well (a) A well that yields water at the land surface without pumping. It is distinguished from a *flowing artesian well* by the possibility that the flow may be due to gas rather than artesian pressure. (b) flowing oil well.
flow joint A joint parallel to the flow layers of a plutonic rock (Tomkeieff, 1943).
flow layer A layer in an igneous rock that is characterized by *flow layering.* Cf: *flow line [petrology].*
flow layering The structure of an igneous rock, characterized by alternating layers of color, mineralogic composition, and/or texture, formed as a result of the flow of magma or lava. Syn: flow banding; *fluxion banding.* See also: *banding [ign].*
flow line [hydraul] (a) The position of the surface of a flowing fluid. (b) A water-level contour around a body of water, e.g. maximum or mean flow line of a lake flowage line. (c) In an open channel, the *hydraulic grade line.* (d) In groundwater, a line that represents the path that a fluid particle would follow through a porous medium (U. S. Army Corps of Engineers, 1991).
flow line [petrology] A lineation of crystals, mineral streaks, or inclusions in an igneous rock, indicating the direction of flow before consolidation. Cf: *flow layer.*
flow lobe In volcanology, a solid or hollow globular extrusion of fluid basalt, generally ~2 m thick and 5-10 m in diameter. Commonly, the "skins" of adjacent flow lobes melt down and disappear, forming homogeneous flow fields (Hon et al., 1994). Cf: *flow-lobe cave.*
flow-lobe cave A cave within a drained *flow lobe* or an interconnected complex of wide, low "pancake rooms" that resulted from such drainage (Halliday, 2003).
flow mark (a) A small channel or gouge cut in a sedimentary surface by a current of water; a flute. (b) A cast of a flow mark, preserved in overlying sediment (Rich, 1950); specif. a *flute cast.* (c) A small ridge formed on the upper surface of a muddy sediment by a current of water (McKee, 1954, p.63). See also: *current mark.*
flow net In the study of seepage phenomena, a graph of flow lines and equipotential lines or net of orthogonal stream lines and equipotential lines.
flow path The subsurface course that a water molecule or solute would follow in a given groundwater velocity field; the direction of movement of water and any contaminants that may be contained therein, as governed principally by the hydraulic gradient.
flow plane The plane along which displacement occurs in both igneous and metamorphic rocks. In the former it is a *flow layer.* In the latter, it is generally subparallel to the foliation visible in hand specimens and corresponds to the average orientation of slip plane of individual crystals, even in conditions of *dynamic recrystallization.*
flow profile The form of the water surface of a *gradually varied flow;* it is commonly known as the *backwater curve.*
flow regime A condition of stream flow defined on the basis of mode of sediment transport, *bed form,* and flow resistance (Middleton, 1965, p.249). The flow *Froude Number* and the phase relationship between the bed and the water surface are also important criteria in determining flow regime (Simons and Richardson, 1961). See also: *lower flow regime; upper flow regime.*
flow roll A rounded, pillow-like body or mass of sandstone occurring within or just above finer-grained sediment or commonly within the basal part of a sandstone overlying shale or mudstone, having a shape approaching that of an elongate, flattened ellipsoid (short axis more or less vertical), and presumed to form by deformation, as by large-scale load casting or mud flowage accompanied by subaqueous slump or foundering of sand channels. The term was used by Pepper et al. (1954, p.88) in reference to the characteristic rolled appearance of the structure and because deformation of strata occurred prior to complete lithification of the rocks. See also: *ball-and-pillow structure; pseudonodule; storm roller.*
flow slide *liquefaction slide.*
flowstone (flow′-stone) A general term for any deposit of calcium carbonate or other mineral formed by flowing water on the walls or floor of a cave. Cf: *dripstone; travertine; cave onyx.*
flow stress A general term for the stress required to sustain *flow [exp struc geol].*
flow structure [ign] A wavy or swirling pattern in which platy or prismatic minerals are oriented along planes of lamellar flowage or in flow lines in fine-grained and glassy igneous.
flow structure [sed struc] A primary sedimentary structure resulting from subaqueous slump or flow (Cooper, 1943, p.190).
flow system In hydrodynamics, a set of flow lines in which any two are always adjacent, and can be intersected in one direction only by an uninterrupted surface across which flow takes place.
flowtill (flow′-till) A superglacial till that is modified and transported by plastic mass flow (Hartshorn, 1958, p.481).
flow tubes Area between two adjacent flow lines of groundwater (Freeze and Cherry, 1979).
flow unit [petroleum] An interval of an oil or gas reservoir whose petrophysical properties determine its flow characteristics. See also: *reservoir compartment.*
flow unit [volc] A group of sheets or beds of lava or pyroclasts that were formed by a single eruption or outpouring.
flow velocity A vector point function used to indicate rate and direction of movement of water through soil, per unit of time and perpendicular to the direction of flow.
fluckite (fluck′-ite) A colorless to rose triclinic mineral related to krautite: $CaMn^{2+}(AsO_3OH)_2{\cdot}2H_2O$.
fluctuation (fluc-tu-a′-tion) The alternate rising and lowering of the water table either regularly or periodically.
fluctus (fluc′-tus) A term established by the International Astronomical Union for a large field of lava flows, often emanating from specific source regions. Originally applied to flow fields observed on Venus, but also applicable to other planetary surfaces. Pl: fluctus.
flue [intrus rocks] A pipelike igneous intrusion. Obsolete.
fluellite (flu′-el-lite) A colorless to yellow orthorhombic mineral: $Al_2F_2(OH)(PO_4){\cdot}7H_2O$.
fluent (flu′-ent) An obsolete term for a stream or other current of water.
fluidal texture (flu′-id-al) Smooth, round surfaces of pyroclasts formed by surface tension within a fluid droplet before it is chilled.
fluid-assisted grain-boundary diffusion *pressure solution [struc petrol].*
fluid decompression The joint-driving mechanism in which the energy release for driving the joint is derived from pore pressure within a flaw or joint (Engelder and Fischer, 1996).
fluid dynamics (flu′-id) That aspect of *fluid mechanics* which deals with motion of fluids.
fluid escape structure A general category of sedimentary feature produced by the escape of fluids from a bed of sediment after deposition. It includes *dish structure, pillar structure* and *vertical sheet structure.* If the escaping fluid is water, the feature may be called a *water escape structure* (Lowe, 1975).
fluid factor [geophys exploration] The difference between the observed fractional change in P-wave velocity and that predicted from the fractional change in S-wave velocity (the fractional change in velocity is the zero-offset reflection coefficient R_P or R_S).
fluid inclusion In a mineral, a tiny cavity, 1s to 100s of micrometers in diameter, containing liquid and/or gas, formed by the entrapment in crystal irregularities of fluid, commonly that from which the rock crystallized.
fluidity index (flu-id′-i-ty) The ratio of sand detritus to the interstitial detrital matrix of a sandstone (Pettijohn, 1954, p.362-363). Rarely used. Syn: fluid factor.
fluidization (flu′-id-i-za′-tion) A process by which a bed of unconsolidated particles behaves as a fluid, as gas is forced upward through the bed. The upward flow of gas supports the solid particles. A fluidized bed will flow down an incline, and objects of density less than the fluidized mass will float on its surface. The concept has been used in attempts to explain the mobility of dense as well as dilute pyroclastic flows.
fluid mechanics The study of the mechanics or behavior of liquids and gases. It is broad in scope, and includes such disciplines as hydraulics and aerodynamics. See also: *fluid dynamics.*
fluid mud Very high concentrations (> 5 g/l) of fine silt and clay suspended in water that creates strong concentration gradient at the base of a water column causing false bottom reflections on an echo-sounder.
fluid potential The mechanical energy per unit mass of a fluid, e.g. water or oil, at any given point in space and time, with respect to an arbitrary state and datum. The fluid potential is proportional to the *total head;* it is the head multiplied by the acceleration due to gravity. Syn: *potential.*
fluke One of two or more recurved components of an *anchor* of a holothurian.

flume [eng] An artificial inclined channel used for conveying water for industrial purposes, such as irrigation, transportation, mining, logging, or power production; or for diverting the water of a stream from its channel for the purpose of washing or dredging the sand and gravel in the dry bed or to aid in engineering construction.

flume [geomorph] A ravine, gorge, or other deep narrow valley, with a stream flowing through it in a series of cascades; e.g. in the White Mountains, N.H.

fluoborite (flu-o-bor'-ite) A colorless hexagonal mineral: $Mg_3(BO_3)(F,OH)_3$. Syn: *nocerite.*

fluocerite-(Ce) A colorless, pale yellow, yellowish brown, or reddish brown trigonal mineral: $(Ce,La)F_3$. Syn: *tysonite.*

fluocerite-(La) A pale greenish yellow trigonal mineral: $(La,Ce)F_3$.

fluor (flu'-or) The original form of *fluorite,* still used chiefly in Great Britain. Etymol: Latin "fluere", "to flow".

fluorannite An iron-black mineral of the *mica* group: $KFe^{2+}{}_3AlSi_3O_{10}F_2$. It represents annite with F>OH.

fluorapatite (flu-or-ap'-a-tite) (a) A common variously colored hexagonal mineral of the *apatite* group: $Ca_5(PO_4)_3F$. It is a common accessory mineral in igneous rocks. Syn: *apatite.* (b) An apatite mineral in which fluorine predominates over chlorine and hydroxyl. Syn: *tavistockite.*

fluorapophyllite (flu-or-a-po'-phyll-ite) A white, pink, yellow, or green platy tetragonal mineral: $KCa_4Si_8O_{20}(F,OH)\cdot 8H_2O$. It represents apophyllite with F>OH.

fluor-arfvedsonite A fibrous light blue-gray mineral of the *amphibole* group: $Na_3(Fe^{2+},Mg)_4Fe^{3+}Si_8O_{22}(F,OH)_2$. It represents arfvedsonite with F>OH.

fluorbritholite-(Ce) A tan or reddish brown hexagonal mineral of the *britholite* group: $(Ca,Ce,La)_5(Si,P)_3O_{12}F$.

fluor-cannilloite A vitreous gray-green mineral of the *amphibole* group: $(Ca,Na)Ca_2(Mg,Al)_3(Si,Al)_8O_{22}(F,OH)_2$. It is regarded as a hyper-calcic amphibole.

fluorcaphite A vitreous pale yellow hexagonal mineral of the *apatite* group: $(Ca,Sr,Na,Ce)_5(PO_4)_3F$.

fluorellestadite (flu-or-el'-les-tad-ite) A colorless, pale bluish, or green hexagonal mineral of the britholite group: $Ca_{10}(SiO_4)_3(SO_4)_3(F,OH)_2$. It represents ellestadite with F>OH.

fluorescence (flu-o-res'-cence) A type of *luminescence* in which the emission of light ceases when the external stimulus ceases; also, the light so produced. Cf: *phosphorescence.*

fluorescence spectrum The *emission spectrum* produced by irradiation of a material with radiation of higher energy, as in *X-ray fluorescence spectroscopy.*

fluoridation Addition of fluorine by means of a chemical reaction between rock and fluid.

fluorine dating (flu'-o-rine) Determination of relative age of Pleistocene or Holocene bones on the basis of fluorine content. The method depends on the gradual combination with time of fluorine in groundwater with the calcium phosphate of bone. In areas where this method has been used, the fluorine content averages 2% in lower Pleistocene bone, 1% in middle Pleistocene bone, 0.5% in upper Pleistocene bone, and 0.3% in Holocene bone.

fluorite (flu'-o-rite) A transparent to translucent mineral: CaF_2. It is found in many different colors (often blue or purple) and has a hardness of 4 on the Mohs scale. Fluorite occurs in veins, usually as a gangue mineral associated with lead, tin, and zinc ores, and is commonly found in crystalline cubes with perfect octahedral cleavage. It is the principal ore of fluorine, and is used as a flux, in the preparation of glass and enamel, in the manufacture of hydrofluoric acid, and for carved ornamental objects. Syn: *fluorspar; fluor; Derbyshire spar.*

fluor-nyböite A monoclinic member of the *amphibole* group: $(Na,Ca,[])_3(Mg,Al,Fe)(Si_7Al)O_{22}(F,OH)_2$. It represents nyböite with F>OH.

fluoro-edenite A vitreous dark brown monoclinic mineral of the *amphibole* group: $(Na,K)Ca_2(Mg,Fe^{2+})_5(Si_7Al)O^{22}(F,OH)$. It represents edenite with F>OH. Syn: fluoredenite.

fluoro-ferroleakeite A vitreous black to bluish black monoclinic mineral of the *amphibole* group: $NaNa_2(Fe^{2+}{}_2Fe^{3+}{}_2Li)Si_80_{22}(F,OH)_2$. It represents leakeite with essential Fe2+ and F>OH.

fluoro-magnesio-arfvedsonitefluoro-magnesio-arfvedsonite A vitreous gray monoclinic mineral of the *amphibole* group: $NaNa_2(Mg_4Fe^{3+})Si_8O_{22}(F,OH)_2$. It represents arfvedsonite with essential Mg and F > OH.

fluoro-magnesiohastingsite A theoretical monoclinic mineral of the *amphibole* group: $(Na,K)Ca_2(Mg,Fe^{2+},Fe^{3+},Al)_5(Si,Al)_8O_{22}(F,OH)_2$. It represents hastingsite with essential Mg and F > OH.

fluoronyböite A bluish gray monoclinic mineral of the *amphibole* group: $NaNa_2(Al_2Mg_3)(Si_7Al)O_{22}F_2$. It represents nyböite with F>OH.

fluororichterite A blue-green, green, or pink mineral of the *amphibole* group: $(Na,K)(Ca,Na)_2(Mg,Fe,Ti)_5Si_8O_{22}(F,OH,O)_2$. It represents richterite with F>OH.

fluor-pargasite A theoretical mineral of the *amphibole* group: $NaCa_2(Mg,Al,Mn)_5(Si_6Al_2)O_{22}F_2$. It represents pargasite with F>OH.

fluorphlogopite A member of the *mica* group: $K(Mg,Fe^{2+})_3(Si_3Al)O_{10}(F,OH)_2$. It represents phlogopite with F>OH.

fluorspar (flu'-or-spar) *fluorite.*

fluor-taramite A theoretical mineral of the *amphibole* group: $(Na,Ca)_3(Mg,Fe,Al)_3(Fe^{3+})_2(Si_6Al_2)O_{22}(F,OH)_2$. It represents taramite with F>OH.

fluorthalénite-(Y) An adamantine colorless monoclinic mineral: $Y_3Si_3O_{10}F$. It represents thalénite with F>OH.

fluortremolite A monoclinic mineral of the *amphibole* group: $Ca_2(Mg,Fe^{2+})_5Si_8O_{22}(F,OH)_2$. It represents tremolite with F>OH.

fluorvesuvianite A silky white or colorless tetragonal mineral of the *vesuvianite* group: $Ca_{19}(Al,Mg)_{13}(SiO_4)_{10}(Si_2O_7)_4O(F,OH)_9$. It represents vesuvianite with F>OH.

flurosion (flu-ro'-sion) A term proposed by Glock (1928, p.477-478) for the work of transportation and erosion carried on by streams. Obsolete.

flush (a) A sudden increase in the volume of a stream; a sudden flow or rush of water, as down a stream and filling the channel or overflowing the banks. Syn: *fresh [water]; freshet.* (b) A British term for a pool or for a low swampy place. Cf: *flash [water].*

flushing period (flush'-ing) The interval of time necessary for an amount of water equal to the volume of a lake to pass through its outlet, computed by dividing lake volume by flow rate (usually mean flow) of the outlet.

flustriform (flus'-tri-form) Said of a flexible, erect colony in cheilostome bryozoans, with lightly calcified zooids and attached by radicles or direct adherence.

flute [geomorph] A small, shallow channel formed by differential weathering and erosion, running nearly vertically down the face of a rock surface.

flute [sed] (a) A primary sedimentary structure, commonly seen as a *flute cast,* consisting of a discontinuous scoop-shaped or lobate depression or groove generally 2-10 cm in length, usually formed by the scouring action of a turbulent, sediment-laden current of water flowing over a muddy bottom, and having a steep or abrupt upcurrent end where the depth of the mark tends to be the greatest. Its long axis is generally parallel to the current. The term was first used by Maxson and Campbell (1935) for flutes cut on boulders in the Colorado River. (b) A term that is sometimes used loosely as a syn. of *flute cast.* (c) A scalloped or rippled rock surface. The term is usually used in the plural. See also: *fluting [sed].*

flute [speleo] Solution grooves formed in the walls of a cave by the solvent action of descending water. Common in the walls of vertical shafts.

flute cast A term suggested by Crowell (1955, p.1359) for a spatulate or lingulate *sole mark* consisting of a raised, oblong, and subconical bulge on the underside of a siltstone or sandstone bed, characterized by a steep or blunt bulbous or beaked upcurrent end from which the structure flattens or flares out in the downcurrent direction and merges with the bedding plane. It is formed by the filling of a flute. See also: *lobate rill mark.* Syn: *fluting [sed]; flute [sed]; flow cast; flow mark; scour cast; vortex cast; linguoid sole mark; lobate plunge structure.*

fluted moraine (flut'-ed) *fluted moraine surface.*

fluted moraine surface Moraine surface in front of a glacier containing parallel ridges that have a more or less constant height over distances of the order of tens or hundreds of meters. The ridge axes are parallel to the flow direction of the glacier (Schytt, 1963). Syn: *fluted moraine.*

fluted till *Lodgment till* that has been eroded into grooves, or in which positive constructional linear ridges have been built.

fluting [geomorph] (flut'-ing) A process of differential weathering and erosion by which an exposed well-jointed coarse-grained rock, such as granite or gneiss, develops a corrugated surface of flutes; esp. the formation of small-scale ridges and depressions by wave action.

fluting [glac geol] (a) The formation by glacial action of smooth deep gutterlike channels or furrows on the stoss side of a rocky hill obstructing the advance of a glacier; the furrows are larger than glacial grooves, and they do not extend around the hill to the lee side (Chamberlin, 1888, p.246). Also, a furrow so formed. (b) Lineations or streamline grooves and ridges parallel to the direction of ice movement, formed in newly deposited till or older drift. They range in height from a few centimeters to 25 m, and in length from a few meters to more than 20 km.

fluting [sed] (a) The process of forming a flute by the cutting or scouring action of a current of water flowing over a muddy surface. (b) Scalloped or rippled rock surfaces. (c) *flute cast.*

Fluvents Entisols that form in recent loamy or clayey alluvial deposits, are usually stratified, and have an organic carbon content that decreases irregularly with depth. Fluvents are not saturated with water for periods long enough to limit their use for most crops. A suborder in the U.S. system of soil taxonomy.

fluvial (flu'-vi-al) (a) Of or pertaining to a river or rivers. (b) Existing, growing, or living in or about a stream or river and its floodplain. (c) Produced by the action of a stream or river. Cf: *fluviatile.* Etymol: Latin "fluvius", "river".

fluvial cycle of erosion *normal cycle.*

fluvial denudation Erosional reduction of a land surface by streams, assisted by weathering, mass wasting, and overland flow.

fluvial deposit A sedimentary deposit consisting of material transported by, suspended in, or laid down by a stream.

fluvialist (flu'-vi-al-ist') An adherent of the doctrine that most widespread surficial deposits and other surface phenomena can be explained by ordinary stream action. Cf: *diluvialist.*

fluvial lake (a) A lake with a perceptible flow of water, e.g. a body of water connecting two larger bodies whose differences in elevation may be sufficient to create a flow from one to another. (b) A slowly moving part of a river, as where its width has been expanded. See also: *fluviatile lake.*
fluvially-dominated delta A river delta built in an environmental setting where fluvial processes of deposition are capable of overwhelming marine processes of sediment reworking. Deltas of this category have elongate distributaries and complex shore line configuration. They have multiple, digitate distributary lobes as a result of considerable fluvial sediment discharge with little reworking by waves or tidal currents. Also called: *bird-foot delta.* Cf: *tide-dominated delta; wave-dominated delta.*
fluviatile (flu'-vi-a-tile') A syn. of *fluvial* (Fowler, 1937, p.184). Geologists tend to use the term for the results of river action (e.g. *fluviatile* dam, or *fluviatile* sands) and for river life (e.g. *fluviatile* fauna).
fluviatile lake A lake formed as a result of the action of a river or stream, or a lake occupying a basin produced by running water capable of erosion or deposition; e.g., an *oxbow lake* on the flood plain of a meandering river, a *varzea lake* on the floodplain of a tropical river, or a lake formed by the damming action of excess sediment at the confluence of a tributary and the main river. Cf: *fluvial lake.*
fluviation (flu-vi-a'-tion) The activities engaged in and the processes employed by streams (Glock, 1928, p.477). Syn: *stream action.*
fluvicoline (flu'-vi-co-line') Said of an animal that lives in or frequents streams.
fluvioclastic rock (flu'-vi-o-clas'-tic) A *hydroclastic rock* containing current- or river-worn fragments (Grabau, 1924, p.295).
fluvioeolian (flu'-vi-o'-e-o'-li-an) Pertaining to the combined action of streams and wind; e.g. a *fluvioeolian* deposit.
fluvioglacial (flu'-vi-o-gla'-cial) A syn. of *glaciofluvial.* The term "glacio-fluvial" is preferred in U.S.
fluvioglacial drift A general term for *glacial drift* transported and deposited by running water emanating from a glacier. Cf: *drift [glacial geol]; glacial drift.* Syn: *glaciofluvial drift.*
fluviograph (flu'-vi-o-graph') A device for measuring and recording automatically the rise and fall of a river. Syn: *fluviometer.*
fluviokarst (flu'-vi-o-karst) A terrain with both fluvial and karstic characteristics. Deranged surface drainage is common; drainage basins usually contain both karstic and nonkarstic rocks.
fluviolacustrine (flu'-vi-o'-la-cus'-trine) Pertaining to sedimentation partly in lake water and partly in streams, or to sediments deposited under alternating or overlapping *lacustrine* and *fluvial* conditions.
fluviology (flu-vi-ol'-o-gy) The science of rivers.
fluviomarine (flu'-vi-o'-ma-rine') Said of marine sediments that contain resorted and redistributed fluvial material along with the remains of marine organisms.
fluviometer (flu-vi-om'-e-ter) *fluviograph.*
fluviomorphology (flu'-vi-o'-mor-phol'-o-gy) *river morphology.*
fluvioterrestrial (flu'-vi-o'-ter-res'-tri-al) Consisting of or pertaining to the land and its streams.
fluviraption (flu-vi-rap'-tion) A term introduced by Malott (1928a) for *hydraulic action* performed by rivers. Obsolete.
flux [cryst] A crystal-growth technique where the crystal solidifies from a cooling molten solution. Used to produce gem materials such as synthetic emerald and other beryls as well as synthetic ruby and sapphire.
flux [geophys] A representation of magnetic, electrical, and other lines of force. The flux density is proportional to field intensity.
flux [metal] *fluxstone.*
flux [radioactivity] The number of particles in a given volume times their mean velocity.
flux [streams] A stream of flowing water; a flood or an outflow.
fluxgate magnetometer (flux'-gate) An instrument that measures the magnetic field component along the axis of its (core) sensor and must be either oriented with the direction of a measured field or arranged as three mutually orthogonal sensors if the total intenisty is to be measured. Capable of detecting changes in the magnetic field on the order of 0.2 nT.
fluxie (flux'-ie) A field term informally used by sedimentologists for *fluxoturbidite.*
fluxion (flux'-ion) Obsolete British usage for "flow". Cf: *flow layering.*
fluxion banding Obsolete British usage for *flow layering.*
flux ore A material containing an appreciable amount of valuable metal but smelted mainly because it contains fluxing agents useful in the reduction of richer ores. Cf: *fluxstone.*
fluxoturbidite (flux-o-tur'-bid-ite) A term proposed by Dzulynski et al. (1959, p.1114) for a sediment produced by a mechanism related both to deposition from turbidity currents and to submarine slumping or sliding. It is characterized by coarse grain, thick bedding, and poor development of grading and of sole marks. Generally speaking, it is a proximal turbidite. Cf: *undaturbidite.* Syn: *fluxie.*
fluxstone (flux'-stone) Limestone, dolomite, or other rock or mineral used in metallurgical processes to lower the fusion temperature of the ore, combine with impurities, and make a fluid slag. Syn: *flux.*
fly ash All particulate matter that is carried in a gas stream, esp. in stack gases at a coal-fired plant for generation of electric power.
flyer (fly'-er) A number of geophones permanently connected at intervals along a short cable. A typical flyer might contain, for example, six geophones connected in series at 20-ft intervals, with clips at one end for joining it to the main cable that carries the signal to the recording equipment.
flying bar (fly'-ing) A looped bar or spit formed on the landward side of an island that is subsequently reduced below sea level by wave erosion before the bar or spit is destroyed. Term originated by Gulliver (1899, p.190).
flying veins A pattern of mineral-deposit veins overlapping and intersecting in a branchlike pattern.
fly leveling Spirit leveling in which some of the restrictions of precise leveling (such as limiting the length of sight, or balancing backsights and foresights) are relaxed in order to obtain elevations of moderate accuracy more rapidly; e.g. running a level line (a line over which leveling operations are accomplished) at the close of a working day in order to check the results of an extended line run in one direction only. Fly leveling may be done with alidade and planetable. Also called "flying levels".
flysch (a) A marine sedimentary facies characterized by a thick sequence of poorly fossiliferous, thinly bedded, graded deposits composed chiefly of marls and sandy and calcareous shales and muds, rhythmically interbedded with conglomerates, coarse sandstones, and graywackes. See also: *wildflysch; macigno.* (b) An extensive, preorogenic sedimentary formation representing the totality of the flysch facies deposited in different troughs, during the later stages of filling of a geosynclinal system, by rapid erosion of an adjacent and rising mountain belt at a time directly previous to the main paroxysmal (diastrophic) phase of the orogeny or when initial diastrophism had already developed interior ridges exposed to erosion; specif. the Flysch strata of late Cretaceous to Oligocene age along the borders of the Alps, deposited in the foredeeps in front of northward-advancing nappes rising from beneath the sea, before the main phase (Miocene) of the Alpine orogeny. (c) A term that has been loosely applied to any sediment with nearly all of the lithologic and stratigraphic characteristics of a flysch, such as almost any turbidite. Etymol: dialectal term of German origin used in Switzerland for a crumbly or fissile material that slides or flows. Pl: flysches. Adj: flyschoid. Cf: *molasse.*
fm *formation [stratig].*
f-number The ratio of the equivalent *focal length* of a photographic lens to the *relative aperture,* or a number expressing the relative aperture of the lens; e.g. the f-number of a lens with a relative aperture of *f*/4.5 is 4.5. The smaller the number, the brighter the image and therefore the shorter the exposure required. Cf: *speed.*
foam crust A snow-surface feature produced during ablation and resembling small overlapping waves, like sea foam on a beach. Cf: *plowshare.*
foaming earth (foam'-ing) Soft or earthy aphrite.
foam mark A surface sedimentary structure consisting of a pattern of almost imperceptible ridges and hollows, formed where foam produced by wind action on seawater is driven over a surface of moist or wet sand.
foam structure A recrystallized grain structure characterized by the presence of grain-boundary triple junctions (forming angles of ~120°). The surface energy associated with the relatively planar grain boundaries is less than that of curved grain boundaries.
focal depth (fo'-cal) *depth of focus.*
focal length A general term for the distance from the principal point or center of a lens to the principal focus. In photogrammetry, the term "equivalent focal length" is the distance measured along the lens axis from the rear nodal point of the lens to the position of the focal plane that provides the best average definition in the aerial negative, and the term "calibrated focal length" is an adjusted value of the equivalent focal length computed to distribute the effect of lens distortion over the entire field of the negative. Symbol: f.
focal mechanism *fault-plane solution.*
focal plane The plane, perpendicular to the axis of a lens, in which images of points in the object field are brought to a focus; a plane that passes through a principal focus.
focal point *principal focus.*
focal sphere An arbitrary reference sphere drawn about the hypocenter or focus of an earthquake, to which body waves recorded at the Earth's surface are projected for studies of earthquake mechanisms.
focus [photo] (fo'-cus) (a) The point at which rays of light converge to form an image after passing through a lens or optical system or after reflection by a mirror. See also: *principal focus.* (b) The condition of sharpest imagery. Pl: foci.
focus [seis] The initial rupture point of an earthquake, where strain energy is first converted to elastic wave energy; the point within the Earth which is the center of an earthquake. Cf: *epicenter.* Syn: *hypocenter; centrum [seis].*
focused-current log (fo'-cused-cur'-rent) The *resistivity log* curves from a multi-electrode *sonde* designed to focus the surveying current radially through the rocks in a horizontal, disk-shaped pattern. This permits sharp definition of bed boundaries and improved measurement of resistivity. Focused-current logs are marketed under several trade names, e.g. Laterolog, Guard Log. Syn: current-focused log; *guard-electrode log.*
focus zone An end-member source for mantle-derived magmas, postulated by Hart et al. (1992). Abbrev: FOZO.
foehn European term for a warm, dry wind that descends the leeward side of a mountain ridge and warms from compression. The type locality is the

Alps but various local names are used elsewhere, e.g. *chinook.* Also spelled: föhn.

fog A cloud at the Earth's surface, formed by condensation of atmospheric water vapor into tiny droplets of water, about 40 µm in diameter, or rarely into crystals of ice. Fog reduces visibility to less than 1 km (5/8 mile). Cf: *advection fog; haze; radiation fog.*

fog desert A west-coast desert having a cold-water marine current just offshore of a warm continental land mass; the coolness and moist air combine to produce fog but little precipitation.

foggara A term used in the Saharan desert region (esp. in Morocco and Mauritania) for a gently inclined, underground conduit or tunnel designed to intercept groundwater near the foot of mountains and to conduct it by gravity to a neighboring lowland for irrigation; a horizontal well. Etymol: Arabic. Cf: *qanat; karez.*

foggite (fogg'-ite) A white platy orthorhombic mineral: $CaAl(PO_4)(OH)_2 \cdot H_2O$.

foid A collective term coined by Johannsen (1917, p.69-70) to denote the *feldspathoid* group of minerals. Etymol: *f*eldspath*oid.*

foid-bearing In the *IUGS classification,* a modifier preceding a rock name (syenite, monzonite, etc.) indicating a plutonic rock with F between 0 and 10. The rock name is determined by P/(A+P) and, for some categories, the plagioclase composition. For a specific rock, the term "foid" is replaced by the feldspathoid that is present. For example, nepheline-bearing gabbro contains nepheline and has F between 0 and 10, P/(A+P) greater than 90, and plagioclase more calcic than An_{50}. Cf: *foid.*

foid diorite In the *IUGS classification,* a plutonic rock with F between 10 and 60, P/(A+P) greater than 90, and plagioclase more sodic than An_{50}. Cf: *theralite.*

foid gabbro In the *IUGS classification,* a plutonic rock with F between 10 and 60, P/(A+P) greater than 90, and plagioclase more calcic than An_{50}. Syn: *theralite.*

foid monzodiorite In the *IUGS classification,* a plutonic rock with F between 10 and 60, P/(A+P) between 50 and 90, and plagioclase more sodic than An_{50}. Syn: *essexite.*

foid monzogabbro In the *IUGS classification,* a plutonic rock with F between 10 and 60, P/(A+P) between 50 and 90, and plagioclase more calcic than An_{50}. Syn: *essexite.*

foid monzosyenite In the *IUGS classification,* a plutonic rock with F between 10 and 60, and P/(A+P) between 10 and 50. Less preferred syn: *foid plagisyenite.*

foidolite (foid'-o-lite) In the *IUGS classification,* a plutonic rock with F greater than 60; a general term that includes *melteigite, missourite, ijolite, fergusite, urtite,* and *italite.*

foid plagisyenite Less preferred synonym of *foid monzosyenite.*

foid syenite In the *IUGS classification,* a plutonic rock with F between 10 and 60, and P/(A+P) less than 10.

foid syenitoid In the *IUGS classification,* a general term in the provisional "field" classification for plutonic rocks thought to contain essential feldspathoids and in which alkali feldspar is thought to be more abundant than plagioclase.

foitite A bluish-black mineral of the *tourmaline* group: $([\,],Na)(Fe^{2+},Al)_3Al_6Si_6O_{18}(BO_3)_3(OH)_4$.

fold [geomorph] A British term for an undulation in the land surface, either a low rounded hill or a shallow depression.

fold [paleont] (a) A major rounded elevation on the surface of a brachiopod valve (generally the *brachial valve*), externally convex in transverse profile and radial from the *umbo,* and usually median in position. It is typically associated with the *sulcus.* Cf: *carina.* (b) A spirally wound ridge on the interior of the wall of a gastropod shell; e.g. *columellar fold* and *parietal fold.*

fold [struc geol] n. A curve or bend of a planar structure such as rock strata, bedding planes, foliation, or cleavage. A fold is usually a product of deformation, although its definition is descriptive and not genetic and may include primary structures.

fold aspect ratio The ratio of the amplitude of a fold to the distance between adjacent inflection points (Twiss, 1988).

fold axis A line which, when moved parallel to itself, traces out a folded surface. It applies only in the case of cylindrical folds.

foldback [geophys exploration] (a) Doubling back a cable on itself such that two or more geophone stations are located at the same position on the ground. Used for making comparisons (e.g., of different types of geophone arrays). (b) Treatment applied to frequencies greater than the *Nyquist frequency,* causing data to *alias.*

fold belt A belt of folds whose hinges are roughly parallel to one another. In general, the folds involved are the product of a single deformation event.

fold breccia A local *tectonic breccia* composed of angular fragments resulting from the sharp folding of thin-bedded, brittle rock layers between which are incompetent ductile beds; e.g. a breccia formed where interbedded chert and shale are sharply folded. Syn: *reibungsbreccia.*

fold coast A coast whose configuration is controlled by folded rocks.

fold facing *facing.*

folding (fold'-ing) The formation of folds in rocks.

folding frequency *Nyquist frequency.*

fold mullion A type of *mullion* formed by the cylindrical undulations of bedding; internal structure displays conformable bedding laminations (Wilson, 1953). Cf: *cleavage mullion.*

fold profile A section of a cylindrical fold on a plane oriented perpendicular to its fold axis.

fold system A group of congruent folds that are produced by the same tectonic episode.

fold-thrust belt A structural domain, usually found along the margins of an orogenic belt, in which crustal shortening occurs by slip on sub-parallel thrust faults that verge toward the exterior of the belt, and the formation of folds associated with thrust faults. Typically, thrust faults merge with a subhorizontal regional décollement at or near the basement-cover contact; in the trailing portions of most fold-thrust belts, basement is involved in the thrusting.

Foleyan (Fo'-ley-an) North American (Gulf Coast) stage: Pliocene (above Clovelly, below Pleistocene).

folgerite (fol'-ger-ite) *pentlandite.*

folia [meta] (fo'-li-a) Thin, leaflike layers or laminae, specif. cleavable folia of gneissic or schistose rocks. Singular: *folium.*

folia [speleo] Downward-sloping, shelf-like, interleaved tiers, commonly of calcite, commonly associated with *calcite rafts;* related to *shelfstone,* but controlled by fluctuating water level surface. Sing: folium, rarely used.

foliar physiognomy (fo'-li-ar phys-i-og'-no-my) *leaf margin analysis.*

foliate (fo'-li-ate) Adj. of *foliation.* Obsolete syn: *parafoliate.* n. A rock showing foliation.

foliated cataclasite A *cataclasite* possessing either bands of distinct mineralogical composition, rock of distinctive grain size, or thin septa aligned roughly parallel to a fault. In some cases, individual mineral or rock fragments in the cataclasite may also define a crude dimensional preferred orientation parallel to the fault (Chester et al., 1985).

foliated ground ice A large mass of ice commonly occupying thermal contraction cracks in permafrost, and characterized by parallel or subparallel structures marked by air bubbles, films of organic or inorganic matter, or boundary surfaces between ice layers of different composition. It is usually, but not always, wedge-shaped. Syn: *wedge ice.*

foliated structure One of various *microstructures* containing sheetlike aggregations of calcitic blades or laths, including regularly foliated, crossed foliated, complex crossed foliated, and calcitic crossed bladed structures (Boggild, 1930, p.249; Kobayashi, 1964, 1969, 1971; Taylor et al., 1969; Grégoire, 1972; Uozumi and Suzuki, 1981; Carter, 1980a).

foliation [glaciol] (fo-li-a'-tion) The planar or layered structure produced in the ice of a glacier, manifest as alternating layers of coarse-grained and fine-grained ice, or bubbly and clear ice, or of dirty and clean ice. Most foliation probably originates as annual layers in the accumulation area. Some of it originates as crevasse fillings or in other ways. After extreme deformation, the origin is difficult or impossible to determine. Syn: *band [glaciol]; glacier band.* Less preferred syn: *banding [glaciol].*

foliation [struc geol] A general term for a planar arrangement of textural or structural features in any type of rock, esp. the locally planar fabric in a rock defined by a fissility, a preferred orientation of crystal planes in mineral grains, a preferred orientation of inequant grain shapes, or from compositional banding. In igneous rocks, planar parallelism of flaky or tabular minerals and mineral aggregates, slabby xenoliths, or flattened vesicles as well as compositional layering. In metamorphic rocks, planar parallelism of flaky minerals and compositional layering. Adj: *foliate.* See also: *cleavage [struc geol].*

foliole (fo'-li-ole) A subordinate element of a *saddle* in the ammonite *suture.*

foliose (fo'-li-ose) Leafy, or resembling a leaf; esp. said of a *corallum* with laminar branches.

Folists Histosols that have an accumulation of organic soil materials mainly as forest litter that is <1m deep to rock or to fragmental materials with interstices filled with organic materials. Folists are not saturated with water for periods long enough to limit their use if cropped. A suborder in the U.S. system of soil taxonomy.

folium Sing. of *folia.*

follicle (fol'-li-cle) A simple dry dehiscent fruit that produces seeds and that has one carpel and splits along one seam only.

following wind (fol'-low-ing) (a) A wind whose direction is the same as that of ocean waves. (b) A tailwind.

fondo (fon'-do) adj. A term applied by Rich (1951, p.2) to the environment of sedimentation that lies on the deep floor of a water body. It may be used alone or as a combining form. See also: *fondoform; fondothem.* Cf: *clino; unda.* Etymol: Spanish *fondo,* "bottom".

fondoform (fon'-do-form) The subaqueous land form constituting the main floor of a water body (Rich, 1951, p.2). It is the site of the *fondo* environment of deposition. Cf: *clinoform; undaform.*

fondothem (fon'-do-them) Rock units formed in the *fondo* environment of deposition (Rich, 1951, p.2). Cf: *clinothem; undathem.*

Foner magnetometer (Fo'-ner) *vibration magnetometer.*

font An archaic term for a stream or a spring, fountain, or source of a stream; it forms part of place names, such as Chalfont.

Fontainebleau sandstone (Fon'-taine-bleau) (a) A desilicified quartz sandstone (or an uncemented quartz sand) whose calcareous cement forms a crystalline aggregate of calcite in which sand grains are embedded; the cement is continuous and the easily fractured surfaces of the rock indicate that the calcite is in crystallographic continuity. Type locality: Fontainebleau, in the Paris Basin of France. Cf: *meulerization.* (b) A name given to a variety of calcite; specif. *sand-calcite.*

fontanite A vitreous yellow orthorhombic mineral: $Ca(UO_2)_3(CO_3)_2O_2{\cdot}6H_2O$.

food chain The passage of energy and materials from producers through a progressive, essentially linear sequence of plant-eating and meat-eating consumers. Cf: *food cycle; food web.*

food cycle All the food chains in an association of organisms; the food relations between the members of a population that make it possible for the population to survive.

food groove (a) An *ambulacral groove* in an echinoderm; e.g. a furrow running along the adoral surface of crinoid ray ossicles and traversing the *tegmen* to the mouth. (b) In a brachiopod, a trough running the length of the frontal surface of the brachia, bounded by a muscular lip and the base of the filaments.

food web The passage of energy and materials from producers through a progressive, many-stranded, anastomosing network of plant-eating and meat-eating consumers. Cf: *food chain; food cycle.*

fool's gold A popular term for pyrites resembling gold in color; specif. pyrite and chalcopyrite.

foordite (foord'-ite) An adamantine brownish yellow monoclinic mineral: $Sn^{2+}(Nb,Ta)_2O_6$.

foot [geol] (a) The bottom of a slope, grade, or declivity. Cf: *head [geomorph].* (b) That portion of the displaced material of a landslide that lies downslope from the toe of the surface of rupture (Varnes, 1978). (c) The lower bend of a fold or structural terrace. Syn: *lower break.*

foot [paleont] (a) The ventral part of the body of a mollusk, consisting chiefly of a muscular surface or process, and used for locomotion; e.g. a broad, flattened muscular sole used for creeping in most gastropods, or a protrusible, tapering, bladelike muscular structure extending from the midline of the body of a bivalve mollusk (anteriorly and ventrally in more typical bivalves), and used for burrowing or locomotion. (b) A limb of an arthropod. (c) One of the radial appendages extending from the ultimate joint of the shell of certain radiolarians (as in the suborders Nassellina and Phaeodarina). Cf: *head [paleont].*

foot cave *cliff-foot cave.*

footeite (foote'-ite) *connellite.*

foot glacier *expanded-foot glacier.*

foothills (foot'-hills) A region of relatively low, rounded hills at the base of or fringing a mountain range; e.g. the low, undulating region along the western base of the Sierra Nevada in California.

footing (foot'-ing) A relatively shallow foundation by which concentrated loads of a structure are distributed directly to the supporting soil or rock through an enlargement of the base of a column or wall. Its ratio of base width to depth of foundation commonly exceeds unity. Cf: *pier [eng].*

foot layer An inward extension of ektexine of a pollen grain, partly surrounded by endexine.

footprint (foot'-print) *track [paleont].*

footprint [geophys] *acquisition footprint.*

foot slope A general term for a hillside surface whose top part is the *wash slope* and that includes "all the slopes of diminishing gradient" (Penck, 1953, p.419). Syn: *fusshang.*

footwall (foot'-wall) The underlying side of a fault, orebody, or mine working; esp. the wall rock beneath an inclined vein or fault. Syn: *heading wall; heading side; lower plate.* Cf: *hanging wall.*

footwall cutoff The intersection between the footwall portion of a once-continuous planar or linear geological feature and a fault surface. In thrust systems, a syn. of *trailing edge.* See: Royse et al. (1975).

footwall flat That portion of a step-shaped fault that parallels strata in the footwall.

footwall imbrication A temporal sequence of formation of imbricates or splays off a thrust fault flat where successively younger imbricates or splays form on the foreland or leading side of preexisting imbricates or splays (Boyer and Elliott, 1982). Syn: *break-forward sequence.* Cf: *hanging-wall imbrication; in-sequence thrusting; out-of-sequence thrusting.*

footwall ramp That portion of a step-shaped fault that cuts across strata in the footwall.

footwall uplift An uplift that occurs below a fault. For large normal faults, a component of footwall uplift is flexural uplift, the buoyancy-induced vertical (isostatic) deformation that decreases in magnitude with distance from the fault (Peacock et al., 2000).

foralite (for'-a-lite) An inorganic structure resembling a worm tube, found in stratified rock.

foram (for'-am) *foraminifer.*

foramen (fo-ra'-men) (a) A small opening, perforation, pore, or orifice, such as the *pedicle foramen* of a brachiopod or the *septal foramen* of a nautiloid; esp. an opening that connects adjacent chambers in the test of a foraminifer, located at the base of septa or areal in position, and often representing a previous aperture or a secondarily formed aperture (but not equivalent to the pore of a perforate test). (b) A small opening in a bone, esp. one that gives passage to a blood vessel or nerve. Pl: *foramina.*

foramina (fo-ram'-i-na) Plural of *foramen.*

foraminifer (for-a-min'-i-fer) Any protozoan belonging to the subclass Sarcodina, order Foraminiferida, characterized by the presence of a test of one to many chambers composed of secreted calcite (rarely silica or aragonite) or of agglutinated particles. Most foraminifers are marine but freshwater forms are known. Range, Cambrian to Holocene. Colloquially shortened to *foram* Pl: foraminifera (fo-ra-mi-nif'-era); informally foraminifers. Etymol: Latin, "foramen (pl: foramina)", small opening + "-fer", carrier.

foraminiferal (fo'-ram-i-nif'-er-al) Pertaining to or derived from the foraminifera or their shells; e.g. "foraminiferal test".

foraminiferal ooze An ooze at least 30% composed of the skeletal remains of the tests of foraminifera; it is a *calcareous ooze.* See also: *globigerina ooze.*

foraminite (fo-ram'-i-nite) A sedimentary rock composed predominantly of the remains of foraminifers.

foram number In quantitative foraminiferal studies, the total number of all foraminiferal shells ("forams") in a one-gram sample (dry weight) of material studied.

foramol A skeletal grain association that characterizes temperate water but extends to the tropics; it includes benthic foraminifera, mollusks, barnacles, bryozoans, and calcareous red algae (Alan Lees, 1975).

forb A noncultivated dicotyledonous herbaceous plant; a herb other than grass; a weed. The term appears in some palynologic literature dealing with Quaternary sediments.

Forbes band An obsolete term for one of a group of bands forming a type of ogive pattern that occurs on valley glaciers below icefalls and is characterized by alternating dark and light curved bands that cross the glacier. These bands normally occur in a regular succession at roughly equal intervals. This type of band was described by James D. Forbes (1809-1868), English physicist, who originally called it a dirt band. Cf: *dirt-band ogive; ogive.*

forbesite (forbes'-ite) A mixture of annabergite and arsenolite.

force [phys] That which tends to put a stationary body in motion or to change the direction or speed of a moving body.

force [streams] A name given in northern England to a waterfall or cascade. Etymol: Old Norse *fors.* Cf: *fors.* Syn: *foss; fosse [streams].*

force couple Two equally intense forces acting in opposite directions but not in the same line on a body, creating a tendency for the body to rotate. Syn: couple; shear couple.

forced convection Convection created by an externally imposed force field, such as groundwater flow caused by differences in water pressure or elevation. Cf: *convection [grd wat]; free convection.*

forced-cut meander A *meander* in which deposition on the inner bank proceeds at the same rate as lateral erosion on the outer bank, thereby maintaining a channel of constant width (Melton, 1936, p.596-597). Cf: *advance-cut meander; scroll meander.*

forced fold A fold whose final overall shape and trend are dominated by the shape of some forcing member below, typically a fault. (Stearns, 1978). See also: *forced folding; drape fold.*

forced folding (a) Deformation of the sedimentary strata above the crystalline basement by movement along faults. (b) A type of *fault-propagation fold* in which the transition from faulting to folding is abrupt. See also: *drape fold; forced fold.*

forced oscillation An oscillation that is imposed on a body, e.g., the Earth, by an external force. Cf: *free oscillation.*

forced regression In sequence stratigraphy, a seaward movement of the shoreline in response to relative sea-level lowering. It is induced by the lack of accommodation space and produces stacking patterns of downward-stepping, prograding clinoforms over the condensed section that was formed during the previous *transgressive* and *highstand systems tracts* (Posamentier et al., 1992).

forceful intrusion Emplacement of magma that forcibly created the space into which it moved. Cf: *passive intrusion.*

force of crystallization *crystallizing force.*

forceps (for'-ceps) A C-shaped siliceous sponge *spicule* (sigma) having the form of tongs, with subparallel extremities.

forcherite (for'-cher-ite) An orange-yellow variety of opal colored with orpiment.

ford (a) A shallow and usually narrow part of a stream, estuary, or other body of water that may be crossed, as by wading or by a wheeled land vehicle. It usually has a firm, level, and relatively boulder-free bottom. Syn: *wath.* (b) An archaic term for a stream or other body of water.

förde A Danish term for a long, narrow straight-sided inlet of the sea in a coastline consisting of till or surrounded by terminal moraines, and typically produced by drowning of a glaciated valley along a low-lying coast, like that of the SE Jutland Peninsula. Pl: förden. Cf: *fjord; bodden.*

fore arc The region between a subduction-related trench and a volcanic arc. Syn: *frontal arc.*

fore-arc basin A sedimentary basin, usually elongate, lying between the volcanic arc and the shelf break in a *convergent plate boundary* zone. It is parallel to the arc and closer to it than to the trench-slope basin and the trench.

foredeep (fore′-deep) (a) An elongate depression bordering an island arc or other orogenic belt. Cf: *trench [marine geol].* (b) A syn. of *foreland basin.*

foredune (fore′-dune) A coastal dune or *dune ridge* oriented parallel to the shoreline, occurring at the landward margin of the beach, along the shoreward face of a beach ridge, or at the landward limit of the highest tide, and more or less completely stabilized by vegetation.

foreland [coast] (fore′-land) (a) An extensive area of land, either high ground or low land, jutting out from the coast into a large body of water (usually the sea); a headland; a *promontory.* See also: *cuspate foreland.* (b) A prograded strip of low, flat land built by waves and currents at the base of a cliff; an initial stage in the development of a *strand plain.* (c) A part of the natural shore, located seaward of an embankment, that receives the shock of sea waves and deadens their force.

foreland [geog] The land lying in front of or adjoining other land and physiographically related to it.

foreland [glac geol] A lowland area, now or formerly covered by piedmont glaciers, or a glacio-isostatically depressed low-elevation area, which is covered with raised beach or marine deposits.

foreland [tect] The exterior area of an orogenic belt where deformation occurs without significant metamorphism. Generally the foreland is closer to the continental interior than other portions of the orogenic belt are.

foreland basin A linear sedimentary basin in a *foreland [tect].* These basins subside in response to flexural loading of the lithosphere by *thrust sheets.* Syn: *foredeep.*

foreland-dipping duplex A *duplex* in which the imbricates dip toward the foreland; structures, and sometimes strata, in the *horses* composing the duplex also dip toward the foreland and usually face downward (Boyer and Elliott, 1982). Cf: *hinterland-dipping duplex.*

foreland facies *shelf facies.*

forelimb (fore′-limb) The steeper of the two limbs of an asymmetrical, anticlinal fold. Cf: *backlimb.*

forelimb thrust A thrust fault that cuts across the forelimb of an asymmetric anticline. Little used syn: *front-limb thrust fault.*

forellenstein (fo-rel′-len-stein) *troctolite.*

Forel scale (Fo-rel′) A color scale of yellows, greens, and blues that is used against a white Secchi disk to measure the color of seawater.

forensic geology (fo-ren′-sic) The application of the earth sciences to the law (Murray, 2004). Syn: *geoforensics; legal geology.*

forepoling (fore-pol′-ing) A method of advancing an underground excavation in loose, caving, or watery earth materials, such as quicksand, by driving sharp-pointed poles, timbers, sections of steel, or slabs into the ground ahead of, or simultaneously with, the excavating operation; a method of supporting a very weak roof. It is useful in tunneling and in extracting coal from under shale or clay.

fore reef The seaward side of a reef; the slope covered with deposits of coarse reef talus. Cf: *back reef; off-reef.* Also spelled: forereef.

forerunner (fore′-run-ner) A low, long-period ocean swell representing the gradual change of water level that commonly begins several hours before the arrival of the main swell from a distant storm, esp. a tropical hurricane.

foreset (fore′-set) n. A *foreset bed.* adj. Pertaining to or forming a steep and advancing frontal slope, or the sediments deposited on such a slope.

foreset bed One of the inclined, internal, and systematically arranged layers of a cross-bedded unit; specif. one of the gently inclined layers of sandy material deposited upon or along an advancing and relatively steep frontal slope, such as the outer margin of a delta or the *lee* side of a dune, and progressively covering the *bottomset bed* and in turn being covered or truncated by the *topset bed.* Foreset beds represent the greater part of the deposits of a delta. Also spelled: fore-set bed. Syn: *foreset.*

foreset bedding A syn. of *cross-bedding;* an internal bedding inclined to the principal surface of accumulation. See also: *compound foreset bedding.*

foreshock (fore′-shock) A small tremor that commonly precedes a larger earthquake or *main shock* by an interval ranging from seconds to weeks and that originates at or near the focus of the larger earthquake. Cf: *aftershock.*

foreshore (fore′-shore) (a) The lower or outer, gradually seaward-sloping, zone of the shore or beach, lying between the crest of the most seaward berm on the *backshore* (or the upper limit of wave wash at high tide) and the ordinary low-water mark; the zone regularly covered and uncovered by the rise and fall of the tide, or the zone lying between the ordinary tide levels. Sometimes referred to as the shore. Syn: *beach face.* (b) A term loosely applied to a strip of ground lying between a body of water, as a lake or stream, and land that is cultivated or otherwise occupied.

foreshortening In radar images, the geometric displacement of the top of objects toward the near range relative to their base. Also called: *layover.*

foresight (fore′-sight) (a) A sight on a new survey point, taken in a forward direction and made in order to determine its bearing and elevation. Also, a sight on a previously established survey point, taken to close a circuit. (b) A reading taken on a level rod to determine the elevation of the point on which the rod rests. Syn: *minus sight.* Abbrev: FS. Ant: *backsight.*

foreslope (fore′-slope) Steeply sloping *fore reef* from outer reef margin to an arbitrary depth of 10 fathoms.

forest bed (for′-est) An interglacial deposit consisting of soil and woody remains of trees and other vegetation. Syn: *black drift; chip yard; woodyard.*

forest floor All organic matter generated by forest vegetation, including litter and unincorporated humus, on the mineral soil surface.

forest-moss peat Peat formed in forested swamps.

forest peat A highmoor peat formed of the remains of trees.

fork (a) A place where two or more streams join to form a larger waterway; a confluence. (b) The smaller of two streams that unite at a fork; any uniting stream at a fork where the resulting stream is much larger; a branch. (c) The land bounded by, adjoining, or lying in the angle made by, a fork. The term is often used in the plural in place names.

form [geomorph] *landform.*

form [mineral] *crystal form.*

form [sed struc] Those aspects of a particle's shape that are not expressed by sphericity or roundness (Sneed and Folk, 1958, p.123). It can be described by the use of ratios of the long, intermediate, and short axes, which can be combined into various "form indices", and by such terms as platy, bladed, elongate, and compact.

forma (for′-ma) The smallest category used in ordinary taxonomic work. It is generally applied to trivial variations occurring among individuals of any population (Lawrence, 1951, p.56).

formal unit (for′-mal) A stratigraphic unit that is "defined and named in accordance with the rules of an established or conventional system of classification and nomenclature" (ISST, 1961, p.18); e.g. a unit that is established in conformance with Article 13 of the Code of Stratigraphic Nomenclature prepared by the ACSN (1961). The initial letter of each word used in forming the name of a formal unit should be capitalized (although a name is not necessarily formal because it is capitalized). Cf: *informal lithostratigraphic unit.*

formanite (for′-man-ite) A brown tetragonal mineral of the *scheelite* group: $Y(Ta,Nb)O_4$. It may contain uranium, thorium, calcium, titanium, and zirconium.

format (for′-mat) An informal lithostratigraphic unit composed of all the rocks between two selected markers; e.g. electric log markers, which can be extended laterally through lithologic changes affecting the enclosed strata (Forgotson, 1957, p.2110).

formation [cart] (for-ma′-tion) A persistent body of igneous, sedimentary, or metamorphic rock, having easily recognizable boundaries that can be traced in the field without recourse to detailed paleontologic or petrologic analysis, and large enough to be represented on a geologic map as a practical or convenient unit for mapping and description; the basic *cartographic unit* in geologic mapping.

formation [drill] A general term applied by drillers without stratigraphic connotation to a sedimentary rock that can be described by certain drilling or reservoir characteristics; e.g. "hard formation", "cherty formation", "porous formation".

formation [geomorph] A naturally formed topographic feature, commonly differing conspicuously from adjacent objects or material, or being noteworthy for some other reason; esp. a striking erosional form on the land surface.

formation [speleo] *speleothem.*

formation [stratig] (a) A body of rock identified by lithic characteristics and stratigraphic position; it is prevailingly but not necessarily tabular, and is mappable at the Earth's surface or traceable in the subsurface (NACSN, 1983, Art. 24). The formation is the fundamental unit in *lithostratigraphic classification.* It may represent a long or short time interval, may be composed of materials from several sources, and may include breaks in deposition. A formation should possess some degree of internal lithic homogeneity or distinctive lithic features such as chemical or mineralogic composition, texture, fossils (viewed as rock-forming particles), or other organic content such as coal or oil shale. A formation must be amenable to being mapped at the scale of geologic mapping practiced in the region when the formation is proposed. Thickness is not a determining factor. Formations may be combined into groups or subdivided into members. A formation name normally consists of a geographic name followed by a descriptive geologic term (usually the dominant rock type) or by the word "formation" if the lithology is so variable that no single term is appropriate, e.g. Dakota Sandstone, Morrison Formation. Abbrev: fm. (b) A lithologically distinctive, mappable body of igneous or metamorphic rock. Syn: *geologic formation.*

formation evaluation The process of evaluating gas- or oil-bearing formations penetrated by a well or wells, and of appraising their commercial significance.

formation factor The ratio of the resistivity of a rock to the resistivity of the electrolyte with which it is completely saturated. Symbol: F. Syn: *resistivity factor; formation resistivity factor.*

formation microscanner A logging tool bearing arrays of microelectrodes that measure conductivity. The data are computer analyzed and displayed as an oriented gray-scale- or color image that resembles a photograph of the walls of the borehole. The log generated by a formation microscanner resembles that of a borehole televiewer (Friedman et al., 1992, p.17).

formation pressure The pressure in a formation resulting only from the fluid it contains. Cf: *bottomhole pressure.*

formation resistivity factor *formation factor.*

formation-volume factor (for-ma′-tion-vol′-ume) The factor applied to convert a barrel of gas-free oil in a stock tank at the surface into an equivalent amount of oil in the reservoir. It generally ranges between 1.14 and 1.60. Cf: *shrinkage factor.*

formation water Water present in a water-bearing formation under natural conditions, as opposed to introduced fluids, such as *drilling mud.* Syn: *native water.* Cf: *connate water.*

Formative (Form'-a-tive) In Mesoamerican archaeology, the cultural stage that follows the *Archaic* and is characterized by the development of agriculture and a settled population. It is followed by the *Classic.* The North American equivalent is called the Woodland. Correlation of relative cultural levels with actual age (and, therefore, with the time-stratigraphic units of geology) varies from region to region. Syn: *Pre-Classic.*

form energy An obsolete term for the potentiality of a mineral to develop its own crystal form against the resistance of the surrounding solid medium (Eskola, 1939). Syn: *power of crystallization; crystalloblastic strength.*

form genus A taxonomic unit used before 2000 for fossil plants not referable to a family. Cf: *morphotaxon; organ genus.*

formicaite A white tetragonal acetate mineral found in a borate deposit: $Ca(CHOO)_2$.

formkohle (form'-kohle) *crumble coal.*

formkreis (form'-kreis) (a) One of a series of related landforms that owe their existence to the same natural agent (such as running water or moving ice). (b) *morphogenetic region.* Etymol: German "Formkreis", "form cycle". Pl: formkreise.

form line A line (usually broken) on a map, sketched or drawn by visual observation, depicting the general surface configuration or shape of the terrain without regard to a true vertical datum and regular spacing and usually without indicating elevations; an uncontrolled or interpolated *contour line,* or one that is not instrumentally or accurately surveyed. Also spelled: formline. Cf: *indefinite contour.*

form ratio The ratio of mean stream depth to width of stream measured from bank to bank (Gilbert, 1914, p.35-36); it expresses the deepness or shallowness of a stream channel. The *hydraulic radius* is sometimes substituted for the mean depth if the stream is broad and shallow. Symbol: R.

form set A set of cross-stratification with an upper surface that preserves the shape of the original bed form (McKee and Weir, 1953). See also: *fossil ripple.*

form symbol A set of *Miller indices* enclosed in braces (as {*100*}), signifying the set of all crystal faces that are symmetrically equivalent to the face with the indicated Miller indices.

fornacite (for'-na-cite) An olive-green monoclinic mineral: $Pb_2Cu(AsO_4)(CrO_4)(OH)$.

fors A Swedish term for a rapids or cataract, or a waterfall of low inclination. Cf: *force [streams].*

forsterite (for'-ster-ite) A whitish, yellowish, or green orthorhombic mineral of the olivine group: Mg_2SiO_4. It is isomorphous with fayalite, and occurs chiefly in metamorphosed dolomites and crystalline limestones. Symbol: Fo. Syn: *white olivine.*

fortification agate (for'-ti-fi-ca'-tion) An agate having angular markings or parallel zigzag lines resembling the plan of a fortification.

fortunite (for'-tu-nite) A *lamproite* that contains phenocrysts of olivine and phlogopite in a very fine-grained groundmass that is megascopically undeterminable but under the microscope can be identified as a combination of ortho- and clinopyroxene, mica, feldspar, and some glass. Its name, given by Osann in 1889, is derived from Fortuna, Spain. Cf: *verite.* Mitchell and Bergman (1991) suggest that this name should be replaced by glassy orthopyroxene-phlogopite *lamproite.*

forward lap (for'-ward) *overlap [photo].*

forward scatter The scattering of radiant energy into the hemisphere of space bounded by a plane normal to the direction of the incident radiation and lying on the side toward which the incident radiation was advancing; the opposite of *backscatter.*

foshagite (fo'-sha-gite) A silky white triclinic mineral: $Ca_4Si_3O_9(OH)_2$.

foshallassite (fos-hal'-las-site) A pearly snow-white mineral: $Ca_3Si_2O_7{\cdot}3H_2O$(?). Also spelled: foschallasite.

foso (fo'-so) A term used in SW U.S. for a stream channel without conspicuous banks or bluffs. Etymol: Spanish, "moat, ditch".

foss (a) *fosse [streams].* (b) *force [streams].*

fossa [paleont] (a) A depression on an articular face of a crinoid *ossicle* for the attachment of muscles or ligaments. (b) A depression in the surface of a bone that serves for the attachment of a muscle or houses a gland, cartilage, or other soft part. Pl: fossae.

fossa [planet] (fos'-sa) A term established by the International Astronomical Union for a linear topographic depression on planetary surfaces (USGS, 2003), similar to a terrestrial graben. Generally used as part of a formal name, such as Claritas Fossae on Mars (Mutch et al., 1976, p.57). Etymol: Latin "fossa", "ditch". Pl: fossae.

fosse [glac geol] A long, narrow depression or troughlike hollow between the edge of a retreating glacier and the wall of its valley, or between the front of a moraine and its outwash plain. It may result from local acceleration of melting due to absorbed or reflected heat from the valley sides. Etymol: French form of Latin "fossa", ditch.

fosse [streams] (a) A long, narrow waterway; a canal, ditch, or trench. Var: *foss.* (b) *force [streams].*

fosse lake A glacial lake occupying a *fosse.*

fossette (fos-sette') (a) One of the slitlike pits, depressions, or grooves paralleling the periphery on the surface of the tests of some foraminifers (such as *Elphidium*). (b) A depression for the *resilium* in the shell of a bivalve mollusk. (c) A complex elaboration of enamel within the molariform teeth of advanced horses or other herbivorous mammals.

fossil (fos'-sil) n. Any remains, trace, or imprint of a plant or animal that has been preserved in the Earth's crust since some past geologic or prehistoric time; loosely, any evidence of past life. adj. Said of any object that existed in the geologic past and of which there is still evidence. Cf: *subfossil.*

fossil assemblage *assemblage [ecol].*

fossil association *association [ecol].*

fossil cave A relict cave no longer linked to its hydrologic mode of origin.

fossil community An *assemblage [ecol]* in which the individuals lived in the same place where their fossils are found, are present in approximately the same numbers and sizes as when alive, and thus have experienced no post-mortem transport. Cf: *winnowed community; disturbed-neighborhood assemblage; transported assemblage; mixed assemblage; paleobiocoenosis.* Syn: *in-place assemblage.*

fossil concentration A densely packed accumulation of *bioclasts* ≥2 mm in size (Kidwell, 1991).

fossil copal *copalite.*

fossil diagenesis The branch of paleoecology concerned with the history of organic remains after burial. Cf: *biostratonomy; taphonomy.*

fossil erosion surface An *erosion surface* that was buried by younger sediments and was later exposed by their removal. Sometimes used as a syn. of *buried erosion surface.*

fossil flood plain A flood plain that is beyond the reach of the highest flood (Bryan, 1923a, p.88). Cf: *living flood plain.*

fossil flour *diatomaceous earth.*

fossil fuel A general term for any hydrocarbon or carbonaceous rock that may be used for fuel: chiefly petroleum, natural gas, and coal.

fossil geochronometry The measurement of growth lines on fossil shells as a means of estimating the length of days and lunar months in *geologic time.* Cf: *geochronometry; lichenometry.*

fossil ice (a) Ice formed in, and remaining from, the geologically recent past. It is preserved in cold regions, such as the coastal plains of northern Siberia, where remains of Pleistocene ice have been found. Cf: *dead ice.* (b) Relatively old *ground ice* in a permafrost region. Also, underground ice in a region where present-day temperatures are not low enough to create it (Huschke, 1959, p.230).

fossil ice wedge *ice-wedge cast.*

fossiliferous (fos-sil-if'-er-ous) Containing fossils.

fossilization (fos'-sil-i-za'-tion) All processes involving the burial of a plant or animal in sediment and the eventual preservation of all, part, or a trace of it.

fossilized brine (fos'-sil-ized) *connate water.*

fossil karst (a) *relict karst.* (b) *paleokarst.*

fossil meteorite crater *astrobleme.*

fossil ore An iron-bearing sedimentary deposit, e.g. *Clinton ore,* in which shell fragments have been replaced and cemented together by hematite and carbonate. Cf: *flaxseed ore.*

fossil patterned ground Patterned ground that is inactive or no longer developing; e.g. ice-wedge casts and involution structures in a region of mild climate that formed during colder periods of the Pleistocene epoch when the region was under periglacial conditions. Ant: *active patterned ground.*

fossil peneplain A peneplain that was buried by younger sediments and was later exposed by their removal. Syn: *stripped peneplain.*

fossil permafrost *relict permafrost.*

fossil pingo The remains of a pingo. See also: *pingo remnant.*

fossil plain A plain that was buried by younger sediments and was later exposed by their removal.

fossil record The record of life in the geologic past as indicated by fossils. Cf: *stratigraphic record.* Syn: *paleontologic record.*

fossil resin Any of various natural resins found in geologic deposits as exudates of long-buried plant life; e.g. amber, retinite, and copal.

fossil ripple A *ripple* preserved as topographic forms on a sedimentary rock surface. See also: *form set.*

fossil time *Geologic time* estimated on the basis of organic evolution (Kobayashi, 1944a, p.476). See also: *marine time; continental time.* Cf: *relative time.*

fossil turquoise *odontolite.*

fossil water *connate water.*

fossil wax *ozocerite.*

fossula (fos'-su-la) (a) An unusually wide or relatively prominent space between septa of a rugose coral, distinguished by its shape and size, and caused by failure of one or more septa to develop as rapidly as others. See also: *cardinal fossula; counter fossula; alar fossula.* (b) A small pitlike depression that may occur in the cephalic axial or posterior border furrow in trilobites.

fossulate (fos'-su-late) Said of sculpture of palynomorphs consisting of grooves that anastomose.

fougerite A green rhombohedral mineral: $Fe^{2+}Fe^{3+}(OH)_5$. It is not approved by the International Mineralogical Association.

fouling (foul'-ing) The attachment and growth of aquatic plants and/or animals on submerged surfaces, usually of man-made structures (ship hulls, submerged pillars).

foundation (foun-da'-tion) (a) The lower, supporting part of an engineering structure, in contact with the underlying soil or rock and transmitting the weight of the structure and its included loads to the underlying earth material (Ireland, 1969). It is usually below ground level. (b) A term that is sometimes applied to the upper part of the soil or rock mass in contact with, and supporting the loads of, an engineering structure; the subsoil.

foundation bed The rock or soil layer immediately beneath the foundation, which receives the load of an engineering structure.

foundation coefficient A coefficient expressing how many times stronger the effect of an earthquake is in a given rock than in an undisturbed crystalline rock under the same conditions (Schieferdecker, 1959, p.197).

founder breccia *collapse breccia.*

fount A *fountain* or spring of water.

fountain (foun'-tain) (a) A spring of water issuing from the Earth. Syn: *fount.* (b) The source or head of a stream.

fountainhead The fountain or spring that is the source of a stream. Syn: *springhead; wellspring.*

fountain head (foun'-tain) According to Tolman (1937, p.559), "The elevation of water surface in a conduit if the overlying confining stratum extends above the water table, or elevation of water table above the upper termination of the confining stratum where the latter is below the water table".

fourchite (fourch'-ite) An olivine-free *monchiquite.* Its name, given by Williams in 1890, is derived from the Fourche Mountains, Arkansas. Obsolete.

four-component recording Sensors that record the three mutually orthogonal components of *particle velocity* plus changes in pressure.

four-dimensional survey A time-lapse survey of three-dimensional data, the fourth dimension being lapsed time. Abbrev: 4-D.

fourfold axis An axis of rotational symmetry by which an object (crystal shape, *crystal structure,* etc.) is invariant to a rotation of 90°. Syn: *tetrad.*

Fourier analysis (Fou'-ri-er) A method for expressing an ordered series or array of measurements as the sum of a series of simple sinusoidal waveforms, each characterized by an amplitude, frequency, and phase angle. The amplitudes and phase angles are shown as spectra. Named after Jean Baptiste Joseph Fourier (1768-1830), French physicist. Syn: *harmonic analysis [math]; spectral analysis.*

Fourier number Ratio of transported heat to stored thermal energy.

fourling (four'-ling) A crystal twin consisting of four individuals. Cf: *twoling; trilling; fiveling; eightling.*

fourmarierite (four-mar'-i-er-ite') An orange-red to brown orthorhombic secondary mineral: $Pb(UO_2)_4O_3(OH)_4{\cdot}4H_2O$.

four-shot method A reversed-refraction method used in engineering studies. Sources at each end of a geophone spread (short shots) are supplemented by additional in-line offset shots (long shots).

fourth-order pinacoid (fourth'-or-der) In a triclinic crystal, the $\{\bar{h}kl\}$, $\{h\bar{k}l\}$, or $\{hk\bar{l}\}$ pinacoid. Cf: *first-order pinacoid; second-order pinacoid; third-order pinacoid.*

fourth-order prism (fourth'-or-der) A crystal form in monoclinic crystals with two pairs of parallel faces, rhombic in cross section and parallel to an inclined direction. Its indices are $\{hkl\}$ or $\{\bar{h}kl\}$. Cf: *first-order prism; second-order prism; fourth-order prism.* Obsolescent syn: *hemipyramid.*

fourth-order sequence In *sequence stratigraphy,* a high-frequency sequence formed during 100,000 to 150,000 years (Mitchum and Van Wagoner, 1991).

four-way dip In seismic prospecting, dip determined by in-line and cross spreads placed at approximately right angles to each other.

foveolate (fo'-ve-o-late') Pitted; e.g. said of sculpture of palynomorphs consisting of pits in the ektexine.

fowlerite (fow'-ler-ite) A variety of rhodonite containing zinc.

foyaite (foy'-a-ite) A *nepheline syenite* containing a predominance of potassium feldspar. Originally described by Blum as synonymous with nepheline syenite, and later applied by Brögger to nepheline syenite with trachytic texture (Johannsen, 1939, p.252). Its name is derived from Foya, Portugal. Cf: *ditroite.* Obsolete.

frac job *hydraulic fracturing.*

fractal A geometric shape that is self-similar at all scales. Fractals are used in the description of complex geometric objects in digital cartography; other geological examples include the self-similar scaling of Pumpelly's Law, rock textures, sedimentary porosity, boudinage structures, drainage systems, topography, mineral concentrations in ore deposits, and the morphology of snow crystals. The term fractal was coined by Benoit Mandelbrot (b. 1924) and is of general applicability in natural science. Cf: *chaos theory.*

fractile In a cumulative distribution, a value at which a specified proportion of the observations lie at or below the value. Cf: *percentile; quartile [math].* Syn: *quantile.*

fractional coordinates The position of an atom in the unit cell, expressed as fractions of the unit-cell axial lengths. Syn: *positional parameters; atomic coordinates; parameter [cryst].*

fractional crystallization [petrology] (frac'-tion-al) A differentiation process whereby previously formed crystals are physically separated from the magma and thus prevented from equilibrating with the liquid from which they grew, resulting in a series of residual liquids of more extreme compositions than would have resulted from *equilibrium crystallization.* Cf: *crystallization differentiation.* Syn: *fractionation [petrology].*

fractional crystallization [salt] Controlled precipitation from a saline solution of salts of different solubilities, as affected by strength ratios, ionic strengths of ions in solution, and temperature.

fractional fusion Fusion in which the liquid produced on heating is isolated from the system as soon as it is formed and is thereby prevented from further reaction with the crystalline residue (Presnall, 1969). Cf: *equilibrium fusion.* Syn: *fractional melting.*

fractional melting *fractional fusion.*

fractional section A *section [surv]* containing appreciably less than 640 acres, usually due to invasion by a body of water or by land which cannot properly be surveyed as part of that section, or due to closing of the public-land survey on the north and west boundaries of the township.

fractional township A *township* containing appreciably less than 36 normal sections, usually due to invasion by a body of water or by land which cannot properly be surveyed as part of that township, or due to closing of the public-land survey on a State boundary or other limiting line.

fractionation [geochem] (frac-tion-a'-tion) Separation of chemical elements in nature, by processes such as preferential concentration of an element in a mineral during magmatic crystallization, or differential solubility during rock weathering.

fractionation [petrology] Change in bulk chemical composition due to physical separation of crystals from magma or physical separation of liquid from a rock undergoing partial melting.

fractionation density The ratio of the gram formula weight to molar volume of the chemical components in the liquid phase that are being removed by fractional crystallization (Sparks and Huppert, 1984). Removal of a crystalline assemblage with a bulk fractionation density greater than or less than one leads to a decrease or increase in the density of the remaining liquid respectively.

fractoconformity (frac'-to-con-form'-i-ty) The relation between conformable strata where faulting of the older beds proceeds contemporaneously with deposition of the newer.

fractography (frac-tog'-ra-phy) The study of the surfaces of fractures and joints, esp. microscopic study.

fracture [exp struc geol] (frac'-ture) Brittle deformation due to a momentary loss of cohesion or loss of resistance to differential stress and a release of stored elastic energy. (Both joints and faults are fractures.) Cf: *flow [exp struc geol].* Syn: *rupture.*

fracture [ice] Any break or rupture through fast ice, a single floe, or highly concentrated pack ice, resulting from deformation caused by tides, temperature changes, currents, or wind. Its length may vary from a few meters to many kilometers, and its width from zero to more than 500 m. Includes: *crack [ice]; flaw [ice]; lead [ice].*

fracture [mineral] The breaking of a mineral other than along planes of cleavage. A mineral can be described in part by its characteristic fracture, e.g. uneven, fibrous, conchoidal, or hackly.

fracture [struc geol] (a) A general term for any surface within a material across which there is no cohesion, e.g. a crack. Fracture includes cracks, joints, and faults. (b) A crack in a rock where the movement of rock separated by the crack is normal to the surface. See also: *extension fracture; extension vein; stylolitic fracture.*

fracture belt Linear to sinuous group of closely spaced, aligned fractures on Venus typically extending for hundreds of kilometers (Bougher et al., 1997).

fracture cleavage An obsolete term for *spaced cleavage* or *disjunctive cleavage,* where a rock tends to break preferentially along the cleavage surface. The nongenetic terms "spaced" or "disjunctive" cleavage are preferred.

fractured deflection (frac'-tured) A marked change in the trend of a mountain range, where arcs meet at large obtuse angles, and from the junction two lineaments appear to cross the ranges and mark major structural changes (Wilson, 1950, p.151).

fracture partitioning The distribution of fractures among different lithologies or beds of different thicknesses (Gross, 1993).

fracture porosity (a) Porosity resulting from the presence of openings produced by the breaking or shattering of an otherwise less pervious rock. (b) The volume of voids formed by fractures divided by the total volume of the porous medium.

fracture skin A coating of a fracture surface and/or the altered zone beneath a fracture surface; fracture skin has different hydrogeological properties from those of the unaltered rock.

fracture spacing index An index used to quantify the relationship between *mechanical layer thickness* and fracture spacing, defined as the slope of the best-fit line on plots of mechanical layer thickness (y-axis) versus median spacing (x-axis), with median spacing as the dependent variable (Narr and Suppe, 1991). A high fracture spacing index indicates closely-spaced fractures. Abbrev: FSI.

fracture spacing ratio The ratio of *mechanical layer thickness* to median fracture spacing (i.e., MLT/spacing) for a given mechanical layer. Abbrev: FSR.

fracture spring A spring whose water flows from joints or other fractures, in contrast to the numerous small openings from which a *filtration spring* flows (Meinzer, 1923, p.50). Cf: *fissure spring; fault spring.*

fracture strength In experimental structural geology, the differential stress at the moment of fracture. Syn: *fracture stress; breaking strength; failure stress.*

fracture stress *fracture strength.*

fracture system A set or group of contemporaneous fractures that formed in the same stress field.

fracture toughness A measure of the tensile strength of rock. The fracture toughness of a rock is equal to the critical stress intensity factor, K_{1c}, at which joint propagation initiates.

fracture zone On the deep-sea floor, an elongate zone of unusually irregular topography that often separates regions of different depths. Such zones commonly follow and extend beyond offsets of the mid-ocean ridge. Fracture zones are scars of *transform faults.*

fracturing (frac'-tur-ing) *hydraulic fracturing.*

fragipan (frag'-i-pan) A natural subsurface horizon with very low organic matter, high bulk density and/or high mechanical strength relative to overlying and underlying horizons; has hard or very hard consistence (seemingly cemented) when dry, but showing a moderate to weak brittleness when moist. The layer typically has redoximorphic features, is slowly or very slowly permeable to water, is considered to be root restricting, and usually has few too many bleached, roughly vertical planes which are faces of coarse or very coarse polyhedrons or prisms.

fragment (frag'-ment) (a) A rock or mineral *particle* larger than a grain. (b) A piece of rock that has been detached or broken from a preexisting mass; e.g., a clast produced by volcanic, dynamic, or weathering processes.

fragmental rock (frag-men'-tal) (a) *clastic rock.* (b) pyroclastic rock. (c) *bioclastic rock.*

fragmental texture (a) A texture of sedimentary rocks, characterized by broken, abraded, or irregular particles in surface contact, and resulting from the physical transport and deposition of such particles; the texture of a clastic rock. The term is used in distinction to a "crystalline" texture. (b) The texture of a pyroclastic rock, such as that of a tuff or a volcanic breccia.

fragmentary Consisting of or characterized by clastic or detrital material; fragmental. The term was formerly applied to rocks composed of fragments of older rocks, or to rocks having an inhomogeneous texture; as used in geology, the term is obsolete.

fragmentation (frag-men-ta'-tion) The act or process of breaking into pieces or fractionating, or the state of being fragmentated or fractionated; esp. the breaking-up of a sponge into several others without concomitant formation of specialized reproductive bodies.

fragmentation index A parameter measuring the grain size of a pyroclastic fall deposit, specifically the percentage of ash finer than 1 mm at the point on the dispersal axis corresponding to $^1/_{10}$ of the maximum thickness of the deposit. Abbrev: F.

fraipontite (frai'-pon-tite) A platy, colorless, yellowish white to bluish monoclinic mineral of the *serpentine* group: $(Zn,Al)_3(Si,Al)_2O_5(OH)_4$.

Fra Mauro basalt (Fra Mau'-ro) A class of basaltic rocks found in the lunar highlands, differing from mare basalts primarily by their higher plagioclase content. A number of varieties, e.g. *KREEP,* are distinguished on the basis of chemical composition (Taylor, 1975, p.234). Named from the Apollo 14 landing site near the crater Fra Mauro, where this basalt was first collected.

framboid (fram'-boid) A microscopic aggregate of pyrite grains in shale, often in spheroidal clusters resembling a raspberry. It was considered to be the result of colloidal processes but is now linked with the presence of organic materials; sulfide crystals fill chambers or cells in bacteria (Park and MacDiarmid, 1970, p.133). Adj: framboidal. Etymol: French "framboise", "raspberry".

frame-builders (frame'-build-ers) Organisms, generally calcareous and sessile-benthonic, that collectively construct a *reef frame* or growth lattice.

frame reef Reef in which essentially in-place skeletons (including calcified microbes) are in contact (Riding, 2002).

framesite (frames'-ite) A South African variety of *bort* showing minute brilliant points.

framestone (frame'-stone) A type of *reef rock* or *boundstone* consisting of colonies, shells, or skeletons attached to each other to form a rigid framework or lattice. Internal cavities are filled with fine sediment, crystalline cement, or coarse skeletal debris (Embry and Klovan, 1971; Cuffey, 1985). Cf: *branchstone; cruststone.*

frame-thicket (frame-thick'-et) A small but proportionately high-standing, open-framework type of *bioherm* or *patch reef,* composed of a branchstone or bafflestone core, in some places flanked by aprons of coarse rudstone and floatstone (Cuffey, 1985). Cf: *crust-mound; mud-mound.*

framework [paleont] (frame'-work) The coarsely porous underlying wall of a double wall in archaeocyathids (TIP, 1972, pt. E, p.40).

framework [sed] (a) The rigid arrangement created in a sediment or sedimentary rock by particles that support one another at their points of contact; e.g. the clasts of a fragmental rock (esp. a sandstone), constituting a mechanically firm structure capable of supporting open pore spaces, although interstices may be occupied by cement or matrix. (b) The rigid, wave-resistant, calcareous structure built by sedentary organisms (such as sponges, corals, and bryozoans) in a high-energy environment.

framework [tect] *tectonic framework.*

framework silicate *tectosilicate.*

francevillite (france'-vil-lite) A yellow, green, or orange orthorhombic secondary mineral: $(Ba,Pb)(UO_2)_2(V_2O_8){\cdot}5H_2O$. It is the barium analogue of curienite.

franciscanite (fran-cis'-can-ite) (a) A red or brownish red trigonal mineral: $Mn^{2+}{}_6V^{5+}(SiO_4)_2(O,OH)_6$. (b) A group name for minerals with the analogous formula, but with V replaced by W or Sb^5.

francisite A bright green orthorhombic mineral: $Cu_3Bi(Se^{4+}O_3)_2O_2Cl$.

franckeite (franck'-e-ite) A dark-gray or black platy triclinic mineral: $Pb_{,5}FeSn_3Sb_2S_{14}$.

francoanellite (fran'-co-a-nel'-lite) A yellow-white rhombohedral mineral: $(K,Na)_3(Al,Fe^{3+})_5(PO_4)_2(PO_3OH)_6{\cdot}12H_2O$.

françoisite-(Nd) (fran-çois'-ite) A yellow monoclinic mineral: $(Nd,Y,Sm,Ce)(UO_2)_3O(PO_4)_2(OH){\cdot}6H_2O$.

francolite (fran'-co-lite) *carbonate-fluorapatite.*

Franconian (Fran-co'-ni-an) North American stage: (Upper Cambrian, above Dresbachian, below Trempealeauan).

franconite (fran'-co-nite) A silky white monoclinic mineral: $Na_2Nb_4O_{11}{\cdot}9H_2O$.

frankamenite A grayish lilac, blue-gray, or light green mineral: $K_3Na_3Ca_5Si_{12}O_{30}(F,OH)_4{\cdot}H_2O$. A dimorph of canasite.

frankdicksonite (frank-dick'-son-ite) A colorless cubic mineral of the *fluorite* group: BaF_2.

frankhawthorneite A leaf-green monoclinic mineral: $Cu_2Te^{6+}O_4(OH)_2$.

franklinfurnaceite (frank'-lin-fur'-nace-ite) A dark brown platy monoclinic mineral of the *chlorite* group: $Ca_2(Fe^{3+},Al)Mn^{3+}Mn^{2+}{}_3Zn_2Si_2O_{10}(OH)_8$.

franklinite (frank'-lin-ite) An iron-black, brown, or red cubic mineral of the *spinel* group: $(Zn,Mn^{2+},Fe^{2+})(Fe^{3+},Mn^{3+})_2O_4$. It resembles magnetite but is less strongly magnetic. It was formerly a major ore of zinc at Franklin, New Jersey.

franklinphilite A dark brown triclinic mineral of the *stilpnomelane* group: $(K,Na)_{1-x}(Mn^{2+},Mg,Zn,Fe^{3+})_8(Si,Al)_{12}(O,OH)_{36}{\cdot}nH_2O$.

fransoletite (fran-so-let'-ite) A colorless monoclinic secondary mineral: $Ca_3Be_2(PO_4)_2(PO_3OH)_2{\cdot}4H_2O$. A dimorph of parafransoletite.

franzinite (fran-zin'-ite) A pearly white trigonal mineral of the *cancrinite* group: $(Na,Ca)_7(Si,Al)_{12}O_{24}(SO_4,CO_3,OH)_3{\cdot}H_2O$. It is the sulfate-analogue of davyne.

Frasch process A process for mining native sulfur, in which superheated water is forced into the deposits for the purpose of melting the sulfur. The molten sulfur is then pumped to the surface.

Frasnian (Fras'-ni-an) A stage of the Standard Global Chronostratigraphic Scale: Upper Devonian (above Givetian, below Famennian). The basal GSSP is at Montagne Noir, southern France (Klapper et al., 1987).

Fraunhofer line (Fraun'-ho-fer) Any of the absorption lines in the spectrum of the sun corresponding to the absorption spectra of the gases around it.

frazil (fra'-zil) (a) A group of individual ice crystals, having the form of small discoids or spicules, which are formed in supercooled turbulent water. Syn: *frazil ice.* (b) *frazil crystal.* Etymol: Canadian French "frasil", from French "fraisil", "forge cinders".

frazil crystal A small discoid or needlelike spicule of ice formed by freezing of supercooled turbulent water. Syn: *frazil; ice crystal.*

frazil ice A spongy, slushy, cinderlike mass or aggregate of frazil crystals collected by adhesion or regelation and suspended in supercooled turbulent water, esp. common in a rapidly flowing stream, but also found in turbulent seawater, where it is called *lolly ice.* Syn: *frazil; needle ice.*

frazil slush An agglomerate of loosely packed *frazil* floating on the water surface which can accumulate under the ice cover. Cf: *slush [snow].*

freboldite (fre'-bold-ite) A metallic copper-red hexagonal mineral of the *nickeline* group: CoSe.

Fredericksburgian (Fred'-er-icks-burg'-i-an) North American (Gulf Coast) stage: Lower Cretaceous (above Trinitian, below Washitan).

fredrikssonite (fred'-riks-son-ite') A reddish-brown orthorhombic mineral of the *ludwigite* group: $Mg_2(Mn^{3+},Fe^{3+})BO_5$.

free Said of a *native element,* e.g. free gold.

free-air anomaly The gravity field values at sea level obtained after applying the *latitude correction* and *free-air correction* to the observed gravity data. Also called: free-air gravity. Shipborne and *satellite gravity* measurements are presented and interpreted as free-air gravity.

free-air correction A correction applied to the observed gravity data to compensate for the elevation of a point of measurement above the survey datum, usually, sea level. It is a correction for the additional distance ("free-air" without rock mass) between the gravimeter and Earth's center of mass. The free-air correction is 0.3086 h (mGal), where h is the elevation above datum in meters and 0.3086 mGal/m is the normal free-air gravity gradient.

free arm The freely mobile part of a crinoid ray not incorporated in the *calyx.*

free blade The portion of a *blade* of a platelike conodont element not flanked by platforms.

freeboard (free'-board) (a) The additional height above the recorded or design high-water mark of an engineering structure, such as a dam, seawall, flume, or culvert, that represents an allowance against overtopping by transient disturbances, including waves induced by winds or landslides. (b) The vertical distance between the water level at a given time and the top of an engineering structure, such as the vertical distance between the normal operating level of a reservoir and the crest of the associated dam.

free-burning coal (free'-burn-ing) *noncaking coal.*

free cheek A lateral part of the *cephalon* of a trilobite outside the facial suture, separated from the *cranidium* at the time of molting and including the visual surface of the eye. See also: *cheek.* Cf: *fixed cheek.* Syn: *librigena.*

free convection Convection created by density (buoyancy) gradients, which can be caused by variations in water temperature, salinity, or suspended matter. Cf: *convection [grd wat]; forced convection.* Syn: *natural convection.*

free corer A type of *gravity corer* that is not attached to a wire but returns tethered to a float.

free degradation Degradation of a slope from the foot of which no debris is removed, e.g. an abandoned cliff (Hutchinson, 1967).

freedite (freed'-ite) A greenish-yellow monoclinic mineral: $Pb_8Cu^{1+}(As^{3+}O_3)_2O_3Cl_5$.

free energy A thermodynamic function of the state of a system, providing a measure of the maximum work obtainable from the system under specified conditions. The functions most commonly used are the *Helmholtz free energy* and the *Gibbs free energy.* Because these functions (among others) also measure the driving forces for processes occurring under specified conditions, they are often referred to as thermodynamic potentials.

free face The part of a hillside surface consisting of an outcrop of bare rock (such as a scarp or cliff) that stands more steeply than the angle of repose of the *constant slope* immediately below (Wood, 1942); a rock wall from which weathered material falls to the slope below.

free flow In hydraulics, flow that is not disturbed by submergence or backwater (Brown and Runner, 1939, p.155).

free-fluid index Proportion of the bulk rock volume that is fluid that is free to flow as measured by the *nuclear-magnetism log,* as distinct from immobile, bound water at solid surfaces.

free groundwater *unconfined groundwater.*

free iron oxides A general term for those iron oxides that can be reduced and dissolved by a dithionite treatment. Generally includes goethite, hematite, ferrihydrite, lepidocrocite, and maghemite, but not magnetite. See also: *iron oxides.*

free margin The peripheral distal border of an ostracode valve, exclusive of the hinge.

free meander A stream meander that displaces itself very easily by lateral *corrasion,* esp. where vertical corrasion is of no importance.

free moisture *free water.*

free operculum The part of a dinoflagellate cyst that is completely surrounded by archeopyle sutures, with no unsutured connection to the rest of the cyst. Cf: *attached operculum.*

free oscillation An oscillation of a body, e.g. the Earth, that occurs without external influence other than the initiating force, and that has its own natural frequency. Such oscillations follow major earthquakes. Cf: *forced oscillation.*

free period The time for one complete swing of the seismograph mass when all damping is removed and no driving force is present.

freestone [rock] Any stone (esp. a thick-bedded, even-textured, fine-grained sandstone) that breaks freely and can be cut and dressed with equal ease in any direction without splitting or tending to split. The ease with which it can be shaped into blocks makes it a good building stone. The term was originally applied to limestone, and is still used for such rock. Cf: *flagstone.*

freestone [water] Water containing little or no dissolved material. Syn: freestone water.

free surface The upper surface of a layer of fluid where the pressure on it is equal to the external atmospheric pressure. Cf: *water table.*

free surface stream In a cave, a stream that does not completely fill its passage.

free water (a) Water in the soil in excess of field capacity that is free to move in response to the pull of gravity. Syn: *gravity water; gravitational water; infiltration water.* Cf: *gravity groundwater.* (b) Water that can be removed from another substance, as in ore analysis, without changing the structure or composition of the substance. Ant: *bound water.* Syn: *free moisture.*

free-water content (free-wa'-ter) *liquid-water content.*

free-water elevation *water table.*

free-water level *free-water surface.*

free-water surface The surface of a body of water at which the pressure is atmospheric and below which the pressure is greater than atmospheric; the surface of any pond, reservoir, etc., that is open to the atmosphere, or a *water table.* Syn: *free-water level.*

freeze-thaw action *frost action.*

freezeup (freeze'-up) (a) The formation of a continuous ice cover, generally restricted to the hardening of locally formed young ice, but sometimes including the freezing together of pieces of drift ice. (b) The period during which a body of water in an area is frozen over, esp. when marking the beginning of winter.

freezing (freez'-ing) The process by which a liquid becomes a solid, involving the removal of heat. Commonly applied to water but also used for solidification on cooling of molten metals and magma.

freezing interval *crystallization interval.*

freezing rain Supercooled raindrops that at least partially freeze on contact with cold surfaces.

freibergite (frei'-berg-ite) A metallic gray cubic mineral of the *tennantite* group: $(Ag,Cu,Fe)_{12}(Sb,As)_4S_{13}$.

freieslebenite (frei-es-le'-ben-ite) A metallic steel-gray to dark-gray monoclinic mineral: $AgPbSbS_3$.

freirinite (frei'-ri-nite) *lavendulan.*

fremontite (fre'-mon-tite) *natromontebrasite.*

French chalk A soft, white variety of talc, steatite, or soapstone, finely ground into powder and used for crayons, as a grease remover in dry cleaning, or for other special purposes.

Frenkel defect (Fren'-kel) In a crystal lattice, the displacement of an atom from its original position to an interstice; it is a type of *point defect.* Cf: *interstitial defect; Schottky defect.*

frenuliniform (fren-u-lin'-i-form) Said of the *loop,* or of the growth stage in the development of the loop, of a dallinid brachiopod (as in the subfamily Frenulininae), marked by lateral resorption gaps (lacunae) occurring in the *hood* but before resorption of the shell occurs posterior to the gaps (TIP, 1965, pt. H, p.145). It is subsequent to the *campagiform* stage.

frenulum (fren'-u-lum) A small cylinder connected to the internal part of the nasal tube of a style near the base of a galea in a phaeodarian radiolarian. Pl: frenula.

frequency A measure of rate of occurence of some phenomenon in time or space.

frequency curve (fre'-quen-cy) A curve that graphically represents a *frequency distribution;* e.g. a smooth line drawn on a histogram if the *class* interval is made smaller and the steps between several bars grow smaller.

frequency distribution A systematic arrangement of statistical data (such as a graphic or tabular display of the number of observations on a variable) in which the variable is divided into mutually exclusive and exhaustive discrete categories or classes and which indicates the frequencies or relative frequencies that correspond to each of the categories or classes. The graphic display of a frequency distribution is a *histogram.*

frequency domain Measurements as a function of frequency, or operations in which frequency is the variable, in contrast to the *time domain.*

frequency-domain method Transmission of a continuous wave (usually sinusoidal) and simultaneous reception of electromagnetic energy as a function of frequency. It is used with induced electrical polarization and electromagnetic methods.

frequency response Attenuation as a function of frequency produced by passage of information through an element such as a geophone or filter.

frequency spectrum The characteristics of a waveform described as a function of frequency. See: *Fourier analysis.*

fresh [water] adj. Said of water that does not contain or is not composed of salt water. n. (a) An increased, sudden flow or rush of water; a *freshet* or a *flush.* (b) A stream, spring, or pool of fresh water. (c) A freshwater stream flowing into the sea, or the part of a stream or its shores above the flow of tidal seawater. (d) The mingling of fresh water and salt water.

fresh [weath] Said of a rock or rock surface that has not been subjected to or altered by surface weathering, such as a rock newly exposed by fracturing. Syn: *unweathered.*

freshening (fresh'-en-ing) Making water less salty; separating water from saline constituents.

freshet (fresh'-et) (a) A great rise in, or a sudden overflowing of, a small stream, usually caused by heavy rains or rapidly melting snow in the highlands at the head of the stream; a rapidly rising flood, usually of minor severity and short duration. See also: *flash flood.* Syn: *fresh [water]; spate.* (b) A small clear freshwater stream or current flowing swiftly into the sea; an area of comparatively fresh water at or near the mouth of a stream flowing into the sea. (c) A small stream flowing swiftly into a lake (as in the spring) and often carrying a heavy silt load during its peak flow.

fresh ice (a) young ice. (b) *freshwater ice.* (c) Ice that was formed on salt water but is now salt-free.

fresh water (a) Water containing less than 1,000 milligrams per liter of dissolved solids; generally, water with more than 500 mg/L is undesirable for drinking and for many industrial uses (Solley et al., 1983). (b) In general usage, the water of streams and lakes unaffected by salt water or salt-bearing rocks. Cf: *potable water; saline water.* Syn: *sweet water.* Also spelled: freshwater; fresh-water.

freshwater estuary (fresh'-wa-ter) (a) An estuary into which river water pours with sufficient volume to exclude salt water. See also: *positive estuary.* (b) In the Great Lakes and other large lakes, the lower reach of a tributary to the lake that has a *drowned river mouth,* shows a zone of transition from stream water to lake water, and is influenced by changes in lake level as a result of seiches or wind tides. It is commonly separated from the adjacent main body of water by a barrier spit or baymouth bar.

freshwater ice Ice formed by the freezing of fresh water in lakes or streams, or in the ground. Syn: *fresh ice.*

fresh-water lens A lenticular fresh groundwater body that underlies an oceanic island. It is underlain by saline water.

freshwater sediment A sediment that accumulates, or has accumulated, in a freshwater environment; e.g. a sediment resulting from lacustrine, fluvial, or glaciofluvial activity.
freshwater swamp A swamp that depends on nontidal fresh water rather than a saltwater source.
Fresnel zone (a) The portion of a reflector from which reflected energy can reach a detector within one-half wavelength of the first reflected energy. The central first Fresnel zone is elliptical (circular for zero offset data) for a horizontal reflector; successive Fresnel zones are annular. Most of the energy reflected from the first Fresnel zone interferes constructively, but the outer portion does not contribute much, and the effective size of the Fresnel zone is smaller by about a factor of v2. Because the size of the Fresnel zone is frequency dependent, the dominant frequency is often used where a spectrum of frequencies is involved. (b) A bundle around a raypath where travel within the bundle differs from that along the raypath by less than a quarter wavelength.
Fresnian (Fres'-ni-an) North American stage: Upper Eocene (above Narizian, below Refugian).
fresnoite (fres'-no-ite) A yellow tetragonal mineral: $Ba_2TiOSi_2O_7$.
fret A spot worn or eroded by *fretting*, as on a limestone surface.
fretted terrain (fret'-ted) A type of erosional landscape on Mars, consisting of high-standing remnants of cratered terrain and lower sparsely cratered plains (R.P. Sharp, 1973).
fretted upland A preglacial upland surface completely consumed by the intersection of cirques from opposite sides; the "ultimate product of cirque sculpture by glaciers" (Hobbs, 1912, p.373). Cf: *grooved upland; scalloped upland.*
fretting (fret'-ting) (a) *honeycomb weathering.* (b) The wearing-away of a rock surface, as by a stream cutting its channel, or by wind-driven (eolian) grains. (c) Agitation or disturbance of running water, such as the rippling of a brook over rocks.
fretwork weathering (fret'-work) *honeycomb weathering.*
freudenbergite (freud'-en-berg-ite') A black monoclinic mineral: $Na_2(Ti,Fe^{3+})_8O_{16}$.
freyalite (frey'-a-lite) A mixture of thorium-bearing melanocerite and its alteration products.
friability (fri-a-bil'-i-ty) The condition of being friable.
friable (fri'-a-ble) (a) Said of a rock or mineral that crumbles naturally or is easily broken, pulverized, or reduced to powder, such as a soft or poorly cemented sandstone. (b) Said of a soil consistency in which moist soil material crushes easily under gentle to moderate pressure (between thumb and forefinger) and coheres when pressed together.
friction (fric'-tion) Mechanical resistance to the relative motion of contiguous bodies one on the other or of a body and a medium; e.g. slip on a fault. Cf: *internal friction.*
frictional (fric'-tion-al) As applied to a soil, a syn. of *cohesionless.*
frictional sliding [exp struc geol] A *brittle-deformation mechanism* involving movement parallel to a preexisting fracture.
frictional sliding [struc geol] A type of brittle shear movement that involves plowing of asperities and/or brittle failure of asperities; it produces frictional wear material, known as *friction breccia* and *gouge.*
friction angle *soil friction angle.*
friction breccia A *breccia* composed of broken or crushed rock fragments resulting from frictional sliding.
friction crack A short crescentic mark that is transverse to the direction of ice movement and that includes a distinct fracture that dips forward into the bedrock, indicating the direction of ice movement (Harris, 1943). It presumably results from local increase in frictional pressure between ice and bedrock.
friction head That head of fluid which is lost because of friction. See also: *friction slope.* Syn: *head loss; friction loss.*
friction layer The top or bottom layer in the ocean where frictional forces due to wind or bottom drag are important. Cf: *Ekman layer.*
friction loss *friction head.*
friction slope The *friction head* or loss per unit length of conduit. For most conditions of flow the friction slope coincides with the energy gradient, but where a distinction is made between energy losses due to bends, expansions, impacts, etc., a distinction must also be made between the friction slope and the energy gradient. In uniform channels, the friction slope is equal to the bed or surface slope only for uniform flow (ASCE, 1962).
friedelite (frie'-del-ite) A pink, dark red, brown, or yellow monoclinic pseudotrigonal mineral of the *pyrosmalite* group: $Mn^{2+}_8Si_6O_{15}(OH,Cl)_{10}$.
Friedel's law The law requiring equivalence of the X-ray diffraction intensities from planes hkl and -h,-k,-l. It is violated only when the effects of *anomalous dispersion* are significant.
friedrichite (fried'-rich-ite) A metallic orthorhombic mineral, a member of the aikinite-bismuthinite series, with a unique superstructure: $Pb_5Cu_5Bi_7S_{18}$.
friendly ice (friend'-ly) A submariner's term for an *ice canopy* containing more than 10 large skylights (or other features that permit a submarine to surface) per 30 nautical miles (56 km) along the submarine's track (U.S. Naval Oceanographic Office, 1968, p. B33). Ant: *hostile ice.*
frigid climate (frig'-id) A type of climate characteristic of a region in which a more or less permanent cover of snow and ice overlies the permanently frozen surface. Cf: *polar climate.*
frigidite (frig'-id-ite) A mixture of tetrahedrite and nickel-bearing minerals.
frigid temperature regime A soil temperature regime in which the mean annual temperature (measured at 50 cm depth) is more than 0°C but less than 8°C, with a summer-winter variation of more than 5°C and with warm summer temperatures (USDA, 1975). Cf: *isofrigid temperature regime; cryic temperature regime.*
frill (a) A relatively large *lamella* projecting well beyond the general contour of a brachiopod valve, deposited by the margin of highly retractile mantle (TIP, 1965, pt. H, p.145). Cf: *growth lamella.* (b) The *velum* in an ostracode. (c) The large, usually fenestrated plate of bones that projects posteriorly from the posterodoral margin of a ceratopsian dinosaur skull.
fringe [glac geol] A thin sprinkling of erratics in front of the end moraine of a glacier.
fringe [paleont] The peripheral pitted bilaminar portion of the *cephalon* in trinucleid and harpetid trilobites.
fringe [seis] (a) The extra area around the area to be surveyed that must be covered because of the inward movement of dipping reflections and focusing of seismic amplitude dispersed over Fresnel zones during 3-D migration. Often referred to as migration fringe or migration halo. (b) The portion of imaged area where coverage is not uniform because of differences in multiplicity, offset or azimuth mixes, or other factors that lessen confidence in data comparisons.
fringe joint A small-scale joint peripheral to a master joint, usually at a 5°-25° angle from the face of the main joint. It is formed by tension. A set of fringe joints forms an en echelon pattern. See also: *cross fracture.*
fringe ore Ore at the outer limits of the mineralization pattern or halo. Syn: *halo ore.*
fringe water Water of the *capillary fringe.*
fringing reef (fring'-ing) An *organic reef* that is directly attached to or borders the shore of an island or landmass, having a rough, table-like surface that is commonly exposed at low tide; it may be more than 1 km wide, and its seaward edge slopes sharply down to the sea floor. There may be a shallow channel or lagoon between the reef and the adjacent mainland. Cf: *barrier reef.* Syn: *shore reef.*
frith *firth.*
fritting (frit'-ting) The partial melting of grains of quartz and other minerals, so that each grain becomes surrounded by a zone of glass. Fritting results from the contact action of basalt and related lavas on other rocks (Johannsen, 1931).
fritzscheite (fritzsch'-e-ite) A platy reddish-brown orthorhombic mineral of the *francevillite* group: $Mn^{2+}(UO_2)_2[(PO_4),(VO_4)]_2\cdot4H_2O$.
frohbergite (froh'-berg-ite) A metallic orthorhombic mineral of the *marcasite* group: $FeTe_2$.
frolovite (fro'-lo-vite) A grayish white triclinic mineral: $CaB_2(OH)_8$.
frond The expanded compound foliage of ferns or a similar leaflike structure.
frondelite (fron-del'-ite) A dark green, olive-green, or greenish black platy orthorhombic mineral: $(Mn^{2+},Fe^{2+})Fe^{3+}_4(PO_4)_3(OH)_5$. It is the Mn analogue of rockbridgeite.
frondescent cast (fron-des'-cent) A term used by Ten Haaf (1959, p.30) for a feather-like *sole mark* resembling certain shrubs or large cabbage leaves, the spreading "foliage" always directed downcurrent; it is usually several decimeters in length. Syn: *cabbage-leaf mark; deltoidal cast.*
frondose (fron'-dose) Said of the flattened, frond-shaped growth habit of erect stenolaemate bryozoan colonies that lack median laminae.
front [geomorph] (a) The more or less linear outer slope of a mountain range that rises above a plain or plateau. (b) Land that faces or abuts, esp. on a body of water.
front [glaciol] (a) *ice front.* (b) *snout [glaciol].*
front [meteorol] The contact at the Earth's surface between two different air masses, commonly cold and warm, that generally moves in an easterly direction. See also: *cold front; warm front; occluded front; stationary front.*
front [paleont] The part of the *carapace* of a brachyuran decapod crustacean (crab) between the orbits.
front [petrology] A metamorphic zone of changing mineralization developed outward from an igneous mass.
frontal (fron'-tal) adj. Pertaining or belonging to the front part; esp. pertaining to the orifice-bearing side of a bryozoan *zooid* or colony. In vertebrates, pertains to the interorbital portion of the skull, to a specific pair of dermal bones anterior to the parietals, or to a horizontal plane of section of the skull at right angles to a sagittal plane. Syn: *obverse.*
frontal apron *apron [geomorph].*
frontal arc *fore arc.*
frontal kame A *kame* that consists of a steep alluvial fan against the edge of an ice sheet.
frontal membrane Uncalcified part of the frontal wall in anascan cheilostome bryozoans, to which parietal muscles are attached. It may be exposed or overarched by a *frontal shield.*
frontal moraine (a) *end moraine.* (b) "A moraine rampart at the front of a former glacier" (Schieferdecker, 1959, term 0918).

frontal plain *outwash plain.*
frontal plate The modified *rostrum* of a brachyuran decapod crustacean, bearing a process that projects ventrally between antennules to unite with the *epistome.*
frontal pore A pore on the front of the lattice shell of a radiolarian and adjacent to the basal ring. It is similar in appearance to the sternal pore.
frontal ramp A thrust ramp that is oriented nearly perpendicular to the transport direction of the thrust sheet.
frontal region The anteromedian part of the *carapace* of some decapod crustaceans, including the *rostrum* and the area behind it (TIP, 1969, pt. R, p.92).
frontal scar The scar on the interior of the carapace of an ostracode, just anterior and dorsal to the adductor muscle scars.
frontal shield The calcareous part of the frontal surface of a cheilostome bryozoan. It is developed in different patterns with varying relationships to the frontal wall of the *zooid.* Cf: *cryptocyst; gymnocyst; pericyst; umbonuloid.*
frontal terrace *outwash terrace.*
frontal wall The exterior zooidal wall in cheilostome and some stenolaemate bryozoans, attached to the wall containing the *orifice* and providing a front side to the *zooid.*
front bay A large, irregular, shallow bay connected with the sea through a pass between barrier islands, as along the coast of Texas. Cf: *back bay.*
front-limb thrust fault A little-used term for a *forelimb thrust.*
front pinacoid In an orthorhombic, monoclinic, or triclinic crystal, the *{100}* pinacoid. Cf: *basal pinacoid; side pinacoid.* Syn: *macropinacoid; orthopinacoid.*
front range The outermost range of a mountain system; e.g. the Front Range of the Rocky Mountains, extending southward from Casper, Wyo., through Colorado and into New Mexico, including the Sangre de Cristo Mountains.
front slope The *scarp slope* of a cuesta.
froodite (frood'-ite) A metallic gray monoclinic mineral: $PdBi_2$. Cf: *michenerite.*
frost (a) A granular or flaky deposit of ice crystals caused by the deposition of water vapor on a surface whose temperature is below the freezing point. Cf: *dew.* Syn: *hoarfrost.* (b) *permafrost.*
frost action (a) The mechanical weathering process caused by alternate or repeated cycles of freezing and thawing of water in pores, cracks, and other openings, usually at the surface. It includes *gelifraction* and *cryoturbation.* (b) The resulting effects of frost action on materials and structures. Syn: *freeze-thaw action.*
frost-active soil (frost'-ac-tive) A fine-grained soil that undergoes changes in volume and bearing capacity due to frost action (Nelson and Nelson, 1967, p.151).
frost belt A ditch designed to assist the early and rapid freezing of the soil in order to obstruct seepage of shallow groundwater. It is commonly placed upslope from foundations in permafrost areas. Syn: *frost dam.*
frost blister A small mound with a solid ice core resembling a laccolith in section. It is formed in the active layer by the freezing of water forced into a plano-convex lens where it is trapped between the downward-freezing active layer and the underlying permafrost. Ice laccoliths are produced in a few days or weeks during fall freeze-up in areas of continuous permafrost. Most are less than a meter high and a few meters to 10 m across. They may not melt for several years. They differ from many hydrolaccoliths that are not active-layer phenomena (Black, 1954, p.848). Syn: *cryolaccolith; soil blister; ice laccolith.*
frost boil (a) An accumulation of excess water and mud liberated from ground ice by accelerated spring thawing, commonly softening the soil and causing a quagmire. (b) A low mound developed by local differential frost heaving at a place most favorable for the formation of segregated ice and accompanied by an absence of an insulating cover of vegetation (Taber, 1943, p.1458-1459). (c) A break in a surface pavement due to swelling frost action; as the ice melts, soupy subgrade materials issue from the break.
frost breaks The effect of repetitive shots at random times following a shot, produced by ice fracturing when shooting in permafrost. Also called ice noise. (Sheriff and Geldart, 1995).
frost bursting *gelifraction.*
frost churning *cryoturbation.*
frost circle A term used by Williams (1936) in referring to a *sorted circle* developed in horizontal thin-bedded limestones in Ontario.
frost crack *thermal contraction crack.*
frost cracking The contraction cracking of frozen ground, and ice on lakes and rivers, at very low temperatures; the formation of frost cracks.
frost-crack polygon A *nonsorted polygon* formed by intersecting frost cracks. It is similar to an *ice-wedge polygon,* but lacks a border underlain by ice wedges and therefore is not necessarily associated with permafrost.
frost creep Soil creep resulting from frost action (Kerr, 1881).
frost dam *frost belt.*
frost drift The movement, by frost action, of debris on a slope (Kerr, 1881). Obsolete.
frost flower A type of *surface hoar,* usually growing on an ice surface, characterized by leafy or dendritic blades oriented at various angles to the surface. Syn: *ice flower.*
frost heave *frost heaving.*
frost heaving The uneven lifting or upward movement, and general distortion, of surface soils, rocks, vegetation, and structures such as pavements, due to subsurface freezing of water and growth of ice masses (esp. ice lenses); any upheaval of ground caused by freezing. Syn: *frost heave.*
frost hillock The marked upward bulging sometimes present in the center of a *mud polygon* (Sharpe, 1938, p.36).
frosting (frost'-ing) (a) A lusterless ground-glass or mat surface on rounded mineral grains, esp. of quartz. It may result from innumerable impacts of other grains during wind action, or from deposition of many microscopic crystals, e.g. fine silica secondarily deposited on quartz grains. (b) The process that produces such a surface.
frost line (a) The maximum depth of frozen ground in areas where there is no permafrost; it may be expressed for a given winter, as the average of several winters, or as the greatest depth on record. Cf: *frost table.* (b) The bottom limit of permafrost. Cf: *permafrost table.* (c) The altitudinal limit below which frost never occurs; applied esp. in tropical regions.
frost mound A general term for a knoll, hummock, or conical mound in a permafrost region, containing a core of ice, and representing a generally seasonal and localized upwarp of the land surface, caused by frost heaving and/or hydrostatic pressure of groundwater. See also: *pingo; palsa; hydrolaccolith; ice laccolith; earth hummock; frost blister; ground-ice mound; ice mound; gravel mound; soil blister; suffosion knob.*
frost-pattern soil A term used by Troll (1944) for what is now known as *patterned ground;* it is a misleading term because patterned ground need not consist of soil, nor need it involve a periglacial origin.
frost point That temperature to which a sample of humid air must be cooled (at constant pressure and water-vapor content) in order to cause the deposition of ice. Cf: *dewpoint.*
frost polygon One of the network polygons forming *polygonal ground.*
frost riving *gelifraction.*
frost scar A *nonsorted circle* or irregular form representing a small patch of bare soil produced by local frost heaving intense enough to disturb the vegetation cover. See also: *mud circle.* Syn: *mud spot; spot medallion.*
frost shattering *gelifraction.*
frost soil *cryoturbate.*
frost splitting *gelifraction.*
frost stirring A syn. of *cryoturbation* involving no mass movement.
frost table An irregular surface that represents, at any given time, the penetration of thawing in seasonally or perennially frozen ground; the upper limit of frozen ground. Cf: *frost line; permafrost table.*
frost weathering *gelifraction.*
frost wedge A term used loosely for any *ice wedge,* whether in perennially or seasonally frozen ground or in fossil form; any wedge-shaped mass whose origin involves cold or freezing conditions.
frost wedging A type of *gelifraction* by which jointed rock is pried and dislodged by ice acting as a wedge.
frostwork A bristly *speleothem,* commonly intricately branched, composed of acicular aragonite crystals. Frequently used synonymously with anthodite, although that usage is controversial.
frost zone *seasonally frozen ground.*
Froude number (Frou'-de) A dimensionless numerical quantity used as an index to characterize the type of flow in a hydraulic structure that has the force of gravity (as the only force producing motion) in conjunction with the resisting force of inertia. It is the ratio of inertia forces to gravity forces, and is equal to the square of a characteristic velocity (mean, surface, or maximum velocity) of the system divided by the product of a characteristic linear dimension (e.g. diameter or depth) and the gravity constant, acceleration due to gravity, all of which are expressed in consistent units in order that the combinations will be dimensionless. The number is used in open-channel flow studies or where the free surface plays an essential role in influencing motion (ASCE, 1962).
frozen [coal] (fro'-zen) *burned.*
frozen [ore dep] Said of the contact between the wall of a vein and the mineral deposit filling it, in which the vein material adheres closely to the wall; also, said of the vein material and of the wall.
frozen ground Ground that has a temperature below freezing and generally contains a variable amount of water in the form of ice. Terms inadvisedly used as syns: *frost; ground frost; permafrost.* Syn: *tjaele; gelisol; merzlota; tele; taele.*
fruchtschiefer (frucht'-schief-er) A type of *spotted slate* characterized by concretionary spots having shapes suggestive of grains of wheat. Etymol: German. Cf: *fleckschiefer; garbenschiefer; knotenschiefer.*
fructification (fruc'-ti-fi-ca'-tion) A reproductive organ or fruiting structure of a plant.
fruit (a) In a strict sense, the pericarp and its seeds, the fertilized and developed ovary. (b) More broadly, the matured pericarp and its contents, with any integral external part (Jackson, 1928, p.153).
frustule (frus'-tule) The siliceous cell wall of a *diatom,* consisting of two halves, the epivalve and the hypovalve. It is ornate, microscopic, and boxlike.
Fry method A type of center-to-center method of strain analysis developed by Fry (1979). The method uses the fact that distances between originally

anticlustered object centers and such centers of their neighbors vary systematically with direction.

FS *foresight.*

F-scale A six-category *tornado*-intensity scale, developed by T. Theodore Fujita, that rates tornadoes from F0 to F5 on the basis of rotational *wind* speed estimated from property damage. Categories are weak (F0, F1), strong (F2, F3), and violent (F4, F5). Also called: Fujita scale.

FSI *fracture spacing index.*

FSR *fracture spacing ratio.*

F test A statistical test of the equality of two population variances, expressed as the ratio between the variances of two random samples taken from the populations. F tests between the variance attributable to a fitted model and the variance of deviations from the model are used in regression, trend-surface analysis, and analysis of variance to determine the statistical significance of fitted models. Syn: *F-distribution test.*

FTP File transfer protocol.

fuchsite (fuchs'-ite) A bright-green, chromium-rich variety of muscovite. Syn: *chrome mica.*

fucoid (fu'-coid) n. (a) An informal name now applied loosely to any indefinite trail-like or tunnel-like sedimentary structure identified as a *trace fossil* but not referred to a described genus. It was once considered to be the remains of the marine alga *Fucus,* and later was regarded as a cylindrical, U-shaped, regularly branching feeding burrow of a marine animal and assigned to the plantlike "genus" *Fucoides.* The term has been broadly applied to crustacean tracks, worm burrows, molluscan trails, marks made by the tide or waves, and rill marks. Fucoids have been defined as being within sedimentary layers and formed of material more or less unlike the matrix (Vassoevich, 1953, p.21); but in common usage fucoids are surface features formed of the same material as the matrix. See also: *chondrite [paleont].* (b) A fossil of an alga, or a fossil resembling an alga or the remains or impression of a seaweed. (c) A seaweed of the order Fucales (brown algae). adj. Pertaining to or resembling a fucoid.

fuel-coolant interaction Referring to an industrial analog to a phreatomagmatic explosion, involving a fuel such as molten iron and a coolant such as water. The interaction of the fuel and coolant has resulted in vapor explosions in industrial environments. Abbrev: FCI.

fuel oil Specifically defined fraction of petroleum distillate.

fuel ratio (fu'-el) In coal, the ratio of fixed carbon to volatile matter. It is sometimes a factor in the analysis or classification of coals. Cf: *carbon ratio [coal].*

fuenzalidaite A colorless to pale yellow trigonal mineral found in a nitrate deposit: $K_3(Na,K)_2Na_3Mg_5(IO_3)_6(SO_4)_6{\cdot}6H_2O$.

fugacity (fu-gac'-i-ty) A thermochemical value (symbolized f) defined by the equation $G_{T,P}=G^{\circ}_{T,1\ bar}+RT\ 1n\ f$ where G is Gibbs free energy, R is the gas constant, and T is temperature (in Kelvins). Fugacity is typically used for gaseous or fluid species or components and is expressed in units of pressure. Fugacities are used in calculations of chemical equilibrium, because they can be calculated in practical applications (based on experimentally derived values) whereas G cannot (Klotz, I. M., 1972).

fugacity coefficient The ratio of the fugacity of a gas to its potential pressure; hence, a measure of the nonideality of the gas.

fugitive (fu'-gi-tive) In ecology, said of an organism, esp. a plant species, that is not permanently established and is likely to disappear.

fugitive constituent A substance that was originally present in a magma but was lost during crystallization, so that it does not commonly appear as a rock constituent (Shand, 1947, p.34). Syn: *volatile component.*

fugitive species *opportunistic species.*

Fujita scale *F-scale.*

fukalite (fu'-ka-lite) A white to pale brown orthorhombic mineral occurring in skarns: $Ca_4Si_2O_6(CO_3)(OH,F)_2$.

fukuchilite (fu-ku'-chi-lite) A metallic pinkish-brown cubic mineral of the *pyrite* group: Cu_3FeS_8.

fulcral plate (ful'-cral) A small plate raised above the floor of the *brachial valve* of a brachiopod, extending between the posterior margin and the *brachiophore base,* and bounding the dental socket anteriorly and laterally (TIP, 1965, pt. H, p.145).

fulcral ridge A linear elevation on an articular face of a crinoid *ossicle,* serving as an axis of differential movement.

fulcrum (ful'-crum) The intersection of the end of a recurved spit with the next succeeding stage in development of a compound spit.

fulgurite (ful'-gu-rite) An irregular, glassy, often tubular or rod-like structure or crust produced by the fusion of loose sand (or rarely, compact rock) by lightning, and found esp. on exposed mountain tops or in dune areas of deserts or lake shores. It may measure 40 cm in length and 5-6 cm in diameter. Etymol: Latin fulgur, "lightning". Syn: *lightning stone; lightning tube; sand tube.*

fulji (ful'-ji) A term used in northern Arabia for a depression between barchans, occurring esp. where the dunes are pressing closely on one another; it has a steep slope on the windward side and a gentle slope on the lee side. Etymol: Arabic. Pl: fuljis. Also spelled: fulje.

full A British term for *beach ridge.*

full-cut brilliant A brilliant-cut diamond or colored stone with the usual total of 58 facets, consisting of 32 facets and a table above the girdle, and 24 facets and a culet below. The girdle is usually polished on colored stones, but seldom on diamonds.

full-depth avalanche A *snow avalanche* that incorporates the entire snow pack to the ground surface; formerly termed *ground avalanche.*

full dip *true dip.*

fullerite A black cubic or tetragonal polymorph of carbon found in carbonaceous shungite: C_{60}. A fullerene. Not approved by the International Mineralogical Association.

fuller's earth (full'-er's) (a) Naturally occurring clay or claylike material possessing a high absorptive capacity. It generally consists largely of hydrous phyllosilicates. Used as early as 5000 B.C. for whitening, degreasing, or fulling (shrinkage, cleaning, and thickening) of wool pelts and woolen fabric. Now extensively used as an adsorbent in refining and decolorizing oils, fats, and wine; it is a natural (but often modified) bleaching agent. (b) Any natural earthy material which will decolorize mineral or vegetable oils to a sufficient extent to be of economic importance. May be composed of attapulgite (= palygorskite), smectite, or kaolinite. (Grim, 1953). (c) In England, fuller's earth refers primarily to Cretaceous beds of calcium montmorillonite in southeast England that have been used for centuries (Robertson, 1986). Known as creta fullonia. Syn: *walker's earth.* Cf: *bentonite; bleaching clay; Florida earth.*

full meander A stream meander consisting of two loops, one in a clockwise direction and the other in a counterclockwise direction.

full-wave sonic log A log similar to the array sonic (q.v.). Halliburton tradename.

fülöppite (fü'-löp-pite) A lead-gray monoclinic mineral with a bluish or bronze tarnish: $Pb_3Sb_8S_{15}$. Also spelled: fuloppite.

fulvic acid (ful'-vic) That pigmented organic matter of indefinite composition that remains in solution when an aqueous alkaline extract of soil is acidified by removal of humic acid. It is separated from the fulvic acid fraction by adsorption on a hydrophobic resin at low pH values.

fulvurite (ful'-vur-ite) An old synonym for *brown coal.*

fumarole (fu'-ma-role) A vent, usually volcanic, from which gases and vapors are emitted; it is characteristic of a late stage of volcanic activity. It is sometimes described by the composition of its gases, e.g. chlorine fumarole. Fumaroles may occur along a fissure or in apparently chaotic clusters or fields. See also: *solfatara; fumarolic stage; fumarole field.*

fumarole field A group of cool fumaroles (Rittmann, 1962, p.10). Cf: *solfatara field.*

fumarole mound A small mound from which fumarole or solfatara gases escape.

fumarolic stage (fu-ma-rol'-ic) A late or decadent type of volcanic activity characterized by the emission of gases and vapors from a vent or *fumarole.* Cf: *solfataric stage.*

fume cloud A vaporous cloud of volcanic gases from a body of molten lava.

functional morphology (func'-tion-al) The study of the form and structure of an organism in relation to its adaptation to a specific environment and/or survival under specific conditions; the morphology of an animal or plant as it responds or responded to environmental changes and conditions.

fundamental n. The lowest frequency of a periodic function. Also called: first harmonic.

fundamental jelly *ulmin.*

fundamental particle Nadeau et al. (1984) showed that illite, smectite, and chlorite particles cleave in sheet thicknesses of integral multiples of 20 Å (two 2:1 layers and interlayer potassium) for illite, 10 Å (one 2:1 layer) for smectite, and 24 Å (two 2:1 layers and an Mg-Fe interlayer) for chlorite, and they referred to each unit (20-, 10-, and 24 Å) as a "fundamental particle."

fundamental strength The maximum stress that a body can withstand, under given conditions but regardless of time, without creep.

fundamental substance *ulmin.*

fungal spore (fun'-gal) A spore of an organism belonging to the Kingdom Fungi. Such spores include a wide variety of types, from simple unicellular to multicellular *sclerotia;* they have a range of Precambrian to Holocene, Cf: *basidiospore; chlamydospore; conidium; dictyospore; phragmospore; teleutospore; urediospore.*

Fungi (Fun'-gi) A *kingdom* of heterotrophic *eukaryotes,* both terrestrial and marine. Fungi are major decomposers in the terrestrial biosphere. Most are multicellular and filamentous, although the yeasts are unicellular. Modes of nutrition range from *saprophytic* to *parasitic* to mutualist. Sing: fungus.

Fungi imperfecti The group of *anamorph* fungi of which sexual stages have not (yet) been recognized or may not exist. Also referred to as Deuteromycetes. Those fungi from which both sexual and asexual spores have been described make up the group of *holomorph* fungi.

funginite Fungal remains in coals and sedimentary rocks; a part of the *inertinite* maceral group (ICCP, 2001).

funicle (fu'-ni-cle) A spirally wound narrow ridge extending upward from the inner lip of a gastropod shell into the *umbilicus.*

funicular regime The condition of high liquid water content in *snow* in which liquid exists in continuous paths; grain-to-grain bonds are weak.

funicular water (fu-nic'-u-lar) *Capillary water* contained in a cluster of rock or soil particles in the unsaturated zone, the interstices of the cluster being completely filled with water bounded by a single closed capillary meniscus

(Smith, W.O., 1961, p.2). Cf: *pendular water; pellicular water; capillary condensation.*

funiculus [bot] (fu-nic'-u-lus) The stalk by which an ovule is attached to the ovary wall or placenta in angiospermous plants.

funiculus [paleont] A strand of tissue connecting the *polypide* with the body wall or septulae in a bryozoan. Pl: funiculi.

funnel filling (fun'-nel) The red-brown to dark-brown, opaque to translucent, coarsely laminated portion of the basal plate of some conodont elements, occupying the cavity in the basal funnel.

funnel intrusion An igneous intrusion with an inverted conical shape; typically layered, and mafic or ultramafic in composition.

furca (fur'-ca) A two-forked last abdominal segment of certain crustaceans; specif. *caudal furca.* Pl: furcae.

furcula (fur'-cu-la) (a) The wishbone, or fused clavicles, of a bird. (b) A wishbone-shaped sponge spicule. Pl: furculae.

furious cross-bedding (fu'-ri-ous) Bedding that is doubly cross-bedded, characterized by foreset beds that are themselves cross-bedded (Reiche, 1937, p.926).

Furongian (Fu-ron'-gi-an) A series of the Standard Global Chronostratigraphic Scale: uppermost Cambrian Series (above unnamed series of Middle Cambrian, below Lower Ordovician Series). The name "Furongian" is from Furong, meaning "lotus," referring to Hunan, the Lotus State. The basal GSSP is near Paibi in Hunan, China, and was ratified in 2003.

furongite (fu-rong'-ite) A yellow triclinic or monoclinic mineral: $Al_{13}(UO_2)_7(PO_4)_{13}(OH)_{14}·58H_2O$. Not approved by the International Mineralogical Association.

furrow [geomorph] (fur'-row) (a) A linear depression produced by the removal of rock material, as by glacial action; e.g. a groove. (b) A term applied in Africa to a natural or artificial watercourse used for drainage or irrigation. (c) *swale.*

furrow [paleont] Used in description of conodont elements for any longitudinal groove, trench, or depression; may be a feature of either *upper* or *under side,* or may describe an elongate depression on one of the faces of a coniform element (see longitudinal furrow); also used to describe a groove on the upper surface of a basal plate or funnel (TIP, 1981, pt. W, supp.2).

furrow [palyn] *colpus.*

furrow [struc geol] A nongenetic term used by Bucher (1933) for a depressed part of the crust of any size with a distinct linear development.

furrow cast A term introduced by McBride (1962, p.58) for a *sole mark* consisting of a cast of a group of closely spaced, parallel, and linear indentations separated by long, narrow, round or flat-topped, slightly sinuous septa which appear as depressions in the cast; it lacks the steep or blunt upcurrent end of a flute cast. The term was suggested hesitantly by Kuenen (1957, p.244) to replace the ambiguous "groove cast". Cf: *furrow flute cast.*

furrow flute cast A *furrow cast* with an upcurrent termination similar to that of a flute cast. Syn: *sludge cast; rill cast.*

furutobeite (fu-ru-to'-be-ite) A metallic gray monoclinic mineral: $(Cu,Ag)_6PbS_4$.

fusain (fu'-sain) A coal lithotype characterized macroscopically by its silky luster, fibrous structure, friability, and black color. It occurs in strands or patches and is soft and dirty when not mineralized. Its characteristic microlithotype is *fusite.* Cf: *vitrain; clarain; durain.* Syn: *mineral charcoal; mother of coal.* Obsolete syn: *motherham.*

fusainisation (fu'-sain-i-sa'-tion) *fusinization.*

fused Said of very closely set conodont denticles, each partly or entirely confluent with adjacent or adjoining denticles; contrasts with discrete denticles (TIP, 1981, pt. W, supp.2). Ant: *discrete [paleont].*

fused cluster A cluster of conodont elements that are fused or cemented to one another (TIP, 1981, pt. W, supp.2).

fusellar fabric (fu-sel'-lar) One of the three major types of materials recognized in electron-microscope study of graptolithine *periderm* as a fundamental structural element of the periderm. Fusellar fabric is formed from slightly wavy, and commonly branching, fibrils that are so interwoven that they form a three-dimensional mesh (Urbanek and Towe, 1974, p.4). Cf: *cortical fabric; sheet fabric.*

fusellar tissue Fusellar tissue in graptolithines is composed of spongy-appearing *fusellar fabric* bounded or enclosed by thin, membranous *sheet fabric;* it forms the inner part of the graptolithine *periderm,* the outer part being formed from *cortical tissue* (Urbanek and Towe, 1974, p.5).

fusibility (fu-si-bil'-i-ty) The property of undergoing fusion in the flame of the blowpipe. See also: *fusibility scale.* Cf: *infusible.*

fusibility scale A temperature scale based on the *fusibility* of a standard group of minerals, with which other minerals may be compared. An analysis that can be made with a burner and a blowpipe is based on the following series: stibnite, 550°C; chalcopyrite, 800°C; almandine garnet, 1,050°C; actinolite, 1,200°C; orthoclase, 1,300°C; enstatite (bronzite), 1,400°C; and quartz, infusible.

fusiform (fu'-si-form) Shaped like a spindle, i.e. tapering toward each end from a swollen middle.

fusiform bomb A type of volcanic bomb that tapers and slightly twists at both ends from an enlarged middle; it includes both rotational bombs and spindle-shaped bombs.

fusil (fu'-sil) A spindle-shaped siliceous concretion.

fusinite (fu'-si-nite) A maceral of coal within the *inertinite* group with intact or broken cellular structure, a reflectance (except in meta-anthracite) well above that of associated vitrinite, and a particle size generally greater than about 50 micrometers except when isolated from other macerals. Cf: *fusain.*

fusinization (fu'-sin-i-za'-tion) A process of *coalification* in which fusain is formed. Cf: *incorporation; vitrinization.* Also spelled: *fusainisation.*

fusinoid (fu'-si-noid) Fusinite that has a reflectance distinctly higher than that of associated xylinoids, vitrinoids, or anthrinoids, and that has well-developed cellular structure (ASTM, 1970, p.175).

fusion [chem] (fu'-sion) The combination, or fusion, of two light nuclei to form a heavier nucleus. The reaction is accompanied by the release of a large amount of energy as in the hydrogen bomb. Cf: *fission [chem].* Syn: *nuclear fusion.*

fusion [petrology] (a) The process whereby a solid becomes liquid by the application of heat; melting. (b) The unification or mixing of two or more substances, as by melting together.

fusion [photo] *stereoscopic fusion.*

fusion crust A thin glassy coating, usually black and rarely exceeding one millimeter in thickness, formed on the surface of a meteorite by the chilling of an oxidized surficial melt produced by frictional heating during atmospheric passage.

fusion tectonite An igneous rock whose alignment of early-formed crystals was caused by continuous movement in an enclosing melt; a type of *primary tectonite* (Turner and Weiss, 1963, p.39).

fusite (fu'-site) A coal microlithotype that contains at least 95% fusinite. It is a variety of *inertite.* Cf: *fusain.*

fusoclarain (fu-so-clar'-ain) A transitional lithotype of coal, characterized by the presence of fusinite and vitrinite with other macerals; fusinite is less abundant than it is in *clarofusain.* Syn: *fusoclarite.*

fusoclarite (fu-so-clar'-ite) *fusoclarain.*

fusodurain (fu-so-du'-rain) A coal lithotype transitional between durain and fusain, but predominantly durain. Cf: *durofusain.*

fusotelain (fu-so-tel'-ain) A coal lithotype transitional between telain and fusain, but predominantly telain. Cf: *telofusain.*

fusovitrain (fu-so-vit'-rain) A coal lithotype transitional between vitrain and fusain, but predominantly vitrain. Cf: *semifusain.* Syn: *fusovitrite.*

fusovitrite (fu-so-vit'-rite) *fusovitrain.*

fusshang (fuss'-hang) A syn. of *foot slope.* Etymol: German "Fusshang", "foot slope".

fusulinacean (fu'-su-li-na'-ce-an) Any fusulinid belonging to the superfamily Fusulinacea, characterized by a spindle-shaped, spheroidal, or discoid test with a complex internal structure. Range, Upper Mississippian to Upper Permian.

fusuline (fu'-su-line) *fusulinid.*

fusulinid (fu-su-lin'-id) Any foraminifer belonging to the suborder Fusulinina, family Fusulinidae, characterized by a multichambered elongate calcareous microgranular test, commonly resembling the shape of a grain of wheat. Range, Middle Pennsylvanian to Upper Permian. Syn: *fusuline.* See also: *alveolinid.*

future ore (fu'-ture) *possible ore.*

fuzzy logic Reasoning involving fuzzy sets, i.e., where elements do not fit within compact boundaries. Fuzzy logic involves probability concepts.

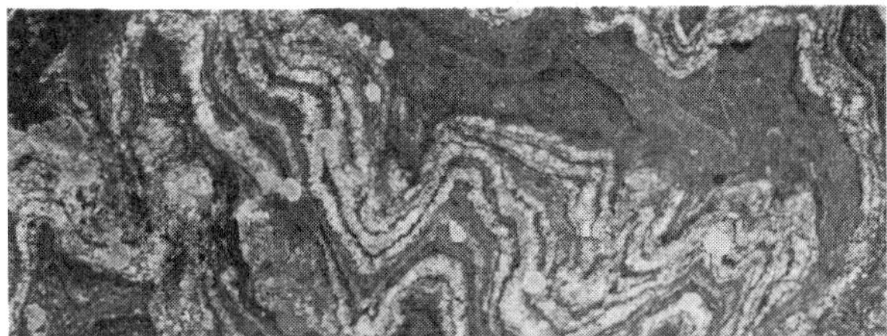

G

Ga Giga-annum, one billion (10^9) years. Informal SI notation, where annum is age in years before present, with "present" fixed as 1950. This term has largely replaced the various abbreviations for "billions of years before present" in geological literature. It is not equivalent to the elapsed time interval or duration in "billions of years" (byr).

gabbro (gab'-bro) (a) In the *IUGS classification,* a plutonic rock with Q between 0 and 5, P/(A+P) greater than 90, and plagioclase more calcic than An_{50}. (b) A group of dark-colored, basic intrusive igneous rocks composed principally of calcic plagioclase (commonly labradorite or bytownite) and clinopyroxene (augite), with or without olivine and orthopyroxene; also, any member of that group. It is the approximate coarse-grained equivalent of basalt. Apatite and magnetite or ilmenite are common accessory minerals. Gabbro grades into *monzonite* with increasing alkali-feldspar content. According to Streckeisen (1967, p.171, 198), plagioclase with more than 50% anorthite distinguishes gabbro from *diorite;* quartz is 0-20% of the light-colored constituents, and the plagioclase/total feldspar ratio is 90/100. The name, introduced by Buch in 1810, is apparently after the town of Gabbro in Tuscany, Italy.

gabbroic layer (gab-bro'-ic) Beneath continents, equivalent to the *basaltic layer.* In marine usage, the lower part of the oceanic crust, known as *Layer 3.* See also: *basaltic layer.*

gabbroid (gab'-broid) (a) In the *IUGS classification,* a preliminary term (for field use) for a plutonic rock with Q less than 20 or F less than 10, P/(A+P) greater than 65, and pl/(pl+px+ol) between 10 and 90. Cf: *melagabbroid.* (b) Said of a rock resembling gabbro.

gabbronorite (gab-bro-nor'-ite) In the *IUGS classification,* a plutonic rock satisfying the definition of *gabbro,* in which pl/(pl+px+ol) and pl/(pl+px+hbl) are between 10 and 90, and ol/(pl+px+ol) and hbl/(pl+px+hbl) are less than 5. Orthopyroxene is between 5 and 95 percent of total px.

gabion (ga'-bi-on) A container, cylinder, box or basket of corrosion-resistant wire, specially designed to hold coarse rock aggregate, e.g., for building a groin or seawall or developing a bar or dike in a harbor. Gabions are also often placed as the toes of slopes for stability, e.g., in gravity retaining walls.

gabrielsonite (ga'-bri-el-son-ite') An adamantine greenish-brown to black orthorhombic mineral of the *adelite* group: $PbFe^{2+}(AsO_4)(OH)$.

gadolinite-(Ce) A black, greenish black, or brown monoclinic mineral: $Be_2Fe^{2+}Ce_2Si_2O_{10}$. It is commonly metamict.

gadolinite-(Y) A black, greenish black, or brown monoclinic mineral: $Y_2Fe^{2+}Be_2Si_2O_{10}$. It is a source of rare earths.

gagarinite-(Y) (ga-ga'-rin-ite-y) A creamy, yellowish, or rosy hexagonal mineral: $NaCaY(F,Cl)_6$.

gagatite (ga'-ga-tite) Coalified woody material, resembling jet. See also: *gagatization.*

gagatization (ga'-ga-ti-za'-tion) In coal formation, the impregnation of wood fragments with dissolved organic substances. See also: *gagatite.*

gage n. In hydraulics, a device for measuring such factors as water-surface elevation, velocity of flow, water pressure, and precipitation. Also spelled: gauge. See also: *staff gage; chain gage.*

gage height *stage [hydraul].*

gageite (gage'-ite) A pink to brown acicular monoclinic pseudo-orthorhombic mineral: $(Mn^{2+},Mg,Zn)_{21}Si_8O_{27}(OH)_{20}$. See the Mg analogue, *balangeroite.*

gage well A sitting well in which stage measurements are performed.

gaging (gag'-ing) *stream gaging.*

gaging station A particular site on a stream, canal, lake, or reservoir where systematic observations of gage height, discharge, or water quality (or any combination of these) are obtained.

gahnite (gahn'-ite) A dark-green to yellowish, gray, or black cubic mineral of the *spinel* series: $ZnAl_2O_4$. It often contains some magnesium. Syn: *zinc spinel.*

gahnospinel (gahn'-o-spi-nel') A blue or greenish variety of spinel containing zinc.

Gaia The Earth as a symbiotic entity in which organisms regulate the climate and the composition of the atmosphere, hydrosphere, and crust. Introduced in 1972 by James Lovelock, British geophysicist.

gaidonnayite (gai-don-nay'-ite) A white to beige, light brown, or pale yellowish green orthorhombic mineral: $Na_2ZrSi_3O_9{\cdot}2H_2O$. It is dimorphous with catapleiite.

gain control In a seismic amplifier, a device to change the amplification with time. It may be automatic by individual channel, ganged for all channels together, programmed prior to the source emission, or otherwise pre-arranged. Syn: *volume control.*

gainesite (gaines'-ite) A pale bluish-lavendar tetragonal mineral: $Na_2(Be,Li)(Zr,Zn)_2(PO_4)_4{\cdot}1.5H_2O$.

gaining stream (gain'-ing) *effluent stream.*

gain-loss diagram [eco geol] A diagram that plots the addition and removal of chemical elements during hydrothermal alteration against distance from the center or core of a mineral deposit.

gaitite (gait'-ite) A colorless triclinic mineral of the *fairfieldite* group: $Ca_2Zn(AsO_4)_2{\cdot}2H_2O$.

gal A unit of acceleration, used in gravity measurements. One gal = 1 $cm/sec^2 = 10^{-2}m/sec^2$. The Earth's normal gravity is 980 gal. The term is not an abbreviation: it was invented to honor the memory of Galileo. See also: *milligal; microgal.*

galactic cluster (ga-lac'-tic) *star cluster.*

galactite (ga-lac'-tite) (a) A variety of white natrolite, occurring in colorless acicular crystals. (b) An obsolete syn. of *novaculite.* (c) An unidentified stone (possibly of calcium nitrate) whose milky solution gave rise to several medieval legends and superstitions.

galaxite (ga'-lax-ite) A black, reddish brown, red, or yellow cubic mineral of the *spinel* series: $(Mn^{2+},Fe^{2+},Mg)(Al,Fe^{3+})_2O_4$.

galaxy (gal'-ax-y) One of billions of large systems of stars, nebulae, star clusters, globular clusters, and interstellar matter that make up the Universe. When the term is capitalized, it refers to the Milky Way stellar system, of which our sun is a member. Syn: *extragalactic nebula.*

galea (ga'-le-a) (a) A conical process in the skeleton of a phaeodarian radiolarian. (b) The spinning tube on the movable finger of the *chelicera* of certain arachnids (pseudoscorpions). (c) The outer distal hoodlike lobe of the second segment of the maxillule of a crustacean, adjacent to the *lacinia* (TIP, 1969, pt. R, p.96).

Gale alidade A syn. of *explorer's alidade.* Named for Hoyt S. Gale (1876-1952), American geologist.

galeate acritarchs (ga'-le-ate) Early Paleozoic *acritarch* taxa (e.g., *Cymatiogalea, Priscogalea*) characterized by a hemispherical vesicle outline and a large *pylome* excystment opening closed by an *operculum* (Servais and Eiserhardt, 1995).

galeite (gale'-ite) A colorless trigonal mineral: $Na_{15}(SO_4)_5F_4Cl$. Cf: *schairerite.*

galena (ga-le'-na) (a) A bluish-gray to lead-gray mineral: PbS. It frequently contains included silver minerals. Galena occurs in cubic or octahedral crystals, in masses, or in coarse or fine grains; it is often associated with sphalerite as disseminations in veins in limestone, dolomite, and sandstone. It has a shiny metallic luster, exhibits highly perfect cubic cleavage, and is relatively soft and very heavy. Galena is the most important ore of lead and one of the most important sources of silver. Syn: *galenite; lead glance; blue lead [mineral].* (b) A group name for cubic minerals with the formula *AX,* where *A* = Mg, Ca,Mn, (AgSb) or (AgBi), and *X* = S, Se or Te.

galenite (ga-len'-ite) *galena.*

galenobismutite (ga-le'-no-bis'-mut-ite) A metallic lead-gray orthorhombic mineral: $PbBi_2S_4$.

galgenbergite-(Ce) A colorless or white triclinic mineral: $Ca(Ce,Nd,La)_2(CO_3)_4{\cdot}H_2O$.

galileiite A waxy pale amber rhombohedral mineral of the *fillowite* group: $NaFe^{2+}{}_4(PO_4)_3$. It is a meteorite mineral.

galkhaite (gal'-kha-ite) A dark orange-red cubic mineral: $(Cs,Tl)(Hg,Cu,Zn)_6(As,Sb)_4S_{12}$.

gall [sed] (a) *clay gall.* (b) A sand pipe.

gall [soil] A small barren or infertile surface spot or area from which the original surface soil has been removed by erosion or excavation.

gallery [grd wat] (gal'-ler-y) *infiltration gallery.*

gallery [paleont] A laterally continuous internal open space between adjacent *laminae* in most stromatoporoid coenostea, partially filled by *pillars* and *dissepiments*.

gallery [speleo] In a cave, a large, more or less horizontal *passage*. Cf: *hall*.

galliard (gal'-liard) A hard, smooth, close-grained, siliceous sandstone; a *ganister*.

gallite (gal'-lite) A metallic gray tetragonal mineral of the *chalcopyrite* group: $CuGaS_2$.

gallobeudantite A pale yellow, greenish, or creamy rhombohedral mineral of the *beudantite* group: $Pb(Ga,Al,Fe)_3(AsO_4,SO_4)_2(OH)_6$.

galloping glacier (gal'-lop-ing) A popular term for *surging glacier*.

Gall projection A stereographic, modified-cylindrical map projection in which the cylinder intersects the globe along the parallels 45°N and 45°S. The scale is preserved along these parallels, but is too small between them and too large poleward of them; there is less distortion of areas and shapes in high latitudes than in the Mercator projection. It was introduced in 1855 by the Rev. James Gall of Scotland.

galmei (gal-mei') *hemimorphite*.

Galperin geophone (Gal'-per-in) A geophone geometry consisting of a triphone with elements making an angle of 54° 35′ with the vertical, with an additional vertical geophone for redundancy (Sheriff, 2002). Named for E. I. Galperin, Russian geophysicist.

galt *gault*.

gamagarite (ga-ma-gar'-ite) A dark-brown monoclinic mineral of the *brackebuschite* group: $Ba_2(Fe^{3+},Mn^{3+})(VO_4)_2(OH)$.

gametangium (gam-e-tan'-gi-um) A gamete-producing structure or sex organ in plants. Plural: gametangia. Cf: *antheridium; archegonium*.

gamete (ga-mete') A haploid reproductive cell whose nucleus and often cytoplasm fuses with that of another gamete (constituting fertilization). The resulting cell (*zygote*) develops into a new individual.

gametophyte (ga-me'-to-phyte) The individual or sexual generation of a plant that produces gametes; e.g. the haploid generation of an embryophytic plant, produced by germination of the spores. In lower vascular plants and bryophytes, the gametophyte is a separate plant, but in seed plants, it is confined to the several cells of the *microgametophyte* in the pollen grain and the multicellular *megagametophyte* in the ovule, with the seed developing from the fertilized ovule. Cf: *sporophyte*. See also: *prothallus*.

gamma [cryst] (gam'-ma) (a) In a biaxial crystal, the largest *index of refraction*. (b) The interaxial angle between the *a* and *b* crystallographic axes. Cf: *alpha [cryst]; beta [cryst]*.

gamma [magnet] The cgs unit of magnetic field intensity: 1 gamma = 10^{-5} gauss = 10^{-9} tesla. It has been replaced in magnetic exploration by the SI preferred name of *nanotesla* (nT).

gamma [mineral] adj. Of or relating to one of three or more closely related minerals and specifying a particular physical structure (esp. a polymorphous modification); specif. said of a mineral that is stable at a temperature higher than those of its *alpha* and *beta* polymorphs (e.g. the gamma (γ) phase of Mg_2SiO_4 is spinel).

gamma* angle The angle of the reciprocal lattice between the *a* axis* and the *b* axis*, which is equal to the dihedral angle between (100) and (010). Cf: *alpha* angle; beta* angle*.

gamma decay De-excitation of an atomic nucleus without a change in atomic number or mass number, usually by emission of *gamma radiation*. Syn: isomeric decay.

gamma diversity (a) The total taxon *richness* of a geographic region (Whittaker, 1972). (b) The taxonomic differentiation between regions (Sepkoski, 1988).

gamma-gamma log *density log*.

gamma-MnO$_2$ *nsutite*.

gamma radiation Electromagnetic radiation from an atomic nucleus, often accompanying emission of alpha particles and beta particles. Cf: *gamma ray*.

gamma ray A photon from an atomic nucleus. See: *gamma radiation; nuclear isomerism; isomer*.

gamma-ray log The *radioactivity log* curve of the intensity of broad-spectrum, undifferentiated natural gamma radiation emitted from the rocks in a cased or uncased borehole. It is used for correlation, and for distinguishing shales (which are usually richer in naturally radioactive elements) from sandstones, carbonates, and evaporites. Cf: *spectral gamma-ray log*.

gamma-ray spectrometer An instrument for measuring the energy distribution, or spectrum, of gamma rays, whether from natural or artificial sources. It is used in airborne remote sensing for potassium, thorium, and uranium. Cf: *scintillation spectrometer*.

gamma-ray spectrometry Radiometric method based on the proportionality between energy of gamma quanta deposited in the detector and pulse amplitudes at the output of the detector that enables qualitative and quantitative analyses of gamma ray emitting sources (IAEA, 2003).

gamma-ray spectroscopy The observation, recording, and quantitative analysis of a gamma-ray spectrum.

gamma-ray surveying Measurement of naturally occuring gamma rays in the search for radioactive materials.

gamma-sulfur (gam'-ma-sul'-fur) *rosickyite*.

gamont The generation of foraminifers that forms gametes in sexual reproduction, commonly with megalospheric test (*A-form*). Cf: *schizont*.

Gamowian A fanciful division of cosmological time from the end of the *Planckian* (first split-second of the Big Bang) to the beginning of the *Hadean* (Emiliani, 1987, p.86). The Gamowian is named after George Gamow (1904-1968).

gananite (ga'-na-nite) A semimetallic brown, black, or greenish black cubic mineral: BiF_3.

gangmylonite (gang-my'-lo-nite) An ultramylonite or mylonite that shows intrusive relations with the adjacent rock with no evidence of fusion (Hammer, 1914).

gangue The valueless rock or mineral aggregates in an ore; that part of an ore that is not economically desirable but cannot be avoided in mining. It is separated from the ore minerals during concentration. Syn: *matrix [ore dep]*. Cf: *ore mineral*.

ganister (gan'-is-ter) (a) A hard, fine-grained quartzose sandstone or quartzite, used in the manufacture of silica brick. It is composed of subangular quartz particles (0.15-0.5 mm in diameter, although some authors use a lower limit of 0.05 mm, thereby extending into the silt-size range), cemented with secondary silica, and possessing a characteristic splintery fracture that gives rise to smooth, subconchoidal surfaces and sharp edges. Ganister is distinguished from chert by its more granular texture and by the relatively small quantity of chalcedonic or amorphous silica. (b) In England, a highly siliceous *seat earth* of coal seams, e.g. the Sheffield ganister of the Lower Coal Measures of Yorkshire. See also: *pencil ganister; bastard ganister; silica rock; galliard*. (c) A mixture of ground quartz and fireclay used as a furnace lining. Also spelled: gannister.

ganoid (gan'-oid) adj. Pertaining to fish scales of rhomboid shape, with a heavy outer coat of enamel. n. (obsolete) Any bony fish with this type of scale, mostly chondrosteans and holosteans but including acanthodians and some crossopterygians.

ganomalite (ga-nom'-a-lite) A colorless to gray hexagonal mineral: $Pb_9Ca_5Mn^{2+}Si_9O_{33}$.

ganophyllite (gan-o-phyl'-lite) A light brown to pale pink platy monoclinic mineral: $(K,Na)(Mn,Al,Mg)_4(Si,Al)_6O_{16}(OH)_2{\cdot}3H_2O$.

ganterite A light gray to silvery monoclinic mineral of the *mica* group: $Ba_{0.5}(Na,K)_{0.5}Al_2(Si_{2.5}Al_{1.5})O_{10}(OH)_2$.

gaotaiite A metallic steel-black cubic mineral of the *pyrite* group: Ir_3Te_8.

gap [coast] (a) A narrow passage or channel between an island and the shore. (b) A break in a levee through which a stream distributary may flow; a tidal inlet.

gap [geomorph] (a) A sharp break or opening in a mountain ridge, or a short pass through a mountain range; e.g. a *wind gap*. Cf: *notch [geomorph]; col*. (b) A ravine or gorge cut deeply through a mountain ridge, or between hills or mountains; e.g. a *water gap*. Cf: *gate [geomorph]*.

gap [marine geol] *abyssal gap*.

gap [stratig] *break [stratig]*.

gap [struc geol] The horizontal component of displacement on a normal or normal/oblique-slip fault that results in a gap in a map of the faulted interval. Syn: *stratigraphic heave*.

gape (a) A localized opening remaining between the margins of a bivalve shell of a mollusk or ostracode when the valves are shut or drawn together by adductor muscles. (b) In brachiopods, the anterior and lateral space between the valves when they are open. (c) In vertebrates, the distance between the opened upper and lower jaws.

gara (ga'-ra) A mushroom-shaped rock occurring in arid or desert regions, resulting from the undercutting of soft rock by wind-driven sand, esp. if the soft rock is overlain by more resistant strata. Pl: gour. Cf: *mushroom rock*.

G/A ratio The principle of the geometric increase in ore tonnage relative to the arithmetic decrease of ore grade, as applied mainly to sedimentary, some residual, and disseminated orebodies (Lasky, 1950).

garavellite (gar-a-vel'-lite) A metallic orthorhombic mineral of the *berthierite* group: $FeSbBiS_4$.

garbage Nonhazardous animal and vegetable wastes from cooking and food processing; garbage contains up to 70% moisture and up to 5% noncombustible solids.

garbenschiefer (gar'-ben-schief'-er) (a) *feather amphibolite*. (b) A type of *spotted slate* characterized by concretionary spots whose shape resembles that of a caraway seed. Etymol: German. Cf: *knotenschiefer; fleckschiefer; fruchtschiefer*.

gardening (gar'-den-ing) A phenomenon in which the lunar regolith is constantly and very slowly churning due to successive impacts whereby bottom material works its way to the top and surface material gets buried.

Gardner's equation The empirical relationship that density is proportional to the $^1/_4$ power of P-wave velocity.

Gargasian (Gar-ga'-si-an) A European substage (France, Switzerland): Lower Cretaceous (middle to upper Aptian; above Bedoulian Substage, below Clansayesian).

gargulho (gar-gu'-lho) A Brazilian term used in the plateau region of Bahia for a comparatively coarse, clay-cemented ferruginous conglomerate in which diamonds are found.

garland (gar'-land) *stone garland*.

garnet (gar'-net) (a) A group of minerals of formula: $A_3B_2(SiO_4)_3$, where A = Ca, Mg, Fe^{2+}, or Mn^{2+}, and B = Al, Fe^{3+}, Mn^{3+}, V^{3+}, or Cr^{3+}. (b) Any of the minerals of the *garnet* group, such as the end members almandine (Fe-Al), andradite (Ca-Fe), grossular (Ca-Al), pyrope (Mg-Al), spessartine (Mn-Al), uvarovite (Ca-Cr), and goldmanite (Ca-V). Garnet is a brittle and transparent to subtransparent mineral, having a vitreous luster, no cleavage, and a variety of colors, dark red being the most common. It occurs as an accessory mineral in a wide range of igneous rocks, but is most commonly found as distinctive euhedral cubic crystals in metamorphic rocks (gneiss, mica schist, marble); it may also be massive or granular. Garnet is used as a semiprecious stone and as an abrasive.

garnetiferous (gar-net-if'-er-ous) Containing garnets; e.g. "garnetiferous peridotite".

garnetite (gar'-net-ite) A metamorphic rock consisting chiefly of an aggregate of interlocking garnet grains. Cf: *tactite.*

garnet jade A light green variety of grossular garnet, closely approaching fine jadeite in appearance, esp. that found in Transvaal, South Africa.

garnet lherzolite A *lherzolite* containing garnet as the dominant aluminum-rich mineral; stable at greater depths in the mantle than *spinel lherzolite* and *plagioclase lherzolite.*

garnetoid (gar'-net-oid) A group name for minerals with structures similar to that of garnet; e.g. katoite and berzeliite.

garnierite (gar'-ni-er-ite') (a) A group name for poorly defined hydrous nickel-magnesium silicates. (b) An apple-green or pale-green substance, probably: $(Ni,Mg)_6(OH)_8Si_4O_{10}$. The ratio Ni:Mg is highly It is sometimes considered to be a nickel-rich antigorite. It is an important ore of nickel and is used as a gemstone. See also: *genthite.* Syn: *noumeite; népouite; pimelite.*

garrelsite (gar'-rel-site) A colorless monoclinic mineral: $Ba_3NaSi_2B_7O_{16}(OH)_4$.

garronite (gar'-ron-ite) A colorless tetragonal zeolite mineral: $Na_2Ca_5Al_{12}Si_{20}O_{64}\cdot 27H_2O$.

Garside's rule The arrangement in tetrads of tri-aperturate pollen grains whereby the apertures occur in groups of three at four points in the tetrad (apparently restricted to Proteaceae). Cf: *Fischer's rule.*

gartrellite (gar-trell'-ite) A greenish-yellow triclinic mineral: $PbCu(Fe^{3+},Cu)(AsO_4)_2(OH,H_2O)_2$.

garyansellite (ga'-ry-an'-sell-ite) A brown orthorhombic mineral of the *phosphoferrite* group: $(Mg,Fe^{3+})_3(PO_4)_2(OH,H_2O)_3$.

gas *natural gas.*

gas barren An area, as large as several acres, that is characterized by a lack of vegetation due to fumarolic activity and acid leaching of the surface rocks.

gas cap Free gas overlying liquid hydrocarbons in a reservoir under trap conditions.

gas-cap drive Energy within an oil pool, supplied by expansion of an overlying volume of compressed free gas. Cf: *dissolved-gas drive; water drive.*

gas chimney [geophys exploration] A region of low-concentration gas escaping and migrating upward from a gas accumulation. Also described as gas cloud. Generally observed as a region of severely deteriorated seismic data quality associated with low velocity and with velocity sags underneath the chimney.

gas chromatography A process for separating gases or vapors from one another by passing them over a solid (*gas-solid chromatography*) or liquid (*gas-liquid chromatography*) phase. The gases are repeatedly adsorbed and released at differential rates resulting in separation of their components. Abbrev: GC. See also: *liquid chromatography; chromatography.*

gas coal Bituminous coal that is suitable for the manufacture of flammable gas because it contains 33-38% volatile matter. Cf: *high volatile bituminous coal.* See also: *coal gas.*

gas-cut mud *Drilling mud,* returned from the open portion of a drill hole, that contains bubbles of natural gas. Most commonly the gas is emitted from strata near the bottom of the hole and is controlled by the use of heavier drilling mud or special surface equipment (e.g., a "gas-buster").

gas depletion *solution gas drive.*

gaseous transfer (gas'-e-ous) Separation from a magma of a gaseous phase that moves relative to the magma and releases dissolved substances, usually in the upper levels of the magma, when it enters an area of reduced pressure. See also: *pneumatolytic differentiation.*

gas field (a) A *gas pool.* (b) Two or more gas pools on a single geologic feature or otherwise closely related.

gas fluxing A rapid upward streaming of free juvenile gas through a column of molten magma in the conduit of a volcano or within a magma chamber. The gas acts as a flux to promote melting of the wall rocks. Syn: *volcanic blowpiping.*

gas giant *giant planet.*

gash breccia A term used in Pembrokeshire, England, for a rock believed to have originated from the collapse of walls and roofs of caves that had been eroded by the solvent action of underground water.

gas heave The distortion of prodelta sediments, produced by the weight of a distributary-mouth bar compacting the underlying sediment and causing carbon dioxide to escape (Moore, 1966, p.101). See also: *air heave.*

gas-heave structure A sedimentary structure, restricted to the smaller distributaries of the Mississippi River delta, produced by gas heave. See also: *air-heave structure.* Cf: *mudlump.*

gash fracture A small-scale tension fracture that occurs at an angle to a fault and tends to remain open. Syn: *open gash fracture.*

gash vein A nonpersistent vein that is wide above and narrow below, and that terminates within the formation it traverses. The term was originally applied to vein fillings of solution joints in limestone.

gas hydrate Hydrated methane in a solid state. A latticelike chemical structure in which one molecular group is surrounded by others, such as H_2O by CH_4. Such crystalline, icelike clathrate compounds can exist at low temperature and high pressure, as in deep water areas, where they often give rise to a *bottom-simulating reflection* (BSR). Sometimes found in permafrost areas. Hydrates also tend to clog and block flow lines and pipelines. (Sheriff and Geldart, 1995).

gasification (gas'-i-fi-ca'-tion) Any process by which a combustible gas is made from coal or other solid hydrocarbons.

gas inclusion A gas bubble within a gemstone, often visible to the unaided eye as in the three-phase inclusions characteristic of some South American emeralds; esp. seen in many melt-grown synthetic stones, often enabling them to be distinguished from their natural counterparts.

gas-liquid chromatography Process in which a gas, such as helium, argon, hydrogen, or nitrogen, carrying a gaseous mixture to be resolved, is passed over a nonvolatile liquid coated on a porous inert solid support (May and Cuttitta, 1967, p.115) where the components are separated by differential mobility rates. Abbrev: GLC. See also: *gas-solid chromatography; gas chromatography.*

gasoclastic sediment (gas-o-clas'-tic) A sediment resulting from sedimentary volcanism, such as mud ejected by enormous volumes of gas (Bucher, 1952, p.87).

gas-oil contact The boundary surface between an accumulation of oil and an overlying accumulation of natural gas. Syn: *gas-oil interface.*

gas-oil interface *gas-oil contact.*

gas-oil ratio (a) The quantity of gas produced with oil from an oil well, usually expressed as the number of cubic feet of gas per barrel of oil. Abbrev: GOR. (b) *reservoir gas-oil ratio.*

gasparite-(Ce) (gas'-par-ite) A light brown-red, monoclinic mineral of the *monazite* group: $(Ce,La,Nd)AsO_4$.

gaspeite (gas-pe'-ite) A light green rhombohedral mineral: $(Ni,Mg)CO_3$. A member of the *calcite* group.

gas phase [volc] That stage in a volcanic eruption that is characterized by the release of large amounts of *volcanic gas.* Cf: *phase [ign].*

gas pit A circular pit, 2.5-30 cm in diameter and from less than 3 cm to more than 30 cm deep, surrounded by a mound of mud and produced by the escape of gas bubbles (as of methane generated during the decomposition of organic matter) rising from the surface of a mud bar (Maxson, 1940a).

gas pool A subsurface accumulation of natural gas that will yield gas in economic quantities. Cf: *gas field.*

gas sag An apparent depression under a gas accumulation resulting from the lowering of velocity associated with the gas. A *hydrocarbon indicator* also called *time sag.*

gas sand A sand or sandstone containing a large quantity of natural gas.

gassi (gas'-si) A term used in the Saharan region for a sand-free interdune passage, in some cases traversing an entire *erg* from end to end. Etymol: Arabic, "neck, closed ground"'. Pl: gassis. Cf: *feidj.*

gas-solid chromatography Process in which a gas, such as helium, argon, hydrogen, or nitrogen, carrying a gaseous mixture to be resolved, is passed over a porous adsorbing solid (May and Cuttitta, 1967, p.114), where the components are separated by differential mobility rates. Abbrev: GSC. See also: *gas-liquid chromatography; gas chromatography.*

gas streaming A process of magmatic differentiation involving the formation of a gaseous phase, usually during a late stage in consolidation of the magma, that results in partial expulsion, by escaping gas bubbles, of residual liquid from the crystal network; crystals are separated from the magma due to their upward transport by streaming gases. Obsolete.

gasteropod (gas'-ter-o-pod') *gastropod.*

gastral (gas'-tral) Pertaining to the surface of the *spongocoel* of a sponge; e.g. "gastral membrane" (endopinacoderm lining the spongocoel) and "gastral cavity" (the spongocoel itself).

gastralium (gas-tral'-i-um) (a) A specialized sponge *spicule* lining the *cloaca.* (b) A dermal ossification ("abdominal rib") in the belly wall of diverse vertebrates; usually used in the plural as a collective term. Pl: gastralia.

gas trap A *trap [petroleum]* that contains gas.

gastric region (gas'-tric) The median part of the carapace in some decapod crustaceans, in front of the *cervical groove* and behind the *frontal region.* It is divided by some authors into epigastric, mesogastric, metagastric, protogastric, and urogastric subregions (TIP, 1969, pt. R, p.92).

gastrolith (gas'-tro-lith) A rounded stone or pebble, commonly highly polished, from the stomach of some reptiles, esp. dinosaurs, plesiosaurs, and crocodilians. Gastroliths are thought to have been used in grinding up food, but marine reptiles may have used them to provide body stability while in the water (Darby and Ojakangas, 1980). Syn: *stomach stone; gizzard stone.*

gastro-orbital groove (gas'-tro-or'-bit-al) A groove in decapod crustaceans that extends forward from the upper part of the *cervical groove* in the direction of the orbit. It forms the upper border of the elevated area behind the antennal spine (Holthuis, 1974, p.733).

gastropod (gas'-tro-pod) Any mollusk belonging to the class Gastropoda, e.g., a snail. Also spelled: *gasteropod.*

Gastropoda A class of *cyrtosome* mollusks have body contained in asymmetric helically coiled shell with apex pointing posteriorly away from head. Head distinct; can be moved independently of rest of body. Foot solelike and used for creeping. Visceral mass torted, that is, rotated up to 180 degrees about vertical axis with respect to foot so that anus and organs of the mangle cavity are above head. Range, Early Cambrian to Holocene.

gastropore (gas'-tro-pore) A relatively large tubular cavity of certain hydrozoans, providing lodgment for a *gastrozooid.* Cf: *dactylopore.*

gastrovascular cavity (gas'-tro-vas'-cu-lar) The interior space (*coelenteron*) of a coral *polyp,* radially partitioned by septa, and functioning in both digestion and circulation.

gastrozooid (gas-tro-zo'-oid) A short, cylindrical feeding and digestive *polyp* housed in a gastropore of a hydrozoan. Cf: *dactylozooid.*

gas-water contact The boundary surface between an accumulation of gas and the underlying bottom water. Syn: *gas-water interface.*

gas-water interface *gas-water contact.*

gas well A well that is capable of producing natural gas or that produces chiefly natural gas. Some statutes define the term on the basis of the *gas-oil ratio.*

gat (a) A natural or artificial opening, as a strait, channel, or other passage, extending inland through shoals, or between sandbanks, or in the cliffs along a coast. (b) A strait or channel from one body of water to another, as between offshore islands or shoals, or connecting a lagoon with the sea. Syn: *gate [coast]; gut.*

gate [coast] (a) An entrance to a bay or harbor, located between promontories; e.g. the Golden Gate in San Francisco, Calif. (b) *gat.*

gate [geomorph] (a) A mountain pass affording an entrance into a country. (b) A broad, low valley or opening between highlands, generally wider than a gap. (c) A restricted passage along a river valley; e.g. the Iron Gate on the River Danube.

gate [geophys] The interval of record time over which a function (such as an autocorrelation or cross-correlation) is evaluated. Also called window. A gate where the boundaries are abrupt is called a *boxcar.*

gate [paleont] A large opening or fissure in the skeleton of a spumellarian radiolarian.

gatehouseite A pale yellow to orange orthorhombic mineral: $Mn^{2+}_5(PO_4,AsO_4)_2(OH)_4$. It is the arsenic analogue of arsenoclasite.

gatelite-(Ce) A colorless monoclinic mineral of the *epidote* group: $(Ca,Ce)_4(Al,Mg,Fe)_4(Si_2O_7)(SiO_4)_3(O,F,OH)_3$.

gather (gath'-er) A display of seismic traces, arranged so that all the traces corresponding to some criterion, such as source-detector distance, are displayed side by side. It is used for checking corrections and evaluating the components of a stack. A *common midpoint* gather displays data for the same midpoint after correction for normal moveout and statics; a "common-range" or "common-offset" gather displays data for the same offset but for a number of nearby reflection points.

gathering ground (gath'-er-ing) (a) An area over which water is collected from precipitation, springs, and surface drainage; esp. an area that supplies water to a reservoir. (b) *drainage basin.*

gathering zone A term suggested for the area between the land surface and the water table.

gating (gat'-ing) In marine seismic profiling and echo sounding, a method of eliminating near-surface background noise from bottom and sub-bottom echoes by restricting the time window of listening.

gatumbaite (ga-tum'-ba-ite) A pearly white monoclinic mineral: $CaAl_2(PO_4)_2(OH)_2{\cdot}H_2O$.

gaudefroyite (gau-de-froy'-ite) A dull black hexagonal mineral: $Ca_4Mn^{3+}_{3-x}(BO_3)_3(CO_3)(O,OH)_3$.

gauge pressure Pressure above one atmosphere.

gault A stiff, firm, compact clay, or a heavy, thick clay soil; specif. the Gault, a Lower Cretaceous clay formation in Great Britain. Also spelled: *galt.*

Gault A European (Great Britain) stage: Lower Cretaceous (above Wealden). Approximately equivalent to Albian and Aptian stages.

gaultite A colorless to pale mauve orthorhombic mineral: $Na_4Zn_2Si_7O_{18}{\cdot}5H_2O$.

Gause's principle In ecology, the statement that two identical species cannot coexist in the same area of space at the same time. Named after G.F. Gause (d. 1855), German geneticist. Var: Gause's rule; Gause's axiom.

gauss The cgs-emu unit of *magnetic induction* (or magnetic flux density). It is a measure of the magnetic force per unit area: 1 gauss = 1 maxwell/cm^2 = 10^5 gamma = 10^{-4} tesla = 10^{-4} weber/m. Also used in the cgs system as a unit of *magnetization,* or dipole moment per unit volume. Named for Karl Friedrich Gauss (1777-1855), German mathematician and natural philosopher.

gaussbergite (gauss'-berg-ite) An extrusive *lamproite* similar to *orendite* but with a glassy groundmass and containing phenocrysts of leucite, and of clinopyroxene and olivine in place of phlogopite. Its name is derived from Gaussberg volcano, Kaiser Wilhelm II Land, Antarctica. Obsolete.

Gauss Chron An interval of geologic time from about 3.58 to about 2.60 Ma, delimited by geomagnetic reversals and dominated by a normal-polarity magnetic field of the present. Named for Karl Friedrich Gauss (1777-1855), German physicist. Symbol: C2An.

Gaussian distribution (Gauss'-i-an) A *bell-shaped distribution* based on two parameters, the *mean* and the *standard deviation*. The distribution characterizes repeated measurements, subject to random, additive errors, and has a fundamental place in *parametric statistics.* Named for Karl Friedrich Gauss (1777-1855), German mathematician. Syn: *normal distribution*.

Gauss projection Any of several conformal map projections used or developed by Karl Friedrich Gauss (1777-1855), German mathematician; esp. the "Gauss-Krüger projection", a special case of the *transverse Mercator projection* derived by the direct conformal representation of the spheroid on a plane.

gauteite (gau'-te-ite) A porphyritic hypabyssal rock, probably *trachyte* or *trachyandesite,* characterized by a bostonitic groundmass predominantly of plagioclase, more than in a typical *bostonite,* along with magnetite, clinopyroxene, hornblende, and biotite and with phenocrysts of hornblende, clinopyroxene, some biotite, and abundant plagioclase. Analcime is also a common constituent. The name, given by Hibsch in 1898, is for Gaute (now Kout), Czech Republic. Less recommended usage.

gaylussite (gay'-lus-site) A yellowish-white to gray monoclinic mineral: $Na_2Ca(CO_3)_2{\cdot}5H_2O$.

Gazdag migration *phase-shift migration.*

GC *gas chromatography.*

geanticline (ge-an'-ti-cline) (a) A mobile upwarping of the crust of the Earth, of regional extent. Ant: *geosyncline.* (b) More specifically, an anticlinal structure that develops in geosynclinal sediments, due to lateral compression. Var: *geoanticline.* Obsolete.

gearksutite (ge-ark'-su-tite) An earthy white triclinic mineral: $CaAl(OH)F_4{\cdot}H_2O$. It occurs with cryolite.

gebhardite (geb'-hard-ite) An adamantine brown monoclinic secondary mineral, originally found at Tsumeb, Namibia: $Pb_8(As^{3+}_2O_5)_2OCl_6$.

gedanite (ged'-a-nite) A brittle, wine-yellow variety of amber with very little succinic acid.

Gedinnian (Ge-din'-ni-an) A European stage: Lower Devonian (above Pridolian of the Silurian, below Siegenian).

gedrite (ged'-rite) A white, gray, green, or brown prismatic monoclinic mineral: $(Mg,Fe^{2+})_2(Mg,Fe^{2+},Al)_5(Si_6Al_2)O_{22}(OH)_2$. A member of the *amphibole* group.

gedroitzite (ged-roit'-zite) A clay substance of the *vermiculite* group. It is characteristic of many alkali soils from the Ukraine. Also spelled: gedroizite.

geerite (geer'-ite) A metallic bluish-white rhombohedral (pseudo-cubic) mineral: $Cu_{8.5}S_5$.

geest (a) Alluvial material that is not of recent origin lying on the surface. An example is the sandy region of the North Sea coast in Germany. (b) *saprolite.*

geffroyite (gef'-froy-ite) A metallic cubic mineral of the *pentlandite* group: $(Ag,Cu,Fe)_9(Se,S)_8$.

gehlenite (geh'-len-ite) A grayish-green to brown tetragonal mineral of the *melilite* group: $Ca_2Al(AlSi)O_7$. It is isomorphous with akermanite. Syn: *velardeñite.*

Geiger counter (Gei'-ger) *Geiger-Müller counter.*

geigerite (gei'-ger-ite) A pale rose-red triclinic mineral: $Mn^{2+}_5(AsO_4)_2(AsO_3OH)_2{\cdot}10H_2O$. It is the Mn analogue of chudobaite.

Geiger-Müller counter (Gei'-ger-Mül'-ler) A radiation detector consisting of a gas-filled tube with a cathode envelope and an axial wire anode. It functions by producing momentary current pulses caused by ionizing radiation. Syn: *G-M counter; Geiger counter.*

geikielite (gei'-kie-lite) A bluish-black or brownish-black rhombohedral mineral of the *corundum* group: $MgTiO_3$. It is isomorphous with ilmenite, and often contains much iron. It is usually found in rolled pebbles.

GEK *geomagnetic electrokinetograph.*

gel (a) A translucent to transparent, semisolid, apparently homogeneous substance in a colloidal state, generally elastic and jellylike, offering little resistance to liquid diffusion, and containing a dispersion or network of fine particles that have coalesced to some degree. (b) A nonhomogeneous gelatinous precipitate; e.g. a coagel. (c) A liquified mud, which became firm and then reabsorbed most of the water released earlier. A gel is in a more solid form than a *sol,* and can sustain limited shear stress.

Gelasian (Gel-as'-i-an) A stage of the Standard Global Chronostratigraphic Scale: upper Pliocene (above Piacenzian, below Pleistocene Series). The basal GSSP is in Sicily, Italy (Rio et al., 1998). Obsolete syn: *Astian.*

gelation [chem] (ge-la'-tion) The formation of a *gel* from a *sol,* as by coagulation or by precipitation with an electrolyte.

gelation [ice] *congelation.*

gelbstoff *yellow substance.*

geli- (gel'-i-) Of or pertaining to cold, frost action, or permafrost; e.g. *geli*vation.

Ge:Li detector *lithium-drifted germanium detector*. Cf: *scintillation counter.*

gelification index A parameter derived from maceral percentages, defined as (vitrinite+macrinite)/(semifusinite+fusinite+inertodetrinite), describing the ratio of gelified material to nongelified material. Developed by Diessel (1986) for Permian coals in the Sydney Basin (New South Wales, Australia) but widely applied to coals of other ages and locations.

gelifluction (gel-i-fluc'-tion) The progressive lateral flow of earth material under periglacial conditions; solifluction in a region underlain by frozen ground. It is commonly used as a modifying or combining term with bench, lobe, sheet, slope, and stream to indicate periglacial origin by soil flow. Also spelled: gelifluxion. Syn: *congelifluction.*

gelifraction (gel-i-frac'-tion) The mechanical disintegration, splitting, or breakup of a rock or soil due to the great pressure exerted by the freezing of water contained in cracks or pores, or along bedding planes; term introduced by Bryan (1946, p.640). Syn: *frost shattering; frost splitting; frost riving; frost bursting; frost weathering; frost wedging; gelivation; congelifraction.*

gelisol (gel'-i-sol) *frozen ground.*

gelisolifluction (gel'-i-so'-li-fluc'-tion) *congelifluction.*

gelite (gel'-ite) A name for opal (or chalcedony?) as a secondary deposit in rocks (Hey, 1962, p.435).

geliturbation (gel'-i-tur-ba'-tion) *cryoturbation.*

gelivation (gel-i-va'-tion) *gelifraction.*

gelivity (ge-liv'-i-ty) The sensitivity or susceptibility of rock to congelifraction; the property of being readily split by frost (Hamelin and Clibbon, 1962, p.219).

gel mineral *mineraloid.*

gelohuminite A subgroup of *huminite* originated from the gelling of humic solutions and not corresponding to specific plant tissues (Pickel, 2004).

gélose (gel'-ose) *ulmin.*

gélosic coal (gé-lo'-sic) Coal that is rich in gélose or ulmin; a syn. of algal coal or *boghead coal.*

gelosite (gel'-o-site) A microscopic constituent of torbanite, consisting of squashed, translucent, pale-yellow spheres of birefringent material (Dulhunty, 1939).

gelovitrinite A subgroup of *vitrinite* originated from the gelling of humic solutions and not corresponding to specific plant tissues (ICCP, 1998).

gem n. (a) A cut-and-polished gemstone that has intrinsic value and possesses the necessary beauty, durability, rarity, and size for use in jewelry as an ornament or for personal adornment; a jewel whose value is not derived from its setting. (b) An especially fine or superlative gemstone specimen (as compared to others cut from the same species), generally of superb color, unusual internal quality, and fine cut; e.g. a gem turquoise having a pure uniform blue of the highest possible intensity. In this usage, the meaning depends on the ethics and the range of qualities handled by the seller. adj. Said of a rough diamond that has the necessary shape, purity, and color to allow it to be used for cutting and polishing into a diamond usable in jewelry. Only about 20% of all diamonds mined meet this requirement. (c) Unfashioned natural gem material; gem rough.

gem color Any color of a gem material that is attractive enough to be used in jewelry. Cf: *perfection color.*

gem crystal A crystal from which a gem can be cut.

gem gravel A gravel *placer* containing an appreciable concentration of gem minerals.

geminite A green triclinic mineral: $Cu^{2+}(AsO_3OH)\cdot H_2O$.

gemma (gem'-ma) An asexual propagule sometimes appearing as, but not homologous with, a vegetative bud (Lawrence, 1951, p.753).

gemmate (gem'-mate) adj. (a) Said of sculpture of palynomorphs consisting of more or less spherical projections. (b) Having gemmae. v. To produce or propagate by a bud.

gem material Any rough, natural material that can be fashioned into a jewel.

gemmation (gem-ma'-tion) Asexual reproduction involving the origination of a new organism as a localized area of growth on or within the body of the parent and subsequently differentiating into a new individual, as in some corals.

gemmiferous (gem-mif'-er-ous) Producing or containing gems.

gem mineral Any mineral species that yields varieties with sufficient beauty and durability to be classed as gemstones.

gemmology (gem-mol'-o-gy) British var. of *gemology.*

gemmule (gem'-mule) An internal resistant asexual reproductive body of a sponge, consisting of a mass of archaeocytes charged with reserves and enclosed in a noncellular protective envelope.

gemmy (gem'-my) Having the characteristics (such as hardness, brilliance, diaphaneity and color) desired in a gemstone.

gemologist (gem-ol'-o-gist) One who has successfully completed recognized courses in *gemology* and has demonstrated competence in the identification and evaluation of gem materials. British spelling: gemmologist.

gemology (gem-ol'-o-gy) The science and study of gemstones, including their source, description, geology, origin, identification, grading, and appraisal. British spelling: *gemmology.*

gem quality A term describing rough gem material which has the weight, shape, clarity, and color to produce a polished stone suitable for use in jewelry. Sometimes used to describe fine diamonds.

gemstone (gem'-stone) Any mineral or rock that, when cut and polished, has sufficient beauty, rarity, and durability for use as a personal adornment or other ornament. Organic materials such as pearl are considered gem materials rather than gemstones.

gem variety The variety of a mineral species that yields gemstones.

gena (ge'-na) A *cheek* of a trilobite. Pl: genae.

genal angle (ge'-nal) The posterior lateral corner of the *cephalon* of a trilobite (typically terminating in a *genal spine,* but may be rounded), or of the *prosoma* of a merostome.

genal spine A spine extending backward from the *genal angle* of the *cephalon* of a trilobite, produced on the outer posterior margin of the *free cheek.*

gendarme (gen'-darme [Pron: zhan-darm']) A sharp rock pinnacle on an arête, such as will retard or prevent progress along the crest of the arête; it is less pointed and more towerlike than an *aiguille,* and is commonly found in the Alps. Etymol: French, "policeman".

gene The fundamental unit governing the transmission and development or determination of hereditary characteristics. Genes occur in a linear sequence on the chromosomes of a cell nucleus, and are now thought to originate in the deoxyribonucleic acid (DNA) component in the chromosomes.

genealogical hierarchy A nested set of entities from genes to organisms to species to clades which are produced by the replication of entities subsumed within them, and which are engaged in the processes of information preservation, promulgation, and expression.

gene complex The system comprising all the interacting genetic factors of an organism.

genera (gen'-er-a) The plural of *genus.*

general age equation (gen'-er-al) The relationship between *radioactive decay* and *geologic time* (Dickin, 1995). Mathematically expressed it is $t = 1/\lambda\ ln[(D_t\text{-}D_o/P)+ 1]$, where t is time, λ is the decay constant for the isotope in question, D_t is the number of daughter atoms at time (t), D_o is the number of daughter atoms initially present, and P is the number of parent atoms at time t, and *ln* is the natural logarithm (logarithm to base e). Syn: *age equation.*

general base level *ultimate base level.*

general equivalent position In a *space group,* a location that does not fall on any symmetry element and thus is mapped into a physically distinct (but symmetrically equivalent) location by each symmetry element of the space group. The *point symmetry* of a general equivalent position is 1. Cf: *special position.*

general form The crystal form in each crystal class that has different intercepts on each crystal axis, and has the largest number of equivalent crystal faces for the symmetry present. The form displays and is characteristic of the point group symmetry; other forms may display more symmetry.

generalist species (gen'-er-al-ist') *opportunistic species.*

generalization *cartographic generalization.*

general-purpose geologic map (gen'-er-al-pur'-pose) A map designed to provide a large amount of geological information for many purposes (land-use planning, petroleum geology, highway location, etc.). In the U.S. its scale is usually 1:24,000 and not smaller than 1:62,500.

generation (gen-er-a'-tion) All the crystals of the same mineral species that appear to have crystallized at essentially the same time; e.g., if there are olivine phenocrysts in a groundmass containing olivine, there are said to be two "generations" of olivine.

generic (ge-ner'-ic) Pertaining to a genus.

generic [geog] The part of a feature name that identifies what kind of feature it is; e.g., "Creek", "Hill", "Trail", etc.

genetic (ge-net'-ic) Pertaining to relationships due to a common origin, or to features involving genes or gene complexes.

genetic algorithms A class of global optimization algorithms which begin with a loose analogy between optimization and a biological system composed of a relatively few organisms that react in a relatively complex way. Algorithms try to evolve a population of trial members in a way mimicking biological evolution (Smith et al., 1992).

genetic drift Gradual change with time in the genetic composition of a continuing population resulting from the elimination of some genetic features and the appearance of others, and appearing to be unrelated to the environmental benefits or detriments of the genes involved.

genetic pan A natural *pan [soil]* of low permeability, with a high concentration of small particles, and differing in certain physical and chemical properties from the soil immediately above or below it (SSSA, 1975, p.12); e.g. claypan or fragipan. Cf: *pressure pan.*

genetic physiography *geomorphogeny.*

genetics (ge-net'-ics) The science that deals with the materials and processes of inheritable characteristics or features from generation to generation.

genetic type A sedimentary body representing a complex of genetically related facies formed in the same environment (on land or in the sea) and mostly as the effect of a single leading process (Krasheninnikov, 1964, p.1245); e.g. alluvial deposits, deltaic deposits, lagoonal deposits, or marine deposits. The concept of genetic types, long established in Russian geology, was introduced by Pavlov (1889) to demonstrate the diversity of continental deposits according to their origin.

genicular spine (ge-nic'-u-lar) A spine originating from the *geniculum* of a graptolite.

geniculate [paleont] (ge-nic'-u-late) Bent abruptly at an angle, as in a knee; specif. said of a brachiopod shell characterized by an abrupt and more or less persistent change in direction of valve growth, producing an angular bend in lateral profile. In vertebrates, describing a ganglion of the facial (seventh cranial) nerve.

geniculate [palyn] adj. In dinoflagellate cysts, a term for the margin of a plate of one latitudinal series when it contacts two plates of an adjacent latitudinal series; the plate margin is thus peaked (=geniculate).

geniculate coniform element Coniform conodont element in which posterior margin of cusp joins upper edge of base to enclose an acute angle (TIP, 1981, pt. W, supp.2).

geniculate twin A type of crystal twin that bends abruptly, e.g. crystals of rutile that are twinned on the second-order pyramid (101). Syn: *elbow twin.*

geniculation point Point of inflection, in lateral view, of the platform margin of planate pectiniform conodont elements of some species of *Polygnathus*; margin slopes upward from its posterior end to geniculation point, then is sharply deflected downward; geniculation point marks anterior limit of ornamented part of platform; anterior of platform consists of adcarinal grooves and their upturned margins (TIP, 1981, pt. W, supp.2).

geniculum [paleont] (ge-nic'-u-lum) A small knee-shaped structure or abrupt bend in an organism; e.g. a distinct change in the direction of growth of the free ventral wall of a graptolithine *theca,* or between two successive segments of erect jointed coralline algae. Pl: genicula.

geniculus [palyn] (ge-nic'-u-lus) A small equatorial peculiarity in the pollen colpus that in compressed specimens resembles a bent knee. It appears to result from a small inflation or evagination at the equator. Pl: geniculi.

genital plate (gen'-i-tal) (a) One of the inner circlet of primordial plates of the *apical system* of an echinoid, arranged in an interradial position and usually perforated by one or more pores for discharge of reproductive products. Cf: *ocular plate.* (b) One of the circlet of five plates on the aboral surface immediately around the *centrale* of an asteroid.

genkinite (gen'-kin-ite) A metallic tetragonal mineral: $(Pt,Pd)_4Sb_3$.

genobenthos (gen-o-ben'-thos) All terrestrial organisms.

genocline (gen'-o-cline) A *cline* caused by hybridization between adjacent populations that are genetically distinct.

genomict breccia A meteorite breccia in which all clasts appear to be from the same parent asteroid but differ in their textures and thus their alteration histories (e.g., because formed or stored under different physical conditions).

genotype (gen'-o-type) (a) *type species.* (b) In genetics, the genetic constitution of an organism or a species in contrast to its observable physical characteristics. See also: *biotype.*

genthelvite (gent-helv'-ite) A white, red, green, or yellow cubic mineral of the *sodalite* group: $Zn_4Be_3(SiO_4)_3S$. It is the zinc end member, isomorphous with helvite and danalite.

genthite (genth'-ite) A soft, amorphous, pale-green or yellowish mixture of hydrous nickel-magnesium silicates, representing a substance near: $(Ni,Mg)_4Si_3O_{10}{\cdot}6H_2O$. See also: *garnierite.*

gentle fold (gen'-tle) A fold with an interlimb angle between 120° and 180° (Fleuty, 1964a, p.470). Obsolete.

genus (ge'-nus) A category in the hierarchy of plant and animal classification intermediate in rank between family and species. Adj: *generic.* Abbrev: gen. Plural: *genera.* Cf: *subgenus.*

genus zone The taxon-range-zone of a genus (ISG, 1976, p.54).

geo- (ge'-o-) A prefix meaning "Earth".

geoacoustics (ge'-o-a-cou'-stics) The study of the sounds of the Earth, e.g., those of a volcanic eruption.

geoanticline (ge-o-an'-ti-cline) Var. of *geanticline.*

geoarchaeology (ge-o-ar-chae-ol'-o-gy) Those disciplines of *archaeology* relating to issues and methods of the *geosciences.*

geobarometer (ge'-o-ba-rom'-e-ter) Any chemical or physical parameter which can be experimentally calibrated and used to estimate the pressure of formation of a rock or mineral.

geobarometry (ge'-o-ba-rom'-e-try) Any method, such as the interpretation of pressure-sensitive chemical or mineral reactions, for the direct or indirect determination of the pressure conditions under which a rock or mineral formed. The term is analogous to *geothermometry.*

geobasin (ge-o-ba'-sin) A structure of geosynclinal type but without folding of its sedimentary filling (Rich, 1938). The term is no longer used.

geobattery (ge-o-bat'-ter-y) An electrical potential generated by a conductive sulfide orebody, with an upper positive cathode and a lower negative anode. Such potentials are readily measured and are sensitive to local conditions. They may be useful in exploration (Sivenas and Beales, 1982).

geobiology (ge'-o-bi-ol'-o-gy) The study of the biosphere, esp. through geologic time. Cf: *biogeology; paleobiology; paleontology.*

geobios (ge-o-bi'-os) That area of the Earth occupied by terrestrial plants and animals. Cf: *hydrobios.*

geobleme (ge'-o-bleme) A term used by Bucher (1963) for a cryptoexplosion structure caused from within the Earth. Cf: *astrobleme.*

geobotanical prospecting (ge'-o-bo-tan'-i-cal) The visual study of plants, their morphology, and their distribution as indicators of such things as soil composition and depth, bedrock lithology, the possibility of ore bodies, and climatic and groundwater conditions. Cf: *biogeochemical prospecting.* See also: *botanical anomaly.*

geocentric (ge-o-cen'-tric) Referring to the center of mass of the Earth, in defining coordinate systems. When combined with latitude, as in "geocentric latitude", it is the angle at the center of the Earth between the plane of the *celestial equator* and a line to a point on the surface of the Earth.

geocentric geodetic datum A *geodetic datum* that specifies the center of the reference ellipsoid shall be located at the Earth's center of mass.

geocentric horizon The plane passing through the center of the Earth, parallel to the apparent horizon.

geocentric latitude (a) The *latitude* or angle at the center of the Earth between the plane of the celestial equator and a line (radius) to a given point on the Earth's surface. It would be identical with *geodetic latitude* only for a truly spherical Earth. Symbol: ψ. (b) The celestial latitude of a body based on or as seen from the Earth's center.

geocentric longitude (a) *geodetic longitude.* (b) The celestial longitude of a body based on or as seen from the Earth's center.

geocerain (ge-o-cer'-ain) *geocerite.*

geocerite (ge-o-cer'-ite) A white, flaky, waxlike resin of approximate composition: $C_{27}H_{53}O_2$. It is found in brown coal. Syn: *geocerain.*

geochemical analysis (ge-o-chem'-i-cal) The application of chemical analysis to geological problems.

geochemical anomaly A concentration of one or more elements in rock, soil, sediment, vegetation, or water that is markedly higher than background. The term may also be applied to hydrocarbon concentrations in soils.

geochemical balance The study of the proportional distribution, and the rates of migration, in the global fractionation of a particular element, mineral, or compound; e.g. the distribution of quartz in igneous rocks, its liberation by weathering, and its redistribution to sediments and, in solution, to terrestrial waters and the oceans.

geochemical cycle The sequence of stages in the exchange of elements among major geochemical reservoirs. Rankama and Sahama (1950) distinguish a major cycle, proceeding from magma to igneous rocks to sediment to sedimentary rocks to metamorphic rocks, and possibly through migmatites and back to magma; and a minor or exogenic cycle proceeding from sediment to sedimentary rocks to weathered material and back to sediments again.

geochemical exploration The search for economic mineral deposits or petroleum by detection of abnormal concentrations of elements or hydrocarbons in surficial materials or organisms, usually accomplished by instrumental, spot-test, or "quickie" techniques that may be applied in the field. Syn: *geochemical prospecting.*

geochemical facies Any areal geological entity that is distinguishable on the basis of trace-element composition, radioactivity, or other geochemical property. A lithofacies could be defined chemically, but a geochemical facies is delimited by features that remain imperceptible lithologically.

geochemical fence In a *fence diagram [geochem],* a boundary between discrete fields that represent different phases.

Geochemical log A trade name for a combination log that records estimates of the concentrations of 10 elements: potassium, thorium, uranium (from the spectral gamma ray log), aluminum (by delayed neutron activation analysis), and silicon, calcium, iron, sulfur, titanium, and gadolinium (from the prompt-capture gamma ray spectrum measure after a 14-Mev neutron burst).

geochemical prospecting *geochemical exploration.*

geochemical survey The mapping of geochemical facies.

geochemistry (ge-o-chem'-is-try) As defined by Goldschmidt (1954, p.1), the study of the distribution and amounts of the chemical elements in minerals, ores, rocks, soils, water, and the atmosphere, and the study of the circulation of the elements in nature, on the basis of the properties of their atoms and ions; also, the study of the distribution and abundance of isotopes, including problems of nuclear frequency and stability in the universe. A major concern of geochemistry is the synoptic evaluation of the abundances of the elements in the Earth's crust and in major classes of rocks and minerals.

geochron (ge'-o-chron) An interval of geologic time corresponding to a *lithostratigraphic unit.* It may, like the age of that unit, vary from place to place (Williams, 1901, p.580, 583.) The term is also used for the meteorite Pb-Pb isotope diagram or list meteorite geochron. Syn: *terrestrial geochron.*

geochrone (ge'-o-chrone) A standard unit of geologic time, now obsolete. It was proposed by Williams in 1893 and was set equal to the duration of the Eocene.

geochronologic (ge'-o-chron'-o-log'-ic) Pertaining or relating to *geochronology.*

geochronologic interval The time span between two geologic events (ISG, 1976, p.14). Syn: *interval [stratig].*

geochronologic unit A division of time traditionally distinguished on the basis of the rock record as expressed by a *chronostratigraphic unit.* It is not a material unit but corresponds to the time span of an established chronostratigraphic unit (NACSN, 1983, Art. 80). Geochronologic units in order of decreasing rank are eon, era, period, epoch, and age. Names of periods and units of lower rank are the same as those of the corresponding chronostratigraphic units; the names of some eras and eons are independently formed. Cf: *chron.* Syn: *geologic-time unit.*

geochronology (ge'-o-chro-nol'-o-gy) The science of dating and determining the time sequence of events in the history of the Earth (ISSC, 1994, p.16). Cf: *chronology; geochronometry.*

geochronometer (ge'-o-chro-nom'-e-ter) A physical feature, material, or element whose formation, alteration, or destruction can be calibrated or related to a known interval of time. Cf: *radioactive clock.*

geochronometric unit (ge'-o-chron'-o-met'-ric) A unit established through the direct division of geologic time, expressed in years (NACSN, 1983,

Art. 96). It is not based on the time span of a designated chronostratigraphic unit, but is simply a time division of convenient magnitude, as in development of a time scale for the Precambrian. Boundaries are arbitrarily chosen or agreed-upon ages in years.

geochronometry (ge'-o-chro-nom'-e-try) The branch of geochronology that deals with the quantitative (numerical) measurement of geologic time in thousands of millions of years (ISSC, 1994, p.16). Cf: *fossil geochronometry; geochronology.*

geocline (ge'-o-cline) A *cline* related to a geographic transition.

geocode A data value assigned to a spatial object that provides information on the geographic location of the object and is used as a key to access data relating to the object.

geocosmogony (ge'-o-cos-mog'-o-ny) The science of the origin of the Earth. Cf: *geogony.*

geocratic (ge-o-crat'-ic) *epeirocratic.*

geocronite (ge-oc'-ro-nite) A metallic lead-gray monoclinic mineral: $Pb_{14}(Sb,As)_6S_{23}$.

geocryology (ge'-o-cry-ol'-o-gy) The study of ice and snow on the Earth, esp. the study of permafrost. See also: *cryolithology.*

geocyclic (ge-o-cy'-clic) (a) Pertaining to or illustrating the rotation of the Earth. (b) Circling the Earth periodically.

geocyclicity (ge'-o-cy-clic'-i-ty) The quality or state of being geocyclic, as in defining all events of cyclic nature recorded in sedimentary rocks for which a specific causal mechanism could not be inferred.

geode (ge'-ode) (a) A hollow or partly hollow and globular or subspherical body, from 2.5 cm to 30 cm or more in diameter, found in certain limestone beds and rarely in shales; it is characterized by a thin and sometimes incomplete outermost primary layer of dense chalcedony, by a cavity that is partly filled by an inner drusy lining of inward-projecting crystals (often perfectly formed and usually of quartz or calcite and sometimes of barite, celestite, and various sulfides) deposited from solution on the cavity walls, and by evidences of growth by expansion in the cavities of fossils or along fracture surfaces of shells. Unlike a *druse,* a geode is separable (by weathering) as a discrete nodule or concretion from the rock in which it occurs and its inner crystals are not of the same minerals as those of the enclosing rock. A geode tends to be slightly flattened with its equatorial plane parallel to the bedding. (b) The crystal-lined cavity in a geode. (c) A term applied to a rock cavity and its lining of crystals that is not separable as a discrete nodule from the enclosing rock. Cf: *vug.*

geodepression (ge'-o-de-pres'-sion) Haarmann's term for a long, narrow depression on the scale of a geosyncline, but not necessarily filled with sediments (Glaessner and Teichert, 1947, p.476). The term is no longer used.

geodesic (ge-o-des'-ic) n. A line of shortest distance between any two points on a mathematically defined surface.

geodesy (ge-od'-e-sy) (a) The science concerned with the determination of the size and shape of the Earth and the precise location of points on its surface. (b) The determination of the gravitational field of the Earth and the study of temporal variations such as Earth tides, polar motion, and rotation of the Earth.

geodetic azimuth (ge-o-det'-ic) The angle at a point A between the tangent to the *meridian* at A and the tangent to the *geodesic* from A to the point B whose geodetic azimuth is wanted. Until 1986 the U.S. National Geodetic Survey had considered a geodetic azimuth to be positive clockwise starting from south. The azimuth is called the "forward azimuth" for the line AB. The angle at B between the tangents to the meridian and to the geodesic is called the "back azimuth" or "reverse azimuth" for the line AB. Because of the convergence of the meridians, the forward and back azimuths of a line do not differ by exactly 180°, except where the two end points have the same geodetic longitude or where the geodetic latitudes of both points are 0°. See also: *Laplace azimuth.*

geodetic control A set of control stations established by geodetic methods.

geodetic coordinates Quantities defining the horizontal position of a point on an ellipsoid of reference with respect to a specific geodetic datum, usually expressed as latitude and longitude. These may be referred to as geodetic positions or *geographic coordinates.* The elevation of a point is also a geodetic coordinate and may be referred to as a height above sea level.

geodetic coordinate system A coordinate system consisting of an ellipsoid, the equatorial plane of the ellipsoid, and a meridional plane through the polar axis. The coordinates of a point in this system are given by the perpendicular distance of the point from the ellipsoid, the angle between that perpendicular (the "normal") and the equatorial plane, and by the dihedral angle between the meridional plane and a plane perpendicular to the equatorial plane and containing the normal.

geodetic datum *datum [geodesy].*

geodetic engineering *geodetic surveying.*

geodetic equator The circle on the spheroid midway between its poles of revolution and connecting points of zero degrees geodetic latitude. Its plane, which contains the center of the spheroid and is perpendicular to its axis, cuts the celestial sphere in a line coinciding with the celestial equator because the axis of the spheroid is by definition parallel to the axis of rotation of the Earth. Cf: *astronomic equator.*

geodetic height The perpendicular distance from an ellipsoid of reference to a point. It is a *geodetic coordinate.*

geodetic inverse Computation of the distance and direction between two points, on an ellipsoid, with known coordinates.

geodetic latitude The *latitude* or angle which the normal at a point on the spheroid makes with the plane of the geodetic equator (the equatorial plane of the spheroid). It is equivalent to the *astronomic latitude* corrected for station error and to true *geographic latitude.* Latitude as shown on topographic maps is geodetic latitude. Symbol: ϕ. Cf: *geocentric latitude.*

geodetic leveling Spirit leveling of a high order of accuracy, usually extended over large areas (and with proper applications of orthometric corrections), to furnish accurate vertical control for surveying and mapping operations.

geodetic longitude The *longitude* or angle between the plane of the geodetic meridian and the plane of an arbitrarily chosen initial meridian (generally the Greenwich meridian). It is equivalent to the *astronomic longitude* corrected for station error and to true *geographic longitude.* Symbol: λ. Syn: *geocentric longitude.*

geodetic meridian A line (ellipse) on the spheroid having the same geodetic longitude at every point. Syn: *geographic meridian.*

geodetic network A network whose nodes are survey stations or gravity stations.

geodetic parallel A line or circle on the spheroid having the same geodetic latitude at every point; a small circle whose plane is parallel with the plane of the geodetic equator.

geodetic position The location of a point on the surface of the Earth expressed in terms of geodetic latitude and longitude. A geodetic position implies that a particular geodetic datum was adopted. In the complete record of a geodetic position, the adopted datum must be stated.

Geodetic Reference System System that defines the geodetic datum. Abbrev: GRS.

geodetic survey (a) A survey of a large land area accomplished by the processes of geodetic surveying and used for the precise location of basic points suitable for controlling other surveys. (b) An organization engaged in making geodetic surveys.

geodetic surveying Surveying in which account is taken of the figure and size of the Earth and corrections are made for earth curvature; the applied science of geodesy. It is used where the areas or distances involved are so great that results of desired accuracy and precision cannot be obtained by *plane surveying.* Syn: *geodetic engineering.*

geodetic zenith The point where the normal to the reference ellipsoid, extended upward, meets the celestial sphere. This term has some use in geodesy, but should be used only in its entirety, because the single word, zenith, is reserved for designating the point determined by the direction of the plumb line.

geodic (ge-od'-ic) Pertaining to a geode; e.g. a "geodic cavity or vein" in which iron sulfide may precipitate.

geodiferous (ge-o-dif'-er-ous) Containing geodes.

Geodimeter (Ge-o-dim'-e-ter) Trade name of an electronic-optical device that measures ground distances precisely by electronic timing and phase comparison of modulated light waves that travel from a master unit to a reflector and return to a light-sensitive tube where an electric current is set up. It is normally used at night and is effective with first-order accuracy up to distances of 5-40 km (3-25 miles). Etymol: acronym for *geo*detic-*di*stance meter.

geodynamic (ge'-o-dy-nam'-ic) Pertaining to physical processes within the Earth as they affect the features of the crust.

geodynamics (ge'-o-dy-nam'-ics) That branch of science which deals with the forces and processes of the interior of the Earth.

Geodynamics Project (Ge'-o-dy-nam'-ics) An international program of research (1971-1977) on the dynamics and dynamic history of the Earth with emphasis on deep-seated geological phenomena, esp. movements and deformations of the lithosphere.

geoecology (ge'-o-e-col'-o-gy) *environmental geology.*

geoelectricity (ge'-o-e'-lec-tric'-i-ty) The Earth's natural electric fields and phenomena. It is closely related to geomagnetism. See also: *telluric current.*

geoevolutionism (ge'-o-ev'-o-lu'-tion-ism) A term introduced by Goode (1969) for the "idea of evolution of geologic processes" in which some geologic changes foster new processes that in turn bring about new changes, i.e., feedback.

geofact (ge'-o-fact) A supposed *artifact* but of geological origin.

geoflex (ge'-o-flex) An obsolete syn. of *orocline.*

geoforensics *forensic geology.*

geogeny (ge-og'-e-ny) *geogony.*

geognosy (ge-og'-no-sy) An eighteenth-century term associated with A.G. Werner for a science accounting for the origin, distribution, and sequence of minerals and rocks in the Earth's crust. It developed out of mineralogy and was distinct from *oryctology.* The term was superseded by *geology* as early ideas were abandoned. It has become restricted to absolute knowledge of the Earth, as distinct from the theoretical and speculative reasoning of geology.

geogony (ge-og'-o-ny) The science or theory of the formation of the Earth, esp. a speculative study of its origin. Syn: *geogeny.* See also: *geocosmogony.*

geogram (ge'-o-gram) A term used by Marr (1905, p. lxii) for a hypothetical geologic column, connoting principally lateral variations in lithology and in organic assemblages that could be traced in a deposit over an area of any width up to that of the circumference of the Earth.

geographic (ge-o-graph'-ic) Pertaining or relating to geography.

geographic center The point on which an area on the Earth's surface would balance if it were a plate of uniform thickness (i.e. the center of gravity of such a plate). The geographic center of the conterminous U.S. is in the eastern part of Smith County, Kansas (lat. 39°50'N, long. 98°35'W); the geographic center of North America is in Pierce County, N.D., a few miles west of Devils Lake.

geographic content The representation of content-worthy features with no generalizations for map symbolization (USGS, 2002b, p.1A-3). Cf: *cartographic content.*

geographic coordinates An inclusive term generally used to designate both geodetic and astronomical coordinates. Also called: *terrestrial coordinates.*

geographic cycle *cycle of erosion.*

geographic grid A system of parallels and meridians used to locate points on the Earth's surface.

geographic horizon The boundary line of that part of the Earth's surface visible from a given point of view; the *apparent horizon.*

geographic information system A computer program and associated data bases that permit cartographic information (including geologic information) to be structured and queried by the geographic coordinates of features. Usually the data are organized in "layers" representing different geographic entities such as hydrology, culture, topography, etc. A geographic information system permits information from different layers to be easily integrated, analyzed, and displayed. Abbrev: GIS.

geographic latitude A general term, applying alike to *astronomic latitude* and *geodetic latitude.*

geographic longitude A general term, applying alike to *astronomic longitude* and *geodetic longitude.*

geographic meridian *geodetic meridian.*

Geographic Names Information System The official repository of domestic geographic names information for the United States. It was developed by the U. S. Geological Survey in cooperation with the U. S. Board on Geographic Names and contains information on almost 2 million physical and cultural geographic features in the United States, its territories, and Antarctica. The federally recognized name of each feature described in the data base is identified, and refernces are made to the feature's location by state, county, and geographic coordinates (adapted from the GNIS website at http://geonames.usgs.gov/index.html). Abbrev: GNIS.

geographic north *true north.*

geographic parallel A general term for a line or circle on the Earth's surface having the same latitude at every point. It is applied to astronomic parallel and geodetic parallel.

geographic pole Either of the two poles or points of intersection of the Earth's surface and its axis of rotation; specif. North Pole and South Pole. Syn: *terrestrial pole.*

geographic position The location of a point on the surface of the Earth, expressed in terms of either geodetic or astronomic latitude and longitude.

geographic position [cart] The position of features as they exist on the ground, as best approximated by high-accuracy sources or geodetic surveys (USGS, 2002b, p.1A-3). Cf: *cartographic position, source position, true position.*

geographic province An extensive region all parts of which are characterized by similar geographic features. Cf: *physiographic province.*

geographic race Part of a biologic species coinciding with a particular geographic area and probably resulting from specific characteristics of the environment; a geographic subspecies or variety.

geography (ge-og'-ra-phy) The study of all aspects of the Earth's surface including its natural and political divisions, the distribution and differentiation of areas and, often, human interactions with the environment. Includes spatial analysis and cartographic sciences. See also: *physical geography.*

geohistory (ge-o-his'-to-ry) *geologic history.*

geohydrology (ge'-o-hy-drol'-o-gy) *hydrogeology.*

geoid [geodesy] (ge'-oid) The *figure of the Earth* considered as a sea-level surface extended continuously through the continents. It is a theoretically continuous surface that is perpendicular at every point to the direction of gravity (the plumb line). It is the surface of reference for astronomical observations and for geodetic leveling. See also: *compensated geoid; datum [geodesy].*

geoid [planet] A surface that conforms to a gravitational-rotational equipotential. For extraterrestrial bodies, typically calculated at the mean planetary radius (Bougher et al., 1997). Syn: *topographic datum.*

geoidal height The distance, taken along a perpendicular to the ellipsoid of reference, from that ellipsoid to the geoid. Also called: geoid height.

geoidal horizon (ge-oid'-al) A circle on the celestial sphere formed by the intersection of the celestial sphere and a plane tangent to the sea-level surface of the Earth at the zenith-nadir line.

geoidal map The contoured development of the separations of the geoid and a specified reference ellipsoid.

geoidal section A section profile of the geoid with relation to the reference ellipsoid.

geoidal separation The distance between the reference ellipsoid and the geoid. Cf: *datum [geodesy].*

geoinformatics The application of modern information technology to organize, preserve, and distribute geoscience data and to facilitate and advance their analysis.

geolith (ge'-o-lith) *lithostratigraphic unit.*

geologese (ge-ol'-o-gese') (a) Literary style or jargon peculiar to geologists. (b) Geological language that is "progressing rapidly" toward the construction of "sentences in such a way that their meaning is not apparent on first reading" (Vanserg, 1952, p.221).

geologic (ge-o-log'-ic) *geological.*

geologic age (a) The age of a fossil organism or of a particular geologic event or feature referred to the geologic time scale and expressed in terms either of years or centuries (*absolute age*) or of comparison with the immediate surroundings (*relative age*); an age datable by geologic methods. (b) The term is also used to emphasize the long-past periods of time in geologic history, as distinct from present-day or historic times. Cf: *age [geochron].*

geologic-age determination Determination of the *relative age* or *numerical age* of a geologic event or feature.

geological (ge-o-log'-i-cal) Pertaining to or related to geology. The choice between this term and *geologic* is optional, and may be made according to the sound of a spoken phrase or sentence. *Geological* is generally preferred in the names of surveys and societies, and in English and Canadian usage.

geological oceanography *marine geology.*

geological ore *possible ore.*

geological science Any of the subdisciplinary specialties that are part of the science of *geology*; e.g. geophysics, geochemistry, paleontology, petrology, etc. The term is commonly used in the plural. See also: *geoscience.* Cf: *Earth science.*

geologic climate *paleoclimate.*

geologic-climate unit (ge-o-log'-ic-cli'-mate) A term used by the ACSN (1961, Art. 39) for "an inferred widespread climatic episode defined from a subdivision of Quaternary rocks"; e.g. glaciation, interglaciation, stade, and interstade. It is strictly not a stratigraphic unit. The different stratigraphic boundaries that define the limits of the geologic-climate unit in different latitudes are not likely to be isochronous. Syn: *climate-stratigraphic unit.*

geologic column (a) A composite diagram that shows in a single column the subdivisions of part or all of geologic time or the sequence of stratigraphic units of a given locality or region (the oldest at the bottom and the youngest at the top, with dips adjusted to the horizontal) so arranged as to indicate their relations to the subdivisions of geologic time and their relative positions to each other. See also: *columnar section.* (b) The vertical or chronologic arrangement or sequence of rock units portrayed in a geologic column. See also: *geologic section.* Syn: *stratigraphic column; column [stratig].*

geologic engineering A specialty of engineering in which the primary materials involved in the planning, design, construction, and maintenance of an engineering project are Earth materials. It includes such projects as highwall stability in open-cast mines; design and construction of tunnels and transportation routes; and design for control and mitigation of hazardous natural conditions. The term is generally considered synonymous with *geotechnical engineering* but is predominantly used to refer to design of rock slopes, tunnels, rock foundations, and reinforcements in rock materials.

geologic erosion A syn. of *normal erosion,* or erosion caused naturally by geologic processes.

geologic formation *formation [stratig].*

geologic hazard A geologic condition or phenomenon, natural or brought about by human activity, that represents a threat to human life, welfare, and property. Such conditions or phenomena can lead to losses through, e.g., landslides, floods, earthquakes, ground subsidence, coastal and beach erosion, faulting, dam leakage and failure, mining disasters, groundwater pollution, and seawater intrusion.

geologic high An oil-field term for a structure on which rocks occur at a higher position than in the surrounding area.

geologic history The history of the Earth and its inhabitants throughout geologic time, often considered for a certain area or duration of time, or in certain aspects. It comprises all chemical, physical, and biologic conditions that have existed on and in the Earth, all processes that have operated to make and modify these conditions, and all events that have affected any part of the Earth, including its inhabitants, from the beginning of the planet to the present, and is circumscribed "in no way by what we think we know about it" (Moore, 1949, p.3). Syn: *geohistory; Earth history.*

geologic horizon *horizon [geol].*

geologic low An oil-field term for a structure on which rocks occur at a lower position than in the surrounding area. Cf: *low [struc geol].*

geologic map A map on which is recorded geologic information, such as the distribution, nature, and age relationships of rock units (surficial deposits may or may not be mapped separately), and the occurrence of structural features (folds, faults, joints), mineral deposits, and fossil localities. It may indicate geologic structure by means of formational outcrop patterns, by conventional symbols giving the direction and amount of dip at certain points, or by structure-contour lines.

geologic norm The condition resulting from *normal erosion* of the land, undisturbed by the activity of man and his agents.

geologic province An extensive region characterized throughout by similar geologic history or by similar structural, petrographic, or physiographic features. Cf: *physiographic province.*

geologic range *stratigraphic range.*

geologic record The "documents" or "archives" of the history of the Earth, represented by bedrock, regolith, and the Earth's morphology; the rocks and the accessible solid part of the Earth. Also, the geologic history based on inferences from a study of the geologic record. See also: *stratigraphic record.*

geologic section (a) Any sequence of rock units found in a given region either at the surface or below it (as in a drilled well or mine shaft); a local *geologic column.* Syn: *stratigraphic section.* (b) *section [geol].*

geologic thermometer *geothermometer.*

geologic time The period of time dealt with by historical geology, or the time extending from the end of the formative period of the Earth as a separate planetary body to the beginning of written history; the part of the Earth's history that is represented by and recorded in the succession of rocks. The term implies extremely long duration or remoteness in the past, although no precise limits can be set. Cf: *time (a).*

geologic time scale An arbitrary chronologic arrangement or sequence of geologic events, used as a measure of the relative or absolute duration or age of any part of geologic time, and usually presented in the form of a chart showing the names of the various rock-stratigraphic, time-stratigraphic, or geologic-time units, as currently understood; e.g. the geologic time scales published by Harland et al. (1982), Odin (1982), Palmer (1983), and Salvador (1985). See also: *atomic time scale; relative time scale; biologic time scale.* Syn: *time scale.*

geologic-time unit *geochronologic unit.*

geologist (ge-ol'-o-gist) One who is trained in and works in any of the geological sciences.

geologize (ge-ol'-o-gize) v. To participate in or talk about geology; to practice geology.

geology (ge-ol'-o-gy) The study of the planet Earth the materials of which it is made, the processes that act on these materials, the products formed, and the history of the planet and its life forms since its origin. Cosmologists speak of a "pre-geologic" stage in planetary history. Geology considers the physical forces that act on the Earth, the chemistry of its constituent materials, and the biology of its past inhabitants as revealed by fossils. Clues on the origin of the planet are sought in a study of the Moon and other extraterrestrial bodies. The knowledge thus obtained is placed in the service of society to aid in discovery of minerals and fuels of value in the Earth's crust, to identify geologically stable sites for major structures, and to provide foreknowledge of some of the dangers associated with the mobile forces of a dynamic Earth. See also: *geological science; Earth science; geoscience; historical geology; physical geology.*

geomagnetic axis (ge'-o-mag-net'-ic) The axis of the dipole magnetic field most closely approximating the actual magnetic field of the Earth.

geomagnetic electrokinetograph An obsolete, ship-towed instrument formerly used to measure ocean currents by measuring the electric field generated as the seawater, a conductor, moves in earth's magnetic field. Abbrev: GEK.

geomagnetic equator The great circle of the Earth whose plane is perpendicular to the geomagnetic axis; the line connecting points of zero geomagnetic latitude.

geomagnetic field *Earth's magnetic field.*

geomagnetic latitude The magnetic latitude that a location would have if the Earth's field were replaced by the dipole field most closely approximating it. It is latitude reckoned relative to the geomagnetic axis instead of to the Earth's rotational axis. Cf: *geomagnetic longitude.*

geomagnetic longitude Longitude reckoned around the geomagnetic axis instead of around the Earth's rotational axis. Cf: *geomagnetic latitude.*

geomagnetic meridian A great circle of the Earth including the geomagnetic poles.

geomagnetic polarity epoch (ge'-o-mag-net'-ic po-lar'-i-ty) *polarity epoch.*

geomagnetic polarity event *polarity event.*

geomagnetic polarity reversal *geomagnetic reversal.*

geomagnetic-polarity time scale A chronology based on counting reversals of the Earth's magnetic field. The four youngest named epochs, from youngest to oldest, are Brunhes normal (now to 0.7 m.y.), Matuyama reversed (0.7 m.y. to 2.5 m.y.), Gauss normal (2.5 m.y. to 3.3 m.y.) and Gilbert reversed (3.3 m.y. to 5 m.y.). See also: *polarity epoch.*

geomagnetic poles The points of emergence at the Earth's surface of the axis of the geocentric magnetic dipole that most closely approximates the Earth's magnetic field. These points are sometimes referred to as the magnetic poles.

geomagnetic reversal A change of the Earth's magnetic field between *normal polarity* and *reversed polarity.* Syn: *field reversal; geomagnetic polarity reversal; magnetic polarity reversal; magnetic reversal; reversal; polarity reversal.*

geomagnetic secular variation *secular variation.*

geomagnetism (ge-o-mag'-net-ism) The magnetic phenomena exhibited by the Earth and its atmosphere; also, the study of such phenomena. Syn: *terrestrial magnetism.*

geomagnetochronology (ge'-o-mag-net'-o-chro-nol'-o-gy) Establishment of a magnetic-polarity time scale of Earth history on the basis of *paleomagnetism.*

geomalism (ge-o-mal'-ism) Equal lateral growth of an organism in response to gravitational force. Cf: *geotaxis; geotropism.*

geomathematics (ge'-o-math'-e-mat'-ics) All applications of mathematics to studies of the Earth. Cf: *mathematical geology.*

geomechanics (ge'-o-me-chan'-ics) A branch of geology that embraces the fundamentals of structural geology and a knowledge of the response, or strain, of natural materials to deformation or changes caused by the application of stress and/or strain energy.

geomedia In engineering geology, a general term for earth materials, including solid, liquid, and gaseous components.

geometric coherence The existence of regular and systematic displacement patterns in a family of faults (Walsh and Watterson, 1991). Cf: *kinematic coherence.*

geometric correction The adjustment of the geometry of a digital image for scaling, skewing, and other spatial distortions (Krzanowski et al., 1993).

geometric factor (ge-o-met'-ric) The factor by which the ratio of voltage to current is multiplied in order to obtain the apparent resistivity. It depends on the electrode array and spacing.

geometric geodesy The part of geodesy which is concerned with positional relationships in the Earth's environment obtained by geometric (rather than physical) methods. Cf: *physical geodesy.*

geometric grade scale A *grade scale* having a constant ratio between size classes; e.g. the Wentworth grade scale, each size class of which differs from its predecessor by the constant ratio 1:2.

geometric horizon A term originally applied to the *celestial horizon,* but now more commonly to the intersection of the celestial sphere and an infinite number of straight lines tangent to the Earth's surface and radiating from the eye of the observer. It would coincide with the *apparent horizon* if there were no terrestrial refraction.

geometric mean An average, calculated as the *n*th root of the product of *n* positive observations; usually determined by taking the antilogarithm of the arithmetic average of the logarithms of the observations. Cf: *mean; harmonic mean.*

geometric mean diameter An expression of the average particle size of a sediment or rock, obtained by taking the antilogarithm of the *phi mean diameter;* the diameter equivalent of the arithmetic mean of the logarithmic frequency distribution.

geometric projection *perspective projection.*

geometrics (ge-o-met'-rics) Measurements of and about the Earth through space and time.

geometric similarity A condition in which all lengths in a *scaled physical model* are reduced by the same amount relative to all lengths in the natural prototype; consequently, all angles in the model are identical to those in the prototype (Hubbert, 1937).

geometric sounding That form of electromagnetic sounding in which separation between transmitting and receiving coils is the variable. Cf: *parametric sounding.*

geometric spreading *spherical divergence.*

geometric transformation (a) Adjustments made in image data to improve geometrical consistency for cartographic purposes. (b) Transformation of image data from its original sensor coordinate system to another selected coordinate system. (Krzanowski et al., 1993). See also: *geometric correction.*

geometry number A dimensionless constant representing the ratio of the product of maximum basin relief and drainage density within a given drainage basin to the tangent of the stream gradient; it summarizes the essentials of landform geometry, particularly the relations between planimetric and relief aspects of a drainage basin (Strahler, 1958, p.287, 295).

geomicrobiology The examination of the role that microbes have played in the past and are currently playing in a number of fundamental geological processes (Ehrlich, 1990).

geomonocline (ge-o-mon'-o-cline) A broad flank of a geosyncline, the beds of which have a uniform dip; a unilateral, marginal geosyncline. Obsolete.

geomorphic (ge-o-mor'-phic) (a) Pertaining to the form of the Earth or of its surface features; e.g. a *geomorphic* province. (b) Geomorphologic. Syn: *physiographic.*

geomorphic cycle *cycle of erosion.*

geomorphic effectiveness The shaping of stream-channel geometry by a *dominant discharge.*

geomorphic geology *geomorphology.*

geomorphogeny (ge'-o-mor-phog'-e-ny) The part of geomorphology that deals with the origin, development, and changes of the Earth's surface features or landforms. Cf: *geomorphography.* Syn: *genetic physiography.*

geomorphography (ge'-o-mor-phog'-ra-phy) The part of geomorphology that deals with the description of the Earth's surface features or landforms. Cf: *geomorphogeny.*

geomorphology (ge'-o-mor-phol'-o-gy) (a) The science that treats the general configuration of the Earth's surface; specif. the study of the classification, description, nature, origin, processes, and development of present landforms and their relationships to underlying structures, and of the history of geologic changes as recorded by these surface features. The term was applied widely in Europe before it was used in the U.S., where it has come to replace the term *physiography* and is usually considered a branch of geology or geography; in the British Commonwealth, Japan and many other countries, it is usually regarded as a branch of geography. Geomorphology as a named science is barely more than a century and a half old (Tinkler, 1985), with the name first appearing in the German language in 1858. A young associate of John Wesley Powell, WJ McGee (indeed "WJ" rather than "W.J." as is often mis-referenced), introduced the term into English in 1888 (McGee, 1888a,b). See also: *physical geography*. Syn: *physiographic geology; geomorphic geology; geomorphy*. (b) Strictly, any study that deals with the form of the Earth (including geodesy, and structural and dynamic geology). This usage is more common in Europe, where the term has even been applied broadly to the science of the Earth. (c) The features dealt with in, or a treatise on, geomorphology; e.g., the *geomorphology* of Texas.

geomorphy (ge'-o-mor'-phy) (a) *geomorphology*. (b) A syn. of *topography* "in the broader application of that term" (Lawson, 1894, p.241). Obsolete.

geomyricite (ge-o-my'-ri-cite) A white, waxy resin of approximate composition: $C_{32}H_{62}O_2$. It is found in brown coal.

geomythology (ge'-o-my-thol'-o-gy) "The geological application of euhemerism" (Vitaliano, 1968; 1973) i.e. of the theory that myths may be interpreted as accounts of historical personages and events. Geomythology was first used for myths or legends embodying the memory of some real geologic event, but has been expanded to include any folklore or myths having a geological basis.

geonomy (ge-on'-o-my) A term variously recommended as a synonym for *geology*, as the science of the dynamic Earth, as the science concerned exclusively with the physical forces relating to the Earth, and to denote the study of the Earth's upper mantle. The word has not been given broad recognition or acceptance, and the variety of proposed definitions suggests that it should be used with care, if at all.

geop (ge'-op) A surface within the gravity field of the Earth in which all points have equal and constant *geopotential*. Syn: *geopotential surface*.

geopetal (ge-o-pet'-al) Pertaining to any rock feature that indicates the relation of top to bottom at the time of formation of the rock; e.g. a "geopetal fabric", the internal structure or organization that indicates original orientation of a stratified rock, such as cross-bedding or grains on a boundary surface. Term introduced by Sander (1936, p.31; 1951, p.2).

geopetal structure A common structure in limestones where a cavity contains sediment in the lower part and cement (usually calcite) in the upper part. These structures are like spirit levels and can be used to deduce the way-up of the rocks and the horizontal at the time of deposition.

geophagous (ge-oph'-a-gous) Said of an organism that feeds on soil.

geophone (ge'-o-phone) A *seismic detector* that produces a voltage proportional to the displacement, velocity, or acceleration of ground motion, within a limited frequency range. Most geophones used in exploration are velocity phones. Syn: *jug; seismometer; pickup*. Cf: *seismograph; transducer; hydrophone*.

geophysical exploration (ge-o-phys'-i-cal) The use of geophysical techniques, e.g. electric, gravity, magnetic, seismic, or thermal, in the search for economically valuable hydrocarbons, mineral deposits, or water supplies, or to gather information for engineering projects. Syn: *applied geophysics; geophysical prospecting*.

geophysical log *well log*.

geophysical prospecting *geophysical exploration*.

geophysical survey The use of one or more geophysical techniques in *geophysical exploration*, such as earth currents, electrical, gravity, or electromagnetic methods, heat flow, magnetics, radioactivity, and seismics. Geochemical techniques are sometimes included.

geophysicist (ge-o-phys'-i-cist) One who studies the physical properties of the Earth, or applies physical measurements to geological problems; a specialist in *geophysics*. Geophysicists also study the Moon and the other planets.

geophysics (ge-o-phys'-ics) Study of the Earth by quantitative physical methods. Basic divisions include solid-earth geophysics, physics of the atmosphere and hydrosphere, and solar-terrestrial physics. There are numerous specialties within the field, e.g. seismology, tectonophysics, engineering geophysics. The term is sometimes used to include instrumental study of the Moon and planets.

geophysiography (ge'-o-phys'-i-og'-ra-phy) The synthesis of all knowledge available about the Earth; "the combination...of relevant parts of geophysics, geochemistry, geomorphology, and geoecology" (Strøm, 1966, p.8). Apparently a unique usage.

geopiezometry (ge'-o-pi'-e-zom'-e-try) Measurement of pressure and compressibility of rocks in metamorphism (Chinner, 1966).

geopolitics (ge-o-pol'-i-tics) A pejorative term suggesting organizational manipulation within government, society, or university circles dealing with earth science.

geopotential (ge'-o-po-ten'-tial) The potential energy of a unit mass relative to sea level, numerically equal to the work that would be done in lifting the unit mass from sea level to the height at which the mass is located. It is commonly expressed in terms of *dynamic height*. Cf: *disturbing potential; spheropotential*.

geopotential height *dynamic height*.

geopotential number The numerical value, C, assigned to a given *geop* or geopotential surface when expressed in geopotential units (g.p.u.), where 1 g.p.u. = 1 m x 1 kGal. Since g = 1.98 kGal, C = gH = 0.98 H. Thus the geopotential number in g.p.u. is almost equal to the height above sea level in meters. Geopotential numbers are generally preferred to dynamic heights because the physical meaning is more correctly expressed.

geopotential surface A surface on which the Earth's gravity potential is constant. Syn: *geop*.

geopressured aquifer (ge-o-pres'-sured) A term used for an aquifer, esp. in the Gulf Coast, in which fluid pressure exceeds normal hydrostatic pressure of 0.465 pound per square inch per foot of depth (Jones, 1969, p.34).

geoprocessing The manipulation and analysis of geographically referenced data (Krzanowski et al., 1993).

georeference v. To establish the relationship between an image coordinate system and a ground coordinate system (USGS, 2001, p.1A-2).

georeference system A coordinate system with which the location of a point on the Earth's surface may be identified (Krzanowski et al., 1993).

georgbokiite An adamantine brown monoclinic mineral: $Cu_5O_2(Se^{4+}O_3)_2Cl_2$.

georgechaoite (george-cha'-o-ite) A colorless or white orthorhombic mineral: $KNaZrSi_3O_9 \cdot 2H_2O$. It is the K analogue of gaidonnayite.

georgeericksenite A vitreous yellow monoclinic mineral: $CaMgNa_6(IO_3)_6(CrO_4)_2 \cdot 12H_2O$.

georgeite (george'-ite) An earthy blue amorphous mineral: $Cu_2(CO_3)(OH)_2$. It is an amorphous analogue of malachite.

georgiadesite (geor-gia'-des-ite) A white or brownish-yellow monoclinic mineral: $Pb_4(As^{3+}O_3)Cl_4(OH)$.

georgiaite (geor'-gia-ite) A greenish *North American tektite* from Georgia, U.S.

Georgian (Geor'-gian) *Waucoban*.

geoscience (ge-o-sci'-ence) (a) A short form, sometimes used in the plural, denoting the collective disciplines of the geological sciences. The term, as such, is synonymous with *geology*. (b) A syn. of *earth science*.

geosere (ge'-o-sere) A series of climax communities that succeed each other as a result of changes in the physical and climatic characteristics of the environment.

geosol (ge'-o-sol) (a) A laterally traceable, mappable, geologic weathering profile that has a consistent position. (b) The fundamental and only *pedostratigraphic unit* (NACSN, 1983, Art. 56).

geosphere (ge'-o-sphere) (a) The *lithosphere*. (b) The lithosphere, hydrosphere, and atmosphere combined. (c) Any of the so-called spheres or layers of the Earth.

geostatic pressure (ge-o-stat'-ic) The vertical pressure at a point in the Earth's crust, equal to the pressure caused by the weight of a column of the overlying rock or soil. Syn: *lithostatic pressure; overburden pressure; rock pressure; stereostatic pressure*.

geostationary Refers to satellites traveling at the angular velocity at which the earth rotates; as a result, they remain above the same point on earth at all times.

geostatistics (ge'-o-sta-tis'-tics) (a) A branch of applied statistics that deals with phenomena that fluctuate in two- or three-dimensional space. Geostatistics is concerned with assessing properties that may be physically continuous (hence cannot be truly random variables) but are too variable to describe by deterministic means; it offers a way of describing the spatial continuity that is an essential feature of many natural phenomena. It adapts classical regression techniques to take advantage of this continuity for interpolation and extrapolation of physical measurements using correlation and probability concepts. Geostatistics has evolved from concepts and techniques for mine evaluation by Georges Matheron (1902-), French mining engineer. (b) Statistics as applied to geology; the application of statistical methods or the collection of statistical data for use in geology. The French spelling "geostatistiques" is sometimes used. Cf: *kriging; cokriging; conditional simulation; variogram*.

geostratigraphic (ge'-o-strat'-i-graph'-ic) Pertaining to worldwide stratigraphy; e.g. "geostratigraphic standards" or "geostratigraphic stage".

geostrome (ge'-o-strome) A term, now obsolete, proposed by Patrin "to denote the strata of the Earth" (Pinkerton, 1811, v. 1, p.542).

geostrophic current (ge-o-stroph'-ic) An ocean current in which the horizontal pressure gradient is exactly balanced by the *Coriolis force*. The flow is along lines of constant pressure in a direction that is clockwise around a high in the northern hemisphere and counterclockwise around a high in the southern hemisphere. The current cannot change with time because there is no unbalanced force acting on the flow. Most currents in the ocean extending over distances of a hundred or more kilometers are very close to geostrophic balance.

geostrophic cycle A term suggested by Tomkeieff (1946, p.326) for the Huttonian concept of one great cycle of dynamic changes occurring in the Earth; the cycle embraces both the organic and the inorganic spheres, and consists alternately of the complementary processes of destruction and construction. Cf: *orogenic cycle*.

geostrophic force The effect of *Coriolis acceleration* on moving fluids such as ocean currents and winds.

geostrophic wind A hypothetical model of the wind that represents an unaccelerated horizontal wind balanced between the horizontal components of the pressure gradient force and the Coriolis effect. Such a wind blows parallel to straight isobars with low pressure to the left in the Northern Hemisphere. The speed is inversely proportional to the isobar spacing.

geosuture (ge-o-su'-ture) In Dietz's use, a boundary zone between contrasting tectonic units of the crust; in many places a fault which probably extends through the entire thickness of the crust. In Wilson's use, a place where two continents have come together. The term is rarely used; the preferred term is *suture*.

geosynclinal (ge-o-syn'-cli-nal) n. The original, now obsolete, term for *geosyncline*, first used by Dana in 1873. adj. Pertaining to a geosyncline.

geosynclinal couple Aubouin's concept of the true geosyncline as miogeosynclinal and eugeosynclinal furrows linked by geanticlinal ridges (1965, p.34); an *orthogeosyncline*. Obsolete.

geosynclinal cycle *orogenic cycle.*

geosynclinal facies A sedimentary facies characterized by great thickness, predominantly argillaceous character, and paucity of carbonate rocks; it consists of uniform, rhythmic, and graded beds of shale or silty shale regularly interbedded with graywackes, deposited rapidly in a strongly subsiding geosyncline of a deep-water marine environment. Cf: *shelf facies.* See also: *graptolitic facies.*

geosynclinal prism The load of sediments that accumulates, often to great thicknesses, in the downwarped part of a geosyncline, having a shape similar to that of a long, plano-convex prism whose convexity is at the floor. Obsolete. A more modern term for this feature is *passive-margin wedge.* Cf: *nepton; clastic wedge.*

geosynclinal trough *trough [tect].*

geosyncline (ge-o-syn'-cline) As originally defined, a mobile downwarping of the crust of the Earth, either elongate or basinlike, measured in scores of kilometers, in which sedimentary and volcanic rocks accumulate to thicknesses of thousands of meters. Formation of a geosyncline, up until the advent of plate tectonics, was thought to be a stage in a tectonic cycle in which orogeny follows. The concept was presented by Hall in 1859, and the term *geosynclinal* was proposed by Dana in 1873. Differing opinions of the origin, mechanics, and essential features of geosynclines are reflected in various schemes that have been used to define their aspects. Some are based on the tectonic relationship of crustal units, some emphasize mountain-building processes, and others are concerned with the relationship of geosynclinal sedimentation to subsidence. Recognition of plate tectonics has led to the appreciation that geosynclines are simply sedimentary basins formed over subsiding lithosphere. The large "geosyncline" recognized by Hall was the passive-margin basin that formed along the eastern margin of North America during the Paleozoic (Wilson, 1968). Cf: *mobile belt.* See also: *synclinorium.* Ant: *geanticline.*

geotaxis (ge-o-tax'-is) *Taxis [ecol]* resulting from gravitational attraction. Cf: *geotropism; geomalism.*

geotechnical (ge-o-tech'-ni-cal) Pertaining to the broad field of *geotechnics.*

geotechnical engineering *geotechnics.*

geotechnics (ge-o-tech'-nics) The application of scientific methods and engineering principles to the acquisition, interpretation, and use of knowledge of materials of the Earth's crust for the solution of engineering problems. It embraces the fields of *rock mechanics* and *soil mechanics* and many of the engineering aspects of geology, geophysics, hydrology, and related sciences. Syn: *geotechnique; geotechnical engineering.*

geotechnique (ge'-o-tech-nique') The French equivalent of *geotechnics.*

geotechnology (ge'-o-tech-nol'-o-gy) A discipline most commonly understood as the application of scientific methods and engineering techniques to the evaluation, exploitation, and conservation of earth resources.

geotectocline (ge-o-tec'-to-cline) The geosynclinal accumulation of sediments formed above a *downbuckle;* the basin between the limbs of a downbuckle (Hess, 1938, p.79). Obsolete. Syn: *tectocline.*

geotectogene (ge-o-tec'-to-gene) *tectogene.*

geotectonic (ge'-o-tec-ton'-ic) *tectonic.* Usually applied when the features being described are of regional scale or larger.

geotectonic cycle A sequence of geosynclinal, orogenic, and cratonic stages. Such a cycle may be repeated (Stille, 1940). The concept is obsolete.

geotectonics (ge'-o-tec-ton'-ics) The term used for *tectonics* when the features being described are of regional scale or larger. Cf: *megatectonics.*

geotextile (ge-o-tex'-tile) Any fabric material used with rock or soil in a human product or structure. Also called filter fabrics, geofabrics, and geosynthetics.

geotexture (ge-o-tex-ture) The texture of the Earth's surface as manifested by the largest features of relief (such as continental massifs and ocean basins), the formation of which is connected with worldwide processes (I.P. Gerasimov and J.A. Mescherikov in Fairbridge, 1968, p.731).

geotherm (ge'-o-therm) Line indicating the increase of temperature with depth within the Earth.

geothermal (ge-o-ther'-mal) Pertaining to the heat of the interior of the Earth. Etymol: Greek "geo", "earth", and "therme", "heat". Syn: *geothermic.*

geothermal anomaly An abnormal temperature condition in relation to the overall temperature field.

geothermal brine A brine that is overheated with respect to its depth. See also: *brine [geol]; geopressured aquifer.*

geothermal energy Heat energy that occurs naturally in the Earth and that can be extracted and used. Geothermal energy is usable for a wide range of temperature and volume, e.g., nonelectric use (direct use of geothermal water, geothermal heat pumps) and electric use of steam or hot water.

geothermal exploration Exploration for sources of *geothermal energy.* Syn: *thermal exploration.*

geothermal gradient The rate of change of temperature in the Earth with depth measured in °C m^{-1} or °C km^{-1}. The gradient differs from place to place depending on the heat flow in the region and the thermal conductivity of the rocks. The average geothermal gradient in the Earth's upper crust approximates 25°C km^{-1} of depth. Symbol: Γ. See also: *temperature gradient.*

geothermal metamorphism A type of deep-seated static metamorphism in which a regular downward increase in temperature attributed to deep burial by overlying rocks is the controlling factor. Cf: *thermal metamorphism; load metamorphism; static metamorphism.*

geothermal reserve Identified geothermal energy that can be extracted legally today at a cost competitive with other energy sources (Muffler and Cataldi, 1978; Muffler, 1985).

geothermal reservoir Underground storage of water trapped in porous rock capable of providing hydrothermal (hot water and steam) resources. Geothermal reservoirs may be classified as being either of low *enthalpy* (<150°C) or high enthalpy (>150°C). The latter term generally is applied to reservoirs that contain water or steam suitable for the generation of electric power, whereas low-enthalpy reservoirs are usually those where the energy is principally of interest for heating purposes (Haenel, Rybach, and Stegena, 1988). Hot-water (liquid-dominated) and dry-steam (vapor-dominated) reservoirs are recognized.

geothermal resource That fraction of the accessible geothermal resource base that might be extracted economically and legally at some reasonable future time (Muffler and Cataldi, 1978; Muffler, 1985). Includes identified and undiscovered components.

geothermal resource base All of the thermal energy in the Earth's crust beneath a specific area, measured from local mean annual temperature (Muffler and Cataldi, 1978; Muffler, 1985).

geothermal system Any regionally localized geological setting where naturally occurring portions of the Earth's thermal energy are transported close enough to the Earth's surface by circulating steam or hot water to be readily harnessed for use; examples are The Geysers region of northern California and the hot brine fields in the Imperial Valley of southern California.

geothermic (ge-o-ther'-mic) *geothermal.*

geothermics A subdiscipline of geophysics that studies the thermal state of the interior of the solid Earth and of the thermal properties of Earth materials. Components of geothermal research are temperature distribution within the Earth, terrestrial heat flow, heat generation, thermal properties of Earth materials, volcanology, geothermal energy, hydrocarbon formation, fluid flow, and permafrost (Jessop, 1990).

geothermobarometry Determination of the temperature and pressure of chemical equilibration of a rock, mineral or fluid.

geothermometer (ge'-o-ther-mom'-e-ter) A mineral or mineral assemblage whose composition, structure, or inclusions are fixed within known thermal limits under particular conditions of pressure and composition and whose presence thus denotes a limit or a range for the temperature of formation of the enclosing rock. Examples are the composition of co-existing minerals that undergo solid solution, the filling temperatures of fluid inclusions, and the thermal discoloration of spores and pollen.

geothermometry (ge'-o-ther-mom'-e-try) (a) Measurement of the temperatures at which geologic processes occur or occurred. (b) Determination of the temperature of chemical equilibration of a rock, mineral, or fluid.

geotomical axis (ge-o-tom'-i-cal) A minor axis with small spines in the shell of an acantharian radiolarian. Cf: *hydrotomical axis.*

geotraverse (ge-o-trav'-erse) A regional-scale profile made by combined geological and deep geophysical methods for structural interpretation of the Earth's crust.

geotropism (ge-ot'-ro-pism) *Tropism* resulting from gravitational attraction. Cf: *geotaxis; geomalism.*

geotumor (ge-o-tu'-mor) An uplift of regional extent (Haarmann, 1930). Obsolete.

geoundation (ge'-o-un-da'-tion) An upward or downward warping of the Earth's crust, on a continental or oceanic scale (Van Bemmelen, 1932). Obsolete.

gerasimovskite (ge-ra'-si-movsk'-ite) A brown or gray amorphous mineral: $(Mn^{2+},Ca)(Nb,Ti)_5O_{12}\cdot 9H_2O$(?).

gerdtremmelite (gerd-trem'-mel-ite) An adamantine yellowish to dark brown triclinic mineral: $(Zn,Fe^{2+})(Al,Fe^{3+})_2(AsO_4)(OH)_5$.

gerenite-(Y) A vitreous white to creamy triclinic mineral: $(Ca,Na)_2(Y,REE)_3Si_6O_{18}\cdot 2H_2O$.

gerhardtite (ger'-hardt-ite) An emerald-green orthorhombic mineral: $Cu_2(NO_3)(OH)_3$.

germ (a) In botany, an obsolete syn. of *embryo.* (b) A common term used in reference to bacteria.

germanite (ger'-man-ite) (a) A metallic reddish-gray cubic mineral: $Cu_{13}Fe_2Ge_2S_{16}$. (b) A group name for minerals with the general formula $Cu_{13}AS16$, where A = Fe, Ge, V, As, Sn or Sb.

germanocolusite A metallic cubic mineral: $Cu_{13}V(Ge,As)_3S_{16}$. It is the Ge analogue of colusite, and a member of the *germanite* group.
germanotype tectonics (ger-man'-o-type) The tectonics of the cratons and stabilized foldbelts, typified by structures in Germany north of the Alps. The milder phases of germanotype tectonics are epeirogenic, but they also include broad folds dominated by vertical uplift and high-angle faults, block-faulted terranes, and sedimentary basins deformed within a frame of surrounding massifs. The term is rarely used. In recent years, North American authors have preferred to use "Laramide-style tectonics" or "thick-skinned tectonics" instead. Cf: *alpinotype tectonics.* Syn: *paratectonics.*
germinal aperture (ger'-mi-nal) An *aperture [palyn],* such as a colpus or germ pore, through which the pollen nuclei emerge on germination of the grain. The term is sometimes used to include the laesura of spores.
germinal furrow *colpus.*
germ pore A *pore [palyn]* or thin area in the exine of a pollen grain through which the pollen tube emerges on germination.
geröllton (ge-röll'-ton) A term used by Pettijohn (1957, p.265) for a nonglacial conglomeratic mudstone (or as a syn. of *tilloid*). It was introduced as "Geröllton" (German for "gravel clay") by Ackermann (1951) who applied it to a pebble-bearing clay in which the pebbles and clay were deposited simultaneously.
gerontic (ge-ron'-tic) The senile or old-age growth stage in ontogeny; the stage following the *ephebic* stage.
gerontomorphosis (ge-ron'-to-mor'-pho-sis) Evolutionary changes involving modifications of the adult characteristics of organisms.
gersdorffite (gers'-dorff-ite) A silver-white to steel-gray cubic or orthorhombic mineral: NiAsS. It closely resembles cobaltite, and may contain some iron and cobalt. Several structurally ordered variants exist. Syn: *nickel glance.*
gerstleyite (gerst'-ley-ite) An adamantine red monoclinic mineral: $Na_2(Sb,As)_8S_{13}{\cdot}2H_2O$.
gerstmannite (gerst'-mann-ite) A white to pale pink orthorhombic mineral: $(Mg,Mn^{2+})_2Zn(SiO_4)(OH)_2$.
getchellite (get'-chell-ite) A dark red monoclinic mineral: $AsSbS_3$.
Getian (Get'-i-an) Eastern European backish-marine substage: lower Pliocene, lower Dacian (above Bosphorian substage, below Parscovian substage).
geversite (ge'-ver-site) A metallic cubic mineral of the *pyrite* group: $Pt(Sb,Bi)_2$.
geyser (gey'-ser) A type of hot spring that intermittently erupts jets of hot water and steam, the result of groundwater coming into contact with rock or steam hot enough to create steam under conditions preventing free circulation; a type of intermittent spring. Syn: *pulsating spring.*
geyser basin A valley that contains numerous springs, geysers, and steaming fissures fed by the same groundwater flow (Schieferdecker, 1959, term 4524).
geyser cone A low hill or mound built up of *siliceous sinter* around the orifice of a geyser. The term is sometimes mistakenly applied to an algal growth on objects (such as wooden snags) occurring along the shores of some Tertiary lakes. Syn: *geyser mound.*
geyser crater The bowl- or funnel-shaped opening of the geyser pipe, which often contains a *geyser pool.* Syn: *crater [grd wat].*
geyserite (gey'-ser-ite) A variety of *siliceous sinter* (opal), used esp. for the compact, loose, concretionary, scaly, or filamentous incrustation of opaline silica deposited by precipitation from the waters of a geyser. Obsolete syn. for all siliceous sinter (White et al., 1964).
geyser jet The plume of water and steam emitted during the eruption of a geyser.
geyser mound *geyser cone.*
geyser pipe The narrow tube or well of a geyser extending downward from the *geyser pool.* Syn: *pipe [grd wat]; geyser shaft.*
geyser pool The comparatively shallow pool of heated water ordinarily contained in a *geyser crater* at the top of a *geyser pipe.*
geyser shaft *geyser pipe.*
GF *grading factor.*
ghat A term used in India originally for a mountain pass, or a path leading down from a mountain, but now commonly and erroneously applied (by Europeans) to a mountain range, specif. the mountain ranges parallel to the east and west coasts of India. Etymol: Hindi. Also spelled: ghaut.
ghost [petrology] A visible outline of a former crystal shape, fossil, or other rock structure that has been partly obliterated (as by diagenesis or replacement) and that is bounded by inclusions and outlined by bubbles or foreign material. Syn: *phantom.*
ghost [seis] Seismic energy that travels upward from a seismic source and then is reflected downward at the base of the weathering or at the surface. It may unite with a downward-traveling wavetrain, or it may be distinguished as a separate wave. Ghosts may also occur where detectors (e.g., in a marine *streamer*) below the water surface respond to events reflected from the surface. Ghosts from earthquakes are indicated by lower-case p or s.
ghost coal A coal that burns with a bright white flame.
ghost crystal *phantom crystal.*
ghost member Any part of the ideal or typical cyclothem that is absent in a particular cyclothem. Cf: *phantom.*
ghost stratigraphy Traces, in highly metamorphosed strata, of the original lithology and stratification.
Ghyben-Herzberg ratio (Ghy'-ben-Herz'-berg) A ratio describing the static relation of fresh groundwater and saline groundwater in coastal areas. For each foot of freshwater head above sea level, the salt-water surface is displaced to 40 feet below sea level, i.e. in a ratio of 1:40. The static relationship is modified by dynamic factors that cause mixing of fresh and salt water, esp. the seaward flow of fresh water and tide-induced fluctuations of the interface. The ratio was formally defined by Badon-Ghyben in 1889 and independently by Herzberg in 1901. Syn: Ghyben-Herzberg relation; Ghyben-Herzberg principle; Ghyben-Herzberg formula. See also: *salt-water encroachment.*
ghyll Obsolescent spelling of *gill.*
gianellaite (gia-nel'-la-ite) A colorless to straw-yellow cubic mineral closely related to mosesite: $Hg_4(SO_4)N_2$.
giannettite A colorless to pale yellow triclinic mineral: $NaCa_2(Ti,Mn,Fe,Ce)Si_2O_7(F,O,OH)_2$. It is probably synonymous with *hainite.*
giant cusp (gi'-ant) A slightly protruding cusp, commonly 300-500 m from adjacent ones, with a submarine ridge continuing seaward as a transverse bar along one or both sides of which is a deep channel. Giant cusps appear to be characteristic of areas or times of relatively strong littoral currents.
giant desiccation polygon A *desiccation polygon* formed on a playa by contraction of muds upon drying, bounded by fissures or cracks measuring several meters in depth and up to 1 m in width, and extending over an area of several hundred square meters (Stone, 1967, p.228). Archaic.
giantism (gi'-ant-ism) The tendency for an evolving *lineage* to develop extremely large body size compared to that in other closely related lineages. Cf: *gigantism.*
giant planet Any one of the large outer planets of the solar system (Jupiter, Saturn, Uranus, and Neptune); also large companions of other stars. Cf: *terrestrial planet.*
giant ripple A ripple that is more than 30 m in length; it usually shows superimposed megaripples (Reineck and Singh, 1973, p.37-39). Called a dune by many workers.
giant's cauldron *giant's kettle.*
giant's kettle A cylndrical hole bored in bedrock beneath a glacier by water falling through a deep *moulin* or by boulders rotating in the bed of a meltwater stream; it may contain a pond or marsh. Syn: *moulin pothole; glacial pothole; pothole [glac geol]; giant's cauldron.*
giant stairway *glacial stairway.*
gibber (gib'-ber) An Australian term for a pebble or boulder; esp. one of the wind-polished or wind-sculptured stones that compose a desert pavement or the lag gravels of an arid region. It is pronounced with a hard "g".
gibber plain A desert plain strewn with wind-abraded pebbles, or *gibbers;* a gravelly desert in Australia.
gibbs A unit of measurement of entropy, heat capacity, and various commonly used thermodynamic functions that is essentially equivalent to *entropy unit.*
Gibbs free energy The Gibbs free energy G is defined by the equation G = H-TS, where H is *enthalpy,* T is absolute temperature, and S is entropy. With respect to any process at constant temperature and pressure, the state of equilibrium is defined by dG=O. Cf: *Helmholtz free energy.* Syn: *Gibbs function.*
Gibbs function *Gibbs free energy.*
gibbsite (gibbs'-ite) A white or tinted monoclinic mineral: $Al(OH)_3$. It is polymorphous with bayerite and nordstrandite. Gibbsite is formed by weathering of igneous rocks and is the principal constituent of bauxite; it occurs in micalike crystals or in stalactitic and spheroidal forms. Syn: *hydrargillite.*
gibbsite-like A term that describes the dioctahedral sheet in 1:1 and 2:1 clay minerals. This sheet is analogous to gibbsite in that gibbsite is two planes of closest packed oxygens with two out of three of the octahedral sites between the two planes occupied by trivalent cations, in this case Al. In the gibbsite-like sheet, some of the oxygens are replaced by hydroxyls.
Gibbs method An analytical method, based on classical thermodynamics, to formalize relationships between gradients in intensive variables (pressure, temperature, and chemical potentials of components) and gradients in the chemical compositions of minerals (Rumble, 1976).
Gibbs phase rule *phase rule.*
gibelite (gi'-bel-ite) A *trachyte* characterized by the presence of abundant large phenocrysts of alkali feldspar, sometimes microperthite, and a small amount of sodic clinopyroxene in a groundmass composed of alkali feldspar and a little quartz, and exhibiting flow texture. Amphibole is present as an accessory. The type locality, described by Washington in 1913, is on the island of Pantelleria in the western Mediterranean. Obsolete.
Gibraltar stone (Gi-bral'-tar) A light-colored *onyx marble* found at Gibraltar and elsewhere.
giessenite (gies'-sen-ite) A metallic grayish-black monoclinic mineral: $Pb_{26}(Cu,Ag)_2(Bi,Sb)_{20}S_{57}$.
gigantism (gi'-gan-tism) (a) In animals, development to abnormally large size as a result of the excessive growth of certain hard parts, often accompa-

nied by structural weakness and sexual impotence. Cf: *nanism.* (b) In plants, excessive vegetative growth.

gigayear (gi'-ga-year) A term proposed by Rankama (1967) for one billion (10^9) years. Cf: *Ga.* Syn: *eon (b).*

Giggenbach bottle An evacuated alkaline-solution-filled bottle (e.g., sodium hydroxide, NaOH), generally connected to a titanium or silica tube which is inserted into a fumarole. The gases bubble through the solution, condense, and are absorbed by the solution, which can be analyzed (William-Jones and Rymer, 2000).

gilalite (gi'-la-lite) A green or blue-green monoclinic mineral: $Cu_5Si_6O_{17}·7H_2O$.

Gilbert Chron An interval of geologic time from about 5.88 to about 3.58 Ma, delimited by geomagnetic reversals, and generally having a magnetic field reversed in polarity from that of the present. Named for William Gilbert, English physicist. Collective symbol: C2Ar, C3n, and C3r.

gilgai (gil'-gai) The microrelief of small basins and knolls or valleys and ridges on a soil surface produced by expansion and contraction during wetting and drying (usually in regions with distinct, seasonal, precipitation patterns) of clayey soils that contain smectite.

gill [paleont] An organ for obtaining oxygen from water; e.g. a *branchia* of a crustacean or mollusk. In aquatic vertebrates, a set of vascularized filaments that function in respiratory, metabolic waste, and ionic exchanges between the bloodstream and surrounding water.

gill [streams] (a) A term used in the English Lake District for a deep narrow rocky valley, esp. a wooded ravine with a rapid stream running through it. (b) A narrow mountain stream or brook flowing swiftly through a gill. Also, a term used in Yorkshire, England, for a stream flowing in a shallow valley, sometimes ending in a pothole. Also spelled: *ghyll;* ghyl; gil.

gill bar One of several serially arranged bars of cartilage or cartilage-replacement bone, jointed and movable in jawed fishes but fixed in Agnatha, which support the gills behind or below the head; collectively, the gill-supporting part of the visceral skeleton.

gill chamber *branchial chamber.*

gillespite (gil-les'-pite) A red tetragonal mineral: $BaFe^{2+}Si_4O_{10}$.

gill slit (a) An opening in an echinoderm, such as a fissure in the disc of an ophiuroid along the side of the base of an arm and leading into the bursa, or an indentation of the peristomial margin of echinoid interambulacra for the passage of the stem of an external branchia. (b) One of several openings behind or below the head of a fish, for the escape of water being passed from the pharynx over the gills. Syn: *branchial slit.*

gillulyite A vitreous to metallic deep red monoclinic mineral: $Tl_2(As,Sb)_8S_{13}$.

gilmarite A vitreous green-blue triclinic mineral: $Cu^{2+}{}_3AsO_4(OH)_3$.

gilpinite (gil'-pin-ite) *johannite.*

gilsonite (gil'-son-ite) *uintahite.*

giniite (gin'-i-ite) A blackish green to brown monoclinic mineral: $Fe^{2+}Fe^{3+}{}_4(PO_4)_4(OH)_2·2H_2O$.

ginkgo (gink'-go) A member of the gymnosperm subclass Ginkgoales, characterized by dichotomous venation, fan-shaped or regularly bifurcating leaves, and terminally borne seeds. Ginkgos range from the Permian.

ginorite (gi'-no-rite) A white monoclinic mineral: $Ca_2B_{14}O_{23}·8H_2O$.

ginzburgite (ginz'-burg-ite) (a) *roggianite.* (b) Formerly, a group name for iron-rich clay minerals of the *kaolin* group.

giobertite (gio'-bert-ite) *magnesite.*

giorgiosite An earthy white mineral of dubious validity: $Mg_5(CO_3)_4(OH)_2·5H_2O$. It may be hydromagnesite.

gipfelflur (gip'-fel-flur) (a) Uniformity or accordance of high mountain summits, presumed to develop independent of geologic structure and of rock type, at a level of uplift limited by isostatic control or because the degrading agencies above that level keep pace with continuing uplift (von Engeln, 1942, p.100). (b) An imaginary, relatively smooth surface touching the tops of the accordant summits of a region. See also: *summit plane; peak plain.* Etymol: German "Gipfelflur", "summit plain". The term was originated by A. Penck (1919).

GIPV *gross in-place value.*

girasol (gir'-a-sol) adj. Said of any gem variety, such as sapphire or chrysoberyl, that exhibits a billowy, gleaming, round or elongated area of light that "floats" or moves about as the stone is turned or as the light source is moved. n. A name that has been applied to many gemstones with a girasol effect, such as moonstone; specif. a translucent variety of *fire opal* with reddish reflections in a bright light and a faint bluish-white floating light emanating from the center of the stone.

giraudite (gi-raud'-ite) A metallic cubic mineral of the *tennantite* group: $(Cu,Zn,Ag)_{12}(As,Sb)_4(Se,S)_{13}$. It is the arsenian analogue of hakite.

girdite (gird'-ite) A chalky white monoclinic mineral: $Pb_3(Te^{4+}O_3)(Te^{6+}O_4)(OH)_2$.

girdle [gem] (gir'-dle) The outer edge or periphery of a fashioned gemstone; the portion that is usually grasped by the setting or mounting; the narrow dividing line between the crown and the *pavilion.*

girdle [paleont] (a) The region of overlap of the two valves of a diatom frustule, consisting of the *connecting band* from each valve. Also, either of the two connecting bands forming the girdle. Syn: *cingulum.* (b) A transverse furrow around the theca or body of a dinoflagellate; the part of the shell lying between epivalve and hypovalve in certain dinoflagellates. (c) The muscular, flexible marginal band of uniform width of the mantle of a chiton, encircling the shell plates and differentiated from the central portion of the back. It belongs with the soft parts of the chiton, although its covering may be studded with needlelike or scalelike calcareous spicules. (d) A spiral or annular shelf in the skeleton of a spumellarian radiolarian. (e) In vertebrates, the ringlike group of bones that connect, muscularly or directly by ligaments or bony fusion, front limbs (through the pectoral girdle) or hind limbs (through the pelvic girdle) to the axial skeleton for purposes of supporting the limbs. See also: *perignathic girdle.*

girdle [struc petrol] On an equal-area projection, a belt of concentration of points representing orientations of fabric elements. If this belt coincides approximately with a great circle of the projection, then it is referred to as a great-circle girdle. If the belt of concentration coincides approximately with a small circle of the projection, then it is called a small-circle or *cleft girdle* (Turner and Weiss, 1963, p.58). Cf: *maximum [struc petrol].*

girdle axis On a *fabric diagram* in which a girdle is produced, the girdle axis coincides with that pole which is normal to the plane best approximating a great-circle girdle and is the axis of the cone corresponding to a small-circle girdle.

girdle facet One of the 32 triangular facets that adjoin the girdle of a round brilliant-cut gem; there are 16 facets above the girdle and 16 below.

girdle list A high membranous ridge perpendicular to the wall and bordering the girdle of a dinoflagellate.

girdle view The side view of a diatom frustule, showing only the edges of the valves.

girdle zone The circular central region with shelves in a radiolarian shell.

Girondian (Gi-ron'-di-an) A European stage: lower Miocene (above Chattian, below Langhian). It includes the Burdigalian and Aquitanian standard stages.

girvanella (gir-va-nel'-la) An *microbial biscuit* characterized by a complex of microscopic filaments.

girvasite A colorless or creamy white monoclinic mineral: $NaCa_2Mg_3(PO_4)_2[PO_2(OH)_2]CO_3(OH)_2·4H_2O$.

GIS *geographic information system.*

gisement (gise'-ment) The angle between the *grid meridian* and the *geographic meridian.* It is a term used primarily in connection with military grids. Cf: *grid azimuth.* Etymol: French, "outcrop, occurrence." Syn: *mapping angle.*

gismondine (gis'-mon-dine) A white monoclinic zeolite mineral: $Ca_2Al_4Si_4O_{16}·8H_2O$. It sometimes contains potassium.

gitology (gi-tol'-o-gy) A term, mainly used in Europe, to describe the study of ore-deposit genesis in the broadest sense, including chemical, thermodynamic, petrological, and economic disciplines. Etymol: French *gisement,* an "ore body".

gittinsite (git'-tins-ite) A chalky white monoclinic mineral of the *thortveitite* group: $CaZrSi_2O_7$.

giumarrite (giu-mar'-rite) An amphibole-bearing *monchiquite,* named for Giumarra, Sicily. Obsolete.

giuseppettite (giu-sep-pet'-tite) A pale violet-blue hexagonal mineral of the *cancrinite* group: $(Na,K,Ca)_{7-8}(Si,Al)_{12}O_{24}(SO_4,Cl)_{1-2}$.

Givetian (Gi-ve'-ti-an) A stage of the Standard Global Chronostratigraphic Scale: Middle Devonian (above Eifelian, below Frasnian). The basal GSSP is near Tafilait, Morocco (Walliser et al., 1995).

gizzard (giz'-zard) (a) The last part of the foregut (anterior part of the alimentary canal) of an arachnid, developed as a pumping organ. Its dorsal dilative muscle is attached to an apodeme visible on the external surface of the carapace (TIP, 1955, pt. P, p.62). (b) The posterior end of the stomach of birds and other parts of the alimentary canal in diverse vertebrates, esp. as characterized by extreme development of muscular walls.

gizzard stone *gastrolith.*

gjá [volc] A gaping, dilation fissure of the Icelandic rift system, from which volcanic eruptions may take place. Etymol: Icelandic, "chasm". Pl: gjár.

gjerdingenite-Fe A pale yellow to orange-yellow monoclinic mineral of the *labuntsovite* group: $K_2(H_2O)_2(Fe,Mn)(Nb,Ti)_4(Si_4O_{12})_2(O,OH)·4H_2O$.

glabella (gla-bel'-la) The raised axial part of the *cephalon* of a trilobite. It represents the anterior part of the axis or axial lobe. In some forms the term includes the *occipital ring.* Pl: glabellae.

glabellar furrow (gla-bel'-lar) A narrow groove extending transversely across the *glabella* of a trilobite. It is commonly incomplete or interrupted. Also called: lateral glabellar furrow.

glabellar lobe A transverse lobe on the *glabella* of a trilobite, more or less bounded by complete or partial glabellar furrows. It represents a remnant of the original segments fused in the *cephalon.* Also called: lateral glabellar lobe.

glacial (gla'-cial) adj. (a) Of or relating to the presence and activities of ice or glaciers, as *glacial* erosion. (b) Pertaining to distinctive features and materials produced by or derived from glaciers and ice sheets, as *glacial* lakes. (c) Pertaining to an ice age or region of glaciation. (d) Suggestive of the extremely slow movement of glaciers. (e) Used loosely as descriptive or suggestive of ice, or of below-freezing temperature. n. A *glacial age,* or *glacial stage,* of a glacial epoch, esp. of the Pleistocene Epoch; e.g. the Wisconsin *glacial.*

glacial action All processes due to the agency of glacier ice, such as erosion, transportation, and deposition. The term sometimes includes the action of meltwater streams derived from the ice. See also: *glacial erosion.*
glacial advance *advance [glaciol].*
glacial age A subdivision of a glacial epoch, esp. of the Pleistocene Epoch. Syn: *glacial.*
glacial basin A *rock basin* caused by erosion of the floor of a glacial valley.
glacial block A large, markedly angular rock fragment that has not been greatly modified during glacial transport.
glacial boulder A *boulder* or large rock fragment that has been moved for a considerable distance by a glacier, being somewhat modified by abrasion but not always "rounded". Cf: *erratic.* Obsolete syn: *ice boulder [glac geol].*
glacial boundary The position occupied, in a given region or during a given glacial stage, by the outer or lower margin of an ice sheet; this may extend beyond the terminal moraine.
glacial canyon A canyon eroded by a glacier, usually occupying the site of an older stream valley and often having a U-shaped cross profile.
glacial chute A term suggested by Harvey (1931, p.231) for one of a group of narrow, closely spaced, steeply plunging glacial troughs, with vertical or nearly vertical walls and gently curving U-shaped bottoms, that acted as valves in passing or stopping snow, ice, and rock material. Examples occur on the flanks of Mount Puy Puy in Peru.
glacial-control theory (gla′-cial-con-trol′) A theory of coral-atoll and barrier-reef formation according to which marine erosion and lowering of the sea level during the Ice Age destroyed existing coral reefs and left an extensive, level rock surface from which coral reefs were built up during the postglacial rise of sea level. Theory was proposed by Reginald A. Daly in 1910. Cf: *antecedent-platform theory; subsidence theory.*
glacial cycle (a) A term used by Davis (1911, p.56) for the ideal case of glaciation operating for a long period of time under fixed climatic conditions such that glacial erosion would be complete and replaced by normal erosion. (b) "A major global climatic oscillation of the order of 10^5 yr, developed within an 'ice age' *sensu lato,* which may last 10^6-10^7 yr and which recurs at widely spaced intervals in geologic time (i.e. 2×10^8yr)" (Fairbridge, 1972).
glacial debris (a) *glacial drift.* (b) *debris [glaciol].*
glacial deposit *glacial drift.*
glacial-deposition coast (gla′-cial-de-po-si′-tion) A coast with partly submerged moraines, drumlins, and other glacial deposits.
glacial dispersal The fan-shaped or wider spread deployment of *erratics* by glacial ice.
glacial drainage (a) The flow system of glacier ice. (b) The system of meltwater streams flowing from a glacier or ice sheet.
glacial drainage channel A "safer expression" for *overflow channel,* because "the exact mode of origin of many channels supposedly due to overflow is uncertain" (Challinor, 1978, p.212); a channel formed by an ice-marginal, englacial, or subglacial stream.
glacial drift A general term for drift transported by glaciers or icebergs, and deposited directly on land or in the sea. Cf: *glaciofluvial drift.* Syn: *glacial deposit; glacial debris.*
glacial epoch Any part of geologic time, from Precambrian onward, in which the climate was notably cold in both the northern and southern hemispheres, and widespread glaciers moved toward the equator and covered a much larger total area than those of the present day; specif. the latest of the glacial epochs, known as the Pleistocene Epoch. Syn: *glacial period; ice age; drift epoch.*
glacial erosion The grinding, scouring, plucking, gouging, grooving, scratching, and polishing effected by the movement of glacier ice armed with rock fragments frozen into it, together with the erosive action of meltwater streams. See also: *glacial action.*
glacial erratic *erratic.*
glacial eustasy *glacio-eustasy.*
glacial flour *rock flour.*
glacial geology (a) The study of the geologic features and effects resulting from the erosion and deposition caused by glaciers and ice sheets, associated melt water, and Earth rheological response. Cf: *glaciology.* (b) The features of a region that has undergone glaciation. Obsolete syn: *glaciogeology.*
glacial groove A deep, wide, usually straight furrow cut in bedrock by the abrasive action of a rock fragment embedded in the bottom of a moving glacier; it is larger and deeper than a *glacial striation,* ranging in size from a deep scratch to a glacial valley. Syn: *groove [glac geol].*
glacialism (gla′-cial-ism) (a) The *glacier theory.* (b) *glaciation.* Obsolete.
glacial lake (a) A lake that derives much or all of its water from the melting of glacier ice, e.g. fed by meltwater, or lying on glacier ice and due to differential melting. (b) A lake occupying a basin produced by glacial deposition, as one held in by a morainal dam. (c) A lake occupying a basin produced in bedrock by glacial erosion (scouring, quarrying), as a *cirque lake* or *fjord lake.* (d) A lake occupying a basin produced by collapse of outwash material surrounding masses of stagnant ice. (e) *glacier lake.*
glacial lobe A short, broad *distributary glacier;* a less preferred syn. for *lobe.* Cf: *tongue [glaciol]; outlet glacier.*
glacial-lobe lake A lake occupying a depression that was excavated by a glacial lobe as it advanced over the drainage basin of a former river.
glacial-marine sedimentation The accumulation of glacially eroded, terrestrially derived sediment in the marine environment. Sediment may be introduced by fluvial transport, by ice rafting, as an ice-contact deposit, or by eolian transport. Deposits can be divided into a variety of glacial-marine (or "glacio-marine") facies (Molnia, 1983).
glacial maximum The time or position of the greatest advance of a glacier, glaciers, and/or ice sheets (such as the greatest extent of Pleistocene glaciation). Ant: *glacial minimum.* Syn: *glaciation limit.*
glacial meal *rock flour.*
glacial milk *glacier milk.*
glacial mill *moulin.*
glacial minimum The time or position of the greatest retreat of a glacier and/or ice sheets. Ant: *glacial maximum.*
glacial pavement A polished, striated, relatively smooth, planed-down rock surface produced by glacial abrasion. Cf: *boulder pavement [glac geol].* Syn: *ice pavement.*
glacial period (a) A syn. of *glacial epoch;* specif. the Pleistocene Epoch. Usage of this term is not strictly correct "as glacial intervals during earth history were not of the period rank, but of shorter duration" (ADTIC, 1955, p.34). (b) A geologic period, such as the Quaternary Period, that embraced an interval of time marked by one or more major advances of glaciers and/or ice sheets.
glacial plain A plain formed by the direct action of the glacier ice itself. Cf: *outwash plain.*
glacial polish A smoothed surface produced on bedrock by glacial abrasion.
glacial pothole A syn. of *giant's kettle.* The term is misleading because the feature is not produced by the direct scouring action of glacier ice, but by a stream of water falling through a *moulin,* by lateral or subglacial meltwater, or perhaps by a viscous mixture of water, ice, and rock fragments.
glacial pressure ridge *ice-pushed ridge.*
glacial recession *recession [glaciol].*
glacial refuge *refugium.*
glacial retreat *retreat.*
glacial scour The eroding action of a glacier, including the removal of surficial material and the abrasion, scratching, and polishing of the bedrock surface by rock fragments dragged along by the glacier. Cf: *grinding [glac geol].* Syn: *scouring.*
glacial scratch *glacial striation.*
glacial spillway *overflow channel.*
glacial stage A major subdivision of a glacial epoch, esp. one of the cycles of growth and disappearance of the Pleistocene ice sheets; e.g. the "Wisconsin Glacial Stage". Syn: *glacial.*
glacial stairway A glacial valley whose floor is shaped like a broad staircase composed of a series of irregular steplike benches (treads) separated by steep risers. Cf: *cirque stairway.* Syn: *giant stairway; cascade stairway; cascade [glaciol].*
glacial stream A flow of water that is supplied by melting glacier ice; a *meltwater* stream.
glacial stria A syn. of *glacial striation.* Pl: *striae.*
glacial striation One of a series of long, delicate, finely cut, commonly straight and parallel furrows or lines inscribed on a bedrock surface by the rasping and rubbing of rock fragments embedded at the base of a moving glacier, and usually oriented in the direction of ice movement; also formed on the rock fragments transported by the ice. Cf: *glacial groove.* Syn: *glacial scratch; glacial stria; drift scratch.*
glacial terrace A terrace formed by glacial action, either by rearranging glacial materials in terrace form (such as a remnant of a valley train), or by cutting into bedrock. Cf: *kame terrace.*
glacial theory *glacier theory.*
glacial till *till.*
glacial trough A deep, steep-sided usually *U-shaped valley* leading down from a cirque, and excavated by an alpine glacier that has widened, deepened, and straightened a preglacial river valley; e.g. Yosemite Valley, Calif.
glacial valley A usually, U-shaped, steep-sided valley showing signs of glacial erosion; a glaciated valley, or one that has been modified by a glacier.
glaciär The Swedish equivalent of *glacier.* Definite form: "glaciären". Example: Storglaciären, the Stor glacier.
glaciated (gla′-ci-at′-ed) Said of a formerly glacier-covered land surface, esp. one that has been modified by the action of a glacier or an ice sheet, as a *glaciated* rock knob. Cf: *glacier-covered.*
glaciated coast A coast whose features were modelled by continental glaciers of the Pleistocene Epoch, or a coast covered by glaciers at the present time.
glaciation (gla-ci-a′-tion) (a) The formation, movement, and recession of glaciers or ice sheets. (b) The covering of large land areas by glaciers or ice sheets. Syn: *glacierization.* (c) The geographic distribution of glaciers and ice sheets. (d) A collective term for the geologic processes of glacial activity, including erosion and deposition, and the resulting effects of such action on the Earth's surface. (e) Any of several minor parts of geologic time during which glaciers were more extensive than at present; a *glacial epoch,* or a *glacial stage.* "A climatic episode during which extensive glaciers developed, at-

tained a maximum extent, and receded" (ACSN, 1961, art.40). Syn: *glacialism.*

glaciation limit (a) The lowest altitude in a given locality at which glaciers can develop; the altitude below the minimum summit altitude of mountains on which glaciers occur but above the maximum summit altitude of mountains, although having topography favorable for glaciers, on which they do not occur. (b) *glacial maximum.*

glacier (gla'-cier) (a) A large mass of ice formed, at least in part, on land by the compaction and recrystallization of snow, moving slowly by creep due to gravitational forces, and surviving from year to year. Included are small mountain glaciers as well as ice sheets of continental size, and ice shelves which float on the ocean but are fed in part by ice formed on land. Cf: *ice stream; ice sheet; ice cap.* (b) Less preferred syn. of *alpine glacier.* (c) A term used in Alaska for floodplain icing or a mass of *ground ice.* (d) A streamlike landform having the appearance of, or moving like, a glacier; e.g. a *rock glacier.* Etymol: French "glace", "ice".

glacier advance *advance [glaciol].*

glacier band *foliation [glaciol].*

glacier bed The surface under a glacier.

glacier berg *glacier iceberg.*

glacier breeze *glacier wind.*

glacier bulb *expanded foot.*

glacier burst *glacier outburst flood.*

glacier cap *ice cap.*

glacier cave A cave that is formed within a glacier. Syn: *ice cave.*

glacier cone *debris cone [glaciol].*

glacier corn Glacier ice broken into irregular crystals of various sizes (Brigham, 1901, p.92).

glacier cornice A projecting mass of glacier ice. It was formerly underlain by ice containing numerous rock fragments which, when warmed by solar radiation, melted the ice around them and caused the ice above them to project like a *cornice* (Hobbs, 1912, p.397).

glacier-covered (gla'-cier-cov'-ered) Said of a land surface overlain by glacier ice at the present time. Cf: *glaciated.* Syn: *glacierized; ice-covered.*

glacière (gla-ci-ere' [Pron: glaciair]) A cave which contains perennial ice. Etymol: French. Cf: *ice cave.*

glacieret (gla-cier-et') (a) A very small glacier on a mountain slope or in a cirque, as in the Sierra Nevada, Calif.; a miniature alpine glacier. Cf: *cirque glacier.* (b) A tiny mass of ice or firn in high mountains, resembling a glacier but defying a precise definition; a *drift glacier.* Also spelled: glacierette.

glacier flood *glacier outburst flood.*

glacier flow The slow downward or outward movement of the ice in a glacier, due to the force of gravity (*gravity flow*). Internal deformation of a glacier by intragranular gliding, grain-boundary migration, and recrystallization, together with sliding of the glacier on its bed in some situations. Usually expressed in meters per day or year. Syn: *ice flow; flow [glaciol].*

glacier grain (a) An individual ice crystal in a glacier. (b) A mechanically separate particle of ice in a glacier. Cf: *grain [glaciol].*

glacier ice Any ice that forms in or was once a part of a glacier, including land ice that is flowing or that shows evidence of having flowed, and glacier-derived ice floating in the sea.

glacier iceberg An iceberg consisting of ice detached from a coastal glacier or ice shelf. Syn: *glacier berg.*

glacierization (gla'-cier-i-za'-tion) A term used in Great Britain for *glaciation* in the sense of the gradual covering or "inundation" of a land surface by glaciers or ice sheets.

glacierized (gla'-cier-ized) A term used in Great Britain for *glacier-covered.*

glacier lake A lake held in place by the damming of natural drainage by the edge or front of a glacier or ice sheet, as a lake ponded by glacier ice advancing across a valley, or a lake occurring along the margin of a continental ice sheet. Cf: *proglacial lake.* Syn: *glacial lake; marginal lake; ice-dammed lake.*

glacier meal *rock flour.*

glacier milk A stream of turbid, whitish meltwater containing *rock flour* in suspension. Syn: *glacial milk.*

glacier mill *moulin.*

glacier outburst flood A sudden, often annual, release of meltwater from a glacier or glacier-dammed lake, sometimes resulting in a catastrophic flood, formed by melting of a drainage channel or buoyant lifting of ice by water or by subglacial volcanic activity. Syn: *jökulhlaup; outburst; glacier burst; glacier flood.*

glacier recession *recession [glaciol].*

glacier remanié *regenerated glacier.*

glacier snout *terminus.*

glacier surge *surge [glaciol].*

glacier table A boulder or large block of rock supported by an *ice pedestal* that rises from the surface of a glacier. It occurs when melting of the glacier is retarded by the insulating effect of the rock.

glacier theory The theory, first propounded about 1840 and now universally accepted, that the drift was deposited through the agency of glaciers and ice sheets that were moving slowly from higher to lower latitudes during the Pleistocene Epoch. Cf: *drift theory [glac geol].* Syn: *glacial theory; glacialism.*

glacier tongue *tongue [glaciol].*

glacier wave A syn. of *ogive;* also used loosely and nonspecifically for *kinematic wave* and the active rapid-flow phase of a *surging glacier.*

glacier well *moulin.*

glacier wind (a) A cold *katabatic wind* blowing off a glacier. (b) A cold wind blowing out of ice caves in a glacier front, due to the difference in density between the colder air inside and the warmer air outside. Syn: *glacier breeze.*

glacioaqueous (gla'-ci-o-a'-que-ous) Pertaining to or resulting from the combined action of ice and water; the term is often used as a syn. of *glaciofluvial.* Syn: *aqueoglacial.*

glacio-eustasy (gla'-ci-o-eu'-sta-sy) The worldwide changes in sea level produced by the successive withdrawal and return of water in the oceans accompanying the formation and melting of ice sheets. Cf: *sedimento-eustasy; tectono-eustasy.* Adj: *glacio-eustatic.* Syn: *glacial eustasy; glacio-eustatism.*

glacio-eustatic (gla'-ci-o-eu-stat'-ic) Pertaining to *glacio-eustatism.*

glacio-eustatism (gla'-ci-o-eu'-stat-ism) *glacio-eustasy.*

glaciofluvial (gla'-ci-o-flu'-vi-al) Pertaining to the meltwater streams flowing from wasting glacier ice and esp. to the deposits and landforms produced by such streams, as kame terraces and outwash plains; relating to the combined action of glaciers and streams. Syn: *fluvioglacial; glacioaqueous.*

glaciofluvial drift *fluvioglacial drift.*

glaciogeology (gla'-ci-o-ge-ol'-o-gy) *glacial geology.*

glacio-isostasy (gla'-ci-o-i-sos'-ta-sy) Crustal adjustment to loading and unloading that is attributed to addition and removal of glacier ice. Cf: *isostasy; hydro-isostasy.*

glaciokarst (gla'-ci-o-karst) Karst developed in glaciated terrains. Cf: *alpine karst.*

glaciolacustrine (gla'-ci-o-la-cus'-trine) Pertaining to, derived from, or deposited in glacial lakes; esp. said of the deposits and landforms composed of suspended material brought by meltwater streams flowing into lakes bordering the glacier, such as deltas, kame deltas, and varved sediments.

glaciology (gla-ci-ol'-o-gy) (a) The study of all aspects of snow and ice; the science that treats quantitatively the whole range of processes associated with all forms of solid water. Syn: *cryology.* (b) The study of existing glaciers and ice sheets, and of their physical properties. This definition is not internationally accepted.

glacionatant (gla'-ci-o-na'-tant) Relating to or derived from floating ice of glacial origin, as "glacionatant till." Rarely used.

glaciosolifluction (gla'-ci-o-sol'-i-fluc'-tion) Gravitational sliding of heterogeneous material mixed with water, on the surface of a melting glacier.

glaciospeleology (gla'-ci-o-spe'-le-ol'-o-gy) The study of glacier caves.

glaciotectonic (gla'-ci-o-tec-ton'-ic) *cryotectonic.*

glacis (gla'-cis) A gently inclined slope or bank, less steep than talus; e.g. a piedmont slope, or an easy slope on a mountain side. Etymol: from its resemblance to a glacis used in fortifications as a defense against attack; originally from Middle French glacer, "to freeze", a British usage.

glade [geog] A term that usually indicates a clearing between slopes; it can be a high meadow, sometimes marshy and forming the headwaters of a stream, or it can be a low, grassy marsh, which is periodically inundated.

glade [karst] (a) A Jamaican term for an elongate depression, having steep sides, in which a generally flat floor is divided into small basins separated by low divides. (b) In Tennessee, *limestone pavement* having extensive growth of cedar trees. Cf: *karst valley.*

gladite (glad'-ite) A lead-gray orthorhombic mineral: $PbCuBi_5S_9$.

gladius (glad'-i-us) The *proostracum,* with rudimentary *rostrum* and *phragmocone,* of the Teuthida (coleoid cephalopods). Etymol: Latin *gladius,* "sword."

gladiusite A vitreous dark green monoclinic mineral: $Fe^{3+}_2(Fe^{2+},Mg)_4PO_4(OH)_{11}\cdot H_2O$.

Gladstone-Dale relationship An empirical equation describing the increase in *index of refraction* with density of a solid. It is $(n-1)/\rho = K_C$, where n is the refractive index, ρ is density, and K_C is the specific refractivity.

glaebule (glae'-bule) A term proposed by Brewer (1964, p.259-260) for a three-dimensional unit, usually prolate to equant in shape, within the matrix of a soil material, recognizable by its greater concentration of some constituent, by its difference in fabric as compared with the enclosing soil material, or by its distinct boundary with the enclosing soil material; e.g. a nodule, concretion, septarium, pedode, or papule. Etymol: Latin "glaebula", "a small clod or lump of earth". Pl: glaebules. See also: *papule.*

glagolevite A vitreous colorless monoclinic mineral of the *chlorite* group: $NaMg_6(Si_3Al)O_{10}(OH,O)_8\cdot H_2O$.

glamaigite (gla-mai'-gite) An intrusive breccia composed of dark patches of *marscoite* in a lighter groundmass. The name, given by Harker in 1904, is for Glamaig, Isle of Skye, Scotland, and was intended for use only at that locality. Obsolete.

glance A mineral that has a *splendent luster;* e.g. chalcocite, or copper glance.

glance coal *pitch coal.*

glance pitch A variety of *asphaltite* with a brilliant conchoidal fracture, sometimes called *manjak.* It is similar to gilsonite, but has a higher specific gravity and fixed-carbon content. It fuses between 230°F and 250°F.

gland duct *setal duct.*

glare ice A smooth, glassy or bright, highly reflective sheet of ice on a surface of water, land, or glacier.
glareous (glar′-e-ous) Said of an organism that lives in gravelly soil.
glaserite (gla′-ser-ite) A white to gray, blue, or greenish trigonal mineral: $K_3Na(SO_4)_2$. It may be synonymous with aphthitalite.
glass [chem] A state of matter intermediate between the close-packed, highly ordered array of a crystal and the poorly packed, highly disordered array of a gas. Most glasses are supercooled liquids, i.e. are metastable, but there is no true break in the change in properties between the metastable and stable states. The distinction between glass and liquid is made solely on the basis of viscosity, and is not necessarily related, except indirectly, to the difference between metastable and stable states.
glass [ign] An amorphous product of the rapid cooling of a magma. It may constitute the whole rock (e.g. obsidian) or only part of a groundmass. Cf: *volcanic glass.*
glass sand A sand that is suitable for glassmaking because of its high silica content (93-99 + %) and its low content of iron oxide, chromium, cobalt, and other colorants.
glass schorl *axinite.*
glass sponge *hyalosponge.*
glass transition temperature In the cooling of a supercooled liquid, the temperature (depending on the cooling rate) below which little structural change occurs and properties such as thermal expansion resemble those of a crystalline solid.
glassy (glass′-y) Said of the texture of certain extrusive igneous rocks, which is similar to that of broken glass or quartz and developed as a result of rapid cooling of the lava, without distinct crystallization. Syn: *hyaline [ign]; vitreous [ign].*
glassy feldspar *sanidine.*
glassy luster *vitreous luster.*
glauberite (glau′-ber-ite) A brittle, lightly colored, monoclinic mineral: $Na_2Ca(SO_4)_2$. It has a vitreous luster and saline taste, and occurs in saline residues.
Glauber's salt (Glau′-ber's salt′) A syn. of *mirabilite.* Named after Johann R. Glauber (1604-1668), German chemist. Also spelled: Glauber salt.
glaucocerinite (glau-co-cer′-i-nite) A sky-blue rhombohedral mineral: $(Zn,CuAl)_9(OH)_{18}(SO_4)_2 \cdot 10H_2O$. Also spelled: *glaucokerinite.*
glaucochroite (glau-co-chro′-ite) A bluish green, white, pinkish, or brown orthorhombic mineral: $CaMn^{2+}SiO_4$. It is a member of the *monticellite* (ordered olivine) group.
glaucodot (glau′-co-dot) A metallic white orthorhombic mineral of the *arsenopyrite* group: (Co,Fe)AsS. Syn: glaucodote.
glaucokerinite (glau-co-ker′-i-nite) *glaucocerinite.*
glauconarenite (glau-co-nar′-e-nite) *glauconitic sandstone.*
glauconite (glau′-co-nite) A group name for a series that comprises dioctahedral interlayer-deficient micas. Often interstratified with smectite as the mixed-layered mineral glauconite/smectite. When mixed with other minerals or when referring to morphological features, the term glauconitic is appropriate. Often assumed to be associated with specific conditions of deposition, but the Nomenclature Committtee of the Clay Minerals Society stated, "Mode of origin is not a criterion..." (Bailey et al., 1979).
glauconitic (glau-co-nit′-ic) Said of a mineral aggregate that contains glauconite, resulting in the characteristic green color, e.g. glauconitic shale or clay.
glauconitic mica One of the "end members of the glauconitic family," the other being *glauconitic smectite* (Odin and Matter, 1981).
glauconitic sand *greensand.*
glauconitic sandstone A sandstone containing sufficient grains of glauconite to impart a marked greenish color to the rock; *greensand; glauconarenite.*
glauconitic smectite One of the "end members of the glauconitic family," the other being *glauconitic mica* (Odin and Matter, 1981).
glaucony (glau′-co-ny) A term introduced by Odin and Letolle (1980) to designate the grains and films formed in the development of *glauconite.* It is a morphological or facies term. Constituents of glaucony are referred to as glauconitic smectite and glauconitic mica. See also: Odin and Matter (1981); Odom (1984). The term "glauconie" was apparently first used by Brongniart (1823). Etymol: Greek "glaukos", "bluish green."
glaucophane (glau′-co-phane) A blue, bluish-black, or grayish-blue monoclinic mineral of the *amphibole* group: $Na_2(Mg_3Al_2)Si_8O_{22}(OH)_2$. It is a fibrous or prismatic mineral that occurs only in certain crystalline schists resulting from regional metamorphism of sodium-rich igneous rocks (such as spilites).
glaucophane schist A type of *amphibole* schist in which glaucophane rather than hornblende is an abundant mineral. Epidote commony occurs, and there are quartz and mica varieties (Holmes, 1928, p.106). Cf: *blueschist.*
glaucophane-schist facies The facies (set of metamorphic mineral assemblages) in which mafic rocks are represented by combinations of sodic amphibole (e.g., glaucophane, crossite), lawsonite, sodic pyroxene, aragonite, epidote, and garnet. The mineral pair jadeite + quartz is also diagnostic. Exact definitions of the facies and its subdivisions vary (Turner, 1968). It represents lower temperatures and higher pressures than the *greenschist facies.* It is characteristic of metamorphism in subduction zones, with their unusually low geothermal gradients. Syn: *blueschist facies.*
glaucopyrite (glau-co-py′-rite) A variety of löllingite containing cobalt.
glaukosphaerite (glau-ko-sphae′-rite) A green monoclinic mineral: $(Cu,Ni)_2(CO_3)(OH)_2$. Cf: *nullaginite; rosasite.*
G layer The seismic region of the Earth below 5,150 km, equivalent to the *inner core.* It is a part of a classification of the Earth's interior made up of layers A to G. Rarely used.
glaze A fired glass surface on ceramic materials.
GLC *gas-liquid chromatography.*
glei *gley.*
gleization (glei-za′-tion) The formation of a *gley soil.* Syn: *gleying.*
glen A narrow, steep-sided secluded valley, usually wooded, often containing a stream or lake at its bottom; esp. a narrow-floored, glaciated mountain valley in Scotland and Ireland. It is narrower and more steep-sided than a strath. Syn: *glyn.*
Glencoe-type caldera (Glen′-coe) *Cauldron subsidence* resulting from the collapse of the roof of a magma chamber along ring fractures, i.e. stoping of a cylindrical block of the crust (Williams, 1941, p.246).
glendonite (glen′-don-ite) A pseudomorph of a carbonate (calcite or esp. siderite) after glauberite.
Glen flow law An empirical relation of the *shear strain* rate (ε) of ice to the *shear stress* (σ), $\varepsilon = k\sigma^n$, where the parameter k varies with temperature, type of ice, and geometry of stress, and n is a number usually taken to be 3 but possibly as low as 1 (under low stresses) or >5 (under deviatoric stresses higher than are found in most naturally occurring ice masses) (Glen, 1955, p.528). This relation is basic to most analyses of glacier flow.
glenmuirite (glen-muir′-ite) Similar to *essexite,* differing only in having analcime, rather than nepheline, as the principal feldspathoid. Named by Johannsen (1938) for Glenmuir Water, Ayrshire, Scotland. Obsolete.
glessite (gles′-site) A brown variety of retinite found on the shores of the Baltic Sea.
gletscherschlucht (glet′-scher-schlucht) A gorge sculptured by meltwater streams, often initiated where a moulin empties onto jointed rock. Etymol: German "Gletscherschlucht", "glacier gorge".
gley A syn. of *gley soil.* Also spelled: *glei.*
gleying (gley′-ing) *gleization.*
gley soil Soil developed under conditions of poor drainage, resulting in reduction of iron and other elements and in gray colors and mottles. The term is obsolete in the U.S.
glide (a) A gently flowing, calm reach of shallow water in a stream. (b) *snow glide.*
glide bedding A variety of *convolute bedding* produced by subaqueous gliding. The term is little used. Cf: *slip bedding.*
glide breccia A breccia formed by subaqueous gravitational movements that deform, shatter, or crush newly formed or partly consolidated strata. It may be produced by overloading, earthquakes, or deformation.
glide direction The direction of translation parallel to glide planes in a crystal structure.
glide fold Rarely used syn. of *shear fold.*
glide horizon *décollement.*
glide plane [cryst] A symmetry element in a *crystal structure* that relates parts on opposite sides by reflection plus translation parallel to the plane. Cf: *a-glide; b-glide; c-glide; d-glide; n-glide.* Syn: *glide reflection; translation plane; gliding plane.*
glide plane [struc geol] *décollement.*
glide plane [struc petrol] The surface along which *dislocations* move. A glide plane is a crystallographic plane across which the bond strength is relatively low. Syn: *slip plane [struc petrol].*
glide reflection *glide plane [cryst].*
glide twin *deformation twin.*
gliding [cryst] (glid′-ing) *crystal gliding.*
gliding [tect] *gravitational sliding.*
gliding flow *Flow [exp struc geol]* involving gliding parallel to the preferred crystallographic orientation, e.g. intragranular deformation in a crystal by twin gliding or translation gliding.
gliding plane *glide plane [cryst].*
gliding surface *slip surface.*
glimmer (glim′-mer) A syn. of *mica.* Etymol: German "Glimmer".
glimmergabbro (glim-mer-gab′-bro) A biotite-bearing *gabbro.* Obsolete.
glimmerite (glim′-mer-ite) *biotite* rock.
glimmerton (glim′-mer-ton) An early name for *illite.*
glinite (glin′-ite) A group name for clay minerals from clay deposits.
glint An escarpment or steep cliff, esp. one produced by erosion of a dipping resistant formation. See also: *klint [coast].* Etymol: Norwegian, "boundary".
glint lake A lake formed along a *glint line,* esp. a long, narrow glacial lake occupying a basin excavated in bedrock where a glacier is dammed by an escarpment ("glint"), e.g. certain lakes in Norway and Scotland.
glint line An extensive erosional escarpment produced by the denudation of a very gently dipping resistant formation, as the Silurian limestone of the Great Lakes region. The term is used specif. for the boundary between an

ancient shield and younger rocks, e.g. in Russia where Paleozoic rocks rise above the Baltic Shield, or at the western margin of the Canadian shield.

Global Boundary Stratotype Section and Point A *boundary stratotype* that designates a globally correlatable chronostratigraphic boundary, following strict procedures set forth by the IUGS International Commission on Stratigraphy. Syn: *marker point.* Abbrev: GSSP.

Global Positioning System The NAVigation Satellite Timing and Ranging (NAVSTAR) GPS is a passive, satellite-based, navigation system operated and maintained by the Department of Defense. Its primary mission is to provide passive global positioning/navigation for land-, air- and sea-based strategic and tactical forces by observation of 24 satellites (4 each in 6 different 55-degree orbital planes) at 20-200 km altitude. It employs two frequencies in the 10-20 GHz band. The radio navigation system, which is used for surveying, navigation, tracking, and field data collection, is continuously available worldwide with 4 satellites observable. From observation of these, a stationary or moving receiver can calculate precise time, three-dimensional position (latitude, longitude, and height), and three-dimensional velocity based on triangulation. Each satellite broadcasts its location. Abbrev: GPS. A GPS receiver is simply a range measurement device: distances are measured between the receiver point and the satellites, and the position is determined from the intersections of the range vectors. These distances are determined by a GPS receiver which accurately measures the time it takes signals to travel from the satellites to the station. This measurement process is similar to that used in conventional pulsing marine navigation systems and in phase comparison electronic distance measurement land surveying equipment. See also: *differential global positioning system.*

global tectonics Tectonics on a global scale, such as tectonic processes related to very large-scale movement of material within the Earth; specif. *plate tectonics.* Cf: *megatectonics.*

globe (a) A body having the form of a sphere; specif. a spherical, typically hollow ball that has a map of the Earth drawn on it and that is usually rotatable at an angle corresponding to the inclination of the Earth's axis. Also, a chart of the celestial sphere, depicted on a sphere. (b) A planet; esp. the Earth.

globigerina ooze (glo'-big-er-i'-na) A deep-sea pelagic sediment containing at least 30% foraminiferal tests, predominantly of the genus *Globigerina.* Archaic synonym for *foraminiferal ooze.*

globigerinid (glo'-big-er-i'-nid) Any planktonic foraminifer belonging to the superfamily Globigerinacea, characterized by a perforate test with bilamellid septa and walls of radial calcite crystals. Range, Middle Jurassic to present. Var: globigerine.

globosphaerite (glo-bo-sphae'-rite) A more or less spherical *cumulite* in which the globulites have a somewhat radial arrangement. Obsolete.

globular (glob'-u-lar) *spherulitic.*

globular projection Generally a non-azimuthal projection, developed before 1700, on which a hemisphere is enclosed in a circle and meridians and parallels are simple curves, such as circular arcs, or are straight lines. The projection is neither conformal nor equal-area. The al-Biruni or Nicolosi form of the globular projection, with all circular arcs, was formerly used for maps of the Eastern and Western Hemispheres.

globulite (glob'-u-lite) A spherical *crystallite* commonly found in volcanic glass.

globulith (glob'-u-lith) An intrusive body, or group of associated bodies, having a globular or botryoidal shape and almost concordant contacts, resulting from the effects of the intrusion(s) on the immediate surroundings (Berthelsen, 1970, p.73). Obsolete.

glockerite (glock'-er-ite) A cryptocrystalline variety of *lepidocrocite* with SO_3 and H_2O.

glomeroclastic (glom'-er-o-clas'-tic) Pertaining to particles grouped together in clusters in a carbonate sedimentary rock. Also said of the texture characterized by lumps. Obsolete.

glomerocryst (glom'-er-o-cryst') An aggregate of crystals of the same mineral. Cf: *polycrystal.*

glomerophyric (glom'-er-o-phyr'-ic) An obsolete term applied to the texture of porphyritic igneous rocks containing closed clusters of equant crystals of the same mineral. Cf: *cumulophyric; gregaritic.* Syn: *glomeroporphyritic.*

glomeroporphyritic (glom'-er-o-por'-phy-rit'-ic) Igneous texture characterized by clusters of phenocrysts scattered throughout the rock. Syn: *glomerophyric.*

glomospirine (glom-o-spi'-rine) Having an irregularly wound coiled tubular chamber; specif. pertaining to the foraminifer *Glomospira.*

glory-hole mining A type of opencut mining in which the orebody is worked from the top down in a conical excavation and the ore is removed by an underground system beneath the orebody. Syn: *mill-hole mining.*

gloss *polish.*

gloss coal The highest ranking lignite. It is deep black, compact, with definite conchoidal fracture and glossy luster. Cf: *subbituminous coal.*

glossopterid (glos-sop'-ter-id) n. The informal name for the fossil gymnosperm genus *Glossopteris* and its allies, whose foliage is common in the Permian of Gondwana. adj. Pertaining to such a plant or plant assemblage.

glossothyropsiform (glos'-so-thy-rops'-i-form) Said of a brachiopod *loop* developed from the cryptacanthiiform stage by final resorption of the posterior part of the *echmidium* and consisting of two descending branches unconnected posteriorly, bearing two broad ascending elements joined by a wide transverse band (TIP, 1965, pt. H, p.145).

glowing avalanche (glow'-ing) *ash flow.*

glowing cloud *nuée ardente.*

glucine (glu'-cine) An acicular colorless or yellow mineral: $CaBe_4(PO_4)_2(OH)_4{\cdot}0.5H_2O$.

glushinskite (glu-shin'-skite) A colorless or white monoclinic oxalate mineral: $Mg(C_2O_4){\cdot}2H_2O$.

glyn A Welsh syn. of *glen.*

glyptogenesis (glyp-to-gen'-e-sis) The sculpture of the Earth's surface by erosion. Obsolete.

glyptolith (glyp'-to-lith) A term proposed by Woodworth (1894a, p.70) for a wind-cut stone or *ventifact.* Etymol: Greek "glyptos", "carved", + "lithos", "stone".

G-M counter *Geiger-Müller counter.*

gmelinite-Ca (gme'-lin-ite) A reddish hexagonal zeolite mineral: $Ca_4(Si_{16}Al_8)O_{48}{\cdot}23H_2O$.

gmelinite-K A colorless to brownish hexagonal zeolite mineral: $K_8(Si_{16}Al_8)O_{48}{\cdot}23H_2O$.

gmelinite-Na A colorless to pale tinted hexagonal zeolite mineral: $Na_8(Si_{16}Al_8)O_{48}{\cdot}21H_2O$.

gnamma hole (gnam'-ma) A term used in the deserts of Western Australia for a rounded hollow eroded or indented in solid rock, usually at the intersections of joints in granite, and frequently containing water; it has a narrow orifice, but widens out below.

gnathal lobe (gnath'-al) The masticatory *endite* of the mandible of a crustacean.

gnathobase (gnath'-o-base) The serrate oral margin on the *coxa* of a eurypterid.

gnathosoma (gnath-o-so'-ma) The anterior part of the body in the arachnid order Acarida, bearing the mouth parts. Obsolete syn: *capitulum.*

gnathostome (gnath'-o-stome) Any vertebrate with anterior visceral skeleton modified into jaws; a member of any vertebrate class except the *Agnatha.*

gnathothorax (gnath-o-tho'-rax) The thorax and the part of the head bearing the feeding organs of an arthropod; e.g. the *tagma* of a crustacean resulting from fusion of mandibular and two maxillary somites with one or more thoracic somites, limbs of which are modified to act as mouth parts (TIP, 1969, pt. R, p.96). Cf: *cephalothorax.*

gneiss A foliated rock formed by regional metamorphism, in which bands or lenticles of granular minerals alternate with bands or lenticles in which minerals having flaky or elongate prismatic habits predominate. Generally less than 50% of the minerals show preferred parallel orientation. Although a gneiss is commonly feldspar- and quartz-rich, the mineral composition is not an essential factor in its definition. Varieties are distinguished by texture (e.g., augen gneiss), characteristic minerals (e.g., hornblende gneiss), or general composition and/or origins (e.g., granite gneiss). Cf: *gneissic; gneissoid; gneissose.*

gneiss dome A domal body of high-grade metamorphic rocks whose surface is typically a shear zone. Cf: *mantled gneiss dome.*

gneissic (gneiss'-ic) Pertaining to the texture or structure typical of gneisses, with foliation that is more widely spaced, less marked, and often more discontinuous than that of a *schistose* texture or structure (Johannsen, 1931). Cf: *gneissoid; gneissose.*

gneissic banding A type of foliation in metamorphic rock defined by compositional banding. Typically, gneissic banding consists of alternating dark (mafic) and light (felsic or silicic) bands of rock. Gneissic banding may reflect original compositional layering in the protolith, or lit-par-lit intrusion, or metamorphic differentiation by diffusion, or transposition. Less used syns: *gneissic structure; gneissose structure; gneissosity.*

gneissic structure *gneissic banding.*

gneissoid (gneiss'-oid) Pertaining to a gneisslike structure or texture that is the result of nonmetamorphic processes, e.g. viscous magmatic flow forming a gneissoid granite. Cf: *gneissic; gneissose.*

gneissose (gneiss'-ose) (a) Said of a rock, or of its structure, that resembles gneiss but that is not the result of metamorphic processes. Cf: *gneissoid.* (b) Said of a rock whose structure is composite, having alternating schistose and granulose bands and lenses which differ in mineral composition and texture. Cf: *gneissic.* Because of its ambiguous nature, this term is not preferred.

gneissose structure *gneissic banding.*

gneissosity (gneiss-os'-i-ty) *gneissic banding.*

GNIS *Geographic Names Information System.*

gnomonic projection (gno-mon'-ic) (a) A perspective azimuthal map projection (of a part of a hemisphere) on a plane tangent to the surface of the sphere, having the point of projection at the center of the sphere. All straight lines on the tangent plane represent arcs of great circles on the Earth's surface; all great circles appear as straight lines. The point of tangency may be at a pole, on the equator, or at any point in between (oblique gnomonic projection). It is used, in conjunction with the Mercator projection, to plot great-circle courses in navigation. Syn: *central projection; great-circle projection; great-circle chart.* (b) A similar projection used in optical mineralogy to plot data obtained by measurements of crystals with a two-circle goniometer, characterized by a plane of projection that is tangent to the north pole of the sphere, with the poles of the faces parallel to the vertical axis of the sphere lying at infinity.

goat beard An informal term for the set of small, closely spaced, mineralized vertical fractures found near the lower end of many small faults in coal. A goat beard in the top coal often warns of the presence of a fault in the roof (Nelson, 1981).

gob (a) That part of a mine from which the coal has been worked away and the space more or less filled up (Fay, 1920). (b) The refuse or waste left in the mine (Fay, 1920). (c) Especially in the Illinois Basin, weak, carbonaceous shale which overlies some seams and is prone to falling.

gobbinsite (gob'-bins-ite) A chalky white orthorhombic mineral: $Na_5(Si_{11}Al_5)O_{32}·11H_2O$. It is a zeolite related to gismondine, merlinoite, phillipsite, garronite and some synthetic Na-P zeolites.

gobi (go'-bi) (a) A Mongolian term introduced by Berkey and Morris (1924, p.105) for a small, open, level-surfaced basin within a *tala.* (b) A lenticular mass of sedimentary deposits occupying a gobi.

goblet valley (gob'-let) *wineglass valley.*

gob pile A heap of mine refuse on the surface.

godlevskite (god-levsk'-ite) A metallic orthorhombic mineral: $(Ni,Fe)_9S_8$.

godovikovite (go-do-vik'-ov-ite) A white trigonal mineral: $(NH_4)(Al,Fe^{3+})(SO_4)_2$. It is a combustion product of a coal heap.

goedkenite (goed'-ken-ite) A colorless to plae yellow monoclinic mineral of the *brackebuschite* group: $(Sr,Ca)_2Al(PO_4)_2(OH)$.

goethite (goe'-thite) A yellowish, reddish, or brownish-black orthorhombic mineral of the *diaspore* group: $\alpha Fe^{3+}O(OH)$. It is trimorphous with lepidocrocite and akaganéite. Goethite is the commonest constituent of many forms of natural rust or of limonite, and it occurs esp. as a weathering product in the gossans of sulfide-bearing ore deposits. Also spelled: *göthite.* Syn: *allcharite; xanthosiderite.*

gold A soft, heavy, yellow, cubic mineral, the native metallic element Au. It is often naturally alloyed with silver or copper and occasionally with bismuth, mercury, or other metals, and is widely found in alluvial deposits (as nuggets and grains) or in veins associated with quartz and various sulfides. Gold is malleable and ductile, and is used chiefly for jewelry and as the international standard for world finance.

goldamalgam (gold-a-mal'-gam) A metallic yellow cubic mineral: (Au,Ag)Hg. Not approved by the International Mineralogical Association.

gold amalgam A variety of native gold containing mercury; a naturally occurring *amalgam* composed of gold, silver, and mercury, the gold averaging about 40%. It is usually associated with platinum, and occurs in yellowish-white grains that crumble readily.

gold beryl A syn. of *chrysoberyl.* Not to be confused with *golden beryl.*

gold dust Fine particles, flakes, or pellets of gold, such as those obtained in placer mining.

golden beryl (gold'-en) A clear, golden-yellow or yellowish-green gem variety of beryl. See also: *heliodor.* Not to be confused with *gold beryl.*

goldfieldite (gold'-field-ite) A dark lead-gray cubic mineral of the *tennantite* group: $Cu_{12}(Te,Sb,As)_4S_{13}$. It is the tellurium analogue of tetrahedrite.

goldichite (gol'-dich-ite) A pale yellowish green monoclinic mineral: $KFe^{3+}(SO_4)_2·4H_2O$.

goldmanite (gold'-man-ite) A green to brownish green cubic mineral of the *garnet* group: $Ca_3(V^{3+},Al,Fe^{3+})_2(SiO_4)_3$.

gold opal A *fire opal* that exhibits only an overall color of golden yellow.

goldquarryite A pale blue to blue-gray triclinic mineral: $Cu(Cd,Ca)_2Al_3(PO_4)_4F_2(H_2O,F)_2·10H_2O$.

gold quartz *Milky quartz* containing small inclusions of gold.

goldschmidtine (gold'-schmidt-ine) *stephanite.*

goldschmidtite (gold'-schmidt-ite) *sylvanite.*

Goldschmidt's phase rule (Gold'-schmidt's) *mineralogical phase rule.*

goldstone (gold'-stone) A translucent, reddish-brown laboratory-grown glass containing a multitude of tiny thin tetrahedral or hexagonal platelets of metallic copper, which exhibit bright reflections and produce a popular but poor imitation of *aventurine.* See also: *sunstone.* Syn: *aventurine glass.*

goletz terrace (go-letz') A *cryoplanation* terrace (Russian).

gompholite (gom'-pho-lite) *nagelfluh.*

gonal spine (go'-nal) In dinoflagellate cysts, a spine situated at the point where three plates join.

gonatoparian (go'-na-to-par'-i-an) adj. Of or concerning a trilobite having facial sutures, the posterior sections of which reach the cephalic margin at the *genal angles;* also, said of the sutures themselves.

gondite (gond'-ite) A metamorphic rock consisting of spessartine and quartz, probably derived from manganese-bearing sediment. It is named after Gonds and the Gondite Series, central India. Rarely used. Cf: *collobrierite.*

Gondwana (Gond-wa'-na) The late Paleozoic continent of the Southern Hemisphere. It was named by Suess for the Gondwana system of India, which has an age range from Carboniferous to Jurassic and contains glacial tillite in its lower part and coal measures higher up. Similar sequences of the same age are found in all the continents of the hemisphere; this similarity, along with much compelling evidence of other sorts, indicates that all these continents were once joined into a single larger mass. The preponderance of modern evidence indicates that the present continents are fragments that have been separated from each other by *continental drift.* The counterpart of Gondwana in the Northern Hemisphere was *Laurasia;* the supercontinent from which both were derived was *Pangea.* Var: *Gondwanaland.*

Gondwanaland (Gond-wa'-na-land) Var. of *Gondwana.* Gondwana is preferred as it means "land of the Gonds" and Gondwanaland is tautological.

gongylodont (gon-gy'-lo-dont) Said of a class of ostracode hinges consisting of three elements wherein the terminal elements are opposites in the same valve, as in the genus *Loxoconcha.*

goniatite (go'-ni-a-tite') Any ammonoid cephalopod belonging to the order Goniatitida, characterized generally by a shell having sutures of angular appearance with eight undivided lobes. Range, Middle Devonian to Upper Permian.

goniatitic suture (go'-ni-a-tit'-ic) A type of *suture* in ammonoids characterized by simple fluting in which most or all of the lobes and saddles are entire or plain (not denticulate or frilled), the only common exception being the ventral lobe, which is subdivided and may be denticulate; specif. a suture in goniatites. Cf: *ammonitic suture; ceratitic suture.*

goniometer (go-ni-om'-e-ter) (a) An instrument used in optical crystallography for measuring the angles between crystal faces. Types are the *contact goniometer,* the *reflection goniometer,* and the *two-circle goniometer.* (b) An instrument that measures X-ray diffractions; a *diffractometer.*

goniometer head A device on which a single crystal may be mounted for diffraction experiments, and which permits translation of the crystal in three mutually perpendicular directions and, with some types, rotation of the crystal about each of those directions.

gonnardite (gon'-nard-ite) A white tetragonal zeolite mineral: $(Na,Ca)_{6-8}(Al,Si)_{20}O_{40}·12H_2O$. It occurs in finely fibrous, radiating spherules. Syn: *tetranatrolite.*

gonopore (gon'-o-pore) (a) A simple opening that serves as an exit from the genital system of an echinoderm, e.g., a stalked echinoderm or an edrioasteroid; a genital pore. In certain cystoids it is combined with a *hydropore.* (b) The outlet of the genital ducts in crustaceans; a sexual pore.

gonozooid (gon-o-zo'-oid) A *zooid* modified as a brood chamber in the stenolaemate bryozoans.

gonyaulacacean (go-ny-au-la-ca'-cean) A *dinoflagellate theca* or *cyst,* or an appearance attributed to them, showing a *tabulation* similar to that of the modern genus *Gonyaulax,* typically with six precingular and six postcingular plates. Syn: gonyaulacoid (Williams et al., 2000, p.92). Cf: *postcingular series; precingular series.*

gonyerite (gon'-yer-ite) A deep-brown orthorhombic trioctahedral Fe,Mn-rich phyllosilicate based on a modulation of the chlorite structure: $(Mn^{2+},Mg)_5Fe^{3+}(Si_3Fe^{3+})O_{10}(OH)_8$.

Goode homolosine projection (ho-mol'-o-sine) An equal-area map projection consisting of a sinusoidal projection (between parallels 40°N and 40°S) combined with a Mollweide homolographic projection (between these parallels and the poles); specif. an *interrupted projection* that allows continental masses to be recentered on several meridians in order that they be shown with a minimum of shape distortion, leaving gaps in the interrupted ocean areas between each section to accommodate errors. Also, a similar projection showing the oceans to best advantage (Snyder and Voxland, 1989, p.66). The homolosine projection was developed in 1923 by J. Paul Goode (1862-1932), of the University of Chicago. Syn: *homolosine projection.*

gooderite (good'-er-ite) A *nepheline syenite* with a predominance of albite rather than potassium feldspar. Named by Johannsen (1938) for Gooderham Township, Ontario. Obsolete.

goodness-of-fit A statistical test used to ascertain agreement of observed data with theoretical distributions, with other observed data, or with some mathematical functions. See also: *chi-square test.*

Goodsprings twin law (Good'-springs) A rare type of normal twin in feldspar, in which the twin plane is ($\bar{1}12$).

goongarrite (goon-gar'-rite) A mixture of cosalite, galena, and heyrovskyite.

gooseberry stone (goose'-ber-ry) A syn. of *grossular,* esp. used for the yellowish-green varieties.

goosecreekite (goose-creek'-ite) A colorless monoclinc mineral belonging to the zeolite family: $Ca(Si_6Al_2)O_{16}·5H_2O$.

gooseneck (goose'-neck) The part of a winding valley resembling in plan the curved neck of a goose; esp. a part formed by an *entrenched meander.*

GOR *gas-oil ratio.*

gorceixite (gor'-ceix-ite) A brown monoclinic mineral of the *crandallite* group: $BaAl_3(PO_4)(PO_3OH)(OH)_6$.

gordaite A micaceous white trigonal mineral: $NaZn_4(SO_4)(OH)_6Cl·6H_2O$.

gordonite (gor'-don-ite) A colorless to lightly tinted triclinic mineral of the laueite group: $MgAl_2(PO_4)_2(OH)_2·8H_2O$. It is isostructural with *laueite.*

gore [cart] One of the series of related and triangular or lune-shaped sections of a map or chart, usually bounded by meridians and tapering to the poles, which can be applied to the surface of a sphere (with a negligible amount of distortion) to form a globe.

gore [surv] A small irregularly shaped tract of land, generally triangular, left between two adjoining surveyed tracts, often because of inaccuracies in the boundary surveys or as a remnant of a systematic survey. It is an officially recognized tract in some States (e.g. Maine and Vermont).

gorge (a) A narrow, deep valley with nearly vertical rocky walls, enclosed by mountains, smaller than a *canyon [geomorph],* and more steep-sided than a *ravine;* esp. a restricted, steep-walled part of a canyon. (b) A narrow defile or passage between hills or mountains. Etymol: French, "throat".

görgeyite (gör′-gey-ite) A colorless or yellowish monoclinic mineral: $K_2Ca_5(SO_4)_6 \cdot H_2O$.

gormanite (gor′-man-ite) A blue-green triclinic mineral that is the Fe^{2+} analogue of souzalite: $Fe^{2+}{}_3Al_4(PO_4)_4(OH)_6 \cdot 2H_2O$.

Gorstian A stage of the Standard Global Chronostratigraphic Scale: lowermost stage of the Ludlovian Series of the Upper Silurian (above the Homerian, below the Ludfordian). The basal GSSP is in southwest England (Holland, 1982).

gortdrumite (gort′-drum-ite) A metallic lead-gray white orthorhombic mineral: $Cu_{18}FeHg_6S_{16}$.

goshenite (go′-shen-ite) A colorless, white, or bluish variety of beryl from Goshen, Mass.

goslarite (gos′-lar-ite) A silky white to lightly tinted orthorhombic mineral of the *epsomite* group: $ZnSO_4 \cdot 7H_2O$. It forms by oxidation of sphalerite and usually occurs massive. Syn: *white vitriol; zinc vitriol; white copperas.*

gossan (gos′-san) An iron-bearing weathered product overlying a sulfide deposit. It is formed by the oxidation of sulfides and the leaching-out of the sulfur and most metals, leaving hydrated iron oxides and rarely sulfates. Syn: *capping; leached capping; iron hat; chapeau de fer.* Adj: gossanous. Also spelled: *gozzan.* Cf: *oxidized zone; false gossan.*

gossany (gos′-san-y) Pertaining to or comprising gossan.

gote A British term for a watercourse.

göthite (gö′-thite) *goethite.*

Gotlandian (Got-land′-i-an) An obsolete syn. of the *Silurian* System and Period, proposed by A. de Lapparent in 1893.

gottardiite A colorless to yellow orthorhombic zeolite mineral: $Na_3Mg_3Ca_5(Si_{117}Al_{19})O_{272} \cdot 93H_2O$.

gottlobite An orange orthorhombic mineral of the *adelite* group: $CaMg(VO_4,AsO_4)OH$.

götzenite (göt′-zen-ite) A colorless to honey-yellow triclinic mineral of the *rosenbuschite* group: $Ca_4(Ca,Na)_2NaTi(Si_2O_7)_2(F,OH,O)_4$.

goudeyite (gou′-dey-ite) A yellow-green hexagonal mineral of the *mixite* group: $(Al,Y)Cu_6(AsO_4)_3(OH)_6 \cdot 3H_2O$.

gouffre A French term for a gulf, chasm, or pit, sometimes applied in English-language publications to a natural gorge, a karstic depression, or a pit cave.

gouge [glac geol] *crescentic gouge.*

gouge [ore dep] (a) A thin layer of soft, earthy fault-comminuted rock material along the wall of a vein or between the country rock and the vein, so named because a miner is able to "gouge" it out and thereby facilitate the mining of the vein itself. Syn: *selvage [ore dep]; pug.* See also: *clay gouge.* (b) A term used in Nova Scotia for a narrow band of gold-bearing slate next to a vein, extractable by a thin pointed stick.

gouge channel A term used by Kuenen (1957, p.242) for a large sole mark (larger than a flute cast) now known as a *channel cast.* Syn: *megaflow mark.*

gouge mark *crescentic gouge.*

gouging [glac geol] (goug′-ing) (a) The formation of crescentic gouges. (b) The local basining of a bedrock surface by the action of glacier ice (Thornbury, 1954, p.48).

gouging [mining] The working of a mine without plan or system, by which only the high-grade ore is mined. Syn: *high-grading.*

gour [geomorph] Plural of *gara.*

gour [speleo] A *rimstone dam* and its product, a *rimstone pool.* Etymol: French. Cf: *microgour.*

gowerite (gow′-er-ite) A colorless monoclinic mineral: $CaB_6O_8(OH)_4 \cdot 3H_2O$.

goyazite (goy′-a-zite) A colorless, pink, or yellow rhombohedral mineral of the *crandallite* group: $SrAl_3(PO_4)(PO_3OH)(OH)_6$.. Syn: *hamlinite.* Cf: *lusungite.*

goyle An English term for a ravine or other steep, narrow valley.

goz A term used in Sudan for a long, gentle, dunelike accumulation of sand ranging in thickness from a few decimeters to tens of meters; also, a large-scale, undulating tract containing such accumulations. Etymol: Arabic. Pl: gozes.

gozzan (goz′-zan) *gossan.*

GPR *ground penetrating radar.*

GPS *Global Positioning System.*

gr *group.*

graben (gra′-ben) An elongate trough or basin, bounded on both sides by high-angle normal faults that dip toward one another. It is a structural form that may or may not be geomorphologically expressed as a *rift valley.* Etymol: German, "ditch". Cf: *horst; half graben.*

grab sampler An ocean-bottom sampler that commonly operates by enclosing material from the seafloor between two jaws upon contact with the bottom. See also: *Shipek bottom sampler; Peterson grab; Van Veen grab sampler; clamshell snapper.* Cf: *dredge [oceanog]; corer.* Syn: *snapper.*

grad The angle subtended at the center of a circle by an arc of $1/400$ of the circumference of the circle.

gradation [geomorph] (gra-da′-tion) (a) The leveling of the land, or the bringing of a land surface or area to a uniform or nearly uniform grade or slope through erosion, transportation, and deposition; specif. the bringing of a stream bed to a slope at which the water is just able to transport the material delivered to it. Cf: *degradation [geomorph]; aggradation [geomorph].* (b) Often used as a syn. of *degradation.*

gradation [part size] The proportion of material of each particle size, or the frequency distribution of various sizes, constituting a particulate material such as a soil, sediment, or sedimentary rock. The limits of each size are chosen arbitrarily. Cf: *sorting; grading [part size].*

gradation period "The entire time during which the base level remains in one position; that is, the interval between two elevations of the Earth's surface of sufficient magnitude to produce a marked change in the position of sea level" (Hayes, 1899, p.22).

gradation zone "A body of rock with upper and lower boundaries defined by selected evolutionary stages of a gradational bioseries of planktonic Foraminifera or of any similarly well distributed bioseries" (Vella, 1964, p.621).

grade [coal] A *coal classification* based on degree of purity, i.e., quantity of inorganic material or ash left after burning. Cf: *type [coal]; rank [coal].*

grade [eng] (a) A degree of inclination, or a rate of ascent or descent, with respect to the horizontal, of a road, railroad, embankment, conduit, or other engineering structure; it is expressed as a ratio (vertical to horizontal), a fraction (such as m/km or ft/mi), or a percentage (of horizontal distance). (b) A graded part of a road, embankment, or other engineering structure that is ascending, descending, or level.

grade [evol] (a) A group of organisms, all at the same or a similar level of organization or advancement; some such groups may be formalized as higher taxa, which then are usually *polyphyletic,* not generally considered desirable in modern systematics. (b) The common level of evolutionary development that has been independently attained by two or more separate but related evolving lineages. (c) A stage of evolution of an animal in which one or more features have undergone a *chronocline.*

grade [meta] *metamorphic grade.*

grade [ore dep] The relative quantity or the percentage of ore-mineral content in an orebody. Syn: *tenor.*

grade [part size] A particular size (diameter), size range, or size class of particles of a soil, sediment, or rock; a unit of a grade scale, such as "clay grade", "silt grade", "sand grade", or "pebble grade".

grade [streams] (a) The condition of balance, achieved by a stream, between erosion and deposition, brought about by the adjustments between the capacity of the stream to do work and the quantity of work that the stream has to do (Davis, 1902, p.86). It is represented by the continuously descending curve (the longitudinal profile of the stream) which everywhere is just steep enough to allow the stream to transport the load of sediment made available to it. Grade involves an equilibrium among slope, load, volume, velocity and channel characteristics. The term was used by Gilbert (1876) but the concept was first formally introduced by Davis (1894) who admits that it "cannot be understood without rather careful thinking"; although a precise definition is difficult, the concept is useful as it implies both an adjustability of the channel to changes in independent variables and a stability in form and profile. (b) A term sometimes used as a syn. of gradient of a given length of a stream. This usage is confusing and not recommended.

grade [surv] (a) A datum level; a level of reference. (b) Height above sea level; actual elevation. Also, the elevation of the finished surface of an engineering project (such as of a canal bed, embankment top, or excavation bottom). (c) Rate of slope; degree of inclination.

grade correction A correction applied to a distance measured on a slope to reduce it to a horizontal distance between the vertical lines through its end points. Syn: *slope correction.*

graded [geomorph] (grad′-ed) Said of a surface or feature when neither degradation nor aggradation is occurring, or when both erosion and deposition are so well balanced that the general slope of equilibrium is maintained. Cf: *in regime.* Syn: *at grade [geomorph].*

graded [part size] (a) A geologic term pertaining to an unconsolidated sediment or to a cemented detrital rock consisting of particles of essentially uniform size or of particles lying within the limits of a single grade. Syn: *sorted [part size].* (b) An engineering term pertaining to a soil or an unconsolidated sediment consisting of particles of several or many sizes or having a uniform or equable distribution of particles from coarse to fine; e.g. a "graded sand" containing coarse, medium, and fine particle sizes. See also: *well-graded.* The term is "rarely used in geology to refer to the sorting of the sediment, although this is common among engineers" (Middleton, 1965, p.249). Ant: *nongraded.*

graded bed A sedimentary bed characterized by distinct vertical gradations in grain size (usually from coarse at the base of the bed to fine at the top).

graded bedding A type of bedding in which each layer displays a gradual and progressive change in particle size, usually from coarse at the base of the bed to fine at the top. It may form under conditions in which the velocity of the prevailing current declined in a gradual manner, as by deposition from a single short-lived turbidity current. Cf: *normal grading; inverse grading.*

graded profile *profile of equilibrium [coast].*

graded reach (a) A part of a stream characterized by a condition of balance between erosion and deposition, as where a stream crossing an outcrop of weak rocks is in equilibrium while its profile across resistant rocks remains for a long time irregular and steep. (b) A reach of a graded stream.

graded shoreline A shoreline that has been straightened or simplified by the formation of barriers across embayments and by the cutting-back of headlands, and that possesses a broadened surface profile so adjusted in slope that the energy of incoming waves is completely absorbed and the shifting of the shoreline is reduced to a very slow rate; a shoreline with a vertical profile of equilibrium, typical of an advanced stage of development. Syn: *equilibrium shoreline.*

graded slope The downstream gradient of a graded stream; it permits the most effective transport of load and is represented by the profile of equilibrium.

graded stream (a) A stream in equilibrium, showing a balance between its transporting capacity and the amount of material supplied to it, and thus between degradation and aggradation in the stream channel. "A graded stream is one in which, over a period of years, slope and channel characteristics are delicately adjusted to provide, with available discharge...just the velocity required for the transportation of the load supplied from the drainage basin. Its diagnostic characteristic is that any change in any of the controlling factors will cause a displacement of the equilibrium in a direction that will tend to absorb the effect of the change" (Mackin, 1948, p.471, revised by Leopold and Maddock, 1953, p.51). A graded stream is not a stream that is loaded to capacity (streams probably never attain this condition), and neither is it a stream that is neither eroding nor depositing (erosion may occur in one part of the channel and deposition in another part). The term is not to be confused with gradient, which is possessed by all streams. Syn: *steady-state stream.* (b) A stream characterized by the absence of waterfalls and rapids (Kesseli, 1941). Cf: *poised stream; regime stream.*

graded unconformity *blended unconformity.*

grade level The level attained by a stream when its "whole course" has been reduced to a uniform gradient, or when its longitudinal profile is a straight line (Park, 1914, p.42).

grade scale A systematic, arbitrary division of an essentially continuous range of particle sizes (of a soil, sediment, or rock) into a series of classes or scale units (or grades) for the purposes of standardization of terms and of statistical analysis; it is usually logarithmic. Examples include: *Udden grade scale; Wentworth grade scale; phi grade scale; Atterberg grade scale; Tyler standard grade scale; Alling grade scale.* See also: *geometric grade scale.*

grade-tonnage model A statistical model or plot showing the relationship between grade and tonnage for a given mineral deposit type.

grade-tonnage relationship The statistical relationship of grade and tonnage for deposits of a given type.

gradient [geomorph] (gra'-di-ent) (a) A degree of inclination, or a rate of ascent or descent, of an inclined part of the Earth's surface with respect to the horizontal; the steepness of a slope. It is expressed as a ratio (vertical to horizontal), a fraction (such as m/km or ft/mi), a percentage (of horizontal distance), or an angle (in degrees). Syn: *slope [geomorph].* (b) A part of a surface feature that slopes upward or downward; a slope, as of a stream channel or of a land surface.

gradient [geophys] The first derivative or change in value of one variable with respect to another such as a change in gravity, temperature, magnetic susceptibility, or electrical potential with respect to horizontal or vertical distance. See also: *gradiometer.*

gradient [hydraul] *hydraulic gradient.*

gradient [streams] A *stream gradient.* Cf: *grade [streams].*

gradient analysis In ecology, the interpretation of the abundance and distribution of taxa in terms of environmental gradients, either through direct gradient analysis, by direct comparison to environmental variables, or through indirect gradient analysis, by multivariate statistical ordinations of faunal abundances within samples (Jongman et al., 1995).

gradient array An electrode array used in resistivity and induced-polarization surveys. Both current electrodes are fixed, while the two potential electrodes are close enough together to measure the gradient of the potential. The potential probes are moved along traverse lines normal to geologic lineation and parallel to a line joining the current electrodes. A square area, of dimensions one-third of the separation of the current electrodes and situated midway between them, is surveyed.

gradienter (gra'-di-ent'-er) An attachment to a surveyor's transit with which an angle of inclination is measured in terms of the tangent of the angle instead of in degrees and minutes. It may be used as a telemeter in observing horizontal distances.

gradient of the head *Hydraulic gradient* for which the specified direction is that of maximum rate of increase in head.

gradient wind A hypothetical horizontal *wind* that blows parallel to curved isobars or height contours above the atmospheric boundary layer. It differs from the *geostrophic wind* in that the path of the gradient wind is curved.

grading [geomorph] (grad'-ing) The reduction of the land to a level surface or equilibrium slope, such as erosion to base level by streams.

grading [part size] The gradual change, in a progressively upward direction within an individual stratification unit, of particle size. Cf: *gradation [part size], coarse-tail grading, distribution grading, normal grading, inverse grading, reverse grading.*

grading factor A *sorting index* developed by Baker (1920, p.368) and defined as the difference between unity and the quotient of mean deviation divided by arithmetic mean diameter (equivalent grade); a measure of how nearly the degree of sorting approaches perfection. Abbrev: GF.

gradiometer (gra-di-om'-e-ter) An instrument that measures the gradient of a physical quantity, e.g., a device consisting of two closely spaced magnetometers, one above the other, that measures the difference in the magnetic field at the two locations.

gradualism The doctrine that the Earth's history is not characterized by sudden and dramatic geologic processes but by everyday processes operating through indefinite spans of time. Gradualism was initiated and championed especially by Charles Lyell. Cf: *catastrophism; uniformitarianism.*

gradualistic evolution (grad'-u-al-is'-tic) That evolutionary pattern in which new taxa develop as the result of gradual continuous changes throughout their history. When applied to the species level, it is known as *phyletic gradualism.* Cf: *episodic evolution.* Ant: *punctuational evolution.*

gradualistic speciation *phyletic gradualism.*

gradually varied flow (grad'-u-al-ly) The flow in an open channel where the velocity changes slowly along the channel and the flow is assumed uniform for an increment of length. See also: *flow profile.*

gradual twist hackle The product of a rupture that grows continuously from the plane of a parent joint into en echelon fringe cracks (Younes and Engelder, 1999). Such fringe cracks are commonly thought to be a product of a spatial rotation of the principal stress axes near the top or bottom edge of a joint-bearing bed. These structures may also form in more isotropic rocks such as granite. Cf: *twist hackle; abrupt twist hackle.*

graduation (grad-u-a'-tion) (a) The method or system of placing degrees or other equally spaced intermediate marks on an instrument or device (such as a thermometer or tape) to represent standard or conventional values. (b) A mark or the marks so placed; one of the equal divisions or dividing lines on a graduated scale.

graemite (gra'-em-ite) An blue-green orthorhombic mineral: $CuTe^{4+}O_3 \cdot H_2O$.

graeserite A metallic black monoclinic mineral: $(Fe,Ti)_4Ti_3AsO_{13}(OH)$.

Graf sea gravimeter A balance-type gravity meter that is heavily damped in order to attenuate shipboard vertical accelerations. It consists of a mass at the end of a horizontal arm, supported by a torsion-spring rotational axis. The mass rises and falls with gravity variation, but is restored to near its null position by a horizontal reading spring, tensioned with a micrometer screw. The difference between actual beam position and null position gives indication of gravity value after the micrometer screw position has been taken into account (U.S. Naval Oceanographic Office, 1966, p.72).

graftonite (graf'-ton-ite) A salmon-pink, reddish brown to dark brown monoclinic mineral: $(Fe^{2+},Mn^{2+},Ca)_3(PO_4)_2$. It is the Fe^{2+} analogue of beusite, and occurs in laminated intergrowths with triphylite and sarcopside.

grahamite [meteorite] (gra'-ham-ite) An obsolete syn. of *mesosiderite.*

grahamite [mineral] A black asphaltite with a variable luster, black streak, high specific gravity, and high fixed-carbon content.

grail Coarse- or medium-grained sediment particles; specif. gravel or sand.

grain [eco geol] A quarrymen's term for a plane of parting in a metamorphic rock, e.g. slate, that is perpendicular to the flow cleavage; or for a direction of parting in massive rock, e.g. granite. Cf: *rift [eco geol].*

grain [gem] (a) A unit of weight commonly used for pearls and sometimes for other gems, equal to 0.3240 *carat,* or 0.0648 gram. (b) A diamond-cutting term referring to the polishing, sawing, or cleavage direction.

grain [geomorph] (a) The broad, linear arrangement of the topographic features (such as mountain ranges and valleys) or underlying geologic structures (such as folds and bedding) of a country or region; e.g. the arrangement of roughly parallel ridges and valleys often displayed in regions of tilted strata. (b) The general direction or trend of such physical or structural features; e.g. the grain of northern Scotland runs NE and SW.

grain [glaciol] An individual particle in snow, ice, or glacier material, consisting of a single ice crystal or a mechanically separate particle of ice. Cf: *snow grain; glacier grain.*

grain [palyn] *pollen grain.*

grain [petrology] (a) A mineral or rock *particle,* smaller than a *fragment,* having a diameter of less than a few millimeters and generally lacking well-developed crystal faces; esp. a small, hard, more or less rounded mineral particle, such as a sand grain. Also, a general term for sedimentary particles of all sizes (from clay to boulders), as used in the expressions "grain size", "fine-grained", and "coarse-grained". (b) The factor of rock texture that depends on the absolute sizes (fineness or coarseness) of the distinct particles composing the rock. Also, the factor of rock texture that is due to the arrangement or trend of constituent particles, such as a lineation or stratification; e.g. the "magnetic grain" in the crustal structure of a region.

grain [struc geol] The second direction of easy splitting in a granite often perpendicular to the rift. Cf: *rift [struc geol].*

grain [water] A unit of hardness of water, expressed in terms of equivalent $CaCO_3$. A hardness of one grain per U.S. gal equals 17.1 ppm by weight as $CaCO_3$. Cf: *Clark degree.*

grain bond The interconnection between grains, usually necklike and narrow.

grain boundary In a polycrystalline solid, the boundary between two crystals. See also: *plane defect.*

grain-boundary diffusion A *diffusional mass transfer* involving the movement of atoms along the boundaries of mineral grains.

grain-boundary diffusion creep Grain-scale, ductile deformation accomplished by the diffusion of atoms along grain boundaries. Syn: *Coble creep.*

grain boundary sliding A high-temperature deformation mechanism in fine-grained rocks whereby grains slip past one another along their grain boundaries, with compatibilities along grain boundaries being accommodated by solid-state diffusion creep or dislocation creep. The grain aggregate changes shape, but individual grains do not.

grain cluster An assemblage of ice crystals with liquid water at crystal boundaries and at triple-crystal junctions. Each crystal is well rounded and easily distinguished. Clusters form without melt-freeze cycles to minimize the surface free energy. They occur only in freely draining snow (Colbeck, 1986). Cf: *melt-freeze grains; melt metamorphism; slush [snow].*

grain density (a) Specific gravity of the grains composing a sediment or sedimentary rock. (b) Syn. of *packing density.*

grain flow *liquefied cohesionless particle flow.*

grain growth (a) The growth of a crystal, as from solution on the walls of a container, in open pore space or in a magma chamber; crystal growth. (b) Applied by Bathurst (1958, p.24) to carbonate sediments, e.g. calcite mud or fibers changing to calcite mosaic with a coarser texture; in this sense it is equivalent to recrystallization; Folk (1965, p.16-20) objects to the usage of this term in carbonate petrology. (c) A metallurgical term for the solid-state growth, coalescence or enlargement of a crystal at the expense of another, occurring between unstrained or undeformed grains.

grain-micrite ratio A ratio that expresses the relative proportion of larger to smaller particles in a carbonate sedimentary rock. It is defined as the sum of the percentages of grains (detrital grains, skeletal grains, pellets, lumps, coated grains, and mineral grains) divided by the percentage of micrite (calcareous mud or its consolidated equivalent). It excludes diagenetic or postdepositional features such as cement, vugs, fractures, vein fillings, and recrystallized areas. Abbrev: GMR.

grain shape *particle shape.*

grain size (a) *particle size.* (b) *granularity.*

grainstone (grain'-stone) A mud-free (less than 1% of material with diameters less than 20 micrometers), grain-supported, carbonate sedimentary rock. It may be current-laid or formed by mud being washed out from previously deposited sediment, or it may result from mud being bypassed while locally produced particles accumulated (Dunham, 1962). Cf: *packstone; mudstone.*

grain-supported A term used to describe a sedimentary carbonate rock with little or no muddy matrix, whose sand-size particles are in three-dimensional contact and able to support one another. Cf: *mud-supported.*

gralmandite (gral'-mand-ite) A variety of garnet intermediate in chemical composition between grossular and almandine (almandite).

gramaccioliite-(Y) A metallic black rhombohedral mineral of the *crichtonite* group: $(Pb,Sr)(Y,Mn)Fe_2(Ti,Fe)_{18}O_{38}$.

gramenite (gra'-men-ite) *nontronite.*

graminivore (gra-min'-i-vore) A *herbivore* that eats primarily grains or grasses.

grandidierite (gran-did'-i-er-ite') A bluish-green orthorhombic mineral: $(Mg,Fe^{2+})Al_3(BO_3)(SiO_4)O_2$.

grandite (grand'-ite) A garnet intermediate in chemical composition between grossular and andradite.

grandreefite (grand-reef'-ite) A colorless subadamantine orthorhombic mineral: $Pb_2(SO_4)F_2$.

granite [eco geol] (gran'-ite) *commercial granite.*

granite [petrology] (a) In the *IUGS classification,* a plutonic rock with Q between 20 and 60 and P/(A+P) between 10 and 65. (b) A plutonic rock in which quartz constitutes 10 to 50 percent of the felsic components and in which the alkali feldspar/total feldspar ratio is generally restricted to the range of 65 to 90 percent. Rocks in this range of composition are scarce, and sentiment has been growing to expand the definition to include rocks designated as *adamellite* or *quartz monzonite,* which are abundant in the U.S. (c) Broadly applied, any holocrystalline, quartz-bearing plutonic rock. Syn: *granitic rock.* Etymol: Latin "granum", "grain".

granite [seis] In early seismologic work, any rock in which velocity of the compressional wave is about 5.5-6.2 km/s.

granite dome A term introduced by Davis (1933) for what he later termed a *desert dome* (Davis, 1938), because the feature was not always developed across granites.

granite gneiss (a) A gneiss derived from a sedimentary or igneous rock and having the mineral composition of a granite. (b) A metamorphosed granite.

granite-greenstone belt A region of Precambrian crust characterized by the occurrence of circular to elliptical bodies of granitoid or high-grade gneiss, surrounded by elongate belts of low-grade metamorphosed mafic and ultramafic volcanic rocks. Most, but not all, major greenstone belts are Archean.

granite-pebble conglomerate (gran'-ite-peb'-ble) A term used by Krumbein and Sloss (1963, p.164) for *arkosic conglomerate.*

granite porphyry A hypabyssal rock differing from a *quartz porphyry* by the presence of sparse phenocrysts of mica, amphibole, or pyroxene in a medium- to fine-grained groundmass.

granite series A sequence of products that evolved continuously during crustal fusion, earlier products tending to be deep-seated, syntectonic, and granodioritic, and later products tending to be shallower, late syntectonic or posttectonic, and more potassic (Turner and Verhoogen, 1960, p.388).

granite tectonics The study of the structural features, such as foliation, lineation, and faults, in plutonic rock masses, and the reconstruction of the movements that created them.

granite wash A driller's term for material eroded from outcrops of granitic rocks and redeposited to form a rock having approximately the same major mineral constituents as the original rock (Taylor and Reno, 1948, p.164); e.g. an *arkose* consisting of granitic detritus. Cf: *basic wash.*

granitic (gra-nit'-ic) (a) Pertaining to or composed of granite. (b) A nonrecommended syn. of granular, or, more restrictively, of subautomorphic. Syn: granitoid; eugranitic.

granitic layer A syn. of *sial,* so named for its supposed petrologic composition. A layer is sometimes called "granitic layer" if it possesses the appropriate seismic velocity (~ 6.0 km/s), although nothing may be known about its composition. Cf: *basaltic layer.*

granitic rock A term loosely applied to any light-colored coarse-grained plutonic rock containing quartz as an essential component, along with feldspar and mafic minerals. When applied to ornamental stone, the terminology is haphazard. Syn: *granite [petrology]; granitoid.*

granitization (gran'-it-i-za'-tion) An essentially metamorphic process or group of processes by which a solid rock is converted or transformed into a granitic rock by the entry and exit of material, without passing through a magmatic stage. Some authors include in this term all granitic rocks formed from sediments by any process, regardless of the amount of melting or evidence of movement. The precise mechanism, frequency, and magnitude of the processes are still in dispute. Cf: *magmatism; transformism.* Syn: *transformation [petrology].*

granitogene (gra-nit'-o-gene) Said of a sediment composed of granitic fragments.

granitoid (gran'-it-oid) n. (a) In the *IUGS classification,* a preliminary term (for field use) for a plutonic rock with Q between 20 and 60. (b) A general term for all *phaneritic* igneous rocks dominated by quartz and feldspars. (c) A *granitic rock.* adj. A syn. of *granitic.*

granoblastic (gran-o-blas'-tic) (a) Pertaining to a *homeoblastic* type of texture in a nonschistose metamorphic rock upon which recrystallization formed essentially equidimensional crystals with normally well *sutured* boundaries (Harker, 1939). Cf: *granuloblastic; sutured.* (b) Pertaining to a secondary texture due to diagenetic change, either by crystallization or recrystallization in the solid state, in which the grains are of equal size (Pettijohn, 1949). This usage is less preferred, because diagenesis is not a metamorphic process.

granodiorite (gran-o-di'-o-rite) (a) In the *IUGS classification,* a plutonic rock with Q between 20 and 60, and P/(A+P) between 65 and 90. (b) A group of coarse-grained plutonic rocks intermediate in composition between *quartz diorite* and *quartz monzonite* (U.S. usage), containing quartz, plagioclase (oligoclase or andesine), and potassium feldspar, with biotite, hornblende, or, more rarely, pyroxene, as the mafic components; also, any member of that group; the approximate intrusive equivalent of *rhyodacite.* The ratio of plagioclase to total feldspar is at least two to one but less than nine to ten. With less alkali feldspar it grades into quartz diorite, and with more alkali feldspar, into granite or quartz monzonite. The term first appeared in print in 1893 in a paper by Lindgren and was applied to all rocks intermediate in composition between granite and *diorite.* The term has the connotation that the rock is a diorite with granitic characteristics, i.e. with quartz and a certain amount of alkali feldspar (Johannsen, 1939, p.254).

granofels (gran'-o-fels) A field name for a medium- to coarse-grained granoblastic metamorphic rock with little or no foliation or lineation (Goldsmith, 1959). Cf: *fels.*

granophyre (gran'-o-phyre) (a) An irregular microscopic intergrowth of quartz and alkali feldspar. (b) As defined by Rosenbusch in 1872, a porphyritic extrusive rock characterized by a micrographic holocrystalline groundmass, or a fine-grained granitic rock having a micrographic texture. (c) As defined by Vogelsang in 1867, a porphyritic rock of granitic composition characterized by a crystalline-granular groundmass. See: Barker, 1970. Adj: *granophyric.* Cf: *vitrophyre.*

granophyric (gran-o-phyr'-ic) (a) An igneous texture characterized by the microscopic intergrowth of quartz and alkali feldspar. Syn: *micrographic.* Cf: *graphic [ign].* (b) As defined by Rosenbusch, a term applied to the texture of a porphyritic igneous rock in which the phenocrysts and groundmass penetrate each other, having crystallized simultaneously; of or pertaining to a *granophyre* (Cross et al., 1906, p.703). (c) As defined by Vogelsang, a term applied to a porphyritic igneous rock having a microgranular groundmass (Johannsen, 1939, p.214).

granoschistose (gran-o-schist'-ose) Pertaining to a structure of a monomineralic metamorphic rock produced by the parallel elongation of grains of a mineral that is normally equidimensional or nearly so.

granosphaerite (gran-o-sphae'-rite) A spherulite composed of radially or concentrically arranged grains. Obsolete.

grantsite (grants'-ite) A dark olive-green to greenish-black monoclinic mineral: $(Na,Ca)_x(V^{5+},V^{4+})_6O_{16}\cdot 4H_2O$(x=~2.3). It is classed as a "vanadium bronze".

granular [geol] (gran'-u-lar) (a) Said of the texture of a rock that consists of mineral grains of approximately equal size. The term may be applied to sedimentary rocks, e.g. sandstones, but is esp. used to describe holocrystalline igneous rocks whose major-phase grain size ranges from 2 to 10 mm. Nonrecommended syn: *granitic.* The syn. *granoblastic* is used for metamorphic rocks. (b) An element in compound adjectives, esp. used to differentiate a texture (e.g. *xenomorphic-granular*) from the corresponding crystal shape (e.g. *xenomorphic*). This usage is obsolescent.

granular [paleont] Covered with very small grains or having numerous small protuberances; e.g. "granular pattern" of ornamentation on the walls of spores and pollen grains, or "granular hyaline wall" representing a perforate and lamellar part of a foraminiferal test composed of minute, equidistant, and variously oriented grains of calcite and seen between crossed nicols as a multitude of tiny flecks of color (TIP, 1964, pt.C, p.60).

granular cementation Chemical deposition of material from solution onto a free surface between detrital grains of a sediment, resulting in outward growth of crystalline material adhering to that surface (Bathurst, 1958, p.14); e.g. growth of calcite in the pores of an unconsolidated sand. Cf: *rim cementation.*

granular chert A compact, homogeneous, hard to soft chert, common in insoluble residues, composed of distinguishable and relatively uniform-sized grains, characterized by an uneven or rough fracture surface and by a dull to glimmering luster (Ireland et al., 1947, p.1486); it may appear saccharoidal. See also: *granulated chert.* Cf: *smooth chert; chalky chert.* Syn: *crystalline chert.*

granular disintegration A type of weathering consisting of grain-by-grain breakdown of rock masses composed of discrete mineral crystals that separate from one another along their natural contacts. It produces a coarse mineral debris, each grain having much the same shape and size as in the original rock. It develops esp. in coarse-grained rocks (such as granite, gneiss, sandstone, and conglomerate) occurring in regions of great temperature extremes. Syn: *mineral disintegration; granular exfoliation.*

granular exfoliation *granular disintegration.*

granular ice Ice made of small crystals of irregular form but having somewhat rounded forms like sand particles.

granularity (gran-u-lar'-i-ty) The quality, state, or property of being granular; specif. one of the component factors of the texture of a crystalline rock, including both grain size and grain-size distribution. Granularity and grain relationships define textures, which, together with orientation, are included in the definition of fabric (in Sander's sense). Friedman (1965, p.647) used the term to refer to the "size and mutual relations of crystals" in a sedimentary rock such as an evaporite, a chemically deposited cement, or a recrystallized limestone or dolomite.

granular snow Cohesionless coarse-grained snow.

granular structure A type of soil structure in which the peds are spheroids or polyhedrons that have little or no accommodation to surrounding peds, are relatively nonporous, and range in size from less than 1.0mm to more than 10.0mm. Cf: *crumb structure.*

granular texture A rock texture resulting from the aggregation of mineral grains of approximately equal size. The term may be applied to a sedimentary or metamorphic rock, but is esp. used to describe an equigranular, holocrystalline igneous rock whose particles range in diameter from 0.05 to 10 mm. See also: *granitic.*

granulated chert (gran'-u-lat'-ed) A type of *granular chert* composed of rough, irregular grains or granules of chert tightly or loosely held together in small masses or fragments (Hendricks, 1952, p.12, 18).

granulation (gran-u-la'-tion) The act or process of being formed into grains, granules, or other small particles; specif. the crushing of a rock under such conditions that no visible openings result. Also, the state or condition of being granulated.

granule [paleont] (gran'-ule) A minute, more or less spherical skeletal element situated on the surface of asterozoan ossicles, generally in a pit or distributed in covering skin.

granule [sed] (a) A term proposed by Wentworth (1922, p.380-381) for a rock fragment larger than a very coarse sand grain and smaller than a pebble, having a diameter in the range of 2-4 mm ($^1/_{12}$ to $^1/_6$ in., or -1 to -2 phi units, or a size between that of the head of a small wooden match and that of a small pea) being somewhat rounded or otherwise modified by abrasion in the course of transport. The term *very fine pebble* has been used as a synonym. (b) A little grain or small particle, such as one of a number of the generally round or oval, nonclastic (precipitated), internally structureless grains of glauconite or other iron silicate in iron formation; a pseudo-oolith.

granule gravel An unconsolidated deposit consisting mainly of granules. Cf: *grit.*

granule ripple A large wind ripple consisting in part of granule-sized particles. Syn: *deflation ripple.*

granule texture A texture of *iron formation* in which precipitated or nonclastic granules are separated by a fine-grained matrix.

granulite [meta] (a) A metamorphic rock consisting of even-sized, interlocking mineral grains, less than 10% of which have any obvious preferred orientation. (b) A relatively coarse, granular rock formed at the high temperatures of the *granulite facies,* which may exhibit a crude gneissic structure due to the parallelism of flat lenses of quartz or feldspar. The texture is typically *granuloblastic.*

granulite [sed] A term that has had so many different uses that it is no longer recommended for use in sedimentology.

granulite facies The facies (set of metamorphic mineral assemblages) in which mafic rocks are represented by diopside+hypersthene+plagioclase, with amphibole generally minor in amount. Almandine is characteristic of mafic and pelitic rocks. Pelitic assemblages show the association of sillimanite or kyanite with perthitic feldspar and almandine, often also with cordierite; muscovite is absent and biotite small in amount. The facies is typical of deep-seated regional dynamothermal metamorphism, at temperatures in excess of 650°C. Cf: *pyroxene-hornfels facies.*

granulitic [meta] (a) Medium- to coarse-grained, equigranular texture and lenticular foliation characteristic of granulites. (b) Pertaining to a *granoblastic* texture having xenoblastic crystal development. (c) Pertaining to a structure resulting from the production of granular fragments in a rock by crushing. Use of this term requires precise definition of its meaning.

granulitic [sed] Said of the rock structure resulting from the production of flattened or granular fragments in a rock by crushing (Stokes and Varnes, 1955, p.66).

granulitization (gran'-u-lit'-i-za'-tion) In *regional metamorphism,* reduction of the components of a solid rock such as a gneiss to grains. The extreme result of the process is the development of *mylonite.*

granuloblastic (gran'-u-lo-blas'-tic) Said of a metamorphic homogranular texture "in which mineral grains largely lack rational faces but have straight or smoothly curving grain boundaries and approximately polygonal shapes" (Harte, 1977). This texture is typically fine-grained (2 mm or less). Granuloblastic and coarser-grained *granoblastic* textures are common in rocks of the granulite facies and in monomineralic rocks (Joplin, 1968). Syn: *homogranular; even-grained.* At times, incorrectly used syn: *equigranular; granulitic [meta].*

granulometric facies (gran'-u-lo-met'-ric) An interpretative term introduced by Rivière (1952) for semilogarithmic cumulative curves representing grain-size analyses of sediments. The facies are subdivided as "linear", "parabolic", "logarithmic", and "hyperbolic" according to the shape of the curve. "The use of the term facies to describe such statistical representations of one single property of a sedimentary rock seems inadmissible" (Teichert, 1958, p.2726).

granulometry (gran-u-lom'-e-try) The measurement of grains, esp. of grain sizes.

granuloreticulose pseudopodia Bifurcating and anastomosing extensions of protoplasm with a relatively solid axis (*stereoplasm*) and granular fluid outer portion (rheoplasm), with granules in continuous movement; they may digest food outside the main body of an animal. A characteristic of foraminifers.

granulose [meta] (gran'-u-lose) Pertaining to the structure that is typical of granulite and that is due to the presence of granular minerals, e.g. quartz, feldspars, garnet, pyroxene, in alternating streaks and bands developed on a megascopic or microscopic scale. No typical foliation is developed due to the absence of lamellar or prismatic minerals.

granulose [paleont] Having a surface roughened with granules; e.g. having very small grains on the tests of certain foraminifers or on the epitheca or tabulae in some corals.

grapestone (grape'-stone) A cluster of small calcareous pellets or other grains, commonly of sand size, stuck together by incipient cementation shortly after deposition (Illing, 1954). The cluster has a lumpy outer surface that resembles a bunch of grapes. Grapestones occur in modern carbonate environments, such as on the Bahama Banks. See also: *bahamite.*

grapevine drainage pattern (grape'-vine) *trellis drainage pattern.*

graphic [cart] In the *GIS* context, the term refers to digital descriptions of map features, logical geographic relationships among features, drawings, and images. A GIS uses the graphic data to generate a map.

graphic [ign] (graph'-ic) Said of the texture of an igneous rock characterized by regular poikilitic intergrowth of triangular or linear-angular quartz grains within larger alkali feldspar grains. Similar intergrowths of other minerals, e.g. ilmenite-pyroxene, are less common. Syn: *runic.*

graphic correlation A method of plotting first and last occurrences of fossils within measured sections to graph the relationships between sections, providing a finer correlation than biostratigraphic techniques relying on biostratigraphic zones. This procedure can help interpret unconformities, sequence boundaries, and rate of rock accumulation (Shaw, 1964; Mann and Lane, 1995).

graphic granite A *pegmatite* characterized by graphic intergrowths of quartz and alkali feldspar. See: Barker, 1970. Syn: *runite.* See also: *pegmatite.*

graphic log *sample log.*

graphic tellurium An old name for sylvanite occurring in monoclinic crystals that are arranged in more or less regular lines, having a fanciful resemblance to writing (such as to runic characters).

graphite (graph'-ite) A hexagonal mineral, a naturally occurring crystalline form of carbon dimorphous with diamond. It is opaque, lustrous, greasy to the touch, and iron black to steel gray in color; it occurs as crystals or as flakes, scales, laminae, or grains in veins or bedded masses or as disseminations in metamorphic rocks. Graphite conducts electricity well, and is soft and unctuous, immune to most acids, and extremely refractory. It is used in "lead" pencils, paints, and crucibles, as a lubricant and an electrode, and as a moderator in nuclear reactors. Syn: *plumbago; black lead.*

graphitic (gra-phit′-ic) Pertaining to, containing, derived from, or resembling graphite; e.g. "graphitic rock".

graphitite (graph′-i-tite) A variety of shungite or type of graphitic rock that does not give the so-called nitric-acid reaction (Tomkeieff, 1954, p.52).

graphitization (graph′-i-ti-za′-tion) The formation of graphitic material from organic compounds.

graphitoid (graph′-it-oid) n. (a) A variety of shungite that will burn in the Bunsen flame. It may represent merely impure graphite. (b) A term applied to meteoritic graphite. adj. Resembling graphite.

graphocite (graph′-o-cite) The end product of coal metamorphism, comparable to *meta-anthracite* and composed mainly of graphitic carbon (ASTM, 1970, p.185).

graphoglypt (graph′-o-glypt) A *trace fossil,* consisting of a presumed worm trail, appearing as a relief on the undersurface of *flysch* beds (mostly sandstones), and having a meandering, spiral, or netlike pattern related to a highly organized foraging behavior; e.g. Paleodictyon. It was interpreted by Fuchs (1895) as a string of spawn of gastropods. Cf: *rhabdoglyph; vermiglyph.*

graptolite (grap′-to-lite) Any colonial marine organism belonging to the class Graptolithina, variously assigned to the phylum Coelenterata or to the Hemichordata, characterized by a cup- or tube-shaped, highly resistant exoskeleton of organic composition, arranged with other individuals along one or more branches (stipes) to form a colony (*rhabdosome*). Graptolites commonly occur in black shales. Range, Middle Cambrian to Carboniferous. Adj: graptolithine; graptolitic.

graptolitic facies (grap-to-lit′-ic) A term applied to a *geosynclinal facies* containing an abundance of graptolites.

graptoloid (grap′-to-loid) Any graptolite belonging to the order Graptoloidea, characterized by a planktonic or epiplanktonic mode of life and by a colony consisting of a few branches with only one kind of theca, the autotheca. Range, Lower Ordovician to Lower Devonian.

grass opal An *opal phytolith* derived from a grass.

grassroots Said of a mineral exploration program, as in "grassroots exploration," where the target region is relatively poorly known geologically and the exploration geologist will likely have to engage in a significant amount of basic mapping and data collecting before identifying specific exploration targets.

grat (Pron: graaht) A term used in the Alps for a small, lateral *arête.* Etymol: German "Grat", "ridge".

graticule [cart] (grat′-i-cule) The network of lines representing meridians of longitude and parallels of latitude on a map or chart, upon which the map or chart was drawn. The graticule can be used to define the position of a point on the Earth's surface with respect to the reference spheroid. Not to be confused with *grid.*

graticule [geophys] A template divided into blocks or cells that is used to integrate graphically a geophysical quantity such as gravity. It is used in computing terrain corrections and gravitational or magnetic effects of bodies of irregular shape.

graticule [optics] An accessory to an optical instrument such as a microscope to aid in measurement of the object under study; it is a thin glass disk bearing a scale which is superimposed upon the object.

grating (grat′-ing) (a) In optical spectroscopy, equidistant and parallel lines that are used in producing spectra by diffraction. Syn: *diffraction grating.* (b) The gratelike pattern of lines observed in some serpentinized hornblende crystals, resulting from the occurrence of the initial alteration along cleavage cracks.

gratonite (grat′-on-ite) A metallic gray rhombohedral mineral: $Pb_9As_4S_{15}$.

grattarolaite A brown to red-brown rhombohedral mineral: $Fe^{3+}_3O_3(PO_4)$.

graulichite-(Ce) A light green to brownish rhombohedral mineral of the *crandallite* group: $(Ce,La,Nd,Ba)(Fe^{3+},Al)_3[(As,Al)O_4]_2(OH)_6$.

graupel (grau′-pel) A soft, usually spherical opaque ice grain 2-5 mm in diameter formed when supercooled water droplets freeze on contact with falling snowflakes or ice crystals. Etymol: German "Graupel", "sleet, soft hail". Syn: *pellet snow; soft hail.*

grauwacke (grau′-wack-e) The German equivalent of *graywacke.*

gravegliaite A vitreous colorless orthorhombic sulfite mineral: $Mn^{2+}S^{4+}O_3{\cdot}3H_2O$.

gravel (a) An unconsolidated, natural accumulation of typically rounded rock fragments resulting from erosion, consisting predominantly of particles larger than sand (diameter greater than 2 mm, or $^1/_{12}$ in.), such as boulders, cobbles, pebbles, granules, or any combination of these fragments; the unconsolidated equivalent of conglomerate. In Great Britain, the range of 2-10 mm has been used. Cf: *rubble; pebble [part size].* (b) A popularly used term for a loose accumulation of rock fragments, such as a detrital sediment associated esp. with streams or beaches, composed predominantly of more or less rounded pebbles and small stones, and mixed with sand that may compose 50-70% of the total mass. (c) A soil term for rock or mineral particles having a diameter in the range of 2-20 mm (Jacks et al., 1960, p.14); in this usage, the term is equivalent to pebbles. The term has also been used in Great Britain for such particles having a diameter in the range of 2-50 mm. In the U.S., the term is used for rounded rock or mineral soil particles having a diameter in the range of 2-75 mm ($^1/_6$ to 3 in.); formerly the term applied to fragments having diameters ranging from 1 to 2 mm. See also: *fine gravel.* (d) An engineering term for rounded fragments having a diameter in the range of 4.76 mm (retained on U.S. standard sieve no. 4) to 76 mm (3 in.). See also: *fine gravel; coarse gravel.* (e) A stratum of gravel. (f) An obsolete term for sand. (g) *volcanic gravel.*

gravel deposit In economic geology, an alluvial deposit consisting mainly of gravel but commonly including sand and clay. The gravel and sand may be used as a construction material, either directly as fill or as aggregate in concrete. See also: *fineness modulus.*

gravel desert *reg.*

gravelly mud An unconsolidated sediment containing 5-30% gravel and having a ratio of sand to mud (silt + clay) less than 1:1 (Folk, 1954, p.346).

gravelly sand (a) An unconsolidated sediment containing 5-30% gravel and having a ratio of sand to mud (silt + clay) greater than 9:1 (Folk, 1954, p.346). (b) An unconsolidated sediment containing more particles of sand size than of gravel size, more than 10% gravel, and less than 10% of all other finer sizes (Wentworth, 1922, p.390).

gravelly soil A soil that contains an abundance of gravel, usually between 35 and 60% by volume.

gravel mound Any mound of gravel; restricted by Muller (1947, p.217) to a low *frost mound* of sand and gravel, formed by hydrostatic pressure of groundwater. Cf: *frost blister.*

gravel pack (a) An envelope of artificially graded coarse material surrounding the *well screen* in a water or monitor well. It provides a more permeable zone around the screen and thus increases the effective diameter of the well. (b) Gravel or coarse sand placed against a loosely consolidated oil-or gas-producing reservoir rock in a well, to prevent or retard the movement of loose sand grains from that interval into the wellbore. It is usually forced through perforations in a casing or liner under pressure.

gravel piedmont A term used by Hobbs (1912, p.214) for a feature now known as a *bajada.*

gravel pipe A *pipe [sed]* filled predominantly with gravel.

gravel rampart A rampart of loosely compacted reef rubble built along the seaward edge of a reef; the rubble pieces average smaller than those in a *boulder rampart.* Syn: *gravel ridge.*

gravel ridge *gravel rampart.*

gravelstone (grav′-el-stone) (a) A rounded rock fragment or constitutent of a gravel. (b) Consolidated gravel; a *conglomerate.*

gravel train A *valley train* composed chiefly of gravel.

gravel wave *coarse-grained ripple.*

gravimeter (gra-vim′-e-ter) An instrument for measuring the *gravitational acceleration* due to the presence of the *Earth's gravity field.* There are two basic types of gravimeter: *absolute-gravity instrument* and *relative-gravity instrument.* Exploration gravimeters are of the second type. The gravitational force (i.e., force of gravitational attraction) on a unit mass in the exploration gravimeter is balanced by a spring arrangement. At each point of measurement, the position of a unit mass is altered by a change in the gravitational attraction of the Earth's mass. Recent developments in gravimeter design include *accelerometer* and magnetic levitation. Syn: *gravity meter.* See also: *astatic gravimeter* and *zero-length spring.*

gravimetric (grav-i-met′-ric) (a) Pertaining or relating to measurement by weight, e.g. *gravimetric analysis.* (b) Pertaining to measurements of variations of the gravitational field.

gravimetric analysis Quantitative chemical analysis in which the different substances of a compound are measured by weight.

gravimetric moisture content The weight of the water divided by the weight of the solids in a porous medium. Cf: *moisture content [soil]; volumetric moisture content.*

gravimetry (gra-vim′-e-try) Methods and instrumentation to collect and process measurements of the *gravitational field,* especially as used in geophysics and geodesy.

gravitation (grav-i-ta′-tion) The mutual attraction between two masses. See also: *Newton's law of gravitation.*

gravitational acceleration (grav-i-ta′-tion-al) The acceleration resulting from the gravitational attraction between any two masses. In gravity exploration, it is measured by the force exerted upon a unit mass in the *gravimeter* due to the attraction of the Earth's mass. The unit of measurement is *gal:* 1 gal = 1 cm/sec^2. The gravitational acceleration at the Earth's surface varies from about 978 gal (9.78 m/sec^2) at the equator to 983 gal at the poles. See also: *mGal* and *Earth's gravity field.*

gravitational constant The constant γ in the law of universal gravitation: its value is $6.670 \pm 0.005 \times 10^{-11}$ newton m^2/kg^2.

gravitational differentiation *Crystal fractionation* brought about through the gravitational settling of floating crystals.

gravitational field A vector field of the attraction that exists between bodies. It is directly proportional to the bodies' masses and inversely proportional to the distance between them. Syn: *gravity field.* See also: *gravitational acceleration* and *Earth's gravity field.*

gravitational gliding *gravitational sliding.*

gravitational intensity The scalar measure of gravitational force exerted on a unit mass at a particular point.
gravitational separation [eco geol] (a) The stratification of gas, oil, and water in a subsurface reservoir according to their specific gravities. (b) The separation of these fluids in a gravity separator after production.
gravitational sliding Downward movement of rock masses on slopes by the force of gravity, e.g. along a bedding weakness. See also: *gravity-driven tectonics; gravity gliding.* Syn: *gravity sliding; gravitational gliding; gliding [tect]; écoulement; sliding.*
gravitational tide *equilibrium tide.*
gravitational water *free water.*
gravitational wave A hypothetical wave that travels at the speed of light, by which gravitational attraction is propagated.
gravitite (grav'-i-tite) A bed of unsorted clastics deposited by a sedimentary flow impelled only by gravitational forces. There is no internal bedding, and particle arrangement is random (Natland, 1976).
gravity [geophys] (grav'-i-ty) The effect on any body of the inverse-square-law attraction between it and all other bodies or the acceleration imparted to a freely falling body in the absence of frictional forces.
gravity [petroleum] A general term for API gravity or *Baumé gravity* of crude oil.
gravity anomaly A difference between the observed (measured) gravity values and those calculated from the assumed Earth's model. The type of gravity anomaly depends on the corrections applied to the observed values, such as *free-air anomaly* or *Bouguer anomaly.* The gravity anomalies are related to the differences in the subsurface mass distribution and are considered to be representing the variations in geological strctures.
gravity compaction Compaction of sediment resulting from pressure of overburden.
gravity corer An oceanographic *corer* that penetrates the ocean floor solely by its own weight, unlike a *piston corer.* Large-diameter covers routinely take undisturbed cores more than 10 m long.
gravity correction *gravity reduction.*
gravity dam A dam so proportioned that by its own weight it will resist overturning moments and sliding forces induced by reservoir water; e.g. Grand Coulee Dam on the Columbia River and Aswan Dam on the Nile. While all *earth dams* are gravity dams, the term is applied principally to concrete and masonry dams.
gravity-driven tectonics Tectonics in which the dominant propelling mechanism is downslope gliding under the influence of gravity. In general, the extent of structures produced mainly by gravity remains controversial. Originally, fold-thrust belt deformation was attributed to it, but this idea is obsolete. See also: *gravitational sliding.*
gravity equipotential surface *equipotential surface.*
gravity erosion *mass erosion.*
gravity fault Obsolete, little used term for a fault with a normal sense of displacement. The term was used as a synonym for *normal fault,* but such usage should be avoided.
gravity field *gravitational field.*
gravity flow The movement of glacier ice as a result of the inclination of the slope of the glacier surface; *glacier flow.* Cf: *extrusion flow.*
gravity fold A fold that is genetically related to isostatic movements.
gravity formula A formula expressing the value of gravity at a point as a function of the coordinates of that point.
gravity gliding Downslope sliding of an overburden sheet or block under its own weight over a weak, ductile décollement (such as salt). The base of the overburden must dip sufficiently that gravitational body force is resolved into a downslope shear stress that is large enough to overcome the shear strength of the décollement layer or the frictional strength of the décollement fault. In salt tectonics, this dip is provided by a continental slope or delta front (Jackson and Talbot, 1991).
gravity gradient The rate of change of the *gravity acceleration* with respect to distance in a particular direction, for which purpose the acceleration of gravity is considered as a scalar.
gravity groundwater The water that would be withdrawn from a body of rock or soil by the influence of gravity should the *saturated zone* and *capillary fringe* be moved downward entirely below that body, remaining there for a specific length of time, no water being lost or received by the body except through the force of gravity (Meinzer, 1923, p.27). Cf: *free water.*
gravity meter *gravimeter.*
gravity orogenesis A concept proposed by Bucher (1965) for mountain building that results entirely from gravitational stresses. Others believe that such forces may account for folding and buckling but not for entire mountains. Cf: *sedimentary tectonics.*
gravity prospecting Measuring, processing, and interpreting gravity data as an aid in locating minerals or other objects.
gravity reduction Applying the *Bouguer correction, free-air correction, isostatic correction, latitude correction, terrain correction,* and other corrections to the observed (measured) gravity values. Airborne and shipborne (marine) gravity reductions include the *Eötvös correction.* Syn: *gravity correction.*
gravity separation Separation of mineral particles, with the aid of water, heavy liquids, or air, according to the differences in their specific gravities. Cf: *heavy liquid.*
gravity sliding *gravitational sliding.*
gravity slope The upper, relatively steep slope of a hillside, commonly lying at the angle of repose of the material eroded from it; it is steeper than the *wash slope* below. Term introduced by Meyerhoff (1940). Cf: *constant slope.* Syn: *steilwand; böschung.*
gravity solution A solution used to separate the different mineral particles of rock by exploiting their differences in specific gravity; e.g. a solution of mercuric iodide in potassium iodide, having a maximum specific gravity of 3.19. Cf: *heavy liquid.*
gravity spreading Vertical collapse and lateral spreading of salt and any overburden under their own weight. A dipping upper surface allows a small component of the gravitational body force to be resolved into a downslope shear stress to drive spreading. The base of the salt can be horizontal, or it can even dip gently opposite to the surface slope. Examples are the spread of salt glaciers or the lateral injection of salt sheets (Jackson and Talbot, 1991).
gravity spring A spring issuing from the point where the water table and the land surface intersect; an outcrop of the water table.
gravity standard *International Gravity Standardization Net 1971.*
gravity survey Measurements of the *Earth's gravity field* at a series of different locations. The object is to associate variations in the measured values with differences in the distribution of rock densities and, hence, rock types. Occasionally the whole *absolute gravity* field is measured (as with a pendulum or free fall), or derivatives of this field (as with a *torsion balance gradiometer*), but usually the difference between the gravity field values at two points is measured (as with a *gravimeter*). Gravity survey data usually are displayed as *Bouguer anomaly* or *free-air anomaly* maps.
gravity unit One tenth of a *milligal.* Abbrev: G unit.
gravity water (a) *free water.* (b) Water delivered in canals or pipelines by gravity instead of by pumping, as for irrigation or a public water supply.
gravity wave A surface wave whose propagation velocity is controlled mainly by gravity, and whose wavelength is greater than 1.7 cm. Cf: *capillary wave.*
gravity well A well that taps an *unconfined aquifer.*
gravity wind *katabatic wind.*
gray The SI unit of radiation absorbed by biota, corresponding to 1 joule (J) of energy per kilogram (kg) of material. 1 Gy = 100 rad. Abbrev: Gy.
gray antimony (a) *stibnite.* (b) *jamesonite.*
grayband (gray'-band) Sandstone used for sidewalks; flagstone.
Gray-Brown Podzolic soil A great soil group of the 1938 classification system, one of a group of zonal soils characterized by a thin A1 horizon, a grayish-brown A2 horizon, and a brown illuvial B horizon in which silicate clays have accumulated. They form under deciduous forest in a humid, temperate climate (USDA, 1938). Not used in current U.S. system of soil taxonomy.
gray cobalt (a) *smaltite.* (b) *cobaltite.*
gray copper ore (a) *tetrahedrite.* (b) *tennantite.* Syn: gray copper.
Gray Desert soil A term used in Russia, and frequently in the United States, synonymously with Desert soil. Not used in current U.S. system of soil taxonomy.
gray hematite *specularite.*
gray ice A type of young ice (10-15 cm thick) that is less elastic than *nilas;* it breaks on swell and usually rafts under pressure. Cf: *gray-white ice.*
grayite (gray'-ite) A dark reddish brown hexagonal mineral of the *rhabdophane* group: $(Th,Pb,Ca)PO_4 \cdot H_2O$.
gray manganese ore (a) *manganite.* (b) *pyrolusite.*
gray mud A type of *mud [marine geol]* that is intermediate in composition between *calcareous ooze* and *red clay.*
gray scale A monochrome strip of continuous tones ranging from white to black with intermediate tones of gray, used to determine the density of a color photograph. Cf: *step wedge.*
graywacke (gray'-wacke) An old rock name that has been variously defined but is now generally applied to a dark gray firmly indurated coarse-grained sandstone that consists of poorly sorted angular to subangular grains of quartz and feldspar, with a variety of dark rock and mineral fragments embedded in a compact clayey matrix and containing an abundance of very fine-grained illite, sericite, and chloritic minerals; e.g. the Jackfork Sandstone (Mississippian) in Oklahoma, parts of the Franciscan Formation (Mesozoic) in western California, and certain Ordovician rocks in the Taconic region of New York and Vermont. This description is similar to Naumann's (1858, p.663) definition of the type graywacke, the Tanner Graywacke (Upper Devonian and Lower Carboniferous) of the Harz Mountains, Germany. It generally reflects an environment in which erosion, transportation, deposition, and burial were so rapid that complete chemical weathering did not occur. Graywackes are typically marine and commonly turbiditic (Pettijohn, 1957, p.313). Selected modern definitions have been given by Allen (1936, p.22); Twenhofel (1939, p.289); Krynine (1948); Folk (1954); Williams et al. (1954, p.293-297); Pettijohn (1957); McBride (1962a); and Krumbein and Sloss (1963, p.171-172). The first recorded use of the term was by Lasius (1789, p.132-152) who referred to "Grauewacke" as a German miner's term for barren country rock of certain ore veins in the Harz Mountains, and who described the rock as a gray or dark quartz "breccia" with mica flakes and fragments of chert or sandstone

in a clay cement (see Dott, 1964). The term "greywacke" was probably first used in English by Jameson (1808). Early usage was wide and vague: "geologists differ much respecting what is, and what is not, Grey Wacce" (Mawe, 1818, p.92), and "it has already been amply shown that this word should cease to be used in geological nomenclature, and...is mineralogically worthless" (Murchison, 1839). Formally defined by Geikie (1885, p.162) as "a compact aggregate of rounded or subangular grains of quartz, feldspar, slate, or other minerals or rocks cemented by a paste...gray, as its name denotes." In view of the diversity of usage, the term "graywacke" should not be used formally without either a specific definition or a reference to a readily available published definition. Folk (1968, p.125) advocates discarding the term for any precise petrographic usage, and relegating it to nonquantitative field usage for a hard, dark, clayey, impure sandstone "that you can't tell much about in the field". Etymol: German *Grauwacke*, "gray stone", probably so named because the original graywackes resembled partly weathered basaltic residues (wackes). See also: *wacke.* Cf: *arkose; subgraywacke.* Also spelled: *greywacke; grauwacke.* Syn: *apogrit.*

graywether (gray'-weth-er) Var. of *greywether.* Also spelled: gray weather.

gray-white ice A type of young ice (15-30 cm thick) that is more likely to be ridged than rafted under pressure. Cf: *gray ice.*

grazing (graz'-ing) The feeding of zooplankton upon phytoplankton, or of tetrapods, esp. mammalian ungulates, on grass and associated plants.

grease ice A soupy layer of *new ice* formed on a water surface (esp. in the sea) by the coagulation of frazil crystals; it reflects little light, giving the sea a matte, greasy appearance. Syn: *ice slush.*

greasy (greas'-y) Said of minerals that appear oily to the touch or to the sight.

greasy quartz A type of *milky quartz* with a greasy luster.

great circle A circle on the Earth's surface, the plane of which passes through the center of the Earth and an arc of which constitutes the shortest distance between any two terrestrial points. Cf: *small circle.*

great-circle belt (a) An alignment of features along a great circle. (b) One in the distribution pattern of primary arcs on the Earth's surface comprising belts of major tectonic activity: the *circum-Pacific belt* and the *Eurasian-Melanesian belt* (Strahler, 1963, p.403).

great-circle chart (a) A chart on a gnomonic projection, on which a great circle appears as a straight line. (b) *gnomonic projection.*

great-circle projection *gnomonic projection.*

great divide A drainage divide between major drainage systems; specif. the Great Divide (Continental Divide) of the North American continent.

great earthquake An earthquake of magnitude greater than 7.5 on the Richter scale.

great elliptic The line of intersection of the ellipsoid and a plane containing two points on the ellipsoid and its center.

Great Ice Age The *Pleistocene* Epoch.

great soil group (a) In the 1938 soil classification system, the most widely used of the higher categories. Soils in each class of this category have common internal characteristics, i.e. kind and sequence of horizons, chemical and physical properties; for example, Podzols and Chernozems. (b) In the current U.S. Dept. of Agriculture soil taxonomy, a category, properly called merely a great group, that divides soil suborders. Great groups are differentiated by properties of the whole soil, not just diagnostic horizons, and by moisture and temperature regimes. Names are formed by prefixing a syllable to suborder names, e.g. Hapludalf, Argiaquoll.

grechishchevite An orange tetragonal mineral: $Hg_3S_2(Br,Cl,I)_2$.

green algae Members of the *phylum* Chlorophyta, owing their grass-green color to the chlorophylls *a* and *b*, the same pigments that are found in land plants and *charophytes.* Both unicellular and multicellular green algae occur in marine, freshwater, and terrestrial environments. Forms that precipitate calcium carbonate (e.g., the *dasycladacean* algae) have an extensive fossil record; modern tropical forms (such as *Halimeda*) contribute to carbonate sands. Green algae range from the Precambrian to the present. Cf: *brown algae; cyanobacteria; red algae; yellow-green algae.*

greenalite (green'-a-lite) An earthy- or pale-green mineral: $(Fe^{2+},Fe^{3+})_{2\text{-}3}Si_2O_5(OH)_4$. It is an iron-rich member of the *serpentine* group. It occurs in small ellipsoidal granules in cherty rock associated with sedimentary iron formations. Greenalite resembles glauconite in appearance, but contains no potassium. Its crystal structure is characterized by a modulation of its tetrahedral sheet to form islands separated by inverted tetrahedra.

greenalite rock A dull, dark-green rock, uniformly fine-grained with conchoidal fracture, containing grains of greenalite in a matrix of chert, carbonate minerals, and ferruginous amphiboles (Van Hise and Leith, 1911, p.165, 474).

green beryl A term applied to the light green or pale green gem variety of beryl, as distinguished from the full green or richly green-colored emerald and the light blue-green aquamarine.

green chalcedony (a) Chalcedony that has been artificially colored green. (b) *chrysoprase.*

green earth Any of various naturally occurring silicates (esp. of iron) used chiefly as bases for green basic dyes and for green pigments; specif. glauconite and celadonite. Syn: *terre verte; terra verde.*

Greene-Kelly test A test used to distinguish smectites whose layer charge originates in the octahedral sheet (e.g., montmorillonite) from smectites whose layer charge originates in the tetrahedral sheet (e.g., beidellite). Li saturation and subsequent heating of the smectite cause migration of Li into the octahedral sheet of a smectite with an octahedral charge, converting the 2:1 layer with a net negative layer charge into a 2:1 layer with a very limited or no layer charge. The change in layer charge affects the physical properties of the smectite, allowing discrimination between smectites with octahedral and tetrahedral charges (Greene-Kelly, 1953).

greenhouse effect (green'-house) The heating of the Earth's surface because outgoing long-wavelength terrestrial radiation is absorbed and emitted by the water vapor and carbon dioxide in the lower atmosphere and eventually returns to the surface as *counterradiation.*

green iron ore *dufrénite.*

green john A green variety of fluorite.

greenlandite (green'-land-ite) *columbite.*

Greenland spar (Green'-land) *cryolite.*

green lead ore *pyromorphite.*

Green Mountains disturbance A name used by Schuchert (1924) for a supposed time of deformation at the end of the Cambrian. Evidence for this disturbance is unconvincing, even in the type area of the Green Mountains of Vermont, and the term should be discarded.

green mud A type of *mud [marine geol]* whose greenish color is due to the presence of chlorite, glauconite, or a reduced state of iron.

greenockite (gree'-nock-ite) A yellow or orange hexagonal mineral of the *wurtzite* group: CdS. It is dimorphous with hawleyite and usually occurs as an earthy incrustation or coating on sphalerite and other zinc ores. Syn: *cadmium blende; cadmium ocher; xanthochroite.*

greenovite (gree'-no-vite) A red, pinkish, or rose-colored variety of sphene containing manganese.

greensand (green'-sand) (a) A sand having a greenish color, specif. an unconsolidated shallow marine sediment consisting largely of dark greenish grains of glauconite, often mingled with clay or sand (quartz may form the dominant constituent). The term is loosely applied to any glauconitic sediment. Syn: *glauconitic sand.* (b) A sandstone consisting of greensand that is commonly little or not at all cemented, having a greenish color when unweathered but an orange or yellow color when weathered, and forming prominent deposits in Cretaceous and Eocene strata of the coastal plain areas of New Jersey and Delaware; specif. either or both of the Greensands (Lower and Upper) of the Cretaceous System in England, whether containing glauconite or not. Syn: *glauconitic sandstone.* Also spelled: green sand.

greensand marl A marl containing sand-size grains of glauconite.

greenschist (green'-schist) A schistose metamorphic rock whose green color is due to the presence of chlorite, epidote, or actinolite. Cf: *greenstone [meta].*

greenschist facies The facies (set of metamorphic mineral assemblages) in which mafic rocks are represented by albite+epidote+chlorite+actinolite (Eskola, 1939). Chlorite, white mica, biotite, and chloritoid are typical minerals in pelitic rocks. It is believed to correspond to temperatures in the range 300°-500°C.

green snow A general name for snow colored by a growth of green microscopic algae, such as *Stichococcus* and *Chlamydomonas.* Cf: *red snow; yellow snow.*

greenstone [ign] (green'-stone) In Scotland, any intrusion of igneous rock in the Coal Measures. Obsolete.

greenstone [meta] A field term applied to any compact dark-green altered or metamorphosed mafic igneous rock (e.g., spilite, basalt, gabbro, diabase) that owes its color to the presence of chlorite, actinolite, or epidote.

greenstone [mineral] (a) *nephrite.* (b) An informal name for a greenish gemstone, such as fuchsite or chiastolite.

greenstone [sed] A compact, nonoolitic, relatively pure chamosite mudstone interbedded with oolitic ironstone in the Lower Jurassic of Great Britain.

greenstone belt Term applied to elongate or beltlike areas within Precambrian shields that are characterized by abundant *greenstone [meta].* An individual belt may contain the deformed and metamorphosed rocks of one or more volcano-sedimentary piles, in each of which there is typically a trend from mafic to felsic volcanics. The resultant volcano-sedimentary complexes are of economic interest as host rocks for presumably volcanogenic metal deposits.

greenstone gold *low-sulfide gold-quartz vein.*

green vitriol *melanterite.*

Greenwich mean time *Universal time.*

Greenwich meridian (Green'-wich) The astronomic meridian that passes through the original site of the Royal Astronomical Observatory at Greenwich, near London, England. Its adoption as the worldwide reference standard, or *prime meridian,* was approved almost unanimously at an International Meridian Conference in Washington, D.C., in 1884. Cf: *national meridian.*

gregaritic (gre-ga-rit'-ic) Said of the texture of a porphyritic igneous rock in which independently oriented grains of the same mineral (esp. augite) in the groundmass occur in clusters. Obsolete. Cf: *synneusis; cumulophyric; glomerophyric.*

gregoryite (greg'-o-ry-ite') A white hexagonal mineral: $(Na_2,K_2,Ca)CO_3$. It is stable above 490°C, and has a disordered structure.

greifensteinite A vitreous olive-green monoclinic mineral of the *roscherite* group: $Ca_2Be_4(Fe^{2+},Mn)_5(PO_4)_6(OH)_4 \cdot 6H_2O$.

greigite (greig'-ite) A pinkish gray cubic mineral of the *linnaeite* group: Fe_3S_4. Syn: *melnikovite.*

greisen (grei'-sen) A pneumatolytically altered granitic rock composed largely of quartz, mica, and topaz. The mica is usually muscovite or lepidolite. Tourmaline, fluorite, rutile, cassiterite, and wolframite are common accessory minerals. Cf: *greisenization.*

greisenization (grei'-sen-i-za'-tion) A process of hydrothermal alteration in which feldspar and muscovite are converted to an aggregate of quartz, topaz, tourmaline, and lepidolite (i.e., *greisen*) by the action of water vapor containing fluorine.

grenatite (gren'-a-tite) (a) *staurolite.* (b) *leucite.*

grennaite (gren'-na-ite) A fine-grained porphyritic *nepheline syenite* containing catapleiite and eudialyte. The name, given by Adamson in 1944, is for Grenna farm, Norra Kärr complex, Sweden. Obsolete.

Grenville Orogeny A name that is widely used for a major plutonic, metamorphic, and deformational event during the Precambrian, dated isotopically as between 880 and 1,100 m.y. ago, which affected a broad province along the southeastern border of the Canadian Shield. Its result, the Grenville orogen, can be traced southwestwards, along the region that later became the Appalachians, and then west through the Llano uplift region of Teaxs. Originally, the name Grenville was used for a metasedimentary series in the southern part of the province, and the name *Laurentian* was used for the associated plutonic rocks. The Grenville Orogeny is thought to be the consequence of a Himalayan-type continental collision during the assembly of a supercontinent (Rodinia). The foreland edge of the Grenville orogen in North America is a shear zone called the Grenville front.

grenz A horizon in coal beds resulting from a temporary halt in the accumulation of vegetal material. It is frequently marked by a bed of clay or sand. Etymol: abbreviated form of grenzhorizont.

grenzhorizont A German term (used especially in European peat) equivalent to *recurrence horizon,* meaning a particular recurrence surface in peat associated with the major change of climate from the Subboreal to the Subatlantic and dated by radiocarbon at 500 B.C., or slightly earlier.

greyite A yellow powdery uncertain mineral: $ThPO_4 \cdot H_2O$.

greywacke (grey'-wacke) *graywacke.*

greywether (grey'-weth-er) Term for silcrete blocks that are spheroidally weathered to develop a fancied resemblance to sheep. Also spelled: *graywether*; gray weather. Cf: *sarsen.*

griceite (grice'-ite) A white cubic mineral of the *halite* group: LiF.

grid (a) A network composed of two sets of uniformly spaced parallel lines, usually intersecting at right angles and forming squares, superimposed on a map, chart, or aerial photograph, to permit identification of ground locations by means of a system of coordinates and to facilitate computation of direction and distance. The term is frequently used to designate a plane-rectangular coordinate system superimposed on a map projection, and usually carries the name of the projection; e.g. "Lambert grid". A grid differs from a coordinate system in being composed of only a finite number of lines. Not to be confused with graticule. A grid differs from a graticule in not necessarily representing lines of longitude and latitude. (b) A data structure composed of points located at the nodes of an imaginary grid, with the spacing between nodes being constant in both the horizontal and vertical directions (Krzanowski et al., 1993). (c) Commonly, raster data structure. (d) A systematic array of points or lines; e.g. a rectangular pattern of pits or boreholes used in alluvial sampling.

grid azimuth The angle in the plane of projection between a straight line and the central meridian of a plane-rectangular coordinate system. In the plane coordinate systems for states established by the former U.S. Coast and Geodetic Survey (now the National Geodetic Survey), grid azimuths are reckoned from south (0°) clockwise through 360°. While essentially a map-related quantity, the grid azimuth may be mathematically transformed into a survey or ground-related quantity. See: *grid convergence.*

grid cell (a) An area enclosed within a network of uniformly spaced horizontal and perpendicular lines. (b) An element of a raster data structure (Krzanowski et al., 1993).

grid convergence (a) The angle, at a point on a map, between the line on the map representing the meridian through that point and the meridional line (through the same point) of a *graticule [cart]* on that map. (b) The angle between a meridional line of a graticule and the central meridional line of that graticule. Syn: *meridional convergence.*

gridiron twinning (grid'-iron) *cross-hatched twinning.*

grid length The distance between two points as obtained by computation from the plane rectangular coordinates of the points. In the plane coordinate systems for states, a grid length differs from the length of a geodetic line by the amount of a correction based on the scale factor for the given line.

grid line One of the lines used to establish a grid.

grid meridian A line through a point parallel to the central meridian or Y axis of a system of plane-rectangular coordinates. Cf: *gisement.*

grid method A method of plotting detail from oblique photographs by superimposing a perspective of a map grid on a photograph and transferring the detail by eye (using the corresponding lines of the map grid and its perspective as placement guides).

grid north The northerly or zero direction indicated by the grid datum of the directional reference. It is coincident with true north only along the meridian of origin.

grid residual The grid (map) after subtracting a "regional" long-wavelength component of the original grid. It emphasizes the anomalies of certain horizontal dimensions in a potential-field map, e.g., gravity or magnetic. See also: *residualizing.*

grid resolution The distance between neighboring grid nodes on the ground (Krzanowski et al., 1993).

grid smoothing A method of smoothing sharp irregularities in potential-field measurements that arise from very shallow disturbances. A grid is drawn on a contour map and the smoothed value at a grid intersection is the average of values a fixed small distance away.

Griesbachian (Gries-bach'-i-an) A Canadian stage and suggested lower substage of the Induan Stage: lowermost Triassic (above Permian System, below Dienerian substage).

Griffith cracks Highly elliptical (in cross section), microscopic to submicroscopic cracks within crystals or along grain boundaries; they concentrate stress and allow crack propagation and fracturing even when the host rock is subject to relatively low remote stress. Named after A.A. Griffith (Griffith, 1924).

griffithite (grif'-fith-ite) An iron-rich clay mineral of the *montmorillonite* group; specif. a variety of saponite containing ferrous iron. It was formerly regarded as identical with nontronite.

grike In England, a vertical or subvertical fissure in a limestone pavement developed by dissolution along a joint. Syn: *solution fissure; kluftkarren.* Cf: *crevice karst.* Also spelled: *gryke.*

grimaldiite (gri-mal'-di-ite) A deep red or pinkish brown rhombohedral mineral polymorphic with guyanaite and bracewellite: CrO(OH).

grimselite (grim'-sel-ite) A yellow hexagonal mineral: $K_3Na(UO_2)(CO_3)_3 \cdot H_2O$.

grinder (grind'-er) A spherical or discus-shaped stone rotated by the force of helical water currents in a stream pothole, the rotation producing a deepening of the pothole.

grinding [geomorph] (grind'-ing) The process of erosion by which rock fragments are abraded, crushed, sharpened, or polished through the frictional effect of continued contact and pressure by other fragments.

grinding [glac geol] Abrasion by rock fragments embedded in a glacier and dragged along the bedrock floor; it produces gouges and grooves, and chips out fragments of bedrock. Cf: *glacial scour.*

griotte (gri-otte') Originally a French quarrymen's term for a marble or fine-grained limestone of red color, commonly variegated with purple and spots or streaks of white or brown. It includes goniatite shells and is commonly nodular. Etymol: French, "morello cherry".

griphite (griph'-ite) A yellow, dark brown, or brownish black cubic mineral: $Ca(Mn^{2+},Na,Li)_6Fe^{2+}Al_2(PO_4)_6(F,OH)_2$. Related to garnet, it occasionally occurs in enormous crystalline masses in pegmatites.

griquaite (gri'-qua-ite) A coarse-grained garnet and clinopyroxene rock that may or may not contain olivine or phlogopite. It occurs as nodular xenoliths in *kimberlite* pipes and dikes; a garnet *pyroxenite.* Cf: *ariegite.* Named by Beck in 1907 for Griqualand, S. W. Africa. Obsolete.

grischunite (grisch'-un-ite) A red-brown orthorhombic mineral: $NaCa_2Mn^{2+}{}_5Fe^{3+}(AsO_4)_6 \cdot 2H_2O$. It is isostructural with wicksite, and occurs intergrown with, and as an alteration product of, sarkinite.

grit (a) A coarse-grained sandstone, esp. one composed of angular particles; e.g. a fine-grained breccia composed of particles ranging in diameter from 2 mm to 4 mm (Woodford, 1925, p.183). (b) A sand or sandstone made up of angular grains that may be coarse or fine. The term has been applied to any sedimentary rock that looks or feels gritty on account of the angularity of the grains. (c) *gritstone.* (d) A sandstone composed of particles of conspicuously unequal sizes (including small pebbles or gravel). (e) A sandstone with a calcareous cement. The term has been applied incorrectly to any nonquartzose rock resembling a grit; e.g. pea grit or a calcareous grit. (f) A small particle of a stone or rock; esp. a hard, angular grain of sand. Also, an abrasive composed of such granules. (g) The structure or "grain" of a stone that adapts it for grinding or sharpening; the hold of a grinding substance. Also, the size of abrasive particles, usually expressed as their *mesh number.* (h) An obsolete term for sand or gravel, and for earth or soil. The term is vague and has been applied widely with many different connotations. Allen (1936, p.22) proposed to restrict the term to a coarse-grained sandstone composed of angular particles varying in diameter from 0.5 mm to 1 mm. Etymol: Old English greot, "gravel, sand".

gritrock *gritstone.*

gritstone A hard, coarse-grained, siliceous sandstone; esp. one used for millstones and grindstones. Syn: *gritrock.* Cf: *grit.*

gritty (a) Said of the feel of a soil or of a loose or cemented sediment containing enough particles of sand to impart a roughness to the touch. The actual quantity of sand in such a soil is usually small. (b) Containing or resembling sand or grit. Syn: *arenose.*

grivation (gri-va'-tion) The angular difference in direction between grid north and magnetic north at any point, measured east or west from grid north.
GRM Generalized reciprocal method of seismic-refraction interpretation.
groin A low, narrow jetty constructed of timber, stone, concrete, or steel, usually extending roughly perpendicular to the shoreline, designed to protect the shore from erosion by trapping sand and littoral drift for the purpose of building up or making a beach. It may be permeable or impermeable. Var. spelling: *groyne.*
groove [glac geol] *glacial groove.*
groove [sed] A long, straight narrow depression, with an almost uniform depth and cross section, on a sedimentary surface (as of mud or shale). It is thought to have been produced by a simultaneous rectilinear advance of objects propelled by a continuous current. Often preserved as a *groove cast,* it is larger and wider than a *striation* but smaller than a channel. See also: *drag mark; slide mark.*
groove [struc geol] One of a series of parallel lineations, often scratches, developed along a fault surface. A groove is a larger structure than a *striation.* Cf: *slickenside; mullion structure; slip-mark.*
groove-and-spur structure *spur-and-groove structure.*
groove cast A term used by Shrock (1948, p.162-163) for a rounded or sharp-crested rectilinear ridge, a few millimeters high and many centimeters in length and width, produced on the underside of a sandstone bed by the filling of a groove on the surface of an underlying mudstone. This structure was called a *drag mark* by Kuenen (1957, p.244) who considered "groove cast" as a general term including drag marks and slide marks. Cf: *striation cast.* See also: *ruffled groove cast.*
grooved lava Lava with grooves or striations made when it was plastic or viscous, by the sliding of one block of lava over another, by the squeezing of viscous lava through cracks in the lava crust, or by impact of volcanic fragments on viscous lava (Nichols, 1938, p.601). See also: *squeeze-up.*
grooved upland An upland surface largely unaffected by feeble cirque-cutting that has left extensive undissected remnants of the preglacial surface (Hobbs, 1911a, p.30). Cf: *fretted upland.* Syn: *channeled upland.*
groove lake A lake occupying a *glacial groove.*
groove spine One of a cluster or row of short, blunt, generally recumbent spines bordering ambulacral grooves in many asteroids.
grooving (groov'-ing) (a) The formation of furrows on a rock surface. (b) A furrow, or a set of furrows.
grorudite (gro'-rud-ite) A hypabyssal rock composed of phenocrysts of microcline or microcline-perthite, aegirine, and less kataphorite in a tinguaitic groundmass of microcline or microperthite, aegirine, and abundant quartz; an aegirine-rich sodic granite. Named by Brögger in 1894 for Grorud, Oslo district, Norway. Obsolete.
grospydite (gros'-py-dite) An ultramafic rock, occurring in nodules in kimberlite pipes of Yakutia, Russia, containing garnets (esp. grossular), plagioclase, pyroxene, and, at high pressures, kyanite, rather than spinel or olivine. Its name, given by Sobolev and others in 1966, is derived from the initial letters of *gros*sular, *py*roxene, and *di*sthene (kyanite).
gross calorific value (a) A *calorific value* calculated on the assumption that the water in the products is completely condensed. Cf: *net calorific value.* (b) For solid fuels and liquid fuels of low volatility, the heat produced by combustion of unit quantity, at constant volume, in an oxygen-bomb calorimeter under specified conditions. Syn: *gross heat of combustion.*
gross heat of combustion *gross calorific value.*
gross in-place value The in-place value of a mineral resource calculated by multiplying the tonnage of contained metals or economic minerals by their market price. No allowance is made for costs of extraction, mineral processing, and marketing. Abbrev: GIPV.
grossite A colorless or white monoclinic mineral: $CaAl_4O_7$.
gross primary production Amount of organic matter produced by living organisms within a given volume or area, including that consumed by the respiratory processes of the organisms (Odum, 1959). Cf: *net primary production.*
grossular (gros'-su-lar) The calcium-aluminum end member of the *garnet* group, usually characterized by a green color: $Ca_3Al_2(SiO_4)_3$. It may be colorless, yellow, orange, brown, rose, or red, and it often occurs in contact-metamorphosed impure limestones. The principal variety is essonite. Syn: *grossularite; gooseberry stone.*
grossularite (gros'-su-lar-ite') *grossular.*
grothite (groth'-ite) *titanite.*
grotto (grot'-to) A small cave, or one of the rooms of a cave.
ground The reference voltage level of an electrical system or instrument.
ground air Air in the ground, principally in the *unsaturated zone,* but including any bubbles trapped in the *saturated zone.* Cf: *subsurface air; included gas.*
ground avalanche An obsolete syn. of *full-depth avalanche.*
ground control [mining] The maintaining of rock mass stability by controlling the movement and support of underground or *openpit mining* excavations.
ground control [remote sensing] Accurate data on the horizontal and/or vertical positions of identifiable ground points so that they may be recognized in aerial photographs.
ground control point (a) A point on the ground, established by survey, and used to determine scale, orientation, etc., for aerial photographs. (b) A point of known location that can be recognized on an IMAGE or a MAP and that can be used to calculate the transformation needed for the registration of images or maps. Ground control points are related to a known projection for use in geometric transformation (Krzanowski et al., 1993). Abbrev: GCP.
ground coupling The mechanical connection of a geophone to the earth. Most often a spike on the geophone base is pressed into the earth.
ground current *earth current.*
ground data Information collected on or near the surface of the Earth, in conjunction with an aerial or remote-sensing survey. Nonrecommended syn: *ground truth.*
grounded hummock (ground'-ed) A hummocked formation of *grounded ice,* appearing singly or in a line or chain. Cf: *stamukha.*
grounded ice Floating ice that is aground in shallow water. Cf: *stranded ice; grounded hummock.*
ground failure A permanent differential ground movement capable of damaging or seriously endangering a structure.
ground fissure *earth fissure.*
ground fog *radiation fog.*
ground frost [meteorol] An occurrence of below-freezing temperature on the surface of the ground, while the air temperature remains above the freezing point.
ground frost [permafrost] Any frozen soil including permafrost; a deprecated syn. of *frozen ground.* See also: *ground ice.*
ground ice (a) All ice, of whatever origin or age, found below the surface of the ground, esp. a lens, sheet, wedge, seam, or irregular mass of clear nonglacial ice enclosed in perennially or seasonally frozen ground, often at considerable depth. Syn: *subsurface ice; subsoil ice; subterranean ice; underground ice.* (b) Glacier ice, sea ice, or lake ice that has been covered with soil (ADTIC, 1955, p.37). (c) Ice formed on the ground by freezing of rain or snow, or by compaction of a snow layer. (d) A deprecated syn. of *anchor ice.*
ground-ice layer *ice layer [permafrost].*
ground-ice mound Any *frost mound, ice laccolith,* or *pingo* containing bodies of ice.
ground-ice wedge *ice wedge.*
grounding (ground'-ing) The temporary dropping and lodgement of sedimentary particles carried in saltation, typically in sandbars, natural levees, or gravel beds (McGee, 1908, p.199).
grounding line The boundary between the grounded and ungrounded (floating) parts of an ice sheet. Example is the *grounded ice* of the ice sheet of West Antarctica and the *floating ice* of the Ross Ice Shelf.
ground magnetometer An instrument for static observations of the magnetic field intensity on the Earth's surface. It is used to monitor and record *daily variations* and *magnetic storm* periods during airborne magnetic surveys. Also used for detailed measurements in mining surveys.
groundmass [ign] (ground'-mass) The finer grained and/or glassy material between the phenocrysts in a porphyritic igneous rock. Syn: *mesostasis; matrix [ign].*
groundmass [sed] A term sometimes used for the matrix of a sedimentary rock.
ground mix The use of a pattern of sources or geophones distributed over a sizable surface area. The objective is to have vertically reflected energy add up in-phase while horizontally traveling energy partially cancels. The term is sometimes reserved for situations where adjacent geophone or source patterns overlap.
ground moraine (a) An accumulation of *till* after it has been deposited or released from the ice during ablation, to form an extensive area of low relief devoid of transverse linear elements. Syn: *bottom moraine.* (b) The rock debris dragged along in and beneath a glacier or ice sheet (a less desirable usage in Europe, where *moraine* is sometimes used as a synonym of *till*).
ground-moraine shoreline (ground'-mo-raine') An irregular shoreline formed where masses of glacial drift abut against the sea as a result of submergence.
ground motion A general term for all seismic motion, including ground acceleration, velocity, displacement, and strain. See also: *strong motion.*
ground noise In exploration seismology, ground motion that is not caused by the source.
ground oscillation (os-cil-la'-tion) The permanent deformation of the ground caused by earthquake-induced liquefaction of susceptible layers or zones, in which the deformed ground surface is more or less in the pre-earthquake position after the ground shaking stops. Also called: *lurching.* Cf: liquefaction-induced *lateral spread.*
ground penetrating radar A means of exploration of the Earth's shallow subsurface with radar energy, commonly in the 10-MHz to 1,000-MHz band. Commonly the two-way traveltime for reflected radar waves defines depth in the Earth where changes in radar propagation occur. Abbrev: GPR.
ground range On *SLAR* images, the distance from the flight path to an object.
ground receiving station A facility that records image data transmitted by *Landsat.*
ground resolution The limit of detail clarity in an image of the Earth's surface collected by a remote sensing device, usually expressed in meters. Each pixel in the image contains a value for a distinct surface area. For example, a

ground resolution of 10 meters means that each pixel contains data to a 10 meter by 10 meter area on the ground.

ground resolution cell The area on the terrain that is covered by the *instantaneous field of view* of a detector. Its size is determined by the altitude of the remote-sensing system and the field of view of the detector.

ground roll A seismic surface wave, generally of low frequency and velocity. Often considered to be a modified *Rayleigh wave.*

ground-surge deposit Thin, fairly well-sorted pyroclastic flow deposits. They occur immediately beneath *ignimbrite* deposits (Sparks and Walker, 1973, p.62) and are intimately associated with the overlying ignimbrite; possibly derived from flow processes within the head of the pyroclastic flow that deposited the ignimbrite.

ground survey A survey made by ground methods, as distinguished from an aerial survey. A ground survey provides coordinates of points on the ground, distances between points on the ground, or angles between points on the ground, or any combination of these quantities. It is used to provide ground control to which photogrammetrically determined coordinates can be adjusted. It may or may not use photography.

ground swell A long, high ocean swell.

ground truth (a) Data obtained on the ground concerning the significance of albedo anomalies observed in remote sensing, to help interpretation. (b) Data from a ground monitor used to show that extraneous events, such as a disturbing magnetic storm, did not occur during the acquisition of airborne data.

groundwater (a) That part of the subsurface water that is in the *saturated zone,* including underground streams. See also: *phreatic water.* (b) Loosely, all *subsurface water* as distinct from surface water. Syn: *subterranean water; underground water.*

groundwater artery A roughly tubular body of permeable material surrounded by impermeable or less permeable material and saturated with water confined under artesian pressure. "The term is especially applicable to deposits of gravel along ancient stream channels that have become buried in less permeable alluvial material under alluvial fans" (Meinzer, 1923, p.42).

groundwater barrier A natural or artifical obstacle, such as a dike or fault gouge, to the lateral movement of ground water, not in the sense of a confining bed. It is characterized by a marked difference in the level of the groundwater on opposite sides. Syn: *barrier [grd wat]; hydrologic barrier; groundwater dam.* Cf: *groundwater cascade; interrupted water table.*

groundwater basin (a) A subsurface structure having the character of a basin with respect to the collection, retention, and outflow of water. (b) An aquifer or system of aquifers, whether basin-shaped or not, that has reasonably well defined boundaries and more or less definite areas of recharge and discharge. Cf: *basin; artesian basin.*

groundwater budget A numerical account, the *groundwater equation,* of the recharge, discharge, and changes in storage of an aquifer, part of an aquifer, or system of aquifers. Syn: *groundwater inventory.*

groundwater cascade The near-vertical or vertical flow of groundwater over a *groundwater barrier.* Cf: *interrupted water table.*

groundwater cement A secondary concentration of calcium carbonate or calcium sulfate, usually in the desert, resulting from evaporation of groundwater at the surface or in shallow soil; a type of *water-table rock.* Syn: *water-table cement.*

groundwater dam *groundwater barrier.*

groundwater discharge (a) Release of water from the *saturated zone.* (b) The water or the quantity of water released.

groundwater divide *divide [grd wat].*

groundwater equation (a) The equation that balances the *groundwater budget;* $P = RO + ET + R + D \pm \Delta S$ where P is rainfall, RO is surface-water runoff, ET is evaporation and transpiration loss, R is groundwater recharge, D is groundwater discharged from the area, and $\pm \Delta S$ is change in groundwater storage. (b) A mathematical statement of groundwater losses and gains in a specified area. (Tolman, 1937, p.560).

groundwater flow (a) *groundwater movement.* (b) *groundwater runoff.*

groundwater geology *hydrogeology.*

groundwater hydraulics Study of the natural or induced movement of groundwater through permeable rock formations (Lohman, 1972, p.1). Cf: *hydrogeology.*

groundwater hydrology The study of the occurrence, movement, and quality of water beneath the Earth's surface (Heath, 1983, p.1). Cf: *hydrogeology.*

groundwater increment *recharge.*

groundwater inventory *groundwater budget.*

groundwater lake A body of surface water that represents an exposure of the upper surface of the *saturated zone,* or of the water table.

Groundwater Laterite soil A great soil group in the 1938 classification system, an intrazonal, hydromorphic group of soils having an A2 horizon containing concretions that is underlain by a hardpan composed of iron and aluminum compounds. These soils are formed in warm-temperate to tropical climates in response to a fluctuating water table (USDA, 1938). They are now classed as *Aquults, Udults,* and *Ustults,* and all contain *plinthite.* Not used in current U.S. system of soil taxonomy.

groundwater level (a) A syn. of *water table.* (b) The elevation of the water table or another potentiometric surface at a particular place or in a particular area, as represented by the level of water in wells or other natural or artificial openings or depressions communicating with the *saturated zone.*

groundwater mining The process, deliberate or inadvertent, of extracting groundwater from a source at a rate so in excess of the replenishment that the ground-water level declines persistently, threatening exhaustion of the supply or at least a decline of pumping levels to uneconomic depths.

groundwater mound A mound-shaped elevation in a water table or other potentiometric surface that builds up as a result of the downward percolation of water, through the *unsaturated zone* or an overlying confining bed, into the aquifer represented by the potentiometric surface. Syn: *water-table mound.*

groundwater movement The movement, or flow, of water in the *saturated zone,* whether naturally or artificially induced. Syn: *groundwater flow.*

groundwater outflow The discharge from a drainage basin, or from any area, occurring as groundwater.

Groundwater Podzol soil A great soil group in the 1938 classification system, an intrazonal, hydromorphic group of soils having a prominent light-colored leached A2 horizon overlain by thin organic material and underlain by a dark brown B horizon, irregularly cemented with iron or organic compounds. It develops under various types of forest vegetation, in humid climates of varying temperature (USDA, 1938). These soils are now classified as *Spodosols.* Not used in current U.S. system of soil taxonomy.

groundwater province An area or region in which geology and climate combine to produce groundwater conditions consistent enough to permit useful generalizations.

groundwater recession curve The part of a stream hydrograph supposedly representing the inflow of groundwater at a decreasing rate after surface runoff to the channel has ceased. Because the base runoff to the stream may include some water that had been stored in lakes and swamps rather than in the ground, the lower recession curve cannot be assumed to represent groundwater only.

groundwater recharge *recharge.*

groundwater replenishment *recharge.*

groundwater reservoir (a) *aquifer.* (b) A term used to refer to all the rocks in the *saturated zone,* including those containing permanent or temporary bodies of *perched groundwater.* Syn: *groundwater zone; reservoir [grd wat].*

groundwater ridge (a) A linear elevation in the water table that develops beneath an influent stream. Cf: *interstream groundwater ridge.* (b) *divide [grd wat].*

groundwater runoff The portion of the runoff that has recharged the groundwater system and has later been discharged into a stream channel or other surface-water body. It is the principal source of base flow or dry-weather flow of streams unregulated by surface storage. Such flow is sometimes called *groundwater flow* (Rogers, 1981, p.174).

groundwater storage (a) The quantity of water in the *saturated zone.* (b) Water available only from storage as opposed to capture.

groundwater surface *water table.*

groundwater table *water table.*

groundwater trench A troughlike depression in the *water table* or other *potentiometric surface,* caused by flow of groundwater into a stream or drainage ditch.

groundwater wave A "high" in the *water table* or other *potentiometric surface* that moves laterally, with a wavelike motion, away from a place where a substantial quantity of water has been added to the *saturated zone* within a brief period. Syn: *phreatic wave.*

groundwater withdrawal The process of withdrawing groundwater from a source; also, the quantity of water withdrawn. Syn: *offtake; recovery [grd wat].*

groundwater zone *groundwater reservoir.*

group (a) A formal lithostratigraphic unit next in rank above a formation. The term is applied most commonly to a sequence of two or more contiguous or associated formations with significant and diagnostic lithologic features in common (ISSC, 1994, p.35). A group name combines a geographic name with the term "group," and no lithic designation is included; for example, San Rafael Group. See also: *subgroup; supergroup; synthetic group.* (b) A stratigraphic sequence that will probably be divided in whole or in part into formations in the future. See also: *analytic group.* (c) A general term for an assemblage or consecutive sequence of related layers of rock, such as of igneous rocks or of sedimentary beds. (d) A term proposed at the 2nd International Geological Congress in Bologna in 1881 as the chronostratigraphic equivalent of an era, and subsequently used quite widely for the rocks now known as an *erathem.* (e) An obsolete term for a chronostratigraphic unit representing a local or provincial subdivision of a system (usually less than a standard series, or the equivalent of "stage" as that term is presently used) and containing two or more formations.

group [geophys] (a) The various geophones or hydrophones which collectively feed a single channel. The number of phones may vary from one to several hundred. A large group is sometimes called a patch. Syn: *array [geophys].* (b) A source group that consists of several sources or source locations that are combined together to form one seismic trace. They may be activated together or combined subsequently.

group velocity (a) The velocity with which seismic energy moves through a medium. Where velocity varies with frequency, individual phases appear to travel at different phase velocities, and the wave shape changes with distance. Cf: *dispersion [seis]; particle velocity; phase velocity.* (b) In anisotropic media, group velocity is the velocity of energy transport radially outward from a point source, called angular dispersion or ray velocity.

grout (a) A cement or bentonite *slurry* of high water content, fluid enough to be poured or injected into spaces and thereby fill or seal them, such as the fissures in the foundation rock of a dam or the space between the lining of a tunnel and the surrounding earth. Many grouts today have fly ash or pozzolan as an additive. (b) The stony waste material, of all sizes, obtained in quarrying.

grouting (grout'-ing) The injection of *grout* into fissured, jointed, or permeable rocks and low-density soils in order to reduce their permeability or increase their strength. Syn: *cementation [eng]; consolidation grouting; compaction grouting.*

groutite (grout'-ite) A jet-black orthorhombic mineral of the *diaspore* group: $Mn^{3+}O(OH)$. It is polymorphous with manganite and feitknechtite.

grovesite (groves'-ite) A brown triclinic mineral of the *chlorite* group: $(Mn,Fe,Al)_6(Si,Al)_4(O,OH)_9$. It is probably a variety of *pennantite.*

growan (gro'-wan) (a) An old English term for a coarse-grained granite, grit, or sandstone. (b) A *grus* developed by the disintegration of a granite.

growler (growl'-er) A small fragment of massive, floating glacier ice or sea ice, extending less than 1 m above sea level and smaller than a *bergy bit.*

growth axis The line formed by tips of lamellae in cusps and denticles of conodont elements and commonly emphasized by concentration of *white matter.*

growth band A wide growth increment on the surface of a bivalve-mollusk shell.

growth band [palyn] An area along plate margins of a dinoflagellate cell where plate growth occurs. The area is usually striated at right angles to the adjacent suture.

growth center Point about which a conodont element developed; the apex of basal cavity (TIP, 1981, pt. W, supp.2).

growth fabric (a) Orientation of fabric elements independent of the influences of stress and deformation, i.e., characteristic of the manner in which the rock was formed. (b) A rock fabric produced by calcifying skeletal organisms that are in situ and in growth position (Insalaco, 1998).

growth fault A fault in sedimentary rock that forms contemporaneously and continuously with deposition, so that the stratigraphic throw of synfaulting sedimentary units increases dramatically with depth and the strata of the downthrown side are thicker than the correlative strata of the upthrown side. The term is most widely used in the U. S. Gulf Coast region. See also: *hinge-line fault.* Syn: *contemporaneous fault; synsedimentary fault.* Less-preferred syn: *depositional fault; Gulf Coast-type fault; progressive fault.* Obsolete syn: *sedimentary fault.*

growth-framework porosity (growth-frame'-work) Primary porosity developed from biotic and/or abiotic processes during the in-place growth of a carbonate-rock framework (Choquette and Pray, 1970, p.246-247). Intraparticle porosity of individual organisms or of particles that were clastic components of the rock is excluded, thus giving the term a more restrictive meaning than that of *constructional void porosity,* which includes these openings.

growth habit The general form, shape and internal structure developed by a bryozoan colony, and its relationship to the substrate.

growth island An irregular layer or patch on a crystal face due to *spiral growth* along an internal screw dislocation.

growth lamella A concentric outgrowth of a brachiopod shell, smaller than a *frill,* deposited by the margin of retractile mantle (TIP, 1965, pt. H, p.145).

growth lattice *reef frame.*

growth layer *growth ring [geochron].*

growth line (a) One of a series of fine to coarse ridges on the outer surface of a brachiopod shell, concentric about the beak and parallel or subparallel to the margins of the valves, and indicating the former positions of the margins when the anterior and lateral growth of the shell temporarily was in abeyance. (b) One of a usually irregularly arranged and more or less obscure series of commarginal lines on the surface of a bivalve-mollusk shell, approximately parallel to the borders of the valve, and representing successive advances of the shell margin at earlier growth stages. Cf: *growth ruga.* (c) One of a series of lines on the surface of a cephalopod conch, denoting periodic increases in size and hence former positions of the aperture. (d) One of a series of collabrally disposed surface markings (low ridges) on the outer surface of a gastropod shell, parallel to and indicating the former positions of the outer lip. (e) An irregular marking on the epitheca of rugose corallites, such as a slight ridge or depression parallel to the upper edge of the corallite, defining a former position of this margin during growth. Syn: *growth ring [geochron].* (f) In conodonts, traces of lamellae in section; also used to refer to striae on an attachment surface (TIP, 1981, pt. W, supp.2). (g) A histologically recognizable increment within a fish scale or bones or teeth of diverse vertebrates that may indicate resumed growth of the tissue following an interval of interrupted growth.

growth ring [geochron] Layer of wood produced in a tree or woody plant during its annual growth period. It is seen in cross section as a ring. Growth rings can be analyzed for chronologic and climatic data based on number and relative sizes. Cf: *dendrochronology; dendroclimatology.* Syn: *annual growth ring; growth layer; tree ring.*

growth ring [paleont] A *growth line* on a rugose coral.

growth ruga An irregular *ruga* or wrinkle on the surface of a bivalve-mollusk shell, having an origin similar to that of a *growth line* but corresponding to a more pronounced halt in growth.

growth twin A twinned crystal that developed as a result of change in lattice orientation during growth.

groyne *groin.*

GRS *Geodetic Reference System.*

grumantite (gru'-mant-ite) A white orthorhombic mineral: $NaSi_2O_4(OH){\cdot}H_2O$.

grumiplucite A metallic gray-black monoclinic mineral: $HgBi_2S_4$.

grumous (gru'-mous) *clotted.*

grunerite (gru'-ner-ite) A green, brown, or gray monoclinic mineral of the *amphibole* group: $Fe^{2+}{}_2(Fe^{2+},Mg)_5Si_8O_{22}(OH)_2$. Cf: *cummingtonite.* Also spelled: grünerite.

grünlingite (grün'-ling-ite) *bismuthinite.*

grus A siliceous sand resulting from weathering and granular disintegration of a parent rock, usually granite. Cf: *residual arkose; slack [weath]; growan.* Etymol: German "Grus" = grit, fine gravel, debris. Pron: rhymes with "goose". Mispronounced and accordingly misspelled "gruss" or "grush."

gruzdevite (gruz'-dev-ite) A metallic white rhombohedral mineral of the *nowackiite* group: $Cu_6Hg_3Sb_4S_{12}$.

gryke *grike.*

gryphaeate (gry'-phae-ate) Shaped like the shell of *Gryphaea* (a genus of fossil bivalve mollusks); i.e. with the left valve strongly convex and its dorsal part incurved, and with the right valve flat.

GSC *gas-solid chromatography.*

GSSP *Global Boundary Stratotype Section and Point.*

guadalcazarite (guad-al-caz-ar'-ite) A variety of metacinnabar containing zinc.

Guadalupian (Gua-da-lu'-pi-an) A series of the Standard Global Chronostratigraphic Scale: Middle Permian (above Cisuralian, below Lopingian). The basal GSSP in the Guadalupe Mountains, Texas, USA, coincides with the lowest occurrence of the conodont *Jinogondolella nankingensis.*

guanajuatite (gua-na-jua'-tite) A metallic bluish-gray orthorhombic mineral of the *stibnite* group: Bi_2Se_3.

guanglinite (guang-lin'-ite) *isomertieite.*

guanine (gua'-nine) A white monoclinic mineral: $C_5H_3(NH_2)N_4O$. It has crystallized from guano.

guano (gua'-no) *cave guano.*

guard The thick hard cigar-shaped calcareous structure that ensheathes the *phragmocone* of a belemnite and is located in the rear portion of the body. Syn: *rostrum.*

guard cells A pair of specialized epidermal cells in land plants that control the opening and closing of stomata.

guard-electrode log (guard-e-lec'-trode) *focused-current log.*

guarinoite A pink hexagonal mineral: $(Zn,Co,Ni)_6SO_4(OH,Cl)_{10}{\cdot}5H_2O$.

guayaquilite (gua-ya-qui'-lite) A soft, pale-yellow, amorphous fossil resin with a high (15%) oxygen content, soluble in alcohol and alkalies, and found near Guayaquil, Ecuador. Its approximate formula: $C_{40}H_{26}O_6$. Also spelled: guayaquillite; *guyaquillite.*

gudmundite (gud'-mun-dite) A steel-gray orthorhombic mineral of the *arsenopyrite* group: FeSbS.

guérinite (gue'-rin-ite) A colorless monoclinic mineral: $Ca_5(AsO_4)_2(AsO_3OH)_2{\cdot}9H_2O$. It is a dimorph of ferrarisite.

guern *khurd.*

guest A mineral introduced into and usually replacing a pre-existent mineral or rock; a *metasome [geol].* Ant: *host.*

guest element *trace element.*

guettardite (guet-tard'-ite) A metallic grayish black monoclinic mineral: $Pb(Sb,As)_2S_4$.

gugiaite (gu'-gia-ite) A colorless tetragonal mineral of the *melilite* group: $Ca_2BeSi_2O_7$.

guhr (a) A white (sometimes red or yellow), loose, earthy, water-laid deposit of a mixture of clay or ocher, occurring in the cavities of rocks. (b) *kieselguhr.*

guidebook (guide'-book) (a) A *road log* of a field trip, summarizing the geology. (b) A guide to the minerals or fossils available in an area.

guided wave (guid'-ed) Any seismic wave that is propagated in a single layer or along some surface or discontinuity, e.g., a surface wave, Stoneley wave, or channel wave.

guide fossil (a) Any fossil that has actual, potential, or supposed value in identifying the age of the strata in which it is found or in indicating the conditions under which it lived; a fossil used esp. as an index or guide in the local correlation of strata. (b) A fossil that is most characteristic of an assemblage-zone, but that is not necessarily restricted to the zone or found throughout every part of it (ACSN, 1961, art. 21e). See also: *zonal guide fossil.* Cf: *index fossil.*

guide meridian A north-south line used for reference in surveying; specif. one of a set of auxiliary governing lines of the U.S. Public Land Survey system, projected north or south from points established on the base line or a standard parallel, usually at intervals of 24 miles east or west of the principal meridian, and on which township, section, and quarter-section corners are established.

guild A group of usually closely related species that are similar in regard to overall morphology, mode of space utilization, and food source (Bambach, 1983).

guildite (guild'-ite) A dark chestnut-brown or yellow monoclinic mineral: $Cu(Fe^{3+},Al)(SO_4)_2(OH)\cdot 4H_2O$.

guilielmite (gui-li-el'-mite) A subaqueous sedimentary structure formed in mud by collapse around a fossil and characterized by small, polished slip surfaces arranged with radial or orthorhombic symmetry around the fossil (Wood, 1935). It is esp. common in the shales of coal-bearing sections.

guilleminite (guil'-le-min-ite') A canary-yellow orthorhombic secondary mineral: $Ba(UO_2)_3O_2(Se^{4+}O_3)_2\cdot 3H_2O$.

gula (gu'-la) A projecting, rather ornate extension of the trilete laesura of fossil megaspores. Pl: gulae. Cf: *apical prominence; acrolamella.*

gulch A term used esp. in the western U.S. for a narrow, deep ravine with steep sides, larger than a gully; esp. a short, precipitous cleft in a hillside, formed and occupied by a torrent, and containing gold (as in California).

gulch placer A placer gold deposit occurring in a *gulch.*

gulf [coast] A relatively large part of an ocean or sea extending far into the land, partly enclosed by an extensive sweep of the coast, and opened to the sea through a strait; the largest of various forms of inlets of the sea. It is usually larger, more enclosed, and more deeply indented than a bay.

gulf [geomorph] A deep, narrow hollow, gorge, or chasm; e.g. one of the long precipitous stream-worn excavations west of the Adirondack Mountains in northern New York State.

gulf [karst] A steep-walled *closed depression* having a flat alluviated bottom; in some gulfs a stream flows across the bottom (Monroe, 1970).

Gulf Coast-type fault *growth fault.*

Gulfian (Gulf'-i-an) North American provincial series: Upper Cretaceous (above Comanchean, below Paleocene of Tertiary).

gull A structure formed by mass-movement processes, consisting of widened, steeply inclined tension fissures or joints, resulting from lateral displacement of a slide mass and filled with debris derived from above. Gulls generally trend parallel to surface contours, and are usually associated with *camber* (Hollingworth et al., 1950). Primarily a British usage.

gullet [paleont] (gul'-let) (a) A variably tubular invagination of the cytoplasm of various protists (such as tintinnids) that sometimes functions in the intake of food. (b) A longitudinal groove present in certain algae (such as some Cryptophyceae and Euglenophyceae). (c) In vertebrates, the esophagus or other, often distensible, modifications of the anterior alimentary canal for taking food into the body.

gullet [streams] A narrow opening or depression, such as a defile or ravine; a gully or other channel for water.

gull hummock A conical or dome-shaped peaty mound, formed by accretion of well-manured grasses, sedges, and other vascular plants near the nest of the great black-backed gull, on islands in the Arctic (ADTIC, 1955, p.61). Owls and other perching birds form similar hummocks in the tundra regions. Cf: *peat mound.*

gully [coast] (gul'-ly) A wave-cut chasm in a cliff, or a minor channel incised in a mud flat below the high-water level (Schieferdecker, 1959, terms 1149, 1233).

gully [geomorph] (a) A very small valley, such as a small *ravine* in a cliff face, or a long, narrow hollow or channel worn in earth or unconsolidated material (as on a hillside) by running water and through which water runs only after a rain or the melting of ice or snow; it is smaller than a *gulch.* Syn: *gullet [streams].* (b) Any erosion channel so deep that it cannot be crossed by a wheeled vehicle or eliminated by plowing, esp. one excavated in soil on a bare slope. (c) A small, steep-sided wooded hollow.

gully erosion Erosion of soil or soft rock material by running water that forms distinct, narrow channels larger and deeper than rills, which usually carry water only during and immediately after heavy rains or following the melting of ice or snow. The term is usually applied to such forms on hillslopes, as opposed to those on valley floors. Cf: *sheet erosion; rill erosion; channel erosion.* Syn: *gullying; ravinement [geomorph].*

gully gravure A term used by Bryan (1940) for the process or processes whereby the steep slopes of hills and mountains retreat by "repeated scoring or graving", each groove (gully) "so disposed as to reduce rather than emphasize inequalities" (p.92); the development of rills into gullies.

gullying (gul'-ly-ing) *gully erosion.*

gum An organic, viscid juice extracted from, or exuded by, certain trees and plants. It hardens in the air and is soluble in water.

gumbo (gum'-bo) A term used locally in the U.S. for a clay soil that becomes sticky, impervious, and plastic when wet.

gumbotil (gum'-bo-til) (a) A gray to dark-colored, leached, deoxidized clay representing the B horizon of fully mature soils, developed from profoundly weathered clay-rich till under conditions of low relief and poor subsurface drainage (as beneath broad, flat uplands). It consists chiefly of expandable clay and/or illite, and may contain altered rock fragments originally mixed with the clay; it is very sticky and plastic when wet, extremely firm when dry. Term introduced by Kay (1916). Cf: *silttil; mesotil.* (b) A term used for a fossilized soil beneath a deposit of later till.

gum copal An inferior resin or amber; copal.

gummite (gum'-mite) A general term for yellow, orange, red, or brown secondary minerals consisting of a mixture of hydrous oxides of uranium, thorium, and lead, and occurring as alteration products of uraninite and not otherwise identified. It includes silicates, phosphates, and oxides; much of the material is probably mixtures or amorphous gels, but some consists perhaps largely of curite. Syn: *uranium ocher.*

gun [seis] (a) An *air gun,* a seismic energy source from which a bubble of highly compressed air is released. (b) A gas gun (gas exploder), a seismic energy source in which an explosive gas mixture is detonated. (c) A *water gun,* a seismic source in which a volume of water is suddenly projected into the water. (d) A so-called *Buffalo gun,* a seismic source used in near-surface studies. (e) A device for obtaining sidewall cores in a borehole. (f) A perforating gun, a device used to perforate or open holes in casing so that fluid can flow into the borehole.

G unit *gravity unit.*

gunite (gun'-ite) n. A mixture of portland cement, sand, and water applied by pneumatic pressure through a specially adapted hose and used as a fireproofing agent and as a sealing agent to prevent weathering of mine timbers and roadways. Etymol: *Gunite,* a trademark. Cf: *shotcrete.* v. To apply gunite; to cement by spraying gunite.

gunningite (gun'-ning-ite) A white monoclinic mineral of the *kieserite* group: $(Zn,Mn^{2+})SO_4\cdot H_2O$.

Gunter's chain (Gun'-ter's) A surveyor's chain that is 66 feet long, consisting of a series of 100 metal links each 7.92 inches long and fastened together with rings. It served as the legal unit of length for surveys of U.S. public lands, but has been superseded by steel or metal tapes graduated in chains and links. Named after Edmund Gunter (1581-1626), English mathematician and astronomer, who invented the device about 1620. Syn: *pole chain.*

Günz (a) A European stage of the Pleistocene, above the Donau gravels, below the Mindel. (b) Long considered the first glacial stage of the Pleistocene Epoch in the Alps; it is now known that there were earlier ones. Cf: *Nebraskan.* Etymol: After the Günz River, Bavaria. Adj: Günzian.

Günz-Mindel (Günz-Min'-del) The term applied in the Alps to the first classical interglacial stage of the Pleistocene Epoch, following the Günz and preceding the Mindel glacial stages. See also: *Aftonian.*

gupeiite (gu-pei'-ite) A metallic steel-gray cubic mineral: Fe_3Si.

gurhofite (gur'-hof-ite) A snow-white variety of dolomite mineral, containing a large proportion of calcium.

gustavite (gus'-tav-ite) A metallic gray monoclinic mineral: $PbAgBi_3S_6$.

gust front Leading edge of a mass of relatively cool gusty air that flows out of the base of a thunderstorm cloud (downdraft) and spreads along the ground well in advance of the parent thunderstorm cell; a mesoscale *cold front.* Uplift along the gust front may produce additional cumulus clouds that may evolve into secondary thunderstorm cells tens of kilometers ahead of the parent cell.

gut (a) A very narrow passage or channel connecting two bodies of water; e.g. a contracted strait, or a small creek in a marsh or tidal flat, or an inlet. Also, "a channel in otherwise shallow water, generally formed by water in motion" (CERC, 1966, p. A14). (b) A tidal stream connecting two larger waterways. (c) A term used in the Virgin Islands and elsewhere for a gully, ravine, small valley, or narrow passage on land. Syn: *gat; gate [coast].*

Gutenberg discontinuity (Gu'-ten-berg) Obsolete term for the seismic-velocity discontinuity at ~2,890 km depth that marks the *core-mantle boundary,* at which the velocities of P waves are reduced and S waves disappear. It probably reflects the change from a solid to a liquid phase and a change in composition. It is named after Beno Gutenberg, seismologist. Also called: *Oldham-Gutenberg discontinuity; Wiechert-Gutenberg discontinuity.* Cf: *core-mantle boundary.*

Gutenberg low-velocity zone The *low-velocity zone* of the upper mantle.

gutkovaite-Mn A pale yellowish pink monoclinic mineral of the *vuoriyarvite* group: $CaK_2Mn(Ti,Nb)_4(Si_4O_{12})_2(O,OH)_4\cdot 5H_2O$.

gutsevichite (gut-se'-vich-ite) An uncertain mineral discredited by the International Mineralogical Association: $(Al,Fe)_3(PO_4,VO_4)_2(OH)_3\cdot 8H_2O$ (?).

guttation (gut-ta'-tion) The process by which water in liquid form is exuded from an uninjured surface of a plant. Cf: *transpiration.*

gutter [ore dep] (gut'-ter) The lowest and usually richest portion of an alluvial placer. The term is used in Australia for the dry bed of a buried Tertiary river. Syn: *bottom [ore dep].*

gutter [streams] (a) A shallow, natural channel, furrow, or gully worn by running water. (b) A shallow, steep-sided valley that drains a marshy upland; it usually marks an area where the drainage is about to be rejuvenated. (c) An artificially paved watercourse, such as a roadside ditch for carrying off excess surface water to a sewer. (d) An archaic term for a brook.

gutter cast A down-bulge on the bottom of a sedimentary bed, of great length (usually one meter or more) compared with its width and depth (a few centimeters to several decimeters). In cross section it has the form of a small channel (Whitaker, 1973, p.405). Cf: *channel [sed struc]; washout [sed struc].*

guyanaite (gu-yan'-a-ite) A reddish brown, golden brown, or green orthorhombic mineral polymorphic with bracewellite and grimaldiite: CrO(OH).
guyaquillite (guy-a-quil'-lite) *guayaquilite.*
guyot (guy-ot' [gee-o']) A type of *seamount* that has a flat top. Etymol: Arnold Guyot, nineteenth-century Swiss-American geologist. Syn: *tablemount; tableknoll.*
G wave A long-period (1-4 min) *Love wave* in the upper mantle, usually restricted to an oceanic path. The G stands for Gutenberg.
gwihabaite A colorless orthorhombic mineral found in guano: $(NH_4,K)NO_3$. It is a dimorph of nitrammite.
Gy *gray.*
gymnite (gym'-nite) *deweylite.*
gymnocyst (gym'-no-cyst) A *frontal shield* in cheilostome bryozoans, formed by calcification of part of the *frontal wall.* In anascan cheilostomes it is principally developed on the proximal and lateral margins of the *frontal membrane.*
gymnodinioid (gym-no-din'-i-oid) Resembling the dinoflagellate genus *Gymnodinium,* used to describe either the unarmored flagellate cell or the *tabulation* pattern consisting of numerous, often hexagonal plates (Williams et al., 2000, p.95).
Gymnophiona (Gym'-no-phi-o'-na) *Caecilia.*
gymnosolen (gym-no-so'-len) A fingerlike or digitate form of stromatolite, splitting off in two or more upward directions from algal structures, resembling a series of stacked inverted thimbles or (if large) of soup bowls similarly arranged (Pettijohn, 1957, p.222). It is produced by blue-green algae of the genus *Gymnosolen.*
gymnosperm (gym'-no-sperm) A *seed plant* in which the seeds are not enclosed by a carpel wall. Seeds may be borne singly or in cones. Examples include *cycads, ginkgos, conifers,* and *seed ferns.* Plants range from Late Devonian to the present.
gymnospore (gym'-no-spore) A naked spore, or one not developing in a sporangium. The term is not in good usage in palynology.
gynoecium *pistil.*
gyp A syn. of *gypsum.* Also spelled: gyps.
gyparenite (gyp-ar'-e-nite) A sandstone composed of discrete, wind-drifted particles of gypsum.
gypcrete (gyp-crete') A *gypsum* deposit that forms cemented desert pavements within the soil profiles developed in the Fayum Depression of Egypt and elsewhere (Aref, 2003). It is the gypsum equivalent of *calcrete.* Also spelled: gypscrete.
gyprock (gyp'-rock) Obsolete syn. of *rock gypsum.*
gypsarenite An evaporative gypsum deposit where crystals or their fragments have been transported by wind or water as sand grains. Contraction of "gypsum arenite."
gypsey (gyp'-sey) A syn. of *bourne.* Also spelled: gipsy; gypsy.
gypsic horizon (gyp'-sic) A mineral soil horizon of secondary $CaSO_4$ enrichment that is >15 cm thick, has at least 50 g kg^{-1} more gypsum than the C horizon, and in which the product of the thickness in centimeters and the amount of $CaSO_4$ is equal to or greater than 1,500g kg^{-1}.
gypsiferous (gyp-sif'-er-ous) Gypsum-bearing, as *gypsiferous* shales.
gypsification (gyp'-si-fi-ca'-tion) Development of, or conversion into, gypsum; e.g. the hydration of anhydrite.
gypsinate (gyp'-si-nate) Cemented with gypsum.
gypsite (gyp'-site) (a) An earthy variety of gypsum containing dirt and sand, found only in arid regions as an efflorescent deposit occurring over the ledge outcrop of gypsum or of a gypsum-bearing stratum. Syn: *gypsum earth.* (b) *gypsum.*
gypsolite (gyp'-so-lite) An *ooid* of gypsum, common in cross-beds (Schreiber, 1986, p.209). Cf: *oolite; halolite.*
gypsrudite A conglomerate composed of coarse, broken gypsum crystals or crystal aggregates.
gypsum (gyp'-sum) A widely distributed mineral consisting of aquated calcium sulfate: $CaSO_4·2H_2O$. It is the commonest sulfate mineral, and is frequently associated with halite and anhydrite in evaporites, forming thick, extensive beds interstratified with limestone, shale, and clay (esp. in rocks of Permian and Triassic age). Gypsum is soft (hardness of 2 on the Mohs scale); it is white or colorless when pure, but commonly has tints of gray, red, yellow, blue, or brown. It occurs massive (alabaster), fibrous (satin spar), or in monoclinic crystals (selenite) Gypsum is used chiefly as a soil amendment, as a retarder in portland cement, and in making Plaster of Paris. Etymol: Greek "gypsos", "chalk". Syn: *gypsite; gyp; plaster stone.*
gypsum cave (a) A cave that is formed in gypsum rock by solution (preferred usage). (b) A cave containing abundant gypsum incrustations. (less preferred usage).
gypsum cavoli Radiating-upward clusters of gypsum crystals that are internally layered across the crystals, appearing much as a cabbage (or cauliflower) in section. These are equivalent to crystal mounds formed at the sediment-water interface in saline water bodies. Best known from the Upper Miocene deposits of Italy, where they still show primary structure. Also recognized from relic structures in the Upper Silurian (Salina) of Indiana.
gypsum earth *gypsite.*
gypsum plate In a polarizing microscope, a plate of clear gypsum (selenite) that gives a first-order red interference color; it is used to determine optical sign with crystals or interference figures and to determine the position of vibration-plane traces in crystal plates. The plate may also be made of quartz.
gyrate (gy'-rate) Winding or coiled round; convolute, like the surface of the brain.
gyre (a) A closed geostrophic circulation in the ocean much larger than an eddy. (b) The closed geostrophic circulation in each ocean basin centered on the subtropical high-pressure region. The water within each subtropical gyre turns clockwise in the Northern Hemisphere and counterclockwise in the Southern Hemisphere. Acceleration causes sea level to fall along mainland coasts; deceleration leads to rise. Cf: *Coriolis force.*
gyrocompass (gy-ro-com'-pass) A nonmagnetic compass that functions by virtue of the couples generated in a rotor when the latter's axis of rotation is displaced from parallelism with that of the Earth and that consists of a continuously driven gyroscope whose supporting ring confines the spring axis to a horizontal plane. It automatically aligns itself in the celestial meridian (thus pointing to the true north) by the Earth's rotation which causes it to assume a position parallel to the Earth's axis. The gyrocompass is used in underground and borehole surveying. Syn: *gyroscopic compass; gyrostatic compass.*
gyrocone (gy'-ro-cone) A loosely coiled cephalopod shell in which the successive whorls are not in contact with each other, or in which only a single whorl is approximately completed. Also called: gyroceracone.
gyrogastric (gy-ro-gas'-tric) Said of a gastropod shell that is coiled toward the posterior of the body; caused by rotation (torsion) of the larval exogastric shell (Pojeta and Runnegar, 1976).
gyrogonite (gy-rog'-o-nite) A dispersed fossil oogonium of charophytes.
gyroid (gy'-roid) An isometric crystal form consisting of 24 crystal faces with indices *{hkl}* and symmetry 432. A gyroidal crystal may be right- or left-handed.
gyroidal class (gy-roid'-al) That crystal class in the isometric system having symmetry 432.
gyrolite (gy'-ro-lite) A white triclinic mineral with a micaceous cleavage: $NaCa_{16}(SiAl)_{24}O_{60}(OH)_8·14H_2O$. Syn: *centrallasite.*
gyroscopic compass (gy-ro-scop'-ic) (a) *gyrocompass.* (b) A magnetic compass whose equilibrium is maintained by the use of gyroscopes.
gyrostatic compass (gy-ro-stat'-ic) *gyrocompass.*
gysinite-(Nd) (gy'-sin-ite) A pink orthorhombic mineral of the *ancylite* group: $Pb(Nd,La)(CO_3)_2(OH)·H_2O$.
gyttja (gyt'-tja [ghytt-ya]) A sediment deposited in *eutrophic* lakes, consisting of microscopic and submicroscopic remains of organisms living in the basin, with an appreciable mineral content added, as found in Sweden. It is only weakly soluble in KOH, yielding an extract that is never dark brown but may be colorless, yellowish, or greenish (Faegri and Iversen, 1975, p.74). Cf: *dy; sapropel.* Etymol: Swedish.
Gzhelian (Gzhel'-i-an) A series or stage of the Standard Global Chronostratigraphic Scale: uppermost Carboniferous (above Kasimovian Stage, below Permian System).

H

haapalaite (haa'-pa-la-ite) A platy metallic bronze-red hexagonal mineral of the *valleriite* group: $4(Fe,Ni)S\cdot 3(Mg,Fe^{2+})(OH)_2$.

HAB *harmful algal bloom.*

habit (hab'-it) (a) The characteristic crystal form or combination of forms of a mineral, including characteristic irregularities. (b) A general term for the outward appearance of a mineral or rock. (c) The characteristic appearance of an organism, esp. those aspects that most affect its mode of life. Syn: *habitus.*

habitat (hab'-i-tat) The particular environment or place where an organism or species tends to live; a more locally circumscribed portion of the total environment.

habitus (hab'-i-tus) *habit.*

hachure (ha'-chure) n. One of a series of short, straight, evenly spaced, parallel lines used on a topographic map for shading and for indicating surfaces in relief (such as steepness of slopes), drawn perpendicular to the contour lines; e.g. an inward-pointing "tick" trending downslope from a depression contour. Hachures are short, broad (heavy), and close together for a steep slope, and long, narrow (light), and widely spaced for a gentle slope, and they enable minor details to be shown but do not indicate elevations above sea level. Etymol: French. Syn: *hatching; hatchure.* v. To shade with or show by hachures.

hackle The irregular outer fringe of a joint and the product of a rupture that is presumably propagating at critical velocities (Bahat, 1991). Cf: *twist hackle.*

hackly fracture (hack'-ly) The property shown by certain minerals or rocks of fracturing or breaking on jagged surfaces.

hackmanite (hack'-man-ite) A variety of sodalite containing a little sulfur and usually fluorescing orange or red under ultraviolet light.

hacksaw structure (hack'-saw) The irregular, *saw-toothed* or saw-shaped termination of a crystal (such as of augite) or mineral particle due to intrastratal solution. Syn: *cockscomb structure.*

hadal (ha'-dal) Pertaining to the deepest oceanic environment, specifically that of oceanic trenches, i.e., over 6,000 m in depth.

hade n. In structural geology, the complement of the dip; the angle that a structural surface makes with the vertical, measured perpendicular to the strike. It is little used. Syn: *underlay.* v. To incline from the vertical.

Hadean A suggested eon of the *Precambrian,* older than the oldest preserved rocks (>~3,800 Ma). Etymol: Allusion to Hades, Greek mythical underworld of the dead.

Hadley cell (Had'-ley) A thermally driven region of atmospheric system circulation that extends in both directions from the equator to about 30° latitude. Air rises at the equator in the vicinity of the Intertropical Convergence Zone, flows poleward, descends in the subtropical high-pressure cells, and then flows toward the equator as surface *trade winds.* Named after G. Hadley who described it in 1735.

Hadrynian (Had-ryn'-i-an) In a three-part division of the Proterozoic of Canada, the latest division, after the *Helikian,* before the Cambrian, from 880 to 600 or 570 Ma.

Haeckel's law (Haeck'-el's) *recapitulation theory.*

haematite (haem'-a-tite) Original spelling of *hematite.*

haff A shallow freshwater coastal lagoon separated from the open sea by a sandspit (*nehrung*) across a river mouth; esp. such a lagoon on the German coast of the Baltic Sea. Pl: haffs; haffe. Etymol: German "Haff", "lagoon".

hafnon (haf'-non) An orange-red, brownish yellow or colorless tetragonal mineral of the *zircon* group: $HfSiO_4$.

hagendorfite (ha'-gen-dorf-ite') A greenish-black monoclinic mineral of the *allaudite* group: $NaCaMn^{2+}(Fe^{2+},Fe^{3+},Mg)_2(PO_4)_3$.

haggertyite A metallic hexagonal mineral of the *magnetoplumbite* group: $BaFe_6Ti_5MgO_{19}$.

häggite (häg'-gite) An opaque black monoclinic mineral: $V_2O_2(OH)_3$.

haidingerite (hai'-ding-er-ite') A white or colorless orthorhombic mineral: $Ca(AsO_3OH)\cdot H_2O$.

haigerachite A vitreous white monoclinic mineral: $KFe^{3+}{}_3(H_2PO_4)_6(HPO_4)_2\cdot 4H_2O$.

hail Precipitation in the form of spheroidal layered balls or lumps of ice that usually fall from cumulonimbus clouds of a thunderstorm. The layered structure is produced by successive accretions of clear and opaque ice.

hail imprint A small, shallow depression or crater-like pit formed by a hailstone falling on a soft sedimentary surface. It is generally larger, deeper, and more irregular than a *rain print.* Syn: *hail pit; hailstone imprint.*

hail pit *hail imprint.*

hailstone imprint (hail'-stone) *hail imprint.*

haineaultite A pale brown, brown, or orange-yellow orthorhombic mineral: $(Na,Ca)_5Ca(Ti,Nb)_5Si_{12}O_{34}(OH,F)_8\cdot 5H_2O$.

hainite A colorless to yellow triclinic mineral of the *rosenbuschite* group: $Na_2Ca_5Ti(Si_2O_7)_2F_2(OH)_2$.

hair ball *lake ball.*

hair copper *chalcotrichite.*

hairpin dune (hair'-pin) A greatly elongated *parabolic dune* that has migrated downwind, its horns or arms drawn out parallel to each other. Hairpin dunes are formed where a constant wind is in conflict with vegetation.

hair pyrites (a) *millerite.* (b) Capillary pyrite.

hair salt (a) *alunogen.* (b) Silky or fibrous epsomite.

hairstone (hair'-stone) A variety of clear crystalline quartz thickly penetrated with fibrous, threadlike, or acicular inclusions of other minerals, usually crystals of rutile or actinolite; esp. sagenitic quartz. See also: *Venus hairstone; Thetis hairstone.* Also spelled: hair stone. Syn: *needle stone.*

hair zeolite A group of fibrous zeolite minerals, including natrolite, mesolite, scolecite, thomsonite, and mordenite. See also: *needle zeolite.* Syn: *feather zeolite.*

haiweeite (hai'-wee-ite) A pale-yellow to greenish-yellow orthorhombic or monoclinic secondary mineral: $Ca(UO_2)_2Si_5O_{12}(OH)_2\cdot 3H_2O$.

hakite (hak'-ite) A metallic grayish brown cubic mineral of the *tennantite* group: $(Cu,Hg)_{12}Sb_4(Se,S)_{13}$. Its arsenian analogue is giraudite.

halbkugelkarst Tropical karst topography containing dome-shaped residual hills surrounding depressions, a kind of *kegelkarst.* Etymol: German "Halbkugel", "hemisphere".

haldenhang (hal'-den-hang) A syn. of *wash slope.* Etymol: German "Haldenhang", "under-talus rock slope of degradation" (Penck, 1924).

Half-Bog soil A great soil group, of the intrazonal order and hydromorphic suborder consisting of soil with dark-brown or black peaty material over grayish and rust mottled mineral soil; formed under conditions of poor drainage under forest, sedge, or grass vegetation in cool to tropical humid climates. Not used in current U.S. system of soil taxonomy.

half breed Nuggets of mixed native copper and native silver found in the copper deposits of the Keweenaw Peninsula, Michigan.

half graben An elongate, asymmetric trough or basin bounded on one side by a *normal fault.* Cf: *graben.*

half-life The time required for one half of a given number of radionuclides to decay. The half-life ($T_{1/2}$) is related to the decay constant (λ) by $T_{1/2} = \ln/2$ where ln is the natural logarithm. Each *radionuclide* has a characteristic half-life.

half-moon cut A style of gem cutting that produces a stone in the shape of a half circle.

half section A half of a normal section of the U.S. Public Land Survey system, representing a piece of land containing 320 acres as nearly as possible; any two quarter sections within a section which have a common boundary. It is usually identified as the north half, south half, east half, or west half of a particular section.

halftone A technique of representing shading with dots produced by photographing the object from behind a fine screen (USGS, 2002a, p.1A-2).

half-tube A remnant or trace of a *tube [speleo]* visible on the roof or walls of a cave.

half-value thickness (half-val'-ue) The thickness of an absorbing medium which will reduce any incident radiation to half its initial intensity.

half width Half the width of a simple *anomaly* (esp. a gravity or magnetic anomaly) at half its maximum value. For simple models the maximum depth at which the body causing the anomaly can lie can be calculated from the half width.

halide (hal'-ide) A mineral compound characterized by a halogen such as fluorine, chlorine, iodine, or bromine as the anion. Halite, NaCl, is an example. Syn: *halogenide*.

halite (hal'-ite) A cubic mineral: NaCl. It is native salt, occurring in massive, granular, compact, or cubic-crystalline forms, and having a distinctive salty taste. Halite is typically colorless, but certain occurrences provide red, yellow, blue, etc. colors owing to impurities and/or "color centers" in the structure. Symbol: Hl. Syn: *common salt; rock salt*.

halite hopper crystals Halite crystals with preferred growth along sides and edges producing depressed faces; includes crystals grown at water-air interface producing an inverted pyramid (Dellwig, 1955), and skeletal halite crystals precipitated in brine-saturated mud (Gornitz and Schreiber, 1981).

halitic (ha-lit'-ic) Pertaining to halite; esp. said of a sedimentary rock containing halite as cementing material, such as "halitic sandstone".

hall In a cave, a lofty chamber which is much longer than it is wide. Cf: *gallery [speleo]*.

hälleflinta (häl'-le-flin'-ta) An obsolete Fennoscandian term for a fine-grained (< 0.25 mm) quartzofeldspathic metavolcanic rock, typically rhyolitic, possibly also meta-arkosic. Its equivalent of higher grade is *leptite*.

hälleflintgneiss *leptite*.

Hallian (Hal'-li-an) North American stage: Pleistocene (above Wheelerian).

hallimondite (hal'-li-mond-ite) A yellow triclinic secondary mineral: $Pb_2(UO_2)(AsO_4)_2$. It is the arsenate analogue of parsonsite.

halloysite (hal-loy'-site) A 1:1 aluminosilicate clay mineral $Al_2Si_2O_5(OH)_4{\cdot}X(H_2O)$ similar to kaolinite but perhaps with some Al(IV) and interlayer cations to compensate for the Al(IV). Probably because of this it is able to incorporate water in the interlayer space (Bailey, 1989). The terms "halloysite (7Å)" and "halloysite (10Å)" were recommended for the anhydrous and dihydrate forms, respectively (Brindley and Pedro, 1976); the term "endellite" should not be used (Bailey et al., 1980).

halmeic (hal-me'-ic) Said of a deep-sea sediment formed directly from solution or around an organic nucleus, e.g. barite, phosphorite, manganese nodules. Cf: *authigenic*. Ant: *chthonic*.

halmyrolysis (hal-my-rol'-y-sis) The geochemical reaction of sea water and sediments in an area of little or no sedimentation. Examples include modification of clay minerals and formation of glauconite and zeolite. Alternate spelling: halmyrosis. Cf: *diagenesis [sed]*. Syn: *submarine weathering*.

halo (ha'-lo) (a) A circular or crescentic distribution pattern about the source or origin of a mineral, ore, mineral association, or petrographic feature. It is encountered principally in magnetic and geochemical surveys. Cf: *dispersion pattern*. (b) Discoloration of a mineral, viewed in thin section, in the form of a ring. Most haloes of this sort are caused by radiation damage by alpha particles emitted from uranium- and thorium-bearing mineral inclusions.

halo [planet] A diffuse, circular radar dark or bright zone surrounding impact craters on the Venusian surface thought to result from impact shocks of disintegrated asteroids and/or comets (Bougher et al., 1997). See also: *splotch, parabola [planet]*.

halocline (hal'-o-cline) A layer in the ocean in which the rate of salinity variation with depth is much larger than layers immediately above or below it. Usually located in the upper 500 m of the ocean and can exhibit a salinity increase with depth (as in a river plume) or a salinity decrease with depth (as in the central gyres of the principal oceans). Cf: *thermocline [oceanog]; pycnocline; mixed layer*.

halocline cave A small cave in coastal karst formed by mixed-water dissolution at the base of a freshwater lens.

halogenic (hal-o-gen'-ic) *halmeic*.

halogenide (hal'-o-gen-ide') *halide*.

halokinesis (hal-o-ki'-ne-sis) (a) The deformation of *halite* by flowage. Mechanisms cited for this process include gravity flow, tectonic thrusting, and diapirism (Kyle and Posey, 1991). Syn: *salt tectonics*. (b) Form of salt tectonics in which salt flow is powered entirely by gravity that is, by release of gravity potential energy alone, in the absence of significant lateral tectonic forces (Jackson and Talbot, 1991).

halolite (hal'-o-lite) An *ooid* of halite, developed in saturated brine under turbulent conditions (Schreiber, 1986, p.209). Cf: *oolite; gypsolite*.

halomorphic soil (hal-o-mor'-phic) A suborder of the intrazonal soil order, consisting of saline and sodic soils formed under imperfect drainage in arid regions and including the great soil groups Solonchak or Saline soils, Solonetz soils, and Soloth soils. Not used in current U.S. system of soil taxonomy.

halo ore *fringe ore*.

halophilic (hal-o-phil'-ic) Said of an organism that prefers a saline environment. Noun: halophile. Cf: *haloxene*.

halophreatophyte (hal'-o-phre-at'-o-phyte) A plant receiving its water supply from saline groundwater.

halophyte (hal'-o-phyte) A plant growing in soil or water with a high content of salts.

halosere (hal'-o-sere) A *sere* that develops in a saline environment.

halotrichite (hal-o-trich'-ite) (a) A mineral: $Fe^{2+}Al_2(SO_4)_4{\cdot}22H_2O$. It occurs in colorless, yellowish, or greenish fibrous crystals. Syn: *feather alum; iron alum; mountain butter; butter rock*. (b) A group name for minerals with the analogous formula, but with Fe^{2+} replaced by Mg, Mn or Zn, and Al sometimes replaced by Fe^{3+}.

haloxene (hal'-ox-ene) Said of an organism that can tolerate saline conditions but does not prefer them. Cf: *halophilic*.

hals A British term for a *pass* or *col*. Cf: *hause*. Also spelled: halse.

halurgite (ha-lur'-gite) A white monoclinic mineral: $Mg_2[B_4O_5(OH)_4]_2{\cdot}H_2O$.

HAM *hypothetical ancestral mollusk*.

hamada (ha-ma'-da) *hammada*.

hambergite (ham'-berg-ite) A grayish-white, yellowish, or colorless orthorhombic mineral: $Be_2BO_3(OH,F)$.

hamlinite (ham'-lin-ite) *goyazite*.

hammada (ham-ma'-da) An extensive, nearly level upland desert surface that is either bare bedrock or bedrock thinly veneered by pebbles, smoothly scoured and polished and generally swept clear of sand and dust by wind action; a *rock desert* of the plateaus, esp. in the Sahara. The term is also used in other regions, as in Western Australia and the Gobi Desert. Etymol: Arabic, "hammadah". Cf: *reg; serir*. Also spelled: *hamada;* hammadah; hammadat; hamadet. Syn: *nejd*.

hammarite (ham'-mar-ite) A metallic steel-gray orthorhombic mineral: $Pb_2Cu_2Bi_4S_9$.

hammer [geophys exploration] A hammer striking a steel plate is used as a seismic source for shallow refraction or reflection measurements. The hammer incorporates a switch which starts a timer when the hammer strikes. Alternatively, closure can be produced by the hammer touching the plate to complete an electrical circuit. In some systems the timer is stopped when energy received by a geophone reaches some threshold value.

Hammer-Aitoff projection (Ham'-mer-Ai'-toff) An equal-area map projection derived from the equatorial aspect of the Lambert azimuthal equal-area projection by doubling the horizontal distances along each parallel from the central meridian until the entire spherical surface can be represented within an ellipse whose major axis (equator) is twice the length of its minor axis (central meridian). It resembles the Mollweide projection, but all parallels (except the equator) are represented by curved lines and there is less angular distortion near the margins. The projection was introduced in 1892 by Ernst H.H. von Hammer (1858-1925), German geodesist, but is often attributed to David Aitoff (1854-1933), Russian geographer, who previously introduced a similar-appearing projection based on the azimuthal equidistant projection. Incorrect syn: *Aitoff projection*.

Hammer chart (Ham'-mer) A template for calculating the *terrain correction* to the gravity data. Named for Sigmund Hammer, American geophysicist.

hammock (ham'-mock) (a) *hummock [geomorph]*. (b) A term applied in the SE U.S. to a fertile area of deep, humus-rich soil, generally covered by hardwood vegetation and often rising slightly above a plain or swamp; esp. an island of dense, tropical undergrowth in the Florida Everglades. Syn: hummock.

hammock structure The intersection of two vein or fracture systems at an acute angle.

hampshirite (hamp'-shir-ite) Steatite pseudomorphous after olivine.

hamrongite (ham-rong'-ite) A dark violet-gray fine-grained *lamprophyre* containing phenocrysts of black mica in a groundmass characterized by intersertal texture and composed of mica, andesine, and some quartz; a quartz *kersantite*. The name, given by Eckermann in 1928, is from Hamrånge, Sweden. Obsolete.

hamulate [palyn] (ham'-u-late) adj. Of spores, having *rugulate* sculpture of irregularly arranged, winding or angularly bent *muri* of various thicknesses, forming a mazelike pattern.

hamulus (ham'-u-lus) (a) A hook-shaped secondary deposit on the chamber floor in foraminifers of the family Endothyridae. The point of the hook is directed toward the aperture of the test. (b) In the vertebrates, any hook-shaped bone or bony process, as in the wrist, that serves as a pulley to modify the direction in which the force of a muscle is exerted. Pl: hamuli.

hanawaltite A metallic black or brown-black monoclinic mineral: $Hg^{1+}{}_6Hg^{2+}Cl_2O_3$.

hancockite (han'-cock-ite) A yellowish brown, yellow-green, or brownish red monoclinic mineral of the *epidote* group: $CaPb(Al,Fe^{3+})_3(SiO_4)(Si_2O_7)O(OH)$.

hand hardness test A layer-parallel strength test which records the penetration of a hand or objects held by a hand.

hand lens A small magnifying glass (usually X6 to X10) for use in the field or in other preliminary investigations of a mineral, fossil, or rock. Syn: *pocket lens*.

hand level A small, hand-held leveling instrument in which the spirit level is so mounted that the observer can view the bubble at the same time that he sights an object through the telescope. The viewing of the bubble is accomplished by means of a prism or mirror in the telescope tube: when the cross hair bisects the bubble and the object in view, that object is on a level with the eye. The hand level is used where a high degree of precision and accuracy is not required, such as in reconnaissance surveys. See also: *Abney level; Locke level*.

hand specimen A piece of rock of a size that is convenient for megascopic study and for preserving in a study collection.

hanger (hang'-er) *hanging wall*.

hanging (hang'-ing) Situated on steeply sloping ground (such as a *hanging* meadow) or on top of other ground (such as a *hanging wall*), or jutting out and downward (such as a *hanging* rock), or situated at or having a discordant junction (such as a *hanging valley*).

hanging cirque A cirque on a mountainside, excavated by a former hanging glacier and not continued in a valley. Cf: *corrie; valley-head cirque.*

hanging delta *step delta.*

hanging drumlin A drumlin on a valley slope, consisting of subglacial debris pushed laterally, molded, or deposited into its present position by an overriding glacier.

hanging glacier A glacier, generally small, in a basin or niche on a mountainside above a cliff or very steep slope. Ice may occasionally break off from such a glacier, resulting in an ice avalanche. Cf: *cliff glacier.*

hanging side *hanging wall.*

hanging tributary A stream whose bed is at a higher elevation than the stream into which it flows. The junction is often the site of a waterfall.

hanging trough A glacial hanging valley.

hanging valley [glac geol] A glacial valley whose mouth is at a relatively high level on the steep side of a larger glacial valley. The larger valley was eroded by a trunk glacier and the smaller one by a tributary glacier, and the discordance of level of their floors, as well as their difference in size, is due to the greater erosive power of the trunk glacier. Syn: *hanging trough; perched glacial valley.*

hanging valley [streams] (a) A tributary valley whose floor at the lower end is notably higher than the floor of the main valley in the area of junction, produced where the more rapid deepening of the main valley results in the creation of a cliff or steep slope over which a waterfall may develop. (b) A coastal valley whose lower end is notably higher than the shore to which it leads, produced where betrunking or rapid cliff recession causes the mouths of streams to "hang" along the cliff front. Syn: *valleuse.*

hanging wall The overlying side of an orebody, fault, or mine working; esp. the wall rock above an inclined vein or fault. Syn: *hanging side; hanger.* Cf: *footwall; upper plate.*

hanging-wall cutoff The intersection between the hanging wall portion of a once-continuous planar or linear geological feature and a fault surface. In thrust systems, a syn. of *leading edge* (Royse et al., 1975).

hanging-wall flat That portion of a step-shaped fault that parallels strata in the hanging wall.

hanging-wall imbrication A temporal sequence of formation of imbricates or splays off a thrust fault flat where successively younger imbricates or splays form on the hinterland or trailing side of preexisting imbricates or splays (Boyer and Elliott, 1982). Syn: *break-back sequence.* Cf: *footwall imbrication; in-sequence thrusting; out-of-sequence thrusting.*

hanging-wall ramp That portion of a step-shaped fault that cuts across strata in the hanging wall.

hanksite (hanks'-ite) A white, gray, or yellow hexagonal mineral: $KNa_{22}(SO_4)_9(CO_3)_2Cl$.

hannayite (han-nay'-ite) A mineral: $(NH_4)_2Mg_3(PO_3OH)_4 \cdot 8H_2O$. It occurs as slender yellowish triclinic crystals in guano.

hannebachite (han'-ne-bach'-ite) A colorless orthorhombic sulfite mineral: $2CaSO_3 \cdot H_2O$.

hanusite (ha-nus'-ite) A mixture of stevensite and pectolite. It was formerly regarded as a mineral: $Mg_2Si_3O_7(OH)_2 \cdot H_2O$.

haplogranite A synthetic mixture of quartz and feldspars having a simplified composition resembling granite, used in experimental petrology.

haplome (hap'-lome) A "more correct", but "not generally accepted", spelling of *aplome* (Hey, 1962, p.446).

haplontic (hap-lon'-tic) Said of a plant life cyle in which the vegetative cells are haploid, the *zygote* being the only diploid cell.

haplopore (hap'-lo-pore) An unbranched pore lying normally within one thecal plate of a cystoid. If connected in pairs, haplopores are designated diplopores.

haploxylonoid (hap-lo-xy'-lo-noid) Said of bisaccate pollen, in which the outline of the sacci in distal-proximal view is more or less continuous with the outline of the body, the sacci appearing more or less crescent-shaped, and the outline of the whole grain presenting generally smooth ellipsoidal forms. Cf: *diploxylonoid.*

haplozoan (hap-lo-zo'-an) Any one of a small group of supposedly free-living echinoderms belonging to the subphylum Haplozoa, comprising only two genera, and characterized by a thick calcareous skeleton composed of a few plates arranged around a median, craterlike depression. Range, Middle Cambrian.

hapteron (hap'-ter-on) An attaching structure in some of the larger seaweeds, esp. in the brown algae; it is usually multicellular, branched, and rootlike.

haptonema (hap-to-ne'-ma) A threadlike to clublike part of a cell in a coccolithophorid, located between, but more rigid than, the flagella. It can be contracted into a spiral or extended and used as an organ of attachment to substrate material.

haptotypic character (hap-to-typ'-ic) A feature of palynomorphs that is a product of its contact with other members of the tetrad in which it was formed; e.g. the laesura and contact areas of spores.

Haq chart A table showing Mesozoic and Cenozoic eustatic cycles, sequence chronostratigraphy, and relationships to other time scales.

haradaite (ha-ra'-da-ite) A green orthorhombic mineral: $SrV^{4+}Si_2O_7$.

harbor (har'-bor) (a) A small bay or a sheltered part of a sea, lake, or other large body of water, usually well protected either naturally or artificially against high waves and strong currents, and deep enough to provide safe anchorage for ships; esp. such a place in which port facilities are furnished. (b) An *estuary* or barred estuary; e.g. Pearl Harbor, Hawaii. British spelling: harbour.

harbor bar A bar built across the exit to a harbor.

hardcap (hard'-cap) A term used in bauxite mining for the uppermost foot or two of a bauxite deposit. Since it is harder and tougher than the material below it, it is usually used as a roof during mining.

hard coal (a) A syn. of *anthracite.* Cf: *soft coal.* (b) Outside the U.S., the term is sometimes used for any coal with a calorific value higher than 23.86 MJ/kg (10,260 BTU/lb) on a moist, mineral-matter-free basis; this includes bituminous and certain subbituminous coals.

hardebank (hard'-e-bank) Unweathered *kimberlite* below the zone of *blue ground.*

hardening [geophys exploration] Increase in acoustic impedance as a result of fluid production and loss of porosity.

hardground (hard'-ground) A limestone which has been cemented on the seafloor. The hardground surface may be encrusted, discolored, case-hardened, bored, and solution-ridden. It implies a temporal gap in sedimentation and may be preserved stratigraphically as an unconformity.

hardground [reef] A generally planar surface of rock, which is generally limestone and is colonized by corals and other attached benthos. Hardgrounds can be pre-existing limestones or those formed by syndepositional cementation of sea floor deposits of skeletal and nonskeletal sediments.

hardhead (hard'-head) *coral head.*

hard linkage The geometry or process whereby two faults are connected by one or more (usually smaller) faults at the scale of observation; the combined fault system exhibits *geometric coherence* (Walsh and Watterson, 1991). Cf: *soft linkage.*

hard magnetization Magnetization that is not easily destroyed; specifically, remanent magnetization with a high *coercive force.* Cf: *soft magnetization; stable magnetization.*

hard mineral A mineral that is as hard as or harder than quartz, i.e. ranking seven or higher on the Mohs scale. Cf: *soft mineral.*

hardness [mineral] (hard'-ness) The resistance of a mineral to scratching; it is a property by which minerals may be described, relative to a standard scale of ten minerals known as the *Mohs scale,* to the *technical scale* of fifteen minerals, or to any other standard.

hardness [water] A property of water causing formation of an insoluble residue when the water is used with soap, and forming a scale in vessels in which water has been allowed to evaporate. It is primarily due to the presence of ions of calcium and magnesium, but also to ions of other alkali metals, other metals (e.g. iron), and even hydrogen. Hardness of water is generally expressed as parts per million as $CaCO_3$ (40 ppm Ca produces a hardness of 100 ppm as $CaCO_3$); also as milligrams per liter; and as the combination of *carbonate hardness* and *noncarbonate hardness.* Syn: *total hardness.* Cf: *soft water.*

hardness points Small, pointed pieces of minerals of different hardness, affixed to handles and used for testing the hardness of another mineral by ascertaining which point will scratch it. Minerals of hardness 6 to 10 on the Mohs scale are usually used as the points for testing gemstones. Because of the potential for damage, this test is not routinely used on faceted gemstones.

hard ore A term used in the Lake Superior region for a compact, massive iron ore mainly composed of specular hematite and/or magnetite and containing more than 58% iron. Cf: *soft ore.*

hardpan (hard'-pan) (a) A general term for a relatively hard, impervious, and often clayey layer of soil lying at or just below the surface, produced as a result of cementation of soil particles by precipitation of relatively insoluble materials such as silica, iron oxide, calcium carbonate, and organic matter. Its hardness does not change appreciably with changes in moisture content, and it does not slake or become plastic when mixed with water. The term is not properly applied to hard clay layers that are not cemented, or to layers that may seem indurated but which soften when soaked in water. See also: *lime pan [soil]; iron pan; ortstein.* Cf: *duricrust; claypan [soil]; fragipan.* (b) A layer of gravel encountered in the digging of a gold placer, occurring one or two meters below the ground surface and partly cemented with limonite. (c) A term commonly applied in NW U.S. to a compact, subglacial till that must be drilled or blasted before removal. Also, a cemented layer of sand or gravel enclosed within till, or a capping of partly cemented material at the top of a water-bearing layer of sand or gravel in till (Wentworth, 1935, p.243). (d) A popular term used loosely to designate any relatively hard layer that is difficult to excavate or drill; e.g. a thin resistant layer of limestone interbedded with easily drilled soft shales. (e) *caliche [soil].* (f) *plow sole.* Legget (1962, p.798) suggests that the term be avoided "in view of its wide and essentially popular local use for a wide range of materials".

hard-rock geology (hard'-rock) A colloquial term for geology of igneous and metamorphic rocks, as opposed to *soft-rock geology.*

hard-rock phosphate A term used in Florida for pebbles and boulders of a hard massive homogeneous light-gray phosphorite, showing irregular cavities that are usually lined with secondary mammillary incrustations of calcium phosphate. It is essentially equivalent to the term *white-bedded phosphate* that is used in Tennessee.

hard shore A shore composed of sand, gravel, cobbles, boulders, or bedrock. Ant: *soft shore.*

hard slab avalanche An avalanche containing a hard slab identified by blocks of slab snow which retain their shape throughout much or all of the path.

hard spar A name applied to corundum and andalusite.

hard water Water that does not lather readily when used with soap, and that forms a scale in containers in which it has been allowed to evaporate; water with more than 60 mg/l of hardness-forming constituents, expressed as $CaCO_3$ equivalent. See also: *hardness [water].* Cf: *soft water.*

hardway A direction that is difficult to split in granite or other crystalline rocks. This direction is usually normal to both *rift* and *grain.*

hardwood (hard'-wood) The wood of an angiospermous tree. Actually, such wood may be either hard or soft. Cf: *softwood.*

hardystonite (har'-dy-ston-ite') A white to pale pink or light brown tetragonal mineral of the *melilite* group: $Ca_2ZnSi_2O_7$.

Harker diagram (Har'-ker) A *variation diagram* on which silica content is shown on the abscissa and other oxides on the ordinate.

harkerite (har'-ker-ite) A colorless rhombohedral mineral: $Ca_{12}Mg_4Al(SiO_4)_4(BO_3)_3(CO_3)_5{\cdot}H_2O$.

Harlechian (Har-lech'-i-an) European stage: Lower Cambrian.

harlequin opal (har'-le-quin) Opal with small, close-set, angular (mosaic-like) patches of play of color of similar size. Cf: *pinfire opal.*

harmful algal bloom *Red tides* and other blooms of toxic algae that kill marine organisms and may cause human paralytic shellfish poisoning when ingested. Abbrev: HAB.

harmomegathus (har-mo-meg'-a-thus) An organ or mechanism, such as the membrane of the pore or colpus, that enables a semirigid pollen wall to accommodate, by expansion or contraction, the changes in its volume resulting from the take-up or loss of water. Pl: harmomegathi. Adj: harmomegathic.

harmonic [geophys] (a) A frequency which is a simple multiple of a fundamental frequency. The third harmonic, for example, has a frequency three times that of the fundamental. (b) Two frequencies are harmonically related if they are each harmonics of a common fundamental. (c) Any component of a Fourier series except the fundamental.

harmonic analysis [math] *Fourier analysis.*

harmonic analysis [struc geol] A method for analyzing the shape of folded layers, usually by determining the wavelength and amplitude of different harmonic components of a Fourier series fit to an expanded quarter wave segment of a fold (Hudleston, 1973).

harmonic fold A fold with uniform wavelength and amplitude. Ant: *disharmonic fold.*

harmonic folding Folding in which the strata remain parallel or concentric, without structural discordances between them, and in which there are no sudden changes in the form of the folds at depth. Ant: *disharmonic folding.*

harmonic mean An average calculated as the reciprocal of the arithmetic average of the reciprocals of the observations. Sometimes used to calculate average permeability. See also: *mean; geometric mean.*

harmonic tremor Long-period volcanic ground vibration that generally lacks a distinct beginning or end and usually indicates the movement of magma or gas. It may precede an eruption by a few hours but also may record fluid movement into an intrusion.

harmotome (har'-mo-tome) A white to variously colored monoclinic zeolite mineral: $(Ba,K)_5(Si,Al)_{16}O_{32}{\cdot}12H_2O$. It forms cruciform twin crystals. Syn: *cross-stone.*

harpolith (harp'-o-lith) A large, sickle-shaped igneous intrusion injected into previously deformed strata and subsequently deformed with the country rock; essentially a *phacolith* with a vertical axis. Obsolete.

harrisite [mineral] (har'-ris-ite) Chalcocite pseudomorphous after galena.

harrisite [petrology] A granular igneous rock composed chiefly of olivine and a smaller amount of calcic plagioclase, and characterized by *harrisitic* texture. Named by Harker in 1908 for Harris, Isle of Rhum. Cf: *troctolite.* Obsolete.

harrisitic (har-ri-sit'-ic) Said of the texture observed in certain olivine-rich rocks (esp. harrisite) in which the olivine crystals are oriented at approximately right angles to the cumulate layering of the rock. This phenomenon is now known to occur with other minerals and is called *crescumulate* texture (Wager, 1968, p.579).

harrisonite A yellow-brown to orange-brown rhombohedral mineral: $Ca(Fe^{2+},Mg)_6(SiO_4)_2(PO_4)_2$.

harrow mark (har'-row) One of a group of parallel fine-grained ridges of sand, silt, and clay, from about 1 to 10 cm high and 5 to 50 cm apart, with intervening trough-like strips of coarser sediments, occurring in stream channels and extending for distances as great as 100 m. It has been ascribed to the action of regular longitudinal helical flow patterns with alternating senses of rotation (Karcz, 1967).

harstigite (har'stig-ite) A colorless orthorhombic mineral: $Ca_6(Mn^{2+},Mg)Be_4(SiO_4)_2(Si_2O_7)_2(OH)_2$.

hartite (hart'-ite) A white, crystalline, triclinic hydrocarbon found in lignites: $C_{20}H_{34}$. Syn: *bombiccite; branchite; hofmannite.*

Hartmann's law (Hart'-mann's) The statement that the acute angle between two sets of intersecting shear planes is bisected by the axis of maximum principal stress, and the obtuse angle by the axis of minimum principal stress.

hartschiefer (hart'-schief-er) A metamorphic rock of compact, dense, cherty, or felsitic texture, having a banded structure in which the bands are of approximately even thickness, have rigid parallelism, and differ considerably in mineral and chemical composition. It is formed by intense dynamic metamorphism from ultramylonites and is associated with other rocks of mylonitic habit (Holmes, 1920, p.116). The term was originated by Quensel in 1916. Etymol: German "Hartschiefer", "hard slate or schist".

harzburgite (harz'-burg-ite) (a) In the *IUGS classification,* a plutonic rock with M equal to or greater than 90, ol/(ol+opx+cpx) between 40 and 90, and cpx/(ol+opx+cpx) less than 5. (b) A *peridotite* composed chiefly of olivine and orthopyroxene. Named by Rosenbusch in 1887 for Harzburg, Germany. See also: *saxonite.*

hashemite (ha'-shem-ite) An adamantine light yellowish brown orthorhombic mineral: $Ba(Cr,S)O_4$.

hassock structure A variety of *convolute bedding* in which the laminae resemble tufts of grass or sedge.

hastate (has'-tate) Said of a leaf that is arrow-shaped, as in leaves with a sharp tip and basal lobes that point away from the petiole.

hastingsite (has'-tings-ite) A dark green monoclinic mineral of the *amphibole* group: $NaCa_2(Fe^{2+}{}_4Fe^{3+})(Si_6Al_2)O_{22}(OH)_2$. It generally contains a little potassium.

hastite (has'-tite) A metallic orthorhombic mineral of the *marcasite* group: $CoSe_2$. It is dimorphous with trogtalite.

hatchettine (hatch'-ett-ine) A soft yellow-white paraffin wax, perhaps $C_{38}H_{78}$, having a melting point of 55-65°C in the natural state and 79°C after purification. It occurs as veinlike masses in ironstone nodules associated with coal-bearing strata (as in south Wales) or in cavities in limestone (as in France). Syn: hatchettite; *adipocire; adipocerite; mineral tallow; mountain tallow; naphthine.*

hatchettolite (hatch-et'-to-lite) *uranpyrochlore.*

hatching (hatch'-ing) (a) The drawing of hachures on a map to give an effect of shading. See also: *cross-hatching.* (b) *hachure.*

hatchite (hatch'-ite) A metallic lead-gray triclinic mineral: $(Pb,Tl)_2AgAs_2S_5$. It is the Ag analogue of wallisite.

hatchure (hatch'-ure) Var. of *hachure.*

hatrurite (ha-tru'-rite) A colorless rhombohedral mineral occurring in the "mottled zone", Israel: Ca_3SiO_5. It corresponds to "alite", a constituent of Portland cement.

hauchecornite (hau-che-cor'-nite) (a) A metallic bronze tetragonal mineral: $Ni_9Bi(Sb,Bi)S_8$. (b) A group name for minerals with the analogous formula, but with Bi replaced by Sb, As or Te.

hauckite (hauck'-ite) A light yellow to orange hexagonal mineral: $(Mg,Mn^{2+})_{24}Zn_{18}Fe^{3+}{}_3(SO_4)_4(CO_3)_2(OH)_{81}$.

hauerite (hau'-er-ite) A reddish-brown or brownish-black mineral of the *pyrite* group: MnS_2. It occurs in octahedral or pyritohedral crystals.

haughtonite (haugh'-ton-ite) A black, iron-rich variety of biotite.

hause An English term for a *pass,* or a ridge connecting two higher elevations, or a narrow gorge (Whitney, 1888, p.137). Cf: *hals.* Syn: *haws.*

hausmannite (haus'-mann-ite) (a) A submetallic, brownish-black, tetragonal mineral: $Mn^{2+}Mn^{3+}{}_2O_4$. Related to the *spinel* group, it is an occasional ore of manganese. (b) A group name for minerals with the analogous formula, but with Mn^{2+} replaced by Fe^{2+} or Zn, and Mn^{3+} replaced by Fe^{3+}.

haustorium (haus-to'-ri-um) A parasitic, food-absorbing outgrowth of a *hypha,* stem, or other fungal or plant organ. Adj: haustorial.

Hauterivian (Hau-te-ri'-vi-an) Stage of the Standard Global Chronostratigraphic Scale: Lower Cretaceous (above Valanginian, below Barremian).

haüyne (ha-uyne' [ah-ween']) A variously colored cubic feldspathoid mineral of the *sodalite* group: $(Na,Ca)_{4\text{-}8}Al_6Si_6(O,S)_{24}(SO_4,Cl)_{1\text{-}2}$. It is related to nosean and occurs in rounded and subangular grains embedded in various volcanic rocks. Also spelled: hauyne. Syn: *hauynite.*

hauyne basanite In the *IUGS classification,* a basanite in which hauyne is the most abundant feldspathoid.

hauyne phonolite In the *IUGS classification,* a *phonolite* in which hauyne is an important feldspathoid.

hauynite [mineral] (ha-uy'-nite) A syn. of *haüyne.* Also spelled: haüynite.

Haüy's law (Ha'-üy's) *law of rational indices.*

haven (ha'-ven) A small bay, recess, or inlet of the sea affording anchorage and protection for ships; a harbor.

Haversian system (Ha-ver'-si-an) The complex network of fine longitudinal canals, each surrounded by concentric layers of apatite in an organic matrix, that is characteristic of the dense cortical bone of many tetrapods, esp. those of large body size. This organization is commonly preserved during fossilization, and when present is fully diagnostic of bony tissue.

havsband (havs′-band) A Swedish term for the outermost or seaward part of a *skerry-guard,* constituting bare skerries and the smallest rock islets (Stamp, 1961, p.230, 419).

Hawaiian-type bomb (Ha-wai′-ian) A type of volcanic bomb formed when a still-plastic clot of lava strikes the ground, so that its shape is controlled by impact, not by its flight through the air. Cf: *pancake bomb; Strombolian-type bomb.*

Hawaiian-type eruption A type of volcanic eruption in which great quantities of extremely fluid basaltic lava are poured out, mainly issuing in lava fountains from fissures on the flanks of a volcano. Explosive phenomena are rare, but much spatter and scoria are piled into cones and mounds along the vents. Characteristic of shield volcanoes. Cf: *Peleean-type eruption; Strombolian-type eruption; Vulcanian-type eruption.*

hawaiite [mineral] (ha-wai′-ite) A pale-green, iron-poor gem variety of olivine from the lavas of Hawaii.

hawaiite [petrology] (a) In the *IUGS classification,* a *trachybasalt* with $Na_2O\text{-}2.0 \geq K_2O$ in wt%. (b) As defined by Iddings in 1913, an olivine "basalt" with andesine as the normative plagioclase (thus differing from true basalt, in which the normative plagioclase is more calcic). (c) Macdonald (1960, p.175) revived the term and extended the definition: "a rock with moderate to high colour index, and frequently basaltic habit, in which the normative and modal feldspar is andesine, and with soda:potash ratio greater than 2:1. It generally, but not always, lacks normative quartz, and commonly contains normative and modal olivine. Hawaiite is intermediate in composition between alkali olivine basalt and *mugearite,* and grades into both."

Haweran (Ha-wer′-an) South Pacific stage: Recent (above Castlecliffian).

hawk's-eye (hawk's′-eye) A transparent to translucent colorless variety of quartz, which contains minute parallel closely packed fibrous crystals of partly replaced crocidolite (from which a pale-blue to greenish-blue sheen is produced by reflection of light) and that resembles the eye of a hawk when cut cabochon; a blue variety of tiger's-eye. Cf: *sapphire quartz.* Syn: *falcon's-eye.*

hawleyite (haw′-ley-ite) A yellow or orange cubic mineral of the *sphalerite* group: CdS. It is dimorphous with greenockite.

haws *hause.*

hawthorneite (haw′-thorne-ite) A metallic black hexagonal mineral of the *magnetoplumbite* group: $Ba(Cr_4Ti_3Fe^{2+}{}_2Fe^{3+}{}_2Mg)O_{19}$.

haxonite (hax′-on-ite) A metallic white cubic meteorite mineral: $(Fe,Ni)_{23}C_6$.

haycockite (hay′-cock-ite) A metallic yellow orthorhombic mineral: $Cu_4Fe_5S_8$.

Hayford zone (Hay′-ford) A subdivision of the globe, used in the calculation of topographic and isostatic reductions around a gravity station. It is named after the U.S. geodesist J.F. Hayford. See also: *Bullard's method.*

haynesite A yellow orthorhombic mineral: $(UO_2)_3(Se^{4+}O_3)_2(OH)_2{\cdot}5H_2O$.

haystack hill (hay′-stack) In the tropics (originally Puerto Rico), a rounded conical hill of limestone developed as a result of *solution.* The now preferred term is *mogote.*

hazardous waste (haz′-ard-ous) Waste that poses a present or potential danger to human beings or other organisms because it is toxic, flammable, radioactive, explosive, or has some other property that produces substantial risk to life.

hazard survey A survey employing several sensors (such as side-scan sonar, magnetometer, and high-resolution seismic) to obtain shallow data in order to identify potential hazards, such as pockets of gas leaking from deeper formations, submarine landslides, gas hydrates, manmade obstacles, etc.

haze (a) Fine particles of dust, salt, smoke, or water dispersed through a part of the atmosphere, diminishing transparency of the air, causing colors to assume a characteristic subdued opalescent appearance, and reducing the horizontal visibility to more than one, but less than two, kilometers. (b) The obscuration, or a lack of transparency, of the atmosphere near the Earth's surface, caused by haze or by heat refraction (shimmering).

H chondrite A member of the high-Fe group of *ordinary chondrites.*

HCI *hydrocarbon indicator.*

HCS *hummocky cross-stratification.*

HDM *humic degradation matter.*

HDR *hot dry rock.*

HE *horizontal equivalent.*

head [coast] (a) A headland, usually coupled with a specific place name; e.g., Diamond Head, Hawaii. (b) The inner part of a bay, creek, or other coastal feature extending farthest inland. (c) The apex of a triangular-shaped delta or fan.

head [geomorph] (a) The influent end of a lake; the end opposite the outlet. (b) The source, beginning, or upper part of a stream. (c) The farthest upstream point reached by vessels; the limit of river navigation. (d) The upper part or end of a slope or valley; *valley head.*

head [hydraul] (a) The height above a standard datum of the surface of a column of fluid that can be supported by the static pressure. (b) Water-level elevation in a well, or elevation to which the water of a flowing artesian well will rise in a pipe extended high enough to stop the flow. (c) When not otherwise specified, it usually refers to *static head.*

head [mass move] (a) The upper parts of landslide material along the contact between the displaced material and the *main scarp* (Varnes, 1978). (b) A term used in southern England for a thick, poorly stratified, compact mass of locally derived angular rubble mixed with sand and clay, formed by solifluction under periglacial conditions, and mantling the high ground or occurring on slopes and in valley bottoms; a *cryoturbate.* See also: *coombe rock.* Syn: *rubble drift.*

head [paleont] (a) The anterior *tagma* of a crustacean, consisting of the *cephalon* alone or comprising the cephalon and one or more anterior thoracomeres (having limbs that are modified as mouth parts) fused to it. See also: *cephalothorax.* (b) The anterior dorsal part of the body of a mollusk, bearing the mouth, sensory organs, and major nerve ganglia. (c) In the vertebrates, the term is often used to designate the convex proximal end of ribs, humerus, and femur. Cf: *foot [paleont].*

head [struc geol] A little-used term for the upper bend of a fold or structural terrace. Cf: *foot [geol].* Syn: *upper break.*

headcut (head′-cut) A vertical face or drop on the bed of a stream channel, occurring at a knickpoint.

head dune A dune that accumulates on the windward side of an obstacle. Cf: *tail dune.*

headed dike (head′-ed) A dike which has a terminal expansion or teardrop shape.

head erosion *headward erosion.*

heading [grd wat] (head′-ing) A horizontal tunnel into an aquifer that taps groundwater penetrating fissures, for the purpose of supplying wells and reservoirs.

heading [surv] The compass direction (azimuth) of the longitudinal axis of a ship or aircraft.

heading side *footwall.*

heading wall *footwall.*

headland [coast] (head′-land) (a) An irregularity of land, esp. of considerable height with a steep cliff face, jutting out from the coast into a large body of water (usually the sea or a lake); a bold *promontory* or a high cape. Syn: *head [coast]; mull [coast].* (b) The high ground flanking a body of water, such as a cove. (c) The steep crag or cliff face of a promontory.

headland [soil] A term used in soil conservation for the source of a stream.

headland beach A narrow beach formed at the base of a *cliffed headland.*

headland mesa A part of a general plateau that projects into a meander loop of a large river (Lee, 1903, p.73). See also: *island mesa.*

head loss *friction head.*

headpool (head′-pool) A pool near the head of a stream.

heads Low-grade material overlying an alluvial *placer.*

headscarp The abrupt *scarp* at the head, or top, of a *landslide* or *slump.*

headspace The empty volume in a sample container between the water level and the cap.

headstream (head′-stream) A stream that is the source or one of the sources of a larger stream or river.

headwall A steep slope at the head of a valley; esp. the rock cliff or steep rock slope at the back of a cirque. Syn: *backwall.*

headwall recession The steepening and backward movement of the headwall of a cirque, caused in part by alternate thawing and refreezing.

headward erosion (head′-ward) The lengthening and cutting upstream of a young valley or gully above the original source of its stream, effected by erosion of the upland at the valley head; it is accomplished by rainwash, gullying, spring sapping, and the slumping of material into the head of the growing valley. Syn: *head erosion; headwater erosion; retrogressive erosion.*

headwater (head′-wa-ter) (a) The source (or sources) and upper part of a stream, esp. of a large stream or river, including the upper drainage basin; a stream from this source. The term is usually used in the plural. Syn: *waterhead.* (b) The water upstream from a structure, as behind a dam.

headwater erosion *headward erosion.*

headwater opposition The position of, or relationship shown by, two valleys facing in opposite directions, each growing upstream by headward erosion and separated from the other by a ridgelike divide (Fenneman, 1909, p.35-36).

head wave A seismic wave traveling downward at the critical angle to a high-velocity layer, moving along the top of that layer, and later emerging at the critical angle. The term is sometimes restricted to the part of such a wave that is also a *first arrival.* Syn: *refracted wave; conical wave; von Schmidt wave.*

healed *crustified.*

heaped dune *star dune.*

heap leaching A process in *hydrometallurgy* by which a lixiviant is added to broken ore heaped on the surface. Cf: *leaching.*

heartwood The central, old, dark-stained wood in most trees, a site of deposition of metabolic wastes. Its tissues are nonliving.

heat Energy in transit from a higher-temperature system to a lower-temperature system; the process ends in thermal equilibrium. Symbol: Q. Measured in joules (J).

heat balance (a) Equilibrium that exists on the average between the radiation received by the Earth and its atmosphere from the Sun and that emitted by the Earth and its atmosphere to space. That the equilibrium does exist in the mean is demonstrated by the observed long-term constancy of the Earth's surface temperature. On the average, regions of the Earth nearer the equator than about 35° latitude receive more energy from the Sun than they are able to radiate, whereas latitudes poleward of 35° receive less. The excess of heat is carried from low latitudes to higher latitudes by atmospheric and oceanic circulations and is radiated. (b) The equilibrium known to exist when all sources of heat gain and heat loss are accounted for in a given region or body. This balance includes advective and evaporative aspects as well as radiation. See also: *heat budget.*

heat budget (a) The amount of heat required to raise the water of a lake from its minimum winter temperature to its maximum summer temperature; it is usually expressed as gram calories of heat per square centimeter of lake surface. (b) The accounting for the total amount of heat received and lost by a particular system, such as a lake, a glacier, or the entire Earth during a specific period. See also: *heat balance.*

heat capacity That quantity of heat required for a unit increase in the temperature of a system. This quantity, which is not completely defined unless the type of change is specified, is expressed in joules per degree Kelvin ($J K^{-1}$) (Haenel et al., 1988, p.11). See also: *specific heat capacity.*

heat conduction The process of heat transfer through a substance, from a higher-temperature to a lower-temperature region, by molecular impact without transfer of the matter itself, i.e., without convection. Syn: *thermal conduction.*

heat content *enthalpy.*

heat convection *thermal convection.*

heated stone (heat′-ed) A gemstone that has been heated to change its color, such as blue zircon and many blue sapphires, or to improve its color, such as many aquamarines. Cf: *stained stone; burnt stone; treated stone.*

heat equator A line that encircles the planet connecting points of highest mean annual surface air temperature. It varies in latitude, ranging from about 20 degrees N in Mexico to about 14 degrees S in Brazil and averaging about 10 degrees N. The mean position of the *intertropical convergence zone* corresponds approximately to the heat equator.

heat flow Rate of heat transfer across a surface measured in watts (W). Symbol: Φ.

heat-flow density Heat-flow divided by area, measured in milliwatt per square meter ($mW\ m^{-2}$). Conversion from SI to CGS units: 1 $mW\ m^{-2}$ is equivalent to 0.239×10^{-17} cal $cm^{-2} s^{-1}$. The average heat-flow density of the Earth is 70 mWm^{-2} (global), 57 mWm^{-2} (continental), and 78 mWm^{-2} (oceanic) (Sclater, et al., 1980). Heat-flow density of the oceanic crust is age-dependent. In some regions, continental heat-flow density is related to the radiogenic heat generation of near-surface basement rocks; empirical relations are defined for particular regions (heat-flow density provinces (Roy et al., 1968)). Abbrev: HFD. Symbol: q. Formerly used term is *heat flow,* and North American scientists and journals still tend to use the term *heat flow* instead of heat-flow density. See also: *terrestrial heat flow.*

heat-flow measurement (heat′-flow) Measurement of the amount of heat leaving the Earth. It involves measuring the *geothermal gradient* of rocks by resistance thermometers in drill holes (preferably more than 300 meters deep), and the *thermal conductivity* of rocks, usually in the laboratory, on samples from the drill holes. Heat-flow measurements on the ocean floors use slightly different techniques.

heat-flow unit A measure of heat-flow density equivalent to 1μcal $cm^{-2}s^{-1}$ or 41.87 $mW\ m^{-2}$. Abbrev: HFU. Obsolete.

heat flux Obsolete term for *heat flow.*

heat flux [oceanog] The transfer of heat through a unit area of the sea surface, usually in watts per square meter. The flux is due to evaporation, solar radiation, infrared radiation, and direct conduction.

heat generation (heat′-gen-er-a′-tion) Rate of heat production (typically symbolized by *H* or A) in 1 m^3 material during 1 second as a result of radioactive decay within the Earth. It is measured in watt per cubic meter (Wm^{-3}). Conversion from SI to CGS units: 1 Wm^{-3} is equivalent to 0.239×10^{-7} cal $cm^{-3}s^{-1}$. In geothermics it is customary to use the unit $\mu W\ m^{-3}$. Radioactive heat generation is a scalar and isotropic petrophysical property independent of in-situ temperature and pressure. The amount of heat generated in a rock can be calculated with the heat generation constants of radioactive isotopes of U, Th, and K if their concentrations and the density of the rock are known (Rybach, 1988). Abbrev: HPR. Syn: *heat-production rate.*

heat-generation unit A measure of the rate of heat production per unit volume of material due to radioactive decay within the Earth. One heat-generation unit is 10^{-13} cal $cm^{-3}s^{-1}$ (= $4.187 \times 10^{-7}\ Wm^{-3}$). Rarely used. Abbrev: HGU.

heath peat *calluna peat.*

heat-production rate *heat generation.*

heat transfer Transfer of heat from points of higher to lower temperature which occurs in one of three ways: heat conduction, heat convection, or radiation.

heave [mining] A rising of the floor of a mine due to the weight on the adjacent pillars. Cf: *creep [mining].*

heave [soil] (a) A predominantly upward movement of a surface, caused by expansion or displacement, as from swelling clay, removal of overburden, seepage pressure, or frost action, esp. *frost heaving.* (b) Upward movement of the bottom of a deep excavation in susceptible soil materials, resulting from removal of confining stress or upward pressure of groundwater. Cf: *air heave; gas heave.* Syn: heaving.

heave [struc geol] The horizontal component of displacement on a dip-slip fault. Cf: *throw.*

heaving shale (heav′-ing) An incompetent or hydrating shale that runs, falls, swells, or squeezes into a borehole.

heavy-bedded (heav′-y-bed′-ded) Said of a shale whose splitting property is intermediate between that of a thin-bedded shale (easy to split) and that of a platy or flaggy shale (hard to split) (Alling, 1945, p.753).

heavy crop (heav′-y) A collective term used in Great Britain for the heavy minerals of a sedimentary rock.

heavy gold Gold occurring as large particles. Cf: *nugget.*

heavy isotope An isotope of an element having a greater than normal mass; e.g. deuterium.

heavy liquid In analysis of minerals, a liquid of high density, such as *bromoform,* in which specific-gravity tests can be made, or in which mechanically mixed minerals can be separated. When a mineral grain is placed in the liquid, the liquid's specific gravity is adjusted by the addition of a lighter or heavier liquid until the mineral neither rises nor sinks; the specific gravity of the liquid and of the mineral are then equal. See also: *Klein solution; Sonstadt solution; Clerici solution; Westphal balance; methylene iodide.* Syn: *specific-gravity liquid.*

heavy metal Any of the metals that react readily with dithizone (C_6H_5N), e.g. zinc, copper, lead, and many others.

heavy mineral [petrology] A rock-forming mineral generally having a specific gravity greater than 2.9; e.g. a mafic mineral.

heavy mineral [sed] A *detrital mineral* from a sedimentary rock, having a specific gravity higher than a standard (usually 2.85), and commonly forming as a minor constituent or *accessory mineral* of the rock (less than 1% in most sands); e.g. magnetite, ilmenite, zircon, rutile, kyanite, garnet, tourmaline, sphene, apatite, biotite. Cf: *light mineral.* See also: *heavy crop.*

heavy-mineral shadow An accumulation of dark *heavy minerals [sed]* on a sand bed that is otherwise dominated by light-colored quartz grains, so that the accumulation stands out as a dark patch; a shadow displays a sharp boundary with the quartz-rich sands on its upstream side, passing gradually downstream from heavy-mineral rich to quartz-rich sands (Cheel, 1984). The structures commonly form on *upper plane bed* surfaces and the flat, stoss sides of a *dune.*

heavy-mineral sheet A lamination, normally dark in color and only a few grain diameters thick, that is formed by the temporary enrichment in dark heavy minerals on an aggrading active *upper plane bed* surface (Cheel and Middleton, 1986).

heavy oil Crude oil that has a low API gravity or *Baumé gravity.* Cf: *light oil.*

heavy soil A soil with a high content of the fine separates, particularly clay, or one with a high drawbar pull and hence difficult to cultivate, especially when wet.

heavy spar *barite.*

heazlewoodite (hea′-zle-wood-ite′) A metallic bronze rhombohedral mineral: Ni_3S_2.

hebronite (he′-bron-ite) *amblygonite.*

hecatolite (hec-at′-o-lite) Orthoclase *moonstone.*

hechtsbergite An adamantine brown monoclinic mineral of the *atelestite* group: $Bi_2O(VO_4)(OH)$.

hectare (hec′-tare) A metric unit of land area equal to 10,000 square meters, 100 ares, or 2.471 acres. Abbrev: ha.

hectorfloresite (hec′-tor-flo′-res-ite) A colorless monoclinic mineral: $Na_9(IO_3)(SO_4)_4$.

hectorite (hec′-tor-ite) A white, creamy, or pale brown monoclinic trioctahedral, lithium-rich clay mineral of the *montmorillonite* group: $Na_{0.3}(Mg,Li)_3Si_4O_{10}(OH)_2$. It represents an end-member, in which the replacement of aluminum by magnesium and lithium in the octahedral sheets is essentially complete.

hedenbergite (hed′-en-berg-ite′) A green to brown or black monoclinic mineral of the pyroxene group: or $CaFe^{2+}Si_2O_6$. It occurs as a *skarn* mineral at the contact of limestones with granitic masses.

hedgehog stone (hedge′-hog) Quartz with needle-shaped inclusions of goethite.

hedleyite (hed′-ley-ite) A metallic white trigonal mineral: Bi_7Te_3.

HED meteorite One of a group of igneous achondrites whose distinctive ages, mineralogy, isotopic compositions, and a tendency to form intermixed breccias suggest an origin on the same parent body, possibly the asteroid Vesta, which shows a basalt-like reflection spectrum. Abbrev. for howardite-eucrite-diogenite.

hedreocraton (hed′-re-o-cra′-ton) A stable, continental craton, including both continental shield and platform. Cf: *thalassocraton.*

hedrumite (hed′-rum-ite) A coarse-grained, light-colored porphyritic hypabyssal rock characterized by trachytoid texture and containing accessory nepheline; a *pulaskite* porphyry. Its name, given by Brögger in 1890, is derived from Hedrum, Norway. Obsolete.

hedyphane (hed′-y-phane) A yellowish-white hexagonal mineral of the *apatite* group: $Pb_3Ca_2(AsO_4)_3Cl$.It may contain barium.

heel Upwardly convex posterior segment of the upper margin of conodont elements of *Belodina* (TIP, 1981, pt. W, supp.2).

Heersian (Heers′-i-an) European stage: Lower Paleocene (above Danian, below Landenian).

heideite (hei'-de-ite) A metallic monoclinic meteorite mineral of the *brezinaite* group: $(Fe,Cr)_{1+x}(Ti,Fe)_2S_4(x=0.15)$.

heidornite (hei'-dorn-ite) A colorless monoclinic mineral: $Na_2Ca_3B_5O_8(SO_4)_2(OH)_2Cl$.

height [geodesy] The distance between an equipotential surface through a point and a reference surface, measured along a line of force or along its tangent. Cf: *dynamic height; orthometric elevation.*

height [geomorph] (a) A landform or area that rises to a considerable degree above the surrounding country, such as a hill or plateau. The term is often used in the plural. (b) The highest part of a ridge, plateau, or other high land.

height [paleont] (a) The maximum distance, measured normal to the length in the plane of symmetry, between a concavo-convex or convexo-concave shell of a brachiopod and the line joining the beak and the anterior margin. Also, the thickness of the shell of a brachiopod. (b) The distance between two planes parallel to the hinge axis of a bivalve-mollusk shell and perpendicular to the plane of symmetry, which just touch the most dorsal and ventral parts of the shell. Cf: *length [paleont].* (c) In coiled cephalopod conchs, the linear distance between the venter and the umbilical seam (TIP, 1964, pt. K, p.22).

height [surv] The vertical distance above a datum (usually the surface of the Earth); *altitude* or elevation above a given level or surface.

height above mean sea level (a) The elevation of a point calculated from the data of a leveling survey, based on the elevations of certain tide-gauge bench marks that have been given specified values. This concept dates from a period when mean sea level and the geoid were thought to be the same or to differ only negligibly from each other.

height of instrument A surveying term used in spirit leveling for the height of the line of sight of a leveling instrument above the adopted datum, in trigonometric leveling for the height of the center of the theodolite above the ground or station mark, in stadia surveying for the height of the center of the telescope of the transit or telescopic alidade above the ground or station mark, and in differential leveling for the height of the line of sight of the telescope at the leveling instrument when the instrument is level. Abbrev: HI.

height of land The highest part of a plain or plateau; specif. a drainage divide, or a part thereof.

Heinrich event A massive release of icebergs from a continental ice sheet, recorded by a layer of sediment in a deep-sea core that is unusually rich in ice-rafted debris. Discovered in the North Atlantic by German marine geologist Hartmut Heinrich (1988) and named by Broecker et al. (1992).

heinrichite (hein'-rich-ite) A yellow to green tetragonal secondary mineral: $Ba(UO_2)_2(AsO_4)_2·10H_2O$. Cf: *metaheinrichite.*

heintzite (heintz'-ite) *kaliborite.*

hejtmanite A brownish yellow to golden yellow monoclinic mineral: $Ba(Mn^{2+},Fe^{2+})_2TiO(Si_2O_7)(OH,F)_2$.

hekistotherm (he-kis'-to-therm) A plant that can grow at low temperatures, esp. in areas where the warmest month has a mean temperature below 10°C.

helatoform (he-lat'-o-form) Shaped like a nail; e.g. "helatoform cyrtolith" having a nail-shaped central structure.

Helderbergian (Hel'-der-berg'-i-an) North American stage: lowermost Devonian (above Upper Silurian, below Deerparkian).

helen (hel'-en) The long narrow curved part of the hyolithid shell that projects laterally from the junction of the operculum and cone; there are right and left helens (Runnegar et al., 1975, p.181). Etymol: the genus *Helenia,* named by Walcott.

helenite (hel'-en-ite) A variety of *ozocerite.*

helical flow (hel'-i-cal) *helicoidal flow.*

helical structure A *microstructure* of nested, dextrally coiled, helical aragonitic rods with helix axes oriented nearly perpendicular to the depositional surface. The helical rods each consist of *basic structural units* (Bé et al., 1972; the "Schraubenschicht" of Bandel, 1977b).

helical tracheid (hel'-i-cal tra'-che-id) A *tracheid* in which secondary cell wall material is deposited in a helical configuration; it is common in protoxylem. Cf: *annular tracheid.* Obsolete syn: *spiral tracheid.*

helicitic (hel-i-cit'-ic) Pertaining to a metamorphic-rock texture consisting of bands of inclusions that indicate original bedding or schistosity of the parent rock and are included in later-formed crystals of the metamorphic rock. The relict inclusions commonly occur in porphyroblasts as curved and contorted strings. The term was originally, but is no longer, confined to microscopic texture. Var. spelling: *helizitic.* Cf: *poikiloblastic.*

helicoid (hel'-i-coid) Forming or arranged in a spiral; specif. said of a gastropod shell having the form of a flat coil or flattened spiral, or sometimes of an ammonoid coiled in regular three-dimensional spiral form with a constant angle, or of the spiraled cochlea of the mammalian inner ear. See also: *torticone.*

helicoidal flow (hel'-i-coid'-al) At the bend of a river, a coiling type of flow motion that results in erosion of the concave, outer bank and deposition along the convex, inner bank. Syn: *helical flow.*

helicoplacoid (hel'-i-co-plac'-oid) Any echinozoan belonging to the class Helicoplacoidea, characterized by a fusiform to pyriform placoid body with a spirally pleated expansible and flexible test (TIP, 1968, pt.U, p.131). Range, Lower Cambrian.

helicospore (hel'-i-co-spore) A cylindric, spiral, or convolute, usually septate, asexual *fungal spore.* Cf: *Saccardoan spore groups.*

helictite (he-lic'-tite) A curved twiglike cave deposit, usually of calcite or aragonite, that grows at the free end by deposition from water emerging from a nearly microscopic central canal. Syn: *eccentric.*

heligmite (he-lig'-mite) A *helictite* that grows from the floor. The preferred term is helictite.

Helikian (He-lik'-i-an) In a three-part division of the Proterozoic of Canada, the middle division, after the *Aphebian* and before the *Hadrynian,* from 1,640 to 880 Ma.

heliodor (he'-li-o-dor') A golden, greenish, or brownish-yellow transparent gem variety of beryl found in southern Africa. See also: *golden beryl.* Obsolete syn: *chrysoberyl.*

heliolite (he'-li-o-lite') *sunstone.*

heliolith (he'-li-o-lith') (a) A *coccolith* constructed of many tiny calcite crystals, commonly in radial arrangement. (b) An individual of the Heliolithae, a subdivision of the family Coccolithophoridae. Cf: *ortholith.*

heliolitid (he'-li-o-lit'-id) Any coral belonging to the family Heliolitidae, characterized by massive coralla with slender tabularia separated by coenenchyme and commonly having 12 equal spinose septa and complete tabulae. Heliolitids are considered by some workers to be tabulates. Range, Middle Ordovician to Middle Devonian.

heliophyllite (he'-li-o-phyl'-lite) A yellow orthorhombic (pseudo-tetragonal) mineral: $Pb_6As^{3+}{}_2O_7Cl_4$. It may be identical to *ecdemite.*

heliotrope [mineral] (he'-li-o-trope') *bloodstone.*

heliotrope [surv] An instrument used in geodetic surveying to aid in making long-distance (up to 320 km) observations and composed of one or more plane mirrors so mounted and arranged that a beam of sunlight may be reflected toward a distant survey station where it can be observed with a theodolite.

heliozoan (he'-li-o-zo'-an) Any actinopod protozoan belonging to the subclass Heliozoa, characterized by pseudopodia that are not stiff or rigid, being strengthened only by an axial rod of fibrils.

helium age method (he'-li-um) Determination of the age of a mineral in years, based on the known radioactive decay rates of uranium and thorium isotopes to helium. Syn: *helium dating.*

helium dating *helium age method.*

helium index An obsolete term for the experimental age obtained by substituting helium and radioactivity values in the age equation.

helizitic (he-li-zit'-ic) *helicitic.*

hellandite-(Y) (hel'-land-ite-y) A reddish-brown monoclinic mineral: $(Ca,Y)_4Y_2(Al,Fe^{3+})B_4Si_4O_{20}(O,OH)_4$.

hellendite-(Ce) A light green monoclinic mineral: $(Ca,Ce)_4(Ce,Ca)_2(Al,Fe^{3+},Ti)(Be,Li)B_4Si_4O_{22}(O,OH,F)_2$.

helluhraun (hell'-u-hraun) An Icelandic term for *pahoehoe.* Cf: *apalhraun.*

hellyerite (hell'-yer-ite) A pale blue monoclinic mineral: $NiCO_3·6H_2O$.

Helmholtz coil (Helm'-holtz) A measuring instrument in which pair of coaxial coils are separated by a distance equal to their radius to provide a nearly constant magnetic field over a large volume between the coils. Used for calibration of magnetometers and for nulling an ambient magnetic field for magnetic measurements of rock samples.

Helmholtz free energy A *thermodynamic potential* that is a function of temperature and volume. It is one of the commonly used functions describing *free energy,* and is useful in determining the course of constant-volume isothermal processes. Cf: *Gibbs free energy.*

Helmholtz waves Waves developed under conditions of *shear* instability at the interface between two fluids, such as warmer and colder layers of water at the *thermocline* of a lake. When Helmholtz waves break, they homogenize the part of the temperature gradient they affect, so that a *metalimnion,* instead of presenting a smooth curve of changing temperature, comes to consist of alternating thick layers of relatively isothermal water and thin sheets in which the temperature gradient is very sharp.

helminthite (hel-min'-thite) An obsolete term for a doubtfully distinguished *trace fossil* consisting of a long, sinuous surface trail or filled-up burrow of a supposed marine worm, without impressions of lateral appendages.

helmutwinklerite (hel-mut-wink'-ler-ite) (a) A light blue triclinic mineral: $PbZn_2(AsO_4)_2·2H_2O$. (b) A group name for minerals with the analogous formula, but with Pb replaced by Ca, Zn replaced by Cu or Fe, and AsO_4 replaced by PO_4.

heloclone (he'-lo-clone) A sinuous, monaxonic sponge spicule of irregular outline, often bearing articulatory notches along its length or at its end.

helophyte (he'-lo-phyte) A perennial marsh plant that has its overwintering buds beneath the water. See also: *hydrophyte.*

helsinkite (hel-sin'-kite) A hypidiomorphic-granular hypabyssal rock composed primarily of albite and epidote. Its name is derived from Helsinki, Finland. Obsolete.

Helvetian (Hel-ve'-ti-an) Obsolete European stage: upper Lower Miocene (above Burdgialian, below Tortonian of Vienna Basin); equiv. lower Vindobonian.

helvite A yellow cubic mineral of the sodalite group: $Mn^{2+}{}_4Be_3(SiO_4)_3S$. It is the manganese end member isomorphous with danalite and genthelvite. Also called: helvine.

hemachate (hem'-a-chate) A light-colored agate spotted with red jasper. Syn: *blood agate.*

hemafibrite (hem-a-fi'-brite) Partly oxidized *synadelphite.*

hematite (hem'-a-tite) A common iron mineral: αFe_2O_3. It is dimorphous with maghemite. Hematite occurs in splendent, metallic-looking, steel-gray or iron-black rhombohedral crystals, in reniform masses or fibrous aggregates, or in deep-red or red-brown earthy forms: it has a distinctive cherry-red to reddish-brown streak and a characteristic brick-red color when powdered. It is found in igneous, sedimentary, and metamorphic rocks, both as a primary constituent and as an alteration product. Hematite is the principal ore of iron. Symbol: Hm. See also: *specularite.* Originally spelled: *haematite.* Syn: *red hematite; red iron ore; red ocher; rhombohedral iron ore; oligist iron; bloodstone.*

hematite schist *itabirite.*

hematolite (he-mat'-o-lite) A blood-red rhombohedral mineral: $(Mn^{2+},Mg,Al)_{15}(As^{3+}O_3)(As^{5+}O_4)_2(OH)_{23}$.

hematophanite (he-ma'-to-phan'-ite) A submetallic red-brown tetragonal mineral: $Pb_4Fe^{3+}{}_3O_8(OH,Cl)$.

hematopore (he-mat'-o-pore) A polymorph that is slender and distally directed on the reverse sides of the colonies of some stenolaemate bryozoans. Cf: *firmatopore.*

hemera (hem'-er-a) (a) The geologic-time unit corresponding to *acme zone;* the time span of the acme or greatest abundance, in a local section, of a taxonomic entity. Also, the period of time during which a race of organisms is at the apex of its evolution. The term was proposed by Buckman (1893, p.481-482) for the time of acme of development of one or more species, but later used by him (Buckman, 1902, 1903) in the sense of *moment* or the time during which a biostratigraphic zone was deposited, and by Jukes-Browne (1903, p.37) for the duration of a subzone. (b) A term sometimes incorrectly applied to a biostratigraphic zone (body of strata) comprising the time range of a particular fossil species. Etymol: Greek, "day". Pl: hemerae. Adj: hemeral.

hemichoanitic (hem'-i-cho'-a-nit'-ic) Said of a *retrochoanitic septal neck* of a nautiloid that extends one-half to three-fourths of the distance to the preceding septum.

Hemichordata (Hem'-i-chor-da'-ta) A subdivision of the *Protochordata* or of the *Chordata,* including animals with a pre-oral notochord and three primary coelom segments in the adult.

hemicolloid A term proposed by Haranczyk (1969) for solutions intermediate between true fluids and colloidal fluids.

hemicone (hem'-i-cone) *alluvial cone.*

hemicyclothem (hem-i-cy'-clo-them) Half of a cyclothem. The term is generally applied either to the lower nonmarine part, or to the upper marine part, of a Pennsylvanian cyclothem.

hemicystal archeopyle (hem-i-cys'-tal) An *archeopyle* formed in a dinoflagellate cyst by separation of the cyst at the *cingulum* (Williams et al., 2000, p.96).

hemidisc (hem'-i-disc) A sponge *spicule* consisting of an unequal-ended *amphidisc.*

hemihedral (hem-i-he'-dral) Said of the *merohedral* crystal class (or classes) in a system, the general form of which has half the number of equivalent faces of the corresponding *holohedral* form. Syn: *hemisymmetrical.* Cf: *ogdohedral; tetartohedral.*

hemihedrite (hem-i-he'-drite) A bright orange, brown, or blackish triclinic mineral: $Pb_{10}Zn(CrO_4)_6(SiO_4)_2F_2$. It is the Zn analogue of iranite.

hemimorph (hem'-i-morph) A crystal having polar symmetry, i.e. displaying *hemimorphism.*

hemimorphism (hem-i-mor'-phism) The characteristic of a crystal that has *polar symmetry,* so that its two ends have different forms. Such a crystal is a *hemimorph.* Adj: hemimorphic.

hemimorphite (hem-i-mor'-phite) (a) A white or colorless to pale-green, blue, or yellow orthorhombic mineral: $Zn_4Si_2O_7(OH)_2 \cdot H_2O$. It is similar to smithsonite, but is distinguished from it by strong pyroelectric properties. Hemimorphite is a common secondary mineral, and is an ore of zinc. Syn: *calamine; electric calamine; galmei.* (b) A term sometimes used (esp. in the gem trade) as a syn. of *smithsonite.*

Hemingfordian (Hem-ing-ford'-i-an) North American land mammal age: Early Miocene (after Arikareean, before Barstovian).

hemiopal (hem-i-o'-pal) *semiopal.*

hemipelagic deposit (hem'-i-pe-lag'-ic) Deep-sea sediment in which more than 25% of the fraction coarser than 5 μm is of terrigenous, volcanogenic, and/or neritic origin. Such deposits usually accumulate near the continental margin and adjacent abyssal plains, so that continentally derived sediment is more abundant than in eupelagic sediments, and the sediment has undergone lateral transport. Cf: *terrigenous deposit; pelagic deposit.*

hemipelagite (hem'-i-pe-lag'-ite) Sediments formed by the slow accumulation on the sea floor of biogenic and fine terrigenous particles. Fossil species are *indigenous* (Natland, 1976, p.702).

hemiperipheral growth (hem'-i-pe-riph'-er-al) Growth of brachiopod shells in which new material is added anteriorly and laterally but not posteriorly.

hemiphragm (hem'-i-phragm) A transverse calcareous shelflike platform extending from the zooecial wall part way across the zooecial chamber in some stenolaemate bryozoans. Hemiphragms commonly alternate in ontogenetic series from opposite sides of chamber walls.

hemipyramid (hem-i-pyr'-a-mid) An old term for *fourth-order prism.*

hemiseptum (hem-i-sep'-tum) (a) A transverse calcareous shelflike platform extending from the zooecial wall part way across the zooecial chamber in some stenolaemate bryozoans. They generally occur singly on the proximal sides of zooecia or in one or two pairs in alternate positions on proximal and distal sides. A "superior hemiseptum" may be present on the proximal wall and an "inferior hemiseptum" on the distal wall. (b) A partial *septum,* between normal ones, subdividing a foraminiferal chamber, as in some Lituolacea. Pl: hemisepta.

hemisphere (hem'-i-sphere) Half a sphere; usually refers to half of the Earth, as divided by the Equator into northern and southern hemispheres, or by the 20°W and 160°E meridians into eastern (Old World) and western (New World) hemispheres. Can also be divided into east and west hemispheres along the Prime Meridian (0°) and 180°.

Hemists Histosols that have an intermediate degree of plant fiber decomposition and a bulk density between about 0.1 and 0.2 g cm^{-3}. Hemists are saturated with water for periods long enough to limit their use for most crops unless they are artificially drained. A suborder in the U.S. system of soil taxonomy.

hemisymmetrical (hem'-i-sym-met'-ri-cal) *hemihedral.*

hemloite (hem'-lo-ite) A metallic black triclinic mineral: $(As^{3+},Sb^{3+})_2(Ti,Fe^{3+},V^{3+},Fe^{2+})_{12}O_{23}(OH)$.

Hemphillian (Hemp-hil'-li-an) North American land mammal age: Late Miocene (after Clarendonian, before Blancan).

hemusite (he'-mus-ite) A metallic gray cubic mineral: Cu_6SnMoS_8.

Henbury glass Glass associated with Henbury craters, Australia, thought to have been produced by melting of local target rocks during the impact event. See also: *impactite.*

hendersonite (hen'-der-son-ite') A dark greenish black orthorhombic mineral: $Ca_{1.3}(V^{5+},V^{4+})_6O_{16} \cdot 6H_2O$. It is classed as a "vanadium bronze".

hendricksite (hen'-dricks-ite) A copper-brown, dark reddish brown, or reddish black monoclinic trioctahedral mica of ideal composition $KZn_3(Si_3Al)O_{10}(OH)_2$.

heneuite (hen'-eu-ite) A pale blue-green triclinic mineral: $CaMg_5(PO_4)_3(CO_3)(OH)$.

henmilite (hen'-mil'-ite) A bluish-violet triclinic mineral: $Ca_2CuB_2(OH)_{12}$.

Hennigian (Hen-nig'-i-an) Referring to *cladism* (or *phylogenetic systematics* in its restricted sense), in honor of its principal proponent, W. Hennig.

hennomartinite A vitreous brown orthorhombic mineral: $SrMn^{3+}{}_2Si_2O_7(OH)_2 \cdot H_2O$.

henritermierite (hen-ri-ter'-mi-er-ite') A brown tetragonal mineral of the *hydrogarnet* group: $Ca_3(Mn^{3+},Al)_2(SiO_4)_2(OH)_4$.

henryite (hen'-ry-ite) A metallic cubic mineral: $Cu_4Ag_3Te_4$.

Henry law constant The equilibrium ratio of the partial pressure of a compound in air to the concentration of the compound in water at a reference temperature. It is sometimes referred to as the air-water partition coefficient (Cohen and Mercer, 1993).

henrymeyerite An adamantine black tetragonal mineral of the *hollandite* group: $BaTi_7Fe^{2+}O_{16}$.

Henry's law States that the equilibrium partial pressure of a compound in the air above the air/water interface is proportional to the concentration of that compound in the water.

hentschelite (hent'-schel-ite) A dark green monoclinic mineral of the *lazulite* group: $CuFe^{3+}{}_2(PO_4)_2(OH)_2$.

hepatic cinnabar (he-pat'-ic) A liver-brown or black variety of cinnabar. Syn: *liver ore.*

hepatic groove A groove in decapod crustaceans that continues from the posterior end of the antennal groove posteriorly, and unites, with a looplike curve, the lower ends of the cervical and postcervical grooves (Holthuis, 1974, p.733).

hepatic region Part of the *carapace* of some decapod crustaceans that may touch the antennal, cardiac, and pterygostomial regions (TIP, 1969, pt. R, p.92).

hepatite (hep'-a-tite) A variety of barite that emits a fetid odor when rubbed or heated.

heptane (hep'-tane) Any of nine colorless liquid isomeric paraffin hydrocarbons of formula C_7H_{16}. One of these, n-heptane, $CH_3(CH_2)_5CH_3$, occurs in crude oils and in some pine oils.

heptaphyllite (hep-ta-phyl'-lite) (a) A group of mica minerals that contain seven cations per ten oxygen and two hydroxyl ions. (b) Any mineral of the *heptaphyllite* group, such as muscovite and other light-colored micas; a dioctahedral clay mineral. Cf: *octaphyllite.*

heptorite (hep'-tor-ite) A dark-colored *lamprophyre* composed of barkevikite, titanian augite, and hauyne phenocrysts in a glassy groundmass containing labradorite microlites. Essentially a hauyne *monchiquite.* Obsolete.

herb Any vascular plant of low stature whose stem does not become woody; either annual or growing from a perennial root or rhizome.

herbaceous (her-ba'-ceous) Said of green, vascular plants of low stature, either annual or perennial.

herbivore (her'-bi-vore) A *heterotrophic* organism that feeds on plants. Cf: *carnivore.*

Hercules stone (Her'-cu-les) A syn. of *lodestone.* Also called: Heraclean stone.

Hercynian Orogeny (Her-cyn'-i-an) By present usage, the late Paleozoic orogenic era of Europe, extending through the Carboniferous and Permian, hence synonymous with the *Variscan Orogeny;* it is roughly equivalent to the Alleghanian orogeny of North America. European usage today is about equally divided between the two terms. Many German geologists regard "Hercynian" as a NW orographic direction without time significance, as proposed by von Buch; hence they prefer "Variscan." Many French and Swiss geologists, following M. Bertrand (1892), prefer "Hercynian" in the time sense; thus, the crystalline massifs of the northern Alps are said to be Hercynian, rather than Variscan. Cf: *Armorican Orogeny.*

Hercynides (Her-cyn'-i-des) A name used for the orogenic belt created by the *Hercynian Orogeny,* extending from southern Ireland and Wales to northern France, Belgium, and northern Germany. Approx. syn: *Variscides.*

hercynite (her'-cy-nite) A black cubic mineral of the *spinel* group: $Fe^{2+}Al_2O_4$. It often contains some magnesium. Syn: *iron spinel; ferrospinel.*

herderite (her'-der-ite) (a) A colorless to pale-yellow or greenish-white monoclinic mineral: $CaBe(PO_4)F$. (b) A group name for minerals with the analogous formula, but with PO_4 replaced by AsO_4, and F replaced by OH.

heredity (he-red'-i-ty) All the qualities and potentialities that an individual has acquired genetically from its ancestors.

heritability The extent to which variations in the phenotype of an organism can be passed to offspring, or are caused by genetic differences.

Herkimer diamond (Her'-ki-mer) A quartz crystal from Herkimer County, N.Y. See also: *Lake George diamond.*

herkomorph (her'-ko-morph) An *acritarch* with a spherical to polygonal vesicle (e.g., *Dictyotidium*); the surface may be divided by crests into polygonal fields (Downie et al., 1963; Mendelson, 1993).

Hermann-Mauguin symbols (Her'-mann-Mau'-guin) An internationally accepted shorthand notation system of the elements of symmetry of crystal classes, which expresses outward symmetry. An example is $4/m\ \bar{3}2/m$ for the hexoctahedral class of the isometric system, in which the numbers with a bar are axes of rotoinversion, *m* is a symmetry plane, and a number over *m* indicates an axis of symmetry with a plane of symmetry perpendicular to it. Cf: *Schoenflies notation; international symbols.*

hermatobiolith (her-mat'-o-bi'-o-lith) An organic *reef rock.*

hermatolith (her-mat'-o-lith) *reef rock.*

hermatopelago (her'-mat-o-pel'-a-go) A submerged *reef cluster.* Etymol: Greek "hermato", "sunken reef", + "pelagos", "sea".

hermatypic coral (her-ma-typ'-ic) A reef-building coral; a coral characterized by the presence within its endodermal tissue of symbiotic algae; a coral incapable of adjusting to aphotic conditions. Ant: *ahermatypic coral.*

heronite (her'-on-ite) A hypabyssal rock composed of spheroidal phenocrysts of alkali feldspar in a groundmass composed of radiating bundles of labradorite and aegirine with interstitial analcime. The rock appears to be an altered *tinguaite* (Johannsen, 1939, p.256). Its name is derived from Heron Bay, Ontario. Obsolete.

herrerite (her'-rer-ite) A blue and green variety of smithsonite containing copper.

herringbone cross-bedding (her'-ring-bone) *chevron cross-bedding.*

herringbone mark *chevron mark.*

herringbone pattern [impact] A characteristic appearance of *secondary crater* chains, in which individual craters overlap and, because they form nearly simultaneously, create dunelike *ejecta blankets* between the craters. First observed among the secondary craters of Copernicus on the Moon (Oberbeck and Morrison, 1974).

herringbone texture In mineral deposits, a pattern of alternating rows of parallel crystals, each row in a reverse direction from the adjacent one. It resembles the "herringbone" textile fabric.

herschelite (her'-schel-ite) *chabazite-Na.*

Hertz The SI unit of frequency, the same as cycles per second (cps). Named after Heinrich Rudolph Hertz (1857-1894), German physicist who discovered electromagnetic waves. Abbrev: Hz.

hervidero (her-vi-de'-ro) A syn. of *mud volcano.* Etymol: Spanish "hervir", "to boil".

herzenbergite (herz'-en-berg-ite') A metallic black orthorhombic mineral: SnS. Syn: *kolbeckine.*

Hesperian The intermediate-aged system of rocks in the geologic stratigraphy of the planet Mars. Named after Hesperia Planum, a broad upland area of plains with numerous wrinkle ridges; the base of the ridged plains defines the base of the Hesperian system (Tanaka, 1986). See also: *Amazonian; Noachian.*

hessite (hess'-ite) A metallic gray monoclinic mineral: Ag_2Te. It is sectile, usually massive, and often auriferous. Cf: *pilsenite.*

hessonite (hes'-son-ite) *essonite.*

hetaerolite (he-tae'-ro-lite) A black tetragonal mineral: $ZnMn^{3+}{}_2O_4$. It is found with chalcophanite. It is a member of the *hausmannite* group, which is related to the *spinel* group.

heteractine (het-er-ac'-tine) A spicule in a heteractinid sponge. It is generally *octactine*-based in more advanced forms, but may be *polyactine* as well.

heteradcumulate (het'-er-ad-cu'-mu-late) A cumulate in which cumulus crystals and unzoned poikilitic crystals have the same composition.

heteradcumulus crystallization (het'-er-ad-cu'-mu-lus) *Adcumulus growth* in which cumulus crystals and poikilitic crystals of the same composition continue to develop until little or no interstitial liquid remains.

hetero- (het'-er-o-) A prefix meaning "different".

heteroblastic (het'-er-o-blas'-tic) Pertaining to a type of *crystalloblastic* texture in a metamorphic rock in which the essential mineral constituents are of two or more distinct sizes. The term was originated by Becke (1903). Cf: *homeoblastic.*

heterochronism (het-er-och'-ro-nism) The phenomenon by which two analogous geologic deposits may not be of the same age although their processes of formation were similar.

heterochronous [evol] (het-er-och'-ro-nous) Said of a fauna or flora appearing in a new region at a time that is quite different from the time it appeared in the region which it previously inhabited.

heterochronous [stratig] Said of a sequence of sediments representing lateral development of a similar lithofacies in successively younger stages. Term introduced by Nabholz (1951).

heterochronous homeomorph A homeomorph from a later geologic time that resembles one from an earlier geologic time. Cf: *isochronous homeomorph.*

heterochrony Differences in the stage of development when a trait in one species appears relative to the stage it appears in a closely related species.

heterochthonous (het-er-och'-tho-nous) (a) Said of a transported rock or sediment, or one that was not formed in the place where it now occurs. Also, said of fossils removed by erosion from their original deposition site and re-embedded. Cf: *allochthonous.* (b) Said of a fauna or flora that is not indigenous.

heterococcolith (het'-er-o-coc'-co-lith) A *coccolith* constructed of differing elements. Cf: *holococcolith.*

heterocolpate (het'-er-o-col'-pate) In *colpate* pollen, having some colpi with pores and others (*pseudocolpi*) without them.

heterocyst (het'-er-o-cyst') A differentiated cell, usually large, produced by some cyanobacteria; its function is uncertain (Fogg et al., 1973).

heterodesmic (het'-er-o-des'-mic) Said of a crystal or other material that is bonded in more than one way. Cf: *homodesmic.*

heterodont (het'-er-o-dont') adj. (a) Said of the dentition of a bivalve mollusk having a small number of distinctly differentiated cardinal and lateral teeth that fit into depressions on the opposed valve. (b) Said of the hingement of ostracode valves effected by a combination of tooth-and-socket and ridge-and-groove types, characterized by pointed or slightly crenulate teeth in one or both valves associated with a ridge in one valve and a groove in the other (TIP, 1961, pt. Q, p.50). (c) Said of vertebrates in which the teeth are markedly nonuniform from front to back and are specialized for particular functions, esp. mammals. n. A heterodont mollusk; specif. a bivalve mollusk of the order Heterodonta, having few hinge teeth but usually with both lateral and cardinal teeth and with unequal adductor muscles. Cf: *taxodont.*

heterogeneous [geophys] Said of a material that has properties such as seismic velocity, density, or anisotropy that vary from point to point in the material. Cf: *homogeneous [geophys].*

heterogeneous equilibrium (het'-er-o-ge'-ne-ous) Equilibrium in a system consisting of more than one phase. Cf: *homogeneous equilibrium.*

heterogenite (het-er-o-gen'-ite) A brownish black to black mineral occurring in mammillary masses: $Co^{3+}O(OH)$. It occurs in two forms: heterogenite-3R and heterogenite-2H. It may contain some copper and iron. Syn: *stainierite.*

heterogony (het-er-og'-o-ny) *alternation of generations.*

heterogranular (het'-er-o-gran'-u-lar) (a) Said of the texture of a rock having crystals of significantly different sizes. In igneous rocks, the grain-size distribution curve may be relatively flat, illustrating a large range of sizes being nearly equally represented. In metamorphic rocks, the grain size is characterized by several overlapping log-normal distribution curves. Cf: *seriate; hiatal [ign].* Ant: *homogranular.* Syn: *inequigranular.*

heterolithic unconformity (het'-er-o-lith'-ic) A term proposed by Tomkeieff (1962, p.412) to replace *nonconformity* in the sense of an unconformity developed between "unlike rocks".

heteromorph (het'-er-o-morph') (a) An organism or part that differs from the normal form, specif. an ammonoid or ammonoid shell that deviates from the normal (planispiral) mode of coiling and/or in which the walls of some or all of the coils are not in contact. (b) In dimorphic ostracodes, the adult form that differs in general shape from the juvenile instars. It is generally presumed to be the female of the species. Cf: *tecnomorph.*

heteromorphic [evol] (het'-er-o-mor'-phic) Deviating from the usual form, or having diversity of form. Cf: *isomorphic.*

heteromorphic [petrology] Said of igneous rocks having similar chemical composition but different mineralogic composition. Obsolete.

heteromorphic processes (het-e-ro-mor'-phic) On *acritarchs* or *dinocysts, processes* of more than one form (e.g., branched and unbranched) on a single specimen (Lister, 1970; Kjellström, 1971). Cf: *acanthomorph; homomorphic processes.*

heteromorphism (het'-er-o-mor'-phism) The crystallization of two magmas of nearly identical chemical composition into two different mineral aggregates as a result of different cooling histories. Obsolete.

heteromorphite (het'-er-o-mor'-phite) A metallic black monoclinic mineral: $Pb_7Sb_8S_{19}$.

heteromorphosis (het'-er-o-mor'-pho-sis) The production by an organism of an abnormal or misplaced part, esp. as the result of regeneration.

heteromyarian (het'-er-o-my-ar'-i-an) adj. Said of a bivalve mollusk, or of its shell, having the anterior adductor muscle conspicuously smaller than the posterior adductor muscle. n. A heteromyarian mollusk. Cf: *anisomyarian; monomyarian.*

hetero-octahedral Said of a mica in which not all octahedral sites are occupied by the same chemical element. Cf: *homo-octahedral.*

heterophragm (het'-er-o-phragm') A small skeletal cystoidal structure that projects from the body wall into the zooecial chamber in some stenolaemate bryozoans.

heteropic (het-er-op'-ic) Said of sedimentary rocks of different facies, or said of facies characterized by different rock types. The rocks may be formed contemporaneously or in juxtaposition in the same sedimentation area or both, but the lithologies are different; e.g. facies that replace one another laterally in deposits of the same age. Also, said of a map depicting heteropic facies or rocks.

heteropod (het'-er-o-pod') Any prosobranch gastropod belonging to the suborder Heteropoda, a group of pelagic forms with shells of aragonite.

heteropolar (het-er-o-po'-lar) Of *pollen,* having opposite poles developed differently. Ant: *isopolar.*

heteropygous (het'-er-o-py'-gous) Said of a trilobite having *cephalon* and *pygidium* of unequal size. Ant: *isopygous.*

heteroscedasticity A condition where the variance is not constant, but changes with the magnitude of the observation. Cf: *homoscedasticity.*

heterosis (het-er-o'-sis) The high capacity for growth and activity frequently displayed by crossbred organisms as compared with those that are inbred.

heterosite (het-er-o'-site) A reddish-purple or brown orthorhombic mineral of the *triphylite* group: $(Fe^{3+},Mn^{3+})PO_4$. It is isomorphous with purpurite.

heterosporous (het'-er-o-spor'-ous) Characterized by *heterospory.* This is said of plants that produce both *microspores* and *megaspores.* Also said of dinoflagellate genera whose constituent species collectively produce several different cyst morphotypes (Wall and Dale, 1968; Williams et al., 2000, p.98). Heterosporous plants date from the Early Devonian.

heterospory (het'-er-o-spo'-ry) The condition in embryophytic plants in which spores are of two types: microspores and megaspores. Cf: *homospory.*

Heterostraci (Het'-er-os'-tra-ci) An order of diplorhinate jawless fishes, characterized by flattened dorsal and ventral head-trunk plates articulated with each other by narrow plates covering the gill apertures. It includes the oldest known vertebrates. Range, Lower Ordovician to Upper Devonian.

heterostrophy (het-er-os'-tro-phy) The quality or state of being coiled in a direction opposite to the usual one; specif. the condition of a gastropod *protoconch* in which the whorls appear to be coiled in a direction opposite to those of the *teleoconch.* Adj: heterostrophic.

heterotactic [stratig] (het'-er-o-tac'-tic) *heterotaxial.*

heterotactic [struc petrol] Said of the symmetry of a fabric in which not all the subfabrics agree in symmetry. Cf: *homotactic [struc petrol].*

heterotaxial (het'-er-o-tax'-i-al) Pertaining to, characterized by, or exhibiting heterotaxy. Syn: heterotactic.

heterotaxis (het'-er-o-tax'-is) An erroneous transliteration of *heterotaxy.*

heterotaxy (het'-er-o-tax'-y) Abnormal or irregular arrangement; specif. the condition of strata that are widely separated and not equivalent as to their relative positions in the geologic sequence, or that are lacking uniformity in stratification or arrangement. Ant: *homotaxy.*

heterotherm (het'-er-o-therm') *poikilotherm.*

heterothermic (het'-er-o-ther'-mic) *poikilothermic.*

heterothrausmatic (het'-er-o-thraus-mat'-ic) A descriptive term applied to igneous rocks with an orbicular texture in which the nuclei of the orbicules are composed of various kinds of rock or mineral fragments. Cf: *allothrausmatic; crystallothrausmatic; isothrausmatic; homeothrausmatic.* Obsolete.

heterotomous (het-er-ot'-o-mous) Said of a crinoid arm characterized by division into unequal branches. Ant: *isotomous.*

heterotrichy (het-er-ot'-ri-chy) The most advanced type of growth habit in filamentous algae, in which both a prostrate or creeping system and a projecting or erect system of filaments are formed (Fritsch, 1961, p.20).

heterotrophic (het'-er-o-troph'-ic) Said of an organism that nourishes itself by using organic material to synthesize its own living matter, a heterotroph. All animals and fungi are heterotrophic. Syn: *metatrophic; allotrophic; zootrophic.* Cf: *autotrophic [ecol].*

heterozooid (het'-er-o-zo'-oid) A bryozoan polymorph differing from an *autozooid* in lacking some or all organs concerned with feeding.

Hettangian (Het-tan'-gi-an) Stage of the Standard Global Chronostratigraphic Scale: lowermost Jurassic (above Rhaetian of Triassic, below Sinemurian).

heubachite (heu'-bach-ite) A nickel-containing variety of heterogenite.

heulandite-Ca (heu'-land-ite) A zeolite mineral: $(Ca,Na)_5(Si,Al)_{36}O_{72}{\cdot}nH_2O$. It often occurs as white to pink foliated masses or as coffin-shaped monoclinic crystals in cavities in decomposed basic igneous rocks. See also: *clinoptilolite; stilbite-Ca.*

heulandite-K A monoclinic zeolite mineral: $(K,Na,Ca)_5(Si,Al)_{36}O_{72}{\cdot}nH_2O$.

heulandite-Na A monoclinic zeolite mineral: $(Na,Ca,K)_5(Si,Al)_{36}O_{72}{\cdot}nH_2O$.

heulandite-Sr A monoclinic zeolite mineral: $(Sr,Ca,Na)_5(Si,Al)_{36}O_{72}{\cdot}nH_2O$.

heumite (heum'-ite) A dark-colored, fine-grained hypabyssal rock characterized by granular texture and composed of alkali feldspar, amphibole, biotite, and smaller amounts of nepheline, sodalite, clinopyroxene, and minor accessories. Its name, given by Brögger, 1898, is derived from Heum, Norway. Obsolete.

heuristic A method of problem-solving in which solutions are discovered by evaluation of progress made toward a solution, resetting parameters in response to new conditions, and iterating in a series of closed loops that converge on the final solution.

hewettite (hew'-ett-ite) A deep-red monoclinic mineral: $CaV^{5+}{}_6O_{16}{\cdot}9H_2O$. It occurs in aggregates of silky, slender orthorhombic crystals. It is classed as a "vanadium bronze".

hexacoral (hex'-a-cor'-al) *scleractinian.*

hexactine (hex-ac'-tine) A siliceous sponge *spicule* having six rays arising from a common center at right angles to one another.

hexactinellid (hex'-ac-ti-nel'-lid) A syn. of *hyalosponge.*

hexad Noun or adj. for a symmetry axis that requires a rotation of 60° to repeat an object's appearance. Cf: *diad; triad; tetrad [cryst].* Syn: *sixfold axis.*

hexadisc (hex'-a-disc) A hexactinellid-sponge spicule (microsclere) composed of three interpenetrating amphidiscs at right angles to one another about a common center. Cf: *staurodisc.*

hexaferrum A steel-gray hexagonal mineral, representing an alloy of Fe, Os, Ru, and Ir.

hexagonal close packing (hex-ag'-o-nal) In a crystal, close packing of atoms by stacking close-packed layers in the sequence ABAB etc. Cf: *cubic close packing.* See also: *close-packed structure.*

hexagonal cross ripple An *oscillation cross ripple* formed by parallel ripples arranged in zigzag fashion and characterized by obtuse angles in adjoining ripples facing in opposite directions, by crossbars connecting apexes on opposite sides of the ripple, and by an enclosed pit that tends to be bounded by six sides. It appears to be formed by waves that oscillate at some angle between 45° and 90° to the direction of the original ripple.

hexagonal cross ripple mark *hexagonal cross ripple.*

hexagonal dipyramid A crystal form of 12 faces consisting of two hexagonal pyramids repeated across a mirror plane of symmetry that is perpendicular to the *sixfold axis.* A cross section perpendicular to the sixfold axis is hexagonal. Its indices are $\{h0l\}$ or $\{hhl\}$ with symmetry $6/m\ 2/m\ 2/m$, and 622, $\{hhl\}$ only with symmetry $\bar{6}m2$, also $\{hkl\}$ with symmetry $6/m$.

hexagonal-dipyramidal class (hex-ag'-o-nal-di-py-ram'-i-dal) That class of the hexagonal system having symmetry $6/m$.

hexagonal indices *Miller-Bravais indices.*

hexagonal prism A crystal form of six faces parallel to the symmetry axis, whose *interfacial angles* are 60°. Its indices are $\{110\}$ or $\{100\}$ in several hexagonal classes, or $\{hk0\}$ in symmetry $6/m$, 6, and $\bar{3}$.

hexagonal pyramid A crystal form consisting of six faces in a pyramid, in which any cross section perpendicular to the *sixfold axis* is hexagonal. Its indices are $\{h01\}$ and $\{hhl\}$ in classes 6mm and 6, and also $\{hkl\}$ in 6.

hexagonal-pyramidal class (hex-ag'-o-nal-py-ram'-i-dal) That crystal class in the hexagonal system having symmetry 6.

hexagonal-scalenohedral class (hex-ag'-o-nal-sca-le'-no-he'-dral) That crystal class in the rhombohedral division of the hexagonal system having symmetry $\bar{3}/m$. Syn: *trigonal-scalenohedral class; ditrigonal-scalenohedral class.*

hexagonal scalenohedron A *scalenohedron* of twelve faces and having symmetry $\bar{3}2/m$. It resembles a ditrigonal pyramid. Cf: *tetragonal scalenohedron.*

hexagonal system One of the six crystal systems, characterized by one unique axis of threefold or sixfold symmetry that is perpendicular and unequal in length to three identical axes that intersect at angles of 120°. This definition includes the *trigonal system* of threefold symmetry; however, the two systems of threefold and sixfold symmetries may be defined separately. Cf: *isometric system; tetragonal system; orthorhombic system; monoclinic system; triclinic system.*

hexagonal-trapezohedral class (hex-ag'-o-nal-tra-pe'-zo-he'-dral) That crystal class in the hexagonal system having symmetry 622.

hexagonal trapezohedron A *crystal form* with 12 faces, a sixfold axis, and three twofold axes, but neither mirror planes nor a center of symmetry. It is composed of top and bottom hexagonal pyramids, one of which is rotated less than 30° about c with respect to the other. It may be right-handed or left-handed, and its indices are $\{hkl\}$ or $\{hk\bar{l}\}$ in symmetry 622.

hexahedral (hex-a-he'-dral) Adj. of *hexahedron.*

hexahedral coordination An atomic structure or arrangement in which an ion is surrounded by eight nearest-neighbor ions of opposite sign, whose centers form the points of a hexahedron (which may or may not be a cube). It may be synonymous with *cubic coordination.*

hexahedrite (hex-a-he'-drite) An *iron meteorite* made up of large single crystals or coarse aggregates of kamacite, usually containing 4-6% nickel in the metal phase, and generally characterized upon etching by the presence of

Neumann bands caused by twinning parallel to the octahedral planes. Symbol: *H.* Cf: *octahedrite [meteorite]; ataxite [meteorite].*

hexahedron (hex-a-he'-dron) A polyhedron of six equivalent faces, e.g. a cube or a rhombohedron. Adj: *hexahedral.*

hexahydrite (hex-a-hy'-drite) A white or greenish-white monoclinic mineral: $MgSO_4 \cdot 6H_2O$. (b) A group name for minerals with the formula $MSO_4 \cdot 6H_2O$, where M = Mg, Fe^{2+}, Mn^{2+}, Co, Ni or Zn.

hexahydroborite (hex'-a-hy'-dro-bo'-rite) A colorless monoclinic mineral: $Ca[B(OH)_4]_2 \cdot 2H_2O$.

hexamethylene (hex-a-meth'-yl-ene) *cyclohexane.*

hexane (hex'-ane) Any of five colorless liquid volatile paraffin hydrocarbons of formula C_6H_{14}. The hexanes, especially n-hexane, $CH_3(CH_2)_4CH_3$, occur in crude oil.

hexarch (hex'-arch) Said of a stele having six strands or origins (Jackson, 1928, p.180).

hexastannite (hex-a-stan'-nite) *stannoidite.*

hexaster (hex'-as-ter) A sponge spicule (microsclere) having the form of a hexactin with anaxial branches or extensions at the ray tips.

hexatestibiopanickelite (hex'-a-te-stib'-i-o-pa-nick'-el-ite) A metallic brownish pink hexagonal mineral: (Ni,Pd)(Te,Sb). Not approved by the International Mineralogical Association.

hexatetrahedron (hex'-a-tet'-ra-he'-dron) *hextetrahedron.*

hexoctahedral class (hex'-oc-ta-he'-dral) That crystal class in the isometric system having symmetry $4/m\ \bar{3}\ 2/m$.

hexoctahedron (hex'-oc-ta-he'-dron) An isometric crystal form of 48 equal triangular faces, each of which cuts the three crystallographic axes at different distances. Its indices are $\{hkl\}$ and its symmetry is $4/m\ \bar{3}\ 2/m$.

hextetrahedral class (hex'-tet-ra-he'-dral) That crystal class in the isometric system having symmetry $\bar{4}\ 3m$.

hextetrahedron (hex'-tet-ra-he'-dron) An isometric crystal form of 24 faces, with indices $\{hkl\}$ and symmetry $\bar{4}\ 3m$. Also spelled: *hexatetrahedron.*

heyite (hey'-ite) A yellow-orange monoclinic mineral: $Pb_5Fe^{2+}{}_2O_4(VO_4)_2$.

heyrovskite (hey-rovsk'-ite) A metallic white orthorhombic mineral: $(Pb,Ag,Bi)_6Bi_2S_9$.

HFSE *high field strength elements.*

HI *height of instrument.*

hiärneite A bright red tetragonal mineral: $Ca_2(Zr,Ti)_5(Sb^{5+},Mn^{3+})_2O_{16}$.

hiatal [ign] (hi-a'-tal) Said of the texture of an igneous rock in which the sizes of the crystals are not in a continuous series but are broken by hiatuses, or in which there are grains of two or more markedly different sizes, as in *porphyritic* rocks (Johannsen, 1939, p.216). Obsolete. Cf: *seriate; heterogranular.*

hiatal [stratig] Pertaining to or involving a stratigraphic hiatus.

hiatus (hi-a'-tus) (a) A break or interruption in the continuity of the geologic record, such as the absence in a stratigraphic sequence of rocks that would normally be present but either were never deposited or were eroded before deposition of the overlying beds. (b) A lapse in time, such as the time interval not represented by rocks at an unconformity. (c) A chronostratigraphic unit representing the space-time value of nondeposition during a regressive-transgressive episode (Wheeler, 1958, p.1057). Cf: *lacuna [stratig].*

hibbingite A colorless to pale green orthorhombic mineral: $(Fe^{2+},Mg)_2(OH)_3Cl$. It is the Fe^{2+} analogue of atacamite.

hibernaculum (hi-ber-nac'-u-lum) A dormant *zooid* with protective walls, produced mainly at the onset of winter by freshwater and brackish-water bryozoans (such as those in the class Gymnolaemata) and developed as part of a colony in the spring. Pl: hibernacula. See also: *statoblast.*

Hibernian Orogeny (Hi-ber'-ni-an) *Erian Orogeny.*

hibonite (hib'-o-nite) A dark brown to black hexagonal mineral of the *magnetoplumbite* group: $(Ca,Ce,La)(Al,Ti,Mg)_{12}O_{19}$.

hibschite (hibsch'-ite) A colorless or pale tinted cubic hydrogarnet: $Ca_3Al_2[SiO_4(OH)_4]_3$..

hidalgoite (hi-dal'-go-ite) A white rhombohedral mineral of the *beudantite* group: $PbAl_3(AsO_4)(SO_4)(OH)_6$.

hidden encruster (hid'-den) A *cryptic* species that is sessile and encrusting in habit, such as a cheilostome bryozoan or serpulid worm.

hiddenite (hid'-den-ite) An intensely green transparant gem variety of spodumene containing chromium.

hidden line removal The process of eliminating from a graphic display, line segments that are obscured from a view in a two-dimensional projection of a three-dimensional object (Krzanowski et al., 1993).

hielmite (hielm'-ite) *hjelmite.*

hieratite (hi-er'-a-tite) A colorless, white, or gray cubic, high-temperature mineral of fumaroles: K_2SiF_6.

hieroglyph (hi'-er-o-glyph') Any sedimentary mark or structure found on a bedding plane; esp. a *sole mark.* A classification of hieroglyphs has been proposed by Vassoevich (1953). The term was first used by Fuchs (1895) for a problematic fossil whose appearance is suggestive of a drawing or ornament.

higginsite (hig'-gins-ite) *conichalcite.*

high n. A general term for such features as a crest, culmination, anticline, or dome. Cf: *low [struc geol].* Syn: *structural high.*

high albite High-temperature albite, stable above 450°C. Natural high albite almost always contains appreciable amounts of potassium and calcium in solid solution. Cf: *low albite.*

high-alumina basalt (high-a-lu'-mi-na) Nonporphyritic basalt distinguished by "higher content of Al_2O_3 (generally higher than 17 percent and rarely as low as 16 percent) than that of the *tholeiite* with the corresponding SiO_2 and total alkalies, and by lower alkali content than that of the *alkali basalt,* provided only aphyric rocks are compared" (Kuno, 1960, p.122). In Japan, high-alumina basalt is transitional between alkaline and tholeiitic basalts geographically, mineralogically, and chemically (except for the higher aluminum content). Kuno's suggestion that it represents a primary magma, independent of alkaline and tholeiitic liquids, has not been widely accepted, and Yoder and Tilley (1962) stated that high-alumina basalt occurs in both the tholeiitic and alkaline basalt suites.

high-angle boundary A grain boundary formed by *rotation recrystallization* across which the crystallographic mismatch exceeds ~6°.

high-angle cross-bedding (high-an'-gle) Cross-bedding in which the cross-beds have an average maximum inclination of 20° or more (McKee and Weir, 1953, p.388). Cf: *low-angle cross-bedding.*

high-angle fault A fault with a dip greater than 45°. Cf: *low-angle fault.*

high chalcocite Hexagonal chalcocite, stable above 105°C.

high-energy coast (high-en'-er-gy) A coast exposed to ocean swell and stormy seas and characterized by average breaker heights of greater than 50 cm. Cf: *moderate-energy coast; low-energy coast.*

high-energy environment An aqueous sedimentary environment characterized by a high *energy level* and by turbulent action (such as that created by waves, currents, or surf) that prevents the settling and accumulation of fine-grained sediment; e.g. a beach or a river channel. Cf: *low-energy environment.*

high field strength elements Elements which, because of the high ratios of their valence charges to ionic radii, have bulk partition coefficients less than unity in the mantle. They include P, Ti, Zr, Hf, Nb, and Ta. Abbrev: HFSE.

high-frequency magnetics Magnetic anomalies of shallow sedimentary origin, often <1 nT, seen in areas of deep magnetic basement. Possible sources include detrital magnetite, diagenetic magnetite, or other minerals possibly related to hydrocarbon microseepage, salt or anhydrite diamagnetism, or fault mineralization. Abbrev: HFMAG.

high-frequency sequence In sequence stratigraphy, a *sequence* formed with cyclicities between 100,000 years and 20,000 years (Mitchum and Van Wagoner, 1991).

high-grade adj. Said of an ore with a relatively high ore-mineral content. Cf: *low-grade.* See also: *grade [ore dep].* v. To extract only the richest ore and leave only low-grade or leave ore behind.

high-grading (high-grad'-ing) (a) Larceny of valuable ore or mineral specimens by employees in a mine. (b) *gouging [mining].*

high island In the Pacific Ocean, a volcanic rather than a coralline island. Cf: *low island.*

high-K In the *IUGS classification,* high-K volcanic rocks plot above a line through two points with the coordinates SiO_2=48, K_2O=1.2 wt% and SiO_2=68, K_2O=2.9 wt%.

highland (high'-land) (a) A general term for a relatively large area of elevated or mountainous land standing prominently above adjacent low areas; a mountainous region. The term is often used in the plural in a proper name; e.g., the Highlands of Scotland or the Hudson Highlands in the U.S. (b) A relative term denoting the higher land of a region; it may include mountains, valleys, and plains. Cf: *upland.* (c) A lofty headland, cliff, or other high landform.

highland glacier A semicontinuous ice cap or glacier system covering the highest or central position of a mountainous area, partly reflecting irregularities of the land surface beneath; e.g. a *plateau glacier.* Cf: *ice field [glaciol].*

highlands (high'-lands) The heavily cratered, relatively high standing regions of the Moon and Mars. The lunar highlands are visually bright and contrast with the lower-lying *maria,* which are visually dark, topographically lower, and geologically younger.

high magnesian calcite Calcite generally containing 5-19 mole % magnesian carbonate. Many bioclasts are composed of high magnesian calcite (e.g., bryozoans, crinoids, red calcareous algae) and some bladed cements in modern carbonate deposits. Abbrev: HMC.

high marsh A syn. of *salting.* The term is a "less correct and rather obsolete" syn. of *salt marsh* (Schieferdecker, 1959, term 1243).

highmoor bog (high'-moor) A bog, often on the uplands, whose surface is largely covered by sphagnum mosses which, because of their high degree of water retention, make the bog more dependent on rainfall than on the water table. The bog often occurs as a *raised peat bog* or *blanket bog.* Cf: *lowmoor bog.*

highmoor peat Peat occurring on high moors and formed predominantly of moss, such as sphagnum. Its moisture content is derived from rain water rather than from groundwater, and is acidic. Mineral matter and nitrogen content is low compared to lowmoor peat, and cellulose content is high. Cf: *domed peat; lowmoor peat.* Syn: *moorland peat; moor peat; sphagnum peat; bog peat; moss peat.*

high oblique n. An *oblique* that includes the apparent horizon. Cf: *low oblique.*

high-pass filter A form of spatial filtering used in image processing systems to emphasize features generally smaller than the size of the averaging area by calculating the average density values within an n by m neighborhood of a pixel and subtracting this average from the value of the pixel being operated on. High-pass filters are defined as any odd number of lines and samples so as to be symmetric around a pixel. Syn: *low-cut filter.*

high plain An extensive area of comparatively level land not situated near sea level; e.g. the High Plains, a relatively undissected section of the Great Plains of the U.S., extending along the eastern side of the Rocky Mountains at elevations above 600 m.

high-polar glacier (high-po'-lar) (a) A less preferred syn. of *polar glacier.* (b) Obsolete: A *polar glacier* that does not melt appreciably even during the summer and in whose accumulation area the firn is at least 100 m thick (Ahlmann, 1933); e.g., most of the glaciers in Antarctica. Cf: *polythermal glacier.*

high quartz High-temperature quartz; specif. beta quartz.

high-rank graywacke A term introduced by Krynine (1945) for a graywacke containing abundant feldspar. The rock is equivalent to *feldspathic graywacke* of Folk (1954) and is regarded as "graywacke proper" by Pettijohn (1957, p.320). Cf: *low-rank graywacke.*

high-rank metamorphism Metamorphism that is accomplished under conditions of high temperature and pressure. Var: high-grade metamorphism. Cf: *low-rank metamorphism.*

high-resolution Use of seismic frequencies above the normal exploration range, recorded with the objective of improving resolution. Often implies frequencies from 80-150 Hz, sometimes to 500 Hz or higher. Abbrev: HR.

high-resolution aeromagnetics High-precision *aeromagnetic survey* flown at low terrain clearance (80-150 m) with close line spacing (100-800 m) recorded at high sample rates (6-14 m) with a high-sensitivity magnetometer (0.001-0.005 nT) using a high-precision positioning system, usually Differential Global Positioning System (DGPS). Abbrev: HRAM.

high-speed layer A subsurface layer in which the speed of seismic-wave propagation is appreciably greater than that in the layers above it. See also: *stringer [seis].*

highstand (high'-stand) The interval of time during one or more cycles of relative change of sea level when sea level is above the shelf edge in a given local area (Mitchum, 1977, p.207). Cf: *lowstand.*

highstand shoreline A *shoreline,* either erosional or constructional, formed at the highest elevation of a *transgression.* Cf: *lowstand shoreline.*

highstand systems tract In sequence stratigraphy, a *systems tract* bounded below by the downlap surface and above by the next sequence boundary; or a unit bounded by a maximum flooding surface below and a slope onlap surface above. The early highstand commonly consists of an aggradational parasequence set; the late highstand is composed of a progradational parasequence set or several. The term is applicable for depositional sequences of type 1, T-R sequences, and genetic stratigraphic sequences (Van Wagoner et al., 1987, 1988; Posamentier and Vail, 1988; Embry and Johannessen, 1992).

high-sulfidation epithermal *acid-sulfate type* deposit.

high-sulfur epithermal *acid-sulfate type* deposit.

high tide The tide at its highest; the accepted popular syn. of *high water* in the sea.

high-velocity layer A layer in which the speed of wave propagation is greater than in the layer above it and which therefore can carry head-wave (refraction) energy.

high volatile A bituminous coal (high-vol'-a-tile) A bituminous coal, commonly agglomerating, that has more than 31% volatile matter (on a dry, mineral-matter-free basis) and 14,000 or more BTU/lb (32.56 or more MJ/kg) (moist, mineral-matter-free). Cf: *high volatile B bituminous coal; high volatile C bituminous coal.*

high volatile B bituminous coal A bituminous coal, characteristically agglomerating, that has 13,000 to 14,000 BTU/lb (30.23 to 32.56 MJ/kg) (moist, mineral-matter-free). Cf: *high volatile A bituminous coal; high volatile C bituminous coal.*

high volatile bituminous coal Bituminous coal that contains more than 31% volatile matter, analyzed on a dry, mineral-matter-free basis. See also: *high volatile A bituminous coal; high volatile B bituminous coal; high volatile C bituminous coal.* Cf: *low volatile bituminous coal; medium volatile bituminous coal; gas coal.*

high volatile C bituminous coal An agglomerating or nonagglomerating bituminous coal that has 11,500 to 13,000 BTU/lb (26.75 to 30.23 MJ/kg) (moist, mineral-matter-free); or an agglomerating, high-volatile bituminous coal that has 10,500 to 11,500 BTU/lb (24.42 to 26.75 MJ/kg) (moist, mineral-matter-free). Cf: *high volatile A bituminous coal; high volatile B bituminous coal.*

highwall (high'-wall) The working face of a surface mine or quarry, esp. of an open-pit coal mine.

high water The height of the maximum vertical elevation of the tide in a 12- or 24-hour cycle relative to an arbitrary datum, usually the tidal benchmark. Thus several high waters may occur in a day. Also, sometimes defined to be the maximum vertical elevation of the tide in a 24-hour period. Abbrev: HW. Cf: *low water.* Syn: *high tide.*

high-water platform (high-wa'-ter) An *abrasion platform* or a *solution platform* developed a little below high-water level, commonly on a rock surface.

highway geology (high'-way) Geology as applied to the planning, design, construction, and maintenance of public roads.

highwoodite (high'-wood-ite) A dark-colored intrusive rock composed of alkali feldspar, labradorite, pyroxene, biotite, iron oxides, apatite, and possibly a small amount of nepheline. It is essentially a *monzonite.* Its name, given by Johannsen in 1938, is derived from the Highwood Mountains, Montana. Obsolete.

hila (hi'-la) Pl. of *hilum.*

hilairite [ign] (hi-lair'-ite) A porphyritic *nepheline syenite* composed of large albite, nepheline, sodalite, aegirine, and eudialyte phenocrysts in a trachytic groundmass of aegirine, nepheline, albite, and orthoclase. The name (Johannsen, 1938) is for Mont St. Hilaire, Quebec. Obsolete.

hilairite [mineral] (hil-air'-ite) (a) A pale brown rhombohedral mineral: $Na_2Zr[Si_3O_9]\cdot3H_2O$. (b) A group name for minerals with the analogous formula, but with Na replaced by Ca, Ba or Y, and Zr sometimes replaced by Ti.

hilate (hi'-late) Said of a spore or pollen grain possessing a *hilum [palyn].*

hilgardite (hil'-gard-ite) A colorless, white, or pale yellow mineral: $Ca_2B_5O_9Cl\cdot H_2O$. It occurs as triclinic and monoclinic forms.

hill (a) A natural elevation of the land surface, rising rather prominently above the surrounding land, usually of limited extent and having a well-defined outline (rounded rather than peaked or rugged), and generally considered to be less than 300 m (1,000 ft) from base to summit; the distinction between a hill and a *mountain* is arbitrary and dependent on local usage. Cf: *mount.* (b) Any slightly elevated ground or other conspicuous elevation in a relatively flat area. (c) An eminence of inferior elevation in an area of rugged relief. (d) Usually used in the plural: A range or group of hills, or a region characterized by hills or by a highland; e.g. the Black Hills of South Dakota. (e) A district whose slope lines run to the same peak (Warntz, 1975, p.213).

hill creep Slow downhill movement, on a steep hillside and under the influence of gravity, of soil and rock waste flowing toward the valleys; it is an important factor in the wasting of hillsides during dissection, as in the Alps. See also: *creep [mass move].* Syn: *hillside creep.*

hillebrandite (hil'-le-brand-ite') A white, colorless, or pale greenish orthorhombic mineral: $Ca_2SiO_3(OH)_2$.

hill-island (hill-is'-land) A glacial moraine, mainly of sand and of variable size, rising as a mature hill from an outwash plain of a later glacial epoch (P.R. Barham in Stamp, 1966, p.235).

hillite A colorless or gray triclinic mineral of the *fairfieldite* group: $Ca_2(Zn,Mg)(PO_4)_2\cdot2H_2O$.

hillock (hil'-lock) A small, low hill; a *mound.* Adj: hillocky.

hillock moraine A moraine consisting of a series of hillocks.

hill of planation A term applied by Gilbert (1877, p.130-131) to a bedrock erosion surface now described as a *pediment* (although such a surface is not a hill in any sense).

hill peat Peat occurring in cold, temperate areas and derived from mosses, heather, pine trees, and related plant forms. Syn: *subalpine peat.*

hillseam Stress-relief joints that have nearly vertical dip, oriented perpendicular to bedding. They occur in shallow overburden where there is high topographic relief, tend to strike parallel to surface contours, and are believed to have formed as a result of local stress relief rather than from tectonic forces (Shea-Albin, 1993).

hill shading (a) A method of showing relief on a map by simulating the appearance of sunlight and shadows, assuming an oblique light from a certain direction (by convention the NW) so that slopes facing away from the light are shaded (the steeper slopes being darker), thereby giving a three-dimensional impression similar to that of a relief model. The method is widely used on topographic maps in association with contour lines. (b) The pictorial effect (of contoured topographic features) emphasized by hill shading, in which the features are shown by the shadows they cast. Syn: *hillwork; relief shading; plastic shading; shading.*

hillside (hill'-side) A part of a hill between its crest and the drainage line at the foot of the hill. Syn: *hillslope.*

hillside creep *hill creep.*

hillside spring *contact spring.*

hillslope (hill'-slope) *hillside.*

hillwash (hill'-wash) The process of *rainwash* operating on a hillslope.

hillwork (hill'-work) *hill shading.*

Hilt's law The generalization that, in a vertical succession at any point in a coal field, coal rank increases with depth.

hilum [bot] (hi'-lum) A scar on a seed coat that marks the place of attachment of the seed stalk to the seed.

hilum [palyn] (a) In *hilate* spores, a laesura that is reduced to a more or less circular, indistinctly delimited aperture, as in certain moss spores. The (proximal) hilum in *Vestispora* is associated with an operculum that commonly is separated from the spore. (b) In *fungal spores,* a scar appearing as a flat spot or porelike mark, indicating the point of attachment to the sporophore or hypha, e.g., the point on a *basidiospore* where the continuity of the outer contour is interrupted. Pl: *hila.* Cf: *ulcus.*

HIMU High μ = high (U + Th)/Pb. An end-member source for mantle-derived magmas, postulated by Zindler and Hart (1986).

hinge [paleont] A collective term for the structures of the dorsal region that function during the opening and closing of the valves of a bivalve shell; esp. a flexible ligamentous joint. The term is often used loosely for hinge line, and for the *cardinal margin* of a brachiopod (TIP, 1965, pt.H, p.145).

hinge [struc geol] The locus of maximum curvature or bending in a folded surface, usually a line. Cf: *hinge line [struc geol].* Syn: *flexure.*

hinge area (a) The flattened area marginal to the hinge of a brachiopod or pelecypod shell. See also: *cardinal area.* (b) The surface involved in the hingement of ostracode valves, commonly differentiated into anterior and posterior areas containing more complex elements and between these an interterminal area with simpler structures (TIP, 1961, pt. Q, p.50-51).

hinge axis (a) An imaginary straight line about which the two valves of a bivalve shell are hinged. Syn: *cardinal axis.* (b) The line joining the points of articulation about which the valves of a brachiopod rotate when opening and closing. Cf: *hinge line [paleont].*

hinge fault An obsolete term for a fault on which the movement of one side hinges about an axis perpendicular to the fault plane; displacement increases with distance from the hinge. It is a questionable term. Cf: *scissor fault; rotational fault; pivotal fault.*

hinge line [paleont] (a) A line along which articulation takes place; e.g. the middorsal line of junction of two valves of a crustacean carapace, permitting movement between them, or the line along which the two valves of an ostracode articulate, seen when the carapace is complete. (b) The straight posterior margin of a brachiopod shell, parallel to the *hinge axis.* The term is also used as a syn. of *cardinal margin.* (c) A term applied loosely to the part of a bivalve-mollusk shell bordering the dorsal margin and occupied by or situated close to the hinge teeth and ligament. The term is sometimes used as a syn. of *hinge axis.* Syn: *hinge [paleont].*

hinge line [struc geol] (a) The axis along which the curvature of a fold is greatest. Syn: fold hinge. (b) A line or boundary between a stable region and a region undergoing upward or downward movement. For example, the hinge line of a sedimentary basin is the boundary between the region that has subsided rapidly and the region that has not. (c) In Pleistocene geology, it is the boundary between regions undergoing postglacial uplift and those of no uplift (for example, in the Great Lakes area).

hinge-line fault A fault that is caused by sedimentary loading along a passive margin. It is used in reference to *growth fault* of the Gulf Coast region of the U.S.

hingement (hinge'-ment) The area of the juncture and articulation of the two halves or valves of the carapace of an ostracode.

hinge node A localized thickened part of the hinge of the right valve of phyllocarid crustaceans, serving to strengthen the hinge.

hinge plate (a) The shelly internal plate bearing the hinge teeth in a bivalve mollusk, situated below the *beak* and the adjacent parts of the dorsal margin of each valve, and lying in a plane parallel to that of commissure. Syn: *cardinal platform.* (b) A plate, simple or divided, typically nearly parallel to the plane between the valves of a brachiopod, lying along the hinge line in the interior of the *brachial valve* and bearing its dental sockets, and joined to crural bases. See also: *inner hinge plate; outer hinge plate.*

hinge surface *axial surface.*

hinge tooth An articulating projection of one valve of a bivalve shell, located near the hinge line or adjacent to the dorsal margin, and fitting into an accompanying socket in the opposite valve for the purpose of holding the valves in position when closed; e.g. a *cardinal tooth* or a *lateral tooth* of a bivalve mollusk, or one of a pair of small or stout, wedge-shaped processes situated at the base of the *delthyrium* of the *pedicle valve* of a brachiopod and articulating with the dental sockets in the *brachial valve.* Syn: *tooth.*

hinge trough A V- or U-shaped depression formed by fusion of the bifurcated median septum with combined socket ridges and crural bases of some terebratellacean brachiopods. Syn: *trough [paleont].*

hingganite-(Ce) A light reddish brown monoclinic mineral of the *datolite* group: $Be_2(Ce,Y,Nd)_2Fe_{0.4}(SiO_4)_2(O,OH)_2$. Not approved by the International Mineralogical Association.

hingganite-(Y) A milky white, light yellow or light green mineral of the *datolite* group: $Be(Y,Ce)SiO_4(OH)$.

hingganite-(Yb) (hing'-gan-ite) A colorless to yellow-green monoclinic mineral of the datolite group: $Be(Yb,Y)SiO_4(OH)$.

hinsdalite (hins'-dal-ite) A colorless or greenish rhombohedral mineral of the *beudantite* group: $(Pb,Sr)Al_3(PO_4)(SO_4)(OH)_6$.

hinterland (hin'-ter-land) An area bordering, or within, an orogenic belt on the internal side, away from the foreland; it is related to the *internides* and to the discredited *borderland* of Schuchert. In the hinterland, rocks are plastically deformed, significant metamorphism accompanies deformation, and the crystalline rocks that compose basement are extensively deformed. Syn: *backland [tect].*

hinterland-dipping duplex A *duplex* in which the imbricates dip toward the hinterland; strata in the horses composing the duplex generally face upward and also dip toward the hinterland (Boyer and Elliott, 1982). Cf: *foreland-dipping duplex.*

hinterland sequence A depositional sequence that consists entirely of nonmarine deposits laid down at a site interior to the coastal area, where depositional mechanisms are controlled only indirectly or not at all by the position of sea level (Mitchum, 1977, p.207).

hintzeite (hin'-tze-ite) *kaliborite.*

hiortdahlite (hiort'-dahl-ite) A pale-yellow or yellowish brown triclinic mineral: $Na_4Ca_8Zr_2(Nb,Mn,Ti,Fe,Mg,Al)_2(Si_2O_7)_4O_3F_5$.

hipotype (hi'-po-type) *hypotype.*

hirst *hurst.*

hisingerite (his'-ing-er-ite') A black or brownish-black monoclinic mineral: $Fe^{3+}{}_2Si_2O_5(OH)_4{\cdot}2H_2O$.

hislopite (his'-lop-ite) A bright grass-green variety of calcite in which the color is due to admixed glauconite.

hispid (his'-pid) Rough, or covered with minute hairlike spines.

histic epipedon (his'-tic) A thin organic soil horizon that is saturated with water at some period of the year unless artificially drained and that is at or near the surface of a mineral soil. The histic epipedon has a maximum thickness depending on the kind of materials in the horizon and the lower limit of organic carbon is the upper limit for the mollic epipedon.

histium (his'-ti-um) The near-ventral ridge confluent with the connecting lobe of the carapace of some heteromorphic ostracodes.

histogram (his'-to-gram) A vertical-bar graph representing a discrete *frequency distribution,* in which the height of bars is proportional to frequency or relative frequency of occurrence within each class interval. The class intervals are mutually exclusive and all-inclusive.

histology The study of tissue anatomy on the microscopic (cellular) level.

histometabasis (his'-to-me-tab'-a-sis) Preservation of structure by the minerals that have replaced organic tissues, as in silicified wood and many other fossils.

historical geology (his-tor'-i-cal) A major branch of *geology* that is concerned with the evolution of the Earth and its life forms from its origins to the present day. The study of historical geology therefore involves investigations into stratigraphy, paleontology, and geochronology, as well as the consideration of paleoenvironments, glacial periods, and plate-tectonic motions. It is complementary to *physical geology.* Not to be confused with *history of geology.*

historical geomorphology A branch of geomorphology concerned with the series of events within a particular geologic period and in a particular geographic region.

history of geology (his'-to-ry) That branch of the history of science that treats the development of geologic knowledge, including the history of observations of geologic features, the history of theories to explain their origin, and the history of the organization and development of geologic institutions and societies. The biographical study of geologists is included. Not to be confused with *historical geology.*

Histosols Organic soils that have organic soil materials in more than half of the upper 80 cm, or that are of any thickness if overlying rock or fragmental materials that have interstices filled with organic soil materials. An order in the U.S. system of soil taxonomy.

hjelmite (hjelm'-ite) A black, often metamict mineral of formula: AB_2O_6 or $A_2B_3O_{10}$, where A = Y, Fe^{2+}, U^{4+}, Mn, or Ca, and B = Nb, Ta, Sn, or W. It is a doubtful mineral, and may be equivalent to pyrochlore + tapiolite. Also spelled: *hielmite.*

hjulstrom effect The relationship whereby fine sediments resist erosion because of their cohesive properties. Thus their erosion requires a faster current than that needed to move larger particles. The "most-movable" sediment is fine sand; both finer and coarser particles are more difficult to move (Friedman et al., 1992, p.20).

***hkl* indices** The *Miller indices,* in general terms that represent integral numbers.

(*hkO*) joint A partial syn. of *diagonal joint.*

H-layer Currently referred to as the Oa horizon.

HLEM Horizontal-loop EM. A configuration in which both transmitter and receiver antennas are in the horizontal plane. Also called: horizontal coplanar.

(*hO1*) joint *longitudinal joint.*

hoa A shallow channel, usually 2-5 m deep traversing the rim of an atoll or barrier reef that is traversible by canoe in calm-conditions. This Polynesian term is derived from the Tuamotu Archipelago, where the hoas typically separate the islets (*motus*). They are attributed to tidal scour (Guilcher, 1988, p.149).

hoarfrost (hoar'-frost) *frost (a).*

hocartite (ho'-cart-ite) A metallic tetragonal mineral of the *stannite* group: Ag_2FeSnS_4.

hochelagaite (ho-che-la'-ga-ite) A vitreous white monoclinic mineral: $(Ca,Na,Sr)Nb_4O_{11}{\cdot}8H_2O$.

hodgkinsonite (hodg'-kin-son-ite') A bright-pink to reddish-brown monoclinic mineral: $Mn^{2+}Zn_2(SiO_4)(OH)_2$.

hodrushite (hod'-rush-ite) A metallic gray monoclinic mineral: $Cu_4Bi_6S_{11}$.

hoe A promontory or a point of land stretching into the sea; a spur of a hill, or a projecting ridge of land; a cliff. Term is obsolete except in English place names; e.g. Plymouth Hoe. Syn: *howe.*

hoegbomite (hoeg'-bom-ite) *högbomite.*

hoelite (hoe'-lite) A yellow monoclinic organic mineral (anthraquinone): $C_{14}H_8O_2$. A product of a burning coal seam.

hoernesite *hörnesite.*

hofmannite (hof'-mann-ite) *hartite.*

hoganite A bluish green monoclinic organic mineral: $Cu(CH_3COO)_2 \cdot H_2O$.
hogback [geomorph] (hog′-back) Any ridge with a sharp summit and steep slopes of nearly equal inclination on both flanks, and resembling in outline the back of a hog; specif. a sharp-crested ridge formed by the outcropping edges of steeply inclined resistant rocks, and produced by differential erosion. The term is usually restricted to ridges carved from beds dipping at angles >20° (Stokes and Varnes, 1955, p.71). Cf: *cuesta; dike wall; razorback; revet-crag.* Syn: *hog's back; stone wall.*
hogback [glac geol] A term applied in New England to a *drumlin* (western Massachusetts) and to a horseback or *esker* (Maine).
högbomite (hög′-bom-ite) A metallic black hexagonal or rhombohedral mineral: $(Mg,Fe^{2+}Al,Ti)_{11}O_{16}$. Many stacking polytypes exist. Also spelled: *hoegbomite.*
hog's back Var. of *hogback [geomorph].* Also spelled: hog's-back; hogsback.
hogtooth spar (hog′-tooth) *dogtooth spar.*
hogtuvaite A subadamantine black triclinic mineral: $(Ca,Na)_2(Fe^{2+},Fe^{3+},Ti)_6(Si,Be,Al)_6O_{20}$. Syn: *makarochkinite.*
högtveitite *thalénite-(Y).*
hog wallow (a) A faintly rolling land surface characterized by many low, coalescent or rounded mounds (such as Mima mounds) that are slightly higher than the basin-shaped depressions between them. (b) A *wallow;* also, a similar depression believed to be formed by heavy rains.
hohmannite (hoh′-mann-ite) An brown, orange, or red triclinic mineral: $Fe^{3+}_2O(SO_4)_2 \cdot 8H_2O$.
holacanth (hol′-a-canth) A *trabecula* of a rugose coral, seemingly consisting of a clear rod of calcite, as in septa of *Tryplasma* (TIP, 1956, pt. F, p.248).
holarctic (hol-arc′-tic) Pertaining to the arctic regions as a whole.
holaspis (hol-as′-pis) A trilobite at the developmental stage having the number of thoracic segments typical of the species. Pl: holaspides.
holdawayite (hol′-da-way-ite) A pink monoclinic mineral: $Mn^{2+}_6(CO_3)_2(OH)_7(Cl,OH)$.
holdenite (hol′-den-ite) A pink to red orthorhombic mineral: $(Mn^{2+},Mg)_6Zn_3(AsO_4)_2(SiO_4)(OH)_8$.
holdfast (hold′-fast) Something that supports or holds in place; e.g. a basal discoid or rootlike structure by which the thallus of many algae is attached to a solid object in water, any structure at the distal extremity of a crinoid column that serves for fixation, or a multiplated cylindrical or globular structure that attaches a primitive crinoid or crinoid calyx to an object on the substrate.
hole [coast] A term used in New England and England for a small bay, cove, or narrow waterway; e.g. Woods Hole, Mass. and Mousehole, Cornwall, U.K.
hole [cryst] *vacancy.*
hole [drill] (a) *borehole.* (b) *drill hole.* (c) A mine, well, or shaft dug in earth material.
hole [geomorph] (a) A term used in the western U.S. for a comparatively level, grassy valley nearly shut in by mountains; e.g. Jackson Hole, Wyoming. Cf: *park.* (b) An abrupt hollow in the ground, such as a pothole, a kettle, or a cave.
hole [streams] (a) A deep place in a stream. (b) A water hole in the bed of an intermittent stream.
hole fatigue A delay in the effective initiation of a seismic impulse from a shot because of changes in the shot environment (usually cavity formation) produced by an earlier shot in the same hole.
holiday *channel [drill].*
holism (ho′-lism) The theory that in nature organisms develop from individual structures acting as "whole" units. Cf: *vitalism.*
hollaite (hol′-la-ite) A hybrid rock produced by interaction between *sövite* and rocks of the ijolite-melteigite series. The name, given by Brögger in 1921, is for Holla Church, Fen Complex, Norway. Obsolete.
hollandite (hol′-land-ite) (a) A silvery-gray to black monoclinic mineral: $Ba(Mn^{4+},Mn^{3+})_8O_{16}$. It is isostructural with coronadite and cryptomelane. (b) A group name for monoclinic or tetragonal minerals with the general formula $AB_6O_{18} \cdot nH_2O$, where A = Ba, Pb, K or Na, and B = Mn, Ti, Fe, V or Cr.
hollingworthite (hol′-ling-worth-ite′) A metallic gray cubic mineral of the *gersdorffite* group: (Rh,Pt,Pd)AsS.
hollow (hol′-low) (a) A low tract of land surrounded by hills or mountains; a small, sheltered valley or basin, esp. in a rugged area. (b) A term used in the Catskill Mountains of New York for a notch or pass. (c) A landform represented by a depression, such as a cirque, a cave, a large sink, or a blowout. (d) A low area of any shape and size in peatland. Cf: *hummock [peat].*
Holmes' classification A classification of igneous rocks based chiefly on the degree of saturation of a rock, with other aspects of the mineralogy as secondary considerations. The system was proposed in 1928 by Arthur Holmes (1890-1965).
Holmes effect The effect leading to overestimation of the relative area of an opaque grain in a thin section of a rock viewed in transmitted light. The apparent area of an opaque grain will always be that of its maximum cross section in the slide, whereas the desired reference is the surface of the thin section. The term was coined by F. Chayes (1956, p.95) for the effect described by the petrographer Arthur Holmes (1890-1965) in 1927.
holmquistite (holm′-quist-ite) A light blue, violet, or bluish-black orthorhombic mineral of the *amphibole* group: $Li_2(Mg,Fe^{2+})_3Al_2Si_8O_{22}(OH)_2$.
holo- (hol′-o-) A prefix meaning "completely, wholly".
holoaxial (hol-o-ax′-i-al) *holohedral.*
holoblast (hol′-o-blast) A *crystalloblast* that is newly and completely formed during metamorphism. The term was first used by Sander (1951). Syn: *neoblast.*
Holocene (Hol′-o-cene) The upper series of the *Quaternary* System of the Standard Global Chronostratigraphic Scale, above the *Pleistocene.* Also the time during which these rocks were formed, the *Holocene* Epoch, covering the time span between 11.5 ka and the present. The term has replaced the earlier "Recent" as a formal name.
Holocephali (Hol-o-ceph′-a-li) A subclass of cartilaginous fishes consisting entirely of the order *Chimaeriformes.* Syn: *ratfish.*
holochoanitic (hol′-o-cho′-a-nit′-ic) Said of a *retrochoanitic septal neck* of a nautiloid that extends backward through the length of one camera.
holochroal eye (hol-och′-ro-al) A trilobite compound eye consisting of numerous adjoining lenses covered by a continuous cornea. Cf: *schizochroal eye.* Syn: *compound eye.*
holoclastic rock (hol-o-clas′-tic) A sedimentary clastic rock, as distinguished from a pyroclastic rock.
holococcolith (hol-o-coc′-co-lith) A *coccolith* consisting entirely of microcrystals of usual crystallographic shape, whether or not they are identical. Cf: *heterococcolith.*
holoconodont *element.*
holocrystalline (hol-o-crys′-tal-line) Said of the texture of an igneous rock composed entirely of crystals, i.e. having no glassy part.
holocyst (hol′-o-cyst) *olocyst.*
holography Recording of the intensity of the interference pattern which results from the scattering of coherent radiation and a reference beam. Optical holography usually involves photographing the interference from laser light (in order to achieve a coherent source); subsequent illumination of the photographic plate allows one to "see" the photographed object in three dimensions. See also: *earth holography.*
holohedral (hol-o-he′-dral) Said of that crystal class having the maximum symmetry possible in each crystal system. Cf: *merohedral; tetartohedral.* Syn: *holosystematic; holosymmetric; holoaxial.*
holohedron (hol-o-he′-dron) Any crystal form in the holohedral class of a crystal system.
holohyaline (hol-o-hy′-a-line) Said of an igneous rock composed entirely of glass.
holokarst (hol′-o-karst) A well-developed karst terrane, formed entirely on soluble rock, which contains no surface drainage (Ford and Williams, 1989).
hololeims (hol′-o-leims) Coalified remains of entire plants (Krishtofovich, 1945, p.138). Cf: *meroleims.* See also: *phytoleims.*
holomictic lake (hol-o-mic′-tic) A lake that undergoes a complete mixing of its waters during periods of *circulation* or *overturn* (Wetzel, 2001, p.83). Cf: *meromictic lake.*
holomixis (hol-o-mix′-is) The process leading to, or the condition of, a *holomictic lake.*
holomorph (hol′-o-morph) In non-lichen-forming ascomycete and basidiomycete fungi, the whole fungus in all its morphs and states, including both the asexual state (*anamorph*) and the corresponding sexual state (*teleomorph*); also called the perfect state.
holoperipheral growth (hol′-o-pe-riph′-er-al) Increase in size of a brachiopod valve all around the margins (in posterior, anterior, and lateral directions). Cf: *mixoperipheral growth.*
holophyletic (hol′-o-phy-let′-ic) In *cladism,* pertaining to a higher taxon that contains all species known to be descended from a given ancestral species. Ant: *paraphyletic.*
holophyly *monophyly.*
holophyte (hol′-o-phyte) A plant that derives its nourishment entirely from its own organs. Adj: *holophytic.*
holophytic (hol-o-phyt′-ic) The obsolescent adj. of *holophyte;* denoting nutrition of the plant type. Cf: *holozoic.* Syn: *phototrophic; photoautotrophic.*
holoplankton (hol-o-plank′-ton) Plankton organisms that are free-floating during their entire life cycles (e.g., copepods), as opposed to *meroplankton* (e.g., fish larvae).
holosome (hol′-o-some) A term introduced by Wheeler (1958, p.1061) for an intertongued chronostratigraphic unit that may be either *depositional* (comprising one or more contiguous holostromes) or *hiatal* (consisting of combined contiguous hiatuses). Cf: *lithosome; biosome.*
Holostei (Ho-los′-tei) An infraclass of ray-finned bony fishes that includes the living bowfin (*Amia*) and gar (*Lepisosteus*), as well as a variety of Mesozoic forms structurally more or less intermediate between the Paleozoic chondrosteans and the Tertiary teleosts. It is probably a grade in the evolution of the *Osteichthyes.*
holostomatous (hol-o-stom′-a-tous) Said of a gastropod shell with a more or less circular apertural margin uninterrupted by a *siphonal canal* or notch. Cf: *siphonostomatous.*
holostratic unit The local part of a stratic unit (Henningsmoen, 1961, p.67).

holostratotype (hol-o-strat′-o-type) The original stratotype designated by the author at the time of proposing a stratigraphic unit or boundary (ISG, 1976, p.26).

holostrome (hol′-o-strome) A term introduced by Wheeler (1958, p.1055-1056) for a chronostratigraphic unit "embodying the space-time value of a complete (restored) transgressive-regressive depositional sequence", including strata that may later have been removed by erosion. Cf: *lithostrome.*

holostylic (hol-o-sty′-lic) Pertaining to a jaw suspension in which the cartilaginous or cartilage-replacement skeleton of the upper jaw is immovably fused to the neurocranium.

holosymmetric (hol′-o-sym-met′-ric) *holohedral.*

holosystematic (hol′-o-sys′-tem-at′-ic) *holohedral.*

holothuroid (hol-o-thu′-roid) Any cylindroid echinozoan, usually free-living, belonging to the class Holothuroidea, characterized by the absence of an articulated test and by the reduction of skeletal elements to microscopic sclerites; e.g. a *sea cucumber.* Var: holothurian.

holotomous (ho-lot′-o-mous) Said of a crinoid arm characterized by a division on each successive *brachial plate;* typically, of a crinoid arm with pinnules.

holotype [bot] A nomenclatural type, which is that element to which the name of a taxon is permanently attached. It is not necessarily the most typical or representative element of the taxon to which it is assigned. The type of a genus is the type (holotype) of a species.

holotype [paleont] The one specimen or other element designated by the author as the nomenclatural type in describing a new species. As long as the holotype is extant, it automatically fixes the application of the name concerned (ICBN, 1972, p.18). Cf: *lectotype; neotype.*

holozoic (hol-o-zo′-ic) Said of an organism that is nourished by the ingestion of organic matter; denoting nutrition of the animal type. Cf: *holophytic.*

holozone A biostratigraphic unit representing the total range of a taxon (Henningsmoen, 1961, p.69).

holozonite A biostratigraphic unit representing an entire range zone (Henningsmoen, 1961, p.68).

Holsteinian (Hol-stein′-i-an) North European climatostratigraphic and floral stage: Middle Pleistocene (above Elsterian, below Saalian). Equivalent in time to Mindel/Riss interglacial.

holtedahlite (hol′-te-dahl-ite′) A colorless trigonal mineral: $Mg_{12}(PO_4)_5(PO_3OH,CO_3)(OH,O)_6$. It is the Mg-analogue of satterlyite.

holtite (holt′-ite) A buff, greenish, or brown orthorhmbic mineral of the *dumortierite* group: $Al_6(Al,Ta)(BO_3)[(Si,Sb,As)O_4]_3(O,OH)_3$.

HOM *homogeneous structure.*

homalographic projection (ho′-mal-o-graph′-ic) *homolographic projection.*

homalozoan (hom′-a-lo-zo′-an) Any echinoderm belonging to the subphylum Homalozoa, characterized by the absence of radial symmetry and having a basically asymmetrical body. The subphylum includes the carpoids and possibly the class Machaeridia.

homeoblastic (ho′-me-o-blas′-tic) Pertaining to a type of *crystalloblastic* texture in a metamorphic rock in which the essential mineral constituents are approximately of equal size. Depending on the habit of the minerals involved, this texture may also be called more specifically *granoblastic, lepidoblastic, nematoblastic,* or *fibroblastic.* The term was originated by Becke (1903). Cf: *heteroblastic.*

homeochilidium (ho′-me-o-chi-lid′-i-um) An externally convex triangular plate closing almost all, or only the apical part, of the *notothyrium* in the brachiopod order Paterinida. Cf: *homeodeltidium.* Also spelled: homoeochilidium.

homeocrystalline (ho′-me-o-crys′-tal-line) *homogranular.*

homeodeltidium (ho′-me-o-del-tid′-i-um) An externally convex triangular plate closing almost all, or only the apical part, of the *delthyrium* in the brachiopod order Paterinida. Cf: *homeochilidium.* Also spelled: homoeodeltidium.

homeomorph [cryst] (ho′-me-o-morph′) A crystal that displays *homeomorphism* with another.

homeomorph [evol] An individual that bears a close resemblance to another organism although the two have different ancestors.

homeomorphic [cryst] (ho′-me-o-mor′-phic) Adj. of *homeomorphism.*

homeomorphic [paleont] Said of a crinoid column consisting of columnals that are all alike in size and shape.

homeomorphism (ho′-me-o-mor′-phism) The characteristic of crystalline substances of dissimilar chemical composition to have similar crystal form and habit; such crystals are known as homeomorphs. Adj: homeomorphic; *homeomorphous.*

homeomorphous (ho′-me-o-mor′-phous) Adj. of *homeomorphism.*

homeomorphy (ho′-me-o-mor′-phy) The phenomenon in which species having superficial resemblance are unlike in structural details; general similarity but dissimilarity in detail. The term is sometimes used as a syn. of *convergent evolution.* Not to be confused with *homomorphy.*

homeostasis (ho-me-ost′-a-sis) The trend toward a relatively stable internal condition in the bodies of the higher animals as a result of a sequence of interacting physiologic processes; e.g., the ability to maintain relatively constant body heat during widely varying external temperatures.

homeothrausmatic (ho′-me-o-thraus-mat′-ic) A genetic term applied to igneous rocks with an orbicular texture in which the nuclei of the orbicules are formed of inclusions of the same generation as the groundmass (Eskola, 1938, p.476). Obsolete. Cf: *isothrausmatic; allothrausmatic; heterothrausmatic; crystallothrausmatic.*

Homerian (Ho-me′-ri-an) A stage of the Standard Global Chronostratigraphic Scale: upper stage of the Wenlockian Series (above the Sheinwoodian, below the Gorstian). The basal GSSP is in southwest England (Holland, 1982).

Homestake-type deposit *arsenical sulfide-silicate iron formation.*

homilite (hom′-i-lite) A black or blackish-brown monoclinic mineral of the *datolite* group: $Ca_2(Fe^{2+},Mg)B_2Si_2O_{10}$.

Hommel's classification (Hom′-mel's) An obsolete classification of igneous rocks in which a rock is represented by a two-part formula, one giving the molecular proportions of the oxides and the other the percentages of the normative minerals. The system was proposed in 1919 by W. Hommel.

homo- (ho′-mo-) A prefix meaning "same".

homoclinal (ho′-mo-cli′-nal) Adj. of *homocline.*

homoclinal shifting A term used by Cotton (1922, p.392) as a syn. of *monoclinal shifting.*

homocline (ho′-mo-cline) A general term for a series of rock strata having the same dip, e.g. one limb of a fold, a tilted fault block, or an isocline. Cf: *monocline.* Adj: *homoclinal.*

homocoelous (ho-mo-coe′-lous) Said of sponges whose flagellated chambers are also spongocoels; specif. pertaining to *asconoid* sponges having calcium-carbonate spicules.

homodesmic (ho-mo-des′-mic) Said of a crystal or other material that is bonded in only one way. Cf: *heterodesmic.*

homogeneity The property of a parameter whose values are unchanged over space, as in hydraulic conductivity of an aquifer.

homogeneous [geophys] Said of a material whose properties are uniform throughout. Not to be confused with *isotropic.* Cf: *heterogeneous [geophys].*

homogeneous equilibrium (ho-mo-ge′-ne-ous) Equilibrium in a system consisting of only one phase, typically liquid or gaseous. Cf: *heterogeneous equilibrium.*

homogeneous fractionation Fractional crystallization wherein the crystals are immediately removed from the system. In this case the crystals being formed are exclusively the liquidus phase(s) of the liquid in the main body of magma and the liquid path is determined solely by the removal of those components contained in the liquidus phase(s) (Nielsen and Delong, 1992).

homogeneous strain A state of *strain* of a deformed body in which all initially straight lines remain straight, because the components of strain are identical at every point. More properly termed a homogeneous strain field. Syn: *uniform strain.*

homogeneous structure An aggregation of more or less equidimensional, irregularly shaped crystallites lacking clear first-order structural arrangement except for possible accretion banding. This excludes fine complex crossed lamellar and crossed acicular structures, which may appear structurally featureless at the optical level but are easily differentiable by SEM. This also excludes diffuse crossed lamellar structure, which shows the general organization of a crossed lamellar structure but may appear similar to homogeneous structure in fractures under the SEM. The present definition of homogeneous structure carries no implication as to mineralogy or crystallographic orientation of the constituent crystallites. Crystallites in fine - and coarse-grained homogeneous structures are generally less than or greater than (respectively) 5 microns in diameter (Carter 1980a,b). Abbrev: HOM.

homogenization temperature (ho-mog′-e-ni-za′-tion) In fluid inclusion geothermometry, the temperature at which a multi-phase fluid inclusion becomes a single, homogeneous phase while being heated on a heating/freezing stage. A minimum temperature of formation of the fluid inclusion.

homogeny (ho-mog′-e-ny) *homology.*

homogranular (ho-mo-gran′-u-lar) Said of the texture of a rock having crystals of the same general size. Obsolete. Ant: *heterogranular.* Syn: *equigranular; even-grained; homeocrystalline.*

homoiothermic (ho′-mo-i′-o-ther-mic) Said of an organism whose body temperature remains relatively uniform and independent of the temperature of the environment; warm-blooded. Cf: *poikilothermic.*

homologous [geol] (ho-mol′-o-gous) (a) Said of strata, in separated areas, that are correlatable (contemporaneous) and that are of the same general lithologic character or facies and/or occupy analogous structural positions along the strike. (b) Said of faults, in separated areas, that have the same relative position or structure.

homologous [paleont] The adj. of *homology.*

homologous character *homologue [phylogen].*

homologous series A group of minerals or other crystalline compounds in which the sequence or the frequency of occurrence of identifiable structural elements is different from one mineral (or compound) to another. Cf: *polysomatic series.*

homolographic projection (ho′-mol-o-graph′-ic) An *equal-area projection.* The term is sometimes given to a particular map projection, such as the "Mollweide homolographic projection". Syn: *homalographic projection.*

homologue [phylogen] (hom'-o-logue) One of several characters having a common origin. Homologues are characters in two or more separate taxa that meet the criteria of homology either by being the same as the character that is present in the ancestor of the two taxa, or by being different characters that have an ancestor-descendant relationship that is described as preexisting-novel (Brooks et al., 1984, p.80; Wiley et al., 1991, p.9). Cf: *plesiomorphic character, apomorphic character.* Ant: *homoplasy.*

homology (ho-mol'-o-gy) (a) In biology, similarity but not identity between parts of different organisms as a result of evolutionary differentiation from the same or corresponding parts of a common ancestor. Syn: *true homology; homogeny.* Cf: *homoplasy.* (b) Similarity of position, proportion, structure, etc. without restriction to common ancestry. Cf: *analogy.* (c) In *cladism,* a character possessed in common by two or more organisms or taxa and also inherited from a common ancestor, remote or immediate. See also: *shared character; shared primitive character; shared derived character.* Adj: *homologous.*

homolosine projection *Goode homolosine projection.*

homomorphic processes (ho-mo-mor'-phic) On *acritarchs* or *dinocysts, processes* that are all of one form on a single specimen (Lister, 1970; Kjellström, 1971). Cf: *heteromorphic processes; acanthomorph.*

homomorphosis (ho-mo-mor'-pho-sis) The regeneration by an organism of a part having a form similar to that of a part that has been lost.

homomorphy (ho'-mo-mor'-phy) Superficial similarity in the morphology of members of different phyla, e.g. the rudistid lamellibranchs and the corals (Whitten, 1972, p.227). Not to be confused with *homeomorphy.*

homomyarian (ho'-mo-my-ar'-i-an) *isomyarian.*

homonym (hom'-o-nym) Any one of two or more identical names used to identify different organisms or objects. In zoologic and botanic nomenclature, it refers to duplication of name for a *taxon* of the same rank based on a different type (McVaugh et al., 1968, p.15). See also: *homonymy.*

homonymy (ho-mon'-y-my) In zoological and botanical nomenclature, identity in spelling of the names applied to different taxa of the same rank. See also: *law of homonymy.*

homo-octahedral Said of a mica in which the same chemical element occupies all octahedral sites. Cf: *hetero-octahedral.*

homoplasy (ho-mop'-la-sy) Similarity or correspondence of parts or organs that developed as a result of convergence or parallelism, rather than from a common ancestry. Cf: *homology.* Adj: homoplastic.

homopolar (ho-mo-po'-lar) Of uniform polarity; not separated or changed into ions; not polar in activity.

homopycnal inflow (ho-mo-pyc'-nal) Flowing water of the same density as the body of water it enters, resulting in easy mixing (Moore, 1966, p.89). Cf: *hyperpycnal inflow; hypopycnal inflow.*

homoscedasticity A condition where the variance remains constant for all magnitudes of a variable. Cf: *heteroscedasticity.*

homospore (ho'-mo-spore) One of the spores of an embryophytic plant that reproduced by homospory; a plant spore functioning as either male or female in reproduction. Its range is Silurian to Holocene. Syn: *isospore.*

homosporous (ho-mo-spor'-ous) Characterized by *homospory.*

homospory (ho-mos'-po-ry) The condition in embryophytic plants in which all spores produced are morphologically of the same kind; the production of homospores, as it occurs in varoious plants. Syn: *isospory.* Cf: *heterospory.*

homotactic [stratig] (ho-mo-tac'-tic) *homotaxial.*

homotactic [struc petrol] Said of the symmetry of a fabric in which all the subfabrics agree in symmetry. Cf: *heterotactic [struc petrol].*

homotaxial (ho-mo-tax'-i-al) Pertaining to, characterized by, or exhibiting homotaxy; e.g. said of rock-stratigraphic units or biostratigraphic units that have a similar order of arrangement in different locations but are not necessarily contemporaneous (ACSN, 1961, Art. 2a). Syn: *homotactic.*

homotaxis (ho-mo-tax'-is) An erroneous transliteration of *homotaxy.*

homotaxy (ho'-mo-tax'-y) Similarity of serial arrangement; specif. taxonomic similarity between stratigraphic or fossil sequences in separate regions, or the condition of strata characterized by similar fossils occupying corresponding positions in different vertical sequences, without connotation of similarity of age. The term was originally proposed as *homotaxis* by Huxley (1862, p. xlvi) to avoid the common fallacy of confusing taxonomic similarity with synchroneity. Etymol: Greek. Cf: *chronotaxy.* Ant: *heterotaxy.*

homothetic (ho-mo-thet'-ic) Said of geomorphologic features that show geometric similarity (similar in shape though perhaps differing in size) and that have corresponding points that are colinear (Strahler, 1958, p.291).

hondo (hon'-do) A term used in the SW U.S. for a broad, low-lying arroyo. Etymol: Spanish, "bottom".

hondurasite (hon-du'-ras-ite) *selen-tellurium.*

honessite (ho-ness'-ite) An orange, yellow, reddish, or green rhombohedral mineral: $(Ni,Fe^{3+})_9(SO_4)_2(OH)_{18}\cdot nH_2O$.

honeycomb structure [ice] (hon'-ey-comb) A sea-ice consisting of soft, spongy ice filled with pockets of meltwater or seawater, and characteristic of *rotten ice.*

honeycomb structure [weath] A rock or soil structure having cell-like forms suggesting a honeycomb; e.g. a *stone lattice.* See also: *tafone.*

honeycomb weathering A type of chemical weathering in which innumerable pits are produced on a rock exposure. The pitted surface resembles an enlarged honeycomb and is characteristic of finely granular rocks, such as tuffs and sandstones, in an arid region. Cf: *cavernous weathering.* Syn: *fretwork weathering; alveolar weathering.*

honeypot An area on a *tidal flat* where very soft mud allows people to sink or become stuck. The feature is common in New England and may be caused by springs.

honey stone (hon'-ey) A syn. of *mellite.* Also spelled: honeystone.

hongquiite (hong-qui'-ite) A discredited metallic white cubic mineral, reported as TiO, but possibly TiC.

hongshiite (hong-shi'-ite) A metallic gray rhombohedral mineral: (Pt,Fe). Not approved by the International Mineralogical Association.

Honkasalo correction (Hon-kas'-a-lo) A term added to the conventional Earth-tide correction formula to reduce observed gravity values at a point to a common average value, instead of eliminating the lunar-solar effect altogether.

hood (a) An arched plate of secondary shell of a brachiopod, arising from the *echmidium* of *Cryptacanthia* or from the median septum of dallinids and terebratellids. (b) The tough fleshy structure located above the head of *Nautilus* and covering the aperture when the head is withdrawn into the living chamber. (c) Ventral flattened part of a *rhyncholite,* commonly of rhomboidal shape (TIP, 1964, pt. K, p.475). (d) Curved lamellose plates of rostroconch mollusks connected to carinae in the Conocardiacea; growing edges of these plates form a tubular extension of the ventral orifice.

hoodoo (hoo'-doo) (a) A fantastic column, pinnacle, or pillar of rock produced in a region of sporadic heavy rainfall by differential weathering or erosion of horizontal strata, facilitated by joints and by layers of varying hardness, and occurring in varied and often eccentric or grotesque forms. Cf: *earth pillar.* Syn: *rock pillar [topog].* (b) *hoodoo column.* Etymol: African; from its fancied resemblance to animals and embodied evil spirits.

hoodoo column A term sometimes applied to an *earth pillar.* Syn: *hoodoo.*

hoodoo rock One of several topographic forms of bizarre shape, developed or modified by differential weathering; e.g. *pedestal rock; earth pillar; hoodoo.*

hook [geomorph] (a) A sandy or gravelly barrier spit or narrow cape turned sharply landward at the outer end, so as to resemble a hook in plan view; e.g., Sandy Hook, N.J.; or a low peninsula or barrier ending in a recurved spit and formed at the end of a bay. Also, a *recurved spit.* (b) A sharp bend, curve, or angle in a stream.

hook [paleont] A holothurian *sclerite* in the form of a fishhook, consisting of an eye, a shank, and a *spear.*

Hookean substance (Hook'-e-an) A material that deforms in accordance with *Hooke's law.* Syn: *elastic material.*

hook echo A distinctive hook-shaped reflectivity pattern in a radar echo that often indicates the presence of a severe *thunderstorm* cell and perhaps tornadic circulation. This pattern is produced by rainfall being drawn around the mesocyclone.

hooked bay A bay similar to a bight but having a headland at only one end.

hooked dune *fishhook dune.*

hooked spit *recurved spit.*

Hooke's law A statement of *elastic deformation,* that the strain is linearly proportional to the applied stress and the strain recovers instantaneously upon removal of the stress. See also: *Hookean substance.*

hook valley The valley containing a *barbed tributary.*

hoop stress Stress acting tangent to the boundary of a rounded opening such as a borehole or elliptical hole.

hope [geog] A piece of dry, arable land surrounded by swamp or marsh.

hope [geomorph] A British term (used esp. in southern Scotland) for a small enclosed valley; esp. the broad upper end of a narrow mountain valley, or a blind valley branching from a larger or wider valley. It is usually rounded and often has a stream flowing through it.

hopeite (hope'-ite) A white, gray, or pale yellow orthorhombic mineral: $Zn_3(PO_4)_2\cdot 4H_2O$. It is dimorphous with parahopeite.

hopper crystal (hop'-per) A cubic crystal, usually of halite, in which the faces of the cube have grown more at the edges than in the center, giving each face a centrally depressed or hopper-shaped form.

horizon [geol] (ho-ri'-zon) An interface indicative of a particular position in a stratigraphic sequence. In practice it is commonly a distinctive very thin bed (ISG, 1976, p.14). Many kinds of stratigraphic horizons occur depending on the stratigraphic characters involved; e.g. lithohorizons, biohorizons, chronohorizons, seismic horizons. See also: *biohorizon.* Syn: *geologic horizon.*

horizon [soil] *soil horizon.*

horizon [surv] One of several lines or planes used as reference for observation and measurement relative to a given location on the Earth's surface and referred generally to a horizontal direction (Huschke, 1959, p.283); esp. *apparent horizon.* The term is also frequently applied to *celestial horizon, actual horizon,* and *artificial horizon.*

horizon circle A circle, in an azimuthal projection, defined by points equidistant from the center of the projection. The maximum horizon circle on a polar projection may be as distant as the opposite pole.

horizon closure The amount by which the sum of a series of adjacent measured horizontal angles around a point fails to equal exactly the theoretical sum of 360 degrees; the *error of closure* of horizon. Also known as "closure of horizon". See also: *closing the horizon.*

horizon slice Display from a three-dimensional data set of the data elements that lie on the same picked reflecting horizon, thus showing areal variations in amplitude or other attributes. Different from a *time slice.*

horizon system of coordinates A set of celestial coordinates, usually altitude and azimuth or azimuth angle, based on the celestial horizon as the primary great circle. Cf: *equator system of coordinates.*

horizontal (hor-i-zon'-tal) adj. In geodesy, said of a direction that is tangent to the *geop* at a given point. Cf: *vertical.*

horizontal accuracy The *positional accuracy* of a dataset with respect to a horizontal datum (Geo-One-Stop, 2003, p.3). See also: *locational accuracy; positional accuracy.*

horizontal angle An angle in a horizontal plane. Cf: *vertical angle.*

horizontal axis The axis about which the telescope of a theodolite or transit rotates when moved vertically. It is the axis of rotation that is perpendicular to the *vertical axis* of the instrument.

horizontal circle A graduated disk affixed to the lower plate of a transit or theodolite by means of which horizontal angles can be measured.

horizontal control A system of points whose horizontal positions and interrelationships have been accurately determined for use as fixed references in positioning and correlating map features.

horizontal control point A control point whose longitude and latitude (but not necessarily its elevation or height) have been defined or determined.

horizontal coordinate (a) One of a pair of coordinates referred to a coordinate system on a level surface, usually the geoid. (b) One pair of coordinates referred to a coordinate system on an ellipsoid taken to represent the Earth. (c) One of a pair of coordinates on a horizontal plane.

horizontal dip slip *horizontal slip.*

horizontal direction A direction in a horizontal plane; an observed horizontal angle at a triangulation station, reduced to a common initial direction.

horizontal displacement *strike slip.*

horizontal drainhole An oil or gas well drilled parallel to the boundaries of a reservoir. It is usually nearly horizontal but not always.

horizontal equivalent The distance between two points on a land surface, projected onto a horizontal plane; e.g. the shortest distance between two contour lines on a map. Abbrev: HE. Cf: *vertical interval.*

horizontal-field balance An instrument that measures the horizontal component of the magnetic field by means of the torque that the field component exerts on a vertical permanent magnet. Cf: *torsion magnetometer; vertical-field balance.*

horizontal fold *nonplunging fold.*

horizontal form index A term used by Bucher (1919, p.154) to express the degree of asymmetry of a current ripple, defined as the ratio of the horizontal length of the steep (downcurrent) side to that of the gentle (upcurrent) side. Twenhofel (1950, p.568) used the ratio of the length of the upcurrent side to that of the downcurrent side. Allen (1963) used the ratio of the span (crest length) to the *chord* (ripple wavelength). Cf: *vertical form index.* See also: *ripple symmetry index.*

horizontal inclined fold A fold with a horizontal axis and an inclined axial plane (Turner and Weiss, 1963, p.119).

horizontal intensity The horizontal component of the vector magnetic-field intensity; it is one of the *magnetic elements,* and is symbolized by H. Cf: *vertical intensity.*

horizontal lamination A flat, horizontal layer of sand or silt that ranges from a few grain diameters to several millimeters thick.

horizontal-loop method An electromagnetic method in which the planes of the transmitting and receiving coils are horizontal.

horizontal separation In faulting, the distance between the two parts of a disrupted unit (e.g. bed, vein, or dike), measured in any specified horizontal direction.

horizontal slip In a fault, the horizontal component of the *net slip.* Cf: *vertical slip.* Syn: *horizontal dip slip.*

horizontal upright fold A fold with a horizontal axis and vertical axial plane (Turner and Weiss, 1963, p.119).

hormite (hor'-mite) A group name suggested (but not approved) for the sepiolite and palygorskite clay minerals.

hormogonium (hor-mo-go'-ni-um) A multicellular trichome of a filamentous blue-green alga, which may become fragmented at random or at a *heterocyst.*

horn [geog] (a) A body of land (such as a spit), or of water, shaped like a horn. (b) The pointed end of a dune or beach cusp; e.g. the forward, outer end of a barchan crescent. Syn: *wing [geomorph].*

horn [glac geol] A high rocky sharp-pointed mountain peak with prominent faces and ridges, bounded by the intersecting walls of three or more cirques that have been cut back into the mountain by headward erosion of glaciers; e.g., the Matterhorn of the Pennine Alps. Cf: *tind; arête.* Syn: *matterhorn; cirque mountain; pyramidal peak; monumental peak; horn peak.*

hornblende (horn'-blende) (a) The commonest mineral of the *amphibole* group: $(Ca,Na)_{2-3}(Mg,Fe^{+2},Fe^{+3},Al)_5(OH)_2[(Si,Al)_8O_{22}]$. It has a variable composition, and may contain potassium and appreciable fluorine. Hornblende is commonly black, dark green, or brown, and occurs in distinct monoclinic crystals or in columnar, fibrous, or granular forms. It is a primary constituent of many acid and intermediate igneous rocks (granite, syenite, diorite, andesite) and less commonly of basic igneous rocks, and it is a common metamorphic mineral in gneiss and schist. (b) A term sometimes used (esp. by the Germans) to designate the *amphibole* group of minerals. The term "Hornblende" is an old German name for any dark, prismatic crystal found with metallic ores but containing no valuable metal (the word "Blende" indicates "a deceiver"). Obsolete syn: *hornstone.*

hornblende gabbro In the *IUGS classification,* a plutonic rock satisfying the definition of *gabbro,* in which pl/(pl+hbl+px) is between 10 and 90, and px/(pl+hbl+px) is less than 5.

hornblende-hornfels facies (horn'-blende-horn'-fels) The facies (set of metamorphic mineral assemblages) in which mafic rocks are represented by hornblende+plagioclase, with epidote and almandine excluded (Turner, 1968). Pelitic assemblages consist of micas accompanied by andalusite, cordierite, or sillimanite; almandine and staurolite are uncommon, kyanite is absent. The facies is typical of the middle grades of thermal (contact) metamorphism and of low-pressure regional dynamothermal metamorphism of the Abukuma or Buchan type. Pressures are generally less than 400 MPa, temperatures in the range 400°-650°C. Cf: *albite-epidote-hornfels facies; amphibolite facies; cordierite-amphibolite facies; pyroxene-hornfels facies.*

hornblende peridotite In the *IUGS classification,* a plutonic rock with M equal to or greater than 90, ol/(ol+hbl+px) between 40 and 90, and px/(ol+hbl+px) less than 5.

hornblende pyroxenite In the *IUGS classification,* a plutonic rock with M equal to or greater than 90, ol/(ol+hbl+px) less than 5, and px/(px+hbl) between 50 and 90.

hornblende schist A schistose metamorphic rock consisting principally of hornblende.

hornblendite (horn'-blend-ite) (a) In the *IUGS classification,* a plutonic rock with M equal to or greater than 90 and hbl/(hbl+px+ol) greater than 90. (b) An igneous rock composed almost entirely of hornblende. The term has been equated incorrectly by some authors with the metamorphic rock *amphibolite.*

horn coral *solitary coral.*

Horner plot The oldest of methods of determining heat flow. See Bullard (1947) and Lachenbruch and Brewer (1959).

hörnesite (hör'-nes-ite) A white monoclinic mineral of the vivianite group: $Mg_3(AsO_4)_2{\cdot}8H_2O$. Its crystals resemble those of gypsum.

hornfels (horn'-fels) A fine-grained rock composed of a mosaic of equidimensional grains without preferred orientation and typically formed by *contact metamorphism.* Porphyroblasts or relict phenocrysts may be present in the characteristically *granoblastic* (or decussate) matrix (Winkler, 1967).

hornfels facies A loosely defined term used to denote the physical conditions involved, or the set of mineral assemblages produced, by thermal (contact) metamorphism at relatively shallow depths in the Earth's crust. It encompasses the *albite-epidote-hornfels facies,* the *hornblende-hornfels facies,* the *pyroxene-hornfels facies,* and the *sanidinite facies.*

hornito (hor-ni'-to) A small mound of spatter built on the back of a lava flow (generally pahoehoe), formed by the gradual accumulation of clots of lava ejected through an opening in the roof of an underlying lava tube. Syn: *driblet cone.*

horn lead *phosgenite.*

horn mercury *calomel.*

horn peak *horn [glac geol].*

horn quicksilver *calomel.*

horn silver *chlorargyrite.*

hornstein (horn'-stein) A syn. of *hornstone [rock].* Etymol: German "Hornstein", "chert".

hornstone [mineral] (horn'-stone) (a) A compact, flinty, brittle variety of chalcedony; "a siliceous mineral substance, sometimes approaching nearly to flint, or common quartz" (Lyell, 1854, p.807). (b) An obsolete name formerly applied to *hornblende.*

hornstone [rock] A general term for a compact tough siliceous rock having a splintery or subconchoidal fracture (Holmes, 1920). It has been used to describe flint or chert as well as hornfels, and has also been confused with hornblende. The term should be abandoned. Syn: *hornstein.*

horny sponge (horn'-y) Any demosponge that possesses spicules but has a skeleton composed of spongin.

horotely (hor'-o-tel'-y) A phylogenetic phenomenon characterized by a normal or average rate of evolution. Cf: *bradytely; tachytely.*

horse [coal] *horseback [coal].*

horse [ore dep] A miners' term for a barren mass of country rock occurring within a vein.

horse [struc geol] A displaced block of rock that is bounded on all sides by faults.

horseback [coal] (horse'-back) (a) A syn. of *cutout.* (b) A bank or ridge of foreign matter in a coal seam. (c) A large roll in a coal seam. (d) A *clay vein* in a coal seam. Syn: *kettleback; horse [coal]; symon fault; washout [mining].*

horseback [glac geol] A low, sharp ridge of sand, gravel, or rock; specif. an *esker* or eskerlike deposit, or a *kame,* in northern New England, esp. Maine. Syn: *hogback [glac geol]; boar's back.*

horseflesh ore (horse'-flesh) Cornish syn. of *bornite.*

horse latitudes Zones of oceanic calms that coincide with the subtropical high-pressure systems at about 30-35°N and S, characterized by warmth and dryness. These belts move north and south about 5°, following the Sun, and are zones of divergence that mark the average region of origin of the *trade winds* and *prevailing westerlies.*

horseshoe (horse'-shoe) A topographic feature, such as a valley or a mountain range, shaped like a horseshoe.

horseshoe bend An *oxbow* in the course of a stream.

horseshoe dissepiment One of a single vertical series of dissepiments of a rugose coral, characterized by a horizontal base and a strongly arched top part. Cf: *lateral dissepiment.*

horseshoe dune *barchan.*

horseshoe flute cast *current crescent.*

horseshoe lake A lake occupying a horseshoe-shaped basin; specif. an *oxbow lake.*

horseshoe moraine A terminal moraine, markedly convex on the down-valley side, usually formed at the end of a valley glacier that never advanced beyond the mountain front.

horseshoe reef A horseshoe-shaped reef that develops from a reef pinnacle or *table reef* parallel to the dominant wave action on a platform reef. Its cusps extend downwind and its interior parts often become densely vegetated to produce a small wooded island of low relief.

horsetail [bot] (horse'-tail) *sphenopsid.*

horsetail [ore dep] adj. Said of a major vein dividing or fraying into smaller fissures; also, said of an ore comprising a series of such veins.

horsetail [struc geol] A fanlike array of fault or tensile fracture splays at the termination of a larger fault. Displacement on the main fault is distributed among the fault splays. See also: *splay [struc geol].*

horsetailing (horse'-tail-ing) A feathery or frondlike fluting, grooving, or other structure developed on the surfaces of shatter-coned rocks by the distinctive striations that radiate from the apex of each *shatter cone* and extend along its length. The presence of multiple nested and parasitic shatter cones produces a distinctive horsetail-like effect.

horsfordite (hors'-ford-ite) A doubtful silver-white mineral: Cu_5Sb. It may be an unstable high-temperature phase.

horst An elongate block that is bounded on both sides by normal faults that dip away from one another. It is a structural form and may or may not be expressed geomorphologically. Etymol: German. Cf: *graben.*

hortonolite (hor-ton'-o-lite) A mineral variety of the *olivine* group: $(Fe,Mg,Mn)_2SiO_4$. It is a variety of fayalite containing magnesium and manganese.

horváthite-(Y) A colorless to pale beige orthorhombic mineral: $NaY(CO_3)F_2$.

hoshiite (ho-shi'-ite) A nickel-containing variety of magnesite.

host A rock or mineral that is older than rocks or minerals introduced into it or formed within or adjacent to it, such as a *host rock,* or a large crystal with inclusions of smaller crystals of a different mineral species; a *palasome.* Ant: *guest.*

host element An essential element replaced by a *guest element* in a mineral.

hostile ice (hos'-tile) A submariner's term for an ice canopy containing no large skylights or other features that permit a submarine to surface (U.S. Naval Oceanographic Office, 1968, p. B33). Ant: *friendly ice.*

host rock A body of rock serving as a *host* for other rocks or for mineral deposits; e.g. a pluton containing xenoliths, or any rock in which ore deposits occur. It is a somewhat more specific term than country rock.

hot As related to radioactivity, said of a highly radioactive substance.

hot brine Warm and very saline water such as is found on the bottom of the Red Sea. Temperature may be as high as 50°C, and salinities may be on the order of 7-8 times average seawater salinity (35).

hot desert An arid area where the mean annual temperature is higher than 18°C (64.4°F) (Stone, 1967, p.230).

hot dry rock A potential source of heat energy within the Earth's crust: rocks at depths less than 10 km and at temperatures above 150°C. They are related to two types of heat source: igneous magmas, and conduction from the Earth's deeper interior. Abbrev: HDR.

Hotelling's T^2 In *geostatistics,* a generalization of *Student's t* distribution to the multivariate case, used in tests of the equalities of multivariate means. Named for Harold Hotelling (1895-1973), American statistician.

hot lahar A flow of hot volcanic materials down the slope of a volcano, produced by heavy rains after an eruption or by intrusion of lava into ice, snow, or water-saturated soil. Cf: *cold lahar.* Syn: *hot mudflow.*

hot mudflow *hot lahar.*

hotsonite (hot'-son-ite) A silky white triclinic mineral: $Al_5(PO_4)(SO_4)(OH)_{10}{\cdot}8H_2O$.

hot spot A volcanic center, 100 to 200 km across and persistent for at least a few tens of millions of years, that is thought to be the surface expression of a persistent rising *plume* of hot mantle material. Hot spots may occur in the interior of a plate. They are not linked with volcanic arcs, and may or may not lie along oceanic ridges. Some 200 late Cenozoic hot spots have been identified (Cloud, 1974, p.879). The largest of these is the Hawaiian hot spot. See also: *melting spot.*

hot-spot track A ridge of volcanic rock formed when a lithospheric plate moves over a *hot spot;* the Hawaiian Ridge is the type example. The active hot-spot volcano lies at the end of the track. Extinct volcanoes lie along the track, with the oldest extinct volcano farthest from the active hot spot.

hot spring A *thermal spring* whose temperature is above that of the human body (Meinzer, 1923, p.54). Cf: *warm spring.*

hot-spring type A *mineral-deposit type* characterized by silica or silica-carbonate sinter deposits and underlying breccias, fractures, and faults mineralized by gold, silver, and/or antimony.

hot-water reservoir *liquid-dominated reservoir.*

hot working Plastic deformation by dislocation glide and climb without significant work hardening. Characteristic of high temperature deformation.

hour angle Angular distance of a body west of the projection of a meridian onto the celestial sphere. Local hour angle is the angle between a body and the projection of the observer's meridian onto the celestial sphere. Greenwich hour angle is with respect to the projection of the Greenwich meridian. Sidereal hour angle is angular distance west (unlike the others which are measured east) of the vernal equinox; it is thus the supplement of right ascension.

hour circle Any great circle on the celestial sphere whose plane is perpendicular to the plane of the celestial equator. Syn: *circle of declination; circle of right ascension.* See also: *celestial meridian.*

hourglass structure (hour'-glass) A type of zoning, especially common in clinopyroxenes and chloritoids, in which a "core", distinguished from the outer part by a difference of color or optical properties, has a cross section resembling that of an hourglass.

hourglass valley (a) A valley whose pattern in plan view resembles an hourglass; e.g. a valley extending without interruption across a former divide, toward which it narrows from both directions (Engeln, 1942, p.377). (b) *wineglass valley.*

hover (hov'-er) A floating island of vegetation.

how An English term for a low, small hill in a valley or dale; a mound or hillock.

howardevansite (how'-ard-ev'-ans-ite) A metallic black or deep red-brown triclinic mineral: $NaCuFe^{3+}_2(VO_4)_3$.

howardite (how'-ard-ite) A group of brecciated meteorites predominantly containing clasts of eucrite and diogenite material; thought to have been produced during impact events on the parent body of those meteorite groups. See also: *HED meteorite; eucrite [meteorite]; diogenite.*

howe (a) A Scottish term for a hollow or depression, esp. one on the Earth's surface, as a basin or a valley. (b) *hoe.*

howieite (how'-ie-ite) A dark-green to black triclinic mineral: $Na(Fe^{2+},Mn,Fe^{3+},Al)_{12}Si_{12}O_{34}(O,OH)_{10}$.

howlite (how'-lite) A white nodular or earthy monoclinic mineral: $Ca_2B_5SiO_9(OH)_5$.

Hoxnian (Hox'-ni-an) British climatostratigraphic stage: Middle Pleistocene (above Anglian, below Wolstonian). Interglacial; probably correlative with Holsteinian.

hoya A very large *closed depression.* The term is used in Puerto Rico for *doline,* in Cuba for *polje.* Etymol: Spanish.

HPC/APC Hydraulic piston corer/advanced (hydraulic) piston corer; coring devices used in the Deep Sea Drilling and Ocean Drilling programs in which a core barrel is lowered inside the drill string, and hydraulically ejected into the sediment and then retrieved on a wire line.

HRAM *high-resolution aeromagnetics.*

HRTEM High resolution transmission electron microscopy, a technique for obtaining nanometer resolution images using the *transmission electron microscope.*

hsianghualite (hsiang'-hua-lite) A white cubic mineral: $Li_2Ca_3Be_3(SiO_4)_3F_2$.

huangho deposit (huang-ho') A general term applied by Grabau (1936, p.253) to a coastal-plain deposit that consists of alluvium that is spread out over a level surface (as a flood plain or a delta) above the normal reach of the sea but that passes laterally into marine beds of equivalent age. Type locality: the loess-derived alluvial deposits at the mouth of the Huang Ho (Yellow River) in northern China. See also: *shantung.*

huanghoite-(Ce) A yellow to yellowish-green rhombohedral mineral: $BaCe(CO_3)_2F$.

huangite A white to pale yellowish rhombohedral mineral of the *alunite* group: $Ca_{0.5}Al_3(SO_4)_2(OH)_6$.

hub The cylindrical or hemispherical projection on the central part (and usually on the lower surface) of a *wheel* of a holothurian.

Hubbert potential Fluid potential that is equal to the sum of a gravity component and a pressure component. Occurs in groundwater where temperature and chemical concentration are constant in space. It was demonstrated by M. King Hubbert, who showed that fluid potential is mechanical energy per unit mass of water.

Hubble constant (Hub'-ble) The amount by which the distance to a galaxy must be multiplied in order to get its velocity of recession or red shift: 75 km sec^{-1} 10^{-6} PSC^{-1}. It represents the present expansion rate of the Universe. It has the unit of reciprocal time and would represent the age of the Universe were there no gravitational deceleration.

hubeite A light to dark brown triclinic mineral: $Ca_2Mn^{2+}Fe^{3+}Si_4O_{12}(OH){\cdot}2H_2O$.

hübnerite (hüb'-ner-ite) A yellowish brown, brownish-red to black monoclinic mineral of the *wolframite* group: $MnWO_4$. It is isomorphous with ferberite, and may contain up to 20% iron tungstate. Also spelled: *huebnerite.*

Hudsonian Orogeny (Hud-son′-i-an) A name proposed by Stockwell (1964) for a time of plutonism, metamorphism, and deformation during the Precambrian in the Canadian Shield (especially in the Churchill, Bear, and Southern provinces), dated isotopically as between 1640 and 1820 m.y. ago.

hudsonite (hud′-son-ite) A syn. of *cortlandtite,* nonpreferred since "hudsonite" had been used earlier for a variety of pyroxene. Obsolete.

hue The attribute of color associated with wavelength or dominant spectral color. In the *IHS* system, represents the dominant wavelength of a color.

huebnerite *hübnerite.*

huemulite (hue′-mu-lite) A reddish-orange triclinic mineral: $Na_4MgV^{5+}{}_{10}O_{28}{\cdot}24H_2O$.

huerfano (huer′-fa-no [ware′-fa-no]) A term used in the SW U.S. for a hill or mountain of older rock entirely surrounded, but not covered, by any kind of later sedimentary material; esp. a solitary eminence separated by erosion from the mass of which it once formed a part. Etymol: Spanish "huérfano", "orphan". Cf: *lost mountain; tejon.*

hügelite (hü′-gel-ite) A brown to orange-yellow monoclinic secondary mineral: $Pb_2(UO_2)_3O_2(AsO_4)_2{\cdot}5H_2O$. It is the arsenate analogue of dumontite.

Hugoniot curve The locus of all shock states that can be achieved by shock-wave compression of variable intensity in any specific material, commonly expressed in the pressure-volume or pressure-particle velocity space. Named after Pierre Henri Hugoniot (1851-1887), French physicist.

hühnerkobelite (hühn′-er-ko′-bel-ite) *ferroalluaudite.*

hullite (hull′-ite) A soft, black, waxy-appearing aluminosilicate of ferric iron, magnesium, calcium, and alkalies, occurring as interstitial matter and amygdaloidal infillings in certain basalts. It is perhaps identical with chlorophaeite.

hulsite (huls′-ite) A submetallic black monoclinic mineral: $(Fe^{2+},Mg)_2(Fe^{3+},Sn)BO_5$.

hum A karst *inselberg*. A residual hill of limestone on a fairly level floor, such as the isolated hills of limestone in *poljes*. In some tropical areas, used loosely as a synonym for *mogote*. Etymol: from a town of that name in Yugoslavia.

human paleontology (hu′-man) That branch of paleontology concerned with the fossil record and evolution of early humans and their immediate predecessors. Syn: *paleoanthropology.*

humanthracite (hum-an′-thra-cite) Humic coal of anthracitic rank; it is the highest stage in the *humolith series.* Cf: *sapanthracite.*

humanthracon (hum-an′-thra-con) Humic coal of bituminous rank; it is the fifth stage in the *humolith series.* Cf: *sapanthracon.*

humate (hu′-mate) A salt or ester of *humic acid.*

humatipore (hu-mat′-i-pore) In cystoids, an exothecal dipore in a single calyx plate having compound calcified canals. These canals carry coelomic fluids for respiration and are found in the rhombiferan family Holocystitidae (Paul, 1972, p.7).

humatirhomb (hu-mat′-i-rhomb) In rhombiferan echinoderms, an exothecal pore structure having calcified thecal canals extending across the suture between adjacent calyx plates. These canals carry coelomic fluids for respiration and are found in the rhombiferan superfamily Caryocystitidae (Paul, 1972, p.2).

humberstonite (hum′-ber-ston-ite′) A colorless rhombohedral mineral: $K_3Na_7Mg_2(SO_4)_6(NO_3)_2{\cdot}6H_2O$. Syn: *Chile-loeweite.*

humboldtine (hum′-boldt-ine) A resinous yellow or amber monoclinic oxalate mineral: $Fe^{2+}C_2O_4{\cdot}2H_2O$. It occurs in capillary or botryoidal forms in brown coal and black shale. Syn: *humboldtite; oxalite.*

humboldtite (hum′-boldt-ite) (a) *datolite.* (b) *humboldtine.*

humic (hu′-mic) Pertaining to or derived from *humus.*

humic acid The dark-colored organic material that can be extracted from soil, low-rank coals, and other decayed plant substances with dilute alkali and other reagents and that is precipitated by acidification to pH 1 to 2. It is insoluble in acids and organic solvents.

humic-cannel coal (hu′-mic-can′-nel) *pseudocannel coal.*

humic coal Coal that is derived from peat by the process of *humification.* Most coal is of this type. Cf: *sapropelic coal.* See also: *humolith series.* Syn: *cahemolith; chameolith; chaemolith; humulith; humus coal; humulite; humolite; humite [coal]; humolith.*

humic decomposition Chemical breakdown of rocks and minerals by the action of vegetable acids.

humic degradation matter Organic degradation matter that is cellulosic and in which the individual particles are still recognizable; it is similar to *anthraxylon.* It is classified according to the type of constituent plant material. See also: *translucent humic degradation matter.* Syn: *cell-wall degradation matter; brown matter.* Abbrev: HDM.

Humic Gley soil Soil of the intrazonal order and hydromorphic suborder that includes Wisenboden and related soils, such as Half-Bog soils, which have a thin muck or peat O2 (Oi) horizon and an A1 (A) horizon. Developed in wet meadow and in forested swamps. Not used in current U.S. system of soil taxonomy.

humidity (hu-mid′-i-ty) The water-vapor content of the atmosphere. The unmodified term often signifies *relative humidity.* See also: *mixing ratio; absolute humidity; specific humidity.*

humification (hu′-mi-fi-ca′-tion) The process whereby the carbon of organic residues is transformed and converted to humic substances through biochemical and abiotic processes.

humin (hu′-min) *ulmin.*

huminite (hu′-mi-nite) (a) A group of macerals in brown coal, consisting of humic matter derived mainly from lignin and cellulose (ICCP, 1971). The group is distinguished microscopically by having a middle level of reflectance and a brownish yellow to reddish brown color in transmitted light. It is the precursor of the *vitrinite* group in bituminous coals. (b) A variety of oxidized bitumen, resembling brown coal, found in a granite-pegmatite vein in Sweden.

humite [coal] (hu′-mite) *humic coal.*

humite [mineral] (a) A white, yellow, brown, or red orthorhombic mineral: $Mg_7(F,OH)_2[SiO_4]_3$. It sometimes contains appreciable iron, and it is found in the masses ejected from volcanoes. (b) A group of homologous magnesium-silicate minerals frequently containing fluorine and closely resembling one another in chemical composition, physical properties, and crystallization. It consists of olivine, humite, clinohumite, chondrodite, and norbergite, and occurs as skarns in high grade marbles.

hummerite (hum′-mer-ite) A bright orange triclinic mineral: $KMgV^{5+}{}_5O_{13}{\cdot}8H_2O$.

hummock [geomorph] (hum′-mock) A rounded or conical knoll, mound, hillock, or other small elevation. Also, a slight rise of ground above a level surface. Syn: *hammock.*

hummock [ice] A mound, hillock, or pile of broken floating ice, either fresh or weathered, that has been forced upward by pressure, as in an ice field or ice floe. Cf: *bummock.* Syn: *ice hummock.*

hummock [peat] Microtopographic elevations usually occurring in ombrotrophic peatlands, consisting of a peat core generally topped with vegetation. Smaller than a *peat mound.*

hummock [permafrost] A small irregular knob of earth or turf. Neither *earth hummock* nor *turf hummock* is diagnostic of permafrost, but both are most common in subpolar and alpine regions. Both require vegetation (Washburn, 1973, p.126).

hummock [volc] Characteristic topographic feature for debris avalanche deposits. The shape of hummocks is variable and irregular.

hummocked ice (hum′-mocked) Sea ice having a rugged, uneven surface due to the formation of hummocks; it has the appearance of smooth hillocks when weathered. A form of *pressure ice.*

hummocking (hum′-mock-ing) Pressure process by which floating ice becomes broken up into hummocks.

hummocky (hum′-mock-y) Said of topographic land or ice forms that are abounding in small hills and depressions meters to tens of meters across (hummocks), such as a hummocky dune or *hummocked ice.*

hummocky cross-stratification A type of cross-stratification in which lower bounding surfaces of sets are erosional and commonly slope at angles less than 10°, though dips can reach 15°; laminae above these erosional set boundaries are parallel to that surface, or nearly so; the hummocks range in scale from 1-5m in diameter. Laminae can systematically thicken laterally in a set so that their traces on a vertical surface are fan-like and dip diminishes regularly; and the dip directions of erosional set boundaries and of the overlying laminae are scattered (Harms et al., 1975, p.87). Abbrev: HCS. Cf: *truncated wave-ripple lamina.*

hummocky moraine An area of *knob-and-kettle topography* that may have been formed either along an active ice front or around masses of stagnant ice (Gravenor and Kupsch, 1959, p.52).

humocoll (hu′-mo-coll) Humic material of the rank of peat; it is the second stage in the *humolith series.* Cf: *saprocol.*

humodil (hu′-mo-dil) Humic coal of lignitic rank; it is the third stage in the *humolith series.* Cf: *saprodil.*

humodite (hu′-mo-dite) Humic coal of subbituminous rank; it is the fourth stage in the *humolith series.* Cf: *saprodite.*

Humods Spodosols that have accumulated organic carbon and aluminum, but not iron, in the upper part of the spodic horizon. Humods are rarely saturated with water or do not have characteristics associated with wetness. A suborder in the U.S. system of soil taxonomy.

humodurite (hu-mo-dur′-ite) *translucent attritus.*

humogelite (hu-mo-gel′-ite) *ulmin.*

humolite (hu′-mo-lite) *humic coal.*

humolith (hu′-mo-lith) *humic coal.*

humolith series Humic material and coals in order of metamorphic rank: *humopel, humocoll, humodil, humodite, humanthracon* and *humanthracite* (Heim and Potonié, 1932, p.146). Cf: *sapropelite series; humosapropelic series; saprohumolith series.* See also: *humic coal.*

humonigritite (hu-mo-nig′-ri-tite) A type of *nigritite* that occurs in sediments. Cf: *polynigritite; exinonigritite; keronigritite.*

humopel (hu′-mo-pel) Organic matter, or *ulmin,* of humic coals; it is the first stage in the *humolith series.* Cf: *sapropel.*

humosapropelic series (hu-mo-sap′-ro-pel-ic) Organic materials and coals intermediate between the *humolith series* and the *sapropelite series,* with humolithic materials predominating. Cf: *saprohumolith series.*

humosite (hu′-mo-site) A dark brownish-red microscopic constituent of torbanite, translucent and isotropic (Dulhunty, 1939).

humovitrinite (hu-mo-vit'-ri-nite) Vitrinite in vitrain of humic coal. Cf: *saprovitrinite.*

humpy (hump'-y) A small morainal mound with a central depression (Gravenor and Kupsch, 1959, p.53).

humulite (hu'-mu-lite) *humic coal.*

humulith (hu'-mu-lith) *humic coal.*

Humults Ultisols that have a high content of organic carbon. Humults are not saturated with water for periods long enough to limit their use for most crops. A suborder in the U.S. system of soil taxonomy.

humus (hu'-mus) The generally dark, more or less stable part of the organic matter of the soil, so well decomposed that the original sources cannot be identified. The term is sometimes used incorrectly for the total organic matter of the soil, including relatively undecomposed material. Syn: soil ulmin.

humus coal *humic coal.*

hunchunite A metallic cubic mineral found in a placer deposit: Au_2Pb.

hungchaoite (hung-chao'-ite) A white triclinic mineral: $MgB_4O_5(OH)_4{\cdot}7H_2O$.

hungry (hun'-gry) (a) Said of a rock, lode, or belt of country that is barren of ore minerals or of geologic indications of ore, or that contains very low-grade ore. Ant: *likely.* (b) Said of a soil that is poor or not fertile.

huntite (hunt'-ite) A chalky white rhombohedral mineral: $CaMg_3(CO_3)_4$.

Hunt-Wentworth recording micrometer (Hunt-Went'-worth) An instrument used for petrographic modal analysis. A series of linear traverses across a thin section are made. The traverse lengths of the various mineral grains are recorded. With a sufficient number of traverses, the total linear distance traversed for a given mineral when divided by the total distance traversed is proportional to its volume in the rock (Heinrich, 1956, p.8). Cf: *Chayes point counter.*

huréaulite (hu'-réau-lite) A monoclinic mineral: $Mn^{2+}_5(PO_4)_2(PO_3OH)_2{\cdot}4H_2O$. It occurs in yellow, reddish, or gray prismatic crystals, or in massive form. Syn: *bastinite; palaite.*

hurlbutite (hurl'-but-ite) A white to greenish white monoclinic mineral: $CaBe_2(PO_4)_2$. It is structurally related to the feldspars.

Huronian (Hu-ro'-ni-an) An obsolete geochronometric subdivision of the *Precambrian,* covering the time span from 2,450 to 2,200 Ma (Harland et al., 1990), therefore approximately equivalent to the *Siderian* Period (earliest *Proterozoic* Eon) of the Standard Global Geochronometric Scale.

hurricane (hur'-ri-cane) A *tropical cyclone,* esp. in the North Atlantic and eastern North Pacific basins, in which the sustained near-surface wind speed equals or exceeds 64 knots (73 mph). See also: *typhoon.*

hurricane delta A delta formed by storm waves carrying sand across a reef or barrier island and depositing it in a lagoon. Cf: *washover fan.*

hurricane surge *storm surge.*

hurricane tide (a) *storm surge.* (b) The height of a storm surge above the astronomically predicted level of the sea.

hurst (a) A wooded knoll, hill, or other small eminence; a grove or a thick wood; a copse; *coppice.* (b) A bank or piece of rising ground; esp. a sandbank in or along a river. The term is very common in place names. Also spelled: *hirst; hyrst.*

husebyite (hu-se'-by-ite) A plagioclase-bearing *nepheline syenite,* named by Brögger in 1933 for Huseby, Oslo district, Norway. Obsolete.

hushing (hush'-ing) *hydraulic prospecting.*

hutchinsonite (hutch'-in-son-ite') A scarlet to deep cherry-red orthorhombic mineral: $(Pb,Tl)_2As_5S_9$.

Huttenlocher intergrowth Plagioclase feldspar consisting of alternating exsolution lamellae of $An_{\sim100}$ and $An_{\sim65}$, where An_x represents percent anorthite composition ($CaAl_2Si_2O_8$) and the implied remainder is albite composition ($NaAlSi_3O_8$). Cf: *Bøggild intergrowth; Peristerite intergrowth.*

Huttonian (Hut-to'-ni-an) Of or relating to James Hutton (1726-1797), Scottish geologist, who proposed a teleological theory of renovation of the Earth to maintain habitable landscapes in the face of obvious decay by erosion. He invoked the principle of uniformity of causes through time, emphasized the indefinite length of geologic time, and developed the theory of *plutonism,* which associated "plutonic" with molten rocks solidified underground. Hutton introduced the concept of a dynamic Earth driven by heat, which challenged the prevailing *neptunian* theory of a static, solid Earth and a receding ocean. His theory of renovation also introduced the *orogenic cycle* (Bailey, 1967; Dean, 1992; 1998). Cf: *neptunism.*

huttonite (hut'-ton-ite) A colorless to pale-cream monoclinic mineral: $ThSiO_4$. It is dimorphous with thorite and isostructural with monazite.

Huygens' principle (Huy'-gens) The statement that any particle excited by wave energy becomes a new point source of wave energy. It is named after the Dutch astronomer and mathematician Christian Huygens (d.1695).

HW *high water.*

H wave *hydrodynamic wave.*

hyacinth (hy'-a-cinth) (a) A transparent orange, red, or reddish-brown variety of *zircon,* sometimes used as a gem. The term has been used interchangeably with "jacinth", and loosely to signify any zircon. (b) Yellow, orange, or brownish essonite used as a gem. (c) A term applied as a syn. of various orange-red to orange minerals, and also of minerals such as harmotome, vesuvianite, and meionite. (d) A precious stone believed by the ancients to be the sapphire.

hyaline [ign] (hy'-a-line) A syn. of *glassy;* sometimes used as a prefix ("hyalo-") to names of volcanic rocks with a glassy texture; e.g., "hyalobasalt".

hyaline [mineral] Said of a mineral that is amorphous.

hyaline [paleont] Said of the glassy clear or transparent outer wall of a foraminifer.

hyaline element A conodont element that lacks *white matter* except, perhaps, along thin growth axes in cusp or denticles; contrasts with albid element; typical fibrous elements are also hyaline, but it is not clear that all hyaline elements are fibrous in structure (TIP, 1981, pt. W, supp.2).

hyalite (hy'-a-lite) A colorless variety of common opal that is sometimes clear as glass and sometimes translucent or whitish and that occurs as globular concretions (resembling drops of melted glass) or botryoidal crusts lining cavities or cracks in rocks. Syn: *water opal; Müller's glass.*

hyalithe (hy'-a-lithe) An opaque glass resembling porcelain and frequently black, green, brown, or red.

hyalo- (hy'-a-lo-) A prefix meaning "glassy". Cf: *vitr-.*

hyaloclastite (hy'-a-lo-clas'-tite) A deposit formed by the flow or intrusion of lava or magma into water, ice, or water-saturated sediment, and its consequent granulation or shattering into small angular fragments. Also includes vitric tuff from shallow-water explosive volcanism or explosive interaction of magma and groundwater. Syn: *aquagene tuff.* Cf: *palagonite tuff.*

hyalocrystalline (hy'-a-lo-crys'-tal-line) Said of the texture of a porphyritic igneous texture in which phenocryst crystals and glassy groundmass are equal or nearly equal in volumetric proportions, the ratio of phenocrysts to groundmass being between 5:3 and 3:5 (Cross et al., 1906, p.694). Obsolete. Cf: *intersertal.* Syn: *semicrystalline.*

hyalomylonite (hy'-a-lo-my'-lo-nite) A glassy rock formed by fusion of granite, arkose, etc. by frictional heat in zones of intense differential movement. Cf: *buchite.*

hyalophane (hy-al'-o-phane) Probably an unnecessary name for a colorless monoclinic mineral of the *feldspar* group intermediate in composition between celsian and orthoclase.

hyalophitic (hy'-a-lo-phit'-ic) Igneous texture in which glass or cryptocrystalline material occurs between, and partly encloses the ends of, feldspar microlites. Cf: *intersertal.*

hyalopilitic (hy-a'-lo-pi-lit'-ic) Igneous texture in which glass occurs between randomly-oriented microlites of feldspar. Cf: *hyalophitic; sporophitic.*

hyalosiderite (hy'-a-lo-sid'-er-ite) A rich olive-green variety of olivine containing considerable iron (30-50 mole percent of Fe_2SiO_4).

hyalosponge (hy'-a-lo-sponge') Any sponge belonging to the class Hyalospongea, characterized chiefly by a skeleton composed of six-rayed siliceous spicules, without calcium carbonate or spongin. Syn: *hexactinellid; glass sponge.*

hyalotekite (hy-al'-o-tek'-ite) A white or gray triclinic (pseudo-monoclinic) mineral: $(Ba,Pb,Ca)_6(B,Si,Al)_2(Si,Be)_{10}O_{28}(F,Cl)$.

hyalotuff Pyroclastic rocks generated by phreatomagmatic explosions when volcanoes erupt in shallow water (Honnorez and Kirst, 1975). The term is applied mainly to the tuffs of nearshore tuff rings.

hybrid [evol] (hy'-brid) An individual having parents belonging to different species.

hybrid [petrology] Rock, or pertaining to such a rock, whose chemical composition is the result of *assimilation.* Syn: *contaminated.* See also: *hybridization.*

hybrid age The radiometric age given by an isotopic system that has partially lost radiogenic isotopes owing to thermal, igneous, or tectonic activity some time after the start of the isotopic system. See also: *overprint [geochron]; mixed ages.*

hybrid computer A computer with combined capabilities of the *analog computer* and *digital computer.*

hybridism (hy'-brid-ism) The process or the result of *hybridization.*

hybridization (hy'-brid-i-za'-tion) Process whereby rocks of composition different from that of the parent magma are formed by *assimilation.* Cf: *contamination [ign].* Syn: *hybridism.*

hybrid shear fractures Shear fractures which form by opening upon shear offset. They are usually characterized by a small conjugate angle ($< 15°$) on the Mohr circle diagram (Hancock, 1985). Theoretically this type of brittle fracture does not exist because in-plane propagation does not occur under combined Mode I and Mode II loading (Engelder, 1999).

hydnophorid (hyd-noph'-o-rid) Said of a scleractinian *corallum* with *corallite* centers arranged around protuberant collines or monticules (TIP, 1956, pt. F, p.248).

hydrarch (hy'-drarch) adj. Said of an ecologic succession (i.e. a *sere*) that develops under *hydric* conditions. Cf: *mesarch; xerarch.* See also: *hydrosere.*

hydrargillite (hy-drar'-gil-lite) (a) *gibbsite.* (b) A name that has been applied to various aluminum-bearing minerals, including aluminite, wavellite, and turquoise.

hydrate (hy'-drate) n. A mineral compound that is produced by hydration, or one in which water is part of the chemical composition. v. To cause the incorporation of water into the chemical composition of a mineral.

hydrated halloysite (hy'-drat-ed) *endellite.*

hydration reaction (hy-dra'-tion) A metamorphic reaction that results in the formation of hydrous minerals. Cf: *dehydration reaction.*

hydration rind dating *obsidian hydration dating.*

hydration shattering The process of grain loosening and rock disintegration by the wedging pressure of water in films of varying thicknesses on silicate mineral surfaces. This water is drawn between the grains by electro-osmosis and exerts differential pressures up to 2,000 kg/cm^2, strong enough to loosen and separate the grains. Such a process may be significant in all climates, without the aid of freezing and thawing. It produces loosened and separated grains, the accumulation being *grus.*

hydration water *water of hydration.*

hydraulic [eng] (hy-drau'-lic) Conveyed, operated, effected, or moved by means of water or other fluids, such as a "hydraulic dredge" using a centrifugal pump to draw sediments from a river channel.

hydraulic [hydraul] Pertaining to a fluid in motion, or to movement or action caused by water.

hydraulic [materials] Hardening or setting under water; e.g. "hydraulic lime" or "hydraulic cement".

hydraulic action The mechanical loosening and removal of weakly resistant material solely by the pressure and *hydraulic force* of flowing water, as by a stream surging into rock cracks or impinging against the bank on the outside of a bend, or by ocean waves and currents pounding the base of a cliff. See also: *fluviraption.*

hydraulic capacitance *specific storage.*

hydraulic conductivity The volume of water at the existing kinematic viscosity that will move in a porous medium in unit time under a unit hydraulic gradient through a unit area measured at right angles to the direction of flow (Lohman, et al., 1972, p.4). In contrast to permeability, it is a function of the properties of the liquid as well as of the porous medium. The term replaces *field coefficient of permeability.* Cf: *permeability coefficient.*

hydraulic current A local current produced by differences in water level at the two ends of a channel, set up by the rising and falling tide at constrictions in a baymouth or in the narrow strait connecting two bodies of water having tides that differ in time or range; e.g. in Hell Gate, where Long Island Sound joins the East River, N.Y.

hydraulic diffusivity The property of an aquifer or confining bed defined as *hydraulic conductivity* divided by *specific storage,* or *transmissivity* divided by *storativity.* It is the conductivity of the saturated medium when the unit volume of water moving is that which is involved in changing the head a unit amount in a unit volume of the medium (Lohman et al., 1972).

hydraulic element A quantity pertaining to a particular stage of flowing water in a particular cross section of a conduit or stream channel, e.g. depth of water, cross-sectional area, hydraulic radius, wetted perimeter, mean depth of water, velocity, energy head, friction factor (ASCE, 1962).

hydraulic fill Earth or waste material that has been excavated, transported, and flushed into place by moving water.

hydraulic-fill dam A dam composed of *hydraulic fill,* in which the sorting of particle sizes into an impervious central core supported by outer zones of coarser material is accomplished through arrangement of peripheral discharge outlets and flow in the central pool. Cores in dams placed by hydraulic-fill methods are called puddled-clay cores.

hydraulic force The eroding and shearing force of flowing water, involving no sediment load and resulting in *hydraulic action.*

hydraulic fracturing A general term, for which there are numerous trade or service names, for the fracturing of rock in an oil or gas reservoir by pumping in water (or other fluid) and sand under high pressure. The purpose is to produce artificial openings in the rock in order to increase permeability. The pressure opens cracks and bedding planes, and sand introduced into these serves to keep them open when pressure is reduced. A natural hydraulic fracture is driven by natural high fluid pressures. Syn: *fracturing; hydrofracturing.*

hydraulic friction The resistance to flow exerted on the perimeter or contact surface between a stream and its containing conduit, due to the roughness characteristic of the confining surface, which induces a loss of energy. Energy losses arising from excessive turbulence, impact at obstructions, curves, eddies, and pronounced channel changes are not ordinarily ascribed to hydraulic friction (ASCE, 1962).

hydraulic geometry The description, at a given cross section of a river channel, of the graphical relationships among plots of hydraulic characteristics (such as width, depth, velocity, channel slope, roughness, and bed particle size, all of which help to determine the shape of a natural channel) as simple power functions of river discharge (Leopold and Maddock, 1953).

hydraulic grade line In a closed channel, a line joining the elevations that water would attain in atmospheric pressure; in an open channel, the free water surface or *flow line [hydraul].* Its slope represents *energy loss.* See also: *hydraulic gradient; hydraulic head.*

hydraulic gradient (a) In an aquifer, the rate of change of *total head* per unit of distance of flow at a given point and in a given direction. Cf: *pressure gradient.* See also: *gradient of the head.* Syn: *potential gradient.* (b) In a stream, the slope of the *hydraulic grade line.* See also: *critical hydraulic gradient.* Syn: *gradient [hydraul].*

hydraulic head (a) The height of the free surface of a body of water above a given subsurface point. (b) The water level at a point upstream from a given point downstream. (c) The elevation of the *hydraulic grade line* at a given point above a given point of a pressure pipe.

hydraulic joints Joints that propagate in response to high pore pressures generated by compaction accompanying overburden loading, the release of water on clay dewatering, the release of gas on the maturation of hydrocarbons, or aquathermal pressuring (e.g. Gulf of Mexcio - relaxed basin) (Engelder, 1985). This is one form of natural hydraulic fracture.

hydraulic jump In fluid flow, a change in flow conditions accompanied by a stationary, abrupt turbulent rise in water level in the direction of flow. It is a type of stationary wave.

hydraulic limestone An impure limestone that contains silica and alumina (usually as clay) in varying proportions and that yields, upon calcining, a cement that will harden under water. See also: *cement rock.*

hydraulic loading rate In groundwater recharge, the average infiltration into a *recharge basin* expressed over time, including flooded, dry, and cleaning cycles for the basin.

hydraulic mean depth *hydraulic radius.*

hydraulic mining The recovery of desired material, generally in open pits, by means of strong jets of water. Cf: *placer mining.*

hydraulic plucking A process of stream erosion by which rock fragments are forcibly removed by the impact of water entering cracks in a rock. Syn: *quarrying [streams].*

hydraulic potential Fluid mechanical energy per unit mass of fluid which is equal to head times gravitational acceleration.

hydraulic profile A vertical section of the *potentiometric surface* of an aquifer.

hydraulic prospecting The use of water to clear away surficial deposits and debris to expose outcrops, for the purpose of exploring for mineral deposits. Syn: *hushing.*

hydraulic radius In a stream, the ratio of the area of its cross section to its *wetted perimeter.* Symbol: R. Syn: *hydraulic mean depth.* See also: *form ratio.*

hydraulic ratio A value expressing the quantity of any given heavy mineral in a sediment, equal to the weight of a heavy mineral in a given size class divided by the weight of light minerals in the hydraulic-equivalent class (Rittenhouse, 1943). The value is commonly multiplied by 100 to reduce the number of decimal places.

hydraulics (hy-drau'-lics) The aspect of engineering that deals with the flow of water or other liquids; the practical application of *hydromechanics.*

hydraulic shrinkage crack A type of joint that forms as a consequence of compressibility of the grains of the rock under increased pore pressure. When grains are compressed under uniaxial strain boundary conditions, the contraction of the rock under increasing pore pressure causes the rock to part. The product of this parting is a natural hydraulic fracture (Miller, 1995). Cf: *natural hydraulic fracture.*

hydraulic wedging Pressure produced in a cavity within a reef or other body of rock by pounding surf (Cloud, 1957, p.1016).

hydric (hy'-dric) Said of a habitat that has or requires abundant moisture; also, said of an organism or group of organisms occupying such a habitat. Cf: *xeric; mesic.* See also: *hydrarch.*

hydric soils Soils that are wet long enough to periodically produce anaerobic conditions, thereby influencing the growth of plants.

hydroamphibole (hy-dro-am'-phi-bole) A mixture of hornblende and chlorite.

hydroastrophyllite (hy'-dro-as'-tro-phyl'-lite) A dark brown triclinic mineral of the *astrophyllite* group: $(H_3O,K,Ca)_3(Fe^{2+},Mn)_{5\text{-}6}Ti_2Si_8(O,OH)_{31}$. Not approved by the International Mineralogical Association.

hydrobasaluminite (hy'-dro-bas'-a-lu'-mi-nite) A white monoclinic mineral: $Al_4(SO_4)(OH)_{10}{\cdot}15H_2O$.

hydrobiology (hy'-dro-bi-ol'-o-gy) The biology of bodies of water, esp. of lakes and other bodies of fresh water. Cf: *biohydrology.*

hydrobios (hy-dro-bi'-os) That area of the Earth occupied by aquatic plants and animals. Cf: *geobios.*

hydrobiotite (hy-dro-bi'-o-tite) A regularly ordered, 1:1 interstratification of biotite and vermiculite (Brindley et al., 1983).

hydroboracite (hy-dro-bor'-a-cite) A silky white monoclinic mineral: $CaMgB_6O_8(OH)_6{\cdot}3H_2O$.

hydrocalcite (hy-dro-cal'-cite) (a) A mineral name applied to material that is perhaps $CaCO_3{\cdot}2H_2O$ or $CaCO_3{\cdot}3H_2O$. (b) A mineral name used by Marschner (1969) for a compound now known as *monohydrocalcite.*

hydrocalumite (hy-dro-cal'-u-mite) A colorless to light green monoclinic mineral: $Ca_2Al(OH)_6[Cl_{1-x}(OH)_x]{\cdot}3H_2O$.

hydrocarbon (hy-dro-car'-bon) Any organic compound, gaseous, liquid, or solid, consisting solely of carbon and hydrogen. They are divided into groups of which those of especial interest to geologists are the paraffin, cycloparaffin, olefin, and aromatic groups. Crude oil is essentially a complex mixture of hydrocarbons.

hydrocarbon fingerprinting Practice of determining the source of an oil spill by looking at the distribution of specific hydrocarbon compounds.

hydrocarbon indicator A measurement which suggests the possibility of a hydrocarbon accumulation. Indicators include local amplitude increase (*bright spot*) or decrese (*dim spot*), polarity reversal, a waveshape change, a change in frequency content (especially a local lowering of frequency), a horizontal event reflected from a gas-water, gas-oil, or oil-water contact (*flat spot*), a decrease in amplitude below the accumulation, lower velocity than laterally equivalent sediments producing an apparent sag in lower reflections because of increased time in transitting the accumulation (*time sag*), an increase in amplitude with offset. See: *amplitude variation with offset.* Abbrev: HCI.

hydrocarbon kitchen The temperature and pressure conditions which generate hydrocarbons without then destroying them.

hydro cast (hy'-dro) *hydrographic cast.*

hydrocerussite (hy'-dro-ce-rus'-site) A colorless or lightly tinted rhombohedral mineral: $Pb_3(CO_3)_2(OH)_2$. It occurs as a secondary product as an encrustation on native lead or on galena.

hydrochemical facies (hy-dro-chem'-i-cal) (a) The diagnostic chemical character of groundwater solutions occurring in hydrologic systems (Back, 1966, p.11). It is determined by the flow pattern of the water and by the effects of chemical processes operating between the groundwater and the minerals within the lithologic framework. (b) A term used by Chebotarev (1955, p.199) to indicate concentration of dissolved solids (facies may be low-, transitional-, or high-saline).

hydrochemistry The study of chemical processes and conditions in groundwater.

hydrochlorborite (hy'-dro-chlor-bor'-ite) A colorless monoclinic mineral: $Ca_2B_4O_4Cl(OH)_7{\cdot}7H_2O$.

hydrochore (hy'-dro-chore) A plant whose seeds or spores are distributed by water.

hydroclast [pyroclast] Volcaniclastic particles formed from *hydrovolcanic* explosions.

hydroclast [sed] A rock fragment that is transported and deposited in an aqueous environment.

hydroclastic rock (hy-dro-clas'-tic) (a) A clastic rock deposited by the agency of water. (b) A rock broken by wave or current action. (c) A volcanic rock broken or fragmented during chilling under water or ice.

hydroclimate (hy-dro-cli'-mate) The physical and often the chemical factors that characterize a particular aquatic environment.

hydrocollapsible soil *collapsible soil.*

hydrocompaction The reduction in volume of a low-density sedimentary deposit in response to soaking by water. Susceptible deposits include loess and alluvial fans derived from fine-grained source materials. Also called: hydroconsolidation. Cf: *collapsible soil.*

hydrocyanite (hy-dro-cy'-a-nite) *chalcocyanite.*

hydrodelhayelite (hy-dro-del-hay'-el-ite) A grayish-white orthorhombic mineral: $KCa_2AlSi_7O_{17}(OH)_2{\cdot}6H_2O$.

hydrodialeima (hy'-dro-di'-a-lei'-ma) A term proposed by Sanders (1957, p.295) for an unconformity caused by subaqueous processes.

hydrodictyacean algae (hy-dro-dic-ty-a'-cean) Colonial *green algae,* primarily of fresh to brackish water, usually occurring as flat colonies (*coenobia*), e.g., *Pediastrum* (Batten, 1996).

hydrodolomite (hy-dro-do'-lo-mite) A mixture of hydromagnesite and calcite.

hydrodresserite (hy-dro-dres'-ser-ite) A colorless triclinic mineral: $BaAl_2(CO_3)_2(OH)_4{\cdot}3H_2O$. It dehydrates to form dresserite.

hydrodynamic dispersion *dispersion [grd wat].*

hydrodynamic jetting (hy'-dro-dy-nam'-ic) The directional ejection of molten or vaporized material at very high velocities as a result of shock-wave interactions at the triple interface between projectile, target, and the free surface, in the early stages of *hypervelocity impact* (Gault et al., 1968, p.90). Syn: *jetting [impact].*

hydrodynamics (hy'-dro-dy-nam'-ics) The aspect of *hydromechanics* that deals with forces that produce motion. Cf: *hydrostatics; hydrokinetics.*

hydrodynamic wave A type of surface wave that is similar to a Rayleigh wave but has an opposite particle motion. Syn: *H wave.*

hydroelectric power (hy'-dro-e-lec'-tric) Electrical energy generated by means of a power generator coupled to a turbine through which water passes. Cf: *waterpower; hydropower; white coal.*

hydroexplosion (hy'-dro-ex-plo'-sion) General term for a volcanic explosion caused by the generation of steam from any body of water. It includes phreatic, phreatomagmatic, submarine, and littoral explosions. Cf: *littoral explosion.*

hydrofracting (hy-dro-fract'-ing) Extension and propagation of fractures in rock by hydraulic pressure exerted by a fluid phase within the fractures, as during crystallization of a water-saturated igneous melt or by water injection into a rock mass (Barnes, 1979, p.113). Not to be confused with *hydrofracturing.*

hydrofracturing (hy-dro-frac'-tur-ing) *hydraulic fracturing.*

hydrogarnet (hy-dro-gar'-net) (a) A group of garnet minerals of the general formula: $A_3B_2[(SiO_4)_{3-x}(OH)_{4x}]$. (b) A mineral of the *hydrogarnet* group, such as hydrogrossular.

hydrogenesis (hy-dro-gen'-e-sis) The natural condensation of moisture in the air spaces of surficial soil or rock material.

hydrogen extraction An analytical technique used to determine the amount of hydrogen in a sample, usually expressed as weight percent of water in the sample. A sample is heated overnight under vacuum to drive off adsorbed water. The sample is then heated to very high temperatures so it breaks down chemically and releases structural water as a gas. The water molecules are separated from other gases, passed through another furnace to release free hydrogen. The hydrogen is then collected in a calibration volume, where it is measured using distilled water for calibration of the volume.

hydrogenic (hy-dro-gen'-ic) Said of a soil whose dominant formative influence is water, as in a cold, humid area.

hydrogenic rock A sedimentary rock formed by the agency of water. The term was restricted by Grabau (1924, p.280) to a *hydrolith* "wholly of chemical origin", such as a precipitate from solution in water. Rarely used.

hydrogen index An expression of the relative abundance of hydrogen with respect to carbon derived from RockEval pyrolysis of sedimentary organic matter. The hydrogen index is derived from measurement of the amount of hydrocarbon-rich material released by RockEval pyrolysis of organic matter divided by the total organic carbon concentration of the sediment sample. It is a proxy for the atomic H/C ratio of organic matter (Last and Smol, 2001, p.465; Talbot and Livingstone, 1989).

hydrogen-index log (hy'-dro-gen-in'-dex) *neutron log.*

hydrogen-ion concentration (hy'-dro-gen-i'-on) *pH [chem].*

hydrogenous (hy-drog'-e-nous) (a) Said of coals high in moisture, such as brown coals. (b) Said of coals high in volatiles, such as sapropelic coals.

hydrogen-sulfide mud (hy'-dro-gen-sul'-fide) *black mud.*

hydrogeochemistry (hy'-dro-ge'-o-chem'-is-try) The chemistry of ground and surface waters, particularly the relationships between the chemical characteristics and quality of waters and the areal and regional geology.

hydrogeologic unit *hydrostratigraphic unit.*

hydrogeology (hy'-dro-ge-ol'-o-gy) (a) The science that deals with subsurface waters and with related geologic aspects of surface waters. Also used in the more restricted sense of *groundwater geology* only. The term was defined by Mead (1919, p.2) as the study of the laws of the occurrence and movement of subterranean waters. More recently it has been used interchangeably with *geohydrology.* (b) The study of the laws governing (1) the movement of groundwater; (2) the mechanical, chemical, and thermal interaction of this water with the porous medium; and (3) the transport of energy and chemical constituents by the flow of groundwater (Domenico and Schwartz, 1990, p.7). Syn: *geohydrology; groundwater geology.* Cf: *groundwater hydraulics; groundwater hydrology.*

hydroglauberite (hy-dro-glau'-ber-ite) A silky white mineral: $Na_{10}Ca_3(SO_4)_8{\cdot}6H_2O$.

hydrograph (hy'-dro-graph) A graph showing stage, flow, velocity, or other characteristics of water with respect to time (Langbein and Iseri, 1960). A stream hydrograph commonly shows rate of flow; a groundwater hydrograph, water level or head. See also: *depletion curve.*

hydrographic basin (hy-dro-graph'-ic) (a) The *drainage basin* of a stream. (b) An area occupied by a lake and its drainage basin.

hydrographic cast The lowering of instruments into the sea by cable from a ship to measure temperature and collect water samples for salinity and water-properties analysis at desired ocean depths. The cast may be made using water bottles and reversing thermometers or using Conductivity, Temperature, Depth *(CTD)* instruments equipped with remotely operated water bottles and other sensors. Cf: *Nansen cast.* Syn: *hydro cast; oceanographic cast.*

hydrographic chart A map used in navigation, showing water depth, bottom relief, tides and currents, adjacent land, and distinguishing surface features. Syn: *nautical chart.*

hydrography (hy-drog'-ra-phy) (a) The science that deals with the physical aspects of all waters on the Earth's surface, esp. the compilation of navigational charts of bodies of water. (b) The body of facts encompassed by hydrography. (c) The geographical description of water bodies on the Earth's surface.

hydrogrossular (hy-dro-gros'-su-lar) (a) A mineral of the *hydrogarnet* group: $Ca_3Al_2[(SiO_4)_{3-x}(OH)_{4x}]$, with x near $\frac{1}{2}$. Syn: *hibschite; plazolite.* (b) A general name for the katoite-hibschite series.

hydrohalite (hy-dro-hal'-ite) A white monoclinic mineral: $NaCl{\cdot}2H_2O$. It is formed only from salty water at or below the freezing temperature of pure water.

hydrohalloysite (hy'-dro-hal-loy'-site) *halloysite.*

hydrohematite (hy-dro-hem'-a-tite) *turgite.*

hydroherderite (hy-dro-her'-der-ite) *hydroxyl-herderite.*

hydrohetaerolite (hy'-dro-het-ae'-ro-lite) A dark brown to black tetragonal mineral of uncertain composition, possibly $Zn_2Mn^{3+}{}_4O_8{\cdot}H_2O$.

hydrohonessite (hy'-dro-hon'-ess-ite) A yellow hexagonal mineral: $(NiFe^{3+})_9(SO_4)_2(OH)_{18}{\cdot}7H_2O$.

hydroid (hy'-droid) Any one of a group of hydrozoans belonging to the order Hydroida, among which the polypoid (usually colonial) generation is dominant, and the skeleton is commonly composed of a hornlike material. Cf: *millepore; stylaster.* Range, Cambrian to Holocene.

hydro-isostasy (hy'-dro-i-sos'-ta-sy) Crustal adjustment to loading and unloading that is attributed directly to water, esp. the addition or removal of the water mass involved in glaciation and deglaciation. Cf: *isostasy; glacio-isostasy.*

hydrokaolin (hy'-dro-ka-o'-lin) (a) *endellite.* (b) A fibrous variety of kaolinite from Saglik in Transcaucasia, former U.S.S.R.

hydrokinetics (hy'-dro-ki-net'-ics) The aspect of *hydromechanics* that deals with forces that cause change in motion. Cf: *hydrodynamics; hydrostatics.*

hydrolaccolith (hy-dro-lac′-co-lith) An approx. syn. of *frost blister.* The term is used often as a syn. of *pingo,* but a hydrolaccolith can be a seasonal mound, whereas a pingo is perennial.

hydrolite (hy′-dro-lite) A term variously applied to enhydros, to siliceous sinter, and to the zeolite mineral gmelinite-Na.

hydrolith (hy′-dro-lith) (a) A term proposed by Grabau (1904) for a rock that is chemically precipitated from solution in water, such as rock salt or gypsum; a *hydrogenic rock.* (b) A rock that is "relatively free from organic material" (Nelson and Nelson, 1967, p.185). (c) A *hydroclastic rock* consisting of carbonate fragments (Bissell and Chilingar, 1967, p.158).

hydrologic balance (hy-dro-log′-ic) *hydrologic budget.*

hydrologic barrier *groundwater barrier.*

hydrologic budget An accounting of the inflow to, outflow from, and storage in a hydrologic unit such as a drainage basin, aquifer, soil zone, lake, or reservoir (Langbein and Iseri, 1960); the relationship between evaporation, precipitation, runoff, and the change in water storage, expressed by the hydrologic equation. Syn: *water balance; water budget; hydrologic balance.*

hydrologic cycle The constant circulation of water from the sea, through the atmosphere, to the land, and its eventual return to the atmosphere by way of transpiration and evaporation from the sea and the land surfaces. Syn: *water cycle.*

hydrologic properties Those properties of a rock that govern the entrance of water and the capacity to hold, transmit, and deliver water, e.g. porosity, effective porosity, specific retention, permeability, and direction of maximum and minimum permeability. Syn: hydraulic properties.

hydrologic regimen *regimen [water].*

hydrologic system A complex of related parts physical, conceptual, or both forming an orderly working body of hydrologic units and their man-related aspects such as the use, treatment and reuse, and disposal of water and the costs and benefits thereof, and the interaction of hydrologic factors with those of sociology, economics, and ecology.

hydrology (hy-drol′-o-gy) (a) The science that deals with global water (both liquid and solid), its properties, circulation, and distribution, on and under the Earth's surface and in the atmosphere, from the moment of its precipitation until it is returned to the atmosphere through evapotranspiration or is discharged into the ocean. In recent years the scope of hydrology has been expanded to include environmental and economic aspects. At one time there was a tendency in the U.S. (as well as in Germany) to restrict the term "hydrology" to the study of subsurface waters (DeWiest, 1965, p.1). Syn: *hydroscience.* (b) The sum of the factors studied in hydrology; the hydrology of an area or district.

hydrolysates (hy-drol′-y-sates) *hydrolyzates.*

hydrolysis (hy-drol′-y-sis) A decomposition reaction involving water. In geology, it commonly indicates reaction between silicate minerals and either pure water or aqueous solution. In such reactions, H^+ ions or OH^- ions are consumed, thus changing the H^+/OH^- ratio. In the hydrolysis of a silicate mineral, an amount of cation chemically equivalent to the quantity of H^+ consumed must be released to the solution. An example is the hydrolysis of potassium feldspar to muscovite plus quartz: $1.5KAlSi_3O_3 + H^+ 0.5KAl_3Si_3O_{10} + K^+ + 3SiO_2$. (See Hemley and Jones, 1964).

hydrolyzates (hy-drol′-y-zates) Sediments characterized by elements that react chemically in water, concentrate in the fine-grained alteration products of primary rocks, and are thus abundant in clays, shales, and bauxites. Hydrolyzate elements are aluminum and associated silicon, potassium, and sodium. It is one of Goldschmidts' groupings of sediments as analogues of differentiation stages in rock analysis. Also spelled: *hydrolysates.* Cf: *resistates; oxidates; reduzates; evaporates.*

hydromagmatic A general term to include all processes, subsurface or surface, involving interaction of magma or magmatic heat with meteoric or connate water in the Earth. Partial syn: *phreatomagmatic* (when describing volcanic eruptions).

hydromagnesite (hy-dro-mag′-ne-site) A white, earthy mineral: $Mg_5(CO_3)_4(OH)_2{\cdot}4H_2O$. It occurs in small monoclinic crystals (as in altered ultrabasic rocks) or in amorphous masses or chalky crusts (as in the temperate caves of eastern U.S.).

hydromagniolite (hy-dro-mag′-ni-o-lite′) A general term for hydrous magnesium silicates.

hydrombobomkulite (hy′-dro-mbo-bom′-ku-lite) A blue monoclinic mineral: $(Ni,Cu)Al_4(NO_3)_2,SO_4(OH)_{12}{\cdot}13\text{-}14H_2O$.

hydromechanics (hy′-dro-me-chan′-ics) The theoretical, experimental, or practical study of the action of forces on water. See also: *hydrodynamics; hydrokinetics; hydrostatics; hydraulics.*

hydrometallurgy The separation of a metal in aqueous solution from the ore and its subsequent precipitation in metallic form.

hydrometamorphism (hy′-dro-met′-a-mor′-phism) Alteration of rock by material that is added, removed, or exchanged by water solutions, without the influence of high temperature and pressure. Syn: *hydrometasomatism.*

hydrometasomatism *hydrometamorphism.*

hydrometeor (hy-dro-me′-te-or) Any product of condensation or deposition in the atmosphere or on the Earth's surface. Precipitation, clouds, dew, and frost are hydrometeors.

hydrometer (hy-drom′-e-ter) An instrument that is used to measure the specific gravity of a liquid such as seawater. Since specific gravity is related to density, a hydrometer may be used to obtain an estimate of the salinity of a seawater sample.

hydrometry (hy-drom′-e-try) (a) The use of the hydrometer to measure specific gravity of a fluid. (b) The study of the flow of water, esp. measurement.

hydromica (hy-dro-mi′-ca) Any of several varieties of muscovite that are less elastic and more unctuous than mica, that have a pearly luster, and that sometimes contain less potash and more water than ordinary muscovite; e.g. a common micaceous clay mineral resembling sericite but having weaker double refraction. The term is practically synonymous with *illite.* Syn: *hydrous mica.*

hydromolysite (hy-dro-mol′-y-site) An ill-defined orange-red monoclinic mineral: $FeCl_3{\cdot}6H_2O$. It is deliquescent.

hydromorphic soils A suborder of intrazonal soils, consisting of seven great soil groups, all formed under conditions of poor drainage in marshes, swamps, seepage areas, or flats. Not used in current U.S. system of soil taxonomy.

hydromuscovite (hy-dro-mus′-co-vite) (a) A term applied loosely to any fine-grained, muscovite-like clay mineral commonly but not always high in water content and deficient in potassium. (b) *illite.*

hydronium jarosite (hy-dro′-ni-um) A yellow to brown rhombohedral mineral of the *alunite* group: $(H_3O)Fe^{3+}{}_3(OH)_6(SO_4)_2$. Cf: *carphosiderite.*

hydroparagonite *brammallite.*

hydrophane (hy′-dro-phane) A white, yellow, brown, or green variety of common opal that becomes more translucent when immersed in water.

hydrophilic (hy-dro-phil′-ic) Said of a mineral, usually of colloidal size, that has a strong affinity for water. Hydrophilic clay minerals or other colloidal-sized minerals are easily dispersed in water (Adamson and Gast, 1997).

hydrophilite (hy-dro-phil′-ite) A white orthorhombic mineral: $CaCl_2$(?). Syn: *chlorocalcite.* It is hygroscopic and deliquescent.

hydrophobic (hy-dro-pho′-bic) Said of a mineral, usually of colloidal size, that lacks a strong affinity for water. Hydrophobic clay minerals or other colloidal-sized minerals are easily flocculated (Adamson and Gast, 1997).

hydrophone (hy′-dro-phone) A pressure-sensitive detector that responds to acoustic or P-waves transmitted through water. It is used in marine seismic surveying. Syn: *pressure detector.* Cf: *geophone.*

hydrophyte (hy′-dro-phyte) (a) A plant growing in water, either submerged, emergent, or floating; esp. a *helophyte.* (b) A plant that requires large quantities of water for its growth. Syn: *hygrophyte.* Cf: *mesophyte; xerophyte.*

hydroplasticity (hy′-dro-plas-tic′-i-ty) Plasticity that results from the presence of pore water and absorbed water films in a sediment, so that it yields easily to changes of pressure.

hydropore (hy′-dro-pore) A pore, slit, or small external opening that serves as an adit to the water-vascular system of an echinoderm. It may be covered by the *madreporite.* Cf: *gonopore.*

hydropore oral An unpaired *oral plate* in edrioasteroids, in the right posterior part of the oral region; it commonly forms the posterior edge of the *hydropore* (Bell, 1976).

hydropore structure A group of plates that externally surround the *hydropore* in edrioasteroids (Bell, 1976).

hydropower (hy′-dro-pow-er) Literally, *waterpower,* but now generally considered a syn. of *hydroelectric power.*

hydroromarchite (hy-dro-ro-march′-ite) A white tetragonal mineral: $Sn^{2+}{}_3O_2(OH)_2$.

hydroscarbroite (hy-dro-scar′-bro-ite) An ill-defined white mineral: $Al_{14}(CO_3)_3(OH)_{36}{\cdot}nH_2O$.

hydroscience (hy-dro-sci′-ence) *hydrology.*

hydroscopic water (hy-dro-scop′-ic) *hygroscopic water.*

hydrosere (hy′-dro-sere) A *sere* that develops in an aquatic environment; a *hydrarch* sere. Cf: *mesosere; xerosere.*

hydrosialite (hy-dro-si′-a-lite) A syn. of *clay mineral.* Also spelled: hydrosyalite.

hydrosilicate inclusion (hy-dro-sil′-i-cate) A fluid inclusion in a crystal representing the late-silicate fraction of magmatic crystallization.

hydrosphere (hy′-dro-sphere) The waters of the Earth, as distinguished from the rocks (lithosphere), living things (biosphere), and the air (atmosphere). Includes the waters of the ocean; rivers, lakes, and other bodies of surface water in liquid form on the continents; snow, ice, and glaciers; and liquid water, ice, and water vapor in both the unsaturated and saturated zones below the land surface. Included by some, but excluded by others, is water in the atmosphere, which includes water vapor, clouds, and all forms of precipitation while still in the atmosphere.

hydrospire (hy′-dro-spire) An infolded, thin-walled elongate calcareous structure in the interior of a blastoid *theca,* beneath and parallel to the ambulacral border. Its function is apparently respiratory.

hydrospire pore One of numerous minute rounded pores between the side plates or outer side plates near the margin of an *ambulacrum,* leading into a *hydrospire* of a blastoid or connecting the space enclosed by a hydrospire with the exterior.

hydrospire slit A longitudinal opening of a *hydrospire* in some fissiculate blastoids, excavated in the surface of *deltoid* and *radial plates.*

hydrostatic equilibrium (hy-dro-stat'-ic) In a fluid, the horizontal coincidence of the surfaces of constant pressure and constant mass; gravity and pressure are in balance.

hydrostatic head The height of a vertical column of water whose weight, if of unit cross section, is equal to the hydrostatic pressure at a given point; *static head* as applied to water. See also: *artesian head.*

hydrostatic level The level to which the water will rise in a well under its full pressure head. It defines the potentiometric surface. Syn: *static level.*

hydrostatic pressure [exp struc geol] Pressure that is uniform in all directions, e.g. beneath a homogeneous fluid, and causes dilation rather than distortion in isotropic materials. Pressure occurs within a fluid.

hydrostatic pressure [hydraul] The pressure exerted by the water at any given point in a body of water at rest. The hydrostatic pressure of groundwater is generally due to the weight of water at higher levels in the saturated zone (Meinzer, 1923, p.37) See also: *artesian pressure.*

hydrostatics (hy-dro-stat'-ics) The aspect of *hydromechanics* that deals with forces that produce equilibrium. Cf: *hydrodynamics; hydrokinetics.*

hydrostatic stress A state of stress in which the normal stresses acting on any plane are equal and where shearing stresses do not exist in the material; applies to a solid but not a liquid. Syn: *lithostatic stress.*

hydrostratigraphic unit (hy'-dro-strat'-i-graph'-ic) A term proposed by Maxey (1964, p.126) for a body of rock having considerable lateral extent and composing "a geologic framework for a reasonably distinct hydrologic system". A body of rock distinguished and characterized by its porosity and permeability. It is unified and delimited on the basis of its observable hydrologic characteristics that relate to its interstices (Seaber, 1988, p.13). Examples are *aquifer* and *confining bed.* Syn: *hydrogeologic unit.*

hydrotalcite (hy-dro-talc'-ite) (a) A pearly white or brownish rhombohedral mineral: $Mg_6Al_2(CO_3)(OH)_{16}·4H_2O$. It is dimorphous with manasseite. (b) A group name for minerals with an analogous formula, but with Mg replaced by Ni, and Al replaced by Cr, Fe^{3+}, Mn^{3+} or Co^{3+}.

hydrotechnical engineering A field similar to *geotechnical engineering* but specializing in the engineering characteristics of flowing water.

hydrothermal [gem] (hy-dro-ther'-mal) A crystal-growth technique in which the crystal solidifies from a water solution at high temperatures and pressures. Used to produce gem materials such as synthetic quartz (including amethyst) and synthetic emerald.

hydrothermal [petrology] (hy-dro-ther'-mal) Of or pertaining to hot water, to the action of hot water, or to the products of this action, such as a mineral deposit precipitated from a hot aqueous solution, with or without demonstrable association with igneous processes; also, said of the solution itself.

hydrothermal alteration Alteration of rocks or minerals by the reaction of hydrothermal water with pre-existing solid phases. Open-system chemical reactions resulting from partial to complete chemical equilibrium between the host rock and hydrothermal fluids of a mineral deposit. Syn: *wall-rock alteration.* See also: *advanced argillic alteration; phyllic alteration; propylitic alteration; silicification [meta].*

hydrothermal deposit A mineral deposit formed by precipitation of ore and gangue minerals in fractures, faults, breccia openings, or other spaces, by replacement or open-space filling, from aqueous fluids ranging in temperature from 50° to 700°C but generally below 400°C, and ranging in pressure from 1 to 3 kilobars. The fluids are of diverse origin. Alteration of host rocks is common.

hydrothermal eruption *phreatic explosion* (Delmelle and Bernard, 2000).

hydrothermal metamorphism A local type of metamorphism caused by the percolation of hot solutions or gases through fractures, causing mineralogic changes in the neighboring rock. Syn: *hydrothermal metasomatism.*

hydrothermal metasomatism *hydrothermal metamorphism.*

hydrothermal stage That stage in the cooling of a magma during which the residual fluid is strongly enriched in water and other volatiles. The exact limits of the stage are variously defined by different authors, in terms of phase assemblage, temperature, composition, and/or vapor pressure; most definitions consider it as the last stage of igneous activity, coming at a later time, and hence at a lower temperature, than the *pegmatitic stage.*

hydrothermal synthesis Mineral synthesis in the presence of water at elevated temperatures.

hydrothermal system A groundwater system that has a source (or area) of recharge, a source (or area) of discharge, and a heat source.

hydrothermal water Subsurface water whose temperature is high enough to make it geologically or hydrologically significant, whether or not it is hotter than the rock containing it. It may include *magmatic water* and *metamorphic water,* water heated by radioactive decay or by energy release associated with faulting, meteoric water that descends slowly enough to acquire the temperature of the rocks in accordance with the normal geothermal gradient but then rises more quickly so as to retain a distinctly above-normal temperature as it approaches the surface, meteoric water that descends to and is heated by cooling intrusive rocks, water of geopressured aquifers, and brine that accumulates in an area of restricted circulation at the bottom of a sea.

hydrotomical axis (hy-dro-tom'-i-cal) A major axis with large spines in the skeleton of an acantharian radiolarian. Cf: *geotomical axis.*

hydrotroilite (hy-dro-troi'-lite) A black, finely divided colloidal material: $FeS·nH_2O$. It is perhaps formed by bacteria on bottoms of marine basins characterized by reducing conditions and restricted circulation; it quickly changes to more stable pyrite.

hydrotungstite (hy-dro-tung'-stite) A dark green monoclinic mineral: $WO_2(OH)_2·H_2O$.

hydrougrandite (hy'-dro-u-grand'-ite) An unnecessary name for hydrous garnets.

hydrous (hy'-drous) Said of a mineral compound containing water.

hydrous mica *hydromica.*

hydrous pyrolysis Breakdown of organic matter at high temperature in the presence of water.

hydrous-silicate nickel laterite A *nickel laterite* composed of hydrous magnesium-nickel silicates.

hydrovolcanic (hy'-dro-vol-can'-ic) Term encompassing all volcanic activity that results from the interaction between lava, magmatic heat, or gases and meteoric or connate water at or near the surface of the Earth (Wohletz and Heiken, 1992, p.378). Syn: *phreatomagmatic.*

hydrowoodwardite A pale blue rhombohedral mineral: $(Cu,Al)_9(SO_4)_2(OH)_{18}·nH_2O$. It corresponds to woodwardite with additional interlayer water.

hydroxide (hy-drox'-ide) A type of *oxide* characterized by the linkage of a metallic element or radical with the ion OH, such as brucite, $Mg(OH)_2$.

hydroxyapatite (hy-drox'-y-ap'-a-tite) *hydroxylapatite.*

hydroxyapophyllite A colorless or white tetragonal mineral of the *apophyllite* group: $KCa_4Si_8O_{20}(OH,F)·8H_2O$. It represents apophyllite with OH>F.

hydroxycancrinite A colorless to light blue hexagonal mineral of the *cancrinite* group: $(Na,Ca,K)_8(AlSiO_4)_6(OH,CO_3)_2·2H_2O$. It represents cancrinite with $OH>CO_3$. Syn: *hydroxylcancrinite, hydroxyl vishnevite.*

hydroxylapatite (hy-drox'-yl-ap'-a-tite) (a) A gray, green, blue, etc. hexagonal mineral of the *apatite* group: $Ca_5(PO_4)_3(OH)$. (b) An apatite mineral in which hydroxyl predominates over fluorine and chlorine. Syn: *hydroxyapatite.*

hydroxyl-bastnäsite-(Ce) A wax-yellow to dark-brown hexagonal mineral of the *bastnäsite* group, with Ce as the dominant REE: $(Ce,La)(OH,F)CO_3$. Also spelled: hydroxyl-bastnaesite-(Ce).

hydroxyl-bastnäsite-(La) A yellow to dark brown hexagonal mineral of the *bastnäsite* group, with La as the dominant REE: $LaCO_3(OH,F)$. Also spelled: hydroxylcarbonate-(La).

hydroxyl-bastnäsite-(Nd) A dull whitish hexagonal mineral of the *bastnäsite* group, with Nd as the dominant REE: $NdCO_3(OH,F)$. Also spelled: hydroxyl-bastnaesite-(Nd).

hydroxylcancrinite *hydroxycancrinite.*

hydroxylclinohumite A yellow-orange monoclinic mineral: $Mg_9Si_4O_{16}(OH,F)_2$. It represents clinohumite with OH>F.

hydroxylellestadite A pale purple to rose-pink monoclinic (pseudo-hexagonal) mineral of the *britholite* group: $Ca_{10}(SiO_4)_3(SO_4)_3(OH,Cl,F)_2$.

hydroxyl-herderite (hy-drox'-yl-her'-der-ite) A colorless to pale yellow monoclinic mineral of the *herderite* group: $CaBe(OH)(PO_4)$. Syn: *hydroherderite.* Also spelled: hydroxylherderite.

hydroxyl vishnevite *hydroxycancrinite.*

hydrozincite (hy-dro-zinc'-ite) A white, grayish, or yellowish monoclinic mineral: $Zn_5(CO_3)_2(OH)_6$. It is a minor ore of zinc and is found in the upper (oxidized) zones of zinc deposits as an alteration product of sphalerite. Syn: *zinc bloom; calamine; earthy calamine.*

hydrozoan (hy-dro-zo'-an) Any coelenterate belonging to the class Hydrozoa, characterized by forms, usually colonial and more specialized than sponges, that are both polypoid and medusoid or exclusively medusoid; and by the absence of nematocysts and a stomodaeum. Range, Precambrian or Lower Cambrian to Holocene.

hyetal (hy'-e-tal) Pertaining to rain, rainfall, or rainy regions; e.g. a *hyetal* interval, or the difference in rainfall between two *isohyets.* Cf: *pluvial [meteorol].*

hyetometer (hy-e-tom'-e-ter) *rain gauge.*

hygrograph (hy'-gro-graph) A self-recording *hygrometer.*

hygrometer (hy-grom'-e-ter) An instrument that is used to measure the humidity of the air; e.g. dewpoint hygrometer and rain hygrometer. See also: *hygrograph.*

hygrophilous (hy-groph'-i-lous) Said of an organism that lives in moist areas.

hygrophyte (hy'-gro-phyte) *hydrophyte.*

hygroscopic capacity (hy-gro-scop'-ic) *hygroscopic coefficient.*

hygroscopic coefficient The ratio of the weight of water that a completely dry mass of soil will absorb if in contact with a saturated atmosphere until equilibrium is reached, to the weight of the dry soil mass, expressed as a percentage. See also: *hygroscopic water.* Syn: *hygroscopic capacity.*

hygroscopicity (hy'-gro-sco-pic'-i-ty) "The quantity of water absorbed by dry soil in a secluded space above 10 percent sulphuric acid at room temperature (about 18°C), expressed as a percentage of the weight of dry soil" (Schieferdecker, 1959, term 0356); the ability of a soil to absorb and retain water.

hygroscopic moisture *hygroscopic water.*

hygroscopic nuclei A special category of *cloud condensation nuclei* that have a special chemical affinity for water molecules, so that condensation may take place on these nuclei at relative humidities under 100%.

hygroscopic water Moisture held in the soil that is in equilibrium with that in the atmosphere to which the soil is exposed. Syn: *hygroscopic moisture; hydroscopic water.* See also: *hygroscopic coefficient.*

hyoid apparatus (hy'-oid) In land vertebrates, that part of the visceral skeleton (second visceral arch) that supports the tongue and related structures; dorsal elements of the hyoid arch serve to brace the upper jaw and gill cover in most fishes and become the stapes (hyomandibular element of fishes) in tetrapods.

hyolitha (hy-o-lith'-a) A high-level taxon of Paleozoic invertebrates, now usually treated as an extinct phylum or class of mollusks. Hyolitha are marine bilaterally symmetrical solitary metazoans, having a generally conical, probably aragonitic, shell with a single aperture. Soft parts consisted of numerous muscles and a long sinuous gut. Range, Lowest Cambrian to Upper Permian (Pojeta, 1986a).

hyostylic (hy-o-sty'-lic) adj. Pertaining to a jaw suspension in fishes in which the cartilaginous or cartilage-replacement skeleton of the upper jaw (mandibular arch) is supported posteriorly by the hyomandibular cartilage or bone which, in turn, is articulated by ligaments with the side of the ear region of the braincase.

hypabyssal (hyp-a-byss'-al) An igneous rock or intrusive body formed at a shallow depth. Term is not used by many petrologists. Syn: *subvolcanic.*

hypautochthony (hyp-au-toch'-tho-ny) (a) Accumulation of plant remains that no longer occur in the exact place of their growth, but still within the same general area, as in a peat bog. Cf: *euautochthony.* (b) A term sometimes used as a syn. of *allochthony.*

hypautomorphic (hyp'-au-to-mor'-phic) A grain partly bounded by crystal faces. Obsolete. Syn: *subhedral; hypidiomorphic.* Cf: *anhedral; xenomorphic; euhedral; idiomorphic; automorphic.*

hypautomorphic-granular (hyp'-au-to-mor'-phic-gran'-u-lar) A phaneritic igneous texture dominated by subhedral, equant grains. Obsolete. Syn: *hypidiomorphic-granular.* Cf: *idiomorphic-granular; xenomorphic-granular.*

hyper- (hy'-per-) A prefix meaning "above, beyond, excessive".

hyperbola An open mathematical curve that corresponds to a conic section with an eccentricity greater than one. An object that follows a hyperbolic orbit relative to a central mass has too much velocity to be gravitationally bound to the central mass, and will escape its attraction. Some comets are perturbed into hyperbolic orbits (with respect to the sun), when they have a close encounter with a large planet such as Jupiter, and exit the solar system. See also: *ellipse; parabola; eccentricity.*

hyperborean (hy-per-bo'-re-an) Pertaining or relating to the far north; of a frigid northern region.

hypercinnabar (hy-per-cin'-na-bar) An adamantine reddish black hexagonal mineral of the *wurtzite* group: HgS. It is a polymorph of cinnabar.

hypercline (hy'-per-cline) Said of the dorsal and anterior inclination of the *cardinal area* in the *brachial valve* of a brachiopod, lying in the top right or second quadrant moving clockwise from the *orthocline* position (TIP, 1965, pt. H, p.60, fig. 61).

hyperconcentrated flow A gravitationally driven, nonuniform mixture of debris and water, having water content larger than that of a debris flow but less than that of a muddy streamflow (Vallance, 2000).

hypercyclothem (hy-per-cy'-clo-them) A term proposed by Weller (1958a, p.203-204) for a great cyclic sequence consisting of four megacyclothems and an alternating detrital sequence "of more than ordinary thickness and complexity".

hyperfusible (hy-per-fu'-si-ble) n. Any substance capable of lowering the melting ranges in end-stage magmatic fluids. Also called: hyperfusible component.

hypergene (hy'-per-gene) *supergene.*

hypergenesis (hy-per-gen'-e-sis) A term introduced by Fersman (1922), and persisting to the present day in Russian geology, for surficial alteration (weathering) of sedimentary rocks. Little used in the English literature. Syn: *retrograde diagenesis; regressive diagenesis; retrodiagenesis.*

hyperite (hy'-per-ite) A plutonic rock composed of orthopyroxene, plagioclase, olivine, and clinopyroxene, being intermediate in composition between *gabbro* and *norite.* See also: *hyperite texture.* Obsolete.

hyperite texture The texture characteristic of *hyperite,* in which a fibrous amphibole reaction rim is formed at the contacts between olivine and plagioclase grains (Johannsen, 1939, p.257). Obsolete.

hypermelanic (hy'-per-me-lan'-ic) Obsolete term for igneous rocks that consist of 90-100% mafic minerals. Cf: *melanocratic; ultramafic.*

hypermorphosis (hy-per-mor'-pho-sis) *anaboly.*

hyperpycnal inflow (hy-per-pyc'-nal) Flowing water that is denser than the body of water it enters, resulting in formation of a turbidity current. Its flow pattern is that of a *plane jet* (Moore, Derek, 1966, p.89). Cf: *hypopycnal inflow; homopycnal inflow.*

hypersaline (hy-per-sal'-ine) Excessively saline; with a salinity substantially greater than that of average seawater (35).

hypersolvus (hy-per-sol'-vus) Refers to the crystallization of a single solid-solution phase in a multicomponent solid solution characterized by a region of immiscibility at lower temperature. In igneous petrology, commonly used to indicate the presence in granites and syenites of a single type of exsolved feldspar (perthite) consisting of lamellae of sodic plagioclase and potassic feldspar and the absence of separate plagioclase crystals. This petrographic phenomenon indicates a temperature gap between the solidus curve and the critical point of the plagioclase-potassic feldspar solvus, implying low water pressure during crystallization (<about 2.5 kbar). Cf: *solvus; subsolvus.*

hyperspectral scanner A special type of multispectral scanner that records many tens of bands of imagery at very narrow bandwidths.

hypersthene (hy'-per-sthene) A common rock-forming mineral of the *orthopyroxene* group: $Mg_4Si_4O_{12}$. It is an unnecessary, but common, name for a member of the enstatite-ferrosilite series. Hypersthene is grayish, greenish, black, or dark brown, and often has a bronze or greenish-brown play of color (*schiller*) on the cleavage surface. It is an essential constituent of many igneous rocks (gabbros, andesites).

hypersthene magma series A term proposed by Kuno for lavas with orthopyroxene in the groundmass; now considered synonymous with calc-alkaline series (Gill, 1981, p.9). Cf: *pigeonitic magma series.*

hyperstomial (hy-per-sto'-mi-al) Said of a cheilostome bryozoan *ovicell* that rests on or is partly embedded in the distal zooid and that opens above the operculum of the maternal zooid.

hyperstrophic (hy-per-stroph'-ic) Said of a rare gastropod shell in which the whorls are coiled on an inverted cone so that the apex points forward rather than backward, and the spire is depressed instead of elevated. A hyperstrophic shell is not easily distinguished from an *orthostrophic* shell unless the aperture shows the siphon pointed in the same direction as the apex (Beerbower, 1968, p.341).

hypertely (hy-per'-tel-y) Evolution to the extreme of being a disadvantage.

hyperthermic temperature regime (hy-per-ther'-mic) A soil temperature regime in which the mean annual temperature (measured at 50cm) is at least 22°C, with a summer-winter variation of at least 5°C (USDA, 1975). Cf: *isohyperthermic temperature regime.*

hypertrophy (hy-per'-troph-y) Excessive growth of a part of an organism, such as a claw, or of a species.

hypervelocity impact (hy'-per-ve-loc'-i-ty) The impact of a projectile onto a surface at a velocity such that the stress waves produced on contact are orders of magnitude greater than the static bulk compressive strength of the target material. The minimum required velocities vary for different materials but are generally 1-10 km/sec, for most crystalline rocks about 4-5 km/sec. In such an impact, about half of the kinetic energy of the projectile is transferred to the target material in the form of a shock wave, whose interactions with the surface produce a crater much larger in diameter than the projectile. Extraterrestrial bodies striking the Earth at speeds in excess of about 5 km/sec give examples of large hypervelocity impacts and produce correspondingly large craters (Dietz, 1959, p.499).

hypha (hy'-pha) One of the individual tubular filaments or threads that make up the *mycelium* of a fungus; e.g. a *conidiophore.* Pl: hyphae.

hypidioblast (hy-pid'-i-o-blast') A mineral grain that is newly formed by metamorphism and is bounded only in part by its characteristic crystal faces. It is a type of *crystalloblast.* Cf: *idioblast; xenoblast.* Syn: *subidioblast.*

hypidioblastic (hy-pid'-i-o-blas'-tic) Pertaining to a *hypidioblast* of a metamorphic rock; also, said of such a texture. It is analogous to the term *hypidiomorphic* in igneous rocks. Cf: *idioblastic.*

hypidiomorphic (hy-pid'-i-o-mor'-phic) *hypautomorphic; subhedral.*

hypidiomorphic-granular (hy-pid'-i-o-mor'-phic-gran'-u-lar) *hypautomorphic-granular.*

hypidiotopic (hy-pid'-i-o-top'-ic) Intermediate between *idiotopic* and *xenotopic;* esp. said of the fabric of a crystalline sedimentary rock in which the majority of the constituent crystals are subhedral. Also, said of the rock (such as an evaporite, a chemically deposited cement, or a recrystallized limestone or dolomite) with such a fabric. The term was proposed by Friedman (1965, p.648).

hypnozygote (hyp-no-zy'-gote) A resting *cyst* resulting from sexual fusion, commonly thick-walled (Williams et al., 2000, p.102).

hypo- (hy'-po-) A prefix meaning "under, nearly". Before a vowel, it is often shortened to "hyp".

hypocenter (hy'-po-cen-ter) *focus [seis].*

hypocotyl (hy-po-cot'-yl) The portion of the axis of a plant embryo or seedling below the attachment of the cotyledons (seed leaves) and above the root.

hypocrystalline (hy-po-crys'-tal-line) An igneous rock composed of a mixture of crystals and glass. Syn: *hypohyaline.*

hypocrystalline-porphyritic (hy-po-crys'-tal-line-por-phy-rit'-ic) The porphyritic texture of an igneous rock having a hypocrystalline groundmass. Obsolete.

hypocyst (hyp'-o-cyst) The part of a dinoflagellate cyst posterior to the *cingulum.* Also the part of an *acritarch* posterior to the equator and containing the excystment opening. Cf: *epicyst.*

hypocystal archeopyle (hyp-o-cys'-tal) An *archeopyle* formed in a dinoflagellate cyst by the loss of one or more plates in the hypocyst (Williams et al., 2000).

hypodeltoid (hy-po-del'-toid) The *anal deltoid* on the aboral side of the anus in many blastoids. In some Pennsylvanian and Permian blastoids, it is either small and loosely attached or completely atrophied.

hypodermalium (hy'-po-der-ma'-li-um) A specialized sponge spicule of the cortex, lying largely beneath the *exopinacoderm.*

hypodermis (hy-po-der′-mis) (a) A reticulate layer beneath the epidermis in the walls of certain foraminifera. (b) The cellular layer that underlies and secretes the external chitinous membrane of arthropods.

hypodigm (hy′-po-digm) A group consisting of all the specimens used by an author as the basis for his description of a species, and all specimens later referred to it (Simpson, 1940, p.18). Cf: *type material.*

hypogastralium (hy′-po-gas-tra′-li-um) A specialized sponge spicule lying largely beneath the *endopinacoderm* of the *spongocoel.*

hypogeal (hy-po-ge′-al) *hypogene.*

hypogeic (hy-po-ge′-ic) *hypogene.*

hypogene (hy′-po-gene) (a) Said of a geologic process, and of its resultant features, occurring within and below the crust of the Earth. Cf: *epigene; endogenetic.* Syn: *hypogenic; hypogeal; hypogeic.* (b) Said of a mineral deposit formed by ascending solutions; also, said of those solutions and of that environment. Cf: *supergene; mesogene.* (c) A rarely used syn. of *plutonic.*

hypogenesis (hy-po-gen′-e-sis) The direct development of an organism without alternation of generations.

hypogenic (hy-po-gen′-ic) *hypogene.*

hypogenic cave A dissolution cave formed by water whose solutional capacity is derived from sources beneath the land surface, most commonly by hydrogen sulfide or cooling of thermal waters (Palmer, 1991). Cf: *epigenic cave.*

hypogynous (hy-pog′-y-nous) Said of flowers in which the ovary surmounts a receptacle, so that the sepals, petals, and stamens radiate from below the locules. Cf: *epigynous; perigynous.*

hypohyaline (hy-po-hy′-a-line) *hypocrystalline.*

hypolimnetic (hy′-po-lim-net′-ic) Pertaining to a *hypolimnion.*

hypolimnion (hy-po-lim′-ni-on) The cold, deep water below the *thermocline* of a thermally stratified lake; the layer below the *metalimnion,* which, in turn, is overlain by the *epilimnion* (Wetzel, 2001, p.74).

hypolithic (hy-po-lith′-ic) Said of a plant that grows beneath rocks.

hypomagma (hy-po-mag′-ma) Obsolete term for relatively immobile, viscous lava that forms at depth beneath a shield volcano, is undersaturated with gases, and gives rise to volcanic activity. Cf: *epimagma; pyromagma.*

hyponome (hy′-po-nome) The muscular tube, nozzle, or swimming funnel just below the head of a cephalopod, extending externally from the mantle cavity, through which water is expelled from the mantle cavity.

hyponomic sinus (hy-po-nom′-ic) The large, concave ventral notch or re-entrant in the middle of the aperture of a cephalopod, marking the location through which the *hyponome* protrudes.

hyponym (hy′-po-nym) The name of a plant *taxon* that cannot be recognized because of the vagueness or inadequacy of its description (Cowan, 1968, p.48).

hypoparian (hy-po-par′-i-an) adj. Of or concerning a trilobite that lacks facial sutures and that is generally blind. n. A hypoparian trilobite; specif. a trilobite of the order Hypoparia (now obsolete).

hypopycnal inflow (hy-po-pyc′-nal) Flowing water that is less dense than the body of water it enters, e.g. a river entering the ocean. Its flow pattern is that of an *axial jet.* (Moore, 1966, p.89). Cf: *hyperpycnal inflow; homopycnal inflow.*

hyporheos In streams, the region adjacent to and below the active channel where water movement is primarily in the downstream direction and the interstitial water is exchanged with the water in the main channel. The boundary of the hyporheos is where 10% of the water has recently been in the stream. Also called: hyporheic zone.

hyposeptal deposit (hy-po-sep′-tal) A distal *cameral deposit* on the convex (adapical) side of *septum* of a nautiloid. Ant: *episeptal deposit.*

hyposome (hy′-po-some) That part of a dinoflagellate theca or cyst posterior to the cingulum. The word "hyposome" encompasses the terms "hypocyst," "hypotheca," and "hypotract" (Fensome et al., 1993; Williams et al., 2000, p.77 and 104). Cf: *episome.*

hypostega (hy-po-ste′-ga) A part of the body cavity of a cheilostome bryozoan zooid between a *cryptocyst* or an umbonuloid *frontal shield* and the overlying membranous wall.

hypostegal coelom (hy-po-ste′-gal) The confluent coelomic spaces beyond the ends of zooidal walls in stenolaemate bryozoans. Also, a syn. of *hypostega.*

hypostoma (hy-po-sto′-ma) A syn. of *hypostome.* Pl: hypostomata.

hypostomal suture (hy-po-sto′-mal) The line of junction in a trilobite between the anterior margin of the *hypostome* and the posterior margin of the frontal *doublure* or *rostral plate.*

hypostome (hy′-po-stome) (a) A ventral plate of the head region behind and above which the mouth of a trilobite is located. Cf: *metastoma.* Syn: *labrum.* (b) A vase-shaped or conical process bearing the mouth of a hydrozoan. Syn: *hypostoma.*

hypostracum (hy-pos′-tra-cum) A term used originally for the inner layer of the shell wall of a bivalve mollusk, secreted by the entire epithelium of the mantle, but also applied in a later sense to the *myostracum.* Cf: *ostracum.*

hypostratotype (hy-po-strat′-o-type) A stratotype proposed after the original designation of the holostratotype (or parastratotype) in order to extend knowledge of the unit or boundary to other geographic areas. It is always subordinate to the *holostratotype* (ISSC, 1994, p.28). See also: *reference section.*

hypotheca (hy-po-the′-ca) (a) The posterior part of a dinoflagellate theca, below the girdle. Cf: *epitheca.* (b) *hypovalve.*

Hypothermal (Hy-po-ther′-mal) n. A term proposed by Cooper (1958, p.944) for a postglacial interval (the last 2,600 years) characterized by a moderate decrease in temperature, some limited glacial expansions, and inferred eustatic lowerings of sea level.

hypothermal [econ geol] Said of a hydrothermal mineral deposit formed at great depth and in the temperature range of 300°-500°C (Park and MacDiarmid, 1970, p.293). Also, said of that environment. Cf: *mesothermal [eco geol]; epithermal; leptothermal; telethermal; xenothermal.* Syn: *katathermal.*

hypothesis (hy-poth′-e-sis) A conception or proposition that is tentatively assumed, and then tested for validity by comparison with observed facts and by experimentation; e.g., the *planetesimal hypothesis* to explain the evolution of the planets. It is less firmly founded than a *theory.*

hypothetical ancestral mollusk (hy-po-thet′-i-cal) A suggested mollusk constructed as a least common denominator of all structures inferred to be primitive in the various classes of living mollusks having shells (Pojeta, 1986b). The concept is no longer popular among workers in the field. Cf: *prototype.* Syn: *schematic mollusk; archetypical mollusk.* Abbrev: HAM.

hypothetical resources Undiscovered mineral resources that we may still reasonably expect to find in known mining districts (Brobst and Pratt, 1973, p.4). Cf: *identified resources; speculative resources.*

hypothyridid (hy-po-thy′-rid-id) Said of a brachiopod *pedicle opening* that is located below or on the dorsal side of the beak ridges with the umbo intact (TIP, 1965, pt. H, p.146).

hypotype (hy′-po-type) A described or figured specimen used in extending or correcting the knowledge of a species, or in other publications regarding it.

hypovalve (hy′-po-valve) The inner valve of a diatom frustule. Cf: *epivalve.* Syn: *hypotheca.*

hypoxenolith (hy-po-xen′-o-lith) Obsolete term for a *xenolith* derived from a source more remote than the adjacent wall rock (Goodspeed, 1947, p.1251). Cf: *epixenolith.*

hypozygal (hy-po-zy′-gal) The proximal brachial plate of a pair joined by *syzygy* in a crinoid.

Hypsithermal (Hyp-si-ther′-mal) n. A term proposed by Deevey and Flint (1957) as a substitute for *Climatic Optimum* and *thermal maximum.* It represents the Holocene interval when "most of the world entered a period when mean annual temperatures exceeded those of the present". The Hypsithermal was taken to include the Boreal, Atlantic, and Subboreal climatic intervals, or from about 9,000 to 2,500 years ago. See also: *Megathermal; Xerothermic.* adj. Pertaining to the postglacial Hypsithermal interval and to its climate, deposits, biota, and events.

hypsographic curve (hyp-so-graph′-ic) A cumulative-frequency profile representing the statistical distribution of the absolute or relative areas of the Earth's solid surface (land and sea floor) at various elevations above, or depths below, a given datum, usually sea level. Syn: *hypsometric curve.*

hypsography (hyp-sog′-ra-phy) (a) A branch of geography dealing with the observation and description of the varying elevations of the Earth's surface with reference to a given datum, usually sea level. Cf: *hypsometry.* (b) Topographic relief. Also, the parts of a map, collectively, that represent topographic relief. (c) The portrayal of topographic relief on maps.

hypsometer (hyp-som′-e-ter) An instrument used in estimating the elevation of a point on the Earth's surface in relation to sea level by determining atmospheric pressure through observation of the boiling point of water at that point. It is useful in mountainous or high-altitude regions. Syn: *thermobarometer.*

hypsometric (hyp-so-met′-ric) Pertaining to hypsometry or to elevation above a datum; esp. relating to elevations above sea level determined with a hypsometer.

hypsometric analysis The measurement of the distribution of ground surface area (or horizontal cross-sectional area) of a landmass with respect to elevation (Strahler, 1952b, p.1118). Syn: *area-altitude analysis.* Cf: *hypsometry.*

hypsometric curve *hypsographic curve.*

hypsometric integral The proportionate area below the *percentage hypsometric curve;* it expresses the relative volume of a landmass at a given contour. In the study of drainage basins, Strahler (1952b, p.1121) used this term to express the ratio of the volume of earth material to the volume of the solid reference figure having a base equal to basin area and a height equal to maximum basin relief. Cf: *erosion integral.* Pike and Wilson (1971) have shown that the hypsometric integral is identical to the *elevation-relief ratio.*

hypsometric map Any map showing relief by means of contours, hachures, shading, tinting, or any other convention.

hypsometric tint A color applied to the area between two selected contour lines on a map of an area whose relief is depicted by layer tinting. Syn: *layer tint.*

hypsometry (hyp-som′-e-try) The science of determining, by any method, height measurements on the Earth's surface with reference to sea level; e.g. "barometric hypsometry" in which elevations are determined by means of mercurial or aneroid barometers. Cf: *hypsography; hypsometric.*

hyrst *hurst.*

hysteresis (hys-ter′-e-sis) (a) A lag in the return of an elastically deformed body to its original shape after the load has been removed. (b) The property of a rock to retain a residual magnetization when the applied magnetic field returns to zero. The retained residual magnetism is called remanent magnetism. Syn: *magnetic hysteresis.* (c) A phase lag of dielectric displacement behind electric-field intensity, due to energy dissipation in polarization processes. (d) The response of a spring gravimeter where the spring maintains the "memory" of its last position.

hystero-brephic stage (hys'-ter-o-breph'-ic) The earliest stage of development of the offset during increase in colonial corals. It commences with the initial modification of the parent *corallite* and normally includes insertion of at least some of the primary septa in the offset (Fedorowski and Jull, 1976, p.42). Cf: *hystero-neanic stage; late neanic stage.*

hysterocrystalline (hys'-ter-o-crys'-tal-line) An obsolete term applied to a mineral produced in an igneous rock as a result of secondary crystallization.

hystero-neanic stage (hys-ter'-o-ne-an'-ic) The stage of development of the offset following the *hystero-brephic stage* during increase in colonial corals. It usually is taken to commence with the appearance of the first *metaseptum* in the offset and commonly includes completion of the partition or dividing wall separating offset from parent *corallite* (Fedorowski and Jull, 1976, p.42). Cf: *hystero-brephic stage; late neanic stage.*

hystero-ontogeny (hys-ter'-o-on-tog'-e-ny) Changes that occur in the offset, or asexually developed *corallite,* during its development (Fedorowski and Jull, 1976, p.40). Cf: *blastogeny; astogeny.*

hysterosoma (hys'-ter-o-so'-ma) That section of the body of an acarid that is behind the second pair of legs.

hystrichosphaerid (hys'-tri-cho-sphae'-rid) A general term formerly used for a great variety of resistant-walled organic microfossils, ranging from Precambrian to Holocene and characterized by spherical to ellipsoidal, usually spinose, remains found among fossil microplankton. These are now divided among the acritarchs and *dinoflagellate* cysts. The term has no formal taxonomic status. Syn: *hystrichosphere.*

hystrichosphere (hys-tri'-cho-sphere) *hystrichosphaerid.*

hyttsjöite A white rhombohedral mineral: $Pb_{18}Ba_2Ca_5Mn^{2+}{}_2Fe^{3+}{}_2Si_{30}O_{90}Cl\cdot 6H_2O$.

Hz *Hertz.*

I

i Symbol to indicate reflection at the boundary between Earth's outer-core and inner-core.
IAB *island-arc basalt.*
IAB iron A member of the non-magmatic iron-meteorite group IAB.
ianthinite (i-an'-thin'-ite) (a) A violet-black orthorhombic secondary mineral: $U^{4+}_2(UO_2)_4O_6(OH)_4 \cdot 9H_2O$. (b) A mineral name that was erroneously given to *wyartite.*
Iapetus Ocean (Ia-pet'-us) A sea that existed east of North America before Europe and Africa collided with North America during the Carboniferous and Permian periods. Iapetus was the father of Atlas, for whom the Atlantic Ocean is named. Syn: *Proto-Atlantic Ocean.*
IASY *International Active Sun Years.*
ICBN *International Code of Botanical Nomenclature.*
ICCL *irregular complex crossed lamellar structure.*
ice (a) Water in the solid state; specif. the solid substance formed in nature by the freezing of liquid water, by the condensation of water vapor directly into ice crystals, or by the recrystallization or compaction of fallen snow. It is colorless to pale blue or greenish blue when pure; white when it includes numerous gas bubbles. At standard atmospheric pressure, it is formed at, and has a melting point of, 0°C; in freezing, it expands about 1/11 in volume and thus floats in water. Ice is technically a mineral and has a hexagonal crystal structure (Cf: *a-axis [ice]; c-axis [ice]*). In large masses, it is a monomineralic rock. (b) A term often substituted for *glacier,* as in "continental ice".
ice age A loosely used syn. of *glacial epoch,* or time of extensive glacial activity; specif. the latest of the glacial epochs, also known as the Pleistocene Epoch.
ice apron The thin mass of snow and ice attached to the headwall of a cirque above the *bergschrund* (but not present at a *randkluft*). Syn: *apron [glaciol].*
ice avalanche A sudden fall, down a steep slope, of ice broken from an ice sheet or glacier (most commonly from a hanging glacier). Syn: *icefall.*
ice bar An ice edge consisting of floes compacted by wind, sea, and swell, and difficult to penetrate.
ice barrier A syn. of *ice front* and *ice shelf.* The term was introduced by Sir James C. Ross in 1841 to designate the high, steep, seaward cliff face or edge of a great ice mass in Antarctica because it obstructed navigation; the syn. "ice front" is now preferred. The term ice barrier was later applied to the entire mass of ice, and was widely adopted for similar polar morphologic features, esp. for an immense ice formation of very great extent; the term "ice shelf" is now used for this type of ice mass. Syn: *barrier [glaciol].*
ice-barrier lake A lake formed in a mountain valley whose lower end is dammed by a glacier descending another valley.
ice bay *bight [ice].*
iceberg (ice'-berg) A piece of floating or grounded (stranded) glacier ice of any shape, detached (calved) from the terminus of a glacier into a body of water. The greatest part of an iceberg's mass (4/5 to 8/9) is located below the water's surface. A unique nomenclature has developed to describe small pieces of calved ice: brash ice is up to 2 m in length; a growler is 2-10 m in length, and a bergy bit is up to 20-25 m in length. Anything larger is termed an iceberg. Icebergs calved from temperate glaciers, such as the glaciers of southern Alaska, rarely exceed 100 m in length. Icebergs calved from polar glaciers, such as the glaciers of Antarctica or Greenland, may reach a length of more than 150 km. Syn: *berg [glaciol].* Cf: *floeberg.*
iceberg tongue "A major accumulation of icebergs projecting from the coast, held in place by grounding, and joined together by fast ice" (U.S. Naval Oceanographic Office, 1968, p.B34).
ice blade A "crest" or "spire" of ice, 0.5-1.5 m high, rising from a surface of firn, and formed by unequal melting (Russell, 1885, p.318).
iceblink [glaciol] (ice'-blink) A cliff extending along the seaward margin of a mass of inland ice. Examples are found on the coast of Greenland. Syn: *isblink.*
iceblink [meteorol] A relatively bright, usually yellowish or whitish glare in the sky near the horizon or on the underside of a cloud layer, produced in a polar region by light reflected from a large ice-covered surface (as an ice sheet) that may be too far away to be visible; not as bright as *snowblink.* Also spelled: ice blink.
ice blister *ice mound.*
ice-block ridge A ridge, either closed or linear, surrounding or separating depressions in a moraine (Deane, 1950, p.14).
ice boulder [glac geol] An obsolete syn. of *glacial boulder.*
ice boulder [ice] A large fragment of sea ice shaped by wave action into a nearly spherical form and then stranded on the shore.
ice boundary "The demarcation at any given time between fast ice and pack ice or between areas of pack ice of different concentrations" (U.S. Naval Oceanographic Office, 1968, p. B34). Cf: *ice edge.*
ice breccia Angular pieces of *sea ice* or *glacier ice* frozen together.
ice cake A *floe* or piece of floating sea ice less than about 10 m across. Syn: *block [ice]; cake [ice].*
ice canopy A submariner's term for *pack ice.*
ice cap A dome-shaped or platelike cover of perennial ice and snow, covering the summit area of a mountain mass so that no peaks emerge through it, or covering a flat landmass such as an Arctic island; spreading due to its own weight outwards in all directions; and having an area of less than 50,000 km². An ice cap is considerably smaller than an *ice sheet.* Cf: *ice field [glaciol]; glacier.* Less preferred syn: *ice carapace; cap [glaciol]; glacier cap.*
ice carapace *ice cap.*
ice cascade *icefall.*
ice cast A shell of ice formed around a beach pebble as a result of the wetting action of spray, tides, and waves, and subsequent freezing; the ice is sometimes separated from the pebble.
ice cauldron A wide area in a valley, upon which glacier ice once piled up so high as to flow radially outward through pre-existing passes that were deepened by glacial scour.
ice cave *glacier cave.*
ice-channel filling An inclusive term for *esker* and *crevasse filling,* to eliminate arbitrary and commonly unfounded distinctions between them. Emphasizes that linear ridges of stratified debris may form in tunnels under the ice, in valleys open to the sky either superglacial or on the substrate, as beaded or continuous deltaic or fluvial forms in standing water at mouths of tunnels ("De Geer eskers"), or as kettle rims between closely spaced kettle holes in outwash (Jahns, 1953).
ice clearing (a) The end phase of *breakup.* (b) *polynya.* Rarely used.
ice cliff Any vertical wall of ice; e.g. a very steep surface bounding a glacier or a mass of shelf ice. Less recommended syn: *ice front; ice wall.* Syn: *ice face.*
ice cluster A concentration of sea ice covering hundreds of square kilometers and found in the same region every summer. See also: *pack ice.* Syn: *ice massif.*
ice column *ice pillar.*
ice concrete A dense frozen mixture of sand, rock fragments, and ice. Syn: *icecrete.*
ice cone *ice pyramid.*
ice-contact delta A delta built by a stream flowing into a lake between a valley slope and the margin of glacier ice. Syn: *delta kame; kame delta.* Obsolete syn: *morainal delta.*
ice-contact deposit Stratified drift deposited in contact with melting glacier ice, such as an esker, a kame, a kame terrace, or a feature marked by numerous kettles.
ice-contact plain *kame plain.*
ice-contact slope The steep slope of sediment that was deposited against a wall of glacier ice, marking the position of the ice margin; an irregular scarp against which glacier ice once rested.
ice-contact terrace *kame terrace.*
ice-contorted ridge *ice-pushed ridge.*
ice core A drilled core taken from a glacier or ice sheet. Ice cores become especially important as source materials for determining the structural properties of glaciers and for obtaining detailed paleoclimate records.
ice cover (a) The extent of *glacier ice* on a land surface at the present time, with special reference to its thickness. (b) The ratio of an area of *sea ice* of any *concentration [ice]* to the total area of sea surface within some large geographic locale that may be global, hemispheric, or specific to a given study

(U.S. Naval Oceanographic Office, 1968, p. B34). (c) The extent of ice on a lake surface.

ice-covered *glacier-covered.*

ice crack *frost crack.*

ice-crack moraine A linear ridge consisting of very sandy, unstratified drift believed to have been deposited between blocks of dead ice in a disintegrating glacier (Sproule, 1939, p.104).

ice creek A narrow inlet or rift in the seaward part of an *ice shelf,* sometimes extending many kilometers in from the *ice front.* Ice creeks are ephemeral, but may recur in the same place (Armstrong et al., 1977).

icecrete (ice-crete') *ice concrete.*

ice crust [glaciol] A type of *snow crust,* formed when meltwater or rainwater freezes to form a continuous layer of ice on the surface of deposited snow. See also: *ice layer [glaciol].*

ice crust [ice] (a) *ice rind.* (b) A thin layer of ice on a rock surface, formed by freezing of water condensed from the air. (c) A layer of ice 2 to 4 mm thick on a snow pack.

ice crystal (a) A single macroscopic particle of ice with a regular structure that is continuous throughout the particle. The structure is hexagonal under pressure and temperature conditions found on Earth. (b) *frazil crystal.*

ice-crystal cast A *crystal cast* formed by the filling of an ice-crystal mark with mud or sand; it commonly appears as a straight, slightly raised ridge on the underside of a sandstone bed.

ice-crystal mark A crack formed on a sedimentary surface by the sublimation of a crystal of ice.

ice dam A river obstruction formed of floating blocks of ice that may cause ponding and widespread flooding during spring and early summer. Syn: *ice barrier.*

ice-dammed lake *glacier lake.*

ice dendrite A thin branching ice crystal. Dendritic ice sometimes forms the first skim of ice over still water, and also may grow under water from an existing ice surface.

iced firn A mixture of ice and firn; firn permeated with meltwater and then refrozen. Syn: *firn ice.*

ice dike A secondary formation of ice, usually made up of columnar crystals, filling a crevasse or other gash in glacier ice.

ice disintegration The process of breaking up a stagnant and wasting glacier into numerous small blocks: it is said to be "controlled" where the blocks are separated along fractures or other lines of weakness to form linear or lobate features, and "uncontrolled" where equal forces break up the glacier along cracks extending in all directions to produce round, oval, or rudely polygonal features (Gravenor and Kupsch, 1959, p.48-49). Syn: *disintegration [glaciol].*

ice dome (a) A rounded, gently sloping elevation in the surface of an inland *ice sheet.* Ice domes do not have precisely defined margins and may cover large areas 100,000 sq km or more (Armstrong et al., 1977). (b) An accumulation of glacier ice in a caldera.

ice edge The boundary at any given time between open water and sea, lake, or river ice of any kind, whether fast or drifting. It may be "compacted" by wind or current, or it may be "diffuse" or "open" when dispersed or poorly defined. Cf: *ice boundary; ice limit.*

ice face *ice cliff.*

icefall (ice'-fall) (a) The part of a glacier that is highly crevassed because the bed is steep. Syn: *ice cascade.* Cf: *cascading glacier.* (b) *ice avalanche.*

ice fan *expanded foot.*

ice field [glaciol] (a) An extensive mass of land ice covering a mountain region, consisting of many interconnected alpine and other types of glaciers, covering all but the highest peaks and ridges. Cf: *ice cap; highland glacier.* (b) A general, but not recommended, designation for a large and irregular body of glacier ice.

ice field [ice] An extensive area of pack ice, consisting of *floes,* and greater than 10 km (6 mi) across; the largest areal subdivision of sea ice. Ice fields are subdivided according to horizontal extent as follows: "large" (over 20 km); "medium" (15-20 km); "small" (10-15 km). Cf: *ice patch.* Syn: *field [ice].*

ice field [permafrost] *icing.*

ice floe (a) A large fragment or extensive sheet of ice, detached and floating freely in open water. (b) *floe.*

ice flow *glacier flow.*

ice flower *frost flower.*

ice foot [glaciol] (a) The ice at the lower end or front of a glacier. (b) A mass or wall of ice formed by the freezing of snow that accumulated along the foot of a mountain slope. It is not formed from converging glaciers.

icefoot [ice] A narrow strip, belt, or fringe of ice formed along and firmly attached to a coast, unmoved by tides, and remaining after the fast ice has broken away; it is usually formed by the freezing of wind-driven spray, or of seawater during ebb tide. A true icefoot has its base at or below the low-water mark. Also used for ice at the foot of lake bluffs. Syn: *kaimoo; shore-ice belt.*

ice foot niche A hollow created at the base of a soft cliff (as of limestone) by floating sea ice; during spring it permits the collapse of the overhanging wall (Hamelin and Cook, 1967, p.101).

ice free "No sea ice present. There may be some ice of land origin" (U.S. Naval Oceanographic Office, 1968, p. B34). Cf: *open water [ice].*

ice fringe (a) A narrow *ice piedmont,* extending less than about one kilometer inland from the sea. (b) A belt of sea ice that extends a short distance offshore.

ice front (a) The floating vertical cliff forming the seaward edge of an ice shelf or other glacier that enters water, ranging in height from 2 to more than 50 m above sea level. Cf: *ice wall.* Syn: *ice barrier; front [glaciol].* (b) Less preferred syn. of *ice cliff* in the sense of "any vertical wall of ice". (c) The *snout* of a glacier.

ice gang A rush of water following a *breakup;* a *debacle.*

ice gland A roughly cylindrical, decimeter-scale, vertical column of ice or iced firn in a *firn field.*

ice gneiss Frozen ground with ice segregated in laminae so as to resemble a gneissic rock; term used by Taber (1943).

ice gorge The vertical-walled opening left after an *ice jam* has broken through.

ice gouge A general term for the processes of ice interacting with the sea floor (Reimnitz et al., 1973, 1977).

ice grass *ice stalk.*

ice grotto *glacier cave.*

ice gruel A type of sludge floating on the sea surface and formed by the irregular freezing together of *frazil* crystals.

ice hummock *hummock [ice].*

ice island A large *tabular iceberg* broken away from an ice shelf, having a thickness of 15-50 m and an area between a few thousand square meters and 500 sq km or even more. The surface of an ice island is usually marked by broad, shallow, regular undulations that give it a ribbed appearance from the air.

ice jam (a) An accumulation of broken river ice lodged in a narrow or obstructed part of the channel; it frequently produces local floods during a spring breakup. Cf: *ice gorge.* (b) An accumulation of large fragments of lake ice or sea ice thawed loose from the shore during early spring and subsequently piled up on or blown against the shore by the wind, often exerting great pressures.

ice keel The submerged mass of broken ice under a *pressure ridge,* forced downward by pressure, and extending as much as 50 m below sea level.

ice laccolith A small mound with a solid ice core resembling a laccolith in section. It is formed in the active layer by the freezing of water forced into a plano-convex lens where it is trapped between the downward-freezing active layer and the underlying permafrost. Ice laccoliths are produced in a few days or weeks during fall freeze-up in areas of continuous permafrost. Most are less than a meter high and a few meters to 10 m across. They may not melt for several years. They differ from many hydrolaccoliths that are not active-layer phenomena (Black, 1954, p.848). Syn: *cryolaccolith; frost blister.*

ice-laid drift *till.*

Iceland agate (Ice'-land) A syn. of *obsidian,* applied to gem-quality varieties.

icelandite (ice'-lan-dite) A term proposed by Carmichael in 1964 for intermediate lavas of Thingmuli volcano, Iceland, that, compared to calc-alkaline andesites, are low in aluminum, high in iron, and have fewer mafic phenocrysts.

Iceland spar A very pure and transparent variety of calcite, the best of which is obtained in Iceland, that cleaves easily and perfectly into rhombohedrons and that exhibits strong double refraction; an *optical calcite.* It occurs in vugs and cavities in volcanic rocks and as nodules in residual clays in limestone regions. Syn: *double-refracting spar.*

ice layer [glaciol] An *ice lens* of particular great lateral extent. It is commonly a remnant *ice crust* that has been covered by snow.

ice layer [permafrost] An approximately horizontal layer of ground ice, sometimes lenticular (see *ice lens [permafrost]*). Syn: *ground-ice layer.*

ice lens [glaciol] A lens of solid ice or iced firn in a mass of snow or firn. The lens is formed by the refreezing of percolating melt water that spreads out along a relative impermeable horizon in the snow or firn and tapers out laterally.

ice lens [permafrost] A discontinuous layer of *ground ice* tapering at the periphery; ice lenses in soil commonly occur parallel to each other in repeated layers.

ice limit (a) The extreme minimum or the extreme maximum extent of the *ice edge* in any given time period, based on observations over several years. The term should be prefaced by "minimum" or "maximum". (b) Obsolete syn. of *mean ice edge.*

ice mantle *ice sheet.*

ice-marginal drainage Stream drainage along the side or front of a glacier.

ice-marginal lake A lake of meltwater dammed between an ice barrier and the valley wall, or between ice lobes. Syn: *proglacial lake.*

ice-marginal terrace *kame terrace.*

ice-marginal valley A valley parallel to the margin of a glacier. It may serve for the draining-away of meltwater.

ice massif *ice cluster.*

ice mound (a) A *frost mound* containing bodies of ice. (b) *icing mound.* Syn: *ice blister.* (c) *moated ice mound.*

ice mountain A popular term for a large *iceberg.*

Icenian (I-cen′-i-an) North Sea stage: upper Pliocene (above Amstelian, in strict sense), based on preglacial molluscan faunas. See: *Baventian.*
I-centered lattice *body-centered lattice.*
ice pack *pack ice.*
ice pan A large flat piece of first-year ice protruding several centimeters to a meter above the sea surface. Syn: *pan [ice].*
ice patch An area of pack ice smaller than an ice field. Syn: *patch [ice].*
ice pavement *glacial pavement.*
ice pedestal A pinnacle, column, or cone of ice projecting from the surface of a glacier and supporting, or formerly supporting, a large rock (*glacier table*) or mass of debris. The rock or debris protected the ice beneath it from solar radiation, so that it ablated less rapidly than the surrounding ice. Cf: *mushroom ice.* Syn: *ice pillar; pedestal [glaciol].*
ice pellets *Precipitation* that consists of frozen raindrops 5mm (0.2 in) or less in diameter that bounce on impact with ground; also called sleet.
ice penitente A *nieve penitente* consisting of glacier ice.
ice period The period of time from freezeup to breakup of ice.
ice piedmont A mass of ice, sloping gently seaward, that covers a coastal strip of low-lying land backed by mountains (Armstrong and Roberts, 1958, p.7). It may be anywhere from one kilometer to 50 km in width, and its outer edge may be marked by a line of ice cliffs. Ice piedmonts frequently merge into ice shelves. Cf: *ice fringe.*
ice pillar Any tall, narrow mass of ice, such as an *ice pedestal* or *mushroom ice.* Syn: *ice column.*
ice pipe An ice mass of cylindrical shape.
ice plateau (a) An ice-covered highland area whose upper surface is nearly level and whose sides descend steeply to lowlands or to the ocean. Cf: *plateau glacier.* (b) Any *ice sheet* with a level or gently rounded surface, as, e.g., the Polar Plateau in Antarctica.
ice pole The approximate center of the most consolidated part of the arctic pack ice, and therefore a difficult point to reach by surface travel; it is near lat. 83°-84°N and long. 160°W. Syn: *pole of inaccessibility.*
ice potential "The potential amount of ice that would be formed in a given water mass if surface heat loss provided the thermohaline circulation" (Baker et al., 1966, p.84).
ice push (a) The lateral pressure exerted by the expansion of shoreward-moving ice, esp. of lake ice. Syn: *ice shove; ice thrust.* (b) The ridge of material formed by an ice push. Syn: *lake rampart; ice-push ridge.*
ice-pushed ridge An asymmetric ridge of local, essentially nonglacial material (such as deformed bedrock, with some drift incorporated in it) that has been pressed up by the shearing action of an advancing glacier. It is typically 10-60 m high, about 150-300 m wide, and as much as 5 km long. Examples are common on the Great Plains of North America, where such ridges occur on the sides of escarpments formed of relatively incompetent rocks that face the direction from which the ice moved. The term is sometimes used "not quite correctly" as a synonym of *push moraine,* which is an accumulation of glacial drift (Schieferdecker, 1959, term 0924). Syn: *ice-contorted ridge; ice-thrust ridge; glacial pressure ridge.*
ice-push ridge *ice push.*
ice-push terrace A terracelike accumulation of coarse material pushed up along a shore; esp. a terrace consisting of successive lake ramparts.
ice pyramid A roughly conical mound of ice on the surface of a glacier, formed by differential ablation; e.g. an ice pedestal whose sides have been melted back, making a cone shape. Syn: *ice cone.*
ice-rafted Said of material such as boulders or till, which is deposited by the melting of floating ice containing it; esp. said of clasts and till distributed widely in marine sediments.
ice-rafting The transporting of rock fragments of all sizes on or within icebergs, ice floes, or other forms of floating ice. Cf: *rafting; dropstone.*
ice ramp An accumulation of snow or ice that forms a slope connecting two levels, including those which may build up on *fast ice* against ice fronts, ice walls, or rock cliffs. Ice ramps can also be artificially constructed or excavated (Armstrong et al., 1977).
ice rampart A syn. of *lake rampart.* The term is misleading because the rampart is a wall of boulders and other coarse material, not of ice.
ice-receiving area The portion of a surging glacier, generally near the terminus, that is periodically refilled by glacier surges.
ice-reservoir area That portion of a glacier that is periodically drained by glacier surges. This reservoir, which is refilled by direct snow accumulation or by normal ice flow between surges, may be located in nearly any part of the glacier system.
ice ribbon A thin, white, curly deposit of ice growing to a length of 10 cm or more, formed by the freezing of moisture exuded from the dead stem of a plant just above the ground in the early frosty period of winter before the ground is thoroughly frozen.
ice-rich permafrost Permafrost containing excess-ice.
ice rind A brittle, thin sea ice formed on a quiet surface by direct freezing or from grease ice, usually in water of low salinity; its thickness is generally less than 5 cm, and it is easily broken into rectangular pieces by wind or swell (U.S. Naval Oceanographic Office, 1968, p. B34). Syn: *ice crust [ice]; crust [ice].*
ice rise An ice mass, usually dome-shaped, resting on unexposed rock, and surrounded by an *ice shelf,* or in part by an ice shelf and in part by sea or ice-free land or both. Cf: *ice rumples.*
ice rumples Irregularities rising a few tens of meters above an *ice shelf* and moving with the shelf. The irregularities form over an area where the ice shelf is grounded but not extensively enough to form an *ice rise.*
ice run (a) Movement of ice *floes* with the current in a river at *breakup.* The ice run may be characterized as thin, close, or compact. (b) A rush of water following a breakup; a *debacle.*
ice-scour lake A glacial lake occupying a rock basin eroded by a glacier; e.g. a *finger lake.*
ice sheet A *glacier* of considerable thickness and more than 50,000 sq km in area, forming a continuous cover of ice and snow over a land surface, spreading outward in all directions and not confined by the underlying topography; a *continental glacier.* Ice sheets are now confined to polar regions (as on Greenland and Antarctica), but during the Pleistocene Epoch they covered large parts of North America and northern Europe. Armstrong and Roberts (1958, p.7) also applied the term to any extensive body of floating sea ice, but it should be restricted to land ice. Not to be confused with *sheet ice.* Cf: *inland ice; ice cap.* Syn: *ice mantle.*
ice shelf A sheet of very thick ice, with a level, gently sloping or gently undulating surface, which is attached to the land along one side but most of which is afloat and commonly bounded on the seaward side by a steep cliff (*ice front*) rising two to more than 50 m above sea level. Ice shelves have been formed along polar coasts (e.g., those of Antarctica, the Canadian Arctic islands, and Greenland), and they are generally of great breadth, some of them extending several hundreds of kilometers seaward from the coastline. They are nourished by annual snow accumulation and by seaward extension of land glaciers; limited areas may be aground. The term was used by Sir Douglas Mawson in 1912. Less recommended syn: *barrier ice; shelf ice.* Cf: *ice barrier.*
ice shove *ice push.*
ice slush *grease ice.*
ice spar A white or colorless, glassy, transparent variety of orthoclase; specif. *sanidine.*
ice spicule A small needlelike ice crystal that grows in water.
ice stalk A fibrous or spiky efflorescence of Taber ice, developed on the surface of freezing sediments. Syn: *ice grass.*
ice stone *cryolite.*
ice stream (a) A current of ice in an ice sheet or ice cap that flows more rapidly than the surrounding ice, usually flowing to the ocean or to an ice shelf and not constrained by exposed rock. Cf: *glacier; outlet glacier.* (b) One component of an *alpine glacier;* e.g. an *inset ice stream.* (c) An obsolete syn. of *outlet glacier* and *alpine glacier.* (d) A term sometimes popularly applied to a glacier of any kind, esp. an alpine glacier. Syn: *stream [glaciol].*
ice table A mass of *level ice.* Cf: *sheet ice.*
ice thrust *ice push.*
ice-thrust ridge (a) *ice-pushed ridge.* (b) *lake rampart.*
ice tongue Less preferred syn. of *tongue [glaciol].*
ice vein *ice wedge.*
ice wall (a) A cliff of ice forming the seaward margin of a glacier that is not afloat, such as of an ice sheet, an ice piedmont, or an ice rise. It is aground, the rock floor being at or below sea level. Cf: *ice front.* (b) Less preferred syn. of *ice cliff* in the sense of "any vertical wall of ice".
ice-walled channel A term used by Gravenor and Kupsch (1959, p.56) for a meltwater channel, either an open trench or a *tunnel valley,* containing a stream that may have flowed beneath a glacier.
ice wedge Wedge-shaped, foliated ground ice produced in permafrost, occurring as a vertical or inclined sheet, dike, or vein tapering downward, and measuring from a few millimeters to as much as 6 m wide and from 1 m to as much as 30 m high. It originates by the growth of hoar frost or by the freezing of water in a narrow crack or fissure produced by thermal contraction of the permafrost. See also: *ice-wedge cast.* Syn: *ground-ice wedge; ice vein.*
ice-wedge cast A sedimentary structure formed by the filling of the space formerly occupied by an *ice wedge* that had melted; the sediment fill may be wedge-shaped or very irregular. Syn: *ice wedge pseudomorph; ice-wedge fill; fossil ice wedge.*
ice-wedge fill *ice-wedge cast.*
ice-wedge polygon A large *nonsorted polygon* characterized by borders of intersecting ice wedges, found only in permafrost regions and formed by contraction of frozen ground. The fissured borders delineating the polygon may be ridges (low-centered polygon in which sediments are being upturned) or shallow troughs (high-centered polygon in which erosion and thawing are prevalent), and are underlain by ice wedges. Diameter: up to 150 m, averaging 10-40 m. In plan, the pattern tends to be three- to six-sided. See also: *fissure polygon.* Cf: *frost-crack polygon.* Syn: *tundra polygon; Taimyr polygon.*
ice-wedge pseudomorph *ice-wedge cast.*
ice-worn Abraded by ice; rubbed, striated, grooved, polished, or scoured by glacial action.
ice yowling A long, high-pitched sound accompanying the formation of contraction cracks in ice.

ichn A combining form signifying preservation of an original feature after alteration, e.g. a "clastichnic rock".
ichnite (ich'-nite) A fossil footprint or track. Syn: *ichnolite.*
ichnocoenosis (ich'-no-coe-no'-sis) An association of trace fossils.
ichnofabric All aspects of the texture and internal structure of a sediment that result from bioturbation and bioerosion at all scales (Ekdale and Bromley, 1983).
ichnofacies (ich-no-fa'-cies) A sedimentary facies characterized by particular trace fossils or by evidences of the life activities of fossil animals.
ichnofossil (ich'-no-fos'-sil) *trace fossil.*
ichnogenus (ich'-no-ge-nus) A genus of *trace fossil.*
ichnolite (ich'-no-lite) (a) *ichnite.* (b) The rock containing an ichnite. Obsolete.
ichnology (ich-nol'-o-gy) The study of trace fossils; esp. the study of fossil tracks. It includes both *palichnology* and *neoichnology.*
ichor (i'-chor) An obsolete term for a fluid thought to be responsible for such processes as *granitization* and *migmatization.*
ichthyodorulite (ich'-thy-o-dor'-u-lite) A spine, commonly a cutwater or support for a fin, of any nontetrapod vertebrate, esp. such a spine used as a basis for a form taxon.
ichthyolith (ich'-thy-o-lith') A microscopic form of fish skeletal debris, esp. denticles, occurring in pelagic sediments (Doyle et al., 1974, p.825). Their time relationships are studied in *stratignathy.*
Ichthyosauria (Ich'-thy-o-saur'-i-a) The sole order of the reptilian subclass Ichthyopterygia, of uncertain ancestry but of porpoiselike or sharklike body form as adaptation for life in the sea. Range, Middle Triassic to Upper Cretaceous.
Ichthyostegalia (Ich'-thy-o-steg-a'-li-a) An order of labyrinthodont amphibians characterized by a fishlike skull pattern and in at least one form by retention of bony supports for tail-fin rays. It includes the earliest known tetrapod. Range, Upper Devonian to Lower Mississippian.
icicle (i'-ci-cle) A narrow cone-shaped spike or shaft of clear ice, hanging with its point downward, formed by the freezing of dripping water; its length ranges from finger length to 8 m.
icing (ic'-ing) (a) A surface ice mass formed during the winter in a permafrost area by successive freezing of sheets of water that may seep from the ground, or from a spring or river. Syn: *aufeis; flood icing; river icing; flooding ice; flood-plain icing; glacier* (colloquial in Alaska); *ice field [permafrost].* (b) The accumulation of an ice deposit on exposed objects.
icing mound A thick, localized surface mound on an *icing;* it may form by the upwarp and breaching of a layer of ice (as in a river) by the hydrostatic pressure of water, which then issues from the mound as a freezing spring. Syn: *ice blister; ice mound.*
IC iron A member of the magmatic iron-meteorite group IC.
ICL *irregular crossed lamellar structure.*
icosacanthic law (i-co'-sa-can'-thic) *Müllerian law.*
icosahedron (i-co'-sa-he'-dron) A solid figure having 20 faces.
icositetrahedron (i-co'-si-tet'-ra-he'-dron) A term for the isometric *trapezohedron.*
Idahoan-type facies series (I'-da-ho'-an) Rocks produced in a type of dynamothermal regional metamorphism characteristic of Boehls Butte, Idaho, in which the polymorphs andalusite, kyanite, and sillimanite occur together with some staurolite and cordierite. Later studies have shown that the aluminum-silicate minerals were not in equilibrium; i.e., they formed at different times and in different conditions. Rarely used. Cf: *Barrovian-type facies series; Saxonian-type facies series.*
idaite (i'-da-ite) A metallic coppery red or brown hexagonal mineral: Cu_5FeS_6(?). It is a doubtful species, and may represent a mixture.
idd A term applied in northern Sudan to a place in the bed of an intermittent stream where water may be obtained in shallow wells for most if not all of the dry season (J.H.G. Lebon in Stamp, 1961, p.254).
Iddings' classification (Id'-dings') An obsolete classification of igneous rocks in which the mineralogic classifications of Rosenbusch and Zirkel are correlated with the CIPW or *normative* classification. The system was proposed in 1913 by J. P. Iddings.
iddingsite (id'-dings-ite) A reddish-brown mixture of silicates (of ferric iron, calcium, and magnesium) formed by the alteration of olivine. It forms rust-colored patches in basic igneous rocks.
ideal cut (i-deal') *Tolkowsky theoretical brilliant cut.*
ideal cyclothem A theoretical *cyclothem* that represents, in a given region and within a given stratigraphic interval, the optimum succession of deposits during a complete sedimentary cycle. It is constructed from theoretical considerations and from accumulated data from modern environments and experimental evidence. An ideal cyclothem of ten members for western Illinois consists of the following sequence (Weller and others, 1942, p.10): (10) marine shale with ironstone concretions; (9) clean marine limestone; (8) black laminated shale with limestone concretions or layers; (7) impure, lenticular, fine-grained marine limestone; (6) gray marine shale with pyritic nodules; (5) coal; (4) underclay; (3) freshwater, usually nonfossiliferous limestone; (2) sandy shale; and (1) fine-grained micaceous sandstone, locally unconformable on underlying beds. See also: *discordance index.*
ideal section A geologic cross section that combines observed evidence on stratigraphy and/or structure with interpretation of what is not present. It may be the summation or average of several successive cross sections.
ideal solution A solution in which the molecular interaction between components is the same as that within each component; a solution that conforms to *Raoult's law.* Cf: *nonideal solution.*
identified resources (i-den'-ti-fied) Resources whose location, grade, quality, and quantity are known or estimated from specific geologic evidence. Identified resources include economic, marginally economic, and subeconomic components. To reflect varying degrees of geologic certainty, these economic divisions can be subdivided into measured, indicated, and inferred (USGS, 1980). Cf: *hypothetical resources; speculative resources.*
identified subeconomic resources Mineral resources that are not *reserves,* but that may become reserves as a result of changes in economic and legal conditions (Pratt and Brobst, 1974, p.2). Syn: *conditional resources.* See also: *identified resources.*
idioblast (id'-i-o-blast') A mineral constituent of a metamorphic rock formed by recrystallization and bounded by its own crystal faces. It is a type of *crystalloblast.* The term was originated by Becke (1903). Cf: *hypidioblast; xenoblast.*
idioblastic (id'-i-o-blas'-tic) Pertaining to an *idioblast* of a metamorphic rock. It is analogous to the term *idiomorphic* in igneous rocks. Cf: *hypidioblastic.*
idioblastic series *crystalloblastic series.*
idiochromatic (id'-i-o-chro-mat'-ic) Said of a mineral whose color is a result of a *chromophore* that is essential to the chemical composition of the mineral. Cf: *allochromatic.*
idiogeosyncline (id'-i-o-ge'-o-syn'-cline) A type of late-cycle geosyncline between stable and mobile areas of the crust, the sediments of which are only weakly folded, such as the marginal basins of the East Indian island arc (Umbgrove, 1933, p.33-43). Obsolete. Cf: *parageosyncline (b).*
idiomorphic (id'-i-o-mor'-phic) A grain bounded by perfect crystal faces; well-formed. Syn: *automorphic; euhedral.*
idiomorphic-granular (id'-i-o-mor'-phic-gran'-u-lar) A phaneritic igneous texture dominated by euhedral, equant grains. Syn: *automorphic-granular.* Cf: *hypidiomorphic-granular; xenomorphic-granular.*
idiophanous (id-i-oph'-a-nous) Said of a crystal that exhibits an interference figure to the naked eye, without the help of optical instruments.
idiotopic (id'-i-o-top'-ic) Said of the fabric of a crystalline sedimentary rock in which the majority of the constituent crystals are euhedral. Also, said of the rock (such as an evaporite, a chemically deposited cement, or a recrystallized limestone or dolomite) with such a fabric. The term was proposed by Friedman (1965, p.648). Cf: *xenotopic; hypidiotopic.*
idocrase (i'-do-crase) *vesuvianite.*
IDP *interplanetary dust particles.*
idrialite (id'-ri-a-lite') An orthorhombic hydrocarbon mineral: $C_{22}H_{14}$. It was previously formulated $C_{24}H_{18}$. Idrialite is often found mixed with cinnabar and clay. Syn: *curtisite.*
IFSAR An airborne or spaceborne interferometer radar system, flown aboard rotary or fixed-wing aircraft or space-based platforms, that is used to acquire three-dimensional coordinates of terrain and terrain features that are both manmade and naturally occurring. IFSAR systems form synthetic aperture images of terrain features from two spatially separated antennae over an imaged swath that may be located to the left, right, or both sides of the imaging platform (Geo-One-Stop, 2003, p.3). Abbreviation for *interferometric synthetic aperture radar.*
igdloite (ig-dlo'-ite) *lueshite.*
IGF *International Gravity Formula.*
igneous (ig'-ne-ous) Said of a rock or mineral that solidified from molten or partly molten material, i.e. from a magma; also, applied to processes leading to, related to, or resulting from the formation of such rocks. Igneous rocks constitute one of the three main classes into which rocks are divided, the others being metamorphic and sedimentary. Etymol: Latin "ignis", "fire". See also: *magmatic; plutonic; hypabyssal; extrusive.*
igneous breccia (a) A breccia that is composed of fragments of igneous rock. (b) Any breccia produced by igneous processes, e.g. volcanic breccia, intrusion breccia. Cf: *agmatite.*
igneous complex An assemblage of intimately associated and roughly contemporaneous igneous rocks differing in form or in petrographic type; it may consist of plutonic rocks, volcanic rocks, or both.
igneous cycle The sequence of events in which volcanic activity is followed by major plutonic intrusions, and then minor intrusions (e.g. dikes). Obsolete.
igneous facies A part or variety of a single igneous rock body, differing in some attribute (structure, texture, or mineralogic or chemical composition) from the "normal" or typical rock of the main mass; e.g. a granite mass may grade into a porphyritic "igneous facies" near its borders.
igneous lamination In plutonic rocks, the arrangement of tabular crystals parallel to each other and to any layering the rocks may have.
igneous metamorphism A high-temperature metamorphic process that includes the effects of magma on adjacent rocks as well as those due to injection pegmatitization (Lindgren, 1933). The term is no longer in common use. Cf: *pyrometamorphism.*
igneous province *petrographic province.*
igneous-rock series An assemblage of temporally and spatially related igneous rocks of the same general form of occurrence (plutonic, hypabyssal, or volcanic), characterized by possessing in common certain chemical,

mineralogic, and textural features or properties so that the rocks together exhibit a continuous variation from one extremity of the series to the other. Syn: *rock series.*

ignimbrite (ig-nim'-brite) The deposit of a pyroclastic flow. The term originally implied dense welding but there is no longer such a restriction, so that the term includes rock types such as *welded tuff* and nonwelded *sillar.* See also: *tufflava; ash-flow tuff.* Syn: *flood tuff.*

ignimbrite shield A *shield volcano* built of rhyolitic ash flows, with a *collapse caldera* at its summit. Esp. well developed in the Tibesti region of Africa (Macdonald, 1972, p.267).

ignispumite (ig-ni-spum'-ite) A type of rhyolite characterized by lenticles and banding, which is believed to have been deposited as an acid, foamy lava and to be transitional with true ignimbrite. Cf: *tufflava.* Obsolete.

IGRF *International Geomagnetic Reference Field.*

IGRF correction *International Geomagnetic Reference Field.*

IGY *International Geophysical Year.*

ihleite (ih'-le-ite) *copiapite.*

IHS Intensity, hue, and saturation system of colors.

IIAB iron A member of the magmatic iron-meteorite group IIAB.

IIC iron A member of the magmatic iron-meteorite group IIC.

IID iron A member of the magmatic iron-meteorite group IID.

IIE iron A member of the non-magmatic iron-meteorite group IIE.

IIF iron A member of the magmatic iron-meteorite group IIF.

IIG iron A member of the magmatic iron-meteorite group IIG.

IIIAB iron A member of the magmatic iron-meteorite group IIIAB.

IIICD iron A member of non-magmatic iron-meteorite group IICD, now thought to be part of the IAB complex.

IIIE iron A member of the magmatic iron-meteorite group IIIE.

IIIF iron A member of the magmatic iron-meteorite group IIIF.

iimoriite (ii-mor'-i-ite) A buff, tan, or light purplish gray triclinic mineral: $Y_2(SiO_4)(CO_3)$.

ijolite (i'-jo-lite) (a) In the *IUGS classification,* a plutonic rock in which F is between 60 and 100, M is between 30 and 70, and sodium exceeds potassium. Cf: *fergusite.* (b) A series of plutonic rocks containing nepheline and 30-60% mafic minerals, generally clinopyroxene, and including titanite, apatite, and andradite; also, any rock of that series. *Melteigite* and *jacupirangite* are more mafic members of the series; *urtite* is a type rich in nepheline. Named by Ramsay in 1891 for Ijola (Iivaara), Finland. Cf: *fergusite; tawite.*

ijussite (i-juss'-ite) A mafic *teschenite* composed of abundant titanian augite and amphibole with smaller amounts of bytownite, anorthoclase, and analcime. The name is for the Ijuss River, Siberia, Russia. Obsolete.

ikaite (i'-ka-ite) A white chalky monoclinic mineral: $CaCO_3 \cdot 6H_2O$.

ikranite A yellow rhombohedral mineral of the *eudialyite* group: $(Na,H_3O)_{15}(Ca,Mn,REE)_6Fe^{3+}{}_2Zr_3Si_{24}O_{66}(O,OH)_6Cl \cdot nH_2O$.

ikunolite (i-ku'-no-lite) A metallic gray rhombohedral mineral: $Bi_4(S,Se)_3$. Cf: *laitakarite.*

ilesite (iles'-ite) A green monoclinic mineral of the *starkeyite* group: $(Mn^{2+},Zn,Fe^{2+})SO_4 \cdot 4H_2O$.

ilimaussite-(Ce) (il-i-mauss'-ite) A brownish-yellow hexagonal mineral: $Ba_2Na_4CeFeNb_2[Si_8O_{28}] \cdot 5H_2O$.

ill-conditioned Said of a situation where a small change in the data results in a large change in the solution.

illidromica (il-lid'-ro-mi'-ca) An unnecessary name for a hydromica low in potassium and high in water; *illite.* Also, a clay mineral intermediate in composition between illite and montmorillonite.

Illinoian (Il-li-noi'-an) Pertaining to the classical third glacial stage of the Pleistocene Epoch in North America, between the Yarmouthian and Sangamonian interglacial stages. Cf: *Riss.* Also spelled: Illinoisan. Obsolete.

illite (il'-lite) The term illite is used in two ways: In the general sense that Grim et al. (1937) introduced it, it is the 2:1 muscovite-like monoclinic or rhombohedral mineral in the clay-size fraction, but which has less K and more water than muscovite and gives a 10 Å d(001) from X-ray diffraction: $(K,H_3O)Al_2(Si_3Al)O_{10}(H_2O,OH)_2$. As a specific mineral, illite is an end-member of a series just as albite is the end-member of the *plagioclase* series. Because less than 5% of interstratified material in illite is difficult to detect by conventional X-ray methods, illite in the sense of a specific mineral, may contain up to 5% of an interstratified component. This component will be the other end of a compositional series. It is most commonly smectite, but can be vermiculite or perhaps chlorite. Srodon et al. (1992) concluded that illite has a layer charge of -0.89. However, others find values as low as -0.70. In the treatment here, the term illitic material covers the original, general intention of Grim et al. (1937), and the term illite should be used when referring to a specific mineral (Moore and Reynolds, 1996). In soil taxonomy, the presence of a 1nm X-ray diffraction peak and $\geq$4% K_2O is used to denote the presence of illite. Syn: *hydromuscovite.*

illite/smectite Probably the most common of the mixed-layered clay minerals formed of interstratified layers of illite and smectite that may be stacked randomly or in several patterns of ordering. Often abbreviated I/S. Cf: *Reichweite (R).* Illite and smectite are two ends of a series. In different geologic circumstances illite may go through a transition to smectite, and in others, smectite may change to illite. Although mechanisms are not yet agreed upon, it is apparently transformation in some situations and neoformation in others. Cf: *smectite-to-illite transition.*

illuvial horizon (il-lu'-vi-al) A soil horizon to which material has been added by the process of *illuviation.* Cf: *eluvial horizon.*

illuviation (il-lu'-vi-a'-tion) The accumulation, in a lower soil horizon, of soluble or suspended material that was transported from an upper horizon by the process of *eluviation.* Adj: illuvial.

Illyrian (Il-lyr'-i-an) A European substage: upper substage of the Anisian Stage, Middle Triassic (above Pelsonian, below Fassanian substage of the Ladinian).

ilmajokite (il-ma-jok'-ite) A yellow monoclinic mineral: $(Na,Ce,Ba)_{10}Ti_5Si_{14}O_{22}(OH)_{44} \cdot nH_2O$.

ilmenite (il'-men-ite) An iron-black, opaque, rhombohedral mineral of the corundum group: $Fe^{2+}TiO_3$. It is the principal ore of titanium. Ilmenite occurs as a common accessory mineral in basic igneous rocks (esp. gabbros and norites), and is also concentrated in mineral sands. See also: *menaccanite.* Syn: *titanic iron ore; mohsite.*

ilmenomagnetite (il'-men-o-mag'-net-ite) (a) Magnetite with microintergrowths of ilmenite. (b) Titanium-bearing maghemite with exsolution ilmenite. Cf: *magnetoilmenite.*

ilmenorutile (il'-men-o-ru-tile') A black tetragonal mineral: $(Ti,Nb,Fe^{2+})O_2$. Probably an unnecessary name for niobium-bearing rutile.

ilsemannite (il'-se-mann-ite') An ill-defined black, blue-black, or blue mineral: $Mo_3O_8 \cdot nH_2O$(?).

iltisite A red to brownish red hexagonal mineral: HgAgS(Cl,Br).

ilvaite (il'-va-ite) A brownish to black orthorhombic mineral: $CaFe^{2+}{}_2Fe^{3+}OSi_2O_7(OH)$. It is related to epidote, and usually contains manganese in small amounts. Syn: *lievrite; yenite.*

image [photo] (im'-age) (a) The recorded representation of a scene or object produced by an optical-mechanical or electronic scanner that converts electromagnetic radiation (EMR) outside the visible part of the spectrum into visible EMR. Cf: *photograph.* (b) The optical counterpart of an object, produced by the reflection or refraction of light when focused by a lens or mirror.

image [sed] (a) A term introduced by Wadell (1932, p.449) for a binomial expression of the shape of a sedimentary particle, expressed as a fraction giving the roundness of the particle in the numerator and the sphericity in the denominator. (b) A two-dimensional (plane) projection or cross section of a sedimentary particle, obtained by photography or by tracing; it is useful in determining *roundness.*

image degradation The loss of image clarity and/or accuracy, due to such things as data transmission, data compression, or data resampling.

image enhancement The application of any of a group of operations that alter or exaggerate the tonal differences in an image. These operations may include *contrast stretching, edge enhancement, filtering,* and/or smoothing to improve the detection or identification of features of interest (Krzanowski et al., 1993).

image map A map composed of satellite or photographic imagery, often cartographically enhanced with lines or symbols. An image map has been processed to be like a map in appearance, scale, geometry, and projection within a specified degree of accuracy.

image motion The smearing or blurring of imagery on an aerial photograph due to the relative movement of the camera with respect to the ground. An "image-motion compensator" installed with the camera intentionally imparts movement to the film at such a rate as to compensate for the forward motion of the aircraft during exposure time.

image processing system A system of software routines for the restoration, enhancement, and computer-assisted interpretation of digital images; most often used on remotely sensed data.

imager (im'-ag-er) A nonrecommended term for a remote-sensing system.

image ray A raypath perpendicular to the Earth's surface. The terminus of the ray on a time-migrated section locates the reflector in the presence of a lateral velocity gradient (Hubral and Krey, 1980, p.106-116).

imagery (im'-ag-ery) (a) The process of producing images; the term is analogous to *photography.* (b) A collective term for the images so produced. (c) The representation of objects on the ground reproduced electronically or by optical means on film, as obtained from satellites or airplanes.

imaging *migration [seis].*

imandrite (im-an'-drite) A yellow orthorhombic mineral: $Na_{12}Ca_3Fe^{3+}{}_2Si_{12}O_{36}$.

imatra stone (i-mat'-ra) *marlekor.*

imbibition [rock] (im-bi-bi'-tion) Formation of feldspathic minerals by the penetration of alkaline solutions of magmatic origin into aluminum-rich metamorphic rocks. Obsolete.

imbibition [water] (a) The absorption of a fluid, usually water, by a granular rock or any other porous material, under the force of capillary attraction, and in the absence of any pressure. Syn: *capillary percolation.* (b) Absorption of water by plants.

imbibometry (im-bi-bom'-e-try) A method of analysis involving measurement of the uptake of water or other fluids in a solid substance.

Imbrian (Im'-bri-an) (a) Pertaining to lunar topographic features and lithologic map units constituting a system of rocks including mare basalts and crater deposits deposited from the time of formation of the Mare Imbrium basin (3.85 Ga) until Imbrian basin flooding was mostly complete (about 3.2 Ga). Imbrian rocks are older than the post-mare craters and associated ejecta of the *Eratosthenian* and *Copernican* Systems and younger than rocks of the *Nectarian* System (Wilhelms, 1987, p.279). The Imbrian is time-equivalent to the late *Eoarchean* and *Paleoarchean* eras of Earth's Precambrian. (b) Said of the stratigraphic period during which the Copernican System was developed.

imbricated (im'-bri-cat'-ed) Overlapping, as tiles on a roof or scales on a bud.

imbricated fault zone A zone of closely spaced faults exhibiting imbricate structure; it is often developed above a décollement, and is underlain by a sole fault.

imbricated texture A texture resembling overlapping plates, seen in certain minerals (such as tridymite) under the microscope.

imbricate fan A series of faults that splay from a thrust fault and separate upward; the individual imbricates in the fan are usually concave-upwards (Boyer and Elliott, 1982). Cf: *imbricate fault array.*

imbricate fault array A series of related, subparallel dip-slip faults with similar displacements that merge with a single detachment horizon at depth. Faults in an imbricate array do not link in an upper detachment. Rock slices between imbricates are arranged in an overlapping shinglelike pattern. Imbricate arrays occur both in thrust systems or in normal-fault systems. Cf: *imbricate fan; duplex fault zone; schuppen zone.*

imbricate structure [sed] (im'-bri-cate) A sedimentary structure characterized by imbrication of pebbles all tilted in the same direction, with their flat sides commonly displaying an upstream dip. Syn: *shingle structure [sed].*

imbricate structure [tect] A tectonic structure displayed by a series of nearly parallel and overlapping minor thrust faults, high-angle reverse faults, or slides, and characterized by rock slices, sheets, plates, blocks, or wedges that are approximately equidistant and have the same displacement and that are all steeply inclined in the same direction (toward the source of stress). See also: *imbricated fault zone.* Syn: *schuppen structure; shingle-block structure.*

imbrication [palyn] (im-bri-ca'-tion) The pattern of overlap of dinoflagellate *thecal* or *cyst plates* along somewhat tapered plate margins (Williams et al., 2000, p.107).

imbrication [sed] (im-bri-ca'-tion) (a) A sedimentary fabric characterized by disk-shaped or elongate fragments dipping in a preferred direction at an angle to the bedding. It is commonly displayed by pebbles on a stream bed, where flowing water tilts the pebbles so that their flat surfaces dip upstream. (b) Formation of imbricate structure. Syn: *shingling.*

imbrication [tect] (a) The steeply inclined, overlapping arrangement of thrust sheets in imbricate structure. (b) Formation of such structure.

Imbrium sculpture A characteristic lunar landform developed radially to the Imbrium basin on the Moon. It consists of radial troughs and valleys and is interpreted primarily as the result of *secondary crater chains* from the impact that created the basin.

imerinite (im-er'-i-nite) *magnesioarfvedsonite.*

imgreite (im'-gre-ite) A metallic gray hexagonal mineral of the nickeline group: NiTe.

imhofite (im'-hof-ite) An adamantine red monoclinic mineral: $Tl_3As_8S_{13}$.

imitation (im-i-ta'-tion) Any material that simulates a genuine, natural gem; specif. glass, plastic, other amorphous material, or certain crystalline materials. Cf: *synthetic stone; assembled stone.* Syn: *simulant.*

imiterite (im-it'-er-ite) A metallic gray monoclinic mineral: Ag_2HgS_2.

immature [geomorph] (im-ma-ture') Said of a topography or region, and of its features, that have not attained maturity; esp. said of a valley or drainage system that is well above base level and whose side slopes descend steeply to the riverbanks, or of a drainage system that is expanding into an incompletely drained upland.

immature [sed] (a) Pertaining to the first stage of textural maturity (Folk, 1951); said of a clastic sediment that has been differentiated or evolved from its parent rock by processes acting over a short time and/or with a low intensity and that is characterized by relatively unstable minerals (such as feldspar), abundance of mobile oxides (such as alumina), presence of weatherable material (such as clay), and poorly sorted and angular grains. Example: an "immature sandstone" containing over 5% clay and commonly occurring in deeper marine, flood-plain, swamp, and mudflow deposits. Cf: *submature [sed]; mature [sed]; supermature.* (b) Said of an *argillaceous* sedimentary material intermediate in character between a clay and a shale; e.g. an "immature shale".

immature region *endozone.*

immature soil *azonal soil.*

immaturity (im-ma-tu'-ri-ty) A stage in the cycle of erosion or the sequential development of land forms that is characterized by immature features.

immediate runoff (im-me'-di-ate) *direct runoff.*

immersed bog (im-mersed') A bog which tends to expand horizontally by growth of plants under water. Cf: *emerged bog.*

immersion cell (im-mer'-sion) A cuplike accessory to a microscope or polariscope, in which a gemstone may be immersed in a liquid, generally of high refractive index, in order to eliminate surface reflection and refraction and thus allow efficient observation of the interior of the stone.

immersion liquid A liquid of known refractive index, used in the *immersion method* of determining a mineral's refractive index. Syn: *index liquid; index oil.*

immersion method A method of determining the refractive index of a mineral, by immersing the sample in a liquid of known refractive index (an *immersion liquid*).

immersion oil *immersion liquid.*

immigrant (im'-mi-grant) In ecology, an organism that becomes established in a region where it was previously unknown (Vermeij, 1991). Cf: *invasive.*

immiscible (im-mis'-ci-ble) Said of two or more phases that, at mutual equilibrium, cannot dissolve completely in one another, e.g. oil and water. Cf: *miscible.*

imogolite (im-o'-go-lite) A semiordered hydrous aluminosilicate with a threadlike morphology in electron microscope images: $Al_2SiO_3(OH)_4$. Closely related to allophane because both seem to be incipient 1:1 minerals both structurally and chemically. The ratio of SiO_2 to Al_2O_3 is usually 1.05 to 1.15, and all Al in imogolite is in sixfold coordination (Wada, 1989). Imogolite is mostly found in soils derived from volcanic ash, and in weathered pumices and spodosols.

impact (im'-pact) A collision of two planetary bodies at or near cosmic velocity, which causes the propagation of a shock wave into both the impactor and the target body.

impact bomb A mass of impact glass formed by shock melting, which can exhibit aerodynamic sculpturing. See also: *Fladen.*

impact cast *prod cast.*

impact crater A generally circular crater formed either by impact of an interplanetary body (projectile) on a planetary surface or by an experimental hypervelocity impact of a projectile into solid matter; craters formed by very oblique impacts may be elliptical. See also: *meteorite crater; penetration funnel; primary crater; secondary crater.*

impact erosion The wearing-away of rocks or fragments through the effect of definite blows of relatively large fragments.

impact glass *impactite.*

impactite (im-pact'-ite) (a) A term used for any shock-metamorphosed rock, including breccias and melt rocks. (b) A term used incorrectly for vesicular, glassy to finely crystalline material produced by fusion or partial fusion of target rock by the heat generated from the impact of a large meteorite and occurring in and around the resulting crater, typically as individual bodies composed of mixtures of melt and rock fragments, often with traces of meteoritic material; a rock (such as suevite) from a presumed impact site. Syn: *impact glass.*

impact law A physical law governing the settling of coarse particles, in which, for a given particle density, fluid density, and fluid viscosity, the settling velocity is directly proportional to the square root of the particle diameter. Cf: *Stokes' law.*

impact mark *prod mark [sed struc].*

impact melt A melt formed by *shock melting* of rocks in *impact craters.*

impact-melt breccia Impact-melt rock containing melt-supported lithic and mineral clasts displaying variable degrees of *shock metamorphism* in a formerly melted matrix that makes up 25-75 volume percent of the rock. Cf: *tagamite.* Syn: *melt-matrix breccia.*

impact-melt rock Crystalline to glassy rock solidified from *impact melt,* containing variable amounts of clastic debris and emplaced in and around an *impact crater* as dikes, sheets, or discrete fragments in breccias.

impact metamorphism Metamorphism of rocks or minerals caused by the passage of a shock wave due to impact of a planetary body (projectile or impactor) on a planetary surface (target). It includes melting and vaporization of the target rock(s). Cf: *shock metamorphism.*

impact plume A hot cloud of debris produced by a large impact event; these plumes are similar to those produced by the explosion of nuclear weapons.

impact slag *impactite.*

impact structure A generally circular or craterlike structure produced by impact (usually extraterrestrial) on a planetary surface. The stage of erosion of the structure and the nature of the impacting body need not be specified.

impact threshold wind velocity The minimum wind velocity required to maintain the saltation of sand (Bagnold, 1941, p.32).

imparipinnate (im-par'-i-pin'-nate) Said of a pinnately compound leaf having alternative leaflets.

impedance [elect] (im-ped'-ance) The complex ratio of voltage to current in an electrical circuit, or the complex ratio of electric-field intensity to magnetic-field intensity in an electromagnetic field. It is the reciprocal of admittance.

impedance [seis] *acoustic impedance.*

imperfect flower (im-per'-fect) A flower having either stamens or carpels but not both. Cf: *perfect flower.*

imperfection (im-per-fec'-tion) *flaw [gem].*

imperforate (im-per'-fo-rate) (a) Not perforated, or lacking a normal opening; esp. descriptive of foraminiferal-test walls without pores. (b) Said of a spiral mollusk shell in which the first whorl is tightly coiled upon itself, leaving no central opening. Cf: *perforate.*

Imperial jade (Im-pe′-ri-al) A translucent to semitransparent variety of jadeite characterized by the finest, highly intense emerald-green color; "true jade". Also spelled: imperial jade.

impermeability (im-per′-me-a-bil′-i-ty) The condition of a rock, sediment, or soil that renders it incapable of transmitting fluids under pressure. Syn: *imperviousness.* Ant: *permeability [geol].* Adj: impermeable.

impervious cover In urban hydrology, the surface of the Earth that is covered with concrete, roofs, or other structures that impede the infiltration of water into the soil (Keller, 2002).

imperviousness (im-per′-vi-ous-ness′) *impermeability.*

impolder (im-pol′-der) *empolder.*

impoverished fauna (im-pov′-er-ished) *depauperate fauna.*

impregnated (im-preg′-nat-ed) Said of a mineral deposit (esp. of metals) in which the minerals are epigenetic and diffused in the host rock. Cf: *disseminated; interstitial.*

impressed area (im-pressed′) The concave dorsum of a coiled cephalopod *conch* in contact with the venter of the preceding (next-older) whorl and tending to overlap it.

impression (im-pres′-sion) (a) The form, shape, or indentation made on a soft sedimentary surface (as of mud or sand) by an organic or inorganic body (usually a harder structure, such as a fossil shell or the strengthened surface of a leaf) that has come in contact with it; a shallow mold. It usually occurs as a negative or concavity found on the top of a bed, and a cast of it may then be found on the base of the overlying bed. (b) A small circular pit formed by rain, hail, drip, or spray. (c) A fossil footprint, trail, track, or burrow. Syn: *imprint.*

imprint (im′-print) An *impression,* esp. one made by a thin object such as a leaf or by a falling hailstone or raindrop.

imprisoned (im-pris′-oned) Said of a boulder or block of rock resting intimately against others with common or closely fitting interfaces, such as those found along a rocky coast like that of Victoria, Australia. Term used by Baker (1959, p.206).

improper symmetry operation Any symmetry operation that changes the handedness of the object on which it operates. *Reflection* and inversion are improper symmetry operations. Cf: *proper symmetry operation.*

impsonite (imp′-son-ite) A dull black, nearly infusible asphaltic pyrobitumen, with a hackly fracture and high fixed-carbon content. It closely resembles albertite, but is almost insoluble in turpentine; it is derived from the metamorphism of petroleum.

impulse (im′-pulse) An energy pulse of such short time-duration that its wave shape is of no consequence. Theoretically it is of infinite magnitude, infinitesimal time duration, and finite energy.

impulsive (im-pul′-sive) Said of the sharp arrival of a seismic phase.

impunctate (im-punc′-tate) Lacking pores, perforations, or punctae; specif. said of a brachiopod shell without endopunctae or pseudopunctae, in which the shell substance is dense. Impunctate brachiopods are by far the most numerous of all brachiopods. Cf: *punctate.*

impure arkose A term commonly applied to a sandstone (esp. a graywacke) that is highly feldspathic but is not an arkose; specif. a sandstone containing 25-90% feldspar and igneous-rock fragments, 10-50% mica and metamorphic-rock fragments, and 0-65% quartz and chert (Folk, 1954, p.354). The term is roughly equivalent to *micaceous arkose* of Hubert (1960), and was used by Krynine (1948, p.137) for a transitional rock between arkose and high-rank graywacke. Cf: *lithic arkose; feldspathic graywacke.* Syn: *dirty arkose.*

IMW *International Map of the World.*

INAA *instrumental neutron activation analysis.*

inactive volcano (in-ac′-tive) A volcano that has not been known to erupt. Cf: *active volcano; dormant volcano; extinct volcano.*

inadunate (in-ad′-u-nate) Any crinoid belonging to the subclass Inadunata, characterized by firmly jointed calyx plates, a mouth concealed by the tegmen, and arms that are free above the radials. Range, Ordovician to Triassic.

inaglyite (in-ag′-ly-ite) A metallic gray hexagonal mineral: $PbCu_3(Ir,Pt)_8S_{16}$. It is the Ir analogue of konderite, and occurs intimately intergrown with other Pt minerals in ultramafic rocks. See also: *konderite.*

in-and-out channel A crescentic valley excavated on a hillside by meltwater flowing around a projecting glacial lobe (Kendall, 1902, p.483).

inaperturate (in-ap-er′-tur-ate) Said of palynomorphs having no germinal, harmomegathic, or other preformed openings. Cf: *acolpate; alete.*

inarticulate (in-ar-tic′-u-late) n. Any inarticulated brachiopod, characterized by valves that are calcareous or composed of chitinophosphate and commonly held together by muscles rather than hinge teeth and dental sockets. adj. Said of a brachiopod having such valves, or of the valves themselves. Cf: *articulate [paleont].*

inarticulated brachiopod Any brachiopod without a distinct hinge and usually with a chitinophosphatic shell, formerly regarded as forming a distinct class, Inarticulata. Cf: *articulated brachiopod.*

inbreak (in′-break) A subsidence of the surface over a mine due to subterranean shattering of rock material. Cf: *crown-in; flash [mining]; creep [mining].*

incaite (in′-ca-ite) A metallic gray monoclinic mineral: $(Pb,Ag)_4Sn_4FeSb_2S_{14}$.

incandescent (in-can-des′-cent) Said of an ash flow, nuée ardente, or any pyroclastic matter that glows as a result of high temperature.

incandescent tuff flow A term essentially synonymous with *ash flow* or *pyroclastic flow;* it was originally used to describe the outbursts of fine-grained fragmental rhyolitic material in the Arequipa region of Peru (Fenner, 1948, p.879).

incarbonization (in-car′-bon-i-za′-tion) *coalification.*

incasion Collapse of a cave ceiling (Ford and Williams, 1989).

inception horizon In speleology, a favorable horizon or zone in sedimentary rock that guides the earliest phase of cave development (Lowe, 2000).

Inceptisols Mineral soils that have one or moer pedogenic horizons in which mineral materials other than carbonates or amorphous silica have been altered or removed but not accumulated to a significant degree. Under certain conditions, Inceptisols may have an ochric, umbric, histic, plaggen or mollic epipedon. Water is available to plants more than half of the year or more than 90 consecutive days during a warm season. An order in the U.S. system of soil taxonomy.

incertae sedis (in-cer′-tae sed′-is) A term applied to a fossil or modern specimen whose classification is regarded as uncertain. Etymol: Latin, "of uncertain place".

inch-scale layering In igneous rocks, a type of rhythmic layering consisting of layers slightly less than one inch thick, which may be repeated hundreds of times (Hess, 1960, p.133).

incidental recharge Water that percolates to the water table by seepage from sources such as water-supply lines, sewage systems, and septic-tank leach fields, which are unintended sources of groundwater recharge.

incidental vein (in-ci-den′-tal) A vein discovered after the original vein on which a claim is based.

incipient peneplain (in-cip′-i-ent) A syn. of *strath [geomorph]* in the restricted sense proposed by Bucher (1932, p.131), of a fluvial-degradation surface consisting of a broad valley floor and extensive valley-floor side strips. Syn: *partial peneplain; local peneplain.*

incised (in-cised′) In geomorphology, said of a stream meander, channel, or notch that has been downcut or entrenched into the surface. Cf: *incision.*

incised meander (a) A generic term for an old stream meander that has become deepened by rejuvenation and that is more or less closely bordered or enclosed by valley walls. Two types are usually recognized: *entrenched meander* and *ingrown meander.* Syn: *inclosed meander.* (b) Used in a more restricted sense as a syn. of *entrenched meander.*

incised valleys In sequence stratigraphy, the channels or valleys formed by fluvial systems that extend their channels basinward and erode into underlying strata in response to a relative fall in sea level. Incised valleys can be up to several hundred feet deep and range in width from half a mile to many tens of miles (Van Wagoner et al., 1990).

incision (in-ci′-sion) (a) The process whereby a downward-eroding stream deepens its channel or produces a narrow, steep-walled valley; esp. the downcutting of a stream during, and as a result of, rejuvenation, whether due to relative movement (uplift) of the crust or to other cause. Also, the product of such a process, e.g. an incised notch or meander. Syn: *downcutting; vertical erosion.* Cf: *entrenchment.* (b) The process whereby a steep-sided trench or notch intersects a plane surface or slope; e.g. current erosion of the continental slope to produce a submarine canyon.

incisor (in-ci′-sor) (a) A process with a biting surface on the gnathal lobe of the *mandible* of a crustacean. (b) One of the premaxillary (anterior) teeth of mammals, generally used for grasping, ripping, or scraping, which are relatively unspecialized morphologically and in most forms are replaced in an individual's lifetime. Cf: *molar.*

inclination [astron] The angle between the plane of an orbit and a reference plane. For planets, the plane of the Earth's orbit (ecliptic) is the usual reference plane. For satellites of planets, the planet's *equatorial plane* is commonly used.

inclination [drill] (in-cli-na′-tion) The angle of the axis of a *wellbore* measured from the vertical at a stated depth. See also: *deviation [drill].* Syn: *drift [drill].*

inclination [magnet] The angle between the horizontal and total magnetic field direction; it is one of the *magnetic elements.* Syn: *dip [magnet]; magnetic dip; magnetic inclination.*

inclination [paleont] The attitude of the *cardinal area* (or pseudointerarea) in either valve of a brachiopod, based on the convention of viewing the specimen in lateral profile with beaks to the left and *brachial valve* uppermost, referring the cardinal area to its position within one of four quadrants defined by the commissural plane and the plane normal to it and the plane of symmetry, touching the base of the cardinal areas (TIP, 1965, pt. H, p.146). See also: *orthocline; catacline; anacline; hypercline; apsacline; procline.* Cf: *obliquity.*

inclination [slopes] (a) A deviation from the true vertical or horizontal. Also, the amount of such deviation; the rate of slope, or grade. (b) An inclined surface; a slope.

inclination [struc geol] A general term for the slope of any geological body or surface, measured in the upward or downward direction and from the horizontal or the vertical. It is often used synonymously with *dip.*

inclinator (in′-cli-na-tor) *inclinometer [magnet].*

inclined bedding (in-clined′) (a) An inclusive term for bedding inclined to the principal surface of deposition. The term is not recommended for use as a syn. of *cross-bedding* because it may "equally refer . . to any initial dip" (Hills, 1963, p.10). See also: *discordant bedding.* Archaic syn: *diagonal bedding;*

oblique bedding. (b) Bedding laid down with primary or initial dip (Dennis, 1967, p.12).

inclined extinction A type of extinction seen in birefringent crystal sections in which the vibration directions are inclined to a crystal axis or direction of cleavage. Cf: *parallel extinction; undulatory extinction.* Syn: *oblique extinction.*

inclined fold A fold whose axial surface is inclined from the vertical, and in which one limb may be steeper than the other. The term sometimes includes the restriction that the steeper of the two limbs not be overturned.

inclinometer [drill] (in-cli-nom'-e-ter) Any of various instruments for measuring *inclination [drill];* a *driftmeter.*

inclinometer [magnet] An instrument that measures magnetic inclination. See also: *earth inductor; dip circle.* Syn: *inclinator.*

inclinometer [surv] Instrument for measuring inclination to the horizontal of a ship or aircraft.

inclosed meander (in-closed') A syn. of *incised meander;* it was proposed as a generic term by Moore (1926). Also spelled: *enclosed meander.*

included gas (in-clud'-ed) Gas in isolated interstices in either the *unsaturated zone* or the *saturated zone* (Meinzer, 1923, p.21). The term may also be applied to bubbles of air or other gas, not in isolated interstices, that are surrounded by water in either zone and that act as obstacles to water flow until the gas disappears by dissolving in the water. Cf: *ground air; subsurface air.*

inclusion (in-clu'-sion) (a) A fragment of older rock within an igneous rock to which it may or may not be genetically related. Syn: *xenolith.* See also: *autolith.* (b) A small foreign body (solid, liquid, or gas) in a gem material, usually seen with magnification. Inclusions often provide important information to help identify gemstones.

inclusion band Inclusions of wall rock material in a vein arranged in bands parallel to the wall of a vein. Inclusion bands are common in antitaxial crystal fiber veins that form by a *crack-seal* mechanism (Ramsay and Huber, 1983).

inclusion hackle The fine-scale roughness within a joint plume or mirror region of a joint surface where microscopic out-of-plane propagation is caused by irregular grain shapes or other small features capable of distorting the local stress field at the propagating crack tip (Kulander et al., 1979). Cf: *plumose surface morphology.*

inclusion trail A trail of small crystals in a porphyroblast commonly arranged in a spiral or sigmoidal pattern.

incoalation (in-coal-a'-tion) *coalification.*

incoherent (in-co-her'-ent) Said of a rock or deposit that is loose or unconsolidated.

incommensurate phase A *modulated structure* in which the modulation takes the form of a wavelike displacement of atoms from the ideal positions of the parent phase, so that the rigid periodicity characteristic of crystalline materials is slightly disrupted. Cf: *continuous transformation.*

incompatible element (a) An element for which the *partition coefficient* for a mineral or assemblage of minerals relative to a liquid is much less than unity; i.e., the element is concentrated in the liquid. (b) A chemical element that tends to remain in the melt when a magma crystallizes. Cf: *compatible element.*

incompetent rock A volume of rock which at a specific time and under specific conditions is not able to support a tectonic force. Cf: *competent rock.*

incomplete caneolith (in-com-plete') A *caneolith* having upper and lower rim elements but lacking a wall. Cf: *complete caneolith.*

incomplete flower A flower which lacks one or more of the four floral appendages (stamens, carpels, corolla or petals, and calyx or sepals). Cf: *complete flower.*

incomplete ripple *starved ripple.*

incomplete tabula A coral *tabula* consisting of several tabellae joined together. Cf: *complete tabula.*

incompressibility modulus (in'-com-pres'-si-bil'-i-ty) *bulk modulus.*

incongruent melting (in-con'-gru-ent) Melting accompanied by decomposition or by reaction with the liquid, so that one solid phase is converted into another; melting to give a liquid different in composition from the original solid. An example is orthoclase melting incongruently to give leucite and a liquid richer in silica than the original orthoclase. Cf: *congruent melting.*

incongruent solution Dissolution accompanied by decomposition or by reaction with the liquid so that one solid phase is converted into another; dissolution to give dissolved material in different proportions from those in the original solid.

incongruous (in-con'-gru-ous) Said of a parasitic fold, the axis and axial surface of which are not parallel to the axis and axial surface of the main fold to which it is related. Ant: *congruent.*

inconsequent (in-con'-se-quent) A syn. of *insequent.* The term was used by Gilbert (1877, p.143-144), but is not now in general use.

incorporation (in-cor'-po-ra'-tion) A process of *coalification* in which there is no modification of material. Cf: *vitrinization; fusinization.*

increase (in'-crease) The addition of corallites to colonies by offset formation resulting from parent polyps. Examples: *axial increase; lateral increase; intermural increase; peripheral increase.*

increment (in'-cre-ment) *recharge.*

incremental strain (in-cre-men'-tal) A finite or infinitesimal strain relating two sequential configurations of a body. The total finite strain of a rock body is the cumulative result of a number of incremental strains (Means, 1976, p.224).

incretion (in-cre'-tion) (a) A term proposed by Todd (1903) for a cylindrical *concretion* with a hollow core; e.g. a rhizocretion. (b) A concretion whose growth has been directed inward from without.

incrustation (in-crus-ta'-tion) *encrustation.*

incumbent (in-cum'-bent) Lying above; said of an overlying or superimposed stratum.

incurrent canal (in-cur'-rent) *inhalant canal.*

indefinite contour A contour used in limited areas where accurate contours are not feasible, usually in areas where the vegetative surface cover precludes economically contouring the ground within accuracy standards. Shown on USGS maps with a dashed line (USGS, 1996, p.7A-5). Syn: *approximate contour.* Cf: *form line.*

indehiscent (in-de-his'-cent) Said of a fruit or other plant structure that does not open spontaneously to distribute its contents (seeds or spores).

indelta (in-del'-ta) A term used in Australia for an inland area where a river subdivides (Taylor, 1951, p.615).

independent ovicell (in-de-pend'-ent) A bryozoan *ovicell* that develops independently of the distal *zooid.*

independent units Stratigraphic units that do not meet each other (Henningsmoen, 1961, p.66).

independent variable The situation in regression analysis and similar procedures where a variable y is treated as a function of other variables x_i plus a random error The x_i variables are independent and the y variable is the dependent variable. Cf: *dependent variable.*

inderborite (in-der-bor'-ite) A colorless monoclinic mineral: $CaMg[B_3O_3(OH)_5]_2{\cdot}6H_2O$.

inderite (in'-der-ite) A colorless to pink monoclinic mineral: $MgB_3O_3(OH)_5{\cdot}5H_2O$. Syn: *lesserite.*

index bed (in'-dex) *key bed.*

index contour A contour line shown on a map in a heavier lineweight than others for ease of identification, and generally labeled with an elevation value along its course. It appears at regular intervals, such as every fifth or sometimes every fourth contour line (depending on the contour interval). Syn: *accented contour.*

index ellipsoid The *indicatrix* of an anisotropic crystal.

index error (a) An instrument error, constant in behavior, caused by the displacement of the zero or index mark of a vernier; e.g. an error resulting from inclination of the upper plate in a transit having a fixed vertical vernier. (b) An instrument error in the magnetic bearing given by readings of the needle of a compass, such as an error arising from oblique magnetization of the needle or from the disturbance of the line of sight.

index factor A constant which, when multiplied by certain measurements made on potential-field anomalies, gives an estimate as to the depth to an anomalous mass (sometimes the maximum depth at which the mass could be located). It is used in magnetic and gravity interpretation. See also: *depth rule.*

index fossil A fossil that identifies and dates the strata or succession of strata in which it is found; esp. any fossil taxon (generally a genus, rarely a species) that combines morphologic distinctiveness with relatively common occurrence or great abundance and that is characterized by a broad, even worldwide, geographic range, and by a narrow or restricted stratigraphic range that may be demonstrated to approach isochroneity. The best index fossils include swimming or floating organisms that evolved rapidly and were distributed widely, such as graptolites and ammonites. The fossil need not necessarily be either confined to, or found throughout every part of, the strata for which it serves as an index. Cf: *characteristic fossil; guide fossil.* Syn: *key fossil; type fossil.*

index horizon A structural surface used as a reference in analyzing the geologic structure of an area. Syn: *index plane.*

index liquid *immersion liquid.*

index map A map showing the location and numbers of flight strips and aerial photographs; a map showing the outline of the area covered by each aerial photograph or by each of the maps in a series. Cf: *photoindex.*

index mineral A characteristic mineral developed under a particular set of temperature and pressure conditions. The first appearance of an index mineral is used to map zones of *progressive metamorphism.* Cf: *critical mineral [meta]; isograd; typomorphic mineral.*

index of refraction In crystal optics, a number that expresses the ratio of the velocity of light in vacuo to the velocity of light within the crystal. Its conventional symbol is n. Modifying factors include wavelength, temperature, and pressure. *Birefringent* crystals have more than one index of refraction. See also: *relative index of refraction; alpha [cryst]; beta [cryst]; gamma [cryst].* Syn: *refractive index.*

index oil *immersion liquid.*

index plane *index horizon.*

index species (a) A species of plant or animal that is characteristic of a particular set of environmental conditions and therefore whose presence in a particular area indicates the existence of those conditions in that area. (b) An *index fossil* of species rank.

index surface A two-shelled geometric surface that represents the indices of refraction in a biaxial crystal in the direction of propagation. Cf: *indicatrix.*
index zone A stratum or body of strata, recognizable by paleontologic or lithologic characters, that can be traced laterally and identifies a reference position in a stratigraphic section.
indialite (in'-di-a-lite') A colorless to blue hexagonal mineral of the *beryl* group: $(Mg,Fe^{2+})_2Al_4Si_5O_{18}$. It is the product of a burning coal seam.
indianaite (in-di-an'-a-ite) A white, porcelainlike clay mineral representing an impure variety of halloysite from Lawrence County, Indiana.
indianite (in'-di-an-ite) A variety of anorthite occurring as gangue of the *corundum* of the Carnatic of India.
Indian ridge (In'-di-an) A term used in New England for a sinuous *esker.* Syn: *serpent kame.*
indicated mineral resource That part of a *mineral resource* for which tonnage, densities, shape, physical characteristics, *grade,* and mineral content can be estimated with a reasonable level of confidence. It is indicated by information that is too widely or inappropriately spaced to confirm geological and/or grade continuity but is spaced closely enough for continuity to be assumed (JORC, 1999). Cf: *inferred mineral resource; measured mineral resource.*
indicated ore (in'-di-cat'-ed) Ore for which there are quantitative estimates of tonnage and grade, made partly from inference and partly from specific sampling. Cf: *inferred ore; possible ore; potential ore.* Syn: *probable ore.*
indicated reserve *probable reserve.*
indicator [eco geol] (in'-di-ca'-tor) A geologic or other feature that suggests the presence of a mineral deposit, e.g. a geochemical anomaly, a carbonaceous shale indicative of coal, or a pyrite-bearing bed that may lead to gold ore at its intersection with a quartz vein.
indicator [ecol] A plant or animal peculiar to a specific environment, which can therefore be used to identify that environment.
indicator [glac geol] *indicator stone.*
indicator fan A pattern formed by the distribution of indicator stones derived from a restricted source. Cf: *boulder train.*
indicator plant (a) A plant whose occurrence is broadly indicative of the soil of an area, e.g. its salinity or alkalinity, level of zone of saturation, and other soil conditions. (b) A plant that grows exclusively or preferentially on soil rich in a given metal or other element. Syn: *plant indicator.*
indicator stone A glacial erratic whose source and direction of transportation are known because of its identity with bedrock in a certain small or restricted area. Syn: *indicator [glac geol].*
indicatrix (in-di-ca'-trix) In optics, a geometric figure that represents the refractive indices of a crystal: it is formed by drawing, from a central point representing the center of the crystal, lines in all directions, whose lengths represent the refractive indices for those vibration directions. The figure for an isotropic crystal is a sphere; for a uniaxial crystal, an ellipsoid of revolution; and for a biaxial crystal, a triaxial ellipsoid (Berry and Mason, 1959, p.192). Cf: *index surface.* Partial syn: *index ellipsoid.* Syn: *optic indicatrix.*
indicatrix [cart] A circle or ellipse having the same shape as that of an infinitesimally small circle (having differential dimensions) on the Earth when it is plotted with finite dimensions on a map projection. Its axes lie in the directions of and are proportional to the maximum and minimum scales at that point. Often called a *Tissot indicatrix* after the originator of the concept (Snyder and Voxland, 1989, p.3).
indices of lattice row (in'-di-ces) Integral numbers, enclosed in square brackets, symbolized by [*uvw*], and determined, with any lattice point on the *lattice row* as origin, by the coordinates ua, vb, and wc, of the next lattice point on the row in terms of the unit cell edges a, b, and c. Indices enclosed in the following manner indicate a symmetrical set of axes: <*uvw*>. See also: *Miller indices; zone symbol.* Syn: *crystal axial indices.*
indicolite (in-dic'-o-lite) An indigo-blue (light-blue to bluish-black) variety of tourmaline, used as a gemstone. Syn: *indigolite.*
indigene (in'-di-gene) An indigenous organism.
indigenous (in-dig'-e-nous) Said of an organism originating in a specific place; native. Syn: *endemic.* Ant: *exotic [ecol].*
indigenous coal Coal formed according to the *in-situ theory;* autochthonous coal.
indigenous limonite Sulfide-derived limonite that remains fixed at the site of the parent sulfide, often as boxworks or other incrustation. Cf: *exotic limonite; relief limonite.*
indigenous stream A stream that lies wholly within its drainage basin. Cf: *exotic stream.*
indigirite (in-dig'-i-rite) A silky white mineral: $Mg_2Al_2(CO_3)_4(OH)_2{\cdot}15H_2O$.
indigo copper (in'-di-go) *covellite.*
indigolite (in-dig'-o-lite) *indicolite.*
indirect effect (in-di-rect') *Bowie effect.*
indirect intake Recharge to an aquifer by way of another body of rock.
indirect leveling A type of *leveling* in which differences of elevation are determined indirectly, as from vertical angles and horizontal distances (trigonometric leveling), from atmospheric pressures (barometric leveling), or from the boiling point of water (thermometric leveling). Cf: *direct leveling.*
indirect linkage A type of linkage in scleractinian corals with one or more couples of mesenteries between each pair of neighboring stomodaea. See also: *trabecular linkage.* Cf: *direct linkage.*
indirect stratification *secondary stratification.*
indite (in'-dite) A metallic black cubic mineral of the *linnaeite* group: $Fe^{2+}In_2S_4$.
indium (in'-di-um) A metallic gray tetragonal mineral, the native metallic element In. It is soft and malleable, and occurs in very small quantities in ores of zinc and other metals.
indochinite (in-do-chi'-nite) A tektite from southeast Asia (Cambodia, Laos, Vietnam, China, Thailand).
Induan (In'-du-an) A stage of the Standard Global Chronostratigraphic Scale: lowermost Triassic (above Changhsingian Stage of the Permian System, below Olenekian Stage).
induced correlation A spurious *correlation [stat]* induced between two or more variables by the nature of their definition or by some mathematical operation. Expressing measurements as a percent will insure spurious negative correlations between some variables because the percentage of one property can increase only if the percentages of others decrease.
induced infiltration (in-duced') Recharge to ground water by *infiltration,* of any type, from a body of surface water as a result of the lowering of the groundwater head below the surface-water level. Syn: *induced recharge.*
induced magnetization The magnetization of a rock in the direction of the applied magnetic field. Its magnitude is directly proportional to the strength of that field and the capacity of a rock to be magnetized, i.e., *susceptibility.* In the absence of *remanent magnetization,* the induced magnetization is the *magnetic moment* per unit volume.
induced meander *advance-cut meander.*
induced permeability *secondary permeability.*
induced polarization The production of electric charge displacement, brought about by application of an electric or magnetic field (induced electrical or magnetic polarization). Induced electrical polarization is manifested either by a slow decay of voltage in the Earth following the cessation of an excitation current pulse, or by a frequency dependence of the apparent resistivity of the Earth. Abbrev: IP.
induced radioactivity Radioactivity that is produced naturally or artificially by the interaction of energetic nuclear particles with stable or long-lived radioactive nuclides. Cf: *artificial radioactivity.*
induced recharge *induced infiltration.*
inductance (in-duct'-ance) That property of an electric circuit by which an electromotive force is induced in it by a current variation, either in it or in a neighboring circuit.
induction [magnet] (in-duc'-tion) (a) *magnetic induction.* (b) *electromagnetic induction.*
induction [philos] Reasoning from the particular to the general, or from the individual to the universal; deriving general principles from the examination of separate facts. Ant: *deduction.*
induction log An electric-log curve obtained, without the use of electrodes, by raising through the uncased borehole transmitting coils (fed with a constant alternating current) that induce, in the rocks surrounding the borehole, concentric eddy currents. These in turn induce fields that are detected by receiver coils. The magnitude of the fields is proportional to the conductivity of the surrounding rocks, and the log gives a continuous record of conductivity with depth. It is appropriate for measurements in empty holes, or in holes drilled with oil or with oil-base or freshwater muds. Syn: *conductivity log.*
induction number A distance in a conductive material measured as a number of radian wave lengths (real wave length/2π). It is a dimensionless quantity used to characterize the dimensions of a conductive body in an electromagnetic field.
inductive coupling (in-duc'-tive) The mutual impedance between a transmitting wire and a potential wire, arising from induction. This effect can lead to fictitious anomalies in induced-electrical-polarization surveys.
inductive method An electrical exploration method in which electric current is introduced into the ground by means of electromagnetic induction and in which the magnetic field associated with the current is determined.
inductura (in-duc'-tu-ra) The smooth shelly layer of a gastropod shell secreted by the general surface of the mantle, commonly extending from the inner side of the aperture over the parietal region, columellar lip, and part or all of the shell exterior (TIP, 1960, pt. I, p.131).
indurated (in'-du-rat'-ed) Said of a rock or soil hardened or consolidated by pressure, cementation, or heat.
indurated soil A very strongly cemented soil horizon. Cf: *cementation [soil].*
induration (in-du-ra'-tion) (a) The hardening of a rock or rock material by heat, pressure, or the introduction of cementing material; esp. the process by which relatively consolidated rock is made harder or more compact. See also: *lithification [sed].* (b) The hardening of a soil horizon by chemical action to form a *hardpan* or *duricrust.* Cf: *calcrete; silcrete; ferricrete; gypcrete; salcrete.*
indusium (in-du'-si-um) The covering of a *sorus* on the leaf of a fern.
industrial diamond (in-dus'-tri-al) A general term for diamonds used in drilling, in wire drawing, and as a general abrasive. See also: *ballas; bort; carbonado.*
industrial lithostratigraphic unit A lithostratigraphic unit recognized primarily for utilitarian purposes, such as an aquifer, oil sand, quarry layer, or ore-bearing bed. It is considered to be an informal unit even if named (ISG, 1976, p.35).

industrial mineral Any rock, mineral, or other naturally occurring substance of economic value, exclusive of metallic ores, mineral fuels, and gemstones; one of the nonmetallics.

inequigranular (in'-e-qui-gran'-u-lar) Texture consisting of grains exhibiting a range in grain size; used primarily for igneous and metamorphic rocks. Syn: *porphyritic; porphyroblastic; heterogranular.*

inequilateral (in'-e-qui-lat'-er-al) Having the two ends unequal; specif. said of a bivalve-mollusk shell whose parts anterior and posterior to the beaks differ appreciably in length. Cf: *inequivalve.* Ant: *equilateral.*

inequilibrium stage (in'-e-qui-lib'-ri-um) In hypsometric analysis of drainage basins, the stage of early development corresponding to youth in the geomorphic cycle (Strahler, 1952b, p.1130); the hypsometric integral is greater than 60%. Cf: *equilibrium stage.*

inequivalve (in'-e-qui-valve') Having valves unequal in size and form; specif. said of a bivalve mollusk or its shell in which one valve is flatter (and often smaller) than the other. Cf: *inequilateral.* Ant: *equivalve.*

inert component (in-ert') A component whose amount in a rock after a metasomatic process depends on its initial concentration rather than on its chemical potential as externally fixed by the environment. Cf: *mobile component; perfectly mobile component.*

inert gas Any of the six elements that have no tendency to react with any of the other elements: helium, neon, argon, krypton, xenon, and radon. They are all gases under usual conditions. Syn: *noble gas.*

inertial current Rotation of the whole ocean surface at the *inertial period,* due to sudden change in wind speed or direction, when only inertia and *Coriolis force* are important. Syn: *inertial wave.*

inertial frame of reference A coordinate system that has no translational acceleration and that is not rotating with respect to the distant galaxies.

inertial period Twelve hours (one half day) divided by the sine of the latitude. Cf: *inertial current.*

inertial surveying instrument (in-er'-tial) A device, mounted in a helicopter or truck, which combines gyroscopes, accelerometers, and electronic computers, and is used to extend positions from a known position or a basic net.

inertial surveying system A complex system of precise gyroscopes and accelerometers combined with electronics to sense and measure movement. Originally designed for navigation and guidance of military aircraft and ships, through continued development and refinement the systems are competitive in remote areas with conventional ground surveying methods for extending horizontal control.

inertial wave *inertial current.*

inertinite (in-ert'-i-nite) A coal maceral group including *micrinite, macrinite, sclerotinite* (term now replaced by *funginite* and *secretinite*), *fusinite, semifusinite,* and *inertodetrinite.* They are characterized by a relatively high carbon content and a reflectance higher than that of vitrinite. They are relatively inert during the carbonization process. Cf: *liptinite; vitrinite.* Syn: *inerts.*

inertite (in-ert'-ite) A microlithotype of coal that contains a combination of inertinite macerals totalling at least 95% (ICCP, 1971). Cf: *fusite; liptite; vitrite.*

inertodetrinite (in-ert'-o-det'-ri-nite) A maceral of coal within the *inertinite* group having a reflectance greater than that of associated vitrinite and occurring as discrete angular clastic fragments surrounded by other macerals or minerals, commonly vitrinite. Fragments are typically between 2 and 50 micrometers in diameter.

inertodetrite (in-ert'-o-det'-rite) A microlithotype of coal consisting of at least 95% of *inertodetrinite* (ICCP, 1971).

inerts (in-erts') An informal term, synonymous with *inertinite.*

inesite (i'-nes-ite) A rose-red, pink, orange-pink, or reddish brown triclinic mineral: $Ca_2Mn^{2+}_7Si_{10}O_{28}(OH)_2{\cdot}5H_2O$.

inface (in'-face) The steeper of the two slopes of a cuesta; the *scarp slope.* The term is an abbreviation of "inward-facing escarpment", referring to the cliff portion (of a cuesta) facing the oldland, as on a coastal plain.

infancy (in'-fan-cy) The hypothetical initial or very early stage in any developmental sequence, esp. in the *cycle of erosion,* commencing when a region is freshly exposed to the action of surface waters. Regional infancy is characterized by smooth, nearly level initial surfaces, imperfectly dissected by narrow stream gorges; numerous original and slight depressions, occupied by marshy lakes and ponds; shallow streams; and imperfect drainage systems. Cf: *youth [geomorph].*

infantile (in'-fan-tile) Pertaining to the initial or very early stage of the cycle of erosion, either of a stream that has just begun its work of erosion, or of a landscape with a smooth surface and numerous shallow lakes. Cf: *youthful.*

infauna (in'-fau-na) Those aquatic animals that live within rather than on the bottom sediment. Adj: infaunal. Cf: *epifauna.*

infaunal Marine organisms living entirely within the sediment, as opposed to epifaunal (living on top of the sediment), or semi-infaunal (living partly infaunally and partly epifaunally).

inferior groove (in-fe'-ri-or) A groove in decapod crustaceans that extends downward from the posterior part of the *hepatic groove;* it might be considered a continuation of the *postcervical groove* (Holthuis, 1974, p.733).

inferior ovary An *epigynous* plant ovary.

inferred mineral resource That part of a *mineral resource* for which tonnage, *grade,* and mineral content can be estimated with a low level of confidence. It is inferred from geological evidence and assumes, but does not verify, geological and/or grade continuity (JORC, 1999). Cf: *indicated mineral resource; measured mineral resource.*

inferred ore (in-ferred') Ore for which there are quantitative estimates of tonnage and grade made in only a general way, based on geologic relationships and on past mining experience, rather than on specific sampling. Cf: *indicated ore; possible ore; potential ore.*

infilling (in'-fill-ing) A process of deposition by which sediment falls or is washed into depressions, cracks, channels, or holes, as the filling-in of crevasses upon the melting of glacier ice.

infiltrated fossils *introduced fossils.*

infiltration (in-fil-tra'-tion) (a) The flow of a fluid into a solid substance through pores or small openings; specif. the movement of water into soil or porous rock. Cf: *percolation.* (b) The process of falling rain or melting snow entering a soil or rock across its interface with the atmosphere, or water from a stream entering its streambed across the stream-streambed interface. Infiltration connotes flow into a material, in contrast to *percolation,* which connotes flow through a material.

infiltration basin A basin in which water is spread for groundwater recharge.

infiltration capacity The maximum or limiting *infiltration rate.* The term is considered an obsolete syn. of *infiltration rate* by the SSSA (1965, p.338). Symbol: f.

infiltration coefficient The ratio of infiltration to precipitation for a specific soil under specified conditions (Nelson and Nelson, 1967, p.192).

infiltration front *pellicular front.*

infiltration gallery A horizontal conduit constructed for the purpose of intercepting groundwater. The galleries often parallel rivers, which provide a perennial water supply to the conduit. Syn: *gallery [grd wat].*

infiltration index The average rate of *infiltration,* expressed in inches per hour. It equals the average rate of rainfall "such that the volume of rainfall at greater rates equals the total direct runoff" (Langbein and Iseri, 1960, p.12).

infiltration metasomatism A process of mass transfer in which chemical components are transported by a stream of aqueous solutions percolating through pores in rocks. Cf: *diffusion metasomatism.*

infiltration rate The amount of *infiltration (b)* per unit of time; expressed in depth of water per unit time (cm/sec; in/hr). Syn: *infiltration velocity.* Obsolete syn: *infiltration capacity.*

infiltration vein An interstitial mineral deposit formed by the action of percolating waters. Cf: *segregated vein.*

infiltration velocity *infiltration rate.*

infiltration water Nonpreferred syn. of *free water.*

infiltrometer (in-fil-trom'-e-ter) An instrument used to measure the infiltration of water into soil.

infinitesimal-strain theory (in'-fin-i-tes'-i-mal-strain') A theory of material deformation in which small displacements and small strains in geology, less than about 1% are analyzed. Cf: *finite-strain theory.*

inflammable cinnabar (in-flam'-ma-ble) A mixture of cinnabar, idrialite, and clay.

inflated lava A local area of flow-field of pahoehoe lava that has swelled as a result of injection of lava beneath its surface crust (Chitwood, 1994).

inflation [paleont] (in-fla'-tion) The distance, measured normal to the plane of symmetry, between the right and left valves of a bivalve mollusk; the length of the middorsal-midventral line, or the "width" of the shell. Syn: *thickness [paleont].*

inflation [volc] *tumescence.*

inflection angle The angle at which a contour line diverges downstream from a stream channel.

inflection line The line created by connecting inflection points in adjacent fold profiles. Syn: inflexion line.

inflection point The point in a fold profile where the curvature of a folded layer changes from concave upward to concave downward.

inflorescence (in-flo-res'-cence) A cluster of flowers. See also: *panicle.*

inflow (in'-flow) (a) The act or process of flowing in; e.g. the flow of water into a lake. Syn: *influx.* (b) Water that flows in; e.g. groundwater and rainfall flowing into the streams of a drainage basin. Also, the amount of water that has flowed in.

inflow cave A cave into which a stream flows, or is known to have flowed. See also: *outflow cave; sinking stream.* Syn: *influent cave.*

influent (in'-flu-ent) adj. Flowing in. n. (a) A surface stream that flows into a lake (e.g. an inlet), or a stream or branch that flows into a larger stream (e.g. a tributary). Ant: *effluent.* Cf: *influent stream.* Syn: *affluent.* (b) A stream that flows into a cave.

influent cave *inflow cave.*

influent flow Flow of water into the ground from a body of surface water; e.g. the flow of water from an *influent stream.*

influent seepage Movement of gravity water in the *unsaturated zone,* from the ground surface toward the water table; seepage of water into the ground.

influent stream (a) A stream or reach of a stream that contributes water to the zone of saturation and develops bank storage; its channel lies above the water table. Syn: *losing stream.* (b) *influent.*

influx (in'-flux) *inflow.*

informal lithostratigraphic unit (in-for'-mal) A lithologic body to which casual reference is made but for which there is insufficient need, insufficient information, or inappropriate basis to justify designation as a *formal unit.* It may be referred to informally as a lithozone, e.g. shaly zone or coal-bearing zone; or as a bed or member, e.g. sandy beds, shaly member (ISSC, 1994, p.36). Names of informal units follow the same rules of capitalization as ordinary common nouns (although failure to capitalize a unit name does not necessarily render the name informal).

infrabasal plate (in'-fra-ba'-sal) Any plate of the proximal circlet in a crinoid dorsal cup that has two circlets of plates (dicyclic) below the radial plates. Syn: infrabasal.

infracrustal (in'-fra-crust'-al) Within the crust (i.e., below the surface) and, in general, below the zone of weathering and diagenesis. Cf: *supracrustal.*

infraglacial (in'-fra-gla'-cial) *subglacial.*

infralaminal accessory aperture (in'-fra-lam'-i-nal) An *accessory aperture* in the test of a planktonic foraminifer that leads to a cavity beneath accessory structures and that is at the margin of these structures (as in *Catapsydrax*). Cf: *intralaminal accessory aperture.*

inframarginal sulcus (in'-fra-mar'-gi-nal) *scrobis septalis.*

infrared (in'-fra-red') Pertaining to or designating that part of the electromagnetic spectrum ranging in wavelength from 0.7 µm to about 1 mm.

infrared absorption spectroscopy The observation, recording, and analysis of an *absorption spectrum* in the infrared frequency region. Absorption of radiant energy for transitions within molecules at the lowest energy levels produces an infrared absorption spectrum. This technique is commonly used to determine the amount and type of water and OH ions present in a mineral crystal, because bonds between the H and O atoms are very polar and move when infrared photons are absorbed.

infrared color photograph A color photograph in which the red-imaging layer is sensitive to photographic *infrared* wavelengths, the green-imaging layer to red light, and the blue-imaging layer to green light. Also known as camouflage detection photographs and false-color photographs.

infrared image An image produced by the *thermal infrared* spectral band.

infrasculpture (in'-fra-sculp'-ture) A kind of structure of palynomorphs consisting of organized internal modifications of exine.

infrastructure (in'-fra-struc'-ture) In tectonics, a structure, produced at a deep crustal level, under conditions of elevated temperature and pressure, that is characterized by plastic folding and the emplacement of granite and other migmatitic and magmatic rocks. This environment occurs in the internal parts of most orogenic belts, but the term is used especially where the infrastructure contrasts with an overlying, less disturbed layer, or superstructure.

infraturma (in'-fra-tur-ma) One of the subdivisions of *turma.* Pl: infraturmae. Cf: *turma.*

infundibulum (in-fun-dib'-u-lum) (a) A deep indentation of the *scrobis septalis* or a basal indentation of the apertural face of a foraminiferal test (as in *Alabamina*) (TIP, 1964, pt.C, p.61). (b) The apertural region of a tintinnid. (c) In the vertebrates, a neural component (neurohypophysis) of the pituitary gland. Pl: infundibula.

infusible (in-fu'-si-ble) Said of a mineral that ranks with quartz on the fusibility scale; i.e., that will not fuse in temperatures up to about 1,500°C. Cf: *fusibility.*

ingenite (in'-ge-nite) A general term, now obsolete, for a rock originating below the Earth's surface; an igneous or metamorphic rock. Cf: *plutonic rock.*

ingersonite (ing'-er-son-ite) A vitreous brownish-yellow trigonal mineral: $Ca_3Mn^{2+}Sb^{5+}{}_4O_{14}$.

ingodite (in'-go-dite) A metallic steel-gray hexagonal mineral: Bi_2TeS.

ingrafted stream (in-graft'-ed) *engrafted stream.*

ingression (in-gres'-sion) The entering of the sea at a given place, as the drowning of a river valley (Schieferdecker, 1959, terms 1260, 1840).

ingroup In cladistic analysis, the group of taxa studied by an investigator (Wiley et al., 1991, p.4). Cf: *outgroup.*

ingrown meander (in'-grown) A term proposed by Rich (1914, p.470) for a continually growing or expanding *incised meander* formed during a single cycle of erosion by the enlargement or accentuation of an initial minor sinuosity while the stream was actively downcutting; a meander that "grows in place". It exhibits a pronounced asymmetric cross profile (a well-developed, steep undercut slope on the outside of the meander, a gentle slip-off slope on the inside) and is produced when the rate of downcutting is slow enough to afford time for lateral erosion. Cf: *entrenched meander.*

ingrown stream A stream that has enlarged its original course by undercutting the outer (concave) banks of its curves.

inhalant canal (in-hal'-ant) Any canal forming part of the *inhalant system* of a sponge. Syn: *incurrent canal; prosochete.*

inhalant system The part of the *aquiferous system* of a sponge between the ostia and the prosopyles, characterized by water flowing inward from the ostia. Cf: *exhalant system.*

inherent ash (in-her'-ent) Ash in coal derived from inorganic material that was structurally part of the original plant material. It cannot be separated mechanically from the coal or peat, of which it commonly constitutes not more than 1%. Cf: *extraneous ash.* Syn: *constitutional ash; fixed ash; intrinsic ash; plant ash.*

inherent mineral matter *Mineral matter* in coal and peat that was structurally part of the original organic material.

inherent moisture (a) In coal, that fraction of the moisture content that is structurally contained in the material. Syn: *bed moisture.* (b) The maximum moisture that a sample of coal will hold at 100% humidity and atmospheric pressure.

inherited (in-her'-it-ed) (a) Said of a geologic structure, feature, or landscape that owes its character to conditions or events of a former period; esp. said of a superimposed stream, valley, or drainage system. (b) Also, said of a soil or sediment characteristic that is directly related to the nature of the parent material rather than to formative processes. See also: *lithomorphic.*

inherited argon Argon-40 that is produced within mineral grains by the decay of potassium-40 before the event being dated. It may have been generated during the premetamorphic history of a rock, which has survived a metamorphic event, or it resulted from the incorporation of older contaminating mineral grains in a dated sample. Cf: *excess argon; extraneous argon; radiogenic argon.*

inherited flow control The control of glacial drainage by the disposition of blocks of dead ice that had broken along thrust planes or open crevasses when the ice was in motion (Gravenor and Kupsch, 1959, p.49).

inherited meander *entrenched meander.*

inherited septum A *septum* originally in the parent *corallite,* which is inherited by the offset during increase in corals (Fedorowski and Jull, 1976, p.41).

inhomogeneity breccia (in-ho'-mo-ge-ne'-i-ty) A term used by Sander (1951, p.2) for a breccia that forms paradiagenetically by the rupture of relatively friable layers occurring within a plastic sediment. It contains sharp fragments with broken borders that sometimes can be matched with each other.

initial bud (in-i'-tial) (a) An outgrowth through a hole in the wall of a graptolithine *sicula,* producing the first *theca* of the *rhabdosome.* (b) The first youthful *polyp* formed asexually from the protopolyp in a colony of corals.

initial dip (a) A syn. of *primary dip.* (b) The dip that a bedded deposit attains due to compaction after sedimentation, but before tectonic deformation.

initial landform A landform that is produced directly by epeirogenic, orogenic, or volcanic activity, whose original features are only slightly modified by erosion; it is dominant in the initial and youthful stages of the erosion cycle. Cf: *sequential landform; ultimate landform.*

initial meridian *prime meridian.*

initial point The point from which any survey is initiated; esp. the point from which a survey within a given area of the U.S. public-land system begins and from which a base line is run east and west and a principal meridian is run north and south. Syn: *point of origin.*

initial production The volume or quantity of oil or gas that a well produces or is capable of producing during the first 24 hours after completion.

initial ratio [radioactivity] The atomic ratio of two isotopes of an element at the time of crystallization of a mineral containing one or more radioactive elements.

initial shoreline A shoreline brought about by regional tectonic activity (subsidence, uplift, faulting, folding), by volcanic accumulation, or by glacial action; it may have any slope, from almost vertical to nearly horizontal, and may be either smooth or irregular.

initiation point The flaw from which cracks and joints start to propagate.

injection [ign] (in-jec'-tion) *intrusion [ign].*

injection [sed] (a) The forcing, under abnormal pressure, of sedimentary material (downward, upward, or laterally) into a pre-existing deposit or rock, either along some plane of weakness or into a crack or fissure; e.g. the transformation of wet sands and silts to a fluid state and their emplacement in adjacent sediments, producing structures such as sandstone dikes or sand volcanoes. See also: *intrusion [sed].* (b) A sedimentary structure or rock formed by injection. Syn: *sedimentary injection.*

injection breccia A fragmental rock formed by the introduction of largely foreign rock fragments into veins and fractures in the host rock (Speers, 1957). Some examples (notably the Sudbury breccias at Sudbury, Canada) are associated with structures of impact origin and have formed as a result of the impact process.

injection complex An assemblage or association of rocks consisting of igneous intrusions in intricate relationship to sedimentary and metamorphic rocks. Cf: *migmatite.*

injection dike A sedimentary dike formed by abnormal pressure of injection from below or above or from the side. Cf: *neptunian dike.*

injection gneiss A composite rock whose banding is wholly or partly caused by *lit-par-lit* injection of granitic magma into layered rock (Holmes, 1928, p.124). Generally considered to form by the process of magmatic injection into high-grade metamorphic rocks; the injections may be concordant with or discordant to the fabric of the rock. Cf: *arterite; phlebite; venite.*

injection ice *intrusive ice.*

injection metamorphism Metamorphism accompanied by intimate injection of sheets and streaks of liquid magma (usually granitic) in zones near deep-seated intrusive contacts. Cf: *lit-par-lit.*

injection well (a) A well used to inject water, steam, or drill cuttings, usually under pressure, into a porous and permeable formation. The injected water may be treated wastewater or surface water injected for aquifer replenishment, or may be waste injected into an aquifer as a means of disposal (Rogers, 1981, p.199). (b) A well in an oil or gas field through which water, gas, steam, or chemicals are pumped into the reservoir formation for *pressure maintenance,* waterflood or enhanced oil recovery. Syn: *input well.*

inland (in'-land) Pertaining to or lying in the interior part of a country or continent, or not bordering on the sea; e.g. an *inland* lake.

inland basin *interior basin.*

inland blue hole A *blue hole* isolated by present topography from surface marine conditions and opening directly onto the land surface or into an isolated pond or lake. It contains tidally-influenced water of a variety of chemistries from fresh to marine (Mylroie et al., 1995). Cf: *ocean hole.*

inland drainage *internal drainage.*

inland ice (a) The ice forming the inner part of a *continental glacier* or large *ice sheet.* The term is applied esp. to the ice on Greenland. (b) A continental glacier or ice sheet in its entirety.

inland sea *epicontinental sea.*

inland waters The *territorial waters* (such as lakes, canals, rivers, inlets, bays) within the territory of a state, but excluding high seas and marginal waters that are subject to sovereign rights of bordering states; waters that are above the rise and fall of the tides.

inland waterway One of a system of navigable inland bodies of water (such as a river, canal, or sound).

inlet (in'-let) (a) A small, narrow opening, recess, indentation, or other entrance into a coastline or a shore of a lake or river, through which water penetrates into the land. Cf: *pass [coast].* (b) A waterway entering a sea, lake, or river; a creek; an inflowing stream. (c) A short, narrow waterway between islands, or connecting a bay, lagoon, or similar body of water with a larger body of water, such as a sea or lake; e.g. a waterway through a coastal obstruction (such as a reef or a barrier island) leading to a bay or lagoon. Syn: *tongue.* (d) *tidal inlet.*

inlier (in'-li-er) An area or group of rocks surrounded by younger rocks, e.g. an eroded anticlinal crest. Cf: *outlier.*

in-line offset The component of the distance from a geophone to the source in the direction of the line of the spread. Cf: *perpendicular offset.*

innate (in-nate') Said of certain igneous rocks that have undergone transformation without intrusion or other change of position, such as rocks formed by simple fusion in place (Medlicott and Blanford, 1879, p.752).

innelite (in'-ne-lite) A pale yellow or brown triclinic mineral: $Na_2CaBa_4Ti_3O_4(Si_2O_7)_2(SO_4)_2$.

inner bar (in'-ner) A bar formed at the upper bend of a flood channel, or where the waters of a river are checked by a flood tide. Ant: *outer bar.*

inner beach The part of a sandy beach that is covered by the wash of gentle waves and is ordinarily saturated. Cf: *foreshore.*

inner channel A deep, narrow, incised portion of a stream channel.

inner core The central part of the Earth's core, extending from a depth of about 5,150 km to the center (6,371 km) of the Earth; its radius is about one third of the whole core. The inner core is probably solid, as evidenced by the observation of *S waves* that are propagated in it, and because compressional waves travel noticeably faster through it than through the outer core. Densities range from 12.8 to 13.1 g/cm^3. It is equivalent to the *G layer.* Cf: *outer core.* Partial syn: *lower core.* Syn: *siderosphere.*

inner hinge plate Either of a pair of subhorizontal hinge plates in the *cardinalia* of some brachiopods (such as rhynchonelloids, spiriferoids, and terebratuloids), located median of the crural bases and fused laterally with them. Cf: *outer hinge plate.*

inner lamella The thin layer covering an ostracode body in the anterior, ventral, and posterior parts of the carapace, chitinous except for calcified marginal parts forming the duplicature (TIP, 1961, pt. Q, p.51). Cf: *outer lamella.*

inner lamina The inner shell layer of a compartmental plate of certain cirripede crustaceans, separated from an outer lamina by parietal tubes.

inner lead An area of calm water between a line of parallel offshore islands (such as a string of skerries) and the mainland.

inner lip The adaxial (inner) margin of the *aperture* of a gastropod shell, extending from the foot of the columella to the suture. It consists of the *columellar lip* and the *parietal lip.* Cf: *outer lip.*

inner lowland The innermost of the lowland belts of a *belted coastal plain,* formed in less-resistant rocks that separate the oldland from the cuesta landscape (the first cuesta scarp descending to the bottom of the lowland). Syn: *inner vale.*

inner mantle *lower mantle.*

inner plate One of a pair of subvertical plates in the *cardinalia* of some pentameracean brachiopods, lying on the ventral side of the base of the *brachial process* and fused dorsally with it. Cf: *outer plate.*

inner reef One of the reefs comprising the landward or shelfward part of a *reef complex* or *reef tract.* These are often smaller and less developed than outer reefs in the same region. Cf: *leeward reef; windward reef.*

inner-shelf shoal An arcuate and linear shelf sand body such as occurs on the U. S. inner continental shelf, which is formed by nearshore or shelf processes (Swift et al., 1972, Ch. 23).

inner side The portion of a *conodont element* on the concave side of the anterior-posterior midline. Ant: *outer side.*

inner vale *inner lowland.*

inner vesicle A membranous expansion from the distal part of a maternal zooid in some cheilostome bryozoans, which partly fills the opening of a brood chamber (Ryland, 1970, p.96).

inner wall The innermost wall in double-walled archaeocyathids, surrounding the central cavity (TIP, 1972, pt. E, p.19-20).

inninmorite (in-nin-mor'-ite) An igneous rock composed of augite and plagioclase (anorthite to labradorite) in a groundmass of sodic plagioclase, augite, and abundant glass. It is similar in composition to *cumbraite.* The name is for Inninmore, Scotland. Obsolete.

inoperculate (in-o-per'-cu-late) adj. Having no operculum; e.g. said of an irregular tear that serves as the opening of a sporangium through which spores are discharged. n. An inoperculate animal or shell; e.g. an inoperculate gastropod shell.

inorganic (in-or-gan'-ic) Pertaining or relating to a compound that contains no carbon. Cf: *organic.*

inosilicate (in-o-sil'-i-cate) A class or structural type of *silicate* characterized by the linkage of the SiO_4 tetrahedra into linear chains by the sharing of oxygens. In a simple chain, e.g. pyroxenes, two oxygens are shared; in a double chain or band, e.g. amphiboles, half the SiO_4 tetrahedra share three oxygens and the other half share two. The Si:O ratio of the former type is 1:3 and for the latter it is 4:11. Cf: *nesosilicate; sorosilicate; cyclosilicate; phyllosilicate; tectosilicate.* Syn: *chain silicate.*

in-phase wave A sinusoidal, *upper flow regime* bed form that normally occur in trains that are more or less in phase with the water surface under upper flow regime conditions (Simons et al., 1961). The structures may migrate upstream or downstream or remain stationary. The breaking form of in-phase waves is termed *antidune.* Some workers consider in-phase waves to be a type of antidune.

in-place assemblage *fossil community.*

in-place value Value of a mineral deposit in-place, prior to mining. Cf: *gross in-place value.*

input well (in'-put) *injection well.*

inquilinism (in'-qui-lin-ism') A form of *commensalism* in which one organism lives inside another, usually in the digestive tract or respiratory chamber. Adj: inquiline. Cf: *parasitism.*

in regime Said of a stream or channel that has attained an average equilibrium or that is capable of adjusting its cross-sectional form or longitudinal profile by means of alterations imposed by the flow, and in which the average values of the quantities that constitute regime show no definite trend over a period of years (such as 10-20 years). Cf: *graded [geomorph].*

in-seam methods Use of channel waves to investigate changes (such as interruptions by faults) in a layer with lower velocity than underlying and overlying beds, such as coal measures. Both source and receivers are in the layer. Both reflection (receivers near the source in the same gallery) and transmission (sources and receivers in different galleries) techniques are used (Sheriff and Geldart, 1995, p.483-486, 510-512).

inselberg (in'-sel-berg) A prominent isolated residual knob, hill, or small *mountain of circumdenudation,* usually smoothed and rounded, rising abruptly from and surrounded by an extensive lowland erosion surface in a hot, dry region (as in the deserts of southern Africa or Arabia), generally bare and rocky although partly buried by the debris derived from and overlapping its slopes; it is characteristic of an arid or semiarid landscape in a late stage of the erosion cycle. The term was originated by W. Bornhardt. Etymol: German "Inselberg", "island mountain". Pl: inselbergs; inselberge. Cf: *monadnock; bornhardt.* Syn: *island mountain.*

in-sequence thrusting Thrusting in which successively younger thrusts form uniformly on the hinterland or on the foreland side of existing thrusts.

insequent (in'-se-quent) adj. Said of a stream, valley, drainage system, or type of dissection that is seemingly uncontrolled by the associated rock structure or surface features, being determined by minor inequalities not falling into any larger-scale pattern. Etymol: in + consequent. Syn: *inconsequent.* n. *insequent stream.*

insequent stream A stream developed on the present surface but not consequent upon it and apparently not controlled or adjusted by the rock structure and surface features; a self-guided stream that develops under accidental or chance controls and whose resulting drainage pattern is dendritic, as a young stream wandering irregularly on a nearly level plain underlain by homogeneous or horizontally stratified rocks. The term was proposed by Davis (1897, p.24). Syn: *insequent.*

insert (in'-sert) adj. Having the ocular plates of an echinoid in contact with the periproctal margin. Ant: *exsert.*

inset [cart] (in'-set) *inset map.*

inset [petrology] A term proposed by Shand (1947) to replace the term *phenocryst.*

inset ice stream Ice from a tributary glacier that is set into the surface of a larger glacier and does not extend to the bed; e.g., a *superimposed ice stream* set into the surface of a *trunk glacier* a short distance downglacier from their confluence. These ice masses do not move appreciably faster than the surrounding ice, so the term is currently misleading. Cf: *juxtaposed ice stream.*

inset map A separate map positioned within the neat line of a larger map. Three forms are recognized: (1) an area geographically outside a sheet but included therein for the convenience of publication, usually at the same scale; (2) a portion of the map or chart shown at an enlarged scale; (3) a smaller map or chart of surrounding areas included for location purposes. Syn: *inset [cart].*

inset terrace A *stream terrace* formed during successive periods of vertical and lateral erosion such that remnants of the former valley floor are left on both sides of the valley (Schieferdecker, 1959, term 1512).

inshore (in'-shore) (a) Situated close to the shore or indicating a shoreward position; specif. said of a zone of variable width extending from the low-water shoreline through the breaker zone. See also: *offshore; nearshore.* (b) In a narrow sense, said of a zone that is equivalent to the *shoreface.*

inshore water (a) Water that is adjacent to land, with physical properties influenced considerably by continental conditions. Ant: *offshore water.* (b) A strip of open water located seaward of an icefoot or land, produced by the melting of *fast ice* along the shore.

in situ [palyn] Of fossil pollen or spores, being prepared from a fructification or sporangium. Ant: *sporae dispersae.*

in-situ bioremediation The breakdown of organic contaminants in the subsurface by the stimulation of existing microorganisms or the injection of new ones that can degrade the contaminants. Stimulation is achieved by the injection of electron acceptors and nutrients for bacterial metabolism.

in-situ combustion (in-si'-tu) A technique used for recovering oil of low gravity and high viscosity from a reservoir when primary methods have failed. The method involves heating the oil in the formation by igniting it (burning it in place), keeping combustion alive by pumping air into the formation. As the front of burning oil advances, the heat breaks down the oil into coke and light oil, and the latter is pushed ahead to producing wells.

in-situ crystallization Magmatic crystallization that occurs "in place". The growing crystals remain stationary and generate a progressively fractionated interstitial liquid that may or may not segregate from the crystalline solids, i.e. compositional convection. Most commonly associated with crystallization in a magmatic boundary layer.

in-situ density The density of a small unit of seawater, measured at its actual depth. Cf: *potential density.*

in-situ mining Removal of the valuable components of a mineral deposit without physical extraction of the rock, e.g. by *solution mining.*

in-situ modeling The interpretation of gravity or magnetic data in which the calculated field points are the same as those of actual field stations, as opposed to points located on a datum or reference plane.

in-situ temperature The temperature of a small volume of seawater measured at its actual depth. Cf: *potential temperature [oceanog].*

in-situ theory The theory that coal originates at the place where its constituent plants grew and decayed. Ant: *drift theory [coal].* See also: *autochthony.* Syn: *swamp theory.*

insizwaite (in-siz'-wa-ite) A metallic cubic mineral of the *pyrite* group: $Pt(Bi,Sb)_2$.

insoak (in'-soak) The absorption of free surface water by unsaturated soil.

insolation [meteorol] (in-so-la'-tion) The combined direct solar and diffuse sky radiation reaching a given body, e.g. the Earth; also, the rate at which it is received, per unit of horizontal surface. Cf: *solar constant.*

insolation [oceanog] The transfer of solar radiation to the ocean, usually in watts per square meter. The radiation may be direct, or indirect after scatter in the atmosphere.

insolation [weath] (a) Exposure to the Sun's rays. (b) The geologic effect of the Sun's rays on the Earth's surficial materials; specif. the effect of changes of temperature on the mechanical weathering of rocks. See also: *shadow weathering.*

insolilith (in-so'-li-lith) A relatively rounded pebble with a rough or cracked surface produced by exfoliation or granular disintegration resulting from insolation.

insoluble residue (in-sol'-u-ble) The material remaining after the more soluble part of a rock sample has been dissolved in, e.g., hydrochloric acid or acetic acid. It is chiefly composed of siliceous material (e.g. chert or quartz) and various detrital minerals (e.g. anhydrite, glauconite, pyrite, and sphalerite). Abbrev: IR. See also: *siliceous residue.*

insonify (in-so'-ni-fy) To suffuse a volume or surface with sound. The term is used in geophysical investigations based on acoustic properties of rocks. It is analogous to the term "illuminate."

inspissation (in-spis-sa'-tion) The alteration of an oil deposit, as by evaporation or oxidation during long exposure, in which gases and lighter fractions escape, and *asphalt* and heavy oils remain.

instant (in'-stant) The geochronologic equivalent of a *chronohorizon* if it has no resolvable time duration on a geologic scale; a "time plane"; the beginning or the end or a moment (ISSC, 1994, p.78). Cf: *moment.*

instantaneous field of view (in-stan-ta'-ne-ous) The solid angle through which a detector is sensitive to radiation. In a scanning system this refers to the solid angle subtended by the detector when the scanning motion is stopped. Cf: *angular field of view.*

instant rock A colloquial term for a fragile rock produced from originally fragmental materials, such as sand, by the shock waves associated with explosions or impacts, i.e. by *shock lithification.*

instar (in'-star) (a) The ontogenetic stage in the life of an arthropod occurring between two successive molts (periods of ecdysis); an immature molted or shed carapace of an ostracode. Also, an individual in a specified instar. (b) A single episode of shell formation in foraminifers, commonly of a single chamber.

instream flow A term used in the western U.S. to refer to flow remaining in a channel, rather than being diverted out of the channel for offstream uses.

instrumental neutron activation analysis An analytical technique in which *neutron activation* in a nuclear reactor is used to transform elements in a sample to isotopes with a higher mass. The energy and quantity of gamma rays given off by the sample (as it decays back to a stable preactivated state) are measured, processed relative to laws of radioactive decay, and the concentrations of elements are determined relative to a standard. Abbrev: INAA.

instrument error (in'-stru-ment) A *systematic error* resulting from imperfections in, or faulty adjustment of, the instrument or device used. Such an error "may be accidental or random in nature and result from the failure of the instrument to give the same indication when subjected to the same input signal" (ASP, 1966, p.1138).

instrument station A station at which a surveying instrument is set up for the purpose of making measurements; e.g. the point over which a leveling instrument is placed for the purpose of taking a backsight or foresight. Syn: *setup.*

insular [clim] (in'-su-lar) Said of a climate exhibiting little seasonal temperature variation, e.g. a marine climate.

insular [ecol] Said of an organism that has a limited or isolated range or habitat, esp. if on an island or group of islands.

insular shelf An area of the ocean floor analagous to the *continental shelf,* but surrounding an island. Syn: *island shelf.*

insular slope An area of the ocean floor analogous to a *continental slope,* but surrounding an island. Syn: *island slope.*

insulated stream (in'-su-lat'-ed) A stream or reach of a stream that neither contributes water to the zone of saturation nor receives water from it (Meinzer, 1923, p.56); it is separated from the zone of saturation by an impermeable bed.

insulosity (in-su-los'-i-ty) The percentage of the area of a lake that is occupied by islands.

insurgence (in-sur'-gence) The sinking of a stream underground in a *karst* area. Cf: *exsurgence; ponor; sinking stream; swallet.*

intaglio (in-ta'-glio [in-tal'-yo]) A carved gem that may be used as a seal, in which the design has been engraved into the stone so that the images are hollowed out and appear as negative relief. Cf: *cameo.* See also: *cuvette [gem].*

intake (in'-take) (a) *recharge.* (b) The openings in water-bearing materials through which water passes into a well.

intake area *recharge area.*

intarsia A lapidary art where small usually flat pieces of gem materials are fitted together and then cemented to a base to form various designs. Syn: inlay work; Pietra dura; mosaic.

intectate (in-tec'-tate) In pollen grains, lacking a *tectum,* but possibly having *sculpture.* Cf: *atectate.*

integrate (in'-te-grate) Said of a type of wall structure in trepostome bryozoans in which the zooecial boundaries appear in tangential section as narrow, well-defined lines. Cf: *amalgamate.*

integrated drainage (in'-te-grat'-ed) Drainage developed during maturity in an arid region, characterized by coalescence (across intervening ridges and mountains) of drainage basins as a result of headward erosion in the lower basins or of spilling over from the upper basins due to aggradation (Lobeck, 1939, p.12-13); drainage developed where various higher local base levels are replaced by a single lower base level.

integrated geophysics The combination of seismic, gravity, magnetic, electrical, and/or well-log data to effect a more accurate and complete interpretation than any one data set could provide if used alone.

Integrated Ocean Drilling Program Successor program to the *ODP,* which terminated drilling in 2003. Abbrev: IODP.

integrated waste management A complex set of procedures for waste management that includes source reduction, recycling, composting, landfill, and incineration (Keller, 2002).

integration [cryst] (in-te-gra'-tion) In petrology, the formation of larger crystals from smaller crystals by recrystallization.

integration [paleont] In colonial animals, the extent to which zooids in combination with extrazooidal parts differ morphologically from solitary animals because of colony control of growth functions.

integripalliate (in-teg'-ri-pal'-li-ate) Said of a bivalve mollusk devoid of a *pallial sinus* in the *pallial line.* Cf: *sinupalliate.*

integument (in-teg'-u-ment) In a flowering plant, the covering layer of an ovule. It contributes to the coat or *testa* of a mature seed.

intensity [remote sensing] (in-ten'-si-ty) In the *IHS* system, brightness ranging from black to white.

intensity [seis] *earthquake intensity.*

intensity [volc] The rate at which magma or ash is disgorged (Walker, 1980). Syn: *eruption intensity.*

intensity scale A standard of measurement of *earthquake intensity* by its effect on people, structures, and the Earth's crust. Three such systems are the *Mercalli scale,* the *modified Mercalli scale,* and the *Rossi-Forel scale.*

intensiveness (in-ten'-sive-ness) In quantification of hydrothermal alteration, the result of fixing intensive variables, which in turn set mineral stabilities. It is described in terms of stable alteration-mineral assemblages, such as "potassic", "quartz-sericite", or "greisen". Cf: *pervasiveness; extensiveness.*

intensive variable (in-ten'-sive) A thermodynamic variable that is independent of the total amount of matter in the system. Examples include temperature, pressure, and mole fraction.

interambulacral (in'-ter-am'-bu-lac'-ral) adj. Situated between ambulacra of echinoderms, such as the *interambulacrum* of an echinoid or the *interray* of a crinoid; esp. referring to the thecal plates making up the area between ambulacra. Cf: *ambulacral.* n. Any plate situated between the ambulacral plates of an echinoderm.

interambulacrum (in'-ter-am'-bu-lac'-rum) One of the areas between two ambulacra in an echinoderm; any of the five interradial sections of most echinoderms.

interantennular septum (in'-ter-an-ten'-nu-lar) A plate that separates two antennular cavities in some malacostracan crustaceans. Syn: *proepistome.*

interarea (in-ter-ar'-e-a) The posterior sector of a brachiopod shell with the growing edge at the hinge line. The term is more commonly used for any plane or curved surface lying between the beak and the posterior margin of a brachiopod valve and bisected by the *delthyrium* or the *notothyrium* (TIP, 1965, pt. H, p.146). The interarea is generally distinguished by a sharp break in angle from the remainder of the valve and by the absence of costae, plicae, or coarse growth lines. Cf: *pseudointerarea; planarea.*

interaxial angle The angle between any two crystallographic axes. The angle between *a* and *b* is γ, that between *a* and *c* is β, and that between *b* and *c* is α.

interbasin area (in-ter-ba'-sin) A roughly triangular area, located between adjacent tributary drainage basins, that has not developed a drainage channel but contributes drainage directly into a higher-order channel (Schumm, 1956, p.608).

interbasin length The maximum horizontal length of the *interbasin area,* measured from the apex of the triangular ground surface to the adjacent channel (Strahler, 1964).

interbed (in'-ter-bed) A bed, typically thin, of one kind of rock material occurring between or alternating with beds of another kind.

interbedded (in-ter-bed'-ded) Said of beds lying between or alternating with others of different character; esp. said of rock material laid down in sequence between other beds, such as a contemporaneous lava flow "interbedded" with sediments. Cf: *intercalated.* Syn: *interstratified.*

interbiohorizon zone *interval zone.*

interbrachial (in-ter-brach'-i-al) adj. Situated between arms; e.g. an "interbrachial margin" of the disc of an asterozoan. n. A crinoid plate occurring in the dorsal cup between brachial plates of adjacent ambulacral rays or between brachs of any single ray.

interburden (in'-ter-bur'-den) A layer of sedimentary rock, of any composition and thickness, that separates two minable coal beds. The term is used mainly in the strip-mining industry of the western United States. It has occasionally been applied in other surface mining, especially for phosphate rock. Cf: *parting [stratig].*

intercalary (in-ter'-ca-lar-y) adj. Interposed or inserted between, or introduced or existing interstitially, e.g. "intercalary apical system" of an echinoid, in which ocular plates II and IV meet at midline so as to separate anterior and posterior portions; hooplike "intercalary bands" located between the valves and connecting bands in a diatom frustule; or "intercalary cuticle" of a cheilostome bryozoan formed by the outermost layers of lateral walls of two contiguous lineal series of zooids. n. One of many thecal plates occurring between radial and basal circlets of some crinoids (e.g. *Acrocrinus*).

intercalary archeopyle (in-ter'-ca-la-ry) An *archeopyle* formed in a dinoflagellate cyst by the loss of one or more plates in the anterior *intercalary* series (Williams et al., 2000, p.110).

intercalary plate A *plate* on a dinoflagellate cyst or theca that lies between the precingular series and the apical series or between the postcingular series and the antapical series (Williams et al., 2000, p.110).

intercalated (in-ter'-ca-lat-ed) Said of layered material that exists or is introduced between layers of a different character; esp. said of relatively thin strata of one kind of material that alternate with thicker strata of some other kind, such as beds of shale that are intercalated in a body of sandstone. Cf: *interbedded.*

intercalation (in-ter'-ca-la'-tion) (a) The existence of one or more layers between other layers; e.g. the presence of sheets of lava between sedimentary strata, the occurrence of a particular fossil horizon between fossil zones of a different character, or the inclusion of lamellar particles of one mineral in another in such a way that the inclusions are oriented in planes related to the crystal structure of the host mineral. (b) The introduction of a layer between other layers already formed, such as the insertion of rock by intrusion or thrusting in a preexisting series of stratified rock. (c) An intercalated body of material, such as a bed in an intertongued zone or a lens of volcanic ash in a sedimentary deposit.

intercamarophorial plate (in'-ter-ca'-ma-ro-pho'-ri-al) A short, low median septum on the posterior midline of the *camarophorium* in brachiopods of the superfamily Stenoscismatacea, extending to the underside of the hinge plate but independent of the median septum duplex (TIP, 1965, pt. H, p.146).

intercameral (in-ter-cam'-er-al) Located between the chambers of a foraminiferal test; e.g. "intercameral foramen" representing a primary or secondary opening between successive chambers.

intercardiophthalmic region (in'-ter-car'-di-oph-thal'-mic) A small rectangular area of the *prosoma* of a merostome, embracing the cardiac lobe and a minor part of the interophthalmic region.

intercellular space (in-ter-cel'-lu-lar) The space resulting from a separation of adjacent plant cell walls from each other along more or less extended areas of contact. In some cases it may result from the splitting of the *middle lamella* in plant cell walls (Esau, 1965, p.62).

intercept [sed] One of the three linear dimensions or diameters of a sedimentary particle: the longest dimension is the "maximum" intercept, and the shorter dimensions are the "intermediate" and "short" intercepts.

intercept [surv] The part of the rod seen between the upper and lower stadia hairs of a transit or telescopic alidade; e.g. a stadia interval.

interception (in-ter-cep'-tion) The process by which water from precipitation is caught and stored on plant surfaces and eventually returned to the atmosphere without having reached the ground. Also, the amount of water intercepted. Cf: *throughfall.*

intercept time The time obtained by extrapolating a seismic refraction alignment on a time-distance curve back to zero source-to-geophone distance; the sum of the delay times at the source and receiver ends of the path.

intercision (in-ter-ci'-sion) (a) A type of capture characterized by sidewise swinging of mature streams (Lobeck, 1939, p.201). (b) A type of diversion accomplished by the cutting back of bluffs along a lake shore such that the lake advances inland and cuts into a bend of a river valley some distance above its mouth (Goldthwait, 1908).

intercolpium (in-ter-col'-pi-um) *mesocolpium.*

intercrater plain (in-ter-cra'-ter) A relatively level terrain between craters in the heavily cratered highlands of terrestrial planets (Carr et al., 1984).

intercretion (in-ter-cre'-tion) A term proposed by Todd (1903) for a *concretion* that grows by accretion (on the exterior) and by irregular and interstitial addition, causing a circumferential expansion and resultant cracking and wedging apart of the interior of the concretion; e.g. a *septarium.*

intercrystal porosity (in-ter-crys'-tal) Porosity between equant, equal-sized crystals (Choquette and Pray, 1970, p.247).

intercumulus (in-ter-cu'-mu-lus) The space between crystals of a *cumulate.*

intercumulus crystals Crystals formed late, during fractional crystallization, from magma trapped between early-formed, settled, cumulus crystals.

intercumulus liquid Magmatic liquid that is trapped between the early-formed, settled, crystals of a *cumulus,* i.e. that occupies the intercumulus. See also: *intercumulus material.*

intercumulus material Material that crystallized from *intercumulus liquid.*

intercutaneous wedge A *thrust sheet* with *passive-roof thrust* above it.

interdigitation (in'-ter-dig'-i-ta'-tion) *intertonguing.*

interdistributary basin (in-ter-dis-trib'-u-tar-y) A relatively low lying depression between two distributary channels.

interdistributary bay (in'-ter-dis-trib'-u-tar-y) A pronounced indentation of the delta front between advancing stream distributaries, occupied by shallow water, and either open to the sea or partly enclosed by minor distributaries.

interdune (in-ter-dune') Pertaining to the relatively flat surface, whether sand-free or sand-covered, between dunes; e.g. said of the long, troughlike, wind-swept passage between parallel longitudinal dunes, such as a *gassi* or a *feidj.*

interestuarine (in-ter-es'-tu-a-rine') Situated between two estuaries.

interface [petroleum] (in'-ter-face) A syn. of *contact (b).*

interface [sed] A depositional boundary separating two different physicochemical regions; specif. the surface separating the top of the uppermost layer of sediment and the medium (usually water) in which the sedimentation is occurring.

interface [seis] *discontinuity [seis].*

interfacial angle (in-ter-fa'-cial) In crystallography, the angle between two faces of a crystal.

interference (in-ter-fer'-ence) The condition occurring when the *area of influence* of a water well comes into contact with or overlaps that of a neighboring well, as when two wells are pumping from the same aquifer or are located near each other.

interference colors In crystal optics, the colors displayed by a *birefringent* crystal in crossed polarized light. Thickness and orientation of the sample and the nature of the light are factors that affect the colors and their intensity.

interference figure The pattern or figure that a crystal displays in polarized light under the conoscope. It is a combination of the *isogyre* and the *isochromatic curve,* and is used to distinguish uniaxial from biaxial crystals and to determine optical sign. See also: *axial figure; biaxial figure.*

interference ripple A *ripple* that is characterized by multiple, symmetrical crests, normally two that are commonly at or close to right angles to each other. The term suggests that interference ripples are formed in response to two sets of waves propagating at right angles to each other. Cf: *cross ripple; compound ripple.*

interference ripple mark *interference ripple.*
interferograms In radar, images that record interference patterns created by superposing images acquired by two antennas that are separated by a short distance.
interferometric synthetic aperture radar *IFSAR.*
interfingering (in-ter-fin'-ger-ing) *intertonguing.*
interflow [ore dep] (in-ter-flow') adj. Occurring in or between volcanic flows, as of an exhalite zone or tuff bed; specif., said of metalliferous sediments or pyroclastics intercalated with flows of Keewatin age at Cobalt, Ontario.
interflow [water] *storm seepage.*
interfluent lava flow (in-ter-flu'-ent) A lava flow that is discharged into and through subterranean fissures and cavities in a volcano and may never reach the surface (Dana, 1890); an obsolete term. Cf: *effluent lava flow; superfluent lava flow.*
interfluve (in'-ter-fluve) The area between rivers (*interstream*), esp. the relatively undissected upland or ridge between two adjacent valleys containing streams flowing in the same general direction. Cf: *doab [streams].*
interfluve hill A relatively flat-topped remnant of an antecedent slope on which gradation was arrested when the adjacent streams and valleys were developed (Horton, 1945, p.360); it occurs along the divides in drainage basins approaching maturity.
interfluvial (in-ter-flu'-vi-al) Lying between streams; pertaining to an *interfluve.* Cf: *interstream.*
interfolding (in-ter-fold'-ing) The simultaneous development of discrete fold systems with different orientations. Rarely used.
interformational (in'-ter-for-ma'-tion-al) Formed or existing between one *formation [stratig]* and another; e.g. "interformational unconformity".
interformational conglomerate A conglomerate that is present within a formation, the constituents of which have a source external to the formation. Cf: *intraformational conglomerate.*
intergelisol (in-ter-gel'-i-sol) *pereletok.*
interglacial (in-ter-gla'-cial) adj. Pertaining to or formed during the time interval between two successive glacial epochs or between two glacial stages. The term implies both the melting of ice sheets to about their present level, and the maintenance of a warm climate for a sufficient length of time to permit certain vegetational changes to occur (Suggate, 1965, 619). n. *interglacial stage.*
interglacial stage A subdivision of a glacial epoch separating two glaciations, characterized by a relatively long period of warm or mild climate during which the temperature rose to at least that of the present day; esp. a subdivision of the Pleistocene Epoch, as the "Sangamon Interglacial Stage". Syn: *interglacial; thermal [glac geol].*
interglaciation (in'-ter-gla'-ci-a'-tion) A climatic episode "during which the climate was incompatible with the wide extent of glaciers that characterized a glaciation" (ACSN, 1961, art. 40).
intergonal (in-ter-go'-nal) In dinoflagellates, having processes or other features that occur along the line of intersection between two plates, in contrast to the junction point of three plates (Williams et al., 2000, p.111).
intergrade (in'-ter-grade) n. A soil that is transitional between two other soils. The subgroup category of the U.S. Dept. of Agriculture soil taxonomy is set up specifically to accommodate soils that are intergrades between two orders, suborders, or great groups.
intergranular (in-ter-gran'-u-lar) Holocrystalline texture consisting of lath-shaped plagioclase grains and interstitial granular pyroxene, olivine, and/or oxide minerals. Syn: *diabasic.* Cf: *intersertal; ophitic; subophitic; hyalophitic; hyalopilitic.*
intergranular movement A process that goes on within a *glacier* when grains of ice rotate and slide over each other. Cf: *intragranular movement.*
intergranular porosity (a) The porosity between the grains or particles of a rock, such as that between the lithoclasts or the bioclasts of a carbonate sedimentary rock. Cf: *interparticle porosity.* (b) The volume of pores between the grains of a sediment or sedimentary rock divided by the total volume of the porous medium.
intergranular pressure *effective stress.*
intergranular stress *effective stress.*
intergrowth (in'-ter-growth) Interlocking of grains of two different minerals as a result of their simultaneous crystallization.
interio-areal aperture (in-te'-ri-o-ar'-e-al) An aperture in the face of the final chamber of a foraminiferal test, not at its base.
interiomarginal aperture (in-te'-ri-o-mar'-gin-al) A basal aperture in a foraminiferal test at the margin of the final chamber and along the final suture. In coiled forms it may be an *equatorial aperture* or an *extraumbilical aperture.* See also: *spiroumbilical aperture.*
interior basin (in-te'-ri-or) (a) A depression entirely surrounded by higher land and from which no stream flows outward to the ocean. Cf: *closed basin.* Syn: *inland basin.* (b) *intracratonic basin.*
interior drainage *internal drainage.*
interior link A link in a channel network emanating from a *fork* (Shreve, 1967). Cf: *link [geomorph]; exterior link.*
interior plain A plain that is situated far from the borders of a continent, as contrasted with a coastal plain.
interior valley A large, flat-floored, *closed depression* in a *karst* area. Its drainage is subsurface, and its floor is commonly covered by alluvium. Interior valleys may become intermittent lakes during periods of heavy rainfall, when the *sinking streams* that drain them cannot manage the runoff. Syn: *polje.* Cf: *karst valley.*
interior wall A body wall in bryozoans that partitions pre-existing body cavity into zooids, parts of zooids, or extrazooidal parts.
interlacing drainage pattern (in-ter-lac'-ing) *braided drainage pattern.*
interlacustrine (in'-ter-la-cus'-trine) Situated between lakes; e.g. an "interlacustrine overflow stream" spilling over from one lake to another.
interlamellar spaces In conodonts, funnel-shaped cavities within lamellae along growth axes (TIP, 1981, pt. W, supp.2).
interlamellar stripes Faint dark bands within, and parallel to the boundaries of lamellae; seen only in thin sections of exceptionally well-preserved conodont elements; thought to represent levels of higher than ordinary concentration of organic matter within lamellae and thus to indicate that formation of a lamella did not take place in a single short episode (TIP, 1981, pt. W, supp.2).
interlaminated (in-ter-lam'-i-nat-ed) Said of laminae occuring between or alternating with others of different character; *intercalated* in very thin layers. Syn: *interleaved.*
interlayer (in-ter-lay'-er) A layer placed between others of a different nature; e.g. an interbed.
interlayering (in-ter-lay'-er-ing) The regular or random arrangement of structural units of clay minerals, each unit differing from the adjacent unit either in composition or in crystallographic orientation.
interlayer sheet A hydroxide sheet, usually in chlorite, that is interleaved in an ordered fashion and of chemical composition $(R^{2+},R^{3+})3(OH)_6$. Interlayer sheets of different composition can form in a less ordered fashion between other 2:1 layers, including smectites, vermiculites, and micas (Bailey, 1980b).
interlayer space The space between 1:1 or 2:1 layers in layer silicates. This is the primary location for exchangeable and nonexchangeable ions that balance layer or permanent charge (Moore and Reynolds, 1997). It also is the locus of attachment of polar molecules and pillars as in pillared clay minerals. Cf: *pillared clay minerals.*
interleaved (in-ter-leaved') *interlaminated.*
inter-level correlation Correlation of two intervals that represent the same span of time (Henningsmoen, 1961, p.65).
interlimb angle The angle between adjacent limbs of a fold, normally determined by drawing tangents to the fold limbs at the inflection points.
interlobate deposit (in-ter-lo'-bate) *Drift* lying between two adjacent glacial lobes.
interlobate moraine A lateral or end moraine formed along the line of junction and roughly parallel to the axes of two adjacent glacial lobes that have pushed their margins together. Syn: *intermediate moraine.*
interlocking (in-ter-lock'-ing) Pertaining to two seismic records made with the end geophone of one record occupying the source location of the other, so that a common raypath and common arrival time are involved.
interlocking spur One of several projecting ridges extending alternately from the opposite sides of the wall of a young, V-shaped valley down which a river with a winding course is flowing, each lateral spur extending into a concave bend of the river so that viewed upstream the spurs seem to "interlock" or "overlap". Syn: *overlapping spur.*
interlocking texture A rock texture in which particles with irregular boundaries interlock by mutual penetration, as in a crystalline limestone.
interloculum (in-ter-loc'-u-lum) A distinct interspace between the ektexine and endexine of certain triporate pollen grains (Kremp, 1965, p.74).
intermediate (in-ter-me'-di-ate) Said of an igneous rock that is transitional between *basic* and *silicic* (or between *mafic* and *felsic*), generally having a silica content of 54 to 65 percent; e.g. syenite and diorite. "Intermediate" is one subdivision of a widely used system for classifying igneous rocks on the basis of their silica content; the other subdivisions are *acidic, basic,* and *ultrabasic.* Obsolete syn: *mediosilicic.*
intermediate belt That part of the *unsaturated zone* that lies between the *capillary fringe* and the *belt of soil water.* Syn: *intermediate zone.*
intermediate coal A type of banded coal defined microscopically as consisting of between 60% and 40% of bright ingredients such as vitrain, clarain, and fusain, with clarodurain and durain composing the remainder. Cf: *semibright coal; semidull coal; bright coal; dull coal.*
intermediate contour A contour line drawn between index contours.
intermediate-focus earthquake (in-ter-me'-di-ate-fo'-cus) An earthquake whose focus occurs between depths of about 60 km and 300 km. Cf: *shallow-focus earthquake; deep-focus earthquake.*
intermediate layer *sima [interior Earth].*
intermediate moraine *interlobate moraine.*
intermediate plain A plain intermediate in altitude between the highest summits of an erosion surface and the bottoms of the deepest valleys (Trowbridge, 1921, p.31-33). The term is misleading because it implies a plain intermediate in position between two other plains.
intermediate-scale map A map at a scale (in the U. S. Geological Survey, 1:50,000 to 1:100,000) that can show most details of roads and culture, together with topography by appropriate contour interval. The U. S. Geological Survey produces intermediate-scale maps on quadrangle or county

format. The Defense Mapping Agency defines intermediate scale as from 1:200,000 to 1:500,000 (Thompson, 1988, p.15). Cf: *large-scale map; small-scale map.*

intermediate vadose water Water of the *intermediate belt.*

intermediate water A water mass generally located below the main thermocline but above deep water masses and centered in the vicinity of 1,000-m depth. Antarctic intermediate water is the best example: It is a cold, low-salinity water layer originating in the Southern Ocean but found through much of the southern parts of the Atlantic, Pacific, and Indian Oceans.

intermediate wave *transitional-water wave.*

intermediate zone *intermediate belt.*

intermineral (in-ter-min'-er-al) Pertaining to a time interval between periods of mineralization (Kirkham, 1971, p.1245-1246); also, pertaining to those features, e.g. dikes, that were emplaced during such an interval. The term has been applied to a time interval within a period of mineralization and to its resulting features, but *intramineral* is preferred for this meaning. Cf: *premineral; postmineral.*

intermittent spring A spring that discharges only periodically. A *geyser* is a special type of intermittent spring (Meinzer, 1923, p.54). Cf: *perennial spring; periodic spring.*

intermittent stream (a) A stream or reach of a stream that flows only at certain times of the year, as when it receives water from springs or from some surface source. The term "may be arbitrarily restricted" to a stream that flows "continuously during periods of at least one month" (Meinzer, 1923, p.68). (b) A stream that does not flow continuously, as when water losses from evaporation or seepage exceed the available streamflow. Cf: *ephemeral stream.* Syn: *temporary stream; seasonal stream.*

intermont (in'-ter-mont) adj .*intermontane.* n. A hollow region between mountains.

intermontane (in-ter-mon'-tane) Situated between or surrounded by mountains, mountain ranges, or mountainous regions; e.g. the Great Basin of western U.S., between the Sierra Nevada and the Wasatch Mountains. Syn: *intermont.*

intermontane glacier A glacier formed by the confluence of several alpine glaciers and occupying a depression between mountain ranges or ridges.

intermontane plateau A plateau that is partly or completely enclosed by mountains, and that is formed in association with them; e.g. the Tibetan plateau.

intermontane trough (a) A subsiding area in an island-arc region of the ocean, lying between stable or uprising regions. (b) A basinlike area between mountain ranges, sometimes occupied by an intermontane glacier.

intermorainal (in'-ter-mo-rain'-al) Situated between moraines, as an *intermorainal* lake occupying a narrow depression between parallel *moraines* of a retreating glacier.

intermural increase (in-ter-mu'-ral) A type of *increase* (offset formation of corallites) in cerioid coralla occurring at or within the intercorallite wall, so that no single corallite is recognizable as the parent.

internal cast (in-ter'-nal) A syn. of *steinkern.* The term should not be used for an *internal mold.*

internal cavity The central opening enclosed in single-walled archaeocyathids, comparable to the central cavity in double-walled forms (TIP, 1972, pt. E, p.40). Cf: *central cavity.*

internal contact The planar or irregular surface between separately emplaced portions of a composite pluton (Compton, 1962, p.277). Cf: *external contact.*

internal drainage Surface drainage whereby the water does not reach the ocean, such as drainage toward the lowermost or central part of an interior basin. It is common in arid and semiarid regions, as in western Utah. Ant: *external drainage.* See also: *centripetal drainage pattern.* Syn: *interior drainage; inland drainage; closed drainage; endorheism.*

internal energy That energy of a system described by the *first law of thermodynamics.*

internal erosion Erosion produced within a compacting sediment by movement of water through the larger pores (Bathurst, 1958, p.33). Cf: *internal sedimentation.*

internal friction That part of the shear strength of a rock or other intact solid that depends on the magnitude of the normal stress on a potential shear fracture. Cf: *cohesion.*

internal lobe The *dorsal lobe* in normally coiled cephalopod conchs. See also: *annular lobe.*

internal mold A mold or impression showing the form and markings of the inner surfaces of a fossil shell or other organic structure; it is made on the surface of the rock material filling the hollow interior of the shell or organism. It is sometimes called incorrectly a "cast of the interior", but can be so called only if the shell or structure itself be regarded as a mold. Cf: *external mold; internal cast.* See also: *steinkern.*

internal oblique muscle One of a pair of muscles in certain brachiopods, originating on the *pedicle valve* between the anterior adductor muscles, and passing posteriorly and laterally to insertions on the *brachial valve* located anteriorly and laterally from the posterior adductor muscles (TIP, 1965, pt. H, p.146). Cf: *lateral oblique muscle.*

internal rotation In structural petrology, a change in the orientation of structural features during deformation, referred to coordinate axes internal to the deformed body.

internal sedimentation Accumulation of clastic or chemical sediments derived from the surface of, or within, a more or less consolidated carbonate sediment (mud or silt), and deposited in secondary cavities formed in the host rock (after its deposition) by bending of laminae or by *internal erosion* or solution (Bathurst, 1958, p.31).

internal seiche A free oscillation of a submerged layer in a stratified body of water occupying an enclosed or semi-enclosed basin; esp. an oscillation of the thermocline in a lake. It is believed to be initiated by the same factors that produce a surface *seiche.*

internal suture The part of a *suture* of a coiled cephalopod *conch* situated on the *dorsum* (or within the impressed area), extending between umbilical seams, and hidden from view unless the conch is broken. Cf: *external suture.*

internal tide Submerged vertical oscillations on density surfaces in the sea with tidal periods.

internal wave A wave occurring on an internal density surface in a stratified ocean, e.g., a wave on the *thermocline.* Because of the small density differences within the sea, compared with the density difference at the sea surface, internal wave heights, periods, and lengths are much greater than surface wave heights, periods, and lengths.

International Active Sun Years (In-ter-na'-tion-al) An international cooperative program of studying solar-terrestrial phenomena during a period of maximum sunspot activity. It is related to the International Geophyscial Year and to the *International Years of the Quiet Sun.* Abbrev: IASY.

international atomic time Time established by the *Bureau International de l'Heure* in accordance with the definition of the second in the International System of weights and measures; set by the General Congress on Weights and Measures, 1971.

International Code of Botanical Nomenclature Only the most current (latest) edition of the Code (currently, the 2000 "St. Louis Code") has legal standing. It provides all the rules and regulations that must be taken into account for validly publishing the name of a modern or fossil plant. A new edition of the Code, often modified, is voted into jurisdiction at the end of each International Botanical Congress, about every six years. Abbrev: ICBN.

international date line (in-ter-na'-tion-al) *date line.*

international ellipsoid The official reference ellipsoid used for the primary geodetic network in Europe (Krzanowski et al., 1993). See also: *North American datum of 1983.*

International Geomagnetic Reference Field A long-wavelength regional magnetic field model representing the Earth's core magnetic field, updated every five years based on data from world-wide magnetic observatories and orbiting satellite-mounted magnetometers. It is subtracted from the observed data to determine a local anomalous field; this procedure is called IGRF correction. Abbrev: IGRF.

International Geophysical Year An international cooperative program of observation of geophysical phenomena from July 1, 1957 to December 31, 1958, a period that was near a maximum in sunspot activity. See also: *International Active Sun Years; International Years of the Quiet Sun.* Abbrev: IGY.

International Gravity Formula According to the 1930 International Gravity Formula (IGF) adopted in Stockholm by the International Association of Geodesy, theoretical (normal) gravity on the Earth's ellipsoid at latitude ϕ is 978,032.68 $(1 + 0.001\ 931\ 851\ 386\ 39 \sin^2\phi)\ (1 - 0.006\ 694\ 379\ 990\ 13 \sin^2\phi)$ mGal. Abbrev: IGF. See also: *latitude correction.*

International Gravity Standardization Net 1971 The world-wide network of the gravity base stations providing a reference frame for the *absolute gravity* measurements using the same type of gravimeters. Supersedes the Potsdam system that was defined by pendulum measurements in 1906.

International Hydrological Decade A ten-year program, 1965-74, patterned after the International Geophysical Year, aimed at training hydrologists and technicians and at establishment of networks for measuring hydrologic data. The idea originated in the United States, but the program was sponsored by UNESCO, and a large proportion of the membership of the United Nations participated.

International Map of the World A map series at a scale of 1:1,000,000 (one inch to 15.78 miles), having a uniform set of symbols and conventional signs, using the metric system for measuring distances and elevations, and printed in modified polyconic projection on 840 sheets, each covering an area of 4°lat. and 6° long. except above the 60th parallel where the longitude covered is 12° on each sheet. The International Map of the World on the Millionth Scale was first suggested at the 5th International Geographical Congress in 1891 and was accepted in principle in 1909. It consists of an incomplete series of map sheets (many needing revision) generally published by national mapping agencies of concerned countries under the auspices of the United Nations. Many countries, including the United States, have discontinued the production, and revision, of this series. The only sheets maintained, but not revised by the U.S. Geological Survey are the ones covering Antarctica. Abbrev: IMW. Syn: *millionth-scale map of the world.*

international nautical mile The *nautical mile* defined as exactly 1,852 m in length.

international symbols An internationally accepted shorthand notation for expressing the symmetry of space groups. These symbols are extensions of the Hermann-Mauguin notation used for crystal classes. Cf: *Hermann-Mauguin symbols; Schoenflies notation.*

International Years of the Quiet Sun An international cooperative program during 1964-1965 of studying solar-terrestrial phenomena during a quiet-sun, i.e. sunspot-minimum, period. It is related to the *International Geophysical Year* and to the *International Active Sun Years.* Abbrev: IQSY.

internides (in-ter'-ni-des) Kober's term for the internal part of an orogenic belt, farthest away from the craton, where high-grade metamorphism, plastic (flow) folding, and plutonism occur. Cf: *externides.* See also: *primary orogeny; hinterland.* Syn: *primary arc.*

internodal (in-ter-nod'-al) A crinoid columnal disposed between nodals, generally smaller than nodals and lacking cirri.

internode (in'-ter-node) An interval or part between two successive nodes; e.g. a segment of a colony of cheilostome bryozoans between articulations, or a segment of a jointed algal thallus.

interophthalmic region (in'-ter-oph-thal'-mic) The space between the cardiac lobe and ophthalmic ridge of a merostome.

interparticle diffraction Nadeau et al. (1984) showed that illite and smectite particles cleave in sheet thicknesses of integral multiples of 20 Å (two 2:1 layers and interlayer potassium) and 10 Å (one 2:1 layer), respectively, which they refer to as "fundamental particles." Diffraction from small (stacked layers, N, of 2, 3, or 4) and independent particles that are stacked upon each other is referred to as "interparticle diffraction." The resulting diffraction pattern is identical to a MacEwan crystallite model of fixed sequences of large (N>20) numbers of layers. Cf: *fundamental particle; MacEwan crystallite.*

interparticle porosity (in-ter-par'-ti-cle) The porosity between particles in a rock; e.g. *breccia porosity.* Choquette and Pray (1970, p.247) recommend use of this term rather than the term *intergranular porosity* which suggests limitation to grain-size particles.

interpenetration twin (in'-ter-pen'-e-tra'-tion) A twinned crystal, in which the individuals appear to have grown through one another. Syn: *penetration twin.*

interplanetary dust Small solid particles that orbit the sun in the vicinity of the planets. These particles, originating from the erosion of asteroids and comets, are most often smaller than about 10 micrometers in diameter, a size where the pressure exerted on them by sunlight can alter their orbits over time. Syn: *zodiacal dust.*

interplanetary dust particles (a) Submillimeter particles in space. (b) Submillimeter interplanetary particles collected for scientific study (particularly in the stratosphere by impactors carried on aircraft). Abbrev: IDP.

interpleural furrow (in-ter-pleu'-ral) A transversely directed furrow on the pleural regions of the *pygidium* of a trilobite, indicating the join between adjacent fused pleurae.

interpluvial (in-ter-plu'-vi-al) adj. Said of an episode of time that was drier than the pluvial periods between which it occurred. n. Such an episode or period of time.

interpolation (in-ter'-po-la'-tion) Estimation of the value of a variate based on two or more known enclosing values; a method used to estimate intermediate values between known points along a line or on a surface. Cf: *extrapolation.*

interpositum (in-ter-pos'-i-tum) In certain receptaculitids, a pentagonal plate which, in conjunction with a triangulum, is associated with an increase in the number of meroms per whorl.

interpretation key Characteristic or combination of characteristics that enable an interpreter to identify an object on an image.

interpretive log (in-ter'-pre-tive) A *sample log* based on rotary well cuttings, in which the geologist attempts to show only the rock encountered by the bit at each sampled depth, ignoring the admixed material from higher levels. Cf: *percentage log.*

interradial [paleont] (in-ter-ra'-di-al) adj. Situated midway between the axes of adjacent rays of an echinoderm; e.g. an "interradial suture" representing a common line or division between adjacent radial plates of a blastoid. n. (a) A plate in the interradial area; e.g. a crinoid plate above a basal. (b) *interray.*

interradial [palyn] Pertaining to areas of the proximal face or equator of a trilete spore, lying between the branches of the laesura. Cf: *radial [palyn].*

interradial loculus A subdivision of intervallar space bounded by septae, rods, pillars, or other radial skeletal elements in double-walled archaeocyathids (TIP, 1972, pt. E, p.13). Pl: loculi.

interray (in'-ter-ray) The area between two adjacent rays of an echinoderm; e.g., the part of a *theca* between any two adjacent crinoid rays. Syn: *interradial [paleont].*

interreef (in'-ter-reef) Situated between reefs; e.g. the "interreef region" characterized by relatively unfossiliferous rock, or "interreef sediments" deposited between reefs. Also spelled: inter-reef. Cf: *off-reef.*

interrupted profile (in-ter-rupt'-ed) A normal profile that has been altered by an *interruption;* e.g. a longitudinal profile of a stream, where, after rejuvenation, the head of the second-cycle valley touches the first-cycle valley.

interrupted projection A map projection lacking continuous outlines, having several central meridians instead of one, or whose origin is repeated, in order to reduce the peripheral shape distortion and the linear scale discrepancy; e.g. a *homolosine projection* split along several meridians. Syn: *recentered projection.*

interrupted stream A stream that contains perennial reaches with intervening intermittent or ephemeral reaches, or a stream that contains intermittent reaches with intervening ephemeral reaches (Meinzer, 1923, p.58). Ant: *continuous stream.*

interrupted water table A water table that slopes steeply over a *groundwater barrier,* with pronounced difference in elevation above and below the barrier, but not as steep as a *groundwater cascade.*

interruption (in-ter-rup'-tion) A break in, or the cutting short of, the cycle of erosion, characterized by a change in the position of base level relative to a landmass or terrain, and resulting in the initiation of a new erosion cycle. It may be dynamic, caused by earth movements (involving deformation, dislocation, or tilting) or fluctuations of sea level; or static, caused by climatic change or piracy. Cf: *accident.*

intersected crossed platy structure A *crossed structure* that has two predominant dip directions and consists of intersected platy crystallites. This structure shows relatively shorter, more rhomboidal second-order structural units than intersected crossed acicular structure; it commonly occurs as a transition from dissected crossed prismatic to scaly structure in aragonitic gastropods (Erben and Krampitz, 1972).

intersecting peneplain (in-ter-sect'-ing) One of two peneplains forming a *morvan* landscape.

intersection (in-ter-sec'-tion) (a) A method in surveying by which the horizontal position of an unoccupied point is determined by drawing lines to that point from two or more points of known position. Cf: *resection.* (b) Determination of positions by triangulation.

intersection lineation A lineation formed by the intersection of two planar surfaces, at least one of which is penetrative, e.g., bedding and cleavage.

intersection shoot An ore shoot located at the intersection of one vein or vein system with another. It is a common type of ore deposit.

intersept (in'-ter-sept) That portion of an archaeocyathid wall between the edges of two adjacent septae (TIP, 1972, pt. E, p.40).

interseptal ridge (in-ter-sep'-tal) A longitudinal elevation on the outer surface of the wall of a *corallite,* corresponding in position to the space between a pair of adjacent septa on the inner surface of the wall. Cf: *septal groove.*

intersequent stream (in-ter-se'-quent) A stream following a consequent course in a depression between the margins of opposing alluvial fans, as on a bajada.

intersertal (in-ter-ser'-tal) Hypocrystalline texture consisting of lath-shaped plagioclase grains and interstitial glass or cryptocrystalline material. Cf: *intergranular; ophitic; subophitic; hyalophitic; hyalopilitic.*

intersilite A yellow monoclinic mineral: $Na_6Mn(Ti,Nb)Si_{10}(O,OH)_{28}{\cdot}4H_2O$.

interstade (in'-ter-stade) A warmer substage of a glacial stage, marked by a temporary retreat of the ice; "a climatic episode within a glaciation during which a secondary recession or a stillstand of glaciers took place" (ACSN, 1961, art. 40). Example: the Alleröd interstade of Denmark. Syn: *interstadial; oscillation [glac geol].*

interstadial (in-ter-sta'-di-al) adj. Pertaining to or formed during an *interstade.* n. *interstade.*

interstice (in-ter'-stice) An opening or space, as in a rock or soil. On the basis of origin, it may be classified as an *original interstice* or a *secondary interstice;* on the basis of size, as a *capillary interstice,* a *subcapillary interstice,* or a *supercapillary interstice.* Syn: *void; pore [geol].* Adj: *interstitial.*

interstitial (in-ter-sti'-tial) Said of a mineral deposit in which the minerals fill the pores of the host rock. Cf: *impregnated.*

interstitial defect In a crystal structure, the filling of a normally void interstice with an extra atom; it is a type of *point defect.* See also: *addition solid solution.* Cf: *Frenkel defect; Schottky defect.*

interstitial solid solution *addition solid solution.*

interstitial water Subsurface water in the voids of a rock. Syn: *pore water.* Cf: *connate water.*

interstratal karst (in-ter-stra'-tal) Karst that forms at depth below a resistant nonsoluble rock. Cf: *subjacent karst; covered karst; paleokarst.*

interstratification (in'-ter-strat'-i-fi-ca'-tion) (a) The state or condition of being interstratified or occurring between strata of a different character. (b) *interlayering.*

interstratification [clay] A clay material composed of two or more kinds of layers in which regularity of interleaving may vary from completely random to prefectly regular. Syn: *interlayering.*

interstratified (in-ter-strat'-i-fied) *interbedded.*

interstratified clay minerals A type of mixed-layer clay mineral.

interstream (in'-ter-stream) Said of an area, divide, or topographic feature situated or lying between streams, such as an *interfluve.* See also: *interfluvial.*

interstream groundwater ridge A residual ridge in the water table that develops between two effluent streams as a result of the percolation of ground water toward the streams. Cf: *groundwater ridge.*

intertentacular (in'-ter-ten-tac'-u-lar) Situated between tentacles; e.g. "intertentacular organ", a flask-shaped, tubular bryozoan structure providing passageway for extrusion of ova between two tentacles on the distal side of the *lophophore* near the midline.

intertextic (in-ter-tex'-tic) Said of an arrangement in a soil fabric whereby the skeleton grains are linked by intergranular braces or are embedded in a porous matrix (Brewer, 1964, p.170).

interthecal (in-ter-the'-cal) Between thecae; e.g. "interthecal septum" separating adjacent thecal cavities in graptoloids.

interthem A minor unconformity-bounded unit comparable in thickness with a formation or in time span with a stage (Chang, 1975, p.1544). Several interthems may by recognized within a *synthem*.

intertidal (in-ter-tid'-al) *littoral [oceanog].*

intertidalite (in-ter-ti'-da-lite) A *tidalite* that is or was known to be deposited by tidal processes in the intertidal zone.

intertidal tombolo A sand body that connects an island to the mainland at low tide but is submerged at high tide; a term commonly used in New England. Cf: *tombolo.*

intertongued lithofacies (in-ter-tongued') A *lithofacies* whose irregular boundaries separate intertonguing stratigraphic bodies of contrasting characteristics (such as shale and sandstone) (Weller, 1958, p.633). It is not a unit that can be mapped in the normal manner. The term *lithosome,* as originally defined, is synonymous. Cf: *statistical lithofacies.*

intertonguing (in-ter-tongu'-ing) The disappearance of sedimentary bodies in laterally adjacent masses owing to splitting into many thin tongues, each of which reaches an independent pinch-out termination; the intergradation of markedly different rocks through a vertical succession of thin interlocking or overlapping wedge-shaped layers. Syn: *interfingering; interdigitation.*

intertrappean (in-ter-trap'-pe-an) Pertaining to a deposit that occurs between two lava flows.

intertropical convergence zone Discontinuous low-pressure belt of *thunderstorms* paralleling the equator and marking the convergence of the Northern and Southern Hemisphere surface *trade winds.* Abbrev: ITCZ.

intertrough (in'-ter-trough) The median, narrowly triangular furrow dividing the pseudointerarea of the *pedicle valve* of some acrotretacean brachiopods.

interval [cart] (in'-ter-val) *contour interval.*

interval [glac geol] An informal term for a subdivision of an *interstade.*

interval [stratig] (a) *stratigraphic interval.* (b) *geochronologic interval.* (c) *polarity interval.*

interval change A lateral increase or decrease of the time interval between two seismic reflection events.

interval correlation Stratigraphic correlation based on well-defined stratigraphic intervals identified by their positions between marker horizons (Krumbein and Sloss, 1963, p.343).

interval density In a well bore, the density of an interval integrated from gamma-gamma log data or determined by a borehole gravity meter.

interval-entropy map (in'-ter-val-en'-tro-py) A multicomponent *vertical-variability map* that expresses the degree of vertical alternations (homogeneity or heterogeneity) of rock types in a given succession of beds or within a given stratigraphic unit. The term was introduced by Forgotson (1960).

intervallar coefficient (in-ter-val'-lar) The ratio of the width of the *intervallum* to the width of the central cavity in double-walled archaeocyathids (TIP, 1972, pt. E, p.11).

intervallum (in-ter-val'-lum) The space between the outer and inner walls of an archaeocyathid (TIP, 1955, pt. E, p.7). It may contain various structures, esp. the septa. Adj: intervallar.

intervalometer (in'-ter-val-om'-e-ter) A timing device on an aerial camera that automatically operates the shutter at predetermined intervals.

interval scale A continuous scale of measurement that has an arbitrary origin and in which the differences between successive units of the scale are constant. The Celsius and Fahrenheit temperature scales are classic examples. Cf: *nominal scale; ordinal scale; ratio scale.*

interval velocity (in'-ter-val) The distance across a stratigraphic interval divided by the time for a seismic wave to traverse it; the *average velocity [seis]* measured over a depth interval, e.g., in a sonic log or borehole survey. It usually refers to P-wave velocity and implies measurement perpendicular to bedding.

interval zone (a) The body of strata, or *biozone,* between two specified, documented lowest and/or highest occurrences of single taxa. Three basic types are recognized: *taxon-range zone, concurrent-range zone,* and *lineage zone* (NACSN, 1983, Art. 50). See also: *partial range zone.* (b) A biozone consisting of the body of fossiliferous strata between two specified *biohorizons* (ISSC, 1994, p.59-61). Syn: *interbiohorizon zone.*

interzonal time (in-ter-zon'-al) *Geologic time* represented by a *diastem* or stratigraphic hiatus (Kobayashi, 1944a, p.745).

interzone (in'-ter-zone) The unfossiliferous beds between fossiliferous ones (Henningsmoen, 1961, p.83).

interzooecial (in'-ter-zo-oe'-cial) Used loosely to indicate skeletal zooecial walls between adjacent zooidal chambers in stenolaemate bryozoans.

interzooidal (in'-ter-zo-oi'-dal) Existing between or among zooids; e.g. "interzooidal avicularia" of cheilostome bryozoans, which are intercalated in lineal series in spaces smaller than those occupied by autozooids.

intexine (in-tex'-ine) A syn. of *endexine.* Also spelled: intextine.

intine (in'-tine) The thin, inner layer of the two major layers forming the wall (sporoderm) of palynomorphs, composed of cellulose and pectates, and situated inside the *exine,* surrounding the living cytoplasm. Syn: *endospore.*

intra-ambulacral extension (in'-tra-am'-bu-lac'-ral) Part of an ambulacral cover plate or oral plate of an edrioasteroid that is produced inward into the *ambulacral tunnel* (Bell, 1976).

intracapsular (in-tra-cap'-su-lar) Said of cell materials within the central capsule of a radiolarian; e.g. "intracapsular layer" consisting of protoplasm exclusive of nucleus.

intraclast (in'-tra-clast) A broad, general term introduced by Folk (1959, p.4) for a component of a limestone, representing a torn-up and reworked fragment of a penecontemporaneous sediment (usually weakly consolidated) that has been eroded within the basin of deposition (such as the nearby sea floor or an exposed carbonate mud flat) and redeposited there to form a new sediment; an *allochem* derived from the same formation. The fragment may range in size from silt to gravel, and is generally rounded but may be equant to discoidal. Cf: *protointraclast; extraclast.*

intracoastal (in-tra-coast'-al) Being within or near the coast; esp. said of inland waters near the coast, e.g., the U.S. Intracoastal Waterway, New York to Texas.

intracontinental geosyncline (in'-tra-con'-ti-nen'-tal) *intrageosyncline.*

intracratonic basin (in'-tra-cra-ton'-ic) A basin formed within the interior region of a continent, away from plate boundaries. It develops where there is subsidence of a portion of a craton, probably due to thermal subsidence of an unsuccessful rift.

intracrystal porosity (in-tra-crys'-tal) The porosity within individual crystals, pores in large crystals of echinoderms, and fluid inclusions (Choquette and Pray, 1970, p.247).

intracyclothem (in-tra-cy'-clo-them) A cyclic sequence of strata resulting from the splitting of a cyclothem. Term introduced by Gray (1955).

intradeep (in'-tra-deep) A geosynclinal trough appearing within a geosynclinal belt at the end of or following uplift of the belt; a type of *secondary geosyncline.* See also: *foredeep; backdeep.* Obsolete.

intradelta (in'-tra-del-ta) *delta plain.*

intrafacies (in'-tra-fa-cies) A term used by Cloud and Barnes (1957, p.169) to denote a minor or subordinate facies occurring within a differing major facies.

intrafolial fold (in-tra-fo'-li-al) A minor fold involving only a few layers in an otherwise unfolded rock.

intraformational (in'-tra-for-ma'-tion-al) (a) Formed within a geologic formation, more or less contemporaneously with the enclosing sediments. The term is esp. used in regard to syndepositional folding or slumping, e.g. "intraformational deformation" or "intraformational breccia". (b) Existing within a formation, with no necessary connotation of time of origin. See also: *intrastratal.*

intraformational breccia A rock formed of angular intraclasts. It is similar in nature and origin to an *intraformational conglomerate* but contains fragments showing greater angularity.

intraformational conglomerate (a) A conglomerate formed by rounded intraclasts; e.g. an *edgewise conglomerate.* Fragmentation is commonly caused by desiccation and mud cracking on low beach ridges and natural levees. (b) A conglomerate occurring in the midst of a geologic formation, such as one formed during a brief interruption in the orderly deposition of strata. It may contain clasts external to the formation. The term is used in this sense esp. in England. Cf: *interformational conglomerate.* Syn: *flat-pebble conglomerate.*

intraformational contortion Intricate and complicated folding, as exhibited in *convolute bedding;* esp. such deformation resulting from the subaqueous slumping or sliding of unconsolidated sediments under the influence of gravity. See also: *intraformational corrugation.* Syn: *intrastratal contortion.*

intraformational corrugation A term applied to *intraformational contortion* on a small scale.

intraformational fold A minor fold confined to a sedimentary layer lying between undeformed beds; it results from processes, such as sliding or slumping, that took place prior to complete lithification.

intrageosyncline (in'-tra-ge'-o-syn'-cline) DuToit's (1937) term for a *parageosyncline.* Obsolete, as well as its Syn.; *intracontinental geosyncline.*

intraglacial (in-tra-gla'-cial) (a) Said of glacial deposits formed on ground actually covered by the ice, or of glacial phenomena pertaining to a region covered by the ice at any given time. Ant: *extraglacial.* (b) *englacial.*

intraglacial volcano A volcano formed by subglacial eruptions (Jones, 1969).

intragranular movement A glacier's gliding movement by which favorably oriented ice crystals are deformed by slip. It is an important mechanism in glacier flow. Cf: *intergranular movement.*

intragranular porosity The porosity existing within individual grains or particles of a rock, esp. within skeletal material of a carbonate sedimentary rock. Cf: *intraparticle porosity.*

intralaminal accessory aperture (in-tra-lam'-i-nal) An *accessory aperture* in the test of a planktonic foraminifer that leads through accessory structures into a cavity beneath them and not directly into the chamber cavity (as in *Rugoglobigerina*). Cf: *infralaminal accessory aperture.*

intramicrite (in-tra-mic′-rite) A limestone containing at least 25% intraclasts and in which the carbonate-mud matrix (micrite) is more abundant than the sparry-calcite cement (Folk, 1959, p.14).
intramicrudite (in′-tra-mic-rud′-ite) An *intramicrite* containing gravel-sized intraclasts.
intramineral (in-tra-min′-er-al) Pertaining to the time interval of a period of mineralization (Kirkham, 1971, p.1245); also, pertaining to those features, e.g. a breccia mineralized during its formation, that were emplaced during such an interval. The term has been applied to a time interval between distinct periods of mineralization and to its resulting features, but *intermineral* is preferred for this meaning.
intramontane space (in-tra-mon′-tane) A rarely used syn. of *Zwischengebirge.*
intramorainal (in′-tra-mo-rain′-al) Said of deposits and phenomena occurring within a lobate curve of a moraine. Ant: *extramorainal.*
intramural budding (in-tra-mu′-ral) A type of *polystomodaeal budding* in which the stomodaea are directly or indirectly linked in a single linear series. Cf: *circummural budding.*
intraparticle porosity (in-tra-par′-ti-cle) The porosity within individual particles of a rock. Choquette and Pray (1970, p.247) recommend use of this term rather than the term *intragranular porosity* which suggests limitation to grain-size particles.
intrapermafrost water (in-tra-per′-ma-frost) Unfrozen groundwater in layers or lenses within permafrost.
intraplate magmatism Igneous activity far from any plate boundary and therefore considered unrelated to subduction or sea-floor spreading processes.
intrapositional deposit (in′-tra-po-si′-tion-al) Sediments deposited by the process of *stratigraphic leakage,* e.g. as crevice or erosional-channel fillings (Foster, 1966).
intrareticulate (in-tra-re-tic′-u-late) In pollen *sculpture,* showing a reticulate pattern formed by the arrangement of the *columellae* below the *tectum.*
intrasediment crystals Saline minerals grown within a sediment host, either by displacing material, incorporating it within the crystal, or both.
intrasparite (in-tra-spar′-ite) A limestone containing at least 25% intraclasts and in which the sparry-calcite cement is more abundant than the carbonate-mud matrix (micrite) (Folk, 1959, p.14). It is common in environments of high physical energy, where the spar usually represents pore-filling cement.
intrasparrudite (in′-tra-spar-rud′-ite) An *intrasparite* containing gravel-sized intraclasts.
intrastratal (in-tra-stra′-tal) Formed or occurring within a stratum or strata; e.g. formation of iron-rich authigenic clay by "intrastratal alteration" of hornblende. See also: *intraformational.*
intrastratal contortion *intraformational contortion.*
intrastratal flow structure A variety of *convolute bedding* formed by flowage.
intrastratal solution Removal by chemical solution of certain mineral species from within a sedimentary bed following deposition. Syn: *differential solution.*
intratabular (in-tra-tab′-u-lar) Said of features of a dinoflagellate cyst that correspond to the central parts of thecal plates rather than to the lines of separation between them. Cf: *nontabular; penitabular.*
intratelluric (in′-tra-tel-lu′-ric) (a) Said of a phenocryst, of an earlier generation than its groundmass, that formed at depth, prior to extrusion of a magma as lava. (b) Said of that period of crystallization occurring deep within the Earth just prior to the extrusion of a magma as lava. (c) Located, formed, or originating deep within the Earth. Obsolete.
intratentacular budding (in′-tra-ten-tac′-u-lar) Formation of new scleractinian coral polyps by invagination of the oral disk of the parent inside the ring of tentacles surrounding its mouth. Cf: *extratentacular budding.*
intrathecal extension (in-tra-the′-cal) Part of an external thecal plate in an edrioasteroid that is produced inward into the interior of the *theca* (Bell, 1976).
intraumbilical aperture (in′-tra-um-bil′-i-cal) An aperture in a foraminiferal test located in the umbilicus but not extending outside of it.
intrazonal soil (a) One of the three orders in soil classification. (b) A soil with more or less well developed soil characteristics that reflect the dominating influence of some local factor of relief, parent material, or age, over the normal effect of climate and vegetation. Not used in current U.S. system of soil taxonomy.
intrazonal time *Geologic time* represented by a *biostratigraphic zone* (Kobayashi, 1944a, p.745).
intrazooidal (in′-tra-zo-oi′-dal) Said of a structure occurring within a *zooid,* e.g. "intrazooidal septulae" that connect the hypostega with the main body cavity in some ascophoran cheilostomes (bryozoans) (Banta, 1970, p.39). Also spelled: intrazoidal.
intrenched meander (in-trenched′) *entrenched meander.*
intrenched stream *entrenched stream.*
intrinsic ash (in-trin′-sic) *inherent ash.*
intrinsic hypothesis In *geostatistics,* an assumption of a form of weak stationarity, in which the differences between successive observations are approximately constant.
intrinsic ionic conduction Electrical conduction arising in transport of charge through a solid as a result of a movement of ions through a crystal lattice. It usually takes place at elevated temperatures and does not depend on the presence of impurities or vacancies.
intrinsic permeability (a) *specific permeability.* (b) Pertaining to the relative ease with which a porous medium can transmit a liquid under a hydraulic or potential gradient. It is a property of the porous medium and is independent of the nature of the liquid or the potential field (Fetter, 1994). The term replaces *coefficient of permeability.* Cf: *permeability [geol]; permeability coefficient.*
introduced fossils Fossils originally in younger rocks later introduced in any one of several ways into the older rocks in which they are now found (ISSC, 1994, p.55).
introgression (in-tro-gres′-sion) The evolutionary process in which a *hybrid [evol]* crosses with a member of one of its parental species, thus introducing some genes from the other parental species into the next generation of the first species. Syn: *introgressive hybridization.*
introgressive hybridization (in-tro-gres′-sive) *introgression.*
intrusion [grd wat] (in-tru′-sion) *salt-water encroachment.*
intrusion [ign] The process of emplacement of magma in pre-existing rock; magmatic activity; also, the igneous rock mass so formed within the surrounding rock. See also: *pluton.* Syn: *injection [ign]; emplacement [intrus rocks]; invasion [ign]; irruption [intrus rocks].*
intrusion [sed] (a) A sedimentary injection on a relatively large scale; e.g. the forcing upward of clay, chalk, salt, gypsum, or other plastic sediment, and its emplacement under abnormal pressure in the form of a diapiric plug. See also: *autointrusion [sed].* (b) A sedimentary structure or rock formed by intrusion. (c) *stone intrusion.* Syn: *sedimentary intrusion.*
intrusion breccia (a) *contact breccia.* (b) *intrusive breccia.*
intrusive (in-tru′-sive) adj. Pertaining to intrusion, both the process and the body so formed. n. An *intrusive* rock or body. Cf: *extrusive.* Syn: irruptive.
intrusive breccia A heterogeneous mixture of angular to rounded fragments in a matrix of clastic material, which has been mobilized and intruded into its present position along pre-existing structures. It is differentiated from *intrusion breccia* because of the absence of igneous matrix and the lack of control by igneous intrusive contacts (Bryant, 1968, p.4). It is commonly hydrothermally altered.
intrusive ice Ice that results from the freezing of injected water, as in a domal or tabular body, forming a raised surface, such as a *hydrolaccolith* or a *pingo.*
intrusive tuff *tuffisite.*
intumescence (in-tu-mes′-cence) The property that some minerals have of swelling or frothing when heated, owing to the release of gases.
inundation (in-un-da′-tion) A rising of water and its spreading over land not normally submerged.
inundite (in-un′-dite) A "sandy flood deposit" consisting of current-rippled sand below a paleosol. It occurs shoreward of tempestite deposits (Einsele and Seilacher, 1982, p.334).
invar (in′-var) An alloy of nickel and iron, containing about 36% nickel, and having an extremely low coefficient of thermal expansion. It is used in the construction of surveying instruments such as pendulums, level rods, first-order leveling instruments, and tapes. Etymol: *Invar,* a trademark.
invariant equilibrium (in-var′-i-ant) Equilibrium of a phase assemblage that has zero degrees of freedom. See also: *invariant point.*
invariant point A point representing the conditions of *invariant equilibrium.*
invasion [ign] (in-va′-sion) *intrusion [ign].*
invasion [stratig] *transgression.*
invasive In ecology, said of an organism that has been introduced into a region where it was not previously found and that has experienced rapid population growth which has negatively affected endemic taxa (Vermeij, 1991). Cf: *immigrant.*
inver (in′-ver) (a) A place where a river flows into the sea or into an arm of the sea; e.g. Inverness, Scotland. (b) The confluence of two streams. Etymol: Gaelic.
inverse estuary An "estuary" in which evaporation exceeds the influx of fresh water (land drainage and precipitation) so that the salinity of its water rises above that of seawater. Ant: *positive estuary.* Syn: *negative estuary.*
inverse filter A filter with characteristics complementary to another filter in such a way that when used in series with the other filter no frequency-selective filtering occurs (except for overall time delay).
inverse grading A systematic particle-size gradation upward through a sedimentary layer in which the finest particles are at the base and the progressively coarser ones, toward the top. Syn: *reverse grading.* Cf: *normal grading* (Friedman et al., 1992, p.21).
inverse position computation On an ellipsoid, computation of the length, the forward azimuth, and the backward azimuth of a geodesic between two points whose geodetic coordinates are known.
inverse problem The problem of gaining knowledge of the physical features of a disturbing body by analysis of its effects, e.g., fields and potentials; finding the model from observed data. It is in contrast to the direct, forward, or normal problem, which is calculating what would have been observed from a given model (Sheriff, 2002, p.194).
inverse projection *transverse projection.*

inverse thermoremanent magnetization An artificial *remanent magnetization* acquired during a temperature increase from below ambient temperature. Cf: *partial thermoremanent magnetization.* Abbrev: ITRM.
inverse zoning *reversed zoning.*
inversion [cryst] (in-ver′-sion) (a) A symmetry transformation through a center of symmetry. (b) A *phase transformation.*
inversion [geomorph] (a) The development of *inverted relief* whereby anticlines are transformed into valleys and synclines into mountains; e.g. the formation of a deep basin in an area formerly occupied by land that produced quantities of sediment. (b) The occupancy by a lava flow of a former ravine or valley in the side of a volcano, thereby producing a divide over the former valley and forcing the stream to develop a new valley on its former divide (Cotton, 1958, p.366-367).
inversion [geophys] Construction of a geophysical model from a set of measurements to determine the values of a property that could be responsible for the measurements; e.g., using numerous gravity measurements to infer subsurface density distributions, calculating a synthetic sonic log from a seismic reflection trace, or using slip vectors and spreading rates to define global plate motions. Without additional constraints, inversion models are inherently ambiguous.
inversion [meteorol] In meteorology, a reversal of the gradient of a meteorologic element, e.g. an increase rather than a decrease of temperature with altitude.
inversion [struc geol] A reversal of tectonic mode. Reverse-reactivation of a normal fault or the contraction of a region that previously underwent extension is termed positive inversion. Alternatively, negative inversion is the normal reactivation of a reverse fault or the extension of a region that previously underwent contraction (Peacock et al., 2000).
inversion center *center of symmetry.*
inversion layer In a body of water, a water layer whose temperature increases rather than decreases with depth.
inversion point (a) A point representing the temperature at which one polymorphic form of a substance, in equilibrium with its vapor, reversibly changes into another under invariant conditions. (b) The temperature at which one polymorphic form of a substance inverts reversibly to another at a specific pressure. (c) More loosely, the lowest temperature at which a monotropic phase inverts at an appreciable rate into a stable phase, or at which a given phase dissociates at an appreciable rate, under given conditions. (d) A single point at which different phases are capable of existing together at equilibrium. Syn: *transition point; transition temperature.*
invert (in′-vert) The floor or bottom of the internal cross section of a closed conduit, such as an aqueduct, tunnel, culvert, or drain. The term originally referred to the inverted arch used to form the bottom of a masonry-lined sewer or tunnel.
invertebrate (in-ver′-te-brate) n. An animal without a backbone, such as the mollusks, arthropods, and coelenterates. adj. Of or pertaining to an animal that lacks a backbone.
invertebrate paleontology The branch of paleontology dealing with fossil invertebrates. Syn: *invertebrate paleozoology.*
invertebrate paleozoology *invertebrate paleontology.*
inverted (in-vert′-ed) *overturned.*
inverted basal cavity *zone of recessive basal margin.*
inverted pendulum *pendulum (b).*
inverted plunge The plunge of folds, or sets of folds, whose inclination has been carried past the vertical, so that the plunge is now less than 90° in a direction opposite from the original attitude. It is a rather common feature in excessively folded or refolded terranes.
inverted relief A topographic configuration that is the inverse of the geologic structure, as where mountains occupy the sites of synclines and valleys occupy the sites of anticlines. See also: *inversion [geomorph].*
inverted siphon A portion of a water conduit that is depressed in a "U-shape". Cf: *siphon [hydraul].*
inverted stream (a) A beheaded stream whose drainage turns back into the capturing stream (Lobeck, 1939, p.199). (b) A term proposed by Davis (1889b, p.210) but later abandoned (Davis, 1895, p.134) in favor of *obsequent stream.*
inverted unconformity (a) An unconformity in which the younger strata end abruptly against the older rocks, such as one produced by intense folding of a complex region (Grabau, 1924, p.826). (b) Truncation of the upper parts of laminae in sediment, shown in some load casts where underlying plastic material has been squeezed upward and intruded laterally into overlying sediment (Kuenen, 1957, p.250).
inverted well A well that takes in water near its top and discharges it at lower levels, into permeable material; e.g. a *drainage well.*
invisible gold (in-vis′-i-ble) A syn. used in the popular press for *Carlin-type gold.*
involucre (in′-vo-lu-cre) One or more whorls of bracts subtending a flower or inflorescence. See also: *cupule.*
involute (in′-vo-lute) Coiled or rolled inward, e.g. said of a foraminiferal *test* having closely coiled and strongly overlapping whorls, in which the inner part of the last whorl extends in toward the center of the coil to cover part of the adjacent inner whorl; or a coiled gastropod shell having the last whorl enveloping the earlier whorls, which are more or less visible in the umbilici; or a coiled cephalopod conch with considerably overlapping whorls and a narrow umbilicus. Cf: *convolute; evolute; advolute.*
involution [sed] (in-vo-lu′-tion) (a) A highly irregular, aimlessly contorted sedimentary structure consisting of local folds and interpenetrations of fine-grained material in clayey strata, and developed by the formation, growth, and melting of ground ice (*cryoturbation*) in the active layer overlying permafrost. Syn: *Brodelboden.* (b) An irregularly contorted and penetrating structure, as a wave or fold, in a soil deposit.
involution [struc geol] The refolding of nappes, resulting in complex patterns.
inwash (in′-wash) Alluvium deposited against the margin of a glacier by a stream of nonglacial origin.
inyoite (in′-yo-ite) A colorless monoclinic mineral: $CaB_3O_3(OH)_5 \cdot 4H_2O$.
IOCG *iron oxide copper-gold deposit.*
iodargyrite (i-o-dar′-gy-rite) A yellowish or greenish hexagonal mineral: AgI. Syn: *iodyrite.*
iodate (i′-o-date) A mineral compound that is characterized by the radical IO_3^-. An example is salesite, $Cu(IO_3)(OH)$.
iodobromite (i-o′-do-bro′-mite) *bromargyrite.*
IODP *Integrated Ocean Drilling Program.*
iodyrite (i-od′-y-rite) *iodargyrite.*
iolanthite (i-o-lan′-thite) A banded reddish jasperlike mineral from Oregon.
iolite A syn. of *cordierite,* esp. the gem variety.
ion exchange (i′-on) Reversible exchange of ions contained in a crystal for different ions in solution without destroying crystal structure or disturbing electrical neutrality. It is accomplished by diffusion and occurs most easily in crystals having one- or two-dimensional channelways where ions are relatively weakly bonded; it also takes place at higher temperatures in network silicates, involving the most weakly bonded cations such as those of potassium and sodium. Ion exchange is also common in resins consisting of three-dimensional hydrocarbon networks to which many ionizable groups are attached. See also: *exchange capacity; anion exchange; cation exchange.*
Ionian (I-on′-ian [Yon′-i-an]) A European stage (Mediterranean): middle Pleistocene (above Calabrian, below Tyrrhenian); base at isotope stage 25. Obsolete syn: *Milazzian.*
ionic substitution (i-on′-ic) The replacement of one or more kinds of ion in a crystal structure by other kinds of generally similar size and charge. Syn: *diadochy; proxying; substitution.*
ionite (i′-o-nite) (a) *anauxite.* (b) An earthy, resinous, brownish-yellow fossil hydrocarbon in the lignite of Ione Valley, Amador County, Calif. It is not a recognized mineral species.
ionium (i-o′-ni-um) An old but still-used name for thorium-230, a member of the uranium series and daughter of uranium-234.
ionium-deficiency method *thorium-230-deficiency method.*
ionium-excess method *ionium-thorium age method; uranium-series age method.*
ionium-thorium age method (i-o′-ni-um-tho′-ri-um) The calculation of an age in years for deep-sea sediments formed during the last 300,000 years, based on the assumption that the initial thorium-230 ratio for accumulating sediments has remained constant for the total section of sediments under study. The age depends on the thorium-230 to thorium-232 ratio, which gradually decreases with the passage of time. Syn: *ionium-excess method; uranium-series age method; thorium-230/thorium-232 age method.*
ionization constant (i′-on-i-za′-tion) As applied to ionization reactions, a syn. of *dissociation constant.*
ionization potential The voltage required to drive an electron from an atom or molecule, leaving a positive ion.
ionizing radiation (i′-on-iz′-ing) Any electromagnetic or particulate radiation that displaces electrons within a medium.
ion microprobe An instrument, similar in principle to an *electron microprobe,* with which a focused beam of ions is made to strike the surface of a sample, with resultant emission of ions from the impacted area. The emitted ions are analyzed in a *mass spectrometer* and are characteristic of the isotopes of the elements present in the area of impact. Also called: *secondary ionization mass spectrometer.*
ionography (i-on-og′-ra-phy) Ion-exhange *electrochromatography* wherein the ions migrate by electrostatic attraction, usually over or through an ion-exchange resin.
ion pairing A property of ionic solutions in which certain anions and cations are not randomly distributed throughout the solution but are in closer proximity to each other due to electrostatic attraction. Can be described by thermodynamic association constants that are similar to other equilibrium constants. Plays important roles in the solubility of calcium carbonate and other minerals in seawater, the partial pressure of dissolved carbon dioxide in seawater, the absorption of sound in seawater, and other oceanic processes.
iowaite (i′-o-wa-ite′) A white, blue-green, or yellowish rhombohedral mineral: $Mg_6Fe^{3+}_2Cl_2(OH)_{16} \cdot 4H_2O$.
Iowan (I′-o-wan) Originally defined as a separate stage between the Illinoian and Wisconsinan, and later as the earliest substage of the Wisconsinan. The area of Iowan drift in northeastern Iowa is now recognized as an erosional surface cut into the Kansan till plain (Ruhe, 1969). Obsolete.

iozite (i'-o-zite) *wüstite.*

IP (a) *induced polarization.* (b) initial production.

IQSY *International Years of the Quiet Sun.*

iquiqueite (i-qui'-que-ite) A yellow trigonal mineral: $K_3Na_4MgCrB_{24}O_{39}(OH)_6 \cdot 9H_2O$.

IR [remote sensing] *infrared.*

IR [sed] *insoluble residue.*

iranite (i-ran'-ite) A saffron-yellow to brown triclinic mineral: $Pb_{10}Cu(CrO_4)_6(SiO_4)_2(F,OH)_2$. Syn: *khuniite.*

iraqite-(La) (i-raq'-ite) A pale greenish-yellow tetragonal mineral: $K(La,Ce,Th)_2(Ca,La,Na)_4Si_{16}O_{40}$.

irarsite (ir-ar'-site) A metallic black cubic mineral of the *gersdorffite* group: (Ir,Ru,Rh,Pt)AsS. It forms a series with hollingworthite.

IR color photograph *infrared color photograph.*

irghizite Glass associated with Zhamanshin crater, Russia, thought to have been produced by melting of local target rocks during the impact event. See also: *impactite.*

irhtemite (irh-tem'-ite) A colorless to pink monoclinic mineral: $Ca_4Mg(AsO_4)_2(HAsO_3OH)_2 \cdot 4H_2O$.

irhzer (ir'-hzer) A term used in northern Africa for a straight groove carved in a mountainside by a stream (Termier and Termier, 1963, p.408). Etymol: Berber.

iridarsenite (ir-id-ar'-sen-ite) A metallic monoclinic mineral: $(Ir,Ru)As_2$.

iridescence (ir-i-des'-cence) The exhibition of prismatic colors (producing rainbow effects) in the interior or on the surface of a mineral, caused by interference of light from thin films or layers of different refractive index.

iridium (i-rid'-i-um) (a) An element of the *platinum* group, symbol Ir. (b) A metallic white cubic mineral consisting of more than 80 percent Ir, the remainder being osmium, palladium, or related elements.

iridosmine (ir-id-os'-mine) A discredited mineral name for iridium-bearing osmium. It usually contains some rhodium, platinum, ruthenium, iron, and copper. Cf: *osmiridium.* Syn: *iridosmium.*

iridosmium (ir-id-os'-mi-um) *iridosmine.* An unnecessary name for iridium-bearing osmium.

iriginite (ir'-ig-in-ite) A canary-yellow orthorhombic secondary mineral: $(UO_2)Mo^{6+}{}_2O_7 \cdot 3H_2O$.

iris (i'-ris) (a) A transparent quartz crystal containing minute air-filled or liquid-filled internal cracks that produce iridescence by interference of light. The cracks may occur naturally or be caused artificially by heating and sudden cooling of the specimen. Syn: *rainbow quartz.* (b) An iridescent mineral; e.g. "California iris" (kunzite).

IRM *isothermal remanent magnetization.*

iron [meteor] *iron meteorite.*

iron [mineral] A heavy, cubic, magnetic, malleable and ductile, and chemically active mineral, the native metallic element Fe. It has a silvery or silver-white color when pure, but readily oxidizes in moist air. Native iron occurs rarely in terrestrial rocks (such as disseminated grains in basalts), but is common in meteorites; it occurs combined in a wide range of ores and in most igneous rocks. Iron is the most widely used of the metals.

Iron Age (I'-ron) In Old World archaeology, a cultural level that is the final age in the *three-age system,* and is characterized by the technology of iron. Correlation of relative cultural levels with actual age (and, therefore, with the time-stratigraphic units of geology) varies from region to region. The Iron Age began in Europe and the Near East about 1100 B.C.

iron alum *halotrichite.*

iron bacteria Anaerobic bacteria that precipitate iron oxide from solution, either by oxidizing ferrous salts or by releasing oxidized metals from organic compounds. Accumulations of iron developed in this way are *bacteriogenic* ore deposits. Cf: *sulfur bacteria.*

iron-bearing formation *iron formation.*

iron cordierite *sekaninaite.*

iron-cross twin law A twin law according to which crystals of the diploidal class of the isometric system are formed by interpenetration twinning of two pyritohedrons. The twin axis is perpendicular to a face of the rhombic dodecahedron.

iron formation A chemical sedimentary rock, typically thin-bedded and/or finely laminated, containing at least 15% iron of sedimentary origin, and commonly but not necessarily containing layers of chert (James, 1954, p.239). Various primary facies (usually not weathered) of iron formation are distinguished on the basis of whether the iron occurs predominantly as oxide, silicate, carbonate, or sulfide. Most iron formation is of Precambrian age. In mining usage, the term refers to a low-grade sedimentary iron ore with the iron mineral(s) segregated in bands or sheets irregularly mingled with chert or fine-grained quartz (Thrush, 1968, p.590). Cf: *ironstone; jaspilite.* See also: *Algoma-type iron formation; Lake Superior-type iron formation; oxide-facies iron formation; carbonate-facies iron formation; silicate-facies iron formation; sulfide-facies iron formation.* Essentially synonymous terms: *itabirite; banded hematite quartzite; taconite; quartz-banded ore; banded ironstone; calico rock; jasper bar; iron-bearing formation.*

iron-formation-hosted gold *arsenical sulfide-silicate iron formation.*

iron froth A fine, spongy or micaceous variety of hematite.

iron glance A variety of hematite; specif. *specularite.*

iron hat *gossan.*

iron hypersthene (a) An iron-rich hypersthene. (b) *ferrosilite.*

iron meteorite A meteorite composed predominantly of iron-nickel metal but commonly containing inclusions of iron sulfide and iron phosphide minerals and sometimes silicates. Iron meteorites are commonly subdivided into IAB, IC, IIAB, IIC, IID, IIE, IIF, IIG, IIIAB, IIICD, IIIE, IIIF, IVA, IVB classes based on their metal compositions. Especially important are the elements Ni, Ga, Ge, Ir, and Au. In the past, iron meteorites were commonly subdivided into structural classes (e.g., hexahedrite, octahedrite, ataxite) based on grain size and crystal structure of the metallic minerals, but these criteria are now known to be inadequate to resolve sets of genetically related irons.

iron mica (a) *lepidomelane.* (b) *biotite.* (c) Micaceous hematite.

iron-monticellite (i'-ron-mon-ti-cel'-lite) *kirschsteinite.*

iron olivine *fayalite.*

iron ore Ferruginous rock containing one or more distinct natural chemical compounds from which metallic iron may be profitably extracted. The chief ores of iron consist mainly of the oxides: hematite (Fe_2O_3); goethite (α-FeO(OH)); magnetite (Fe_3O_4); and the carbonate, siderite or chalybite ($FeCO_3$).

iron oxide copper-gold deposit A group of mineral deposits characterized by magnetite and/or hematite breccia zones and veins forming pipelike and tabular orebodies hosted by subaerial volcanic, sedimentary, and intrusive rocks (Lefebure, 1995). These deposits vary from apatite-iron (*Kiruna-type*) to polymetallic iron, copper, uranium, gold, and rare earth element-bearing (*Olympic Dam-type*) subtypes. Abbrev: IOGC.

iron oxides Group name for the oxides and hydroxides of iron. Includes the minerals goethite, hematite, lepidocrocite, ferrihydrite, maghemite, and magnetite. Sometimes referred to as "sesquioxides," or "iron hydrous oxides." See also: *free iron oxides.*

iron pan A general term for a *hardpan* in a soil in which iron oxides are the principal cementing agents; several types of iron pans are found in dry and wet areas and in soils of widely varying textures. Also spelled: ironpan. See also: *moorpan.* Cf: *claypan [soil].*

iron pyrites (a) *pyrite.* (b) *marcasite [mineral].* Sometimes incorrectly spelled: iron pyrite.

iron range A term used in the Great Lakes region of the U.S. and Canada for a productive belt of iron formations. The term implies a linear region rather than a topographic elevation.

iron sand A sand containing particles of iron ore (usually magnetite), as along a coastal area.

iron shale A material, usually with a laminated structure, consisting of iron oxides and produced by the weathering of an iron meteorite.

ironshot (i'-ron-shot) adj. (a) Said of a mineral that is streaked, speckled, or marked with iron or an iron ore. (b) Containing small nodules or oolitic bodies of limonite or hematite; e.g. an "ironshot rock" in which the ooliths are essentially composed of limonite. n. A limonitic oolith in an ironshot rock.

iron spar *siderite [mineral].*

iron spinel *hercynite.*

ironstone (i'-ron-stone) (a) Any rock containing a substantial proportion of an iron compound, or any iron ore from which the metal may be smelted commercially; specif. an iron-rich sedimentary rock, either deposited directly as a ferruginous sediment or resulting from chemical replacement. The term is customarily applied to a hard, coarsely banded or nonbanded, and noncherty sedimentary rock of post-Precambrian age, in contrast with *iron formation.* The iron minerals may be oxides (limonite, hematite, magnetite), carbonate (siderite), or silicate (chamosite); most ironstones containing iron oxides or chamosite are oolitic. (b) *clay ironstone.* (c) *banded ironstone.*

ironstone cap A surficial or near-surface sheet or cap of concretionary *clay ironstone.*

iron-stony meteorite (i'-ron-ston-y) *stony-iron meteorite.*

iron vitriol *melanterite.*

irradiance (ir-ra'-di-ance) The radiant energy per unit time per unit area incident upon a surface.

irradiated stone A gemstone that has been exposed to radiation to change its color, such as some colored diamonds.

irreducible liquid content The liquid content held by capillarity against the pull of gravity.

irregular (ir-reg'-u-lar) Pertaining to an echinoid of the order Irregularia displaying an *exocyclic* test in which the *periproct* is located outside of the *oculogenital ring* or in a posterior or oral position. Cf: *regular.*

irregular complex crossed lamellar structure A *complex crossed lamellar structure* in which the first-order lamellae consist of irregularly shaped, laterally interdigitating aggregations of parallel second-order structural units, the latter consisting of fibers, rods, or planar lamellae. This is comparable to simple lamellar and rod-type crossed lamellar structure but with more than two dip directions. Calcitic irregular CCL structure is also called "irregular complex crossed foliated and is characterized by much larger first-order lamellae with much lower dip angles than the aragonitic irregular CCL structure (Carter and Tevesz, 1978, p.863; Carter, 1980b). Abbrev: ICCL.

irregular crossed lamellar structure A *crossed lamellar structure* with irregularly shaped first-order lamellae as seen on the depositional surface and in horizontal sections. This structure should not be confused with *irregular complex crossed lamellar structure,* which shows three or more rather than just two predominant dip directions of its elongated structural units (Carter, 1990, Fig. 25D). Abbrev: ICL.

irregular fibrous prismatic structure A *fibrous prismatic structure* in which each prism has an irregular form, commonly showing nodular subunits along the length of the prism (Carter, 1990, Fig. 2E).

irregular simple prismatic structure A *simple prismatic structure* in which the prism cross sections appear highly variable along their lengths (Carter, 1980a). Abbrev: ISP.

irregular spherulitic prismatic structure A *spherulitic prismatic structure* in which the first-order prisms are more or less conical and, in some instances, strongly interdigitating along their mutual boundaries (Carter, 1990, Fig. 5). Abbrev: ISphP.

irreversibility (ir'-re-vers'-i-bil'-i-ty) In evolution, the theory that an evolving group of organisms, or part of an organism, does not return to the ancestral condition. Syn: *Dollo's law.*

irreversible process (ir-re-vers'-i-ble) Any process which proceeds spontaneously, without external influence, and produces a reduction in energy and increase in entropy of a system.

irrotational strain (ir-ro-ta'-tion-al) Strain at a point, in which the orientation of the principal axes of strain remains unchanged before and after deformation. Cf: *rotational strain.* Syn: *nonrotational strain.* Not to be confused with *coaxial progressive deformation.*

irrotational wave *P wave.*

irruption [ecol] (ir-rup'-tion) An abrupt, sharp, but temporary increase in a natural population, usually connected with exceptionally favorable environmental conditions. Adj: *irruptive [ecol].*

irruption [intrus rocks] An obsolete syn. of intrusion.

irruptive [ecol] Pertaining to *irruption [ecol].*

irruptive [intrus rocks] *intrusive.*

irtyshite (ir'-tysh-ite) An adamantine colorless hexagonal mineral: $Na_2(Ta,Nb)_4O_{11}$.

Irvingtonian (Ir-ving-ton'-i-an) North American land mammal age: early Pleistocene (after Blancan, before Rancholabrean).

isallobar (is-al'-lo-bar) A line on a weather map connecting points of equal barometric tendencies (time rate of change of air pressure). Cf: *isobar.*

isallotherm (is-al'-lo-therm) A line connecting points of equal temperature variation in a given time interval. Cf: *isotherm.*

isanomaly (is-a-nom'-a-ly) *isoanomaly.*

isarithm (is'-a-rithm) An isopleth, esp. one drawn through points on a graph at which a given quantity has the same numerical value. An isarithm is created at the intersection of the mapped surface with a plane parallel to the datum. Syn: *isoline.*

isarithmic line *isarithm.*

isarithmic map A map that uses isarithms to represent a statistical surface.

isblink (is'-blink) A Danish term used in Greenland for *iceblink [glaciol]* or seaward cliff of ice.

ischium (is'-chi-um) (a) The posteroventral of the three major bones on each side of the vertebrate pelvic girdle. (b) The third pereiopodal segment from the body of a malacostracan crustacean, distal to the basis and proximal to the merus. It comprises the first segment of the endopod. Pl: ischia.

isentrope (is'-en-trope) A line or surface that represents the locus of points of a given entropy.

isentropic (is-en-trop'-ic) Said of a process that is at constant entropy.

iserine (is'-er-ine) A variety of ilmenite found as loose rounded crystals or grains in the sands at Iserwiese in Bohemia. It was formerly regarded as probably a ferruginous rutile. Syn: *iserite.*

iserite (is'-er-ite) (a) A doubtful variety of rutile with considerable amounts of FeO. (b) *iserine.*

ishikawaite (ish-i-ka'-wa-ite) A black metamict mineral of the *samarskite* group: $(U,Fe,Y,Ca)(Nb,Ta)O_4$(?).

ishkulite (ish-ku'-lite) A chromium-bearing variety of magnetite.

ishkyldite (ish-kyl'-dite) A variety of chrysotile high in silica. Also spelled: ishkildite.

isinglass (i'-sin-glass) A syn. of *mica,* esp. muscovite in thin transparent sheets.

island [cart] A polygon within another, not sharing a common boundary. Islands require special handling in polygon oriented spatial data bases (Krzanowski et al., 1993).

island [geog] (is'-land) (a) A tract of land smaller than a continent, surrounded by the water of an ocean, sea, lake, or stream. The term has been loosely applied to land-tied and submerged areas, and to land cut off on two or more sides by water, such as a peninsula. (b) An elevated piece of land surrounded by a swamp, marsh, or alluvial land, or isolated at high water or during floods. (c) Any isolated and distinctive tract of land surrounded by terrain with other characteristics; e.g. a woodland surrounded by prairie or flat open country.

island arc An offshore *volcanic arc.*

island-arc basalt The "common mafic members (<53% SiO_2) of the island arc orogenic lava suites developed near convergent plate margins" (Basaltic Volcanism Study Project, 1981, p.204). Abbrev: IAB.

island mesa A *headland mesa* that has been cut off from the main plateau by a river, so that it stands as an isolated mass (Lee, 1903, p.73).

island mountain (a) A mountain more or less completely encircled by valleys that separate it from other mountains or drainage-divide ridges. (b) *inselberg.*

island shelf *insular shelf.*

island slope *insular slope.*

island-tying The process of *tombolo* formation.

isle An island, generally but not necessarily of small size; e.g. the British Isles. Etymol: Old French "isle" (modern "île"), from Latin "insula", "island".

islet (is'-let) A small or minor island.

iso- (i'-so-) A prefix meaning "equal", e.g. in *isopach,* equal thickness, or *isotherm,* equal temperature.

isoanomaly (i'-so-a-nom'-a-ly) A line connecting points of equal geophysical anomalies. Syn: *isanomaly.*

isoanomaly curve A curve on a map representing equal potential field (e.g., gravity or magnetic) anomaly values.

isoanthracite line (i-so-an'-thra-cite) (a) On a map or diagram, a line connecting points of equal carbon-hydrogen ratio in anthracite. (b) A nonpreferred synonym of *isovol* as applied to anthracite.

isobar (i'-so-bar) A line on a weather map or chart connecting points of equal pressure. Cf: *isallobar.*

isobaric surface (i-so-bar'-ic) A surface, all points of which have equal pressure; it is not necessarily horizontal.

isobase (i'-so-base) A term used for a line that connects all areas of equal uplift or depression; it is used especially in Quaternary geology as a means for expressing crustal movements related to postglacial uplift.

isobath [grd wat] (i'-so-bath) An imaginary line on a land surface along which all points are the same vertical distance above the upper or lower surface of an aquifer or above the water table.

isobath [oceanog] A line on a map or chart that connects points of equal water depth. Syn: *bathymetric contour; depth contour.*

isobiolith (i-so-bi'-o-lith) A *para-time-rock unit* defined by its contained fossils (Wheeler et al., 1950, p.2362).

isocal (i'-so-cal) On a map or diagram, a line connecting points of equal calorific value in coal. Cf: *isocarb; isodeme; isohume; isovol.*

isocarb (i'-so-carb) On a map or diagram, a line connecting points of equal fixed-carbon content in coal. Cf: *isocal; isodeme; isohume; isovol.* See also: *isocarbon map.*

isocarbon map (i'-so-car'-bon) A coal-deposit map showing points of equal fixed-carbon content by contour lines, or isocarbs.

isocenter (i'-so-cen-ter) (a) The unique point common to the principal plane of a tilted photograph and the plane of an assumed truly vertical photograph taken from the same camera station and having an equal principal distance. It is the center of radial displacement of images due to tilt. (b) The point on an aerial photograph intersected by the bisector of the angle between the plumb line and the perpendicular to the photograph. (c) The point of intersection (on a photograph) of the principal line and the isometric parallel.

isochalcopyrite (i'-so-chal'-co-py'-rite) A metallic cubic mineral found in Red Sea mud: (Fe,Cu)S. Not approved by the International Mineralogical Association.

isochela (i-so-che'-la) A sponge *chela* having equal or similar ends. Cf: *anisochela.*

isochemical metamorphism (i-so-chem'-i-cal) Metamorphism that involves no change in bulk chemical composition (Eskola, 1939). It is a theoretical concept that is probably only approached in nature. Obsolete syn: *treptomorphism.*

isochemical series Rocks displaying the same bulk chemical composition throughout a sequence of mineralogic or textural changes, as in a sequence of metamorphic rocks of varying grade.

isochore [chem] (i'-so-chore) In a phase diagram, a line connecting points of constant volume.

isochore [stratig] A line drawn on a map through points of equal drilled thickness for a specified subsurface unit. Thickness figures are uncorrected for dip in vertical wells, and corrected for hole angle, but not for dip, in deviated wells. Cf: *isopach.*

isochore map (a) A map showing drilled thickness of a given stratigraphic unit by means of isochores. Syn: *convergence map.* (b) A map showing by contours the thickness of the pay section of an oil pool between the oil-water contact and the cap rock. It is used for making calculations of reservoir volume (Levorsen, 1967, p.616). Cf: *isopach map.*

isochromatic curve (i'-so-chro-mat'-ic) In optics of biaxial and uniaxial crystals, a band of color indicating the emergence of those components of light having equal path difference. It is a part of the *interference figure.* Cf: *isogyre.* See also: *Cassinian curve.*

isochrome map (i'-so-chrome) A contour map that depicts the continuity and extent of color stains on geologic formations.

isochron [geochron] (i'-so-chron) In isotopic age measurements, a straight line that is yielded by a plot of D_r/D_n versus P/D_n, when isotope ratios are determined separately on several systems with the same history (for instance, different minerals in a rock) and the ratio of a daughter isotope (D_r) to a related nonradiogenic isotope (D_n) was initially the same for all samples, with P representing the parent of D_r. The slope of such an isochron increases with the age of the systems investigated. Cf: *isochrone.*

isochron [seis] A line on a map connecting points at which a characteristic time or interval has the same value; e.g., in seismology, a line passing through points at which the difference between arrival times of seismic waves from two reflecting surfaces is equal. Syn: *isotime line.*

isochronal (i-soch'-ro-nal) (a) *isochronous.* (b) *isochronic.*

isochrone (i'-so-chrone) A line, on a map or chart, connecting all points that indicate simultaneous occurrences of events or phenomena or represent the same time value or time difference; e.g., a line along which duration of travel is constant, or a line indicating the places at which rain begins at a specified time. Cf: *isochron [geochron].*

isochroneity (i'-so-chro-ne'-i-ty) The state or quality of being *isochronous;* equivalence in duration. Syn: *isochronism.*

isochronic (i-so-chron'-ic) Having isochrones; e.g., an isochronic map. Syn: *isochronal.*

isochronism (i-soch'-ro-nism) *isochroneity.*

isochronous (i-soch'-ro-nous) (a) Equal in duration or uniform in time; e.g., an "isochronous interval" between two synchronous surfaces, or an "isochronous unit" of rock representing the complete rock record of an isochronous interval. Mann (1970, p.750) has proposed that the word *coetaneous* would be more appropriate for use in this sense. (b) A term frequently applied in the sense of *synchronous,* such as an "isochronous surface" having everywhere the same age or time value within a body of strata (Hedberg, 1958, p.1890; and ACSN, 1961, Art. 28c). Syn: *isochronal.*

isochronous homeomorph A homeomorph that develops at the same geologic time as another and therefore poses special problems in identification. Cf: *heterochronous homeomorph.*

isoclasite (i-so-clas'-ite) A white monoclinic mineral: $Ca_2(OH)(PO_4){\cdot}2H_2O$.

isoclinal (i-so-cli'-nal) Adj. of *isocline.*

isocline (i'-so-cline) A fold whose limbs are parallel. Adj: *isoclinal.*

isoclinic line (i-so-clin'-ic) An *isomagnetic line* connecting points of equal magnetic inclination.

isocommunity (i'-so-com-mu'-ni-ty) A natural community that closely resembles another community in morphology and ecology.

isocon (i'-so-con) A line connecting points of equal geochemical concentration, e.g. salinity.

isocubanite (i'-so-cu'-ban-ite) A metallic bronze cubic mineral: $CuFe_2S_3$. Syn: cubic cubanite, chalcopyrrhotite.

isodeme (i'-so-deme) On a map or diagram, a line connecting points of equal swelling characteristics. Cf: *isocal; isocarb; isohume; isovol.*

isodesmic (i-so-des'-mic) Said of a crystal or other material having ionic bonding of equal strength, e.g. NaCl. Cf: *anisodesmic.*

isodiff (i'-so-diff) A line on a map or chart connecting points of equal correction or difference in datum; e.g. *isolat* and *isolong.*

isodimorphism (i'-so-di-mor'-phism) The characteristic of two crystalline substances to be both dimorphous and isomorphous, e.g. calcite and aragonite. Adj: *isodimorphous.*

isodimorphous (i'-so-di-mor'-phous) Said of two crystalline substances displaying *isodimorphism.*

isofacial [petrology] (i-so-fa'-cial) Pertaining to rocks belonging to the same metamorphic facies and having reached equilibrium under the same set of physical conditions. Cf: *allofacial.* Syn: *isogradal.*

isofacial [stratig] Pertaining to rocks belonging to the same facies; e.g. an "isofacial line" on a map, along which the thickness of stratum of the same lithologic composition is constant.

isofacies map (i-so-fa'-cies) A map showing the distribution of one or more facies within a designated stratigraphic unit. See also: *facies map.*

isoferroplatinum (i'-so-fer'-ro-plat'-i-num) A metallic cubic mineral: $(Pt,Pd)_3(Fe,Cu)$.

isofract (i'-so-fract) A graphic representation of the locus of all compositions in a system having a given value for the index of refraction.

isofrigid temperature regime (i-so-frig'-id) A soil temperature regime having the same temperature range as the *frigid temperature regime,* but with a summer-winter variation of less than 5°C (USDA, 1975).

isogal (i'-so-gal) A contour line of equal gravity values. Cf: *gal.*

isogam (i'-so-gam) An *isomagnetic line* connecting points of equal magnetic-field intensity. It is used for maps of total, horizontal, or vertical magnetic intensity.

isogeolith (i-so-ge'-o-lith) A *para-time-rock unit* defined by its lithology (Wheeler et al., 1950, p.2362).

isogeotherm (i-so-ge'-o-therm) A line or surface within the Earth connecting points of equal temperature. Syn: *isotherm.*

isogon (i'-so-gon) An *isomagnetic line* connecting points of equal magnetic declination. See also: *agonic line.*

isograd (i'-so-grad) A line on a map joining points at which metamorphism proceeded at similar values of pressure and temperature as indicated by rocks containing a diagnostic mineral or mineral assemblage. Such a line represents the intersection of a reaction surface with the Earth's surface corresponding to the boundary between two contiguous zones of metamorphic grade, as defined by the appearance of specific index minerals, e.g., garnet isograd, staurolite isograd.

isogradal (i-so-grad'-al) Pertaining to rocks which have reached the same grade of metamorphism irrespective of their initial compositions. Cf: *isophysical series.* Syn: *isograde; isofacial [petrology].*

isograde (i'-so-grade) *isogradal.*

isogram (i'-so-gram) A general term proposed by Galton (1889, p.651) for any line on a map or chart connecting points having an equal numerical value of some physical quantity (such as temperature, pressure, or rainfall); an isopleth.

isogriv (i'-so-griv) A line on a map or chart connecting points of equal grivation.

isogyre (i'-so-gyre) In crystal optics, a black or shadowy part of an *interference figure* that is produced by extinction and indicates the emergence of those components of light having vibration directions parallel to the privileged directions of the nicols. Cf: *isochromatic curve.* Syn: *polarization brush.*

isohaline (i-so-hal'-ine) adj. Of equal or constant salinity. n. A line on a chart that connects points of equal salinity in the ocean.

isoheight (i'-so-height) *isohypse.*

isohume (i'-so-hume) On a map or diagram, a line connecting points of equal moisture content in coal. Cf: *isocal; isocarb; isodeme; isovol.*

isohyet (i-so-hy'-et) A line connecting points of equal precipitation on a map.

isohyperthermic temperature regime (i'-so-hy'-per-ther'-mic) A soil temperature regime having the same temperature range as the *hyperthermic temperature regime,* but with a summer-winter variation of less than 5°C (USDA, 1975).

isohypse (i'-so-hypse) An isopleth for height or elevation; a *contour* on a topographic map. Syn: *isoheight.*

isokite (i'-so-kite) A colorless, buff, pinkish monoclinic mineral of the *tilasite* group: $CaMg(PO_4)F$. Its hydroxyl analogue is panasqueiraite.

isolat (i'-so-lat) An *isodiff* connecting points of equal latitude correction.

isolated crystal morphotype One of various sparsely distributed, irregularly oriented crystallites or crystalline aggregates that do not comprise persistent, major shell layers and show morphologies typical of inorganically precipitated crystal forms, such as calcitic rhombohedra, aragonitic or calcitic spindles, aragonitic spherules, isolated aragonitic spherulites, aragonitic acicular crystallites, etc. (Carter, 1980b). Based on "crystal morphotypes" of Wind and Wise (1976).

isolated fault A fault that was not affected by interaction with other faults during its propagation. An isolated fault is usually characterized by a displacement maximum near the center of the fault trace, with displacement decreasing towards the tips (Peacock et al., 2000). Cf: *hard linkage; soft linkage.*

isolated porosity (i'-so-lat'-ed) The property of rock or soil of containing noncommunicating interstices, e.g. vesicles in lava, expressed as the percent of bulk volume occupied by such interstices; the numerical difference between *total porosity* and *effective porosity.*

isolation (i-so-la'-tion) In biology, any process or condition by which a group of individuals is separated for a considerable length of time from other groups, as a result of geographic, behavioral, or ecologic factors.

isolation of outcrops A method of geologic mapping that outlines all areas of exposed rock to distinguish them from areas where the rock is buried or otherwise concealed. Syn: *multiple-exposure method.*

isoline (i'-so-line) *isopleth [cart]; isarithm.*

isolith (i'-so-lith) (a) An imaginary line connecting points of similar lithology and separating rocks of differing nature, such as of color, texture, or composition (Kay, 1945a, p.427). The term "isolithic boundary" was used by Grossman (1944, p.48) for a zone of lithofacies change separating rocks of different grain sizes. (b) An imaginary line of equal aggregate thickness of a given lithologic facies or particular class of material within a formation, measured perpendicular to the bedding at selected points (which may be on outcrops or in the subsurface).

isolith map A map that depicts isoliths; esp. a *facies map* showing the net thickness of a single rock type or selected rock component in a given stratigraphic unit.

isolong (i'-so-long) An *isodiff* connecting points of equal longitude correction.

isolueshite A brownish black cubic mineral: $NaNbO_3$. It is a dimorph of lueshite.

isomagnetic line (i'-so-mag-net'-ic) A line connecting points of equal value of some magnetic element, e.g., *isoclinic line; isogon; isopor.*

isomegathy (i-so-meg'-a-thy) A term introduced by Shepard and Cohee (1936) for a line, on a map, connecting points of equal median size of sedimentary particles.

isomer A nuclide whose nucleus can remain in an excited state for a measurable interval of time following radioactive decay of its parent.

isomeric decay *gamma decay.*

isomeric transition Reduction of the energy of the nucleus of an isomer to the groundstate by the emission of a gamma ray.

isomertieite (i-so-mer'-tie-ite) A metallic cubic mineral: $Pd_{11}(Sb,As)_4$. It is polymorphous with mertieite.

isomesic temperature regime (i-so-me'-sic) A soil temperature regime having the same temperature range as the *mesic temperature regime,* but with a summer-winter variation of less than 5°C (USDA, 1975).

isomesobath A line of equal average depth, used to calculate isostatic response to water loading.
isometric (i-so-met′-ric) *equant.*
isometric line A term introduced by Wright (1944) for a line, drawn on a map, representing a constant value obtained from measurement at a series of points along its course; an isopleth.
isometric projection A projection in which the plane of projection is equally inclined to the three spatial axes of a three-dimensional object, so that equal distances along the axes are drawn equal. It gives a bird's-eye view, combining the advantages of a ground plan and elevation; e.g. as in a block diagram showing three faces.
isometric system One of the six crystal systems, characterized by four threefold axes of symmetry as body diagonals in a cubic unit cell of the lattice. It comprises five crystal classes or point groups. Cf: *hexagonal system; tetragonal system; orthorhombic system; monoclinic system; triclinic system.* Syn: *cubic system.*
isometry (i-som′-e-try) The constancy of shape or proportions of a system (or organism) as the magnitude of the system changes. Cf: *allometry [biol].*
isomicrocline (i-so-mi′-cro-cline) An optically positive variety of microcline.
isomodal layering (i-so-mod′-al) Layering in a cumulate in which the layers are characterized by a uniform proportion of one or more cumulus minerals.
isomorph (i′-so-morph) An organism, or a part of an organism, that is similar to another but unrelated to it.
isomorphic (i-so-mor′-phic) Having identical or similar form. Cf: *heteromorphic [evol].*
isomorphism [cryst] (i-so-mor′-phism) The characteristic of two or more crystalline substances to have similar chemical composition, axial ratios, and crystal forms, and to crystallize in the same crystal class. Such substances form an *isomorphous series.* Adj: *isomorphous.* Cf: *isostructural.* Syn: *allomerism.*
isomorphism [evol] The similarity that develops in organisms of different ancestry as a result of *convergence [evol].*
isomorphous (i-so-mor′-phous) Adj. of isomorphism. Syn: *isomorphic; allomeric.*
isomorphous mixture *isomorphous series.*
isomorphous series Two or more crystalline substances that display isomorphism; their physical properties vary along a smooth curve. An example is olivine, usually found in nature as a solid solution of Mg_2SiO_4 and Fe_2SiO_4, i.e. an isomorphous series between forsterite and fayalite. The exact lattice dimensions and other physical properties vary with change of the Mg/Fe ratio. Syn: *solid-solution series.*
isomyarian (i′-so-my-ar′-i-an) adj. Said of a bivalve mollusk or its shell having two adductor muscles of equal or nearly equal size. Syn: *homomyarian.* n. An isomyarian mollusk.
isontic line (i-son′-tic) Obsolete syn. of isopleth. Term proposed by Lane (1928, p.37).
iso-orthoclase (i-so-or′-tho-clase) An optically positive variety of orthoclase. It has been found in granitic gneiss. Also called: isorthoclase; isorthose.
isopach (i′-so-pach) A line drawn on a map through points of equal true thickness of a designated stratigraphic unit or group of stratigraphic units. Cf: *isochore [stratig].* Syn: *isopachyte; thickness line; thickness contour.*
isopach map A map that shows the thickness of a bed, formation, sill, or other tabular body throughout a geographic area by means of isopachs at regular intervals. Cf: *isochore map.* Syn: *thickness map.* Nonrecommended syn: *isopachous map.*
isopachous (i-so-pach′-ous) Of, relating to, or having an isopach; e.g. an "isopachous contour". Not recommended usage.
isopachous cement A crust of carbonate cement, commonly fibrous, of equal thickness around all grains; indicates precipitation of cement in a phreatic environment.
isopachous map A nonrecommended syn. of *isopach map.*
isopach strike The compass direction of an isopach at a given point on a map.
isopachyte (i-so-pach′-yte) British term for *isopach.*
isopag (i′-so-pag) An *equiglacial line* connecting points where ice is present for approximately the same number of days per year.
isopectic (i-so-pec′-tic) An *equiglacial line* connecting points where ice begins to form at the same time in winter. Cf: *isotac.*
isoperimetric curve (i′-so-per′-i-met′-ric) A line on a map or map projection (as on an equal-area projection) along which there is no variation from exact scale.
isoperthite (i-so-perth′-ite) A variety of alkali feldspar consisting of perthitic intergrowths of the same kind of feldspar or of two kinds of feldspar belonging to the same isomorphous series.
isophysical series (i-so-phys′-i-cal) A series of rocks of different chemical composition that were metamorphosed under identical physical conditions. Obsolete. Cf: *isogradal.*
isopleth [cart] (a) A general term for a line, on a map or chart, along which all points have a numerically specified constant or equal value of any given variable, element, or quantity (such as abundance or magnitude), with respect to space or time; esp. a *contour.* Etymol: Greek "isos", "equal", + "plethos", "fullness, quantity, multitude". Syn: *isogram; isoline; isontic line; isometric line.* (b) A line drawn through points on a graph at which a given quantity has the same numerical value (or occurs with the same frequency) as a function of two coordinate variables. It is often used in connection with a meteorologic element that varies with the time of the year (month) and the time of day (hour). Syn: *isarithm.*
isopleth [geochem] (i′-so-pleth) In a strict sense, a line or surface on which some mathematical function has a constant value. It is sometimes distinguished from a *contour* by the fact that an isopleth need not refer to a directly measurable quantity characteristic of each point in the map area, e.g. maximum temperature of a particular point. More generally, the term is used as a synonym of "isocompositional section".
isopleth map A general term for any map showing the areal distribution of some variable quantity in terms of lines of equal or constant value; e.g. an *isopach map.*
isopod (i′-so-pod) Any malacostracan crustacean belonging to the order Isopoda, characterized generally by the absence of a carapace and the presence of sessile eyes and a compressed body. Range, Triassic to Holocene. Cf: *amphipod.*
isopolar (i-so-po′-lar) Of *pollen,* having opposite poles developed equally. Ant: *heteropolar.*
isopoll (i-so-poll′) A line on a map connecting locations with samples having the same percentage or amount of pollen of a given kind.
isopor (i′-so-por) An *isomagnetic line* of equal secular change, e.g., equal annual change of isogonic or isoclinic lines.
isopotal (i-so-pot′-al) Having equal infiltration capacities; e.g. an isopotal area in a watershed.
isopotential line (i′-so-po-ten′-tial) *equipotential line.*
isopotential surface A surface on which points of equal *fluid potential* lie.
isoprenoid A branched molecule found in chlorophyll as well as in plants and bacteria; consists of long-chained hydrocarbons with repeating 5 carbon units.
isopycnal (i-so-pyc′-nal) adj. Of constant or equal density, measured in space or in time. n. A line on a chart that connects points of equal density. Cf: *isohaline.* See also: *isostere.* Also called isopycnic.
isopycnal surface A surface of which all points have equal density.
isopygous (i-so-py′-gous) Said of a trilobite having *cephalon* and *pygidium* of subequal size. Ant: *heteropygous.*
isorad (i′-so-rad) A line connecting points of equal radioactivity or radiation exposure.
isorat (i′-so-rat) Line connecting points of equal isotope ratios.
isoseismal (i-so-seis′-mal) n. A syn. of *isoseismal line.*
isoseismal line A line connecting points on the Earth's surface at which earthquake intensity is the same. It is usually a closed curve around the *epicenter.* Cf: *coseismal line.* Syn: *isoseismal.*
isosinal map (i-so-si′-nal) A *slope map* whose contour lines are lines of equal slope represented by sines of slope angles read from a topographic map.
isospore (i′-so-spore) *homospore.*
isospory (i-so-spo′-ry) *homospory.*
isostannite (i-so-stan′-nite) *kësterite; ferrokësterite.*
isostasy (i-sos′-ta-sy) The condition of equilibrium, comparable to floating, of the units of the lithosphere above the asthenosphere. Crustal loading, as by ice, water, sediments, or volcanic flows, leads to isostatic depression or *downwarping;* removal of load, to isostatic uplift or *upwarping.* Two differing concepts of the mechanism of isostasy are the *Airy hypothesis* of constant density and the *Pratt hypothesis* of differing root depth and differing density and constant root depth. See also: *isostatic compensation; depth of compensation; glacio-isostasy; hydro-isostasy.*
isostatic adjustment (i-so-stat′-ic) *isostatic compensation.*
isostatic anomaly A gravity anomaly calculated on a hypothesis that the gravitational effect of masses extending above sea level is compensated by a deficiency of density of the material beneath those masses; the effect of deficiency of density in ocean waters is compensated by an excess of density in the material under the oceans.
isostatic compensation The adjustment of the lithosphere of the Earth to maintain equilibrium among units of varying mass and density; excess mass above is balanced by a deficit of density below, and vice versa. See also: *depth of compensation; isostasy.* Syn: *isostatic adjustment.*
isostatic correction A correction applied to gravity data to compensate for lateral density or thickness variations between large blocks of the Earth's crust. It is based on a hypothesis that the gravitational effect of rock masses above sea level is compensated by a deficiency of density of the material beneath those masses; the density deficiency in ocean waters is compensated (balanced) by an excess of density of rock masses under oceans. This correction is based on the isostatic model obtained from elevation and water-depth data, often using zone charts. See *isostasy.*
isostere (i′-so-stere) A line connecting points of equal density of the Earth's atmosphere on a chart.
isostratification map (i′-so-strat′-i-fi-ca′-tion) A map that shows the number or thickness of beds in a stratigraphic unit by means of contour lines representing equal stratification indices (Kelley, 1956, p.299).

isostructural (i-so-struc′-tur-al) Said of two or more chemical compounds with similar crystal structures but with little tendency to show isomorphism.
isotac (i′-so-tac) An *equiglacial line* connecting points where ice melts at the same time in spring. Cf: *isopectic.*
isotach (i′-so-tach) A line connecting points of equal wind speed on a map.
isotangent map (i-so-tan′-gent) A *slope map* whose contour lines are lines of equal slope represented by tangents of slope angles read from a topographic map.
isotaque (i′-so-taque) In crystal optics, one of several curves representing equal wave-normal velocities; in a uniaxial crystal it is a circle concentric with the optic axis, and in a biaxial crystal, it is a spherical ellipse.
isotherm (i′-so-therm) (a) A line connecting points of equal temperature on a map or chart. Isotherm maps are often used to portray surface temperature patterns of water bodies. Cf: *isallotherm.* (b) *isogeotherm.*
isothermal (i-so-ther′-mal) Pertaining to the process of changing the thermodynamic state of a substance, e.g., its pressure and volume, while maintaining the temperature constant.
isothermal remanent magnetization Remanent magnetization acquired by application of a direct magnetic field, without change of temperature. Abbrev: IRM.
isothermic temperature regime (i-so-ther′-mic) A soil temperature regime that has the characteristics of a *thermic temperature regime* except for a summer-winter variation of less than 5°C (USDA, 1975).
isothrausmatic (i′-so-thraus-mat′-ic) A descriptive term applied to igneous rocks with an orbicular texture in which the nuclei of the orbicules are composed of the same rock as the groundmass (Eskola, 1938, p.476). Obsolete. Cf: *allothrausmatic; crystallothrausmatic; homeothrausmatic; heterothrausmatic.*
isotime line (i′-so-time) *isochron [seis].*
isotomous (i-sot′-o-mous) Said of a crinoid arm characterized by division into equal branches. Ant: *heterotomous.*
isotope (i′-so-tope) One of two or more species of the same chemical element; i.e., having the same number of protons in the nucleus, but differing from one another by having a different number of neutrons. The isotopes of an element have slightly different physical and chemical properties, owing to their mass differences, by which they can be separated. See also: *radioisotope.*
isotope dilution An analytical method in which a known quantity of an element with an isotopic composition different from that of the natural element (a *spike* or *tracer*) is mixed with the sample being analyzed. Measurement of the isotopic composition of the mixture allows calculation of the amount of the natural element in the sample.
isotope effect *Isotopic fractionation* in a variety of chemical and physical processes, resulting from slight differences in the properties (heat capacity, vapor pressure, density, free energy, etc.) of the various isotopic forms of the elements involved.
isotope geochemistry *isotope geology.*
isotope geology The application of the study of radioactive and stable isotopes, especially their abundances, to geology. It includes the calculation of geologic time, and the determination of the origin, mechanisms, and conditions of geologic processes by isotopic means. Syn: *isotope geochemistry; nuclear geology; nuclear geochemistry; radiogeology.*
isotope ratio The ratio of abundance of any two isotopes of a given element; e.g., ^{18}O to ^{16}O. It is conventionally written as the ratio of the heavy isotope to the light isotope. Cf: *delta value.*
isotopic age An age expressed in years and calculated from the quantitative determination of radioactive elements and their decay products. Imprecise and less preferred syn: *absolute age; radiometric age.*
isotopic age determination *isotopic dating.*
isotopic anomaly An excess of deficiency on a particular isotope (e.g., a ^{26}Mg anomaly produced by the decay of extinct ^{26}Al).
isotopic dating Calculating an age in years for geologic materials by measuring the presence of a short-lived radioactive element (e.g., carbon-14) or by measuring the presence of a long-lived radioactive element plus its decay product (e.g., potassium-40/argon-40). The term applies to all methods of age determination based on nuclear decay of naturally occurring radioactive isotopes. Syn: *isotopic age determination; nuclear age determination; radioactive age determination; radioactive dating; radiogenic age determination; radiogenic dating.* Imprecise and less preferred syn: *radiometric age determination; radiometric dating.*
isotopic fractionation The relative enrichment of one isotope of an element over another, owing to slight variations in their physical and chemical properties. It is proportional to differences in their masses. See also: *isotope effect.*
isotopic number The number of excess neutrons, i.e. the number of neutrons minus the number of protons, in an atomic nucleus. It is usually an indication of the radioactivity of the nucleus.
isotropic [cryst] (i-so-trop′-ic) Said of a medium whose properties are the same in all directions; in crystal optics, said of a crystal whose physical properties do not vary according to crystallographic direction, e.g. one in which in which light travels with the same speed in any direction. Cubic crystals and amorphous substances are usually *isotropic.* Ant: *anisotropic [cryst].*
isotropic [grd wat] A condition in a geologic formation in which a hydraulic property, e.g., *hydraulic conductivity,* is independent of the direction of measurement at a point in the formation (Freeze and Cherry, 1979, p.32). Ant: *anisotropic [grd wat].* See also: *isotropy.*
isotropic displacement parameters *displacement parameters.*
isotropic temperature factor *temperature factor.*
isotropization (i-sot′-ro-pi-za′-tion) The solid-state conversion of an originally birefringent mineral such as quartz or feldspar into a more or less isotropic phase at temperatures below the melting point, as a result of destruction of the crystallinity by such processes as shock-wave action or neutron bombardment.
isotropy (i-sot′-ro-py) (a) The condition of having properties at a point that do not depend on direction. (b) The condition of having properties that are uniform in all directions. Adj: isotropic.
isotypic (i-so-typ′-ic) Said of crystalline substances that have analogous crystal structures and chemical compositions, e.g. zircon and xenotime.
isovite A metallic iron-gray cubic mineral: $(Cr,Fe)_{23}C_6$. It is the Cr analogue of haxonite.
isovol (i′-so-vol) On a map or diagram, a line connecting points of equal volatile content in coal. Cf: *isocarb; isocal; isodeme; isohume.* See also: *isoanthracite line.*
ISP *irregular simple prismatic structure.*
ISphP *irregular spherulitic prismatic structure.*
issue (is′-sue) The place where a stream flows out into a larger body of water.
isthmus (isth′-mus) A narrow strip or neck of land, bordered on both sides by water, connecting two larger land areas, such as a peninsula and the mainland (e.g., the Isthmus of Suez) or two continents (e.g., the Isthmus of Panama). Etymol: Greek.
Isuan (Is′-u-an) An obsolete geochronometric subdivision of the *Precambrian,* covering the time span from 3,800 to 3,500 Ma (Harland et al., 1990), therefore approximately equivalent to the *Eoarchean* Era of the Standard Global Chronostratigraphic Scale.
itabirite (it-a-bi′-rite) A laminated, metamorphosed *oxide-facies iron formation* in which the original chert or jasper bands have been recrystallized into megascopically distinguishable grains of quartz and in which the iron is present as thin layers of hematite, magnetite, or martite (Dorr and Barbosa, 1963, p.18). The term was originally applied in Itabira, Brazil, to a high-grade massive specular-hematite ore (66% iron) associated with a schistose rock composed of granular quartz and scaly hematite. The term is now widely used outside Brazil. Cf: *jacutinga; canga.* Syn: *banded quartz-hematite; hematite schist.*
itacolumite (it-a-col′-u-mite) A micaceous sandstone or a schistose quartzite that contains interstitial, loosely interlocking grains of mica, chlorite, and talc, and that exhibits flexibility when split into thin slabs. Type locality: Itacolumi Mountain in the state of Minas Gerais, Brazil. Syn: *flexible sandstone.* Rarely used syn: *articulite.*
italite (it′-a-lite) (a) In the *IUGS classification,* a plutonic rock in which F is between 60 and 100, M is 10 or less, and potassium exceeds sodium. Cf: *urtite.* (b) A volcanic rock rich in leucite and containing up to 30 percent mafic minerals, such as melilite, biotite, and apatite. Its name, given by Washington in 1920, is derived from Italy. Cf: *fergusite; missourite.* Obsolete.
ITCZ *intertropical convergence zone.*
iterative (a) A procedure that repeats with an improved model until some condition is satisfied. (b) Processing by successive approximations, each based on changes concluded from the preceding iteration, in such a way as to converge onto the desired solution, such as used in computing a satellite-navigation fix.
iterative evolution (it′-er-a-tive) Repeated development of new forms from the same ancestral stock; repeated, independent evolution.
itoigawaite A vitreous blue orthorhombic mineral of the *lawsonite* group: $SrAl_2Si_2O_7(OH)_2{\cdot}H_2O$.
itoite (i′-to-ite) A silky white orthorhombic mineral: $Pb_3Ge^{4+}O_2(SO_4)_2(OH)_2$.
ITRM *inverse thermoremanent magnetization.*
itsindrite (it-sin′-drite) A potassium-rich hypabyssal *nepheline syenite* containing microcline, nepheline, biotite, aegirine, and zoned andradite. Its name, given by Lacroix in 1922, is derived from the Itsindra Valley, Malagasy. Obsolete.
I-type granitoid A granitic rock produced by magma that was generated by partial fusion of mafic or intermediate igneous or meta-igneous rocks. The term was proposed by Chappell and White in 1974 and is thoroughly discussed by Clarke (1992).
IUGS classification The classification of igneous rocks established by the International Union of Geological Sciences Subcommission on the Systematics of Igneous Rocks, working from 1973 through 2002. Two methods of classification were adopted. One, for plutonic rocks and volcanic rocks that are sufficiently crystalline that the *mode [petrology]* may be estimated, is based on the proportions of minerals in five groups: Q=quartz and other polymorphs of SiO_2; A=alkali feldspars; P=plagioclase more calcic than An_5, and scapolite; F=feldspathoids (foids); M=all other minerals (mafites). If the percentage of mafites M (*color index*) is less than 90, a rock is classified according to its position in a double triangle QAPF. *Ultramafic* rocks (M=90 to 100) are classified according to proportions of mafites. The second method, for volcanic rocks, does not use the mode and requires major-element analysis of the rock. The analysis is recalculated to sum 100 after H_2O and CO_2 are deleted. SiO_2 is plotted on the horizontal axis and the sum of Na_2O+K_2O on

the vertical axis of a total alkali silica (TAS) diagram. The TAS diagram is divided into fields approximately corresponding to rock types defined on the QAPF modal plot. See Le Maitre (1989 and 2002) and Woolley et al. (1996).
IVA iron A member of magmatic group IVA *iron meteorite.*
IVB iron A member of magmatic group IVB *iron meteorite.*
ivorite (i′-vo-rite) An *Ivory Coast tektite.*
ivory (i′-vo-ry) The fine-grained creamy-white dentine forming the tusks of elephants, and the teeth or tusks of certain other large animals such as the walrus; it has long been esteemed for a wide variety of ornamental articles. It is illegal to import ivory into many countries. "Vegetable ivory" is produced from the corozo nut and the Doum palm nut.
Ivory Coast tektite A tektite from one of the four main tektite fields, formed about 1.1 Ma ago.
iwakiite (i-wak′-i-ite) A metallic greenish-black tetragonal mineral of the *hausmannite* group: $Mn^{2+}(Fe^{3+},Mn^{3+})_2O_4$. It is a dimorph of jacobsite.
I wave A longitudinal or *P wave* in the Earth's inner core. Cf: *K wave.*
ixiolite (ix′-i-o-lite) A submetallic gray orthorhombic mineral of the *samarskite* group: $(Ta,Nb,Sn,Fe,Mn)O_2$. It was previously considered to be a mixture of cassiterite with columbite or tapiolite, and to be a manganese-tantalate isomorph of tapiolite and of mossite. Syn: *ashanite.*
izoklakeite A metallic lead-gray orthorhombic mineral: $(Pb,Ag)_{26.5}(Cu,Fe)_2(Sb,Bi)_{19.5}S_{57}$.

J

jáchymovite A vitreous yellow monoclinic mineral: $(UO_2)_8(SO_4)(OH)_{14}\cdot 13H_2O$.

jack [coal] (a) *Cannel coal* interstratified with shale. (b) Coaly, often canneloid, shale. (c) A large ironstone nodule in the coal measures of Wales.

jack [mineral] A zinc ore; specif. *sphalerite.*

jack iron A term used in the zinc-mining area of Missouri for a solid flint rock containing disseminated *sphalerite,* or *blackjack [mineral].*

jackknife A class of statistical procedures for estimating *bias* and *standard errors* from the available observations without invoking a theoretical distribution. The procedures involve iteratively recalculating the statistic in question using random subsets of the data, and using the scatter in the statistic to estimate the standard error. Cf: *bootstrap.*

jackrock A hard flint claystone parting in coal. Hard and structurally competent enough to set roof support jacks on.

Jacksonian (Jack-so'-ni-an) North American (Gulf Coast) stage: Eocene (above Claibornian, below Vicksburgian).

jackstraw texture (jack'-straw) A texture of metamorphic rocks characterized by a criss-cross arrangement of an elongate mineral. It is often exhibited by olivine in a matrix of talc in metamorphosed ultramafic rocks (Snoke and Calk, 1978).

jacobsite (ja'-cobs-ite) A black cubic magnetic mineral of the *spinel* group: $(Mn^{2+},Fe^{2+},Mg)(Fe^{3+},Mn^{3+})_2O_4$. Also spelled: jakobsite.

Jacob's staff (Ja'-cob's) A single, straight rod, staff, or pole, pointed and shod with iron at its lower end for insertion in the ground, and fitted with a ball-and-socket joint at its upper end for adjustment to a level position, used instead of a tripod for mounting and supporting a surveyor's compass or other instrument. Named after Jacob (St. James), symbolized in religious art by a pilgrim's staff.

jacquesdietrichite A vitreous blue orthorhombic mineral: $Cu_2BO(OH)_5$.

jacupirangite (jac'-u-pi-ran'-gite) An ultramafic plutonic rock that is part of the *ijolite* series, composed chiefly of titanian augite and magnetite, with a smaller amount of nepheline; a nepheline-bearing *clinopyroxenite.* Its name, given by Derby in 1891, is derived from Jacupiranga, Brazil.

jacutinga (jac-u-tin'-ga) A term used in Brazil for disaggregated, powdery *itabirite,* and for variegated thin-bedded high-grade hematite iron ores associated with and often forming the matrix of gold ore. Etymol: from its resemblance to the colors of the plumage of *Pipile jacutinga,* a Brazilian bird.

jade (a) A hard, extremely tough, compact gemstone consisting of either the pyroxene mineral jadeite or the amphibole variety nephrite, and having an unevenly distributed color ranging from dark or deep green to dull or greenish white. It takes a high polish, and has long been used for jewelry, carved articles, and various ornamental objects. Syn: *jadestone.* (b) A term that is often applied to various hard green minerals; e.g. "California jade" (or californite, a green compact variety of vesuvianite), "Mexican jade" (or tuxtlite, and also green-dyed calcite), saussurite, and green varieties of sillimanite, pectolite, garnet, and serpentine.

jade-albite (jade-al'-bite) *Maw-sit-sit.*

jadeite (jade'-ite) A high-pressure monoclinic mineral of the *clinopyroxene* group, essentially: $Na(Al,Fe^{3+})Si_2O_6$. It occurs in various colors (esp. green) and is found chiefly in Burma; when cut, it furnishes the most valuable and desirable variety of jade and is used for ornamental purposes.

jadeitite (jade'-i-tite) A metamorphic rock consisting principally of jadeite, commonly associated with small amounts of feldspar or feldspathoids. It is probably derived from an alkali-rich igneous rock by high-pressure metamorphism.

jadestone (jade'-stone) *jade.*

jaffeite (jaf'-fe-ite) A colorless trigonal mineral: $Ca_6Si_2O_7(OH)_6$.

jager (ja'-ger) A high-quality bluish-white diamond.

jagoite (ja'-go-ite) A yellow-green hexagonal platy mineral. The structure presents extensive solid solution: $(Pb,Na,Ca)_9(Fe^{3+},Mg,Mn)_2(Si,Fe,Pb)_{17}O_{41}(Cl,OH)_3$.

jagowerite (ja-gow'-er-ite) A light green triclinic mineral: $BaAl_2(PO_4)_2(OH)_2$.

jahnsite-(CaMnMg) (jahns'-ite) A greenish-yellow to brown monoclinic mineral of the *whiteite* group: $CaMn^{2+}Mg_2Fe^{3+}{}_2(PO_4)_4(OH)_2\cdot 8H_2O$.

jahnsite-(CaMnMn) (jahns'-ite) A brownish yellow monoclinic mineral of the *whiteite* group: $CaMn^{2+}{}_3Fe^{3+}{}_2(PO_4)_4(OH)_2\cdot 8H_2O$.

jahnsite-(MnMnMn) (jahns'-ite) A brown, yellow or greenish yellow monoclinic mineral of the *whiteite* group: $Mn^{2+}{}_4Fe^{3+}{}_2(PO_4)_4(OH)_2\cdot 8H_2O$.

jahresringe (jah'-res-rin'-ge [yah'-ress-ring-uh]) An annual lamination, e.g., a varve; also used in the more general sense of any sedimentary couplet that may have specific time significance, though this is generally implied rather than documented. Etymol: German, Pl. of "Jahresring" = year-ring.

jaipurite (jai'-pur-ite) A metallic gray hexagonal mineral of the *nickeline* group: CoS.

jalpaite (jal'-pa-ite) A metallic light gray tetragonal mineral: Ag_3CuS_2.

jama [Pron: ya-ma] (a) A Slavic word for a vertical or steeply inclined shaft in limestone (Cvijic, 1924), known as *abime* or *aven* in France, and as *pothole* in England. (b) Any cave.

jamborite (jam'-bor-ite) A green hexagonal mineral: $(Ni^{2+},Ni^{3+},Fe)(OH)_2(OH,S,H_2O)$(?).

jameo A large collapse sink formed by structural failure of the roof of more than one level of a multi-level lava tube cave. The term originated in the Canary Islands and is commonly used in the Spanish literature.

jamesite (james'-ite) A reddish-brown subadamantine triclinic mineral: $Pb_2ZnFe^{3+}{}_2(Fe^{3+},Zn)_4(AsO_4)_4(OH,O)_{10}$.

jamesonite (jame'-son-ite) A lead-gray to gray-black monoclinic mineral: $Pb_4FeSb_6S_{14}$. It is a minor ore of lead and antimony, and sometimes contains copper and zinc. Jamesonite has a metallic luster and commonly occurs in acicular crystals with fibrous or featherlike forms. Syn: *feather ore; gray antimony.*

Jamin effect (Ja'-min) The restrictive force exerted upon the flow of fluids through narrow tubes or passages by successive bubbles of air or other gas. If a narrow tube is expanded at several places, and each bulb or enlargement contains a gas, the liquid can support a pressure of several atmospheres before it begins to flow.

Jänecke diagram (Jä'-neck-e) A square phase diagram whose corners represent two reciprocal salt pairs (e.g., NaCl - KCl - NaBr - KBr) and on which is plotted the configuration of the surface representing the aqueous solution saturated with the salts. It is particularly useful in the study of phase equilibria relevant to evaporites. Syn: *reciprocal salt-pair diagram.*

janggunite (jang'-gun-ite) A dull black orthorhombic mineral: $(Mn^{4+},Mn^{2+},Fe^{3+})_6O_8(OH)_6$.

janhaugite (jan'-haug-ite) A reddish-brown monoclinic mineral: $Na_3Mn^{2+}{}_3Ti_2Si_4O_{15}(OH,F,O)_3$.

jankovicite A metallic black triclinic mineral: $Tl_5Sb_9(As,Sb)_{4S.22}$.

Japanese twin law (Jap'-a-nese) A twin law in quartz that governs twinning of two individuals with a composition plane of (1122); four varieties are possible.

Jaramillo Subchron An interval of geologic time from about 1.07 to about 0.99 Ma, delimited by geomagnetic reversals, and having a normal magnetic field polarity. Named for Jaramillo Creek, Jemez Mountains, N.M. It is part of the *Matuyama Chron.* Symbol: C1r.1n.

jardang (jar'-dang) *yardang.*

jargoon (jar-goon') A colorless, pale-yellow, or smoky gem variety of zircon from Ceylon. Also spelled: jargon.

jarlite (jar'-lite) A colorless to gray monoclinic mineral: $Na(Sr,Na)_7MgAl_6F_{32}(OH,H_2O)_2$.

jarosewichite (jar-o'-se-wich-ite) A dark red to black orthorhombic mineral: $Mn^{2+}{}_3Mn^{3+}(AsO_4)(OH)_6$.

jarosite (ja'-ro-site) (a) An ocher-yellow or brown rhombohedral mineral of the *alunite* group: $KFe^{3+}{}_3(SO_4)_2(OH)_6$. Syn: *utahite.* (b) A sub-group of minerals consisting of basic iron sulfates, including jarosite, natrojarosite, ammoniojarosite, argentojarosite, plumbojarosite, and hydronium jarosite.

jars A mechanical device included in the *drillstring* or *fishing* string to allow application of a sharp vertical stress to be applied if the string should become stuck while drilling or fishing for "stuck tools".

jaskolskiite (jas-kol'-ski-ite) A metallic gray orthorhombic mineral: $Pb_{2.2}Cu_{0.2}(Sb,Bi)_{1.8}S_5$. It is a member of the meneghinite homologous series.

jasmundite (jas'-mund-ite) A greenish-brown tetragonal mineral: $Ca_{11}(SiO_4)_4O_2S$.
jaspachate (jasp-a-chat'-e) *jaspagate.*
jaspagate (jasp-a'-gate) A syn. of *agate jasper,* esp. that in which jasper predominates. Also spelled: *jaspachate.*
jasper (jas'-per) A variety of *chert* associated with iron ores and containing iron-oxide impurities that give it various colors, characteristically red, although yellow, green, grayish-blue, brown, and black cherts have also been called jasper. The term has also been applied to any red chert or chalcedony irrespective of associated iron ore. Syn: *jasperite; jaspis; jasperoid.*
jasper bar A term used in Australia for *iron formation.* Syn: *bar [eco geol]; jaspilite.*
jasperine (jas'-per-ine) Banded jasper of varying colors.
jasperite (jas'-per-ite) *jasper.*
jasperization (jas'-per-i-za'-tion) The conversion or alteration of igneous or sedimentary rocks into banded rocks like jaspilite by metasomatic introduction of iron oxides and cryptocrystalline silica.
jasperoid (jas'-per-oid) n. (a) A dense, usually gray, chertlike siliceous rock, in which chalcedony or cryptocrystalline quartz has replaced the carbonate minerals of limestone or dolomite; a silicified limestone. It typically develops as the gangue of metasomatic sulfide deposits of the lead-zinc type, such as those of Missouri, Oklahoma, and Kansas. Jasperoid is commonly associated with epithermal gold deposits. (b) *jasper.* adj. Resembling jasper.
jasper opal A yellow or yellow-brown, almost opaque common opal containing iron oxide and other impurities, having the color of yellow jasper but the luster of common opal. Some varieties are almost reddish brown to red. Syn: *jaspopal; opal jasper.*
jaspery (jas'-per-y) Resembling or containing jasper; e.g. "jaspery iron ore" (impure hematite interbedded with jasper), or "jaspery chert" (a silicified radiolarian ooze associated with volcanic rocks in Ordovician strata of southern England). Syn: *jaspidean.*
jaspidean (jas-pid'-e-an) Resembling or containing jasper; jaspery.
jaspilite (jas'-pi-lite) (a) A banded compact siliceous rock containing at least 25% iron, occurring with iron ores, and resembling jasper; e.g. the rock of the Precambrian iron-bearing district of the Lake Superior region. (b) A general term (used esp. in Australia) for banded *iron formation.* Syn: *jasper bar.* Also spelled: jaspilyte. See also: *jasperization.*
jaspis (jas'-pis) A syn. of *jasper.* Etymol: German *Jaspis.*
jasponyx (jasp-on'-yx) An opaque onyx, part or all of whose bands consist of jasper.
jaspopal (jasp-o'-pal) A syn. of *jasper opal.* Also spelled: jasp-opal.
javaite (ja'-va-ite) An Australasian tektite from Java. Also spelled: javanite.
jeanbandyite (jean-band'-y-ite) A brown-orange tetragonal mineral of the *stottite* group: $(Fe^{3+},Mn^{2+})Sn^{4+}(OH)_6$.
jebel (je'-bel) A hill, mountain, or mountain range in northern Africa. Etymol: Arabic. Also spelled: jabal; djebel.
jedwabite A metallic grayish yellow hexagonal mineral: $Fe_7(Ta,Nb)_3$.
jefferisite (jef'-fer-is-ite') A variety of vermiculite.
jeffersonite (jef'-fer-son-ite') A dark-green or greenish-black zinc- or manganese-bearing augite or diopside.
jeffreyite (jef'-frey-ite) A colorless orthorhombic mineral related to the *melilite* group: $(Ca,Na)_2(Be,Al)Si_2(O,OH)_7$.
jelly (jel'-ly) *ulmin.*
jenkinsite (jen'-kins-ite) A variety of antigorite containing iron.
jennite (jen'-nite) A white triclinic mineral: $Ca_9Si_6O_{16}(OH)_{10}·6H_2O$.
jensenite An emerald-green triclinic mineral: $Cu^{2+}{}_3Te^{6+}O_6·2H_2O$.
jentschite A metallic black monoclinic mineral: $TlPb(As,Sb)_3S_6$.
jeppeite (jep'-pe-ite) A submetallic black or brown monoclinic mineral: $(K,Ba)_2(Ti,Fe^{3+})_6O_{13}$.
jeremejevite (jer-e-me'-jev-ite) A colorless to pale yellowish-brown or blue hexagonal mineral: $Al_6(BO_3)_5(F,OH)_3$. It is structurally related to fluoborite. Syn: *eremeyevite.*
jeromite (je-ro'-mite) A cherry-red amorphous mineral of doubtful validity: $As(S,Se)_2$(?).
jerrygibbsite (jer'-ry-gibbs'-ite) A violet-pink orthorhombic mineral: $Mn^{2+}{}_9(SiO_4)_4(OH)_2$.
jervisite (jer'-vis-ite) A light-green monoclinic mineral of the *pyroxene* group: $(Na,Ca,Fe^{2+})(Sc,Mg,Fe^{2+})Si_2O_6$.
Jervois-type deposit *Broken Hill-type deposit.*
jet [coal] A hard, lustrous pure black variety of lignite; it has a conchoidal fracture and will take a high polish. It occurs as isolated masses in bituminous shale and is probably derived from waterlogged pieces of driftwood. Jet is used for jewelry and other ornamentation. Syn: *black amber.* See also: *jet shale; pitch coal.*
jet [hydraul] A sudden and forceful rush or gush of fluid through a narrow or restricted opening, either in spurts or in a continuous flow; e.g. a stream of water or air used to flush cuttings from a borehole.
jet flow A type of streamflow characterized by water moving in plunging, jetlike surges, produced where a stream reaches high velocity along a sharply inclined stretch or moves swiftly over a waterfall, or where a turbulent stream enters a body of standing water. Syn: *shooting flow.*
jetonized wood (jet'-o-nized) Lamellae of vitrain in coal.
jet rock *jet shale.*
jetsam Debris, such as objects thrown overboard, that sinks. Cf: *flotsam.*
jet shale Bituminous shale containing jet [coal]. Syn: *jet rock.*
jet streak An area of accelerated air flow within a jet stream; wind speeds may increase by as much as 100 km (62 mi) per hr. Jet streaks occur where surface horizontal temperature gradients are particularly steep and play an important role in the generation and maintenance of *extratropical cyclones.* The strongest jet streaks develop during winter in the *polar front* jet stream along the east coasts of North America and Asia.
jetted well (jet'-ted) A shallow water well, constructed by a high-velocity stream of water directed downward into the ground.
jetting [drilling] (a) The process of sinking a *borehole* in relatively unconsolidated sediments by using a directed, forceful stream (jet) of drilling mud, air, or water from the bit. (b) The process of removing excess drill solids from surface mud tanks in order to maintain desireable *drilling mud* properties.
jetting [impact] *hydrodynamic jetting.*
jetty (jet'-ty) (a) An engineering structure (such as a *groin* or small pier) extending more or less perpendicularly from the shore into a body of water and designed to direct and confine the current or tide, to protect a harbor, or to prevent shoaling of a navigable inlet by littoral materials. Generally, a jetty is larger than a groin. Jetties are often built in pairs on either side of a harbor entrance, inlet, or at the mouth of a river (Bush et al., 1996). (b) A British term for a landing wharf or pier used as a berthing place for vessels.
jezekite (jez'-ek-ite) *morinite.*
jheel *jhil.*
jhil [Pron: jeel] A term applied in the Ganges flood plain of India to a backwater, such as a pool, marsh, or lake, remaining from inundation, existing during the cold weather at about the same level as that of the river and rising with the river during the rainy season. Etymol: Hindi. Cf: *bhil.* Also spelled: *jheel.*
jianshuiite A submetallic brown to brownish- black triclinic mineral: $(Mg,Mn^{2+},Ca)Mn^{4+}{}_3O_7·3H_2O$.
jimboite (jim'-bo-ite) A light purplish-brown orthorhombic mineral: $Mn^{2+}{}_3(BO_3)_2$. It is the Mn^{2+} analogue of kotoite.
jimthompsonite (jim-thomp'-son-ite) A colorless to pale pinkish-brown orthorhombic pyribole mineral: $(Mg,Fe^{2+})_5Si_6O_{16}(OH)_2$.
jinshajiangite (jin-sha-jiang'-ite) A blackish, brownish, or golden red monoclinic mineral: $Na_2KBaCa(Fe^{2+},Mn^{2+})_8Ti_4Si_8O_{32}(O,F,H_2O)_6$. It is isomorphous with perraultite.
jixianite (ji-xian'-ite) A red or brownish-red cubic mineral of the *pyrochlore* group: $Pb(W,Fe^{3+})_2(O,OH)_7$.
joaquinite-(Ce) (joa-quin'-ite) (a) A brown, honey-yellow or amber monoclinic mineral: $NaBa_2Ce_2Fe^{2+}(Ti,Nb)_2O_2(OH)([Si_4O_{12}])_2·H_2O$. (b) A group name for minerals with the analogous formula, but with Ce replaced by Sr or Ba, and Fe replaced by Mn.
Job's tears Rounded grains of olivine (peridot) found associated with garnet in Arizona and New Mexico.
joch [Pron. yoch, as in Scottish "Loch"] A mountain pass with a long, approximately level summit between two parallel slopes (Stamp, 1961, p.491); a *col.* Etymol: German "Joch", "yoke". Syn: *yoke-pass.*
joesmithite (joe-smith'-ite) A deep brown-black monoclinic amphiboloid mineral: $PbCa_2(Mg,Fe^{3+},Fe^{2+})_5(Si_6Be_2)O_{22}(OH)_2$.
jog *overstep [struc geol].*
johachidolite (jo-ha-chid'-o-lite) A hard colorless orthorhombic mineral: $CaAlB_3O_7$.
johannite (jo'-han-nite) A secondary green or yellowish-green triclinic mineral: $Cu(UO_2)_2(SO_4)_2(OH)_2·8H_2O$. Syn: *gilpinite.*
Johannsen classification A quantitative mineralogic classification of igneous rocks developed by the petrographer Albert Johannsen (1939). Cf: *Johannsen number.*
johannsenite (jo-hann'-sen-ite) A colorless, clove-brown, grayish, blue, or greenish monoclinic mineral of the *pyroxene* group: $CaMn^{2+}Si_2O_6$.
Johannsen number (Jo-hann'-sen) A number, composed of three or four digits, that defines the position of an igneous rock in the *Johannsen classification.* The first digit represents the class, the second the order, and the third and fourth the family.
johillerite (jo-hil'-ler-ite) A violet monoclinic mineral of the *alluaudite* group: $NaCu(Mg,Zn)_3(AsO_4)_3$.
johnbaumite (john-baum'-ite) A colorless to gray hexagonal member of the *apatite* group: $Ca_5(AsO_4)_3(OH,F)$.
johninnesite (john-in'-nesite) A light yellowish-brown triclinic mineral: $Na_2Mn^{2+}{}_9(Mg,Mn^{2+})_7(AsO_4)_2(Si_6O_{17})_2(OH)_8$.
johnsomervilleite (john-som'-er-ville-ite') A dark brown rhombohedral minera of the *fillowite* group: $Na_2Ca(Fe^{2+},Mg,Mn^{2+})_7(PO_4)_6$.
johnstrupite (john'-strup-ite) *mosandrite.*
johntomaite A greenish-black monoclinic mineral of the *bjarebyite* group: $Ba(Fe^{2+},Ca)_2Fe^{3+}{}_2(PO_4)_3(OH)_3$.
johnwalkite (john-walk'-ite) A dark reddish-brown orthorhombic mineral: $K(Mn^{2+},Fe^{3+},Fe^{2+})_2(Nb,Ta)O_2(PO_4)_2(H_2O,OH)_2$. It is the Mn^{2+} analogue of johnwalkite.
JOIDES *Joint Oceanographic Institutions for Deep Earth Sampling.*

join The line or plane drawn between any two or three composition points in a phase diagram. There is no special phase significance to a join; it need not be a limiting binary or ternary subsystem.
joint [paleont] (a) An articulation in a crustacean; commonly, the movable connection of an individual *segment* of an appendage with its neighbors or with the body, or the movable connection of body parts. (b) A connection between any pair of contiguous crinoid ossicles. (c) A segment of the shell of a nasselline radiolarian. (d) For vertebrates, any arthrosis between adjacent bones, whether highly mobile or almost fixed. See: *articular; articulation.*
joint [struc geol] A planar fracture, crack, or parting in a rock, without shear displacement; the surface is usually decorated with a plumose structure. Often occurs with parallel joints to form part of a *joint set.* See also: *jointing.*
joint block A body of rock that is bounded by joints; the rock that occurs between adjacent joints.
joint-block separation A type of mechanical weathering in which the rock breaks down or comes apart along well-defined joint planes. Syn: *block disintegration.*
joint cavity A solutional hollow whose position is controlled by a joint on the inner surface of a cave. See also: *ceiling cavity.*
joint density The number of joints per length of scanline (1-D sample). The ratio of joint length per unit surface area (2-D sample). The surface area of joints per unit volume (3-D sample). Cf: *joint frequency.*
joint frequency The number of joints per distance; inverse of *joint spacing.*
jointing (joint'-ing) n. The condition or presence of joints in a body of rock.
joint-normal stretching The joint-driving mechanism in which the energy release for driving the joint is derived from extension of a rock containing the joint (Engelder and Fischer, 1996).
Joint Oceanographic Institutions for Deep Earth Sampling A consortium first formed in 1964 to provide overall scientific policy and management guidance for the Deep Sea Drilling Project and later for the *Ocean Drilling Program.* It now includes 14 university departments and marine-research institutions in the United States, and four European agencies, one Canadian, and one Japanese. Abbrev: JOIDES.
Joint Oceanographic Institutions, Inc. A corporation formed by the U.S. members of *JOIDES* for the purpose of managing major oceanographic research projects, such as the *Ocean Drilling Program.*
joint plane The surface of a joint.
joint-plane cave A cavity high in relation to width developed along steeply dipping joint planes.
joint-plane fall A waterfall whose crest is irregularly broken by the falling away of joint blocks (Tarr and Von Engeln, 1926, p.83).
joint probability The probability that two or more events will occur together.
joint set A group of more or less parallel joints. See also: *joint system.*
joint spacing The interval between joints of a particular joint set, measured on a line perpendicular to the joint planes; inverse of *joint frequency.*
joint spectrum A collection of joints that exhibits a fan-like pattern emanating from a local position in a bed (Hancock, 1985).
joint system Two or more joint sets that intersect. They may be of the same age or of different ages.
joint valley A valley whose drainage pattern is controlled by master joint systems, e.g. in a *rectangular drainage pattern.*
jokokuite (jo-ko'-ku-ite) A pale pink triclinic mineral of the *pentahydrite* group: $Mn^{2+}SO_4{\cdot}5H_2O$.
jökulhlaup (jök'ul-hlaup) An Icelandic term for *glacier outburst flood.*
jökull (jök'-ull) An Icelandic term meaning *glacier* or *ice sheet.* In geologic use it is best considered only as a compounding element in a specific name, e.g. Vatnajökull. Pl: jöklar.
joliotite (jo'-li-o-tite) A yellow orthorhombic mineral: $(UO_2)CO_3{\cdot}2H_2O$.
jolliffeite A metallic cubic mineral of the *gersdorffite* group: NiAsSe.
Jolly balance (Jol'-ly) In mineral analysis, a delicate spring balance used to measure specific gravity.
jonesite (jones'-ite) A colorless monoclinic mineral: $Ba_4(K,Na)_2Ti_4Al_2Si_{10}O_{36}{\cdot}6H_2O$.
Joplin-type lead (Jop'-lin-type) *J-type lead.*
jordanite (jor'-dan-ite) A lead-gray monoclinic mineral: $Pb_{14}(As,Sb)_6S_{23}$.
Jordan's law (Jor'-dan's) A theory in evolutionary biology stating that closely related organisms tend to occupy adjacent rather than identical or distant ranges. Named after the American biologist David Jordan (1851-1931). Not to be confused with *Jordan's rule.*
Jordan's rule The empirical relationship, noted by the biologist David Jordan, that the number of vertebrae in a freshwater fish is larger in coldwater than in warm water. (Orr, 1961, p.167). Not to be confused with *Jordan's law.*
jordisite (jor'-dis-ite) A black amorphous mineral: MoS_2. Cf: *molybdenite.*
jørgensenite A colorless or white monoclinic mineral of the *jarlite* group: $Na_2(Sr,Ba)_{14}Na_2Al_{12}F_{64}(OH,F)_4$.
joséite (jo-sé'-ite) A metallic gray rhombohedral mineral: Two different compositions are given: joseite-A: Bi_4TeS_2, and joseite-B: Bi_4Te_2S. A further joseite-C with a composition $Bi_{16}Te_3S_9$ has been reported, but is probably not a valid species.
josephinite (jo'-se-phin-ite') A natural alloy of iron and nickel occurring in stream gravel from Josephine County, Oregon; nickel-iron with nickel greater than 50%.
jotunite (jo-tun'-ite) A plutonic rock of the *charnockite series,* intermediate between *monzonite* and *norite,* containing orthopyroxene, plagioclase, and microperthite (Streckeisen, 1967, p.169). The name, given by Goldschmidt in 1916, is for Jotunheim, Norway.
jouravskite (jou-rav'-skite) A greenish-yellow to greenish-orange hexagonal mineral: $Ca_3Mn^{4+}(SO_4)(CO_3)(OH)_6{\cdot}12H_2O$. It possesses the thaumasite structure.
J-type lead *Anomalous lead* that gives model ages younger than the age of the enclosing rock, in some cases even negative model ages. Cf: *B-type lead.* Syn: *Joplin-type lead.*
juabite An emerald-green triclinic mineral: $CaCu_{10}(TeO_3)_4(AsO_4)_4(OH)_2{\cdot}4H_2O$.
juanitaite An olive-green to grass-green tetragonal mineral: $(Cu,Ca,Fe)_{10}Bi(AsO_4)_4(OH)_{11}{\cdot}2H_2O$.
juanite (juan'-ite) A white to pistachio-green orthorhombic mineral: $Ca_{10}Mg_4Al_2Si_{11}O_{39}{\cdot}4H_2O$ (?).
jug A colloquial syn. of *geophone.*
jugum (ju'-gum) (a) A medially placed connection of secondary shell between two primary lamellae of brachiopod spiralia; a more or less complex skeletal crossbar linking the right and left halves of the *brachidium* of certain brachiopods. (b) A transverse structure crossing the center of a heterococcolith and connecting one side of the cycle with the other. Pl: juga or jugums. Adj. jugal.
julgoldite-(Fe²⁺) (jul-gold'-ite) A submetallic green to greenish-black monoclinic mineral of the *pumpellyite* group: $Ca_2Fe^{2+}Fe^{23+}(OH)_2(SiO_4)(Si_2O_7){\cdot}H_2O$. The compositional suffix was added later to the original name "julgoldite".
julgoldite-(Fe³⁺) A submetallic green, greenish-black to black monoclinic mineral of the *pumpellyite* group: $Ca_2Fe^{3+}Fe^{3+}{}_2(Si_2O_7)(SiO_4)O(OH){\cdot}H_2O$.
julgoldite-(Mg) A green to black mineral of the *pumpellyite* group: $Ca_2MgFe^{3+}{}_2(Si_2O_7)(SiO_4)(OH)_2{\cdot}H_2O$.
Julian (Ju'-li-an) A European substage: lower substage of Carnian Stage, Upper Triassic (above Longobardian substage of the Ladinian, below Tuvalian).
Julian date A 4-digit way of expressing the date, e.g., the 33rd day of the year 1999 is expressed as 3399. An instant is identified by giving the fractional part of the day from the preceding noon (12th) U.T. to the instant.
julienite (ju'-lien-ite) A blue monoclinic thiocyanate mineral occurring in needlelike crystals: $Na_2Co(SCN)_4{\cdot}8H_2O$.
jumillite (ju-mil'-lite) An extrusive rock, commonly fine-grained, composed of phenocrysts of barium-bearing sanidine, olivine, and phlogopite in a fine-grained groundmass of altered leucite, sanidine, and iron-rich diopside mantled with aegirine-augite, with interstitial kataphorite. A variety of *lamproite,* named by Osann in 1906 for Jumilla, Spain. Cf: *orendite.* Mitchell and Bergman (1991) suggest that this name should be replaced by olivine-diopside-richterite *madupitic lamproite.* Obsolete.
jump correlation Identification of events on noncontiguous seismic records as involving the same interfaces in the Earth.
junction [streams] (junc'-tion) The meeting of two or more streams; also, the place of such a meeting; a confluence. Examples: *accordant junction; deferred junction.*
junction [surv] A point common to two or more survey lines.
junction closure The amount by which a new survey line into a junction fails to give the previously determined position or elevation for the junction point.
jungite (jung'-ite) A dark green platy orthorhombic mineral: $Ca_2Zn_4Fe^{3+}{}_8(PO_4)_9(OH)_9{\cdot}[16]H_2O$.
jungle (jun'-gle) A tropical region of wild, tangled, dense vegetation.
junitoite (jun-i'-to-ite) A colorless orthorhombic mineral: $CaZn_2Si_2O_7{\cdot}H_2O$.
junk *fish [drill].*
junoite (ju'-no-ite) A metallic monoclinic mineral: $Pb_3Cu_2Bi_8(S,Se)_{16}$.
juonniite A colorless, light yellow to yellowish-brown orthorhombic mineral of the *overite* group: $CaMgSc(PO_4)_2(OH){\cdot}4H_2O$.
Jura (Ju'-ra) *Jurassic.*
Jurassian relief (Ju-ras'-si-an) A type of relief found in young mountains that consist of many parallel anticlines and synclines, characterized by primary structural forms or by features upon which erosion has had relatively little influence (Schieferdecker, 1959, term 1944). Type example: the relief of the Jura Mountains in Switzerland. Cf: *Appalachian relief.*
Jurassic (Ju-ras'-sic) The middle system of the *Mesozoic* Erathem of the Standard Global Chronostratigraphic Scale, above the *Triassic* and below the *Cretaceous.* Also the time during which these rocks were formed, the Jurassic Period, covering the time span between 203 and 145 Ma. It is named after the Jura Mountains between France and Switzerland, in which rocks of this age were first studied. Informally called: *age of cycads.* Syn: *Jura.*
Jura-type fold *detachment fold.*
jurbanite (jur'-ban-ite) A colorless monoclinic mineral: $Al(SO_4)(OH){\cdot}5H_2O$.
jurupaite (ju-ru'-pa-ite) A variety of xonotlite containing magnesium.
juvenarium (ju-ve-na'-ri-um) The *proloculus* and first few chambers of a foraminifer. See also: *embryonic apparatus.*
juvenile [geomorph] (ju'-ve-nile) *youthful.*

juvenile [ore dep] Said of an ore-forming fluid or mineralizer that is derived from a magma, via fractional crystallization or other plutonic mechanism, as opposed to fluids of surface, connate, or meteoric origin. Cf: *assimilated; filtrational.*

juvenile [volc] In the classification of pyroclastics, the equivalent of *essential;* derived directly from magma reaching the surface.

juvenile [water] A term applied to water and gases that are known to have been derived directly from magma and are thought to have come to the Earth's surface for the first time. Cf: *resurgent [petrology].*

juvite (ju'-vite) A light-colored *nepheline syenite* in which the feldspar is exclusively or predominantly potassium feldspar and the potassium-oxide content is higher than the sodium oxide. Obsolete.

juxta-epigenesis (jux'-ta-ep'-i-gen'-e-sis) Post-diagenetic changes that affect sediments while they are near the original environment of deposition, either under a relatively thin overburden or exposed above sea level (Chilingar et al., 1967, p.316).

juxtaposed ice stream (jux'-ta-posed) Ice from a tributary glacier that is set into the surface of a glacier and extends to the bed. Cf: *inset ice stream.*

juxtaposition twin (jux'-ta-po-si'-tion) *contact twin.*

J wave A transverse or *S wave* traveling through the Earth's inner core.

K

ka Kilo-annum, one thousand (10^3) years. Informal SI notation, where "annum" is age in years before present, with "present" fixed as 1950. This term has largely replaced the various abbreviations for "thousands of years before present" in geological literature. It is not equivalent to the elapsed time interval or duration in "thousands of years" (kyr).

kaatialaite (kaa-ti-a'-la-ite) A greenish-blue monoclinic mineral: $Fe^{3+}(H_2AsO_4)_3{\cdot}3H_2O$.

Kachkanar-type PGE deposit *zoned ultramafic PGE deposit.*

kadryelite (kad'-ry-el-ite) An adamantine orange cubic mineral: $Hg^{1+}{}_6(Br,Cl)_3O_{1.5}$. It is the Br analogue of eglestonite.

kaersutite (kaer-su'-tite) A dark-brown to black monoclinic member of the *amphibole* group: $NaCa_2(Mg_4Ti)(Si_6Al_2)O_{23}(OH)$.

kafehydrocyanite (ka'-fe-hy'-dro-cy'-a-nite) A yellow tetragonal ferrocyanide mineral: $K_4[Fe(CN)_6]{\cdot}3H_2O$.

kahlerite (kahl'-er-ite) A yellow to yellow-green tetragonal secondary mineral of the *autunite* group: $Fe^{2+}(UO_2)_2(AsO_4)_2{\cdot}12H_2O$.

kaimoo (kai-moo') A stratified ice and sediment rampart built during the autumn on an Arctic beach by wave action. Etymol: Eskimo.

kainite (kain'-ite) A white and variously colored monoclinic mineral: $KMg(SO_4)Cl{\cdot}3H_2O$. It is a natural salt occurring in irregular granular masses, and is used as a source of potassium and magnesium compounds.

kainosite-(Y) (kai'-nos-ite) A white, yellow or yellowish-brown orthorhombic mineral: $Ca_2(Y,Ce)_2Si_4O_{12}(CO_3){\cdot}H_2O$. Syn: *cenosite.*

Kainozoic (Kai-no-zo'-ic) *Cenozoic.*

kaiwekite (kai'-wek-ite) A trachytic extrusive rock composed of phenocrysts of amphibole, small aegirine-augite crystals, and anorthoclase with inclusions of aegirine-augite and other minerals, and with a few pseudomorphs of serpentine after olivine, in a groundmass composed chiefly of oligoclase with some augite and magnetite; the approximate extrusive equivalent of *larvikite.* Essentially a *trachyte.* Named by Marshall in 1906 for Kaiweke, New Zealand. Obsolete.

kakirite (ka-kir'-ite) A megascopically sheared and brecciated rock in which fragments of original material are surrounded by gliding surfaces along which intense granulation and some recrystallization has occurred. Rarely used. Named by Svenonius after Lake Kakir, Swedish Lapland.

kakortokite (ka-kor'-tok-ite) A layered *nepheline syenite* of varied composition, having light-colored layers rich in feldspar and nepheline (white) or in eudialyte and nepheline (red) and dark-colored layers rich in aegirine and arfvedsonite (black). The rock was originally described by Ussing in 1912 from Kakortok in the Ilimaussaq complex, SW Greenland. Obsolete.

kalborsite (kal-bor'-site) A pale rose tetragonal zeolite mineral: $K_6BAl_4Si_6O_{20}Cl(OH)_4$.

kali- (kal'-i-) A prefix which, when in an igneous rock name, signifies an absence of plagioclase or a plagioclase content of less than 5.0%.

kaliborite (kal-i-bor'-ite) A colorless to reddish-brown monoclinic mineral: $KHMg_2B_{12}O_{16}(OH)_{10}{\cdot}4H_2O$. Syn: *heintzite; hintzeite; paternoite.*

kalicinite (ka-lic'-i-nite) A colorless to white or yellowish monoclinic mineral: $KHCO_3$. Syn: kalicine; kalicite.

kalifersite A fibrous pink-brownish triclinic mineral: $(K,Na)_5Fe^{3+}{}_7Si_{20}O_{50}(OH)_6{\cdot}12H_2O$. It is a member of the palygorsksite-sepiolite polysomatic series.

kalininite (kal-in'-in-ite) A metallic black cubic mineral of the *linnaeite* group: $ZnCr_2S_4$.

kalinite (kal'-i-nite) A white monoclinic mineral: $KAl(SO_4)_2{\cdot}11H_2O$. It is the K analogue of mendozite.

kaliophilite (kal-i-o-phil'-ite) An colorless hexagonal mineral of volcanic origin: $KAlSiO_4$. It is dimorphous with kalsilite. Syn: *facellite; phacellite.*

kalipyrochlore (kal-i-py'-ro-chlore) A greenish cubic mineral of the *pyrochlore* group: $(H_2O)(Nb,Ti)_2(O,OH)_6{\cdot}H_2O$.

kalistrontite (kal-i-stront'-ite) A colorless rhombohedral mineral: $K_2Sr(SO_4)_2$. It is the Sr analogue of palmierite.

kalkowskite (kal-kow'-skite) A very rare, brownish or black mineral of doubtful validity: $Fe_2Ti_3O_9$(?). It may be ilmenite. It usually contains small amounts of rare-earth elements, niobium, and tantalum. Cf: *arizonite [mineral].*

kalmafite (kal-ma'-fite) A general term for igneous rocks composed of *kal*silite (or, presumably, any other polymorph of $KAlSiO_4$) and *maf*ic minerals, proposed by Hatch, et al. (1961). Syn: *mafurite; katungite.*

kalsilite (kal'-sil-ite) A colorless mineral of the *nepheline* group: $KAlSiO_4$. It exists in hexagonal and trigonal polytypes, and is dimorphous with kaliophilite.

kaluginite A greenish-yellow orthorhombic mineral: $(Mn,Ca)MgFe^{3+}(PO_4)_2(OH){\cdot}4H_2O$. Not approved by the International Mineralogical Association.

kalyptra (ka-lyp'-tra) A formless outer covering of adherent, flocculent organic material, structureless or granular, within which certain dinoflagellate cysts are enclosed. Pl: kalyptrae.

kamacite (kam'-a-cite) A metallic gray meteorite mineral consisting of the body-centered cubic alpha-phase of a *nickel-iron* alloy, with a fairly constant composition of 5-7% nickel: (Fe,Ni). It occurs in iron meteorites as bars or "girders" flanked by lamellae of taenite.

kamafugite (ka-ma-fu'-gite) A general term for potassium-rich, silica-undersaturated igneous rocks, proposed by Sahama in 1974. The term, embracing *katungite, mafurite,* and *ugandite,* is derived from the first letters of these. Obsolete.

kamaishilite (ka-ma-ish'-i-lite) A colorless tetragonal mineral dimorphous with bicchulite: $Ca_2Al_2SiO_6(OH)_2$.

kamarezite (ka-mar'-e-zite) *brochantite.*

kambaldaite (kam-bal'-da-ite) An acicular green hexagonal mineral: $NaNi_4(CO_3)_3(OH)_3{\cdot}3H_2O$.

kamchatkite (kam-chat'-kite) A greenish-brown orthorhombic mineral: $KCu_3(SO_4)_2OCl$.

kame A low mound, knob, hummock, or short irregular ridge, composed of stratified sand and gravel deposited by a subglacial stream as a fan or delta at the margin of a melting glacier; by a superglacial stream in a low place or hole on the surface of the glacier; or as a ponded deposit on the surface or at the margin of stagnant ice. The term has undergone several changes in meaning, but can still be usefully applied to a deposit of glaciofluvial and glaciolacustrine sand and gravel whose precise mode of formation is uncertain (Thornbury, 1954, p.378-379). Etymol: a Scottish variant of "comb", a long, steep-sided ridge. Cf: *esker.*

kame-and-kettle topography *knob-and-kettle topography.*

kame complex An assemblage of *kames,* constituting a hilly landscape.

kame delta *delta kame.*

kame field A group of closely spaced *kames,* interspersed in places with *kettles* and *eskers,* and having a characteristic hummocky topography.

kame moraine (a) An *end moraine* that contains numerous kames. (b) A group of kames along the front of a stagnant glacier, commonly comprising the slumped remnants of a formerly continuous outwash plain built up over the foot of rapidly wasting or stagnant ice. Cf: *moraine kame.*

kamenitza (ka-me-nit'-za) *solution pan.*

kame plain A flat-topped outwash plain originally entirely bounded by ice-contact slopes. Syn: *ice-contact plain.*

kame terrace A terracelike ridge consisting of stratified sand and gravel formed as a *glaciofluvial* or *glaciolacustrine* deposit between a melting glacier or a stagnant ice lobe and a higher valley wall or lateral moraine, and left standing after the disappearance of the ice; a filling of a fosse. A kame terrace terminates a short distance downstream from the terminal moraine; it is commonly pitted with kettles and has an irregular ice-contact slope. Cf: *glacial terrace.* Syn: *ice-contact terrace; ice-marginal terrace.*

kamiokite (ka-mi-o'-kite) A metallic black hexagonal mineral: $Fe^{2+}{}_2Mo^{4+}{}_3O_8$.

kamitugaite (ka-mi-tu'-ga-ite) A yellow triclinic mineral: $PbAl(UO_2)_5(PO_4,AsO_4)_2(OH)_9{\cdot}9.5H_2O$.

kämmererite (käm'-mer-er-ite') (a) A reddish variety of clinochlore containing chromium. Its formula is near: $(Mg,Al,Cr)_6(OH)_8AlSi_3O_{10}$. (b) A hypothetical end-member of the *chlorite* group: $(Mg_4Cr_2)(OH)_8(Cr^{3+}{}_2Si_2)O_{10}$.

kamotoite-(Y) (ka-mo′-to-ite-y) A yellow monoclinic mineral: $Y_2(UO_2)_4(CO_3)_3O_4 \cdot 14.5H_2O$.

kamperite (kam′-per-ite) A fine-grained black hypabyssal rock composed of small euhedral alkali feldspar crystals and a small amount of oligoclase in a groundmass of dark-colored mica. The name, given by Brögger in 1921, is for Kamperhaug in the Fen complex, Norway. Obsolete.

kampfite A light blue-gray hexagonal mineral: $Ba_6(Si,Al)_8O_{16}(CO_3)_2Cl_2(Cl,H_2O)_2$.

kamphaugite-(Y) A colorless tetragonal mineral: $Ca_2(Y,Dy,Gd,Nd)_2(CO_3)_4(OH)_2 \cdot 3H_2O$.

kanat (ka-nat′) *qanat.*

kandic horizon Subsoil diagnostic horizon having a clay increase relative to overlying horizons and has low activity clays i.e., < 160 $cmol_c\ kg^{-1}$ clay.

kandite (kan′-dite) A name suggested (but not approved) for the *kaolin* group of clay minerals, including kaolinite, nacrite, dickite, and halloysite.

kanemite (kan′-e-mite) A colorless orthorhombic mineral: $NaHSi_2O_5 \cdot 3H_2O$.

kankan-ishi (kan′-kan-ish′-i) A black resinous flinty *andesite,* composed of orthopyroxene, oligoclase, and hornblende microphenocrysts in a groundmass of colorless glass and a network of acicular crystals of colorless orthopyroxene (Johannsen, 1939, p.259). Obsolete.

kankar (kan′-kar) A term used in India for *calcrete.* Also spelled: *kunkur; kunkar.*

kankite (kank′-ite) A yellowish-green monoclinic mineral originally found as an alteration product of arsenopyrite: $Fe^{3+}AsO_4 \cdot 3.5H_2O$.

kanoite (ka′-no-ite) A pinkish-brown monoclinic mineral of the *pyroxene* group: $(Mn^{2+},Mg)SiO_3$.

kanonaite (ka-no′-na-ite) A greenish-black orthorhombic mineral, isotypic with andalusite: $(Mn^{3+},Al)AlSiO_5$.

kanonerovite A colorless monoclinic triphosphate mineral: $Na_3MnP_3O_{10} \cdot 12H_2O$.

Kansan (Kan′-san) Pertaining to the classical second glacial stage of the Pleistocene Epoch in North America, after the Aftonian interglacial stage and before the Yarmouthian. Cf: *Mindel.* Obsolete.

kansite (kan′-site) *mackinawite.*

kantography (kan-tog′-ra-phy) The depiction of *edge lines* on relief maps. Etymol: German "Kantographie".

kaolin [eco geol] A soft, relatively nonplastic but dispersible, usually white or nearly white claystone composed primarily of minerals of the kaolin group, principally kaolinite but often containing a variable proportion of, e.g. mica or quartz. Kaolin is white or nearly white on firing; a porcelain clay or natural (unwashed) china clay; used in the manufacture of ceramics, refractories, and paper. Type locality: Kao-ling (meaning "high hill"), a hill in Jiang Xi province, SE China. Syn: *white clay; bolus alba.*

kaolin [mineral] (ka′-o-lin, ka-o′-lin) (a) A group of clay minerals characterized by a two-layer crystal structure in which each silicon-oxygen sheet is alternately linked with one aluminum-hydroxyl sheet and having approximate composition: $Al_4(OH)_8[Si_4O_{10}]$. The kaolin minerals include kaolinite, nacrite, dickite, and anauxite; although the minerals halloysite, endellite, and allophane are structurally and chemically different, they are sometimes included. The kaolin minerals are generally derived from alteration of alkali feldspars and micas. They have lower base-exchange capacities than montmorillonite and illite, and they absorb less water and thus have lower plasticity indexes, lower liquid limits, and less shrinkage when drying from a wet state. See also: *kandite.* Syn: *kaolinite.* (b) A mineral of the *kaolin* group; specif. *kaolinite.* The term was once applied to a single clay mineral which later was known to include at least four minerals of the kaolin group.

kaolinic (ka-o-lin′-ic) Pertaining to or resembling kaolin.

kaolinite (ka-o′-lin-ite) (a) A common earthy white, grayish, yellowish, etc. triclinic clay mineral of the *kaolin* group: $Al_2Si_2O_5(OH)_4$. It is the characteristic mineral of most kaolins, and is polymorphous with dickite and nacrite. Kaolinite consists of sheets of tetrahedrally coordinated silicon joined by an oxygen shared with octahedrally coordinated aluminum; it also occurs as a disordered monoclinic variant. Kaolinite is a high-alumina clay mineral that does not appreciably expand under varying water content and does not exchange iron or magnesium. The mineral was formerly known as kaolin. (b) A name sometimes applied to the kaolin group of clay minerals, and formerly applied to individual minerals of that group (such as to dickite and nacrite).

kaolinite-alunite epithermal *acid-sulfate type* deposit.

kaolinite/smectite A mixed-layered clay mineral formed of interstratified layers of kaolinite and smectite or a layer that expands. These seem to be stacked randomly with very few reports of ordering or any kind. Typically found in paleosols. Often abbreviated K/S or may be referred to as kaolinite/expandable K/E (Hughes et al., 1993).

kaolinitic shale (ka′-o-lin-it′-ic) *feldspathic shale.*

kaolinton (ka-o-lin′-ton) An obsolescent term used by ceramists (esp. in Europe) for the portion of a clay that is soluble in sulfuric acid but not soluble in hydrochloric acid. Syn: *allophaneton.* Etym: German "kaolin" + "ton" (clay).

Kapitean (Kap-i-te′-an) South Pacific stage: uppermost Miocene (above Tongaporutuan, below Pliocene Opoitian).

kapitsaite-(Y) A white to light pink triclinic mineral: $(Ba,K,Pb)_4(Y,Ca)_2Si_8(B,Si)_4O_{28}F$. It is the Y analogue of hyalotekite.

kar A Swiss-German equivalent of *cirque [glac geol].*

K-Ar age method *potassium-argon age method.*

karang (ka-rang′) A term used in Indonesia for an emerged terrace composed of ancient fringing-reef material, and also for the coral limestone itself (Termier and Termier, 1963, p.408). Etymol: Malay, "reef, coral reef".

karasugite A colorless monoclinic mineral: $SrCaAl(F,OH)_7$.

karat (kar′-at) The proportion of pure gold in an alloy. Pure or fine gold is 24 karat; 10-karat gold is $^{10}/_{24}$ pure, or 10 parts of pure gold by weight mixed with 14 parts of other metals. Not to be confused with *carat.* Abbrev: k.

Karatau An obsolete geochronometric subdivision of the *Precambrian,* covering the time span from 1,050 to 800 Ma (Harland et al., 1990), therefore approximately equivalent to the *Tonian* Period (early *Neoproterozoic* Era) of the Standard Global Geochronometric Scale.

karelianite (ka-re′-lian-ite) A black rhombohedral mineral of the *corundum* group: V_2O_3. It may contain some iron, chromium, and manganese.

karewa (ka-re′-wa) A term applied in Kashmir to the level surface between the incised streams dissecting a terrace. Etymol: Kashmiri.

karez (ka′-rez) A term used in Pakistan for a gently inclined, underground channel dug so as to conduct groundwater by gravity from alluvial gravels and the foot of hills to an arid lowland or basin; a horizontal well. Etymol: Baluchi. Pl: karezes. Cf: *qanat; foggara.*

karibibite (kar′-i-bib-ite) A brownish-yellow orthorhombic mineral: $Fe^{3+}{}_2As^{3+}{}_4(O,OH)_9$.

karling (kar′-ling) (a) A high, dissected region, characterized by cirques, as Mount Anne in Tasmania. (b) A cluster or group of cirques.

karlite (kar′-lite) A silky white to light green orthorhombic mineral: $Mg_7(BO_3)_3(OH)_4Cl$.

Karlsbad twin law (Karls′-bad) *Carlsbad twin law.*

kärnäite (kär′-nä-ite) A rock found on the island of Kärnä in Lake Lappajärvi, central Finland, originally described as a volcanic rock with glassy groundmass and numerous inclusions, consisting of agglomeratelike tuff, and having a composition similar to dacite, the principal feldspar phenocrysts being monoclinic (probably sanidine). The rock has also been interpreted as an *impactite* with bedrock fragments (Svensson, 1968). Obsolete.

karnasurtite-(Ce) (kar-na-sur′-tite) A honey to pale-yellow metamict mineral: $(Ce,etc.)(Ti,Nb)Al(OH)_4[Si_2O_7] \cdot 3H_2O$. It gives a monazite-like X-ray pattern when heated, but it may be equivalent to rhabdophane. Syn: kozhanovite.

Karnian (Kar′-ni-an) *Carnian.*

karoo (ka-roo′) *karroo.*

karooite A hypothetical end-member of the *armalcolite* series: $MgTi^{4+}{}_2O_5$. Not approved by the International Mineralogical Association.

Karpatian (Kar-pa′-ti-an) Central European stage: upper Lower Miocene, planktonic zone N.7/8 (above Ottnangian, below Badenian).

karpatite (kar′-pa-tite) *carpathite.*

karpinskite (kar-pin′-skite) A greenish-blue mineral of doubtful validity: $(Mg,Ni)_2Si_2O_5(OH)_2$(?). Not to be confused with *karpinskyite.*

karpinskyite (kar-pin′-sky-ite) A mixture of leifite and Zn-bearing clay, formerly thought to be a distinct mineral species. Not to be confused with *karpinskite.*

karren (kar′-ren) Channels or furrows (always plural) caused by *solution* on massive, bare limestone surfaces. They vary in depth from a few millimeters to more than a meter and are separated by ridges. In modern usage, the terms are general, describing the total complex of superficial solution forms found on compact pure limestone. Classified by Bögli (1960) into several kinds, of which the most common are *deckenkarren; kluftkarren; mäanderkarren; rillenkarren; rinnenkarren; rundkarren; spitzkarren; trittkarren.* Etymol: German, "wheel tracks". Cf: *karrenfeld.* Syn: *lapiaz; lapiés; solution grooves.*

karrenfeld (kar′-ren-feld) A karstic surface on limestone, characterized by *solution grooves.* Cf: *limestone pavement; karren.*

karroo (kar-roo′) A tableland, found esp. in South Africa, that commonly rises to a considerable height in terraces. It does not support vegetation in the dry season, but becomes a grassy plain or pastureland during the wet season. Also spelled: *karoo.*

karst A type of topography that is formed on limestone, gypsum, and other soluble rocks, primarily by *dissolution.* It is characterized by *sinkholes, caves,* and underground drainage. First published in 1774 on a topographic map, *Ducatus Carnioliae.* Type locality: a limestone plateau in the Dinaric Alps. Adj: *karstic.* Syn: *karst topography.* Cf: *causse; pseudokarst; thermokarst.* Etymol: German, from the Slovenian region of Krs; Slavic "kras" or "krš" "a bleak, waterless place".

karst breccia *collapse breccia; solution breccia.*

karst bridge A *natural bridge* or arch in limestone or other soluble rock.

karst cone A karst tower with rounded or slanted sides. See also: *karst tower; fengcong; fenglin.*

karst corridor *solution corridor.*

karst feature A geologic feature formed directly or indirectly by *solution,* including *caves.* The term is often used to describe features that are not large enough to be considered caves but have some probable relation to subsurface drainage or groundwater movement. These features typically include, but are not limited to, *sinkholes,* enlarged fractures, noncavernous springs and seeps, soil pipes, and epikarstic solution cavities.

karst fen A marsh developed in *sinkhole* terrain; a swampy *solution* fen.

karst fenster The German equivalent of *karst window.*

karst-fill *residual and karst-fill nonsulfide zinc deposit.*

karst hydrology (a) The drainage phenomena of karstified limestones, dolomites and other slowly soluble rocks. (b) The surface and groundwater hydraulics in karst drainage systems.

karstic (karst′-ic) Pertaining to or having the characteristics of *karst.*

karstification (karst′-i-fi-ca′-tion) The action of water, mainly solutional but also mechanical, that produces features of a karst topography, including such surface features as *dolines, karren,* and *mogotes* and such subsurface features as *caves* and *shafts.*

karst lake A large area of standing water in extensive *closed depression* in limestone. Cf: *karst pond.*

karst-margin plain (karst′-mar-gin) A plain, generally of limestone, between higher country of limestone on one side and less pervious rocks on the other, but having a cover of impervious detritus which allows surface drainage. Syn: *marginal karst plain.*

karst plain A plain, usually of limestone, on which karst features are developed. Cf: *marginal karst plain; labyrinth karst.* Syn: *karst plateau.*

karst plateau A plateau, usually of limestone, on which karst features are developed.

karst pond A body of standing water in a *closed depression* of a karst region. Syn: *karst lake; solution lake.*

karst river A river that originates from a *karst spring.*

karst spring A spring emerging from karstified limestone. Cf: *emergence [karst]; exsurgence; resurgence; rise [karst].*

karst street *solution corridor.*

karst topography *karst.*

karst tower An isolated hill with steep vertical sides on a karst plain that is commonly alluviated. Syn: *haystack hill; hum; mogote; pepino; pepino hill.* Cf: *fengcong; fenglin; karst cone; kegelkarst.*

karst valley (a) An elongate *solution valley.* (b) A valley produced by collapse of a cavern roof. (c) A *closed depression* formed by the coalescence of several *sinkholes.* Its drainage is subsurface, its size is measured in hundreds of meters to a few kilometers, and it usually has an irregular floor and a scalloped margin inherited from the sinkholes. Syn: *nested sinkholes; solution valley; uvala.* Cf: *glade [karst]; interior valley; sinkhole; solution corridor.*

karst window A *collapse sinkhole* opening into a cave. Syn: *karst fenster.* Cf: *cenote.*

karupmøllerite-Ca A white monoclinic mineral of the *labuntsovite* group: $(Na,Ca,K)_2Ca(Nb,Ti)_4(Si_4O_{12})_2(O,OH)_4{\cdot}7H_2O$.

karyogamy (kar-y-og′-a-my) The second phase of fungal sexual reproduction, in which the zygote is formed.

kåsenite (kå′-se-nite) A pyroxene-rich, nepheline-bearing carbonatite, named by Brögger in 1921 at the Fen complex, Norway. Obsolete.

kashinite A metallic grayish-black monoclinic mineral: $(Ir,Rh)_2S_3$. It is the Ir analogue of bowieite.

Kasimovian (Kas-i-mo′-vi-an) A stage of the Standard Global Chronostratigraphic Scale: middle of Pennsylvanian Subsystem of Carboniferous (above Moscovian, below Gzhelian).

kasoite (ka′-so-ite) A variety of celsian containing potassium.

kasolite (ka′-so-lite) A yellow to brown monoclinic mineral: $Pb(UO_2)SiO_4{\cdot}H_2O$. Syn: *droogmansite.*

kassaite (kas′-sa-ite) A fine-grained hypabyssal rock composed of phenocrysts of hauyne, labradorite with oligoclase rims, amphibole, and augite in a holocrystalline tinguaitic groundmass of acicular hastingsite crystals and andesine with oligoclase and orthoclase rims. Described by Lacroix in 1918 from the Los Archipelago, Guinea. Obsolete.

kassite (kass′-ite) An adamantine yellow orthorhombic mineral: $CaTi_2O_4(OH)_2$.

kastningite A colorless, white or beige triclinic mineral of the *laueite* group: $(Mn^{2+},Fe^{2+},Mg)Al_2(PO_4)_2(OH)_2{\cdot}8H_2O$.

kata- (kat′-a-) A prefix meaning "down", "beneath".

katabatic wind (kat-a-bat′-ic) A local wind that moves down a slope, driven by density differences, e.g. as a result of surface cooling during the night. Ant: *anabatic wind.* See also: *mountain wind; glacier wind.* Syn: *gravity wind.*

katagenesis (kat-a-gen′-e-sis) *catagenesis [evol].*

katamorphic zone (kat-a-mor′-phic) The shallow zone in the Earth's crust in which *katamorphism* takes place. The term was originated by Van Hise. Obsolete. Cf: *anamorphic zone.*

katamorphism (kat-a-mor′-phism) Destructive metamorphism in the *katamorphic zone,* at or near the Earth's surface, in which complex minerals are broken down and altered through oxidation, hydration, solution, and allied processes to produce simpler and less dense minerals. The term was introduced by Van Hise in 1904. Obsolete. Also spelled: *catamorphism.* Cf: *anamorphism.*

kataphorite (kat-a-phor′-ite) *katophorite.*

kataseism (kat′-a-seism) Earth movement toward the focus of an earthquake. Cf: *anaseism.*

katatectic layer (kat-a-tec′-tic) A sedimentary layer that is "built downward"; specif. a distinct, generally horizontal or slightly dipping layer of solution residue (gypsum and/or anhydrite) formed by intermittent compaction of sulfate accumulating on top (in the cap rock) of a salt stock by solution of salt. The term was introduced by Goldman (1933, p.84). Etymol: Greek "kata", "down", + "tekton", "builder".

katatectic surface A surface separating two katatectic layers (Goldman, 1952, p. v).

Katathermal (kat-a-ther′-mal) n. The interval of gradually cooling climates since the *Climatic Optimum.* See also: *Little Ice Age.* adj. Pertaining to such a time interval.

katathermal [paleoclim] *Hypothermal.*

katayamalite (ka-ta-ya′-ma-lite) A vitreous white monoclinic mineral: $(K,Na)Li_3Ca_7(Ti,Fe^{3+},Mn)_2(OH,F)_2(SiO_3)_{12}$. It is a cyclosilicate with similarities to baratovite.

katazone (kat′-a-zone) According to Grubenmann's classification of metamorphic rocks (1904), the lowermost depth zone of metamorphism, which is characterized by high temperatures (500°-700°C), mostly strong hydrostatic pressure, and low or no shearing stress. Long-continued reconstitution and recrystallization, often without deformation, and deep-seated metamorphism associated with igneous action, produce such rocks as high-grade schists and gneisses, granulites, eclogites, and amphibolites. The concept was modified by Grubenmann and Niggli (1924) to include effects of high-temperature contact metamorphism and metasomatism. Modern usage stresses temperature-pressure conditions (highest metamorphic grade) rather than the likely depth of zone. Cf: *mesozone; epizone.* Also spelled: *catazone.* See also: *katamorphism.*

Katmaian-type eruption (Kat-mai′-an-type) The violently explosive ejection of huge amounts of pumice and ash, followed by an *ash flow* and extensive fumarole activity (Macdonald, 1972, p.241). Type area: vicinity of Mount Katmai, Alaska, including Valley of Ten Thousand Smokes. Cf: *Peleean-type eruption; Vulcanian-type eruption.*

katogene (kat′-o-gene) Pertaining to the breaking-down of a rock by atmospheric or other agents, or to shallow-depth replacement. Obsolete.

katoite (ka′-to-ite) A colorless cubic member of the katoite-hibschite-grossular solid-solution series, with end-member composition $Ca_3Al_2(OH)_{12}$ and generally less than 50% of the grossular component.

katophorite A black, greenish black or bluish black monoclinic member of the *amphibole* group: $Na_2Ca[Fe^{2+}{}_4(Al,Fe^{3+})](Si_7Al)O_{22}(OH)_2$.

katoptrite (kat-opt′-rite) A metallic brilliant black monoclinic mineral with perfect cleavage: $(Mn^{2+},Mg)_{13}(Al,Fe^{3+})_4Sb^{5+}{}_2Si_2O_{28}$. Also spelled: *catoptrite.*

katungite (ka-tung′-ite) An extrusive rock composed chiefly of melilite, with a smaller amount of olivine and magnetite and minor leucite, kalsilite, nepheline, and perovskite; a pyroxene-free melilitite. Named by Holmes in 1937 for Katunga, Uganda. Obsolete. Woolley et al. (1996) replace this name with kalsilite-leucite-olivine melilitite.

katzenbuckelite (katz-en-buck′-el-ite) A hypabyssal rock with tinguaitic texture and composed of phenocrysts of nepheline, biotite, olivine, nosean, leucite, and apatite in a fine-grained groundmass of nepheline, leucite, and aegirine. Named by Osann in 1903 for Katzenbuckel, Germany. Obsolete.

kauaiite (kau-ai′-ite) An orthoclase-bearing olivine-augite *diorite* in which the feldspar is zoned, with calcic labradorite in the inner zones grading outward into alkali feldspar. Its name, given by Iddings in 1913, is derived from the Hawaiian island of Kauai. Obsolete.

kauri (kau′-ri) A light-colored, whitish-yellow, or brown *copal,* usually found as a fossil resin from the kauri pine (a tree of the genus *Agathis*), esp. from *Agathis australis,* a tall timber tree of New Zealand. Etymol: Maori "kawri". Also called: kauri resin; kauri gum; kauri copal; agathocopalite.

kavir (ka-vir′) (a) A term used in Iran for a *salt desert,* specif. the Great Kavir of inner Iran, a series of closed basins noted for marshy conditions and high salinities. (b) A *playa* on a kavir. Also spelled: kevir; kewire.

kawazulite (ka-wa-zu′-lite) A metallic white rhombohedral mineral of the tetradymite group: Bi_2Te_2Se.

kaxtorpite (kax′-torp-ite) A *nepheline syenite* containing pectolite, eckermannite, and sodic clinopyroxene. The name, given by Adamson in 1944, is for Kaxtorp, Norra Kärr complex, Sweden. Obsolete.

kay A variant of *key [coast]* and *cay.*

kazakhstanite (ka-zakh′-stan-ite) An adamantine or dull black monoclinic mineral: $Fe^{3+}{}_5V^{4+}{}_3V^{5+}{}_{12}O_{39}(OH)_9{\cdot}9H_2O$.

kazakovite (ka′-za-kov′-ite) A pale yellow rhombohedral mineral: $Na_6Mn^{2+}TiSi_6O_{18}$.

Kazanian (Ka-zan′-i-an) A Russian stage: Middle Permian (above Kungurian, below Tatarian).

KB *kelly bushing.*

K-bentonite (K-ben′-ton-ite) A potassium-bearing clay-rich rock formed by the alteration of volcanic ash and consisting dominantly of randomly or regularly interstratified layers of illite and smectite, with illite generally consistituting between 60% and 95% of the mixture (Weaver, 1953). Syn: *potassium bentonite.* Cf: *metabentonite; potash bentonite.*

KBO Abbrev. for Kuiper Belt Object, one of the small bodies that populate the *Kuiper Belt.*

keatite (keat′-ite) A tetragonal polymorph of SiO_2, synthesized hydrothermally at high pressures.

keckite (keck′-ite) A yellow-brown monoclinic mineral of the *whiteite* group: $Ca(Mn^{2+},Zn)_2Fe^{3+}{}_3(PO_4)_4(OH)_3{\cdot}2H_2O$.

kedabekite (ke-da-bek′-ite) A rock, possibly of igneous origin, composed of bytownite, calcium-iron garnet, and hedenbergite. The name, given by Fedorov in 1901, is for Kedabek, Caucasus. Obsolete.

keel (a) Ridgelike longitudinal structure on *under side* of planate pectiniform conodont elements; main keel is situated beneath blade-carina; secondary keels extend from basal pit beneath lateral processes. Term has also been used to describe longitudinal ridges or flangelike projections along anterior or posterior margins of coniform elements (TIP, 1981, pt. W, supp.2). (b) A continuous sharp ridge along the venter of a coiled nautiloid or ammonoid conch. (c) A *carina* of a bryozoan; also that part of a distal zooecial wall in the endozone of a stenolaemate that is between sinuses, and proximally is the recumbent basal zooecial wall and distally is the flattened median part of the vertical zooecial wall. (d) A vertical sail-like plate in a radiolarian; a keel-like ridge along the outer margin of the test of a foraminifer. (e) A canal or cleft in the valve of some pennate diatoms. (f) A spiral ridge on a gastropod shell. (g) The *carina* of the *sternum* of most birds.

keel angle In conodont morphology, angle opening anteriorly between main keel and secondary keel (TIP, 1981, pt. W, supp.2).

keeleyite (kee′-ley-ite) *zinkenite.*

Keewatin (Kee-wa′-tin) A provincial division of the *Precambrian* in the Lake Superior area. Also called: *Keewatinian.*

Keewatinian (Kee-wa-tin′-i-an) *Keewatin.*

kegelite (ke′-gel-ite) A colorless monoclinic pseudohexagonal mineral: $Pb_4Al_2SiO_{10}(SO_4)(CO_3)_2(OH)_4$.

kegelkarst (ke′-gel [Pron: keh-gul]) A general term used to describe several types of tropical humid karst characterized by numerous, closely spaced conical or tower-shaped hills having intervening *closed depressions* and narrow steep-walled karst valleys. Cf: *cockpit karst; halbkugelkarst; tower karst.* Etymol: German "Kegel", "cone".

kehoeite (ke′-hoe-ite) A discredited mixture of phosphate minerals, originally described as a hydrated zinc aluminum phosphate.

keilhauite (keil′-hau-ite) A radioactive variety of sphene containing aluminum, iron, and yttrium and other rare earths.

keilite A metallic cubic meteorite mineral of the *galena* group: (Fe,Mg)S.

keithconnite (keith-con′-nite) A metallic rhombohedral mineral: $Pd_{20}Te_7$.

keiviite-(Y) (kei′-vi-ite) A colorless monoclinic mineral of the *thortveitite* group: $Y_2Si_2O_7$.

keiviite-(Yb) A colorless monoclinic mineral of the *thortveitite* group: $Yb_2Si_2O_7$.

keldyshite (kel′-dysh-ite) A white triclinic mineral: $Na_2ZrSi_2O_7$.

kellerite (kel′-ler-ite) A cuprian variety of pentahydrite.

Kellerwasser event A distinctive horizon of black shales and limestones marking the Frasnian/Famennian boundary, particularly in Europe, and the peak of the Late Devonian biodiversity crisis.

kelly (kel′-ly) A steel pipe of square or hexagonal cross section, 10-50 ft (3-15 m) long, forming the top section of the rotary *drillstring.* It is fitted into and passes through the *rotary table* and is turned by it during drilling, thereby transmitting the rotary motion of the table to the drill pipe.

kelly bushing The journal-box insert on the *rotary table* of a rotary drilling rig through which the *kelly* passes. Its upper surface is commonly used as the zero-depth reference for well-log and other downhole measurements in a wellbore. Abbrev: KB.

kellyite (kel′-ly-ite) A yellow mineral of the *serpentine* group: $(Mn^{2+},Mg,Al)_3(Si,Al)_2O_5(OH)_4$. It occurs in hexagonal and rhombohedral polymorphs, and is the Mn^{2+}-analogue of *amesite,* with Mn in place of octahedral Mg.

Kelvin wave [lake] A wave component that develops in a lake because of the effect of Coriolis acceleration on a *seiche.* It rotates around the lake at right angles to the shore, clockwise in the northern hemisphere, counterclockwise in the southern. At the air-water interface, the Kelvin component of lake waviness is easily detected only in very large lakes, but at a thermocline or chemocline, where the density difference between upper and lower layers is smaller, the Kelvin wave can be significant, even in lakes of moderate size (Wetzel, 2001, p.115).

Kelvin wave [oceanog] A surface gravity wave with a period greater than a few hours, such that the wave is influenced by the *Coriolis force* and a boundary such as a coast. Each tidal constituent is a Kelvin wave. The direction of propagation is such that the wave moves poleward along west coasts and equatorward along east coasts.

kelyanite (kel′-yan-ite) A reddish-brown monoclinic mineral: $Hg_{36}Sb_3O_{28}(Cl,Br)_9$.

kelyphitic A variety of corona texture involving radial-fibrous concentric shells of one or two minerals on grains of another mineral. Core grains are commonly garnet or olivine and rims commonly amphibole and/or pyroxene.

kelyphitic rim (a) In some igneous rocks, a peripheral zone of pyroxene or amphibole developed around olivine where it would otherwise be in contact with plagioclase, or around garnet where it would otherwise be in contact with olivine or other magnesium-rich minerals. Cf: *reaction rim; corona [petrology].* (b) A secondary *reaction rim.* (c) *reaction rim.* Syn: kelyphytic border.

kemmlitzite (kemm′-litz-ite) A colorless to light grayish-brown rhombohedral mineral of the *beudantite* group: $(Sr,Ce)Al_3(AsO_4)(SO_4)(OH)_6$.

kempite (kemp′-ite) An emerald-green orthorhombic mineral: $Mn^{2+}{}_2Cl(OH)_3$. It is isotypic to atacamite.

kenhsuite A vitreous yellow orthorhombic mineral: $Hg_3S_2Cl_2$. It is photosensitive and blackens on exposure to light.

kennedyite (ken′-ne-dy-ite′) *armalcolite.*

Kennedy's critical velocity (Ken′-ne-dy′s) *critical velocity (d).*

kennel coal *cannel coal.*

Kennicott-type copper Carbonate-hosted copper deposits that form in dissolution cavities above *volcanic redbed copper* occurrences in greenstone. Named for the Kennicott copper deposits in Alaska.

Kenoran Orogeny (Ke-nor′-an) A name proposed by Stockwell (1964) for a time of plutonism, metamorphism, and deformation during the Precambrian of the Canadian Shield (especially in the Superior and Slave provinces), dated radiometrically at 2,390-2,600 Ma, or near the end of the Archean. It is synonymous with *Algoman Orogeny* of Minnesota.

kenozooid (ke-no-zo′-oid) A bryozoan *heterozooid* without a polypide and usually without either orifice or muscles.

kentallenite (ken-tal′-len-ite) A dark-colored *monzonite* composed of approximately equal amounts of augite, olivine, potassium feldspar, and plagioclase, with smaller amounts of biotite, apatite, and opaque oxides; an olivine-bearing monzonite. Its name is derived from Kentallen, Argyllshire, Scotland. Not recommended usage.

kentbrooksite A yellow-brown rhombohedral mineral of the *eudialyte* group: $(Na,REE)_{15}(Ca,REE)_6Mn_3Zr_3NbSi_{25}O_{73}(O,OH,H_2O)_3(F,Cl)_2$.

kentrolite (ken′-tro-lite) A submetallic dark reddish-brown to black orthorhombic mineral: $Pb_2Mn^{3+}{}_2O_2Si_2O_7$.

kentsmithite (kent-smith′-ite) A local name used in the Paradox Valley, Colorado, for a black vanadium-bearing sandstone.

kenyaite (ken′-ya-ite) A white friable monoclinic mineral: $Na_2Si_{22}O_{41}(OH)_8{\cdot}6H_2O$.

kenyte (ke′-nyte) An olivine-bearing *phonolite* composed of anorthoclase, nepheline, aegirine-augite, sodic amphibole, olivine, apatite, and opaque oxides. The groundmass may have a trachytic or hyalopilitic texture. Its name, given by Gregory in 1900, is derived from Mount Kenya, Kenya. Obsolete.

Kepler's laws of planetary motion (Kep′-ler′s) The statements that each planet moves in an elliptical orbit with the Sun at one focus of the ellipse; that the line from the Sun to any planet sweeps out equal areas of space in equal intervals of time; and that the squares of the sidereal periods of the several planets are proportional to the cubes of their mean distances from the Sun. Although Kepler's laws are a mathematical consequence of Newton's laws, which are more fundamental, they preceded Newton's laws, were empirically based on the observations of Tycho Brahe, and were a significant extension of Copernican philosophy.

kerabitumen (ker′-a-bi-tu′-men) *kerogen.*

keralite (ker′-a-lite) A quartz-biotite *hornfels.* The term was originated by Cordier in 1868. Rarely used.

kerargyrite (ker-ar′-gyr-ite) *cerargyrite.*

keratophyre (ker-a′-to-phyre) A name originally applied by Gümbel (1874, p.43) to trachytic rocks containing highly sodic feldspars, but now more generally applied to all salic extrusive and hypabyssal rocks characterized by the presence of albite or albite-oligoclase and chlorite, epidote, and calcite, all of metamorphic origin. Originally the term was restricted to lavas of pre-Tertiary age but this distinction is not recognized in current usage. Some varieties of keratophyre contain sodic orthoclase and sodic amphiboles and pyroxenes. Keratophyres commonly are associated with spilitic rocks and interbedded with marine sediments.

keratose (ker′-a-tose) Said of a horny sponge in which the skeleton consists entirely of organic fibers without spicules (although it sometimes may contain foreign particles including spicules of other sponges).

keriotheca (ke′-ri-o-the′-ca) The relatively thick shell layer with honeycomblike structure in the wall of some fusulinids (such as schwagerinids), occurring next below the tectum, and forming part of the spirotheca. It may be divisible into *lower keriotheca* and *upper keriotheca.*

kermesite (ker′-mes-ite) A cherry-red triclinic mineral: Sb_2S_2O. It usually occurs as tufts of capillary crystals resulting from the alteration of stibnite. Syn: *antimony blende; red antimony; purple blende; pyrostibite.*

kernbut (kern′-but) A projecting ridge or buttress created by displacement on a fault traversing a hillslope and separated from the hill by a *kerncol;* the outer ridgelike edge of a fault terrace or fault bench. The term was introduced by Lawson (1904, p.332) for a primary feature occurring in Kern Canyon, Calif., but the type locality has since been shown to be one of fault-line forms (Webb, 1936). Etymol: Kern Canyon + buttress.

kerncol (kern′-col) A low sag or trough separating a *kernbut* from the hillside, occurring where a faulted block joins the hill. The term was introduced by Lawson (1904, p.332) for a primary feature occurring in Kern Canyon, Calif., but the type locality has since been shown to be one of fault-line forms (Webb, 1936). Etymol: Kern Canyon + *col.*

kernite (kern′-ite) A colorless to white monoclinic mineral: $Na_2B_4O_6(OH)_2{\cdot}3H_2O$. It is an ore of boron. Syn: *rasorite.*

kerogen (ker′-o-gen) Fossilized insoluble organic material found in sedimentary rocks, usually shales, which can be converted to petroleum products by distillation. Syn: *kerabitumen.*

kerogenite (ke-rog′-e-nite) *oil shale.*

kerogen shale *oil shale.*

kerolite (ker′-o-lite) A disordered variety of talc, regarded by some as a mixture of serpentine and saponite. Syn: *cerolite.*

keronigritite (ker-o-nig′-ri-tite) A type of *nigritite* that is derived from kerogen. Cf: *polynigritite; humonigritite; exinonigritite.*

kerosene shale (ker′-o-sene) *kerosine shale.*

kerosine shale (ker′-o-sine) (a) A syn. of *torbanite.* (b) Any bituminous oil shale. Also spelled: *kerosene shale.*

kersantite (ker-san′-tite) (a) In the *IUGS classification,* a *lamprophyre* in which the modal proportion of plagioclase exceeds that of alkali feldspar, feldspathoids are lacking, and biotite forms phenocrysts. (b) A *lamprophyre* containing biotite and plagioclase (usually oligoclase or andesine), with or without clinopyroxene and olivine. Defined by Delesse in 1851, named for the village of Kersanton, France.

kerstenite (ker′-sten-ite) A yellow orthorhombic mineral of the *barite* group: $PbSeO_4$. It is inadequately characterized.

kertschenite (kertsch′-e-nite) An amorphous oxidation product of vivianite.

kerzinite (kerz′-i-nite) In the Urals, lignite that is impregnated with hydrated nickel silicate and thus is mined for nickel.

kess-kess An Arabic term used in Morocco for a fossil reef knoll isolated by erosion. Cf: *klint [reef].*

kësterite (kes′-ter-ite) A metallic greenish-black tetragonal mineral: $Cu_2(Zn,Fe)SnS_4$. It is the zinc analogue of ferrokesterite. Also spelled: kesterite. Syn: *isostannite.*

kettle [glac geol] (ket′-tle) A steep-sided, usually basin- or bowl-shaped hole or depression, commonly without surface drainage, in glacial-drift deposits (esp. outwash and kame fields), often containing a lake or swamp; formed by the melting of a large, detached block of stagnant ice (left behind by a retreating glacier) that had been wholly or partly buried in the glacial drift. Kettles range in depth from about a meter to tens of meters, and in diameter to as much as 13 km. Thoreau's Walden Pond is an example. Cf: *pothole [glac geol].* Syn: *kettle hole; kettle basin; potash kettle.*

kettle [streams] *pothole [streams].*

kettleback (ket′-tle-back) *horseback [coal].*

kettle basin *kettle [glac geol].*

kettle bottom *caldron bottom.*

kettle drift *Drift* within and adjacent to a *kettle [glac geol].*

kettle hole *kettle [glac geol].*

kettle lake A body of water occupying a *kettle [glac geol],* as in a *pitted outwash plain* or in a *kettle moraine.* Cf: *cave-in lake.* Syn: *pit lake.*

kettle moraine A *moraine* whose surface is marked by many kettles; e.g., the Kettle Moraine district of SE Wisconsin.

kettle plain A *pitted outwash plain* marked by many kettles.

kettnerite (kett′-ner-ite) A brown to yellow orthorhombic (pseudotetragonal) mineral: $CaBiO(CO_3)F$.

Keuper (Keu′-per) A term used (esp. in Germany) to what is now approx. the Upper Triassic (above Muschelkalk, below Jurassic). Essentially a lithostratigraphic term.

kevir (ke-vir′) *kavir.*

Keweenawan (Ke-wee-naw′-an) A provincial series of the *Precambrian* (middle *Proterozoic*) in the Lake Superior area.

kewire (ke-wire′) *kavir.*

key [cart] A *legend* on a map.

key [coast] A *cay,* esp. one of the coral islets or barrier islands off the southern coast of Florida. Not to be confused with *quay.* See also: *sand key.* Syn: *kay.*

key [taxon] An artificial analytic device or arrangement for use in identification of plants or animal forms whereby a choice is provided between two contradictory propositions resulting in the acceptance of one and rejection of the other (Lawrence, 1951, p.225).

key bed (a) A well-defined, easily identifiable stratum or body of strata that has sufficiently distinctive characteristics (such as lithology or fossil content) to facilitate correlation in field mapping or subsurface work. (b) A bed the top or bottom of which is used as a datum in making structure-contour maps. Syn: *key horizon; index bed; marker horizon [stratig].*

keyed *sutured.*

key fossil *index fossil.*

keyhole passage In a cave, a *passage* with a keyhole-shaped cross section; initially a *tube,* then experiencing downcutting that formed a *canyon [speleo].* It is thought to reflect the transition from initial pipe-full flow to vadose conditions.

key horizon (a) The top or bottom of an easily recognized, extensive bed or formation that is so distinctive as to be of great value in stratigraphy and structural geology; e.g. a *datum horizon.* (b) A term that is used interchangeably with *key bed.*

keyite (key′-ite) A blue monoclinic mineral: $Cu_3(Zn,Cu)_4Cd_2(AsO_4)_6{\cdot}2H_2O$.

keystone graben A graben developed on the crest on an anticline as a consequence of the outer-arc extension that accompanied bending of beds to form the anticline.

keystoneite (key′-stone-ite) An adamantine yellow hexagonal mineral of the *zemannite* group: $Mg_{0.5}NiFe^{3+}(Te^{4+}O_3)_3{\cdot}4.5H_2O$.

keystone vug A millimetric-size cavity occurring in carbonate grainstones deposited in a beach or foreshore environment; formed by the trapping of air within the sand.

K-feldspar (K-feld′-spar) *potassium feldspar.*

K-feldspar-cordierite-hornfels facies (K-feld′-spar-cor′-di-er-ite-horn-fels) A name given to the *pyroxene-hornfels facies* (Winkler, 1967) in order to indicate a more diagnostic mineral assemblage that would distinguish it from the hornblende-hornfels facies. Characteristic orthopyroxenes form in the higher-temperature parts of this facies (in excess of about 700°C).

khadar (kha′-dar) A term used in India for a low-lying area, e.g. an alluvial plain, that is liable to be flooded by the waters of a river. Etymol: Urdu-Hindi. Cf: *bhangar.*

khademite (kha′-de-mite) A colorless orthorhombic mineral: $Al(SO_4)F{\cdot}5H_2O$. It is the F analogue of rostite.

khagiarite (kha′-gi-a-rite) A black *pantellerite* characterized by a glassy microlitic groundmass exhibiting flow texture. Defined by Washington in 1913. Obsolete.

khaidarkanite A vitreous blue monoclinic mineral: $Cu_4Al_3(OH)_{14}F_3{\cdot}2H_2O$.

khal (a) A term used in East Pakistan for a narrow stream channel. (b) A sluggish creek on the lower delta of the Ganges. Etymol: Bengali.

khamrabaevite (kham-ra-ba′-ev-ite) A metallic gray cubic mineral of the *tantalcarbide* group occurring with native iron and magnetite: (Ti,V,Fe)C.

khanneshite (khan-nesh′-ite) A pale yellow hexagonal mineral of the *burbankite* group: $(Na,Ca)_3(Ba,Sr,Ce,Ca)_3(CO_3)_5$.

Khaprovian (Khap-ro′-vi-an [Hap-ro′-vi-an]) Eastern European mammal age ("complex"): late Pliocene (after Moldavian, before Odessan); equiv. Villanyian.

kharaelakhite (kha-rae′-lakh-ite) A metallic grayish orthorhombic mineral of the *pentlandite* group: $(Pt,Cu,Pb,Fe,Ni)_9S_8$.

kharafish (kha′-ra-fish) A limestone plateau in the Libyan desert, formed by wind erosion.

khari (kha′-ri) A term used in East Pakistan for a small, deep stream of local origin. Etymol: Bengali.

khatyrkite (kha′-tyrk-ite) A metallic gray-yellow mineral: $(Cu,Zn)Al_2$.

kheneg (khe′-neg) A term used in the Atlas Mountains of northern Africa for a canyon. Etymol: Arabic.

Khersonian (Kher-son′-i-an [Hair-son′-i-an]) Ponto-Caspian substage: middle upper Miocene, upper Sarmatian (above Bessarabian substage, below Maeotian).

khibinite [mineral] (khib′-i-nite) *mosandrite.*

khibinskite (khi-bin′-skite) A pale yellow monoclinic, pseudotrigonal mineral: $K_2ZrSi_2O_7$.

khinite (khin′-ite) A green orthorhombic mineral: $PbCu_3Te^{6+}O_4(OH)_6$.

khlopinite (khlo′-pi-nite) A tantalian variety of samarskite.

khmaralite A dark green monoclinic mineral: $(Al,Mg,Fe^{2+},Fe^{3+})_4(Al,Si,Be)_3O_{10}$.

khoharite (kho-har′-ite) A hypothetical end member of the *garnet* group: $Mg_3Fe_2(SiO_4)_3$.

khomyakovite An orange to orange-red rhombohedral mineral of the *eudialyte* group: $Na_{12}Ca_6Sr_3Fe_3WZr_3(Si_{25}O_{73})(O,OH,H_2O)_3(Cl,OH)_2$.

khondalite (khon′-dal-ite) A group of metamorphosed aluminous sediments consisting of garnet-quartz-sillimanite rocks with garnetiferous quartzites, graphite schists, and marbles (Walker, 1902, p.11). Rarely used. Named after Khonds and the Khondalite series, India.

khor (a) A term used in Sudan for an intermittent stream. (b) A term used in northern Africa for a watercourse or ravine, esp. one that is dry. Etymol: Arabic "khawr", "wadi, dry wash".

khristovite-(Ce) A dark grayish-brown monoclinic mineral of the *epidote* group: $(Ca,La)Ce(Mg,Fe)MnAl(Si_2O_7)(SiO_4)(OH)(F,O)$.

khud A term used in India for a ravine or precipice. Etymol: Hindi "khad".

khuniite (khun′-i-ite) *iranite.*

khurd A term used in Algeria for a pyramid-shaped sand dune, 80 to 150 m high, with curved slopes, formed by the intersection of seif dunes (Capot-Rey, 1945, p.393). Cf: *rhourd.* Syn: *guern.*

kick [drill] (a) A flow of fluid from a subsurface reservoir surrounding a wellbore during drilling, caused by the formation pressure in the well exceeding the pressure exerted by the drilling mud. (b) A quick snap of the drillstring caused by the core breaking in a blocked core barrel. (c) A small sidewise displacement in a borehole, caused by the deflection of the drill bit when entering a hard, dipping stratum underlying softer rock.

kick [seis] *arrival.*

kick off To start a drill a deviated hole out from a straight hole.

kick off point The point where a well deviates from a straight hole.

kickout (kick′-out) The lateral distance from a drilling site reached by a *directional well.*

kiddcreekite (kidd-creek′-ite) A metallic gray cubic mineral: Cu_6SnWS_8. It is the tungsten analogue of hemusite.

kidney ore (kid'-ney) A variety of hematite occurring in compact kidney-shaped masses, concretions, or nodules, together with clay, sand, calcite, or other impurities; concretionary ironstone. Also called: kidney iron ore.

kidneys A miner's term for a mineral zone that contracts, expands, and again contracts downwards.

kidney stone [mineral] *actinolite.*

kidney stone [sed] A pebble or nodule roughly resembling the shape of a kidney; e.g. a small hard red-coated ironstone nodule common in the Oxford Clay of England.

kidwellite (kid'-well-ite) A yellow to greenish-yellow monoclinic mineral: $NaFe^{3+}_9(PO_4)_6(OH)_{11}\cdot 6H_2O$.

kieftite A metallic cubic mineral of the *skutterudite* group: $CoSb_3$.

kies A general term for the sulfide ores. Etymol: German *Kies,* "finer gravel".

kieselguhr (kie'-sel-guhr [Pron: keezelgoor]) *diatomite; guhr.* Etymol: German, "Kieselgu(h)r".

kieserite (kie'-ser-ite) (a) A white, grayish-white, or yellowish monoclinic mineral: $MgSO_4\cdot H_2O$. It occurs in saline residues. (b) A group name for minerals of composition $ASO_4\cdot H_2O$, where A = Mg, Fe^{2+}, Mn, Ni, Cu or Zn.

Kikuchi lines (Ki-ku'-chi) Pairs of parallel lines that appear in transmission-electron-microscope diffraction patterns and are formed as a result of coherent scattering of inelastically scattered electrons in the upper parts of the sample foil.

kilchoanite (kil-cho'-a-nite) A colorless orthorhombic mineral: $Ca_3Si_2O_7$. It is dimorphous with rankinite.

kilkenny coal (kil-ken'-ny) *anthracite.*

kill A creek, channel, stream, or river. The term is used chiefly in place names in Delaware and New York State; e.g. Peekskill, N.Y. Etymol: Dutch "kil".

killalaite (kil-la'-la-ite) A colorless monoclinic mineral: $Ca_3Si_2O_7\cdot H_2O$.

Killarney Revolution (Kil-lar'-ney) A name proposed by Schuchert (1924) for a supposed major orogeny at the end of Precambrian time in North America; based on the Killarney Granite north of Lake Huron in Ontario, supposed to be of post-Keweenawan age. Radiometric data now indicate that the Keweenawan is 1,100 m.y. old, and that the Killarney Granite is older, and probably equivalent to the Penokean Granite (see *Penokean Orogeny*) of Michigan and Minnesota. Actually, no notable tectonic events are now known to have occurred in this part of North America at the end of the Precambrian. The term Killarney Revolution is obsolete, and should be abandoned.

killas (kil'-las) A name used in Devon and Cornwall for any rock that has been metamorphosed by contact with granite.

kimberlite (kim'-ber-lite) An ultramafic igneous rock containing at least 35% olivine, with one or more of the following in the groundmass; monticellite, phlogopite, carbonate, serpentine, diopside. No leucite is allowed in the definition of kimberlite (Woolley et al., 1996). The name, proposed by Lewis in 1888, is for the Kimberley district, South Africa, where kimberlite is a host for diamonds. Two types of kimberlite have long been distinguished. *Basaltic kimberlite* or Type 1 kimberlite is more widespread, and constitutes the classic diatreme-filling diamond-bearing rocks of South Africa. Texturally, most Type 1 kimberlites are serpentinite microbreccias that have xenolithic fragments. Micaceous kimberlite, also called lamprophyric kimberlite, orangeite, or Type 2 kimberlite, is apparently restricted to southern Africa, and differs so strongly from Type 1 that some authorities do not consider it a variety of kimberlite. Cf: *orangeite.*

Kimmerian (Kim-me'-ri-an) *Cimmerian.*

Kimmerian Orogeny *Cimmerian Orogeny.*

Kimmeridgian (Kim-me-ridg'-i-an) A stage of the Standard Global Chronostratigraphic Scale: Upper Jurassic (above Oxfordian, below Tithonian). The spelling "Kimeridgian" was used by Arkell (1956, p.20) on the basis that the type locality, the village of Kimmeridge in the Isle of Purbeck, southern England, was spelled with one "m" until 1892.

kimrobinsonite (kim-ro'-bin-son-ite) A white cubic mineral: $(Ta,Nb)(OH)_3(O,CO_3)$.

kimuraite-(Y) (ki-mur'-a-ite-y) A silky light purplish orthorhombic mineral: $CaY_2(CO_3)_4\cdot 6H_2O$.

kimzeyite (kim'-zey-ite) A brown cubic mineral of the *garnet* group: $Ca_3(Zr,Ti)_2(Si,Al,Fe^{3+})_3O_{12}$.

kin A headland. Also, a term used in Ireland for the highest point of anything. Etymol: Gaelic.

kindchen (kind'-chen) A nodule or concretion that resembles the head of a child; specif. *loess kindchen.* Etymol: German "Kindchen", "little child, infant, baby".

Kinderhookian (Kin-der-hook'-i-an) North American series: lowermost Mississippian (above Conewangoan of Devonian, below Osagean).

K index A measure of the average intensity of a magnetic disturbance in time, such as *magnetic storm..* K index value varies from zero to nine, indicating a range of magnetic intensity in a three-hour interval, after subtraction of normal *daily variation.*

kindly (kind'-ly) *likely.*

kindred (kin'-dred) *rock association.*

kinematically admissible cross section A restorable cross section for which there is a sequence of intermediate stages between the undeformed and deformed states, each related by a kinematically reasonable deformation increment.

kinematic analysis The part of structural analysis that deals with the movements that occurred during deformation. These movements include rigid-body translation and rotation, and non-rigid movements of distortion and dilation.

kinematic coherence The existence of synchronous slip rates and slip distributions that are arranged such that *geometric coherence* is maintained (Walsh and Watterson, 1991).

kinematic indicator A feature in a deformed rock body that indicates the shear sense or movement direction during the deformation. Kinematic indicators include asymmetric folds, asymmetric porphyroclast systems, S-C fabrics, and asymmetric inclusion trails. Syn: *shear-sense indicator*.

kinematic similarity The condition in which the scaled physical model and the natural prototype are geometrically similar throughout the deformation history (Hubbert, 1937). Specifically, the time of deformational events (e.g., the onset of faulting) in the model must be proportional to the timing of those same events in the prototype.

kinematic viscosity (kin-e-mat'-ic) The ratio of the viscosity coefficient (in poises) to density at room temperature (in q/cu cm). See also: *eddy viscosity.* Syn: *coefficient of kinematic viscosity.*

kinematic wave A zone of increased mass flux, commonly expressed on a glacier surface as a very slight bulge that travels down the glacier with a velocity three to four times faster than the surface velocity of the glacier. Cf: *glacier wave.* Syn: *traveling wave.*

kinetic growth form (ki-net'-ic) The form of faceted ice crystals in a dry snow cover undergoing rapid recrystallization due to a high-temperature gradient (Colbeck, 1986). Syn: *faceted form.*

kinetic metamorphism A type of metamorphism that produces deformation of rocks without chemical reconstitution or recrystallization to form new minerals (Turner and Verhoogen, 1951, p.370). Obsolete. Cf: *cataclastic metamorphism.* Syn: *mechanical metamorphism.*

kingdom (king'-dom) (a) The level of classification of organisms below domain. Depending on the classification system used, the number of kingdoms recognized varies from five to seven, sometimes even more (ICZN, 1964; ICBN, 2000). (b) Any one of the three major divisions into which all natural objects are traditionally classified, viz. animal kingdom, plant kingdom, mineral kingdom.

kingeniform (kin-gen'-i-form) Said of the *loop* of an adult dallinid brachiopod (as in the subfamily Kingeninae) in which the tendency to retain the *campagiform* hood during loop development leads to a broad sheetlike transverse band with connecting bands leading to the septum "in addition to normal ones joining descending branches with septum" (TIP, 1965, pt. H, p.147).

kingite (king'-ite) A white triclinic mineral: $Al_3(PO_4)_2(OH,F)_2\cdot 8H_2O$.

kingle (king'-le) A Scottish term for a very hard rock, esp. a siliceous or calcareous sandstone occurring in oil shales but without bituminous matter.

kingsmountite (kings-mount'-ite) A white to light brown monoclinic mineral, of the *montgomeryite* group: $(Ca,Mn^{2+})_4(Fe^{2+},Mn^{2+})Al_4(PO_4)_6(OH)_4\cdot 12H_2O$.

kinichilite (ki-nich'-i-lite) A dark brown hexagonal mineral that of the *zemannite* group: $Mg_{0.5}(Fe^{2+},Zn,Mn^{2+})_{1.7}(Te^{4+}O_3)_3\cdot 3.2H_2O$.

kink [paleont] A sharp lateral bend in the carina of some platformed pectiniform conodont elements (TIP, 1981, pt. W, supp.2).

kink [struc geol] In joint propagation, the tip of a crack that tilts abruptly out of plane because of a change in orientation of the local stress field. This abrupt tilt is thought to be a consequence of reinitiation of joint propagation following the rotation of the remote stress field after the arrest of the parent crack (Cruikshank et al., 1991). Cf: *tail crack.*

kink band A type of *deformation band* occurring microscopically in crystals and megascopically in foliated and thin-bedded rocks, in which the orientation of the lattice or of the foliation is changed or deflected by gliding or slippage. Kink bands are associated with shock-wave action as well as with normal deformation. They commonly occur as conjugate systems. Cf: *conjugate fold system.* Syn: *knick band; knick zone.*

kink fold A fold with planar limbs and a sharp angular hinge. Syn: *concertina fold; accordion fold; angular fold.*

kinoite (kin'-o-ite) A blue monoclinic mineral: $Ca_2Cu_2Si_3O_8(OH)_4$.

kinoshitalite (ki-nosh'-i-ta-lite') A yellow-brown platy monoclinic trioctahedral brittle mica, of ideal composition $BaMg_3(Si_2Al_2)O_{10}(OH)_2$.

kinradite (kin'-rad-ite) A name used in California and Oregon for jasper containing spherical inclusions of colorless or nearly colorless quartz.

kintoreite A cream to brownish-yellow or yellowish-green rhombohedral member of the *crandallite* group: $PbFe^{3+}_3(PO_4)_2(OH,H_2O)_6$.

kinzigite (kin'-zig-ite) A coarse-grained metamorphic rock of pelitic composition occurring in the granulite facies. Essential minerals are garnet and biotite, with which occur varying amounts of quartz, K-feldspar, oligoclase, muscovite, cordierite, and sillimanite. The term was originated in 1860 by Fischer, who named it after the Kinzig River, Black Forest, Germany.

kipuka (ki-pu'-ka) An area surrounded by a lava flow. Etymol: Hawaiian, "opening".

Kipushi copper-lead-zinc A *mineral-deposit type* characterized by massive base-metal sulfides and arseno-sulfosalts in dolomite breccias. Named for the Kipushi deposit in Zaire.

kipushite (ki-push'-ite) A green monoclinic mineral: $(Cu,Zn)_6(PO_4)_2(OH)_6{\cdot}H_2O$. It is the phosphate analogue of philipsburgite.
kirchheimerite (kirch-hei'-mer-ite) A pink tetragonal mineral of the *autunite* group: $Co(UO_2)_2(AsO_4)_2{\cdot}12H_2O$.
Kirchhoff migration Migration accomplished by integration along diffraction curves.
Kirkfeldian North American stage: Upper Ordovician, above Rocklandian, below Shermanian.
kirkiite (kir'-ki-ite) A metallic white hexagonal mineral: $Pb_{10}Bi_3As_3S_{19}$.
kirovite (ki'-rov-ite) A mineral: $(Fe,Mg)SO_4{\cdot}7H_2O$. It is a variety of melanterite containing magnesium.
kirschsteinite (kirsch'-stein-ite) A greenish orthorhombic mineral: $CaFe^{2+}SiO_4$. It is the Fe^{2+} analogue of monticellite. Syn: *iron-monticellite.*
Kiruna-type iron *Apatite iron* mineral deposit, a subtype of *iron oxide copper-gold deposits.* Named for the Kiruna iron deposits in Sweden.
kitaibelite (ki-ta-i'-bel-ite) An ill-defined gray metallic mineral: $Ag_{10}PbBi_{30}S_{51}$.
kitkaite (kit'-ka-ite) A metallic gray trigonal mineral of the *melonite* group: NiTeSe.
kittatinnyite (kitt-a-tin'-ny-ite) A yellow hexagonal mineral: $Ca_4Mn^{2+}{}_2Mn^{3+}{}_4Si_4O_{16}(OH)_8{\cdot}18H_2O$. It is isostructural with wallkilldellite.
kittlite (kitt'-lite) An ill-defined metallic gray cubic mineral: (Hg,Ag)(S,Se) (?).
kivite (kiv'-ite) A dark-colored olivine-bearing leucite basanite. Its name, given by Lacroix in 1923, is derived from Lake Kivu in east-central Africa. Obsolete.
kjelsasite (kjel'-sas-ite) A *monzodiorite* similar to *larvikite* but with more calcium and less alkalis. Brögger in 1933 derived the name from Kjelsås, Oslo district, Norway. Obsolete.
kladnoite (klad'-no-ite) A mineral corresponding to phthalimide: $C_6H_4(CO)_2NH$. It occurs as monoclinic crystals formed in burning waste heaps in the Kladno coal basin of Bohemia.
klapperstein (klap'-per-stein) A *rattle stone* that results from the weathering of a box-stone. Etymol: German *Klapperstein.*
Klauder wavelet The autocorrelation of a *vibroseis* sweep. In vibroseis processing, the Klauder wavelet is effectively the seismic source waveform. Named for John Rider Klauder (1932-), American mathematician.
klebelsbergite (kle'-bels-berg-ite') A pale yellow to orange-yellow orthorhombic mineral: $Sb^{3+}{}_4O_4(SO_4)(OH)_2$.
kleemanite (klee'-man-ite) A yellow to colorless monoclinic mineral: $ZnAl_2(PO_4)_2(OH)_2{\cdot}3H_2O$.
kleinite (klein'-ite) A yellow to orange hexagonal mineral: $Hg_2N(Cl,SO_4){\cdot}nH_2O$. Cf: *mosesite.*
Klein solution A solution of cadmium borotungstate that is used as a *heavy liquid;* its specific gravity is 3.6. Cf: *bromoform; Clerici solution; Sonstadt solution; methylene iodide.*
kliachite (kli'-ach-ite) A colorless ill-defined colloidal mineral: $Al_2O_3{\cdot}nH_2O$.
klimakotopedion (kli-ma'-ko-to-ped'-i-on) A term introduced by Schwarz (1912, p.95) as a syn. of *stepped plain.* Etymol: translation into Greek of "stepped plain".
klingstein (kling'-stein) An obsolete syn. of *phonolite.*
klint [coast] A term used in Denmark and Sweden for a vertical mountain wall or abrasion precipice, several meters high and 100 m or more long; esp. a steep cliff along the shore of the Baltic Sea. Pl: klintar. Cf: *glint.*
klint [reef] An exhumed fossil *bioherm* or coral reef, its surrounding rocks having been eroded, leaving the reef core standing in relief as a prominent knob, ridge, or hill. Pl: klintar. Cf: *tepee butte.* Not to be confused with clint.
klippe (klip'-pe) (a) An erosional remnant of a thrust sheet that is completely surrounded by exposure of the footwall. (b) An isolated rock unit that is an erosional remnant or outlier of a nappe. The original sense of the term was merely descriptive, i.e. included any isolated rock mass such as an erosional remnant. Plural: klippen. Etymol: German, "cliff, crag, reef". Syn: *thrust outlier.*
klockmannite (klock'-mann-ite) A metallic reddish-violet to slate-gray hexagonal mineral: CuSe. It tarnishes blue-black and is found in granular aggregates.
kloof A term used in South Africa for a deep rugged gorge, ravine, glen, or other short steep-sided valley, and also for a mountain pass. In some place names, the term may refer to a wide, open valley. Etymol: Afrikaans/Dutch.
kluftkarren (kluft'-kar-ren) A group or series of solution fissures.
klyuchevskite (kly-u-chevsk'-ite) A dark green to olive-green monoclinic mineral: $K_3Cu_3(Fe^{3+},Al)O_2(SO_4)_4$.
k-means A form of hierarchical *cluster analysis* in which the similarities between every object and *k* arbitrary points are calculated rather than the similarities between all pairs of objects. This reduces the magnitude of the clustering process from n^2 to $n{\cdot}k$, and because *k* is usually smaller than *n*, there is a great savings in computation time and memory requirements.
knap (a) A crest or summit of a hill. (b) A small hill or slight rise of ground.
kneaded (knead'-ed) (a) Said of a vague sedimentary structure resembling kneaded dough, such as a variety of flow roll or ball-and-pillow structure, or a structure formed by intrastratal slippage. (b) Said of a sediment or sedimentary particles transported by mudflows; e.g. "kneaded gravel".
knee fold A *zigzag fold* occurring in gravity-collapse structures. Rarely used.
knick (a) A *knickpoint;* esp. the place of junction where a gently inclined pediment and the adjacent mountain slope meet at a sharp angle. Syn: *knickpunkt.* (b) *nick.*
knick band *kink band.* Rarely used.
knickline (knick'-line) A line formed by the angle of a *knick* in a slope, esp. in a desert region where there is an abrupt transition from a pediment surface to the mountain slope.
knickpoint (knick'-point) Any interruption or break of slope; esp. a point of abrupt change or inflection in the longitudinal profile of a stream or of its valley, resulting from rejuvenation, glacial erosion, or the outcropping of a resistant bed. Etymol: German "Knickpunkt", "bend point". Syn: *knick; nick; nickpoint; knickpunkt; break [geomorph]; rejuvenation head; rock step.*
knickpunkt (knick'-punkt) (a) A *knickpoint* in a stream profile, esp. one resulting from rejuvenation or from an uplift. (b) The sharp angle made by the *haldenhang* and the *steilwand;* a *knick.* Pl. knickpunkte. Etymol: German "Knickpunkt", "bend point".
knick zone *kink band.* Rarely used.
knife edge (a) A narrow ridge of rock or sand. (b) *feather edge.*
knipovichite (knip'-o-vich-ite) A chromium-containing variety of *alumohydrocalcite.*
knitted texture (knit'-ted) A texture that is typical of the mineral *serpentine* in a rock when it replaces a clinopyroxene. Cf: *lattice texture [meta].*
knob (a) A rounded eminence, as a knoll, hillock, or small hill or mountain; esp. a prominent or isolated hill with steep sides, commonly found in the southern U.S. See also: *knobs.* (b) A peak or other projection from the top of a hill or mountain. Also, a boulder or group of boulders or an area of resistant rocks protruding from the side of a hill or mountain.
knob-and-basin topography (knob-and-ba'-sin) *knob-and-kettle topography.*
knob-and-kettle topography An undulating landscape in which a disordered assemblage of knolls, mounds, or ridges of glacial drift is interspersed with irregular depressions, pits, or kettles that are commonly undrained and may contain swamps or ponds. Cf: *hummocky moraine.* Syn: *knob-and-basin topography; kame-and-kettle topography.*
knob and trail A structure, found in glaciated areas, that is made up of a protruding mass of resistant rock (the "knob") and a ridge of softer rock (the "trail") extending from the lee side of the knob (Chamberlin, 1888, p.244-245). Cf: *crag and tail.*
knobs An area marked by a group of rounded, isolated hills (knobs).
knock A hill in the English Lake District or in Scotland. Etymol: Gaelic.
knocker A colloquial field term denoting a resistant, rounded monolith, a few feet to several hundred feet across, that stands out prominently above the level of the surrounding *mélange* terrain. It is a convenient term for both *tectonic blocks* and *exotic blocks* (Berkland et al., 1972, p.2296).
knoll [geomorph] (a) A small, low, rounded hill; a hillock or mound. (b) The rounded top of a hill or mountain. Syn: *knowe.*
knoll [marine geol] A mound-like relief form of the sea floor, less than 1000 m in height. Syn: *seaknoll.*
knoll reef *reef knoll.*
Knoop hardness test A test of resistance to deformation of metals or minerals in which a rhombus-shaped diamond is forced into a polished surface of the specimen to be tested under various static loads. The result is a function of the length of the long diagonal of the resulting indentation. Cf: *Vickers hardness test; Brinell hardness test.*
knopite (knop'-ite) A variety of perovskite containing cerium.
knorringite (knor'-ring-ite) A bluish-green cubic mineral of the *garnet* group: $Mg_3Cr_2(SiO_4)_3$. It forms a series with pyrope.
knot [geomorph] (a) A term used in the English Lake District for a hill of moderate height; esp. one having a bare-rock surface. (b) An elevated land area formed by the meeting of two or more mountainous regions; e.g. the structural-junction area of ridges of folded mountains.
knot [mining] A miner's term for small concretions, e.g. galena in sandstone, or for segregations of darker minerals in granite and gneiss.
knotenschiefer (knot'-en-schief'-er) A type of *spotted slate* characterized by conspicuous subspherical or polyhedral clots that are often individual minerals, e.g. cordierite, biotite, chlorite, andalusite (Holmes, 1928). Etymol: German. Cf: *garbenschiefer; fruchtschiefer; fleckschiefer.*
knotted (knot'-ted) With reference to metamorphic rocks, a syn. of *maculose.*
knotted-hornfels facies Metamorphic rocks formed in the lowest grades of thermal (contact) metamorphism at temperatures between 200° and 350°C and at pressures not exceeding 250 MPa (Hietanen, 1967). Obsolete. Syn: *albite-epidote-hornfels facies.*
knotted schist *spotted slate.*
knotted slate *spotted slate.*
knowe A Scottish syn. of knoll.
knoxvillite (knox'-vil-lite) *copiapite.*
Knudsen's tables Older version of hydrographic tables used to calculate salinity from chlorinity as well as the density of seawater under various conditions. Densities may be expressed as sigma (s), where s = [(density - 1)/1,000]. Uses an approximate relationship of salinity to chlorinity [Salinity = 1.8050 Chlorinity + 0.030]

which has now been superseded by the *salinity- chlorinity ratio.* Cf: *salinity-chlorinity ratio.*

koashvite (ko-ash'-vite) A pale yellow orthorhombic mineral: $Na_6(Ca,Mn)(Fe^{3+},Ti)Si_6O_{18}{\cdot}H_2O$.

kobeite-(Y) (ko'-be-ite) A metamic orthorhombic brown or black mineral of the *euxenite* group: $(Y,U)(Ti,Nb)_2(O,OH)_6$.

kobellite (ko'-bell-ite) A metallic blackish-gray orthorhombic mineral: $Pb_{22}Cu_4(Bi,Sb)_{30}S_{69}$.

kochite A colorless to pale brown triclinic mineral of the *rosenbuschite* group: $Na_2(Na,Ca)_4Ca_4(Mn,Ca)_2Zr_2Ti_2(Si_2O_7)_4(F,O)_8$.

kochkarite A metallic silver-gray trigonal mineral; $PbBi_4Te_7$.

koechlinite (koech'-lin-ite) A greenish-yellow orthorhombic mineral: Bi_2MoO_6.

koenenite (koe'-nen-ite) A very soft colorless, yellow, or red platy rhombohedral mineral: $Na_4Mg_9Al_4Cl_{12}(OH)_{22}$.

Koenigsberger ratio (Koe'-nigs-berg'-er) The ratio of the *remanent magnetization* to the *induced magnetization.* Abbrev: Q.

koenlinite (koen'-lin-ite) *könlite.*

koettigite (koet'-tig-ite) *köttigite.*

köfelsite (kö'-fels-ite) A frothy, pumiceous, high-silica glass occurring as small veins in fractured gneisses in the Köfels structure, Austria, and apparently formed by vesiculation of a melt. The material is extremely heterogeneous, ranging in color from white to dark brown. Controversy presently exists as to whether the rock is of impact or frictional (landslide) origin.

köflachite (kö'-flach-ite) A dark-brown variety of retinite found in brown coal at Köflach in Styria, Austria.

Kofoid tabulation system The method of describing the number and arrangement of plates in dinoflagellates developed by Kofoid (1907, 1910, 1911) and modified by later workers. Plates are considered to lie in successive series, apical ('), precingular (''), cingular (g or c), postcingular ('''), and antapical ('''') and are numbered from left to right starting at the sulcus.

kogarkoite (ko-gar'-ko-ite) A colorless to pale blue monoclinic mineral: $Na_3(SO_4)F$.

kohalaite (ko-hal'-a-ite) An *andesite* that contains normative oligoclase and may or may not contain modal olivine. Its name, given by Iddings in 1913, is derived from Kohala volcano, Hawaii. Cf: *mugearite.* Obsolete.

koktaite (kok'-ta-ite) A colorless monoclinic mineral: $(NH_4)_2Ca(SO_4)_2{\cdot}H_2O$. It is the NH_4 analogue of syngenite.

kolarite (ko'-lar-ite) A metallic gray orthorhombic mineral: $PbTeCl_2$.

kolbeckine (kol'-beck-ine) *herzenbergite.*

kolbeckite (kol'-beck-ite) A colorless, yellow, green, blue, or bluish gray monoclinic mineral of the metavariscite group: $ScPO_4{\cdot}2H_2O$. It was formerly described as a hydrous phosphate and silicate of aluminum, beryllium, and calcium. Syn: *sterrettite.*

kolfanite (kol'-fa-nite) An adamantine red, orange, or yellow monoclinic mineral: $Ca_2Fe^{3+}{}_3O_2(AsO_4)_3{\cdot}2H_2O$. It is related to arseniosiderite.

kolicite (ko'-lic-ite) A yellow-orange orthorhombic mineral: $Mn^{2+}{}_7Zn_4(AsO_4)_2(SiO_4)_2(OH)_8$.

kolk n. A deep, isolated hole or depression, scoured out by eddying water in soft rock. Etymol: German "Kolk", "deep pool, eddy, scour". v. To quarry jointed rock by violent swirling currents.

kolm Nodular or concretionary bodies of coal found in the Paleozoic alum shales of Sweden and that contain rare metals, especially uranium. Syn: *culm.*

Kolmogorov-Smirnov statistic A nonparametric test of goodness-of-fit of an observed sample distribution to a theoretical distribution such as the normal. The test is performed by ranking the sample, then determining the largest difference between the ranked values and the corresponding theoretical values. Named for A.N. Kolmogorov (1903-1987) and V.I. Smirnov (1887-1974), Russian mathematicians.

kolovratite (ko-lov'-ra-tite) An incompletely studied greenish-yellow to dark citrine mineral which is a hydrous vanadate of nickel and zinc.

kolskite (kolsk'-ite) A mixture of lizardite and sepiolite.

kolwezite (kol-wez'-ite) A dull beige to black triclinic mineral: $(Cu,Co)_2(CO_3)(OH)_2$. It is isostructural with rosasite and glaukosphaerite.

kolymite (ko'-ly-mite) A metallic white cubic mineral that is identical with a synthetic amalgam: Cu_7Hg_6.

komarovite (ko-ma'-rov-ite) A pale rose orthorhombic mineral: $(Ca,Mn)Nb_2(Si_2O_7)(O,F)_3{\cdot}3.5H_2O$.

komatiite (ko-mat'-i-ite) (a) In the *IUGS classification,* an ultramafic lava with SiO_2<52, Na_2O+K_2O<1, MgO>18, and TiO_2<1 wt%. Cf: *meimechite; picrite.* (b) An igneous suite, analogous to the tholeiitic, calc-alkaline, and alkaline suites, distinguished by the presence of ultramafic lavas. (c) Ultramafic lava. The term was originally applied to basaltic and ultramafic lavas by Viljoen and Viljoen (1969), but was extended to include the associated rocks by Arndt et al. (1977). Their definition of the komatiite suite includes "noncumulate rocks ranging in composition from peridotite (≈30 percent MgO, 44 percent SiO_2) to basalt (8 percent MgO, 52 percent SiO_2) or andesite (12 percent MgO, 56 percent SiO_2), and cumulate rocks ranging from peridotite (up to 40 percent MgO) to mafic gabbro (≈12 percent MgO)." The lavas commonly exhibit *spinifex texture.* All rocks of the series have low Ti and Fe/(Fe+Mg), and high Mg, Ni, and Cr. The name is for the Komati River, Barberton Mountain Land, Transvaal, South Africa.

kombatite A yellow monoclinic mineral: $Pb_{14}O_9(VO_4)_2Cl_4$. It is the vanadate analogue of sahlinite.

komkovite A vitreous brown rhombohedral mineral of the *hilairite* group: $BaZrSi_3O_9{\cdot}3H_2O$.

kona (ko'-na) A term used in Hawaii for the leeward side, i.e. one away from the trade winds (Stamp, 1966, p.282).

konderite (kon'-de-rite) A metallic gray hexagonal mineral closely related to inaglyite: $PbCu_3(Rh,Pt,Ir)_8S_{16}$.

kongsbergite (kongs'-berg-ite) A silver-rich variety of native amalgam, containing about 95% silver and 5% mercury.

koninckite (ko'-ninck-ite) A yellow tetragonal mineral: $(Fe^{3+},Al)PO_4{\cdot}3H_2O$.

Konkian (Kon'-ki-an) Ponto-Caspian stage: upper middle Miocene (above Chokrakian, below Sarmatin); equiv. Serravallian.

könlite (kön'-lite) A brown to yellow hydrocarbon found in brown coal and having an approximate composition of 91.75% carbon, 7.50% hydrogen, and 0.75% oxygen. Syn: könleinite; koenlinite.

konyaite (kon'-ya-ite) A white monoclinic mineral of the *picromerite* group: $Na_2Mg(SO_4)_2{\cdot}5H_2O$.

kop A mountain or large hill that stands out prominently. Etymol: Dutch, "head".

kopje *koppie.*

Köppen's classification (Köp'-pen's) A *climate classification,* formulated by W. Köppen in 1918, that is based on the climatic requirements of certain types of vegetation. Cf: *Thornthwaite's classification.*

koppie (kop'-pie) A small but prominent hill occurring on the veld of South Africa, sometimes reaching 30 m above the surrounding land; esp. an isolated, elongate, scrub-covered hillock or knob representing an erosion remnant, such as a small *inselberg.* Cf: *castle koppie.* Etymol: Afrikaans, from Dutch *kopje,* "small head". Syn: *kopje.*

koppite (kop'-pite) A variety of pyrochlore containing iron, potassium, and cerium and lacking titanium.

koragoite A metallic red to dark reddish-brown mineral: $Mn^{2+}{}_2Mn^{3+}Nb_2(Nb,Ta)_3W_2O_{20}$.

koris (kor'-is) A term used in northern Africa for a dry valley.

koritnigite (ko-rit'-nig-ite) A colorless triclinic mineral: $Zn(AsO_3OH){\cdot}H_2O$.

kornelite (kor'-nel-ite) A colorless, pink, violet to brown monocinic mineral: $Fe^{3+}{}_2(SO_4)_3{\cdot}7\text{-}8H_2O$.

kornerupine (kor-ne-rup'-ine) A colorless, yellow, brown, or sea-green hard orthorhombic mineral: $(Mg,Fe^{2+})_4Al_6(Si,Al,B)_5O_{21}(OH)$. It resembles sillimanite in appearance and typically occurs in granulite facies rocks.

kornite A dark red or brownish-lilac monoclinic mineral of the *amphibole* group: $(K,Na)(Na,Li)_2(Mg,Mn^{3+},Fe^{3+},Li)_5Si_8O_{22}(OH)_2$.

korobitsynite A colorless orthorhombic mineral: $(Na,[])_8Ti_4(Si_4O_{12})_2(O,OH)_4{\cdot}8H_2O$. It is the titanium analogue of nenadkevichite.

korshunovskite (kor-shu-nov'-skite) A colorless triclinic mineral: $Mg_2Cl(OH)_3{\cdot}3.5\text{-}4.0H_2O$.

korzhinskite (kor-zhin'-skite) A colorless monoclinic mineral: $CaB_2O_4{\cdot}0.5H_2O$. Not approved by the International Mineralogical Association.

kosmochlor (kos'-mo-chlor) A dark green monoclinic meteorite mineral of the *pyroxene* group: $NaCr^{3+}Si_2O_6$. Syn: *ureyite.*

kosnarite A pale blue to blue-green rhombohedral mineral: $KZr_2(PO_4)_3$.

kostovite (kos'-tov-ite) A metallic creamy-white orthorhombic mineral: $CuAuTe_4$.

kostylevite (kos'-tyl-ev-ite) A colorless monoclinic mineral: $K_2ZrSi_3O_9{\cdot}H_2O$. It is a cyclosilicate and a dimorph of umbite.

kotoite (ko'-to-ite) A colorless orthorhombic mineral: $Mg_3(BO_3)_2$. It is the Mg analogue of jimboite.

kotschubeite (ko-tschu'-be-ite) A rose-red variety of clinochlore containing chromium.

köttigite (köt'-tig-ite) A monoclinic colorless to carmine-red mineral of the *vivianite* group: $Zn_3(AsO_4)_2{\cdot}8H_2O$. Its dimorph is metaköttigite. Also spelled: *koettigite.*

kotulskite (ko-tul'-skite) A metallic cream hexagonal mineral: $Pd(Te,Bi)_{2-x}(x=0.4)$.

koum A *sandy desert* or continuous tract of sand dunes in central Asia, equivalent to an *erg.* Etymol: French. See also: *kum.*

koutekite (kou'-tek-ite) A metallic orthorhombic or hexagonal mineral: Cu_5As_2.

kovdorskite (kov-dor'-skite) A colorless to pale rose monoclinic mineral: $Mg_2(PO_4)(OH){\cdot}3H_2O$.

Kozachurian Or Kotsakhurian; see: *Takaraulian.*

kozoite-(La) A white to pale pinkish-purple orthorhombic mineral of the *ancylite* group: $LaCO_3(OH)$.

kozoite-(Nd) A white to pale pinkish-purple orthorhombic mineral of the *ancylite* group: $NdCO_3(OH)$.

kôzulite (ko'-zu-lite) A reddish-black to black monoclinic mineral of the *amphibole* group: $NaNa_2[Mn^{2+}{}_4(Fe^{3+},Al)]Si_8O_{22}(OH)_2$.

krablite (krab'-lite) A rhyolitic *crystal tuff* containing plagioclase grains enclosed in orthoclase phenocrysts, along with smaller amounts of augite and quartz. The rock was originally identified as the mineral feldspar. It occurs as ejecta from Krafla, Iceland. Obsolete.

krageröite (kra-ge-rö'-ite) A rutile-bearing albite *aplite,* with minor amounts of quartz, potassium feldspar, and ilmenite. Its name, given by Brögger in 1904, is derived from Kragerö, Norway. Also spelled: kragerite. Not recommended usage.

kraisslite (krais'-sl-ite) A coppery-brown platy hexagonal mineral: $Zn_3(Mn^{2+},Mg,Fe^{3+})_{25}(AsO_4)_4(SiO_4)_8(OH)_{12}$.

Krakatoan caldera (Kra-ka-to'-an) A type of caldera formed in the summit region of a volcano following evacuation of the underlying magma chamber by a voluminous outpouring of pyroclasts, normally of silicic composition.

kramerite (kra'-mer-ite) *probertite.*

krans *krantz.*

krantz A term used in southern Africa for a precipitous rock face or sheer cliff. Etymol: Afrikaans *krans,* "wreath". Pl: krantzes. Syn: *krans.*

krantzite (krantz'-ite) A variety of retinite found in small yellowish grains disseminated in brown coal.

kras A Slavic word meaning "bleak, waterless place," from which the term *karst* is derived.

krasnovite A pale blue orthorhombic mineral: $Ba(Al,Mg)(PO_4,CO_3)(OH)_2{\cdot}H_2O$.

Krasnozem (Kras'-no-zem) A Russian term for a zonal red soil developed in a Mediterranean climate.

kratochvilite (kra-toch'-vil-ite) A colorless orthorhombic hydrocarbon mineral: $C_{13}H_{10}$.

kratogen (krat'-o-gen) *craton.*

kraton (kra'-ton) *craton.*

kraurite (krau'-rite) *dufrénite.*

krausite (kraus'-ite) A yellowish-green monoclinic mineral: $KFe^{3+}(SO_4)_2{\cdot}H_2O$.

krauskopfite (kraus'-kopf-ite) A colorless monoclinic mineral: $BaSi_2O_4(OH)_2{\cdot}2H_2O$.

krautite (kraut'-ite) A pale rose monoclinic mineral: $Mn^{2+}(AsO_3OH){\cdot}H_2O$.

KREEP An acronym for a basaltic lunar rock type, first found in Apollo 12 fines and breccias, characterized by an unusually high content of potassium (K), rare-earth elements (REE), phosphorus (P), and other trace elements, in comparison to other lunar rock types. The material, which is found in a variety of crystalline and glassy (shock-melted?) rock types, is distinctly different from the iron-rich mare basalts. Syn: *nonmare basalt.*

kreittonite (kreit'-to-nite) A black variety of gahnite containing ferrous iron or ferric iron, or both.

kremersite (kre'-mers-ite) A ruby-red to brownish-red orthorhombic mineral: $(NH_4,K)_2Fe^{3+}Cl_5{\cdot}H_2O$. It is the ammonium analogue of erythrosiderite.

krennerite (kren'-ner-ite) A silver-white to pale-yellow metallic orthorhombic mineral: $(Au,Ag)Te_2$. It often contains silver. Syn: *white tellurium.*

krettnichite A dark brown to black monoclinic mineral: $PbMn^{3+}{}_2(VO_4)_2(OH)_2$.

kribergite (kri'-berg-ite) A white, chalklike triclinic mineral: $Al_5(PO_4)_3(SO_4)(OH)_4{\cdot}4H_2O$.

kriging (krig'-ing) (a) A statistical technique employed in calculating grade and tonnage of ore reserves from sampling data. The data are handled by computer. Named for Danie Krige, a South African geostatistician. (b) A technique for interpolating which honors data points exactly. An output point is calculated as a linear combination of known data points. Kriging attempts to produce the best linear unbiased estimate. Used to interpolate between boreholes. Cf: *cokriging.*

krinovite (kri'-nov-ite) A green triclinic meteorite mineral of the *aenigmatite* group: $NaMg_2CrSi_3O_{10}$.

kristiansenite A colorless triclinic mineral: $Ca_2ScSn(Si_2O_7)(Si_2O_6OH)$.

kröhnkite (kröhnk'-ite) An azure-blue or greenish-blue monoclinic mineral: $Na_2Cu(SO_4)_2{\cdot}2H_2O$. Also spelled: kroehnkite.

krokidolite (kro-kid'-o-lite) *crocidolite.*

krokydite (kro'-kyd-ite) *crocydite.*

krotovina (kro-to-vi'-na) An irregular tubular or tunnel-like structure in soil, made by a burrowing animal and subsequently filled with material from another horizon. Also spelled: *crotovina.* Etymol: Russian.

krummholz (a) *timberline* (area with stunted trees). Etymol: German, "crooked, twisted wood".

krupkaite (krup'-ka-ite) A metallic gray orthorhombic mineral: $PbCuBi_3S_6$.

krutaite (kru'-ta-ite) A metallic gray cubic mineral of the *pyrite* group: $CuSe_2$.

krutovite (kru'-tov-ite) A metallic grayish-white cubic mineral: $NiAs_2$. A polymorph of rammelsbergite and pararammelsbergite.

kryokonite (kry-ok'-o-nite) *cryoconite [glaciol].*

kryomer (kry'-o-mer) A relatively cold period within the Pleistocene Epoch, such as a *glacial stage* (Lüttig, 1965, p.582). Ant: *thermomer.* Rarely used.

kryoturbation (kry'-o-tur-ba'-tion) *cryoturbation.*

krystic (krys'-tic) Pertaining to ice in all its forms, as a surface feature of the Earth. Rarely used.

kryzhanovskite (kry-zha-ovsk'-ite) A reddish-brown to greenish-brown orthorhombic mineral of the *phosphferrite* group: $(Fe^{3+},Mn^{2+})_3(PO_4)_2(OH,H_2O)_3$.

K selection The evolutionary process favoring development of K strategists or *equilibrium species* (Wilson and Bossert, 1971, p.110-111).

K-spar (a) *potassium feldspar.* (b) *potash spar.*

K strategist *equilibrium species.*

ktenasite (kte'-na-site) A blue-green monoclinic mineral: $(Cu,Zn)_5(SO_4)_2(OH)_6{\cdot}6H_2O$.

ktypéite (kty'-pé-ite) A mineral substance intermediate between calcite and aragonite.

kuannersuite-(Ce) A vitreous light rose-colored trigonal mineral of the *apatite* family: $Ba_6NaCe(PO_4)_3(F,Cl)$.

kuemmerform (kuem'-mer-form) n. An organism that is small and stunted as a result of environmental stress. The term is commonly applied to planktonic foraminifers but may also refer to other forms. Erroneous spelling: kummerform. Etymol: German, "Kümmerform", stunted, retarded, undersized variant.

kugdite In the *IUGS classification*, an obsolete syn. for *olivine melilitolite*.

Kuiper Belt A zone of distant small bodies orbiting beyond the orbit of Neptune, analogous to the asteroid belt between Mars and Jupiter. The objects presumably are leftovers from the accretion of planets. Named for Dutch-American astronomer Gerard Kuiper. Syn: Trans-Neptunian Objects (TNOs). See also: *KBO.*

kukersite (ku'-ker-site) An organic sediment rich in the alga *Gloexapsamorpha prisca,* found in the Ordovician of Estonia.

kukharenkoite-(Ce) A yellow monoclinic mineral: $Ba_2Ce(CO_3)_3F$.

kukharenkoite-(La) A colorless to pale green monoclinic mineral: $Ba_2(La,Th,Ce)(CO_3)_3F$.

kukisvumite A colorless orthorhombic mineral: $Na_6ZnTi_4O_4(SiO_3)_8{\cdot}4H_2O$.

kuksite A gray orthorhombic mineral: $Pb_3Zn_3TeO_6(PO_4)_2$. It is the phosphate analogue of cheremnykhite.

kulanite (ku'-lan-ite) A bluish-green monoclinic mineral of the *bjarebyite* group: $Ba(Fe^{2+},Mn,Mg)_2Al_2(PO_4)_3(OH)_3$.

kuliokite-(Y) (ku'-li-o-kite-y) A colorless triclinic mineral: $(Y,Yb)_4Al(SiO_4)_2(OH)_2F_5$.

kulkeite (kul'-ke-ite) A colorless monoclinic regularly ordered, 1:1 mixed-layer talc/chlorite interstratification.

Kullenberg corer (Kul'-len-berg) The original type of *piston corer* used by the Swedish Deep Sea Expedition (1947-1949) and later modified into many varieties. Named for Borje Kullenberg.

kullerudite (kul'-le-rud-ite') A metallic orthorhombic mineral of the *marcasite* group: $NiSe_2$.

kum A Turkish term for "sand", applied to the sandy deserts of central Asia; e.g. Kizil Kum. See also: *koum.*

Kungurian (Kun-gur'-i-an) A stage of the Standard Global Chronostratigraphic Scale: Lower Permian (above Artinskian, below Roadian).

kunkar (kun'-kar) *kankar; calcrete.*

kunkur (kun'-kur) *kankar.*

kunzite (kunz'-ite) A pinkish, light-violet, or lilac-colored transparent gem variety of spodumene. An occasional semiprecious gemstone, its chromophore is Mn^{3+}.

kupcíkite A metallic gray monoclinic mineral: $Cu_{3.4}Fe_{0.6}Bi_5S_{10}$.

kupfernickel (kup'-fer-nick'-el) *nickeline.*

kupferschiefer-type Stratiform and stratabound, sedimentary Cu-Pb-Zn sulfide deposits associated with carbonaceous, metalliferous shales. Assumed by most modern authors to be sygenetic. From the kupferschiefer, literally "copper-shale," unit, a one-meter thick, dolomitic shale at the base of the Middle Permian Zechstein formation in the Harz Mountain, Germany. Cf: *sediment-hosted copper deposit.*

kupletskite (kup-let'-skite) A yellow, dark brown to black platy triclinic mineral of the *astrophyllite* group: $(K,Na)_3(Mn,Fe^{2+})_7(Ti,Nb)_2Si_8O_{26}(OH)_4F$.

kuramite (kur'-am-ite) A metallic gray tetragonal mineral of the *stannite* group: Cu_3SnS_4.

kuranakhite (kur'-a-nakh-ite) A brownish-black orthorhombic mineral: $PbMn^{4+}Te^{6+}O_6$.

kurchatovite (kur'-cha-tov'-ite) A pale gray orthorhombic mineral: $Ca(Mg,Mn^{2+},Fe^{2+})B_2O_5$. It has a monoclinic dimorph.

kurgantaite (kur-gan'-ta-ite) A colorless mineral triclinic mineral of the *hilgardite* group: $Ca,SrB_5O_9{\cdot}H_2O$.

kurilite A metallic cubic mineral: $(Ag,Au)_2(Te,Se,S)$. Not approved by the International Mineralogical Association.

kurnakovite (kur'-na-kov'-ite) A white triclinic mineral: $MgB_3O_3(OH)_5{\cdot}5H_2O$. It is a dimorph of inderite.

kuroko deposit (ku-ro'-ko) A type of massive polymetallic base-metal sulfide deposit in Japan. Kuroko deposits are typically zoned and strata-bound. They are volcanogenic deposits of Miocene age, precipitated on the sea floor adjacent to fumaroles or hot springs on the flanks of submarine dacite domes during the late stages of explosive felsic volcanic cycles. Kuroko-type refers to polymetallic, submarine, volcanogenic massive sulfide deposits hosted by intermediate to felsic volcanic tuffs (Matsukuma and Horikoshi, 1970). Cf: *Cyprus-type deposit.*

kurskite (kurs'-kite) *carbonate-fluorapatite.*

kurtosis (kur-to'-sis) (a) A statistic expressing the departure of a frequency distribution from normality, as the degree of peakedness or concentration around the mode. Kurtosis is defined as the ratio of the fourth moment over the square of the second moment, and is equal to 3 for a normal distribution, determined as a ratio between percentile differences $(P_{75}P_{25})/2(P_{90}-P_{10})$. Cf: *skewness.*

kurumsakite (kur'-um-sak-ite) An ill-defined yellowish-green to bright yellow mineral, possibly related to montmorillonite: $(Zn,Ni,Cu)_8Al_8V^{5+}{}_2Si_5O_{35}\cdot 27H_2O(?)$.

kusachiite A metallic black tetragonal mineral: $Cu^{2+}Bi^{3+}{}_2O_4$.

kusuite (ku'-su-ite) *wakefieldite-(Ce).*

kutinaite (ku-ti'-na-ite) A metallic gray cubic mineral: $Cu_{14}Ag_6As_7$.

kutnahorite (kut-na-hor'-ite) A white to pink rhombohedral mineral of the *dolomite* group: $Ca(Mn,Mg,Fe)(CO_3)_2$. Also spelled: kutnohorite.

Kutter's formula (Kut'-ter's) A formula that expresses the value of the Chézy coefficient in the *Chézy equation* in terms of the friction slope, hydraulic radius, and a roughness coefficient (Brown and Runner, 1939, p.199).

kuzelite A colorless trigonal mineral: $Ca_4Al_2(OH)_{12}(SO_4)\cdot 6H_2O$.

kuzmenkoite-Mn A yellow monoclinic mineral of the *labuntsovite* group: $K_4Mn_2Ti_8(Si_4O_{12})_4(OH,O)_8\cdot 10\text{-}12H_2O$.

kuzmenkoite-Zn A colorless, white, gray, pink or pinkish brown monoclinic mineral of the *labuntsovite* group: $K_2Zn(Ti,Nb)_4(Si_4O_{12})_2(OH,O)_4\cdot 6\text{-}8H_2O$.

kuzminite (kuz'-min-ite) A colorless or bluish-gray tetragonal mineral of the calomel group: $Hg^{1+}(Br,Cl)$.

kuznetsovite (kuz'-net-sov-ite) An adamantine pale brown to yellow cubic mineral: $Hg^{1+}{}_2Hg^{2+}(AsO_4)Cl$.

kvanefjeldite (kva-ne-fjeld'-ite) A violet-pink orthorhombic mineral: $Na_4(Ca,Mn)Si_6O_{14}(OH)_2$.

kvellite (kvel'-lite) A very dark-colored hypabyssal rock containing phenocrysts of biotite, olivine, amphibole, apatite, ilmenite, and magnetite in a groundmass of lath-shaped anorthoclase and nepheline. The name, given by Brögger in 1898, is derived from Kvelle, Oslo district, Norway. Obsolete.

K wave A longitudinal or *P wave* in the Earth's outer core. Cf: *I wave.*

kX unit A unit in which X-ray wave lengths were generally given prior to World War II. 1 kX = 1.002056 ± 0.000005 A.

kyanite (ky'-a-nite) A mainly blue but variously colored triclinic mineral: Al_2SiO_5. It is trimorphous with andalusite and sillimanite. Kyanite occurs in long, thin, bladed crystals and crystalline aggregates in schists, gneisses, and granite pegmatites, and has a hardness of 4-5 along the length of the crystal and 6-7 across it. It forms at medium temperatures and high pressures in regionally metamorphosed sequences. Also spelled: *cyanite.* Syn: *sappare; disthene.*

kyanophyllite (ky'-a-no-phyll'-ite) A mixture of paragonite and muscovite.

kyle A Scottish term for a narrow channel, sound, or strait between two islands or an island and the mainland, or for a narrow inlet into the coast. Etymol: Gaelic.

kylite (ky'-lite) An olivine-rich *theralite,* named by Tyrrell in 1912 for the Kyle district, Ayrshire, Scotland. Obsolete.

kymoclastic rock (ky-mo-clas'-tic) A hydroclastic rock containing marine or other wave-formed fragments (Grabau, 1924, p.295).

kyphorhabd (ky'-pho-rhabd) A curved monaxonic sponge spicule with transverse swellings on the convex side.

kyr A term used in central Asia for flat land, a plateau, or the top of a small hill or mountain, and indiscriminately for a low hill or small mountain, but applied specif. in Turkmenia to stony, hard ground as contrasted to *adyr* (Murzaevs and Murzaevs, 1959, p.131).

kyriosome (ky'-ri-o-some') A term used by Niggli (1954, p.191) for the fundamental mass or framework fraction of a complex rock; the major part of a *migmatite.* Cf: *akyriosome.* Rarely used.

kyrtome (kyr'-tome) A triradiate, more or less thickened area associated with and bordering the laesura of a trilete spore. Cf: *torus.*

kyzylkumite (ky-zyl-kum'-ite) A black monoclinic mineral: $BeV^{3+}2Ti^{4+}O_6$. The originally reported chemical formula was $V^{3+}{}_2Ti^{4+}{}_3O_9$.

L

laagte (laag′-te [loch′-ta]) A broad, almost level drainage course in the veld of southwestern Africa, less well-defined than a valley, and dry for most of the year. Etymol: Afrikaans. Syn: *leegte*.

laavenite (laa′-ven-ite) *låvenite*.

labes (la′-bes) A term established by the International Astronomical Union for a landslide feature on planetary surfaces (USGS, 2003). Pl: labes.

labia (a) Syn. of *labium*. Pl: labiae. (b) Pl. of *labium*.

labial aperture (la′-bi-al) An *accessory aperture* formed in a foraminiferal test by the free parts of the apertural lip, not leading directly to a chamber.

labial pore A median or submedian pore in the proximal wall of the *peristome* in ascophoran cheilostomes (bryozoans), resulting from closure of a notch in the secondary orifice.

labiate (la′-bi-ate) Having lips; e.g. said of an exaggerated marginate foramen of a brachiopod in which the dorsal edge is prolonged liplike (TIP, 1965, pt. H, p.147).

labiatiform cyrtolith (la-bi-at′-i-form) A cyrtolith *coccolith* with a central structure shaped like a double lip (as in *Anthosphaera robusta*).

labile (la′-bile) (a) Said of rocks and minerals that are mechanically or chemically unstable; e.g. a "labile sandstone" or "labile graywacke" containing abundant unstable fragments of rocks and minerals and less than 75% matrix of fine silt and clay (Packham, 1954), or "labile constituents" (such as feldspar and rock fragments in a sandstone) that are easily decomposed. Cf: *unstable [sed]*. (b) Said of protobitumen that represents easily decomposable plant and animal products (such as fat, oil, or protein) in peat and sapropel. Ant: *stabile*.

labilizing force (la′-bi-liz′-ing) In an unstable gravimeter, a force acting in the same direction as the force being measured, and therefore opposite to the direction of an ordinary restoring force.

labite (la′-bite) *chrysotile*.

labium (la′-bi-um) (a) The lower lip of an arthropod (such as of an insect); the *metastoma* of a crustacean. Cf: *labrum*. (b) The columellar part of the aperture of a gastropod shell. Pl: labia.

labium [palyn] (la′-bi-um) *labrum*. Pl: *labia*. Syn: *labia* (a singular).

labor (la-bor′ [la-bore′]) A Spanish term used in early land surveys in Texas for unit of area equal to about 177.14 acres (representing a tract 100 varas square).

labra Pl. of *labrum*.

labradorescence (lab′-ra-dor-es′-cence) An optical phenomenon consisting of flashes of a laminated iridescence of a single bright hue that changes gradually as a mineral or gemstone is moved about in reflected light, caused by internal structures that selectively reflect only certain colors; specif. the light-interference effect exhibited by labradorite and set up in thin plates of feldspar (produced by repeated twinning or by exsolution), resulting in a series of vivid colors (usually brilliant blue or green) spread over large areas. Syn: *change of color*.

labradorite A colorless to dark mineral of the plagioclase feldspar group with composition ranging from $Ab_{50}An_{50}$ to $Ab_{30}An_{70}$. It commonly shows a rich, beautiful play of colors (commonly blue or green), and is therefore much used for ornamental purposes. Labradorite is common in igneous rocks of intermediate to low silica content. Syn: *Labrador spar*.

Labrador spar (Lab′-ra-dor) A syn. of *labradorite*. Also called: Labrador stone; Labrador rock.

labrum (la′-brum) (a) An unpaired outgrowth of an arthropod, consisting of a single median piece or flap immediately in front of or above the mandibles and more or less covering the mouth; e.g. the upper lip in front of the mouth of a crustacean or of a merostome, or the *hypostome* of a trilobite. Cf: *labium*. (b) The external margin of a gastropod shell. (c) A more or less enlarged and modified liplike primordial plate of an echinoid, bordering the peristome in the interambulacrum 5. Adj: labral.

labrum [palyn] (la′-brum) The more or less elevated and/or thickened margin or lip of a spore, edging the *laesura*. Pl. labra. Etym: Originally preferred for the single (upper or lower) lip, while the syn. *labium* was preferably used in the plural, referring to a pair of lips (thus leading to the use of *labia* as a singular noun).

labuntsovite (la-bunt′-sov-ite) A group name for monoclinic minerals of the general formula $A_8B_2C_8(Si_4O_{12})_4(O,OH)_8{\cdot}nH2O$, where A = Na, K , Ca or Ba; B = Mg, Mn, or Fe; and C = Ti or Nb.

labuntsovite-Fe An orange monoclinic mineral of the *labuntsovite* group: $Na_4K_4Fe^{2+}{}_2Ti_8(Si_4O_{12})_4(O,OH)_8{\cdot}10\text{-}12H_2O$.

labuntsovite-Mg A colorless, yellowish, pink or light orange monoclinic mineral of the *labuntsovite* group: $Na_4K_4Mg_2Ti_8(Si_4O_{12})_4(O,OH)_8{\cdot}10\text{-}12H_2O$.

labuntsovite-Mn An orange monoclinic mineral of the *labuntsovite* group: $Na_4K_4Mn^{2+}{}_2Ti_8(Si_4O_{12})_4(O,OH)_8{\cdot}10\text{-}12H_2O$.

labyrinth *maze cave*.

labyrinthic (lab-y-rin′-thic) Said of some agglutinated foraminifers having a complex spongy wall with interlaced dendritic channels perpendicular to the surface.

labyrinth karst (lab′-y-rinth) A *karst plain* characterized by intersecting *solution corridors* (Brook and Ford, 1978, p.493).

labyrinthodont (lab-y-rin′-tho-dont) adj. Pertaining to conical teeth in which the outer layer of enamel is complexly folded to form a labyrinthine pattern in cross section. Such teeth occur in crossopterygian fish and labyrinthodont amphibians. n. A member of the amphibian subclass *Labyrinthodonta*.

Labyrinthodonta (Lab′-y-rin′-tho-don′-ta) A subclass of amphibians, characterized by teeth with labyrinthine infolding of the enamel and by composite vertebral centra.

labyrinthus (lab-y-rin′-thus) A term established by the International Astronomical Union for a region of intersecting linear depressions on planetary surfaces (USGS, 2003). These depressions are generally considered to be graben (Mutch et al., 1976, p.57, 77).

lac The French word for lake. It appears in proper names in parts of the U.S. where the influence of early French settlement remains, and in French-speaking regions of Canada.

LaCasitan (La-Ca-si′-tan) North American (Gulf Coast) stage: Upper Jurassic (above Zuloagan, below Durangoan of Cretaceous; it is equivalent to European Portlandian and Kimmeridgian) (Murray, 1961).

laccolite (lac′-co-lite) The original term, now superseded, for a *laccolith*. Obsolete.

laccolith (lac′-co-lith) A concordant igneous intrusion with a convex-up roof and known or assumed flat floor. Syn: *laccolite*.

Lacian (La′-cian) A European substage: lower substage of the Norian Stage, Upper Triassic (above Tuvalian, below Alaunian).

lacine (la-cine′) One of a series of detached, tongue-shaped, ridgelike meander scrolls, frequently found spread apart like the rays of a fan. Lacines are not as long, as smoothly curved, or as closely spaced as the more symmetric meander scrolls. Term introduced by Melton (1936, p.599). Etymol: Latin "lacinia", "flap, tongue".

lacine meander A detached *scroll meander* in which lateral erosion of the outer bank is somewhat retarded, thereby producing a low-water channel of unequal width (Melton, 1936, p.599-600). Examples occur in the Mississippi River between Cairo, Ill., and Baton Rouge, La.

lacinia (la-cin′-i-a) The inner distal spiny lobe of the second segment of the maxillule of a crustacean, adjacent to the *galea* (TIP, 1969, pt. R, p.97).

lacinia mobilis (la-cin′-i-a mo′-bi-lis) A small, generally toothed process articulated with the incisor process of the *mandible* of a malacostracan crustacean.

laciniate (la-cin′-i-ate) Deeply incised, as in a leaf having narrow, pointed lobes, as if slashed.

LaCoste-Romberg gravimeter (La-Coste′-Rom′-berg) A spring-type gravimeter where the gravity sensor design is based on the use of a unit mass at the end of a lever arm with quartz or metal *zero-length spring* suspension, which provides a linear deflection of a suspended weight (unit mass) and extreme sensitivity to small changes in *gravitational acceleration*. The most commonly used type of exploration gravimeter.

lacroixite (la-croix′-ite) A colorless to pale yellowish-green monoclinic mineral of the tilasite group: $NaAl(PO_4)F$. It is isotypic with titanite.

lacullan (la-cul′-lan) *anthraconite*.

lacuna [geochron] A *chronostratigraphic unit,* representing the interpreted space-time value of both nondeposition (hiatus) and the erosionally removed part of the subjacent transgressive-regressive succession (Wheeler, 1958, p.1058; 1964, p.599).

lacuna [paleont] (la-cu'-na) (a) A pore or slitlike opening between costae making up the frontal shield in cribrimorph cheilostomes (bryozoans). (b) A space, in the *lorica* of a tintinnid, lacking reticulation or other surface marking. (c) A lateral hole or gap in the *hood* of some terebratellacean brachiopods, produced by resorption during loop development. (d) In receptaculitids, a circular area devoid of meroms, in the apical portion of the thallus. (e) Any one of the tiny spaces within the microscopic structure of bone that house the living cells (osteocytes) of that tissue. Pl: lacunae or lacunas.

lacuna [palyn] A depressed space surrounded by ridges on the outer surface of pollen grains, particularly those in Compositae. Pl: lacunae.

lacuna [stratig] A period during which sedimentation was either nil or, more likely, was replaced by erosion (Gignoux, 1950, p.19; 1955, p.15-16).

lacus (la'-cus) A term established by the International Astronomical Union for a small plain. Used only on the Moon (USGS, 2003). Pl: lacus.

lacustrine (la-cus'-trine) (a) Pertaining to, produced by, or formed in a lake or lakes; e.g., "lacustrine sands" deposited on the bottom of a lake, or a "lacustrine terrace" formed along its margin. (b) Growing in or inhabiting lakes; e.g., a "lacustrine fauna". (c) Said of a region characterized by lakes; e.g., a "lacustrine desert" containing the remnants of numerous lakes that are now dry. Cf: *limnic [lake].*

lacy residue (lac'-y) An insoluble residue containing irregular openings and having constituent material comprising less than 25% of the volume (Ireland et al., 1947, p.1482). Cf: *skeletal residue.*

LAD *last appearance datum.*

ladder lode (lad'-der) *ladder vein.*

ladder reef *ladder vein.*

ladder vein One of a series of mineral deposits in transverse, roughly parallel fractures that have formed along foliation planes perpendicular to the walls of a dike during its cooling, or along shrinkage joints in basaltic rocks or dikes. Syn: *ladder lode; ladder reef.*

lade (a) The mouth of a river. (b) A watercourse.

Ladinian (La-din'-i-an) Stage of the Standard Global Chronostratigraphic Scale: upper Middle Triassic (above Anisian, below Carnian).

ladu (la-du') The basal avalanchelike part of a pyroclastic flow.

laesura (lae-su'-ra) The line or scar on the proximal face of an embryophytic spore, marking the contact with other members of the tetrad. It may be trilete, monolete, or rarely, dilete. Some palynologists consider each branch of a trilete mark a separate laesura; then, a trilete spore has three laesurae. Pl: laesurae. Syn: *tetrad mark; tetrad scar.* See also: *suture [palyn]; Y-mark.*

laevigate (lae'-vi-gate) Smooth, i.e., lacking sculpture. The term is applied more often to spores than to pollen. Sometimes spelled levigate. Syn: *psilate.*

laffittite (laf-fit'-tite) A metallic bluish-white monoclinic mineral: $AgHgAsS_3$.

laflammeite A metallic monoclinic mineral: $Pd_3Pb_2S_2$.

laforêtite A metallic brown tetragonal mineral of the *chalcopyrite* group: $AgInS_2$.

lag [geophys] (a) *Time lag.* Cf: *time lead.* (b) The phase angle by which the current is behind the emf in an inductive circuit. (c) The distance a sensor is behind the point where location is determined, as the distance an aerosurvey bird is behind the aircraft or a fish behind a boat. Also called: layback.

lag [sed] (a) *lag gravel.* (b) More general than (a): Concentration of the coarsest fraction of a sedimentary deposit on a surface after dispersion of finer-grained fractions.

lag deposit *lag gravel.*

Lagerstätte In North America, layers of exceptionally well-preserved biotas of whole fossils (for example, *Burgess Shale fauna*).

lagg A moatlike area of shallow water around the central peat mat of a *raised peat bog* (Wetzel, 2001, p.819-820). Etymol: Swedish. Syn: *bog moat.*

lag gravel (a) A residual accumulation of coarse, usually hard rock fragments remaining on a surface after the finer material has been blown away by winds. See also: *desert pavement.* (b) Coarse-grained material that is rolled or dragged along the bottom of a stream or sea floor at a slower rate than the finer material, or that is left behind after currents have winnowed or washed away the finer material. Syn: *lag; lag deposit.*

lag mound A remnant of thin, unconsolidated surface material left on a limestone pavement when the cover is partly removed through underlying solution fissures by *piping.*

lagoon [coast] (la-goon') (a) Typically, a narrow water body that is parallel to the shore and is between the mainland and a barrier and parallel to the shore. Little or no fresh water in flux and limited tidal flux cause elevated salinities (Davis, 1983). (b) A shallow freshwater pond or lake near or communicating with a larger lake or a river; a stretch of freshwater cut off from a lake by a barrier, as in a depression behind a shore dune; a barrier lake. (c) A shallow body of water enclosed or nearly enclosed within an atoll. (d) The term has been widely applied to other coastal features, such as an estuary, a slough, a bayou, a marsh, and a shallow pond or lake into which the sea flows. Etymol: Latin lacuna, "pit, pool, pond". Syn: *lagune; laguna [coast].*

lagoon [geog] A closed depression in a high, grass-covered tableland of the cordilleras of the western U.S.

lagoon [water] (a) A body of water along the seashore, either separated from the sea, as a *beach pond,* or joined to it by a breach in the sandbar or atoll that partly encloses it. (b) The basin of a hot spring; also, the pool formed by a hot spring in such a basin. (c) A perennial brine pool near the margin of an alkaline lake; e.g., near Lake Magadi in southern Kenya. (d) Any shallow artificial pond or other water-filled excavation for the natural oxidation of sewage or disposal of farm manure, or for some decorative or aesthetic purpose. Etymol: Anglicized version of Italian/Spanish "laguna" (Latin "lacuna"), "small/minor lake".

lagoonal (la-goon'-al) Pertaining to a lagoon, esp. *lagoonal* deposition.

lagoon atoll *pseudoatoll.*

lagoon channel (a) The stretch of deep water separating a reef from the neighboring land (mainland or island). (b) A pass through a reef, and into and through a lagoon.

lagoon cliff A very steep to vertical lagoon slope. Syn: *lagoon scarp.*

lagoon island (a) One of many scattered islets rising from within the lagoon of a composite *atoll* or a large *barrier reef,* generally marking former fringing reefs that grew up with the postglacial eustatic rise. (b) *atoll.*

lagoon phase The strata or stratigraphic facies formed by accumulation of sediment in a shallow coastal water body that is separated from the open sea by a barrier.

lagoon plain A flat landform produced by the filling of lagoons with sediments.

lagoon scarp *lagoon cliff.*

lagoon shelf The part of a *reef* that borders the lagoon side of a reef island; the sand-covered, lagoonward-sloping shelf commonly found where sedimentation conspicuously exceeds organic growth.

lagoonside (la-goon'-side) The land bordering on a lagoon.

Lagorio's rule (La-go'-ri-o's) An approximate rule according to which quartz usually begins to crystallize early from highly siliceous rhyolites and porphyries, and late from the less siliceous ones. The rule was proposed in 1887 by A. Lagorio.

Lagrangean (La-gran'-ge-an) A way of describing fluid motion whereby the motion and properties of the fluid are calculated following fluid particles as the particles move through space. This is the most common way for describing particle dynamics, e.g., the trajectory of a bullet, but not fluid dynamics. Named in honor of Joseph Louis Lagrange (1736-1813), French mathematician. Cf: *Eulerian.*

laguna [coast] (la-gu'-na) A var. of *lagoon [coast].* Etymol: Spanish and Italian.

laguna [lake] (la-gu'-na [la-goo-na]) The Spanish term for "minor lake" used in the SW U.S. for a lake or lagoon; esp. a shallow ephemeral lake in the lower part of a *bolson,* fed by streams rising in the neighboring mountains and flowing only as a result of rainstorms.

lagunar (la-gu'-nar) Var. of *lagoonal,* sometimes used in Britain.

lagune (la-gune') (a) Var. of *lagoon [coast].* Etymol: French. (b) A term used in the SW U.S. for a small lake.

lahar (la-har') A *mudflow [mass move]* composed chiefly of volcaniclastic materials on the flank of a volcano. The debris carried in the flow includes pyroclasts, blocks from primary lava flows, and epiclastic material. Etymol: Indonesian. Syn: *mudflow [volc].*

laihunite (lai-hun'-ite) A submetallic black monoclinic mineral related to fayalite, having Fe^{3+} and vacancies in place of Fe^{2+}: $Fe^{2+}Fe^{3+}_2(SiO_4)_2$. Syn: *ferrifayalite.*

laitakarite (lai'-ta-kar'-ite) A metallic rhombohedral mineral: $Bi_4(Se,S)_3$. Cf: *ikunolite.* Syn: *selenjoséite.*

lakarpite (la-karp'-ite) A subsolvus *nepheline syenite* composed of orthoclase or microcline, calcic plagioclase, and an amphibole resembling arfvedsonite, accessory aegirine, altered nepheline, rosenbuschite, and secondary natrolite. The name, proposed by Törnebohm in 1906, is from Lakarp, Norra Kärr complex, Sweden. Obsolete.

lake [coast] A term loosely applied to a sheet of water lying along a coast and connected with the sea; e.g. one of the shallow (1-2 m deep), interconnected bodies of water in the Florida Bay area, Fla.

lake [ice] A submariner's term for a *polynya* during the summer. Cf: *skylight.*

lake [lake] Any inland body of standing water occupying a depression in the Earth's surface, generally of appreciable size (often larger than a *pond,* in common usage; but there is no limnological distinction between them) and too deep to permit vegetation (excluding subaqueous vegetation) to take root completely across the expanse of water; the water may be fresh or saline. The term includes an expanded part of a river, a reservoir behind a dam, or a lake basin intermittently or formerly covered by water.

lake [streams] An English term for a brook or small stream; also, a channel.

lake asphalt Soft asphalt, rich in bitumen, from the pitch lake of Trinidad. See also: *land asphalt.* Syn: *lake pitch.*

lake ball A spherical mass of tangled, waterlogged fibers and other filamentous material of living or dead vegetation (such as blue-green algae, moss, spruce needles, and fragments of peat, grass, or twigs), produced mechanically along a lake bottom by wave action, and usually impregnated with sand and fine-grained mineral fragments. Lake balls may range in size up to that of a man's head. Cf: *peat ball; sea ball.* Syn: *hair ball; burr ball.*

lake basin (a) The depression in the Earth's surface occupied or formerly occupied by a lake and containing its shore features. (b) The area from which a lake receives drainage (Wetzel, 2001, p.133).

lake bed (a) The flat to gently undulating ground underlain by fine-grained sediments deposited in a former lake. (b) The ground on which a lake rests at present; the bottom of a lake; a *lake basin.*

lake biscuit *microbial biscuit.*

lake district A region marked by the grouping together of lakes, e.g., the Lake District of England, the Sand Hills lake district of Nebraska.

lake gage A gage for measuring the elevation of the water surface of a lake.

Lake George diamond Colorless, doubly terminated quartz crystal from Herkimer County, N.Y. See also: *Herkimer diamond.*

lake gun A term used on Seneca Lake, New York, for a phenomenon that produces sounds like distant thunder or gunfire. It is known to occur on several European lakes and has a variety of names (Hutchinson, 1957, p.361-362).

lake ice Ice formed on a lake, regardless of observed location; it is usually *freshwater ice.*

lakelet (lake'-let) A small lake.

lake loam A term applied to loess that may have been formed by deposition in lakes (Veatch and Humphrys, 1966, p.171).

lake marl *bog lime.*

lake ocher Ocherous deposits formed on the bottom of a lake by bacteria capable of precipitating ferric hydroxide, or found in a marsh or swamp that was formerly the site of a lake.

lake ore (a) A disklike or irregular concretionary mass of ferric hydroxide less than a meter thick, or a layer of porous, yellow bedded limonite, formed along the borders of certain lakes. Cf: *bog ore.* (b) *bog iron ore.*

lake peat *sedimentary peat.*

lake pitch *lake asphalt.*

lake plain (a) The nearly level surface marking the floor of an extinct lake, filled in by well-sorted deposits from inflowing streams. (b) A flat lowland or a former lake bed bordering an existing lake. See also: *lake terrace.*

lake rampart A conspicuous wall-like ridge composed of unconsolidated coarse material along a lake shore. It is produced by shoreward movement of lake ice, as by winds, waves, or currents, and esp. by expansion of ice against yielding lake-shore deposits; or by pushing and then stranding bottom-lake deposits as it overrides the shore. Lake ramparts may reach 2 m in height. Examples occur along the shores of the Great Lakes. See also: *walled lake.* Syn: *ice rampart; ice push.*

lakescape (lake'-scape) All or part of a lake, including water surface, islands, and shoreline features, that can be viewed from an observation point.

Lake Superior-type iron formation Siliceous *iron formation* consisting of chamosite-siderite-goethite with appreciable silica, clay, and detritus. It is commonly oolitic. Associated rocks are quartzite, dolomite, and black shale. It formed in continental-shelf environments remote from volcanism, chiefly in Proterozoic-Paleozoic time (Gross, 1980). Cf: *Algoma-type iron formation.*

lake terrace A narrow shelf, partly cut and partly built, produced along a lake shore in front of a nip or line of low cliffs, and later exposed when the water level falls (Cotton, 1958, p.489). See also: *lake plain.*

lalondeite A pearly colorless triclinic mineral: $(Na,Ca)_6(Ca,Na)_3Si_{16}O_{38}(F,OH)_2{\cdot}3H_2O$.

lalongate (la-lon'-gate) Of a *colpus,* elongated transversally ("latitudinally"). Cf: *lolongate.*

lamalginite An *alginite* with indistinct (microscopic) algal forms; a part of the *liptinite* maceral group.

Lamarckism (La-marck'-ism) A 19th-century theory of evolution stating that changes in the environment cause structural changes in an organism especially by inducing new or increased use of organs or parts as a result of adaptive modification or greater development, and also cause disuse and eventual atrophy of other parts, and that these changes are passed on to offspring. This theory is named after the French naturalist J.B. de Monet Lamarck (1744-1829).

lambda (lamb'-da) A polymorphic transformation characterized by a cusp in a graph of heat capacity against temperature. It is typical of order-disorder transformations and magnetic transitions (Thompson and Perkins, 1981).

lambda transformation A *phase transformation* in which the second derivatives of *Gibbs free energy* with respect to temperature and pressure tend to infinity at the transformation temperature.

Lambert azimuthal equal-area projection (Lam'-bert) An azimuthal map projection having the pole of the projection at the center of the area mapped, the azimuths of great circles radiating from this pole (center) and being truly represented on the map but the scale along such great-circle lines so varying with distance from the center that an equal-area projection is produced. The pole (center) of the projection may be at the pole of the sphere, on its equator, or at any point in between. The projection is useful for representing a single hemisphere or continental masses, but extreme distortion of areas is encountered near the periphery if extended to a world map. See also: *Schmidt projection.* Syn: *azimuthal equal-area projection.*

Lambert conformal conic projection A conformal conic map projection on which all meridians are represented by equally spaced straight lines that radiate from a common point outside the map limits and the parallels (of which one or two are standard parallels along which the scale is exact) are represented by circular arcs having this common point for a center and intersecting the meridians at right angles. The scale is the same in every direction at any point on the map, but increases north and south from the standard parallel(s); where there are two standard parallels, the scale is too small between them and too large beyond them. The projection is used for maps of middle latitudes (for maps of the conterminous U.S., smallest distortion occurs when the standard parallels represent latitudes 33°N and 45°N) and as a base for sets of large-scale aeronautical charts produced by the National Geodetic Survey. Named after Johann H. Lambert (1728-1777), German physicist, who introduced the projection in 1772. Cf: *conformal projection.*

Lambert's law (Lam'-bert's) The statement that the intensity of blackbody radiation emerging from an aperture is greatest in the direction perpendicular to the plane of the aperture, and decreases with the cosine of the angle between the perpendicular and the direction of observation. Such a reflective body is called a perfectly diffuse reflector; real bodies seldom approach this condition.

Lamb's problem An investigation in seismology that is concerned with disturbances initiated at a point or along a line on the surface of a semi-infinite, perfectly elastic medium.

Lamé constants (La-me') Two elastic constants or parameters, λ and μ, which express the relationships between the components of stress and strain for linear elastic behavior of an isotropic solid; λ is identical with rigidity, and μ is equivalent to the bulk modulus minus $2\mu/3$.

lamella (la-mel'-la) (a) A thin scale, leaf, lamina, or layer, e.g. one of the units of a polysynthetically twinned mineral, such as plagioclase. See also: *deformation lamella; exsolution lamella.* (b) An organ, process, or part of an organism resembling a leaf or thin plate, e.g. a *primary lamella* of a brachiopod. Pl: lamellae. (c) One of the many thin layers of minute phosphatic crystallites of which a conodont element is composed (TIP, 1981, pt. W, supp.2).

lamellar (la-mel'-lar) Composed of or arranged in lamellae; disposed in layers like the leaves of a book.

lamellar columella A platelike coral *columella.* In rugose corals, it is generally in the plane of the *cardinal septum* and the *counter septum;* in scleractinian corals, it is oriented parallel with the longer axis of the *calice.*

lamellar conodont element A *conodont element* consisting of numerous thin layers or sheaths, being most obvious in specimens that also contain opaque *white matter.* See also: *fibrous conodont element.*

lamellar flow Flow of a liquid in which layers glide over one another. Cf: *laminar flow [hydraul].*

lamellar layer The *primary layer* of a brachiopod.

lamellar ligament The part of a *ligament* of a bivalve mollusk characterized by lamellar structure and containing no calcium carbonate. It is secreted at the edge of the mantle and is elastic to both compressional and tensional stresses. Cf: *fibrous ligament.*

lamellar linkage The joining, by lamellar septal plates, of corallite centers in scleractinian corals, corresponding to *direct linkage* of stomodaea.

lamellar pyrites *marcasite [mineral].*

lamellar wall A foraminiferal test constructed of thin platelike layers of aragonite or calcite, one layer being formed with addition of each new chamber, and covering the whole previously formed test.

lamellibranch (la-mel'-li-branch) *bivalve; pelecypod.*

lamello-fibrillar structure A plywood-type laminar structure consisting of sheets of more or less parallel horizontal fibers that differ in orientation by irregularly varying angles in successive laminae (Erben, 1972, p.28). The "Spirula nacre" of Mutvei (1964).

lamina [bot] (lam'-i-na) The blade or expanded portion of a leaf.

lamina [paleont] A thin platelike, scalelike, or sheetlike structure in an organism, including bones, certain teeth, and layers of the brain; e.g. a uniform thin sheet of wall substance in the *lorica* of a tintinnid, or a sheetlike structure in some corals, formed by the juxtaposition of two layers of skeletal material in septa and the column. Cf: *lamella.* Pl: laminae.

lamina [sed] The thinnest recognizable unit layer of original deposition in a sediment or sedimentary rock, differing from other layers in color, composition, or particle size; specif. such a sedimentary layer less than 1 cm in thickness (commonly 0.05-1.00 mm thick) (Otto, 1938, p.575; Campbell, 1967, p.16-20). It may be parallel or oblique to the general stratification. Several laminae may constitute a *bed [stratig]* (Payne, 1942, p.1724) or a *stratum* (McKee and Weir, 1953, p.382). Pl: laminae. Syn: *lamination [sed]; straticule.*

laminar (lam'-i-nar) Consisting of, arranged in, or resembling laminae; e.g. "laminar structure" produced by alternation of thin sedimentary layers of differing composition.

laminar flow [glaciol] The normal type of *glacier flow* in which flow vectors are nearly parallel.

laminar flow [hydraul] Water flow in which the stream lines remain distinct and in which the flow direction at every point remains unchanged with time. It is characteristic of the movement of groundwater. Cf: *turbulent flow; mixed flow; lamellar flow.* Syn: *streamline flow; sheet flow [hydraul].*

laminarian (lam-i-nar'-i-an) Pertaining to a large family of kelps, Laminariaceae, of the order Laminariales (brown algae).

laminar layer The secondary shell layer in the majority of strophomenid brachiopods, consisting of flat-lying blades or laths of calcite amalgamated laterally and disposed subparallel with the surface of deposition (Williams, 1970).

laminar structure A *microstructure* of rods, laths, blades, or tablets comprising sheets that are oriented parallel or nearly parallel to the depositional surface. This includes nacreous, seminacreous, plywood, matted, regularly foliated, and semifoliated structures (Carter, 1980a,b).

laminar velocity That velocity of water in a stream below which the flow is laminar and above which it may be either laminar or turbulent. Cf: *turbulent velocity.*

laminar wall A single layer of silica in the wall of a diatom frustule. It may be either of uniform thickness or have local thickenings that form ribs or costae.

laminaset A relatively conformable succession of genetically related laminae bounded by surfaces (called laminaset surface) of erosion and non-deposition (Campbell, 1967).

laminate (lam'-i-nate) adj. Consisting of or containing laminae.

laminated [petrology] (lam'-i-nat'-ed) Consisting of very thin compositional layers. In sedimentary rocks, layers less than 1 cm. thick.

laminated [sed struc] (a) Said of a rock (such as shale) that consists of laminae or that can be split into thin (<1 cm thick) layers. Syn: *laminate.* (b) Said of a substance that exhibits lamination; e.g. "laminated clay" formed in a lake. (c) Said of the sedimentary structure possessed by a laminated rock.

laminated quartz Vein quartz containing slabs, blades or laminar films of other material.

lamination [ign] (lam-i-na'-tion) The spreading-out of the constituents of a lava parallel to the underlying rocks.

lamination [sed] (a) *lamina [sed].* (b) The formation of a lamina or laminae. (c) The state of being laminated; specif. the finest stratification or bedding, typically exhibited by shales and fine-grained sandstones. (d) A laminated structure.

laminite (lam'-i-nite) (a) A term used by Lombard (1963) for a finely laminated detrital rock of the flysch lithofacies, frequently occurring in geosynclinal successions in natural sequences complementary to typical turbidites. It is finer-grained and thinner-bedded than a turbidite, ranging in thickness from a few millimeters to 30 cm, and is believed to form seaward from turbidites as a bottomset bed of a large delta. (b) A term suggested by Adolph Knopf (in Sander, 1951, p.135) to replace *rhythmite* in order to avoid the positive implication of perfect periodicity in the recurrence of laminae.

laminoid (lam'-i-noid) Laterally elongate parallel to stratification; e.g. "laminoid-fenestral fabric" of a limestone, characterized by particulate carbonate interrupted by horizontally elongate gaps (fenestrae) that tend to outline lamination (Tebbutt et al., 1965, p.4).

laminoid-fenestral structure (lam'-i-noid-fe-nes'-tral) A sedimentary fabric in limestone or dolomite, consisting of irregular patches or laminae, once voids, now filled by clear calcite. Patches are crudely oriented parallel with bedding. The structure is formed by desiccation and separating of microbial mat layers or other tissues, burrowing, gas escape, etc. Syn: *bird's-eye; dismicrite.*

lammerite (lam'-mer-ite) A dark green monoclinic mineral: $Cu_3(AsO_4,PO_4)_2$.

lammuchangite A white to pale yellow cubic mineral of the alum group: $TlAl(SO_4)_2 \cdot 12H_2O$. It is water-soluble.

lampadite (lam'-pa-dite) A variety of wad containing as much as 18% copper oxide and often containing cobalt. The term is often used for all hydrous manganese oxides containing copper.

lamprobolite (lam-pro'-bo-lite) A discredited name for an oxygen-rich hornblende. Syn: *basaltic hornblende.*

lamproite (lam'-pro-ite) A group name for hypabyssal or extrusive rocks rich in potassium and magnesium; also, any rock in that group, such as *madupite, orendite, fitzroyite, verite, cedricite,* or *wyomingite.* The term was proposed by Niggli in 1923. Modified in an addendum to the *IUGS classification* by Wooley et al. (1996). See also: Le Maitre (2002).

lamprophyllite (lam-pro-phyl'-lite) A golden-brown platy orthorhombic or monoclinic mineral: $(Na,Mn,K)_3(Sr,Ba)_2Ti_3(Si_2O_7)_2(O,OH,F)_4$.

lamprophyre (lam'-pro-phyre) (a) In the *IUGS classification,* a group of porphyritic igneous rocks in which mafic minerals form the phenocrysts; feldspars, if present, are restricted to the groundmass. Varieties of lamprophyre are camptonite, kersantite, minette, monchiquite, sannaite, spessartite, and vogesite. (b) A group of dark-colored, porphyritic, hypabyssal igneous rocks characterized by panidiomorphic texture, a high percentage of mafic minerals (esp. biotite, hornblende, and pyroxene), which form the phenocrysts, and a fine-grained groundmass with the same mafic minerals in addition to feldspars and/or feldspathoids; also, any rock in that group, e.g. *minette, vogesite, kersantite, spessartite, camptonite, monchiquite, fourchite, alnoite.* Many lamprophyres are altered. Adj: *lamprophyric.*

lamprophyric (lam-pro-phyr'-ic) Said of the holocrystalline-porphyritic texture exhibited by lamprophyres, in which phenocrysts of mafic minerals occur in a fine-grained crystalline groundmass.

lamproschist (lam-pro-schist') Metamorphosed lamprophyre with a schistose structure containing brown biotite and green hornblende.

lamp shell A syn. of *brachiopod.*

Lampson scaling A scaling relation between impact or explosion crater diameter or depth and the cube root of the energy that formed it. Syn: *yield scaling.*

lanarkite (lan'-ark-ite) A white, greenish, or gray monoclinic mineral: $Pb_2O(SO_4)$.

lanceolate (lan'-ce-o-late') Spear-shaped, or shaped like a lance head, such as a leaf or prism that is much longer than broad, widening above the base and tapering to a point at the apex; e.g. said of the form of a lobe of an ammonoid suture, or of a nautiloid whorl section with an acute periphery, or of a bryozoan colony.

lancet plate (lan'-cet) An elongate and spear-shaped or triangular plate located along the midline of the *ambulacrum* of blastoids.

land In a general sense, that part of the Earth's surface that stands above mean sea level. The inclusion of Antarctica's permanent ice in calculating the land surface of the Earth is controversial. Not to be confused with the term "soil".

land accretion Reclamation of land from the sea or other low-lying or flooded areas by draining and pumping, dumping of fill, or planting of marine vegetation.

land air gun An impulsive seismic energy source which involves injecting highly compressed air into a water filled expandable tank that is held against the surface.

land application An alternative for disposal of certain types of hazardous chemical waste in which the waste is applied to the soil and degraded by natural biological activity in the soil (Keller, 2002).

land asphalt Hard asphalt, containing less bitumen and more impurities than *lake asphalt,* from areas outside the pitch lake of Trinidad. It is divided into several varieties depending on the depth at which it is found.

landauite (lan'-dau-ite) A semi-metallic black or brownish-green rhombohedral mineral of the *crichtonite* group: $NaMn^{2+}Zn_2(Ti,Fe^{3+})_6Ti_{12}O_{38}$.

landblink (land'-blink) A yellowish reflection on the underside of a cloud layer over snow-covered land in a polar region; yellower than *iceblink [meteorol].* Also spelled: land blink.

land boundary A line of demarcation between adjoining parcels of land that are politically or legally distinct. The parcels of land may be of the same or of different ownership, but are distinguished at some time in the history of their ownership by separate legal descriptions. A land boundary may be marked on the ground by material monuments placed primarily for this purpose by fences, hedges, ditches, roads and other structures along the line or may be specified by astronomically described points and lines; or by coordinates on a survey system whose position on the ground is witnesses by material monuments which are established without reference to the boundary line. Various other methods are also in use. Although the land boundary is defined to be a line and is determined as such by the surveyor, it is usually considered by law to be a vertical surface extending upwards and downwards from the line.

land breeze A relatively cool surface wind, directed from land to sea (or land to lake) in response to differential cooling between land and sea (Lake); develops at night when the land surface cools more than the adjacent water surface.

land bridge A land connection between continents or landmasses, often subject to temporary or permanent submergence, that permits the migration of organisms; e.g. the Bering Land Bridge. See also: *neck [geog]; filter bridge.*

land compass *surveyor's compass.*

land description The exact location of a parcel of land stated in terms of lot, block, and tract; metes and bound; or section, township and range.

Landenian (Lan-den'-i-an) Obs. European stage: Upper Paleocene (above Montian-Heersian, below Ypresian of Eocene). It includes Thanetian and Sparnacian.

landerite (lan'-der-ite) A pink to rose-pink variety of grossular garnet. Syn: *rosolite; xalostocite.*

landesite (lan'-des-ite) A brown orthorhombic mineral of the *phosphoferrite* group: $(Mn^{2+},Mg)_9Fe^{3+}_3(PO_4)_8(OH)_3 \cdot 9H_2O$. It is an oxidation product of ferroan reddingite. Cf: *kryzhanovskite.*

landfill (land'-fill) *sanitary landfill.*

landflood (land'-flood) An overflowing of inland water onto the land.

landform (land'-form) Any physical, recognizable form or feature of the Earth's surface, having a characteristic shape, and produced by natural causes; it includes major forms such as plain, plateau, and mountain, and minor forms such as hill, valley, slope, esker, and dune. Taken together, the landforms make up the surface configuration of the Earth. Cf: *physiographic form; topographic form.* Syn: *relief feature.*

landform map *physiographic diagram.*

land hemisphere That half of the Earth containing the bulk (about six-sevenths) of the dry land surface; it is mostly north of the Equator, with Paris as its approx. center. Cf: *water hemisphere.*

land ice Any ice mass formed from snow, rain, or other freshwater on land, as a glacier, even though it may be floating in the sea, as an iceberg. Ant: *sea ice.*
land information system A spatial information system containing data on land and land use (Krzanowski et al., 1993).
landlocked (land'-locked) Said of a body of water that is nearly or entirely enclosed by land; e.g. a landlocked bay separated from the main body of water by a bar, or a landlocked lake having no surface outlet.
landmark (land'-mark) (a) Any conspicuous object, natural or artificial, located near or on land and of sufficient interest or prominence in relation to its surroundings to make it outstanding or useful in determining a location or a direction. (b) Any monument, material mark, or fixed object (such as a river, tree, or ditch) used to designate the location of a land boundary on the ground.
landmass (land'-mass) A land area studied as a unit, without regard necessarily to size or relief, on the basis of the sediments derived from it or the paleogeographic evidence indicated by the change in shorelines (Eardley, 1962, p.6).
landmass volume The volume, beneath a land surface, of a body with vertical sides and a base equal to *basin area* at the elevation of the stream's mouth (Strahler, 1952b, p.1120).
land pebble *land-pebble phosphate.*
land-pebble phosphate (land'-peb-ble) A term used in Florida for a *pebble phosphate* occurring as pellets, pebbles, and nodules in gravelly beds a few feet below the ground surface. It is extensively mined. Cf: *river-pebble phosphate.* Syn: *land pebble; land rock; matrix [ore dep] (b).*
land rock A syn. used in South Carolina for *land-pebble phosphate.*
Landsat (Land'-sat) The series of American satellites that collect multispectral images of the Earth's surface in the visible, reflected, and thermal infrared bands. The sensors flown on the satellites include Return Beam Vidicon (RBV), a Multispectral scanner (MSS), and the Thematic Mapper (TM) scanner. The spectral bands and pixel resolution have varied with the different sensors. It was formerly called Earth Resource Technology Satellite, or ERTS. Similar systems include AVHRR and SPOT. See also: *Landsat TM.*
Landsat TM Thematic mapper system that collects electromagnetic radiation from a ground resolution cell of 30 meters in the following seven channels: Channel 1 (0.45-0.52 micrometers), Channel 2 (0.52-0.60 micrometers), Channel 3 (0.63-0.69 micrometers), Channel 4 (0.76-0.90 micrometers), Channel 5 (1.55-1.75 micrometers), Channel 6 (10.4-12.5 micrometers), Channel 7 (2.08-2.46 micrometers) (Krzanowski et al., 1993).
landsbergite (lands'-berg-ite) *moschellandsbergite.*
landscape (land'-scape) The distinct association of landforms, esp. as modified by geologic forces, that can be seen in a single view, e.g. glacial landscape.
land sculpture *sculpture [geomorph].*
landside (land'-side) (a) That part of a near-water feature that is facing toward the land. (b) An obsolete term for shore.
land sky Dark or gray streaks or patches in the sky near the horizon or on the underside of low clouds, caused by the absence of reflected light from ground that is not snow-covered; not as dark as *water sky.* See also: *blink.*
landslide (land'-slide) A general term covering a wide variety of mass-movement landforms and processes involving the downslope transport, under gravitational influence, of soil and rock material en masse. Usually the displaced material moves over a relatively confined zone or surface of shear. The wide range of sites and structures, and of material properties affecting resistance to shear, result in a great range of landslide morphology, rates, patterns of movement, and scale. Landsliding is usually preceded, accompanied, and followed by perceptible creep along the surface of sliding and/or within the slide mass. Terminology designating landslide types generally refers to the landform as well as the process responsible for it, e.g. *rockfall, translational slide, block glide, avalanche, mudflow, liquefaction slide,* and *slump.* Syn: *landsliding; slide [mass move]; landslip.*
landslide breccia A *breccia* that is largely fragmented and wholly assembled by the force of gravity, as by a rockfall or a rockslide.
landslide lake (a) A lake resulting from the damming of a stream valley by the material in a landslide. (b) A long, narrow lake between the back slope of a landslide terrace and a valley wall.
landslide sapping The process of causing landslides by a stream undermining a canyon wall (Freeman, 1925, p.78).
landslide scar A bare or relatively bare surface or niche on the side of a mountain or other steep slope, left by the removal of earth material from the place where a landslide started.
landslide shear surface *slip surface.*
landslide terrace A short, rough-surfaced terrace resulting from a landslide.
landslide track The exposed path in rock or earth formed by a landslide. Syn: *slide [mass move].* Not to be confused with *avalanche track.*
landsliding (land'-slid-ing) The downward movement of a landslide.
landslip (land'-slip) A British syn. of *landslide.*
land survey A survey made to determine boundaries and areas of tracts of land, esp. of privately owned parcels of land. Cf: *cadastral survey; boundary survey.*
land-tied island *tied island.*
lane (a) A narrow, not necessarily navigable, fracture or channel of water through sea ice; it may widen into a lead. (b) A syn. of *lead [ice].* Rarely used.
lane [geophys] A unit of measuring position with standing-wave radio-positioning systems. In phase-comparison (CW) systems, a lane is one cycle of the standing-wave interference pattern resulting from two radiated waves and it depends on the position within the network. Phase-comparison systems yield location within a lane but do not necessarily determine in which lane; lane ambiguity has to be resolved by counting lanes or by some additional measurement.
långbanite (lång'-ban-ite) An iron-black monoclinic (pseudo-trigonal) mineral: originally from Långban, Varmland, Sweden: $(Mn^{2+},Ca)_4(Mn^{3+},Fe^{3+})_9Sb^{5+}Si_2O_{24}$. Not to be confused with *langbeinite.*
langbeinite (lang'-bein-ite) (a) A colorless and variously colored cubic mineral: $K_2Mg_2(SO_4)_3$. It is much used in the fertilizer industry as a source of potassium compounds. Not to be confused with *långbanite.* (b) A group name for minerals with an analogous formula, but with K replaced by NH_4, and Mg replaced by Mn^{2+}.
Langhian (Lan'-ghi-an) A stage of the Standard Global Chronostratigraphic Scale: lower middle Miocene (above Burdigalian, below Serravallian).
langisite (lang'-i-site) A metallic pinkish hexagonal mineral of the *nickeline* group: (Co,Ni)As.
langite (lang'-ite) A blue to green monoclinic mineral: $Cu_4(SO_4)(OH)_6{\cdot}2H_2O$.
Langmuir cell (Lang'-muir) A unit of water circulation, at or adjacent to the surface, driven by the wind, having its long axis essentially parallel with the wind direction, having opposite senses of spiral flow in adjacent cells, and having pronounced ellipticity in shallow water, where it may influence sediment transport (Wetzel, 2001, p.106).
La Niña (Ni'-ña [neenya]) A term introduced by oceanographers to describe a situation, within the ENSO system, that is opposite to *El Niño.* It is marked by anomalously cold sea surface temperatures and often accompanies strong *trade winds.* Cf: *El Niño; ENSO.* Etymol: Spanish, "the girl"; feminine equivalent of *El Niño.*
lanmuchangite A white to pale yellow cubic mineral of the *alum* group: $TlAl(SiO_4)_2{\cdot}12H_2O$. It is water-soluble.
lannonite (lan'-non-ite) A white tetragonal mineral occurring as efflorescences on mine walls and in a breccia zone: $HCa_4Mg_2Al_4(SO_4)_8F_9{\cdot}32H_2O$.
lansfordite (lans'-ford-ite) A colorless monoclinic mineral: $MgCO_3{\cdot}5H_2O$. It alters to nesquehonite on exposure to air.
lantern (lan'-tern) *Aristotle's lantern.*
lanthanides *rare earths.*
lanthanite-(Ce) A white to yellowish orthorhombic mineral: $(Ce,La,Nd)_2(CO_3)_3{\cdot}8H_2O$.
lanthanite-(La) A colorless, white, pink, or yellow orthorhombic mineral: $(La,Ce)_2(CO_3)_3{\cdot}8H_2O$.
lanthanite-(Nd) A pink orthorhombic mineral: $(Nd,La)_2(CO_3)_3{\cdot}8H_2O$.
laphamite (lap'-ham-ite) A dark red monoclinic mineral: $As_2(Se,S)_3$. It is the Se analogue of orpiment.
lapiaz (la-piaz') *karren.* Etymol: Pyrennees dialect. Also spelled: *lapiés.*
lapidary (lap'-i-dar-y) (a) A cutter, grinder, and polisher of colored stones, or of precious stones other than diamonds. Syn: lapidist. (b) The art of cutting, polishing, and engraving gemstones other than diamonds. (c) An obsolete term for a short treatise on metals, stones, and gems, describing their supposed medicinal, magical, or mythical characteristics.
lapidification (la-pid'-i-fi-ca'-tion) An obsolete term signifying conversion into stone or stony material, such as the process of petrifaction or lithification.
lapidofacies (lap'-i-do-fa'-cies) Facies related to diagenesis (Vassoevich, 1948).
lapieite (la-pie'-ite) A metallic greenish-gray orthorhombic mineral: $CuNiSbS_3$.
lapiés (la-pies') *lapiaz.*
lapilli (la-pil'-li) Pyroclastic materials that may be either essential, accessory, or accidental in origin, of a size range that has been variously defined within the limits of 2 and 64 mm. The fragments may be either solidified or still viscous when they land (though some classifications restrict the term to the former); thus there is no characteristic shape. An individual fragment is called a *lapillus.* Cf: *volcanic gravel; block [volc]; cinder.*
lapillistone (la-pil'-li-stone) A pyroclastic rock composed largely of *lapilli.*
lapillite (la-pil'-lite) *lapilli tuff.*
lapilli tuff An indurated deposit that is predominantly lapilli, with a matrix of ash. Syn: *lapillite.*
lapillus (la-pil'-lus) The singular form of *lapilli.*
lapis lazuli (lap'-is laz'-u-li) (a) A blue, semitranslucent to opaque, granular crystalline rock used as a semiprecious stone for ornamental purposes and composed essentially of lazurite and calcite but also containing hauyne, sodalite, pyrite inclusions, and other minerals. It usually has a rich azure-blue color, but may be other shades of blue, depending on the amount of inclusions. It is probably the original sapphire of the ancients. Syn: *lazuli.* (b) An old name for lazurite, still used esp. for the gem variety. (c) An ultramarine-colored serpentine from India.

Laplace azimuth (La-place') A *geodetic azimuth* derived from an astronomic azimuth by means of the *Laplace equation,* expressing the relationship between astronomic and geodetic azimuths in terms of astronomic and geodetic longitudes and geodetic latitude.

Laplace equation An equation, used to derive the *Laplace azimuth,* that expresses the relationship between astronomic and geodetic azimuths in terms of astronomic and geodetic longitudes and geodetic latitude. See also: *Laplace station.*

Laplace station A triangulation or traverse station at which the *Laplace azimuth* is observed. At a Laplace station, both astronomic longitude and astronomic azimuth are observed.

laplandite (lap'-land-ite) A light gray, yellowish, or bluish orthorhombic mineral: $Na_4CeTiPSi_7O_{22}·5H_2O$.

lapout The lateral termination of strata at their depositional pinchout. Lapout may occur at the upper boundary of a sequence, where it is called *toplap,* or at the lower boundary, where it is called *onlap* or *downlap* (Mitchum, 1977, p.208).

lap-out map A map showing the areal distribution of formations immediately overlying an unconformity. Syn: *worm's-eye map.*

lapparentite (lap-pa-rent'-ite) *tamarugite.*

lappet (lap'-pet) Projection of the *peristome* on whorl sides or venter of nautiloid and ammonoid conchs (TIP, 1957, pt. L, p.4).

lapse rate The rate at which some atmospheric property, usually temperature, decreases with altitude; the vertical gradient. For temperature, the average in the troposphere is about 0.6°C/100 meters; however, it varies according to time of day, season, and latitude.

laqueiform (la-que'-i-form) Said of the *loop* pattern in a dallinid brachiopod (as in the family Laqueidae) in which posterior connecting bands from the ascending branches to the descending branches "are retained during enlargement and proportional thinning during change from frenuliniform to terebrataliiform loop" (TIP, 1965, pt. H, p.147).

Laramian Orogeny (La-ram'-i-an) *Laramide Orogeny.*

Laramic Orogeny (Lar'-a-mic) *Laramide Orogeny.*

Laramide Orogeny (Lar'-a-mide) A time of deformation, typically recorded in the eastern Rocky Mountains of the United States, whose several phases extended from late Cretaceous until the end of the Paleocene (approximately 80 to 50 Ma). Intrusives and accompanying ore deposits emplaced about this time in the mountain states are commonly called Laramide (e.g. the Boulder Batholith, Montana). The term is also used for orogenic time in western Canada that produced the Canadian Rockies. In Canada, the foreland edge of the Laramide structures is part of the thin-skinned foreland fold-thrust belt that started forming during the *Sevier Orogeny.* In the U.S., Laramide structures include the thick-skinned basement-cored uplifts of the Rockies and the monoclines of the Colorado Plateau. These structures lie east of the Sevier fold-thrust belt. Arc volcanism swept eastward during Laramide time, probably due to shallowing of the dip angle of the subducting Farallon plate, beneath North America. It is named for the Laramie Formation of Wyoming and Colorado, probably a synorogenic deposit. Also called: *Laramic Orogeny; Laramian Orogeny.* Obsolete syn: *Laramide Revolution.*

Laramide Revolution *Laramide Orogeny.*

larderellite (lar-de-rel'-lite) A white monoclinic mineral: $(NH_4)B_5O_7(OH)_2·H_2O$.

lardite (lar'-dite) (a) White hydrated silica, probably a variety of opal, occurring in clay in central Russia. (b) Massive talc; steatite. (c) *agalmatolite.*

lard stone Massive talc; steatite.

large boulder A *boulder* having a diameter in the range of 1,024-2,048 mm (40-80 in., or -10 to -11 phi units).

large cobble A term for a cobble having a diameter in the range of 128-256 mm (5-10 in., or -7 to -8 phi units).

large igneous province A region characterized by massive crustal emplacements of predominantly mafic (Mg- and Fe-rich) extrusive and intrusive rocks which originated through processes other than "normal" seafloor spreading (Coffin and Eldholm, 1994); includes continental flood basalt provinces (e.g., Deccan), oceanic plateaus (e.g., Ontong, Java), and North Atlantic volcanic passive margins. Abbrev: LIP.

large-ion lithophile elements Elements which, because of their large ionic radii, have *bulk partition coefficients* < 0.5 for basaltic magma equilibrated with upper mantle lherzolite (Gill, 1981, p.10). They include K, Rb, Sr, Y, Cs, Ba, the rare-earth elements, Pb, Th, and U. Abbrev: LILE.

larger foraminifera (larg'-er) An informal term generally used to designate those foraminifers studied with the aid of thin-sectioning. Cf: *smaller foraminifera.*

large-scale map A map at a scale (in the U. S., 1:25,000 and larger) such that a small area can be shown in fine detail and with great accuracy; a map whose representative fraction has a small denominator (such as 1:10,000). The U.S. Geological Survey defines large-scale maps as those at a scale larger than 1:50,000. The Defense Mapping Agency defines large-scale maps as those at 1:75,000 or larger. Cf: *intermediate-scale map; small-scale map.*

larisaite A vitreous yellow monoclinic mineral: $Na(H_3O)(UO_2)_3(SeO_3)_2·4H_2O$.

larnite (larn'-ite) A white to gray mineral: βCa_2SiO_4. It is a metastable monoclinic phase of calcium orthosilicate, stable from 520° to 670°C, and tending to break down or "fall" to the stable calcio-olivine at standard pressure and temperature. Cf: *bredigite.* Syn: *belite.*

larosite (la'-ro-site) A metallic orthorhombic mineral: $(Cu,Ag)_{21}(Pb,Bi)_2S_{13}$.

larsenite (lar'-sen-ite) A colorless or white acicular orthorhombic mineral: $PbZnSiO_4$.

Larsen method (Lar'-sen) *Lead-alpha age method.* This method was suggested and developed under the guidance of E. S. Larsen, Jr. (1879-1961), U.S. mineralogist and petrologist.

Larsen variation diagram A diagram in which the weight percent of each oxide constituent of a rock is plotted as the ordinate against the abscissa, which is one-third of the SiO_2+K_2O-FeO-MgO-CaO. The diagram was devised by E. S. Larsen (1938).

larvikite (lar'-vik-ite) An alkalic *syenite,* grading to *monzonite,* composed of phenocrysts of two feldspars (esp. oligoclase and alkali feldspar), often intimately intergrown, which comprise up to 90% of the rock, with diopsidic augite and titanian augite as the chief mafic minerals, and accessory apatite (generally abundant), ilmenite, and titaniferous magnetite, and less commonly olivine, orthopyroxene, biotite, and quartz or feldspathoids (less than 10 percent by volume). Its name, given by Brögger in 1890, is derived from Larvik, Norway.

laser leveling instrument A leveling rod having as target a photodetector in a mounting which can be slid up or down on the rod. Also referred to as a laser-type leveling-rod. It is used with a laser-type leveling-instrument to indicate differences of elevation above or below the plane described by rotation of the laser beam about the standing axis of the laser-type leveling instrument.

Lasky's law A general principle, proposed by Lasky (1950), that, given the graduation from relatively rich to relatively lean material in many ore deposits, the tonnage of the deposit will increase at a constant geometric rate as successively lower *cutoff grades* are selected.

last appearance datum Commonly used in its acronym form, LAD; mostly applied to fossils in cored sedimentary rock sequences. Cf: *FAD.*

late Pertaining to or occurring near the end of a segment of time. The adjective is applied to the name of a geologic-time unit (era, period, epoch) to indicate relative time designation and corresponds to *upper* as applied to the name of the equivalent time-stratigraphic unit; e.g. rocks of an Upper Jurassic batholith were intruded in Late Jurassic time. The initial letter of the term is capitalized to indicate a formal subdivision (e.g. "Late Devonian") and is set in lower case indicate an informal subdivision (e.g. "late Miocene"). The informal term may be used for eras and epochs, and for periods where there is no formal subdivision. Cf: *middle [geochron]; early.*

late diagenesis Deep-seated diagenesis, occurring a long time after deposition, when the sediment is more or less compacted into a rock, but still in the realm of pressure-temperature conditions similar to those of deposition; it represents a transition from diagenesis to metamorphism. Syn: *epigenesis; epidiagenesis; metharmosis.*

late glacial Pertaining to the time of the waning of the last glaciation; specif. the "Late Glacial period" of the Pleistocene Epoch, immediately preceding the Preboreal phase.

late neanic stage During increase in colonial corals, the stage following the hystero-neanic stage. It is taken to start when the offset possesses all or nearly all the characters of the adult stage except that it has smaller dimensions and fewer septa (Fedorowski and Jull, 1976, p.42). Cf: *hystero-brephic stage; hystero-neanic stage.*

latent magma (la'-tent) A highly viscous magma that exists under high pressure beneath the Earth's crust and reacts as a solid body, e.g. with respect to the propagation of earthquake waves. With a decrease in pressure, the magma becomes sufficiently fluid to flow (Schieferdecker, 1959, term 3827). Obsolete.

latera (lat'-er-a) Plural of *latus.*

lateral [drill] A relatively short branch, usually horizontal, drilled out from an oil or gas well.

lateral [paleont] (lat'-er-al) n. (a) A compartmental plate in certain cirripede crustaceans, bounded by a carinolateral and a rostrolateral. In other cirripedes, the term is synonymous with *latus.* (b) One of a series of ossicles along the side of an arm in an ophiuroid. Also, one of a circlet of five plates in certain cystoids. (c) A lateral part; e.g. a lateral tooth. (d) In coiled cephalopod conchs, those parts of the shell wall lying between the ventral and dorsal sides.

lateral [streams] n. A *lateral stream.*

lateral [volc] *parasitic [volc].*

lateral accretion Sediment accumulation at the side of a channel or bank. Lateral accretion deposits or sidebars grow towards the channel axis whereas cross bedding dips parallel to channel axis. Cf: *vertical accretion.*

lateral accretion surface A sedimentary structure preserved in deposits laid down by *lateral accretion* on the side of a channel or by bars migrating toward the channel axis. Commonly observed in vertical section, a *set [stratig]* of such surfaces makes up a form of large-scale *low-angle cross-bedding* termed *epsilon cross-stratification* by Allen (1963); internal cross-strata drape the former depositional surface and are visible due to variations in texture within the deposit.

lateral bud In seed plants, an axillary bud or an *adventitious* bud on the side of a stem.
lateral channel A channel formed by a meltwater stream flowing laterally away from a glacier through a notch in bordering hills (Rich, 1908, p.528).
lateral consequent stream A *secondary consequent stream* flowing down the flank of an anticline or syncline.
lateral corrasion Corrasion of the banks of a stream.
lateral crevasse *marginal crevasse.*
lateral depression series A group of small, basinlike depressions on the exterior lateral edges of the cover plates of lebetodiscid edrioasteroids. They flank the suture lines between contiguous ambulacral cover plates, and are associated with the external foramina of the cover-plate passageways (Bell, 1976).
lateral depressor pit A small hollow near one or both basal angles of the *scutum* of a cirripede crustacean, serving for attachment of the lateral muscle that depresses or draws down.
lateral dissepiment A *dissepiment* of a rugose coral, characterized by a blisterlike form and developed in isolated manner on the sides of septa. Cf: *horseshoe dissepiment.*
lateral dune A sand dune flanking a larger dune, formed around an obstacle.
lateral erosion The wearing-away of its banks by a meandering stream as it swings from side to side, impinging against and undercutting the banks as it flows downstream; it results in *lateral planation.*
lateral face One of the sides of a conodont element flanking the anteroposterior midline (TIP, 1981, pt. W, supp.2).
lateral fault A fault along which there has been strike separation. Cf: *strike-slip fault; dip-separation fault.* See also: *wrench fault; right-lateral separation; left-lateral separation.* Obsolete syn: *strike-separation fault.*
lateral increase A type of *increase* (offset formation of corallites) in fasciculate and massive coralla characterized by sideward outgrowth.
lateral lake A *fluviatile lake* formed in the valley of a tributary stream by the silting-up of the channel of the main stream, thereby producing embankments or levees that impound the water of the tributary (Wetzel, 2001, p.31).
lateral levee lake A lake occupying a depression behind a *natural levee* (Wetzel, 2001, p.31).
lateral lobe Any adapical inflection of a *suture* of a nautiloid or ammonoid shell between the ventral and dorsal lobes; in coiled conchs, the lateral lobes may be external or internal according to whether they are on the flanks or dorsal areas (TIP, 1964, pt. K, p.57).
lateral log An obsolete 3-electrode *resistivity log,* designed to record resistivity in a porous formation well beyond the radius of invasion by drilling mud. Cf: *focused-current log.*
lateral migration Movement of oil or gas through permeable zones parallel to the stratification.
lateral moraine (a) A low ridgelike moraine carried on, or deposited at or near, the side margin of a mountain glacier. It is composed chiefly of rock fragments loosened from the valley walls by glacial abrasion and plucking, or fallen onto the ice from the bordering slopes. (b) An end moraine built along the side margin of a glacial lobe occupying a valley. Cf: *flanking moraine.* Syn: *side moraine; valley-side moraine.*
lateral oblique muscle One of a pair of muscles in certain brachiopods, originating on the pedicle valve anteriorly and laterally from the posterior adductor muscles, and passing anteriorly and dorsally to insertions either on the brachial valve and anterior body wall against the anterior adductor muscles (as in the family Discinidae) or entirely on the anterior body wall (as in the family Craniidae) (TIP, 1965, pt. H, p.147). Cf: *internal oblique muscle.*
lateral planation The reduction of the land in an interstream area to a plain or a nearly flat surface by the *lateral erosion* of a meandering stream; the creation and development by a stream of its floodplain.
lateral process Process projecting from one of the faces of a conodont element flanking the anteroposterior midline (TIP, 1981, pt. W, supp.2).
lateral ramp A thrust ramp that strikes generally parallel to the transport directon of the thrust sheet.
lateral rupture An *acritarch excystment* form consisting of a more or less straight *suture* in the vesicle wall that does not divide the vesicle into two equal halves, e.g. in *Visbysphaera* (Colbath and Grenfell, 1995). Syn: *lateral split.*
lateral saddle An adoral inflection of a *suture* of a nautiloid or ammonoid shell between the ventral and dorsal lobes; in coiled conchs, lateral saddles may be external or internal according to whether they are on the flanks or dorsal areas (TIP, 1964, pt. K, p.57).
lateral search *profiling.*
lateral secretion A theory of ore genesis formulated in the 18th century and passing in and out of use since. It postulates the formation of ore deposits by the leaching of adjacent wall rock. In current usage, convectively driven fluids associated with cooling plutons are thought to have removed metals from adjacent host rocks and transported them to new sites of deposition, as in the formation of certain porphyry base-metal deposits. See also: *lithogene; segregated vein.*
lateral shared cover plate One of a pair of oral plates in edrioasteroids that flank the transverse oral midline. They are proximal to the lateral ambulacral bifurcation plate and distal to the central primary orals (Bell, 1976).
lateral sinus A notch or re-entrant in the lateral part of the apertural margin (peristome) of a cephalopod.
lateral split *lateral rupture.*
lateral spread (a) Lateral movements in a fractured mass of rock or soil, which result from liquefaction or plastic flow of subjacent materials. (b) Lateral movement of intact blocks of soil or rock material which slide on a gently inclined planar surface, such as a bedding plane or a liquefied sand layer.
lateral storage *bank storage.*
lateral stream A stream situated on, directed toward, or coming from the side; e.g. a stream flowing along the edge of a lava flow that recently filled part of a valley. Syn: *lateral.*
lateral tooth A *hinge tooth* located some distance from the *beak* of a bivalve mollusk, situated anterior or posterior to the middle of the hinge, lying ahead of or behind the *cardinal teeth,* and separated from them by an edentulous space. Its long axis is subparallel to the *hinge line.*
lateral valley A *longitudinal valley* developed parallel to the regional structure.
lateral variation A change in the sedimentary characteristics of a formation in a horizontal direction. It is particularly significant in petroleum geology when porosity and permeability are affected, as in a *facies change* from sandstone to siltstone or shale (a *shale-out*).
lateral wall One of a pair of oppositely placed *vertical walls* bounding the sides of a cheilostome bryozoan zooid and commonly developed as an *exterior wall.*
laterite (lat'-er-ite) (a) An older term for a highly weathered red subsoil or material rich in secondary oxides of iron, aluminum, or both, nearly devoid of bases and primary silicates, and commonly with quartz and kaolinite. It develops in a tropical or forested warm to temperate climate, and is a residual product of weathering. Laterite is capable of hardening after a treatment of wetting and drying, and can be cut and used for bricks; hence its etymology: Latin, latericius, "brick". See also: *laterite soil; plinthite.* (b) In modern usage, the iron oxide rich, silica-poor upper soil horizon of intensely weathered *regolith* found in tropical climates (Eggleton, 2001).
laterite soil A great soil group in the 1938 classification system, characterized by a thin A horizon and a reddish leached B horizon overlying laterite (USDA, 1938). This group was dropped in the 1949 revision and these soils were included with Latosols. Laterite soils are now classified as *Orthox.*
lateritic residuum Ferruginous *duricrust.*
lateritization (lat'-er-it'-i-za'-tion) *laterization.*
laterization (lat'-er-i-za'-tion) A general term for the process that converts a rock or soil to *laterite.* Also spelled: *lateritization.*
Laterolog (Lat'-er-o-log') A trade name for a *focused-current log.* See also: *Microlaterolog.*
late wood Xylem formed in late stages of a growth zone, which is more dense than the wood produced earlier. Cf: *early wood.* Syn: *summerwood.*
lath [cryst] A crystal that is *lath-shaped.*
lath [paleont] The part of a *heterococcolith* with one large dimension, one intermediate, and one very small.
lath-shaped Said of the habit of a crystal that is long and thin, and of moderate to narrow width. In thin section, lath-shaped crystals are often cross sections of platy or tabular crystals.
lath-type fibrous prismatic structure A *fibrous prismatic structure* in which the fibers show generally rectangular cross sections with shapes intermediate between rods and blades (Carter, 1990, Fig. 2B).
latilamina (lat-i-lam'-i-na) A layer, 1-10 mm thick, composed of several laminae or dissepiments in some stromatoporoids, the boundaries of which are marked as planes of weakness in the coenosteum and by crowding of the laminae. Pl: latilaminae.
Latin square (Lat'-in) An array of size $n \chi n$, containing *n* letters displayed *n* times in such a manner that each letter occurs once only in each row and in each column of the array. Used in experimental design to control noninteracting extraneous factors in an efficient manner.
latite (la'-tite) (a) In the *IUGS classification,* a volcanic rock defined in the QAPF diagram by Q/(Q+A+P) between 0 and 5% and P/(P+A) between 35 and 65%, and in the TAS diagram as a *trachyandesite* with $Na_2O-2<K_2O$. (b) A porphyritic extrusive rock having phenocrysts of plagioclase and potassium feldspar (probably mostly sanidine) in nearly equal amounts, little or no quartz, and a finely crystalline to glassy groundmass, which may contain obscure potassium feldspar; the extrusive equivalent of *monzonite.* Latite grades into *trachyte* with an increase in the alkali feldspar content, and into *andesite* or basalt, depending on the presence of sodic or calcic plagioclase, as the alkali feldspar content decreases. It is usually considered synonymous with *trachyandesite* and *trachybasalt,* depending on the color. The name, given by Ransome in 1898, is derived from Latium, Italy.
latitude (lat'-i-tude) (a) The angle that the normal to a given point on the surface of a sphere or ellipsoid makes with the plane of the equator; specif. angular distance of a point on the Earth's surface north or south of the equator, measured along a meridian through 90 degrees (the equator being latitude zero degrees, the North Pole lat. 90°N, and the South Pole lat. 90°S). A degree of latitude on the Earth's surface ranges in length from 68.704 statute miles at the equator to 69.407 statute miles at the poles. Abbrev: lat. Symbol: ϕ. See also: *astronomic latitude; geodetic latitude;*

geocentric latitude; geographic latitude; celestial latitude. Cf: *parallel.* (b) The projection on the meridian of a given course in a plane survey equal to the length of the course multiplied by the cosine of its bearing. (c) A linear coordinate distance measured north or south from a specified east-west line of reference; e.g. *northing* and *southing.* Cf: *longitude.*

latitude correction (a) A correction of gravity data with latitude, because of variations in centrifugal force owing to the Earth's rotation and because of differences in the radius owing to polar flattening. The correction for latitude ϕ amounts to 1.3049 sin 2 ϕ mgal/mi = 0.8108 sin 2 ϕ mgal/km. See also: *international gravity formula.* (b) The north-south sorrection made to observed magnetic field intensities in order to remove the Earth's normal field (leaving, as the remainder, the anomalous field). Cf: *longitude correction.*

latitude difference The length of the projection of a line onto a meridian of reference in a plane survey, being equal to the length of the line multiplied by the cosine of its bearing; e.g. *northing* (positive difference) and *southing* (negative difference). Cf: *meridional difference; departure.*

latiumite (lat'-i-um-ite') A colorless monoclinic mineral: $(Ca,K)_4(Si,Al)_5O_{11}(SO_4,CO_3)$.

Latosol (Lat'-o-sol) A suborder of zonal soils including soils formed under forested, tropical, humid conditions and characterized by low silica-sesquioxide ratios of the clay fractions, low base-exchange capacity, low activity of the clay, low content of most primary minerals, low content of soluble constituents, a high degree of aggregate stability, and usually having a red color. Not used in current U.S. system of soil taxonomy.

latrappite (la-trap'-pite) A submetallic black orthorhombic mineral of the *perovskite* group: $(Ca,Na)(Nb,Ti,Fe)O_3$.

lattice [cryst] (lat'-tice) *crystal lattice.*

lattice [reef] *growth lattice.*

lattice bar A bar composing the *lattice shell* of a radiolarian.

lattice constant In a crystal lattice, the length of one edge of its unit cell; also, the angle between two edges of the cell. Syn: *lattice parameter; parameter [cryst]; cell dimension.*

lattice defect *crystal defect.*

lattice drainage pattern *rectangular drainage pattern.*

lattice parameter *lattice constant.*

lattice plane In a *crystal lattice,* a plane whose orientation may be expressed by *Miller indices.* In a *crystal structure,* a lattice plane need not contain atoms. Cf: *atomic plane.*

lattice point One of the points constituting a *crystal lattice.*

lattice pore One of the open spaces surrounded by lattice bars in a radiolarian skeleton of the subfamily Trissocyclinae.

lattice-preferred orientation (lat'-tice-pre-ferred') The preferred orientation of crystallographic axes or planes. In metamorphic rocks, it results from ductile deformation (crystal gliding and/or dynamic recrystallization) and is dependent on the mineral structure and on the physical conditions (pressure, temperature, stress) during deformation. In igneous rocks, it is related mainly to the original shape of the crystals during settling or flow. Cf: *shape-preferred orientation.*

lattice row A row of lattice points. See also: *crystal axis.*

lattice shell A porous sheath that surrounds all or a part of the sagittal ring of some radiolarians, or is divided into symmetric halves by the sagittal ring in other radiolarians; a meshwork radiolarian skeleton.

lattice spine One of the spines that project from the lattice bar in a radiolarian skeleton, either distributed randomly or confined to the junctions of two or more lattice bars.

lattice texture [eco geol] In mineral deposits, a texture produced by exsolution in which elongate crystals are arranged along structural planes.

lattice texture [meta] A texture that is typical of the mineral serpentine in a rock when it replaces an amphibole. Cf: *knitted texture.*

Lattorfian (Lat-torf'-i-an) North European stage: upper Eocene, in type section (below Rupelian). Syn: Latdorfian; approx. equiv. *Priabonian.*

latus (la'-tus) (a) Any of paired plates forming part of the shell in certain cirripede crustaceans, but not including tergum and scutum; e.g. "carinal latus" (carinolateral) or "rostral latus" (rostrolateral). Syn: *lateral [paleont].* (b) The surface of a crinoid *columnal* or *cirral,* exclusive of articular facets. Pl: *latera.*

laubmannite (laub'-mann-ite) Discredited as a mixture of dufrénite, kidwellite and beraunite.

Laue camera (Lau'-e) The instrument used in the *Laue method* of X-ray diffraction analysis. A pinhole defines an X-ray beam perpendicular to a flat film. When the back-reflection method is used, a pinhole is made through the film which is then mounted between the X-ray tube and the crystal.

Laue class *Laue group.*

Laue equations Three simultaneous equations that represent the necessary conditions for radiation to be diffracted by a three-dimensional crystal structure.

Lauegram (Lau'-e-gram) The diagram of X-ray diffraction made according to the *Laue method.* See also: *Laue spot.* Syn: *Laue pattern.*

Laue group One of eleven subsets of the 32 crystal classes, obtained by adding a *center of symmetry* to each class. This is the symmetry of the *weighted reciprocal lattice,* except in cases where *Friedel's law* is violated by the effects of *anomalous dispersion.*

laueite (lau'-e-ite) (a) A honey-brown to orange triclinic mineral: $Mn^{2+}Fe^{3+}_2(PO_4)_2(OH)_2{\cdot}8H_2O$. (b) A group name for minerals with the analogous formula, but with Mn^{2+} replaced by Mg, Fe^{2+}, and Fe^{3+} replaced by Al.

Laue method A technique of X-ray diffraction analysis using a single fixed crystal irradiated by a beam of a continuous spectrum of X-rays. The patterns, or *Lauegram,* are observed after X-ray transmission or by reflection back to their source, called back reflection. See also: *Laue camera.*

Laue pattern *Lauegram.*

Laue spot A single spot on a *Lauegram.*

Laue symmetry The symmetry that results when a *center of symmetry,* if absent, is added to the class of a crystal. There are eleven Laue symmetries, corresponding to the eleven Laue groups.

laumontite (lau'-mont-ite) A white or variously colored monoclinic zeolite mineral: $CaAl_2Si_4O_{12}{\cdot}4H_2O$. It sometimes contains appreciable sodium, and on exposure to air it loses water, becomes opaque, and crumbles. It occurs as prismatic crystals in veins in schist and slate, and in cavities in igneous rocks. Also spelled: *lomonite; lomontite.*

launayite (lau-nay'-ite) A metallic monoclinic mineral: $Pb_{22}Sb_{26}S_{61}$.

launder (laun'-der) A trough, channel, gutter, flume, or chute by which water or powdered ore is conveyed in a mining operation.

lauoho o pele *Pele's hair.*

Laurasia (Lau-ra'-sia) The protocontinent of the Northern Hemisphere, corresponding to *Gondwana* in the Southern Hemisphere, from which the present continents of the Northern Hemisphere have been derived by separation and continental drift. The supercontinent from which both Laurasia and Gondwana were derived is *Pangea.* The protocontinent Laurasia included most of North America, Greenland, and most of Eurasia, excluding India. Etymol: a combination of "Laurentia", a paleogeographic term for the Canadian Shield and its surroundings, and "Eurasia".

laurdalite (laur'-dal-ite) An alkalic *syenite* containing more than 10 percent modal feldspathoids and characterized by porphyritic texture. The name, given by Brögger in 1890, is for Laurdal, Norway.

laurelite (lau'-rel-ite) A silky colorless hexagonal mineral: $Pb_7F_{12}Cl_2$.

Laurentian (Lau-ren'-tian) A name that is widely and confusingly used for granites and orogenies of Precambrian age in the Canadian Shield. It is named for the Laurentian Highlands northwest of the St. Lawrence River in eastern Canada (a part of the Grenville province of current usage), where Logan (1863) recognized the Laurentian granites, now dated radiometrically at about 1,000 Ma. The term was misapplied by Lawson (1885) to the oldest granites near the U.S.-Canadian border northwest of Lake Superior, from which Schuchert subsequently derived his Laurentian Revolution, or orogeny, that was supposed to have terminated the Archeozoic. Modern work shows that Lawson's Laurentian is older than the 2,400-m.y.-old Algoman Orogeny and granites, at the end of the Archean of the present Canadian classification, but no radiometric dates for it survive, and its significance and extent are uncertain. It has been suggested that the term "Laurentian" be restored to Logan's original meaning. See also: *Grenville Orogeny.*

laurionite (lau'-ri-on-ite') A colorless orthorhombic mineral: PbCl(OH). It is dimorphous with paralaurionite.

laurite (lau'-rite) An iron-black mineral of the *pyrite* group: RuS_2. It often contains osmium. Laurite is found in association with platinum in placer deposits and usually occurs in minute octahedrons resembling those of magnetite.

lausenite (lau'-sen-ite) An ill-defined white silky or fibrous monoclinic mineral: $Fe^{3+}_2(SO_4)_3{\cdot}6H_2O$.

lautarite (lau'-ta-rite) A colorless or yellowish monoclinic mineral: $Ca(IO_3)_2$.

lautenthalite A vitreous blue monoclinic mineral: $PbCu_4(SO_4)_2(OH)_6{\cdot}3H_2O$. It is the Pb analogue of devilline.

lautite (lau'-tite) A metallic gray or black orthorhombic mineral: CuAsS.

lava (la'-va) A general term for a molten *extrusive;* also, for the rock that is solidified from it.

lava ball A globular mass of lava that is scoriaceous inside and compact on the outside; it is formed by the coating of a fragment of scoria by fluid lava. Syn: *pseudobomb; volcanic ball.*

lava blister *blister [volc].*

lava breccia (a) Autoclastic breccia produced by fragmentation during movement of the upper and lower crusts of lava flows. (b) A type of *volcanic breccia.*

lava cascade A cascade of fluid, incandescent lava, formed when a lava river passes over a cliff or steep part of its course.

lava cave Any cave formed by volcanic processes, including *lava tubes, blister* caves, caves formed from fractured solidified flows, and occasionally dormant volcanic *vents* (Larson, 1990).

lava channel A long, open trough in a lava flow that carries a river of lava to a flow front (Larson, 1990).

lavacicle (la'-va-ci-cle) A general term applied to nearly anything that protrudes into a *lava tube* (Larson, 1990).

lava column (a) The column of fluid or solidified lava in a volcanic conduit. (b) A formation of lava dripstone extending to the floor from a ledge in a lava tube. Syn: *magma column.*

lava dam A *levee [speleo]* across a lava tube (Larson, 1990, p.27).

lava dome A steep-sided, rounded accumulation of lava extruded from a volcano to form a dome-shaped or bulbous mass of congealed lava above and around the vent (Peterson and Tilling, 2000). Commonly parasitic on the flanks of, or within the crater of, larger edifices, e.g., Unzen, Showa Sin-Zan. Syn: *volcanic dome.* Cf: *coulee [volc]; endogenous dome; exogenous dome; low dome; Peleean dome; upheaved plug; torta.*

lava eruption A volcanic phase that is characterized by the emission of lava, with few if any explosive phenomena. Cf: *explosive eruption; mixed eruption.*

lava fall A precipitous drop in the floor of a lava tube over which lava flowed (Larson, 1990).

lava field A more or less well-defined area that is covered by lava flows. Cf: *ash field; volcanic field.*

lava flow A lateral, surficial outpouring of molten lava from a vent or a fissure; also, the solidified body of rock that is so formed. Syn: *flow [volc].*

lava flow-unit A separate, distinct lobe of lava that issues from the main body of a lava flow (Nichols, 1936). Obsolete.

lava fountain A jet of incandescent lava, shot into the air as the magma reaches the surface by the hydrostatic pressure on the liquid and the expansion of gas bubbles forming in it. Fountains usually range from about 10 to 100 m in height, but occasionally reach 300 m. They are characteristic of Hawaiian-type eruptions. See also: *curtain of fire.*

lava-inflation cleft A deep axial or star-shaped crack characteristic of *tumulus*, commonly invaded by squeeze-ups (Walker, 1991).

lava lake A lake of molten lava, usually basaltic, in a volcanic crater or depression. The term refers to solidified and partly solidified stages as well as to the molten, active lava lake.

lava levee The scoriaceous sheets of lava that overflowed their natural channels and solidified to form a levee, similar to levees formed by an overflowing stream of water.

lava plain A broad stretch of level or nearly level land, usually many hundreds of square kilometers in extent, underlain by a relatively thin succession of lava flows, most of which are basaltic and the product of fissure eruptions.

lava plateau A broad, elevated tableland or flat-topped highland, usually many hundreds or thousands of square kilometers in extent, underlain by a thick succession of lava flows, most of which are tholeiitic basalts and the product of fissure eruption. Syn: *basaltic plateau.* Cf: *flood basalt.*

lava rag An irregular-shaped bit of scoriaceous material ejected from a volcano.

lava rise A more or less flat-topped portion of a pahoehoe lava flowfield that was bodily uplifted by injection of lava beneath the surface crust. Hollow rises are a form of subsidence cave (Walker, 1991). Syn: *pressure plateau.*

lava rise pit A negative topographic feature of pahoehoe lava flowfields, surrounded by lava rises and/or tumuli, the site of comparatively little or no subcrustal injection of lava (Walker, 1991). Syn: *sink [volc].*

lava shield A *shield volcano* of the basaltic type.

lava stalactite A *stalactite* composed of lava (Larson, 1990). Cf: *lavacicle.*

lava stalagmite A vertically oriented accretion of droplets and dribbles of semisolid lava occurring in a wide variety of shapes (Larson, 1990).

lava toe One of a series of small, bulbous projections that develop at the front of a moving pahoehoe flow, formed by the breaking-open of the crust and the emergence of fluid lava.

lava tree A half-cylindrical projection above the surface of a lava flow, formed when fluid lava flows against the upstream side of a tree; the tree then burns, forming a *lava-tree mold.* An unusual form, seen only on recent flows.

lava-tree mold A cylindrical hollow in a lava flow formed by the envelopment of a tree by the flow, solidification of the lava in contact with the tree, and disappearance of the tree by burning and subsequent removal of the charcoal and ash. The inside of the mold preserves the surficial features of the tree. See also: *lava tree.*

lava trench (a) A lava channel that never had a roof. (b) A lava tube which has lost its roof through crustal sagging or collapse.

lava tube [speleo] A conduit formed of hardened lava, on or within a lava *flow* through which lava flows to an advancing flow front. Also, a cavernous segment of the conduit remaining after flow ceases (Larson, 1990). Lava tubes have been surveyed that extend for tens of kilometers, although they are generally fragmented into shorter stretches.

lava tube [volc] A roofed conduit of molten lava flowing from an eruptive vent or locus of subcrustal injection of lava to a depositional site, formed by one or more of the following processes: (1) growth of flat, rooted crusts across lava streams within confined channels; (2) overflow and accretion of spatter to levees, producing a roof arched across a lava stream; (3) coalescence of plates of solidified crust floating downstream, forming a roof over a lava stream; (4) extension of pahoehoe lobes through injection of lava beneath a solidified crust (Peterson et. al., 1994). Cf: *lava cave; lava trench.*

lava tube system A distributive network of *lava tubes* that is characteristic of tube-fed pahoehoe flows and the principal means by which such flows are so widely and thinly spread. It is usually broadly dendritic in pattern with an identifiable master tube (Larson, 1990).

lava tunnel *lava tube [volc].*

lavendulan (la-ven'-du-lan) (a) A lavender-blue orthorhombic mineral: $NaCaCu_5(AsO_4)_4Cl{\cdot}5H_2O$. Syn: *freirinite.* (b) A group name for minerals with an analogous formula, but with Cu replaced by Zn and AsO_4 replaced by PO_4.

låvenite (lå'-ven-ite) (a) A light yellow, reddish-black, and brownish-black monoclinic mineral: $(Na,Ca)_2(Mn^{2+},Fe^{2+})(Zr,Ti,Nb)Si_2O_7(O,OH,F)_2$. An orthorhombic dimorph has also been reported. Cf: *wöhlerite.* Also spelled: *laavenite.* (b) A group name for monoclinic minerals with an analogous composition, but with (Mn^{2+},Fe^{2+}) replace by Ca, with variations in the Na:Ca ratio, or in the O:OH:F ratio.

lavialite (la'-vi-a-lite') A metamorphosed basaltic rock with relict phenocrysts of labradorite in an amphibolitic groundmass. The term was originated by Sederholm in 1899, who named it after Lavia, Finland. Rarely used.

lavrentievite (lav-ren'-tiev-ite) An adamantine colorless, yellow, or light olive-green monoclinic mineral: $Hg_3S_2(Cl,Br)_2$.

lavrovite (lav'-rov-ite) A green chromium-bearing diopside.

law [geomorph] A Scottish term for a more or less rounded or conical hill or mound. Syn: *low [geomorph].*

law [philos] A formal statement of the invariable and regular manner in which some natural phenomena occur under given conditions; e.g., the "law of superposition" or a "law of thermodynamics".

law of acceleration A theory in biology stating that the order of development of a structure or organ is directly related to its importance to the organism.

law of accordant junctions *Playfair's law.*

law of basin areas A general law expressing the direct geometric relation between stream order and the mean basin area of each order in a given drainage basin, originally stated by Schumm (1956, p.606). The law is expressed as a linear regression of logarithm of mean basin area on stream order, the positive regression coefficient being the logarithm of the basin-area ratio.

law of constancy of interfacial angles The statement in crystallography that the angles between corresponding faces on different crystals of one substance are constant. It was first noted by the Danish scientist Nicolaus Steno in 1669. Syn: *constancy of interfacial angles.*

law of constancy of relative proportions *constancy of relative proportions.*

law of correlation of facies A ruling principle in stratigraphy, enunciated by Walther (1893-1894): within a given sedimentary cycle, the same succession of facies that occurs laterally is also present in vertical succession.

law of crosscutting relationships A stratigraphic principle whereby relative ages of rocks can be established: a rock (esp. an igneous rock) is younger than any other rock across which it cuts.

law of equal declivities Where homogeneous rocks are maturely dissected by consequent streams, all hillside slopes of the valleys cut by the streams tend to develop at the same slope angle, thereby producing symmetric profiles of ridges, spurs, and valleys. The principle was formulated by Gilbert (1877, p.141).

law of equal volumes *Lindgren's volume law.*

law of extralateral rights *apex law.*

law of faunal assemblages A general law of geology: Similar assemblages of fossil organisms (faunas and floras) indicate similar geologic ages for the rocks that contain them.

law of faunal succession A general law of geology: Fossil organisms (faunas and floras) succeed one another in a definite and recognizable order, each geologic formation having a different total aspect of life from that in the formations above it and below it; or, the relative age of rocks can be determined from their fossil content.

law of homonymy A principle in taxonomy stating that any name that is a junior homonym of another name must be rejected and replaced. See also: *homonymy; law of priority.*

law of nature *natural law.*

law of original continuity A general law of geology: A water-laid stratum, at the time it was formed, must continue laterally in all directions until it thins out as a result of nondeposition or until it abuts against the edge of the original basin of deposition. The law was first clearly stated by Steno (1669).

law of original horizontality A general law of geology: Water-laid sediments are deposited in strata that are horizontal or nearly horizontal, and parallel or nearly parallel to the Earth's surface. The law was first clearly stated by Steno (1669).

law of priority A principle in taxonomy stating that nomenclature of a taxonomic group is based on priority of publication (ICBN, 1972, p.16). See also: *priority.*

law of rational indices The statement in crystallography that crystal faces make simple rational intercepts on suitable crystal axes, i.e. the axes of reference or the three axes forming the edges of the unit cell of each crystal lattice. Syn: *Haüy's law; law of rational intercepts.*

law of rational intercepts *law of rational indices.*

law of reflection The statement in physics that the angle between the reflected ray and the normal to the reflecting surface is the same as the angle between this normal and the incident ray, provided the wave travels with the same velocity as the incident wave. See also: *reflection.*

law of refraction The statement in physics that when a wave crosses a boundary between two isotropic substances, the wave normal changes direction in such a manner that the sine of the angle of incidence between wave normal and boundary normal divided by the velocity in the first medium equals the angle of refraction divided by the velocity in the second medium. Syn: *Snell's law.*

law of stream gradients A general law expressing the inverse geometric relation between stream order and the mean stream gradient of a given order in a given drainage basin, originally stated by Horton (1945, p.295).

law of stream lengths A general law expressing the direct geometric relation between stream order and the main stream lengths of each order in a given drainage basin, originally stated by Horton (1945, p.291). The law is expressed as a linear regression of logarithm of mean stream length on stream order, the positive regression coefficient being the logarithm of the stream-length ratio.

law of stream numbers A general law expressing the inverse geometric relation between stream order and the number of streams of each order in a given drainage basin, originally stated by Horton (1945, p.291). The law is expressed as a linear regression of logarithm of number of streams on stream order, the negative regression coefficient being the logarithm of the bifurcation ratio. See also: *number of streams.*

law of superposition A general law upon which all geologic chronology is based: In any sequence of layered rocks, sedimentary or extrusive volcanic, that has not been overturned, the youngest stratum is at the top and the oldest at the base; i.e., each bed is younger that the bed beneath, but older than the bed above it. The law was first clearly stated by Steno (1669).

law of surface relationships A principle developed by Wheeler (1964, p.602-603): "time as a stratigraphic dimension has meaning only to the extent that any given moment in the Earth's history may be conceived as precisely coinciding with a corresponding worldwide lithosphere surface and all simultaneous events either occurring thereon or directly related thereto". The "lithosphere surfaces" (surfaces of deposition or surfaces of erosion) are envisioned as "the only universal physical geologic 'datum' surfaces with direct stratigraphic implication".

law of unequal slopes A stream flowing down the steeper slope of an asymmetric ridge or divide erodes its valley more rapidly than one flowing down the gentler slope, thereby causing the crest of the divide to migrate away from the more actively eroding stream toward the less actively eroding one. The principle was first recognized by Gilbert (1877, p.140).

law of universal gravitation *Newton's law of gravitation.*

lawrencite (law'-renc-ite) A white, green, or brown rhombohedral meteorite mineral: $(Fe^{2+},Ni)Cl_2$. It has been reported as an abundant accessory mineral in iron meteorites, but its natural occurrence is unproven. It is deliquescent.

lawsonbauerite (law'-son-bau'-er-ite) A colorless monoclinic mineral: $(Mn^{2+},Mg)_9Zn_4(SO_4)_2(OH)_{22}\cdot 8H_2O$.

lawsonite (law'-son-ite) (a) A colorless to grayish-blue orthorhombic mineral: $CaAl_2Si_2O_7(OH)_2\cdot H_2O$. (b) A group name for minerals with an analogous formula, but with Ca replaced by Ba or Sr, and Al replaced by Mn^{3+}.

lawsonite-albite facies A term introduced by Winkler (1967) for rocks formed by *burial metamorphism* at the same temperatures as those of the *lawsonite-glaucophane-jadeite facies* but at lower pressures, e.g., 600-750 MPa. Obsolete.

lawsonite-glaucophane-jadeite facies A term introduced by Winkler (1967) for that part of the *glaucophane-schist facies* formed at 250° to 400°C. The rocks are formed by *burial metamorphism* at very high pressures (> 750 MPa), and the coexistence of lawsonite and glaucophane as index minerals is required. Obsolete. Cf: *lawsonite-albite facies.*

laxite (lax'-ite) An old name for unconsolidated fragmental rocks.

layer [cart] A term referring to a set of homogeneous map features registered positionally to other database layers through a common coordinate system. Each layer contains data separated by logical relationships and the graphic portrayal of sets of features. Syn: *data layer.*

layer [sed] A portion of a *bed* that is separated from the remainder of the bed by a minor but distinct discontinuity in texture or composition; e.g., separated by an *amalgamation surface* (Blatt et al., 1980).

layer [seis] (lay'-er) One of a series of concentric zones or belts of the Earth, delineated by seismic discontinuities. A classification of the interior of the Earth designates layers A to G from the surface inward.

layer [stratig] A general term for any tabular body of rock (igneous, metamorphic, or sedimentary), of ice, or of unconsolidated material, lying in a position essentially parallel to the surface or surfaces on or against which it was formed, and more or less distinctly limited above and below; specif. a bed or *stratum* of rock, with no limitation as to thickness (Calkins, 1941).

Layer 1 The uppermost layer of oceanic crust, corresponding to sediments. Typically less than 500 m thick, and characterized by seismic P wave velocities of 1.5-3.5 km/s.

Layer 2 A layer of oceanic crust, originally identified from seismic refraction measurements. It is a 1-2 km thick layer typified by seismic P wave velocities of 4.5-5.5 km/s and densities near 2.7 g/cm^3. The upper part, Layer 2a, is often associated with a basaltic pillow and sheet flow zone, and the lower part, Layer 2b, is associated with basalt dikes.

Layer 3 The lower part of oceanic crust, originally identified from seismic refraction measurements. It is typically 4.5-5 km thick, and characterized by seismic P velocities of 6.5-7 km/s and densities near 3.0 g/cm^3. Often, this layer is assumed to be gabbroic in composition.

layer-cake Said of the geologic concept of successive layers of strata, each separated by an unconformity and completely independent of or structurally different from other layers above and below (Levorsen, 1943, p.907-912).

layer charge The charge on individual 1:1 or 2:1 layers in phyllosilicates generated by cation substitution in tetrahedral and/or octahedral sheets. Apart from neutral-layer species such as talc and pyrophyllite, the layer charge in phyllosilicates is generally negative and is balanced by the presence of positively charged interlayer species or hydroxide layers. The layer charge in layered double hydroxide minerals is generally positive and is balanced by the presence of negatively charged interlayered species.

layer depth In the ocean, the depth to the top of the *pycnocline;* i.e., to the bottom of the *mixed layer.*

layered intrusion (lay'-ered) An intrusive body in which there are layers, centimeters to many meters thick, of varying mineralogical composition, e.g. the Bushveld Complex, Stillwater Complex, and Skaergaard Intrusion. Syn: *stratiform intrusion.*

layered permafrost Ground consisting of perennially frozen layers alternating with unfrozen layers or taliks (Muller, 1947, p.218).

layered series A body of igneous rocks showing banding that simulates the stratification of a sedimentary sequence and having a parallelism of platy minerals that resembles the bedding of sediments (Wager and Deer, 1939, p.36).

layered tektite Tektites having irregular shapes but generally pronounced layering defined by differences in color and/or bubble content. Layered tektites are identified with certainty only in the Southeast Asian part of the Australasian field. They are chemically similar to splash-form tektites but have higher contents of some volatile metals such as Zn, Ga, As, and Sb.

layering [cart] (lay'-er-ing) *layer tinting.*

layering [petrology] A succession of tabular units exhibiting distinct variation in mineralogic, textural or structural characteristics within igneous, sedimentary, or metamorphic rocks; or the formation of tabular bodies at different rock type, one upon the other, in a particular rock; e.g. the phenomenon in plutonic rocks resulting from *crystal settling* in magma. Describes high-temperature sedimentation features of igneous rocks and tabular mineral segregation in metamorphic rocks. Bedding and stratification refer to layering in sedimentary rocks. Cf: *rhythmic layering; phase layering.*

layering [stratig] *bedding [stratig].*

layer silicate *phyllosilicate.*

layer stripping A procedure that removes the effects of layers one at a time. It endeavors to modify seismic or gravity data to make the data appear as though the seismic source and receivers, or the gravity stations, were at the base of the stripped layer and unaffected by it. Several layers may be stripped off in succession. Syn: *stripping.*

layer structure A type of crystal structure built up by distinct layer units, as in the micas, clay, and graphite.

layer tint *hypsometric tint.*

layer tinting A method of depicting relief on a map by the distinctive shading or coloring of the areas between contour lines in a manner suggestive of progressive change, so that the pattern of distribution of high and low areas is revealed or emphasized at a glance. Syn: *layering [cart].*

layer tomography *tomography.*

lay of the land *topography.*

layover (lay'-o-ver) In *SLAR* images, the geometric displacement of the top of objects toward the near infrared range, relative to their base.

lazarenkoite (laz-a-ren'-ko-ite) A silky bright orange orthorhombic mineral: $(Ca,Fe^{2+})Fe^{3+}As^{3+}{}_3O_7\cdot 3H_2O$.

lazarevicite (laz-a-re'-vi-cite) *arsenosulvanite.*

Lazarus taxa Taxa, usually genera, which disappear from the fossil record before a mass extinction, only to reappear later (Jablonski, 1986).

lazuli (laz'-u-li) A syn. of *lapis lazuli.* Also spelled: lazule.

lazulite (laz'-u-lite) An azure-blue to violet-blue or bluish-green mineral: $MgAl_2(PO_4)_2(OH)_2$. It is isomorphous with scorzalite, and occurs in small masses or in monoclinic crystals. Syn: *blue spar; false lapis; berkeyite.* Not to be confused with *lazurite.*

lazurite (laz'-u-rite) An intense blue, greenish-blue or violet-blue cubic feldspathoid mineral of the sodalite group: $Na_3CaSi_3Al_3O_{12}S$. It also occurs as orthorhombic, monoclinic and triclinic variants. It is the principal constituent of *lapis lazuli.* See also: *ultramarine.* Not to be confused with *lazulite.*

L chondrite A member of the low-Fe group of *ordinary chondrites.*

LCL *linear crossed lamellar structure.*

leachate (leach'-ate) (a) A solution obtained by leaching; e.g. water that has percolated through soil containing soluble substances and that contains certain amounts of these substances in solution. Syn: *lixivium.* (b) Groundwater containing pollutants leached from buried solids, esp. from waste in landfills.

leached capping *gossan.*

leaching (leach'-ing) (a) The separation, selective removal, or dissolving-out of soluble constituents from a rock or orebody by the natural action of percolating water. (b) The removal in solution of nutritive or harmful constituents (such as mineral salts and organic matter) from an upper to a lower soil horizon by the action of percolating water, either naturally (by rainwater) or artificially (by irrigation). Cf: *eluviation.* (c) The extraction of soluble metals or salts from an ore by means of slowly percolating solutions; e.g. the separation of gold by treatment with a cyanide solution. Syn: *lixiviation.* (d) The chemical process of the removal or extraction of soluble compounds from solid wastes that are exposed to weathering; leached fluids can contaminate groundwater and surface water systems.

lead [eco geol] (a) A syn of *lode [eco geol].* (b) A placer deposit. See also: *back lead; blue lead [ore dep]; deep lead.* Pron: leed.

lead [geophys] (a) *time lead.* Cf: *time lag.* (b) The phase angle by which current is ahead of emf in a capacitive electrical circuit.

lead [ice] (Pron: leed) Any *fracture [ice], water opening,* or long narrow strip of ocean water through *sea ice* (esp. *pack ice*), navigable by surface vessels, and sometimes covered by young ice; wider than a *lane.* Cf: *polynya.* Syn: *channel [ice]; lane.*

lead [mineral] (a) A soft, heavy, malleable and ductile cubic mineral, the native metallic element Pb. It is silvery when freshly cut, but tarnishes readily in moist air to dull gray. Lead rarely occurs in native form, being found mostly in combination (as in galena, cerussite, and anglesite). (b) A term sometimes applied to graphite.

lead [seis] A syn. of *time lead.* Pron: leed.

lead [streams] An open watercourse, usually artificial, leading to or from a mill, mine, reservoir, etc. Syn: *leat.*

lead-210 age method A method of age determination based on the activity of ^{210}Pb (half-life = 22.2 years), and comparing the measured activity with the activity of present-day samples. It has been applied to studies of past precipitation rates in Antarctica and to sedimentation rates in lakes.

lead-alpha age method (lead-al'-pha) A method of calculating an age in years by spectrographically determining the total lead content and the alpha-particle activity of a zircon, monazite, or xenotime concentrate, the alpha-particle activity representing the uranium-thorium content. This age method is not reliable and obsolete. Syn: *Larsen method; lead-alpha dating.*

lead-alpha dating *lead-alpha age method.*

leadamalgam (lead-a-mal'-gam) A metallic white tetragonal mineral: $HgPb_2$.

lead glance *galena.*

leadhillite (lead'-hill-ite) A variously colored monoclinic mineral: $Pb_4(SO_4)(CO_3)_2(OH)_2$. It is polymorphous with macphersonite and susannite.

leading edge (lead'-ing) The frontal edge of a thrust sheet (Dahlstrom, 1970) with respect to a given stratigraphic unit, i.e. as identified by the map-view of the *hanging-wall cutoff..* Partial syn: *hanging-wall cutoff.* Cf: *trailing edge.*

leading stone *lodestone.*

lead-isotope age (lead-i'-so-tope) *lead-lead age.*

lead-lead age An age in years calculated from the ratio of lead-207 to lead-206, a by-product of the *uranium-thorium-lead age method.* Syn: *lead-isotope age.*

lead line A weighted line of wire or cord formerly used in sounding.

lead ocher A yellow or red, scaly or earthy lead monoxide; specif. massicot and litharge. Syn: *plumbic ocher.*

lead ratio The ratio of one lead isotope to another. The ratios normally encountered are $^{206}Pb/^{204}Pb$, $^{207}Pb/^{204}Pb$, $^{208}Pb/^{204}Pb$, and $^{207}Pb/^{206}Pb$.

lead spar (a) *cerussite.* (b) *anglesite.*

lead-uranium age method (lead'-u-ra'-ni-um) *uranium-lead age method.*

lead-uranium ratio The ratio of lead-206 to uranium-238 and/or lead-207 to uranium-235, formed by the radioactive decay of uranium within a mineral. The ratios are essential to the *uranium-thorium-lead age method.*

lead vitriol *anglesite.*

leaf The principal photosynthetic appendage of many higher plants. Typically leaves have a *petiole* or stalk and a veined widened portion or lamina.

leaf clay *book clay.*

leaf gap A parenchymatous opening into a stele, left by the departure of a leaf trace (Cronquist, 1961, p.877).

leaf gold Gold occurring naturally as thin flakes or sheets; not to be confused with man-made gold leaf.

leaflet (leaf'-let) One of the parts of a *compound leaf.*

leaf margin analysis A method of studying the size, shape, and margin of leaves in order to estimate paleoclimate, especially paleotemperature and paleoprecipitation (Wolfe, 1993; Wolfe and Spicer, 1999). Syn: *leaf physiognomy; foliar physiognomy; Climate Leaf Analysis Multivariate Program.*

leaf peat *paper peat.*

leaf physiognomy *leaf margin analysis.*

leaf primordium The *meristem* within a bud which develops into a leaf.

leaf scar A scar on a twig following the abscission of a leaf. See also: *bundle scar.*

leaf trace Vascular tissue extending from a stem into a leaf. Cf: *branch trace.*

leafy (leaf'-y) Pertaining to a sedimentary structure resembling a leaf, or said of a rock containing such a structure; e.g. "leafy post", a thinly laminated sandstone containing micaceous layers (as in Durham, England).

league (a) Any of various linear units of distance, ranging from about 2.42 to 4.6 statute miles; esp. "land league" (an English land unit equal to 3 statute miles) and "marine league" (a marine unit equal to 3 nautical miles). (b) Any of various units of land area equal to a square league; esp. an old Spanish unit for the area of a tract 5,000 varas square, equal to 4,428.4 acres (1,792.1 hectares) in early Texas land descriptions or equal to 4,439 acres (1,796 hectares) in old California surveys.

leaked fossil A fossil deposited as a result of a *stratigraphic leak.*

leakeite A vitreous dark red monoclinic mineral of the *amphibole* group: $(Na,K)Na_2Mg_2(Fe^{3+},Li)_3(Si,Al)_8O_{22}(OH,F)_2$.

leaking mode (leak'-ing) A seismic wave that is imperfectly trapped, so that its energy leaks or escapes across a layer boundary, causing some attenuation.

leaking polygon A topological error denoting a polygon with a boundary that is not closed (Krzanowski et al., 1993).

leaky aquifer (leak'-y) A confined aquifer whose confining beds conduct significant quantities of water into or out of the aquifer. Cf: *leaky confining bed.*

leaky confining bed A *confining bed* through which water can move into or out of the adjacent aquifer. Cf: *aquitard; leaky aquifer.*

lean *low-grade.*

lean cannel coal Cannel coal that is low in hydrogen and transitional to bituminous coal in rank. Cf: *subcannel coal.* Syn: *semicannel coal.*

lean clay A clay of low to medium plasticity owing to a relatively high content of silt or sand. Ant: *fat clay.*

leaping (leap'-ing) The sudden and radical *shifting [streams]* of a *divide* from one position to another. It occurs when the valley system drained by a captured stream is "transferred and added in a moment to that of the master stream" (Cotton, 1958, p.69). Cf: *creeping.*

least squares *method of least squares.*

least-time path *minimum-time path.*

leat An English dialectal syn. of *lead [streams].*

leatherstone (leath'-er-stone) *mountain leather.*

lebensspur (le'-bens-spur) A sedimentary structure left by a living organism; a fossil form is a *trace fossil.* The term is also applied to a Holocene track or burrow. Etymol: German "Lebensspur", "life mark". Pl: lebensspuren.

lechatelierite (le-cha'-tel-ier-ite) Naturally fused amorphous silica, occurring in fulgurites and impact craters as a vitreous or glassy product formed by the melting of quartz sand as a result of lightning or of the heat generated by the impact of a meteorite; a natural *silica glass* formed at high temperatures.

Le Chatelier's rule (Le Cha-te'-lier's) The statement in chemistry that, if conditions of a system that is initially at equilibrium are changed, the equilibrium will shift in a direction toward the original conditions.

lechosos opal (le-cho'-sos) A variety of precious opal exhibiting a deep-green play of color; esp. a Mexican opal exhibiting emerald-green play of color and flashes of carmine, dark violet, dark blue, and purple.

lecontite (le-cont'-ite) A colorless orthorhombic mineral found in bat guano: $(NH_4,K)Na(SO_4)\cdot 2H_2O$.

lectoparatype (lec-to-par'-a-type) *paralectotype.*

lectostratotype (lec-to-strat'-o-type) A stratotype for a previously described stratigraphic unit selected later in the absence of an adequately designated original stratotype (holostratotype) (ISG, 1976, p.26). Etymol: Latin "lectus", "bed", + "stratotype".

lectotype (lec'-to-type) A specimen, selected from a syntypic series, subsequent to the original description, to serve as holotype (Schenk and McMasters, 1956, p.7). Cf: *holotype [paleont]; neotype.*

ledge (a) A narrow shelf or projection of rock, much longer than wide, formed on a rock wall or cliff face, as along a coast by erosion of rocks, usually limestone; erosion is by combined biological and chemical action. Cf: *bioerosion; notch [geomorph].* The ledge may not be amplified by encrusting growth of calcareous alga annelids or coral, to form a *trottoir.* (b) A rocky outcrop; solid rock. (c) An underwater ridge of rocks, esp. near the shore; also, a nearshore reef. (d) A quarry exposure or natural outcrop of a mineral deposit.

Ledian (Led'-i-an) European stage: Middle Eocene (equivalent to Middle Lutetian).

ledikite (led'-i-kite) Interstratified biotite and vermiculite.

ledmorite (led'-mor-ite) An andradite-bearing *nepheline syenite* or *malignite* with more pyroxene and less andradite than *borolanite* with which it occurs (Johannsen, 1939, p.262). Its name, given by Shand in 1910, is derived from the Ledmore River, Scotland. Obsolete.

lee [glac geol] n. The side or slope of a hill or knob that faces away from an advancing glacier or ice sheet. adj. facing the downstream side of a glacier and relatively protected from its abrasive action. Ant: *stoss.*

lee [glaciol] adj. Said of a side or slope of a hill or knob that faces away from an advancing glacier or ice sheet; facing the downstream side of a glacier and relatively protected from its abrasive action. Ant: *stoss.*

lee [meteorol] The part or side of a hill or prominent object that is sheltered or turned away from the wind.

lee [sed struc] n. or adj. The term used for the relatively steep, downstream side of an asymmetrical *bed form* such as a *ripple* or *dune.* The lee surface, or slope, extends from the *brinkpoint* into the deepest point of the downstream trough of the *bed form.*

lee dune A general term for a dune formed to the leeward of a source of loose sand or of an obstacle of any kind, and generally under a wind of constant direction. See also: *umbracer dune; umbrafon dune.*

leegte (leeg'-te) *laagte.*

lee shore A shore crossed by wind from the land and thereby protected from strong wave action. Ant: *weather shore.*

lee-source dune *umbrafon dune.*

leeward (lee'-ward) adj. Said of the side or slope (as of a hill or prominent rock) sheltered or located away from the wind; downwind. n. The *lee* side, or the lee direction. Ant: *windward.*

leeward reef In a *reef complex* or *reef tract,* a reef on the side opposite that from which the prevailing winds blow. See also: *outer reef; inner reef.*

left bank The bank of a stream situated to the left of an observer who is facing downstream.

left-handed [cryst] (left-hand'-ed) Said of an optically active crystal that rotates the plane of polarization of light to the left. Cf: *right-handed [cryst].* Syn: *levorotatory.*

left-handed [paleont] *sinistral.*

left-handed separation *left-lateral separation.*

left-lateral fault (left-lat'-er-al) A fault on which the displacement is *left-lateral separation.* Syn: *sinistral fault.*

left-lateral separation Displacement along a fault such that, in plan view, the side opposite the observer appears displaced to the left. See also: *left-lateral fault.* Cf: *right-lateral separation.* Syn: *left-handed separation.*

left-slip fault A strike-slip fault on which the side opposite the observer has been displaced to the left.

left valve The valve lying on the left-handed side of a bivalve mollusk when the shell is placed with the anterior end pointing away from the observer, the commissure being vertical and the hinge being uppermost. Abbrev: LV. Ant: *right valve.*

leg On a seismogram, a single cycle in a wave train. Cf: *leggy.*

legal geology (le'-gal) *forensic geology.*

legend (leg'-end) A brief explanatory list of the symbols, cartographic units, patterns (shading and color hues), and other cartographic conventions appearing on a map, chart, or diagram. On a geologic map, it shows the sequence of rock units, the oldest at the bottom and the youngest at the top. The legend formerly included a textual inscription of, and the title on, the map or chart. Syn: *explanation; key [cart].*

leggy (leg'-gy) Said of a wavetrain that contains a number of cycles or legs. Syn: *tailing.*

legrandite (le-grand'-ite) A yellow to nearly colorless monoclinic mineral: $Zn_2(AsO_4)(OH){\cdot}H_2O$.

legume (leg-ume') A dry, dehiscent fruit that is produced from a single carpel that splits along two seams, as in pods of peas and beans.

lehiite (le'-hi-ite) *crandallite.*

lehm A term used in Alsace, France, for *loess.* Etymol: German *Lehm,* "loam".

Lehmann discontinuity A discontinuity often associated with the transition between asthenosphere and mesosphere, at about 200-250 km depth.

lehnerite (leh'-ner-ite) A bronze-yellow monoclinic mineral: $Mn^{2+}(UO_2)_2(PO_4)_2{\cdot}8H_2O$.

leidleite (leid'-le-ite) A glassy variety of *dacite* or *rhyodacite* containing microlites, not phenocrysts, of calcic plagioclase and pyroxene, with accessory apatite and opaque oxides. Named by Thomas and Bailey in 1915 for Glen Leidle, Mull, Scotland. Obsolete.

leifite (leif'-ite) A colorless trigonal mineral: $Na_7Be_2Al_3Si_{15}O_{39}(F,OH)_2$.

leightonite (leigh'-ton-ite) A pale-blue to greenish-blue orthorhombic mineral: $K_2Ca_2Cu(SO_4)_4{\cdot}2H_2O$. It is the Cu analogue of polyhalite.

Leintwardinian (Leint-war-din'-i-an) European stage: Upper Silurian (above Bringewoodian, below Whitecliffian).

leiolite Microbial sediment with a relatively structureless, aphanitic macrofabric lacking clear lamination, clots, or dendritic fabrics (Braga et al., 1995).

leiosphaerid (lei-o-sphae'-rid) Having the characteristics of a *leiosphere;* or a syn. of *leiosphere.*

leiosphere (lei'-o-sphere) A name usually referring to simple *sphaeromorphs* with a smooth wall, mostly Ordovician to Silurian in age. Derived from the *acritarch* generic name *Leiosphaera* (Williams et al., 1978). Cf: *cryptarch; leiosphaerid.*

leisingite A pale yellow to pale orange hexagonal mineral: $Cu(Mg,Cu,Fe,Zn)_2Te^{6+}O_6{\cdot}6H_2O$.

leiteite (lei'-te-ite) A colorless to brown monoclinic mineral: $ZnAs^{3+}{}_2O_4$.

lekolith (lek'-o-lith) A term proposed by Coats (1968, p.71) for "a mass of extrusive igneous rock more or less equant in plan, with a nearly level upper surface, commonly a lower surface determined by the shape of the basin that it filled, and a diameter greater than its depth"; e.g. a mass formed by a congealed lava lake. Etymol: Greek "lekos", "dish", + "lithos", "stone".

Lemberg's stain (Lem'-berg's) A test used to distinguish calcite from dolomite. A solution of logwood (dye) in an aqueous solution of aluminum chloride is used to stain the minerals by boiling; calcite and aragonite become violet, whereas dolomite does not change color. Cf: *Meigen's reaction.*

lemmleinite-Ba An orange to reddish-orange monoclinic member of the *labuntsovite* group: $Na_4K_4Ba_{2+x}Ti_8(Si_4O_{12})_4(O,OH)_8{\cdot}8H_2O$. It was originally called labuntsovite.

lemmleinite-K A vitreous colorless monoclinic member of the *labuntsovite* group: $Na_4K_8Ti_8(Si_4O_{12})_4(O,OH)_8{\cdot}8H_2O$.

lemoynite (le-moyn'-ite) A yellowish-white monoclinic mineral: $(Na,K)_2CaZr_2Si_{10}O_{26}{\cdot}5\text{-}6H_2O$.

Lemuria (Le-mu'-ri-a) An imaginary continent, also known as *Amosnuria,* beloved by science-fiction writers, that is alleged to have occupied most of the central Pacific Ocean until historic time, when it sank, leaving only the Pacific islands as tiny remnants. The dispersal of Polynesian peoples and cultures is supposed to have been facilitated by the existence of Lemuria, but this dispersal is easily explained otherwise. Geologically, the existence of such a continent, either modern or ancient, is impossible.

lenaite A metallic tetragonal mineral of the *chalcopyrite* group: $AgFeS_2$.

lengaite *natrocarbonatite.*

lengenbachite (leng'-en-bach-ite') A metallic-gray triclinic mineral: $Pb_6(Ag,Cu)_2As_4S_{13}$.

length [lake] The shortest distance, through the water or on the water surface, between the most distant points on a lake shore.

length [paleont] (a) The distance from the most posterior part (normally the umbo) of a brachiopod valve to the farthest point on the anterior margin, measured on or parallel with the commissural plane in the plane of symmetry. (b) The distance between two planes perpendicular to the hinge axis of a bivalve-mollusk shell and just touching the anterior and posterior extremities of the shell. Cf: *height [paleont].*

length distortion The ratio of scaled map distance to true ground distance.

length of overland flow The distance along the ground surface, projected to the horizontal, of nonchannel flow from a point on the drainage divide to a point of contact with a definite stream channel; the length is always measured at right angles to the contour lines in the drainage basin. Cf: *critical length.* Syn: *slope length.*

Lenian Russian stage: Lower Cambrian (above Atdabanian), below Solvan (Harland et al., 1990).

leningradite A vitreous red orthorhombic mineral: $PbCu_3(VO_4)_2Cl_2$.

lennilenapeite (len'-ni-len'-a-pe-ite') A black, brown, or yellowish-green triclinic mineral of the *stilpomelane* group: $K_{6\text{-}7}(Mg,Mn,Fe^{2+},Fe^{3+})_{48}(Si,Al)_{72}(O,OH)_{216}{\cdot}16H_2O$.

lennilite (len'-ni-lite) (a) A green variety of feldspar (orthoclase) from Lenni Mills, Delaware County, Penna. (b) A vermiculite mineral.

lenoblite (le-nob'-lite) A blue orthorhombic mineral: $V^{4+}{}_2O_4{\cdot}2H_2O$. It alters to a greenish color in air.

lens n. A geologic deposit bounded by converging surfaces (at least one of which is curved), thick in the middle and thinning out toward the edges, resembling a convex lens. A lens may be double-convex or plano-convex. See also: *lentil.* v. To disappear laterally in all directions; e.g. a unit is said to "lens out" within a mapped area.

lensing (lens'-ing) The thinning-out of a stratum in one or more directions; e.g. the disappearing laterally of a stratum.

lentelliptical (len-tel-lip'-ti-cal) Lenticular and elliptical; e.g. said of a lens-shaped radiolarian shell with elliptical outline.

lentic Pertaining to standing-water systems, e.g., lakes, ponds, pools (Kalff, 2002, p.5; Wetzel, 2001, p.17). Cf: *limnal; lotic.*

lenticel (len'-ti-cel) In stems and other plant parts, a mass of thin-walled cells or a pore through which the exchange of gases occurs. In a woody stem, lenticels occur in the bark (Fuller and Tippo, 1954, p.962).

lenticle (len'-ti-cle) (a) A large or small lens-shaped stratum or body of rock; a *lentil.* (b) A lens-shaped rock fragment of any size.

lenticula (len-tic'-u-la) A term established by the International Astronomical Union for small, dark spots on Europa (USGS, 2003). Pl: lenticulae.

lenticular (len-tic'-u-lar) (a) Resembling in shape the cross section of a lens, esp. of a double-convex lens. The term may be applied, for example, to a body of rock, a sedimentary structure, or a mineral habit. (b) Pertaining to a stratigraphic lens or lentil. Syn: *lentiform.*

lenticular bedding A form of interbedded mud and ripple cross-laminated sand, in which "the ripples or lenses are discontinuous not only in the vertical but also more or less in the horizontal direction" (Reineck and Wunderlich, 1968, p.102). Cf: *flaser structure [sed]; wavy bedding.*

lenticule (len'-ti-cule) A small *lentil.*

lentiform (len'-ti-form) *lenticular.*

lentil (len'-til) (a) A minor rock-stratigraphic unit of limited geographic extent, being a subdivision of a formation and similar in rank to a *member,* and thinning out in all directions; "a geographically restricted member that terminates on all sides within a formation" (ACSN, 1961, Art. 7). Term originated by Keith (1895). Cf: *tongue [stratig].* (b) A lens-shaped body of rock, enclosed by strata of different material; a geologic *lens.* See also: *lenticule; lenticle.*

lentil ore *liroconite.*

leogangite A vitreous bluish-green monoclinic mineral: $Cu_{10}(AsO_4)_4SO_4(OH)_6{\cdot}8H_2O$.

Leonardian (Leo-nar'-di-an) North American series: Lower Permian (above Wolfcampian, below Guadalupian).

leonardite (leon'-ard-ite [len'-ard-ite]) A weathering product of subbituminous coal or lignite, rich in humic and fulvic acids and soluble in alkaline water. It is a byproduct of mining near-surface coal seams, and is used as a soil conditioner, additive to drilling mud, and binder for taconite iron ore (Fowkes and Frost, 1960). Not to be confused with *leonhardite* or *leonhardtite.*

leonhardite (leon′-hard-ite [len′-hard-ite]) A variety of laumontite altered by partial loss of water. Not to be confused with *leonhardtite* or *leonardite.*

leonhardtite (leon-hardt′-ite [len′-hardt-ite]) A synonym of *starkeyite.* Not to be confused with *leonhardite* or *leonardite.*

leonite (le′-on-ite) A colorless, white, or yellowish monoclinic mineral: $K_2Mg(SO_4)_2{\cdot}4H_2O$. It is the Mg analogue of mereiterite.

leopardite (leop′-ard-ite) An igneous rock composed of small quartz phenocrysts in a microgranitic groundmass of quartz, orthoclase, albite, and mica. Iron and manganese hydroxide stains give the rock a characteristic streaked or spotted appearance. Obsolete.

leopoldite (le′-o-pold-ite′) *sylvite.*

leperditiid (lep-er-di′-ti-id) Any ostracode belonging to the order Leperditicopida, characterized by a large, strongly calcified, thick-walled shell that is usually smooth, but sometimes finely ornamented to nodose, and that has a long straight hinge, large muscle-scar pattern, and secondary shell layers. Leperditiids are commonly four or five times larger than other ostracodes. Range, Lower Ordovician (or possibly Upper Cambrian) to Upper Devonian.

lepersonnite-(Gd) A yellow orthorhombic mineral: $Ca(Gd,Dy)_2(UO_2)_{24}(OH)_{32}(CO_3)_8[Si_4O_{12}]{\cdot}44H_2O$.

lepidoblastic (lep′-i-do-blas′-tic) Pertaining to a *homeoblastic* type of texture of a foliated or schistose rock that is due to the parallel orientation during recrystallization of minerals with a flaky or scaly habit, e.g. mica, chlorite.

lepidocrocite (lep′-i-do-cro′-cite) A ruby-red or blood-red to reddish-brown orthorhombic mineral: $\gamma Fe^{3+}O(OH)$. It is trimorphous with akaganéite and goethite, and is associated with limonite in iron ores. Syn: *glockerite.*

lepidodendrid (lep′-i-do-den′-drid) n. An arborescent lycopsid of the family Lepidodendraceae, well known from Carboniferous deposits. adj. Pertaining to the genus *Lepidodendron* or to related genera. Cf: *sigillarian.*

lepidolite (le-pid′-o-lite) Group name for trioctahedral Li-rich micas, including the species polylithionite, trilithionite, taeniolite, and zinnwaldite. It commonly occurs in rose or lilac-colored masses made up of small scales, as in pegmatites. Syn: *lithium mica; lithia mica; lithionite.*

lepidolith (le-pid′-o-lith) A thin, apparently homogeneous, elliptical *coccolith;* e.g. a surface plate of the coccolithophorid *Thorosphaera flagellata.*

lepidomelane (lep′-i-do-mel′-ane, lep-i-do′-mel-ane) A black variety of biotite with a high content of ferric iron. Syn: *iron mica.*

lepidote (lep′-i-dote) Said of a plant part that is covered with fine scales. Cf: *squamose.*

lepisphere (lep′-i-sphere) A micrometer-sized spheroidal diagenetic body, usually composed of a silica mineral, with radial crystal orientation and scaly crystal terminations on the outer surface.

Lepospondyli (Lep-o-spon′-dy-li) A subclass of amphibians, characterized by unitary vertebral centra and distinctive skull pattern. Range, Paleozoic.

leptite (lep′-tite) An obsolete Fennoscandian term for a fine-grained (< 1 mm) quartzofeldspathic metavolcanic rock, roughly equivalent to metamorphosed rhyolite or felsite. Granulite rank is no longer implied. Its lower-rank equivalent is *hälleflinta.* The term was first used by Hummel in 1875. It has been dropped from usage. Obsolete syn: *hälleflintgneiss.*

leptochlorite (lep-to-chlo′-rite) (a) A group name for chlorites of indistinct crystallization. (b) A group name for chlorites with a composition corresponding to: $(Mg,Fe^{+2},Al)_n(OH)_8(Si,Al)_4O_{10}$, where *n* is less than 6 (Hey, 1962, p.495). Cf: *orthochlorite.*

leptogeosyncline (lep′-to-ge′-o-syn′-cline) An oceanic trough containing only a minor sedimentary accumulation and associated with volcanism (Trümpy, 1955). Obsolete.

leptokurtic (lep-to-kur′-tic) (a) Said of a frequency distribution that has a concentration of values about its mean greater than for the corresponding normal distribution. (b) Said of a narrow frequency distribution curve that is more peaked than the corresponding normal distribution curve. Cf: *platykurtic; mesokurtic.*

leptoma (lep-to′-ma) A thin region of exine situated at the distal pole of a pollen grain and usually functioning as the point of emergence of the pollen tube. See also: *pseudopore [palyn].*

leptopel (lep′-to-pel) Finely particulate, mainly colloidal organic and inorganic matter (such as silicates, hydrous oxides, or insoluble carbonates) occurring suspended in natural waters (Fox, 1957, p.383). Cf: *pelogloea.*

leptosporangiate (lep′-to-spo-ran′-gi-ate) A fern of the subclass Leptosporangiatae, in which the sporangium wall is only one cell-layer thick and the sporangial structure arises from only the outer segment of an initial cell (Melchior and Werdermann, 1954, p.287).

leptothermal (lep-to-ther′-mal) Said of a hydrothermal mineral deposit formed at temperature and depth conditions intermediate between *mesothermal [eco geol]* and *epithermal;* also, said of that environment. Cf: *hypothermal; xenothermal; telethermal.*

leptynolite (lep-tyn′-o-lite) A fissile or schistose variety of *hornfels* containing mica, quartz, and feldspar, with or without accessories such as andalusite and cordierite. The term was originated by Cordier in 1868 (Holmes, 1928, p.139). Cf: *cornubianite.* Rarely used.

lermontovite (ler-mon′-tov-ite) A silky grayish-green orthorhombic mineral: $U^{4+}(PO_4)(OH){\cdot}H_2O(?)$.

lesserite (les′-ser-ite) *inderite.*

lessingite-(Ce) (les′-sing-ite) A reddish-yellow, greenish, or reddish-brown monoclinic mineral of the *britholite* group: $(Ce,Ca)_5(SiO_4)_3(OH,F)$.

lestiwarite (les-ti-war′-ite) A syenite-aplite composed chiefly of microperthite, aegirine, arfvedsonite, and accessory titanite. Its name, given by Brögger in 1898, is derived from Lestiware, Kola Peninsula, Russia. Obsolete.

lesukite A yellow-orange cubic mineral: $Al_2(OH)_5Cl{\cdot}2H_2O$.

letdown (let′-down) The natural lowering (in the stratigraphic section) of slabs and fragments of a resistant formation by weathering and erosion of more vulnerable underlying rock; e.g. the stage of brecciation in which bedding is but little disturbed (Landes, 1945, p.125).

letovicite (let-o′-vic-ite) A colorless monoclinic or triclinic mineral: $H(NH_4)_3(SO_4)_2$.

lettsomite (lett′-som-ite) *cyanotrichite.*

lettucestone (let′-tuce-stone) A type of *reef rock, boundstone,* or *framestone* composed of erect foliaceous sheetlike colonies forming a skeletal framework enclosing numerous cavities, which may remain open or be filled with fine sediment, coarse skeletal debris, or crystalline cement (Cuffey, 1985).

leucaugite (leuc-au′-gite) *diopside.*

leuchtenbergite (leuch′-ten-berg-ite′) A variety of chlorite, often resembling talc and containing little or no iron.

leucite (leu′-cite) A white or gray tetragonal mineral of the *feldspathoid* group: $KAlSi_2O_6$. It is an important rock-forming mineral in alkalic rocks (esp. lavas), and usually occurs in trapezohedral crystals with a glassy fracture. Syn: *amphigene; grenatite; white garnet; Vesuvian garnet; vesuvian.*

leucite basanite In the *IUGS classification,* a basanite in which leucite is the most abundant feldspathoid. No unambiguous chemical criteria have been found to distinguish this group of rocks (Le Maitre, 2002).

leucite phonolite In the *IUGS classification,* a *phonolite* in which leucite is the most abundant feldspathoid. No unambiguous chemical criteria have been found to distinguish this group of rocks (Le Maitre, 2002).

leucite tephrite In the *IUGS classification,* a *tephrite* in which leucite is the most abundant feldspathoid. No unambiguous chemical criteria have been found to distinguish this group of rocks (Le Maitre, 2002).

leucitite (leu′-ci-tite) (a) In the *IUGS classification,* a volcanic rock defined in the QAPF diagram by F/(F+A+P)>90%, and in the TAS diagram as a foidite. Leucite is the most abundant feldspathoid. (b) A fine-grained or porphyritic extrusive or hypabyssal igneous rock chiefly composed of pyroxene (esp. titanian augite) and leucite, with little or no feldspar and without olivine.

leucitohedron (leu′-ci-to-he′-dron) *trapezohedron.*

leuco- (leu′-co-) A prefix meaning "light-colored"; in the *IUGS classification,* it is used to designate a rock that is more felsic than the specified range, e.g. leucogabbro. Cf: *mela-.*

leucochalcite (leu-co-chal′-cite) *olivenite.*

leucocratic (leu-co-crat′-ic) Light-colored; applied to light-colored igneous rocks that are relatively poor in mafic minerals. The percentage of mafic minerals necessary for a rock to be classified as leucocratic varies among petrologists, but is usually given as less than 30 to 37.5 percent. Cf: *melanocratic; mesocratic.* Noun: leucocrate. Syn: *light-colored.*

leucon (leu′-con) A sponge or sponge larva in which the flagellated chambers are connected to both exhalant and inhalant canals and do not open directly either to the *spongocoel* or to the exterior except through a canal. Cf: *ascon; sycon.* See also: *rhagon.* Adj: leuconoid.

leucophanite (leu′-co-phan′-ite) A vitreous, variously colored orthorhombic mineral: $NaCaBeSi_2O_6F$.

leucophoenicite (leu-co-phoe′-ni-cite) A brown, red, or pink monoclinic mineral: $Mn^{2+}{}_7(SiO_4)_3(OH)_2$.

leucophosphite (leu-co-phos′-phite) (a) A variously colored monoclinic mineral: $KFe^{3+}{}_2(PO_4)_2(OH){\cdot}2H_2O$. (b) A group name for minerals with an analogous formula, but with K replaced by NH4, and Fe^{3+} replaced by Al.

leucopyrite (leu-co-py′-rite) *löllingite.*

leucosome (leu′-co-some) The light-colored part of a migmatite, usually rich in quartz and feldspar (Mehnert, 1968, p.355). Cf: *melanosome; paleosome; mesosome.*

leucosphenite (leu-co-sphe′-nite) A variously colored monoclinic mineral: $BaNa_4Ti_2B_2Si_{10}O_{30}$.

leucoxene (leu′-cox-ene) A general term for fine-grained, opaque, whitish alteration products of ilmenite, commonly consisting mostly of rutile and partly of anatase or sphene, and occurring in some igneous rocks. The term has also been applied to designate a variety of sphene.

leurodiscontinuity (leu′-ro-dis′-con-ti-nu′-i-ty) A term proposed by Sanders (1957, p.295) for an unconformity characterized by a regular surface. Cf: *trachydiscontinuity.* Etymol: Greek "leuros", "smooth, even", + discontinuity.

levee [marine geol] (lev′-ee) An embankment of sediment, bordering one or both sides of a submarine canyon, fan valley, or deep-sea channel. It is similar to a river-channel levee in the subaerial environment.

levee [speleo] A retaining wall of hardened lava along the side of a *lava channel,* built up incrementally by overflow, overthrusting of lava crusts of blocks, or spatter; also, the freestanding cooled edge of a lava tongue or flow left after evacuation of the molten lava (Larson, 1990). Cf: *lava dam.*

levee [streams] (a) *natural levee.* (b) An artificial embankment built along the bank of a watercourse or an arm of the sea, to protect land from inundation or to confine streamflow to its channel. Cf: *floodwall.* Syn: *earth dike.* (c) A landing place along a river; a pier or quay. (d) *mudflow levee.* Etymol: French "levée".
levee [volc] *lava levee.*
levee delta A delta having the form of a long narrow ridge, resembling a natural levee (Dryer, 1910).
levee-flank depression *backswamp depression.*
levee lake A lake formed by a *natural levee* that acts as a barrier or enclosure for holding water. See also: *lateral levee lake; delta levee lake.*
levee ridge The elevated strip of land on which a river flows, produced by the building-up of the stream bed and the natural levees on each side.
level [geog] (lev'-el) n. Any large expanse of relatively flat, usually low-lying country, unbroken by noticeable elevations or depressions; specif. any alluvial tract of recent formation, such as the Bedford Level in Lincolnshire, England.
level [speleo] (a) Within a cave, a group of *passages* developed in the same horizontal plane. (b) The altitudinal relation of a cave floor to an outside surface. Syn: *story; tier.*
level [surv] n. (a) A *leveling instrument.* (b) A device or attachment for finding a horizontal line or plane or for adjusting an instrument to the horizontal; specif. a *spirit level.* (c) A measurement of the difference of altitude of two points on the Earth's surface by means of a level. v. To find the heights of different points by means of a level.
level [water] (a) An open reach of water in a stream or canal, such as between two canal locks. (b) The elevation of the surface of a body of water; a water table.
level fold *nonplunging fold.* Rarely used.
level ice *Sea ice* that has not been deformed, displays a flat surface, and typically occurs in undisturbed waters. Ant: *pressure ice.*
leveling (lev'-el-ing) The operation of determining the comparative altitude of different points on the Earth's surface, usually by sighting through a leveling instrument at one point to a level rod at another point. Also, the finding of a horizontal line or the establishing of grades (such as for a railway roadbed) by means of a level. See also: *spirit leveling; direct leveling; indirect leveling.* Also spelled: levelling.
leveling instrument An instrument for establishing a horizontal line of sight, usually by means of a spirit level or a pendulum device; e.g. a *surveyor's level* and a *pendulum level.* It is used, with a level rod, to determine differences in elevation between two widely separated points on the Earth's surface. Syn: *level [surv].*
leveling rod A syn. of *level rod.* Also known as a "leveling pole" or "leveling staff".
leveling screw One of three or more adjusting screws for bringing an instrument (such as a surveyor's level) to the horizontal.
level of compensation *depth of compensation.*
level of saturation *water table.*
level of zero amplitude The maximum depth below the Earth's surface reached by seasonal temperature changes.
levelperson (lev'-el-per-son) A surveyor who operates a leveling instrument.
level rod A straight rod or bar, with a flat face graduated in plainly visible linear units with zero at the bottom, used in measuring the vertical distance between a point on the Earth's surface and the line of sight of a leveling instrument that has been adjusted to a horizontal position. It is usually made of metal or well-seasoned wood. See also: *target rod; self-reading leveling rod.* Syn: *rod [surv]; leveling rod; surveyor's rod.*
level slicing The mapping of a continuous valued function into discrete classes (Krzanowski et al., 1993).
level surface *equipotential surface.*
level trier An apparatus for use in measuring the angular value of the divisions of a spirit level.
leverrierite (le-ver'-rier-ite) A discredited name for a clay mineral known to be kaolinite, or a mixture of alternating plates of kaolinite and muscovite, or a mixture of kaolinite and illite.
levigate (lev'-i-gate) *laevigate.*
levinsonite-(Y) A vitreous colorless monoclinic oxalate mineral: $(Y,Nd,Ce)Al(SO_4)_2(C_2O_4)\cdot 12H_2O$.
levorotatory (le'-vo-ro'-ta-to'-ry) *left-handed [cryst].*
lévyclaudite A metallic gray monoclinic mineral: $Pb_8Cu_3Sn_7(Bi,Sb)_3S_{28}$.
levyne-Ca (le-vyn') A white or light-colored rhombohedral zeolite mineral: $(Ca,Na_2,K_2)Al_2Si_4O_{12}\cdot 6H_2O$. Also called: levynite; levyite.
levyne-Na A white or light-colored rhombohedral zeolite mineral: $(Na_2,Ca)Al_2Si_4O_{12}\cdot 6H_2O$.
lewisite (lew'-is-ite) A brown cubic mineral of the *stibiconite* group: $(Ca,Fe^{2+},Na)_2(Sb,Ti)_2O_7$. It may be a titanian roméite.
lewistonite (lew'-is-ton-ite') A discredited syn. of *carbonate-fluorapatite.*
lexicon (lex'-i-con) An alphabetic compilation of geologic names, accompanied by formal definitions that state the lithology, thickness, age, underlying and overlying formations, type locality, and original reference; e.g. the "Wilmarth lexicon" containing 13,090 names (Wilmarth, 1938) and the "Keroher lexicon" containing 14,634 names (Keroher et al., 1966).
Lg wave A short-period, higher-mode surface wave, with a group velocity of about 3.5 km/s, that travels over long paths in the continental crust only. The "g" refers to the granitic layer. Cf: *Rg wave.*
lherzite (lherz'-ite) A *hornblendite* composed chiefly of brown hornblende, with minor amounts of biotite, ilmenite, and occasionally garnet. The name, given by Lacroix in 1917, is from Lherz in the Pyrenees. Obsolete.
lherzolite (lherz'-o-lite) (a) In the *IUGS classification,* a plutonic rock with M equal to or greater than 90, ol/(ol+opx+cpx) between 40 and 90, and both opx/(ol+cpx+opx) and cpx/(ol+cpx+opx) greater than 5. (b) *Peridotite* composed chiefly of olivine, orthopyroxene, and clinopyroxene, in which olivine is generally most abundant; a two-pyroxene peridotite. The name, dating from 1797, is for Lherz in the Pyrenees.
liandratite (li-an'-dra-tite) A yellow to yellowish-brown metamict mineral that converts to a rhombohedral form on heating: $U^{6+}(Nb,Ta)_2O_8$.
lias (li'-as) A bluish or whitish, compact, argillaceous limestone or cement rock, typically interbedded with shale or clay; esp. such a limestone quarried in Somerset and other parts of SW England. Syn: lyas.
Lias (Li'-as) Lower part of the Jurassic; corresponds approximately with the standard Lower Jurassic (above Triassic, below Dogger). Syn: *Liassic.*
Liassic (Li-as'-sic) *Lias.*
liberite (li'-ber-ite) A pale yellow to brown monoclinic mineral: Li_2BeSiO_4.
libethenite (li'-beth-en-ite) (a) An olive-green to dark-green orthorhombic mineral: $Cu_2(PO_4)(OH)$. (b) A group name for minerals with the formula $A_2(XO_4)(OH)$, where A = Cu, Mn or Zn, and X = P or As.
libolite (lib'-o-lite) A variety of albertite from Angola.
libration (li-bra'-tion) The small, periodic angular change in the face that the Moon presents toward the Earth. The effect is due primarily to the fact that while the Moon's orbital motion and rotation are synchronous, its rotation is uniform while its rate of orbital motion varies because of the orbital eccentricity. The same effect would apply to other spin-orbit-coupled planets and satellites.
libriform (lib'-ri-form) Said of wood fibers that are thick-walled and elongate, and have simple pits.
librigena (li-bri-ge'-na) The *free cheek* of a trilobite.
Libyan Desert glass (Lib'-yan) Tektite-like objects from the Libyan Desert (more exactly, the Great Sand Sea of western Egypt), probably a form of *layered tektite.* Thus these tektites from a fifth field of tektites. Formed about 29 Ma ago. Cf. *tektite.*
lichen (li'-chen) A symbiotic association between a fungus (as *mycobiont*) and one or more members of the green algae and/or Cyanobacteria (as *photobionts*), forming a *thallus* different from either of the bionts.
lichenometry (li-che-nom'-e-try) Measurement of the diameter of lichens growing on exposed rock surfaces as a method of dating geomorphic features. Cf: *fossil geochronometry.*
lichen polygon A highly specialized *vegetation polygon,* in which the pattern appears to be confined to thick reindeer moss, as in northern Quebec (Rousseau, 1949, p.50). The polygon sides are 30-50 cm in length.
lick *salt lick.*
LIDAR *light detection and ranging.*
liddicoatite (lid'-di-coat-ite) A pink to brown rhombohedral mineral of the *tourmaline* group: $Ca(Li,Al)_3Al_6(BO_3)_3Si_6O_{18}(O,OH,F)_4$.
lido (li'-do) (a) An Italian term for a barrier beach; e.g. the one protecting the lagoon of Venice. (b) A bathing beach at a seaside resort, but now extended to include those at freshwater and artificial-lake resorts. Type example: the Lido near Venice. Syn: *plage.*
lie A British term for the disposition of topographic features or for the slope of the land surface.
liebauite A bluish-green monoclinic mineral: $Ca_3Cu_5Si_9O_{26}$.
liebenbergite (lie'-ben-berg-ite') A yellow-green orthorhombic mineral of the *olivine* group: $(Ni,Mg)_2SiO_4$.
liebenerite (lie'-be-ner-ite') A variety of pinite containing alkalies, iron, and calcium.
liebigite (lie'-big-ite) An apple-green or yellow-green orthorhombic mineral: $Ca_2(UO_2)(CO_3)_3\cdot 11H_2O$. It occurs as secondary concretions or coatings. Syn: *uranothallite.*
Liesegang banding (Lie'-se-gang) *Liesegang rings.*
Liesegang rings Secondary, nested rings or bands caused by rhythmic precipitation within a fluid-saturated rock. Syn: *Liesegang banding.*
lievrite (liev'-rite) *ilvaite.*
life assemblage *biocoenosis.*
life cycle The phases, changes, or stages that an individual organism passes through during its lifetime. Syn: *ontogeny.*
life zone A *biostratigraphic unit* based on the range of a fossil form (Henningsmoen, 1961, p.68).
lift force In streams, a force normal to the streambed, caused by velocity gradients and turbulence.
lift joint A horizontal tension joint in massive rock such as granite, probably formed as a result of reduction of overburden load pressure during quarrying; a type of strain break. Syn: *sheet fracture.*
ligament (lig'-a-ment) (a) A tough structure of connecting tissue in an animal; esp. a horny elastic band in bivalve mollusks connecting the valves of the shell dorsally along a line adjacent to the umbones and acting as a spring to open the valves when the adductor muscles relax. (b) In the vertebrates, a

strong fibrous band or cord of movable tissue connecting collagenous skeletal elements, usually without direct involvement of muscle.

ligamentary articulation (lig-a-men'-ta-ry) A general type of articulation between crinoid ossicles effected solely by ligaments but sometimes supplemented by calcareous deposition. This articulation type may have many different facet topographies. Cf: *muscular articulation.*

ligament fossa A concave or flat part of a crinoid articular facet for the attachment of ligaments.

ligament groove A narrow elongate depression in the *cardinal area* of a bivalve mollusk for the attachment of the fibers of a ligament. Cf: *ligament pit.*

ligament pit (a) A relatively broad depression in the *cardinal area* of a bivalve mollusk for the attachment of the ligament. Cf: *ligament groove.* (b) A generally steep-sided small depression in a crinoid dorsal-ligament fossa adjoining the center of the transverse ridge.

ligand In crystallography, a term for a group, molecule, or atom directly coordinated to a central atom (Sposito, 1989). See: *coordination number.*

light-colored (light-col'-ored) Said of a rock-forming mineral that is light in color and generally also light in weight; also, said of the rock such minerals form; *leucocratic.* By convention, light-colored aphanites include those that are white, light and medium gray, yellow, light and medium green, red, purple, and brown. Cf: *dark-colored.*

light detection and ranging A method and instrument that measure distance to a reflecting object by emitting timed pulses of light and measuring the time between emission and reception of reflected pulses. The measured time interval is converted to distance (Geo-One-Stop, 2003, p.3). Shortened forms: lidar, LiDAR. Abbrev: LIDAR.

lighter-than-water nonaqueous-phase liquid A polluting liquid with a density less than that of water, which accumulates or "floats" on the top of the water table or *capillary fringe.* Abbrev: LNAPL.

lighthouse (light'-house) A term used in Kentucky for a *natural bridge.*

light metal Those metals prized and marketed for their metallic properties and relatively low density. These include aluminum, lithium, and titanium.

light mineral (a) A rock-forming mineral of a detrital sedimentary rock, having a specific gravity lower than a standard (usually 2.85); e.g. quartz, feldspar, calcite, dolomite, muscovite, feldspathoids. Cf: *heavy mineral [sed].* (b) A *light-colored* mineral.

lightning A brilliant flash of light produced by an electrical discharge in response to the buildup of an electrical potential between a cloud and Earth's surface, between different clouds, or between different portions of the same cloud.

lightning stone (light'-ning) *fulgurite.*

lightning tube A tubular *fulgurite.*

light oil Crude oil that has a high API gravity or *Baumé gravity.* Cf: *heavy oil.*

light red silver ore *proustite.*

light ruby silver *proustite.*

lightweight aggregate (light'-weight) An *aggregate* with a relatively low specific gravity, e.g. pumice, volcanic cinders, expanded shale, foamed slag, or expanded *perlite* or *vermiculite.*

lightweight concrete A concrete made with *lightweight aggregate.*

lignin (lig'-nin) A complex amorphous polysaccharide containing methoxyl and phenyl propane units, always occurring with cellulose and forming an important chemical constituent of wood (Treiber, 1957, p.446).

lignite (lig'-nite) A brownish-black coal that is intermediate in coalification between peat and subbituminous coal; consolidated coal with a calorific value less than 8,300 BTU/lb (19.31 MJ/kg), on a moist, mineral-matter-free basis. Cf: *brown coal; lignite A; lignite B.*

lignite A *Lignite* that contains 6,300 or more BTU/lb (14.65 MJ/kg) but less than 8,300 BTU/lb (19.31 MJ/kg) (moist, mineral-matter-free). Cf: *lignite B.* Syn: *black lignite.*

lignite B *Lignite* that contains less than 6,300 BTU/lb (14.65 MJ/kg) (moist, mineral-matter-free); essentially synonymous with *brown lignite* or *brown coal.*

ligule (lig'-ule) A term used for various straplike plant structures, e.g. a membranous structure internal to the leaf base in the heterosporous lycopsids *Isoëtes, Selaginella,* and certain fossil *lepidodendrids* or the limb of the ray flowerets in a member of the family Compositae.

ligurite (lig'-u-rite) An apple-green variety of sphene.

likasite (li-kas'-ite) A sky-blue orthorhombic mineral: $Cu_3(NO_3)(OH)_5{\cdot}2H_2O$.

likely (like'-ly) Said of a rock, lode, or belt of ground that gives indications of containing valuable minerals. Syn: *kindly.* Ant: *hungry.*

LIL An acronym for large ion lithophile elements, such as potassium, fluorine, boron, rubidium, and uranium, which have large ionic radii. First coined to describe the composition of lunar samples (Taylor, 1975).

Lillburnian (Lill-bur'-ni-an) South Pacific stage: middle Miocene (above Clifdenian, below Waiauan).

lillianite (lil'-li-an-ite') A steel-gray mineral: $Pb_3Bi_2S_6$.

lily pad (lil'-y) A colloquial term for a type of *shelfstone* that resembles water lily leaves.

lily-pad ice A term used for *pancake ice* consisting of circular pieces of ice that are not more than about 50 cm in diameter. Rarely used.

liman (li'-man) (a) A shallow muddy lagoon, bay, or marshy lake, formed at the mouth of a river behind the seaward deposits of a delta and protected by a barrier or a spit; an *estuary* or broad freshwater bay of the sea. Etymol: Russian, from Greek "limen", "harbor". (b) An area of mud or slime deposited near the mouth of a river.

liman coast A coast with many lagoons (limans) and drowned valleys, protected from the open sea by a barrier or a spit; e.g. the northern coast of the Black Sea.

limb [astron] The outer edge of a lunar or planetary disk.

limb [palyn] *equatorial limb.*

limb [struc geol] That area of a fold between adjacent fold hinges. It generally has a greater radius of curvature than the hinge region and may be planar. Syn: *flank [struc geol].* Obsolete syn: *shank [struc geol].*

limb [surv] (a) The graduated margin of an arc or circle in an instrument for measuring angles, such as the part of a marine sextant carrying the altitude scale. (b) The graduated staff of a leveling rod.

limbate (lim'-bate) Said of a foraminiferal chamber with a thickened border or edge, which commonly occurs at the suture but is sometimes elevated.

limb darkening A photometric function of celestial bodies in which, at zero phase angle, the center of the disk is much brighter than the limbs, evidently because of surface roughness or atmospheric scattering at optical dimensions.

limburgite (lim'-burg-ite) A dark-colored, porphyritic extrusive igneous rock having olivine and clinopyroxene as phenocryst minerals in an alkali-rich glassy groundmass that may have microlites of clinopyroxene, olivine, and opaque oxides; some nepheline and/or analcime may be present, and feldspars are typically absent. Its name, given by Rosenbusch in 1872, is derived from Limburg, Kaiserstuhl, Germany. In the *IUGS classification,* an approved synonym for a *basanite* with a glassy groundmass.

limbus (lim'-bus) A crease or hemlike thickening in a pollen grain, seen in polar view along the outline or edge of a saccus or pseudosaccus, where the proximal and distal layers are more or less fused.

lime (a) Calcium oxide, CaO; specif. quicklime and hydraulic lime. The term is used loosely for calcium hydroxide (as in hydrated lime) and incorrectly for calcium carbonate (as in *agricultural lime*). (b) A white cubic mineral of the periclase group: CaO. (c) A term commonly misused for calcium in such unfortunate expressions as "carbonate of lime" or "lime feldspar". (d) A limestone. The term is sometimes used by drillers for any rock consisting predominantly of calcium carbonate. (e) A product produced from the calcination of limestone, and used in the manufacture of cement and other industrial processes. Lime is one of the oldest chemicals known to man; primitive kilns have been discovered by archeologists which are believed to have been used during the Stone Age for burning lime, and the use of lime as a cementing and plastering material is probably almost as old as the history of fire.

limeclast (lime'-clast) A lithoclast derived by erosion from an older limestone; also, an intraclast disrupted from partly consolidated calcareous mud on the bottom of a sea or lake.

lime concretion A concretion in soil, having a variable shape and size and consisting of an aggregate of precipitated calcium carbonate or of other material cemented by it.

lime feldspar A misnomer for *calcium feldspar.*

lime mica *margarite [mineral].*

lime mud The unconsolidated micritic component of a limestone.

lime mudstone A fairly pure (93-99% calcium carbonate), mainly nonporous and impermeable, texturally uniform limestone whose main constituent (75-85%) is calcite mud (micrite) (Dunham, 1962). See also: *micritic limestone.*

lime olivine *calcio-olivine.*

lime pan [geomorph] A playa with a smooth, hard surface of calcium carbonate, commonly tufa.

lime pan [soil] A type of *hardpan* cemented chiefly with calcium carbonate. Also spelled: limepan. Better term may be caliche.

lime-silicate rock *calc-silicate rock.*

lime-soda feldspar (lime-so'-da) A misnomer for *sodium-calcium feldspar.*

limestone (lime'-stone) (a) A sedimentary rock consisting chiefly (more than 50% by weight or by areal percentages under the microscope) of calcium carbonate, primarily in the form of the mineral calcite, and with or without magnesium carbonate; specif. a carbonate sedimentary rock containing more than 95% calcite and less than 5% dolomite. Common minor constituents include silica (chalcedony), feldspar, clays, pyrite, and siderite. Limestones are formed by either biotic or abiotic processes, and may be detrital, chemical, oolitic, earthy, crystalline, or recrystallized; many are highly fossiliferous and clearly represent ancient shell banks or coral reefs. Limestones include chalk, calcarenite, coquina, and travertine, and they effervesce freely with any common acid. Abbrev: lst. (b) A general term used commercially (in the manufacture of lime) for a class of rocks containing at least 80% of the carbonates of calcium or magnesium and which, when calcined, gives a product that slakes upon the addition of water.

limestone buildup *carbonate buildup.*

limestone log An obsolete *resistivity log* device of short *lateral log* type that used 5 electrodes.

limestone pavement (a) A limestone bedding-plane or glaciated surface in a karst area that is divided into clints by solution fissures. See also: *crevice karst.* (b) A solution-grooved surface on limestone. See also: *karrenfeld.*

lime uranite *autunite.*

limewater (lime'-wa-ter) Natural water with large amounts of dissolved calcium bicarbonate or calcium sulfate.

limited-mobility displacement grout *compaction grout.*

limiting beds (lim'-it-ing) The oldest strata immediately above and the youngest strata immediately below an angular unconformity; they are used to date the folding and erosion (Spieker, 1956).

limiting sphere In X-ray crystallography, an imaginary sphere of radius $2/\lambda$, where λ is the wavelength of the X-rays, within which a *reciprocal lattice point* must fall for the corresponding lattice planes to cause *X-ray diffraction.*

Limmasol Forest cobalt-nickel deposit A *mineral-deposit type* characterized by irregular veins, pods, and lenses of cobalt-nickel sulfides associated with serpentinized peridotite and dunite in ophiolite complexes. Named for a deposit in Cyprus.

limnal (lim'-nal) Pertaining to a body or bodies of fresh water, esp. to a lake or lakes. Cf: *lentic.*

limnetic (lim-net'-ic) (a) Relating to the pelagic or open part of a body of fresh water. (b) Said of lake-dwelling organisms and communities that are free from direct dependence on the bottom or shore. Syn: *limnic [lake].*

limnic [coal] (lim'-nic) (a) Said of coal deposits formed inland in freshwater basins, peat bogs, or swamps, as opposed to paralic coal deposits. (b) Said of peat formed beneath a body of standing water. Its organic material is mainly planktonic.

limnic [lake] (a) Pertaining to a body of fresh water. Cf: *lacustrine.* (b) *limnetic.*

limnic material One of the common components of organic soils and includes both organic and inorganic materials that were either (1) deposited in water by precipitation or through the action of aquatic organisms, or (2) derived from underwater and floating aquatic plants and aquatic animals.

limnic peat Peat formation occurring on or in deep water from free-floating or deeply rooted plants. Cf: *telmatic peat; terrestric peat.*

limnite (lim'-nite) *bog iron ore.*

limnogenic rock (lim-no-gen'-ic) A sedimentary rock formed by precipitation from fresh water, esp. that of a lake (Grabau, 1924, p.329).

limnogeology (lim'-no-ge-ol'-o-gy) The geology of lakes. Cf: *paleolimnology.*

limnogram (lim'-no-gram) A record of lake-level variations as recorded by a water-level gage, such as a record made on a limnimeter. Syn: *limnograph.*

limnograph (lim'-no-graph) *limnogram.*

limnology (lim-nol'-o-gy) The scientific study of the physical, chemical, meteorological, biological, and ecological characteristics and interactions of all inland waters, such as pools, streams, bogs, ponds, lakes, rivers, and wetlands (Wetzel, 2001, p.4). Etymol: Greek "limne", "marsh, lake, pool".

limnoquartzite (lim-no-quartz'-ite) A fine-grained siliceous rock of fresh-water origin (Kraus, 1979); a lacustrine chert.

limon (li-mon') (a) Viscous mud deposited during floods by rivers of the Mediterranean basin, the Atlantic coast of Morocco, and western Africa, and characterized by a binder of fine iron-hydroxide grains. (b) A widespread, fine-grained, surficial deposit of periglacial loam in northern France, from which brown loamy soils have developed. It is probably of windblown and wind-deposited origin, but different from loess in that it is formed under a more humid climate. (c) A term sometimes used as a French syn. of *loess.* Etymol: French, "loam, silt, ooze, mud".

limonite (li'-mo-nite) (a) A general field term for a group of brown, amorphous, to cryptocrystalline naturally occurring hydrous ferric oxides whose real identities are unknown in absense of determinitive study. Limonite was formerly thought to be a distinct mineral ($2Fe_2O_3{\cdot}3H_2O$), but is now considered to have a variable composition (and variable chemical and physical properties) and to consist of any of several iron oxyhydroxides (commonly goethite) or of a mixture of several minerals (such as hematite, goethite, and lepidocrocite) with or without presumably adsorbed additional water. It is a common secondary substance formed by oxidation (weathering) of iron or iron-bearing minerals, and it may also be formed as an inorganic or biogenic precipitate in bogs, lakes, springs, or marine deposits; it occurs as coatings (such as ordinary rust), as loose or dense earthy masses, as pseudomorphs after other iron minerals, and in a variety of stalactitic, fibrous, reniform, botryoidal, or mammillary forms, and it represents the coloring material of yellow clays and soils. Limonite is commonly dark brown or yellowish brown, but may be yellow, red, or nearly black; it is a minor ore of iron. See also: *bog iron ore.* Syn: *brown iron ore; brown hematite; brown ocher.* (b) Informal term for the iron oxide-rich upper portion of *saprolite* soil profiles.

limonite nickel laterite *oxide nickel laterite.*

limpid dolomite (lim'-pid) Mineral dolomite that occurs as exceptionally clear, perfectly formed euhedra with mirrorlike faces and gemlike luster; it is most easily recognized under the binocular microscope on acid-etched surfaces. Thought to form by slow crystallization in fresh or brackish water (Folk and Siedlecka, 1974).

limu fragment One of the parts of glass bubbles that were formed by expansion of vapor through the surface of fluid lava.

limurite (lim'-u-rite) A *metasomatic* rock found at the contact of calcareous rocks with intruded granite and consisting of over 50% axinite. Other minerals include diopside, actinolite, zoisite, albite, and quartz. The term was originated by Zirkel in 1879. Rarely used.

lin *linn.*

linarite (li-na'-rite) A deep-blue monoclinic mineral: $PbCu(SO_4)(OH)_2$.

lindackerite (lin-dack'-er-ite) A light-green or apple-green triclinic mineral: $(Cu,Co)_5(AsO_4)_2(AsO_3OH)_2{\cdot}10H_2O$. It may contain a little nickel or cobalt.

Lindeman efficiency A measure of the calories passing through one level of a food chain, expressed as a percentage of the calories passing through the level below it.

lindgrenite (lind'-gren-ite) A green to yellowish-green monoclinic mineral: $Cu_3(MoO_4)_2(OH)_2$.

Lindgren's volume law (Lind'-gren's) The principle that during formation of ore by replacement, there is no change in rock volume or form (Lindgren, 1933). Syn: *law of equal volumes.*

lindinosite (lin-din'-o-site) A dark granite composed of more than 50 percent sodic amphibole, with quartz and microcline. The name was derived by Lacroix in 1922 from Lindinosa, Corsica. Obsolete.

lindoite (lin'-do-ite) A light-colored hypabyssal rock characterized by bostonitic texture and similar in composition to *sölvsbergite,* but being rich in quartz and poor in dark-colored minerals; the extrusive equivalent of an alkalic granite. The name, given by Brögger in 1894, is for the island of Lindö, Oslo district, Norway. Obsolete.

lindqvistite A submetallic black hexagonal mineral of the *magnetoplumbite* group: $Pb_2(Mn^{2+},Mg)Fe^{3+}{}_{16}O_{27}$.

lindsleyite (lind'-sley-ite) A metallic black rhombohedral mineral of the *crichtonite* group: $(Ba,Sr)(Ti,Cr,Fe,Mg,Zr)_{21}O_{38}$.

lindströmite (lind'-ström-ite) A metallic lead-gray orthorhombic mineral: $Pb_3Cu_3Bi_7S_{15}$. Also spelled: lindstromite.

line [cart] A shape having length and direction but no area, connecting at least two (x,y) coordinates. Lines represent geographic features too narrow to be displayed as an area at a given scale, such as contours, street centerlines, or streams, or linear features with no area, such as state and county boundary lines (Kennedy et al., 2001, p.60).

line [seis] A linear array of seismic observation points.

linea [paleont] (lin'-e-a) A linear marking on the *carapace* of some crustaceans, typically weakly calcified (TIP, 1969, pt. R, p.97).

linea [planet] A term for a dark or bright elongate marking on planetary surfaces, which may be curved or straight (USGS, 2003). An irregular linear topographic or albedo feature of unknown origin on solid surfaces, typically several hundred km long (Anonymous, 1984). Pl: lineae.

lineage (lin'-e-age) In evolution, a line of descent, usually expressed as a chronological succession of ancestor-descendant species (or genera). Although it is sometimes used synonymously with *evolutionary series,* it usually refers to a particular line of descent within the evolutionary plexus.

lineage boundary (line'-age) The surface along which plane defects in a crystal occur.

lineage zone (a) The *interval zone* between successive documented lowest occurrences within an evolutionary lineage (NACSN, 1983, Art. 50). Cf: *taxon-range zone; concurrent-range zone.* (b) A type of *range zone* consisting of the body of strata containing specimens that represent a segment of an evolutionary or developmental line or trend (ISG, 1976, p.58-59). (c) A body of strata containing specimens representing a specific segment of an evolutionary lineage. It may represent the entire range of a taxon within a lineage or only that part of the range of a taxon below the appearance of a descendent taxon (ISSC, 1994, p.61-62). Syn: *evolutionary zone; morphogenetic zone; phylogenetic zone.*

lineal series (lin'-e-al) A single line of connected zooids in a cheilostome bryozoan colony, sequentially related by direct asexual descent, and commonly in contact with zooids of adjacent series along exterior walls (Boardman and Cheetham, 1969, p.233).

lineament (lin'-e-a-ment) (a) An extensive linear surface feature. (b) A linear topographic feature of regional extent that is believed to reflect crustal structure (Hobbs et al., 1976, p.267). Examples are fault lines, aligned volcanoes, and straight stream courses. Nonrecommended syn: *linear.*

linear (lin'-e-ar) adj. Arranged in a line or lines; pertaining to the linelike character of some object or objects. n. A nonrecommended syn. of *lineament.*

linear correlation The strength of the linear relationship between two variables, measured by the *correlation coefficient* (ratio of the *covariance* to the product of the standard deviations of the two variables). The correlation coefficient varies from +1 (perfect linear relationship) through 0 (no linear relationship) to -1 (perfect inverse linear relationship).

linear crossed lamellar structure A *crossed lamellar structure* in which the first-order lamellae have rectilinear shapes as observed on the depositional surface and in horizontal sections (Carter, 1976). Abbrev: LCL.

linear elastic fracture mechanics The analysis of fracture strength based on the crack-tip stress fields developed as a consequence of stress concentration within an elastic material (Pollard and Aydin, 1988).

linear element A fabric element having one dimension that is much greater than the other two. Mineral lineations are common linear elements. Cf: *planar element; equant element.*

linear flow structure An igneous rock structure of aligned elongate minerals, mineral aggregates, xenoliths, or vesicles. Cf: *platy flow structure.*

linear scale *scale [cart].*

linear scale ratio In model analysis, a ratio of the length in the prototype to the length in the model (Strahler, 1958, p.291).
linear selection Natural selection favoring variation in a particular direction.
lineation [sed] (lin-e-a'-tion) Any linear structure, of megascopic or microscopic nature, on or within a sedimentary rock, and esp. characterizing a bedding plane; e.g. a ripple, a sole mark, or a linear parallelism in fabric caused by preferred alignment of long axes of clasts or fossils at the time of deposition. It is largely the product of current action. See also: *parting lineation.*
lineation [struc geol] A general, nongeneric term for a locally linear structure or fabric in a rock, e.g. flow lines, scratches, striae, slickensides or slickenfibers on a single surface; linear arrangements of components in sediments; or axes of folds. Lineation in metamorphic rocks includes aligned rod-shaped and/or elongate minerals grains, crenulation fold axes, and the lines of intersection between bedding and cleavage or any two sets of oriented surfaces (O'Leary et al., 1976; El-Etr, 1976).
line coverage A coverage containing information on geographic entities represented by lines and related attributes (Krzanowski et al., 1993). Cf: *point coverage.*
line defect A type of *crystal defect* occurring along certain lines in the crystal structure. Cf: *plane defect; point defect.* See also: *screw dislocation [cryst]; edge dislocation [cryst].* Syn: *dislocation [cryst]; crystal dislocation.*
line map A map composed of lines as opposed to photographic imagery. See also: *planimetric map.*
line of collimation The *line of sight* of the telescope of a surveying instrument, defined as the line through the rear nodal point of the objective lens of the telescope and the center of the *reticle* when they are in perfect alignment. Syn: *collimation line.*
line of concrescence In an ostracode, the proximal line of junction or fusion of the duplicature with the outer lamella.
line of dip The direction of the angle of dip, measured in degrees by compass direction. It generally refers to true dip, but can be said of apparent dip as well. Syn: *direction of dip.*
line of force [magnet] *magnetic-field line.*
line of force [phys] In a *field of force,* a line that is perpendicular to every equipotential surface it intersects; a line designating the direction of the force at every point along it.
line of induction *magnetic-field line.*
line of section A line on a map, indicating the position of a *profile section.* It is the *profile line* of the section as seen in plan.
line of seepage *seepage line.*
line of sight (a) A line extending from an observer's eye or an observing instrument to a distant point (such as on the celestial sphere) toward which the observer is looking or directing the instrument; e.g. *line of collimation.* (b) The straight line between two points. It is in the direction of a great circle but does not follow the curvature of the Earth. (c) A line joining the Earth or the Sun and a distant astronomic body.
line-of-sight map *viewshed map.*
line of strike *strike [struc geol].*
liner Heavy, large diameter, steel pipe that is installed in an oil or gas well. It is very similar to *casing* but, unlike casing, a liner string does not run all the way up the well to the surface.
line rod *range rod.*
line scanner A nonrecommended syn. of *scanner.*
line source A straight current element of infinite extent but infinitesimal cross section.
line spectrum The array of intensity values in the spectrum that occurs in very short, distinct ranges (i.e. only certain wavelengths) of the ordering variable, so that the spectrum appears to be a number of discrete lines with spaces between. An optical line spectrum results from electron transitions within atoms. Cf: *band spectrum.*
line-up n. On a seismic trace, alignment in phase; an *event [seis].*
lingaitukuang *brabantite.*
linguiform process A conodont process that is tongue shaped when viewed from its *upper* or *under side* (TIP, 1981, pt. W, supp.2).
lingulacean (lin-gu-la'-ce-an) n. Any inarticulate brachiopod belonging to the superfamily Lingulacea, characterized by subequal, generally phosphatic valves, with the pedicle valve being slightly larger. Range, Lower Cambrian to Holocene. adj. Said of a brachiopod having subequal phosphatic valves, or of the valves themselves. Var: *linguloid.*
lingulid (lin'-gu-lid) Any lingulacean brachiopod belonging to the family Lingulidae, characterized mainly by an elongate oval to spatulate outline and a biconvex shell. Range, Silurian (possibly Ordovician) to Holocene. The genus *Lingula* belongs to this family and has frequently been used loosely for any Ordovician species in the family.
linguloid (lin'-gu-loid) *lingulacean.*
linguloid ripple *linguoid ripple.*
linguoid current ripple (lin'-gu-oid) *linguoid ripple.*
linguoid ripple An aqueous *current ripple* characterized by a tongue-shaped outline or having a barchanlike shape whose horns point into the current; it is best developed on the bottoms of shallow streams where it shows a highly irregular pattern with a wide variety of forms. The term "linguoid" applied to a ripple was introduced by Bucher (1919, p.164). See also: *cusp-ripple.* Syn: *linguoid current ripple; linguloid ripple; cuspate ripple.*
linguoid ripple mark *linguoid ripple.*
linguoid sole mark *flute cast.*
lining (lin'-ing) (a) A casing of brick, concrete, *gunite* or *shotcrete,* cast iron, or steel, placed in a tunnel or shaft to provide support. (b) A layer of clay, concrete, synthetic film, or other material, placed under or over all or part of the perimeter of a conduit, reservoir, or landfill to resist erosion, minimize seepage losses or the escape of gases, withstand pressure, and improve flow.
lining pole *range rod.*
link [geomorph] An unbroken section of stream channel between two nodes in a *drainage network* (Shreve, 1966). The upstream node may be either a source or *fork.* The downstream node may be a fork or the mouth of the network. Cf: *bifurcating link; cis link; trans link; exterior link; interior link.*
link [paleont] A radial lath of skeletal material connecting the walls or vertical pillars in cups of dictyocyathid archaeocyathids (TIP, 1972, pt. E, p.40).
link [surv] (a) One of the 100 standardized divisions of a surveyor's chain, each consisting of iron rods or heavy steel wire looped at both ends and joined together by three oval rings, and measuring 7.92 inches in length. (b) A unit of linear measure equal to 7.92 inches or one one-hundredth of a chain.
linkage [geomorph] (link'-age) The joining at a sharp angle of two branches of a mountain range, as shown in plan view. Cf: *deflection [geomorph].*
linkage [paleont] A type of intratentacular budding in scleractinian corals, characterized by development of two or more mouths with stomodaea inside the same tentacular ring. See also: *direct linkage; indirect linkage.*
linkage [struc geol] The process by which, or condition in which, two originally separate faults become connected (Peacock et al., 2000). Cf: *hard linkage; isolated fault; soft linkage.*
link distance A dimensionless integer assigned to each link in a channel network, denoting the number of links in the flow path from the link to the mouth of the network (Jarvis, 1972). It is a measure of the relative position of the link in the drainage network. Syn: *topologic path length.*
linked veins An ore-deposit pattern in which adjacent, more or less parallel veins are connected by diagonal veins or veinlets.
links [coast] A Scottish term for a narrow area of flat or undulating land built up along a coast by drifting sand, and covered with turf or coarse grass; in Scotland, such land is often used as a golf course.
links [streams] A winding course of a river. Also, the ground along such a winding course.
linn (a) A pool of water, esp. a deep one below a fall. (b) A torrent running over rocks; a waterfall, cataract, or cascade. (c) A precipice or a steep ravine. The term is used chiefly in Scotland and northern England. Etymol: Gaelic "linne", "pool". Also spelled: *lin; lyn;* lynn.
Linnaean (Lin-nae'-an) Conforming to the principles of *binominal nomenclature* as advocated by the Swedish botanist Carl von Linné, who Latinized his name to Carolus Linnaeus.
Linnaean species A species defined, usually in a broad sense, on the basis of its morphology. Also spelled: Linnean species.
linnaeite (lin-nae'-ite) (a) A pale steel-gray cubic mineral: Co_3S_4. It has a coppery-red tarnish and constitutes an ore of cobalt. Syn: *cobalt pyrites.* Also spelled: linneite. (b) A group name for thiospinel minerals with the general formula A_3X_4, where A = Fe, Co, Ni, Cu. Cr. Zn, In, Rh, Ir, Pt or Rh; and X = S or Se.
linotolypa (li-no-tol'-y-pa) Black, reticulate microfossils of unknown affinity (Eisenack, 1962); that are now known as *cenospheres.* Sing: linotolypum.
linsey (lin'-sey) A term used in Lancashire, NW England, for a strong, striped shale and a streaky, banded sandstone or siltstone, interbedded in such a manner as to resemble a mixed linen and woolen fabric ("linsey-woolsey").
lintisite A colorless to pale yellow monoclinic mineral: $Na_3LiTi_2O_2(SiO_3)_4{\cdot}2H_2O$.
lintonite (lin'-ton-ite) A greenish, agate-like variety of thomsonite from the Lake Superior region.
liottite (li-ott'-ite) A colorless hexagonal mineral of the *cancrinite* group: $(Na,K)_{16}Ca_8Si_{18}Al_{18}O_{72}(SO_4)_5Cl_4$.
LIP *large igneous province.*
lip [eng] A low parapet erected on the downstream edge of a millrace or dam apron, to direct discharged water so that scouring of the river bottom is minimized.
lip [geomorph] (a) A projecting or overhanging edge, rim, or margin, such as of a rock on a mountainside. (b) A steep slope or abyss. (c) crater lip.
lip [paleont] (a) A margin of the aperture of a gastropod shell; e.g. *inner lip* and *outer lip.* (b) An elevated border of the aperture of a foraminiferal test. It may be small and at one side of the aperture, or completely surround it. (c) The labrum (upper lip) or labium (lower lip) of an arthropod. (d) Small lateral expansion flanking a conodont basal cavity; commonly projects downward (TIP, 1981, pt. W, supp.2).
liparite (lip'-a-rite) An obsolete syn. of *rhyolite* used by German and Russian authors. Its name, given by Roth in 1861, is derived from the Lipari Islands, in the Tyrrhenian Sea. Not recommended usage; the much more

widely used synonym *rhyolite* has priority by one year and overwhelmingly by usage.

lipid (lip′-id) Any of several saponifiable oxygenated fats or fatty-acid-containing substances such as waxes, exclusive of hydrocarbon and certain other nonsaponifiable ether-soluble compounds, which in general are soluble in organic solvents, but barely soluble in water. They, along with proteins and carbohydrates, are the principal structural components of living cells. Also spelled: lipide.

lipogenesis (li-po-gen′-e-sis) In evolution, accelerated development as a result of the omission of certain ancestral stages. Cf: *bradytely; tachytely.*

lipotexite (li-po-tex′-ite) Nonliquified mafic material within anatectic magma (Dietrich and Mehnert, 1961). Rarely used. Var: lipotectite.

lipscombite (lips′-comb-ite) An olive-green to black tetragonal mineral: $(Fe^{2+},Mn^{2+})Fe^{3+}_2(PO_4)_2(OH)_2$.

liptinite (lip′-ti-nite) A coal maceral group including *sporinite, cutinite, alginite, resinite,* and *liptodetrinite,* derived from spores, cuticular matter, resins, and waxes. Exinite is relatively rich in hydrogen. It is a common component of attrital coal. Cf: *inertinite; vitrinite.* Old syn: *exinite.*

liptite (lip′-tite) A microlithotype of coal that contains a combination of liptinite macerals totalling at least 95%. Cf: *vitrite; inertite.* Syn: *exite [coal].*

liptobiolite (lip-to-bi′-o-lite) (a) A resistant plant material that is left behind after the less resistant parts of the plant have wholly decomposed and that is characterized by relative stability of composition; e.g. resin, gum, wax, amber, copal, and pollen. (b) *liptobiolith.*

liptobiolith (lip-to-bi′-o-lith) A combustible organic rock formed by an accumulation of liptobiolites; e.g. spore coal and pollen peat. Syn: *liptobiolite.*

liptocoenosis (lip′-to-coe-no′-sis) In paleontology, an assemblage of dead organisms together with the traces and products of their life prior to burial. Its syn., *necrocoenosis,* is used more commonly in biology.

liptodetrinite (lip-to-det′-ri-nite) A maceral of coal within the *liptinite* group, having no recognizable structure and low reflectance and fluorescence; because of its finely detrital condition it cannot be assigned with certainty to any of the other macerals of the group (ICCP, 1971).

liquation (li-qua′-tion) (a) In a magma, the separation of the residual liquid from earlier formed crystals. Rarely used. (b) In Russian usage, a syn. of *liquid immiscibility.* Obsolete.

liquefaction [sed] (liq-ue-fac′-tion) The transformation of loosely packed sediment into a more tightly packed fluid mass as a result of sudden shocking preliminary to movement of a turbidity current by subaqueous slumping or sliding or in situ.

liquefaction [soil] In saturated, cohesionless soil, the transformation from a solid to a liquid state as a result of increased *pore pressure* and reduced *effective stress.* It is typically caused by rapid loading of the soil during earthquake shaking.

liquefaction slide The rapid and often catastrophic failure of a loose mass of predominantly cohesionless material which is generally at or near full saturation. The essential mechanism of such a slide is the sudden transfer of load from the particle contacts to the pore fluid, with resultant high transient pore-fluid pressures and consequent loss of strength. Liquefaction slides usually follow upon a disturbance (e.g. by earthquake or conventional slide) and can occur both subaqueously and subaerially (Koppejan et al., 1948). Syn: *flow slide.*

liquefied cohesionless particle flow (liq′-ue-fied) A body of sediment particles that flows as a liquid because it has become dilated past its critical density (Friedman and Sanders, 1978, p.514-515). Syn: *grain flow.*

liquefied natural gas Gas (mainly methane) that has been liquefied by refrigeration and pressure. Abbrev: LNG. Cf: *liquefied petroleum gas.*

liquefied petroleum gas A compressed hydrocarbon gas obtained through distillation and usable as a motor fuel, for heating, or in certain industrial processes. Abbrev: LPG.

liquid chromatography (liq′-uid) A process for separating components in a liquid phase from one another by passing them over a solid or liquid stationary phase where the components are separated by their differential mobility rates. The technique used, based on the nature of the stationary phase, is often *column chromatography, paper chromatography,* or *thin-layer chromatography.* Cf: *gas chromatography.* See also: *chromatography.*

liquid-dominated reservoir A geothermal reservoir in which the fluid at depth is a single phase, liquid water (Muffler, 1985). Syn: hot-water reservoir.

liquid flow Movement of a liquid that is usually of low viscosity, involving laminar and/or turbulent flow. Cf: *viscous flow; solid flow.*

liquid fractionation Generation of a compositionally graded column of magma through the buoyancy-driven gravitational segregation of differentiated liquids along the margins of an active body (McBirney et al., 1985). Liquids may be generated by boundary layer crystallization or wall-rock melting. See also: *sidewall crystallization; compositional convection.*

liquid immiscibility A process of magmatic differentiation involving division of the magma into two or more liquid phases.

liquidity index (li-quid′-i-ty) An expression of the consistency of a soil at its natural moisture content: its water content minus the water content at the plastic limit, all divided by the plasticity index at the liquid limit (Nelson and Nelson, 1967). Syn: *water-plasticity ratio; relative water content.*

liquid limit The water-content boundary between the semiliquid and the plastic states of a sediment, e.g. a soil. It is one of the *Atterberg limits.* Cf: *plastic limit.*

liquid-line of descent The compositional path followed by a liquid generated through fractional crystallization.

liquidus (liq′-ui-dus) The locus of points in a temperature-composition (T-X) diagram representing the maximum solubilities of solid components in the liquid (or melt) phase. In a binary system this is a curve; in a ternary system it is a curved surface. The intersection of a compositional isopleth and a liquidus indicates the temperature at which crystallization (generally of a single solid phase) begins for that composition.

liquid-water content (liq′-uid-wa′-ter) The fraction of the total mass (or volume) of wet snow that is liquid. Not to be confused with *water equivalent.* Syn: *snow moisture; free-water content; water saturation.* Cf: *water content [snow].*

lira [paleont] (li′-ra) A fine raised line or linear elevation on the surface of some shells, resembling a thread or a hair; e.g. one of the parallel fine ridges on the surface of a nautiloid conch separated by *striae* and not easily discernible with the naked eye, or a fine linear elevation within the outer lip or on the shell surface of a gastropod. Pl: lirae or liras.

lira [palyn] (li′-ra) One of the ridges separating the striae in a striate spore. Pl: lirae. Syn: *murus.*

liroconite (lir′-o-con-ite′) A blue to green monoclinic mineral: $Cu_2Al(AsO_4)(OH)_4{\cdot}4H_2O$. It usually contains some phosphorus. Syn: *lentil ore.*

lisetite A colorless orthorhombic mineral: $Na_2CaAl_4Si_4O_{16}$.

lishizhenite A vitreous pale violet triclinic mineral: $ZnFe^{3+}_2(SO_4)_4{\cdot}14H_2O$.

lisitsynite A vitreous colorless orthorhombic mineral: $KBSi_2O_6$.

liskeardite (lis-keard′-ite) An ill-defined white, greenish, bluish, or brownish acicular mineral: $(Al,Fe^{3+})_3(AsO_4)(OH)_6{\cdot}5H_2O$.

lissamphibian (lis-sam-phib′-i-an) Member of a subclass of amphibians characterized by loss or extreme reduction of dermal armor. The subclass was established primarily to accommodate living forms. Cf: *Anura.*

list [palyn] In *dinoflagellate cysts,* a high, membranous flange or crest perpendicular to the cell or cyst wall and arising from the margins of plates, most notably along the cingular and sulcal edges in dinophysialeans.

listric fault (lis′-tric) A curved downward-flattening fault, generally concave upward. Listric faults may be characterized by *normal* or *reverse separation* (Biddle and Christie-Blick, 1985), but the term is used most frequently in normal faulting and is often a characteristic of *growth faults.*

listrium (lis′-tri-um) A plate closing the anterior end of the pedicle opening that has migrated posteriorly in some discinacean brachiopods.

listwaenite (list-wae′-nite) A carbonatized and variably silicified serpentinite, occurring as dikes in ophiolite complexes in the Arabian shield. Rarely used. Var. spelling: listwänite.

litchfieldite (litch′-field-ite) A *nepheline syenite* composed of albite, with smaller amounts of potassium feldspar, nepheline, biotite, cancrinite, and sodalite. Its name, given by Bayley in 1892, is derived from Litchfield, Maine. Obsolete.

-lite A rock-name suffix derived from the Greek word "lithos", "stone".

lith A combining form, prefix or suffix, which means stone or rock.

litharenite (lith-ar′-e-nite) (a) A term introduced by McBride (1963, p.667) as a shortened form of *lithic arenite* and used by him for a sandstone containing more than 25% fine-grained rock fragments, less than 10% feldspar, and less than 75% quartz, quartzite, and chert. See also: *sublitharenite.* (b) A general term used by Folk (1968, p.124) for a sandstone containing less than 75% quartz and metamorphic quartzite and more than 25% fine-grained volcanic, metamorphic, and sedimentary rock fragments, including chert; or whose content of such rock fragments is at least three times that of feldspar and plutonic-rock fragments.

litharge (lith′-arge) A red or yellow tetragonal mineral: PbO. Cf: *massicot.* Syn: *lead ocher.* It is the Pb analogue of romarchite.

lithia mica (lith′-i-a) *lepidolite.*

lithia water Mineral water containing lithium salts (e.g. lithium bicarbonate, lithium chloride).

lithic (a) A syn. of *lithologic,* as in "lithic unit". (b) Said of a medium-grained sedimentary rock or of a pyroclastic deposit if either one contains abundant fragments of previously formed rocks; also said of those fragments. (c) Pertaining to or made of stone; e.g., "lithic artifacts" or "lithic architecture". Also a stone implement made by chipping or grinding a rock or mineral.

Lithic (Lith′-ic) In Mesoamerican archaeology, the basal prehistoric cultural stage, characterized by the migration of humans into the New World and the hunting of big game. It is followed by the *Archaic.* In North America, the equivalent period is called Paleo-Indian. Correlation of relative cultural levels with actual age (and, therefore, with the time-stratigraphic units of geology) varies from region to region.

lithic arenite (a) A term used by Williams et al. (1954, p.294, 304) for a sandstone containing abundant quartz, chert, and quartzite, less than 10% argillaceous matrix, and more than 10% feldspar, and characterized by an abundance of unstable materials in which the fine-grained rock fragments exceed feldspar grains. It is better sorted and more porous and permeable, and contains better-rounded grains, than lithic wacke. The rock is roughly equivalent to "subgraywacke" as redefined by Pettijohn (1957). See also: *litharenite; subgraywacke.* (b) A term used by Pettijohn (1954, p.364) as a syn. of *lithic sandstone* and by Krynine (1945) as a syn. of *low-rank graywacke.*

lithic arkose (a) A term used by McBride (1963, p.667) for an arkose containing appreciable rock fragments; specif. a sandstone containing 10-50% fine-grained rock fragments, 25-90% feldspar, and 0-65% quartz, quartzite, and chert. (b) A term used by Folk (1968, p.124) for a sandstone containing less than 75% quartz and metamorphic quartzite and having an "F/R ratio" between 1:1 and 3:1, where "F" signifies feldspars and fragments of gneiss and granite, and "R" signifies all other fine-grained rock fragments. Cf: *feldspathic litharenite; impure arkose.*

lithic arkosic wacke A graywacke in which feldspar exceeds rock particles (Pettijohn et al., 1972, p.167). Syn: *feldspathic graywacke.*

lithic clast A *pyroclast* consisting of previously formed rocks, fragments of igneous, metamorphic, or sedimentary rock or pieces of earlier lavas in the vent area that are included in a pyroclastic deposit.

lithic contact A boundary between soil and continuous, coherent, underlying material. The underlying material must be sufficiently coherent to make hand-digging with a spade impractical. If a single mineral, its hardness is 3 (Mohs scale), and gravel size chunks that do not disperse with 15 hours shaking in water or sodium hexametaphosphate solution.

lithic-crystal tuff (lith'-ic-crys'-tal) A tuff that is intermediate between *crystal tuff* and *lithic tuff* but is predominantly the latter. Cf: *crystal-lithic tuff.*

lithic graywacke A graywacke characterized by abundant unstable materials; specif. a sandstone containing a variable content (generally less than 75%) of quartz and chert and 15-75% clay matrix, and having rock fragments (primarily of sedimentary or low-rank metamorphic origin) in greater abundance than feldspar grains (Pettijohn, 1957, p.304). Example: some of the gray sandstones of the Siwalik Series (India), with little or no feldspar and 40-45% metamorphic-rock fragments (mainly phyllite or schist). The rock is equivalent to *low-rank graywacke* of Krynine (1945) and to *subgraywacke* as originally defined by Pettijohn (1949). The term was introduced by Pettijohn (1954, p.364) and by Williams et al. (1954, p.294). Cf: *feldspathic graywacke.*

lithic sandstone A sandstone containing rock fragments in greater abundance than feldspar grains. The term was used by Pettijohn (1954, p.364) for such a sandstone with less than 15% detrital clay matrix (e.g., subgraywacke and protoquartzite), by Williams et al. (1954, p.310) to include lithic arenite and lithic wacke, and by Hatch and Rastall (1965, p.111-112) to include the sublitharenite of McBride (1963). Cf: *lithic arenite.*

lithic subarkose A term used by McBride (1963, p.667) for a sandstone composed of subequal amounts of feldspar and rock fragments; specif. a sandstone containing 10-25% feldspar, 10-25% rock fragments, and 50-80% quartz, quartzite, and chert. Syn: *feldspathic sublitharenite.*

lithic subarkosic wacke A wacke with subequal proportions of feldspar and rock fragments but no more than 25 percent of either (Pettijohn et al., 1972, p.167).

lithic tuff An indurated deposit of volcanic ash in which the fragments are composed of previously formed rocks, e.g. accidental particles of sedimentary rock, accessory pieces of earlier lavas in the same cone, or small bits of new lava (essential ejecta) that first solidify in the vent and are then blown out. Cf: *crystal-lithic tuff; crystal tuff; lithic-crystal tuff.*

lithic wacke (a) A sandstone containing abundant quartz, chert, quartzite, >10% argillaceous matrix, and >10% feldspar, and characterized by an abundance of unstable materials in which the fine-grained rock fragments exceed feldspar grains (Williams et al., 1954, p.291-292, 301). (b) A quartz wacke containing abundant (up to 40-50%) fine-grained rock fragments (bits of shale, coal, etc.) (Krumbein and Sloss, 1963, p.172-173). Syn: *lithic graywacke* of Pettijohn (1954) in most cases.

lithidionite (li-thid'-i-o-nite') *litidionite.*

lithifaction (lith-i-fac'-tion) *lithification [sed].*

lithification [coal] (lith'-i-fi-ca'-tion) A compositional change in a coal seam from coal to bituminous shale or other rock; the lateral termination of a coal seam due to a gradual increase in impurities.

lithification [sed] (a) The conversion of a newly deposited, unconsolidated sediment into a coherent, solid rock, involving processes such as cementation, compaction, desiccation, crystallization. It may occur concurrent with, soon after, or long after deposition. (b) A term sometimes applied to the *solidification* of a molten lava to form an igneous rock. See also: *consolidation [geol]; induration.* Syn: *lithifaction.*

lithify (lith'-i-fy) To change to stone, or to petrify; esp. to consolidate from a loose sediment to a solid rock.

lithiomarsturite (lith'-i-o-mar'-stur-ite) A light pinkish-brown to light yellow triclinic mineral: $LiCa_2Mn^{2+}{}_2Si_5O_{14}(OH)$.

lithionite (lith'-i-o-nite') *lepidolite.*

lithiophilite (lith'-i-o-phil'-ite) A salmon-pink, yellow, or clove-brown orthorhombic mineral of the *triphylite* group: $Li(Mn^{2+},Fe^{2+})PO_4$. It is isomorphous with triphylite. Both are ordered olivine structures.

lithiophorite (lith'-i-o-phor'-ite) A metallic bluish-black trigonal mineral: $(Al,Li)Mn^{4+}O_2(OH)_2$. It has also been reported as monoclinic.

lithiophosphate (lith'-i-o-phos'-phate) A white, colorless, or light rose orthorhombic mineral: Li_3PO_4. It is a hydrothermal alteration product of montebrasite. Also called: lithiophosphatite.

lithiotantite (lith'-i-o-tan'-tite) A colorless monoclinic mineral: $Li(Ta,Nb)_3O_8$.

lithiowodginite (lith-i-o-wod'-gin-ite) An adamantine dark pink to red monoclinic mineral of the *wodginite* group: $(Li,Mn)(Ta,Nb,Sn)_3O_8$.

lithistid (li-this'-tid) adj. Said of a stonelike or stony sponge whose rigid skeletal framework consists of interlocking or fused siliceous spicules (desmas). n. Any demosponge belonging to the order Lithistida and characterized by the presence of desmas, interlocked and cemented to form a rigid framework.

lithium-drifted germanium detector (lith'-i-um-drift'-ed) A *semiconductor radiation detector* containing germanium rather than silicon, in which lithium is diffused into the semiconductor to compensate for impurities. Syn: *Ge:Li detector.* Cf: *scintillation counter.*

lithium mica *lepidolite.*

lithizone (lith'-i-zone) A *para-time-rock unit* representing a zone or succession of strata possessing common lithologic characteristics (Wheeler et al., 1950, p.2364). Cf: *monothem.* Syn: *lithozone.*

litho- (lith'-o-) A prefix meaning "rock" or "stone".

lithochronostratic zone A chronostratic unit based on a lithostratic unit (Henningsmoen, 1961, p.66, 73).

lithochronozone (lith-o-chron'-o-zone) A *chronozone* based on a lithostratigraphic unit, e.g. Woodbend Lithochronozone (NACSN, 1983, Art. 75). Cf: *biochronozone; polarity chronozone.*

lithoclast (lith'-o-clast) A mechanically formed and deposited fragment of a carbonate rock, normally larger than 2 mm in diameter, derived from an older, lithified limestone or dolomite within, adjacent to, or outside the depositional site.

lithocorrelation (lith'-o-cor'-re-la'-tion) *lithostratigraphic correlation.*

lithodeme (lith'-o-deme) The fundamental unit in lithodemic classification. A lithodeme is a body of intrusive, pervasively deformed, or highly metamorphosed rock, generally nontabular and lacking primary depositional structures, and characterized by lithic homogeneity. It is mappable at the Earth's surface and traceable in the subsurface (NACSN, 1983, Art. 33). For cartographic and hierarchical purposes, it is comparable to a *formation.* The name of a lithodeme combines a geographic term with a lithic or descriptive term, e.g. Duluth Gabbro.

lithodemic unit (lith-o-dem'-ic) A body of predominantly intrusive, highly deformed, and/or highly metamorphosed rock, distinguished and delimited on the basis of rock characteristics. Unlike lithostratigraphic units, a lithodemic unit does not conform to the Law of Superposition. Its contacts with other rock units may be sedimentary, extrusive, intrusive, tectonic, or metamorphic (NACSN, 1983, Art. 31). The fundamental unit in lithodemic classification is the *lithodeme.* Lithodemic units are the practical units of geological work in terranes in which rocks generally lack primary stratification. See also: *complex (b).*

Lithodensity log A trade name for a *density log* that measures both *bulk density* and photoelectric absorption cross-section (Pe). The Pe measurement is a direct function of the aggregate atomic number of elements in the formation and is therefore a good lithology indicator.

lithodesma (lith-o-des'-ma) A small calcareous plate reinforcing the internal *ligament* (resilium) in some shells of bivalve mollusks. Pl: lithodesmata. Syn: *ossiculum.*

lithodomous (li-thod'-o-mous) *lithotomous.*

lithoecozone A body of strata characterized by a certain paleoenvironment that is inferred from the lithology and can be distinguished from the paleoenvironment of adjacent strata (Poag and Valentine, 1976, p.188).

lithofacies (lith-o-fa'-cies) (a) A lateral, mappable subdivision of a designated stratigraphic unit, distinguished from adjacent subdivisions on the basis of lithology, including all mineralogic and petrographic characters and those paleontologic characters that influence the appearance, composition, or texture of the rock; a *facies* characterized by particular lithologic features. Laterally equivalent lithofacies may be separated by vertical arbitrary-cutoff planes, by intertonguing surfaces, or by gradational changes. Cf: *statistical lithofacies; intertongued lithofacies; operational facies.* (b) A term used by Moore (1949, p.17, 32) to signify any particular kind of sedimentary rock or distinguishable rock record formed under common environmental conditions of deposition, without regard to age or geologic setting or without reference to designated stratigraphic units, and represented by the sum total of the lithologic characteristics of the rock. This usage closely parallels Wells' (1947) definition of *lithotope.* (c) A term that has been applied to "lithology", "lithologic type", and the "manifestation" of lithologic characters.

lithofacies map A *facies map* based on lithologic attributes, showing areal variation in the overall lithologic character of a given stratigraphic unit. The map may emphasize the dominant, average, or specific lithologic aspect of the unit, and it gives information on the changing composition of the unit throughout its geographic extent.

lithofraction (lith-o-frac'-tion) The fragmentation of rocks during transportation in streams or by wave action on beaches.

lithogene (lith'-o-gene) adj. Said of a mineral deposit formed by the process of mobilization of elements from a solid rock and their transportation and redeposition elsewhere. On a local scale the process may be called a product of *lateral secretion;* on a larger scale, the deposit may be called a product of regional metamorphism (Lovering, 1963, p.315-316).

lithogenesis (lith-o-gen'-e-sis) The origin and formation of rocks, esp. of sedimentary rocks. Also, the science of the formation of rocks. Cf: *petrogenesis.* adj: lithogenetic. Syn: *lithogeny.*

lithogenetic unit (lith'-o-ge-net'-ic) A term used by Schenck and Muller (1941) for a local mappable assemblage of rock strata (such as a formation, member, or bed), considered without regard to time; a cartographic unit. Cf: *lithostratigraphic unit.*

lithogenous (li-thog'-e-nous) Said of stone-secreting organisms, such as a coral polyp.

lithogeny (li-thog'-e-ny) *lithogenesis.*

lithogeochemical (lith'-o-ge'-o-chem'-i-cal) Said of "geochemical exploration techniques dealing with sampling and studying the mineral fraction of the lithosphere (rock, soil, stream and lake sediments)" (Beus, 1978, p.110).

lithogeochemistry (lith'-o-ge'-o-chem'-is-try) The chemistry of the mineral fraction of the lithosphere, i.e. rocks, soils, and stream and lake sediments (Beus, 1978, p.110; Govett, 1978, p.109). Cf: *biogeochemistry; hydrogeochemistry.*

lithographic limestone (lith-o-graph'-ic) A compact, dense, homogeneous, exceedingly fine-grained limestone having a pale creamy yellow or grayish color and a conchoidal or subconchoidal fracture; a *micritic limestone.* It was formerly much used in lithography for engraving and the reproduction of colored plates. An example is the *Solenhofen stone.*

lithographic texture A sedimentary texture of certain calcareous rocks, characterized by uniform particles of clay size and by an extremely smooth appearance resembling that of the stone used in lithography.

lithoherm (lith'-o-herm) A deep-water mound (up to several hundred meters long by 50 m high) of limestone, apparently formed by submarine lithification of carbonate mud, sand, and skeletal debris; e.g., in the Florida Straits (James, 1977). Cf: *bioherm; bank [sed].*

lithohorizon (lith'-o-ho-ri'-zon) A surface of lithostratigraphic change or of distinctive lithostratigraphic character, pre-eminently valuable for correlation (not necessarily time-correlation); commonly the boundary of a lithostratigraphic unit, though also often a lithologically distinctive horizon or very thin marker bed within a lithostratigraphic unit (ISG, 1976, p.32). Cf: *biohorizon; chronohorizon.* Syn: *lithostratigraphic horizon.*

lithoid (lith'-oid) Pertaining to or resembling a rock or stone, e.g. lithoid tufa.

lithoidal (lith-oid'-al) Said of the texture of some dense, microcrystalline igneous rocks, or of devitrified glass, in which individual constituents are too small to be distinguished with the unaided eye. Obsolete.

lithoid tufa Gray, compact, bedded tufa, locally containing gastropod shells, occurring in the core of domelike masses in the desert basins of NW Nevada, as along the shore of the extinct Lake Lahontan. It is older and more stonelike than the overlying *thinolitic tufa* and *dendroid tufa.*

lithologic (lith-o-log'-ic) Adj. of *lithology.* Syn: *lithic.*

lithologic correlation A kind of *stratigraphic correlation* based on the correspondence in lithologic characters such as particle size, color, mineral content, primary structures, thickness, weathering characteristics, and other physical properties. Cf: *biocorrelation.* Syn: *lithocorrelation; lithostratigraphic correlation.*

lithologic guide In mineral exploration, a kind of rock known to be associated with an ore. Cf: *stratigraphic guide.* See also: *ore guide.*

lithologic log *sample log.*

lithologic map A type of geologic map showing the rock types of a particular area.

lithologic unit *lithostratigraphic unit.*

lithology (li-thol'-o-gy) (a) The description of rocks, esp. in hand specimen and in outcrop, on the basis of such characteristics as color, mineralogic composition, and grain size. As originally used, "lithology" was essentially synonymous with *petrography* as currently defined. (b) The physical character of a rock. Adj: *lithologic.* Cf: *petrology.*

lithomarge (lith'-o-marge) A smooth, indurated variety of common kaolin, consisting at least in part of a mixture of kaolinite and halloysite. Rarely used.

lithomorphic (lith-o-mor'-phic) Said of a soil whose characteristics are mainly *inherited.*

lithophagous (li-thoph'-a-gous) Said of an organism that ingests rock material.

lithophile (lith'-o-phile) (a) Said of an element that is concentrated in the silicate rather than in the metal or sulfide phases of meteorites. Such elements concentrate in the Earth's silicate crust in Goldschmidt's tripartite division of elements in the solid Earth. Cf: *chalcophile; siderophile.* (b) Said of an element with a greater free energy of oxidation per gram of oxygen than iron. It occurs as an oxide and more often as an oxysalt, esp. in silicate minerals. (Goldschmidt, 1954, p.24). Examples are: Se, Al, B, La, Ce, Na, K, Rb, Ca, Mn, U. Syn: *oxyphile.*

lithophilous (li-thoph'-i-lous) *rupestral.*

lithophysa (lith-o-phy'-sa) A hollow, bubblelike structure composed of concentric shells of finely crystalline alkali feldspar, quartz, and other minerals; found in silicic volcanic rocks, e.g. rhyolite and obsidian. Pl: lithophysae.

lithophyte (lith'-o-phyte) A plant living on the surface of a rock. adj: lithophytic.

lithorelic (lith-o-rel'-ic) Said of a soil feature that is derived from the parent material. Cf: *pedorelic.*

lithosere (lith'-o-sere) A *sere* that develops on a rock surface.

lithosiderite (lith-o-sid'-er-ite) An obsolete syn. of *stony-iron meteorite.*

lithosite (lith'-o-site) A colorless monoclinic zeolite mineral: $K_3Al_2Si_4O_{12}(OH)$.

Lithosols A great soil group of azonal soils characterized by an incomplete solum or no clearly expressed soil morphology and consisting of freshly and imperfectly weathered rock or rock fragments. Not used in current U.S. system of soil taxonomy.

lithosome (lith'-o-some) (a) A rock mass of essentially uniform or uniformly heterogeneous lithologic character, having intertonguing relationships in all directions with adjacent masses of different lithologic character. The term was introduced by Wheeler and Mallory (in Fischer et al., 1954, p.929) and defined by them (1956, p.2722) as a lithostratigraphic body or vertico-laterally segregated unit that is "mutually intertongued with one or more bodies of differing lithic constitution". It is essentially identical with Caster's (1934) *magnafacies.* Cf: *biosome; holosome; intertongued lithofacies.* (b) The sedimentary record of a physicochemical environment or of a more or less uniform lithotope; a body of sediment deposited under uniform physicochemical conditions (Sloss, in Weller, 1958, p.624). (c) A term defined by Moore (1957a, p.1787-1788) as "an independent body of genetically related sedimentary deposits of any sort", or, alternatively, "a spatially segregated part of any genetically related body of sedimentary deposits". Cf: *lithostrome.*

lithospar (lith'-o-spar) A naturally occurring mixture of spodumene and feldspar.

lithosphere (lith'-o-sphere) The outer, relatively rigid layer of the Earth that responds to the emplacement of a load by flexural bending. The lithosphere consists of the entire crust, plus the uppermost mantle (i.e., the mantle that is cool enough to behave flexurally). The lithosphere has been broken into about 20 plates. According to the theory of plate tectonics, motion and interaction of lithosphere plates is responsible for most geologic activity. Cf: *tectosphere.*

lithospheric mantle (lith-o-spher'-ic) The outer part of the mantle, above the asthenosphere, which with the crust constitutes tectonic plates.

lithostatic pressure (lith-o-stat'-ic) *geostatic pressure.*

lithostatic stress A state of stress in which the normal stresses acting on any plane are equal. Sometimes used as the term for overburden or vertical stress in the Earth's crust (Engelder, 1993).

lithostratic unit (lith-o-strat'-ic) A protostratic unit based on lithologic data (Henningsmoen, 1961, p.72-74). Syn: *lithostratigraphic unit.*

lithostratigraphic classification (lith'-o-strat'-i-graph'-ic) The organization of rock strata into units on the basis of their lithologic character (ISG, 1976, p.30). The fundamental unit is the *formation [stratig].*

lithostratigraphic correlation Similarity or correspondence in lithologic properties or lithostratigraphic position between two stratigraphic sections (ISSC, 1994, p.15). Syn: *lithocorrelation; lithologic correlation.*

lithostratigraphic horizon *lithohorizon.*

lithostratigraphic unit (a) A defined body of sedimentary, extrusive igneous, metasedimentary, or metavolcanic strata that is distinguished and delimited on the basis of lithic characteristics and stratigraphic position. It generally conforms to the Law of Superposition and commonly is stratified and tabular in form (NACSN, 1983, Art. 22). (b) A body of rocks, sedimentary, igneous or metamorphic, that is defined and recognized on the basis of its observable and distinctive lithologic properties or combination of lithologic properties and its stratigraphic relations (ISSC, 1994, p.31-43). Boundaries of lithostratigraphic units are placed at positions of lithic change, either at distinct contacts or arbitrarily within zones of gradation. The fundamental unit is the *formation.* A lithostratigraphic unit has a binomial designation, consisting of a geographic name, derived from the *type locality,* combined with a descriptive lithologic term; both are capitalized. See also: *lithogenetic unit.* Syn: *rock-stratigraphic unit; lithostratic unit; lithologic unit; rock unit; geolith.*

lithostratigraphic zone *lithozone.*

lithostratigraphy (lith'-o-stra-tig'-ra-phy) The element of stratigraphy that deals with the description and systematic organization of the rocks of the Earth's crust into distinctive named units based on the lithologic character of the rocks and their stratigraphic relations (ISSC, 1994, p.31). Syn: *rock stratigraphy.*

lithostrome (lith'-o-strome) A term introduced by Wheeler and Mallory (1956, p.2720-2722) for a sedimentary unit "consisting of one or more beds of essentially uniform or uniformly heterogeneous lithologic character" and representing the "three-dimensional counterpart of a lithotope"; esp. an individual tongue projecting from a lithosome. The term is regarded as essentially synonymous with *lithosome* as defined by some, and with *lithotope* as defined by Wells (1947). Weller (1958, p.636) would reject the term because it represents "nothing more than a rock-stratigraphic unit" such as "bed", "member", "tongue", "stratum", or "layer". Cf: *holostrome.*

lithotectonic unit (lith'-o-tec-ton'-ic) An assemblage of rocks that is unified on the basis of structural or deformational features, mutual relations, origin, or historical evolution. It may be igneous, sedimentary, or metamorphic.

lithothamnion (lith-o-tham'-ni-on) A plant of the genus *Lithothamnion,* an encrusting or nodular red calcareous alga of the family Corallinaceae, abundant in post-Jurassic rocks, and reported as a living form from considerable depths and very cold waters. It is most abundant on the seaward edge of reef flats, where it acts as a cementing medium of some coral reefs.

lithothamnion ridge An *algal ridge* built by *Lithothamnion* and other crustose coralline algae, rising about 1 m above the surrounding reef and extending to depths of 6-7 m below sea level. See also: *algal ridge, cup reef; algal reef; boiler; breaker [reef].*

lithotomous (li-thot'-o-mous) Said of an organism that bores into rock. Also spelled: *lithodomous.*

lithotope (lith'-o-tope) (a) An area or surface of uniform sedimentation or sedimentary environment, including associated organisms (Krumbein and Sloss, 1951, p.194). (b) A paleoecologic term originally proposed by Wells (1944, p.284) for "the sedimentary rock record of a biotope" whose life community or biocoenosis was preserved; it was later defined by Wells (1947, p.119) as "the rock record of the environment" (including both its physical and biologic expressions). The term has subsequently been used for a stratigraphic unit, a part of a stratigraphic section, a particular kind of sediment or rock, and a body of uniform sediments formed by persistence of the depositional environment, and also in an intangible sense for a sedimentary rock environment and a physical environment. Cf: *lithofacies; lithostrome; biotope [stratig].*

lithotype (lith'-o-type) A macroscopically visible band in humic coals, analyzed by physical characteristics rather than by botanical origin. The four lithotypes of banded bituminous coal are *vitrain, clarain, durain,* and *fusain.* These were originally described by Stopes in 1919. Beds or bands of cannel and boghead coal are sometimes included as lithotypes of coal. See also: *banded ingredients.*

lithoxyl (li-thox'-yl) A term applied to *wood opal* in which the original woody structure is observable. The term is also used to designate petrified (opalized) wood.

lithozone (lith'-o-zone) (a) An informal term to indicate a body of strata that is unified in a general way by lithologic features but for which there is insufficient need or information to justify its designation as a formal unit, e.g. the shaly zone in the lower part of the Parker Formation, the coal-bearing zone exposed south of Ravar, the Burgan oil-producing zone (ISG, 1976, p.30). Syn: *lithostratigraphic zone.* (b) A "more euphonious" syn. of *lithizone* (P. F. Moore, 1958, p.449).

litidionite (lit-id'-i-on-ite) A blue triclinic mineral of the *fenaksite* group: $KNaCu[Si_4O_{10}]$.

lit-par-lit adj. Having the characteristic of a layered rock, the laminae of which have been penetrated by numerous thin, roughly parallel sheets of igneous material, usually granitic. The term denotes a layered *migmatite* texture. Etymol: French, "bed-by-bed". Cf: *injection gneiss.* Syn: *stromatic.*

litter (lit'-ter) In forestry, a general term for the layer of loose organic debris, composed of freshly fallen or only slightly decayed material, that accumulates in wooded areas. It is a constituent of the *forest floor.*

Little Ice Age (Lit'-tle) A cool, brief interval in an otherwise warm interglacial stage. Originally employed for a mid-Holocene event in the Yosemite area, California, about 3,000 years B.P. (Matthes, 1930), it is now used only for the 16th- through 19th-century cool phases; in some areas, this cool phase began earlier (mid-13th century) (Bradley and Jones, 1992, 1993; Grove, 1988, 2001).

littoral [lake] Refers to the shoreward region of a lake, where rooted vegetation may be present.

littoral [oceanog] (lit'-to-ral) Pertaining to the benthic ocean environment or depth zone between high water and low water; also, pertaining to the organisms of that environment. Syn: *intertidal.* See also: *sublittoral.* Cf: *supralittoral.*

littoral cone A mound of hyaloclastic debris constructed by steam explosions at the point where lava enters the sea. Littoral cones lack feeding vents connected to subsurface magma supplies (Fisher and Schmincke, 1984, p.263).

littoral current An ocean current caused by the approach of waves to a coast at an angle. It flows parallel to and near to the shore. See also: *littoral drift.* Syn: *longshore current.*

littoral drift Material (such as shingle, gravel, sand, and shell fragments) that is moved along the shore by a *littoral current.* Syn: *longshore drift; shore drift.*

littoral explosion An explosion that is the result of the contact of a flow of molten lava with the edge of a body of water; a *hydroexplosion.*

littoral shelf A shallow, nearshore, terracelike part of a submerged lake bed, produced by the combined effects of wave erosion and current deposition, and often extending a considerable distance lakeward from the beach.

lituicone (lit'-u-i-cone') A nautiloid *conch* (like that of *Lituites*) that is coiled in the early stages of development and then becomes straight in the mature stages.

lituid (lit'-u-id) An internal bifid lobe in some ammonoids (e.g., Phyllocerida) which has straight sides, not folded or serrated.

litvinskite A red to reddish-brown monoclinic mineral: $Na_2([],Na,Mn)ZrSi_6O_{12}(OH,O)_6$.

liujinyinite *uytenbogaardtite.*

live cave A cave in which there is moisture and growth of *speleothems* associated with the presence of moisture. Cf: *dead cave.* Syn: *active cave.*

liveingite (live'-ing-ite) A bluish-gray metallic monoclinic mineral: $Pb_9As_{13}S_{28}$. Syn: *rathite-II.*

live oil Oil that is saturated with gas.

liverite (liv'-er-ite) Name used locally in Utah for *elaterite.*

liver opal (liv'-er) *menilite.*

liver ore *hepatic cinnabar.*

liverwort (liv'-er-wort) A member of the bryophyte class Hepaticae, characterized by a creeping or dichotomously branched thallus, no roots or leaves, and unicellular rhizoids. Liverworts range from the Devonian. Cf: *moss [bot].*

livesite (liv'-es-ite) A clay mineral intermediate between kaolinite and halloysite; a disordered kaolinite.

live speleothem A moist-cave mineral deposit that is still forming by crystal precipitation. Syn: *active speleothem.*

live stream *perennial stream.*

living chamber (liv'-ing) (a) The *body chamber* housing the soft parts of a cephalopod. (b) The outermost part of the zooidal body cavity in stenolaemate bryozoans, which contains the functional organs.

living flood plain A flood plain that is overflowed in times of high water (Bryan, 1923a, p.88). Cf: *fossil flood plain.*

living fossil An animal or plant that lives at the present time, is also known as a fossil from an earlier geologic time, and has undergone relatively little modification since that earlier time.

livingstonite (liv'-ing-ston-ite') A metallic lead-gray monoclinic mineral: $HgSb_4S_8$.

lixiviation (lix-iv'-i-a'-tion) *leaching.*

lixivium (lix-iv'-i-um) *leachate.*

lizardite (liz'-ard-ite) The most abundant form of the trioctahedral serpentine minerals. It crystallizes as flat platelets. Variable amounts of Al substitute for both Mg and Si in the ideal serpentine formula of $Mg_3Si_2O_5(OH)_4$ to create a better lateral fit between the component octahedral and tetrahedral sheets than found in antigorite and chrysotile. Several polytypes exist: rhombohedral, trigonal, hexagonal, or monoclinic.

L-joint A syn. of *primary flat joint.* Rarely used.

Llandeilian A European series or stage: Middle Ordovician (above Llanvirnian, below Caradocian).

Llandoverian (Llan-do-ve'-ri-an) A series of the Standard Global Chronostratigraphic Scale: Lower Silurian (above Ordovician System, below Wenlockian Series). Syn: Llandovery.

llanite (lla'-nite) A hypabyssal porphyritic *rhyolite* composed of phenocrysts of red alkali feldspar and blue quartz in a fine-grained groundmass of quartz, microcline, albite, and biotite. The name was derived by Iddings in 1904 from Llano County, Texas. Obsolete.

llano (lla'-no [ya'-no]) A term for an extensive subtropical plain, with or without vegetation, applied esp. to the generally treeless plains of northern South America and the southwestern U.S. Etymol: Spanish.

Llano (Lla'-no) A provincial series of the *Precambrian* in Texas.

Llanoria (Lla-nor'-i-a) One of the borderlands proposed by Schuchert (1923), in this case south of North America, between the Ouachita geosyncline and the Gulf of Mexico. Obsolete.

Llanvirnian (Llan-vir'-ni-an) A British series or stage: Middle Ordovician (above Arenigian, below Llandeilian).

LL chondrite A member of the low-Fe, low-metal group of *ordinary chondrites.*

llyn A Welsh term for a pool or lake.

LNAPL *lighter-than-water nonaqueous-phase liquid.*

LNG *liquefied natural gas.*

load (a) The material that is moved or carried by a natural transporting agent, such as a stream, a glacier, the wind, or waves, tides, and currents; specif. *stream load.* (b) The quantity or amount of such material at any given time. Syn: *sediment load.* (c) The power which a device consumes or delivers. (d) An impedance connected across an output.

load [drill] The weight pressing on a *drill bit.*

load cast A *sole mark,* usually measuring less than a meter in any direction, consisting of a swelling in the shape of a slight bulge, a deep or shallow rounded sack, a highly irregular protuberance, or a bulbous, mammillary, or papilliform protrusion of sand or other coarse clastics, extending downward into finer-grained, softer, and originally hydroplastic underlying material, such as wet clay, mud, or peat, that contained an initial depression. It is produced by the exaggeration of the depression as a result of unequal settling and compaction of the overlying material and by the partial sinking of such material into the depression, as during the onset of deposition of a turbidite on unconsolidated mud. A load cast is more irregular than a flute cast (it is usually not systematically elongated in the current direction), and is characterized by an absence of a distinction between the upcurrent and downcurrent ends. The term was proposed by Kuenen (1953, p.1058) to replace *flow cast* used by Shrock (1948, p.156), although Kuenen excluded the phenomenon of warping of underlying laminae and applied the term to a feature resulting from vertical adjustment only. See also: *load-flow structure.* Syn: *load casting.*

load-casted (load'-cast-ed) Said of a current mark (such as a groove or flute) that is exaggerated, misshapen, or obscured by the development of a load cast. Also said of a sole mark (such as a flute cast or groove cast) that is similarly modified by load casting.

load-casted ripple A term apparently first used by Dzulynski and Kotlarczyk (1962) for a structure formed by sinking of ripples into the underlying mud during deposition of the ripple. See also Dzulynski and Walton (1965, p.146). Syn: *ripple load cast.*

load casting (a) The formation or development of a load cast or load casts; also, the configuration of the underside of a stratum characterized by load casts. (b) *load cast.*

load-cast lineation A small-scale, poorly defined, irregular linear structure that appears as a cast on the underside of a sandstone bed, and that is attributed to a dense, sluggish turbidity current moving over soft mud (Crowell, 1955, p.1358).

load-cast striation A rill-like sedimentary structure of uncertain origin (Pettijohn and Potter, 1964, p.319).

loaded stream (load'-ed) A stream that has all the sediment it can carry. A partly loaded stream is one carrying less than full capacity. See also: *overloaded stream; underloaded stream.*

load-flow structure A term sometimes used for *load cast* because the structure forms by downsinking of overlying material and not by infilling of a depression (as implied by the term "cast").

load fold A plication of an underlying stratum, believed to result from unequal pressure and settling of overlying material (Sullwold, 1959).

load metamorphism A type of *static metamorphism* in which pressure due to deep burial has been a controlling influence, along with high temperature. Cf: *geothermal metamorphism; thermal metamorphism.*

load mold The mold of a load cast; the depression in an underlying stratum, occupied by a load pocket, such as the sea-floor surface beneath a depositing turbidite (Sullwold, 1960).

load pocket The material within a load cast, consisting of a "bulge of sand" pressing into an underlying stratum (Sullwold, 1959, p.1247).

loadstone (load'-stone) *lodestone.*

load wave The "salient curved unevenness" of underlying material that appears to have been "squirted up" into a superjacent turbidity-current deposit as a result of unequal settling of the overlying material (Sullwold, 1959); it resembles a ripple or other wavelike structure. The term "refers to smooth upward bulges as well as to tenuous breaking wave or flame shapes" (Sullwold, 1960, p.635). See also: *flame structure [sed].*

loam (a) A rich, permeable soil composed of a friable mixture of relatively equal and moderate proportions of clay, silt, and sand particles, and usually containing organic matter (humus); specif. a soil consisting of 7-27% clay, 28-50% silt, and 23-52% sand. It has a somewhat gritty feel yet is fairly smooth and slightly plastic. Loam may be of residual, fluvial, or eolian origin, and includes many loesses and many of the alluvial deposits of flood plains, alluvial fans, and deltas. It usually implies fertility, and is sometimes called *topsoil* in contrast to the subsoils that contain little or no organic matter. (b) A term used in the old English literature for a mellow soil rich in organic matter, regardless of texture. (c) An obsolete term formerly used in a broad sense for clay, impure clay, clayey earth, and mud.

loaming (loam'-ing) A method of geochemical prospecting in which samples of soil or other surficial material are tested for traces of the metal desired, its presence presumably indicating a near-surface orebody.

loamy sand (loam'-y) A soil containing 70-90% sand, 0-30% silt, and 0-15% clay, or a soil containing 85-90% sand at the upper limit and having the percentage of silt plus 1.5 times the percentage of clay not less than 15, or a soil containing 70-85% sand at the lower limit and having the percentage of silt plus twice the percentage of clay not exceeding 30 (SSSA, 1965, p.347); specif. such a soil containing at least 25% very coarse sand, coarse sand, and medium sand, and less than 50% fine sand or very fine sand. It is subdivided into loamy coarse sand, loamy fine sand, and loamy very fine sand. Cf: *sandy loam.*

loamy soil A soil (such as a clay loam, silt loam, or sandy loam) whose texture and properties are intermediate between those of a coarse-textured or sandy soil and a fine-textured or clayey soil.

LO-analysis A palynological method for analyzing the sculpture and structure of the *sexine* by comparing the elements in high and low focus. In high focus, raised elements are bright ("lux"), holes are dark ("obscuritas"). At low focus, this pattern is reversed.

lobate delta (lo'-bate) *arcuate delta.*

lobate lava A submarine lava type consisting of elongated, flattish lobes with smooth outer glassy skins (Batiza and White, 2000).

lobate plunge structure *flute cast.*

lobate rill mark A term used by Clarke (1918) and Shrock (1948, p.131) for a spatulate or lingulate sedimentary structure (cast) resembling the bowl of an inverted spoon and believed to develop on a beach by ebbing tidal currents or retreating storm waves of the intertidal zone. The structure is now considered to be a *flute cast* formed by current action. See also: *rill mark.*

lobate soil *step [pat grd].*

lobe [bot] Any protuberant part or segment of a plant organ; specifically a segment of a petal or calyx or leaf that represents division to about the middle (Lawrence, 1951, p.759).

lobe [glaciol] (a) A rounded, tonguelike projection of glacial ice or drift; e.g., the Michigan Lobe which occupied and extended southward from Lake Michigan.(b) *glacial lobe.*

lobe [paleont] (a) An element or undulation of a suture line in a cephalopod shell that forms an angle or curve whose convexity is directed backward or away from the aperture (or toward the apex). Ant: *saddle.* (b) One of the longitudinal divisions of the body, or one of the lateral divisions of the glabella, in a trilobite. (c) A rounded major protuberance of the valve surface of an ostracode, generally best developed in the dorsal part of the carapace (TIP, 1961, pt.Q, p.52); e.g. *connecting lobe.* (d) A conodont process flanked by flanges or platforms; commonly formed by lateral expansion of a lateral process; may bifurcate (TIP, 1981, pt. W, supp.2). (e) In vertebrates, any rounded projection of a body part or organ (brain, liver, lung, etc.).

lobe [streams] (a) *meander lobe.* (b) *floodplain lobe.*

lobefin (lobe'-fin) *Sarcopterygii.*

local base level (lo'-cal) *temporary base level.*

local current A natural earth current of any origin, e.g. from the oxidation of a sulfide deposit.

local fauna *faunule.*

local-gravity map (lo'-cal-grav'-i-ty) A *residual gravity* map.

local horizon (a) *apparent horizon; visible horizon.* (b) The actual lower boundary of the observed sky or the upper outline of terrestrial objects including nearby obstructions or irregularities.

localized deformation A type of heterogeneous deformation taking place within a *deformation band* (such as a shear zone) in an otherwise homogeneously deforming or rigid body (Rutter, 1986). Cf: *distributed deformation.*

local magnitude A measure of the strain energy released by an earthquake based on measurements within 100 km of the epicenter. See: *earthquake magnitude.*

local metamorphism Metamorphism caused by a local process, e.g. contact metamorphism or metasomatism near an igneous body, hydrothermal metamorphism, or dislocation metamorphism in a fault zone. Cf: *regional metamorphism.*

local peat Peat developed by groundwater. Syn: *basin peat; azonal peat.*

local peneplain (a) *incipient peneplain.* (b) *partial peneplain.*

local-range zone (a) The *range zone* of a specified taxon in some particular section or local area (ACSN, 1957, p.1883). The sum of all the local range zones is the true range zone of the taxon. (b) A body of strata in a specific geographical section or area characterized by the occurrence of a specified fossil group (Geological Society of London, 1967, p.85). Syn: *teilzone; topozone; partial range zone.*

local relief The vertical difference in elevation between the highest and the lowest points of a land surface within a specified horizontal distance or in a limited area. Syn: *relative relief.* Cf: *available relief.*

local sorting The action responsible for the size-sorting of sedimentary particles at a given place (Pettijohn, 1957, p.540-541). Cf: *progressive sorting.*

local unconformity An unconformity that is strictly limited in geographic extent and that usually represents a relatively short period, such as one developed around the margins of a sedimentary basin or along the axis of a structural trend that rose intermittently while continuous deposition occurred in an adjacent area. It may be similar in appearance to, but lacks the regional importance of, a disconformity. Cf: *regional unconformity.*

location (lo-ca'-tion) The spot or place where a borehole is to be drilled; e.g. a *well site.*

locational accuracy The accuracy with which spatial objects are positioned in spatial data bases (Krzanowski et al., 1993). See also: *horizontal accuracy; positional accuracy.*

loch A Scottish term for a lake or arm of the sea.

lochan (loch'-an) A Scottish term for a small lake (*loch*).

Lochkovian (Loch-kov'-i-an) A stage of the Standard Global Chronostratigraphic Scale: lowermost Devonian (above Silurian System, below Pragian Stage). The basal GSSP is near Prague, Czech Republic (Martinsson, 1977).

lock A stretch of water in a canal, stream, or dock, enclosed by gates at each end, and used in raising or lowering boats as they pass from one water level to another.

Locke level A *hand level* with a fixed bubble tube that can be used only for approximate horizontal sighting.

Lockportian (Lock-port'-i-an) North American provincial stage: Upper Silurian (above Cliftonian, below Cayugan).

locomorphic stage (lo-co-mor'-phic) A term introduced by Dapples (1962) for the middle geochemical stage of diagenesis characterized by prominent mineral replacement (without reactions). It is typical of lithification of a clastic sediment, and is more advanced than the *redoxomorphic stage* and precedes the *phyllomorphic stage.* Rarely used.

locular dimorphism (loc'-u-lar) Development of a *loculus* in certain ostracodes.

locular wall (a) A wall of a diatom frustule having separate inner and outer laminae connected by vertical partitions that form areolae. (b) The wall of a seed or spore chamber (locule) in an ovary or sporangium.

locule (loc′-ule) A compartment, cavity or chamber; in flowering plants, a cavity in the ovary containing ovules, or in the anther containing pollen grains (Scagel et al., 1965, p.622). Syn: *loculus [bot].*

loculus [bot] (loc′-u-lus) *locule.*

loculus [paleont] One of the chambers in a foraminiferal test. (b) The space between two adjacent septa in the *intervallum* of archaeocyathids. (c) In certain dimorphic ostracodes, a pocketlike structure developed in the presumed female (heteromorph) between the frill and the marginal ridge, usually several in each valve (as in *Ctenoloculina, Abditoloculina, Tetrasacculus,* and others). Pl: loculi. Adj: locular.

lode [eco geol] A mineral deposit consisting of a zone of veins, veinlets, disseminations, or planar breccias; a mineral deposit in consolidated rock as opposed to *placer* deposits. Syn: *lead [eco geol].* Cf: *vein [ore dep]; vein system.*

lode [streams] A local English term for a channel or watercourse, usually partly artificial and embanked above the surrounding country.

lode claim A mining claim on an area containing a known vein or lode. Cf: *placer claim.*

lode country *ore channel.*

Lode's parameter A parameter that describes the shape of the strain ellipsoid in terms of the logarithmic values of its semi-axis lengths (stretches) (Hossack, 1968).

lodestone (lode′-stone) (a) A magnetic variety of natural iron oxide (Fe_3O_4) or of the mineral magnetite; specif. a piece of magnetite possessing polarity like a magnet or magnetic needle and hence one that, when freely suspended, will attract iron objects. Also spelled: *loadstone.* Syn: *leading stone; Hercules stone.* (b) An intensely magnetized rock or ore deposit.

lodestuff (lode′-stuff) Both the gangue and the economically valuable minerals of a lode; the contents of an *ore channel.*

lode tin Cassiterite occurring in veins, as distinguished from *stream tin.*

lodge moraine A terminal moraine of billowy relief, consisting of subglacial debris lodged under a thin margin of a glacier; widespread in North America. Syn: *submarginal moraine.*

lodgment (lodg′-ment) The plastering beneath a glacier of successive layers of basal till upon bedrock or other glacial deposits. Cf: *plastering-on.* Also spelled: lodgement.

lodgment till A *basal till* commonly characterized by compact fissile structure and containing stones oriented with their long axes generally parallel to the direction of ice movement. Cf: *fluted till.*

lodranite (lod′-ra-nite) A group of primitive achondrites rich in olivine and pyroxene, and containing significant metal and sulfides; thought to have formed from chondritic materials as refractory residues from which melts have been lost; similar in composition to acapulcoites.

loellingite *löllingite.*

loess A widespread, homogeneous, commonly nonstratified, porous, friable, slightly coherent, usually highly calcareous, fine-grained blanket deposit (generally less than 30 m thick), consisting predominantly of silt with secondary grain sizes ranging from clay to fine sand. It covers areas extending from north-central Europe to eastern China as well as the Mississippi Valley and Pacific Northwest of the U.S. Loess is generally buff to light yellow or yellowish brown, often contains shells, bones, and teeth of mammals, and is traversed by networks of small narrow vertical tubes (frequently lined with calcium-carbonate concretions) left by successive generations of grass roots, which allow the loess to stand in steep or nearly vertical faces. Loess is now generally believed to be windblown dust of Pleistocene age, carried from desert surfaces, alluvial valleys, and outwash plains, or from unconsolidated glacial or glaciofluvial deposits uncovered by successive glacial recessions but prior to invasion by a vegetation mat. The mineral grains, composed mostly of silica and associated heavy minerals, are fresh and angular, and are generally held together by calcareous cement. In some regions, e.g. Moravia and China, more than 10 successive loess formations are separated by red to dark brown paleosols. A "lee-desert loess" commonly found in the Middle East on the downwind side of the northeastern Sahara is reddish in color and commonly noncalcareous. Etymol: German "Löss", from dialectal (Switzerland) "lösch", "loose", so named by peasants and brickworkers along the Rhine valley where the deposit was first recognized. Pron: luehss. Cf: *limon; adobe.* Syn: *löss; lehm.*

loessal (loess′-al) Pertaining to or consisting of *loess.*

loess doll A compound nodule or concretion of calcium carbonate found in loess and resembling a doll, a potato, or a child's head. It is often hollow but may contain a loose stone. Syn: *loess nodule; loess kindchen; puppet.*

loess flow A fluid suspension of dry porous silt in air, moving downslope, such as occurred following the 1920 earthquake in Kansu Province, China (Close and McCormick, 1922).

loessification (loess′-i-fi-ca′-tion) Formation and development of *loess.*

loess kindchen A *loess doll* resembling the head of a child. Etymol: German "Lösskindchen".

loessland (loess′-land) Land whose surface is underlain by *loess.*

loess nodule *loess doll.*

loessoïde (loess-oï′-de) A Dutch term for deposits in southern Limburg (a province in southern Netherlands), believed to be of loessal origin, but reworked and redeposited by streams, possibly with an admixture of residual material from in-situ decomposition.

loeweite *löweite.*

Loewinson-Lessing classification (Loe′-win-son-Les′-sing) A chemical classification of igneous rocks (into the four main types acid, intermediate, basic, and ultrabasic) based on silica content.

loferite (lo′-fer-ite) A limestone or dolomite riddled by shrinkage pores, such as the carbonate sediments in the Triassic Dachstein Formation (Lofer facies) in Salzburg, Austria (Fischer, 1964, p.124). The term is partly synonymous with *bird's-eye limestone.*

log A continuous record as a function of depth, usually graphic and plotted to scale on a narrow paper strip, of observations made on the rocks and fluids of the geologic section exposed in a well bore; e.g., *well log, sample log, strip log, drilling-time log.*

logan [geog] (lo′-gan) Shortened form of *pokelogan.*

logan [geomorph] *logan stone.*

logan stone An English name for a *rocking stone* consisting of a large mass of granite or gneiss chemically weathered along horizontal joints and so balanced on its base as to "log" or rock from side to side; e.g. the stone weighing about 80 tons near Land's End in Cornwall. Syn: *logan [geomorph]; loggan stone; logging stone.*

logarithmic mean diameter (log-a-rith′-mic) An expression of the average particle size of a sediment or rock, obtained by taking the arithmetic mean of the particle-size distribution in terms of logarithms of the class midpoints. Cf: *phi mean diameter.*

loggan stone (log′-gan) *logan stone.*

logging (log′-ging) (a) The act or process of making or recording a *log.* (b) The method or technique by which subsurface formations are characterized relative to depth by measurements or observations on the rocks of a borehole.

logging stone *logan stone.*

logging tool *sonde.*

logistic growth (lo-gis′-tic) The term for a model describing the growth of resource-limited populations. Such populations initially grow at a rate proportional to their size (*Malthusian principle*); but their growth is damped to an equilibrium state as the *carrying capacity* of the environment is approached. It is also used to model biologic diversification (Sepkoski, 1978).

lognormal distribution (log-nor′-mal) A *frequency distribution* of a variable whose logarithm follows a normal distribution.

loipon (loi′-pon) A term proposed by Shrock (1947) for a residual surficial layer produced by intense and prolonged chemical weathering and composed largely of certain original constituents of the source rock. Typical accumulations of loipon are the gossans over orebodies, bauxite deposits in Arkansas, terra rossa deposits of Europe, and duricrust of Australia. Etymol: Greek, "residue". Adj: loiponic.

lokkaite-(Y) (lok′-ka-ite) A white orthorhombic mineral: $CaY_4(CO_3)_7{\cdot}9H_2O$.

löllingite (löl′-ling-ite) (a) A metallic bright gray orthorhombic mineral: $FeAs_2$. Also spelled: *loellingite.* Syn: *leucopyrite.* (b) A group name for minerals with the formula AX_2, where A = Fe, Co, Ni or platinum-group elements, X = As, or Sb.

lolly ice (lol′-ly) Soft *frazil ice* formed in turbulent seawater. Rarely used.

lolongate (lo-lon′-gate) Of a *colpus,* elongated longitudinally. Cf: *lalongate.*

loma (lo′-ma) A term used in the SW U.S. for an elongated, gentle swell or rise of the ground (as on a plain), or a rounded, broad-topped, inconspicuous hill. Etymol: Spanish, "hillock, rising ground, slope".

lomita (lo-mi′-ta) A small, low *loma.*

lomonite (lo′-mon-ite) Original spelling of *laumontite.*

lomonosovite (lo-mo-no′-sov-ite′) A dark cinnamon-brown to black or rose-violet triclinic mineral: $Na_5Ti_2O_2(Si_2O_7)(PO_4)$. Cf: *murmanite.*

lomontite (lo′-mont-ite) *laumontite.*

lonchiole (lonch′-i-ole) A *sceptrule* with a single spine opposite the single ray.

londonite A colorless to yellow cubic mineral: $(Cs,K)Al_4Be_4(B,Be)_{12}O_{28}$. It is the Cs analogue of rhodizite.

lonecreekite (lone-creek′-ite) A colorless cubic mineral of the *alum* group: $(NH_4)(Fe^{3+},Al)(SO_4)_2{\cdot}12H_2O$.

lonestone (lone′-stone) An oversized exotic stone in finer sediment, with or without clear evidence of having been dropped in from above (Ojakangas, 1985). Cf: *dropstone (a).*

Longaxones (Long′-ax-ones) A group of mid-Cretaceous and younger tricolpate angiosperm pollen in which the polar axis is as long as, or longer than, the equatorial diameter. Cf: *Brevaxones.*

long clay A highly plastic clay; a *fat clay.*

longicone (lon′-gi-cone) A long, slender, conical, gradually tapering shell characteristic of certain orthoconic cephalopods. Also, a fossil animal having such a shell. Cf: *brevicone.*

longitude (lon′-gi-tude) (a) An angular distance between the plane of a given meridian through any point on a sphere or spheroid and the plane of an arbitrary meridian selected as a line of reference, measured in the plane of a great circle of reference or in a plane parallel to that of the equator; specif. the length of the arc or portion of the Earth's equator or of a parallel of latitude intersected between the meridian of a given place and the prime meridian (or sometimes a national meridian), expressed either in time or in degrees east or west of the prime or national meridian (which has longitude zero degrees) to a maximum value of 180 degrees. A degree of longitude on the Earth's surface varies in length approximately as the cosine of the latitude, being 69.95 statute miles at the equator, 53.43 miles at lat. 40°, and zero at the poles; it represents 4 minutes of time, so that 15 degrees of longitude is equivalent to a difference of one hour of local time. Longitude may also be measured as the angle at the poles lying between the two planes that intersect along the Earth's axis to produce the two meridians. Abbrev: long. Symbol λ. See also: *astronomic longitude; geodetic longitude; geographic longitude; celestial longitude.* Cf: *meridian.* (b) A linear coordinate distance measured east or west from a specified north-south line of reference; e.g. *easting* and *westing.* Cf: *latitude.*

longitude correction A correction applied to the observed magnetic field data to compensate for the Earth's internal (core) magnetic field, i.e., IGRF. Obsolete, replaced by IGRF correction.

longitude difference *departure.*

longitudinal (lon-gi-tu'-di-nal) Said of an entity that is extended lengthwise; esp. said of a topographic feature that is oriented parallel to the general strike or topographic trend of a region. Ant: *transverse [geomorph].*

longitudinal band *Foliation,* or a zone of foliation in a glacier that is parallel to the direction of ice movement.

longitudinal coastline *concordant coastline.*

longitudinal conductance The product of the average conductivity and thickness of a rock layer, measured in siemens units.

longitudinal consequent stream A consequent stream whose direction is determined by the plunge of a fold; esp. a stream flowing in a synclinal trough.

longitudinal crevasse A crevasse roughly parallel to the direction of ice movement. In a valley glacier, this type of crevasse is longitudinal only in the center of the glacier; away from the center it becomes a *splaying crevasse* (the preferred term).

longitudinal dispersion Dispersion that is parallel to the direction of groundwater flow.

longitudinal drift A long, tapered, sharp-crested *sand drift* formed on the lee side of a narrow gap in a ridge or scarp oriented transversely to the prevailing wind, esp. in a desert or steppe region where ridges interrupt flat plains or plateaus; it may be 500 m in length.

longitudinal dune A long, narrow sand dune, usually symmetrical in cross profile, oriented parallel with the direction of the prevailing wind; it is wider and steeper on the windward side but tapers to a point on the leeward side, and commonly forms behind an obstacle in an area where sand is abundant and the wind is strong and constant. Such dunes may be a few meters high and up to 100 km long. See also: *seif.*

longitudinal fault A fault whose strike is parallel with that of the general structural trend of the region.

longitudinal flagellum A thread-shaped flagellum in a dinoflagellate, trailing after the body and arising from the posterior pore in the sulcus if two are present, its proximal part lying in the ventral sulcus near the major axis. It is used mainly for steering (Fensome et al., 1993; Williams et al., 2000, p.126).

longitudinal fold (a) A fold whose axis trends in accordance with the general strike of the area's structures. (b) A fault-related fold whose axis is parallel to the strike of the associated fault (Schlische, 1985). Cf: *discordant fold; transverse fold.* Syn: *strike fold.*

longitudinal furrow [paleont] A deep longitudinal groove formed by infolding of the surface on one side of conodont elements proper of *Panderodus, Neopanderodus, Belodina,* and related genera of the Panderodontacea (TIP, 1981, pt. W, supp.2).

longitudinal furrow [streams] One of many closely spaced continuous furrows elongated parallel to the current and developed rhythmically on all or part of a stream bed. Furrows are separated by regularly spaced longitudinal ridges, the separation ranging from 3 mm to as much as 5 cm (Dzulynski and Walton, 1965, p.61). The furrows are rounded in section; they begin upstream in a convex "beak" and are broken along their length by occasional cuspate bars. They are commonly preserved as sole marks. The term was first defined by Dzulynski and Walton (1963, p.285). Cf: *furrow cast; furrow flute cast; rib and furrow.*

longitudinal joint A steeply dipping joint plane in a pluton that is oriented parallel to the lines of flow. Syn: *S-joint; (hO1) joint; bc-joint.*

longitudinal moraine A moraine rampart consisting of a medial moraine and an englacial moraine of a former glacier (Schieferdecker, 1959, term 0920).

longitudinal profile (a) The profile of a stream or valley, drawn along its length from the source to the mouth of the stream; it is the straightened-out, upper edge of a vertical section that follows the winding of the stream or valley. See also: *thalweg [streams].* Cf: *cross profile.* Syn: *long profile; valley profile; stream profile; river profile.* (b) A similar profile of a landform, such as a pediment.

longitudinal resistivity Resistivity of rock measured along the direction of bedding. Cf: *transverse resistivity.*

longitudinal ridge One of many closely spaced continuous ridges elongated parallel to the current and developed rhythmically on all or part of a stream bed. Ridges are separated by regularly spaced longitudinal furrows, the separation ranging from 3 mm to as much as 5 cm (Dzulynski and Walton, 1965, p.61). Cf: *rib and furrow.*

longitudinal ripple A *ripple* with a relatively straight crest, formed parallel to the direction of the current, such as one related to oscillatory wave action (Van Straaten, 1951); its profile may be asymmetric or symmetric. See also: *corrugated ripple; mud-ridge ripple.*

longitudinal ripple mark *longitudinal ripple.*

longitudinal section A diagram drawn on a vertical or inclined plane and parallel to the longer axis of a given feature; e.g. a section drawn parallel to the strike of a vein, the length of a valley, or the axis of a fossil. Cf: *cross section [geol].*

longitudinal septum A *septum* in certain cirripede crustaceans disposed normal to the inner and outer laminae of a compartmental plate and separating the parietal tubes. Syn: *parietal septum.*

longitudinal splitting crack The crack that forms along the axis of an experimental cylinder of rock that is subject to an unconfined compression test.

longitudinal strain Any one of several measures of the change in length of a material line in a rock body.

longitudinal stream A *subsequent stream* that follows the strike of the underlying strata.

longitudinal valley (a) A subsequent valley developed along or in the same direction as the general strike of the underlying strata; a valley at right angles to a consequent stream. This is the current usage of the term, as used by Powell (1873, p.463). (b) A term originally applied by Conybeare and Phillips (1822, p. xxiv) to a long valley developed parallel to the general trend of a ridge, range, or chain of mountains or hills. According to current usage, the term is correctly used only "where the mountain or hill ranges are parallel to the strike" (Stamp, 1961, p.300). Syn: *lateral valley.* Cf: *transverse valley.*

longitudinal wave *P wave.*

Longobardian (Lon-go-bar'-di-an) A European substage: upper substage of the Ladinian Stage, Middle Triassic (above Fassanian, below Julian substage of the Carnian).

long period A period of seismic activity that is more than six seconds in duration. Cf: *short period.*

long profile *longitudinal profile.*

long-range order (a) A strong tendency for the random atoms in a random solid solution to become ordered as the solution cools from the elevated temperature at which it was formed. (b) That state of a crystal structure in which a given atomic species occupies the same specific site in each unit cell, so that the probability of finding atoms on wrong sites is zero. Cf: *short-range order.*

longshore (long'-shore) Pertaining or belonging to the shore or coast, or a seaport; littoral. Syn: *alongshore.*

longshore bar A low, elongate sand ridge, built chiefly by wave action, occurring at some distance from, and extending generally parallel with, the shoreline, being submerged at least by high tides, and typically separated from the beach by an intervening trough. Syn: *ball [coast]; offshore bar; submarine bar.*

longshore current A current that is formed by waves approaching the coast at an angle. Generally confined to the surf zone. Syn: *littoral current; littoral drift.*

longshore drift *littoral drift.*

longshore trough Elongate depression between longshore bars.

Longtanian Chinese stage of the Upper Permian (Harland et al., 1990). Above Capitanian, below Changxingian.

longulite (lon'-gu-lite) Obsolete term for a cylindrical or conical belonite thought to have formed by the coalescence of globulites. See also: *bacillite.*

longwall (long'-wall) Said of a method of underground mining in flat-lying strata, esp. of coal. Parallel entries are driven into the seam, to the limit of the block to be mined; from the end of these entries, workings are driven at right angles in both directions. A longwall face is produced as these workings are widened back toward the point of entry. Working space is provided by timbers or other supports; the roof caves as mining progresses. Almost all of the coal or other desired mineral is recovered, in contrast to the *room-and-pillar* method.

long wave [seis] An obsolete syn. of *surface wave [seis].*

long wave [water] *shallow-water wave.*

lonsdaleite (lons'-dale-ite) A meteorite mineral consisting of a form of carbon. It is hexagonal and is polymorphous with diamond, graphite, and chaoite.

lonsdaleoid dissepiment (lons'-dale-oid) In rugose corals, a *dissepiment* lying between the peripheral ends of septa and the outer wall, as in *Lonsdaleia.* They are usually larger than other types of dissepiments (Jull, 1967, p.622). See also: *lonsdaleoid septum.*

lonsdaleoid septum A rugose corallite *septum* characterized by discontinuity toward the peripheral edge of the septum, as in *Lonsdaleia.* See also: *lonsdaleoid dissepiment.*

look direction The direction in which pulses of microwave energy are transmitted by a *SLAR* system. It is normal to the flight direction. Syn: *range direction.*

loop [geophys] A pattern of field observations that begins and ends at the same point with a number of intervening measurements. Such a pattern is useful in correcting for drift in gravity-meter observations and for diurnal variation in magnetometer surveys, and in detecting faults or other causes of misclosure in seismic shooting.

loop [glac geol] *loop moraine.*

loop [paleont] (a) A support (*brachidium*) for a brachiopod *lophophore,* composed of secondary shell and extending anteriorly from crura as a closed apparatus, variably disposed and generally ribbonlike, with or without a supporting septum from the floor of the brachial valve (TIP, 1965, pt. H, p.147). (b) Elliptical segment of conodont attachment surface enclosing basal pit or segminate or segminiplanate pectiniform elements. Closed posteriorly; continuous anteriorly with attachment surface beneath anterior process; and commonly expressed as raised area of recessive basal margin around basal pit (TIP, 1981, pt. W, supp.2).

loop [surv] A pattern of measurements in the field, such that the final measurement is made at the same place as the first measurement.

loop bedding Bedding characterized by small groups of laminae that are sharply constricted or that end abruptly at intervals, giving the effect of long, thin loops or links of a chain; it is found in fine calcareous sediments and in oil shale.

looped bar A curved bar on the leeward or landward side of an offshore island undergoing wave erosion, formed by the union of two separate spits that have trailed off behind and joined together to form a loop that encloses or nearly encloses a body of water. Cf: *cuspate bar.*

loop lake *oxbow lake.*

loop moraine An end moraine of a valley glacier, shaped like an arc or half-loop, concave toward the direction from which the ice approached; it is usually steep on both sides and extends across the valley. Syn: *valley-loop moraine; moraine loop; loop [glac geol].*

loose ice A rarely used syn. for broken ice.

loose line In cross-section balancing, a line drawn normal to layering in either the deformed or restored cross section that is free to change its orientation relative to layering during the restoration. Loose lines are used to identify and estimate the amount of layer-parallel slip needed to generate a given fold-thrust belt structure (Geiser, 1988; Woodward et al., 1989).

loose-snow avalanche A snow avalanche that starts at a point and widens downhill, in snow lacking cohesion. Cf: *wind-slab avalanche; slab avalanche.* Syn: *point release avalanche.*

lopadolith (lo-pad'-o-lith) A basket-shaped *coccolith* opening distally.

loparite-(Ce) (lo-pa'-rite) A brown to black cubic mineral: $(Ce,Na,Ca)_2(Ti,Nb)_2O_6$. It was formerly regarded as a variety of perovskite containing alkalies and cerium.

lopezite (lo'-pez-ite) An orange-red triclinic mineral: $K_2Cr_2O_7$.

lophophore (loph'-o-phore) A feeding organ in bryozoans and brachiopods, usually consisting of a circular or horseshoe-shaped fleshy ridge surrounding the mouth and bearing the tentacles that serve to engulf food particles and provide a respiratory current. In brachiopods, it is a feeding organ with filamentous appendages, symmetrically disposed about the mouth, suspended from the anterior body wall or attached to the dorsal mantle, and occupying the mantle cavity (TIP, 1965, pt. H, p.147). See also: *brachia.*

lophophytous (loph-o-phyt'-ous) Said of a sponge that is fastened to the substrate by a tuft of spicules.

lophotrichous (lo-phot'-ri-chous) Said of a bacterial cell having a tuft of flagella at one or both ends. Cf: *monotrichous; peritrichous.*

Lopingian (Lo-pin'-gi-an) A series of the Standard Global Chronostratigraphic Scale: Upper Permian (above Guadalupian Series, below Triassic System). The basal GSSP is in China.

lopolith (lop'-o-lith) A large, concordant, typically layered igneous intrusion whose floor is convex-down and whose roof may be convex-down or flat.

Lorac (Lo'-rac) A hyperbolic radio location system of four radio transmitters operating at four different frequencies between 1.650 and 1.700 MHz, forming two families of hyperbolas (iso-phase lines) along each of which the difference of phase of the signals from a pair of stations is a constant. A receiver's location can be determined by measuring the differences of phase in signals from the two pairs of stations and plotting the location at the intersection of a pair of hyperbolas. A plotting accuracy of 1 meter is attainable. The accuracy with which a location can be determined depends on many factors such as the location of the receiver within the pattern of iso-phase lines. A trade name. Etymol: *lo*ng-range *ac*curacy.

loran (lo'-ran) Any of various long-range radio position-fixing systems by which hyperbolic lines of position are determined by measuring the difference in arrival times of synchronized pulse signals from two or more fixed transmitting radio stations of known geographic position. Loran fixes may be obtained at a range of 1400 nautical miles at night. Cf: *shoran.* Etymol: *lo*ng-*ra*nge *n*avigation.

lorandite (lor'-an-dite) A metallic adamantine vivid red or dark lead-gray monoclinic mineral: $TlAsS_2$.

loranskite-(Y) (lo-ran'-skite) An ill-defined submetallic black mineral: $(Y,Ce,Ca,Zr)TaO_4$ (?).

Lorentz factor In calculating or correcting X-ray diffraction intensities, a factor that accounts for the time required for a *reciprocal lattice point* to pass through the *sphere of reflection.* It is often combined with the *polarization factor.*

Lorentz-polarization factor Two factors almost always used together to explain the very high background at very low angles of X-ray diffraction tracings. The Lorentz factor is a combination of two geometrical factors. The first is a formulation for the volume of the crystal that is exposed to primary irradiation. The second relates the number of crystals favorably oriented for diffraction at any Bragg angle Θ_b. The Lorentz factor is different for random powders and single crystals (Reynolds, 1976). The polarization factor accounts for increases in peak and background scattering from a maximum at 0° 2Θ to a minimum at 90° 2Θ. As radiation comes from the tube it is unpolarized, but the process of scattering causes a degree of polarization related to the angle of the incident and diffracted beams. The X-rays are polarized as they are diffracated from a plane of atoms for the same reason a beam of light is polarized by reflecting off of a smooth body of water, and polarized in the same way, i.e., the vibrations surviving the reflection are those parallel to the water surface or the plane of the atoms. As amplitude in directions other than that of the plane of polarization is lost, the intensity decreases (Klug and Alexander, 1974, p.142).

lorenzenite (lo-ren'-zen-ite) A pale purple-brown, pale pink, pale mauve, brown, or black orthorhombic mineral: $Na_2Ti_2Si_2O_9$. Syn: *ramsayite.*

lorettoite (lo-ret'-to-ite) A discredited honey-yellow to reddish-yellow mineral: $Pb_7O_6Cl_2$. Its natural occurrence has not been established.

lorica (lo'-ri-ca) (a) A hard, protective covering, commonly tubular or vase-like, secreted or built with agglutinated foreign matter by tintinnids, thecamoebians, certain algae, and other protists, and having an organic, calcareous, or siliceous composition. (b) The cell wall or two valves of a diatom. Pl: loricae.

loseyite (lo'-sey-ite) A bluish-white to brownish monoclinic mineral: $(Mn^{2+},Zn)_7(CO_3)_2(OH)_{10}$.

losing stream (los'-ing) *influent stream.*

löss Var. of *loess.* Etymol: German *Löss.*

lost circulation The condition during *rotary drilling* when excessive amounts of drilling mud flows into porous, fractured, or cavernous rocks penetrated by the borehole and very little or none return to the surface.

lost mountain An isolated mountain standing in a desert and so far removed from the main mass of mountains as to have no apparent connection with them; e.g. an outlier or a monadnock that has resisted erosion more effectively than the surrounding land. A smaller feature is called a "lost hill". Cf: *huerfano.*

lost stream (a) A dried-up stream in an arid region. (b) A *sinking stream* in a karst region.

Lotharingian (Lo-tha-rin'-gi-an) A European substage: upper substage of Sinemurian Stage, Lower Jurassic (below Carixian substage of Pliensbachian).

lotharmeyerite (lo-thar-mey'-er-ite) A reddish-orange monoclinic mineral of the *tsumcorite* group: $Ca(Mn^{3+},Zn)_2(AsO_3OH)_2(OH)_3$.

lotic Pertaining to running-water systems, e.g., rivers, brooks, springs (Kalff, 2002, p.5; Wetzel, 2001, p.17). Cf: *lentic.*

lotrite (lot'-rite) *pumpellyite.*

lottal (lot'-tal) A field term used by King (1962, p.179) for the aqueous clayey mixtures formed by mass movement down hillslopes. Etymol: Richard Armour's verse, "Careful with that catsup bottle; none'll come, and then a lot'll."

louderback (lou'-der-back) A remnant of a lava flow appearing in a tilted fault block and bounded by a dip slope. It is named after George D. Louderback, a North American geologist, who used it as evidence of block faulting in basin-and-range topography.

louderbackite (lou'-der-back-ite') *römerite.*

loudounite (lou'-dou-nite) A white to light green mineral: $NaCa_5Zr_4Si_{16}O_{40}(OH)_{11}·8H_2O$.

loughlinite (lough'-lin-ite) A pearly-white, asbestiform mineral of the *sepiolite* group: $Na_2Mg_3Si_6O_{16}·8H_2O$. It is the Na analogue of sepiolite with Na in place of octahedral Mg.

loupe Any small magnifying glass mounted for use in the hand or so that it can be held in the eye socket or attached to spectacles, and used to study gemstones.

lourenswalsite (lou'-rens-wal'-site) A silvery gray to light brownish-gray hexagonal mineral: $(K,Ba)_2Ti_4(Si,Al)_6O_{14}(OH)_{12}$.

louver (lou'-ver) A transverse wall plate in archaeocyathids, commonly developed between the edges of adjacent septa or longitudinal ribs and usually tilted with reference to the wall surface (TIP, 1972, pt. E, p.41).

lovchorrite (lov-chor'-rite) *mosandrite.*

lovdarite (lov-dar'-ite) A colorless to yellowish orthorhombic mineral: $K_2Na_6(Be,Al)_4Si_{14}O_{36}·9H_2O$.

love arrow *flèche d'amour.*

Lovenian system (Lo-ve'-ni-an) A numbering system in which the individual ambulacral areas of the tests of echinoids and edrioasteroids are designated by Roman numerals (I-V), and the interambulacral areas by Arabic numerals (1-5). It is based on bilateral symmetry with respect to a plane passing through the apical system, peristome, and periproct in irregular echinoids, and chiefly by the position of the madreporite in regular echinoids. Named after Sven L. Lovén (1809-1895), Swedish zoologist.

loveringite A metallic black rhombohedral mineral of the *crichtonite* group: $(Ca,Ce,La)(Ti,Fe,Cr)_{21}O_{38}$.

Love wave A type of surface wave having a horizontal motion that is shear or transverse to the direction of propagation, that is, SH motion. It is named after A.E.H. Love, the English mathematician who discovered it. See also: *G wave.* Syn: *Q wave; Querwellen wave;* L_q.

lovozerite (lo-vo-zer'-ite) A reddish-brown, brown, or black monoclinic or rhombohedral mineral: $Na_2Ca(Zr,Ti)Si_6(O,OH)_{18}$.

low [beach] (a) *swale.* (b) *trough [beach].*

low [geomorph] *law [geomorph].*

low [meteorol] *depression [meteorol].*

low [struc geol] n. A general term for such features as a structural basin, a syncline, a saddle, or a sag. Cf: *high.* Syn: *structural low.*

low albite Low-temperature albite common in nature, stable below 450°C. It takes almost no calcium or potassium into solid solution, and has a completely ordered structure. Cf: *high albite.*

low-angle boundary A *tilt boundary* formed by *recovery* across which the crystallographic mismatch is <6°.

low-angle cross-bedding (low-an'-gle) Cross-bedding in which the cross-beds have an average maximum inclination of less than 20° (McKee and Weir, 1953, p.388). Cf: *high-angle cross-bedding.*

low-angle fault A fault with a dip of 45° or less. Cf: *high-angle fault.*

low-angle thrust *overthrust.*

low chalcocite Orthorhombic chalcocite, stable below 105°C.

low-cut filter *high-pass filter.*

low dome A flat-topped, roughly symmetrical type of *lava dome* that has grown by internal expansion from the successive influx of new lava (Blake, 1989; Francis, 1993).

löweite (lö'-we-ite) A colorless to reddish-yellow rhombohedral mineral: $Na_{12}Mg_7(SO_4)_{13}·15H_2O$. Also spelled: *loeweite.*

low-energy coast (low-en'-er-gy) A coast protected from strong wave action by headlands, wide gently sloping bottom, dominance by winds from the land, or other factors, and characterized by average breaker heights of less than 10 cm. Cf: *high-energy coast; moderate-energy coast; zero-energy coast.*

low-energy environment An aqueous sedimentary environment characterized by a low kinetic energy and by standing water or a general lack of wave or current action, thereby permitting very fine-grained sediment to settle and accumulate; e.g. a coastal lagoon or an alluvial swamp. Cf: *high-energy environment.*

lower (low'-er) Pertaining to rocks or strata that are normally below those of later formations of the same subdivision of rocks. The adjective is applied to the name of a chronostratigraphic unit (system, series, stage) to indicate position in the geologic column and corresponds to *early* as applied to the name of the equivalent geologic-time unit; e.g. rocks of the Lower Jurassic System were formed during the Early Jurassic Period. The initial letter of the term is capitalized to indicate a formal subdivision (e.g. "Lower Devonian") and is lowercased to indicate an informal subdivision (e.g. "lower Miocene"). The informal term may be used where there is no formal subdivision of a system or of a series. Cf: *upper; middle [stratig].*

lower beach Portion of the beach submerged at all times, extending out as far as ordinary high waves entrain and transport sediment from and along the bottom, commonly extending inland as far as the step (if present) dividing it from the upper beach.

lower break *foot [geol].*

lower core A term that includes the *inner core* and the transitional zone of the *outer core,* i.e. the equivalent of the *F layer* and the *G layer.* Archaic.

lower crustal layer *Conrad layer.*

lower flow regime The *flow regime* of a *unidirectional current* that is characterized by relatively low sediment transport rates (largely as *bed load*) and *bed forms* that are out of phase with the water surface, including *ripple, dune,* and *lower plane bed;* flow *Froude number* is less than unity (Simons and Richardson, 1961).

lower flow regime plane bed *lower plane bed.*

lower keriotheca The adaxial (lower) part of *keriotheca* in the wall of a fusulinid, characterized by coarse alveolar structure (as in *Schwagerina*). Cf: *upper keriotheca.*

lower mantle That part of the mantle that lies below a depth of about 660 km. With increasing depth, density increases from ~4.4 g/cm^3-to ~5.6 g/cm^3, and velocity of compressional seismic waves increases from ~10.7 km/s to ~13.7 km/s (Dziewonski and Anderson, 1981). It is equivalent to the *D layer.* Syn: *inner mantle.*

lower Paleolithic In archaeology, the first and oldest division of the *Paleolithic,* characterized by *Australopithecus* and *Homo erectus.* Correlation of cultural levels with actual age (and, therefore, with the time-stratigraphic units of geology) varies from region to region. Cf: *middle Paleolithic; upper Paleolithic.*

lower plane bed A bed configuration of the *lower flow regime* that is characterized by a flat, almost featureless surface and very low rates of sediment transport. Syn: *lower flow regime plane bed; lower stage plane bed.*

lower plate The footwall of a fault. Cf: *upper plate.*

lower stage plane bed *lower plane bed.*

lower tectorium The adaxial secondary dark layer of spirotheca in the wall of a fusulinid, next below the diaphanotheca or tectum, as in *Profusulinella.* Cf: *tectorium; upper tectorium.*

low flow The lowest sustained flow during base runoff conditions of a river.

low-flow frequency curve A graphic illustration of both the magnitude and frequency of minimum flows in a given time span.

low-fluorine porphyry molybdenum A sub-group of *porphyry molybdenum deposit* associated with intrusive rocks having low fluorine contents (generally less than 0.1 percent) compared to mineralizing intrusions associated with *Climax-type* porphyry molybdenum deposits.

low-grade Said of an ore with a relatively low ore-mineral content. Syn: *lean.* Cf: *high-grade.* See also: *grade [ore dep].*

low island In the Pacific Ocean, a coralline rather than a volcanic island. Cf: *high island.*

low-K In the *IUGS classification,* low-K volcanic rocks plot below a line through two points with the coordinates SiO_2=48, K_2O=0.3 wt% and SiO_2=68, K_2O=1.2 wt%.

lowland (low'-land) (a) A general term for low-lying land or an extensive region of low land, esp. near the coast and including the extended plains or country lying not far above tide level. (b) The low and relatively level ground of a region, in contrast with the adjacent, higher country; e.g. a vale between two cuestas. (c) A low or level tract of land along a watercourse; a bottom. The term is usually used in the plural. Ant: *upland.*

low-latitude desert (low-lat'-i-tude) *tropical desert.*

low magnesian calcite Calcite containing 0-4 mole % magnesian calcite. Abbrev: LMC.

lowmoor bog (low'-moor) A bog that is at or only slightly above the water table, on which it depends for accumulation and preservation of peat (chiefly the remains of sedges, reeds, shrubs, and various mosses). Cf: *highmoor bog.*

lowmoor peat Peat occurring on low-lying moors or swamps and containing little or no sphagnum. Its moisture is standing surface water and is low in acidity. Mineral matter and nitrogen content is high compared to highmoor peat; cellulose content is low. Cf: *highmoor peat; planar peat; topogenous peat.* Syn: *fen peat.*

low oblique n. An *oblique* that does not include the apparent horizon. Cf: *high oblique.*

low-pass filter An operation used in image processing systems to enhance features that are generally larger than the averaging neighborhood. The enhancement is accomplished by a smoothing operation which removes most of the high-frequency variations and emphasizes the low frequencies.

low quartz Low-temperature quartz; specif. *alpha quartz.*

low-rank graywacke A term introduced by Krynine (1945) for a graywacke in which feldspar is almost absent. The rock is equivalent to *subgraywacke* as originally defined by Pettijohn (1949) and to *lithic graywacke* of Pettijohn (1954). Cf: *high-rank graywacke.*

low-rank metamorphism Metamorphism that is accomplished under conditions of low to moderate temperature and pressure. Var: low-grade metamorphism. Cf: *high-rank metamorphism.*

lowstand (low'-stand) The interval of time during one or more cycles of relative change of sea level when sea level is below the shelf edge (Mitchum, 1977, p.208). Cf: *highstand.*

lowstand shoreline A *shoreline,* either erosional or constructional, formed at the lowest elevation of a *regression* (Kelley et al., 2003). Cf: *highstand shoreline.*

lowstand systems tract In sequence stratigraphy, a *systems tract* bounded below by the sequence boundary and above by the first major flooding surface, called the *transgressive surface;* or a unit bounded by a slope onlap surface below and a conformable transgressive surface above. It may consist of a basin-floor fan, a slope fan, and a lowstand wedge. The proximal part of the lowstand systems tract is commonly an incised valley filled with fluvial or estuarine strata. The term is applicable for depositional sequences of type 1, T-R sequences, and genetic stratigraphic sequences (Van Wagoner et al., 1987, 1988; Posamentier and Vail, 1988; Embry and Johannessen, 1992).

lowstand wedge In sequence stratigraphy, one or more progradational to aggradational parasequence sets making up a sedimentary wedge that is restricted seaward of the shelf break and that onlaps the slope of the preceding sequence. The most proximal part of the wedge consists of incised-valley fills and their associated lowstand-shoreline deposits on the shelf or upper slope. In basins with shelf-slope breaks, the major part of the wedge is composed of a thick, mostly shale-prone, prograding wedge-shaped unit that downlaps onto the slope fan, and consists of shallowing-upward lowstand deltas and associated sediments that prograde basinward and pinch out landward (Van Wagoner et al., 1987, 1988; Posamentier and Vail, 1988).

low-sulfidation epithermal *adularia-sericite epithermal* deposit.

low-sulfide gold-quartz vein A *mineral-deposit type* characterized by gold-bearing quartz veins with minor base-metal sulfide minerals hosted by regional fault systems. Syn: *Archean lode gold; greenstone gold; Mother Lode vein; mesothermal gold-quartz vein; shear-hosted lode gold.*

low-sulfur epithermal *adularia-sericite epithermal* deposit.

low-sun-angle photograph Aerial photograph acquired in the morning, evening, or winter when the sun is at a low elevation above the horizon.

low tide The tide at its lowest; the accepted popular syn. of *low water* in the sea.

low-tide delta A delta formed at the step by drainage of water from the beach onto the tidal flat. It is associated with enlarged and accentuated rills.

low-tide platform *solution platform.*

low-tide terrace A relatively horizontal zone of the foreshore near the low-water line.

low-velocity layer (a) A near-surface belt of very low-velocity material, also called *weathering* or *weathered layer*. This layer, which commonly varies fairly rapidly in thickness, lithology, density, velocity, and attenuation effects, has a marked effect on seismic reflections. (b) A layer whose velocity is lower than that of shallower refractors. See: *blind zone.* (c) Any layer bounded on both sides by layers of higher velocity, which can carry channel waves. The layer boundaries may be gradational. (d) The B-layer in the upper mantle from 60 to 250 km deep, where velocities are about 6 percent lower than in the outermost mantle. Abbrev: *LVL*. See also: *low-velocity zone.*

low-velocity-layer correction (low-ve-loc'-i-ty-lay'-er) *weathering correction.*

low-velocity zone (a) The zone in the upper mantle, variously defined as from 60 to 250 km in depth, in which velocities are about 6% lower than in the outermost mantle. It is probably caused by the near-melting-point temperature of the material. Syn: *Gutenberg low-velocity zone; B layer.* (b) A region inside the core boundary below a depth of 2,900 km which produces a *shadow zone [seis]* at the Earth's surface. Abbrev: LVZ. See also: *low-velocity layer*.

low volatile bituminous coal (low-vol'-a-tile) Bituminous coal, characteristically agglomerating, that contains 15-22% volatile matter, analyzed on a dry, mineral-matter-free basis. Cf: *high volatile bituminous coal; medium volatile bituminous coal.*

low water Water at the minimum level reached during a tide cycle. The height of the minimum vertical elevation of the tide in a 12- or 24-hour cycle relative to an arbitrary datum, usually the tidal benchmark. Abbrev: LW. Cf: *high water.* Syn: *low tide.*

low-water datum (low-wa'-ter) An approximation to the plane of *mean low water*, adopted as a standard reference plane for a specific area (as the Atlantic coast of the U.S.).

loxochoanitic (lox'-o-cho'-a-nit'-ic) Said of a short, straight *retrochoanitic septal neck* of a nautiloid that points obliquely toward the interior of the *siphuncle.*

loxoclase (lox'-o-clase) A variety of orthoclase containing considerable sodium: $(K,Na)AlSi_3O_8$. It has a green tinge due to small inclusions of diopside. The loxoclase series ranges from Or_1Ab_1 to Or_1Ab_4, with K_2O in the range of 4-7%. Syn: *soda orthoclase.*

loxodrome (lox'-o-drome) *rhumb line.*

loxodromic curve (lox-o-drom'-ic) *rhumb line.*

LPG *liquefied petroleum gas.*

Lq *Love wave.*

L_q *Love wave.*

LS-tectonite Tectonite whose fabric has both planar and linear fabric elements caused by deformation. Cf: *L-tectonite; S-tectonite.*

L-tectonite (L-tec'-ton-ite) A tectonite whose fabric is dominated by the presence of lineations, such as a deformed conglomerate in which the pebbles are strongly elongate. Cf: *S-tectonite; B-tectonite; LS-tectonite.*

LTP *lunar transient phenomenon.*

luanheite (lu-an'-he-ite) A metallic black hexagonal mineral: Ag_3Hg.

luberoite A metallic dark bronze monoclinic mineral: Pt_5Se_4.

lublinite (lub'-lin-ite) A very soft, cheesy, or spongy mixture of calcite and water. Cf: *moonmilk.* Syn: rock milk; mountain milk.

lubricating layer (lu'-bri-cat'-ing) In a *décollement*, that stratum which acted as a lubricant for the gliding of the overthrust. Often used erroneously in cases where the décollement has ordinary frictional strength.

lucasite-(Ce) (lu'-cas-ite) A brown or gray monoclinic mineral: $(Ce,La)Ti_2(O,OH)_6$.

lucinite (lu'-ci-nite) *variscite.*

lucinoid (lu'-ci-noid) Said of *heterodont* dentition of a bivalve mollusk with two cardinal teeth in each valve, the anterior tooth in the left valve occupying a median position below the beaks. Cf: *corbiculoid.*

luddenite (lud'-den-ite) A green monoclinic mineral: $Pb_2Cu_2Si_5O_{14}{\cdot}14H_2O$.

Lüder's lines (Lü'-der's) Planar deformation features, wider than ordinary shear fractures, inclined along planes of high shear stress, on which plastic or cataclastic deformation is concentrated.

Ludfordian A stage of the Standard Global Chronostratigraphic Scale: upper stage of Ludlovian Series of the Upper Silurian (above the Gorstian Stage, below the Pridolian Series). The basal GSSP is in southwest England (Holland, 1982).

Ludhamian (Lud-ha'-mi-an) English continental stage: middle Pliocene (above Waltonian, below Baventian); based on cool-climate deposits. Syn: *Norwich Crag* (lower).

Ludian (Lud'-i-an) Obsolete European stage: upper Eocene (above Bartonian, below Tongrian of Oligocene); equiv. Lattorfian.

ludjibaite A blue-green triclinic mineral: $Cu_5(PO_4)_2(OH)_4$. It is one of the polymorphs of pseudomalachite.

ludlamite (lud'-lam-ite) A white to green monoclinic mineral: $(Fe^{2+},Mg,Mn^{2+})_3(PO_4)_2{\cdot}4H_2O$.

ludlockite (lud'-lock-ite) A red triclinic mineral: $PbFe^{3+}_4As^{3+}_{10}O_{22}$.

Ludlovian (Lud-lo'-vi-an) A series or stage of the Standard Global Chronostratigraphic Scale: Upper Silurian (above Wenlockian, below Pridolian Series). The basal GSSP is in southwest England (Holland, 1982).

ludwigite (lud'-wig-ite) (a) A blackish-green orthorhombic mineral: $Mg_2Fe^{3+}BO_5$. It is isomorphous with vonsenite. Syn: *magnesioludwigite.* (b) A group name for minerals with the formula $A^{2+}{}_2B^{3+}BO_5$, where A = Mg, Fe2+ or Ni, and B = Fe^{3+} or Mn^{3+}.

lueshite (lu'-e-shite) A black orthorhombic mineral: $NaNbO_3$. It has structure of the perovskite type and is dimorphous with natroniobite. Syn: *igdloite.*

luetheite (lue'-the-ite) A blue or pale green monoclinic mineral, the Al-analogue of chenevixite: $Cu_2Al_2(AsO_4)_2(OH)_4{\cdot}H_2O$.

lugarite (lu'-gar-ite) A coarse-grained porphyritic *ijolite* that contains analcime in place of nepheline. Phenocrysts of amphibole, titanian augite, and zoned labradorite occur in an analcime groundmass. The name, given by Tyrrell in 1912, is for Lugar, Ayrshire, Scotland. Obsolete.

luhite (lu'-hite) A calcite-rich dike rock with olivine and titanian augite in a groundmass of clinopyroxene, melilite, hauyne, nepheline, and calcite; a hauyne-melilite *damkjernite* or hauyne-nepheline *alnoite.* The rock was named by Scheumann in 1913 for Luh, Czech Republic. Obsolete.

Luisian (Lu-is'-i-an) A Pacific Coast provincial stage: middle Miocene (above Relizian, below Mohnian).

lujavrite (lu-ja'-vrite) A coarse-grained trachytic eudialyte-bearing *nepheline syenite,* containing thin parallel feldspar crystals with interstitial nepheline grains and acicular aegirine crystals. The rock was originally described by Brögger in 1890 from Luijaur (Lujavr), now Lovozero, Kola Peninsula, Russia. Also spelled: lujaurite; luijaurite; lujauvrite. Cf: *chibinite.* Obsolete.

lukechangite-(Ce) A colorless to pale beige hexagonal mineral: $Na_3Ce_2(CO_3)_4F$. It is the Na analogue of cordylite-(Ce).

lukrahnite A yellow triclinic mineral of the *helmutwinklerite* group: $Ca(Cu,Zn)(Fe^{3+},Zn)(AsO_4)_2(OH,H_2O)_2$.

lulzacite A pale blue to dark yellow-green triclinic mineral: $Sr_2Fe^{2+}(Fe,Mg)_2Al_4(PO_4)_4(OH)_{10}$.

lumachelle (lu-ma-chelle') (a) A compact limestone or marble, composed chiefly of fossil mollusk shells, and characterized by a brilliant iridescence or chatoyant reflection from within. (b) Any accumulation of shells (esp. oysters) in stratified rocks. Etymol: French, "coquina, oyster bed", from Italian "lumachella", "little snail".

lumen (lu'-men) (a) A small, round central open space through a columnal of a crinoid, blastoid, or cystoid; e.g. the wide cavity in a short, ringlike, proximal columnal of many cystoids. (b) Central invagination in a conodont basal funnel (TIP, 1981, pt. W, supp.2). (c) One of the spaces between muri of pollen and spores exhibiting reticulate sculpture. (d) The internal space within any saclike or tubular structure, organ, or organelle. Pl: *lumina.* Adj: luminal.

lumen pore A pseudopore on the frontal surface of a costa in some cribrimorph cheilostomes (bryozoans), opening into the costal lumen. Cf: *pelma; pelmatidium.*

lumina (lu'-mi-na) Plural of *lumen.*

luminescence (lu-mi-nes'-cence) The emission of light of a different wavelength by a substance that has received energy or electromagnetic radiation from an external stimulus; also, the light so produced. It occurs at temperatures lower than those required for incandescence. See also: *phosphorescence; fluorescence.*

lump (a) A descriptive term applied to a composite, lobate grain in recent carbonate sediments, believed to have formed by aggregation, flocculation, or clotting of two or more pellets, ooids, skeletons, etc., or fragments thereof, or by disruption of newly deposited or partly indurated carbonate mud. It typically possesses surficial re-entrants. See also: *megalump.* (b) *mudlump.*

lump graphite Cryptocrystalline or very finely crystalline natural graphite from vein deposits, occurring in particle sizes ranging from that of walnuts to finer than 60-mesh.

lumping (lump'-ing) In taxonomy, the practice of ignoring minor differences in the recognition or definition of species and genera. A taxonomist known for his frequent lumping of taxa is called a "lumper". Cf: *splitting [paleont].*

lumpy (lump'-y) Term used in the gem trade to describe a gemstone characterized by a thick cut (too great a depth in proportion to its width).

lunar (lu'-nar) (a) Pertaining to or occurring on the Moon, such as a "lunar probe" designed to pass close to the Moon, or "lunar dust" consisting of fine-grained material produced by meteoritic bombardment. (b) Resembling the surface of the Moon, such as the "lunar landscape" of certain glaciers.

lunar day The time required for the Earth to rotate once with respect to the Moon, or the interval between two successive upper transits of the Moon across a local meridian; it is about 24.84 hours (24 hours and 50 minutes) of solar time.

lunar geology A science that applies geologic principles and techniques to the study of the Moon, esp. its composition and the origin of its surface features. See also: *selenology.*

lunarium (lu-nar'-i-um) A skeletal structure in some stenolaemate bryozoans that is part of the zooecial wall, occurs on the proximal or lateral side of a feeding zooid, and generally produces a troughlike depression along the length of the zooidal chamber. It commonly projects beyond the remainder of the zooecial wall. Pl: lunaria.

lunar meteorite A group of largely brecciated basaltic or anorthitic achondrites whose petrology and chemical compositions indicate that they are lunar, based on similarity to rocks returned by lunar sample return missions.

lunar playa A relatively small, level area of the Moon's surface, as much as a few kilometers long, occupying a low place in the ejecta blankets surrounding lunar craters such as Tycho and Copernicus. It is believed to be either a *fallback* deposit or a small lava flow. Lunar playa does not imply any association with water, as in the use of "playa" on Earth.

lunar regolith A thin, gray layer on the surface of the Moon, perhaps several meters deep, consisting of partly cemented or loosely compacted fragmental material ranging in size from microscopic particles to blocks more than a meter in diameter. It is believed to be formed by repeated meteoritic and secondary fragment impact over a long period of time. Syn: *lunar soil; regolith [planet].*

lunar soil *lunar regolith.*

lunar tide The part of the tide caused solely by the tide-producing force of the Moon. Cf: *solar tide.*

lunar transient phenomenon A temporarily abnormal appearance of a small area of the Moon, generally involving brightening, darkening, obscuration of surface features, or significant color changes, especially in the red and blue. Durations range from a few seconds to about a day, with a typical duration of about 30 minutes. The areas involved are generally part of a single crater or an isolated mountain but may occasionally include a whole crater or larger areas. These phenomena are strongly distributed around mare margins, as in the occurrence of dark-floored and dark-haloed craters, domes, and sinous rilles. They are generally thought to arise from some type of internal volcanic action involving outgassing or extrusion of lava (Cameron, 1972). Abbrev: LTP.

lunate bar (lu′-nate) A crescent-shaped bar commonly found off a pass between barrier islands, at the entrance to a harbor, or at a stream mouth.

lunate fracture A *crescentic mark [glac geol]* that is similar to a *crescentic fracture* but consists of two fractures from between which rock has been removed.

lunate mark *crescentic mark [glac geol].*

lunate ripple A type of *ripple* in which the crest lines are strongly curved and open out downcurrent: the form is similar to that of a *barchan* and has the opposite orientation to that of a *linguoid ripple.* Cf: *cuspate ripple.*

lunate ripple mark *lunate ripple.*

lunate sandkey A *lunate bar* that has been built up above the water surface to form a crescent-shaped island, as along the west coast of Florida.

lundyite (lun′-dy-ite) An intrusive granite or *quartz syenite* characterized by orthophyric texture, a high content of alkali minerals, and a kataphoritelike amphibole. The name, given by Hall in 1914, is for Lundy Island in the Bristol Channel, England. Obsolete.

lüneburgite (lue′-ne-burg′-ite) A colorless to brownish-white triclinic mineral: $Mg_3(PO_4)_2[B(OH)_3]_2 \cdot 6H_2O$. Also spelled: lueneburgite.

lunette (lu-nette′) A term proposed by Hills (1940) for one of the broad low even-crested crescentic mounds or ridges, rarely more than 6-9 m high, of clay loam or silty clay bordering the leeward (eastern) shore of almost every lake and swamp in the plains of northern Victoria, Australia; it is produced by dust-laden winds. Cf: *clay dune.*

lungfish (lung′-fish) *Dipnoi.*

lunijianlaite A colorless regular interstratification of cookeite and pyrophyllite: $Li_{0.7}Al_{6.2}(Si_7Al)_{20}(OH,O)_{10}$.

lunitidal interval (lu-ni-tid′-al) The interval between the Moon's transit over the local meridian and the time of the following high water. Syn: *retardation [tides].*

lunker (lun′-ker) A Scottish term for a lenticular mass of sandstone or clay ironstone; a big nodule.

lunoid furrow (lu′-noid) *crescentic gouge.*

lun'okite A colorless orthorhombic mineral: $(Mn^{2+},Ca)(Mg,Fe^{2+},Mn^{2+})Al(PO_4)_2(OH) \cdot 4H_2O$. It is the Mn-analogue of segelerite and a member of the *overite* group.

lunula (lu′-nu-la) A crescentic marking on the *selenizone* of certain gastropods, formed by growth increments.

lunule (lu′-nule) (a) A flat or curved, commonly cordate, area below the beak on the outside of some bivalve (pelecypod) shells, corresponding to the anterior part of the cardinal area, and distinguished from the remainder of the shell surface by a sharp change in angle. (b) One of the openings in an echinoid test from the aboral surface through the oral surface at a perradial or interradial suture.

lunulitiform (lu-nu-lit′-i-form) Said of conical or discoid colonies of cheilostome bryozoans in late stages of growth.

lurching *ground oscillation.*

lusakite (lu′-sa-kite) A variety of staurolite containing cobalt.

luscladite (lus-clad′-ite) An olivine *theralite* or *essexite* having hyperite texture and characterized by the absence of hornblende and by the presence of biotite as well as olivine. Alkali feldspar forms the reaction rim around the plagioclase. Nepheline is not abundant and fills the interstices. Cf: *berondrite; kylite.* Named by Lacroix in 1920 for Ravin de Lusclade, Mont Dore, France. Obsolete.

Lusitanian (Lu-si-ta′-ni-an) European stage: Upper Jurassic (above Oxfordian, below Kimmeridgian). It includes the Argovian, Rauracian, and Sequanian substages.

lusitanite (lu-si-tan′-ite) A dark-colored albite *syenite* composed of sodic amphibole, aegirine, alkali feldspar and minor amounts of quartz. Its name, given by Lacroix in 1916, is derived from Lusitania (i.e. Portugal). Obsolete.

luster (lus′-ter) The reflection of light from the surface of a mineral, described by its quality and intensity; the appearance of a mineral in reflected light. Terms such as metallic or resinous refer to general appearance; terms such as bright or dull refer to intensity.

luster mottling [ign] The macroscopic appearance of *poikilitic* rocks. The term was originated by Raphael Pumpelly (Johannsen, 1939, p.183). Obsolete.

luster mottling [sed] The shimmering appearance of a broken surface of a sandstone cemented with calcite, produced by the brilliant reflection of light from the cleavage faces of conspicuously large and independently oriented calcite crystals, a centimeter or more in diameter, incorporating colonies of detrital sand grains; e.g. the luster displayed by *Fontainebleau sandstone.* It may also develop locally in barite, gypsum, or dolomite cements.

lusungite (lu-sung′-ite) *goyazite.*

lutaceous (lu-ta′-ceous) Said of a sedimentary rock formed from mud or having the fine texture of powder or rock flour; pertaining to a lutite. Also said of the texture of such a rock. Term introduced by Grabau (1904, p.242). Cf: *argillaceous; pelitic.*

lutecite (lu′-te-cite) Fibrous chalcedony characterized by inclined extinction and by fibers that are seemingly elongated about 30° to the *c*-axis.

Lutetian (Lu-te′-tian) A stage of the Standard Global Chronostratigraphic Scale: lower Middle Eocene (above Ypresian, below Bartonian).

lutetium-hafnium age method (lu-te′-ti-um-haf′-ni-um) The determination of an age in years based on the known radioactive decay rate of lutetium-176 (half-life approximately 2.2×10^{10} years) to hafnium-176. The method can be used, under favorable conditions, for dating minerals containing rare earths.

lutite (lu′-tite) A general name used for consolidated rocks composed of silt and/or clay and of the associated materials which, when mixed with water, form mud; e.g., shale, mudstone, and calcilutite. The term is equivalent to the Greek-derived term *pelite* and was introduced as *lutyte* by Grabau (1904, p.242) who used it with appropriate prefixes in classifying fine-grained rocks (e.g., "anemolutyte", "anemosilicilutyte", "hydrolutyte", and "hydrargillutyte"). Var: *lutyte.* Etymol: Latin "lutum", "mud". See also: *rudite; arenite.*

lutyte (lu′-tyte) Var. of *lutite.*

luxullianite (lux-ul′-li-an-ite′) A granite characterized by phenocrysts of potassium feldspar and quartz which enclose clusters of radially arranged acicular tourmaline crystals in a groundmass of quartz, tourmaline, alkali feldspar, brown mica, and cassiterite. Its name is derived from Luxulyan, Cornwall. Also spelled: luxulianite; luxulyanite. Var: luxuliane. Obsolete.

luzonite (lu-zon′-ite) (a) A metallic pinkish-brown tetragonal mineral: Cu_3AsS_4. It is dimorphous with enargite, and was formerly regarded as a variety of famatinite containing arsenic. Cf: *sinnerite.* (b) A group name for minerals of composition A_3XY_4, where A = Cu, Fe, Zn or Cd, X = As, Sb or Ge, and Y = S or Se.

LV *left valve.*

LVL *low-velocity layer.*

LVZ *low-velocity zone.*

LW *low water.*

L wave *surface wave [seis].*

lycophyte (ly′-co-phyte) A *pteridophyte* characterized by dense, simple, spirally arranged leaves (*microphylls*) with sporangia borne in the axils of the leaves. Range extends from the Late Silurium to the present.

Lycophytina (Ly-co-phy-ti′-na) A *clade* containing the *lycophytes* and the *zosterophylls.*

lycopod (ly′-co-pod) (a) an old synonym for *lycophyte* in general, Late Silurian to present. (b) A member of the Lycopodiales, an order of *homosporous lycophytes,* ranging from the Pennsylvanian to the present and including the modern *club mosses.*

lycopsid (ly-cop′-sid) A member of the subdivision Lycopsida; a *lycopod.*

Lydian stone (Lyd′-i-an) A *touchstone* consisting of a compact, extremely fine-grained, velvet- or gray-black variety of *jasper.* Etymol: Greek *Lydia,* ancient country in Asia Minor. Syn: *lydite; basanite [sed].*

lydite (lyd′-ite) A syn. of *Lydian stone.* Also spelled: lyddite.

lyn *linn.*

lynchet (lyn′-chet) A bank of earth that accumulates on the downhill side of an ancient ploughed field as the disturbed soil moves downslope under gravity; terraced field characteristic of the Early Iron Age common in the southern half of England.

lyndochite (lyn′-doch-ite) A black metamict mineral, generally regarded as a variety of aeschynite-(Y), relatively high in calcium and thorium.

lyonsite (ly′-ons-ite) A metallic black orthorhombic mineral: $Cu_3Fe^{3+}_4(VO_4)_6$.

lyrula (ly'-ru-la) In some acrophoran cheilostomes (bryozoans), a median skeletal protuberance projecting over the operculum from the proximal lip of the orifice. Cf: *mucro.*

lysimeter (ly-sim'-e-ter) A structure used to measure quantities of water used by plants, evaporated from soil, and lost by *percolation*. It consists of a basin, having closed sides and a bottom fitted with a drain, in which soil is placed and plants are grown. Quantities of natural and/or artificial precipitation are measured, the deep percolate is measured and analyzed, water taken up by plants is weighed, etc.

lysocline (ly'-so-cline) (a) The depth range in the oceanic water column wherein the rate of solution of solid particles of calcium carbonate (i.e. mainly calcitic plankton fossils) increases markedly compared to the rate in overlying layers. Situated above the *calcite compensation depth.* (b) The depth, in any one region, where dissolution tends to remove a substantial portion of the assemblage of foraminifers accumulating on the sea floor, before burial. (c) The depth where a substantial portion of carbonate is removed. (d) The depth of the switchover from saturated to undersaturated waters (that is, appearance of potential dissolution rate from the chemistry of the bottom water).

lyssacine (lys'-sa-cine) adj. Said of a hexactinellid sponge whose megascleres (spicules) are unfused or separate, incompletely fused, or so fused that their individual boundaries are apparent. Ant: *dictyonine.* n. A lyssacine sponge.

M

M In the *IUGS classification* of plutonic rocks, the symbol for mafic and related minerals (micas, amphiboles, pyroxenes, olivines, opaque minerals, epidote, garnets, primary carbonates, etc.) (IUGS, 1973).

Ma Mega-annum, one million (10^6) years. Informal SI notation, where annum is age in years before present, with "present" fixed as 1950. This term has largely replaced the various abbreviations for "millions of years before present" in geological literature. It is not equivalent to the elapsed time interval or duration in "millions of years" (myr).

mäanderkarren Small winding or meandering solution channels in limestone. See: *karren*. Etymol: German.

maar A low-relief, broad volcanic crater formed by multiple shallow explosive eruptions. It is surrounded by a *crater ring*, and may be filled by water. Type occurrence is in the Eifel area of Germany.

Maastrichtian (Maas-tricht'-i-an) A stage of the Standard Global Chronostratigraphic Scale: uppermost Cretaceous (above Campanian, below Danian Stage of the Tertiary). The basal GSSP is in southwest France (Odin and Lamaurelle, 2001). Occasionally spelled: Maestrichtian.

macallisterite (ma-cal'-lis-ter-ite') *mcallisterite*.

macaluba (ma-ca-lu'-ba) A syn. of *mud volcano*. The name is taken from that of a low mud volcano, Macaluba, in Sicily.

macaulayite (ma-cau'-lay-ite) An earthy red monoclinic mineral with an expandable layer structure consisting of silicate and hematite-like sheets: $(Fe^{3+},Al)_{24}Si_4O_{43}(OH)_2$.

macconnellite (mac-con'-nel-lite) *mcconnellite*.

macdonaldite (mac-don'-ald-ite) A colorless orthorhombic mineral: $BaCa_4Si_{16}O_{36}(OH)_2 \cdot 10H_2O$.

macedonite [mineral] (mac'-e-don-ite') A black tetragonal mineral: $PbTiO_3$.

macedonite [rock] A fine-grained rock composed of orthoclase, sodic plagioclase, biotite, olivine, and rare pyriboles; an olivine-biotite *trachyte* or *trachyandesite*. Named for Mt. Macedon, Victoria, Australia, by Skeats and Summers in 1909. Obsolete.

maceral (mac'-er-al) One of the organic constituents that comprise the coal mass; all petrologic units seen in polished or thin sections of coal. Macerals are to coal as minerals are to inorganic rock. Maceral names bear the suffix "-inite" (*vitrinite, liptinite*, etc.). Cf: *phyteral*. Syn: *micropetrological unit*.

maceration (mac-er-a'-tion) The act or process of disintegrating sedimentary rocks (such as coal and shale) by various chemical and physical techniques in order to extract and concentrate acid-insoluble microfossils (including palynomorphs). It includes mainly chemical treatment by oxidants and alkalies and use of other separating techniques that will remove extraneous mineral and organic constituents. Maceration is widely used in palynology.

MacEwan crystallite A term synonymous with "mixed-layered crystal" used in clay mineralogy as a name for nonperiodic crystals, i.e., crystals that are not perfectly ordered in the direction perpendicular to the stacking direction of two or more types of layers. Observed in HRTEM images and used in modeling one-dimensional (001) X-ray diffraction patterns of mixed-layer clays (in these examples equivalent to coherent scattering domain of such minerals). Cf: *interparticle diffraction*.

macfallite (mac-fall'-ite) A reddish-brown to maroon monoclinic mineral of the *epidote* group: $Ca_2Mn^{3+}{}_3(SiO_4)(Si_2O_7)(OH)_3$.

macfarlanite (mac-far'-lan-ite) A silver ore consisting of a mixture of sulfides, arsenides, etc., and containing cobalt, nickel, and lead. Cf: *animikite*.

macgovernite *mcgovernite*.

Machaeridia (Mach-ae-rid'-i-a) A class questionably assigned to the homalozoans and characterized by an elongate, bilaterally symmetrical test composed of an even number of longitudinal columns of plates. They have been variously classified as mollusks, annelids, and arthropods.

machatschkiite A colorless rhombohedral mineral: $Ca_6(AsO_4)(AsO_3OH)_3PO_4 \cdot 15H_2O$.

macigno (ma-ci'-gno [mah-cheen'-yo]) The classical *flysch* facies in the northern Apennines, consisting of alternating strata of sandstone and mudstone, and showing graded bedding. Etymol: Italian, "millstone".

mackayite (mac-kay'-ite) A green to brownish-green tetragonal mineral: $Fe^{3+}Te^{4+}{}_2O_5(OH)$.

mackelveyite *mckelveyite-(Y)*.

Mackereth sampler (Mack'-er-eth) A variety of *piston corer* that is operated by compressed air.

mackinawite (mack'-i-naw-ite') A metallic reddish-gray tetragonal mineral: $(Fe,Ni)_9S_8$. It occurs as a corrosion product on iron pipes. Syn: *kansite*.

mackinstryite *mckinstryite*.

mackintoshite (mack'-in-tosh-ite') *thorogummite*.

macle [cryst] (ma'-cle) A twinned crystal; esp. a flat, often triangular, diamond composed of two flat crystals. Etymol: French, "wide-meshed net".

macle [mineral] (a) A dark or discolored spot in a mineral. (b) *chiastolite*. (c) A twin crystal, esp. of diamond.

macled [cryst] (ma'-cled) Said of a crystal having a twin structure.

macled [mineral] (a) Said of a mineral that is marked like chiastolite. (b) Said of a mineral that is spotted.

maconite (ma'-con-ite) A vermiculite from North Carolina.

macphersonite (mac-pher'-son-ite) A pale amber, colorless or white orthorhombic mineral: $Pb_4(SO_4)(CO_3)_2(OH)_2$. It is a polymorph of susannite and leadhillite.

macquartite (mac-quart'-ite) An orange monoclinic mineral: $Pb_3Cu(CrO_4)SiO_3(OH)_4 \cdot 2H_2O$.

macrinite (mac'-ri-nite) A maceral of coal within the *inertinite* group having a reflectance higher than that of associated vitrinite, without recognizable plant-cell structure, with a non-angular shape and a particle diameter greater than about 2 micrometers. Cf: *micrinite*.

macro- (mac'-ro-) A prefix meaning "large" or "great". Cf: *micro-*. Syn: *mega-*.

macro-axis (mac'-ro-ax'-is) The longer lateral axis of an orthorhombic or triclinic crystal; it is usually the *b* axis. Cf: *brachy-axis*.

macrochoanitic (mac'-ro-cho'-a-nit'-ic) Said of a *retrochoanitic septal neck* of a nautiloid that reaches backward beyond the preceding septum and is invaginated into the preceding septal neck.

macroclastic (mac-ro-clas'-tic) Said of coal that contains many recognizable fragments. Cf: *microclastic*.

macroclastic rock A clastic rock whose constituents are visible to the unaided eye. Ant: *cryptoclastic rock*.

macroclimate (mac-ro-cli'-mate) The general climate of an extensive region or a country. See also: *mesoclimate; microclimate*.

macrococcolith (mac-ro-coc'-co-lith) One of the larger coccoliths in coccolithophores exhibiting dimorphism but with the dimorphic coccoliths irregularly placed. Cf: *micrococcolith*.

macroconch (mac'-ro-conch) A mature *conch* of a chambered cephalopod which, in all respects except size and occasional modification of the aperture, resembles smaller conchs (microconchs) found in the same fossil association. Macroconchs are now generally regarded as representing females. Macroconchs and microconchs have in the past been described as different species or even placed in different genera (Callomon, 1963).

macrocrystalline (mac-ro-crys'-tal-line) Said of the texture of a rock consisting of or having crystals that are large enough to be distinctly visible to the unaided eye or with the use of a simple lens; also, said of a rock with such a texture. Howell (1922) applied the term to the texture of a recrystallized sedimentary rock having crystals whose diameters exceed 0.75 mm, and Bissell and Chilingar (1967, p.103) to the texture of a carbonate sedimentary rock having crystals whose diameters exceed 1.0 mm. Syn: *megacrystalline; eucrystalline*. See also:*phaneritic*.

macrodome (mac'-ro-dome) An obsolete term for a {*h0l*} prism in an orthorhombic crystal. A macrodome with four faces is a rhombic prism, or a *second-order prism* of the orthorhombic system.

macroevolution (mac'-ro-ev'-o-lu'-tion) (a) The evolution or origin of higher taxa (i.e., above species rank; esp. orders or classes), as contrasted to *microevolution*. Syn: *megaevolution*. (b) Evolution occurring in large, complex stages, such as the development of one species from another. Cf: *microevolution*.

macrofabric (mac'-ro-fab'-ric) *megafabric*.

macrofacies (mac'-ro-fa'-cies) *facies tract.*
macrofauna (mac'-ro-fau'-na) (a) Living or fossil animals large enough to be seen with the naked eye; benthic animals larger than 0.5 mm. (b) An obsolete term for the animals occupying a broad area of uniform characteristics; a large or widespread group of animals. Cf: *microfauna; megaflora.* Syn: *megafauna.*
macroflora (mac-ro-flo'-ra) *megaflora.*
macrofossil (mac'-ro-fos'-sil) A fossil large enough to be studied without the aid of a microscope. Cf: *microfossil.* Syn: *megafossil.*
macrofragmental (mac'-ro-frag-men'-tal) Said of a coal composed of recognizable fragments or lenses of vegetal matter. Cf: *microfragmental.*
macrograined (mac'-ro-grained) Said of the texture of a carbonate sedimentary rock having clastic particles whose diameters are greater than one millimeter (Bissell and Chilingar, 1967, p.103). See also: *megagrained.*
macrohabitat (mac-ro-hab'-i-tat) A *habitat* large enough for a human observer to move around in, i.e. tens or hundreds of feet in dimensions. Cf: *microhabitat.*
macroite (mac'-ro-ite) A microlithotype of coal, a variety of *inertite,* consisting of 95% or more of macrinite (ICCP, 1971).
macrolinear (mac-ro-lin'-e-ar) "Any lineation....which is two to less than ten kilometers long" (El-Etr, 1976, p.485).
macrolithology (mac'-ro-li-thol'-o-gy) The study of rocks considered as a part of the stratigraphic column in a given area; also, the collective characteristics of such rocks. Cf: *microlithology.*
macromodel A velocity model for use in ray-tracing and/or depth migration. Usually made up of a few thick layers with variable thickness.
macromolecule High molecular (>1,000 daltons) weight bio- or geopolymer ribonucleic acid (RNA).
macromutation (mac'-ro-mu-ta'-tion) Mutation that is large and easily observed.
macronucleus (mac-ro-nu'-cle-us) A relatively large vegetative nucleus that is believed to exert a controlling influence over the trophic activities in the body of a tintinnid. Cf: *micronucleus.*
macrophagous (mac-roph'-a-gous) Said of an organism that feeds on relatively large particles. Cf: *microphagous.*
macrophyte (mac'-ro-phyte) A megascopic plant, esp. in an aquatic environment.
macropinacoid (mac-ro-pin'-a-coid) *front pinacoid.*
macroplankton (mac-ro-plank'-ton) Plankton organisms of the size range 2-20 cm. They are larger than *ultraplankton, nannoplankton,* and *microplankton,* but smaller than *megaplankton.*
macropolyschematic (mac'-ro-pol'-y-sche-mat'-ic) Said of mineral deposits having megascopically distinguishable textural elements. The term is little used.
macropore (mac'-ro-pore) A pore too large to hold water by capillarity. Syn: *megapore.*
macroporphyritic (mac'-ro-por'-phy-rit'-ic) *megaphyric.*
macropygous (mac-ro-py'-gous) Said of a trilobite with a *pygidium* larger than the *cephalon.* Cf: *isopygous; micropygous.*
macropylome (mac-ro-py'-lome) A circular to subpolygonal *acritarch excystment* structure with free *operculum,* having a total diameter of >25 % of the vesicle diameter, e.g. *Priscogalea, Cymatiogalea* (Miller, 1987). Cf: *micropylome; pylome [palyn].*
macrorelief (mac'-ro-re-lief') Regional, major irregularities of a land surface, including such features as mountains and valleys. (b) Relief features that are dominant landforms in topography represented at scales of 1:100,000 or larger. Cf: *microrelief.*
macrosclere (mac'-ro-sclere) *megasclere.*
macroscopic (mac-ro-scop'-ic) (a) *megascopic.* (b) According to Dennis (1967, p.152), a term introduced to describe tectonic features that are too large to be observed directly in their entirety. Cf: *mesoscopic.*
macroseism (mac'-ro-seism) A syn. of *earthquake,* as opposed to *microseism.*
macrospore (mac'-ro-spore) An obsolete syn. of *megaspore;* or a spore >200 µm in diameter, in contrast to *miospore.*
macrotectonics (mac'-ro-tec-ton'-ics) A rarely used term introduced by Tomkeieff (1943, p.348) as a syn. of *megatectonics.*
macrotherm (mac'-ro-therm) *megatherm.*
macrotidal (mac'-ro-tid'-al) Said of a coast with a tidal range greater than 4 meters (Davies, 1963). Cf: *mesotidal.*
macula [intrus rocks] (mac'-u-la) A local pocket of magma, formed by the fusion of shale, that acts as a type of *magma chamber.* Pl: maculae.
macula [paleont] In many stenolaemate bryozoans, a cluster of a few polymorphs, an *extrazooidal skeleton,* or a combination. They are more or less evenly spaced among feeding zooids. Pl: maculae.
macula [planet] (mac'-u-la) A term established by the International Astronomical Union for a dark spot on planetary surfaces (USGS, 2003). Pl: maculae.
maculose (mac'-u-lose) Said of a group of contact-metamorphic rocks, e.g. spotted slates, that have a spotted or knotted character; also, said of the structure itself. Syn: *spotted; knotted.*
macusanite A peraluminous *rhyolite* lava with phenocrysts of andalusite and sillimanite in a glassy groundmass. Named by Martin and de Sitter-Koomans (1955) for Macusani, Peru. Obsolete.
made land Constructed land; an area artificially filled with earth materials that may be mixed with a variety of waste material. Made land is common along marshy shorelines and at many former *sanitary landfill* sites. Cf: *fill [eng].*
madocite (ma'-doc-ite) A metallic gray orthorhombic mineral: $Pb_{18}(Sb,As)_{15}S_{41}$.
madreporite (mad'-re-por-ite') A porous or sievelike structure that is situated at the distal end of the *stone canal* in an echinoderm and that provides access to the water-vascular system from the exterior; e.g. a conspicuous plate on the aboral surface of an asteroid or in the right anterior genital plate of an echinoid.
madupite (ma-du'-pite) An extrusive rock composed of phlogopite, clinopyroxene, and perovskite phenocrysts in a brown glassy groundmass with the composition of leucite and nepheline; a *lamproite.* The name was coined by Cross in 1897 from an Indian word for "sweetwater", after Sweetwater County, Wyoming, where the Leucite Hills are located. Mitchell and Bergman (1991) suggest that this name should be replaced by diopside *madupitic lamproite.* Obsolete.
madupitic lamproite A group name for lamproites containing poiklilitic phlogopite in the groundmass (Mitchell and Bergmann, 1991). Varieties are *jumillite, madupite,* and *wolgidite.* Obsolete.
maelstrom (mael'-strom) A rapid, confused, and often destructive current, formed by the combination of strong wind-generated waves and a strong opposing tidal current; it may display *eddy*-type or *whirlpool*-type characteristics. It typically occurs along the south shore of the Lofoten Islands of Norway.
maenaite (mae'-na-ite) A hypabyssal plagioclase-bearing *bostonite* differing from normal bostonite in being richer in calcium and poorer in potassium; an altered *trachyte.* It was named by Brögger in 1894 for Lake Maena, Oslo district, Norway. Obsolete.
Maeotian (Mae-o'-tian [My-o'-tian]) Ponto-Caspian stage: middle upper Miocene (above Sarmatian, below Pontian).
Maestrichtian (Maes-tricht'-i-an) *Maastrichtian.*
mafelsic (ma-fel'-sic) Said of an igneous rock in which the felsic and mafic minerals are present in approximately equal amounts. Obsolete.
mafic (maf'-ic) Said of an igneous rock composed chiefly of one or more ferromagnesian, *dark-colored* minerals in its mode; also, said of those minerals. The term was proposed by Cross, et al. (1902, p.561) to replace the term *femag,* which they did not consider to be euphonious. Etymol: a mnemonic term derived from *magnesium* + *ferric* + *ic.* It is the complement of *felsic.* Cf: *femic; salic; basic.* Partial syn: *ferromagnesian.*
mafic front A term preferred by some petrologists to its synonym *basic front.*
mafic index A chemical parameter of igneous rocks, equal to $100\times(FeO+Fe_2O_3)/MgO+FeO+Fe_2O_3$. It is most commonly plotted as the ordinate on variation diagrams, on which the abscissa represents the *felsic index.* It reflects changes produced by fractional crystallization of the mafic minerals (Simpson, 1954). Abbrev: MI.
mafic margin *basic border.*
mafite (maf'-ite) (a) A mafic mineral. (b) A dark-colored *aphanite.* Obsolete.
mafraite (maf'-ra-ite) A hypabyssal *theralite* containing labradorite with alkali feldspar rims, pyroxene, magnetite, and euhedral hornblende, and without modal nepheline although nepheline does occur in the norm. Cf: *berondrite.* Named by Lacroix in 1920 for Mafra, Portugal. Obsolete.
mafurite (ma-fur'-ite) A variety of olivine *leucitite* in which kalsilite is present instead of leucite. The name, proposed by Holmes in 1945, is for Mafura, Uganda. Woolley et al. (1996) replace this name with olivine-clinopyroxene kalsilitite.
magadiite (ma-ga'-di-ite) A white monoclinic (pseudo-tetragonal) mineral: $Na_2Si_4O_{29}\cdot 11H_2O$. It is found in lake beds at Lake Magadi, Kenya.
magadiniform (mag-a-din'-i-form) Said of the *loop,* or of the growth stage in the development of the loop, of a terebratellid brachiopod (as in the subfamily Magadinae), marked by completed descending branches from the cardinalia to the median septum, with a ringlike structure on the septum representing an early ascending portion of the loop (TIP, 1965, pt. H, p.147). Cf: *premagadiniform.*
magbasite (mag'-ba-site) A rose violet orthorhombic mineral: $KBa(Mg,Fe^{2+})_6(Al,Sc)Si_6O_{20}F_2$.
magellaniform (mag-el-lan'-i-form) Said of the free *loop,* or of the growth stage in the development of the loop, of a terebratellid brachiopod (as in *Magellania*), consisting of long descending branches recurved into ascending branches that meet in transverse band (TIP, 1965, pt. H, p.147). The magellaniform loop is morphologically similar to the *dalliniform* loop.
magelliform (ma-gel'-li-form) Said of the *loop,* or of the growth stage in the development of the loop, of a terebratellid brachiopod (as in *Magella*), in which the bases of the septal ring on the median septum meet and fuse with the attachments of completed descending branches (TIP, 1965, pt. H, p.147).
maghagendorfite (mag-hag'-en-dorf-ite') A greenish-black monoclinic mineral of the *alluaudite* group: $NaMn^{2+}(Mg,Fe^{3+},Fe^{2+})_2(PO_4)_3$.
maghemite (mag-hem'-ite) A strongly magnetic brown to bluish-black mineral possessing cation vacancies in the magnetite series of the *spinel* group: $Fe_{2.67}O_4$. It is dimorphous with hematite. Syn: *oxymagnite.*

magin (ma-gin') In Argentina, a damp, grassy concavity on an otherwise sparcely vegetated slope.
magma (mag'-ma) Naturally occurring molten or partially molten rock material, generated within the Earth and capable of intrusion and extrusion, from which igneous rocks are derived through solidification and related processes. It may or may not contain suspended solids (such as crystals and rock fragments) and/or gas phases. Adj: *magmatic.*
magma blister A pocket of magma whose formation has raised the overlying land surface.
magma chamber A reservoir of magma in the shallow part of the lithosphere (to a few km. or tens of km.), from which volcanic materials are derived; the magma has ascended into the crust from a deeper source. See also: *macula [intrus rocks].* Syn: *magma reservoir.*
magma column *lava column.*
magma province *petrographic province.*
magma reservoir *magma chamber.*
magma series Magmas that are genetically related to each other by some differentiation process(es) or by being separate partial melts of a common source under similar conditions (Gill, 1981, p.7). Examples are the alkaline, calcalkaline, and tholeiitic series.
magmatic (mag-mat'-ic) Of, pertaining to, or derived from *magma.* Obsolete syn: *orthotectic.*
magmatic arc An arcuate line of plutons, volcanic rocks, or active volcanoes formed at a convergent plate boundary; most of the magma forms by melting in the asthenosphere above the downgoing plate.
magmatic deposit *magmatic ore deposit.*
magmatic differentiation *differentiation [intrus rocks].*
magmatic dissolution The dissolving of country rock by magma; *assimilation.*
magmatic emanation A combination of gases and liquids given off by a magma, e.g. aqueous, pegmatitic, or hydrothermal fluids. Cf: *mineralizer.*
magmatic evolution The continuing change in composition of a magma as a result of magmatic differentiation, assimilation, or mixing of magmas.
magmatic iron meteorite An iron meteorite showing evidence of fractional crystallization and a general absence of chondritic or subchondritic silicates.
magmatic ore deposit An ore deposit formed by *magmatic segregation,* generally in mafic rocks and layered intrusions, as crystals of metallic oxides or from an immiscible sulfide liquid. Syn: *magmatic segregation deposit; magmatic deposit.*
magmatic segregation Concentration of crystals of a particular mineral (or minerals) in certain parts of a magma during its cooling and crystallization. Some economically valuable ore deposits (i.e. magmatic ore deposits) are formed in this way. See also: *differentiation [intrus rocks].* Syn: *segregation [petrology].*
magmatic segregation deposit *magmatic ore deposit.*
magmatic stoping A process of magmatic emplacement or intrusion that involves detaching and engulfing pieces of the country rock which sink downward, thereby providing space above for the magma. See also; *piecemeal stoping; ring-fracture stoping; cauldron subsidence.*
magmatic water Water contained in or expelled from magma. Cf: *juvenile [water]; plutonic water.*
magmation (mag-ma'-tion) Conversion of pre-existing rocks to magma; esp. the in situ formation of granite magma (Dunn, 1942). Obsolete.
magmatism (mag'-ma-tism) (a) The development and movement of magma, and its solidification to igneous rock. (b) The theory that much granite has formed through crystallization from magma rather than through *granitization;* opposed to *transformism.*
magmatite (mag'-ma-tite) A rock formed from magma.
magma type Categorization of magma having a distinctive chemical composition.
magmosphere (mag'-mo-sphere) *pyrosphere.*
magnacycle (mag'-na-cy'-cle) A term proposed by Merriam (1963, p.106) for a large, complex rock unit that follows a repetitious pattern and that can be considered cyclic in nature.
magnacyclothem (mag-na-cy'-clo-them) A magnacycle that is larger than a *megacyclothem* (Merriam, 1963, p.106).
magnafacies (mag-na-fa'-cies) A term proposed by Caster (1934, p.19) for a major, continuous, and homogeneous belt of deposits that is distinguished by similar lithologic and paleontologic characters and that extends obliquely across time planes or through several defined chronostratigraphic units; a complete "lithic member" or perfect lithostratigraphic unit of the same facies but formed at different times; an isopic facies of European usage. It represents a distinct depositional environment that persisted with more or less shifting of geographic placement during time, and it may be divisible into, or assignable to, several noncontemporaneous *parvafacies.* The term is very nearly synonymous with *lithosome* as defined by Wheeler and Mallory (1956). Etymol: Latin "magna", "great", + facies. See also: *megafacies.*
magnesia alum (mag-ne'-sia) *pickeringite.*
magnesia mica (a) *phlogopite.* (b) *biotite.*
magnesian calcite (mag-ne'-sian) A variety of calcite: $(Ca,Mg)CO_3$. It consists of randomly substituted magnesium ions in solid solution for calcium in the calcite structure. Low-magnesian calcite has less than 4% $MgCO_3$ in solid substitution, and is essentially the common form of calcite. High-magnesian calcite has 4-19% $MgCO_3$ in solid substitution; it is metastable and during limestone formation converts to low-magnesian calcite or to dolomite. Syn: *magnesium calcite.*
magnesian dolomite A dolomite rock with an excess of magnesium; specif. a dolomite rock whose Ca/Mg ratio ranges from 1.0 to 1.5 (Chilingar, 1957), or a dolomite rock containing 50-75% dolomite and 25-50% magnesite (Bissell and Chilingar, 1967, p.108).
magnesian limestone (a) An obsolete term for limestone that contains appreciable magnesium; specif. a limestone having at least 90% calcite, no more than 10% dolomite, an approximate MgO equivalent of 1.1-2.1%, and an approximate magnesium-carbonate equivalent of 2.3-4.4% (Pettijohn, 1957, p.418); or a limestone whose Ca/Mg ratio ranges from 60 to 105 (Chilingar, 1957); or a limestone containing 5-15% magnesium carbonate but in which dolomite cannot be detected (Holmes, 1928, p.149). Some petrographers use the term for a limestone with some MgO but no dolomite; others for a rock with all possible mixtures of dolomite and calcite. Cf: *dolomitic limestone.* (b) Formerly, dolomitic limestone; specif. the Magnesian Limestone, a facies of the Permian of NE England. (c) A term commonly but loosely used to indicate dolomite rock.
magnesian marble A type of metamorphosed magnesian limestone containing some dolomite (generally less than 15%). Cf: *dolomitic marble.*
magnesian spar *dolomite [mineral].*
magnesioarfvedsonite (mag-ne'-si-o-arf'-ved-son-ite) A green, bluish green, greenish-black to black monoclinic mineral of the *amphibole* group: $Na,Na_2(Mg_4Fe^{3+})Si_8O_{22}(OH)_2$.
magnesioaubertite A vitreous blue triclinic mineral of the *aubertite* group: $(Mg,Cu)Al(SO_4)_2Cl{\cdot}14H_2O$.
magnesioaxinite (mag-ne'-si-o-ax'-i-nite) A pale blue brown or pink triclinic mineral of the *axinite* group: $Ca_2MgAl_2BSi_4O_{15}(OH)$.
magnesiocarbonatite A chemically defined carbonatite in which wt% $CaO/(CaO+MgO+FeO+Fe_2O_3+MnO)<0.8$ and $MgO>(FeO+Fe_2O_3+MnO)$ (Woolley and Kempe, 1989).
magnesiocarpholite A light green to grayish orthorhombic mineral of the *carpholite* group: $(Mg,Cu)Al(SO_4)_2Cl{\cdot}14H_2O$.
magnesiochloritoid A dark blue monoclinic mineral of the *chloritoid* group: $(Mg,Fe^{2+})Al_2O(SiO_4)(OH)_2$.
magnesiochromite (mag-ne'-si-o-chro'-mite) (a) A mineral of the *spinel* group: $MgCr_2O_4$. It is isomorphous with chromite. Syn: *magnochromite.* (b) *picrochromite.*
magnesiocopiapite (mag-ne'-si-o-co-pi-a'-pite) An orange or greenish-yellow triclinic mineral of the *copiapite* group: $MgFe^{3+}{}_4(SO_4)_6(OH)_2{\cdot}20H_2O$.
magnesiocoulsonite A black cubic mineral of the *spinel* group: $Mg(V,Cr)_2O_4$.
magnesiodumortierite A pink or red orthorhombic mineral of the *dumortierite* group: $(Mg,Ti)(Al,Mg)_2Al_4BSi_3(O,OH)_{18}$.
magnesio-ferri-fluor-katophorite A monoclinic amphibole representing magnesiokatophorite with essential ferric iron and fluorine: $(Na,K)_{1.5}CaMg_4Fe^{3+}(Si,Al)_8O_{22}(F,OH)_2$. It has not been approved by the International Mineralogical Association.
magnesio-ferri-fluor-oxy-katophorite A monoclinic amphibole representing an oxygen-rich magnesiokatophorite with essential ferric iron and fluorine: $Na_2CaMg_4Fe^{3+}(Si_7Al)O_{22}(F,O,OH)_2$. It has not been approved by the International Mineralogical Association.
magnesio-ferri-katophorite A monoclinic amphibole representing magnesiokatophorite with essential ferric iron: $Na_2CaMg_4Fe^{3+}(Si_7Al)O_{22}(OH)_2$.
magnesioferrite (mag-ne'-si-o-fer'-rite) A black strongly magnetic cubic mineral of the *spinel* group: $MgFe^{3+}{}_2O_4$. Syn: *magnoferrite.*
magnesiofoitite A dull bluish-gray rhombohedral mineral of the *tourmaline* group: $(Mg_2Al)Al_6Si_6O_{18}(BO_3)_3(OH)_4$.
magnesiohastingsite A green monoclinic mineral of the *amphibole* group, representing hastingsite with essential Mg: $NaCa_2(Mg_4Fe^{3+})(Si_6Al_2)O_{22}(OH)_2$.
magnesiohögbomite-2N2S A metallic black trigonal mineral of the *högbomite* group: $(Al,Mg,Fe,Ti)_{22}(O,OH)_{32}$. The structure is interpreted in terms of two nolanite modules and two spinel modules.
magnesiohögbomite-2N3S A brown trigonal mineral of the *högbomite* group: $(Mg,Fe,Zn,Ti)_{9.6}Al_{18.3}O_{38}(OH)_2$. The structure is interpreted in terms of two nolanite modules and three spinel modules.
magnesiohögbomite-6N6S A metallic black trigonal mineral of the *högbomite* group: $(Al,Mg,Ti)_{66}(O,OH)_{96}$. The structure is interpreted in terms of six nolanite modules and six spinel modules.
magnesiohornblende A green to brown monoclinic member of the *amphibole* group, representing hornblende with essential Mg: $Ca_2(Mg_4Al)(Si_7Al)O_{22}(OH)_2$.
magnesiohulsite A submetallic black monoclinic mineral of the *pinakiolite* group: $(Mg,Fe^{2+})_2(Fe^{3+},Sn,Mg)O_2(BO_3)$.
magnesiokatophorite A black, greenish black or reddish brown monoclinic mineral of the *amphibole* group, representing katophorite with essential Mg: $Na_2Ca(Mg_4Al)(Si_7Al)O_{22}(OH)_2$.
magnesioludwigite (mag-ne'-sio-lud'-wig-ite) *ludwigite.*
magnesionigerite-2N1S A yellowish-brown, light yellow or colorless trigonal mineral of the *nigerite* group: $(Mg,Al,Zn)_4(Sn,Fe)_2Al_{10}O_{22}(OH)_2$. The

structure is interpreted as comprising two nolanite modules and one spinel module. Also known as: pengzhizhongite-6T.

magnesionigerite-6N6S A yellowish-brown, light yellow or colorless trigonal mineral of the *nigerite* group: $(Mg,Al,Zn)_4(Sn,Fe)_2Al_{10}O_{22}(OH)_2$. The structure is interpreted as comprising six nolanite modules and six spinel modules. Also known as: pengzhizhongite-24R.

magnesioriebeckite (mag-ne'-si-o-rie'-beck-ite) A blue or black monoclinic mineral of the *amphibole* group: $Na_2Mg^{2+}{}_3Fe^{3+}{}_2Si_8O_{22}(OH)_2$.

magnesiosadanagaite A dark brown to black monoclinic mineral of the *amphibole* group, representing sadanagaite with essential Mg: $NaCa_2(Mg_3Fe^{3+}{}_2)(Si_5Al_3)O_{22}(OH)_{22}$.

magnesiostaurolite A monoclinic member of the *staurolite* group: $Mg_4Al_{17}(Si,Al)_8O_{45}(OH)_3$.

magnesiotaaffeite-2N'2S A variously colored hexagonal member of the *taaffeite* group: $Mg_3BeAl_8O_{16}$. The structure is interpreted as comprising two nolanite-like modules and one spinel module. Also known as: taprobanite.

magnesiotaaffeite-6N'3S A pale olive-green member of the *taaffeite* group: $(Mg,Fe,Zn)_2BeAl_6O_{12}$. The structure is interpreted as comprising six nolanite-like modules and three spinel module. Syn: musgravite.

magnesiotantalite A metallic black orthorhombic mineral of the *columbite* group: $(Mg,Fe)(Ta,Nb)_2O_6$.

magnesiotaramite A monoclinic member of the *amphibole* group, representing taramite with essential Mg: $Na_2Ca(Mg_3AlFe^{3+})(Si_6Al_2)O_{22}(OH)_2$.

magnesite (mag'-ne-site) A white to grayish, yellow, or brown rhombohedral mineral: $MgCO_3$. It is isomorphous with siderite. Magnesite is generally found as earthy masses or irregular veins resulting from the alteration of dolomite rocks, or of rocks rich in magnesium silicates, by magmatic solutions. It is used chiefly in making refractories and magnesia. Syn: *giobertite.*

magnesium astrophyllite (mag-ne'-si-um) A yellow platy monoclinic mineral of the *astrophyllite* group: $K_2Na_2Mg_2(Fe^{2+},Mn)_4Ti_2Si_8O_{26}(OH)_4$.

magnesium calcite *magnesian calcite.*

magnesium-chlorophoenicite (mag-ne'-si-um-chlo-ro-phoe'-ni-cite) A colorless or white monoclinic mineral: $(Mg,Mn)_3Zn_2AsO_4(OH,O)_6$. It is the Mg analogue of chlorophoenicite.

magnesium front *basic front.*

magnesium number A ratio of magnesium to magnesium plus iron in a rock or mineral, variously defined as $100Mg/(Mg+Fe^{2+})$, $(Mg/Mg+Fe_{total})$, etc., either by weight or atomic proportions. Because there are so many forms for this carelessly named but useful ratio, the method of calculation should be explicitly stated in every publication using the term. Also called: mg-number; mg-value; Mg'-value.

magnesium-zippeite (mag-ne'-si-um-zip'-pe-ite) A yellow monoclinic mineral of the *zippeite* group: $Mg_2(UO_2)_4(SO_4)_2O_2{\cdot}7H_2O$.

magnet (mag'-net) A magnetized body, especially a *permanent magnet.*

magnetic aftereffect (mag-net'-ic) *magnetic viscosity.*

magnetic anisotropy (a) *susceptibility anisotropy.* (b) *magnetocrystalline anisotropy.*

magnetic azimuth The *azimuth [surv]* measured clockwise from magnetic north through 360 degrees; the angle at the point of observation between the vertical plane through the observed object and the vertical plane in which a freely suspended magnetized needle, influenced by no transient artificial magnetic disturbance, will come to rest.

magnetic balance An instrument in which the translational force on a magnetic moment in a nonuniform magnetic field is balanced against a spring, torsional, or gravitational force. See also: *Curie balance.*

magnetic basement The upper surface of extensive heterogeneous rocks having relatively large magnetic susceptibilities compared with those of sediments; often but not necessarily coincident with the geologic *basement.* It generally excludes magnetic sediments and thin volcanic and other high-susceptibility rocks intruded into the sedimentary section, but thick volcanic rocks in the sedimentary section would be classed as magnetic basement where the magnetic effects of deeper bodies would not be resolvable.

magnetic bearing The *bearing* expressed as a horizontal angle between the local magnetic meridian and a line on the Earth; a bearing measured clockwise from magnetic north. It differs from a *true bearing* by the amount of magnetic declination at the point of observation.

magnetic chron The time interval of constant polarity of the Earth's magnetic field, generally on the order of a few 10^5 to a few 10^6 yrs.

magnetic cleaning Partial or selective demagnetization of natural remanent magnetization, by the removal of less geologically stable, secondary components of magnetization or viscous remanent magnetization. Syn: magnetic washing; *demagnetization.*

magnetic compass A compass whose operation depends upon an element that senses the Earth's magnetic field; e.g. an instrument having a magnetic needle that turns freely on a pivot in a horizontal plane and that always swings to such a position that one end points to magnetic north. See also: *prismatic compass.*

magnetic declination The angular direction, east or west from the north branch of the celestial meridian, of magnetic north as determined by the positive pole of a freely suspended magnetic needle that is not subject to any transient, artificial disturbance. In nautical and aeronautical navigation, the term variation is used instead of declination, and the angle is called variation of the compass or magnetic variation. Except for usage in navigation, magnetic declination is not synonymous with magnetic variation, which refers to regular or irregular change with time of the magnetic declination, dip, or intensity.

magnetic dip *inclination [magnet].*

magnetic domain *domain [magnet].*

magnetic elements The characteristics of a magnetic field that can be expressed numerically. The seven magnetic elements are *declination* D, *inclination* I, *total intensity* F, *horizontal intensity* H, *vertical intensity* Z, north component X, east component Y. Only three elements are needed to give a complete vector specification of the magnetic field.

magnetic epoch *polarity epoch.*

magnetic equator The line on the Earth's surface at which magnetic inclination is zero; the locus of points with zero magnetic latitude. Syn: *aclinic line; dip equator.*

magnetic exploration A technique of applied geophysics: a survey, conducted with a magnetometer or gradiometer, on the ground, in the sea, in the air, or in orbiting satellites, that yields local variations, or anomalies, of the *Earth's magnetic field* and its gradients. These anomalies are interpreted as to the depth, size, shape, and magnetization of geologic features causing them. Syn: *magnetic prospecting.* See also: *high-resolution aeromagnetics.*

magnetic field A vector field of force exerted by a *magnetic dipole* upon any magnetized bodies or electric currents present; the field of influence of a magetized body or an electric current. Its intensity is inversely proportional to the distance from a magnetic field source. See also: *Earth's magnetic field.*

magnetic-field intensity The force exerted by the magnetic field on a magnetic material at a point in space. Commonly used as a syn. for total intensity. It is expressed in SI as teslas and in cgs units as *gauss* or gammas. Syn: *magnetic field; magnetic-field strength.* Nonrecommended syn: *magnetic force.*

magnetic-field line A curve whose tangent at any point is in the magnetic-field direction at that point. Syn: *line of force [magnet]; line of induction; magnetic-flux line.*

magnetic-field strength *magnetic-field intensity.*

magnetic flux An integral over the surface of the normal component of *magnetic induction* B; the number of magnetic-field lines crossing the surface of a given area. Expressed in webers in SI or in maxwells in the cgs system: 1 weber = 10^8 maxwell.

magnetic-flux line *magnetic-field line.*

magnetic force (a) The physical force experienced by a magnetic substance when placed in a *magnetic field* or between magnetized bodies and electric currents. (b) A nonrecommended syn. of *magnetic-field intensity.*

magnetic hysteresis *hysteresis (b).*

magnetic inclination *inclination [magnet].*

magnetic induction (a) Magnetic-flux density, symbolized by B. In a magnetic medium, it is the vector sum of the inducing field H and the magnetization M. B is expressed in teslas in SI and in *gauss* or gammas in the cgs system. Syn: *magnetic field.* (b) A nonrecommended syn. of *electromagnetic induction.* (c) The process of magnetizing a body by applying a magnetic field. This usage is not recommended.

magnetic iron ore A syn. of *magnetite.* Var: magnetic iron.

magnetic latitude The angle whose tangent is one-half the tangent of the magnetic inclination. It would equal geographic latitude if the Earth's actual magnetic field were an axial dipole field.

magnetic lineation Relatively strong alignments and close spacings in magnetic contours. Sequences of magnetic lineations, identified by ship or aircraft surveys, were instrumental in working out the sea-floor spreading history of the world's oceans.

magnetic meridian *magnetic north.*

magnetic moment A vector quantity characteristic of a magnetized body or an electric-current system; it is proportional to the magnetic-field intensity produced by this body and also to the force experienced in the magnetic field of another magnetized body or electric current. The magnetic moment per unit volume is the *magnetization.*

magnetic needle A short, slender, wirelike length of magnetic material (such as a bar magnet) that is used as a compass and that is so suspended at its midpoint as to indicate the direction of the magnetic field in which it is placed by orienting itself toward the Earth's magnetic north. Usually referred to as needle. Syn: *needle [surv].*

magnetic north The uncorrected direction indicated by the north-seeking end of the needle of a magnetic compass; the direction from any point on the Earth's surface of the horizontal component of the Earth's magnetic lines of force connecting the observer with the north magnetic pole; the northerly direction of the magnetic meridian at any given point. It is the common zero-degree (or 360-degree) reference in much of navigational practice. Cf: *true north.* Syn: *magnetic meridian.*

magnetic order A repetitive arrangement of the magnetic moments of ions in mineral crystals, analogous to the repetitive arrangement of the positions of the ions. It is applicable only for ions with an intrinsic magnetic moment, such as Fe^{+3}, Fe^{+2}, or Mn^{+2}. See also: *exchange force; ferromagnetism; ferrimagnetism; antiferromagnetism.*

magnetic permeability The ratio of the magnetic induction B to the inducing field strength H. It is dimensionless.

magnetic polarity reversal *geomagnetic reversal.*
magnetic poles (a) Two areas near opposite ends of a magnet toward which the magnetic lines of force are oriented and concentrated. If the magnet is permitted to rotate about its center, the pole which points in the direction of the Earth's north magnetic pole is the north-seeking or *positive pole;* the other pole is the south-seeking or *negative pole.* The magnetic lines of force leave the magnet at the positive pole and enter at the negative pole. Always occur as part of a magnetic dipole, the elementary source of a magnetic field. (b) An approx. syn. for *geomagnetic poles.*
magnetic potential A mathematical function that describes, through its derivatives, the *magnetic field* at any space point. The magnetizing force H is the negative gradient of the magnetic potential, a scalar quantity representing the work done in moving a fictitious unit magnetic pole against the magnetic field from "infinity" to a given point.
magnetic prospecting *magnetic exploration.*
magnetic pyrites *pyrrhotite.*
magnetic quiet epoch A long interval when the Earth's magnetic field did not reverse its polarity, and no sea-floor magnetic lineations were created; esp. the long Cretaceous normal magnetic epoch.
magnetic resonance An interaction between the magnetic moments (electron spin and/or nuclear spin) of atoms with an external magnetic field. It is basic to the operation of the *optically pumped magnetometer*.
magnetic reversal *geomagnetic reversal.*
magnetic signature The shape of a magnetic anomaly, useful for comparison with known or model anomalies.
magnetic spherule A black *cosmic spherule* consisting of magnetite, sometimes including a metal core. Often collected by dragging magnets on the sea floor.
magnetic stability The ability of a magnetization to remain constant in direction over geologic time regardless of the direction or polarity of the geomagnetic field, as proven by field-based tests such as fold, conglomerate, contact, or consistency tests.
magnetic storm A period of rapid, irregular fluctuations of the Earth's external magnetic field, commonly with amplitudes of 50 to 200 nanoteslas, occasionally thousands of nanoteslas. Duration is often several days. Magnetic storms are caused by bombardment of the Earth by high-energy charged particles from the sun. Magnetic prospecting must be suspended during such periods.
magnetic stratigraphy *paleomagnetic stratigraphy.*
magnetic survey Measurement of the total magnetic intensity or rarely, a component or element of the geomagnetic field at different locations. See also: *aeromagnetic survey; magnetic prospecting.*
magnetic susceptibility *susceptibility [magnet].*
magnetic variation (a) Changes of the magnetic field in time or in space. (b) Magnetic *declination.*
magnetic viscosity A systematic change in the magnetization of an assemblage of grains towards the direction of the ambient magnetic field. See also: *viscous magnetization.* Syn: *magnetic aftereffect.*
magnetism (mag'-net-ism) Physical phenomena associated with moving electricity in conductive rocks or the permanent magnetic effects of naturally magnetic rocks.
magnetite (mag'-net-ite) (a) A black, cubic, strongly magnetic, opaque mineral of the spinel group: $(Fe^{2+}Fe^{3+})Fe^{3+}O_4$. More properly it is an inverse spinel, with ferric iron in the bracketed tetrahedral site. It often contains variable amounts of titanium oxide, and it constitutes an important ore of iron. Magnetite commonly occurs in octahedrons and also granular or massive; it is a very common and widely distributed accessory mineral in rocks of all kinds (in orebodies as a magmatic segregation, in lenses enclosed in schists and gneisses, in igneous rocks as a primary mineral or as an alteration product, in placer deposits, and as a heavy mineral in sands). Syn: *magnetic iron ore; octahedral iron ore.* (b) A name applied to a series of isomorphous or near-isomorphous minerals in the spinel group, consisting of magnetite, magnesioferrite, franklinite, jacobsite, trevorite, and maghemite. Symbol: Mt.
magnetization (mag'-ne-ti-za'-tion) The *magnetic moment* per unit volume, a vector quantity symbolized by M or I. Magnetization of a rock is the vector sum of its two types: *induced magnetization* and *remanent magnetization*. Syn: *volume magnetization.* Non-recommended syn: *polarization [magnet].*
magnetocrystalline anisotropy (mag'-ne-to-crys'-tal-line) Dependence of the electronic energy of a magnetically ordered crystal upon the direction in which the atomic magnetic moments are aligned. Those crystallographic directions for which the energy is lowest are called "easy" directions. In magnetite these are the [111] directions. Syn: *magnetic anisotropy.*
magnetogram *magnetograph.*
magnetograph (mag-ne'-to-graph) An instrument to record, automatically and continuously, temporal variations in the magnetic elements; the record it produces is a *magnetogram.*
magnetohydrodynamics (mag-ne'-to-hy'-dro-dy-nam'-ics) The study of the relationship between a magnetic field and the motion of an electrically conducting fluid. It is relevant to studies of the Earth's core.
magnetoilmenite (mag-ne'-to-il'-men-ite) A high-temperature solid solution of magnetite in ilmenite. Cf: *ilmenomagnetite.*
magnetometer (mag-ne-tom'-e-ter) An instrument that measures the Earth's magnetic field and its changes, or the magnetic field of a particular rock (from which its magnetization is deduced).
magnetometric resistivity method (mag-ne'-to-met'-ric) A method of electrical surveying in which the ground is energized with direct current through a pair of electrode contacts, and the behavior of the current is surveyed by measuring the resulting magnetic field. Abbrev: MMR.
magnetoplumbite (mag-ne'-to-plum'-bite) (a) A black magnetic hexagonal mineral: $Pb(Fe^{3+},Mn^{3+})_{12}O_{19}$. Cf: *plumboferrite.* It is isotypic with yimengite. (b) A group name for hexagonal minerals with formula $AB_{12}O19$, where A = Pb, Ca, Ce, Na, Ba, or K, and B = Fe^{3+}, Al, Ti, Cr or Mn.
magnetopolarity unit (mag-ne'-to-po-lar'-i-ty) A body of rock unified by its remanent magnetic polarity and distinguished from adjacent rock that has different polarity (NACSN, 1983, Art. 44). Its upper and lower limits are defined by boundaries marking a change of polarity. Syn: *polarity rock-stratigraphic unit; magnetostratigraphic polarity unit.*
magnetopolarity zone This term, rather than *polarity zone,* "should be used where there is risk of confusion with other kinds of polarity" (NACSN, 1983, Art. 46).
magnetosphere (mag-ne'-to-sphere) The confines of a planet's magnetic field, established by interaction between the solar wind and the planetary magnetic field. On the sunlit side of Earth, the magnetosphere is approximately hemispherical, with a radius of about ten Earth radii under quiet conditions; it may be compressed to about six Earth radii by magnetic storms. Opposite the sunlit side, the magnetosphere extends in a "tail" of several hundred Earth radii.
magnetostratigraphic classification The organization of rock bodies into units based on differences in magnetic character (ISSC, 1994, p.20).
magnetostratigraphic polarity classification The organization of rock bodies into units based on changes in the polarity of their remanent magnetization, related to reversals in the polarity of the Earth's magnetic field (ISSC, 1994, p.71).
magnetostratigraphic polarity-reversal horizon A surface or very thin transition interval separating sequences of rock strata of opposite magnetic polarity (ISSC, 1994, p.71). Syn: *polarity-reversal horizon*.
magnetostratigraphic polarity-transition zone A substantial interval in a succession of rock strata (of the order of 1 m in thickness) in which the magnetic polarity is reversed gradually (ISSC, 1994, p.71). Syn: *polarity-transition zone.*
magnetostratigraphic polarity unit A body of rocks characterized by its magnetic polarity (ISSC, 1994).
magnetostratigraphic polarity zone The basic formal unit of magnetostratigraphic polarity classification. Polarity zones may be subdivided into polarity subzones and grouped into polarity superzones (ISSC, 1994, p.71). Syn: *polarity zone.*
magnetostratigraphic unit A body of rocks unified by specified remanent-magnetic properties and distinct from underlying and overlying magnetostratigraphic units having different magnetic properties (NACSN, 1983, Art. 43). Syn: *magnetozone.*
magnetostratigraphy (mag-ne'-to-stra-tig'-ra-phy) The element of stratigraphy that deals with the magnetic characteristics of rock units (ISSC, 1994, p.70).
magnetostriction (mag-ne'-to-stric'-tion) Elastic strain or deformation accompanying magnetization. Cf: *piezomagnetism.*
magnetotelluric method (mag-ne'-to-tel-lu'-ric) An electromagnetic method of surveying in which natural electric and magnetic fields are measured. Usually the two horizontal electric-field components (E) plus the three magnetic-field components (H) are recorded. Ratios of the orthogonal E:H pairs yield the principal elements of the tensor impedance of the Earth. This impedance is measured at frequencies of 10-5 Hz to 10 Hz.
magnetotellurics Electromagnetic methods that measure natural electrical and magnetic fields. Abbrev: MT.
magnetozone *magnetostratigraphic unit.*
magnioborite (mag'-ni-o-bo'-rite) *suanite.*
magniophilite (mag'-ni-o-phil'-ite) *beusite.*
magniotriplite (mag'-ni-o-trip'-lite) A reddish-brown monoclinic mineral of the *triplite* group: $(Mg,Fe^{2+},Mn^{2+})_2(PO_4)(F,OH)$.
magnioursilite A silky yellow orthorhombic mineral: $(Mg,Ca)_4(UO_2)_4(Si_2O_5)_5(OH)_6{\cdot}20H_2O$. It is the Mg-dominant analogue of ursilite, but has not been approved by the International Mineralogical Association.
magnitude [geomorph] (mag'-ni-tude) A dimensionless integer assigned to each link in a channel network denoting the number of sources ultimately tributary to the link. The magnitude of a network is given by the magnitude of its outlet link, and thus denotes the number of exterior links in the network (Shreve, 1967). Cf: *bifurcating link.*
magnitude [seis] *earthquake magnitude.*
magnitude [volc] The total volume of erupted material (Walker, 1980). Syn: *eruption magnitude.*
magnocalcite (mag-no-cal'-cite) An unnecessary name for dolomitic calcite; a mixture of dolomite and calcite.
magnochromite (mag-no-chro'-mite) *magnesiochromite.*

magnocolumbite (mag'-no-co-lum'-bite) A black or brownish-black orthorhombic mineral of the *columbite* group: $(Mg,Fe^{2+},Mn^{2+})(Nb,Ta)_2O_6$.

magnoferrite (mag-no-fer'-rite) *magnesioferrite.*

magnoliid complex (mag-no'-li-id) An informal grouping of woody angiosperms with pollen that is neither tricolpate nor tricolpate derived. The term includes the orders Magnoliales, Laurales, Winterales, and Piperales and refers especially to *angiosperm* evolution.

magnolite (mag'-nol-ite) A colorless, creamy, brown, or yellow-green orthorhombic mineral: $Hg^{1+}_2Te^{4+}O_3$.

magnophorite (mag'-no-phor'-ite) A discredited name for a titanium- and potassium-bearing richterite.

magnussonite (mag'-nus-son-ite') A green to blue-green cubic mineral: $Mn^{2+}_{18}(As^{3+}_6Mn^{1+}O_{18})_2Cl_2$. It may contain some magnesium and copper.

Mahalanobis distance A multivariate measure of the dissimilarity (or "distance" in variable space) between two populations or samples. It is a multivariate extension of the standardized difference between two samples, $D^2=d'S^{-1}d$, where d is the vector of differences between the means and S is the pooled covariance matrix. The measure is used in *discriminant function analysis* and similar procedures. Named for P. C. Mahalanobis (1893-1972), Indian statistician.

mahlmoodite A creamy-white monoclinic mineral: $Fe^{2+}Zr(PO_4)_2{\cdot}4H_2O$.

mahnerite A blue or emerald-green tetragonal mineral: $(Na,Ca,K)Cu_3(AsO_4)_2Cl{\cdot}5H_2O$.

maikainite A metallic cubic mineral of the *germanite* group: $Cu_{10}(Fe,Cu)_3MoGe_3S_{16}$.

main carina Portion of axis on upper side of planate or scaphate pectiniform conodont elements posterior to apex of basal cavity or growth center; consists of a row of fused or discrete denticles, commonly nodelike (TIP, 1981, pt. W, supp.2).

main joint *master joint.*

mainland (main'-land) A continuous body of land constituting the chief part of a country; e.g. a continent, or a main island relative to an adjacent smaller island. Syn: *fastland.*

main partition A radial wall of a foraminiferal test, extending from the marginal zone toward the center of the chamber (as in Orbitolinidae). It may be a simple transverse septum.

main scarp A steep surface on the undisturbed ground at the edge of a landslide, caused by movement of the slide material away from the undisturbed ground. See also: *minor scarp; surface of rupture.*

main shock The largest earthquake in a sequence. See also: *aftershock; foreshock.*

main stem The principal course of a stream.

main stream The principal, largest, or dominating stream of any given area or drainage system, having an axial or central position. Syn: *master stream; stem stream; trunk stream.*

main trough Furrow or depression parallel to anteroposterior midline on upper side of planate or scaphate pectiniform conodont elements (TIP, 1981, pt. W, supp.2). Syn: *trough [paleont]; sulcus.*

maitlandite (mait'-land-ite) *thorogummite.*

majakite (ma-jak'-ite) A metallic grayish-white hexagonal mineral: PdNiAs.

major earthquake (ma'-jor) An earthquake having a *surface-wave magnitude* of seven or greater on the Richter scale. Such a limit is arbitrary, and may vary according to the user. Cf: *microearthquake; ultramicroearthquake.*

major element (a) One of the eight elements, O, Si, Al, Fe, Ca, Na, K, Mg, that are present in amounts greater than one percent in the Earth's crust. (b) An element that comprises more than five percent of a rock; it is not quantitatively defined. Cf: *minor element.*

major elements The ions that comprise the bulk of the salinity of seawater and have the highest concentrations in seawater. Essentially the nonreactive or *conservative elements* in seawater. Cf: *minor elements.*

major fold A large-scale or dominant fold in an area, with which minor folds are usually associated.

majorite (ma'-jor-ite) A purple to pale yellowish-brown cubic meteorite mineral of the *garnet* group: $Mg_3(Fe,Al,Si)_2(SiO_4)_3$. It possesses silicon in both tetrahedral and octahedral coordinations by oxygen.

major joint *master joint.*

major septum One of the initial or secondary septa of a *corallite;* specif. a protoseptum or a metaseptum. Major septa are of subequal length and extend most of the distance from the wall to the axis. Cf: *minor septum.*

makarochkinite A black triclinic mineral of the *aenigmatite* group: $(Ca,Na)_2(Fe^{2+},Fe^{3+},Ti)_6(Si,Be,Al)_6O_{20}$. It is not approved by the International Mineralogical Association, and may be høgtuvaite.

makatea (ma-ka-tea') A Polynesian term used in the South Pacific for a raised rim of a *coral reef,* for a broad uplifted coral reef surrounding an island, or for an *atoll* uplifted so that its lagoon waters drained away and left the atoll exposed as a large carbonate island.

makatite (ma'-ka-tite) A white monoclinic mineral: $Na_2Si_4O_8(OH)_2{\cdot}4H_2O$.

make n. (a) A formation or accumulation of ore in a vein; esp. the wide or thick part of a lode or orebody. Cf: *pinch.* (b) The output, actual yield, or amount produced by an oil or gas well or a mine over a specified period. The term is colloquial.

makhtésh (makh-tesh') A term used in Israel for a huge, cirque-like hollow somewhat resembling an elongated meteorite crater, produced by erosion of a structural dome (Amiran, 1950-1951). Etymol: Hebrew, "mortar". Pl: makhtéshim. Syn: *erosion crater.*

mäkinenite (mäk'-in-en-ite) A metallic rhombohedral mineral: gamma-NiSe. It is the Se analogue of millerite.

making hole (mak'-ing) The act of drilling a well.

makovickyite A metallic gray monoclinic mineral: $(Cu,Ag)_{1.5}(Bi,Pb)_{5.5}S_9$. It is a member of the pavonite homologous series.

malachite (mal'-a-chite) A bright-green monoclinic mineral: $Cu_2(CO_3)(OH)_2$. It is an ore of copper and is a common secondary mineral associated with azurite in the upper (oxidized) zones of copper veins. Malachite occurs in masses having smooth mammillated or botryoidal surfaces, and it is often concentrically banded in different shades of colors. It is used to make ornamental objects.

malacolite (mal'-a-co-lite) A syn. of *diopside.* The term originally designated a light-colored (pale-green or yellow) translucent variety of diopside from Sweden.

malacology (mal-a-col'-o-gy) The study of mollusks. Cf: *conchology.*

malacoma (mal-a-co'-ma) Collective name for the soft parts of radiolarians.

malacon (mal'-a-con) A brown altered or hydrated variety of zircon. Also spelled: malakon; malacone.

malacostracan (mal-a-cos'-tra-can) (a) Any crustacean belonging to the class Malacostraca, characterized by compound eyes, a thorax composed of eight somites, typically with a carapace, and by an abdomen composed of six or seven somites. Range, Lower Cambrian to Holocene. (b) In very early usage, a soft-shelled crustacean. Cf: *entomostracan.*

malanite (mal'-a-nite) A metallic white cubic mineral of the *linnaeite* group: $Cu(Pt,Ir)_2S_4$.

malayaite (ma-lay'-a-ite) A pale yellow to orange monoclinic mineral of the *titanite* group: $CaSnSiO_5$.

malchite (mal'-chite) A fine-grained *lamprophyre,* generally porphyritic, with small phenocrysts of hornblende, labradorite, and sometimes biotite, in a groundmass of hornblende, andesine, and a small amount of quartz. Its name, given by Osann in 1892, is derived from Malchen, Germany. Obsolete.

maldonite (mal'-don-ite) A cubic mineral: approximately Au_2Bi. It is a pinkish to silvery-white alloy of gold and bismuth. Syn: *black gold; bismuth gold.*

maleevite A vitreous white orthorhombic mineral: $BaB_2Si_2O_8$. It is the Ba analogue of danburite.

malenclave (mal'-en-clave) A body of contaminated or unusable groundwater surrounded by uncontaminated water. Classification of malenclaves depends on whether their volume expands, diminishes, or is constant with time (Legrand, 1965, p.88).

malezal swamp (ma-le'-zal) A swamp due to drainage of water over an extensive plain which has only a slight, almost imperceptible slope.

malignite (ma-lig'-nite) (a) In the *IUGS classification,* a plutonic rock in which F is between 10 and 60, P/A+P is 10 or less, and M is between 30 and 60. (b) A mafic *nepheline syenite* which has more than 5% nepheline and roughly equal amounts of pyroxene and potassium feldspar. The name, given by Lawson in 1896, is derived from the Maligne River, Ontario, Canada.

malinkoite A colorless, pale green, or greenish-blue hexagonal mineral: $NaBSiO_4$.

malinowskite (mal-i-now'-skite) A variety of tetrahedrite containing lead.

malladrite (mal'-la-drite) A pale rose or white trigonal, low-temperature mineral of fumaroles: Na_2SiF_6. Not to be confused with *mallardite.*

mallardite (mal'-lard-ite) A white or pale rose monoclinic mineral of the *melanterite* group: $Mn^{2+}SO_4{\cdot}7H_2O$. Not to be confused with *malladrite.*

Mallard's constant (Mal'-lard's) In *Mallard's law,* the constant for any combination of lenses on a given microscope; it is written as K.

Mallard's law (a) A statement in optics that relates to the determination of 2V, the optic axial angle. The formula is D=K sinE, in which D equals half the distance between the points of emergence of the optic axes, E equals one half the optic axial angle in air, and K equals *Mallard's constant.* (b) An empirical relationship in twinning, which states that if a unit cell has a symmetry or pseudosymmetry greater than that of the crystal structure, then twinning is likely to occur, with the additional symmetry or pseudosymmetry acting as the twin element. Named after the French crystallographer and mineralogist of the nineteenth century, Ernest Mallard.

malleable (mal'-le-a-ble) Said of a mineral, e.g. gold, silver, copper, platinum, which can be plastically deformed under compressive stress, e.g. hammering.

mallestigite An adamantine colorless hexagonal mineral of the *fleischerite* group: $Pb_3Sb(SO_4)(AsO_4)(OH)_6{\cdot}3H_2O$.

Malm Upper part of the Jurassic. Corresponds approximately with the standard Upper Jurassic (above Dogger, below Cretaceous).

malmstone (malm'-stone) (a) A hard, cherty, grayish-white sandstone whose matrix contains minute opaline globules derived from sponge spicules that once filled now-empty molds; specif. the Malmstone from the upper part of the Upper Greensand (Cretaceous) of Surrey and Sussex in

England, used as a building and paving material. (b) A marly or chalky rock.

malpais (mal'-pais) A term used in the southwestern U.S. and Mexico for a region of rough and barren lava flows. The connotation of the term varies according to the locality. Etymol: Spanish, "mal país", "bad land".

Malthusian principle (Mal-thu'-si-an) The concept that all animals, including man, potentially outbreed the food supply; conversely, the food supply is the primary limiting factor on population. Thus most populations, if allowed a free breeding range, maintain themselves at the point of starvation.

mamelon [paleont] (mam'-e-lon) (a) A raised, rounded top of an echinoid tubercle, on which the spine articulates. (b) A member of a group of similarly shaped and regularly arranged domelike elevations on the surface of some stromatoporoids, the summits of which commonly mark the points of divergence of astrorhizae. Cf: *stromatoporoid; astrorhiza.*

mamelon [volc] A small, rounded volcano formed over a vent by slow extrusion of viscous, siliceous lava (Cotton, 1958).

mamlahah (mam'-la-hah) A term used on the Arabian Peninsula for an interior, salt-encrusted *playa* or region from which salt is extracted. Etymol: Arabic derivation from "milh" = salt. Cf: *sabkha.*

mammal (mam'-mal) Any vertebrate of the class Mammalia: endothermic, clothed in hair, bringing forth their young alive and nursing them. Range, Jurassic to Holocene.

mammillary [mineral] Forming smoothly rounded masses resembling breasts or portions of spheres; said of the shape of some mineral aggregates, as malachite or limonite. Cf: *botryoidal; colloform; reniform.*

mammillary [speleo] (a) A rounded *speleothem* consisting of carbonate layers coating bedrock. (b) A variety of rounded, subaqueous coating. Cf: *cloud [speleo].*

mammillary hill A smooth, rounded, more or less elongate drumlin having an elliptical base.

mammillary structure *pillow structure [sed].*

mammillated surface (mam'-mil-lat'-ed) A hummocky rock surface characterized by smoothed and rounded mounds alternating with hollows, esp. a streamlined surface formed by glacial erosion in mountainous areas, as in the Adirondack Mountains, N.Y.

mammoth (mam'-moth) Extinct Pleistocene ancestor of the elephant, with a hairy skin and long tusks curving upward. Mammoths were nearly worldwide in distribution but underwent a severe wave of extinction in the late Pleistocene.

mammothite (mam'-moth-ite) A blue monoclinic mineral: $Pb_6Cu_4AlSb^{5+}O_2(SO_4)_2Cl_4(OH)_{16}$.

manaccanite (ma-nac'-ca-nite) *menaccanite.*

manaksite A colorless, creamy, or rosy monoclinic mineral of the *fenaksite* group: $KNaMn^{2+}Si_4O_{10}$.

manandonite (ma-nan'-do-nite) A white or yellow triclinic mineral of the *serpentine* group: $LiAl_4(Si_2AlB)O_{10}(OH)_8$.

manasseite (ma-nas'-se-ite) (a) A white or pale colored hexagonal mineral: $Mg_6Al_2(CO_3)(OH)_{16}\cdot 4H_2O$. It is dimorphous with hydrotalcite. Syn: *quintinite-2H.* (b) A group name for hexagonal minerals of an analogous composition, but with Al replaced by Fe^{3+} or Cr, and CO_3 replaced by Cl_2.

mandarinoite (man-da-ri'-no-ite) A light green monoclinic mineral: $Fe^{3+}_2(Se^{4+}O_3)_3\cdot 6H_2O$.

mandchurite (mand-chur'-ite) A glassy nepheline basanite, named by Lacroix in 1923 for Manchuria. Obsolete.

mandible (man'-di-ble) (a) In cheilostome bryozoans, the relatively enlarged, generally intricately reinforced but uncalcified, movable part of an *avicularium.* It is generally hinged on condyles or a crossbar. (b) Any of various invertebrate mouth parts serving to hold or bite into food materials and/or to move food into the mouth; e.g. one of the third pair of cephalic appendages of a crustacean. (c) An obsolete term used by some arachnologists for chelicera and by others for pedipalpal coxa. (d) The lower jaws of a gnathostome, whether fused into a single element or composed of ligamentously connected individual bones.

mandibular joint covering (man-dib'-u-lar) A point near the end of the *cervical groove* in decapod crustaceans (Holthuis, 1974, p.735).

mandibular muscle scar The place of attachment of the muscle leading to the mandible of an ostracode from the inner surface of the carapace just anterior and ventral to the adductor muscle scars.

mandibular palp The distal articulated part of the *mandible* of a crustacean, which aids in feeding and cleaning.

Manebach-Ala twin law (Man'-e-bach-Al'-a) A complex twin law in triclinic feldspar according to which the twin axis is perpendicular to {001} and the composition plane is {001}. Cf: *Ala-A twin law.* Syn: *acline-A twin law.*

Manebach pericline twin law A complex twin law in feldspars, in which the twin axis is at right angles to {010}, and the composition plane is {001}.

Manebach twin law A twin law in feldspars, both monoclinic and triclinic, usually simple, with the twin plane and composition plane of {001}.

mangan (man'-gan) A *cutan* consisting of manganese oxides or hydroxides (Brewer, 1964, p.215).

manganandalusite (man'-gan-an'-da-lu'-site) *viridine.*

manganapatite (man-gan-ap'-a-tite) A variety of apatite containing managnese in solid solution for calcium.

manganarsite (man'-gan-ars'-ite) A pinkish-brown trigonal mineral: $Mn^{2+}_3As^{3+}_2O_4(OH)_4$.

manganaxinite (man-gan-ax'-i-nite) A honey-yellow or dark grayish-brown triclinic mineral of the *axinite* group: $Ca_2Mn^{2+}Al_2BSi_4O_{15}(OH)$.

manganbabingtonite (man'-gan-bab'-ing-ton-ite') A greenish-black or yellowish-green triclinic mineral: $Ca_2(Mn^{2+},Fe^{2+})Fe^{3+}Si_5O_{14}(OH)$. It is the Mn analogue of babingtonite.

manganbelyankinite (man'-gan-bel-yan'-kin-ite) A brownish-black amorphous mineral of the *belyankinite* group: $(Mn^{2+},Ca)(Ti,Nb)_5O_{12}\cdot 9H_2O$.

manganberzeliite (man'-gan-ber-ze'-li-ite) A brown, yellow, or orange cubic mineral of the *berzeliite* group: $NaCa_2(Mn^{2+},Mg)_2(AsO_4)_3$. It is isostructural with garnet.

manganblende (man'-gan-blende) *alabandite.*

manganese The element Mn, which has been reported as a naturally-occuring native metal.

manganese alum (man'-ga-nese) *apjohnite.*

manganese epidote *piemontite.*

manganese-hörnesite (man'-ga-nese-hör'-nes-ite) A white monoclinic mineral: $(Mn,Mg)_3(AsO_4)_2\cdot 8H_2O$.

manganese nodule An irregular, black to brown, friable, laminated concretionary mass consisting primarily of manganese-oxide minerals (Mn content is 15-30%), alternating with iron oxides. These nodules are abundant on the floors of the world's oceans (and also of the Great Lakes) as a result of precipitation, esp. in an area of slow deposition, and occur on or in sediments (esp. red clay and sometimes organic ooze). Manganese nodules range from a few mm to 25 cm in diameter (generally 3-5 cm) and have an average weight of about 100 grams, although larger ones exist (a nodule weighing 770 kg has been found).

manganese-shadlunite (man'-ga-nese-shad'-lun-ite) A metallic grayish-yellow cubic mineral of the *pentlandite* group: $(Mn,Pb,Cd)(Fe,Cu)_8S_8$. Not approved by the International Mineralogical Association.

manganese spar (a) *rhodonite.* (b) *rhodochrosite.*

mangangordonite A colorless to pale yellow triclinic mineral of the *laueite* group: $(Mn^{2+},Fe^{2+},Mg)Al_2(PO_4)_2(OH)_2\cdot 8H_2O$.

manganhumite (man-gan-hu'-mite) A pale to deep brownish-orange orthorhombic mineral, the Mn-analogue of humite: $(Mn^{2+},Mg)_7(SiO_4)_3(OH)_2$.

manganite (man'-ga-nite) A brilliant steel-gray or iron-black monoclinic mineral: $Mn^{3+}O(OH)$. It is trimorphous with groutite and feitknechtite, and is a minor ore mineral of manganese. Syn: *gray manganese ore.*

mangankhomyakovite An orange rhombohedral mineral of the *eudialyte* group: $Na_{12}Ca_6Sr_3Mn_3WZr_3Si_{25}O_{73}(O,OH,H_2O)_3(Cl,OH)_2$.

manganlotharmeyerite A brown-red to dark reddish-orange monoclinic mineral: $Ca(Mn^{3+}, Mg)_2(AsO_4)_2(OH,H_2O)_2$.

mangan-neptunite (man'-gan-nep'-tun-ite) A dark cherry-red, orange, or black monoclinic mineral: $KNa_2Li(Mn,Fe)_2^{2+}Ti_{12}Si_{18}O_{24}$. Cf: *neptunite.* It is the Mn analogue of neptunite.

manganocalcite (man'-ga-no-cal'-cite) (a) A variety of rhodochrosite containing calcium. (b) A variety of calcite containing manganese.

manganochromite (man'-ga-no-chro'-mite) A submetallic cubic mineral of the *spinel* group: $(Mn^{2+},Fe^{2+})(Cr^{3+},V^{3+})_2O_4$.

manganocolumbite (man'-ga-no-co-lum'-bite) A brownish-black to black orthorhombic mineral of the *columbite* group: $(Mn^{2+},Fe^{2+})(Nb,Ta)_2O_6$.

manganocummingtonite A gray, greenish-gray, pink, yellow, orange, or pale green monoclinic member of the *amphibole* group, representing cummingtonite with essential Mn: $Mn_2Mg_5Si_8O_{22}(OH)_2$. Syn: *parvowinchite.*

manganogrunerite A yellowish brown, greenish gray, or colorless monoclinic member of the *amphibole* group, representing grunerite with essential Mn: $Mn_2Fe^{2+}_5Si_8O_{22}(OH)_2$.

manganokukisvumite A vitreous colorless orthorhombic mineral: $Na_6MnTi_4Si_8O_{28}\cdot 4H_2O$. It is the Mn analogue of kukisvumite.

manganolangbeinite (man'-ga-no-lang'-bein-ite) A rose-red cubic mineral of the *langbeinite* group: $K_2Mn^{2+}_2(SO_4)_3$.

manganolite [mineral] (man-gan'-o-lite) *rhodonite.*

manganolite [rock] A general term for rocks composed of manganese minerals, esp. manganese oxides such as wad and psilomelane.

manganomelane (man-ga-nom'-e-lane) A field term used synonymously for *psilomelane* to designate hard, massive, botryoidal, colloform manganese oxides not specifically identified. The term was rejected by the International Mineralogical Association in favor of romanèchite.

manganonaujakasite A vitreous blue monoclinic mineral: $Na_6(Mn,Fe^{2+})Al_4Si_8O_{26}$. It is the Mn analogue of naujakasite.

manganonordite-(Ce) A colorless to light brown orthorhombic mineral of the *nordite* group: $Na_3Sr(Ce,La)(Mn,Zn,Fe)Si_6O_{17}$.

manganoparawollastonite A colorless monoclinic mineral: $(Ca,Mn)SiO_3$. This is probably an unnecessary name for manganese-bearing wollastonite. Not approved by the International Mineralogical Association.

manganophyllite (man'-ga-no-phyl'-lite) (a) A coppery-black monoclinic platy manganoan variety of biotite: $K(Mg,Mn)_3(OH)_2[AlSi_3O_{10}]$. (b) A hypothetical biotite end member: $KMn_3^{2+}(OH)_2[AlSi_3O_{10}]$.

manganosegelerite A vitreous yellow or yellow-green orthorhombic mineral of the *overite* group: $(Mn^{2+},Ca)(Mn^{2+},Fe^{2+},Mg)Fe^{3+}(PO_4)_2(OH)\cdot 4H_2O$.

manganosiderite (man'-ga-no-sid'-er-ite) A variety of siderite containing manganese. It is an intermediate member of the isomorphous series siderite-rhodochrosite.
manganosite (man-ga-no'-site) A cubic mineral of the *periclase* group: $Mn^{2+}O$. It occurs in small emerald-green octahedrons that turn black on exposure.
manganostibite (man'-ga-no-stib'-ite) A black orthorhombic mineral: $(Mn^{2+},Fe^{2+})_7Sb^{5+}As^{5+}O_{12}$.
manganotantalite (man'-ga-no-tan'-ta-lite) A black or brownish-black orthorhombic mineral of the *columbite* group: $(Mn^{2+},Fe^{2+})(Ta,Nb)_2O_6$. Cf: *tantalite.*
manganotapiolite (man'-ga-no-tap'-i-o-lite) A submetallic dark brown tetragonal mineral of the *tapiolite* group: $(Mn^{2+},Fe^{2+})(Ta,Nb)_2O_6$.
manganotychite A pale pink cubic mineral of the *northupite* group: $Na_6(Mn^{2+},Fe^{2+},Mg)_2(CO_3)_4(SO_4)$.
manganpyrosmalite (man'-gan-py-ro-smal'-ite) A green, brown, or gray trigonal trigonal mineral of the pyrosmalite group: $(Mn^{2+},Fe^{2+})_8Si_6O_{15}(OH,Cl)_{10}$. Cf: *pyrosmalite.*
manganvesuvianite A deep maroon-red tetragonal mineral of the *vesuvianite* group: $Ca_{19}Mn^{3+}(Al,Mn^{3+})_{10}(Mg,Mn^{2+})_2(SiO_4)_{10}(Si_2O_7)_4O(OH)_9$.
Mangapanian (Man-ga-pa'-mi-an) South Pacific stage: middle Pliocene (above Waipipian, below Nakamaruan).
mangerite (man'-ger-ite) (a) A member of the *charnockite series*, equivalent to orthopyroxene-bearing *monzonite*. (b) A plutonic rock of the *charnockite series*, corresponding to *monzonite.* Typically it contains microperthite as the dominant feldspar, with varying amounts of mafic minerals, esp. orthopyroxene; an orthopyroxene-bearing alkalic *monzonite* containing a predominance of perthitic feldspars. It is the intrusive equivalent of *doreite* (Streckeisen, 1967, p.209). See also: *pyroxene monzonite.* The name was given by Kolderup in 1903 for Manger, Norway.
mangrove coast (man'-grove) A tropical or subtropical low-energy coast with a shore zone overgrown by mangrove vegetation. Such coasts are common in Indonesia, Papua New Guinea, and other tropical regions. The marine coast of southern Florida is the only significant U.S. example.
mangrove swamp A tropical or subtropical *marine swamp* characterized by abundant mangrove trees.
maniculifer (man-i-cu'-li-fer) Said of brachiopod crura derived from the *radulifer* type, with handlike processes at the end of straight, ventrally directed crura.
manjak (man'-jak) A variety of asphaltite found in Barbados, which contains 0.7% to 0.9% sulfur and 1% to 2% mineral matter. See also: *glance pitch.*
manjiroite (man-ji'-ro-ite) A brownish-gray tetragonal mineral of the *hollandite* group: $(Na,K)(Mn^{4+},Mn^{2+})_8O_{16}\cdot[n]H_2O$.
man-made shoreline A shoreline consisting of constructed works, such as harbor areas, breakwaters, causeways, piers, seawalls, and docks.
mannardite (man-nard'-ite) An adamantine black tetragonal mineral of the *hollandite* group: $BaTi_6(V^{3+},Cr^{3+})_2O_{16}\cdot H_2O$.
Manning equation (Man'-ning) An equation used to compute the velocity of uniform flow in an open channel: $V=1.486/n\ R^{2/3}\ S^{1/2}$, where V is the mean velocity of flow (in cfs units), R is the hydraulic radius in feet, S is the slope of the channel or sine of the slope angle, and n is the Manning roughness coefficient. Cf: *Chézy equation.*
mansfieldite (mans'-field-ite) A white to pale-gray orthorhombic mineral of the *variscite* group: $AlAsO_4\cdot 2H_2O$. It is isomorphous with scorodite.
mantiennéite (man-ti-enne'-ite) (a) A yellowish-brown orthorhombic mineral: $KMg_2Al_2Ti(PO_4)_4(OH)_3\cdot 15H_2O$. (b) A group name for minerals with the analogous formula, but with Mg replaced by Fe^{2+} or Mn^{2+}, and Al replaced by Fe^{3+}.
mantle [cryst] (man'-tle) The outer zone in a zoned crystal; an overgrowth.
mantle [geol] A general term for an outer covering of material of one kind or another, such as a *regolith;* specif. *waste mantle.*
mantle [interior Earth] The zone of the Earth below the crust and above the core, which is divided into the *upper mantle* and the *lower mantle,* with a transition zone between.
mantle [paleont] (a) The fold, lobe, or pair of lobes of the body wall in a mollusk or brachiopod, lining the shell and bearing the shell-secreting glands, and usually forming a mantle cavity; e.g. a prolongation of the body wall of a brachiopod, such as the two folds of ectodermal epithelium lying above and below the viscera and lining the inner surface of each valve, or the integument surrounding the vital organs of a bivalve mollusk. Syn: *pallium.* (b) The fleshy structure of cirripede crustaceans, strengthened by five calcified plates (carina, terga, and scuta) (TIP, 1969, pt.R, p.98). (c) Variously formed covering or coat in a radiolarian.
mantle canal Any of the flattened tubelike branching extensions of the body cavity into the mantle of a brachiopod, through which fluids circulate in the mantle. Syn: *pallial sinus.*
mantle cavity The cavity, between the mantle and the body proper, holding the viscera and respiratory organs of a mollusk or brachiopod; e.g. the anterior space between brachiopod valves, bounded by the mantle and the anterior body wall, and containing the lophophore. Syn: *pallial chamber.*
mantle-crust mix Rock whose properties are between those of the crust and those of the mantle, e.g. having *P wave* velocities between 7.4 and 7.7 km/s.
mantled (man'-tled) Covered, as by an *ash fall* that conforms to the underlying surface.
mantled gneiss dome A term used by Eskola (1948) for a dome in metamorphic terranes that has a core of gneiss that was remobilized from an original basement and has risen through a cover of younger rocks, also metamorphosed. The gneiss is surrounded by a concordant sheath of the basal part of the overlying metamorphic sequence.
mantle plume (a) A rising pipe-shaped volume of mantle that is either abnormally hot or wet or both, such that during decompression is partially melts more than "normal" mantle (in the oceans, often taken to be MORB-type mantle). The geochemically distinct (from MORB) basalt that segregates from the plume is called an *oceanic island basalt* (OIB) and forms oceanic island chains like the Hawaiian islands (Morgan, 1971; Hart, 1988). (b) A vertical cylindrical part of the Earth's mantle, hotter than its surroundings, within which larger-than-normal amounts of heat are conducted upward to form a "hot spot" at the Earth's surface (Friedman et al., 1992, p.24).
mantle rock A syn. of *regolith.*
mantle source volume The region in the asthenosphere in which basalt is melted from its parent rock (Dalrymple et al., 1974, p.30).
manto (man'-to) A flat-lying, bedded deposit; either a sedimentary bed or a replacement strata-bound orebody. Etymol: Spanish, "cloak, cloak". Cf" *blanket deposit [ore dep].*
manto-type copper deposit Strata-bound, volcanic-rock-hosted copper deposit found in northern Chile.
manus (ma'-nus) The broad proximal part of a cheliped propodus (i.e., this *propodus* minus the fixed *finger*) (TIP, 1969, pt. R, p.98).
map n. A diagram, drawing, or other graphic representation, usually on a flat surface, of selected features (natural, artificial, or both) of a part or the whole of the surface of the Earth, some other planet, the Moon, or any desired surface or subsurface area, by means of signs and symbols and with the means of orientation indicated, so that the relative position and size of each feature on the map corresponds to its correct geographic situation according to a definite and established scale and projection. The type of information that a map is primarily designed to convey is frequently designated by a descriptive adjective, e.g. "geologic map", "topographic map", or "structure map". Etymol: Latin "mappa", "napkin, cloth". Cf: *chart; plan.* v. To produce or prepare a map; to represent or delineate on a map; to engage in a mapping operation.
map algebra A set of operations defined on spatial data sets for the analysis and synthesis of spatial information (Krzanowski et al., 1993).
map collar *marginalia [cart].*
map convolution A method of making grid residuals.
map face The area on a map, enclosed by the *neat line.*
map grid A grid superposed on a map to provide a coordinate system more convenient than that provided by the graticule.
mapimite (map'-i-mite) A blue-green monoclinic mineral: $Zn_2Fe^{3+}_3(AsO_4)_3(OH)_4\cdot 10H_2O$.
map measurer *chartometer.*
map migration Constructing a map from unmigrated seismic data and then migrating the map so as to show features in their correct locations.
mappability The quality or state of being mappable, determined by such considerations as the scale of base maps, purpose of the project and time assigned for completing the mapping, kind and number of exposures of the strata, the experience and skill of the mapper(s), and extent of the previous geologic study and mapping of surrounding areas (Friedman et al., 1992, p.24).
mapping (map'-ping) The process of making a map of an area; esp. the field work necessary for the production of a map.
mapping angle *gisement.*
map projection (a) Any orderly system or arrangement of lines drawn on a plane surface and representing a corresponding system of imaginary lines on an adopted terrestrial or celestial datum surface; esp. a graticule formed by two intersecting systems of lines (representing parallels of latitude and meridians of longitude) that portray upon a flat surface the whole or any part of the curved surface of the Earth, or a grid based on such parallels and meridians. It is frequently referred to as a *projection.* (b) Any systematic method by which a map projection is made; the process of transferring the outline of surface features of the Earth onto a plane. (c) The mathematical concept of such a system.
map reading The interpretation of the information shown on a map.
map scale *scale [cart].*
map separate *separate [cart].*
map series A group of maps generally conforming to the same cartographic specifications or having some common unifying characteristic, such as the same scale or the same size of area covered. It usually has a uniform format and is identified by a name, number, or a combination of both. Examples are the National Topographic Map Series and the Geologic Quadrangle Map Series published by the U.S. Geological Survey. Syn: *series.*
map sheet An individual map, including *marginalia [cart],* either complete in itself or part of a map series.
mar A Swedish term for a bay or creek whose entrance is filled with silt so that the water is almost fresh (Stamp, 1961, p.308). Pl: marer.

marais (ma-rais') A French term for *swamp* used in place names in certain localities of the U.S.

marble (mar'-ble) (a) A metamorphic rock consisting predominantly of fine- to coarse-grained recrystallized calcite and/or dolomite, usually with a granoblastic, saccharoidal texture. (b) In commerce, any crystallized carbonate rock, including true marble and certain types of limestone (*orthomarble; travertine*), that will take a polish and can be used as architectural or ornamental stone. (c) *verd antique.*

marcasite [gem] (mar'-ca-site) A popular term used in the gemstone trade to designate any of several minerals with a metallic luster (esp. crystallized pyrite, as used in jewelry) and also polished steel and white metal.

marcasite [mineral] A common metallic light yellow or grayish orthorhombic mineral: FeS_2. It is dimorphous with pyrite and resembles it in appearance, but marcasite has a lower specific gravity, less chemical stability, and usually a paler color. Marcasite often occurs in sedimentary rocks (such as chalk) in the form of nodules or concretions with a radiating fibrous structure. Syn: *white iron pyrites; iron pyrites; white pyrite; white pyrites; cockscomb pyrites; spear pyrites; lamellar pyrites.*

March model A mathematical model that describes the rotation path of perfectly linear and perfectly planar deformable markers, called passive markers, that deform along with their matrix in three-dimensional strain (March, 1932).

mare (ma'-re) (a) One of the several dark, low-lying, level, relatively smooth, plainslike areas of considerable extent on the surface of the Moon, having fewer large craters than the *highlands,* and composed of mafic or ultramafic volcanic rock; e.g. Mare Imbrium (a circular mare) and Mare Tranquillitatis (a mare with an irregular outline). It is completely waterless. Cf: *terra.* (b) A dark area on the surface of a planet, whose origin is not definitely known. Etymol: Latin, "sea", from Galileo's belief that lunar maria represented great seas of water. Pron: mah-rey. Pl: *maria.* Syn: *sea [planet].*

mare basalt Dark, relatively smooth, heavily cratered igneous rock, chiefly of mafic or ultramafic composition, comprising the lunar maria.

mare basin A large, approximately circular or elliptical topographic depression in the lunar surface, filled or partly filled with mare material; e.g. the Imbrium basin. See also: *thalassoid.*

marecottite A yellow-orange triclinic mineral of the *zippeite* group: $Mg_3O_6(UO_2)_8(SO_4)_4(OH)_2{\cdot}28H_2O$.

maremma (ma-rem'-ma) A low, marshy or swampy tract of coastland. Etymol: Italian.

Maremmian (Ma-rem'-mi-an) *Astaracian.*

mare ridge *wrinkle ridge.*

mareugite (ma-reu'-gite) A bytownite- and hauyne-bearing plutonic rock; a hauyne *gabbro.* Its name, given by Lacroix in 1917, is derived from Mareuges, Auvergne, France, where the rock forms inclusions in *ordanchite.* Obsolete.

margaritasite (mar-gar-i'-tas-ite) A yellow monoclinic mineral of the *carnotite* group: $(Cs,K,H_3O)_2(UO_2)_2(V_2O_8){\cdot}H_2O$.

margarite [ign] (mar'-ga-rite) A beadlike string of globulites, commonly found in glassy igneous rocks. Obsolete.

margarite [mineral] A dioctahedral monoclinic brittle mica of ideal composition $CaAl_2(Si_2Al_2)O_{10}(OH,F)_2$. It has a pale pink, green, or yellowish color, and is marked by a pearly luster. Syn: *lime mica; calcium mica; pearl mica.*

margarodite (mar-gar-o'-dite) A pearly variety of muscovite, resembling talc and giving a small percentage of water on ignition.

margarosanite (mar'-ga-ro-san'-ite) A colorless or snow-white platy triclinic mineral: $Pb(Ca,Mn^{2+})_2Si_3O_9$. It is the Pb analogue of walstromite.

marginal basin A loose term for a basin marginal to a continent, usually between the continent and a volcanic arc. Syn: *back-arc basin.*

marginal carina (mar'-gin-al) The ridge forming the posterior and lateral margins of the *carapace* on some decapod crustaceans (Holthuis, 1974, p.735).

marginal chamberlet A simple subdivision of a primary chamber of a foraminiferal test, located in the marginal zone of the chamber, and formed by main partitions only (as in Orbitolinidae).

marginal channel A channel formed by a meltwater stream flowing along the margin of a glacier or an ice sheet (Rich, 1908, p.528).

marginal conglomerate A conglomerate that forms along a shore, on the landward margins of sediments of other types into which it grades (Twenhofel, 1939, p.30). It lies at different stratigraphic levels in the section (as seen over a large area) and thereby diagonally transects time intervals. If sea level is rising, the conglomerate is a *basal conglomerate.*

marginal cord A thick spiral structure beneath the surface at the periphery of a foraminiferal test (as in Nummulitidae) (TIP, 1964, pt. C, p.61).

marginal crevasse A crevasse near the margin of a glacier. Initially, it normally extends obliquely upstream from either side toward the center of the glacier at an angle of about 45° (as seen in plan). The flow of the glacier tends to rotate such crevasses toward and through the perpendicular to the margin and eventually closes them. Cf: *transverse crevasse; splaying crevasse.* Syn: *lateral crevasse.*

marginal fault *boundary fault.*

marginal granule A dotlike body in a lamella of a tintinnid.

marginal groove The groove on decapod crustaceans that extends along the inner side of the marginal carina of the carapace (Holthuis, 1974, p.733).

marginalia [cart] (mar-gi-nal'-i-a) All printed or other material outside the *neat line* of a map.

marginalia [paleont] Sponge spicules (prostalia) around or on an oscular margin.

marginal karst plain *karst-margin plain.*

marginal lake *glacier lake.*

marginal moraine A term formerly used as a syn. of *terminal moraine* (Hobbs, 1912, p.279).

marginal plain An obsolete term for an *outwash plain* flanking the margin of a terminal moraine. Also, a vague term loosely applied to various topographic features around the margins of glaciers.

marginal plate A large elongate plate forming part of the massive marginal rim of some flattened early echinoderms, such as stylophorans, ctenocystoids, and some eocrinoids (TIP, 1968, pt. S, p.538).

marginal plateau A relatively flat shelf adjacent to a continent and similar topographically to, but deeper than, a *continental shelf.* The Blake Plateau is an example.

marginal polje A flat limestone plain that is surrounded by higher country but is bordered on one side by impervious rock. Cf: *blind valley.*

marginal probability In *Bayesian statistics,* the unconditional probability that an event will occur. Cf: *joint probability; conditional probability.*

marginal reserves That part of the reserve base which, at the time of determination, borders on being economically producible. Its essential characteristic is economic uncertainty. Included are resources that would be producible, given postulated changes in economic or technologic factors (USGS, 1980).

marginal ring The prominent circlet of thick plates exposed on both the oral and aboral sides of *cyclocystoids.*

marginal salt pan A natural salt pan along a coast, such as the Great Rann of Kutch in the Gujarat region of western India; a salt marsh along a coast; a *saline pan.*

marginal sea A sea distinctly smaller than the Arctic, and with a restricted exchange with the ocean. The prime example is the Mediterranean. Also called "adjacent sea," following the usage of Sverdrup et al., 1942.

marginal spine In some anascan cheilostomes (bryozoans), one of a series of spines, some uncalcified at the base, placed on the inner margin of the *gymnocyst* around the frontal membrane; e.g. a *scutum.*

marginal suture (a) The ecdysial (molting) junction between exoskeleton elements at the prosomal margin in a merostome. (b) A *suture* running along the edge of the cephalon of certain trilobites (TIP, 1959, pt. O, p.122).

marginal trench *trench [marine geol].*

marginal zone The peripheral portion of foraminiferal chambers where chamberlets are subdivided by primary and secondary partitions (as in Orbitolinidae).

marginarium (mar-gi-nar'-i-um) The peripheral part of the interior of a *corallite,* characterized by generally abundant dissepiments or by a dense deposit of skeletal tissue producing a *stereozone.* Adj: marginarial. Cf: *tabularium.*

marginate chorate cyst (mar'-gin-ate) *marginate cyst.*

marginate cyst A dinoflagellate cyst whose outgrowths are characteristically located on the lateral margins, leaving the dorsal and more often the ventral surfaces free of outgrowths. Syn: *marginate chorate cyst.*

margo (mar'-go) (a) A modified margin of the colpus of a pollen grain, consisting of a thickening or thinning in the ektexine. Cf: *annulus [palyn].* (b) A term sometimes used for similar marginal features associated with the laesura of spores.

maria (ma'-ri-a) Plural of *mare.*

marialite (ma-ri'-a-lite) A variously colored tetragonal mineral of the *scapolite* group: $3NaAlSi_3O_8{\cdot}NaCl$ (or three albite plus sodium chloride). It is isomorphous with meionite. Symbol: Ma.

maricite (mar'-i-cite) A colorless, gray, or pale brown orthorhombic mineral: $NaFe^{2+}PO_4$.

maricopaite (ma-ri-co'-pa-ite) A colorless orthorhombic zeolite mineral: $Pb_7Ca_2Al_{12}Si_{36}(O,OH)_{100}{\cdot}32(H_2O,OH)$.

marienbergite (ma'-ri-en-berg'-ite) A plagioclase-bearing *phonolite* containing natrolite instead of nepheline. Named by Johannsen in 1938 for Marienberg (now Mariánske Lázne), Czech Republic. Obsolete.

marignacite (mar-ig'-na-cite) *ceriopyrochlore-(Ce).*

marigraph (mar'-i-graph) A self-registering *tide gage,* usually actuated by a float in a tube or pipe communicating with the sea through a small hole that filters out short-period waves.

marine abrasion (ma-rine') (a) Erosion of the ocean floor by sediment that is moved by wave energy. Syn: *wave erosion.* (b) Erosion of submarine canyons by downslope movement of sediments under the influence of gravity.

marine biology The study of marine organisms.

marine climate The climate of islands and of land areas bordering the ocean, characterized by only moderate diurnal and annual temperature ranges and by the occurrence of maximum and minimum temperatures longer after the summer and winter solstices, respectively, than in a *continental climate.* Syn: *oceanic climate; maritime climate.*

marine delta plain A nearly flat plain built in a bay by stream deposits at the place where the current is checked upon entering quiet water (Tarr, 1902, p.73-74); it is built a slight distance above sea level. Cf: *delta plain.*

marine-deposition coast (ma-rine'-de-po-si'-tion) A coast whose configuration results chiefly from marine deposition, such as one straightened by the formation of spits or bars, or prograded by wave and current deposits.

marine ecology The study of the relationships between marine organisms and their environment, including associated organisms.

marine-erosion coast (ma-rine'-e-ro'-sion) A coast whose configuration results chiefly from marine erosion, as in the straightening of sea cliffs by waves.

marine erratic A sedimentary particle of anomalous size or lithology, transported and deposited in marine sediments by ice rafting, plants, or animals.

marine-flooding surface *flooding surface [stratig].*

marine geodesy The precise determination of positions at sea and the establishing of boundaries and boundary markers at sea. It also includes the measurement of gravity at sea and the study of all the physical characteristics of the sea environment that effect such measurements.

marine geology That aspect of the study of the ocean that deals specifically with the ocean floor, ocean crust, and the ocean-continent border, including submarine relief features, the geochemistry and petrology of the sediments and rocks of the ocean floor, and the influence of seawater and waves on the ocean bottom and its materials. Syn: *geological oceanography; submarine geology.*

marine ice sheet An ice sheet that is grounded below sea level. This applies to most of the West Antarctic ice sheet.

marine invasion The spreading of the sea over a land area.

marine limit The present or former limit of the sea; a shoreline. Commonly used in areas of postglacial isostatic uplift for the highest record of late-glacial submergence. Cf: *marin gräns; highstand shoreline.*

marinellite A colorless trigonal mineral of the *cancrinite* group: $(Na,K)_{42}Ca_6Al_{36}Si_{36}O_{144}(SO_4)_8Cl_2{\cdot}6H_2O$.

marine magnetic anomaly The magnetic pattern on the sea floor, symmetric about the spreading center, and reflecting the accomodation, within the basalt, of the prevailing magnetic signal at the time the sea floor was created and cooled. The anomalies were discovered in the late 1950s and emerged several years later as the chief evidence for seafloor spreading.

marine marsh A flat vegetated savannalike land surface at the edge of the sea, usually covered by water during high tide. Cf: *salt marsh.*

Marine Observation Satellites Satellites launched by the Space Development Agency of Japan (MOS-1 on February 19, 1987 and MOS-1b on February 7, 1990) (Krzanowski et al., 1993).

marine onlap A term proposed by Melton (1947, p.1869) for *onlap* in connection with marine strata that are progressively pinched out landward above an unconformity. Example: the relations of the Cambrian rocks of the Grand Canyon.

marine peneplain A hypothetical *abrasion platform* of large areal extent, uplifted above the reach of the waves before wave erosion had succeeded in perfecting a smooth plane; an almost plane surface of uncompleted marine denudation. Cf: *plain of marine erosion.*

marine placer *beach placer.*

marine plain (a) *plain of marine erosion.* (b) A coastal plain of marine sediments.

marine plane A hypothetical wave-cut surface produced during the ultimate stage of marine erosion; a *plain of marine erosion.*

marine salina A body of salt water along an arid coast, separated from the sea by a sand or gravel barrier through which seawater enters, and having little or no inflow of fresh water; e.g. at Larnaca on Cyprus; a *saltern.* Some salt may be deposited in it.

marine snow The aggregated organic detritus sinking in the ocean. Syn: *plankton snow; sea snow.*

marine stack *stack [coast].*

marine swamp A low area of salty or brackish water along the seashore, characterized by an abundant growth of grass, reeds, mangrove trees, and similar types of vegetation. See also: *mangrove swamp.* Syn: *paralic swamp.*

marine terrace (a) A narrow constructional coastal strip, sloping gently seaward, veneered by a marine deposit (typically silt, sand, fine gravel). Cf: *wave-built terrace.* (b) A narrow coastal plain whose margin has been strongly cliffed by marine erosion. (c) Loosely, a wave-cut platform that has been exposed by uplift along a seacoast or by the lowering of the sea level, and from 3 m to more than 40 m above mean sea level. Cf: *raised beach.* (d) A terrace formed along a seacoast by the merging of a wave-built terrace and a wave-cut platform. Syn: *shore terrace.*

marine time A term used by Kobayashi (1944b, p.477) for *fossil time* as indicated by marine organisms. Cf: *continental time.*

marine transgression *transgression.*

marin gräns Any maximum stand of the sea against the coast; esp. the highest *marine limit* or coastline of the postglacial sea. Etymol: Swedish, "marine border" (or "limit"). Abbrev: MG.

mariposite (mar-i-po'-site) A discredited name for a bright green, chromium-rich variety of muscovite (or phengite), having a high silica content.

maritime (mar'-i-time) Bordering on the sea, as a *maritime* province.

maritime climate *marine climate.*

maritime plant A plant growing in salty conditions of the foreshore.

maritime polar air mass Cool moist air derived from oceanic areas at high latitudes. Abbrev: mP.

maritime tropical air mass Warm moist air derived from oceanic areas at low latitudes. Abbrev: mT.

mariupolite (ma-ri-u'-po-lite) An albite-nepheline *syenite* containing aegirine and biotite, with zircon and beckellite as the main accessories. Named by Morozewicz in 1902 for Mariupol (now Oktj' abv), Ukraine. Obsolete.

mark A sedimentary structure along a bedding plane. The term usually signifies a mold, such as a slide mark or a tool mark, but is also frequently applied to a cast (filling), such as a sole mark or a drag mark. Sometimes applied to *bed forms* such as a ripple. Syn: *marking.*

marker [seis] (mark'-er) (a) A layer that accounts for a characteristic segment of a seismic-refraction time-distance curve and can be followed over reasonably extensive areas. (b) A layer that yields characteristic reflections over a more or less extensive area.

marker [stratig] (a) An easily recognized stratigraphic feature having characteristics distinctive enough for it to serve as a reference or datum or to be traceable over long distances, esp. in the subsurface, as in well drilling or in a mine working; e.g. a stratigraphic unit readily identified by characteristics recognized on an electric log, or any recognizable rock surface such as an unconformity or a corrosion surface. See also: *format.* Syn: *marker bed; marker horizon [stratig].* (b) A term used in South Africa for an outcrop.

marker band An identifiable thin bed that has the same stratigraphic position throughout a considerable area (Wills, 1956, p.14).

marker bed (a) A geologic formation serving as a marker. (b) *key bed.*

marker horizon [stratig] (a) A marker represented by a rock surface or stratigraphic level, such as a vertical or lateral boundary based on electric or other mechanically recorded logs, that may serve to delineate lithostratigraphic units. Also called: *marker band.* (b) *key bed.*

marker horizon [struc geol] A distinctive surface or layer in a body of rock that can be used to identify slip on a fault, folding, or used to measure strain. Syn: reference horizon.

marker point *Global Boundary Stratotype Section and Point.*

markfieldite (mark'-field-ite) A hypabyssal granite containing plagioclase phenocrysts in a granophyric groundmass. Its name, given by Hatch in 1909, is derived from Markfield, England. Obsolete.

marking (mark'-ing) *mark.*

Markov process (Mar'-kov) A *stochastic process* in which the state of a system at time *t(n)* depends in part on the state of the system at time *t(n-1).* It assumes that in a sequence of random events, the outcome or probability of each event is influenced by or depends upon the outcome of the immediately preceding event. Named for A.A. Markov (1856-1922), Russian mathematician.

marl (a) A term loosely applied to a variety of materials, most of which occur as loose, earthy deposits consisting chiefly of an intimate mixture of clay and calcium carbonate, formed under marine or esp. freshwater conditions; specif. an earthy substance containing 35-65% clay and 65-35% carbonate (Pettijohn, 1957, p.410). Marl is usually gray; it is used esp. as a fertilizer for acid soils deficient in lime. In the Coastal Plain area of SE U.S., the term has been used for calcareous clays, silts, and sands, esp. those containing glauconite (greensand marls); and for newly formed deposits of shells mixed with clay. The term has also been used to designate a soft, friable clay with very little calcium carbonate, and a very fine, loose, almost pure calcium carbonate with little clay or silt. Syn: *calcareous clay.* (b) A soft, grayish to white, earthy or powdery, usually impure calcium carbonate precipitated on the bottoms of present-day freshwater lakes and ponds largely through the chemical action of aquatic plants, or forming deposits that underlie marshes, swamps, and bogs that occupy the sites of former (glacial) lakes. The calcium carbonate may range from 90% to less than 30%. Syn: *bog lime.* (c) A term occasionally used (as in Scotland) for a compact, impure, argillaceous limestone. (d) A term loosely applied to any soil that falls readily to pieces on exposure to air. (e) A literary term for clay or earthy material. Etymol: French marle.

marlaceous (mar-la'-ceous) Resembling or abounding with marl.

marl ball *marl biscuit.*

marl biscuit A *microbial biscuit* found on the shore or shallow bottom of a lake (esp. in northern U.S. and southern Canada), consisting of a hard, flattish, rounded concretion of marl formed around a shell fragment or other nucleus. Syn: *marl ball; marl pebble.*

marlekor (mar'-le-kor) A calcareous concretion of certain glacial clays, as of the varved lake clays of Scandinavia and in the Connecticut River valley of New England. Syn: *imatra stone.*

marlite (marl'-ite) (a) A hardened marl resistant to the action of air; *marlstone.* (b) A semi-indurated sheet or crust formed on the bottoms and shores of lakes by the intergrowth or cementation of a considerable number of marl biscuits. Rarely used.

marl lake (a) A lake whose bottom deposits contain large quantities of marl. (b) A lake that has been mined or dredged as a commercial source of marl, esp. for the manufacture of portland cement.

marloesite (mar'-loes-ite) A pale-gray, fine-grained extrusive rock, apparently an altered *trachyte,* composed of phenocrysts of plagioclase and lath-shaped pseudomorphs of mica after olivine in a groundmass characterized by glomerophyric texture and composed of augite, sodic plagioclase, and iron oxides. Its name, given by Thomas in 1911, is derived from Marloes, Pembrokeshire, Wales. Cf: *skomerite.* Obsolete.

marl pebble *marl biscuit.*

marl slate An English term for fissile calcareous rock (shale); it is not a true slate.

marlstone (marl'-stone) (a) An indurated rock of about the same composition as *marl,* called an earthy or impure argillaceous limestone. It has a blocky subconchoidal fracture, and is less fissile than shale. Syn: *marlite.* (b) A hard ferruginous rock (ironstone) of the Middle Lias in England, worked as an iron ore; specif. the Marlstone, a calcareous and sideritic oolite made up of ooliths, shell chips, and crinoid ossicles, set in a carbonate cement. (c) A term originally applied by Bradley (1931) to slightly magnesian calcareous mudstones or muddy limestones in the Green River Formation of the Uinta Basin, Utah, but subsequently applied to associated rocks (including conventional shales, dolomites, and oil shales) whose lithologic characters are not readily determined. Picard (1953) recommends abandonment of the term as used in the Uinta Basin.

marly (marl'-y) Pertaining to, containing, or resembling marl; e.g. "marly limestone" containing 5-15% clay and 85-95% carbonate, or "marly soil" containing at least 15% calcium carbonate and no more than 75% clay (in addition to other constituents).

marmatite (mar'-ma-tite) A dark-brown to black, iron-rich variety of sphalerite. Syn: *christophite.*

marmolite (mar'-mo-lite) A thinly laminated, usually pale-green serpentine mineral; a variety of chrysotile.

marne A French term for a marl or calcareous clay containing more than 50% clay and not less than 15% calcium carbonate.

marokite (mar'-o-kite) A black orthorhombic mineral: $CaMn^{3+}_2O_4$.

Marplatan (Mar-plat'-an) South American mammal age: late Pliocene (after Chapadmalalan, before Ensenadan). Syn: *Uquian.*

marquise cut (mar-quise') A doubly pointed, elongated variation of the *brilliant cut* in which the girdle outline is boat-shaped.

marrite (marr'-ite) A metallic gray monoclinic mineral: $PbAgAsS_3$. It is the As analogue of freieslebenite.

marscoite (mar'-sco-ite) A hybrid intrusive rock containing quartz and feldspar phenocrysts in a gabbroid groundmass. The term, proposed by Harker in 1904, was intended for use only in the Marsco area of Skye, Scotland. Obsolete.

Marsden chart (Mars'-den) A map used to give the position of oceanographic data. It is based on a Mercator map projection divided into rectangles called squares, each square being 10 degrees on a side. The squares are systematically numbered, with number 1 having its southeast corner on the equator at the *prime meridian.* Each Marsden square can be further subdivided into 100 squares each one degree on a side.

Marsden square One of a system of numbered areas each 10 degrees latitude by 10 degrees longitude, based on the Mercator projection, and used chiefly for identifying geographic positions and showing distribution of worldwide oceanographic and meteorologic data on a chart. Each square is subdivided into 100 one-degree subsquares which are numbered from 00 to 99 starting with 00 nearest the intersection of the equator and the Greenwich meridian. The system was introduced in 1831 by William Marsden (1754-1836), Irish orientalist.

marsh A water-saturated, poorly drained area, intermittently or permanently water-covered, having aquatic and grasslike vegetation, essentially without the formation of peat. Cf: *bog; fen; swamp.*

Marshall line (Mar'-shall) A syn. of *andesite line,* named after the New Zealand geologist P. Marshall. Obsolete.

marsh bar A narrow ridge of sand piled up at the seaward edge of a marsh undergoing wave erosion, as along the Delaware Bay shores of New Jersey. Cf: *chenier.*

marsh creek A drainage channel developed on a salt marsh.

marsh gas *Methane* produced during the decay of vegetable substances in stagnant water.

marshite (marsh'-ite) A colorless, pink or reddish-brown cubic mineral: CuI.

marsh lake (a) An area of open water in a marsh, surrounded by wide expanses of marshland. (b) A lake covered completely or nearly so by emergent aquatic plants, esp. sedge and grasses.

marsh ore *bog iron ore.*

marsh pan A *salt pan* in a marsh.

marsh peat Peat that is derived from both plant debris and sapropelic matter. Cf: *banded peat.*

marsh shore A lake shore consisting of marsh vegetation, which often merges with the emergent aquatic vegetation of the lake.

marsturite (mar'-stur-ite) A colorless to light pink triclinic mineral: $NaCaMn^{2+}_3Si_5O_{14}(OH)$.

Marsupialia (Mar-su'-pi-a'-li-a) An order of mammals, characterized by lack of placenta and consequent birth of young at a very immature state, with later development taking place in a specialized pouch. See also: *Eutheria.*

martensitic transformation A *phase transformation* that is dependent on temperature and independent of time.

marthozite (mar'-tho-zite) A yellow-green or greenish-brown orthorhombic mineral: $Cu(UO_2)_3O_2(Se^{4+}O_3)_2{\cdot}8H_2O$..

martinite (mar'-tin-ite) (a) A colorless or lavender triclinic mineral: $(Na,Ca)_{11}Ca_4(Si,S,B)_{14}B_2O_{40}F_2{\cdot}4H_2O$. (b) A variety of whitlockite containing carbonate.

martite (mar'-tite) Hematite occurring in iron-black octahedral crystals pseudomorphous after magnetite.

masafuerite (ma'-sa-fu-e'-rite) A hypabyssal *picrite* consisting of olivine phenocrysts, comprising over 50 percent of the rock, in a groundmass of pleochroic augite and calcic plagioclase, ilmenite, and magnetite. The name, given by Johannsen in 1937, is for Masafuera in the Juan Fernandez islands, Chile. Obsolete.

masanite (ma'-san-ite) A *quartz monzonite* containing phenocrysts of zoned plagioclase and corroded quartz in a granophyric groundmass. The name, given by Koto in 1909, is from Masan-po, Korea. Obsolete.

mascagnite (mas-cagn'-ite) A colorless to yellowish-gray orthorhombic mineral: $(NH_4)_2SO_4$. It is the NH_4 analogue of arcanite, and occurs as powdery crusts in volcanic districts and with other ammonium sulfates in guano deposits.

mascon (mas'-con) A large-scale, high-density, planetary mass concentration below an impact basin (Muller and Sjogren, 1968, p.680). Etymol: mass + concentration.

maskeeg (mas'-keeg) *muskeg.*

maskelynite (mas'-ke-lyn-ite') Thetomorphic plagioclase glass; a colorless meteorite substance consisting of a shock-formed noncrystalline phase that results from vitrification of plagioclase in rocks transfigured by shock waves and that retains the external features of crystalline plagioclase.

masking A process of excluding of a part of an image or map from further processing (Krzanowski et al., 1993).

maslovite (mas'-lov-ite) A metallic gray cubic mineral: (Pt,Pd)BiTe. Its powder diffraction pattern is similar to those of insizwaite and michenerite.

masonite (ma'-son-ite) A variety of chloritoid occurring in broad dark-green plates.

massa (mas'-sa) The spongy mass of aborted spores and tapetal material covering the trilete mark of certain megaspores.

mass balance The change in mass of a glacier over some defined interval of time (usually winter, summer, or year), determined either at a point on the glacier (specific balance) or over the entire glacier. The units normally used are millimeters, meters, or cubic meters of water equivalent, but kilograms per square meter or kilograms are used in some situations. Syn: *balance; mass budget; regimen [glaciol]; regime.* Cf: *net balance; annual balance.*

mass budget *mass balance.*

mass defect The difference between the sum of the atomic weights of the constituent particles of an isotope and its atomic weight as a whole.

mass erosion A term which includes all processes by which soil and rock materials fail and are transported downslope predominantly en masse by the direct application of gravitational body stresses. Syn: *gravity erosion.*

mass heaving The all-sided, general expansion of the ground during freezing, involving significant horizontal forces over a considerable area (Washburn, 1956, p.840). Syn: mass heave.

massicot (mas'-si-cot) A yellow orthorhombic mineral: PbO. Cf: *litharge.* Syn: *lead ocher.*

massif (mas-sif') A massive topographic and structural feature, especially in an orogenic belt, commonly formed of rocks more rigid than those of its surroundings. These rocks may be protruding bodies of basement rocks, consolidated during earlier orogenies, or younger plutonic bodies. Examples are the crystalline massifs of the Helvetic Alps, whose rocks were deformed mainly during the Hercynian Orogeny, long before the Alpine Orogeny.

massive [eco geol] (mas'-sive) Said of a mineral deposit (esp. of sulfides) characterized by a great concentration of ore in one place, as opposed to a disseminated or veinlike deposit.

massive [ign] Igneous rocks possessing a more or less homogeneous texture (fabric) over wide areas and lacking layering, foliation, cleavage, or similar features.

massive [meta] Said of a metamorphic rock whose constituents are neither oriented in parallel position nor arranged in layers; that is, a rock that does not have schistosity, foliation, or any similar structure.

massive [mineral] (a) Said of a mineral that is physically isotropic, e.g. lacking a platy, fibrous, or other structure. (b) Said of an *amorphous* mineral, or one without apparent crystalline structure. This usage is not recommended.

massive [paleont] Said of a *corallum* composed of corallites with or without intercorallite walls closely in contact with one another, or of a bryozoan colony consisting of a thick heavy *zoarium,* generally hemispherical or subglobular in shape.

massive [rock mech] (a) Said of a durable rock that is considered to be essentially isotropic and homogeneous and is free of fissures, bedding, foliation, and other discontinuities. Massive rock possesses a strength that does not vary appreciably from point to point. (b) A descriptive term for a lack of field-observable structure, characteristic of unstratified sediments or pedogenic soil horizons exposed in trenches typically excavated for neotectonic assessments.

massive [sed] (a) Said of a stratified rock that occurs in very thick, homogeneous beds, or of a stratum that is imposing by its thickness; specif. said of a bed that is more than 10 cm (4 in.) in thickness (Payne, 1942) or more than 1.8 m (6 ft) in thickness (Kelley, 1956, p.294). (b) Said of a stratum or stratified rock that is obscurely bedded, or that is or appears to be without internal structure (such as a rock free from minor joints, fissility, or lamination), regardless of thickness. The massive appearance may be deceptive, as many "massive" beds display laminae and other structures when X-rayed. See also: *unstratified.* (c) Descriptive of a sedimentary rock that is difficult to split, or that splits into layers greater than 120 cm (4 ft) in thickness (McKee and Weir, 1953, p.383).

massive sulfide deposit Any mass of unusually abundant metallic sulfide minerals, e.g. a *kuroko deposit.*

mass movement A unit movement of a portion of the land surface; specif. *mass wasting* or the gravitative transfer of material down a slope. Cf: *mass transport [sed].*

mass number The sum of protons and neutrons in the nucleus of an atom. Abbrev: A.

mass point One in a number of discrete points usually represented by a collection of irregularly or nonuniformly organized points. Each point is identified by a coordinate pair (x,y) and an elevation value (z). While mass points are not necessarily present in a uniform or structured pattern, they are often acquired in a systematic distribution, which is intended to facilitate accurate representation of the surface or features the data are intended to model. When generated manually, mass points are ideally chosen so that subtle terrain characteristics, such as gradual variations in slope or aspect, or distinct features, such as a levee or a river embankment, are adequately represented in the data. However, when generated automatically, such as through the use of an active sensing system (LIDAR), mass point distribution depends upon the characteristics of the sensors used to acquire the data, their performance in different terrain, and land cover types (Geo-One-Stop, 2003, p.3).

mass property A characteristic of a sediment considered as an aggregate, e.g. porosity, color, density, plasticity.

mass spectrograph Strictly, a recording *mass spectrometer,* but commonly used only for those instruments that record on a photographic plate as contrasted to those that record numerically or graphically. The latter are usually simply called mass spectrometers.

mass spectrometer An instrument for producing and measuring, usually by electrical means, a *mass spectrum.* It is especially useful for determining molecular weights and relative abundances of isotopes within a compound. See also: *mass spectrograph.*

mass spectrometry The art or process of using a *mass spectrometer* to study mass spectra.

mass spectroscopy The observation of a *mass spectrum* and all processes of recording and measuring that go with it.

mass spectrum The pattern of relative abundances of ions of different atomic or molecular mass (mass-to-charge ratio) within a sample. It frequently refers to the measured relative abundances of various isotopes of a given element.

mass susceptibility *specific susceptibility.*

mass transfer The redistribution of matter, typically involving chemical interaction between water and rock, and generally also including diffusion. It is now used chiefly to analyze major and minor element redistribution in hydrothermal alteration systems and in modelling of those systems.

mass transport [oceanog] The mass of water passing through a vertical plane one meter wide that is perpendicular to the transport and extending from the surface to some depth, measured in kg per second. Cf: *volume transport.*

mass transport [sed] The carrying of material in a moving medium such as water, air, or ice. Cf: *mass movement.*

massula (mas'-su-la) A more or less irregular, coherent mass of many pollen grains shed from the anther while fused together. Cf: *pollinium.*

mass wasting A general term for the dislodgement and downslope transport of soil and rock material under the direct application of gravitational body stresses. In contrast to other erosion processes, the debris removed by mass wasting is not carried within, on, or under another medium. The mass properties of the material being transported depend on the interaction of the soil and rock particles and on the moisture content. Mass wasting includes slow displacements, such as *creep* and *solifluction,* and rapid movements such as *rockfalls, rockslides,* and *debris flows.* Cf: *mass erosion.* Syn: *mass movement.*

master cave (mas'-ter) An area in or a portion of a cave that seems to be the largest and to which the auxiliary passages seem to lead.

master curve One of a set of theoretical curves calculated from models, against which an observed curve is matched in an effort to find a fit sufficiently close that the model parameters can be considered applicable to the actual situation. Used in resistivity, gravity, and magnetic interpretation. Syn: *type curve.*

master joint A persistent joint plane of greater-than-average extent. Syn: *main joint; major joint.*

master map An original map, usually of large scale, containing all the information from which other maps showing specialized information can be compiled; a primary source map. Syn: *base map.* Cf: *derivative map [cart].*

master stream *main stream.*

mastigoneme (mas'-ti-go-neme') One of the delicate, hairlike lateral threads, filaments, or processes along the length of some flagella. Syn: *flimmer.*

mastodon (mas'-to-don) One of a group of extinct, elephantlike mammals widely distributed in the Northern Hemisphere. It differs from mammoths and other true elephants in that teeth are low-crowned, with closed roots. Range, Oligocene to Pleistocene.

masutomilite (ma-su-to'-mil-ite) A purplish-pink monoclinic mineral of the trioctahedral mica group: $K(Li,Al,Mn^{2+})_3(Si,Al)_4O_{10}(F,OH)_2$.

masuyite (ma-su'-y-ite) An orange to reddish-brown orthorhombic mineral: $Pb(UO_2)_3O_3(OH)_2 \cdot 3H_2O$.

matched terrace *paired terrace.*

maternal zooid (ma-ter'-nal) A *zooid* in cheilostome bryozoans that deposits its eggs in a brood chamber enclosed by body walls of the same or adjacent zooids (Woollacott and Zimmer, 1972, p.165).

mathematical geography (math-e-mat'-i-cal) That branch of geography that is concerned with the representation of the Earth on maps and charts using various projection methods.

mathematical geology Geologic studies that principally employ mathematics, especially statistics and probability theory. Cf: *geomathematics.*

mathewrogersite (ma'-thew-ro'-gers-ite) A colorless to pale greenish-yellow rhombohedral mineral: $Pb_7(Fe^{2+},Cu)Al_3GeSi_{12}O_{36}(OH,H_2O)_6$.

mathiasite (ma-thi'-as-ite) A metallic black rhombohedral mineral of the *crichtonite* group: $(K,Ca,Sr)(Ti,Cr,Fe,Mg)_{21}O_{38}$.

matildite (ma-til'-dite) (a) A metallic gray trigonal mineral: $AgBiS_2$. Syn: *schapbachite; plenargyrite.* (b) A group name for minerals with the analogous composition, but with S replaced by Se or Te.

matlockite (mat'-lock-ite) (a) A variously colored tetragonal mineral: PbFCl. (b) A group name for minerals with the analogous composition, but with Bi replaced by Ca, and F or Cl replaced by O or OH.

matraite (mat'-ra-ite) A rhombohedral polymorph of wurtzite: ZnS.

matric potential The absolute value of the gage pressure head of water in an unsaturated material. It is inversely proportional to water saturation. By convention, the gage is set to read zero at atmospheric pressure.

matrix [gem] (ma'-trix) A gemstone cut from material consisting of a mineral and the surrounding rock material; e.g., opal matrix or turquoise matrix.

matrix [ign] *groundmass [ign]; mesostasis.*

matrix [meteorite] In a chondrite, the fine (<1 µm) material between chondrules.

matrix [ore dep] (a) A syn. of *gangue.* (b) A local term for the phosphate-bearing gravel in the land-pebble deposits of Florida.

matrix [paleont] The natural rock or earthy material in which a fossil is embedded, as opposed to the fossil itself.

matrix [sed] The finer-grained material enclosing, or filling the interstices between, the larger grains or particles of a sediment or sedimentary rock; the natural material in which a sedimentary particle is embedded. The term refers to the relative size and disposition of the particles, and no particular particle size is implied. In carbonate sedimentary rocks, the matrix usually consists of clay minerals or micritic components surrounding coarser material; although the term should be used in a descriptive, nongenetic, and noncompositional manner, it has been applied inappropriately as a syn. of *micrite.* Syn: *groundmass [sed].*

matrix [stat] A rectangular array of numbers, referred to as the elements of the matrix, arranged in *n* row and *m* columns. Using matrix algebra, a matrix can be mathematically manipulated as a single entity, greatly simplifying the notation that otherwise would be necessary to indicate manipulation of the individual elements.

matrix block Unaltered block or core in a breccia matrix bounded by mineralized fractures.

matrix porosity The porosity of the matrix or finer part of a carbonate rock, as opposed to the porosity of the coarser constituents (Choquette and Pray, 1970, p.247).

matrix rock *matrix [ore dep] (b).*

matrix support A sedimentary texture in which the larger particles are not in contact but are separated by finer particles. Cf: *clast support.*

matrosite (mat'-ro-site) Black, opaque microscopic material forming the matrix of torbanite (Dulhunty, 1939).

matsubaraite An adamantine gray monoclinic mineral of the *chevkinite* group: $Sr_4Ti_5O_8(Si_2O_7)_2$.

mattagamite (mat-ta-gam'-ite) A metallic violet-gray orthorhombic mineral of the *marcasite* group: $CoTe_2$.

matted structure A *microstructure* of, in general, mutually isolated, predominantly horizontal, elongated crystallites oriented in more or less random directions in the plane of their lamina (Peyer, 1968; Carter, 1990). The "Glanzschicht" or "shiny layer" of Preuschoft et al. (1974) for the dense vitrodentine or enameloid covering of shark fangs and cutting teeth.

matterhorn (mat'-ter-horn) A glacial *horn* resembling the Matterhorn, a peak in the Pennine Alps.

matteuccite (mat-te-uc′-cite) A white monoclinic mineral: $NaHSO_4 \cdot H_2O$.

matthedleite A creamy-white hexagonal mineral of the *britholite* group: $Pb_5(SiO_4,SO_4)_3(Cl,OH)$.

matulaite (ma-tu′-la-ite) A pearly-white platy monoclinic mineral: $CaAl_{18}(PO_4)_{12}(OH)_{20} \cdot 28H_2O$.

Matura diamond (Ma-tu′-ra) Colorless to faintly smoky gem-quality zircon from the Matara (Matura) district of southern Ceylon. When it does not occur colorless, the color may be removed by heating.

mature [geomorph] (ma-ture′) Pertaining to the hypothetical stage of maturity of the cycle of erosion; esp. said of a topography or region and of its landforms (such as a plain or plateau) that have undergone maximum development and accentuation of form; also, of a stream (and its valley) with a fully developed profile of equilibrium; or of a coast that is relatively stable.

mature [sed] Pertaining to the third stage of textural maturity (Folk, 1951); said of a clastic sediment that has been differentiated or evolved from its parent rock by processes acting over a long time and with a high intensity and that is characterized by stable minerals (such as quartz), deficiency of the more mobile oxides (such as soda), absence of weatherable material (such as clay), and well-sorted but subangular to angular grains. Example: a clay-free "mature sandstone" on a beach, with a standard deviation of less than 0.5 phi units (a range of less than 1 phi unit between the 16th and 84th percentiles of the particle-size distribution). Cf: *immature [sed]; submature [sed]; supermature.*

matureland (ma-ture′-land) The land surface of the mature stage of the cycle of erosion, ranging from surfaces having attained maximum relief to those of reduced "but not low" relief (Maxson and Anderson, 1935, p.90). The term was introduced by Willis (1928, p.493) in a broader sense to include eroded surfaces "qualified as vigorous, advanced, or subdued, according to the stage of development", a subdued matureland approaching the flatness of a peneplain. Davis (1932, p.429), noting that a "subdued" surface is neither mature nor old, but senescent, proposed the term "senesland" for this kind of matureland.

mature region *exozone.*

mature soil A soil with well-developed soil horizons produced by the natural processes of soil formation and essentially in equilibrium with its present environment. See also: *zonal soil.*

maturity [coast] (ma-tu′-ri-ty) A hypothetical stage in the development of a shore, shoreline, or coast that begins when a profile of equilibrium is attained, and that is characterized by decrease of wave energy; creation of beaches; disappearance of lagoons and marshes; straightening of the shoreline by bridging of bays and cutting back of headlands so as to produce a smooth, regular shoreline consisting of sweeping curves; and, eventually, retrogradation of the shore beyond the bayheads so that it lies against the mainland as a line of eroded cliffs throughout its course. This process does not necessarily occur at the same rate everywhere, owing to varying rock resistance. Cf: *secondary [coast].* Obsolete.

maturity [geomorph] The second of the three hypothetical principal stages of the *cycle of erosion* in the topographic development of a landscape or region, intermediate between *youth* and *old age* (or following *adolescence*), lasting through the period of greatest diversity of form or maximum topographic differentiation, during which nearly all the gradation resulting from operation of existing agents has been accomplished. It is characterized by numerous, closely spaced mature streams; disappearance of initial level surfaces, as the land is completely dissected and reduced to slopes; large, well-defined drainage systems with numerous and extensive tributaries and sharp, narrow divides and an absence of swamps or lakes on the uplands; greatest degree of ruggedness possible, with a new plain of erosion just beginning to appear; and *pedimentation* (in an arid cycle).

maturity [sed] The extent to which a clastic sediment texturally and compositionally approaches the ultimate end product to which it is driven by the formative processes that operate upon it (Pettijohn, 1957, p.508, 522). See also: *textural maturity; mineralogic maturity; compositional maturity.*

maturity [streams] The hypothetical stage in the development of a stream at which it has reached its maximum efficiency, having attained a profile of equilibrium and a velocity that is just sufficient to carry the sediment delivered to it by tributaries. It is characterized by: a load that is just about equal to the ability of the stream to carry it; lateral erosion predominating over downcutting, with the formation of a broad, open, flat-floored valley having a regular and moderate or gentle gradient and gently sloping, soil-covered walls with few outcrops; absence of waterfalls, rapids, and lakes; a steady but deliberate current, and muddy water; numerous and extensive tributaries, some of whose headwaters may still be in the youthful stage; development of floodplains, alluvial fans, deltas, and meanders, as the stream begins to deposit material; and a graded bed.

maturity index A measure of the progress of a clastic sediment in the direction of chemical or mineralogic stability; e.g. a high ratio of alumina/soda, of quartz/feldspar, or of quartz + chert/feldspar + rock fragments, indicates a highly mature sediment (Pettijohn, 1957, p.509).

Matuyama Chron An interval of geologic time from about 2.60 to about 0.78 Ma, delimited by geomagnetic reversals, and generally having a magnetic field reversed from that of the present. Named for Motonori Matuyama, Japanese geophysicist. Collective symbol: C1r, C2n, and C2r.

matveevite A white to yellowish monoclinic mineral: $KTi(Mn,Mg)_2(Fe^{3+},Al)_2(PO_4)_4(OH)_3 \cdot 15H^2O$. Not approved by the International Mineralogical Association.

maucherite (mau′-cher-ite) A metallic gray tetragonal mineral: $Ni_{11}As_8$. It tarnishes to coppery-red. Syn: *temiskamite.*

maufite (mau′-fite) A green incompletely investigated mineral, possibly an interstratified lizardite-chlorite: $(Mg,Ni)Al_4Si_3O_{13} \cdot 4H_2O$(?).

Maui-type well (Mau′-i-type) A type of *basal tunnel* characterized by a vertical or inclined shaft dug from the land surface to the *basal water table,* and by one or more tunnels dug along the water table to skim off the uppermost *basal groundwater* to avert possible *salt-water encroachment* (Stearns and Macdonald, 1942, p.126). This procedure was first used on the island of Maui, Hawaii.

mauzeliite (mau-ze′-li-ite) A variety of roméite containing Pb.

mawbyite (maw′-by-ite) An adamantine pale brown, orange-brown, or reddish monoclinic mineral of the *tsumcorite* group: $PbFe^{3+}_2(AsO_4)_2(OH)_2$.

Maw-sit-sit A rock used as a jade simulant from Myanmar, consisting of an intimate mixture of chrome-rich jadeite Kosmochlor (ureyite) and albite amphibole, and chromite. Syn: *jade-albite.*

mawsonite (maw′-son-ite) A metallic orange tetragonal mineral: $Cu_6Fe_2SnS_8$. It is isotypic with chatkalite.

maxilla (max-il′-la) (a) The main tooth-bearing bone in the upper jaws of Osteichthyes and tetrapods. (b) One of the first or second pairs of mouth parts posterior to the mandibles in various arthropods; e.g. the last cephalic appendage of a crustacean, following the maxillule and serving for feeding and respiration, or the *coxa* of a pedipalpus of an arachnid. Pl: maxillae. Adj: maxillary.

maxilliped (max-il′-li-ped) One of the three pairs of appendages of a crustacean, situated next behind the maxillae; an anterior *thoracopod* modified to act as a mouth part, its somite usually fused to the cephalon.

maxillule (max′-il-lule) The fourth cephalic appendage of a crustacean, between the *mandible* and the *maxilla,* serving as a mouth part. Syn: *first maxilla.*

maximum [geophys] (max′-i-mum) n. An anomaly characterized by values greater than those at neighboring locations; e.g., a gravity maximum or a geothermal maximum. Cf: *minimum [geophys].*

maximum [struc petrol] On a fabric diagram, a single area of concentration of poles representing the orientations of fabric elements (Turner and Weiss, 1963, p.58). Cf: *girdle [struc petrol]; cleft girdle.* Syn: *point maximum.*

maximum contaminant level One of a group of drinking-water standards set by the U.S. Environmental Protection Agency under the Safe Drinking Water Act (SDWA). Abbrev: MCL. Primary MCLs are health-based criteria, and secondary MCLs are based on aesthetic criteria such as color, odor, and taste.

maximum dry density In engineering geology, the largest mass of soil mixed with water that can be compacted into a standard mold with a standard amount of energy. Cf: *compaction test.*

maximum entropy filtering Filtering which attempts to produce an output which is as unpredictable as possible but which has the same autocorrelation as the input (Sheriff and Geldart, 1995, p.299, 561-2).

maximum-flooding interval In sequence stratigraphy, an interval that corresponds with the time of maximum transgression. It is typically associated with sediment starvation in deep water and with the development of *downlap.* It is also called "maximum flooding surface" but is usually an interval, not a surface (Miller et al., 1998).

maximum-flooding surface In sequence stratigraphy, a surface that separates underlying, upward-deepening, transgressive strata from overlying, upward-shallowing, regressive strata. There may or may not be a significant depositional hiatus associated with the surface (Embry and Johannessen, 1992).

maximum likelihood filtering Filtering which attempts to determine the most probable value(s) (Sheriff and Geldart, 1995, p.342, 559).

maximum projection sphericity The maximum projection area of a sphere of the same volume as a sedimentary particle, divided by the maximum projection area of the particle (Sneed and Folk, 1958, p.118). It is a measure of shape more closely related to the hydraulic resistance of the particle than is *sphericity* as defined by Waddell.

maximum slope A slope that is steeper than the slope units above or below it.

maximum unit weight The *dry unit weight* defined by the peak of a compaction curve.

maximum water-holding capacity The average moisture content of a disturbed soil sample, one centimeter high, after equilibration with a water table at its lower surface (Jacks et al., 1960, p.45). The retained water represents the lower part of the capillary fringe.

maxwell (max′-well) The cgs (centimeter-gram-second) unit of magnetic flux. One maxwell = 10^{-8} *weber*, or the flux through one square centimeter normal to a field of magnetic induction of one *gauss.*

maxwellite A vitreous red monoclinic mineral of the *tilasite* group: $NaFe^{3+}AsO_4F$.

Maxwell liquid (Max′-well) A model of *elasticoviscous behavior.* During the application of stress the body deforms both elastically and viscously. When the stress is released the elastic strain is recovered, releasing the stored energy. If the body is retained in a strained condition, the stress is relaxed as the elastic strain is slowly recovered.

Mayaian European stage: Middle Cambrian (above Amgaian, below Dresbachian) (Cowie and Bassett, 1989).

mayaite (ma'-ya-ite) (a) A white to gray-green or yellow-green material grading from tuxtlite to a nearly pure albite, found in the ancient tombs of the Mayans and elsewhere in Central America. (b) *omphacite.*

mayenite (may'-e-nite) A colorless cubic mineral: $Ca_{12}Al_{14}O_{33}$.

mayingite A steel-black cubic mineral of the *gersdorffite* group: IrBiTe. Syn: telluromayingite.

Maysvillian (Mays-vil'-li-an) North American stage: Upper Ordovician (above Edenian, below Richmondian).

maze cave Any cave consisting of repeatedly rejoined passages. Syn: *network cave; labyrinth.* Cf: *branchwork cave.*

mazzite (maz'-zite) A white hexagonal zeolite mineral: $(Mg_{2.5}K_2Ca_{1.5})(Si_{26}Al_{10})O_{72}\cdot 30H_2O$.

mbobomkulite (m-bo-bom'-ku-lite) A blue monoclinic mineral occurring in a cave deposit: $(Ni,Cu)Al_4[(NO_3)_2,SO_4]_2(OH)_{12}\cdot 3H_2O$. It is the dehydrated equivalent of hydrombobomkulite and is the Ni-NH_4 analogue of chalcoalumite.

mboziite (m-bo'-zi-ite) A black monoclinic mineral of the *amphibole* group discredited as being potassium-bearing taramite.

mbuga (mbu'-ga) A term used in SW Africa for a temporary swamp or "black claypan" (playa) marking the last stand of a now desiccated lake.

mcallisterite (mc-al'-lis-ter-ite') A white to amber rhombohedral mineral: $Mg_2B_{12}O_{14}(OH)_{12}\cdot 9H_2O$. Also spelled: *macallisterite.*

mcalpineite An emerald-green cubic mineral: $Cu_3Te^{6+}O_6\cdot H_2O$.

mcauslanite (mc-aus'-lan-ite) A yellowish-white triclinic mineral: $HFe^{2+}{}_3Al_2(PO_4)_4F\cdot 18H_2O$.

mcbirneyite (mc-bir'-ney-ite) A metallic black triclinic mineral: $Cu_3(VO_4)_2$. It is the vanadium analogue of stranskiite.

mcconnellite (mc-con'-nel-lite) A red rhombohedral mineral first described as an intergrowth with grimaldiite: $Cu^{1+}Cr^{3+}O_2$. It is the Cr analogue of delafossite.

mccrillisite A colorless tetragonal mineral of the *gainesite* group: $NaCs(Be,Li)Zr_2(PO_4)_4\cdot 1\text{-}2H_2O$.

mcgillite (mc-gill'-ite) A pearly light to dark pink monoclinic mineral of the *pyrosmalite* group: $(Mn^{2+},Fe^{2+})_8Si_6O_{15}Cl_2(OH)_8$.

mcgovernite (mc-gov'-ern-ite) A coppery-bronze foliated rhombohedral mineral: $(Mn,Mg)_{19}Zn_3(AsO_3)(AsO_4)_3(SiO_4)_3(OH)_{21}$. Syn: *macgovernite.*

mcguinessite (mc-gui'-ness-ite) A pale blue-green monoclinic mineral of the *malachite* group: $(Mg,Cu)_2(OH)_2(CO_3)$.

McKelvey box A diagram that portrays the formal classification of mineral reserves and resources proposed by McKelvey (1973). Reserves and resources are classified on one axis according to degree of certainty and along the other according to feasibility of economic recovery.

mckelveyite-(Nd) A pale green triclinic mineral of the *donnayite* group: $NaCa(Ba,Sr)_3Nd(CO_3)_6\cdot nH_2O$.

mckelveyite-(Y) (mc-kel'-vey-ite) A green triclinic mineral of the *donnayite* group: $NaBa_3CaY(CO_3)_6\cdot 3H_2O$. Syn: mackelveyite.

mckinstryite (mc-kin'-stry-ite) A metallic gray orthorhombic mineral: $(Ag,Cu)_2S$.

MCL *maximum contaminant level.*

mcnearite (mc-near'-ite) A white triclinic mineral: $NaCa_5(AsO_4)(AsO_3OH)_4\cdot 4H_2O$.

MCS *multichannel seismic profiling.*

md *millidarcy.*

M-discontinuity (M-dis-con'-ti-nu'-i-ty) Syn. of *Mohorovicic discontinuity,* suggested by Vening Meinesz (1955, p.321). Also spelled: *M discontinuity.*

M-domains *mica-rich domains.*

meadow ore (mead'-ow) *bog iron ore.*

meadow peat Peat derived from grasses.

mealy (meal'-y) *farinaceous.*

mean An arithmetic average of a series of values; esp. *arithmetic mean.* See also: *geometric mean; harmonic mean.* Cf: *mode [stat]; median.*

mean depth The cross-sectional area of a stream divided by its width at the surface. Cf: *mean hydraulic depth.*

meander [streams] (me-an'-der) n. (a) One of a series of regular freely developing sinuous curves, bends, loops, turns, or windings in the course of a stream. It is produced by a mature stream swinging from side to side as it flows across its flood plain or shifts its course laterally toward the convex side of an original curve. Etymol: Greek maiandros, from Maiandros River in western Asia Minor (now known as Menderes River in SW Turkey), proverbial for its windings. (b) *valley meander.* v. To wind or turn in a sinuous or intricate course; to form a meander.

meander [surv] v. To survey on or along a meander line. n. *meander line.*

meander amplitude The distance between points of maximum curvature of successive meanders of opposite phase, measured in a direction normal to the general course of the meander belt (Langbein and Iseri, 1960, p.14).

meander bar A deposit of sand and gravel located on the inside of, and extending into the curve of, a meander; specif. a *point bar.*

meander belt The zone along a valley floor across which a meandering stream shifts its channel from time to time; specif. the area of the flood plain included between two lines drawn tangentially to the extreme limits of all fully developed meanders. It may be from 15 to 18 times the width of the stream.

meander breadth The distance between the lines used to define the *meander belt* (Langbein and Iseri, 1960, p.14).

meander core (a) A central hill encircled or nearly encircled by a stream meander. Syn: *rock island.* (b) *cutoff spur.*

meander cusp A projection on the eroded edge of a meander-scar terrace, formed by the intersection of two or more meander scars. See also: *two-swing cusp; three-swing cusp; two-sweep cusp.* Syn: *terrace cusp.*

meander cutoff A cutoff formed when a stream cuts through a meander neck.

meandering stream (me-an'-der-ing) A stream having a pattern of successive meanders. Syn: *snaking stream.*

meandering valley A valley having a pattern of successive windings broadly resembling the trace of a meandering stream. The windings, or valley meanders, are of the same general order of size. Many are relict former channels of much larger rivers, such as glacial meltwater rivers. Cf: *underfit stream.*

meander length (a) The distance between corresponding parts of successive meanders of the same phase, measured along the general course of the meanders (Langbein and Iseri, 1960, p.14). (b) Twice the distance between successive points of inflection of the meander (Leopold and Wolman, 1957, p.55). Also called: meander wavelength.

meander line A surveyed line, usually of irregular course, that is not a boundary line; esp. a metes-and-bounds traverse of the margin or bank of a permanent natural body of water, run approximately along the mean-high-water line for the purpose of defining the sinuosities of the bank or shoreline and as a means of providing data for computing the area of land remaining after the water area has been segregated. Syn: *meander [surv].*

meander lobe The more or less elevated, tongue-shaped area of land enclosed within a stream meander. Syn: *tongue.*

meander neck The narrow strip of land, between the two limbs of a meander, that connects a meander lobe with the mainland.

meander niche On the wall of a cave, a crescentic opening formed by stream erosion. Syn: *wall niche.*

meander plain A term introduced by Melton (1936, p.594) for a plain built by the meandering process, or a plain of lateral accretion; it is seldom or never subject to overbank floods and thus lacks any alluvial cover. Cf: *covered plain; bar plain.*

meander scar (a) A crescentic, concave mark on the face of a bluff or valley wall, produced by the lateral planation of a meandering stream which undercut the bluff, and indicating the abandoned route of the stream. See also: *floodplain meander scar.* Syn: *meander scarp.* (b) An abandoned meander, often filled in by deposition and vegetation, but still discernible (esp. from the air).

meander scarp *meander scar.*

meander-scar terrace A local terrace formed by the shifting of meanders during the slow and continuous excavation of a valley (Schieferdecker, 1959, term 1519). Syn: *alternate terrace.*

meander scroll (a) One of a series of long, parallel, closely fitting, arcuate ridges and troughs formed along the inner bank of a stream meander as the channel migrated laterally down-valley and toward the outer bank. Cf: *lacine; point bar.* (b) A small, elongate lake occurring on a floodplain in a well-defined part of an abandoned stream channel, commonly in an oxbow.

meander spur An undercut projection of high land extending into the concave part of, and enclosed by, a meander.

meander terrace A small, relatively short-lived *stream terrace* formed by a freely swinging meander cutting into a former and higher floodplain; an *unpaired terrace.*

mean deviation The arithmetic mean of the absolute deviations of observations from their mean. Syn: *average deviation.*

mean diameter (a) *arithmetic mean diameter.* (b) *geometric mean diameter.* (c) *logarithmic mean diameter.* (d) *phi mean diameter.*

meandroid (me-an'-droid) Said of a *corallum* characterized by meandering rows of confluent corallites with walls only between the rows.

mean Earth ellipsoid The hypothesized ellipsoid that coincides most closely with the actual figure of the Earth at sea level, i.e., with the geoid.

mean higher high water A *chart datum,* the average of the higher high-water heights of each tidal day, averaged over a specific 19-year period. Cf: *National Tidal Datum Epoch.* Abbrev: *MHHW.*

mean high water Average of high water values. Used in England, Wales, and some states to determine ownership of coastal property. Abbrev: *MHW.*

mean high-water spring Average spring-tide high-water level averaged over a sufficiently long period. Abbrev: *MHWS.*

mean hydraulic depth The cross-sectional area of a stream divided by the length of its wetted perimeter. Cf: *mean depth.*

mean ice edge The average position of the *ice edge* in any given time period (usually a month), based on observations over several years. Formerly known as *ice limit.*

mean life In a *radionuclide* with a large number of atoms, the average of the lives of the individual atoms (each life ending in *radioactive decay*). For a single atom, the mean life is the time interval for which the probability of decay within the interval is 1/2. Mean life is the reciprocal of the decay constant.

mean lower low water A *chart datum,* the average of the lower low-water heights of each tidal day, averaged over a specific 19-year period. Cf: *National Tidal Datum Epoch.* It is a *chart datum* along the west and Gulf coasts of the United States. Abbrev: *MLLW.*

mean low water Average of low-water values. Used in some states to determine ownership of coastal property. Abbrev: *MLW.*

mean low water spring The *chart datum,* the average spring-tide low-water level averaged over a sufficiently long period. Abbrev: *MLWS.*

mean range The difference in height between mean high water and mean low water. Cf: *tide range.*

mean refractive index (a) The median *index of refraction* for any crystalline substance, with variation due to zoning. (b) The median index of refraction in any microcrystalline substance for which specific index values related to crystal directions are not determinable. (c) In a biaxial crystal, the beta, N_y or N_m index of refraction (in which y or m = mean). This is not the average index.

mean sea level A *chart datum,* the arithmetic mean of hourly heights observed over some specified time, usually 19 years. Abbrev: *MSL.*

mean stress The algebraic average of the three principal stresses.

mean tide level The arithmetic mean of mean high water and mean low water.

mean velocity *average velocity [hydraul].*

mean velocity curve *vertical-velocity curve.*

measured mineral resource That part of a *mineral resource* for which tonnage, densities, shape, physical characteristics, *grade,* and mineral content can be estimated with a high level of confidence. It is measured using information from locations spaced closely enough to confirm geological and/or grade continuity (JORC, 1999). Cf: *indicated mineral resource; inferred mineral resource.*

measured ore (meas'-ured) *developed reserves.*

measured reserve *proven reserve.*

measurement while drilling Obtaining log information while drilling operations are going on without making special logging runs. Abbrev: MWD.

measures (meas'-ures) A group or series of sedimentary rocks having some characteristic in common, specif. *coal measures.* The term apparently refers to the old practice of designating the different seams of a coalfield by their "measure" or thickness.

mechanical analysis Determination of the particle-size distribution of a soil, sediment, or rock by screening, sieving, or other means of mechanical separation; "the quantitative expression of the size-frequency distribution of particles in granular, fragmental, or powdered material" (Krumbein and Pettijohn, 1938, p.91). It is usually expressed in percentage by weight (and sometimes by number or count) of particles within specific size limits. See also: *particle-size analysis.*

mechanical clay A clay formed from the products of abrasion of rocks.

mechanical erosion *corrasion.*

mechanical layer thickness The joint parallel or bedding normal dimension defined by systematic joints of uniform height. Abbrev: MLT.

mechanically layered rocks A volume of rocks consisting of beds of varying lithologies and mechanical properties.

mechanical metamorphism *kinetic metamorphism.*

mechanical sediment *clastic sediment.*

mechanical stage A microscope stage that allows exact recording of the position of the object, e.g. a thin section, and that has a device for moving the object sideways, forward, and backward.

mechanical twin *deformation twin.*

mechanical twinning A deformation mechanism that reorients a crystal lattice by a discrete angle, so that the twin boundary becomes a crystallographic mirror plane. It results from the movement of *partial dislocations.*

mechanical weathering The process of weathering by which frost action, salt-crystal growth, absorption of water, and other physical processes break down a rock to fragments, involving no chemical change. Cf: *chemical weathering.* Syn: *physical weathering; disintegration [weath]; disaggregation.*

medaite (me'-da-ite) A sub-adamantine brownish-red monoclinic mineral: $(Mn^{2+},Ca)_6(V^{5+},As^{5+})Si_5O_{18}(OH)$.

medano (me-da'-no) A Spanish term for a *sand dune,* esp. one occurring along a seashore, as in Chile or Peru.

medenbachite A yellow to brownish-yellow triclinic mineral: $Bi_2Fe^{3+}(Cu,Fe^{2+})O(AsO_4)_2(OH)_3$.

medial (me'-di-al) *middle [geochron].*

medial moraine (a) An elongate moraine carried in or upon the middle of a glacier and parallel to its sides, usually formed by the merging of adjacent and inner lateral moraines below the junction of two coalescing valley glaciers. (b) A moraine formed by glacial abrasion of a rocky protuberance near the middle of a glacier and whose debris appears at the glacier surface in the ablation area. (c) The irregular ridge left behind in the middle of a glacial valley, when the glacier on which it was formed has disappeared. Syn: *median moraine.*

median (me'-di-an) The value of the middle item in a set of data arranged in rank order; the 50th percentile. If the set of data has an even number of items, the median is the arithmetic mean of the middle two ranked items. Cf: *mean; mode [stat].*

median carina In some decapod crustaceans, a longitudinal ridge extending over the full length of the middorsal area of the *carapace* from the tip of the *rostrum* to the middle of the posterior margin (Holthuis, 1974, p.735).

median diameter An expression of the average particle size of a sediment or rock, obtained graphically by locating the diameter associated with the midpoint of the particle-size distribution; the middlemost diameter that is larger than 50% of the diameters in the distribution and smaller than the other 50%.

median dorsal plate An elongate plate that posteriorly and dorsally separates the *carapace* valves of a phyllocarid crustacean.

median lamina In many stenolaemate bryozoans, the erect median colony wall from which zooids grow in two back-to-back layers to form *bifoliate* colonies. Syn: *mesotheca.*

median mass A rarely used syn. of *Zwischengebirge.*

median moraine *medial moraine.*

median muscle An anterior or posterior pedal retractor muscle inserted across the dorsal midline of the shells of primitive rostroconchs (Pojeta and Runnegar, 1976, p.15).

median rupture *median split.*

median section A slice in the central sagittal part and perpendicular to the axis of coiling of a foraminiferal test.

median septum (a) A calcareous ridge built along the midline of the interior of a brachiopod valve (Beerbower, 1968, p.284). (b) In *biserial scandent* graptoloid graptolithines, a partition that separates two rows of thecae.

median split An *acritarch excystment* structure in which the *vesicle* is sutured into roughly equal halves e.g., *Orthosphaeridium, Riculasphaera* (Strother, 1996). Syn: *median rupture.*

median sulcus A prominent vertical depression in the anterior and median surface of an ostracode valve.

median suture A suture along the sagittal line on the ventral cephalic *doublure* of a trilobite, developed in many forms where no rostral plate is present. Cf: *connective suture.*

median valley *rift valley [oceanog].*

medical geology (med'-i-cal) The application of geology to medical and health problems, involving such studies as the occurrence of toxic elements in unusual quantities in parts of the Earth's crust, the distribution of trace elements as related to nutrition, or the geographic patterns of disease. The medical syn. is "regional pathology". See also: *environmental geochemistry.*

medicinal spring (me-dic'-i-nal) A spring of reputed therapeutic value due to the substances contained in its waters. Cf: *spa.*

Medieval Warm Period A relatively mild episode of the Holocene bewteen about A.D. 950 and 1250.

mediglacial (me-di-gla'-cial) Relating to or formed between glaciers, or situated in the midst of glaciers.

Medinan (Me-di'-nan) (a) Obsolete syn. of *Alexandrian.* (b) North American stage: Lower Silurian (above Richmondian of Upper Ordovician, below Clintonian).

mediosilicic (me'-di-o-si-lic'-ic) An obsolete term proposed by Clarke (1908, p.357) to replace *intermediate.* Cf: *persilicic; subsilicic.*

mediterranean (med'-i-ter-ra'-ne-an) A rarely used term for *mesogeosyncline.*

Mediterranean climate (Med'-i-ter-ra'-ne-an) A climate characterized by hot, dry summers and mild, rainy winters. Typical of the climate surrounding the Mediterranean Sea.

mediterranean delta A delta built out into a landlocked sea that is tideless or has a low tidal range (Lyell, 1840, v.1. p.422).

mediterranean sea A restricted but deep body of water surrounded by continents connected with the world ocean by a one or more-narrow openings.

Mediterranean suite Obsolete name for a major group of igneous rocks, characterized by high potassium content. This suite was so named because of the predominance of potassium-rich lavas around the Mediterranean Sea; specif. those of Vesuvius and Stromboli. Cf: *Atlantic suite; Pacific suite.*

Medithermal (Med-i-ther'-mal) A term used by Antevs (1948, p.176) for a period of time in the late Holocene marked by decreasing temperatures.

medium bands (me'-di-um) In banded coal, vitrain bands from 2.0 to 5.0 mm thick (Schopf, 1960, p.39). Cf: *thin bands; thick bands; very thick bands.*

medium-bedded (me'-di-um-bed'-ded) A relative term applied to a sedimentary bed whose thickness is intermediate between *thin-bedded* and *thick-bedded.* See also: *stratification index.*

medium boulder A *boulder* having a diameter in the range of 512-1,024 mm (20-40 in., or -9 to -10 phi units).

medium clay A geologic term for a clay particle having a diameter in the range of 1/1024 to 1/512 mm (1-2 μm or 10 to 9 phi units). Also, a loose aggregate of clay consisting of medium clay particles.

medium-crystalline (me'-di-um-crys'-tal-line) Descriptive of an interlocking texture of a carbonate sedimentary rock having crystals whose diameters are in the range of 0.062-0.25 mm (Folk, 1959) or 0.1-0.2 mm (Carozzi and Textoris, 1967, p.5) or 1-4 mm (Krynine, 1948, p.143). Cf: *medium-grained.*

medium-grained (a) Said of an igneous rock, and of its texture, in which the individual crystals have an average diameter in the range of 1-5 mm (0.04-0.2 in.). Johannsen (1931, p.31) earlier used the range of 1-10 mm. (b) Said of a sediment or sedimentary rock, and of its texture, in which the individual particles have an average diameter in the range of 1/16 to 2 mm (62-2,000 μm or sand size). Cf: *medium-crystalline.* The term is used in a relative sense to describe rocks that are neither *coarse-grained* nor *fine-grained.*

medium-K In the *IUGS classification,* medium-K volcanic rocks plot above a line through two points with the coordinates SiO_2=48, K_2O=0.3wt% and SiO_2=68, K_2O=1.2 wt%, but below a line through two points with the coordinates SiO_2=48, K_2O=1.2 wt% and SiO_2=68, K_2O=3.1 wt%.

medium pebble A geologic term for a pebble having a diameter in the range of 8-16 mm (0.3-0.6 in., or -3 to -4 phi units).

medium sand (a) A geologic term for a sand particle having a diameter in the range of 0.25-0.5 mm (250-500 micrometers, or 2 to 1 phi units). Also, a loose aggregate of sand consisting of medium sand particles. (b) An engineering term for a sand particle having a diameter in the range of 0.42 mm (retained on U.S. standard sieve no. 40) to 2 mm (passing U.S. standard sieve no. 10). (c) A soil term used in the U.S. for a sand particle having a diameter in the range of 0.25-0.5 mm.

medium silt A geologic term for a silt particle having a diameter in the range of 1/64 to 1/32 mm (16-31 μm or 6 to 5 phi units). Also, a loose aggregate of silt consisting of medium silt particles.

medium volatile bituminous coal (me'-di-um-vol'-a-tile) Bituminous coal, characteristically agglomerating, that contains 23-31% volatile matter, analyzed on a dry, mineral-matter-free basis. Cf: *high volatile bituminous coal; low volatile bituminous coal.*

medmontite (med'-mont-ite) A mixture of chrysocolla and mica.

medulla (me-dul'-la) (a) The central zone of certain octocorals; e.g. the central chord of the axis of the Holaxonia (TIP, 1956, pt. F, p.174). (b) The internal part of some protozoans. (c) The central part of an organ or tissue (surrounded by the cortex).

medullary shell (med'-ul-lar'-y) The internal concentric shell of spumellarian radiolarians.

Medullosales (Me-dul-lo-sa'-les) *medullosans.*

medullosans (me-dul-lo'-sans) A group of Paleozoic *seed ferns* with large, complex fronds, massive petioles, synangiate (fused) pollen organs, and typically large, radially symmetrical seeds. Medullosans are generally considered to be the ancestors of the *cycads* and ranged from the Early Carboniferous to the Permian.

medusa (me-du'-sa) A basic *cnidarian* (coelenterate) body form, shaped like a bell or inverted bowl, with tentacles on the margin. It is adapted to free swimming.

meerschaum (meer'-schaum) A fine, compact, usually white claylike rock of hydrous magnesium silicate found in the Mediterranean area and used in fashioning tobacco pipes and as a building stone. Consists of the mineral sepiolite.

mega- (meg'-a-) A prefix signifying "large" or "great". Syn: *macro-.*

megabarchan (meg-a-bar'-chan) A giant *barchan,* up to 100 m or more in height (Stone, 1967, p.232).

megabias (meg'-a-bi-as) Long-term changes in the quality of the fossil record due to changes in styles of preservation, tectonic history, sea level, paleolatitudinal position, ocean chemistry, and other such geologic changes.

megabreccia (meg-a-brec'-cia) (a) A term used by Landes (1945) for a rock produced by brecciation on a very large scale, containing blocks that are randomly oriented and invariably inclined at angles from 6° to 25° and that range from a meter to more than 100 m in horizontal dimension. (b) A term used by Longwell (1951) for a coarse breccia containing individual blocks as much as 400 m long, inferred to develop downslope from large thrusts by gravitational sliding. It may be partly tectonic and partly sedimentary in origin, containing blocks that are shattered but little rotated. Cf: *chaos [geol].* (c) A *caldera collapse breccia* in which many clasts are so large that the brecciated nature of the rock is obscure. It is probably most abundant deep in caldera fills, where it may be confused with floor of the caldera (Lipman, 1984). Cf: *mesobreccia.*

megacanthopore (meg-a-can'-tho-pore) The larger of two sizes of *acanthopore* in stenolaemate bryozoans, which increases in size through the exozone until comparable in diameter to mesozooecia and feeding zooecia. Cf: *micracanthopore.*

megaclast (meg'-a-clast) (a) One of the larger fragments in a variable matrix of a sedimentary rock (Crowell, 1964). Cf: *phenoclast.* (b) A constituent of a *mixtite* (Schermerhorn, 1966).

megaclone (meg'-a-clone) A large, smooth monaxonic desma (of a sponge), having branches that bear cuplike articular facets, mostly terminal.

megacryst (meg'-a-cryst) A nongenetic term introduced by Clarke (1958, p.12) for "any crystal or grain" in an igneous or metamorphic rock that is "significantly larger" than the surrounding groundmass or matrix; e.g. a large microcline crystal in porphyritic granite. It may be a phenocryst, a xenocryst, a porphyroblast, or a porphyroclast.

megacrystalline (meg-a-crys'-tal-line) *macrocrystalline.*

megacycle (meg'-a-cy'-cle) An informal term applied by Mathur (1981b) to strata in the Vindhyan Supergroup of India. Each megacycle "is composed of an orderly and regular superposition of well-defined lithologic units that are of the status of formations and/or members, and the cycle is measured in tens or hundreds of metres." Cf: *magnacycle; supercycle (b).*

megacyclite A colorless monoclinic mineral: $KNa_8Si_9O_{18}(OH)_9{\cdot}19H_2O$.

megacyclothem (meg-a-cy'-clo-them) A term introduced by Moore (1936, p.29) to designate a combination of related cyclothems, or a cycle of cyclothems, such as in the Pennsylvanian of Kansas. Also, a cyclothem on a large scale, comprising minor cyclothems. Cf: *hypercyclothem; magnacyclothem.*

megaevolution (meg'-a-ev'-o-lu'-tion) *macroevolution.*

megafabric (meg'-a-fab'-ric) The fabric of a rock as seen in hand specimen or outcrop, without the aid of a microscope. Cf: *microfabric.* Syn: *macrofabric.*

megafacies (meg'-a-fa'-cies) (a) A term used by Cooper and Cooper (1946, p.68) apparently for a large intertonguing lithologic body. (b) A term used mistakenly for *magnafacies.*

megafauna (meg'-a-fau'-na) *macrofauna.*

megaflora (meg'-a-flo'-ra) (a) Plants large enough to be seen with the naked eye. (b) An obsolete term for the plants of a large habitat; a large, widespread group of plants. Cf: *microflora; macrofauna.* Syn: *macroflora.*

megaflow mark (meg-a-flow') A term used by Kuenen (1957, p.243) as a syn. of *gouge channel.*

megafossil (meg'-a-fos'-sil) *macrofossil.*

megagametophyte (meg'-a-ga-me'-to-phyte) The multicellular female *gametophyte* or haploid generation that develops from the megaspore of a heterosporous embryophytic plant. In lower vascular plants, it may be a small free-living plant bearing archegonia, but in seed plants it is contained within the ovule, and the egg is produced in it. Cf: *microgametophyte.*

mega-geomorphology The study of landforms and landscapes on large spatial and temporal scales, e.g. large regions, continents, or planets (Gardner and Scoging, 1983).

megagrained (meg'-a-grained) The texture of a carbonate sedimentary rock having particles whose diameters are greater than one millimeter (DeFord, 1946). See also: *macrograined.*

megagroup (meg'-a-group) A term used by Swann and Willman (1961, p.471, 475-476) for a litho-stratigraphic unit that is next higher in rank than group and that represents a major event in the course of geologic history. It is not recognized as a formal unit by the ACSN (1961) or by the NACSN (1983). Cf: *supergroup.*

megakalsilite A colorless hexagonal mineral: $KAlSiO_4$. It is a polymorph of kalsilite.

megalineament (meg-a-lin'-e-a-ment') "Any lineation more than 100 kilometers long" (El-Etr, 1976, p.485).

megalospheric (meg'-a-lo-spher'-ic) Said of a foraminiferal test or shell produced asexually and characterized by a large initial chamber (proloculus), relatively few chambers, small size of the adult test, and incomplete ontogeny. Cf: *microspheric.*

megalump (meg'-a-lump) A gravel-sized *lump* in a limestone. It usually originates by disruption (by high-energy waves or currents or possibly by turbidity currents) of newly deposited or partly indurated carbonate mud, which is then incorporated within the sedimentary unit from which it was derived.

megaphyric (meg-a-phyr'-ic) Said of the texture of a porphyritic igneous rock in which the greatest dimension of the phenocrysts is more than 2 mm (Cross et al., 1906, p.702); also, said of a rock having such texture. Obsolete. Cf: *microphyric.* Syn: *macroporphyritic.*

megaplankton (meg'-a-plank'-ton) The largest plankton organisms; they are more than 20 cm in size. Cf: *ultraplankton; nannoplankton; microplankton; macroplankton.*

megapore (meg'-a-pore) (a) *macropore.* (b) In the pore-size classification of Choquette and Pray (1970, p.233), an equant to equant-elongate pore or a tubular or platy pore with an average diameter or thickness greater than 4 mm. Cf: *mesopore [petrology]; micropore.*

megaregolith On heavily impacted planetary surfaces this comprises rock comminuted by impact that dates from the early history of bombardment. It is formed of overlapping layers of ejecta blankets grading downward into fractured crust (Kieffer et al., 1992). See also: *regolith [planet].*

megarhizoclone (meg-a-rhi'-zo-clone) A large *rhizoclone* approaching the form of a megaclone.

megaripple (meg-a-rip'-ple) A large bedform or ripplelike feature having a wavelength greater than 1 m (Van Straaten, 1953) or a ripple height greater than 10 cm (Imbrie and Buchanan, 1965, p.155), composed of sand, and formed in a subaqueous environment; also called dune. See also: *giant ripple; sand wave; dune.*

megasclere (meg'-a-sclere) A large *sclere;* specif. one of the primary spicules forming the principal skeletal support in a sponge. It usually differs in form from a *microsclere.* Syn: *macrosclere.*

megascopic (meg-a-scop'-ic) Said of an object or phenomenon, or of its characteristics, that can be observed with the unaided eye or with a hand lens. Syn: *macroscopic.*

megashear (meg'-a-shear) (a) A strike-slip fault with a horizontal displacement measured in tens to hundreds of kilometers. (b) A regional fault or fault zone whose trace length is greater than about 700 km. Infrequently used term, typically in reference to an ancient plate boundary scale fault.

megasporangium (meg'-a-spo-ran'-gi-um) A *sporangium* that develops or bears megaspores; e.g. the nucellus in a gymnospermous seed plant. Cf: *microsporangium.*

megaspore (meg'-a-spore) One of the spores of a heterosporous embryophytic plant that produces a *megagametophyte* and that is ordinarily larger than the *microspore.* Its range is mid-Devonian to Holocene. Cf: *gametophyte; miospore.* Syn: *macrospore.*

megatectonics (meg'-a-tec-ton'-ics) A rarely used term for the tectonics of the very large structural features of the Earth, or of the whole Earth. Similar terms are *geotectonics* and *global tectonics,* but all these large, vague words seem superfluous since the subject of *tectonics* itself differs from the subject of *structural geology* in dealing only with the very large structural features. Cf: *microtectonics.* Syn: *macrotectonics.*

megatherm (meg'-a-therm) A plant requiring high temperatures and large quantities of water for its existence. Syn: *macrotherm.* Cf: *microtherm; mesotherm.*

megathermal (meg-a-ther'-mal) Pertaining to a climate characterized by high temperature. Cf: *mesothermal [clim]; microthermal.*

Megathermal [paleoclim] n. A term proposed by Judson (1953, p.59) for *Altithermal.* See also: *Hypsithermal.*

megayear (meg'-a-year) A term proposed by Rankama (1967) for one million (10^6) years. Cf: *Ma.*

megazone (meg'-a-zone) *superzone.*

megerliiform (me-ger'-lii-form) Said of the *loop* of a terebratellacean brachiopod with descending branches joining anterior projections from the large ring on a low median septum, differing from similar dallinid and terebratellid loops by the appearance of a well-developed ring before growth of descending branches (TIP, 1965, pt. H, p.148).

Meigen's reaction (Mei'-gen's) A test used to distinguish calcite from aragonite. A cobalt-nitrate solution is used to stain the minerals by boiling; aragonite becomes lilac and retains this color in thin section, whereas calcite and dolomite become pale blue but do not show the color in thin section. Cf: *Lemberg's stain.*

meimechite (mei'-mech-ite) In the *IUGS classification,* an ultramafic lava with SiO_2<53, Na_2O+K_2O<1, MgO>18, and TiO_2>1 wt%. Kotulsky in 1943 derived the name from the Meimecha River, Siberia. Also spelled: meymechite. Cf: *komatiite; picrite.* Obsolete.

meinzer (mein'-zer) A syn. of *permeability coefficient,* named for O.E. Meinzer (1876-1948), a hydrogeologist with the U.S. Geological Survey. Syn: *Meinzer.*

meiofauna (mei'-o-fau'-na) Benthic animals ranging in size from 63-500 µm that live on and within sediments; size class between *macrofauna* and *microfauna.*

meionite (mei'-o-nite) A variously colored tetragonal mineral of the *scapolite* group: $3CaAl_2Si_2O_8{\cdot}CaCO_3$ (or three anorthite plus calcium carbonate). It is isomorphous with marialite, and may contain other anions (sulfate, chloride). Symbol: Me.

meiosis (mei-o'-sis) The division of chromosomes that results in the reduction (halving) of their number in the production of haploid egg and sperm cells.

meixnerite (meix'-ner-ite) (a) A colorless rhombohedral mineral: $Mg_6Al_2(OH)_{18}{\cdot}4H_2O$. It has the hydrotalcite structure, but with CO_3 replaced by $(OH)_2$. (b) A group name for rhombohedral minerals with the analogous composition, but with Al replaced by Cr or Fe^{3+}, and OH replaced by Cl.

mela- (me'-la-) A prefix meaning "dark-colored"; in the *IUGS classification,* it is used to designate a rock that is more mafic than the specified range, e.g. melagranite. Cf: leuco-.

melaconite (mel'-a-con-ite) *tenorite.*

melagabbroid (me-la-gab'-broid) In the *IUGS classification,* a plutonic rock satisfying the definition of *gabbroid,* and in which pl/(pl+px+ol) is between 10 and 35.

melanchym (mel'-an-chym) A complex humic substance separated into two fractions by alcohol: insoluble *melanellite* and soluble *rochlederite.* It is found in the brown coal of Bohemia.

melanellite (mel-a-nel'-lite) The insoluble portion remaining when *melanchym* is treated with alcohol. See also: *rochlederite.*

mélange [gem] (me-lange') An assortment of mixed sizes and/or qualities of diamonds larger than those of a *mêlée,* i.e. weighing more than 0.2 carat.

mélange [sed] A body of rock mappable at a scale of 1:24000 or smaller, characterized by a lack of internal continuity of contacts or strata and by the inclusion of fragments and blocks of all sizes, both exotic and native, embedded in a fragmental matrix of finer-grained material (Raymond, 1984). Criteria of matrix composition and fabric are not employed in the definition, and no genetic significance is implied. Cf: *tectonic mélange; allolistostrome.* See also: *dismembered formation; chaos [geol].* The term was introduced by Greenly (1919, p.980). Etymol: French, "mixture."

melanite (mel'-a-nite) A black variety of andradite garnet containing titanium. Cf: *schorlomite.* Syn: *pyreneite.*

melanocerite-(Ce) A brown or black rhombohedral metamict mineral: $(Ce,Ca)_5(OH,F)(Si,B)_3O_{12}{\cdot}nH_2O$ (?).

melanocratic (mel'-a-no-crat'-ic) Dark-colored; applied to dark-colored igneous rocks rich in mafic minerals. The percentage of mafic minerals required for a rock to be classified as melanocratic varies among petrologists; the lower limit ranges from 60 to 67%. n. melanocrate. Cf: *leucocratic; mesocratic.* Syn: *dark-colored.*

melanoidin Insoluble, brown organic compounds precipitated from sugars and amino acids.

melanophlogite (mel'-a-no-phlo'-gite) A colorless, white, yellow, reddish, or brown tetragonal mineral consisting of silicon dioxide (SiO_2) and containing carbon and sulfur. It was formerly believed to be a partly oriented pseudomorph of alpha quartz after cristobalite containing H_2SO_4.

melanophyre (me-lan'-o-phyre) A broad term used in the field for any dark-colored porphyritic igneous rock having a fine-grained groundmass. Obsolete.

melanosclerite (mel-a-no-scle'-rite) A problematic organic (chitinous) microfossil with a dense wall and a variety of forms ranging in size from approximately 50 µm to several millimeters. It is present in organic residues from the Paleozoic of Laurentia, Baltica, and Laurasia and thought to have an affinity with hydrozoans (Cashman, 1996).

melanosome (me-lan'-o-some) The dark-colored part of a migmatite, rich in mafic minerals (Mehnert, 1968, p.355). Cf: *leucosome; mesosome; paleosome.* Var: *melasome.*

melanostibite (mel'-a-no-stib'-ite) A deep reddish-black platy rhombohedral mineral of the *corundum* group: $Mn^{2+}(Sb^{5+},Fe^{3+})O_3$.

melanotekite (mel'-a-no-tek'-ite) A black or dark gray-green orthorhombic mineral: $Pb_2Fe^{3+}{}_2O_2Si_2O_7$. It is the Fe^{3+} analogue of kentrolite.

melanothallite (mel'-an-o-thall'-ite) A bluish-black orthorhombic mineral: Cu_2OCl_2.

melanovanadite (mel'-a-no-van-a'-dite) A submetallic black triclinic mineral: $Ca(V^{5+},V^{4+})_4O_{10}{\cdot}5H_2O$. It is classed as a "vanadium bronze".

melanterite (me-lan'-ter-ite) (a) A colorless to green or greenish-blue monoclinic mineral: $Fe^{2+}SO_4{\cdot}7H_2O$. It usually results from the decomposition of iron sulfides. Syn: *copperas; green vitriol; iron vitriol.* (b) A group name for minerals with the analogous formula, but with Fe^{2+} replaced by Mn^{2+}, Co, Cu, or Zn.

melaphyre (mel'-a-phyre) A term originally applied to any dark-colored porphyritic igneous rock but later restricted to altered basalt, esp. of Carboniferous and Permian age. Obsolete.

melasome (mel'-a-some) *melanosome.*

melatope (mel'-a-tope) In an *interference figure,* a point indicating the crystal's optic axis.

mêlée (a) A collective term for small round faceted diamonds, such as those mounted in jewelry. The term is sometimes applied to colored stones of the same size and shape as the diamonds. (b) A small diamond cut from a fragment of a larger size. (c) In diamond classification, a term for small round cut diamonds weighing less than 0.2 carat. (d) Rough sorting term for regular, sawable stones weighing up to 1 ct. from which polished mêlée can be manufactured. French, meaning mixed. Also spelled melée, melee (Liddicoat et al., 1993). Cf: *mélange [gem].* Etymol: French, "mixed".

melikaria (mel-i-kar'-i-a) (a) Skeletal structures of quartz formed in place by deposition of silica from rising waters in the bottoms of deep shrinkage cracks in septaria or other concretions, the enclosing rock having been removed by solution (Burt, 1928). They resemble septarian veins in form, and may be as large as 45 x 20 x 10 cm (as in the Quaternary alluvial deposits of Brazos County, Tex.). Cf: *septarium.* (b) A term applied to the vein skeletons of septaria (Twenhofel, 1939, p.552). Etymol: Greek, "honeycombs".

melilite (mel'-i-lite) (a) A group of pale-colored tetragonal minerals of general formula: $(Na,Ca)_2(Mg,Al)(Si,Al)_2O_7$. It consists of an isomorphous solid-solution series, and may contain some iron. (b) A tetragonal, often honey-yellow mineral of the melilite group, such as the end members gehlenite and åkermanite. It occurs as a component of certain recent basic volcanic rocks. The melilites of volcanic rocks are usually classed as feldspathoids (even though they do not possess framework structures), but have also been considered as "undersaturated pyroxenes". Also spelled: *mellilite.*

melilite nephelinite In the *IUGS classification,* a volcanic rock falling in the foidite and tephrite + basanite fields in the TAS diagram, and containing more than 0 but less than 10 wt% calcium orthosilicate in the *CIPW norm.*

meliphanite (mel'-i-phan'-nite) A yellow or red tetragonal mineral: $(Ca,Na)_2Be(Si,Al)_2(O,OH,F)_7$. Syn: meliphane; *gugiaite.*

melissopalynology (me-lis'-so-pal-y-nol'-o-gy) A subset of *palynology* concerned with the study of pollen in honey.

melkovite (mel'-kov-ite) A yellow or brownish-yellow monoclinic mineral: $CaFe^{3+}{}_2Mo_5O_{10}(PO_4)(OH)_{12}{\cdot}8H_2O$.

mellilite (mel'-li-lite) (a) *melilite.* (b) *mellite.*

mellite (mel'-lite) A yellow, reddish, brownish, or white tetragonal mineral: $Al_2(COO)_6{\cdot}16H_2O$. It has a resinous luster, usually occurs as nodules in brown coal, and is in part a product of vegetable decomposition. It is a salt of mellitic acid (=benzene hexacarboxylic acid). Syn: *honey stone.*

mellorite (mel'-lor-ite) A name suggested for a poorly crystallized material of the *kaolin* group of clay minerals in which randomness of stacking of the layer packets in the *c*-axis direction is present. Because there is considerable range of disorder in the less well-crystallized kaolinites, there is no need for a specific mineral name such as "mellorite"; the term in general use is *fireclay mineral.*

melnikovite (mel'-ni-kov-ite') *greigite.*

melonite (mel'-o-nite) (a) A reddish-white trigonal mineral: $NiTe_2$. (b) a group name for trigonal minerals with the general formula AX_2, where A = Ni, Sn, Ir, Pd, or Pt, and X = Te, S, Se, or Bi.

mélonjosephite (mel-on-jo'-se-phite) A sub-adamantine dark-green orthorhombic mineral: $CaFe^{2+}Fe^{3+}(PO_4)_2(OH)$.

melt [gem] A crystal-growth technique in which the crystal solidifies from a cooling liquid of the same chemical composition. Used to produce gem materials such as synthetic ruby and sapphire.

melt [petrology] n. A liquid, fused rock.

melteigite (mel'-teig-ite) (a) In the *IUGS classification,* a plutonic rock in which F is between 60 and 100, M is between 70 and 90, and sodium exceeds potassium. Cf: *missourite.* (b) A dark-colored plutonic rock that is part of the *ijolite* series and contains nepheline and 60-90% mafic minerals, esp. green pyroxene. Cf: *turjaite; urtite; algarvite.* The name is from Melteig farm, Fen complex, Norway.

melt-fragment breccia *suevite.*

melt-freeze grains The amorphous form of *grain cluster* subjected to repeated melt-freeze cycles. Individual ice crystals are not distinguishable and internal water inclusions are frozen. Melt-freeze grains may be restored to grain clusters by absorption of solar radiation. They occur only in freely draining snow (Colbeck, 1986). Cf: *melt metamorphism; slush [snow].*

melting column A partly molten permeable rock column (most frequently peridotite) in which the melt (most frequently basalt) is flowing upward, e.g., the rising and decompressing mantle under mid-ocean ridges; the segregated basalt forms the oceanic crust (Klein and Langmuir, 1987).

melting spot (melt'-ing) A region in the mantle within which basalt magma is generated and whose vertical projection on the Earth's surface is an area within which eruptions occur repeatedly. Example: Hawaii. (Dalrymple et al., 1974, p.31). See also: *hot spot.*

melt-matrix breccia *impact-melt breccia.*

melt metamorphism The process of snow-grain growth and rounding under melting conditions, caused by an inverse relationship between grain radius and melting temperature. It usually results in an increase in density and in modification of other snow properties. Cf: *grain cluster; melt-freeze grains.*

melt-out till Till derived from slow melting of thick masses of debris-rich stagnant ice buried beneath sufficient overburden to inhibit deformation under gravity, thus preserving structures derived from the parent ice (Boulton, 1970).

meltwater (melt'-water) Water derived from the melting of snow or ice, esp. the stream flowing in, under, or from melting glacier ice.

member (mem'-ber) A formal lithostratigraphic unit next in rank below a formation, comprising some specially developed part of a *formation.* A named member may extend laterally from one formation into another. A member name combines a geographic name followed by the word "member"; where a lithologic designation is useful, it should be included (e.g. the Wedington Sandstone Member of the Fayetteville Shale). It is higher in rank than a bed. Abbrev: mbr. Cf: *lens; lentil; tongue [stratig].*

membranate chorate cyst (mem'-bra-nate) *membranate cyst.*

membranate cyst A dinoflagellate cyst with a prominent membrane, e.g., *Membranilarnacia.* Syn: *membranate chorate cyst.*

membrane (mem'-brane) The thinned, generally delicate and elastic exinous floor of a pore or colpus of a pollen grain; e.g. *harmomegathus.*

membranelle (mem-bra-nelle') A flattened, bladelike vibrating organ in a tintinnid, consisting of a row of fused cilia and fringed with lamellae, and used for locomotion.

membranimorph (mem-bran'-i-morph) adj. Pertaining to generally simple anascan cheilostomes (bryozoans), characterized by extensive frontal membranes, slightly to moderately developed gymnocysts with or without marginal spines, and small or no cryptocysts. n. An anascan cheilostome having such a structure.

membraniporiform (mem'-bra-ni-por'-i-form) Said of a generally unilaminate encrusting colony in cheilostome bryozoans, firmly attached by calcified or membranous walls of zooids (Lagaaij and Gautier, 1965, p.51).

membranous sac (mem'-bra-nous) In living stenolaemate bryozoans, the membrane surrounding the digestive and reproductive organs in retracted position.

menaccanite (me-nac'-can-ite) A variety of *ilmenite* found as a sand near Manaccan (Menachan) in Cornwall, England. Also spelled: *manaccanite.*

Menap (Me'-nap) North European glacial stage: Pleistocene (below Elster). Equivalent to the Günz of the Alpine sequence. Replaces *Elbe* in glacial chronology.

Menapian (Men-ap'-i-an) European cold-climate paleofloral stage: lower Pleistocene (above Waalian, below Cromerian). Equiv. Günz glacial system.

mendeleyevite (men-de-ley'-ev-ite) An unnecessary name for titanium- and rare-earth-bearing betafite.

mendip (men'-dip) (a) A buried hill that is exposed (by the cutting of a valley across a *cuesta*) as an inlier. (b) A coastal-plain hill that at one time was an offshore island. Type locality: Mendip Hills in Somerset, England.

mendipite (men'-dip-ite) An adamantine colorless, white, brownish cream, or gray orthorhombic mineral: $Pb_3O_2Cl_2$.

mendozavilite (men-do'-za-vil'-ite) A yellow or orange mineral of undetermined crystallography: $NaCa_2Fe_6^{3+}(OH)_{10}(PO_4)_2(PMo_{11}^{6+}O_{39})\cdot 33H_2O$.

mendozite (men-do'-zite) A white monoclinic mineral: $NaAl(SO_4)_2\cdot 11H_2O$. Cf: *soda alum.* It is the Na analogue of kalinite.

meneghinite (men-e-ghi'-nite) A blackish lead-gray orthorhombic mineral: $Pb_{13}CuSb_7S_{24}$.

Menevian (Me-ne'-vi-an) British stage: Middle Cambrian (above Solvan, below Maentwrogian) (Harland et al., 1990).

mengxianminite (meng-xian'-min-ite) A brownish-green monoclinic mineral: $(Ca,Na)_4(Mg,Fe,Zn)_5Sn_4Al_{16}O_{41}$. Not approved by the International Mineralogical Association.

menilite (men'-i-lite) An opaque, impure, dull-grayish or brown variety of opal found in rounded or flattened concretions at Menilmontant near Paris, France. Syn: *liver opal.*

meniscus (me-nis'-cus) A liquid surface curved by capillarity. Pl: menisci.

meniscus cement A particular type of cement, generally of calcite, precipitated in the meteoric vadose zone (above the watertable), occurring between grains and having a meniscus shape.

mensa A term established by the International Astronomical Union for a flat-topped prominence with cliff-like edges on planetary surfaces (USGS, 2003). Pl: mensae.

menshikovite A metallic hexagonal mineral: $Pd_3Ni_2As_3$.

Meotian (Me-o'-tian) Eastern European stage (Black Sea area): Upper Miocene (above Sarmatian, below Pontian). It has also been regarded as lowermost Pliocene, and is equivalent in age to Pannonian.

Meramecian (Mer-a-mec'-i-an) North American series: Upper Mississippian (above Osagean, below Chesterian).

Merapian Said of volcanic activity caused by the disintegration and collapse of a silicic dome or lava flow, producing pyroclastic flows.

meraspid (me-ras'-pid) Adj. of *meraspis.*

meraspis (me-ras'-pis) A juvenile trilobite that has a distinct *cephalon* and *pygidium* but does not yet have the number of thoracic segments typical of the species; a late trilobite larva in which the *thorax* progressively develops. Pl: meraspides.

Mercalli scale (Mer-cal'-li) An arbitrary scale of *earthquake intensity,* ranging from I (detectable only instrumentally) to XII (causing almost total destruction). It is named after Giuseppi Mercalli (d.1914), the Italian geologist who devised it in 1902. Its adaptation to North American conditions is known as the *modified Mercalli scale.*

mercallite (mer-cal'-lite) A colorless to sky-blue orthorhombic mineral: $KHSO_4$.

Mercator chart (Mer-ca'-tor) A chart or map drawn on the Mercator projection. It is commonly used for marine navigation.

Mercator equal-area projection *sinusoidal projection.*

Mercator projection An equatorial, cylindrical, conformal map projection derived by mathematical analysis (not geometrically) in which the equator is represented by a straight line normally true to scale, the meridians by parallel straight lines perpendicular to the equator and equally spaced according to their distance apart at the equator, and the parallels by straight lines perpendicular to the meridians and parallel with (and the same length as) the equator. True scale may also be placed along a symmetrical pair of other parallels. The parallels are spaced so as to achieve conformality, their spacing increasing rapidly with their distance from the equator so that at all places the degrees of latitude and longitude have the same ratio to each other as to the sphere itself, resulting in great distortion of distances, areas, and shapes in the polar regions (above 80° lat.), the scale increasing poleward as the secant of the latitude. Because any line of constant direction (azimuth) on the sphere is truly represented on the projection by a straight line, the Mercator projection is of great value in navigation and is used for hydrographic charts, and also to show geographic variations of some physical property (such as magnetic declination) or to plot trajectories of Earth satellites in oblique orbits. Named after Gerhardus Mercator (1512-1594), Flemish mathematician and geographer, whose world map of 1569 used this projection. See also: *transverse Mercator projection.*

Mercator track A *rhumb line* constructed on a Mercator projection.

mercury (mer'-cu-ry) A heavy, silver-white to tin-white mineral, the native metallic element Hg. It is the only metal that is liquid at ordinary temperatures. It solidifies to a rhombohedral solid at -38.9°C. Native mercury is found as minute fluid globules disseminated through cinnabar or deposited from the waters of certain hot springs, but it is unimportant as a source of the metal. It usually contains small amounts of silver. Mercury combines with many metals to form alloys or amalgams. Syn: *quicksilver.*

mercury barometer A type of *barometer* that measures air pressure and its changes by balancing the weight of the atmosphere on a unit area against the weight of a column of liquid mercury supported by the atmosphere. Cf: *aneroid barometer.*

mer de glace A French term meaning "sea of ice" and applied to any of the large glaciers or ice sheets of the Pleistocene Epoch. The term refers to Mer de Glace, the largest glacier on the Mont Blanc massif in the Alps.

mere [coast] An obsolete term for an estuary, creek, inlet, or other arm of the sea, and for the sea itself.

mere [lake] (a) A sheet of standing water, a lake, pond, or pool; now chiefly British regional usage. (b) A *marsh* or *fen.* (c) A *levee lake* behind a barrier consisting of sediment carried upstream by the tide.

mereheadite A pale yellow to reddish-orange monoclinic mineral: $Pb_2O(OH)Cl$.
mereiterite A colorless to light yellow monoclinic mineral: $K_2Fe^{2+}(SO_4)_2 \cdot 4H_2O$. It is the Fe^{2+} analogue of leonite.
merenskyite (me-ren'-sky-ite) A metallic white trigonal mineral of the *melonite* group: $(Pd,Pt)(Te,Bi)_2$. Syn: *biteplapallidite.*
Merensky Reef-type deposit *stratiform PGE deposit.*
mergifer (mer'-gi-fer) Said of a variant of the *radulifer* type of long brachiopod crura, very close together and parallel, arising directly from the swollen edge of a high dorsal *median septum.*
meridian (me-rid'-i-an) (a) An imaginary great circle on the surface of the Earth passing through the poles and perpendicular to the equator, connecting all points of equal *longitude;* a north-south line of constant longitude, or a plane, normal to the geoid or spheroid and passing through the Earth's axis, defining such a line. Also, a half of such a great circle included between the Earth's poles. Syn: *terrestrial meridian.* (b) Any one of a series of lines, corresponding to meridians, drawn on a globe, map, or chart at intervals due north and south and numbered according to the degrees of longitude east or west from the prime meridian. (c) *celestial meridian.* Cf: *parallel.*
meridian hole A term introduced by Agassiz (1866, p.293-294) for a shallow, crescent-shaped *dust well* that accurately registers on the surface of a glacier the position of the Sun during the day. In the northern hemisphere it has a steeper wall on its southern side than on its northern side.
meridian line A line running accurately north and south through any given point on or near the Earth's surface; specif. a line used in plane surveying and defined by the intersection of the plane of the celestial meridian and the plane of the horizon.
meridional (me-rid'-i-o-nal) (a) Pertaining to a movement or direction between the poles of an object, e.g. the Earth's north-south water or air circulation patterns, or the alignment of colpi on a pollen grain. (b) Southern. (c) n. In a north-south direction, along meridians, a line of constant longitude.
meridional convergence *grid convergence.*
meridional difference The difference (distance) between the meridional parts of any two given parallels of latitude. It is found by subtraction if the two parallels are on the same side of the equator, and by addition if they are on opposite sides. Cf: *latitude difference; departure.* Also called: *meridional difference of latitude.*
meridional difference of latitude *meridional difference.*
meridional part The linear length of the arc of a meridian between the equator and a given parallel of latitude on a Mercator chart, expressed in units of one minute of longitude at the equator.
meridional projection Former name for an azimuthal map projection centered on the equator. It is now usually called an equatorial projection or aspect.
merismite (mer'-is-mite) A *chorismite* in which there is irregular penetration of the diverse units (Dietrich and Mehnert, 1961). Rarely used.
meristele (mer'-i-stele) A strand of vascular tissue enclosed in a sheath of endodermis forming part of a *dictyostele* (Swartz, 1971, p.289).
meristem (mer'-i-stem) A plant tissue consisting of undifferentiated formative or generative cells that give rise to daughter cells capable of differentiation, as found in the *cambium* and other plant tissues and organs (Swartz, 1971, p.289).
merlinoite (mer-lin'-o-ite) A white orthorhombic zeolite mineral: $K_5Ca_2(Si_{23}Al_9)O_{64} \cdot 22H_2O$.
mero- (mer'-o-) A prefix signifying "part" or "portion".
merodont (mer'-o-dont) Said of a class of ostracode hinges having three elements and characterized by crenulate terminal elements with either a positive or a negative, crenulate or smooth, median element.
merohedral (mer-o-he'-dral) Said of crystal classes in a system, the general form of which has only one half, one fourth, or one eighth the number of equivalent faces of the corresponding form in the *holohedral* class of the same system. This condition is known as *merohedrism.* Cf: *tetartohedral; ogdohedral.* Syn: *merosymmetric.* See also: *hemihedral.*
merohedral twinning *twinning by merohedry.*
merohedrism (mer-o-he'-drism) The condition of being *merohedral.* Syn: *merohedry.*
merohedry (mer-o-he'-dry) *merohedrism.*
merokarst (mer'-o-karst) Imperfect karst topography as found on thin, impure, or chalky limestone, where surface drainage and dry valleys are present in addition to some karstic features. Cf: *holokarst.*
meroleims (mer'-o-leims) Coalified remains of plant debris (Krishtofovich, 1945, p.138). Cf: *hololeims.* See also: *phytoleims.*
merom (mer'-om) (a) In receptaculitids, an elongate calcified branch (lateral) originating on the central axis and consisting of shaft and head. The head has a four-ribbed stellate structure and a terminal plate. (b) In cyclocrinitids, similar structure but without the stellate structure and terminating with an often porous lid. Also spelled: merome.
meromictic lake (mer-o-mic'-tic) A lake that undergoes incomplete mixing of its waters during periods of circulation; specif. a lake in which the bottom, noncirculating water mass (*monimolimnion*) is adiabatically isolated from the upper, circulating layer (*mixolimnion*) (Wetzel, 2001, p.83). Cf: *holomictic lake.*
meromixis (mer-o-mix'-is) The process leading to, or the condition of, a *meromictic lake.*
meroplankton (mer-o-plank'-ton) Plankton organisms that represent life stages of otherwise nonplankton organisms; e.g., eggs and larvae of benthic and nektonic species. Cf: *holoplankton.* Syn: *temporary plankton.*
meropod (mer'-o-pod) *merus.*
merostome (mer'-o-stome) Any aquatic *arthropod* belonging to the class Merostomata, characterized by the presence of one pair of preoral appendages with three, possibly four, joints. Cf: *eurypterid; arachnid.*
merostratic unit Any stratic unit based on a partial range (Henningsmoen, 1961, p.67).
merosymmetric (mer'-o-sym-met'-ric) *merohedral.*
merosyncline (mer-o-syn'-cline) Bubnoff's term for that part of a geosynclinal belt having independent mobility (Glaessner and Teichert, 1947, p.588). Obsolete.
meroxene (mer'-o-xene) A variety of biotite with its axial plane parallel to the crystallographic *b*-axis.
merozone A *biostratic unit* representing the partial range of a taxon (Henningsmoen, 1961, p.69).
merozonite A biostratic unit representing part of a range-zone; equivalent to partial-range zone (Henningsmoen, 1961, p.68).
Merriam effect (Mer'-ri-am) The relationship between mountain mass and the vertical distribution of animals and plants. The term was designated by Lowe (1961, p.45-46) for the influence of factors such as the elevation of a mountain, its size or mass, and the elevation of the basin or plain from which it rises, on the vertical placement of species and communities of plants and animals. Named after Clinton Hart Merriam (1855-1942), U.S. biologist, who first recognized the relationship (Merriam, 1890).
merrihueite (mer-ri-hue'-ite) A greenish-blue hexagonal meleorite mineral of the *milarite* group: $(K,Na)_2(Fe^{2+},Mg)_5Si_{12}O_{30}$.
merrillite (mer'-ril-lite) A colorless rhombohedral meteorite mineral of the *whitlockite* group: $Ca_9Na(Mg,Fe)(PO_4)_7$.
Mersey yellow coal (Mer'-sey) *tasmanite [coal].*
mertieite (mer'-tie-ite) A metallic brassy-yellow mineral. Two crystallographic and compositional variants have been reported; the hexagonal and rhombohedral variants have the composition $Pb_{11}As_2Sb_2$ and $Pd_8(Sb,As)_3$, respectively.
merumite (mer'-u-mite) A mixture of eskolaite and other minerals.
merus (me'-rus) The fourth pereiopodal segment from the body of a malacostracan crustacean, bounded proximally by the *ischium* and distally by the *carpus.* Pl: meruses. Syn: *meropod.*
merwinite (mer'-win-ite) A colorless to pale green monoclinic mineral: $Ca_3Mg(SiO_4)_2$.
merzlota (mer-zlo'-ta) A Russian term for *frozen ground.*
mesa (me'-sa) (a) An isolated, nearly level landmass standing distinctly above the surrounding country, bounded by abrupt or steeply sloping erosion scarps on all sides, and capped by layers of resistant, nearly horizontal rock (often lava). Less strictly, a very broad, flat-topped, usually isolated hill or mountain of moderate height bounded on at least one side by a steep cliff or slope and representing an erosion remnant. A mesa is similar to, but has a more extensive summit area than, a *butte,* and is a common topographic feature in the arid and semiarid regions of the U.S. See also: *table mountain.* (b) A broad terrace or comparatively flat plateau along a river valley, marked by an abrupt slope or escarpment on one side. See also: *bench [geomorph].* Etymol: Spanish, "table".
mesabite (me-sa'-bite) An ocherous variety of goethite from the Mesabi Range in Minnesota.
mesa-butte A *butte* formed by the erosion and reduction of a mesa. Cf: *volcanic butte.*
mesa plain The flat summit of a hilly mountain or plateau (Hill, 1900, p.6). Cf: *plateau plain.*
mesarch (mes'-arch) adj. Said of an ecologic succession (i.e. a sere) that develops under *mesic* conditions. Cf: *hydrarch; xerarch.* See also: *mesosere.*
mesa-terrace (me'-sa-ter'-race) An obsolete term used by Lee (1900, p.504-505) for an alluviated, planate rock surface contained within a valley, lying between the flood plain of a nearby stream and the steeper slope leading up to a mesa.
mesenchyme (mes'-en-chyme) (a) The *mesohyle* of a sponge. (b) A term used by zoologists for the fleshy connective tissue in coelenterates, but applied by paleontologists to the stony skeletal structures between corallites secreted by the common fleshy connective tissue (Shrock and Twenhofel, 1953, p.133). Cf: *sclerenchyme.* Also spelled: mesenchyma.
mesentery (mes'-en-ter'-y) One of several radially disposed fleshy laminae or sheets of soft tissue that are attached to the inner surface of the oral disk and column wall of a coral *polyp,* and that partition the internal body cavity by extending inward from the body wall. Adj: mesenterial.
meseta (me-se'-ta) (a) A small *mesa.* (b) An extensive plateau or flat upland, often with an uneven or eroded surface, forming the central physical feature of a region; e.g. the high, dissected tableland of the interior of Spain. Etymol: Spanish, "tableland". Syn: *mesita; mesilla.*
mesh [part size] One of the openings or spaces between the wires of a sieve or screen. See also: *mesh number.*

mesh [pat grd] The unit component of patterned ground (excepting steps and stripes), as a circle, a polygon, or an intermediate form (Washburn, 1956, p.825).

mesh number The size of a sieve or screen, or of the material passed by a sieve or screen, in terms of the number of meshes per linear inch; e.g. mesh number 20 indicates that the sieve or screen has 20 holes per linear inch (this takes no account of the diameter of the wire, so that the mesh number does not have a simple relation to the size of the hole).

mesh texture A rock texture that is reticulate.

mesic (me′-sic) Said of a habitat receiving a moderate amount of moisture; also, said of an organism or group of organisms occupying such a habitat. Cf: *hydric; xeric.* See also: *mesarch.*

mesic temperature regime A soil temperature regime in which the mean annual temperature (measured at 50cm) is at least 8°C but less than 15°C, with a summer-winter variation of more than 5°C (USDA, 1975). Cf: *isomesic temperature regime.*

mesilla *meseta.*

mesistele (mes′-i-stele) The intermediate part of a crinoid column, between *proxistele* and *dististele;* it is doubtfully distinguishable in pluricolumnals. Also, the medial part of the stele in certain homalozoans.

mesita *meseta.*

mesitine *mesitite.*

mesitine spar *mesitite.*

mesitis (me-si′-tis) Obsolete term for a transformation tending to promote homogenization between chemically different rocks under the same temperature and pressure (Dietrich and Mehnert, 1961, p.61).

mesitite (mes′-i-tite) A white variety of magnesite containing 30-50% iron carbonate. Syn: *mesitine; mesitine spar.*

meso- (mes′-o-) A prefix meaning "middle".

Mesoarchean (Mes-o-ar′-che-an) An era of the Standard Global Geochronometric Scale (third-oldest era of the *Archean* Eon). Defined as the time span between 3,200 Ma, the end of the *Paleoarchean* Era, and 2,800 Ma, the beginning of the *Neoarchean* Era (Plumb, 1991).

mesobreccia (mes-o-brec′-cia) A *caldera collapse breccia* in which numerous small clasts are visible within a single outcrop. It is commonly found as thin tabular layers in the middle and upper parts of caldera fills (Lipman, 1984). Cf: *megabreccia (c).*

mesoclade (mes′-o-clade) One of the median clades or skeletal branches that connect the bifurcating parts of the actines in an ebridian skeleton.

mesoclimate (mes′-o-cli′-mate) The climate of a small area, for example a valley or a densely forested area, that may differ from the general climate of the region. See also: *macroclimate; microclimate.*

mesocolpium (mes-o-col′-pi-um) The meridional segment between two colpi of a pollen grain, extending to the pole and bordered by the lines demarking the *apocolpia.* Syn: *intercolpium.*

mesoconch (mes′-o-conch) The part of a *dissoconch* of a bivalve mollusk formed at an intermediate stage of growth and separated from earlier- and later-formed parts by pronounced discontinuities.

mesocratic (mes-o-crat′-ic) Composed of almost equal amounts of light and dark constituents; applied to igneous rocks intermediate in color between *leucocratic* and *melanocratic.* The percentage of mafic minerals required for a rock to be classified as mesocratic varies among petrologists; the lower limit ranges from 30 to 37%, the upper limit from 60 to 67%.

mesocrystalline (mes-o-crys′-tal-line) Said of the texture of a rock consisting of or having crystals whose diameters are intermediate between those of a microcrystalline and a macrocrystalline rock; also, said of a rock with such a texture. Howell (1922) applied the term to the texture of a recrystallized sedimentary rock having crystals whose diameters are in the range of 0.20-0.75 mm, and Bissell and Chilingar (1967, p.103) to the texture of a carbonate sedimentary rock having crystals whose diameters are in the range of 0.05-1.0 mm.

mesocumulate (mes-o-cu′-mu-late) A cumulate containing a small amount of intercumulus material; a cumulate intermediate between an *orthocumulate* and an *adcumulate.*

mesocumulate texture An igneous texture consisting of early-formed, accumulated crystals and relatively small amounts of interstitial minerals formed later by crystallization of magma trapped between the early crystals. Cf: *adcumulate texture; orthocumulate texture.*

mesoderm (mes′-o-derm) The middle primary germ layer of a triploblastic embryo, which in the adult gives rise to muscular, vascular, and connective tissues and lines the coelomic cavity.

mesofossil (mes′-o-fos-sil) A term defined by Hughes (1994) as a plant fossil of intermediate size. Ranging between megafossils (e.g., leaves) and spores or pollen, it still requires microscopic study, such as megaspores and small seeds.

Mesogea (Mes-o-ge′-a) A name, used mainly in France, for the sea usually called *Tethys.* Also spelled: Mesogaea.

mesogene (mes′-o-gene) Said of a mineral deposit or enrichment of mingled *hypogene* and *supergene* solutions; also, said of such solutions and environment.

mesogenetic (mes′-o-ge-net′-ic) A term for the period between the time when newly buried deposits are affected mainly by processes related to the depositional interface (*eogenetic* stage) and the time when long-buried deposits are affected by processes related to the erosional interface (*telogenetic* stage). Also applied to the porosity that develops during the mesogenetic stage (Choquette and Pray, 1970, p.220).

mesogeosyncline (mes′-o-ge′-o-syn′-cline) A sedimentary basin, on oceanic crust between two convergent continents, which receives clastics from both of them. Obsolete.

mesogloea (mes-o-gloe′-a) A gelatinous substance between *endoderm* and *ectoderm* of certain invertebrates; e.g. an extracellular gel, containing proteins and carbohydrates, found in the *mesohyle* of many sponges, or a noncellular jellylike middle layer of the outer walls and mesenteries of coral polyps. Also spelled: mesoglea.

mesograined (mes′-o-grained) The texture of a carbonate sedimentary rock having clastic particles whose diameters are in the range of 0.05-1.0 mm (Bissell and Chilingar, 1967, p.103) or 0.1-1.0 mm (DeFord, 1946).

mesogyrate (mes-o-gy′-rate) Said of the *umbones* (of a bivalve mollusk) curving toward the center. Cf: *orthogyrate.*

mesohyle (mes′-o-hyle) Loosely organized material constituting a sponge between the *pinacoderm* and the *choanoderm,* commonly consisting of spongin, spicules, and various types of cells (mainly amoebocytes), embedded in *mesogloea,* although one or more of these elements may be missing. Syn: *mesenchyme;* parenchyma.

mesokurtic (mes-o-kur′-tic) Closely resembling a normal frequency distribution; a distribution curve that is neither *leptokurtic* (very peaked) nor *platykurtic* (flat across the top).

mesolimnion (mes-o-lim′-ni-on) *metalimnion.*

mesolite (mes′-o-lite) An orthorhombic zeolite mineral: $Na_2Ca_2Al_6Si_9O_{30}{\cdot}8H_2O$. It is intermediate in chemical composition between natrolite and scolecite, and is usually found in white or colorless tufts of very delicate acicular crystals in amygdaloidal basalts. Syn: *cotton stone.*

Mesolithic (Mes-o-lith′-ic) In archaeology, the middle division of the *Stone Age,* characterized by the change from glacial to postglacial climate and the absence of agriculture. Correlation of relative cultural levels with actual age (and, therefore, with the time-stratigraphic units of geology) varies from region to region. Cf: *Paleolithic; Neolithic.* Syn: *middle Stone Age.*

mesolithion (mes-o-lith′-i-on) A grouping of animals that live in cavities in rock.

mesomicrocline (mes-o-mi′-cro-cline) A pseudomonoclinic mineral of the alkali *feldspar* group: $KAlSi_3O_8$. It is intermediate in degree of ordering between microcline and orthoclase.

mesonorm (mes′-o-norm) Theoretical calculation of normative minerals in metamorphic rocks of the *mesozone* from chemical analyses (Barth, 1959). Cf: *catanorm; epinorm.* See also: *Niggli molecular norm.*

mesopelagic (mes′-o-pe-lag′-ic) Pertaining to the pelagic environment of the ocean between 200 and 1000 meters. Cf: *epipelagic.*

mesopeltidium (mes′-o-pel-tid′-i-um) A *sclerite* (commonly one of a pair) of the segmented carapace of an arachnid, situated immediately behind the *propeltidium* and in front of the *metapeltidium.*

mesoperthite (mes-o-perth′-ite) A variety of perthitic feldspar consisting of an intimate mixture of about equal amounts of potassium feldspar and plagioclase (usually albite, sometimes oligoclase). It is intermediate in composition between perthite and antiperthite. Syn: *eutectoperthite.*

mesophilic (mes-o-phil′-ic) Said of an organism that prefers a moderate environment; e.g. mesothermal conditions. n. mesophile.

mesophragm (mes′-o-phragm) The name of the wall layer between the *endophragm* and *periphragm* in certain dinoflagellate cysts with three wall layers.

mesophyll (mes′-o-phyll) The chlorophyllous tissues in the interior of a leaf.

mesophyte (mes′-o-phyte) A plant that cannot survive extreme conditions of temperature or water supply. Cf: *hydrophyte; xerophyte.* Adj: mesophytic.

Mesophytic (Mes-o-phyt′-ic) A paleobotanic division of geologic time, signifying the time between the first occurrence of gymnosperms and that of angiosperms. Cf: *Aphytic; Archeophytic; Eophytic; Paleophytic; Cenophytic.*

mesophytic flora The collection of seed plants that dominated terrestrial ecosystems from the mid-Permian until the expansion of the angiosperms in the Cretaceous, and including cycads, conifers, ginkgoes and a variety of other early seed plants.

Mesophyticum (Mes-o-phyt′-i-cum) A paleobotanic division of geologic time, corresponding approximately to, and characterized by the plant life of, the Mesozoic. Cf: *Palaeophyticum; Cainophyticum.*

mesoplankton (mes-o-plank′-ton) (a) Plankton organisms of the size range 0.2-20 mm; a type of *microplankton.* (b) Plankton that live at middle depths. The term is rarely used because it is confusing.

mesopore [paleont] (mes′-o-pore) *mesozooecium.*

mesopore [petrology] In the pore-size classification of Choquette and Pray (1970, p.233), an equant to equant-elongate pore or a tubular or platy pore with an average diameter or thickness between 4 and 1/16 mm. Cf: *megapore; micropore.*

mesoporium (mes-o-po′-rium) The area between two adjacent *pores* of a *pollen grain,* bordered by tangential lines connecting the upper and lower margins of both pores. Pl: mesoporia.

Mesoproterozoic (Mes-o-prot-er-o-zo′-ic) An era of the Standard Global Geochronometric Scale: the middle era of the *Proterozoic* Eon. Defined as the time span between 1,600 Ma, the end of the *Paleoproterozoic* Era, and 1,000 Ma, the beginning of the *Neoproterozoic* Era (Plumb, 1991).

mesopsammon (mes-o-psam'-mon) A grouping of animals that live in cavities in sand.

Mesosauria (Mes-o-saur'-i-a) An order of presumably anapsid reptiles, aquatic fish catchers, known only from black shales of Early Permian age in South Africa and Brazil.

mesoscale eddy (mes'-o-scale) The most common geostrophic eddy found in the ocean, having a diameter of 40-200 km.

mesoscopic (mes-o-scop'-ic) According to Dennis (1967, p.152), a term introduced to describe a tectonic feature large enough to be observed without the aid of a microscope yet small enough that it can still be observed directly in its entirety. Cf: *macroscopic.*

mesosere (mes'-o-sere) A *sere* that develops in an environment having a moderate amount of moisture, i.e. in a mesic environment; a *mesarch* sere. Cf: *hydrosere; xerosere.*

mesosiderite (mes-o-sid'-er-ite) A *stony-iron meteorite* in which the silicates are mainly pyroxene (usually orthopyroxene) and calcic plagioclase. The silicates are chemically and isotopically similar to those in HED meteorites, and they appear to have originated on similar asteroids (or the same asteroid). Mesosiderites often appear to be breccias made up of small fragments, cemented together by a fine nickel-iron matrix. Olivine is sometimes present.

mesosoma (mes-o-so'-ma) The middle region of the body of some invertebrates, esp. when this cannot be readily analyzed into its primitive segmentation (as in arachnids and most mollusks); specif. the anterior part of a merostome *opisthosoma* carrying appendages. Cf: *metasoma.*

mesosome "A body of that part of a migmatite complex that is not neosome; generally having the appearance of an ordinary metamorphic rock" (Henkes and Johannes, 1981; Ashworth, 1985). A mesosome may not be identical to the parent rock of the migmatite. Cf: *leucosome; melanosome; paleosome.*

mesosphere (mes'-o-sphere) The region of the mantle beneath the asthenosphere, having a greater rigidity.

mesosphere [meteorol] The atmospheric layer above the *stratopshere;* average altitudes of 50 to 80 km (31 to 50 mi).

mesostasis (me-sos'-ta-sis) Groundmass in an igneous rock. Syn: *groundmass [ign]; matrix [ign].*

mesotheca (mes-o-the'-ca) *median lamina.*

mesothem A time-significant stratigraphic unit forming major cycles of deposition which are normally bounded above and below by unconformities on shelf areas, but which have their limits defined by marker points at the bases of chronozones in continuously deposited sediments in basins (Ramsbottom, 1977, p.282; 1978, p.307).

mesotherm (mes'-o-therm) A plant that requires moderate temperatures for successful growth. Cf: *microtherm; megatherm.*

mesothermal [clim] (mes-o-ther'-mal) Pertaining to a climate characterized by moderate temperature. Cf: *megathermal; microthermal.*

mesothermal [eco geol] Said of a hydrothermal mineral deposit formed at considerable depth and in the temperature range of 200°-300°C (Park and MacDiarmid, 1970, p.317). Also, said of that environment. Cf: *hypothermal; epithermal; leptothermal; telethermal; xenothermal.*

mesothermal [ecol] Said of an organism that prefers moderate temperatures, i.e. in the 25-37°C range.

mesothermal gold-quartz vein *low-sulfide gold-quartz vein.*

mesothyridid (mes-o-thy'-ri-did) Said of a brachiopod *pedicle opening* when the *foramen* is located partly in the ventral *umbo* and partly in the *delthyrium,* with the beak ridges appearing to bisect the foramen (TIP, 1965, pt. H, p.148).

mesotidal (mes'-o-tid'-al) Said of a coast with a tidal range of 2 to 4 meters (Davies, 1963). Cf: *macrotidal.*

mesotil (mes'-o-til) A semiplastic or semifriable derivative of chemically weathered till, developed beneath a partially drained area, and intermediate in texture between *gumbotil* and *silttil* (Leighton and MacClintock, 1930, p.42-43).

mesotrophic lake (mes-o-troph'-ic) A lake that is characterized by a moderate supply of nutrient matter, neither notably high nor low in its total production; it is intermediate between a *eutrophic lake* and an *oligotrophic lake.*

mesotrophic peat Peat containing a moderate amount of plant nutrients. Cf: *oligotrophic peat; eutrophic peat.*

mesotrophy (me-sot'-ro-phy) The quality or state of a *mesotrophic lake.*

mesotype (mes'-o-type) (a) A group of zeolite minerals, including natrolite, mesolite, and scolecite. Syn: *needle zeolite.* (b) A term used, mainly in France, in the restricted meaning of *natrolite,* because its form is intermediate between those of stilbite and analcime.

Mesozoic (Mes-o-zo'-ic) The middle erathem of the *Phanerozoic* Eonothem of the Standard Global Chronostratigraphic Scale, above the *Paleozoic* and below the *Cenozoic.* Also the time during which these rocks were formed, the Mesozoic Era, covering the time span between 250 and 65 Ma.

mesozonal pluton An intrusive body that formed at intermediate depth (6.5-14 km); typically with sharp contacts and internal flow structure; associated with low- to medium-grade metamorphic rocks. Cf: *epizonal pluton; catazonal pluton.*

mesozone (mes'-o-zone) According to Grubenmann's classification of metamorphic rocks (1904), the intermediate depth zone of metamorphism, which is characterized by temperatures of 300°-500°C, moderate hydrostatic pressure and shearing stress. Chemical and regional metamorphism predominate; association of some epizone and katazone minerals is characteristic. The concept was modified by Grubenmann and Niggli (1924) to include effects of intermediate-temperature contact metamorphism. Modern usage stresses temperature-pressure conditions (medium to high metamorphic grade) rather than the likely depth of zone. Cf: *katazone; epizone.*

mesozooecium (mes'-o-zo-oe'-ci-um) In some Paleozoic bryozoans, a space-filling polymorph in the *exozone* between feeding zooids. It is closely tabulated out to the distal end, so that little or no room is available for functional organs. Pl: mesozooecia. Syn: *mesopore [paleont].*

messelite (mes'-sel-ite) A colorless, pale green, white or greenish-gray triclinic mineral of the *fairfieldite* group: $Ca_2(Fe^{2+},Mn^{2+})(PO_4)_2{\cdot}2H_2O$. Syn: *neomesselite.*

messenger (mes'-sen-ger) A sliding metal weight on the cable of an oceanographic vessel, which activates an oceanographic device at depth.

Messinian (Mes-sin'-i-an) A stage of the Standard Global Chronostratigraphic Scale: uppermost Miocene (above Tortonian, below the Zanclean Stage of the Pliocene). The basal GSSP is near Rabat, Morocco (Hilgen et al., 2000).

meta- (met'-a-) A prefix that, when used with the name of a sedimentary or igneous rock, indicates that the rock has been metamorphosed, e.g. metabasalt, metaquartzite.

meta-aluminite (met-a-a-lu'-mi-nite) A white monoclinic mineral: $Al_2(OH)_4(SO_4){\cdot}5H_2O$. It is the lower hydrate of aluminite.

meta-alunogen (met-a-a-lu'-no-gen) A colorless orthorhombic mineral: $Al_2(H_2O)_{12}(SO_4)_3{\cdot}2H_2O$. It is the lower hydrate of alunogen.

meta-ankoleite (met-a-an-ko'-le-ite) A yellow tetragonal secondary mineral of the *natroautunite* group: $K(UO_2)(PO_4){\cdot}3H_2O$.

meta-anthracite (met-a-an'-thra-cite) Coal having a fixed-carbon content of 98% or more; the highest rank of anthracite. Cf: *graphocite.* Syn: *superanthracite; subgraphite.*

meta-argillite (met-a-ar'-gil-lite) An *argillite* that has been metamorphosed.

meta-arkose (met-a-ar'-kose) Arkose that has been "welded" or recrystallized by metamorphism so that it resembles a granite or a granitized sediment (Pettijohn, 1957, p.325). Cf: *recomposed granite.*

meta-autunite (met-a-au'-tun-ite) (a) A yellow secondary tetragonal mineral: $Ca(UO_2)_2(PO_4)_2{\cdot}2\text{-}6H_2O$. It is apparently not formed directly in nature, but most field and museum specimens of autunite have been partly dehydrated to this phase. Cf: *para-autunite.* (b) A group name for tetragonal minerals of analogous composition, but with Ca replaced by Mg, Ba, Mn, Fe^{2+}, Co, Cu, Zn or Pb, and PO_4 replaced by AsO_4.

metabasite (met-a-ba'-site) A collective term, first used by Finnish geologists, for metamorphosed mafic rock.

metabentonite (met-a-ben'-ton-ite) Metamorphosed, altered, or somewhat indurated bentonite, characterized by clay minerals (esp. illite) that no longer have the property of absorbing or adsorbing large quantities of water; nonswelling bentonite, or bentonite that swells no more than do ordinary clays. The term has been applied to certain Ordovician clays of the Appalachian region and upper Mississippi River valley. See also: *potassium bentonite.*

metaberyllite (me-ta-ber'-yll-ite) An incompletely investigated mineral: $Be_3SiO_5{\cdot}2H_2O$.

metabituminous coal (met'-a-bi-tu'-mi-nous) Coal that contains 89-91.2% carbon, analyzed on a dry, ash-free basis. Cf: *semibituminous coal.*

metablastesis (met-a-blas'-te-sis) (a) Recrystallization and growth of a preferred mineral or group of minerals. (b) Essentially isochemical recrystallization without evidence of a separate mobile phase. (Dietrich and Mehnert, 1961, p.61). Obsolete.

metaboghead coal (met-a-bog'-head) High-rank torbanite.

metabolite [ecol] (me-tab'-o-lite) An excretion or external secretion (e.g., an enzyme, hormone, or vitamin) of an organism that affects the associated organisms by inhibiting their activities or even killing them.

metaboly (me-tab'-o-ly) The capability of an organism to change its shape.

metaborite (met-a-bo'-rite) A white to brownish mineral: HBO_2. It is the cubic modification of metaboric acid.

metacalciouranoite (met-a-cal'-ci-o-u-ran'-o-ite) An orange incompletely investigated mineral: $(Ca,Na,Ba)U_2O_7{\cdot}2H_2O$. Syn: *metacaltsuranoite.*

metacaltsuranoite (met'-a-calts'-u-ran'-o-ite) *metacalciouranoite.*

metacannel coal (met-a-can'-nel) Cannel coal of high metamorphic rank. Cf: *subcannel coal.*

metacarbonatite (met'-a-car-bon'-a-tite) Rock formed from metamorphism of a carbonatite.

metacinnabar (met-a-cin'-na-bar) A black cubic mineral of the sphaelerite group: HgS. It is dimorphous with cinnabar and represents an ore of mercury. Syn: *saukovite.*

metaclase (met'-a-clase) Obsolete term used by Leith (1905, p.12) for a rock possessing *secondary cleavage,* or *cleavage [struc geol]* in its modern meaning. Cf: *protoclase.*

metacolloid (met-a-col'-loid) An originally colloidal substance that has become crystalline, e.g. serpophite.

metacommunity (met-a-com-mu′-ni-ty) The set of all taxa in a regional association of communities in which, however, not all taxa may exist in one place at a given moment; or a community of many discrete patches.
metacryst (met′-a-cryst) Any large crystal developed in a metamorphic rock by recrystallization, such as garnet or staurolite in mica schist; a syn. of *porphyroblast.* Rarely used. Syn: *metacrystal.*
metacrystal (met-a-crys′-tal) *metacryst.*
metadata Data about data. Metadata provide information on the identification, sources, accuracy, spatial reference system, attributes, and other characteristics of a data set.
metadelrioite (met-a-del-ri′-o-ite) A pale yellow-green triclinic mineral: $CaSrV^{5+}{}_2O_6(OH)_2$. It is a dimorph of delrioite.
metadiagenesis (met′-a-di′-a-gen′-e-sis) *epigenesis [sed].*
metagenesis [evol] (met-a-gen′-e-sis) *alternation of generations.*
metagenesis [sed] A term applied by Russian geologists to *epigenesis* (changes occurring in a more or less compact sedimentary rock) or to late epigenesis.
metagenic (met-a-gen′-ic) Said of a sediment or sedimentary rock formed through diagenetic alteration of other sediments (Grabau, 1920, p.2). Rarely used.
metahaiweeite (met-a-hai′-wee-ite) A pale yellow to greenish-yellow secondary mineral: $Ca(UO_2)_2[Si_6O_{15}]{\cdot}nH_2O$, where *n* is less than 5. It is a lower hydrate of haiweeite.
metahalloysite (met′-a-hal-loy′-site) A name used in Europe for the less hydrous form of halloysite. It is synonymous with *halloysite* of U.S. authors. The term has also been used to designate the nonhydrated form of halloysite.
metaharmosis (met-a-har′-mo-sis) Var. of *metharmosis.*
metaheinrichite (met-a-hein′-rich-ite) A yellow to green tetragonal secondary mineral: $Ba(UO_2)_2(AsO_4)_2{\cdot}8H_2O$. Cf: *heinrichite.* It is a lower hydrate of heinrichite.
metahewettite (met-a-hew′-ett-ite) A red monoclinic mineral: $CaV^{5+}{}_6O_{16}{\cdot}3H_2O$. It resembles hewettite but differs slightly from it in its behavior during hydration; it is found in highly oxidized ore as coatings and fracture fillings. It is classed as a "vanadium bronze".
metahohmannite (met-a-hoh′-mann-ite) An orange mineral: $Fe^{3+}{}_2(SO_4)_2O{\cdot}4H_2O$. It is a lower hydrate hohmannite.
metajennite (met-a-jen′-nite) A synthetic product: $Na_2Ca_8Si_5O_{19}{\cdot}7H_2O$.
metakahlerite (met-a-kahl′-er-ite) A yellow to yellowish-green tetragonal secondary mineral: $Fe^{2+}(UO_2)_2(AsO_4)_2{\cdot}8H_2O$. It is a lower hydrate of kahlerite.
metakaolinite (met′-a-ka-o′-lin-ite) An intermediate product obtained when kaolinite is heated between about 500°C and 850°C; artificially dehydrated kaolinite. Syn: metakaolin.
metakirchheimerite (met-a-kirch′-heim-er-ite) A pale rose tetragonal mineral: $Co(UO_2)_2(AsO_4)_2{\cdot}8H_2O$.
metaköttigite (met-a-köt′-tig-ite) A bluish-gray triclinic mineral of the *symplesite* group: $(Zn,Fe^{3+},Fe^{2+})_3(AsO_4)_2{\cdot}8(H_2O,OH)$. It is the dimorph of köttigite.
metal (met′-al) (a) Any of a class of chemical elements, such as iron, gold, and aluminum, that have a characteristic luster, are good conductors of heat and electricity, and are opaque, fusible, and generally malleable or ductile. (b) *road metal.*
metal factor A derived parameter used to represent induced polarization anomalies. Abbrev: MF.
metalignitous coal (met-a-lig′-ni-tous) Coal that contains 80-84% carbon, analyzed on a dry, ash-free basis. Not listed by ASTM as a rank classification. Cf: *subbituminous coal.*
metalimnion (met-a-lim′-ni-on) The horizontal layer of a thermally stratified lake in which the temperature decreases rapidly with depth. The metalimnion lies between the *epilimnion* and the *hypolimnion,* and includes the *thermocline.* Syn: *mesolimnion.* Less preferred syn: *thermocline [lake] (b).*
metallic (me-tal′-lic) (a) Pertaining to a metal. (b) Said of a type of luster that is characteristic of metals. Cf: *nonmetallic; submetallic luster.*
metalliferous (me-tal-lif′-er-ous) Metal-bearing; specif., pertaining to a mineral deposit from which a metal or metals can be extracted by metallurgical processes. Also used as a modifier for brines rich in base metals.
metallization (met′-al-li-za′-tion) The process or processes by which metals are introduced into a rock, resulting in an economically valuable deposit; the mineralization of metals.
metallized hood (met′-al-lized) The upper shell or roof of a batholith, which is the first area to solidify after intrusion and may contain virtually all the hydrothermal metalliferous lodes of the intrusion (Emmons, 1933). Syn: *cupola [eco geol].*
metallogenetic (me-tal′-lo-ge-net′-ic) *metallogenic.*
metallogenic (me′-tal-lo-gen′-ic) Adj. of *metallogeny.* Syn: *metallogenetic; minerogenic; minerogenetic.*
metallogenic element An element that occurs as a native element or that occurs in sulfides, selenides, tellurides, arsenides, antimonides, or sulfosalts. It is one of H.S. Washington's bipartate groupings of elements of the lithosphere, now obsolete. Cf: *petrogenic element.*
metallogenic epoch A unit of geologic time favorable for the deposition of ores, or characterized by a particular assemblage of mineral deposits. Several metallogenic epochs may be represented within a single area, or *metallogenic province.*
metallogenic map A map, usually on a regional scale, on which is shown the distribution of particular assemblages, belts, or provinces of mineral deposits and their relationship to such geologic features as tectonic trends and petrographic types.
metallogenic province An area characterized by a particular assemblage of mineral deposits, or by one or more characteristic types of mineralization. A metallogenic province may have had more than one episode of mineralization, or *metallogenic epoch.* Syn: *metallographic province.*
metallogeny (me-tal-log′-e-ny) The study of the genesis of mineral deposits, with emphasis on their relationship in space and time to regional petrographic and tectonic features of the Earth's crust. The term has been used for both metallic and nonmetallic mineral deposits. Adj: *metallogenic.*
metallographic province (me′-tal-lo-graph′-ic) A little-used syn. of *metallogenic province.*
metallo-organic (me-tal′-lo-or-gan′-ic) Said of a compound in which an atom of a metal is bound to an organic compound through an atom other than carbon, such as oxygen, nitrogen, or sulfur, to form a coordination compound. Cf: *organometallic.*
metallotect (me-tal′-lo-tect) A term used in metallogenic studies for any geologic feature (tectonic, lithologic, geochemical, etc.) considered to have influenced the concentration of elements to form mineral deposits; an *ore control,* but without the implication of economic value.
metallotectonic (me-tal′-lo-tec-ton′-ic) Said of concentrations of ore, or of metal anomalies, in specific tectonic environments or regional structural contexts.
metallurgy (met′-al-lur′-gy, me-tal′-lur-gy) The science and technology of metals and alloys. Includes *extractive metallurgy,* the science and art of separating metals and metallic minerals from their ores by mechanical and chemical processes; and *physical metallurgy.*
meta-lodèvite (met-a-lo′-de-vite) A pale yellow to olive tetragonal mineral of the *meta-autunite* group: $Zn(UO_2)_2(AsO_4)_2{\cdot}10H_2O$.
metaluminous (met-a-lu′-mi-nous) Said of an igneous rock in which the molecular proportion of aluminum oxide is greater than that of sodium and potassium oxides combined but generally less than of sodium, potassium, and calcium oxides combined; one of Shand's (1947) groups of igneous rocks, classified on the basis of the degree of aluminum-oxide saturation. Cf: *peralkaline; peraluminous; subaluminous.*
metamarble (met-a-mar′-ble) A term proposed by Brooks (1954) for metamorphic carbonate rock that is commercially valuable because it will take a polish, e.g. the Vermont metamarble. Cf: *orthomarble.*
metamict (met′-a-mict) Said of a mineral containing radioactive elements in which lattice disruption and changes have taken place as a result of radiation damage while its original external morphology has been retained. Internal radiation damage in crystals containing radioactive nuclides depends on the duration of the exposure, the concentration of radionuclides, and the energy released by the decay of the radionuclides. Examples occur in zircon, thorite and several other minerals. Not all minerals containing radioactive elements are metamict; e.g., xenotime and apatite.
metamictization The loss of crystallinity that occurs when radiation from radioactive isotopes in a crystal disrupts the atomic arrangement.
metamorphic (met-a-mor′-phic) Pertaining to the process of *metamorphism* or to its results.
metamorphic assemblage (a) A metamorphic *mineral assemblage.* (b) *metamorphic complex.*
metamorphic aureole *aureole.*
metamorphic complex The metamorphic rocks constituting a whole group closely related on a regional or stratigraphic basis; e.g., the Dalradian metamorphic complex of Scotland. Syn: *metamorphic assemblage.*
metamorphic core complex [meta] Any *metamorphic complex* that occurs in the "core" (internal region) of an orogen. Syn: *core complex.*
metamorphic core complex [struc geol] A generally domal or archlike uplift of deformed metamorphic and plutonic rocks overlain by tectonically detached and distended relatively unmetamorphosed cover rocks (Peacock et al., 2000).
metamorphic differentiation (a) Development of layers or segregations, e.g. in a gneiss. (b) A collective term for the various processes by which minerals or mineral assemblages are locally segregated from an initially uniform parent rock during metamorphism, e.g. garnet porphyroblasts in fine-grained mica schist.
metamorphic facies A set of metamorphic mineral assemblages, repeatedly associated in space and time, such that there is a constant and therefore predictable relation between mineral composition and chemical composition (Turner, 1968). The concept was introduced by Eskola (1915, 1939). It is generally assumed that the metamorphic facies represent the results of equilibrium crystallization of rocks under a restricted range of externally imposed physical conditions, e.g. temperature, lithostatic pressure, H_2O-pressure. Syn: *mineral facies; densofacies.* See also: *metamorphic facies series; metamorphic subfacies.*
metamorphic facies series A group of *metamorphic facies* characteristic of an individual area or terrane, and represented by a curve or a group of curves in a pressure-temperature diagram illustrating the range of the different types

of metamorphism and metamorphic facies (Hietanen, 1967). The term was introduced by Miyashiro (1961).

metamorphic grade The intensity or rank of metamorphism, measured by the amount or degree of difference between the original parent rock and the metamorphic rock. It indicates in a general way the P-T environment in which the metamorphism took place. For example, conversion of shale to slate or phyllite would be low-grade dynamothermal metamorphism (greenschist facies), whereas its continued alteration to a garnet-sillimanite schist would be high-grade metamorphism (almandine-amphibolite facies). Syn: *grade [meta]; metamorphic rank; rank [meta].*

metamorphic overprint *overprint [struc petrol].*

metamorphic rank *metamorphic grade.*

metamorphic rock (a) In its original usage (Lyell, 1833), the group of gneisses and crystalline schists. (b) In current usage, any rock derived from preexisting rocks by mineralogical, chemical, or structural changes, essentially in the solid state, in response to marked changes in temperature, pressure, shearing stress, and chemical environment, generally at depth in the Earth's crust. Rarely used syn: *metamorphite.*

metamorphic subfacies A subdivision of a *metamorphic facies* based on minor but significant differences in mineral assemblages. Such subdivisions lead to a misunderstanding of the facies concept, and the trend is away from their use.

metamorphic terrane A terrane characterized by a regional penetrative metamorphic fabric and development of metamorphic minerals to such a degree that original stratigraphic features and relations are obscured. Protolithic contrasts with adjoining terranes must be demonstrable (D.L. Jones et al., 1983).

metamorphic water Water that is driven out of rocks by the process of metamorphism (Tolman, 1937).

metamorphic zone In a metamorphic terrain, a region characterized by a diagnostic index mineral or mineral assemblage; e.g., garnet zone, kyanite zone. Var: *zone [meta].*

metamorphic zoning *zoning [meta].*

metamorphism (met-a-mor′-phism) The mineralogical, chemical, and structural adjustment of solid rocks to physical and chemical conditions that have generally been imposed at depth, below the surface zones of weathering and cementation, and differ from the conditions under which the rocks in question originated. In an older and now obsolete sense, the scope of the term included *katamorphism,* i.e., the processes of cementation and weathering (Van Hise, 1904).

metamorphite (met-a-mor′-phite) *metamorphic rock.*

metamorphosis (met-a-mor′-pho-sis) (a) In biology, a process involving marked or abrupt reorganization of an animal during post-embryonic development, such as the transformation of a larva into a succeeding stage of development and growth. (b) Any change in form, structure, substance, etc.

metamunirite A colorless orthorhombic mineral: $NaV^{5+}O_3$.

metanatroautunite A yellow tetragonal mineral: $Na_2(UO_2)_2(PO_4)_2{\cdot}6\text{-}8H_2O$. Syn: natroautunite, sodium meta-autunite.

metanauplius (met-a-nau′-pli-us) A postnaupliar crustacean larva with the same general body and limb morphology as a *nauplius,* but having additional limbs (about seven pairs).

metanovacekite (met-a-no′-va-cek-ite) A yellow tetragonal mineral: $Mg(UO_2)_2(AsO_4)_2{\cdot}4\text{-}8H_2O$. It is a lower hydrate of novacekite.

metaparian (met-a-par′-i-an) Of or concerning a trilobite that appears to have nonfunctional facial sutures both beginning and ending at the posterior margin of the *cephalon.*

metapeltidium (met′-a-pel-tid′-i-um) The last *sclerite* (usually single, rarely one of a pair) of a segmented carapace of an arachnid, following upon the *mesopeltidium.*

metaphyte (met′-a-phyte) A general term for a multicellular plant. Cf: *protophyte.*

metapodosoma (met′-a-po′-do-so′-ma) A section of the body of an acarid arachnid, bearing the third and fourth pairs of legs. Cf: *propodosoma.*

metapopulation (met-a-pop-u-la′-tion) A set of populations that lie in discrete patches and are linked through migration and extinction.

metaprotaspis (met′-a-pro-tas′-pis) A large trilobite *protaspis* in which the *protopygidium* is well defined. Cf: *anaprotaspis.* Pl: metaprotaspides.

metaquartzite (met-a-quartz′-ite) A quartzite formed by metamorphic recrystallization, as distinguished from an *orthoquartzite,* whose crystalline nature is of diagenetic origin.

metarossite (met-a-ross′-ite) A light-yellow or pale greenish-yellow triclinic mineral: $CaV^{5+}{}_2O_6{\cdot}2H_2O$. It is a lower hydrate of rossite.

metasaléeite A yellow or yellow-green tetragonal mineral: $Na_2(UO_2)_2(PO_4)_2{\cdot}6\text{-}8H_2O$. It is a lower hydrate of saléeite.

metaschoderite (met-a-scho′-der-ite) A yellow-orange monoclinic mineral: $Al_2(PO_4)(VO_4){\cdot}6H_2O$. It is a lower hydrate of schoderite.

metaschoepite (met-a-schoep′-ite) A yellow-orange orthorhombic mineral: $(UO_2)O(OH)_6{\cdot}5H_2O$. It is a lower hydrate of schoepite.

metasediment (met-a-sed′-i-ment) A sediment or sedimentary rock that shows evidence of having been subjected to metamorphism.

metaseptum (met-a-sep′-tum) One of the main septa of a corallite other than a *protoseptum,* generally distinguished by its extension axially much beyond that of minor septa (TIP, 1965, pt. F, p.249).

metasicula (met-a-sic′-u-la) The distal part of the *sicula* of a graptolithine, formed of normal growth increments of fusellar tissue overlain by cortical tissue. Cf: *prosicula.*

metasideronatrite (met-a-sid′-er-o-nat′-rite) A silky yellow orthorhombic mineral: $Na_2Fe^{3+}(SO_4)_2(OH){\cdot}2H_2O$. It is a lower hydrate of sideronatrite.

metasilicate (met-a-sil′-i-cate) According to the now obsolete classification of silicates as oxyacids of silicon, a salt of the hypothetical metasilicic acid, H_2SiO_3. Cf: *orthosilicate.*

metasom (met′-a-som) *metasome [geol].*

metasoma (met-a-so′-ma) The hind region of the body of some invertebrates, esp. when this cannot be readily analyzed into its primitive segmentation (as in some mollusks and arachnids); specif. the posterior part of a merostome *opisthosoma* lacking appendages, or the *metasome* of a copepod crustacean. Cf: *mesosoma.*

metasomasis (met-a-so′-ma-sis) *metasomatism.*

metasomatic (met′-a-so-mat′-ic) Pertaining to the process of metasomatism and to its results. The term is especially used in connection with the origin of ore deposits.

metasomatic rock A rock whose chemical composition, exclusive of volatile components, has been substantially changed by the metasomatic alteration of its original constituents; occasionally called a *metasomatite.*

metasomatism (met-a-so′-ma-tism) The open-system metamorphic process in which the original chemical composition of a rock is changed by reaction with an external source. The process is commonly thought to occur in the presence of a fluid medium flowing through the rock (*infiltration metasomatism*). Metasomatism may also occur by grain-boundary diffusion or by diffusion through a static fluid medium (*diffusion metasomatism*). Skarn, tactite, and serpentinite are examples of metasomatic rocks. Cf: *allochemical metamorphism; pyrometasomatism.* Syn: *metasomasis.*

metasomatite (met-a-so′-ma-tite) *metasomatic rock.*

metasome [geol] (met′-a-some) (a) A replacing mineral, which grows in size at the expense of another mineral (the host or *palasome*); a mineral grain formed by metasomatism. Syn: *guest.* (b) The newly formed part of a migmatite or composite rock, introduced during metasomatism. Cf: *neosome.* Also spelled: *metasom.*

metasome [paleont] The posterior part of the *prosome* of a copepod crustacean, consisting of free thoracic somites in front of the major articulation.

metaspondyle (met-a-spon′-dyle) In dasycladacean algae, a regular whorled arrangement of branches (laterals) in clusters.

metastable (met-a-sta′-ble) (a) Said of a phase that is stable with respect to small disturbances but that is capable of reaction with evolution of energy if sufficiently disturbed. (b) Said of a phase that exists in the temperature range in which another phase of lower vapor pressure is stable. A vapor phase need not be present. The metastable phase is not to be confused with instability. In general, metastability is due to the reluctance of a system to initiate the formation of a new, stable phase.

metastable relict *unstable relict.*

metastasis [meta] (me-tas′-ta-sis) Changes of a paramorphic character, such as the recrystallization of a limestone or the devitrification of a glassy rock (Bonney, 1886).

metastasis [tect] *metastasy.*

metastasy (me-tas′-ta-sy) A term used by Gussow (1958) for lateral adjustments of the Earth's crust, as opposed to vertical movements (*isostasy*). Syn: *metastasis [tect].*

metaster (me-tas′-ter) The portion of a *migmatite* that remained solid (immobile or less mobile) during migmatization. Cf: *paleosome; restite; stereosome.* Rarely used.

metastibnite (met-a-stib′-nite) A red noncrystalline mineral: Sb_2S_3. It is an amorphous form of stibnite.

metastoma (met-a-sto′-ma) A median platelike process behind the mouth in certain arthropods; e.g. a plate at the posterior edge of the mouth of a merostome. It is possibly represented in some trilobites by a plate posterior to the *hypostome.* Also, the lower lip behind the mandibles of a crustacean, usually cleft into paragnaths. Pl: metastomata. Cf: *hypostome.* Syn: *labium.* Also spelled: *metastome.*

metastome (met′-a-stome) *metastoma.*

metastrengite (met-a-streng′-ite) *phosphosiderite.*

metastudtite (met-a-studt′-ite) A yellow orthorhombic mineral: $(UO_2)O_2{\cdot}2H_2O$. It is a lower hydrate of studtite.

metaswitzerite (met-a-switz′-er-ite) A pale pink to golden brown monoclinic mineral: $Mn^{2+}{}_3(PO_4)_2{\cdot}4H_2O$. It was originally called switzerite.

metatarsus (met-a-tar′-sus) (a) Collectively, the foot bones of a tetrapod distal to the tarsal elements and proximal to the phalanges. (b) The proximal (typically the sixth) segment of a leg of an arachnid, following the tibia and preceding the tarsus. Pl: metatarsi.

metatect (met′-a-tect) The fluid or more mobile part of a *migmatite.* Cf: *chymogenic; mobilizate; neosome.* Rarely used.

metatectite (met-a-tec′-tite) A *lipotexite* whose mineralogy and texture have been changed mainly through *metasomatism* accompanying *anatexis.* Synonymous with the *metatexite* of some workers (Dietrich and Mehnert, 1961). Rarely used.

metatexis (met-a-tex′-is) "Moderate degree of *anatexis* (partial melting); leucosome is subordinate to nonleucosome; pre-migmatization structures not disrupted" (Ashworth, 1985, p.3). Cf: *anamigmatization; diatexis.*

metatexite (met-a-tex′-ite) The rock resulting from *metatexis.* Synonymous with the *metatectite* of some workers (Dietrich and Mehnert, 1961). Rarely used.

metatheca (met-a-the′-ca) The distal part of a graptoloid *theca.* It is equivalent to the *autotheca* in those graptolites with more than one type of theca.

metathenardite (met′-a-the-nard′-ite) A trigonal mineral stable above 271°C, occurring in fumaroles on Martinique Island: Na_2SO_4. It is the high-temperature dimorph of thenardite.

metatorbernite (met-a-tor′-bern-ite) A tetragonal pale to dark green secondary mineral: $Cu(UO_2)_2(PO_4)_2 \cdot 8H_2O$. It is a lower hydrate of torbernite.

metatrophic (met-a-troph′-ic) *heterotrophic.*

metatyuyamunite (met-a-tyu-ya-mun′-ite) A yellow to greenish-yellow orthorhombic secondary mineral: $Ca(UO_2)_2(V_2O_8) \cdot 3H_2O$. It is a lower hydrate of tyuyamunite.

meta-uranocircite (met-a-u-ran′-o-circ′-ite) A yellow-green monoclinic mineral of the meta-*autunite* group: $Ba(UO_2)_2(PO_4)_2 \cdot 6H_2O$. It exists in two crystallographic variants, called I and II.

meta-uranopilite (met-a-u-ran′-o-pil′-ite) A yellow, grayish, brown, or green mineral: $(UO_2)_6(OH)_{10}(SO_4) \cdot 5H_2O$. It is lower hydrate of uranopilite. Syn: β-uranopilite.

meta-uranospinite met-a-u-ran′-o-spin′-ite) A yellow tetragonal secondary mineral: $Ca(UO_2)_2(AsO_4)_2 \cdot 8H_2O$. It is a lower hydrate of uranospinite.

metavandendriesscheite (met-a-van′-den-dries′-sche-ite) An yellow-orange orthorhombic mineral: $PbU_7O_{22} \cdot nH_2O$. Syn: vandendriesschite II.

metavanmeersscheite (met-a-van-meers′-sche-ite) A yellow orthorhombic mineral: $U(UO_2)_3(PO_4)_2(OH)_6 \cdot 2H_2O$. It is a lower hydrate of vanmeersscheite.

metavanuralite (met-a-van-ur-al′-ite) A yellow triclinic mineral: $Al(UO_2)_2(V_2O_8)(OH) \cdot 8H_2O$. It is a lower hydrate of vanuralite.

metavariscite (met-a-var′-is-cite) (a) A pale green monoclinic mineral: $AlPO_4 \cdot 2H_2O$. (b) A group name for minerals with the analogous composition, but with Al replaced by Fe^{3+} or Sc.

metavauxite (met-a-vaux′-ite) A colorless to pale green monoclinic mineral: $Fe^{2+}Al_2(PO_4)_2(OH)_2 \cdot 8H_2O$. It has more water than vauxite but less than paravauxite.

metavivianite (met-a-viv′-i-an-ite′) A green platy triclinic mineral: $Fe^{2+}_{3-x}Fe^{3+}_x(PO_4)_2(OH)_x \cdot 8\text{-}xH_2O$. It is a dimorph of vivianite.

metavolcanics (met′-a-vol-can′-ics) An informal term for volcanic rocks that show evidence of having been subjected to *metamorphism.*

metavoltine (met-a-vol′-tine) A yellow, green, orange, or brown trigonal mineral: $K_2Na_6Fe^{2+}Fe^{3+}_6O_2(SO_4)_{12} \cdot 18H_2O$.

metaxite [mineral] (me-tax′-ite) A fibrous serpentine mineral; a variety of chrysotile.

metaxite [sed] *micaceous sandstone.*

metaxylem (met-a-xy′-lem) Primary xylem which matures after the *protoxylem,* concomitantly with or after the surrounding tissues (Cronquist, 1961, p.877).

metazellerite (met-a-zel′-ler-ite) A yellow orthorhombic secondary mineral: $Ca(UO_2)(CO_3)_2 \cdot 3H_2O$. It is a lower hydrate of zellerite.

metazeunerite (met-a-zeu′-ner-ite) A tetragonal green secondary mineral: $Cu(UO_2)_2(AsO_4)_2 \cdot 8H_2O$. It is a lower hydrate of zeunerite.

Metazoa (Met-a-zo′-a) The large group of multicellular animals in which the cells are arranged in two layers in the embryonic gastrula stage.

meteor (me′-te-or) (a) The visible streak of light resulting from the entry into the atmosphere of a solid particle from space. (b) In popular jargon, any physical object or relatively small fragment of solid material associated with a meteor and made luminous as a result of friction during its passage through the Earth's atmosphere; a *meteoroid.* Syn: *shooting star.*

meteor crater *meteorite crater.*

meteoric [meteorite] (me-te-or′-ic) Relating to or composed of meteors or meteoroids.

meteoric [water] (a) Pertaining to water of recent atmospheric origin. (b) Pertaining to, dependent on, derived from, or belonging to the Earth's atmosphere; e.g. "meteoric erosion" caused by rain, wind, or other atmospheric forces.

meteoric iron (a) Iron of meteoric origin. (b) An *iron meteorite.*

meteoric stone (a) A stone of meteoric origin; a *stony meteorite.* (b) A meteorite having the appearance of a stone.

meteorite (me′-te-or-ite′) A *meteoroid* that has survived atmospheric passage and fallen to the Earth's surface in one or more recoverable fragments. Immediately recovered meteorites are called falls; those later recovered are called finds.

meteorite crater An *impact crater* formed by the impact of a large meteorite onto a surface; e.g., Barringer Crater (Meteor Crater) in Coconino County, Arizona, and Chubb Crater in Quebec, Canada. Cf: *penetration funnel.* Syn: *meteor crater.*

meteorite flux The rate at which meteorites fall to Earth, measured in terms of the number of specimens of a specific minimum size or total mass to fall over a unit area during a given period of time. Syn: meteorite influx rate.

meteoritic dust (me′-te-or-it′-ic) Particles produced by the fragmentation or crusting of meteorites during atmospheric passage. Cf: *cosmic dust.*

meteoritics (me′-te-or-it′-ics) A science that deals with meteors and meteorites and related fields such as impact studies.

meteorogram A graphical depiction of the simultaneous varation in two or more weather elements, e.g. temperature, pressure, cloud cover, over a given time interval at a particular locale; used for studying frontal passages.

meteoroid (me′-te-or-oid′) Solid objects in interplanetary space that would form meteorites if they are captured by the Earth and survive atmospheric passage. Meteoroids are distinguished from asteroids and planets by their smaller size. Cf: *meteor; meteorite.*

meteorologic tide (me′-te-or′-o-log′-ic) A change in water level due to periodic forcing by the atmosphere at the same frequency as the solar tides, e.g., daily changes in sea level due to the sea breeze.

meteorology (me′-te-o-rol′-o-gy) The study of the Earth's atmosphere, including its movements and other phenomena, especially as they relate to weather forecasting.

meteor shower A large concentration of falling meteors observed when members of a meteor swarm encounter the Earth's atmosphere and their luminous paths appear to diverge from a single point in the sky (the radiant). Meteor showers result from the infall of dust associated with comets, and following similar orbits. The dust was entrained into interplanetary space by cometary comas; showers occur when the Earth passes near the orbit of a comet.

meteor swarm A group of dust particles that have been expelled from a comet and thus have closely similar orbits around the sun. Swarms of larger meteoroids have not been documented.

meter rod (me′-ter) A precise leveling rod graduated in whole and fractional meters.

metes and bounds The boundaries or limits of a tract of land; esp. the boundaries of irregular pieces of land (such as claims, grants, and reservations) in which the bearing and length of each successive line is given and in which the lines may be described by reference to local natural or artificial monuments along it (such as a stream, ditch, road, or fence). Such boundaries have been established for much of the land in non-public-land surveys, and are distinguished from those established by beginning at a fixed starting point and running therefrom by stated compass courses and distances.

methane (meth′-ane) A colorless odorless inflammable gas, the simplest paraffin hydrocarbon, formula CH_4. It is the principal constituent of natural gas and is also found associated with crude oil. See also: *marsh gas; firedamp.*

methane hydrate An icelike compound made up of molecules of methane gas that is trapped within cages of frozen water beneath the seafloor at water depths of about 1,000 m (Keller, 2002).

methane series The homologous series of saturated aliphatic hydrocarbons, empirical formula C_nH_{2n+2}, of which methane is the lowest and representative member, followed by ethane, propane, the butanes, etc. Syn: *paraffin series.*

methanogens Bacteria that produce methane gas from carbon dioxide, acetate, or methanol.

metharmosis (me-thar′-mo-sis) The changes occurring in a sediment after its burial (after uplift or consolidation) but before weathering begins; in this usage, the term is equivalent to *late diagenesis* or *epigenesis.* The term was proposed by Kessler (1922) in a less restricted sense to designate all changes that a sediment may undergo, including diagenesis proper and metamorphism. Syn: *metaharmosis.*

method of least squares (meth′-od) Any of several statistical procedures for estimating a parameter or fitting a line, curve, or higher-degree surface to a set of data so that the sum of the squared differences between the observed values and the estimates is minimized. Syn: *least squares.*

methylene iodide (meth′-yl-ene) A liquid compound that is used as a *heavy liquid;* its specific gravity is 3.33. Cf: *Clerici solution; Sonstadt solution; Klein solution; bromoform.*

metric carat (met′-ric) *carat.*

meulerization (meu′-ler-i-za′-tion) Local cementation, and replacement (in part) by opaline or chalcedonic silica carried by groundwater, of a carbonate sandstone or a limestone, such as the reaction occurring in certain sedimentary rocks of the Paris Basin. Etymol: French "meule", "millstone". Cf: *Fontainebleau sandstone.*

meurigite A yellow, creamy white, or pale orange monoclinic mineral: $KFe^{3+}_7(PO_4)_5(OH)_7 \cdot 8H_2O$.

Mexican onyx (Mex′-i-can) Yellowish-brown or greenish-brown *onyx marble,* found chiefly in Tecali, Mexico.

meyerhofferite (mey′-er-hoff′-er-ite) A colorless triclinic mineral: $CaB_3O_3(OH)_5 \cdot H_2O$. It is an alteration product of inyoite.

meymacite (mey′-mac-ite) A resinous, yellow-brown amorphous mineral: $WO_3 \cdot 2H_2O$.

MF *metal factor.*

MG *marin gräns.*

mGal *milligal.*

mgd Million gallons per day: a rate of flow of water.

mgriite (m-gri′-ite) A metallic cubic mineral: Cu_3AsSe_3.

MHHW *mean higher high water.*

MHW *mean high water.*

MHWS *mean high-water spring.*

MI *mafic index.*

mianthite (mi-an'-thite) Dark-colored enclosures, patches, or streaks in an *anatexite* (Dietrich and Mehnert, 1961). Rarely used.

miargyrite (mi-ar'-gyr-ite) An iron-black to steel-gray monoclinic mineral: $AgSbS_2$. It has a cherry-red powder.

miarolithite (mi'-a-ro-lith'-ite) A *chorismite* having *miarolitic* cavities or remnants thereof; a variety of *ophthalmite [meta]*. Rarely used.

miarolitic (mi'-a-ro-lit'-ic) Small irregular cavities in phaneritic igneous rocks, esp. "granites", into which small crystals of the rock-forming minerals protrude.

miarolitic pegmatite Gem-bearing pegmatite formed under conditions of greenschist metamorphism (~1 to 2 kilobars pressure) within or peripheral to subvolcanic granitic plutons.

miaskite (mi-as'-kite) (a) In the *IUGS classification*, a special term for a *nepheline monzosyenite*. (b) A biotite-bearing *nepheline syenite* containing oligoclase and microperthite. Its name, given by Rose in 1839, is derived from Miask, in the Urals. Also spelled: miascite. Obsolete.

miassite A metallic cubic mineral: $Rh_{17}S_{15}$. Syn: *prassoite*. It is the Rh analogue of palladseite.

mica (mi'-ca) (a) A group of minerals of general formula: $(K,Na,Ca)(Mg,Fe,Li,Al)_{2-3}(OH,F)_2[(Si,Al)_4O_{10}]$. It consists of complex phyllosilicates that crystallize in forms apparently orthorhombic or hexagonal (such as tabular six-sided prisms) but really monoclinic; that are characterized by low hardness and by perfect basal cleavage, readily splitting into thin, tough, somewhat elastic laminae or plates with a splendent pearly luster; and that range in color from colorless, silvery white, pale brown, or yellow to green or black. Micas are prominent rock-forming constituents of igneous and metamorphic rocks, and commonly occur as flakes, scales, or shreds. Sheet muscovite is used in electric insulators; ground mica in paint and as a dusting agent. Cf: *brittle mica*. Syn: *isinglass; glimmer*. (b) Any mineral of the mica group, including muscovite, biotite, lepidolite, phlogopite, zinnwaldite, roscoelite, paragonite, and sericite.

mica book A crystal of mica, often large and irregular. It is so named because of the resemblance of its cleavage plates to the leaves of a book. Syn: *book*.

micaceous (mi-ca'-ceous) (a) Consisting of, containing, or pertaining to mica; e.g. a "micaceous sediment". (b) Resembling mica; e.g., a "micaceous mineral" capable of being easily split into thin sheets, or a "micaceous luster".

micaceous arkose A term used by Hubert (1960, p.176-177) for a sandstone containing 25-90% feldspars and feldspathic crystalline-rock fragments, 10-50% micas and micaceous metamorphic-rock fragments, and 0-65% quartz, chert, and metamorphic quartzite. The term is roughly equivalent to *impure arkose* of Folk (1954). Cf: *feldspathic graywacke*.

micaceous iron ore A soft, unctuous variety of hematite having a foliated structure resembling that of mica.

micaceous quartzite A term used by Hubert (1960, p.176-177) for a sandstone containing 70-95% quartz, chert, and metamorphic quartzite, 5-15% micas and micaceous metamorphic-rock fragments, and 0-15% feldspars and feldspathic crystalline-rock fragments. Cf: *feldspathic quartzite*.

micaceous sandstone A sandstone containing conspicuous layers or flakes of mica, usually muscovite. Syn: *metaxite [sed]*.

micaceous shale A gray or brownish-gray shale, usually well-laminated, containing abundant muscovite flakes along its lamination planes and finer-grained sericite in its clay matrix; it is commonly associated with subgraywacke.

mica fish Lenticular mica porphyroclasts with long axes inclined to mylonitic foliation. The direction of inclination can be used to determine shear sense.

mica layer In phyllosilicates, a structural unit consisting of two offset *tetrahedral sheets* stacked together to form an intervening *octahedral sheet*.

mica plate In a polarizing microscope, a *phase plate* consisting of a sheet of muscovite (or other *birefringent* material) that is used to determine optical sign from interference figures. Its interference color in white light is a light, neutral gray, and it provides 147 nm of retardation. Syn: *quarter-wave plate*.

micarelle (mi-ca-relle') Mica pseudomorphous after scapolite.

mica-rich domains Cleavage domains composed of oriented micas.

mica schist A schist whose essential constituents are mica and quartz, and whose schistosity is mainly due to the parallel arrangement of mica flakes.

micheelsenite A colorless hexagonal mineral: $(Ca,Y)_3Al(PO_3OH)CO_3(OH)_6 \cdot 12H_2O$.

michenerite (mich'-e-ner-ite') A metallic lead-gray cubic mineral of the *gersdorffite* group: PdBiTe. Cf: *froodite*.

micracanthopore (mi-cra-can'-tho-pore) An *acanthopore* in stenolaemate bryozoans that belongs to the smaller of two distinct sizes occurring in the same *zoarium*. Cf: *megacanthopore*.

micrinite (mic'-ri-nite) A maceral of coal within the *inertinite* group having a reflectance higher than that of associated vitrinite, without recognizable plant-cell structure, and occurring as discrete particles of size less than 2 micrometers, commonly about 1 micrometer. Cf: *macrinite; residuum [coal]*.

micrinoid (mic'-ri-noid) A maceral group that includes the macerals in the micrinite series.

micrite (mic'-rite) (a) A descriptive term used by Folk (1959) for the semiopaque crystalline matrix of limestones, consisting of carbonate mud with crystals less than 4 μm in diameter, and interpreted as a lithified ooze. The term is now commonly used in a descriptive sense without genetic implication. Leighton and Pendexter (1962) used a diameter limit of 31 μm. Chilingar et al. (1967, p.317) and Bissell and Chilingar (1967, p.161) extended usage of the term to include unconsolidated material that may be of either chemical or mechanical origin (and possibly biogenic, biochemical, or physicochemical). Micrite is finer-textured than *sparite*. See also: *matrix [sed]*. (b) A limestone with less than 1% allochems and consisting dominantly of micrite matrix (Folk, 1959, p.14); e.g. lithographic limestone. See also: *micritic limestone*.

micrite envelope A thin coating of micrite around allochems, particularly skeletal grains. It is produced by coating or the boring of algae or other organisms, or perhaps by mechanical adhesion of carbonate mud (Bathurst, 1966).

micritic limestone (mic-rit'-ic) A limestone consisting of more than 90% micrite (Leighton and Pendexter, 1962, p.60) or less than 10% allochems (Wolf, 1960, p.1415); a *micrite*. See also: *calcilutite; lithographic limestone; lime mudstone*.

micritization (mic'-rit-i-za'-tion) Decrease in the size of sedimentary carbonate particles, due to boring activities of certain organisms (Bogas, 1987, p.225).

micro- (mi'-cro-) A prefix meaning "small". When modifying a rock name, it signifies fine-grained hypabyssal, as in microgranite. Cf: *macro-*.

microaerophilic (mi'-cro-aer'-o-phil'-ic) Said of an organism that can exist with very little free oxygen present. Noun: microaerophile.

microanalyzer (mi-cro-an'-a-lyz-er) *electron microprobe*.

microaphanitic (mi'-cro-aph'-a-nit'-ic) *cryptocrystalline*.

microatoll (mi-cro-at'-oll) (a) A ring-shaped growth of corals or serpulids, surrounding a central dead area or depression, with a width of 1-6 m. They are commonly found in the intertidal belt of relatively warm seas or scattered across a reef flat. Cf: *cup reef*. Syn: *miniature atoll*. (b) A small atoll-like reef or knoll, developed within the lagoon or shallows of a reef complex like an atoll or platform reef, characterized by a rim of coral growth surrounding a central sandy depression. Smaller than a *faro*, such a microatoll is on the order of 100 m across and 10 m high, with a depression of 1-3 m in the center (Kornicker and Boyd, 1962).

microbial biscuit Any of various hemispherical or disk-shaped calcareous masses, up to 20 cm in diameter, produced in freshwater as a result of precipitation by various cyanobacteria: e.g. a deposit of marl formed around a piece of algal material or other nucleus as a result of photosynthesis and found on the shallow bottoms of hard-water lakes of the temperate region (as in Wisconsin). Syn: *oncolite; water biscuit; marl biscuit; girvanella; pycnostromid*.

microbialite *microbial laminite*.

microbial laminite Laminated limestone where the laminae have formed by the trapping of grains and precipitation of lime mud through the effects of eubacteria, cyanobacteria, and fungi in a microbial mat (formerly called algal mat). A lithified microbial laminite is also called a stromatolite. Syn: *microbialite*.

microbial mat A layer of eubacteria, cyanobacteria, and fungi growing over the surface of the sediment. The microbially induced precipitation of $CaCO_3$ plus the trapping and binding of sediment by these mats gives rise to microbial laminites and stromatolites. Formerly called: *algal mat*.

microbiofacies (mi'-cro-bi'-o-fa'-cies) The biologic aspect of a *microfacies* (Fairbridge, 1954, p.683).

microbiostratigraphy (mi'-cro-bi'-o-stra-tig'-ra-phy) Biostratigraphy based on microfossils.

microbreccia (mi-cro-brec'-cia) (a) Poorly sorted sandy rocks with sharply angular grains (Williams et al., 1954, p.283). (b) A breccia withhin fragments of a coarser breccia (Sander, 1951, p.28). (c) A well-indurated, massive rock that has been crushed to very fine grain size through *cataclastic flow*, commonly in detachment faults (Davis, 1984).

microchemical test (mi-cro-chem'-i-cal) A chemical test made on minute grains or polished surfaces under a microscope. It is often combined, in identifying a substance, with observations on form, color, and optical properties.

microclastic (mi-cro-clas'-tic) Said of coal that is composed mainly of fine particles, e.g. cannel coal. Cf: *macroclastic*.

microclastic rock A clastic rock whose constituents are minute. Cf: *cryptoclastic rock*.

microclimate (mi-cro-cli'-mate) The climatic structure close to the Earth's surface, affected by the character of the surface materials; for example, over a snow surface, lake, or cornfield. See also: *macroclimate; mesoclimate*.

microcline (mi'-cro-cline) A clear, white to gray, brick-red, or green mineral of the alkali *feldspar* group: $KAlSi_3O_8$. It is the fully ordered, triclinic modification of potassium feldspar and is dimorphous with orthoclase, being stable at lower temperatures; it usually contains some sodium in minor amounts. Microcline is a common rock-forming mineral of granitic rocks and pegmatites, and is often secondary after orthoclase. It is generally characterized by cross-hatch twinning.

microcline-perthite (mi'-cro-cline-perth'-ite) A perthite consisting of an intergrowth of microcline and plagioclase.

micrococcolith (mi-cro-coc′-co-lith) One of the smaller coccoliths in coccolithophores exhibiting dimorphism but with the dimorphic coccoliths irregularly placed. Cf: *macrococcolith.*

microcodium A type of bacteria which lives off limestone and carbonate grains and produces spherulitic aggregates of calcite crystals.

microconch (mi′-cro-conch) A mature *conch* of a chambered cephalopod which, in all respects except size and occasional modification of the aperture, resembles larger conchs (macroconchs) found in the same fossil association. Microconchs are now generally regarded as representing males (Callomon, 1963).

microconglomerate (mi'-cro-con-glom′-er-ate) A poorly sorted sandstone containing relatively large rounded particles of sand set in a very fine silty or clayey matrix.

microcontinent (mi-cro-con′-ti-nent) A submarine plateau that is an isolated fragment of continental crust. Cf: *aseismic ridge.*

microcoquina (mi'-cro-co-qui′-na) (a) A detrital limestone composed wholly or chiefly of shell detritus of sand size (2 mm in diameter) or less. (b) A variety of chalk (Bissell and Chilingar, 1967, p.153). Cf: *coquina.*

microcosmic salt (mi-cro-cos′-mic) *stercorite.*

microcracks Cracks forming within individual grains or at grain boundaries. These form either upon the removal of the high lithostatic stress and temperature present at the time of rock formation or as the precursors to macroscopic brittle failure. Syn: *microfracture.*

microcrater (mi-cro-cra′-ter) Small impact crater, ranging from micrometers to several centimeters in diameter produced by hypervelocity impact of primary micrometeorite particles on exposed surfaces of any airless body. Such craters on the Moon are typically less than a few millimeters in diameter and are characterized by a central glass-lined pit, a concentric lightened area of shock-fractured minerals, and a roughly circular spall area approximately 4.5 times larger in diameter than the central pit. Syn: *micrometeorite crater, zap crater.*

micro cross-lamination A small but distinctive cross-lamination, similar to a small-scale trough cross-bedding. See also: *rib and furrow.*

microcryptocrystalline (mi'-cro-cryp'-to-crys′-tal-line) *cryptocrystalline.*

microcrystal (mi-cro-crys′-tal) A crystal, the crystalline nature of which is discernible only under the microscope; such crystals form a *microcrystalline* substance.

microcrystalline (mi-cro-crys′-tal-line) Said of the texture of a rock consisting of or having crystals that are small enough to be visible only under the microscope; also, said of a rock with such a texture. In regard to carbonate sedimentary rocks, various diameter ranges are in use: 0.01-0.20 mm (Pettijohn, 1957, p.93); less than 0.01 mm (Carozzi and Textoris, 1967, p.6); and 0.001-0.01 mm (Bissell and Chilingar, 1967, p.161, who note that some petrographers use 0.004-0.062 mm). Cf: *cryptocrystalline; felsophyric.* See also: *microcrystal.*

microdelta (mi-cro-del′-ta) A small-scale delta or bar, generally not more than a few meters across, with a slip face on which foreset beds are deposited. It resembles a dune or megaripple, but does not show a repetitive, wavelike form, and lacks a well-developed inclined *stoss* side. Also spelled: micro-delta.

microdolomite A specific type of dolomite occurring as small crystals (1-20 micrometers) in calcite. They occur within cements, ooids, and bioclasts which were originally composed of high magnesian calcite (HMC).

microearthquake (mi-cro-earth′-quake) An earthquake having a *body-wave magnitude* of two or less on the Richter scale. Such a limit is arbitrary, and may vary according to the user. Cf: *major earthquake; ultramicroearthquake.*

microelement (mi-cro-el′-e-ment) *trace element.*

microeutaxitic (mi'-cro-eu′-tax-it′-ic) The texture of many welded tuffs that are microscopically *eutaxitic.*

microevolution (mi'-cro-ev′-o-lu′-tion) (a) The evolution or origin of species, as contrasted to that of higher taxa. (b) Evolution that occurs within a continuous population but does not result in the development of genetic discontinuities; the changes, brought about by selective accumulation of minute variations, are thought to be chiefly responsible for evolutionary differentiation. Cf: *macroevolution,* from which it probably differs only in degree.

microfabric (mi′-cro-fab′-ric) The fabric of a rock as seen under the microscope. Cf: *megafabric.*

microfacies (mi′-cro-fa′-cies) Those characteristic and distinctive aspects of a sedimentary rock that are visible and identifiable only under the microscope (low-power magnification) (Brown, 1943, p.325; Cuvillier, 1951). See also: *microbiofacies; microlithofacies.*

microfauna (mi′-cro-fau′-na) (a) Living or fossil animals too small to be seen with the naked eye; benthic animals smaller than 0.1 mm. (b) An obsolete term for a very localized or small group of animals; animals occupying a small habitat. Cf: *microflora; macrofauna.*

microfelsitic (mi′-cro-fel-sit′-ic) *cryptocrystalline.*

microflora (mi′-cro-flo′-ra) (a) An assemblage of extant microscopic organisms. The term is commonly misapplied to the *microfossil* remains of higher plants. (b) An obsolete term for a very localized or small group of plants; plants occupying a very small habitat. Cf: *microfauna; megaflora.*

microforaminifera (mi-cro-fo′-ram-i-nif′-er-a) (a) The chitinous inner tests of certain foraminifers, almost always spiral, frequently found in palynologic preparations of marine sediments; they are generally much smaller than "normal" whole foraminifers but display recognizable characteristics of "normal" species. (b) Foraminifers much smaller than those generally observed and studied.

microfossil (mi′-cro-fos′-sil) A fossil too small to be studied without the aid of a microscope, e.g. an invertebrate such as a *foraminifer* or an *ostracode.* It may be the remains of a microscopic organism or a part of a larger organism. Cf: *macrofossil; nannofossil.*

microfracture A crack within a single grain or cutting several grains. Most commonly observed with a microscope. Syn: *microcracks.*

microfragmental (mi′-cro-frag-men′-tal) Said of a coal composed of macerated vegetal matter. Cf: *macrofragmental.*

microgal (mi′-cro-gal) A unit of acceleration commonly used in borehole gravity work; 10^{-6} gal. Abbrev: µGal.

microgametophyte (mi′-cro-ga-me′-to-phyte) The male *gametophyte* or haploid generation that develops from the microspore of a heterosporous embryophytic plant. In lower vascular plants, a few-celled microgametophyte as well as the sperm cells are produced entirely within the microspore; in seed plants, the microgametophyte plus the surrounding microspore wall is the pollen grain, in which the microgametophyte is further reduced, consisting of only three cells in the angiosperms. Cf: *megagametophyte.*

micro-gas survey Soil analysis to determine the presence of hydrocarbon gases that have presumably seeped upwards into the overburden from buried sources.

microgeography (mi′-cro-ge-og′-ra-phy) The detailed analysis of the natural features of a very limited area.

microgeology (mi′-cro-ge-ol′-o-gy) (a) Study of the geologic and geochemical role of microorganisms (Ehrenberg, 1854). (b) Study of microscopic features of rocks.

microgour (mi′-cro-gour) A *gour* or *rimstone dam* or *rimstone pool* on the scale of a few millimeters.

micrograined (mi′-cro-grained) (a) The texture of a carbonate sedimentary rock having particles whose diameters are in the range of 0.001-0.01 mm (Bissell and Chilingar, 1967, p.103) or 0.001-0.004 mm (DeFord, 1946). Some petrographers use the limits of 0.004-0.062 mm. (b) The texture of a carbonate sedimentary rock wherein the particles are mostly 0.01-0.06 mm in diameter, are poorly sorted, and are admixed with clay-sized calcareous mud (Thomas, 1962). Also said of a sedimentary rock with such a texture. Cf: *microgranular [sed].*

microgranular [ign] (mi-cro-gran′-u-lar) Said of the texture of a *microcrystalline, xenomorphic-granular* igneous rock. Syn: *fine-granular.*

microgranular [paleont] Said of a foraminiferal wall (as in Endothyracea) composed of minute calcite crystals, probably originally granular but possibly recrystallized. The granules may be aligned in rows perpendicular to the outer wall, resulting in fibrous structure.

microgranular [sed] Minutely granular; specif. the texture of a carbonate sedimentary rock wherein the particles are mostly 10-60 micrometers in diameter and are well-sorted, and the finer clay-sized matrix is absent (Thomas, 1962). Also said of a sedimentary rock with such a texture. Cf: *micrograined.*

micrograph (mi′-cro-graph) A graphic recording of something seen through the microscope, e.g. a *photomicrograph* of a petrologic thin section.

micrographic (mi-cro-graph′-ic) Said of the graphic texture of an igneous rock, distinguishable only with a microscope. Syn: *granophyric.*

microgroove cast (mi′-cro-groove) A term used by McBride (1962, p.56) for a *striation cast* of a striation less than 2.5 cm in length.

microhabitat (mi-cro-hab′-i-tat) A very small *habitat* (dimensions measurable in mm or cm). Cf: *macrohabitat.*

microhardness Mineral hardness determined by applying a static load with an indenter to a polished surface. The two most commonly used indenters are those of Vickers and Knoop.

microhill (mi-cro-hill′) A very rough, miniature sand column raised by the formation of pipkrakes, ranging from a few millimeters to several centimeters high and having a height-diameter ratio of 2:5 (Otterman and Bronner, 1966, p.56).

microindentation hardness *microhardness.*

microkarst Karst topography in which all surficial features are small; an area dominated by minor *karst features.* The term is not easily applied, because it lacks numerical limits.

Microlaterolog (Mi′-cro-lat′-er-o-log′) Trade name for a *microresistivity log,* obtained with a miniaturized *focused-current log* electrode arrangement, designed to measure a shallow volume of rock at the borehole face. Cf: *Microlog.* Syn: *trumpet log.*

microlinear (mi-cro-lin′-e-ar) "Any lineation that is invisible to the unaided eye" (El-Etr, 1976, p.485).

microlite [cryst] (mi′-cro-lite) A microscopic crystal that polarizes light and has some determinable optical properties. Cf: *crystallite; crystalloid.* Syn: *microlith.*

microlite [mineral] (mic′-rol-ite) A pale-yellow, reddish, brown, or black cubic mineral of the *pyrochlore* group: $(Na,Ca)_2(Ta,Nb)_2O_6(O,OH,F)$. It is isomorphous with pyrochlore, with Ta greater than Nb, and it often contains small amounts of other elements (including uranium and titanium). Microlite occurs in granitic pegmatites and in pegmatites related to alkalic igneous rocks, and it constitutes an ore of tantalum. Syn: *djalmaite;* neotantalite.

microlith (mi′-cro-lith) *microlite [cryst].*

microlithofacies (mi′-cro-lith′-o-fa′-cies) The lithologic aspect of a *microfacies* (Fairbridge, 1954, p.683).

microlithology (mi′-cro-li-thol′-o-gy) The study, or characteristics, of rocks as they appear under the microscope. Cf: *macrolithology.*

microlithon (mi-cro-lith′-on) The tabular to lenticular regions of rock material between surfaces in a spaced cleavage, either disjunctive or domainal cleavage; they either lack cleavage or have poorly developed cleavage (DeSitter, 1954; Powell, 1979).

microlithotype (mi-cro-lith′-o-type) A typical association of macerals in coals, occurring in bands at least 50 micrometers wide. Microlithotype names bear the suffix "-ite". Cf: *lithotype.*

microlitic (mi-cro-lit′-ic) Volcanic groundmass texture in which tiny fibrous crystals (microlites) occur with interstitial glass or cryptocrystalline material. Hyalopilitic, pilotaxitic, orthophyric, and trachytic are microlitic textures.

Microlog (Mi′-cro-log) Trade name for a *well log* consisting of two microresistivity curves: the micronormal curve (see *normal log*) and a very short lateral curve (see *lateral log*). Response is dominated by the presence of mud cake, which causes separation between otherwise virtually coincident curves and thus indicates porous zones. Use of logs of this type for quantitative estimation of porosity is obsolete, although such logs with accompanying caliper curves are still valued for precise location of filterable rock layers of implied reservoir quality. Cf: *Microlaterolog.*

micromeritics (mi′-cro-me-rit′-ics) The study of the characteristics and behavior of small particles. It is applicable to soil physics.

micrometeorite (mi-cro-me′-te-or-ite′) A meteorite or meteoritic particle with a diameter generally less than a millimeter; a meteorite so small that it undergoes atmospheric entry without vaporizing or becoming intensely heated and hence without disintegration. Cf: *cosmic dust.*

micrometeorite crater *microcrater.*

micrometer A unit of length equal to one one-millionth of a meter. Symbol: μm. Obsolete syn: *micron.*

micron (mi′-cron) An obsolete syn. of *micrometer.*

micronucleus (mi-cro-nu′-cle-us) A small nucleus that is concerned with reproductive functions in the body of a tintinnid. Cf: *macronucleus.*

micro-oil A term used by Vernadskiy for hydrocarbons occurring in a diffused state in sedimentary rock; the nascent oil, still within and sorbed to its source rock (Vassoevich, 1965, p.510).

micropaleontology (mi′-cro-pa′-le-on-tol′-o-gy) A branch of paleontology that deals with the study of fossils too small to be observed without the aid of a microscope; the study of microfossils.

micropegmatite (mi-cro-peg′-ma-tite) A less-preferred syn. of *granophyre.*

micropegmatitic (mi-cro-peg′-ma-tit′-ic) A nonrecommended syn. of graphic and *micrographic.*

micropellet (mi-cro-pel′-let) A pellet or pelletlike sedimentary particle of a fine to very fine grade size, "possibly smaller than 0.01 mm in diameter" (Bissell and Chilingar, 1967, p.161). Adj: micropeloid.

microperthite (mi-cro-perth′-ite) A variety of perthite in which the lamellae (5-100 micrometers wide) are visible only with the aid of the microscope. Cf: *cryptoperthite.*

micropetrological unit (mi′-cro-pet′-ro-log′-i-cal) *maceral.*

microphagous (mi-croph′-a-gous) Said of an organism that feeds on relatively minute particles. Cf: *macrophagous.*

microphotograph (mi-cro-pho′-to-graph) A less-preferred syn. of *photomicrograph.*

microphyll (mi′-cro-phyll) (a) A leaf with a single trace (vein) that does not leave a gap in the stele (vascular tissue) of the stem when it is produced; characteristic of the *lycophytes.* Although modern forms are small, some Carboniferous microphylls reached 1 m in length. (b) A leaf size within the Raunkiaer-Webb system of vegetation classification for leaves with a surface area of 225-2,025 mm^2. Xerophytes generally have microphylls. Adj: microphyllous. Cf: *enation; leaf physiognomy; sclerophyll.*

microphyric (mi-cro-phyr′-ic) Said of the texture of a porphyritic igneous rock in which the phenocrysts are of microscopic size, i.e. their longest dimension does not exceed 0.2 mm (Cross et al., 1906, p.702); also, said of a rock having such texture. Cf: *megaphyric.* Syn: *microporphyritic.*

micropiracy (mi-cro-pi′-ra-cy) The overtopping and breaking-down of the narrow ridge between adjacent rill channels, and diversion of flow from the higher, shorter, and shallower channel to the lower, longer, and deeper one closer to the initial rill (Horton, 1945, p.335). See also: *cross-grading.*

microplankton (mi-cro-plank′-ton) Plankton organisms of the size range 0.02-0.2 mm, e.g. most phytoplankton. They are larger than *nannoplankton,* but smaller than *macroplankton.*

microplate A small lithospheric plate.

micropore (mi′-cro-pore) (a) A pore small enough to hold water against the pull of gravity and to inhibit the flow of water. (b) In the pore-size classification of Choquette and Pray (1970, p.233), an equant to equant-elongate pore or a tubular or platy pore with an average diameter or thickness of less than 1/16 mm. Cf: *mesopore [petrology]; megapore.*

microporosity Pores ranging in size from 0.1 to 0.5 micrometers.

microporphyritic (mi′-cro-por′-phy-rit′-ic) *microphyric.*

micropulsation (mi′-cro-pul-sa′-tion) An oscillatory geomagnetic variation in the frequency range from 0.01 to 3 Hz, commonly with amplitudes less than 10 nanotesla (Sumner, 1976, p.243).

micropygous (mi-cro-py′-gous) Said of a trilobite with a *pygidium* smaller than the *cephalon.* Cf: *isopygous; macropygous.*

micropyle (mi′-cro-pyle) The minute opening in the integument of an ovule or seed, through which a pollen tube grows to reach the female gametophyte (Fuller and Tippo, 1954, p.963).

micropylome (mi-cro-py′-lome) A circular to subpolygonal *acritarch excystment* structure with free *operculum,* having a total diameter of <25 % of the vesicle diameter, e.g. *Polyancistrodorus* (Miller, 1987). Cf: *macropylome; pylome [palyn].*

microquartz (mi-cro-quartz′) Nonclastic anhydrous crystalline silica occurring in sediments and having particle diameters usually less than 20 micrometers.

microradiograph (mi′-cro-ra′-di-o-graph′) A picture produced by X-rays or rays from a radioactive source showing the minute internal structure of a substance.

microrelief (mi′-cro-re-lief′) (a) Local, slight irregularities of a land surface, including such features as low mounds, swales, and shallow pits, generally ~1 m in diameter and causing variations amounting to no more than 3 m. (b) Relief features that are too small to show on a topographic map; e.g., gullies, mounds, boulders, pinnacles, or other features <60 m in diameter and <6 m in elevation, in an area for which the topographic map has a scale of 1:50,000 or smaller and a contour interval of 3 m (10 ft) or larger. Cf: *macrorelief.*

microresistivity log (mi′-cro-re′-sis-tiv′-i-ty) Generic term for any *resistivity log* curve obtained from measurements between electrodes spaced a few inches apart and held in direct contact with the wall of the borehole. *Microlog* and *Microlaterolog* are examples of such logs used to determine porosity in potential reservoir rocks flushed by mud filtrate.

microrhabd (mi′-cro-rhabd) A rod-shaped monaxonic sponge spicule (microsclere).

Microsauria (Mi-cro-saur′-i-a) An order of lepospondylous amphibians characterized by small size and salamanderlike or snakelike body form. The older literature includes (in error) early and primitive cotylosaurian reptiles in this group. Range, Lower Mississippian to Lower Permian.

microsclere (mi′-cro-sclere) A small *sclere;* specif. one of the minute secondary spicules scattered throughout a sponge or concentrated in the cortex or elsewhere. It usually differs in form from a *megasclere.*

microscope (mi′-cro-scope) An optical instrument that is used to produce an enlarged image of a small object; it consists of the lens (or lenses) of the objective and of the eyepiece set into a tube, and held by an adjustable arm over a stage on which the object is placed. Types of microscopes vary according to intended use and to type of energy used, e.g. natural light, polarized light, transmitted light, reflected light, electrons, or X-rays.

microscopic (mi-cro-scop′-ic) (a) Said of an object or phenomenon or of its characteristics that cannot be observed without the aid of a microscope. (b) Of or pertaining to a microscope.

microsection (mi-cro-sec′-tion) (a) Any thin section used in microscopic analysis. (b) A *polished section.* (c) A *polished thin section.*

microseism (mi′-cro-seism) A collective term for small motions in the Earth that are unrelated to an earthquake and that have a period of 1.0-9.0 s. They are caused by a variety of natural and artificial agents, esp. atmospheric events. Cf: *macroseism.* Syn: *seismic noise.*

microsere (mi′-cro-sere) A *sere* of a very small habitat, usually failing to attain climax and ending with the loss of identity of the habitat. Syn: *serule.*

microsolifluction (mi′-cro-sol′-i-fluc′-tion) The frost movements that produce patterned ground (Troll, 1944).

microsommite (mi-cro-som′-mite) A colorless hexagonal mineral of the *cancrinite* group: $(Na,Ca,K)_{7-8}(Si,Al)_{12}O_{24}(Cl,SO_4)_{2-3}$.

microspar (mi′-cro-spar) Calcite matrix in limestones, occurring as uniformly sized and generally loaf-shaped crystals ranging from 5 to more than 20 micrometers in diameter. It develops by recrystallization or neomorphism of carbonate mud (micrite) (Folk, 1959). Not to be used as a synonym for fine pore-filling spar (cement). Cf: *microsparite.*

microsparite (mi-cro-spar′-ite) (a) A term used by Folk (1959, p.32) for a limestone whose carbonate-mud matrix has recrystallized to microspar. (b) A term used by Chilingar et al. (1967, p.320) for a sparry crystal of calcite whose diameter ranges from 5 to 20 micrometers. Cf: *microspar.*

microspheric (mi-cro-spher′-ic) Said of a foraminiferal test or shell produced sexually and characterized by a very small initial chamber (proloculus), many chambers, often large size of the adult test, and more complete ontogeny. Cf: *megalospheric.*

microspherulitic (mi′-cro-spher′-u-lit′-ic) Said of the spherulitic texture distinguishable only with the aid of a microscope.

microsphorite (mi-cros′-pho-rite) Primary in situ microcrystalline phosphorite of the central Florida phosphate district. Disrupted and fragmented, it forms clastic particles that are the most common form of phosphorite in the sediments (Freas and Riggs, 1968).

microsporangium (mi'-cro-spo-ran'-gi-um) A *sporangium* that develops or bears microspores; e.g. the anther in an angiosperm or the pollen sac in all other seed plants. Cf: *megasporangium.*

microspore (mi'-cro-spore) One of the spores of a heterosporous embryophytic plant that germinates to produce a microgametophyte and that is ordinarily smaller than the *megaspore* of the same species. In seed plants, pollen grains consist of an exinous microspore wall or exine with a microgametophyte contained inside. See also: *miospore; microgametophyte; small spore.*

microstriation (mi'-cro-stri-a'-tion) A microscopic scratch developed on the polished surface of a rock or mineral as a result of abrasion.

microstructure (mi'-cro-struc'-ture) (a) The internal structure and character of plant and animal tissues, esp. skeletal tissues, as revealed by the microscope. (b) Structural features of rocks that can be discerned only with the aid of a petrographic, scanning electron, and/or transmission electron microscope. Cf: *Pumpelly's Rule.*

microstylolite (mi-cro-sty'-lo-lite) A *stylolite* in which the relief along the surface is less than a millimeter, such as one indicating differential solution between two mineral grains.

microtectonics (mi'-cro-tec-ton'-ics) A syn. of *structural petrology.*

microtektite (mi-cro-tek'-tite) A small glassy object, less than one millimeter in diameter and usually spherical, found in some deep-sea sediments. Most recovered microtektites are associated with the Australasian tektite event 0.8 Ma ago.

microtherm (mi'-cro-therm) A plant that requires low temperatures for successful growth. Cf: *mesotherm; megatherm.*

microthermal (mi-cro-ther'-mal) Pertaining to a climate characterized by low temperature. Cf: *mesothermal [clim]; megathermal.*

microtopography (mi'-cro-to-pog'-ra-phy) (a) Topography on a small scale. The term has been applied to features having relief as small as 1-10 cm as well as to those involving amplitudes of 50-100 m and wavelengths of a few km. (b) The term is also applied to the microscopic relief on the surface of particles.

microtwinning *Twinning* on a submicroscopic scale.

microvitrain (mi-cro-vit'-rain) Vitrain bands occurring in clarain, 0.05-2.0 mm thick.

microwave (mi'-cro-wave) The region of the electromagnetic spectrum in the approximate wavelength range from 1 mm to beyond 1 m. *Passive remote sensing* systems operating at these wavelengths are called microwave systems; *active remote sensing* systems are called radar.

microXANES A technique using XANES methodology combined with micrometer scale incident X-ray beam to provide spatial resolution comparable with modern analytical techniques.

mictite (mic'-tite) Coarsely composite rock formed as the result of contamination of a magma, by the incorporation and partial or complete assimilation of country-rock fragments, under conditions of relatively low temperature and probably at relatively high levels in the crust (Dietrich and Mehnert, 1961). Rarely used.

mid-bay bar A bar built across a bay at some point between its mouth and its head.

midcontinent fault-and-fold zone A zone of faulting and associated folding that occurs within a continental interior craton. In most cases, these zones have been active during the Phanerozoic, and movement within them has generated monoclinal folds. The zones may represent reactivation of long-lived crustal weaknesses that originated during unsuccessful Proterozoic rifting. Also called: continental-interior fault zone; intracratonic fault zone.

midden [arch] (mid'-den) A heap or stratum of refuse (broken pots and tools, ashes, shells and other food remains, etc.). Normally found on the site of an ancient settlement.

midden [sed] A moundlike accumulation of calcareous sediment trapped or bound together by microbial growth.

midden [soil] A mass of highly organic soil formed by an earthworm around its burrow; also, any organic debris on soil, deposited by an animal.

middle [geochron] (mid'-dle) Pertaining to a segment of time intermediate between *late* and *early.* The adjective is applied to the name of a geologic-time unit (era, period, epoch) to indicate relative time designation and corresponds to middle as applied to the name of the equivalent time-stratigraphic unit; e.g., rocks of a Middle Jurassic batholith were intruded in Middle Jurassic time. The initial letter of the term is capitalized to indicate a formal subdivision (e.g., "Middle Devonian") and is lowercased to indicate an informal subdivision (e.g., "middle Miocene"). The informal term may be used for eras and epochs, and for periods where there is no formal subdivision. Syn: *medial.*

middle [stratig] Pertaining to rocks or strata that are intermediate between *upper* and *lower.* The adjective is applied to the name of a chrono-stratigraphic unit (system, series, stage) to indicate position in the geologic column and corresponds to middle as applied to the name of the equivalent geologic-time unit; e.g., rocks of the Middle Jurassic Series were formed during the Middle Jurassic Epoch. The initial letter of the term is capitalized to indicate a formal subdivision (e.g., "Middle Devonian") and is lowercased to indicate an informal subdivision (e.g., "middle Miocene"). The informal term may be used where there is no formal subdivision of a system or of a series.

middle diagenesis *anadiagenesis.*

middle ground A bar deposit or shoal formed in the middle of a channel or fairway at the entrance and exit of a constricted passage (as a strait) by the rise and fall of the tide, and characterized by a flow of water on either side of the deposit.

middle lamella In plants, the intercellular substance between the primary walls of two contiguous cells, composed chiefly of calcium pectate (Esau, 1965, p.34). See also: *intercellular space.*

middle lateral muscle One of a pair of muscles in some lingulid brachiopods, originating on the *pedicle valve* between the central muscles, and diverging slightly posteriorly before insertion on the *brachial valve* (TIP, 1965, pt. H, p.148). Cf: *outside lateral muscle.*

middle latitude n. The latitude of the point situated midway on a north-south line between two parallels; half the arithmetic sum of the latitudes of two places on the same side of the equator.

middle Paleolithic In archaeology, the second division of the *Paleolithic.* Correlation of cultural levels with actual age (and, therefore, with the time-stratigraphic units of geology) varies from region to region. Cf: *lower Paleolithic; upper Paleolithic.*

middle Stone Age *Mesolithic.*

midfan (mid'-fan) The area between the fanhead and the outer, lower margins of an alluvial fan.

midfan mesa A much eroded, islandlike remnant of an old upfaulted alluvial fan, commonly a primary product of piedmont faulting (Eckis, 1928, p.243-246).

mid-latitude desert A vast desert area occurring within lat. 30°-50° north or south of the Equator in the interior of a large continental mass, usually situated in the lee of high mountains that stand across the path of prevailing winds (thus, a *rain-shadow desert*), and commonly characterized by a cold, dry climate.

mid-latitude rain forest A *biome* associated with cool winters and abundant moisture, and very huge biomass; e.g. spruce-hemlock forest of the NW United States and *temperate rain forest.*

mid-ocean canyon (mid-o'-cean) *deep-sea channel.*

mid-ocean ridge (mid'-o-ce-an') A continuous, seismically active, median mountain range extending through the North and South Atlantic Oceans, the Indian Ocean, and the South Pacific Ocean. It is a broad, fractured swell with a central rift valley and usually extremely rugged topography; it is 1-3 km in height, about 1,500 km in width, and over 84,000 km in length. The mid-ocean ridge is the source of new crustal material. See also: *rift valley [oceanog]; sea-floor spreading.* Syn: *mid-ocean rise; oceanic ridge; spreading center.*

mid-ocean-ridge basalt Basalt erupted at a sea-floor spreading axis. Abbrev: MORB.

mid-ocean rift *Rift valley [oceanog]* associated with the mid-ocean ridge.

mid-ocean rise *mid-ocean ridge.*

midpoint The point midway between a source and receiver.

midrange (mid'-range) The arithmetic average of the smallest and largest values in a sample. Syn: *range midpoint.*

midrib (mid'-rib) The central rib of leaf venation. It is a continuation of the petiole.

midstream (mid'-stream) (a) The part of a stream well removed from both sides or from the source and the mouth. (b) A line along a stream course, midway between the sides of the stream.

midwater trawl (mid'-wa-ter) A towed net with anterior depressor panels that is used for catching marine organisms anywhere between the bottom and the water surface.

Midwayan (Mid'-way-an) North American (Gulf Coast) stage: Paleocene (above Navarroan of Cretaceous, below Sabinian).

miemite (mi'-em-ite) A yellowish-brown, fibrous dolomite occurring at Miemo in Tuscany, Italy.

miersite (mi'-ers-ite) A canary-yellow adamantine cubic mineral of the *nantokite* group: (Ag,Cu)I.

Mie scattering Multiple reflection of light waves by atmospheric particles with the general dimensions of the wavelength of light. Cf: *Mie theory.*

Mie theory A theory of the scattering of electromagnetic radiation by spherical particles, developed by G. Mie in 1908. In contrast to *Rayleigh scattering,* the Mie theory embraces all ratios of diameter to wavelength. Mie theory is important in meteorological optics, where diameter-to-wavelength ratios of the order of unity and larger are characteristic of many problems regarding haze and cloud scattering.

migma (mig'-ma) A mobile, or potentially mobile, mixture of solid rock material(s) and magma, the magma having been injected into or melted out of the rock material (Dietrich and Mehnert, 1961). Etymol: Greek, "mixture".

migmatite (mig'-ma-tite) "A rock found in medium-grade to high-grade metamorphic areas, that is pervasively inhomogeneous on a macroscopic scale, one part being pale colored and consistently of quartzofeldspathic or feldspathic composition" (Ashworth, 1985, p.3). A composite rock composed of igneous or igneous-appearing or metamorphic materials, which are generally distinguishable megascopically (Dietrich, 1960, p.50). The term was introduced by Sederholm (1907, p.88-89). Cf: *chorismite.*

migmatitization *migmatization.*

migmatization (mig'-ma-ti-za'-tion) The formation of a *migmatite.* The more mobile, typically light-colored, part of a migmatite may be formed as the result of anatexis, metamorphic differentiation, metasomatism, or injection. Var: *migmatitization.*

migrating dune *wandering dune.*

migrating inlet A tidal inlet, such as that connecting a coastal bay with the open sea, that shifts its position laterally in the direction in which the dominant longshore current flows. It results from deposition on one side of the inlet, accompanied by erosion on the other.

migration [ecol] (mi-gra'-tion) A broad term applied to the movements of plants and animals from one place to another over long periods of time.

migration [geomorph] The movement of a topographic feature from one locality to another by the operation of natural forces; specif. the movement of a dune by the continual transfer of sand from its windward to its leeward side. Cf: *shifting [streams].*

migration [petroleum] The movement of liquid and gaseous hydrocarbons from their source or generating beds through permeable formations into reservoir rocks.

migration [seis] The process by which events on a reflection seismogram are mapped to their true spatial positions. It requires knowledge of the velocity distribution along the raypath. Also called: *imaging.*

migration [streams] (a) *shifting [streams].* (b) The slow downstream movement of a system of meanders, accompanied by enlargement of the curves and widening of the meander belt.

migration recrystallization *boundary migration recrystallization.*

miharaite (mi-ha'-ra-ite) A metallic orthorhombic mineral: $PbCu_4FeBiS_6$.

mikasaite A white to light brown rhombohedral mineral: $(Fe^{3+},Al)_2(SO_4)_3$. It is the Fe^{3+} analogue of millosevichite.

Milankovitch curve (Mi-lan'-ko-vitch) That curve drawn for any latitude (commonly 65°N) and any geologic time that represents the amount of extra atmospheric solar energy received by that latitude. It is computed from an analysis of the earth's orbital variations to support the Milankovitch theory (Berger and Loutre, 1991).

Milankovitch theory An astronomical theory of glaciation, formulated by Milutin Milankovitch (1879-1958), Yugoslav mathematician, in which climatic changes result from fluctuations in the seasonal and geographic distribution of insolation, determined by variations of the Earth's orbital elements, namely eccentricity, tilt of rotational axis, and longitude of perihelion with periods on the order of 100,000 years, 41,000 years, and 23,000 years respectively. (Milankovitch, 1941). It is supported by recent isotopically dated reconstructions of ocean temperature and ice-core records (Imbrie and Imbrie, 1979).

milarite (mi-lar'-ite) (a) A colorless to greenish, glassy, hexagonal mineral: $(K,Na)Ca_2(Be,Al)_3Si_{12}O_{30}{\cdot}H_2O$. (b) A group name for minerals with the analogous formula, but with Ca replaced by Mg, Mn, Y, Zr, Ti Na or a vacant structural site; and (Be,Al) by Li, B, Mg, Zn or Fe.

Milazzian (Mi-laz'-zi-an) Obs. European stage: Upper Pleistocene (above Sicilian, below Tyrrhenian). See: *Ionian.*

mile Any of various units of distance that were derived from the ancient Roman marching unit of 1,000 double paces (a double pace = 5 ft) and that underwent many changes as the term came into use among the western nations (e.g., a mile = 1,620 English yards or 1,482 meters); specif. *statute mile* and *nautical mile.* Etymol: Latin "mille", "thousand".

milieu (mi'-lieu) A French term used in paleontology, sedimentation, and stratigraphy for environment, surroundings, or setting; e.g. the environment characteristic of a stratigraphic facies.

miliolid (mil'-i-o-lid') A foraminifer belonging to the family Miliolidae, characterized by a test that usually has a porcelaneous and imperforate wall and has two chambers to a whorl variably arranged about a longitudinal axis.

milioline (mil'-i-o-line') Pertaining or belonging to or resembling the foraminiferal order Miliolida; e.g. formed as in the foraminiferal tests of the superfamily Miliolacea, commonly with narrow elongate chambers (two to a whorl) added in differing planes of coiling (TIP, 1964, pt. C, p.61).

miliolite (mil'-i-o-lite') A fine-grained limestone of eolian origin, consisting chiefly of the tests of *Miliola* and other foraminifers.

military geology (mil'-i-tar'-y) (a) Those branches of the earth sciences, especially geomorphology, soil science, and climatology, that are applied to such military concerns as *terrain analysis,* water supply, cross-country movement, location of construction materials, and the building of roads and airfields. (b) The application of geological sciences to decision-making processes required by the military command (Rosenbaum, 1994).

milk opal A translucent and milk-white to green, yellow, or blue variety of common opal.

milky quartz (milk'-y) A milk-white, nearly opaque variety of crystalline quartz often having a greasy luster. The milkiness is usually due to the presence of innumerable very small cavities containing fluids. Syn: *greasy quartz.*

Milky Way The *galaxy* in which the sun and the solar system reside.

milled ring A flange near the base of an echinoid spine for the attachment of muscles controlling the movement of the spine.

millepore (mil'-le-pore) Any one of a group of hydrozoans belonging to the order Milleporina, characterized by a calcareous skeleton and free-swimming sexual individuals. Cf: *hydroid; stylaster.*

Miller-Bravais indices (Mil'-ler-Bra'-vais) A four-index type of *Miller indices,* useful but not necessary in order to define planes in crystal lattices in the hexagonal system; the symbols are hkīl, in which ī=-(h+k). Syn: *hexagonal indices.*

Miller cylindrical projection A *cylindrical projection* for world maps based on modifying the *Mercator projection* so that the poles appear as straight lines. There is no distortion along the equator. The parallels are gradually spaced farther apart toward the two poles, but not as rapidly as on the Mercator projection. Introduced in 1942 by O.M. Miller of the American Geographical Society. Resembles the *Gall projection.*

Miller indices (Mil'-ler) A set of three or four symbols (letters or integers) used to define the orientation of a crystal face or internal crystal plane. The indices are determined by expressing, in terms of lattice constants, the reciprocals of the intercepts of the face or plane on the 3 crystallographic axes, and reducing (clearing fractions) if necessary to the lowest integers retaining the same ratio. When the exact intercepts are unknown, the general symbol (*hkl*) is used for the indices, where *h, k,* and *l* are respectively the reciprocals of rational but undefined intercepts along the *a, b,* and *c* crystallographic axes. In the hexagonal system, the Miller indices are (*hkīl*); these are known as the *Miller-Bravais indices.* Indices designating individual crystal faces are enclosed in parentheses; complete crystal forms, in braces; crystal zones, in square brackets; and crystallographic lines, in greater-than/less-than symbols. To denote the interception at the negative end of an axis, a line is placed over the appropriate index, as (*1̄11*). The indices were proposed by William H. Miller (1801-1880), English mineralogist. See also: *indices of lattice row.* Cf: *crystal form; diffraction indices.* Syn: *crystal indices; hkl indices.*

millerite (mil'-ler-ite) A brass-yellow to bronze-yellow rhombohedral mineral: NiS. It usually has traces of cobalt, copper, and iron, and is often tarnished. Millerite generally occurs in fine hairlike or capillary crystals of extreme delicacy, chiefly as nodules in oxidized nickel sulfide ores or clay ironstone. Syn: *capillary pyrites; nickel pyrites; hair pyrites.*

millet-seed sand Sand that consists essentially of smoothly and conspicuously rounded grains about the size of a millet seed; specif. a desert sand whose grains have a surface like that of ground glass and are very perfectly rounded as a result of wind action.

mill-hole mining *glory-hole mining.*

millidarcy (mil'-li-dar'-cy) The customary unit of measurement of fluid permeability, equivalent to 0.001 *darcy.* Abbrev: md.

milligal (mil'-li-gal) A unit of acceleration used with gravity measurements; 10^{-3} gal = 10^{-5}m/sec^2. Abbrev: mGal.

milling ore (mill'-ing) *second-class ore.*

millionth-scale map of the world (mil'-lionth-scale) *International Map of the World.*

millisite (mil'-lis-ite) A white, light gray, or greenish tetragonal mineral of the *wardite* group: $(Na,K)CaAl_6(PO_4)_4(OH)_9{\cdot}3H_2O$.

mill ore Var. of *milling ore.*

millosevichite (mil-lo-sev'-ich-ite) A violet-blue rhombohedral mineral: $(Al,Fe^{3+})_2(SO_4)_3$. It is the Al analogue of mikasaite.

mill-rock (a) A type of coarse acidic pyroclastic breccia found in or close to the volcanic units in which Canadian massive sulfide ore deposits occur (Sangster, 1972, p.3). (b) More generally, any proximal, typically explosive rhyolite pyroclastic breccia.

Mill's knife A mechanical downhole device used inside a water-well casing to create or enlarge perforations.

millstone (mill'-stone) A buhrstone; e.g. a coarse-grained sandstone or a fine-grained quartz conglomerate. Also, one of two thick disks of such material formerly used for grinding grain and other materials, which were fed through a center hole in the upper stone.

millstone grit Any hard, siliceous rock suitable for use as a material for millstones; specif. the Millstone Grit of the British Carboniferous, a coarse conglomeratic sandstone.

Mima mound (Mi'-ma [my'-ma]) A term used in the NW U.S. for one of numerous low, circular or oval domes composed of loose, unstratified, gravelly silt and soil material, built upon glacial outwash on a hog-wallow landscape; the basal diameter varies from 3 m to >30 m, and the height from 30 cm to about 2 m. Named after the Mima Prairie in western Washington State. Cf: *pimple mound.*

mimetene (mim'-e-tene) *mimetite.*

mimetesite (mi-met'-e-site) *mimetite.*

mimetic [cryst] (mi-met'-ic) Pertaining to a twinned or malformed crystal that appears to have higher symmetry than it actually does.

mimetic [evol] Said of an organism that exhibits or is characterized by *mimicry.*

mimetic [struc petrol] Said of a tectonite whose deformation fabric, formed by recrystallization or neomineralization, reflects and is influenced by pre-existing anisotropic structure; also, said of the fabric itself.

mimetic crystallization Recrystallization or neomineralization in *metamorphism* that reproduces any preexistent anisotropy, bedding, schistosity, or other structures (Knopf and Ingerson, 1938). Obsolete. Syn: *facsimile crystallization.*

mimetite (mim'-e-tite) A yellow to yellowish-brown hexagonal mineral of the *apatite* group: $Pb_5(AsO_4)_3Cl$. It is isomorphous with pyromorphite, and commonly contains some calcium or phosphate. Clinomimetite has a closely related structure with monoclinic symmetry. Mimetite usually occurs in the oxidized zone of lead veins, and is a minor ore of lead. Syn: *mimetene; mimetesite.*

mimicry (mim'-ic-ry) A superficial similarity between organisms, or between an organism and its surroundings, as a means of concealment, protection, or other advantage. See also: *mimetic [evol].*

minable (min'-a-ble) Said of a mineral deposit for which extraction is technically feasible and economically worthwhile.

minal (min'-al) *end member.*

minamiite (mi'-na-mi'-ite) A colorless rhombohedral mineral: $(Na,Ca,K)Al_3(SO_4)_2(OH)_6$. It is a member of the *alunite* group.

minasgeraisite-(Y) (mi'-nas-ger-ais'-ite-y) A lavender or purple monoclinic mineral of the *datolite* group: $Ca(Y,Bi)_2Be_2Si_2O_{10}$.

minasragrite (mi-nas-rag'-rite) A blue monoclinic efflorescent mineral: $V^{4+}O(SO_4)\cdot 5H_2O$.

Mindel (Min'-del) (a) European stage: Pleistocene (above Günz, below Riss). (b) The second classical glacial stage of the Pleistocene Epoch in the Alps, after the Günz-Mindel interglacial stage. See also: *Kansan; Elster.* Etymol: Mindel River, Bavaria. Adj: Mindelian.

Mindel-Riss The term applied in the Alps to the second classical interglacial stage of the Pleistocene Epoch, after the Mindel glacial stage and before the Riss. Cf: *Yarmouth.*

mine n. (a) An underground excavation for the extraction of mineral deposits, in contrast to surficial excavations such as quarries. The term is also applied to various types of open-pit workings. (b) The area or property of a mineral deposit that is being excavated; a mining claim. v. To excavate for and extract mineral deposits or building stone.

mineevite-(Y) A light green to yellowish-green hexagonal mineral: $Na_{25}Ba(Y,Gd,Dy)_2(CO_3)_{11}(HCO_3)_4(SO_4)_2F_2Cl$.

minehillite (mine-hill'-ite) A colorless trigonal mineral: $(K,Na)_{2-3}Ca_{28}Zn_5Al_4Si_{40}O_{112}(OH)_{16}$.

mineragraphy (min-er-ag'-ra-phy) An obsolescent syn. of *ore microscopy.*

mineral (min'-er-al) (a) A naturally occurring inorganic element or compound having a periodically repeating arrangement of atoms and characteristic chemical composition, resulting in distinctive physical properties. (b) An element or chemical compound that is crystalline and that has formed as a result of geologic processes. Materials formed by geologic processes from artificial substances are no longer accepted (after 1995) as new minerals (Nickel, 1995). Mercury, a liquid, is a traditional exception to the crystallinity rule. Water is not a mineral (although ice is), and crystalline biologic and artificial materials are not minerals. Cf: *mineraloid.* (c) Any naturally formed inorganic material, i.e. a member of the mineral kingdom as opposed to the plant and animal kingdoms.

mineral aggregate An *aggregate* or assemblage of more than one crystal grain (which may be of one or several mineral species) and containing more than one crystal lattice. It can occur as sediment if loosely bound, or as rock if tightly bound.

mineral assemblage (a) The minerals that compose a rock, esp. an igneous or metamorphic rock. The term includes the different kinds and relative abundances of minerals, but excludes the texture and fabric of the rock. See also: *metamorphic assemblage.* (b) *mineral association.*

mineral association A group of minerals found together in a rock, esp. in a sedimentary rock. Syn: *mineral assemblage.*

mineral belt An elongated region of mineralization; an area containing several mineral deposits.

mineral blossom drusy quartz.

mineral caoutchouc *elaterite.*

mineral charcoal *fusain.*

mineral deposit A mass of naturally occurring mineral material, e.g. metal ores or nonmetallic minerals, usually of economic value, without regard to mode of origin. Accumulations of coal and petroleum may or may not be included; usage should be defined in context. Cf: *mineral occurrence.*

mineral-deposit model A conceptual model that describes the essential characteristics of a group of similar mineral deposits or of a *mineral-deposit type.*

mineral-deposit type A collective term for mineral deposits that (a) share a set of geologic attributes, and (b) contain a particular mineral commodity or combination of commodities such that (a) and (b) together distinguish them from other types of mineral deposits (Eckstrand et al., 1995). Mineral deposits of the same type are likely to have a common or similar genesis.

mineral disintegration *granular disintegration.*

mineral economics The subdiscipline of economics which studies the pricing, supply and demand, and marketing of mineral commodities.

mineral endowment *endowment.*

mineral facies (a) An approx. syn. of *metamorphic facies.* (b) Rocks of any origin whose constituents have been formed within the limits of a certain pressure-temperature range characterized by the stability of certain index minerals.

mineral filler A finely pulverized inert mineral or rock that is included in a manufactured product, e.g. paper, rubber, and plastics, to impart certain useful properties, such as hardness, smoothness, or strength. Common mineral fillers include asbestos, kaolin, and talc.

mineralization [ore dep] (min'-er-al-i-za'-tion) The process or processes by which a mineral or minerals are introduced into a rock, resulting in a valuable or potentially valuable deposit. It is a general term, incorporating various types, e.g. fissure filling, impregnation, replacement.

mineralization [paleont] A process of fossilization whereby the organic components of an organism are replaced by inorganic ("mineral") material.

mineralize (min'-er-al-ize') To convert to a mineral substance; to impregnate with mineral material. The term is applied to the processes of ore formation and also to the process of fossilization.

mineralizer (min'-er-al-iz'-er) (a) A gas or fluid that dissolves, receives by fractionation, transports, and precipitates ore minerals. A mineralizer is typically aqueous, with various hyperfusible gases (CO_2, CH_4, H_2S, HF), simple ions (H^+, HS, Cl^-, K, Na, Ca), complex ions (esp. chloride complexes), and dissolved base and precious metals. Syn: *ore-forming fluid.* (b) A gas that is dissolved in a magma and that aids in the concentration, transport, and precipitation of certain minerals and in the development of certain textures as it is released from the magma by decreasing temperature and/or pressure. Cf: *fugitive constituent; volatile component.*

mineral lands Legally, areas considered more valuable for their ore deposits or mineral potential than for agriculture or other purpose. Cf: *stone land.*

mineral lineation A linear fabric element defined by the preferred orientation of inequant mineral grains (e.g. prismatic grains of horneblende) or elongated grain aggregates.

mineral matter The inorganic material in coal and peat. See also: *inherent mineral matter.*

mineral occurrence Any ore or economic mineral in any concentration found in bedrock or as float; esp. a valuable mineral in sufficient concentration to suggest further exploration. Cf: *mineral deposit.*

mineralogic (min'-er-a-log'-ic) Adj. of *mineralogy.*

mineralogical (min'-er-a-log'-i-cal) Adj. of *mineralogy.*

mineralogical phase rule Any of several modifications of the fundamental Gibbs *phase rule,* taking into account the number of degrees of freedom consumed by the fixing of physical-chemical variables in the natural environment. The most famous such rule, that of Goldschmidt, assumes that two variables (taken as pressure and temperature) are fixed externally and that consequently the number of phases (minerals) in a system (rock) will not generally exceed the number of components. The Korzhinskiy-Thompson version takes into account the external imposition of chemical potentials of perfectly mobile components, and thereby reduces the maximum expectable number of minerals in a given rock to the number of inert components. Syn: *Goldschmidt's phase rule.*

mineralogic maturity A type of sedimentary maturity in which a clastic sediment approaches the mineralogic end product to which it is driven by the formative processes that operate upon it (Pettijohn, 1957). The ultimate sand is a concentration of pure quartz, and the mineralogic maturity of sandstones is commonly expressed by the quartz/feldspar ratio; this ratio is not so appropriate for sand derived from feldspar-poor rocks and the ratio of quartz + chert/feldspar + rock fragments may be substituted as more generally applicable. Cf: *maturity [sed]; compositional maturity; textural maturity.*

mineralogist (min-er-al'-o-gist) One who studies the formation, occurrence, properties, composition, and classification of minerals; a geologist whose field of study is *mineralogy.*

mineralography (min'-er-a-log'-ra-phy) A syn. of *mineragraphy;* both are obsolescent terms for *ore microscopy.*

mineralogy (min-er-al'-o-gy) (a) The study of minerals: formation, occurrence, properties, composition, and classification. See also: *mineralogist.* Adj: *mineralogic; mineralogical.* Obsolete syn: *oryctology; oryctognosy.* (b) An obsolete use of the term is for the general geology of a region.

mineraloid (min'-er-al-oid') A naturally occurring, usually inorganic substance that is not considered to be a *mineral* because it is amorphous and thus lacks a periodically repeating arrangement of atoms; e.g., opal. Syn: *gel mineral.*

mineral pathology Study of the changes undergone by unstable minerals in an environment whose conditions of temperature, pressure, and composition are different from those under which the minerals originally formed (Pettijohn, 1957, p.502).

mineral pigment An inorganic pigment, either natural or synthetic, used to give color, opacity, or body to a paint, stucco, plaster, or similar material. See also: *ocher; sienna.*

mineral pitch An obsolete syn. of *asphalt.*

mineral reserve *ore reserve.*

mineral reserves *reserves.*

mineral resin Any of a group of resinous, usually fossilized, mineral hydrocarbon deposits; e.g. bitumen and asphalt. See also: *resin.*

mineral resource A concentration or occurrence of material of intrinsic economic interest in or on the Earth's crust in such form and quantity that there are reasonable propspects for eventual economic extraction (JORC, 1999). Cf: *indicated mineral resource; inferred mineral resource; measured mineral resource.*

mineral resources *resources.*

mineral rod *divining rod.*

mineral sands *beach placer.*

mineral sequence *paragenetic sequence.*

mineral soap *bentonite.*
mineral soil A soil consisting predominantly of, and having its properties determined predominantly by, mineral matter. Usually contains < 200 g kg^{-1} organic carbon (<120-180 g kg^{-1} if saturated with water), but may contain an organic surface layer up to 30 cm thick.
mineral spring A spring whose water contains enough mineral matter to give it a definite taste, in comparison to ordinary drinking water, esp. if the taste is unpleasant or if the water is regarded as having therapeutic value. This type of spring is often described in terms of its principal characteristic constituent; e.g. *salt spring.*
mineral streaking A type of lineation in metamorphic rock defined by alignment of elongate or stretched crystals or esp. by linear streaks of microcrystalline mica in the plane of foliation.
mineral survey The marking of legal boundaries of ore deposits or mineralized formations on public land, where the boundaries are to be determined by lines other than those of the normal subdivision of the public lands.
mineral tallow *hatchettine.*
mineral time *Geologic time* estimated on the basis of radioactive minerals (Kobayashi, 1944b, p.476). Cf: *numerical time.*
mineral water Water that contains naturally or artificially supplied mineral salts or gases (e.g. carbon dioxide).
mineral wax *ozocerite.*
mineral wool A generic term for felted or matted fibers manufactured by blowing or spinning threads of molten rock, slag, or glass. The material is used for thermal insulation. Syn: *rock wool.*
mineral zone An informal term for a stratigraphic unit classified on the basis of mineral content (usually detrital minerals) and usually named from characteristic minerals (ISST, 1961, p.29).
mineral zoning *zoning of ore deposits.*
minerocoenology (min'-er-o-coe-nol'-o-gy) The study of mineral associations in the broadest sense, such as the correlation of igneous rocks or magmatic provinces with their ore deposits (Thrush, 1968, p.712). Obsolete.
minerogenetic (min'-er-o-ge-net'-ic) *metallogenic.*
minerogenic (min'-er-o-gen'-ic) *metallogenic.*
minerotrophic peat Peat whose water is originally derived from mineral soils or rocks but sometimes via lakes or rivers as intermediates; it may be eutrophic, mesotrophic, or oligotrophic. Cf: *topogenous peat; ombrotrophic peat.*
miner's inch (min'-er's) A measure of water flow equal to 1.5 cu ft/min.
minette (mi-nette') (a) In the *IUGS classification,* a variety of *lamprophyre* with alkali feldspar more abundant than plagioclase, no feldspathoid, and generally containing biotite and clinopyroxene phenocrysts. (b) A *lamprophyre* primarily composed of biotite phenocrysts in a groundmass of alkali feldspar and biotite.
Minette-type ironstone *Ironstone* deposits principally composed of goethite. Syn: *brown ironstone.*
minguzzite (min-guz'-zite) A green or yellow-green monoclinic oxalate mineral: $K_3Fe^{3+}(C_2O_4)_3{\cdot}3H_2O$.
miniature atoll (min'-i-a-ture') *microatoll.*
minibasin A small basin subsiding into relatively thick allochthonous or autochthonous salt. Such basins merge over time into larger, composite basins.
minimum [geophys] (min'-i-mum) n. An anomaly characterized by values smaller than those at neighboring locations; e.g., a gravity minimum or a geothermal minimum. Cf: *maximum [geophys].*
minimum-delay wavelet *minimum-phase wavelet.*
minimum detectable power In infrared detector technology, the incident power that will give a signal-to-noise ratio equal to unity at the output of the detector (Smith et al., 1968, p.250).
minimum pendulum A pendulum used in gravity measurements, so designed that changes in period resulting from small changes in length are at a minimum. Among factors that may tend to change the length are temperature, creep, and knife-edge wear.
minimum phase A characteristic of a waveform that has its energy concentrated in its front portion and has no anticipatory component.
minimum-phase wavelet A wavelet whose energy is concentrated in the front end of the wavelet, sometimes called front-loaded. Most seismic sources produce wavelets which are nearly minimum-phase. Also called: *minimum-delay wavelet.*
minimum slope A slope that is flatter in gradient than the slope units above or below it.
minimum-time path The path between two points along which the time of travel is less than on neighboring paths. See also: *Fermat's principle.* Syn: *least-time path; brachistochrone.*
mining (min'-ing) The process of extracting metallic or nonmetallic mineral deposits from the Earth. The term may also include preliminary treatment, e.g. cleaning or sizing. Cf: *mining geology; mining engineering.*
mining claim A *claim* on mineral lands.
mining engineering The planning and design of mines, taking into account economic, technical, and geologic factors; also supervision of the extraction, and sometimes the preliminary refinement, of the raw material. Cf: *mining; mining geology.*
mining geology The study of the geologic aspects of mineral deposits, with particular regard to problems associated with mining. Cf: *mining; mining engineering.*
minipermeameter (min-i-per-me-am'-e-ter) Gas-flow device used to measure permeability of flat rock faces at the millimeter scale (Goggin, 1993).
minipermeametry Measurement of permeability of flat rock faces at the millimeter scale (Goggin, 1993).
minium (min'-i-um) A bright-red, scarlet, or orange-red tetragonal mineral: $Pb^{2+}{}_2Pb^{4+}O_4$. Syn: *red lead.*
minnesotaite (min-ne-so'-ta-ite) A green to brown triclinic silicate mineral commonly found in metamorphosed Fe,Mg-rich sediments: $(Fe^{2+},Mg)_3Si_4O_{10}(OH)_2$. Most specimens are mixtures of two phases, differing in Fe/Mg ratio.
minor element (mi'-nor) (a) A syn. of *trace element.* (b) A term that is occasionally used for an element that normally comprises between one and five percent of a rock; it is not quantitatively defined.
minor elements The solutes that are present in seawater in abundances ranging from low to trace. Essentially the reactive or *nonconservative elements* in the ocean. Cf: *major elements.*
minor fold A small-scale fold that is associated with or related to the *major fold* of an area. Cf: *parasitic fold.*
minor metal Said of a metal with a very small annual market demand. Typically, such metals are by-products of base, ferrous, or precious metal mining or produced in small quantities on demand from a few mines. Examples are arsenic, selenium, the rare earth metals, or zirconium.
minor planet *asteroid [astron].* The term *minor planet* is preferred as the official name by the International Astronomical Union, although asteroid is more common in general usage.
minor scarp A steep surface on the displaced material of a landslide, produced by differential movements within the sliding mass. Cf: *main scarp.*
minor septum One of the relatively short, third-cycle septa of a *corallite,* commonly inserted between, and much shorter than, adjacent major septa.
minrecordite (min-re'-cord-ite) A colorless rhombohedral mineral of the *dolomite* group: $CaZn(CO_3)_2$.
minus-blue photograph A black-and-white photograph, acquired using a filter that removes blue wavelengths to produce higher spatial resolution.
minus-cement porosity (mi'-nus-ce-ment') The porosity that a sedimentary material would have if it contained no chemical cement.
minus sight *foresight.*
minute (min'-ute) (a) A unit of time equal to $^1/_{60}$ of an hour and containing 60 seconds. Abbrev: min; m (in physical tables). (b) A unit of angular measure equal to 1/60 of a degree and containing 60 seconds of arc. Symbol: '.
minyulite (min-yu'-lite) A white orthorhombic mineral: $KAl_2(PO_4)_2(OH,F){\cdot}4H_2O$.
Miocene (Mi'-o-cene) A series of the *Neogene* Subsystem of the *Tertiary* System of the Standard Global Chronostratigraphic Scale, above the *Oligocene* and below the *Pliocene.* Also the time during which these rocks were formed, the Miocene Epoch, covering the time span between 23 and 5 Ma.
miogeocline (mi-o-ge'-o-cline) A prograding wedge of shallow-water sediment at the continental margin (Dietz and Holden, 1966) or along a geosynclinal seaway. A shortened version of *miogeosyncline.* Cf: *eugeocline.*
miogeosyncline (mi'-o-ge'-o-syn'-cline) A geosyncline in which volcanism is not associated with sedimentation; the nonvolcanic aspect of an orthogeosyncline, located near the craton (Stille, 1940). Syn: *miomagmatic zone.* Cf: *eugeosyncline.* See also: *ensialic geosyncline.* Obsolete.
miomagmatic zone (mi'-o-mag-mat'-ic) *miogeosyncline.*
miomirite (mi-o-mir'-ite) A variety of davidite containing lead.
miospore (mi'-o-spore) A term arbitrarily defined in paleopalynology as a spore or pollen grain <200 µm in maximum diameter, regardless of its possible biological status. Cf: *megaspore; microspore; small spore.*
miosynthem A relatively small, minor *synthem* within a large synthem but not a component of a hierarchy of unconformity-bounded units (ISSC, 1987a, p.236; the term was not used in ISSC, 1994).
mirabilite (mi-rab'-i-lite) A white or yellow monoclinic mineral: $Na_2SO_4{\cdot}10H_2O$. It occurs as a residue from saline lakes, playas, and springs, and as an efflorescence. Syn: *Glauber's salt.*
mire (a) A small piece of marshy, swampy, or boggy ground; wet spongy earth. (b) Soft, heavy, often deep mud or slush. Obsolete syn: *slough [geog].* (c) Commonly used in the sense of *peatland,* particularly in Europe, New Zealand, and the former Soviet Union. Now an internationally accepted term that includes the generally ombrotrophic (e.g. bog) and minerotrophic (e.g. fen) types of peat (Gore, 1983). The collective term for all peat-forming ecosystems (Immirzi et al., 1992).
mirror The planar portion of a joint characterized by either a smooth surface (e.g., in glass or in shale) or by plumose surface morphology (e.g., in granite or in sandstone) (Bahat, 1991). Some interpret this region as a manifestation of subcritical joint growth.
mirror glance (mir'-ror) *wehrlite [mineral].*
mirror plane of symmetry *plane of mirror symmetry.*
mirror stone *muscovite.*
mischungskorrosion (misch'-ungs-kor-ro'-sion) The German equivalent of *mixing dissolution* or *mixture dissolution.*

miscibility gap (mis-ci-bil'-i-ty) An intermediate compositional range within a solid solution where stable single-phase compositions do not exist. Compositions within the gap are represented by a coexistence of two solid-solution phases with individual compositions at the limits of the gap. Miscibility gaps commonly widen with decreasing temperature, leading to progressive exsolution. Common examples of miscibility gaps include coexisting sodic plagioclase and potassic feldspar (perthite) and coexisting dolomite and magnesian calcite.

miscible (mis'-ci-ble) Said of two or more phases that, when brought together, have the ability to mix and form one phase. Cf: *immiscible.*

misclosure (mis-clo'-sure) *error of closure.*

mise a la masse A drill-hole resistivity or induced-polarization survey technique in which a buried conductor is directly energized and serves as a large buried electrode. Potentials are measured on the surface, in bore holes, or underground. Etymol: French.

misenite (mi'-sen-ite) A white or grayish-white monoclinic mineral: $K_8(SO_4)(SO_3OH)_6$.

miserite (mi'-ser-ite) A pink, lavender, or reddish brown coarsely fibrous triclinic mineral: $K(Ca,Ce)_6Si_8O_{22}(OH,F)_2$.

misfit stream (mis'-fit) (a) A stream whose meanders are obviously not proportionate in size to the meanders of the valley or to the meander scars preserved in the valley wall; a stream that is either too large (an *overfit stream*) or too small (an *underfit stream*) to have eroded the valley in which it flows. (b) A term that is often incorrectly used as a syn. of *underfit stream.*

mispickel (mis'-pick-el) *arsenopyrite.*

Mississippian (Mis-sis-sip'-pi-an) The lower subsystem of the *Carboniferous* System of the Standard Global Chronostratigraphic Scale, above the *Devonian* System and below the *Pennsylvanian* Subsystem. Also the time during which these rocks were formed, the Mississippian Subperiod, covering the time span between 355 and 320 Ma. It is named after the Mississippi River valley, in which there are good exposures of rocks of this age and is the approximate equivalent of the Lower Carboniferous of some chronostratigraphic scales.

Mississippi bentonite *sodium bentonite.*

Mississippi Valley-type deposit (Mis-sis-sip'-pi) A strata-bound deposit of lead and/or zinc minerals in carbonate rocks, together with associated fluorite and barite. These deposits characteristically have relatively simple mineralogy, occur as veins and replacement bodies, are at moderate to shallow depths, show little post-ore deformation, are marginal to sedimentary basins, and are without an obvious source of the mineralization. Examples: Wisconsin-Illinois lead deposits; Kentucky-Illinois fluorspar deposits; Appalachian zinc and barite deposits. Abbrev: MVT.

Missourian (Mis-sou'-ri-an) A North American stage or series: Middle Pennsylvanian (above Desmoinesian, below Virgilian).

missourite (mis-sou'-rite) (a) In the *IUGS classification,* a plutonic rock in which F is between 60 and 100, M is between 70 and 90, and potassium exceeds sodium. Cf: *melteigite.* (b) A plutonic rock containing a potassium feldspathoid (leucite) and 60 to 90 percent mafic minerals, such as pyroxene and olivine. Its name, proposed by Weed and Pirsson in 1896, is derived from the Missouri River in Montana. Cf: *fergusite; italite.* Obsolete.

mist (a) Very thin *fog* in which visibility is greater than 1.0 km (0.62 mi). Mist is also known as light *drizzle.* (b) In glass, the transition from a *mirror* surface to *hackle* (Bahat, 1991). This term does not apply to a joint with plumose surface morphology in the mirror region.

mis-tie A term used in surveying for the failure of the first and final observations around a closed loop to be identical, or for the failure of the values at identical points on intersecting loops to be the same. See also: *error of closure.*

misy (mi'-sy) A term for various poorly defined iron sulfates.

mitosis (mi-to'-sis) The somatic cell nuclear division involving no reduction in chromosome number. Adj: *mitotic.*

mitotic (mi-to'-tic) Relating to *mitosis.*

mitridatite (mit-ri-da'-tite) An earthy variously colored monoclinic mineral: $Ca_2Fe^{3+}{}_3O_2(PO_4)_3 \cdot 3H_2O$. It characteristically forms dull bile-green stains on the walls of abandoned pegmatites.

mitryaevaite A dull white triclinic mineral: $Al_5(PO_4)_2[(P,S)O_3(OH,O)]_2F_2(OH)_2 \cdot 14.5H_2O$.

mitscherlichite (mitsch'-er-lich-ite') A greenish-blue tetragonal mineral: $K_2CuCl_4 \cdot 2H_2O$.

mixed ages *discordant ages.*

mixed assemblage An *assemblage [ecol]* composed of some specimens representing a *fossil community* or a *winnowed community,* plus others representing one or more transported assemblages brought into the locality where found. See also: *disturbed-neighborhood assemblage; thanatocoenosis.* Syn: *remanié assemblage.*

mixed-base crude A crude oil in which both paraffinic and naphthenic hydrocarbons are present in approximately equal proportion. Cf: *paraffin-base crude; asphalt-base crude.*

mixed breccia Breccia containing multiple rock types that may either be melt bearing or melt free. Frequently found in a *breccia lens.* Syn: *polymict breccia.*

mixed crystal *solid solution.*

mixed cut A combination of *brilliant cut* above the girdle, usually with 32 facets, and *step cut* below, with the same number of facets or vice versa. It is often used for colored stones, esp. fancy sapphires, to improve color and retain brilliancy. Now also used for diamonds.

mixed dislocation A dislocation that is oblique to the direction of slip and has both screw and edge components.

mixed eruption A volcanic phase that includes both the emission of lava and the explosive ejection of pyroclasts. Cf: *explosive eruption; lava eruption.*

mixed flow Water flow that is partly *turbulent flow* and partly laminar flow.

mixed gneiss *composite gneiss.*

mixed layer The layer of ocean water above the pycnocline; it is mixed by wind and wave action. It is equivalent to the *epilimnion* in a lake. In the ocean, vertical mixing is commonly so intense that both nonconservative and *conservative elements* have constant or nearly constant concentrations throughout the layer. It frequently contains much or all of the *euphotic* zone.

mixed-layered clay minerals Phyllosilicate structures in which two or more kinds of layers occur in a stacking sequence along one direction, usually along a line normal to the layers. Some interstratified clay minerals, most notably those containing interstratified 10 Å layers and smectite layers, have orderly stacking in 1:1 or 1:3 ratios. Phyllosilicates with regular interstratifications may be designated as mineral species, for which criteria are given in Bailey (1982). Interstratified clay minerals without sufficient stacking order are usually designated by having the component names of the layers separated by either a slash(/) or a hyphen (-) and the component with the smaller layer value named first; e.g., illite/smectite or illite-smectite. Syn: *interstratified clay minerals.* Cf: *Reichweite (R); illite/smectite.*

mixed-layer mineral A mineral whose structure consists of alternating layers of clay minerals and/or mica minerals; e.g. chlorite, made up of alternating biotite and brucite sheets.

mixed-mode crack A crack whose tip is subject to any combination of opening (mode I) displacement and shear (modes II and III) displacements.

mixed ore An ore of both oxidized and unoxidized minerals.

mixed peat Peat that is stratified according to plant associations. Cf: *banded peat.*

mixed tide Tide that is a mixture of diurnal and semidiurnal periods of about equal amplitude, which causes two *high waters* and two *low waters* each day with marked differences in *tidal range.*

mixed topostratic unit A topostratic unit with the lower boundary based on lithologic data and the upper boundary based on paleontologic data, or vice versa (Henningsmoen, 1961, p.74).

mixed volatile Pertaining to more than one species or component in the fluid phase of natural, experimental, or hypothetical rock-fluid systems. In metamorphic petrology, mixed volatile reactions are considered in particular in the progressive metamorphism of impure limestones and dolostones.

mixed water A term used by White (1957, p.1639) for any mixture of volcanic and meteoric waters in any proportion. White recommended discontinuing use of the term for chloride- and sulfide-rich acid waters.

mixing (mix'-ing) Summing the output of different channels to attenuate noise. Syn: *compositing.*

mixing coefficient *austausch.*

mixing dissolution The dissolution of carbonate rock by the mixing of two or more waters of contrasting content of carbon dioxide or hydrogen sulfide or degree of salinity. Although all waters are originally at or near saturation with respect to the local bedrock, the mixture becomes undersaturated because of the nonlinear increase in solubility of carbonate rocks with increasing carbon dioxide, hydrogen sulfide, or salinity (Bogli, 1964). Syn: *mischungskorrosion.* Cf: *mixture dissolution.*

mixing length The length, normal to the flow direction, over which a small volume of fluid is assumed to retain its identity in the mass exchange process in turbulent flow. It is related to the coefficient of eddy viscosity and the rate of change of velocity normal to the line of flow (Middleton, 1965, p.250).

mixing ratio The ratio of the mass of water vapor to the mass of dry air with which it is associated. It is an important atmospheric quantity and is often expressed in units of g per kg. Cf: *humidity; absolute humidity; relative humidity; specific humidity.*

mixing zone cave A dissolution cave formed by water whose solutional capacity is derived from the mixing of waters of contrasting chemical properties.

mixite (mix'-ite) (a) An emerald-green or blue-green hexagonal mineral: $BiCu_6(AsO_4)_3(OH)_6 \cdot 3H_2O$. (b) A name for a group of minerals with an analogous composition, but with Bi replaced by Al, Y or REE, and PO_4 replaced by AsO_4.

mixolimnion (mix-o-lim'-ni-on) The upper, low-density, freely circulating layer of a *meromictic lake.* (Wetzel, 2001, p.83). Cf: *monimolimnion; chemocline.*

mixoperipheral growth (mix'-o-pe-riph'-er-al) Growth of a brachiopod valve in which the posterior part increases in size anteriorly and toward the other valve. Cf: *holoperipheral growth.*

mixotrophic (mix-o-troph'-ic) Said of an organism that is nourished by both *autotrophic* and *heterotrophic* mechanisms.

mixstone Scleractinian reef growth framework in which no single form dominates (Insalaco, 1998).

mixtite (mix'-tite) A descriptive group term proposed by Schermerhorn (1966, p.834) for a coarsely mixed, nonsorted or poorly sorted, clastic sedimentary rock, without regard to composition or origin; e.g. a tillite. Syn: *diamictite.*

mixton *diamicton.*

mixtum (mix'-tum) A term proposed by Schermerhorn (1966, p.834) for an unconsolidated *mixtite.*

mixture dissolution (mix'-ture) The ability of two calcite-saturated waters of different carbon-dioxide content to dissolve additional calcite when mixed. The CO_2 limits the solubility at low CO_2 content, whereas the water limits it at high CO_2 content, and the two factors together provide optimal solubility at an intermediate content of CO_2. Consequently, despite the saturated condition of the two initial waters, their mixture can dissolve more calcite (Moore and Sullivan, 1978, p.147). Syn: *mischungskorrosion.*

mizzonite (miz'-zo-nite) An unnecessary name for a variety of scapolite intermediate between meionite and marialite, and containing 54-57% silica; esp. such a variety of scapolite occurring in clear crystals in ejected masses on volcanoes. Syn: *dipyre.*

MLLW *mean lower low water.*

MLT *mechanical layer thickness.*

MLW *mean low water.*

MLWS *mean low water spring.*

MMcf A thousand thousand (million) cubic feet of gas.

MMR *magnetometric resistivity method.*

MM scale *modified Mercalli scale.*

mo A Swedish term for "glacial silts or rock flour having little plasticity" (Stokes and Varnes, 1955, p.93).

moat [glac geol] A glacial channel resembling a moat; e.g. a deep trench in glacier ice, surrounding a *nunatak,* and produced by ablation; or a channel at the margin of a dwindling glacier.

moat [marine geol] A ringlike depression around the base of many seamounts and islands. It may be discontinuous. Syn: *sea moat.*

moat [reef] An elongate water-filled channel, on or adjacent to a reef flat, and only a few meters deep and wide. Cf: *boat channel.*

moat [streams] A syn. of *oxbow lake.* The term is used in New England and was also applied by Shaler (1890, p.277) to the waters in abandoned channels in the Mississippi River flood plain.

moat [volc] A valleylike depression around the inner side of a crater or caldera, between its rim and a resurgent dome or cone constructed within it.

moated ice mound (moat'-ed) A mass of ice several feet in diameter, surrounded by a pebble-studded moat, found on the ice of Greely Fiord, Ellesmere Island. It is interpreted as the result of the melting of an *armored snowball* (Ferguson, 1970).

moat lake A lake of annular shape such as a *caldera lake* lying between the caldera wall and a resurgent dome (Heiken et al., 2000). (b) A lake in a basin of which the central part is occupied by a bog or fen (Veatch and Humphrys, 1966).

móberg An Icelandic term for hyaloclastite and basalt breccias formed by subglacial eruptions during the Pleistocene.

mobile belt (mo'-bile) A long, relatively narrow crustal region of tectonic activity, measured in scores of miles. See also: *orogenic belt; orogenic cycle.*

mobile component A component whose amount in a system may change during a given process because its chemical potential is fixed externally to the system being considered. In petrologic processes, such components are typically those with gaseous or fluid properties such as water, carbon dioxide, fluorine, or chlorine. In metasomatic processes, normally inert components such as alkalis, calcium, or magnesium may become mobile. Cf: *perfectly mobile component; inert component.*

mobility (mo-bil'-i-ty) A term used by W. Penck (1924) for the concept that the relative rate of uplift of the Earth's crust primarily determines the nature of the landforms produced by erosional processes.

mobilizate (mo'-bi-li-zate') The English equivalent of the German word "Mobilisat", introduced to refer to the mobile phase, of any consistency, that existed during migmatization. Cf: *chymogenic; metatect.* Cf: *neosome.*

mobilization (mo'-bi-li-za'-tion) (a) Any process that renders a solid rock sufficiently plastic to permit it to flow or to permit geochemical migration of the mobile components. Cf: *rheomorphism.* (b) Any process that redistributes and concentrates the valuable constituents of a rock into an actual or potential ore deposit.

Mocha stone (Mo'-cha) A white, gray, or yellowish form of moss agate containing brown to red iron-bearing or black manganese-bearing dendritic inclusions. The term is also used as a syn. of *moss agate.* Named for the city of Mocha (Al Mukha) in Yemen. Also spelled: mocha stone; mochastone. Syn: Mocha pebble.

mock lead *sphalerite.*

mock ore *sphalerite.*

moctezumite (moc-te-zu'-mite) A bright orange monoclinic mineral: $Pb(UO_2)(Te^{4+}O_3)_2$.

modal analysis A statement of the composition of a rock in terms of the relative amounts of minerals present; also, the procedure (usually point counter analysis or *Rosiwal analysis*) that yields such a statement (Chayes, 1956, p.1).

modal cycle A term proposed by Duff and Walton (1962) for a particular group of beds that occurs most frequently through a succession displaying cyclic sedimentation.

modal diameter An expression of the average particle size of a sediment or rock, obtained graphically by locating the highest point of the frequency curve or by finding the point of inflection of the cumulative curve; the diameter that is most frequent in the particle-size distribution.

modderite (mod'-der-ite) A metallic black orthorhombic mineral of the *westerveldite* group: (Co,Fe)As.

mode [petrology] The actual mineral composition of a rock, usually expressed in weight or volume percentages. Adj: modal. Cf: *norm.*

mode [stat] The value or group of values that occurs with the greatest frequency in a set of data; the most typical observation. Cf: *mean; median.*

mode I crack A crack whose walls have displaced normal to the crack plane.

mode II crack That portion of a crack tip whose walls have displaced parallel to the crack plane with slip in the direction of crack growth. A buried crack whose tip line curves back around itself must have a portion of its wall subject to mode III displacement.

mode III crack That portion of a crack tip whose walls have displaced parallel to the crack plane with slip normal to the direction of crack growth. A buried crack whose tip line curves back around itself must have a portion of its wall subject to mode II displacement.

model (mod'-el) A working hypothesis or precise simulation, by means of description, statistical data, or analogy, of a phenomenon or process that cannot be observed directly or that is difficult to observe directly. Models may be derived by various methods, e.g. by computer, from stereoscopic photographs, or by scaled experiments. Syn: *conceptual model.* See: Wolf, 1976, vol. 1, chap.1.

modeling materials Materials used in *scaled physical models* that have strengths or viscosities scaled down relative to those of the natural prototype (e.g., Withjack and Callaway, 2000). Sand and wet clay are commonly used to simulate faulting in the upper crust, where *strength* increases with depth and where the rocks obey *Coulomb's failure criterion.* Silicone putty is widely used to model *ductile behavior.*

model scale The relationship existing between a distance measured in a model (such as in a stereoscopic image) and the corresponding distance on the Earth.

moder (mod'-er) Plant material in a state intermediate between living and decayed.

moderate-energy coast (mod'-er-ate-en'-er-gy) A coast protected from strong wave action by headlands, wide gently sloping bottom, dominance of winds from the land, or other factors, and characterized by average breaker heights of less than 30 cm. Cf: *high-energy coast; low-energy coast.*

moderately sorted (mod'-er-ate-ly) Said of a sorted sediment that is intermediate between a well-sorted sediment and a poorly sorted sediment and that has a *sorting coefficient* (Trask's So) in the range of 2.5 to 4.0. On the basis of the phi values associated with the 84 and 16 percent lines, Folk (1954, p.349) suggests sigma phi limits of 0.50-1.00 for moderately sorted material.

modern carbon (mod'-ern) Carbon of living organisms. Due to changes in the production rate in the atmosphere, it has a carbon-14 activity of about 16 disintegrations per minute per gram carbon. Carbon-14 activity may differ among organisms because they obtain carbon from different reservoirs (oceans, biosphere), and among different parts of the same organism because of isotopic fractionation. Syn: *contemporary carbon.*

modern evolutionary fauna The classes of marine invertebrates which account for the bulk of marine diversity during the Mesozoic and Cenozoic, includes bivalves, gastropods, malacostracan arthropods, bony fish, and echinoids.

modification stage The last of the three stages of impact cratering, in which the *transient impact crater* collapses to form either a *simple crater,* a *complex impact crater,* or a *multi-ring basin.* Cf: *contact-and-compression stage; excavation stage.*

modified Mercalli scale (mod'-i-fied) An earthquake *intensity scale,* having twelve divisions ranging from I (not felt by people) to XII (damage nearly total). It is a revision of the *Mercalli scale* made by Wood and Neumann in 1931. Cf: *Rossi-Forel scale.* Abbrev: MM scale.

modified polyconic projection A projection used for many of the maps of the International Map of the World series at a scale of 1:1,000,000. It is derived from the regular *polyconic projection* by so altering the scale along the central meridian that the scale is exact along two standard meridians, one on either side of the central meridian and equidistant therefrom. Scale is preserved along these two meridians.

modulated structure An atomic arrangement in which the rigid periodicity required of a *crystal structure* has been violated by structural fluctuations that are caused either by incipient *exsolution* or by *microtwinning.* Cf: *continuous transformation.*

modulus The geometric scale that defines a cave space; cave passages of smaller scale are not considered included in a cave of specified modulus (Curl, 1966). See also: *proper modulus.*

modulus of compression (mod'-u-lus) *compressibility.*

modulus of elasticity The ratio of stress to its corresponding strain under given conditions of load, for materials that deform elastically, according to Hooke's law. It is one of the elastic constants. See also: *Young's modulus; modulus of rigidity; static modulus; bulk modulus.* Syn: *elastic modulus; modulus of volume elasticity.*

modulus of incompressibility *bulk modulus.*

modulus of rigidity A *modulus of elasticity* in shear. Symbol: µ or G. Syn: *shear modulus; rigidity modulus; Coulomb's modulus.*

modulus of volume elasticity *modulus of elasticity.*

modumite (mo'-dum-ite) A light-colored *essexite* or *anorthosite* containing bytownite, pyroxene, amphibole, and biotite. The name, given by Brögger in 1933, is from Modum in the Oslo district, Norway. Obsolete.

moel (mo'-el) A term used in Wales for a rounded hill with a vegetation-clad summit (Marr, 1901). Etymol: Welsh, "bare field".

moëloite A metallic orthorhombic mineral: $Pb_6Sb_6S_{17}$.

mofette (mo-fette') The exhalation of carbon dioxide in an area of late-stage volcanic activity; also, the small opening from which the gas is emitted. Examples are in Yellowstone National Park in the U.S. Etymol: French, "noxious gas".

mofettite (mof'-et-tite) A natural carbon-dioxide gas.

moganite A white, gray, or creamy monoclinic mineral: $SiO_2{\cdot}nH_2O$. Syn: *lutecite.*

mogote (mo-go'-te) A steep-sided hill of limestone, generally surrounded by nearly flat alluviated plains; karst inselberg. Cf: *karst tower.* Less preferred syn: *haystack hill.*

mohavite (mo-hav'-ite) *tincalconite.*

Mohawkian (Mo-hawk'-i-an) North American stage: Middle Ordovician (above Chazyan, below Cincinnatian). See also: *Trentonian.*

mohite (moh'-ite) A metallic greenish-gray triclinic mineral: Cu_2SnS_3.

Mohnian (Mohn'-i-an) A Pacific Coast provincial stage: middle and upper Miocene (above Luisian, below Repettian).

Moho (Mo'-ho) Abbreviated form of *Mohorovicic discontinuity,* suggested by Birch (1952, p.229).

Mohorovicic discontinuity (Mo'-ho-ro-vi'-cic) The seismic-velocity discontinuity or boundary surface that separates the Earth's crust from the subjacent mantle. It marks the level in the Earth at which *P wave* velocities change somewhat abruptly from 6.7-7.2 km/sec (in the lower crust) to 7.6-8.6 km/sec or average 8.1 km/sec (at the top of the upper mantle); its depth ranges from 5-10 km beneath the ocean floor to about 40 km below the continents, although it reaches more than 70 km under some mountain ranges. The discontinuity probably represents a chemical change from basaltic or simatic materials above to peridotitic or dunitic materials below, rather than a phase change (basalt to eclogite); however, the discontinuity is defined by seismic velocities alone and it is variously estimated to be 0.2-3 km thick. It is named in honor of its discoverer, Andrija Mohorovicic (1857-1936), Croatian seismologist. Syn: *Moho; M-discontinuity.*

Mohr circle A graphic representation of the state of stress at a particular point at a particular time. The coordinates of each point on the circle are the shear stress and the normal stress on a particular plane. See also: *Mohr envelope.*

Mohr envelope An envelope of a series of Mohr circles; the locus of points whose coordinates represent the normal and shear stresses at failure. Syn: *rupture envelope; Coulomb-Mohr envelope.*

mohrite (mohr'-ite) A pale green monoclinic mineral of the *picromerite* group: $(NH_4)_2Fe^{2+}(SO_4)_2{\cdot}6H_2O$.

Mohr-Knudsen method (Mohr-Knud'-sen) In oceanography, a chemical method for measuring the chlorinity of seawater involving the precipitation of halide ions with a titrant solution of silver nitrate. Cf: *chlorinity.*

mohsite (mohs'-ite) *crichtonite.*

Mohs scale A standard of ten minerals by which the hardness of a mineral may be rated. The scale includes, from softest to hardest and numbered one to ten: talc; gypsum; calcite; fluorite; apatite; orthoclase; quartz; topaz; corundum; and diamond. Cf: *technical scale.*

moire [cart] A wavy pattern on a map resulting from the overprinting of screens having inadequate angular separation.

moire [mineral] Said of feldspars having the appearance of watered silk.

moissanite (mois-san'-ite) A metallic green-black meteorite mineral: SiC. It is identical with the manufactured product Carborundum. It crystallizes in various hexagonal or rhombohedral polytypes.

moist playa *wet playa.*

moisture (mois'-ture) Water diffused in the atmosphere or the ground, including soil water.

moisture capacity The dewatering component of *matric potential.* The derivative of volumetric moisture content with reference to *matric potential.*

moisture content [coal] In coal, both the surface or free moisture that can be removed by natural drying, and the *inherent moisture* that is structurally contained in the substance. Also, the total of air-dried-loss moisture and residual moisture.

moisture content [soil] The amount of moisture in a given soil mass, expressed as weight of water divided by weight of oven-dried soil, multiplied by 100 to give a percentage. See also: *water content [sed].*

moisture deficiency The amount of water required to restore moisture to field capacity in desiccated soil.

moisture-density curve (mois'-ture-den'-si-ty) *compaction curve.*

moisture-density test *compaction test.*

moisture equivalent The ratio of weight of water that a saturated soil will retain against a centrifugal force 1,000 times the force of gravity to the weight of dry soil (Meinzer, 1923, p.25). Syn: *centrifuge moisture equivalent.*

moisture index A means for classifying climates devised by Thornthwaite; wet-season surplus minus 0.6 times dry-season deficiency, divided by total need all expressed in the same unit, such as inches multiplied by 100 to get a percentage. Cf: *Thornthwaite's classification.*

moisture meter An instrument for determining the percentage of moisture in a substance such as timber or soil, usually by measuring its electrical resistivity.

molar (mo'-lar) n. (a) One of the posterior grinding or shearing teeth of mammals, which are not normally replaced in an individual's lifetime. (b) A process with a grinding surface on the gnathal lobe of the *mandible* of a crustacean. Cf: *incisor.*

molasse (mo-lasse') (a) A paralic (partly marine, partly continental or deltaic) sedimentary facies consisting of a very thick sequence of soft, ungraded, cross-bedded, fossiliferous conglomerates, sandstones, shales, and marls, characterized by primary sedimentary structures and sometimes by coal and carbonate deposits. It is more clastic and less rhythmic than the preceding flysch facies. (b) An extensive, postorogenic sedimentary formation representing the totality of the molasse facies resulting from the wearing down of elevated mountain ranges during and immediately succeeding the main paroxysmal (diastrophic) phase of an orogeny, and deposited considerably in front of the preceding flysch; specif. the Molasse strata, mainly of Miocene and partly of Oligocene age, deposited on the Swiss Plain and Alpine foreland of southern Germany subsequent to the rising of the Alps. Etymol: French "mollasse", "soft". Adj: molassic. Cf: *flysch.*

molasse sandstone A sandstone of the *molasse* facies, characterized by Cayeux as a poorly rounded, poorly sorted, coarse sand rich in rock fragments and generally calcareous; generally a lithic arenite, in places arkosic (Pettijohn et al., 1972, p.168). The weathered and eroded product of a newly uplifted orogenic chain.

mold [paleont] (a) An *impression* made in the surrounding earth or rock material by the exterior or interior of a fossil shell or other organic structure. A complete mold would be the hollow space with its boundary surface. Cf: *cast [paleont].* See also: *external mold; internal mold.* (b) *natural mold.* (c) A cast of the inner surface of a fossil shell or other organic structure. Also spelled: *mould.*

mold [sed] An original mark or primary depression made on a sedimentary surface; e.g. a flute, striation, or groove. The filling of such a depression produces a cast; unfortunately "some authors reverse the usage and regard the structures on the bottoms of beds as molds" (Middleton, 1965, p.247), and others regard "cast" and "mold" as synonymous. Syn: *mark.*

Moldavian (Mol-da'-vi-an) Eastern European mammal age ("complex"): early Pliocene (before Khaprovian); equiv. Csarnotian, Ruscinian.

moldavite [astron] (mol'-da-vite) A tektite from the Central European field, one of the four main tektite fields, deposited about 15 Ma ago. Most moldavites are spin-shaped tektites, mainly found in the states of Bohemia and Moravia of the Czech Republic, characterized by marked sculpturing on their surfaces due to solution etching. Named after the Bohemian river Moldau (German name for Vltava), in whose valley many moldavites are found. Also spelled: *vltavite.*

moldavite [mineral] A variety of *ozocerite* from Moldavia.

moldering (mold'-er-ing) An obsolete term for decomposition of organic matter under conditions of insufficient oxygen.

moldic porosity (mold'-ic) Porosity resulting from the removal, usually by dissolution, of an individual constituent of a rock, such as a shell (Choquette and Pray, 1970, p.248-249).

mole In coastal context, a massive solid-fill protective structure extending from the shore into deep water, formed of masonry and earth or large stones and serving as a *breakwater* or a *pier.*

molecular clock Hypothesis that the rate of mutation in DNA is relatively constant over time.

molecular paleontology (mo-lec'-u-lar) The study of the molecular-scale aspects of fossils, such as skeletal mineralogy and geochemistry, or chemical traces of fossils that have been largely destroyed by diagenesis. Cf: *chemical fossil.*

molecular proportion The ratio of the weight percentage of a particular rock component, esp. an oxide, to its molecular weight.

mole fraction The number of moles of a given component in a phase, divided by the total number of moles of all components in the phase. Mole fractions are thus useful in defining the composition of a phase.

mole track A small, geologically short-lived ridge, 30-60 cm high, formed by the humping up and cracking of the ground where movement along a large strike-slip fault occurred in heavily alluviated terrain. It resembles the track of a gigantic mole, or a line of disturbed earth turned by a great plowshare.

mollic epipedon (mol'-lic) A surface horizon of mineral soil that is dark colored and relatively thick, contains at least 5.8 g kg^{-1} organic carbon, is not massive and hard or very hard when dry, has a base saturation of >50% when measured at pH 7, has <110 mg P kg^{-1} soluble in 0.05 M citric acid, and is dominantly saturated with divalent cation.

mollisol (mol'-li-sol) *active layer [permafrost].*

Mollisols Mineral soils that have a mollic epipedon overlying mineral material with a base saturation of 50% or more when measured at pH 7. Mollisols may have an argillic, natric, albic, cambic, gypsic, calcic, or petrocalcic horizon, a histic epipedon, or a duripan, but not an oxic or spocid horizon. An order in the U.S. system of soil taxonomy.

molluscoid (mol-lus′-coid) In older classifications, any invertebrate animal possessing a lophophore; i.e. a brachiopod or bryozoan.

mollusk (mol′-lusk) A solitary invertebrate belonging to the phylum Mollusca, characterized by a nonsegmented body that is bilaterally symmetrical and by a radially or biradially symmetrical mantle and shell. Among the classes included in the mollusks are the gastropods, bivalves, and cephalopods. Also spelled: mollusc. Adj: molluscan.

Mollweide projection (Moll′-wei-de) An equal-area map projection on which the entire surface of the Earth is enclosed within an ellipse whose major axis (the equator, representing 360° of longitude) is twice the length of the minor axis (the central meridian, representing 180° of latitude). All parallels are represented by straight lines at right angles to the central meridian and more widely spaced at the equator than at the poles, and all meridians are represented by equally spaced elliptical arcs with the exception of the central meridian (a straight line) and the meridian 90° from the center (a full circle, representing the hemisphere centered at the origin of the projection). The meridional curvature increases away from the central meridian. There is excessive angular distortion (shearing) at the margins of the map. Named after Karl B. Mollweide (1774-1825), German mathematician and astronomer, who introduced the projection in 1805. Also known as "Mollweide homolographic projection".

moluranite (mol-u-ran′-ite) A black amorphous incompletely investigated mineral: $H_4U^{4+}(U^{6+}O_2)_3(MoO_4)_7{\cdot}18H_2O$.

molybdate (mo-lyb′-date) A mineral compound characterized by the radical MoO_4. An example of a molybdate is wulfenite, $PbMoO_4$. Cf: *tungstate.*

molybdenite (mo-lyb′-de-nite) A soft platy lead-gray mineral crystallizing in hexagonal aand rhombohedral polytypes: MoS_2. It is the principal ore of molybdenum. Molybdenite generally occurs in foliated masses or scales, and is found in pegmatite dikes and quartz veins or disseminated in porphyry; it resembles graphite in appearance and to the touch, but has a bluer color. Cf: *jordisite.*

molybdenum The metal Mo. It has been reported as a mineral from a lunar sample but has not been approved by the International Mineralogical Association.

molybdic ocher (mo-lyb′-dic) (a) *ferrimolybdite.* (b) *molybdite.*

molybdite (mo-lyb′-dite) An adamantine greenish-yellow orthorhombic mineral: MoO_3. Much so-called molybdite is *ferrimolybdite.* Syn: *molybdic ocher.*

molybdofornacite (mo-lyb′-do-for′-nac-ite) An adamantine light-green monoclinic mineral: $Pb_2Cu(AsO_4,PO_4)(MoO_4,CrO_4)(OH)$. It is the Mo analogue of fornacite.

molybdomenite (mo-lyb′-do-me-nite′) A colorless to yellowish-white monoclinic mineral: $PbSe^{4+}O_3$.

molybdophyllite (mo-lyb′-do-phyl′-lite) A platy colorless, white, or pale-green hexagonal mineral: $Pb_2Mg_2Si_2O_7(OH)_2$.

molysite (mol′-y-site) A brownish-red, purple-red, or yellow rhombohedral mineral: $Fe^{3+}Cl_3$. It is deliquescent.

moment (mo′-ment) (a) The geochronologic equivalent of a *chronohorizon* (ISG, 1976, p.68). (b) A term recommended by Renevier et al., (1882) and Teichert (1958a, p.113-115, 117) for the time interval during which a biostratigraphic zone was deposited; the geologic-time unit corresponding to Oppel's (1856-1858) "zone". It is, for all practical purposes, the shortest perceptible interval into which geologic time can be subdivided. Cf: *instant; hemera.* Syn: *phase [geochron]; secule; zone time.*

moment magnitude Earthquake size calculated from the seismic moment of the earthquake. This measure is the most valid size calculation for earthquakes above Richter or body wave magnitude 7-7.5. Symbol: Mw. See also: *seismic moment.*

moment map A stratigraphic map that expresses the positional relations of beds as a continuous variable (Krumbein and Libby, 1957, p.200); e.g., a *center-of-gravity map* and a *standard-deviation map.*

moment measure The expected value of each of the powers of a random variable that has a given distribution; a weighted measure of central tendency. It is used to describe the character of a distribution curve. The first moment measure is the *mean;* the second is the *standard deviation;* the third is the *skewness;* and the fourth is the *kurtosis* (Cole and King, 1968, p.660). In sedimentology, moment measures are related to the center of gravity of the particle-size distribution curve and are defined about the mean value of the variable.

moments Moments about the origin are means of the *k*th powers of a set of observations. Only the first moment about the origin, the *mean,* is widely used. Moments about the mean are the *k*th powers of deviations from the mean; *variance* is the second moment about the mean; *skewness* is the third; and *kurtosis* is the fourth. In sedimentology, graphical approximations of the moments are used as decriptors of the shape of a grain-size distribution.

moment tensor An earthquake parameter that depends on source strength and fault orientation. It characterizes all the information about the source that can be learned from observing waves whose wavelengths are much longer than the linear dimensions of the fault area that broke during the quake.

monacanth (mon′-a-canth) A *trabecula* in a rugose coral in which the fibers are related to a single center of calcification and radiate upward and outward from the axis formed by upward shifting of the center (TIP, 1956, pt. F, p.235). Cf: *rhabdacanth; rhipidacanth.*

monactine (mon-ac′-tine) A sponge spicule having a single ray.

monadnock (mo-nad′-nock) An upstanding rock, hill, or *mountain of circumdenudation* rising conspicuously above the general level of a peneplain in a temperate climate, representing an isolated remnant of a former erosion cycle in a mountain region that has been largely beveled to its base level. Type locality: Mount Monadnock in New Hampshire. Cf: *catoctin; unaka; inselberg.* Syn: *torso mountain.*

monadnock phase In hypsometric analysis of drainage basins, the transitory stage characterized by abnormally low hypsometric integrals (less than 35%): removal of the monadnock by fluvial erosion will restore the distorted hypsographic curve to equilibrium form (Strahler, 1952b, p.1130). Cf: *equilibrium stage.*

monalbite (mon-al′-bite) Monoclinic albite; a monoclinic, high-temperature modification of sodium feldspar. It forms a complete solid-solution series with sanidine. Formerly called: *barbierite.*

monaxial (mon-ax′-i-al) Having one axis.

monaxon (mon-ax′-on) A simple uniaxial sponge *spicule* with a single axial filament or canal, or one developed by growth along a single axis. It may be curved or straight and may bear expansions at one or both ends. Obsolete syn: *rhabd.*

monazite (mon′-a-zite) A yellow, brown, or reddish-brown monoclinic mineral: $(Ce,La,Nd,Th)(PO_4,SiO_4)$. It is a rare-earth phosphate with appreciable substitution of thorium for rare earths and silicon for phosphorus; thorium-free monazite is rare. It is widely disseminated as an accessory mineral in granites, gneisses, and pegmatites, and it is often naturally concentrated in detrital sand, gravel, and alluvial tin deposits. Monazite is a principal ore of the rare earths and the main source of thorium. Several end-members of the lanthanide elements are known, and they are acknowledged by the Levinson notation: a hyphenated suffix of the principal element. Syn: *cryptolite.*

monazite-(Ce) A monoclinic mineral of the *monazite* group, with end-member composition $CePO_4$.

monazite-(La) A monoclinic mineral of the *monazite* group, with end-member composition $LaPO_4$.

monazite-(Nd) A monoclinic mineral of the *monazite* group, with end-member composition $NdPO_4$.

monazite-(Sm) A monoclinic mineral of the *monazite* group, with end-member composition $SmPO_4$.

moncheite (mon′-che-ite) A metallic white trigonal mineral of the *melonite* group: $(Pt,Pd)(Te,Bi)_2$. Syn: *biteplatinite; chengbolite.*

monchiquite (mon-chi′-quite) (a) In the *IUGS classification,* a variety of *lamprophyre* with glass or feldspathoid in the groundmass, no feldspars, and generally containing amphibole phenocrysts. (b) A *lamprophyre* containing phenocrysts of olivine, clinopyroxene, and typically biotite or amphibole, in a groundmass of glass or analcime, often highly altered. Nepheline or leucite may be present. Its name, given by Hunter and Rosenbusch in 1890, is derived from Serra de Monchique, Portugal. Cf: *fourchite.*

mondmilch (mond′-milch) The German equivalent of *moonmilk.*

monetite (mon′-e-tite) A colorless or yellowish-white triclinic mineral: $CaHPO_4$. It is the phosphate analogue of weilite.

mongolite (mon′-go-lite) A lilac tetragonal mineral: $Ca_4Nb_6Si_5O_{24}(OH)_{10}{\cdot}6H_2O$.

mongshanite (mong-shan′-ite) An hexagonal mineral: $(Mg,Cr)_2(Ti,Zr)_5O_{12}$. Not approved by the International Mineralogical Association.

monheimite (mon′-heim-ite) A variety of smithsonite with iron in solid solution for zinc.

moniliform (mo-nil′-i-form) Beadlike, or jointed at regular intervals so as to resemble a string of beads; e.g. "moniliform antennae". Syn: *nummuloidal.*

moniliform wall A vertical zooecial wall in stenolaemate bryozoans that is thickened in transverse annular ridges so that in longitudinal section it looks beaded.

monimolimnion (mo′-ni-mo-lim′-ni-on) The deep, high-density water of a perennially stratified lake. Cf: *mixolimnion; chemocline.*

monimolite (mon-i-mol′-ite) A yellowish, brownish, or greenish cubic mineral of the *stibiconite* group: $(Pb,Ca)_3Sb_2O_8$(?). It may contain ferrous iron.

monk rock *penitent rock.*

monocentric (mon-o-cen′-tric) Said of a *corallite* formed by a monostomodaeal polyp.

monochromatic illuminator (mon′-o-chro-mat′-ic) *monochromator.*

monochromatic light Electromagnetic radiation of a single wavelength or frequency. It is used in crystal optics to determine indices of refraction.

monochromator (mon-o-chro′-ma-tor) An instrument for selecting a narrow portion of a spectrum. In optics, a variable filter, grating, or prism which can isolate light of only one wavelength (color) or of a very narrow range of wavelengths. Syn: *monochromatic illuminator.*

monoclinal (mon′-o-cli′-nal) Adj. of *monocline.*

monoclinal scarp A scarp resulting from a steep downward flexure between an upland block and a tectonic basin (Cotton, 1958, p.174).

monoclinal shifting The downdip migration of a divide (and of a stream channel) resulting from the tendency of streams in a region of inclined strata to flow along the strike of less resistant strata, as where differential erosion proceeds more rapidly along the steeper slope of a cuesta or monoclinal ridge. The process was first noted by Gilbert (1877, p.135-140). See also: *shifting [streams].* Syn: *homoclinal shifting; uniclinal shifting.*

monocline (mon'-o-cline) A local steepening in an otherwise uniform gentle dip. Cf: *homocline.* Adj: *monoclinal.* Obsolete syn: *unicline.*

monoclinic system (mon-o-clin'-ic) One of the six crystal systems, characterized by either a single twofold axis of symmetry, a single plane of symmetry, or a combination of the two. Of the three nonequivalent axes, one is perpendicular to the plane formed by the other two. Cf: *isometric system; hexagonal system; tetragonal system; orthorhombic system; triclinic system.*

monocolpate (mon-o-col'-pate) Said of pollen grains having a single, normally distal colpus. Syn: *monosulcate.*

monocot (mon'-o-cot) *monocotyledon.*

monocotyledon (mon'-o-cot'-y-le'-don) An angiosperm whose seeds contain a single, parallel-veined embryonic leaf. Such a plant usually has flowering parts in threes, parallel leaf venation, and monocolpate pollen. Examples include grasses, palms, and lilies. Monocotyledons range from the Cretaceous. Cf: *dicotyledon.* Syn: *monocot.*

monocyclic (mon-o-cy'-clic) (a) Said of a crinoid having only a single circlet of plates proximal to the radial plates. (b) Said of the apical system of an echinoid in which genital plates and ocular plates are arranged in a single ring around the *periproct.* Cf: *dicyclic.*

monoecious (mo-noe'-cious) Said of a plant that has both staminate and pistillate flowers, or both male and female gametangia.

monofacies (mon'-o-fa'-cies) A term used by Bailey and Childers (1977, p.27) for roll-front orebodies of uranium that "occur in formations which are uniformly reducing". Cf: *bifacies.*

monogene (mon'-o-gene) adj. (a) *monogenetic.* (b) A term applied specif. by Naumann (1850, p.433) to an igneous rock (such as dunite) composed essentially of a single mineral. Obsolete. Cf: *polygene; monomineralic; monomictic [sed].* Syn: *monogenic.*

monogenetic (mon'-o-ge-net'-ic) (a) Resulting from one process of formation or derived from one source, or originating or developing at one place and time; e.g. said of a volcano built up by a single eruption. (b) Consisting of one element or type of material, or having a homogeneous composition; e.g. said of a gravel composed of a single type of rock. Cf: *polygenetic.* Syn: *monogene; monogenic.*

monogenetic volcano A volcano that erupts only once (Walker, 2000). Cf: *polygenetic volcano.*

monogenic (mon-o-gen'-ic) (a) *monogenetic.* (b) *monogene.*

monogeosyncline (mon'-o-ge'-o-syn'-cline) A single geosynclinal trough along the continental margin and receiving sediments from a borderland on its oceanic side (Schuchert, 1923). Obsolete. Cf: *polygeosyncline.*

monoglacial theory (mon-o-gla'-cial) The obsolete belief that the Pleistocene ice sheet made only one general advance and one general recession, without any substantial "interglacial" recession and readvance (Wright, 1914, p.124-125).

monograptid (mon-o-grap'-tid) n. Any *graptoloid* belonging to the family Monograptidae, characterized by scandent uniserial rhabdosomes with thecae of variable form. Range, Silurian to Lower Devonian. adj. Said of the upward direction of growth in graptoloid thecae.

monohydrocalcite (mon'-o-hy'-dro-cal'-cite) A rare white or gray trigonal rhombohedral mineral: $CaCO_3 \cdot H_2O$. It was first observed in lake-bottom sediments, and it may be formed by precipitation from cold water in contact with air. Cf: *hydrocalcite.*

monolete (mon'-o-lete) adj. Said of an embryophytic spore having a laesura consisting of a single line or mark. Cf: *trilete.* n. A monolete spore. The usage of this term as a noun is improper.

monolith (mon'-o-lith) (a) A piece of unfractured bedrock, generally more than a few meters across; e.g. an unweathered joint block moved by a glacier. (b) A large upstanding mass of rock, such as a volcanic spine. (c) One of many large blocks of stone or concrete forming the component parts of an engineering structure, such as a dam. (d) A vertical soil section, taken to illustrate the soil profile.

monomaceral (mon-o-mac'-er-al) Said of a coal microlithotype consisting of a single maceral. Cf: *bimaceral; trimaceral.*

monomict breccia (mon'-o-mict) A brecciated meteorite in which all the fragments have essentially the same composition and metamorphic grade. Cf: *polymict breccia; genomict breccia.*

monomictic [lake] (mon-o-mic'-tic) Said of a lake with only one yearly *overturn.* Tropical lakes overturn in the winter and polar lakes in the summer. Cf: *amictic; dimictic; oligomictic [lake]; polymictic [lake].*

monomictic [sed] Said of a clastic sedimentary rock composed of a single mineral or rock species. Cf: *oligomictic [sed]; polymictic [sed].* Syn: monomict.

monomineralic (mon'-o-min'-er-al'-ic) Said of a rock composed wholly or almost wholly of a single mineral; esp. said of an igneous rock (such as anorthosite or dunite) consisting of one essential mineral. The amounts of other minerals tolerated under the definition vary with different authors. Cf: *polymineralic; anchimonomineralic; monogene.*

monomorphic (mon-o-mor'-phic) Said of a bryozoan colony in which all zooids in zones of astogenetic repetition are feeding zooids of similar morphology.

monomyarian (mon'-o-my-ar'-i-an) adj. Said of a bivalve mollusk or its shell with only the posterior *adductor muscle.* Cf: *dimyarian; anisomyarian.* n. A monomyarian mollusk, such as an oyster or scallop.

Monongahelan (Mo-non'-ga-he'-lan) North American provincial stage: Upper Pennsylvanian (above Conemaughian, below Dunkardian).

monophyletic (mon'-o-phy-let'-ic) Evolving from a single ancestral stock. Cf: *nonmonophyletic; polyphyletic.*

monophyletic group An evolutionary lineage (group of species) that includes an ancestral species and all of its descendants; a clade (Wiley et al., 1991, p.3; Brooks et al., 1984, p.80). Cf: *paraphyletic group, polyphyletic group, natural taxon, grade.*

monophyly Condition of or relating to a *monophyletic group.*

Monoplacophora (mon'-o-pla-coph'-o-ra) A class of pseudosegmented cyrtosome mollusk with a cap-shaped to helical shell. Soft parts not twisted (untorted) and anus not over the head, but some with gut bent into U-shape. Most bilaterally symmetrical, with one to several pairs of muscles (pedal retractors) for clamping shell over soft parts. Apex of shell points anteriorly and overhangs head. Range, Lower Cambrian to Holocene; about 135 genera.

monoplacophoran Any mollusk belonging to the class Monoplacophora.

monopleural (mon-o-pleu'-ral) Said of the arrangement of the two rows of thecae in the biserial *rhabdosome* of a scandent graptoloid in which the rows are in side-by-side contact. Cf: *dipleural.*

monopodial (mon-o-po'-di-al) Having one main axis of growth.

monoporate (mon-o-por'-ate) Said of pollen grains provided with a single pore, as in grasses.

monopyroxene (mon-o-py'-rox-ene, mon'-o-py-rox'-ene) *clinopyroxene.*

Monorhina (Mon-o-rhi'-na) A subclass of the Agnatha characterized by a single midline nostril associated with an externalized hypophysis; bone, if present, is cellular and histologically comparable to that of gnathostomes.

monosaccate (mon-o-sac'-cate) Said of pollen with a single *saccus* that extends all around the grain more or less at the equator.

monoschematic (mon'-o-sche-mat'-ic) Said of mineral deposits having a uniform texture. Cf: *polyschematic.*

monosomatic chondrule (mon'-o-so-mat'-ic) Obsolete term for a chondrule consisting of a single crystal. Cf: *polysomatic chondrule.*

monostomodaeal (mon'-o-sto-mo'-dae-al) Said of the stomodaea of a scleractinian coral *polyp,* each having its own tentacular ring after originating by distomodaeal or tristomodaeal budding.

monostratum (mon-o-stra'-tum) A simple layer, as in a first-order *laminite* (Lombard, 1963, p.14).

monostromatic (mon'-o-stro-mat'-ic) Said of foliaceous tissue that is composed of only one layer of cells, such as leaves of mosses or algal thalli that are only one cell thick. Cf: *oligostromatic; polystromatic.*

monosulcate (mon-o-sul'-cate) *monocolpate.*

monothalamous (mon-o-thal'-a-mous) *unilocular.*

monothem (mon'-o-them) A term proposed by Caster (1934, p.18) for a noncyclic, or not obviously cyclic, chronostratigraphic unit of genetically related strata, representing a "more ordinary, and perhaps more normal major subdivision" of a stage; but interpreted by Moore (1949, p.19) as a "local deposit having essentially uniform lithologic character" and corresponding to formation or member of lithostratigraphic classification. Weller (1958, p.636) regarded the term as "superfluous because its meaning is the same as 'substage'". Cf: *lithizone.*

monothermite (mon-o-ther'-mite) A clay-mineral material that shows a single high-temperature endothermal reaction at about 550°C. It appears to be a mixture in which illite and kaolinite are important components (Grim, 1968, p.48).

monothetic (mon-o-thet'-ic) Said of a taxonomic classification in which all members of a taxon possess a characteristic or a combination of characteristics unique to that taxon. Cf: *polythetic.*

monothetic classification [phylogen] System in which one or a few essential characters are used in classification (see Wiley, 1981, p.80). Ant: *polythetic classification [phylogen].*

Monotremata (Mon'-o-tre-ma'-ta) The sole order of the mammalian subclass Prototheria, characterized by egg-laying reproduction, lack of nipples, and near or complete loss of teeth.

monotrichous (mo-not'-ri-chous) Said of a bacterial cell with a single flagellum occurring at one pole. Cf: *lophotrichous; peritrichous.*

monotrophic (mon-o-troph'-ic) Said of an organism that feeds on one kind of food only.

monotropy (mo-not'-ro-py) The relationship between two different forms of the same substance, e.g. pyrite and marcasite, that have no definite transition point, since only one of the forms, e.g. pyrite, is stable; and in which the change from the unstable to the stable form is irreversible. Cf: *enantiotropy.*

monotypic (mon-o-typ'-ic) Said of a taxon that includes only one taxon of the next lower rank, e.g. a genus or subgenus with only one originally included species.

monrepite *tetra-ferri-annite.*

monroe (mon-roe') A small mound of mud occurring on tidal flats in cold regions. They generally occur in groups, have a rounded conical shape; have steep slopes, and are 5 to 25 cm in diameter. The name is "a descriptive term referring to a world-known American movie star that may easily be used in many languages" (Dionne, 1973, p.848). They are thought to form under an *ice foot* as a result of load pressure that expels air and water trapped under mud and ice.

mons A term established by the International Astronomical Union for a large isolated mountain on planetary surfaces (USGS, 2003). Most are of volcanic origin. Generally used as part of a formal name for a landform, such as Olympus Mons on Mars (Mutch et al., 1976, p.57). Etymol: Latin "mountain". Pl: montes.

monsmedite (mons-med'-ite) A dark green to black cubic mineral of the *voltaite* group: $KTl^{3+}(SO_3OH)_4{\cdot}6H_2O$.

monsoon (mon-soon') A type of wind system whose direction changes with the seasons, for example over the Indian subcontinent where an onshore southwesterly flow occurs during the high-sun season (April-Sept.) and an offshore northeasterly flow occurs during the low-sun season (Oct.-March).

monster (mon'-ster) An organism with extreme departure in form or structure, generally pathologic, from the usual type of its species.

monstrosity (mon-stros'-i-ty) A part of an organism exhibiting considerable deviation in structure or form, which may be injurious or useless to the species and is usually not propagated.

montane (mon'-tane) Of, pertaining to, or inhabiting cool upland slopes below the timber line, characterized by the dominance of evergreen trees. Cf: *alpine [ecol].* Syn: *subalpine; alpestrine.*

montanite (mon-tan'-ite) An ill-defined earthy white, yellowish, or greenish mineral: $Bi_2Te^{6+}O_6{\cdot}2H_2O$.

montan wax (mon'-tan) A solid bitumen that may be extracted by solvents from certain lignites or brown coals. It is white to brown and melts at 77° to 93°C.

montbrayite (mont-bray'-ite) A metallic tin-white triclinic mineral: $(Au,Sb)_2Te_3$.

montdorite (mont'-dor-ite) A green to brownish-green monoclinic mica of ideal composition $K(Fe^{+2},Mn,Mg)_{2.5}Si_4O_{10}(OH,F)_2$, thus approximately midway between dioctahedral and trioctahedral structure.

montebrasite (mon-te-bra'-site) A white to variously colored triclinic mineral of the *amblygonite* group: $LiAl(PO_4)(OH,F)$.

Monte Carlo method (Mon-te Car'-lo) A computer-implemented random sampling process that generates uniformly distributed random numbers and used these to select values from known frequency distributions.

monteponite (mon-te-pon'-ite) A black cubic mineral of the *periclase* group: CdO.

monteregianite-(Y) (mon-te-re'-gian-ite) A white, mauve, pale green, or gray monoclinic mineral: $KNa_2Y\ Si_8O_{19}{\cdot}5H_2O$.

montesite (mon'-te-site) A Pb-bearing herzenbergite.

montesommaite A colorless orthorhombic zeolite mineral: $(K,Na)_9(Si_{23}Al_9)O_{64}{\cdot}10H_2O$.

montgomeryite (mont-gom'-er-y-ite') (a) A green to colorless monoclinic mineral: $Ca_4MgAl_4(PO_4)_6(OH)_4{\cdot}[12]H_2O$. (b) A name for a group of minerals with analogous compositions, but with Mg replaced by Mn^{2+} or Fe^{2+}, and Al replaced by Fe^{3+}.

Montian (Mon'-ti-an) Obsolete European stage: equivalent to Selandian Stage of the Paleocene.

monticellite (mon-ti-cel'-lite) A colorless, greenish-gray, or yellowish-gray orthorhombic mineral related to olivine: $CaMgSiO_4$. It is isomorphous with kirschsteinite and glaucochroite, and usually occurs in contact-metamorphosed limestones.

monticle (mon'-ti-cle) *monticule [geomorph].*

monticule [geomorph] (mon'-ti-cule) (a) A little mound; a hillock, mound, knob, or other small elevation. (b) A small, subordinate volcanic cone developed on the flank or about the base of a larger volcano. Etymol: French. Syn: *monticle.*

monticule [paleont] (a) A protuberant part of the corallum surface of a scleractinian coral, produced in circummural budding. Cf: *colline.* (b) One of the small rounded nodes or swellings of a brachiopod shell, commonly bearing spines. (c) One of the clusters of small, modified zooecia that may be regularly spaced throughout a stenolaemate bryozoan colony, appearing at the surface as a small protuberance or elevation. Cf: *macula [paleont].*

montiform (mon'-ti-form) Having the shape of a mountain; mountainlike.

montmartrite (mont-mar'-trite) A variety of gypsum from the Montmartre section of Paris, France.

montmorillonite (mont-mo-ril'-lon-ite) (a) A white, yellow, or green monoclinic mineral that has the origin of the layer charge primarily in the octahedral sheet. An ideal formula is: $(Na,Ca)_{0.3}(Al,Mg)_2Si_4O_{10}(OH)_2{\cdot}nH_2O$. Cf: *smectite.* (b) A name for a group of monoclinic dioctahedral micaceous minerals of analogous composition, but with (Na,Ca) replaced by Ca, and (Al,Mg) replaced by Fe^{3+} or Cr. Syn: *smectite.*

montrealite (mon-tre-al'-ite) An ultramafic rock (*hornblende peridotite*) containing, in order of decreasing abundance, clinopyroxene, amphibole, and olivine, with little or no feldspar or nepheline. The name, given by Adams in 1913, is for Montreal (Mount Royal), Quebec. Obsolete.

montroseite [mineral] (mont'-rose-ite) A black orthorhombic mineral of the *diaspore* group: $(V^{3+},Fe^{2+},V^{4+})O(OH)$.

montroseite [sed] A uranium-bearing sandstone.

montroyalite (mont-roy'-al-ite) A white triclinic mineral: $Sr_4Al_8(CO_3)_3(OH,F)_{26}{\cdot}10H_2O$.

montroydite (mon-troy'-dite) A red or reddish-brown orthorhombic mineral: HgO.

monument [geomorph] (mon'-u-ment) (a) An isolated pinnacle, column, or pillar of rock resulting from erosion and resembling a man-made monument or obelisk, usually extremely regular in form and of grand dimensions. (b) *tind.*

monument [surv] A natural or artificial physical structure that marks the location on the ground of a *corner* or other survey point; e.g. a pile of stones indicating the boundary of a mining claim, or a road or fence marking the boundary of real property. See also: *boundary monument.*

monumental peak (mon-u-men'-tal) *horn [glac geol].*

monumented upland (mon'-u-ment-ed) A term proposed by Hobbs (1921, p.373) for "the extreme type of mountain sculpture ... believed to be due to continued glacial action upon a fretted upland like that of the Alps", characterized by enlargement of cirques and reduction of horns. Example: Glacier National Park, Mont.

monzodiorite (mon-zo-di'-o-rite) In the *IUGS classification,* a plutonic rock with Q between 0 and 5, P/(A+P) between 65 and 90, and plagioclase more sodic than An_{50}.

monzogabbro (mon-zo-gab'-bro) In the *IUGS classification,* a plutonic rock with Q between 0 and 5, P/(A+P) between 65 and 90, and plagioclase more calcic than An_{50}.

monzogranite In the *IUGS classification,* a plutonic rock with Q/(Q+A+P) between 20 and 60, and P/(A+P) between 35 and 65; in other words, a variety of granite in which alkali feldspar and plagioclase are in approximately equal proportions. Cf: *syenogranite.*

monzonite (mon'-zo-nite) (a) In the *IUGS classification,* a plutonic rock with Q between 0 and 5, and P/(A+P) between 35 and 65. (b) A group of plutonic rocks intermediate in composition between *syenite* and *diorite,* containing approximately equal amounts of alkali feldspar and plagioclase, little or no quartz, and commonly augite as the main mafic mineral; also, any rock in that group; the intrusive equivalent of *latite.* With a decrease in the alkali feldspar content, monzonite grades into diorite or *gabbro,* depending on the composition of the plagioclase; with an increase in alkali feldspar, it grades into syenite. Syn: *syenodiorite.* The name, given by de Lapparent in 1864, is for Monzoni, Tyrolean Alps.

monzonorite (mon-zo-nor'-ite) In Tobi's classification (1971, p.202) of the *charnockite series,* a quartz-poor member containing more plagioclase than microperthite.

Moody diagram (Mood'-y) A diagram showing the variation of the Darcy-Weisbach coefficient against the Reynolds number.

mooihoekite (moo-i-hoek'-ite) A metallic yellow tetragonal mineral related to chalcopyrite: $Cu_9Fe_9S_{16}$.

moolooite (moo'-loo-ite) A blue or green orthorhombic oxalate mineral: $Cu(C_2O_4){\cdot}0.4H_2O$.

Moon The Earth's only known natural satellite and next to the Sun the most conspicuous object in the sky, deriving its light from the Sun and reflecting it to the Earth. The Moon revolves about the Earth from west to east in about 29.53 days with reference to the Sun (interval from new moon to new moon) or about 27.32 days with reference to the stars; it has a mean diameter of 3,475.9 km (2,160 miles, or about 27% that of the Earth), a mean distance from the Earth of about 384,400 km (238,857 miles), a mass of 7.354×10^{25} g (about $^1/_{81}$ that of the Earth), a volume about $^1/_{49}$ that of the Earth, and a mean density of 3.34 g/cm^3. The Moon rotates once on its axis during each revolution in its orbit and therefore it always presents nearly the same face (59%) to the Earth; it has essentially no atmosphere, no water, and no life forms or organic matter.

moon [planet] Any natural satellite of a planet.

moonmilk (moon'-milk) (a) A soft white, initially plastic deposit that occurs on the walls of caves. It may consist of calcite, hydromagnesite, nesquehonite, huntite, aragonite, magnesite, dolomite, or other minerals. The term was introduced in 1555 by Conrad Gesner, Swiss naturalist. Etymol: from a 16th-Century belief that the Moon's rays produce dew at the surface and white material ("elves' milk") underground. Syn: mountain milk; *mondmilch.* Cf: *lublinite.*

moonquake (moon'-quake) An agitation or disturbance of the Moon's surface, analogous to a terrestrial earthquake.

moonscape (moon'-scape) The surface of the Moon, as observed in photographs or through a telescope or as delineated on the basis of photographic or telescopic evidence.

moonstone (moon'-stone) (a) A semitransparent to translucent alkali feldspar (adularia) or cryptoperthite that exhibits a bluish to milky-white pearly or opaline luster; an opalescent variety of orthoclase. Flawless moonstones are used as gemstones. Cf: *sunstone.* Syn: *hecatolite.* (b) A name incorrectly applied to peristerite or to opalescent varieties of plagioclase (esp. albite). (c) A name incorrectly applied (without proper prefix) to milky or *girasol* varieties of chalcedony, scapolite, corundum, and other minerals.

moor A German term meaning peatland. In England the term "moor" is used in the sense of peatland.

moor coal A lignite or brown coal that is friable.

mooreite (moore′-ite) A glassy white monoclinic mineral: $Mg_9Zn_4Mn^{2+}_2(SO_4)_2(OH)_{26}{\cdot}8H_2O$. Cf: *torreyite.*
moorhouseite (moor′-house-ite) A pink monoclinic mineral of the *hexahydrite* group: $(Co,Ni,Mn^{2+})SO_4{\cdot}6H_2O$.
moorland pan (moor′-land) *moorpan.*
moorland peat *highmoor peat.*
moorpan (moor′-pan) An *iron pan* occurring in a peaty soil or forming at the bottom of a bog, containing compact redeposited iron and humus compounds. Syn: *moorland pan.*
moor peat *highmoor peat.*
mopungite (mo-pung′-ite) A white tetragonal mineral of the *stottite* group occurring as an alteration product of stibnite: $NaSb^{5+}(OH)_6$.
mor A type of forest humus characterized by an accumulation or organic matter on the soil surface in matted Oe(F) horizons, reflecting the dominant mycogenous decomposers. The boundary between the organic horizon and the underlying mineral soil is aburpt. Sometimes differentiated into the following groups: Hemimor, Humimor, Resimor, Lignomor, Hydromor, Fibrimor, and Mesimor.
moraesite (mo-ra′-es-ite) A white fibrous monoclinic mineral: $Be_2(PO_4)(OH){\cdot}4H_2O$.
morainal (mo-rain′-al) Of, relating to, forming, or formed by a *moraine.* Also called: *morainic.*
morainal apron *outwash plain.*
morainal channel A meltwater-stream channel formed during the construction of a moraine (Rich, 1908, p.528).
morainal-dam lake A glacial lake impounded by a *drift dam* left in a preexisting valley by a retreating glacier.
morainal delta An obsolete syn. of *ice-contact delta.*
morainal lake A glacial lake occupying a depression resulting from irregular deposition of drift in an end moraine or ground moraine of a continental glacier.
morainal plain *outwash plain.*
morainal stuff An obsolete term for the material carried upon the surface of a glacier.
morainal topography An irregular landscape produced by deposition of drift and characterized by irregularly scattered hills and undrained depressions, or *kettles.*
moraine [glac geol] (mo-raine′) A mound, ridge, or other distinct accumulaion of unsorted, unstratified glacial drift, predominantly till, deposited chiefly by direct action of glacier ice, in a variety of topographic landforms that are independent of control by the surface on which the drift lies. The term was probably used originally, and is still often used in European literature, as a petrologic name for *till* that is being carried and deposited by a glacier; but it is now more commonly used as a geomorphologic name for a landform composed mainly of till that has been deposited by either a living or an extinct glacier. Etymol: French, a term used by Alpine peasants in the 18th century for any heap of earth and stony debris. The term moraine was introduced in print in glaciology by Saussure in 1779 (Carozzi, 1996).
moraine [volc] Solidified volcanic debris carried on the surface of a lava flow.
moraine kame A term applied by Salisbury et al. (1902, p.118) to a kame that forms one of a group having the characteristics of, and "the same general significance" as, a terminal moraine. Cf: *kame moraine.*
moraine loop *loop moraine.*
moraine plateau A relatively flat area within a hummocky moraine, generally at the same elevation as, or a little higher than, the summits of surrounding knobs (Gravenor and Kupsch, 1959, p.50).
moraine rampart An elongated ridge or row of lateral and terminal moraines, sometimes forming an amphitheatrical arrangement (Schieferdecker, 1959, terms 0916 and 0922).
morainic (mo-rain′-ic) *morainal.*
moralla (mo-ral′-la) (a) Poorly crystallized or massive opaque-appearing greenish material from Colombian emerald mines. (b) Any of the poorer grades of emerald. Syn: morallion; morallon.
morass ore (mo-rass′) *bog iron ore.*
MORB *mid-ocean-ridge basalt.*
mordenite (mor′-den-ite) A white, yellowish, or pinkish orthorhombic zeolite mineral: $(Na_2,Ca,K_2)Al_2Si_{10}O_{24}{\cdot}7H_2O$. Syn: *ashtonite; flokite; arduinite; ptilolite.*
moreauite (mo-reau′-ite) A greenish-yellow monoclinic mineral: $Al_3(UO_2)(PO_4)_3(OH)_2{\cdot}13H_2O$.
morelandite (more′-land-ite) A colorless, light yellow, or gray hexagonal mineral of the *apatite* group: $(Ba,Ca,Pb)_5(AsO_4,PO_4)_3Cl$.
morencite (mo-ren′-cite) *nontronite.*
morenosite (mo-ren′-o-site) An apple-green or light green orthorhombic mineral of the *epsomite* group: $NiSO_4{\cdot}7H_2O$. It may contain appreciable magnesium, and it occurs in secondary incrustations. Syn: *nickel vitriol.*
morganite (mor′-gan-ite) *vorobievite.*
morimotoite An adamantine black cubic mineral of the *garnet* group: $Ca_3(Ti,Fe^{2+},Fe^{3+})_2(Si,Fe^{3+})_3O_{12}$.
morinite (mo′-rin-ite) A colorless or pink prismatic monoclinic mineral: $NaCa_2Al_2(PO_4)_2(F,OH)_5{\cdot}2H_2O$. Syn: *jezekite.* It is the phosphate analogue of esperanzaite.
morion (mo′-ri-on) A nearly black, opaque variety of smoky quartz or cairngorm.
morlop (mor′-lop) A mottled variety of jasper found in New South Wales, Australia. It often occurs as pebbles associated with diamonds.
morozeviczite (mo-ro-ze′-vicz-ite) A metallic white cubic mineral with an X-ray pattern resembling that of germanite: $(Pb,Fe)_3Ge_{1-x}S_4$(x=0.18 to 0.69). It is the Pb-analogue of polkovicite, with which it forms a solid-solution series.
morphocline (mor′-pho-cline) In *cladism,* the ordered spectrum of variation displayed by a homologous character. Syn: *transformation series.*
morphogenesis (mor-pho-gen′-e-sis) The origin and subsequent growth or development of landforms or of a landscape.
morphogenetic (mor′-pho-ge-net′-ic) Pertaining to the origin of morphological features.
morphogenetic region A climatic zone in which the predominant geomorphic processes produce distinctive regional landscape characteristics that contrast with those of other areas developed under different climatic conditions. Peltier (1950) postulated nine morphological regions based on temperature and moisture conditions. Cf: *formkreis.*
morphogenetic zone *lineage zone.*
morphogeny (mor-phog′-e-ny) The interpretative morphology of a region; specif. *geomorphogeny.*
morphographic map (mor-pho-graph′-ic) *physiographic diagram.*
morphography (mor-phog′-ra-phy) The descriptive morphology of a region, or the phenomena so described; specif. *geomorphography.*
morphologic region (mor-pho-log′-ic) A region delimited according to its distinctive landforms, rock structure, and evolutionary history. Cf: *physiographic province.*
morphologic sequence concept *morphosequence concept.*
morphologic series A graded series of fossils showing variation either in individuals as a whole or in some particular variable feature.
morphologic species A species based solely on morphologic characteristics.
morphologic unit A surface, either depositional or erosional, that is recognized by its topographic character.
morphology [geomorph] (mor-phol′-o-gy) (a) The shape of the Earth's surface; *geomorphology,* or "the morphology of the Earth" (King, 1962). (b) The external structure, form, and arrangement of rocks in relation to the development of landforms.
morphology [paleont] (a) A branch of biology or paleontology that deals with the form and structure of animals and plants or their fossil remains; esp. a study of the forms, relations, and phylogenetic development of organs apart from their functions. See also: *paleomorphology; functional morphology.* (b) The features included in the form and structure of an organism or any of its parts.
morphology [soil] The study of the distribution patterns of horizons in a soil profile, and of the soil's properties.
morphometrics (mor-pho-met′-rics) Statistical analysis of morphological measurements in organisms.
morphometry [geomorph] (mor-phom′-e-try) "The measurement and mathematical analysis of the configuration of the Earth's surface and of the shape and dimensions of its landforms. The main aspects examined are the area, altitude, volume, slope, profile, and texture of the land as well as the varied characteristics of rivers and drainage basins" (Clarke, 1966, p.235).
morphometry [lake] (a) The measurement of the form characteristics (area, depth, length, width, volume, bottom gradients) of lakes and their basins. (b) The branch of limnology dealing with such measurements. (Wetzel, 2001, p.34).
morphosculpture (mor-pho-sculp′-ture) A topographic feature smaller than, and often developed within or on, a *morphostructure;* e.g. a ripple mark, ledge, or knoll on the ocean floor.
morphosequence concept (mor-pho-se′-quence) The recognition that the retreatal positions of the last ice sheet in New England can be traced by careful mapping of landforms composed of meltwater deposits such as eskers and kames, termed sequences (Koteff, 1974; Koteff and Pessl, 1981). The concept was originally developed by Jahns (1941, 1953). Syn: *morphologic sequence concept.*
morphosequent (mor-pho-se′-quent) Said of a surface feature that does not reflect the underlying geologic structure. Ant: *tectosequent.*
morphospace The range of morphology exhibited by a group depicted as a volume within a statistically defined space following quantitative analysis.
morphospecies (mor-pho-spe′-cies) Living or fossil species that are recognized only on the basis of morphology.
morphostratigraphic unit (mor′-pho-strat′-i-graph′-ic) A distinct stratigraphic unit, defined by Frye and Willman (1960, p.7) as "a body of rock that is identified primarily from the surface form it displays"; it may or may not be distinctive lithologically from contiguous units, and may or may not transgress time throughout its extent. The term is used in stratigraphic classification of surficial deposits such as glacial moraines, alluvial fans, beach ridges, and other such deposits where landforms serve to give identity to a body of clastic sediments.

morphostructure (mor-pho-struc′-ture) A major topographic feature that coincides with or is an expression of a geologic structure (e.g. a trench or ridge on the ocean floor) or that is formed directly by tectonic movements (e.g. a basin or dome). It is produced by the interaction of endogenic and exogenic forces, the former being predominant. Cf: *morphosculpture.*

morphotaxon (mor′-pho-tax-on) A taxon (morphogenus, morphospecies, etc.) of fossil plants comprising only the parts, life-history stages, or preservational states represented by the nomenclatural type of the taxon concerned (ICBN).

morphotectonics (mor′-pho-tec-ton′-ics) *tectonic geomorphology.*

morriner (mor′-ri-ner) *esker.*

morro (mor′-ro) A term used in Latin America for an isolated hill or ridge, which may or may not be on a coastal plain near the present shoreline; esp. a headland or bluff. Etymol: Spanish and Portuguese.

Morrowan (Mor′-row-an) A North American stage or series: lowermost Pennsylvanian (above Chesterian of the Mississippian, below Atokan).

Morsumian (Mor-sum′-i-an) North Sea stage: upper Miocene or lower Pliocene (above Syltian, below Scaldisian).

mortar bed (mor′-tar) A valley-flat deposit, occurring in Nebraska and Kansas, consisting of sand, or of a mixture of clay, silt, sand, and gravel, firmly cemented by calcium carbonate and resembling hardened mortar; a type of caliche.

mortar structure A structure in crystalline rocks characterized by a mica-free aggregate of small grains of quartz and feldspar occupying the interstices between, or forming borders on the edges of, much larger and rounded grains of the same minerals. Long considered a product of cataclasis, the structure may actually be the result of plastic deformation and dynamic recrystallization (Harte, 1977). Cf: *pseudogritty structure.* Syn: *cataclastic structure; murbruk structure; porphyroclastic structure.*

mortar texture A texture found in crystalline sedimentary rocks in which relatively large crystalline grains are separated by a microcrystalline mosaic.

Mortensnes A suggested geochronometric subdivision equivalent to the early part of the *Ediacaran* Period (latest *Precambrian*), covering the time span from 600 to 590 Ma (Harland et al., 1990).

mortlake (mort′-lake) A British syn. of *oxbow lake.* The term is "practically obsolete" in Britain (Stamp, 1961, p.327). Etymol: probably from Mortlake, a parish in a SW suburb of London, situated near a drained oxbow lake of the River Thames.

morvan (mor′-van) (a) The intersection of two peneplains, as where an exhumed, tilted peneplain is cut across obliquely by a younger surface that has more nearly retained its original horizontal attitude; e.g. the intersection between the stripped and distinctly sloped Fall Zone peneplain (along the eastern Piedmont Plateau of the U.S.) with the late Tertiary Harrisburg peneplain. Also, the "problem" of the intersection of two peneplains. (b) A region that exhibits a *morvan* relationship, marked by a hard-rock upland bordered by a sloping land of older rock. The term was introduced by Davis (1912, p.115); "a region of composite structure, consisting of an older undermass, usually made up of deformed crystalline rocks, that had been long ago worn down to small relief and that was then depressed, submerged, and buried beneath a heavy overmass of stratified deposits, the composite mass then being uplifted and tilted, the tilted mass being truncated across its double structure by renewed erosion, and in this worn-down condition rather evenly uplifted into a new cycle of destructive evolution". Type locality: Morvan region of central France. Syn: *skiou.*

MOS *Marine Observation Satellites.*

mosaic [geomorph] (mo-sa′-ic) *desert mosaic.*

mosaic [paleont] (a) A pattern formed on the interior of a brachiopod valve by outlines of the adjacent fibers of the secondary layer of shell. (b) Arrangement of plates in edrioasteroids and cyclocystoids, more or less in plane and not imbricating, and presumably rather rigid.

mosaic [petrology] A textural subtype in which individual mineral grains are approximately equant (Harte, 1977).

mosaic [photo] An assembly of aerial or space photographs or images whose edges have been feathered and matched to form a continuous photographic representation of a part of the Earth's surface; e.g. a composite photograph formed by joining together parts of several overlapping vertical photographs of adjoining areas of the Earth's surface. See also: *controlled mosaic; uncontrolled mosaic.* Syn: *aerial mosaic; photomosaic.*

mosaic breccia A breccia having fragments that are largely but not wholly disjointed and displaced.

mosaic evolution The pattern within an evolving *lineage,* in which various morphologic characters of the organisms change at different rates; as a result, a particular species within the lineage appears to possess some advanced, some primitive, and some intermediate characters simultaneously.

mosaic structure Slight irregularity of orientation of small, angular, and granular fragments of varying sizes in a crystal, the fragments appearing in polarized light like pieces of a mosaic.

mosaic texture [meta] A granoblastic texture in a dynamically metamorphosed rock in which the individual grains meet with straight or only slightly curved, but not interlocking or sutured, boundaries. Obsolete syn: *cyclopean texture.*

mosaic texture [sed] A texture in a crystalline sedimentary rock characterized by more or less regular grain-boundary contacts; e.g. a texture in a dolomite in which the mineral dolomite forms rhombs of uniform size so that in section contiguous crystals appear to dovetail, or a texture in an orthoquartzite in which secondary quartz is deposited in optical continuity on detrital grains.

mosandrite (mo-san′-drite) A colorless, greenish, reddish-brown, or yellowish-brown monoclinic mineral: $(Na,Ca)_3(Ca,Ce)_3(Ti,Nb,Al,Zr)(Si_2O_7)_2(O,F)_4$. Cf: *johnstrupite.* Syn: *rinkite; rinkolite; lovchorrite; khibinite [mineral].*

moschelite An adamantine yellow tetragonal mineral of the calomel group: HgI. It changes to dark green on exposure to light.

moschellandsbergite (mo-schel′-lands-berg-ite′) A metallic white cubic mineral: Ag_2Hg_3. It consists of a naturally occurring alloy of silver with mercury, and was formerly included with *amalgam.* Syn: *landsbergite.*

Moscovian (Mos-co′-vi-an) A stage of the Standard Global Chronostratigraphic Scale: middle of the Pennsylvanian Subsystem of the Carboniferous (above Bashkirian, below Kasimovian).

moscovite (mos′-co-vite) *muscovite.*

mosesite (mo′-ses-ite) An adamantine yellow to yellow-green cubic mineral: $Hg_2N(Cl,SO_4,MoO_4,CO_3)\cdot H_2O$. Cf: *kleinite.*

moskvinite-(Y) A vitreous white monoclinic mineral: $Na_2K(Y,REE)Si_6O_{15}$.

mosor (mo′-sor) A *monadnock* that has survived because of remoteness from the main drainage lines; esp. a *hum* in a karstic region. Etymol: originally a German term named by Penck (1900) after the Mosor Mountains in Dalmatia, Yugoslavia. Pl: mosore.

moss [bot] A bryophyte of the class Musci, characterized by a leafy upright gametophyte, which bears the sporophyte in haustorial connection. Cf: *liverwort.*

moss [eco geol] adj. A syn. of *capillary [mineral];* e.g., moss gold.

moss [gem] An inclusion in a gemstone, having the appearance of moss; specif. chlorite inclusions in quartz, i.e. moss agate.

moss agate (a) A general term for any translucent chalcedony containing inclusions of any color arranged in dendritic patterns resembling trees, ferns, leaves, moss, and similar vegetation; specif. an agate containing brown, black, or green mosslike markings due to visible inclusions of oxides of manganese and iron. See also: *Mocha stone; tree agate.* (b) A moss agate containing green inclusions of actinolite or of other green minerals.

moss animal *bryozoan.*

Mössbauer effect (Möss′-bau-er) The almost recoil-free emission and absorption of nuclear gamma rays by atoms tightly bound in a solid. It is a special case of *nuclear resonance* characterized by an extremely sharply defined resonant frequency as the atom is tightly bound in a crystal lattice, atomic recoil is minimized, and the emitted gamma ray is limited to a narrow frequency range. The Mössbauer effect was named in honor of Rudolf L. Mössbauer (1929-), German physicist. So sharp is the resonance that very slight Doppler shifts and even gravitational shifts of the emitted frequency can be detected.

Mössbauer spectrometer Instrument used to conduct *Mössbauer spectroscopy.*

Mössbauer spectrometry The art or process of using a spectrometer to analyze a *Mössbauer spectrum,* mainly for determining chemical structure. Measurement is made of the nuclear resonant absorption of gamma rays passing from a radioactive source to an absorber, usually the material being studied (DeVoe and Spijkerman, 1966, p.382R).

Mössbauer spectroscopy An analytical technique that uses a spectrometer to analyze a Mössbauer spectrum; mainly for determining oxidation states of elements in order to decipher their chemical structure. The technique involves measurement of the nuclear resonant absorption of gamma rays as they pass from a radioactive source to the absorbing medium being studied.

Mössbauer spectrum The spectrum seen when Mössbauer gamma-ray intensity is plotted as a function of relative velocity between the radioactive source and the absorber, usually the material being studied (DeVoe and Spijkerman, 1966, p.382R).

moss coral *bryozoan.*

mossite (moss′-ite) A discredited name for tantalum-bearing *ferrocolumbite.*

moss land An area with abundant moss, yet not wet enough to be a bog.

moss peat *highmoor peat.*

moss polyp *bryozoan.*

mother cell (moth′-er) A cell from which new, particularly haploid, cells are formed; e.g., *spore mother cell; pollen mother cell.*

mother crystal A mass of raw quartz (faced or rough) as found in nature.

mother geosyncline Stille's term for a geosyncline that matured by evolving into a folded mountain system (Glaessner and Teichert, 1947, p.588). Obsolete. See also: *orogeosyncline.*

motherham (moth′-er-ham) An old syn. of *fusain.*

mother lode (a) A main mineralized unit that may not be economically valuable in itself but to which workable veins are related, e.g. the Mother Lode of California. (b) An ore deposit from which a placer is derived; the mother rock of a placer.

Mother Lode vein *low-sulfide gold-quartz vein.*

mother of coal A syn. of *fusain.* Also spelled: mother-of-coal.

mother-of-emerald (moth′-er-of-em′-er-ald) (a) *prase.* (b) Green fluorite.

mother-of-pearl The *nacre* of a pearl-bearing mollusk, extensively used for making small ornamental objects.

mother rock [eco geol] A general term for the rock in which a secondary or transported ore deposit originated; *mother lode.*

mother rock [sed] *parent rock [sed].*
motile (mo'-tile) Exhibiting or capable of movement, as by cilia; e.g. the flagellate stage or "motile phase" in the life cycle of a coccolithophorid. Ant: *nonmotile.*
mottanaite-(Ce) A vitreous brownish monoclinic mineral: $Ca_4(Ce,Ca)_2AlBe_2O_2Si_4B_4O_{22}$.
mottle (mot'-tle) (a) A spot, blotch, or patch of color occurring on the surface of a sediment or soil. (b) A small, irregular body of material in a sedimentary matrix of different texture (difference in color not being essential) (Moore and Scruton, 1957, p.2727).
mottled [sed] (mot'-tled) Said of a sediment or sedimentary rock marked with spots of different colors, usually as a result of oxidation of iron compounds. Cf: *variegated.*
mottled [soil] Said of a soil that is irregularly marked with spots or patches of different colors, usually indicating poor aeration or seasonal wetness.
mottled limestone Limestone with color variations or grain/crystal size variations in branching cylindrical masses; may be biotic or abiotic in origin.
mottled structure Discontinuous lumps, tubes, pods, and pockets of a sediment, randomly enclosed in a matrix of contrasting textures, and usually formed by the filling of animal borings and burrows (Moore and Scruton, 1957, p.2727). Syn: *mottling.*
mottling (mot'-tling) (a) Variation of color in sediments and soils, as represented by localized spots, patches, or blotches of color or shades of color. Also, the formation of mottles or of a mottled appearance. (b) *mottled structure.* (c) *luster mottling [sed].* (d) *dolomitic mottling.*
mottramite (mot'-tram-ite) A variously colored orthorhombic mineral of the *descloizite* group: $PbCu(VO_4)(OH)$. Syn: *cuprodescloizite; psittacinite.*
motu (mo-tu') A Polynesian term for a small coral island with vegetation. Pl: motu; motus.
motukoreaite (mo-tu'-ko-rea'-ite) A white to pale yellow-green rhombohedral mineral related to hydrotalcite: $[Mg_6Al_3(OH)_{18}][Na_{0.6}(SO_4,CO_3)_2]\cdot 12H_2O$.
mould *mold [paleont].*
moulin (mou-lin' [moo-lanh']) A roughly cylindrical, nearly vertical, well-like opening, hole, or shaft in the ice of a glacier, scoured out by swirling meltwater as it pours down from the surface. Etymol: French, "mill". (The feature is called this because the loud roaring noise made by the falling water is reminiscent of a water-driven mill). Syn: *glacier mill; glacial mill; glacier well.*
moulin kame A conical hill of glaciofluvial material formed in a large circular hole (*moulin*) in glacier ice.
moulin pothole *giant's kettle.*
mounanaite (mou'-na-na-ite) A brownish-red monoclinic mineral of the *tsumcorite* group: $PbFe^{3+}{}_2(VO_4)_2(OH)_2$.
mound [arch] A small, human-made hill, composed either of debris accumulated during successive occupations of the site (syn: *tell; teppe*) or of earth heaped up to mark a burial site.
mound [geomorph] A low, rounded hill, generally of earth, a *knoll [geomorph].*
mound [paleont] An organic structure built by fossil colonial organisms, such as crinoids.
mound spring A spring characterized by a mound at the place where it flows onto the land surface. According to Meinzer (1923, p.55) "mound springs may be produced, wholly or in part, by the precipitation of mineral matter from the spring water; or by vegetation and sediments blown in by the wind a method of growth common in arid regions". Cf: *pool spring.* See also: *spring mound.*
mount (a) An abbreviated form of the term *mountain,* esp. used preceding a proper name and usually referring to a particular summit within a group of elevations; e.g. Mount Marcy in the Adirondack Mountains. Abbrev: mt. (b) A high *hill;* esp. an eminence rising abruptly above the surrounding land surface, such as Mount Vesuvius. (c) *seamount.*
mountain (moun'-tain) (a) Any part of the Earth's crust higher than a *hill,* sufficiently elevated above the surrounding land surface of which it forms a part to be considered worthy of a distinctive name, characterized by a restricted summit area (as distinguished from a plateau), and generally having comparatively steep sides and considerable bare-rock surface; it can occur as a single, isolated eminence, or in a group forming a long chain or range, and it may form by earth movements, erosion, or volcanic action. Generally, a mountain is considered to project at least 300 m (1,000 ft) above the surrounding land, although older usage refers to an altitude of 600 m (2,000 ft) or more above sea level. When the term is used following a proper name, it usually signifies a group of elevations, such as a range (e.g., the Adirondack Mountains) or a system (e.g., the Rocky Mountains). Abbrev: mt.; mtn. Syn: *mount.* (b) Any conspicuous or prominent elevation in an area of low relief, esp. one rising abruptly from the surrounding land and having a rounded base. (c) A region characterized by mountains; term usually used in the plural.
mountain and bolson desert A desert area made up of elongated mountain ranges and intervening alluvium-filled fault basins or bolsons (Stone, 1967, p.233).
mountain apron *bajada.*
mountain blue A blue copper mineral; specif. azurite and chrysocolla.
mountain building *orogeny.*
mountain butter A term used for various salts; esp. *halotrichite.*
mountain chain A complex, connected series of several more or less parallel mountain ranges and mountain systems grouped together without regard to similarity of form, structure, and origin, but having a general longitudinal arrangement or well-defined trend; e.g. the Mediterranean mountain chain of southern Europe. See also: *cordillera.*
mountain climate A climate of high elevations, characterized by extremes of surface temperature, low atmospheric temperature, strong winds, and rarefied air (Swayne, 1956, p.98).
mountain cork (a) A white or gray variety of *asbestos* consisting of thick interlaced fibers and resembling cork in texture and lightness (it floats on water). Syn: *rock cork.* (b) A fibrous clay mineral such as sepiolite or palygorskite.
mountain flax A fine silky *asbestos.*
mountain glacier *alpine glacier.*
mountain green A green mineral; specif. malachite, green earth, and chrysocolla.
mountain group An assemblage of several mountain peaks or of short mountain ridges; e.g. the Catskill Mountains, N.Y.
mountainite (moun'-tain-ite) A silky white monoclinic zeolite mineral: $(Ca,Na_2,K_2)_2Si_4O_{10}\cdot 3H_2O$. Cf: *rhodesite.*
mountain leather (a) A tough variety of *asbestos* occurring in thin flexible sheets made of interlaced fibers. Syn: *rock leather; mountain paper.* (b) A fibrous clay mineral such as sepiolite or palygorskite. Syn: *leatherstone.*
mountain of accumulation A symmetric mountain, frequently of great height, formed by the accretion of material on the Earth's surface, esp. by the ejection of material from a volcano; it tends to occur as an isolated peak. Syn: *accumulation mountain.*
mountain of circumdenudation A mountain consisting of resistant rock that remains after the surrounding, less resistant rock has been worn away, or a mountain representing the remains of a preexisting plateau; e.g. a *monadnock* or an *inselberg.* Syn: *relict mountain; remainder mountain; circumdenudation mountain.*
mountain of denudation A remnant of "undisturbed and otherwise continuous strata, that have been in part removed by erosion" (Gilbert, 1875, p.21).
mountain of dislocation A mountain "due to the rearrangement of strata, either by bending or fracture" (Gilbert, 1875, p.21); a "fold mountain" or "fault mountain".
mountainous (moun'-tain-ous) (a) Descriptive of a region characterized by conspicuous peaks, ridges, or mountain ranges. (b) Resembling a mountain, such as a *mountainous* dome that is strongly elevated and around whose flanks the strata are steeply dipping.
mountain paper A paperlike variety of *asbestos* occurring in thin sheets; specif. *mountain leather.*
mountain pediment (a) A term introduced by Bryan (1923a. p.30, 52-58, 88) for a plain of combined erosion and transportation at the foot of a desert mountain range, similar in form to an alluvial plain, and surrounding a mountain in such a manner that at a distance the plain appears to be a broad triangular mass (resembling a pediment or gable of a low-pitched roof) above which the mountain projects. This usage is similar to *piedmont pediment.* (b) A pediment occurring within a mountain mass as a relatively high-altitude surface truncating a mountain structure (Tator, 1953, p.51).
mountain range A single, large mass consisting of a succession of mountains or narrowly spaced mountain ridges, with or without peaks, closely related in position, direction, formation, and age; a component part of a *mountain system* or of a *mountain chain.*
mountainside (moun'-tain-side) A part of a mountain between the summit and the foot. Syn: *mountain slope.*
mountain slope *mountainside.*
mountain soap A dark clay mineral having a greasy feel and streak; specif. *saponite.* Syn: *rock soap.*
mountain system A group of mountain ranges exhibiting certain unifying features, such as similarity in form, structure, and alignment, and presumably originating from the same general causes; esp. a series of ranges belonging to an *orogenic belt.* Cf: *mountain chain.*
mountain tallow *hatchettine.*
mountain-top detritus *block field.*
mountain wall A very steep mountainside.
mountain wind A nighttime *katabatic wind,* flowing down a mountain slope. It often alternates with a daytime *valley wind.*
mountain wood (a) A compact, fibrous, gray to brown variety of *asbestos* resembling dry wood in appearance. Syn: *rock wood.* (b) A fibrous clay mineral such as sepiolite or palygorskite.
mountkeithite (mount-keith'-ite) A pale pink or white hexagonal mineral related to pyroaurite: $(Mg,Ni)_{11}(Fe^{3+},Cr,Al)_3(SO_4,CO_3)_{3.5}(OH)_{24}\cdot 11H_2O$.
mourite (mour'-ite) A violet monoclinic mineral: $(UO_2)Mo^{6+}{}_5O_{16}\cdot 5H_2O$.
mouth [geol] (a) The place of discharge of a stream, as where it enters a larger stream, a lake, or the sea; debouchment (b) *baymouth.* (c) The entrance or opening of a geomorphic feature such as a cave, valley, or canyon. (d) The surface outlet of an underground conduit, as of a volcano.
mouth [paleont] The entrance to the digestive tract through which food and water passes into the body of an animal; e.g. the central opening at the summit of the theca leading to the alimentary system of a blastoid or the external

opening of the body cavity of a coelenterate through which indigestible material is also discharged.
mouth frame The angulated girdle of ossicles surrounding the mouth of an asterozoan.
moutonnée A French adjective meaning "fleecy" or "curled", but often used (incorrectly) as a shortened form of *roche moutonnée.* The term was introduced into geologic literature by Saussure (1786, par. 1061, p.512-513) in describing an assemblage of rounded Alpine hills whose contiguous and repeated curves, taken as a whole and as seen from a distance, resemble a thick fleece and also a curly or wavy wig ("perruque moutonnée") that was fashionable in the late 18th century. The term later implied a fancied resemblance between the general form of a roche moutonnée and that of a grazing sheep whose head is represented by the stoss side.
movable bed (mov'-a-ble) A stream bed consisting of readily transportable materials.
movement horizons In flexural slip folds, zone of interlayer slip on several layer surfaces linked by faults cutting obliquely across the layering (Tanner, 1989).
movement picture (move'-ment) *deformation plan.*
movement plan *deformation plan.*
moveout (move'-out) The difference in arrival times of a reflection event on adjacent traces of a seismic record, esp. resulting from the dip of the reflecting interface. Cf: *normal moveout.* Syn: *stepout.*
moving average A smoothing procedure applied to a time series by replacing each observation with the average of the surrounding observations. An analogous procedure is used to smooth map data.
moydite-(Y) A vitreous yellow orthorhombic mineral: $YB(OH)_4CO_3$.
mozarkite (mo'-zar-kite) The state rock of Missouri: a varicolored, easily polished Ordovician chert.
mozartite A reddeish-brown orthorhombic mineral: $CaMn^{3+}SiO_4(OH)$. It is the Mn^{3+} analogue of vuagnatite.
mozgovaite A metallic silver-gray orthorhombic mineral: $PbBi_4(S,Se)_7$.
MPa Symbol for megapascal(s), 10^6 pascals.
mpororoite (m-po-ro'-ro-ite) A greenish-yellow monoclinic mineral: $(Al,Fe)_2O(WO_4)_2{\cdot}6H_2O$.
M position In locational notation for conodont skeletal apparatuses, *M* designates the position occupied by pick-shaped dolabrate elements or by similar bipennate, digyrate, or coniform elements (TIP, 1981, pt. W, supp.2).
mrázekite A bright blue monoclinic mineral: $Bi_2Cu_3(PO_4)_2O_2(OH)_2{\cdot}2H_2O$.
mroseite (mrose'-ite) An adamantine colorless orthorhombic mineral: $CaTe^{4+}O_2(CO_3)$.
MSL *mean sea level.*
MSS *multispectral scanner.*
MT *magnetotellurics.*
M-type granitoid A granitic rock produced from mantle-derived magmas that was generated by partial fusion of subducted oceanic crust or fractionation of peridotite followed by mafic magma. The term was proposed by A. J. R. White in 1979. See: Clarke (1992).
muchuanite A metallic black trigonal mineral: $MoS_2{\cdot}0.5H_2O$. It has not been approved by the International Mineralogical Association, and may be altered molybdenite.
muck [eng] A general term applied to soil and rock material excavated from a tunnel.
muck [mining] n. Rock or ore broken in mining v. To remove broken rock or ore.
muck [sed] Dark fine-textured well-decomposed organic material, intermixed with a high percentage of mineral matter, usually silt; it forms surface deposits in some poorly drained areas, e.g. areas of permafrost and lake bottoms.
mückeite (mü'-cke-ite) A metallic gray orthorhombic mineral: $CuNiBiS_3$. It is the Bi analogue of lapieite.
muckite (muck'-ite) A yellow variety of retinite found in minute particles in coal in a region of central Europe about the upper valley of the Oder River. Named after its discoverer, H. Muck, 19th-century German mineralogist.
muck soil An organic soil in which the plant residues have been altered beyond recognition. The sum of the thicknesses of organic layers is usually greater than the sum of the thicknesses of mineral layers.
mucro (mu'-cro) An abrupt, sharp terminal point or process of an animal part or a plant part; e.g. a projection of the frontal shield over the proximal lip of the orifice in some ascophoran cheilostome bryozoans (cf: *lyrula*), the terminal segment of the springing appendage of certain arthropods, a short abrupt spur or spiny tip of some leaves, or the elevation at the apex of the tail valve of a polyplacophoran. Pl: mucrones or mucros. Syn: *mucron.*
mucron (mu'-cron) (a) A perforate central scar or buttonlike projection on the aboral end of a chitinozoan test, serving for attachment. (b) *mucro.*
mucronate (mu'-cro-nate) Ending in an abrupt, sharp terminal point or process, such as by a distinct and obvious mucro; e.g. said of the cardinal margin of a brachiopod in which the posterolateral extremities extend into sharp points.
mud [drill] *drilling mud.*
mud [marine geol] A sticky, fine-grained, (predominantly clay- and silt-sized) marine detrital sediment, either pelagic or terrigenous. Muds are usually described by color: *blue mud; black mud; gray mud; green mud; red mud.*
mud [sed] (a) A slimy, sticky, or slippery mixture of water and silt- or clay-sized earth material, with a consistency ranging from semifluid to soft and plastic; a wet, soft soil or earthy mass; mire, sludge. (b) An unconsolidated sediment consisting of clay and/or silt, together with material of other dimensions (such as sand), mixed with water, without connotation as to composition; e.g. a recently exposed lake-bottom clay in a soft, oozelike condition. (c) A mixture of silt and clay; the silt-plus-clay portion of a sedimentary rock, such as the finely divided calcareous matrix of a limestone.
mud aggregate An aggregate of mud grains, commonly having the size of a sand or silt particle, and usually mechanically deposited.
mud ball (a) A spherical mass of mud or mudstone in a sedimentary rock, developed by weathering and breakup of clay deposits. It may measure as much as 20 cm in diameter. (b) *armored mud ball.*
mudbank (mud'-bank) (a) A submerged or partly submerged ridge of mud along a shore or in a river, usually exposed during low tide. (b) A *bank [coast]* consisting largely of carbonate mud (*micrite*); some mudbanks preserve traces of plants or sessile animals that helped to stabilize the loose sediment. Cf: *bank [sed].*
mud breccia A term used by Ransome and Calkins (1908, p.31) for a *desiccation breccia* containing angular or slightly rounded fragments of "fine-grained argillite embedded in somewhat coarser-grained and more arenaceous material".
mud-buried ripple A *ripple* covered by mud settling out of water, characterized by the filling up of troughs and little or no accumulation on the crests (Shrock, 1948, p.109-110). Cf: *flaser structure [sed].*
mud-buried ripple mark *mud-buried ripple.*
mud cake [drill] In *rotary drilling,* a layer of concentrated solids adhering to the walls of a well or borehole, formed where the *drilling mud* lost its fluid phase (e.g., water filtrate) into a permeable formation. Syn: *filter cake; cake [drill].*
mud cake [sed] A clast formed by desiccation and occurring in an *intraformational breccia.*
mud circle A *nonsorted circle* characterized by a central core of upwardly injected clay, silt, or sometimes fine sand, surrounded by vegetation; the center is round and generally 10 cm to 2 m in diameter. See also: *frost scar; plug [pat grd].* Syn: *clay boil; tundra ostiole.*
mud clast A *clast* that is commonly eroded from a soft, muddy substrate and transported only a relatively short distance before being deposited. Cf: *rip-up.*
mud column The height, measured from the bottom of a borehole, of *drilling mud* standing in the hole.
mud cone A small cone of sulfurous mud built around the opening of a *mud volcano* or mud geyser.
mud crack (a) An irregular fracture in a crudely polygonal pattern, formed by the shrinkage of clay, silt, or mud, generally in the course of drying under the influence of atmospheric surface conditions. Also referred to as a *sun crack,* a *shrinkage crack,* and a *desiccation crack.* (b) *mud-crack cast.* Also spelled: mudcrack.
mud-crack cast A mud crack after it has been filled and the filling material (generally sand) has been hardened into rock; often occurs on the underside of a bed immediately overlying a mudstone. Syn: *mud crack.*
mud-crack polygon *desiccation polygon.*
mudding off (mud'-ding) The blockage of the flow of reservoir fluids, e.g. oil, into a wellbore that results from the formation of *mud cake* and/or the loss of permeability from *drilling mud* invasion. It may require remedial treatment to establish production.
mud drape A thin layer of clay and silt deposited on top of coarser sediment (typically rippled) during waning flow. It is common in fluvial and tidal environments. Cf: *flaser structure [sed].*
muddy [sed] adj. Pertaining to or characterized by mud; esp. said of water made turbid by sediment, or of sediment consisting of mud.
muddy gravel An unconsolidated sediment containing 30-80% gravel and having a ratio of sand to mud (silt + clay) less than 1:1 (Folk, 1954, p.346).
muddy sand An unconsolidated sediment containing 50-90% sand and having a ratio of silt to clay between 1:2 and 2:1 (Folk, 1954, p.349).
mud engineer A specialist who studies and describes the materials, chemicals, and proprietary additives to make up and maintain the properties of the *drilling mud* used in rotary drilling.
mud field An area saturated with groundwater owing to the presence of fumaroles (Schieferdecker, 1959, term 4511).
mud flat A relatively level area of fine silt along a shore (as in a sheltered estuary) or around an island, alternately covered and uncovered by the tide, or covered by shallow water; a muddy *tidal flat* barren of vegetation. Cf: *sand flat; slikke.* Syn: *flat [geomorph].*
mud-flat polygon *desiccation polygon.*
mudflow [mass move] (mud'-flow) A general term for a mass-movement landform and a process characterized by a flowing mass of predominantly fine-grained earth material possessing a high degree of fluidity during movement. The degree of fluidity is revealed by the observed rate of movement or by the distribution and morphology of the resulting deposit. If

more than half of the solid fraction of such a mass consists of material larger than sand size, the term *debris flow* is preferable (Sharp and Nobles, 1953; Varnes, 1958). Mudflows are intermediate members of a gradational series of processes characterized by varying proportions of water, clay, and rock debris. The water content of mudflows may range up to 60%. The degree of water bonding, determined by the clay content and mineralogy, critically affects the viscosity of the matrix and the velocity and morphology of the flow. With increasing fluidity, mudflows grade into loaded and clear streams; with a decrease in fluidity, they grade into earthflows. Also spelled: mud flow.

mudflow [sed] A minor sedimentary structure found in fine-grained rocks and indicative of local flowage while the material was still soft. Not in common usage.

mudflow [volc] *lahar.*

mudflow fan Splay resulting from a mudflow, or in particular its lower part if spread across flat territory.

mudflow levee A sharp linear ridge marking the edge of a narrow debris flow or mudflow, and consisting of boulders shoved aside by the force of the flow.

mud geyser A geyser that erupts sulfurous mud; a type of mud volcano.

mud glacier A viscous mass of surficial material moving slowly downslope by solifluction.

mud lava [grd wat] The sulfurous and sometimes carbonaceous material contained in mud pots or erupted from mud volcanoes or mud geysers.

mud lava [volc] A term that has been used to describe some materials in Japan that are now known to be *ash flow* deposits (Ross and Smith, 1961, p.5).

mud log A continuous analysis of the drilling mud and well cuttings during *rotary drilling,* for entrained oil or gas. Visual observation, ultraviolet fluoroscopy, partition gas chromatograph, and hydrogen-flame ionization analyzer may be used. A *drilling-time log* is kept concurrently.

mudlump (mud'-lump) A diapiric sedimentary structure that forms a small short-lived island, some 4000 square meters in area, near the mouth of a major distributary of the Mississippi River; it consists of a broad mound or swelling of silt or thick plastic clay that stands 2 to 4 m above sea level. It is created by the loading of rapidly deposited delta-front sands upon lighter-weight prodelta clays, causing the clays to be intruded or thrust upward into and through the overlying sandbar deposits. Cf: *gas-heave structure.* Also spelled: mud lump.

mud motor *turbine motor.*

mud-mound (a) A small to large mound consisting chiefly of micrite, which may contain layers or pockets of more fossiliferous carbonate rock types. Mud-mounds commonly contain the cavity structure, *stromatactis.* (b) Carboniferous through Permian carbonate build-ups with abundant crinoids, but lacking framework builders. Cf: *crust-mound; frame-thicket.*

mud pellet A small, flattened to rounded, irregularly shaped mass of mud or mudstone, 3 to 13 mm in diameter, in a sedimentary rock. Pellets are pieces of compacted mud produced by breakup, transported short distances, and redeposited.

mud pit In drilling, an excavated pit or diked area or metal tub, often divided into two or more sections, used to impound water or *drilling mud* or to retain fluids discharged from a well or to slow the flow of the drilling mud to allow *cuttings* to settle out. Syn: *slush pit.*

mud polygon (a) A *nonsorted polygon* whose center is bare of vegetation but whose outlining reticulate fissures contain peat and plants. The term was suggested by Elton (1927, p.165) to replace *fissure polygon,* but it is ambiguous because forms without a stone border do not invariably consist of "mud" but may consist of sand, gravel, or a nonsorted mixture of sand, clay, and silt with stones (Washburn, 1956, p.825-826). See also: *frost hillock.* (b) *desiccation polygon.*

mud pot A type of hot spring containing boiling mud, usually sulfurous and often multicolored, as in a *paint pot.* Mud pots are commonly associated with geysers and other hot springs in volcanic areas, esp. Yellowstone National Park, Wyo. Syn: *sulfur-mud pool.*

mud pump The reciprocating pump used to circulate *drilling mud* through the circulating system used in rotary drilling. Syn: *slush pump.*

mud-ridge ripple A *longitudinal ripple* with a regular profile, a usually symmetric crest, and a narrow and angular ridge, that is situated between much wider and relatively flat troughs and that frequently branches (always converging downcurrent). Cf: *corrugated ripple.*

mud-ridge ripple mark *mud-ridge ripple.*

mud rock A syn. of *mudstone.* Also spelled: mudrock.

mud-rock flood A violent and destructive rush of water generated by a cloudburst and laden with rocks, mud, and debris engulfed along its path.

mudrush (mud'-rush) The sudden inflow of mud into shallow mine workings. See also: *running ground.*

mud shale A consolidated sediment consisting of no more than 10% sand and having a silt/clay ratio between 1:2 and 2:1 (Folk, 1954, p.350); a fissile mudstone.

mudslide (mud'-slide) A relatively slow-moving type of mudflow in which movement occurs predominantly by sliding upon a discrete boundary shear surface (Hutchinson and Bhandari, 1971). Cf: *earthflow.*

mud spot *frost scar.*

mud stalagmite A *stalagmite* that is composed of mud and may be cemented by calcite.

mudstone (mud'-stone) (a) An indurated mud having the texture and composition of shale, but lacking its fine lamination or fissility; a blocky or massive, fine-grained sedimentary rock in which the proportions of clay and silt are approximately equal; a nonfissile mud shale. Shrock (1948a) regards mudstone as a partly indurated mud that slakes upon wetting. See also: *claystone [sed]; siltstone.* (b) A general term that includes clay, silt, claystone, siltstone, shale, and argillite, and that should be used only when the amounts of clay and silt are not known or specified or cannot be precisely identified; or "when a deposit consists of an indefinite mixture of clay, silt, and sand particles, the proportions varying from place to place, so that a more precise term is not possible" (Twenhofel, 1937, p.98); or when it is desirable to characterize the whole family of finer-grained sedimentary rocks (as distinguished from sandstones, conglomerates, and limestones). Syn: *mud rock.* (c) A term used by Dunham (1962) for a mud-supported carbonate sedimentary rock containing less than 10% grains (particles with diameters greater than 20 micrometers); e.g. a calcilutite or micstone. The term specifies neither mineralogic composition nor mud of clastic origin. Cf: *wackestone; packstone; grainstone.* The term was apparently first used by Murchison (1839) for certain massive dark-gray fine-grained Silurian shales of Wales, which on exposure and wetting rapidly disintegrate into mud.

mudstone conglomerate (a) A conglomerate containing mudstone clasts. (b) *desiccation conglomerate.* Cf: *conglomeratic mudstone.*

mudstone ratio A uranium prospector's term, esp. on the Colorado Plateau, for the ratio of the total thickness of red mudstone to that of green mudstone within an assumed stratigraphic interval. "Its value is based upon the premise that uranium-bearing solutions will bleach red mudstone containing ferric iron to green mudstone containing ferrous iron" (Ballard and Conklin, 1955, p.194), in the course of depositing uranium minerals.

mudstream A well-defined coastal band of turbid water and fluid mud characterized by unidirectional transport in response to fluvial and/or marine processes; e.g., the mudstream trailing west from the Atchafalaya River to the west Louisiana chenier plain.

mud-supported (mud'-sup-port'-ed) A term used to describe a sedimentary carbonate rock whose sand-size particles (at least 10% of the total bulk) are embedded or "floating" in, and supported by, the muddy matrix. Cf: *grain-supported.*

mud tuff *dust tuff.*

mud volcano An accumulation, usually conical, of mud and rock ejected by volcanic gases; also, a similar accumulation formed by escaping petroliferous gases. The term has also been used for a *mud cone* not of eruptive origin. Syn: *hervidero; macaluba.* Cf: *air volcano; salinelle.*

mud wave A bank of fine sediment that occurs along some coasts, such as Suriname and on shelves and in the deep sea. Lateral dimensions can be a few to tens of kilometers and heights range from a few to a few tens of meters.

mud weight The density of drilling fluid (*mud*), usually expressed in pounds per gallon; 9.0 lb/gallon mud produces hydrostatic or normal pressure. Generally, mud weight must be kept between normal (hydrostatic) and fracture pressures to avoid risking a blowout; if it is smaller than the normal pressure, formation fluids are apt to flow into the borehole, if it is too large, the formation is apt to fracture.

mugearite (mu-gear'-ite) An extrusive or hypabyssal igneous rock, a member of the *alkali basalt* suite, consisting of oligoclase with subordinate alkali feldspar and mafic minerals. In many examples, olivine is more abundant than clinopyroxene. Although generally nepheline-normative, mugearite may contain normative hypersthene, or even quartz. It has been defined (Baker et al., 1974) as having a differentiation index between 45 and 65 and normative plagioclase more sodic than An_{30}. Cf: *hawaiite [petrology]; benmoreite; basalt [petrology].* The term, proposed by Harker in 1904, is from the village of Mugeary, Isle of Skye, Scotland. Usage was revived by Macdonald (1960), and by Muir and Tilley (1961). In the *IUGS classification,* a *basaltic trachyandesite* with $Na_2O-2 \geq K_2O$ in wt%. Cf: *shoshonite.*

muirite (mu'-ir-ite) An orange tetragonal mineral: $Ba_{10}Ca_2Mn^{2+}TiSi_{10}O_{30}(OH,Cl,F)_{10}$.

mukhinite (mukh'-in-ite) A brownish-black or reddish monoclinic mineral of the *epidote* group: $Ca_2Al_2V^{3+}(SiO_4)(Si_2O_7)O(OH)$.

mull [coast] A Scottish term for a *headland [coast];* e.g. Mull of Galloway.

mull [soil] A type of humus, usually developed in the forest, that is incorporated with underlying mineral matter. Cf: *mor.*

Müllerian law (Mül-le'-ri-an) The law that expresses the regularity in distribution of 20 radial spines on the shells of radiolarians of the suborder Acantharina (four spines on each of the five circles that are comparable to the equatorial, two tropical, and two polar circles of the terrestrial globe): "between two poles of a spineless axis are regularly disposed five parallel zones, each with four radial spines; the four spines of each zone are equidistant one from another, and also equidistant from each pole; and the four spines of each zone are so alternating with those of each neighboring zone, that all twenty spines together lie in four meridian planes, which transect one another at an angle of 45°" (translated from Haeckel, 1862, p.40). Named by Haeckel in honor of Johannes Müller (1801-1858), German physiologist and zoologist, who first recognized the regularity in the disposition of the 20 radial spines (Müller, 1858, p.12, 37). Syn: *icosacanthic law.*

Müller's glass (Mül'-ler's) *hyalite.*

mullicite (mul'-li-cite) A variety of vivianite occurring in cylindrical masses.

mullion (mul'-lion) A type of lineation in metamorphic rock formed where the interface between lithologies with differing ductilities becomes corrugated into a surface of alternating pointed peaks (pointing into the stiffer layers) and intervening smoothly curving cusps as a consequence of deformation. In profile, mullion structure looks like waves or ripples. In three dimensions, mullions are linear, rodlike structures.

mullion structure A wavelike pattern of parallel V-shaped grooves and rounded ridges. The distance between adjacent grooves or between adjacent ridges, which may be millimeters to several meters, is fairly constant. Mullion structure typically forms as a consequence of layer-parallel shortening where there is significant ductility contrast between adjacent layers. Etymol: Old French "moienel", "medial". Cf: *striation; groove [struc geol]; slickenside; slip-mark.*

mullite (mull'-ite) A white or variously colored orthorhombic mineral: $Al_{4+2x}Si_{2-2x}O_{10-x}$ (x~0.4). Synthetic mullite is a valuable refractory material. Syn: *porcelainite.*

mullitization (mull'-i-ti-za'-tion) The formation of mullite from minerals of the sillimanite group by heating.

mullock (mul'-lock) An Australian term for *waste rock.*

multiband (mul'-ti-band) A nonrecommended syn. of multispectral.

multibeam mapping A method of making contoured bathymetric maps both of the seafloor directly beneath the surveying ship and of the off-track region. The hull-mounted array of narrow-beam transducers receives sideways-reflected signals, recording their ranges and bearings. The data are processed to compute the off-track ranges and depths, and the results plotted, in near-real time, as contour charts of the swath of seafloor ensonified by the transducers. Also called: swath mapping.

multichannel amplitude analyzer The part of a gamma-ray spectrometer that sorts input pulses into channels according to the amplitude (energy) of the input pulses (IAEA, 2003).

multichannel seismic profiling A method of reflection seismic profiling in which the hydrophone receivers are grouped into separate sections, each section being recorded on a separate channel. The data can then be processed by stacking and other techniques for enhancing the primary reflections. Abbrev: MCS.

multicycle (mul-ti-cy'-cle) adj. Said of a landscape or landform produced during or passing through more than one cycle of erosion, and bearing the traces of the former condition(s); e.g., a *multicycle* coast with a series of elevated sea cliffs separated from each other in stair-like fashion by narrow wave-cut benches, each sea cliff representing a separate shoreline cycle (Cotton, 1922, p.426); or a multicycle valley showing on its sides a series of *straths* resulting from successive uplifts. Syn: *multiple-cycle; polycyclic.*

multidomains In a *ferromagnetic, ferroelastic,* or *ferroelectric* crystal, a region consisting of many *single domains.* Cf: *single domains.*

multifossil-range zone A biostratigraphic unit, a special type of *range zone,* marked by a concentration of the range zones of a number of different fossils (ACSN Rept. 5, 1957, p.1884). See also: *concurrent-range zone.*

multigelation (mul'-ti-ge-la'-tion) Often-repeated freezing and thawing (Washburn, 1956, p.838). See also: *regelation.*

multilocular (mul-ti-loc'-u-lar) Divided into or composed of many small chambers or vesicles; specif. said of a many-chambered *test* of a unicellular organism such as a foraminifer. See also: *polythalamous.*

multimembrate apparatus A general term for conodont skeletal apparatuses consisting of more than one morphologically distinct type of element, regardless of the total number of elements of all types in the complete apparatus; see bimembrate, trimembrate, quadrimembrate, quinquimembrate, seximembrate, septimembrate (TIP, 1981, pt. W, supp.2).

multipartite map (mul-ti-par'-tite) A *vertical-variability map* that shows the degree of distribution of one lithologic type within certain parts (such as the top, middle, and bottom thirds) of a given stratigraphic unit. The map was introduced by Forgotson (1954).

multiple [seis] n. Seismic energy which has been reflected more than once. While virtually all seismic energy involves some multiples, the important distinction is between long-path and short-path multiples: A long-path multiple arrives as a distinct event, whereas a short-path multiple arrives so soon after the primary that it merely adds tail to the primary (i.e., changes the wave shape). Short-path multiples may obscure stratigraphic detail even where structural aspects are not affected significantly. The attitude of long-path multiples is apt to not represent the portion of the section associated with their arrival time.

multiple biseries (mul'-ti-ple) Two or more sets of coverplates in edrioasteroids, each of which forms an alternating biseries of pairs of plates, the pairs of each set alternating with those of the other sets to form an integrated system (Bell, 1976).

multiple-cycle (mul'-ti-ple-cy'-cle) *multicycle.*

multiple detectors Two or more seismic detectors whose combined outputs are fed into a single amplifier-recorder channel in order to reduce undesirable noise.

multiple-exposure method (mul'-ti-ple-ex-po'-sure) *isolation of outcrops.*

multiple glaciation The alternating advance and recession of glacier ice during the Pleistocene Epoch.

multiple intrusion An igneous intrusion composed of two or more kinds of rock relatively close in chemical and mineralogic composition, formed by separate injections. Cf: *composite intrusion.*

multiple reflection A seismic wave that has been reflected more than once. Syn: *secondary reflection.*

multiple retrogressive slide A type of mass movement associated with shear failure in unfrozen sediment underlying permafrost, leading to detachment of blocks of frozen ground that move downslope.

multiple tunnel One of a series of openings in the chamber of a fusulinid test, produced by resorption of the lower (adaxial) parts of septa.

multiple twin A twinned crystal that is formed by *repeated twinning.*

multiple working hypotheses The name given by Chamberlin (1897) to a method of "mental procedure" applicable to geologic studies, in which several rational and tenable explanations of a phenomenon are developed, coordinated, and evaluated simultaneously in an impartial manner.

multiplex (mul'-ti-plex) n. A stereoscopic (anaglyphic) plotting instrument used in preparing topographic maps from aerial photographs. v. To transmit several channels of seismic information over a single channel without crossfeed. Usually different input channels are sampled in sequence at regular intervals and the samples are fed into a single output channel. Digital tapes are sometimes multiplexed in this way (Sheriff, 1973, p.148).

multiplicity [cryst] (a) The number of lattice points in a *unit cell.* (b) Any position in a unit cell is said to have a multiplicity equal to the number of different, symmetrically equivalent positions in the unit cell that are generated by all of the symmetry operations of the *space group.*

multiplicity [seis] The number of common-midpoint traces which sample essentially the same portion of a reflector but with different offsets, hence the number of independent raypaths which add together to provide a single output trace.

multiply primitive cell A *unit cell* possessing more than one *lattice point.*

multiply primitive lattice *centered lattice.*

multiramate element A ramiform conodont element with more than four basic processes (TIP, 1981, pt. W, supp.2).

multi-ring basin An *impact crater* with a relatively low depth/diameter ratio and with at least one asymmetric concentric ring outside the crater rim. The two basic types of multi-ring basins are (1) the Orientale type (after the Orientale basin on the Moon), with just a few inward-facing concentric scarps, and (2) the Valhalla type (after the Valhalla basin on Callisto), with dozens of concentric scarps whose distal members appear to be grabens. Syn: *multi-ring crater.*

multi-ring crater *multi-ring basin.*

multisaccate (mul-ti-sac'-cate) Describes pollen with more than two sacci.

multiserial (mul-ti-se'-ri-al) Arranged in or consisting of several or many rows or series; e.g. "multiserial ambulacrum" of an echinoid with pore pairs arranged in more than two longitudinal series, or said of a protist composed of numerous rows of cells or other structural features.

multispectral data set A data set containing a record of more than one spectral band (Krzanowski et al., 1993).

multispectral images Images optically acquired in more than one spectral band. Each individual image covers the same physical area and is the same scale but of a different spectral band.

multispectral scanner (mul-ti-spec'-tral) An electro-optical sensor carried on the Landsat satellites, that acquires images in the visible and reflected infrared bands by oscillating a flat mirror between a field of view and a set of optical scanners to gather data on several bands simultaneously. The Landsat MSS system gathered data on four bands roughly equivalent to the wavelengths for green, red and two infrared bands. A MSS pixel covers a ground area approximately 79 meters on a side that is rectified to about 57 meters on a side. Abbrev: *MSS.*

multispectral system Framing or scanning system that simultaneously acquires images of the same scene at different wavelengths.

multisystem (mul-ti-sys'-tem) A set of phases more numerous than can coexist stably under any set of conditions, thus formally possessing, in the phase-rule sense, a negative number of degrees of freedom; the equilibrium relationships among the phases of a multisystem can be represented by an array of invariant points connected by univariant, bivariant, etc., equilibria, some of which will in general be stable and some metastable.

multitemporal data set A data set containing information pertaining to more than one time or period (Krzanowski et al., 1993).

Multituberculata (Mul'-ti-tu-ber'-cu-la'-ta) *Allotheria.*

multivariate (mul-ti-var'-i-ate) Pertaining to, having, or involving two or more properties; e.g. "multivariate statistics" utilizes methods that consider the simultaneous variation in two or more variables.

multivariate analysis of variance Multivariate generalization of analysis of variance for testing the equality of mean vectors of several populations. Acronym: MANOVA.

multivincular (mul-ti-vin'-cu-lar) Said of a type of *ligament* of a bivalve mollusk (e.g. *Isognomon*) consisting of serially repeated elements of *alivincular* type.

multi-year ice *Old ice* up to 3 m thick that has survived at least two summers' melt. The hummocks are smooth, the ice is almost salt-free, and the color when bare is usually blue. The melt pattern consists of interconnecting puddles and a well-developed drainage system.

multopost (mul'-to-post) Said of a process involving an igneous rock that occurs some time after consolidation of the magma, i.e. later than a *deuteric* process. Obsolete.

mummeite (mum'-me-ite) A metallic gray monoclinic mineral: $Ag_{2.8}Cu_{0.7}Pb_{1.3}Bi_{6.5}S_{13}$. It is a pavonite homologue.

mundic (mun'-dic) A syn. of *pyrite.* Drillers often use "mundick" to designate pyrite.

mundite (mund'-ite) A pale yellow orthorhombic mineral that is structurally related to phosphuranylite: $Al(UO_2)_3(PO_4)_2(OH)_3{\cdot}5.5H_2O$.

mundrabillaite (mun-dra-bil'-la-ite) An earthy colorless monoclinic mineral occurring in cave deposits: $(NH_4)_2Ca(PO_3OH)_2{\cdot}H_2O$.

municipal solid waste Discarded solid materials, such as used paper, packaging, or food.

muniongite (mu-niong'-ite) A hypabyssal rock resembling *tinguaite,* being composed of alkali feldspar, nepheline, aegirine, and sometimes cancrinite; a nepheline-rich *phonolite.* The name is for Muniong, New South Wales, Australia. Obsolete.

munirite (mu-nir'-ite) A pearly white monoclinic mineral with a green tinge: $NaV^{5+}O_3{\cdot}1.9H_2O$.

Munsell color system (Mun-sell') A system of color classification that is applied to rocks and soils. Color is defined by its hue, value (brilliance), and chroma (purity).

muntenite (mun'-ten-ite) A variety of amber from Rumania.

Muong-Nong type tektite The name derives from the fact that the first scientifically characterized layered tektites were recovered near Muong-Nong, Laos. Syn: *layered tektite.*

mural deposit (mu'-ral) A *cameral deposit* along the wedgelike extension of each septum attached to the wall of a nautiloid conch.

mural escarpment A rocky cliff with a face nearly vertical, like a wall (Lee, 1840, p.350).

muralite (mu'-ral-ite) A *phyteral* of coal that represents the structure of plant cell walls and that occurs in some types of vitrain.

mural plate *compartmental plate.*

mural pore (a) A small hole in the wall between adjoining corallites, as in some tabulates. (b) An opening in the shell wall of a foraminifer, as distinguished from a *septal pore.* (c) A *communication pore* in a bryozoan.

mural ridge Elevation of the inner prismatic layer of the shell along the edge of the last septum in ectocochleate cephalopods.

mural rim The line of attachment of the *frontal membrane* to the vertical walls or the inner margin of the gymnocyst in anascan cheilostomes (bryozoans). It commonly forms a ridge, bears marginal spines, or borders the outer margin of a cryptocyst.

murambite (mu-ram'-bite) A leucite basanite containing abundant mafic minerals, named by Holmes in 1936. Obsolete.

murataite-(Y) (mu-rat'-a-ite) A sub-metallic black cubic mineral: $(Y,Na)_6(Zn,Fe^{3+})_5(Ti,Nb)_{12}O_{29}(O,F)_{14}$.

murbruk structure (mur'-bruk) *mortar structure.*

murchisonite (mur'-chi-son-ite') (a) A flesh-red perthitic variety of orthoclase with good cleavage and often gold-yellow reflections in a direction perpendicular to (010). (b) A name applied to moonstone and the iridescent feldspar from Frederiksvaern, Norway.

Murderian (Mur-der'-i-an) Stage in New York State: Upper Silurian (upper Cayugan; above Canastotan).

murdochite (mur'-doch-ite) A black cubic mineral: $Cu_{12}Pb_2O_{15}(Cl,Br)_2$.

muri (mu'-ri) Pl. of *murus.*

muriform (mu'-ri-form) Of fungal *conidia,* being divided into multiple cells by septa in more than one plane (usually transverse and longitudinal), like bricks in a wall. Cf: *murus.*

murite (mu'-rite) A dark-colored feldspathoid-rich *phonolite* in which the mafic minerals comprise about 50 percent of the rock. Named by Lacroix in 1927 for Cape Muri, Rarotonga, Cook Islands. Obsolete.

murmanite (mur'-man-ite) A lilac, bright pink, yellow, or brown triclinic mineral: $Na_3(Ti,Nb)_4(Si_2O_7)_2{\cdot}4H_2O$. Cf: *lomonosovite.*

muromontite (mu-ro-mon'-tite) A doubtful mineral: $Be_2FeY_2(SiO_4)_3$ (?). It is perhaps identical with gadolinite or is a member of the clinozoisite group.

murram (mur'-ram) Deposits of *bog iron ore* in the tropics of Africa.

murunskite (mu-run'-skite) A metallic reddish-brown tetragonal mineral: $K_2Cu_3FeS_4$.

murus (mu'-rus) The walls of a positive *reticulate* sculpture on the surface of palynomorphs. Pl: *muri.* Cf: *intrareticulate; negative reticulum.*

murus reflectus (mu'-rus re-flec'-tus) A sutural indentation of the apertural face of a foraminiferal test, longitudinally and obliquely folded below the aperture (as in *Osangularia*). Etymol: Latin.

Muschelkalk (Mu'-schel-kalk) A term used esp. in Germany to refer to a carbonate section approx. in what is now the Middle Triassic (above Bunter, below Keuper). Essentially a lithostratigraphic term.

muscle field (mus'-cle) An area of a brachiopod valve where muscle scars are concentrated.

muscle fossa A concave or flat area on the ventral (inner) side of an articular face of a muscularly articulated plate of a crinoid ray, serving for the attachment of muscle fibers.

muscle platform A relatively broad and solid or undercut elevation of the inner surface on either valve of some brachiopods, to which muscles are attached. Syn: *platform [paleont] (b).*

muscle scar (a) One of the differentiated, more or less well-defined impressions or elevations on the inner surface of a bivalve shell (as in an ostracode, rostroconch, brachiopod, or pelecypod), or on a bone, marking the former place of attachment of a muscle; e.g. an adductor muscle scar. Syn: *scar [paleont].* (b) Smooth or slightly depressed paired areas in the external surface of the axial region of a trilobite exoskeleton, interpreted as areas of muscle attachment (TIP, 1959, pt. O, p.123).

muscle track The path of successive muscle impressions formed in a brachiopod, rostroconch, or pelecypod shell by movement of the muscle base during growth. Syn: *track [paleont] (b).*

muscovado (mus-co-va'-do) A term applied in Minnesota to rusty-colored outcropping rocks, such as gabbros and quartzites, that resemble brown sugar. Etymol: Spanish, "brown sugar".

muscovite (mus'-co-vite) (a) A mineral of the *mica* group: $KAl_2(Si_3Al)O_{10}(OH,F)_2$. It is colorless to yellowish or pale brown, and is a common mineral in gneisses and schists, in most acid igneous rocks (such as granites and pegmatites), and in many sedimentary rocks (esp. sandstones). Several monoclinic, triclinic and trigonal polytypes are recognized. Also spelled: *moscovite.* Syn: *white mica; potash mica; common mica; Muscovy glass; mirror stone.* (b) A term applied in clay mineralogy to *illite.* See also: *sericite; talcite.*

muscovite pegmatite Muscovite-feldspar-uraninite pegmatite formed under conditions of upper amphibolite metamorphism (~5 to 8 kilobars pressure).

Muscovy glass (Mus'-co-vy) *muscovite.*

muscular articulation (mus'-cu-lar) A type of articulation of crinoid ossicles effected by muscle fibers in addition to ligaments. Cf: *ligamentary articulation.*

musgravite (mus'-grav-ite) *magnesiotaaffeite-6N'3S.*

mush *brash ice.*

mush frost *pipkrake.*

mushistonite (mush-i'-ston-ite) A brownish-green, yellow-brown, or green cubic mineral of the *schoenfliesite* group: $(Cu,Zn,Fe^{2+})Sn^{4+}(OH)_6$.

mushroom diapir Salt diapir having a broad *bulb* fringed by one or more laterally flattened, pendant *skirts* (Jackson and Talbot, 1991).

mushroom ice (mush'-room) An *ice pedestal* with a round and expanded top. Cf: *ice pillar.*

mushroom rock A tablelike rock mass formed by wind abrasion or differential weathering in an arid region, consisting of an upper layer of resistant rock underlain by a softer, partially eroded layer, thereby forming a thin "stem" supporting a wide mass of rock, the whole feature resembling a mushroom in shape. Cf: *pedestal rock; zeuge; gara.* Syn: *toadstool rock.*

musical sand A *sounding sand* that emits a definite musical note or tone when stirred, trodden on, or otherwise disturbed; esp. *whistling sand.*

muskeg (mus'-keg) (a) A *bog,* usually a *sphagnum bog,* frequently with grassy tussocks, growing in wet, poorly drained boreal regions, often areas of permafrost. Tamarack and black spruce are commonly associated with muskeg areas. Also spelled: *maskeeg.* (b) A term sometimes used in Michigan for a *bog lake.*

muskoxite (musk'-ox-ite) A dark reddish-brown rhombohedral mineral: $Mg_7Fe^{3+}{}_4(OH)_{26}{\cdot}H_2(?)$.

mussel (mus'-sel) (a) Any of the common freshwater bivalves belonging to the superfamily Unionacea. (b) *mytilid.*

mustard gold (mus'-tard) A spongy type of free gold found in the gossan above gold-silver-telluride deposits.

mutant (mu'-tant) The offspring bearing a *mutation.*

mutation (mu-ta'-tion) A fundamental change in heredity, which results in the development of new individuals that are genetically unlike their parents and, therefore, can be acted upon by natural selection to effect changes and eventually to establish new species. Mutations are now thought to be chemical changes in the DNA of a chromosome; some are visible, but most are not; some are deleterious. Mutations are the raw material of evolution. See also: *mutant.*

mute v. To exclude the relative contribution of the components of a record stack with record time. The early portions of long-offset traces may be muted or excluded because they are dominated by refraction arrivals or because their frequency content after correction for normal moveout is appreciably lower than that of other traces. Muting may also be done over certain time intervals to keep ground roll, air waves, or noise bursts out of the stack. Cf: *diversity stack.*

muthmannite (muth'-mann-ite) A metallic yellow orthorhombic mineral: $Ag,AuTe_2$.

mutinaite A white orthorhombic mineral: $Na_3Ca_4Al_{11}Si_{85}O_{192}{\cdot}60H_2O$.

mutual inductance (mu'-tu-al) The complex ratio of voltage in one circuit to current in another to which it is inductively coupled.

mutualism (mu'-tu-al-ism') A relationship between two organisms in which both are benefitted. Cf: *commensalism; symbiosis.*

MVT *Mississippi Valley-type deposit.*

mycelium (my-ce'-li-um) The entire filamentous vegetative growth of fungi, composed of hyphae.

mycobiont (my-co-bi'-ont) The fungal partner, or component, of a lichen. Cf: *photobiont.*

mycorrhiza (my-cor-rhi'-za) A symbiotic relationship between a fungus (Zygomycota) and roots of a land plant. Members of this group occur in approximately 90 percent of modern land plants and range from Early Devonian to the present. Pl: micorrhizae.

mylonite (my'-lo-nite) A fine-grained, foliated rock, commonly with poor fissility and possessing a distinct lineation. Usually found in narrow, planar zones of localized ductile deformation, although kilometer-scale zones exist. Mylonites are often inferred to indicate extensive simple shear, but they may also record pure shear or volume loss or both. Introduced by Lapworth in 1885 and originally thought to have formed by the crushing and milling of minerals, it is now understood to form by one or more of the crystal plastic deformation mechanisms (Holmes, 1920). Cf: *blastomylonite; cataclasite; microbreccia; protomylonite; ultramylonite.*

mylonite gneiss A metamorphic rock that is intermediate in character between mylonite and schist. Felsic minerals show cataclastic phenomena with little or no recrystallization, and commonly occur as *augen* surrounded by and alternating with schistose streaks and lenticles of recrystallized mafic minerals (Holmes, 1928, p.164).

mylonitic structure (my-lo-nit'-ic) A structure characteristic of mylonites, produced by intense microbrecciation and shearing which gives the appearance of a flow structure. Cf: *flaser structure [meta].*

mylonitization (my'-lo-nit'-i-za'-tion) Deformation of a rock by extreme microbrecciation, due to mechanical forces applied in a definite direction, without noteworthy chemical reconstitution of granulated minerals. Characteristically the mylonites thus produced have a flinty, banded, or streaked appearance, and undestroyed augen and lenses of the parent rock in a granulated matrix (Schieferdecker, 1959). Var: *mylonization.*

mylonization (my'-lo-ni-za'-tion) *mylonitization.*

myocyte (my'-o-cyte) A fusiform contractile cell in sponges.

myodocope (my-o'-do-cope) Any ostracode belonging to the order Myodocopida, characterized by a shell with subequal valves that may be ornamented or smooth and by a well developed rostrum. Most planktonic marine ostracodes are myodocopes. Range, Ordovician to Holocene.

myophore (my'-o-phore) A part of a shell adapted for the attachment of a muscle; e.g. a process for attachment of an *adductor muscle* of a pelecypod, or the distal expanded part of the differentiated *cardinal process* of a brachiopod to which the diductor muscles were attached.

myophragm (my'-o-phragm) A median ridge of secondary shell of a brachiopod, secreted between muscles and not extending beyond the muscle field.

myostracum (my-os'-tra-cum) The part of the shell wall of a mollusk secreted at the attachments of the muscles or *mantle.* Syn: *hypostracum.*

myriapod (myr'-i-a-pod') Any arthropod belonging to the superclass Myriapoda, which includes millipeds, centipeds, pauropods, symphylids, and some related extinct forms. They are rarely preserved as fossils. Range, Upper Silurian to Holocene.

myrickite (my'-rick-ite) (a) A white or gray chalcedony, opal, or massive quartz unevenly colored by or intergrown with pink or reddish inclusions of cinnabar, the color of which tends to become brown. The opal variety is known as *opalite.* (b) Cinnabar intergrown with common white opal or translucent chalcedony.

myrmekite (myr'-me-kite) An intergrowth of plagioclase feldspar (generally oligoclase) and vermicular quartz, generally replacing potassium feldspar, formed during the later stages of consolidation in an igneous rock or during a subsequent period of plutonic activity (Barker, 1970). The quartz occurs as blobs, drops, or vermicular shapes within the feldspar.

myrmekite-antiperthite (myr'-me-kite-an-ti-perth'-ite) A myrmekite-like intergrowth of predominant plagioclase and vermicular orthoclase (Schieferdecker, 1959, term 5177).

myrmekite-perthite (myr'-me-kite-perth'-ite) A myrmekite-like intergrowth of microcline and vermicular plagioclase (Schieferdecker, 1959, term 5176).

myrmekitic (myr-me-kit'-ic) Microscopic igneous texture involving a wormy intergrowth of optically continuous quartz bodies with sodic plagioclase.

mytilid (myt'-i-lid) Any bivalve mollusk belonging to the family Mytilidae, characterized by an equivalve, inequilateral shell with prosogyrate umbones.

mytiliform (my-til'-i-form) Said of a slipper-shaped shell of a bivalve mollusk; specif. shaped like a marine mussel shell, such as the elongated and equivalve shell of *Mytilus* (a genus of marine pelecypods).

myxomycete (myx-o-my'-cete) An organism of the class Myxomycetes, commonly called the slime molds, of uncertain systematic position but usually associated with the fungi. It exists as complex, mobile plasmodia and reproduces by spores. Cf: *eumycete; schizomycete.*

myxosponge (myx'-o-sponge) Any *demosponge* whose only skeleton is mesogloea. It is without spicules or spongin.

N

N *neutron number.*

nab A British term for a projecting part of an eminence; e.g., a headland or promontory (a *ness*), or a spur of an escarpment.

nabaphite (na′-ba-phite) A colorless cubic mineral: $NaBaPO_4 \cdot 9H_2O$. It is the Ba analogue of nastrophite.

Nabarro-Herring creep *volume-diffusion creep.*

nabesite A colorless orthorhombic mineral: $Na_2BeSi_4O_{10} \cdot 4H_2O$.

nabiasite An adamantine dark red cubic mineral: $BaMn_9[(V,As)O_4]_6(OH)_2$.

nabokoite A vitreous yellowish-brown tetragonal mineral: $Cu_7Te^{4+}O_4(SO_4)_5 \cdot KCl$.

nacaphite (na′-ca-phite) A colorless triclinic or orthorhombic mineral: $Na_2Ca(PO_4)F$.

nacareniobsite-(Ce) (na-ca-re-ni′-ob-site) A colorless monoclinic mineral of the *rinkite* group: $Na_3Ca_3CeNbO(Si_2O_7)_2F_3$.

NaCl equivalent In fluid inclusion studies, the unit used to estimate the salinity of the fluid in the inclusion. Since the specific ionic content of the fluid is typically unknown, salinity is expressed as the weight precent NaCl that would account for the freezing point depression observed on a freezing stage.

nacre (na′-cre) The hard, iridescent internal layer of various mollusk shells, having unusual luster and consisting chiefly of calcium carbonate in the form of aragonite deposited as thin tablets normal to the surface of the shell and interleaved with thin sheets of organic matrix. Cf: *calcitostracum; treppen.* Syn: *mother-of-pearl.*

nacreous luster (na′-cre-ous) A type of mineral luster resembling that of mother-of-pearl. Syn: *pearly luster.*

nacreous structure A *microstructure* of polygonal to rounded aragonitic tablets arranged in broadly continuous, regular, mutually parallel sheets. This includes sheer nacreous, row stack nacreous, and columnar nacreous varieties (Boggild, 1930, p.250).

nacrite (na′-crite) A variously colored dioctahedral kaolin mineral, polymorphic with kaolinite and dickite: $Al_2Si_2O_5(OH)_4$. It is characterized by a 6-layer monoclinic stacking sequence and an ordered distribution of octahedral vacancies.

NAD *North American datum.*

NAD27 North American Datum 1927; a datum with its origin at Meades Ranch, Kansas (Krzanowski et al., 1993).

NAD83 North American Datum 1983; an Earth-centered datum defined on satellite and terrestrial data (Krzanowski et al., 1993).

nadir (na′-dir) (a) The point on the celestial sphere that is directly beneath the observer and directly opposite the *zenith.* (b) The point on the ground vertically beneath the perspective center of an aerial-camera lens. (c) A low point in any fluctuating record, e.g. an economic nadir such as a great depression.

nadorite (nad′-or-ite) A brownish-yellow orthorhombic mineral: $PbSb^{3+}O_2Cl$. It is the Sb analogue of perite.

naëgite (naëg′-ite) A variety of zircon containing thorium and uranium. Also spelled: naegite.

nafertisite A light green monoclinic mineral: $(Na,K)_3(Fe^{2+},Fe^{3+},Mg)_6Ti_2(Si,Fe^{3+})_{12}O_{30}(OH,O)_{11} \cdot 2H_2O$.

nafud (na′-fud) *nefud.*

nagashimalite (na-ga-shi′-ma-lite) A submetallic greenish-black orthorhombic mineral of the *taramellite* group: $Ba_4(V^{3+},Ti)_4B_2Si_8O_{27}Cl(O,OH)_2$.

nagatelite (nag-a-tel′-ite) A phosphatian variety of allanite.

nagelfluh (na′-gel-fluh [nahgulflue]) A massive and variegated Miocene conglomerate accompanying the *molasse* of the Alpine region in Switzerland. Etymol: German "Nagel", "nail", + "Fluh", "mass of rock, stratum, layer". Syn: *gompholite.*

nagelschmidtite (nag′-el-schmidt-ite′) A white to brownish hexagonal or trigonal mineral occurring in the "mottled zone", Israel: $Ca_7(SiO_4)_2(PO_4)_2$. Not approved by the International Mineralogical Association.

nagyágite (nagy′-ag-ite) A metallic dark lead-gray monoclinic (pseudo-tetragonal) mineral: $[Pb(Pb,Sb)S_2][Te,Au]$. Syn: *black tellurium; tellurium glance.*

nahcolite (nah′-co-lite) A colorless, white, gray, or buff monoclinic mineral: $NaHCO_3$.

nahpoite (nah′-po-ite) A white monoclinic mineral: Na_2HPO_4.

naif (na-if′) (a) Natural, unpolished surface or skin of a rough diamond. Small pieces of naif are sometimes left on polished stones, usually on the girdle, to indicate that maximum yield has been obtained. (b) Well-formed, undistorted diamond crystal with bright faces. (c) Diamond which is pointed rather than flat (Liddicoat et al., 1993). Also spelled: naife; naive; nyf.

nailhead spar (nail′-head) A variety of calcite in crystals showing a combination of hexagonal prisms with flat rhombohedrons.

nailhead striation A glacial striation with a definite or blunt head or point of origin, generally narrowing or tapering in the direction of ice movement and coming to an indefinite end.

Nain Province A Precambrian province in the Canadian Shield (Stockwell, 1964).

nakaséite A discredited metallic light gray monoclinic mineral: $Ag_3CuPb_4Sb_{12}S_{24}$. It may be a variety of andorite.

nakauriite (na-kau-ri′-ite) A blue orthorhombic mineral first described as occurring in serpentinite: $Cu_8(SO_4)_4(CO_3)(OH)_6 \cdot 48H_2O$.

naked flagellar field (na′-ked) The area around the flagella in which no coccoliths are present in coccolithophores lacking a complete cover of coccoliths. Cf: *covered flagellar field.*

naked karst Karst that is developed in a region without soil cover, so that its topographic features are well exposed. Syn: *bare karst.* Cf: *covered karst.*

naked pole The end in a nonflagellate coccolithophore that is free of coccoliths.

nakhlite (nakh′-lite) One of a group of augite-rich cumulate achondrites whose young age, petrologic, chemical and isotopic composition suggest origin on the same parent body as the chassignites and shergottites, thought by many to be the planet Mars. See also: *SNC meteorite; chassignite; shergottite.*

Na-komarovite A white or yellowish orthorhombic mineral: $Na_6Ca(Nb,Ti)_6Si_4O_{26}F_2 \cdot 4H_2O$. It is the Na analogue of komarovite, but has not been approved by the International Mineralogical Association.

naled A Russian term for *icing; aufeis.* Pl: naledi.

nalipoite A white, pale yellow or pale blue orthorhombic mineral: $NaLi_2PO_4$.

nallah (nal′-lah) *nullah.*

namakier *salt glacier.*

namansilite An adamantine red-violet or dark red monoclinic mineral of the *pyroxene* group: $Na(Mn^{3+},Fe^{3+})Si_2O_6$.

nambulite (nam′-bu-lite) A reddish-brown triclinic mineral: $(Li,Na)Mn^{2+}_4Si_5O_{14}(OH)$. It is a pyroxenoid mineral, related to rhodonite.

namibite (na-mib′-ite) A dark green triclinic mineral: $Cu(BiO)_2(VO_4)(OH)$.

Nammalian (Nam-ma′-li-an) A term used to combine the Dienerian and Smithian substages of the Lower Triassic (Harland et al., 1990).

Namurian (Na-mu′-ri-an) A European series: middle Carboniferous (above Visean, below Westphalian). The term is poorly defined, however, and has been used with a variety of meanings.

namuwite (na-mu′-wite) A hexagonal pale green trigonal mineral: $(Zn,Cu)_4(SO_4)(OH)_6 \cdot 4H_2O$.

nanism (na′-nism) The development of abnormally or exceptionally small size; dwarfishness. Cf: *gigantism.*

nanlingite (nan-ling′-ite) A brownish-red rhombohedral mineral: $CaMg_4(As^{3+}O_3)_2F_4$. Not approved by the International Mineralogical Association.

nannobacteria Spheroidal forms, generally 0.03 to 0.2 micrometers; best seen in acid-etched limestone and thought to be dwarf forms of bacteria; highly active geochemically (Folk, 1993). Also spelled: nanobacteria.

nannocrystalline Descriptive of an interlocking texture of a carbonate sedimentary rock having crystals whose diameters are in the range of 0.001-0.004 mm.

nannofossil (nan′-no-fos′-sil) (a) A collective term for fossil discoasters and coccoliths, both primarily calcareous microfossils, mostly rather near the limit of resolution of the light microscope and hence best studied with electron microscopy. (b) A term sometimes used in a more general sense for other extremely small marine (usually algal) fossils, smaller than microfossils.

nanno ooze (nan′-no) Unconsolidated pelagic deposit, consisting of more than 69% $CaCO_3$, in which nannoplankton makes up more than 30% of the recognizable skeletal remains.

nannoplankton (nan-no-plank′-ton) Passively floating unicellular organisms ranging in size from 0.002-0.02 mm. They are larger than *ultraplankton* but smaller than *microplankton.*

nanotesla (na-no-tes′-la) A unit of magnetic field intensity (*magnetic flux* density): 1 nanotesla = 10^{-9} tesla = 1 gamma. It is a preferred unit of measurement in *magnetic exploration.* Abbrev: nT.

nanozooid (na-no-zo′-oid) A small polymorph with a single *tentacle* and no digestive system in a few stenolaemate bryozoans.

nanpingite (nan-ping′-ite) A white monoclinic platy mineral of the *mica* group: $Cs(Al,Mg,Fe^{2+},Li)_2(Si_3Al)O_{10}(OH,F)_2$.

Nansen bottle (Nan′-sen) A device used in oceanography that is attached to a *wireline* and lowered into the sea to obtain subsurface seawater samples. The open Nansen bottle is tripped at depth by a *messenger.* When tripped, it closes its end valves and upends, causing an attached pair of reversing thermometers to record the water temperature. Nansen bottles are used in a series fastened to a hydro-wire to collect samples at desired depths. Named after the Norwegian explorer and oceanographer Fridtjof Nansen. Cf: *thermometric depth.*

Nansen cast The use of a series of Nansen bottles to obtain seawater samples and measurements. Cf: *hydrographic cast.*

nant A little valley with a stream. Etymol: Celtic, "brook".

nantokite (nan′-to-kite) A colorless, white, greenish, or grayish cubic mineral: CuCl.

naotic septum (na-ot′-ic) A rugose corallite *septum* characterized by development peripherally in a series of closely spaced dissepimentlike plates, as in *Naos.*

NAP *nonarboreal pollen.*

Na-phlogopite *aspidolite.*

naphtha (naph′-tha) An archaic term for liquid petroleum. It is now used to designate those hydrocarbons of the lowest boiling point (under 250°C) that are liquid at standard conditions, but easily vaporize and become inflammable. They are used as cleaners and solvents.

naphthalene (naph′-tha-lene) A white crystalline bicyclic aromatic hydrocarbon, formula $C_{10}H_8$, that has a characteristic odor and occurs in coal tar and some crude oils.

naphthene (naph′-thene) *cycloparaffin.*

naphthene-base crude *asphalt-base crude.*

naphthine (naph′-thine) A syn. of *hatchettine.* Also spelled: naphtine; naphtein.

Napoleonville (Na-po′-le-on-ville′) North American (Gulf Coast) stage: Miocene (above Anahauc, below Duck Lake).

nappe [hydraul] A sheet of water overflowing a dam.

nappe [struc geol] A sheetlike, allochthonous rock unit, which has moved on a predominantly horizontal surface. The mechanism may be thrust faulting, recumbent folding, or both. The term was first used as "nappe de recouvrement" (Schardt, 1893) for the large allochthonous sheets of the western Alps, and it has been adopted into English. The German equivalent, *Decke,* is also sometimes used in English. Etymol: French, "cover sheet, tablecloth". See also: *klippe.*

nappe outlier *klippe.*

nari A variety of caliche that forms by surface or near-surface alteration of permeable calcareous rocks (dissolution and redeposition of calcium carbonate) and that occurs in the drier parts of the Mediterranean region. It is characterized by a fine network of veins surrounding unreplaced remnants of the original rock, and it often contains clastic particles (rocks and shells). Etymol: Arabic "nar", "fire", in allusion to its use in limekilns.

Narizian (Na-riz′-i-an) North American stage: Middle Eocene (above Ulatisian, below Refugian).

narrow (nar′-row) A constricted section of a mountain pass, valley, or cave; a gap or narrow passage between mountains. Commonly used in the plural, e.g. the Narrows of New River.

narsarsukite (nar-sar-suk′-ite) A yellow brownish-gray, greenish-yellow, or dark green tetragonal mineral: $Na_2(Ti,Fe^{3+})Si_4(O,F)_{11}$. Syn: gouréite.

NASA National Aeronautics and Space Administration.

nasal tube (na′-sal) A curved cylinder or prismatic tube in a phaeodarian radiolarian, embracing the central capsule on one side and a galea on the other side. Syn: *rhinocanna.*

nase *naze.*

nasinite (nas′-i-nite) A colorless orthorhombic mineral: $Na_2B_5O_8(OH){\cdot}2H_2O$.

nasledovite (nas′-le-dov-ite) An ill-defined silky white mineral: $PbMn^{2+}{}_3Al_4O_5(CO_3)_4(SO_4){\cdot}5H_2O$.

nasonite (na′-son-ite) A white, yellow, blue, or green hexagonal mineral: $Pb_6Ca_4(Si_2O_7)_3Cl_2$.

Na-spar (a) *sodium feldspar.* (b) *soda spar.*

nasselline (nas′-sel-line) Any radiolarian belonging to the suborder Nassellina, characterized by a central capsule perforated only at one pole and enclosed by a single membrane.

nastrophite (na-stro′-phite) A colorless cubic mineral: $Na(Sr,Ba)PO_4{\cdot}9H_2O$. It is the Sr analogue of nabaphite.

nasturan (na′-stu-ran) *pitchblende.*

natalyite (na′-ta-ly-ite) A green monoclinic mineral of the *pyroxene* group: $Na(V^{3+},Cr^{3+})Si_2O_6$.

natanite (nat′-a-nite) A greenish-brown cubic mineral of the *schoenfliessite* group: $Fe^{2+}Sn^{4+}(OH)_6$.

National Map *The National Map.*

National Map Accuracy Standards Define accuracy standards for published maps, including horizontal and vertical accuracy, accuracy testing method, accuracy labeling on published maps, labeling when a map is an enlargement of another map, and basic information for map construction as to latitude and longitude boundaries (USGS, 1947). Abbrev: NSSDA.

national meridian (na′-tion-al) A meridian chosen in a particular nation as the reference datum for determining longitude for that nation. It is commonly defined in several European countries with respect to a key point in the capital city. Cf: *Greenwich meridian.*

National Standard for Spatial Data Accuracy Implements a statistical and testing methodology for estimating the positional accuracy of points on maps and in digital geospatial data, with repsect to georeferenced ground positions of higher accuracy (FGDC, 1998b, p.3-1). Abbrev: NSSDA.

National Tidal Datum Epoch A changing 19-year period over which data are collected to determine the datum for nautical charts produced by the National Oceanic and Atmospheric Administration's National Ocean Survey. Cf: *chart datum.*

natisite (na′-ti-site) A vitreous yellow-green or greenish-gray tetragonal mineral: $Na_2TiO(SiO_4)$. It is a dimorph of paranatisite.

native (na′-tive) *endemic.*

native asphalt Liquid or semiliquid asphalt in exudations or seepages including surface flows and lakes. Syn: *natural asphalt.*

native coke *natural coke.*

native element Any element found uncombined in a nongaseous state in nature. Nonmetallic examples are carbon, sulfur, and selenium; semimetal examples are antimony, arsenic, bismuth, and tellurium; native metals include silver, gold, copper, iron, mercury, iridium, lead, palladium, and platinum.

native metal A metallic *native element.*

native mud The aqueous suspension consisting of pulverized *drill cuttings* without additives and produced by the bit drilling a borehole through the "native" strata". Thus, native mud often does not contain the most desirable lubricating and cooling properties of drilling muds.

native paraffin *ozocerite.*

native water (a) *connate water.* (b) *formation water.*

natramblygonite (na′-tra-am-blyg′-on-ite) *natromontebrasite.*

natric horizon (na′-tric) A diagnostic subsurface soil horizon that has the same properties as an argillic horizon but also displays a blocky, columnar, or prismatic structure and has a subhorizon with an exchangeable-sodium percentage of over 15 (USDA, 1975).

natrite (na′-trite) A colorless, rose, or yellow monoclinic mineral: Na_2CO_3. It is the low-temperature dimorph of gregoryite.

natroalunite (na-tro-al′-u-nite) A colorless to variously colored rhombohedral mineral of the alunite group: $NaAl_3(SO_4)_2(OH)_6$. Syn: *almeriite.*

natroapophyllite (na′-tro-a-po′-phyll-ite) A colorless to yellowish-brown orthorhombic mineral of the *apophyllite* group: $NaCa_4Si_8O_{20}F{\cdot}8H_2O$.

natroautunite *metanatroautunite.*

natrobistantite (na′-tro-bis-tan′-tite) A colorless, blue-green, or yellow-green cubic mineral of the *pyrochlore* group: $(Na,Cs)Bi(Ta,Nb,Sb)_4O_{12}$.

natroborocalcite (na′-tro-bo′-ro-cal′-cite) A syn. of *ulexite.* Also spelled: natronborocalcite.

natrocarbonatite A variety of carbonatite, so far known only from the active volcano Oldoinyo Lengai in Tanzania, composed mostly of Na-K-Ca carbonates (*nyerereite* and *gregoryite*). Less preferred syn. *lengaite.*

natrochalcite (na-tro-chal′-cite) An emerald-green monoclinic mineral: $NaCu_2(SO_4)_2(OH){\cdot}H_2O$.

natrodufrénite (na-tro-du-frén′-ite) A pale blue-green monoclinic mineral of the *dufrénite* group: $NaFe^{2+}(Fe^{3+},Al)_5(PO_4)_4(OH)_6{\cdot}2H_2O$.

natrofairchildite (na-tro-fair′-child-ite) A white hexagonal mineral: $Na_2Ca(CO_3)_2$. It has not been approved by the International Mineralogical Association, and may be identical with nyerereite.

natroglaucocerinite A pale blue trigonal mineral: $Zn_{8-x}Al_x(OH)_{16}(SO_4)_{x/2+y/2}Na_y(H_2O)_6$. It is a glaucocerinite-type mineral containing essential Na.

natrojarosite (na-tro-ja′-ros-ite) A yellowish-brown to golden-yellow rhombohedral mineral of the *alunite* group: $NaFe^{3+}{}_3(SO_4)_2(OH)_6$. Syn: *utahite.*

natrolemoynite A colorless monoclinic mineral: $Na_3Zr_2Si_{10}O_{26}{\cdot}9H_2O$. It is the Na analogue of lemoynite.

natrolite (nat′-ro-lite) A variously colored monoclinic zeolite mineral: $Na_2Al_2Si_3O_{10}·2H_2O$. It sometimes contains appreciable calcium, and usually occurs in slender, acicular or prismatic crystals. Partial syn: *mesotype; needle zeolite.*

natromontebrasite (na′-tro-mon′-te-bra′-site) A white to variously colored triclinic mineral of the *amblygonite* group: $(Na,Li)Al(PO_4)(OH,F)$. Syn: *fremontite; natramblygonite.*

natron (na′-tron) A white, yellow, or gray monoclinic mineral: $Na_2CO_3·10H_2O$. It is very soluble in water, and occurs mainly in solution (as in the soda lakes of Egypt and the western U.S.) or in saline residues. Syn: *soda.*

natronambulite A pinkish-orange triclinic mineral: $(Na,Li)Mn^{2+}{}_4Si_5O_{14}(OH)$. It is a pyroxenoid mineral related to rhodonite, and is the Na analogue of nambulite.

natroniobite (na′-tro-ni-o′-bite) A yellowish, brownish, or blackish monoclinic mineral: $NaNbO_3$. It is dimorphous with lueshite.

natron lake *soda lake.*

natrophilite (na-tro-phil′-ite) A lustrous yellow orthorhombic mineral of the *triphylite* group: $NaMn^{2+}PO_4$.

natrophosphate (na-tro-phos′-phate) A vitreous white cubic mineral: $Na_7(PO_4)_2(F,Cl)·19H_2O$.

natrosilite (na-tro-si′-lite) A colorless monoclinic mineral: $Na_2Si_2O_5$.

natrotantite (na-tro-tan′-tite) An adamantine colorless to yellowish rhombohedral mineral: $Na_2Ta_4O_{11}$.

natroxalate A pale yellow monoclinic oxalate mineral: $Na_2C_2O_4$.

natural arch (nat′-u-ral) (a) A *natural bridge* resulting from erosion. (b) A landform similar to a natural bridge but not formed by erosive agencies (Cleland, 1910, p.314). (c) *sea arch.* Syn: *arch [geomorph].*

natural area (a) An area of land or water that has retained its wilderness character, although not necessarily completely natural and undisturbed, or that has rare or vanishing flora, fauna, archaeological, scenic, historical, or similar features of scientific or educational value (Ohio Legislative Service Commission, 1969, p.3); e.g., a "Research Natural Area" where "natural processes are allowed to predominate and which is preserved for the primary purposes of research and education" (U.S. Federal Committee on Research Natural Areas, 1968, p.2). (b) Any outdoor site that contains an unusual biologic, geologic, or scenic feature or that illustrates "common principles of ecology uncommonly well" (Lindsey et al., 1969, p.4). Cf: *wilderness area.*

natural asphalt *native asphalt.*

natural assemblage *assemblage [ecol].*

natural attenuation The net result of those processes in soil or groundwater environments that act to reduce the mass, toxicity, mobility, volume, or concentration of contaminants. These processes may include dilution, dispersion, *sorption, ion exchange,* precipitation, *redox* reactions, *radioactive decay,* and biodegradation.

natural avalanche Any avalanche which occurs without being triggered by active control procedures or without accidental triggering by a snow traveler.

natural bridge (a) Any archlike rock formation created by erosive agencies and spanning a ravine or valley; an opening found where a stream broke through the narrow meander neck, as at Rainbow Bridge, Utah. Syn: *bridge [geomorph].* (b) In a limestone terrane, the remnant of the roof of an underground cave or tunnel that has collapsed. Syn: *karst bridge.* (c) *sea arch; natural arch.* Cf: *lighthouse.*

natural brine *brine [geol].*

natural classification A biological classification system containing only monophyletic groups or species (Wiley, 1981, p.7). Ant: *artificial classification.*

natural coke Coal that has been naturally carbonized by contact with or proximity to an igneous intrusion, or by natural combustion. Syn: *carbonite; coke coal; cokeite; native coke; finger coal; blind coal; cinder coal.* Cf: *clinker [coal]; coke.*

natural convection *free convection.*

natural gas (a) Hydrocarbons that exist as a gas or vapor at ordinary pressures and temperatures. Methane is the most important, but ethane, propane, and others may be present. Common impurities include nitrogen, carbon dioxide, and hydrogen sulfide. Natural gas may occur alone or associated with oil. Syn: *gas.* (b) Gaseous hydrocarbons trapped in the zone of groundwater saturation, under pressure from, and partially dissolved in, underlying water or petroleum (Meinzer, 1923, p.21). Cf: *subsurface air; included gas.*

natural-gas liquids Hydrocarbons that occur naturally in gaseous form or in solution with oil in the reservoir, and that are recoverable as liquids by condensation or absorption; e.g. *condensate* and *liquefied petroleum gas.*

natural gasoline *condensate.*

natural glass A vitreous, amorphous, inorganic substance that has solidified from magma too quickly to crystallize. Granitic or acid natural glass includes pumice and obsidian; an example of a basaltic natural glass is tachylyte.

natural history The study of the nature and history of all animal, vegetable, and rock and mineral forms. The term is sometimes considered old-fashioned except when applied to the animal world; but it is still used by some in its geologic sense, e.g. in environmental geology.

natural horizon *apparent horizon.*

natural hydraulic fracture Either hydraulic joint or tectonic joint in which fluid pressure is essential for crack propagation. Such joints are driven by fluid pressures working against a compressive crack-normal stress (Engelder and Lacazette, 1990).

natural landscape A landscape that is unaffected by the activities of man (in contrast to the "cultural landscape" resulting from man's settlement); it includes landforms and their natural plant cover, and the contrast between land and water. Syn: *physical landscape.*

natural law A generalization of science, representing an intrinsic orderliness of natural phenomena or their necessary conformity to reason; e.g., the laws of conservation of matter and energy. Syn: *law of nature.*

natural levee (a) A long broad low ridge or embankment of sand and coarse silt, built by a stream on its flood plain and along both banks of its channel, esp. in time of flood when water overflowing the normal banks is forced to deposit the coarsest part of its load. It has a gentle slope (about 60 cm/km) away from the river and toward the surrounding flood plain, and its highest elevation (about 4 m above the flood plain) is closest to the river bank, at or near normal flood level. Syn: *levee [streams]; raised bank; spill bank.* (b) Any naturally produced low ridge resembling a natural levee; e.g. a lava levee, or a sediment ridge bordering a fan-valley.

natural load The quantity of sediment that a stable stream carries.

natural mold The empty space or cavity left after solution of an original fossil shell or other organic structure, bounded by the external impression (external mold) and the surface of the internal filling (*steinkern*) (Shrock and Twenhofel, 1953, p.19). See also: *mold [paleont]; cast [paleont].*

natural radiation Radiation originating from the decay of naturally occurring radionuclides and cosmic radiation (IAEA, 2003).

natural region (a) A part of the Earth's surface characterized by relatively uniform and distinctive physical features (relief, structure, climate, vegetation) within its borders, and therefore possessing to a certain extent a uniformity in human activities. (b) A region that possesses a unity based on significant geographic characteristics (physical, biological, cultural), in contrast to an area marked out by boundaries imposed for political or administrative purposes. The term provides a convenient regional basis for nomenclature and integration of the whole landscape. Syn: *ecoregion.*

natural remanence *natural remanent magnetization.*

natural remanent magnetism *natural remanent magnetization.*

natural remanent magnetization The entire *remanent magnetization* of a rock as measured in the absence of an inducing magnetic field. Abbrev: NRM. Syn: *natural remanence; natural remanent magnetism.*

natural resin An unmodified *resin* from a natural source (such as a tree) and distinguished from synthetic resin; e.g. a copal.

natural scale (a) The scale of a map, expressed in the form of a fraction or ratio, independent of the linear units of measure; specif. *representative fraction.* The term is not recommended. (b) True scale, as it exists in nature, without magnification or reduction.

natural selection The process by which less vigorous or less well-adapted individuals tend to be eliminated from a population, so that they leave fewer descendants to perpetuate their stock. See also: *struggle for existence.* Syn: *selection.*

natural slope (a) The slope assumed by a mass of loose heaped-up material, such as earth. (b) *angle of repose.*

natural stone A gemstone that occurs in nature, as distinguished from a man-made substitute.

natural target An existing natural or cultural feature, on the ground, which serves as a photographic target. Also called substitute point; photocontrol point.

natural taxon A group of organisms that exists in nature as the result of evolution; either a species or a *monophyletic group* (Wiley et al., 1991, p.3). Ant: *artificial taxon.*

natural tunnel A cave that is nearly horizontal and that is open at both ends. It may contain a stream. Syn: *tunnel cave; tunnel [speleo].*

natural well A sinkhole or other natural opening resembling a well that extends below the water table and from which groundwater can be withdrawn.

nature The vertical direction of easiest splitting in a granite (Engelder, 1993).

naujaite (nau′-ja-ite) A coarse, hypidiomorphic-granular sodalite-rich *nepheline syenite* that contains microcline and small amounts of albite, analcime, aegirine, and sodic amphiboles and is characterized by poikilitic texture; the sodalite is surrounded by the other phases. The rock was first described by Ussing in 1912 from Naujakasik, Ilimaussaq complex, on the southwest coast of Greenland. Obsolete.

naujakasite (nau-ja-ka′-site) A pearly white or grayish monoclinic mineral: $Na_6(Fe^{2+},Mn^{2+})Al_4Si_8O_{26}$.

naumannite (nau′-mann-ite) An metallic iron-black cubic mineral: Ag_2Se.

naupliar eye (nau′-pli-ar) An unpaired median eye appearing in the *nauplius* and retained in some mature stages.

nauplius (nau′-pli-us) A crustacean larva in the early stage after leaving the egg, having only three pairs of limbs (corresponding to antennules, antennae, and mandibles), a median (naupliar) eye, and little or no segmentation of the body. Pl: nauplii. Cf: *metanauplius.*

nautical chart (nau′-ti-cal) *hydrographic chart.*

nautical distance The length in nautical miles of the *rhumb line* joining any two places on the Earth's surface.
nautical mile A unit of distance used principally in navigation; for practical consideration it is usually considered the length of 1 minute of any *great circle* of the Earth, the meridion being the great circle most commonly used; the International Hydrographic Bureau in 1929 proposed a standard length of 1,852 meters, which is known as the *international nautical mile* (McGraw-Hill, 1984).
nautilicone (nau-til'-i-cone) A strongly involute nautiloid *conch* (like that of *Nautilus*) coiled in a plane spiral with the outer whorls embracing the inner whorls.
nautiloid (nau'-ti-loid) Any cephalopod belonging to one of the subclasses Nautiloidea, Endoceratoidea, or Actinoceratoidea, characterized by a centrally located siphuncle and by a straight, curved, or coiled chambered external shell with less elaborate sutural flexures than in ammonoids. Nautiloids, known today only from the genus *Nautilus*, reached their peak in the Ordovician and Silurian. Range, Upper Cambrian to Holocene. adj. Pertaining to Nautiloidea.
navajoite (nav'-a-jo-ite') A silky dark brown monoclinic mineral: $(V^{5+},Fe^{3+})_{10}O_{24}\cdot 12H_2O$.
Navarroan (Na-var'-ro-an) North American (Gulf Coast) stage: Upper Cretaceous (above Tayloran, below Midwayan).
Navier-Stokes equation *Eulerian* form of Newton's second law applied to a viscous fluid. Newton's second law states that the time-rate of change of momentum is proportional to force. The transformation from *Lagrangean* to *Eulerian* coordinates changes Newton's linear differential equation into a nonlinear partial differential equation.
Navier-Stokes equations (Na'-vier-Stokes) Equations of motion for a viscous fluid.
NAVSTAR system *Global Positioning System.*
naze A promontory or headland; a *ness.* Etymol: perhaps from the Naze, a promontory in Essex, England. Syn: *nase.*
nchwaningite A vitreous yellowish-brown orthorhombic mineral: $Mn_2SiO_3(OH)_2\cdot H_2O$.
NDCP *nondenticular composite prismatic structure.*
NEA *Near-Earth Asteroid.*
nealite (neal'-ite) An adamantine orange triclinic mineral: $Pb_4Fe^{2+}(As^{3+}O_3)_2Cl_4\cdot 2H_2O$.
neanic (ne-an'-ic) Said of a youthful or immature growth stage of an organism; the stage following the *nepionic* stage and preceding the ephebic stage.
neap tide A tide occurring at the first and third quarters of the Moon when the gravitational pull of the Sun opposes (or is at right angles to) that of the Moon, and having an unusually small or reduced tide range (usually 10-30% less than the mean range). Cf: *spring tide.*
Near-Earth Asteroid An asteroid in an orbit that approaches the Earth's orbit, specifically with perihelion less than 1.3 astronomical units. Such asteroids are responsible for occasional impacts on the Earth, possible causing biologic extinctions and posing a hazard to current life, including human. Abbrev: NEA.
near earthquake An earthquake whose epicenter is within 1,000-1,200 km of the detector.
nearest neighbor interpolation Resampling a source raster cell to yield a new raster with a different cell size, raster orientation, and/or internal geometry by computing the distance between the centers of each cell in the output raster and the four nearest cells in the input raster. The data value for the closest input cell is assigned without modification as the data value of the output cell. Since this technique does not mathematically calculate the value for the output raster, it is a suitable choice for rasters with categorical data such as land cover.
near-field processes Defining the intensity of waves (emissions) from a point source involves two terms which describe waves that decrease with distance at different rates; near field implies that the one of these that falls off most rapidly is dominant, far field that the other dominates.
near infrared *reflected infrared.*
nearshore (near'-shore) Extending seaward or lakeward an indefinite but generally short distance from the shoreline; specif. said of the indefinite zone extending from the low-water shoreline well beyond the breaker zone, defining the area of nearshore currents, and including the *inshore* zone and part of the *offshore* zone. Nearshore is sometimes defined as extending across the area of longshore bars. Depths are generally less than 5 fathoms (10 m).
near-surface geophysics Generally means investigating the uppermost 50-100 m of the Earth's surface. The objects and objectives of near-surface geophysics are often noninvasive investigations of pollution problems, obtaining engineering information, locating and defining archaeological sites, near-surface structural features such as paleochannels and faults, cavities and tunnels, minerals, groundwater, pipelines, etc. Methods include surveys using gravity and magnetics, electrical resistivity, self potential (SP), electromagnetics, ground-penetrating radar (GPR), seismic reflection, seismic refraction, and other methods.
neat line The line that surrounds a map, separating the map from the margin. Neat lines usually consist of parallels and meridians, but conventional or arbitrary grid lines may be used. Cf: *extent.* Also spelled: neatline.
nebkha (neb'-kha) *shrub-coppice dune.*
Nebraskan (Ne-bras'-kan) Pertaining to the first classical glacial stage of the Pleistocene Epoch in North America, followed by the Aftonian interglacial stage. Cf: *Günz.* Obsolete.
nebula (neb'-u-la) Historically, any faintly luminous, diffuse object seen in the heavens. In modern usage, an interstellar cloud of gas or dust. The other diffuse objects seen by early astronomers are clusters of stars, or galaxies. Galaxies are still often referred to as extragalactic nebulae (Stokes and Judson, 1968, p.512). Cf: *solar nebula.*
nebular hypothesis (neb'-u-lar) A model for the origin of the Solar System (by Laplace in 1796) which supposes a rotating, primeval nebula of gas and dust which increased its rotation as it contracted. This led to a flattening of the mass and, as centrifugal forces exceeded gravity, ejections of matter from its equator. This castoff matter formed planets around the original mass, the Sun. The model has been largely abandoned in favor of the *dust-cloud hypothesis.*
nebulite (neb'-u-lite) A *migmatite* characterized by indistinct inclusions (*schlieren* or *skialiths*) (Dietrich and Mehnert, 1961). Rarely used. Adj: nebulitic.
neck [bot] The tapering apical portion of an *archegonium.*
neck [currents] The narrow band or "rip" of water forming the part of a rip current where feeder currents converge and flow swiftly through the incoming breakers or surf.
neck [geog] (a) A narrow stretch or strip of land connecting two larger areas; e.g. the lowest part of a level mountain pass between two ridges, or a narrow isthmus joining a peninsula with the mainland. See also: *land bridge.* (b) Any narrow strip of land such as a cape, promontory, peninsula, bar, or hook. (c) *meander neck.*
neck [ore dep] *pipe [ore dep].*
neck [paleont] (a) The constricted anterior part of the body chamber in specialized brevicones between the flared aperture and the inflated portion (TIP, 1964, pt. K, p.57). (b) The often constricted, usually flexible part of the tetrapod body between head and shoulders; neck vertebrae are morphologically distinct from trunk vertebrae in all except snakelike forms.
neck [volc] A vertical, pipelike intrusion that represents a former volcanic *vent.* The term is usually applied to the form as an erosional remnant. Cf: *plug [volc].*
neck cutoff A meander cutoff formed where a stream breaks through or across a narrow meander neck, as where downstream migration of one meander has been slowed and the next meander upstream has overtaken it. River processes such as erosion, scour, and caving produce neck cutoffs. The term cutoff is used when they are the result of human activity. Cf: *chute cutoff.*
neck ring *occipital ring.*
necrocoenosis (nec'-ro-coe-no'-sis) *liptocoenosis.*
necronite (nec'-ro-nite) A blue pearly variety of orthoclase that emits a fetid smell upon hammering. It occurs in limestone near Baltimore, Md.
necrophagous (ne-croph'-a-gous) Said of an organism that feeds on dead matter.
Nectarian A geochronometric subdivision of the Precambrian Eon. A period of the Hadean Era covering the time span from 3,950 to 3,850 Ma (Harland et al., 1990).
Nectarian [planet] (Nec-tar'-i-an) (a) Pertaining to lunar topographic features and lithologic map units constituting a system of rocks including basin and crater deposits beginning with the formation of the Nectaris basin at about 3.92 Ga and ending with the Imbrium basin impact at 3.85 Ga. Nectarian rocks are older than the mare rocks of the Imbrian System and postdate older impact-basin rocks of the pre-Nectarian (Wilhelms, 1987). The Nectarian is time-equivalent to the early Eoarchean Era of Earth's Precambrian. (b) Said of the stratigraphic period during which the Nectarian system was developed.
Nectridia (Nec-trid'-i-a) An order of lepospondylous amphibians characterized by fan-shaped neural and haemal spines on caudal vertebrae, elongate body form, and limb reduction. Range, Lower Pennsylvanian to Upper Permian.
needle [cryst] (nee'-dle) A needle-shaped or acicular mineral crystal.
needle [geol] A pointed, elevated, and detached mass of rock formed by erosion, such as an *aiguille* or a stack.
needle [surv] *magnetic needle.*
needle ice (a) *pipkrake.* (b) *frazil ice.* (c) *candle ice.*
needle ironstone A variety of goethite occurring in fibrous aggregates of acicular crystals.
needle ore (a) Iron ore of very high metallic luster, found in small quantities, which may be separated into long slender filaments resembling needles. (b) *aikinite.*
needle stone (a) *needle zeolite.* (b) *hairstone.* Also spelled: needlestone.
needle tin ore A variety of cassiterite with acute pyramidal forms.
needle zeolite A syn. of *mesotype;* specif. a syn. of *natrolite.* See also: *hair zeolite.* Syn: *needle stone.*
Néel point (a) The temperature at which the susceptibility of an antiferromagnetic mineral has a maximum. Above this point, thermal agitation prevents antiferromagnetic ordering. (b) The temperature at which

thermal agitation overcomes magnetic order in a ferrimagnetic mineral. *Curie point* is also used with this meaning. Syn: *Néel temperature.*

Néel temperature *Néel point.*

nefedovite (ne-fe'-dov-ite) A colorless triclinic, pseudotetragonal mineral: $Na_5Ca_4(PO_4)_4F$.

neftdegil (neft'-de-gil) *neft-gil.*

neft-gil (neft'-gil) A mixture of paraffins and a resin found in the Caspian area on Cheleken Island; it is related to *pietricikite*. Syn: *neftdegil.*

nefud (a) A deep or large *sandy desert* in Arabia, equivalent to an *erg*. Also spelled: *nafud.* (b) A high *sand dune* in the Syrian desert (Stone, 1967, p.267).

negative [optics] (neg'-a-tive) (a) Said of anisotropic crystals: of a uniaxial crystal, in which the extraordinary index of refraction is greater than the ordinary index; and of a biaxial crystal in which the intermediate index of refraction β is closer to γ than to α. Cf: *positive [optics].* (b) Said of a crystal containing a cavity, the form of which is one of the possible crystal forms of the mineral.

negative [photo] (a) A photographic image on film, plate, or paper that reproduces the bright parts of the subject as dark areas and the dark parts as light areas. Cf: *positive [photo].* (b) A film, plate, or paper containing such an image.

negative area *negative element.*

negative crystal A *fluid inclusion* whose walls have crystallized to conform with the lattice of the host mineral and display inward-facing crystal morphology (Roedder, 1979, p.701-704).

negative delta A term used by Playfair (1802, p.430) for an *estuary*.

negative element A large structural feature or portion of the Earth's crust, characterized through a long period of geologic time by frequent and conspicuous downward movement (subsidence, submergence) or by extensive erosion, or by an uplift that is considerably less rapid or less frequent than those of adjacent positive elements. Syn: *negative area.*

negative elongation In an elongated section of an anisotropic crystal, a term signifying that the long dimension is parallel to the faster of the two plane-polarized rays. Cf: *positive elongation*. See also: *sign of elongation.*

negative estuary *inverse estuary.*

negative flower structure A *flower structure* formed along a transtensional fault in which the upward diverging splays have predominantly normal separation. Development of a negative flower structure usually results in the formation of a narrow depression along the fault-zone trace. Partial syn: *releasing bend.*

negative landform A relatively depressed or low-lying topographic form, such as a valley, basin, or plain, or a volcanic feature formed by a lack of material (as a caldera). Ant: *positive landform.*

negative movement (a) A downward movement of the Earth's crust relative to an adjacent part of the crust, such as produced by subsidence; a negative movement of the land may result in a *positive movement* of sea level. (b) A relative lowering of the sea level with respect to the land, such as produced by a *positive movement* of the Earth's crust or by a retreat of the sea.

negative pole The south-seeking member of the *magnetic poles*. Cf: *positive pole.* See also: *dipole field.*

negative reticulum A reticuloid pattern simulated by anastomosing grooves or *fossulae* that cut into a thicker outer wall layer of *palynomorphs* or separate closely set *verrucate* formations.

negative shoreline *shoreline of emergence.*

negative skin friction Frictional forces developed on the sides of a pile, or other structure embedded in soil, that tends to cause it to be dragged downward. It is often developed in a consolidating soil mass. See also: *skin friction [eng].*

negative snowflake *Tyndall figure.*

nehrung (neh'-rung) A long, narrow sandspit, sandbar, or barrier beach enclosing or partially enclosing a lagoon (*haff*), formed across a river mouth by longshore drifting of sand; esp. such a feature along the East German coast of the Baltic Sea. Pl: nehrungs; nehrungen. Etymol: German "Nehrung", "sandbar, spit".

neighborhood analysis The analysis of the relationships between an object and similar surrounding objects, used in image processing, etc. (Krzanowski et al., 1993).

neighborite (neigh'-bor-ite) A colorless, pink, brown, or cream orthorhombic mineral: $NaMgF_3$.

nejd A syn. of *hammada.* Var: *nijd.*

nek A term used in South Africa for a low place in a mountain range; a saddle or col. Etymol: Afrikaans, "neck".

nekoite (nek'-o-ite) A white triclinic mineral: $Ca_3Si_6O_{15}{\cdot}7H_2O$. Cf: *okenite.*

nekrasovite (ne'-kra-sov-ite) A metallic pale brown cubic mineral of the *germanite* group: $Cu_{26}V_2(Sn,As,Sb)_6S_{32}$. It is the Sn-analogue of colusite.

nektic (nek'-tic) *nektonic.*

nektobenthic (nek'-to-ben-thic) Said of the type of pelagic organism that actively swims but whose ecology and life habitat are intimately associated with the benthos.

nekton (nek'-ton) Aquatic animals that are actively free-swimming, e.g. cephalopods, fish. Adj: *nektonic.*

nektonic (nek-ton'-ic) Referring to that type of pelagic organism which actively swims; adj. of *nekton.* Cf: *planktonic.* Syn: *nektic.*

nelenite (nel'-e-nite) A brown monoclinic mineral: $(Mn^{2+},Fe^{2+})_{16}As^{3+}{}_3Si_{12}O_{36}(OH)_{17}$. It is polymorphous with schallerite.

nelsonite (nel'-son-ite) A group of hypabyssal rocks composed chiefly of ilmenite and apatite, with or without rutile. The ratio of ilmenite to apatite varies widely. It was named by Watson in 1907 after Nelson County, Virginia.

neltnerite (nelt'-ner-ite) A submetallic black tetragonal mineral of the *braunite* group: $CaMn^{3+}{}_6SiO_{12}$.

nema (ne'-ma) A hollow threadlike prolongation of the apex of the *prosicula* of a graptolite. The term is used where the prolongation is "exposed", as in all except scandent rhabdosomes. Cf: *virgula.*

nemalite (ne'-ma-lite) A fibrous variety of brucite containing ferrous oxide.

nematath (ne'-ma-tath) A term used by Carey (1958) for a submarine ridge across an Atlantic-type ocean basin, which is not an orogenic structure but is composed of continental crust that has been stretched across a *sphenochasm* or *rhombochasm*. Carey cites as an example the Lomonosov Ridge that extends across the Arctic Ocean basin from North America to Asia. Obsolete.

nematoblastic (nem'-a-to-blas'-tic) Pertaining to a *homeoblastic* type of texture of a metamorphic rock, a texture that is due to the development of slender parallel prismatic crystals during recrystallization. Cf: *fibroblastic.*

nematocyst (nem'-a-to-cyst') (a) One of the minute stinging cells or organs of hydrozoans, scyphozoans, and anthozoans; e.g. a "thread cell" formed within a *cnidoblast* of a coral. (b) A large, ejectile organelle found in some dinoflagellates (Williams et al., 2000, p.140).

nematomorph (nem'-a-to-morph) One of the *cryptarchs*, consisting of septate or aseptate filaments or chains of vesicles. The filaments may be branched or unbranched, uniseriate or multiseriate.

Nematophytales (Nem-a-to-phy-ta'-les) *nematophytes.*

nematophytes (nem'-a-to-phytes) A group of enigmatic early land plants that have a *thallus* plant body (e.g., *Protosalvinia, Nematothallus*). All appear to have some type of *cuticle;* some consist of a series of tubes.

nenadkevichite (ne-nad'-ke-vich-ite) A rose, brown, or dark brown orthorhombic mineral: $(Na,[])_8Nb_4(Si_4O_{12})_2(O,OH)_4{\cdot}8H_2O$. It is the Nb analogue of korobitsynite.

nenadkevite (ne-nad'-kev-ite) A mixture of several uranium minerals.

n-en log *epithermal-neutron log.*

Neoarchean (Ne-o-ar'-che-an) An era of the Standard Global Geochronometric Scale: the youngest era of the *Archean* Eon. Defined as the time span between 2,800 Ma, the end of the *Mesoarchean* Era, and 2,500 Ma, the beginning of the *Paleoproterozoic* Era of the Paleoproterozoic Eon (Plumb, 1991).

neoautochthon (ne'-o-au-toch'-thon) A stable basement or autochthon formed where a nappe has ceased movement and has become defunct. Cf: *paleoautochthon.*

neoblast (ne'-o-blast) *holoblast.*

Neocomian (Ne-o-co'-mi-an) A European series: Lower Cretaceous. The term has never been properly defined and has been used in different contexts, but it generally is considered equivalent to the *Berriasian* through *Barremian* stages.

neocrystallization (ne'-o-crys'-tal-li-za'-tion) Crystallization or recrystallization that involves the development of new minerals among the fabric elements.

neo-Darwinism (ne-o-Dar'-win-ism) *Darwinism* modified or rephrased in accordance with modern genetics.

neoformation (ne'-o-for-ma'-tion) *neogenesis.*

Neogene (Ne'-o-gene) The upper subsystem of the *Tertiary* System of the Standard Global Chronostratigraphic Scale, above the *Paleogene* Subsystem and below the *Quaternary* System. Also the time during which these rocks were formed, the Neogene Subperiod, covering the time span between 23 and 1.75 Ma.

neogenesis (ne-o-gen'-e-sis) The formation of new minerals, as by diagenesis or metamorphism. Cf: *authigenesis.* Syn: *neoformation.*

neogenic (ne-o-gen'-ic) Said of newly formed minerals; pertaining to neogenesis.

neoglaciation (ne'-o-gla'-ci-a'-tion) The readvance of mountain glaciers during the late Holocene (Moss, 1951, p.62). There have been several neoglacial episodes, the most recent of which is referred to as the *Little Ice Age* (Porter and Denton, 1967; Grove, 1988).

Neognathae (Ne-o-gnath'-ae) A superorder of the avian subclass Neornithes which includes, in some 25 orders, all birds except ground-dwelling forms and toothed birds of the Mesozoic. See also: *Paleognathae.*

neoichnology (ne'-o-ich-nol'-o-gy) The study of Holocene tracks, burrows, and other structures left by living organisms, as opposed to *palichnology*. See also: *ichnology.*

neokaolin (ne-o-ka'-o-lin) Kaolinite artificially produced from nepheline.

Neolithic (Ne-o-lith'-ic) In Old World archaeology, the last division of the *Stone Age,* characterized by the development of agriculture and the domestication of farm animals. Correlation of relative cultural levels with actual age (and, therefore, with the time-stratigraphic units of geology) varies from region to region. Obsolete syn: *New Stone Age.*

neomesselite (ne-o-mes'-sel-ite) *messelite.*

neomineralization (ne′-o-min′-er-al-i-za′-tion) A chemical reaction within a rock whereby its mineral constituents are converted into new mineral species; a type of *recrystallization.*
neomorphism (ne-o-mor′-phism) An inclusive term suggested by Folk (1965, p.20-21) for all transformations between one mineral and itself or a polymorph, whether the new crystals are larger or smaller or simply differ in shape from the previous ones, or represent a new mineral species, e.g. fibrous calcite altering to equant calcite or aragonite mud altering to calcite microspar. It includes the processes of inversion, recrystallization, and strain recrystallization, in which the gross composition remains essentially constant. The term is appropriate where it is not possible to distinguish between recrystallization and inversion, or where the mechanism of change is not known. See also: *aggrading neomorphism; degrading neomorphism.*
neontology (ne-on-tol′-o-gy) The study of existing organisms, as opposed to *paleontology.* Approx. syn: *biology.* Cf: *cenozoology.*
Neophytic (Ne-o-phyt′-ic) *Cenophytic.*
Neoproterozoic (Ne-o-prot-er-o-zo′-ic) An era of the Standard Global Geochronometric Scale: the youngest era of the *Proterozoic* Eon, covering the time span between 1,000 Ma, the end of the *Mesoproterozoic* Era, and ~540 Ma, the age of the base of the *Paleozoic* Era (Plumb, 1991).
Neornithes (Ne-or′-ni-thes) A subclass of birds that includes all except the Upper Jurassic genus *Archaeopteryx.*
neosome (ne′-o-some) A geometric element of a composite rock or mineral deposit, appearing to be younger than the main rock mass (or *paleosome*); e.g. an injection in country rock, or the introduced or newly formed material of a migmatite. Sometimes used in place of the term *metatect, mobilizate,* or *chymogenic.* Cf: *metasome [geol].*
neostratotype (ne-o-strat′-o-type) A stratotype established after the holostratotype has been destroyed or is otherwise not usable (Sigal, 1964).
neotectonic joint A joint that propagates under the influence of the neotectonic stress field where the joint driving stress develops as a consequence of layer-parallel contraction by strain relaxation and thermal cooling upon *exhumation.* Such a joint is a useful tool for mapping the present stress field (Hancock and Engelder, 1989).
neotectonic map (ne′-o-tec-ton′-ic) A map portraying post-Miocene geologic structures. Cf: *neotectonics.*
neotectonics (ne′-o-tec-ton′-ics) The study of the post-Miocene structures and structural history of the Earth's crust.
neoteny (ne-ot′-e-ny) (a) Arrested development such that youthful characteristics are retained by the adult organism. Syn: *paedomorphism.* (b) Acceleration in the attainment of sexual maturity relative to general body development. Syn: *paedogenesis.*
Neotian (Ne-o′-tian) *Pannonian.*
neotocite (ne-o-toc′-ite) A resinous brown, black, or olive-green amorphous mineral consisting of a hydrous silicate of manganese and iron of uncertain formula. It may be an alteration product of rhodonite, possibly an opal with disseminated manganese and iron oxides.
neotype (ne′-o-type) A single specimen designated as the *type specimen* of a species or subspecies when the *holotype* (or *lectotype*) and all paratypes or all syntypes have been lost or destroyed (ICZN, 1964, p.150).
neovolcanic (ne′-o-vol-can′-ic) Said of extrusive rocks that are of Tertiary or younger age. Cf: *paleovolcanic.*
nepheline (neph′-e-line) A hexagonal feldspathoid mineral: $(Na,K)AlSiO_4$. It occurs as glassy crystals or colorless grains, or as coarse crystals or green to brown masses of greasy luster without cleavage, in alkalic igneous rocks; it is an essential constituent of some sodium-rich rocks. Syn: *nephelite; eleolite.*
nepheline basalt *olivine nephelinite.*
nepheline basanite In the *IUGS classification,* the variety of basanite in which nepheline is the dominant feldspathoid.
nepheline diorite In the *IUGS classification,* the variety of *foid diorite* in which nepheline is the dominant feldspathoid.
nepheline gabbro In the *IUGS classification,* the variety of *foid gabbro* in which nepheline is the dominant feldspathoid.
nepheline monzodiorite In the *IUGS classification,* the variety of *foid monzodiorite* in which nepheline is the dominant feldspathoid.
nepheline monzogabbro In the *IUGS classification,* the variety of *foid monzogabbro* in which nepheline is the most abundant feldspathoid.
nepheline monzosyenite In the *IUGS classification,* the variety of *foid monzosyenite* in which nepheline is the most abundant feldspathoid. Syn: *nepheline plagisyenite.*
nepheline plagisyenite In the *IUGS classification,* the variety of *foid plagisyenite* in which nepheline is the most abundant feldspathoid. Syn: *nepheline monzosyenite.*
nepheline syenite (a) In the *IUGS classification,* the variety of *foid syenite* in which nepheline is the most abundant feldspathoid. (b) A plutonic rock composed essentially of alkali feldspar and nepheline. It may contain an alkali ferromagnesian mineral, e.g. an amphibole (riebeckite, arfvedsonite, barkevikite) or a pyroxene (aegirine or aegirine-augite); the intrusive equivalent of *phonolite.* Sodalite, cancrinite, hauyne, and nosean, in addition to apatite, titanite, and opaque oxides, are common accessories. Rare minerals are also frequent accessories. Cf: *foyaite; ditroite; foid syenite.*
nepheline tephrite In the *IUGS classification,* the variety of *tephrite* in which nepheline is the most abundant feldspathoid.
nephelinite (neph′-e-lin-ite′) (a) In the *IUGS classification* of volcanic rocks, a foidite in which nepheline is the dominant feldspathoid. (b) A fine-grained or porphyritic extrusive or hypabyssal rock, of basaltic character, but primarily composed of nepheline and clinopyroxene, esp. titanian augite, and lacking olivine and feldspar. Cf: *olivine nephelinite.*
nephelinitoid (neph-e-lin′-i-toid) A nepheline-rich groundmass in an igneous rock; the glassy groundmass in nepheline rocks. Obsolete.
nephelinolite In the *IUGS classification* of plutonic rocks, a *foidolite* in which nepheline is the dominant feldspathoid; the plutonic equivalent of *nephelinite.* Nephelinolite is subdivided into *urtite, ijolite,* and *melteigite* on the basis of *color index.*
nephelite (neph′-e-lite) *nepheline.*
nepheloid layer (neph′-e-loid) A layer of water in the deep ocean basin that contains significant amounts of suspended sediment. It is from 200 m to 1,000 m thick.
nephelometer (neph-e-lom′-e-ter) An instrument used in nephelometry, designed to measure the amount of cloudiness of a medium.
nephelometry (neph-e-lom′-e-try) The measurement of the cloudiness of a medium; esp. the determination of the concentration or particle sizes of a suspension by measuring, at more than one angle, the scattering of light transmitted or reflected by the medium. Cf: *turbidimetry.*
nephrite (neph′-rite) An exceptionally tough, compact, fine-grained, greenish or bluish amphibole (specif. tremolite or actinolite) constituting the less rare or valuable kind of jade and formerly worn as a remedy for kidney diseases. Syn: *greenstone [mineral].*
nepioconch (nep′-i-o-conch′) The earliest-formed part of a *dissoconch* of a bivalve mollusk, separated from the later part by a pronounced discontinuity. Cf: *mesoconch.*
nepionic (nep-i-on′-ic) Said of the stage or period in which the young shell of an invertebrate does not yet show distinctive specific characteristics, i.e. of the stage following the *embryonic* stage and preceding the *neanic* stage.
népouite (né-pou′-ite) A green orthorhombic mineral: $Ni_3Si_2O_5(OH)_4$. It is a member of the kaolinite-serpentine group of two-layer phyllosilicates.
nepskoeite A yellowish-white orthorhombic mineral: $Mg_4Cl(OH)_7{\cdot}6H_2O$.
nepton (nep′-ton) A term used by Makiyama (1954) for a body of sedimentary rock filling a basin; e.g. a *geosynclinal prism.* Rarely used.
neptunian (nep-tu′-ni-an) (a) Pertaining to *neptunism* and the rocks whose origin was explained by it. (b) Formed by the agency of water.
neptunian dike A sedimentary dike formed by infilling of sediment, generally sand, in an undersea fissure or hollow. Cf: *injection dike.*
neptunian joint A joint that forms in the seabed commonly as a consequence of *syneresis.*
neptunic rock (nep-tu′-nic) (a) A rock formed in the sea. (b) A general term proposed by Read (1944) for all sedimentary rocks. Cf: *plutonic rock; volcanic rock.*
neptunism (nep′-tun-ism) A concept of the eighteenth and early nineteenth centuries that considered the origin of all rocks, including the crystalline rocks such as granite and metamorphic rocks as well as basalt, as a succession of deposits from a primeval sea. In the context, as propounded by A.G. Werner and his students, volcanoes and "true" volcanic rocks-lavas for example-were the result of fires in coal beds within the underlying sedimentary rocks (Dean, 1998). Ant: *plutonism [hist].* Cf: *Wernerianism.*
neptunist (nep′-tun-ist) An adherent to the theory of neptunism. Ant: *plutonist.*
neptunite (nep′-tun-ite) A black to deep reddish-brown monoclinic mineral: $KNa_2Li(Fe^{2+},Mn^{2+},Mg)_2Ti_2Si_8O_{24}$. Cf: *mangan-neptunite.*
nereite (ne′-re-ite) A *trace fossil* of the "genus" *Nereites,* consisting of a meandering feeding trail 1 to 2 cm wide with a narrow central axis and regularly spaced lateral, leaf-shaped, or lobelike projections, and perhaps formed by a worm or gastropod.
neritic (ne-rit′-ic) Pertaining to the ocean environment or depth zone between low-tide level and 200 meters, or between low-tide level and approximately the edge of the continental shelf; also, pertaining to the organisms living in that environment. Syn: *sublittoral.*
Nernst distribution law The statement that the ratio of molar concentrations of a component dissolved in two immiscible phases is constant and depends only on temperature. The ratio is called the *partition coefficient.*
nesaster (nes′-as-ter) The only skeletal element of radiocyathids, consisting of shafts and one or two many-ribbed rosettes. Two rosettes, when present, are similar and at opposite ends of shafts.
neskevaarite-Fe A pale brown to yellowish-brown monoclinic mineral of the *vuoriyarvite* group: $NaK_3Fe(Ti,Nb)_4(Si_4O_{12})_2(O,OH)_4{\cdot}6H_2O$.
nesosilicate (nes-o-sil′-i-cate) A class or structural type of *silicate* characterized by isolated SiO_4 tetrahedra, rather than by linkage of tetrahedra by the sharing of common oxygens. An example of a nesosilicate is olivine, $(Mg_2SiO_4\text{-}Fe_2{}^{+2}SiO_4)$. Cf: *sorosilicate; cyclosilicate; inosilicate; phyllosilicate; tectosilicate.* Syn: *orthosilicate; single island silicate.*
nesquehonite (nes-que-ho′-nite) A colorless or white monoclinic mineral: $Mg(HCO_3)(OH){\cdot}2H_2O$. It occurs in radiating groups of prismatic crystals.

ness A British term used esp. in Scotland for a *promontory,* headland, or cape, or any point or projection of the land into the sea; commonly used as a suffix to a place name, e.g. Fife*ness.* Syn: *naze; nose [geomorph]; nore; nab.*
nest A concentration of some relatively conspicuous element of a geologic feature, such as a "nest" of pebbles in a sand layer or inclusions in an igneous rock; esp. a small, pocketlike mass of ore or mineral within another formation.
nest cup A carbonate-lined, bowl-shaped depression containing *cave pearls.*
nested (nest'-ed) (a) Said of volcanic cones, craters, or calderas that occur one within another, i.e. show *cone-in-cone structure [volc].* (b) Said of two or more calderas that intersect, having been formed at different times or by different explosions.
nested semivariogram In *geostatistics,* a *semivariogram* composed of two or more models, such as a *nugget effect* model plus a spherical model.
nested sinkholes *karst valley; uvala.*
net [pat grd] A form of horizontal patterned ground whose mesh is intermediate between a circle and a polygon. See: *sorted net; nonsorted net.*
net [struc petrol] In structural petrology, a stereographic or an equal-area projection of a sphere in which the network of meridians and parallels forms a coordinate system. The meridians and parallels are normally projected at 2° intervals, and the net is used to plot points that represent linear and/or planar elements. Syn: *projection net; stereographic net.*
net [surv] A series of surveying or gravity stations that have been interconnected in such a manner that closed loops or circuits have been formed or that are so arranged as to provide a check on the consistency of the measured values; e.g. a *base net* and a *triangulation net.* Syn: *network.*
net ablation A less preferred term with various meanings, such as *summer balance* and *net balance* of the *ablation area* of a glacier.
net accumulation A less preferred term with various meanings, such as *winter balance* and *net balance* of the *accumulation area* of a glacier.
net balance The change in mass of a glacier from the time of minimum mass in one year to the time of minimum mass in the succeeding year (*balance year*); the mass change between one *summer surface* and the next. It can be determined at a point, in which case it is referred to as a "specific net balance", or as a total-mass change for the glacier. Units of millimeters, meters, or cubic meters of water equivalent are generally used. Syn: *net budget.* Cf: *annual balance.*
net budget *net balance.*
net calorific value A *calorific value* calculated from *gross calorific value* under conditions such that all the water in the products remains in the form of vapor. Syn: *net heat of combustion.*
net fabric Seams of replacement dolomite, concentrated along stylolites and microfractures, that enclose remnants of micrite in which are disseminated large baroque cross-shaped dolomite crystals (Zenger, 1983). Cf: *styloreactate.*
net heat of combustion *net calorific value.*
net plankton Any planktonic organism filterable with towed nets.
net primary production The amount of organic matter produced by living organisms within a given volume or area, minus that which is consumed by the respiratory processes of the organisms. NPP represents organic material available for consumers and decomposers. Cf: *gross primary production.*
net profits interest The amount payable to an interested party from the net profits produced by a mine or petroleum well. Abbrev: NPI.
netromorph (net'-ro-morph) An *acritarch* having an elongate to fusiform *vesicle* (e.g., *Dactylofusa, Leiofusa*). It may have *processes* or spines at one or both poles (Downie et al., 1963; Mendelson, 1993).
net slip On a fault, the distance between two formerly adjacent points on either side of the fault, measured on the fault surface or parallel to it, and in the direction of the slip vector. It defines both the direction and relative amount of *displacement.* Syn: *total displacement; total slip.*
net smelter royalty A royalty calculated on the basis of the actual sale price of a mineral product, net of smelting, refining, and related transportation charges. Abbrev: NSR.
net texture A network system of nickel sulfides in peridotite, originally a heavy interstitial melt in which lighter solid elements floated, or at least levitated (Naldrett, 1973).
net-transfer reaction (net-trans'-fer) A balanced reaction among phases or components of phases in which progress of the reaction results in a change in the modal amounts of the phases (Thompson et al., 1982). Cf: *exchange reaction.*
net venation In a leaf, a type of *venation* in which the veins branch repeatedly to form a network. Cf: *parallel venation.* See also: *pinnate venation; palmate venation.*
network (net'-work) *net [surv].*
network cave (a) Any cave consisting of repeatedly rejoined passages, commonly guided by joints. Syn: *maze cave.* See also: *branchwork cave.* (b) A cave in which the passages intersect as a grid. See also: *anastomotic cave; spongework cave.*
network deposit *stockwork.*
neudorfite (neu'-dorf-ite) A waxy, pale-yellow variety of retinite containing a little nitrogen, found in coal at Neudorf in Moravia, Czechoslovakia.
Neumann bands (Neu'-mann) Fine straight lines observed on etched surfaces of the low-Ni (A or kamacite) phase of iron meteorites, caused by mechanical twinning on (211) planes in cubic kamacite. Named after Franz E. Neumann (1798-1895), German mineralogist. Syn: Neumann lines; Neumann lamellae.
Neumann's principle The statement that the macroscopic properties of a crystal have at least the symmetry of the class of the crystal.
Neumann's problem (Neu'-mann's) A well-known problem in geodesy: to determine a function that is harmonic outside of a given surface and whose normal derivatives assume prescribed boundary values on the surface. Cf: *boundary-value problem; Dirichlet's problem.*
neural network An information-processing array utilizing a large number of simple modules; stored information components affect connections between the modules.
neurocranium (neu-ro-cra'-ni-um) The cartilage or cartilage-replacement component of the vertebrate skull. Syn: *braincase.*
neuromotorium (neu'-ro-mo-to'-ri-um) A ganglionlike granular body forming the dynamic center of ciliates (as in tintinnids).
neustädtelite An adamantine brownish triclinic mineral: $Bi_2Fe^{3+}(Fe^{3+},Co)_2(O,OH)_4(AsO_4)_2$. It is isomorphous to medenbachite.
neuston (neus'-ton) Animals floating on or living attached to surface film of water either on the air side or water side. Most animals are on the water side and are called hyponeuston.
neutral axis (neu'-tral) In a two-dimensional structural model, the equivalent of a *neutral surface.*
neutral density A value assiged to all parcels of water on the same *neutral surface.* It has units of density. It is a function of salinity, *in-situ temperature,* pressure, latitude and longitude.
neutral depth *normal depth.*
neutral dune A small, irregular sand dune (Wolfe et al., 1966, p.614).
neutral estuary An estuary in which neither freshwater inflow nor evaporation dominates.
neutral model In ecology, a model that seeks to explain diversity-abundance patterns in a given system purely as a function of stochastic processes of birth, death, dispersal, and speciation, without assuming any innate differences among the individuals or species comprising the system (Hubbell, 2001).
neutral pressure (a) *neutral stress.* (b) The lateral earth pressure when the soil is *at rest.*
neutral shoreline A *shoreline* whose essential features are independent of either the submergence of a former land surface or the emergence of a former underwater surface (Johnson, 1919, p.172;187); a shoreline resulting without a change in the relative level of land and water. It includes shorelines of deltas, alluvial and outwash plains, volcanoes, and coral reefs, as well as those produced by faulting. Obsolete.
neutral soil A soil whose pH value is 7.0. In practice, the pH value of a neutral soil ranges from 6.6 to 7.3.
neutral stress The stress transmitted by the fluid that fills the voids between particles of a soil or rock mass; e.g. that part of the total normal stress in a saturated soil caused by the presence of interstitial water. Syn: *pore pressure; pore-water pressure; neutral pressure.*
neutral surface *surface of no strain.*
neutral surface [oceanog] A surface in the ocean such that a parcel of water moving along the surface is always below less dense water and above denser water. No work is done moving along the surface if friction is neglected, because there is no buoyancy force acting on the parcel as it moves. Cf: *neutral density.*
neutron activation (neu'-tron) *Activation analysis* that relies on neutrons inside a nuclear reactor to irradiate a sample. See also: *instrumental neutron activation analysis.*
neutron-activation log (neu'-tron-ac-ti-va'-tion) A *radioactivity log* of neutron-spectral gamma type, usually run in cased wells, in which high-energy neutrons (about 14 Mev) bombard well-bore rocks and transmute natural elements to gamma-ray-emitting isotopes of characteristic identity. Behavior of calcium versus silicon permits lithology interpretation, and that of carbon versus oxygen may distinguish oil from water. See also: *spectral gamma-ray log.*
neutron-capture cross-section The probability that a thermal neutron will be captured by the nucleus of a target atom. Measured in units of barns, where 1 barn = 10^{-24} cm^2.
neutron diffraction (a) Diffraction of neutrons by a crystalline material. (b) A method of crystal analysis that operates on the same principles as *X-ray powder diffraction* analysis but uses beams of neutrons instead of X-rays. It is used when a crystal contains a large proportion of light elements, to determine all of the atomic positions of those elements in the crystal structure.
neutron-gamma log (neu'-tron-gam'-ma) The *well log* curve of induced gamma radioactivity that results from bombardment of rocks near the well bore by fast neutrons. A low count rate implies near-source dissipation in high-porosity rocks, esp. capture by chlorine. See also: *neutron log.* Syn: *n-g log.*
neutron log A *radioactivity log* curve that indicates the intensity of radiation (neutrons or gamma rays) produced when the rocks in a borehole are bombarded by neutrons from a *sonde.* It indicates the presence of fluids (but does not distinguish between oil and water) in the rocks, and is used with the *gamma-ray log* to differentiate porous from nonporous formations. See also:

neutron-gamma log; neutron-neutron log; epithermal-neutron log. Syn: *hydrogen-index log; nuclear log.*

neutron-neutron log Any of the several *neutron log* curves that measure the abundance of neutrons of a discrete energy range. Neutrons arrive at the detector after "random walk" scattering and slowing, most effectively by hydrogen nuclei. Depending on the neutron-energy selectivity level of the detector, these curves may be subdivided into *epithermal-neutron log* and *thermal-neutron log* types. See also: *neutron log.* Syn: *n-n log.*

neutron number The number of neutrons in the nucleus of an atom. Abbrev: N.

neutron soil-moisture meter An instrument for measuring water content of soil and rocks as indicated by the scattering and absorption of neutrons emitted from a source, and resulting gamma radiation received by a detector, in a probe lowered into an access hole.

Nevadan Orogeny (Ne-vad'-an) A time of deformation, metamorphism, and plutonism during Jurassic and Early Cretaceous time in the western part of the North American Cordillera, typified by relations in the Sierra Nevada, California. In that area, deformation of the supracrustal rocks can be closely dated by limiting fossiliferous strata as late in the Jurassic (between the Kimmeridgian and Portlandian Stages), but earlier and later Nevadan deformation occurs elsewhere. In the Sierra Nevada itself, the emplacement of granite and other plutonic activity were more prolonged than the deformation, and have been dated radiometrically between 180 Ma and 80 Ma, or from Early Jurassic to Early Cretaceous. Geologists differ as to whether to restrict the Nevadan closely in time and space, or to use it broadly; it can most properly be considered as an orogenic era, in the sense of Stille. Also spelled: *Nevadian Orogeny; Nevadic Orogeny.* Cf: *Coast Range Orogeny.*

Nevada twin law (Ne-vad'-a) A rare, parallel twin law in feldspar, with a twin axis of [112].

Nevadian Orogeny (Ne-vad'-i-an) *Nevadan Orogeny.*

Nevadic Orogeny (Ne-vad'-ic) *Nevadan Orogeny.*

nevadite (ne-vad'-ite) A term, now obsolete, that was applied to *rhyolite* containing abundant large phenocrysts of quartz, feldspar, biotite, and hornblende in a small amount of groundmass. The name, given by Richtofen in 1868, was derived from Nevada.

névé A French term meaning a mass of hardened snow at the source or head of a glacier; it refers to the overall snow cover that exists during the melting period and sometimes from one year to another. The term was originally used in English as an exact equivalent of *firn* (the material), and is still frequently so used, but it is perhaps best to restrict it, as proposed by Bristish glaciologists, to a geographic meaning, such as an area covered with perennial snow or an area of firn (a *firn field*), or more generally the *accumulation area* above or at the head of a glacier.

nevskite (nev'-skite) A metallic gray trigonal mineral: Bi(Se,S) It is the Se analogue of tsumoite.

nevyanskite (nev-yansk'-ite) A tin-white variety of iridosmine containing 35-50% osmium or more than 40% iridium and occurring in flat scales.

newberyite (new'-ber-y-ite') A white orthorhombic mineral: $Mg(PO_3OH){\cdot}3H_2O$. It occurs in guano caves.

new global tectonics A general term introduced by Isacks et al. (1968) for *global tectonics* based on the related concepts of continental drift, sea-floor spreading, transform faults, and underthrusting of the lithosphere (crust and uppermost mantle) at island arcs, as they are jointly applied to an integrated global analysis of the relative motions of crustal segments delineated by the major seismic belts. Contemporary authors prefer to use the term *plate tectonics.*

new ice A general term for recently formed ice (esp. floating sea ice) less than 5 cm thick, composed of ice crystals that may be weakly frozen together and that have a definite form only while they are afloat; e.g. *frazil ice, grease ice, sludge [ice], shuga, ice rind, nilas,* and *pancake ice.*

New Red Sandstone The red sandstone facies of the Permian and Triassic systems, well-developed in NW England. See William Buckland in Phillips (1818, p.71-79).

new snow (a) Fallen snow in which the original crystalline structure is still recognizable. Ant: *old snow.* (b) Snow that has fallen very recently, as in the past 24 hours.

New Stone Age *Neolithic.*

Newtonian flow (New-to'-ni-an) In experimental structural geology, flow in which the rate of shear strain is directly proportional to the shear stress; flow of a *Newtonian liquid.* Cf: *non-Newtonian flow.* Syn: *viscous flow.*

Newtonian fluid A fluid in which *shear stress* is directly proportional to the gradient velocity in the direction perpendicular to flow.

Newtonian liquid A substance in which the rate of shear strain is proportional to the shear stress. This constant ratio is the *viscosity* of the liquid. See also: *Newtonian flow.*

Newton's law of gravitation (New'-ton's) The statement that every particle of matter attracts every other particle with a force whose magnitude is proportional to the product of their masses and inversely proportional to the square of the distance between them. Syn: *law of universal gravitation.* See also: *gravitation.*

nexine (nex'-ine) The inner division of the exine of pollen, more or less equivalent to *endexine* (but includes the *foot layer*). Cf: *endexine; perine; sexine.*

neyite (ney'-ite) A metallic gray monoclinic mineral: $Ag_2Cu_6Pb_{25}Bi_{26}S_{68}$.

nezilovite A metallic black hexagonal mineral of the *magnetoplumbite* group: $PbZn_2(Mn^{4+},Ti^{4+})_2Fe^{3+}{}_8O_{19}$.

***n*-glide** A *glide plane* involving reflection across a mirror plane combined with translation parallel to the mirror plane of one-half the vector sum of two or three of the *basis vectors* of the *crystal lattice.* Syn: *diagonal glide plane.*

n-g log *neutron-gamma log.*

n'hangellite (n'han'-gel-lite) A green, elastic bitumen, similar to *coorongite,* that represents deposits of the alga *Coelosphaerium.*

Niagaran (Ni-ag'-a-ran) North American series: Lower and Upper Silurian (above Alexandrian, below Cayugan).

niahite (ni'-a-hite) A pale orange orthorhombic mineral occurring in cave deposits: $(NH_4)(Mn^{2+},Mg)PO_4{\cdot}H_2O$. It is the Mn analogue of dittmarite.

niccolite (nic'-co-lite) A syn. of *nickeline.* Also spelled: *nicolite.*

niche [ecol] The position of an organism or a population in the environment as determined by its mode of life, needs, contributions, potential, and interaction with other organisms or populations. Syn: *ecologic niche.*

niche [geomorph] A shallow cave or re-entrant produced by weathering and erosion near the base of a rock face or cliff, or beneath a waterfall.

niche glacier A common type of small mountain glacier, occupying a funnel-shaped hollow or irregular recess in a mountain slope. Cf: *cirque glacier.*

nichromite (ni-chro'-mite) A metallic black cubic mineral of the *spinel* group: $(Ni,Co,Fe^{2+})(Cr,Fe^{3+},Al)_2O_4$. Not approved by the International Mineralogical Association.

nick (a) A place of abrupt inflection in a stream profile; a *knickpoint.* (b) A sharp angle cut by waves, currents, or ice at the base of a cliff. Syn: *knick.*

nickel (nick'-el) A nearly silver-white cubic mineral, the metallic element Ni. It occurs native esp. in meteorites and also alloyed with iron in meteorites. Nickel is used chiefly in alloys and as a catalyst.

nickelalumite (nick'-el-al'-um-ite) A pale blue monoclinic mineral: $(Ni,Cu)Al_4(OH)_{12}(SO_4){\cdot}3H_2O$. It is the Ni analogue of chalcoalumite.

nickel-antimony glance (nick'-el-an'-ti-mo-ny) *ullmannite.*

nickelaustinite (nick'-el-aus'-tin-ite) A pale yellowish-green to grass-green orthorhombic mineral of the *adelite* group: $Ca(Ni,Zn)(AsO_4)(OH)$.

nickelbischofite (nick-el-bisch'-of-ite) A green monoclinic mineral: $NiCl_2{\cdot}6H_2O$.

nickelblödite (nick-el-blöd'-ite) A pale yellowish-green to green monoclinic mineral of the *blödite* group: $Na_2Ni(SO_4)_2{\cdot}4H_2O$. Also spelled: nickelbloedite.

nickel bloom A green hydrated and oxidized patina or incrustation on outcropping rocks, indicating the existence of primary nickel minerals; specif. annabergite (a nickel arsenate). The term is also applied to zaratite (a nickel carbonate) and to morenosite (a nickel sulfate).

nickelboussingaultite A greenish-blue monoclinic mineral of the *picromerite* group: $(NH_4)_2(Ni,Mg)(SO_4)_2{\cdot}6H_2O$.

nickel glance *gersdorffite.*

nickelhexahydrite (nick'-el-hex'-a-hy'-drite) A blue-green monoclinic mineral of the *hexahydrite* group: $(Ni,Mg,Fe^{2+})SO_4{\cdot}6H_2O$.

nickeline (nick'-el-ine) (a) A pale copper-red hexagonal mineral: NiAs. It is an important ore of nickel, and may contain antimony, cobalt, iron, and sulfur. Syn: *niccolite; arsenical nickel; copper nickel; kupfernickel.* (b) A group name for hexagonal minerals with the general formula *AX,* where *A* = Ni, Co, Fe, Pd, or Pt; and *X* = As, Se, Te, Sb, Bi, or S.

nickel-iron (nick'-el-i'-ron) An alloy of nickel and iron (Ni,Fe) occurring in pebbles and grains (as in stream gravel), and also in meteorites. See also: *kamacite; taenite.* Syn: *awaruite; josephinite.*

nickel laterite Any *regolith* that contains economically recoverable concentrations of nickel. Usually applied loosely to all significant nickel-bearing regolith layers, which may or may not include any true *laterite* (Gleeson et al., 2003).

nickellotharmeyerite A yellow or brown monoclinic mineral: $Ca(Ni,Fe)_2(AsO_4)_2(H_2O,OH)_2$. It is the Ni analogue of lotharmeyerite and a member of the *tsumcorite* group.

nickel ocher *annabergite.*

nickelphosphide A metallic tetragonal meteorite mineral: $(Ni,Fe)_3P$. It is the Ni analogue of schreibersite.

nickel pyrites *millerite.*

nickelschneebergite An adamantine brown or beige monoclinic mineral: $Bi(Ni,Co)_2(AsO_4)_2(OH,H_2O)_2$. It is the Ni analogue of schneebergite, and a member of the *tsumcorite* group.

nickel-skutterudite (nick'-el-skut-te-rud'-ite) A tin-white to steel-gray cubic mineral of the *skutterudite* group: $(Ni,Co)As_3$. It may contain iron, and it represents a valuable ore of nickel, often associated with smaltite and skutterudite. Syn: *chloanthite; white nickel.*

nickel vitriol *morenosite.*

nickel-zippeite (nick'-el-zip'-pe-ite) An orange or tan monoclinic mineral of the *zippeite* group: $Ni(UO_2)_2(SO_4)O_2{\cdot}3.5H_2O$.

nickenichite A vitreous blue monoclinic mineral of the *alluaudite* group: $(Na,Ca,Cu)_{1.6}(Mg,Fe^{3+},Al)_3(AsO_4)_3$.

nickpoint *knickpoint.*

nicol (nic'-ol) (a) *Nicol prism.* (b) Any apparatus that produces polarized light, e.g. Nicol prism or Polaroid; a *polarizer.*

nicolite (nic'-o-lite) *niccolite.*

nicolo (nic′-o-lo) A variety of onyx with a black or brown base and a bluish-white or faint-bluish top layer.

Nicol prism (Nic′-ol) In a polarizing microscope, a pair of prisms that polarize and analyze the light used for illumination of the thin section under study. The lower nicol, or *polarizer,* is located below the stage; it consists of a rhombohedron of optically clear calcite so cut and recemented that the ordinary ray produced by double refraction in the calcite is totally reflected and the extraordinary ray transmitted. The upper nicol, or *analyzer,* is located above the objective and receives the polarized light after it has passed through the object under study. Its vibration direction is normally set at right angles to that of the polarizer. Partial syn: *nicol.* Syn: *polarizing prism.*

nicopyrite (nic-o-py′-rite) *pentlandite.*

niedermayrite A vitreous bluish-green monoclinic mineral: $Cu_4Cd(SO_4)_2(OH)_6{\cdot}4H_2O$.

nierite A brownish-red or colorless trigonal meteorite mineral: Si_3N_4.

nieve penitente (ni-e′-ve pen-i-ten′-te) (a) A jagged pinnacle or spike of snow or firn, typically a decimeter or two in height but possibly up to several meters in height, resulting from differential ablation under conditions of strong insolation, especially in high-altitude/low-latitude environments; an advanced stage of sun cup development. The top is typically bent over, giving the illusion of figures hanging their heads in penitence. (b) An assemblage of nieve penitentes. Etymol: Spanish, "penitent snow", shortened from "nieve de los penitentes"; first used in South America. Syn: *penitent [glaciol]; ice penitente; snow penitente; sun spike.* Cf: *serac.*

nifontovite (ni-fon-tov′-ite) A colorless monoclinic mineral: $Ca_3B_6O_6(OH)_{12}{\cdot}2H_2O$.

nigerite (ni′-ger-ite) The original name for a dark-brown trigonal mineral: $(Fe^{2+},Zn)_4Sn_2(Al,Fe^{3+})_{15}O_{30}(OH)_2$. The name was later changed to *ferronigerite-2N1S.*

niggliite (nig′-gli-ite) A metallic silver-white hexagonal mineral: PtSn.

Niggli molecular norm (Nig′-gli) A *norm* in which the reported content of a mineral represents the percentage of the total cations present in the rock that are tied up in that particular mineral. Three variant mineral associations are possible: the *catanorm,* the *mesonorm,* and the *epinorm.* Syn: molecular norm.

Niggli's classification (Nig′-gli's) A chemical classification of igneous rocks that is essentially a modification and simplification of *Osann's classification.* This system was proposed in 1920 by the Swiss mineralogist Paul Niggli (1888-1953).

Niggli value A parameter produced from the chemical analysis of a rock reported as molecular proportions. The proportions of oxides other than SiO_2 are summed, and this sum is divided into SiO_2 to produce *si;* into Al_2O_3 to produce *al;* into the sum $FeO+2Fe_2O_3+MgO$ to produce *fm;* into CaO to produce *c;* and into the sum Na_2O+K_2O to produce *alk.* Also called: Niggli number.

night emerald *evening emerald.*

nigrine (nig′-rine) A black variety of rutile containing iron.

nigritite (nig′-ri-tite) Coalified, carbon-rich bitumens. See also: *polynigritite; humonigritite; exinonigritite; keronigritite.*

niigataite A pale gray monoclinic mineral of the *epidote* group: $CaSrAl_3O(Si_2O_7)(SiO_4)(OH)$.

nijd *nejd.*

nikischerite A green rhombohedral mineral: $NaFe^{2+}{}_6Al_3(SO_4)_2(OH)_{18}(H_2O)_{12}$. It is isostructural with shigaite and motukoreaite.

nilas (ni′-las) A thin elastic crust of gray-colored ice formed on a calm sea, having a matte surface, and easily bent by waves and thrust into a pattern of interlocking "fingers"; it is subdivided by color into "dark nilas" (less than 5 cm thick) and "light nilas" (5-10 cm thick). Etymol: Russian.

niligongite (nil-i-gong′-ite) A plutonic foidite intermediate in composition between *fergusite* and *ijolite,* containing approximately equal amounts of nepheline and leucite and 30 to 60 percent mafic minerals. Its name, given by Lacroix in 1933, is derived from Niligongo, Zaire. Obsolete.

nimite (nim′-ite) A yellowish-green monoclinic trioctahedral member of the *chlorite* group in which Ni is the dominant divalent octahedral cation: $(Ni,Mg,Fe^{2+})_5Al(Si_3Al)O_{10}(OH)_8$.

ningyoite (ning′-yo-ite) A brownish-green to brown orthorhombic mineral: $(U,Ca,Ce)_2(PO_4)_2{\cdot}1\text{-}2H_2O$. It occurs as coatings or cavity fillings in uranium ore.

niningerite (nin′-in-ger-ite′) A metallic gray cubic meteorite mineral of the *galena* group: $(Mg,Fe^{2+},Mn)S$.

niobite (ni-o′-bite) *columbite.*

niobo-aeschynite-(Ce) A red-brown to dark brown orthorhombic mineral of the *aeschynite* group: $(Ce,Ca,Th)(Nb,Ti)_2(O,OH)_6$.

niobo-aeschynite-(Nd) An orthorhombic mineral of the *aeschynite* group: $(Nd,Ce)(Nb,Ti)_2(O,OH)_6$. Not approved by the International Mineralogical Association.

niobocarbide A metallic bronze cubic mineral of the *tantalcarbide* group: (Nb,Ta)C.

niobokupletskite A beige to brown triclinic mineral of the *astrophyllite* group: $K_2Na(Mn,Zn,Fe)_7(Nb,Zr,Ti)_2Si_8O_{26}(OH,O,F)_5$. It is the Nb analogue of kupletskite.

nioboloparite (ni-o′-bo-lo′-par-ite) Niobium-bearing *loparite-(Ce).*

niobophyllite (ni-o′-bo-phyl′-lite) A brown platy triclinic mineral of the *astrophyllite* group: $K_2Na(Fe^{2+},Mn^{2+})_7(Nb,Ti)_2Si_8O_{26}(O,OH,F)_5$.

niocalite (ni-o-cal′-ite) A pale-yellow monoclinic mineral: $Ca_7Nb(Si_2O_7)_2O_3F$. It is closely related to wöhlerite.

nip [coal] A *pinch* or thinning of a coal seam, esp. as a result of tectonic movements. Cf: *want.*

nip [coast] (a) A small, very low cliff or break in slope produced at the high-water mark by wavelets, and often cited as an initial feature in the development of a shoreline of emergence. The term has also been applied in a broader sense to the small notch resulting from the formation of such a cliff. (b) An undercutting notch in rock, particularly limestone, along a seacoast between high and low tide levels, produced by erosion or dissolution.

nisbite (nis′-bite) A metallic orthorhombic mineral of the *löllingite* group: $NiSb_2$.

nissonite (nis′-son-ite) A bluish-green monoclinic mineral: $Cu_2Mg_2(PO_4)_2(OH)_2{\cdot}5H_2O$.

niter (ni′-ter) (a) A colorless, white, or gray orthorhombic mineral: KNO_3. It is a soluble crystalline salt that occurs as a product of nitrification in most arable soils in hot, dry regions, and in the loose earth forming the floors of some natural caves. Cf: *soda niter.* Syn: *saltpeter.* (b) A term that was formerly used for a variety of saline efflorescences, including natron and soda niter. Also spelled: *nitre.*

nitrammite (ni-tram′-mite) A colorless orthorhombic mineral: NH_4NO_3. Its natural occurrence is in doubt; it may be gwihabaite. Syn: ammonia-niter.

nitrate (ni′-trate) A mineral compound characterized by a fundamental anionic structure of NO_3^-. Soda niter, $NaNO_3$, and niter, KNO_3, are nitrates. Cf: *carbonate [mineral]; borate.*

nitratine (ni′-tra-tine) A colorless, white, or yellowish rhombohedral mineral: $NaNO_3$. Syn: *soda niter.*

nitre (ni′-tre) *niter.*

nitride (ni′-tride) A mineral compound that is a combination of nitrogen with a more positive element. An example is osbornite, TiN.

nitrification (ni′-tri-fi-ca′-tion) The formation of nitrates by the oxidation of ammonium salts to nitrites (usually by bacteria) followed by oxidation of nitrites to nitrates. It is one of the processes of soil formation.

nitrobarite (ni-tro-bar′-ite) A colorless cubic mineral: $Ba(NO_3)_2$.

nitrocalcite (ni-tro-cal′-cite) A white or gray monoclinic mineral: $Ca(NO_3)_2{\cdot}4H_2O$. It occurs as an efflorescence, as on walls and in limestone caves. Syn: *wall saltpeter.*

nitrogen fixation (ni′-tro-gen) In a soil, the conversion of atmospheric nitrogen to a combined form by the metabolic processes of some algae, bacteria, and actinomycetes.

nitroglauberite (ni-tro-glau′-ber-ite) A mixture of darapskite and soda niter.

nitromagnesite (ni-tro-mag′-nes-ite) A colorless or white monoclinic mineral: $Mg(NO_3)_2{\cdot}6H_2O$. It occurs as an efflorescence in limestone caverns.

nitrophyte (ni′-tro-phyte) A plant that requires nitrogen-rich soil for growth.

nival (ni′-val) Characterized by or living in or under snow, or pertaining to a snowy environment; e.g. *nival* fauna or climate.

nival karst *Alpine karst* in which snowmelt accounts for much of the dissolution.

nivation (ni-va′-tion) (a) The process of excavation of a shallow depression or *nivation hollow* in a mountainside by removal of fine material around the edge of a shrinking snow patch or snowbank, chiefly through sheetwash, rivulet flow, and solution in meltwater (Thorn, 1976). Freeze-thaw action is apparently insignificant. Syn: *snow-patch erosion.* (b) More generally, the work of snow and ice beyond the limits of glacier action.

nivation cirque *nivation hollow.*

nivation glacier A small, "new-born" glacier, representing the initial stage of glaciation. Syn: *snowbank glacier.*

nivation hollow A shallow depression or hollow in a mountainside, permanently or intermittently occupied by a snowbank or snow patch and produced by *nivation.* If the snow completely melts each summer the hollow is deepened; otherwise not (Thorn, 1976). It has been suggested that deepening of a nivation hollow produces a cirque, but this is not proven. Syn: *nivation cirque; snow niche.*

nivation ridge A low convex accumulation of fine sediment downslope from a *nivation hollow,* consisting of fine material carried by sheetwash and rivulet flow from beneath the melting edge of a snow patch or snowbank. Cf: *nivation.*

niveal (niv′-e-al) Said of features and effects "due to the action of snow and ice" (Scheidegger, 1961, p.24). See also: *niveoglacial.*

niveau surface (ni-veau′) *equipotential surface.*

nivenite (niv′-en-ite) A velvet-black variety of uraninite containing rare earths (cerium and yttrium).

niveo-eolian (niv′-e-o-e-o′-li-an) *niveolian.*

niveoglacial (niv′-e-o-gla′-cial) Pertaining to the combined action of snow and ice. See also: *niveal.*

niveolian (niv-e-o′-li-an) Pertaining to simultaneous accumulation and intermixing of snow and airborne sand at the side of a gentle slope; e.g. said of material deposited by snowstorms under periglacial conditions. Syn: *niveo-eolian.*

nivo-karst (ni′-vo-karst) "A characteristic of periglacial areas" (Hamelin and Cook, 1967, p.73) whereby differential chemical weathering beneath snowbanks produces a karstlike topography, as the solution of limestone fragments by snowmelt containing carbonic acid.

NMAS *National Map Accuracy Standards.*

NML *nuclear-magnetism log.*

NMO *normal moveout.*

NMO stretch The shift toward lower frequencies that results from applying the *normal-moveout correction* to offset seismic traces. Measured by the change in a period as a result of applying NMO. Muting is often done at 50-100 percent NMO stretch.

NMR *nuclear magnetic resonance.*

n-n log *neutron-neutron log.*

Noachian The oldest system of rocks in the geologic stratigraphy of the planet Mars. Named after Noachis Terra, a portion of the southern highlands near the prime meridian that is characterized by rugged, densely cratered materials typical of the oldest exposed surfaces on Mars (Tanaka, 1986). See also: *Amazonian; Hesperian.*

Noachian flood (No-a′-chi-an) The flood described in the Bible (Genesis, chap.6-8), during which the patriarch Noah was said to have saved his family and representative creatures. Early writers believed that the waters of this flood deposited material now known as *drift* (drifted in by the flood as in icebergs). Also known as "The Deluge".

noble gas (no′-ble) *inert gas.*

nobleite (no′-ble-ite) A colorless monoclinic mineral: $CaB_6O_9(OH)_2{\cdot}3H_2O$. It is the Ca analogue of tunellite.

noble metal Any metal or alloy of comparatively high economic value, or one that is superior in certain desired properties, e.g. gold, silver, or platinum. Cf: *base metal.*

nocerite (no-cer′-ite) *fluoborite.*

nodal (nod′-al) A crinoid *columnal* that is generally larger than adjacent columnals and that commonly bears cirri.

nodal point *amphidrome.*

node [bot] The place on a plant stem from which a leaf and bud normally emerge.

node [cart] A zero-dimensional object, and a special type of point, that is a topological junction or end point and that may specify a geographic location.

node [evol] A branching point on a *dendrogram*, especially on a *cladogram* or *phylogenetic tree.*

node [paleont] (a) The uncalcified proximal extremity of a branch forming an articulation in a jointed colony of cheilostome bryozoans. (b) A knob, protuberance, or thickened or swollen body part of an animal, such as a small boss at the end of a foraminiferal pillar.

node [phylogen] The representation of a speciation event in a cladogram (Brooks et al., 1984, p.80).

node [waves] That point on a standing wave at which the vertical motion is least and the horizontal velocity is greatest. It is also associated with seiches.

nodular (nod′-u-lar) (a) Composed of nodules; e.g., "nodular bedding" consisting of scattered to loosely packed nodules in matrix of like or unlike character. (b) Having the shape of a nodule, or occurring in the form of nodules; e.g. "nodular ore" such as a colloform mineral aggregate with a bulbed surface. Syn: *nodulated.* (c) *orbicular.*

nodular anhydrite *chickenwire anhydrite.*

nodular chert (a) *Chert* in the form of small, rounded lumps. (b) A term used in Missouri for *chalky chert* containing small irregular grains (Grohskopf and McCracken, 1949, p.3).

nodular texture An intrasediment occurrence of finely crystalline material as displacive, ellipsoidal to nodular, irregular masses. Nodular texture is common for gypsum, anhydrite, and other saline minerals such as ulexite, glauberite, and magadiite. See also: *chickenwire anhydrite.*

nodulated (nod′-u-lat-ed) Occurring in the form of nodules; *nodular.*

nodule [ign] (nod′-ule) A rounded fragment of a coarse-grained igneous rock, apparently crystallized at depth, occurring as an *inclusion* in an extrusive rock; e.g. a "peridotite nodule" in a flow of olivine basalt. Cf: *xenolith.* Syn: *plutonic nodule.*

nodule [sed] (a) A small, irregularly rounded knot, mass, or lump of a mineral or mineral aggregate, normally having a warty or knobby surface and no internal structure, and usually exhibiting a contrasting composition from the enclosing sediment or rock matrix in which it is embedded; e.g. a nodule of pyrite in a coal bed, a *chert nodule* in limestone, or a *phosphatic nodule* in marine strata. Most nodules appear to be secondary structures: in sedimentary rocks they are primarily the result of postdepositional replacement of the host rock and are commonly elongated parallel to the bedding. Nodules can be separated as discrete masses from the host material. (b) One of the widely scattered concretionary lumps of manganese, cobalt, iron, and nickel found on the floors of the world's oceans; esp. a *manganese nodule.* Etymol: Latin "nodulus", "small knot". Cf: *concretion.*

noélbensonite A dark brown orthorhombic mineral of the *lawsonite* group: $BaMn^{3+}{}_2Si_2O_7(OH)_2{\cdot}H_2O$.

noise (a) Any undesired disturbance; everything except the signal. (b) An erratic, intermittent, or statistically random oscillation. (c) That portion of the unwanted signal that is statistically random, as distinguished from "hum", which is an unwanted signal occurring at multiples of the power-supply frequency.

noise analysis A seismic profile made by moving the source to progressively larger offsets while keeping geophones fixed, or keeping the source stationary while the geophones are moved to progressively larger offsets. Used to examine coherent noise trains. Syn: *walkaway.*

nolanite (no′-lan-ite) A black hexagonal mineral: $(V^{3+},Fe^{3+},Fe^{2+},Ti)_{10}O_{14}(OH)_2$.

nomen ambiguum (no′-men am-big′-u-um) A name used in different senses (e.g. the same name applied to different taxa) so that it has become "a long-persistent source of error" (McVaugh et al., 1968, p.18). Etymol: Latin, "uncertain name". Pl: nomina ambigua.

nomenclature (no′-men-cla′-ture) The practice of naming allied groups of plants and animals (*taxa*) according to the hierarchical system and formal procedure prescribed by accepted authoritative codes, i.e. the International Code of Botanical Nomenclature and the International Code of Zoological Nomenclature.

nomen confusium (no′-men con-fu′-si-um) A name applied to a *taxon* of dubious integrity, e.g. one based on a type consisting of discordant elements (Cowan, 1968, p.70; McVaugh et al., 1968, p.19). Etymol: Latin, "confused name". Pl: nomina confusia.

nomen conservandum (no′-men con-ser-van′-dum) Any name that must be adopted as a correct name in accordance with a special regulation, although otherwise contrary to nomenclatural rules (McVaugh et al., 1968, p.10; ICBN, 1972, appendix III). Etymol: Latin, "name to be preserved". Pl: nomina conservanda.

nomen nudum (no′-men nu′-dum) A scientific name published without description or *diagnosis* (McVaugh et al., 1968, p.10). Etymol: Latin, "nude name, mere name". Pl: nomina nuda.

nominal (nom′-i-nal) In zoological nomenclature, a term applied to a particular *taxon* whose rank has been objectively defined by its *type;* e.g. "the nominal genus *Musca* is always that to which its type species, *Musca domestica,* belongs" (ICZN, 1964, p.153).

nominal diameter The computed diameter of a hypothetical sphere having the same volume as that calculated for a given sedimentary particle; it is a true measure of particle size independent of either the shape or the density of the particle. Cf: *equivalent radius; sedimentation diameter.*

nominal scale A classification of observation data into mutually exclusive categories of equal rank.

nomogenesis (no-mo-gen′-e-sis) A theory of evolution stating that evolutionary change is governed by predetermined natural processes and is independent of environmental influences.

nomogram (nom′-o-gram) A type of line chart that graphically represents an equation of three variables, each of which is represented by a graduated straight line. It is used to avoid lengthy calculations; a straight line connecting values on two of the lines automatically intersects the third line at the required value. Syn: *nomograph.*

nomograph (nom′-o-graph) *nomogram.*

nonangular unconformity (non-an′-gu-lar) *disconformity.*

nonarboreal pollen (non-ar-bo′-re-al) Pollen of herbs and shrubs. Abbrev: NAP. Syn: *nonarborescent pollen; nontree pollen.*

nonarborescent pollen (non′-ar-bo-res′-cent) *nonarboreal pollen.*

nonartesian groundwater (non-ar-te′-sian) *unconfined groundwater.*

nonasphaltic pyrobitumen (non-as-phal′-tic) Any of a group of pyrobitumens, including peat, coal, and nonasphaltic pyrobituminous shales, that are dark-colored relatively hard nonvolatile solids, composed of hydrocarbons containing oxygenated bodies. They are sometimes associated with mineral matter, the nonmineral constituents being infusible and largely insoluble in carbon disulfide (Abraham, 1960, p.57).

nonassociated gas (non-as-so′-ci-at-ed) Natural gas that occurs in a reservoir without oil. Cf: *associated gas.*

nonbanded coal (non-band′-ed) Coal without bands of vitrain or lustrous material, consisting mainly of clarain, durain, or intermediate material. Cf: *sapropelic coal.*

nonbridging oxygen An oxygen atom that is bonded to only one tetrahedrally coordinated atom in a silicate *crystal structure;* that is, it is not shared by two tetrahedra. Cf: *bridging oxygen.*

noncaking coal (non-cak′-ing) Coal that does not cake or agglomerate when heated; it is usually a hard or dull coal. Syn: *free-burning coal.*

Noncalcic Brown soil (Non-cal′-cic) A great soil group in the 1938 classification system, for a group of zonal soils having a slightly acidic, light pink or reddish brown A horizon and a light brown or dull red B horizon. It is developed under a mixture of grass and forest vegetation, in a subhumid climate (USDA, 1938). Most of these soils are now classified as *Xeralfs.* Syn: *Shantung soil.*

noncapillary porosity (non-cap′-il-lar′-y) The volume of large interstices in a rock or soil that do not hold water by capillarity (Jacks et al., 1960). Cf: *aeration porosity.*

noncarbonate hardness (non-car′-bon-ate) Hardness of water, expressed as $CaCO_3$, that is in excess of the $CaCO_3$ equivalent of the carbonate and bicarbonate alkalinity. It cannot be removed by boiling and hence is sometimes called *permanent hardness,* although this synonym is becoming obsolete. Cf: *carbonate hardness; hardness [water].*

nonclassical biopyribole One of a *polysomatic* series of minerals consisting of pyroxenelike and micalike modules (exclusive of amphiboles, which consist of a simple 1:1 alternation of pyroxenelike and micalike modules). These are chain silicates with finite chain widths in excess of two.

nonclassical pyribole *nonclassical biopyribole.*

nonclastic (a) Said of a sedimentary texture showing no evidence that the sediment was derived from a preexisting rock or was deposited mechanically. (b) Pertaining to a chemically or organically formed sediment or sedimentary rock. Syn: *nonmechanical.*

noncoaxial (non-co-ax'-i-al) Antonym of *coaxial* (Hsu, 1966, p.217).

noncoaxial progressive deformation A deformation in which the principal strain directions of the incremental and finite strain ellipsoids are not parallel during all stages of deformation. Cf: *coaxial progressive deformation.*

noncognate (non-cog'-nate) *accidental.*

noncohesive (non-co-he'-sive) *cohesionless.*

nonconformable (non-con-form'-a-ble) Pertaining to a nonconformity or to the stratigraphic relations shown by a nonconformity.

nonconformity (non-con-form'-i-ty) (a) An *unconformity* developed between sedimentary rocks and older rocks (plutonic igneous or massive metamorphic rocks) that had been exposed to erosion before the overlying sediments covered them. The restriction of the term to this usage was proposed by Dunbar and Rodgers (1957, p.119). Although the term is "well known in the classroom", it is "not commonly used in practice" (Dennis, 1967, p.160). Syn: *heterolithic unconformity.* (b) A term that formerly was widely, but now less commonly, used as a syn. of *angular unconformity,* or as a generic term that includes angular unconformity. Term proposed by Pirsson (1915, p.291-293).

nonconservative elements (non-con-serv'-a-tive) (a) In seawater, those elements whose total amounts and average concentrations vary considerably over short time scales (day to years) and distance scales (meters to hundreds of km). Their abundances relative to each other, and relative to salinity or chlorinity, are thus not constant. (b) Elements in seawater that participate in internal chemical reactions in the ocean on short time scales (tens to thousands of years). Cf: *conservative elements; salinity; chlorinity.*

noncryotic ground Any ground (soil or rock) having a temperature above 0°C.

noncyclic terrace (non-cy'-clic) A terrace or bench along a stream valley that does not represent a former cycle of erosion. It may be made of resistant rocks that are unrelated to former valley floors or may be formed by continuous valley deepening accompanied by lateral stream migration and erosion, in which cases terraces on opposite sides of the valley are *unpaired terraces.* Cf: *cyclic terrace.*

noncylindrical fold A fold whose hinge line itself is markedly curved.

nondarcy flow Flow of water under conditions in which the volumetric rate of flow is not directly proportional to *hydraulic gradient.* It occurs when flow is either nonlaminar or is in transition between nonlaminar and laminar flow. See also: *Darcy's law.*

nondenticular composite prismatic structure A *composite prismatic structure* in which divergence of the second-order prisms occurs independently of curvature and denticulation of the depositional surface. This includes both vertical and reclined nondenticular composite prisms. In addition, nondenticular composite prisms constitute the second-order structural units of compound composite prismatic and crossed composite prismatic structures (Carter, 1980b). Abbrev: NDCP.

nondepositional unconformity (non'-dep-o-si'-tion-al) A term used by Tomkeieff (1962, p.412) for a surface of nondeposition in marine sediments. It is equivalent to *paraconformity.*

nondetrital (non-de-tri'-tal) Pertaining to sedimentary material derived from solution by physicochemical or biochemical means, including authigenic minerals formed in the sediment after deposition.

nonequilibrium system A geomorphic system that does not tend toward a stable steady state. It is distinct from a *disequilibrium system,* which tends toward equilibrium but has not attained a steady state (Renwick, 1992).

noneroding velocity (non-e-rod'-ing) The velocity of water in a channel that will maintain silt in movement but will not scour the bed. Cf: *transporting erosive velocity.*

nonfaradaic path (non'-far-a-da'-ic) One of the two available paths for transfer of energy across an electrolyte-metal interface. Energy is carried by capacitive transfer, i.e. charging and discharging of the double-layer capacitance. Cf: *faradaic path.*

nonferrous (non-fer'-rous) Said of metals other than iron, usually the base metals.

nonflowing artesian well (non-flow'-ing) An *artesian well* that taps an artesian aquifer in which the head is not sufficient to raise the water in the well above the land surface. Cf: *flowing artesian well; nonflowing well.*

nonflowing well A well that yields water at the land surface only by means of a pump or other lifting device. It may be either a *water-table well* or a *nonflowing artesian well.*

nonfoliate (non-fo'-li-ate) Pertaining to a metamorphic rock lacking foliation on the scale of hand specimens.

nongeniculate coniform element A coniform *conodont element* in which there is a smooth transition from posterior cusp margin to upper basal edge; in lateral view, posterior margin and upper basal edge form a straight or smoothly arcuate line (TIP, 1981, pt. W, supp.2).

nongraded (non-grad'-ed) (a) A rarely used geologic term pertaining to an unconsolidated sediment or to a cemented detrital rock consisting notably of particles of more than one size or of particles lying within the limits of more than one grade; e.g. a loam or a till. Syn: *poorly sorted.* (b) An engineering term pertaining to a soil or an unconsolidated sediment consisting of particles of essentially the same size. See also: *poorly graded.* Ant: *graded [part size].*

nongraphic data Digital representations of the characteristics, qualities, or relationships of map features and geographic locations usually stored in alphanumeric formats. Attributes are nongraphic data which are linked to the features they describe.

nonideal solution (non-i-deal') A solution in which the atomic or molecular interactions between components are not the same as those within each component, leading to chemical activities (or fugacities) that are different from mole fractions. Cf: *ideal solution.*

nonlinear interaction *wave-wave interaction.*

nonlinear sweep A vibroseis sweep where more time is spent sweeping high frequencies than low frequencies, to compensate for the frequency-dependent absorption in the signal's travel through the earth.

nonlinear system A geomorphic system where the output or response is not proportional to the input or stimulus across the entire range of the inputs. Due to thresholds, storage effects, *hysteresis,* and other phenomena, most geomorphic systems are nonlinear. While nonlinear systems are not necessarily complex, and complex systems are not necessarily nonlinear, the study of nonlinearity in the geosciences has often focused on complex nonlinear dynamics (Phillips, 2003).

nonmare basalt (non-ma'-re) *KREEP.*

nonmechanical *nonclastic.*

nonmetal (non-met'-al) (a) A naturally occurring substance that does not have metallic properties, such as high luster, conductivity, opaqueness, and ductility. (b) In economic geology, any rock or mineral mined for its nonmetallic value, such as stone, sulfur, or salt. Syn: *nonmetallic; industrial mineral.*

nonmetallic (non-me-tal'-lic) adj. (a) Of or pertaining to a nonmetal. (b) Said in general of mineral lusters other than *metallic* luster. Cf: *submetallic luster.* n. A *nonmetal* or *industrial mineral;* usually used in the plural.

nonmimetic dolomitization A process by which dolomite crystals grow and do not preserve the shapes of the original sediment particles (Friedman et al, 1992, p.26).

nonmonophyletic (non-mon'-o-phy-let'-ic) In *cladism,* referring to a taxon which is either paraphyletic or polyphyletic.

nonmotile (non-mo'-tile) Not *motile;* e.g. the nonflagellate stage or "nonmotile phase" in the life cycle of a coccolithophorid.

non-Newtonian flow (non-New-to'-ni-an) Flow in which the relationship of the shear stress to the rate of shear strain is nonlinear, i.e. flow of a substance in which viscosity is not constant. Cf: *Newtonian flow.*

non-occupation surface A level in an archaeological excavation or site that was not inhabited by humans. Cf: *occupation surface.*

nonparametric statistics (non'-par-a-met'-ric) Statistics that do not assume specific distributions. Cf: *parametric statistics.*

nonpenetrative (non-pen'-e-trat'-ive) Said of a texture of deformation that affects only part of a rock, where some subdivisions of a rock lack the texture, e.g., *kink bands.* Cf: *spaced cleavage.*

nonplunging fold (non-plung'-ing) A fold whose hinge line is horizontal. Cf: *plunging fold.* Syn: *horizontal fold; level fold.*

nonpoint-source pollution Pollution from sources that cannot be defined as discrete points, such as areas of crop production, timber, surface mining, disposal of refuse, and construction. Cf: *point-source pollution.*

nonrotational fault A fault whose slip does not result in rotation of the hanging wall block relative to the footwall block around an axis parallel to the strike of the fault. Imprecise term referring to displacement along a planar fault.

nonrotational strain (non-ro-ta'-tion-al) *irrotational strain.*

nonsaline alkali soil (non-sa'-line) *nonsaline sodic soil.*

nonsaline sodic soil A soil with a content of exchangeable sodium greater than 15%, which gives a conductance of less than 4 mmhos/cm. The pH values are usually between 8.5 and 10.0. Cf: *saline-sodic soil.* Syn: *nonsaline alkali soil.*

nonsequence (non-se'-quence) A term used in Great Britain for a *diastem,* or for a break or gap in the continuity of the geologic record, representing a time during which no permanent deposition took place. A nonsequence usually can be detected only by a study of successive fossil contents. Cf: *paraconformity.*

nonsilting velocity (non-silt'-ing) The velocity of water in a channel that maintains silt in movement. Syn: *transportation velocity.*

nonsorted (non-sort'-ed) Said of a nongenetic group of *patterned ground* features that do not have a border of stones surrounding or alternating with finer material, as in sorted patterned ground; often there is a border of vegetation between areas of relatively bare ground or finer material.

nonsorted circle A form of patterned ground "whose mesh is dominantly circular and has a nonsorted appearance due to the absence of a border of stones" (Washburn, 1956, p.829); developed singly or in groups. Vegetation characteristically outlines the pattern by forming a bordering ridge. When well-developed, it has a distinctly domed central area. Diameter: commonly 0.5 to 3 m. Examples: *mud circle; frost scar; peat ring; tussock ring.*

nonsorted crack A rare form of patterned ground representing the boulder-free variant of a *sorted crack.*

nonsorted net A form of patterned ground "whose mesh is intermediate between that of a nonsorted circle and a nonsorted polygon and has a nonsorted appearance due to the absence of a border of stones" (Washburn, 1956, p.830).

nonsorted polygon A form of patterned ground "whose mesh is dominantly polygonal and has a nonsorted appearance due to the absence of a border of stones" (Washburn, 1956, p.831-832); never developed singly. Its borders commonly, but not invariably, are marked by wedge-shaped fissures narrowing downward; it typically results from infilling of these fissures. Diameter: a few centimeters to tens of meters. See also: *fissure polygon; mud polygon; ice-wedge polygon; vegetation polygon; sand-wedge polygon; frost-crack polygon; desiccation polygon.*

nonsorted step A form of patterned ground "with a steplike form and a nonsorted appearance due to a downslope border of vegetation embanking an area of relatively bare ground upslope" (Washburn, 1956, p.834); formed in groups. See also: *sorted step; turf-banked terrace.*

nonsorted stripe One of the alternating bands comprising a form of patterned ground characterized by "a striped pattern and a nonsorted appearance due to parallel lines of vegetation-covered ground and intervening strips of relatively bare ground oriented down the steepest available slope" (Washburn, 1956, p.837). Vegetation characteristically outlines the pattern, as the absence of lines of stones is an essential feature; the bare ground consists of finer-grained material or a nonsorted mixture of fines and stones. See also: *solifluction stripe; vegetation stripe; stripe hummock.*

nonstationary Said of a variable whose statistical characteristics, especially the *mean,* change with time or location. In *geostatistics,* the *drift* component of a spatial variable.

nonsteady flow (non-stead'-y) *unsteady flow.*

nonstrophic (non-stroph'-ic) Said of a brachiopod shell in which the posterior margin is not parallel with the hinge axis. Cf: *strophic.*

nonsulfide zinc deposit Any or all types of zinc deposits composed of nonsulfide zinc minerals, principally carbonates, silicates, but less commonly, oxides. Cf: *direct-replacement nonsulfide zinc deposit; wall-rock-replacement nonsulfide zinc deposit; residual and karst-fill nonsulfide zinc deposit.*

nonsystematic fractures Typically irregular, nonplanar fractures that are not part of a set and cannot be correlated from one outcrop to the next. However, they may play a significant role in enhancing fracture connectivity (Hodgson, 1961). Syn: *nonsystematic joints.*

nonsystematic joints (non'-sys-tem-at'-ic) Joints that are not part of a set. They do not cross other joints, they often terminate at bedding surfaces, and their surfaces may be strongly curved. Cf: *systematic joints.*

nontabular (non-tab'-u-lar) Said of surface features of a dinoflagellate cyst or theca that are neither sutural nor *intratabular* and have a random arrangement. Cf: *penitabular.*

nontectonite (non-tec'-ton-ite) Any rock whose fabric shows no influence of movement of adjacent grains, e.g. a rock formed by mechanical settling. Some rocks are transitional between a tectonite and a nontectonite (Turner and Weiss, 1963, p.39).

nonthermal spring (non-ther'-mal) A spring in which the temperature of the water is not appreciably above the mean atmospheric temperature in the vicinity. A spring whose temperature approximates the mean annual temperature, or a *cold spring,* is considered a nonthermal spring (Meinzer, 1923, p.55).

nontree pollen (non-tree') A syn. of *nonarboreal pollen.* Abbrev: NTP.

nontronite (non'-tron-ite) A member of the dioctahedral *montmorillonite* group of minerals in which the layer charge is the result of substitution primarily in the tetrahedral sheet. An ideal formula is: $(Ca_{0.5},Na)_{0.3}Fe^{3+}{}_2(Si,Al)_4O_{10}(OH)_2{\cdot}nH_2O$. One of the dominant minerals along mid-ocean ridges. Cf: *smectite.*

nonuniform flow (non-u'-ni-form) In hydraulics, a type of steady flow in an open channel in which velocity varies at different points along the channel.

nonuniformist (non'-u-ni-form'-ist) One who believes that past changes in the Earth have proceeded from cataclysms or violent processes not now operating; a believer in the doctrine of *catastrophism.* Cf: *catastrophist.* Also called: nonuniformitarian.

nonvascular plant (non-vas'-cu-lar) A plant without a vascular system or well differentiated roots, stems, and leaves, e.g. a thallophyte or bryophyte.

nonwetting sand Sand that resists infiltration of water, consisting of angular particles of varying sizes, and occurring as a tightly packed mass, generally lenticular.

nook An obsolete syn. of *promontory.*

norbergite (nor'-berg-ite) A tan, yellow yellow-orange, orange-brown, or purplish pink orthorhombic mineral related to humite: $Mg_3(SiO_4)(F,OH)_2$.

nordenskiöldine (nord'-en-skiöld'-ine) A colorless or yellow rhombohedral mineral: $CaSn^{4+}(BO_3)_2$. It is the Ca analogue of tusionite.

nordfieldite (nord'-field-ite) *esmeraldite.*

nordite (nord'-ite) A group name for minerals with the general formula $Na_3(Sr,Ca)REEASi_6O_{17}$, where *REE* is a rare-earth element and *A* = Zn, Mg, Fe^{2+} or Mn^{2+}.

nordite-(Ce) A brown, black, or white orthorhombic mineral of the *nordite* group: $Na_3(Sr,Ca)(Ce,La)(Zn,Mg)Si_6O_{17}$.

nordite-(La) (nord'-ite) A pale brown orthorhombic mineral of the *nordite* group: $Na_3(Sr,Ca)(La,Ce)(Zn,Mg)Si_6O_{17}$.

nordmarkite [mineral] (nord'-mark-ite) A variety of staurolite containing manganese.

nordmarkite [rock] A quartz-bearing alkali *syenite* that has microperthite as its main constituent with smaller amounts of oligoclase, quartz, and biotite, and is characterized by granitic or trachytoid texture. The name, given by Brögger in 1890, is for Nordmark, Oslo district, Norway. Obsolete.

nordsjoite (nord'-sjo-ite) A *nepheline syenite* that contains andradite and calcite. Named by Johannsen in 1938 for Nordsjö, the lake bordering the Fen complex, Norway. Obsolete.

nordstrandite (nord'-strand-ite) A colorless, white, cream, pink, beige, or pale green triclinic mineral: $Al(OH)_3$. It is trimorphous with gibbsite and bayerite.

nordströmite (nord'-ström-ite) A metallic gray monoclinic mineral: $Pb_3CuBi_7(S,Se)_{14}$.

nore *ness.*

Norian (Nor'-i-an) Stage of the Standard Global Chronostratigraphic Scale: Upper Triassic (above Carnian, below Rhaetian).

norilskite (nor-ilsk'-ite) An alloy consisting of platinum with a high content of iron and nickel.

Noril'sk-type A class of magmatic Ni-Cu sulfide deposits enriched in PGE's associated with flood basalts in cratonic rift environments. Mineralization may be both disseminated within picritic basalts and massive at the basal contact and in fractures in the country rock. Named for deposits in the Noril'sk-Talnakh region, Siberia.

Noril'sk-type copper-nickel-PGE deposit A *mineral-deposit type* characterized by disseminated to massive nickel-copper and PGE-bearing sulfides in mafic-ultramafic intrusive rocks associated with continental flood basalt. Named for the Noril'sk mining camp in Siberia, Russia.

norite (nor'-ite) (a) In the *IUGS classification,* a plutonic rock satisfying the definition of *gabbro,* in which pl/(pl+px+ol) is between 10 and 90 and opx/(opx+cpx) is greater than 95. (b) A coarse-grained plutonic rock containing basic plagioclase (labradorite) as the chief constituent and differing from gabbro by the presence of orthopyroxene as the dominant mafic mineral. The name was first used by Esmark in 1823.

norm The theoretical mineral composition of a rock expressed in terms of *normative mineral* molecules that have been determined by specific chemical analyses for the purpose of classification and comparison; "the theoretical mineral composition that might be expected had all chemical components crystallized under equilibrium conditions according to certain rules" (Stokes and Varnes, 1955, p.94). Adj: *normative.* Cf: *mode [petrology].* See also: *CIPW classification.*

normal [meteorol] n. The average value of a meteorological element (such as pressure, temperature, rainfall, or duration of sunshine) over any fixed period of years that is recognized as standard for a given country or element. The period 1901-1930 was set by the International Meteorological Organization at a Warsaw conference in 1935 as the international standard period for climatological normals. This interval is updated every 10 years to include more recent data, so that commencing with 2001, temperature and precipitation normals are computed from the period 1971-2000. adj. Approximating the statistical norm or average, such as the 'normal rainfall' of a region for a definite time.

normal [struc geol] Said of an anticlinorium in which the axial surfaces of the subsidiary folds converge downwards; said of a synclinorium in which the axial surfaces of the subsidiary folds converge upwards. Cf: *abnormal.*

normal color film Film in which the colors are essentially true representations of the colors of the terrain.

normal consolidation Consolidation of sedimentary material in equilibrium with overburden pressure. Cf: *overconsolidation.*

normal curve A symmetrical, bell-shaped graphical representation of a *normal distribution.*

normal cycle A *cycle of erosion* in which the complete reduction or lowering of a region to base level is effected largely by running water, specif. the action of rivers as the dominant erosion agent. Cf: *arid cycle.* Syn: *fluvial cycle of erosion.*

normal depth (a) Water depth in an open channel that corresponds to uniform velocity for a given flow. It is the hypothetical depth in a steady, nonuniform flow; the depth for which the surface and bed are parallel. Syn: *neutral depth.* (b) Water depth measured perpendicular to the bed. (ASCE, 1962).

normal dip *regional dip.*

normal displacement Component of displacement in which the hanging wall of a fault moves down-dip relative to the footwall. Syn: *normal separation.* See also: *dip slip.* Cf: *total displacement.*

normal distribution A *frequency distribution* defined by two parameters, the mean and the variance, and which forms the cornerstone of much of statistical inference. It has the form of a continuous, infinite, bell-shaped curve that is symmetrical about its arithmetic mean, mode, and median, which coincide. Sometimes referred to as the normal distribution of error. Syn: *Gaussian distribution; bell-shaped distribution*.

normal drag An outdated term for folding adjacent to a fault where the fold appears to have formed in response to resistance to sliding along the fault. Most geologists now believe that such folding actually formed prior to fault surface formation and represents strain that accumulates beyond the propagating *tip line*.. Thus, the term "drag" is misleading.

normal erosion (a) The wearing-away of topographic features that is effected by prevailing agencies and that is mainly responsible for the present modification of the habitable land surface; specif. subaerial erosion by running water, rain, and certain physical and organic weathering processes. The term, used originally for stream erosion in a temperate climate, is open to criticism because erosion as found in temperate areas may in fact be "abnormal" (esp. in regard to past geologic conditions) or because one mode of erosion is just as "normal" as another. Cf: *special erosion*. (b) Erosion of rocks and soil under natural environmental conditions, undisturbed by human activity. It includes erosion by running water, rain, wind, ice, waves, gravity, and other geologic agents. Cf: *accelerated erosion*. See also: *geologic norm*. Syn: *geologic erosion*.

normal fault A fault in which the hanging wall has moved downward relative to the footwall. The angle of the fault is usually 45-90°, and in most cases close to 60°. Cf: *reverse fault; thrust fault*.

normal geopotential number *spheropotential number*.

normal gradient *normal gravity*.

normal grading A systematic particle-size gradation upward through a sedimentary layer in which the coarsest particles are at the base and the progressively finer ones, toward the top (Friedman et al., 1992, p.26). Cf: *inverse grading*.

normal gravity The gravity caused by the attraction of the equipotential *mean earth ellipsoid* combined with the centrifugal force due to the Earth's rotation. Syn: *normal gradient*.

normal hydrostatic pressure In porous strata or in a well, pressure at a given point that is approximately equal to the weight of a column of water extending from that point to the surface.

normal incidence A raypath impinging on an interface at right angles. In isotropic media, equivalent to a wavefront striking an interface broadside, i.e., so that the angle between the wavefront and the interface (angle of incidence) is zero.

normalized histogram A histogram whose distribution has been adjusted to have as close to a normal (bell-shaped) distribution as possible. The data value that occurs most frequently will be near the center of the histogram.

normal log The *resistivity log* curve derived from a simple 2-electrode array, with spacing that ranges from 2 inches in the micronormal (see *Microlog*) to 16 and 64 inches for the short and long normal curves (see *electric log*) and to as much as 1,000 feet in the *ultra-long-spaced electric log*. See also: *lateral log; focused-current log; induction log*.

normal magnetic field A smooth, long wavelength component of the *Earth's magnetic field*, that represents its main (core) component and does not contain anomalies of exploration interest. Compensated by applying the IGRF correction.

normal mode propagation *channel wave*.

normal modes Patterns of standing-wave free vibrations. Syn: *eigenmodes*.

normal moisture capacity *field capacity*.

normal moveout The increase in arrival time of a seismic-reflection event resulting from an increase in the distance from source to detector. Seismic data must be corrected for normal moveout. Abbrev: NMO. See also: *moveout*.

normal-moveout correction The time correction applied to reflection times because of the variation of reflection arrival time with source-to-geophone distance. See: *offset [seis]*. Abbrev: NMO correction.

normal-moveout velocity (nor'-mal-move'-out) *stacking velocity*.

normal polarity (a) A natural *remanent magnetization* closely parallel to the present ambient geomagnetic field direction. (b) A configuration of the Earth's magnetic field with the magnetic negative pole, where field lines enter the Earth, located near the geographic south pole. Cf: *reversed polarity*.

normal pore canal A tubule or *pore canal* piercing an ostracode carapace at right angles, and believed to serve as a receptor of sensory setae. Cf: *radial pore canal*.

normal pressure *normal hydrostatic pressure*.

normal-pressure surface (nor'-mal-pres'-sure) A potentiometric surface that coincides with the upper surface of the zone of saturation (Meinzer, 1923, p.39). It is usually the same as the water table. Cf: *subnormal-pressure surface; artesian-pressure surface*.

normal projection (a) A projection in which a three-dimensional object is projected onto two mutually perpendicular planes. (b) A projection whose surface axes coincide with those of the sphere.

normal ripple An aqueous *current ripple* consisting of a "simple asymmetrical ridge" that may have "various ground plans" (Shrock, 1948, p.101).

normal ripple mark *normal ripple*.

normal sandstone A term used by Shrock (1948a) for a sandstone composed almost exclusively of quartz, with subordinate amounts of other minerals.

normal section A line between two points on the surface of an ellipsoid, formed by the intersection of the ellipsoid and plane containing the normal at one point and the other point.

normal separation *Separation* measured parallel to the dip of a fault across which the hanging wall is apparently lowered with respect to the footwall (Biddle and Christie-Blick, 1985). Syn: *normal displacement*. See also: *dip slip*. Cf: *reverse separation*.

normal shift In a fault, the horizontal component of the shift, measured perpendicular to the strike of the fault. Obsolete and little used. Cf: *offset [struc geol]*.

normal slip fault Obsolete term for a *normal fault*. Cf: *reverse slip fault*.

normal strain Change of length per unit length in a given direction.

normal stress That component of *stress* which is perpendicular to a given plane. It may be either *tensile stress* or *compressive stress*. Symbol: σ_n. Cf: *shear stress*.

normal twin A twinned crystal, the twin axis of which is perpendicular to the *composition surface*. Cf: *parallel twin; complex twin*.

normal water A standardized seawater; its chlorinity is between 19.30 and 19.50 parts per thousand and it has been analyzed to within 0.001 part per thousand. Syn: *Copenhagen water; standard seawater*.

normal zoning Zoning in a crystal of plagioclase, in which the zones become progressively more sodic outward. Cf: *reversed zoning*.

normandite An orange-brown or yellow monoclinic mineral of the *låvenite* group: $NaCa(Mn,Fe)(Ti,Nb,Zr)(Si_2O_7)OF$.

Normapolles (Nor-ma-pol'-les) A group of Cretaceous and lower Paleogene porate pollen with a complex pore apparatus (e.g. an *oculus*) and sometimes other peculiarities such as double Y-marks. Cf: *Postnormapolles*.

normative (nor'-ma-tive) The adj. of *norm*.

normative mineral A mineral whose presence in a rock is theoretically possible on the basis of certain chemical analyses. A normative mineral may or may not be actually present in the rock. See also: *norm*.

norm system *CIPW classification*.

norrishite (nor'-rish-ite) A black monoclinic micaceous mineral: $KLiMn^{3+}_2Si_4O_{12}$ It is a di-trioctahedral mica.

norsethite (nor'-seth-ite) A colorless or white rhombohedral mineral: $BaMg(CO_3)_2$.

North American datum First known as the United States standard datum, in 1930 it was adopted by Canada and Mexico and renamed North American datum of 1927 (NAD27). It was defined by the data of the station at Meades Ranch in Kansas: latitude 39°13'28.686"N., longitude 98°32'30.506"W., azimuth to Waldo 75°28'14.52"; and by the *Clarke ellipsoid of 1866*. When the entire triangulation network of the United States was readjusted between 1925 and 1930, the azimuth to Waldo was corrected to 75°28'09.64".

North American Datum of 1983 The horizontal control datum for the United States, Canada, Mexico and Central America, based on a geocentric origin and the Geodetic Reference System 1980. This datum, designated as NAD 83, is the new geodetic reference system.

North American tektite A tektite from one of the four main tektite fields, formed about 36 Ma ago.

northing (north'-ing) A *latitude difference* measured toward the north from the last preceding point of reckoning; e.g. a linear distance northward from the east-west line that passes through the origin of a grid.

north pole [astron] The north *celestial pole*, representing the zenith of the heavens as viewed from the north geographic pole.

north pole [geog] The *geographic pole* in the northern hemisphere of the Earth at lat. 90°N, representing the northernmost point of the Earth or the northern extremity of its axis of rotation. Also spelled: North Pole.

northupite (north'-up-ite) (a) A colorless, white, yellow, or gray cubic mineral: $Na_3Mg(CO_3)_2Cl$. (b) A group name for minerals with the analogous composition, but with Mg replaced by Fe^{2+} or Mn^{2+}, and Cl replaced by $0.5SO_4$.

Norwegian cyclone model The original description of the structure and life cycle of an extratropical low-pressure system based mostly on surface observations, first proposed during World War I by researchers at the Norwegian School of Meteorology at Bergen.

Norwich Crag *Baventian; Ludhamian*.

nose [geomorph] (a) A projecting and generally overhanging buttress of rock. (b) The projecting end of a hill, spur, ridge, or mountain. (c) The central forward part of a parabolic dune. (d) *ness*.

nose [sed] The forward part of a turbidity current, which is more dense than the tail and carries coarser material.

nose [struc geol] A short, plunging anticline without closure. Syn: *structural nose; anticlinal nose*.

nosean (no'-se-an) A cubic feldspathoid mineral of the *sodalite* group: $Na_8Al_6Si_6O_{24}(SO_4)\cdot H_2O$. It is gray, blue, or brown, and is related to hauyne. Syn: *noselite*.

noseanite (no'-se-an-ite') In the *IUGS classification*, a foidite in which nosean is the dominant feldspathoid.

noselite (no'-se-lite) *nosean*.

notch [coast] A deep, narrow cut or hollow along the base of a sea cliff near the high-water mark, formed by undercutting due to wave erosion and/or chemical solution, and above which the cliff overhangs. See also: *nip [coast].*

notch [geomorph] (a) A narrow passageway or short defile between mountains or through a ridge, hill, or mountain; a deep, close *pass.* Also, the narrowest part of such a passage. Cf: *gap [geomorph]; col.* (b) A breached opening in the rim of a volcanic crater.

notite (no'-tite) A variety of porphyry with quartz, feldspar, and mica phenocrysts, named by Jurino in 1806. Obsolete.

notochord (no'-to-chord) A flexible or elastic rod that provides a supporting and stiffening structure in a chordate animal's body; in the subphylum Vertebrata it is replaced or enclosed by a vertebral column composed of bone or cartilage.

notothyrial chamber (no-to-thy'-ri-al) The cavity in the *umbo* of the *brachial valve* of a brachiopod, bounded laterally by brachiophore bases (or homologues) or by posterior and lateral shell walls if brachiophore bases are absent. It corresponds to the *delthyrial chamber* of the pedicle valve.

notothyrial platform Umbonal thickening of the floor of the *brachial valve* of a brachiopod between brachiophore bases (or homologues).

notothyrium (no-to-thy'-ri-um) The median subtriangular opening in the brachial valve of a brachiopod, bisecting the dorsal *cardinal area* or pseudointerarea. Pl: notothyria. Cf: *delthyrium.*

noumeite (nou'-me-ite) A syn. of *garnierite,* esp. of a dark-green unctuous variety.

nourishment [beach] (nour'-ish-ment) The replenishment of a beach, either naturally (as by littoral transport) or artificially (as by the deposition of dredged materials).

nourishment [glaciol] *accumulation.*

novácekite (no'-va-ce-kite) A yellow or greenish-yellow tetragonal secondary mineral of the *autunite* group: $Mg(UO_2)_2(AsO_4)_2 \cdot 9\text{-}12H_2O$.

novaculite (no-vac'-u-lite) (a) A dense hard even-textured light-colored cryptocrystalline siliceous sedimentary rock, similar to chert but characterized by dominance of microcrystalline quartz over chalcedony. It was formerly believed to be the result of primary deposition of silica, but in the type occurrence (Lower Paleozoic of the Ouachita Mountains, Arkansas and Oklahoma) it appears to be a thermally metamorphosed *bedded chert,* distinguished by characteristic polygonal triple-point texture (Keller et al., 1977). The origin of novaculite has also been ascribed to crystallization of opaline skeletal material during diagenesis. The rock is used as a *whetstone.* See also: *Arkansas stone; Washita stone.* Syn: *razor stone; Turkey stone; galactite.* (b) A term used in southern Illinois for an extensive *bedded chert* (J.E. Lamar, in Tarr, 1938, p.19). (c) A general name formerly used in England for certain argillaceous stones that served as whetstones.

novaculitic chert (no-vac'-u-lit'-ic) A generally gray chert that breaks into slightly rough, splintery fragments; it is less vitreous and somewhat coarser-grained than *chalcedonic chert.*

novákite (no'-va-kite) A metallic gray monoclinic mineral: $(Cu,Ag)_{21}As_{10}$.

novgorodovaite A vitreous colorless monoclinic oxalate mineral: $Ca_2C_2O_4C_{12} \cdot 2H_2O$.

nowackiite (no-wack'-i-ite) (a) A metallic black rhombohedral mineral: $Cu_6Zn_3As_4S_{12}$. It is isotypic with aktashite and gruzdevite. (b) A group name for rhombohedral minerals with the general formula $Cu_6A_3X_4S_{12}$, where A = Zn or Hg, and X = As or Sb.

NPI *net profits interest.*

NPP *net primary production.*

NQR *nuclear quadrupole resonance.*

NRA *nuclear reaction analysis.*

NRM *natural remanent magnetization.*

NSR *net smelter royalty.*

NSSDA *National Standard for Spatial Data Accuracy.*

nsutite (n'-sut-ite) A metallic gray or brown hexagonal mineral: $Mn^{4+}_{1-x}Mn^{2+}_xO_{2-2x}(OH)_{2x}$(x is small).

n-tn log *thermal-neutron log.*

NTP *nontree pollen.*

nubbin (nub'-bin) (a) An isolated bedrock knob or small hill forming the last remnant of the crest of a mountain that has succumbed to desert erosion. The term was introduced by Lawson (1915) and extended by Cotton (1942) to include small remnants of spurs and ridges. (b) A residual boulder, commonly granitic, occurring on a *desert dome* or broad *pediment* (Stone, 1967, p.235).

nuclear age determination (nu'-cle-ar) *isotopic dating.*

nuclear clock *radioactive clock.*

nuclear fission *fission [chem].*

nuclear fusion *fusion [chem].*

nuclear geochemistry *isotope geology.*

nuclear geology *isotope geology.*

nuclear isomerism Delayed de-excitation of the nucleus of radiogenic product atoms by emission of gamma rays.

nuclear log (a) *neutron log.* (b) *radioactivity log.*

nuclear-magnetic resonance The selective absorption of electromagnetic radiation at the appropriate resonant frequency by nuclei undergoing precession in a strong magnetic field. Abbrev: NMR.

nuclear-magnetic-resonance spectrometer An instrument for scanning and measuring the *nuclear magnetic resonance* spectrum of nuclei.

nuclear-magnetism log (nu'-cle-ar-mag'-net-ism) An open-hole *well log* that measures the free protons present in a few inches of rock near the well bore. The log gives a measure of *free-fluid index* and can be used to estimate permeability in sandstones and effective porosity in carbonates. Abbrev: NML. Syn: *free-fluid index* log.

nuclear quadrupole resonance Resonance of an atomic nucleus whose electric charge distribution deviates from a spherical distribution. Abbrev: NQR.

nuclear radiation Radiation from an atomic nucleus, commonly alpha or beta particles and gamma rays.

nuclear reaction A change in identity of an atomic nucleus, brought about by interaction with an elementary particle or with another nucleus.

nuclear reaction analysis A group of analytical techniques based on the energies of particle recoil, including Rutherford backscattering and elastic recoil detection analysis. Abbrev: NRA.

nuclear resonance *Resonance* occurring when a nucleus is irradiated with gamma rays of exactly the same frequency as those that the nucleus naturally tends to radiate.

nuclear-resonance magnetometer (nu'-cle-ar-res'-o-nance) A type of magnetometer that measures total magnetic-field intensity by means of the precession of magnetic nuclei, precession frequency being proportional to field intensity. In practice, only the *proton-precession magnetometer* has been used.

nucleation (nu-cle-a'-tion) The beginning of crystal growth at one or more points in a system (usually a liquid or vapor).

nucleic acid (nu-cle'-ic) A complex organic substance that is the genetic material in all known organisms. Cf: *DNA.*

nucleoconch (nu'-cle-o-conch') *embryonic apparatus.*

nucleogenesis (nu'-cle-o-gen'-e-sis) The origin of the chemical elements of the universe.

nucleosynthesis (nu'-cle-o-syn'-the-sis) The generation of elements from hydrogen nuclei or protons by nuclear processes under the high-temperature, high-pressure conditions common in the life of a star.

nucleus (nu'-cle-us) The earliest-formed part of the shell or operculum of a gastropod. The term should not be used synonymously with *protoconch.*

nuclide (nu'-clide) A species of atom characterized by the number of neutrons and protons in its nucleus. See also: *radionuclide; cosmogenic nuclide.*

nucule (nu'-cule) (a) In fossil seeds, the central cavity formerly occupied by nucellus and gametophyte. (b) The female reproductive structure of a charophyte. It includes the oogonium and the outer protective cells.

nuculoid (nu'-cu-loid) Any bivalve mollusk belonging to the order Nuculoida, characterized by a taxodont, equivalve, isomyarian shell with closed margins.

nudibranch (nu'-di-branch) Any *opisthobranch* belonging to the order (or suborder) Nudibranchia, characterized chiefly by the absence of a shell in the adult stage and by the absence of gills, or their replacement by secondary gills.

nuée ardente (nu-ee ar-dente') A swiftly flowing, turbulent gaseous cloud, sometimes incandescent, erupted from a volcano and containing ash and other pyroclastic materials in its lower part; a *density current* of pyroclastic flow. Etymol: French, "glowing cloud". Syn: *Peleean cloud; glowing cloud; fiery cloud.*

nuevite (nu-e'-vite) *samarskite.*

Nuevoleonian (Nu-e'-vo-le-on'-i-an) North American (Gulf Coast) stage: Lower Cretaceous (above Durangoan, below Trinitian).

nuffieldite (nuf'-field-ite) A metallic gray orthorhombic mineral: $Pb2Cu_{1.4}(Pb,Bi,Sb)_2S_7$.

nugget (nug'-get) A large lump of placer gold or other metal. Cf: *heavy gold.*

nugget effect In *geostatistics,* said of a *semivariogram* whose value is greater than zero at the origin. It indicates variation at a spatial scale smaller than the distance between observations, or variation among replicate measurements.

Nukumaruan (Nu-ku-ma-ru'-an) South Pacific stage: upper Pliocene (above Mangapanian, below Pleistocene Castlecliffian).

nukundamite (nu-kun'-da-mite) A metallic reddish-orange or pale gray trigonal mineral related to idaite: $Cu_{3.4}Fe_{0.6}S_4$.

nullaginite (nul'-lag-in-ite) A silky-green monoclinic mineral related to rosasite and glaukosphaerite: $Ni_2(CO_3)(OH)_2$.

nullah (nul'-lah) (a) A term used in the desert regions of India and Pakistan for a sandy river bed or channel, or a small ravine or gully, that is normally dry except after a heavy rain. (b) The small, intermittent, generally torrential stream that flows through a nullah. Etymol: Hindi "nala". See also: *wadi; arroyo.* Also spelled: nulla; nallah; nalla.

null hypothesis A statistical statement of no real difference between the true value and the hypothesized value; a null hypothesis and an *alternative hypothesis* are required in any statistical analysis.

nullipore (nul'-li-pore) A coralline alga, formerly thought to be an animal.

null point In *inversion* tectonics, the point along a partially inverted fault at which normal displacement changes to reverse displacement.

number of streams (num'-ber) Total number of stream segments of a specified order or orders in a given drainage basin. The symbol N_u refers to the total number of stream segments of a given order *u* within a specified drainage basin. See also: *law of stream numbers.*

numerical age The *geologic age* of a fossil organism, rock, or geologic feature or event given in units of time, usually years. Commonly used as a syn. of *isotopic age* or *radiometric age,* but may also refer to ages obtained from tree rings, varves, etc. Cf: *actual age; relative age.* Less preferred syn: *absolute age.*

numerical age determination The measurement of *numerical age,* usually but not always on the basis of radioactive isotopes. The ratio of decay products to parent products in the sample is calibrated to a certain number of years as in the *age equation.* Less preferred syn: *absolute age determination.*

numerical aperture (nu-mer'-i-cal) A measurement or indicator of a microscope's resolving power.

numerical chronology *Geochronology* in which the time-order is based on measured numerical age, usually measured in years by radiometric dating, rather than on superposition and/or fossil content as in *relative chronology.* Less preferred syn: *absolute chronology.*

numerical date The date of an event usually expressed in years (*numerical age*) and related to a specific time scale. Less preferred syn: *absolute date.*

numerical model A model whose solution must be approximated by varying the values of controlling parameters and using computers to solve approximate forms of the model's governing equations. See also: *analytical model.*

numerical model [oceanog] A computer program that uses an algebraic form of the equations of motion and *equation of state of seawater* to calculate oceanic variables at a fixed grid of points as a function of time. Grid points may be as close as one kilometer in a model of a small region to 200 km in a climate model. The model is usually forced by heat fluxes and wind stress. It provides no information about the variables between the grid points. Typically used to calculate currents in the ocean, and the interaction of the ocean with the atmosphere.

numerical taxonomy The use of statistics in classifying and analyzing fossils and their paleoecologic implications. Syn: *phenetics.*

numerical time *Geologic time* measured in years; specif. time as determined by radioactive decay of elements. Cf: *mineral time; physical time; relative time.* Less preferred syn: *absolute time.*

nummulite (num'-mu-lite) Any foraminifer belonging to the family Nummulitidae, characterized by a test that is usually planispiral. Range, Upper Cretaceous to present. Adj: nummulitic. Var: nummulitid.

nummuloidal (num-mu-loid'-al) *moniliform.*

nunakol (nun'-a-kol) A *nunatak* rounded by glacial erosion; a rounded "island" of rock in a glacier. Etymol: Eskimo. Syn: *rognon.*

nunatak (nun'-a-tak) An isolated hill, knob, ridge, or peak of bedrock that projects prominently above the surface of a glacier and is completely surrounded by glacier ice. Nunataks are common along the coast of Greenland. Etymol: Eskimo, "lonely peak". The plural is usually given in Swedish: "nunatakker". Cf: *rognon; nunakol.* Also spelled: nunatag.

Nusselt number The ratio of total heat transmitted to that transmitted by conduction alone, and therefore an index of the magnitude of convective heat transfer.

nut An indehiscent, one-celled, and one-seeded hard and bony fruit, even if resulting from a compound ovary (Lawrence, 1951, p.762).

nutation (nu-ta'-tion) The motion of the axis of rotation of the Earth about its mean position with a principal term of about 18.6 years.

nutrient [oceanog] (nu'-tri-ent) (a) A substance dissolved in seawater that is required for growth by phytoplankton which can become limiting to that growth due to depletion. Usually applied to simple dissolved inorganic ions, e.g. nitrate, nitrite, phosphate, ammonia, urea, and silicate but can also be applied to dissolved organic forms of nitrogen and phosphorus or to certain trace elements, e.g. zinc or iron. (b) Any substance required for the maintenance or growth of a marine organism, e.g., silica for diatoms.

nutrient-like profile A vertical distribution of a solute in the oceanic water column in which the concentration of the solute in the upper ocean is markedly less than that in the deeper ocean. So named because of the proclivity of oceanic phytoplankton to deplete the euphotic zone of inorganic nutrients such as phosphate during photosynthesis (carbon fixation). Found for other elements in seawater; e.g. cadmium, barium, and other trace metals. Cf: *nutrient [oceanog]; carbon fixation.*

nyböite (ny'-bö-ite) A gray-green monoclinic member of the *amphibole* group: $NaNa_2[Mg_3Al_2](Si_7Al)O_{22}(OH)_2$.

nyerereite (nye-re'-re-ite) A colorless orthorhombic (pseudo-hexagonal) mineral occurring in carbonate lavas: $Na_2Ca(CO_3)_2$.

nymph (a) One of the narrow, thickened lunate processes or platforms of many bivalve mollusks extending posteriorly from the beak along the dorsal margin and serving for attachment of the *ligament.* Also called: nympha. (b) An immature stage in the life cycle of an acarid arachnid; e.g. protonymph, deutonymph, and tritonymph.

Nyquist frequency (Ny'-quist) The maximum unaliased frequency for a given sample interval. If Δt is the sample interval, the Nyquist frequency is ($\frac{1}{2}\Delta t$). Cf: *alias.*

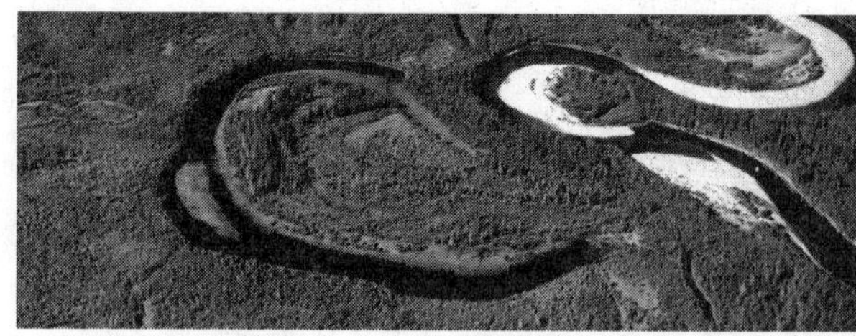

Oa horizon A layer occuring in mor humus consisting of well-decomposed organic matter of unrecognizable origin (sapric material). Syn: *H-layer.*

oasis A fertile, vegetated area in the midst of a desert, where the water table has come close enough to the surface for wells and springs or seepages to exist, thus making it suitable for human habitation.

obcordate (ob-cor'-date) Inversely heart-shaped, as in a leaf that is deeply lobed at the tip. Cf: *cordate.*

obduction (ob-duc'-tion) The overriding or overthrusting of oceanic crust onto the leading edges of continental lithospheric plates; plate accretion (Coleman, 1971).

obertiite A pink monoclinic mineral of the *amphibole* group: $Na_3(Mg_3Fe^{3+}Ti^{4+})Si_8O_{22}O_2$.

object glass (ob'-ject) *objective.*

objective (ob-jec'-tive) The lens (or lenses) that gives an image of an object in the focal plane of a microscope's or telescope's *eyepiece.* Syn: *objective lens; object glass.*

objective lens *objective.*

objective synonym In taxonomy, one of two or more names based on the same type. Cf: *subjective synonym.*

oblate [palyn] In pollen, being distinctly flattened or depressed at the poles, thus having the polar axis significant shorter than the equatorial diameter. Ant: *prolate.*

oblate ellipsoid An ellipsoid that is flattened at its poles.

oblate-prolate index (ob'-late-pro'-late) A numerical index devised by Dobkins and Folk (1970, p.1188) to describe the shape of a particle. It is based mainly on the value (L-I/L-S), which defines whether the intermediate axis I is closer in length to the short axis S or to the long axis L.

oblate strain *flattening strain.*

obligate (ob'-li-gate) adj. Said of an organism that can grow only under certain restricted conditions. Cf: *facultative.*

oblique (ob-lique') n. An aerial photograph taken with the camera axis intentionally inclined (not vertical). It combines the ground view with the pattern obtained from a height. See also: *high oblique; low oblique.*

oblique aerial photograph An aerial photograph taken with the camera access directed between the horizontal and vertical planes.

oblique bedding An archaic syn. of *inclined bedding,* or bedding oblique to the principal surface of deposition; specif. *cross-bedding.*

oblique extinction *inclined extinction.*

oblique illumination method *van der Kolk method.*

oblique joint *diagonal joint.*

oblique lamination (a) *cross-lamination.* (b) *transverse lamination.*

oblique projection A projection that is not centered on a pole or on the equator and that does not use the equator as a center line of orientation, or that has an axis inclined at an oblique angle to the equatorial plane; e.g. "oblique stereographic projection" or "oblique Mercator projection".

oblique ramp A thrust ramp that is neither perpendicular nor parallel to the transport direction of the thrust sheet.

oblique section A slice through a foraminiferal test cut in a direction neither parallel nor normal to the axis of coiling.

oblique slip In a fault, movement or net slip that has *dip slip* and *strike slip* components.

oblique-slip fault A fault on which the net slip has dip slip and strike slip components. Obsolete syn: *diagonal-slip fault.*

oblique subduction The relative displacement of one lithospheric plate beneath another plate, such that in map view the displacement vector is oblique to the plate boundary (Biddle and Christie-Blick, 1985).

obliquity A measure of the angle formed by the midumbonal line and the dorsal margin, or hinge axis of a pelecypod. Cf: *prosocline; acline [paleont]; opisthocline.*

obliquity [astron] (ob-liq'-ui-ty) The tilt of the Earth's axis to the plane of the orbit, which gives rise to the seasons. Currently the obliquity is 23° 27'.

obliquity cycle One of the three "Milankovitch parameters" (also known as "tilt cycle") that contribute to the major ice-age fluctuation of climate and related geologic processes which lead to change in paleotemperatures, sedimentation rates and glacier dynamics. Modeling shows that a minimum obliquity confirms a minimal seasonal contrast but a maximum meridional energy gradient (Berger, 1995). Cf: *Milankovitch theory; ecliptic.*

obovate (ob-o'-vate) Inversely egg-shaped, as in a leaf whose terminal end is broader than its basal end. Cf: *ovate.*

obovoid (ob-o'-void) Inversely ovoid, as in a fruit whose terminal portion is broader than its basal portion. Cf: *ovoid.*

oboyerite (o-boy'-er-ite) A white triclinic mineral: $H_6(Pb,Ca)_6(Te^{4+}O_3)_3(Te^{6+}O_6)_2 \cdot 2H_2O$.

obradovicite (o-brad'-o-vic-ite) A green orthorhombic mineral: $H_4(K,Na)CuFe^{3+}{}_2(AsO_4)(MoO_4)_5 \cdot 12H_2O$.

obruchevite (ob-ru-chev'-ite) *yttropyrochlore-(Y).*

obsequent [geomorph] (ob'-se-quent) Said of a geologic or topographic feature that does not resemble or agree with a consequent feature from which it developed at a later date; esp. said of a tilt-block mountain (or of a rift-block mountain) that was formerly the floor of the original valley (or graben) but that was left standing as a result of differential erosion, or said of a tilt-block valley (or of a rift-block valley) that occupies the site of the former mountain (or horst) after the original topography was modified by differential erosion or of a stream that flows in a direction opposite to that of the initial consequent drainage. Ant: *resequent [geomorph].* Cf: *subsequent [geomorph]; consequent [geomorph].*

obsequent [streams] adj. Said of a stream, valley, or drainage system whose course or direction is opposite to that of the original consequent drainage. The term was proposed by Davis (1895, p.134). Etymol: "opposite to consequent". n. *obsequent stream.*

obsequent fault-line scarp A *fault-line scarp* that faces in the opposite direction from the original fault scarp (i.e. faces the upthrown block) or in which the structurally downthrown block is topographically higher than the upthrown block. Cf: *resequent fault-line scarp.*

obsequent stream A stream that flows in a direction opposite to that of an original consequent stream and that is a tributary to a subsequent stream developed along the strike of weak beds; e.g. a short stream flowing down the scarp slope of a cuesta, or a stream flowing in a direction opposite to that of the dip of the local strata or the tilt of the land surface. See also: *scarp stream; antidip stream; reversed consequent stream.* Syn: *obsequent [streams]; anticonsequent stream; inverted stream.*

observation (a) The act of deliberately sensing an event or thing and noting the circumstances. An observation is distinguished from a sensing by being deliberate and involving the noting (as by making a record) of the circumstances. It differs from a measurement in that a number, the measurement, need not be given to the thing or event observed. That is, a measurement is an observation and the act of assigning a numerical value to the observation. (b) The notes or records of the circumstances accompanying the sensing of a thing or event.

observation well (ob-ser-va'-tion) (a) A well used for the purpose of observing parameters such as fluid levels and pressure changes, e.g. within an oil reservoir as production proceeds. (b) A nonpumping well that is used to observe the elevation of the water table or other potentiometric surface (Fetter, 1994). It may be used temporarily during a pumping test or as a long-term monitoring well. Cf: *aquifer test.*

observed gravity (ob-served') Gravity value obtained by either relative or absolute measurements, usually with gravity corrections applied.

obsidian (ob-sid'-i-an) A black or dark-colored volcanic glass, usually of rhyolite composition, characterized by conchoidal fracture. It is sometimes banded or has microlites. Usage of the term goes back as far as Pliny, who described the rock from Ethiopia. Obsidian has been used for making arrowheads, other sharp implements, jewelry, and art objects. Syn: *Iceland agate.*

obsidian dating *obsidian hydration dating.*

obsidian hydration dating A method of calculating an age in years for an obsidian artifact or Holocene volcanic glass by determining the thickness of the hydration rim which has been produced by water vapor slowly diffusing into a freshly chipped surface and producing a hydrated layer or rind. It is applicable to glasses 200 to 200,000 years old. Syn: *hydration rind dating; obsidian dating.*

obsidianite (ob-sid'-i-an-ite') An obsolete term, proposed by Walcott (1898) for a small rounded glassy object now known as a *tektite.* Most stones originally described as "obsidianite" were later shown to be true obsidian and not tektites.

obstacle mark (ob'-sta-cle) *obstacle scour.*

obstacle scour A term introduced by Dzulynski and Walton (1965, p.40) to refer to a scour mark produced by the interaction of an obstacle on the bed and a current flowing around it. Cf: *scour mark; current crescent; current mark; current scour.* Syn: *obstacle mark.*

obstructed stream (ob-struct'-ed) A stream whose valley has been blocked, as by a landslide, glacial moraine, sand dune, or lava flow; it frequently consists of a series of ponds or small lakes.

obstruction cave (ob-struc'-tion) A *glacier cave* formed as the result of interrupted ice flow due to a bedrock or other subglacial protuberance. Entrance to such caves is generally impossible without tunneling.

obstruction moraine A moraine formed where the movement of ice is obstructed, as by a ridge of bedrock.

obtuse (ob-tuse') Blunt or rounded at one end, as in a leaf or tooth of a leaf. Sometimes, the term is defined more exactly as a basal or terminal leaf or leaf-tooth feature that appears as flat-rounded or has an angle of >90° between the two sides.

obtuse bisectrix The *bisectrix* of the obtuse angle between the optic axes of a biaxial crystal. Cf: *acute bisectrix.*

obverse (ob-verse') Pertaining to the *frontal* side of a bryozoan *zooid* or colony. Cf: *reverse.*

Occam's razor (Oc'-cam's) A traditional term for the principle of *parsimony [philos].*

occidental cat's-eye A syn. of quartz cat's-eye.

occipital condyle (oc-cip'-i-tal) A prominently convex articular surface, capped with cartilage during life, on the bone along the back part of the skull of a tetrapod by which the skull articulates with the first cervical vertebra (atlas). See also: *condyle [paleont].*

occipital furrow (a) The transverse groove on the *cephalon* of a trilobite running from axial furrow to axial furrow and forming the posterior boundary of the *glabella.* (b) A groove in front of the rim along the posterior border of the *prosoma* of a merostome (TIP, 1955, pt. P, p.8).

occipital ring The axial region of the most posterior element of the cephalon of a trilobite, generally delimited frontally by a prominent *occipital furrow.* Syn: *neck ring.*

occluded front A boundary between two air masses that forms when a cold front overtakes a warm front.

occludent margin (oc-clud'-ent) The margin of *scutum* and *tergum* forming the aperture in a cirripede crustacean and occluding it with comparable margins of opposed scutum and tergum.

occlusion (oc-clu'-sion) A syn. of absorption. Adj: *occluded.*

occlusor lamina (oc-clu'-sor) In some cheilostome bryozoans, one of a pair of oppositely placed calcareous partitions arising from the lateral walls to provide the origin for the occlusor muscles.

occlusor muscle One of a pair of oppositely placed muscles that originate on the walls of a cheilostome bryozoan zooid and insert on the *operculum* (Ryland, 1970, p.34).

occult mineral (oc'-cult) A mineral that might be expected to be present in a rock (as from the evidence of chemical analysis) but which is not identifiable, even with the aid of a microscope. Common examples are quartz and orthoclase in the glassy or cryptocrystalline groundmasses of certain lavas.

occupation surface A level in an archaeological excavation or site that was once inhabited by humans. Cf: *non-occupation surface.*

occupy (oc'-cu-py) To set a surveying instrument over a point for the purpose of making observations or measurements.

ocean (o'-cean) The continuous salt-water body that surrounds the continents and fills the Earth's great depressions; also, one of its major geographic divisions. See also: *sea [oceanog].*

ocean-basin floor (o'-cean-ba'-sin) The area of the sea floor between the base of the continental margin, usually the foot of the continental rise, and the mid-ocean ridge (Heezen and Menard, 1963, p.236).

ocean current (a) A permanent predominantly horizontal movement of the surface water of the ocean, constituting part of its general circulation. (b) Broadly, any current in the ocean tidal or nontidal, permanent or seasonal, horizontal or vertical characterized by regularity, either as a continuous stream flowing along a definable path, or less commonly of a cyclic nature.

Ocean Drilling Program Successor to the Deep Sea Drilling Program. The Ocean Drilling Program is a multinationally funded program of exploration of the oceanic crust, with scientific advice and guidance by *JOIDES.* It operated the drilling vessel JOIDES Resolution and is managed by Texas A&M, College Station, TX. Drilling commenced in January 1984 and terminated in September 2003. Results are reported in "Proceedings of the Ocean Drilling Program." Abbrev: *ODP.*

ocean-floor spreading *sea-floor spreading.*

ocean hole A *blue hole* that opens directly into the present marine environment and usually contains marine water with tidal flow (Mylroie et al., 1995). Cf: *inland blue hole.*

oceanic (o-ce-an'-ic) (a) Pertaining to those areas of the ocean that are deeper than the littoral and neritic zones. (b) Pertaining to the open ocean.

oceanic climate *marine climate.*

oceanic crust That type of the Earth's crust which underlies the ocean basins. The oceanic crust is 5-10 km thick; it has a density of 2.9 g/cm^3, and compressional seismic-wave velocities travelling through it at 4-7.2 km/sec. Cf: *continental crust; Layer 1; Layer 2; Layer 3.*

oceanic formation Name proposed for a formal or informal ocean-basin lithostratigraphic unit; the word "oceanic" should precede the formal or informal name (Wilson, J. A., 1971).

oceanic island Island either composed of basalt or of biogenic origin (coral reef, etc.), as distinguished from islands having rocks characteristic of continents. The Hawaiian and most islands in the Pacific Ocean are oceanic, whereas Japan, the Philippines, New Zealand, and most of the larger islands toward the western side of the basin are continental. Beaches of true oceanic islands consist of rock fragments or of coral and shell debris, and have a dark, white, or reddish appearance. They lack white quartz sand, the most characteristic component of continental beaches.

oceanic island basalt Mafic lavas of *intraplate magmatism* on oceanic portions of lithospheric plates, as distinguished from mid-ocean-ridge or island-arc magmatism. Abbrev: OIB.

oceanic ridge *mid-ocean ridge.*

oceanic tholeiite The principal igneous rock of the ocean basins (Engel et al., 1965, p.720). Syn: *abyssal tholeiite.* Cf: *mid-ocean-ridge basalt.*

oceanic trench *trench [marine geol].*

oceanite (o'-cean-ite) A picritic basalt with more than 50 percent olivine in the mode. Named by Lacroix in 1923 for its occurrence at Piton de la Fournaise, Réunion Island, Indian Ocean. Obsolete.

oceanization (o'-cean-i-za'-tion) The supposed conversion of continental crust into oceanic crust. Archaic.

oceanographic cast (o'-cean-o-graph'-ic) *hydrographic cast.*

oceanography (o-cean-og'-ra-phy) (a) The study of the ocean, including its physical, chemical, biologic, and geologic aspects. (b) In a narrower sense, the study of the marine environment. Syn: *oceanology.*

oceanology (o-cean-ol'-o-gy) *oceanography.*

oceanus A term established by the International Astronomical Union for a very large dark area on the Moon (USGS, 2003).

ocellar (o-cel'-lar) An igneous rock texture, esp. one with nepheline, in which the "phenocrysts" consist of aggregates of smaller crystals (e.g. of biotite or acmite) arranged radially or tangentially around larger, euhedral crystals (e.g. of leucite or nepheline) or form rounded eyelike branching forms.

ocellus [paleont] (o-cel'-lus) (a) A minute simple eye in an arthropod; e.g. the only type of eye found in an arachnid, the median visual organ located on the *prosoma* of a merostome, or an unpaired median eye common in some branchiopod and copepod crustaceans. Also, one of the elements of a *compound eye.* (b) A short hyaline process on the frustule in some diatoms (as in *Auliscus*). Pl: ocelli.

ocher (o'-cher) (a) An earthy, pulverulent, red, yellow, or brown iron oxide that is used as a pigment; e.g. "yellow or brown ocher" (limonite) and "red ocher" (hematite). Also, any of various clays strongly colored by iron oxides. (b) A similar earthy and pulverulent metallic oxide used as a pigment; e.g. "antimony ocher" (stibiconite and cervantite), "lead ocher" (massicot and litharge), and "tungstic ocher" (tungstite and ferritungstite). Cf: *umber.* Also spelled: *ochre.*

ocherous (o'-cher-ous) Pertaining to, containing, or resembling ocher; e.g. "ocherous iron ore", a red, powdery or earthy hematite. Var: *ochreous; ochrous.*

Ochoan (O-cho'-an) North American series: uppermost Permian (above Guadalupian, below Triassic).

ochre (o'-chre) *ocher.*

ochreous (o'-chre-ous) *ocherous.*

Ochrepts Inceptisols formed in cold or temperate climates and that commonly have an ochric epipedon and a cambic horizon. They may have an umbric or mollic epipedon <25 cm thick or a fragipan or duripan under certain conditions. These soils are not dominated by amorphous materials and are not saturated with water for periods long enough to limit their use for most crops. A suborder in the U.S. system of soil taxonomy.

ochric epipedon (och'-ric) A surface horizon of mineral soil that is too light in color, too high in chroma, too low in organic carbon, or too thin to be a plaggen, mollic, umbric, anthropic or histic epipedon, or that is both hard and massive when dry.

ochrous (och'-rous) *ocherous.*

Ockham's razor *Occam's razor.*

ocrite (oc'-rite) A group name for powdery ochers.

octactine (oc-tac'-tine) A sponge spicule having six equidistant rays in one plane and two rays at right angles to them.

octahedral (oc-ta-he'-dral) Pertaining to an *octahedron.*

octahedral borax A variety of tincalconite occurring in crystals that simulate octahedrons, as from the lagoons of Tuscany, Italy.

octahedral cleavage Mineral cleavage parallel to the faces of the octahedron (111); e.g. in diamond.

octahedral coordination An atomic arrangement in which an ion is surrounded by six ions of opposite sign, whose centers form the corners of an octahedron. Cf: *six coordination.*

octahedral copper ore *cuprite.*

octahedral iron ore *magnetite.*

octahedral sheet A structural component of phyllosilicates and some hydroxides formed by two closest-packed anion planes in which the six-coordinated (i.e., octahedral) sites are occupied by cations, commonly Mg, Al, Fe. The individual octahedra share edges to form a sheet. Cf: *tetrahedral sheet.*

octahedral structure *Widmanstätten structure.*

octahedrite [meteorite] (oc-ta-he'-drite) The commonest structural class of *iron meteorites.* Octahedrites contain 6-18% nickel in the metal phase and, on etching, showing Widmanstätten structure due to the presence of oriented plates of kamacite (with narrow borders of taenite) of thickness t, oriented parallel to four octahedral planes. Octahedrites are commonly divided into finest (Off, $t<0.2$ mm), fine (Of, $0.2\leq t>0.5$ mm), medium (Om, $0.5\leq t<1.3$ mm), coarse (Og, $1.3\leq t<3$ mm) and coarsest (Ogg, $t\geq 3$ mm) octahedrites. Symbol: O. Cf: *hexahedrite; ataxite [meteorite].*

octahedrite [mineral] A syn. of *anatase.* The term is a misnomer because anatase crystallizes in tetragonal dipyramids and not in octahedrons.

octahedron (oc-ta-he'-dron) An isometric crystal form of eight faces that are equilateral triangles. Its indices are {*111*} and its symmetry is $4/m\ \bar{3}\ 2/m$. Adj: *octahedral.*

octane (oc'-tane) Any of the several isomeric liquid paraffin hydrocarbons having the formula C_8H_{18}, including n-octane $CH_3(CH_2)_6CH_3$ which is found in petroleum.

octant search In *computer contouring,* an algorithm to find the points used to estimate the value at a location by dividing the surrounding area into eight equal radial sectors, then locating the nearest *k* observations within each sector.

octaphyllite (oc-ta-phyl'-lite) (a) An unnecessary name for mica minerals that contain eight cations per ten oxygen and two hydroxyl ions. (b) Any mineral of the *octaphyllite* group, such as biotite; a trioctahedral clay mineral. Cf: *heptaphyllite.*

octocoral (oc'-to-co'-ral) Any anthozoan belonging to the subclass Octocorallia, characterized by exclusively polypoid forms with pinnate tentacles and colonial growth. Range, Silurian (questionably) to Holocene. See also: *alcyonarian.*

ocular [optics] (oc'-u-lar) *eyepiece.*

ocular [paleont] n. *ocular plate.*

ocular plate One of the outer circlet of primordial plates of the *apical system* of an echinoid, located at the aboral terminus of an ambulacrum and perforated by an ocular pore. Cf: *genital plate.* Syn: *ocular [paleont].*

ocular pore A perforation in an *ocular plate* of an echinoid for the passage of a *terminal tentacle.*

ocular sinus One of a pair of small, shallow sinuses at the sides of the aperture in the position of the eyes in *Nautilus.*

Oculata (Oc-u-la'-ta) A group of Late Cretaceous to earliest Paleogene pollen with one pair of pores on one side superimposed over another pair on the other side; e.g., *Wodehouseia.*

oculi Pl. of *oculus.*

oculogenital ring (oc'-u-lo-gen'-i-tal) A ring formed in echinoids by a circlet of ocular plates surrounding a circlet of genital plates in the center of the aboral surface at the apical end of ambulacral and interambulacral areas. It surrounds the *periproct* in regular echinoids, and it represents the initial plates of an echinoid skeleton. See also: *apical system.*

oculus (oc'-u-lus) A much enlarged part of the pore structure of triporate pollen (typical in *Normapolles*), consisting of a thick protrusion of ektexine. Pl: *oculi.*

OD *ordnance datum.*

o'danielite (o'-dan'-iel-ite) A pale violet monoclinic mineral of the *alluaudite* group: $H_2Na(Zn,Mg)_3(AsO_4)_3$.

Oddo-Harkins rule (Od'-do-Har'-kins) A statement in geochemistry that, with four exceptions, the cosmic abundances of elements of even atomic number exceed those of adjacent elements of odd atomic number. This relationship was perceived by both Oddo and Harkins.

odenite (o'-den-ite) A variety of biotite supposed to contain a new element (odenium).

Odessan (O-des'-san) Eastern European mammal age ("complex"): earliest Pleistocene (after Khaprovian, before Tamanian); equiv. Biharian; late Villafranchian.

odinite [mineral] A silky green, 1:1 serpentine type clay mineral that is intermediate between dioctahedral and trioctahedral: $(Fe^{3+},Mg,Al,Fe^{2+})_{2.4}(Si,Al)_2O_5(OH)_4$. It crystallizes in trigonal and monoclinic polytypes. Apparently it forms in association with organic material on shallow marine shelves and reef lagoons in tropical latitudes (Bailey, 1988).

odintsovite A pink, dark-red, or white orthorhombic mineral: $K_2(Na,Li)_4Ca_3Ti_2Be_4Si_{12}O_{38}$.

odograph (o'-do-graph) An instrument that automatically plots the course and distance traveled by a vehicle and that draws directly on paper by electronic or photoelectric methods a continuous map of the route taken. The instrument containing a rotating part that moves or turns by an amount proportional to the actual distance traveled, a compass giving a reference direction with respect to the distance which is resolved into orthogonal components and an integrator for determining the components of the distance.

odometer (o-dom'-e-ter) An instrument attached to a wheel of a vehicle to count the number of turns made by the wheel and used to measure the approximate distance traveled as a function of the number of revolutions and the circumference of the wheel.

odometry (o-dom'-e-try) Mechanical measurement of distances.

Odontognathae (O-don-tog'-nath-ae) A superorder of the avian subclass Neornithes that is restricted to the toothed forms of the upper Mesozoic.

odontolite (o-don'-to-lite) A fossil bone or tooth colored deep blue by iron phosphate (vivianite), and rarely green by copper compounds, resembling turquoise, such as that from the tusks of mammoths found in Siberia. It is cut and polished for jewelry. Syn: *bone turquoise; fossil turquoise.*

odontology (o-don-tol'-o-gy) The study of teeth, including their structure, development, and diseases.

ODP *Ocean Drilling Program.*

oeciopore (oe'-ci-o-pore') The skeletal aperture of a brood chamber of stenolaemate bryozoans through which larvae escape.

Oe horizon A layer of partially decomposed litter with portions of plant structures still recognizable (hemic material). Occurs below the L layer on the forest floor in forest soils. It is the fermentation layer. Syn: *F layer.*

oenite A metallic silver-white orthorhombic mineral of the *löllingite* group: CoSbAs.

oersted (oer'-sted) (a) Commonly used as the cgs (centimeter-gram-second) unit of *magnetic-field intensity.* Except in magnetized media, a magnetic field with an intensity H of one oersted has an induction B of one gauss. (b) The cgs unit of *magnetic force.*

offlap (off'-lap) (a) The progressive offshore regression of the updip terminations of the sedimentary units within a conformable sequence of rocks (Swain, 1949, p.635), in which each successively younger unit leaves exposed a portion of the older unit on which it lies. Also, the successive contraction in the lateral extent of strata (as seen in an upward sequence) due to their being deposited in a shrinking sea or on the margin of a rising landmass. Ant: *onlap.* Syn: *regressive overlap.* (b) The progressive withdrawal of a sea from the land. Cf: *regression [stratig].* (c) A term commonly used by seismic interpreters for reflection patterns generated from strata prograding into deep water (Mitchum, 1977, p.208).

off-lying (off'-ly-ing) *offshore.*

off-reef Pertaining to the sea floor away from the margin of a reef; e.g. the "off-reef facies" or the "off-reef sediments" deposited immediately surrounding a reef. Cf: *fore reef; interreef.* Also spelled: offreef.

offrétite (of'-fret-ite) A colorless hexagonal zeolite mineral: $CaKMg(Si_{13}Al_5)O_{36}\cdot 6H_2O$.

offscraping The addition of material to the toe of an *accretionary prism* by formation of new thrusts at its toe.

offset [cart] (off'-set) The small distance added (during construction of a map projection) to the length of meridians on each side of the central meridian in order to determine the top latitude of the constructed chart.

offset [coast] The migration of an upcurrent part of a shore to a position a little farther seaward than a downcurrent part; esp. the offset of a spit across a coastal inlet. Cf: *overlap [coast].*

offset [geomorph] (a) A spur or minor branch from a range of hills or mountains. (b) A level terrace on a hillside.

offset [paleont] A new *corallite* formed in a *corallum* by budding; a corallite formed directly or indirectly from a protocorallite.

offset [seis] n. (a) The horizontal distance from energy source to detector, esp. to the nearest detector; *in-line offset.* (b) The horizontal distance from a source to the line of profile, measured perpendicularly to the line; *perpendicular offset.* (c) The horizontal displacement, measured from the detector, of a point for which a calculated depth applies. v. To make such an adjustment of position or depth.

offset [struc geol] (a) An informal term for the separation on a fault as observed in the plane of an outcrop or cross section. Cf: *normal shift.* (b) *Overstep [struc geol].*

offset [surv] (a) A short line measured perpendicular to a traverse course or a surveyed line or principal line of measurement, for the purpose of locating a point with respect to a point on the course or line; e.g. a perpendicular distance measured from a great-circle line to a parallel of latitude in order to locate a section corner on that parallel in the U.S. Public Land Survey system. (b) A jog in a survey line which has approximately the same direction both before and after passing the jog.

offset deposit (a) A mineral deposit, esp of sulfides, formed partly by magmatic segregation and partly by hydrothermal solution, near the source rock. (b) At Sudbury, Ontario, the term refers to dikelike bodies radiating from the Sudbury Complex, thought to have been filled from above by xenolithic rock fragments and massive pyrrhotite-chalcopyrite-pentlandite.

offset line A supplementary line established close to and roughly parallel with the main survey line to which it is referenced by measured offsets. Offset lines are used where it is convenient to avoid obstructions, over which it would be difficult to make measurements, located along the main line.

offset source In seismic surveying, a source at a small distance from the line of active detectors. See also: *perpendicular offset; in-line offset.*

offset stream A stream displaced laterally or vertically by faulting.

offset vertical seismic profiling A vertical seismic profile where the source point is an appreciable horizontal distance from the wellhead.

offset well (a) An oil well drilled near the boundary of a property and opposite to a producing or completed well on an adjoining property, for the purpose of preventing the drainage of oil or gas by the earlier well. (b) Any *development well.*

offshore (off'-shore) (a) Situated off or at a distance from the shore; specif. said of the comparatively flat, always submerged zone of variable width extending from the breaker zone to the seaward edge of the continental shelf. Ten meters is a suggested minimal depth. The offshore zone is seaward of the *inshore* or *nearshore* zone or the *shoreface* (CERC, 1966, p. A43; and Johnson, 1919, p.161), although it is often regarded (e.g. Shepard, 1967, p.43) as the zone extending seaward from the low-water shoreline. (b) Pertaining to a direction seaward or lakeward from the shore; e.g. an *offshore* wind or one that blows away from the land, or an *offshore* current or one moving away from the shore. Ant: *onshore.* See also: *nearshore.* Syn: *off-lying.*

offshore bar (a) *longshore bar.* (b) A catchall term used by Johnson (1919) for features now known as a *barrier beach* and a *barrier island.* The term is undesirable as it has been applied both to a submerged feature (a bar) and an emergent feature (a barrier).

offshore barrier *barrier beach.*

offshore beach A syn. of *barrier beach.* The term was used by Gilbert and Brigham (1902, p.306) for a long narrow sandy beach with a belt of quiet water separating it from the mainland.

offshore permafrost *subsea permafrost.*

offshore slope The frontal slope below the outer edge of the wave-built terrace.

offshore terrace A wave-built terrace in the offshore zone, composed of gravel and coarse sand. See also: *shoreface terrace.*

offshore water Water lying seaward of *inshore water,* whose physical properties are influenced only slightly by continental conditions. Ant: *inshore water.*

offtake (off'-take) *groundwater withdrawal.*

ogdensburgite (og'-dens-burg-ite') A dark brownish-red to reddish-orange platy orthorhombic mineral:$Ca_2(Zn,Mn^{2+})Fe^{3+}_4(AsO_4)_4(OH)_6{\cdot}6H_2O$.

ogdohedral Said of a *crystal class* in which the number of faces in the *general form* is one eighth that of the general form of the *holohedral* class in that *crystal system.* Cf: *merohedral; hemihedral; tetartohedral.*

oghurd (og'-hurd) A term used in the Saharan region for a massive, mountainous dune, formed by some underlying rocky topographic feature, and rising considerably above the general dune level.

ogive (o'-give [Pron: o'-jive]) One of a series of alternating bands of light and dark ice, extending across a glacier below an *icefall* and bowed (ogive-fashion, i.e., resembling the pointed arch or rib across a Gothic vault) downglacier by faster flow in the center of the glacier. As ice passes through the icefall, more melting occurs in the summer than in the winter, and dirt is concentrated on the surface by the summer melting. Thus the dark bands are summer bands, the ice in them is typically thinner due to the melting, and they lie lower than the thicker, lighter bands of winter, so that the ogives form a series of waves (Nye, 1958). Cf: *dirt band [glaciol]; dirt-band ogive; Forbes band.*

ohmilite (oh'-mil-ite) A brownish-pink monoclinic mineral: $Sr_3(Ti,Fe^{3+})(Si_2O_6)_2(O,OH){\cdot}2H_2O$.

O horizons Layers dominated by organic material. These horizons incorporate one or more of the following subdivisions: Oa horizon (H layer) A layer occuring in mor humus consisting of well-decomposed organic matter of unrecognizable origin; Oe horizon (F layer) A layer of partially decomposed litter with portions of plant structures still recognizable. Occurs below the L layer on the forest floor in forest soils. It is the fermentation layer; Oi horizon [L layer (litter)] A layer of organic material having undergone little or no decomposition. On the forest floor this layer consists of freshly fallen leaves, needles, twigs, stems, bark, and fruit. This layer may be very thin or absent during the growing season.

OIB *oceanic island basalt.*

-oid A suffix (derived from the Greek) meaning "like, having the form of". A rock name or geologic feature thus qualified (e.g. granitoid, gneissoid) resembles but is not the same as the name or feature to which it is attached.

Oi horizon A layer of organic material having undergone little or no decomposition (fibric material). On the forest floor this layer consists of freshly fallen leaves, needles, twigs, stems, bark, and fruits. This layer may be very thin or absent during the growing season. Syn: L layer; *litter.*

oikocryst (oi'-ko-cryst) In poikilitic fabric, the enclosing crystal.

oil *petroleum.*

oil accumulation *oil pool.*

oil-base mud A *drilling mud* with clay particles suspended in oil. The oil may be either crude oil or a refined product such as "mineral oil". Abbrev: OBM.

oil-cut mud *Drilling mud* with some crude oil. Such mud could result from oil entering it during drilling or from a *drill-stem test* of an oil reservoir.

oil expulsion The process whereby hydrocarbons are pushed out of a source rock.

oil field (a) An *oil pool.* (b) Two or more oil pools on a single geologic feature or otherwise closely related.

oil-field brine Water found by the drill in rocks penetrated at depth. It usually has a high concentration of dissolved salts.

oil mining Recovery of oil by surface extraction, as in tar sands or oil shales, or from tunnels in a shallow reservoir.

oil pool A subsurface accumulation of petroleum that will yield crude oil in economic quantities. Cf: *oil field.* Syn: *oil accumulation.*

oil sand A term applied loosely to any porous stratum containing petroleum or impregnated with hydrocarbons; specif. a sandstone or unconsolidated sand from which oil is obtained by drilled wells. The term is also applied to productive limestone and dolomite. See also: *gas sand; tar sand; sand [drill].* Cf: *water sand.*

oil seep The emergence of liquid petroleum at the surface as a result of slow upward migration from its buried source. Syn: *seepage [petroleum].*

oil shale A *kerogen*-bearing, finely laminated brown or black sedimentary rock that will yield liquid or gaseous hydrocarbons on distillation. Syn: *kerogen shale; kerogenite.*

oil show *show (a).*

oil trap A *trap [petroleum]* that contains oil.

oil-water contact (oil-wa'-ter) The boundary surface between an accumulation of oil and the underlying bottom water. Syn: *oil-water interface.*

oil-water interface *oil-water contact.*

oil well A well from which petroleum is obtained by pumping or by natural flow. Some statutes define the term on the basis of the gas-oil ratio.

oil window The temperature and pressure in the Earth where oil is formed from the kerogen in shales.

ojo (o'-jo [o-ho]) A term used in SW U.S. for a very small lake or a pond. Etymol: Spanish, "eye". An artesian spring in limestone regions, especially one forming a small pond. See: *vauclusian spring.* Also called: ojo de agua.

ojuélaite A pale green monoclinic mineral of the whitmoreite group: $ZnFe^{3+}_2(AsO_4)_2(OH)_2{\cdot}4H_2O$.

okanoganite-(Y) (ok'-an-og'-an-ite-y) A tan or pale pink rhombohedral mineral: $(Na,Ca)_3(Y,Ce)_{12}B_2Si_6O_{27}F_{14}$.

okayamalite A creamy-white tetragonal mineral of the *melilite* group: $Ca_2B_2SiO_7$.

okenite (o'-ken-ite) A white, yellowish, or bluish triclinic mineral: $Ca_5Si_9O_{23}{\cdot}9H_2O$. Cf: *nekoite.*

okhotskite (o-khots'-kite) An orange monoclinic mineral of the *pumpellyite* group: $Ca_2(Mn^{2+},Mg)(Mn^{3+},Al,Fe^{3+})_2Si_3O_{10}(OH)_4$.

old age [coast] A hypothetical stage in the development of a shore, shoreline, or coast, characterized by a wide wave-cut platform, a faintly sloping sea cliff pushed far inland, and a coastal region approaching peneplanation. The stage is probably a theoretical abstraction, since it is doubtful whether stability of sea level is maintained long enough for the land to be so reduced (Dietz, 1963).

old age [geomorph] The hypothetical final stage of the *cycle of erosion* of a landscape or region, in which the surface has been reduced almost to base level and the landforms are marked by simplicity of form and subdued relief. It is characterized by a few large meandering streams flowing sluggishly across broad floodplains, separated by faintly swelling hills, and having dendritic distributaries; and by *peneplanation.* Cf: *senescence [geomorph]; senility.*

old age [streams] The hypothetical stage in the development of a stream at which erosion is decreasing in vigor and efficiency, and aggradation becomes dominant as the gradient is greatly reduced. It is characterized by: a load that exceeds the stream's ability to carry it, and is therefore readily deposited; a very broad, shallow, open valley with gently sloping sides and a nearly level floor (flood plain) that may be 15 times the width of the meander belt; numerous oxbows, meander scars, levees, yazoos, bayous, and swamps and lakes on valley floors; a sluggish current; graded or mature tributaries, few in number; and slow erosion, effected chiefly by mass-wasting at valley sides. Note that many of these characteristics are most common on geologically recent alluvial plains as a result of postglacial hydrologic adjustments or rising sea level and are not evidence of old age in the *cycle of erosion.*

Older Dryas (Old'-er Dry'-as) n. A term used primarily in Europe for an interval of late-glacial time (centered about 14,500 calendar years ago) following the Bølling and preceding the Allerød. It was characterized by tundra vegetation, and the climate, as inferred from stratigraphic and pollen data in Scandinavia (Iversen, 1954), deteriorated so as to favor either expansion or retarded retreat of glaciers. adj. Pertaining to the late-glacial Older Dryas interval and to its climate, deposits, biota, and events.

Oldest Dryas (Old'-est Dry'-as) n. A term used primarily in Europe for a late-glacial interval (about 15,000 calendar years ago) preceding the Bølling, characterized by tundra vegetation. The climate as inferred from stratigraphic and pollen data in Scandinavia (Iversen, 1954) was colder than in the succeeding Bølling. adj. Pertaining to the late-glacial Oldest Dryas interval and to its climate, deposits, biota, and events.

old-from-birth peneplain A term used by Davis (1922) for a peneplain presumably formed during an uplift of such extreme slowness over a long period of time that vertical corrasion was outpaced by valley-side grading and by general downwearing of the interstream uplands, thereby producing a landscape that will at once be "old" or that lacks any features characterizing youth or maturity; it is essentially a *primärrumpf.*

Oldham-Gutenberg discontinuity (Old'-ham-Gu'-ten-berg) *Gutenberg discontinuity.*

oldhamite (old'-ham-ite) A pale-brown cubic meteorite mineral of the *galena* group: (Ca,Mg)S.

old ice (a) Floating *sea ice* more than two years old (Armstrong et al., 1966, p.30). It may be more than 3 m thick and shows features that are smoother than those in *second-year ice.* (b) A term formerly applied to sea ice that has survived at least one summer's melt and that shows features that are smoother than those in first-year ice; e.g., second-year ice and *multi-year ice.* (c) A term loosely applied to a deposit of ice in permafrost.

oldland (old'-land) (a) Any ancient land; specif. an extensive area (such as the Canadian shield) of ancient crystalline rocks, reduced to low relief by long-continued erosion, from which the materials of later sedimentary deposits were derived. (b) A region of older land, projected above sea level behind a coastal plain, that supplied the material of which the coastal-plain strata were formed; the land adjoining a new land surface that has just been brought above sea level. (c) A term proposed by Maxson and Anderson (1935, p.90) for the land surface of the old-age stage of the *cycle of erosion,* characterized by subdued relief. Maxson later adopted (1950, p.101) the earlier term "senesland" for this feature.

old mountain A mountain that was formed prior to the Tertiary period, esp. a fold mountain produced before the Alpine orogeny. Ant: *young mountain.*

Old Red Sandstone A thick sequence of nonmarine, predominantly red sedimentary rocks, chiefly sandstones, conglomerates, and shales, representing the Devonian System in parts of Great Britian and elsewhere in NW Europe. See Miller (1841).

old snow Fallen snow in which most traces of the original snow-crystal shapes have been lost due to metamorphism, as *firn* or *settled snow.* Ant: *new snow.* Syn: *firn snow.*

Old Stone Age *Paleolithic.*

olefin (o'-le-fin) An unsaturated *aliphatic hydrocarbon,* empirical formula C_nH_{2n}, which contains at least one double bond. The olefins form a series, analogous to the methane series, of which ethylene, C_2H_4, a sweet-smelling gas present in common gas, is the lowest member. Also spelled: *olefine.*

olefine (o'-le-fine) *olefin.*

olekminskite A vitreous white trigonal mineral: $Sr(Sr,Ca,Ba)(CO_3)_2$.

Olenekian (O-le-nek'-i-an) A stage of the Standard Global Chronostratigraphic Scale: Lower Triassic (above Induan, below Anisian).

olenellid (o-le-nel'-lid) Any trilobite belonging to the family Olenellidae, characterized generally by a subovate to elongate exoskeleton, the absence of dorsal sutures on the cephalon, numerous segments in the thorax, and well-developed pleural spines or acutely terminating, falcate distal portions (TIP, 1959, pt.O, p.191). Range, Lower Cambrian.

olenite (ol'-en-ite) A pink rhombohedral mineral of the *tourmaline* group: $Na_{0.5}Al_9(BO_3)_3(Si_6O_{18})(O,OH)_4$.

olgite (ol'-gite) A bright blue to bluish green trigonal mineral: $Na(Sr,Ba,Na)_3(PO_4)_{1.8}$.

oligist iron (ol'-i-gist) A syn. of *hematite.* Also spelled: oligiste iron.

oligo- (ol'-i-go-) A prefix meaning "small", "a little".

Oligocene (Ol'-i-go-cene') The upper series of the *Paleogene* Subsystem of the *Tertiary* System of the Standard Global Chronostratigraphic Scale, above the *Eocene* and below the *Miocene* Series of the *Neogene* Subsystem. Also the time during which these rocks were formed, the *Oligocene* Epoch, covering the time span between 34 and 23 Ma.

oligoclase (ol'-i-go-clase') A triclinic mineral of the plagioclase *feldspar* group with composition ranging from $Ab_{90}An_{10}$ to $Ab_{70}An_{30}$. It is common in igneous rocks of intermediate to high silica content.

oligomictic [lake] (ol'-i-go-mic'-tic) Said of a lake that circulates only at rare intervals when abnormally cold spells occur; e.g. a lake of small or moderate area or of very great depth, or in a region of high humidity, in which a small temperature difference between surface and bottom suffices to maintain stable stratification (Hutchinson, 1957, p.462; Wetzel, 2001, p.81). Cf: *polymictic [lake].*

oligomictic [sed] Said of a clastic sedimentary rock composed of a single rock type, such as an *orthoquartzitic conglomerate;* also, said of the clasts of such a rock. Cf: *monomictic [sed]; polymictic [sed].*

oligonite (ol'-i-go-nite') A variety of siderite containing up to 40% manganese carbonate. Syn: oligon spar.

oligopelic (ol'-i-go-pel'-ic) Said of a lake-bottom deposit that contains very little clay (Veatch and Humphrys, 1966, p.218).

oligostromatic (o-lig'-o-stro-mat'-ic) Said of a plant part that is composed of only a few layers of cells. Cf: *monostromatic; polystromatic.*

oligotrophic lake (ol'-i-go-troph'-ic) A lake that is characterized by a deficiency in plant nutrients and usually by abundant dissolved oxygen in the *hypolimnion;* its bottom deposits have relatively small amounts of organic matter and its water is often deep (Wetzel, 2001, p.273-274). Cf: *dystrophic lake; mesotrophic lake; eutrophic lake.*

oligotrophic peat Peat containing a small amount of plant nutrients. Cf: *mesotrophic peat; eutrophic peat.*

oligotrophic water A layer of seawater at the ocean surface in which biochemical processes associated with carbon fixation have reduced inorganic nutrient concentrations to very low levels. Essentially the ocean equivalent of deserts on land. Found in the central-gyre areas of all major oceans and in many coastal waters. Ant: *eutrophic* water. Cf: *carbon fixation; nutrient [oceanog].*

oligotrophy (o-li-got'-ro-phy) The quality or state of an *oligotrophic lake.*

olistolith (o-lis'-to-lith) A large *exotic block* or other rock mass (usually >10m) transported by submarine gravity sliding or slumping and included within the binder of an olistostrome. Term introduced by G. Flores in Beneo (1955, p.122). See also: Abbate et al., (1970) and Reedins (1986) p.402.

olistostrome (o-lis'-to-strome) A debris-flow deposit consisting of a chaotic mass of intimately mixed heterogeneous materials (such as blocks and muds) that accumulated by submarine gravity sliding or slumping of unconsolidated sediments. It is a mappable, lens-like stratigraphic unit lacking true bedding but intercalated among normally bedded sequences, as in the Tertiary basin of central Sicily. Raymond (1978) gives "olistostrome" as a general term for either a *broken formation* or a *mélange* of sedimentary origin. Cf: *allolistostrome; endolistostrome.* Term introduced by G. Flores in Beneo (1955, p.122). Etymol: Greek "olistomai", "to slide", + "stroma", "bed".

olivenite (o-liv'-en-ite, ol'-i-ven-ite) An olive-green, dull-brown, gray, or yellowish orthorhombic mineral of the *libethenite* group: $Cu_2(AsO_4)(OH)$. Syn: *leucochalcite; wood copper.*

olivine (ol'-i-vine) (a) An olive-green, grayish-green, or brown orthorhombic mineral: $(Mg,Fe)_2SiO_4$. It consists of the isomorphous solid-solution series forsterite-fayalite. Olivine is a common rock-forming mineral of basic, ultrabasic, and low-silica igneous rocks (gabbro, basalt, peridotite, dunite); it crystallizes early from a magma, weathers readily at the Earth's surface, and metamorphoses to serpentine. (b) A name applied to a group of minerals with the general formula *A*2SiO4, where *A* = Mg, Fe^{2+}, Mn^{2+} or Ni. See also: *peridot; chrysolite [mineral].* Syn: *olivinoid.*

olivine basalt (a)-(normative) Basalt that contains normative olivine and diopside with no nepheline or quartz (Yoder and Tilley, 1962). (b)-(modal) A basalt with olivine in the mode.

olivine clinopyroxenite In the *IUGS classification,* a plutonic rock with M ≥90, ol/(ol+opx+cpx) between 5 and 40, opx/(ol+opx+cpx) less than 5, and cpx/(ol+opx+cpx) less than 90.

olivine gabbro In the *IUGS classification,* a *gabbro* containing more than 5% olivine.

olivine gabbronorite In the *IUGS classification,* a plutonic rock satisfying the definition of *gabbro* and in which pl/(pl+px+ol) is between 10 and 90, px/(pl+px+ol) is greater than 5, and ol/(pl+px+ol) is greater than 5.

olivine-hornblende pyroxenite (ol'-i-vine-horn'-blende) In the *IUGS classification,* a plutonic rock with M equal to or greater than 90, ol(ol+hbl+px) between 5 and 40, and pyroxene more abundant than amphibole.

olivine hornblendite In the *IUGS classification,* a plutonic rock with M equal to or greater than 90, ol/(ol+hbl+px) between 5 and 40, px/(ol+hbl+px) less than 5, and hbl/(ol+hbl+px) less than 90.

olivine leucitite *ugandite.*

olivine melilitite In the *IUGS classification,* a volcanic rock containing at least 10% olivine and at least 10% melilite, most of the remainder being clinopyroxene.

olivine melilitolite In the *IUGS classification,* a plutonic rock containing at least 10% melilite, at least 10% olivine, and less than 10% clinopyroxene. Obsolete syn: *kugdite.*

olivine nephelinite An extrusive igneous rock differing in composition from nephelinite only by the presence of olivine. Syn: *nepheline basalt; ankaratrite.*

olivine norite In the *IUGS classification,* a plutonic rock containing at least 10% plagioclase and at least 10% olivine, with orthopyroxene making up at least 95% of the pyroxenes.

olivine orthopyroxenite In the *IUGS classification,* a plutonic rock with M ≥90, ol/(ol+opx+cpx) between 5 and 40, cpx/(ol+opx+cpx) <5, and opx/(ol+opx+cpx) <90.

olivine-pyroxene hornblendite (ol'-i-vine-py'-rox-ene) In the *IUGS classification,* a plutonic rock with M equal to or greater than 90, ol/(ol+hbl+px) between 5 and 40, and amphibole more abundant than pyroxene.

olivine pyroxene melilitolite In the *IUGS classification,* a plutonic rock consisting mostly of melilite, olivine, and clinopyroxene, with clinopyroxene > olivine.

olivine pyroxenite In the *IUGS classification,* a plutonic rock with M equal to or greater than 90, ol/(ol+hbl+px) between 5 and 40, hbl/(ol+hbl+px) less than 5, and px/(ol+hbl+px) less than 90.

olivine tholeiite Basalt that is silica-undersaturated, containing normative olivine, hypersthene, and diopside, with neither quartz nor nepheline (Yoder and Tilley, 1962).

olivine websterite In the *IUGS classification,* a plutonic rock with M equal to or greater than 90, ol/(ol+opx+cpx) between 5 and 40, opx/(ol+opx+cpx) greater than 5, and cpx/(ol+opx+cpx) greater than 5.

olivinoid (ol'-i-vin-oid') (a) An olivine-like substance found in meteorites. (b) *olivine.*

olkhonskite A metallic black monoclinic mineral: $(Cr,V^{3+})_2Ti_3O_9$. It is the Cr analogue of schreyerite.

ollenite (ol'-len-ite) A rarely used term for type of *hornblende schist* characterized by abundant epidote, sphene, and rutile. Garnet is one of the accessories (Holmes, 1928, p.170).

olmsteadite (olm'-stead-ite) A sub-adamantine dark brown, reddish-brown, or black orthorhombic mineral: $KFe^{2+}_2(Nb,Ta)O_2(PO_4)_2 \cdot 2H_2O$. It is the Fe^{2+} analogue of johnwalkite.

olocyst (o'-lo-cyst) A term used in the older literature for a thin, smooth calcareous layer thought to form the initial deposit in frontal shields of certain ascophoran cheilostomes (bryozoans). Syn: *holocyst*.

olsacherite (ol-sach'-er-ite) A colorless orthorhombic mineral: $Pb_2(Se^{6+}O_4)(SO_4)$.

olshanskyite (ol-shan'-sky-ite) A colorless triclinic mineral: $Ca_3[B_3O_3(OH)_6]OH \cdot 3H_2O$.

Olympic Dam-type deposit A polymetallic iron, copper, uranium, gold, and rare earth element-bearing subtype of *iron oxide copper-gold deposit*. Named for the Olympic Dam polymetallic deposit in South Australia.

olympite (o-lym'-pite) A colorless orthorhombic mineral: $LiNa_5(PO_4)_2$.

olynthus (o-lyn'-thus) (a) The first stage in the development of a sponge, in which the initial functional aquiferous system has a single flagellated chamber. (b) Newly attached sponge larva resembling a vase in form and having a simple and asconoid body wall.

ombrogenous peat (om-brog'-e-nous) Peat that receives essentially all of its moisture from precipitation. It is rich in cellulose and low in plant nutrients, mineral matter, and nitrogen. Ombrogenous peat is highly acidic and is generally well preserved. Cf: *topogenous peat; soligenous peat; highmoor peat; domed peat.*

ombrophilous (om-broph'-i-lous) Said of a plant adapted to extremely rainy conditions. Cf: *ombrophobous.* Noun: ombrophile.

ombrophobous (om-broph'-o-bous) Said of an organism that cannot tolerate extremely rainy conditions. Cf: *ombrophilous.* Noun: ombrophobe.

ombrotrophic peat Peat whose nutrient supply is exclusively from rain water (including snow and atmospheric fallout) therefore making nutrients extremely oligotrophic. Cf: *minerotrophic peat; ombrogenous peat.*

omeiite (o'-mei-ite) A metallic gray orthorhombic mineral of the *löllingite* group: $(Os,Ru)As_2$.

ominelite A vitreous blue orthorhombic mineral: $(Fe^{2+},Mg)Al_3O_2(BO_3)SiO_4$. It is the Fe^{2+} analogue of grandidierite.

omission solid solution A crystal in which there is incomplete filling of particular atomic sites. Cf: *substitution solid solution.* Syn: *defect-lattice solid solution.*

omission surface A *discontinuity surface* of the most minor nature, which marks a temporary halt in deposition but little or no erosion (Bromley, 1975). The term was first used by Heim in 1924.

ommatidium (om-ma-tid'-i-um) One of the basic visual units of the *compound eye* of an arthropod. Pl: ommatidia.

omphacite (om'-pha-cite) A grass-green to pale-green granular monoclinic of the pyroxene group, found as a common constituent in the rock *eclogite*; $(Ca,Na)(Fe^{2+},Al)Si_2O_6$. In thin section it is colorless, superficially resembling olivine. Syn: *tuxtlite.*

omuramba (om-u-ram'-ba) A term used in central and NE Namibia for the clearly defined dry bed of an intermittent stream, carrying water sometimes only as a series of shallow lakes and *vleis.* Etymol: Bantu (Herero). Pl: omirimbi. Cf: *oshana.*

Oncequethwan (On-ce-queth'-wan) A North American stage: Middle Devonian (above Onondagan, below Cazenovian).

oncoid (on'-coid) A microbial biscuit that resembles an ancient oncolite.

oncolite (on'-co-lite) A small, variously shaped, concentrically laminated, calcareous sedimentary structure, resembling an oolith, and formed by the accretion of successive layered masses of gelatinous sheaths of blue-green algae. It is smaller than a *stromatolite* and generally does not exceed 10 cm in diameter. Also spelled: *onkolite.* Cf: *catagraphite.*

oncus (on'-cus) In *pollen grains,* a lens-shaped structure located beneath the apertures of many kinds, not resistant to *acetolysis.*

one-face-centered lattice A type of *centered lattice* in which the unit cell has one pair of faces centered, i.e. there are two lattice points per unit cell. If the (100) plane is centered, the symbol A is used; if the (010) plane is centered, the symbol B; and if the (001) plane is centered, the symbol C. In orthorhombic crystal lattices, all types are possible; in monoclinic crystal lattices, only the C and A type of centering is possible. Syn: *base-centered lattice; end-centered lattice.*

onegite (o-ne'-gite) A pale amethyst-colored sagenitic quartz penetrated by needles of goethite.

oneillite A yellowish-brown rhombohedral mineral of the *eudialyte* group: $Na_{15}Ca_3Mn_3Fe_3Zr_3NbSi_{25}O_{73}(O,OH,H_2O)_3(OH,Cl)_2$.

one-M (1M) polytype Polytypes are special types of polymorphs, where stacking of identical or nearly identical layers occurs in one direction. A 1M polytype is a crystal structure modification showing overall monoclinic symmetry and a unit repeat of one layer along the stacking direction (Bailey, 1980a). This polytype is a common form in the trioctahedral mica minerals but is considerably less common in the serpentine minerals. Cf: *two-M (2M) polytype.*

one-to-one (1:1) layer A component of a hydrous silicate composed of a continuous tetrahedral sheet of composition T_2O_5 (T commonly Si, Al, or Fe^{3+}) and an octahedral sheet where the octahedral sheet is composed of (a) medium-sized cations (e.g., Mg, Al, Fe^{2+}, Fe^{3+}) or vacancies and (b) anions of O and OH (Bailey, 1980a). The ratio of tetrahedral to octahedral sheets is 1:1. Minerals with 1:1 layers include the serpentine minerals and the kaolin minerals. Cf: *octahedral sheet; tetrahedral sheet.*

one-to-one (1:1) phyllosilicate A hydrous silicate composed of 1:1 layers. Adjacent unit components (adjacent 1:1 layers) are linked by hydrogen bonding and electrostatic forces. Minerals with 1:1 layers include the serpentine minerals and the kaolin minerals. (Bailey, 1980a). In the older literature, the 1:1 layer was referred to as "two-layer clay," and such usage is no longer acceptable. Cf: *octahedral sheet; tetrahedral sheet.*

one-year ice Sea ice of not more than one winter's growth, and a thickness of 70 cm to 2 m; it includes the "medium" and "thick" subdivisions of *first-year ice.*

onion-skin weathering (on'-ion-skin) *spheroidal weathering.*

onkilonite (on-kil'-o-nite) An olivine-leucite *nephelinite* that also contains augite and perovskite, but no feldspar. Backlund in 1915 named the rock for the Onkilones, the inhabitants of the type locality, Wilketski Island, Russia. Obsolete.

onkolite (on'-ko-lite) *oncolite.*

onlap (on'-lap) (a) An overlap characterized by the regular and progressive pinching out, toward the margins or shores of a depositional basin, of the sedimentary units within a conformable sequence of rocks (Swain, 1949, p.635; Mitchum, 1977, p.208-209), in which the boundary of each unit is transgressed by the next overlying unit and each unit in turn terminates farther from the point of reference. Also, the successive extension in the lateral extent of strata (as seen in an upward sequence) due to their being deposited in an advancing sea or on the margin of a subsiding landmass. Ant: *offlap.* Cf: *overstep [stratig].* See also: *marine onlap; apparent onlap; proximal onlap; distal onlap.* Syn: *transgressive overlap; coastal onlap.* (b) The progressive submergence of land by an advancing sea. Cf: *transgression.*

onofrite (o-nof'-rite) A mineral variety: Hg(S,Se). It is metacinnabar containing selenium, and is a source of selenium.

onokoid (on'-o-koid) A small or microcrystalline, dense, nodular, pealike body in ophthalmitic rocks (Niggli, 1954, p.191). Rarely used.

Onondagan (On-on-da'-gan) A North American provincial stage: Middle Devonian (above Deerparkian, below Oncequethwan).

onoratoite (on-o-ra'-to-ite) A colorless monoclinic mineral: $Sb_8O_{11}Cl_2$.

onset (on'-set) *arrival.*

onset-and-lee topography *stoss-and-lee topography.*

onshore (on'-shore) (a) Pertaining to a direction toward or onto the shore; e.g. an *onshore* wind or one that blows landward from a sea or lake, or an *onshore* current or one moving toward the shore. (b) Situated on or near the shore, as *onshore* oil reserves. Ant: *offshore.*

Ontarian (On-tar'-i-an) (a) Stage in New York State: Middle Silurian (middle and lower parts of Clinton Group). (b) An obsolete name for the Middle and Upper Ordovician in New York State.

ontogenetic stage (on'-to-ge-net'-ic) Developmental stage in the growth of an individual organism.

ontogeny (on-tog'-e-ny) Development of an individual organism in its various stages from initiation through maturity. Adj: ontogenetic. Cf: *phylogeny.* Syn: *life cycle.*

ontozone (on'-to-zone) A biostratigraphic unit based on the range of a fossil form; equivalent to "biozone" (Henningsmoen, 1961, p.68-69).

onychium (o-nych'-i-um) The distal subsegment of the *tarsus* carrying the claws, found in some arachnids but lacking in others. Pl: onychia.

onyx (on'-yx) (a) A variety of chalcedony that is like *banded agate* in consisting of alternating bands of different colors but unlike it in that the bands are always straight and parallel. Onyx is used esp. in making cameos. Cf: *agate; sardonyx; jasponyx.* (b) A name applied incorrectly to dyed, unbanded, solid-colored chalcedony; esp. *black onyx.* (c) *onyx marble.* adj. (a) Parallel-banded; e.g. "onyx marble" and "onyx obsidian". (b) Jet black.

onyx agate A banded agate with straight parallel bands of white and different tones of gray.

onyx marble A compact, usually banded, generally translucent variety of calcite (or rarely of aragonite) resembling true onyx in appearance; esp. parallel-banded *travertine* capable of taking a good polish, and used as a decorative or architectural material or for small ornamental objects. It is usually deposited from cold-water solutions, often in the form of stalagmites and stalactites in caves. See also: *cave onyx.* Syn: *onyx; Mexican onyx; alabaster; oriental alabaster; Gibraltar stone; Algerian onyx.*

onyx opal Common opal with straight parallel markings.

oocast (o'-o-cast) *oolicast.*

ooecium (o-oe'-ci-um) The *ovicell* in cheilostome bryozoans. Pl: ooecia.

oogonia Pl. of *oogonium.*

oogonium (o-o-go'-ni-um) The female sex organ in the thallophytes, containing one or more oospheres (Swartz, 1971, p.324). Pl: ogonia. Cf: *antheridium.*

ooid (o'-oid) (a) An individual spherite of an oolitic rock; an *oolith.* The term has been used in preference to "oolith" to avoid confusion with "oolite". (b)

A general, nongeneric term for a particle that resembles an oolith in outer appearance and size (Henbest, 1968, p.2). Cf: *pseudo-oolith.* Adj: ooidal.

oolicast (o-ol′-i-cast) One of the small, subspherical openings found in an oolitic rock, produced by the selective solution of ooliths without destruction of the matrix. The term is inappropriate unless the opening is subsequently filled. See also: *oomold.* Syn: *oocast.*

oolicastic porosity (o′-ol-i-cas′-tic) The porosity produced in an oolitic rock by removal of the ooids and formation of oolicasts (Imbt and Ellison, 1947, p.369-370).

oolite (o′-o-lite) (a) A sedimentary rock, usually a limestone, made up chiefly of ooliths cemented together. The rock was originally termed *oolith.* Syn: *roestone; eggstone.* (b) A term often used for *oolith,* or one of the ovoid particles of an oolite. Etymol: Greek "oon", "egg". Cf: *pisolite [sed].* Also spelled: oölite.

oolite bar *oolite shoal.*

oolite shoal A shallow wave-agitated marine shoal consisting primarily of sand-sized (0.25-2 mm diameters) accretionary particles (ooliths) composed most commonly of aragonite, but may be of calcite, dolomite, or less common mineralogies. Syn: *oolite bar.*

oolith (o′-o-lith) One of the small round or ovate accretionary bodies in a sedimentary rock, resembling the roe of fish, and having diameters of 0.25 to 2 mm (commonly 0.5 to 1 mm). It is usually formed of calcium carbonate, but may be of dolomite, silica, or other minerals, in successive concentric layers, commonly around a nucleus such as a shell fragment, an algal pellet, or a quartz-sand grain, in shallow, wave-agitated water; it often shows an internal radiating fibrous structure indicating outward growth or enlargement at the site of deposition. Ooliths are frequently formed by inorganic precipitation, although many noncalcareous ooliths are produced by replacement, in which case they are less regular and spherical, and the concentric or radial internal structure is less well-developed, than in accretionary oolites. The term was originally used for a rock composed of ooliths (an *oolite*), and is sometimes so used today. Cf: *pisolith.* Also spelled: oölith. Syn: *ooid; oolite; ovulite.*

oolitic (o-o-lit′-ic) Pertaining to an oolite, or to a rock or mineral made up of ooliths; e.g. an "oolitic ironstone", in which iron oxide or iron carbonate has replaced the calcium carbonate of an oolitic limestone. Also spelled: oölitic.

oolitic texture The texture of a sedimentary rock consisting largely of ooliths showing tangential contacts with one another.

oolitization (o-o′-li-ti-za′-tion) The act or process of forming ooids or an oolitic rock. Also, the result of such action or process.

oolitoid (o-o-lit′-oid) A sedimentary particle similar in size and shape to an oolith, but lacking its internal structure (Bissell and Chilingar, 1967, p.162). Cf: *pseudo-oolith.*

ooloid (o′-o-loid) A term used by Martin (1931, p.15) for a tiny, elliptically shaped, concretionlike siliceous form constructed of thin concentric layers around a central siliceous mass; found singly or cemented together in irregularly shaped clusters embedded in silicified shells of bryozoans, brachiopods, etc.; may compose the entire pseudomorphic shell. See also: *beekite.*

oomold (o′-o-mold) A spheroidal opening in a sedimentary rock or insoluble residue, produced by solution of an oolith. Adj: oomoldic. See also: *oolicast.*

Oort Cloud The swarm of comets whose members occasionally enter the inner solar system but most of the time remain in orbits about 10,000 astronomical units distant from the sun. They were ejected into these distant orbits by the giant planets at the time of solar system formation. Named for Dutch astronomer Jan Oort.

oosparite (o-o-spar′-ite) A limestone containing at least 25% ooids and no more than 25% intraclasts and in which the sparry-calcite cement is more abundant than the carbonate-mud matrix (micrite) (Folk, 1959, p.14). It is common in environments of high wave or current energy, where the spar represents pore-filling cement.

oosterboschite (oos′-ter-bosch-ite′) A metallic yellowish-white orthorhombic mineral: $(Pd,Cu)_7Se_5$.

oovoid (o-o′-void) A void in the center of an incompletely replaced oolith.

ooze [geog] A piece of soft, muddy ground, such as a mudbank; a marsh, fen, or bog resulting from the flow of a spring or brook.

ooze [marine geol] A pelagic sediment consisting of at least 30% skeletal remains of pelagic organisms (either calcareous or siliceous), the rest being clay minerals. Grain size is often bimodal (partly in the clay range, partly in the sand or silt range). Oozes are further defined by their characteristic organisms: *diatom ooze; foraminiferal ooze; globigerina ooze; pteropod ooze; radiolarian ooze.* See also: *calcareous ooze; siliceous ooze.*

ooze [sed] (a) A soft, soupy mud or slime, typically found covering the bottom of a river, estuary, lake or ocean. (b) Wet earthy material that flows gently, or that yields easily to pressure.

oozy (ooz′-y) Pertaining to or composed of ooze; e.g. "oozy fraction" of soils in which mineral grains are less than one micrometer in diameter.

opacite (o′-pa-cite) An obsolete general term applied to swarms of opaque, microscopic grains in rocks, esp. as rims that develop mainly on biotite and hornblende phenocrysts in volcanic rocks, apparently as a result of post-eruption oxidation and dehydration. Opacite is generally supposed to consist chiefly of magnetite dust. Cf: *viridite; ferrite [ign].*

opal (o′-pal) A mineral or mineral gel: $SiO_2 \cdot nH_2O$. It has been shown by electron diffraction to consist of packed spheres of silica; some so-called opal gives weak X-ray patterns of cristobalite or tridymite. Opal has a varying proportion of water (as much as 20% but usually 3 to 9%); it occurs in nearly all colors, is transparent to nearly opaque and typically exhibits a marked iridescent *play of color.* It differs from quartz in being isotropic, and has a lower refractive index than quartz and is softer and less dense. Opal usually occurs massive and frequently pseudomorphous after other minerals, and is deposited at low temperatures from silica-bearing water. It is found in cracks and cavities of igneous rocks, in flintlike nodules in limestones, in mineral veins, in deposits of thermal springs, in siliceous skeletons of various marine organisms (such as diatoms and sponges), in serpentinized rocks, in weathering products, and in most chalcedony and flint. The transparent colored varieties exhibiting opalescence are valued as gemstones. Syn: *opaline.*

opal-agate (o′-pal-ag′-ate) A variety of banded opal having different shades of color and an agate-like structure, consisting of alternate layers of opal and chalcedony. Cf: *agate opal.*

opalescence (o-pal-es′-cence) A milky or somewhat pearly appearance or luster of a mineral, such as that shown by opal and moonstone. Cf: *play of color.*

opaline (o′-pal-ine) n. (a) Any of several minerals related to or resembling opal; e.g. a pale blue to bluish-white opalescent or girasol corundum, or a brecciated impure opal pseudomorphous after serpentine. (b) *opal.* (c) An earthy form of gypsum. (d) A rock with a groundmass or matrix consisting of opal. adj. Resembling opal, esp. in appearance; e.g. "opaline feldspar" (labradorite) or "opaline silica" (tabasheer).

opalite (o′-pal-ite) An impure, colored variety of common opal; e.g. *myrickite.*

opalized wood (o′-pal-ized) *silicified wood.*

opal jasper *jasper opal.*

opal phytolith A discrete, distinctively shaped *phytolith [paleont],* or solid body of isotropic silica, usually less than 80 micrometers in diameter, precipitated by terrestrial plants (sedges, reeds, some woods, and esp. grasses) as excess material or as reinforcement of cell structures. Such bodies may represent recent or fossil forms, can be transported by wind, and may be deposited in the ocean. Syn: *plant opal; grass opal.*

opaque (o-paque′) Said of a material that is impervious to visible light, or of a material that is impervious to radiant energy other than visible light, e.g. X-rays. Cf: *transopaque; translucent; transparent.*

opaque attritus Attrital material in coal, consisting of abundant particles of *inertinite,* opaque in transmitted light. Cf: *translucent attritus.*

opaque zone A zone of subsurface rocks from or below which no seismic reflections are obtained. Cf: *transparent zone.*

opdalite (op′-dal-ite) (a) A member of the *charnockite series,* equivalent to orthopyroxene-bearing *granodiorite.* (b) A orthopyroxene-biotite *granodiorite.* It was named by Goldschmidt in 1916 after Opdal, Norway. Cf: *farsundite.* Obsolete.

open bay (o′-pen) An indentation between two capes or headlands, so broad and open that waves coming directly into it are nearly as high near its center as on adjacent parts of the open sea; a bight.

opencast mining (o′-pen-cast) *opencut mining.*

open-cavity ice (o′-pen-cav′-i-ty) Ice that results by condensation from water vapor in an open cavity or crack in the ground. It is similar to hoarfrost, except that the ice crystals grow in cavities rather than on the surface. It is common in thermal contraction cracks and mine workings in permafrost.

open channel A conduit in which water flows with a free surface (ASCE, 1962).

open coast A coast exposed to the full action of waves and currents.

open-coast marsh A *salt marsh* found along an open coast. Cf: *coastal marsh.*

opencut mining (o′-pen-cut) Surficial mining, in which the valuable rock is exposed by removal of overburden. Coal, numerous nonmetals, and metalliferous ores (as of iron and copper) are worked in this way. Cf: *quarrying [mining].* Syn: *strip mining; opencast mining; openpit mining; contour mining.*

open fold A fold with an inter-limb angle between 70° and 120° (Fleuty, 1964, p.470).

open form A crystal form whose faces do not enclose space, e.g. a trigonal prism. Cf: *closed form.*

open gash fracture *gash fracture.*

open hole An uncased well or borehole, or that portion extending below the depth at which *casing* has been set. Cf: *cased hole.*

open ice (a) Ice that is sufficiently broken up to permit passage of vessels. (b) broken ice.

open joint *fissure [geol].*

open lake (a) A lake that has an effluent; e.g. a *drainage lake.* Ant: *closed lake.* (b) A lake having open water, free of ice or emergent vegetation (Wetzel, 2001, p.47).

open-packed structure In crystal structure, a pattern of stacking of equal spheres in an orthogonal arrangement such that each sphere is in contact with six others. Cf: *close-packed structure.*

open pack ice *Pack ice* in which the concentration is 4/10 through 6/10 with many *leads* and *polynyas;* the *floes* are generally not in contact with one another. Cf: *scattered ice.*

open packing The manner of arrangement of uniform solid spheres packed as loosely as possible so that the porosity is at a maximum; e.g. *cubic packing.* Ant: *close packing.*

openpit mining (o'-pen-pit) *opencut mining.*

open rock Any stratum sufficiently open or porous to contain a significant amount of water or to convey it along its bed.

open sound A *sound [coast]* with large openings between the protecting islands.

open-space structure A structure in a carbonate sedimentary rock, formed by the partial or complete occupation by internal sediments and/or cement (Wolf, 1965).

open structure A geologic structure which, when represented on a map by contour lines, is not surrounded by closed contours. Ant: *closed structure.*

open system [chem] A chemical system in which, during the process under consideration, both energy (thermal and mechanical) and mass may be transferred into or out of the system. Cf: *closed system [chem].*

open system [permafrost] A condition of freezing of the ground in which additional groundwater is available through either free percolation or capillary movement (Muller, 1947, p.219), exemplified by the pingos of East Greenland. Ant: *closed system [permafrost].*

open traverse A surveying traverse that starts from a station of known or adopted position but does not terminate upon such a station and therefore does not completely enclose a polygon. Cf: *closed traverse.*

open valley (a) A broad band of lowland, with relatively straight and parallel valley sides, through which a stream swings from side to side in broad, open curves (Rich, 1914, p.469). (b) A strath produced by progressive widening of a valley by lateral stream cutting (Bucher, 1932, p.131).

open venation A vein pattern in which the lateral (secondary) veins do not anastomose.

open water [ice] A relatively large area of freely navigable water in an ice-filled region; specif. water in which the *concentration* of *floating ice* is less than $^1/_8$ (or $^1/_{10}$). Cf: *ice free; polynya.*

open water [lake] (a) Lake water that remains unfrozen or uncovered by ice during the winter. (b) Lake water that is free of emergent vegetation or artificial obstructions and of dense masses of submerged vegetation at very shallow depths.

open well (a) A well large enough (one meter or more in diameter) for a person to descend to the water level. (b) An artificial pond formed where a large excavation into the zone of saturation has been filled with water to the level of the water table (Veatch and Humphrys, 1966, p.351).

openwork (o'-pen-work) Said of a gravel with unfilled voids.

operational facies (op-er-a'-tion-al) A term used by Krumbein and Sloss (1963, p.328) for stratigraphic *facies* designating lateral variations of any characteristic of a defined stratigraphic unit, occupying mutually exclusive areas bounded by arbitrarily (or preferably, quantitatively) determined limits, and usually comprising one or several lithosomes and biosomes that occur in vertical succession or are intertongued.

operational unit A term used by Sloss et al. (1949, p.109-110) for an arbitrary stratigraphic unit that is distinguished by objective criteria for some practical purpose (such as regional facies mapping or analysis); e.g. a unit delimited by easily recognizable and traceable markers, or a unit defined by the velocity of transmission of seismic or sonic energy. Its boundaries do not necessarily correspond with those of any conventional stratigraphic unit. Syn: *parastratigraphic unit.*

operator error An effect in all experimental procedures conducted by human operators that arises from a constant *bias* characteristic of an operator and inconsistent values produced by one operator or among a group of operators. Also called: "operator variance" or "operator variation".

opercula Pl. of *operculum.*

opercular (o-per'-cu-lar) Relating to an *operculum [palyn].*

operculate [paleont] adj. Having an *operculum.* n. An operculate gastropod.

operculate [palyn] (o-per'-cu-late) In *pollen grains,* having a cover of *ektexine* over an *aperture.* In *archeopyles,* covered by a lid.

operculum [bot] (o-per'-cu-lum) A lid or cover, as in a protistan, a moss capsule, or the fungi; it may be part of a cell wall.

operculum [paleont] (a) A corneous or calcareous plate that develops on the posterior dorsal surface of the foot of a gastropod and that serves to close the aperture. (b) A generally uncalcified lamina or flap, hinged or pivoting on condyles, that closes the zooidal orifice in cheilostome bryozoans. (c) The valves (terga or scuta) and associated membranes forming an apparatus that guards the aperture of cirripede crustaceans. (d) A lid, usually disklike and composed of one or two parts, that closes an opening (such as the anus or genital opening) of an arachnid; a plate adjoining the appendages of the genital segment of a merostome. (e) A lidlike covering of the calice in some solitary corals, formed of one or more independent plates. (f) A structure that may serve to close the pseudostome of chitinozoans. It may be external in position or sunken within the neck. (g) The flat pore-bearing base of the podoconus in nassellarian radiolarians; the central part of the astropyle of phaeodarian radiolarians. (h) In the vertebrates, a movable, flat, scalelike complex of bones or a flap of soft tissues that serves as a controllable cover of an opening, e.g. in Osteichthyes the gill cover, in frogs and toads the small bone that partially closes the lateral opening of the middle ear. Pl: opercula.

operculum [palyn] (a) A lid consisting of the plate or plates that originally closed the archeopyle of a dinoflagellate or the pylome of an acritarch. (b) Ektexinous structure that covers (part of) an ectoaperture of a pollen grain. (c) For certain hilate spores, an exinous lid associated with formation of the hilum. Pl: *opercula.* Adj: operculate.

opesiula (o-pes'-iu-la) One of the small notches or pores in a *cryptocyst* for the passage of parietal muscles attached to the frontal membrane of some anascan cheilostomes (bryozoans). Pl: opesiulae.

opesium (o-pe'-si-um) In some anascan cheilostomes (bryozoans), an opening defined by the inner margin of the *cryptocyst,* serving as a passage for the *lophophore.* It also serves as a passageway for parietal muscles in some anascans lacking opesiulae. Pl: opesia.

opferkessel (op'-fer-kes-sel) A *solution pan* formed on silicate rocks and commonly stained by reddish-brown iron minerals. Etymol: German, "sacrificial basin".

Ophiacodontia (O-phi'-a-co-dont'-i-a) A suborder of pelycosaurian synapsid reptiles, characterized by generally conservative structure; later forms appear to be more highly aquatic than most pelycosaurs. Range, Lower Pennsylvanian to Lower Permian.

ophicalcite (o-phi-cal'-cite) A recrystallized metamorphic rock composed of calcite and serpentine, commonly formed by dedolomitization of a siliceous dolostone. Some ophicalcites are highly veined and brecciated and are associated with serpentinite. Cf: *ophicarbonate.*

ophicarbonate (o-phi-car'-bon-ate) A metamorphic rock composed of serpentine and a carbonate mineral (calcite, dolomite, or magnesite). Cf: *ophicalcite.*

ophiocistioid (o-phi-o-cis'-ti-oid) Any quinqueradiate, free-living echinozoan belonging to the class Ophiocistioidea, having a depressed, dome-shaped body covered entirely or on one side only by plates. Range, Lower Ordovician to Upper Silurian (possibly Middle Devonian).

ophiolite (o'-phi-o-lite') An assemblage of *ultramafic* and *mafic* intrusive and extrusive rocks, widely believed to represent oceanic crust (Coleman, 1977). In a complete ophiolite, sheared and serpentinized ultramafic rocks are overlain by less deformed ultramafic and mafic cumulates, which are overlain by a sheeted dike complex topped by pillow lavas and deep marine sediments. The entire sequence is rarely preserved. Originally defined as a group of mafic and ultramafic igneous rocks ranging from spilite and basalt to gabbro and peridotite, including rocks rich in serpentine, chlorite, epidote, and albite derived from them by later metamorphism, whose origin was associated with an early phase of the development of a geosyncline. The term was originated by Steinmann in 1905.

ophiolitic suite (o'-phi-o-lit'-ic) The association of ultramafic rocks, coarse-grained gabbro, coarse-grained diabase, volcanic rocks, and radiolarian chert that characteristically occurs in eugeosynclinal sequences.

ophirhabd (o'-phi-rhabd) A sinuous oxea (sponge spicule). Cf: *eulerhabd.*

ophite (o'-phite) A general term for diabases that have retained their ophitic structure although the pyroxene is altered to uralite. The term was originated by Palasson in 1819. Rarely used.

ophitic (o-phit'-ic) Igneous texture characterized by plagioclase laths largely or entirely enclosed by pyroxene grains. Cf: *subophitic.* The term *diabasic* was distinguished from "ophitic" by Kemp (1900, p.158-159), who considered the latter as requiring an excess of augite over plagioclase, and the former as having a predominance of plagioclase, with augite filling the interstices. Cf: *poikilitic; poikilophitic.* Syn: *doleritic.*

ophiuroid (o'-phi-ur-oid') Any asterozoan echinoderm belonging to the subclass Ophiuroidea, characterized by slender, elongate arms that are distinct from the disc in almost all cases; e.g. starfishlike animals such as brittle stars and basket stars. Var: ophiurid; ophiuran.

ophthalmic ridge (oph-thal'-mic) A longitudinal ridge above the *compound eye* of a merostome and extending forward and backward from it.

ophthalmite [meta] (oph-thal'-mite) A *migmatite* characterized by augen or other lenticular aggregates of minerals (Dietrich and Mehnert, 1961). Rarely used.

ophthalmite [paleont] *eyestalk.*

opisometer (op-i-som'-e-ter) A *chartometer* consisting of a small toothed wheel, geared to a pointer moving over a graduated recording dial, used for measuring distances on a map. The wheel is run along a given line, which may be curved or irregular (such as one representing a stream, road, or railway).

opisthobranch (op-is'-tho-branch) Any marine gastropod belonging to the subclass Opisthobranchia, characterized by the reduction or absence of the shell. Range, Mississippian to Holocene.

opisthoclade (op-is'-tho-clade) A clade or bar in the ebridian skeleton that arises from an upper actine and is directed toward the posterior. In the triaene ebridian skeleton, it may rejoin the distal extremity of the rhabde. Cf: *proclade.*

opisthocline (op-is'-tho-cline) (a) Said of the body of the shell (and in some genera, of the hinge teeth) of a bivalve mollusk, sloping (from the lower end) in the posterior or backward direction, so that the midumbonal line intersects the dorsal margin at an angle of more than 90°. Syn: *backward obliquity.* (b) Said of the growth lines that incline backward relative to the growth direction of a gastropod shell. Cf: *prosocline; acline [paleont]; obliquity.*

opisthodetic (op-is'-tho-det'-ic) Said of a *ligament* of a bivalve mollusk situated wholly posterior to (or behind) the beaks. Cf: *amphidetic.*

opisthogyrate (op-is'-tho-gy'-rate) Said of the *umbones* (of a bivalve mollusk) curved so that the beaks point in the posterior or backward direction. Ant: *prosogyrate.*

opisthoparian (op-is'-tho-par'-i-an) adj. Of or concerning a trilobite whose facial sutures extend backward from the eyes to the posterior margin of the cephalon; e.g. an "opisthoparian facial suture" that crosses a cheek, passes along the medial edge of the eye, and intersects the posterior border of the cephalon medial to the genal angle. Cf: *proparian.* n. An opisthoparian trilobite; specif. a trilobite of the order Opisthoparia including those in which the genal angles or genal spines are borne by the free cheeks.

opisthosoma (op-is'-tho-so'-ma) The posterior part of the body of an arthropod; esp. the *abdomen* behind the *prosoma* of a merostome or following the fourth pair of legs of an arachnid. Syn: *thoraceton.*

opisthosome (op-is'-tho-some) A dark fusiform body at the base of the body chamber of a chitinozoan, usually convex upward or even spherical, and commonly having a ragged appearance (as if burst open) and a longitudinally striate surface. It is not always present, and may at times be mistaken for a fold of the body-chamber wall. Cf: *prosome.*

Opoitian (O-po-i-ti'-an) South Pacific stage: lowermost Pliocene (above Miocene Kapitean, below Waipipian).

opoka (op-o'-ka) A porous, flinty, and calcareous sedimentary rock, with conchoidal or irregular fracture, consisting of fine-grained opaline silica (up to 90%), and hardened by the presence of silica of organic origin (silicified residues of radiolaria, sponge spicules, and diatoms). Etymol: Polish.

Oppel zone (Op'-pel-zone) The Oppel zone, named after the German biostratigrapher Albert Oppel, has been considered as a type of assemblage zone or as a multi-taxon concurrent-range zone (ISG, 1976, p.57-58). However, neither Oppel nor subsequent biostratigraphers have precisely defined the biozones used by Oppel, which, in any case, do not appear to correspond consistenly to any one kind of biozone. For this reason, the Oppel zone was not included as a distinct kind of biozone in the second edition of the International Stratigraphic Guide (1994).

opportunistic species (op'-por-tun-is'-tic) A species that tends to be generalized in its adaptations, disperses and reproduces rapidly (a high "r" in the logistic population-growth equation), but remains at small population size because of the unstable, unpredictable, or temporary nature of the environment occupied. Syn: *fugitive species; generalist species; r strategist.* Ant: *equilibrium species.*

opt Abbreviation for troy ounces per ton, used to report grade of precious metal ore.

optical absorption spectroscopy *visible region spectroscopy.*

optical activity (op'-ti-cal) The property or ability of a mineral, e.g. quartz, to rotate the plane of polarization of light. Such a mineral is said to be optically active. Syn: *rotary polarization.*

optical axis In an optical system, the line passing through the nodal points of a lens.

optical calcite Crystalline calcite so clear that it has value for optical use. It is usually *Iceland spar.*

optical center That point on the axis of an optical system at which light rays cross.

optical character In optical crystallography, the designation positive or negative, depending on the values of the different indices of refraction of a mineral. For uniaxial crystals with two indices of refraction, if the index of the extraordinary ray exceeds that of the ordinary ray, the mineral has a positive optical character. For biaxial crystals with three indices of refraction, the intermediate index is nearer in value to the smaller index than to the larger one for optically positive crystals.

optical constant Any characteristic optical property of a crystal, e.g. index of refraction, optic angle.

optical crystallography That branch of crystallography that deals with the optical properties of crystals; *crystal optics.* Cf: *optical mineralogy.*

optical emission spectrometry Chemical analysis performed by heating the sample to high temperatures whereupon the atoms emit light of definite wavelengths characteristic of the specific elements or molecules present (May and Cuttitta, 1967, p.130). See also: *emission spectrum; arc spectrum; flame spectrum; spark spectrum.*

optical emission spectroscopy The observation of an optical emission spectrum and all processes of recording and measuring that go with it.

optical glass Glass that is suitable for use as prisms, lenses, and other optical items.

optically pumped magnetometer (op'-ti-cal-ly) A type of magnetometer that measures total magnetic-field intensity by means of the precession of magnetic atoms, with precession frequency proportional to field intensity. The magnetic atoms are usually gaseous rubidium, cesium, or helium, which are magnetized by optical pumping, i.e., irradiation by circularly polarized light of suitable wavelength. See also: *cesium-vapor magnetometer; rubidium-vapor magnetometer.*

optical microscope A microscope that utilizes visible light for illumination.

optical mineralogy That branch of science dealing with the optical properties of minerals. Cf: *optical crystallography.*

optical oceanography That aspect of physical oceanography which deals with the optical properties of seawater and natural light in seawater.

optical path The path along which light rays travel through the *optical system* of a microscope or other optical apparatus. Syn: *path [optics].*

optical pyrometer A type of *pyrometer* that measures high temperature by comparing the color and intensity of light from the hot material with that of a filament of known temperature. It is used to determine the temperature of incandescent lavas.

optical rotation The angle of rotation, measured in degrees, of plane-polarized light as it passes through an optically active crystal. Cf: *specific rotation.*

optical section The image seen when a microscope's plane of focus is half-way through a palynomorph.

optical square A small hand instrument used in surveying for accurately setting off a right angle by means of two plane mirrors placed at an angle of 45 degrees to each other or by means of a single plane mirror so placed that it makes an angle of 45 degrees with a sighting line.

optical system The lenses, prisms, and mirrors of an optical apparatus such as a microscope, through which goes the *optical path.* Syn: *optical train.*

optical train *optical system.*

optical twinning A type of twinning in quartz, the individuals of which are alternately right-handed and left-handed, e.g. Brazil twinning. Syn: *chiral twinning.*

optical wedge (a) A refracting prism of very small angle, inserted in an optical train to introduce a small bend in the ray path. It is used in the eyepiece of certain stereoscopes. (b) A strip of film or a glass plate used to reduce the intensity of light or radiation (gradually or in steps, as in determining the density of a photographic negative), and having a layer of neutral or colored substance varying progressively in transmittance with distance along the wedge; e.g. a *step wedge.*

optic angle (op'-tic) The acute angle between the two optic axes of a biaxial crystal; its symbol is 2V. See also: *apparent optic angle.* Syn: *axial angle; optic axial angle.*

optic axial angle (op'-tic-ax'-i-al) *optic angle.*

optic axis A direction in an anisotropic crystal along which there is no double refraction. In tetragonal and hexagonal crystals it is parallel to the threefold, fourfold, or sixfold symmetry axis; in orthorhombic, monoclinic, and triclinic crystals there are two optic axes, which are determined by the indices of refraction. See also: *primary optic axis; secondary optic axis.*

optic ellipse Any noncircular section of an *index ellipsoid.*

optic indicatrix *indicatrix.*

optic normal The principal vibration direction of a biaxial crystal that is perpendicular to the plane of the optic axes.

optic plane In optics, the plane in a biaxial *indicatrix* containing both optic axes.

optimal channel network A fluvial network that develops so as to minimize energy expenditure in any link while simultaneously equalizing the energy expenditure per unit channel anywhere in the network (Rigon et al., 1993).

optimum moisture content (op'-ti-mum) The water content at which a specified force can compact a soil mass to its maximum dry unit weight.

optimum water content The water content in a soil sample at which the *maximum dry density* is achieved. Cf: *compaction test.*

OQQ *orthophoto quarter-quadrangle.*

ora (o'-ra) Pl. of *os [palyn].* Cf: *ossa.*

oral (o'-ral) adj. (a) Said of the surface on which the mouth of an invertebrate is situated, such as the upward-directed *actinal* surface of the theca of an edrioasteroid. Also, relating to or located on an oral surface, or situated at, near, or toward the mouth or peristome (such as of an echinoderm); e.g. an "oral pole" representing the end theca containing the mouth in a cystoid. Ant: *aboral.* (b) Pertaining to the *orifice* (not the mouth) of a bryozoan zooid. (c) Toward the upper side of a conodont element. (d) In vertebrates, a general term related to any aspect of the mouth. n. An *oral plate* of an echinoderm.

oral disk The fleshy, more or less flattened wall closing off the upper or free end of the cylindrical column that forms the sides of a scleractinian coral polyp, its center containing the mouth. Cf: *basal disk.* Also spelled: oral disc.

oral frame A structure in edrioasteroids underlying the external oral area, formed by proximal ambulacral floorplates and commonly other elements. It surrounds the central lumen or stomial chamber (Bell, 1976).

oral margin The trace of the oral side of a *conodont element* in lateral (side) view. The term has also been used for the oral side itself.

oral membrane A sheet of cilia in the gullet of a tintinnid.

oral pinnule Any proximal pinnule of a crinoid, differentiated from distal pinnules in function, structure, or both.

oral plate (a) Any of five interradially disposed plates forming a circlet surrounding or covering the mouth of an echinoderm. Syn: *oral.* (b) One of the elements in edrioasteroids that form the external covering of the oral area and roof the central lumen; distally continuous with the ambulacral cover-plate series (Bell, 1976).

oral pole (a) The open end of a flask-shaped *chitinozoan* that includes the neck and the mouth. (b) The center of its circular operculum.

oral side The upper side of a *conodont element* opposite that toward which the basal cavity opens. It commonly supports denticles, nodes, and ridges in compound and platelike conodont elements. Cf: *aboral side.*

oral tooth One of the sharp triangular projections around the basal shell opening in phaeodarian radiolarians.

orangeite A name proposed by Wagner in 1928 for a rock type that he had previously named micaceous kimberlite. R. H. Mitchell (1995) revived the name orangeite for micaceous or Type 2 kimberlites. Orangeite contains phlogopite as a major mineral; smaller grains in the groundmass may vary from phlogopite to tetraferriphlogopite. Other groundmass minerals are diopside mantled by titanian aegirine, spinels (chromite to magnetite), perovskite, Sr-rich apatite, carbonates, and serpentine. Unlike Type 1 kimberlites, orangeites lack monticellite, and olivine is commonly subordinate to phlogopite in abundance. Syn: Type 2 kimberlite.

orangite (or′-an-gite) A bright orange-yellow variety of thorite.

oranite (or′-an-ite) A lamellar intergrowth of a potassium feldspar and a plagioclase near anorthite.

orate (o′-rate) Said of a porate pollen grain having an internal opening in the endexine.

O ray In uniaxial crystals, the ray that vibrates perpendicular to the optic axis; the *ordinary ray.* Cf: *E ray [cryst].*

orbicular (or-bic′-u-lar) Igneous texture characterized by numerous orbicules. Cf: *centric; nodular; spheroidal; spherulitic.*

orbiculate (or-bic′-u-late) Said of a circular or disk-shaped leaf.

orbicule (or′-bi-cule) A more or less spherical body, from microscopic size to several centimeters or more in diameter, commonly composed of concentric shells of alternating light and dark minerals. Their centers may or may not exhibit xenolithic nuclei. Cf: *spherulite [petrology].*

orbicule [palyn] (or′-bi-cule) A distinctive orbicular granule of *sporopollenin* produced by the *tapetum,* scattered on the surface of pollen; e.g., on TCT pollen grains. Syn: *ubish bodies.*

orbit [paleont] (or′-bit) (a) A circular opening in the anterior part of the carapace of a decapod crustacean, enclosing the eyestalk. (b) In the vertebrates, a circular opening between skull roof and cheek that houses the eye.

orbit [waves] The path of a water particle affected by wave motion, being almost circular in deep-water waves and almost elliptical in shallow-water waves. Orbits are generally slightly open in the direction of wave motion, giving rise to *wave drift.*

orbital carina (or′-bit-al) The ridge forming the orbital margin on some decapod crustaceans (Holthuis, 1974, p.735).

orbital region Part of the carapace of some decapod crustaceans behind the eyes, bordered by the frontal and antennal regions (TIP, 1969, pt. R, p.92).

orbitoid (or′-bi-toid) Any foraminifer belonging to the superfamily Orbitoidacea, characterized by large discoidal saddle-shaped or stellate tests with walls composed of radially arranged calcite crystals and with bilamellid septa. Range, Cretaceous to Holocene.

orbitolinid (or′-bi-to-lin′-id) Any foraminifer belonging to the family Orbitolinidae, characterized by a relatively large conical test ranging from a high pointed cone to a broad shield or disc. Range, Lower Cretaceous to Eocene.

orbitolite (or-bit′-o-lite) Any foraminifer belonging to the genus *Orbitolites* of the suborder Miliolina, characterized by a discoidal test containing numerous small chambers in annular series. Range, Upper Paleocene to Eocene.

orcelite (or-cel′-ite) A metallic hexagonal mineral: $Ni_{5-x}As_2(x=0.23)$.

ordanchite (or-danch′-ite) An extrusive rock containing phenocrysts of sodic plagioclase, hauyne, hornblende, augite, and some olivine; an olivine-bearing hauyne *trachyandesite.* Named by Lacroix in 1917 for Banne d'Ordanche, Auvergne, France. Obsolete.

order [geomorph] (a) *stream order.* (b) *basin order.*

order [petrology] In the CIPW classification of igneous rocks, the basic unit of the *class [petrology].*

order [taxon] A category in the hierarchy of classification of plants and animals intermediate between class and family. In botany, the name of an order characteristically ends in -ales; e.g. Filicales. Cf: *suborder.*

order-disorder inversion (or′-der-dis′-or-der) *substitutional transformation.*

order-disorder polymorphs Two crystal substances of the same composition but of different atomic arrangement with respect to *ordering.* In the higher-temperature or disordered form, two or more elements are randomly distributed over a particular set of atom sites; in the lower-temperature or ordered form, the atoms become ordered with respect to the same sites. The ordered form usually has lower symmetry.

order-disorder transformation A *transformation [cryst]* between two polymorphic forms, one of which has a more ordered structure than the other. In general, if the ordered, low-symmetry, low-temperature form is heated, a point is reached at which some portion of the structure becomes disordered, or random, usually with an increase in crystal symmetry, to produce the high-temperature form. Cf: *substitutional transformation; lambda transformation.*

ordering The phenomenon of *order in minerals.*

order in minerals The segregation of certain species of atoms into a subset of all occupied sites in a *crystal structure.* The term is usually applied to the segregation of cations that are otherwise randomly distributed (disordered) over those sites. Cf: *disorder in minerals.* See also: *short-range order; long-range order.*

order of crystallization The apparent chronologic sequence in which crystallization of the various minerals of an assemblage takes place, as evidenced mainly by textural features.

ordinal scale A scale of measurement in which observation data are ranked in a hierarchy of states. The ranks are monotonic but not uniform in magnitude.

ordinary chert (or′-di-nar′-y) A generally homogeneous *smooth chert.* It has an even fracture surface, is nearly opaque, has slight granularity or crystallinity, and may be of any color (chiefly white, gray, or brown, or sometimes mottled) (Ireland et al., 1947, p.1485).

ordinary chondrite A stony meteorite characterized by abundant mafic chondrules, metal, and troilite (FeS) embedded in a finely crystalline matrix. On a plot of $\delta^{17}O$ vs. $\delta^{18}O$ they plot 0.5-1.3% above the terrestrial fractionation line. Ordinary chondrites are the most abundant type of meteorite and are divided into H (high-Fe), L (low-Fe), LL (low-Fe, low metal) groups. The classes show distinctive abundances of nonvolatile lithophile and siderophile elements and degrees of oxidation. Ordinary chondrites are also classified into petrographic (also called petrologic) types from 3 (low grade) to 7 (high grade) according to the presence of features suggesting various amounts of secondary thermal metamorphism and recrystallization. Thus an L3.0 chondrite has experienced negligible minimal thermal metamorphism, while an L6 or H6 chondrite has experienced enough thermal metamorphism to make it difficult to recognize chondrules.

ordinary coccolith One of the unmodified coccoliths in a coccolithophore exhibiting dimorphism.

ordinary lead *common lead.*

ordinary ray *O ray.*

ordination In *geostatistics,* the placing of multivariate observations into a logical sequence, based on the similarities between the observations.

ordnance datum (ord′-nance) A name given to several horizontal datums to which heights have been referred on official maps of the British Ordnance Survey; specif. in Great Britain (but not Ireland) the mean sea level at Newlyn in Cornwall. Abbrev: OD.

ordoñezite (or-don′-ez-ite) A colorless, gray, or brown tetragonal mineral of the tapiolite group: $ZnSb^{5+}_2O_6$. Also spelled: ordonezite.

ordosite (or′-do-site) A dark-colored *syenite,* containing about 60 percent sodic clinopyroxene. The name, given by Lacroix in 1925, is for the Ordos Plateau, China. Obsolete.

Ordovician (Or-do-vi′-cian) A system of the *Paleozoic* Erathem of the Standard Global Chronostratigraphic Scale, above the *Cambrian* and below the *Silurian.* Also the time during which these rocks were formed, the Ordovician Period, covering the time span between 500 and 435 Ma. It is named after a Celtic tribe called the Ordovices.

ore (a) The naturally occurring material from which a mineral or minerals of economic value can be extracted at a reasonable profit. Also, the mineral(s) thus extracted. The term is generally but not always used to refer to metalliferous material, and is often modified by the name of the valuable constituent; e.g., "iron ore". See also: *mineral deposit; orebody; ore mineral.* (b) The term "ores" is sometimes applied collectively to opaque accessory minerals, such as ilmenite and magnetite, in igneous rocks.

ore beds Metal-rich layers in a sequence of sedimentary rocks.

ore block A section of an orebody, usually rectangular, that is used for estimates of overall tonnage and quality. See also: *blocking out.*

ore blocked out *developed reserves.*

orebody (ore′-bod-y) A continuous, well-defined mass of material of sufficient *ore* content to make extraction economically feasible. See also: *mineral deposit.*

örebroite (ö-re-bro′-ite) A dark brown trigonal mineral of the *welinite* group: $Mn^{2+}_3(Sb^{5+},Fe^{3+})(SiO_4)(O,OH)_3$.

ore channel A little-used term for the orebody or lode, including both gangue and economically valuable minerals. See also: *lodestuff.* Syn: *lode country.*

ore chimney *pipe [ore dep].*

ore cluster A genetically related group of orebodies that may have a common root or source rock but that may differ structurally or otherwise.

ore control Any tectonic, lithologic, or geochemical feature considered to have influenced the formation and localization of ore. Cf: *metallotect.*

ore-forming fluid (ore′-form-ing) *mineralizer.*

oregonite (or′-e-gon-ite′) A metallic white hexagonal mineral: Ni_2FeAs_2.

ore guide Any natural feature, such as alteration products, geochemical variations, local structures, or plant growth, known to be indicative of an orebody or mineral occurrence. See also: *lithologic guide; stratigraphic guide.*

ore in sight *developed reserves.*

Orellan (O-rel′-lan) North American land mammal age: Upper Oligocene (after Chadronian, before Whitneyan).

ore magma A term proposed by Spurr (1923) for a magma that may crystallize into an ore; the sulfide, oxide, or other metallic facies of a solidified magma.

ore microscopy The study of opaque ore minerals in polished section with a reflected-light microscope. Syn: *mineragraphy; mineralography.*

ore mineral The part of an *ore*, usually metallic, which is economically desirable, as contrasted with the *gangue*.

Orenburgian (O'-ren-burg'-i-an) Stage in Russia: uppermost Upper Carboniferous (above Gzhelian, below Permian Asselian).

orendite (o-ren'-dite) A porphyritic leucite *lamproite* containing phlogopite phenocrysts in a nepheline-free reddish-gray groundmass of leucite, sanidine, phlogopite, amphibole, and diopside. Its name, given by Cross in 1897, is derived from Orenda Butte, Leucite Hills, Wyoming. Mitchell and Bergman (1991) suggest that this name should be replaced by diopside-sanidine-phlogopite *lamproite*.

ore of sedimentation *placer.*

oreography (ore-og'-ra-phy) *orography.*

ore pipe *pipe [ore dep].*

ore reserve The economically minable part of a *measured mineral resource* or *indicated mineral resource.* It includes diluting materials and allowances for losses that may occur when the material is mined. Appropriate assessments, which may include feasibility studies, have been carried out, and include consideration of and modification by realistically assumed mining, metallurgical, economic, marketing, legal, environmental, social, and governmental factors. These assessments demonstrate at the time of reporting that extraction could reasonably be justified (JORC, 1999). Cf: *reserve.*

ore roll *roll orebody.*

ore shoot An elongate pipelike, ribbonlike, or chimneylike mass of ore within a deposit (usually a vein), representing the more valuable part of the deposit. Syn: *shoot [ore dep].*

organ genus (or'-gan) A generic name applied to a fossil plant fragment which, even though detached and isolated, is assignable to a family. Since 2000, the term and concept are no longer used, according to the ICBN. Cf: *form genus; morphotaxon.*

organic (or-gan'-ic) adj. Pertaining or relating to a compound containing carbon, especially as an essential component. Organic compounds usually have hydrogen bonded to the carbon atom. Cf: *inorganic.* n. A substance containing carbon, as in such expressions as "organic-rich shale".

organic bank *bank [sed].*

organic evolution *evolution.*

organic geochemistry That branch of chemistry concerned with naturally occurring carbonaceous and biologically derived substances of geological interest.

organic hieroglyph *bioglyph.*

organic lattice *growth lattice.*

organic mound *bioherm.*

organic reef A *bioherm* of sufficient size to develop associated facies. It is erected by, and composed mostly of the remains of, sedentary, solitary, colonial, and sediment-binding organisms, generally marine: chiefly corals and algae, less commonly crinoids, bryozoans, sponges, mollusks, and other forms that live their mature lives near but below the surface of the water (although they may have some exposure at low tide). Their exoskeletal hard parts remain in place after death, and the deposit is firm enough to resist wave erosion. An organic reef may also contain still-living organisms. See also: *coral reef; algal reef.* Cf: *bank [sed].*

organic rock A sedimentary rock consisting primarily of the remains of organisms (plant or animal), such as of material that originally formed part of the skeleton or tissues of an animal. Cf: *biogenic rock.*

organic soil A soil in which the sum of the thicknesses of layers containing organic soil materials is generally greater than the sum of the thicknesses of mineral layers.

organic soil materials Soil materials that are saturated with water and have 174 g kg^{-1} or more organic carbon if the mineral fraction has 500 g kg^{-1} or more clay, or 116 g kg^{-1} organic carbon if the mineral fraction has no clay, or has proportional intermediate contents, or if never saturated with water, have 203 g kg^{-1} or more organic carbon.

organic texture A sedimentary texture resulting from the activity of organisms, e.g., *bioturbation.*

organic weathering Biologic processes and changes that assist in the breakdown of rocks; e.g. the penetrating and expanding force of roots, the presence of moss and lichen causing humic acids to be retained in contact with rock, and the work of animals (worms, moles, rabbits) in modifying surface soil. Syn: *biologic weathering.*

organogenic (or'-ga-no-gen'-ic) Said of a rock or sediment made up of products of organic activity; e.g. a crinoidal limestone (termed an "organogenic conglomerate" by Hadding, 1933).

organolite (or-gan'-o-lite) Any rock consisting mainly of organic material, esp. one derived from plants; e.g. coal, resin, and bitumen.

organometallic (or-ga'-no-me-tal'-lic) Said of a compound in which an atom of a metal is bound to an organic compound directly through a carbon atom. Cf: *metallo-organic.*

organosedimentary (or-ga'-no-sed'-i-men'-ta-ry) Pertaining to sedimentation as affected by organisms; e.g. said of a stromatolite, a sedimentary structure produced by the life processes of cyanobacteria.

organotrophic (or'-ga-no-troph'-ic) Relating to the development and nourishment of living organs.

organovaite-Mn A vitreous brown or pink monoclinic mineral of the *labuntsovite* group: $K_2MnNb_4(Si_4O_{12})_2O_4{\cdot}5\text{-}7H_2O$.

organovaite-Zn A vitreous pink, pinkish-brown or white monoclinic mineral of the *labuntsovite* group: $K_2Zn(Nb,Ti)_4(Si_4O_{12})_2(O,OH)_4{\cdot}6H_2O$.

orickite (or'-ick-ite) A metallic yellow hexagonal mineral: $CuFeS_2{\cdot}nH_2O$.

orido An Italian term for a gorge cut through a rock barrier holding a lake in a glaciated region, like those around Lago di Como in Italy.

orient [gem] (o'-ri-ent) (a) The minute play of color on, or just below, the surface of a gem-quality pearl, caused by diffraction and interference of light from the irregular edges of the overlapping crystals or platelets of aragonite that comprise the nacre of the pearl. (b) A pearl of great luster.

orient [surv] (a) To place or set a map so that the map symbols are parallel with their corresponding ground features. (b) To turn a plane table in a horizontal plane until all lines connecting positions on the plane-table sheet have the same azimuths as the corresponding lines connecting ground objects. (c) To turn a transit so that the direction of the zero-degree line of its horizontal circle is parallel to the direction it had in the preceding (or in the initial) setup or parallel to a standard line of reference. Etymol: Latin "oriens", "rising", used originally in connection with the rising of the Sun in the east.

oriental alabaster *onyx marble.*

oriental amethyst (a) Violet to purple variety of sapphire. (b) Any amethyst of exceptional beauty.

oriental cat's-eye A syn. of chrysoberyl *cat's-eye.*

oriental chrysolite Greenish-yellow chrysoberyl.

oriental jasper *bloodstone.*

orientation [cryst] (o'-ri-en-ta'-tion) In describing crystal form and symmetry, the placing of the crystal so that its crystallographic axes are in the conventional position.

orientation [photo] The direction in which an aerial photograph is turned with respect to observer or map. A single photo is best oriented for study when turned so that the shadows are cast toward the observer.

orientation [surv] The assignment or imposition of a definite direction in space; the act of establishing the correct relationship in direction, usually with reference to the points of the compass. Also, the state of being in such relationship.

orientation diagram In structural petrology, a general term for a fabric diagram.

oriented (o'-ri-ent-ed) (a) Said of a specimen or thin section that is so marked as to show its original orientation in place. (b) In *spatial statistics*, possessing a sense of bearing with respect to spatial axes, but not having a sense of direction. A fault line may be oriented north-south. See: *directional data.*

oriented core A full-diameter *core [drill]* whose azimuth was marked as it was being cut.

orientite (o'-ri-en-tite') A brownish-red to dark brown orthorhombic mineral: $Ca_8Mn^{3+}{}_{10}(SiO_4)_3(Si_3O_{10})_3(OH)_{10}{\cdot}4H_2O$.

orifice (or'-i-fice) (a) An opening in the *zooid* wall through which the lophophore and tentacles of a bryozoan are protruded. Cf: *aperture [paleont].* (b) An opening in the upper part of a crustacean shell, containing the *operculum.* (c) Any major opening through the outer covering of an echinoderm. (d) An aperture or other opening in a foraminiferal test. (e) A general term for mouth or aperture.

origin (or'-i-gin) (a) A point in a coordinate system that serves as an *initial point* in computing its elements or in prescribing its use; esp. the point defined by the intersection of coordinate axes, from which the coordinates are reckoned. The term has also been applied to the point to which the coordinate values of zero and zero are assigned (regardless of its position with reference to the axes) and to the point from which the computation of the elements of the coordinate system, or projection, proceeds. Syn: *origin of coordinates.* (b) Any arbitrary zero or starting point from which a magnitude is reckoned on a scale or other measuring device.

original dip (o-rig'-i-nal) *primary dip.*

original interstice An *interstice* that formed contemporaneously with the enclosing rock. Cf: *secondary interstice.* Syn: *primary interstice.*

original stream *consequent stream.*

original valley A valley formed by *hypogene* action, or by *epigene* action other than that of running water (Geikie, 1898, p.347).

origin of a datum The point in a geodetic network whose coordinates are fixed by definition. Not necessarily the same as the origin of the coordinate system involved.

origin of coordinates *origin.*

origofacies (o'-ri-go-fa'-cies) Facies of the primary sedimentary environment (Vassoevich, 1948); the sedimentary "facies" of most western authors (Teichert, 1958, p.2736).

Oriskanyan (O-ris-kan'-y-an) A North American provincial stage: Lower Devonian (above Helderbergian, below Onondagan).

orizite (or'-i-zite) *epistilbite.*

orlandiite A colorless or white triclinic mineral: $Pb_3Cl_4(Se^{4+}O_3){\cdot}H_2O$.

Orleanian (Or-le-an'-i-an) European mammal age: early Miocene (after Agenian, before Astaracian).

orlymanite A vitreous dark brown trigonal mineral: $Ca_4Mn^{2+}{}_3Si_8O_{20}(OH)_6{\cdot}2H_2O$.

ornament (or'-na-ment) A pattern of diagonal lines, plus-signs, curlicues, or the like, printed in black, gray, a contrasting color, or negative (white), over a color hue on a geologic map, distinguishing one cartographic unit from another of basically the same hue. The pattern may suggest lithology and/or internal arrangement of rock-unit constituents.

ornamental stone (or-na-men'-tal) An attractive natural stone, usually opaque, that is not practical for jewelry but is useful for fashioning into ornamental and decorative objects, such as figurines, ash trays, and lamp bases; e.g. onyx marble, agate, or malachite. Cf: *decorative stone.*

ornamentation [paleont] (or'-na-men-ta'-tion) The characteristic markings or patterns on the body of an animal; e.g. the external surface features of preserved hard parts (ridges, grooves, granules, spines, etc.) that may interrupt the smooth surface of a shell. See also: *sculpture [paleont]; prosopon.*

ornamentation [palyn] *sculpture [palyn].*

Ornithischia (Or-nith-is'-chia) One of the two orders of archosaurian reptiles commonly treated as dinosaurs, characterized by a bifurcate or reduced pubis, a tendency toward elaboration of the dental battery and development of a beak, and herbivorous habit. Range, Upper Triassic to Upper Cretaceous. Cf: *Saurischia.*

orocline (or'-o-cline) An orogenic belt with an imposed curvature or sharp bend, formed by rotation of fold and fault traces around a vertical axis; the trend of structures changes progressively during formation of an orocline. Obsolete syn: *geoflex.*

oroclinotath (or-o-cli'-no-tath) An orogenic belt, interpreted by Carey (1958) as having been subjected both to substantial horizontal bending and to stretching along the strike. Obsolete.

orocratic (or-o-crat'-ic) Pertaining to a period of time in which there is much *diastrophism.* Cf: *pediocratic.*

orogen (or'-o-gen) *orogenic belt.*

orogenesis (or-o-gen'-e-sis) *orogeny.*

orogenetic (or'-o-ge-net'-ic) Adj. of *orogenesis.*

orogenic (or-o-gen'-ic) Adj. of *orogeny.* Cf: *orographic [tect].*

orogenic andesite A suite of volcanic rocks that are hypersthene-normative, contain 53 to 63 wt% SiO_2, and have TiO_2<1.75 wt% and K_2O<(0.145SiO_2-5.135) (Gill, 1981, p.6).

orogenic belt A linear or arcuate region that has been subjected to folding and other deformation during an *orogenic cycle.* Orogenic belts are mobile belts during their formative stages, and most of them later became mountain belts by postorogenic processes. Syn: *fold belt; orogen.*

orogenic collapse Extensional deformation within an orogenic belt in response to gravitational loading of the thickened crust in the orogen.

orogenic cycle The interval of time during which an originally *mobile belt* evolved into a stabilized *orogenic belt,* passing through a *preorogenic phase,* an *orogenic phase,* and a *postorogenic phase.* The concept has been rendered obsolete by the recognition of the plate structure of the Earth (Coney, 1970). Syn: *tectonic cycle.* Cf: *geotectonic cycle.*

orogenic phase The median part of an *orogenic cycle,* characterized by a climax of crustal mobility and orogenic activity. It is commonly shorter than the *preorogenic phase* and the *postorogenic phase,* and may be less than a geologic period in length, although it is commonly prolonged by a succession of pulsations.

orogenic sediment Any sediment that is produced as the result of an *orogeny* or that is directly attributable to the orogenic region in which it later becomes involved; e.g. a clastic sediment such as *flysch* or *molasse.*

orogeny (o-rog'-e-ny) Literally, the process of formation of mountains. The term came into use in the middle of the 19th Century, when the process was thought to include both the deformation of rocks within the mountains, and the creation of the mountainous topography. Only much later was it realized that the two processes were not always closely related, either in origin or in time. Today, most geologists regard the formation of mountainous topography as postorogenic. By present geological usage, orogeny is the process by which structures within fold-belt mountainous areas were formed, including thrusting, folding, and faulting in the outer and higher layers, and plastic folding, metamorphism, and plutonism in the inner and deeper layers. Only in the very youngest, late Cenozoic mountains is there any evident causal relation between rock structure and surface landscape. Little such evidence is available for the early Cenozoic, still less for the Mesozoic and Paleozoic, and virtually none for the Precambrian yet all the deformational structures are much alike, whatever their age, and are appropriately considered as products of orogeny. See also: *diastrophism.* Cf: *epeirogeny; tectogenesis; cymatogeny; tectonic geomorphology.* Syn: *orogenesis; mountain building; tectogenesis.* Adj: *orogenic; orogenetic.*

orogeosyncline (or'-o-ge'-o-syn'-cline) Kober's term for a geosyncline that later became an area of orogeny (Glaessner and Teichert, 1947, p.588). Obsolete. See also: *mother geosyncline.*

orographic [geog] (or-o-graph'-ic) (a) Pertaining to mountains, esp. in regard to their location and distribution. (b) Said of the precipitation that results when moisture-laden air encounters a high barrier and is forced to rise over it, such as the precipitation on the windward slopes of a mountain range facing a steady wind from a warm ocean. Also, said of the lifting of an air current caused by its passage up and over a mountain. (c) Pertaining to a *rain-shadow desert.*

orographic [tect] A term, now little used, for features relating to mountain structure and topography. More explicit adjectives such as *diastrophic, epeirogenic,* and *orogenic,* are now preferred.

orographic desert *rain-shadow desert.*

orographic lifting The forced rising of air up the slopes of a hill or mountain. Air that is forced to ascend the slopes facing the oncoming *wind* (windward slopes) expands and cools, which increases its *relative humidity.* With sufficient cooling, *clouds* and *precipitation* develop.

orographic snowline The lower elevation limit of patches of perennial snow between glaciers. Term was introduced by Ratzel in 1886 (Flint, 1971, p.64). Cf: *regional snowline.*

orography (o-rog'-ra-phy) (a) The branch of physical geography that deals with the disposition, character, formation, and structure of mountains and of chains, ranges, and systems of mountains. (b) Broadly, the description or depiction of the relief of the Earth's surface or of a part of it, or the representation of such relief on a map or model; the land features of a specified region. Etymol: Greek "oros", "mountain". Syn: *orology; oreography.*

orohydrography (or'-o-hy-drog'-ra-phy) A branch of hydrography dealing with the relations of mountains to drainage.

orology (o-rol'-o-gy) A syn. of *orography;* esp. the study of mountain building and mountain formation.

orophilous (o-roph'-i-lous) Said of an organism that lives in *montane* conditions.

orophyte (or'-o-phyte) A plant growing in *montane* regions.

Orosirian A period of the Standard Global Geochronometric Scale: the third-oldest period of the *Paleoproterozoic* Era, characterized by a global orogenic event ("orosira", "mountain range"). It is defined as the time span between 2,050 Ma, the end of the *Rhyacian* Period, and 1,800 Ma, the beginning of the *Statherian* Period (Plumb, 1991).

orotath (or'-o-tath) An orogenic belt, interpreted by Carey (1958) as having been substantially stretched in the direction of its length.

orotvite (o-rot'-vite) A *diorite* composed of hornblende, biotite, plagioclase, nepheline, and cancrinite, with accessory titanite, ilmenite, and apatite. The name was given by Streckeisen in 1939 for the Orotva Valley, Ditro, Romania. Obsolete.

orpheite (or'-phe-ite) A colorless, gray, pale blue, yellowish, or green rhombohedral mineral of the *beudantite* group: $H_6Pb_{10}Al_{20}(PO_4)_{12}(SO_4)_5(OH)_{40}\cdot 11H_2O$.

orpiment (or'-pi-ment) A lemon-yellow to orange monoclinic mineral: As_2S_3. It is generally foliated or massive, and is frequently associated with realgar. Orpiment occurs as a deposit from some hot springs and as a sublimate from some volcanoes. Syn: *yellow arsenic.*

orschallite A vitreous colorless rhombohedral sulfite mineral: $Ca_3(S^{4+}O_3)_2SO_4\cdot 12H_2O$.

orthembadism (or-them'-bad-ism) A term used in cartography as a synonym of equivalence. An "orthembadic projection" is an equal-area projection.

Orthents Entisols that have either textures of very fine sand or finer in the fine earth fraction, or textures of loamy fine sand or coarser and a coarse fragment content of 35% or more and that have an organic carbon content that decreases regularly with depth. Orthents are not saturated with water for periods long enough to limit their use for most crops. A suborder in the U.S. system of soil taxonomy.

Orthids Previous to 1994 this term was used to indicate Aridisols that have a cambic, calcic, petrocalcic, gypsic, or salic horizon or a duripan but that lack an argillic or natric horizon. The term was dropped as a suborder in the 1994 revision of the USDA, Soil Taxonomy.

orthite (or'-thite) A syn. of *allanite,* applied esp. when occurring in slender prismatic or acicular crystals.

ortho- (or'-tho-) In petrology, a prefix that, when used with the name of a metamorphic rock, indicates that it was derived from an igneous rock, e.g. orthogneiss, orthoamphibolite; it may also indicate the primary origin of a crystalline, sedimentary rock, e.g. "orthoquartzite" as distinguished from "metaquartzite".

orthoamphibole (or-tho-am'-phi-bole) (a) A group name for amphiboles crystallizing in the orthorhombic system. (b) Any orthorhombic mineral of the *amphibole* group, such as anthophyllite, gedrite, and holmquistite. Cf: *clinoamphibole.*

orthoantigorite (or'-tho-an-tig'-o-rite) A mineral of the *serpentine* group: $Mg_6(OH)_8[Si_4O_{10}]$. It is a six-layer orthorhombic form of antigorite.

orthoapsidal projection (or-tho-ap'-si-dal) A map projection produced by means of the orthographic projection of a graticule from some solid body other than the sphere or spheroid.

ortho-arenite (or-tho-ar'-e-nite) An arenite with detrital matrix under 15 percent (Pettijohn et al., 1972, p.168).

orthoaxis (or'-tho-ax'-is) In a monoclinic crystal, the lateral axis that has twofold symmetry and/or is perpendicular to the mirror plane of symmetry; it is the *b* axis in the *second setting.* Cf: *clinoaxis.*

orthobituminous (or'-tho-bi-tu'-mi-nous) Said of bituminous coal containing 87-89% carbon, analyzed on a dry, ash-free basis. Not listed by ASTM as a rank classification. Cf: *parabituminous; perbituminous.*

orthobrannerite (or-tho-bran'-ner-ite) An adamantine black metamict mineral: $U^{4+}U^{6+}Ti_4O_{12}(OH)_2$.

orthoceratite (or-tho-cer'-a-tite) Any nautiloid belonging to the genus *Orthoceras,* characterized by the presence of three longitudinal furrows on the body chamber.

orthochamosite (or-tho-cham'-o-site) A green platy mineral of the *chlorite* group: $(Fe^{2+},Mg,Fe^{3+})_5Al(Si_3Al)O_{10}(OH,O)_8$. It is the orthorhombic dimorph of chamosite.

orthochem (or'-tho-chem) An essentially normal precipitate formed by direct chemical action within a depositional basin or within the sediment itself, as distinguished from material transported in a solid state (Folk, 1959, p.7). Obsolete. Cf: *allochem*.

orthochlorite (or-tho-chlo'-rite) (a) A group name for distinctly crystalline forms of chlorite (such as clinochlore and penninite). (b) A group name for chlorites conforming to the general formula: $(M^{2+},M^{3+})_6(OH)_8[(Si,Al)_4O_{10}]$. (Hey, 1962, p.546). Cf: *leptochlorite*.

orthochoanitic (or'-tho-cho'-a-nit'-ic) Said of a straight, cylindrical, *retrochoanitic septal neck* of a nautiloid that extends only a short distance to the preceding septum.

orthochronology (or'-tho-chro-nol'-o-gy) *Geochronology* based on a standard succession of biostratigraphically significant faunas or floras or on irreversible evolutionary processes (Schindewolf, 1950, p.85). Ideally, it is based on a stratigraphic succession of species "where each successive species is the descendent of the one which immediately precedes it stratigraphically" (Teichert, 1958a, p.106). Cf: *biochronology; parachronology*.

orthochrysotile (or-tho-chrys'-o-tile) A pale green mineral of the *serpentine* group: $Mg_3Si_2O_5(OH)_4$. It is an orthorhombic form of chrysotile. Cf: *clinochrysotile*.

orthoclase (or'-tho-clase) (a) A colorless, white, cream-yellow, flesh-pink, or gray mineral of the alkali *feldspar* group: $KAlSi_3O_8$. It is the partly ordered, monoclinic modification of potassium feldspar and is dimorphous with microcline, being stable at higher temperatures; it usually contains some sodium in minor amounts. Ordinary or common orthoclase is a common rock-forming mineral; it occurs esp. in granites, acid igneous rocks, and crystalline schists, and is usually perthitic. Syn: *common feldspar; orthose; pegmatolite*. (b) A general term applied to any potassium feldspar that is or appears to be monoclinic; e.g. sanidine, submicroscopically twinned microcline, adularia, and submicroscopically twinned analbite. Cf: *plagioclase; anorthoclase*.

orthocline (or'-tho-cline) (a) Said of the inclination of the *cardinal area* in either valve of a brachiopod, lying on the continuation of the commissural plane. (b) Said of the body of the shell (and in some genera, of the hinge teeth) of a bivalve mollusk, oriented perpendicular or nearly perpendicular to the hinge axis. Syn: *acline [paleont]*. (c) Said of the growth lines that traverse the whorl at right angles to the growth direction of a gastropod shell.

orthocone (or'-tho-cone) A straight, slender nautiloid *conch*, resembling that of *Orthoceras*.

orthoconglomerate (or'-tho-con-glom'-er-ate) A term used by Pettijohn (1957, p.256) for a conglomerate with an intact gravel framework (stones in contact), characterized by a mineral cement, and containing <15% matrix (Pettijohn, 1975, p.165, Table 6.2); e.g., orthoquartzite conglomerate and arkosic conglomerate. It is strongly current bedded and is associated with coarse cross-bedded sandstones. Cf: *paraconglomerate*.

orthocontour An isoline of the gradient of a spatial function. Orthocontours show a rate and direction of change in a mapped phenomena (Krzanowski et al., 1993).

orthocumulate (or-tho-cu'-mu-late) A cumulate composed chiefly of one or more cumulus minerals plus the crystallization products of the intercumulus liquid. Cf: *mesocumulate*.

orthocumulate texture Igneous texture consisting of early-formed, accumulated crystals and relatively large amounts of interstitial minerals formed later by crystallization of magma trapped between the early crystals. Cf: *adcumulate texture; mesocumulate texture*.

orthodome (or'-tho-dome) An old term for a monoclinic crystal form whose faces parallel the orthoaxis. Its indices are {*h*01}.

orthodromic projection (or-tho-drom'-ic) A map projection, derived from the gnomonic projection, in which angles are correct at two points and all great circles are straight lines.

Orthods Spodosols that have less than six times as much free iron (elemental) than organic carbon in the spodic horizon but the ratio of iron to carbon is 0.2 or more. Orthods are not saturated with water for periods long enough to limit their use for most crops. A suborder in the U.S. system of soil taxonomy.

orthoericssonite (or-tho-er'-ics-son-ite') A reddish-black mineral: $BaMn^{2+}_2Fe^{3+}OSi_2O_7(OH)$. It is the orthorhombic dimorph of ericssonite.

orthoferrosilite (or'-tho-fer'-ro-sil'-ite) *ferrosilite*. Cf: *clinoferrosilite*.

orthogenesis (or-tho-gen'-e-sis) Evolution that follows a single direction or specific trend continuously for many generations of an evolving *lineage*, and commonly appears independent of the effects of natural selection or other external factors. Cf: *rectilinear evolution*. Syn: *straight-line evolution*.

orthogeosyncline (or'-tho-ge'-o-syn'-cline) A geosyncline between continental and oceanic cratons, containing both volcanic (eugeosynclinal) and nonvolcanic (miogeosynclinal) belts (Stille, 1935, p.77-97). Obsolete. Syn: *primary geosyncline; geosynclinal couple*. See also: *eugeosyncline; miogeosyncline*.

orthogneiss (orth'-o-gneiss) A *gneiss* formed from igneous rocks. Cf: *paragneiss*.

orthogonal (or-thog'-o-nal) The path along which wave energy travels. For *gravity waves*, it is everywhere perpendicular to the wave crest and in the horizontal plane. Syn: *wave ray*.

orthogonal joint sets Two sets of systematic joints that have propagated at right angles. They may not crosscut. Abutting relationships and joint surface morphology are often useful for sorting out the sequence of joint propagation.

orthogonal projection (a) A projection in which the projecting lines (straight and parallel) are perpendicular to the plane of the projection; e.g. an *orthographic projection*. (b) A projection with all meridians and parallels intersecting at right angles; e.g. normal azimuthal and cylindrical projections, conformal projections.

orthograde (or'-tho-grade) Pertaining to the uniform distribution of dissolved oxygen in the hypolimnion of a lake, dependent only "on conditions at circulation and on subsequent physical events" (Hutchinson, 1957, p.603; Wetzel, 2001, p.154). Cf: *clinograde*.

orthographic projection (or-tho-graph'-ic) (a) A perspective azimuthal map projection produced by straight parallel lines from a point at an infinite distance from the sphere to points on the sphere and perpendicular to the plane of projection. The largest area depicted is that of a hemisphere, and the projection is true to scale only at the center. The plane of projection may be oblique to the Earth's axis of rotation, perpendicular to it (polar orthographic projection, with the center at a pole), or parallel to the plane of some selected meridian (meridional orthographic projection, with the center on the equator). It is used for star charts and for pictorial world maps. (b) A similar projection used in optical mineralogical study of the origin of interference phenomena under the polarizing microscope, obtained by dropping perpendiculars from the poles (in the projection of the sphere) to the plane of projection which is normal to the north-south axis of the sphere. (c) *orthogonal projection*.

orthogyrate (or-tho-gy'-rate) Said of the umbones (of a bivalve mollusk) curved so that each *beak* points neither anteriorly nor posteriorly but directly toward the other valve. Cf: *mesogyrate*.

orthohexagonal (or'-tho-hex-ag'-o-nal) Pertaining to an orthorhombic lattice having a hexagonal array of points. Such a lattice is c-centered and has $b=2\sqrt{3}a$.

orthohydrous (or-tho-hy'-drous) (a) Said of coal containing 5-6% hydrogen, analyzed on a dry, ash-free basis. (b) Said of a maceral of normal hydrogen content, e.g. vitrinite. Cf: *subhydrous; perhydrous*.

orthojoaquinite-(Ce) A brown orthorhombic mineral of the joaquinite group: $NaBa_2Ce_2FeTiO_2(Si_4O_{12})_2(OH)\cdot H_2O$.

orthojoaquinite-(La) A silky brown orthorhombic mineral of the joaquinite group: $NaBa_2(La,Ce)_2Fe^{2+}TiO_2(Si_4O_{12})_2(OH,O,F)\cdot H_2O$.

ortholignitous (or-tho-lig'-ni-tous) Said of coal containing 75-80% carbon, analyzed on a dry, ash-free basis. Not listed by ASTM as a rank classification.

ortholith (or'-tho-lith) (a) A *coccolith* composed of one or very few crystals, as in the coccolithophorid *Braarudosphaera*. (b) An individual of the Ortholithae, a subdivision of the family Coccolithophoridae. Cf: *heliolith*.

orthomagmatic stage (or'-tho-mag-mat'-ic) Obsolete term for the main stage in the crystallization of silicates from a typical magma; the stage during which as much as 90 percent of a magma may crystallize. Syn: *orthotectic stage*.

orthomarble (or-tho-mar'-ble) A term proposed by Brooks (1954) for sedimentary carbonate rock that is commercially valuable because it will take a polish, e.g. the Holston orthomarble of Tennessee. Cf: *metamarble*.

orthomatrix (or-tho-ma'-trix) A term introduced by Dickinson (1970, p.702) for recrystallized detrital clay or *protomatrix* in graywackes and arkoses.

orthometric correction (or-tho-met'-ric) A systematic correction that must be applied to a gravity-oriented measured difference of elevation because level surfaces at different elevations are not exactly parallel.

orthometric elevation The distance of a point above the geoid expressed in linear units measured along the plumb line at the point. Orthometric corrections are applied to measurements of precise leveling because level surfaces at different elevations are not parallel. Cf: *height [geodesy]*.

orthomimic feldspar (or-tho-mim'-ic) A group of feldspars that by repeated twinning simulate a higher degree of symmetry with rectangular cleavages. Also spelled: orthomic feldspar.

orthominasragrite A pale to bright blue orthorhombic mineral: $V^{4+}O(SO_4)\cdot 5H_2O$. It is the orthorhombic dimorph of minasragrite.

orthomorphic projection (or-tho-mor'-phic) *conformal projection*.

orthomorphism (or-tho-mor'-phism) *conformality*.

orthophotograph (or-tho-pho'-to-graph) A photographic copy, prepared from a photograph formed by a perspective projection, in which the displacements due to tilt and relief have been removed; a photograph that has been transformed to an orthographic projection.

orthophotomap (or-tho-pho'-to-map) An orthophotograph, or a mosaic of orthophotographs, in standard quadrangle format, printed in colors to approximate ground conditions and enhanced with cartographic symbols including contours, elevations, boundaries, roads, and drainage. Syn: *photomap*.

orthophotomosaic (or'-tho-pho'-to-mo-sa'-ic) A uniform-scale photographic mosaic consisting of an assembly of orthophotographs.

orthophotoquad (or-tho-pho'-to-quad) *orthophoto quadrangle.*

orthophoto quadrangle An *orthophotograph* in standard quadrangle format of 7.5 minutes of longitude by 7.5 minutes of latitude at 1:24,000 scale with minimum cartographic treatment. Displacement of images in the source caused by ground relief and camera tilt must be corrected within specified limits. The photoimage must be brought to a defined common scale and related to a geodetic reference system (USGS, 1993, p.1-2). Shortened form: *orthophotoquad.*

orthophoto quarter-quadrangle An *orthophoto quadrangle* in a standard quadrangle format of 3.75 minutes of longitude by 3.75 minutes of latitude at 1:12,000 scale (USGS, 1991, p.1-1, 1-2). Abbrev: OQQ.

orthophyric (or-tho-phyr'-ic) Texture of the groundmass in certain holocrystalline porphyritic igneous rocks in which the feldspar crystals have quadratic or short, stumpy rectangular cross sections, rather than the lath-shaped outline observed in *trachytic* texture. Obsolete.

orthopinacoid (or-tho-pin'-a-coid) *front pinacoid.*

orthopinakiolite (or-tho-pi-na-ki'-ol-ite) A metallic black orthorhombic mineral: $(Mg,Mn^{2+})_2Mn^{3+}BO_5$. It is the orthorhombic dimorph of pinakiolite.

orthopyroxene (or-tho-py'-rox-ene) (a) A group name for pyroxenes crystallizing in the orthorhombic system and usually containing no calcium and little or no aluminum. (b) Any orthorhombic mineral of the *pyroxene* group, such as enstatite, bronzite, hypersthene, and orthoferrosilite. Cf: *clinopyroxene.*

orthopyroxene gabbro In the *IUGS classification,* a *gabbro* consisting mostly of calcic plagioclase and clinopyroxene, with orthopyroxene exceeding 5%.

orthopyroxenite (or'-tho-py-rox'-e-nite) In the *IUGS classification,* a plutonic rock with M equal to or greater than 90, and opx/(ol+opx+cpx) greater than 90.

orthoquartzite (or-tho-quartz'-ite) A clastic sedimentary rock that is made up almost exclusively of quartz sand (with or without chert), that is relatively free of or lacks a fine-grained matrix, a quartzite of sedimentary origin, or a "pure quartz sandstone". The term generally signifies a sandstone with more than 90-95% quartz and detrital chert grains that are well-sorted, well-rounded, and cemented primarily with secondary silica (sometimes with carbonate) in optical and crystallographic continuity with the grains. The rock is characterized by stable but scarce heavy minerals (zircon, tourmaline, rutile, magnetite), by lack of body fossils, and by prominence of cross-beds and ripple marks. It commonly occurs as thin but extensive blanket deposits associated with widespread unconformities (e.g. an epicontinental transgressive deposit). Example: St. Peter Sandstone (Middle Ordovician) of midwestern U.S. The term was introduced by Tieje (1921, p.655) for a quartz sandstone whose interlocking particles were cemented by infiltration and pressure (in contrast to paraquartzite), and was used by Krynine (1948, p.149) in contrast to *metaquartzite,* but the term in this sense is objectionable because it is an exception to the use of "ortho-" for a metamorphic rock indicating an igneous origin and because "quartzite" is traditionally applied to quartzose rocks that break across instead of between grains. See also: *quartzose sandstone.* The term is essentially equivalent to *quartzarenite* and *quartzitic sandstone.* Syn: *sedimentary quartzite; orthoquartzitic sandstone.*

orthoquartzitic conglomerate (or'-tho-quartz-it'-ic) A well-sorted, lithologically homogeneous, light-colored *orthoconglomerate* consisting of mature or supermature quartzose residues (chiefly vein quartz, chert, and quartzite, in fine to medium pebble size) that represent relatively stable material typically derived from eroded granitic or metamorphic terrain, with removal of finer material and less-stable lithologic types. It is commonly interbedded with pure quartz sandstone. Syn: *quartz-pebble conglomerate.*

orthoquartzitic sandstone *orthoquartzite.*

orthorhombic system (or-tho-rhom'-bic) One of the six crystal systems, characterized by three axes of symmetry that are mutually perpendicular and of unconstrained relative lengths. Cf: *isometric system; tetragonal system; hexagonal system; monoclinic system; triclinic system.* Syn: *rhombic system.*

orthorostrum (or-tho-ros'-trum) The proximal part of the *rostrum* in some belemnites, consisting of primary aragonitic lamellae.

orthoscope (or'-tho-scope) A polarizing microscope in which light is transmitted by the crystal parallel to the microscope axis, in contrast to the *conoscope,* in which a converging lens and *Bertrand lens* are used.

orthose (or'-those) (a) A syn. of *orthoclase,* esp. yellow orthoclase. (b) An obsolete term introduced by Haüy (1801) for the *feldspar* group of minerals. Etymol: French.

orthoselection (or-tho-se-lec'-tion) The continuous action of *natural selection* in the same direction over a long period of time.

orthoserpierite (or'-tho-ser'-pier-ite) A blue orthorhombic mineral: $Ca(Cu,Zn)_4(SO_4)_2(OH)_6{\cdot}3H_2O$. It is the orthorhombic dimorph of serpierite.

orthosilicate (or-tho-sil'-i-cate) According to the now obsolete classification of silicates as oxyacids of silicon, a salt of the hypothetical orthosilicic acid, H_4SiO_4. Although the basis of this classification system is obsolete, the term *orthosilicate* is still widely used as a synonym for *nesosilicate.* Cf: *metasilicate.*

orthosite (or'-tho-site) A light-colored coarse-grained *syenite* composed almost entirely of orthoclase, described by Turner in 1900. Obsolete.

orthostratigraphy (or'-tho-stra-tig'-ra-phy) Standard or "main" stratigraphy based on fossils identifying recognized biostratigraphic zones (such as trilobites in the Cambrian and graptolites in the Silurian) (Schindewolf, 1950, p.85; 1957, p.397). Cf: *parastratigraphy.*

orthostrophic (or-tho-stroph'-ic) Having harmonious coiling throughout; specif. said of the common gastropod shell in which the whorls are coiled on an erect cone so that the apex points backward rather than forward, and the spire is slightly to strongly elevated. Cf: *hyperstrophic.*

orthotectic (or-tho-tec'-tic) Obsolete syn. of *magmatic.*

orthotectic stage Obsolete syn. of *orthomagmatic stage.*

orthotectonics (or'-tho-tec-ton'-ics) A syn. of *alpinotype tectonics.* It is used by Dewey (1969) for orogenic belts of the Andean type.

orthotill (or'-tho-till) A till formed by immediate release from the transporting ice, as by ablation and melting (Harland et al., 1966, p.231). Ant: *paratill.* Rarely used.

orthotriaene (or-tho-tri'-aene) A sponge triaene in which the cladi are oriented close to 90 degrees to the rhabdome. Cf: *protriaene.*

orthotropous Said of a plant ovule with a straight axis, i.e., with the micropylar opening farthest away from the basal attachment. Cf: *amphitropous; anatropous; campylotropous.*

orthotropy A condition of a porous material in which the channels of preferred flow are arranged sets that are mutually perpendicular, thereby resulting in *anisotropy.* Cf: *isotropy.*

orthowalpurgite A yellow orthorhombic mineral: s $(UO_2)Bi_4O_4(AsO_4)_2{\cdot}2H_2O$. It is the orthorhombic dimorph of walpurgite.

Orthox (Or'-thox) Oxisols that are moist all or most of the time, and that have a low to moderate content of organic carbon within the upper 1 m or a mean annual soil temperature of 22°C or more. A suborder in the U.S. system of soil taxonomy.

orthozone (or'-tho-zone) A term suggested by Kobayashi (1944a, p.742) to replace zone as defined by Oppel (1856-1858).

ortstein (ort'-stein) A cemented *spodic horizon* in which the cementing material consists of illuviated sesquioxides, mostly iron and organic matter; the hardened B horizon of a Podzol. See also: *hardpan.*

orvietite (or-vi-e'-tite) An extrusive *trachybasalt* composed of approximately equal amounts of plagioclase and sanidine, along with leucite, augite, minor biotite and olivine, and accessory apatite and opaque oxides. The name, proposed by Niggli in 1923, is for Orvieto, Italy. Obsolete.

oryctocoenosis (o-ryc'-to-coe-no'-sis) That part of a *thanatocoenosis* that has been preserved as fossils.

oryctognosy (o-ryc-tog'-no-sy) An obsolete syn. of *mineralogy.*

oryctology (o-ryc-tol'-o-gy) The study of and about the objects of mines and excavations; esp. fossils, also figured stones or petrifactions including minerals; synonymous with *mineralogy* in the eighteenth century and the first half of the nineteenth century. Etymol: Greek "oryktos", "dug up, mined".

os [glac geol] Anglicized spelling of the Swedish term *ås,* meaning *esker.* Pl: *osar.* Var. spelling: *ose.*

os [paleont] (a) bone. Pl: *ossa.* Adj: osseous Cf: *ossicle; osteolith..* (b) mouth, orifice. Pl: *ora.* Adj: *oral; orate.* Syn: *endopore [paleont]; endoaperture.*

os [palyn] An aperture in the inner layer of the exine. Syn: *endoaperture; endopore [palyn].* Pl: *ora.* Adj: *orate.* As a component, "or-" survives in a few terms, e.g., *colporate* (colpus + orate).

Osagean (O-sag'-e-an) A North American series: Lower Mississippian (above Kinderhookian, below Meramecian).

Osann's classification (O-sann's') A purely chemical classification of igneous rocks. "The system is based on certain definite characteristics of the mineral combinations formed from the magmas, namely, on the combination of the alkalies with Al_2O_3 in definite proportions in the feldspars and feldspathoids, and on the union of lime with alumina in the anorthite molecule of the plagioclase and with iron and magnesia in the ferromagnesian minerals. The rock is classified from the amounts of these combinations, the percentage of silica, the silica coefficient, and the ratio of soda to the sum of the alkalies" (Johannsen, 1931, v.1, p.68). Obsolete. See also: *Niggli's classification.*

osar (os'-ar) Plural of *os [glac geol].* The term is often mistakenly used as a singular noun.

osarizawaite (o'-sar-i-za'-wa-ite) A greenish-yellow to plae green rhombohedral mineral of the *alunite* group: $PbCuAl_2(SO_4)_2(OH)_6$.

osarsite (os-ars'-ite) A metallic gray monoclinic mineral of the *arsenopyrite* group: (Os,Ru)AsS.

osbornite (os'-born-ite) A golden-yellow cubic meteorite mineral: TiN. It is the Ti analogue of carlsbergite.

oscillation [glac geol] (os-cil-la'-tion) *interstade.*

oscillation [stratig] A term used by Ulrich (1911) for the repeated transgressions and regressions of the seas in constantly shifting patterns, bringing about changes in the character of the sediments being deposited. Cf: *pulsation [stratig].*

oscillation cross ripple A *cross ripple* resulting from the concurrent or successive action of two sets of waves or from the intersection of a set of waves with a pre-existing current ripple. Cf: *interference ripple.*

oscillation ripple A *symmetric ripple* with a sharp, narrow, relatively straight crest between broadly rounded troughs, formed by the orbital or to-and-fro motion of water agitated by oscillatory waves. Cf: *current ripple.* Syn: *oscillatory ripple; wave ripple.*

oscillation ripple mark *oscillation ripple.*

oscillation theory A theory, proposed by Haarmann (1930), that cosmic energy produces the Earth's major tectonic features, and that secondary features are the result of gravitational sliding, compressional settling, or subsidence.

oscillatory extinction (os-cil'-la-to-ry) *undulatory extinction.*

oscillatory ripple *oscillation ripple.*

oscillatory twinning Repeated, parallel twinning.

oscillatory zoning Compositional zoning in plagioclase wherein zones from core to rim alternate between sodic and calcic compositions. May also be used for other minerals displaying analogous zonation, but of different compositions than plagioclase. Cf: *normal zoning; reversed zoning.*

osculum (os'-cu-lum) A large opening from the internal cavity of a sponge to the exterior, through which water leaves the sponge. Pl: oscula. Cf: *ostium; pore [paleont].*

ose Var. of *os.* Pl: oses.

oshana (o-sha'-na) A poorly defined stream channel in the flat-lying Ovamboland region of Namibia, containing water only during the highest floods and usually in the form of a chain of standing pools that quickly dry away. Cf: *omuramba.* Etymol: Afrikaans.

osmiridium (os-mi-rid'-i-um) A discredited name for osmium-bearing iridium. It is a native alloy containing 25-40% osmium and 50-60% iridium, and is often found with platinum. The name has also been used as a syn. of *iridosmine.*

osmium (os'-mi-um) (a) An element of the *platinum* group, symbol Os. (b) A metallic white hexagonal mineral, Os.

osmosis (os-mo'-sis) The movement at unequal rates of a solvent through a semipermeable membrane, which usually separates the solvent and a solution, or a dilute solution and a more concentrated one, until the solutions on both sides of the membrane are equally strong. Cf: *dialysis.* See also: *electro-osmosis.*

ossa Pl. of *os [paleont].*

osseous amber (os'-se-ous) Opaque or cloudy amber containing numerous minute bubbles. Syn: *bone amber.*

osseous breccia *bone breccia.*

ossicle (os'-si-cle) (a) Any of the numerous individual calcified elements or pieces of the skeleton of many echinoderms; e.g. a plate. The term is normally used for the larger of such elements. (b) A tiny bone, esp. of the middle ear. Formal Latin form: *ossiculum.*

ossiculum (os-sic'-u-lum) (a) *ossicle.* (b) *lithodesma.* Pl: ossicula.

Osteichthyes (Os-te-ich'-thy-es) A class of vertebrates, the bony fishes. Range, Devonian to Holocene.

osteolite (os'-te-o-lite') A massive, earthy mineral (apatite) consisting of an impure, altered calcium phosphate.

osteolith (os'-te-o-lith') (a) A femur-shaped heterococcolith built up of lamellae as in the coccolithophorid *Ophiaster hydroideus.* (b) A fossil bone.

Osteostraci (Os-te-os'-tra-ci) An order of monorhinate jawless vertebrates characterized by expanded head and trunk shield and flattened body form. Range, Upper Silurian to Upper Devonian.

ostiole [geomorph] (os'-ti-ole) *tundra ostiole.*

ostiole [paleont] One of the small inhalant openings of a sponge; an *ostium.*

ostiole [palyn] An exitus or pore, often at the end of a necklike structure, in an *ascocarp* or *pycnidium.*

ostium (os'-ti-um) Any opening through which water enters a sponge. The term is sometimes applied only to an opening larger than a pore, and it was used in the older literature as a synonym of *posticum.* Pl: ostia. Cf: *osculum.* Syn: *ostiole [paleont].*

ostracode (os'-tra-code) Any aquatic crustacean belonging to the subclass Ostracoda, characterized by a bivalve, generally calcified carapace with a hinge along the dorsal margin. Most ostracodes are of microscopic size (0.4-1.5 mm long), although freshwater forms up to 5 mm long and marine forms up to 30 mm long are known. Range, Lower Cambrian to Holocene. Also spelled: ostracod.

ostracum (os'-tra-cum) A term used originally for the outer part of the calcareous wall of the shell of a bivalve mollusk, secreted at the edge of the mantle, but also applied by some later authors to the entire calcareous wall. Cf: *hypostracum; periostracum.* Pl: ostraca.

Ostwald ripening A recrystallization process in which, within a solution containing the material of the crystals with which it is in contact, the smallest crystals are dissolved and their material is added to the larger crystals of the same phase. The driving mechanism is a shift to lower surface free energy (Ostwald, 1900 in Baronnet, 1982).

Ostwald's rule (Ost'-wald's) The statement in phase studies that an unstable phase does not necessarily transform directly to the truly stable phase, but rather it may first pass through successive intermediate phases, presumably due to lower activation energy barriers via that route.

osumilite (o-su'-mi-lite) A variously colored hexagonal mineral of the *milarite* group: $(K,Na)(Fe^{2+},Mg)_2(Al,Fe^{3+})_3(Si,Al)_{12}O_{30}$. It is commonly mistaken for cordierite.

osumilite-(Mg) A colorless, pink, blue, or black mineral of the *milarite* group: $(K,Na)(Mg,Fe^{2+})_2(Al,Fe)_3(Si,Al)_{12}O_{30}$.

oswaldpeetersite A vitreous yellow monoclinic mineral: $(UO_2)_2CO_3(OH)_2{\cdot}4H_2O$.

Otaian (O-tai-an) South Pacific stage; lower Miocene (above Waitakian, below Altonian).

otavite (o-ta'-vite) A white, yellow-brown, or reddish rhombohedral mineral of the *calcite* group: $CdCO_3$.

otjisumeite (ot'-ji-su'-me-ite) A colorless triclinic mineral: $PbGe_4O_9$.

otolith (o'-to-lith) Ear bone of a vertebrate animal; esp. a fish.

ottajanite (ot-ta-jan'-ite) A leucite *tephrite* or *trachybasalt* having the chemical, but not mineralogic, composition of a *sommaite,* being composed of augite and leucite phenocrysts in a groundmass of calcic plagioclase, leucite, and augite, with some sanidine, nepheline, olivine, opaque oxides, hornblende, biotite, and apatite. Its name, given by Lacroix in 1917, is derived from Ottajano, Vesuvius, Italy. Obsolete.

ottemannite (ot'-te-mann-ite') A metallic gray orthorhombic mineral: Sn_2S_3.

Ottnangian (Ott-nang'-ian) Central European stage, lower Miocene, planktonic zone N.6 (above Eggenburgian, below Karpatian). Obs. syn: *Helvetian* (part).

ottoliniite A monoclinic member of the *amphibole* group: $NaLi(Mg_3Fe^{3+}Al)Si_8O_{22}(OH)_2$.

ottrélite (ot'-tre-lite) A yellow-green, gray to black mineral of the *chloritoid* group: $(Mn^{2+},Fe^{2+},Mg)Al_2SiO_5(OH)_2$.

otwayite (ot'-way-ite) A silky-green orthorhombic mineral: $Ni_2(CO_3)(OH)_2{\cdot}H_2O$.

Ouachita stone (Ouach'-i-ta) *Washita stone.*

ouachitite (ouach'-i-tite) An olivine-free biotite *lamprophyre* having a glassy or analcime-bearing groundmass. The name, given by Kemp in 1891, is for the Ouachita River, Arkansas. Obsolete.

oued A var. of *wadi.* Pl: oueds; ouadi. Etymol: French.

ouenite (ou'-en-ite) A fine-grained *diabase* resembling *eucrite* and containing chrome diopside, anorthite, and smaller amounts of orthopyroxene and olivine. The name, given by Lacroix in 1911, is for Ouen, New Caledonia. Obsolete.

ouklip (ouk'-lip) A term used in southern Africa for a conglomerate. Etymol: Afrikaans, "old rock".

oulankaite A metallic tetragonal mineral: $(Pd,Pt)_5(Cu,Fe)_4SnTe_2S_2$.

oule A term used in the Pyrenees for *cirque [glac geol].* Etymol: Spanish "olla", "pot, kettle".

ourayite (ou'-ray-ite) A metallic gray orthorhombic mineral: $Ag_3Pb_4Bi_5S_{13}$(?). It is a member of the lillianite homologous series. A polymorph with a primitive unit cell has been called ourayite-P.

oursinite (our'-sin-ite) A yellow orthorhombic mineral: $(Co,Mg)(UO_2)_2(SiO_3OH)_2{\cdot}6H_2O$.

outburst (out'-burst) *glacier outburst flood.*

outcrop (out'-crop) n. That part of a geologic formation or structure that appears at the surface of the Earth; also, bedrock that is covered only by surficial deposits such as alluvium. Cf: *exposure [geol].* Syn: *crop* (deprecated); *cropping* (deprecated); *outcropping.*

outcrop area The area occupied by a particular rock unit.

outcrop curvature *settling [mass move].*

outcrop map A type of geologic map that shows the distribution and shape of actual outcrops, leaving blank those areas without outcrops. It often includes measured data for specific places, such as specimen or fossil collections, or strike and dip of beds.

outcropping (out'-crop-ping) n. *outcrop.*

outcrop spring *contact spring.*

outcrop water Rain and surface water that seeps downward through outcropping porous and fissured rock, fault planes, old shafts, or surface drifts.

outer bar (out'-er) A bar formed at the mouth of an ebb channel of an estuary. Ant: *inner bar.*

outer bark For stems and roots of dicotyledons and gymnosperms, a nontechnical term incorporating the rough corky tissue developed from the *cork cambium* (Swartz, 1971, p.328). The technical term for outer bark is *rhytidome.*

outer core The outer or upper zone of the Earth's core, extending from a depth of 2,900 km to 5,100 km, and including the *transition zone;* it is equivalent to the *E layer* and the *F layer.* It is presumed to be liquid because it sharply reduces compressional-wave velocities and does not transmit shear waves. Its density ranges from 9.9-12.2 g/cm^3. The outer core is the source of the principal geomagnetic field. Cf: *inner core.*

outer epithelium The ectodermal epithelium adjacent to the shell of a brachiopod and responsible for its secretion.

outer hinge plate Either of a pair of concave or subhorizontal hinge plates in the *cardinalia* of some brachiopods, separating inner socket ridges and crural bases. Cf: *inner hinge plate.*

outer lamella The relatively thick mineralized shell layer of an ostracode, enclosed between thin chitinous layers, and serving to conceal and protect the soft parts of the body and appendages (TIP, 1961, pt. Q, p.53). Cf: *inner lamella.*

outer lip The abaxial (lateral) margin of the aperture of a gastropod shell, extending from the suture to the foot of the *columella.* Cf: *inner lip.*

outer mantle *upper mantle.*

outer mantle lobe The outer peripheral part of the mantle of a brachiopod, separated by a mantle groove from an inner lobe, and responsible (in articulate brachiopods) for the secretion of the primary shell layer (TIP, 1965, pt. H, p.149).

Outer planets Planets beyond the orbit of Mars, which marks the furthest member of the terrestrial planets. With the exception of icy Pluto, the planets beyond Mars are largely gaseous and comprise the largest members of the solar system. See also: *giant planet.* Cf: *terrestrial planet.*

outer plate One of a pair of subvertical plates in the *cardinalia* of pentameracean brachiopods, with the ventral surface fused to the base of the brachial process and the dorsal edge attached to the floor of the valve (TIP, 1965, pt. H, p.149). Cf: *inner plate.*

outer reef One of the reefs comprising the seaward or basinward part of a *reef complex* or *reef tract.* They tend to be larger and better developed than inner reefs in the same region. Cf: *leeward reef; windward reef; bank-barrier reef; barrier reef.*

outer side The portion of a conodont element on the convex side of the anterior-posterior midline. Ant: *inner side.*

outer wall The exterior layer in the skeletion of double-walled archaeocyathids. It is homologous with the single wall in one-walled forms (TIP, 1972, pt. E, p.7).

outface (out′-face) *dip slope.*

outfall (out′-fall) (a) The mouth of a stream or the outlet of a lake; esp. the narrow end of a watercourse or the lower part of any body of water where it drops away into a larger body. (b) The vent or end of a drain, pipe, sewer, ditch, or other conduit that carries waste water, sewage, storm runoff, or other effluent into a stream, lake, or ocean.

outflow (out′-flow) (a) The act or process of flowing out; e.g. the discharge of water from a river into the sea. Syn: *efflux.* (b) Water that flows out; e.g. groundwater seepage and stream water flowing out of a drainage basin. Also, the amount of water that has flowed out. (c) An *outlet* where water flows out of a lake.

outflow cave A cave from which a stream issues, or is known to have issued. See also: *inflow cave; through cave.* Syn: *effluent cave.*

outflow channel A broad valley on Mars consisting of irregular, occasionally braided networks, thought to have been formed by sudden release of water (Carr et al., 1984).

outgassing (out′-gas-sing) The escape of absorbed gases, usually by heating; e.g. the process involving the release of gases and water vapor from molten rocks, leading to the formation of the Earth's atmosphere and oceans.

outgroup In cladistic analysis, a group of organisms (*species, genera,* etc.) that is related to, but removed from, the group of study taxa. One or more outgroups are examined to determine which character states are evolutionary novelties (Brooks et al., 1984, p.80). Cf: *sister group, ingroup.*

outlet (out′-let) (a) The relatively narrow opening at the lower end of a lake through which water is discharged into an outflowing stream or other body of water. Syn: *outflow.* (b) A stream flowing out of a lake, pond, or other body of standing water; also, the channel through which such a stream flows. (c) The lower end of a watercourse where its water flows into a lake or sea; e.g. a channel, in or near a delta, diverging from the main river and delivering water into the sea. (d) A crevasse in a levee.

outlet glacier A glacier issuing from an *ice sheet* or *ice cap* through a mountain pass or valley and constrained to a channel or path by exposed rock. Cf: *ice stream; glacial lobe; distributary glacier.*

outlet head The place where water leaves a lake and enters an effluent.

outlier (out′-li-er) An area or group of rocks surrounded by older rocks, e.g. an isolated hill or butte. Cf: *inlier.*

outline map (out′-line) A map that presents minimal geographic information, usually only coastlines, principal streams, major civil boundaries, and large cities, leaving as much space as possible for the reception of additional particular data. See also: *base map.*

out-of-sequence thrusting Thrusting in which successively younger thrusts do not form uniformly on the hinterland or foreland side of existing thrusts.

outpost well (out′-post) A hole drilled for oil or gas with the thought that it will probably extend, by a considerable distance, a pool already partly developed. It is far enough from the limits of the pool to make its outcome uncertain, but not far enough to be designated a *wildcat well* (Lahee, 1962, p.133).

outside lateral muscle (out′-side) One of a pair of muscles in some lingulid brachiopods, originating on the *pedicle valve* laterally to the central muscles, and extending posteriorly to insertions behind the middle lateral muscles on the *brachial valve* (TIP, 1965, pt. H, p.149).

outwash [glac geol] (out′-wash) (a) Stratified detritus (chiefly sand and gravel) removed or "washed out" from a glacier by meltwater streams and deposited in front of or beyond the end moraine or the margin of an active glacier. The coarser material usually is deposited nearer to the ice. Syn: *outwash drift.* (b) The meltwater from a glacier.

outwash [sed] Soil material washed down a hillside by rainwater and deposited upon more gently sloping land, e.g., *colluvium.*

outwash apron *outwash plain.*

outwash cone A steeply sloping, cone-shaped accumulation of outwash deposited by meltwater streams at the margin of a shrinking glacier. Syn: *wash cone.*

outwash drift A deposit of *outwash.*

outwash fan A fan-shaped accumulation of outwash deposited by meltwater streams in front of the end moraine of a glacier. Coalescing outwash fans form an *outwash plain.*

outwash plain A broad, gently sloping sheet of outwash deposited by meltwater streams flowing in front of or beyond a glacier, and formed by coalescing outwash fans; the surface of a broad body of outwash. Cf: *valley train; glacial plain.* See also: *sand plain [glac geol].* Syn: *outwash apron; morainal apron; frontal plain; wash plain; marginal plain; sandur; morainal plain.*

outwash-plain shoreline A prograding shoreline formed where the outwash plain in front of a glacier is built out into a lake or the sea.

outwash terrace A dissected and incised valley train or benchlike deposit extending along a valley downstream from an outwash plain or terminal moraine; a flat-topped bank of outwash with an abrupt outer face. Syn: *frontal terrace.*

outwash train *valley train.*

ouvala *uvala.*

ouvarovite (ou-va′-rov-ite) *uvarovite.*

oval cut (a) Brilliant cut with an elliptical girdle outline; also called an oval brilliant cut. (b) Obsolete barrel-shaped cut, circular in section and covered with triangular facets.

ovamboite A metallic cubic mineral: $Cu_{10}(Fe,Zn,Cu)_3WGe_3S_{16}$.

ovary (o′-va-ry) In a flower, the basal, enlarged part of the pistil, in which seeds develop.

ovate (o′-vate) Shaped like an egg in sectional view, as a leaf whose basal end is broader than its terminal end. Cf: *obovate.*

oven (ov′-en) (a) A rounded, sacklike pit or hollow in a rock, esp. in a granitic rock, having an arched roof and resembling an oven (Bell, 1894, p.358). It is produced by chemical weathering. Cf: *weather pit.* (b) *spouting horn.*

oven-dry soil Soil that has been dried at 105°C until it reaches constant mass.

overbank deposit (o′-ver-bank) Fine-grained sediment (silt and clay) deposited from suspension on a flood plain by floodwaters that cannot be contained within the stream channel. See also: *flood-plain deposit.*

overbank splay *floodplain splay.*

overburden [eco geol] (o′-ver-bur′-den) Barren rock material, either loose or consolidated, overlying a mineral deposit, which must be removed prior to mining. Syn: *top [ore dep]. Cap rock* and *capping* are synonyms that are usually used for consolidated material.

overburden [sed] (a) The upper part of a sedimentary deposit, compressing and consolidating the material below. (b) The loose soil, silt, sand, gravel, or other unconsolidated material overlying bedrock, either transported or formed in place; *regolith.* (c) In salt tectonics, strata younger than the salt substratum or source layer (Jackson and Talbot, 1991).

overburden pressure *geostatic pressure.*

overconsolidation (o′-ver-con-sol′-i-da′-tion) Consolidation (of sedimentary material) greater than that normal for the existing overburden; e.g. consolidation resulting from desiccation or from pressure of overburden that has since been removed by erosion, e.g. a *lodgment till.* Ant: *underconsolidation.* Cf: *normal consolidation.*

overdeepened valley (o-ver-deep′-ened) The degraded channel or valley of an alpine glacier, now occupied by an aggrading stream.

overdeepening (o-ver-deep′-en-ing) The process by which an eroding glacier excessively deepens and broadens an inherited preglacial valley to a level below that of the original base level. Undrained depressions and rock-floored lakes are common. Cf: *oversteepening.*

overdip slope (o′-ver-dip) A land surface sloping in approximately the same direction as, but more steeply than, the dip of the rock layers that crop out on that surface (Briggs, 1974).

overdraft (o′-ver-draft) Withdrawal of groundwater in excess of replenishment.

overedge Any portion of a map extending outside the neatline, usually done for convenience of publication, especially in a map series.

overfall [eng] A place provided on a dam or weir for the overflow of surplus water. Syn: *spillway [eng].*

overfall [streams] An obsolete term for a waterfall.

overfit stream (o′-ver-fit) A *misfit stream* that is too large to have eroded the valley in which it flows, or whose flood plain is too small for the size of the stream. There is some doubt as to whether such a stream exists.

overflow (o′-ver-flow) v. To flow over the margin of; to cover with water. n. A flowing over the banks of a stream or river; an inundation.

overflow channel A channel or notch cut by the overflow waters of a lake, esp. the channel draining meltwater from a glacially dammed lake; an outlet of a proglacial lake. Cf: *glacial drainage channel.* Syn: *spillway [glac geol]; glacial spillway; sluiceway.* Cf: *crease.*

overflow ice Ice formed during high spring tides by water rising through cracks in the surface ice and then freezing (Swayne, 1956, p.104).

overflow spring A type of *contact spring* that develops where a permeable deposit dips beneath an impermeable mantle. Groundwater overflows onto the land surface at the edge of the impermeable stratum.

overflow stream (a) A stream containing water that has overflowed the banks of a river. Syn: *spill stream.* (b) An effluent from a lake, carrying water to a stream, sea, or another lake.

overgrown denticles Conodont denticles that formed in early stages of growth, but later fused completely with adjacent denticles; outlines of such denticles can commonly be seen through the thin, subsequently secreted lamellae by which they are enveloped; commonly (but inappropriately) termed "germ denticles"; may also be termed suppressed denticles (TIP, 1981, pt. W, supp.2).

overgrowth [cryst] (o'-ver-growth) (a) Secondary material deposited in optical and crystallographic continuity around a crystal grain of the same composition, as in the diagenetic process of secondary enlargement. (b) A deposit of one mineral growing in oriented crystallographic directions on the surface of another mineral; e.g. hematite on quartz, or chalcopyrite on galena. See also: *mantle [cryst].*

overgrowth [paleont] A subsequent encrusting growth of zooids and possible extrazooidal parts separated from the supporting stenolaemate bryozoan colony by a basal lamina. The overgrowth can be intracolonial, intercolonial and conspecific, or interspecific.

overhand stoping (o'-ver-hand) The working of a block of ore from a lower level to a level above.

overhang (o'-ver-hang) (a) *cliff overhang.* (b) A part of the mass of a *salt dome* that projects out from the top of the dome much like the cap of a mushroom.

overhanging ripple (o-ver-hang'-ing) *rhomboid ripple.*

overite (o'-ver-ite) (a) A pale green to colorless orthorhombic mineral: $CaMgAl(PO_4)_2(OH)\cdot4H_2O$. (b) A group name for minerals with the analogous composition, but with Ca replaced by Mg or Mn, Mg replaced by Mn^{2+}, and Al replaced by Fe^{3+} or Sc.

overland flow (o'-ver-land) That part of surface runoff flowing over land surfaces toward stream channels; specif. *sheet flow [geomorph].* After it enters a stream, it becomes a part of the total runoff (Langbein and Iseri, 1960). Syn: *unconcentrated flow.* Cf: *channel flow; streamflow.*

overlap [coast] (o'-ver-lap) The migration of an upcurrent part of a shore to a position that extends seaward beyond a downcurrent part; esp. the lapping-over of an inlet by a spit. Cf: *offset [coast].*

overlap [paleont] In ostracodes, the closure of the two valves in such a manner that the contact margin of the larger valve extends over the margin of the smaller.

overlap [photo] The area common to two successive aerial or space photographs or images along the same flight strip, expressed as a percentage of the photo area. Cf: *sidelap.* Syn: *end lap; forward lap.*

overlap [stratig] A general term referring to the extension of marine, lacustrine, or terrestrial strata beyond underlying rocks whose edges are thereby concealed or "overlapped", and to the unconformity that commonly accompanies such a relation; esp. the relationship among conformable strata such that each successively younger stratum extends beyond the boundaries of the stratum lying immediately beneath. The term is often used in the sense of *onlap,* and sometimes in the sense of *overstep* (as by De la Beche, 1832); because of such conflicting usage, Melton (1947, p.1869) and Swain (1949, p.634) urged that the term be abandoned. See also: *replacing overlap.*

overlap fault (a) An obsolete term for a *thrust fault.* (b) A fault structure in which the displaced strata are doubled back upon themselves. Little used.

overlapping faults Two subparallel, noncollinear faults with traces that partly extend past each other (Peacock et al., 2000). Cf: *underlapping faults.*

overlapping pair (o-ver-lap'-ping) Two photographs taken at different camera stations in such a manner that part of one photograph shows the same terrain as shown on a part of the other photograph; e.g. *stereoscopic pair.*

overlapping spur *interlocking spur.*

overlap sequence A depositional sequence that unconformably lies across the contacts of two or more terranes (Howell, 1995).

overlap zone (o'-ver-lap) *concurrent-range zone.*

overlay (o'-ver-lay) Graphic data on a transparent or translucent sheet to be superimposed on another sheet (such as a map or photograph) to show details not appearing, or requiring special emphasis, on the original; a template. Also, the medium or sheet containing an overlay.

overload (o'-ver-load) The amount of sediment that exceeds the ability of a stream to transport it and that is therefore deposited.

overloaded stream *transport-limited stream.*

overpressure (o'-ver-pres'-sure) The amount by which the formation-fluid pressure exceeds hydrostatic or normal pressure. Overpressure, which can have several causes, generally indicates that fluid pathways to the surface were cut off at some time so that the interstitial fluid could not escape to restore equilibrium. Normal pressure may exist below overpressured formations.

overpressured system A hydrogeologic system in which fluid pressure exceeds hydrostatic pressure. Cf: *underpressured system.*

overprint [geochron] (o'-ver-print) A complete or partial disturbance of an isolated radioactive system, by thermal, igneous, or tectonic activities, which results in loss or gain of radioactive or radiogenic isotopes, and hence a change in the isotopic age that will be given by the disturbed system. Cf: *mixed ages; hybrid age.*

overprint [struc petrol] The superposition of a new set of structural features on an older set. Syn: *superprint; metamorphic overprint.*

overprinting (o'-ver-print-ing') Successive episodes of hydrothermal alteration acting on a volume of rock and on alteration minerals generated by previous episodes. Phyllic overprinting is a common late hypogene effect in porphyry copper deposits (Guilbert and Park, 1986, p.416-418). Cf: *phyllic alteration.*

overriding plate At a *convergent plate boundary,* the plate that is not being subducted. A *volcanic arc* develops along the edge of the overriding plate. The overriding plate can be composed either of oceanic or continental lithosphere.

oversaturated (o-ver-sat'-u-rat-ed) Said of an igneous rock or magma that contains silica (or occasionally alumina) in excess of the amount required to form saturated minerals from the bases present. Syn: *silicic.* Cf: *undersaturated.*

overshoot A topological error denoting a *dangle* (Krzanowski et al., 1993).

oversteepened valley (o-ver-steep'-ened) "An ice-free valley in which one side is higher and steeper than the other, a condition caused by the swing of a former glacier directed against that side" (Swayne, 1956, p.105).

oversteepened wall A *trough end* having an almost vertical slope due to glacial action.

oversteepening (o-ver-steep'-en-ing) The erosive process by which an alpine glacier excessively steepens the sides of an inherited preglacial valley. Cf: *overdeepening.*

overstep [stratig] n. (a) An *overlap [stratig]* characterized by the regular truncation of older units of a complete sedimentary sequence by one or more later units of the sequence (Swain, 1949, p.635). The term, which is more commonly used in Great Britain than in U.S., refers to the progressive burial of truncated edges of underlying strata below an unconformity (esp. when an unconformity is not very obvious but is made evident by detailed mapping). Cf: *onlap.* See also: *strike-overlap; complete overstep; regional overstep.* (b) A stratum laid down on the upturned edges of underlying strata. v. To transgress; e.g. an unconformable stratum that truncates the upturned edges of the underlying older rocks is said to "overstep" each of them in turn.

overstep [struc geol] The area between two subparallel, noncollinear faults. Syn: *jog* (Biddle and Christie-Blick, 1985).

overthrust (o'-ver-thrust) A low-angle *thrust fault* of large scale, with displacement generally measured in kilometers. Cf: *underthrust fault.* Syn: *low-angle thrust; overthrust fault.*

Overthrust Belt An ill-defined term widely used for promoting oil and gas development in western North America. "Because of wide disagreement about its meaning and geomorphic extent, the term is not generally acceptable in scientific literature" (Hunt, 1982).

overthrust block *overthrust nappe.*

overthrust fault *overthrust.*

overthrust nappe The body of rock that forms the hanging wall of a large-scale overthrust; a *thrust nappe.* Syn: *overthrust block; overthrust sheet; overthrust slice.*

overthrust sheet *overthrust nappe.*

overthrust slice *overthrust nappe.*

overturn (o'-ver-turn) The *circulation,* esp. in the fall and spring, of the layers of water in a lake or sea, whereby surface water sinks and mixes with bottom water; it is caused by changes in density differences due to changes in temperature, and is esp. common wherever lakes are icebound in winter. See also: *turnover [lake]; circulation [lake].*

overturned (o-ver-turned') Said of a fold, or the limb of a fold, that has tilted beyond the perpendicular. Sequence of strata thus appears reversed. Syn: *inverted; reversed.*

overturned flap An inversion of the normal stratigraphy that occurs in the *ejecta blanket* near the rim of a *simple crater* of either impact or explosion origin. First described by Shoemaker (1963) at Meteor Crater, Arizona.

overturning (o-ver-turn'-ing) Seasonal vertical mixing in lakes, upon cooling in fall, and analogous processes in restricted basins in the sea.

overwash (o'-ver-wash) (a) A mass of water representing the part of the uprush that runs over the berm crest (or other structure) and that does not flow directly back to the sea or lake. (b) The flow of water in restricted areas over low parts of barriers or spits, esp. during high tides or storms.

overwash fan *washover fan.*

overwash mark A narrow, tongue-like ridge of sand formed by overwash on the landward side of a berm.

overwash pool A *tide pool* between a berm and a beach scarp, which water enters only at high tide.

ovicell (o'-vi-cell) (a) The structure for brooding embryos present in most cheilostome bryozoans. It generally consists of a fold of body wall, one or both walls of which are calcified, enclosing the water-filled *brood chamber.* Syn: *ooecium.* (b) A term used loosely for any skeletal structure that houses bryozoan larvae during their development.

ovoid (o'-void) Egg-shaped, as in a fruit whose basal portion is broader than its terminal portion. Cf: *obovoid.*

ovulate (o'-vu-late) Containing or bearing an egg or ovule (Swartz, 1971, p.329).

ovule (o'-vule) The unfertilized young seed in the ovary of a seed plant (Swartz, 1971, p.329).

ovulite (o'-vu-lite) *oolith.*

ovum (o'-vum) The female gamete of an *embryophytic* plant; a nonmotile gamete which can fuse with a sperm to form a *zygote* (Cronquist, 1961, p.874).

owensite A metallic cubic mineral: $(Ba,Pb)_6(Cu^{1+},Fe,Ni)_{25}S_{27}$.

owyheeite (o-wy-hee'-ite) A metallic steel-gray orthorhombic mineral: $Pb_{10-2x}Ag_{3+x}Sb_{11+x}S_{28}$(x=-0.13 to 0.20). It occurs in fibrous masses and acicular crystals. Syn: *silver jamesonite.*

oxalate A mineral compound containing the anion $(C_2O_4)^{-2}$. An example is whewellite, $Ca(C_2O_4)\cdot H_2O$.

oxalite (ox'-a-lite) *humboldtine.*

oxammite (ox-am'-mite) A colorless or yellowish-white orthorhombic oxalate mineral: $(NH_4)_2C_2O_4\cdot H_2O$.

oxbow (ox'-bow) (a) A closely looping stream meander resembling the U-shaped frame embracing an ox's neck, having an extreme curvature such that only a neck of land is left between two parts of the stream. Syn: *horseshoe bend.* (b) A term used in New England also for the land enclosed, or partly enclosed, within an oxbow (bend of a stream). (c) The abandoned, bow- or horseshoe-shaped channel of a former meander, left when the stream formed a cutoff across a narrow meander neck. See also: *cutoff meander.* Syn: *abandoned channel.* (d) *oxbow lake.*

oxbow lake The crescent-shaped, often ephemeral, body of standing water situated by the side of a stream in the abandoned channel (oxbow) of a meander after the stream formed a neck cutoff and the ends of the original bend were silted up. Examples are common along the banks of the Mississippi River, where they are often known as bayous. See also: *billabong.* Syn: *oxbow; loop lake; mortlake; moat [streams]; horseshoe lake; cutoff lake; crescentic lake.*

oxea (ox'-e-a) A needle-shaped monaxonic sponge *spicule* tapering to a sharp point at each end. Pl: oxeas or oxeae.

Oxfordian (Ox-for'-di-an) A stage of the Standard Global Chronostratigraphic Scale: lowermost Upper Jurassic (above Callovian, below Kimmeridgian).

oxic horizon (ox'-ic) A mineral soil horizon that is at least 30 cm thick and characterized by the virtual absence of weatherable primary minerals or 2:1 layer silicate clays, the presence of 1:1 layer silicate clays and highly insoluble minerals such as quartz sand, the presence of hydrated oxides of iron and aluminum, the absence of water-dispersible clay, and the presence of low cation exchange capacity and small amounts of exchangeable bases.

oxidates (ox'-i-dates) Sediments composed of the oxides and hydroxides of iron and manganese, crystallized from aqueous solution. It is one of Goldschmidt's groupings of sediments or analogues of differentiation stages in rock analysis. Cf: *resistates; evaporates; reduzates; hydrolyzates.*

oxidation potential (ox-i-da'-tion) *Eh.*

oxide (ox'-ide) A mineral compound characterized by the linkage of oxygen with one or more metallic elements, such as cuprite, Cu_2O, rutile, TiO_2, or spinel, $MgAl_2O_4$. See also: *hydroxide.*

oxide-facies iron formation (ox'-ide-fa'-cies) An *iron formation* in which the principal iron-rich minerals are oxides, typically hematite or magnetite (James, 1954, p.256-263). It is thought to be the shoreward facies of sedimentary iron formation. See also: *specular schist; itabirite.*

oxide nickel laterite A *nickel laterite* composed of nickel-bearing iron oxyhydroxides, principally goethite.

oxide zinc deposit *nonsulfide zinc deposit.*

oxidite (ox'-i-dite) Shale-like iron oxides formed by the oxidation of iron meteorites. Cf: *shale-ball.*

oxidized zone (ox'-i-dized) An area of mineral deposits modified by surface waters, e.g. sulfides altered to oxides and carbonates. See also: *supergene enrichment.* Cf: *sulfide zone; gossan; protore.*

oxidizing flame (ox'-i-diz'-ing) In blowpiping, the outer, almost invisible, and less intense part of the flame, from which oxygen may be added to the compound being tested. Cf: *reducing flame.*

Oxisols Mineral soils that have an oxic horizon within 2 m of the surface or plinthite as a continuous phase within 30 cm of the surface, and that do not have a spodic or argillic horizon above the oxic horizon. An order in the U.S. system of soil taxonomy.

oxoferrite (ox-o-fer'-rite) A variety of native iron with some FeO in solid solution.

oxyanion In an *anisodesmic* compound, a group of atoms consisting of three or four anions (usually oxygen) surrounding a central cation, and having an overall negative charge (e.g. CO_3^{-2}, SiO_4^{-4}). A "radical" in chemical terms. Cf: polyanion.

oxyaquic conditions Pertaining to soils that are saturated but are not reduced and do not contain redoximorphic features.

oxyaster (ox'-y-as'-ter) A stellate sponge spicule (aster) having acute, sharp rays.

oxycone (ox'-y-cone) A laterally compressed, coiled cephalopod *conch* with an acute periphery and a usually narrow or occluded *umbilicus,* as in *Oxynoticeras.*

oxyferropumpellyite *pumpellyite-(Fe^{3+}).*

oxygen deficit (ox'-y-gen) The difference between the actual amount of dissolved oxygen in lake or sea water and the saturation concentration at the temperature of the water mass sampled.

oxygen demand *chemical oxygen demand.*

oxygen-isotope fractionation (ox'-y-gen-i'-so-tope) Fractionation of oxygen isotopes (oxygen-18/oxygen-16) in oxygen-bearing geologic materials, e.g. carbonate shells of marine organisms, which may be used as an indication of the temperature of formation of the materials. See also: *carbonate thermometer.*

oxygen minimum A zone in a body of water where the value of dissolved oxygen is less than it is either above or below. In extreme cases, anoxia results from depletion of all oxygen, marked by deposition of laminated black mud.

oxygen ratio Obsolete term for the ratio of the number of atoms of oxygen in the basic oxides of a mineral or rock to the number of atoms of oxygen in SiO_2 (Johannsen, 1939, v.1, p.164). Syn: *acidity coefficient; acidity quotient; coefficient of acidity.*

oxyhexaster (ox-y-hex'-as-ter) A hexaster whose simple terminal rays end in sharp points.

oxyhornblende (ox-y-horn'-blende) *basaltic hornblende.*

oxylophyte (ox-yl'-o-phyte) A plant preferring or restricted to acid soil.

oxymagnite (ox-y-mag'-nite) *maghemite.*

oxyphile (ox'-y-phile) *lithophile.*

oxysphere (ox'-y-sphere) A term that was proposed as a replacement for *lithosphere;* that zone or layer of the Earth whose constituent rocks are 60% oxygen. Archaic.

oxysulfide A *sulfide* mineral containing oxygen. An example is kermesite, Sb_2OS_2.

oxytylote (ox-y-ty'-lote) A sponge spicule shaped like a common pin.

oyelite (oy'-e-lite) A white orthorhombic mineral: $Ca_{10}B_2Si_8O_{29}\cdot 12H_2O$.

oyster reef (oys'-ter) An *organic reef* or *bank [sed]* composed mostly of oyster shells attached upon one another in growth position; living examples tend to be small (a hundred meters or so across, by a few meters high) and to occur in estuarine waters.

ozalid (oz'-a-lid) *diazo print.*

ozarkite (o'-zark-ite) White massive thomsonite from Arkansas.

ozocerite (o-zo'-ce-rite) A brown to jet black paraffin wax. It occurs in irregular veins, is soluble in chloroform, has a variable melting point, and yields *ceresine* on heating with a 20-30% solution of concentrated H_2SO_4 at 120° to 200° C. Varieties: *baikerite; celestialite; helenite; moldavite; pietricikite.* Also spelled: *ozokerite.* Syn: *ader wax; earth wax; fossil wax; mineral wax; native paraffin.*

ozokerite (o-zo'-ke-rite) Original spelling of *ozocerite.*

P

pääkkönenite (pääk′-kön-en-ite) A metallic gray monoclinic mineral: Sb_2AsS_2.

paar A depression produced by the moving-apart of crustal blocks rather than by subsidence within a crustal block. It is floored with upper-mantle igneous rocks and is essentially devoid of crustal material. Examples are the Gulf of California and the Dead Sea. Rarely used. Etymol: Hebrew. Cf: *rift [tect].*

paarite A metallic gray orthorhombic mineral: $Cu_{1.6}Pb_{1.6}Bi_{6.4}S_{12}$.

pabstite (pabst′-ite) A colorless or pinkish hexagonal mineral of the *benitoite* group: $Ba(Sn,Ti)Si_3O_9$.

paceite A deep sky-blue tetragonal acetate mineral: $CaCu(CH_3COO)_2{\cdot}6H_2O$.

pachnolite (pach′-no-lite) A colorless to white monoclinic mineral: $NaCaAlF_6{\cdot}H_2O$.

Pacific Decadal Oscillation (Pa-cif′-ic Dec′-a-dal Os-cil-la′-tion) A 20- to 30-year cycle of variability of sea-surface temperature of the North Pacific associated with major changes in the average distribution of atmospheric pressures. Abbrev: *PDO.* PDO influences weather patterns over the northern hemisphere.

Pacific suite (Pa-cif′-ic) Obsolete name for one of two large groups of igneous rocks, characterized by calcic and calc-alkalic rocks. Harker (1909) divided all Tertiary and Holocene igneous rocks of the world into two main groups, the *Atlantic suite* and the Pacific suite, the latter being so named because of the predominance of calcic and calc-alkalic rocks in the area of the circum-Pacific orogenic belt. Because there is such a wide variation in tectonic environments and associated rock types in the areas of Harker's Atlantic and Pacific suites, the terms are now seldom used to indicate kindred rock types. Cf: *Mediterranean suite.* Syn: *anapeirean.* See also: *andesite line.*

Pacific-type coastline A *concordant coastline,* esp. one as developed around the Pacific Ocean (e.g. the coastline of British Columbia), reflecting the continuous linear trends of the circum-Pacific fold-mountain system. A "modified" Pacific-type coastline develops behind festoons of island arcs and adjacent foredeep trenches, such as the coastline of Asia. Ant: *Atlantic-type coastline.*

pack *pack ice.*

packed ice *close pack ice.*

packer A short cylinder of expandable, rubberlike material that can be compressed to block a cased or uncased *wellbore* and prevent upward or downward fluid movement. It is generally limited to temporary use.

pack ice (a) A term used in a broad sense to include any area of *sea ice* (other than *fast ice*) regardless of its form or disposition, composed of a heterogeneous mixture of ice of varying sizes and ages, and formed by the jamming or crushing together of pieces of *floating ice;* the mass may be either loosely or tightly packed but it covers the sea surface with little or no open water. Cf: *drift ice.* (b) The ice material in an area of pack ice, or forming an ice pack. Syn: *ice canopy; ice pack; pack.* "The terms 'pack ice' and 'ice pack' have been used indiscriminately for both the sea area containing floating ice, and the material itself" (Huschke, 1959, p.410). Cf: *ice cluster.*

packing (pack′-ing) The spatial distribution and arrangement of the solid particles in a sediment or sedimentary rock, or of the atoms or ions in a crystal lattice; specif. the arrangement of clastic grains, entirely apart from any authigenic cement that may have crystallized between them. Cf: *fabric [sed]; compaction [sed].*

packing density A measure of the extent to which the grains of a sedimentary rock occupy the gross volume of the rock in contrast to spaces between the grains, equal to the cumulated grain-intercept length along a traverse in a thin section (Kahn, 1956). Syn: *grain density.*

packing index The ratio of the ion volume to the volume of the unit cell in a crystal (Fairbairn, 1943).

packing proximity An estimate of the number of grains (in a sedimentary rock) that are in contact with their neighbors, equal to the total percentage of grain-to-grain contacts along a traverse measured on a thin section (Kahn, 1956).

packing radius Half the distance of closest approach of like atoms or ions in a crystal.

packsand A very fine-grained sandstone that is so loosely consolidated by a little calcareous cement as to be readily cut by a spade.

packstone A term used for a sedimentary carbonate rock whose granular material is arranged in a self-supporting framework, yet also contains some matrix of calcareous mud (Dunham, 1962). Cf: *mudstone; grainstone; wackestone.*

paddle (pad′-dle) (a) The flat distal part of the last prosomal appendage (toward the rear) in a merostome. (b) A tetrapod limb modified into a rowing organ for life in the sea, as in plesiosaurs and sea turtles.

paderaite (pa-de′-ra-ite) A metallic gray monoclinic mineral: $AgPb_2Cu_6Bi_{11}S_{22}$.

padmaite A metallic pale yellow cubic mineral of the *gersdorffite* group: PdBiSe.

padmaragaya A light orange or reddish-yellow variety of sapphire; a synthetic corundum of various shades of yellow or orange. Etymol: Sinhalese, "lotus color". Syn: *padparadscha.*

padparadscha A syn. of *padmaragaya.* Also spelled: padparadschah.

paedogenesis (pae-do-gen′-e-sis) *neoteny (b).*

paedomorphism (pae-do-mor′-phism) *neoteny (a).*

paedomorphosis (pae-do-mor′-pho-sis) Evolution as a result of modification in the immature growth stages.

paganoite An orange-brown to deep golden-brown triclinic mineral: $NiBi^{3+}OAsO_4$.

pagoda halite A type of *halite* in which edges and corners (crystal coigns) are extended by skeletal growth. This is a three-dimensional mechanism, so that random cuts through a cubic crystal give pagoda (extended-corner) structures. Similar skeletal extensions may also occur along horizontal depositional surfaces, creating a two-dimensional network of right-angle crystal traceries (reticulate ridge halite). In depositional-plane growth, similar extended corners are also termed "pagoda" structures (Southgate, 1982).

pagoda stone (pa-go′-da) (a) A Chinese limestone showing in section fossil orthoceratites arranged in pagoda-like designs. (b) An agate whose markings resemble pagodas. (c) *pagodite.* Syn: *agalmatolite.*

pagodite (pa-go′-dite) Massive pinite or *agalmatolite* carved by the Chinese into miniature pagodas. Syn: *pagoda stone.*

paha (pa-ha′) A low, elongated, rounded glacial ridge or hill consisting mainly of drift, rock, or windblown sand, silt, or clay but capped with a thick cover of loess; found esp. in NE Iowa. Height varies between 10 and 30 m. Etymol: Dakota "pahá", "hill". Pl: paha; pahas.

pahasapaite (pa-ha-sa′-pa-ite) A colorless to pale pink cubic mineral with zeolitic properties: $Li_8(Ca,Li,K,Na)_{11}Be_{24}(PO_4)_{24}{\cdot}38H_2O$.

pahoehoe (pa-ho′-e-ho′-e) A Hawaiian term for a type of basaltic lava flow typified by a smooth, billowy, or ropy surface. Varieties include corded, elephant-hide, entrail, festooned, filamented, sharkskin, shelly, and slab pahoehoe. Cf: *aa.* Obsolete syn. *dermolith.* Syn: *ropy lava.*

PAHs *polyaromatic hydrocarbons.*

Paibian (Pai′-bi-an) A stage of the Standard Global Chronostratigraphic Scale: lowest stage of Furongian Series of the Cambrian. The basal GSSP is near Paibi in Hunan, China.

paigeite (paige′-ite) *vonsenite.*

painite (pain′-ite) A hard, ruby-red, orange-red, or brownish hexagonal mineral: $CaZrAl_9O_{15}(BO_3)$. It is structurally related to *fluoborite.*

paint A term used in SW U.S. for an earthy, pulverulent variety of cinnabar (Thrush, 1968).

paint pot A type of *mud pot* containing multicolored mud. Also spelled: paintpot.

paired terrace One of two stream terraces that face each other at the same elevation from opposite sides of the stream valley and that represent the remnants of the same flood plain or valley floor. Cf: *unpaired terrace.* Syn: *matched terrace.*

pair production The interaction of a gamma-ray photon with the nucleus of an atom in which the photon is absorbed and energy, $E > 1.02$ MeV, is transformed into an electron-positron pair (IAEA, 2003).

paisanite (pai′-sa-nite) A light-colored peralkaline hypabyssal or extrusive *rhyolite* characterized by small sanidine and quartz phenocrysts, few in number, and aggregates of sodic amphibole in a groundmass of quartz and alkali feldspar, frequently intergrown. Its name, given by Osann in 1893, is derived from Paisano Pass, Texas. Cf: *comendite; pantellerite*. Obsolete.

pakihi (pa-ki'-hi) A term used in New Zealand for a waterlogged gravel flat (Stamp, 1966, p.352). Etymol: Maori.

palaeo- (pa'-lae-o-) *paleo-*.

palaeocope (pa'-lae-o-cope') Any ostracode belonging to the order Palaeocopida, characterized by a shell with a long straight dorsal margin, commonly with lobes, sulci, and ventral structures. Range, Lower Ordovician to Middle Permian, with some questionably identified in present-day waters.

palaeoethnobotany (pa'-lae-o-eth'-no-bot'-a-ny) The study of plants used by ancient human societies.

Palaeophyticum (Pa'-lae-o-phy'-ti-cum) A paleobotanic division of geologic time, corresponding approximately to, and characterized by the plant life of, the Paleozoic. Cf: *Mesophyticum; Cainophyticum*. Also spelled: Paleophyticum.

palaetiology (pa'-lae-ti-ol'-o-gy) The explanation of past changes in the Earth's condition as being governed by the laws of cause and effect. Also such a type of explanation that invokes design or purpose, as in the phrase "intelligent design". Var. spelling: paletiology.

palagonite (pa-lag'-o-nite) An altered *basaltic glass*, brown to yellow or orange and found in pillow lavas as interstitial material or in amygdules. Named by Waltershausen in 1845, for Palagonia, Sicily.

palagonite tuff A pyroclastic rock consisting of angular fragments of hydrothermally altered or weathered basaltic glass. Cf: *hyaloclastite*.

palagonitization (pa-lag'-o-nit'-i-za'-tion) Formation of palagonite by hydration of volcanic (usually basaltic) glass.

palaite (pa'-la-ite) *huréaulite*.

palarstanide (pal-ar'-stan-ide) A metallic gray trigonal mineral: $Pd_5(Sn,As)_2$.

palasome (pal'-a-some) A syn. of *host*, used in economic geology. Also spelled: *palosome*. Cf: *paleosome; metasome [geol]*.

palate (pal'-ate) (a) In the vertebrates, esp. Osteichthyes and tetrapods, the bony roof of the mouth and its soft covering. (b) In some *cheilostome* bryozoans, the membranous or partly calcified wall of an *avicularium*, on which the mandible occludes.

Palatinian Orogeny (Pal-a-tin'-i-an) *Pfalzian Orogeny*.

paleic surface (pa-le'-ic) A smooth preglacial erosion surface, such as widely developed in Scandinavia.

palenzonaite (pal-en-zo'-na-ite) An adamantine red cubic mineral with the garnet structure: $NaCa_2Mn^{2+}_2(VO_4)_3$.

paleo- (pa'-le-o-) (a) A combining form denoting the attribute of great age or remoteness in regard to time (*Paleo*zoic), or involving ancient conditions (*paleo*climate), or of ancestral origin, or dealing with fossil forms (*paleo*anthropic). Sometimes given as *pale-* before vowels (paleoceanography; palevent). Also spelled: *palaeo-;* palaio-. (b) A prefix indicating pre-Tertiary origin, and generally altered character, of a rock to the name of which it is added, e.g. *paleo*picrite; by some the prefix has been applied to pre-Carboniferous rocks or features, e.g. the *Paleo*atlantic Ocean.

paleoagrostology (pa'-le-o-ag'-ros-tol'-o-gy) The study of fossil grasses.

paleoaktology (pa'-le-o-ak-tol'-o-gy) Study of ancient nearshore and shallow-water environments.

paleoalgology (pa'-le-o-al-gol'-o-gy) The study of fossil algae. Syn: *paleophycology*.

paleoanthropology (pa'-le-o-an'-thro-pol'-o-gy) *human paleontology*.

paleoaquifer (pa'-le-o-aq'-ui-fer) A stratigraphic unit or subterranean channel system that functioned as an aquifer at some time in the geologic past.

Paleoarchean (Pa-le-o-ar'-che-an) An era of the Standard Global Geochronometric Scale: the second-oldest era of the *Archean* Eon. Defined as the time span between 3,600 Ma, the end of the *Eoarchean* Era, and 3,200 Ma, the beginning of the *Mesoarchean* Era (Plumb, 1991).

paleoautochthon (pa'-le-o-au-toch'-thon) The original autochthon or basement of a tectonic region, esp. where overlain successively by *neoautochthon* and allochthon. It may be folded and faulted by later movements, but it has not been greatly displaced in a horizontal direction.

paleobiochemical (pa'-le-o-bi'-o-chem'-i-cal) Relating to ancient biochemical products, such as amino acids, fatty acids, and sugars, isolated from geological specimens, that have undergone little change since they were produced.

paleobiocoenosis (pa'-le-o-bi'-o-coe-no'-sis) An assemblage of organisms that lived together in the geologic past as an interrelated community. Syn: *paleocoenosis*. Cf: *fossil community*.

paleobiogeography (pa'-le-o-bi'-o-ge-og'-ra-phy) That branch of paleontology that treats the geographic distribution of plants and animals in past geologic time, esp. with regard to the ecologic, climatologic, and evolutionary factors involved. See also: *paleophytogeography; paleozoogeography*.

paleobiology (pa'-le-o-bi-ol'-o-gy) A branch of paleontology dealing with the study of fossils as organisms rather than as features of historical geology. Cf: *biogeology; geobiology; cenozoology; paleozoology*.

paleobiotope (pa'-le-o-bi'-o-tope) A term sometimes used in paleoecology to designate a region of unspecified size that is characterized by essentially uniform environmental conditions and by a correspondingly uniform population of animals or plants or both. See also: *biotope [ecol]*.

paleoblast (pa'-le-o-blast') A crystal, or a remnant of a crystal, esp. in a metamorphic rock, that is older than other grains of the same or other mineral species in the rock. Such relicts represent former conditions of equilibration of the rock. Because of the kinetics of solid-state *recrystallization*, paleoblasts have high dislocation densities and narrow subgrain-boundary spacings. Cf: *holoblast; porphyroclast*.

paleobotanic province (pa'-le-o-bo-tan'-ic) A large region characterized by a similar fossil flora over the whole area.

paleobotany (pa'-le-o-bot'-a-ny) The study of the plant life of the geologic past (Arnold, 1947, p.1). Syn: *phytopaleontology; paleophytology*.

paleoceanography (pa'-le-o'-cean-og'-ra-phy) The study of the physical, chemical, biologic, and geologic aspects of past oceans, as deduced from the geologic record.

Paleocene (Pa'-le-o-cene') The lower series of the *Paleogene* Subsystem of the *Tertiary* System of the Standard Global Chronostratigraphic Scale, above the *Cretaceous* System and below the *Eocene* Series. Also the time during which these rocks were formed, the Paleocene Epoch, covering the time span between 65 and 53 Ma.

paleochannel (pa'-le-o-chan'-nel) A remnant of a stream channel cut in older rock or sediment and filled by the younger overlying rock or sediment; a buried stream channel.

paleoclimate (pa'-le-o-cli'-mate) The climate of a given interval of time in the geologic past. Syn: *geologic climate*.

paleoclimatology (pa'-le-o-cli'-ma-tol'-o-gy) The study of climates prior to the period of instrumental records, and of the causes of their variation, on either a local or a worldwide basis. It involves the interpretation of natural phenomena that are climate dependent and incorporate into their structure a measure of this dependency; e.g., land and marine sediments, loess, tree rings, ice cores, speleotherms, corals (Bradley, 1999).

paleocoenosis (pa'-le-o-coe-no'-sis) *paleobiocoenosis*.

paleocrystic ice (pa'-le-o-crys'-tic) Old *sea ice*, esp. well-weathered *polar ice*, generally considered to be at least 10 years old; it is often found in *floebergs* and in the pack ice of the central Arctic Ocean.

paleocurrent (pa'-le-o-cur'-rent) An ancient current (generally of water) that existed in the geologic past, whose direction is inferred from the sedimentary structures and textures of the rocks formed at that time.

paleocurrent structure *directional structure*.

paleodepth (pa'-le-o-depth') The depth at which an ancient organism or group of organisms lived.

paleodrainage pattern (pa'-le-o-drain'-age) A drainage pattern representing the distribution of a valley system as it existed at a given moment of geologic time (Andresen, 1962).

paleoecology (pa'-le-o-e-col'-o-gy) The study of the relationships between ancient organisms and their environments, the death of organisms, and their burial and postburial history in the geologic past, based on fossil faunas and floras and their stratigraphic position. See also: *ecology*.

paleoenvironment (pa'-le-o-en-vi'-ron-ment) An environment in the geologic past.

paleoequator (pa'-le-o-e-qua'-tor) The position of the Earth's equator in the geologic past as defined for a specific geologic period and based on geologic evidence such as paleomagnetic measurements, oxygen-isotope ratios, fauna and flora, distribution of evaporites, reefs, coal deposits, and tillites; e.g., the Ordovician paleoequator for North America, running from the southern tip of Baja California to the north end of Greenland. Paleoequators are great circles that were formerly normal to the axis of rotation but are now displaced and vary from continent to continent and plate to plate.

paleoflood Any flood occurring before systematic discharge records were kept for a stream.

paleofloristics (pa'-le-o-flo-ris'-tics) The composition and distribution of ancient floras.

paleofluminology (pa'-le-o-flu'-mi-nol'-o-gy) The study of ancient stream systems. Cf: *paleohydrology*.

Paleogene (Pa'-le-o-gene') The lower subsystem of the *Tertiary* System of the Standard Global Chronostratigraphic Scale, above the *Cretaceous* and below the *Neogene* Subsystem. Also the time during which these rocks were formed, the Paleogene Subperiod, covering the time span between 65 and 23 Ma.

paleogeographic event (pa'-le-o-ge'-o-graph'-ic) *palevent*.

paleogeographic map A map that shows the reconstructed physical geography at a particular time in the geologic past, including such information as the distribution of land and seas, geomorphology of the land, depth of the sea, directions of currents in water and air, distribution of bottom sediments, and climatic belts. Cf: *paleotectonic map*.

paleogeographic stage *palstage*.

paleogeography (pa'-le-o-ge-og'-ra-phy) The study and description of the physical geography of the geologic past, such as the historical reconstruction of the pattern of the Earth's surface or of a given area at a particular time in the geologic past, or the study of the successive changes of surface relief during geologic time. Syn: *paleophysiography*.

paleogeologic map (pa'-le-o-ge'-o-log'-ic) A map that shows the areal geology of an ancient surface at some time in the geologic past; esp. such a map of the surface immediately below an unconformity, showing the geology as it existed at the time the surface of unconformity was completed and before the overlapping strata were deposited. Paleogeologic maps were introduced by Levorsen (1933). Cf: *subcrop map.* Syn: *peel map.*

paleogeomorphology (pa'-le-o-ge'-o-mor-phol'-o-gy) A branch of geomorphology concerned with the recognition of ancient surfaces and materials and with the study of ancient topographies and topographic features that are now concealed beneath the surface or have been removed by erosion. Syn: *paleophysiography.*

Paleognathae (Pa'-le-o-gnath'-ae) A superorder of the avian subclass Neornithes which encompasses the ground-dwelling orders of birds: ostriches, cassowaries, rheas, emus, tinamous, moas, elephant birds, and kiwis. See also: *Neognathae.*

paleohydrology (pa'-le-o-hy-drol'-o-gy) (a) The study of the earliest uses and management of water. (b) The study of ancient hydrologic features preserved in rock, e.g. paleokarst. Cf: *paleofluminology.*

paleoichnology (pa'-le-o-ich-nol'-o-gy) *palichnology.*

paleointensity The intensity of the ancient Earth's magnetic field, as determined by experiments on geologic materials that attempt to duplicate the magnetization acquisition process, using the correct ancient field intensity. Paleointensity values may be expressed as "virtual" dipole moments for the Earth.

paleoisotherm (pa'-le-o-i'-so-therm) The locus of points of equal temperature at some time in the geologic past.

paleokarst (pa'-le-o-karst') A *karst* formed under an earlier erosion cycle, preserved by burial or suspension of karstification processes. Caves and sinkholes are usually filled with sediment. Cf: *buried karst; covered karst; fossil karst; interstratal karst; relict karst; resurrected karst.*

paleokarstic surface An irregular, hummocky, pot-holed surface on a limestone formed by exposure and dissolution by meteoric (fresh) water.

paleolatitude (pa'-le-o-lat'-i-tude) The latitude of a specific area on the Earth's surface in the geologic past; specif. the distance measured in degrees from the *paleoequator* or *paleopole.*

paleolimnology (pa'-le-o-lim-nol'-o-gy) (a) The study of the conditions and processes in ancient lakes; interpretation of the accumulated sediments and the geomorphology and geologic history of ancient lake basins. (b) The study of the sediments and history of existing lakes (Cohen, 2003; Wetzel, 2001, p.785).

Paleolithic (Pa'-le-o-lith'-ic) In archaeology, the first division of the *Stone Age,* characterized by the appearance of humans and their implements. Correlation of relative cultural levels with actual age (and, therefore, with the time-stratigraphic units of geology) varies from region to region; however, the age generally given for the Paleolithic more or less coincides with the Pleistocene. Cf: *Mesolithic; Neolithic; lower Paleolithic; middle Paleolithic; upper Paleolithic; Eolithic.* Obsolete syn: *Old Stone Age.*

paleolithologic map (pa'-le-o-lith'-o-log'-ic) A *paleogeologic map* that shows lithologic variations at some buried horizon or within some restricted zone at a particular time in the geologic past.

paleomagnetic pole (pa'-le-o-mag-net'-ic) A conventional means of expressing the direction of a geologically stable remanence thought to adequately average the geomagnetic field. The pole location of the geocentric dipole magnetic field representing the field direction, at the rock's location, is parallel to its remanent magnetization. See also: *virtual geomagnetic pole.*

paleomagnetic stratigraphy The use of natural remanent magnetization to identify stratigraphic sequences, depending on the polarity and/or paleosecular variation of the ambient magnetic field. Syn: *magnetic stratigraphy.*

paleomagnetism (pa'-le-o-mag'-net-ism) The study of natural remanent magnetization of Earth materials in order to determine the direction, and indirectly intensity, of the Earth's magnetic field in the geologic past.

paleomorphology (pa'-le-o-mor-phol'-o-gy) The morphology or study of form and structure of fossil remains (hard parts) in order to determine the original anatomy (soft parts) of an organism; e.g. the study of a brachiopod muscle scar whose depth may indicate the strength of the muscle.

paleomycology (pa'-le-o-my-col'-o-gy) The study of fossil fungi.

paleontography (pa'-le-on-tog'-ra-phy) The formal, systematic description of fossils. Adj: paleontographic.

paleontological resource (pa'-le-on'-to-log'-i-cal) A locality, esp. one of such limited size that it can be destroyed by human activity, that has yielded type fossils or described fossils of unusually high quality, or is documented for its preservation of paleoenvironmental features. Also, the fossils, described or not, from such a locality.

paleontologic facies (pa'-le-on'-to-log'-ic) A term recommended by Teichert (1958, p.2734) to replace *biofacies* as used in stratigraphy, signifying the paleontologic characteristics of a sedimentary rock.

paleontologic record A syn. of *fossil record.* Cf: *stratigraphic record.*

paleontologic species A *morphologic species* based on fossil specimens. It may include specimens that would be considered specifically distinct if living individuals could be observed.

paleontologist (pa'-le-on-tol'-o-gist) One who studies the fossilized remains of animals and/or plants.

paleontology (pa'-le-on-tol'-o-gy) The study of life in past geologic time, based on fossil plants and animals and including phylogeny, their relationships to existing plants, animals, and environments, and the chronology of the Earth's history. Cf: *neontology.* See also: *historical geology; paleobiology; paleobiogeography.*

paleopalynology (pa'-le-o-pal'-y-nol'-o-gy) The study of robust organic-walled microfossils. Usually this means fossil plant pollen and spores and single-celled algae (especially dinoflagellates and acritarchs). It also includes other organic, decay-resistant microfossils such as various algal tissues, coenobia, cysts and phycomata, and fungal spores; animal microfossils such as scolecodonts and chitinozoans; and the organic linings of foraminifera (microforaminifera). Also included are studies of fragmented, more or less degraded organic particles (palynodebris) (MacRae, 1994).

paleopathology The study of the effects of disease in fossils.

paleopeat An ancient peat deposit that was buried and is now transformed into coal. Useful to denote and discuss the conditions during peat formation millions of years ago for what is now coal.

paleopedology (pa'-le-o-pe-dol'-o-gy) The study of soils of past geologic ages, including determination of their ages.

paleophycology (pa'-le-o-phy-col'-o-gy) *paleoalgology.*

paleophysiography (pa'-le-o-phys'-i-og'-ra-phy) (a) *paleogeomorphology.* (b) *paleogeography.*

Paleophytic (Pa'-le-o-phyt'-ic) A paleobotanic division of geologic time, signifying that time during which pteridophytes were abundant, between the development of algae and the appearance of the first gymnosperms. Cf: *Aphytic; Archeophytic; Eophytic; Mesophytic; Cenophytic.* Syn: *Pteridophytic.*

paleophytic flora The collection of spore-bearing plants that dominated terrestrial ecosystems before the expansion of seed plants in the Permian, including ferns, mosses, horsetails, and lycophytes.

paleophytogeography (pa'-le-o-phy'-to-ge-og'-ra-phy) The branch of *paleobiogeography* dealing with the distribution of plants in past geologic time. Cf: *paleozoogeography.*

paleophytology (pa'-le-o-phy-tol'-o-gy) An obsolescent syn. of *paleobotany.*

paleoplacer A lithologic unit which hosts a mineral deposit formed in a *placer* environment in the geological past, e.g. the Witwatersrand auriferous quartzite.

paleoplain (pa'-le-o-plain') A term introduced by Hill (1900, p.5) for an ancient degradational plain that is now buried beneath later deposits.

paleopole (pa'-le-o-pole') A pole of the Earth, either magnetic or geographic, at some former geologic time.

Paleoproterozoic (Pa-le-o-prot-er-o-zo'-ic) An era of the Standard Global Geochronometric Scale: the oldest era of the *Proterozoic* Eon. Defined as the time span between 2,500 Ma, the end of the *Neoarchean* Era of the *Archean* Eon, and 1,600 Ma, the beginning of the *Mesoproterozoic* Era (Plumb, 1991).

paleosalinity (pa'-le-o-sa-lin'-i-ty) The salinity of a body of water in the geological past, as evaluated on the basis of chemical analyses of sediment or formation water.

paleoseismicity Prehistoric earthquake events resulting in slip on faults (Engelder, 1974).

paleosere (pa'-le-o-sere') A sequence of ecologic communities in the geologic past that led to a *climax community;* a *sere* in the geologic past.

paleoslope (pa'-le-o-slope') The direction of initial dip of a former land surface; esp. the regional slope of a large, ancient physiographic unit, such as a flood plain or a continental slope.

paleosol (pa'-le-o-sol') A soil that formed on a landscape in the past with distinctive morphological features resulting from a soil-forming environment that no longer exists at the site. The former pedogenic process was either altered because of external environmental change or interrupted by burial. A paleosol (or component horizon) may be classed as relict if it has persisted in a land-surface position without major alteration of morphology by processes of the prevailing pedogenic environment. An exhumed paleosol is one that formerly was buried and has been re-exposed by erosion of the covering mantle.

paleosome (pa'-le-o-some') A geometric element of a composite rock or mineral deposit, (or *neosome*); e.g., wall rock in a vein or replacement deposit, or the unaltered and relatively immobile pre-existing part of a migmatite. Var. spelling: palaeosome (British). Commonly used in place of the terms *stereosome, metaster,* or *restite.* Cf: *host; leucosome; melanosome; mesosome.*

paleostage indicator Any erosional or depositional feature that represents past river stages.

paleostress analysis Application of one of several techniques used to estimate the stress tensor for a particular deformation episode. Commonly used techniques include measuring the mean diameter of subgrains or recrystallized grains in rocks deformed by dislocation creep; measuring the diameters or crystallographic orientations of twinned carbonate grains; and calculating a tensor that conforms with the inferred directions of shear stresses on numerous planes with different attitudes. In the latter case, most workers infer that the *rake* of a slip vector on a fault surface gives the rake of the maximum resolved shear stress on that fault plane.

paleostress tensor An estimate of the stress tensor that acted on rocks during a particular deformation episode, usually defined by the orientation and relative magnitudes of the three principal stresses; in some cases the estimate will give absolute magnitudes of stresses.

paleostructure (pa'-le-o-struc'-ture) The geologic structure of a region or sequence of rocks at some time in the geologic past; the structure of a paleogeologic area.

paleostructure map A map that shows, by thickness contour lines, the geologic structure that existed at the time when a surface of unconformity was completed and before the overlapping strata were deposited (Levorsen, 1960, p.4).

paleotectonic map (pa'-le-o-tec-ton'-ic) A map intended to show geologic and tectonic features as they existed at some time in the geologic past, rather than the sum of all the tectonics of the region, as portrayed on a *tectonic map.* It is similar to a *paleogeographic map* but more emphasis is placed on the tectonic features than on the distribution of lands and seas. See also: *neotectonic map.*

paleotemperature (pa'-le-o-tem'-per-a-ture') The mean temperature at a given time or place in geologic history.

paleothanatocoenosis (pa'-le-o-than'-a-to-coe-no'-sis) A group of organisms buried together in the geologic past.

paleo-thanatotope The total of all sediments in which a fossil assemblage occurs (excluding reworked occurrences) (van Morkhoven, 1966, p.312).

paleothermal (pa'-le-o-ther'-mal) Pertaining to or characteristic of warm climates of the geologic past; e.g. a "paleothermal fauna". Syn: paleothermic.

paleothermometry (pa'-le-o-ther-mom'-e-try) Measurement or estimation of paleotemperatures by a variety of methods.

paleotopographic map (pa'-le-o-top'-o-graph'-ic) A map that shows the relief of a surface of unconformity, or the relief believed to have existed at some time in the geologic past.

paleotopography (pa'-le-o-to-pog'-ra-phy) The topographic relief of an area at a particular time in the geologic past; the topography of a paleogeologic area, such as the configuration of the surface of an unconformity at the time it was overlapped.

paleotoxicology (pa'-le-o-tox'-i-col'-o-gy) Monitoring of biologically active compounds in the biosphere of former ages, e.g. information preserved in permafrost or old glacier ice.

paleotypal (pa'-le-o-typ'-al) Said of a fine-grained porphyritic igneous rock having the characteristics of altered extrusive or hypabyssal rocks such as those of pre-Tertiary age. This term and the term *cenotypal* were introduced to distinguish Tertiary (*neovolcanic*) and pre-Tertiary (*paleovolcanic*) fine-grained igneous rocks; both are little used.

paleovolcanic (pa'-le-o-vol-can'-ic) Said of extrusive rocks that are of pre-Tertiary age. Cf: *neovolcanic.*

paleovolcanology (pa'-le-o-vol'-can-ol'-o-gy) The study of the processes and products of volcanic activity in the geologic past.

paleowind (pa'-le-o-wind') A wind of the geologic past. Its direction is recorded by distribution of volcanic ash falls, growth patterns of coral reefs, orientation of sand dunes, etc.

Paleozoic (Pa'-le-o-zo'-ic) The lowest erathem of the *Phanerozoic* Eonothem of the Standard Global Chronostratigraphic Scale, above the *Precambrian* and below the *Mesozoic.* Also the time during which these rocks were formed, the Paleozoic Era, covering the time span between 540 and 250 Ma.

paleozoogeography (pa'-le-o-zo'-o-ge-og'-ra-phy) The branch of *paleobiogeography* dealing with the distribution of animals in past geologic time. Cf: *paleophytogeography.*

paleozoology (pa'-le-o-zo-ol'-o-gy) That branch of paleontology dealing with the study of *subfossil* and *fossil* animals, both invertebrate and vertebrate. See also: *paleobiology.*

palermoite (pa-ler'-mo-ite) A colorless orthorhombic mineral: $(Li,Na)_2(Sr,Ca)Al_4(PO_4)_4(OH)_4$. It is the Sr analogue of bertossaite.

palette (pa-lette') *shield [speleo].*

palevent (pal-e-vent') A relatively sudden and short-lived paleogeographic happening, such as the brief existence of a particular depositional environment, or a rapid geographic change separating two palstages (Wills, 1956, p.14). Syn: *paleogeographic event.* Rarely used.

pali [geog] (pa'-li) An Hawaiian term for a steep slope; e.g. the Nuuanu Pali, a steep-faced scarp on the NE side of Oahu.

pali [paleont] Plural of *palus.*

palichnology (pal-ich-nol'-o-gy) The study of trace fossils, as opposed to *neoichnology.* Also spelled: *paleoichnology.* See also: *ichnology.*

paliform (pal'-i-form) Resembling a *palus;* specif. "paliform lobes" of the septa in corals, formed by detached trabecular offsets from the inner edges of the septa, appearing in vertical succession, and differing from pali in not being formed as a result of substitution.

palimpsest [marine geol] (pal'-imp-sest) Said of sediments of the *continental shelf,* reworked several times by physical or biological processes.

palimpsest [meta] adj. Said of a structure or texture in a metamorphic rock in which remnants of some pre-existing structure or texture, perhaps even the original one, are preserved, and sometimes even megascopically visible. The term was first used by Sederholm (1891). Cf: *relict [meta].*

palimpsest [streams] Said of a kind of drainage in which a modern, anomalous drainage pattern is superimposed upon an older one, clearly indicating different topographic and possibly structural conditions at the time of development.

palimpsest landscape A subaerial landscape from which previous processes and stages of development can be recognized within the landforms that are now being shaped.

palimpsest surface *Continental shelf* morphology that results from shore-face retreat and modification (but not obliteration) by subsequent processes (Swift et al., 1972, p.537).

palingenesis [paleont] (pal-in-gen'-e-sis) Recapitulation, without change, in the young stages of an organism of the characteristics of its ancestors. See also: *recapitulation theory.*

palingenesis [petrology] Obsolete term for the formation of a new magma by the melting of pre-existing magmatic rock in situ. Considered incorrectly by some workers as a syn. of *anatexis.* Adj: palingenic.

palingenetic (pal'-in-ge-net'-ic) *resurrected.*

palinspastic map (pal-in-spas'-tic) A name coined by Kay (1937) for a paleogeographic or paleotectonic map in which the features represented have been restored as nearly as possible to their original geographic positions, before the rocks of the crust were deformed.

palinspastic restoration *palinspastic section.*

palinspastic section A cross section in which the features represented have been restored as nearly as possible to their original geographic positions, before the rocks of the crust were deformed (Kay, 1937). Syn: *palinspastic restoration.*

palintrope (pal'-in-trope) A term used initially for the morphologically posterior sector of either valve of some brachiopod shells, which was reflexed to grow anteriorly (mixoperipheral growth), but more recently for the curved surface of the shell, bounded by beak ridges and cardinal margin of nonstrophic shells (TIP, 1965, pt. H, p.149). It differs from a *planarea* in being curved in all directions.

palisade (pal-i-sade') A picturesque rock cliff or line of bold cliffs, rising precipitously from the margin of a stream or lake; esp. one consisting of basalt with columnar structure, such as the Palisades along the Hudson River of New York and New Jersey. Term is usually used in the plural.

palisade mesophyll *palisade tissue.*

Palisades disturbance (Pal-i-sades') A time of deformation, or orogeny, supposed by Schuchert (1924) to have closed the Triassic Period in eastern North America and elsewhere. It is based on the block-faulted structure of the Upper Triassic Newark series in the Appalachian area, which was truncated before younger Mesozoic (mainly Cretaceous) strata were laid over it. The concept of a distinct orogeny at this time is dubious and has only local application at most. Named for the Palisades of New York and New Jersey, the edge of a diabase sill intruded at this time. Essentially, this event is related to the rifting that led to formation of the Atlantic Ocean.

palisade tissue Tissue composed of long cylindrical chlorophyllous cells oriented normal to the lamina beneath the upper epidermis of angiospermous leaves. Syn: *palisade mesophyll.*

palladinite An ill-defined tetragonal mineral: $(Pd,Cu)O$.

palladium (pal-la'-di-um) A soft silver-white or steel-gray cubic mineral, the native metallic element Pd. It is one of the platinum metals, and it resembles and occurs with platinum, usually occurring in grains and frequently alloyed with platinum and iridium.

palladium amalgam *potarite.*

palladium gold *porpezite.*

palladoarsenide (pal'-la-do-ar'-se-nide) A metallic gray monoclinic mineral: Pd_2As.

palladobismutharsenide (pal'-la-do-bis'-muth-ar'-se-nide) A metallic cream-colored orthorhombic mineral: $Pd_2(As,Bi)$.

palladodymite A metallic orthorhombic mineral: $(Pd,Rh)_2As$. It is the Pd analogue of rhodarsenide.

palladseite (pal-lad'-se-ite) A metallic cubic mineral: $Pd_{17}Se_{15}$.

pallasite (pal'-la-site) A *stony-iron differentiated meteorite* composed essentially of large crystals of olivine embedded in a network of nickel-iron. These are two main kinds of pallasites, the main group and the Eagle-Station grouplet. Pallasites are believed to have been formed at the interface of the stony mantle and metal core of one or more differentiated asteroids.

pallial chamber (pal'-li-al) *mantle cavity.*

pallial line A line or narrow band on the inner surface of a valve of a bivalve-mollusk or rostroconch shell, close to and more or less parallel with the margin, and marking the line of attachment of the marginal muscles of the mantle. It is typically distinguished by a groove or ridge and by a change in texture of shell material.

pallial sinus (a) An often conspicuous embayment or inward bend in the posterior and ventral part of the *pallial line* of a bivalve mollusk, marking the point of attachment of the siphonal retractor muscles. See also: *sinus [paleont].* (b) A *mantle canal* of a brachiopod.

pallite (pal'-lite) A ferrian variety of millisite.

pallium (pal'-li-um) The *mantle* of a mollusk or brachiopod. Pl: pallia.

pallomancy (pal'-lo-man'-cy) A form of *dowsing* using a pendulum. Cf: *rhabdomancy.*

palmate (pal'-mate) Having lobes or divisions radiating from a common center in a handlike fashion; digitate.

palmate gypsum Naturally occurring habit of primary gypsum developed at the bottom of a shallow saline water body. The 112 crystal prisms are

elongated away from a twin plane with an obtuse angle of juncture, creating broad, leaflike crystal projections. A variation of swallowtail twinning.
palmate venation In a leaf, a type of *net venation* in which the main veins branch out from the stalk apex like the fingers of a hand. Cf: *pinnate venation.*
palmierite (pal-mie'-rite) A white rhombohedral mineral: $(K,Na)_2Pb(SO_4)_2$. It is the Pb analogue of kalistrontite.
palmitic acid (pal-mit'-ic) A long-chain waxlike fatty acid, formula $C_{16}H_{32}O_2$, present in numerous plant and animal fats as glycerides.
palm-tree structure *flower structure.*
palosome (pal'-o-some) *palasome.*
palp A reduced distal portion of the limb of a crustacean, usually only one of its rami, but sometimes comprising both rami and basis (TIP, 1969, pt. R, p.99). See also: *palpus.*
palpebral lobe (pal-pe'-bral) (a) An elevated portion of the *fixed cheek* of a trilobite, extending laterally from the axial furrow to the upper and inner margin of the visual surface of an eye. (b) The elevated portion of the lateral eye of a eurypterid.
palpi (pal'-pi) Plural of *palpus.*
palpus (pal'-pus) A term applied either to a *pedipalpus* (including pedipalpal coxae) or more properly to one of the five segments following the coxa in an arachnid (TIP, 1955, pt. P, p.62). Pl: *palpi.* Adj: palpal. See also: *palp.*
pals *palsa.*
palsa (pal'-sa) An elliptical domelike *frost mound* containing ice lenses in peat, commonly 3-6 m high and 2-25 m long, occurring in subarctic bogs of the tundra, esp. in Scandinavia, and often surrounded by shallow open water. Etymol: Swedish, "elliptical". Pl: *palsen.* See also: *frost mound; peat mound; peat hummock.* Also spelled: *pals.*
palsen (pal'-sen) Plural of *palsa.*
palstage (pal'-stage) A period of time when paleogeographic conditions were relatively static, or were changing gradually and progressively, with relation to such factors as sea level, surface relief, or distance from shore (Wills, 1956, p.14). Cf: *palevent.* Syn: *paleogeographic stage.* Rarely used.
paludal (pa-lu'-dal) Pertaining to a marsh. Syn: *paludous; palustral.*
paludification (pa-lu'-di-fi-ca'-tion) Peat formation over previously forested land, grassland, or even bare rock as a result of a rise in the local water table due to changes in climatic conditions, rise in base level, or autogenic processes. Cf: *terrestrialization.*
paludous (pa-lu'-dous) *paludal.*
palus (pal'-us) Any of several slender vertical calcareous lamellae, plates, or pillars, developed along the inner edge of certain entosepta of a coral and comprising the remnant part of a pair of exosepta joined at their inner margins. Pl: pali. See also: *paliform.*
palus [planet] A term established by the International Astronomical Union for a small plain on the Moon (USGS, 2003). Pl: paludes.
palustral *paludal.*
palustrine (pa-lus'-trine) Pertaining to material growing or deposited in a marsh or marsh-like environment.
palygorskite (pal-y-gor'-skite) (a) A white, grayish, yellowish, or grayish-green chain-structure clay mineral: $(Mg,Al)_2Si_4O_{10}(OH){\cdot}4H_2O$. It crystallizes in several monoclinic and orthorhombic polytypes. (b) A group name for monoclinic minerals with an analogous composition, but with Mg replaced by Mn or Na, and Al replaced by Fe^{3+} or Mn^{3+}. (c) The term has also been used as a group name for lightweight fibrous clay minerals showing a considerable amount of substitution of aluminum for magnesium and characterized by distinctive rodlike shapes under the electron microscope. Syn: *attapulgite.*
palyniferous (pal-y-nif'-er-ous) Of rocks and other samples, yielding pollen, spores, or other palynomorphs on maceration.
palynivore (pal'-y-ni-vore) An animal that feeds on pollen (Labandeira, 2000).
palynobiofacies (pa-ly-no-bi'-o-fa-cies) The aspect of *palynofacies* referring to organic composition.
palynodebris (pa-ly-no-de-bris') The resistant-walled, palynomorph-size organic particles other than palynomorphs in a sediment or sedimentary rock: wood fragments, charcoal, amorphous organic matter, etc. Cf: *palynofacies; phytoclast; particulate organic matter; POM.*
palynofacies (pal'-y-no-fa'-cies) The assemblage of *palynomorphs* and *palynodebris* in a sediment, studied from the point of view of expressing the extent of local difference from the average for an area, which can be a difference in taxonomic composition (= *palynobiofacies*) or in kind of particle, e.g., abundance of charcoal (*palynolithofacies*).
palynoflora (pal'-y-no-flo'-ra) The whole suite of fossil palynomorphs from a given rock sample. The term *microflora* is sometimes incorrectly used as a synonym; however, it refers to extant, not fossil, microscopic organisms.
palynolithofacies (pa-ly-no-lith'-o-fa-cies) The aspect of *palynofacies* referring to inorganic composition.
palynology (pal-y-nol'-o-gy) Originally, the study of pollen and spores, extant and fossil, including stratigraphic and paleoecological applications. Now the term includes also a wider range of other, resistant-walled microscopic remains of various plants, animals, fungi, and protists. Cf: *paleopalynology; pollen analysis.*
palynomorph (pal'-y-no-morph') A microscopic, resistant-walled organic body found in palynologic maceration residues; a palynologic study object. Palynomorphs include pollen, spores of many sorts, acritarchs, chitinozoans, dinoflagellate thecae and cysts, certain colonial algae, and other acid-insoluble microfossils. Cf: *sporomorph.*
palynostratigraphy (pal'-y-no-stra-tig'-ra-phy) The stratigraphic application of palynologic methods.
pamet (pam'-et) A dry valley formed in glacial deposits on the outer part of Cape Cod, Mass.
pampa (pam'-pa) A vast treeless grassy plain of temperate regions, esp. as used in Argentina and adjacent parts of Uruguay. It is comparable to the *prairies* of North America, the *steppes* of Russia, and the *veld* of South Africa.
pan [geomorph] (a) A shallow, natural depression or basin, esp. one containing a lake, pond, or other body of standing water; e.g. a shallow depression holding a temporary or permanent pool in a tidal marsh along the Atlantic coast of the U.S. (b) A term used in South Africa for a hollow in the ground where the neck of a volcano formerly existed. (c) A *salt pan,* esp. in South African usage. Cf: *vloer.* (d) An artificial basin for producing salt by evaporation of salt water or brine. Also, a vessel for evaporating salt water or brine.
pan [ice] (a) A shortened form of *pancake ice.* (b) An individual piece of pancake ice. (c) *ice pan.* (d) A large fragment of flat, relatively thin ice, having a diameter about 60 m, formed in a bay or fiord or along the shore and subsequently loosened to drift about the sea.
pan [soil] A hard, cementlike layer, crust, or horizon within or just beneath the surface soil, being strongly compacted, indurated, or high in clay content, and usually impeding the movement of water and air and the growth of plant roots; specif. *hardpan.* See also: *genetic pan; pressure pan.*
panabase (pan'-a-base) *tetrahedrite.*
panasqueiraite (pa-na-squeir'-a-ite) A pink monoclinic mineral of the *tilasite* group: $CaMg(PO_4)(OH,F)$.
pancake bomb (pan'-cake) A type of volcanic bomb whose flattened shape is due to impact. Cf: *Hawaiian-type bomb.*
pancake dome Steep-sided, flat-topped volcanic feature on Venus; typically a few hundred meters high, ranging from 10 to almost 100 km in diameter (Bougher et al., 1997). Syn: *farrum.*
pancake ice One or more small, predominantly circular pieces of newly formed sea ice (diameter ranging from about 30 cm to about 3 m) with slightly raised rims caused by the pieces rotating and striking against one another; it often forms during the early fall in polar regions. Cf: *lily-pad ice.* Syn: *pan [ice].*
pandaite (pan'-da-ite) *bariopyrochlore.*
pandasutural (pan-da-su'-tur-al) Of bands or zones on a *dinoflagellate theca* or *cyst,* being located along the outer margins of plates and related to zones of incremental plate growth (Williams et al., 2000, p.151).
pandemic (pan-dem'-ic) Said of conditions that occur over a broad geographic area and affect a major part of the population; also said of a widely dispersed population.
panderian opening (pan-der'-i-an) A small opening in the fixigenal or thoracic *doublure* of trilobites, situated immediately behind the *panderian protuberance.* Its function is not known, but it may have been a pore leading to some internal organ. Cf: *panderian protuberance.*
panderian protuberance A small protuberance on the fixigenal or thoracic *doublure* of trilobites, which functioned as a stopping device to prevent segments overlapping too far on enrollment. Cf: *panderian opening.*
pandermite (pan'-der-mite) *priceite.*
panethite (pan'-e-thite) A pale amber monoclinic meteorite mineral: $(Na,Ca)(Mg,Fe^{2+})PO_4$.
panfan (pan'-fan) A graded bedrock surface consisting of a series of coalescing pediments and representing the penultimate stage of an arid cycle of erosion. The synonymous term *pediplain* is preferred because the feature does not involve alluvial fans, although the term "panfan" was proposed by Lawson (1915, p.33) for a vast alluvial fan representing the end stage in the process of geomorphic development in a desert region.
Pangaea (Pan-gae'-a) *Pangea.*
Pangea (Pan-ge'-a) A supercontinent that existed from about 300 to about 200 Ma ago and included most of the continental crust of the Earth, from which the present continents were derived by fragmentation and *continental drift.* During an intermediate stage of the fragmentation, between the existence of Pangea and that of the present continents, Pangea is believed to have split into two large fragments, *Laurasia* on the north and *Gondwana* on the south. The proto-ocean around Pangea has been termed *Panthalassa.* Also spelled: *Pangaea.*
Pang Kiang Hollows Very large depressions in Pliocene continental sediments of the Mongolian Desert; formed by a combination of Pleistocene deflation and arid scarp retreat. Described by Berkey and Morris (1927, p.336-37, 347).
Pang-Yang depression A large erosional basin with a flat bottom and steep sides, developed on a rocky plain or plateau (Stone, 1967, p.236). Type locality: Pang Yang, Burma.
panicle (pan'-i-cle) A compound *inflorescence* with several main branches, each of which bears pedicelled flowers arranged along its axis; in grass pani-

cles, the flowers are borne in spikelets on the pedicels (Fuller and Tippo, 1954, p.965).

pan lake A lake occupying a shallow natural depression, or pan. See also: *ephemeral lake.*

panmixis (pan-mix'-is) The free interchange of genes within an interbreeding population.

panning (pan'-ning) A technique of prospecting for heavy metals, e.g. gold, by washing placer or crushed vein material in a pan. The lighter fractions are washed away, leaving the heavy minerals behind in the pan.

Pannonian (Pan-no'-ni-an) Eastern European stage: Upper Miocene (above Sarmatian, below Pontian). Equivalent to lower Pontian in Ponto-Caspian basin.

panplain (pan'-plain) (a) A term introduced by Crickmay (1933, p.344-345) for a very broad plain formed by the coalescence of several adjacent floodplains, each resulting from long-continued lateral erosion by meandering streams; it represents the end stage of an erosion cycle. Cf: *peneplain; plain of lateral planation.* Syn: *panplane.* (b) A very level plain with a general seaward inclination (Engeln, 1942).

panplanation (pan-pla-na'-tion) The action or process of formation and development of a *panplain.*

panplane (pan'-plane) *panplain.*

panspermia A theory that life, or the precursor materials necessary for life, can be exchanged between planetary bodies by the impact-generated dispersal of meteorites.

pantellerite (pan-tel'-ler-ite) (a) In the *IUGS classification,* a peralkaline *rhyolite* in which Al_2O_3<1.33 (total iron as FeO)+4.4. Cf: *comendite.* (b) A peralkaline *rhyolite* or *quartz trachyte* with normative quartz exceeding 10%. It is more mafic than *comendite* (Macdonald and Bailey, 1973). Its name, given by Förstner in 1881, is derived from Pantelleria, an island in the Mediterranean Sea south of Sicily.

Panthalassa (Pan-tha-las'-sa) The ocean surrounding *Pangea.*

pantocolpate (pan-to-col'-pate) Of *pollen grains,* having more than three colpi with some, or all, not meridionally arranged. Syn: *pericolpate.*

pantocolporate (pan-to-col-por'-ate) Of *pollen grains,* having more than three colpi with none of them meridionally arranged. Syn: *pericolporate.*

pantograph (pan'-to-graph) An instrument for copying a map or drawing on any predetermined scale of reduction or enlargement. It consists of four bars hinged to form an adjustable parallelogram, so that as one tracing stylus is moved over the material to be copied the other makes the desired copy. Specif: a mechanical attachment to a stereo-plotting instrument to permit tracing map detail at model scale while reproducing on the compilation manuscript at a predetermined reduction.

pantonematic (pan'-to-ne-mat'-ic) Said of a featherlike *flagellum* (as in Euglenophyta) provided throughout its length with a single row of tiny cilia.

pantoporate (pan-to-por'-ate) Of *pollen grains,* having apertures distributed over the entire surface. Syn: *periporate; polyporate.*

panunzite (pan-unz'-ite) A colorless hexagonal mineral related to nepheline: $(K,Na)AlSiO_4$. Syn: *tetrakalsilite.*

paolovite (pao'-lo-vite) A metallic lilac-rose orthorhombic mineral: Pd_2Sn.

papagoite (pa-pa'-go-ite) A blue monoclinic mineral: $CaCuAlSi_2O_6(OH)_3$.

paper chromatography (pa'-per) A chromatographic technique for separating components of a sample by moving it in a mixture of solution by gravity or capillarity through a paper substrate in such a way that the different components have different mobilities and thus become separated. The technique usually involves partition procedures (May and Cuttitta, 1967, p.116). See also: *chromatography.*

paper clay A fine-grained, white, kaolinic clay with high retention and suspending properties and a very low content of free silica, used for coating or filling paper.

paper coal Coal that is composed dominantly of cuticles, with some intercalated, *vitrain* bands.

paper peat Thinly laminated peat. Syn: *leaf peat.*

paper shale A shale that easily separates on weathering into thin layers or laminae suggesting sheets of paper; it is commonly highly carbonaceous.

paper spar A crystallized variety of calcite occurring in thin lamellae or paperlike plates.

papery (pa'-per-y) Descriptive of a fine-grained sedimentary rock that splits into laminae less than 2 mm in thickness (McKee and Weir, 1953, p.383).

papilla [paleont] (pa-pil'-la) (a) A surficial mound associated with a pore in cystoids. (b) A minute scalelike *ossicle* or projection in ophiuroids. (c) *apical papilla.* Pl: papillae.

papilla [palyn] A small protuberance; the term is mostly used in referring to the knoblike projection carrying the germinal site in pollen of some Taxodiaceae.

papillate (pa-pil'-late) Of pollen *sculpture,* having low (<1 µm) elements with rounded tops. Syn: *scabrate.*

papillus polaris (pa-pil'-lus po-lar'-is) A swelling or protuberance in the polar region ("polpapillus") of pollen grains (Tschudy and Scott, 1969, p.27).

papula (pap'-u-la) A short protuberance of integument between ossicles of the aboral or oral surface of an asteroid and functioning as an external gill. Pl: papulae.

papule (pap'-ule) A prolate to equant, somewhat rounded *glaebule* composed dominantly of clay minerals with a continuous lamellar fabric and having sharp external boundaries (Brewer, 1964, p.274-275); e.g. a clay gall in soil material.

para- (par'-a-) (a) A prefix that, when used with a metamorphic rock name, indicates that the rock was derived from a sediment; e.g., *paragneiss.* (b) When used with the name of a clastic sedimentary rock, a prefix signifying "matrix-rich" (Pettijohn et al., 1972, p.168), as, e.g., in *pararenite* ("para-arenite").

para-alumohydrocalcite (par'-a-a-lu'-mo-hy'-dro-cal'-cite) A white mineral: $CaAl_2(OH)_4(CO_3)_2 \cdot 6H_2O$. It is a higher hydrate of alumohydrocalcite.

para-autochthon A body of rock that is autochthonous with respect to the fault block above it, but is allochthonous relative to the fault block below it.

para-autunite (par-a-au'-tun-ite) An artificial compound: $Ca(UO_2)_2(PO_4)_2$. It represents the complete dehydration product of autunite. Cf: *meta-autunite.*

parabariomicrolite A white to pale pink rhombohedral mineral: $BaTa_4O_{10}(OH)_2 \cdot 2H_2O$.

parabiont (par-a-bi'-ont) Any one of the organisms involved in *parabiosis.*

parabiosis (par'-a-bi-o'-sis) The condition in which members of two or more species maintain colonies close to one another without conflict. See also: *parabiont.*

parabituminous (par'-a-bi-tu'-mi-nous) Said of bituminous coal containing 84-87% carbon, analyzed on a dry, ash-free basis. Not listed by ASTM as a rank classification. Cf: *perbituminous; orthobituminous.*

parabola [astron] An open mathematical curve that corresponds to a conic section with an eccentricity of one. Objects that follow a parabolic orbit will not return again past the central gravitational mass they are passing; many newly discovered comets from the Oort Cloud are in nearly parabolic orbits that will not bring them back to inner solar system again. See also: *hyperbola; ellipse; eccentricity.*

parabola [planet] Parabolic, radar dark or bright patches on the Venusian surface thought to result from impact shocks of disintegrated asteroids and/or comets without formation of an attendant impact crater (Bougher et al., 1997). See also: *halo [planet], splotch.*

parabolic dune (par-a-bol'-ic) (a) A sand dune with a long, scoop-shaped form, convex in the downwind direction so that its horns point upwind, whose ground plan, when perfectly developed, approximates the form of a parabola. It is characteristically covered with sparse vegetation, and is often found along the coast where strong onshore winds are supplied with abundant sand. (b) A term used loosely as a syn. of *upsiloidal dune.* Cf: *barchan.*

parabolic equation The partial differential equation that governs the transient diffusion-type process in groundwater.

parabrandtite (par-a-brandt'-ite) A colorless triclinic mineral: $Ca_2Mn^{2+}(AsO_4)_2 \cdot 2H_2O$.

parabutlerite (par-a-but'-ler-ite) An pale orange to orange-brown orthorhombic mineral: $Fe^{3+}(SO_4)(OH) \cdot 2H_2O$. It is a dimorph of butlerite.

paracelsian (par-a-cel'-si-an) A white to pale-yellow monoclinic mineral: $BaAl_2Si_2O_8$. It is dimorphous with celsian.

parachoma (par-a-cho'-ma) A ridge of dense calcite developed between adjacent foramina in some fusulinacean foraminiferal tests having multiple foramina (as in Verbeekinidae and Neoschwagerininae). Pl: parachomata. Cf: *choma.*

parachronology (par'-a-chro-nol'-o-gy) (a) Practical dating and correlation of stratigraphic units. (b) Geochronology based on fossils that supplement, or are used instead of, biostratigraphically significant fossils (Schindewolf, 1950, p.84-85). Cf: *orthochronology.*

parachrysotile (par-a-chrys'-o-tile) A yellow, white, gray, or green orthorhombic mineral of the *serpentine* group: $Mg_3Si_2O_5(OH)_4$. It is a polymorph of chrysotile, distinct from clinochrysotile and lizardite. It may be the same as orthochrysotile.

paraclase (par'-a-clase) An obsolete term for a *fault.*

paraclavule (par-a-clav'-ule) An apparently monaxonic sponge spicule (microsclere) consisting of a short straight shaft pointed at one end and bearing an umbel at the other end. It resembles an amphidisc with one umbel missing.

paraclinal (par-a-cli'-nal) Said of a stream or valley that is oriented in a direction parallel to the fold axes of a region. Also said of a region having paraclinal streams. Term introduced by Powell (1874, p.50). Ant: *diaclinal.*

paracme (par-ac'-me) The period in the phylogeny of a group of organisms that follows the *acme* and is marked by decadence or decline.

paraconformable (par'-a-con-form'-a-ble) Not really or not quite conformable; esp. said of strata exhibiting paraconformity.

paraconformity (par'-a-con-form'-i-ty) A term introduced by Dunbar and Rodgers (1957, p.119) for an obscure or uncertain *unconformity* in which no erosion surface is discernible or in which the contact is a simple bedding plane, and in which the beds above and below the break are parallel. This type of unconformity was formerly classed by Pirsson (1915, p.291-293) as a kind of *disconformity,* and is recognized in Great Britain as a *nonsequence* "of major time-significance" rather than as an unconformity (Challinor, 1978, p.321). Cf: *diastem.* Syn: *nondepositional unconformity.*

paraconglomerate (par'-a-con-glom'-er-ate) A term proposed by Pettijohn (1957, p.261) for a conglomerate that is not a product of normal aqueous flow but is deposited by such modes of mass transport as debris flows and glacier ice; it is characterized by a disrupted gravel framework (stones not generally in contact), is often unstratified, and is notable for containing more matrix than gravel-sized fragments (pebbles may form less than 10% of the rock). Examples include tillites and pebbly mudstones. Cf: *orthoconglomerate.* Syn: *conglomeratic mudstone.*

paraconodont Microscopic phosphatic sclerites, usually conical, consisting of lamellae that were accreted around the inner and outer surfaces, but were not continuous around the apex (tip). Many workers include paraconodonts among the conodonts, as forms immediately ancestral to euconodont grade of structure, differing by having less highly mineralized structure than euconodonts. Range: Cambrian and Ordovician. See also: *conodont.* Cf: *protoconodont; euconodont.*

paracontinuity (par'-a-con'-ti-nu'-i-ty) A small-scale *disconformity* traceable over a wide geographic area. It exhibits slight, but distinct, channeling and is associated with a slight, but significant, faunal discontinuity (Conkin and Conkin, 1973). A small-scale disconformity that is also a small-scale *faunal break.* It is "a kind of geographically widespread diastem" and "offers a reliable means by which significant time-stratigraphic boundaries may be determined precisely" (Conkin and Conkin, 1975, p.1-2).

paracoquimbite (par'-a-co'-quim-bite) A pale-violet rhombohedral mineral: $Fe^{3+}_2(SO_4)_3 \cdot 9H_2O$. It is dimorphous with coquimbite.

paracostibite (par'-a-co-stib'-ite) A metallic orthorhombic mineral: CoSbS. It is dimorphous with costibite.

paracycle In sequence stratigraphy, a small-scale relative rise and stillstand of sea level, followed by another relative rise with no intervening fall (Vail et al., 1977). Paracycle can also refer to an autocyclically controlled change in water depth such as those produced by compaction following delta-lobe shifts.

paracycle of relative change of sea level (par'-a-cy-cle) The interval of time occupied by one regional or global relative rise and stillstand of sea level, followed by another relative rise, with no intervening relative fall (Mitchum, 1977, p.209). See also: *cycle of relative change of sea level; supercycle.*

paradamite (par-ad'-a-mite) A pale yellow triclinic mineral: $Zn_2(AsO_4)(OH)$. It is the arsenate analogue of tarbuttite and dimorphous with adamite.

paradiagenetic (par'-a-di'-a-ge-net'-ic) Signifying a close relation with sedimentary diagenesis; e.g. "paradiagenetic movement", or deformation that is precrystalline in relation to spathization (Sander, 1951, p.52).

paradocrasite (par'-a-do-cra'-site) A metallic white monoclinic mineral: Sb_3As.

paradoublural line (par'-a-dou-blu'-ral) Furrow, flexure, or weak ridge developed on the dorsal exoskeleton of a trilobite directly above the inner margin of the *doublure.*

paradox of anisotropy (par'-a-dox) In a sequence of rocks that exhibits anisotropy because of layering, a measurement of apparent resistivity made with a four-terminal electrode array laid out in the direction of maximum resistivity (that is, perpendicular to the bedding) will yield the minimum value of resistivity, i.e. for current flowing along the bedding planes. This is known as the "paradox of anisotropy".

para-ecology (par'-a-e-col'-o-gy) *taphonomy.*

paraffin-base crude (par'-af-fin-base') Crude oil which will yield large quantities of paraffin in the process of distillation. Cf: *asphalt-base crude; mixed-base crude.*

paraffin coal (par'-af-fin) A type of light-colored bituminous coal from which oil and paraffin are produced.

paraffin hydrocarbon Any of the hydrocarbons of the *methane series.*

paraffinic (par-af-fin'-ic) Pertaining or relating to a *paraffin hydrocarbon* or paraffin wax.

paraffin series *methane series.*

paraffin wax A colorless, odorless, tasteless, amorphous solute of complex hydrocarbons with a high methane-series composition.

paraflagellar boss (par'-a-fla-gel'-lar) A swelling near the base of a *flagellum* in some Euglenophyta. It possibly serves as a photoreceptor.

parafoliate *foliate.*

parafransoletite A white triclinic mineral: $Ca_3Be_2(PO_4)_2(PO_3OH)_2 \cdot 4H_2O$. It is a dimorph of fransoletite.

paragaster (par'-a-gas'-ter) The *spongocoel* of a sponge.

paragenesis [eco geol] (par-a-gen'-e-sis) A characteristic association or occurrence of minerals or mineral assemblages in ore deposits, connoting contemporaneous formation. Cf: *paragenetic sequence.*

paragenesis [speleo] The upward solution of a cave ceiling due to nonsoluble fluvial fill on the floor (Ford and Williams, 1989).

paragenetic (par'-a-ge-net'-ic) (a) Pertaining to paragenesis. (b) Pertaining to the genetic relations of sediments in laterally continuous and equivalent facies.

paragenetic sequence The sequential order of mineral deposition, as individual phases or assemblages, in an ore deposit. The temporal ordering of minerals or mineral assemblages in contrast to zoning which is the spatial ordering of minerals or mineral assemblages. Cf: *paragenesis [eco geol].* Syn: *mineral sequence.*

parageosyncline (par'-a-ge'-o-syn'-cline) (a) A geosyncline within a craton or stable area; an epeirogenic basin rather than an orogenic belt (Stille, 1936). Syn: *intrageosyncline.* (b) A contemporary oceanic depression marginal to the craton (Schuchert, 1923, p.151-260). Obsolete. Cf: *idiogeosyncline.*

paraglacial (par-a-gla'-cial) *periglacial.*

paragnath (par'-a-gnath) One of a pair of leafy lobes of the *metastoma* lying behind the mandibles in most crustaceans.

paragneiss A *gneiss* formed from sedimentary rocks. Cf: *orthogneiss.*

paragonite (par-ag'-o-nite) A yellowish or greenish mineral of the *mica* group: $NaAl_2(Si_3Al)O_{10}(OH)_2$. It crystallizes in several monoclinic and trigonal polytypes, and is the Na-analogue of muscovite, with Na in place of K in the interlayer position. It usually occurs in metamorphic rocks. Syn: *soda mica.*

paraguanajuatite (par'-a-gua'-na-jua'-tite) A metallic gray rhombohedral mineral of the *tetradymite* group: $Bi_2(Se,S)_3$. It is the Se analogue of tetradymite.

parahilgardite (par-a-hil'-gard-ite) A mineral of the *hilgardite* group. Also called: hilgardite-3A.

parahopeite (par-a-hope'-ite) A colorless triclinic mineral: $Zn_3(PO_4)_2 \cdot 4H_2O$. It is dimorphous with hopeite.

parajamesonite (par-a-jame'-son-ite) A metallic gray mineral: $Pb_4FeSb_6S_{14}$. It is dimorphous with jamesonite.

parakeldyshite (par-a-kel'-dysh-ite) A white triclinic mineral that is dimorphous with keldyshite: $Na_2ZrSi_2O_7$.

parakhinite A dark green trigonal mineral: $Cu^{2+}_3PbTe^{6+}O_6(OH)_2$. It is a dimorph of khinite.

parakuzmenkoite-Fe An orange to reddish-orange monoclinic mineral of the *labuntsovite* group: $(K,Ba)_8Fe_4Ti_{16}(Si_4O_{12})_8(OH,O)_{16} \cdot 20\text{-}28H_2O$.

paralabuntsovite-Mg A monoclinic mineral of the *labuntsovite* group: $Na_8K_8Mg_4Ti_{16}(Si_4O_{12})_8(O,OH)_{16} \cdot 20\text{-}24H_2O$.

paralaurionite (par-a-lau'-ri-on-ite') A white monoclinic mineral: PbCl(OH). It is dimorphous with laurionite.

paralectotype (par-a-lec'-to-type) Any of the syntypes other than the one designated as *lectotype.* Syn: *lectoparatype.*

paraliageosyncline (pa-ral'-i-a-ge'-o-syn'-cline) A geosyncline developing along a present-day continental margin, e.g. the Gulf Coast geosyncline (Kay, 1945). Obsolete.

paralic [coal] (pa-ral'-ic) Said of coal deposits formed along the margin of the sea, as opposed to limnic coal deposits.

paralic [sed] By the sea, but nonmarine; esp. pertaining to intertongued marine and continental deposits laid down on the landward side of a coast or in shallow water subject to marine invasion, and to the environments (such as lagoonal or littoral) of the marine borders. Also said of basins, platforms, marshes, swamps, and other features marked by thick terrigenous deposits intimately associated with estuarine and continental deposits, such as deltas formed on the heavily alluviated continental shelves. Etymol: Greek "paralia", "seacoast".

paralic swamp *marine swamp.*

paralimnion (par-a-lim'-ni-on) (a) The entire spaciotemporal collection of objects and events that constitute the environment of a lake and act on it. (b) The *littoral* part of a lake (Hutchinson, 1967, p.239). Because of the ambiguity of the term, littoral is the preferred usage.

paralithic contact Similar to a lithic contact except that it is softer, can be dug with difficulty with a spade, if a single mineral has a hardness <3 (Mohs scale), and gravel size chunks will partially disperse within 15 hours shaking in water or sodium hexametaphosphate solution.

parallax [surv] (par'-al-lax) (a) The apparent displacement of the position of an object, with respect to a reference point or system, caused by an actual shift in the point of observation; e.g. "instrument parallax" in which an imperfect adjustment of a surveying instrument or a change in the position of the observer causes a change in the apparent position of an object with respect to the reference mark(s) of the instrument. (b) The difference in the apparent direction of an object as seen from two different points not on a straight line with the object (such as the apparent difference in position of a point on two consecutive photographs, or the apparent difference in direction between objects on the Earth's surface due to their difference in elevation); the angular distance between two straight lines drawn to an object from two different points of view.

parallax bar *stereometer.*

parallel (par'-al-lel) (a) One of the imaginary circles on the surface of the Earth, parallel to the equator and to one another and connecting all points of equal *latitude;* a circle parallel to the primary great circle of a sphere or spheroid, or a closed curve approximating such a circle; an east-west line of constant latitude. Each parallel is a small circle except for the equator. (b) A line, corresponding to a parallel, drawn on a globe, map, or chart. Cf: *meridian.* Syn: *parallel of latitude.*

parallel bedding *concordant bedding.*

parallel cleavage An obsolete syn. of *bedding-plane cleavage.*

parallel-displacement fault (par'-al-lel-dis-place'-ment) An obsolete term for a fault on which the linear features that were parallel before displacement are still parallel afterwards.

parallel drainage pattern A drainage pattern in which the streams and their tributaries are regularly spaced and flow parallel or subparallel to one another over a considerable area. It is indicative of a region having a pronounced, uniform slope and a homogeneous lithology and rock structure.

parallelepiped (par'-al-lel'-e-pi'-ped) A closed crystal form bounded by three pairs of parallelograms.

parallel evolution The development of similar forms by related but distinct phylogenetic lineages. See also: *parallelism.* Cf: *convergent evolution.*

parallel extinction A type of extinction in anisotropic crystals that occurs when crystal outlines or traces of cleavage planes are parallel to the privileged directions of the nicols. Cf: *inclined extinction; undulatory extinction.*

parallel fold A fold in which the orthogonal thickness of the layers is constant. Syn: *concentric fold.*

parallel growth *parallel intergrowth.*

parallel intergrowth Intergrowth of two or more crystals in which one or more crystallographic axes in each crystal are almost parallel. Syn: *parallel growth.*

parallelism (par'-al-lel-ism') (a) The development or possession of similar characteristics by two or more related organisms in separate lineages, often as a result of similar environmental conditions acting upon similar heredities derived from a long-distant common ancestor. See also: *parallel evolution.* Cf: *convergence [evol].* (b) In *cladism,* a character shared by two taxa but mistakenly considered a *derived character.* Syn: *false synapomorphy.* Cf: *convergence [phylogen]; homoplasy.*

parallelkanter (par'-al-lel-kan'-ter) An elongated *ventifact* having parallel faces or edges. Etymol: German *Parallelkanter,* "one having parallel edges".

parallel of latitude A *parallel* or line of latitude. Cf: *circle of latitude.*

parallel retreat of slope (a) The recession of a scarp or of the side of a hill or mountain (once the angle of slope is established) without change in declivity, the slope at any given time retreating parallel to its former positions. (b) The concept or principle of backwearing of a slope as proposed by W. Penck (1924).

parallel ripple A *ripple* with a relatively straight crest and an asymmetric profile; specif. a *current ripple.*

parallel roads A series of horizontal beaches or wave-cut terraces occurring parallel to each other at different levels on each side of a glacial valley, as those at Glen Roy in Scotland. Each beach, or *road [glac geol],* represents a former shoreline that corresponds with a temporary level of overflow from a proglacial lake formed by ice-damming.

parallel section A slice through a foraminiferal test in a plane normal to the axis of coiling but not through the *proloculus.*

parallel shot In seismic prospecting, a test made with all the amplifiers connected in parallel and activated by a single geophone, in order to check for lead, lag, polarity, and phasing in the geophone-to-recording circuits. Syn: *bridle.*

parallel twin A twinned crystal, the twin axis of which is parallel to the composition surface and to some rational direction [uvw] within that surface. Cf: *normal twin; complex twin.*

parallel unconformity *disconformity.*

parallel venation In a leaf, a type of *venation* in which the main veins are parallel with each other and with the longitudinal axis of the leaf. Cf: *net venation.*

parallochthon (par-al-loch'-thon) Rocks that were brought from intermediate distances and deposited on or near an allochthonous mass during transit.

paralstonite (par-al'-ston-ite) A colorless or grayish-white trigonal mineral: $BaCa(CO_3)_2$. It is a dimorph of alstonite.

paramagnetic (par'-a-mag-net'-ic) Having a positive magnetic susceptibility. A paramagnetic mineral such as olivine, pyroxene, or biotite contains magnetic cations that will partially align along an applied magnetic field but do not have a spontaneous magnetic order. Cf: *diamagnetic.* See also: *superparamagnetism.*

paramagnetic resonance *electron spin resonance.*

paramarginal resources (par-a-mar'-gi-nal) Low-grade *resources* that are recoverable at prices as much as 1.5 times those prevailing now (Brobst and Pratt, 1973, p.3).

paramelaconite (par'-a-mel'-a-con-ite) A lustrous black tetragonal mineral: $Cu^{1+}{}_2Cu^{2+}{}_2O_3$.

paramendozavilite A pale yellow mineral: $NaAl_4Fe_7(PO_4)_5(PMo_{12}O_{40})(OH)_{16}{\cdot}56H_2O$. It is a higher hydrate of mendozavilite.

parameter [cryst] (pa-ram'-e-ter) (a) Any of the axial lengths or interaxial angles that define a unit cell. Syn: *lattice constant.* (b) On a crystal face, the rational multiple of the axial length intercepted by a plane, which determines the position of the plane relative to the crystal lattice. (c) The proportions (x, y, z) of the unit-cell axial lengths that define the position of an atom relative to any lattice point. Syn: *fractional coordinates; positional parameters; atomic coordinates.*

parameter [stat] A numerical quantity that characterizes a *population.* Parameters are estimated by the corresponding *statistics* calculated from samples taken from the population.

parametric hydrology (par-a-met'-ric) That branch of hydrology dealing with "the development and analysis of relationships among the physical parameters involved in hydrologic events and the use of these relationships to generate, or synthesize, hydrologic events" (Hofmann, 1965, p.120). Cf: *stochastic hydrology; synthetic hydrology.*

parametric latitude *reduced latitude.*

parametric sounding That form of electromagnetic sounding in which frequency is the variable. Cf: *geometric sounding.*

parametric statistics Statistical tests which assume that the sampled population follows a specified form such as a normal distribution. Cf: *nonparametric statistics.*

paramontroseite (par-a-mont'-rose-ite) A submetallic grayish-black orthorhombic mineral: VO_2. It is isomorphous with montroseite.

paramorph (par'-a-morph) A *pseudomorph* with the same composition as the original crystal, as calcite after aragonite.

paramorphism (par-a-mor'-phism) The property of a mineral to change its internal structure without changing its external form or chemical composition. Cf: *paramorph.* Syn: *allomorphism [cryst].*

paramoudra (par-a-mou'-dra) A flint nodule of exceptionally large size (up to a meter in length and one-third meter in diameter), shaped like a barrel, pear, or cylinder, standing erect in the chalk beds of NE Ireland and the eastern coast of England. It appears to be a gigantic fossil zoophyte allied to the sponges. Term introduced by Buckland (1817). Etymol: vernacular Irish. Pl: paramoudras. Syn: *potstone [mineral].*

paranatisite A yellow, orange to brown orthorhombic mineral: $Na_2TiO(SiO_4)$. It is a dimorph of natisite.

paranatrolite (par-a-nat'-ro-lite) A white, colorless, grayish, pale pink or yellow orthorhombic zeolite mineral: $Na_2Al_2Si_3O_{10}{\cdot}3H_2O$. It is a higher hydrate of natrolite.

paraniite-(Y) A creamy-yellow tetragonal mineral: $(Ca,Y,Dy)_2Y(WO_4)_2AsO_4$.

paraotwayite A silky emerald-green monoclinic mineral: $Ni(OH)_{2-x}(SO_4,CO_3)_{0.5x}$. It is similar in appearance to otwayite.

parapet Wall-like longitudinal structure on platform of pectiniform conodont element, or on flange of ramiform element (TIP, 1981, pt. W, supp.2).

paraphreatic (par-a-phre-at'-ic) In speleology, said of a *passage* that has an air surface under relatively low flow conditions, when drainage is within the capacity of its downstream continuation, but reverts to being water filled (phreatic) under conditions of high flow or when the downstream drainage is temporarily impeded.

paraphyletic group (par'-a-phy-let'-ic) In *cladism,* a taxon that contains some but not all of the species descended from an ancestor.

paraphyly (pa-raph'-y-ly) Of or relating to a paraphyletic group. Adj: paraphyletic.

parapierrotite (par'-a-pier-rot'-ite) A semi-metallic black monoclinic mineral: $Tl(Sb,As)_5S_8$. It is a dimorph of pierrotite.

paraplate (par'-a-plate) The area on a dinoflagellate *cyst* that corresponds to the area of a *plate* on the dinoflagellate *theca.* The prefix "para-" may be used to indicate cyst features as opposed to the presumed thecal counterparts, e.g., *cingulum* on a theca-"paracingulum" on a cyst (Williams et al., 2000, p.151-152).

parapyla (par-a-py'-la) An accessory tubular aperture of the central capsule (in addition to the astropyle) of a phaeodarian radiolarian. Pl: parapylae.

pararammelsbergite (par-a-ram'-mels-berg-ite') A metallic white orthorhombic mineral: $NiAs_2$. It is dimorphous with rammelsbergite.

pararealgar (par'-a-re-al'-gar) An orange-yellow monoclinic mineral: AsS. It is a dimorph of realgar.

pararenite (par-ar'-e-nite) An arenite with detrital and/or diagenetic matrix between 15% and 70% (Pettijohn et al., 1987, p.202).

para-ripple (par-a-rip'-ple) A term introduced by Bucher (1919, p.262-263) for a large symmetric or nearly symmetric ripple having gentle surface slopes and "showing no assortment of grains".

pararobertsite (par-a-rob'-erts-ite) A vitreous red platy monoclinic mineral: $Ca_2Mn^{3+}{}_3O_2(PO_4)_3{\cdot}3H_2O$. It is a dimorph of robertsite.

pararsenolamprite One of the two natural occurrences of elemental arsenic. It is a dimorph of arsenolamprite.

paraschachnerite (par-a-schach'-ner-ite) A metallic gray orthorhombic mineral: Ag_3Hg_2. It is a dimorph of schachnerite.

paraschoepite (par-a-schoep'-ite) An ill-defined yellow orthorhombic mineral: $UO_3{\cdot}1\text{-}2H_2O$. It is closely related to *schoepite* (schoepiite III).

parascholzite (par-a-scholz'-ite) A colorless monoclinic mineral: $CaZn_2(PO_4)_2{\cdot}2H_2O$. It is a dimorph of scholzite.

parascorodite A yellowish-white hexagonal mineral: $Fe^{3+}AsO_4{\cdot}2H_2O$. It is a dimorph of scorodite.

parasequence In sequence stratigraphy, a relatively conformable succession of genetically related beds or bedsets bounded by flooding surfaces or their correlative surfaces. In special positions within a sequence, parasequences may be bounded either above or below by sequence boundaries (Van Wagoner, 1985; Van Wagoner et al., 1987, 1988, 1990).

parasequence boundary In sequence stratigraphy, a flooding surface and its correlative surfaces. It is a planar surface of local to basinal extent and exhibits only minor topographic relief over large areas (Van Wagoner et al., 1987, 1988, 1990).

parasequence set In sequence stratigraphy, a succession of genetically related parasequences forming a distinctive stacking pattern bounded by major flooding surfaces and their correlative surfaces. In special positions within the sequence, parasequence sets may be bounded either above or below by sequence boundaries (Van Wagoner 1985, Van Wagner et al., 1987, 1988, 1990).

parasequence set boundary In sequence stratigraphy, a type of flooding surface with all of the attributes of a parasequence boundary, bounding a parasequence set (Van Wagoner et al., 1987, 1988, 1990).

parasibirskite A pearly white monoclinic mineral: $Ca_2B_2O_5 \cdot H_2O$. It is a dimorph of sibirskite.

parasitic [ecol] (par-a-sit′-ic) Said of an organism that lives by *parasitism.*

parasitic [volc] Said of a volcanic cone, crater, or lava flow that occurs on the side of a larger cone; it is a subsidiary form. Syn: *lateral [volc]; adventive; satellite.*

parasitic ferromagnetism *weak ferromagnetism.*

parasitic fold A fold on the limb or hinge of a larger fold with which it is *congruent.* The small fold is said to be parasitic on the larger. Syn: *subsidiary fold.*

parasitism (par′-a-sit-ism′) The relationship that exists when one organism derives its food, and usually other benefits, from another living organism without killing it, but usually causing it some harm. Cf: *commensalism; inquilinism; mutualism; symbiosis.* Adj: *parasitic [ecol].*

paraspurrite (par-a-spurr′-ite) A colorless monoclinic mineral: $Ca_5(SiO_4)_2(CO_3)$. It is a dimorph of spurrite.

parastratigraphic rock unit A grouping of strata identified by objective lithologic criteria but lacking in either mappability or, most commonly, in lithologic homogeneity and constancy (Krumbein and Sloss, 1963, p.333).

parastratigraphic unit (par′-a-strat′-i-graph′-ic) *operational unit.*

parastratigraphy (par′-a-stra-tig′-ra-phy) (a) Supplemental stratigraphy based on fossils other than those governing the prevalent *orthostratigraphy* (Schindewolf, 1957, p.397). (b) Stratigraphy based on operational units.

parastratotype (par-a-strat′-o-type) A supplementary stratotype used in the original definition by the original author to illustrate the diversity or heterogeneity of the defined stratigraphic unit or some critical feature not evident or exposed in the holostratotype (ISSC, 1994, p.28).

parasutural (par-a-su′-tur-al) Of the surface features of a dinoflagellate cyst, lying at the plate margins. Cf: *sutural.*

parasymplesite (par-a-sym′-ple-site) A green to greenish-blue monoclinic mineral: $Fe^{2+}_3(AsO_4)_2 \cdot 8H_2O$. It is a dimorph of symplesite.

parasyncolpate (pa-ra-syn-col′-pate) A variation of *syncolpate* pollen grains in which the apices of the colpi divide into two branches that anastomose with the neighboring branches, delimiting small areas referred to as *apocolpial fields.*

paratabulation (pa-ra-tab-u-la′-tion) The pattern or arrangement of the constituent *paraplates* of a dinoflagellate cyst; the cyst's equivalent of the *theca tabulation* (Williams et al., 2000).

paratacamite (par-at-a-cam′-ite) A vitreous green rhombohedral mineral: $Cu_2Cl(OH)_3$. It is trimorphous with atacamite and clinoatacamite.

parataxitic (par′-a-tax-it′-ic) Said of an extremely well layered *eutaxitic* texture (Whitten, 1972). Obsolete.

parataxon (par-a-tax′-on) An informal designation for those fossilized remains, esp. of animals, that are only part of the whole individual and that usually occur widely separated from other parts; e.g. a particular conodont element. Cf: *form genus.* See also: *taxon.*

paratectonic (par′-a-tec-ton′-ic) A rarely used term pertaining to orogenic belts characterized by steep cleavages in low-grade metamorphic rocks (Dewey, 1969).

paratectonics (par′-a-tec-ton′-ics) *germanotype tectonics.*

paratellurite (par-a-tel′-lu-rite) A grayish-white tetragonal mineral: TeO_2. It is dimorphous with tellurite.

paratheca (par-a-the′-ca) A wall of a scleractinian *corallite,* formed by closely spaced rows of dissepiments. Cf: *septotheca; synapticulotheca.*

paratill (par′-a-till) A till formed by ice-rafting in a marine or lacustrine environment; it includes deposits from ice floes and icebergs (Harland et al., 1966, p.232). Ant: *orthotill.* Rarely used.

para-time-rock unit A term introduced by Wheeler et al. (1950, p.2364) for a working *time-stratigraphic unit* that is biostratigraphic and lithostratigraphic in character and that therefore is intrinsically transgressive with respect to time; e.g. zone (lithizone, radiozone), stage, isobiolith, and isogeolith. It "approaches synchrony", whereas a true chronostratigraphic unit (such as system and series) expresses "absolute synchrony".

paratsepinite-Ba A vitreous brown monoclinic mineral of the *labuntsovite* group: $(Ba,Na,K)_{2-x}(Ti,Nb)_2Si_4O_{12}(OH,O)_2 \cdot 4H_2O$.

paratype (par′-a-type) Any of the specimens, other than the *holotype,* on which the original description of a species or subspecies is based (Frizzell, 1933, p.660).

paraumbite (par-a-um′-bite) A colorless orthorhombic mineral: $K_3Zr_2HSi_6O_{18} \cdot 3H_2O$. It is crystallographically similar to umbite.

parautochthon (par-au-toch′-thon) A tectonic unit only slightly displaced with respect to the autochthon and usually lying between autochthon and allochthon.

parautochthonous (par-au-toch′-tho-nous) Said of a rock unit that is intermediate in tectonic character between *autochthonous* and *allochthonous.*

parautochthonous assemblage A fossil assemblage composed of *autochthonous* specimens that have been reworked to some degree but not transported out of the original life habitat (Kidwell et al., 1986).

paravane (par′-a-vane) A device with vanes, used in marine seismic surveying, that is towed through the water. The force of the flowing water on the vanes causes the device to dive, maintain a particular orientation, or move to one side.

paravauxite (par-a-vaux′-ite) A colorless or pale greenish-white triclinic mineral of the *laueite* group: $Fe^{2+}Al_2(PO_4)_2(OH)_2 \cdot 8H_2O$. It has more water than *vauxite* and *metavauxite,* and is isostructural with *laueite* and *gordonite.*

paraveatchite (par-a-veatch′-ite) *p-veatchite.*

paravinogradovite A white or colorless triclinic mineral: $(Na,[])_2(Ti^{4+},Fe^{3+})_4(Si_2O_6)_2(Si_3Al)O_{10}(OH)_4DH_2O$. The composition is similar to that of vinogradovite.

parawollastonite A dimorphous variant of *wollastonite:* wollastonite-2M.

parenchyma [bot] (pa-ren′-chy-ma) A plant tissue composed of thin-walled cells that are relatively undifferentiated or unspecialized, vacuolate, and isodiametric or polyhedral. Adj: parenchymatous.

parenchyma [paleont] (a) The *mesohyle* of a sponge. (b) The endoplasm of a protozoan.

parenchymella (pa-ren′-chy-mel′-la) Sponge larva composed of an envelope of uniflagellate cells surrounding more or less completely an internal mass of cells. See also: *parenchymula.*

parenchymula (pa-ren-chy′-mu-la) (a) The *planula* of a coelenterate. (b) An alternate, but not recommended, spelling of *parenchymella.*

parent (par′-ent) A *radionuclide* regarded in relation to the nuclide or nuclides into which it is transformed by decay. Cf: *daughter; end product.*

parental magma (pa-ren′-tal) The magma from which a particular igneous rock solidified or from which another magma was derived. It is sometimes used incorrectly as a syn. of *primary magma.*

parent bodies The planetary objects (asteroids, Mars and the Moon) from which meteorites were originally derived. Studies suggest that at least 70-80 parent bodies are required to account for the chemical and petrologic diversity seen among meteorites.

parent material The unconsolidated material, mineral or organic, from which the *solum* develops. See also: *parent rock [soil]; residual material; transported soil material.*

parent rock [sed] A rock from which other sediments or rocks are derived. Syn: *mother rock [sed]; source rock [sed].*

parent rock [soil] The rock mass from which a soil's *parent material* is derived.

parfacies (par-fa′-cies) A subfacies of a *diagenetic facies,* based on pH-Eh limits (Packham and Crook, 1960, p.400).

pargasite (par′-gas-ite) (a) A monoclinic mineral of the *amphibole* group: $NaCa_2(Mg_4Al)(Si_6Al_2)O_{22}(OH)_2$. Cf: *edenite.* (b) A green or blue-green variety of hornblende containing sodium and found in contact-metamorphosed rocks.

parichno (pa-rich′-no) One of a pair of scars located adjacent to the leaf-trace scar of a lepidophytic leaf cushion that reflects the presence of aerenchymatous tissue extending into the cortex.

paries (pa-ri′-es) The triangular middle part of a compartmental plate of a cirripede crustacean. Pl: *parietes.*

parietal (pa-ri′-e-tal) (a) Pertaining to the walls of a part or cavity of an organism. (b) Said of a plant part that is peripheral in position or orientation.

parietal fold A fold or spirally wound ridge on the parietal region of a gastropod, projecting into the shell interior.

parietal gap An opening from the *spongocoel* to the exterior of a lyssacine hexactinellid sponge, extending completely through the body wall and interrupting the regular skeletal framework.

parietal lip The part of the *inner lip* of a gastropod shell situated on the parietal region.

parietal muscle One of a pair or of multiple pairs of oppositely placed muscles that originate on the lateral or basal walls of a cheilostome bryozoan zooid and insert on the *frontal membrane* or the floor of the *ascus* (Ryland, 1970, p.34).

parietal pore (a) A *parietal tube* in a cirripede crustacean. (b) A *gonopore* in a cystoid.

parietal region (a) The basal surface of a gastropod shell just within and immediately outside the aperture. (b) In gnathostomes, the roof of the skull behind the orbits.

parietal septum (a) A *longitudinal septum* in a cirripede crustacean. (b) A seldom preserved longitudinal wall extending inward in the posterior region of certain echinoderms.

parietal tube One of the longitudinal tubes in certain cirripede crustaceans situated between the inner and outer laminae of a compartmental plate and separated by longitudinal septa. Syn: *parietal pore.*

parietes (pa-ri′-e-tes) Plural of *paries.*

pariety (pa-ri′-e-ty) A name formerly applied to the *septum* of an archaeocyathid.

paripinnate Said of a pinnately compound leaf having leaflets. Cf: *imparipinnate.*

parisite-(Ce) A brownish-yellow rhombohedral secondary mineral: $(Ce,La)_2Ca(CO_3)_3F_2$. It crystallizes in numerous rhombohedral, hexagonal, and monoclinic polytypes. It is related to synchysite.

parisite-(Nd) A pale yellowish-brown Nd analogue of parasite-(Ce). It has not been fully characterized, and has not been approved by the International Mineralogical Association.

parivincular (pa-ri-vin'-cu-lar) Said of a longitudinally elongated type of *ligament* of a bivalve mollusk, located posterior to the beaks and comparable to a cylinder split on one side with severed edges attached respectively along the dorsal margin of two valves.

park (a) A term used in the Rocky Mountain region of Colorado and Wyoming for a wide, grassy open valley lying at a high altitude and walled in by wooded mountains; e.g. South Park in central Colorado. It is more extensive than a *hole.* Also, a level valley between mountain ranges. (b) A relatively large, open, grassy area surrounded by woodland, or interrupted by scattered clumps of trees and shrubby vegetation; e.g. a tropical grassland in Africa.

parkerite (park'-er-ite) A metallic creamy-white orthorhombic or monoclinic mineral: $Ni_3(Bi,Pb)_2S_2$.

parkinsonite A red to purplish-red tetragonal mineral: $(Pb,Mo,[])_8O_8Cl_2$.

parmal pore (par'-mal) One of the pores piercing the shield of an acantharian radiolarian and bordered only by united branches of apophyses. Cf: *sutural pore.*

parna (par'-na) A term used in SE Australia for an eolian clay occurring in sheets. Rarely used. Etymol: an aboriginal word for "sandy and dusty ground".

parnauite (par-nau'-ite) A blue-green orthorhombic mineral: $Cu_9(AsO_4)_2(SO_4)(OH)_{10}{\cdot}7H_2O$.

parogenetic (par'-o-ge-net'-ic) Formed before the enclosing rock; esp. said of a concretion formed in a different (older) rock from its present (younger) host. Term introduced by Bates (1938, p.91). Etymol: Greek "paros", "before, formerly".

paroxysm (par'-ox-ysm) Any sudden and violent action of physical forces occurring in nature, such as the explosive eruption of a volcano or the convulsive "throes" of an earthquake; specif. the most violent and explosive action during a volcanic eruption, sometimes leading to the destruction of the volcano and generally preceded and followed by smaller explosions. Cf: *catastrophe.*

paroxysmal eruption (par-ox-ys'-mal) An eruption of the Katmaian, Peléean, Plinian, or Vulcanian type.

parricidal budding (par'-ri-cid-al) Formation of a new scleractinian coral *polyp* from the inner surface of a wedge-shaped fragment split off lengthwise from the parent.

parrot coal (par'-rot) A syn. of *cannel coal,* so named because of the crackling noises it makes while burning.

Parscovian (Pars-co'-vi-an) Eastern European brackish-marine substage: lower Pliocene, upper Dacian (above Bosphorian, below Romanian).

parsettensite (par-set'-tens-ite) A copper-red or tan monoclinic mineral: $(K,Na,Ca)_{7.5}(Mn,Mg)_{49}Si_{72}O_{168}(OH)_{50}{\cdot}nH_2O$. It often contains appreciable aluminum.

parsimony [philos] (par'-si-mo-ny) The principle of parsimony, which holds that the most economical or parsimonious explanation of natural phenomena is to be preferred at least until new evidence requires otherwise; also use of the minimum number of assumptions. It was set forth by William of Occam (or Ockham), English philosopher (~1300-1349), thus the nickname "Occam's razor". Also called the principle of *simplicity.*

parsimony [phylogen] Economy of assumption in reasoning. In a cladistic analysis, it requires choosing the cladogram postulating the least number of character state changes (Brooks et al., 1984, p.80).

parsonsite A sub-adamantine pale yellow to pale brown triclinic mineral: $Pb_2(UO_2)(PO_4)_2{\cdot}2H_2O$. It is the phosphate analogue of hallimondite.

parthéite (par'-thé-ite) A white monoclinic zeolite mineral: $Ca_2(Si_4Al_4)O_{15}(OH)_2{\cdot}4H_2O$.

partial dislocation A *dislocation [struc petrol]* in which the glide, or slip, distance is not equal to the length of one unit lattice distance (the length of one atomic bond) or multiples thereof. The movement of partial dislocations is responsible for *mechanical twinning.*

partial-duration flood (par'-tial-du-ra'-tion) A flood peak that exceeds a given base stage or discharge. Syn: basic-stage flood; flood above a base.

partial pediment (a) A term proposed by Mackin (1937, p.877) for a broadly planate gravel-capped interstream bench or terrace. (b) A broad, planate erosion surface formed by the coalescence of contemporaneous, valley-restricted benches developed at the same elevation in proximate valleys, which would produce a pediment if uninterrupted planation were to continue at this level (Tator, 1953, p.52-53).

partial peneplain (a) *incipient peneplain.* (b) A planation surface intermediate in development between a berm (or a strath terrace) and a peneplain; a base-leveled area that need not be limited to the confines of a valley. It can cross divides on rocks of medium resistance or on decayed resistant rocks (Bascom, 1931, p.173). Syn: *local peneplain.*

partial pressure The pressure of a gas in the atmosphere which would be in solubility equilibrium with the dissolved form of the same gas at the temperature of the seawater. Used to evaluate the directions and magnitudes of the net fluxes of gases such as carbon dioxide, oxygen, or fluorocarbons across the air-sea interface.

partial range zone (a) The *interval zone* between the documented lowest occurrence of one taxon and the documented highest occurrence of another taxon, that does not result in stratigraphic overlap but is used to partition the range of a third taxon (NACSN, 1983, Art. 50). Cf: *taxon-range zone; concurrent-range zone; lineage zone.* (b) A body of strata within the range of a taxon above the uppermost documented occurrence of one taxon and below the lowermost documented occurrence of another taxon (Geological Society of London, 1967, p.85).

partial thermoremanent magnetization The thermoremanent magnetization acquired or blocked in cooling in an ambient field over a distinct temperature interval, as opposed to the entire temperature range from *Curie point* to room temperature. Cf: *inverse thermoremanent magnetization.* Abbrev: PTRM.

partial tide *tidal constituent.*

particle (par'-ti-cle) A general term, used without restriction as to shape, composition, or internal structure, for a separable or distinct unit in a rock; e.g. a "sediment particle", such as a *fragment* or a grain, usually consisting of a mineral.

particle diameter The length of a straight line through the center of a sedimentary particle considered as a sphere; a common expression of *particle size.*

particle-induced gamma-ray emission An analytical technique based on the use of an incident proton beam to cause changes in the nuclear structure of sample atoms, generating emission of gamma-rays. Most commonly used to analyze elements with weights from H up to Si. Abbrev: PIGE.

particle-induced X-ray emission A type of scanning proton microprobe using an incident proton beam to generate characteristic secondary X-rays, generally used for analysis of elements heavier than Si. Analogous to electron excitation from an *electron microprobe,* but with higher spacial resolution, lower X-ray background, and far lower detection limits. Abbrev: PIXE.

particle shape The spatial or geometric form of the particles in a sediment or rock; a fundamental property of a particle that determines the relation between its mass and surface area. It is expressed as the *sphericity* and *roundness* of the particle, although the term is frequently applied to sphericity as distinguished from roundness. Syn: *grain shape.*

particle size The general dimensions (such as average diameter or volume) of the particles in a sediment or rock, or of the grains of a particular mineral that make up a sediment or rock, based on the premise that the particles are spheres or that the measurements made can be expressed as diameters of equivalent spheres. It is commonly measured by sieving, by calculating settling velocities, or by determining areas of microscopic images. See also: *particle diameter.* Syn: *grain size.*

particle-size analysis Determination of the statistical proportions or distribution of particles of defined size fractions of a soil, sediment, or rock; specif. *mechanical analysis.* Syn: *size analysis; size-frequency analysis.*

particle-size distribution The percentage, usually by weight and sometimes by number or count, of particles in each size fraction into which a disaggregated sample of a soil, sediment, or rock has been classified, such as the percentage of sand retained on each sieve in a given size range. It is the result of a *particle-size analysis.* It can be expressed by a curve derived by mathematical differentiation of the cumulative size distribution. Syn: *size distribution; size-frequency distribution.* Cf: *cumulative-size distribution.*

particle velocity The velocity with which an individual particle of a medium moves under the influence of seismic wave motion. Cf: *group velocity; phase velocity.*

particulate organic carbon Carbon present in seawater as particles of organic matter. Abbrev: POC.

particulate organic matter (a) Organic matter in the water of an ocean or lake that is retained by a 0.45-µm sieve. It consists mostly of the debris of aquatic organisms and also that of land plants that is transported by rivers and winds to the ocean or lake. Abbrev: POM. (b) Microscopic fragments of organic matter, a constituent of *palynodebris,* used in analysis of *palynofacies.* It is a subset of sedimentary organic matter.

particulate organic nitrogen Nitrogen present in seawater in the form of organic compounds within particulate matter. Abbrev: PON.

particulate organic phosphorus Phosphorus present in seawater in the form of organic compounds within particulate matter. Abbrev: POP.

partiform tabulation (par'-ti-form) A dinoflagellate *tabulation* pattern in which there are six postcingular plates and two relatively large plates that make up the antapical area, e.g., in the fossil genus *Microdinium* (Williams et al., 2000, p.156).

parting [cryst] (part'-ing) The breaking of a mineral along planes of weakness caused by deformation or exsolution; e.g. in garnet. Cf: *cleavage [mineral].*

parting [metal] In metallurgy, the process of separating (purifying) different metal ores.

parting [ore dep] (a) A band or bed of waste material between veins or beds of ore. (b) *clay parting.*

parting [stratig] A lamina or very thin sedimentary layer, following a surface of separation between thicker strata of different lithology; e.g. a *shale break* in sandstone, or a thin bed of shale or slate in a coal bed.

parting [struc geol] A joint or fissure; specif. a plane or surface along which a rock is readily separated or is naturally divided into layers; e.g., a *bedding-plane parting.* See also: *splitting [sed].*

parting cast A sand-filled tension crack produced by creep along the sea floor (Birkenmajer, 1959, p.111). Syn: *pseudo mud crack.*

parting lineation A term introduced by Crowell (1955, p.1357) for "faint irregularities or streaks on lamination planes", parallel to the direction in which the current flowed. It consists of parallel ridges and grooves a few millimeters wide and many centimeters long, esp. in thin-bedded sandstone. This feature has been termed *current lineation* by Stokes (1947), *parting-plane lineation* by McBride and Yeakel (1963), and *streaming lineation* by Conybeare and Crook (1968).

parting-plane lineation *current lineation.*

parting-step lineation A *parting lineation* characterized by subparallel, step-like ridges where the parting surface cuts across several adjacent laminae (McBride and Yeakel, 1963).

partition [paleont] (par-ti'-tion) The wall separating an offset from its parent corallite during *increase* in corals. It is constructed of some combination of the thickened ends of septa, septal and/or pseudoseptal pinnacles, sclerenchyme, and horizontal skeletal elements. An epitheca is absent. It is a permanent structure in some species, and temporary in others (Fedorowski and Jull, 1976, p.41). Cf: *dividing wall.*

partition [speleo] In a cave, a solutional remnant of rock that spans a passage from floor to ceiling. See also: *bridge [speleo]; wall [speleo].*

partition coefficient The ratio of the molar concentration of a substance dissolved in two immiscible liquids, as described by the *Nernst distribution law.*

partition gas chromatograph A device for quantitative analysis of hydrocarbon constituents. A fixed quantity of sample is carried with a stream of sweep gas through a partition column packed with an inert solid coated with a nonvolatile organic liquid. The lighter fractions traverse the column faster than the heavier fractions so that the components appear separately at the column exit, where their amounts can be measured.

partiversal (par-ti-ver'-sal) Said of a series of local dips in different directions ranging through about 180° in compass direction, occurring at or near the end of a plunging anticlinal axis.

partridgeite (par'-tridge-ite) *bixbyite.*

partzite (partz'-ite) An olive-green, yellowish-green, blackish-green, or black cubic mineral of the *stibiconite* group: $Cu_2Sb_2(O,OH)_7$(?).

parvafacies (par-va-fa'-cies) A term proposed by Caster (1934, p.19) for a body of rock that comprises the part of any *magnafacies* lying between designated chronostratigraphic horizons or key beds traced across the magnafacies; a laterally limited or grading chronostratigraphic unit of different facies but formed at the same time; a *heteropic* facies of European usage. Etymol: Latin "parva", "small", + *facies.*

parvowinchite A vitreous yellow monoclinic mineral of the *amphibole* group: $Na_2Mn(Mg_4Fe^{3+})Si_8O_{22}(OH)_2$. Syn: *tirodite; manganocummingtonite.*

parwelite (par'-wel-ite) A translucent yellowish-brown, orange, or tan monoclinic mineral: $(Mn^{2+},Mg)_5Sb^{5+}As^{5+}SiO_{12}$.

Pasadenan Orogeny (Pas-a-de'-nan) The youngest of 30 or more short-lived orogenies during Phanerozoic time recognized by Stille (1936), in this case in the middle of the Pleistocene, based on relations in southern California between Pliocene and lower Pleistocene strata, and the unconformably overlying upper Pleistocene; named for Pasadena, California. Syn: *Coast Range Orogeny.*

pascoite (pas'-co-ite) An orange monoclinic mineral: $Ca_3V^{5+}_{10}O_{28}{\cdot}17H_2O$.

pass [coast] (a) A relatively permanent channel through which a distributary on a delta flows to the sea; specif. a navigable channel in the Mississippi River delta. (b) A navigable channel connecting a body of water with the sea; e.g. a narrow opening between two closely adjacent islands or through a coastal obstruction such as a barrier reef, a barrier island, a bar, or a shoal. Cf: *inlet.* Syn: *passage [coast].* (c) An expanse of open water in a marsh.

pass [geomorph] A natural passageway through high, difficult terrain; e.g. a break, depression, or other relatively low place in a mountain range, affording a passage across, or an opening in a ridge between two peaks, usually approached by a steep valley. Cf: *col; gap [geomorph]; notch [geomorph].*

pass [streams] A river crossing; a ford. Syn: *passage [streams].*

passage [coast] (pas'-sage) *pass [coast].*

passage [speleo] A cave opening, greater in length than in width or height, that is traversible by a person. Cave passages interconnect to form the overall pattern of a cave; e.g., branchwork, network, anastomotic, ramiform, or spongework caves. Cf: *conduit.*

passage [streams] *pass [streams].*

passage zone The subhorizontal zone of transition between steeply dipping hyaloclastite breccias and overlying sheet lava in intraglacial volcanoes. It is believed to represent the water level or ice level at the time of eruption (Jones, 1969).

pass band In seismic profiling, the range of frequencies transmitted without attenuation.

passing (pass'-ing) Transportation of sediment; e.g. *bypassing* and *total passing.*

passive diapirism Syndepositional diapir growth. The diapir increases relief by growing downward relative to the sedimentary surface. The diapir base subsides, together with surrounding strata, as the basin fills with sediment. The diapir crest remains at or just below a thin roof that is continually thickened by sedimentation and thinned by erosion or extensional thinning (Barton, 1933; Nelson, 1989; Vendeville and Jackson, 1991). Syn: *downbuilding; passive piercement.*

passive earth pressure (pas'-sive) The maximum resistance of a vertical earth face to deformation by a horizontal force. Also considered to be the maximum value of lateral *earth pressure* exerted by soil on a structure, which occurs when the soil is compressed laterally, thereby causing its internal shearing resistance along a potential failure surface to be completely mobilized. Cf: *active earth pressure.*

passive fold A fold model in which the folded rocks are believed to have behaved in a purely passive way during folding.

passive glacier A glacier with sluggish movement, generally occurring in a continental environment at a high latitude, where accumulation and ablation are both small. Ant: *active glacier.*

passive intrusion Emplacement of magma by a stoping mechanism, involving no force. Cf: *forceful intrusion.*

passive margin A continental boundary formed by rifting and continental rupture and without plate-boundary tectonism. Syn: *trailing margin.* Cf: *active margin.*

passive-margin basin The subsided region along a continental *passive margin.* The underlying continental crust stretched and subsided during the rifting that preceded establishment of a new mid-ocean ridge. Subsidence is due to thermal cooling.

passive-margin wedge *geosynclinal prism.*

passive method (a) A seismic method that involves monitoring naturally produced ground motions. (b) A construction method in permafrost areas, in which the frozen ground near the structure is not disturbed or altered, and the foundations are provided with additional insulation to prevent thawing of the underlying ground. Cf: *active method.*

passive permafrost *relict permafrost.*

passive piercement *passive diapirism.*

passive remote sensing Remote-sensing methods that utilize energy naturally reflected or radiated from the terrain, e.g. photography which records available light. Cf: *active remote sensing.*

passive-roof duplex A *duplex* in which the roof thrust is blind and terminates just in front of the duplex. Strata in front of the duplex are pinned to their footwall, so strata above the roof thrust cannot move forward. They must deform passively by layer-parallel shortening, or translate backwards above the roof thrust (Banks and Warburton, 1986).

passive-roof thrust The thrust at the top of a *passive-roof duplex* or above an *intercutaneous wedge.* Passive-roof thrusts usually are nearly parallel to bedding in their hanging walls, have a tip line on the foreland side of the passive-door duplex or intercutaneous wedge, and dip toward the foreland.

passive seismic methods Seismic investigations using listening capabilities only; no seismic energy is added by the investigator. Earthquake seismology and natural noise studies are examples. Such techniques are used for thermal exploration, studies of micro-earthquakes, amplitude spectra of ground noise, fracturing, etc.

pass point A point whose horizontal and/or vertical position is determined from photographs by photogrammetric methods and which is intended for use as a supplemental *control point* in the orientation of other photographs.

paste The claylike matrix of a "dirty" sandstone; e.g. the microcrystalline matrix of a graywacke, consisting of quartz, feldspar, clay minerals, chlorite, sericite, and biotite.

pastinate element A pectiniform *conodont element* with three primary processes, which are anterior, posterior, and lateral (TIP, 1981, pt. W, supp.2).

pastiniplanate element A pastinate *conodont element* with lateral platform extensions and the attachment surface characteristic of planate pectiniform elements (TIP, 1981, pt. W, supp.2). See also: *pastinate element; planate element.*

pastiniscaphate element A pastinate *conodont element* with lateral platform extensions and the attachment surface characteristic of scaphate pectiniform elements (TIP, 1981, pt. W, supp.2). See also: *pastinate element; scaphate element.*

pastplain (past'-plain) A plain that has been uplifted and dissected; thus "it is no longer a true plain" (Davis, 1890, p.88).

pat (a) A term used in Pakistan for an arid plain formed by deposits of fine light-colored muddy clay that accumulates on evaporation of shallow pools of water. (b) A term used in Chota Nagpur, India, for a small, steep-sided plateau. Etymol: Sindhi.

patagium (pa-ta'-gi-um) A spongy veil between the arms in the skeleton of a spumellarian radiolarian. Pl: patagia.

patch [geophys] (a) A jumper or a temporary connection, expecially one which can be changed easily, as a connection on a "patch panel". (b) More-or-less uniform distribution of geophones over an area (template), which gives single-fold subsurface coverage for a single source over an area of one-half the dimensions. (c) A large geophone group feeding a single channel, occasionally several hundred feet across containing several hundred geophones.

patch [ice] (a) *ice patch.* (b) An irregular small mass of floating sea-ice fragments of any concentration.

patch reef A moundlike or flat-topped *organic reef,* generally less than a kilometer across, isolated from other *bioherms,* less extensive than a *platform* *reef*, and frequently forming a part of a larger reef complex. Cf: *reef patch; table reef, chapeiro.*

patchy gas Small accumulations of gas that present potential hazards but of no commercial value.

patella (pa-tel′-la) (a) The fourth segment in the pedipalpus or in the leg of an arachnid, following upon and forming the "knee" articulation with the femur (TIP, 1955, pt. P, p.62). (b) A joint forming the "knee" in the prosomal appendage of a merostome. (c) In the tetrapoda, a sesamoid (heterotopic) bone commonly present at the anterior surface of the joint between femur and tibia. Pl: patellae or patellas.

patellate (pa-tel′-late) (a) Pertaining to the tetrapod kneecap. (b) Said of a low solitary *corallite* with sides expanding from the apex at an angle of about 120 degrees. Cf: *trochoid; turbinate.*

patera (pa-te′-ra) A term established by the International Astronomical Union for an irregular crater with scalloped edges on planetary surfaces (USGS, 2003). Most are believed to be calderas. Generally used as part of a formal name for a landform, such as Tyrrhenum Patera on Mars (Mutch et al., 1976, p.57, 66). Etymol: Latin *patera*, saucer. Pl: paterae.

paternoite (pa-ter′-no-ite) *kaliborite.*

paternoster lake (pa′-ter-nos′-ter) One of a chain or series of small circular lakes occupying rock basins, usually at different levels, in a glacial valley, separated by morainal dams or riegels, but connected by streams, rapids, or waterfalls, to resemble a rosary or a string of beads. Syn: *rock-basin lake; step lake; beaded lake.*

path [optics] *optical path.*

path [seis] *raypath.*

pathfinder (path′-find-er) In geochemical exploration, a relatively mobile element or gas that occurs in close association with an element or commodity being sought, but can be more easily found because it forms a broader halo or can be detected more readily by analytical methods. A pathfinder serves to lead investigators to a deposit of a desired substance.

pathline (path′-line) In computer modeling of hydrothermal systems, the path along which a fluid packet moves in response to physicochemical gradients near an intruded stock (Norton, 1978). See also: *sourceline.*

patina [geol] (pat′-i-na) (a) A colored film or thin outer layer produced on the surface of a rock or other material by weathering after long exposure; e.g. a desert varnish, or a case-hardened layer on a chert nodule. (b) Strictly, the greenish film formed naturally on copper and bronze after long exposure to a moist atmosphere, and consisting of a basic carbonate. Etymol: Italian.

patina [palyn] A thickening of the exine of a spore, extending over essentially the whole distal surface.

patination (pat-i-na′-tion) The quality or state of being coated with a patina, or the act or process of coating with a patina.

patrónite (pat′-ro-nite) A gray-black monoclinic mineral: VS_4. It is mined as an ore of vanadium at Minasragra, Peru.

patterned ground (pat′-terned) A group term suggested by Washburn (1950, p.7-8) for certain well-defined, more or less symmetrical forms, such as circles, polygons, nets, steps, and stripes, that are characteristic of, but not necessarily confined to, surficial material subject to intensive frost action. It is classified according to type of pattern and presence or absence of sorting. Patterned ground occurs principally in polar, subpolar, and arctic regions, but also includes features in tropical and subtropical areas. Previous terms more or less synonymous: *structure ground; soil structure [pat grd]; Strukturboden; frost-pattern soil;* patterned soil.

patterned sedimentation Sedimentation characterized by a systematic sequence of beds; e.g. recurrent sedimentation (interbedding, interdigitation, etc.), repetition of beds, or rhythmic or cyclic sedimentation.

pattern recognition (a) Identifying characteristic patterns on images. (b) A branch of computer science and allied disciplines concerned with the analysis of images, including their characterization, classification, and the extraction of features.

pattern shooting (pat′-tern) The use of a number of seismic energy sources arranged in a definite geometric pattern.

Patterson function (Pat′-ter-son) *Patterson synthesis.*

Patterson map *Patterson projection.*

Patterson projection A projection of the *Patterson synthesis* on a section through a crystal. See also: *Patterson vectors.*

Patterson synthesis A type of Fourier synthesis whose coefficients are the squares of the structure factors; it is used in direct determination of crystal structure. See also: *Patterson vectors.* Syn: *Patterson function.*

Patterson vectors In analysis of crystal structure, the vectors of peaks relative to the origin in a *Patterson synthesis* or *Patterson projection.*

paulingite-Ca A colorless or pale yellow cubic zeolite mineral with Ca as the dominant exchangeable cation: $(Ca,K,Na,Ba)_7(Si,Al)_{42}O_{84}\cdot nH_2O$.

paulingite-K (paul′-ing-ite) A colorless cubic zeolite mineral with K as the dominant exchangeable cation: $(K,Na,Ca)_7(Si,Al)_{42}O_{84}\cdot 34H_2O$.

paulkellerite (paul-kel′-ler-ite) An adamantine greenish-yellow monoclinic mineral: $Bi_2Fe^{3+}O_2(PO_4)(OH)_2$.

paulkerrite (paul-kerr′-ite) A light yellow-brown orthorhombic mineral of the mantienneite group: $K(Mg,Mn^{2+})_2(Fe^{3+},Al)_2Ti(PO_4)_4(OH)_3\cdot 15H_2O$.

paulmooreite (paul-moore′-ite) An adamantine colorless to light orange monoclinic mineral: $Pb_2As^{3+}{}_2O_5$.

paulopost (paul′-o-post) Obsolete syn. of *deuteric.*

pavement (pave′-ment) A rocky surface that suggests a paved road in smoothness, hardness, horizontality, surface extent, or close packing of its units. Examples: *boulder pavement [geomorph]; glacial pavement; desert pavement; limestone pavement; erosion pavement.*

pavement [streams] A coarse surface layer on the streambed that is rarely disrupted.

pavement prismatic structure A regular *simple prismatic structure* in which the prisms are wider (in the horizontal plane) than they are tall (in the vertical direction), thereby comprising a thin layer of tablet-shaped prisms (Carter, 1990, Fig. 7). The "pavimental prismatic structure" of Schein-Fatton (1988, Pl. 2, Fig. 6).

pavilion (pa-vil′-ion) The portion of a faceted gemstone below the girdle that extends from the girdle to the culet. Cf: *crown [gem].* Syn: *base [gem].*

pavilion facet A main facet on the pavilion of any fashioned gemstone; e.g. a large facet extending from the girdle to the culet of a brilliant-cut gem, or a facet in the center row of facets on the pavilion of a step-cut gem.

pavonite (pav′-o-nite) A metallic gray monoclinic mineral: $(Ag,Cu)(Bi,Pb)_3S_5$.

paxilla (pax-il′-la) A pillarlike *ossicle* of an asterozoan, having a flattened summit bearing a tuft of spinelets or granules. Pl: paxillae.

paxillose (pax′-il-lose) (a) Resembling a little stake. (b) Bearing paxillae.

paxite (pax′-ite) A metallic gray monoclinic mineral: $CuAs_2$.

pay adj. Said of a structure or stratum that contains a mineral deposit (pay gravel, pay dirt, pay streak) or oil and gas (pay sand); also, said of a mineral deposit or part of it that is especially profitable, e.g. pay ore. n. A reservoir rock containing oil or gas. The term is colloquial.

pay zone The vertical interval(s) of the stratigraphic section in an oil or gas field that will yield oil or gas in economic quantities.

PBM *permanent bench mark.*

PCA *principal components analysis.*

PDB standard Carbon dioxide prepared from belemnites (*Belemnitella americana*) collected from the Peedee Formation (Upper Cretaceous) of South Carolina. It is used as a standard of comparison in determining the isotopic composition of carbon and of oxygen. It was first used by H. C. Urey in 1951 to determine paleotemperatures on the basis of the fractionation of oxygen isotopes. The initials stand for "Peedee belemnite".

PDO *Pacific Decadal Oscillation.*

PDR *precision depth recorder.*

peachblossom ore (peach′-blos-som) *erythrite.*

peacock coal A thin layer of iron oxides on the cleat surfaces of coal that takes on a purple-blue color.

peacock copper (pea′-cock) *peacock ore.*

peacock ore Informal name for an iridescent copper mineral having a lustrous, tarnished surface exhibiting variegated colors, such as chalcopyrite and esp. bornite. Syn: *peacock copper.*

pea gravel Clean gravel, the particles of which are similar in size to that of peas.

pea iron ore *pea ore.*

peak [coast] A headland or promontory; a jut of land.

peak [geomorph] (a) The more or less conical or pointed top of a hill or mountain; one of the crests of a mountain; a prominent summit or the highest point. (b) An individual mountain or hill taken as a whole, esp. when isolated or having a pointed, conspicuous summit.

peak [paleont] A skeletal extension of the wall, shaped like the brim of a baseball cap, from the upper half of the rim in skeletal pores in archaeocyathids (TIP, 1972, pt. E, p.41).

peak diameter The dominant or modal particle diameter as determined on a particle-size distribution curve.

peak discharge *flood peak.*

peak plain A high-level plain formed by a series of accordant summits, often explained as an uplifted and fully dissected peneplain. See also: *gipfelflur.* Syn: *summit plain.*

peak-ring crater A *complex impact crater* with a central mountainous ring, instead of a central peak, surrounding its center. Peak-ring craters are generally larger than central-peak craters on a given planet or satellite. The radial profile of the ring is roughly symmetric, as opposed to that of a *multi-ring crater.*

peak runoff The maximum rate of runoff at a given point or from a given area, during a specified period.

peak zone *abundance zone.*

pea ore A variety of pisolitic limonite or *bean ore* occurring in small, rounded grains or masses. Syn: *pea iron ore.*

Pearce element ratio Simple or compound molar concentration ratio where the denominator is an element that is conserved (i.e. fully partitioned into the liquid) during the crystallization or melting process (Pearce, 1968). Two-variable Pearce Element Ratio Diagrams (PERDs) eliminate the *constant sum effect* and are a powerful tool for quantitatively evaluating melting, fractional crystallization, and magma mixing processes. Obsolete.

pearceite (pearce′-ite) A metallic black monoclinic mineral: $(Ag,Cu)_{16}As_2S_{11}$. Cf: *polybasite.*

pearl A dense spherical calcareous concretion, usually white or light-colored, consisting of occasional layers of *conchiolin* and predominant nacreous layers of aragonite (or rarely calcite), deposited concentrically about a foreign particle within or beneath the mantle of various marine and freshwater mollusks, either free from or attached to the shell.

pearl mica *margarite [mineral].*

pearl opal *cacholong.*

pearl spar A crystalline carbonate mineral, such as ankerite, having a pearly luster; specif. *dolomite [mineral].* Also spelled: pearlspar.

pearlstone (pearl'-stone) *perlite.*

pearly luster (pearl'-y) *nacreous luster.*

pear-shaped cut A variation of the *brilliant cut,* usually with 58 facets, having a pear-shaped girdle outline.

Pearson product-moment correlation A measure of the linear relationship between two quantitative variables, the *correlation coefficient.* Named for Karl Pearson (1857-1936), English mathematician. Also called Pearsonian correlation.

peat An unconsolidated deposit of semicarbonized plant remains in a watersaturated environment, such as a bog or fen, and of persistently high moisture content (at least 75%). It is an early stage or rank in the development of coal; carbon content is about 60% and oxygen content is about 30% (moisture-free). Structures of the vegetal matter may or may not be seen depending on the degree of organic matter degradation. When dried, peat burns freely.

peat ball A *lake ball* containing an abundance of peaty fragments.

peat bed *peat bog.*

peat bog A *bog* in which peat has developed, under conditions of acidity, from the characteristic vegetation, esp. sphagnum. Syn: *peat moor; peat bed.*

peat breccia Peat that has been broken up and then redeposited by water. Syn: *peat slime.*

peat coal (a) A coal transitional between peat and brown coal or lignite. (b) Artificially carbonized peat that is used as a fuel.

peat deposit Term usually referring to an exploitable peat resource.

peat flow A flow of peat produced in a peat bog by a *bog burst.* Cf: *bog flow.*

peat formation The partial decomposition of vegetable matter under waterlogged and oxygen-deficient conditions. Cf: *humification; putrefaction; terrestrialization; paludification.* Syn: *ulmification; peatification.*

peat hummock A hummock of peat rather than soil. Cf: *earth hummock.*

peatification (peat'-i-fi-ca'-tion) *peat formation.*

peatland Applies to areas having peat-forming vegetation on peat and includes all peat originating from that vegetation. Cf: *bog; fen; wetland.* Syn: *mire; moor.*

peat moor *peat bog.*

peat moss Peat that has formed from moss, usually *sphagnum moss.*

peat mound Any mound composed largely of peat, ranging in size from a small hummock to a *palsa.* Cf: *gull hummock.*

peat plateau A low, generally flat-topped expanse of peat, rising one or more meters above the general surface of the peatland. A layer of permafrost exists in the peat plateau, which may extend into the peat below the general peatland surface and even into the underlying mineral soil (Stanek and Worley, 1983). Cf: *bog expanse; palsa.*

peat ring A *nonsorted circle* in peat.

peat-sapropel (peat-sap'-ro-pel) Organic degradation matter that is transitional between peat and sapropel. Syn: *sapropel-peat.*

peat slime *peat breccia.*

peat soil An organic soil in which the plant residues are recognizable. The sum of the thicknesses of the organic layers are usually greater than the sum of the thicknesses of the mineral layers.

peat swamp An area of wet, spongy ground in which the dominant vegetation is trees and the surface is markedly convex and not subject to flooding. The peat that accumulates is low in plant nutrients, has a pH below 4.0, and a content of mineral matter lower than about 25 percent on a dry-weight basis (Anderson, 1964). Cf: *peat bog.*

peat-to-anthracite theory (peat'-to-an'-thra-cite) A theory of coal formation as a process in which the progressive ranks of coal are indicative of the degree of *coalification* and, by inference, of the relative geologic age of the deposit. Peat, as the initial stage of coalification, is of recent geologic age; lignite, as an intermediate stage, is usually Tertiary or Mesozoic, and bituminous coal and anthracite, as the more advanced stages of coalification, are usually Carboniferous (Nelson and Nelson, 1967, p.271).

pebble [gem] (peb'-ble) (a) A rough gem occurring in the form of a pebble, as in a stream. (b) Transparent, colorless quartz, or *quartz crystal;* e.g. "Brazilian pebble".

pebble [part size] (a) A general term for a small, roundish, esp. waterworn stone; specif. a rock fragment larger than a granule and smaller than a cobble, having a diameter in the range of 4-64 mm (1/6 to 2.5 in., or -2 to -6 phi units, or a size between that of a small pea and that of a tennis ball), being somewhat rounded or otherwise modified by abrasion in the course of transport. In Great Britain, the range of 10-50 mm has been used. The term has been used to include fragments of cobble size; it is frequently used in the plural as a syn. of *gravel.* See also: *very coarse pebble; coarse pebble; medium pebble; fine pebble.* Syn: *pebblestone.* (b) A rock or mineral fragment in the soil, having a diameter in the range of 2-20 mm (Atterberg, 1905). The U.S. Bureau of Soils has used a range of 2-64 mm. Cf: *gravel.*

pebble armor A *desert armor* consisting of rounded pebbles, as on a *serir.*

pebble bed Any *pebble conglomerate,* esp. one in which the pebbles weather conspicuously and fall loose; e.g. the Bunter pebble beds of Devon and Somerset in England. Syn: *popple rock.*

pebble coal *ball coal.*

pebble conglomerate A consolidated rock consisting mainly of pebbles.

pebble dent A depression formed by a pebble on an unconsolidated sedimentary surface, represented by a downward curvature of laminae beneath the pebble.

pebble dike (a) A *clastic dike* composed largely of pebbles. (b) A tabular body containing sedimentary fragments in an igneous matrix, as from the Tintic district in Utah (Farmin, 1934); e.g. one whose fragments were broken from underlying rocks by gaseous or aqueous fluids of magmatic origin and injected upward into country rock, becoming rounded due to the milling and/or corrosive action of the hydrothermal fluids.

pebble gravel An unconsolidated deposit consisting mainly of pebbles.

pebble mosaic A *desert mosaic* consisting of pebbles.

pebble peat Peat that is formed in a semiarid climate by the accumulation of moss and algae, no more than 1/4 inch in thickness, under the surface pebbles of well-drained soils.

pebble phosphate A secondary phosphorite of either residual or transported origin, consisting of pellets, pebbles and nodules of phosphatic material mixed with sand and clay, as in Florida; e.g. *land-pebble phosphate* and *river-pebble phosphate.*

pebble pup (a) A geologist's assistant. (b) A student of geology. (c) An inexperienced *rock hound.*

pebble size A term used in sedimentology for a particle size greater than that of a sphere with a diameter of 4 mm (1/6 in.) and less than that of a sphere with a diameter of 64 mm (2.5 in.).

pebblestone *pebble [part size].*

pebbly mudstone (peb'-bly) A *conglomeratic mudstone* in which thinly scattered pebbles are embedded among somewhat distorted bedding planes. The term is advocated by Crowell (1957, p.1003) as a descriptive name, without regard to manner of origin, for a poorly sorted, till-like rock composed of dispersed pebbles in an abundant mudstone matrix. See also: *tilloid; pseudotillite.*

pebbly sand An unconsolidated sediment containing at least 75% sand and a conspicuous number of pebbles that do not exceed 25% of the total aggregate (Willman et al., 1942, p.343-344). Cf: *sandy gravel.*

pebbly sandstone (a) A consolidated *pebbly sand.* (b) A sandstone containing 10-20% pebbles (Krynine, 1948, p.141). Cf: *conglomeratic sandstone.* (c) A term used in Scotland for a conglomerate.

Péclet number (a) Ratio of heat transferred by convection to heat transferred by conduction in the bulk flow far from any boundaries. (b) A dimensionless number that relates *diffusion coefficient,* average linear velocity, and average particle diameter. It is the ratio of advectance to conductance and is an index to the degree of spreading out of a concentration profile of a tracer or contaminant in groundwater (Freeze and Cherry, 1979, p.593). The larger it is, the stronger the role of advection in comparison to *diffusion.*

pecoraite (pe-cor'-a-ite) A green monoclinic mineral of the *serpentine* group: the Ni-analogue of clinochrysotile, with Ni in place of octahedrally coordinated Mg: $Ni_3Si_2O_5(OH)_4$.

pectinacean (pec-ti-na'-cean) Any bivalve mollusk assigned to the superfamily Pectinacea, characterized by an orbicular, monomyarian, subequilateral shell with winglike extensions (auricles) from the hinge margin; e.g. a *scallop [paleont].*

pectinate (pec'-ti-nate) Comblike, with close narrow divisions or parts; also said of spine connections in cacti when small lateral spines radiate like comb teeth from the areole.

pectinate tabula A transverse, porous skeletal element in archaeocyathids, shaped like the teeth of two combs, each projecting toward the other from adjoining septa (TIP, 1972, pt. E, p.41).

pectiniform elements A category of basically blade- or comb-shaped conodont elements, with two, three, or four processes that are commonly much higher than they are wide; one or more of the processes may develop lateral flanges or platforms; a general designation for the shape categories termed "blades" and "platforms" in much of the conodont literature (TIP, 1981, pt. W, supp.2).

pectinirhomb (pec-tin'-i-rhomb) A specialized type of *pore rhomb* in rhombiferan echinoderms, consisting of a compact rhomboidal structure of closely spaced comblike grooves. It is typically set in a distinct depressed area on thecal plates.

pectolite (pec'-to-lite) A whitish or grayish triclinic mineral: $NaCa_2Si_3O_8(OH)$. It occurs in compact masses of divergent or parallel fibers, commonly in cavities in basalts and scoriaceous lavas. Cf: *sérandite.*

pectolite-M2abc A white, pale tan, pinkish, or pale blue monoclinic dimorph of pectolite: $NaCa_2Si_3O_8(OH)$. Also called: parapectolite.

ped A naturally formed unit of soil structure, e.g. granule, block, crumb, aggregate. Cf: *clod [soil].*

pedal elevator muscle (ped'-al) A thin bundle of muscle fibers attached to the bivalve-mollusk shell in the umbonal cavity and serving to raise the foot.

pedalfer (pe-dal′-fer) A subdivision of a soil order comprising a large group of soils in which sesquioxides increased relative to silica during soil formation. Not used in current U.S. system of soil taxonomy.

pedal gape An opening between margins of the shell of a bivalve mollusk for the protrusion of the foot.

pedality (pe-dal′-i-ty) The physical nature of a soil as expressed by the features of its constituent peds.

pedal levator muscle A pedal muscle serving to retract the foot of a bivalve mollusk.

pedal muscle One of a pair or several pairs of muscles connecting the foot of a mollusk to the interior surface of the shell; e.g. "pedal protractor muscle" serving to extend the foot of a bivalve mollusk or gastropod, and "pedal retractor muscle" serving to retract it.

pedcal (ped′-cal) *pedocal.*

pedestal [geomorph] (ped′-es-tal) A relatively slender neck or column of rock capped by a wider mass of rock and produced by undercutting as a result of wind abrasion (as in the SW U.S.) or by differential weathering. See also: *pedestal rock.* Syn: *rock pedestal.*

pedestal [glaciol] *ice pedestal.*

pedestal rock (a) An isolated and residual or erosional mass of rock supported by or balanced on a *pedestal.* The term is also applied to the entire feature. Cf: *balanced rock; mushroom rock.* (b) *perched block.*

pedia (ped′-i-a) Plural of *pedion.*

pedial class (ped′-i-al) That class in the triclinic system having symmetry 1 (no symmetry).

pedicel [bot] (ped′-i-cel) In an inflorescence, the stalk of an individual flower.

pedicel [paleont] (a) The greatly modified first segment of the abdomen in arachnids of the subclass Caulogastra, reaching its extreme development in spiders. (b) A small foot or footlike organ of an invertebrate, such as a *tube foot* of an echinoderm. (c) The area of attachment of the body of a tintinnid to the *lorica.* (d) A small or short stalk or stem in an animal; esp. a narrow basal part by which a larger part or body is attached, such as the *pedicle* of a brachiopod. (e) In many tetrapods, that part of a vertebra lateral to the neural canal that connects the centrum to the bases of the zygapophyses and lamina (neural arch).

pedicellaria (ped′-i-cel-lar′-i-a) Any of various minute organs resembling forceps that are borne in large numbers on certain echinoderms, e.g. a minute stalked organ in the external integument of an echinoid, used for grasping or defending; or a minute pincerlike or valvate calcareous appendage on or in the skin, ossicles, or spines of an asteroid. Pl: pedicellariae.

pedicle (ped′-i-cle) A variably developed, cuticle-covered, fleshy or muscular appendage of a brachiopod, commonly protruding from the *pedicle valve* and serving to attach the animal to the substratum. See also: *pedicel [paleont].*

pedicle callist A localized thickening of secondary-shell layer in the apex of the *pedicle valve* of a brachiopod, representing the track of anterior migration of junction between pedicle epithelium and outer epithelium.

pedicle collar The complete or partial ringlike thickening of the inner surface of the ventral beak of a brachiopod, "continuous laterally with internal surface of deltidial plates, sessile, with septal support, or free anteriorly and secreted by anteriorly migrating outer epithelium at its junction with pedicle epithelium" (TIP, 1965, pt. H, p.149).

pedicle epithelium The ectodermal epithelium investing the *pedicle* of a brachiopod.

pedicle foramen The subcircular to circular perforation of the shell, adjacent to the beak of the pedicle valve, through which the *pedicle* of a brachiopod passes. Cf: *pedicle opening.* Syn: *foramen.*

pedicle groove A commonly subtriangular groove dividing the ventral pseudointerarea medially and affording passage for the *pedicle* in many lingulid brachiopods.

pedicle opening *pedicle foramen.*

pedicle sheath An externally directed tube projecting posteriorly and ventrally from the ventral *umbo* of a brachiopod, probably enclosing the *pedicle* in young stages of development of some shells with supra-apical pedicle opening (TIP, 1965, pt. H, p.150).

pedicle tube An internally directed tube of the secondary shell of a brachiopod, continuous with the margin of the pedicle foramen and enclosing the proximal part of the *pedicle* (TIP, 1965, pt. H, p.150).

pedicle valve The valve of a brachiopod through which the pedicle commonly emerges. It is usually larger than the *brachial valve,* and it contains the teeth by which the valves are hinged. Syn: *ventral valve.*

pediment (ped′-i-ment) A broad gently sloping rock-floored erosion surface or plain of low relief, typically developed by subaerial agents (including running water) in an arid or semiarid region at the base of an abrupt and receding mountain front or plateau escarpment, and underlain by bedrock (occasionally by older alluvial deposits) that may be bare but are more often partly mantled with a thin discontinuous veneer of alluvium derived from the upland masses and in transit across the surface. The longitudinal profile of a pediment is normally slightly concave upward, and its outward form may resemble a *bajada* (which continues the forward inclination of a pediment). The term was first applied to a landform by McGee (1897, p.92), although Gilbert (1877, p.130-131) recognized and described the feature as a *hill of planation.* Etymol: from an architectural pediment, a triangular feature crowning a portico of columns in front of a Grecian-style building; in this sense the term is not appropriate for a gently sloping surface commonly forming a broad approach to a mountain range. Cf: *rock fan; plain of lateral planation; peripediment.* Syn: *piedmont interstream flat; conoplain.*

pedimentation (ped′-i-men-ta′-tion) The action or process of formation and development of a *pediment* or pediments. The two processes recognized as being most active in pediment formation are lateral planation by steep-gradient streams, and backwearing and removal of debris by rill wash and unconcentrated flow; the latter process appears to be the more widely accepted. Cf: *pediplanation.*

pediment dome *desert dome.*

pediment gap A term applied by Sauer (1930) to a broad opening formed by the enlargement of a *pediment pass.*

pediment pass A term applied by Sauer (1930) to a narrow, flat, rock-floored tongue extending back from a pediment and penetrating sufficiently along a mountain to join another pediment extending into the mountain front from the other side; the pediments are frequently at different levels. Cf: *pediment gap.*

pediocratic (ped′-i-o-crat′-ic) Pertaining to a period of time in which there is little *diastrophism.* Cf: *orocratic.*

pedion (ped′-i-on) An open crystal form having only a single face, with no symmetrically equivalent face. Pl: *pedia.*

pedipalpus (ped-i-pal′-pus) One of the second pair of cephalothoracic appendages that lie on each side of the mouth of an arachnid and that are subject to many variations in structure, such as being the largest and most conspicuous appendages in scorpions (ending in a powerful chela), stout and conspicuous in whip scorpions and the order Phrynichida (but ending in a pointed joint), and the least conspicuous appendages in the order Architarbida. Pl: pedipalpi. See also: *palpus.* Syn: pedipalp.

pediplain (ped′-i-plain) A term proposed by Maxson and Anderson (1935, p.94) for an extensive thinly alluviated erosion surface formed in a desert region by the coalescence of two or more adjacent pediments and occasional desert domes, and representing the end result (the "peneplain") of the mature stage of the arid erosion cycle. Howard (1942) objected to the term because the surface to which it is applied is not wholly at the base of a slope and is not a "plain" in the true geomorphic sense. Cf: *pediplane; coalescing pediment.* Syn: *panfan; desert peneplain; desert plain.*

pediplanation (ped′-i-pla-na′-tion) (a) A general term for all the processes by which pediplanes are formed (Howard, 1942, p.11). (b) The action or process of formation and development of a *pediplain* or pediplains; *pedimentation* of regional magnitude, assisted by slope retreat.

pediplane (ped′-i-plane) (a) A general term proposed by Howard (1942, p.11) for any planate erosion surface, such as a pediment or a peripediment, produced in the piedmont area of an arid or semiarid region, either exposed or covered with a veneer of alluvium no greater than the depth of effective scour (the thickness that can be moved during floods). (b) A term sometimes used as a syn. of *pediplain.*

pedocal (ped′-o-cal) A subdivision of a soil order comprising a large group of soils in which calcium accumulated during soil formation. Not used in current U.S. system of soil taxonomy.

pedoderm A mappable unit mantle of soil, entirely or partially truncated, at the Earth's surface or partially or wholly buried, which has physical characteristics and stratigraphic relationships that permit its consistent recognition and mapping.

pedogenesis (ped-o-gen′-e-sis) *soil genesis.*

pedogenic (ped-o-gen′-ic) Pertaining to soil formation.

pedogeography (ped′-o-ge-og′-ra-phy) The study of the geographic distribution of soils.

pedography (pe-dog′-ra-phy) The systematic description of soils; an aspect of *soil science.*

pedologic age (ped-o-log′-ic) The relative maturity of a soil profile.

pedological features Recognizable units within a soil material which are distinguishable from the enclosing material for any reason such as origin (deposition as an entity), differences in concentration of some fraction of the plasma, or differences in arrangement of the constituents (fabric).

pedologic horizon (a) *soil horizon.* (b) The product of soil development (pedogenesis), which occurred subsequent to formation of the lithostratigraphic, allostratigraphic, or lithodemic unit or units on which the buried soil was formed (NACSN, 1983, p.864).

pedology (pe-dol′-o-gy) One of the disciplines of soil science, the study of soil morphology, genesis, and classification. It is sometimes used as a syn. of *soil science.*

pedometer (pe-dom′-e-ter) A pocket-size, watch-shaped instrument that registers the linear distance a pedestrian covers by responding to his body motion at each step. It is carried in an upright position attached to the body or to a leg, and it can be adjusted to the length of the pace of the person carrying it.

pedon (ped′-on) A three-dimensional body of soil with lateral dimensions large enough to permit the study of horizon shapes and relations. Its area ranges from 1 to 10 square meters. Where horizons are intermittent or cyclic, and recur at linear intervals of 2 to 7 m, the pedon includes one-half of the cycle. Where the cycle is < 2m, or all horizons are continuous and of uniform thickness, the pedon has an area of approximately 1 m^2. If the horizons are cyclic, but recur at intervals > 7 m, the pedon reverts to the 1 m^2, and more than one soil will usually be represented in each cycle.

pedorelic (ped-o-rel'-ic) Said of a soil feature that is derived from a pre-existing soil horizon. Cf: *lithorelic.*

pedosphere (ped'-o-sphere) That shell or layer of the Earth in which soil-forming processes occur.

pedostratigraphic unit (ped'-o-strat'-i-graph'-ic) A buried, traceable, three-dimensional body of rock that consists of one or more differentiated soil horizons. It is developed in one or more lithostratigraphic, allostratigraphic, or lithodemic units, and is overlain by one or more formally defined lithostratigraphic or allostratigraphic units (NACSN, 1983, Art. 55). The fundamental unit in pedostratigraphic classification is the *geosol.*

pedrizite A monoclinic mineral of the *amphibole* group: $Li_2(Li,Mg,Fe^{2+},Al)_5(Si,Al)_8O_{22}(OH)_2$.

pedrosite (pe-dro'-site) A peralkaline *hornblendite.* The name, given by Osann in 1923, is for Pedrosa, Portugal. Obsolete.

peduncle [bot] (pe'-dun-cle) A plant stalk that bears an inflorescence or a strobilus.

peduncle [paleont] (a) A narrow part by which some larger part or the whole body of an animal is attached, such as the *pedicle* of a brachiopod, the *column* of an echinoderm, or the basal portion of certain crustacean appendages. (b) The fleshy stalklike portion of the body of certain cirripedes. (c) The mass of cyptoplasm projecting from a thecamoebian-test aperture, giving rise to pseudopodia; pseudopodial trunk.

peel A thin film of acetate or other plastic substance, obtained in the *peel technique* for rock study. In some fields, e.g. paleobotany, peels are superior to thin sections for microscopic study, esp. because they cover much larger areas.

peel map A *paleogeologic map* on which the formations overlying an unconformity are, in effect, "peeled off" and the pre-unconformity distribution of the underlying formations is uncovered and mapped (Levorsen, 1960, p.4).

peel-off time The time correction to be applied to observed seismic data to adjust them to a depressed reference datum.

peel technique A method of preparing rock sections for study. A smooth surface is differentially etched with acid, and then covered with cellulose acetate or a similar liquid that will dry to a transparent film. When this film is peeled off, it brings with it a thin layer of rock substance, preserving this material in its original space relations. It may then be studied microscopically.

peel thrust A sedimentary sheet peeled off a sedimentary sequence, essentially along a bedding plane. A series of peel thrusts may be imbricated above a décollement (Bucher, 1955). Little used.

peepsight alidade An *alidade* used with a plane table, consisting of a rear (open) sight mounted on a straightedge.

peg adjustment The adjustment of a spirit-leveling instrument of the dumpy-level type in which the line of collimation is made parallel with the axis of the spirit level by means of two stable marks (pegs) the length of one instrument sight apart.

peg-leg multiple A multiple reflection involving successive reflection at different interfaces so that its travel path is not symmetrical. Sometimes refers to short-path multiples within thin beds, which result in transferring energy from the front of a wavetrain and adding it back later, and sometimes to long-path multiples, which lead to separate events.

pegma (peg'-ma) In rostroconchs, a plate connecting right and left valves in the umbonal part of the shell; it supports a large muscle in primitive rostroconchs (Pojeta and Runnegar, 1976, p.47). Etymol: Greek, "fastened".

pegmatite (peg'-ma-tite) An exceptionally coarse-grained igneous rock, with interlocking crystals, usually found as irregular dikes, lenses, or veins, esp. at the margins of batholiths. Most grains are one cm or more in diameter. Although pegmatites having gross compositions similar to other rock types are known, their composition is generally that of granite; the composition may be simple or complex and may include rare minerals rich in such elements as lithium, boron, fluorine, niobium, tantalum, uranium, and rare earths. Pegmatites represent the last and most hydrous portion of a magma to crystallize and hence contain high concentrations of minerals present only in trace amounts in granitic rocks. The first use of the term "pegmatite" is attributed to Haüy, who used it in 1822 as a syn. of *graphic granite.* Haidinger introduced the present meaning in 1845. Cf: *pegmatoid; symplectite.* Adj: *pegmatitic.* Syn: giant granite.

pegmatitic (peg-ma-tit'-ic) Said of the texture of very coarse-grained (crystal diameter greater than 3 cm) igneous rocks.

pegmatitic stage A stage in the normal sequence of crystallization of a magma containing volatiles, at which time the residual fluid is sufficiently enriched in volatile materials to permit the formation of coarse-grained rocks (i.e. pegmatites). Cf: *hydrothermal stage.*

pegmatoid (peg'-ma-toid) n. An igneous rock that has the coarse-grained texture of a *pegmatite* but lacks graphic intergrowths and/or typically granitic composition. adj. Said of the texture of a pegmatitic rock lacking graphic intergrowths and/or typical granitic composition. Introduced by Shand in 1910. Obsolete.

pegmatolite (peg-mat'-o-lite) *orthoclase.*

peg test In surveying, a test for the magnitude of the collimation error, carried out by leveling around a small loop which exaggerates the difference between backsight and foresight distances and produces a misclosure proportional to the collimation error of the leveling instrument.

pehrmanite (pehr'-man-ite) *ferrotaaffeite-6N'3S.*

peisleyite (peis'-ley-ite) A white monoclinic mineral: $Na_3Al_{16}(PO_4)_{10}(SO_4)_2(OH)_{17}{\cdot}20H_2O$.

pekoite (pe'-ko-ite) A metallic gray orthorhombic mineral: $PbCuBi_{11}(S,Se)_{18}$.

pekovite A vitreous colorless orthorhombic mineral: $SrB_2Si_2O_8$. It is the Sr analogue of danburite.

pelagic [lake] (pe-lag'-ic) Pertaining to the deeper part of a lake (10-20 m or more), characterized by deposits of mud or ooze and by the absence of aquatic vegetation (Wetzel, 2001, p.131).

pelagic [oceanog] (a) Pertaining to the water of the ocean as an environment. See also: *epipelagic; mesopelagic.* (b) Said of marine organisms whose environment is the open ocean, rather than the bottom or shore areas. Pelagic organisms may be either *nektonic* or *planktonic.*

pelagic deposit Marine sediment that has formed by settling of material (either biogenic or terrigenous) through the deep water column. Generally consists of fine-grained terrigenous and sand-sized biogenic material. Cf: *terrigenous deposit; hemipelagic deposit; deep-sea deposit.*

pelagic limestone A fine-textured limestone formed chiefly by the accumulation of the calcareous tests of floating organisms. It is characteristic of sedimentary basins, but can also occur as a shelf sediment.

pelagochthonous (pe-la-goch'-tho-nous) Said of coal derived from a submerged forest or from driftwood.

pelagosite (pe-lag'-o-site) A term used by Revelle and Fairbridge (1957, p.258) for a superficial calcareous crust a few millimeters thick, generally white, gray, or brownish, with a pearly luster, formed in the intertidal zone by ocean spray and evaporation (alternate dissolution and evaporation), and composed of calcium carbonate (See Purser and Loreau, 1973). Its first mention appears to have been by Charles Darwin in 1839 as a "white incrustation" seen on Ascension and St Helena (Fairbridge, 1995, p.94).

pelecypod (pe-lec'-y-pod) Any benthic aquatic mollusk belonging to the class Bivalvia. Syn: *lamellibranch.* Partial syn: *bivalve.*

Pelecypoda A class of diasome mollusk with bivalved shell. Dorsal margins of valves typically with crenulations known as teeth and sockets. Shell opened by doral ligament; closed by adductor muscles. Head and radula lacking. Range, Lower Cambrian to Holocene.

Peleean cloud A syn. of *nuée ardente,* so named because it is a characteristic type of eruption of Mt. Pelée on the island of Martinique in the West Indies.

peleean dome A type of lava dome characterized by craggy topography and lava spines that form jagged battlements along the crest of the dome (Francis, 1993), e.g., Mount Pelée.

Peleean-type eruption A type of volcanic eruption characterized by gaseous clouds (*nuées ardentes*) and/or the development of *volcanic domes.* Etymol: Mont Pelée, island of Martinique. Cf: *Hawaiian-type eruption; Strombolian-type eruption; Vulcanian-type eruption.*

Pele's hair (Pe'-le's) A natural spun glass formed by blowing-out during quiet fountaining of fluid lava, cascading lava falls, or turbulent flows, sometimes in association with *Pele's tears.* A single strand, with a diameter of less than half a millimeter, may be as long as two meters. Etymol: Pele, Hawaiian goddess of fire. Syn: *lauoho o pele; filiform lapilli; capillary ejecta.*

Pele's tears Small, solidified drops of volcanic glass behind which trail fibers of *Pele's hair.* They may be tear-shaped, spherical, or nearly cylindrical. Etymol: Pele, Hawaiian goddess of fire. Cf: *tear-shaped bomb.*

pelhamine (pel'-ham-ine) A light gray-green variety of serpentine from Pelham, Mass. It may be altered chrysotile.

pelionite (pel'-i-o-nite') A name proposed by W.F. Petterd for a bituminous coal resembling English cannel coal, occurring near Monte Pelion in Tasmania (Thrush, 1968, p.802).

pelite (pe'-lite) (a) A sediment or sedimentary rock composed of the finest detritus (clay- or mud-size particles); e.g. a *mudstone,* or a calcareous sediment composed of clay and minute particles of quartz. The term is equivalent to the Latin-derived term, *lutite.* (b) A fine-grained sedimentary rock composed of more or less hydrated aluminum silicates with which are mingled small particles of various other minerals (Twenhofel, 1937, p.90); an aluminous sediment. (c) A term regarded by Tyrrell (1921, p.501-502) as the metamorphic derivative of lutite, such as the metamorphosed product of a siltstone or mudstone. "As commonly used, a pelite means an aluminous sediment metamorphosed, but if used systematically, it means a fine-grained sediment metamorphosed" (Bayly, 1968, p.230). Etymol: Greek "pelos", "clay mud". See also: *psammite; psephite.* Also spelled: *pelyte.*

pelitic (pe-lit'-ic) (a) Pertaining to or characteristic of pelite; esp. said of a sedimentary rock composed of clay, such as a "pelitic tuff" representing a consolidated volcanic ash consisting of clay-size particles. (b) Said of a metamorphic rock derived from a pelite; e.g. a "pelitic hornfels" or a "pelitic schist", derived by metamorphism of an argillaceous or a fine-grained aluminous sediment. Cf: *argillaceous; lutaceous.*

pellet (pel'-let) A small, usually rounded aggregate of accretionary material, such as a *fecal pellet;* specif. a spherical to elliptical homogeneous clast made up almost exclusively of clay-sized calcareous (micritic) material, devoid of internal structure, and contained in a well-sorted carbonate rock. Folk (1959; 1962) suggested that the term apply to allochems less than

0.15-0.20 mm in diameter. Pellets appear to be mainly the feces of mollusks and worms. It is also likely that may pellets are the product of microbially induced precipitation around a bacterial colony. See also: *peloid*.

pelletization *pellet.*

pellet snow *graupel.*

pellicle (pel'-li-cle) An imperforate *pellis.*

pellicular envelope (pel-lic'-u-lar) The delicate outer covering of soft parts in a tintinnid.

pellicular front The even front, developed only in pervious granular material, on which pellicular water depleted by evaporation, transpiration, or chemical action is regenerated by influent seepage (Tolman, 1937, p.593). Syn: *infiltration front; wetting front.*

pellicular water Water in layers more than one or two molecules thick that adheres to the surfaces of soil and rock particles in the *unsaturated zone.* Layers more than a few micrometers in thickness are short-lived, owing to the requirement that free energy and capillary surface be at a minimum when moisture is at equilibrium (Smith, W.O., 1961, p.11). Syn: *adhesive water; film water; sorption water.* Cf: *pendular water; funicular water; attached groundwater.*

pellis (pel'-lis) A thin, commonly *imperforate* sheet or sheath outside the cup in archaeocyathids (TIP, 1972, pt. E, p.41).

pellodite (pel'-lo-dite) A term used by Schuchert (1924, p.441) for a water-laid sandy varved clay, and by Pettijohn (1957, p.273) for the lithified equivalent of a varved clay. It is apparently an equivalent or variant spelling of *pelodite.*

pelloid (pel'-loid) *peloid.*

pellouxite A metallic black monoclinic mineral: $(Cu,Ag)Pb_{10}Sb_{12}S_{27}(Cl,S)_{0.6}O$.

pellyite (pel'-ly-ite) A colorless to pale yellow orthorhombic mineral: $Ba_2Ca(Fe^{2+},Mg)_2Si_6O_{17}$.

pelma (pel'-ma) (a) An entire crinoid *column* with attached cirri and holdfast structure, if present. (b) A relatively large *lumen pore* in a costa of a cribrimorph cheilostome (bryozoan). Pl: pelmata.

pelmatidium (pel-ma-tid'-i-um) A relatively small *lumen pore* in a costa of a cribrimorph cheilostome (bryozoan).

pelmatozoan (pel'-ma-to-zo'-an) n. Any echinoderm, with or without a stem, that lives attached to a substrate. adj. Said of an echinoderm having an attached mode of life. Var: pelmatozoic. Cf: *eleutherozoan.*

pelmicrite (pel-mic'-rite) A limestone consisting of a variable proportion of pellets and carbonate mud (micrite); specif. a limestone containing less than 25% intraclasts and less than 25% ooids, with a volume ratio of pellets to bioclasts greater than 3 to 1, and the carbonate-mud matrix more abundant than the sparry-calcite cement (Folk, 1959, p.14).

pelodite (pe'-lo-dite) A term proposed by Woodworth (1912, p.78) for a lithified glacial rock flour, composed of glacial pebbles in a silty or clayey matrix, formed by redeposition of the fine fraction of a till. Cf: *pellodite.*

pelogloea (pel-o-gloe'-a) Organic matter, mainly colloidal, adsorbed on sedimentary particles in natural waters (Fox, 1957, p.384). Cf: *leptopel.*

peloid (pel'-oid) (a) An *allochem* composed of micrite, irrespective of size or origin, for which exact origin is unknown; e.g., completely micritized fossils or ooids. Other types of peloid include pseudo-ooids and aggregates produced by gas bubbling, by microbial precipitation, or by other intraformational reworking of lithified or semilithified carbonate mud. Some peloids in reefs are the products of microbially induced precipitation around clumps of bacteria. (b) A small, rounded aggregate (0.1-0.3 mm in diameter) of clay minerals and fine quartz found in some shales and clays, separated from a matrix of the same materials by a shell of organic material, and ascribed to the action of water currents (Allen and Nichols, 1945). Also spelled: pelloid. See also: *pellet.*

peloidal limestone (pel'-let-ed) A limestone characterized by abundant peloids.

pelolithic (pe-lo-lith'-ic) *argillaceous.*

Pelsonian (Pel-so'-ni-an) A European substage: middle of the Anisian Stage, Middle Triassic (above Bithynian, below Illyrian).

pelsparite (pel-spar'-ite) A limestone consisting of a variable proportion of pellets and clear calcite (spar); specif. a limestone containing less than 25% intraclasts and less than 25% ooids, with a volume ratio of pellets to bioclasts greater than 3 to 1, and the sparry-calcite cement more abundant than the carbonate-mud matrix (micrite) (Folk, 1959, p.14).

pelta (pel'-ta) A lidlike flap, dorsally closing or partly closing the internal cavity in one-walled archaeocyathids; it may or may not be porous (TIP, 1972, pt. E, p.41).

peltasperms (pel'-ta-sperms) A group of Permian-Triassic *seed ferns* (Peltaspermaceae or Peltaspermales) with bi- to tripinnate leaves and seeds borne on the lower surface of flattened, *peltate* megasporophylls. Species are known from Eurasia, Greenland, and Gondwana.

peltate (pel'-tate) Shield-shaped, as in a leaf that is attached to its stalk inside the margin (Lawrence, 1951, p.764).

pelvis (pel'-vis) (a) The proximal or dorsal part of a blastoid *theca* from the aboral tips of the ambulacra to the proximal pole. (b) In vertebrates, a term applied to the combined bones of the pelvic girdle or to the expansion of the ureter at the joint of its junction with the kidney.

Pelycosauria (Pel'-y-co-sau'-ri-a) An order of synapsid reptiles of generally reptilian grade, as indicated by crocodiliform or lacertiform habitus. Range, Lower Pennsylvanian to ?Upper Permian.

pelyte (pe'-lyte) *pelite.*

pen A British term variously used for a hill, mountain, highland, or headland. Etymol: Celtic.

peña (pe'-ña) A rock; a rocky point; a needlelike eminence; a cliff. Term is used in the SW U.S. Etymol: Spanish, "rock".

peñasco (pe-ña'-sco) A term used in the SW U.S. for a projecting rock, esp. one isolated by the recession of a cliff or of a mountain slope. Etymol: Spanish, "large rock".

pencil cleavage (pen'-cil) Cleavage whose presence causes a rock body to break into long, slender (pencil-like) pieces. Typically, it is produced by the intersection of cleavage planes with bedding planes in a fissile rock. Pencil cleavage may represent an intermediate stage in the development of slaty cleavage, and thus generally occurs only in weakly metamorphosed rocks. Syn: *pencil structure.*

pencil ganister A variety of *ganister* characterized by fine carbonaceous streaks or markings, and so called from the likeness of these to pencil lines. The carbonaceous traces are often recognizable as roots and rootlets of plants.

pencil gneiss A gneiss that breaks into roughly cylindrical, pencil-like quartz-feldspar crystal aggregates, often mantled by mica flakes. Rarely used. Syn: *stengel gneiss.*

pencil ore Hard, fibrous masses of hematite that can be split up into thin rods.

pencil stone A compact pyrophyllite once used for making slate pencils.

pencil structure *pencil cleavage.*

pendant [intrus rocks] (pen'-dant) *roof pendant.*

pendant [speleo] A solutional remnant hanging from the ceiling or wall of a cave. Syn: *rock pendant; solution pendant.* Cf: *deckenkarren.*

pendeloque (pen-de-loque') A modification of the round *brilliant cut,* having a drop-shaped outline similar to that of the *pear-shaped cut,* but with the narrow end longer and more pointed. Frequently drilled near the apex for use as a pendant for a necklace or a brooch. Cf: *briolette.*

pendent [geomorph] (pen'-dent) Said of a landform that slopes steeply down (as a hillside) or overhangs (as a cliff).

pendent [paleont] Said of a graptoloid *rhabdosome* with approximately parallel stipes that hang below the *sicula.*

pendletonite (pen'-dle-ton-ite') *carpathite.*

pendular water (pen'-du-lar) *Capillary water* ringing the contact points of adjacent rock or soil particles in the unsaturated zone (Smith, W.O., 1961, p.2). Cf: *funicular water; pellicular water; capillary condensation; attached groundwater.*

pendulum (pen'-du-lum) (a) A body suspended from a fixed point so as to swing freely under the combined action of gravity and momentum. Syn: *physical pendulum.* (b) A vertical bar supported from below by a stiff spring so as to vibrate under the combined action of gravity and the restoring force of the spring (Dobrin, 1952, p.50-51). Syn: *inverted pendulum.*

pendulum level A *leveling instrument* in which the line of sight is automatically maintained horizontal by means of a built-in pendulum device (such as a horizontal arm and a plumb line at right angles to the arm).

penecontemporaneous (pe'-ne-con-tem'-po-ra'-ne-ous) Formed or existing at almost the same time; e.g. said of a structure or mineral that was formed immediately after deposition of a sediment but before its consolidation into rock. Cf: *contemporaneous.*

penecontemporaneous deformation A type of deformation, usually in soft-sedimentary layers, that occurs almost at the same time as deposition (e.g., folding related to slumping of a sedimentary mass). Sometimes called (less precisely) contemporaneous deformation.

penecontemporaneous faulting A deformation occurring in sediment or partially lithified rock, soon after the deposition of the strata involved, and caused by gravitational sliding or slumping.

penecontemporaneous fold A fold that develops in sediment shortly after deposition.

peneloken (pen-e-lo'-ken) *pokelogan.*

peneplain (pe'-ne-plain) n. A term introduced by Davis (1889a, p.430) for a low, nearly featureless, gently undulating land surface of considerable area, which presumably has been produced by the processes of long-continued subaerial erosion (primarily mass-wasting of and sheetwash on interstream areas of a mature landscape, assisted by stream erosion) almost to base level in the penultimate stage of a humid, fluvial geomorphic cycle; also, such a surface uplifted to form a plateau and subjected to dissection. A peneplain may be characterized by gently graded and broadly convex interfluves sloping down to broad valley floors, by truncation of strata of varying resistance and structure, by accordant levels, and by isolated erosion remnants (monadnocks) rising above it. Etymol: Latin "pene-", "almost", + plain. Cf: *endrumpf; base-level plain; marine peneplain; panplain; plain of denudation; rumpffläche.* Syn: *peneplane; base-level peneplain.* v. *peneplane.*

peneplanation (pe'-ne-pla-na'-tion) The act or process of formation and development of a *peneplain;* esp. the decline and flattening out of hillsides during their retreat and the accompanying downwasting of divides and residual hills.

peneplane (pe′-ne-plane) n. A term suggested by Johnson (1916) to replace *peneplain* since the latter, when first introduced by Davis (1889a), was not intended to signify "almost a plain" (i.e. a region of nearly horizontal structure) but a region with almost a flat surface; the suggestion has received little support. v. To erode to a peneplain. Syn: *peneplain.*

penesaline (pe-ne-sa′-line) A condition intermediate between normal marine and hypersaline. Having a salinity too high for normal marine organisms, yet below saturation with halite (Sloss, 1953).

penesaline environment An environment with *penesaline* water, characterized by sedimentary carbonates with restricted fauna or no fauna, and often interbedded with gypsum or anhydrite.

penetration funnel (pen-e-tra′-tion) An *impact crater* that is generally funnel-shaped and was formed by a small meteorite striking the Earth at a relatively low velocity; it contains nearly all the impacting mass within it (Cassidy, 1968, p.117). Cf: *meteorite crater.*

penetration test A test to determine the relative densities of noncohesive soils, sands, or silts; e.g., the *standard penetration test (SPT)* that determines the number of blows required by a standard weight when dropped from a standard height (30 in. [75 cm]) to drive a standard sampling device a standard penetration (12 in. [30 cm]); or the "dynamic penetration test" that determines the relative densities of successive layers by recording the penetration per blow or a specified number of blows. Cf: *cone penetration test; standard penetration test.*

penetration twin *interpenetration twin.*

penetrative Said of a feature or texture if all subdivisions of the rock at some scale possess that feature. Features or textures may be penetrative at one scale, but may be nonpenetrative at a different scale.

penetrometer (pen-e-trom′-e-ter) An instrument for measuring the consistency of materials (such as soil, snow, asphalt, or coal) by indicating the pressure necessary to inject a rigid weight-driven rod or needle of specified shape to a specific depth.

penfieldite (pen′-field-ite) A white, yellowish, or bluish hexagonal mineral: $Pb_2Cl_3(OH)$.

penginite (pen-gin′-ite) *penzhinite.*

penikisite (pe-nik′-i-site) A blue or green triclinic (pseudo-monoclinic) mineral of the *bjarebyite* group: $Ba(Mg,Fe^{2+})_2Al_2(PO_4)_3(OH)_3$.

penikkavaarite (pe-nik′-ka-vaar′-ite) An intrusive rock, probably an *essexite,* composed chiefly of augite and amphibole in a groundmass of andesine. Named by Johannsen (1938) for Penikkavaara, Finland. Obsolete.

peninsula (pe-nin′-su-la) (a) An elongated body or stretch of land nearly surrounded by water and connected with a larger land area, usually by a *neck [geog]* or an *isthmus.* (b) A relatively large tract of land jutting out into the water, with or without a well-defined isthmus; e.g. the Italian peninsula. Etymol: Latin, "almost-island".

penitabular (pe-ni-tab′-u-lar) Of the surface features of a dinoflagellate cyst, lying immediately interior to the margins of reflected plate areas (as in *Areoligera* and *Eisenackia*) (Williams et al., 2000, p.159). Cf: *intratabular; nontabular.* Obsolete syn: *peritabular.*

penitent [geomorph] (pen′-i-tent) A term used in the French Alps for an *earth pillar.*

penitent [glaciol] n. A syn. of *nieve penitente.* adj. A term used to refer to a nieve penitente; e.g. "penitent ice" is a nieve penitente consisting mainly of ice, and "penitent snow" is one consisting mainly of snow.

penitent rock A variety of *tor* formed on rock with a dipping foliation, joint pattern, or bedding (Ackermann, 1962). Syn: *monk rock.*

penkvilksite (penk-vilk′-site) A silky white monoclinic or orthorhombic mineral: $Na_4Ti_2Si_8O_{22}·4H_2O$.

pennantite (pen′-nant-ite) An orange, reddish brown, brown, dark red, or dark green monoclinic trioctahedral member of the *chlorite* group of minerals, in which Mn is the dominant divalent octahedral cation: $(Mn^{2+},Al)_6(Si,Al)_4O_{10}(OH)_8$.

pennate diatom (pen′-nate) A diatom having elongate form and bilateral symmetry; a member of the diatom order Pennales. Cf: *centric diatom.*

pennine (pen-nine′) *penninite.*

penninite (pen-nin′-ite) *clinochlore.*

Pennsylvanian (Penn-syl-va′-ni-an) The upper subsystem of the *Carboniferous* System of the Standard Global Chronostratigraphic Scale, above the *Mississippian* Subsystem and below the *Permian* System. Also the time during which these rocks were formed, the Pennsylvanian Subperiod, covering the time span between 320 and 295 Ma. It is named after the State of Pennsylvania, in which rocks of this age are widespread and yield much coal. It is the approximate equivalent of the Upper Carboniferous of some chronostratigraphic scales.

penobsquisite A pale yellow monoclinic mineral: $Ca_2Fe^{2+}[B_9O_{13}(OH)_6]Cl·4H_2O$.

Penokean Orogeny (Pe-no′-ke-an) A time of deformation and granite emplacement during the Precambrian in Minnesota and Michigan, dated radiometrically at about 1,800 Ma, which occurred between the formation of the Animikie (Huronian) Series and the Keweenawan Series. The Penokean Orogeny probably resulted from collision of crustal blocks with the southern edge of the Superior Province. It is perhaps the same as the *Hudsonian Orogeny* of the Canadian Shield. See also: *Killarney Revolution.*

peñon (pe-ñon′) A high, rocky point. Etymol: Spanish, "large rock, rocky mountain".

penroseite (pen′-rose-ite) A metallic lead-gray cubic mineral of the *pyrite* group: $(Ni,Co,Cu)Se_2$. It may contain some lead and silver. It occurs in radiating columnar masses. Syn: *blockite.*

penstock A surface or underground conduit through which water moves into a hydroelectric turbine or electricity generator.

pentactine (pen-tac′-tine) A sponge spicule having five rays; specif. a hexactine with one ray suppressed.

pentagonal dodecahedron (pen-tag′-o-nal) *pyritohedron.*

pentagonite (pen-tag′-o-nite) A greenish-blue orthorhombic mineral: $Ca(V^{4+}O)Si_4O_{10}·4H_2O$. It is polymorphous with cavansite.

pentahydrite (pen-ta-hy′-drite) (a) A colorless or pale blue triclinic mineral: $MgSO_4·5H_2O$. It is isostructural with chalcanthite. Syn: *allenite.* (b) A group name for triclinic minerals with the general composition $ASO_4·5H_2O$, where A = Mg, Fe^{2+}, Mn^{2+}, or Cu^{2+}.

pentahydroborite (pen′-ta-hy′-dro-bor′-ite) A colorless triclinic mineral: $CaB_2O(OH)_6·2H_2O$.

pentalith (pen′-ta-lith) A *coccolith* formed of five crystal units diverging at 72 degrees.

pentameracean (pen-tam′-e-ra′-ce-an) Any articulate brachiopod assigned to the superfamily Pentameracea; characterized in general by a large, strongly biconvex shell with a smooth, costellate, or costate exterior, and by a spondylium in the pedicle valve. Range, Upper (possibly Middle) Ordovician to Upper Devonian. Var: pentamerid.

pentamerous (pen-tam′-er-ous) *quinqueradiate.*

pentane (pen′-tane) Any of three isomeric, low-boiling paraffin hydrocarbons, formula C_5H_{12}, found in petroleum and natural gas.

pentlandite (pent′-land-ite) (a) A pale bronze to light-brown cubic mineral: $(Fe,Ni)_9S_8$. It is commonly intergrown with pyrrhotite, from which it is distinguished by its octahedral cleavage and lack of magnetism. Pentlandite is the principal ore of nickel. Syn: *folgerite; nicopyrite.* (b) A group name for cubic minerals with the general composition A_9S_8, where A = Fe, Ni, Co, Ag, Pb, Mn or Cu.

penumbra (pe-num′-bra) (a) The partly shadowed region of an eclipse. (b) The outer, lighter region of a sunspot. Cf: *umbra.*

Penutian (Pe-nu′-tian) North American stage: Lower Eocene (above Bulitian, below Ulatisian).

penzhinite (pen-zhin′-ite) A metallic hexagonal mineral: $(Ag,Cu)_4Au(S,Se)_4$. Syn: *penginite.*

Peorian (Pe-or′-i-an) A term previously used for an interglacial stage between the now-discarded Iowan (earlier) and the Wisconsinan (later) glacial stages. Named for exposures near Peoria, Ill.

peperino (pe-pe-ri′-no) (a) An unconsolidated, gray tuff of the Italian Albano hills, containing crystal fragments of leucite and other minerals. (b) An indurated pyroclastic deposit containing fragments of various sizes and types.

peperite (pe′-pe-rite) A breccialike material in marine sedimentary rock, interpreted by some as a mixture of lava with sediment, and by others as shallow intrusions of magma into wet sediment.

pepino (pe-pi′-no) *karst tower.*

pepino hill *karst tower.*

pepper-and-salt texture (pep′-per-and-salt′) Said of disseminated ores, esp. with dark grains in a light matrix.

peprossiite-(Ce) A pale yellow hexagonal mineral: $(Ce,La)Al_2B_4O_{10}$.

peptide Amino acids linked together by a specific bond that releases water during its formation; the backbone of proteins.

peracidite (per-ac′-id-ite) *silexite [ign].*

peraeopod (pe-rae′-o-pod) *pereiopod.*

peralkaline (per-al′-ka-line) Said of an igneous rock in which the molecular proportion of aluminum oxide is less than that of sodium and potassium oxides combined; one of Shand's (1947) groups of igneous rocks, classified on the basis of the degree of aluminum-oxide saturation. Cf: *peraluminous; metaluminous; subaluminous.*

peraluminous (per-a-lu′-mi-nous) Said of an igneous rock in which the molecular proportion of aluminum oxide is greater than that of sodium and potassium oxides combined; one of Shand's (1947) groups of igneous rocks, classified on the basis of the degree of aluminum-oxide saturation. Cf: *peralkaline; metaluminous; subaluminous.*

perbituminous (per-bi-tu′-mi-nous) Said of bituminous coal containing more than 5.8% hydrogen, analyzed on a dry, ash-free basis. Not listed by ASTM as a rank classification. Cf: *orthobituminous; parabituminous.*

percentage hypsometric curve (per-cent′-age) A *hypsographic curve* using dimensionless parameters independent of an absolute scale of topographic features by relating the area enclosed between a given contour and the highest contour to the height of the given contour above a basal plane.

percentage log A *sample log* in which the percentage of each type of rock (except obvious *cavings*) present in each sample of well cuttings is estimated and plotted. Cf: *interpretive log.*

percentage map A *facies map* that depicts the relative amount (thickness) of a single rock type in a given stratigraphic unit.

percent frequency effect The basic polarization parameter measured in frequency-domain resistivity surveys. Abbrev: PFE.

percentile In a *cumulative frequency distribution,* the *fractile* values at or below which lie the specified percentage of the observations.
percent slope (per-cent') The direct ratio between the vertical distance and the horizontal distance for a given slope; e.g. a 3-meter rise in 10 meters horizontal distance would be a 30 *percent slope.*
perch A unit of length, varying locally in different countries, but by statute in Great Britain and U.S. equal to 16.5 ft. It was used extensively in the early public-land surveys and is equivalent in length to a rod or pole.
perched aquifer An aquifer containing *perched groundwater.*
perched block (a) A large, detached rock fragment, generally of boulder size, believed to have been transported and deposited by a glacier, and lying in a conspicuous and relatively unstable or precariously poised position on a hillside. Cf: *erratic.* Syn: *perched boulder; perched rock; balanced rock; pedestal rock.* (b) A rock forming a glacier table in a glacier. (c) A rock capping an earth pillar.
perched boulder *perched block.*
perched glacial valley *hanging valley [glac geol].*
perched groundwater Unconfined groundwater separated from an underlying main body of groundwater by an unsaturated zone. See also: *perched aquifer; perching bed.* Syn: *perched water.*
perched lake A perennial lake whose surface level lies at a considerably higher elevation than those of other bodies of water, including aquifers, directly or closely associated with it; e.g. a lake on a bench that borders the shore of a larger lake.
perched rock *perched block.*
perched spring A spring whose source of water is a body of perched groundwater.
perched stream (a) A stream or reach of a stream whose upper surface is higher than the water table and that is separated from the underlying groundwater by an impermeable bed in the zone of aeration (Meinzer, 1923, p.57). (b) A stream flowing on an antecedent hillside along a graded valley of a higher-order stream into which it flows at nearly right angles (Horton, 1945, p.352).
perched water *perched groundwater.*
perched water table The water table of a body of perched groundwater. Syn: *apparent water table.*
perching bed (perch'-ing) A body of relatively impermeable rock that supports a body of *perched groundwater.* At a given place there may be two or more perching beds and bodies of perched groundwater, separated from each other and from the main zone of saturation by unsaturated zones.
percleveite-(Ce) A colorless or white tetragonal mineral: $(Ce,La,Nd)_2Si_2O_7$.
percolating water (per'-co-lat'-ing) (a) A legal term for water that oozes, seeps, or filters through the soil without a definite channel in a course that is unknown or not discoverable. Cf: *underground stream.* (b) Water involved in percolation.
percolation (per-co-la'-tion) (a) Slow laminar movement of water through small openings within a porous material. Also used incorrectly as a syn. of "infiltration". Flow in large openings such as caves is not included. (b) In the unsaturated zone, the flow of water that has infiltrated and is moving downward or laterally toward the water table. Cf: *infiltration; seepage [grd wat].*
percolation rate The rate, expressed as either velocity or volume, at which water percolates through a porous medium.
percolation test A term used in sanitary engineering for a test to determine the suitability of a soil for the installation of a domestic sewage-disposal system. A hole is dug and filled with water and the rate of water-level decline measured; the dimensions of the hole and acceptable rate of decline differ from one jurisdiction to another.
percolation zone The area on a glacier or ice sheet where a limited amount of surface melting occurs, but the meltwater refreezes and the snow layer is not completely soaked or brought up to the melting temperature. The percolation zone may be bounded at higher altitudes by the *dry-snow line* and is bounded at lower altitudes by the *wet-snow line.* Cf: *soaked zone.*
percussion figure (per-cus'-sion) A pattern of radiating lines produced on a section of a crystal by a blow from a sharp point. Cf: *pressure figure.*
percussion mark A crescentic scar produced on a hard, dense pebble (esp. one of chert or quartzite) by a sharp blow, as by the violent impact of one pebble on another; it may be indicative of high-velocity flow. Syn: *percussion scar.*
percussion scar *percussion mark.*
percylite (per'-cy-lite) A pale-blue cubic mineral: $PbCu(OH)_2Cl_2$. It is a doubtful species, and may be a mixture.
pereion (pe-rei'-on) The *thorax* of a malacostracan crustacean, exclusive of the somites bearing maxillipeds. It is usually provided with locomotory appendages (pereiopods). Pl: pereia. Also spelled: *pereon.*
pereionite (pe-rei'-o-nite) A *somite* of the pereion of a malacostracan crustacean. Also spelled: *pereonite.*
pereiopod (pe-rei'-o-pod) A locomotory *thoracopod* of a malacostracan crustacean; an appendage of the *pereion.* Also spelled: *peraeopod; pereopod.* Syn: *walking leg.*
pereletok A frozen layer of ground, between the active layer above and the *permafrost* below, that remains unthawed for one or several years. Etymol: Russian, "survives over the summer". Syn: *intergelisol.*
perennial lake (pe-ren'-ni-al) A lake that retains water in its basin throughout the year, and that usually is not subject to extreme fluctuations in level.
perennially frozen ground (pe-ren'-ni-al-ly) *permafrost.*
perennial spring A spring that flows continuously, as opposed to an *intermittent spring* or a *periodic spring.*
perennial stream A stream or reach of a stream that flows continuously throughout the year and whose upper surface generally stands lower than the water table in the region adjoining the stream. Syn: *permanent stream; live stream.*
pereon (pe-re'-on) *pereion.*
pereonite (pe-re'-o-nite) *pereionite.*
pereopod (pe-re'-o-pod) *pereiopod.*
peretaite (per-et'-a-ite) A colorless monoclinic mineral: $CaSb^{3+}{}_4O_4(SO_4)_2(OH)_2{\cdot}2H_2O$.
peretite A pearly yellow orthorhombic mineral: $Ca(UO_2)_3(Se^{4+}O_3)_2(OH)_4{\cdot}4H_2O$.
perfect crystal (per'-fect) A crystal without lattice defects. It is an unattained ideal or standard.
perfect dislocation A *dislocation [struc petrol]* in which the glide, or slip, distance is equal to the length of one unit lattice distance (the length of one atomic bond) or multioples thereof.
perfect flower A flower having both stamens and carpels. Cf: *imperfect flower.*
perfect fractionation path On a phase diagram, a line or a path representing a crystallization sequence and path followed by residual melt in which any crystal that has been formed remains inert, i.e. does not have its composition altered.
perfection color (per-fec'-tion) As applied to gem diamonds, the finest color grade; i.e. a total absence of color. As applied to colored stones, the finest color for that particular variety of a gemstone. Cf: *gem color.*
perfectly mobile component A component whose amount in a system is determined by its externally imposed chemical potential rather than by its initial amount in the system. Cf: *inert component; mobile component.* Syn: *boundary-value component.*
perfect stone (a) A *flawless* gemstone. (b) A colored stone in which small inclusions or structural faults are less undesirable than in a flawless gemstone and may even be desirable.
perfemic (per-fem'-ic) One of five classes in the *CIPW classification* of igneous rocks, in which the ratio of salic to femic minerals is less than one to seven. Cf: *dofemic.*
perfoliate (per-fo'-li-ate) Said of a sessile leaf or bract that surrounds the stem so that the stem seems to pass through it.
perforate (per'-fo-rate) (a) Said of the wall between corallites of some colonies, characterized by the presence of many irregularly arranged small openings. Also, pertaining to Perforata, a division of corals whose skeleton has a porous texture. (b) Descriptive of foraminiferal-test walls punctured or pierced by numerous pores or small openings that are distinct from apertures, foramina, and canals. Perforate walls are esp. characteristic of calcareous hyaline tests. Also, pertaining to Perforata, a division of foraminifers whose shells have small perforations for the protrusion of pseudopodia. (c) Said of an echinoid *tubercle* with a small depression in the top for the ligament connecting a spine with the tubercle. (d) Said of a spiral mollusk shell having an opening in the center of the umbilicus. Cf: *imperforate.*
perforated crust (per'-fo-rat-ed) A type of *snow crust* containing pits and hollows produced by *ablation.*
perforation (per-fo-ra'-tion) Puncturing of well casing opposite an oil- or gas-bearing zone to permit oil or gas to flow into a cased borehole.
perforation deposit A term suggested by Cook (1946) for an isolated kame consisting of material that accumulated in a vertical shaft piercing a glacier and affording no outlet for water at the bottom.
pergelation (per-ge-la'-tion) The formation of permafrost in the present or in the past (Bryan, 1946, p.640).
pergelic temperature regime (per-gel'-ic) A soil temperature regime in which the mean annual temperature is less than 0°C, and there is permafrost (USDA, 1975).
pergelisol (per-gel'-i-sol) A term introduced by Bryan (1946) for *permafrost.* Also, the French term for permafrost.
perhamite (per'-ham-ite) A white or tan platy hexagonal mineral: $Ca_3Al_7(SiO_4)_3(PO_4)_4(OH)_3{\cdot}16.5H_2O$.
perhyaline (per-hy'-a-line) In the *CIPW classification* of igneous rocks, those rocks in which the ratio of crystals to glassy material is greater than one to seven. Obsolete. Cf: *dohyaline.*
perhydrous (per-hy'-drous) (a) Said of coal containing more than 6% hydrogen, analyzed on a dry, ash-free basis. (b) Said of a maceral of high hydrogen content, e.g. liptinite, resinite. Cf: *orthohydrous; subhydrous.*
peri- (per'-i-) A prefix meaning "around", "near".
perianth (per'-i-anth) A collective term for the corolla and the calyx of a flower when considered together or when structurally undifferentiated.
periblain (per'-i-blain) A kind of *provitrain* in which the cellular structure is derived from cortical material. Cf: *suberain; xylain.*
periblinite (pe-rib'-li-nite) A variety of provitrinite characteristic of periblain and consisting of cortical tissue. Cf: *xylinite; suberinite; telinite.*

pericarp (per'-i-carp) (a) The wall of a matured ovary. (b) The wall of the capsule in mosses. (c) Improperly used for the protective husks surrounding certain fruits (Jackson, 1928, p.273).

periclase (per'-i-clase) (a) A colorless cubic mineral with the rocksalt structure: MgO. It alters easily to brucite. Syn: periclasite. (b) A group name for cubic minerals with the rocksalt structure and the general formula *AO*, where *A* = Mg, Fe^{2+}, Ni, Mn^{2+}, Cd, or Ca.

periclinal [bot] (per-i-cli'-nal) Parallel to the surface or circumference of a plant organ. Cf: *anticlinal [bot].*

periclinal [geol] Said of strata and structures that dip radially outward from, or inward towards, a center, to form a dome or a basin. Cf: *quaquaversal; centroclinal.*

pericline [mineral] A variety of albite elongated in the direction of the *b*-axis and often twinned with this as the twinning axis. It occurs in veins as large milky-white opaque crystals. Pericline is probably an albitized oligoclase.

pericline [struc geol] (per'-i-cline) A general term for a fold in which the dip of the beds has a central orientation; beds dipping away from a center form a dome, and beds dipping towards a center form a *basin.* The term is generally British in usage. See also: *centrocline; quaquaversal.*

pericline ripple A term used by Ten Haaf (1959, p.22) for a ripple arranged in an orthogonal pattern either parallel or transverse to the current direction and having a wavelength up to 80 cm and amplitude up to 30 cm.

pericline twin law A parallel twin law in triclinic feldspars, in which the twin axis is the crystallographic *b* axis and the composition surface is a rhombic section. It occurs alone or with the albite twin law.

pericoel (per'-i-coel) The space between the periphragm and endophragm in a cavate dinoflagellate cyst. Cf: *autocoel; endocoel.*

pericolpate (per-i-col'-pate) *pantocolpate.*

pericolporate (per-i-col'-por-ate) *pantocolporate.*

pericycle (per'-i-cy-cle) In roots and stems, a layer (or layers) of cells immediately outside the phloem and inside the endodermis, from which branch roots develop (Fuller and Tippo, 1954, p.966).

pericyst (per'-i-cyst) The *frontal shield* of a cribrimorph cheilostome (bryozoan), composed of costae united medially and commonly laterally.

perideltaic (per'-i-del-ta'-ic) Adjacent to or surrounding a delta.

perideltidium (per'-i-del-tid'-i-um) One of a pair of slightly raised triangular parts of the interarea of a brachiopod, flanking the *pseudodeltidium* or lateral to it, and characterized by vertical striae in addition to horizontal growth lines parallel to the posterior margin (TIP, 1965, pt. H, p.150).

periderm [bot] (per'-i-derm) A collective name for *cork, cork cambium,* and *phelloderm.* It constitutes a major portion of what is commonly called bark in woody plants.

periderm [paleont] (a) The protein substances that compose the *rhabdosome* of a graptolite, consisting of an inner part (fusellar tissue) with growth increments and an outer part (cortical tissue) of layers of material (cortical fabric). (b) In fetal amniotes, a transitory cellular or keratinized tissue (generally lost shortly after hatching or birth) that protects immature openings of the eyes, external ears, and margins of the mouth from dessication and attack by microorganisms (in mammals, commonly known as epitrichium).

periderre (per'-i-derre) A thinner layer located outside the *ectoderre* in the wall of a chitinozoan. Cf: *endoderre.*

peridiniacean (per-i-di-ni-a'-cean) A dinoflagellate theca or cyst with a tabulation similar to that of the modern genus *Peridinium,* with typically seven precingular and five postcingular plates (Williams et al., 2,000, p.161-162). Cf: *peridinioid.*

peridinioid (pe-ri-din'-i-oid) A dinoflagellate theca or cyst that resembles the modern genus *Peridinium.* The term is sometimes used synonymously with *peridiniacean* (implying tabulation pattern) but may also be used simply to imply a shape similar to modern *Peridinium* (Williams et al., 2000, p.161-162).

peridot (per'-i-dot) (a) A transparent to translucent green gem variety of olivine. Syn: peridote. (b) A yellowish-green or greenish-yellow variety of tourmaline, approaching olivine in color. It is used as a semiprecious stone. Syn: peridot of Ceylon.

peridotite (pe-rid'-o-tite) (a) In the *IUGS classification,* a plutonic rock with M equal to or greater than 90 and ol/(ol+opx+cpx) greater than 40. (b) A general term for a coarse-grained plutonic rock composed chiefly of olivine with or without other mafic minerals such as pyroxenes, amphiboles, or micas, and containing little or no feldspar. Peridotites encompass the more specific terms saxonite, harzburgite, lherzolite, wehrlite, dunite. Accessory minerals of the spinel group are commonly present. Peridotite is commonly altered to *serpentinite.*

peridotite shell A syn. of the outer or *upper mantle,* so named because its composition may be peridotitic. Rarely used.

periembryonic chamber (per'-i-em'-bry-on'-ic) An immature (nepionic) part of a foraminiferal test formed on the ventral side and partly surrounding the proloculus (as in Orbitolinidae).

perigean tide (per-i-ge'-an) A tide of increased range (e.g. a spring tide) occurring monthly when the Moon is at or near the perigee of its orbit. Ant: *apogean tide.*

perigee (per'-i-gee) That point on the orbit of an Earth satellite, natural or manufactured, which is nearest to the Earth. Cf: *apogee.*

perigenic (per-i-gen'-ic) Said of a rock constituent or mineral formed at the same time as the rock of which it constitutes a part "but not at the specific location in which it is now found in that rock" (Lewis, 1964, p.875); e.g. said of a glauconite grain formed from an agglutinated clay pellet and subjected to "short" transportation prior to its final incorporation into the sediment.

periglacial (per-i-gla'-cial) (a) Said of the processes, conditions, areas, climates, and topographic features at the immediate margins of former and existing glaciers and ice sheets, and influenced by the cold temperature of the ice. (b) By extension, said of an environment in which frost action is an important factor, or of phenomena induced by a periglacial climate beyond the periphery of the ice. Syn: *cryergic; cryonival; paraglacial; subnival.* Term introduced by Lozinski (1909).

periglacial geomorphology "The study of all processes and phenomena found in cold regions" (Hamelin and Cook, 1967, p.11). See also: *cryomorphology.*

perignathic girdle (per-i-gnath'-ic) A continuous or discontinuous ring of internal processes around the peristomial opening of an echinoid, serving for the attachment of muscles supporting and controlling *Aristotle's lantern.* See also: *girdle [paleont].*

perigynous (pe-rig'-y-nous) Said of a plant in which the stamens and petals are borne on a ring of the receptacle surrounding a pistil and usually adnate to the calyx. Cf: *epigynous; hypogynous.*

perihelion (per-i-he'-li-on) The point at which any orbiting object passes nearest to the sun. Cf: *aphelion.*

perilith (per'-i-lith) *cored bomb.*

perilumen (per-i-lu'-men) A raised inner border of a columnal *articulum* of a crinoid, developed as a crenulate, smooth-topped, granulose, tuberculate or vermiculate ridge or field surrounding a *lumen.*

perimarine (per'-i-ma-rine') Said of the sedimentary facies of a lowland area protected by barrier islands, consisting of fluvial, lagoonal, and peat-swamp deposits (Hageman, 1969). Cf: *paralic [sed].*

perine (per'-ine) A *sporoderm* layer that is situated around the exine of many pollen grains. It is not always resistant to acetolysis or acid. The term is used in conjunction with *sexine* and *nexine* for pollen, whereas *perispore* is used in conjunction with *exospore* and *endospore* for spores.

perinium *perine.* Pl: perinia. Adj: perinate.

period [geochron] (pe'-ri-od) (a) The *geochronologic unit* lower in rank than *era* and higher than *epoch [geochron],* during which the rocks of the corresponding *system* were formed. It is the fundamental unit of the worldwide geologic time scale. (b) A term used informally to designate a length of geologic time; e.g. *glacial period.*

period [phys] The interval of time required for the completion of a cyclic motion or recurring event, such as the time between two consecutive like phases of a wavetrain, the tide, or a current.

periodicity hypothesis The proposal, based on the analysis of the stratigraphic records of marine families and genera by David Raup and John Sepkoski Jr. of the University of Chicago, that mass extinctions have occurred approximately every 26 million years from about 250 million years ago to the present.

periodic spring A spring that ebbs and flows, owing to natural siphon action. Such springs issue mainly from carbonate rocks, in which solution channels form the natural siphons. It is distinguished from a geyser by its temperature that of ordinary groundwater and general lack of gas emission. Syn: *ebbing-and-flowing spring.* Cf: *perennial spring; intermittent spring.*

periodite (pe'-ri-od-ite') A gradually and periodically changing sediment succession, most common in limestone-marl sequences. Periodites occur in environments free of tempestites and turbidites, as on an outer shelf or shallow basin below storm wave base, an isolated platform, or a deeper basin floor beyond the range of turbidites (Einsele and Seilacher, 1982, p.532).

periostracum (per-i-os'-tra-cum) The thin organic layer covering the exterior of the shell of brachiopods and many mollusks, such as the thin coat covering the calcareous part of the shell of a bivalve mollusk, or the outer horny shell layer of a gastropod, composed dominantly of *conchiolin.* Cf: *ostracum.*

peripediment (per-i-ped'-i-ment) A term proposed by Howard (1942, p.11) for the segment of a *pediplane* extending across the younger rocks or alluvium of a basin which is always beyond but adjacent to the segment (termed *pediment* by Howard) developed on the older upland rocks.

peripheral counter (pe-riph'-er-al) In structural petrology, an instrument used to prepare density contours for the marginal areas of a fabric diagram. It consists of a strip of plastic or similar material with a circular hole in either end, the area of each of which is equivalent to 1.0% of the area of the total diagram. Cf: *counter.*

peripheral depression *ring depression.*

peripheral fault An *arcuate fault* bounding an elevated or depressed area such as a *diapir.*

peripheral increase A type of *increase* (offset formation of corallites) characterized by offsets that arise in marginarial or coenenchymal tissue. A number of offsets are usually formed, and increase characteristically involves termination of development of the parietal corallite.

peripheral moraine A term proposed by Chamberlin (1879, p.14) and now considered an obsolete syn. of *recessional moraine.*

peripheral rim A quasi-regularized series of plates in edrioasteroids that form the margin of the oral surface; it includes several circlets of plates, which diminish in size distally (Bell, 1976).

peripheral-rim transition plate One of the plates in the *peripheral rim* of edrioasteroids that separate externally elongate plates concentric with the thecal margin from those that are radial (Bell, 1976).

peripheral sink Locally thickened, synkinematic strata accumulating in a *rim syncline* as a result of *salt withdrawal.* Peripheral sinks may have any shape in plan and are classified according to thickness variations within them (Jackson and Talbot, 1991).

peripheral stream A stream that flows parallel with the edge of a glacier, usually just beyond the moraine (Todd, 1902, p.39).

periphery (pe-riph'-er-y) The part of a gastropod shell or any particular whorl that is farthest from or most lateral to the axis of coiling; also, the outer margin of a coiled foraminiferal test.

periphract (per'-i-phract) A continuous band, composed of muscles and of fibrous tissues (aponeuroses) providing means of linear attachment to the muscles, that encircles the body of a nautiloid (TIP, 1964, pt. K, p.57). Adj: periphractic. Syn: *annulus [paleont].*

periphragm (per'-i-phragm) The outer layer of a dinoflagellate cyst, carrying extensions in the form of surface sculpture, flanges, and processes. The periphragm is not the outermost layer if processes or other positive features on the periphragm merge distally to form an additional layer (Evitt et al., 1977; Tappan, 1980; Edwards, 1993; Williams et al., 2000, p.164). Cf: *ectophragm; endophragm.*

periphyton (per-i-phy'-ton) Micro-organisms, primarily algae and heterotrophic microbes, that coat rocks, plants, and other surfaces on the water bottom. Cf: *aufwuchs.*

peripolar space (per-i-po'-lar) A three-sided pyramidal space resulting from formation of *basal leaf cross* in an acantharian radiolarian. Cf: *perizonal space.*

periporate (per-i-po'-rate) *pantoporate.*

periproct (per'-i-proct) The membranous or irregularly polyplated area surrounding the anal opening of an echinoderm, such as the space in the CD interray of an echinoid containing the anus and covered in life by skin in which small plates are embedded; or the irregular circlet of plates around the anus of lebetodiscid edrioasteroids. Adj: periproctal.

perirheic zone The portion of a floodplain on which different water types mix, e.g., river water, groundwater, and direct precipitation.

perisome (per'-i-some) The *body wall* of an invertebrate, esp. of an echinoderm. Formal Greek form: perisoma.

perispore (per'-i-spore) Spore term, equivalent to *perine,* but applied to spores.

perisporia Pl. of *perisporium.*

perisporium (per-i-spo'-ri-um) *perispore.* Pl: *perisporia.*

peristerite (pe-ris'-te-rite) A gem variety of albite with blue or bluish-white luster characterized by sharp internal reflections of blue, green, and yellow; an inhomogeneous, unmixed sodic plagioclase with a composition ranging between An_2 and An_{24}. It resembles moonstone, and is falsely called *moonstone* by jewelers.

Peristerite intergrowth Plagioclase feldspar consisting of alternating exsolution lamellae of An_{-2} and An_{-24}, where An_x represents percent anorthite composition ($CaAl_2Si_2O_8$) and the implied remainder is albite composition ($NaAlSi_3O_8$). Cf: *Bøggild intergrowth; Huttenlocher intergrowth.*

peristome (per'-i-stome) (a) The region around the mouth in various invertebrates; e.g. the space containing the oral plates at the summit of the theca of a blastoid, the edge of the aperture of the body chamber of a cephalopod, the frontal depression above the mouth of a tintinnid, or the raised rim around the aperture of a foraminiferal test. (b) An elevated rim surrounding the primary orifice in a cheilostome bryozoan, or the raised terminal portion of the zooid in a stenolaemate bryozoan. Syn: *peristomie.*

peristomial (per-i-stom'-i-al) Pertaining to the *peristome;* e.g. "peristomial ovicell" formed as a dilatation of the peristome of a bryozoan.

peristomice (per-i-sto'-mice) *secondary orifice.*

peristomie (per-i-sto'-mie) *peristome.*

peritabular (per-i-tab'-u-lar) Obsolete syn. of *penitabular* (Williams et al., 2000, p.165).

perite (per'-ite) An adamantine yellow orthorhombic mineral: $PbBiO_2Cl$.

peritectic point (per-i-tec'-tic) *reaction point.*

perithecium (per-i-the'-ci-um) A rounded, oval, pyriform (pear-shaped) or beaked *ascocarp* that opens by a pore or slit and in which *asci* are borne.

peritidal (per-i-tid'-al) Referring to depositional environments in a zone from somewhat above highest storm or spring tides to somewhat below lowest tides; a broader term than "intertidal" (Folk, 1973).

peritreme (per'-i-treme) The edge or margin of the aperture of a shell; e.g. the peristome of a gastropod.

peritrichous (pe-rit'-ri-chous) Said of a bacterial cell having flagella uniformly distributed over all the surface. Cf: *lophotrichous; monotrichous.*

perizonal space (per-i-zon'-al) A four-sided region resulting from the formation of *basal leaf cross* in an acantharian radiolarian. Cf: *peripolar space.*

perizonium (per-i-zo'-ni-um) An outer, silicified membrane on a diatom frustule, formed during auxospore development, from which the new hypovalve and epivalve are produced.

perknite (perk'-nite) Any of a group of igneous rocks containing as their main constituents clinopyroxene and amphibole, with accessory orthopyroxene, biotite, iron oxides, and little or no feldspar. Included in this group are pyroxenite and hornblendite. Obsolete.

perlialite (per'-li-a-lite') A pearly white hexagonal zeolite mineral: $K_9Na(Ca,Sr)Al_{12}Si_{24}O_{72}\cdot15H_2O$. The X-ray diffraction pattern resembles those of the synthetic zeolites L and K,Ba-G.

perlite (per'-lite) A volcanic glass having the composition of *rhyolite,* a perlitic texture, and a generally higher water content than *obsidian.* Syn: *pearlstone.*

perlitic (per-lit'-ic) Said of the texture of glassy volcanic rocks characterized by numerous curving cracks roughly concentric around closely spaced centers.

perloffite (per'-loff-ite) A sub-adamantine dark brown, greenish-brown, or black monoclinic mineral of the *bjarebyite* group: $Ba(Mn^{2+},Fe^{2+})_2Fe^{3+}{}_2(PO_4)_3(OH)_3$.

permafrost (per'-ma-frost) Any soil, subsoil, or other surficial deposit, or even bedrock, occurring in arctic, subarctic, and alpine regions at a variable depth beneath the Earth's surface in which a temperature below freezing has existed continuously for a long time (from two years to tens of thousands of years). This definition is based exclusively on temperature, and disregards the texture, degree of compaction, water content, and lithologic character of the material. The thickness of permafrost ranges from over 1,000 m in the north to 30 cm in the south of the northern permafrost region; it underlies about one-fifth of the world's land area. Etymol: permanent + frost. See also: *tjaele.* Cf: *pereletok.* Term introduced by Muller (1947) who included as synonyms the terms "frozen ground" or "frozen soil" preceded by any of the following modifiers: "constantly", "eternally", "ever", "perennially", "permanently", "perpetually", and "stable". Syn: *pergelisol; perennially frozen ground.*

permafrost base The lower boundary of permafrost, above which the temperature is perennially below 0°C, and below which the termperature is perennially above 0°C.

permafrost island A small, shallow, isolated patch of permafrost surrounded by unfrozen ground; occurs on protected north-facing slopes in regions of *sporadic permafrost.*

permafrost line A line on a map representing the geographic limits of permafrost.

permafrost table The upper limit of permafrost, represented by an irregular surface dependent on local factors. Cf: *frost table; frost line.* Syn: *pergelisol table.*

permanence of continents (per'-ma-nence) A hypothesis, propounded as a virtual dogma by many North American geologists during the 19th and early part of the 20th centuries, that the continents (and by implication the intervening ocean basins) have been fixed in their present positions throughout geologic time. Even these geologists found it necessary, in order to explain at least some of the intercontinental resemblances, to resort to borderlands, isthmian links, and other supposed land features, now foundered beneath the oceans. The hypothesis was severely challenged later in the 20th Century by opposing hypotheses, such as *continental drift* and *oceanization,* and is now out of favor in anything like its original form.

permanent axis (per'-ma-nent) The axis of the greatest moment of inertia of a rigid body, about which it can rotate in equilibrium.

permanent bench mark A readily identifiable, relatively permanent, recoverable *bench mark* that is intended to maintain its elevation with reference to an adopted datum without change over a long period of time and located where disturbing influences are believed to be negligible. Abbrev: PBM. Cf: *temporary bench mark.*

permanent charge The net negative (or positive) charge of clay particles inherent in the crystal structure of the particle; not affected by changes in pH or by ion-exchange reactions.

permanent hardness *noncarbonate hardness.*

permanent icefoot An *icefoot [ice]* that does not melt completely during the summer.

permanently frozen ground (per'-ma-nent-ly) *permafrost.*

permanent magnet A *magnet* having a large, hard remanent magnetization.

permanent set The amount of permanent deformation of a material that has been stressed beyond its elastic limit. Syn: *set [exp struc geol].*

permanent stream A syn. of *perennial stream.* The term should be avoided because a stream is not permanent (unchanging) in course, volume, or velocity.

permanent water A source of water that remains constant throughout the year.

permanent wilting A degree of wilting from which a plant can recover only if water is added to the soil. Cf: *wilting point; temporary wilting.*

permanganogrunerite A monoclinic member of the *amphibole* group. It is a member of the cummingtonite-grunerite series with 3-5 Mn atoms per formula unit.

permeability [geol] (per'-me-a-bil'-i-ty) The property or capacity of a porous rock, sediment, or soil for transmitting a fluid; it is a measure of the relative ease of fluid flow under unequal pressure and is a function only of the medium. The customary unit of measurement is the *millidarcy.* Cf: *absolute permeability; effective permeability; relative permeability; hydraulic conductivity.* Syn: *perviousness.* Adj: permeable. Ant: *impermeability.*

permeability [magnet] The ratio of magnetic induction B to inducing field strength H. Syn: *magnetic permeability.*

permeability coefficient The rate of flow of water in gallons per day through a cross section of one square foot under a unit hydraulic gradient, and adjusted for a temperature of 60°F (Stearns, 1927, p.148). It is a function of the properties of the porous meedium and is independent of the properties of the liquid. Cf: *capillary conductivity; hydraulic conductivity; intrinsic permeability.* Syn: *coefficient of permeability; meinzer.*

permeameter (per-me-am'-e-ter) An instrument for measuring permeability.

permeation [grd wat] (per-me-a'-tion) Penetration by passing through the interstices, as of a rock or soil, without causing physical change.

permeation [meta] The intimate penetration of country rock by metamorphic solutions, particularly of an already metamorphosed rock so that it becomes more or less completely recrystallized.

permeation gneiss A gneiss formed as a result of or modified by the passage of geochemically mobile materials through or into solid rock. Rarely used.

permeation grout (per-me-a'-tion) Grout injected into a porous soil or rock without disturbing the original structure. Typically, the grout is a water-based slurry of cement, fly ash, lime, or other finely ground solids that hardens with time. Permeation grout that is used to fill pores and joints in soil and rock has an effective particle diameter of approximately 20% of the pore or joint aperture.

permesothyridid (per'-mes-o-thy'-ri-did) Said of a brachiopod *pedicle foramen* located mostly within the ventral umbo (TIP, 1965, pt. H, p.150). Cf: *submesothyridid.*

Permian (Per'-mi-an) The uppermost system of the *Paleozoic* Erathem of the Standard Global Chronostratigraphic Scale, above the *Carboniferous* and below the *Triassic* System of the *Mesozoic* Erathem. Also the time during which these rocks were formed, the Permian Period, covering the time span between 295 and 250 Ma. It is named after the province of Perm, Russia, where rocks of this age were first studied.

per mille Per thousand.

permineralization (per'-min-er-al-i-za'-tion) A process of fossilization whereby the original hard parts of an animal have additional mineral material deposited in their pore spaces.

permingeatite (per-min-geat'-ite) A metallic brownish-pink tetragonal mineral: Cu_3SbSe_4. It is the Se analogue of famatinite.

permissive area A contiguous area with geologic characteristics permissive of the presence and future discovery of mineral deposits of a particular type.

permissive intrusion (per-mis'-sive) Passive intrusion. Obsolete. Cf: *forceful intrusion.*

permutation (per-mu-ta'-tion) In *geostatistics,* any differently ordered subset, or arrangement, of a given set of objects. Cf: *combination.*

perofskite (pe-rof'-skite) *perovskite.*

peroikic (per-oi'-kic) In the *CIPW classification* of igneous rocks, those rocks in which the ratio of oikocrysts to chadacrysts is greater than seven to one. Obsolete. Cf: *domoikic.*

perovskite (pe-rov'-skite) (a) A yellow, brown, or grayish-black cubic mineral: $CaTiO_3$. It sometimes has cerium and other rare-earth elements. Cf: *latrappite.* Also spelled: *perofskite.* (b) A group name for cubic minerals with an analogous composition, but with Ca replaced by Na, and Ti replaced by Nb.

Perox Oxisols that have a perudic soil moisture regime. A suborder in the U.S. system of soil taxonomy.

perpatic (per-pat'-ic) In the *CIPW classification* of igneous rocks, those rocks in which the ratio of groundmass to phenocrysts is greater than seven to one. Obsolete. Cf: *dopatic.*

perpendicular (per-pen-dic'-u-lar) n. A very steep slope or precipitous face, as on a mountain.

perpendicular offset In seismic surveying, the perpendicular distance from a source to the spread line. Cf: *in-line offset.*

perpendicular separation The separation of a fault as measured at right angles to the fault plane.

perpendicular slip The component of the slip of a fault that is measured perpendicular to the trace of the fault on any intersecting surface (Dennis, 1967, p.138).

perpendicular throw In a faulted bed, vein, or other planar feature, the distance between two formerly adjacent points, measured perpendicular to the surface. Obsolete.

perpetual frost climate (per-pet'-u-al) A type of *polar climate* having a mean temperature in the warmest month of less than 0°C. This zone coincides with the polar ice caps where ablation does not exceed accumulation. Cf: *tundra climate.*

perpetually frozen ground (per-pet'-u-al-ly) *permafrost.*

perpetual spouter A hot spring resembling a geyser but whose eruptive activity is continuous (Bryan, 2001, p.449).

perpotassic rock A rare igneous rock in which the molar proportion of K_2O exceeds that of Al_2O_3 (Johannsen, 1939). Cf: *potassic rock; ultrapotassic rock.* Obsolete.

perradial (per-ra'-di-al) Having a meridional position; e.g. a "perradial suture" situated at the midline between two columns of an ambulacrum in an echinoid, a "perradial position" precisely along one of the radii of a crinoid, or a "perradial plane" occupying a meridional position in acantharian radiolaria.

perradial line The junction in edrioasteroids between closed opposing members of cover-plate pairs or series; it extends approximately along the midline of each ambulacrum (Bell, 1976).

perraultite An orange-brown monoclinic mineral: $(Na,Ca)_2(Ba,K)_2(Mn^{2+})_8(Ti,Nb)_4O_4(Si_2O_7)_4(OH,F)_6$.

Perret phase (Per-ret') That stage of a volcanic eruption characterized by the emission of much high-energy gas that may significantly enlarge the volcanic conduit.

perrierite-(Ce) (per'-ri-er-ite') A black, brownish-black, or brownish-red monoclinic mineral of the *chevkinite* group: $(Ce,La,Ca)_4(Fe^{2+},Mg)_2(Ti,Fe^{3+})_3Si_4O_{22}$.

perrierite-(La) A vitreous black monoclinic mineral of the *chevkinite* group: $(La,Ce,Ca,Th)_4(Fe^{2+},Mg)_2(Fe^{3+},Ti)_3Si_4O_{22}$.

perroudite (per-roud'-ite) An adamantine red orthorhombic mineral: $Hg^{2+}{}_4Ag_4S_5(Cl,I,Br)_4$.

perryite (per'-ry-ite) A metallic rhombohedral meteorite mineral: $(Ni,Fe)_8(Si,P)_3$.

persalic (per-sal'-ic) One of five classes in the *CIPW classification* of igneous rocks, in which the ratio of salic to femic minerals is greater than seven to one. Obsolete. Cf: *dosalic.*

persemic (per-sem'-ic) In the *CIPW classification* of igneous rocks, those rocks in which the ratio of groundmass to phenocrysts is less than one to seven. Rarely used. Cf: *dosemic.*

persilicic (per-si-lic'-ic) An obsolete term proposed by Clarke (1908, p.357) to replace "acidic". Syn: *silicic.* Cf: *subsilicic; mediosilicic.*

personal error (per'-son-al) An *error,* either random or systematic, caused by an observer's personal habits in making observations, by his mental or physical reactions, or by his inability to perceive or measure dimensional values exactly.

perspective (per-spec'-tive) The appearance to the eye of objects in respect to their relative distance and positions. Also, a picture (or other representation) in perspective.

perspective center The point of origin or termination of bundles of rays directed to a point object such as a photographic image.

perspective plane Any plane containing the perspective center; its intersection with the ground always appears as a straight line on an aerial photograph.

perspective projection A projection of points by straight lines drawn through them from some given point to an intersection with the plane of projection; e.g. a photograph is formed by a perspective projection of light rays from a point within the lens. The point of projection (unless otherwise indicated) is understood to be within a finite distance from the plane of projection. Examples include: stereographic projection, orthographic projection, and gnomonic projection. Syn: *geometric projection.*

perthite (perth'-ite) A variety of alkali feldspar consisting of parallel or subparallel intergrowths in which the potassium-rich phase (usually microcline) appears to be the host from which the sodium-rich phase (usually albite) exsolved. The exsolved areas are visible to the naked eye, and typically form strings, lamellae, blebs, films, or irregular veinlets. Cf: *antiperthite; mesoperthite.*

perthitic (per-thit'-ic) An igneous texture consisting of parallel or subparallel intergrowths of sodic plagioclase (typically albite) occurring as small stringers or irrgular veinlets in potassium-rich feldspar (typically microcline). Cf: *perthitoid.*

perthitoid (per'-thi-toid) A perthitelike texture produced by minerals other than the feldspars. Obsolete. Cf: *perthitic.*

perthophyte (per'-tho-phyte) A plant living on a dead plant or on the decaying portions of a live plant.

perthosite (per'-tho-site) A light-colored *syenite* composed almost entirely of *perthite,* with less than three percent mafic minerals. The name was introduced by Phemister in 1926. Obsolete.

pertsevite A colorless orthorhombic mineral: Mg_2BO_3F.

perudic A udic soil moisture regime in which water moves through the soil in all months when it is not frozen.

Peru saltpeter (Pe-ru') Naturally occurring sodium nitrate; *soda niter* occurring in Peru. Cf: *saltpeter.*

pervalvar axis (per-val'-var) The axis connecting the midpoints of the two valves in a diatom frustule. Cf: *apical axis.*

pervasiveness (per-va'-sive-ness) In quantification of hydrothermal alteration, the degree to which alteration is strictly veinlet-controlled at one extreme and generally distributed without regard to veinlets at the other. Cf: *extensiveness; intensiveness.*

perviousness (per'-vi-ous-ness') *permeability [geol].*

perxenic (per-xen'-ic) In the *CIPW classification* of igneous rocks, those rocks in which the ratio of oikocrysts to chadacrysts is less than one to seven. Obsolete. Cf: *doxenic.*

petal [bot] (pet'-al) A member of the second set of floral leaves, i.e. the set just internal to the sepals (Cronquist, 1961, p.879).

petal [paleont] An expanded, differentiated, petal-shaped segment of the *ambulacrum* situated toward the apical system of an echinoid, characterized by tube feet more or less specialized for respiration and by typically unequal or enlarged pore pairs.

petalite (pet'-al-ite) A white, gray, pink, or green monoclinic mineral: $LiAlSi_4O_{10}$.

petaloid process (pet'-al-oid) An *acritarch process* that has a characteristic, broad petal or leaf shape in lateral view and consists of three laminae, e.g., *Cyclosphaeridium* (Loeblich and Tappan, 1969).

petarasite (pe-tar'-a-site) A monoclinic mineral in shades of yellow, green, or brown, with zeolite-like physical properties: $Na_5Zr_2Si_6O_{18}(OH,Cl){\cdot}2H_2O$.

petedunnite (pete-dunn'-ite) A dark green monoclinic mineral of the *pyroxene* group: $Ca(Zn,Mn^{2+},Fe^{2+},Mg)Si_2O_6$.

peteinoid process (pet'-ei-noid) One of discrete solid *processes* linking a cohesive group of *acritarchs* (e.g., *Peteinosphaeridium, Liliosphaeridium*) and having longitudinally and radially oriented bladelike elements (Playford et al., 1995). Etymol: Greek "peteinos," winged.

peterbaylissite A black or dark red-brown orthorhombic mineral: $Hg_3CO_3(OH){\cdot}2H_2O$.

petersenite-(Ce) A vitreous gray monoclinic mineral: $Na_4(Ce,La,Nd)_2(CO_3)_5$.

petersite-(Y) (pe'-ters-ite) A yellow-green hexagonal mineral of the *mixite* group: $(Y,Ca)Cu_6(OH)_6(PO_4)_3{\cdot}3H_2O$. It is related to mixite.

Peters' length (Pe'-ters) A horizontal distance between the points on the side of the potential-field (usually, magnetic) anomaly profile, where the slope line is drawn at half of the maximum slope line. This measurement is made with the objective of determining the depth of an anomalous mass (magnetized body). For tabular body or dike models, the Peters' rule gives a depth as $^5/_8$ of the Peters' length (Peters, 1949).

Peterson grab (Pe'-ter-son) *Van Veen grab sampler.*

petiole (pet'-i-ole) The stalk of a leaf.

petitjeanite A white, yellow, pale pink, or dark brown triclinic mineral of the *preisingerite* group: $Bi_3O(PO_4)_2(OH)$.

Petoskey stone (Pe-tos'-key) A waterworn fragment of Devonian colonial coral from the beach of Lake Michigan at Petoskey, Mich. It is the "state rock" of Michigan.

petraliiform (pe-tra'-lii-form) Said of a unilamellar encrusting colony in cheilostome bryozoans, loosely attached by protuberances from basal walls of zooids or by radicles.

petricole (pet'-ri-cole) *petrocole.*

petricolous (pe-tric'-o-lous) *endolithic.*

petrifaction (pet-ri-fac'-tion) A process of fossilization whereby organic matter is converted into a stony substance by the infiltration of water containing dissolved inorganic matter (e.g. calcium carbonate, silica) which replaces the original organic materials, sometimes retaining the structure. Syn: *petrification.*

petrification (pet'-ri-fi-ca'-tion) *petrifaction.*

petrified moss (pet'-ri-fied) A moss-like coating of *tufa* deposited on growing plants.

petrified rose *barite rosette.*

petrified wood Wood that has become lignified by impregnation with mineral matter, most often silica. Cf: *silicified wood.*

petro- (pet'-ro-) A prefix meaning "rock".

petroblastesis (pet-ro-blas'-te-sis) Formation of rocks chiefly as the result of crystallization of diffusing ions (Dietrich and Mehnert, 1961). Rarely used.

petrocalcic horizon (pet-ro-cal'-cic) A continuous, indurated calcic horizon that is cemented by calcium carbonate and, in some places, with magnesium carbonate. It cannot be penetrated with a spade or auger when dry, dry fragments do not slake in water, and it is impenetrable to roots.

petrochemistry (pet-ro-chem'-is-try) (a) The study of the chemical composition of rocks. (b) The study of the chemistry of petroleum and its products.

petroclastic rock (pet-ro-clas'-tic) *detrital rock.*

petrocole (pet'-ro-cole) An organism that lives in rocky areas. Also spelled: *petricole.*

petrofabric (pet-ro-fab'-ric) *fabric [struc geol].*

petrofabric analysis An equivalent term for *structural petrology*, used by Knopf and Ingerson (1938, p.13).

petrofabric diagram *fabric diagram.*

petrofabrics (pet-ro-fab'-rics) *structural petrology.*

petrofacies (pet-ro-fa'-cies) *petrographic facies.*

petroferric contact A boundary between soil and a continuous layer of indurated soil in which iron is an important cement. Contains little or no organic matter.

petrofilaments (pet'-ro-fil-a-ment) Artifacts created by the physico-chemical reaction between in situ bitumen and a specific combination of palynological embedding and mounting media. In palynological preparations they appear as angular to subangular pieces of solid hydrocarbon (usually between 20 and 30 µm in diameter) with protruding, corkscrew-shaped filaments (30-100 µm in length). The presence of these artifacts can be used as a "hydrocarbon show indicator," which can be used to identify possible carrier beds (Graham et al, 2000; Wood et al., 2002).

petrogenesis (pet-ro-gen'-e-sis) A branch of petrology that deals with the origin and formation of rocks. Cf: *lithogenesis.* Adj: petrogenetic; petrogenic. Syn: *petrogeny.*

petrogenetic grid (pet'-ro-ge-net'-ic) A diagram whose coordinates are intensive parameters characterizing the rock-forming environment (e.g. pressure, temperature) on which may be plotted equilibrium curves delimiting the stability fields of specific minerals and mineral assemblages.

petrogenic element (pet-ro-gen'-ic) An element that occurs mainly as an oxide, silicate, fluoride, or chloride, and is therefore a characteristic occurrence in ordinary rocks. It is one of H.S. Washington's two major groupings of elements of the lithosphere, now obsolete. Cf: *metallogenic element.*

petrogeny (pe-trog'-e-ny) *petrogenesis.*

petrogeny's primitive system The system in whose constituents the residual liquids produced by crystallization differentiation are invariably depleted: $CaAl_2Si_2O_8$-$CaMgSi_2O_6$-Mg_2SiO_4. (Poldervaart and Parker, 1964). Cf: *crystallization index; petrogeny's residua system.*

petrogeny's residua system The system $NaAlSiO_4$ - $KAlSiO_4$ - SiO_2, whose constituents are invariably concentrated in the final residual melts produced by crystallization differentiation. It includes as phases quartz, albite, the potassium feldspars, leucite, nepheline, and kaliophilite (kalsilite). The term was first used by N. L. Bowen. Cf: *petrogeny's primitive system; differentiation index.*

petroglyph (pet'-ro-glyph) Literally, a rock carving; it usually excludes writing and therefore is of prehistoric or protohistoric age.

petrographer (pe-trog'-ra-pher) One versed in the science of petrography.

petrographic (pet-ro-graph'-ic) Adj. of *petrography.*

petrographic facies Facies distinguished primarily on the basis of appearance or composition without respect to form, boundaries, or mutual relations (Weller, 1958, p.627). They consist of actual large bodies of rock occurring in certain areas and in more or less restricted parts of the stratigraphic section (e.g. "red-bed facies", "paralic facies", "geosynclinal facies", "evaporite facies"), or they may consist of all rocks of a single kind (e.g. "black-shale facies", "graywacke facies"). See also: *facies.* Cf: *stratigraphic facies.* Syn: *petrofacies.*

petrographic microscope *polarizing microscope.*

petrographic period The extension in time of a rock association. Obsolete. Cf: *petrographic province.*

petrographic province A broad area in which similar rocks are considered to have been formed during the same period of igneous activity. Cf: *petrographic period.* Syn: *magma province; comagmatic region; igneous province.*

petrography (pe-trog'-ra-phy) That branch of geology dealing with the description and systematic classification of rocks, esp. igneous and metamorphic rocks and esp. by means of microscopic examination of thin sections. Petrography is more restricted in scope than *petrology.* Adj: *petrographic.* See also: *lithology; sedimentary petrography.*

petrogypsic horizon A continuous, strongly cemented, massive, gypsic horizon that is cemented by calcium sulfate. It can be chipped with a spade when dry. Dry fragments do not slake in water and it is impenetrable to roots.

petroleum (pe-tro'-le-um) (a) A naturally occurring complex liquid hydrocarbon, which after distillation and removal of impurities yields a range of combustible fuels, petrochemicals, and lubricants. Syn: *crude oil; oil.* (b) A general term for all naturally occurring hydrocarbons, whether gaseous, liquid, or solid (Levorsen, 1967, p.3).

petroleum coke (a) The residue obtained in the distillation of petroleum. (b) A cokelike substance found in cavities of igneous intrusions into carbonaceous sediments (Tomkeieff, 1954).

petroleum geologist A geologist engaged in exploration for, or production of, oil or gas. See also: *petroleum geology.*

petroleum geology The study of the origin, migration, and accumulation of oil and gas. See also: *petroleum geologist.*

petroleum migration The process by which hydrocarbons travel through rocks to reach a reservoir.

petroliferous (pet-ro-lif'-er-ous) Bearing crude oil or natural gas. The term may be applied to a province, a geologic structure, or a geologic formation or unit.

petrologic (pet-ro-log'-ic) Adj. of *petrology.*

petrologic or petrographic type Chondrites are assigned to textural classifications according to the degree of secondary thermal or aqueous alteration experienced, with 3.0 chondrites showing the least evidence of aqueous or thermal alteration. The type number is listed immediately after the group symbol and just before the name (e.g., LL3.0 Semarkona). The suffixes 2 and 1 denote progressive aqueous alteration, and the suffixes 4, 5, or 6 denote progressive thermal alteration. For example, a CM2 chondrite has experienced mild aqueous alteration, while a CK5 meteorite has experienced significant thermal metamorphism. Cf: *ordinary chondrite.*

petrologic province *sedimentary petrologic province.*

petrologist (pe-trol'-o-gist) One who is engaged in the study of *petrology.*

petrology (pe-trol'-o-gy) That branch of geology dealing with the origin, occurrence, structure, and history of rocks. Petrology is broader in scope than *petrography.* Adj: *petrologic.* See also: *sedimentary petrology.*

petromictic (pet-ro-mic'-tic) Said of a sedimentary deposit characterized by an assortment of metastable rock fragments; e.g. a "petromictic conglomerate" containing a mixture of pebbles or cobbles of plutonic, eruptive, sedimentary, and/or metamorphic rocks. McElroy (1954, p.151) proposed the term "petromictic sandstone" to replace "greywacke" for certain Permian and Triassic sedimentary rocks of New South Wales, having the general granular composition of classical graywackes but being light-colored, well-sorted, and mildly indurated, with a matrix that may consist of an introduced mineral cement. Pettijohn (1975, p.165, Table 6.2) proposed a "petromict conglomerate" to contain >10% metastable clasts and to be composed of clasts of mixed lithology and <15% matrix. Syn: petromict.

petromorph (pet'-ro-morph) A feature, such as a calcite vein, that stands in relief on a cave wall because limestone was dissolved from around it. Etymol: Greek, "rock form".

petrophilous (pe-troph'-i-lous) *epilithic.*

petrophysics (pet-ro-phys'-ics) The study of the physical and chemical properties of rocks which relates to the distribution of the pore system and its contained hydrocarbons and water.

petrosilex [petrology] (pet-ro-si'-lex) (a) An old name for an extremely fine crystalline porphyry or quartz porphyry, and for the groundmass of such porphyries; also, a finely crystalline aggregate now known to be devitrified glass. "It was practically a confession by the older petrographers that they did not know of what the rock consisted" (Kemp, 1896, p.156). (b) A term applied by Lyell (1839, p.99) to igneous rocks. (c) An obsolete term formerly applied to clinkstone, to fusible hornstone, to felsite, and to compact feldspar.

petrous (pet'-rous) Said of a material that resembles stone in its hardness; e.g. petrous phosphates. Little used.

petrovicite (pet'-ro-vic-ite) A metallic orthorhombic mineral: $PbHgCu_3BiSe_5$.

petrovskaite (pet-rov'-ska-ite) A metallic dark gray to black monoclinic mineral: AuAg(S,Se).

petrukite A metallic orthorhombic mineral: $(Cu,Fe,Sn)_3SnS_4$. It is isomorphous with enargite.

Petschau twin law (Pet'-schau) A rare type of parallel twin law in feldspar, having a twin axis of $[\bar{1}10]$.

petschekite (pet'-schek-ite) A submetallic dark brown trigonal mineral: $UFe^{2+}Nb_2O_8$. It is isomorphous with liandratite.

petterdite A pale gray to pinkish-violet orthorhombic mineral: $PbCr_2(CO_3)_2(OH)_4{\cdot}H_2O$. It is the Cr analogue of dundasite.

petuntse (pe-tun'-tse) *china stone [ign].*

petzite (petz'-ite) A steel-gray to iron-black mineral: Ag_3AuTe_2.

pezzottaite A purplish-pink rhombohedral mineral related to beryl: $CsLiBe_2Al_2Si_6O_{18}$.

Pfalzian Orogeny (Pfalz'-i-an) One of the 30 or more short-lived orogenies during Phanerozoic time identified by Stille, in this case at the end of the Permian. Syn: *Palatinian Orogeny.*

p-form A smooth erosional form of *glacier bed* produced by subglacial meltwater during the summer but then commonly striated by glacier ice that extends into the eroded form in winter, when meltwater from the surface no longer reaches the bed. A shortened version of "plastically molded form", a term applied to the feature before its origin was understood. Cf: *current scour.*

PGE *platinum group element.*

pH [chem] The negative $\log_{10}$ of the hydrogen-ion activity in solution; a measure of the acidity or basicity of a solution.

pH [soil] The pH of a solution in equilibrium with soil. It is determined by means of a glass, quinhydrone, or other suitable electrode or indicator at a specified soil-solution ratio in a specified solution, usually distilled water, 0.01 M $CaCl_2$, or 1M KCl.

phacellite (pha-cel'-lite) *kaliophilite.*

phaceloid (phac'-e-loid) Said of a *fasciculate corallum* having subparallel corallites. Also spelled: phacelloid.

phacoidal structure (pha-coid'-al) An infrequently used term for a lenticular metamorphic structure, e.g., *augen structure; flaser structure [meta].*

phacolite (phac'-o-lite) A variety of chabazite, characterized by colorless lenticular crystals.

phacolith (phac'-o-lith) A concordant, lenticular, sill-like intrusion emplaced along a fold axis. Cf: *harpolith.*

phaeodarian (phae-o-da'-ri-an) Any radiolarian belonging to the suborder Phaeodarina, characterized mainly by a central capsule enclosed by a double-walled membrane.

Phaeophyta (Phae-o-phy'-ta) *brown algae.*

phagotrophic (phag-o-troph'-ic) Said of an organism that is nourished by the ingestion of solid matter.

phanerite (phan'-er-ite) An igneous rock having the grains of its essential minerals large enough to be seen macroscopically.

phaneritic (phan-er-it'-ic) Said of the texture of an igneous rock in which the grains are large enough to be distinguished with the unaided eye, i.e. megascopically crystalline. Cf: *aphanitic [ign].*

phanerogam (phan'-er-o-gam') A seed plant, as opposed to a spore-bearing plant or *cryptogam.* Cf: *spermatophyte.*

phanerogenic (phan'-er-o-gen'-ic) Of known origin; e.g. said of a species that is proved to descend from a known species found in an older geologic formation.

phaneromphalous (phan-er-om'-pha-lous) Said of a gastropod shell with a completely open *umbilicus.* Cf: *anomphalous.*

phanerophyte (phan'-er-o-phyte') A perennial plant whose overwintering buds are above the ground surface.

Phanerozoic (Phan'-er-o-zo'-ic) The uppermost eonothem of the Standard Global Chronostratigraphic Scale. It comprises the *Paleozoic, Mesozoic* and *Cenozoic* erathems, which include rocks with abundant evidence of life. Also the time during which these rocks were formed, the Phanerozoic Eon, covering the time span between 540 Ma and the present.

phanoclastic rock (phan-o-clas'-tic) An "even-grained or uniformly sized" clastic rock (Pettijohn, 1949, p.30).

phantom (phan'-tom) (a) A bed or member that is missing from a given stratigraphic section although it elsewhere occupies a characteristic position in a sequence of similar age. Cf: *ghost member.* (b) *ghost [petrology].*

phantom bottom *deep scattering layer.*

phantom crystal A crystal within which an earlier stage of crystallization or growth is outlined by dust, tiny inclusions, mineral coatings, or bubbles, e.g. serpentine containing a ghost or phantom of original olivine. Syn: *ghost crystal.*

phantom horizon A line drawn on seismic sections so that it is parallel to nearby event segments thought to indicate structural attitude. It is used where events are not sufficiently continuous.

pharaonite (phar'-a-o-nite) *davyne.*

pharetrone (phar'-e-trone) Any calcisponge having spicules that have the shape of a tuning fork.

pharmacolite (phar-ma-col-ite') A colorless monoclinic mineral: $Ca(HAsO_4){\cdot}2H_2O$. It occurs in white or grayish silky fibers. Syn: *arsenic bloom.*

pharmacosiderite (phar'-ma-co-sid'-er-ite) (a) A variously colored cubic mineral: $KFe^{3+}_4(AsO_4)_3(OH)_4{\cdot}6\text{-}7H_2O$. It commonly occurs in cubic crystals. Cf: *alumopharmacosiderite; sodium-pharmacosiderite.* Syn: *cube ore.* (b) A group name for minerals with an analogous composition, but with K replaced by Na or Ba, and Fe^{3+} replaced by Al.

pharynx (phar'-ynx) (a) A differentiated part of the alimentary canal in many invertebrates; e.g. the tubular passageway between the mouth and gastrovascular cavity of an octocoral, or the internal oral tube in a phaeodarian radiolarian. (b) In the vertebrates, that part of the alimentary tract that connects the buccal cavity to the esophagus. Its walls support the gills in gilled vertebrates and, although perforated in many lower chordates, the pharyngeal walls may be devoid of gills. In tetrapods, the lateral walls of the pharynx are perforated in the embryo but variously elaborated into ducts and specialized tissues in adults.

phase [chem] A part of a chemical system that is homogeneous, physically distinct and at least hypothetically separable, and which has continuously variable chemical and mechanical properties.

phase [geochron] (a) A term approved by the 8th International Geologic Congress in Paris in 1900 for the geologic-time unit next in order of magnitude below *age,* during which the rocks of a *substage* (then referred to as a "zone") were formed; a *subage.* The term was seldom used and is now obsolete in this usage. Syn: *episode; time (d).* (b) moment.

phase [glac geol] An informal subdivision of a *stage.* The term was by Flint (1957) for the deposits of various glacial Great Lakes at different levels and dates.

phase [ign] An interval in the development of a given process; esp. a chapter in the history of the igneous activity of a region, such as the "volcanic phase" and major and minor "intrusive phases".

phase [phys] (a) A stage in periodic motion (rotation, oscillation, etc.), measured with respect to a given initial point and expressed in angular measure. (b) A peak, trough, zero-crossing, or other identifiable point of a travelling wave. See also: *Airy phase; T phase.*

phase [sed] (a) A product of "deposition during a single fluctuation in the competency of the transporting agent" (Apfel, 1938). Such a subunit is probably a lamina. (b) A transitory or minor fluctuation in the velocity of a depositing current, resulting in the formation of a lamina.

phase [soil] A utilitarian grouping of soils defined by soil or environmental features that are not class differentia used in U.S. system of soil taxonomy, e.g., surface texture, surficial rock fragments, rock outcrops, substratum, special soil water conditions, salinity, physiographic position, erosion, thickness, etc. Phase identifications are introduced into soil names by adding them to a taxon name as modifiers.

phase [stratig] (a) A lithologic facies, esp. on a small scale, such as a minor variety within a dominant or normal facies, or a facies of short duration or local occurrence; e.g. the "marine phase" and the "fluviatile phase" of the Pocono Formation (Barrell, 1913, p.465). The term was used by McKee (1938, p.13-14) for a lateral subdivision (or facies) of a formation. (b) A term defined by Fenton and Fenton (1930, p.150) as "a local or regional aspect or condition of a stratum or group of strata, as determined both by original nature and secondary change; the latter being the determining factor"; e.g., a change arising from faulting, folding, secondary dolomitization, or erosion. (c) Characteristic strata repeated at various positions in a stratigraphic section that record the recurrence of a particular kind of environment (Allan, 1948, p.8). (d) The formal, hierarchical *diachronic unit* next lower in rank than *episode* and higher than span (NACSN, 1983, Art. 93). The term has been used with numerous, diverse, and often vague stratigraphic meanings; best avoided in stratigraphic nomenclature (Weller, 1958, p.619-620, 633).

phase [surv] The apparent displacement of a surveying object or signal caused by one side being more strongly illuminated than the other and resulting in an error in sighting.
phase boundary *boundary line [geochem].*
phase diagram A graphical representation of phase equilibria that is designed to show stability relations of the various phases of the system. Plotting coordinates are commonly two or more of the intensive variables (e.g., pressure, temperature, composition) but may also include extensive parameters such as entropy or volume for specific representational purposes. Syn: *equilibrium diagram.*
phase equilibria In physical chemistry, the study of those phases which, under specified conditions, may exist in equilibrium in a given system.
phase layering (a) The mineralogical layering of plutonic rocks caused by the generally sudden appearance and progressive disappearance (from bottom to top) of some mineral phase. Cf: *cryptic layering; rhythmic layering.* (b) A large-scale structure in metamorphic rocks produced by subparallel layers of various origins that are mineralogically distinct from the host rocks such as gneisses and peridotites. Syn: *compositional layering; primary layering.*
phase microscope A microscope that utilizes the phase differences of light rays transmitted by different portions of an object to create an image in which the details of the object are distinct, despite their near uniformity of refractive index.
phase plate In a polarizing microscope, a plate of doubly refracting material, e.g. mica of a *quarter-wave plate,* that changes the relative phase of the polarized light's components.
phase response *phase spectrum.*
phase rule The statement that, for any system in equilibrium, the number of *degrees of freedom,* or *variance [exp petrol]* (independently variable intensive parameters) is two greater than the difference between the number of components and the number of phases. It is symbolically stated as: F (or V) = (C-P) + 2. Syn: *Gibbs phase rule.* See also: *mineralogical phase rule.*
phase-shift migration A method of migration in the frequency-domain by phase-shifting at each step in downward continuation. The process can be used to discriminate against dipping coherent noise (Sheriff and Geldart, 1995). Syn: *Gazdag migration.*
phase spectrum The variation of phase with frequency (or period), such as produced by passage through a filter. Syn: *phase response.*
phase transformation The change of a crystal from one polymorph to another (i.e., a change in *crystal structure* not accompanied by a change in composition), usually driven by changes in external conditions such as temperature or pressure. Syn: *phase transition; inversion [cryst].*
phase transition *phase transformation.*
phase velocity The velocity with which an observable, individual wave or wave crest is propagated through a medium; the velocity of a point of constant phase. It is the product of wavelength and frequency. Symbol: c. Cf: *group velocity; particle velocity.* The reciprocal of phase velocity is *slowness.*
phasor A rotating line used to represent a sinusoidally varying quantity. The length of the line represents the magnitude and the angle with the x-axis at any instant represents the phase. Used for quantities that are complex numbers. Master curves for the electromagnetic method are often presented as phasor diagrams.
phassachate (phas'-sa-chate) A lead-colored agate.
phaunouxite (phau-noux'-ite) A colorless triclinic mineral: $Ca_3(AsO_4)_2 \cdot 11H_2O$. It dehydrates to rauenthalite at room temperature.
phelloderm (phel'-lo-derm) Secondary tissue of woody plant stems, produced by the cork cambium on its inner surface, as opposed to the cork, which is produced by the cork cambium on its outer surface (Cronquist, 1961, p.879). See also: *periderm [bot].*
phellogen (phel'-lo-gen) *cork cambium.*
phenacite (phen'-a-cite) *phenakite.*
phenakite (phen'-a-kite) (a) A colorless, white, or pale varicolored glassy rhombohedral mineral: Be_2SiO_4. It is sometimes confused with quartz. Phenakite is used as a minor gemstone. Not to be confused with *fenaksite.* Also spelled: *phenacite.* (b) A group name for minerals with the analogous composition, but with Be replaced by Zn or (LiAl).
phenanthrene *ravatite.*
phenetics A set of numerical techniques that group taxa based on overall morphologic similarity. Although common in the 1960s and 70s, it is now rarely used. Syn: *numerical taxonomy.*
phenetic system (phe-net'-ic) A classification system for organisms based on degrees of overall similarity, using morphologic, anatomic, physiologic, or any other available criteria. By definition the system does not have evolutionary implications.
phengite (phen'-gite) (a) A series name for dioctahedral micas of composition $K(AlMg)_2(OH)_2(SiAl)_4O_{10}$. (b) A transparent or translucent stone (probably crystallized gypsum) used by the ancients for windows.
phenhydrous (phen-hy'-drous) Said of vegetal matter deposited under water.
phenicochroite (phe'-ni-co-chro'-ite) *phoenicochroite.*
phenoclast (phe'-no-clast) One of the larger and more conspicuous fragments in a sediment or sedimentary rock composed of various sizes of material, such as a cobble or pebble (*spheroclast*) embedded in a fine-textured matrix of a conglomerate, or a fragment (*anguclast*) of a breccia.
phenoclastic rock (phe'-no-clas'-tic) A "nonuniformly sized" clastic rock containing phenoclasts (Pettijohn, 1949, p.30).
phenocryst (phe'-no-cryst) A term suggested by J.P. Iddings, and widely used, for a relatively large, conspicuous crystal in a porphyritic rock. The term *inset [petrology]* has been suggested as an alternative.
phenogenesis (phe-no-gen'-e-sis) The development of the *phenotype.*
phenogram A *dendrogram* expressing the hierarchical relationships among a group of organisms, as based on observed morphologic similarities alone rather than inferred phylogenetic relationships. Cf: *cladogram.*
phenomenal gem (phe-nom'-e-nal) A gemstone exhibiting an optical phenomenon, such as asterism, chatoyancy, or play of color.
phenomenology (phe-nom'-e-nol'-o-gy) The science that treats of the description and classification of phenomena.
phenoplast (phe'-no-plast) A large rock fragment (in a *rudaceous* rock) that was plastic at the time of its incorporation in the matrix.
phenotype (phe'-no-type) The visible characters of an organism that reflect the interaction of *genotype* and environment.
phi A logarithmic transformation of the ratio of a grain diameter in mm to a standard grain diameter of 1 mm. It is not a dimension but a ratio. Symbol: ϕ.
phialine (phi'-a-line) Said of an everted apertural rim (as on the neck of a vial or bottle) of some foraminiferal tests.
phi deviation measure A graphic measure of dispersion of particle size shown on a plot of phi units.
phi grade scale A logarithmic transformation of the *Wentworth grade scale* in which the negative logarithm to the base 2 of the particle diameter (in millimeters) is substituted for the diameter value (Krumbein, 1934); it has integers for the class limits, increasing from -5 for 32 mm to +10 for $^1/_{1024}$ mm. The scale was developed specifically as a statistical device to permit the direct application of conventional statistical practices to sedimentary data. Syn: *phi scale.*
Philadelphia leveling rod A two-piece leveling rod having a target but with graduations so styled that the rod may also be used as a self-reading leveling rod. Also called a Philadelphia rod. If a length greater than 7 ft is needed, the target is clamped at 7 ft and raised by extending the rod. When the target is used, the rod is read by vernier to 0.001 ft. When the rod is used as a self-reading leveling rod, the rod is read to 0.005 ft. In practice, the target is seldom used.
philippinite (phil'-ip-pin-ite') An *Australasian tektite* from the Philippine Islands. See also: *rizalite.*
philipsbornite (phil-ips-born'-ite) A grayish-green rhombohedral mineral of the *crandallite* group: $PbAl_3(AsO_4)_2(OH)_5 \cdot H_2O$.
phillipsburgite (phil'-lips-burg-ite) A green monoclinic mineral: $(Cu,Zn)_6(OH)_6[(As,P)O_4]_2 \cdot H_2O$. It is the arsenate analogue of kipushite.
phillipsite-Ca A colorless or white monoclinic zeolite mineral with Ca as the dominant exchangeable cation: $(Ca,K,Na)_2(Si,Al)_8O_{16} \cdot 6H_2O$.
phillipsite-K A white or reddish monoclinic zeolite mineral with K as the dominant exchangeable cation: $(K,Na,Ca)_{1\text{-}2}(Si,Al)_8O_{16} \cdot 6H_2O$. It sometimes contains no sodium, but always contains considerable potassium. It commonly occurs in complex (often cruciform) fibrous crystals, and makes up an appreciable part of the red-clay sediments in the Pacific Ocean.
phillipsite-Na A colorless or white monoclinic zeolite mineral with Na as the dominant exchangeable cation: $(Na,Ca,K)_2(Si,Al)_8O_{16} \cdot 6H_2O$.
philolithite A vitreous green tetragonal mineral: $Pb_{12}O_6(Mn,Mg)_7(SO_4)(CO_3)_4Cl_4(OH)_{12}$.
phi mean diameter A *logarithmic mean diameter* obtained by using the negative logarithms of the class midpoints to the base 2. See also: *geometric mean diameter.*
phi scale *phi grade scale.*
phi standard deviation The *standard deviation* of a grain-size distribution expressed in phi units. It can be determined by calculation or graphically. Syn: *phi deviation measure.*
phi unit The unit interval on the curve of values in phi transformation. It should be used in the same context as intervals on the *Wentworth grade scale* (McManus, 1962, p.673).
phlebite (phle'-bite) A metamorphic rock or *migmatite* with roughly banded or veined appearance (Dietrich and Mehnert, 1961). Originally proposed, without genetic connotation, to replace the term *veined gneiss* (Mehnert, 1968, p.17). Cf: *venite; arterite; composite gneiss; injection gneiss; diadysite.* Rarely used.
Phleger corer (Phleg'-er) A type of *gravity corer* that has a check valve and/or core catcher to retain the sample. Named for Fred Phleger.
phloem The food-conducting tissue of vascular plants, consisting of various types of cells, such as sieve tubes, phloem parenchyma, fibers, and companion cells.
phlogopite (phlog'-o-pite) A magnesium-rich mineral of the *mica* group: $KMg_3Si_3AlO_{10}(F,OH)_2$. It is yellowish brown to brownish red or copper-colored, and usually occurs in crystalline limestones as a result of dedolomitization. It crystallizes in monoclinic, trigonal and orthorhombic polytypes. Phlogopite is near biotite in composition, but contains little iron. Cf: *sodium phlogopite.* Syn: *magnesia mica; amber mica; brown mica.*

phobotaxis (pho-bo-tax'-is) *Taxis [ecol]* in which an organism avoids a concentration or intensity of something. Cf: *strophotaxis; thigmotaxis.*

phoenicochroite (phoe'-ni-co-chro'-ite) A dark red monoclinic mineral: $Pb_2O(CrO_4)$. Also spelled: *phenicochroite.* Syn: *berezovite; scheibeite.*

pholad (pho'-lad) Any bivalve mollusk belonging to the family Pholadidae, characterized by an equivalve shell of variable size, commonly gaping open at the posterior end.

pholerite (pho'-ler-ite) A discredited name for a clay mineral identical with kaolinite or one of the other kaolin minerals.

phonolite (pho'-no-lite) (a) In the *IUGS classification,* a volcanic rock defined in the QAPF diagram by F/(F+A+P) between 10 and 60 and P/(P+A)<10, and in the TAS diagram by a field open at the top but defined in its lower boundaries by points with the SiO_2 and total alkali coordinates: 52.5, 14; 57.6, 11.7; 61, 13.5. (b) In the strictest sense, a group of fine-grained extrusive rocks primarily composed of alkali feldspar (esp. anorthoclase or sanidine), and with nepheline as the main feldspathoid (Streckeisen, 1967, p.185); also, any rock in that group; the extrusive equivalent of *nepheline syenite.* (c) In the broadest sense, any extrusive rock composed of alkali feldspar, mafic minerals and any feldspathoid, such as nepheline, leucite, or sodalite. Syn (all obsolete): *clinkstone; klingstein.* Streckeisen (1967, p.186) suggests that "phonolite" be preceded by the name of the main feldspathoid mineral (e.g. "leucite phonolite", "analcime phonolite", etc.). Etymol: Greek "phone", "sound", in reference to the allegedly characteristic ringing sound emitted by a phonolite when struck with a hammer.

phonolitic basanite In the *IUGS classification,* a volcanic rock with F between 10 and 60, P/(P+A) between 50 and 90, and modal olivine exceeding 10%.

phonolitic foidite In the *IUGS classification,* a volcanic rock with F between 60 and 90, and P/(P+A)<50. Preferably, the root "foidite" should be replaced to indicate the most abundant feldspathoid, as in phonolitic nephelinite or phonolitic leucitite.

phonolitic tephrite In the *IUGS classification,* a volcanic rock with F between 10 and 60, P/(P+A) between 50 and 90, and modal olivine less than 10 percent.

phonolitoid In the *IUGS classification,* a general term in the provisional "field" classification for a volcanic rock thought to contain feldspathoid and in which alkali feldspar is estimated to exceed plagioclase in modal proportions. Obsolete.

phorogenesis (phor-o-gen'-e-sis) A raely used term for the shifting or slipping of the Earth's crust relative to the mantle.

phosgenite (phos'-gen-ite) A white, yellow, brown, or grayish tetragonal mineral with adamantine luster: $Pb_2(CO_3)Cl_2$. Syn: *horn lead; cromfordite.*

phosinaite-(Ce) (pho-sin'-a-ite) A pale rose or brownish-rose orthorhombic mineral: $Na_{13}(Ca,Mn)_2(Ce,La)(SiO_3)_4(PO_4)_4$.

phosphammite (phos-pham'-mite) A colorless monoclinic mineral: $(NH_4)_2HPO_4$.

phosphate (phos'-phate) A mineral compound containing tetrahedral PO_4^{-3} groups. An example is fluorapatite, $Ca_5(PO_4)_3F$. Cf: *arsenate; vanadate.*

phosphate rock Any rock that contains one or more phosphatic minerals of sufficient purity and quantity to permit its commercial use as a source of phosphatic compounds or elemental phosphorus. About 90% of the world's production is sedimentary phosphate rock, or *phosphorite;* the remainder is igneous rock rich in apatite.

phosphatic (phos-phat'-ic) Pertaining to or containing phosphates or phosphoric acid; said esp. of a sedimentary rock containing phosphate minerals, such as a "phosphatic limestone" produced by secondary enrichment of phosphatic material, or a "phosphatic shale" representing mixtures of primary or secondary phosphate and clay minerals.

phosphatic nodule A black, gray, or brown rounded mass or "pebble", ranging in diameter from a few millimeters to more than 30 cm, consisting of coprolites, corals, shells, bones, sand grains, mica flakes, or sponge spicules, more or less enveloped in *collophane* (calcium phosphate). They occur in marine strata (as in Permian beds of western U.S. and in the Cretaceous chalk of England), and are forming at present on the sea floor (as off the coast of California). See also: *coprolite.*

phosphatization (phos'-pha-ti-za'-tion) Conversion to a phosphate or phosphates; e.g. the diagenetic replacement of limestone, mudstone, or shale by phosphate-bearing solutions, producing phosphates of calcium, aluminum, or iron. Cf: *phosphorization.*

phosphide (phos'-phide) A mineral compound that is a combination of phosphorus with a metal. An example is schreibersite, $(Fe,Ni)_3P$.

phosphochalcite (phos-pho-chal'-cite) *pseudomalachite.*

phosphoellenbergerite A vitreous blue hexagonal mineral: $(Mg,[])_2Mg_{12}(PO_4,PO_3OH)_6(PO_3OH,CO_3)_2(OH)_6$.

phosphoferrite (phos-pho-fer'-rite) (a) A white, pale green, yellow, or dark reddish-brown orthorhombic mineral: $(Fe^{2+},Mn^{2+})_3(PO_4)_2{\cdot}3H_2O$. It is isomorphous with reddingite. (b) A group name for minerals with the analogous composition, but with Fe^{2+} replaced by Mg or Fe^{3+} and H_2O replaced by OH where necessary to retain charge balance.

phosphofibrite (phos-pho-fi'-brite) A yellow or yellow-green orthorhombic mineral: $KCuFe^{3+}_{15}(PO_4)_{12}(OH)_{12}{\cdot}12H_2O$.

phosphogartrellite A vitreous to adamantine green triclinic mineral: $PbCuFe^{3+}(PO_4)_2(OH,H_2O)_2$. It is the phosphate analogue of gartrellite.

phosphophyllite (phos-pho-phyl'-lite) A colorless or pale blue-green monoclinic mineral with perfect micaceous cleavage: $Zn_2(Fe^{2+},Mn^{2+})(PO_4)_2{\cdot}4H_2O$.

phosphorescence (phos-pho-res'-cence) A type of *luminescence* in which the stimulated substance continues to emit light after the external stimulus has ceased; also, the light so produced. The duration of the emission is temperature-dependent, and has a characteristic rate of decay. Cf: *fluorescence.*

phosphorite (phos'-pho-rite) A sedimentary rock with a high enough content of phosphate minerals to be of economic interest. Most commonly it is a bedded primary or reworked secondary marine rock composed of microcrystalline carbonate fluorapatite in the form of laminae, pellets, oolites, nodules, and skeletal, shell, and bone fragments. See also: *brown rock; bone phosphate of lime; microsphorite; pebble phosphate; guano.*

phosphorization (phos'-pho-ri-za'-tion) Impregnation or combination with phosphorus or a compound of phosphorus; e.g. the diagenetic process of *phosphatization.*

phosphorochalcite (phos'-pho-ro-chal'-cite) *pseudomalachite.*

phosphorrösslerite (phos-phor-ross'-ler-ite) A colorless or yellowish monoclinic mineral: $Mg(PO_3OH){\cdot}7H_2O$. It is the phosphate analogue of rösslerite. Also spelled: phosphorroesslerite.

phosphosiderite (phos-pho-sid'-er-ite) A pinkish-red, reddish-violet, or green monoclinic mineral: $Fe^{3+}PO_4{\cdot}2H_2O$. It is dimorphous with strengite and isomorphous with metavariscite. Syn: *clinostrengite; metastrengite.*

phosphovanadylite A light greenish-blue cubic mineral: $(Ba,Ca,K,Na)_{0.7}(V,Al)_4P_2(O,OH)_{16}{\cdot}12H_2O$.

phosphuranylite (phosph-u-ran'-yl-ite) A golden-yellow orthorhombic secondary mineral: $Ca(UO_2)_7(PO_4)_4(OH)_4{\cdot}12H_2O$. It exhibits phosphorescence upon exposure to radium emanations. It is structurally related to mundite.

photic zone (pho'-tic) *euphotic [oceanog].*

photoautotrophic (pho'-to-au'-to-troph'-ic) *holophytic.*

photobase (pho'-to-base) The length of the *air base* as represented on a photograph.

photobiont (pho-to-bi'-ont) The photosynthetic partner in a *lichen;* either a member of the *green algae* (Chlorophyta) or the *Cyanobacteria* or sometimes both. Obsolete syn: *phycobiont.*

photoclinometry (pho'-to-cli-nom'-e-try) A technique for ascertaining slope information from a distribution of image brightness; it is used esp. for studying the amount of slope of a lunar crater wall or ridge by measuring the density of its shadow.

photoeffect The interaction of a photon with an orbital electron of an atom in which the proton is absorbed and its energy is used for the release of the orbital electron (kinetic energy) (IAEA, 2003).

photoelasticity (pho'-to-e-las-tic'-i-ty) The property of a transparent, isotropic solid to become doubly refracting under nonhydrostatic states of stress. This makes it possible to study stress-distribution patterns under the polariscope.

photoelectric absorption cross-section A logging measurement in units of barns per electron that reflects the absorption of low-energy gamma rays and is a function of the aggregate atomic number of the formation elements. Recorded together with *bulk density* on the *Lithodensity log.* Symbol: Pe.

photogeologic map A geologic map based on information derived from the interpretation of aerial photographs.

photogeology (pho'-to-ge-ol'-o-gy) The identification, recording, and study of geologic features and structures by means of photography; specif. the geologic interpretation of aerial and space photographs and images and the presentation of the information so obtained. It now includes the interpretation of second-generation photographs obtained by photographing images recorded on television-type tubes (the images recording wavelengths outside the visible spectrum).

photogrammetry (pho-to-gram'-me-try) The art and science of obtaining reliable measurements from photographic images. Measurements relate not only to size, shape, and position, but also to color or tone, texture, and patterns of distribution of these elements.

photograph (pho'-to-graph) The recorded representation of a scene or object made by the action of light on a base material coated with a sensitized solution that is chemically treated to fix the image points at the desired density. The term is generally taken to mean the direct action of electromagnetic radiation on the sensitized material. Cf: *image [photo].*

photograph center The center of a photograph as indicated by the images of the fiducial marks of the camera. For a perfectly adjusted camera, it is identical to the *principal point* of the photograph.

photographic interpretation (pho-to-graph'-ic) *photointerpretation.*

photographic target A specially-made, natural or already existing, man-made mark or feature, on the ground, which can be identified on an aerial photograph and which indicates the location of a station such as a control point. Also called a signal or target. A photographic target made by assembling large strips of cloth or plastic into a geometric figure with the control point at the center of symmetry is called a panel or paneling.

photography (pho-tog'-ra-phy) (a) The process of producing photographs. Cf: *imagery.* (b) A collective term for the photographs so produced.

photohydrology (pho'-to-hy-drol'-o-gy) The science involving extraction of hydrologic data from aerial photographs and Landsat imagery.

photoindex (pho-to-in′-dex) A mosaic made by assembling individual photographs (with accompanying designations) into their proper relative positions and copying the assembly photographically at a reduced scale. It is not an *index map.* Also spelled: photo index.

photointerpretation (pho′-to-in-ter′-pre-ta′-tion) The extraction of information from aerial photographs and images for a particular purpose, such as mapping the geologic features of an area. Also spelled: *photo interpretation.* Syn: photographic interpretation.

photoluminescence The emission of light by a substance when it is exposed to ultraviolet radiation. The two types of photoluminescence are *fluorescence* and *phosphorescence.*

photoluminescent decay The gradual decrease in the intensity of light emitted by a photoluminescent substance after the source of excitation is eliminated. Cf: *fluorescence; phosphorescence.*

photomap (pho′-to-map) An aerial photograph or a controlled mosaic of rectified photographs to which have been added a reference grid, scale, place names, marginal information, and other pertinent data or map symbols; e.g. an *orthophotomap.*

photometer (pho-tom′-e-ter) An instrument for measuring the intensity of light. See also: *spectrophotometer; flame photometer.*

photometry (pho-tom′-e-try) (a) Study of ways and means to measure the intensity of light. (b) The art or process of using a *photometer.*

photomicrograph (pho-to-mi′-cro-graph) A photograph of an image made with the aid of a microscope such as a petrologic thin section; a type of *micrograph.* See also: *photomicrography.* Less-preferred syn: *microphotograph.*

photomicrography (pho′-to-mi-crog′-ra-phy) The preparation of photomicrographs, performed by the projection of the image through the eyepiece of the microscope onto the photographic recording medium.

photomosaic (pho′-to-mo-sa′-ic) *mosaic [photo].*

photon A single quantum of electromagnetic energy, the smallest amount by which electromagnetic energy can change and be measured. See also: *electromagnetic radiation.*

photopeak A local maximum in the gamma energy spectrum representing the emission energy of photons of a source (IAEA, 2003).

photoperiod (pho-to-pe′-ri-od) The relative number of alternating daylight and dark hours in a 24-hour period. The photoperiod has a significant effect on the development of certain organisms, esp. flowering plants.

photoreceptor (pho′-to-re-cep′-tor) A molecule, part of a cell, sensory cell, or organ of special sense that undergoes specific stimulation when exposed to light, such as (perhaps) a paraflagellar body in some Euglenophyta or the rods and cones of the vertebrate retina.

photorelief map (pho′-to-re-lief′) (a) A map consisting of a photograph of a relief model of the area under study and showing salient physical features. (b) A diagrammatic map that simulates or gives the impression of a photograph of a relief model of the area under study. Also spelled: photo-relief map.

photosynthesis Biological process whereby CO_2 is converted to organic matter in the presence of light; oxygen gas can also be produced.

phototaxis (pho-to-tax′-is) Taxis resulting from stimulation by light.

phototheodolite (pho′-to-the-od′-o-lite) A ground-surveying instrument used in terrestrial photogrammetry, combining the functions of a theodolite and a camera mounted on the same tripod.

phototriangulation (pho′-to-tri-ang′-u-la′-tion) The addition of horizontal and/or vertical control points by photogrammetric methods, whereby "the measurements of angles and/or distances on overlapping photographs are related into a spatial solution using the perspective principles of the photographs" (ASP, 1966, p.1148); esp. aerotriangulation.

phototrophic (pho-to-troph′-ic) *holophytic.*

phragmites peat (phrag-mi′-tes) Peat that is derived mainly from the reed genus *Phragmites.*

phragmocone (phrag′-mo-cone) (a) The chambered part of the *conch* of a nautiloid or an ammonoid. (b) The conical, chambered internal shell of a belemnite or aulacocerid, fitted into a deep anterior cavity (the *alveolus*) in the *rostrum* or *telum.*

phragmospore (phrag′-mo-spore) A plant spore having two or more septa. Cf: *Saccardoan spore groups.*

phreatic cycle (phre-at′-ic) The period of time during which the water table rises and falls. It may be a daily, annual, or other cycle. Syn: *cycle of fluctuation.*

phreatic decline *water-table decline.*

phreatic explosion A volcanic eruption or explosion of steam, mud, or other material that is not incandescent; it is caused by the heating and consequent expansion of groundwater due to an underlying igneous heat source. Cf: *phreatomagmatic explosion.* Syn: *hydrothermal eruption.*

phreatic gas A gas formed by the contact of atmospheric or surface water with ascending magma. Cf: *juvenile [water]; resurgent gas.*

phreatic line *seepage line.*

phreaticolous (phre-a-tic′-o-lous) Said of an organism or of the fauna inhabiting the interstices of mixtures of sand and gravel.

phreatic passage A cave passage developed primarily beneath a stable water table, or during flood events, as indicated by passage morphology characteristic of tube-full conditions.

phreatic solution The solution action by groundwater below the water table. Cf: *vadose solution.*

phreatic surface *water table.*

phreatic water A term that originally was applied only to water that occurs in the upper part of the saturated zone under water-table conditions (syn. of *unconfined groundwater,* or *well water*), but has come to be applied to all water in the saturated zone, thus making it an exact syn. of *groundwater* (Meinzer, 1923, p.5).

phreatic-water discharge (phre-at′-ic-wa′-ter) *groundwater discharge.*

phreatic wave *ground-water wave.*

phreatic zone *saturated zone.*

phreatomagmatic *hydrovolcanic.*

phreatomagmatic eruption (phre-a′-to-mag-ma′-tic) Explosive volcanic eruption caused by interaction of magma with groundwater or shallow surface water. Syn: *hydrovolcanic.*

phreatomagmatic explosion (phre-at′-o-mag-mat′-ic) A volcanic explosion that extrudes both magmatic gases and steam; it is caused by the contact of magma with groundwater or shallow surface water. Cf: *phreatic explosion.*

phreatophyte (phre-at′-o-phyte) A plant that obtains its water supply from the *saturated zone* or through the *capillary fringe* and is characterized by a deep root system.

phreatoplinian eruption Plinian eruption during which interaction of vesiculating magma and near-surface water produce very fine ash and abundant steam in an eruption column that reaches great heights and very energetic pyroclastic flows.

phrenotheca (phren-o-the′-ca) One of the thin, dense, diaphragmlike partitions that extend across the chamber of a foraminiferal test at various angles and in various parts of the chamber (as in *Pseudofusulina*).

phthanite (phthan′-ite) Siliceous shale. The term is used esp. by European geologists.

Phthinosuchia (Phthin-o-such′-i-a) A suborder of predaceous synapsid reptiles, the order Therapsida but showing many features transitional between sphenacodont pelycosaurs and synapsids proper. Range, early Upper Permian.

phuralumite (phur-al-um′-ite) A yellow monoclinic mineral: $Al_2(UO_2)_3(PO_4)_2(OH)_6 \cdot 10H_2O$.

phurcalite (phur′-cal-ite) A yellow orthorhombic mineral: $Ca_2(UO_2)_3O_2(PO_4)_2 \cdot 7H_2O$.

phycobiont (phy-co-bi′-ont) An obsolete syn. of *photobiont.* Cf: *mycobiont.*

phycoma (phy-co′-ma) A highly resistant, apparently sporopollenin-containing, nonmotile stage in the life cycle of *prasinophyte* algae, e.g., *Pterospermopsis, Pterospermella, Tasmanites* (Tappan, 1980). It has sometimes been referred to as a cyst. Pl: phycomata.

phyla (phy′-la) The plural of *phylum.*

phyletic (phy-let′-ic) *phylogenetic.*

phyletic evolution Evolution involving changes in lineages but little or no increase in the number of taxonomic groups.

phyletic gradualism That evolutionary process or pattern in which morphologic change occurs gradually and continuously throughout the time duration of several successive species within an evolutionary *lineage,* so that an ancestral species appears to grade imperceptibly into its immediately descendent species. Cf: *punctuated gradualism.* Syn: *gradualistic speciation.* Ant: *punctuated equilibria.*

phyllarenite (phyl-lar′-e-nite) A term used by Folk (1968, p.124) for a *litharenite* composed chiefly of foliated, phyllosilicate-rich metamorphic-rock fragments (as of slate, phyllite, and schist). It may have any particle size from silt through gravel, and any clay content, sorting, or rounding. See also: *subphyllarenite.*

phyllic alteration (phyl′-lic) Hydrothermal alteration typically resulting from removal of sodium, calcium, and magnesium from calc-alkalic rocks, with pervasive replacement of silicates by micas, muting the original rock texture. It is a common style of alteration in porphyry base-metal systems around a central zone of *potassic alteration* (Guilbert and Park, 1986, p.183). See also: *overprinting; propylitization.* Syn: *sericitic alteration.*

phyllic overprint In many *porphyry copper deposits,* the superposition of younger sericite-pyrite alteration on older *potassic alteration* and *propylitic alteration.*

phyllic zone The zone in porphyry ore deposits next outward from the *potassic zone,* characterized by the assemblage quartz-sericite-pyrite, with less than 5 percent kaolin, biotite, or potassium feldspar (Lowell and Guilbert, 1970). Cf: *argillic zone; potassic zone; propylitic zone.*

phyllite [mineral] (phyl′-lite) (a) A general term used by some French authors for the scaly minerals, such as the micas, chlorites, clays, and vermiculites. (b) A general term for minerals with a layered crystal structure.

phyllite [petrology] A metamorphosed rock, intermediate in grade between slate and mica schist. Minute crystals of graphite, sericite, or chlorite impart a silky sheen to the surfaces of cleavage (or schistosity). Phyllites commonly exhibit corrugated cleavage surfaces. Cf: *phyllonite.*

phyllite-mylonite *phyllonite.*

phyllocarid (phyl-lo-car′-id) Any malacostracan belonging to the subclass Phyllocarida, characterized by a relatively large bivalve carapace that may or may not be hinged along the dorsal margin. Range, Lower Cambrian to Holocene.

phylloclade (phyl′-lo-clade) A somewhat flattened branch or stem that functions as a leaf, e.g. in Christmas cactus. Cf: *phyllode [bot].*

phyllode [bot] (phyl′-lode) A flattened and expanded petiole that functions as a leaf. Cf: *phylloclade.*

phyllode [paleont] The more or less depressed area of enlarged pores in the adoral part of an *ambulacrum* in an echinoid. It bears specialized podia. Cf: *bourrelet.*

phyllofacies (phyl-lo-fa′-cies) A facies differentiated on the basis of stratification characteristics, esp. the *stratification index* (Kelley, 1956, p.299).

phylloid (phyl′-loid) Leaf-shaped, or resembling a leaf; esp. said of the minor elements or endings of the saddles of an ammonoid suture (as in *Phylloceras*).

phyllomorphic stage (phyl-lo-mor′-phic) A term introduced by Dapples (1962) for the latest (most advanced) geochemical stage of diagenesis characterized by authigenic development of micas, feldspars, and chlorites at the expense of clays (unidirectional reactions). It follows the *locomorphic stage.* See also: *redoxomorphic stage.*

phyllonite (phyl′-lon-ite) A rock that macroscopically resembles *phyllite [petrology]* but that is formed by mechanical degradation (mylonization) of initially coarser rocks (e.g. graywacke, granite, or gneiss). Characteristic features are silky films of recrystallized mica or chlorite, smeared out along schistosity surfaces; and formation by *dislocation metamorphism.* The term was originated by Sander (1911). Syn: *phyllite-mylonite.*

phyllonitization (phyl-lon′-i-ti-za′-tion) The processes of *mylonitization* and *recrystallization* to produce a *phyllonite.*

phyllopodium (phyl-lo-po′-di-um) A broad flat leaflike thoracic appendage of a crustacean. Pl: phyllopodia. Cf: *stenopodium.*

phylloretene A colorless orthorhombic hydrocarbon mineral: $C_{18}H_{18}$=1-methyl1-1-isopropylphenanthrene.

phyllosilicate (phyl-lo-sil′-i-cate) A class or structural type of *silicate* characterized by the sharing of three of the four oxygens in each tetrahedron with neighboring tetrahedra, to form flat sheets; the Si:O ratio is 2:5. An example is the micas. Cf: *nesosilicate; sorosilicate; cyclosilicate; inosilicate; tectosilicate.* See also: *structural terms.* Syn: *layer silicate; physil; sheet mineral; sheet silicate.*

phyllosilicate cement A term applied by Dickinson (1970, p.702) to a cement in graywackes and arkoses that is clear, monomineralic, and crystalline, and has certain other features that distinguish it from inhomogeneous matrix.

phyllotaxy (phyl′-lo-tax′-y) The arrangement of leaves or floral parts on an axis. It is generally expressed numerically by a fraction, the numerator representing the number of revolutions of a spiral made in passing from one leaf past each successive leaf to reach the leaf directly above the initial leaf, and the denominator representing the number of leaves passed in the spiral thus made (Lawrence, 1951, p.765).

phyllotriaene (phyl-lo-tri′-aene) A sponge *triaene* in which the cladi are expanded into flattened, sometimes digitate, leaflike structures.

phyllotungstite (phyl-lo-tung′-stite) A platy yellow orthorhombic mineral: $CaFe^{3+}{}_3W_6O_{23}(OH)\cdot10H_2O$.

phyllovitrinite (phyl-lo-vit′-ri-nite) *provitrinite.*

phylogenetic (phy′-lo-ge-net′-ic) The adj. of *phylogeny.* Syn: *phyletic.*

phylogenetic evolution Evolution within a single *lineage.*

phylogenetics *cladistic analysis.*

phylogenetic systematics (a) The study of *systematics* with the ultimate goal of inferring *phylogeny* of the organisms investigated. (b) In more restricted usage, the study of systematics by means of the methods of *cladism;* this usage results from the work of W. Hennig, and thus may be termed *Hennigian* systematics. (c) An approach to systematics that attempts to recover the phylogenetic (genealogical) relationships among groups of organisms, and produces classifications that reflect exactly those genealogical relationships (Wiley, 1981, p.6). Cf: *cladistic analysis.* Syn: *phylogenetic taxonomy.*

phylogenetic taxonomy *phylogenetic systematics.*

phylogenetic tree (a) A *dendrogram* or other diagrammatic portrayal expressing ancestor-descendant or genealogical relationships among several taxa, as inferred from various types of evidence. Syn: *phylogram; cladogram.* (b) More specifically, in *cladism,* a *dendrogram* showing phylogenetic relationships among taxa in terms of inferred genealogical history; identities of nodes (ancestors) are specified, connecting lines represent ancestral taxa, and not all the included taxa are terminal in position. Cf: *cladogram.* (c) In cladistic analysis, a graphic representation of the genealogical relationships between taxa as the relationships are understood by an investigator; an hypothesis of genealogical relationships among taxa (Wiley et al., 1991, p.6).

phylogenetic zone *lineage zone.*

phylogeny (phy-log′-e-ny) (a) The line, or lines, of direct descent in a given group of organisms, as opposed to the development of an individual organism. Cf: *ontogeny.* (b) The study or history of such relationships. Adj: *phylogenetic.*

phylogerontism (phy-lo-ger′-on-tism) The condition of apparent deterioration and probable approaching extinction of an evolving *lineage.* Syn: *racial senescence.* The term has been overapplied.

phylogram (phy′-lo-gram) *phylogenetic tree (a).*

phylozone (phy′-lo-zone) (a) A belt of rock formed during a biochron (Van Hinte, 1969, p.271; ISSC, 1994, p.62). (b) lineage zone.

phylum (phy′-lum) The primary taxonomic division of the animal kingdom, characterized by organisms that share a major type of body plan implying common ancestry. A phylum is generally divided into classes, e.g. the phylum Mollusca includes the classes Gastropoda, Cephalopoda, and Pelecypoda. In botanical nomenclature, the equivalent of *phylum* has traditionally been *division,* but the 2000 *ICBN* officially recognizes *phylum* as an acceptable, optional synonym of *division.* Pl: *phyla.* Cf: *subphylum.*

-phyre A suffix which, in a rock name, signifies "porphyry".

physical exfoliation (phys′-i-cal) A type of *exfoliation* caused by physical forces, such as by the freezing of water in fine cracks in the rock or by the removal of overburden concealing deeply buried rocks.

physical geodesy The part of geodesy concerning the study and determination of the Earth's gravity field. Physical geodesy deals with those portions of geodesy that involve the principles and methods of physics, as opposed to geometric geodesy, which deals primarily with geometrical relationships. When positional relationships can be determined by physical methods (e.g., via satellite orbits), the procedures are considered physical geodesy.

physical geography That branch of *geography* which is the study of the Earth's surface as the physical environment of humans, dealing with the classification, form, processes, and extent of the natural phenomena directly related to the exterior physical features and changes of the Earth, including land, water, and air. It differs chiefly from *geology* in that it considers conditions of the Earth primarily over shorter time scales (e.g. Quaternary and/or modern rates). It is more inclusive than *geomorphology,* dealing not only with landforms but also climate, oceans, atmosphere, soils, geologic processes, natural resources, and the biogeographical distribution of animal and plant life. In the 18th century, the term was applied in a broader sense, commonly including the races of men and their physical works on the Earth. Cf: *physiography.*

physical geology A broad division of *geology* that concerns itself with the processes and forces involved in the inorganic evolution of the Earth and its morphology, and with its constituent minerals, rocks, magmas, and core materials. Cf: *historical geology.*

physical landscape *natural landscape.*

physical metallurgy The science and technology of metals as applied to the mechanical and physical properties of metals.

physical oceanography The study of physical processes in the ocean, such as the interactions of the ocean with the atmosphere, the ocean's influence on weather and climate, the transport of heat and momentum, the propagation of sound and light, currents, waves, tides, and the distribution of temperature and density.

physical pendulum *pendulum* (a).

physical properties Those characteristics, processes, or reactions of a soil which are caused by physical forces and which can be described by, or expressed in, physical terms or equations. Examples of physical properties are bulk density, hydraulic conductivity, porosity, pore-size distribution, etc.

physical residue A *residue* formed by mechanical weathering in place; e.g. a deposit of gravel resulting from the removal of finer particles by water or wind, as on the floor of a desert valley.

physical stratigraphy Stratigraphy based on the physical aspects of rocks (esp. the sedimentologic aspects); e.g. *lithostratigraphy.*

physical time A term used by Jeletzky (1956, p.682) to designate time as measured by any physical phenomenon or process (such as by radioactive decay of elements), and proposed by him to replace "absolute time" as used in the geologic sense. Cf: *numerical time.*

physical weathering *mechanical weathering.*

physil (phy′-sil) An abbreviated form of *phyllosilicate,* proposed to apply to "all sheet silicate minerals regardless of grain size" (Weaver, 1980).

physilite (phy′-si-lite) A rock with a high content of physils (Weaver, 1980).

physiofacies (phys′-i-o-fa′-cies) A term suggested by Moore (1949, p.17) for "the total inorganic characteristics of a sedimentary rock", or that part of lithofacies not represented by biofacies. The term is essentially identical with *lithofacies* as that term has been interpreted by some. Moore (1957a, p.1784-1785) later wrote that "the concept of physiofacies...may well be forgotten". Cf: *physiotope.*

physiographic *geomorphic.*

physiographic cycle (phys′-i-o-graph′-ic) *cycle of erosion.*

physiographic diagram A small-scale map showing landforms by the systematic application of a standardized set of simplified pictorial symbols that represent the appearances such forms would have if viewed obliquely from the air at an angle of about 45°. The first major map of this kind was published by Lobeck (1921). Syn: *morphographic map; landform map.*

physiographic feature A prominent or conspicuous *physiographic form* or noticeable part thereof (Mitchell, 1948, p.64). Cf: *topographic feature.*

physiographic form A *landform* considered with regard to its origin, cause, or history (Mitchell, 1948, p.64). Cf: *topographic form.*

physiographic geology A branch of geology that deals with topography; *geomorphology.* The term was previously used as a syn. of *physiography.*

physiographic pictorial map *trachographic map.*

physiographic province A region of which all parts are similar in geologic structure and climate and which has consequently had a unified geomorphic history; a region whose pattern of relief features or landforms differs significantly from that of adjacent regions (see Fenneman,

1914). Examples: the Valley and Ridge, Blue Ridge, and Piedmont provinces in eastern U.S., and the Basin and Range, Rocky Mountains, and Great Plains provinces in western U.S. Cf: *geologic province; geographic province; morphologic region; structural province.*

physiography (phys-i-og'-ra-phy) Originally, a description of the physical nature (form, substance, arrangement, changes) of objects, esp. of natural features; the term was introduced into geography in 1869 by Huxley for the study or description of "natural phenomena in general". The term later came to mean, esp. in the U.S., "a description of the surface features of the Earth, as bodies of air, water and land" (Powell, 1895), with an emphasis on mode of origin; i.e. it became synonymous with *physical geography,* and embraced geology, meteorology, and oceanography. Still later, esp. in the U.S., the term was restricted to a part of physical geography, namely the description and origin of landforms; in this sense, it is obsolescent and is replaced by *geomorphology,* although there is a general tendency to regard "physiography" as the descriptive, and "geomorphology" as the interpretative, study of landforms. See also: *physiographic geology.* Etymol: Greek "physis", "nature", "graphein", "to write".

physiotope (phys'-i-o-tope') A term defined by Moore (1949, p.17) as "designation of all purely physiochemical elements of an environment", but intended by him to represent the sedimentary environment of a *physiofacies.* Weller (1958, p.616) notes that if the term is to be accepted as having a meaning similar to "biotope" or "lithotope", it should be defined as an "area".

phytal (phy'-tal) Said of those parts of a body of water that are shallow enough for the growth of rooted green plants. Cf: *aphytal.*

phytane The long-chained isoprenoid hydrocarbon derived from phytol.

phytem (phy'-tem) A major Precambrian biostratigraphic unit defined largely on the basis of stromatolites (Keller et al., 1968, p.190).

phyteral (phy'-ter-al) Vegetal matter in coal and peat that is recognizable as morphologic forms, e.g. cuticle, spore coats, or wax, as distinguished from the macerals, or organic material forming the coal mass. See also: *muralite.*

phytoclast (phy'-to-clast) An organic particle of roughly the same size as a mineral clast of the containing rock. Phytoclasts make up 0.1-0.5% of most shale and are less abundant in sandstone; they also occur widely in metasediments (Bostick, 1970, p.74).

phytocoenosis (phy'-to-coe-no'-sis) The plant population of a particular habitat.

phytocollite (phy-to-col'-lite) A black gelatinous nitrogenous humic body occurring beneath or within peat deposits. Cf: *dopplerite.*

phytoecology (phy'-to-e-col'-o-gy) The branch of ecology concerned with the relationships between plants and their environment. Cf: *zooecology.*

phytogenic dam (phy-to-gen'-ic) A natural dam consisting of plants and plant remains. Such dams may account for ponds and lakes in tundra regions.

phytogenic dune Any dune in which the growth of vegetation influences the form of the dune, as by arresting the drifting of sand; e.g. a *foredune.*

phytogenic rock A *biogenic rock* produced by plants or directly attributable to the presence or activities of plants; e.g. algal deposits, peat, coal, some limestones, and lithified ooze containing diatoms. Cf: *phytolith [sed].* Syn: *phytogenous rock.*

phytogenous rock (phy-tog'-e-nous) *phytogenic rock.*

phytogeography (phy'-to-ge-og'-ra-phy) The branch of *biogeography* dealing with the geographic distribution of plants. Cf: *zoogeography.*

phytokarst (phy'-to-karst) A type of *solution* landscape in which the major morphology is produced by the attack of boring algae and/or fungi on limestone. It is characterized by jagged, grotesque sculpture; it differs from ordinary solution karst by a random orientation of the sculpture with respect to gravity, and a black algal coat (Folk et al., 1973). Syn: *biokarst.*

phytol Long-chained, isoprenoid alcohol connected to the porphyrin group of chlorophyll a.

phytoleims (phy'-to-leims) Coalified remains of plants (Krishtofovich, 1945, p.138). See also: *meroleims; hololeims.*

phytolith [paleont] (phy'-to-lith) A stony or mineral structure, generally microscopic, secreted by a living plant; often composed of calcium oxalate or opaline silica. Cf: *opal phytolith.*

phytolith [sed] A *biolith* formed by plant activity or composed of plant remains and silica; specif. *phytogenic rock.*

phytopaleontology (phy'-to-pa'-le-on-tol'-o-gy) *paleobotany.*

phytophagous (phy-toph'-a-gous) Said of an organism that feeds on plants.

phytoplankton (phy-to-plank'-ton) The chlorophyll-bearing forms of *plankton,* e.g. diatoms. Phytoplankton are the dominant primary producers in the sea, and the base of the marine food web. These single-celled organisms are the principal agents for photosynthetic *carbon fixation* in the ocean along with all the reactions associated with it; e.g., oxygen production, carbon dioxide consumption, inorganic nutrient consumption. Cf: *zooplankton.*

phytozoan (phy-to-zo'-an) *zoophyte.*

Piacenzian (Pia-cen'-zi-an) A stage of the Standard Global Chronostratigraphic Scale: Middle Pliocene (above Zanclean, below Gelasian). The basal GSSP is in Sicily, Italy (Castradori et al., 1998).

pi axis *pi pole.*

picacho (pi-ca'-cho) A term used in the SW U.S. for a large, sharply pointed, isolated hill or mountain; a peak. Etymol: Spanish.

pi circle On a *pi diagram,* a girdle of points representing poles to folded surfaces.

pick In the interpretation of seismic records, the selection of an event; also, a selected event on a seismic section.

pickeringite (pick'-er-ing-ite') A colorless, white, yellowish or reddish monoclinic mineral of the halotrichite group: $MgAl_2(SO_4)_4{\cdot}22H_2O$. It occurs in white to faintly colored fibrous masses. Syn: *magnesia alum.*

pickup (pick'-up) *geophone.*

picoplankton *ultraplankton.*

picotite (pic'-o-tite) A dark-brown, chromium-bearing variety of hercynite (spinel). Much so-called picotite is ceylonite or magnesiochromite. Syn: *chrome spinel.*

picotpaulite (pi-cot-paul'-ite) A metallic creamy-white orthorhombic mineral: $TlFe_2S_3$.

picrite (pic'-rite) (a) In the *IUGS classification,* a volcanic rock containing (in wt%)<52 SiO_2, >18 MgO, and between 2 and 3 Na_2O+K_2O. Additionally, if MgO is between 12 and 18, then alkalis must be <3. (b) A dark-colored, generally hypabyssal rock containing abundant olivine along with pyroxene, biotite, possibly amphibole, and less than 10 percent plagioclase. The term was first used by Tschermak who applied it to a rock composed chiefly of olivine, titanian augite, and amphibole, with or without biotite; later the term was used by Rosenbusch for a rock composed chiefly of olivine and augite, with or without amphibole and biotite (Streckeisen, 1967, p.176).

picrobasalt In the *IUGS classification,* a volcanic rock defined in the TAS diagram by a field bounded by points with the SiO_2, total alkali coordinates: 41, 0; 41, 3; 45, 3; 45, 0.

picrochromite (pic-ro-chro'-mite) *magnesiochromite.*

picrocollite (pic-ro-col'-lite) A hypothetical end-member of the *palygorskite* group: $Mg_2(OH)_8Si_6O_{10}{\cdot}4H_2O$.

picrolite (pic'-ro-lite) A fibrous or columnar variety of serpentine.

picromerite (pic-rom'-er-ite) (a) A white monoclinic mineral: $K_2Mg(SO_4)_2{\cdot}6H_2O$. Syn: *schoenite.* (b) A group name for minerals with the analogous composition, but with K replaced by NH_4 or Na, and Mg replaced by Fe^{2+}, Ni or Cu.

picropharmacolite (pic'-ro-phar'-mac-o-lite) A silky white triclinic mineral: $Ca_4Mg(AsO_4)_2(HAsO_3OH)_2{\cdot}11H_2O$.

pictograph (pic'-to-graph) A picture painted on a rock by primitive peoples and used as a sign.

picture element (pic'-ture) *pixel.*

picurite (pic'-u-rite) *pitch coal.*

pi diagram In structural petrology, a fabric diagram in which the poles representing normals to fabric planes have been plotted. In an area of simple cylindrical folding, the pi diagram for bedding would be a great-circle girdle at right angles to the fold axis. Also written: π diagram. Cf: *beta diagram.*

piecemeal stoping (piece'-meal) *Magmatic stoping* involving the sinking, into a magma body, of relatively small blocks of country rock. Cf: *cauldron subsidence.*

piedmont (pied'-mont) n. An area, plain, slope, glacier, or other feature at the base of a mountain; e.g., a foothill or a bajada. In the U.S., the Piedmont is a gently rolling plateau extending from New Jersey to Alabama and lying east of the Appalachian Mountains. adj. Lying or formed at the base of a mountain or mountain range; e.g., a piedmont terrace or a piedmont pediment. Etymol: from Piemonte, a region of NW Italy at the foot of the Alps.

piedmont alluvial plain *bajada.*

piedmont angle The sharp break of slope between a hill and a plain, such as the angle at the junction of a mountain front and the pediment at its base. Cf: *knick.*

piedmont bench (a) An upfaulted alluvial fan or pediment surface at the base of a mountain, bounded on its outer side by a scarplet (piedmont scarp) (Sharp, 1954, p.23). Such features are displayed along the south sides of the San Gabriel and San Bernardino ranges in southern California. (b) *piedmont step.*

piedmont benchland One of several successions or systems of piedmont steps. Syn: *piedmonttreppe; piedmont stairway.*

piedmont bulb *expanded foot.*

piedmont flat *piedmont step.*

piedmont glacier A thick continuous sheet of ice at the base of a mountain range, resting on land, formed by the spreading out and coalescing of valley glaciers from the higher elevations of the mountains. Cf: *expanded-foot glacier.*

piedmont gravel Gravel derived from mountainous terrain and spread out on relatively flat ground where the velocity of the water is decreased.

piedmont interstream flat A term used by Tator (1949) for a planate rock surface along the east flank of the Colorado Front Range, and regarded by him (1953, p.47) as a syn. of *pediment.*

piedmontite (pied'-mont-ite) *piemontite.*

piedmont lake An oblong lake occupying a partly overdeepened basin excavated in rock by, or dammed by a moraine of, a *piedmont glacier.*

piedmont pediment A term used by Davis (1930, p.154) for a pediment peripheral to, and along the base of, a mountainous area. Cf: *mountain pediment.*

piedmont plain *bajada.*

piedmont plateau A plateau lying between the mountains and the plains or the ocean; e.g. the plateau of Patagonia in southern Argentina and southern Chile, between the Andes and the Atlantic Ocean.

piedmont scarp A small, low cliff occurring in alluvium on a piedmont slope at the foot of, and essentially parallel to, a steep mountain range (as in the western U.S.), resulting from dislocation of the surface, esp. by faulting; term proposed by Gilbert (1928, p.34). See also: *fan scarp.* Syn: *scarplet.*

piedmont slope (a) *bajada.* (b) A gentle slope at the base of a mountain in a semiarid or desert region, composed of a pediment (upper surface of eroded bedrock) and a bajada (lower surface of aggradational origin).

piedmont stairway *piedmont benchland.*

piedmont step An extensive or regional terrace- or bench-like feature sloping outward or down-valley (as in the Black Forest or Schwarzwald region of SW Germany), assumed by W. Penck (1924) to develop in response to a continually accelerated uplift of a rising or expanding dome. See also: *treppen concept; piedmont benchland.* Syn: *piedmont bench; piedmont flat.*

piedmonttreppe (pied-mont-trep'-pe) A syn. of *piedmont benchland.* Etymol: German *Piedmonttreppe,* "piedmont staircase".

piemontite (pie'-mont-ite) A reddish-brown, deep red, purplish-red or black monoclinic manganese-bearing mineral of the *epidote* group: $(Ca,Pb,Ce)_2(Mn^{3+},Fe^{3+})Al_2(Si_2O_7)(SiO_4)(O,OH)_2$. Cf: *withamite.* Also spelled: *piedmontite.* Syn: *manganese epidote.*

pienaarite (pie-naar'-ite) A titanite-rich *malignite* in which the feldspar is anorthoclase. The name, given by Brouwer in 1910, is for Pienaar Creek, Transvaal, South Africa. Obsolete.

pier [coast] A breakwater, groin, mole, or other structure used to protect a harbor or shore, and serving also as a promenade or as a landing place for vessels.

pier [eng] (a) An underground structural member that transmits a concentrated load to a stratum capable of supporting it without danger of failure or unacceptable settlement. Its ratio of base width to depth of foundation is usually less than 1:4. Cf: *footing.* (b) A rectangular or circular column, usually of concrete or masonry, designed to support heavy, concentrated loads from arches or the superstructure of a bridge.

piercement dome (pierce'-ment) *diapir [struc geol].*

piercing fold (pierc'-ing) *diapir [struc geol].*

piercing point [chem] In a quaternary chemical system, the point at which a univariant curve (representing the compositions of liquids that can exist in equilibrium with three particular solid phases) and a ternary *join* intersect at some point other than a ternary univariant point.

piercing point [struc geol] A point of intersection of a geologically defined line with a fault surface that can be used to determine the *net slip* along the fault.

pier dam An engineering structure, such as a groin, built from shore to deepen a channel, to divert logs, or to direct the water flow. Syn: *wing dam.*

pierre-perdue (pi-erre'-per-due') Blocks of stone or concrete heaped loosely in the water to make a foundation. Etymol: French, "lost stone".

pierrepontite (pierre'-pont-ite) An iron-rich variety of tourmaline.

pierrotite (pier-rot'-ite) A metallic grayish-black orthorhombic mineral: $Tl_2Sb_6As_4S_{16}$.

Pierson-Moskowitz wave spectrum An idealized form of the *wave spectrum,* which represents the ocean waves generated by a constant, homogeneous wind blowing for many days over the ocean.

pietricikite (pie-tri'-ci-kite) A variety of *ozocerite.* Originally incorrectly spelled *zietrisikite.* See also: *neft-gil.*

piezocrystallization (pi-e'-zo-crys'-tal-li-za'-tion) Crystallization of a magma under pressure, such as pressure associated with orogeny. Obsolete concept.

piezoelectric crystal (pi-e'-zo-e-lec'-tric) A crystal, e.g. of quartz or tourmaline, that displays the *piezoelectric effect.* Any nonconducting crystal lacking a center of symmetry may be piezoelectric.

piezoelectric effect In certain crystals, the development of an electric potential in certain crystallographic directions when mechanical strain is applied, or, the development of a mechanical strain, hence vibration, when an electric potential is applied. Quartz and tourmaline are examples of naturally piezoelectric crystals.

piezomagnetism (pi-e'-zo-mag'-net-ism) Stress dependence of magnetic properties. It is the inverse of *magnetostriction.*

piezometer (a) An instrument for measuring *pressure head* in a conduit, tank, or soil. It usually consists of a small pipe or tube tapped into the side of the container, with its inside end flush with, and normal to, the water face of the container, and connected with a manometer pressure gage, or mercury or water column. (b) A device or technique that measures stress. In structural geology, piezometers are usually techniques that use measurable geometric features in deformed rocks, such as the diameter of subgrains or recrystallized grains in plastically deformed minerals or the orientations of slickensides on fault surfaces, to estimate the relative or absolute magnitudes and orientations the principal stresses (Fetter, 1994).

piezometric limit (pi-e'-zo-met-ric) The point in a cave *passage* where vadose morphology changes downstream to phreatic morphology (Palmer, 1972). In a relict cave, it shows the former position of the water table.

piezometric surface A less preferred synonym for *potentiometric surface* (Lohman, 1972).

PIGE *particle-induced gamma-ray emission.*

pigeonite (pi'-geon-ite) A brown, greenish-brown, or black monoclinic mineral of the *pyroxene* group: $(Mg,Fe^{2+},Ca)SiO_3$. It is intermediate in composition between clinoenstatite and diopside, and has little calcium, little or no aluminum or ferric iron, and less ferrous iron than magnesium. Pigeonite is characterized optically by a small and variable axial angle (2V = 0-30°). It is found in basic igneous rocks at Pigeon Point in Minnesota. Cf: *augite.*

pigeonitic magma series A term proposed by Kuno for lavas containing low-calcium monoclinic pyroxene in the groundmass; this series is equivalent to the tholeiitic series. Obsolete.

pigeon's-blood ruby (pi'-geon's-blood) A gem variety of ruby of the finest color: intense, clear, dark red to slightly purplish red. It is found almost exclusively in upper Burma. Also spelled: pigeon-blood ruby.

pigment mineral (pig'-ment) A mineral having economic value as a coloring agent. The most important are hematite and limonite.

pigotite (pi-got'-ite) An ill-defined salt of a humic acid: $Al_4[C_6H_5O_{10}]\cdot 13H_2O(?)$.

pike A term used in England for any summit or top of a mountain or hill, esp. one that is peaked or pointed. Also, a mountain or hill having a peaked summit.

pikeite (pike'-ite) An obsolete name for an augite-bearing phlogopite *peridotite.* Its name (Johannsen, 1938) is derived from Pike County, Arkansas. Cf: *kimberlite.* Obsolete.

pila (pi'-la) Plural of *pilum.*

pilandite (pi-lan'-dite) A hypabyssal *syenite* containing abundant anorthoclase phenocrysts in a groundmass of the same mineral. The name, given by Henderson in 1898, is for the Pilanesberg complex, Transvaal, South Africa. Obsolete.

pilar (pi-lar') A term used in the SW U.S. for a large pillarlike or projecting rock. Etymol: Spanish, "pillar".

pilate (pi'-late) This term ideally refers to *clavate*-like sculpture of sporomorphs, but it and such terms as "piliferous," "pilose," "pilum," "pila," and "pilosus" have been defined and applied so differently that they are preferably avoided.

pile A long, relatively slender structural foundation element, usually made of timber, steel, or reinforced or prestressed concrete, that is driven or jetted into the ground or cast in place in a borehole. Piles are used to support vertical or lateral loads, to form a wall to exclude water or soft material or to resist their pressure, to compact the surrounding ground, or rarely to restrain the structure from uplift forces. Cf: *battered pile; sheet pile.*

pile cap A structural member at the top of and connected to a group of *piles* that improves the lateral stiffness of the piles.

piliferous (pi-lif'-er-ous) A less-preferred syn. of *pilate.*

piling (pil'-ing) A structure or group of *piles.*

pilite (pi'-lite) (a) Actinolite pseudomorphous after olivine. (b) *tinder ore.*

pill An English term for a pool, and for a small stream or creek.

pillaite A metallic black monoclinic mineral: $Pb_9Sb_{10}S_{23}Cl_{0.5}$.

pillar [geomorph] (pil'-lar) A natural formation shaped like a pillar; specif. an *earth pillar* and a *rock pillar.*

pillar [paleont] (a) A tiny rodlike structure, larger and straighter than a *trabecula,* connecting discrete layers of sclerite in holothurians. (b) An elongate peglike structure produced near the center or axis of coiling in certain foraminifera by thickening of the wall. The ends of the pillars appear as small bosses or nodes on the ventral side of the test. (c) In archaeocyathids, longitudinal cylindrical to laterally flattened rodlike skeletal elements, which may bridge upward and outward from the inner wall to the outer wall. (d) A rodlike internal structure, 0.02 to 0.25 mm in diameter, between adjacent laminae, which partly fills the gallery space in most stromatoporoid coenostea.

pillar [speleo] Bedrock support remaining after removal of surrounding rock by solution and/or collapse.

pillar [struc geol] A joint block produced by columnar jointing.

pillared clay minerals Clay minerals modified by the intercalation of complex cations, such as the Keggin cation, that prop open the interlayer space. Choice of groups to serve as props provides custom-size openings. Such modified clay minerals have enhanced catalytic activity or serve as nano-scale sieves (Rausell-Colom and Serratosa, 1987). Cf: *interlayer space.*

pillar reef A *coral reef* consisting of large massive corals grown on one another to form tall sturdy pillars, separated by spacious caverns that tend to become filled by secondary internal encrustations and fine carbonate sediments. Cf: *thicket reef.*

pillarstone Scleractinian reef growth framework dominated by organisms that have a dominant vertical component of growth and relatively restricted lateral growth (for example, all types of branching colonies and rod and tabular solitary forms). These growth forms constitute >60% of the skeletal volume (Insalaco, 1998).

pillar structure A columnar structure, normal or oblique to the bedding, consisting of massive or "swirled" sand that cuts across laminae or other primary structures in a sand bed. It is thought to be formed by the escape of water after deposition of the bed. The term was first published by Lowe and LoPiccolo (1974). Cf: *vertical sheet structure.*

pillow breccia (pil'-low) A deposit of pillows and fragments of lava in a matrix of tuff.

pillow lava A general term for those lavas displaying pillow structure and considered to have formed in a subaqueous environment; such lava is usually basaltic or andesitic. Syn: *ellipsoidal lava.*

pillow structure [ign] A structure, observed in certain extrusive igneous rocks, that is characterized by discontinuous bun-shaped masses ranging in size from a few centimeters to a meter or more in greatest dimension (commonly between 30 and 60 cm). The pillows are close-fitting, the concavities of one matching the convexities of another. Grain size within pillows tends to decrease toward the exterior. Pillow structures are considered to be the result of subaqueous extrusion, as evidenced by their association with sedimentary deposits, usually of deep-sea origin. See also: *pillow lava.*

pillow structure [sed] A primary sedimentary structure resembling the size and shape of a pillow; it is most characteristic of the basal parts of a sandstone overlying shale. See also: *ball-and-pillow structure.* Syn: *mammillary structure.*

pilotaxitic (pi'-lo-tax-it'-ic) Volcanic groundmass texture in which tightly-packed, weakly-aligned, lath-shaped microlites (typically plagioclase) occur with interstitial micro- or cryptocrystalline material. Cf: *trachytic.* Syn: *felty.*

pilot channel (pi'-lot) One of a series of cutoffs for converting a meandering stream into a straight channel of greater slope. It is built only large enough to start flow along the new course, since erosion during floods is expected to create channels of adequate capacity.

pilsenite (pil'-sen-ite) A metallic white rhombohedral mineral occurring as one component in a phase formerly known as wehrlite: Bi_4Te_3. It occurs as a mixture with hessite.

pilum (pi'-lum) One of the small, spinelike rods on the outer surface of certain sporomorphs, dinoflagellates, and acritarchs, characterized by rounded or swollen knoblike ends. Pl: *pila.* See also: *pilate.*

pimelite (pi'-me-lite) An apple-green hexagonal clay mineral with a talc-like structure: $(Ni,Mg)_3(OH)_2Si_4O_{10}{\cdot}nH_2O$. Syn: *desaulesite; garnierite.*

pimple (pim'-ple) *pimple mound.*

pimple mound A term used along the Gulf Coast of eastern Texas and SW Louisiana for one of hundreds of thousands of low, flattened, rudely circular or elliptical domes composed of sandy loam that is coarser than, and distinct from, the surrounding soil; the basal diameter ranges from 3 m to more than 30 m, and the height from 30 cm to more than 2 m. Cf: *Mima mound.* Syn: *pimple.*

pimple plain A plain characterized by numerous conspicuous pimple mounds.

pin [geomorph] An Irish term for a mountain peak. Etymol: Gaelic "beann" or "beinn", "peak". Cf: *ben.*

pin [sed] (a) A thin, irregular bed, band, or seam of ironstone or other hard rock in the coal measures of south Wales. (b) A cylindrical nodule, usually of clay ironstone, in the coal measures of south Wales.

pin [surv] A sharp pointed metal rod with a ring at the end, used for marking taped measurements on the ground. A chaining pin set has 11 pins with two carrying rings. After 10 tape lengths are measured, the forward chainperson has an empty ring and rear chainperson has a ring with 10 pins. The last pin in the ground marks the end of the "tally" (Brinker and Wolf, 1977, p.59).Syn: *arrow.*

pinacocyte (pi-nac'-o-cyte) One of the cells, generally flat, of the *pinacoderm* of a sponge.

pinacoderm (pi-nac'-o-derm) An unstratified layer of cells (pinacocytes), other than the choanoderm, constituting the soft parts of a sponge and delimiting it from the external milieu; e.g. *endopinacoderm* and *exopinacoderm.*

pinacoid (pin'-a-coid) An open crystal form consisting of two parallel symmetrically equivalent faces. Adj: pinacoidal.

pinacoidal class (pin-a-coid'-al) That crystal class in the triclinic system having only a center of symmetry.

pinacoidal cleavage Mineral cleavage parallel to one of the crystal's pinacoidal surfaces; e.g. the (010) cleavage of gypsum.

pinakiolite (pi-na-ki'-ol-ite) A metallic black monoclinic mineral: $(Mg,Mn^{2+})_2(Mn^{3+},Sb^{5+})BO_5$. It is polymorphous with orthopinakiolite.

pinalite (pin-al'-ite) An adamantine yellow orthorhombic mineral: $Pb_3W^{6+}O_5Cl_2$.

pinch n. (a) A marked thinning or squeezing of a rock layer; e.g. a coming-together of the walls of a vein, or of the roof and floor of a coal seam, so that the ore or coal is more or less completely displaced. See also: *nip [coal].* (b) A thin place in, or a narrow part of, an orebody; the part of a mineral zone that almost disappears before it widens out in another place to form an extensive orebody, or swell. Cf: *make.* v. *pinch out.*

pinch-and-swell structure A structural condition commonly found in quartz veins and pegmatites in metamorphosed rocks, in which the vein is pinched and thinned at frequent intervals, leaving expanded parts between (Ramberg, 1955). Cf: *boudinage.*

pinchite (pinch'-ite) A dark-brown or black orthorhombic mineral: $Hg_5O_4Cl_2$.

pinch-out The termination or end of a stratum, vein, or other body of rock that narrows or thins progressively in a given horizontal direction until it disappears and the rocks it once separated are in contact; esp. a *stratigraphic trap* formed by the thinning-out of a porous and permeable sandstone between two layers of impermeable shale. The lithologic character of the stratum is typically maintained to the feather edge. Cf: *shale-out.* See also: *wedge-out.*

pinch out To taper or narrow progressively to extinction; to *thin out.* See also: *pinch-out.*

pinchout trap The updip loss of porosity causing the entrapment of oil or gas.

pinfire opal (pin'-fire) Opal in which the patches of *play of color* are very small and close together and usually less regularly spaced than in *harlequin opal.*

pinger (ping'-er) (a) An instrument, lowered over the side of the ship, that emits sound pulses used to determine the height of the instrument above the sea floor by observing the time for the pulse to travel directly to the ship relative to the time for the pulse to travel from the pinger to the bottom and to the ship. Used to position other instruments close to the sea floor. (b) An instrument that emits sound pulses, used to locate other instruments attached to it, e.g., used to locate the "black box" on sunken aircraft. (c) A low-frequency sound source towed from a ship and used to map reflections from below the sea floor.

pingguite A yellowish-green orthorhombic mineral: $Bi_6(Te^{4+})_2O_{13}$.

pingo (pin'-go) (a) A large *frost mound,* esp. a relatively large conical mound of soil-covered ice (commonly 30-50 m high and up to 400 m in diameter), raised in part by hydrostatic pressure of water within or below the permafrost of Arctic regions (esp. Canada), and of more than one year's duration; an intrapermafrost ice-cored hill or mound. Its crest is sometimes ruptured or collapsed due to melting of the ice, thus forming a star-shaped crater; the term has also been applied to such a depression (Monkhouse, 1965, p.237). The mound itself often resembles a small volcano. The term was introduced for this feature by Porsild (1938, p.46). Pl: pingos. See also: *ground-ice mound; ice laccolith.* Syn: *boolgoonyakh.* (b) The term has been used in several related senses, as a conical hill or mound, or as a hill completely covered by an ice sheet but revealing its presence by surface indications (ADTIC, 1955, p.61). Etymol: Eskimo, "conical hill". Syn: *pingok.*

pingok (pin'-gok) *pingo.*

pingo remnant A rimmed depression, as in the northern Netherlands, where it was previously regarded as a kettle (Gravenor and Kupsch, 1959, p.62). It is formed by rupturing of a pingo summit resulting in exposure of the ice core to melting followed by partial or total collapse. See also: *fossil pingo.* Syn: *pseudokettle.*

pinguite (pin'-guite) *nontronite.*

pinhole chert (pin'-hole) Chert containing weathered pebbles pierced by minute holes or pores.

pinite (pi'-nite) A compact, fine-grained, usually amorphous mica (chiefly muscovite) of a dull-gray, green, or brown color, derived from the alteration of other minerals (such as cordierite, nepheline, scapolite, spodumene, and feldspar).

pink snow *red snow.*

pin line In cross section balancing, a line drawn with a given orientation relative to layering in a deformed-state cross section that is envisioned to have the same orientation relative to layering in the undeformed state, or restored, cross section. Pin lines are placed where rocks lack evidence for interlayer slip, and they are used to determine the amount of interlayer slip at other locations during deformation (Geiser, 1988; Woodward et al., 1989).

pinna (pin'-na) A primary subdivision of a pinnately compound leaf or frond; a leaflet. See also: *pinnule [bot].* Pl: pinnae.

pinnacle [geomorph] (pin'-na-cle) (a) A tall slender tapering tower or spire-shaped pillar of rock, either isolated or at the summit of a mountain or hill; esp. a lofty peak. (b) A hill or mountain with a pointed summit.

pinnacle [reefs] A small, isolated spire or column of rock or coral, either slightly submerged or awash; specif. a small reef patch, consisting of coral growing sharply upward (with slopes ranging from 45° to nearly vertical), usually within an atoll lagoon, often rising close to the water surface. Syn: *pinnacle reef; reef pinnacle; coral pinnacle; patch reef; bommy.*

pinnacled iceberg (pin'-na-cled) An irregular iceberg, shaped and weathered in such a way as to be topped with spires and pinnacles.

pinnacle karst A high-relief landscape of naked, reticulated, saw-toothed ridges having almost vertical slopes. The ridges rise above forest-covered corridors and depressions. Cf: *tower karst.* Syn: *arete and pinnacle karst.*

pinnacle reef A syn. of *pinnacle [reefs].*

pinnate (pin'-nate) (a) Having leaflets or veins on each side of a common axis in a featherlike arrangement (Swartz, 1971, p.358). (b) Referring to an erect growth habit of colonies of stenolaemate bryozoans, in which lateral branches grow in the same plane from opposite sides of a main axial branch.

pinnate drainage pattern A *dendritic drainage pattern* in which the main stream receives many closely spaced, subparallel tributaries that join it at acute angles, resembling in plan a feather; it is believed to indicate unusually steep slopes on which the tributaries developed.

pinnate joints A set of en echelon cracks propagating away from mesofaults at small angles. Often pinnate joints propagate only within the two tensile quadrants immediately in the vicinity of the fault. The term, pinnate joint, is preferred over feather joint.

pinnate venation A type of *venation* in which the secondary veins branch from the midrib in parallel pattern. Cf: *palmate venation.*

pinnoite (pin'-no-ite) A yellow to greenish-yellow tetragonal mineral: $MgB_2O(OH)_6$.

pinnular (pin'-nu-lar) A plate forming part of a *pinnule* of a crinoid.

pinnule [bot] (pin′-nule) A subdivision of a *pinna,* or a secondary subdivision of a pinnately compound leaf or frond; a secondary leaflet.

pinnule [paleont] (a) One of several generally slender, unbifurcated, uniserial branches of the food-gathering system of a crinoid arm, borne on alternate sides of successive brachial plates. (b) A secondary branch of a plumelike organ, such as a digitate lateral branch of a tentacle of an octocoral polyp or one of the biserial branches of the cystoid *Caryocrinites.* Also spelled: *pinule.*

pinolite (pin′-o-lite) A metamorphic rock containing magnesite (*breunnerite*) as crystals and granular aggregates in a matrix of phyllite or talc schist. It is so named because the magnesite inclusions commonly resemble pine cones in shape (Holmes, 1928, p.184). Rarely used.

pintadoite (pin-ta′-do-ite) An ill-defined green mineral: $Ca_2V^{5+}{}_2O_7 \cdot 9H_2O$.

pinule (pin′-ule) Var. of *pinnule.*

pinwheel garnet *rotated garnet.*

pioneer (pi-o-neer′) In ecology, a community, species, flora, fauna, or individual that establishes itself in a barren area, initiating a new ecologic cycle or *sere.* Cf: *climax.*

piotine (pi′-o-tine) *saponite.*

pipe [grd wat] *geyser pipe.*

pipe [intrus rocks] A discordant pluton of tubular shape.

pipe [ore dep] A cylindrically shaped, more or less vertical orebody. The ore may be a vein deposit, a breccia column, or a diamond-bearing volcanic breccia. Syn: *ore pipe; ore chimney; chimney [ore dep]; neck [ore dep]; stock [ore dep].*

pipe [sed] (a) A tubular cavity from several centimeters to a few meters in depth, formed esp. in calcareous rocks, and often filled with sand and gravel; e.g., a vertical joint or sinkhole in chalk, enlarged by solution of the carbonate material and filled with clastic material. See also: *sand pipe.* (b) *clastic pipe.*

pipe [volc] A vertical conduit through the Earth's crust below a volcano, through which magmatic materials have passed. It is usually filled with volcanic breccia and fragments of older rock. As a zone of high permeability, it is commonly mineralized. Cf: *plug [volc]; chimney [volc]; neck [volc]; vent; diatreme.* Syn: *breccia pipe.*

pipe amygdule An elongate amygdule that occurs in a lava, towards the base of the flow, probably formed by the generation of gases or vapor from the underlying material.

pipe clay (a) A white to grayish-white, highly plastic clay, practically free from iron, suitable for use in making tobacco pipes. The term has been extended to include any white-burinng clay of considerable plasticity. Syn: *ball clay.* (b) A mass of fine clay, generally of lenticular form, forming the surface of bedrock, on which the gravel of deep leads (old river beds) frequently rests.

piperno (pi-per′-no) A welded tuff characterized by conspicuous *fiamme,* or flame structure. Such a rock is said to be pipernoid. Etymol: Italian. Syn: *welded tuff.* See also: *eutaxitic.*

pipernoid (pi-per′-noid) Said of the eutaxitic texture of certain extrusive igneous rocks, e.g. *piperno,* in which dark patches and stringers occur in a light-colored matrix. Obsolete.

pipe-rock A marine sandstone containing abundant *Scolithus* structures.

pipe-rock burrow *scolithus.*

pipestone (pipe′-stone) A collective archeological term applied to claystones and massive, fine-grained, carvable metamorphic rocks (e.g., argillite) that early people made into utilitarian and ceremonial objects. Common in North America are platform and tubular pipes (Hughes et al., 1998). Cf: *catlinite; flint clay.*

pipette analysis (pi-pette′) A kind of particle-size analysis of fine-grained sediment, made by removing samples from suspension with a pipette.

pipe vesicle Slender vertical cavities a few centimeters or tens of centimeters in length extending upward from the base of a lava flow. Most are formed by water vapor, derived from the underlying wet ground, that streamed upward into the lava. Cf: *spiracle [volc].*

piping (pip′-ing) Erosion or solution by percolating water in a layer of subsoil, resulting in caving and in the formation of narrow conduits, tunnels, or "pipes" through which soluble or granular soil material is removed; esp. the movement of material, from the permeable foundation of a dam or levee, by the flow or seepage of water along underground passages. Cf: *water creep.* Syn: *tunnel erosion.*

pipkrake (pip′-krake) (a) A small, thin spike or needlelike crystal of ground ice, from 2.5 to 6 cm in length, formed just below, and growing perpendicular to, the surface of the soil in a region where the daily temperatures fluctuate across the freezing point. It is common in periglacial areas, where it contributes to the sorting of material in patterned ground and to downslope movement of surface material. (b) A bundle, cluster, or tuft of pipkrakes. Etymol: Swedish, "needle ice". Syn: *needle ice; feather ice; mush frost; spew frost.*

pi pole In structural petrology, that pole which represents the normal to a fabric plane. Commonly written: π pole. There is confusion in the literature, and some workers have used the term as equivalent to *beta axis;* but the π pole for a particular plane is always at right angles to the β axis defined by the intersection of that plane with some other plane.

piracy (pi′-ra-cy) *capture [streams].*

pirate (pi′-rate) (a) *capturing stream.* (b) *pirate valley.*

pirated stream (pi′-rat-ed) *captured stream.*

pirate stream *capturing stream.*

pirate valley A valley that appropriates the waters of another valley; a valley containing a capturing stream. Syn: *pirate.*

pirquitasite (pir-qui′-ta-site) A metallic brownish-gray tetragonal mineral having a stannite-like structure and related to hocartite: Ag_2ZnSnS_4.

pirssonite (pirs′-son-ite) A white or grayish orthorhombic mineral: $Na_2Ca(CO_3)_2 \cdot 2H_2O$.

pisanite (pi-san′-ite) A blue triclinic mineral: $(Fe,Cu)(H_2O)_6(SO_4) \cdot H_2O$. It is a variety of melanterite containing copper.

pisekite-(Y) An ill-defined yellow-green, green, dark brown, or black metamict mineral: $(Y,As,Ca,Fe,U)(Nb,Ti,Ta)O_4$. It is probably samarskite-(Y).

pisiform (pi′-si-form) Resembling the size and shape of a pea; e.g. a "pisiform concretion" (or pisolith).

pisoid A round or ellipsoidal accretionary body commonly formed of calcium carbonate. A pisoid or pisolith is larger and less regular in form than an *ooid,* although it may have the same concentric and/or radial internal structure. Syn: *pisolith; pisolite [sed].*

pisolite [mineral] (pi′-so-lite) A variety of calcite or aragonite.

pisolite [sed] (a) A sedimentary rock, usually a limestone, made up chiefly of pisoids cemented together. (b) A term often used for a *pisolith,* or one of the spherical particles of a pisolite. Etymol: Greek "pisos", "pea". Cf: *oolite.*

pisolite [volc] An individual in a mass of *accretionary lapilli.* Obsolete.

pisolith *pisoid.*

pisolitic [ign] (pi-so-lit′-ic) Said of a tuff composed of accretionary lapilli. Obsolete.

pisolitic [sed] Pertaining to *pisolite [sed],* or to the texture of a rock made up of pisoliths; e.g. "pisolitic bauxite" or "pisolitic limestone".

pistacite (pis-ta′-cite) A syn. of *epidote,* esp. the pistachio-green variety rich in ferric iron. Also spelled: pistazite.

pistil (pis′-til) In a flower, the female reproductive organ consisting of an ovary, style (when present), and stigma. It may consist of a single *carpel* (simple pistil) or of two or more carpels (compound pistil). Syn: *gynoecium.*

pistillate (pis′-til-late) Said of a flower which has a pistil but no stamens. Cf: *staminate.*

piston corer (pis′-ton) An oceanographic *corer* containing a piston inside the cylinder which facilitates the filling of the corer with sediment by creating suction. There are several varieties, usually based on the design of the *Kullenberg corer.* Cf: *Ewing corer, Mackereth sampler, gravity corer.*

piston organelle A moundlike structure rising from the floor of peristome in a tintinnid.

pit [bot] A thin place in a cell wall. See also: *simple pit; bordered pit.*

pit [geol] (a) A small indentation or depression left on the surface of a rock or particle (esp. of a clastic particle) as a result of some eroding or corrosive process, such as etching, differential solution, or impact. (b) A shallow excavation, up to several meters depth, dug to expose a soil or geologic section of relative soft sediments.

pit [speleo] A *vertical shaft* in a cave which may or may not be open to the surface. Cf: *pothole [speleo].*

pit-and-mound structure A sedimentary structure consisting of a small blister-like mound, 1 mm high and 3-12 mm in diameter, that surrounds or contains at its summit a tiny crater-like central pit up to 1 mm in diameter simulating a rain print. It is formed during rapid settling of low-viscosity mud, by gas bubbles or water currents emerging at the surface. The term was introduced by Kindle (1916).

pitch [petroleum] *asphalt.*

pitch [photo] The angular deviation from the horizontal of a craft about the horizontal axis normal to its direction of movement. The pitch is positive if it causes the craft's front end to rise. In photogrammetry, the term is applied to a rotation of the camera, or of the coordinate system of the photograph, about either the y-axis of the photograph's coordinate system or about the y-axis of the ground coordinate system (Maune, 2001, p.493). Cf: *roll [photo], yaw [photo].*

pitch [slopes] A steep place; a declivity.

pitch [speleo] A *vertical shaft* or abrupt drop in floor level in a cave.

pitch [struc geol] *rake.*

pit chamber The space between the pit chamber and the over-arching border of a *bordered pit* (Swartz, 1971, p.359).

pitchblende (pitch′-blende) A massive brown to black variety of uraninite, found in hydrothermal sulfide-bearing veins. It is fine-grained (colloform), amorphous, or microcrystalline, and has a distinctive pitchy to dull luster. Pitchblende contains a slight amount of radium; thorium and the rare earths are generally absent. Syn: *pitch ore; nasturan.*

pitch coal (a) A brittle, lustrous bituminous coal or lignite, with conchoidal fracture. Syn: *bituminous lignite; bituminous brown coal; glance coal; picurite; specular coal.* (b) A kind of *jet [coal].*

Pitcher sampler A double-walled soil sampler in which a spring-loaded Shelby tube is placed inside a steel sampling barrel. The Shelby tube extends below the bottom of the outer barrel in soft soil but retracts relative to the outer barrel in stiff soil.

pitches and flats (pitch′-es) *flats and pitches.*

pitching fold (pitch'-ing) A fold of which the hinge line is inclined to the horizontal. Cf: *rake.*
pitch opal A yellow to brown inferior quality of common opal displaying a pitchy luster.
pitch ore (a) *pitchblende.* (b) *pitchy copper ore.*
pitch peat Peat that resembles asphalt.
pitchstone (pitch'-stone) A volcanic glass (usually silicic) with a waxy dull resinous luster. Its color and composition vary widely; it contains a higher percentage of water than *obsidian.* Crystallites are detectable in thin section.
pitchy copper ore (pitch'-y) A dark, pitchlike oxide of copper; a mixture of chrysocolla and limonite. Syn: *pitch ore.*
pitchy iron ore (a) *pitticite.* (b) *triplite.*
pit crater A *sink [volc]* or a small *caldera.*
pith The spongy central tissue of an exogenous stem, consisting chiefly of parenchyma (Jackson, 1928, p.287).
pith rays Primary bands of parenchyma cells extending from the pith to the pericycle in herbaceous and young woody stems (Fuller and Tippo, 1954, p.967).
pitiglianoite A vitreous colorless hexagonal mineral of the *cancrinite* group: $K_2Na_6(AlSiO_4)_6(SO_4){\cdot}2H_2O$.
pit lake *kettle lake.*
piton [karst] A limestone hill having a sharply pointed peak. Etymol: French.
piton [volc] (pi-ton') A term commonly used for volcanic peaks, especially steep-sided domes, in the West Indies and other French-speaking regions.
pit run *bank gravel.*
pitted outwash (pit'-ted) Outwash with pits or kettles, produced by the partial or complete burial of glacial ice by outwash and the subsequent thaw of the ice and collapse of the surficial materials. Cf: *pitted outwash plain.*
pitted outwash plain An *outwash plain* marked by many irregular depressions such as kettles, shallow pits, and potholes; many are found in Wisconsin and Minnesota. Cf: *pitted outwash; kettle plain.* Syn: *pitted plain.*
pitted pebble A pebble having marked concavities not related to the texture of the rock in which it appears or to differential weathering (Kuenen, 1943). The depressions range in size from minute pits caused by sand particles to cups a few centimeters across and a centimeter deep; they are common at the contacts between adjacent pebbles, and have been explained as the result of pressure-induced solution at points of contact. The term has also been applied to cobbles. Cf: *cupped pebble.* Syn: *scarred pebble.*
pitted plain *pitted outwash plain.*
pitticite (pit'-ti-cite) A generic name for ill-defined yellowish or reddish hydrous ferric arsenates found in reniform masses. Syn: *pitchy iron ore.* Also spelled: pittizite.
pivotability (piv'-ot-a-bil'-i-ty) A measure of roundness of sedimentary particles, expressed by the ease with which a particle can be dislodged from a surface or by the tendency of a particle to start rolling on a slope. The term was introduced by Shepard and Young (1961, p.198) who assigned the highest values of roundness to the particles that "could be most easily pivoted".
pivotal fault (piv'-o-tal) An obsolete partial syn. of *hinge fault.* Cf: *scissor fault.*
PIXE *particle-induced X-ray emission.*
pixel (pix'-el) In a digitized image, the area on the ground represented by each digital value. Because the analog signal from the detector of a scanner may be sampled at any desired interval, the size of the pixel may be different from the *ground resolution cell* of the detector. Etymol: Contraction of *picture element.*
piypite (pi'-yp-ite) A green to black tetragonal mineral that occurs as a volcanic sublimate: $K_4Cu_4O_2(SO_4)_4(Na,Cu)Cl$. Syn: *caratiite.*
placanticline (plac-an'-ti-cline) A gentle, anticlinelike uplift of the continental platform, usually asymmetric and without a typical outline. There is no corresponding synclinelike structure. The term is used mainly in the Russian literature of the Volga-Urals region (Shatsky, 1945).
placental (pla-cen'-tal) A member of the mammalian subclass Eutheria, characterized by bearing young in a relatively advanced state of development by means of a placenta. Range, Cretaceous to Holocene.
placer (plac'-er [plass'-er]) A surficial mineral deposit formed by mechanical concentration of mineral particles from weathered debris. The common types are beach placers and *alluvial* placers. The mineral concentrated is usually a heavy, durable mineral such as gold, cassiterite, or rutile. Cf: *lode [eco geol].* Syn: *lead [eco geol]; ore of sedimentation.*
placer claim A *claim* on a placer deposit in which a discovery has been made. Cf: *lode claim.*
placer mining The extraction and concentration of heavy metals or minerals from placer deposits by various methods, generally using running water. Cf: *hydraulic mining; drift mining; dredging.*
place value The value that a mineral deposit has by virtue of its location. Not to be confused with *in-place value.*
placic horizon (plac'-ic) A black to dark reddish mineral soil horizon that is usually thin but that may range from 1 mm to 25 mm in thickness. The placic horizon is commonly cemented with iron and is slowly permeable or impenetrable to water and roots.
Placodermi (Plac-o-der'-mi) A class of jawed vertebrates characterized by development of dermal armor, very heavy in primitive forms, in which it occurs as elaborately jointed head and trunk shields; ossification of the internal skeleton is poor. Range, Lower to Upper Devonian.
placolith (plac'-o-lith) A perforate coccolith having two shields connected by a central tube. See also: *tremalith.* Syn: *cyatholith.*
pladdy (plad'-dy) A term used in Northern Ireland for a "residual island drumlin awash at high tide" (Stamp, 1961, p.365).
plage A French term for a sandy beach, esp. at a seaside resort. The term is being supplanted by *lido.*
plaggen epipedon (plag'-gen) A diagnostic surface soil horizon that is manmade by manuring and mixing, and that is more than 50cm in thickness (USDA, 1975).
Plaggepts Inceptisols that have a plaggen epipedon. A suborder in the U.S. system of soil taxonomy.
plagiaplite (pla-gi-ap'-lite) An *aplite* composed chiefly of plagioclase (oligoclase to andesine), possibly green hornblende, and accessory quartz, biotite, and muscovite; a fine-grained *diorite.* Obsolete.
plagioclase (pla'-gi-o-clase') (a) A group of triclinic feldspars of general formula: $(Na,Ca)[Al(Si,Al)Si_2O_8]$. At high temperatures it forms a complete solid-solution series from Ab $Na[AlSi_3O_8]$ to An $Ca[Al_2Si_2O_8]$. The plagioclase series is arbitrarily subdivided and named according to increasing mole fraction of the An component: albite (An 0-10), oligoclase (An 10-30), andesine (An 30-50), labradorite (An 50-70), bytownite (An 70-90), and anorthite (An 90-100). The Al/Si ratio ranges with increasing An content from 1:3 to 1:1. Plagioclase minerals are among the commonest rock-forming minerals, have characteristic twinning, and commonly display zoning. (b) A mineral of the plagioclase group; e.g. albite, anorthite, peristerite, and aventurine feldspar. The term was introduced by Breithaupt (1847, p.490) who applied it to all feldspars having an oblique angle between the two main cleavages. Cf: *alkali feldspar; orthoclase.* Syn: *sodium-calcium feldspar.*
plagioclase arenite A term used by McBride (1963, p.668) for an arkose containing more than 25% plagioclase, and by Folk (1968, p.124) for an arkose in which plagioclase is the main feldspar.
plagioclase arkose An arkose of which the chief feldspar is plagioclase (Folk, 1968, p.130).
plagioclase-bearing hornblende pyroxenite In the *IUGS classification,* a plutonic rock in which plagioclase/(plagioclase+pyroxenes+hornblende) <10 and pyroxenes>hornblende.
plagioclase-bearing hornblendite In the *IUGS classification,* a plutonic rock in which plagioclase/(plagioclase+pyroxenes+hornblende)<10 and hornblende/(plagioclase+pyroxenes+hornblende)>90.
plagioclase-bearing pyroxene hornblendite In the *IUGS classification,* a plutonic rock in which plagioclase/(plagioclase+pyroxenes+hornblende) <10, pyroxenes/(plagioclase+pyroxenes+hornblende)<90 and pyroxenes<hornblende.
plagioclase-bearing pyroxenite In the *IUGS classification,* a plutonic rock in which plagioclase/(plagioclase+pyroxenes+hornblende)<10, and pyroxenes/(plagioclase+pyroxenes+hornblende)>90.
plagioclase lherzolite A *lherzolite* containing plagioclase as the dominant aluminum-rich mineral, stable at shallower depths in the mantle than either *spinel lherzolite* or *garnet lherzolite.*
plagioclase rock *anorthosite.*
plagiogranite (pla'-gi-o-gran'-ite) (a) In the *IUGS classification,* a syn. of *trondhjemite* and leucocratic *tonalite.* (b) A term commonly used by Russian petrologists for a *granitoid* having a low potassium content. A low-K *granitoid* associated with mid-ocean-ridge (or ophiolite) basaltic rocks as either a partial melt or differentiate. It includes rocks ranging from *quartz diorite* to *trondhjemite.*
plagionite (pla'-gi-o-nite') A blackish lead-gray monoclinic mineral: $Pb_5Sb_8S_{17}$.
plagiostome (pla'-gi-o-stome') An asymmetrically placed aperture or pseudostome in a thecamoebian test (as in *Centropyxis* or *Plagiopyxis*).
plain (a) Any flat area, large or small, at a low elevation; specif. an extensive region of comparatively smooth and level or gently undulating land, having few or no prominent surface irregularities but sometimes having a considerable slope, and usually at a low elevation with reference to surrounding areas. A plain may be either forested or bare of trees, and may be formed by deposition or by erosion. (b) An extensive tract of level or rolling, almost treeless country with a shrubby vegetation; a *prairie.* In Australia, "plain" implies treelessness. The term is usually used in the plural. (c) A region underlain by horizontal strata or characterized by horizontal structure, which may be dissected into hills and valleys by stream erosion; Davis (1885) introduced this concept of a "plain" but the term should be used without regard to the underlying geologic structure. Cf: *plateau [geomorph].*
plain of denudation A surface that has been reduced to or nearly to sea level by the agents of erosion (usually considered to be of subaerial origin); it is relatively flat but may be marked by residual hills of resistant rock rising somewhat above the general level. See also: *plain of marine denudation.* Cf: *peneplain.*
plain of lateral planation An extensive smooth apronlike surface developed at the base of a mountain or escarpment (as at Book Cliffs, Utah) by the widening of valleys and the coalescence of flood plains as a result of *lateral*

planation. It resembles the landscape form of a *pediment* and is often so called. Cf: *panplain.*

plain of marine denudation A hypothetical plane or nearly plane surface worn down by the gradual encroachment of ocean waves upon the land; or an imaginary plane representing such a surface after uplift and partial subaerial erosion. The concept was first recognized by Ramsay (1846); however, the term has often been applied to a plain produced by subaerial or nonmarine agents, which was subsequently submerged beneath the sea, only a minor role being assigned to wave erosion. Syn: *plain of submarine denudation.*

plain of marine erosion A hypothetical *platform,* representing a plane surface of unlimited width, produced below sea level by the cutting-away of the land by marine processes acting over a very long period of stillstand; the ultimate abrasion platform. Cf: *plain of marine denudation; marine peneplain.* Syn: *marine plain; marine plane; submarine plain.*

plain of submarine denudation *plain of marine denudation.*

plain tract The lower part of a stream, characterized by a low gradient and a wide floodplain. Cf: *valley tract.* See also: *deltaic tract.*

Plaisancian (Plai-san'-ci-an) A European stage: Lower Pliocene (above Zanclean, below Astian).

plaiting (plait'-ing) A texture seen in some schists that results from the intersection of relict bedding planes with well developed cleavage planes.

plan A drawing, sketch, or diagram of any object or structure, made by horizontal projection upon a plane or flat surface; esp. a very large-scale and considerably detailed *map* of a small area, such as one showing underground mine workings.

planaas (pla-naas' [pla-nohs']) A Danish term for an outwash plain formed as a flat-topped delta in standing water between two walls of stagnant ice.

planaperturate (plan-a-per'-tu-rate) In reference to a pollen grain's angular *amb,* having the apertures situated in the middle of the sides, e.g., in *Tilia.* Ant: *angulaperturate.*

planar (pla'-nar) Lying or arranged as a plane or in planes, usually implying more or less parallelism, as in bedding or cleavage. It is a two-dimensional arrangement, in contrast to the one-dimensional *linear* arrangement.

planar cross-bedding (a) Cross-bedding in which the lower bounding surfaces are planar surfaces of erosion (McKee and Weir, 1953, p.385); it results from beveling and subsequent deposition. (b) Cross-bedding characterized by planar foreset beds.

planar deformation features Submicroscopic amorphous lamellae occurring in shocked minerals as multiple sets of planar lamellae (optical discontinuities under the petrographic microscope) parallel to rational crystallographic planes; they are indicative of *shock metamorphism.* Less preferred syn: planar elements; *shock lamellae.*

planarea (pla-nar'-e-a) One of two flattened areas developed on either side of the posterior part of a brachiopod shell in place of the more common single median *interarea.* Cf: *palintrope.*

planar element A fabric element having two dimensions that are much greater than the third. Examples are bedding, cleavage, and schistosity. Cf: *linear element; equant element.*

planar features Closely spaced parallel microscopic planes, distinct from cleavage planes, that occur in shock-metamorphosed minerals (particularly quartz and feldspar) and are regarded as unique and important indicators of shock metamorphism. The structures are characteristically multiple (often more than five distinct sets per grain) and are oriented parallel to specific planes in the host-crystal lattice. They have been produced experimentally by shock pressures of 80 to 250 kb. Syn: *shock lamellae.*

planar flow structure *platy flow structure.*

planar irregular spherulitic prismatic structure An *irregular spherulitic prismatic structure* in which the second-order prisms are horizontally flattened (Carter, 1990, Fig. 8). This corresponds in part with the "planar spherulitic" structure of Sandberg (1977).

planar peat Peat that fills topographic depressions and has a nearly flat upper surface. The water table in the peat is generally at or above the surface and at or below the local water table (Cecil et al., 1985). Cf: *topogenous peat; lowmoor peat; domed peat.*

planar spherulitic structure A *microstructure* of horizontally flattened spherulites, with second-order structural units radiating more or less equally in all horizontal directions from a single crystal aggregate or point of origin. This is the "planar spherulitic ultrastructure" in the basal wall of the calcitic bryozoan Megapora ringens described by Sandberg (1983).

planate (pla'-nate) adj. Said of a surface that has been flattened or leveled by *planation;* e.g. a pediment is a nearly planate erosion surface. v. To reduce to a plain or other flat surface.

planate element A pectiniform *conodont element* with conspicuous lateral ledges, brims, or platforms flanking one or more of its processes; and with an attachment surface on its *under side* distinguished by a zone of recessive basal margin, which at least partially surrounds a basal pit that has groovelike extensions beneath at least the primary processes (TIP, 1981, pt. W, supp.2).

planation (pla-na'-tion) (a) The process or processes of erosion whereby the surface of the Earth or any part of it is reduced to a fundamentally even, flat, or level surface; specif. *lateral planation* by a meandering stream. The term also includes erosion by waves and currents, and abrasion by glaciers or wind, in producing a flat surface. The term was originated by Gilbert (1877, p.126-127) who considered alluviation of the flattenend surface as part of the planation process; however, this condition is not necessary. (b) A broad term for the general lowering of the land; e.g. *peneplanation; panplanation; pediplanation; cryoplanation; altiplanation.*

planation stream piracy Capture effected by the lateral planation of a stream invading and diverting the upper part of a smaller stream.

planation surface *erosion surface.*

planchéite (plan-ché'-ite) A blue orthorhombic mineral: $Cu_8Si_8O_{22}(OH)_4{\cdot}H_2O$. It possesses amphibole inosilicate ribbons. Cf: *shattuckite.*

Planckian The Planck "era" is a fanciful term for the first split-second of the Big Bang, ranging from cosmological time $t = 0$ to cosmological time $t = 5.390 \times 10^{-44}$ seconds (Emiliani, 1987, p.163). It is the time light takes to cross the "Planck length", a fundamental unit of length equal to $(Gh/2\pi^3)^{1/2} = 1/616 \times 10^{-35}$ m, where G = universal gravitational constant, h = Planck's constant, c = speed of light (Emiliani, 1992, p.66). The Planckian is named after Max Planck (1858-1947).

plane A two-dimensional form that is without curvature; ideally, a perfectly flat or smooth surface. In geology the term is applied to such features as a *bedding plane* or a planation surface. Adj: *planar.* Cf: *surface.*

plane bed A flat, almost featureless surface that is produced by the interaction of a *unidirectional current* flowing over a mobile sediment bed (e.g., *upper plane bed; lower plane bed*).

plane coordinates (a) Two coordinates that represent the perpendicular distances of a point from a pair of axes that intersect at right angles, reckoned in the plane of those axes. (b) A coordinate system in a horizontal plane, used to describe the positions of points with respect to an arbitrary origin by means of plane coordinates. It is used in areas of such limited extent that the errors introduced by substituting a plane for the curved surface of the Earth are within the required limits of accuracy. Syn: *rectangular coordinates; plane-rectangular coordinates.*

plane correction A correction applied to observed surveying data to reduce them to a common reference plane.

plane defect A type of *crystal defect* that occurs along the boundary plane (*lineage boundary*) between two regions of a crystal, or along the *grain boundary* between two crystals. Cf: *line defect; point defect.* See also: *stacking fault.*

plane group One of 17 kinds of two-dimensional patterns that can be produced by one asymmetric motif that is repeated by symmetry operations to produce a unit of pattern, which then is repeated by translation to build up an ordered pattern that fills any two-dimensional area. Cf: *space group.* Syn: *plane symmetry group.*

plane jet A flow pattern characteristic of *hyperpycnal inflow,* in which the inflowing water spreads as a parabola whose width is about three times the square root of the distance downstream from the mouth (Moore, 1966, p.87). Cf: *axial jet.*

plane of composition A term used by Cullison (1938, p.983) for the plane of contact between the part of an internal mold composed of material similar to the matrix in which the fossil shell was embedded and the material (such as secondary calcite) that partially filled the hollow interior of the shell. It defines the horizontal plane at the time the filling became hardened. Not to be confused with *composition plane.* Cf: *geopetal.*

plane of contemporaneity A term used by Caster (1934, p.19, 24) for the horizontal or nearly horizontal surface between stratigraphic units (primarily formations) as seen in section; e.g. the surface separating parvafacies belonging to the same magnafacies. Cf: *facies plane.*

plane of incidence A plane that contains an incident ray and the normal to the surface at the point of incidence.

plane of maximum shear stress Either of two planes that lie on opposite sides of the maximum principal stress axis at angles of 45° to it and that are parallel to the intermediate principal stress axis.

plane of mirror symmetry A symmetry element that is a plane dividing an object (crystal, array of atoms, etc.) into halves, one of which is the mirror image of the other. Syn: *mirror plane of symmetry; plane of symmetry [cryst]; symmetry plane; reflection plane.*

plane of polarization *vibration plane.*

plane of stratification *bedding plane.*

plane of symmetry [cryst] *plane of mirror symmetry.*

plane of symmetry [paleont] The plane that bisects the body of an organism symmetrically.

plane of vibration *vibration plane.*

plane-polarized (plane-po'-lar-ized) Said of a moving wave, e.g. of light, that has been polarized so that it vibrates in a single plane.

plane-rectangular coordinates *plane coordinates.*

planerite (pla'-ner-ite) A chalky white, pale blue or pale green triclinic mineral of the turquoise group: $Al_6(PO_4)_2(PO_3OH)_2(OH)_8{\cdot}4H_2O$.

plane strain A state of strain in which all displacements that arise from deformation are parallel to one plane, and the longitudinal strain is zero in one principal direction.

plane stress A state of stress in which one of the principal stresses is zero.

plane surveying Ordinary field and topographic surveying in which earth curvature is disregarded and all measurements are made or reduced parallel to a plane representing the surface of the Earth. The accuracy and precision of results obtained by plane surveying will decrease as the area surveyed increases in size. Cf: *geodetic surveying.*

plane symmetry group *plane group.*

planet (plan'-et) (a) One of the nine large celestial bodies of the solar system that revolve around the Sun in elliptical orbits and in the same direction. A planet shines only by reflected light. (b) A similar body in another solar system.

planetable (plane'-ta-ble) A simple surveying instrument for graphically plotting the lines of a survey directly from field observations. A drawing board mounted on a tripod, together with a straight-edge having a sighting device such as a telescope attached for pointing the straight-edge at the object observed. The planetable allows an observed direction to be plotted directly on a sheet fastened to the drawing boart. The straight-edge is frequently graduated so that the distance to the object can be laid off to scale also. The term planetable is usually applied to the assemblage of drawing board, tripod and alidade (straight-edge plus sighting device) but may also be applied to the drawing board alone or to the combination of drawing board and tripod. Also spelled: plane table.

planetable map A map made by planetable surveying methods. It includes maps made by complete field mapping on a base projection and by field contouring on a planimetric base map.

planetabling (plane'-ta-bling) Plotting with, or making use of, a planetable.

planetary (plan'-e-tar-y) (a) Pertaining to the planets of the solar system. Cf: *terrestrial.* (b) Pertaining to the Earth as a whole.

planetary geology A science that applies principles and techniques of geology, geochemistry, and geophysics to the study of the nature, origin, and history of the condensed matter and gases in the solar system (usually excluding the Earth). It includes: remote-sensing observations and in situ manned exploration of other planetary bodies, such as the Moon and Mars, Venus, Mercury, and "earthlike" or "rocky" satellite moons of the giant planets (Shirley and Fairbridge, 1997); the study of the chemistry, mineralogy, and history of objects that occur on the Earth but are of known or possible extraterrestrial origin (such as meteorites and tektites) or that are returned to the Earth (such as lunar samples); and the study of the effects of extraterrestrial processes (such as meteorite impact, solar energy changes, and tides) on the Earth in the present and past. The term was first used by Lesevich (1877) for a branch of astronomy based primarily on the study of meteorites and secondarily on telescopic spectroscopy (see Milton, 1969). Syn: *astrogeology, planetology.*

planetary vorticity The spin imparted to all elements of the Earth, including seawater, by Earth's rotation. It is equal to the *Coriolis parameter;* e.g., the planetary vorticity of the ocean at the North Pole is 4 pi radians per sidereal day.

planetary wave A major, prominent atmospheric wave characterized by a long wavelength, a significant amplitude, and a velocity directed always to the west. Also, a similar free progressive wave in the model ocean, not yet convincingly demonstrated to exist there, but believed to be about 1,600 km (1,000 mi) long and to be caused by the gravitational attraction of the Sun and Moon on the Earth but largely governed by the depth of the water and by the Earth's rotational effects. Also known as a *Rossby wave,* for Carl Gustav Rossby, a Swedish American meteorologist.

planetary wind Any major part of the worldwide system of large-scale wind circulation, e.g. the *trade winds.*

planetesimal A hypothetical solid celestial body that accumulated during the first stages of planet formation. Planetesimals aggregated into increasingly larger bodies, ultimately forming the terrestrial planets.

planetesimal hypothesis (plan-e-tes'-i-mal) *dust-cloud hypothesis.*

planetoid (plan'-et-oid) *asteroid [astron].* Planetoid is sometimes used in popular literature, but is now uncommon.

planetology (plan-e-tol'-o-gy) A term originally applied to the study and interpretation of surface markings of planets and their natural satellites, and later to the study of the condensed matter of the solar system, including planets, satellites, asteroids, meteorites, and interplanetary material. The term is frequently used as a syn. of *planetary geology,* and was redefined by Rankama (1962, p.519) as "the universal science that studies the configuration and movements of matter and the accompanying energy transformations in planets, their natural satellites, and other cosmic bodies of a similar nature" in our solar system and other possible planetary systems. Syn: *planetary geology.*

plane wave A wave with planar wavefronts.

plane-wave decomposition *tau-p mapping.*

planèze (pla-neze') An erosional relief form consisting of a lava flow protecting the underlying volcanic cone. It may be a wedge-shaped unit on the slope of an erosionally dissected volcano, or a lava-capped plateau. Etymol: French, "lava plateau". Also spelled: planeze.

planform The shape of a feature as seen in two dimensions, horizontally, as on a map.

planimeter (pla-nim'-e-ter) A mechanical instrument for measuring the area of any plane figure by means of a pointer or moving arm that traces its boundary or perimeter. It is used esp. for measuring irregular areas on a chart or map.

planimetric map (pla-ni-met'-ric) A map that presents only the relative horizontal positions of natural or cultural features, by lines and symbols. It is distinguished from a *topographic map* by the omission of relief in measurable form. Cf: *contour map.*

planimetry (pla-nim'-e-try) (a) The measurement of plane surfaces; e.g. the determination of horizontal distances, angles, and areas on a map. (b) The plan details of a map; the natural and cultural features of a region (excluding relief) as shown on a map.

planispiral (pla-ni-spi'-ral) adj. Having the shell coiled in a single plane; esp. said of gastropod and cephalopod shells formed by a spiral coiled in a single plane and ideally symmetric in that plane, and said of a coiled foraminiferal test with whorls of the coil in a single plane. Also spelled: *planospiral.* n. A planispiral shell or test.

planitia (pla-nit'-ia) A term established by the International Astronomical Union for a plain lower than the surrounding terrain on a planet, generally of regional extent (USGS, 2003). Generally used as part of a formal name for a landform, such as Acidalia Planitia on Mars (Mutch et al., 1976, p.57). Pl: planitiae.

plankter (plank'-ter) An individual planktonic organism.

planktic (plank'-tic) *planktonic.*

planktivorous (plank-tiv'-o-rous) Said of an organism that feeds on plankton.

plankton (plank'-ton) Aquatic organisms that drift, or swim weakly. See also: *phytoplankton; zooplankton.* Adj: *planktonic.*

plankton bloom An aquatic growth of algae in such concentration as to discolor the water and possible cause fish kill. See also: *red tide, harmful algal bloom.* Syn: *water bloom, bloom [oceanog].*

plankton equivalent A quantitative chemical relationship between one aspect of plankton and another, e.g. in phytoplankton, 1 mg carbon = 2.3 mg dry organic matter.

planktonic (plank-ton'-ic) Refers to that type of pelagic organism which drifts; adj. of *plankton.* Cf: *nektonic.* Syn: *planktic.*

plankton snow *marine snow.*

planoconformity (pla'-no-con-form'-i-ty) A term used by Crosby (1912, p.297) for the relation between conformable strata that are approximately uniform in thickness and sensibly parallel throughout.

plano-convex (pla'-no-con-vex') Flat on one side and convex on the other; e.g. said of a brachiopod shell having a flat *brachial valve* and a convex *pedicle valve.* Cf: *convexo-plane.*

planophyre (plan'-o-phyre) A porphyritic rock characterized by planophyric texture. Obsolete.

planorasion (pla-no-ra'-sion) The process by which wind, working in conjunction with other erosional agents in a desert, "acts as an abrading and eroding agent" that works uphill (Hobbs, 1917, p.48); the process may grade a slope as steep as 4 degrees. Obsolete. Cf: *antigravitational gradation.*

Planosol (Plan'-o-sol) A great soil group of the intrazonal order and hydromorphic suborder consisting of soils with eluviated surface horizons underlain by B horizons more strongly eluviated, cemented, or compacted than associated normal soil. Not used in current U.S. system of soil taxonomy.

planospiral (pla-no-spi'-ral) *planispiral.*

plan-position indicator (plan-po-si'-tion) A radar-display device that consists of an oscilloscope (radarscope) equipped with a cathode-ray tube and a rotating antenna for scanning horizontally all or part of a complete circle, and that presents visually or graphically in plan (map) position the range and direction (azimuth or bearing) of an object (such as a ship, building, cliff, or mountain), from which echoes are reflected in the form of spots of light whose brightness corresponds to the strength of the target signal detected by radar. The position of the radar itself is displayed in the center of the indicator. Abbrev: PPI.

plant [bot] Any member of the vegetable group (plant kingdom) of living organisms.

plant [seis] n. (a) The manner in which a geophone is placed on or in the earth. Syn: *seismometer plant.* (b) The coupling to the ground. v. To place a geophone in its proper place on the ground.

plant ash *inherent ash.*

plant indicator *indicator plant.*

plant opal *opal phytolith.*

plant residues A term used in archaeology for organic remains preserved within archaeological materials, usually pottery.

planula (plan'-u-la) The very young free-swimming larva of a coelenterate (such as of a coral polyp), consisting of an outer layer of ciliated ectoderm cells and an internal mass of endoderm cells. Pl: planulae. Syn: *parenchymula.*

planulate (plan'-u-late) Said of a moderately *evolute* and compressed cephalopod shell with an open *umbilicus* and a bluntly rounded venter.

planum (pla'-num) A term established by the International Astronomical Union for a plateau or high plain on planetary surfaces (USGS, 2003). Generally used as part of a formal name for a landform or region, such as Lunae Planum on Mars (Mutch et al., 1976, p.57). Etymol: Latin *planum,* level. Pl: plana.

plash A shallow short-lived pool or small pond resulting from a flood, heavy rain, or melting snow; a puddle.

plasma [mineral] (plas'-ma) A semitranslucent green variety of chalcedony, sometimes having white or yellowish spots. The green color is attributed to chlorite. Cf: *bloodstone.*

plasma [soil] The part of a soil material that is capable of being, or has been, moved, reorganized, and/or concentrated by soil-forming processes (Brewer and Sleeman, 1960); e.g. all mineral or organic material of colloidal size, and the relatively soluble material that is not bound up in skeleton grains. See also: *floc.*

plaster A hemihydrate of calcium sulfate: $CaSO_4 \cdot \frac{1}{2}H_2O$. A construction material made by the calcining of *gypsum.* Also known as plaster of Paris.

plaster conglomerate (plas′-ter) A conglomerate composed entirely of boulders derived from a partially exhumed monadnock and forming a wedgelike mass on its flank.

plastering-on (plas′-ter-ing-on′) The addition of material to a ground moraine by the melting of ice at the base of a glacier (Gravenor and Kupsch, 1959, p.60). Cf: *lodgment.*

plaster stone *gypsum.*

plastic [biol] (plas′-tic) Having the capability of variation and phylogenetic change.

plastic [struc geol] Said of a body in which strain produces continuous, permanent deformation without rupture. Cf: *elastic.*

plastic behavior The response of a rock to stress such that permanent strain does not accumulate until the yield stress is exceeded.

plastic deformation (a) Permanent deformation of the shape or volume of a substance, without rupture. It is mainly accommodated by *crystal gliding* and/or *dynamic recrystallization.* (b) Deformation by one or both of two grain-scale mechanisms: slip, and twinning. This is a metallurgical definition, increasingly used by geologists. Sometimes called "crystal plasticity." (c) Rheological term for deformation characterized by a *yield stress,* which must be exceeded before flow begins.

plastic deformation mechanism The mechanism by which permanent deformation occurs, including *mechanical twinning, dislocation glide, dislocation creep, diffusional mass transfer,* and *grain boundary sliding.*

plastic equilibrium State of stress within a soil mass or a portion thereof that has been deformed to such an extent that its ultimate shearing resistance is mobilized (ASCE, 1958, term 263). Plastic equilibrium is "active" if it is obtained by lateral expansion of a soil mass, and "passive" if obtained by lateral compression of a soil mass.

plasticity index (plas-tic′-i-ty) The water-content range of a soil at which it is plastic, defined numerically as the liquid limit minus the plastic limit.

plastic limit The water-content boundary of a sediment, e.g. a soil, between the plastic and semisolid states. It is one of the *Atterberg limits.* Cf: *liquid limit.*

plastic relief map A topographic map printed on plastic and then molded by heat and pressure into a three-dimensional form to emphasize the relief.

plastic shading Archaic syn. of *hill shading* (BNCG, 1966, p.31).

plastic soil A soil capable of being molded or deformed continuously and permanently, by relatively moderate pressure, into various shapes.

plastic strain *plastic deformation.*

plastic zone (a) The region of inelastic deformation surrounding a crack tip in elastic-plastic materials such as metals. (b) A region adjacent to the rupture zone of an explosion crater and at an increased distance from the shock site, differing from the rupture zone by having less fracturing and only small permanent deformations.

plastotype (plas′-to-type) An artificial specimen molded or cast directly from a type specimen.

plastron (plas′-tron) (a) The more or less inflated and enlarged adoral segment of the posterior interambulacral area of certain echinoids. (b) A firmly sutured, rigid belly-shield, formed from flattened and expanded girdle bones (in plesiosaurs) or from dermal scales combined with gastralia and girdle bones (in turtles). Adj: plastral.

plat [cart] (a) A diagram drawn to scale, showing boundaries and subdivisions of a tract of land as determined by survey, together with all essential data required for accurate identification and description of the various units shown and including one or more certificates indicating due approval. It differs from a map in that it does not necessarily show additional cultural, drainage, and relief features. (b) A precise and detailed plan or map representing a township, private land claim, mineral claim, or other surveyed area, and showing the actual or proposed divisions, special features, or uses of the land. Syn: *cadastral map.* See also: *plot.*

plat [geog] An obsolete term for a plateau, tableland, or other expanse of open level land.

platarsite (plat-ars′-ite) A metallic cubic mineral of the *gersdorffite* group: (Pt,Rh,Ru)AsS.

plate [geol] (a) A thin flat fragment of rock, such as a slab or flagstone. (b) A torsionally rigid thin segment of the Earth's lithosphere, which may be assumed to move horizontally and adjoins other plates along zones of seismic activity (Dennis and Atwater, 1974, p.1032). See also: *plate tectonics.* (c) A snow crystal in the form of a flat hexagonal plate.

plate [paleont] (a) Any discrete, normally flat or tabular *ossicle* in the skeleton of an echinoderm, composed of a single crystal of calcium carbonate. The term is sometimes used only for external plates, "but all calcareous bodies formed serve as framework of support for soft parts and constitute plates" (TIP, 1967, pt. S, p.113). (b) A structure consisting of inner and outer platforms and adjoining a portion of the axis of a platelike conodont. The term is used incorrectly when referring to a platform. (c) A *lamina* that forms part of an animal body, such as a valve of a mollusk or crustacean, or a flat octocorallian *sclerite* too thick to be called a scale. (d) A flat, calcified, generally rhomboidal structure, forming the most distal portion of a receptaculitid merome.

plate [palyn] A flat, polygonal, separable unit of the *theca* of a *dinoflagellate.* The corresponding polygonal area of a dinoflagellate *cyst* may also be called a plate or may be called a plate-equivalent or *paraplate* (Williams et al., 2000, p.168-169, 152).

plateau [geomorph] (pla-teau′) (a) Broadly, any comparatively flat area of great extent and elevation; specif. an extensive land region considerably elevated (>150 m in altitude) above the adjacent country or above sea level. It is commonly limited on at least one side by an abrupt descent; has a flat or nearly smooth surface but is often dissected by deep valleys and surmounted by high hills or mountains; and has a large part of its total surface at or near the summit level. A plateau is usually higher and has more noticeable relief than a *plain* (it often represents an elevated plain), and is usually higher and more extensive than a *mesa;* it may be tectonic, residual, or volcanic in origin. Cf: *tableland.* (b) A flat, upland region underlain by horizontal strata or characterized by horizontal structure, which may be highly dissected. Davis (1885) introduced this concept of a "plateau", but the term should be used without regard to the underlying geologic structure. Etymol: French. Pl: plateaus; plateaux.

plateau [marine geol] A broad, more or less flat-topped elevation of the sea floor, generally over 200 m in height. Syn: *submarine plateau.*

plateau age In the *argon-40/argon-39 age method* and the *xenon-xenon age method,* an age calculated by averaging results from gas fractions released continuously that have similar apparent ages, within analytical uncertainties.

plateau basalt *flood basalt.*

plateau eruption Successive lava flows that spread in sheets over a large area. Cf: *fissure eruption.*

plateau glacier A *highland glacier* overlying a relatively flat mountain tract, and usually overflowing its edges, forming *hanging glaciers.* Cf: *ice plateau.*

plateau gravel A sheet or patch of surficial gravel commonly occupying a flat area on a hilltop, plateau, or other high region at a height above that normally occupied by a stream-terrace gravel. It may represent a formerly extensive deposit that has been tectonically raised and largely removed by erosion.

plateau mountain A *pseudomountain* produced by the dissection of a plateau; e.g. the Catskill Mountains, N.Y.

plateau plain An extensive plain surmounted by a sublevel summit area and bordered by escarpments (Hill, 1900, p.8). Cf: *mesa plain.*

plate boundary Zone of seismic and tectonic activity along the edges of lithosphere plates, due to relative motion between plates (Dennis and Atwater, 1974, p.1033). Syn: *plate margin.* Rarely used syn: *plate juncture.*

plate carrée projection A simple cylindrical projection with an evenly spaced network of horizontal parallels (spaced at their correct meridional distance from the equator) and vertical meridians (spaced at their correct equatorial distances). Only cardinal directions are true, and scale is true on all meridians and the standard parallel but is greatly distorted away from the center. The equator is the standard parallel, and the network consists of squares. Any other standard parallel will produce rectangles with the north-south dimension the longer. The projection is neither equal-area nor conformal, and is used in geologic mapping for small areas and in geographic referencing for large-scale city maps. It is also commonly used for the distribution (formatting) of global raster data. A modified version has the scale preserved along two parallels of latitude other than the equator. Etymol: French, "regular-square projection". Syn: *equirectangular projection.*

plate-equivalent (plate-e-quiv′-a-lent) Said of that wall part of the dinoflagellate-cyst wall judged to occupy a position equivalent to that occupied by a plate of the theca.

plate juncture *plate boundary.*

platelet (plate′-let) A small *ice crystal* which, when united with other platelets, forms a layer of floating ice, esp. sea ice, and serves as a seed crystal for further thickening of the ice cover. Platelets in sea ice retain their identity for some time because they are bounded by rows or layers of brine cells.

platelike conodont element (plate′-like) A *conodont element* having platforms or a greatly expanded basal cavity.

plate margin *plate boundary.*

plate-scale One of the oval to elliptic organic scales embedded in the surface layer of the periplast (cell membrane) of a coccolithophorid.

platestone Scleractinian reef growth framework dominated by platy to tabular colonies where calcification in the horizontal plane dominates over that of the vertical plane. Width-to-height ratio of dominant organisms is 5:1 to 30:1, with these growth forms constituting >60% of the skeletal volume (Insalaco, 1998).

plate subsidence caldera A *caldera* bounded by a complete arcuate, steeply dipping fault that outlines a coherent subsided plate (or piston) (Lipman, 2000).

plate tectonics A theory of *global tectonics* in which the lithosphere is divided into a number of plates whose pattern of horizontal movement is that of torsionally rigid bodies that interact with one another at their boundaries, causing seismic and tectonic activity along these boundaries (Dennis and Atwater, 1974, p.1031).

platform [coast] (plat'-form) A flat or gently sloping underwater erosional surface extending seaward or lakeward from the shore; specif. a *wave-cut platform* or an *abrasion platform*. Cf: *wave-built platform*.

platform [geomorph] (a) A general term for any level or nearly level surface; e.g. a terrace or bench, a ledge or small space on a cliff face, a flat and elevated piece of ground such as a tableland or plateau, a peneplain, or any beveled surface. (b) A small plateau.

platform [paleont] (a) A laterally broadened structure or shelf along the inner side or the outer side of the anterior-posterior axis of a conodont. The term is also used (incorrectly) for plate. (b) The *muscle platform* of a brachiopod. (c) The flat bottom or floor of the calyx of a coral.

platform [tect] That part of a continent that is covered by flat-lying or gently tilted strata, mainly sedimentary, which are underlain at varying depths by a *basement* of rocks that were consolidated during earlier deformations. A platform is a part of the *craton*.

platform facies *shelf facies*.

platform reef An *organic reef*, more extensive than a *patch reef* (i.e., several km across), with a flat upper surface, and sometimes forming an island. Platform reefs are common off the coast of Australia. Cf: *table reef*.

platidiiform (pla-tid'-ii-form) Said of a brachiopod *loop* consisting of descending branches from the *cardinalia* to the *median septum*, with only rudimentary prongs on the septum representing the ascending part of the loop (TIP, 1965, pt. H, p.150).

platina (pla'-ti-na) Crude native platinum.

platiniridium (plat'-in-i-rid'-i-um) An unnecessary name for platinum-bearing iridium.

platinite (plat'-i-nite) *platynite*.

platinum (plat'-i-num) A very heavy, steel-gray to silvery-white, cubic mineral, the native metallic element Pt, commonly containing palladium, iridium, iron, and nickel. It occurs as grains and nuggets in alluvial deposits (often associated with nickel sulfide and gold ores), and disseminated in basic and ultrabasic igneous rocks. Platinum is a highly corrosion-resistant, ductile, and malleable metal, and is the most abundant metal of the *platinum* group. Syn: *polyxene*.

platinum group element Refers specifically to ruthenium, rhenium, palladium, osmium, iridium, and platinum. Abbrev: PGE.

platte A resistant knob of rock in a glacial valley or rising in the midst of an existing glacier, often causing a glacier to split near its snout. Etymol: German "Platte", "slab". Pl: platten.

platting (plat'-ting) The action or process of mapping a surveyed area; the making of a plat.

plattnerite (platt'-ner-ite) A metallic iron-black tetragonal mineral of the *rutile* group: PbO_2.

platy (plat'-y) (a) Said of a sedimentary particle whose length is more than three times its thickness (Krynine, 1948, p.142). Cf: *acicular [sed]*. (b) Said of a sandstone or limestone that splits into laminae having thicknesses in the range of 2 to 10 mm (McKee and Weir, 1953, p.383).

platycone (plat'-y-cone) A coiled cephalopod *conch* with a flattened form, without implication as to the width of the umbilicus or the shape of the venter.

platy flow structure An igneous rock structure of tabular sheets suggesting stratification. It is formed by contraction during cooling; the structure is parallel to the surface of cooling, and is commonly accentuated by weathering. Syn: *platy structure; planar flow structure*.

platykurtic (plat-y-kur'-tic) (a) Said of a *frequency distribution* that has a concentration of values about its mean less than for the corresponding normal distribution. (b) Said of a broad, flat-topped frequency-distribution curve that is less peaked than the corresponding normal curve. Cf: *leptokurtic; mesokurtic*.

platynite (plat'-y-nite) A metallic gray or iron-black trigonal mineral: $PbBi_2(Se,S)_4$. It occurs in thin metallic plates resembling graphite. Syn: *platinite*.

platy structure *platy flow structure*.

play The active exploration or leasing of land for petroleum or natural gas, based on a concept of the petroleum geology of an area.

playa [coast] (pla'-ya [ply'-ah]) (a) A small, generally sandy land area at the mouth of a stream or along the shore of a bay. (b) A flat, alluvial coastland, as distinguished from a beach. Etymol: Spanish, "beach, shore, strand, coast".

playa [geomorph] (a) A term used in SW U.S. for a dry, vegetation-free, flat area at the lowest part of an undrained desert basin, underlain by stratified clay, silt, or sand, and commonly by soluble salts. The term is also applied to the basin containing an expanse of playa, which may be marked by ephemeral lakes. See also: *salina; alkali flat; salt flat; salt pan; salar; salada; saline mudflat; saline pan; dry mudflat*. Syn: *dry lake; vloer; sabkha; kavir; takir*. (b) A term that is often used for *playa lake*.

playa basin *bolson*.

playa lake A shallow, intermittent lake in an arid or semiarid region, covering or occupying a playa in the wet season but subsequently drying up; an *ephemeral lake* that upon evaporation leaves or forms a playa. Syn: *playa*.

playback [seis] (a) The procedure of creating a new form of record from magnetic tapes (or other reproducible recording) by filtering, gain adjustment, time shifting, mixing, stacking, migrating, etc. (b) The result of such processing, as opposed to the original recording.

playfairite (play'-fair-ite) A metallic gray or black monoclinic mineral: $Pb_8(Sb,As)_{10}S_{23}$.

Playfair's law (Play'-fair's) A generalized statement about the relation of stream systems to their valleys in areas of uniform bedrock and structure that have been subject to stream erosion for a long period of time; viz. that streams cut their own valleys, which are proportional in size to the streams they contain, and that the stream junctions in these valleys are accordant in level. For a quantitative statement of a major part of Playfair's law, see Horton (1945, p.293, eq. 17). The law was enunciated by John Playfair (1747-1819), professor of natural philosophy at the University of Edinburgh: "Every river appears to consist of a main trunk, fed from a variety of branches, each running in a valley proportioned to its size, and all of them together forming a system of valleys, communicating with one another, and having such a nice adjustment of their declivities, that none of them join the principal valley, either on too high or too low a level; a circumstance which would be infinitely improbable, if each of these valleys were not the work of the stream which flows in it" (Playfair, 1802, p.102). Syn: *law of accordant junctions*.

play of color An optical phenomenon consisting of flashes of a variety of prismatic colors, seen in rapid succession as certain minerals (esp. opal) or cabochon-cut gems are moved about; e.g. *opalescence*. In the case of opal, it is caused by diffraction of light from innumerable minute regularly arranged optically transparent uniform spherical particles of amorphous silica, and from the spaces between them, stacked in an orderly three-dimensional pattern that behaves like a diffraction grating. Cf: *fire; change of color*. Syn: *schiller*.

plaza (pla'-za) A term used in the SW U.S. for the exceptionally wide floor of a flat, open valley; the flat bottom of a shallow canyon. Etymol: Spanish, "square, marketplace".

plazolite (pla'-zo-lite) *hydrogrossular*.

pleat A longitudinal fold of retractile muscle fibers with associated mesogloea on the side of a coral *mesentery*.

plectogyral *streptospiral*.

plectolophe (plec'-to-lophe) A brachiopod *lophophore* in which each *brachium* consists of a U-shaped side arm bearing a double row of paired filamentary appendages "but terminating distally in medially placed plano-spire normal to commissural plane and bearing single row of paired appendages" (TIP, 1965, pt. H, p.150). Cf: *deuterolophe; spirolophe*.

Pleistocene (Pleis'-to-cene) The lower series of the *Quaternary* System of the Standard Global Chronostratigraphic Scale, above the *Pliocene* Series of the *Tertiary* System and below the *Holocene* Series. Also the time during which these rocks were formed, the Pleistocene Epoch, covering the time span between 1.75 Ma and 11.5 ka.

Pleistogene (Pleis'-to-gene) A term suggested for the latest period of the *Cenozoic* Era, proposed as an alternative for the term *Quaternary* (Harland, 1982).

plenargyrite (ple-nar'-gyr-ite) *matildite*.

pleniglacial (plen-i-gla'-cial) The full glacial phase of a paleoclimate cycle.

pleochroic (ple-o-chro'-ic) Said of a mineral that displays *pleochroism*.

pleochroic formula An expression of a crystal's pleochroism, or the color of transmitted light.

pleochroic halo A spherical shell or shells of discoloration around an inclusion of a radioactive mineral in another mineral caused by the emission of alpha particles.

pleochroic halo dating Determination of geologic age by measuring the increase in color darkening of the *pleochroic halo* of alpha-particle radiation damage around a zircon, monazite, xenotime, or apatite crystal as a function of time and alpha-particle activity. So many variables (e.g., mica sensitivity to alpha radiation, thermal annealing, color reversal) have been discovered that this dating method has only limited application. The radius of a halo measures the kinetic energy, therefore the probability of emission of the α-particle, and thus the half-life. Identity of the radii of pleochroic haloes in biotites of all ages implies constancy of decay rates.

pleochroism (ple-och'-ro-ism) The ability of an anisotropic crystal to differentially absorb various wavelengths of transmitted light in various crystallographic directions, and thus to show different colors in different directions. This property is more easily seen under polarized light than by the unaided eye. A mineral showing pleochroism is said to be *pleochroic*. Cf: *bireflectance*. Syn: *polychroism*. See also: *dichroism; trichroism*.

pleomere (ple'-o-mere) A *somite* of the abdomen of a malacostracan crustacean. Syn: *pleonite*.

pleomorph (ple'-o-morph) *polymorph [evol]*.

pleomorphism (ple-o-mor'-phism) *polymorphism [evol]*.

pleomorphous (ple-o-mor'-phous) *polymorphic*.

pleon (ple'-on) The *abdomen* of a malacostracan crustacean.

pleonaste (ple'-o-naste) *ceylonite*.

pleonite (ple'-o-nite) *pleomere*.

pleopod (ple'-o-pod) An abdominal limb of a crustacean; specif. any appendage of the *pleon* of a malacostracan crustacean, excluding *caudal ramus* and *uropod.* Syn: pleopodite.

pleosponge (ple'-o-sponge) *archaeocyathid.*

pleotelson (ple-o-tel'-son) A structure of a malacostracan crustacean resulting from fusion of one or more abdominal somites (pleomeres) with the *telson.*

plesiomorphic character Ancestral character in an ancestor-descendant relationship that is described as preexisting-novel (Wiley et al., 1991, p.9). Cf: *homologue; apomorphic character.*

plesiomorphy (ple'-si-o-mor'-phy) (a) *Primitive character.* (b) In cladistic analysis, the original, preexisting character from which a homologous character was derived; character exhibiting generalized or relatively primitive traits (Brooks et al., 1984, p.80). Adj: *plesiomorphic.* Cf: *symplesiomorphy, apomorphy.*

plesion A name of variable rank accorded a fossil species or a monophyletic group of fossil species when classified with one or more Holocene species or groups of species. Plesion substitutes for a categorical rank; categorical ranks within a plesion can be no higher than rank within that plesion's Holocene sister group (Wiley, 1981, p.205).

plesiostratotype (ple'-si-o-strat'-o-type) A complementary stratotype (Sigal, 1964).

plessite (ples'-site) A fine-grained intergrowth of kamacite and taenite. It occurs as triangular or polygonal areas in iron meteorites exhibiting Widmanstätten structure.

plessite [meteorite] (ples'-site) In meteorites, fine mixtures of the Fe-Ni phases kamacite and taenite. In octahedrites plessite occupies the regions between kamacite lamellae.

pleura (pleu'-ra) (a) A laterally located part of the body of an invertebrate; e.g. a lateral part of the opisthosoma of a merostome. (b) One of the two lateral parts of each exoskeletal segment of a trilobite that extend outward from its axis; the part of the thoracic segment or pygidium that is lateral to the axial lobe. In this use, the term has been used as both a syn. and a plural of *pleuron.* (c) An *epimere* of a crustacean. Pl: pleurae.

Pleuracanthodii (Pleu'-rac-an-tho'-di-i) An order of elasmobranch fishes, almost exclusively freshwater, characterized by lobate fins, amphistylic jaw suspension, and distinctive double-cusped teeth. Range, Upper Devonian to Upper Triassic.

pleural angle (pleu'-ral) The angle between two straight lines lying tangential to the last two whorls on opposite sides of a gastropod shell.

pleural furrow (a) A groove along the surface of a *pleura* of a trilobite. (b) A groove crossing the pleura of a merostome.

pleuralia (pleu-ral'-i-a) Sponge spicules (prostalia) on the sides of the body.

pleural spine A pointed or sharply rounded extension of the distal end of a *pleura* of a trilobite. It is narrower than the medial part of the pleura.

pleural suture The line of splitting-apart in molting of the *carapace* of a decapod crustacean. It is present in all brachyurans.

pleurite (pleu'-rite) *epimere.*

pleurocyst (pleu'-ro-cyst) A term used in the older literature for a generally granular imperforate calcareous layer thought to form a secondary deposit over an *olocyst* in frontal shields of certain ascophoran cheilostomes (bryozoans).

pleurodont (pleu'-ro-dont) Pertaining to vertebrate teeth that have been fused (see: *ankylosis*) to the medial bony surface (often within a defined groove) in the upper and lower jaws; contrasted with being fused to the biting edges of the jaws (acrodont) or being set in sockets (thecodont).

pleuromyarian (pleu'-ro-my-ar'-i-an) Said of a nautiloid in which the retractor muscles of the head-foot mass are attached to the shell along the lateral areas of the interior of the body chamber (TIP, 1964, pt. K, p.57). Cf: *dorsomyarian; ventromyarian.*

pleuron (pleu'-ron) A term that has been used as both a synonym and a singular form of *pleura.*

pleuston (pleus'-ton) Animals living in both air and water simultaneously, as do certain jellyfish, e.g., *Physalia.*

plexus [evol] (plex'-us) *evolutionary plexus.*

plica [paleont] (pli'-ca) (a) One of the strong ridges and depressions involving the entire thickness of a bivalve shell (mollusk or brachiopod), extending radially from beak to shell margin, and appearing as corrugations on the inner as well as the outer surface of the shell; e.g. a major undulation of the commissure of a brachiopod, with crest directed dorsally, commonly but not invariably associated with the dorsal *fold* and the ventral *sulcus* (TIP, 1965, pt. H, p.150). Syn: *plication [paleont].* (b) A term used, irrespective of commissure, for a small *carina* or fold in the surface of a brachiopod valve. (c) A small projection on the surface of an opisthosomal segment of a eurypterid. Pl: plicae.

plica [palyn] A ridgelike fold as, e.g., in the exine of *Ephedra* pollen. Pl: plicae. Adj: plicate. Cf: *polyplicate.*

plicate (pli'-clate) Adj. of *plication.* Syn: *crumpled.*

plication [paleont] (pli-ca'-tion) A coarse radial corrugation in the surface of a bivalve-mollusk or brachiopod shell; specif. a *plica [paleont].*

plication [struc geol] Intense, small-scale folding. Adj: *plicate.* Cf: *crenulation.*

Pliensbachian (Pliens-bach'-i-an) Stage of the Standard Global Chronostratigraphic Scale: Lower Jurassic (above Sinemurian, below Toarcian).

Plinian eruption (Plin'-i-an) An explosive eruption in which a steady, turbulent stream of fragmented magma and magmatic gas is released at a high velocity from a vent. Large volumes of *tephra* and tall eruption columns are characteristic (Wilson, 1976, p.543). Etymol: Pliny the Younger, A.D. 79.

plinth A term suggested by Bagnold (1941, p.229) for the lower and outer part of a *seif dune,* beyond the slip-face boundaries, that has never been subjected to sand avalanches.

plinthite (plin'-thite) In a soil, a material consisting of a mixture of clay and quartz, that is rich in sesquioxides and poor in humus and is highly weathered. It occurs as red mottles in a platy, polygonal, or reticulate pattern. Repeated wetting and drying changes plinthite to ironstone hardpan or irregular aggregates (USDA, 1975). See also: *laterite.*

Pliocene (Pli'-o-cene) The uppermost series of the *Neogene* Subsystem of the *Tertiary* System of the Standard Global Chronostratigraphic Scale, above the *Miocene* and below the *Pleistocene* Series of the *Quaternary* System. Also the time during which these rocks were formed, the Pliocene Epoch, covering the time span between 5 and 1.75 Ma.

pliomagmatic zone (pli'-o-mag-mat'-ic) *eugeosyncline.*

plocoid (plo'-coid) Said of a massive scleractinian *corallum* in which corallites have separated walls and are united by costae, dissepiments, or coenosteum.

plombièrite (plom-bier'-ite) A white to pink orthorhombic mineral: $Ca_5H_2Si_6O_{18}{\cdot}6H_2O$(?).

plot To place survey data upon a map or plat; to draw to scale. The term was formerly used in noun form as a syn. of *plat.*

plotter A computer output device capable of presenting data in graphic form, such as raster images or line drawings.

plowshare (plow'-share) In glaciology, a wedge-shaped feature developed on a snow surface by deposition of windblown snow, and sometimes accentuated by *ablation.* Cf: *sastrugi; foam crust.*

plow sole A *pressure pan* representing a layer of soil compacted by repeated plowing to the same depth. Also spelled: plowsole. Syn: *hardpan.*

PLSS *public land survey system.*

plucking [glac geol] (pluck'-ing) The process of glacial erosion by which sizable rock fragments, such as blocks, are loosened, detached, and borne away from bedrock by the freezing of water along joints and stratification surfaces with resulting removal of rock as the ice advances. See also: *sapping [glac geol].* Syn: *quarrying [geomorph].*

plucking [streams] The *entrainment* of sediment from the bed of a stream as a result of *lift forces* within the flow. Cf: *quarrying [streams].*

pluck side The downstream, or lee, side of a *roche moutonnée,* roughened and steepened by glacial plucking. Ant: *scour side.*

plug [drill] n. A watertight or gastight seal, such as an interval of cement or a *bridge plug,* installed in a *borehole* or well to prevent movement of fluids. v. To set a plug in a borehole; to fill in or seal off fractures, cavities, or other pores in the walls of a borehole.

plug [paleont] *umbilical plug.*

plug [pat grd] (a) A cohesive, commonly vertical column of gravelly material with considerable fines, representing the continuance at depth of a *sorted circle* in a gravel beach, as on Victoria Island, Canada (Washburn, 1956, p.844). (b) A similar columnlike feature occurring with a *mud circle* (Bird, 1957, p.194).

plug [sed] A mass of sediment filling the part of a stream channel abandoned by the formation of a cutoff; e.g. a *clay plug* or a *sand plug.* See also: *valley plug.*

plug [volc] (a) A vertical, pipelike body of magma that represents the conduit to a former volcanic *vent.* Cf: *neck [volc].* (b) A crater filling of lava, the surrounding material of which has been removed by erosion.

plug dome A volcanic dome characterized by an upheaved, consolidated conduit filling (Williams, 1932).

plugging (plug'-ging) The sealing of a *dry hole* or depleted well. It is usually accomplished by pumping cement into the well, setting a surface plug, and capping the hole with a metal plate.

plugging back The act or process of cementing off a lower section of a *wellbore* and completing in a section higher in the well.

plug reef A small triangular reef that grows with its apex pointing seaward through openings between linear shelf-edge reefs (Maxwell, 1968, p.101). Its outline is analogous with that of a sand ridge formed in the lower reach of a large river. Plug reefs are found off the coast of Australia where high tide range results in strong currents.

plum A clast embedded in a matrix of a different kind; esp. a pebble in a conglomerate.

plumalsite (plum-al'-site) An ill-defined mineral: $Pb_4Al_2(SiO_3)_7$(?).

plumasite (plu'-mas-ite) A coarsely xenomorphic-granular hypabyssal rock of variable composition, composed chiefly of corundum crystals enclosed in oligoclase grains. The name, given by Lawson in 1903, is for Plumas County, California. Obsolete.

plumbago (plum-ba'-go) A syn. of *graphite.* The term has also been applied to graphitic rock, to an impure graphite, and to graphitoid minerals such as molybdenite.

plumb bob A conical metal weight suspended by a cord, used to project a point vertically in space for short distances. Also called: plumb.

plumbic ocher (plum′-bic) *lead ocher.*

plumb line A line perpendicular to all equipotential surfaces of the Earth's gravity field that intersect with it. The line of force (in the direction of gravity) in the geopotential field; a continuous curve to which the direction of gravity is everywhere tangential.

plumbobetafite An adamantine yellowish metamict mineral of the *pyrochlore* group: $(Pb,U,Ca,[])_2(Ti,Nb)_2(O,OH,F)_7$.

plumboferrite (plum-bo-fer′-rite) A metallic black hexagonal mineral of the *magnetoplumbite* group: $Pb_2(Fe^{3+},Mn^{2+},Mg)_{11}O_{19}$.

plumbogummite (a) A variously colored rhombohedral mineral of the *crandallite* group: $PbAl_3(OH)_6(PO_4)(PO_3OH)$. (b) A group of isostructural minerals consisting of plumbogummite, gorceixite, goyazite, crandallite, florencite, and dussertite, and related to alunite and other sulfates isostructural with it.

plumbojarosite (plum-bo-ja′-ro-site) A brown rhombohedral mineral of the *alunite* group: $Pb_{0.5}Fe^{3+}{}_3(SO_4)_2(OH)_6$.

plumbomicrolite (plum-bo-mi′-cro-lite) A greenish-yellow or orange cubic mineral of the *pyrochlore* group: $(Pb,Ca)_2(Ta,Nb)_2O_6(OH)$.

plumbonacrite (plum-bo-na′-crite) A white hexagonal mineral: $Pb_5O(CO_3)_3(OH)_2$.

plumboniobite (plum′-bo-ni-o′-bite) A dubious dark-brown to black mineral of complex composition, consisting of a niobate of yttrium, uranium, lead, iron, and rare earths. It resembles samarskite, and may be a lead-bearing variety of samarskite.

plumbopalladinite (plum′-bo-pal-lad′-i-nite) A metallic hexagonal mineral: Pd_3Pb_2.

plumbopyrochlore (plum-bo-py′-ro-chlore) A greenish-yellow, brown, or red cubic mineral of the *pyrochlore* group: $(Pb,Y,U,Ca)_{2-x}Nb_2O_6(OH)$.

plumbotellurite (plum-bo-tel′-lu-rite) A gray, yellow-gray, or brown orthorhombic mineral: $PbTe^{4+}O_3$. It is a dimorph of fairbankite.

plumbotsumite (plum-bo-tsu′-mite) A colorless orthorhombic mineral: $Pb_5Si_4O_8(OH)_{10}$.

plume [env geol] (a) A featherlike flow, as of polluted water entering a river. (b) A zone of dissolved-phase contaminants in groundwater migrating down the *hydraulic gradient* from a surface or subsurface waste source below the water table.

plume [mineral] A featherlike inclusion in a gem, as in a "plume agate".

plume [volc] A localized body of volcanic rock rising into the crust from the mantle and thought to be the causal mechanism of a *hot spot.* Cf: *mantle plume.*

plume structure On the surface of a master joint or fringe crack, a ridgelike tracing in a plumelike pattern, usually oriented parallel to the upper and lower surfaces of the containing rock unit. Syn: *plumose structure.* Less preferred syn: *feather fracture.*

plumicome (plu′-mi-come) A sponge spicule (hexaster) in which the terminal branches are S-shaped and arranged in several tiers, forming a plumelike structure.

plumose mica (plu′-mose) A feathery variety of muscovite mica.

plumose spiculofiber A spiculofiber of an axinellid or ectyonine sponge skeleton, in which some or all of the component spicules face obliquely outward.

plumose structure *plume structure.*

plumose surface morphology The delicate irregularity on the surface of joints which gives the impression of a feather.

plumosite (plu′-mo-site) A dubious antimony-sulfide mineral having a feathery form; e.g. jamesonite and boulangerite. Syn: plumose ore.

plum-pudding stone *puddingstone.*

plumule (plu′-mule) *epicotyl.*

plunge [struc geol] The angle between a linear structure, e.g. fold hinge, slip lineation, mineral lineation, and the horizontal as measured in the vertical plane containing the line. Cf: *dip [struc geol].*

plunge [surv] v. (a) To set the horizontal cross wire of a theodolite in the direction of a grade when establishing a grade between two points of known level. (b) *transit.*

plunge basin A deep, relatively large hollow or cavity scoured in the bed of a stream at the foot of a waterfall or cataract by the force and eddying effect of the falling water. It is often called a *plunge pool.* Cf: *pothole [streams].*

plunge line Line on a map that joins points of known elevation, marked with spot heights to indicate its plunge.

plunge pool (a) The water in a plunge basin. (b) A deep, circular lake occupying a plunge basin after the waterfall has ceased to exist or the stream has been diverted. Syn: *waterfall lake.* (c) A small, deep *plunge basin.*

plunge step Subtle step at base of the foreshore on a beach. Also marked by coarsest sediments on the beach profile. Cf: *step [coast]; step zone.*

plunging breaker (plung′-ing) A type of breaker whose crest curls over and collapses suddenly, with complete disintegration of the wave. Cf: *surging breaker; spilling breaker.*

plunging cliff A sea cliff bordering directly on deep water, having a base that lies well below water level. Waves are generally deflected rather than caused to break.

plunging fold A fold of which the hinge line is inclined to the horizontal. Cf: *nonplunging fold; doubly plunging fold; pitching fold.* See also: *plunging inclined fold; plunging upright fold.*

plunging inclined fold A fold with a plunging axis and inclined axial plane (Turner and Weiss, 1963, p.119).

plunging upright fold A fold with an inclined axis and vertical axial plane (Turner and Weiss, 1963, p.119).

pluricolumnal (plu′-ri-co-lum′-nal) Two or more crinoid columnals attached to one another.

plush copper ore *chalcotrichite.*

plus sight *backsight.*

plutology (plu-tol′-o-gy) The study of the interior of the Earth. Rarely used.

pluton (plu′-ton) A deep-seated igneous *intrusion.* Cf: *plutonism [ign].*

plutonic (plu-ton′-ic) Pertaining to an igneous rock or intrusive body formed at great depth. See also: *plutonic rock.* Cf: *hypabyssal.* Syn: *abyssal [intrus rocks]; deep-seated; hypogene.*

plutonic cognate ejecta Fragments that were solidified at depth but were brought to the surface by eruption of the magma.

plutonic nodule *nodule [ign].*

plutonic rock An igneous rock formed at considerable depth; it is characteristically medium- to coarse-grained and of granitoid texture. Syn: *plutonite.*

plutonic water Juvenile water in, or derived from, magma at a considerable depth, probably several kilometers. Obsolete. Cf: *magmatic water.*

plutonism [hist] (plu′-to-nism) The concept of the formation of the Earth by solidification of a molten mass underground. The theory was promulgated by Hutton in the eighteenth century (Dean, 1998).

plutonism [ign] A general term for the phenomena associated with the formation of *plutons.* Cf: *pluton; plutonic rocks.*

plutonist (plu′-to-nist) A believer in the theory of plutonism as promulgated by Hutton. Ant: *neptunist.* Obsolete.

plutonite (plu′-ton-ite) An obsolete syn. of *plutonic rock.*

pluvial [clim] (plu′-vi-al) Said of a climate characterized by relatively high precipitation, or of the time interval during which such a climate prevailed. Formerly equated with the glacial stage of the Quaternary glacial/interglacial sequence, pluvial intervals are now regarded more as transitional, or, in low latitudes, as typical of interglacials.

pluvial [geomorph] Said of a geologic episode, change, process, deposit, or feature resulting from the action or effects of rain; e.g. pluvial denudation, a landslide, or gully erosion and the consequent spreading-out of the eroded material below. The term sometimes includes the fluvial action of rainwater flowing in a stream channel, esp. in the channel of an ephemeral stream.

pluvial [meteorol] (a) Pertaining to rain, or more broadly, to precipitation. (b) Characterized by or regularly receiving abundant rain. Cf: *hyetal.*

pluvial lake A lake formed in a period of exceptionally heavy rainfall; specif. a lake formed in the Pleistocene epoch during a time of glacial advance, and now either extinct or existing as a remnant. Example: Lake Bonneville, a prehistoric lake in present Utah, eastern Nevada, and southern Idaho.

pluviofluvial (plu′-vi-o-flu′-vi-al) Pertaining to the combined action of rainwater and streams; e.g. *pluviofluvial* denudation.

pluviometer (plu-vi-om′-e-ter) *rain gauge.*

plywood structure A *microstructure* of horizontal laminae of more or less parallel, horizontal, elongate structural units oriented in different directions in adjacent laminae. This includes orthogonal-plywood, twisted-plywood, and lamello-fibrillar structures in invertebrates. (Carter, 1990, Figs. 12, 40). In elasmoid scales, where this structure is also called "isopedin," the layers reach 30 microns in thickness, and the collagen fibrils reach 30-190 nanometers in diameter. In lamellar bone, the layers and collagen fibrils are thinner. In some primitive teleost fishes, vertical fibers (TC fibers, transverse collagen fibers) cross the plywood layers perpendicularly. In some vertebrate plywood structures, mineralization may be delayed or completely lacking (Weiss and Ferris, 1954; Meunier and Géraudie, 1980; Meunier, 1987; Carter, 1990).

plywood structure with transverse fibers A *plywood structure* in which the horizontal laminae are traversed by vertical fibers.

Pn A seismic phase that is a P wave, of velocities 7.7-8.3 km/s, associated with propagation in the uppermost mantle. Also, the velocity of that wave. Also called: Moho headwave.

pneumatocyst (pneu-ma′-to-cyst) A hollow inflated structure of a stipe that serves to keep some of the *brown algae* afloat.

pneumatogenic (pneu′-ma-to-gen′-ic) *pneumatolytic [ore dep].*

pneumatolysis (pneu-ma-tol′-y-sis) Alteration of a rock or crystallization of minerals by gaseous emanations derived from solidifying magma. Obsolete. Adj: *pneumatolytic [petrology].*

pneumatolytic [ore dep] (pneu′-ma-to-lyt′-ic) Said of a rock or mineral deposit formed by a gaseous agent, for example, lithium- or tin-bearing pegmatite deposits.

pneumatolytic [petrology] (a) Formed by *pneumatolysis.* (b) Sometimes applied to the surface products of gaseous emanations near volcanoes. (c) Applied to the stage of magmatic differentiation between the pegmatitic and hydrothermal stages. (d) Said of the effects of contact metamorphism adjacent to deep-seated intrusions. Obsolete.

pneumatolytic differentiation Magmatic differentiation by the process of *gaseous transfer.*

pneumatolytic metamorphism *Contact metamorphism,* accompanied by strong *metasomatism,* resulting from the chemical action of magmatic gases on both country rock and intrusion.

pneumatolytic stage That stage in the cooling of a magma during which the solid and gaseous phases are in equilibrium.

pneumotectic (pneu-mo-tec'-tic) Said of processes and products of magmatic consolidation affected to some degree by gaseous constituents of the magma. Obsolete.

POC *particulate organic carbon.*

pocket [coast] (pock'-et) An enclosed or sheltered place along a coast, such as a reentrant between rocky cliffed headlands or a bight on a lee shore.

pocket [eco geol] (a) A small, discontinuous occurrence or patch of ore, e.g. a mineralized cavity or crevice. (b) A localized enrichment of an ore deposit. Syn: *belly.*

pocket [geog] (a) A *water pocket* in the bed of an intermittent stream. (b) A *hollow* or *glen* in a mountain.

pocket [soil] The downfolded or sagging convex part of a layer in a *cryoturbate.* Ant: *festoon.*

pocket [speleo] A solutional concavity in the ceiling, walls, or floor of a cave. See also: *ceiling cavity.*

pocket beach A small, narrow beach formed in a *pocket [coast],* commonly crescentic in plan and concave toward the sea, and generally displaying well-sorted sand or gravel; a *bayhead beach.*

pocket lens *hand lens.*

pocket penitente A *nieve penitente* on the north side of which there is a water-filled depression in the ice, 30 to 60 cm deep, rounded or oval in outline, and with perpendicular walls; at the bottom is a thin layer of debris that absorbed solar heat and melted the ice (Workman, 1914, p.306). Such features are formed in late summer in the Himalayas.

pocket rock A term used in the SW U.S. for a desert boulder with a hard case of desert varnish.

pocket transit A small, compact surveyor's transit that fits in a pocket; specif. a *Brunton compass.*

pocket valley A valley whose head is enclosed by steep walls at the base of which underground water emerges as a spring. Syn: *blind valley.* See also: *steephead.*

pockmark (pock'-mark) A concave crater-like depression of the type that occurs in profusion on mud bottoms across the Scotian Shelf (King and MacLean, 1970). They range in diameter from 15 to 45 meters and in depth from 5 to 10 meters. Pockmarks have also been found in the North Sea (Fannin, 1981) and elsewhere. Their origin is debatable but may be due to gas escape.

pocosen (po-co'-sen) *pocosin.*

pocosin (po-co'-sin) A local term along the Atlantic coastal plain south of Virginia for a swamp or marsh on a flat upland, bordering on or near the sea, in many places enclosing knobs or hummocks. Etymol: American Indian (Delaware). Also spelled: *pocoson; pocosen.* Syn: *dismal.*

pocoson (po-co'-son) *pocosin.*

pod [eco geol] An orebody of *podiform* shape.

pod [geomorph] A term used in the steppes of southern Russia for a very shallow depression as much as 10 km in diameter, containing an intermittent lake or lakes; it may indicate uneven loess deposition, preloess topography, deflation, or solution.

pod [meta] A term formerly used to describe certain bodies that are long in one dimension and short in two dimensions and are enclosed in schist with the long axis parallel to the schistosity.

podial pore (po'-di-al) A pore admitting the passage of a *tube foot* between the ambulacral plates of an echinoderm (as of an asteroid). See also: *tentacle pore.*

podiform (pod'-i-form) Said of an orebody, either diffuse or sharply demarcated, of an elongate lenticular or rodlike shape, e.g. chromite in alpine-type peridotite.

podite (pod'-ite) A limb segment of an arthropod; e.g. a joint of a biramous appendage of a trilobite.

podium (po'-di-um) The cylindrical outer part of a *tube foot* of an echinoderm. The term has also been applied to the tube foot itself. Pl: podia.

podoconus (po-do-co'-nus) Internal cone within the central capsule of a nasselline radiolarian.

podocope (po'-do-cope) Any ostracode belonging to the order Podocopida, characterized by a calcified shell with a curved dorsal margin or a straight dorsal margin shorter than the total length of the shell, and by a muscle scar pattern usually consisting of a few secondary scars. Range, Lower Ordovician to Holocene.

podolite (po'-do-lite) *carbonate-apatite.*

podomere (po'-do-mere) An individual leg *segment* of an arthropod, connected by articulation with adjoining segments.

podophthalmite (pod-oph-thal'-mite) The distal segment of the eyestalk of a decapod crustacean, bearing the corneal surface of the eye. Cf: *basiophthalmite.*

podostyle (po'-do-style) A mass of cytoplasm that projects from the test aperture of monothalamous foraminifers and gives rise to pseudopodia (TIP, 1964, pt. C, p.62).

Podsol (Pod'-sol) *Podzol.*

Podzol (Pod'-zol) A great soil group in the 1938 classification system, a group of zonal soils having an organic mat and a very thin organic-mineral layer overlying a gray, leached A2 horizon and a dark brown, illuvial B horizon enriched in iron oxide, alumina, and organic matter. It develops under coniferous or mixed forests or under heath, in a cool to temperate moist climate. Not used in current U.S. system of soil taxonomy. Also spelled: *Podsol.* Etymol: Russian *podsol,* "ash soil".

podzolization (pod'-zol-i-za'-tion) The process by which a soil becomes more acid owing to depletion of bases, and develops surface layers that are leached of clay and develop illuvial B horizons; the development of a *podzol.*

poëchore (po'-ë-chore) A climatic term for the part of the Earth's surface represented by steppes.

poeciloblast (poe'-ci-lo-blast') *poikiloblast.*

poeciloblastic (poe'-ci-lo-blas'-tic) *poikiloblastic.*

poikilitic (poi-ki-lit'-ic) Said of the texture of an igneous rock in which small grains of one mineral (e.g. plagioclase) are irregularly scattered within a typically anhedral larger crystal of another mineral (e.g. pyroxene). The enclosing crystal is an *oikocryst.* In hand specimen, this texture produces lustrous patches (luster mottling) due to reflection from cleavage planes of the oikocrysts. Originally spelled poecilitic. Cf: *ophitic.* Nonrecommended syn: semipegmatitic.

poikilo- (poi'-ki-lo-) A prefix meaning "mottled", "spotted".

poikiloblast (poi'-ki-lo-blast') A large crystal formed by recrystallization during metamorphism and containing numerous inclusions of small grains. Adj: *poikiloblastic.* Var. spelling: *poeciloblast.*

poikiloblastic (poi'-ki-lo-blas'-tic) (a) Pertaining to a *poikiloblast.* (b) A metamorphic texture in which small grains of one constituent lie within larger metacrysts. Modern usage favors this meaning. Syn: *sieve texture.* (c) Said of a metamorphic texture due to the development, during recrystallization, of a new mineral around numerous relicts of the original minerals, thus simulating the poikilitic texture of igneous rocks. Cf: *helicitic.* Var. spelling: *poeciloblastic.*

poikilocrystallic (poi'-ki-lo-crys'-tal-lic) A syn. of *poikilotopic.* The term was introduced by Phemister (1956, p.74).

poikilophitic (poi'-ki-lo-phit'-ic) Said of *ophitic* texture characterized by lath-shaped plagioclase crystals completely included in large, anhedral pyroxene crystals; an intermediate texture between *ophitic* and *sporophitic.* Obsolete. Cf: *poikilitic.*

poikilotherm (poi'-ki-lo-therm') A poikilothermic organism; e.g. a frog. Syn: *heterotherm.*

poikilothermic (poi'-ki-lo-ther'-mic) Said of an organism having no internal mechanism for temperature regulation; having a body temperature that varies with the temperature of the environment; cold-blooded. Syn: *heterothermic.* Cf: *homoiothermic.*

poikilotope (poi'-ki-lo-tope') A large crystal enclosing smaller crystals of another mineral in a sedimentary rock showing poikilotopic fabric; e.g. a large calcite crystal enclosing smaller relics of incompletely replaced dolomite crystals in a dedolomitized rock, or a large gypsum crystal enclosing numerous grains of quartz and/or feldspar.

poikilotopic (poi'-ki-lo-top'-ic) Said of the fabric of a recrystallized carbonate rock or a chemically precipitated sediment in which the constituent crystals are of more than one size and in which the larger crystals enclose smaller crystals of another mineral. The term was proposed by Friedman (1965, p.651). Cf: *porphyrotopic.* Syn: *poikilocrystallic.*

point [coast] A tapering tract of relatively low land, such as a small cape, projecting from the shore into a body of water; specif. the tip section or extremity of such a projection, or the sharp outer end of any land jutting out into a body of water. Syn: *tongue [coast].*

point [gem] A unit of weight for diamonds and other gemstones, equal to $^{1}/_{100}$ *carat,* or 2 mg. A stone weighing $^{32}/_{100}$ carat is called a 32-point stone or a 32 pointer.

point [geomorph] A sharp projecting rocky prominence; esp. a peak of a mountain range.

point [surv] (a) One of the 32 precisely marked equidistant spots about the circumference of a circular card attached to a compass, which indicate the direction in which the various parts of the horizon lie; e.g. a *cardinal point.* The term is also applied to the angular distance of 11.25 degrees between two such successive points, and to the part of the horizon indicated precisely or approximately by a point of a compass card. (b) A *position* on a reference system determined by a survey and represented by a fix; e.g. a "point of observation".

point bar One of a series of low, arcuate ridges of sand and gravel developed on the inside of a growing meander by the slow addition of individual accretions accompanying migration of the channel toward the outer bank. Cf: *channel bar; meander scroll.* Syn: *meander bar.*

point-bar deposit A deposit consisting of a series of alternating point bars and intervening troughs.

point-counter analysis (point'-count-er) A statistical method involving the estimation of the frequency of occurrence of an object, such as a fossil or mineral species, in a sample, determined by counting the number of times it occurs at specified intervals throughout the sample. The analysis is commonly made with an automatic point counter attached to a microscope.

point coverage A coverage containing information on geographic entities represented by points and related attributes (Krzanowski et al., 1993). See also: *line coverage; polygon coverage*.

point defect A type of *crystal defect* occurring at a particular point and involving a particular atom in a lattice. Cf: *line defect; plane defect.* See also: *interstitial defect; Frenkel defect; Schottky defect.*

point diagram A *fabric diagram* in which poles representing lineations, normals to fabric planes, or crystallographic directions have been plotted. Syn: *scatter diagram.*

point group A set of compatible *symmetry elements* that intersect at a common point, and whose operations leave any object they describe invariant. There are an infinite number of point groups, but only 32 that describe crystal symmetry; these are the 32 crystal classes.

point mass Any mass whose geophysical response is equivalent to the theoretical response of the same mass if concentrated at a point. In gravity, a uniform sphere can be treated as if all its mass were concentrated at its center. At large distances, nonspherical masses can be approximated by point masses.

point maximum *maximum [struc petrol].*

point mode A mode of digitizing in which each digitized point must by explicitly entered by the operator (Krzanowski et al., 1993). See also: *stream mode.*

point of origin *initial point.*

point of percussion In archaeology, the impact point on a *flake.*

point release avalanche *loose-snow avalanche.*

point sample A sample of the sediment contained at a single point in a body of water. It is obtained either by an instantaneous sampler or a time-integrating sampler.

point source The origin of a tracer, in the shape of a point within a three-dimensional framework wherein the tracer is distributed by currents and through diffusion.

point-source pollution Pollution resulting from any confined, discrete source, such as a pipe, ditch, tunnel, well, container, concentrated animal-feeding operation, or floating craft. Cf: *nonpoint-source pollution.*

point symmetry (a) The *point group* that describes the symmetry about any particular point of interest. (b) *site symmetry.*

poised stream (a) A stream that is neither eroding nor depositing sediment. Cf: *graded stream.* (b) A stream that possesses stability from an engineering viewpoint.

Poiseuille's law (Poi-seuille's') A statement in physics that the velocity of flow of a liquid through a capillary tube varies directly as the pressure and the fourth power of the diameter of the tube and inversely as the length of the tube and the coefficient of viscosity.

Poisson distribution (Pois-son') A discrete frequency distribution, the limiting form of the binomial distribution when the probability of an event approaches zero, the number of trials approaches infinity, and the product of the two is a constant. The probability that an event will occur is proportional to the length of time preceding the event, and the probability that two events will occur at the same time is vanishingly small.

Poisson's equation The partial differential equation that governs steady-state diffusion processes in groundwater flow systems with sources and sinks. Cf: *Laplace equation.* In a space where the source density is ρ, the Laplacian of a potential *U* is: $\nabla^2 U = 4\pi\rho K$, where ∇ is the operator del and *K* is a constant (the gravitational constant in case of mass and gravitational potential). The constant 4π is deleted in some systems. In empty space where π = 0, this becomes Laplace's equation. Named for Simeon Denis Poisson (1781-1840). French mathematician.

Poisson's number (Pois-son's') The reciprocal of *Poisson's ratio.*

Poisson's ratio The ratio of the lateral unit strain to the longitudinal unit strain in a body that has been stressed longitudinally within its elastic limit. It is one of the elastic constants. See also: *Poisson's number; Poisson's relation.*

Poisson's relation In experimental structural geology, a model of elastic behavior that takes *Poisson's ratio* as equal to 0.25; i.e., approximating that of many solids.

poitevinite (poit-ev'-in-ite) A salmon-colored triclinic mineral: $(Cu,Fe^{2+})SO_4 \cdot H_2O$.

pokelogan (po-ke-lo'-gan) A term of Algonquian origin for a marshy *cove* or inlet of a stream or lake. Sometimes the term, used primarily in Wisconsin, is shortened to *logan [geog].* Syn: *bogan; peneloken.*

pokrovskite (po-krov'-skite) A pinkish-white monoclinic mineral: $Mg_2(CO_3)(OH)_2 \cdot 0.5H_2O$.

polar (po'-lar) Relating or pertaining to the region of either or both of the two poles of the Earth. Region poleward of the Arctic and Antarctic circles.

polar air mass An *air mass* originating over oceans or continents between 50° and 60° north and 50° and 60° south of the Equator; characterized by cold air. Cf: *tropical air mass.*

polar anisotropy *transverse isotropy.*

polar area The part of a pollen grain poleward from the ends of the colpi.

polar-area index (po'-lar-ar'-e-a) The ratio between the width of the polar area of a pollen grain and its equatorial diameter. Syn: *apocolpium index.*

polar axis (a) The primary axis of direction or the fixed reference line from which the angle coordinate is measured in a system of polar coordinates; e.g. the axis of rotation of the Earth. (b) An axis of symmetry that has different crystal faces at opposite ends. (c) An imaginary line connecting the two poles of spores and pollen grains.

polar cap [glaciol] (a) An ice sheet centered at the South Pole, i.e. Antarctica. (b) A term incorrectly applied to the sea ice of the Arctic Ocean.

polar cap [planet] An area at the pole of a planet, covered with condensed volatiles, and often varying in extent with the seasons.

polar climate A type of climate of polar latitudes (above 66°33'), characterized by temperatures of 10°C and below, which are too cold to support tree growth. The two types of polar climate in Köppen's classification are *tundra climate* and *perpetual frost climate.* Cf: *frigid climate.*

polar convergence The line of convergence of polar and subpolar water masses in the ocean. It is indicated by a sharp change in water-surface temperatures.

polar coordinate (a) One of two coordinates that represent the distance from a central point of reference (the pole or origin) along a line to a point whose position is being defined, and the direction (angle) this line makes with a fixed line. (b) A coordinate used to define the position of a point in space with respect to an arbitrarily chosen origin by means of three polar coordinates (two directions or angles and one distance).

polar coordinates A graphical coordinate system in which locations are specified by (a) the angular direction of a radius drawn from a center, and (b) the distance from the center along the line.

polar desert A high-latitude desert where the moisture present is frozen in ice sheets, and thus is unavailable for plant growth (Stone, 1967, p.239). Syn: *arctic desert.*

polar firn Firn formed at low temperatures with no melting or liquid water present.

polar front The line of discontinuity, developed under favorable conditions between air originating in polar regions and air from low latitudes, on which the majority of the depressions of temperate latitudes develop.

polar glacier A glacier whose temperature is below the *pressure-melting temperature* throughout, except possibly at the base. The two types have been distinguished as "type I polar glacier" for a glacier frozen to its bed and "type II polar glacier" for one that is at the melting point at the base (Hooke, 2005). Cf: *high-polar glacier; temperate glacier; polythermal glacier.* Syn: *cold glacier.*

polar ice Any sea ice more than one year old and more than 3 m thick. It is heavily hummocked and usually the thickest form of sea ice. Syn: *arctic pack.*

polarimeter (po-la-rim'-e-ter) An instrument for measuring the amount of polarization of light or the proportion of polarized light in a partially polarized ray caused by some property of a substance. Sometimes called a *polariscope.*

polarimetry (po-la-rim'-e-try) The art or process of using a *polarimeter.*

polariscope (po-lar'-i-scope) (a) Any of several optical instruments for observing the properties of polarized light or the effects produced on polarized light by various materials, e.g. the refraction of a crystal. It consists of a polarizer and an analyzer. (b) A *polarimeter.*

polarite (po'-la-rite) A metallic white orthorhombic mineral: Pd(Bi,Pb).

polarity [elect] (po-lar'-i-ty) The orientation of charge separation.

polarity [evol] In *cladism,* the direction from primitive characters toward derived characters within a *morphocline.*

polarity-change horizon *polarity-reversal horizon.*

polarity chron The fundamental unit of geologic time designating the time span of a *polarity chronozone* (NACSN, 1983, Art. 89). It is the polarity-chronologic unit of middle rank, between *polarity superchron* and *polarity subchron.* Obsolete syn: *polarity epoch (b).*

polarity-chronologic unit (po-lar'-i-ty-chron-o-log'-ic) A division of time distinguished on the basis of the record of magnetopolarity as expressed by polarity-chronostratigraphic units. Polarity-chronologic units correspond to the time spans represented by polarity chronozones (NACSN, 1983, Art. 88).

polarity-chronostratigraphic unit A body of rock that contains the primary magnetic-polarity record imposed when the rock was deposited, or crystallized, during a specific interval of geologic time. The *polarity chronozone* is the fundamental unit (NACSN, 1983, Art. 83).

polarity chronozone The fundamental unit of worldwide polarity-chronostratigraphic classification, consisting of rocks of a specified primary polarity. Individual polarity zones are the basic building blocks of polarity chronozones. A formal polarity chronozone has a compound name beginning with that of a geographic feature; the second component indicates the normal, reversed, or mixed polarity of the unit, and the third is the term chronozone. If the same geographic name is used for both a polarity zone and a polarity chronozone, the latter is distinguished by an -an or -ian ending, e.g. Tetonian Reversed-Polarity Chronozone (NACSN, 1983, Art. 85, 86). Divisions of polarity chronozones are designated polarity subchronozones; assemblages may be termed polarity superchronozones. The time span represented by a polarity chronozone is a *polarity chron*. See also: *polarity-chronostratigraphic unit.*

polarity epoch (a) A period of time during which the Earth's magnetic field was predominantly or entirely of one polarity. The chronology of the epochs and shorter polarity events defines the geomagnetic polarity time scale. (b) *polarity chron.*

polarity event *polarity subchron.*

polarity excursion (a) A magneto-stratigraphic polarity unit that may not be clearly classifiable as normal or reversed but may be designated as intermediate in direction. (b) A swing in the direction of the Earth's dipole magnetic field, with typical amplitude of 15 degrees and period of 10^2 to 10^4 years (Harland et al., 1982, p.63; 1990, p.140-141).

polarity interval The fundamental unit of worldwide polarity-chronostratigraphic classification. The term is applied to rock, not time; it is used in a spatial sense (Oriel et al., 1976, p.276).

polarity period *polarity superchron.*

polarity reversal *geomagnetic reversal.*

polarity-reversal horizon (po-lar'-i-ty-re-ver'-sal) Either a single clearly definable surface or a body of strata no more than 1 m thick, across which a change in magnetic polarity is recorded. Polarity-reversal horizons or polarity transition-zones form the upper and lower boundaries of polarity zones (NACSN, 1983, Art. 45). Syn: *polarity-change horizon; magnetostratigraphic polarity-reversal horizon.*

polarity rock-stratigraphic unit *magnetopolarity unit.*

polarity subchron The shortest *polarity-chronologic unit* (NACSN, 1983, Art. 89). Obsolete syn: *polarity event.* Syn: *event [magnet].*

polarity subchronozone A division of a *polarity chronozone.*

polarity subzone When continued work at the stratotype for a *polarity zone,* or new work in correlative rocks elsewhere, reveals smaller polarity units, these may be recognized formally as polarity subzones (NACSN, 1983, Art. 46).

polarity superchron The longest *polarity-chronologic unit* (NACSN, 1983, Art. 89). Syn: *polarity period.*

polarity superchronozone An assemblage of polarity chronozones.

polarity superzone When continued work at the stratotype for a *polarity zone,* or new work in correlative rocks elsewhere, makes it necessary or desirable to group polarity zones, these should be termed polarity superzones (NACSN, 1983, Art. 46).

polarity-transition zone A body of strata greater than 1 m in thickness constituting a transitional interval across which a change in magnetic polarity is recorded. Polarity-transition zones or polarity-reversal horizons form the upper and lower boundaries of polarity zones (NACSN, 1983, Art. 45). Syn: *magnetostratigraphic polarity-transition zone*.

polarity zone The fundamental unit of magnetopolarity classification; a unit of rock characterized by the polarity of its magnetic signature. The formal name consists of a geographic name and the term polarity zone; the name may be modified to indicate the polarity of the unit, e.g. Deer Park Reversed Polarity Zone (NACSN, 1983, Art. 46). See also: *polarity chronozone.* Syn: *magnetopolarity zone; magnetostratigraphic polarity zone.*

polarization [elect] (po'-lar-i-za'-tion) (a) The production of dipoles or higher-order multipoles in a medium. (b) The polarity or potential near an electrode.

polarization [magnet] A nonrecommended syn. of *magnetization.*

polarization [optics] The modification of light so that its vibration directions are restricted. Plane polarized light is used in the study of thin sections of minerals and rocks, by the polarizing microscope.

polarization [phylogen] *character polarization.*

polarization [seis] The sense of displacement for a wave (such as an S-wave). A preferential direction of wave motion, as the component of S-waves whose motion is confined to a horizontal plane. A preferred orientation may also be imposed by anisotropy, for example, S-waves involving displacement parallel (or perpendicular) to vertical fractures in rocks.

polarization brush *isogyre.*

polarization factor In calculating or correcting X-ray diffraction intensities, a factor that accounts for the reduction in intensity of a diffracted beam owing to its state of polarization. It is often combined with the *Lorentz factor.*

polarized light (po'-lar-ized) Light that has been changed by passage through a prism or other polarizer so that its transverse vibrations occur in a single plane, or in a circular or elliptical pattern. (In general, the term polarized light is taken to mean plane-polarized light). It is used in the polarizing microscope for optical analysis of minerals or rocks in thin section.

polarized S wave *S wave; SH wave; SV wave.*

polarized transformation series A transformation series in which the relative apomorphy and plesiomorphy of characters has been determined (Wiley et al., 1991, p.9). Ant: *unpolarized transformation series.*

polarizer (po'-lar-iz'-er) An apparatus for polarizing light; in a polarizing microscope, it may be the lower *Nicol prism* or the *Polaroid.* Cf: *analyzer.* Partial syn: *nicol.*

polarizing angle (po'-lar-iz'-ing) The angle at which unpolarized light is incident upon a surface so that it acquires the maximum degree of plane polarization. Syn: *Brewster angle.* See also: *Brewster's law.*

polarizing microscope A microscope that uses polarized light and a revolving stage for analysis of petrographic thin sections. Two nicols, one above and the other below the stage, polarize and analyze the light; the stage rotates about the line-of-sight axis. Syn: *petrographic microscope.*

polarizing prism *Nicol prism.*

polar lake A lake whose surface temperature never exceeds 4°C. Cf: *tropical lake.*

polar layered deposits On Mars, deposits approximately centered on each pole extending up to 20° from the poles and displaying nearly horizontal layers (Kieffer et al., 1992).

polar migration *polar wandering.*

polarograph (po-lar'-o-graph) An instrument for analyzing solutions by electrolysis using a cathode consisting of falling mercury drops. See also: *polarography.*

polarography (po-lar-og'-ra-phy) An electrolytic technique for chemical analysis based on diffusion rates of ions to an electrode as a measure of the concentration of ions in the solution. A readily polarized electrode consisting of falling mercury drops is used to keep the ion concentration in the solution uniform. See also: *polarograph.*

Polaroid (Po'-lar-oid) A trademark name for a *polarizer* that consists of a sheet of cellulose that is impregnated with crystals of quinine iodosulphate. The crystals are aligned so that their optical axes are parallel, and they polarize light in two directions at right angles: in one direction most visible light is absorbed, and in the other, essentially white light is transmitted. The term is often used in the noncapitalized form.

polar-orbiting satellite A satellite in relatively low orbit that travels near the geographical poles on meridional trajectories. Earth rotates through the plane of the satellite's orbit. The altitude of the satellite's orbit is about 800 to 1,000 km (500 to 620 mi).

polar projection One of a group of projections that are centered on a pole of a sphere. Examples include any of several azimuthal map projections (polar stereographic projection, polar gnomonic projection, and polar orthographic projection).

polar space A four-sided region resulting from formation of *basal leaf cross* in an acantharian radiolarian.

polar spine (a) One of the modified coccoliths located at the ends of nonmotile fusiform coccolithophores (such as *Calciosolenia*). (b) A spine normal to the equator, and defining one axis, of the shell in a radiolarian; e.g. a radial spine disposed according to the *Müllerian law* and marking a zone in an acantharian comparable to the polar zone of the terrestrial globe.

polar stereographic projection A *stereographic projection* generally on an equatorial plane, having its center located at one of the poles of the sphere. It is suitable for maps of the polar regions on planes cutting the Earth north of 60°, and it serves as the base of the Universal Polar Stereographic (UPS) Military Grid System for latitudes between 80° and 90°. It is also widely used in optical mineralogy and in structural geology. See also: *Wulff net.*

polar symmetry A type of crystal symmetry in which the two ends of the central crystallographic axis are not symmetrical. Such a crystal is said to display *hemimorphism;* hemimorphite is the characteristic example.

polar tubule One of the external cylinders occurring at opposed poles in the main axis of an elliptic shell of a spumellarian radiolarian.

polar view The view of a spore or pollen grain from more or less directly above one of the poles. See also: *amb.*

polar wandering (a) Short-period movement of the Earth's poles, resulting from wobbling of its axis. (b) Long-period, more or less systematic displacement of the Earth's poles, which may have occurred during the passage of geologic time. Syn: *polar migration; Chandler motion.*

polder (pol'-der) A generally fertile tract of flat, low-lying land (as in Netherlands and Belgium) reclaimed and protected from the sea, a lake, a river, or other body of water by the use of embankments, dikes, dams, or levees; e.g. a marsh that has been drained and brought under cultivation, or a lake that has been dried out by pumping. The term is usually reserved for coastal areas that are at or below sea level and that are constantly protected by an organized system of maintenance and defense. Etymol: old Flemish poelen, "to dig out". Syn: *polderland.*

polderization (pol'-der-i-za'-tion) The creation of a polder or polders; esp. the draining and bringing under cultivation of a low-lying area reclaimed from the sea. See also: *empolder.* Syn: *empoldering.*

polderland (pol'-der-land) *polder.*

poldervaartite A vitreous colorless orthorhombic mineral: $Ca(Ca,Mn)(SiO_3OH)(OH)$.

pole [cryst] In crystallography, a line that is perpendicular to a crystal face and that passes through the center of the crystal. See also: *face pole.*

pole [geog] Either extremity of an axis of a sphere or spheroid, or one of the two points of intersection of its surface and its axis; specif. a *geographic pole* of the Earth.

pole [paleont] (a) An end of the axis of coiling in planispirally coiled shells or tests, as in the fusulinids. (b) An end of the theca in cystoids; e.g. "oral pole" containing the mouth, or "aboral pole" opposite the mouth and usually marking the end to which the column is attached. (c) A general point of anatomical reference (such as the animal versus vegetal poles of early embryos).

pole [palyn] (a) Either termination of the axis of a pollen grain or spore running from the center of the original tetrad to the center of the distal side of the grain, hence the center of both distal and proximal surfaces. The term is esp. useful for angiosperm pollen in which it is not apparent which is the proximal and which the distal surface. (b) One of the two ends of a dinoflagellate cell when oriented in the direction of swimming (Fensome et al., 1993, p.259). (c) In spherical or elongate acritarchs, either of the opposite ends of the central body along the axis perpendicular to the equator (Cramer and Diez Rodriguez, 1979; Willliams et al., 2000, p.129).

pole [struc petrol] A point on a stereographic or equal-area projection that represents the projection of a lineation, the normal to a fabric plane, or a crystallographic direction. The term is also used for a point which represents the normal to any plane on a fabric diagram; thus the normal to a great-circle girdle would be called the pole of that distribution.

pole [surv] (a) A bar, staff, or rod used as a target in surveying; e.g. a *range rod.* (b) A unit of length measuring 16.5 ft or equivalent to a *perch* or rod. Also, a unit of area measuring 30.25 square yards or equal to a square perch or square rod. (c) The origin of a system of polar coordinates.

pole chain *Gunter's chain.*

pole coccolith A modified coccolith found at the flagellar and tail ends in flagellate coccolithophores exhibiting dimorphism (such as *Acanthoica*). Cf: *tail coccolith.*

pole-dipole array (pole-di′-pole) An electrode array in which one current electrode is placed at infinity while one current electrode and two potential electrodes in close proximity are moved across the structure to be investigated. The separation between the near current electrode and the closest potential electrode is an integral number times the spacing between the potential electrodes. The array is used in resistivity and induced-polarization surveys and in drill-hole logging.

pole-fleeing force (pole′-flee-ing) A component of forces resulting from the Earth's rotation that carry the crust away from the poles, toward the equator. It was supposed by Wegener and Staub to have been sufficiently great to have displaced free-moving continents, an example cited being the closing-up of Tethys between the northern and southern continents of the Eastern Hemisphere. Jeffreys showed its force to be negligible. It is not considered to be a cause of plate motion. Etymol: German, "Polflucht". Cf: *Coriolis force.*

pole of inaccessibility (a) A syn. of *ice pole.* (b) Point in Antarctica or Siberia most remote from the sea.

pole-pole array An electrode array, used in profiling or in logging, in which one current electrode and one potential electrode are removed to infinity while the other current electrode and the other potential electrode are kept in close proximity and are traversed across the structure.

polhemusite (pol-he′-mus-ite) An adamantine black tetragonal mineral: (Zn,Hg)S.

polianite (po′-li-a-nite′) A syn. of *pyrolusite,* esp. in well-formed tetragonal crystals.

polish (pol′-ish) An attribute of surface texture of a rock or particle, characterized by high luster and strong reflected light, produced by various agents; e.g. *desert polish, glacial polish,* or the coating formed on a gastrolith. Syn: *gloss.*

polished section (pol′-ished) A section of rock or mineral that has been highly polished. It is used for study of opaque minerals by plane or polarized reflected light. Cf: *polished thin section.* Partial syn: *microsection.*

polished surface *slickenside.*

polished thin section A thin section similar to that used in petrography but finished with a polished surface and not covered with a cover glass. It is useful for study by both transmitted and reflected light and by the electron microprobe. Cf: *polished section.* Partial syn: *microsection.*

polje (pol′-je) In areas of *karst topography,* a very large *closed depression,* in some places several kilometers long and wide, having a flat floor either of bare limestone or covered by alluvium, and surrounded by generally steep walls of limestone. Typically, the floor is flooded annually. Syn: *interior valley.* Etymol: Serbo-Croatian, "field".

polkanovite A metallic hexagonal mineral: $Rh_{12}As_7$.

polkovicite (pol′-ko-vic-ite) A metallic white cubic mineral that forms a solid-solution series with morozeviczite: $(Fe,Pb)_3(Ge,Fe)_{1-x}S_4$ (x=0.18 to 0.69).

pollen (pol′-len) The several-celled microgametophyte of seed plants, enclosed in the microspore wall. Fossil pollen consists entirely of the microspore wall or exine, from which the microgametophyte itself was removed during or before lithification. The term "pollen" is a collective plural noun, and it is incorrect to say "a pollen". See also: *pollen grain.*

pollen analysis (a) A subset of palynology dealing with the study of Pleistocene and Holocene sediments and paleoenvironments, employing *pollen diagrams* and *isopoll* maps. (b) Prior to 1944, the term was used for what is now called *palynology.* Syn: *pollen statistics.*

pollen diagram Any diagram of pollen abundance showing stratigraphic fluctuation; strictly, the graphical presentation of relative abundances of various genera of palynomorphs at successive levels of cores of Pleistocene and Holocene sediment studied in pollen analysis. Cf: *pollen profile.*

pollen grain One of the dustlike particles of which *pollen* is made up; a single unit of pollen. Syn: *grain [palyn].*

pollenite (pol′-len-ite) An olivine-bearing *tephritic phonolite,* also containing phenocrysts of sanidine, plagioclase, nepheline, clinopyroxene, amphibole, and biotite, in a glassy groundmass. The name, given by Lacroix in 1907, is from Vallone di Pollena, Monte Somma, Italy. Not recommended usage.

pollenkitt (pol′-len-kitt) A sticky material, produced by the *tapetum,* that may hold pollen grains together during dispersal. Etymol: German "Kitt," putty, cement.

pollen mother cell A special case of *spore mother cell.*

pollen profile A vertical section diagram of an organic deposit (such as a peat bog) showing the sequence of buried or fossil pollen; a *pollen diagram.* Syn: *profile [palyn].*

pollen rain The total deposit of palynomorphs in a given area and period of time, as estimated by study of sediment samples and by pollen-trapping devices.

pollen sac One of the pouchlike organs in a seed plant that contain the pollen; e.g. the *anther* of an angiosperm or flowering plant.

pollen spectrum One of the characteristic horizontal lines in a pollen diagram, showing the relative abundances of the various sorts of pollen and spores diagrammed in a single sample analyzed from a single level.

pollen statistics *pollen analysis.*

pollen sum In pollen analysis, a portion of the total pollen count from which certain sorts of pollen are excluded by definition. The most usual pollen sum excludes all nonarborescent pollen and some arborescent pollen as well. Where pollen sums are used, pollen abundances are calculated as ratios of given sorts of pollen to the pollen sum, rather than to the raw total count.

pollen symbol An arbitrary sign used in Quaternary pollen diagrams, representing a genus or other group of plants, and serving as an internationally understood identification of a line in the pollen diagram.

pollen tube A more or less cylindric extension that emerges from the wall of a pollen grain and that protrudes through one of its apertures when the grain germinates on contact with the stigmatic surface of flowering plants or the megasporangium of gymnosperms. The tube acts as a food-absorbing organ in lower seed plants such as cycads, but in flowering plants its primary function is to conduct the male nuclei to the vicinity of the female gametophyte to effect fertilization.

pollination (pol-li-na′-tion) The fertilization of a seed plant; specif. the transfer of pollen from a stamen or anther to an ovule or megasporangium.

pollinium (pol-lin′-i-um) A large, coherent mass of pollen, usually the contents of a whole locule of an anther, shed in the mature stage as a unit (as in the milkweed *Asclepias*). Pl: pollinia. Cf: *polyad; massula.*

pollucite (pol′-lu-cite) A colorless, pinkish, or bluish cubic mineral of the analcime group: $(Cs,Na)(AlSi_2)O_6 \cdot nH_2O$. It occurs massive or in cubes, and is used as a gemstone and as an occasional ore of cesium. Obsolete syn: *pollux.*

pollute (pol′-lute) (a) To make physically impure or unclean. (b) Often considered synonymous with "contaminate", but in some environmental regulations *pollute* represents a more advanced state of degradation. To pollute is to introduce any substance that alters some aspect of the environment in a manner to make it unfit for a particular use.

pollution (pol-lu′-tion) *contamination [water].*

pollux (pol′-lux) An obsolete syn. of *pollucite.*

polster (pol′-ster) A stunted, perennial plant, usually isolated, that grows in a dense cushiony hummock capped with concentric layers of moss or lichen which incorporate sandy silt and small pebbles as they grow. It is esp. abundant on the snout of Matanuska Glacier, Alaska. Etymol: German "Polster", "cushion".

polyactine (pol-y-ac′-tine) A sponge *spicule* having many rays diverging from a common center along more than four axes.

polyad (pol′-y-ad) A group of more than four mature pollen grains shed from the anther as a unit (as in *Acacia*). The grains within the polyad are usually in multiples of four. Cf: *pollinium; dyad; tetrad [palyn].*

polyakovite-(Ce) A vitreous black metamict mineral of the *chevkinite* group: $(Ce,Ca)_4(Mg,Fe^{2+})(Cr,Fe^{3+})_2(Ti,Nb)_2Si_4O_{22}$.

polyanion In an *anisodesmic* crystal structure, an insular fragment with overall negative charge, consisting of several atoms involved in relatively strong bonds; e.g., the $[B_3O_3(OH)_5]^{-2}$ polyanion in meyerhofferite or the $[V_{10}O_{28}]^{6-}$ polyanion in hummerite. Cf: *oxyanion.*

polyannulate (pol-y-an′-nu-late) Of certain Normapolles pollen (e.g., *Basopollis*), having a structure in which the sexine of the outer aperture has multiple layers.

polyargyrite (pol-y-ar′-gyr-ite) A dubious gray to black mineral, possibly a mixture of tetrahedrite and acanthite: $Ag_{24}Sb_2S_{15}$ (?).

polyaromatic hydrocarbons Various hydrocarbons consisting of multiple aromatic rings and formed principally as a byproduct of combustion. Found in atmospheric particles and used to identify the source of oils and oil spills. Abbrev: PAHs.

polyaxon (pol-y-ax′-on) A sponge *spicule* in which the rays grow along many axes of development emanating from a central point.

polybasite (pol-y-ba′-site) An metallic iron-black to steel-gray monoclinic mineral: $(Ag,Cu)_{16}Sb_2S_{11}$. Cf: *pearceite.*

polycentric (pol-y-cen′-tric) Said of a *corallite* formed by a polyp retaining polystomodaeal condition permanently.

polychroism (pol′-y-chro-ism) *pleochroism.*

polyclinal fold (pol'-y-cli'-nal) One of a group of adjacent folds, the axial surfaces of which have various orientations.

polyconic projection (pol-y-con'-ic) (a) A map projection (neither conformal nor equal-area) in which a series of right circular cones are each tangent to the Earth's surface at successive latitudes, each parallel thus constructed serving as if it were the chosen standard parallel for a simple *conic projection.* All parallels (developed from the bases of the cones) are arcs of nonconcentric circles with their centers on the straight line generally representing the central meridian, all other meridians being curved lines drawn through the true divisions of the parallels. The scale along each parallel and along the central meridian is true, but it increases on the meridians with increasing distance from the central meridian. The projection is suitable for maps of small areas and for areas of great longitudinal extent (such as Chile). (b) *modified polyconic projection.*

polycrase-(Y) (pol'-y-crase) A submetallic black or brownish-black orthorhombic mineral of the *euxenite* group: $(Y,Ca,Ce,U,Th)(Ti,Nb,Ta)_2(O,OH)_6$.

polycrystal (pol-y-crys'-tal) An assemblage of crystal grains of a mineral, of unspecified number, shape, size, orientation, or bonding, that together form a solid body. Cf: *glomerocryst.*

polycyclic (pol-y-cy'-clic) A term favored by many geomorphologists in place of the syn. *multicycle,* esp. for a stream whose course reflects base-leveling to more than one former sea level.

polydemic (pol-y-dem'-ic) Said of an organism that is native to several regions.

polydymite (pol-y-dym'-ite) A metallic gray cubic thiospinel mineral of the *linnaeite* group: $NiNi_2S_4$. Much so-called polydymite is really violarite.

polygene (pol'-y-gene) adj. (a) *polygenetic.* (b) A term applied specif. by Naumann (1850, p.433) to an igneous rock composed of two or more minerals. Obsolete. Cf: *monogene; polymineralic; polymictic [sed].* Syn: *polymere; polygenic.*

polygenetic (pol'-y-ge-net'-ic) (a) Resulting from more than one process of formation, derived from more than one source, or originating or developing at various places and times; e.g. said of a mountain range resulting from several orogenic episodes. (b) Consisting of more than one type of material, or having a heterogeneous composition; e.g. said of a conglomerate composed of materials from several different sources. Cf: *monogenetic.* Syn: *polygene; polygenic.*

polygenetic volcano A volcano that erupts repeatedly, often in an episodic manner (Walker, 2000). Cf: *monogenetic volcano.*

polygenic (pol-y-gen'-ic) (a) *polygenetic.* (b) *polygene.*

polygeosyncline (pol'-y-ge'-o-syn'-cline) A geosynclinal-geoanticlinal belt along the continental margin and receiving sediments from a borderland on its oceanic side (Schuchert, 1923). Obsolete. Cf: *monogeosyncline.* See also: *sequent geosyncline.*

polygon [geog] A closed, two-dimensional figure with three or more sides and intersections. A spatial object having nonzero area and perimeter, and representing a closed boundary region of uniform characteristics. Syn: *area.*

polygon [pat grd] (pol'-y-gon) A unit of horizontal patterned ground whose mesh is tetragonal, pentagonal, or hexagonal. Its formation is favored by intensive frost action. See also: *sorted polygon; nonsorted polygon.*

polygonal ground (po-lyg'-o-nal) A form of patterned ground marked by polygonal arrangements of rock, soil, and vegetation, produced on a level or gently sloping surface by frost action; esp. a ground surface consisting of a large-scale network of ice-wedge polygons. Syn: polygon ground; polygonal markings; *polygonal soil; Polygonboden; cellular soil.*

polygonal karst A karst area where the surface is completely pitted with *closed depressions,* the divides of which form a crudely polygonal pattern. Cf: *cockpit karst; tower karst.*

polygonal soil *polygonal ground.*

polygon coverage A coverage containing information on geographic entities represented by polygons and related attributes (Krzanowski et al., 1993). Cf: *line coverage; point coverage.*

polygonization (po-lyg'-o-ni-za'-tion) A process through which crystals reach an equilibrium number of dislocations per unit volume during deformation. Newly created dislocations that do not have the time to glide to a crystal boundary may climb into walls. Both dislocation density and wall ("subgrain boundaries") spacing are proportional to the applied differential stress.

polygonomorph (pol-y-gon'-o-morph) Of an *acritarch vesicle,* having a polygonal outline (e.g., *Veryhachium, Stellinium, Estiastra*) and typically simple processes (Downie et al., 1963). Also such a vesicle.

polygon retrieval A function of spatial information systems by which polygons falling within a query window or meeting a specific set of search conditions are selected (Krzanowski et al., 1993).

polyhalite (pol-y-hal'-ite) A colorless, gray, pink, or red triclinic mineral: $K_2Ca_2Mg(SO_4)_4{\cdot}2H_2O$. The pink or red color is due to iron oxide.

polyharmonic fold A fold train that contains shorter wavelength, usually lower amplitude, components and longer wavelength, usually higher amplitude, components. The component with the longest wavelength is the first order fold; components with successively shorter wavelengths are successively higher order folds.

polyhedric projection (pol-y-he'-dric) A projection for large-scale topographic maps in which a small quadrangle on the sphere or spheroid is projected onto a plane trapezoid, the rectilinear parallels and meridians corresponding closely to arc distances on the sphere or spheroid.

polykinematic mélange (pol'-y-kin'-e-mat'-ic) A mélange that includes elements derived from an earlier mélange.

polylitharenite (pol'-y-lith-ar'-e-nite) A lithic arenite with a compositional diversity of sand-sized rock particles volcanic, sedimentary, and metamorphic (Folk, 1968, p.135).

polylithionite (pol-y-lith'-i-o-nite') A pink, cream, white, or colorless monoclinic trioctahedral lepidolite mica of ideal composition: $KLi_2AlSi_4O_{10}F_2$.

polymere (pol'-y-mere) An obsolete adjective applied by Rosenbusch (1898, p.17) to an igneous rock composed of two or more minerals. Syn: *polygene; polymineralic.*

polymetallic (pol'-y-me-tal'-lic) Said of deposits that contain economically important quantities of three or more metals.

polymetamorphism (pol'-y-met'-a-mor'-phism) Polyphase or multiple metamorphism, whereby two or more successive metamorphic events have left their imprint upon the same rocks. The superimposed metamorphism may be of a higher or lower grade than the earlier type. Cf: *prograde metamorphism; retrograde metamorphism.* Syn: *superimposed metamorphism.*

polymict breccia (pol'-y-mict) A brecciated meteorite containing fragments of differing composition and origin (e.g., CM chondrite clasts in an H chondrite). Cf: *monomict breccia; genomict breccia.*

polymictic [lake] (pol-y-mic'-tic) Said of a lake that is continually mixing and has no persistent thermal stratification; e.g. a lake of great area, moderate or little depth, in a region of low humidity or at great altitude (Wetzel, 2001, p.82). Cf: *oligomictic [lake].*

polymictic [sed] (a) Said of a clastic sedimentary rock composed of many mineral or rock types, such as an arkose or graywacke, or a conglomerate with more than one variety of pebble; also, said of the clasts of such a rock. Cf: *oligomictic [sed].* (b) Said of a clastic sedimentary rock composed of more than one mineral species. Cf: *monomictic [sed]; polygene.* Syn: polymict.

polymignyte (pol-y-mig'-nyte) *zirconolite.*

polymineralic (pol'-y-min'-er-al'-ic) Said of a rock composed of two or more minerals; esp. said of an igneous rock consisting of more than one essential mineral. Cf: *monomineralic.* Syn: *polymere; polygene.*

polymodal distribution (pol-y-mod'-al) A frequency distribution characterized by two or more modes, each of which has a higher frequency of occurrence than other immediately adjacent individuals or classes. Cf: *bimodal distribution.*

polymodal sediment A sediment whose particle-size distribution shows one or more secondary maxima (modes).

polymorph [cryst] (pol'-y-morph) A crystal form of a substance that displays polymorphism. Syn: *polymorphic modification; allomorph.*

polymorph [evol] An organism exhibiting *polymorphism [evol];* also, one of the forms of such an organism. Syn: *pleomorph.*

polymorphic (pol-y-mor'-phic) Said of a chemical substance that displays polymorphism, and said of the different crystal forms so displayed. Syn: *polymorphous; pleomorphous; allomorphic; allomorphous.*

polymorphic modification *polymorph [cryst].*

polymorphism [cryst] (pol-y-mor'-phism) The characteristic of a chemical substance to crystallize in more than one form, e.g. rhombic and monoclinic sulfur. Such forms are called polymorphs. Adj: *polymorphic.* See also: *dimorphism [cryst]; trimorphism; tetramorphism; polytypism; allotropy.*

polymorphism [evol] The existence of a species in several forms independent of sexual variations; esp. referring to different types of individuals within a colony, as in bryozoans or corals. Adj: *polymorphic.* See also: *polymorph [evol]; polytypy [evol].* Syn: *pleomorphism.*

polymorphous (pol-y-mor'-phous) *polymorphic.*

polynigritite (pol-y-nig'-ri-tite) A type of *nigritite* that occurs finely dispersed in argillaceous rocks. Cf: *exinonigritite; humonigritite; keronigritite.*

polynomial A linear equation of the form $y=b_0+b_1x+b_2x^2+...+b_kx^k$. Trend surfaces are polynomial equations based on two variables, the geographic coordinates of the observations.

polynya (po-lyn'-ya) Any nonlinear opening enclosed in ice, esp. a large expanse of water, other than a *lead [ice],* surrounded by sea ice, but not large enough to be called *open water [ice];* commonly found off the mouth of a large river. Pl: polynyas; polynyi. Etymol: Russian. Cf: *shore polynya; recurring polynya.* Syn: *clearing; ice clearing; pool [ice].*

polyp (pol'-yp) A typical coelenterate individual, with a hollow and tubular or columnar body terminating anteriorly in a central mouth surrounded by tentacles directed upward. It is posteriorly closed, and attached to the bottom (as in *Hydra*) or more or less directly continuous with other individuals of a compound animal (as in most corals).

polypary (pol'-yp-ar'-y) The common investing structure or tissue in which coral polyps are embedded; a coral colony as a whole. Also called: polyparium.

polypedon A group of contiguous similar pedons. The limits of a polypedon are reached at a place where there is no soil or where the pedons have characteristics that differ significantly.

polyphite A light brown triclinic mineral: $Na_{17}Ca_3Mg(Ti,Mn)_4O_2(Si_2O_7)_2(PO_4)_6F_6$.

polyphyletic (pol'-y-phy-let'-ic) (a) Evolving from more than one ancestral stock. Cf: *monophyletic.* (b) In *cladism,* pertaining to a higher taxon that contains species descended from different ancestral stocks but incorrectly united on the basis of one or more misinterpreted shared derived characters.

polyphyletic group An artificial group of organisms in which the common ancestor is classified in another taxon (Wiley et al., 1991, p.4). Cf: *paraphyletic group, monophyletic group, grade [evol].*

polyphyly Condition of or relating to a polyphyletic group. Adj: *polyphyletic.*

polypide (pol'-y-pide) The parts of a bryozoan *autozooid* that undergo periodic replacement, viz. tentacles, tentacle sheath, alimentary canal, associated musculature, and nerve ganglion.

Polyplacophora A class of amphineurans with head, elongated body, and dorsal shell of eight articulated valves surrounded by muscular mantle girdle covered by cuticle with embedded spicules. Range, Upper Cambrian to Holocene.

polyplacophoran (pol'-y-pla-coph'-o-ran) A marine mollusk, considered a class of the *amphineurans.*

polyplicate (pol-y-pli'-cate) Said of pollen grains (such as those of *Ephedra*) with multiple longitudinal linear thinnings in the exine that resemble, but are not, true colpi.

polyporate (pol-y-po'-rate) *pantoporate.*

polyquartz (pol'-y-quartz) A group term for $Al(PO_4)$, $Al(AsO_4)$, $B(PO_4)$, etc.

polyschematic (pol'-y-sche-mat'-ic) Said of mineral deposits having more than one textural element. Cf: *monoschematic.*

polysomatic Said of a series of minerals or other crystalline compounds that display *polysomatism.*

polysomatic chondrule (pol'-y-so-mat'-ic) Obsolete term referring to a chondrule consisting of several crystals. Cf: *monosomatic chondrule.*

polysomatic series A group of minerals or other crystalline compounds that are related by *polysomatism.* Cf: *homologous series.*

polysomatism The phenomenon, displayed by certain groups of minerals or other crystalline compounds, of consisting of different proportions and arrangements of two or more identical modules. The proportions and arrangement of the modules give each member of a *polysomatic series* a different, but related, chemical composition and *crystal structure*.

polysome One of a series of minerals or other compounds whose structures consist of modules of different compositions stacked together in such a way as to yield a structure and chemical composition different from those of other minerals or compounds formed of the same modules. A member of a *polysomatic series.* Polysomes are structures of minerals that can be created by combining two or more structurally and stoichiometrically distinct types of slab modules. Polysomatic series are groups of structures that are made up of different ratios of the same types of slabs (Veblen, 1991).

polystomodaeal budding (pol'-y-sto-mo'-dae-al) A type of budding in scleractinian corals in which more than three stomodaea are developed within a common tentacular ring. Examples: *intramural budding; circummural budding; circumoral budding.*

polystromatic (pol'-y-stro-mat'-ic) Said of a plant part, e.g. an algal thallus, composed of many layers of cells. Cf: *monostromatic; oligostromatic.*

polysynthetic twinning (pol'-y-syn-thet'-ic) *Repeated twinning* of three or more individuals according to the same twin law and on parallel composition planes; e.g. albite twinning of plagioclase. It is often revealed megascopically by striated surfaces. Cf: *cyclic twinning.*

polythalamous (pol-y-thal'-a-mous) Many-chambered; esp. said of a foraminifer or foraminiferal test composed of numerous chambers. See also: *multilocular.*

polythermal glacier A glacier composed of large volumes of ice at the *pressure-melting temperature* and similarly large volumes below that point. The polythermal glacier is transitional between a *polar glacier* and a *temperate glacier.* Less preferred syn: *subpolar glacier.*

polythetic (pol-y-thet'-ic) Said of a taxonomic classification in which clustering of individuals, populations, or taxa is based on possession of characteristics common to a majority of the members of a cluster. Cf: *monothetic.*

polythetic classification [phylogen] System in which many characters are correlated for use in classification (see Wiley, 1981, p.80). Ant: *monothetic classification [phylogen].*

polytype (pol'-y-type) A type of *polymorph [cryst]* whose different possible forms result from different stackings of similar atomic structural units. For example, in any particular metal, hexagonal close-packed (ABABAB) sequences and cubic close-packed (ABCABC) sequences would be called polytypes. Polytypism is a special case of polymorphism: the two-dimensional translations within the layers are (essentially) preserved whereas the spacings normal to the layers vary between polytypes and are indicative of the stacking period. No such restrictions apply to polymorphism (Guinier et al., 1984). The 1M and 2M polytypes are the most commonly found polytypes of the many possible stacking patterns. Polytypes can occur in a variety of minerals and manufactured compounds. Cf: *one-M (1M) polytype; two-M (2M) polytype*. Adj: *polytypic.*

polytypic (pol-y-typ'-ic) (a) Said of a species consisting of subspecies that replace each other geographically. (b) Said of a taxon that contains two or more taxa of the next lower rank. Cf: *monotypic.* (c) Adj. of *polytype.*

polytypism (pol-y-typ'-ism) The property of a mineral to crystallize in more than one form, owing to more than one possible mode of stacking similar modules, such as layers; a form of one-dimensional *polymorphism [cryst].* Such a mineral is a *polytype.* Syn: *polytypy [cryst].*

polytypy [cryst] (pol-y-typ'-y) *polytypism.*

polytypy [evol] (a) The existence, within a species, of multiple morphologic types. (b) The existence of a species in several geographically separated forms; the existence of a species containing several geographic subspecies. Adj: *polytypic.* Cf: *polymorphism [evol].*

polyvalent (pol-y-va'-lent) Said of foraminiferal specimens or individuals forming a vegetative and accidental association (probably due to crowding) with two or more embryonic apparatuses always of the same generation (microspheric or megalospheric) and of approximately the same age (TIP, 1964, pt.C, p.62).

polyxene (pol'-y-xene) An unnecessary name for an alloy of platinum with other metals.

polyzoan (pol-y-zo'-an) *bryozoan.*

polyzoic (pol-y-zo'-ic) Said of a habitat that supports a wide variety of animals.

polzenite (pol'-zen-ite) (a) In the *IUGS classification* an olivine-bearing melilitite with a glassy or feldspathoidal groundmass and no clinopyroxene (Woolley et al. 1996). (b) A group of lamprophyres characterized by the presence of olivine and melilite; also, any rock in that group, e.g. *luhite, vesecite.* The name, given by Scheumann in 1913, is for Polzen, Czech Republic. Obsolete.

POM *particulate organic matter.*

PON *particulate organic nitrogen.*

pond (a) A natural body of standing fresh water occupying a small surface depression, usually smaller than a *lake* and larger than a *pool.* (b) A term frequently used interchangeably with lake and pool and applied indiscriminately to water bodies in various sections of the U.S. (c) A body of water formed in a stream by *ponding.* (d) A small, artificial body of water, used as a source of water. In Great Britain, the term usually refers only to a small body of standing water of artificial formation.

pondage land (pond'-age) Land on which water is stored as dead water during flooding. It does not contribute to the downstream passage of flow. Syn: *flood fringe.*

ponding (pond'-ing) (a) The natural formation of a pond in a stream by an interruption of the normal streamflow, either by a transverse uplift whose rate of elevation exceeds that of the stream's erosion, or by a dam caused by landsliding, glacial deposition, volcanism, or strong flow of water from a side valley. (b) The artificial impoundment of stream water to form a pond.

pongo (pon'-go) A term used in South America (esp. in Peru) for a canyon or gorge, esp. one cutting through a ridge or mountain range; also, a narrow and dangerous ford. Etymol: Quechua "puncu", "door".

ponomarevite (po-no-mar'-ev-ite) A red monoclinic mineral: $K_4Cu_4OCl_{10}$.

ponor (po'-nor) Hole in the bottom of side of a *closed depression* water disappears into a cave. Etymol: Serbo-Croatian, "insurgence". Cf: *insurgence; swallet; swallow hole.*

Pontian (Pon'-ti-an) (a) Eastern European stage: uppermost Miocene (above Pannonian, below Dacian). Equiv. Messinian. (b) Ponto-Caspian stage: upper Miocene (above Maeotian, below Cimmerian).

pontic (pon'-tic) Pertaining to sediments or facies deposited in comparatively deep and motionless water, such as an association of black shales and dark limestones deposited in a stagnant basin. Rarely used. Etymol: Greek "pontos", "sea". Cf: *euxinic.*

ponzaite (pon'-za-ite) A term proposed by Reinisch in 1912 embracing feldspathoidal and feldspathoid-free trachytes. Cf: *ponzite.* Obsolete.

ponzite (pon'-zite) A feldspathoid-free *trachyte* containing augite and pyroxenes that may be rimmed with aegirine or aegirine-augite. Little or no biotite or amphibole occurs as phenocrysts. Its name, given by Washington in 1913, is derived from the Ponza Islands, Italy. Cf: *ponzaite.* Obsolete.

pool [coast] (a) *tide pool.* (b) *beach pool.*

pool [ice] (a) *polynya.* (b) A large *puddle.*

pool [petroleum] A subsurface accumulation of oil and/or gas in porous and permeable rock. See also: *oil pool; gas pool.*

pool [water] (a) A small, natural body of standing water, usually fresh; e.g. a stagnant body of water in a marsh, or a transient puddle in a depression following a rain, or a still body of water within a cave. (b) A small, quiet, rather deep reach of a stream, as between rapids or where there is little current. See also: *plunge pool.* (c) A body of impounded water, artificially confined above a dam or the closed gates of a lock.

pool-riffle sequence A downstream alternation of *pools* and *riffles* on the bed of a stream.

pool spring A spring fed from a deep source, sometimes related to a fault, and forming a pool. A pool spring may develop the shape of a jug, because a peripheral platform is developed over the water by vegetation and sediments blown in by the wind (Meinzer, 1923, p.55). Cf: *mound spring.*

poop shot *weathering shot.*

poorly graded (poor'-ly) (a) A geologic term for *poorly sorted.* (b) An engineering term pertaining to a *nongraded* soil or unconsolidated sediment in which all the particles are of about the same size or in which a continuous distribution of particle sizes from the coarsest to the finest is lacking. Ant: *well-graded.*

poorly sorted Said of a clastic sediment or of a cemented detrital rock that is not sorted or that shows a wide spread of sizes around the average size, or that consists of particles of many sizes mixed together in an unsystematic manner so that no one size class predominates and that has a sorting coefficient in the range of 3.5 to 4.5 and higher. Based on the phi values associated with the 84 and 16 percent lines, Folk (1954, p.349) suggests *sigma phi* limits of 1.00-2.00 for poorly sorted material. Ant: *well-sorted.* Syn: *unsorted; assorted; nongraded; poorly graded.*

poort A term used in southern Africa for a mountain pass, esp. a water gap or a gorge cut by a river through a ridge or a range of hills or mountains. Etymol: Afrikaans, "gate".

POP *particulate organic phosphorus.*

popcorn In speleology, a colloquial name for *coralloids.*

popple rock (pop'-ple) An English term for *pebble bed.*

popular name (pop'-u-lar) *vernacular name.*

population [ecol] (pop-u-la'-tion) (a) All the individuals of the same species, or of a group of closely related species. (b) In former usage, all organisms occupying a certain area or environment.

population [stat] Any theoretical group of items or samples, all of which are capable of being measured statistically in one or more respects; all possible values of a variable, either finite or infinite, continuous or discrete. Syn: *universe.*

pop-up A mass of rock uplifted by reverse slip on two faults that dip toward a common point beneath the rock mass. In fold-thrust belts, pop-ups occur where synthetic and antithetic splays together lift part of a thrust hanging wall.

porate (por'-ate) Of pollen grains, germinating through a pore or pores in the exine. Some fungal spores show similar preformed *hila* or germination pores (although germination may take place outside these structures).

porcelain clay (por'-ce-lain) A clay suitable for use in the manufacture of porcelain; specif. *kaolin.* Cf: *china clay.*

porcelainite (por'-ce-lain-ite) (a) *porcellanite.* (b) *mullite.*

porcelain jasper A hard, naturally baked, impure clay or *porcellanite* which, because of its red color, was long considered a variety of jasper.

porcelain stone *china stone [ign].*

porcelaneous (por-ce-la'-ne-ous) Resembling unglazed porcelain; e.g. said of a foraminiferal test having a calcareous wall with a dull white luster, or said of a rock consisting of chert and carbonate impurities or of clay and opaline silica. Also spelled: porcellaneous; porcelanous; porcelainous.

porcelaneous chert A hard, opaque to subtranslucent *smooth chert,* having a smooth fracture surface and a typically china-white appearance resembling chinaware or glazed porcelain (Ireland et al., 1947, p.1485).

porcelanite (por'-ce-la-nite') *porcellanite.*

porcellanite (por'-cel-la-nite') A dense siliceous rock having the texture, dull luster, hardness, conchoidal fracture, and general appearance of unglazed porcelain; it is less hard, dense, and vitreous than chert. The term has been used for an impure chert, in part argillaceous (see also *siliceous shale*); for an indurated or baked clay or shale often found in the roof or floor of a burned-out coal seam (see also *porcelain jasper*); and for a fine-grained, acidic tuff compacted by secondary silica (see also *hälleflinta*). Etymol: Italian *porcellana,* "porcelain". Also spelled: *porcelanite; porcelainite.* Syn: *thermuticle.*

pore [geol] A small to minute opening or passageway in a rock or soil; an *interstice.*

pore [paleont] (a) A small opening in an echinoderm; e.g. a *hydrospire pore* of a blastoid, or an opening from the exterior through the thecal plates of a cystoid. The term has also been used for a horizontal perforation (tube, canal, or slit) occupying parts of two adjoining thecal plates of a cystoid (TIP, 1967, pt. S, p.113). (b) One of numerous small openings from the exterior of a sponge; e.g. the terminus of a canal at any surface, an opening surrounded by a single cell, or a smaller-sized opening serving for inward flow of water. Cf: *osculum; ostium; skeletal pore.* (c) A hole through the wall, *septum,* or *tabula* in archaeocyathids; it may be round, oval, slitlike, rectangular, or polygonal (TIP, 1972, pt. E, p.41).

pore [palyn] (a) In the exine of an angiosperm pollen grain, an external, more or less circular to oval preformed thinning or opening that may occur by itself or in conjunction with a colpus. (b) a perforation or depression in the outer surface of a wall layer in dinoflagellates and acritarchs (Williams et al., 1973, p.129). Cf: *colpus; germ pore.* In fungal spores, a preformed opening or thin-walled structure which is commonly marked by a surrounding thickening. In dinoflagellate cysts, the flagellar pores may be reflected in the ventral area.

pore canal (a) A minute tubular passageway extending through the shell of an ostracode; e.g. *normal pore canal* and *radial pore canal.* (b) A perforation in a thecal plate of an echinoderm. (c) A pore in relatively thick skeletal elements of archaeocyathids, in which the length of the opening is longer than its diameter. They may be straight or curved, circular or hexagonal in cross section, and normal or oblique to the wall surface. (d) A duct passing through the exoskeleton of a trilobite, ranging from about 1 to 75 μm in diameter. Some authors restrict the term to those ducts of about 1 to 2 μm diameter, using the term *setal duct* for the larger ones.

pore chamber A *dietella* of a cheilostome bryozoan.

pore closeoff The closing of passageways between air pockets in *firn.* The air pockets then become bubbles in ice, the point at which firn becomes glacier ice. It occurs at a density of approximately 830 kg m^{-3} and at depths ranging from a few meters to over 100 m, depending largely on temperature.

pore diameter The diameter of a pore in a rock; it is measured as the diameter of the largest sphere that may be contained within the pore.

pore frame The raised edge around the area enclosing a minute opening in a radiolarian.

pore ice Ground ice that fills or partially fills pore spaces in the ground; it is formed by freezing of pore water in situ with no addition of water.

pore pair An ambulacral pore of an echinoid, divided by a wall of stereom and through which a single *tube foot* passes.

pore plate (a) The flat, pore-bearing base of the podoconus in a nassellarian radiolarian. (b) In cheilostome bryozoans, a thin part of a calcareous wall forming the skeletal part of a septula. It is perforated by one or more pores. Syn: *rosette plate.*

pore plug Minute, single, organic, microporous plates lying at the base of external openings in certain foraminifers (TIP, 1964, pt. C, p.62).

pore pressure *neutral stress.*

pore rhomb One of the diamond-shaped structures on the surface of thecal plates of rhombiferan echinoderms, consisting of a group of parallel, laterally directed perforations (tubes, grooves, slits) each end of which occupies parts of two adjacent plates (so that each plate of a pair bears one half of the rhomb). The ends may be exposed to the outside or covered by thin calcareous layers. Cf: *diplopore.* See also: *pectinirhomb.* Syn: *rhomb [paleont].*

pore-size distribution The volume fractions of the various size ranges of pores in a soil, expressed as percentages of the soil bulk volume (soil particles plus pores).

pore space The open spaces in a rock or soil, considered collectively (Stokes and Varnes, 1955, p.112). Syn: *pore volume.*

pore system All the openings in a rock or sediment, considered as a unit (Choquette and Pray, 1970, p.214).

pore throat A constricted opening connecting pores in a pore system (Choquette and Pray, 1970, p.214).

pore-tube In archaeocyathids, a commonly hexagonal thin-walled tube, formed by horizontal, oblique, or curved wall plates, or by scooplike bracts or peaks (TIP, 1972, pt. E, p.41).

pore volume *pore space.*

pore water *interstitial water.*

pore-water pressure (pore'-wa-ter) *neutral stress.*

poriferan (po-rif'-er-an) *sponge.*

porocyte (po'-ro-cyte) One of the large tubular cells that constitute the wall of the inhalant canals of some sponges, completely enclosing or surrounding a pore and capable of regulating its size by expansion or contraction.

porolith (po'-ro-lith) A *coccolith* in the form of a polygonal prism with an axial perforation; an axially perforated *prismatolith.* The term was introduced for the elements of the coccolithophorid *Thoracosphaera,* which electron-microscopic studies have shown to be both perforate and imperforate.

porosimeter (po-ro-sim'-e-ter) An instrument that measures *porosity.*

porosity (po-ros'-i-ty) The percentage of the bulk volume of a rock or soil that is occupied by interstices, whether isolated or connected. Cf: *effective porosity.* Syn: *total porosity.* See also: *primary porosity; secondary porosity; porous.*

porosity log A generic term for *well log* curves whose measurements relate easily to formation porosity (*sonic log, density log, neutron log, epithermal-neutron log*).

porospore (por'-o-spore) A deciduous *fungal spore* that originates through a small pore dissolved in the beak of a preceding spore, e.g., in *Alternaria.*

porous (po'-rous) Having numerous interstices, whether connected or isolated. "Porous" usually refers to openings of smaller size than those of a *cellular* rock. Cf: *cavernous [volc].* See also: *porosity.*

porpezite (por'-pe-zite) An unnecessary name for a mineral consisting of a native alloy of gold and 5-10% of palladium. Syn: *palladium gold.*

porphyrin (por'-phy-rin) A large complex organic ring compound made up of, in addition to other rings, four substituted pyrrole rings. Chlorophyll is a porphyrin with magnesium coordinated in the center of the ring; heme (of hemoglobin) is a porphyrin with iron coordinated in the center. Porphyrins are found not only in plants, but also in carbonaceous shale, crude oil, and coal.

porphyrite (por'-phy-rite) An obsolete term synonymous with *porphyry.* The term was originally used to distinguish porphyries that contain plagioclase phenocrysts from those that contain alkali feldspar phenocrysts. Obsolete.

porphyritic (por-phy-rit'-ic) (a) Said of the texture of an igneous rock in which larger crystals (phenocrysts) are set in a finer-grained groundmass, which may be crystalline or glassy or both; a rock with such texture.

porphyro-aphanitic (por'-phy-ro-aph-a-nit'-ic) Said of the texture of a porphyritic igneous rock, esp. an extrusive rock, consisting of phenocrysts in an aphanitic groundmass. Obsolete.

porphyroblast (por′-phy-ro-blast′) A large crystal (relative to average matrix minerals) in a rock produced by metamorphic recrystallization. Adj: *porphyroblastic.* Syn: *metacryst.* Rarely used syn: *pseudophenocryst.*

porphyroblastic (por′-phy-ro-blas′-tic) Pertaining to the texture of a recrystallized metamorphic rock having large idioblasts of minerals (e.g., garnet, andalusite) in a finer-grained crystalloblastic matrix. Cf: *pseudoporphyroblastic.*

porphyroclast (por′-phy-ro-clast′) A relict, partly crushed protolith grain within a finer-grained (often recrystallized) matrix in a metamorphic rock. See also: *porphyroclastic.*

porphyroclastic (por′-phy-ro-clas′-tic) Said of a heterogranular metamorphic texture characterized by volumetrically significant amounts of both porphyroclasts and neoblasts. Also, said of a rock with such a texture. This follows "a longstanding descriptive usage in geology, and does not imply the earlier genetic and etymological connotation of breakage or fracture. Indeed, it is believed that porphyroclastic rocks are the product of plastic deformation and dynamic recrystallization rather than brittle deformation or cataclasis" (Harte, 1977). Semantically correct, though less used, syn: *blastogranular. Blastolaminar* applies to the most strongly laminated facies of porphyroclastic rocks.

porphyroclastic structure *mortar structure.*

porphyrocrystallic (por′-phy-ro-crys′-tal-lic) A syn. of *porphyrotopic.* The term was introduced by Phemister (1956, p.74).

porphyroid (por′-phy-roid) n. A *blastoporphyritic* or sometimes *porphyroblastic* metamorphic rock of igneous origin, or a feldspathic metasedimentary rock having the appearance of a porphyry. It occurs in the lower grades of regional metamorphism. adj. Said of or pertaining to such a rock.

porphyroid neomorphism A term introduced by Folk (1965, p.22) for *aggrading neomorphism* in which small crystals are converted to large ones by growth of a few large crystals in and replacing a static matrix; e.g. the replacement of an aragonite shell by calcite mosaic. Cf: *coalescive neomorphism.*

porphyrotope (por′-phy-ro-tope′) A large crystal enclosed in a finer-grained matrix in a sedimentary rock showing porphyrotopic fabric; e.g., a large dolomite crystal in finer-grained calcitic matrix.

porphyrotopic (por′-phy-ro-top′-ic) Said of the fabric of a recrystallized carbonate rock or a chemically precipitated sediment in which the constituent crystals are of more than one size and in which the larger crystals are enclosed in a finer-grained matrix. The term was proposed by Friedman (1965, p.649). Cf: *poikilotopic.* Syn: *porphyrocrystallic.*

porphyry (por′-phy-ry) An igneous rock of any composition that contains conspicuous phenocrysts in a fine-grained groundmass; a porphyritic igneous rock. The term (from a Greek word for a purple dye) was first applied to a purple-red rock quarried in Egypt and characterized by phenocrysts of alkali feldspar. The rock name descriptive of the groundmass composition usually precedes the term, e.g. diorite porphyry. Obsolete syn: *porphyrite.*

porphyry copper deposit A large body of rock, typically porphyry, that contains disseminated chalcopyrite and other sulfide minerals. Such deposits are mined in bulk on a large scale, generally in open pits, for copper and by-product molybdenum. Most deposits are 3 to 8 km across, and of low grade (less than 1% Cu). They are always associated with intermediate to felsic hypabyssal porphyritic intrusive rocks. Distribution of sulfide minerals changes outward from dissemination to veinlets and veins. *Supergene enrichment* has been historically very important at most deposits, as without it the grade would be too low to permit mining. Cf: *porphyry molybdenum deposit; porphyry tin deposit.*

porphyry gold A *mineral-deposit type* characterized by narrow (< 1 cm) gold-bearing veins and veinlets hosted by subvolcanic porphyry intrusions.

porphyry molybdenum deposit A large, low-grade molybdenite deposit, in which sulfide minerals are distributed as stockworks and disseminated grains in and near siliceous porphyritic intrusive rocks with quartz-latite to rhyolite composition; associated hydrothermal alteration is characterized by potassium feldspathization, sericitization, argillization and silicification. These deposits are characteristically found in the western North American Cordillera (Clark, 1972), and appear more continentally affiliated than porphyry copper deposits. See also: *porphyry tin deposit; Climax-type.* Syn: *stockwork molybdenum deposit.*

porphyry tin deposit A center of tin mineralization in which a tin-bearing mineral, usually cassiterite, is distributed in porphyritic igneous stocks of intermediate composition as disseminated grains, stockworks, or intrusive-breccia fillings, with associated hydrothermal alteration, especially sericitization-silicification, of the wallrocks (Sillitoe et al., 1975). The term is considered to convey the same implications as *porphyry copper deposit* and *porphyry molybdenum deposit,* and is applied to many of the tin occurrences in Bolivia.

porphyry tungsten Large, low-grade deposits of tungsten, in which scheelite- and/or wolframite-bearing quartz stockwork veinlets are distributed throughout a felsic intrusion and associated country rock (Sinclair, 1995).

porphyry uranium deposit A uranium deposit in granitic rocks cut by aplite and pegmatite. The ore minerals, which include primary uraninite and uranothorite, are disseminated and also occur in joints and openings; they are microscopic to submicroscopic in size. Most of the uranium is contained in "felsic and varietal minerals" of the rock (Armstrong, 1974, p.629). Examples are cited in Africa, Canada, Australia, and the U. S., but the designation "porphyry" remains provisional.

portal (por′-tal) The mouth of an adit or tunnel. Syn: *entry.*

porticus (por′-ti-cus) A distinctly asymmetrical apertural flap in the tests of some planktonic foraminifers (such as *Ticinella* and *Praeglobotruncana*). It was originally defined as imperforate. Pl: portici.

portland cement (port′-land) A *cement* made by grinding a mixture of limestone and shale (or equivalent raw materials), heating the mixture to incipient fusion in a rotary kiln, and fine-grinding the resulting clinker. Practically all cement produced today is portland cement. First produced in England, it was named for its resemblance to Portland stone. See also: *concrete.*

Portlandian (Port-land′-i-an) A European stage: uppermost Jurassic (above Kimmeridgian, below Purbeckian-Tithonian).

portlandite (port′-land-ite) A colorless trigonal mineral of the *brucite* group: $Ca(OH)_2$. It occurs as hexagonal plates in contact-metamorphic rocks and also in portland cement.

Portland stone (Port′-land) (a) A yellowish-white, oolitic limestone from the Isle of Portland (a peninsula in southern England), widely used for building purposes. (b) A purplish-brown sandstone (*brownstone*) from Portland, Conn.

pošepnyte (po-šep′-nyte) A light-green to red-brown resin with a high oxygen content (18%), found in plates and nodules in the Great Western mercury mine, Lake County, Calif. Also spelled: pošepnyite; posepnyte.

position (po-si′-tion) (a) Data that define the location of a point with respect to a reference system in surveying. (b) The place occupied by a point on the surface of the Earth or in space. (c) The coordinates that define the location of a point on the geoid or spheroid. (d) A prescribed reading of the graduated horizontal circle of a direction instrument theodolite to be used for the observation on the initial station of a series of stations which are to be observed.

positional accuracy Reliability of the locating of cartographic features relative to their *true position* (Krzanowski et al., 1993). See also: *horizontal accuracy; locational accuracy.*

positional parameters *fractional coordinates.*

positive [optics] (pos′-i-tive) Said of anisotropic crystals: of a uniaxial crystal in which the ordinary index of refraction is greater than the extraordinary index; and of biaxial crystal in which the intermediate index of refraction β is closer in value to α, and in which Z is the *acute bisectrix.* Cf: *negative [optics].*

positive [photo] A photographic transparency or print having approximately the same or similar rendition of tones or colors as that of the original subject. Cf: *diapositive; negative [photo].*

positive [tect] n. An area of a craton that tends to stand higher than the surrounding area; a *positive element.*

positive area *positive element.*

positive birefringence *Birefringence* in which the velocity of the ordinary ray is greater than that of the extraordinary ray.

positive element A structural feature or area, characterized by conspicuous upward movement (uplift, emergence), by relative stability, or by subsidence that is less rapid or less frequent than that of adjacent negative elements. Syn: *positive area; positive [tect].* Obsolete syn: *archibole.* Cf: *negative element.*

positive elongation In an elongated section of an anisotropic crystal, a term signifying that the long dimension is parallel to the slower of the two plane-polarized rays. Cf: *negative elongation.*

positive estuary An estuary in which there is a measurable dilution of seawater by land drainage. Ant: *inverse estuary.* See also: *freshwater estuary.*

positive flower structure A *flower structure* in which the upward-diverging fault splays are predominantly of reverse separation and commonly associated with a prominent antiformal structure in strata above, or cut by, the faults (Biddle and Christie-Blick, 1985). Cf: *negative flower structure.* See also: *restraining bend.*

positive landform An upstanding topographic form, such as a mountain, hill, or plateau, or a volcanic feature formed by an excess of material (as a cinder cone). Ant: *negative landform.*

positive movement (a) An upward movement of the Earth's surface relative to an adjacent part of the surface, such as produced by an uplift or by isostatic recovery; a positive movement of the land may result in a *negative movement* of sea level. (b) A relative rise of sea level with respect to the land, as produced by a *negative movement* of the Earth's crust or by an advance of the sea.

positive ore An orebody that has been exposed and developed on four sides, and for which tonnage and quality estimates have been made. Cf: *developed reserves; proved reserves.*

positive pole The north-seeking member of the *magnetic poles.* Cf: *negative pole; dipole field.*

posnjakite (pos′-n-jak-ite) A blue monoclinic mineral: $Cu_4(SO_4)(OH)_6{\cdot}H_2O$.

possible ore (pos′-si-ble) A mineral deposit whose existence and extent is postulated on the basis of past geologic and mining experience. Syn: *future ore; geological ore.* Cf: *inferred ore; indicated ore; potential ore.* See also: *extension ore.*

post (a) An old English term, now largely obsolete, for a thick bed of sandstone or limestone. (b) A mass of slate traversed by so many joints as to be useless for building purposes.

postabdomen (post-ab′-do-men) (a) The slender, attenuated posterior part of the *abdomen* of a scorpion, composed of five segments and a *telson* modified as a poison gland; the narrow posterior part of the abdomen of a merostome. (b) The *telson* of a crustacean. (c) A joint succeeding the third segment (abdomen) of the shell of a nassellarian radiolarian.

postadaptation (post′-ad-ap-ta′-tion) More perfect adjustment to an adaptive zone after an organism has entered it.

postcervical groove (post-cer′-vi-cal) One of the most conspicuous of the grooves on the *carapace* of decapod crustaceans. It extends from about the middle of the dorsum of the carapace, or somewhat behind it, downward and slightly forward, its end meeting the *hepatic groove* (Holthuis, 1974, p.733).

postcingular series (post-cin′-gu-lar) The series of plates immediately posterior to the girdle of a dinoflagellate theca or cyst, usually fewer in number and often larger in size than those of the *precingular series*.

Post-Classic (Post-Clas′-sic) In Mesoamerican archaeology, the final cultural stage before the arrival of the European colonists; it follows the *Classic*.

postcollarette (post′-col-lar-ette′) A fine membrane prolonging the collar and appearing to close the *pseudostome* in some chitinozoans. It may have served for temporary closure or be a remnant of an attachment; it is commonly ragged or folded back and ornamented by a filamentous network of clear lines.

poster (post′-er) Part of the *orifice* in ascophoran cheilostomes (bryozoans) that is proximal to the condyles and leads to the ascus. Cf: *anter*.

posterior (pos-te′-ri-or) adj. Situated toward the back (tail) of an animal, or at or toward the hinder part of the body, as opposed to *anterior*; e.g. in a direction (in the plane of symmetry or parallel to it) toward the *pedicle* and away from the mantle cavity of a brachiopod, or (in the plane of bilateral symmetry) opposite the position of the mouth of a bivalve mollusk, or in a typically apical direction along the midline axis of a gastropod and opposite the head. n. The hinder part or end of an animal, e.g. the part of a brachiopod shell occupied by the viscera and including the area nearest to the pedicle opening (the side defined by the position of the hinge line); the end defined by the position of the anus of a bivalve mollusk; or the end opposite the *aperture* of a gastropod.

posterior face Back side of a *conodont element* or feature of an element in conventional orientation (TIP, 1981, pt. W, supp.2). Obsolete syn: *posterior side*.

posterior margin (a) The posterior part of the junction between the edges of brachiopod valves. It may be a hinge line or a cardinal margin. (b) In conodonts, trace of posterior side of an element or feature of an element in lateral view.

posterior oblique muscle One of a pair of muscles in discinacean brachiopods, originating posteriorly and laterally on the *pedicle valve*, and converging dorsally to insertions on the *brachial valve* between posterior adductor muscles (TIP, 1965, pt. H, p.150).

posterior side *posterior face.*

postglacial (post-gla′-cial) Pertaining to the time interval since the total disappearance of continental glaciers in middle latitudes or esp. from a particular area; e.g. "postglacial rebound".

posthumous fold (post′-hu-mous) A kind of recurrent folding that occurs in younger sedimentary rocks overlying a buried fold belt. The term is little used.

posthumous movement Movement on a preexisting structure. Rarely used.

posticum (pos′-ti-cum) An obsolete term for the opening of an *exhalant canal* on the external or spongocoel surface of a sponge. Pl: postica. See also: *ostium*.

postkinematic (post′-kin-e-mat′-ic) *posttectonic.*

postkinematic layer Strata interval overlying the *synkinematic layer* and recording sedimentation after salt flow or any other deformation has ceased. Basal postkinematic strata can onlap or truncate an underlying, uneven, deformed surface but show no thickness changes ascribed to local deformation (Jackson and Talbot, 1991).

postmagmatic (post-mag-mat′-ic) An indefinite term applied generally to reactions or events occurring after crystallization of the bulk of a magma, and usually including the hydrothermal stage. Cf: *deuteric*.

postmineral (post-min′-er-al) adj. In economic geology, said of a structural or other feature formed after mineralization. Cf: *premineral; intermineral.*

Postnormapolles (Post′-nor-ma-pol′-les) A group of Cretaceous and Cenozoic porate pollen without the usual pore apparatus or other features of the *Normapolles* group, from which it presumably derived.

postorogenic (post′-or-o-gen′-ic) Said of a geologic process or event occurring after a period of orogeny; or said of a rock or feature so formed. Cf: *posttectonic*.

postorogenic phase The final phase of an orogenic event, following the climactic orogeny. Cf: *preorogenic phase; orogenic phase; orogenic cycle.*

postorogenic pluton An intrusive igneous body emplaced after a period of orogenic activity. Cf: *anorogenic pluton.*

postseptal passage (post-sep′-tal) An opening that connects all chamberlets of the same chamber of a foraminiferal test (as in the Alveolinidae), located between wall and septum at the rear of the chamber. Cf: *preseptal passage.*

post stone An English term for any fine-grained sandstone or limestone. Also spelled: poststone.

posttectonic (post-tec-ton′-ic) Said of a geologic process or event occurring after any kind of tectonic activity; or said of a rock or feature so formed. Cf: *postorogenic*. Syn: *postkinematic.*

post-uplift joints Joints forming upon exhumation and controlled by local conditions developed coincident with local geomorphology (Bahat, 1991). Cf: *unloading joint; release joint.*

pot [coal] (a) *pot bottom.* (b) *caldron bottom.*

pot [geomorph] A general term for any hole, pit, or depression produced naturally in the ground (and often containing water) that suggests the shape or form of a pot or kettle; specif. any of various kinds of *pothole*.

pot [permafrost] A sedimentary deposit in the shape of a pot in distinctively different host materials. Typical pots of the Upland Gravels of Maryland and Virginia are 2 m to 8 m wide; they are composed of sandy-gravelly silt in gravel. A seasonal frost origin is proposed (Conant et al., 1976).

pot [seis] A colloquial syn. of *geophone*.

pot [speleo] A British term for a *vertical cave*.

potable water (po′-ta-ble) Water that is safe and palatable for human use; *fresh water* in which any concentrations of pathogenic organisms and dissolved toxic constituents have been reduced to safe levels, and which is, or has been treated so as to be, tolerably low in objectionable taste, odor, color, or turbidity and of a temperature suitable for the intended use.

potamic (po-tam′-ic) Pertaining to rivers or river navigation; e.g. "potamic transport", or transportation of sediments by river currents.

potamography (po-ta-mog′-ra-phy) The description of rivers. Etymol: Greek potamos, "river".

potamology (po-ta-mol′-o-gy) The scientific study of rivers. Etymol: Greek potamos, "river".

potarite (po-tar′-ite) A metallic silver-white tetragonal mineral: PdHg. It is a natural alloy of palladium and mercury. Syn: *palladium amalgam*.

potash (pot′-ash) (a) Potassium carbonate, K_2CO_3. (b) A term that is loosely used for potassium oxide, potassium hydroxide, or even for potassium, in such informal expressions as *potash spar.* (c) an industry term for a group of potassium-bearing salts that includes potassium chloride, and may also refer to potassium sulfate, nitrate, and oxide.

potash alum (a) *alum.* (b) *kalinite.*

potash bentonite *potassium bentonite.*

potash feldspar A misnomer for *potassium feldspar*.

potash kettle *kettle [glac geol].*

potash lake An *alkali lake* whose waters are rich in dissolved potassium salts. Examples occur in the western part of the sandhills region of north-central Nebraska.

potash mica A misnomer for potassium-rich mica; specif. *muscovite*.

potash spar An informal term for potassium feldspar, i.e. *orthoclase* or *microcline*, or for a feldspar mixture assaying at least 10% K_2O (Rogers and Neal, 1975, p.638). Syn: *K-spar.* Cf: *soda spar.*

potassic (po-tas′-sic) Said of a rock or mineral containing a significant amount of potassium.

potassic alteration Hydrothermal alteration resulting from potassium metasomatism, commonly accompanied in calc-alkalic rocks by removal of calcium and sodium. Characteristic major minerals are orthoclase, biotite, sericite, and quartz, with common albite, anhydrite, Fe-Mg carbonate, apatite, and chalcopyrite. It is a common style of alteration at the deep, central cores of porphyry base-metal systems (Guilbert and Park, 1986, p.181). See also: *phyllic alteration.*

potassic-carpholite A silky white or straw-yellow orthorhombic mineral of the *carpholite* group: $K(Mn^{2+},Li)_2Al_4Si_4O_{12}(OH,F)_8$.

potassic-chlorohastingsite A green monoclinic mineral of the *amphibole* group: $(K,Na)Ca_2(Fe^{2+},Mg)_4Fe^{3+}(Si_6Al_2)O_{22}(Cl,OH)_2$. It represents hastingsite with $Cl>(OH)$, and with K as the dominant alkali cation.

potassic-chloropargasite A vitreous black monoclinic mineral of the *amphibole* group: $(K,Na)Ca_2(Mg,Fe^{2+})_4Al(Si_6Al_2)O_{22}(Cl,OH)_2$. It represents pargasite with $Cl>(OH)$, and with K as the dominant alkali cation.

potassicferrisadanagaite A vitreous black monoclinic mineral of the *amphibole* group: $(K,Na)Ca_2Fe^{2+}{}_3Fe^{3+}{}_2(Si,Al)_8O_{22}(OH)_2$. It represents ferrisadanagaite with K as the dominant alkali cation.

potassic-fluorrichterite A light gray monoclinic mineral of the *amphibole* group: $(K,Na)(Ca,Na)_2Mg_5Si_8O_{22}(F,OH)_2$. It represents richterite with $F>(OH)$, and K as the dominant alkali cation.

potassicleakeite A vitreous reddish-brown monoclinic mineral of the *amphibole* group: $KNa_2(Mg_2Fe^{3+}{}_2Li)Si_8O_{22}(OH)_2$. It represents leakeite with K as the dominant alkali cation.

potassicpargasite A vitreous black monoclinic mineral of the *amphibole* group: $KNa_2(Mg_2Fe^{3+}{}_2Li)Si_8O_{22}(OH)_2$. It represents pargasite with K as the dominant alkali cation.

potassicrichterite A pale to bright blue monoclinic mineral of the *amphibole* group: $K(Na,Ca,Sr)_2(Mg,Fe)_5Si_8O_{22}(OH)_2$. It represents richterite with K as the dominant alkali cation. Not approved by the International Mineralogical Association.

potassic rock In the *IUGS classification,* an igneous rock in which the molecular ratio $K_2O/Na_2O>1$, or one in which wt% $K_2O>Na_2O-2$.

potassic trachybasalt In the *IUGS classification,* the potassic analogue of hawaiite; defined in the TAS diagram as rocks with wt% $Na_2O-2<K_2O$ and

falling in the triangular field with SiO_2 and total alkali coordinates 45 and 5, 49.4 and 7.3, and 52 and 5, respectively.

potassic zone The innermost, core zone of porphyry ore deposits, characterized by introduced or recrystallized potassium feldspar and biotite, with minor sericite and variable but persistent minor amounts of anhydrite (Lowell and Guilbert, 1970). Cf: *argillic zone; phyllic zone; propylitic zone.*

potassium alum (po-tas'-si-um) A colorless or white cubic mineral of the *alum* group: $KAl(SO_4)_2 \cdot 12H_2O$. Syn: *potash alum.*

potassium-argon age method (po-tas'-si-um-ar'-gon) Determination of the age of a mineral or rock in years, based on measurement of the ratio of radiogenic argon-40 to potassium-40 and the known radioactive decay rate of potassium-40 to argon-40. Cf: *argon-40/argon-39 age method.* Abbrev: K-Ar age method. Syn: *potassium-argon dating.*

potassium-argon dating *potassium-argon age method.*

potassium bentonite A potassium-bearing clay of the illite group, formed by alteration of volcanic ash; a *metabentonite* consisting of randomly interstratified layers of illite and montmorillonite with a ratio of 4 to 1 (potassium occupying about 80% of the exchangeable-cation positions of the mica portion). Syn: *K-bentonite; potash bentonite.*

potassium-calcium age method (po-tas'-si-um-cal'-ci-um) The determination of the age of a mineral or rock in years based on measurement of the ratio of radiogenic calcium-40 to potassium-40 and the known radioactive decay rate of potassium-40 to calcium-40. The method is not in common use, as initially there is apt to be a significant amount of calcium-40 present.

potassium feldspar An alkali feldspar containing the Or molecule, $KAlSi_3O_8$; e.g. orthoclase, microcline, sanidine, and adularia. See also: *potash spar.* Syn: *K-feldspar; K-spar.*

potassium fixation The process of converting exchangeable or water-soluble potassium to that occupying the position of K^+ in the micas. They are counter-ions entrapped in the ditrigonal voids in the plane of basal oxygen atoms of some phyllosilicates as a result of contraction of the interlayer space. The fixation may occur spontaneously with some minerals in aqueous suspensions or as a result of heating to remove interlayer water in others. Fixed K^+ ions are exchangeable only after expansion of the interlayer space.

pot bottom A large boulder or concretion in the roof of a coal seam, having the rounded appearance of the bottom of an iron pot and easily detached. Cf: *caldron bottom; bell; kettle bottom; camel back; tortoise.* Syn: *pot [coal]; potstone [coal].*

potch An Australian term for an opal of inferior quality that does not exhibit play of color; it may be found in association with precious opal.

pot clay (a) A refractory clay (fireclay) suitable for the manufacture of the melting pots in which glass is produced. (b) A clay bed associated with coal measures. (c) A kaolin-rich residual clay.

pot earth *potter's clay.*

potential (po-ten'-tial) Any of several different scalar quantities, each of which involves energy as a function of position or of condition; e.g. the *fluid potential* of groundwater.

potential [geophys] (a) A scalar representation of a gravitational, magnetic, or electric effect, the spatial derivative of which is the corresponding field (Telford et al., 1990). (b) The amount of work required to place a unit charge, unit pole, or unit mass at a given position, usually with respect to infinity. (c) A function from which a quantity can be determined by specified mathematical operations, as a potential field from which seismic displacement, velocity, etc. can be ascertained by differentiation. See also: *field [geophys]; field intensity.*

potential barrier The resistance to change from one energy state to another in a chemical system, which must be overcome by *activation energy.*

potential density The density of a small parcel of water at the sea surface after it has been raised adiabatically (without gain or loss of heat) from some depth in the ocean. It is calculated from the *in-situ temperature,* pressure and salinity. More generally, the density can be referenced to any depth, not just the surface.

potential difference The difference in electric potential between two points that represents the work involved or the energy released in the transfer of a unit amount of electricity between them.

potential disturbance *disturbing potential.*

potential drop The difference in pressure between two equipotential lines.

potential electrode One of two electrodes between which potential is measured.

potential evaporation Maximum amount of water lost under natural climatic conditions when water in surplus is available to the root system of plants.

potential gradient *hydraulic gradient.*

potential of random masses *disturbing potential.*

potential ore (a) As yet undiscovered mineral deposits. (b) A known mineral deposit for which recovery is not yet economically feasible. Cf: *possible ore; inferred ore; indicated ore.* See also: *resources; reserves.*

potential temperature [meteorol] The temperature that a given unit of air would attain if it were reduced to a pressure of 1,000 hectopascals (mb) without any heat transfer to or from it.

potential temperature [oceanog] The temperature of a small parcel of water at the sea surface after it has been raised adiabatically (without gain or loss of heat) from some depth in the ocean. It is calculated from the *in-situ temperature* and pressure.

potential vorticity (a) The *absolute vorticity* of a layer of seawater in the ocean divided by the thickness of the layer. (b) More generally, at any depth in the ocean, it is the *Coriolis parameter* divided by density times the vertical density gradient at the point.

potential well In a field of force, a sharply defined area of minimum potential.

potentiometer (po-ten'-ti-om'-e-ter) An electrical instrument for the precise measurement of low-level direct-current voltages.

potentiometric contour *equipotential line.*

potentiometric map (po-ten'-ti-o-met'-ric) A map showing the elevation of a *potentiometric surface* of an aquifer by means of contour lines or other symbols. Syn: *pressure-surface map.*

potentiometric surface A surface representing the total head of groundwater and defined by the levels to which water will rise in tightly cased wells. The water table is a particular potentiometric surface. Syn: *piezometric surface; pressure surface.*

pothole [coast] (pot'-hole) A small, rounded, steep-sided depression or pit in a coastal marsh, containing water at or below low-tide level (Veatch and Humphrys, 1966, p.245). Syn: *rotten spot.*

pothole [geomorph] Any pot-shaped pit or hole. A term used in Death Valley, Calif., for a circular opening, about a meter in diameter, filled with brine and lined with halite crystals. Archaic.

pothole [glac geol] (a) *giant's kettle.* (b) A term applied in Michigan to a small pit depression (1-15 m deep), generally circular or elliptical, occurring in an outwash plain, a recessional moraine, or a till plain (Veatch and Humphrys, 1966, p.244). Cf: *kettle [glac geol].* (c) *moulin.*

pothole [lake] A shallow depression, generally less than 10 acres in area, occurring between dunes or on subdued morainic relief on a *prairie* (as in Minnesota and the Dakotas), often containing an intermittent pond or marsh and serving as a nesting place for waterfowl.

pothole [speleo] A term used in England for a vertical or steeply inclined *shaft* in limestone. See also: *pit [speleo]; vertical cave.*

pothole [streams] A smooth, bowl-shaped or cylindrical hollow, generally deeper than wide, formed in the rocky bed of a stream by the grinding action of a stone or stones, or of coarse sediment (sand, gravel, pebbles, boulders), whirled around and kept in motion by eddies or the force of the stream current in a given spot, as at a strong rapid or the foot of a waterfall. Cf: *plunge basin.* Syn: *pot [geomorph]; kettle [streams]; eversion hollow; rock mill; churn hole; eddy mill; kolk.*

pothole erosion *eversion.*

potholer (pot'-hol-er) British term for *caver.*

potholing (pot'-hol-ing) A British term for *caving [speleo].*

pot lead Graphite used on the bottoms of racing boats.

potosiite (po-to'-si-ite) A metallic triclinic mineral of the cylindrite-franckeite group: $Pb_{48}Fe_7Sn_{18}Sb_{16}S_{115}$.

potrero (po-tre'-ro) An elongate, island-like beach ridge, surrounded by mud flats and separated from the coast by a lagoon and barrier island, and made up of a series of accretionary dune ridges (Fisk, 1959, p.113); e.g. Potrero Lopeno, rising 10 m above the Laguna Madre Flats along the southern Texas coast.

Potsdam ellipsoid (Pots'-dam) An obsolete geodetic reference ellipsoid used in most of Europe, Africa, the Middle East, and the Soviet Union.

Potsdam gravity A former *gravity standard,* the absolute gravity value determined at the Pendelstaal of the Geodetic Institute in Potsdam, Germany, in 1906.

Potsdam system A system of gravity values based on the determination of absolute gravity at Potsdam, Germany in 1906. Recent determinations show that the Potsdam value is about 13 milligals too high. Replaced by *International Gravity Standardization Net 1971.*

potstone [coal] (pot'-stone) *pot bottom.*

potstone [mineral] (a) A dark-green or dark-brown impure steatite or massive talc, used in prehistoric times to make cooking pots and vessels. (b) A term used in Norfolk, England, for *paramoudra.* Also spelled: *pot stone.*

potter's clay (pot'-ter's) A plastic clay free from iron and devoid of fissility, suitable for modeling or making of pottery or adapted for use on a potter's wheel. It is white after burning. Syn: potter's earth; pot earth.

pottsite (potts'-ite) A yellow tetragonal mineral: $HPbBi(VO_4)_2 \cdot 2H_2O$.

Pottsvillian (Potts-vil'-li-an) North American provincial stage: Lower Pennsylvanian (above Mauch Chunk, below Allegheny).

poubaite (pou'-ba-ite) A metallic rhombohedral mineral: $PbBi_2Se_2(Te,S)_2$.

poudretteite (pou-drette'-ite) A colorless or pale pink hexagonal mineral of the *milarite* group: $KNa_2B_3Si_{12}O_{30}$.

poughite (pough'-ite) A dark yellow, brownish-yellow, or greenish-yellow orthorhombic mineral: $Fe^{3+}_2(Te^{4+}O_3)_2(SO_4) \cdot 3H_2O$.

Poulter method (Poul'-ter) *air shooting.*

Poundian A suggested geochronometric subdivision equivalent to the later part of the *Ediacaran* Period (latest *Precambrian*), covering the time span from 580 to 570 Ma (Harland et al., 1990).

Pourtalès plan (Pour'-ta-les) The arrangement of septa in some scleractinian corals (notably in the family Dendrophylliidae) characterized by much greater development of exosepta than of entosepta. Named after Louis F. de Pourtalès (1824-1880), Swiss naturalist.

povondraite A black rhombohedral mineral of the *tourmaline* group: $(Na,K)(Fe^{3+},Mg)_3Fe^{3+}_6(BO_3)_3Si_6O_{18}(O,OH)_4$.

pow A Scottish term for a slow-moving stream, or for a ditch or canal.

powder (pow′-der) *powder snow.*

powder avalanche *dry-snow avalanche.*

powder diffraction X-ray or neutron diffraction by a powdered, crystalline sample, commonly observed by the Debye-Scherrer camera method or by a recording diffractometer.

powder method *Debye-Scherrer method; powder diffraction.*

powder pattern In the *powder method* of X-ray or neutron diffraction analysis, the display of lines made on film by the Debye-Scherrer method or on paper by a recording *diffractometer.* See also: *powder photograph.*

powder photograph The *powder pattern* made on film in the Debye-Scherrer method of X-ray diffraction analysis.

powder snow Dry snow of low density, composed of loose crystals that accumulated under conditions of low temperature and no wind and that have not been compacted. Cf: *sand snow.* Syn: *powder.*

powdery efflorescence Saline mineral efflorescent crust comprised of very fine, loosely cemented crystals that readily disaggregate to powder. Common habit of alkali sufates and carbonates.

powellite (pow′-ell-ite) A sub-adamantine variously colored tetragonal mineral of the *scheelite* group: $CaMoO_4$. It is a minor ore of molybdenum.

power In statistical testing, the probability of rejecting a *null hypothesis* when it is false, or the probability of not committing a *type II error* (power is the complement of β, the probability of committing a type II error). Syn: *power efficiency.*

power efficiency *power.*

power of crystallization *form energy.*

power spectrum In a Fourier transform, a plot of one-half the square of the amplitudes versus frequencies, showing the distribution of variance among the frequencies. Syn: variance spectrum.

poyarkovite (po-yar′-kov-ite) An adamantine red to black monoclinic mineral: Hg_3ClO.

Poynting-Robertson Effect The motion of a particle in orbit causes radiant energy from the sun, which travels at a finite velocity, to strike the particle slightly in front of the radial direction, thereby imparting a miniscule retarding drag. As a result, small particles drift slowly inward. A 1 cm diameter body of density 3 gm/cm^3 would spiral from an orbit at the distance of the Earth into the sun in about ten million years. The effect is probably important for removing particles from centimeters in size down to fine dust from the inner solar system. Particles smaller than about a micron in size are blown directly out of the solar system by radiation pressure. The effect was first suggested by J.H. Poynting in 1903 and elaborated upon by H.P. Robertson in 1937. Cf: *Yarkovsky effect.*

Poynting's law (Poyn′-ting's) A special case of the *Clapeyron equation,* in which the fluid is removed as fast as it forms, e.g. under metamorphic stress, so that its volume may be ignored.

pozzolan (poz′-zo-lan) Siliceous material such as diatomaceous earth, opaline chert, and certain tuffs, which can be finely ground and combined with *portland cement* (in a proportion of 15 to 40 percent by weight). The pozzolan reacts with calcium hydroxide that is liberated as concrete hardens, forming compounds with cementitious properties. Pozzolans also counteract the adverse effects of certain undesirable aggregates that may have to be used in concrete. Portland-pozzolan cements are highly resistant to penetration and corrosion by salt water. The name comes from the town of Pozzuoli, Italy, near which occurs a leucite tuff that was used in cement in Roman times. Also spelled: pozzolana; puzzolan; puzzuolana.

ppb parts per billion.

PPI *plan-position indicator.*

ppm parts per million.

P position In locational notation for conodont skeletal apparatuses, *P* designates the positions occupied by pectiniform elements; characteristically, two types of pectiniform elements are in fully developed apparatuses and the positions occupied by these are designated *Pa* and *Pb* (TIP, 1981, pt. W, supp.2).

pradetite A green triclinic mineral discredited in favor of *lindackerite.*

pradolina The Polish equivalent of *urstromtal.* Pl: pradoliny.

Praetiglian (Prae-ti′-gli-an) A North European climatostratigraphic and floral stage: middle Pliocene (above Reuverian, below Tiglian). Equivalent to Lower *Villafranchian.* Also spelled: Pretiglian.

praevestibulum (prae-ves-tib′-u-lum) A chamber within the ektexine of palynomorphs but not contiguous to the endexine (Tschudy and Scott, 1969, p.27).

Pragian A stage of the Standard Global Chronostratigraphic Scale: Lower Devonian (above Lochkovian, below Emsian). The basal GSSP is near Prague, Czech Republic (Chlupac and Oliver, 1989).

prairie (prai′-rie) An extensive tract of level to rolling grassland, generally treeless, in the temperate latitudes of the interior of North America (esp. in the Mississippi Valley region). The North American prairie consisted of tall-grass prairie in the humid east characterized by a deep, fertile soil and by a covering of tall, coarse grass and herbaceous plants; and short-grass prairie in the semiarid west characterized by little biomass and thin soils. In addition, there is a low, wet grass-grown tract or sink, often water-covered, in the pinewoods of Florida. Etymol: French, "meadow, grassland". Cf: *steppe; black prairie.*

Prairie soils A zonal great soil group consisting of soils formed under temperate to cool-temperate, humid regions under tall grass vegetation. Not used in current U.S. system of soil taxonomy.

prairillon (prai′-ril-lon) A small *prairie.*

Prandtl number In fluid mechanics, a nondimensional parameter that is the ratio of kinematic viscosity to thermometric conductivity. It is named after Ludwig Prandtl, German physicist (d. 1953).

prase (a) A translucent dull green or yellow-green variety of chalcedony. (b) Crystalline quartz containing a multitude of green hairlike crystals of actinolite. Syn: *mother-of-emerald; chrysoprase.*

prasinite (pra′-si-nite) A *greenschist* in which the proportions of the hornblende-chlorite-epidote assemblage are more or less equal (Holmes, 1928, p.189). Rarely used.

prasinophyte (pra-sin′-o-phyte) A phylogenetically primitive, primarily marine *green alga* with motile cells having *flagella* with scales and a nonmotile stage that produces highly resistant phycomata (Mendelson, 1993). Also a term descriptive of such algae. See also: *sphaeromorph; tasmanites.*

prasopal (pras-o′-pal) A green variety of common opal containing chromium. Also spelled: prase opal.

prassoite (prass′-o-ite) A metallic gray cubic mineral: $Rh_{17}S_{15}$. Syn: *miassite.*

pratincolous (pra-tin′-co-lous) Said of an organism that lives in meadows or low grassy areas.

Pratt hypothesis *isostasy.*

praya (pra′-ya) A beach or waterfront. Etymol: Portuguese "praia".

preabdomen (pre-ab′-do-men) The enlarged anterior part of the abdomen of a scorpion, composed of seven segments; the broad anterior part of the *opisthosoma* of a merostome.

preadaptation (pre′-ad-ap-ta′-tion) Appearance of nonadaptive or inadaptive characters that later prove to be adaptive in a different or changed environment.

preapical plate (pre-ap′-i-cal) Any one of the plates, typically small, in the region of the apical pore of a dinoflagellate theca; or a corresponding plate on a dinoflagellate cyst (Williams et al., 2000, p.178).

prebiotic Originating before life began.

Preboreal (Pre-bo′-re-al) n. A term used primarily in Europe for an interval of Holocene time (centered around 10,500 calendar years ago) following the Younger Dryas and preceding the Boreal, during which the inferred climate was somewhat colder and wetter than during the Boreal; a unit of the *Blytt-Sernander sequence,* characterized by birch and pine vegetation in Scandinavia. Also spelled: Pre-Boreal. adj. Pertaining to the postglacial Preboreal interval and to its climate, deposits, biota, and events.

Precambrian (Pre-cam′-bri-an) A commonly used term to designate all rocks older than the *Cambrian* Period of the Standard Global Chronostratigraphic Scale. It includes the *Archean* and *Proterozoic* eons and represents 90 percent of geologic time. Syn: *Archeozoic.*

precession camera (pre-ces′-sion) An X-ray camera used to register the diffraction from a single crystal showing individual layers of the reciprocal lattice without distortion. Cf: *Buerger precession method.*

precession cycle A 19,000 to 23,000 yr astronomic periodicity, the Precession of Equinoxes (one of the three "Milankovitch parameters" of the Sun's effective insolation at the Earth's surface), that gives rise to important climatic and geologic cycles, notably in sedimentation rates and composition. A second cycle, the axial precession, 25,694 yr rotating in the opposite direction (clockwise), also affects climate through the ecliptic angle.

precession method *Buerger precession method.*

precession of the equinoxes A consequence of the precession of the Earth's spin axis wherein the intersection of the ecliptic with the celestial equator advances along the equator. A full precessional cycle takes approximately 26,000 years. It produces approximately a twenty-second difference in length between the calendar year and the sidereal year.

precingular archeopyle (pre-cin′-gu-lar) An *archeopyle* formed in a dinoflagellate cyst by loss of one or more of the plates of the precingular series.

precingular series The series of plates immediately anterior to the cingulum in dinoflagellate theca. Cf: *postcingular series.*

precious (pre′-cious) Said of the finest variety of a gem or mineral; e.g., "precious jade" (true jadeite that is wholly or partly deep green) or "precious scapolite" (gem-quality scapolite).

precious garnet (a) An unusually purple and brilliant almandine. (b) An unusually red and brilliant pyrope.

precious metal A general term for gold, silver, or any of the minerals of the platinum group.

precious opal A gem variety of opal that exhibits a brilliant play of delicate colors; e.g. *white opal* and *black opal.* Cf: *common opal.*

precious serpentine A green massive translucent variety of the mineral serpentine.

precious stone (a) A term with little effect, meaning that in the past it was confined to a gemstone of a small group, owing to its hardness, beauty, durability, and rarity, such as a diamond, emerald, ruby, sapphire, chrysoberyl, perhaps black opal and pearl (which is not a mineral). All other gemstones were classified as semiprecious stones. The term semiprecious is now discouraged (Manutchehr-Danai, 2000). (b) An imprecise term, now little used, which encompassed such high-value gems as diamond, emerald, ruby, and sapphire (Raed, 1988). Cf: *semiprecious stone.* See also: *precious.*

precipice (prec'-i-pice) A very steeply inclined, vertical, or overhanging wall or surface of rock; e.g. the high, steep face of a cliff. Syn: *sheer.*

precipitable water The depth of water that would be produced if all the water vapor in a vertical column of air were condensed; usually the column of air extends from Earth's surface to the top of the *troposphere.*

precipitation (pre-cip'-i-ta'-tion) Water that falls to the surface from the atmosphere as rain, snow, hail, or sleet. It is measured as a liquid-water equivalent regardless of the form in which it fell. Cf: *rainfall.*

precipitation excess The volume of water from precipitation that is available for direct runoff. Cf: *rainfall excess; abstraction [water]; effective precipitation.*

precipitation facies Facies characteristics that provide evidence of depositional conditions, as revealed mainly by sedimentary structures (such as cross-bedding and ripple marks) and by primary constituents (esp. fossils) (Sonder, 1956). Cf: *alimentation facies.*

precision (pre-ci'-sion) (a) The degree of agreement or uniformity of repeated measurements of a quantity; the degree of refinement in the performance of an operation or in the statement of a result. It is exemplified by the number of decimal places to which a computation is carried and a result stated. Precision relates to the quality of the operation by which a result is obtained, as distinguished from *accuracy,* but it is of no significance unless accuracy is also obtained. (b) The deviation of a set of estimates or observations from their mean. (c) A term applied in surveying to the degree of perfection in the methods and instruments used when making measurements and obtaining results of a high order of accuracy.

precision depth recorder An *echo sounder* having an accuracy better than 1 in 3000. Abbrev: PDR.

Pre-Classic (Pre-Clas'-sic) *Formative.*

preconsolidation pressure (pre'-con-sol'-i-da'-tion) The greatest effective stress to which a soil or sediment has been subjected; the pressure exerted on unconsolidated sediment by present or former overlying material, or by desiccation of a silt or clay, resulting in compaction. Syn: *prestress.*

precoxa (pre-cox'-a) An occasionally occurring limb segment proximal to the *coxa* of a crustacean.

precurrent mark (pre-cur'-rent) A structure produced on the surface of an unconsolidated sediment before the arrival of a turbidity current; e.g. an animal track.

precursor *foreshock.*

predazzite (pre-daz'-zite) A recrystallized limestone containing brucite and calcite, with calcite predominating.

prediagenesis (pre'-di-a-gen'-e-sis) A term used by Chilingar et al. (1967, p.322) for that part of *syngenesis* responsible for "those parts that were introduced subsequently" to *syndeposition* but "before the principal processes of diagenesis began"; e.g. internal sedimentation of clastic material.

predictive deconvolution Use of information from the earlier part of a seismic trace to predict and deconvolve the latter part of that trace. Some types of systematic noise, such as reverberations and multiples can be predicted. Predictive deconvolution may also be used in a multitrace sense, where one tries to predict a trace from neighboring traces (Sheriff and Geldart, 1995).

prediluvian (pre-di-lu'-vi-an) *antediluvian.*

pre-feasibility [eco geol] An informal evaluation of the anticipated profitability of mining a particular mineral deposit. Such studies are significantly less detailed and less reliable than a *feasibility study [eco geol].*

preferred orientation [cryst] In *X-ray powder diffraction,* the nonrandom alignment of crystallites in the powder specimen, yielding statistically incorrect relative intensities of the peaks or lines of the *X-ray powder pattern.*

preferred orientation [struc geol] (pre-ferred') In structural geology, nonrandom orientation of planar or linear fabric elements, including crystallographic directions (*lattice-preferred orientation*) or elongation/flattening axes of crystals (*shape-preferred orientation*). See also: *foliation [struc geol].*

pregeologic (pre'-ge-o-log'-ic) (a) Antedating reliable geologic data or theory. (b) Before the time when the surface of the Earth became generally similar to what it is today; e.g. "pregeologic time", or the part of geologic history that antedates the oldest rocks (about 4-4.5 Ga).

preglabellar field (pre-gla-bel'-lar) That part of a trilobite *cranidium* between the glabella and the anterior border furrow.

preglacial (pre-gla'-cial) (a) Pertaining to the time preceding a period of glaciation; specif. that immediately before the Pleistocene Epoch. (b) Said of material underlying glacial deposits; e.g. the loose sand and gravel lying beneath till in Iceland, where the term "preglacial drift" is (incorrectly) used.

prehistoric (pre-his-tor'-ic) (a) Said of or pertaining to something in the past that is prior to the written records. (b) Pertaining to prehistory, i.e., the study of human activities during the time prior to written records.

prehnite (prehn'-ite) A pale-green, yellow-brown, or white orthorhombic mineral: $Ca_2Al_2Si_3O_{10}(OH)_2$. It usually occurs in crystalline aggregates having a botryoidal or mammillary and radiating structure, and is commonly associated with zeolites in geodes, druses, fissures, or joints in altered igneous rocks.

prehnite-pumpellyite facies *pumpellyite-prehnite facies.*

preisingerite (prei'-sin-ger-ite) (a) A white or gray triclinic mineral: $Bi_3O(AsO_4)_2(OH)$. (b) A group name for minerals with an analogous composition, but with AsO_4 replaced by PO_4 or VO_4.

preiswerkite (preis'-werk-ite) A pale greenish monoclinic trioctahedral mineral of the *mica* group: $Na(Mg_2Al)(OH)_2(Si_2Al_2)O_{10}$.

prekinematic layer A strata interval whose initial stratigraphic thickness is constant (or no more variable than is typical for the region) above a salt structure or its adjacent *rim syncline* (but below any *synkinematic layer*). The prekinematic layer records sedimentation before salt movement or any other deformation began (Jackson and Talbot, 1991).

preliminary waves (pre-lim'-i-nar-y) The body waves of an earthquake, including both P waves and S waves.

premagadiniform (pre'-mag-a-din'-i-form) Said of the *loop,* or of the early stages in the development of the loop, of a terebratellid brachiopod, marked by growth and completion of descending branches from both the cardinalia and the median septum and by the appearance of a tiny hood developing into a ring on the septum (TIP, 1965, pt. H, p.151). Cf: *magadiniform.*

premineral (pre-min'-er-al) adj. In economic geology, said of a structural or other feature extant before mineralization. Cf: *intermineral; postmineral.*

pre-Nectarian (pre-Nec-tar'-i-an) (a) Pertaining to lunar topographic features and lithologic map units constituting the earliest system of rocks beginning with the origin of the Moon at 4.55 Ga, including crustal solidification, and ending with the formation of the Nectaris basin at about 3.92 Ga. Pre-Nectarian rocks are older than the rocks of the Noachian System on Mars (Wilhelms, 1987). (b) Said of the stratigraphic period during which pre-Nectarian rocks developed.

preobrazhenskite (preo-bra-zhens'-kite) A colorless, yellow, or dark gray orthorhombic mineral: $Mg_3B_{11}O_{15}(OH)_9$.

preoccupied name (pre-oc'-cu-pied) In taxonomy, a name that is a junior, i.e. a later, *homonym,* unavailable for use because it was given previously to a different taxon.

preoral cavity (pre-o'-ral) The depression above the gullet in a tintinnid.

preorogenic phase (pre'-or-o-gen'-ic) The initial phase of an *orogenic cycle,* prior to the climactic orogeny. This phase was considered to be the time of formation of geosynclines, most of which are clearly divisible into eugeosynclinal (internal) and miogeosynclinal (external) parts, the first characterized by abundant submarine volcanism, the second by little magmatism and by carbonate-quartzite sedimentation. Preorogenic plutonic rocks include ultramafic bodies and rare early granitic plutons. Cf: *orogenic phase; postorogenic phase; orogenic cycle.*

prepollen (pre-pol'-len) Functional pollen grains, the fossil examples showing *haptotypic* features like those of spores, usually a trilete *laesura* on the proximal side. They may also have a *colpus* on the distal side and other pollenlike features such as *sacci.* They are typical of extinct primitive gymnosperms, mostly Mississippian to Permian in age. They are also found in the modern gymnosperms, *ginkgo* and *cycads,* in which the distally emergent pollen tube functions haustorially, not for the delivery of nuclear material, which instead occurs proximally (Proot et al., 1996).

preprismatic enamel A term for various enamel structures organizationally intermediate between "nonprismatic" and "prismatic" enamel, in which enamel crystallites diverge and converge in a chevron pattern within mutually parallel columns. Some preprismatic enamel structures resemble the nondenticular composite prismatic structure of bivalve molluscs, i.e., with each first-order prism showing a wide, central core of mutually parallel, elongated crystallites and a peripheral region of outward-flaring crystallites (S. Carlson *in* Carter, 1990, Fig. 3B). The "pseudo-prismatic" of Poole (1956, 1967a,b) and "prismatic without prism sheaths" of Osborn and Hillman (1979); Frank et al. (1984) for Triassic Haramiyidae.

preraft A fault block still resting on its original footwall during thin-skinned extension above a salt detachment.

present (pres'-ent) As a convention to standardize isotopic dating and discussions of past geologic events, the "present" is defined as the year 1950, which marks the advent of carbon isotope dating. An age before the year 1950 is expressed as "years before present" (yrBP). Cf: *Ga; ka; Ma.*

preseptal passage (pre-sep'-tal) An opening that connects all chamberlets of the same chamber of a foraminiferal test (as in the Alveolinidae), located in the anterior part of the chamber. Cf: *postseptal passage.*

presque isle A promontory or peninsula extending into a lake, nearly or almost forming an island, its head or end section connected with the shore by a sag or low gap only slightly above water level, or by a strip of lake bottom exposed as a land surface by a drop in lake level (Veatch and Humphrys, 1966, p.246). Type examples: Presque Isle (Mich.), extending into Lake Huron; Presque Isle (Penn.), extending into Lake Erie. Etymol: French "presque", "almost".

pressed amber *amberoid.*

pressolution (pres-so-lu'-tion) *pressure solution [sed].*

pressolved (pres'-solved) Said of a sedimentary bed or rock in which the grains have undergone *pressure solution;* e.g. "pressolved quartzite" whose toughness and homogeneity is due to a tightly interlocked texture of quartz grains subjected to pressure solution. Term was introduced by Heald (1956, p.22).

pressure (pres′-sure) As applied to a fluid but not a solid: (a) The force exerted across a real or imaginary surface divided by the area of that surface; the force per unit area exerted on a surface by the fluid in contact with it. (b) An incorrectly used short form for *lithostatic stress*. Cf: *stress*.
pressure altimeter *barometric altimeter*.
pressure bulb The zone in a loaded soil mass bounded by an arbitrarily selected isobar of stress (ASCE, 1958, term 277).
pressure burst *rock burst*.
pressure compartment Volume of rock hydraulically sealed from the fluids in surrounding rock whose pressure is above or below the normal static water pressure.
pressure cone Obsolete term for *shatter cone*.
pressure decay The decline, usually gradual, from a temporary, abnormal pressure condition toward a pressure that is more nearly in balance with permanent or steady-state environmental conditions.
pressure detector *hydrophone*.
pressure dissolution The process of dissolution under the effects of overburden pressure. Pressure dissolution results in the formation of stylolites, flasers, and pressure dissolution seams.
pressure dome *tumulus [volc]*.
pressure drag *pressure resistance*.
pressure figure A pattern resembling a six-rayed star, produced by intersecting lines of parting due to gliding, when certain minerals, esp. mica, are compressed by a blunt point. The rays are similar in character, but not necessarily in position, to percussion figures.
pressure fringe *pressure shadow*.
pressure gradient (a) The rate of variation of pressure in a given direction in space at a fixed time. In the ocean, pressure gradients are caused by the vertical distribution of density (which depends on water temperature and salinity), by the slope of the sea surface with respect to the level surface, and by the difference at atmospheric pressure at the sea surface. (b) Loosely, the magnitude of the pressure gradient.
pressure head The height of a column of liquid supported, or capable of being supported, by pressure *p* at a point in the liquid. See also: *static head; total head*.
pressure ice A general term for ice, esp. sea ice, whose surface in places has been deformed by stresses generated by wind, currents, or waves. It includes pieces of ice squeezed against the shore or each other, or forced upwards or downwards. Pressure ice may be "rafted", "hummocked", or "tented". Cf: *deformed ice; rough ice; screw ice*. Ant: *level ice*.
pressure maintenance The introduction of a fluid into an oil or gas reservoir to maintain the pressure and thereby improve *ultimate recovery*.
pressure melting The melting of ice at temperatures below 0°C due to increased pressure.
pressure-melting temperature (pres′-sure-melt′-ing) The temperature at which ice can melt at a given pressure. Cf: *warm ice*.
pressure pan An induced soil pan having a higher bulk density and a lower total porosity than the soil directly above or below it, produced as a result of pressure applied by normal tillage operations or by other artificial means (SSSA, 1965, p.341). Cf: *genetic pan; plow sole*. Syn: *traffic pan*.
pressure penitente A *nieve penitente* composed of brilliantly white ice, shaped into a slender ridge by lateral pressure of converging morainal streams and by melting of the adjacent debris-covered ice (Workman, 1914, p.316-317). Such features have broken upper surfaces and sharply inclined sides (resembling pointed cones, wedges, or pyramids), and usually occur on the lower parts of glaciers where morainal streams are strongly developed.
pressure plateau *lava rise*.
pressure-release jointing (pres′-sure-re-lease′) *Exfoliation* that occurs in once deeply buried rock that erosion has brought nearer the surface, thus releasing its confining pressure. See also: *lift joint*.
pressure resistance In fluid dynamics, a normal stress caused by acceleration of the fluid, which results in a decrease in pressure from the upstream to the downstream side of an object, and acting perpendicular to the boundary (Chow, 1957). Cf: *shear resistance*. Syn: *pressure drag*.
pressure ridge [ice] A rugged, irregular wall of broken floating ice buckled upward by the lateral pressure of wind or current forcing or squeezing one floe against another; it may be fresh or weathered, and extend many kilometers in length and up to 30 m in height. Cf: *ice keel*. Syn: *ridge [ice]*.
pressure ridge [seis] A seismic feature due to transverse pressure and shortening of the land surface; a *slice ridge*.
pressure ridge [volc] An elongate uplift of the congealing crust of a lava flow, buckled upward by laterally directed pressure (Walker, 1991). Cf: *tumulus [volc]*.
pressure shadow In structural petrology, aggregates of new grains growing on opposed sides of a host porphyroblast or detrital grain, thereby producing an elongate structure. This structure is generally aligned parallel to a foliation and may define a lineation (Hobbs, Means and Williams, 1976, p.274). Syn: *pressure fringe; strain shadow; stress shadow*.
pressure solution [sed] Solution (in a sedimentary rock) occurring preferentially at the contact surfaces of grains (crystals) where the external pressure exceeds the hydraulic pressure of the interstitial fluid. It results in enlargement of the contact surfaces and thereby reduces pore space and tightly welds the rock. See also: *solution transfer*. Syn: *pressolution*.
pressure solution [struc petrol] A *grain-boundary diffusion* assisted by the presence of fluid along the grain boundaries. Syn: *fluid-assisted grain-boundary diffusion*.
pressure solution cleavage *solution cleavage*.
pressure solution slip A type of aseismic fault slip that occurs by local interpenetration and removal of asperities, and concomitant growth of mineral fibers, by diffusive mass transfer. It results in slickenfiber coatings on fault surfaces (Elliott, 1976).
pressure surface A less-preferred syn. of *potentiometric surface*.
pressure-surface map (pres′-sure-sur′-face) A less-preferred syn. of *potentiometric map*.
pressure tendency The net rise or fall of barometric pressure within a specified time (usually three hours) before a particular observation. Syn: *barometric tendency*.
pressure texture *cataclastic texture*.
pressure-transient analysis The study and interpretation of time-dependent changes of fluid pressures in groundwater flow systems to estimate the hydraulic properties of the system.
pressure vessel *bomb [geochem]*.
pressure wave *P wave*.
prestack migration Migration of the components of a CMP stack before stacking. This permits proper migration where velocity varies and avoids the reflection-point smearing of dipping reflections.
prestress (pre-stress′) *preconsolidation pressure*.
presuppression (pre-sup-pres′-sion) Holding the gain of a seismic amplifier low until after the appearance of the strong first arrivals.
Pretiglian *Praetiglian*.
pretulite A colorless or pale pink mineral of the *xenotime* group: $ScPO_4$.
prevailing current (pre-vail′-ing) The ocean current most frequently observed during a given period, as a month, season, or year.
prevailing westerlies The belt of westerly winds between 35° and 60° north and south of the Equator. These winds blow from the subtropical anticyclones toward the subarctic lows, from the southwest in the Northern Hemisphere and from the northwest in the Southern Hemisphere.
previtrain (pre-vit′-rain) The woody lenses in lignite that are equivalent to vitrain in coal of higher rank (Schopf, 1960, p.30).
Priabonian (Pri-a-bo′-ni-an) A stage of the Standard Global Chronostratigraphic Scale: Upper Eocene (above Bartonian, below Rupelian Stage of the Oligocene).
priceite (price′-ite) A snow-white earthy monoclinic mineral: $Ca_2B_5O_7(OH)_5{\cdot}H_2O$. Syn: *pandermite*.
priderite (pri′-der-ite) A black or reddish tetragonal mineral of the *hollandite* group: $(K,Ba)(Ti,Fe^{3+})_8(O,OH)_{16}$.
Pridolian A series or stage of the Standard Global Chronostratigraphic Scale: Uppermost Silurian (above Ludlovian, below Devonian System). The basal GSSP is near Prague, Czech Republic (Holland, 1985).
prill An English term for a running stream.
primanal (pri-ma′-nal) The proximal anal plate in the posterior interray of camerate crinoids.
primärrumpf (pri-mär′-rumpf) A term proposed by W. Penck (1924) for a low, convex, rather featureless erosional landscape or plain produced by waxing uplift that proceeded so slowly with respect to the rate of denudation that there was no net rise of the surface or increase in its relief; an "expanding dome" that represents the universal and initial geomorphic unit. Etymol: German *Primärrumpf*, "primary torso". Cf: *endrumpf; old-from-birth peneplain*. Syn: *primary peneplain*.
Primary (Pri′-ma-ry) A term applied in the early 19th century as equivalent to *Primitive* or the period of time and associated rocks now referred to as Precambrian. It was later extended to include the lower Paleozoic, and still later restricted to the whole of the Paleozoic Era. The term was abandoned in the late 19th century in favor of *Paleozoic*. See also: *Secondary*.
primary [coast] (pri′-ma-ry) Said of a youthful coast or shoreline where waves have not had time to produce notable effects and having features that are produced chiefly by nonmarine agencies (Shepard, 1937, p.605); e.g. coasts shaped by diastrophism, volcanism, subaerial deposition, or land erosion. Cf: *secondary [coast]*.
primary [eco geol] (a) Said of a mineral deposit unaffected by supergene enrichment. (b) Said of fluid inclusion when it formed at the same time and in equilibrium with the same fluids that precipitated the host mineral.
primary [metal] Said of a metal obtained from ore rather than from scrap. Ant: *secondary [eco geol]*. Syn: *virgin*.
primary allochthony In coal formation, accumulation of plant remains in a region that does not correspond to that in which the plants grew. Cf: *secondary allochthony*.
primary ambulacral radius One of the three radii of edrioasteroids: anterior, right lateral, and left lateral. They extend out from the center of the oral area and are marked by the anterior and transverse oral midlines. The two lateral radii bifurcate to form the two lateral pairs of ambulacra, I-II and IV-V; the anterior radius continues as ambulacrum III (Bell, 1976).
primary aperture *protoforamen*.

primary arc Obsolete term for a *volcanic arc* that is convex outward from a continent and overlies a deep-seated tectonic feature; a curved orogen or fold-thrust belt whose structures originated with nearly their present curvature (Marshak and Mitra, 1988).

primary axial septulum A *primary septulum* in a foraminiferal test, representing an *axial septulum* observable in sagittal (equatorial) section (as in *Lepidolina* and *Yabeina*).

primary clay A clay found in the place where it was formed; a *residual clay.* Cf: *secondary clay.*

primary colors The basic colors from which other hues can be produced. The additive primaries (red, green, and blue) are used with light, whereas the subtractive primaries (yellow, magenta, and cyan) are used with printing inks.

primary crater An *impact crater* produced directly by the high-velocity impact of a projectile; e.g. any of the lunar craters formed by collision of the Moon with objects from space. Cf: *secondary crater.*

primary creep Elastic deformation that is time-dependent and results from a constant differential stress acting over a long period of time. Cf: *secondary creep.* Syn: *transient creep.*

primary current lineation *current lineation.*

primary dip The slight dip of a bedded deposit assumed at its moment of deposition. Syn: *original dip; depositional dip.* Cf: *initial dip; regional dip.*

primary dolomite A dense, finely textured dolomite rock, made up of grains less than 0.01 mm in diameter, formed in place by direct chemical or biochemical precipitation from seawater or lake water. Characteristically it is well-stratified, unfossiliferous, and interbedded with anhydrite, clay, and micritic limestone. Also, a similarly textured dolomite rock made up of clastic particles formed by direct accumulation. Some authors consider the rock to be syndiagenetic (Fairbridge, 1967, p.66-67). See also: *dolomicrite.* Syn: orthodolomite.

primary flat joint An approximately horizontal joint plane in igneous rocks. Syn: *L-joint.*

primary flowage Movement within an igneous rock that is still partly fluid (Cloos, 1946).

primary fluid inclusion A *fluid inclusion* containing fluid trapped during original crystallization of its host mineral (Roedder, 1979).

primary foliation A foliation that develops during the formation of a rock. Includes bedding in sedimentary rocks and flow layering in igneous rocks.

primary fumarole A *fumarole* formed over a volcanic fissure and fed directly from the main source of activity, thus giving a true index of internal conditions. Cf: *rootless fumarole.*

primary geosyncline Peyve and Sinitzyn's term for an *orthogeosyncline* (1950). Obsolete.

primary gneiss A rock that exhibits planar or linear structures characteristic of metamorphic rocks but lacks observable granulation or recrystallization and is therefore considered to be of igneous origin. Cf: *protoclastic.*

primary interstice *original interstice.*

primary lamella The first half-whorl of each brachiopod *spiralium* distal from its attachment to a *crus* (TIP, 1965, pt. H, p.151).

primary layer The outer shell layer immediately beneath the *periostracum* of a brachiopod, deposited extracellularly by columnar outer epithelium of the outer mantle lobe. It forms a well-defined calcareous layer, devoid of cytoplasmic strands, in most rhynchonelliform (formerly articulate) brachiopods. Cf: *secondary layer.* Syn: *lamellar layer.*

primary layering *phase layering; compositional layering; rhythmic layering.*

primary ligament The part of a *ligament* of a bivalve mollusk representing the original condition of structure, consisting of *periostracum* and *ostracum,* but excluding secondary additions.

primary magma A magma that has not been chemically modified from the time of melt extraction from the source region to final solidification. It is sometimes used incorrectly as a syn. of *parental magma.*

primary mineral A mineral formed at the same time as the rock enclosing it, by igneous, hydrothermal, or pneumatolytic processes, and that retains its original composition and form. Cf: *secondary mineral.*

primary optic axis One of two optic axes in a crystal that are perpendicular to the circular sections of the *indicatrix* and along which all wave normals travel with equal velocity. Cf: *secondary optic axis.*

primary oral In edrioasteroids, one of a number of large oral plates, with intrathecal extensions, that participate in formation of the underlying *oral frame* and have a fixed position within the theca relative to ambulacral and other oral plates (Bell, 1976).

primary orogeny Orogeny that is characteristic of the *internides* and that involves deformation, regional metamorphism, and granitization. Obsolete.

primary peneplain A syn. of *primärrumpf,* originally a German word with no adequate English equivalent. The term is unsatisfactory because the Davisian peneplain is developed by different processes and has different characteristics.

primary phase The only crystalline phase capable of existing in equilibrium with a given liquid; it is the first to appear on cooling from a liquid state, and the last to disappear on heating to the melting point.

primary phase region On a phase diagram, the locus of all compositions having a common primary phase.

primary pollutants In air pollution, those pollutants that are emitted directly into the atmosphere, including particulates, sulfur oxides, carbon monoxide, nitrogen oxides, and hydrocarbons (Keller, 2002).

primary porosity The *porosity* that developed during the final stages of sedimentation or that was present within sedimentary particles at the time of deposition. "Primary porosity includes all predepositional and depositional porosity of a particle, sediment, or rock" (Choquette and Pray, 1970, p.249). Cf: *secondary porosity.*

primary precipitate A precipitate formed directly; e.g. an evaporite formed by evaporation of a saline solution, or a sediment formed as a result of a reaction between dissolved material and suspended clay or as a result of a change in acidity or a shift in the oxidation-reduction potential.

primary process A conodont process that projects from the cusp; *under side* occupied by an extension of the basal cavity or basal pit (TIP, 1981, pt. W, supp.2).

primary productivity A measure of *carbon fixation* by chlorophyll-bearing organisms in a body of water; commonly expressed as gC/m^2 per unit time.

primary rocks (a) Rocks of which the constituents are newly formed particles that have never been constituents of previously formed rocks and that are not the products of alteration or replacement, esp. igneous rocks formed directly by solidification from a magma. Cf: *secondary rocks.* (b) A "more appropriate" syn. of *primitive rocks* (Humble, 1843, p.210). The term in this usage is now obsolete.

primary sedimentary structure A syngenetic *sedimentary structure* determined by the conditions of deposition (mainly current velocity and rate of sedimentation) and developed before lithification of the rock in which it is found. It includes bedding in the broad sense (esp. the external form of the beds and their continuity and uniformity of thickness), bedding-plane markings such as ripples and sole marks, and those deformational structures produced by preconsolidation movement due to unequal loading or to downslope sliding or slumping. Syn: *primary structure [geol].*

primary septulum A major partition of a chamberlet in a foraminiferal test; e.g. *primary axial septulum* and *primary transverse septulum.* Cf: *secondary septulum.*

primary septum (a) In rugose and scleractinian corals, one of the six first-formed septa in the corallite. Syn: *protoseptum.* Cf: *alar septum; cardinal septum; counter septum; counter-lateral septum.* (b) The septum immediately following the proseptum in ammonoid conchs.

primary spine The first-formed and usually largest spine of a plate of the *corona* of an echinoid. It is situated over the growth center of the plate except on a *compound plate.* Cf: *secondary spine.*

primary stratification Stratification developed when the sediments were first deposited. Syn: *direct stratification.*

primary structure [geol] (a) A structure in an igneous rock that originated contemporaneously with the formation or emplacement of the rock, but before its final consolidation; e.g. pillow structure developed during the eruption of a lava, or layering developed during solidification of a magma. (b) *primary sedimentary structure,* e.g. bedding or ripple marks. (c) The structure pre-existing the deformation and re-equilibration associated with the emplacement at shallow depth of a metamorphic body of deep origin during an orogeny (e.g. Alpine lherzolitic massifs). Cf: *secondary structure [geol].*

primary structure [paleont] Fine vacuoles or spaces in the wall of a tintinnid *lorica.* Cf: *secondary structure [paleont]; tertiary structure.*

primary suture The line of junction between the primary septum and the wall of an ammonoid *conch.*

primary tectogenesis Vertical uplift over a *geotumor* (Haarmann, 1930). Obsolete.

primary tectonite A tectonite whose fabric is *depositional fabric.* Most tectonites, however, are secondary tectonites. See also: *fusion tectonite.*

primary tissue Plant tissue derived directly by differentiation from an apical or intercalary meristem (Cronquist, 1961, p.880).

primary transverse septulum A *primary septulum* in a foraminiferal test that has a plane approximately normal to the axis of coiling and that is observable in axial section (as in *Lepidolina* and *Yabeina*).

primary type A specimen on which the description of a new species is based, wholly or in part; e.g. a *holotype, syntype,* or *lectotype* (Frizzell, 1933, p.662).

primary-type coal *banded ingredients.*

primary wall The first wall formed in a developing plant cell. It is the only wall in many types of cells (Esau, 1965, p.36-37).

primary wastewater treatment The first steps in *wastewater* treatment; screens and sedimentation tanks are used to remove most materials that float or will settle. Primary treatment removes about 30 percent of carbonaceous biochemical oxygen demand from domestic sewage (USEPA, 1994). Cf: *advanced wastewater treatment; secondary wastewater treatment.*

primary wave *P wave.*

primary zooid A single *zooid* (ancestrula), or one of the multiple, simultaneously formed zooids resulting from the metamorphosis of a single larva to found a cheilostome bryozoan colony (Boardman and Cheetham, 1973, p.173).

primatology (pri-ma-tol'-o-gy) The study of fossil and living primates.

prime meridian An arbitrary meridian selected as a reference line having a longitude of zero degrees and used as the origin from which other longitudes are reckoned east and west to 180 degrees; specif. the *Greenwich meridian.* Local or national prime meridians are occasionally used. Syn: *zero meridian; initial meridian; first meridian.*

prime movers The large diesel engines used to supply power to a rotary drilling rig.

primeval (pri-me'-val) Pertaining to the earliest ages of the Earth; e.g. said of lead that is associated with so little uranium (as in some meteorites) that the Pb-isotope composition has not changed appreciably in five billion years. See also: *primordial.*

primeval-fireball hypothesis (pri-me'-val-fire'-ball) *"big bang" hypothesis.*

primeval lead *primordial lead.*

prime vertical (a) A vertical circle perpendicular to the plane of the celestial meridian. The plane of the prime vertical cuts the horizon in the east and west points. (b) At a given point on an ellipsoid, the geodesic perpendicular to the meridian through that point.

primibrachial (pri-mi-brach'-i-al) A plate of the proximal brachitaxis of a crinoid. It may or may not be an axillary, and it may or may not be fixed into the *calyx.*

Primitive (Prim'-i-tive) A name applied from the teachings of A.G. Werner in the 1790's to the group or series of rocks that were considered the first chemical precipitates derived from the ocean before emergence of land areas and that were believed to extend uninterrupted around the world. The rocks included the larger intrusive igneous masses, all highly metamorphosed rocks, and roughly the rocks that later came to be known as Precambrian in age. See also: *Primary; Transition.*

primitive achondrite A *stony meteorite* of igneous texture (essentially free of chondrules) whose mineral and bulk composition differ little from those of chondritic matter, suggesting appreciable melting but without significant phase segregation. See also: *acapulcoite.*

primitive area (prim'-i-tive) Land belonging to the U.S. government that is to be preserved in its natural state. The only changes permitted are those for fire prevention.

primitive character In *cladism,* a character or character state possessed by an ancestral species. Cf: *homology (c).* Syn: *plesiomorphy.*

primitive circle That circle on a *stereographic projection* which is the intersection of the stereographic plane with the sphere of reflection; it is the sphere's equatorial circle.

primitive lattice A *crystal lattice* with lattice points only at the corners of the unit cell. Cf: *centered lattice.* Syn: *simple lattice.*

primitive magma Magma that approaches the compositional criteria for primary magma (Basaltic Volcanism Study Project, 1981, p.399). These criteria are high *magnesium number,* the presence of olivine on the liquidus, and high concentrations of Ni and Cr, sufficient to indicate equilibrium between the magma and a mantle source rock.

primitive mantle The hypothetical composition of the Earth's mantle after formation of the core but before separation of the crust (Sun, 1982). Also called: *bulk silicate Earth.*

primitive rocks A term applied by Lehmann (1756) to crystalline rocks devoid of fossils and rock fragments, and believed to be of chemical origin, having formed prior to the advent of life; also, a term for the rocks believed to have been first formed, being irregularly crystallized and aggregated without cement. They include gneiss, schist, primary limestone, and plutonic rocks such as granite. The term is obsolete as many of these rocks are found in all ages and formations. Cf: *secondary rocks.* Syn: *primary rocks.*

primitive unit cell A *unit cell* that contains only one *lattice point.* Cf: *conventional unit cell; reduced unit cell.*

primordial (pri-mor'-di-al) Original, first in development, earliest, or existing from the beginning; e.g. "primordial dust", "primordial ocean basin" or a "primordial magma". See also: *primeval.*

Primordial (Pri-mor'-di-al) An obsolete term formerly applied to what is now called *Cambrian.* It was used by Joachim Barrande (1799-1883), French paleontologist, for the oldest or lowest fossiliferous strata as developed in Bohemia.

primordial lead Lead with isotopic ratios unaffected by addition of lead from radioactive decay of uranium and thorium since the formation of the Earth (and solar system). It is often taken to be lead with an isotopic composition the same as that found in the *troilite* phase of meteorites. Syn: *primeval lead.*

primordial plate One of the first plates formed following metamorphosis in each plate system of an echinoid.

primordial valve A chitinous plate in certain cirripede crustaceans, having a distinctive honeycomb appearance, and developing at incipient umbones of the terga, scuta, and carina during metamorphosis (TIP, 1969, pt. R, p.100).

primordium (pri-mor'-di-um) The rudiment or earliest trace of development of any plant structure or organ (Swartz, 1971, p.379).

principal axis [cryst] (prin'-ci-pal) That crystallographic axis which is the most prominent. In the tetragonal and hexagonal systems, it is the vertical or *c* axis; in the orthorhombic, monoclinic, or triclinic systems, it is also usually the *c* axis, although in monoclinic minerals such as epidote it may be the *b* axis.

principal axis [exp struc geol] In experimental structural geology, a *principal axis of stress* or a *principal axis of strain.*

principal axis of strain One of the three mutually perpendicular axes corresponding to the three axes of the body that were also mutually perpendicular before deformation; also described as the axes of the strain ellipsoid. The longest or greatest is the axis of elongation, and the shortest or least is the axis of shortening. Syn: *strain axis; principal axis.*

principal axis of stress One of the three mutually perpendicular axes that are perpendicular to the principal planes of stress. In a body subjected to unequal tractions in different directions, a line normal to one of three planes across which there are no shear stresses. One of the three principal axes will indicate the direction of the most compressive normal stress, and another will indicate the direction of the most tensile normal stress. Syn: *stress axis; principal axis.*

principal components analysis A projection of multivariate observations onto new orthogonal axes, termed principal components, that are linear combinations of the original variables. The principal components are the *eigenvectors* of the *variance-covariance matrix.* Syn: *components analysis.*

principal coordinates analysis A Q-mode procedure in which observations are projected onto axes defined by the *eigenvectors* of a centered *matrix* of similarities between the observations. Cf: *Q-mode factor analysis.*

principal distance The perpendicular distance from the internal perspective center to the plane of a particular finished negative or print.

principal focus The focus for a beam of incident rays parallel to the axis of a lens or optical system. Syn: *focal point.*

principal layer The main layer of a trilobite cuticle, commonly showing some internal lamination. Some authors recognize three distinct zones: outer, central, and inner. Cf: *prismatic layer.*

principal line [geochem] The *spectral line* that is most easily excited or observed.

principal line [photo] The trace of the principal plane upon a photograph; e.g. the line through the principal point and the nadir of a tilted photograph.

principal meridian (a) The 0° longitude line that passes through Greenwich, England, and from which longitude is measured. (b) *central meridian* on which a rectangular grid is based; specif. one of a pair of coordinate axes (along with the *base line*) used in the U.S. Public Land Survey system to subdivide public lands in a given region. It consists of a line extending north and south along the astronomic meridian passing through the initial point and along which standard township, section, and quarter-section corners are established. The principal meridian is the line from which the survey of the township boundaries is initiated along the parallels.

principal plane [cryst] One of the three planes of a biaxial *indicatrix* containing two of the principal vibration directions, X, Y, and Z. Cf: *principal section.*

principal plane [photo] (a) The vertical plane through the internal perspective center and containing the perpendicular from that center to the plane of a tilted photograph. (b) Any plane perpendicular to the axis of an optical system and passing through its principal points.

principal plane of strain One of three planes in the *strain ellipsoid* containing any two of the principal axes of strain. Shear strain will be zero across each of the principal planes.

principal plane of stress One of three mutually perpendicular planes, upon each of which the resultant stress is normal, i.e. on which shear stress is zero. See also: *principal axis of stress.*

principal point (a) The foot of the perpendicular from the interior perspective center of a lens to the plane of the photograph; the geometric center of an aerial photograph, or the point where the optical axis of the lens meets the film plane in an aerial camera. Symbol: p. See also: *fiducial mark; photograph center.* Syn: *center point.* (b) Either of two points on the axis of a lens such that a ray from any point of the object directed toward one principal point will emerge from the lens in a parallel direction but directed through the other principal point. (c) The point at which a principal visual ray intersects a perspective plane.

principal ray (a) The one ray within a bundle of incident rays that, upon entering an optical instrument from any given point of the object, passes through the optical center of the lens. (b) *principal visual ray.*

principal section In a uniaxial *indicatrix,* any plane containing the optic axis. Cf: *principal plane [cryst].*

principal spine One of the large regularly placed spikes or needles in acantharian and spumellarian radiolarians.

principal strain The magnitude of the strain in the direction of a *principal axis of strain.* Equal to the stretch parallel to a principal axes.

principal stress A stress that is perpendicular to one of three mutually perpendicular planes that intersect at a point in a body on which the shear stress is zero; a stress that is normal to a principal plane of stress. The magnitude of stress in the direction of a *principal axis of stress.* The three principal stresses are identified as least or minimum, intermediate, and greatest or maximum. See also: *mean stress.*

principal visual ray A perpendicular extending from a station point to a perspective plane and theoretically passing exactly along the visual axis of a viewing eye. Syn: *principal ray.*

principle of uniformity *uniformity.*

pringleite A colorless, pale yellow, or orange triclinic mineral: $Ca_9B_{26}O_{34}(OH)_{24}Cl_4{\cdot}13H_2O$.

print A photographic copy made from a negative or from a positive transparency by photographic means. Prints may be positive (the usual case) or negative. Cf: *positive [photo].*
priorite (pri'-or-ite) *aeschynite-(Y).*
priority (pri-or'-i-ty) State of being earlier or first; see *law of priority.*
Prior's rules (Pri'-or's) A redox relationship among the constituents of chondritic meteorites, first recognized by George T. Prior (1862-1936), an English mineralogist, and generally stated: The less the amount of nickel-iron in a chondritic stone, the richer in nickel is the metal content, and the richer in iron are the magnesium silicates.
Priscoan A suggested geochronometric subdivision of the *Precambrian,* a pre-*Archean* time unit earlier than 4,000 Ma, prior to the age of the oldest known preserved rocks (Harland et al., 1990).
prisere (pri'-sere) A *sere* that takes place in a barren area undisturbed by man's activities.
prism [cryst] A crystal form having three, four, six, eight, or twelve faces, with parallel intersection edges, and which is open only at the two ends of the axis parallel to the intersection edges of the faces.
prism [sed] (a) A long, narrow, wedge-shaped sedimentary body whose width/thickness ratio is greater than 5 to 1 but less than 50 to 1 (Krynine, 1948, p.146); e.g. an alluvial fan adjacent to an escarpment, or one of the great conglomerates of the geologic record. It is typical of orogenic sediments formed during periods of intense crustal deformation, such as the arkoses found in fault troughs. Cf: *tabular; shoestring.* Syn: *wedge [stratig].* (b) *geosynclinal prism.*
prismatic (pris-mat'-ic) (a) Said of a sedimentary particle whose length is 1.5 to 3 times its width (Krynine, 1948, p.142). Cf: *tabular.* (b) Pertaining to a sedimentary prism. (c) Pertaining to a crystallographic prism. (d) Said of a crystal that shows one dimension markedly longer than the other two. (e) Said of a metamorphic texture in which a large proportion of grains are prismatic and have approximately parallel orientation, so that a lineated appearance is usually visible in hand specimen or thin section.
prismatic class That crystal class in the monoclinic system having symmetry *2/m.* Prisms of this system have four faces, are rhombic in cross section, and have as their axis either the *c* axis, the *a* axis, or any lattice row perpendicular to the *b* axis.
prismatic cleavage Mineral cleavage parallel to the faces of a prism, e.g. the (110) cleavage of amphibole.
prismatic compass A small *magnetic compass* held in the hand when in use and equipped with peep sights and a glass prism so arranged that the magnetic bearing or azimuth of a line can be read (through the prism) from a circular graduated scale at the same time that the line is sighted over.
prismatic joints *columnar joints.*
prismatic layer (a) The middle layer of the shell of a mollusk, consisting essentially of prisms of calcium carbonate (calcite or aragonite). (b) Part of the *secondary layer* in some articulate brachiopods, secreted extracellularly as prismatic calcite. (c) The most common type of outer layer of trilobite cuticle, generally about 20 to 30 μm thick. In some species the outer layer is pigmented or phosphatic rather than prismatic. Cf: *principal layer.*
prismatic structure (a) A *microstructure* of mutually parallel, adjacent, polygonal columns that do not strongly interdigitate along their mutual boundaries and may be separated by organic matrix (Boggild, 1930, p.246; Taylor et al., 1969, p.4; Carter, 1980a). (b) *columnar joints.*
prismatine A gold-brown orthorhombic mineral of the *kornerupine* group: $(Mg,Al,Fe)_6Al_4(Si,Al)_4(B,Si,Al)(O,OH,F)_{22}$.
prismatolith (pris-mat'-o-lith) A *coccolith* constructed of polygonal prisms. It may be solid or axially perforated. See also: *porolith.*
prismatomorph (pris-mat'-o-morph) Of *acritarchs,* having prismatic to polygonal *vesicles* (e.g., *Polyedryxium*) with more or less sharp edges that may be extended into a flange, entire or serrate, with or without *processes* at the angles (Downie et al., 1963). Also such an *acritarch.*
prism crack A mud crack that develops in polygonal patterns on the surface of drying mud puddles and that breaks the sediment into prisms standing normal to bedding (Fischer, 1964, p.114).
prism level A type of *dumpy level* in which the level bubble can be viewed from the eyepiece end by means of an attached prism at the same time the rod is being read.
prismoid (pris'-moid) A solid body resembling a prism, having similar but unequal parallel polygonal ends. It is a textural term used for sedimentary particles. Adj: *prismoidal.*
prismoidal (pris-moi'-dal) Adj. of *prismoid.* It is a term used in sedimentary petrology, and is not to be confused with *prismatic,* which is a crystallographic term.
prism twin law A rare, normal twin law in monoclinic or triclinic feldspars, having a twin plane of (110) or ($\bar{1}$10).
pristane Another isoprenoid hydrocarbon, pristane/phytane ratios are used to determine the source of organic matter.
pristine rocks (pris-tine') Rocks from the lunar highlands thought to have been produced by early igneous activity, which have retained their original chemical characteristics (S.R. Taylor, 1982).
privileged direction In optics, the direction in which light waves vibrate as they pass through a crystal along any particular path. Syn: *vibration direction.*
proancestrula (pro-an-ces'-tru-la) The first-formed or basal part of the *ancestrula* of stenolaemate bryozoans. Syn: *basal disk; protoecium.*
probability (prob-a-bil'-i-ty) A scale indicating the degree to which an event is expected to occur; 0 indicates the event is impossible, 1 indicates the event is certain to occur. Probabilities may be based on mathematical logic, as in the discrete probabilities arising in fixed games of chance; on the past record of occurrences or relative frequencies; or on subjective belief.
probable ore (prob'-a-ble) (a) A syn. of *indicated ore.* (b) A mineral deposit adjacent to developed ore but not yet proven by development. Cf: *extension ore.*
probable ore reserve The economically minable part of an *indicated mineral resource,* and in some circumstances *measured mineral resource* (JORC, 1999). Cf: *ore reserve.*
probable reserve A *reserve* for which quantity and *grade* and/or quality are computed from information similar to that used for a *proven reserve,* but the sites for inspection, sampling, and measurement are farther apart or are otherwise less adequately spaced. The degree of assurance, although lower than that for a proven reserve, is high enough to assume continuity between points of observation (SEC, 2003).
probe n. Any measuring device that is placed in the environment to be measured, e.g. a potential electrode, a density probe in a drill hole, or oceanographic instruments that are lowered into the sea.
probertite (prob'-ert-ite) A colorless monoclinic mineral: $NaCaB_5O_7(OH)_4{\cdot}3H_2O$. Syn: *kramerite.*
problematic fossil (prob-lem-at'-ic) A natural object, structure, or marking in a rock, resembling a fossil but having a doubtful organic nature or origin. Cf: *pseudofossil.* Syn: *dubiofossil.*
problematicum (prob-lem-at'-i-cum) A marking, object, structure, or other feature in a rock whose nature presents a problem, such as a doubtful "fossil" that may be of inorganic origin or whose organic nature is uncertain; esp. an undoubted organic remain (such as a trace fossil) with a more or less obscure origin. Pl: problematica.
proboscis (pro-bos'-cis) A distal cylindrical tube extending from an astropyle of a phaeodarian radiolarian.
procephalic (pro-ce-phal'-ic) Pertaining to, forming, or situated on or near the front of the head; e.g. the "procephalic lobe", or anterior preoral part, of a merostome embryo.
process [paleont] Anterior, posterior, or lateral projections of ramiform and pectiniform conodont elements; commonly, but not invariably denticulate (TIP, 1981, pt. W, supp.2).
process [palyn] An essentially columnar or spinelike projection arising from the surface of a *dinoflagellate cyst* or a projection of >5 μm arising from the surface of an *acritarch* or *sporomorph* (Lentin and Williams, 1976; Williams et al., 2000, p.183).
process zone Region of inelastic deformation surrounding a crack tip in brittle materials like rock. The inelastic deformation is largely a consequence of the high stress concentration near the crack tip.
prochlorite (pro-chlo'-rite) *ripidolite.*
prochoanitic (pro'-cho-a-nit'-ic) Said of a *septal neck* of a cephalopod directed forward (adorally, or toward the aperture). Ant: *retrochoanitic.*
prochronic (pro-chron'-ic) Before time or creation. Ant: *diachronic.*
proclade (pro'-clade) A clade or bar in the ebridian skeleton that arises from the end of an upper actine and is directed toward the anterior or nuclear pole. Cf: *opisthoclade.*
procline (pro'-cline) Said of the ventral and anterior inclination of the *cardinal area* in the pedicle valve of a brachiopod, lying in the bottom right or second quadrant moving counterclockwise from the *orthocline* position (TIP, 1965, pt. H, p.60, fig. 61).
proclined When used to describe the cusp or denticles of a conodont element, this term implies that the axis of the cusp or denticles forms an angle of more that 90 degrees with the upper edge of the base or of a posterior process (TIP, 1981, pt. W, supp.2).
Proctor (Proc'-tor) Pertaining to or determined by a procedure designed by Ralph R. Proctor (1894-1962), U.S. civil engineer, to establish water content-density relationships of a remolded soil by application of compactive effort under standardized conditions; e.g., the "Proctor curve" (or *compaction curve*), the "Proctor compaction test", and the "Proctor density" of soil.
procumbent (pro-cum'-bent) Said of a stem that trails or lies flat on the ground but does not take root.
prod cast The cast of a *prod mark,* consisting of a short ridge that rises downcurrent and ends abruptly. Originally defined by Dzulynski and Slaczka (1958, p.232). Syn: *impact cast.*
prodelta (pro-del'-ta) The part of a delta that is below the effective depth of wave erosion, lying beyond the *delta front,* and sloping gently down to the floor of the basin into which the delta is advancing and where clastic river sediment ceases to be a significant part of the basin-floor deposits; it is entirely below the water level. Cf: *delta plain.*
prodelta clay The fine-grained river-borne material (very fine sand, silt, and clay) deposited as a broad fan on the floor of a sea or lake beyond the main body of a delta; the material in a *bottomset bed.*
prodissoconch (pro-dis'-so-conch) The rudimentary or earliest-formed shell of a pelecypod preserved as the beak of some adult shells. The terms

prodissoconch I and prodissoconch II may be used to refer to earlier-formed and later-formed parts of prodissoconch, respectively.

prod mark [sed struc] A short *tool mark* oriented parallel to the current of a stream and produced by an object that plowed into and was then raised above the bottom. Its longitudinal profile is asymmetric. The mark deepens gradually downcurrent where it ends abruptly (unlike a *flute*). Cf: *bounce mark.* Syn: *impact mark.*

prod mark [struc geol] An indicator of slip direction on a slickensided fault surface, consisting of a groove made by a clast.

producer [ecol] (pro-duc'-er) An organism (e.g., most plants and many bacteria) that can form new organic matter from inorganic matter (e.g., carbon dioxide, water, soluble salts). Cf: *consumer.*

producer [petroleum] A well that produces oil or gas.

producing zone (pro-duc'-ing) The rock stratum of an oil field that will produce petroleum or gas when penetrated by a well. Often incorrectly referred to as "producing horizon".

productid (pro-duc'-tid) *productoid.*

production [lake] In a lake, the increase in biomass of new organic material over a period, including any losses attributable to *respiration, excretion, secretion,* injury, death, or *grazing* (Wetzel, 2001, p.143).

production [oceanog] (pro-duc'-tion) Rate of potential increase of a given biological group within the marine carbon cycle (e.g., phytoplankton, zooplankton, fish).

productivity [lake] The rate of *production [lake]* per unit time.

productivity [oceanog] A general term for the organic fertility of a body of water. Also used synonymously with *primary productivity* and *production.*

productoid (pro-duc'-toid) Any articulate brachiopod belonging to the suborder Productidina, characterized by a pseudopunctate shell having a flat or concave, rarely convex, brachial valve and a convex pedicle valve. This group includes the largest and most aberrant brachiopods yet known. Range, Lower Devonian to Upper Permian. Var: *productid.*

proepistome (pro-ep'-i-stome) *interantennular septum.*

profile [geomorph] (pro'-file) (a) The outline produced where the plane of a vertical section intersects the surface of the ground; e.g. the *longitudinal profile* of a stream, or the profile of a coast or hill. (b) *profile section.*

profile [geophys] A graph or drawing that shows the variation of one property such as elevation or gravity, usually as ordinate, with respect to another property, such as distance.

profile [palyn] *pollen profile.*

profile [seis] In seismic prospecting, the data recorded from one source point by a number of groups of detectors. Syn: *seismic profile.*

profile [soil] A vertical section of the soil through all its horizons and extending into the C horizon.

profile [struc geol] Cross section of a region of cylindrical folds drawn perpendicular to the fold axes. Syn: *tectonic profile; right section.*

profile [water] A vertical section of a water table or other potentiometric surface, or of a body of surface water.

profile line The top line of a *profile section,* representing the intersection of a vertical plane with the surface of the ground. Cf: *line of section.*

profile method *two-dimensional method.*

profile of equilibrium [coast] The slightly convex slope of the floor of a sea or lake, taken in a vertical plane and extending away from and transverse to the shoreline, being flattest near the shore, and having a gradient such that wave energy is expended at a uniform rate across the shore zone; the transverse slope of a *graded shoreline.* The profile is easily disturbed by strong winds, large waves, and exceptional high tides. The concept is hypothetical; see, for example, Bloom, 1978. Syn: *equilibrium profile; graded profile.*

profile of equilibrium [streams] The longitudinal profile of a graded stream or of one whose gradient at every point is just sufficient to enable the stream to transport the load of sediment made available to it. It has long been regarded as a smooth, parabolic curve, gently concave to the sky, practically flat at the mouth and steepening toward the source; but current thought is that it need not be smooth.

profiler (pro'-fil-er) A marine seismic system that achieves limited penetration into the sedimentary section. It usually employs a low-energy source, often a *sparker,* and only one or a few recording channels.

profile section A diagram or drawing that shows along a given line the configuration or slope of the surface of the ground as it would appear if intersected by a vertical plane. The vertical scale is often exaggerated. See also: *line of section; profile line.* Syn: *profile [geomorph].*

profiling (pro'-fil-ing) Exploration wherein sensors are moved along a line and a profile is developed. Cf: *sounding [elect].* Syn: *lateral search.*

profluent stream (pro'-flu-ent) A stream that is flowing copiously or smoothly.

profundal (pro-fun'-dal) Pertaining to or existing in the deeper part of a lake, below the limit of well developed zones of vegetation. Cf: *littoral [lake]; tropholytic.*

progenesis pro-gen'-e-sis *Paedomorphosis* resulting from the early onset of reproductive maturity.

progenitor (pro-gen'-i-tor) In biology, an ancestor.

proglacial (pro-gla'-cial) Immediately in front of or just beyond the outer limits of a glacier or ice sheet, generally at or near its lower end; said of lakes, streams, deposits, and other features produced by or derived from the glacier ice.

proglacial lake A lake formed just beyond the frontal margin of an advancing or retreating glacier, generally in direct contact with the ice. Cf: *glacier lake.* Syn: *ice-marginal lake.*

progradation (pro-gra-da'-tion) Outward or basinward building of a shoreline occurring when the rate of sediment supply at the shoreline exceeds the rate of relative rise in sea level. Ant: *retrogradation [coast].* Cf: *advance [coast]; regression [stratig].*

progradational parasequence set In sequence stratigraphy, a parasequence set in which successively younger parasequences are deposited farther basinward; overall, the rate of deposition is greater than the rate of accommodation (Van Wagoner et al., 1990).

prograde metamorphism (pro'-grade) Metamorphic changes in response to a higher pressure or temperature than that to which the rock last adjusted itself. Cf: *polymetamorphism; retrograde metamorphism.*

prograde motion The predominant eastward direction of angular motion in the solar system, applying both to planetary orbital motion and the axial rotation of the Sun and most planets, and also to the orbital motion of most satellites of planets. The direction is counterclockwise as viewed from the north. Cf: *retrograde motion.*

prograding shoreline (pro'-grad-ing) A shoreline that is being built forward or outward into a sea or lake by deposition and accumulation. Ant: *retrograding shoreline.*

progression (pro-gres'-sion) *advance [coast].*

progressionism (pro-gres'-sion-ism) The special belief that accompanied *catastrophism,* that each successive creation consisted of animals of a higher order than the previous one.

progressive deformation (pro-gres'-sive) A continuous sequence of configurations through which a body passes. Distinguished from a *deformation,* which involves just two discrete states of the body (commonly the initial and final states) (Flinn, 1962).

progressive fault *growth fault.*

progressive metamorphism Progressive change in the degree of metamorphism from lower to higher grade across a metamorphic terrane. The term may be applied to rocks in contact aureoles or to rocks traced through the different isograds or facies of regional metamorphism. It usually implies sequential passage of rocks up through all the lower grade zones.

progressive overlap A general term used by Grabau (1906, p.569) for a "regular progressive" onward movement or spreading of the "zones of deposition", and including what are now known as onlap, offlap, and continental transgression.

progressive sorting Sorting in the downcurrent direction, resulting in a systematic downcurrent decrease in the mean grain size of the sediment (Pettijohn, 1957, p.541).

progressive wave A water wave, the wave form of which appears to move progressively. Cf: *standing wave.*

progymnosperm (pro-gym'-no-sperm) One of a group of arboreal Devonian *vascular plants* with fernlike leaves and coniferlike wood. Progymnosperms represent the first arboreal plants.

projectate (pro-jec'-tate) Of *pollen grains,* having the *apertures* borne on the ends of strongly projecting arms, e.g., in *Aquilapollenites.*

projected profile (pro-ject'-ed) A diagram that includes only those features of a series of profiles, usually drawn along several regularly spaced and parallel lines on a map, that are not obscured by higher intervening ground (Monkhouse and Wilkinson, 1952); it gives a panoramic effect with a distant skyline, a middleground, and a foreground, and it represents an outline landscape-drawing showing only summit detail. Cf: *superimposed profile; composite profile.*

projection (pro-jec'-tion) (a) A systematic, diagrammatic representation on a plane (flat) surface of three-dimensional space relations, produced by passing lines from various points to their intersection with a plane; esp. a *map projection.* (b) Any orderly method by which a projection is made; the process or operation of transferring a point from one surface to a corresponding position on another surface by graphical or analytical means, so that each point of one corresponds to one and only one point of the other.

projection change The transfer of digital map data from one map projection to another (Krzanowski et al., 1993).

projection net *net [struc petrol].*

prokaryote (pro-kar'-y-ote) A member of a group of organisms that lack a membrane-bound nucleus or other organelles. The group includes the true *bacteria* (Eubacteria), *Cyanobacteria,* and the thermophilic and extremophilic bacteria (*Archaea*). Procaryotes range from the Archean to the present. Cf: *eukaryote.*

prokaryote [paleont] Any organism, usually single celled, of the kingdom Bacteria (also referred to as Prokaryotae, Procaryotae, and Monera) characterized by having no nucleus, typically a smaller size than *eukaryotes,* and a thin cell wall; formerly referred to in part as *protists.* Range, Archean to Holocene.

prolapsed bedding (pro-lapsed') A term used by Wood and Smith (1958, p.172) for bedding characterized by a series of flat folds with near-horizontal axial planes contained entirely within a bed having undisturbed boundaries.

prolate (pro′-late) Extended or elongated in the direction of a line joining the poles; e.g. "prolate pollen" whose equatorial diameters are much shorter than the dimensions from pole to pole. Ant: *oblate.*

prolate strain Three-dimensional strain defined by a strain tensor in which the magnitude of the intermediate and minimum strains are approximately equal and significantly less than that of the maximum strain. Cf: *oblate strain; plane strain.*

proloculus (pro-loc′-u-lus) The initial or first-formed chamber of a foraminiferal test, typically at the small end of a series or at the center of a coil. Pl: proloculi.

proloculus pore A single circular opening in a *proloculus,* leading to the next-formed chamber of a foraminiferal test (as in fusulinids).

promontorium A term established by the International Astronomical Union for a cape, or headland promontory on the Moon (USGS, 2003). Pl: promontoria.

promontory (prom′-on-to′-ry) (a) A high, prominent projection or point of land, or cliff of rock, jutting out boldly into a body of water beyond the coastline; a headland. Syn: *cobb; reach; ness; nook.* (b) A cape, either low-lying or of considerable height, with a bold termination. (c) A bluff or prominent hill overlooking or projecting into a lowland.

prong [geomorph] *spur [geomorph].*

prong [streams] A term applied in the southern Appalachian Mountains to a fork or branch of a stream or inlet.

prong reef A *wall reef* that has developed irregular buttresses normal to its axis in leeward and (to a smaller degree) seaward directions (Maxwell, 1968, p.99, 101).

proof roll In engineering geology, a procedure in which a loaded dump truck or other heavy vehicle with pneumatic tires systematically passes over a prepared soil site to detect soft or loose zones where compactive effort is needed to provide a uniform consistency or density prior to placing compacted fill or structural elements of a building.

proostracum (pro-os′-tra-cum) The anterior (adoral) horny or calcareous bladelike prolongation of the dorsal border of the *phragmocone* of belemnites and related cephalopods, forming a protecting shield over the visceral mass of the animal. Pl: proostraca.

propagule (prop′-a-gule) The minimum number of individuals of a species required for the successful colonization of a habitable island (MacArthur and Wilson, 1967, p.190).

propane (pro′-pane) An inflammable gaseous hydrocarbon, formula C_3H_8, of the methane series. It occurs naturally in crude petroleum and natural gas. It is also produced by cracking and is used primarily as a fuel and in the making of chemicals.

proparea (pro-par′-e-a) One of a pair of flattened subtriangular halves of the pseudointerarea of a brachiopod, divided medially by various structures (such as homeodeltidium, intertrough, or pedicle groove).

proparian (pro-par′-i-an) adj. Of or concerning a trilobite whose facial sutures extend outward from the eyes to the lateral margin of the *cephalon;* e.g. a "proparian facial suture" that crosses the dorsal surface of the cephalon, passes along the medial edge of the eye, and intersects the lateral border of the cephalon in front of or at the genal angle. Cf: *opisthoparian.* n. A proparian trilobite; specif. a trilobite of the order Proparia (now obsolete) in which the posterior branch of the *facial suture* cuts the lateral margin of the cephalon.

propeltidium (pro-pel-tid′-i-um) An anterior *sclerite* of a segmented carapace of an arachnid, situated in front of the *mesopeltidium.*

proper cave A *cave* large enough for a person to enter (Curl, 1964). The term was introduced to remove anthropomorphic attributes from the term "cave", which then could refer to all solutional openings. Proper caves were analyzed as geometrically three-dimensional fractals, and the results were used for extrapolation of some of their geometric properties to caves smaller than proper ones (Curl, 1986).

proper entrance A cave entrance large enough for a person to enter (Curl, 1964).

proper modulus The modulus defining a proper cave; e.g. 0.6 m (Curl, 1966).

proper symmetry operation A symmetry operation which leaves the handedness of the objects upon which it operates unchanged. Rotation axes are proper symmetry operations. Cf: *improper symmetry operation.*

prophylaxis layer A relatively thick organic shell layer deposited in the umbonal region of some freshwater bivalves in anticipation of deep corrosion of their shell. Unlike some other interior conchiolin layers, these are not deposited in response to irritation of the mantle epithelium and are not directly underlain by prismatic structure (compare with "intranacreous periostracal extensions" defined by Grégoire, 1974, p.867).

propodosoma (pro′-po-do-so′-ma) A section of the body of an acarid arachnid, bearing the first and second pairs of legs. Cf: *metapodosoma.*

propodus (prop′-o-dus) The sixth or penultimate segment of the *pereiopod* of a malacostracan crustacean, bounded proximally by the carpus and distally by the dactylus. Pl: propodi.

proportional counter (pro-por′-tion-al) A radiation detector consisting of a gas-filled tube in which the amplitude of the discharge pulses is proportional to the energy of the ionizing particles.

proportional limit The highest value of stress that a material can undergo before it loses its linear relationship between stress and strain, i.e. before it ceases to behave according to Hooke's law.

propylite (prop′-y-lite) A hydrothermally altered *andesite* resembling a greenstone and containing calcite, chlorite, epidote, serpentine, quartz, pyrite, and iron oxides. The term was first used by Richthofen in 1868. See also: *propylitization.*

propylitic alteration (pro-py-lit′-ic) *propylitization.*

propylitic zone The zone in porphyry ore deposits characterized by a chlorite-calcite-epidote-adularia-albite alteration assemblage (Lowell and Guilbert, 1970). Cf: *argillic zone; phyllic zone; potassic zone.*

propylitization (prop′-y-lit′-i-za′-tion) The result of low-pressure-temperature alteration around many ore bodies. The propylitic assemblage consists of epidote, chlorite, Mg-Fe-Ca carbonates, and sometimes albite-orthoclase, all involved in partial replacement of wall-rock minerals (Guilbert and Park, 1986, p.173). Syn: *propylitic alteration.*

proration (pro-ra′-tion) Restriction of oil and gas production by a regulatory commission, usually in anticipation of market demand. It is the basis on which allowables are assigned.

prorsiradiate (pror-si-ra′-di-ate) Said of an ammonoid rib inclined forward (adorally) from the umbilical side toward the venter. Cf: *rursiradiate; rectiradiate.*

proseptum (pro-sep′-tum) The septum closing the *protoconch* in an ammonoid shell.

prosicula (pro-sic′-u-la) The proximal, initially formed part of the *sicula* of a graptolite, secreted as a single conical unit. Cf: *metasicula.*

prosiphon (pro-si′-phon) A small, threadlike structure extending from the adapical part of the caecum to the wall of the *protoconch* of an ammonoid shell.

prosobranch (pros′-o-branch) Any gastropod belonging to the subclass Prosobranchia, characterized in most cases by the presence of a shell, commonly with an operculum, and by the anterior position of the auricle with respect to the ventricle. Range, Cambrian to Holocene.

prosochete (pros′-o-chete) An *inhalant canal* of a sponge.

prosocline (pros′-o-cline) (a) Said of the hinge teeth or of the body of the shell of a bivalve mollusk, sloping from the lower end in the anterior or forward direction, such that the angle formed by the midumbronal line and the hinge axis is acute. (b) Said of the growth lines that incline forward relative to the growth direction of a gastropod shell. Cf: *opisthocline; acline; obliquity.*

prosodus (pros′-o-dus) A small canal of uniform diameter in a sponge, leading from an *inhalant canal* to a *prosopyle* of approximately the same cross-sectional area. Pl: prosodi. Cf: *aphodus.*

prosogyrate (pros-o-gy′-rate) Said of the umbones (of a bivalve mollusk) curved so that the beaks point in the anterior or forward direction. Ant: *opisthogyrate.*

prosoma (pro-so′-ma) (a) The anterior part of the body of various invertebrates; esp. the *cephalothorax* of an arachnid or merostome. See also: *opisthosoma.* (b) The *prosome* of a copepod.

prosome (pro′-some) (a) The anterior region of the body of a copepod crustacean, commonly limited behind by major articulation. See also: *urosome.* Formal Greek form: *prosoma.* (b) A structure within the neck of the body of a chitinozoan, extended to various intermediate positions, or even projecting beyond the collar. Its upper surface may be flat, convex, conical, or truncate, and an upper flange may lie against the pseudostome; its top may be marked by dark radial fibers, and the tubular area commonly has many dark annular rings. Cf: *opisthosome.*

prosopite (pro′-sop-ite) A colorless, gray, or blue monoclinic mineral: $CaAl_2(F,OH)_8$.

prosopon (pro-so′-pon) Sculpture on the external surface of a trilobite *exoskeleton,* with presumed but unknown functional significance. This term is preferred by some authors to *ornamentation [paleont].*

prosopore (pros′-o-pore) The entrance opening of an inhalant canal of a sponge. Cf: *apopore.*

prosopyle (pros′-o-pyle) Any aperture through which water enters a *flagellated chamber* of a sponge. Cf: *apopyle.*

prospect (pros′-pect) n. (a) An area that is a potential site of mineral deposits, based on preliminary exploration. (b) Sometimes, an area that has been explored in a preliminary way but has not given evidence of economic value. (c) An area to be searched by some investigative technique, e.g. geophysical prospecting. (d) A geologic or geophysical anomaly, especially one recommended for additional exploration. A prospect is distinct from a *mine* in that it is nonproducing. See also: *prospecting.*

prospect hole A general term for any shaft, pit, adit, drift, tunnel, or drill hole made for the purpose of prospecting mineral-bearing ground. More specific terms, such as *prospect shaft* and *prospect pit,* are generally used. Informal term is *dog hole.*

prospecting (pros′-pect-ing) (a) Searching for economically valuable deposits of fuel or minerals. Cf: *exploration.* (b) *geophysical exploration.*

prospecting seismology *applied seismology.*

prospection (pros-pec′-tion) *prospecting.*

prospective ecospace (pro-spec′-tive) As proposed by Valentine (1969, p.687), "the total ecospace that an organism or other ecological unit may utilize if it is physically available". Cf: *realized ecospace.*

prospector (pros′-pec-tor) An individual engaged in prospecting for valuable mineral deposits, generally working alone or in a small group, and on foot with simple tools or portable detectors. The term implies an individual searching on his own behalf, rather than an employee of a mining company.

prospect pit *prospect hole.*

prospect shaft *prospect hole.*

prosperite (pros′-per-ite) A colorless monoclinic mineral: $CaZn_2(AsO_4)_2{\cdot}H_2O$.

prostratigraphy (pro-stra-tig′-ra-phy) A term proposed by Schindewolf (1954, p.25) for "preliminary stratigraphy" including lithologic and paleontologic studies without consideration of the time factor; the "raw material" for stratigraphy, consisting of local observation, description, and arrangement of strata, but not yet methodically linked together by the concept of time. See also: *protostratigraphy; topostratigraphy.*

prosuture (pro-su′-ture) The line of junction of a proseptum with the walls of an ammonoid shell.

protactinium-231/thorium-230 age method *protactinium-ionium age method.*

protactinium-ionium age method (pro-tac-tin′-i-um-i-o′-ni-um) Calculation of an age in years for deep-sea sediments formed during the last 150,000 years, based on the assumption that the initial protactinium-231 to ionium (thorium-230) ratio for newly formed sediments has remained constant for the total section of sediments under study. The age depends on the gradual change with time of the protactinium-231 to ionium ratio because of the difference in half-lives. See also: *uranium-series age method.* Syn: *protactinium-231/thorium-230 age method; thorium-230/protactinium-231 excess method.*

protalus rampart (pro-ta′-lus) An arcuate ridge of coarse, angular blocks of rock derived by single rockfalls from a cliff or steep rocky slope above, marking the downslope edge of an existing or melted snowbank. The blocks roll and slide across the snowbank but no fine material reaches its edge. After the snowbank melts, the rampart or ridge stands some distance beyond any talus near the base of the cliff (Bryan, 1934, p.656). Cf: *winter protalus ridge.*

protasite (pro-tas′-ite) An orange monoclinic mineral: $Ba(UO_2)_3O_3(OH)_2{\cdot}3H_2O$.

protaspis (pro-tas′-pis) An early juvenile trilobite whose small, oval exoskeleton is not yet divisible into *cephalon, thorax,* and *pygidium.* Cf: *anaprotaspis; metaprotaspis.* Pl: protaspides.

protaxis (pro-tax′-is) An antique term for the central axis of a mountain chain, supposedly consisting of the oldest rocks and structures; for example, an "Archean protaxis".

protected thermometer (pro-tect′-ed) A *reversing thermometer* that is protected against hydrostatic pressure by a glass shell. Cf: *unprotected thermometer.*

protectite (pro-tec′-tite) A rock formed by the crystallization of a primary magma. Obsolete. See also: *anatexite, syntectite.*

protegulal node (pro-teg′-u-lal) The apical, commonly raised portion of an adult brachiopod shell, representing the site of the *protegulum* and later growth up to the brephic stage (TIP, 1965, pt. H, p.151).

protegulum (pro-teg′-u-lum) The smooth biconvex first-formed shell of organic material (chitin or protein) of a brachiopod, secreted simultaneously by both mantles.

proteroglacial (prot′-er-o-gla′-cial) Pertaining to the earlier of two postulated great ice ages (Hansen, 1894, p.128). Cf: *deuteroglacial.*

proterosoma (prot′-er-o-so′-ma) The anterior section of the body of an acarid arachnid, ending behind the second pair of legs.

Proterozoic (Prot′-er-o-zo′-ic) The upper eon of the *Precambrian* of the Standard Global Chronostratigraphic Scale, above the *Archean* and below the *Phanerozoic* Eon. The lower boundary has been established geochronometrically at 2,500 Ma; the upper boundary is at 540 Ma.

prothallus (pro-thal′-lus) The *gametophyte* of a fern or other pteridophyte, usually a flattened thalluslike structure attached to the soil. Pl: prothalli.

protheca (pro-the′-ca) (a) The proximal part of graptoloid theca before it is differentiated from the succeeding theca. It is considered equivalent to the *stolotheca* in those graptolites with more than one type of theca. (b) A primary element of the wall of a fusulinid, comprising diaphanotheca and tectum.

protist (pro′-tist) (a) Any organism assigned to the kingdom Protista, also called *Protoctista,* which includes forms with both plant and animal affinities, e.g. protozoans, and some algae. No agreement exists on the limits of nomenclature of the Protista. (b) A term formerly applied to any single-celled organisms including both prokaryotes and eukaryotes but now best used informally. Var: protistan.

protoanthophyllite A vitreous white orthorhombic mineral of the *amphibole* group: $(Mg,Fe)_7Si_8O_{22}(OH)_2$. It is a dimorph of anthophyllite.

Proto-Atlantic Ocean (Pro′-to-At-lan′-tic) *Iapetus Ocean.*

protobitumen (pro′-to-bit-u′-men) Any of the fats, oils, waxes, or resins present as unaltered or nearly unaltered plant and animal products from which fossil bitumens are formed (Tomkeieff, 1954). See also: *labile; stabile.*

Protochordata (Pro′-to-chor-da′-ta) A phylum or subphylum of animals that possess a notochord during some part of their life history but lack a bony skeleton or spinal column. They occupy a position intermediate between the invertebrates and vertebrates; they are included by some in the phylum Chordata.

protoclase (pro′-to-clase) Obsolete term for a rock possessing *primary cleavage,* e.g., bedding planes in sedimentary rock (Leith, 1905, p.12). Cf: *metaclase.*

protoclastic (pro-to-clas′-tic) An igneous texture characterized by earlier formed crystals that have been broken or deformed due to differential flow of the magma before complete solidification.

protoconch (pro′-to-conch) (a) The first portion of the embryonic shell of a cephalopod, its preservation in fossil and in living forms being uncertain. The term is sometimes applied to the first camera (chamber) of the shell, located at the apex of the *phragmocone* or at the center of the coil, and closed in an ammonoid by the *proseptum.* (b) The apical, often smooth whorl of a gastropod shell, usually well-demarcated from the *teleoconch.* The term applies to the fully formed embryonic shell of a gastropod and should not be used synonymously with *nucleus,* although it has been restricted by some authors to the simple cap-shaped plate that constitutes the first shell rudiment (see Knight, 1941).

protoconodont Microscopic phosphatic sclerites, usually long and conical, consisting of weakly mineralized (apatitic) and organic layers accreted basally from the inner surface. Many workers do not include the group among the Conodonta, but either as ancestral to the paraconodont structural grade or as a non-related group. Some consider protoconodonts to be Chaetognatha sclerites because of compositional, structural, and morphological similarity. Range: latest Precambrian to Ordovician. See also: *conodont.* Cf: *euconodont; paraconodont.*

protocorallite (pro-to-cor′-al-lite) The first-formed *corallite* of a colony.

protoctista A kingdom of *eukaryotic* (nucleated) microorganisms and their descendants, usually but not always single celled, that are either meiotic or nonmeiotic, a group formerly referred to as *protists.* Range, Proterozoic to Holocene.

protodolomite (pro-to-do′-lo-mite) (a) A crystalline calcium-magnesium carbonate with a disordered structure in which the magnesium ions occur in the same crystallographic layers instead of in alternate layers as in the dolomite mineral. (b) An imperfectly crystallized artificial material of composition near $CaMg(CO_3)_2$.

protoecium (pro-to-e′-ci-um) (a) *proancestrula.* (b) *ancestrula.*

protoenstatite (pro-to-en′-sta-tite) An artificial, unstable modification of $MgSiO_3$, produced by decomposition of talc by heating, and convertible to enstatite by grinding or by heating to a high temperature.

protoferro-anthophyllite (pro-to-fer′-ro-an-tho′-phyll-ite) A light brownish-yellow orthorhombic mineral of the *amphibole* group: $(Fe^{2+},Mn^{2+})_7Si_8O_{22}(OH)_2$. It is a dimorph of ferro-anthophyllite.

protoforamen (pro′-to-fo-ra′-men) The *primary aperture* of a foraminiferal test associated with a fully developed or rudimentary tooth plate. Cf: *deuteroforamen.*

protogene (pro′-to-gene) An old term for a primary rock. Adj: *protogenous.* Cf: *deuterogene.* Syn: *protogine.*

protogenesis (pro-to-gen′-e-sis) Reproduction by budding.

protogenous (pro-tog′-e-nous) Adj. of *protogene.*

protogine (pro′-to-gine) A granitic rock, occurring in the Alps, that has gneissic structure, contains sericite, chlorite, epidote, and garnet, and shows evidence of a composite origin or crystallization (or partial recrystallization) under stress after consolidation. Also spelled: *protogene.* The term, dating from 1806, is obsolete.

protogranular (pro-to-gran′-u-lar) Said of the xenomorphic, granoblastic texture of a rock, characterized by sinuous mineral boundaries, intergrowths, and other features suggesting previous equilibrium with a melt (Mercier and Nicolas, 1975).

protointraclast (pro-to-in′-tra-clast) A genetic term suggested by Bosellini (1966) for a limestone component that resulted from a premature attempt at resedimentation while still being in an unconsolidated and viscous or plastic state and that never existed as a free, clastic entity. Cf: *intraclast.*

protojoséite (pro-to-jo-sé-ite) A metallic hexagonal mineral: Bi_3TeS. It has not been approved by the International Mineralogical Association.

protolith (pro′-to-lith) (a) The unmetamorphosed rock from which a given metamorphic rock was formed by metamorphism. Syn: *parent rock.* (b) The parent or unweathered rock from which *regolith* is formed.

protomangano-ferro-anthophyllite A light brownish-yellow orthorhombic mineral of the *amphibole* group: $(Mn,Fe^{2+})_2Fe_5Si_8O_{22}(OH)_2$.

protomatrix (pro-to-ma′-trix) A term introduced by Dickinson (1970, p.702), for unrecrystallized clayey material in weakly consolidated graywackes and arkoses.

protomylonite (pro-to-my′-lo-nite) (a) A mylonitic rock produced from contact-metamorphosed rock, with granulation and flowage caused by overthrusts following the contact surfaces between intrusion and country rock (Holmes, 1920). (b) A coherent crush breccia whose characteristically lenticular, megascopic particles faintly retain primary structures. It is a lower grade in the development of *mylonite* and *ultramylonite* (Waters and Campbell, 1935, p.479).

proton Principle particle of an atom nucleus. The proton is a positively charged nucleon (IAEA, 2003).

protonema (pro-to-ne′-ma) The green filamentous growth from a germinating spore that develops into the gametophyte of mosses. Has anticlinal cross walls.

proton-precession magnetometer (pro'-ton-pre-ces'-sion) A type of nuclear resonance magnetometer that accurately measures absolute total magnetic intensity by the use of the precession of protons in a hydrogen-rich liquid (such as water) about the magnetic field direction. Precession frequency is proportional to field strength. See also: *proton-vector magnetometer.*

proton-vector magnetometer (pro'-ton-vec'-tor) A type of *proton-precession magnetometer* with a system of auxiliary coils that permits measurement of horizontal intensity H or vertical intensity Z as well as total intensity F.

protonymph (pro'-to-nymph) The first postembryonic stage in the arachnid order Acarida.

protophyte (pro'-to-phyte) An obsolete term for a unicellular or primitive plant.

protopod (pro'-to-pod) The proximal portion of a limb of a crustacean, consisting of *coxa, basis,* and sometimes *precoxa,* often fused to each other. Its distal edge generally bears the *endopod* and *exopod.* Syn: *sympod.*

protopore (pro'-to-pore) A single fine opening or perforation in a foraminiferal test, rounded at least on the inner wall. Cf: *deuteropore.*

protopygidium (pro'-to-py-gid'-i-um) The postcephalic portion of a trilobite *protaspis.* Pl: protopygidia.

protoquartzite (pro-to-quartz'-ite) A well-sorted, quartz-enriched sandstone that lacks the well-rounded grains of an *orthoquartzite;* specif. a lithic sandstone intermediate in composition between subgraywacke and orthoquartzite, containing 75-95% quartz and chert, less than 15% detrital clay matrix, and 5-25% unstable materials in which the rock fragments exceed the feldspar grains in abundance (Pettijohn, 1954, p.364). It commonly forms shoestring sands. Examples: Venango Formation (Upper Devonian) of New York and Pennsylvania, and Hartshorne Sandstone (Pennsylvanian) of Oklahoma and Arkansas. The term was used by Krynine (1951) for a "cleaned-up" graywacke (matrix washed out), intermediate in composition between quartzose graywacke and orthoquartzite. Syn: *quartzose subgraywacke.*

protore (prot'-ore) The rock below the sulfide zone of *supergene enrichment;* the primary, subeconomic material. See also: *oxidized zone; sulfide zone.*

protosaccus (pro-to-sac'-cus) A *saccus* filled with *alveoli.* This structure occurs in pollen of the Perm-Trias interval (e.g., *Lueckisporites virkkiae*) and is considered to be primitive.

protoscience (pro-to-sci'-ence) Attempts to explain the natural world preceding or leading up to systematic studies.

protoseptum (pro-to-sep'-tum) *primary septum.*

protostele (pro'-to-stele) A *stele* with a solid xylem core (Swartz, 1971, p.384).

protostome Coelomate animal where the blastopore becomes the mouth, e.g. molluscs, arthropods, annellids.

protostratic classification Stratigraphic classification embracing biostratic and lithostratic classification (Henningsmoen, 1961, p.63).

protostratic unit A stratigraphic unit whose boundaries delimit and embrace beds with a common physical property, such as a particular lithology or a particular fossil content (Henningsmoen, 1961, p.66).

protostratigraphy (pro'-to-stra-tig'-ra-phy) A term proposed by Henningsmoen (1961) for preliminary or introductory stratigraphy, including lithostratigraphy and biostratigraphy; *prostratigraphy.*

prototheca (pro-to-the'-ca) The roughly conical or cup-shaped structure constituting the embryonic exoskeleton of a coral.

prototype (pro'-to-type) An ancestral form; the most primitive form in a group of related organisms. Syn: *archetype.*

protoxylem (pro-to-xy'-lem) The first-formed, primary xylem of a plant. Cf: *metaxylem; annular tracheid; helical tracheid.*

protozoan (pro-to-zo'-an) An organism of the kingdom Protoctista that is predominantly heterotrophic and typically but not always single celled. In the two-kingdom classification of organisms, protozoans were classified in the kingdom Animalia as phylum Protozoa. The term is now best used informally to encompass a number of phyla.

protractor (pro-trac'-tor) An instrument used in drawing and plotting, designed for laying out or measuring angles on a flat or curved surface, and consisting of a plate marked with units of circular measure.

protractor muscle A muscle that extends an organ or part; e.g. an *outside lateral muscle* or *middle lateral muscle* in some lingulid brachiopods, or a longitudinal fibril in connective tissue of the pedicle of some articulate brachiopods (TIP, 1965, pt.H, p.151). Cf: *retractor muscle.*

protriaene (pro-tri'-aene) A sponge triaene in which the cladi curve away from the rhabdome, making an angle to the rhabdome that is noticeably greater than that of a normal tetraxon. Cf: *orthotriaene.*

protrusion (pro-tru'-sion) A proposed term for a rock mass that has been tectonically intruded in the solid state; it is in contrast to an igneous *intrusion* (Lockwood, 1971).

proudite (proud'-ite) A metallic silver-gray monoclinic mineral: $Pb_{7.5}Cu_{0-1}Bi_{9.3-9.7}(S,Se)_{22}$.

proustite (proust'-ite) A cochineal-red rhombohedral mineral: Ag_3AsS_3. It is isomorphous with pyrargyrite, and is a minor ore of silver. Cf: *xanthoconite.* Syn: *light ruby silver; light red silver ore.*

prove v. In economic geology, to establish, by drilling, trenching, underground openings, or other means, that a given deposit of a valuable substance exists, and that its grade and dimensions equal or exceed some specified amounts. See also: *proved reserves.*

proved ore *proved reserves.*

proved ore reserve The economically minable part of a *measured mineral resource* (JORC, 1999). Cf: *ore reserve.*

proved reserves Reserves of metallic and nonmetallic minerals, and of oil and gas, for which reliable quantity and quality estimates have been made. Cf: *developed reserves; positive ore.* Syn: *proved ore.*

provenance (prov'-e-nance) A place of origin; specif. the area from which the constituent materials of a sedimentary rock or facies are derived. Also, the rocks of which this area is composed. Cf: *distributive province.* Syn: *provenience; source area; sourceland.*

provenancial linkage A tectonic linkage among terranes is inferred if one terrane was the source of the detritus deposited on another adjoining terrane (Howell, 1995).

provenience (pro-ve'-ni-ence) *provenance.*

proven reserve A *reserve* for which (a) quantity is computed from dimensions revealed in outcrops, trenches, workings, or drill holes; *grade* and/or quality are computed from the results of detailed sampling and (b) the sites for inspection, sampling, and measurement are spaced so closely and the geologic character is so well defined that size, shape, depth, and mineral content of the reserve is well-established (SEC, 2003). Cf: *probable reserve.*

proven reserves (prov'-en) Oil that has been discovered and determined to be recoverable but is still in the ground.

province [ecol] (prov'-ince) (a) Part of a region, isolated and defined by its climate and topography and characterized by a particular group of organisms. (b) A group of temporally and spatially associated plant or animal communities.

province [geog] Any large area or region considered as a whole, all parts of which are characterized by similar features or by a history differing significantly from that of adjacent areas; specif. a *geologic province* or a *physiographic province.*

provincial alternation (pro-vin'-cial) The overlapping of sedimentary petrologic provinces, caused by oscillation of the boundary between two provinces during time (Pettijohn, 1957, p.573-574).

provinciality A measure of the number of large, discrete biogeographic units present on the Earth at any one time. Provinciality changes due to continental drift, climatic change, modification of sea level, and biotic crises.

provincial series A *series [stratig]* recognized only in a particular region and involving a major division of time within a period; e.g. the Wolfcampian Series within the Permian System in west Texas and New Mexico.

provincial succession A succession of sedimentary petrologic provinces, produced by changes in provenance leading to mineral associations that change with time (Pettijohn, 1957, p.574).

provinculum (pro-vin'-cu-lum) A taxodont hinge composed of minute teeth developed in some bivalve mollusks before the permanent dentition.

provitrain (pro-vit'-rain) *Vitrain* in which some plant structure is microscopically visible. Cf: *euvitrain.* See also: *periblain; suberain; xylain.* Syn: *telain.*

provitrinite (pro-vit'-ri-nite) A variety of the maceral *vitrinite* characteristic of provitrain and including the varieties periblinite, suberinite, and xylinite. Plant cell structure is visible under the microscope. The term *telinite* has been proposed as a preferable synonym. Cf: *euvitrinite.* Syn: *phyllovitrinite.*

prowersite (prow'-ers-ite) An orthoclase- and biotite-rich *minette.* Its name, given by Rosenbusch in 1908, is derived from Prowers County, Colorado. Obsolete.

proximal [eco geol] (prox'-i-mal) Said of an ore deposit formed immediately adjacent to, perhaps part of, a volcanic hearth, pile, or fumarole to which it is genetically related and from which its constituents have been derived. Cf: *distal [eco geol].*

proximal [paleont] Next to or nearest the point of attachment or place of reference, a point conceived of as central, or the point of view. Examples in invertebrate morphology: "proximal direction" toward the dorsal pole or mouth of a crinoid; "proximal ray" of a sponge *spicule,* directed inward from a bounding surface of the sponge; and "proximal direction" toward the *ancestrula* or origin of growth of a bryozoan colony. Ant: *distal [paleont].*

proximal [palyn] Said of the parts of pollen grains or spores nearest or toward the center of the original tetrad; e.g. said of the side of a monocolpate pollen grain opposite the colpus, or said of the side of a trilete spore provided with contact areas. Ant: *distal [palyn].*

proximal [sed] Said of a sedimentary deposit consisting of coarse clastics and formed nearest the source area; e.g. a "proximal turbidite" consisting of thick sandy varves. Cf: *distal [sed].*

proximale (prox'-i-male) The noncirriferous topmost *columnal* of a crinoid, typically distinguished by enlargement and permanent attachment to the *dorsal cup.*

proximal onlap *Onlap* in the direction of the source of clastic supply (Mitchum, 1977, p.208). Cf: *distal onlap.*

proximate admixture (prox'-i-mate) A term applied by Udden (1914) to an *admixture* (in a sediment of several size grades) whose particles are most similar in size to those of the dominant or maximum grade; material in one of the two classes adjacent to the maximum histogram class.

proximate analysis The determination of compounds contained in a mixture; for coal, the determination of moisture, volatile matter, ash, and fixed carbon (by difference) (ASTM). Cf: *ultimate analysis.*

proximate cyst A dinoflagellate cyst of nearly the same size as, and closely resembling, the motile theca. The ratio of the diameter of the main body to the total diameter of the cyst exceeds 0.8. The term refers to the supposed proximity of the main cyst wall to the theca at the time of encystment. See also: *chorate cyst; proximochorate cyst.*

Proximegerminantes (Prox-i-me-ger-mi-nan'-tes) The name of an *anteturma* in the turmal system for the classification of dispersed fossil palynomorphs. Cf: *turma; Variegerminantes.*

proximity A measure of inter-object distance (Krzanowski et al., 1993).

proximity analysis The analysis of inter-object distances.

proximochorate cyst (prox'-i-mo-chor'-ate) A dinoflagellate cyst having sutured outgrowths that readily indicate the tabulate character. The ratio of the diameter of the main body to the total diameter of the cyst is between 0.6 and 0.8. See also: *chorate cyst; proximate cyst.*

proxistele (prox'-i-stele) The proximal region of a crinoid *column* near the theca, generally not clearly delimited from the *mesistele.* Cf: *dististele.*

proxy (prox'-y) v. To substitute one ion or atom for another in a crystal structure. adj. Said of such a substituted ion or atom.

proxying (prox'-y-ing) *ionic substitution.*

pruning The deletion of extraneous points in stream digitizing (Krzanowski et al., 1993). Cf: *stream mode; weeding.*

przhevalskite (pr-zhe-val'-skite) A bright greenish-yellow tetragonal mineral: $Pb(UO_2)_2(PO_4)_2{\cdot}4H_2O$.

Psamments Entisols that have textures of loamy fine sand or coarser in all parts, have <35% coarse fragments, and that are not saturated with water for periods long enough to limit their use for most crops. A suborder in the U.S. system of soil taxonomy.

psammite (psam'-mite) (a) A clastic sediment or sedimentary rock composed of sand-size particles; a sandstone. The term is equivalent to the Latin-derived term, *arenite.* (b) A term formerly used in Europe for a fine-grained, fissile, clayey sandstone (as distinguished from a more siliceous and gritty one) in which "the component grains are scarcely distinguishable by the unassisted eye" (Oldham, 1879, p.44). (c) A term regarded by Tyrrell (1921, p.501-502) as the metamorphic derivative of arenite. Etymol: Greek "psammos", "sand". See also: *psephite; pelite.* Also spelled: *psammyte.*

psammitic (psam-mit'-ic) (a) Pertaining to or characteristic of psammite; *arenaceous.* Cf: *sandy.* (b) Said of a metamorphic rock derived from a psammite; e.g. a "psammitic gneiss" or a "psammitic schist".

psammobiotic (psam'-mo-bi-ot'-ic) Said of an organism that lives in sand or sandy areas.

psammofauna (psam'-mo-fau'-na) The animals associated with sandy substrates.

psammogenic dune (psam-mo-gen'-ic) A dune "caused by the effect of sand surfaces in trapping more sand" (Schieferdecker, 1959, term 0148).

psammon (psam'-mon) The interstitial organisms found between sand grains.

psammophilic (psam-mo-phil'-ic) Said of an organism or of the fauna found in sand. Noun: psammophile.

psammophyte (psam'-mo-phyte) A plant preferring sand or very sandy soil for growth.

psammosere (psam'-mo-sere) A *sere* that develops in a sandy environment.

psammyte (psam'-myte) *psammite.*

psephicity (pse-phic'-i-ty) A term used by Mackie (1897, p.301) for the "coefficient of roundability" of a pebble- or sand-size mineral fragment, expressed as the ratio of specific gravity to hardness (as measured in air) or the quotient of specific gravity minus one divided by hardness (as measured in water).

psephite (pse'-phite) (a) A sediment or sedimentary rock composed of large fragments (coarser than sand) set in a matrix varying in kind and amount; e.g. rubble, talus, breccia, glacial till, tillite, shingle, gravel, and esp. conglomerate. The term is equivalent to the Latin-derived term, *rudite.* (b) A term regarded by Tyrrell (1921, p.501-502) as the metamorphic derivative of rudite. Etymol: Greek "psephos", "pebble". See also: *psammite; pelite.* Also spelled: psephyte.

psephitic (pse-phit'-ic) (a) Pertaining to or characteristic of psephite. (b) Said of a metamorphic rock derived from a psephite. Cf: *rudaceous.*

psephonecrocoenosis (pse'-pho-nec'-ro-coe-no'-sis) A necrocoenosis of dwarf individuals.

psephyte [sed] *psephite.*

pseudo- (pseu'-do-) A prefix meaning "false" or "spurious".

pseudoactine (pseu-do-ac'-tine) A raylike arm or branch of a sponge spicule that contains no axial filament or axial canal.

pseudoallochem (pseu-do-al'-lo-chem) An object resembling an *allochem* but produced in place within a calcareous sediment by a secondary process such as recrystallization (Folk, 1959, p.7).

pseudoaquatic (pseu'-do-a-quat'-ic) Said of an organism living in moist or wet but not truly aquatic conditions.

pseudoatoll (pseu-do-at'-oll) (a) With the prefix "pseudo-", implying a false characteristic, a pseudoatoll is a ring-shaped reef of non-coralline nature (e.g. the rim of a submerged volcanic crater). (b) A term sometimes used for *bank atoll* (q.v.) or *shelf atoll* (q.v.), on the mistaken assumption that only the Darwinian volcanic basement model represented a "true" atoll. Also spelled: pseudatoll.

pseudo-autunite (pseu-do-au'-tun-ite) A discredited pale yellow to white mineral: $(H_3O)_4Ca_2(UO_2)_2(PO_4)_4{\cdot}5H_2O$(?). It is not a member of the *autunite* group.

pseudobed (pseu-do-bed') A group of nearly parallel plane surfaces that dip upcurrent in climbing-ripple laminae and that are formed either by nondeposition or by erosion on the upcurrent sides of migrating superimposed ripple laminae: "between successive pseudobeds are sets of laminae that dip steeply in the opposite direction, formed by deposition on the *lee* side of each ripple crest and resembling, in general, the foresets of tabular planar cross-beds" (McKee in Middleton, 1965, p.250). See also: *cross-bedding.*

pseudobedding [sed] Bedding developed by concentration or combining of ripple laminae representing the approach slopes of ripple deposits (McKee, 1939, p.72); i.e. bedding produced by the bounding surfaces between sets of cross-beds deposited by climbing ripples. See also: *false bedding; pseudo cross-bedding.*

pseudobivalved (pseu-do-bi'-valved) Said of rostroconch mollusks in which the larval shell is univalved and the adult shell bivalved. In this group there are always one or more shell layers continuous across the dorsal margin, and a dorsal commissure such as occurs in pelecypods is lacking; thus the two valves cannot be separated from one another without breaking them.

pseudoboléite (pseu-do-bo'-lé-ite) A blue tetragonal mineral: $Pb_{31}Cu_{24}Cl_{62}(OH)_{48}$. Also spelled: pseudoboleite.

pseudobomb (pseu-do-bomb') *lava ball.*

pseudobreccia (pseu-do-brec'-cia) A limestone, characterized by a mottled appearance that gives the rock a texture mimicking that of a breccia, or by a weathered surface that appears deceptively fragmental. It may be produced diagenetically by selective grain growth in which localized, patchy, and irregularly shaped recrystallized masses of coarse calcite (usually visible to the naked eye: 1-20 mm in diameter) are embedded in a lighter-colored and less-altered matrix of calcareous mud. The boundaries between the "clasts" and the matrix are indistinct or gradational.

pseudobrookite (pseu-do-brook'-ite) An adamantine reddish-brown or black orthorhombic mineral: $Fe^{3+}{}_2TiO_5$. It resembles brookite.

pseudocannel coal (pseu-do-can'-nel) Cannel coal that contains much humic matter. Syn: *humic-cannel coal.*

pseudocarina (pseu'-do-ca-ri'-na) A perforate, ridgelike thickening of the peripheral part of a chamber wall of a foraminiferal test, situated approximately in the plane of coiling.

pseudoceratite (pseu-do-cer'-a-tite) A Jurassic and Cretaceous ammonoid cephalopod having a suture similar to that of a ceratite. It is explained as a reversionary or atavistic modification of a normal ammonite.

pseudoceratitic suture (pseu'-do-ce'-ra-tit'-ic) A type of *suture* in ammonoids that approximates a *ceratitic suture* in form but is not related to ceratites; specif. a suture in pseudoceratites.

pseudochamber (pseu-do-cham'-ber) A partly subdivided cavity of a foraminiferal test (as in the family Tournayellidae), indicated by a slight protuberance or an incipient septum.

pseudochitin (pseu-do-chi'-tin) A resistant organic substance, the exact chemical structure of which is uncertain, though it apparently consists of compounds of C-H-O-N. The behavior of pseudochitin is similar to that of *chitin,* but by definition it does not yield a positive chitin staining reaction. Various fossils, including graptolites and chitinozoans, contain or consist mostly of this substance.

pseudochitinous (pseu-do-chi'-tin-ous) Consisting of *pseudochitin.*

pseudochlorite (pseu-do-chlo'-rite) (a) *swelling chlorite.* (b) *septechlorite.* (c) An artificial product obtained by adsorbing magnesium salts on montmorillonite and precipitating magnesium hydroxide between the layers of the mineral (Youell, 1960).

pseudocirque (pseu-do-cirque') A term used by Freeman (1925) and recommended by Charlesworth (1957, p.244) for a feature that is similar but not homologous to a glacial cirque. See also: *cirque [geomorph].*

pseudocol (pseu-do-col') A term proposed by Chamberlin (1894a) for a landform represented by a constriction of the valley of a stream diverted by glacial ponding, formed by the cutting through of a cover of drift and subsequent exposure of a former col; the feature occurs in regions of reversed drainage along the border of ancient glacial formations, as along several segments of the Ohio River valley.

pseudocolpus (pseu-do-col'-pus) A colpuslike modification of the exine of pollen grains, differing from a true *colpus* in that it is not a site of pollen-tube emergence. Pl: pseudocolpi.

pseudoconcretion (pseu'-do-con-cre'-tion) A subspherical sedimentary structure resembling a concretion, but not formed by orderly precipitation of mineral matter in the pores of a sediment; e.g. an armored mud ball or certain algal structures.

pseudoconformity (pseu'-do-con-form'-i-ty) A term used by Fairbridge (1946, p.88) for a stratigraphic relationship that appears conformable but is characterized by nonaccumulation or deficiency of sediment, such as a slump gap in which an entire formation slipped away off the crest of a rising anticline or in which no trace of a hiatus is immediately apparent from the structure.

pseudoconglomerate (pseu'-do-con-glom'-er-ate) A rock that resembles, or may easily be mistaken for, a normal sedimentary conglomerate; e.g. a *crush conglomerate* consisting of cemented fragments that have been rolled and rounded nearly in place by orogenic forces; a sandstone packed with many rounded concretions; or an aggregate of rounded boulders produced in place by *spheroidal weathering* and surrounded by clayey material. Term introduced by Van Hise (1896, p.679). Cf: *pseudopsephite.*

pseudocotunnite (pseu'-do-co-tun'-nite) A colorless, yellow, or greenish-yellow orthorhombic mineral of uncertain composition: K_2PbCl_4(?).

pseudo cross-bedding (a) An inclined bedding produced by deposition in response to ripple migration, and characterized by foreset beds that appear to dip into the current. See also: *pseudobedding [sed].* (b) A structure resembling cross-bedding, caused by distortion-free slumping and sliding of a semiconsolidated mass of sediments. Also spelled: pseudocross-bedding. Syn: pseudo cross-stratification.

pseudocruralium (pseu'-do-cru-ral'-i-um) An excessive thickening of the secondary shell of a brachiopod, bearing dorsal adductor impressions, and elevated anteriorly above the floor of the valve.

pseudocrystal (pseu-do-crys'-tal) A substance that appears to be crystalline but does not give a diffraction pattern that confirms it as truly crystalline.

pseudodeltidium (pseu'-do-del-tid'-i-um) A single convex or flat plate affording variably complete cover of the *delthyrium* of a brachiopod, but always closing the apical angle when the pedicle foramen is supra-apical or absent, and dorsally enclosing the apical foramen (TIP, 1965, pt. H, p.151). Cf: *deltidium.*

pseudo-diffusion (pseu'-do-dif-fu'-sion) Mixing of thin layers of slowly accumulated marine sediments by the action of water motion and/or subsurface organisms. This phenomenon can lead to serious errors in determining the rate of sedimentation if the disturbed sediments are dated by carbon-14 or other isotopic methods (Bowen, 1966, p.208).

pseudoextinction Apparent extinction due to the change in nature of a lineage, but unaccompanied by any lineage termination.

pseudofault (pseu-do-fault') A term coined by Palmer (1920, p.851) for a faultlike feature resulting from weathering along joints, shrinkage cracks, or bedding planes.

pseudofibrous peat (pseu-do-fi'-brous) Peat that is fibrous in texture but that is plastic and incoherent. Cf: *fibrous peat; amorphous peat.*

pseudofossil (pseu-do-fos'-sil) A natural object, structure, or mineral of inorganic origin that may resemble or be mistaken for a fossil. Cf: *problematic fossil.*

pseudogalena (pseu'-do-ga-le'-na) *sphalerite.*

pseudogradational bedding (pseu'-do-gra-da'-tion-al) A structure in metamorphosed sedimentary rock in which the original textural gradation (coarse at the base, finer grained at the top) appears to be reversed because of the formation of porphyroblasts in the finer portion of the rock.

pseudograndreefite (pseu'-do-grand-reef'-ite) A sub-adamantine colorless orthorhombic mineral: $Pb_6(SO_4)F_{10}$.

pseudogravity The assumed gravity field calculated from the magnetic-field measurements by means of *Poisson's relation.* Calculation involves conversion of *magnetic susceptibility* to *density* and vertical integration of the reduced-to-pole magnetic data. See *Reduction to the pole.*

pseudogravity anomaly The gravity anomaly that would be observed if a magnetic body were replaced by mass in equal proportions. Magnetic anomalies are transformed to pseudogravity anomalies by application of a linear filter in the Fourier domain, equivalent to calculating magnetic potential (Blakely, 1995). See also: *potential [geophys].*

pseudogritty structure (pseu-do-grit'-ty) A type of *mortar structure* in which the larger relics are angular, due to fracture along cleavage planes. The term is not in common use.

pseudohexagonal (pseu'-do-hex-ag'-o-nal) Said of a crystal form, e.g. some orthorhombic forms, that simulate the hexagonal form.

pseudointerarea (pseu'-do-in'-ter-ar'-e-a) The somewhat flattened posterior sector of the shell of some brachiopods "secreted by posterior sector of mantle not fused with that of opposite valve" (TIP, 1965, pt. H, p.151). Cf: *interarea.*

pseudokame (pseu-do-kame') *residual kame.*

pseudokarst (pseu-do-karst') (a) A karstlike terrain having *closed depressions, sinking streams,* and *caves,* but produced by a process other than the dissolving of rock. Processes and forms involving *piping* and *thermokarst* are included (Otvos, 1976); some authors also include terrain characterized by *lava tubes, sea caves,* and *blowouts.* The term has been applied to *covered karst* and to karst produced by the dissolution of rocks that are relatively insoluble, such as quartzite and granite, but more general usage regards these as varieties of true karst. The term was first used by von Knebel in 1906.

pseudokeel Raised flat area on *under side* of planate conodont elements of some species of *Siphonodella;* contains deep pit near anterior end and is bordered in elements of mature individuals by narrow zones of recessive basal margin (TIP, 1981, pt. W, supp.2).

pseudokettle (pseu-do-ket'-tle) *pingo remnant.*

pseudolagoon (pseu'-do-la-goon') The shallow pool of water encircled by a *pseudoatoll.* Syn: miniature lagoon.

pseudolaueite (pseu-do-lau'-e-ite) A yellow-orange monoclinic mineral: $Mn^{2+}Fe^{3+}{}_2(PO_4)_2(OH)_2{\cdot}8H_2O$. It is a trimorph of *laueite* and *stewartite.*

pseudoleucite (pseu-do-leu'-cite) A pseudomorph after leucite, consisting of a mixture of nepheline, orthoclase, and analcime, such as occur in certain syenites in Arkansas, Montana, and Brazil.

pseudomalachite (pseu-do-mal'-a-chite) A bright-green to blackish-green monoclinic mineral: $Cu_5(PO_4)_2(OH)_4$. It resembles malachite and occurs in the oxidized zone of hydrothermal copper deposits. Syn: *dihydrite; phosphochalcite; phosphorochalcite; tagilite.*

pseudomatrix (pseu-do-ma'-trix) A term introduced by Dickinson (1970, p.702) for "a discontinuous interstitial paste formed by the deformation of weak detrital grains" in graywackes and arkoses.

pseudomicrite (pseu-do-mic'-rite) A genetic term applied to calcareous micrite that has formed by secondary changes such as "degenerative" recrystallization (crystal diminution) of faunal and floral material (Chilingar et al., 1967, p.319). Obsolete.

pseudomonoclinic (pseu'-do-mon'-o-clin'-ic) Said of a triclinic crystal form, e.g. that of microcline, that simulates the monoclinic form.

pseudomorph (pseu'-do-morph) A mineral whose outward crystal form is that of another mineral species; it has developed by alteration, substitution, incrustation, or paramorphism. A pseudomorph is described as being "after" the mineral whose outward form it has, e.g. quartz after fluorite. See also: *pseudomorphism; paramorph.* Adj: *pseudomorphous.* Syn: *false form; allomorph.*

pseudomorphism (pseu-do-mor'-phism) The process of becoming, and the condition of being, a *pseudomorph.*

pseudomorphous (pseu-do-mor'-phous) Adj. of *pseudomorph.*

pseudomountain (pseu-do-moun'-tain) A term used by Tarr (1902) for a mountain formed by differential erosion, as contrasted with one produced by uplift; e.g. a *plateau mountain.*

pseudo mud crack A term used by Ksiazkiewicz (1958, pl.16, fig.2) for a sedimentary structure now known as a *parting cast.* See also: *false mud crack.*

pseudonodule (pseu-do-nod'-ule) A primary sedimentary structure consisting of a ball-like mass of sandstone enclosed in shale or mudstone, characterized by a rounded base with upturned or inrolled edges, and resulting from the settling of sand into underlying clay or mud which welled up between isolated sand masses. The term was introduced by Macar (1948) who attributed the structure to horizontal displacement or vertical foundering. See also: *ball-and-pillow structure; flow roll.* Syn: *sand roll.*

pseudo-oolith (pseu-do-o'-o-lith) A spherical or roundish pellet or particle (generally less than 1 mm in diameter) in a sedimentary rock, externally resembling an oolith in size or shape but of secondary origin and amorphous or crypto- or micro-crystalline, and lacking the radial or concentric internal structure of an oolith; e.g. a fecal pellet, a worn calcite grain, a shell fragment, a glauconite granule, or an oolith whose peripheral layers have been resorbed or replaced. Cf: *oolitoid; ooid.* Also spelled: pseudoolith. Syn: *false oolith.*

pseudo-ophitic (pseu'-do-o-phit'-ic) Said of a texture of rock gypsum that is formed by a diagenetic rather than a metamorphic process, and that is characterized by large, platy selenite crystals enclosing small, well formed euhedra. The large crystals are probably of later origin than the matrix in which they are found (Pettijohn, 1957, p.479).

pseudo-orthorhombic (pseu'-do-or'-tho-rhomb'-ic) Said of a monoclinic or triclinic crystal that approximates an orthorhombic crystal in lattice geometry or crystal form.

pseudophenocryst *porphyroblast.*

pseudophite (pseu'-do-phite) A general name for compact, massive chlorite resembling serpentine, in part clinochlore and in part penninite (Hey, 1962, p.569).

pseudoplankton (pseu-do-plank'-ton) *epiplankton.*

pseudopodium (pseu-do-po'-di-um) A temporary or semipermanent projection or retractile process of the protoplasm of a cell (such as a unicellular organism) that serves for locomotion, attachment, and food gathering and that changes in shape, character, and position with the activity of the cell. It may be lobose, filamentous, bifurcating, or anastomosing. Examples: *axopodium; reticulopodium; rhizopodium.* Pl: pseudopodia. Informally: pseudopod.

pseudopore [paleont] (pseu'-do-pore) (a) A tissue-filled space in the calcified layer of a body wall closed by an outer cuticle in many bryozoans. (b) A pore in the outer covering of various calcisponges, the covering being formed by outgrowth from the peripheral part of the inhalant canals.

pseudopore [palyn] A thinning of the palynomorph *exine,* superficially resembling a pore but presumed not to function as an *exitus.*

pseudoporphyroblastic (pseu'-do-por'-phy-ro-blas'-tic) Pertaining to a structure resembling *porphyroblastic* texture but due to processes other than growth, e.g. to differential granulation.

pseudopsephite (pseu-do-pse'-phite) The equivalent of *pseudobreccia* or *pseudoconglomerate* (Read, 1958).

pseudopuncta (pseu-do-punc'-ta) A conical deflection of the secondary shell of a brachiopod, with or without a *taleola,* pointing inwardly and commonly anteriorly so as to appear on the internal surface of the valve as a *tubercle.* It may weather out in fossil shells, leaving a tiny opening that may be mistaken for a *puncta* in punctate shells. Pl: pseudopunctae.

pseudopylome (pseu-do-py′-lome) A prominent thickening of the wall at the antapical end of the vesicle in some acritarchs, resembling the rim of a pylome. In some species, such as *Axisphaeridium* and *Polyancistrodorus*, a central depression or notch seemingly continues as a canal.

pseudoraphe (pseu-do-raphe′) On the frustule of some pennate diatoms, a clear area on the valve between striae or costae.

pseudo ripple mark A term used by Kuenen (1948, p.372) for a bedding-plane feature resembling a ripple but attributed to lateral pressure caused by slumping (such as a mudflow structure imitating a ripple) or by local, small-scale tectonic deformation (such as a corrugation on the cleavage face of slate). See also: *crinkle mark; creep wrinkle.*

pseudorostrum (pseu-do-ros′-trum) The anterior part of the *gnathothorax* of a malacostracan crustacean, formed by a pair of anterior and lateral parts of the cephalic shield projecting forward and meeting in front of the true *rostrum.*

pseudorutile (pseu-do-ru′-tile) A gray, black, brown, or red hexagonal mineral: $Fe^{3+}_2Ti^{4+}_3O_9$. It is an oxidation product of ilmenite, and is common in beach sands. Cf: *arizonite [mineral].*

pseudosaccus (pseu-do-sac′-cus) An extensive separation of the wall layers of a spore; it resembles a *saccus* but lacks alveolate intrastructure. Cf: *cavate.*

pseudoscience (pseu-do-sci′-ence) A term applied to procedures that use the appearance and lingo of science to obtain dubious and often fraudulent results.

pseudosecondary inclusion (pseu-do-sec′-ond-ar-y) A fluid inclusion formed by healing of a fracture occurring during growth of the host crystal and thus appearing to be a primary inclusion while, in facet, being secondary (Roedder, 1967, p.522).

pseudosection [geophys] A display of resistivity and induced-polarization data, obtained with the pole-dipole or dipole-dipole array, in which the observed data values are plotted in section at the intersections of lines drawn at 45 degrees from the mid-point of the current and potential electrode pairs (midpoint of potential pair and through the near current electrode for pole-dipole array); an artifice used to present all of the data from a sounding-profiling in one section. The vertical dimension of the pseudo section bears no simple relationship to the geologic section.

pseudosection [meta] In metamorphic studies, a pressure-temperature plot of the stable mineral assemblages in a metamorphic rock. The diagram is specific to a rock composition and represents the combination of continuous and discontinuous reactions that affect the mineral assemblage in the rock (Hensen, 1971).

pseudoseptal pinnacle (pseu-do-sep′-tal) One of the sclerenchymal pillars constructed on the upper surface of a *tabula* in the zone between parent corallite and offset during the early stages of *increase* in corals (Fedorowski and Jull, 1976, p.37). Cf: *septal pinnacle.*

pseudoseptum (pseu-do-sep′-tum) (a) A spinelike or toothlike skeletal projection in octocorals of the order Coenthecalia. Pseudosepta "bear no constant relationship with soft septa of polyps" (TIP, 1956, pt.F, p.174). (b) The plane of junction in a nautiloid conch between hyposeptal deposits of one *septum* and episeptal deposits on the preceding septum (TIP, 1964, pt. K, p.58), probably representing remnants of the cameral mantle.

pseudosinhalite A colorless monoclinic mineral: $Mg_2Al_3B_2O_9(OH)$.

pseudoskeleton (pseu-do-skel′-e-ton) A sponge skeleton consisting of foreign bodies not secreted by the sponge. Cf: *autoskeleton.*

pseudosparite (pseu-do-spar′-ite) Relatively large, clear calcite crystals that have developed by recrystallization (Folk, 1959, p.33). Syn: neomorphic spar.

pseudospicule (pseu-do-spic′-ule) *spiculoid.*

pseudospondylium (pseu′-do-spon-dyl′-i-um) A cup-shaped chamber accommodating the ventral muscle field of a brachiopod and comprising an undercut *callus* contained between discrete dental plates. Cf: *spondylium.*

pseudostome (pseu′-do-stome) (a) An aperture in a thecamoebian test from which pseudopodia protrude. It may be a simple opening or have definite structure. (b) An opening at the end of a chitinozoan neck. It may be simple, bordered by a small lip, or have a tubular collar.

pseudostratification [glac geol] (pseu′-do-strat′-i-fi-ca′-tion) A concentric structure, resembling stratification, that occurs in till deposits overridden by ice, formed partly by plastering-on of layers of debris and partly by shearing of the till due to pressure of superincumbent ice.

pseudosymmetry (pseu-do-sym′-me-try) Apparent symmetry of a crystal, resembling that of another system; it is often due to twinning.

pseudotachylyte (pseu-do-tach′-y-lyte) (a) A dense rock produced in the compression and shear associated with intense fault movements, involving extreme mylonitization or partial melting. Similar rocks, such as the Sudbury breccias, contain shock-metamorphic effects and may be injection breccias emplaced in fractures formed during meteoritic impact. Cf: *ultramylonite.* (b) A dark gray or black rock that externally resembles *tachylyte* and that typically occurs in irregularly branching veins. The material carries fragmental enclosures, and shows evidence of having been at high temperature. Miarolitic and spherulitic crystallization has sometimes taken place in the extremely dense base. Some pseudotachylyte has behaved like an intrusive and has no structures obviously related to local crushing.

pseudotachylyte [struc geol] A glass or devitrified glass along or adjacent to a fault; the melt is produced by frictionally generated heat, which generally requires sliding velocities consistent with seismic slip.

pseudo telescope structure A term proposed by Blissenbach (1954, p.181) for an alluvial-fan structure resulting from slumping of unconsolidated deposits, such as that formed in a fan that has been cut by a series of small normal faults.

pseudotheca (pseu-do-the′-ca) The false wall of a coral, formed by the thickening and fusion of the outer ends of septa.

pseudotill (pseu-do-till′) A nonglacial deposit resembling a glacial till.

pseudotillite (pseu-do-till′-ite) A term proposed by Schwarzbach (1961) for a definitely nonglacial tillitelike rock, such as a *pebbly mudstone* formed on land by flow of nonglacial mud or deposited by a subaqueous mud flow; an indurated *pseudotill.* Harland et al. (1966, p.233) urge the use of "pseudotillite" as an "unambiguous" and "negative" term for tillitelike rocks found to be nonglacial: "a deposit so named is ... likely to show positive characters which will lead to different nomenclature". The term is equivalent to *tilloid* as used by Pettijohn (1957, p.265).

pseudoumbilicus (pseu′-do-um-bil′-i-cus) A deep depression, either wide or narrow, between the inner umbilical chamber walls of a trochospirally enrolled foraminiferal test where the sharply angled umbilical shoulder occurs (as in *Globorotalites*).

pseudounconformity (pseu′-do-un′-con-form′-i-ty) A term used by Fairbridge (1946, p.88) for a stratigraphic relationship that appears unconformable but is characterized by superabundance or excess accumulation of sediment, such as due to submarine slumping penecontemporaneous with sedimentation off the sides of a rising anticline or dome.

pseudoviscous flow (pseu-do-vis′-cous) *secondary creep.*

pseudovitrinite (pseu-do-vit′-ri-nite) A maceral of coal that is superficially similar to *vitrinite* but is higher in reflectance and has slitted structure, remnant cellular structures, uncommon fracture patterns, higher relief, and paucity or absence of pyrite inclusions (Benedict et al., 1968, p.125).

pseudovitrinoid (pseu-do-vit′-ri-noid) Pseudovitrinite that occurs in bituminous coals (Benedict et al., 1968, p.126).

pseudovolcano (pseu′-do-vol-ca′-no) A large crater or circular hollow believed not to be associated with volcanic activity; e.g. a crater that is possibly meteoritic in origin but may be the result of phreatic explosion or cauldron subsidence. Adj: pseudovolcanic.

pseudowavellite (pseu-do-wa′-vell-ite) *crandallite.*

pseudowollastonite A colorless monoclinic mineral: $CaSiO_3$. It is a high-temperature polymorph of wollastonite. Syn: *cyclowollastonite.*

psi A negative logarithmic transformation (to the base 2) of settling velocity in cm/sec, analogous to the *phi* transformation of grain size. Proposed by Middleton (1967, p.484).

psilate (psi′-late) Smooth. This term may be applied to exines with pits <1 μm in diameter. Syn: *laevigate.*

psilomelane (psi-lo′-mel-ane) (a) A general field term for mixtures of manganese oxide minerals, or for a botryoidal, colloform manganese oxide whose mineral composition is not specifically determined. Cf: *wad [mineral].* Syn: *manganomelane.* (b) A manganese-oxide mineral; specif. *romanèchite.*

psilophyte (psi′-lo-phyte) *rhyniophyte.*

psilopsid (psi-lop′-sid) A term that was formerly used for leafless, rootless vascular plants with terminal sporangia, whether fossil or modern; now it is preferably restricted to the Psilotaceae, a fern-related family of modern plants that have such morphology, presumably by reduction. Cf: *psilophyte; rhyniophyte.*

psittacinite (psit-tac′-i-nite) *mottramite.*

psychrophilic (psy-chro-phil′-ic) A syn. of *cryophilic.* Noun: psychrophile.

psychrophyte (psy′-chro-phyte) A plant adapted to arctic or alpine conditions.

psychrosphere (psy′-chro-sphere) That part of the marine realm characterized by cold, temperature-stable hydroclimates. It lies below the *thermocline* in low and intermediate latitudes, and at all depths in polar regions. Etymol: Greek "psychros", "cold".

psychrotolerant (psy-chro-tol′-er-ant) Said of an organism that lives at 0°C and tolerates temperatures above 20°C.

p-tau mapping *tau-p mapping.*

pterate chorate cyst (pte′-rate) A dinoflagellate *chorate cyst* characterized by a pronounced equatorial outgrowth in the form of solid processes linked distally or in meshlike fashion, as in *Wanaea.*

pterate cyst A dinoflagellate cyst characterized by a pronounced equatorial outgrowth in the form of solid processes linked distally or in meshlike fashion, as in *Wanaea* (Sarjeant, 1969; Williams et al., 2000, p.188). Syn: *pterate chorate cyst.*

pteridophyte (pte-rid′-o-phyte) An informal term (a *grade*) for a member of a group of vascular plants that reproduce spores. The group includes the *lycophytes* (club mosses), *ferns,* and *equisetophytes* (horsetails, scouring rushes). Cf: *bryophyte; spermatophyte; thallophyte.*

Pteridophytic (Pte-rid′-o-phyt′-ic) *Paleophytic.*

pteridosperm (pte-rid′-o-sperm) A *gymnosperm* with fernlike foliage and true seeds borne on leaves, not in cones; a *seed fern.* Its range is Late Devonian to Early Cretaceous.

pterocavate (pte-ro-ca′-vate) Said of a dinoflagellate cyst having a pronounced equatorial pericoel, as in *Stephodinium.*

pterodactyl (pte-ro-dac′-tyl) (a) Strictly, any member of the Pterodactyloidea, the more advanced of two suborders into which the archosaurian order Pterosauria is divided, characterized by a reduced tail and a tendency toward loss of teeth and increase in size. Range, Middle Jurassic to Upper Cretaceous. (b) More loosely, any *pterosaur.*

pteromorph (pte'-ro-morph) An *acritarch* with a sphaeroidal *vesicle* (e.g., *Pterospermella, Pterospermopsis*) surrounded by an equatorial flange. Most acritarchs of this type are believed to represent *prasinophyte* algae (Downie et al., 1963). See also: *ala [palyn]; phycoma.*

pteropod (pte'-ro-pod) Any opisthobranch gastropod belonging to the order Pteropoda, which includes pelagic forms sometimes with shells. The shells are generally conical and composed of aragonite. Range, Cretaceous to Holocene.

pteropod ooze A pelagic ooze containing at least 30% remains of pteropods. The shells are made of *aragonite*, generally found at much shallower depth than other oozes.

pteropsid (pte-rop'-sid) A vascular plant (of the class Pteropsida) that has leaf gaps, as in ferns and seed plants. Rarely used.

pterosaur (pte'-ro-saur) A member of the order Pterosauria, archosaurian reptiles highly adapted to flight. They were characterized by extreme elongation of the fourth digit of the hand for support of a membranous wing, and by reduction of the hind limbs. Range, Upper Triassic to Upper Cretaceous. Partial syn: *pterodactyl.*

pterygostomial region (pter'-y-go-sto'-mi-al) The anterolateral part of the carapace on the ventral surface of some decapod crustaceans, located on opposite sides of the buccal cavity (TIP, 1969, pt. R, p.92).

ptilolite (pti'-lo-lite) *mordenite.*

PTRM *partial thermoremanent magnetization.*

ptycholophe (pty'-cho-lophe) A brachiopod *lophophore* in which the brachia are folded into one or more lobes in addition to median indentation (TIP, 1965, pt. H, p.151).

ptychopariid (pty-cho-pa'-ri-id) Any trilobite belonging to the order Ptychopariida, characterized generally by opisthoparian sutures, more than three segments in the thorax, and a simple glabella. Range, Lower Cambrian to Middle Permian.

ptygma (ptyg'-ma) Granitic material within migmatite or gneiss, having the appearance of disharmonic folds (Dietrich, 1959, p.358). Rarely used. The term is also applied to veins, e.g., quartz veins. Syn: *ptygmatic folds.*

ptygmatic folds Folds with rounded hinges and fold amplitudes nearly equal to fold wavelength; they develop in isolated layers, usually in metamorphic rocks. Syn: *ptygma.*

pubescent (pu-bes'-cent) Said of a plant that is covered with soft, downy hairs.

public domain (pub'-lic) (a) Land owned, controlled, or heretofore disposed of by the U.S. Federal government. It includes the land that was ceded to the government by the original thirteen States, together with certain subsequent additions acquired by cession, treaty, and purchase. At its greatest extent, the public domain occupied more than 1,820,000,000 acres. See also: *public land.* (b) The status of products, publications, and processes unprotected by copyright or patent (Webster's, 1984, p.951).

public land Land owned by a government, esp. a national government; specif. the part of the U.S. *public domain* to which title is still vested in the Federal government and that is subject to appropriation, sale, or disposal under the general laws.

public land survey A survey of public lands; specif. the U.S. Public Land Survey system (USPLS) by which much of the United States was surveyed and divided into a rectangular grid system using townships, sections, and fractions of sections (USBLM, 1973).

public land survey system See *public land survey.* Abbrev: PLSS.

pucherite (pu'-cher-ite) An adamantine dark reddish-brown or yellowish-brown orthorhombic mineral: $BiVO_4$.

puddingstone (a) A popular name applied chiefly in Great Britain to a conglomerate consisting of well-rounded pebbles whose colors are in such marked contrast with the abundant fine-grained matrix or cement that the rock suggests an old-fashioned plum pudding. Example: the Hertfordshire Puddingstone (lower Eocene) in England, composed of black or brown flint pebbles cemented by white silica, with or without brown iron hydroxide. Syn: *plum-pudding stone.* (b) A siliceous rock cut into blocks for furnace linings. Also spelled: pudding stone.

puddle (pud'-dle) A small accumulation of meltwater in a depression or hollow on the surface of any form of ice, produced mainly by the melting of snow and ice, and in most cases fresh and potable.

puddle-core dam An earth dam in which the impervious core is constructed of puddled clay deposited by *hydraulic fill* methods.

puddle wall The impervious core of a dam, or an impervious cutoff wall in natural materials, made of puddled clay.

Puercan (Puer'-can) North American land mammal age: lowermost Paleocene (after Cretaceous, before Dragonian or Torrejorian).

puerto (puer'-to) A term used in the SW U.S. for a pass over or through an escarpment or mountain range. Etymol: Spanish.

puff n. A high spot or elevation in *gilgai.*

puffing hole (puff'-ing) *blowhole [coast].*

puffy ground A surface formed on *playa* mudflats by efflorescent crystal growth that causes sediment to be pushed apart, creating mounds and polygons that are easily collapsed. Syn: *powdery efflorescence.*

pug *gouge [ore dep].*

puglianite (pu'-gli-an-ite') A coarse-grained *foid monzogabbro* composed of euhedral augite, leucite, anorthite, sanidine, amphibole, and biotite. It occurs as fragments in the lavas of Monte Somma, Italy. Lacroix proposed the term in 1917. Cf: *sebastianite.* Obsolete.

pulaskite (pu-las'-kite) A li●ht-colored, feldspathoid-bearhng, granular or trachytoid alkali *syenite* composed chiefly of alkali feldspar, sodic pyroxene, arfedsonite, and nepheline. The term has also been applied to quartz-bearing syenites. The name, given by Williams in 1891, is from Pulaski County, Arkansas. Obsolete.

pull-apart (a) n. A precompaction sedimentary structure resembling *boudinage,* consisting of beds that have been stretched and torn apart into relatively short slabs, the intervening cracks being filled in from the top (or in some cases possibly from below) (Natland and Kuenen, 1951, p.89-90); e.g. stiff clay embedded in more mobile, water-soaked sand, or compact sandstone embedded in hydroplastic clayey rock. (b) In ultramafic complexes, blocks or slabs of chromite cut by dikes of serpentinized dunite. adj. Said of a structure or bed characterized by pull-aparts.

pull-apart basin A topographic depression created by an extensional bend or extensional overstep along a strike-slip fault (Aydin and Nur, 1982). Syn: *sag pond; rhombochasm.*

Pulmonata A subclass of presumably a detorted gastropod with shell commonly present, sometimes reduced and concealed in mantle. Gills absent; mantle cavity vascularized and altered into lung and fused with the neck except for small air hole. Operculum rarely present. The vast majority live in freshwater or terrestrial environments. Range, Pennsylvanian to Holocene; about 2,500 genera.

pulmonate (pul'-mo-nate) Any terrestrial or freshwater gastropod belonging to the subclass Pulmonata.

pulp A fluid mixture (slurry) of coal and water.

pulp cavity A cavity inside the base of any vertebrate tooth, or one of several in the base of an unreduced scale, which in life contains vascular, nervous, and connective tissue. The term has been incorrectly used for the *basal cavity* of a conodont.

pulpit rock (pul'-pit) *chimney rock.*

pulps A term used by Allen and Day (1935, p.65) for "a fine, mealy, opaline silica", much like sand.

pulpy peat (pulp'-y) *sedimentary peat.*

pulsating spring (pul'-sat-ing) *geyser.*

pulsation [stratig] (pul-sa'-tion) (a) A term used by Grabau (1936a) for a long rhythm, conceived to be nearly the length of a geologic-time period, representing a eustatic movement of sea level that resulted in simultaneous transgression and regression of widespread and semipermanent seas over whole continents. Cf: *oscillation [stratig].* (b) A distinct step or change in a series of rhythmical or regularly recurring movements.

pulsation [tect] A minor time of deformation, or a subdivision of a more prolonged epoch of orogeny. Cf: *event [tect]; disturbance.*

pulsation tectonics A theory of tectonics involving the cyclic eruption of plumes of hot mantle material from the lowermost mantle (Sheridan, 1983).

pulse In ecology, a sudden increase in the number of organisms or kinds of organisms, usually at regularly recurring intervals.

pulsed-neutron-capture log (pulsed-neu'-tron-cap'-ture) A specialized *radioactivity log* of *neutron-gamma log* type. A neutron generator produces short bursts of high-energy neutrons (about 14 Mev), which are slowed in the borehole and nearby rocks to thermal-energy level (about 0.025 ev); capture by nuclei (especially chlorine) results in gamma-ray emissions. The log is used in cased oil wells to determine changes in fluid saturation during oil production. It is marketed under several trade names, e.g. Neutron Lifetime Log, Thermal Decay Time Log.

pulverite (pul'-ver-ite) A sedimentary rock composed of silt- or clay-sized aggregates of constructional (nonclastic) origin, simulating in texture a *lutite* of clastic origin; e.g. a rock formed of diatom frustules. The term was introduced by Grabau (1911, p.1007). Syn: pulveryte.

pulverization (pul'-ver-i-za'-tion) *comminution.*

pulverulent (pul-ver'-u-lent) Said of a mineral that is easily powdered.

pumice (pum'-ice) Highly vesicular pyroclasts with very low bulk density and thin vesicle walls.

pumice fall The descent of pumice from an eruption cloud; pumice fallout. Cf: *ash fall.*

pumice flow A type of *pyroclastic flow* in which a large proportion of the fragments are of pumice. Cf: *ash flow.*

pumiceous (pu-mi'-ceous) Volcanic igneous vesicular texture involving tiny gas holes such as in pumice. Finer than scoriaceous.

pumicite (pum'-i-cite) Obsolete syn. of *ash [volc].*

pumpage (pump'-age) (a) The quantity of water or other liquid pumped, as of groundwater. (b) The act of pumping.

pumpellyite (pum-pel'-ly-ite) A group name for monoclinic minerals with the general formula $Ca_2AB^{3+}_2(SiO_4)(Si_2O_7)(OH,O)_2{\cdot}H_2O$, where A = Mg, Mn^{2+}, Fe^{2+}, Fe^{3+}, or Al and B = Al, Fe^{3+}, Mn^{3+}, or Cr. See also: *chlorastrolite.* Syn: *zonochlorite; lotrite.*

pumpellyite-(Al) A theoretical Al end-member pumpellyite: $Ca_2AlAl_2(SiO_4)(Si_2O_7)(OH,O)_2{\cdot}H_2O$.

pumpellyite-(Fe^{2+}) A greenish-black monoclinic mineral of the *pumpellyite* group: $Ca_2Fe^{2+}Al_2(SiO_4)(Si_2O_7)(OH)_2{\cdot}H_2O$. Syn: *ferropumpellyite.*

pumpellyite-(Fe^{3+}) A greenish-black monoclinic mineral of the *pumpellyite* group: $Ca_2(Fe^{3+},Mg)Al_2(SiO_4)(Si_2O_7)(OH,O)_2{\cdot}H_2O$. Syn: *oxyferropumpellyite.*

pumpellyite-(Mg) (pum-pel'-ly-ite) A bluish-green, lath-shaped, monoclinic mineral of the *pumpellyite* group: $Ca_2(H_2O)MgAl_2(SiO_4)(Si_2O_7)(OH)_2{\cdot}H_2O$. It is related to clinozoisite. Its Fe^{3+} iron analogue is julgoldite-(Mg).

pumpellyite-(Mn^{2+}) A brownish-pink monoclinic mineral of the *pumpellyite* group: $Ca_2Mn^{2+}Al_2(SiO_4)(Si_2O_7)(OH)_2{\cdot}H_2O$.

pumpellyite-prehnite facies The facies (set of metamorphic mineral assemblages) in which metagraywackes contain albite+quartz+prehnite+pumpellyite+chlorite+sphene (Coombs et al., 1959). It is generally believed to represent pressure-temperature conditions between those of the *zeolite facies* and the *greenschist facies.* Var: *prehnite-pumpellyite facies.*

Pumpelly's rule (Pum-pel'-ly's) The generalization that the axes and axial surfaces of minor folds of an area are congruent with those of the major fold structures of the same phase of deformation (Pumpelly, 1894, p.158).

pumping test A test made by pumping a well, generally at a constant rate, and observing the change in hydraulic head in the aquifer. It may be used to determine the capacity of the well and the hydraulic properties of the aquifer (Fetter, 1994). Cf: *aquifer test.*

punch register To make punch holes at predetermined positions in multiple pieces of cartographic copy which are in perfect registry for the insertion of pins or studs so that registry can be maintained in subsequent cartographic steps and in the preparation of negatives to be used in making color printing plates. Cf: *register punch.*

puncta (punc'-ta) (a) One of the minute, closely spaced pores, perforations, or tubules extending perpendicularly a variable distance from the inner or outer surface of a brachiopod shell. See also: *endopuncta; exopuncta; pseudopuncta.* Often used interchangeably with *punctum.* (b) Any of various thin places arranged in characteristic pattern in the frustule of pennate diatoms, being smaller and simpler than an *areola;* specif. the smallest structure on a diatom valve, such as one of the pores having diameters as small as 0.037 µm (but commonly 0.5 to 1.0 µm), occurring either scattered or in rows, and sometimes having fine porous plates at their inner extremity. (c) A hole in the external wall of a foraminiferal chamber. Pl: punctae. Etymol: Latin, stab, prick (mark).

punctate (punc'-tate) Minutely pitted, or having minute dots, punctae, or depressions, such as a "punctate leaf"; specif. said of a brachiopod or brachiopod shell possessing endopunctae. Cf: *impunctate.*

punctation (punc-ta'-tion) The condition of being *punctate.*

punctual kriging In *geostatistics,* estimates of values at point locations, based on observations which themselves are measured at point locations.

punctuated equilibria (punc'-tu-at-ed) The pattern of evolution that predominates in the fossil record, in which the morphology of a species remains constant for long periods of time, and descendant species evolve in allopatry and appear relatively suddenly, with the ancestral species generally persisting without changing morphologically.

punctuated gradualism The evolutionary pattern shown by some species lineages in which intervals of morphologic change by *phyletic gradualism* alternate with intervals of morphological constancy or *stasis* (Malmgren et al., 1983). Cf: *punctuated equilibria.*

punctuational evolution (punc-tu-a'-tion-al) That evolutionary pattern in which new taxa appear relatively suddenly via major shifts in characteristics within very short time spans and then remain essentially constant in those characters for long periods. When applied to the species level, it is known as *punctuated equilibria.* Cf: *episodic evolution.* Ant: *gradualistic evolution.*

punctum (punc'-tum) A small area marked off in any way from a surrounding surface; specif. a minute pit on the shell surface of a gastropod (not a tubule penetrating the shell substance), or *puncta* in the shell of a brachiopod. Pl: puncta. Etymol: Latin, result of stinging or stabbing.

punky (punk'-y) Said of a semi-indurated rock, such as a leached limestone; esp. said of a tuff that is weakly welded or consolidated.

pup A term used in Alaska for a small tributary stream.

puppet (pup'-pet) *loess doll.*

Purbeckian (Pur-beck'-i-an) A European regional stage (Great Britain): lowermost Cretaceous (above Portlandian, below Wealden).

pure coal An informal syn. of *vitrain.*

pure rotation *rotational strain.*

pure shear A particular example of *irrotational strain* in which the body is elongated in one direction and shortened at right angles to this. Cf: *simple shear.*

purga (pur'-ga) A violent arctic snowstorm. Etymol: Karelian "purgu", "snowstorm".

purgatory (pur'-ga-to-ry) (a) A term used in New England for a long, deep, narrow, steep-sided cleft or ravine along a rugged coast, into which waves rush during a storm with great noise and violence; a rock chasm without a stream, often covered at the bottom with large, angular rocks, and difficult to traverse. (b) A swamp that is dangerous or difficult to traverse.

purge well A well designed to control contaminant movement in an aquifer and remove contaminants for treatment.

purl A swirling or eddying stream or rill, moving swiftly around obstructions; a stream making a soft, murmuring sound.

purple blende *kermesite.*

purple copper ore *bornite.*

purpurite (pur'-pu-rite) A deep-brown to reddish-purple orthorhombic mineral of the *triphyllite* group: $(Mn^{3+},Fe^{3+})PO_4$. It is isomorphous with heterosite. Both are derived from the alkali leaching of triphylite-lithiophilite.

push [hydraul] A force exerted directly by the wind upon the exposed sides of wave crests (Strahler, 1963, p.308). Cf: *drag [hydraul].*

pushcharovskite A light green or colorless triclinic mineral: $K_{0.6}Cu_{18}[AsO_2(OH)_2]_4[AsO_3OH]_{10}(AsO_4)(OH)_{9.6}{\cdot}\ 18.6H_2O$.

push moraine A broad, smooth, arc-shaped morainal ridge consisting of material mechanically pushed or shoved along or deformed by an advancing glacier. Examples are common in the Netherlands and NW Germany. Cf: *ice-pushed ridge.* Syn: *shoved moraine; upsetted moraine; thrust moraine.*

push-pull wave *P wave.*

pustule (pus'-tule) A minute boss on an asterozoan *ossicle,* having a central depression in which a spine articulates.

pusule apparatus (pu'-sule) The sacklike vacuole in a dinoflagellate connected with the exterior by a slender canal opening into a flagellar pore.

putoranite (pu-to'-ra-nite) A metallic yellow cubic mineral that is the ordered equivalent of a disordered high-temperature phase: $Cu_{1.1}Fe_{1.2}S_2$.

putrefaction (pu-tre-fac'-tion) The decomposition of organic matter by slow distillation, in the presence of water, without air. Methane and other gaseous products (H_2, NH_3, H_2S) are formed. Cf: *disintegration [coal]; moldering; peat formation.*

puy A small, remnant volcanic cone; it is the French word for such structures in the Auvergne district of central France.

puzzolan (puz'-zo-lan) *pozzolan.*

p-veatchite A colorless monoclinic mineral: $Sr_2[B_5O_8(OH)]_2[B(OH)_3]{\cdot}H_2O$. It is dimorphous with veatchite. Syn: *paraveatchite.*

P wave A seismic *body wave* that involves particle motion (alternating compression and expansion) in the direction of propagation. It is the fastest of the seismic waves, traveling 1.5 to 6.0 km/s in sedimentary rocks, 5.5-7.2 km/s in the crust, 7.8-8.5 km/s in the upper mantle, up to 14 km/s in the mantle, and 8 to 12 km/s in the core. Sound waves are P waves. The P stands for primary; it is so named because it is the *first arrival* from earthquakes. Syn: *longitudinal wave; irrotational wave; pressure wave; dilatational wave; primary wave; compressional wave; push-pull wave.* Cf: *S wave; surface wave [seis].*

pyatenkoite-(Y) A vitreous colorless rhombohedral mineral of the *hilairite* group: $Na_5(Y,Dy,Gd)(Ti,Nb)Si_6O_{18}{\cdot}6H_2O$.

pycnidium (pyc-nid'-i-um) An asexual, hollow, fungal fruiting body lined with *conidiophores* bearing *conidia.* It may be fully enclosed or have an opening (*ostiole*).

pycnite (pyc'-nite) A variety of topaz occurring in massive columnar aggregations.

pycnocline (pyc'-no-cline) (a) The depth of the maximum rate of increase of density with depth, the depth of the maximum of the vertical density gradient. (b) More generally, the region of large vertical density gradients below the *mixed layer.*

pycnogonid (pyc-nog'-o-nid) Any marine arthropod belonging to the subphylum Pycnogonida, resembling the chelicerates in having one pair of chelae but lacking a well-developed abdomen. Range, Lower Devonian to Holocene.

pycnometer A glass vessel of known volume used in the measurement of specific gravity of fine grained and powdered substances.

pycnostromid (pyc-nos'-tro-mid) A *microbial biscuit* produced by *Pycnostroma.*

pycnotheca (pyc-no-the'-ca) The dense, nonalveolate inner layer of the test wall of schwagerinid fusulinids, penetrated by septal pores, and wedged between tectum and keriotheca or antetheca.

pygidium (py-gid'-i-um) A caudal structure or terminal body region of various invertebrates; esp. the posterior part or tail piece of an exoskeleton of a trilobite, consisting of several fused segments. Pl: pygidia. Adj: pygidial. Cf: *abdomen.*

pylome [paleont] A large opening in spumellarian radiolarians, commonly only in the outermost of concentric shells.

pylome [palyn] A circular to subcircular *excystment* opening of an *acritarch.* It differs from an *archeopyle* by lacking any indication of relation to *paratabulation* (Evitt, 1961; Williams et al., 2000). Cf: *macropylome; micropylome.*

pyralmandite (pyr-al'-man-dite) A garnet intermediate in chemical composition between pyrope and almandine.

pyralspite (pyr-al'-spite) A group of garnets of formula: $M_3Al_2(SiO_4)_3$, where M = Mg, Fe^{+2}, or Mn^{+2}. It includes pyrope, almandine, and spessartine, and their intermediate forms.

pyramid [cryst] (pyr'-a-mid) An open crystal form consisting of three, four, six, eight, or twelve nonparallel faces that meet at a point. Cf: *dipyramid.* Adj: *pyramidal.*

pyramid [paleont] A large beaklike or winglike element of *Aristotle's lantern* in the interambulacral position of an echinoid. See also: *demipyramid.*
pyramidal (py-ram'-i-dal) Having the symmetry of a pyramid.
pyramidal cleavage Mineral cleavage parallel to the faces of a pyramid, e.g. the (101) cleavage of scheelite.
pyramidal dune *star dune.*
pyramidal peak *horn [glac geol].*
pyramidal system *tetragonal system.*
pyramid pebble *dreikanter.*
pyranometer The standard instrument for measuring solar radiation incident on a horizontal surface; calibrates the temperature response of a special sensor in units of radiation flux, such as watts per square meter.
pyrargyrite (pyr-ar'-gy-rite) A dark-red, gray, or black rhombohedral mineral: Ag_3SbS_3. It is isomorphous with proustite and polymorphous with pyrostilpnite, and is an important ore of silver. Syn: *dark ruby silver; dark red silver ore.*
pyrene (py'-rene) The small stone of a drupe or similar fruit.
Pyrenean Orogeny (Py-re-ne'-an) One of the 30 or more short-lived orogenies during Phanerozoic time identified by Stille, in this case during the late Eocene, between the Bartonian and Ludian stages.
Pyrenean-type facies series Rocks produced in a type of dynamothermal regional metamorphism characteristic of the Pyrenees, in which the pressure range is 350-500 MPa. The mineral sequence, in order of rising temperature, is staurolite - andalusite - sillimanite - cordierite (Hietanen, 1967, p.193). Rarely used. Cf: *Idahoan-type facies series.*
pyreneite (py-re-ne'-ite) *melanite.*
pyribole (pyr'-i-bole) A mnemonic term coined by Johannsen in 1911 in his classification of igneous rocks to indicate the presence of either or both a pyroxene and/or an amphibole. Also spelled: pyrabole; pyrabol; pyrobol. Obsolete. Etymol: *pyr*oxene + amph*ibole.*
pyric pond (pyr'-ic) A pool of water that collects in a shallow hole or sink formed as a result of fires and subsequent subsidence in peat deposits, lignite, and coal beds.
pyrite (py'-rite) (a) A common, pale-bronze or brass-yellow, cubic mineral: FeS_2. It is dimorphous with marcasite, and often contains small amounts of other metals. Pyrite has a brilliant metallic luster and an absence of cleavage, and has been mistaken for gold (which is softer and heavier). It commonly crystallizes in cubes (whose faces are usually striated), octahedrons, or pyritohedrons, and it also occurs in shapeless grains and masses. Pyrite is the most widespread and abundant of the sulfide minerals and occurs in all kinds of rocks, such as in nodules in sedimentary rocks and coal seams or as a common vein material associated with many different minerals. Pyrite is an important ore of sulfur, less so of iron, and is burned in making sulfur dioxide and sulfuric acid; it is sometimes mined for the associated gold and copper. Cf: *pyrites.* Syn: *iron pyrites; fool's gold; mundic; common pyrites.* (b) A group name for minerals isomorphous with pyrite, with the general formula AX_2, where A = Fe, Co, Ni, Mn, Cu, Au, Ru, Os, Ir or Pt, and X = S and, more rarely, Se, Te, Sb or As.
pyrites (py-ri'-tes) (a) Any of various metallic-looking sulfides of which pyrite ("iron pyrites") is the commonest. The term is used with a qualifying term that indicates the component metal; e.g. "copper pyrites" (chalcopyrite), "tin pyrites" (stannite), "white iron pyrites" (marcasite), "arsenical pyrites" (arsenopyrite), "cobalt pyrites" (linnaeite), and "nickel pyrites" (millerite). When used popularly and without qualification, the term usually signifies *pyrite.* (b) An obsolete term for a stone that may be used for striking fire.
pyrite-type pressure shadow A *pressure shadow* containing crystal fibers growing from the walls towards a rigid object (Ramsay and Huber, 1983). Fibers generally are a different material than those in the rigid object. Cf: *crinoid-type pressure shadow.*
pyritohedral (py'-ri-to-he'-dral) Adj. of *pyritohedron.*
pyritohedron (py'-ri-to-he'-dron) A crystal form that is a *dodecahedron* consisting of 12 pentagonal faces that are not regular. Its symmetry is $2/m\bar{3}$ and its indices are *{210}*. It is named after pyrite, which characteristically has this crystal form. Adj: *pyritohedral.* Cf: *rhombic dodecahedron.* Syn: *pentagonal dodecahedron; regular dodecahedron; pyritoid.*
pyritoid (py'-ri-toid) As a noun, a syn. of *pyritohedron.*
pyroaurite (py-ro-au'-rite) A colorless, golden-yellow or brownish platy rhombohedral mineral of the *hydrotalcite* group: $Mg_6Fe^{3+}{}_2(CO_3)(OH)_{16}\cdot 4H_2O$. It is dimorphous with sjögrenite, and may contain up to 5% MnO.
pyrobelonite (py-ro-bel'-o-nite) A fire-red to deep brilliant-red orthorhombic mineral: $PbMn^{2+}(VO_4)(OH)$.
pyrobiolite (py-ro-bi'-o-lite) An organic rock containing organic remains that have been altered by volcanic action.
pyrobitumen (py'-ro-bi-tu'-men) Any of the dark-colored, fairly hard, nonvolatile native substances composed of hydrocarbon complexes, which may or may not contain oxygenated substances and are often associated with mineral matter. The nonmineral constituents are infusible, insoluble in water, and relatively insoluble in carbon disulfide. On heating, pyrobitumens generally yield bitumens, i.e. decompose rather than melt.
pyrobituminous (py'-ro-bi-tu'-mi-nous) Pertaining to substances which yield bitumens upon heating.
pyrochlore (py'-ro-chlore) (a) A pale-yellow, red, brown, or black cubic mineral: $(Na,Ca)_2(Nb,Ta)_2O_6(OH,F)$. It is isomorphous with microlite, with Nb greater than Ta, and it usually contains cerium and titanium. Pyrochlore occurs in pegmatites derived from alkalic igneous rocks and constitutes an ore of niobium. Syn: *pyrrhite.* (b) A group of minerals of the general formula: $A_2B_2O_6(O,OH,F)$, where A = Na, Ca, K, Fe^{2+}, U^{4+}, Sb^{3+}, Pb, Th, Ce, or Y, and B = Nb, Ta, Ti, Sn, Fe^{3+}, or W. It includes minerals such as pyrochlore, microlite, betafite, obruchevite, and pandaite.
pyrochroite (py-ro-chro'-ite) A trigonal mineral of the *brucite* group: $Mn^{2+}(OH)_2$. It is white when fresh, but darkens on exposure; it is similar to brucite in appearance, to which it is isomorphic.
pyroclast (py'-ro-clast) An individual particle ejected during a volcanic eruption. It is usually classified according to size. Cf: *pyroclastic deposit.*
pyroclastic (py-ro-clas'-tic) Pertaining to clastic rock material formed by volcanic explosion or aerial expulsion from a volcanic vent; also, pertaining to rock texture of explosive origin. It is not synonymous with the adjective "volcanic".
pyroclastic breccia *explosion breccia.*
pyroclastic density current A gravity-controlled, laterally moving mixture of pyroclasts and gas. It is a more general term that includes *pyroclastic flow* and *pyroclastic surge.*
pyroclastic deposit A general term for a deposit of *pyroclasts.* Syn: *tephra.*
pyroclastic flow A density current of pyroclastic material, usually very hot and composed of a mixture of gases and particles. A syn. of ash flow used in a more general sense in that an ash flow is composed of ash-sized *pyroclasts.* Cf: *pyroclastic surge.*
pyroclastic surge Low-density, dilute, turbulent *pyroclastic flow.* Types of pyroclastic surges include base surges, ash-cloud surges, and ground-surges.
pyroelectricity (py'-ro-e'-lec-tric'-i-ty) The simultaneous development, in any crystal lacking a center of symmetry, of opposite electric charges at opposite ends of a crystal axis, due to certain changes in temperature.
pyrogeology (py'-ro-ge-ol'-o-gy) An obsolete synonym of *volcanology* that was proposed by Grabau (1924).
pyrolite (py'-ro-lite) A model proposed by Ringwood (Green and Ringwood, 1963) for the material of the upper mantle, composed of one part basalt to three parts dunite and consisting mainly of olivine and pyroxenes. It is so designed that a partial melt will yield a basaltic magma.
pyrolith (py'-ro-lith) A term proposed by Grabau (1904) for igneous rock. Obsolete.
pyrolusite (py-ro-lu'-site) A soft iron-black or dark steel-gray tetragonal mineral of the *rutile* group: $Mn^{4+}O_2$. It is the most important ore of manganese and is dimorphous with ramsdellite. Pyrolusite is generally massive or reniform, sometimes with a fibrous or radiate structure. Syn: *polianite; gray manganese ore.*
pyromagma (py-ro-mag'-ma) Obsolete term for a highly mobile lava, oversaturated with gases, that exists at shallower depths than *hypomagma.* Cf: *epimagma.*
pyromelane (py-ro-mel'-ane) *brookite.*
pyrometallurgy The chemical separation of a metal from its ores and concentrates by using heat at high temperatures by means of roasting, smelting, and refining.
pyrometamorphism (py'-ro-met'-a-mor'-phism) (a) Metamorphic changes taking place without the action of pressure or water vapor, at temperatures near the melting points of the component minerals; it is a local, intense type of *thermal metamorphism,* resulting from the unusually high temperatures at the contact of a rock with magma, e.g. in xenoliths (Turner, 1948). Cf: *igneous metamorphism.* (b) "Metamorphism" caused by lightning strikes and burning coal seams.
pyrometasomatic (py'-ro-met'-a-so-mat'-ic) Formed by *metasomatic* changes in rocks, principally in limestone, at or near intrusive contacts, under influence of magmatic emanations and high temperature and pressure.
pyrometasomatism (py'-ro-met'-a-som'-a-tism) The formation of contact-metamorphic mineral deposits at high temperatures by emanations issuing from the intrusive and involving replacement of enclosing rock with addition or subtraction of materials; *skarn* formation. Cf: *metasomatism.*
pyrometer (py-rom'-e-ter) An instrument that measures high temperature, e.g. of molten lavas, by electrical or optical means. See also: *optical pyrometer; pyrometry.*
pyrometric cone (py-ro-met'-ric) *Seger cone.*
pyrometry (py-rom'-e-try) The measurement of high temperatures by electrical or optical means, using a *pyrometer.* Its geologic application is to incandescent lavas.
pyromorphite (py-ro-mor'-phite) A green, yellow, brown, gray, or white hexagonal mineral of the *apatite* group: $Pb_5(PO_4)_3Cl$. It is isomorphous with mimetite and vanadinite, and may contain arsenic or calcium. Pyromorphite is found in the oxidized zone of lead deposits, and is a minor ore of lead. Syn: *green lead ore.*
pyrope (py'-rope) (a) The magnesium-aluminum end-member of the garnet group, characterized by a deep fiery-red color: $Mg_3Al_2(SiO_4)_3$. It rarely occurs in crystals, but is found in detrital deposits as rounded and angular fragments, or associated with olivine and serpentine in basic igneous rocks

such as kimberlite. See also: *Cape ruby; Bohemian garnet.* Syn: *rock ruby.* (b) An obsolete name for a bright red gem, such as a ruby.

pyrophane (py'-ro-phane) (a) *fire opal.* (b) An opal (such as hydrophane) artificially impregnated with melted wax.

pyrophanite (py-ro-phan'-ite) A metallic blood-red rhombohedral mineral of the *corundum* group: $Mn^{2+}TiO_3$. It is isomorphous with ilmenite.

pyrophosphite A vitreous white monoclinic mineral occurring in a cave deposit: $K_2CaP_2O_7$. Not approved by the International Mineralogical Association.

pyrophyllite (py-ro-phyll'-ite) A white, gray, blue, green, or brown mineral occurring in monoclinic and triclinic polytypes: $Al_2Si_4O_{10}(OH)_2$. It resembles talc and occurs in a foliated form or in compact masses in quartz veins, granites, and esp. metamorphic rocks. Syn: *pencil stone.*

pyropissite (py-ro-pis'-site) An earthy, nonasphaltic pyrobitumen made up primarily of water, humic acids, wax (it is a source of *montan wax*), and silica. It is frequently found associated with *brown coal,* which is then called pyropissitic brown coal.

pyroretinite (py-ro-ret'-i-nite) A type of retinite found in the brown coals of Aussig (Usti nad Labem), Czech Republic.

pyroschist (py-ro-schist') A schist or shale that has a sufficiently high carbon content to burn with a bright flame, or to yield volatile hydrocarbons, when heated.

pyrosmalite (py-ro-smal'-ite) A group name for rhombohedral or monoclinic minerals with the general formula $(Fe^{2+},Mn^{2+})_8Si_6O_{15}(OH,Cl)_{10}$, and including *ferropyrosmalite, manganpyrosmalite, friedelite* and *mcgillite.*

pyrosphere (py'-ro-sphere) An archaic term for the zone of the Earth below the lithosphere, possibly partly molten; it is equivalent to the *barysphere.* Syn: *magmosphere.*

pyrostibite (py-ro-stib'-ite) *kermesite.*

pyrostilpnite (py-ro-stilp'-nite) A hyacinth-red monoclinic mineral: Ag_3SbS_3. It is polymorphous with pyrargyrite. Syn: *fireblende.*

pyroxene (py'-rox-ene, py-rox'-ene) (a) A group of dark rock-forming silicate minerals, closely related in crystal form and composition and having the general formula: $A_2M_2Si_4O_{12}$, where A = Ca, Na, Mg, or Fe^{2+}, and B = Mg, Fe^{2+}, Fe^{3+}, Cr, Mn, or Al, with silicon sometimes replaced in part by aluminum. It is characterized by a single chain of tetrahedra with a silicon:oxygen ratio of 1:3; by short, stout prismatic crystals; and by good prismatic cleavage in two directions parallel to the crystal faces and intersecting at angles of about 87° and 93°. Colors range from white to dark green or black. Pyroxenes may crystallize in the orthorhombic or monoclinic systems; they constitute a common constituent of igneous rocks, and are similar in chemical composition to the amphiboles (except that the pyroxenes lack hydroxyls). (b) A mineral of the pyroxene group, such as enstatite, hypersthene, diopside, hedenbergite, acmite, jadeite, pigeonite, and esp. augite. Etymol: Greek "pyros", "fire", + "xenos", "stranger", apparently so named from the mistaken belief that the pyroxenes "were only accidentally caught up in the lavas that contain them" (Challinor, 1978, p.250). Pron: pie-rok-seen or peer-ok-seen.

pyroxene alkali syenite In Tobi's classification of the *charnockite series* (1971, p.202), a member with less than 20% quartz and characterized by the presence of microperthite.

pyroxene-hornblende gabbronorite (py'-rox-ene-horn'-blende) In the *IUGS classification,* a plutonic rock satisfying the definition of *gabbronorite* with hornblende between 5 and 85 percent.

pyroxene-hornblende peridotite In the *IUGS classification,* a plutonic rock with M equal to or greater than 90, ol/(ol+hbl+px) between 40 and 90, px/(ol+hbl+px) greater than 5, and hbl/(ol+hbl+px) greater than 5.

pyroxene hornblendite In the *IUGS classification,* a plutonic rock with M equal to or greater than 90, ol/(ol+hbl+px) less than 5, and hbl/(px+hbl) between 50 and 90.

pyroxene-hornfels facies The facies (set of metamorphic mineral assemblages) in which mafic rocks are represented by diopside+hypersthene+ plagioclase, with amphibole typically absent. Pelitic assemblages exhibit the association of sillimanite (or andalusite) and cordierite with potassium feldspar; muscovite is absent. Marbles contain wollastonite and calcite+forsterite+periclase (Turner, 1968). The facies is typical of high-grade thermal metamorphism, as in the inner parts of contact aureoles. It corresponds to temperatures in excess of about 550°C, and to relatively low pressures.

pyroxene melilitolite In the *IUGS classification,* a plutonic rock consisting mostly of melilite and clinopyroxene, with less than 10% olivine.

pyroxene monzonite In Tobi's classification of the *charnockite series* (1971, p.202), a quartz-poor member containing approximately equal amounts of microperthite and plagioclase; *mangerite.*

pyroxene olivine melilitolite In the *IUGS classification,* a plutonic rock consisting mostly of melilite, olivine, and clinopyroxene, with olivine> clinopyroxene.

pyroxene peridotite In the *IUGS classification,* a plutonic rock with M equal to or greater than 90, ol/(ol+hbl+px) between 40 and 90, and hbl/(ol+hbl+px) less than 5.

pyroxene-perthite (py'-rox-ene-perth'-ite) Lamellar intergrowths of any of several pyroxenes, as with the feldspars.

pyroxene syenite In Tobi's classification of the *charnockite series* (1971, p.202), a quartz-poor member having more microperthite than plagioclase; a mangerite-syenite.

pyroxenide (py-rox'-e-nide) An informal term, used in the field, for any holocrystalline, medium- to coarse-grained igneous rock composed chiefly of pyroxene; e.g. a pyroxenite (Johannsen, 1931). Obsolete.

pyroxenite (py-rox'-e-nite) (a) In the *IUGS classification,* a plutonic rock with M equal to or greater than 90 and ol/(ol+opx+cpx) less than 40. (b) An ultramafic plutonic rock chiefly composed of pyroxene, with accessory hornblende, biotite, or olivine. Syn: *pyroxenolite.*

pyroxenoid (py-rox'-e-noid) Any mineral chemically analogous to pyroxene but with the SiO_4-tetrahedra connected in chains with a repeat unit of 3, 5, 7, or 9; e.g. wollastonite and rhodonite.

pyroxenolite (py-ro-xen'-o-lite) A term proposed by Lacroix in 1894 as a synonym for *pyroxenite* of English-speaking petrologists, as the French usage of *pyroxenite* was confined to metamorphic rocks. Obsolete.

pyroxferroite (py-rox-fer'-ro-ite) A yellow triclinic mineral of the *pyroxenoid* group found in Apollo 11 lunar samples: $(Fe^{2+},Mn^{2+},Ca)SiO_3$. It is the iron analogue of pyroxmangite.

pyroxmangite (py-rox-man'-gite) A red or brown triclinic mineral of the *pyroxenoid* group: $Mn^{2+}SiO_3$. It is the Mn analogue of pyroxferroite.

pyrrhite (pyr'-rhite) *pyrochlore.*

pyrrhotine (pyr'-rho-tine) *pyrrhotite.*

pyrrhotite (pyr'-rho-tite) A common red-brown to bronze hexagonal to pseudohexagonal mineral: $Fe_{1-x}S$(x=0 to 0.17). Many ordered superstructures are known for discrete values of x. It has a defect structure with a lattice of hexagonal close-packed S ions in which some of the Fe ions are lacking. Some pyrrhotite is magnetic. The mineral is darker and softer than pyrite; it is usually found massive and commonly associated with pentlandite, sometimes containing as much as 5% nickel, in which case it is mined as an ore of nickel. Syn: *pyrrhotine; magnetic pyrites; dipyrite.*

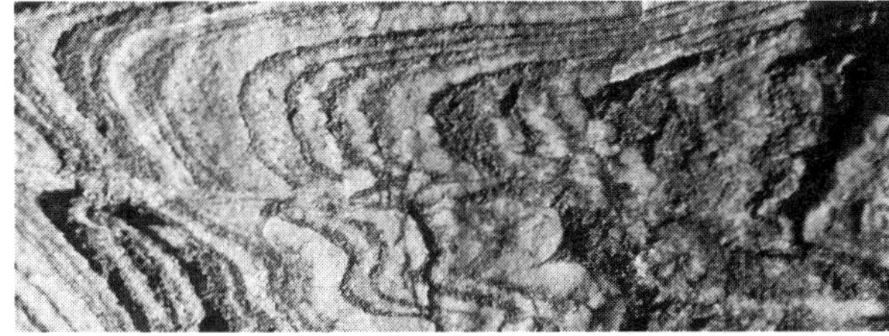

Q [part size] Abbrev. of *quartile [part size].*

Q [seis] A measure of the loss of energy by absorption. Q is 2π times the ratio of the peak energy in a cycle to the energy dissipated in that cycle. The larger the value of Q, the less the absorption. The dimensionless parameter Q is used to describe the intrinsic attenuation of seismic waves. Syn: *quality factor; attenuation coefficient.*

QAL *quartz-alunite epithermal* deposits.

qanat (qa-nat') A term used in Iran for an ancient, gently inclined, underground channel or conduit dug so as to conduct groundwater by gravity from alluvial gravels and the foot of hills to an arid lowland; a horizontal well. Etymol: Arabic. Cf: *foggara; karez.* Syn: *kanat.*

qandilite (qan'-dil-ite) A metallic black cubic mineral of the *spinel* group: $(Mg,Fe^{2+})_2(Ti,Fe^{3+},Al)O_4$.

QF-domains *Microlithons* composed of lenticular quartz and feldspar aggregates.

qilianshanite A vitreous colorless monoclinic mineral: $NaH_4(CO_3)(BO_3)\cdot 2H_2O$.

qingheiite (qing-hei'-ite) A green monoclinic mineral of the *wyllieite* group: $Na_2NaMn_2Mg_2(Al,Fe)_2(PO_4)_6$.

qitianlingite (qi'-ti-an'-ling-ite) A metallic black orthorhombic mineral: $(Fe^{2+},Mn^{2+})_2(Nb,Ta)_2W^{6+}O_{10}$.

Q-joint A partial syn. of *cross joint,* used for a cross joint that is perpendicular to flow structure.

Q-mode factor analysis *Factor analysis* concerned with relationships among observations, as opposed to the relationships between variables. Cf: *R-mode factor analysis.*

quad (a) Shortened form of *quadrangle.* (b) 1,015 BTU, approximately equal to 109 ft³ natural gas, or approximately equal to 300 x 109 kWh.

quadrangle (quad'-ran-gle) (a) A rectangular area bounded by parallels of latitude and meridians of longitude, used as a unit in systematic mapping. The dimensions of a quadrangle are not necessarily the same in both directions, and its size and the scale at which it is mapped are determined by the prime purpose of the map. (b) A sheet representing a quadrangle. Syn: *quad.*

quadrangle map A rectangular map bounded by parallels of latitude and meridians of longitude and generally published in a series with prescribed scale.

quadrant [paleont] (quad'-rant) The space in the interior of a rugose corallite, bounded by the *cardinal septum* and an *alar septum* or by the *counter septum* and an alar septum.

quadrant [surv] (a) An instrument formerly used in surveying and astronomy for measuring angles and altitudes, consisting of a graduated arc of 90 degrees (180 degrees in range) equipped with a sighting device and a movable index or vernier and usually a plumb line or spirit level for fixing the vertical or horizontal direction. It is now largely superseded by the *sextant.* (b) A quarter of a circle, an arc of 90 degrees, or an arc subtending a right angle at the center. Also, the area bounded by a quadrant and two radii.

quadrat (quad'-rat) In ecology, a sample area (usually a square) chosen as the basis for studying a particular assemblage of organisms. Cf: *transect.*

quadrat analysis A procedure for studying the spatial distribution of points on a map or in an area by dividing the area into equal-sized subareas (quadrats) and counting the number of points per quadrat.

quadratic elongation (qua-drat'-ic) A measure of the change in the length of a line, specifically the square of the ratio of its final length to its initial length (Ramsay, 1967, p.52). Cf: *stretch [exp struc geol].*

quadratite A metallic brownish-red tetragonal mineral: $Ag(Cd,Pb)AsS_3$.

quadrature [geophys] A component of a vector that has a phase difference of one-quarter cycle as compared to the primary quantity.

quadridavyne A vitreous colorless hexagonal mineral of the *cancrinite* group: $(Na,K)_6Ca_2(AlSiO_4)_6Cl_4$.

quadriramate element A ramiform conodont element with anterior and posterior processes, and a lateral process on either side of the midplane (TIP, 1981, pt. W, supp.2).

quadruphite A light brown triclinic mineral: $Na_{14}Ca_2Ti_4O_4(Si_2O_7)_2(PO_4)_4F_2$.

quadtree Hierarchical data structures based on the principle or recursive decomposition of space into square tiles, resulting in a balanced tree structure of degree 4 (Krzanowski et al., 1993). Abbrev: Q-tree.

quagmire (quag'-mire) (a) A soft marsh or bog that gives under pressure. (b) *quaking bog.*

quake n. A syn. of *earthquake;* also, a *seismic event* on another planetary body.

quake sheet A well-defined bed resembling a *slump sheet* but produced by an earthquake and resulting in load casting without horizontal slip (Kuenen, 1958, p.20).

quaking bog (quak'-ing) A peat bog that is either floating or is growing over water-saturated ground, so that it shakes or trembles when walked on. *Quagmire* is sometimes used as a synonym. Syn: *floating bog.*

quality assurance Programs and sets of procedures including but not limited to quality control that are used to assure product quality or data reliability.

quality factor A measure of 2π times the ratio of the peak energy to the energy dissipated in a cycle. Q is related to other measures of absorption (Sheriff and Geldart, 1995, p.59-60). Symbol: Q.

quantile *fractile.*

quantitative geomorphology The application of mathematical or statistical numerical analysis to the study of landform geometry, geomorphic processes, and interrelationships in Earth surface systems.

quantum evolution A now discredited mode of *macro-* and/or *megaevolution* whereby new higher taxa evolve when an ancestral population goes through an inadaptive and preadaptive phase and comes to occupy a new adaptive peak.

quaquaversal (qua-qua-ver'-sal) adj. An obsolete term describing strata and structures that dip outward in all directions away from a central point. Ant: *centroclinal.* n. A geologic structure, such as a dome or ridge, having a quaquaversal dip. Cf: *pericline [struc geol].* Ant: *centrocline.*

quaquaversal fold *dome [struc geol].* Rarely used.

quar ice A term used in Labrador for ice formed during the spring by meltwater running off the land onto an icefoot or fast ice, where it refreezes (ADTIC, 1955, p.64).

quarry (quar'-ry) Open workings, usually for the extraction of stone.

quarrying [geomorph] (quar'-ry-ing) (a) *plucking [glac geol].* (b) *hydraulic plucking.*

quarrying [mining] The extraction of building stone or other valuable nonmetallic constituent from a surficial mine, or quarry. See also: *opencut mining.*

quarrying [streams] The erosion of chunks of cohesive material such as jointed bedrock from the bed of a stream as a result of *lift forces* within the flow. Cf: *plucking [streams].*

quarry sap *quarry water.*

quarry sap [sed] The trace liquids found in some *freestones* which, on freezing, allow the rock to be split along its bedding planes (Winchester, 2002, p.308).

quarry water Subsurface water retained in freshly quarried rock. Syn: *quarry sap.*

quarter post (quar'-ter) A post marking a corner of a *quarter section* of the U.S. Public Land Survey system. It is located midway between the controlling section corners, or 40 chains (0.5 mi) from the controlling section corner, depending on location within the township.

quarter-quarter section A sixteenth of a normal section of the U.S. Public Land Survey system, representing a piece of land normally a quarter mile square and containing 40 acres as nearly as possible; a quarter section divided into four parts. It is usually identified as the northeast quarter, northwest quarter, southeast quarter, or southwest quarter of a particular quarter section and section.

quarter section A fourth of a normal section of the U.S. Public Land Survey system, representing a piece of land normally a half mile square and containing 160 acres nearly as possible. It is usually identified as the northeast quarter, northwest quarter, southeast quarter, or southwest quarter of a particular section.

quarter-wave plate *mica plate.*
quartile [math] A *fractile* division of a cumulative distribution into four equal parts. The first quartile includes the smallest 25% of the observations, the second quartile (up to the median) includes the smallest 50% of the observations, and the third quartile includes the smallest 75% of the observations.
quartile [part size] (quar'-tile) (a) Any one of three particle-size values (diameters) dividing a frequency distribution into four classes, obtained graphically from a cumulative curve by following the 25, 50, or 75 percent line to its intersection with the curve and reading the value on the diameter scale directly below the intersection; e.g. the first quartile (the 25 percentile) is the size such that 25% of the particles are larger than itself and 75% smaller, this size being larger than the third quartile (the 75 percentile) which is the size such that 75% of the particles are larger than itself and 25% smaller. Abbrev: Q.
quartz (a) Crystalline silica, an important rock-forming mineral: SiO_2. It is, next to feldspar, the commonest mineral, occurring either in transparent hexagonal crystals (colorless, or colored by impurities) or in crystalline or cryptocrystalline masses. Quartz is the commonest gangue mineral of ore deposits, forms the major proportion of most sands, and has a widespread distribution in igneous (esp. granitic), metamorphic, and sedimentary rocks. It has a vitreous to greasy luster, a conchoidal fracture, an absence of cleavage, and a hardness of 7 on the Mohs scale (scratches glass easily, but cannot be scratched by a knife); it is composed exclusively of silicon-oxygen tetrahedra with all oxygens joined together in a three-dimensional network. It is polymorphous with cristobalite, tridymite, stishovite, coesite and keatite. Symbol: Q. Abbrev: qtz; qz. Etymol: German provincial "Quarz". Cf: *tridymite; cristobalite; coesite; stishovite.* (b) A general term for a variety of noncrystalline or cryptocrystalline minerals having the same chemical composition as that of quartz, such as chalcedony, agate, and opal.
quartz-adularia epithermal *adularia-sericite type* deposit.
quartz alkali feldspar syenite In the *IUGS classification,* a plutonic rock in which Q/(Q+A+P) is between 5 and 20, and P/(A+P) is less than 10.
quartz alkali feldspar trachyte In the *IUGS classification,* a volcanic rock in which Q/(Q+A+A) is between 5 and 20, and P/(A+P) is less than 10. Not defined in the TAS diagram.
quartz-alunite epithermal *acid-sulfate type* deposit.
quartz anorthosite In the *IUGS classification,* a plutonic rock with Q between 5 and 20, P/(A+P) greater than 90, and color index less than 10.
quartzarenite (quartz-ar'-e-nite) A sandstone that is composed primarily of quartz; specif. a sandstone containing more than 95% quartz framework grains (excluding detrital chert grains) and having little clay matrix and any sorting, rounding, texture, or hardness (Folk, 1968). McBride (1963, p.667), who included chert and quartzite in the 95% quartz content, coined the term as a contracted form of "quartz arenite", a term used by Williams et al. (1954, p.294, 316) for a mature sandstone containing more than 80% quartz, chert, and quartzite and less than 10% each of argillaceous matrix, feldspars, and unstable fine-grained rock fragments. The term is essentially equivalent to *orthoquartzite.*
quartz-banded ore A term used in Scandinavia for a metamorphosed *iron formation.*
quartz-bearing diorite A syn. of *quartz diorite,* although Streckeisen (1967, p.157) suggests that the term be restricted to diorite in which quartz constitutes 5 to 20 percent of the light-colored components.
quartz-bearing monzonite As recommended by Streckeisen (1967, p.157), any *monzonite* in which quartz constitutes from 5 to 20 percent of the light-colored components. In the most recent Russian classification, it is a syn. of *quartz monzonite.*
quartz crystal Quartz that is transparent or nearly so, is usually colorless, and has a low refractive index resulting in low brilliancy. It is used for lenses, wedges, and prisms in optical instruments and for frequency control in electronics, or it is fashioned into beads or other ornamental objects. It may or may not be in distinct crystals. Syn: *rock crystal; pebble [gem].*
quartz diorite (a) In the *IUGS classification,* a plutonic rock with Q between 5 and 20, P/(A+P) greater than 90, and plagioclase more sodic than An_{50}. (b) A group of plutonic rocks having the composition of *diorite* but with an appreciable amount of quartz, i.e. between 5 and 20 percent of the light-colored constituents, according to Streckeisen (1967, p.157); also, any rock in that group; the approximate intrusive equivalent of *dacite.* Quartz diorite grades into *granodiorite* as the alkali feldspar content increases. Syn: *quartz-bearing diorite.*
quartz dolerite A *diabase* or *dolerite* of silica-oversaturated composition, generally containing interstitial *granophyre.*
quartz felsite An obsolete syn. of *quartz porphyry.*
quartz-free wacke A wacke with more than 90% unstable materials (Fischer, 1934).
quartz gabbro In the *IUGS classification,* a plutonic rock with Q between 5 and 20, P/(A+P) greater than 90, and plagioclase more calcic than An_{50}.
quartz graywacke A term used by Williams et al. (1954, p.294) for a graywacke containing abundant grains of quartz and chert and less than 10% each of feldspar and rock fragments. See also: *quartzose graywacke.*
quartzic (quartz'-ic) *quartziferous.*
quartziferous (quartz-if'-er-ous) Quartz-bearing. The term is applied to a rock (such as a limestone or syenite) that contains a minor proportion of quartz, to distinguish it from a variety (usually commoner) of the same rock that contains no quartz. Cf: *quartzose.* Syn: *quartzic.*
quartz index [petrology] A derived quantity (qz) in the Niggli system of rock classification, which may be either positive or negative, and is an indicator of a rock's degree of silica saturation. Obsolete.
quartz index [sed] A term used by Dapples et al. (1953, p.294, 304) to indicate the mineralogic maturity of a sandstone by measuring the percentage of detrital quartz. It is expressed as the ratio of quartz and chert to the combined percentage of sodic and potassic feldspar, rock fragments, and clay matrix. The index is used as a basis for evaluating the degree of weathering of the source rock and the degree to which the sediment has been transported. Values for sandstones range between 3 and 19.
quartzine (quartz'-ine) Chalcedony characterized by fibers having a positive crystallographic elongation (parallel to the *c*-axis). Also spelled: quartzin.
quartzite [meta] (quartz'-ite) A granoblastic metamorphic rock consisting mainly of quartz and formed by recrystallization of sandstone or chert by either regional or thermal metamorphism; *metaquartzite.* Cf: *orthoquartzite.*
quartzite [sed] A very hard but unmetamorphosed sandstone, consisting chiefly of quartz grains that have been so completely and solidly cemented with secondary silica that the rock breaks across or through the grains rather than around them; an *orthoquartzite.* The cement may grow in optical and crystallographic continuity around each quartz grain, thereby tightly interlocking the grains as the original pore spaces are filled. Skolnick (1965) believed that most sedimentary quartzites are compacted sandstones developed by pressure solution of quartz grains. The term is also used for a metamorphic rock produced by recrystallization of a sandstone.
quartzitic arkose (quartz-it'-ic) *arkosite.*
quartzitic sandstone A term used by Krynine (1940, p.51) for a sandstone that contains 100% quartz grains cemented with silica. The term is essentially equivalent to *orthoquartzite.* Cf: *quartzose sandstone.*
quartz latite In the *IUGS classification*, a volcanic rock in which Q/(Q+A+P) is between 5 and 20, and P/(A+P) is between 35 and 65. Not defined in the TAS diagram.
quartz mengwacke A wacke with 10-33% unstable materials (Fischer, 1934).
quartz mine A miner's term for a mine in which the valuable constituent, e.g. gold, is found in siliceous veins rather than in placers. It is so named because quartz is the chief accessory mineral.
quartz monzodiorite In the *IUGS classification,* a plutonic rock with Q between 5 and 20, P/(A+P) between 65 and 90, and plagioclase more sodic than An_{50}.
quartz monzogabbro In the *IUGS classification,* a plutonic rock with Q between 5 and 20, P/(A+P) between 65 and 90, and plagioclase more calcic than An_{50}.
quartz monzonite (a) In the *IUGS classification,* a plutonic rock with Q between 5 and 20 and P/(A+P) between 35 and 65. (b) In former U.S. usage, granitic rock in which quartz comprises 10-50% of the felsic constituents, and in which the alkali feldspar/total feldspar ratio is between 35% and 65%; the approximate intrusive equivalent of *rhyodacite.* With an increase in plagioclase and femic minerals, it grades into *granodiorite,* and with more alkali feldspar, into a granite. As introduced by Brögger in 1895, the term designated monzonite containing only small amounts of quartz, and it is still used in this sense by Russian geologists. According to Tröger (1935, p.47), Lindgren, in 1900, changed the definition to apply to andesine-bearing granites. Now the term is applied by most British petrologists to granites with quartz constituting 20-60% of the light-colored components and with a plagioclase/total feldspar ratio of 35/65. Streckeisen (1967, p.167) recommends replacing the term with *quartz-bearing monzonite.* Syn: *adamellite.*
quartz norite In Tobi's classification of the *charnockite series* (1971, p.202), a member that contains plagioclase but no potassium feldspar.
quartzolite (quartz'-o-lite) In the *IUGS classification,* a plutonic rock with Q greater than or equal to 90: preferred over *silexite [ign].*
quartzose (quartz'-ose) Containing quartz as a principal constituent; esp. applied to sediments and sedimentary rocks (such as sands and sandstones) consisting chiefly of quartz. Cf: *quartziferous.* Syn: quartzous; quartzy.
quartzose arkose A term used by Hubert (1960, p.176-177) for a sandstone containing 50-85% quartz, chert, and metamorphic quartzite, 15-25% feldspars and feldspathic crystalline-rock fragments, and 0-25% micas and micaceous metamorphic-rock fragments. Cf: *quartzose graywacke.*
quartzose chert A vitreous, sparkly, shiny chert, which under high magnification shows a heterogeneous mixture of pyramids, prisms, and faces of quartz (Grohskopf and McCracken, 1949, pl.3), but also including chert in which the secondary quartz is largely anhedral. Also known as "drusy chert".
quartzose graywacke (a) A term used by Hubert (1960, p.176-177) for a sandstone containing 50-85% quartz, chert, and metamorphic quartzite, 15-25% micas and micaceous metamorphic-rock fragments, and 0-25% feldspars and feldspathic crystalline-rock fragments. Cf: *quartzose arkose.* (b) A term used by Krynine (1951) for a graywacke that has lost its micaceous con-

stituents through abrasion and thus tends to approach an orthoquartzite. It is equivalent to *subgraywacke* of Folk (1954). See also: *quartz graywacke.*

quartzose sandstone (a) A well-sorted sandstone that contains (if pure) more than 95% clear quartz grains and 5% or less of matrix and cement (Krumbein and Sloss, 1963, p.169-170). (b) A sandstone that contains at least 95% quartz, but is not cemented with silica (Krynine, 1940, p.51). Cf: *quartzitic sandstone.* (c) A sandstone that contains 99% quartz and quartz cement (Shrock, 1948a). (d) A sandstone that contains 90% quartz grains (Dunbar and Rodgers, 1957). See also: *orthoquartzite.* Syn: *quartz sandstone.*

quartzose shale A shale composed dominantly of rounded quartz grains of silt size. It is commonly associated with highly mature sandstones (orthoquartzites).

quartzose subgraywacke *protoquartzite.*

quartz-pebble conglomerate A term used by Krumbein and Sloss (1963, p.163) for *orthoquartzitic conglomerate.*

quartz plate *quartz wedge.*

quartz porphyry A porphyritic extrusive or hypabyssal rock containing phenocrysts of quartz and alkali feldspar in a microcrystalline or cryptocrystalline groundmass; a *rhyolite.* European petrologists called pre-Tertiary and Tertiary extrusive equivalents of granite "quartz porphyry", and post-Tertiary equivalents "rhyolite" (Streckeisen, 1967, p.189). Obsolete syn: *quartz felsite.* Cf: *granite porphyry.*

quartz-rich granitoid In the *IUGS classification,* a plutonic rock with Q between 60 and 90. Cf: *tarantulite.*

quartz sandstone *quartzose sandstone.*

quartz schist A schist whose foliation is due mainly to streaks and lenticles of nongranular quartz. Mica is present but in lesser quantities than in mica schist. A syn. of "schistose quartzite" of some petrologists.

quartz syenite In the *IUGS classification,* a plutonic rock with Q between 5 and 20 and P/(A+P) between 10 and 35.

quartz topaz A frequently used but incorrect syn. of *citrine.* Cf: *topaz quartz.*

quartz trachyte (a) In the *IUGS classification,* a volcanic rock in which Q/(Q+A+P) is between 5 and 20, and P/(A+P) is between 10 and 35. Not defined in the TAS diagram. (b) A fine-grained igneous rock consisting mostly of alkali feldspar, with normative quartz between 5 and 20 percent; the volcanic equivalent of *quartz syenite* (Williams et al., 1954, p.100).

quartz wacke A gray to buff, moderately well-sorted, commonly fine-grained sandstone containing up to 90% quartz and chert, and with more than 10% argillaceous matrix (largely sericite and chlorite), less than 10% feldspar, and less than 10% rock fragments (bits of coal, shale, etc.) (Williams et al., 1954, p.292-293). Krumbein and Sloss (1963, p.172-173) give a lower limit of 15-20% matrix, and regard the rock as a "washed" graywacke. The rock is equivalent to *subgraywacke* as originally defined by Pettijohn (1949, p.227, 256). Term introduced by Fischer (1934) for a wacke with less than 10% unstable materials. Examples include many coal-measures sandstones of Pennsylvanian age (such as those of the Atokan Series). Also spelled: quartzwacke.

quartz wedge In an optical system such as a polarizing microscope, an elongate wedge of clear quartz that is used in analysis of a mineral's fast and slow vibration-plane traces, optical sign, and interference colors. Syn: *quartz plate.*

quartzy (quartz'-y) *quartzose.*

quasicratonic (qua'-si-cra-ton'-ic) Said of semiconsolidated regions in which paratectonic deformation (*germanotype tectonics*) tends to dominate (Stille, 1940). Obsolete. Syn: *semicratonic.*

quasi-equilibrium (qua'-si-e'-qui-lib'-ri-um) The state of balance or grade in a stream's cross section, whereby "conditions of approximate equilibrium tend to be established in a reach of the stream as soon as a more or less smooth longitudinal profile has been established in that reach even though downcutting may continue" (Leopold and Maddock, 1953, p.51).

quasi-geoid (qua-si-ge'-oid) A nonequipotential surface in the vicinity of the geoid that is defined as the locus of points whose distances below the terrain are the normal heights.

quasi-instantaneous (qua'-si-in'-stan-ta'-ne-ous) Geologically instantaneous; occurring within an interval of geologic time too small to be subdivided (Termier and Termier, 1956).

Quaternary (Qua-ter'-na-ry) The upper system of the *Cenozoic* Erathem of the Standard Global Chronostratigraphic Scale, above the *Tertiary* System. Also the time during which these rocks were formed, the Quaternary Period, covering the time span between 1.75 Ma and the present.

quaternary system A chemical system having four components.

quay (Pron: key) A wharf of solid construction, built roughly parallel to the shoreline and accommodating vessels on one side only. Not to be confused with *key* or *cay.* Cf: *pier [coast].*

quebrada (que-bra'-da) A term used in Latin America for a ravine or gorge, esp. one that is usually dry but is filled by a torrent during a rain; a barranco. Also, a stream or brook. Etymol: Spanish.

queitite (queit'-ite) A pale yellow or colorless monoclinic mineral: $Pb_4Zn_2(SiO_4)(Si_2O_7)(SO_4)$.

quenching (quench'-ing) In experimental petrology, the very rapid cooling of a heated or pressurized charge in order to preserve the physical (textural) or chemical characteristics of the high-temperature equilibrium state that would be changed by slow cooling. For example, quenching of a melt crystallization experiment would typically result in crystals suspended in a homogeneous glass of the composition of the high-temperature melt. Also refers to rapidly cooled magmas in volcanic eruptions.

quenselite (quen'-sel-ite) An adamantine black monoclinic mineral: $PbMn^{3+}O_2(OH)$.

quenstedtite (quen'-stedt-ite) A pale violet or reddish-violet triclinic mineral: $Fe^{3+}{}_2(SO_4)_3{\cdot}10\text{-}11H_2O$.

Querwellen wave (Quer-wel'-len) *Love wave.*

quetzalcoatlite (quet'-zal-co-at'-lite) A blue trigonal mineral: $Zn_6Cu_3(Te^{6+}O_6)_2(OH)_6{\cdot}0.24AgCl{\cdot}0.30PbCl_2$.

quick [mineral] A local term used in the western U.S. for *quicksilver.*

quick [ore dep] Said of an economically valuable or productive mineral deposit, in contrast to a *dead* ground or area. An ore is said to be quickening as its mineral content increases. Syn: *alive.*

quick [sed] Said of a sediment that, when mixed with water, becomes extremely soft and incoherent, and is capable of flowing easily under load or by force of gravity; e.g. "quick clay" of glacial or marine origin, which, if disturbed, loses practically all its shear strength and flows plastically.

quick clay Clay-sized (<0.004 μm) minerals, commonly composed of rock flour containing quartz, feldspar, and less commonly clay minerals, that lose all or nearly all shear strength after being disturbed; i.e., the material is thixotropic. Such material does not gain in strength on remolding. Cf: *clay, clay minerals.*

quicklime Calcium oxide: CaO. Produced by calcination of limestone in kilns. Syn: *lime.*

quicksand (quick'-sand) (a) A mass or bed of fine sand, as at the mouth of a river or along a seacoast, that consists of smooth rounded grains with little tendency to mutual adherence and that is usually thoroughly saturated with water flowing upward through the voids, forming a soft, shifting, semiliquid, highly mobile mass that yields easily to pressure and hence is apt to engulf persons and animals coming upon it. Capable of injection into fissures. (b) An area marked by the presence of one or more such beds. (c) The sand found in a bed of quicksand. Cf: *liquefaction [sed].*

quicksilver (quick'-sil-ver) A term applied to *mercury* where it occurs as a native mineral or has been mined but not yet used (as in "flasks of quicksilver").

quickstone (quick'-stone) A consolidated rock that had flowed under the influence of gravity before lithification; a quick sediment that has become lithified.

quickwater (quick'-wa-ter) The part of a stream characterized by a strong current.

quiet reach *still water.*

quiet zone A subsurface zone that does not give rise to effects; e.g., a quiet zone is not associated with magnetic or gravity anomalies.

quilted surface (quilt'-ed) A land surface characterized by broad, rounded, uniformly convex hills separating valleys that are comparatively narrow "like the seams by which a quilt is furrowed" (Davis, 1918, p.124).

quinary system (qui'-na-ry) *five-component system.*

quinqueform antapical plate (quin'-que-form) A pentagonal *antapical plate* on a dinoflagellate theca or cyst that is in contact with three postcingular plates and two additional plates near the *sulcus,* as in the modern genus *Triadinium* (Williams et al., 2000, p.194).

quinqueloculine (quin-que-loc'-u-line) Said of a foraminiferal test having five externally visible chambers, as a result of growth in varying planes about an elongate axis; specif. pertaining to the foraminifer *Quinqueloculina.*

quinqueradiate (quin-que-ra'-di-ate) Said of radial symmetry of certain echinoderms characterized by five rays extending from the mouth. Syn: *pentamerous.*

quintinite-2H An orange-brown or pale brown hexagonal mineral: $Mg_4Al_2(OH)_{12}CO_3{\cdot}3H_2O$. It is dimorphous with quintinite-3T. Syn: *manasseite.*

quintinite-3T A yellow trigonal mineral, dimorphous with quintinite-2H: $Mg_4Al_2(OH)_{12}CO_3{\cdot}3H_2O$.

quisqueite (quis'-que-ite) A highly sulfurous asphaltum; a black, brittle, lustrous substance mostly composed of sulfur (37%) and carbon (43%) and accompanying the vanadium ores of Peru.

Q wave *Love wave.*

R

ra A Norwegian term, used esp. in southern Norway, for a morainal ridge covered with a surface layer of large stones (Stamp, 1961, p.383); most of them are in or near the sea.
raadeite A pearly colorless monoclinic mineral: $Mg_7(PO_4)_2(OH)_8$. It is the Mg analogue of allactite.
rabbittite (rab'-bitt-ite) A pale greenish-yellow monoclinic secondary mineral: $Ca_3Mg_3(UO_2)_2(CO_3)_6(OH)_4 \cdot 18H_2O$.
rabdolith (rab'-do-lith) *rhabdolith.*
rabejacite A vitreous yellow orthorhombic mineral: $Ca(UO_2)_4(SO_4)_2(OH)_6 \cdot 6H_2O$.
race [paleont] A group of organisms with similar characteristics but not sufficiently distinctive to be classified as a species or *subspecies.*
race [water] (a) A strong or rapid current of water flowing through a narrow channel or river, e.g. a *tide race.* (b) The constricted channel or river in which such a current flows. It may occur by the meeting of two tides, as near a headland separating two bays, or it may be artificial and used for an industrial purpose, such as conveying water to or away from the waterwheel of a mill. Syn: *water race.*
racemization (ra'-ce-mi-za'-tion) A process in which an optically active stereoisomer is converted to a mixture of two isomers which possesses no optical activity (Cram and Hammond, 1959, p.131). Cf: *racemization age method.*
racemization age method A method of geochronology based on the chemical *racemization* (or *epimerization*) of amino acids. Amino acids of living organisms consist virtually entirely of the L-enantiomer (or diastereomer). After death the L-enantiomer (or diastereomer) for each amino acid is slowly racemized and eventually forms an equilibrium mixture consisting of equal amounts of the D- and L-enantiomers (or diastereomers). The increase in D/L ratio can be used to obtain a measure of the time that has elapsed since the organism died. The method is sensitive to temperature, pH, and other environmental factors (Schroeder and Bada, 1976). See also: *racemization; epimerization.* Syn: *amino-acid racemization age method.*
rachis (ra'-chis) The axis of a large compound leaf, or of a flowering spike or raceme (Swartz, 1971, p.395).
racial senescence (ra'-cial) A now discredited view that as a species or higher taxon becomes older its overall vigor decreases due to increasing specialization. Syn: *phylogerontism.*
rad A dose of ionizing radiation of 100 ergs of energy per gram of absorber. One rad is equivalent to 0.01 grays (Gy).
radar (ra'-dar) (a) An electronic detection device or active system for locating or tracking a distant object by measuring elapsed circuit time of travel of ultrahigh-frequency radio waves of known propagation velocity emitted from a transmitter and reflected back by the object to or near the point of transmission in such a way that range, bearing, height, and other characteristics of the object may be determined. Electromagnetic radiation, in the spectral range from 1 mm to 0.8 m in wavelength, is beamed out to illuminate target areas, and the intensity and time delay of the returning energy are measured. Radar operation is unaffected by darkness, but moisture in the form of fog, snow, rain, or heavy clouds may cause varying degrees of attenuation or reflection of the radio energy. The term is usually prefixed by a code letter indicating the frequency band for certain wavelength ranges; e.g. K-band (about 2 cm), X-band (about 4 cm), and P-band (about one meter). (b) A name applied to the method or technique of locating or tracking objects by means of radar, such as the observation and analysis of minute radio signals reflected from the objects and displayed in a radar system. Etymol: *ra*dio *d*etecting *a*nd *r*anging. See also: *bistatic radar; ground penetrating radar.*
radar imagery Imagery provided by scanning devices using microwave radiation.
radar shadow A dark area of no return on a radar image that extends in the far-range direction from an object on the terrain that intercepts the radar beam.
radhakrishnaite (rad-ha-krish'-na-ite) A metallic gray tetragonal mineral: $PbTe_3(Cl,S)_2$.

radial [paleont]
radial [paleont] (ra'-di-al) adj. (a) Belonging to or in the direction of a ray of an echinoderm. (b) A syn. of *ambulacral;* e.g. referring to the position of a line extending from the centrally placed mouth to the aboral end of any *ambulacrum* of a blastoid. (c) Directed outward from the center of the *umbilicus* of an ammonoid and at right angles to the axis of coiling and growth; transverse. (d) Said of the microstructure of hyaline calcareous foraminiferal tests consisting of calcite or aragonite crystals with *c*-axes perpendicular to the surface and exhibiting between crossed nicols a black cross with concentric rings of color mimicking a negative uniaxial interference figure (TIP, 1964, pt. C, p.63). (e) Pertaining to the *radius* (bone of the forearm in tetrapods). n. A *radial plate* together with all structures borne by it.
radial [palyn] Pertaining to features beyond the ends of the laesura branches in a trilete spore. Cf: *interradial [palyn].*
radial [photo] A line or direction from the center (principal point, isocenter, nadir point, or substitute center) to any point on a photograph.
radial array *azimuthal survey.*
radial beam An internal rod usually connecting concentric lattice shells of spumellarian radiolarians.
radial canal (a) A tube extending radially from the *ring canal* beneath or along an *ambulacrum* of an echinoderm. It is closed at its outer end and bears rows of podia. The ambulacral system of echinoderms contains five radial canals. (b) One of the numerous minute canals, lined with choanocytes, radiating from the *spongocoel* in some sponges and ending just below the surface of the sponge.
radial canal [palyn] A distinctive pore that partially or completely penetrates the *vesicle* wall of a tasmanitid (see: *tasmanites*). See also: *phycoma; prasinophyte.*
radial coefficient A ratio expressing the number of radial elements divided by the diameter of the cup in archaeocyathids (TIP, 1972, pt. E, p.13). Abbrev: RK.
radial drainage pattern A drainage pattern in which consequent streams radiate or diverge outward, like the spokes of a wheel, from a high central area; it is best developed on the slopes of a young, unbreached domal structure or of a volcanic cone. Cf: *centripetal drainage pattern.* Syn: *centrifugal drainage pattern.*
radiale (ra'-di-ale) (a) A *radial plate* of a crinoid. Pl: radialia. (b) A proximal carpal bone on the radial (thumb) side of the wrist of tetrapods.
radial facet A smooth or sculptured distal facet of a radial plate of a crinoid that supports the arm. It may have only ligamentary connective tissue or muscular and ligamentary connective tissue connecting radial facet to first primibrachial. It is lacking in radial plates that bear no arms.
radial fault One of a group of faults that radiate from a central point. Typically used in reference to faulting near a diapir, impact crater, or volcanic center.
radially elongate simple prismatic structure A *simple prismatic structure* in which the prisms are distinctly elongated in the radial direction. Unlike lathic simple prismatic structure, the prisms are generally taller (in the vertical direction) than they are wide (in the horizontal plane) (Carter, 1980a). Abbrev: RESP.
radial plate An echinoderm ray plate that, in crinoids, lies between the basal and brachial plates and bears an arm; in blastoids, rhombiferan cystoids, and parablastoids, lies between the basal and deltoid plates and bears part or all of an ambulacrum; in asteroids, is a prominent aboral plate at the base of each arm.
radial pore canal One of a series of tubules or pore canals in an ostracode leading from the line of concrescence through the area of the duplicature to the free margin of the valve, usually housing sensory setae protecting the gape of the open carapace. Cf: *normal pore canal.*
radial ridge Ridge on upper surface of platformed pectiniform *conodont element* trending diagonally from longitudinal axis of element toward margin of platform; contrasts with longitudinal and transverse ridges (TIP, 1981, pt. W, supp.2).
radial shield In many ophiuroids, one of a pair of relatively large ossicles adjacent to the base of an arm on the aboral surface of the disc.

radial spine A tangential rod or needle in the skeleton of an acantharian or phaeodarian radiolarian.
radial suture A suture in a *heterococcolith* corresponding to a radius in a circular coccolith, or to a straight line drawn through the nearest focus or connecting the foci of an elliptic coccolith.
radial symmetry The condition, property, or state of having similar parts of an organism regularly arranged about a common central axis (as in a starfish); e.g. a type of symmetry exemplified in a flower that can be separated into two approximate halves by a longitudinal cut in any plane passing through the center of the flower. Cf: *bilateral symmetry.*
radial triangulation A triangulation procedure in which direction lines from the centers of overlapping vertical or oblique photographs to control points imaged on the photographs are measured and used for horizontal-control extension by successive intersection and resection.
radial tube A centrifugal cylinder in an acantharian radiolarian.
radial zone The chamber portion of a foraminiferal test with essentially radial elements, situated between the marginal zone and the central complex (as in Orbitolinidae).
radianal (ra-di-a'-nal) In a cladid, disparid, or flexible crinoid, the proximal C-ray plate or the proximal CD-interray plate; in either case, the proximal plate of the posterior interray.
radiance (ra'-di-ance) Radiant flux per unit solid angle per unit area.
radiant (ra'-di-ant) n. An organism or group of organisms, such as a species, that has arrived at its present geographic location as the result of dispersal from its main place of origin. Cf: *radiation [evol].* Obsolete.
radiante (ra-di-ante') That point in the skeletal end of an anthaspidellid sponge from which the trabs radiate (Rigby and Bayer, 1971, p.609).
radiant emittance Radiant flux emitted per unit area of a source.
radiant energy Energy transferred by electromagnetic waves, measured in joules or ergs.
radiant flux The rate of transfer of radiant energy, measured in watts or ergs per second. See also: *spectral radiant flux.* Syn: *radiant power.*
radiant intensity Radiant flux per unit solid angle.
radiant power *radiant flux.*
radiant-power peak (ra'-di-ant-pow'-er) The wavelength at which the maximum electromagnetic energy is radiated at a particular temperature.
radiate aperture (ra'-di-ate) A foraminiferal-test opening consisting of numerous diverging slits (as in the superfamily Nodosariacea).
radiate mud crack A mud crack that displays an incomplete radiate pattern and that lacks normal polygonal development (Kindle, 1926, p.73).
radiation [evol] (ra-di-a'-tion) (a) The dispersal of a group of organisms into different environments often accompanied by divergent change in morphology. (b) *adaptive radiation.* Cf: *convergence [evol]; radiant.*
radiation [surv] A method of surveying in which points are located by a knowledge of their distances and directions from a central point.
radiation damage The damage done to a crystal lattice (or glass) by passage of fission particles or alpha-particles from the nuclear decay of a radioactive element residing in the lattice. The damage paths (*alpha-particle recoil tracks* or *fission tracks*) can be enlarged to microscopic size by suitable etching techniques and used to determine an age for the material.
radiation detector *detector.*
radiation fog A type of *fog* that occurs at night when the air cools by radiation to a temperature below the dewpoint. Syn: *ground fog.* Cf: *advection fog.*
radiation log *radioactivity log.*
radiaxial (ra-di-ax'-i-al) Radially axial; e.g. "radiaxial calcite" in sedimentary rocks, occurring as cavity linings composed of subparallel individual crystals elongated normal to the cavity wall (Fischer, 1964, p.148).
radiaxial fibrous calcite A type of calcite cement forming a crust of crystals elongated perpendicular to the substrate; crystals have curved cleavage and twinning planes (concave outward), and optic axes of the crystals converge outward (Bathurst, 1959).
radiciform (ra-dic'-i-form) Said of a rootlike epithecal process (outgrowth) of a corallite wall, serving for fixation.
radicle [bot] (rad'-i-cle) The root-primordium of an embryo, which grows into the primary root of a seedling (Fuller and Tippo, 1954, p.969).
radicle [paleont] (a) A bryozoan rootlike structure composed of one or more kenozooids. Syn: *rhizoid [paleont].* (b) An individual rootlike branch of a crinoid *radix.*
radii (ra'-di-i) Plural of *radius.*
radioactivation analysis (ra'-di-o-ac'-ti-va'-tion) *activation analysis.*
radioactive (ra'-di-o-ac'-tive) Pertaining to or exhibiting *radioactivity.*
radioactive age determination *isotopic dating.*
radioactive chain *radioactive series.*
radioactive clock A *geochronometer* consisting of a radioactive isotope, e.g., carbon-14, rubidium-87, or potassium-40, whose decay constant is known and is low enough to be calibrated to time units, usually years. Radioactive clocks are the basis of isotopic age determinations and the specific element being used is sometimes designated as a clock, e.g., carbon clock. Syn: *atomic clock; nuclear clock.*
radioactive dating *isotopic dating.*
radioactive decay Spontaneous emission of particles (alpha or beta) and gamma rays from the nucleus of an unstable nuclide. The resulting product nucleus may be stable or unstable, in which case it decays until at last a stable nuclide is formed. Radioactive decay involves the emission of alpha particles, beta particles, and other energetic particles and usually is followed by emission of gamma rays and by atomic de-excitation phenomena. It always results in the generation of heat. Cf: *radioactivity.* Syn: *disintegration [radioactivity]; radioactive disintegration.*
radioactive disintegration *radioactive decay.*
radioactive equilibrium A relationship between a long-lived *parent* and one or more short-lived radioactive daughters in which the ratio of daughter activity to parent activity is equal to unity.
radioactive heat Heat produced within a medium as a result of the decay of its constituent radioactive elements.
radioactive series A series or succession of nuclides, each of which becomes the next by radioactive decay, until a stable nuclide is formed. There are three important natural radioactive series, the *actinium series, thorium series,* and *uranium series.* See also: *parent; daughter; end product.* Syn: *radioactive chain.*
radioactive spring A spring whose water has a high and readily detectable radioactivity.
radioactive waste "Equipment and materials from nuclear operations that are radioactive and for which there is no further economic use. Waste is generally classified as high-level (having concentrations of radioactivity of hundreds to thousands of curies per gallon or cubic foot or cubic meter), low-level (averaging in the range of 1 microcurie per cubic foot or so), or intermediate-level (between these extremes). Waste is classified by its level of radioactivity at the time of burial. As radioactive decay proceeds, the level diminishes according to known and predictable mathematical expressions, depending on the radioactive nuclides present" (National Research Council Panel on Land Burial, 1976, p.7).
radioactivity (ra'-di-o-ac-tiv'-i-ty) (a) The emission of energetic particles and/or radiation from the nucleus during *radioactive decay.* (b) A particular radiation component from a radioactive source, such as gamma radioactivity. Adj: *radioactive.*
radioactivity anomaly A deviation from expected results found when making a radioactivity survey. Such anomalies are important in mineral exploration.
radioactivity log The generic name for well logs that measure the reactions of atomic nuclei involving the behavior of gamma rays and/or neutrons as a function of drillhole depth. Except for the natural *gamma-ray log* and the *spectral gamma-ray log,* they record the response of rocks very near the well bore to bombardment by gamma rays or neutrons from a source in the logging *sonde.* Most can be obtained in cased, empty, or fluid-filled well bores. See also: *density log; neutron log; neutron-activation log; epithermal-neutron log; pulsed-neutron-capture log.* Syn: *radiation log; nuclear log.*
radioassay (ra'-di-o-as'-say) An assay procedure involving the measurement of alpha, beta, or gamma radiation intensity of a sample.
radiocarbon (ra'-di-o-car'-bon) Radioactive carbon, esp. *carbon-14,* but also carbon-10 and carbon-11.
radiocarbon age *carbon-14 age.*
radiocarbon clock A popular syn. of *carbon-14,* used in radiometric dating of rocks. The term is used more commonly today than its syn. *carbon clock.*
radiocarbon dating *carbon-14 dating.*
radiochemistry (ra'-di-o-chem'-is-try) The chemical study of irradiated and naturally occurring radioactive materials and their behavior. It includes their use in tracer studies and other chemical problems.
radiocyathid (ra'-di-o-cy-ath'-id) Any of a group of Lower Cambrian problematic calcareous fossils belonging to the order Radiocyatales and characterized by an apically open globose body consisting of nesasters which form one or two body walls. Radiocyathids have been classified with archaeocyathids, sponges, receptaculitids, echinoderms, and as members of an uncertain phylum.
radio-echo sounding A form of radar sounding used to measure the thickness of glaciers and to examine their internal structure and basal-ice and bedrock characteristics.
radioecology (ra'-di-o-e-col'-o-gy) The branch of ecology concerned with the relationship between natural communities and radioactive material.
radiogenic (ra'-di-o-gen'-ic) Said of a product of a radioactive decay process, e.g., heat, lead.
radiogenic age determination *isotopic dating.*
radiogenic argon (a) Argon-40 formed by the decay of potassium-40. (b) Argon-40 in a rock or mineral formed by the decay of potassium-40 in situ since the rock or mineral formed. Cf: *atmospheric argon; excess argon; inherited argon.*
radiogenic dating *isotopic dating.*
radiogenic isotope An isotope that was produced by the decay of a *radionuclide,* but which itself may or may not be radioactive. See also: *radioisotope.*
radiogenic lead (a) Lead formed as a result of radioactive decay of uranium or thorium. (b) Lead in a rock or mineral formed by radioactive decay of uranium or thorium in situ after formation of the rock or mineral. (c) Lead which has $^{207}Pb/^{204}Pb$ and $^{206}Pb/^{204}Pb$ ratios greater than they would be if the lead had evolved in a single stage, because it evolved in an integrated system with $^{238}U/^{204}Pb$ about 9. Cf: *common lead; primeval lead.*

radiogenic strontium (a) Strontium-87 formed from decay of rubidium-87. (b) Strontium-87 occurring in rocks and minerals that is the direct result of decay of rubidium-87 in situ since the rock or mineral formed. Cf: *common strontium.*

radiogeology (ra'-di-o-ge-ol'-o-gy) An obsolete syn. of *isotope geology.*

radiograph (ra'-di-o-graph') A less-preferred syn. of *autoradiograph.*

radiohydrology (ra'-di-o-hy-drol'-o-gy) The study of the hydrologic relationships of extraction, processing, and use (including use in hydrologic investigations) of radioactive materials and disposal of the associated waste products.

radioisotope (ra'-di-o-i'-so-tope) A radioactive *isotope* of an element. Also incorrectly used as a syn. of *radionuclide.* See also: *radiogenic isotope.* Syn: *unstable isotope.*

radiolarian (ra'-di-o-lar'-i-an) Any actinopod belonging to the subclass Radiolaria, characterized mainly by a siliceous skeleton and a marine pelagic environment. Range, Cambrian to Holocene. In some classifications the radiolarians are grouped with the rhizopods.

radiolarian chert A well-bedded, microcrystalline radiolarite that has a well-developed siliceous cement or groundmass.

radiolarian earth A *siliceous earth* composed predominantly of the remains (lattice-like skeletal framework) of Radiolaria; the unconsolidated equivalent of radiolarite.

radiolarian ooze A deep-sea pelagic sediment containing at least 30% opaline-silica tests of radiolarians. It is a *siliceous ooze.*

radiolarite [paleont] (ra'-di-o-lar'-ite) A fossil shell of the Radiolaria.

radiolarite [sed] (a) The comparatively hard fine-grained chertlike homogeneous consolidated equivalent of radiolarian earth. (b) Indurated radiolarian ooze. (c) A syn. of *radiolarian earth.*

radiolitic [ign] (ra'-di-o-lit'-ic) Said of the texture of an igneous rock characterized by radial, fanlike groupings of acicular crystals, resembling sectors of spherulites.

radiolitic [sed struc] Said of limestones in which the components radiate from central points, with the cement comprising less than 50 percent of the total rock (Krumbein and Sloss, 1963, p.179).

radioluminescence (ra'-di-o-lu'-mi-nes'-cence) Luminescence that is stimulated by the impact of radioactive particles.

radiolysis (ra-di-ol'-y-sis) Chemical decomposition caused by radiation.

radiometer (ra-di-om'-e-ter) A nonimaging device for measuring radiant energy, especially thermal radiation.

radiometric (ra'-di-o-met'-ric) Pertaining to the measurement of geologic time performed with a radiometer. Often imprecisely used to refer to *isotopic dating.*

radiometric age *isotopic age.*

radiometric age determination *isotopic dating.*

radiometric dating *isotopic dating.*

radionuclide (ra'-di-o-nu'-clide) A radioactive *nuclide.* The term *radioisotope* is incorrectly used as a synonym. See also: *radioactivity.*

radiosonde (ra'-di-o-sonde') An instrument package, carried aloft by a balloon, that is used for the gathering and radio transmission of meteorologic data from the free atmosphere to an altitude of approximately 30 km.

radio-wave method Any electromagnetic exploration method wherein electromagnetic waves transmitted from radio broadcast stations are used as an energy source in determining the electrical properties of the Earth.

radiozone (ra'-di-o-zone') A *para-time-rock unit* representing a zone or succession of strata established on common radioactivity criteria (Wheeler et al., 1950, p.2364).

radius (ra'-di-us) (a) A *ray* of an echinoderm, such as any of five radiating ossicles in *Aristotle's lantern* of an echinoid; esp. a *radial plate.* (b) The lateral part of a compartmental plate of a cirripede crustacean, overlapping the *ala* of an adjoining plate, and differentiated from the *paries* by a change in direction of the growth lines. (c) An imaginary radial line dividing the body of a radially symmetrical animal into similar parts. (d) The more medial of the two bones of the forearm of tetrapods; its distal end is in line with the first digit (thumb). Pl: *radii.*

radius [palyn] In a trilete spore, one arm of the *laesura.*

radius of influence The radial distance from the center of a well bore to the edge of its area of influence. See also: *area of influence.*

radius ratio The radius of a cation divided by that of an ion. Relative ionic radii are pertinent to coordinations in ionic crystal structures.

radix (ra'-dix) Rootlike distal anchorage of a crinoid *column.*

radon A noble gas having radioactive isotopes ^{222}Rn, ^{220}Rn, ^{219}Rn (radon, thoron, and actinon) (IAEA, 2003).

radon-220 (ra'-don) A radioactive, gaseous isotope of radon; it is a member of the *thorium series* and a daughter of radium-224.

radon background The component of background gamma radiation originating in disintegration and gamma radiation of short-lived decay products of ^{222}Rn in air, particularly ^{214}Pb and ^{214}Bi (IAEA, 2003).

radon emanometry Measurement of the radon content of the atmosphere.

Radon transform *tau-p mapping.*

radovanite A pistachio-green orthorhombic mineral: $Cu_2Fe^{3+}AsO_4AsO_2(OH)_2{\cdot}H_2O$.

radtkeite An adamantine orange or yellow-orange monoclinic mineral: Hg_3S_2ICl. It is photosensitive and darkens on exposure to light.

radula (rad'-u-la) A chitinous band or strip of horny material in nearly all univalve mollusks that bears numerous transverse rows of filelike or rasplike and usually very minute teeth on its dorsal surface, that can be protruded through the mouth from its position on the floor of the digestive canal, and that serves to gather and tear up food and draw it into the mouth. Pl: radulae.

radulifer (ra-du'-li-fer) Said of hook-shaped or rodlike brachiopod crura that arise on the ventral side of the hinge plates and project toward the pedicle valve.

rafaelite (ra'-fa-el-ite') A nepheline-free orthoclase-bearing hypabyssal *syenite* that also contains analcime and calcic plagioclase. The name (Johannsen, 1938) is from the San Rafael Swell, Utah. Obsolete.

raft [coal] *float coal.*

raft [ign] A rock fragment within a magma body and drifting freely. Cf: *xenolith.*

raft [streams] An accumulation or jam of floating logs, driftwood, dislodged trees, or other debris, formed naturally in a stream by caving of the banks, and acting as an impedance to navigation. See also: *raft lake.*

raft [tect] A fault block that has separated from its original footwall and lies entirely on a décollement layer, which typically consists of salt. Rafts may themselves consist of smaller, older rafts, which became yoked together by later sedimentation before being ruptured again. Rafts are separated by troughlike depocenters of younger, synkinematic strata (Burollet, 1975; Duval et al., 1990).

raft breccia A breccia having fragments that remained unworn during transportation, as by an iceberg or floating vegetation such as trees or seaweed (Norton, 1917, p.172).

raft cone A conical *speleothem* composed of a pile of sunken *calcite rafts.* The apex of the cone corresponds to a drip point; dripping causes rafts to sink at that point.

rafted block (raft'-ed) *dropstone.*

rafted ice A form of *pressure ice* in which one floe overrides another.

rafting (raft'-ing) (a) The transporting of land-derived rocks, soil, and the like, by floating organic material, as seaweed or logs, or by floating ice (*ice-rafting*). (b) A form of deformation of floating ice wherein one piece overrides another, thus creating *rafted ice.* It is most common in new ice and young ice.

raft lake A relatively short-lived body of water impounded along a stream by a raft; examples are commonly found in the Red River of Louisiana during times of high water.

raft tectonics Extreme extension characterized by opening of deep, syndepositional grabens and separation of intervening fault blocks of overburden into rafts, which slide downslpe on a décollement (typically of thin salt), like a block-glide landslide. The décollement faults may commonly be cryptic fault welds if all the salt is removed or thinned beyond seismic recognition. The type area for raft tectonics is the continental margin of Angola (Burollet, 1975; Duval et al., 1990). Cf: *plate tectonics.*

rag In British usage, any of various hard, coarse, rubbly or shelly rocks that weather with a rough irregular surface; e.g. a flaggy sandstone or limestone used as a building stone. The term appears in certain British stratigraphic names, as the Kentish Rag (a Cretaceous sandy limestone in East Kent). Syn: *ragstone.*

raggioni (rag-gio'-ni [rah-joe'-nee]) Large ray-crystals of former aragonite, now converted to a calcite mosaic. They are typically 1-10 cm long, generally with squared-off ends and hexagonal cross-sections; they occur in some sabkha-type carbonates. Plural form commonly used; sing: raggione. Etymol: Italian, "large rays".

raglanite (rag'-lan-ite) A *nepheline diorite* composed of oligoclase, nepheline, and corundum with minor amounts of mica, calcite, magnetite, and apatite. Cf: *craigmontite.* The name, given by Adams and Barlow in 1910, is from Raglan Township, Ontario. Obsolete.

ragstone (rag'-stone) *rag.*

raguinite (ra-guin'-ite) A metallic bronze monoclinic mineral: $TlFeS_2$.

rain A falling of numerous particles, such as the long unending deposition of pelagic matter to the bottom of the ocean, or of micrometeorite dust from interplanetary space. Also, the falling particles themselves, such as *pollen rain.*

rain [meteorol] Form of *precipitation* consisting of liquid water drops having diameters generally bewteen 0.5 and 6.0 mm (0.02 and 0.2 in); falls mostly from nimbostratus and cumulonimbus clouds.

rainbeat (rain'-beat) *raindrop impact.*

rainbow (rain'-bow) Chromatic iridescence observed in drilling fluid that has been circulated in a well, indicating contamination or contact with fresh hydrocarbons.

rainbow quartz An *iris* quartz that exhibits the colors of the rainbow.

rain crust A type of *snow crust* formed by refreezing of surface snow that has been melted or wetted by liquid precipitation.

rain desert A desert in which rainfall is sufficient to maintain a sparse general vegetation. Cf: *runoff desert.*

raindrop impact (rain'-drop) The action of raindrops striking or falling upon the surface of the ground. Syn: *rainbeat.*

raindrop impression *rain print.*

raindrop imprint *rain print.*

rainfall (rain'-fall) (a) The quantity of water that is precipitated out in the atmosphere as rain, in a given period of time. (b) The liquid product of precipitation in whatever form. In this sense the term is synonymous with *precipitation.*
rainfall excess The volume of water from rainfall that is available for direct runoff (Langbein and Iseri, 1960). Cf: *abstraction [water]; precipitation excess.* Syn: *excess water.*
rainfall penetration The depth below the soil surface to which water from a given rainfall event is able to infiltrate.
rain forest A forest where the annual rainfall is very high and with no dry season. The region is characterized by tall, lush evergreen trees.
rain gauge A device used to measure precipitation (melted snow, sleet, or hail as well as rain). It consists of a receiving funnel, a collecting vessel, and a measuring cylinder. Also spelled: rain gage. Syn: *pluviometer; hyetometer.*
rain pillar A minor landform consisting of a column of soil or soft rock capped and protected by pebbles or concretions, produced by the differential erosion effected by the impact of falling rain (Stokes and Varnes, 1955, p.118).
rain print A small, shallow craterlike pit surrounded by a slightly raised rim, formed in soft fine sand, silt, or clay, or in the mud of a tidal flat, by the impact of a falling raindrop, and sometimes preserved on the bedding planes of sedimentary rocks or as casts on the underside of overlying sandstone beds. See also: *hail imprint; spray print.* Syn: *raindrop imprint; raindrop impression.*
rain shadow A very dry region on the lee side of a topographic obstacle, usually a mountain range, where the rainfall is noticeably less than on the windward side. The White Mountains in east-central California are in the rain shadow of the Sierra Nevada.
rain-shadow desert (rain'-shad-ow) A desert occurring on the lee side of a mountain or mountain range that deflects moisture-laden air upward on the windward side. See also: *mid-latitude desert; topographic desert.* Syn: *orographic desert.*
rainwash (rain'-wash) (a) The washing-away of loose surface material by rainwater after it has reached the ground but before it has been concentrated into definite streams; specif. *sheet erosion.* Also, the movement downslope (under the action of gravity) of material loosened by rainwater. It occurs esp. in semiarid or scantily vegetated regions. Syn: *hillwash.* (b) The material that originates by the process of rainwash; material transported and accumulated, or washed away, by rainwater. (c) The rainwater involved in the process of rainwash.
rainwater (rain'-wa-ter) Water that has fallen as rain and has not yet collected soluble matter from the soil, thus being quite soft.
rain-wave train *Overland flow* in the form of wave trains or series of uniformly spaced waves and involving nearly all the runoff. Rain-wave trains are usually associated with heavy rains, esp. cloudbursts (Horton, 1945, p.313).
raise A mine shaft driven upward from a lower to a higher level.
raised bank *natural levee.*
raised beach An ancient beach occurring above the present shoreline and separated from the present beach, having been elevated above the high-water mark either by local crustal movements (uplift) or by lowering of sea level, and often bounded by inland cliffs. Cf: *marine terrace; elevated shoreline.* See also: *strandline.*
raised peat bog A *highmoor bog* with a thick accumulation of peat in the center, giving it a convex surface.
raised reef An *organic reef* elevated and standing above sea level.
raite (ra'-ite) A brown or gold orthorhombic mineral: $Na_3Mn^{2+}{}_3Ti_{0.25}Si_8O_{20}(OH)_2{\cdot}10H_2O$.
rajite (raj'-ite) A bright green monoclinic mineral: $CuTe^{4+}{}_2O_5$.
rake The angle between the horizontal (strike line) and any linear structure, e.g., a slickenline, measured in the plane containing the linear feature (Phillips, 1954, p.10). Syn: *pitch [struc geol].*
Raleigh-Taylor instability A situation that develops when a warmer, less dense layer underlies a cooler, denser layer. If the buoyancy force exceeds the strength of materials, diapiric masses of the less dense layer rise upwards through the denser layer.
ralstonite (ral'-ston-ite) A colorless, white, or yellowish cubic mineral: $Na_{0.5}(Al,Mg)_2(F,OH)_6{\cdot}H_2O$. It occurs in octahedral crystals.
ram An underwater ledge or projection from an ice wall, ice front, iceberg, or floe, caused by more intensive melting of the part above the water level. Syn: *apron [ice]; spur [ice].*
Raman effect (Ra'-man) The resonant absorption and re-emission of radiant energy by specific interatomic valence bonds in a sample. When intense monochromatic radiation (as from a laser) is scattered by a sample, *Raman lines* appear in addition to the normal Rayleigh scattering of the exciting radiation. These lines are shifted relative to the exciting wavelength (Raman shifts) by amounts characteristic of the vibrational and rotational frequencies of the specific bonds causing the effect. Named in honor of C. V. Raman (1883-1970), Indian physicist.
Raman lines Shifted lines in the *Raman spectrum.* They are typically at longer wavelengths than the exciting radiation (Stokes lines), but shorter wavelengths (anti-Stokes lines) may occasionally appear.
Raman spectroscopy The observation of a *Raman spectrum* and all processes of recording and measuring that go with it.
Raman spectrum The characteristic spectrum observed when monochromatic light is scattered by a transparent substance. See also: *Raman effect.*
ramassis (ra-mas'-sis) A local term in southern Louisiana for a mass of decomposed plant debris, dried plant remains, driftwood, and other flotsam occurring in coastal marshes or flotants (Russell, 1942, p.96-97).
rambergite A dark brown to black hexagonal mineral of the *wurtzite* group: MnS. It is a dimorph of alabandite.
rambla (ram'-bla) A dry ravine, or the dry bed of an ephemeral stream. Etymol: Spanish, from Arabic "ramlah", "sand".
Ramblian (Ramb'-li-an) European continental stage: lower Miocene (below Aragonian).
ramdohrite (ram'-dohr-ite) A dark gray monoclinic mineral: $Ag_3Pb_6Sb_{11}S_{24}$. It is closely related to andorite and fizelyite.
rameauite (ra-meau'-ite) An orange monoclinic mineral: $K_2CaO_8(UO_2)_6{\cdot}9H_2O$.
ramentum (ra-men'-tum) A thin, chaffy scale on a leaf, e.g. on some ferns.
rami (ra'-mi) Plural of *ramus.*
ramiform cave A dissolution cave consisting of irregular rooms and galleries extending outward in various directions from the main area of cave development; a typical pattern of *hypogenic caves* formed by rising hydrogen sulfide (Palmer, 1991).
ramiform element A *conodont element* in which at least one of the sides or edges of the base is drawn out laterally, anteriorly, or posteriorly from the cusp into a process that is serrate on its upper edge (TIP, 1981, pt. W, supp.2).
rammelsbergite (ram'-mels-berg-ite') A metallic white orthorhombic mineral of the *löllingite* group: $NiAs_2$. It is dimorphous with pararammelsbergite. Syn: *white nickel.*
rammesonde A cone penetrometer driven vertically into the snowpack by a weighted hammer on a rule guide rod.
ramose (ra'-mose) Consisting of or having branches; e.g. said of a bryozoan colony consisting of erect, round, or moderately flattened branches. Syn: *dendroid.*
ramp [oceanog] An area of no significant bathymetric changes between the shoreline and the deepest parts of a sedimentary basin.
ramp [paleont] (a) The abapically sloping surface of a gastropod *whorl* next below a *suture.* (b) Sloping surface on some platformed pectiniform conodont elements, between low- and higher-level areas of the platform (TIP, 1981, pt. W, supp.2).
ramp [snow] A drift of snow that forms an inclined plane between land or land ice and sea or shelf ice.
ramp [struc geol] A portion of a step or thrust fault that cuts across formational contacts in a short distance (Woodward et al., 1985, p.3). Thrust fault and normal fault ramps commonly have dips of 30° and 60°, respectively. Cf: *flat [struc geol].* Syn: *riser (b).*
ramp anticline An *anticline* formed in a thrust sheet as a result of movement up a *ramp.* Partial syn: *fault-bend fold.*
rampart [geomorph] (ram'-part) (a) A narrow, wall-like ridge, 1-2 m high, built up by waves along the seaward edge of a reef flat, and consisting of boulders, shingle, gravel, or reef rubble, commonly capped by dune sand. (b) A wall-like ridge of unconsolidated material formed along a beach by the action of strong waves and currents.
rampart [volc] A crescentic or ringlike deposit of pyroclastics around the top of a volcano.
rampart crater On Mars, impact craters whose ejecta deposits are terminated by a distal rampart (Kieffer et al., 1992).
rampart wall A *rimming wall* formed along the outer or seaward margin of a terrace, as on various "high limestone" Pacific islands (Flint et al., 1953, p.1258).
ram penetrometer *ramsonde.*
ramp-on-flat geometry The situation where a *ramp* in the *hanging wall* of a fault is juxtaposed against a *flat* in the *footwall* of the fault.
ramp trough *ramp valley.*
ramp valley A valley that is bounded by high-angle thrust faults, or ramps. Syn: *ramp trough.* Cf: *graben.*
ram resistance The resistance of a snow layer to penetration by a *ramsonde.* The energy per unit of distance penetrated gives a ram resistance index in units of force.
ramsayite (ram'-say-ite) *lorenzenite.*
ramsbeckite (rams'-beck-ite) A green monoclinic mineral: $(Cu,Zn)_{15}(SO_4)_4(OH)_{22}{\cdot}6H_2O$.
ramsdellite (rams'-dell-ite) A gray to black submetallic orthorhombic mineral: $Mn^{4+}O_2$. It is dimorphous with pyrolusite.
ramsonde (ram'-sonde) A cone-tipped metal rod that is driven downward into snow by repeated blows from a drop-weight in order to measure its penetration resistance. Syn: *ram penetrometer.* Also spelled: ramsond; Rammsonde.
ramule (ram'-ule) A minor branch of a crinoid arm, differing from a *pinnule* in less regular occurrence; in some crinoids it bears pinnules.
ramus (ra'-mus) (a) In the vertebrates, a branch or process, esp. of a bone, nerve, or blood vessel. (b) A projecting part or elongated process of an in-

vertebrate; e.g. a branch of a crustacean limb, or the main branch of a crinoid arm. Pl: *rami.*

raña (ra'-ña) A Spanish term for a consolidated mudflow deposit containing angular blocks of rock of all sizes.

Rancholabrean (Ran'-cho-la-bre'-an) North American land mammal age: Late Pleistocene (after Irvingtonian).

ranciéite (ran'-cie-ite) A metallic black, brownish, or violet hexagonal mineral: $(Ca,Mn^{2+})Mn^{4+}{}_4O_9{\cdot}3H_2O$.

rand (a) An English term for the low, marshy border of a lake or of a river overgrown with reeds. (b) A term used in South Africa for a long, low, rocky ridge or range of hills often covered with scrub; e.g. Witwatersrand (popularly contracted to "The Rand"), a ridge containing rich gold-bearing rocks, 100 km long, situated near Johannesburg.

randannite (ran-dan'-nite) (a) A dark variety of diatomaceous earth containing humic material, occurring in the Puy-de-Dôme (Randan) region of France. (b) An earthy form of opal. Also spelled: randanite.

Randian A geochronometric subdivision of the early *Precambrian,* covering the time span between 2,800 and 2,450 Ma (Harland et al., 1990), therefore approximately equivalent to the *Neoarchean* Era of the Standard Global Chronostratigraphic Scale.

randkluft (rand'-kluft) A *crevasse* at the head of a *mountain glacier,* separating the moving ice and snow from the surrounding rock wall of the valley where no *ice apron* is present. It may be enlarged where heat radiating from the rock wall causes the ice to melt. Etymol: German "Randkluft", "rim crevice". Cf: *bergschrund.*

random error (ran'-dom) Variance that cannot be attributed to known sources of variation, such as differences between samples, operators, or experimental conditions.

randomization A process in which samples are selected or experiments performed in an order that is determined solely by chance.

random line (a) A trial surveying line that is directed as closely as circumstances permit toward a fixed terminal point that cannot be seen from the initial point. (b) *random traverse.* (c) *arbitrary line [surv].*

random powder mount A sample preparation for powder X-ray or powder neutron diffractometry in which particles are ideally oriented randomly. Grain size is preferably <5 µm, and sufficient particle numbers (for X-ray diffractometry this number is 10^6) are required so that all possible diffracting hkl surfaces are presented to the analyzing beam in true proportions to their populations in each grain and a diffraction pattern is obtained that is true in both reflection position and reflection intensity.

random process *stochastic process.*

random sample A subset of a population in which each observation has an equal and independent probability of being included in the sample. A sample that results from a selection procedure that operates by pure chance.

random traverse A survey traverse run from one survey station to another station which cannot be seen from the first station in order to determine their relative positions. Syn: *random line.*

random variable A real-valued mathematical function that is defined over a sample space. A random variable may be either continuous or discrete. Syn: *variate.*

randpolje An enclosed plain at the edge of a karst area receiving surface water from the nonkarstic area. The water drains out through underground passages in the karst area. The plain is thus completely enclosed by higher ground. Cf: *blind valley; karst-margin plain.*

range [eco geol] An area in which a mineral-bearing formation crops out, e.g. the "iron range" and "copper range" of the Lake Superior region; a mineral belt.

range [ecol] The geographic area over which an organism or group of organisms is distributed. Syn: *distribution [ecol].*

range [geomorph] (a) A *mountain range.* Also, a line of hills if the heights are comparatively low. (b) A term sometimes used in Australia for a single mountain. (c) Mountainous country; term usually used in the plural.

range [hydrogr] (a) An established or well-defined line or course whose position is known and along which soundings are taken in a hydrographic survey. (b) The distance to a radar target.

range [radioactivity] The distance that radiation penetrates a medium before its velocity becomes no longer detectable.

range [sed] A measure of the variability between the largest and smallest particle sizes of a sediment or sedimentary rock.

range [stat] The numerical difference between the highest and lowest values in any series.

range [surv] (a) Any series of contiguous townships (of the U.S. Public Land Survey system) aligned north and south and numbered consecutively east and west from a principal meridian to which it is parallel (e.g. "range 3 east" indicates the third range or row of townships to the east from a principal meridian). Also, any series of contiguous sections similarly situated within a township. Abbrev (when citing specific location): R. Cf: *tier [surv].* (b) The distance between any two points, usually an observation point and an object under observation; also, two points in line with the point of observation. Two or more objects in line are said to be "in range".

range chart A chart that records for a given area the local range zone (and often the peak zone) of each significant fossil taxon encountered in terms of genera and species.

range direction *look direction.*

range finder A *tachymeter* designed for finding the distance from a single point of observation to other points at which no instruments are placed. It uses the parallax principle, and is usually constructed to give a rapid mechanical solution of a triangle having the target at its apex and the range finder at one corner of its base. See also: *telemeter.*

range line One of the imaginary boundary lines running north and south at six-mile intervals and marking the relative east and west locations of townships in a U.S. public-land survey; a meridional township boundary line. Cf: *township line.*

range midpoint *midrange.*

range-overlap zone (range-o'-ver-lap) *concurrent-range zone.*

range pole A *range rod.* Syn: *ranging pole.*

range rod A wooden or metal *level rod,* rounded or octagonal in section, 6-8 ft long and one inch or less in diameter, fitted with a sharp-pointed metal shoe, usually painted in one-foot bands of alternate red and white, and used for sighting points and lines in surveying or for showing the position of a ground point. Syn: *range pole; lining pole; line rod; sight rod.*

range zone A body of strata representing the known stratigraphic and geographic range of occurrences of any selected element or elements of the assemblage of fossils in a stratigraphic sequence. (ISSC, 1994, p.57). See also: *taxon-range zone; concurrent-range zone; partial range zone; lineage zone.* Syn: *acrozone; zonite.*

ranging pole *range pole.*

rank [bot] (a) A vertical row of leaves. Cf: *decussate; distichous.* (b) The position of a category in the taxonomic hierarchy of plants.

rank [coal] Degree of metamorphism. It is the basis of *coal classification* into a natural series from lignite to anthracite. Cf: *type [coal]; grade [coal].*

rank [meta] *metamorphic grade.*

rank [stat] An ordering by relative magnitude. The rank of an observation indicates only the position of the observation in a sequence from smallest to largest.

rankachite (ran-kach'-ite) A sub-adamantine dark brown or brownish-yellow orthorhombic mineral: $CaFe^{2+}V^{5+}{}_4W^{6+}{}_8O_{36}{\cdot}12H_2O$.

rankamaite (ran'-ka-ma-ite) A white to creamy-white orthorhombic mineral: $(Na,K,Pb,Li)(Ta,Nb,Al)_4(O,OH)_{10}$.

rankinite (ran'-kin-ite) A colorless monoclinic mineral: $Ca_3Si_2O_7$. It is dimorphous with kilchoanite.

Ranney collector (Ran'-ney) *collector well.*

ranquilite (ran'-qui-lite) Possibly = *haiweeite.*

ransomite (ran'-som-ite) A sky-blue monoclinic mineral: $CuFe^{3+}{}_2(SO_4)_4{\cdot}6H_2O$.

ranunculite (ra-nun'-cu-lite) A yellow monoclinic mineral: $HAl(UO_2)PO_4(OH)_3{\cdot}4H_2O$.

Raoult's law (Ra-oult's') In its original sense, the statement that the partial vapor pressure of a solvent liquid is proportional to its mole fraction. It is now usually used, however, in a more general form to specify a model for the ideal solution: The statement that the activity of each component in a solution is equal to its mole fraction. It is obeyed by all solutions for the major component in sufficiently concentrated regions, and approximately by many solutions over large compositional ranges.

rapakivi (ra-pa-ki'-vi) n. In the U.S., granite or *quartz monzonite* that is characterized by alkali feldspar phenocrysts (commonly ellipsoidal) that are mantled with plagioclase (oligoclase, albite, or andesine). The term was introduced in 1694 by Urban Hjarne to denote crumbly stone in certain weathered outcrops in Finland. Etymol: Finnish, "rotten stone". Syn: *wiborgite.* adj. Said of volcanic as well as plutonic rocks having alkali feldspar phenocrysts that are mantled with plagioclase.

rapakivi texture A texture of igneous and metamorphic rocks in which relatively large crystals of potassium feldspar are surrounded by a rim of sodic plagioclase, set in a finer-grained matrix.

raphe (a) That portion of the funiculus of an ovule that is adnate to the integument, usually represented by a ridge. It is present in most anatropous ovules (Lawrence, 1951, p.767). (b) A vertical, unsilicified groove or cleft in the valve of some pennate diatoms (Scagel et al., 1965, p.630).

rapidcreekite (ra'-pid-creek'-ite) A colorless orthorhombic mineral: $Ca_2(SO_4)(CO_3){\cdot}4H_2O$.

rapid flow *supercritical flow.*

rapids (rap'-ids) (a) A part of a stream where the current is moving with a greater swiftness than usual and where the water surface is broken by obstructions but without a sufficient break in slope to form a waterfall, as where the water descends over a series of small steps. It commonly results from a sudden steepening of the stream gradient, from the presence of a restricted channel, or from the unequal resistance of the successive rocks traversed by the stream. The singular form "rapid" is rarely used. See also: *cascade [streams]; cataract.* (b) A swift, turbulent flow or current of water through a rapids.

rappoldite A vitreous red to red-brown triclinic mineral: $Pb(Co,Ni,Zn)_2(AsO_4)_2{\cdot}2H_2O$.

raqqaite (raq'-qa-ite) An extrusive rock having the composition of a *pyroxenite* (Streckeisen, 1967, p.188). Cf: *komatiite.* Obsolete.

rare In the description of coal constituents, less than 5% of a particular constituent occurring in the coal (ICCP, 1963). Cf: *common; very common; abundant; dominant [coal].*
rare earths Oxides of a series of fifteen metallic elements, from lanthanum (atomic number 57) to lutetium (71), and of three other elements: yttrium, thorium, and scandium. These elements are not especially rare in the Earth's crust, but concentrations are. The rare earths are constituents of certain minerals, esp. monazite, bastnaesite, and xenotime. Syn: *lanthanides.* Abbrev: REE.
rare element [eco geol] Any element of which the concentration in the Earth's crust is less than 0.01 percent.
rare-element pegmatite Rare-earth- and rare-metal-bearing pegmatite formed under conditions of lower amphibolite metamorphism (~2 to 4 kilobars pressure) peripheral to granitic intrusions.
rarefaction (rar-e-fac'-tion) *dilatation [seis].*
rarefaction [paleont] Statistical technique that assesses how large a sample is required to obtain different degrees of sampling completeness (Raup, 1975).
rarefaction wave Expansion wave that initially moves in the opposite direction from an associated shock wave, e.g., down a volcanic conduit during an explosive eruption.
rare metal A loose, informal term for the less common and more expensive metallic elements, such as cadmium, cobalt, gallium, molybdenum, and titanium.
ras A cape or headland. Etymol: Arabic "ra's", "head".
rash Very impure coal, so mixed with noncoal material (clay, shale, or other argillaceous substances from the top or bottom of the coal seam) as to be unsalable; a dark substance intermediate in character between coal and shale; a dirty coal. Not to be confused with *rashing.*
rashing (rash'-ing) A soft, friable, and flaky or scaly shale or clay immediately benenath a coal seam, often containing much carbonaceous material (numerous slickensided surfaces and streaks of coal), and readily mixed with the coal in mining. It may also overlie or be interstratified with the coal. The term often used in the plural. It is not to be confused with *rash.*
raslakite A vitreous red rhombohedral mineral of the *eudialyte* group: $Na_{15}Ca_3Fe_3(Na,Zr)_3Zr_3(Si,Nb)Si_{25}O_{73}(OH,H_2O)_3(Cl, OH)$.
rasorite (ra'-sor-ite) *kernite.*
raspberry spar (rasp'-ber-ry) (a) *rhodochrosite.* (b) Pink tourmaline.
raspite (rasp'-ite) A yellow, brown, or gray monoclinic mineral: $PbWO_4$. It is dimorphous with stolzite.
Rassenkreis (Ras'-sen-kreis) A *polytypic* species. Etymol: German "Rasse", "race", plus "Kreis", "circle". Plural: Rassenkreise.
rasskar (ras'-skar) A Norwegian term for a cirque which has served as "an old scree channel" and has been "carved upward by weathering" (Termier and Termier, 1963, p.412).
raster A format for storing, processing, and displaying graphic data in which graphic data are stored as values for uniform grid cells or pixels.
raster to vector The process of converting an image made of cells into one composed of lines and polygons.
rastrate In conodont morphology, an element that is rake-shaped, posteriorly denticulate or serrate, and deeply excavated, and whose surface is longitudinally striate and bears a deep groove or furrow on one side.
rasvumite (ras-vu'-mite) A metallic gray orthorhombic mineral: KFe_2S_3.
rate-of-change map A derived stratigraphic map that shows the rate of change of structure, thickness, or composition of a given stratigraphic unit (Krumbein and Sloss, 1963, p.484). It is based on analysis of the contour lines on an initial map (structure-contour map, isopach map, facies map, etc.).
rate of sedimentation The amount of sediment accumulated in an aquatic environment over a given period of time, usually expressed as thickness of accumulation per unit time. Syn: *sedimentation rate.*
ratfish (rat'-fish) *Holocephali.*
rathite (rath'-ite) A dark gray orthorhombic or monoclinic mineral: $(Pb,Tl)_{11}Ag_2As_{20}S_{40}$. Other compositions have been reported. It is a sartorite homologue.
rathite-II *liveingite.*
rating curve (rat'-ing) *stage-discharge curve.*
ratio map (ra'-tio) A *facies map* that depicts the ratio of thicknesses between rock types in a given stratigraphic unit; e.g. a "sand-shale ratio map" showing the ratio of sandstone thickness to shale thickness in a given unit.
rational face (ra'-tion-al) A crystal face naturally suggested by and peculiar to the internal molecular structure of the mineral species to which the crystal belongs. Such faces usually have low *Miller indices.*
rational formula In hydraulics, the expression of peak discharge (in cfs units) as equal to rainfall (in inches/hr) times drainage area (in acres) times a runoff coefficient depending on drainage-basin characteristics (Chow, 1957).
rational horizon (a) A *celestial horizon;* e.g. a great circle 90 degrees from the zenith and constituting the equator of the horizon system of coordinates. (b) *actual horizon.*
ratio scale A quantitative scale of measurement in which the unit intervals are equal in magnitude and there is a true origin for the scale. An example is the millimeter scale of length, as "negative lengths" are physically impossible. Cf: *interval scale.*
rattlesnake ore (rat'-tle-snake) A gray, black, and yellow mottled ore of carnotite and vanoxite, its spotted appearance resembling that of a rattlesnake.
rattle stone (rat'-tle) A *concretion* composed of concentric laminae of different composition, in which the more soluble layers have been removed by solution, leaving the central part detached from the outer part, such as a concretion of iron oxide filled with loose sand that rattles on being shaken. Syn: *klapperstein.* Also spelled: rattlestone; rattle-stone.
rauenthalite (rau'-en-thal-ite') A colorless triclinic mineral: $Ca_3(AsO_4)_2{\cdot}10H_2O$.
rauhaugite (rau'-haug-ite) A carbonatite that contains ankerite or dolomite. The name, given by Brögger in 1921, is from Rauhaug in the Fen complex, Norway. Cf: *beforsite.*
rauk A Swedish term for a stack. Pl: raukar.
Rauracian (Rau-ra'-cian) Substage in Great Britain: Upper Jurassic (middle Lusitanian: above Argovian, below Sequanian).
rauvite (rau'-vite) An ill-defined purplish-black to bluish-black mineral: $Ca(UO_2)_2V^{5+}{}_{10}O_{28}{\cdot}16H_2O$.
ravatite A colorless, white or pale gray monoclinic hydrocarbon mineral: $C_{14}H_{10}$. Syn: *phenanthrene.*
raveling ground (rav'-el-ing) A term for ground that begins to drop out of the roof or sides of a tunnel after being exposed. The term is also applied to steep natural and cut slopes that may produce isolated or small groups of falling rock fragments or gravel. Cf: *firm ground; flowing ground; running ground; squeezing ground; swelling ground.*
ravinated (rav'-i-nat-ed) Said of a landform or landscape having or cut by ravines.
ravine (ra-vine') (a) A small narrow deep depression, smaller than a *gorge* or a canyon but larger than a gully, usually carved by running water; esp. the narrow excavated channel of a mountain stream. (b) A stream with a slight fall between rapids. Etymol: French, "mountain torrent".
ravinement [geomorph] (ra-vine'-ment) (a) The formation of a ravine or ravines. (b) *gully erosion.*
ravinement [stratig] A term introduced by Stamp (1921, p.109) for "an irregular junction which marks a break in sedimentation", such as an erosion line occurring where shallow-water marine deposits have "scooped down into" (or "ravined") slightly eroded underlying beds; a small-scale disconformity caused by periodic invasions of the sea over a deltaic area. Etymol: French, "hollowing out (by waters), gullying".
ravinement surface *ravinement [stratig].*
raw Said of a mineral, fuel, or other material in its natural, unprocessed state, as mined.
ray [bot] (a) In a composite inflorescence, the corolla of a marginal flower. (b) A *vascular ray.* (c) In dasycladacean algae, a branch.
ray [paleont] (a) Any of the radiating divisions of the body of an echinoderm together with all structures borne by it; e.g. a segment of an echinoderm body that includes one ambulacral axis, or a radial plate or an arm of a crinoid. Also, a radial direction established by the position of an *ambulacrum.* (b) One of the primary subdivisions of a sponge *spicule* containing an axial filament or an axial canal. (c) In Osteichthyes and Chondrichthyes, one of many fine rods, respectively bony or horny, that support a fin.
ray [phys] A vector normal to wavefront surfaces in isotropic media, indicating the direction and sometimes the velocity of propagation.
ray [planet] One of the long, relatively bright, almost white streaks, loops, or lines observed on a solid body's surface and appearing to radiate from a large, well-formed crater, in some examples extending for hundreds of kilometers. Rays are brightest at high Sun angles and nearly invisible at low Sun angles except on rough ground. They are believed to be formed in some way by fine-grained debris explosively ejected from craters either by impact or by volcanic activity.
ray crater A large, relatively young crater with visible rays; e.g. Copernicus on the Moon.
ray-crystal shrub Subparallel to fan-shaped array of coarse to extremely coarsely crystalline blades of calcite or aragonite. Ray-crystal shrubs are dominantly, but not completely, abiotically produced features (Chafetz and Guidry, 1999).
rayite (ray'-ite) A metallic gray monoclinic mineral: $(Ag,Tl)_2Pb_8Sb_8S_{21}$. It is structurally related to semseyite.
Rayleigh criterion (Ray'-leigh) The relationship between surface roughness, depression angle, and wavelength that determines whether a surface will respond in rough or smooth fashion to a radar pulse. Cf: *rough criterion; smooth criterion.*
Rayleigh scattering Selective scattering of electromagnetic radiation by particles in the atmosphere that are small relative to the wavelength. The scattering is inversely proportional to the fourth power of the wavelength. See also: *Mie theory.*
Rayleigh wave A type of surface wave having a retrograde, elliptical motion at the free surface. It is named after Lord Rayleigh, the English physicist who predicted its existence. Abbrev: L_R. See also: *Rg wave.* Syn: *R wave.*
ray parameter A function *p* that is constant along a given seismic ray when there is no horizontal change of velocity. For planar geometry, it is defined

as $p = v^{-1} \sin i$, where v is the velocity and i is the angle that the ray makes with the vertical. For spherical geometry, it is defined as $p = (r/v) \sin i$, where r = distance from the center of the sphere.

raypath (ray′-path) The imaginary line along which wave energy travels. A raypath is always perpendicular to the wavefront in isotropic media. Syn: *path [seis]; trajectory [seis].*

ray tracing Determining the arrival time at detector locations by following raypaths which obey Snell's law through a model for which the velocity distribution is known. Ray tracing represents a high-frequency approximation to the wave equation.

razorback (ra′-zor-back) A sharp, narrow ridge, resembling the back of a razorback hog. "There is little or no implication as to geologic structure, hence the term is not quite so specific as hogback" (Stokes and Varnes, 1955, p.119).

razor stone (ra′-zor) *novaculite.*

RBS *Rutherford backscattering.*

Rb-Sr age method *rubidium-strontium age method.*

R chondrite A member of a small chondrite group having exceptionally high mean $\Delta^{17}O$ values (about +3 per mil); possibly related to ordinary chondrites.

RE In chemical formulas, a symbol for a *rare earth* element.

reach [coast] (a) An arm of the sea extending up into the land; e.g. an estuary or a bay. (b) *promontory.*

reach [geog] (a) A continuous and unbroken expanse or surface of water or land. Syn: *stretch [geog].* (b) An unstated but specific distance; an interval.

reach [hydraul] (a) The length of a channel, uniform with respect to discharge, depth, area, and slope. (b) The length of a channel for which a single gage affords a satisfactory measure of the stage and discharge. (c) The length of a stream between two specified gaging stations. See also: *test reach.*

reach [lake] (a) A relatively long, straight section of water along a lake shore; also, a narrow arm of a lake, reaching into the land. (b) A straight, narrow expanse of shore or land extending into a lake.

reach [streams] (a) A straight, continuous, or extended part of a stream, viewed without interruption (as between two bends) or chosen between two specified points; a straight section of a restricted waterway, much longer than a narrows. Cf: *sea reach.* (b) The level expanse of water between locks in a canal. (c) A length of stream with consistent morphology, e.g., *pools* and *riffles.*

reactance (re-ac′-tance) That part of the impedance of an alternating-current circuit that is due to capacitance and/or inductance. It is expressed in ohms.

reaction line In a ternary system, a special case of the boundary line, along which one of the two crystalline phases reacts with the liquid, as the temperature is decreased, to form the other crystalline phase.

reaction pair Any two minerals, one of which is formed at the expense of the other by reaction with liquid; esp., any two adjacent minerals in a *reaction series.*

reaction point An invariant point on a liquidus diagram in which the composition of the liquid cannot be stated in terms of position quantities of all the solid phases in equilibrium at the point. In a binary system it is equivalent to an incongruent melting point, or *peritectic point.*

reaction rim A peripheral zone around a mineral; it is composed of another mineral species and represents the reaction of the earlier solidified mineral with the surrounding magma. Cf: *corrosion border; corona [petrology]; kelyphytic rim.*

reaction series A series of minerals in which any early-formed mineral phase tends to react with the melt, later in the differentiation, to yield a new mineral further down in the series; e.g. early-formed crystals of olivine react with later liquids to form pyroxene crystals, and these in turn may further react with still later liquids to form amphiboles. There are two different series, a *continuous reaction series* and a *discontinuous reaction series.* This concept is frequently referred to as *Bowen's reaction series,* after N.L. Bowen, who first proposed it, or as the reaction principle. See also: *reaction pair.*

reactivation surface (re-ac′-ti-va′-tion) An inclined bedding surface, separating otherwise conformable cross-beds; it is formed by slight erosion of the *lee* side of a dune, sand wave, megaripple, or bar, during a period when deposition is temporarily interrupted. Originally described by Collinson (1970) from braided river sands, but more abundant in sands deposited by the migration of megaripples or dunes in the tidal environment.

reactive diapirism Emplacement of an elongated, sharp-crested diapir of salt or shale into the space created by regional extensional thinning during rifting or gravity spreading or gliding.

readvance (re-ad-vance′) (a) A new advance made by a glacier after receding from the marginal position reached in an earlier advance. (b) A time interval during which a readvance occurred.

real-aperture radar (real-ap′-er-ture) A *SLAR* system in which azimuth resolution is determined by the length of the antenna and by the wavelength. The radar returns are recorded directly to produce images. Syn: *brute-force radar.*

realgar (re-al′-gar) A bright-red to orange-red monoclinic mineral: AsS. It occurs as nodules in ore veins and as a massive or granular deposit from some hot springs, and it is frequently associated with orpiment. Pararealgar is a polymorph. Syn: *red arsenic; sandarac; red orpiment.*

realized ecological hyperspace (re′-al-ized) *biospace.*

realized ecospace That portion of the *ecospace* actually utilized by an organism (Valentine, 1969, p.687). Cf: *prospective ecospace.*

realm (a) A portion of the Earth consisting of several regions. (b) A large *region [ecol].*

real time Processing of data such that results are available immediately, allowing influence on further processing of data.

reamer (ream′-er) A rotary-drilling tool with cutters used for enlarging, smoothing, or straightening a well, or making the hole circular or meet a minimum diameter.

rebound (re′-bound) *upwarping.*

rebulite (reb′-ul-ite) A submetallic gray monoclinic mineral: $Tl_5Sb_5As_8S_{22}$.

recapitulation theory (re′-ca-pit′-u-la′-tion) A theory in biology stating that an organism passes through successive stages resembling its ancestors so that the ontogeny of the individual is a recapitulation of the phylogeny of its group. See also: *palingenesis [paleont].* Syn: *Haeckel's law.*

recemented glacier (re-ce-ment′-ed) *regenerated glacier.*

Recent (Re′-cent) *Holocene.*

recentered projection (re-cen′-tered) A term preferred by the BNCG (1966, p.33) to the synonym *interrupted projection.*

receptacle (re-cep′-ta-cle) The apex of a pedicel (or peduncle) from which the organs of a flower grow out; also, the inflated tip of certain brown algae within which gametangia are borne (Fuller and Tippo, 1954, p.969).

receptaculitid (re′-cep-tac′-u-li′-tid) Any of a group of Paleozoic fossils of uncertain systematic position, belonging to the order Receptaculitales and characterized by an ovoid, globose calcareous body with the lower nuclear area closed and the upper lacunal area open. Meroms, always with stellate structures of four ribs, are borne on the central axis in whorls, circlets, or spirals. Receptaculitids have been classified as calcareous algae and as members of various animal phyla. Range, Ordovician to Permian.

recess [geomorph] An indentation in a surface; e.g. a cleft in a steep rock bank. See also: *reentrant.*

recess [struc geol] (a) An area in which the axial traces of folds are concave toward the outer edge of the folded belt. (b) A map-view curve in a mountain belt that is concave-to-the-foreland. Recesses may occur at the endpoints of a salient, or may form where different, non-parallel, segments of an orogen overlap. Ant: *salient [struc geol].*

recession [coast] (re-ces′-sion) A continuing landward movement of a shoreline or beach undergoing erosion or submergence. Also, a net landward movement of the shoreline or beach during a specified period of time. Ant: *advance [coast].* Cf: *retrogradation [coast].* Syn: *retrogression [coast].*

recession [geomorph] (a) The backward movement or retreat of an eroded escarpment; e.g. the slow wasting-away of a cliff under the influence of weathering and erosion. (b) The moving-back of a slope from a former position without a change in its angle. (c) The gradual upstream retreat of a waterfall or *knickpoint.*

recession [glaciol] An overall decrease in the volume of a glacier; a *retreat.*

recession [grd wat] The decline of a system's output in the absence of any inputs, e.g., the decline of the water table following a rise due to a recharge event.

recessional moraine (re-ces′-sion-al) An end or lateral moraine built during a temporary but significant pause in the final retreat of a glacier. Also, a moraine built during a slight or minor readvance of the ice front during a period of general recession. Syn: *peripheral moraine; retreatal moraine; stadial moraine.*

recession curve A hydrograph showing the decrease of the runoff rate after rainfall or the melting of snow. Direct runoff and base runoff are usually given separate curves as they recede at different rates. The use of the term *depletion curve* in reference to the base-runoff recession is considered incorrect (Langbein and Iseri, 1960).

recessive (re-ces′-sive) Said of a characteristic of an organism that must be inherited from both parents if it is to be exhibited by offspring. A recessive character can be passed on to offspring without being exhibited by a parent.

recharge The processes involved in the addition of water to the saturated zone, naturally by precipitation or runoff, or artificially by spreading or injection; also, the amount of water added. Syn: *intake; replenishment [grd wat]; groundwater replenishment; groundwater recharge; groundwater increment; increment.*

recharge area An area beneath which water reaches the saturated zone following *infiltration* and *percolation.* Beneath it, downward components of hydraulic head exist and groundwater moves downward into deeper parts of the aquifer (Fetter, 1994). Cf: *catchment area [grd wat]; discharge area.* Syn: *intake area.*

recharge basin A basin constructed in sandy material to collect water, as from storm drains, for the purpose of replenishing groundwater supply.

recharge well A well used to inject water into one or more aquifers in the process of *artificial recharge.* Syn: *injection well.*

recharge zone Region of a lake bottom where water from the lake recharges the shallow groundwater.

reciprocal bearing (re-cip′-ro-cal) *back bearing.*

reciprocal lattice A lattice array of points formed by drawing perpendiculars to each plane (hkl) in a crystal lattice through a common point as origin. Points are located on each perpendicular at a distance from the origin (000) inversely proportional to spacing of the specific lattice planes (hkl). The axes of the reciprocal lattice are the *a* axis,* the *b* axis,* and the *c* axis,* which are perpendicular, respectively, to (100), (010), and (001) of the crystal lattice. The coordinates of each reciprocal lattice point are (*hkl*) or whole multiples (n*h*, n*k*, n*l*) in terms of the unit lengths a*, b*, and c*. Cf: *direct lattice.*

reciprocal lattice point One of the points making up a *reciprocal lattice.*

reciprocal leveling Trigonometric leveling in which vertical angles have been observed at both ends of the line in order to eliminate instrumental errors; e.g. leveling across a wide river by establishing a turning point on each bank of the river from one side and taking a backsight on each to determine the height of instrument on the other side. The mean of the differences in level represents the true difference.

reciprocal salt-pair diagram *Jänecke diagram.*

reciprocal space A mathematical space defined by the coordinate system of the *reciprocal lattice.* Cf: *direct space.*

reciprocal strain ellipsoid In elastic theory, an ellipsoid of certain shape and orientation which under homogeneous strain is transformed into a sphere. Cf: *strain ellipsoid.*

reciprocity principle (a) The rule that the potential at a recording point M with respect to a current source point A is the same as if the points of measurement and source were reversed. Applied in electrical exploration (Sheriff, 2002). (b) The rule that the seismic trace from a source at A to a geophone at B is the same as from a source at B to a geophone at A, if sources and receivers are similarly coupled to the Earth (Sheriff, 2002). For converted waves, reciprocity requires interchanging the motion senses of the sources/receivers as well as their locations (Dellinger and Noble, 1997).

reclaimed water Wastewater made fit for reuse for potable or nonpotable purposes.

reclined (re-clined') (a) Said of a graptoloid rhabdosome with stipes extending above the sicula and enclosing an angle less than 180 degrees between their dorsal sides. Cf: *reflexed; declined; deflexed.* (b) Said of a tabulate corallite growing and opening obliquely with respect to the surface of corallum (TIP, 1956, pt.F, p.250). (c) When used to describe the cusp or denticles of a conodont element, this term implies that the axis of the cusp or of the denticles is straight and forms an angle of less than 90 degrees with the upper edge of the base or of a posterior process (TIP, 1981, pt. W, supp.2).

reclined fold A fold whose hingeline plunges parallel to the direction of dip of the axial surface (Turner and Weiss, 1963, p.119).

recomposed granite (a) An arkose consisting of consolidated feldspathic residue (produced by surface weathering of an underlying granitic rock) that has been so little reworked and so little decomposed that upon cementation the rock looks very much like the granite itself. It may have a faint bedding, an unusual range of particle sizes (unlike the even-grained or porphyritic texture of true granite), and a greater percentage of quartz than is normal for granite. Syn: *reconstructed granite.* (b) A conglomerate that has been recrystallized by strong metamorphism into a rock that simulates granite, as in the Lake Superior region. Cf: *meta-arkose.*

recomposed rock A rock produced in place by the cementation of the fragmental products of surface weathering; e.g. a *recomposed granite.* The term has been applied to a rock of intermediate character straddling an unconformable surface between the breccia of the lower formation and the conglomeratic base of the upper formation (Leith, 1923).

reconnaissance (re-con'-nais-sance) (a) A general, exploratory examination or survey of the main features (or certain specific features) of a region, usually conducted as a preliminary to a more detailed survey; e.g. an engineering survey in preparing for triangulation of a region. It may be performed in the field or office, depending on the extent of information available. (b) A rapid geologic survey made to gain a broad, general knowledge of the geologic features of a region.

reconnaissance map A map based on the information obtained in a reconnaissance survey and on data obtained from other sources.

reconnaissance survey A preliminary survey, usually executed rapidly and at relatively low cost, prior to mapping in detail and with greater precision.

reconnoiter (re-con-noi'-ter) To make a reconnaissance of; esp. to make a preliminary survey of an area for military or geologic purposes.

reconsequent (re-con'-se-quent) *resequent [streams].*

reconstructed glacier (re-con-struct'-ed) *regenerated glacier.*

reconstructed granite *recomposed granite.*

reconstructed stone A gem material made by the fusing or sintering of small particles of the genuine stone to make a larger stone (Manutchehr-Danai, 2000).

reconstructed vein An idealized composite vein model proposed by Emmons (1936) to illustrate typical mineral assemblages in veins formed under different pressure and temperature conditions.

reconstructive transformation (re-con-struc'-tive) A type of *phase transformation* that involves the breaking of either first- or second-order coordination bonds. It is usually a slow transformation. An example is quartz-tridymite. Cf: *dilatational transformation; displacive transformation; rotational transformation; substitutional transformation.*

record (re'-cord) (a) *geologic record.* (b) *stratigraphic record.* (c) *seismic record.* (d) A collection of related date elements, items, or codes, treated as a unit. Records my be of fixed or variable lengths.

record section A display of seismic traces side-by-side to show the continuity of events along a line of profile.

recovery [grd wat] (re-cov'-er-y) (a) The rise in static water level in a well that occurs when discharge from that well or a nearby well is stopped. (b) *groundwater withdrawal.*

recovery [mining] In mining, the percentage of valuable constituent derived from an ore, or of coal from a coal seam; a measure of mining or extraction efficiency.

recovery [struc petrol] Any of the processes through which the number of grain dislocations (i.e. strain energy) produced during rock deformation can be reduced. It includes *polygonization,* by which dislocations guide and climb into walls or "subgrain boundaries", and *recrystallization,* by which new strain-free material is formed at the expense of paleoblasts.

recovery [surv] A visit to a survey station to identify its mark or monument as authentic and in its original location and to verify or revise its description.

recovery test The part of an *aquifer test* that involves measurement of the rate of fluid-pressure or water-level return toward static conditions in a well following a period of pumping or injection. Syn: build-up test.

recrystallization (re-crys'-tal-li-za'-tion) The formation, essentially in the solid state, of new crystalline mineral grains in a rock. The new grains are generally larger than the original grains and may have the same or a different mineralogical composition. Specif., it is the way in which a deformed crystal aggregate releases stored strain energy due to deformation. It consists of several discrete phases: annihilation of crystal defects and their rearrangement into stable arrays, which may lead to the formation of polygonal subgrains; primary crystallization, involving the nucleation of new grains and their growth by the migration of high-angle boundaries, producing an aggregate of strain-free grains; and normal grain growth, in which the average grain size increases, driven by a tendency to decrease the interfacial energies of adjacent grains. Secondary recrystallization may follow, in which some grains grow very large by consuming neighboring grains. Cf: *Riecke's principle.*

recrystallization flow *Flow [exp struc geol]* in which there is molecular rearrangement by solution and redeposition, solid diffusion, or local melting.

rectangular coordinates (rect-an'-gu-lar) Two- or three-dimensional coordinates on any system in which the axes of reference intersect at right angles; *plane coordinates.* Also, a coordinate system using rectangular coordinates. Syn: *rectilinear coordinates.*

rectangular cross ripple *interference ripple.*

rectangular drainage pattern A drainage pattern in which the main streams and their tributaries display many right-angle bends and exhibit sections of approximately the same length; it is indicative of streams following prominent fault or joint systems that break the rocks into rectangular blocks. It is more irregular than the *trellis drainage pattern,* as the side streams are not perfectly parallel and not necessarily as conspicuously elongated, and secondary tributaries need not be present (Zernitz, 1932, p.503). Examples are well developed along the Norwegian coast and in parts of the Adirondack Mountains. See also: *angulate drainage pattern; joint valley.* Syn: *lattice drainage pattern.*

rectangular speciation *punctuated equilibria.*

rectification [coast] (rec'-ti-fi-ca'-tion) The simplification and straightening of the outline of an initially irregular and crenulate shoreline by marine erosion cutting back headlands and offshore islands, and by deposition of waste resulting from erosion or of sediment brought down by neighboring rivers.

rectification [data] (a) A set of techniques for the elimination of errors in data. (b) A process in which aerial photographs, remotely sensed data, or analog maps are tied precisely to a geographic reference system. Cf: *rubber sheeting.*

rectification [eng] A new alignment to correct a deviation of a stream channel or bank.

rectification [photo] The process of projecting a tilted or oblique aerial photograph onto a horizontal reference plane, the angular relation between photography and plane being determined by ground reconnaissance. Cf: *transformation [photo].*

rectilinear coordinates (rec-ti-lin'-e-ar) *rectangular coordinates.*

rectilinear evolution Continued change of the same kind and in the same direction within a line of descent over a considerable length of time; similar to *orthogenesis,* but without implying how the direction is determined and maintained.

rectilinear shoreline A long, relatively straight shoreline, or one with nearly right-angle bends, such as are caused by intersecting joints.

rectimarginate (rec-ti-mar'-gin-ate) Said of a brachiopod having a plane (straight) anterior commissure; also, said of such a commissure.

rectiradiate (rec-ti-ra'-di-ate) Said of an ammonoid rib in straight radial position, bending neither forward nor backward. Cf: *prorsiradiate; rursiradiate.*

rectorite (rec'-tor-ite) A regularly ordered, 1:1 mixed-layer dioctahedral mica/dioctahedral smectite (Brown and Weir, 1963): $(Na,Ca)Al_4(Si,Al)_8O_{20}(OH)_4 \cdot 2H_2O$. The name rectorite has priority over *allevardite* (Bailey, 1982). A prefix of Na-, K-, or Ca- may be used to indicate the dominant interlayer cation in the mica component.

recumbent fold (re-cum'-bent) An overturned fold, the axial surface of which is horizontal or nearly so (Turner and Weiss, 1963, p.119).

recurrence horizon (re-cur'-rence) In peat bogs, the demarcation between older, more decomposed peat and younger material; a parting or horizon marking an abrupt change in lithology of a peat bog, reflecting climatic change. Cf: *grenz.*

recurrence interval (a) The average time interval between occurrences of a hydrological event of a given or greater magnitude. (b) In an annual flood series, the average interval in which a flood of a given size recurs as an annual maximum. (c) In a partial duration series, the average interval between floods of a given size, regardless of their relationship to the year or any other period of time. This distinction holds even though for large floods recurrence intervals are nearly the same on both scales. (ASCE, 1962).

recurrent (re-cur'-rent) Said of an organism or group of organisms that reappears in an area from which it had been previously expelled; e.g., a fossil present in two different rock units separated by a unit or units in which it is absent.

recurrent folding A type of folding that results from periodic deformation or subsidence and is characterized by thinning or disappearance of formations at the crest of uplifts. Cf: *supratenuous fold.* Syn: *revived folding.* Rarely used.

recurring polynya (re-cur'-ring) A *polynya* that is found in the same region every year.

recurve (re'-curve) A feature produced by the successive landward extension of a spit.

recurved When used to describe the cusp or denticles of a conodont element, this term implies that the cusp or denticle axis is not straight, but arcuate, and that the apex of the cusp or denticles points posteriorly (TIP, 1981, pt. W, supp.2).

recurved spit (re-curved') A spit whose outer end is turned landward by current deflection, by the opposing action of two or more currents, or by wave refraction. Syn: *hook [geomorph]; hooked spit.*

recycled grain (re-cy'-cled) A grain derived from a pre-existing sedimentary rock and incorporated into a new sediment or sedimentary rock; as contrasted to a grain derived directly from an igneous or metamorphic rock.

recycling (re-cy'-cling) The processes by which particles are weathered from an existing sedimentary rock, transported, deposited, and incorporated into a new sediment or sedimentary rock.

red algae A group of *algae,* members of the *phylum* Rhodophyta, with red color from the pigment phycoerythrin. Red algae are predominantly tropical; unicellular forms exist, but most are multicellular with very complex life cycles and reproduction. The *coralline algae* deposit calcium carbonate within their cell walls, forming extensive carbonate deposits; they also form *rhodoliths* and contribute to *marls.* The range of this group is from the Proterozoic to the present.

red antimony *kermesite.*

red arsenic *realgar.*

red-bed copper Stratabound copper sulfide and native copper deposits associated with nearshore, red-bed sedimentary sequences, often associated with evaporites, e.g. Creta, Oklahoma, and Nacimiento, New Mexico.

red beds Sedimentary strata composed largely of sandstone, siltstone, and shale, with locally thin units of conglomerate, limestone, or marl, that are predominantly red in color due to the presence of ferric oxide (hematite) usually coating individual grains; e.g. the Permian and Triassic sedimentary rocks of western and eastern U.S., and the Old Red Sandstone facies of the European Devonian. At least 60% of any given succession must be red before the term is appropriate, the interbedded strata being of any color (Hatch and Rastall, 1965, p.371). Also spelled: redbeds. Syn: *red rock.*

red clay A pelagic deposit that is fine grained and bright to reddish brown or chocolate colored, formed by the slow accumulation of material a long distance from the continents and at depths generally greater than 3,500 meters. It contains relatively large proportions of windblown particles, meteoric and volcanic dust, pumice, shark teeth, whale earbones, and manganese nodules. The content of $CaCO_3$ is low, generally <10%. Syn: *brown clay.*

red cobalt *erythrite.*

red copper ore *cuprite.*

Red Desert soil A zonal great soil group consisting of soils formed under warm-temperate to hot, dry regions under desert-type vegetation, mostly shrubs. Not used in current U.S. system of soil taxonomy.

reddingite (red'-ding-ite) A pink-white or yellow-white orthorhombic mineral of the *phosphoferrite* group: $Mn^{2+}_3(PO_4)_2{\cdot}3H_2O$.

Reddish-Brown Lateritic soil (Red'-dish-Brown') A great soil group in the 1938 classification system, a group of zonal soils developed from a mottled red lateritic parent material and characterized by a reddish-brown surface horizon and an underlying red-clay B horizon. These soils are now classified as *Humults* and *Udults.*

Reddish Brown soil A great soil group in the 1938 classification system, a group of zonal soils having a reddish, light brown surface horizon overlying a heavier, more reddish horizon and a light-colored horizon with calcium-carbonate accumulation. These soils are developed in warm, temperate to tropical, semiarid climate under shrub and short-grass vegetation (USDA, 1938). Most are now classified as *Ustalfs, Orthids,* and *Argids.* Cf: *Red Desert soil.*

Reddish Chestnut soil A great soil group in the 1938 classification system, a group of zonal soils having a thick surface horizon that ranges from dark brown to reddish or pinkish, below which is a heavier, reddish-brown horizon and a carbonate accumulation. They are developed under mixed grasses with some shrubs, in a warm to temperate, semiarid climate (USDA, 1938). These soils are now classified as *Ustalfs* and *Ustolls.* Cf: *Chestnut soil.*

reddle (red'-dle) *red ocher.*

red earth Highly leached, red clayey soils of the humid tropics, usually with very deep profiles that are low in silica and high in sesquioxides. Not used in current U.S. system of soil taxonomy.

redeposition (re'-dep-o-si'-tion) Formation in a new accumulation, such as the deposition of sedimentary material that has been picked up and moved (reworked) from the place of its original deposition, or the solution and reprecipitation of mineral matter. See also: *resedimentation.*

redgillite An emerald-green monoclinic mineral: $Cu_{12}(SO_4)_2(OH)_{20}{\cdot}2H_2O$. Not approved by the International Mineralogical Association.

red hematite A syn. of *hematite.* Cf: *brown hematite; black hematite.*

redikortsevite A light yellow or colorless orthorhombic mineral: $(NH_4)MgCl_3{\cdot}6H_2O$. It is a product of a burning coal dump, and has not been approved by the International Mineralogical Association.

redingtonite (red'-ing-ton-ite') A silky white or purple mineral of the *halotrichite* group: $(Fe^{2+},Mg,Ni)(Cr,Al)_2(SO_4)_4{\cdot}22H_2O$. It has not been adequately characterized.

red iron ore *hematite.*

red lead *minium.*

red lead ore *crocoite.*

redledgeite (red'-ledge-ite) A black monoclinic mineral of the *hollandite* group: $BaTi_6Cr^{3+}_2O_{16}$. Syn: *chromrutile.*

red loam *red earth.*

red manganese A reddish manganese mineral; specif. rhodonite and rhodochrosite.

Red Mediterranean soil An obsolete term for a reddish soil that forms in a Mediterranean climate.

red mud A type of *mud [marine geol]* that is terrigenous and contains as much as 25% calcium carbonate. Its color is due to the presence of ferric oxide.

red ocher A red clayey or earthy hematite used as a pigment. Syn: *reddle; ruddle.*

redondite An ill-defined colorless or green mineral: (Al,Fe) (PO4)$\cdot2H_2O$. It may be the same mineral as Messbach-type variscite. Syn: *tangaite.*

Redonian (Re-do'-ni-an) European stage: Upper Pliocene (above Plaisancian, below Calabrian). Equivalent to Astian.

red ore A red-colored ore mineral; specif. hematite or metahewettite.

red ore [eco geol] Residually enriched iron-rich slate composed of earthy red hematite and goethite found in the Knob Lake iron range of Quebec and Labrador, Canada. Cf: *blue ore; yellow to brown ore; rubble ore; treat rock.*

red orpiment *realgar.*

redox Reduction-oxidation.

redox concentrations Zones of apparent accumulation of Fe-Mn oxides in soils.

redox depletions Zones of low chroma (2 or less) where Fe-Mn oxides alone or both Fe-Mn oxides and clay have been stripped out of the soil.

red oxide of copper *cuprite.*

red oxide of zinc *zincite.*

redoximorphic features Soil properties associated with wetness that result from the reduction and oxidation of iron and manganese compounds in the soil after saturation with water and desaturation, respectively. See: *redox concentrations; redox depletions.*

redoxomorphic stage (re-dox'-o-mor'-phic) A term introduced by Dapples (1962) for the earliest geochemical stage of diagenesis, characterized by mineral changes primarily due to oxidation and reduction reactions (reversible reactions). It is typical of unlithified sediment and preceeds the *locomorphic stage.* See also: *phyllomorphic stage.*

red rock (a) *red beds.* (b) A driller's term for any reddish sedimentary rock.

redruthite (red'-ruth-ite) *chalcocite.*

red schorl (a) *rubellite.* (b) *rutile.*

red silver ore A red silver-sulfide mineral; specif. "dark red silver ore" (pyrargyrite) and "light red silver ore" (proustite).

red snow A general name for snow colored by the presence of various red or pink microscopic algae (such as *Sphaerella* and *Protococcus*) in the upper layers of snow in arctic and alpine regions. Cf: *green snow; yellow snow.* Syn: *pink snow.*

redstone (red'-stone) A reddish sedimentary rock; specif. a deep-red, clayey sandstone or siltstone, as in the Triassic deposits of Connecticut. Cf: *brownstone.*

red tide A type of *plankton bloom* that is caused by dinoflagellates, commonly toxic.

reduced cell *reduced unit cell.*

reduced latitude The angle at the center of a sphere tangent to a reference ellipsoid along the equator, between the plane of the equator and a radius to the point intersected on the sphere by a straight line perpendicular to the plane of the equator. Syn: *parametric latitude.*

reduced major axis A form of line-*fitting* in which the sum of products of the deviations from the fitted line in the *x*-direction and the *y*-direction is minimized.

reduced matrix A soil matrix which has a low chroma in situ, but undergoes a change in hue or chroma within 30 minutes after the soil material is exposed to air. The color change is due to the oxidation of iron.

reduced mud *black mud.*

reduced ore Sulfide-rich *protore* that has not been oxidized by exposure to groundwater. Typically used to refer to unoxidized ore beneath and oxidized blanket produced by weathering and erosion.

reduced unit cell The *unit cell* constructed from the three shortest, noncoplanar lattice vectors that form a right-handed basis. It is not constrained by symmetry requirements, and thus may not correspond to the *conventional unit cell.* Apart from the possibility of different orientations arising from lattice symmetry, the reduced unit cell is unique for a lattice. Cf: *conventional unit cell; primitive unit cell.* Syn: *reduced cell.*

reducing flame (re-duc'-ing) In blowpiping, the blue part of the flame, in which oxygen in the compound being tested is partly burned away. Cf: *oxidizing flame.*

reduction [geomorph] (re-duc'-tion) The lowering of a land surface by erosion.

reduction [geophys] *gravity reduction.*

reduction body A multicellular mass resulting from the disorganization of a sponge and capable of reorganizing into a sponge with a functional aquiferous system.

reduction index The rate of wear of a sedimentary particle subject to abrasion in the course of transportation, expressed as the difference between the mean weight of the particle before and after transport divided by the product of mean weight before transport and the distance traveled (Wentworth, 1931, p.25). Abbrev: RI. Cf: *durability index.*

reduction sphere A white, leached spheroidal mass produced in a reddish or brownish sandstone by a localized reducing environment, commonly surrounding an organic nucleus or a pebble and ranging in size from a poorly defined speck to a large perfect sphere more than 25 cm in diameter (Hamblin, 1958, p.24-25); e.g. in the Jacobsville Sandstone of northern Michigan.

reduction to sea level The application of a correction to a measured distance or other quantity on the Earth's surface to reduce it to its corresponding value at sea level.

reduction to the equator Converting the magnetic data recorded at low latitudes to what they would be at the *magnetic equator.* This procedure severley distorts shapes of magnetic anomalies and is usually not applied in magnetic exploration. Abbrev: RTE. See also: *reduction to the pole.*

reduction to the pole Converting magnetic data that have been recorded in the inclined Earth's magnetic field to what they would be if the magnetic field had been vertical. It removes the asymmetry of magnetic anomalies caused by *inclination* and locates anomalies above their sources, assuming that *remanent magnetization* is small compared to *induced magnetization.* The procedure becomes unstable at low latitudes. Abbrev: RTP.

reduzates (re-du'-zates) Sediments accumulated under reducing conditions and thus characteristically rich in organic carbon and in iron sulphide; coal and black shale are principal examples. It is one of Goldschmidt's groupings of sediments or analogues of differentiation stages in rock analysis. Cf: *resistates; evaporates; hydrolyzates; oxidates.*

red vitriol *bieberite.*

Red-Yellow Podzolic soils A combination of the zonal great soil groups, Red Podzolic and Yellow Podzolic, consisting of soils formed under warm-temperature to tropical, humid climates, under deciduous or coniferous forest vegetation and usually, except for a few members of the Yellow Podzolic Group, under conditions of good drainage. Not used in current U.S. system of soil taxonomy.

red zinc ore *zincite.*

REE Rare-earth elements.

reed cast A vertical and cylindrical cast of sand presumably representing the filling of a mold left by a reed.

reederite-(Y) A yellow or orange-brown hexagonal mineral: $(Na,Mn)_{15}(Y,Ce)_2(CO_3)_9(SO_3F)Cl$.

reedmergnerite (reed-merg'-ner-ite) A colorless triclinic mineral of the *feldspar* group: $NaBSi_3O_8$. It is the boron analogue of albite.

reed peat A type of *telmatic peat* derived from predominantly reed plants.

reef (a) A ridgelike or moundlike structure, layered or massive, built by sedentary calcareous organisms, esp. corals, and consisting mostly of their remains; it is wave-resistant and stands above the surrounding contemporaneously deposited sediment. Also, such a structure built in the geologic past and now enclosed in rock, commonly of differing lithology. See also: *bank [oceanog]; bioherm; biostrome.* Syn: *organic reef.* (b) A mass or ridge of rocks, esp. coral or shells and sometimes sand, gravel, boulder conglomerate, hogback ridges, dikes, or sills, rising above the surrounding sea or lake bottom to or nearly to the surface, and forming an obstruction to navigation; specif. such a feature at 10 fathoms (formerly 6) or less. See also: *shoal.* (c) A provincial term for a metalliferous mineral deposit, esp. gold-bearing quartz (e.g. *saddle reef*).

reefal (reef'-al) Pertaining to a reef, its integral parts, organisms frequenting reefs in general, and sediments accumulating on or around reefs.

reef apron The gently sloping surface of sediment accumulation behind or surrounding a reef. Cf: *reef flank.*

reef breccia A rock formed of limestone fragments broken off from a reef by the action of waves and tides.

reef buttress A long, narrow sloping ridge or vertical promontory, standing out or projecting seaward beyond the steep fore-reef slope or vertical *reef wall,* and flanked by shallow valleys or reentrants. Cf: *spur-and-groove structure.*

reef cap A deposit of fossil-reef material overlying or covering an island or mountain. Cf: *coral cap.*

reef cluster A group of reefs of wholly or partly contemporaneous growth, found within a circumscribed area or geologic province. See also: *hermatopelago.*

reef complex A solid reef and the heterogeneous and contiguous fragmentary material derived from it by abrasion; the aggregate of reef, fore-reef, back-reef, and interreef deposits, bounded on the seaward side by basin sediments and on the landward side by lagoonal sediments (Nelson et al., 1962, p.249). Term introduced by Henson (1950, p.215-216) to include the reef and "all genetically(?) associated sediments". Cf: *reef tract.*

reef conglomerate *reef talus.*

reef core Within an *organic reef,* the centrally located solid rock mass constructed in place by reef-building organisms; the solid reef proper. See also: *reef flank.*

reef crest The sharp break in slope at the seaward margin or edge of the *reef flat,* located at the top of the *reef front;* marked by dominance of a particular coral species (such as *Acropora palmata* throughout the Caribbean) or by an algal ridge and/or surge channels. Cf: *reef edge; reef front.*

reef debris *reef detritus.*

reef detritus Fragmental material derived from the erosion of an organic reef; some is produced by mechanical breakage by waves or surf, some by accumulation of shells, and some by boring animals detaching solid materials. The finer particles tend to be carried away by waves, while the coarser fragments often form a *talus apron* around the reef. See also: *reef talus.* Syn: *reef debris.*

reef edge The seaward margin of the reef flat, commonly marked by surge channels. Cf: *reef crest; reef front.*

reef flank The part of a reef that surrounds, interfingers with, and locally overlies the *reef core,* often indicated by beds of *reef detritus* dipping away from the core. Cf: *reef apron.*

reef flat A stony platform of *reef rock,* landward of the *reef crest* at or above the low tide level, occasionally with patches of living coral and associated organisms, and commonly strewn with coral fragments and coral sand. It may include shallow pools, irregular gullies, low islands of sand or rubble (often vegetated, esp. by palms), and scattered colonies of the more hardy species of coral.

reef frame The rigid, reef-building, in-situ framework of an *organic reef,* consisting of the skeletons of sessile calcareous organisms and excluding reef-flank and associated fragmental deposits (MacNeil, 1954, p.390). Syn: *organic lattice; growth lattice; lattice [reef].*

reef front The upper part of the outer or seaward slope of a reef, extending to the *reef edge* from above the depth limit of abundant living coral and coralline algae. Cf: *reef crest.*

reef-front terrace A shelflike or benchlike eroded surface, sometimes veneered with organic growth, often sloping gently seaward; may be developed at various depths, as at 6, 8, 20, and 35 m on reefs off northern Eleuthera. Syn: *reef terrace.*

reef knoll (a) A bioherm or fossil coral reef now represented by a small, prominent hill, up to 100 m high; specif. a small, pinnaclelike or conical mass of coralline limestone, more or less circular in ground plan and commonly surrounded by rock of different lithology, as in the type area of the Craven district in Yorkshire, England. (b) A present-day reef in the form of a knoll; a small *reef patch* developed locally and upward rather than outward or laterally. The term was first used by Tiddeman (1890, p.600) for a reef feature that originated as a knoll. Syn: *knoll reef.*

reef limestone A limestone consisting of the remains of active reef-building organisms, such as corals, sponges, and bryozoans, and of sediment-binding biotic constituents, such as calcareous algae.

reef milk A very fine-grained matrix material of the back-reef facies, consisting of white, opaque microcrystalline calcite and/or aragonite derived from bioerosion and abrasion of the reef core and reef flank.

reefoid (reef'-oid) Resembling a reef; e.g. "reefoid rocks".

reef patch A growth of coral formed independently on a shelf of less than 70 m depth, often in the lagoon of a barrier reef or atoll, ranging from an expanse several kilometers across down to that of a single large colony. See also: *reef knoll.* Cf: *patch reef; bommy.*

reef pinnacle *pinnacle [reefs].*

reef ring *atoll.*

reef rock A resistant massive unstratified rock consisting of the calcareous remains of reef-building organisms, often intermingled with carbonate sand and shingle, the whole cemented by calcium carbonate. Also spelled: reefrock. Cf: *biolithite; boundstone.* Syn: *hermatolith; hermatobiolith.*

reef segment A part of an organic reef lying between passes, gaps, or channels.
reef slope The face or flank of a reef, rising from the sea floor (Maxwell, 1968, p.106-107).
reef talus Massive or thick-bedded inclined strata consisting of coarse *reef detritus,* usually deposited along the seaward margin of an organic reef; it is one type of *reef flank* deposit. Syn: *reef conglomerate.*
reef terrace *reef-front terrace.*
reef tract An extensive offshore area in which many reefs are found, such as the Florida reef tract. Cf: *reef complex.*
reef tufa Drusy, prismatic, fibrous calcite deposited directly from supersaturated water upon the void-filling internal sediment of the calcite mudstone of a reef knoll (Bissell and Chilingar, 1967, p.165). See also: *stromatactis.*
reef wall (a) A wall-like upgrowth of living coral and the skeletal remains of dead coral and other reef-building organisms, reaching intertidal level where it acts as a partial barrier between adjacent environments (Henson, 1950, p.227); an elongate *reef core.* See also: *wall reef.* (b) Near-vertical to overhanging wall developed on sea-facing reef margins below the zone of shallow reef growth (ca. 30 m) and extending to depths of ca. 150 m in the Caribbean. The irregular and cavernous surface is veneered with luxurious growths of platey corals and associated skeletal organisms to depths of some 100 m. Cavities become infilled with sand and mud sized skeletal debris, which becomes cemented by non-skeletal aragonite and high-Mg calcite to form an armor at least a few meters thick. Repeated episodes of this process during the Quaternary may have produced some walls in the Caribbean, others may be drowned Pleistocene sea cliffs veneered with Holocene accretion.
reefy (reef'-y) (a) Containing reefs, as a "reefy harbor". (b) Containing sedimentary material that resembles the material of an organic reef.
reentrant (re-en'-trant) adj. Reentering or directed inward; e.g. a *reentrant* angle in a coastline or on a twinned crystal. n. A prominent, generally angular indentation in a landform; e.g. an inlet between two promontories along a coastline, or a transverse valley extending into an escarpment. Ant: *salient [geomorph].* Cf: *recess [geomorph].*
reentrant angle The angle between two plane surfaces on a solid, in which the external angle is less than 180°.
reevesite (reeves'-ite) A yellow or greenish-yellow rhombohedral mineral: $Ni_6Fe^{3+}_2(CO_3)(OH)_{16}·4H_2O$.
reference axis (ref'-er-ence) *fabric axis.*
reference ellipsoid An ellipsoid of specified dimensions and associated with a geodetic reference system or a geodetic datum. Coordinates given in this system are said to be "with respect to the reference ellipsoid." Reference ellipsoids are most commonly ellipsoids of revolution and are sometimes called reference spheroids.
reference frame A coordinate system associated with a physical system.
reference level A datum plane; e.g. a standard level (in the study of underwater sound) to which sound levels can be related.
reference line Any line that serves as a reference or base for the measurement of other quantities; e.g. a *datum line.*
reference locality A locality containing a reference section, established to supplement the *type locality.*
reference plane *datum plane [surv].*
reference section A rock section, or group of sections, designated to supplement the *type section,* or sometimes to supplant it (as where the type section is no longer exposed), and to afford a standard for correlation for a certain part of the geologic column; e.g. an auxiliary section of particular regional or facies significance, established through correlation with the type section, and from which lateral extension of the boundary horizons may be made more readily than from the type section (ISSC, 1994, p.28). See also: *standard section; hypostratotype.*
reference seismometer In seismic prospecting, a detector placed on the Earth's surface to record the effects of successive sources under similar conditions, to permit overall time comparisons. It is used in connection with well velocity surveys.
reference station A place where tidal constants previously have been determined and which is used as a standard for the comparison of simultaneous observations at a second station. Also, a place where independent daily predictions are given in the tide and tidal-current tables, from which corresponding predictions are obtained for other stations by means of differences or factors (CERC, 1966, p. A26). Cf: *tide station.* British syn: *standard port.*
referencing (ref'-er-enc-ing) The process of measuring the horizontal (or slope) distances and directions from a survey station to nearby landmarks, reference marks, and other permanent objects which can be used in the recovery or relocation of the station.
refikite (re'-fik-ite) A white, very soft orthorhombic mineral occurring in modern resins: $C_{20}H_{32}O_2$. Also spelled: reficite.
reflectance (re-flec'-tance) The ratio of the energy reflected by a body to that incident upon it. See also: *spectral reflectance.*
reflected infrared (re-flect'-ed) Wavelengths from 0.7 to about 3 µm that are primarily reflected solar radiation. Syn: *near infrared.*
reflected refraction (a) Head-wave energy which has been reflected or diffracted back from a discontinuity in the refractor, such as a fault. (b) A refraction multiple, involving head-wave energy which has been multiply reflected between reflectors. The extra travel may occur at any time during the travel along the refractor. The effect is to add cycles to the refraction arrival. (c) Any of a number of travel paths which involve reflection at an interface as well as travel through a high-velocity refractor. (Sheriff, 2002).
reflected wave An elastic wave that has been reflected at an interface between media with different elastic properties. It is indicated by symbols such as PP, SS, SP, PS, PKP, etc., each successive letter indicating a leg of the travel path. Cf: *converted wave.*
reflecting goniometer (re-flect'-ing) *reflection goniometer.*
reflecting horizon In seismic profiling of the ocean floor, a prominent layer of *reflection* caused by an acoustic impedance contrast between adjacent layers, sedimentary or igneous.
reflection (re-flec'-tion) (a) The return of a wave incident upon a surface to its original medium. See also: *law of reflection; total reflection.* (b) In seismic prospecting, the indication on a record of such reflected energy. (c) In X-ray diffraction, a synonym for diffraction peak, diffraction spot, or diffraction line. (d) In crystallography, a symmetry transformation by a *plane of mirror symmetry.* Cf: *refraction [cryst]; diffraction [cryst].*
reflection angle *Bragg angle.*
reflection coefficient The ratio of the amplitude of the reflected wave to that of the incident wave, often taken at normal incidence. The ratio of the reflected energy to the incident energy is the reflection coefficient squared.
reflection goniometer A *goniometer* that measures the angles between crystal faces by reflection of a parallel beam of light from the successive crystal faces. Cf: *contact goniometer; two-circle goniometer.*
reflection plane *plane of mirror symmetry.*
reflection pleochroism *bireflectance.*
reflection profile Seismic recordings from a number of individual seismometer groups arranged in a line and at relatively short distances from the source, in which the data displayed primarily result from reflected seismic waves. Cf: *refraction profile.*
reflection shooting A type of seismic survey based on measurement of the travel times of waves that originate from an artificially produced disturbance and are reflected back at near-vertical incidence from subsurface boundaries separating media of different densities and/or elastic-wave velocities. Cf: *refraction shooting.*
reflection signature The aspect of a waveshape which makes it distinctive; character. A waveshape may distinguish a particular source, transmission path, or reflecting sequence.
reflection spectrum The spectrum seen when incident waves are selectively altered by a reflecting substance. It is analyzed with a *spectroreflectometer.*
reflection tomography *tomography.*
reflection twin A crystal twin whose symmetry is formed by apparent mirror image across a plane. Cf: *rotation twin.*
reflectometer (re-flec-tom'-e-ter) An apparatus for measuring the reflectivity of a substance, using some form of radiant energy such as light.
reflector (re-flec'-tor) An interface between media of different densities and/or elastic properties that reflects seismic waves.
reflexed (re-flexed') Said of a graptoloid *rhabdosome* with stipes extending above the *sicula,* their initial parts enclosing an angle less than 180 degrees between their dorsal sides, and their distal parts tending to the horizontal. Cf: *reclined; deflexed; declined.*
reflux (re'-flux) A return flow, especially the return flow of concentrated brine through the floor or across the barrier sill of an evaporite basin. Because such brines may be enriched in magnesium compared to seawater, reflux is believed to contribute to the dolomitization of carbonate rocks in some basinal sequences (Adams and Rhodes, 1960).
refolded fold A complex three-dimensional structure created by the interference of two or more folding events, or by a progressive rotating deformation event (Thiessen, 1986). Cf: *superposed fold.*
refoliation (re'-fo-li-a'-tion) A foliation that is subsequent to and oriented differently from an earlier foliation.
refracted cleavage (re-fract'-ed) Cleavage that changes orientation from layer to layer where layers are of different rock types.
refracted wave *head wave.*
refraction [cryst] (re-frac'-tion) The deflection of a ray of light or of an energy wave (such as a seismic wave) due to its passage from one medium to another of differing density, which changes its velocity. Cf: *reflection; diffraction [cryst].* See also: *single refraction; birefringence [cryst].*
refraction angle *angle of refraction.*
refraction profile A seismic profile obtained by designing the spread geometry in such a manner as to enhance the recording of energy that has traveled more-or-less horizontally through a medium with large velocity. Cf: *reflection profile.*
refraction shooting A type of seismic survey based on the measurement of the travel times of seismic waves that have travelled nearly parallel to the bedding in high-velocity layers, in order to map such layers. Cf: *reflection shooting.*
refractive index (re-frac'-tive) *index of refraction.*
refractive power *refractivity.*

refractivity (re-frac-tiv′-i-ty) The power of a substance to refract light. Such ability can be quantitatively expressed by the *index of refraction.* See also: *specific refractivity.* Syn: *refractive power; refringence.*

refractometer (re-frac-tom′-e-ter) An apparatus for measuring the indices of refraction of a substance, either solid or liquid; e.g., can be used to measure salinity of seawater since the refractive index of seawater is a function of salinity. Various types are designed for various substances; the chief type used for analysis of gems and minerals is the *Abbe refractometer.*

refractometry (re-frac-tom′-e-try) The measurement of indices of refraction, by means of a *refractometer.*

refractory (re-frac′-to-ry) (a) Said of an ore from which it is difficult or expensive to recover its valuable constituents. (b) Exceptionally resistant to heat.

refractory clay *fireclay.*

refractory inclusions *calcium aluminum-rich inclusions.*

refringence (re-frin′-gence) *refractivity.*

Refugian (Re-fu′-gi-an) A Pacific Coast provincial stage: Eocene and Oligocene (above [Fresnian]-Narizian, below Zemorrian).

refugium (re-fu′-gi-um) (a) An isolated area that underwent little environmental change, permitting a *biota* to persist locally long after it had been exterminated elsewhere. Syn: *asylum.* (b) A restricted area in which plants and animals persisted during a period of continental climatic change that made surrounding areas uninhabitable; esp. an ice-free or unglaciated area within or close to a continental ice sheet or upland ice cap, where hardy biotas eked out an existence during a glacial phase. Later, it might serve as a center of dispersal for the repopulation of surrounding areas after climatic readjustment. Pl: refugia. Syn: *glacial refuge.*

refuse An approximate equal mix of non-hazardous garbage (food wastes) and rubbish (combustible solid wastes); contains up to 50% moisture and 7% non-combustible solid wastes.

reg An extensive desert plain from which fine sediment has been removed by the wind, leaving a sheet of coarse, smoothly angular, wind-polished gravel and small stones lying on a broad *desert pavement;* a *stony desert* of the plains, as in the Algerian Sahara and parts of American deserts. Etymol: Hamitic. Pl: regs. See also: *serir; hammada.* Syn: *gravel desert.*

regelation (re-ge-la′-tion) A two-fold process involving the melting of ice in a region of high pressure (*pressure melting*) and the refreezing of the derived meltwater in a region of lower pressure. For the process to operate, the latent heat must be able to flow from the region where water freezes to the region where ice melts. The term is sometimes restricted to the refreezing part of the process, but in some European literature it has been applied to often-repeated freezing and thawing (or *multigelation*).

regelation layer Ice at the bottom of a glacier or an ice sheet that has formed by the *regelation* process (Paterson, 1969, p.129).

regenerated anhydrite (re-gen′-er-at′-ed) Anhydrite produced by the dehydration of gypsum that was itself formed by the hydration of anhydrite (Goldman, 1961).

regenerated flow control In glacial geology, control of glacial drainage by modified morainal features, resulting from the readvance of a previously stagnant glacier (Gravenor and Kupsch, 1959, p.56).

regenerated glacier (a) A glacier formed by accumulated ice blocks brought down by avalanches and icefalls from the terminus of a glacier at a higher level. Syn: *recemented glacier; reconstructed glacier; glacier remanié.* (b) A glacier that becomes active after a period of stagnation.

regenerated rock A clastic rock. The term "regenerirte Gesteine" was used by Zirkel (1866, p.3).

regeneration (re′-gen-er-a′-tion) The renewal, regrowth, or restoration of a body or of a part, tissue, or substance of a body, following injury or as a normal bodily process.

regime (re-gime′) (a) A regular or systematic pattern of occurrence or action, or a condition or style having widespread influence, as a sedimentation regime. (b) The existence in a stream channel of a balance or grade between erosion and deposition over a period of years. (c) The condition of a stream with respect to the rate of its average flow as measured by the volume of water passing different cross sections in a specified period of time. In this sense the term is used incorrectly as a syn. of regimen. (d) In glaciology, a syn. of *mass balance.*

regimen [glaciol] (reg′-i-men) *mass balance.*

regimen [streams] The flow characteristics of a stream; specif. the habits of an individual stream (including low flows and floods) with respect to such quantities as velocity, volume, form of and changes in the channel, capacity to transport sediment, and amount of material supplied for transportation. Cf: *regime.*

regimen [water] The characteristic behavior and the total quantity of water involved in a drainage basin, determined by measuring such quantities as rainfall, surface and subsurface storage and flow, and evapotranspiration. Syn: *hydrologic regimen; water regimen.*

regime stream A stream with a mobile (erodible) boundary, making at least part of its boundary from its transported load and part of its transported load from its boundary, carrying out the process at different places and times in a balanced or alternating manner that prevents unlimited growth or removal of the boundary (Blench, 1957, p.2). Cf: *graded stream.*

regime theory A theory of the formation of a channel in material carried and deposited by its stream.

regio (re′-gio) A term established by the International Astronomical Union for a large area on planetary surfaces marked by reflectivity or color distinctions from adjacent areas. Also a broad geographic region (USGS, 2003). Pl: regiones.

region [ecol] (re′-gion) A major division of the Earth having distinctive climatic and topographic features and floral and faunal provinces. Cf: *realm.*

region [geog] A very large expanse of land usually characterized or set apart by some aspect such as its being a political division or area of similar geography.

regional (re′-gion-al) n. In geophysics, the general field, attitude, or configuration, disregarding features smaller than a given size. Cf: *residual [geophys].*

regional correlation Correlation of rock units, major structures, or other geologic features over or across wide areas of the Earth's surface.

regional dip The nearly uniform inclination of strata over a wide area, generally at a low angle, as in the Atlantic and Gulf coastal plains and parts of the Midcontinent region. Cf: *homocline; primary dip.* Syn: *normal dip.*

regional geology The geology of any relatively large region, treated broadly and primarily from the viewpoint of the spatial distribution and position of stratigraphic units, structural features, and surface forms. Cf: *areal geology.*

regional-gravity map (re′-gion-al-grav′-i-ty) A gravity map showing gradual changes of gravity associated with regional geologic features.

regionalized variable In *geostatistics,* a spatial property that is continuous, smooth at a large scale, but so erratic at a small scale that it cannot be modeled by a deterministic function.

regional metamorphism A general term for metamorphism affecting an extensive region, as opposed to *local metamorphism* that is effective only in a relatively restricted area. As introduced in the nineteenth century, the term covered only those changes due to deep burial metamorphism; today it is used almost synonymously with *dynamothermal metamorphism* (Holmes, 1920). Cf: *dynamic metamorphism.*

regional metasomatism Metasomatic processes affecting extensive areas. Cf: *contact metasomatism.*

regional overstep A term proposed by Swain (1949, p.634) for an *overstep* in which an unconformity occurs "widespread, but not universally, over very large parts of a craton (platform, shelf)".

regional snowline The level above which, averaged over a large area, snow accumulation exceeds ablation year after year. Cf: *climatic snowline; snowline; orographic snowline.*

regional unconformity An unconformity that extends continuously throughout an extensive region. It may be nearly continent-wide and usually represents a relatively long period. Cf: *local unconformity.*

register mark (reg′-is-ter) A small figure, cross, circle, or other pattern on the cartographic copy of a map that is to be printed in more than one color. The accuracy of printing of each color is checked by synchronization of the register marks on each printing plate. Also called: *corner mark; corner tick; register tick.*

register punch The equipment used in *punch register* systems for making multiple copies of cartographic material. There are many designs, from a simple hand-operated punch to elaborate automated systems with material held down by vacuum during simultaneous punching of numerous precisely located holes. The punched holes may be round or slotted, and many sizes and spacing arrangements are possible. Cf: *punch register.*

register tick *register mark.*

registration (reg-is-tra′-tion) (a) The process of superposing two or more images or photographs so that equivalent points coincide. (b) The correct position of one component of a composite map image in relation to other components. (c) The process of geometrically aligning two or more cartographic data sets or digital images, in vertical juxtaposition, while maintaining true geographic referencing (Krzanowski et al., 1993).

regmagenesis (reg-ma-gen′-e-sis) The production of global strike-slip displacements (Sonder, 1956). Also spelled: *rhegmagenesis.* Adj: *regmatic.* Obsolete.

regmaglypt (reg′-ma-glypt) Any of various small, well-defined, characteristic indentations or pits on the surface of meteorites produced by selective erosion during atmospheric passage. They are polygonal, round, almond-shaped, or elliptic; their diameters range from a few millimeters to many centimeters.

regmatic (reg-mat′-ic) Adj. of *regmagenesis;* said of global strike-slip displacements related to a presumed simple stress pattern.

regolith [planet] (reg′-o-lith) Soil formed on planetary surfaces by weathering, impact, and aeolian activity (Bougher et al., 1997).

regolith [weath] (reg′-o-lith) A general term for the layer or mantle of fragmental and unconsolidated rock material, whether residual or transported and of highly varied character, that nearly everywhere forms the surface of the land and overlies or covers the bedrock. It includes rock debris of all kinds, volcanic ash, glacial drift, alluvium, loess and eolian deposits, vegetal accumulations, and soil. The term was originated by Merrill (1897, p.299). Var. spelling: *rhegolith.* Etymol: Greek "rhegos", "blanket", +

"lithos", "stone". Cf: *lunar regolith.* Syn: *mantle [geol]; soil [eng]; mantle rock; rock mantle; overburden.*

Regosol (Reg'-o-sol) Any soil of the azonal order without definite genetic horizons and developing from or on deep, unconsolidated, soft mineral deposits such as sands, loess, or glacial drift. Not used in current U.S. system of soil taxonomy.

regradation (re-gra-da'-tion) The formation by a stream of a new profile of equilibrium, as when the former profile, after gradation, became deformed by crustal movements, climatic change, or piracy.

regrading stream (re-grad'-ing) A stream that is simultaneously upbuilding (aggrading) and downcutting (degrading) along different parts of its profile.

regression [evol] (re-gres'-sion) (a) A hypothetical reversal in the direction of evolution that is sometimes used to explain certain paleontologic phenomena such as the extinction of the graptolites. (b) The trend exhibited by offspring, in respect to their inherited characteristics, away from advanced or specialized characters exhibited by their parents and toward a simpler state.

regression [stratig] The retreat or contraction of the sea from land areas, and the consequent evidence of such withdrawal (such as enlargement of the area of deltaic deposition). Also, any change (such as fall of sea level or uplift of land) that brings nearshore, typically shallow-water environments to areas formerly occupied by offshore, typically deep-water conditions, or that shifts the boundary between marine and nonmarine deposition (or between deposition and erosion) toward the center of a marine basin. Ant: *transgression.* Cf: *offlap.*

regression [streams] The name given to the theory that some rivers have their sources on the rainier sides of mountain ranges and gradually erode their heads backward until the ranges are cut through.

regression analysis The linear relationship between a dependent y variable and one or more independent x variables, estimated by fitting a function of the x variable(s) that minimizes the sum of the squared deviations of y from the function.

regression coefficient A coefficient in a *regression equation;* in linear regression, the intercept and slope of the regression line.

regression conglomerate A coarse (>2mm) sedimentary deposit formed during a retreat of the sea.

regression curve A graphical representation of a *regression equation* that relates a dependent variable to a polynomial or other expansion of an *independent variable.*

regression equation A model equation of the relationship between a *dependent variable y* and a function of one or more *independent variables x.* The coefficients of the equation are estimated by the *method of least squares,* in which the sum of the squared deviations of the values of y from the fitted model is minimized.

regression line A *regression curve* that is a straight line; the line or curve from a family of curves that best fits the empirical relation between a dependent variable and an independent variable.

regressive diagenesis (re-gres'-sive) *hypergenesis.*

regressive overlap *offlap.*

regressive reef One of a series of nearshore reefs or bioherms superimposed on basinal deposits during the rising of a landmass or the lowering of the sea level, and developed more or less parallel to the shore (Link, 1950). Cf: *transgressive reef.*

regressive ripple A term used by Jopling (1961) for an asymmetric ripple formed by a locally reversed current (e.g. in the *lee* of a sand wave or small delta) and therefore oriented in a direction opposite to the general movement of current flow.

regressive sand wave A term proposed by Bucher (1919, p.165) to replace *antidune* as used by Gilbert (1914, p.31). Most authors prefer the original term.

regressive sediments Sediments deposited during the retreat or withdrawal of water from a land area or during the emergence of the land, and characterized by an *offlap* arrangement.

regressive surface of marine erosion In sequence stratigraphy, an abrupt, scoured surface that is interpreted to have been formed by the submarine erosive action of waves and currents on the marine shelf during seaward movement of the shoreline. Marine strata, which in general display a shallowing-upward trend, must directly overlie the surface (Embry and Johannessen, 1992).

regressive systems tract In sequence stratigraphy, a unit bounded below by a maximum flooding surface and above by a subaerial unconformity or a *shoreface ravinement surface* that has eroded through a subaerial unconformity. Applicable for *T-R sequences* and genetic stratigraphic sequences (Embry and Johannessen, 1992).

regular (reg'-u-lar) In paleontology, pertaining to an echinoid of the Regularia division having a more or less globular symmetrical shell with 20 meridional rows of plates and displaying an *endocyclic* test in which the *periproct* is located within the *oculogenital ring.* Cf: *irregular.*

regular dissepimentarium A *dissepimentarium* in rugose corals in which the dissepiments are developed only in spaces between major septa and minor septa.

regular dodecahedron *pyritohedron.*

regularly foliated structure A *laminar structure* in which parallel calcitic laths or blades are arranged in sheets dipping at the same angle and in the same general direction over a large portion of the depositional surface (MacClintock, 1967; Carter, 1980a). This is the "regularly foliated structure" of Boggild (1930, p.249), with the exclusion of his example of *Tentaculites,* which is irregularly crossed bladed (cf. Towe, 1978).

regular simple prismatic structure A *simple prismatic structure* in which each first-order prism is columnar and has a more or less equidimensional (non-elongated) polygonal cross-sectional shape; the prism diameters and their shapes appear rather uniform in sections cut perpendicular to the prism length axes, although prism diameters may increase toward the inner part of the layer through geometric selection (Carter, 1980b). This application of "regular simple prismatic" corresponds generally with "calcitic simple prismatic" of Taylor et al. (1969), Suzuki and Uozumi (1981), Uozumi and Suzuki (1981), and Watabe (1984), but it also includes aragonitic simple prismatic structures, e.g., in the outer prismatic shell layer of the bivalve *Parapholas californica.* Vertically oriented regular spherulitic prismatic and nondenticular composite prismatic structures may appear superficially similar to regular simple prisms, but they have radiating substructures. The substructure of regular simple prisms may vary from homogeneous (see *Pinna muricata* in Keller and Dauphin, 1983) to fine complex crossed lamellar (see *Pinna* in Cuif et al., 1980), irregular complex crossed lamellar (see *Pinctada fucata martensii* in Suzuki and Uozumi, 1981, Pl. 3, Fig. 5), and obliquely oriented acicular crystallites (see *Malleus albus* in Suzuki and Uozumi, 1981, Pl. 3, Fig. 2), but it does not consist of a spherulite sector or composite prism. Abbrev: RSP.

regulation (reg-u-la'-tion) Artificial manipulation of stream flow.

Regur (Reg'-ur) An intrazonal group of dark calcareous soils high in clay, which is mainly montmorillontic, and formed mainly from rocks low in quartz; occurring extensively on the Deccan Plateau of India. Not used in current U.S. system of soil taxonomy.

reibungsbreccia (rei'-bungs-brec'-cia) A syn. of *fold breccia.* Etymol: German "Reibung", "rubbing, friction." Rarely used.

reichenbachite (reich'-en-bach-ite) A dark green monoclinic mineral: $Cu_5(PO_4)_2(OH)_4$. It is polymorphous with pseudomalachite and ludjibaite.

Reichenbach lamellae (Rei'-chen-bach) Thin platy inclusions of *troilite* occurring in iron meteorites. Named after Karl von Reichenbach (1788-1869), German chemist.

Reichweite (R) A parameter used in the Markovian statistic description of layered systems. For clay mineralogy, Reichweite involves the ordering of stacking of layers in phyllosilicates (Moore and Reynolds, 1997). For a given set of layers in a two-component system containing A-type and B-type layers, the terminology expresses the probability of finding the next layer to be B after an A layer; it addresses the question how much influence A has on what the next layer will be or the next layer after that, i.e., what the "reach back" of A is. For example, flipping a coin is R = 0; there is no influence at all of one flip on another. The probability of obtaining a "heads" depends only on the proportion of heads and tails, in this case 0.5. Therefore, for perfectly ordered 50/50 illite/smectite (I/S), given an S or I layer, the other must be next in line. The S layer, then, has a reach of one unit along the line of stacking or R = 1. For an ordered I/S of ISII (i.e., ISIIISIIISII), then R = 3. The most common use of R in mineralogy is for characterizing interstratification in mixed-layered clay minerals. Cf: *mixed-layered clay minerals; illite/smectite.* Etymol: German, "reichen" = reach + "Weite" = distance, range.

reidite An adamantine white tetragonal mineral: $ZrSiO_4$. It is a high-pressure polymorph of zircon.

Reid mechanism *elastic-rebound theory.*

reinerite (rei'-ner-ite) A blue or pale yellow-green orthorhombic mineral: $Zn_3(As^{3+}O_3)_2$. Not to be confused with renierite.

reinhardbraunsite (rein-hard-brauns'-ite) A light pink monoclinic mineral that is the natural equivalent of synthetic "calcio-chondrodrite": $Ca_5(SiO_4)_2(OH,F)_2$.

rejected recharge (re-ject'-ed) Water that infiltrates to the water table but then discharges because the aquifer is full and cannot accept it.

rejuvenated (re-ju'-ve-nat'-ed) Said of a structural feature, e.g. a fault scarp, along which the original movement has been renewed. Syn: *revived.*

rejuvenated fault scarp A fault scarp freshened by renewed movement along an old fault line after the initial scarp had been partly dissected or eroded. Syn: *revived fault scarp.*

rejuvenated stream A stream that, after having developed to maturity or old age, has had its erosive ability renewed as a result of rejuvenation. It may be characterized by entrenched meanders, stream terraces, and meander cusps. Syn: *revived stream.*

rejuvenated water Water returned to the terrestrial water supply as a result of compaction and metamorphism. See also: *water of compaction.*

rejuvenation (re-ju'-ve-na'-tion) (a) The action of stimulating a stream to renewed erosive activity, as by uplift or by a drop of sea level; the renewal or restoration of youthful vigor in a stream that has attained maturity or old age. The causes of rejuvenation may be dynamic, eustatic, or static. (b) The development or restoration of youthful features of a landscape or landform in an area previously worn down nearly to base level, usually caused by regional uplift or eustatic movements, followed by renewed downcutting by streams; a change in conditions of erosion, leading to the initiation of a new cycle of erosion. (c) The renewal of any geologic process, such as the reactivation of a fissure. Syn: *revival.*

rejuvenation head A *knickpoint* resulting from *rejuvenation* or from an uplift.

relational data base A set of relations whose structure is specified by the relational scheme. Commonly, a data base where information is arranged into tables and the dependencies between information is mapped by dependencies between two or more tables (Krzanowski et al., 1993).

relative abundance (rel'-a-tive) The number of individuals of a taxon in comparison with the number of individuals of other taxa in a certain area or volume. See also: *abundance [ecol]; absolute abundance.*

relative age The *geologic age* of a fossil organism, rock, geologic feature, or event, defined relative to other organisms, rocks, features, or events rather than in terms of years. Cf: *numerical age.*

relative aperture The diameter of the stop, diaphragm, or other physical element that limits the size of the bundle of rays traversing an optical instrument from a given point. It is expressed as a fraction of the focal length of the camera lens, with the symbol *f* being used instead of 1 as the numerator; e.g., a lens whose relative aperture is 1/45 of its focal length has a relative aperture of *f*/4.5 or *f*:4.5. See also: *f-number; speed.*

relative change in sea level In sequence stratigraphy, a change in the sea level measured relative to a predeterminated datum on the sea floor or within the sediment below the sea floor (Mitchum, 1977). For example, if the distance between the sea surface and the datum, selected in this case to be the sea floor, increased, a relative rise in sea level occurred. If the distance between the sea surface and the sea floor decreased, a relative fall in sea level occurred. Depositing sediment on the datum, in this case the sea floor, does not change the distance between the datum and the sea surface, hence deposition alone cannot cause a relative fall in sea level. Deposition alone causes water-depth decrease, which is different from a relative fall in sea level. A relative change in sea level is caused either by subsidence, uplift, or eustasy. The term is used to encompass any of the three, especially when the mechanism of sea level change cannot be determined.

relative chronology *Geochronology* in which the time-order is based on superposition and/or fossil content rather than on an age expressed in years. Cf: *numerical chronology.*

relative consistency *consistency index.*

relative dating The proper chronological placement of a feature, object, or event in the *geologic time scale* without reference to its absolute age.

relative density The ratio of the difference between the void ratio of a cohesionless soil in the loosest state and any given void ratio to the difference between its void ratios in the loosest and in the densest states (ASCE, 1958, term 296).

relative dispersion *dispersive power.*

relative fall in sea level In sequence stratigraphy, an apparent fall of sea level with respect to the underlying initial surface of deposition or some other predetermined datum. It may result from tectonic uplift or eustatic fall. The term relative fall in sea level is used when the process of sea-level fall in unknown or in doubt (Mitchum, 1977).

relative fugacity The ratio of the fugacity in a given state to the fugacity in a defined standard state.

relative-gravity instrument (rel'-a-tive-grav'-i-ty) A device for measuring the difference in the gravity field at two or more points. There are two principal types: (a) static type, in which a linear or angular displacement is observed or nulled by an opposing force, and (b) dynamic type, in which the period of oscillation is a function of gravity and is the quantity directly observed (as with a *vibrating-string gravimeter*). Exploration gravimeters are relative-gravity instruments. Cf: *absolute-gravity instrument.* See also: *gravimeter.*

relative humidity The ratio, expressed as a percentage, of the actual amount of water vapor in a given volume of air to the amount that would be present if the air were saturated at the same temperature. See also: *saturation.* Cf: *absolute humidity; specific humidity; mixing ratio.*

relative index of refraction An *index of refraction* that is the ratio of the velocity of light in one crystal to that in another crystal. Syn: *relative refractive index.*

relative permeability (a) The ratio between the *effective permeability* to a given fluid at a partial saturation and the permeability at 100% saturation (the *absolute permeability*). It ranges from zero at a low saturation to 1.0 at a saturation of 100% (Levorsen, 1967, p.110). (b) In a hydrogeologic system, the ratio of the effective conductivity of a fluid at a given saturation to its conductivity at full saturation. In a system containing more than one fluid, the facility with which each fluid moves in response to its own field of fluid potential is lessened because each fluid competes for the limited number of available flow paths.

relative permittivity Dielectric constant normalized by dividing by the permittivity of free space (8.85 times 10^{-12} farad per meter) so as to give a dimensionless quantity.

relative position The location of a point with respect to other points. The relative positions of two points whose locations are given in the same coordinate system are expressed as the differences of their coordinates or by the direction and distance from one point to the other.

relative refractive index *relative index of refraction.*

relative relief (a) *local relief.* (b) Within a drainage basin, the ratio of *basin relief* to *basin perimeter.* Cf: *relief ratio.*

relative rise in sea level In sequence stratigraphy, an apparent rise in sea level with respect to the underlying initial surface of deposition or some other predetermined datum. It may result from tectonic subsidence or eustatic rise. The term relative rise in sea level is used when the process of sea-level rise is unknown or in doubt (Mitchum, 1977).

relative roughness In streams, a ratio of flow depth to grain size, designed to characterize resistance to flow. Several version exist.

relative sea level The vertical position of global sea level with respect to an arbitrary scale, for example, extending from a low of 0 (position 30 million years ago) to a high of 1 (position 65 million years ago) (Vail et al. 1977, p.83).

relative time *Geologic time* determined by the placing of events in a chronologic order of occurrence; esp. time as determined by organic evolution or superposition. Cf: *fossil time; numerical time.*

relative time scale An uncalibrated *geologic time scale,* based on layered rock sequences and the paleontologic evidence contained therein, giving the relative order for a succession of events. Cf: *biologic time scale; atomic time scale.*

relative vorticity The local rate of rotation of a fluid relative to the Earth. Mathematically, it is the vertical component of the curl of the horizontal component of the local current. It can be due to rotation, such as an eddy, or to the horizontal shear of a current, e.g., the shear at the edge of the Gulf Stream.

relative water content *liquidity index.*

relaxation [exp struc geol] (re-lax-a'-tion) In experimental structural geology, the release of applied stress with time, due to any of various creep processes.

relaxation [geophys] (a) A process in which a system returns to a steady state after being suddenly changed by a disturbing force. (b) In an elastic medium, the decrease of elastic restoring force under applied stress, resulting in permanent deformation. (c) In nuclear magnetic resonance experimentation, returning to a previous more-or-less random state after being forced into alignment of spins.

relaxation time The time required for a substance to return to its normal state after release of stress. See also: *rheidity.*

relay An area along a segmented fault array where displacement is transferred from one fault segment to another.

relay geometry An arrangement of overlapping or underlapping subparallel faults in which the strike of the faults is the same as the trend of the fault zone (Biddle and Christie-Blick, 1985).

relay ramp In normal faulting, the ramplike zone between two overlapping fault segments, connecting the footwall of one segment to the hanging wall of the other. A relay ramp is the expression of *soft linkage* between faults. First called "relais des failles" (relay faults) by Goguel (1952).

release adiabat (re-lease') A curve or locus of points that defines the succession of states through which a material, shocked to a high-pressure state, passes while monotonically returning to ambient pressure. The process operates over a short time interval, compared with the characteristic time for heat flow in the material.

release fracture A fracture developed as a consequence of the relief of stress in one particular direction. The term is generally applied to a fracture formed when the maximum principal stress decreases sufficiently that it becomes the minimum principal stress; the fracture is an extension fracture oriented perpendicular to the then-minimum principal-stress direction.

release joint An exhumation-related joint whose orientation is controlled by a rock fabric; e.g. cleavage, schistosity, etc. (Engelder, 1985).

releasing bend (re-leas'-ing) A bend or step in a fault where fault movement results in the development of local extensional strain. In strike slip faulting, a syn. of *negative flower structure.*

relic [geomorph] (rel'-ic) A landform that has survived decay or disintegration, such as an *erosion remnant;* or one that has been left behind after the disappearance of the greater part of its substance such as a remnant island. The term is sometimes used adjectively as a synonym of relict, but this usage is not recommended.

relic [meta] *relict [meta].*

relic [sed] A vestige of a particle in a sedimentary rock, such as a trace of skeletal material in a carbonate rock or an incompletely recrystallized mineral in a diagenetic rock.

relict [geomorph] (rel'-ict) adj. Said of a topographic feature that remains after other parts of the feature have been removed or have disappeared; e.g. a "relict beach ridge" or a "relict hill". Cf: *relic [geomorph]; residual [geomorph].* n. A relict landform. Also, a landform made by processes no longer operative, e.g. glaciated forms in the northern United States or sand dunes in rain forests.

relict [meta] adj. Pertaining to a mineral, structure, or feature of an earlier rock that has persisted in a later rock in spite of processes tending to destroy it. n. Such a mineral, structure, or other feature. As a noun only, var: *relic.* Cf: *palimpsest [meta]; stable relict; unstable relict.*

relict [paleont] n. A remnant of an otherwise extinct flora, fauna, or kind of organism that has persisted since the extinction of the rest of the group. adj. (a) Said of a remnant of an extinct group. (b) Showing traces of a former structure or microstructure, now largely obscured by *diagenesis;* e.g., relict aragonitic shell microstructure in a calcitized molluscan shell.

relict aperture One of the short radial slits around the umbilicus of a planktonic foraminiferal test that remain open when the umbilical parts of the equatorial aperture are not covered by succeeding chambers (as in Planomalinidae) or that, even when secondarily closed, allow the elevated apertural lips or flanges to remain visible around the umbilicus (as in *Planomalina* and *Hastigerinoides*) (TIP, 1964, pt.C, p.63).

relict glacier The remnant of an older and larger glacier.

reliction (re-lic'-tion) The slow and gradual withdrawal of the water in the sea, a lake, or a stream, leaving the former bottom as permanently exposed and uncovered dry land; it does not include seasonal fluctuations in water levels. Legally, the added land belongs to the owner of the adjacent land against which it abuts. Also, the land left uncovered by reliction. Cf: *dereliction; accretion [sed].*

relict karst Old karst that was never covered by deposits after its original formation. Cf: *paleokarst.* Syn: *fossil karst.*

relict lake A lake that survives in an area formerly covered by the sea or a larger lake, or a lake that represents a remnant resulting from a partial extinction of an original body of water; a lake that has become separated from the sea, e.g., by gentle uplift of the sea bottom.

relict mountain *mountain of circumdenudation.*

relict permafrost (a) Permafrost left from colder times and commonly below the average depth reached by seasonal frost. (b) Permafrost that, having formed during an earlier colder period, will not, under present climatic conditions, reform after it is once disturbed or destroyed. (c) Permafrost that was formed in the past and persists in places where it could not form today (Hopkins et al., 1955). Ant: *active permafrost.* Syn: *passive permafrost; fossil permafrost.*

relict sediment A sediment that had been deposited in equilibrium with its environment, but that is now unrelated to its present environment even though it remains unburied by later sediments; e.g. a land-laid or shallow-marine sediment occurring in deep water (as near the seaward edge of the continental shelf).

relict texture In mineral deposits, an original texture that remains after partial or total replacement.

relief [geomorph] (re-lief') (a) A term used loosely for the physical shape, configuration, or general unevenness of a part of the Earth's surface, considered with reference to variations of height and slope or to irregularities of the land surface; the elevations or differences in elevation, considered collectively, of a land surface. The term is frequently confused with *topography,* although the use of the two terms in the sense of surface configuration is "thoroughly established both in general speech and in technical geomorphological literature" in the U.S. (C.D. Harris, in Stamp, 1961, p.454). (b) The vertical difference in elevation between the hilltops or mountain summits and the lowlands or valleys of a given region. A region showing a great variation in elevation has "high relief", and one showing little variation has "low relief". See also: *local relief; available relief.*

relief [geophys] The range of values over an anomaly or within an area, e.g. the "gravity relief" for the magnitude of a gravity anomaly.

relief [optics] The degree of visibility of a crystal section or fragment relative to its surroundings under the microscope. High relief indicates a great difference in *index of refraction* between the crystal and its mounting medium. The relief is positive if the refractive index of the mineral is greater than that of the medium, and negative in the reverse case. Syn: *shagreen.*

relief displacement The geometric distortion on vertical aerial photographs. The tops of objects are located on the photograph radially outward from the base.

relief feature *landform.*

relief limonite *Indigenous limonite* that is porous and cavernous in texture, commonly botryoidal after chalcocite.

relief map A map that depicts the surface configuration or relief of an area by any method, such as by use of contour lines (contour map) and hachures, by hill shading (shaded-relief map), by photography (photorelief map), by layer tinting, by pictorial symbols (physiographic diagram), by molding plastic in three dimensions (plastic relief map), or by a combination of these methods. Cf: *relief model.*

relief model A three-dimensional representation of the physical features or relief of an area, in any size or medium, but not necessarily constructed to true scale (the vertical scale is generally exaggerated to accentuate the relief). Cf: *relief map.*

relief ratio Within a drainage basin, the ratio of *basin relief* to *basin length;* it is a measure of the overall steepness of the basin and the intensity of erosion on its slopes. Cf: *relative relief.*

relief shading *hill shading.*

relief well A well used to relieve excess hydrostatic pressure, as to reduce waterlogging of soil or to prevent blowouts on the land side of levees or dams at times of high water. Cf: *drainage well.*

Relizian (Re-liz'-i-an) A Pacific Coast provincial stage: middle Miocene (above Saucesian, below Luisian).

rem A dose of ionizing radiation that deposits 100 ergs of energy in biological tissue. One rem is equivalent to 0.001 sieverts (Sv).

remainder mountain (re-main'-der) *mountain of circumdenudation.*

remanent magnetization (rem'-a-nent) That component of a rock's *magnetization* that has a fixed direction relative to the rock and is independent of moderate, applied magnetic fields such as the Earth's magnetic field. The natural remanent magnetization (NRM) may consist of more that one remanence. Cf: *induced magnetization.* See also: *hysteresis; natural remanent magnetization.*

remanié (re-ma-nie') adj. A French word meaning "reworked" or "rehandled", applied in geology to fragments or entities derived from older materials, esp. to fossils removed from or washed out of an older bed and redeposited in a new one ("remanié assemblage"). The term is also applied to boulders in a glacial till, country rock engulfed in a batholith, and glaciers that have been "reconstructed" ("glacier remanié"). Cf: *reworked; derived.* Anglicized sometimes as remanie. n. A fragment of an older formation incorporated in a younger deposit.

remanié assemblage *mixed assemblage.*

remineralization The process, mainly owing to bacterial activity, by which inorganic nutrients are released to seawater during the consumption of organic matter by the processes involved in *respiration.*

remnant (rem'-nant) *erosion remnant.*

remnant arc A fossil volcanic arc commonly in the back-arc position of an active arc.

remolded soil (re-mold'-ed) Soil that has had its natural internal structure modified or disturbed by manipulation so that it loses shear strength and gains compressibility.

remolding (re-mold'-ing) (a) The disturbance of the naturally occurring internal structure of a soil. (b) The placement and compaction of soil in a standard steel cylinder, called a mold, in a *compaction test* or in a brass ring in a direct shear test. Cf: *Proctor.*

remolding index The ratio of the modulus of deformation of a soil in the undisturbed state to that of a soil in the remolded state.

remolding sensitivity *sensitivity ratio.*

remolinite (re-mo'-lin-ite) *atacamite.*

remondite-(Ce) A vitreous orange-red monoclinic mineral: $Na_3(Ca,Ce,La,Na,Sr)_3(CO_3)_5$.

remondite-(La) A vitreous orange-yellow monoclinic mineral: $Na_3(La,Ce,Ca)_3(CO_3)_5$.

Remotely Operated Vehicle An unmanned undersea vehicle that performs underwater tasks in response to directions from an operator in another vessel at the surface. Abbrev: ROV.

remote sensing (re-mote') The science of collecting, processing, and interpreting images that record the interaction between electro-magnetic energy and matter. Acquiring information about an object or phenomenon by a recording device that is not in physical contact with the subject being studied. Aerial photographs and imagery acquired from a satellite are examples of remotely sensed information.

remote stress A homogeneous stress arising from a source, generally a traction, at significant distances from some region of interest.

renardite (re-nard'-ite) A yellow orthorhombic mineral: $Pb(UO_2)_4(OH)_4(PO_4)_2{\cdot}7H_2O$. Probably an unecessary name for an intermediate member of the dewindtite-phosphuranylite series.

Rendolls Mollisols that have no argillic or calcic horizon but that contain material with $CaCO_3$ equivalent > 400 g kg^{-1} within or immediately below the mollic epipedon. Rendolls are not saturated with water for periods long enough to limit their use for most crops. A suborder in the U.S. system of soil taxonomy.

Rendzina A great soil group of the intrazonal order and calcimorphic suborder consisting of soils with brown or black friable surface horizons underlain by light gray to pale yellow calcareous material developed from soft, highly calcareous parent material under grass vegetation or mixed grasses and forest in humid and semiarid climates. Not used in current U.S. system of soil taxonomy.

renewed consequent stream (re-newed') *resequent stream.*

rengite A dark greenish-brown monoclinic mineral of the *chevkinite* group: $Sr_4Ti_4ZrO_8(Si_2O_7)_2$.

renierite (re-nier'-ite) A metallic orange tetragonal mineral: $(Cu,Zn)_{11}(Ge,As)_2Fe_4S_{16}$. Not to be confused with reinerite.

reniform (ren'-i-form) Kidney-shaped. Said of a crystal structure in which radiating crystals terminate in rounded masses; also said of mineral deposits having a surface of rounded, kidneylike shapes. Cf: *colloform; botryoidal.*

Renninger effect *double reflection.*

rensselaerite (rens-se-laer'-ite) A compact fibrous talc, pseudomorphous after pyroxene, found in Canada and northern New York. It is harder than talc, takes a good polish, and is often made into ornamental articles.

repeated twinning (re-peat'-ed) Crystal twinning that involves more than two simple crystals; it may be *cyclic twinning* or *polysynthetic twinning.* See also: *multiple twin.*

Repettian (Re-pet'-ti-an) Western North American stage: lower Pliocene (above Mohnian, below Venturian).

replacement [meta] (re-place'-ment) Change in a mineral, or composition of a mineral, or mineral aggregate, presumably by diffusion of new material in and old material out.

replacement [paleont] A process of fossilization involving substitution of inorganic matter for the original organic constituents of an organism.

replacement [stratig] The gradual movement of the sea either toward or away from land areas, such as "marine replacement" (or transgression) and "continental replacement" (or regression).

replacement magnetite [eco geol] A *mineral-deposit type* characterized by veins and lenticular masses of magnesite and/or brucite replacing carbonate and calc-silicate rocks. These deposits are an important source of commercial magnesite.

replacement talc [eco geol] A *mineral-deposit type* characterized by lensoid bodies of massive talc and other magnesium silicates replacing dolostone.

replacing overlap (re-plac'-ing) A term, now obsolete, used by Grabau (1920, p.398) for a nonmarine overlap involving a receding shoreline, occurring where continental sediments are deposited and progressively "replace" the corresponding and all but contemporaneous marine sediments into which they grade (as described by Grabau, 1906, p.628-629). The misuse of overlap in this sense for a facies change from marine to continental sediments of the same age is confusing and "ungeological" (Lovely, 1948, p.2295).

replat (re-plat') (a) A French term for a horizontal surface (such as a bench, shelf, or shoulder), wider than a ledge, occurring along the steep side of a U-shaped valley (Stamp, 1961, p.391). (b) A French term for a stretch of relatively horizontal land interrupting a slope.

replenishment [coast] The *nourishment* of a beach with introduced sand; it often refers to the first nourishment of a series.

replenishment [grd wat] (re-plen'-ish-ment) *recharge.*

repose imprint (re-pose') A term used by Kuenen (1957, p.232) for a *sole mark* formed by an animal lying on or taking cover in bottom sediment.

repose period The interval of solfataric or fumarolic activity between volcanic outbursts.

reppiaite An orange-red monoclinic mineral: $Mn^{2+}_5(VO_4)_2(OH)_4$.

representative fraction (rep-re-sent'-a-tive) The scale of a map, expressed in the form of a numerical fraction that relates linear distances on the map to the corresponding actual distances on the ground, measured in the same unit (centimeters, inches, feet); e.g. "1/24,000" indicates that one unit on the map represents 24,000 equivalent units on the ground. Abbrev: RF. Syn: *natural scale.*

represo (re-pre'-so) *charco.*

reproduction (re-pro-duc'-tion) *synthetic stone; reconstructed stone.*

reptant (rep'-tant) (a) Creeping or prostrate; esp. said of a *corallite* with a creeping habit, growing attached along one side to some foreign body. (b) Said of a bryozoan colony consisting of largely separate recumbent zooids attached to the substrate.

reptation (rep-ta'-tion) A syn. of *surface creep* (Scheidegger, 1961, p.290).

reptile (rep'-tile) Any vertebrate of the class Reptilia; cold-blooded tetrapods that are air-breathing at all stages of development. Range, Pennsylvanian to Holocene.

resaca (re-sa'-ca) A term applied in the SW U.S. to a long, narrow, meandering lake occupying the bed of a former stream channel; a series of connected *bancos.* Also, the dry channel or the former marshy course of a stream, now containing a resaca. Etymol: American Spanish, from Spanish "resacar", "to redraw".

resampling In digital image processing, the interpolation of pixels on a source digital image to new locations of transformed pixels, usually coinciding with a georegistered grid (Krzanowski et al., 1993).

resection (re-sec'-tion) (a) A method in surveying by which the horizontal position of an occupied point is determined by drawing lines from the point to two or more points of known position. The most usual problem in resection is the *three-point problem* when three known positions are observed to locate the occupied station. Cf: *intersection.* (b) A method of determining a plane-table position by orienting along a previously drawn foresight line and drawing one or more rays through the foresight from previously located stations.

resedimentation (re'-sed-i-men-ta'-tion) (a) Sedimentation of material dervied from a pre-existing sedimentary rock; *redeposition* of sedimentary material. (b) Mechanical deposition of material in cavities of postdepositional age, such as the deposition of carbonate muds and silts by internal mechanical erosion or solution of a limestone. (c) The general process of subaqueous, downslope movement of sediment under the influence of gravity, such as the formation of a turbidity-current deposit.

resedimented rock (re-sed'-i-ment-ed) (a) A rock consisting of reworked sediments. (b) A turbidity-current deposit; e.g. a flysch or other similar graywacke, showing graded bedding, and alternating with shales in a thick sequence.

resequent [geomorph] (re'-se-quent) Said of a geologic or topographic feature that resembles or agrees with a consequent feature but that developed from such a feature at a later date; esp. said of a block mountain that is similar in form to the original tilted block or horst but that is shaped by differential erosion after the original topography was destroyed and uplifted or of a river that flows in the same direction as the original consequent drainage: a "renewal consequent" stream. Ant: *obsequent [geomorph].* Cf: *consequent [geomorph]; subsequent [geomorph].*

resequent [streams] adj. Said of a stream, valley, or drainage system whose course or direction follows an earlier pattern but on a newer and lower surface, as in an area of ancient folding subjected to long-continued erosion. Etymol: *re* + consequent. Syn: *reconsequent.* n. *resequent stream.*

resequent fault-line scarp A *fault-line scarp* that faces in the same direction as the original fault scarp (i.e. facing the downthrown block), or in which the downthrown block is topographically lower than the upthrown block. Cf: *obsequent fault-line scarp.*

resequent stream A stream that flows down the dip of underlying strata in the same direction as an original consequent stream but developed later at a lower level than the initial slope (as on formerly buried resistant strata) and generally tributary to a subsequent stream; e.g. a stream flowing down the back slope of a cuesta. Syn: *resequent [streams]; renewed consequent stream.*

reserve That part of a *mineral deposit* which could be economically and legally extracted or produced at the time of the reserve determination (SEC, 2003).

reserve base That part of an identified resource that meets specified minimum physical and chemical criteria related to current mining and production practices, including those for grade, quality, thickness, and depth. The reserve base is the in-place demonstrated (measured plus indicated) resources from which reserves are estimated. It may encompass those parts of the resources that have a reasonable potential for becoming economically available within planning horizons beyond those that assume proven technology and current economics. The reserve base includes those resources that are currently economic (reserves) marginally economic (marginal reserves), and some of those that are currently subeconomic (subeconomic resources) (USGS, 1980).

reserves (re-serves') (a) An estimate within specified accuracy limits of the valuable metal or mineral content of known deposits that may be produced under current economic conditions and with present technology (Schanz and Ellis, 1983). (b) That part of the reserve base that could be economically extracted or produced at the time of determination. The term reserves need not signify that extraction facilities are in place and operative. Reserves include only recoverable materials; thus, terms such as extractable reserves and recoverable reserves are redundant and are not a part of this classification system (USGS, 1980). (c) *Identified resources* of mineral- or fuel-bearing rock from which the mineral or fuel can be extracted profitably with existing technology and under present economic conditions (Brobst and Pratt, 1973, p.2). The concept can be used in global, regional, or local senses, or applied as a measure of the remaining effective life of an individual mine. See also: *resources.* Syn: *mineral reserves.*

reservoir [grd wat] (res'-er-voir) *groundwater reservoir.*

reservoir [paleont] The enlarged posterior part of the gullet in some motile cells in protists such as Cryptophyceae and Euglenophyta.

reservoir [petroleum] In petroleum geology, any porous and permeable rock that yields oil or gas. Sandstone, limestone and dolomite are the most common reservoir rocks, but accumulation in fractured igneous and metamorphic rocks is not unknown.

reservoir [water] An artificial or natural storage place for water, such as a lake, pond, or aquifer, from which the water may be withdrawn for such purposes as irrigation, municipal water supply, or flood control.

reservoir compartment Subdivision of an oil or gas reservoir consisting of a pore system that responds as a unit when fluid is withdrawn.

reservoir energy The energy or "drive" within a petroleum reservoir. See: *dissolved-gas drive; gas-cap drive; water drive.*

reservoir gas-oil ratio The number of cubic feet of gas per barrel of oil in the reservoir. See also: *gas-oil ratio.*

residence time [geol] (res'-i-dence) "The average amount of time a particular substance spends within a designated earth system. The residence time is inversely proportional to the rate of movement within the system and directly proportional to the size of the system" (Laporte, 1975, p.175).

residence time [lake] The length of time to renew the water in a lake by inflow streams, seepage, or rainfall, which can be calculated by dividing the lake volume by the outflow from the lake (Wetzel, 2001, p.17).

residual [cryst] *R-factor.*

residual [geomorph] (re-sid'-u-al) adj. Said of a topographic or geologic feature, such as a rock, hill, or plateau, that represents a small part or trace of a formerly greater mass or area, and that remains above the surrounding surface which has been reduced by erosion. Cf: *relict [geomorph].* n. *erosion remnant.*

residual [geophys] n. In geophysics, what is left after the *regional* has been subtracted; a field from which gross effects have been removed, in order to emphasize local anomalies. Cf: *regional.*

residual [ore dep] adj. Said of a mineral deposit formed by mechanical concentration, e.g. a placer, or by chemical alteration in the zone of weathering, e.g. kaolinite from feldspar or bauxite after syenite.

residual [stat] *residual error.*

residual [weath] adj. Pertaining to or constituting a *residue;* esp. said of material left after the weathering of rock in place, such as a *residual deposit* or a *residual soil.* Syn: *eluvial [weath].*

residual and karst-fill nonsulfide zinc deposit Zinc deposits formed by mechanical erosion and concentration of zinc oxides and silicated minerals as residual concentrations and karst fillings (Hitzman et al., 2003).

residual anticline In salt tectonics, a relative structural *high* that is created as the result of the depression of two adjacent rim synclines. Syn: *residual dome.*

residual arkose An arkose formed in situ by disintegration of a granite; an untransported arkose, commonly grading into the underlying granite (Barton, 1916, p.447). Also called: sedentary arkose. See also: *grus.*

residual boulder *boulder of weathering.*

residual clay Clay material formed in place by the weathering of rock, derived either from the chemical weathering of feldspar and other rock minerals or from the removal of nonclay-mineral constituents by solution from a clay-bearing rock (such as an argillaceous limestone); a soil or a product of the soil-forming processes. Cf: *secondary clay.* Syn: *primary clay.*

residual community *winnowed community.*

residual compaction The difference between the amount of compaction that will occur ultimately for a given increase in applied stress, and that which has occurred at a specified time (Poland et al., 1972).

residual deposit (a) The *residue* formed by weathering in place. (b) An ore deposit formed in clay by the conversion of metallic compounds (as of manganese, iron, lead, or zinc) into oxidized forms by weathering at or near the Earth's surface.

residual dome *residual anticline.*

residual error The difference between any measured value of a quantity in a series of observations (corrected for known systematic errors) and the computed value of the quantity obtained after the adjustment of that series. In practice, it is the residual errors that enter into a computation of probable error. Syn: *residual [stat].*

residual geosyncline *autogeosyncline.*

residual gravity Short- and mid-wavelength components of the Bouguer gravity or free-air gravity, attributed to density contrasts within the high-density geological basement and the lower density sedimentary section. Anomalies in the residual gravity field are of primary exploration interest. See also: *residualizing.*

residualizing (re-sid'-u-a-liz'-ing) In applied geophysics, separating a curve or surface into its broad components (the "regional") and its local components (the "residual"). Residualizing is an attempt to sort out of the total field those anomalies that result from local structure; that is, to find local anomalies by subtracting gross (regional) effects (Sheriff, 2002). See also: *grid residual.*

residual kame A ridge or mound of sand or gravel formed by the denudation of glaciofluvial material that had been deposited in glacial lakes or on the flanks of hills of till (Gregory, 1912, p.175). Syn: *pseudokame.*

residual liquid The still-molten part of a magma that remains in the magma chamber after some crystallization has taken place during a series of differentiations. Obsolete. Syn: *residual liquor; residual magma; rest magma; ichor.*

residual liquor *residual liquid.*

residual magma *residual liquid.*

residual map A stratigraphic map that displays the small-scale variations (such as local features in the sedimentary environment) of a given stratigraphic unit (Krumbein and Sloss, 1963, p.486). It is superimposed on the underlying pattern of a *trend map.*

residual material Unconsolidated or partly weathered *parent material* of a soil, presumed to have developed in place, by weathering, from the consolidated rock on which it lies; it is the material from which soils are formed. See also: *residual soil.* Cf: *transported soil material.*

residual oil Oil that is left in the reservoir rock after the pool has been depleted.

residual soil A soil formed from, or resting on, consolidated rock of the same kind as that from which it was formed, and in the same location. Not used in current U.S. system of soil taxonomy.

residual strength The *ultimate strength* that develops along a surface in soil or a discontinuity in rock, usually at high strain or large displacement. For soil or rock not previously sheared, a marked decrease in strength usually occurs with increasing displacement until the residual value is reached.

residual stress *residual strength.*

residual swelling "The difference between the original prefreezing level of the ground and the level reached by the settling after the ground is completely thawed" (Muller, 1947, p.221).

residual valley A trough intervening between uplifted mountains, as in the Basin and Range Province of western U.S. (Gilbert, 1875, p.63).

residue (res'-i-due) An accumulation of rock debris formed by weathering and remaining essentially in place after all but the least soluble constituents have been removed, usually forming a comparatively thin surface layer concealing the unweathered or partly altered rock below. See also: *chemical residue; physical residue.* Syn: *residuum [weath]; residual deposit; eluvium [weath].*

residuite (re-sid'-u-ite) The translucent residuum that occurs in *clarain.*

residuo-aqueous sand (re-sid'-u-o-a'-que-ous) A term used by Sherzer (1910, p.627) for a sand containing water-rounded particles that were subsequently subjected to weathering. Cf: *aqueo-residual sand.*

residuum [coal] (re-sid'-u-um) The structureless groundmass of microscopically unresolvable constituents, consisting of particles of one to two micrometers or less, usually opaque, and of a dark color. It is the same as the lower range of fine *micrinite.* See also: *desmite; residuite.*

residuum [weath] *residue.*

resilience (re-sil'-i-ence) The ability of a material to store the energy of elastic strain. This ability is measured in terms of energy per unit volume.

resilifer (re-sil'-i-fer) A spoon-shaped recess or process on the hinge plate of some bivalve mollusks (as in *Mactra*) to which the *resilium* is attached or by which it is supported. See also: *chondrophore.*

resilium (re-sil'-i-um) The internal *ligament* within the hinge line of a bivalve mollusk, compressed by the hinge plate when the valves are closed. Pl: resilia.

resin (res'-in) Any of various hard, brittle, transparent or translucent substances formed esp. in plant secretions and obtained as exudates of recent or fossil origin (as from pine or fir trees, or from certain tropical trees) by the condensation of fluids on the loss of volatile oils. Resins are yellowish to brown, with a characteristic luster; they are fusible and flammable, are soluble in ether and other organic solvents but not in water, and represent a complex mixture of terpenes, resin alcohols, and resin acids and their esters. See also: *amber; fossil resin; mineral resin.* Syn: *natural resin.*

resin canal *resin duct.*

resin duct A long, narrow intercellular canal in wood, surrounded by one or more layers of parenchyma cells of the epithelium (Record, 1934, p.72). See also: *resin rodlet.* Syn: *resin canal.*

resinite (res'-in-ite) A maceral of coal within the *liptinite* group, consisting of resinous compounds, often in elliptical or spindle-shaped bodies representing cell-filling matter or resin rodlets. Cf: *alginite; cutinite; sporinite.*

resin jack *rosin jack.*

resinoid (res'-in-oid) A group term for the macerals in the resinite series.

resin opal A wax-yellow or honey-yellow variety of common opal with a resinous luster.

resinous coal (res'-in-ous) Coal, usually younger coal, in which the attritus contains a large proportion of resinous material.

resinous luster The luster on the fractured surfaces of certain minerals (such as opal, sulfur, amber, and sphalerite) and rocks (such as pitchstone) that resembles the appearance of resin.

resin rodlet A fossil resinous secretion that may be isolated from coal. It was presumably deposited in a *resin duct* by a secretory epithelium.

resin tin *rosin tin.*

resistance (re-sis'-tance) The ratio of an applied constant electromotive force to the current it produces. Cf: *conductance.*

resistates (re-sis'-tates) Sediments composed of chemically resistant minerals, enriched in weathering residues; thus highly quartzose sediments characteristically rich also in zircon, ilmenite, rutile, and, more rarely, cassiterite, monazite, and gold. It is one of Goldschmidt's groupings of sediments or analogues of differentiation stages in rock analysis. Cf: *hydrolyzates; oxidates; reduzates; evaporates.*

resistivity (re-sis-tiv'-i-ty) (a) *electrical resistivity.* (b) *thermal resistivity.*

resistivity factor *formation factor.*

resistivity log A *well log* consisting of one or more resistivity curves, which may be of the *normal log, lateral log,* or *focused-current log* types (or their *microresistivity log* equivalents). As a spontaneous-potential curve is commonly also present, the term resistivity log is often used as a syn. of *electric log.*

resistivity method Any electrical exploration method in which current is introduced into the ground by two contact electrodes and potential differences are measured between two or more other electrodes.

resistivity profile A survey by the resistivity method in which an array of electrodes is moved along profiles to determine lateral variations in resistivity.

resistivity sounding *electrical-resistivity sounding.*

resistivity-thickness product The product of resistivity and thickness of a sequence of layers. Units are ohm-meters^2.

resolution [remote sensing] (a) A measure of the ability of individual components, and of remote-sensing systems, to distinguish detail or to define closely spaced targets. (b) The minimum size of a feature that can be detected. Syn: *resolving power; spatial resolution.* (c) The measure of the accuracy or detail of a graphic display expressed as dots per inch, pixels per line, lines per millimeter, etc.

resolution [speleo] The solution of *speleothems* that come into contact with undersaturated water, often during flooding; sometimes called "redissolving".

resolving power (re-solv'-ing) *resolution [remote sensing].*

resonance (res'-o-nance) A buildup of amplitude in a physical system when the frequency of an applied oscillatory force is close to the natural frequency of the system. Cf: *nuclear resonance; electron spin resonance; nuclear magnetic resonance.*

resonant frequency (res'-o-nant) The frequency at which maximum response of a system occurs.

resorbed reef (re-sorbed') A reef characterized by embayed margins and by numerous isolated patches of reef that are closely distributed about the main mass (Maxwell, 1968, p.106-107). Resorbed reefs frequently rise from larger, submerged platforms, and suggest restrictive growth or degeneration of the reef mass.

resorption (re-sorp'-tion) The act or process of reabsorption or readsorption; specif. the partial or complete re-fusion or solution, by and in a magma, of previously formed crystals or minerals with which the magma is not in equilibrium or, owing to changes of temperature, pressure (depth), or chemical composition, with which it has ceased to be in equilibrium. "The term is often wrongly applied to immature crystals, and to crystals which have decomposition borders through change of pressure or otherwise" (Holmes, 1928, p.198).

resorption rim A *corrosion rim* representing partial resorption and recrystallization by a molten magma of previously crystallized minerals.

resource base The entire quantity of a particular metal or mineral in the Earth's crust in a given region.

resources (re-sourc'-es) (a) A concentration of naturally occurring solid, liquid, or gaseous materials in or on the Earth's crust in such form and amount that economic extraction of a commodity form the concentration is currently or potentially feasible (USGS, 1980). (b) *Reserves* plus all other mineral deposits that may eventually become available: either known deposits that are not economically or technologically recoverable at present, or unknown deposits, rich or lean, that may be inferred to exist but have not yet been discovered (Brobst and Pratt, 1973, p.2). They represent the mineral endowment, global, regional, or local, ultimately available for man's use. See also: *identified subeconomic resources; hypothetical resources; speculative resources.* Syn: *mineral resources.*

RESP *radially elongate simple prismatic structure.*

respiration A process in which organic matter is oxidized with the result that dissolved oxygen in seawater is consumed and both carbon dioxide and inorganic forms of nutrients are released to seawater; usually the result of oxidative processes in *phytoplankton* or marine bacteria. The opposite of *photosynthesis* or photosynthetic *carbon fixation*.

response surface A form of regression analysis that seeks to determine an optimal combination of experimental conditions by fitting a second-degree trend surface and seeking a minimum or maximum on the surface.

responsivity (re-spon-siv'-i-ty) In infrared detector terminology, the ratio of signal output to incident radiant flux, usually expressed as volts/watt (Bernard, 1970, p.56).

rest-hardening The increase of strength, with time, of a clay subsequent to its deposition, remolding, or modification by the application of shear stress. It appears to be an electrochemical bonding phenomenon.

resting spore (rest'-ing) A spore that remains dormant for a period before germination; e.g. a *chlamydospore*, or a desmid *zygospore* having thick cell walls and able to withstand adverse conditions such as heat, cold, or drying out. See also: *statospore; cyst [palyn].*

restite (rest'-ite) An essentially nongenetic designation for all immobile or less mobile parts of *migmatites*. Cf: *melanosome; metaster; paleosome; stereosome.*

rest magma *residual liquid.*

restored cross section A cross section, constructed with all effects of deformation removed, that shows the distribution of rock prior to deformation.

restraining bend (re-strain'-ing) A bend in a fault where fault movement results in the development of local contractional strain. Syn: contractional bend; convergent bend.

restraining overstep *contractional overstep.*

restricted basin (re-strict'-ed) A depression in the ocean floor characterized by topographically restricted water circulation, often resulting in oxygen depletion. Syn: *silled basin; barred basin.*

restricted resources That part of any resource or reserve category that is restricted from extraction by laws or regulations. For example, restricted reserves meet all the criteria of reserves except that they are restricted from extraction by laws or regulation (USGS, 1980).

reststrahlen (rest'-strahl-en) Narrow bands of enhanced reflectance that occur in transparent materials in which the refractive index is high or the absorption coefficient is large. Etymol: German, "residual rays".

resultant The sum of a set of directional vectors, expressed as the sum of their sines and cosines.

resupinate (re-su'-pi-nate) Inverted or reversed in position; esp. referring to reversal in relative convexity of postbrephic brachiopod shells in which the convex *pedicle valve* becomes concave and the concave *brachial valve* becomes convex during successive adult stages of growth.

resurgence (re-sur'-gence) In a *karst* region, the reappearance of a stream by discharge from a cave, the water having entered the cave as a *sinking stream.* Cf: *exsurgence;* also, the point where this happens, the *emergence [karst].* Cf: *insurgence.*

resurgent [petrology] (re-sur'-gent) Said of magmatic water or gases that were derived from sources on the Earth's surface, from the atmosphere, or from country rock adjacent to the magma. Cf: *juvenile [water].*

resurgent [pyroclast] In the classification of pyroclastics, the equivalent of accessory. Cf: *juvenile [volc].* Obsolete.

resurgent caldera A caldera in which the downdropped block is uplifted by magmatic intrusion following crater formation.

resurgent cauldron A *cauldron [volc]* in which the cauldron block, following subsidence, has been uplifted, usually in the form of a structural dome (Smith and Bailey, 1968, p.613).

resurgent gas Obsolete syn. of *resurgent vapor.*

resurgent vapor (a) Groundwater volatilized by contact with hot rock. (b) Gas in magma, possibly derived from dissolved or assimilated country rock. Obsolete. Syn: *resurgent gas.* Cf: *phreatic gas; juvenile [water].*

resurrected (res-ur-rect'-ed) (a) Said of a surface, landscape, or feature (such as a mountain, peneplain, or fault scarp) that has been restored by *exhumation* to its previous status in the existing relief. Syn: *exhumed.* (b) Said of a stream that follows an earlier drainage system after a period of brief submergence had slightly masked the old course by a thin film of sediments. Syn: *palingenetic.*

resurrected karst A *paleokarst* that has been reactivated.

resurrected-peneplain shoreline (res-ur-rect'-ed-pe'-ne-plain) A hypothetical *shoreline of submergence,* formed where the sea rests against an inclined resurrected peneplain (Johnson, 1925, p.27); it may be remarkably straight for long distances. Obsolete.

retained water (re-tained') Water retained in a rock or soil after the *gravity groundwater* has drained out. It is no longer ground water but has become *vadose water.* Most of it is held by molecular attraction, but part may be in isolated interstices or held by other, more or less obscure forces, and part remains as water vapor in interstices from which water has drained (Meinzer, 1923, p.27-28).

retaining wall (re-tain'-ing) A type of engineered retaining structure, such as a *crib* wall, bin wall, *gabion* wall, or mechanically stabilized earth (MSE) wall. Typically, it is designed to resist the lateral pressure of the material behind it; e.g., a *bulkhead* for the prevention of earth movement.

retard (re'-tard) A permeable bank-protection structure situated at and parallel to the toe of a slope and projecting into a stream channel, designed to reduce stream velocity and induce silting or accretion.

retardation [cryst] (re-tar-da'-tion) In crystal optics, the amount by which the slow wave falls behind the fast wave during passage through an anisotropic crystal plate. Retardation depends on the plate's thickness and the difference in refractive indices of its two privileged directions.

retardation [env geol] In the movement of contaminants through a groundwater flow system, the delay resulting from partitioning (adsorption and/or precipitation) of the contaminants to the aquifer solids.

retardation [tides] *lunitidal interval.*

retention (re-ten'-tion) The amount of water from precipitation that has not escaped as runoff or through evapotranspiration; "the difference between total precipitation and total runoff on a drainage area" (Nelson and Nelson, 1967).

reteporiform (re-te-po'-ri-form) Said of a rigid, erect *fenestrate* colony in cheilostome bryozoans, firmly attached by a calcified base.

retgersite (ret'-ger-site) A bluish-green tetragonal mineral: $NiSO_4 \cdot 6H_2O$. It is dimorphous with nickel hexahydrite.

reticle (ret'-i-cle) A system of wires, cross hairs, threads, dots, or very fine etched lines, placed in the eyepiece of an optical instrument (such as a surveyor's telescope) perpendicular to its principal focus, to define the line of sight of the telescope or to permit a specific pointing to be made on a target or signal or a reading to be made on a rod or scale. Syn: *reticule.*

reticulate [ore dep] (re-tic'-u-late) Said of a vein or lode with netlike texture, e.g. *stockwork.*

reticulate [paleont] (a) Said of a netted pattern of an invertebrate, or of one resembling a network, e.g. a "reticulate layer" consisting of ornamental ridges at the surface of a foraminiferal test, or a "reticulate ornamentation" on the exterior of a brachiopod shell, commonly involving the intersection of concentric rugae with radial costae or costellae. (b) Said of evolutionary change that involves repeated intercrossing between a number of lines; specif. a change involving the complex recombination of genes from varied strains of a diversified interbreeding population.

reticulate [palyn] Said of palynomorphs having sculpture consisting of a network of ridges or muri. Cf: *intrareticulate; negative reticulum.*

reticulate [petrology] Said of a rock texture in which crystals are partially altered to a secondary mineral, forming a network that encloses remnants of the original mineral. Cf: *mesh texture.*

reticulated bar (re-tic'-u-lat-ed) One of a group of slightly submerged sandbars in two sets, both of which are diagonal to the shoreline, forming a crisscross pattern (Shepard, 1952, p.1909). Reticulated bars are observed in bays and lagoons on the inside of barrier islands.

reticulate speciation Evolutionary process resulting in the origin of one new species from two previously existing species (Wiley, 1981, p.8). Cf: *cladogenesis, anagenesis.*

reticule (ret'-i-cule) *reticle.*

reticulite (re-tic'-u-lite) *thread-lace scoria.*

reticulopodium (re-tic'-u-lo-po'-di-um) A bifurcating and anastomosing ectoplasmic pseudopodium that is characteristic of foraminifers. Pl: reticulopodia.

reticulum (re-tic'-u-lum) A structure or pattern suggesting a net or network. Most commonly used in derivative terms such as reticle, reticular, reticulate, reticulated, reticule, reticuloid. Pl: reticula. Etymol: Latin "reticulum" (diminutive of "rete"), small net.

retiform wall (ret'-i-form) A wall in archaeocyathids in which the pores are so closely spaced that the entire wall appears netlike. The pores may be of various shapes (TIP, 1972, pt. E, p.9).

retinalite (re-tin'-a-lite) A massive, honey-yellow or greenish serpentine mineral with a waxy or resinous luster; a variety of chrysotile.

retinasphalt (ret-in-as'-phalt) A light-brown variety of retinite usually found with lignite.

retinite (ret'-i-nite) (a) A general term for a large group of fossil resins of variable composition (oxygen content generally 6-15%), characterized by the absence of succinic acid, and found in the younger (brown) coals or in peat. (b) Any fossil resin of the retinite group, such as glessite, krantzite, muckite, and ambrite. (c) A general name applied to fossil resins.

retinosite (ret-i-no'-site) A microscopic constituent of torbanite, consisting of translucent orange-red discs (Dulhunty, 1939).

retipilate (re-tip'-i-late) Of palynomorphs, having a *reticulum* formed by rows of *columellae (pila)* instead of *muri.*

retractor muscle (re-trac'-tor) A muscle that draws in an organ or part, e.g. a "siphonal retractor muscle" serving to withdraw the *siphon* of a bivalve mollusk partly or wholly into the shell, a "pallial retractor muscle" withdrawing marginal parts of *mantle* into a bivalve-mollusk shell where there is no distinct line of muscle attachment, or an *anterior lateral muscle* in a lingulid brachiopod. Cf: *protractor muscle.*

retral (re'-tral) Posterior, or situated at or toward the back; e.g. "retral processes" in foraminiferal tests, consisting of backward-pointing extensions of chamber cavity and enclosed protoplasm, located beneath external ridges on the chamber wall, and ending blindly at the chamber margins (as in *Elphidium*).

retreat (re-treat') The decrease in length of a *glacier,* resulting in a displacement upvalley or upslope of the position of the terminus, caused when processes of *ablation* (usually melting and/or calving) exceed the speed of *ice flow;* normally measured in meters per year. Syn: *recession [glaciol]; glacial recession; glacial retreat; glacier recession.*

retreatal moraine (re-treat'-al) *recessional moraine.*

retreat path The track of movement defined by erosional surfaces and associated deposits formed under local-to-regional transgressive conditions.

retrochoanitic (ret'-ro-cho'-a-nit'-ic) Said of a septal neck of a cephalopod directed backward (adapically). Ant: *prochoanitic.*

retrodiagenesis (ret'-ro-di'-a-gen'-e-sis) *hypergenesis.*

retrodirective prism *corner reflector.*

retrogradation [coast] (ret'-ro-gra-da'-tion) The backward (landward) movement or retreat of a shoreline or of a coastline by wave erosion; it produces a steepening of the beach profile at the breaker line. Ant: *progradation.* Cf: *recession [coast].*

retrogradation [stratig] In sequence stratigraphy, landward stepping of the shoreline occurring when the overall rate of sediment supply at the shoreline is less than the overall rate of relative rise in sea level. As a result of retrogradation, the distal toes of successively younger parasequences are deposited farther landward. Typically, retrogradation occurs because discrete prograding parasequences are stacked in a backstepping pattern.

retrogradational parasequence set In sequence stratigraphy, a parasequence set in which successively younger parasequences are deposited farther landward, in a backstepping pattern. Overall, the rate of deposition is less than the rate of accommodation (Van Wagoner et al., 1990).

retrograde boiling (ret'-ro-grade) The separation of a gas phase in a cooling magma as a result of its residual enrichment in the dissolved gaseous components by progressive crystallization of the magma.

retrograde condensation Appearance of a *condensate* in deep formations as the reservoir pressure is reduced through production of natural gas. The gas condenses to form a liquid, instead of the usual pattern of liquid changing to gas; hence the term retrograde.

retrograde diagenesis *hypergenesis.*

retrograde metamorphism A type of *polymetamorphism* by which metamorphic minerals of a lower grade are formed at the expense of minerals characteristic of a higher grade of metamorphism; it is a readjustment necessitated by a change in physical conditions, e.g., the lowering of temperature. Cf: *prograde metamorphism.* Syn: *retrogressive metamorphism.* Rarely used syn: *diaphthoresis.*

retrograde motion Rotation or orbital motion in the opposite direction of *prograde motion,* that is, westward or clockwise as viewed from the north.

retrograding shoreline (ret'-ro-grad'-ing) A shoreline that is being moved backward by wave attack. Ant: *prograding shoreline.* Syn: *abrasion shoreline; transgressive shoreline.*

retrogression [coast] (ret-ro-gres'-sion) *recession [coast].*

retrogression [evol] The passage from a higher to a lower or from a more to a less specialized state or type of organization or structure during the development of an organism.

retrogressive erosion (ret-ro-gres'-sive) *headward erosion.*

retrogressive metamorphism *retrograde metamorphism.*

retrogressive thaw A slope failure or slump resulting from thawing of *ice-rich permafrost.*

return (re-turn') In *SLAR,* a pulse of microwave energy reflected by the terrain and received at the radar antenna. The strength of a return is referred to as "return intensity".

return flow The amount of water that reaches a ground- or surface-water source after release from the point of use and thus becomes available for further use (Solley et al., 1983). Syn: *wastewater; return water.*

return water *return flow.*

retusoid (re-tu'-soid) Of spores, mostly of Devonian age, having prominent contact areas and "curvaturae perfectae" as, e.g., in *Retusotriletes.* Cf: *curvatura.*

retzian (ret'-zi-an) A group name for orthorhombic minerals with the general formula $Mn_2REE(AsO_4)(OH)_4$, where REE are rare-earth elements.

retzian-(Ce) A dark brown orthorhombic mineral of the *retzian* group: $Mn^{2+}_2CeAsO_4(OH)_4$.

retzian-(La) A dark reddish-brown orthorhombic mineral of the *retzian* group: $Mn^{2+}LaAsO_4(OH)_4$.

retzian-(Nd) A dark reddish brown orthorhombic mineral of the *retzian* group: $Mn^{2+}_2NdAsO_4(OH)_4$.

retzian-(Y) A mineral of the *retzian* group with a compostion reported as $Mn^{2+}_2YAsO_4(OH)_4$. It was later shown to be retzian-(Nd).

Reuverian (Reu-ver'-i-an [Roy-ver'-ian]) European warm-climate paleofloral stage: lower Pliocene (above Brunsummian, below Pretiglian).

revdite (rev'-dite) A colorless monoclinic mineral: $Na_{16}Si_{16}O_{27}(OH)_{26}·28H_2O$.

reverberation (re-ver'-ber-a'-tion) *singing.*

reversal (re-ver'-sal) *geomagnetic reversal.*

reverse (re-verse') (a) Pertaining to the basal side of an incrusting or freely growing bryozoan colony. (b) Aspect of graptoloid rhabdosome (especially early growth stages or biserial forms) in which *sicula* is more or less concealed by *crossing canals.* Cf: *obverse.*

reverse bearing *back bearing.*

reverse branch A seismic event where right-to-left orientation on the event corresponds to the reverse orientation (left-to-right) on the reflector. Such an event could originate from a sharp syncline, which makes it appear to have anticlinal curvature on an unmigrated seismic section. For zero-offset seismic data, this occurs where the radius of curvature for the concave-upward reflector is less than the reflector depth. See also: *buried focus.*

reversed (re-versed') *overturned.*

reversed chevron mark A *chevron mark* that points upstream (Dzulynski and Walton, 1965, p.110).

reversed consequent stream A consequent stream whose direction of flow is contrary to that normally consistent with the geologic structure; e.g. the part of a captured consequent stream between the escarpment slope and the elbow of capture. See also: *obsequent stream.*

reversed fault *reverse fault.*

reversed gradient A local gradient opposite to the general gradient; esp. a valley gradient at the downstream side of a glacially overdeepened valley.

reversed magnetic field *reversed polarity.*

reversed magnetization A *natural remanent magnetization* that is opposite to the present (normal polarity) ambient geomagnetic field.

reversed polarity (a) A *natural remanent magnetization* opposite to the present ambient geomagnetic-field direction. See also: *geomagnetic reversal.* Syn: *reversed magnetic field.* (b) A configuration of the Earth's magnetic field with the magnetic positive pole, where field lines leave the Earth, located near the geographic north pole. Cf: *normal polarity.*

reverse drag Deformation along a fault that creates a fold or set of folds whose curvature is opposite that which is formed by normal drag folding. Reverse drag is a common feature of listric normal faults where hanging-wall folds are concave toward the slip direction (Biddle and Christie-Blick, 1985). Also used for the hanging-wall subsidence and *footwall uplift* adjacent to a planar normal fault (Barnett et al., 1987) that result from a decrease in displacement away from the fault plane. Because drag is a misleading term, "reverse drag" should be avoided; better terms are *rollover* and *fault-bend fold.* Cf: *drag fold; normal drag.*

reversed stream A stream whose direction of flow has been reversed, as by glacial action, landsliding, gradual tilting of a region, or capture.

reversed zoning Zoning in a plagioclase crystal in which the core is more sodic than the rim. Cf: *normal zoning.* Syn: *inverse zoning.*

reverse fault A fault on which the hanging wall has moved upward relative to the footwall. The dip of the fault is usually greater than 45°. Cf: *normal fault.* Partial syn: *thrust fault.* Obsolete syn: *reversed fault.*

reverse-flowage fold (re-verse'-flow'-age) A fold in which flow from deformation has thickened the anticlinal crests and thinned the synclinal troughs, contrary to the normal flow pattern of a *flow fold.*

reverse grading *inverse grading.*

reverse saddle A mineral deposit associated with the trough of a synclinal fold and following the bedding plane. Syn: *trough reef.* Cf: *saddle reef.*

reverse scarplet An *earthquake scarplet* facing in toward the mountain slope and enclosing a trench, produced by reversal of earlier movement along a fault (Cotton, 1958, p.165-166); examples are numerous in New Zealand. Syn: *earthquake rent.*

reverse separation Separation measured parallel to the dip of a fault across which the hanging wall is apparently elevated with respect to the footwall (Biddle and Christie-Blick, 1985). Cf: *normal separation.*

reverse similar fold A fold in which the strata of which are thickened on the limbs and thinned in the hinges, contrary to the pattern of a *similar fold.*

reverse slip fault Obsolete term for *reverse fault.* Cf: *normal slip fault.*

reverse VSP A vertical seismic profile made with the source in the borehole and geophones on the surface.

reversible pendulum (re-vers'-i-ble) A pendulum that can swing around either of two knife edges placed in such a way that the period is the same for both; it is used in absolute-gravity determinations.

reversible process A thermodynamic process in which an infinitesimal change in the variables characterizing the state of the system can change the direction of the process. A process proceeding at equilibrium where the free energy of the product (G_P) is equal to the reactant (G_R).

reversing current (re-vers'-ing) A tidal current that flows in an alternating pattern of opposite directions for approximately equal lengths of time, with a slack period of no movement at each reversal. Reversing currents occur in estuaries, restricted channels, and inland bodies of water. Cf: *rotary current.*

reversing dune A dune that tends to develop unusual height but migrates only a limited distance "because seasonal shifts in direction of dominant wind cause it to move alternately in nearly opposite directions" (McKee, 1966, p.10). Its general shape may resemble that of a barchan or a transverse dune, but it differs in the complexity of its internal structural orientation due to reversals in direction of the slip face.

reversing thermometer A mercury-in-glass thermometer used to measure temperatures of the sea at depth. The temperature is recorded when the thermometer is inverted; and the recording is maintained until it is once again upright. A *protected thermometer* and an *unprotected thermometer* are usually used as a pair, attached to a *Nansen bottle,* to measure both in-situ temperature and pressure.

reversion (re-ver'-sion) A return toward an ancestral type or condition, such as the reappearance in an organism of an ancestral characteristic. Syn: *atavism.*

revet-crag (re-vet'-crag) A term proposed by Gilbert (1877, p.26) for one of a series of narrow, pointed *outliers* or ridges of eroded strata inclined like a *revetment* against a mountain spur. Cf: *hogback [geomorph].*

revetment (re-vet'-ment) A facing of stone, riprap, concrete, or other material, built to protect an embankment or a shore structure from wave erosion.

revier (re-vier') A term applied in SW Africa to a deeply cut riverbed that usually remains dry.

revival (re-viv'-al) *rejuvenation.*

revived (re-vived') *rejuvenated.*

revived fault scarp *rejuvenated fault scarp.*

revived folding *recurrent folding.*

revived stream *rejuvenated stream.*

revolution (rev-o-lu'-tion) A term formerly popular among geologists for a time of profound orogeny and other crustal movements, on a continentwide or even worldwide scale, the assumption being that such revolutions produced abrupt changes in geography, climate, and environment. Schuchert (1924) classed all orogenies at the close of geologic eras as revolutions, in contrast to disturbances, or orogenies within the eras. The basic premises of these concepts are dubious, and the term revolution is little used today.

reworked (re-worked') Said of a sediment, fossil, rock fragment, or other geologic material that has been removed or displaced by natural agents from its place of origin and incorporated in recognizable form in a younger formation, such as a "reworked tuff" carried by flowing water and redeposited in another locality. Cf: *derived; remanié.*

reworked fossil Fossils from rocks of one age that have been eroded, transported, and redeposited in sediments of a younger age.

reworked tuff *sedimentary tuff.*

reyerite (rey'-er-ite) A colorless or pale green trigonal mineral: $(Na,K)_2Ca_{14}Al_2Si_{22}O_{58}(OH)_8{\cdot}6H_2O$.

Reynolds critical velocity (Reyn'-olds) *critical velocity (b).*

Reynolds number A numerical quantity used as an index to characterize the type of flow in a hydraulic structure in which resistance to motion depends on the viscosity of the liquid in conjunction with the resisting force of inertia. It is the ratio of inertia forces to viscous forces, and is equal to the product of a characteristic velocity of the system (e.g. the mean, surface, or maximum velocity) and a characteristic linear dimension, such as diameter or depth, divided by the kinematic viscosity of the liquid; all expressed in consistent units in order that the combinations will be dimensionless. The number is chiefly applicable to closed systems of flow, such as pipes or conduits where there is free water surface, or to bodies fully immersed in the fluid so the free surface need not be considered (ASCE, 1962).

Reynolds stress The transfer of momentum by action of turbulence in a fluid. Turbulent *eddies* in the ocean transfer momentum horizontally over tens to hundreds of kilometers.

rézbányite (réz-bán'-yite) A mixture of bismuthinite derivatives.

RF *representative fraction.*

R-factor A numerical index that provides a gauge of the quality of a crystal-structure refinement and of the data on which the refinement is based. It is generally formulated as the sum, over all reflections, of the differences between observed and calculated structure amplitudes divided by the sum of the observed structure amplitudes. Syn: *residual [cryst]; discrepancy index.*

R_f/phi (R_f/ϕ) method A method of stain analysis based on the shape and orientation of deformed markers (Dunnet, 1969). The method uses a plot of the ratio of the long axis to the short axis (R_f) and long axis direction (ϕ) of a number of elliptical grains to determine the ellipticity and orientation of the strain ellipse.

Rg wave A slow, short-period *Rayleigh wave* that travels only along a nonoceanic path. The "g" refers to the granitic layer. Cf: *Lg wave.*

rhabd (a) An obsolete term for *monaxon.* (b) *rhabdome.*

rhabdacanth (rhab'-da-canth) A *trabecula* of a rugose coral in which the fibers are related to any number of separate, transient (shifting) centers of growth grouped around a main one (TIP, 1956, pt. F, p.235). Cf: *monacanth; rhipidacanth.*

rhabde (rhab'-de) The lower or axial branch in the triaene spicule of an ebridian skeleton.

rhabdite (rhab'-dite) A syn. of *schreibersite,* esp. occurring in rods or needle-shaped crystals.

rhabdodiactine (rhab'-do-di-ac'-tine) A seemingly monaxonic sponge spicule, formed by suppression of two of the axes of a hexactine which are preserved internally as an axial cross. See also: *diactine.*

rhabdoglyph (rhab'-do-glyph) A collective term used by Fuchs (1895) for a *trace fossil* consisting of a presumed worm trail appearing on the undersurface of flysch beds (sandstones) as a nearly straight bulge with little or no branching. Cf: *graphoglypt; vermiglyph.*

rhabdolith (rhab'-do-lith) A calcareous, spinose, rodlike or clublike, supposedly perforate *coccolith,* averaging 3 μm in diameter, having a shield surmounted by a long stem. Rhabdoliths are found both at the surface and on the bottom of the ocean, and they have been classed as protozoans and as algae. Also spelled: *rabdolith.*

rhabdomancy (rhab'-do-man'-cy) A form of *dowsing* using a rod or twig. Cf: *pallomancy.*

rhabdome (rhab'-dome) The long ray of a triaene sponge spicule. Syn: *rhabd.*

rhabdophane (rhab'-do-phane) A group name for hexagonal mineral with the general formula $REEPO_4{\cdot}H_2O$, where REE stands for "rare-earth elements".

rhabdophane-(Ce) A brown, pinkish, or yellowish-white hexagonal mineral of the *rhabdophane* group: $(Ce,La,Ca)PO_4{\cdot}H_2O$.

rhabdophane-(La) A brown, pinkish, or yellowish-white hexagonal mineral of the *rhabdophane* group: $(La,Ce)PO_4{\cdot}H_2O$.

rhabdophane-(Nd) A brown, pinkish, or yellowish-white hexagonal mineral of the *rhabdophane* group: $(Nd,Ce,La)PO_4{\cdot}H_2O$.

rhabdosome (rhab'-do-some) The skeleton of a graptolithine colony.

Rhaetian (Rhae'-tian) Stage of the Standard Global Chronostratigraphic Scale: uppermost Triassic (above Norian, below Hettangian of Jurassic). Syn: Rhaetic.

rhagon (rha'-gon) (a) The earliest developmental stage of a sponge with a functional aquiferous system having several flagellated chambers. Also, a sponge or sponge larva in such a stage. (b) A term used incorrectly as a syn. of *leucon.* Adj: rhagonoid.

rhegmagenesis (rheg-ma-gen'-e-sis) *regmagenesis.*

rhegolith (rheg'-o-lith) *regolith [weath].*

rheid (rhe'-id) A substance (below its melting point) which deforms by viscous flow during the time of applied stress at an order of magnitude at least three times that of the elastic deformation under similar conditions.

rheid fold A fold in which the strata have deformed by flow as if they were fluid. Cf: *flow fold.* Syn: *rheomorphic fold.*

rheidity (rhe-id'-i-ty) *Relaxation time* of a substance, multiplied by 1,000.

rheniite A ill-defined mineral reported as Rh_2S_3. Not approved by the International Mineralogical Association.

rhenium-osmium age method (rhe'-ni-um-os'-mi-um) The determination of an age in years based on the known radioactive decay rate of rhenium-187 to osmium-187. The low crustal abundance of rhenium limits the application of this method to problems related to core-mantle evolution.

rheoignimbrite (rhe'-o-ig-nim'-brite) A densely welded ignimbrite that flowed after welding due to high temperatures and low viscosities.

rheologic settling (rhe-o-log'-ic) Failure of a sediment under a stress load by plastic deformation or flow.

rheology (rhe-ol'-o-gy) The study of the deformation or *flow* of matter and the relationship between stress and strain and the responses of rock to deformation, including *elastic, brittle, viscous,* and *plastic behaviors.*

rheomorphic (rhe-o-mor'-phic) Said of a rock whose form and internal structure indicate that it was subjected to ductile flow; also, the phenomena causing such a rock. See also: *rheomorphism.*

rheomorphic fold *rheid fold.*

rheomorphic intrusion A body of country rock rheomorphically injected into the igneous intrusion that caused the rheomorphism.

rheomorphism (rhe-o-mor'-phism) The process by which a rock becomes mobile as a result of at least partial fusion, commonly accompanied by, if not promoted by, addition of new material by diffusion. The term was created by H.G. Backlund in 1937. Cf: *mobilization.*

rheopexy (rhe'-o-pex-y) The accelerated gelation of a thixotropic sol by agitating it in some manner, e.g. stirring.

rheophile (rhe'-o-phile) adj. Said of an organism that lives in or prefers flowing water.

rheoplasis (rhe-o-pla'-sis) A process of plastic flow-molding at the interface between two immiscible sediments during liquefaction (Van de Poll and Patel, 1981).

rheoplast (rhe'-o-plast) A sedimentary structure resembling common *sole mark,* produced as a result of liquefaction, diapirism, and rheoplasis during compaction (Van de Poll and Patel, 1981).

rheotaxis (rhe-o-tax'-is) *Taxis [ecol]* resulting from mechanical stimulation by a stream of fluid, such as water. Cf: *rheotropism.*

rheotropism (rhe-ot'-ro-pism) *Tropism* resulting from mechanical stimulation by a stream of fluid, such as water. Cf: *rheotaxis.*

rhexistasy (rhex-is'-ta-sy) The mechanical breaking-up and transport of old soils or other surface residual materials (Erhart, 1955). Etymol: Greek "rhexis", "act of breaking", + "stasis", "condition of standing". Adj: rhexistatic. See also: *biorhexistasy; biostasy.*

rhinestone (rhine'-stone) Originally a colorless quartz pebble from the Rhine Valley, now loosely applied to foilbacked and colorless lead-glass imitations of diamond (Liddicoat et al., 1993).

rhinocanna (rhi-no-can'-na) *nasal tube.*

rhipidacanth (rhi-pid'-a-canth) A compound *trabecula* in Paleozoic rugose corals with spindle-shaped septa assigned to the Phillipsastraeidae. Each is composed of a central primary trabecula from which short secondary trabeculae radiate at varying angles perpendicular to the long axis of the septum. Cf: *monacanth; rhabdacanth.*

Rhipidistia (Rhip-i-dis'-ti-a) A suborder of crossopterygian fish which remained primarily freshwater during its entire history; it includes the best candidates for tetrapod ancestry. Range, Lower Devonian to Permian.

rhizoclone (rhi'-zo-clone) A desma (of a sponge) consisting of a straight or curved body bearing branching outgrowths along its entire length. See also: *megarhizoclone.*

rhizoconcretion (rhi'-zo-con-cre'-tion) A small concretionlike structure in a sedimentary rock, cylindrical or conical, usually branching or forked, resembling a root of a plant, usually a tree. It may consist of material such as caliche or chert. Cf: *rhizocretion.* See also: *root sheath.* Syn: *rhizomorph; root cast; rhizolith.*

rhizocretion (rhi-zo-cre'-tion) A term used by Kindle (1923, p.631) for a hollow concretionlike mass that had formed around the root of a living plant. Cf: *rhizoconcretion.*

rhizoid [bot] (rhi'-zoid) A unicellular or multicellular, rootlike filament that attaches some nonvascular plants and gametophytes of some vascular plants to the substrate and serves to absorb water and minerals from the soil (Scagel et al., 1965, p.630). Cf: *rhizome; rhizophore.*

rhizoid [paleont] adj. Resembling a root; e.g. "rhizoid spine" of a brachiopod, resembling a plant rootlet and serving for attachment either by entanglement or by extending along and cementing itself to a foreign surface. n. A radicle of a bryozoan.

rhizolith *rhizoconcretion.*

rhizome (rhi'-zome) An underground stem that lies horizontally and that is often enlarged in order to store food. Not to be confused with rhizoid. Cf: *rhizophore.*

rhizomorph (rhi'-zo-morph) A term used by Northrop (1890) for a structure now known as a *rhizoconcretion.*

rhizophore (rhi'-zo-phore) A naked branch that grows down into the soil and develops roots from the apex (Swartz, 1971, p.405). Cf: *rhizoid [bot]; rhizome.*

rhizophytous (rhi-zo-phy'-tous) Said of a sponge that is fastened to the substrate by branching extensions of the body.

rhizopod (rhi'-zo-pod) A protozoan belonging to the class Sarcodina, subclass Rhizopoda, generally characterized by lobose pseudopodia and by zoned protoplasm in shelled forms and protoplasm differentiated into endo- and ectoplasm in nonshelled forms. Range, Ordovician to Holocene. Cf: *actinopod.*

rhizopodial (rhi-zo-po'-di-al) Said of a morphologic type or growth form in which the cell is somewhat amoeboid.

rhizopodium (rhi-zo-po'-di-um) A bifurcating and anastomosing ectoplasmic *pseudopodium* that is characteristic of foraminifers. Pl: rhizopodia. Syn: *reticulopodium.*

rhizosphere (rhi'-zo-sphere) The soil in the immediate vicinity of plant roots, in which the abundance or composition of the microbial population is affected by the presence of the roots. Syn: *root zone [soil].*

Rhodanian Orogeny (Rho-dan'-i-an) One of the 30 or more short-lived orogenies during Phanerozoic time identified by Stille, in this case at the end of the Miocene.

rhodarsenide A metallic orthorhombic mineral: $(Rh,Pd)_2As$.

rhodesite (rhodes'-ite) A silky white orthorhombic mineral: $KHCa_2Si_8O_{19}{\cdot}5H_2O$. Cf: *mountainite.*

rhodite (rho'-dite) A rhodium-bearing (about 40%) variety of gold.

rhodium (rho'-di-um) (a) An element of the *platinum* group, symbol Rh. (b) A white cubic mineral, Rh, commonly alloyed with other platinum-group elements.

rhodizite (rho'-di-zite) A colorless, white, grayish, or yellowish cubic hard refractory mineral: $(K,Cs)Al_4Be_4(B,Be)_{12}O_{28}$.

rhodochrosite (rho-do-chro'-site) A rose-red or pink to gray rhombohedral mineral of the calcite group: $Mn^{2+}CO_3$. It is isomorphous with calcite and siderite, and commonly contains some calcium and iron; it is a minor ore of manganese. Syn: *dialogite; manganese spar; raspberry spar.*

rhodochrostone (rho-do-chro'-stone) A sedimentary rock consisting mainly of rhodochrosite, with or without other carbonate minerals (Kim, 1975).

rhodolite (rho'-do-lite) A pink, rose, or purple to violet variety of garnet that is intermediate in chemical composition between pyrope and almandine, characterized by a lighter tone and a higher degree of transparency than either of the other two, and used as a gem.

rhodolith (rho'-do-lith) A nodule of red (coralline) algae, concentrically encrusted, often rolled by bottom currents (Bosellini and Ginsburg, 1971). Term proposed as "rhodolite", but this was already a mineral name. Cf: *oncolite.*

rhodonite (rho'-do-nite) A pale-red, rose-red, or flesh-pink to brownish-red or red-brown triclinic pyroxenoid mineral: $(Mn^{2+},Fe^{2+},Mg,Ca)SiO_3$. It sometimes contains zinc, and is often marked by black streaks and veins of manganese oxide. Rhodonite is used as an ornamental stone, esp. in Russia. Syn: *manganese spar; manganolite [mineral]; rhodarsenide.*

Rhodophyta (Rho'-do-phy-ta) *red algae.*

rhodostannite (rho-do-stan'-nite) A metallic reddish tetragonal mineral: $Cu_2FeSn_3S_8$. It may contain appreciable Ag.

rhodplumsite (rhod'-plum-site) A metallic rhombohedral mineral: $Pb_2Rh_3S_2$. It is the Rh analogue of shandite.

rhodusite (rho'-du-site) *magnesioriebeckite.*

rhohelos (rho-he'-los) A stream-crossed, nonalluvial marsh typical of filled lake areas.

rhomb [cryst] An oblique, equilateral parallelogram; in crystallography, a *rhombohedron.*

rhomb [paleont] (a) *pore rhomb.* (b) A six-sided, roughly equidimensional crystal composing some heterococcoliths.

rhombic (rhom'-bic) (a) Adj. of rhomb. (b) Adj. of orthorhombic.

rhombic-dipyramidal class (rhom'-bic-di-py-ram'-i-dal) That crystal class in the orthorhombic system having symmetry $2/m\ 2/m\ 2/m$.

rhombic-disphenoidal class (rhom'-bic-di-sphe-noi'-dal) That crystal class in the orthorhombic system having symmetry 222.

rhombic dodecahedron A crystal form in the cubic system that is a *dodecahedron,* the faces of which are equal rhombs. Cf: *pyritohedron.*

rhombic-pyramidal class (rhom'-bic-py-ram'-i-dal) That crystal class in the orthorhombic system having symmetry mm2.

rhombic section That plane in a triclinic feldspar containing both the b axis and the normal to b lying in (010).

rhombic system A syn. of *orthorhombic system.* It is an undesirable term because it may be confused with rhombohedral.

rhombochasm (rhom'-bo-chasm) *pull-apart basin.*

rhomboclase (rhom'-bo-clase) A white, gray, or pale yellow orthorhombic mineral: $HFe^{3+}(SO_4)_2{\cdot}4H_2O$.

rhombohedral class (rhom-bo-he'-dral) That crystal class in the rhombohedral division of the hexagonal system having symmetry $\bar{3}$.

rhombohedral cleavage Mineral cleavage parallel to the faces of the rhombohedron, e.g. the $(10\bar{1}1)$ cleavage of calcite.

rhombohedral iron ore (a) *hematite.* (b) *siderite [mineral].*

rhombohedral lattice A centered lattice of the trigonal system in which the unit cell is a rhombohedron. It may occur in crystal classes having one threefold axis. The hexagonal unit cell contains three lattice points: one at the corners and two equally spaced along one long diagonal. The rhombohedral unit cell is primitive, having all three crystal axes equal and all interaxial angles equal, but not constrained by symmetry to any particular value.

rhombohedral packing The "tightest" manner of systematic arrangement of uniform solid spheres in a clastic sediment or crystal lattice, characterized by a unit cell of six planes passed through eight sphere centers situated at the corners of a regular rhombohedron (Graton and Fraser, 1935). An aggregate with rhombohedral packing has the minimum porosity (25.95%) that can be produced without distortion of the grains. Cf: *cubic packing.* See also: *close packing.*

rhombohedral system A division of the *trigonal system* in which the unit cell is a rhombohedron.

rhombohedron (rhom-bo-he'-dron) A trigonal crystal form that is a parallelepiped whose six identical faces are rhombs. Syn: *rhomb [cryst].*

rhomboidal lattice structure (rhom-boi'-dal) A common feature on sand beaches, resulting from two sets of small grooves that trend diagonally along the beach slope, and formed immediately after or during the last stages of backwash. "The grooves are probably the surface expression of planar zones of high shear stress in which the grains become more closely packed than elsewhere in the beach sand" (Stauffer et al., 1976, p.1667). Cf: *rhomboid ripple; rhomboid rill mark.*

rhomboid current ripple (rhom'-boid) *rhomboid ripple.*

rhomboid reef Diamond or polygonal shaped reef up to four nautical miles in long dimension developed in the lagoon of the Belize Barrier Reef. Rhomboid reefs consist of narrow reef-built rims with steep slopes rising from depths of 20-40 m to near sea level and surrounding lagoons ranging from 10-30 m deep. Some lagoons are segmented by spitlike coral walls. The unusual rhomboid morphology is controlled by pre-Holocene foundations and either structurally controlled karsted blocks or the interfluves of sinuous streams developed during Pleistocene lowstands of sea level.

rhomboid rill mark A sandy-beach surface pattern consisting of shallow grooves, incised by breaker swash or backwash. The grooves form a fairly uniform rhomboid patterned network; the long diagonal of the rhombs parallels the direction of the creating current (Otvos, 1965, p.271). They are distinguished from rhomboid ripple marks by lower relief and absence of slip faces; they are not a type of ripple, but a network of lineations. Cf: *rhomboid ripple; rhomboidal lattice structure.*

rhomboid ripple An aqueous *current ripple* characterized by diamond-shaped tongues of sand arranged in a reticular pattern resembling the scales of certain fish. Tongues range from 12 to 25 mm in width and 25 to 50 mm in length; each has one acute angle (formed by two steep sides) pointing downcurrent, and another acute angle (formed by the gentle side extending into the reentrant angle of the steep sides of two tongues of the following) pointing upcurrent; it is common on sand beaches where it forms during the final stages of backwash of each retreating wave. The sides are not more than 1 mm high. The term "rhomboid" applied to a *ripple* was introduced by Bucher (1919, p.153). Syn: *rhomboid current ripple; overhanging ripple.*

rhomboid ripple mark *rhomboid ripple.*

rhombolith (rhom'-bo-lith) *scapholith.*

rhomb-porphyry (rhomb-por'-phy-ry) Porphyritic alkalic *syenite* or *trachyte* containing phenocrysts of augite, sparse olivine, and anorthoclase or potassium oligoclase with rhombohedron-shaped cross sections, in a groundmass composed chiefly of alkali feldspars. Var: rhombenporphyry; rhombenporphyr. Named by Buch in 1810. Usually applied to lavas and dikes of the Oslo district, Norway, but also used elsewhere as a textural term. Obsolete.

rhomb spar A dolomite mineral in rhombohedral crystals.

rhönite (rhön'-ite) A dark brown to black triclinic mineral of the *aenigmatite* group: $Ca_2(Mg,Fe^{2+},Fe^{3+},Ti)_6(Si,Al)_6O_{20}$. Syn: rhoenite.

rhopaloid septum (rho'-pa-loid) A rugose corallite *septum* characterized by distinctly thickened axial edge, appearing club-shaped in cross section (TIP, 1956, pt. F, p.250).

rhourd A pyramid-shaped sand dune, formed by the intersection of other dunes (Aufrère, 1931, p.376). Cf: *khurd.*

Rhuddanian (Rhud-da'-ni-an) A stage of the Standard Global Chronostratigraphic Scale: lowermost stage of the Silurian Llandoverian Series (above the Ordovician System, below the Aeronian Stage). The basal GSSP is at Dob's Lin, Scotland (Cocks, 1985).

rhumb line A curved line on the surface of the Earth that crosses successive meridians at a constant oblique angle and that spirals around and toward the poles in a constant true direction but theoretically never reaches them; a straight line on a Mercator projection, representing a line of constant bearing or direction. Syn: *loxodrome; loxodromic curve; Mercator track.*

Rhyacian (Rhy-a'-ci-an) A period of the Standard Global Geochronometric Scale: the second-oldest period of the *Paleoproterozoic* Era, characterized by injection of layered complexes ("rhyax", "stream of lava"). It is defined as the time span between 2,300 Ma, the end of the *Siderian* Period, and 2,050 Ma, the beginning of the *Orosirian* Period (Plumb, 1991).

rhyacolite (rhy-ac'-o-lite) *sanidine.*

rhyncholite (rhyn'-cho-lite) A fossil beak or part of a cephalopod jaw; specif. the calcified tip of a jaw of fossil cephalopods.

rhynchonelloid (rhyn-cho-nel'-loid) Any articulated brachiopod belonging to the order Rhynchonellida, characterized by a rostrate shell, a functional pedicle, and a delthyrium partially closed by deltidial plates. Range, Middle Ordovician to present. Var: rhynchonellid.

Rhynchosauria (Rhyn'-cho-saur'-i-a) An order of archosaurian reptiles that includes large beaked animals with crushing dentitions, of Triassic age. Not to be confused with the living Sphenodon and its extinct relatives, which are lepidosaurs.

rhyniophyte (rhy'-ni-o-phyte) A member of the Rhyniophyta, one of the three major groups of early vascular plants as defined by Banks (1975), originally called Rhyniophytina. The simplest vascular plants are leafless and dichotomously branched and bear terminal sporangia. Named after the genus *Rhynia* from the Early Devonian. Cf: *psilopsid; trimerophyte; zosterophylls.* Obsolete syn: *psilophyte.*

rhyocrystal (rhy-o-crys'-tal) One of a group of idiomorphs arranged in "stream-lines" (Lane, 1903, p.386). Obsolete.

rhyodacite (rhy-o-da'-cite) A volcanic rock intermediate between rhyolite and dacite. Winchell proposed the name in 1913, apparently not knowing that Brögger had earlier proposed *dellenite* for the same composition. The *IUGS classification* does not include a category between rhyolite and dacite. Obsolete.

rhyolite (rhy'-o-lite) (a) In the *IUGS classification,* a volcanic rock defined in the QAPF diagram as having Q/(Q+A+P) between 20 and 60 and P/(A+P) between 10 and 65, and in the TAS diagram by a field, extending to higher silica and alkali contents from a vertical line with its lower end at 69 wt% SiO_2, 8% total alkalis, and a line sloping from 69 SiO_2, 8 alkalis to 77 SiO_2, 0 alkalis. (b) A group of igneous rocks, typically porphyritic and commonly exhibiting flow texture, with phenocrysts of quartz and alkali feldspar in a glassy to cryptocrystalline groundmass; also, any rock in that group; the fine-grained equivalent of granite. Rhyolite grades into *rhyodacite* with decreasing alkali feldspar content and into *trachyte* with a decrease in quartz. The term was coined in 1860 by Baron von Richthofen (grandfather of the World War I aviator). Etymol: Greek "rhyo-", from "rhyax", "stream of lava". Syn: *liparite.* Cf: *quartz porphyry.*

rhyolitoid In the *IUGS classification,* a general term in the provisional "field" classification for a rock tentatively identified as *rhyolite.* Obsolete.

rhythmic crystallization (rhyth'-mic) A phenomenon, observed in igneous rocks, in which different minerals crystallize in concentric layers, giving rise to *orbicular* structure. See also: *cyclic crystallization.*

rhythmic layering Readily observable structure in an igneous intrusion involving the repetition of gravity-stratified layers (more mafic minerals along the bottom, more plagioclase along the top, typically), resulting from periodic crystallization and accumulation (settling) of crystals. See also: *cyclic crystallization; cryptic layering; zebra layering; phase layering; compositional layering.*

rhythmic sedimentation The consistent repetition, through a sedimentary succession, of a regular sequence of two or more rock units organized in a particular order and indicating a frequent and predictable recurrence or pattern of the same sequence of conditions. It may involve only two components (such as interbedded laminae of silt and clay), or broad changes in sediment character spanning whole systems (or longer intervals) and units up to hundreds of meters thick, or any sequence intermediate between these two extremes. See also: *cyclic sedimentation.*

rhythmic succession A succession of rock units showing continual and repeated changes of lithology. The term was used by Hudson (1924) for a continual repetition of a more or less complete suite comprising successive beds of certain kinds of sediments accompanied by an equally marked variation in the kind of fossils they contain.

rhythmic topography *beach rhythm.*

rhythmic unit (a) *rhythmite.* (b) A layer or band of a rhythmically layered intrusive igneous rocks.

rhythmite (rhyth'-mite) (a) An individual unit of a rhythmic succession or of beds developed by rhythmic sedimentation; e.g. a *cyclothem.* The term was used by Bramlette (1946, p.30) for the couplet of distinct types of sedimentary rock, or the graded sequence of sediments, that forms a unit in rhythmically bedded deposits. The term implies no limit as to thickness or complexity of bedding and it carries no time or seasonal connotation. See also: *laminite.* Syn: *rhythmic unit.* (b) An orthorhombic mineral formed in a burning coal heap: $Ca_4(SiO_4)\cdot 3CaCl_2$. Not approved by the International Mineralogical Association.

rhytidome (rhyt'-i-dome) *outer bark.*

ria (ri'-a) (a) Any long, narrow, sometimes wedge-shaped inlet or arm of the sea (but excluding a fjord) whose depth and width gradually and uniformly diminish inland and which is produced by drowning due to submergence of the lower part of a narrow river valley or of an estuary; it is shorter and shallower than a *fjord.* Originally, the term was restricted to such an inlet produced where the trend of the coastal rock structure is at right angles to the coastline; it was later applied to any submerged land margin that is dissected transversely to the coastline. (b) Less restrictedly, any broad or estuarine river mouth, including a fjord, and not necessarily an embayment produced by partial submergence of an open valley. Cf: *estuary.* Etymol: Spanish "ría", from "río", "river".

ria coast A coast having several long parallel *rias* extending far inland and alternating with ridge-like promontories; e.g. the coasts of SW Ireland and NW Spain. It is especially developed where the trend of the coastal structures is transverse to that of the coastline.

ria shoreline A shoreline characterized by numerous *rias* and produced by drowning due to partial submergence of a land margin subaerially dissected by numerous river valleys (Johnson, 1919, p.173).

rib [bot] In a leaf or similar organ, the primary *vein [bot].*

rib [geomorph] A layer or dike of rock forming a small ridge on a steep mountainside.

rib [mining] Sidewall of an underground mine working. Cf: *footwall, hanging wall.*

rib [paleont] (a) A radial or transverse fold upon a shell; e.g. any radial ornament on a brachiopod, pelecypod, or rostroconch shell, or a raised radial ridge on the coiled conch of a nautiloid or ammonoid. (b) One of the arcuate, paired bones of the axial skeleton that connects dorsally to the vertebral column and ventrally to the sternum and helps support the body wall in Osteichthyes and tetrapods.

rib and furrow A term used by Stokes (1953, p.17-21) for the bedding-plane expression of micro cross-bedding. It consists of small transverse arcuate markings, convex upcurrent, occurring in sets and confined to relatively long grooves, 3-5 cm wide, oriented parallel to the current flow and separated from one another by narrow discontinuous ridges. It represents the eroded edges of upturned arcuate laminae.

riband jasper (rib'-and) *ribbon jasper.*

ribbed moraine One of a group of irregularly subparallel, locally branching, generally smoothly rounded and arcuate ridges that are convex in the downstream direction of a glacier but that curve upstream adjacent to eskers (J.A. Elson in Fairbridge, 1968, p.1217). They are most common in the continental ice sheets, and are abundant in the Arctic.

ribbeite (rib'-be-ite) A pink orthorhombic mineral allied to the *leucophoenicite* group: $(Mn^{2+},Mg)_5(SiO_4)_2(OH)_2$.

ribble (rib'-ble) *ripple till.*

ribbon [ore dep] (rib'-bon) adj. Said of a vein having alternating streaks of ore with gangue or country rock, or simply of varicolored ore minerals. Cf: *banded; book structure.*

ribbon [petrology] One of a set of parallel bands or streaks in a mineral or rock, e.g. ribbon jasper; when the lines of contrast are on a larger scale, the term banding is used. When occurring in slate, the structure is known as *slate ribbon.* Syn: *stripe [meta].*

ribbon banding A banding produced in the bedding of a sedimentary rock by thin strata of contrasting colors, giving the rock an appearance suggesting bands of ribbons.

ribbon bomb A type of volcanic bomb that is elongate and flattened, and derived from a rope of lava.

ribbon diagram A single, continuous geologic cross section drawn in perspective along a curved or sinuous line.

ribbon injection A tonguelike igneous intrusion along the cleavage planes of a foliated rock. Obsolete.

ribbon jasper Beautifully banded jasper with parallel, ribbon-like stripes of alternating colors or shades of color (as of red, green, and esp. brown). Syn: *riband jasper.*

ribbon reef A linear reef within the Great Barrier Reef off the NE coast of Australia, having inwardly curved extremities, and forming a festoon along the precipitous edge of the continental shelf. They are variable in length (3-24 km), less so in width (300-470 m).

ribbon rock A rock characterized by a cyclic or rhythmic succession of thin layers of contrasting composition or color; e.g. gray shale interspersed with thin seams of brown dolomite and lighter-colored limestone (Goldring, 1943), or a vein rock with narrow quartz bands separated by stripes of altered wall rock. Applies also to the Precambrian *banded iron formation* or *banded ironstone.*

ribbon slate Slate produced by incomplete metamorphism, with still visible bedding that cuts across the cleavage surface; slate characterized by varicolored ribbons.

ribonucleic acid An essential component of all cells; a nucleic acid composed of a long, usually single-stranded chain of nucleotide units that contain the sugar ribose. Abbrev: RNA.

Richard's equation The non-linear parabolic equation that governs flow of water under unsaturated conditions.

richellite (ri-chell'-ite) A reddish-brown tetragonal mineral: $(Ca,Fe^{2+})(Fe^{3+},Al)_2(PO_4)_2(OH)_2$.

richelsdorfite (rich'-els-dorf-ite') A turquoise-blue monoclinic mineral: $Ca_2Cu_5Sb^{5+}(AsO_4)_4(OH)_6Cl \cdot 6H_2O$.

richetite (rich'-e-tite) An adamantine black triclinic mineral: $Pb_9(UO_2)_{36}(O,OH)_{60} \cdot 41H_2O$.

Richmondian (Rich-mon'-di-an) North American stage: Uppermost Ordovician (above Maysvillian, below Alexandrian of Silurian).

richness In ecology, the total number of taxa in a sample, location, or habitat, without regard to differences in the relative abundance of those taxa. It is equivalent to *diversity* in its strict sense (Hayek and Buzas, 1997).

richterite (rich'-ter-ite) A brown, yellow, or rose-red monoclinic mineral of the *amphibole* group: $Na_2CaMg_5Si_8O_{22}(OH)_2$. Cf: *soda tremolite.*

Richter scale (Rich'-ter) A numerical scale of *earthquake magnitude,* devised in 1935 by the seismologist C.F. Richter. Very small earthquakes, or microearthquakes, can have negative magnitude values. The strength of Earth materials produces an actual upper limit of slightly less than 9. Usually the scale refers to *local magnitude,* but for large earthquakes it often refers to *surface-wave magnitude.*

Richter's Rule Named for the German paleontologist Rudolf Richter (1881-1957), who noticed that bottom currents tend to tilt clam and trilobite shells to a convex-up position, which thus indicates sedimentation polarity in complexly folded sequences (Shrock, 1948).

rickardite (rick'-ard-ite) A deep-purple orthorhombic mineral: Cu_7Te_5.

Ricker wavelet A zero-phase wavelet used in seismic modeling. It is the second derivative of the Gaussian function. Named for Norman H. Ricker (1896-1980), American geophysicist.

rideau (ri-deau') A small ridge or mound of earth, or a slightly elevated piece of ground. Etymol: French.

ridge [beach] (a) *beach ridge.* (b) A low sandbar that is sometimes found above the water level on the foreshore of a sand beach during low tide. See also: *runnel [beach].*

ridge [geomorph] (a) A general term for a long, narrow elevation of the Earth's surface, usually sharp-crested with steep sides, occurring either independently or as part of a larger mountain or hill; e.g. an extended upland between valleys. A ridge is generally less than 8 km long (Eardley, 1962, p.6). (b) A term occasionally applied to a range of hills or mountains. (c) The top or upper part of a hill; a narrow, elongated crest of a hill or mountain.

ridge [ice] *pressure ridge [ice].*

ridge [marine geol] An elongate, steep-sided elevation of the ocean floor, having rough topography. Syn: *submarine ridge.*

ridge [meteorol] An elongated region of relatively high atmospheric pressure at a given altitude, with anticyclonic curvature of the winds and isobars. If a closed circulation occurs, the ridge is called an anticyclone. Ant: *trough [meteorol].*

ridge [paleont] An elevated body part of an animal, projecting from a surface; e.g. a relatively long narrow elevation of secondary shell of a brachiopod, or a *transverse ridge* on a crinoid. Also, an area separating adjacent pairs of ambulacral pores of a regular echinoid.

ridge-and-groove slickensides Slickensides, normally found on shiny fault surfaces, in which the opposite walls of a fault possess matched ridges on one wall and grooves on the opposite wall. They are thought to be caused by the combined mutual plowing and mutual deformation of opposing wall rocks (Means, 1987).

ridge-and-ravine topography Hack's (1960) term for landscapes in dynamic equilibrium, undergoing active regional erosion in a humid temperate climate, such as the *ridge-and-valley topography* of the Appalachians.

ridge-and-swale topography Long subparallel ridges and swales aligned obliquely across the regional trend of the contours (Swift et al., 1972, p.501); e.g., the inner-shelf topography of the Middle Atlantic Bight.

ridge-and-valley topography A land surface characterized by a close succession of parallel or nearly parallel ridges and valleys, and resulting from the differential erosion of highly folded strata of varying resistances. Type region: Ridge and Valley region in the Appalachian Mountains, lying to the west of the Blue Ridge. Cf: *ridge-and-ravine topography.*

ridge belt Narrow to broad belt of mostly parallel ridges in the plains of Venus.

ridged ice Sea ice having readily observed surface features in the form of one or more pressure ridges; it is usually found in first-year ice. See also: *ropak; ridging.*

ridge-push force In plate tectonics, the force, outward directed from the axis of a mid-ocean ridge, that is due to the gravitational potential energy of the elevated ridge area. It is probably a key plate-driving force.

ridge-ridge transform fault A *transform fault* that offsets mid-ocean ridges.

ridge-top trench A trench, occasionally found at or near the crest of high, steep-sided mountain ridges, formed by the creep displacement of a large slab of rock along shear surfaces more or less parallel with the side slope of the ridge. Trenches are usually parallel with the crest of the ridge. See also: *sackungen.*

ridging (ridg'-ing) A form of deformation of *floating ice,* caused by lateral pressure, whereby ice is forced or piled haphazardly one piece over another to form *ridged ice.* Cf: *tenting.*

riebeckite (rie'-beck-ite) A dark blue or black monoclinic mineral of the *amphibole* group: $Na_2Fe^{2+}{}_3Fe^{3+}{}_2Si_8O_{22}(OH)_2$. It occurs as a primary constituent in some acid or sodium-rich igneous rocks. See also: *crocidolite.*

Riecke's principle (Rieck'-e's) The statement in thermodynamics that solution of a mineral tends to occur most readily at points where external pressure is greatest, and that crystallization occurs most readily at points where external pressure is least. It is applied to recrystallization in metamorphic rocks with attendant change in mineral shapes. It is named after the German physicist E. Riecke (1845-1915) although it was actually discovered and described by H.C. Sorby in 1863.

Riedel shear (rie'-del) A secondary slip surface which develops during the early stage of shearing. Such shears are typically arranged en échelon, usually at inclinations of between 10° and 30° to the direction of relative movement (Riedel, 1929). Riedel shears are subsidiary sets of faults inclined to the trace of a throughgoing fault. The most common sets are conjugate faults whose acute bisectrix is inclined roughly 45° to the throughgoing fault.

riedenite (rie'-den-ite) A *foidolite* composed of large tabular biotite crystals in a granular groundmass of nosean, biotite, pyroxene, and small amounts of titanite and apatite. The name, given by Brauns in 1922, is from Rieden, Laacher See district, Germany. Obsolete.

riegel (rie'-gel) A low, transverse ridge or barrier of bedrock on the floor of a glacial valley, esp. common in the Alps; it separates a rock basin from the gently sloping valley bottom farther downstream. Cf: *rock step.* Etymol: German "Riegel", "crossbar". Syn: *rock bar; threshold [glac geol]; verrou.*

Riel discontinuity A seismic-velocity discontinuity noted in Alberta that may be equivalent to the *Conrad discontinuity.*

Rietveld refinement A procedure for carrying out a *crystal-structure refinement* using X-ray or neutron *powder diffraction* data, in which an entire *powder pattern* is simulated for a trial structure and matched against the observed powder pattern. *Atomic parameters* and other variables are modified to achieve an acceptable fit between the calculated and observed powder patterns.

riffle (rif'-fle) (a) A natural shallows or other expanse of shallow bottom extending across a stream bed over which the water flows swiftly and the water surface is broken into waves by obstructions wholly or partly submerged or a shallow rapids of comparatively little fall. It has a larger downstream channel-bed gradient and coarser sediment than adjacent portions of the stream. See also: *rift [streams].* (b) An expanse of shallow water flowing over a riffle or at the head of a rapids. (c) A low bar or bedrock irregularity in a stream, resembling a riffle. (d) A wave of a riffle. Syn: *ripple.*

riffle hollow A shallow depression in a stream bed, commonly 8 to 30 cm deep, produced by differential erosion of alternate layers of hard and soft rock (Bryan, 1920, p.192).

riffler (rif'-fler) *sample splitter.*

rift [eco geol] A quarrying term for a direction of parting in a massive rock, e.g. granite, at approximately right angles to the *grain [eco geol].*
rift [geomorph] A narrow cleft, fissure, or other opening in rock, made by cracking or splitting.
rift [speleo] A narrow, high *passage* in a cave, the shape of which is controlled by a joint or by a bedding or fault plane.
rift [streams] A shallow or rocky place in a stream, forming either a ford or a rapids. The term is used in NE U.S. as a syn. of *riffle.*
rift [struc geol] The direction of easiest splitting in granite or other crystalline rocks often enhanced by a microcrack fabric that forms as a result of either cooling or decompression (Dale, 1923).
rift [tect] (a) A long, narrow continental trough that is bounded by normal faults; a *graben* of regional extent. It marks a zone that has undergone extension. Cf: *paar.* (b) A belt of strike-slip faulting of regional extent.
rift-drift transition The stage in the evolution of a *passive margin* at which rift faulting ceases and sea-floor spreading begins. After the rift-drift transition, the stretched lithosphere of the newly formed continental margin undergoes thermal subsidence and the regional principal stress state changes.
rift fault A fault that bounds a rift valley, graben, or half-graben. The term has been applied to normal faults and less commonly to large strike-slip faults. Cf: *border fault; boundary fault.*
rift lake An elongate lake that forms in a rift valley.
riftogenic Applied to fracture belts of great lineal extent and traversing the depth of the craton; introduced by Russian geologists. Cf: *taphrogeny.*
rift trough *rift valley [tect].*
rift valley [oceanog] The deep central cleft in the crest of the *mid-ocean ridge.* Syn: *central valley; median valley; mid-ocean rift.*
rift valley [tect] A valley that has developed along a *rift [tect].* Syn: *rift trough.*
rift-valley lake A lake formed in a *rift valley,* esp. a *graben* lake such as Lake Malawi (Nyasa), Lake Rudolf (Turkana), or Lake Tanganyika in Africa. Cf: *sag pond.* Syn: *rift lake.*
rift zone A region of crust where extension results in formation of an array of kinematically related normal faults, along with associated grabens, half grabens, and horsts. Some active rift zones have associated volcanic activity. Some rift zones evolve into troughs filled by very thick sequences of sediment. Some are broad with distributed faults, whereas some are narrow with localized faulting.
rig *drilling rig.*
right bank The bank of a stream situated to the right of an observer who is facing downstream.
right-handed [cryst] Said of an optically active crystal that rotates the plane of polarization of light to the right. Cf: *left-handed [cryst].* Syn: *dextrorotatory.*
right-handed [paleont] *dextral.*
right-handed separation Obsolete term for *right-lateral separation.*
right-lateral fault (right-lat'-er-al) A strike-slip fault on which the side opposite the observer has been displaced to the right. Syn: *dextral fault.*
right-lateral separation Displacement along a fault such that, in plan view, the side opposite the observer appears displaced to the right. Cf: *left-lateral separation.* Obsolete syn: *right-handed separation.*
right reading image An image that is not reversed, one that reads correctly when viewed normally.
right section *profile [struc geol].*
right-slip fault An obsolete term for a *right-lateral fault.*
right valve The valve lying on the right-hand side of a bivalve mollusk when the shell is placed with the anterior end pointing away from the observer, the commissure being vertical and the hinge being uppermost. Abbrev: RV. Ant: *left valve.*
rigid-body rotation A type of deformation where *rotation* is the only non-zero component. Deformation where the body suffers no distortion, dilation, or translation. Cf: *rigid-body translation.*
rigid-body translation A type of deformation where *translation* is the only non-zero component. Deformation where the body suffers no distortion, dilation, or rotation. Cf: *rigid-body rotation.*
rigidity (ri-gid'-i-ty) The property of a material to resist applied stress that would tend to distort it. A fluid has zero rigidity.
rigidity modulus *modulus of rigidity.*
rigolet (ri-go-let') A term applied in the Mississippi River valley to a small stream, creek, or rivulet. Etymol: French "rigole", "trench, small ditch, channel".
Rigsby Stage A 4-axis *universal stage* constructed especially for measuring the orientations of *c*-axes of ice crystals. Named after George Rigsby who designed it.
rijkeboerite (rijk-e-boer'-ite) *bariomicrolite.*
rilandite (ri'-land-ite) An ill-defined brownish-black mineral: $(Cr^{3+},Al)_6SiO_{11}\cdot 5H_2O(?)$.
rill [beach] (a) A small, transient runnel carrying to the sea or a lake the water of a wave after it breaks on a beach, esp. one formed following an outgoing tide. It may be 2-10 mm wide, 0.5 m or more long, and about 1 mm deep. (b) The minute stream or thin sheet of water flowing in a rill.
rill [planet] *rille.*
rill [streams] (a) A very small brook or trickling stream of water; a streamlet or rivulet, usually without any tributaries. (b) A small channel eroded by a rill, esp. in soil; one of the first and smallest channels formed by runoff, such as a *shoestring rill.*
rill cast A term used by Dzulynski and Slaczka (1958, p.230) for a sole mark that is a type of *flute cast.* See also: *rill mark.*
rille One of several trenchlike or cracklike valleys, up to several hundred km long and 1-2 km wide, commonly occurring on planetary surfaces subjected to plains volcanism. Rilles may be extremely irregular with meandering courses (sinuous rilles), or they may be relatively straight (normal rilles); they have relatively steep walls and usually flat bottoms. Rilles apparently represent fractures originating in brittle material. Syn: *rima.* Also spelled: rill.
rillenkarren (ril'-len-kar'-ren) Downslope solution channels separated by sharp intergroove crests typically several centimeters apart. Etymol: German, "rill tracks". Cf: *karren; rinnenkarren; deckenkarren.*
rillenstein (ril'-len-stein) Microsolution grooves and pitting on rock surface. Etymol: German, "rilled rock".
rill erosion The development of numerous minute closely spaced channels resulting from the uneven removal of surface soil by running water that is concentrated in streamlets of sufficient discharge and velocity to generate cutting power. It is an intermediate process between *sheet erosion* and *gully erosion.* Cf: *channel erosion.* Syn: *rill wash; rilling; rillwork.*
rill flow Surface runoff flowing in small, irregular channels too small to be considered rivulets.
rilling (rill'-ing) *rill erosion.*
rill mark (a) A small, dendritic channel, groove, or furrow formed on the surface of beach mud or sand by a wave-generated rill or by a retreating tide; esp. one formed on the lee side of a half-buried pebble, shell, or other obstruction, and usually branching upstream. (b) A dendritic channel formed by a small stream or spring debouching on a sand flat or a mud flat; it shows a downslope bifurcation. Cf: *lobate rill mark.*
rill wash A syn. of *rill erosion.*
rillwork (rill'-work) *rill erosion.*
rim [geomorph] The border, margin, edge, or face of a landform, such as the curved brim surrounding the top part of a crater or caldera; specif. the *rimrock* of a plateau or canyon.
rim [glac geol] A ridge of morainal material, generally unbroken and of uniform height, surrounding a central depression (Gravenor and Kupsch, 1959, p.52).
rim [paleont] (a) One of the two flanges of a caneolith coccolith peripheral to the wall; e.g. the distal "upper rim" and the proximal "lower rim". (b) The outer, usually flangelike component of a *wheel* of a holothurian. It may be recurved, and the inner margin of its upper side is commonly denticulate or dentate. The rim is inclined to or within the plane of the wheel.
rima (ri'-ma) A term established by the International Astronomical Union for a long narrow aperture, cleft, or fissure on planetary surfaces (USGS, 2003); specif. a lunar *rille.* Pl: rimae.
rim cementation A term used by Bathurst (1958, p.21) for *secondary enlargement* in detrital sediments; e.g. the chemical deposition of calcium carbonate forming a single, completely enveloping rim on a grain of the same composition, as in a crinoidal limestone where each grain (or crinoidal fragment) is a single crystal and is permeated by the calcite cement in lattice or optical continuity. Cf: *granular cementation.*
rime A deposit of white opaque ice crystals formed as a result of contact between supercooled droplets of fog and a solid object at a temperature below 0°C.
rim gypsum Gypsum in thin films between anhydrite crystals, believed to have been introduced in solution rather than produced by replacement (Goldman, 1952, p.2).
rim height The maximum height of the rim of a crater above the original ground surface.
rimkorolgite A yellow-brown to light pink monoclinic mineral: $(Ba,Sr,Ca)(Mg,Mn^{2+})_5(PO_4)_4\cdot 8H_2O$.
rimmed kettle A morainal depression with raised edges (Gravenor and Kupsch, 1959, p.53); a kettle in stratified drift with raised edges.
rimming wall (rim'-ming) A steep, ridgelike erosional remnant of continuous layers of porous, permeable, poorly cemented, detrital limestones, believed to form under tropical or subtropical conditions (as on Okinawa and other Pacific islands) by surface-controlled secondary cementation on an original steep slope followed by differential erosion that brings the cemented zone into relief (Flint et al., 1953). See also: *rampart wall.*
rimpylite (rim'-py-lite) A group name for several green and brown hornblendes having high contents of $(Al,Fe)_2O_3$.
rim ridge A minor ridge of till defining the edge of a *moraine plateau* (Hoppe, 1952, p.5).
rimrock [eco geol] (rim'-rock) The bedrock forming or rising above the margin of a placer or gravel deposit. Also spelled: rim rock.
rimrock [geomorph] (a) An outcrop of a horizontal layer of resistant rock, such as a lava flow, exposed at the edge of a plateau, butte, or mesa; it generally forms a cliff or ledge. (b) The edge or face of an outcrop of rimrock, esp. a cliff or a relatively vertical face of rock in the wall of a canyon. Syn: *rim [geomorph].*

rimrocking (rim′-rock-ing) Prospecting for carnotite, specifically on the Colorado Plateau, where the favorable beds, more or less flat-lying, crop out in cliffs, or rims.
rimstone (rim′-stone) A calcium carbonate deposit which forms around the periphery of a cave pool.
rimstone barrier *rimstone dam.*
rimstone dam A barrier of calcite, aragonite, or other material that obstructs a cave stream to form a pool. It is the product of carbonate precipitation, which occurs at a break in slope in the stream. Syn: *gour [speleo]; rimstone barrier; travertine dam.*
rimstone pool The pool or basin of water that is formed of and bounded by a *rimstone dam.* Syn: *gour [speleo].*
rim syncline A fold having an arcuate or subcircular axial tract on the outer margin of a salt upwelling. Rim syncline is a nongenetic term, but in the context of salt tectonics a rim syncline typically results from salt withdrawal in the source layer. Peripheral sinks of sediments accumulate within rim synclines (Nettleton, 1934).
rim terrace An outward-dipping bench on the wall of a *complex impact crater,* generally interpreted to be the top of a *rotational slump* block that formed as the crater rim collapsed.
rimula (rim′-u-la) The subequatorial ring-furrow characteristic of *Classopollis* and related fossil pollen.
rimule (rim′-ule) A *sinus* in the operculum-bearing *orifice* of an ascophoran cheilostome (bryozoan).
rincon (rin-con′) (a) A term used in the SW U.S. for a square-cut recess or hollow in a cliff or a reentrant in the borders of a mesa or plateau. Cf: *cove [geomorph].* (b) A term used in the SW U.S. for a small, secluded valley, and for a bend in a stream. Etymol: Spanish "rincón", "inside corner, nook".
rindle (rin′-dle) An English syn. of *runnel.*
ring [geol] *ring structure.*
ring [paleont] The precursor to the ascending branches of the premagadiniform *loop* of terebratellid brachiopods, "consisting of thin circular ribbon, narrow ventrally and broadening dorsally to its attachment on median septum" (TIP, 1965, pt. H, p.152).
ring [remote sensing] A closed sequence of nonintersecting chains or arcs (Krzanowski et al., 1993).
ring canal An internal tube encircling the oral region of an echinoderm, from which five radial canals branch.
ring complex An association of ring dikes and cone sheets.
ring depression The annular, structurally depressed area surrounding the central uplift of an *impact structure.* Faulting and folding may be involved in its formation. Syn: *ring syncline; peripheral depression.*
ring dike A dike that is arcuate or roughly circular in plan and is vertical or inclined away from the axis of the arc. Ring dikes are commonly associated with cone sheets to form a *ring complex.*
ringer (ring′-er) A thin bed of tough, tightly cemented, fine-grained sandstone that gives out a clear, resonant sound when struck with a hammer.
ring fault A circular fault pattern that is associated with *cauldron subsidence, diapirism,* and meteorite impacts. Syn: *ring fracture.*
ring fracture *ring fault.*
ring-fracture stoping Large-scale *magmatic stoping* associated with *cauldron subsidence.* Cf: *ring dike; piecemeal stoping.*
ring hill An isolated, till-covered hill in Scandinavia, which remained above the marine limit and is surrounded by a very pronounced ring of bedrock washed clean of material (Stephens and Synge, 1966, p.28).
ringing (ring′-ing) *singing.*
ringite (ring′-ite) A carbonatite containing aegirine and alkali feldspar. The name, given by Brögger in 1921, is from Ringe in the Fen complex, Norway. Obsolete.
ring mark A *skip mark* produced by a fish vertebra, consisting of a ring-like ridge whose higher side is upcurrent; often the ring is incomplete, forming a semicircle that is concave downcurrent.
ring moor A *string bog* with concentric ridges.
ring ore *cockade ore.*
ring plain [volc] The region surrounding a volcano beyond its lower topographic flanks over which tephra and mass wasting products are radially distributed (Davidson and De Silva, 2000).
ring-porous wood (ring′-po-rous) Wood in which the pores (vessels) of one part of an annual ring are distinctly different in size or number (or both) from those in the other part of the ring (Fuller and Tippo, 1954, p.969).
ring reef *atoll.*
ring septum In some stenolaemate bryozoans, a perforated skeletal *diaphragm* that forms a washerlike transverse structure in living chambers of feeding zooecia.
ring silicate *cyclosilicate.*
ring structure A general term for an epigenetic structure with a ring-shaped trace in plan; e.g. a ring dike, or a lunar crater resulting from a meteorite impact. Syn: *ring.*
ring syncline *ring depression.*
ringwoodite (ring′-wood-ite) A colorless, pale purple, or smoky gray cubic meteorite mineral of the *spinel* group: $(Mg,Fe^{2+})_2SiO_4$. It is a cubic dimorph of olivine.
rinkite (rink′-ite) *mosandrite.*
rinkolite (rin′-ko-lite) *mosandrite.*
rinmanite A submetallic black hexagonal mineral: $Mg_2Fe_4Zn_2Sb_2O_{14}(OH)_2$. It is isomorphous with nolanite.
rinneite (rin′-ne-ite) A colorless, pink, violet, or yellow rhombohedral mineral: $K_3NaFe^{2+}Cl_6$. It is isomorphous with chlormanganokalite.
rinnenkarren (rin′-nen-kar′-ren) Small, flat-bottomed *solution grooves,* several centimeters wide and with sharp intergroove crests, that occur on sloping rock surfaces. Etymol: German, "channel tracks". Cf: *karren; rillenkarren.*
Rinnental (Rin′-nen-tal) A syn. of *tunnel valley.* Etymol: German, "channel valley".
rio (ri′-o) A term used in SW U.S. for a river or stream, usually a permanent stream. Etymol: Spanish "río".
rip (a) A turbulent agitation of water, generally caused in the sea by the meeting of water currents or the interaction of currents and wind, or in a river or a nearshore region by currents flowing rapidly over an irregular bottom; an overfall. See also: *tide rip; current rip.* (b) An abbreviated form of *ripple [currents],* often used in the plural.
ripa (ri′-pa) A legal term for the bank of a stream or lake (Veatch and Humphrys, 1966, p.268). Etymol: Latin.
riparian (ri-par′-i-an) Pertaining to or situated on the bank of a body of water, esp. of a watercourse such as a river; e.g. "riparian land" situated along or abutting upon a stream bank, or a "riparian owner" who lives or has property on a riverbank. Cf: *riverain.* Syn: riparial; riparious.
riparian water loss Discharge of water through evapotranspiration by vegetation growing along a watercourse. Discharged water may be derived from the watercourse, adjacent groundwater, and/or soil moisture.
rip channel A channel, often >2 m deep, carved on the shore by a *rip current.*
rip current A strong, narrow surface or near-surface current, of short duration (a few minutes to an hour or two) and high velocity (up to 2 knots), flowing seaward from the shore through the breaker zone at nearly right angles to the shoreline, appearing as a visible band of agitated water returning to the sea after being piled up on the shore by incoming waves and wind; it consists of a *feeder current,* a neck, and a head. Cf: *undertow.* Often miscalled a *rip tide.*
ripe Said of peat that is in an advanced state of decay. Cf: *unripe.*
ripe snow Snow that has reached the melting point and has its minimum capillary requirements satisfied, so that additional melting can produce meltwater runoff. Cf: *ripe-snow area.*
ripe-snow area The area of a drainage basin where coarsely crystalline snow is in a condition to discharge meltwater upon the addition of heat (as by rain); expressed in percent of drainage basin or in square kilometers. Abbrev: RSA. See also: *ripe snow.*
Riphean (Riph′-e-an) An obsolete geochronometric subdivision of the late *Precambrian,* covering the time span from 1,650 to 800 Ma (Harland et al., 1990), therefore approximately equivalent to the *Mesoproterozoic* Era and the earliest *Neoproterozoic* Era of the Standard Global Chronostratigraphic Scale.
ripidolite (ri-pid′-o-lite) A varietal mineral of the *chlorite* group, intermediate between clinochlore and chamosite in composition: $(Mg,Fe^{2+},Al)_6(OH)_8[Al_2Si_2O_{10}]$. The name is sometimes applied to clinochlore. Syn: *prochlorite; aphrosiderite.*
rippability (rip-pa-bil′-i-ty) In engineering geology, the mechanical power needed to excavate cemented soil and rock material in an open excavation. It is related to the strength of the cemented soil or rock mass, including fracture orientation and spacing, and is commonly estimated by a seismic-refraction survey and indexed to heavy construction equipment required for the particular excavation.
ripple [currents] (rip′-ple) (a) A syn. of *capillary wave.* (b) The light ruffling of the surface of the water by a breeze. Syn: *rip.*
ripple [sed struc] (a) This term includes a variety of relatively small *bed forms* that are exposed on bed surfaces of modern sediments and bedding plane surfaces of sedimentary rocks. All ripples are roughly triangular in transverse cross-section and develop due to the interaction of a moving fluid (air or water) with a mobile substrate (most commonly sand-size sediment). The overall morphology of a ripple is determined by the nature of the fluid that acts to form it. Cf: *current ripple; oscillation ripple; wind ripple [sed].* (b) A general term for all bedforms with shapes similar to small-scale ripples, no matter what the scale (e.g. Reineck and Singh, 1973, p.14-47). Syn: *sedimentary ripple.*
ripple [streams] (a) A shallow reach of running water in a stream, roughened or broken into small waves by a rocky or uneven bottom. (b) *riffle.*
ripple amplitude The height of a ripple measured as the vertical distance between the crest of the *ripple* and the adjacent *trough;* it is generally a centimeter or less. This use of the term "amplitude" is at variance with the convention used in physics and mathematics in which "amplitude" refers to displacement relative to a mean or equilibrium value. A better term is *ripple height.*
ripple bedding (a) A bedding surface characterized by ripples. (b) A term preferred by Hills (1963, p.10-11) to *current bedding* when used for "the small-scale ripple-like bedding of rapidly deposited sand".
ripple biscuit A bedding structure produced by lenticular lamination of sand in a bay or lagoon (Moore, 1966, p.99).

ripple cross-lamination Small-scale cross-lamination formed by migrating current ripples developed during deposition, characterized by individual laminae whose thicknesses range between 0.08 cm (1/32 in.) and 0.3 cm (1/8 in.) (McKee, 1939, p.72). See also: *ripple lamina.* Syn: *rolling strata.*
ripple drift A term used by Sorby (1852, p.232; see also 1857, p.278) for small-scale cross-lamination formed by migrating ripples. See also: *climbing ripple.*
ripple height The vertical distance from crest to trough of a ripple. If the ripple is asymmetric, the height is measured from the trough adjacent to the steeper (downcurrent) slope. The term was used by Allen (1963, p.192). See also: *ripple amplitude.*
ripple index The ratio of ripple wavelength to ripple amplitude. The ratio usually ranges from 6 to 22 for ripples produced by water currents or waves and from 20 to 50 for ripples produced by wind. Syn: *ripple-mark index; vertical form index.*
ripple lamina An internal sedimentary structure formed in sand or silt by currents or waves, as opposed to a ripple preserved on a surface. The term, as commonly used in the plural, "includes sets of laminae in incomplete ripple profiles and isolated ripple lenses, as well as series of superposed rippled layers" (McKee, 1965, p.66). McKee proposed: "ripple laminae-in-rhythm", a general term for all ripple structures superimposed in an orderly sequence; and "ripple laminae-in-phase", a general term for ripple laminae in which the crests of vertically succeeding laminae (as seen in sections parallel to the direction of current or wave motion) are directly above one another. See also: *ripple cross-lamination.*
ripple load cast A *load cast* of a ripple that shows signs of penecontemporaneous deformation (caused by unequal loading, settling, and compaction) in the accentuation of its trough and crest and in the oversteepening of component laminae (Kelling, 1958, p.120-121). Now generally termed a *load-casted ripple.*
ripple mark A *ripple.* The addition of the word "mark" was common in the past and remains a relatively common practice among geologists; particularly, but not exclusively, nonspecialists in the area of sedimentology. The use of the word "mark" provides no additional information about the bed form itself and is therefore not recommended.
ripple-mark index *ripple index.*
ripple-mark wavelength *chord.*
ripple ridge *ripple [sed struc].*
ripple scour A large shallow linear trough with superimposed *transverse ripple* (Potter and Glass, 1958, pl. 5).
ripple spacing *chord.*
ripple symmetry index A term used by Tanner (1960, p.481) to express the degree of symmetry of a *ripple* defined as the ratio of the horizontal length of the gentle (upcurrent) side to that of the steep (downcurrent) side; an *asymmetric ripple* has an index greater than 1. Abbrev: RSI. See also: *horizontal form index.*
ripple till A till sheet containing low, winding, smooth-topped ridges, 6-15 m high, 200-3,000 m long, lying at right angles to the direction of ice movement, and grouped into narrow belts up to 80 km long that are generally parallel to the direction of ice movement (F.K. Hare in Stamp, 1961, p.395); found in parts of northern Ontario. Syn: *ribble.*
ripple wavelength *chord.*
rippling (rip'-pling) A surface characterized by ripple mark; a collective term for a series of occurrences of ripples.
riprap (rip'-rap) (a) A layer of large, durable fragments of broken rock, specially selected and graded, thrown together irregularly (as offshore or on a soft bottom) or fitted together (as on the upstream face of a dam). Its purpose is to prevent erosion by waves or currents and thereby preserve the shape of a surface, slope, or underlying structure. It is used for irrigation channels, river-improvement works, spillways at dams, and sea walls for shore protection. (b) The stone used for riprap.
rip tide A popular, but improper, term used as a syn. of *rip current.* The usage is erroneous because a rip current has no relation to the tide. Also spelled: riptide.
rip-up (a) Said of a sedimentary structure formed by mud clasts (usually of flat shape) that have been "ripped up" by currents from a semiconsolidated mud deposit and transported to a new depositional site. (b) Said of a clast in a *rip-up* structure.
rip-up clast A *mud clast* within a *rip-up* structure.
rischorrite (ri-schor'-rite) A variety of *nepheline syenite* in which nepheline is poikilitically enclosed in microcline perthite. The name, given by Kupletsky in 1932, is for Rischorr in the Khibina complex, Kola Peninsula, Russia (Sørensen, 1974, p.572). Obsolete.
rise [geomorph] (a) An upward slope in the land. (b) The top part of a hill or other landform that is higher than the surrounding ground.
rise [karst] A Jamaican term for *resurgence.*
rise [marine geol] A broad, elongate, smooth elevation of the ocean floor. Syn: *swell [marine geol].*
rise pit A pit through which an underground stream rises to the surface.
riser (ris'-er) (a) The vertical or steeply sloping surface of one of a series of natural steplike landforms, as those of a glacial stairway or of successive stream terraces. Cf: *tread.* (b) *ramp [struc geol].*
rise time The time history of the fault dislocation at the focus of an earthquake.
rising (ris'-ing) *resurgence.*
rising dune *climbing dune.*
rising tide That part of a tide cycle between low water and the following high water, characterized by landward or advancing movement of water. Also, an inflowing tidal river. Ant: *falling tide.* Syn: *flood tide.*
risk The probability of undesirable consequences arising from possible events. Risk management is the systematic estimation of risk and implementation of risk controls. Risk differs from *uncertainty,* which is the precision with which a measurement or value is known; while risk can be managed, uncertainty requires more or better data.
Riss (a) A European stage: Pleistocene (above Mindel, below Würm). (b) The third classical glacial stage of the Pleistocene Epoch in the Alps, after the Mindel-Riss interglacial stage. Cf: *Illinoian; Saale.* Etymol: After the river Riss, Germany. Adj: Rissian.
Riss-Würm A term applied in the Alps to the third classical interglacial stage of the Pleistocene Epoch, after the Riss glacial stage and before the Würm. See also: *Sangamon.*
rithe An English term for a small stream. Syn: *rive.*
rittmanite (ritt'-man-ite) A pale-yellow monoclinic mineral of the *whiteite* group: $(Mn^{2+},Ca)Mn^{2+}(Fe^{2+},Mn^{2+},Mg)_2(Al,Fe^{3+})_2(PO_4)_4(OH)_2{\cdot}8H_2O$.
Rittmann norm (Ritt'-mann) A norm calculated after the manner of the *Niggli molecular norm* but in which greater variability is introduced into the composition of the minerals in order to bring the resulting mineral percentages into closer agreement with the mode of the rock (Rittmann, 1973).
rivadavite (ri'-va-dav'-ite) A colorless monoclinic mineral: $Na_6Mg[B_6O_7(OH)_6]_4{\cdot}10H_2O$.
rive *rithe.*
river [coast] (riv'-er) A term used in place names for an estuary, lagoon, tidal river, inlet, or strait; e.g. York River, Va., and Indian River, Fla.
river [gem] An obsolete term for a pure-white diamond of very high grade. Cf: *water [gem].*
river [streams] (a) A general term for a natural freshwater surface stream of considerable volume and a permanent or seasonal flow, moving in a definite channel toward a sea, lake, or another river; any large stream, or one larger than a brook or a creek, such as the trunk stream and the larger branches of a drainage system. (b) A term applied in New England to a small watercourse which elswhere in the U.S. is known as a creek.
riverain (riv'-er-ain) Pertaining to a riverbank; situated on or near a river. The term has a wider meaning than *riparian.*
river bar A ridgelike accumulation of alluvium in the channel, along the banks, or at the mouth, of a river. It is commonly emergent at low water and constitutes a navigational obstruction.
river-bar placer *bench placer.*
river basin The entire area drained by a river and its tributaries. Cf: *drainage basin.*
riverbed The *channel* containing or formerly containing the water of a river.
river bluff A *bluff* or steep hillslope or line of slopes above a river bank. Cf: *river cliff.*
river bottom The low-lying alluvial land along a river.
river breathing Fluctuation of the water level of a river (ASCE, 1962).
river cliff The steep *cutbank* formed by the lateral erosion of a river. Cf: *river bluff.*
river-delta marsh (riv'-er-del'-ta) A brackish or freshwater marsh bordering the mouth of a distributary stream.
river-deposition coast (riv'-er-de-po-si'-tion) A deltaic coast characterized by lobate seaward bulges crossed by river distributaries and bordered by lowlands (Shepard, 1948, p.72).
river-dominated delta *fluvially-dominated delta.*
river drift An old term for *riverwash.*
river end The lowest point of a river with no outlet to the sea, situated where its water disappears by percolation or evaporation (Swayne, 1956, p.121).
river engineering A branch of civil engineering that deals with the control of rivers, their improvement, training, and regulation; and with flood mitigation.
river flat An *alluvial flat* adjacent to a river; a bottom.
river forecasting Forecasting the river stage and discharge, by hydrology and meteorology, including research into forecasting methods. In some countries, the term hydrometeorology is used with this limited meaning (ASCE, 1962).
riverhead (riv'-er-head) The source or beginning of a river.
river ice (a) Ice formed on a river. (b) Ice carried by a river.
river icing A common *icing* in stream courses. Syn: *flood icing; aufeis.*
riverine (riv'-er-ine) (a) Pertaining to or formed by a river; e.g. a "riverine lake" created by a dam across a river. (b) Situated along the banks of a river; e.g. a "riverine ore deposit".
river morphology The study of the *channel pattern* and the *channel geometry* at several points along a river channel, including the network of tributaries within the drainage basin. Syn: *channel morphology; fluviomorphology; stream morphology.*
river pattern *channel pattern.*

river-pebble phosphate (riv'-er-peb'-ble) A term used in Florida for a transported, dark variety of *pebble phosphate* obtained from bars and flood plains of rivers. Cf: *land-pebble phosphate.* Syn: *river rock.*
river plain *alluvial plain.*
river profile The *longitudinal profile* of a river.
river rock A syn. used in South Carolina for *river-pebble phosphate.*
river-run gravel Natural gravel as found in deposits that have been subjected to the action of running water (Nelson, 1965, p.373).
rivershed (riv'-er-shed) The drainage basin of a river.
riversideite (riv'-er-side-ite') A silky white orthorhombic mineral: $Ca_5Si_6O_{16}(OH)_2{\cdot}2H_2O$.
river system A river and all of its tributaries. Syn: *water system.*
river terrace *stream terrace.*
river valley An elongate depression of the Earth's surface, carved by a river during the course of its development.
riverwash (riv'-er-wash) (a) Soil material that has been transported and deposited by rivers. (b) An alluvial deposit in a riverbed or flood channel, subject to erosion and deposition during recurring flood periods.
riviera (riv-i-er'-a) A resort coastline much frequented by tourists, usually having extensive sandy beaches and a mild climate. Type locality: the Riviera along the coast of the Mediterranean Sea between Marseilles, France, and La Spezia, Italy.
riving (riv'-ing) The splitting off, cracking, or fracturing of rock, esp. by frost action. See also: *gelifraction.*
rivotite (riv'-o-tite) A mixture of malachite and stibiconite.
rivulet (riv'-u-let) (a) A small stream; a brook or a runnel. (b) A small river. (c) A streamlet developed by rills running down a steep slope.
rizalite (ri'-za-lite) An *Australasian tektite* from the Rizal region of the Philippines.
RK *radial coefficient.*
R layers Hard bedrock including granite, basalt, quartzite and indurated limestone or sandstone that is sufficiently coherent to make hand digging impractical.
R-mode factor analysis *Factor analysis* concerned with relationships among variables. Cf: *Q-mode factor analysis.*
RMR *rock mass rating.*
rms velocity Root-mean-square velocity. For a series of parallel layers of velocity V_i, where the traveltime for seismic energy perpendicularly through each is t_i, the rms velocity is $V_{rms} = [(\Sigma V_i^2 t_i)/(\Sigma t_i)]^{1/2}$. Rms velocity is often used erroneously for *stacking velocity*, the velocity determined from velocity analysis based upon normal-moveout measurements.
RNA *ribonucleic acid.*
road [coast] A *roadstead.* Term is usually used in the plural; e.g. Hampton Roads, Va.
road [glac geol] One of a series of erosional terraces in a glacial valley, formed as the water level dropped in an ice-dammed lake. See also: *parallel roads.*
Roadian A stage of the Standard Global Chronostratigraphic Scale: lowest stage of the Guadalupian Series, middle Permian (above Kungurian, below Wordian). The basal GSSP is in the Guadalupe Mountains, Texas, USA.
road log A descriptive record of the route taken on a field trip and of the geology observed along it. Syn: *guidebook.*
road metal Rock suitable for surfacing macadamized roads, and for foundations for asphalt and concrete roadways; also used without asphaltic binder as the traffic-bearing surface, generally on secondary roads. Also called: road aggregate.
roadstead (road'-stead) An area of water near a shore, sheltered by a reef, sandbank, or island, or an open anchorage, usually a shallow indentation in the coast, where vessels may lie in relative safety from winds and heavy seas; it is often outside, and less sheltered than, a harbor. An "open roadstead" is unprotected from the weather. Syn: *road [coast].*
roaldite (ro'-ald-ite) A metallic white cubic mineral occurring in meteorites: $(Fe,Ni)_4N$.
roaring sand (roar'-ing) A *sounding sand,* found on a desert dune, that sets up a low roaring sound that sometimes can be heard for distances of up to 400 m. See also: *booming sand.*
robbery (rob'-ber-y) *capture [streams].*
robertsite (rob'-erts-ite) A red, reddish-brown or black platy monoclinic mineral: $Ca_2Mn^{3+}{}_3O_2(PO_4)_3{\cdot}3H_2O$. It is a dimorph of *pararobertsite,* and isomorphic to *arseniosiderite.*
robinsonite (rob'-in-son-ite') A metallic gray monoclinic mineral: $Pb_4Sb_6S_{13}$.
Robinson projection A compromise projection for world maps on which the parallels of latitude are straight lines parallel with the equator and the meridians are curved arcs. The poles are straight lines about half the length of the equator. Scale is correct along the 38th parallels, but no point is completely free of shape, area, or distance distortion. Presented in 1963 by Arthur H. Robinson of the University of Wisconsin and used especially by Rand McNally and the National Geographic Society.
robust Said of a statistical procedure when its performance is relatively unaffected by violations of its underlying assumptions.
rocdrumlin (roc-drum'-lin) *rock drumlin.*

Roche limit The distance from a primary planetary body inside which a secondary (satellite) will be disrupted by tidal force. The classical Roche limit, defined by the French mathematician Edouard Roche, is the distance inside which a fluid body is no longer contained within a closed equipotential surface (sometimes called a Roche lobe) and thus would flow out into an orbiting ring. The classical Roche limit is mainly applicable to contact binary stars, which behave as fluids. For solid planetary bodies, other limits, sometimes also referred to as a Roche limit, can be defined where tidal stresses exceed material strength and result in disruption of the body. Even a body with only compressive strength and no tensile strength (e.g., a rubble pile) can maintain itself as a coherent body inside the classical Roche limit. The Martian satellite Phobos is an example of a body inside the classical Roche limit, but requiring only "rubble pile" cohesion to maintain itself as a coherent body.
roche moutonnée (roche' mou-ton-nee' [rosh mootonnay']) A small elongate protruding knob or hillock of bedrock, so sculptured by a large glacier as to have its long axis oriented in the direction of ice movement, an upstream (stoss or scour) side that is gently inclined, smoothly rounded, and striated, and a downstream (lee or pluck) side that is steep, rough, and hackly. It is usually a few meters in height, length, and breadth. In 1776, Saussure used "moutonnée", a French adjective meaning "fleecy, ruffled, or curled", in describing an assemblage of rounded knobs in the Alps. *Roche moutonnée* came to connote a resemblance between a single knob of the character described and a grazing sheep. Much later, the term was applied to a single glaciated knob so roughened by plucking as to allow comparison to a sheep's back in surface texture as well as in general form, but this kind of surface is not generally regarded as essential. Pl: roches moutonnées. Syn: *sheepback rock; sheep rock; whaleback; embossed rock.*
rochlederite (roch-le'-der-ite) The soluble resin extracted from *melanchym* by alcohol. See also: *melanellite.*
rock [geol] An aggregate of one or more minerals, e.g., granite, shale, marble; or a body of undifferentiated mineral matter, e.g., obsidian; or a lithified organic material, e.g., coal.
rock [topog] Any prominent peak, cliff, or promontory, usually bare, when considered as a mass, e.g., the Rock of Gibraltar. Also, a rocky mass lying at or near the surface of a body of water, or along a jagged coastline, esp. where dangerous to shipping.
rockallite (rock'-al-lite) A coarse-grained, mafic, alkalic granite composed of quartz, aegirine, albite, and microcline. The name, given by Judd in 1897, is for Rockall Bank in the North Atlantic Ocean 350 km west of Scotland. Obsolete.
rock asphalt *asphalt rock.*
rock association A group of igneous rocks within a petrographic province that are related chemically and petrographically, generally in a systematic manner such that chemical data for the rocks plot as smooth curves on variation diagrams. Syn: *rock kindred; kindred; association [petrology].*
rock avalanche The very rapid downslope flowage of rock fragments, during which the fragments may become further broken or pulverized. Rock avalanches typically result from large rockfalls and rockslides, and their patterns of displacement have led to the term *rock-fragment flow* (Varnes, 1958). Characteristic features include chaotic distribution of large blocks, flow morphology and internal structure, relative thinness in comparison to large areal extent, high porosity, angularity of fragments, and lobate form. Cf: *debris flow.* Preferred syn: *sturzstrom.*
rock baby An odd-shaped, protruding hill of sandy bedrock, produced by differential erosion in the desert region of the Henry Mountains, Utah (Hunt et al. 1953, p.175).
rock bar *riegel.*
rock basin A depression in solid rock, sometimes of great extent; esp. one formed by local erosion of the uneven floor of a cirque or glacial valley in a mountainous region, and usually containing a lake. See also: *glacial basin.*
rock-basin lake (rock-ba'-sin) A glacial lake occupying a rock basin; e.g. a *paternoster lake.*
rock bench (a) A narrow valley-side niche developed during backweathering of weaker beds in a section of stratified rocks; a *structural bench* cut in solid rock. (b) A *wave-cut bench* produced on a rock surface.
rock bind An English term for a sandy shale or a banded or nonbanded siltstone.
rock bit *drill bit.*
rock bolt A bar, usually of steel, used in *rock-bolting.* It is generally at least one meter in length and about 2 cm in diameter, and it is provided with a device for expanding the leading end so that it may be anchored firmly in rock. Rock bolts are classified according to the means by which they are secured or anchored: expansion, wedge, grouted, and explosive. Cf: *cable bolt; roof bolt.*
rock-bolting A procedure to secure or strengthen joints in fissured rock in mine workings, tunnels, or rock abutments by inserting and anchoring *rock bolts* in predrilled holes that range in length from less than one meter to about 12 m. Some rock bolts are anchored firmly to prevent rock masses from moving; whereas others are installed somewhat loosely and take on load as the rock mass moves, thereby becoming "self-anchoring" bolts.

rock borer Any of certain bivalve mollusks that live in cavities they have bored into soft rock, concrete, or other material, usually by rotating the shell. Cf: *saxicavous.*

rockbridgeite (rock'-bridge-ite) A dark green, olive-green, brown, or yellow-brown columnar to platy orthorhombic mineral: $(Fe^{2+},Mn^{2+})Fe^{3+}_4(PO_4)_3(OH)_5$. It is the Fe^{2+} analogue of frondelite.

rock burst A sudden and often violent breaking of a mass of rock from the walls of a tunnel, mine, or deep quarry, caused by failure of highly stressed rock and the rapid or instantaneous release of accumulated strain energy. It may result in closure of a mine opening, or projection of broken rock into it, accompanied by ground tremors, rockfalls, and air concussions. Syn: *pressure burst.*

rock city An area of large blocks of rock formed in situ, usually on a hillcrest or summit, as a result of weathering and creep along several joint systems. The blocks and open joint fissures resemble the buildings and streets of a city.

rock cork *mountain cork.*

rock crystal *quartz crystal.*

rock cycle A sequence of events involving the formation, alteration, destruction, and reformation of rocks as a result of such processes as magmatism, erosion, transportation, deposition, lithification, and metamorphism. A possible sequence involves the crystallization of magma to form igneous rocks that are then broken down to sediment as a result of weathering, the sediments later being lithified to form sedimentary rocks, which in turn are altered to metamorphic rocks.

rock-defended terrace (rock'-de-fend'-ed) (a) A river terrace protected from later undermining by a projecting ledge or outcrop of resistant rock at its base (or at successively lower levels of the river). (b) A marine terrace protected from wave erosion by a mass of resistant rock at the base of the wave-cut cliff formed in the overlying coastal-plain sediments. Syn: *rock-perched terrace.*

rock desert An upland desert area in which bedrock has been exposed after the removal of sand and dust particles by wind, or in which bedrock is covered by a thin veneer of coarse rock fragments; e.g. a *hammada.* Cf: *stony desert.* Syn: *rocky desert.*

rock doughnut A raised annular ridge encircling a *weather pit,* such as occur on certain granite domes of central Texas (Blank, 1951). Syn: *doughnut.*

rock drift *creep [mass move].*

rock drumlin A smooth, streamlined hill, having a core of bedrock usually veneered with a layer of till; it is modelled by glacial erosion, and its long axis is parallel to the direction of ice movement. It is similar in outline and form to a true drumlin, but is generally less symmetrical and less regularly shaped. Syn: *false drumlin; drumlinoid.* Also spelled: *rocdrumlin.*

rockery (rock'-ery) A low, steep excavation faced with boulder-sized fragments of natural rock. Often erroneously called "retaining walls", rockeries are decorative, erosion-resistant slope facings common in the Pacific Northwest of the United States.

rock face An exposed surface of rock in a wall or cliff.

rock failure *failure.*

rockfall (rock'-fall) (a) The relatively free falling or precipitous movement of a newly detached segment of bedrock (usually massive, homogeneous, or jointed) of any size from a cliff or other very steep slope; it is the fastest form of mass movement and is most frequent in mountain areas and during spring when there is repeated freezing and thawing of water in cracks in the rock. Movement may be straight down, or in a series of leaps and bounds down the slope; it is not guided by an underlying slip surface. Syn: *sturzstrom.* (b) The mass of rock moving in or moved by a rockfall; a mass of fallen rocks. Also spelled: rock fall.

rockfall avalanche A rockfall that has turned into a flow, occurring only when large rockfalls and rockslides, involving millions of metric tons, attain extremely rapid speeds; most common in a rugged mountainous region, as that which occurred in 1903 at Frank, Alberta (McConnell and Brock, 1904). Cf: *debris flow.* Preferred syn: *sturzstrom.*

rockfall talus An accumulation of coarse, angular rock fragments, derived by falling from a cliff or steep rocky slope above. Blocks may be derived by pressure release and freeze-thaw action in previously formed cracks; heavy rain also helps release blocks. The rocks may fall and shatter, roll, or bounce; large sizes collect at the bottom. The angle of slope is greater than 32° and may reach 42° (White, 1967, p.237). Syn: *scree.*

rock fan An eroded, convex, fan-shaped bedrock surface having its apex at the point where a mountain stream debouches upon a piedmont slope, and occupying the zone where a pediment meets the mountain slope (assuming that the mountain front retreats as a result of lateral planation). According to Johnson (1932), a *pediment* evolves from a coalescence of rock fans, although the term "rock fan" is often considered an equivalent of "pediment".

rock-fill dam A dam composed primarily of large, broken, loosely placed or pervious rocks, with either an impervious core or an upstream facing.

rock-floor robbing A form of *sheetflood erosion* in which sheetfloods remove crumbling debris from rock surfaces in desert mountains (Cotton, 1958, p.258).

rock flour Finely comminuted, chemically unweathered material, consisting of silt- and clay-sized angular particles of rock-forming minerals, chiefly quartz, formed when rock fragments are pulverized while being transported or are crushed by the weight of superincumbent material. The term is most commonly applied to the very fine powder that is formed when stones embedded in a glacier or ice sheet abrade the underlying rocks, and that is deposited as the matrix in till or in outwash deposits. Syn: *glacier meal; glacial meal; glacial flour; rock meal.*

rock flowage *flow [exp struc geol].*

rock-forming (rock'-form-ing) Said of those minerals that enter into the composition of rocks, and determine their classsification. The more important rock-forming minerals include quartz, feldspars, micas, amphiboles, pyroxenes, olivine, calcite, and dolomite.

rock-fragment flow (rock'-frag-ment) *sturzstrom.*

rock glacier A mass of poorly sorted angular boulders and fine material, with interstitial ice a meter or so below the surface (ice-cemented) or containing a buried ice glacier (ice-cored). It occurs in high mountains in a permafrost area, and is derived from a cirque wall or other steep cliff. Rock glaciers have the general appearance and slow movement of small valley glaciers, ranging from a few hundred meters to several kilometers in length, and having a distal area marked by a series of transverse arcuate ridges. When active, they may be 50 m thick with a surface movement (resulting from the flow of interstitial ice) of 0.5-2 m/yr. Rock glaciers are classified in plan as lobate, tongue-shaped, or spatulate. Cf: *block stream; chrystocrene.*

rock-glacier creep (rock'-gla-cier) The slow creep of tongues of rock glaciers.

rock gypsum A sedimentary rock composed chiefly of the mineral gypsum; it is generally massive, and ranges from coarsely crystalline to finely granular. It often shows disturbed bedding owing to expansion during hydration of the parent anhydrite. Obsolete syn: *gyprock.*

rock hole An Australian term for a *rock tank.*

rock hound (a) An amateur mineralogist. (b) A petroleum exploration geologist. See also: *pebble pup.*

rocking stone (rock'-ing) A stone or boulder, often of great size, so finely poised on its foundation that it can be rocked or moved backward and forward with little force. It may be a glacial erratic or a rounded residual block formed in place by weathering. See also: *logan stone; balanced rock; elephant rock.* Syn: *roggan.*

rock island (a) *meander core.* (b) A bedrock hill surrounded by alluvium in an aggraded stream valley.

rock kindred *rock association.*

Rocklandian North American stage: Upper Ordovian, above Blackriverian, below Kirkfeldian.

rock leather *mountain leather.*

rock magnetism The study of the origins and characteristics of magnetization in rocks and minerals.

rock mantle *regolith [weath].*

rock mass A term, used especially for engineering purposes, that includes a volume of earth material, including rock, soil, water, and air-filled voids, and the discontinuities that divide it (Bell, 1992, p.54). The behavior of a rock mass is determined, to a large extent, by the type, spacing, orientation, and characteristics of the discontinuities present.

rock mass rating A rating system for rock masses based on five parameters, (1) strength of intact rock material, (2) RQD, (3) rock discontinuity spacing, (4) rock discontinuity condition, and (5) groundwater condition; it is also adjusted for rock discontinuity orientation with respect to a tunnel or cut-slope geometry. RMR values range from 0 to 100 and indicate very poor rock (RMR ≤20), poor rock (40 ≤RMR >20), fair rock (60 ≤RMR >40), good rock (80 ≤RMR >60), and very good rock (100 ≤RMR >80) (Bieniawski, 1993). Abbrev: RMR.

rock meal (a) *moonmilk.* (b) *rock flour.*

rock mechanics The theoretical and applied science of the physical behavior of rocks, representing a "branch of mechanics concerned with the response of rock to the force fields of its physical environment" (NAS-NRC, 1966, p.3).

rock mill *pothole [streams].*

rock pedestal *pedestal [geomorph].*

rock pediment A *pediment* developed on a bedrock surface.

rock pendant *pendant [speleo].*

rock-perched terrace *rock-defended terrace.*

rock phosphate *phosphate rock.*

rock physics *petrophysics.*

rock pillar [speleo] In a cave, a columnlike structure that is residual bedrock rather than a *speleothem.*

rock pillar [topog] A column of rock produced by differential weathering or erosion, as along a joint plane; a *hoodoo.*

rock plane A term used by Johnson (1932) as a syn. of *pediment;* the term is not recommended in this usage as there are many approximately planate rock surfaces that lack the areal extent and climatic restriction of a pediment.

rock platform (a) A *wave-cut platform* eroded on a rock surface. (b) A *high-water platform* eroded on a rock surface.

rock pool A *tide pool* formed along a rocky shoreline.

rock pressure (a) The pressure exerted by surrounding solids on the support system of underground openings, including that caused by the weight of the overlying material, residual unrelieved stresses, and pressures associated with swelling clays (Stokes and Varnes, 1955, p.125). (b) The compressive stress within the solid body of underground geologic material. (c) *geostatic pressure.*

rock quality designation The cumulative length of core pieces longer than 100 mm in a run, divided by the total length of the core run, expressed as a percentage. Mechanical breaks caused by the drilling process or extracting the core from the core barrel are ignored, but lost or missing core is included in the total core-run length (Deere and Deere, 1988). Abbrev: RQD.

rock ruby A fine red variety of garnet; specif. *pyrope.*

rock salt (a) Coarsely crystalline *halite* occurring as a massive, fibrous, or granular aggregate, and constituting a nearly pure sedimentary rock that may occur in domes or plugs or as extensive beds resulting from evaporation of saline water. It is frequently stained by iron or mixed with fine-grained sediments. (b) Artificially prepared salt in the form of large crystals or masses.

rock sea *block field.*

rock series *igneous-rock series.*

rockshelter (rock-shel'-ter) A cave, commonly formed in nonsoluble rock, that extends only a short way underground, with a roof of overlying rock that usually extends beyond its sides. Syn: *shelter cave.*

rock silk A silky variety of asbestos.

rockslide (rock'-slide) (a) A slide involving a downward and usually sudden and rapid movement of newly detached segments of bedrock sliding or slipping over an inclined surface of weakness, as a surface of bedding, jointing, or faulting, or other pre-existing structural feature. The moving mass is greatly deformed and usually breaks up into many small independent units. Rockslides frequently occur in high mountain ranges, as the Alps or Canadian Rockies. (b) The mass of rock moving in or moved by a rockslide. Also spelled: rock slide. Syn: *rock slip.*

rockslide avalanche An obsolescent syn. of *rockslide.*

rock slip *rockslide.*

rock soap *mountain soap.*

rock stack *stack [coast].*

rock step (a) A *knickpoint* produced by the outcrop of a resistant rock. (b) One of a series of ledges or other irregularities of gradient in the upper reaches of a hanging valley; an abrupt descent in the floor of a glacial valley. Cf: *riegel.*

rock-stratigraphic unit (rock'-strat-i-graph'-ic) *lithostratigraphic unit.*

rock stratigraphy *lithostratigraphy.*

rock stream *block stream.*

rock stripe *stone stripe.*

rock tank A natural *tank* formed in rock by differential weathering or differential erosion. See also: *sand tank.* Syn: *rock hole.*

rock terrace A *stream terrace* produced on the side of a valley by erosion in horizontal beds of unequal resistance; it is composed of strong bedrock that is worn back less rapidly than the weaker beds above and below. Cf: *alluvial terrace.* Syn: *erosion terrace; stream-cut terrace.*

rock train A term suggested by Kendall and Wroot (1924, p.448) for the rock material in "process of transport at the sides and in the middle of a glacier" and "subject to the dynamic forces of the glacier".

rock type [coal] *banded ingredients.*

rock type [petrology] (a) One of the three major groups of rocks: igneous, sedimentary, metamorphic. (b) A particular kind of rock having a specific set of characteristics. It may be a general classification, e.g. a basalt, or a specific classification, e.g. a basalt from a particular area and having a unique description.

rock unit *lithostratigraphic unit.*

rock varnish A thin dark shiny film or coating composed of iron oxide accompanied by traces of organic matter, manganese oxide, and silica. It is formed on the surfaces of pebbles, boulders, and other rock fragments. Commonly forms on rocks in desert regions after long exposure, as well as on ledges and other rock outcrops. It is believed to be caused by exudation of mineralized solutions from within and deposition by evaporation on the surface. A similar appearance produced by wind abrasion is properly known as *desert polish.* Syn: *desert varnish; desert patina; desert lacquer; desert crust; desert rind; varnish.*

rock waste *debris [geol].*

rock weathering The chemical decomposition and mechanical disintegration of rocks in place, at or near the Earth's surface.

rock wood *mountain wood.*

rock wool *mineral wool.*

rock wreath *sorted circle.*

rocky desert *rock desert.*

Rocky Mountain Orogeny (Rock'-y Moun'-tain) A name proposed by W.H. White (1959) for a time of major folding and thrusting during Late Cretaceous and Paleocene time in the Rocky Mountains of eastern British Columbia and adjacent Alberta; it is broadly equivalent to the *Laramide Orogeny* of western United States. Obsolete.

rod [paleont] (a) An elongate holothurian *sclerite* having a circular cross section, one or more axes, and an eye at its end. (b) A part of a heterococcolith having one dimension large and two much smaller. (c) A thin cylindrical or prismatic skeletal element in archaeocyathids, commmonly radially oriented. (d) A kind of photoreceptive sensory cell in the vertebrate retina that is sensitive under conditions of low light and functions in non-color vision.

rod [sed] A rod-like or prolate shape of a sedimentary particle, defined in *Zingg's classification* as having a width/length ratio less than 2/3 and a thickness/width ratio greater than 2/3. Syn: *roller [sed].*

rod [surv] (a) A bar or staff for measuring, such as a graduated pole used as a target in surveying; specif. a *level rod.* (b) A unit of length equal to 16.5 ft. Also called a *perch* or pole.

rodalquilarite (ro-dal-qui'-lar-ite) A green triclinic mineral: $H_3Fe^{3+}{}_2(Te^{4+}O_3)_4Cl$.

rødberg An altered carbonatite in which ferrous iron, formerly held in solid solution within ankerite, has become oxidized and formed disseminated hematite. The name is derived from the Norse, "red rock", for the occurrence at the Fen complex, Norway. Obsolete.

rodding (rod'-ding) In metamorphic rocks, a linear structure in which the stronger parts, such as vein quartz or quartz pebbles, have been shaped into parallel rods. Whether the structure is formed parallel to the direction of transport or parallel to the fold axes has been debated. Cf: *mullion.*

roddon (rod'-don) A term used in East Anglia, England, for a natural levee built of sediment carried upstream by the tide rather than downstream by a river.

rodingite (ro'-ding-ite) A massive dense buff to pink metamorphic rock typically rich in grossular garnet and calcic pyroxene, and enveloped in serpentinite. Epidote, vesuvianite, and other calcium-rich minerals are commonly present. It is formed by metasomatic alteration of a protolith which in many cases was a dike rock, as shown by preservation of structures. The name was applied by Bell in 1911.

rodite (ro'-dite) An obsolete syn. of *diogenite.*

rod level A spirit level attached to a level rod or stadia rod to assure a vertical position of the rod prior to instrument reading.

rodolicoite A brown to red-brown trigonal mineral of the *berlinite* group: $Fe^{3+}PO_4$.

rod person One who uses or carries a surveyor's level rod; a *chainperson.*

rod-type crossed lamellar structure A *crossed lamellar structure* in which the second-order structural units consist of ovoid to rectangular rods rather than planar lamellae (Carter, 1990, Fig. 24).

rod-type fibrous prismatic structure A *fibrous prismatic structure* in which the fibers are generally equidimensionally rounded to polygonal in cross section (Carter, 1990, Fig. 2A).

roeblingite (roe'-bling-ite) A white to pale pink porcellaneous to platy monoclinic mineral: $Pb_2Ca_6Mn^{2+}Si_6O_{18}(SO_4)_2(OH)_2{\cdot}4H_2O$.

roedderite (roed'-der-ite) A colorless hexagonal meteorite mineral of the *milarite* group: $(Na,K)_2(Mg,Fe^{2+})_5Si_{12}O_{30}$.

Roelof's prism A prism made to be attached to the front of the objective lens-system of a theodolite's telescope and permitting precise pointing at the Sun's center by creating four overlapping images of the Sun. The images overlap in such a way that all four images intersect at a common point on the optical axis when the telescope is pointed at the center of the Sun. Also called: Roelof's solar prism; *solar prism.*

roemerite *römerite.*

roentgen A dose of ionizing radiation that generates 2.58×10^{-4} coulomb (C) of charge per kilogram of air. The effect of one roentgren on dry air is roughly equal to the *rad.* Obsolete.

roentgenite *röntgenite-(Ce).*

roentgen meter (roent'-gen) A device that measures the cumulative quantity of X-rays or gamma-rays.

roepperite (roep'-per-ite) A variety of fayalite with Mn^{2+} and Zn replacing some Fe^{2+}.

roesslerite *rösslerite.*

roestone (roe'-stone) *oolite.*

rofla (ro'-fla) A term used by E. Desor for an extremely narrow, tortuous gorge, formed by meltwater streams flowing from a glacier (Marr, 1900, p.172 and 314); e.g. the gorge of the Trient, near Vernayaz, Switzerland. Pl: roflas.

rogenstein (ro'-gen-stein) An oolite in which the ooliths are united by argillaceous cement. Etymol: German "Rogenstein", "roestone". Occasionally misspelled: roggenstein.

roggan (rog'-gan) *rocking stone.*

roggianite (rog'-gi-a-nite') A white to yellowish-white tetragonal mineral: $Ca_2BeAl_2Si_4O_{13}(OH)_2{\cdot}2.5H_2O$.

rognon (ro'-gnon) (a) A small rocky peak or ridge surrounded by glacier ice in a mountainous region. Also, a similar peak projecting above the bed of a former glacier (ADTIC, 1955, p.67). Cf: *nunatak.* (b) A rounded nunatak (Lliboutry, 1958, p.264); *nunakol.*

rogue wave An abnormally large open-ocean wave caused by the chance interaction of large peaks in intersecting wave trains. Waves with heights greater than 100 feet (30 meters) have been reported, and large vessels have been damaged or (presumed) sunk by such waves.

rohaite (ro'-ha-ite) A submetallic yellow-cream to bluish-gray orthorhombic mineral: $TlCu_5SbS_2$.

roho system A system of major, listric, basinward-dipping growth faults that gradually flatten with depth into conformity with an allochthonous *salt sheet* or *salt weld.* Sediment wedges in the fault blocks dip and thicken landward but become younger seaward (Schuster, 1995). Cf: *stepped counter-regional system.*

roil A small section of a stream, characterized by swiftly flowing, turbulent water.

roily (roil′-y) (a) Said of muddy or sediment-filled water. Cf: *turbid.* (b) Said of turbulent, agitated, or swirling water.

rokühnite (ro-kühn′-ite) A colorless monoclinic mineral occurring in evaporite deposits: $Fe^{2+}Cl_2 \cdot 2H_2O$.

roll [coal] (a) An elongate protrusion of shale, siltstone, or sandstone (locally limestone) from the roof into a coal seam, causing a thinning of the seam and sometimes replacing it almost entirely; cf: *cutout.* A roll is commonly overlain by a thin coal stringer. (b) An elongate upheaval of the floor material into a coal seam, causing thinning of the seam (Woolnough, 1910). Syn: *horseback [coal].* (c) Various minor deformations or dislocations of a coal seam, such as washouts, small monoclinal folds, or faults with little displacement.

roll [ore dep] *roll orebody.*

roll [photo] The angular deviation from the horizontal of a craft about its longitudinal axis. In photogrammetry, the angular deviation of an aerial camera or coordinate system of a photograph about either the photograph's x-axis or about the x-axis of the ground coordinate system (Maune, 2001, p.495). Cf: *pitch [photo], yaw [photo].*

roll [sed] A primary sedimentary structure produced by deformation involving subaqueous slump or vertical foundering; e.g. a flow roll or a pseudonodule.

roll-along switch A switch which permits connecting different geophone groups to the recording instruments, used in common-midpoint recording.

rollandite A vitreous green orthorhombic mineral: $Cu_3(AsO_4)_2 \cdot 4H_2O$.

rollback At a convergent plate boundary, the downgoing plate sinks into the mantle. The bend between the subducted slab and the yet to be subducted slab gradually migrates away from the volcanic arc. This movement is rollback.

rolled garnet *rotated garnet.*

roller [sed] *rod [sed].*

roller [waves] A general term, usually meaning one of a series of massive, long-crested waves that roll in upon a coast (as after a storm), usually retaining its form until it reaches the beach or shoal. Cf: *comber.*

roller-compacted concrete Low-strength concrete used for mass applications such as gravity dams. It is typically transported by haul truck or conveyor, spread with earth-moving equipment, and compacted with smooth-tire rollers used for compacting asphalt concrete pavement.

roll-front deposit *roll-front orebody.*

roll-front orebody A *roll orebody* of the Wyoming type, which is bounded on the updip concave side by oxidized altered rock typically containing hematite or limonite, and on the convex side by relatively reduced altered rock typically containing pyrite and organic matter. Cf: *bifacies.*

rolling prairie A term used in Texas for "a plain of undulating or rounded hilly relief" (Hill, 1900, p.7).

rolling strata (a) *ripple cross-lamination.* (b) *wavy bedding.*

rolling topography Any land surface having a gradual succession of low, rounded hills or undulations that impart a wave effect to the surface; esp. a land surface much varied by many small hills and valleys.

roll mark One of a series of similar tool marks following each other in a line parallel to the direction of the current, produced by an object that was rolled along the bottom. Originally proposed (as "roll-spuren") by Krejci-Graf in 1932; first usage in English was by Dzulynski and Slaczka (1958, p.234). Cf: *skip mark.*

roll-off The rate at which seismic attenuation changes with frequency in a filter.

roll orebody A uranium and/or vanadium orebody in a sandstone lens or layer, which cuts across bedding in sharply curving forms, commonly C-shaped or S-shaped in cross section. Two types can be distinguished: the Colorado Plateau type, named in 1956, and the Wyoming type, named in 1962. Roll orebodies of the Colorado Plateau type are of highly variable geometry, with their longest dimension in plan view parallel to the axes of buried sandstone lenses representing former stream channels, and surrounded by a wide halo of reduced (altered) rock. Orebodies of the Wyoming type are crescent-shaped in cross section and typically form in relatively thick, tabular, or elongate sandstone bodies, with the tips of the crescent thinning and becoming tangent to mudstone layers above and below (Bailey and Childers, 1977). See also: *roll-front orebody.* Syn: *ore roll; roll-type orebody; roll [ore dep].*

rollover (roll′-o-ver) An anticlinal feature adjacent to listric normal faults, in which the beds of the downthrown block dip toward the fault surface. Partial syn: *reverse drag; fault-bend fold.*

roll-type orebody *roll orebody.*

roll-up structure *convolutional ball.*

romanèchite (ro-man′-è-chite) An iron-black to steel-gray monoclinic mineral: $(Ba,H_2O)_2(Mn^{4+},Mn^{3+})_5O_{10}$. Calcium, potassium, sodium, cobalt, and copper are sometimes present. Romanèchite has a brownish-black streak, and commonly occurs massive, botryoidal, reniform, or stalactitic. It is an important ore of manganese. Syn: *psilomelane; black hematite.*

Romanian Eastern European stage: Mid- to Upper Pliocene (above Dacian).

romanzovite (ro-man′-zov-ite) A dark-brown variety of grossular garnet.

romarchite (ro′-march-ite) A black tetragonal mineral: SnO. It is isomorphous to litharge.

roméite (romé′-ite) A honey-yellow to yellow-brown cubic mineral occurring in minute octahedrons: $(Ca,Fe^{2+},Mn^{2+},Na)_2(Sb,Ti)_2O_6(O,OH,F)$.

römerite (rö′-mer-ite) A rust-brown, yellow or violet-brown triclinic mineral: $Fe^{2+}(H_2O)_6Fe^{3+}{}_2(H_2O)_8(SO_4)_4$. Also spelled: *roemerite.* Syn: *louderbackite.*

rond A British term for a narrow *washland* that separates a broad from the river; it is connected with the river by an artificial passageway.

rondorfite An orange-brown or amber mineral: $Ca_8Mg(SiO_4)_4Cl_2$.

rongstockite (rong′-stock-ite) A medium- to fine-grained plutonic rock composed of zoned plagioclase, orthoclase, some cancrinite, augite, mica, hornblende, magnetite, titanite, and apatite. The rock resembles *essexite* but contains less nepheline and has sodic rather than calcic plagioclase. The name, given by Tröger in 1935, is from Rongstock (Roztoky), Czech Republic.

ronneburgite A reddish-brown monoclinic mineral: $K_2MnV_4O_{12}$.

röntgenite-(Ce) (rönt′-gen-ite) A wax-yellow to brown rhombohedral mineral: $Ca_2(Ce,La)_3(CO_3)_5F_3$. Also spelled: roentgenite.

roof [intrus rocks] The country rock bordering the upper surface of an igneous intrusion. Cf: *floor [intrus rocks].*

roof [ore dep] The rock above an orebody; the *back.*

roof bolt Usually, a *rock bolt* that is short and mechanically anchored (wedge or expansion) and used to support the roof of a tunnel, mine or mine shaft.

roof collapse *roof foundering.*

roof control The engineering study of the behavior of unsupported rocks in mining operations, the systematic measurement of the movement of roof strata and the stresses involved, and the most effective measures to prevent or reduce roof movements (Nelson, 1965, p.378). Syn: *strata control.*

roof crystallization Boundary layer crystallization beneath the top (or roof) or a magma body. Typically accompanied by *thermal convection* of the underlying liquid magma.

roof foundering Collapse of rocks into an underlying reservoir or magma, usually following the evacuation of a large quantity of the magma. Less preferred syn: *roof collapse.*

roof pendant A downward projection of country rock into an igneous intrusion. Cf: *cupola [intrus rocks].* Syn: *pendant [intrus rocks].*

roof rock *cap rock.*

roof thrust The upper boundary of a *duplex fault zone* (Dahlstrom, 1970, p.418).

room In speleology, an exceptionally wide portion of a cave, often at the junction of *passages.* Cf: *chamber [speleo].*

room-and-pillar (a) Said of a system of mining in which the ore is mined in rooms separated by pillars of undisturbed rock left for roof support. (b) Said of a coral-reef structure characterized by interconnected and roofed-over surge channels or caverns.

room-and-pillar [mining] A type of underground mining involving the excavation of room-like openings and the leaving of intervening blocks as pillars to support the roof or back.

room-and-pillar [reef] Said of a coral-reef structure characterized by interconnected and roofed-over surge channels or caverns. Examples: *abrolhos* made of *chapeiros.*

rooseveltite (roo′-se-velt-ite′) A gray monoclinic mineral of the *monazite* group: $BiAsO_4$.

rooster tail (roost′-er) A plumelike form of water and sometimes spray that occurs at the intersection of two crossing waves.

root [ore dep] (a) Syn. of *bottom [ore dep].* (b) The conduit leading up through the basement to an ore deposit in the superjacent rocks.

root [paleont] *radix.*

root [struc geol] The basal part of a fold *nappe* that was originally linked to its source, or root zone.

root [tect] According to the *Airy hypothesis,* the downward extension of lower-density crustal material as isostatic compensation for its greater mass and high topographic elevation. Cf: *antiroot.*

root cap A thimblelike cellular tissue that fits over the growing tip of a rootlet and protects it.

root cast (a) A slender, nearly vertical, commonly downward-branching sedimentary structure, formed by the filling of a tubular opening left by a root. (b) *rhizoconcretion.*

root clay An *underclay* characterized by the occurrence of fossil roots of coal plants. See also: *rootlet bed.*

rootless fumarole (root′-less) A *fumarole* that derives its gases from the lava flow or ash flow on which it occurs, rather than from some deep source. Syn: *secondary fumarole.*

rootless vent A source of lava that is not directly connected to a volcanic vent or magma source; it may be an accumulation of overflow or an outflow from an otherwise solidified lava flow.

rootlet bed (root′-let) A stratum characterized by the occurrence of fossil rootlets of plants; e.g. a *root clay* beneath a coal bed.

root level The place within a sediment at which plant roots are found in the living position.

root-mean-square deviation *standard deviation.*

root scar *root zone [struc geol].*
root sheath A hollow *rhizoconcretion.*
root-tuft A tuft of subparallel, elongate spicules protruding from the base of a sponge and serving to fix it in the substrate.
root zone [soil] *rhizosphere.*
root zone [struc geol] (a) The zone in an orogenic belt in which thrust faults appear to emerge, i.e. where fault dips increase and faults may cut into basement. (b) The source or original attachment of the root of a nappe. Syn: *root scar.*
ropak (ro'-pak) A pinnacle or slab of sea ice standing vertically, rising as high as 8 m above the surrounding ice, and representing an extreme formation of *ridged ice.* Etymol: Russian.
ropy lava (rop'-y) *pahoehoe.*
roquesite (ro'-que-site) A metallic gray tetragonal mineral of the *chalcopyrite* group: $CuInS_2$.
rorisite A vitreous colorless tetragonal mineral of the *matlockite* group: CaClF.
rosasite (ro'-sa-site) A pale-green or sky-blue monoclinic mineral: $(Cu,Zn)_2(CO_3)(OH)_2$. Cf: *glaukosphaerite; nullaginite.*
roscherite (rosch'-er-ite) A dark brown or olive-green mineral occurring in both monoclinic and triclinic forms: $Ca(Mn^{2+},Mg,Fe^{2+})_2Be_2(Fe^{3+},[])(PO_4)_3(OH)_3{\cdot}2H_2O$.
roscoelite (ros'-coe-lite) The V^{3+}-analogue of muscovite, with V in place of octahedral Al: $K(V^{3+},Al,Mg)_2AlSi_3O_{10}(OH)_2$. It is tan to greenish brown and occurs in minute scales or flakes in the cement of certain sandstones and in some gold-quartz deposits.
rose [gem] n. (a) *rose cut.* (b) *rose diamond.* adj. Said of a gem having a rose, pink, or lilac color; e.g. "rose topaz".
rose [sed] *rosette [sed].*
rose [surv] *compass rose.*
rose cut Historic gemstone cut characterized by a flat base, a circular girdle outline, a pointed, dome-shaped crown, and a varying number of triangular facets. Thought to have originated in India, and to have been introduced into Europe by Venetian polishers in the fifteenth century; it is sometimes called a rosette cut or rosie (Liddicoat et al., 1993). Obsolete. Syn: *rose [gem].*
rose diagram A circular or semicircular star-shaped graph indicating values or quantities in several directions of bearing, consisting of radiating rays drawn proportional in length to the value or quantity; e.g. a current rose, a structural diagram for plotting strikes of planar features, or a "histogram" of orientation data.
rose diamond A rose-cut diamond. Syn: *rose [gem].*
roselite (ro'-se-lite) (a) A rose-red or pink monoclinic mineral: $Ca_2(Co,Mg)(AsO_4)_2{\cdot}2H_2O$. (b) A group name for minerals with the analogous composition, but with (Co,Mg) replaced by Zn or Mn.
rosemaryite (rose'-mar-y-ite') A green, dark brown, or reddish-brown monoclinic member of the *wyllieite* group: $(Na,Mn^{2+},Ca,[])(Mn^{2+},Fe^{3+},Fe^{2+})_2(Fe^{3+}Al)(PO_4)_3$.
rosenbergite A colorless tretragonal mineral: $AlF[F_{0.5}(H_2O)_{0.5}]_4{\cdot}H_2O$.
rosenbuschite (ro'-sen-busch-ite') A gray, orange, or yellowish triclinic mineral: $(Ca,Na)_3(Zr,Ti)Si_2O_7(O,F)_2$.
Rosenbusch's law (Ro'-sen-busch's) A statement of the sequence and crystallization of minerals from magmas, proposed in 1882 by the German petrologist Harry Rosenbusch, to which there are many exceptions. Obsolete.
rosenhahnite (ro'-sen-hahn-ite') A colorless, buff, or pale yellow triclinic mineral: $Ca_3Si_3O_8(OH)_2$.
rose opal An opaque variety of common opal having a fine red color.
rose quartz A pink to rose-red and commonly massive variety of crystalline quartz often used as a gemstone or ornamental stone. The color is perhaps due to titanium, and is destroyed or becomes paler on exposure to strong sunlight. Syn: *Bohemian ruby.*
rosette [paleont] (a) A delicate calcareous plate formed of metamorphosed basal plates, centrally located within the radial pentagon in some free-swimming crinoids. (b) The group of five petal-shaped ambulacra on certain echinoids. (c) A cluster of parts in circular form; e.g. a *discoaster.* (d) A flower-shaped button within a hexagonal pore frame in a radiolarian skeleton. (e) The cellular part of a *septula* in cheilostome and ctenostome bryozoans.
rosette [sed] A radially symmetric, sand-filled crystalline aggregate or cluster with a fancied resemblance to a rose, formed in sedimentary rocks by barite, marcasite, or pyrite. See also: *barite rosette.* Syn: *rose [sed].*
rosette plate (a) A large complex linguiform radially lobate structure anterior to the base of each limb of the giant Late Carboniferous arthropod Arthropleura. These unique and distinctive structures often occur as isolated, well-preserved fossils. (b) *pore plate.*
rosette texture A flowerlike or scalloped pattern of a mineral aggregate.
roshchinite A metallic gray orthorhombic mineral: $(Ag,Cu)_{19}Pb_{10}(Sb,As)_{51}S_{96}$.
rosiaite A colorless or pale yellow trigonal mineral: $PbSb_2O_6$.
rosickyite (ro-sic'-ky-ite) An adamantine yellow monoclinic mineral: S. It consists of native *sulfur* in the gamma crystal form. Syn: *gamma-sulfur.*
rosiérésite (ro-si-er'-e-site) An incompletely characterized yellow, greenish-yellow, or brown amorphous mineral consisting of a hydrous phosphate of lead, copper, and aluminum.
rosin jack (ros'-in) A yellow variety of sphalerite. Syn: *resin jack.*
rosin tin A reddish or yellowish variety of cassiterite. Syn: *resin tin.*
Rosiwal analysis (Ro'-si-wal) In petrography, a quantitative method of estimating the volume percentages of the minerals in a rock. Thin sections of a rock are examined with a microscope fitted with a micrometer which is used to measure the linear intercepts of each mineral along a particular set of lines. This method "is based on the assumption that the area of a mineral on an exposed surface is proportional to its volume in the rock mass" (Nelson and Nelson, 1967, p.320).
rosolite (ro'-so-lite) *landerite.*
ross A *promontory.* Etymol: Celtic.
Rossby wave (a) A wave in a nearly *geostrophic current* resulting from the North-South variability of the *Coriolis force* tending to restore the current to its original position after the current has been perturbed. (b) *planetary wave.*
Rossi-Forel scale (Ros'-si-Fo-rel') An earthquake *intensity scale,* devised by the Italian geologist Michele Stefano de Rossi and the Swiss naturalist Francois Alphonse Forel in 1883. It has a range of one to ten. It has been replaced by the *modified Mercalli scale.* Cf: *Richter scale.*
rossite (ross'-ite) A yellow triclinic mineral: $CaV^{5+}{}_2O_6{\cdot}4H_2O$.
rösslerite (röss'-ler-ite) A colorless monoclinic mineral: $Mg(AsO_3OH){\cdot}7H_2O$. It is isomorphous with phosphorroesslerite. Also spelled: *roesslerite.*
rossmanite A colorless to pale pink rhombohedral member of the *tourmaline* group: $([],Na)(Al,Li)_3Al_6(Si_6O_{18})(BO_3)_3(OH)_4$.
rostellum (ros-tel'-lum) A low projection between the anterior adductor muscle scars of the pedicle valve of some craniacean brachiopods to which the internal oblique muscles are attached.
rosterite (ros'-ter-ite) *vorobievite.*
rosthornite (ros'-thorn-ite) A brown to garnet-red variety of retinite with a low oxygen content (4.5%), found in lenticular masses in coal.
rostite (rost'-ite) A chalky white orthorhombic mineral: $Al(SO_4)(OH,F){\cdot}5H_2O$.
rostral carina (ros'-tral) The ridge forming the lateral margin of the *rostrum* on some decapod crustaceans, often passing into the orbital carina (Holthuis, 1974, p.735).
rostral incisure The anterior opening in the carapace of many ostracodes having a rostrum, which persists when the valves are closed, serving for the permanent protrusion of the antennae; sometimes referred to as the "permanent opening" of the carapace.
rostral notch The indentation, as seen in lateral view, below the rostrum in certain marine ostracodes (such as *Cyridina*).
rostral plate (a) An anteriorly projecting movable median extension of the carapace of a malacostracan crustacean. (b) A small median ventral plate of the head region in a trilobite, immediately anterior to the *hypostome.* Syn: *epistome.*
rostral ridge In conodont morphology, a ridge of more or less fused nodelike denticles extending posteriorly from anterior portion of platform; rostral ridges outline the rostrum and adcarinal grooves; they form parallel, collar- or rufflelike structures on pectiniform elements of some species of *Siphonodella* and *Polygnathus* (TIP, 1981, pt. W, supp.2). Syn: *cristula.*
rostral suture That portion of the *facial suture* of a trilobite along the join of the anterior of the *rostral plate* and the *cranidium.*
rostrate (ros'-trate) Having a *rostrum;* specif. said of a brachiopod with a prominent beak of the pedicle valve projecting over a narrow cardinal margin.
rostroconch (ros'-tro-conch) Any benthic marine mollusk belonging to the class Rostroconchia, characterized by an uncoiled univalved larval shell that straddles the dorsal midline and a bivalved adult shell with one or more shell layers continuous across the dorsal margin so that a dorsal commissure is lacking (Pojeta and Runnegar, 1976).
rostroconcha A diasome mollusk with pseudobivalved shell; radula probably present. In primitive rostroconchs an anterodorsal plate called the pegma connects the two valves and was a site of muscle attachment. Range, Lower Cambrian to Upper Permian; about 35 genera.
rostrolateral (ros-tro-lat'-er-al) One of a pair of plates in certain cirripede crustaceans lying between the rostrum and the lateral.
rostrum (ros'-trum) (a) A part of an arachnid suggesting a bird's bill, e.g. the tubelike "beak" in the order Solpugida, or the anterior spike of the carapace of the order Eophrynidae, similar to the rostrum of a lobster (TIP, 1955, pt. P, p.62). (b) The anteriorly projecting spinelike median extension of the carapace of a crustacean; e.g. an unpaired plate adjacent to the scuta of a cirripede. See also: *rostral plate.* (c) An elevation of the secondary shell on the inner surface of the brachial valve of some craniacean brachiopods, in front of the anterior adductor muscles, consisting of a pair of low club-shaped protuberances forming the seat of attachment for the brachial protractor muscles. Also, the *beak* of an articulate brachiopod. (d) The grooved extension of a gastropod shell protecting the *siphon;* the attenuated extremity of the last whorl other than the siphonal canal, as in *Tibia.* Also, the snout of a gastropod when nonretractile. (e) A pointed projection of *peristome* on the venter of an ammonoid. Also, the *guard* of a belemnite. (f) A tubular extension of the posterodorsal part of the shell of some rostroconchs. (g) In cheilostome bryozoans, the rounded or pointed skeletal rim around the palate on which the mandible occludes in an *avicularium.*

Used as a syn. of *palate.* Syn: *beak.* Pl: rostra or rostrums. (h) Narrow neckline or beaklike structure in anterior part of platform in planate pectiniform conodont elements of most *Siphonodella* and some *Polygnathus* species (TIP, 1981, pt. W, supp.2). (i) May be applied to the beak of birds or, among vertebrates generally, to any anteriorly extended nasal or prenasal region.

rotaliid (ro-tal'-i-id) Any trochospirally, rather than planispirally, coiled foraminifer, excluding trochospirally coiled planktonic genera.

rotary current (ro'-ta-ry) A tidal current of variable velocity that flows continually (generally clockwise in the Northern Hemisphere) and changes direction progressively through 360° during a tide cycle. Cf: *reversing current.*

rotary drilling The most common method of drilling wells, for oil and gas or water, also used to drill monitor wells in hard rock terrains. A hard-toothed *drill bit* at the bottom of a rotating *drillstring* made primarily of drill pipes grinds a hole in the rock. Lubrication and cooling are provided by continuously circulating *drilling mud,* which also brings the *well cuttings* to the surface. Cf: *cable-tool drilling.*

rotary polarization *optical activity.*

rotary table In *rotary drilling,* a power-driven circular platform on the derrick floor that rotates the *kelly, drillstring,* and *drill bit.* It is sometimes used as the zero-depth reference for downhole measurements. Cf: *kelly bushing.* Abbrev: RT.

rotate (ro'-tate) Wheel-shaped, as in a flower whose parts are flat and radiating.

rotated factor In *factor analysis,* one of the factors subjected to *varimax rotation.* Factors are rotated in a manner that maximizes the variance of the factor loadings. This tends to make the factors correspond more closely to some of the original variables.

rotated garnet (ro'-tat-ed) A garnet crystal that shows evidence of having been rotated during metamorphism (Knopf and Ingerson, 1938). Syn: *pinwheel garnet; rolled garnet; snowball garnet; spiral garnet.*

rotating dipole An energy source consisting of two fixed dipoles, either magnetic or grounded electric current dipoles, oriented with their axes mutually perpendicular and energized by sinusoidal current with a 90° phase shift between sources, so that the resultant magnetic and electric fields appear to rotate in space. It produces elliptically polarized waves.

rotation (ro-ta'-tion) (a) *internal rotation.* (b) *external rotation.* (c) In crystallography, a symmetry transformation by a *symmetry axis.*

rotational bomb (ro-ta'-tion-al) A pyroclastic bomb whose shape is formed by spiral motion or rotation during flight; rotation produces such types as spheroidal, tear-shaped, and spindle-shaped bombs. See also: *fusiform bomb.*

rotational cylindroidal fold A cylindrical fold, the axial surface of which has been distorted by a subsequent or cross fold (Whitten, 1959).

rotational deflection The deflection of currents of air or water by rotation of the Earth, as stated in *Ferrel's law.*

rotational fault A fault on which rotational movement is exhibited; a partial syn. of *hinge fault.* Cf: *scissor fault.* Often used erroneously, as rock units can be tilted and hence rotated by different amounts across planar faults.

rotational flow Flow in which each fluid element rotates about its own mass center.

rotational landslide A slide in which shearing takes place on a well defined, curved shear surface, concave upward, producing a backward rotation in the displaced mass (Hutchinson, 1968). It may be single, successive (repeated up- and down-slope), or multiple (as the number of slide components increases). Syn: *rotational slump.* See also: *Toreva block.*

rotational movement Apparent fault-block displacement in which the blocks have rotated relative to one another, so that alignment of formerly parallel features is disturbed. Cf: *translational movement.* See also: *rotational fault.*

rotational slump *rotational landslide.*

rotational stacking disorder Layered minerals may be visualized as stacked with or without rotation of successive layers in relation to the layer below. Phyllosilicates can have stacking disorder described by n60° and n120°. Smectites have turbostratic stacking which is simply n° rotation.

rotational strain *Strain* in which the orientation of the principal axes of strain is different before and after deformation. Not to be confused with *noncoaxial progressive deformation.* Cf: *irrotational strain.* Syn: *pure rotation.*

rotational transformation A type of phase transformation that is a change from an ordered phase to a partially disordered phase by rotation of groups of atoms. It is usually a rapid process. Cf: *dilatational transformation; displacive transformation; reconstructive transformation; substitutional transformation.*

rotational wave *S wave.*

rotation axis *symmetry axis.*

rotation method A method of X-ray diffraction analysis using a rotating single crystal, monochromatic radiation, and a cylindrical film coaxial with the rotation axis of the crystal.

rotation recrystallization The mechanism by which *dislocations* preferentially accumulate in a *tilt boundary,* resulting in a rotation of the bounded crystal latttice.

rotation twin A crystal twin whose symmetry is formed by apparent axial rotation of 180°. Cf: *reflection twin.*

rotatory dispersion (ro'-ta-to'-ry) In crystal optics, the breaking-up of white light into colors by passing it through an optically active substance, such as quartz.

rotatory reflection axis *rotoreflection axis.*

Rotliegendes (Rot'-lie-gen-des) A European lithostratigraphic unit (esp. in Germany): Lower and Middle Permian (below Zechstein). It contains the Autunian and Saxonian stages. According to German usage, the term used with a definite article is changed to (das/the) Rotliegende.

rotoinversion axis (ro'-to-in-ver'-sion) A type of crystal symmetry element that combines a rotation of 60°, 90°, 120°, or 180° with inversion through the center. Syn: symmetry axis of rotoinversion; *symmetry axis of rotary inversion.*

rotoreflection axis (ro'-to-re-flec'-tion) A type of symmetry element that combines a rotation of 60°, 90°, 120°, or 180° with reflection across the plane perpendicular to the axis. Syn: *rotatory reflection axis.*

rotten ice (rot'-ten) Ice in which the grains or crystals are loosened, one from the other, by melting along grain boundaries. It may form a *honeycomb structure [ice].* Cf: *candle ice.*

rotten spot *pothole [coast].*

rottenstone (rot'-ten-stone) Any highly decomposed but still coherent rock; specif. a soft, friable, lightweight, earthy residue consisting of fine-grained silica and resulting from the decomposition of siliceous limestone (or of a highly shelly sandstone) whose calcareous material has been removed by the dissolving action of water. Cf: *tripoli.* Syn: *terra cariosa.*

rotula (rot'-u-la) One of the five massive radial elements in ambulacral position at the top of *Aristotle's lantern* of an echinoid. Pl: rotulae.

rouaite A dark green monoclinic mineral: $Cu_2NO_3(OH)_3$. It is a dimorph of gerhardtite.

roubaultite (rou-bault'-ite) A green triclinic mineral: $Cu_2(UO_2)_3(CO_3)_2O_2(OH)_2{\cdot}4H_2O$.

rougemontite (rouge'-mont-ite) A coarse-grained *gabbro* composed of anorthite, titanian augite, and small amounts of olivine and iron oxides. Its name, given by O'Neill in 1914, is derived from Rougemont, Quebec. Obsolete.

rough n. An uncut gemstone. adj. Pertaining to an uncut gemstone; e.g. a *rough* diamond.

rough cleavage A *disjunctive cleavage* characterized by short, discontinuous concentrations of oriented platy minerals that bound larger grains. Common cleavage type in micaceous sandstones. Cf: *smooth cleavage.*

rough criterion In radar, the relationship between surface roughness, depression angle, and wavelength that determines whether a surface will scatter the incident radar pulse in rough or intermediate fashion. Cf: *smooth criterion; Rayleigh criterion.*

rough ice An expanse of ice having an uneven surface caused by formation of *pressure ice* or by *growlers* frozen in place (ADTIC, 1955, p.68).

roughneck (rough'-neck) A general workman in a drilling crew; also, one who builds and repairs oil derricks. Cf: *roustabout.*

roughness coefficient (rough'-ness) A factor in formulas for computing the average velocity of flow of water in a conduit or channel which represents the effect of roughness of the confining material upon the energy losses in the flowing water.

rounded Round or curving in shape; specif. said of a sedimentary particle whose original edges and corners have been smoothed off to rather broad curves and whose original faces are almost completely removed by abrasion (although some comparatively flat surfaces may be present), such as a pebble with a roundness value between 0.40 and 0.60 (midpoint at 0.500) and few (0-5) and greatly subdued secondary corners that disappear at roundness 0.60 (Pettijohn, 1957, p.59). The original shape is still readily apparent. Also, said of the *roundness class* containing rounded particles. Cf: *well-rounded.*

rounded form *equilibrium form.*

roundness The degree of abrasion of a clastic particle as shown by the sharpness of its edges and corners, expressed by Wadell (1932) as the ratio of the average radius of curvature of the several edges or corners of the particle to the radius of curvature of the maximum inscribed sphere (or to one-half the nominal diameter of the particle). The value is more conveniently computed from a plane figure (a projection or cross section); thus, roundness may be defined as the ratio of the average radius of curvature of the corners of the particle image to the radius of the maximum inscribed circle. A perfectly rounded particle (such as a sphere) has a roundness value of 1.0; less-rounded particles have values less than 1.0. The term has been used carelessly and should not be confused with *sphericity*: a nearly spherical particle may have sharp corners and be angular, while a flat pebble, far from spherical in shape, may be well-rounded. Cf: *flatness.* See also: *angularity; roundness class.*

roundness class An arbitrarily defined range of *roundness* values for the classification of sedimentary particles. Pettijohn (1957, p.58-59) recognizes five classes: *angular; subangular; subrounded; rounded; well-rounded.* Powers (1953) adds a sixth class: *very angular.* Syn: *roundness grade.*

roundness grade *roundness class.*

roundstone (round'-stone) (a) A term proposed by Fernald (1929) for any naturally rounded rock fragment of any size larger than a sand grain (diameter greater than 2 mm), such as a boulder, cobble, pebble, or granule. Cf: *sharpstone.* (b) *cobblestone.*

rouseite (rouse'-ite) An orange-yellow triclinic mineral: $Pb_2Mn^{2+}(As^{3+}O_3)_2 \cdot 2H_2O$.

roustabout (roust'-a-bout) A common laborer called upon to do any of the unskilled jobs in an oil field or refinery, or around a mine. Cf: *roughneck.*

routhierite (rou'-thier-ite) A violet-red tetragonal mineral: $(Tl,Cu)Hg(As,Sb)S_3$.

routivarite (rou-ti-var'-ite) A fine-grained garnet-bearing *anorthosite,* named by Sjögren in 1893 for Routivara, Norrbotten, Sweden. Obsolete.

rouvilleite A colorless or tan monoclinic mineral: $Na_3Ca(Mn,Ca)(CO_3)_3F$.

rouvillite (rou'-vil-lite) A light-colored *theralite* composed predominantly of labradorite or bytownite and nepheline, with small amounts of titanian augite, hornblende, pyrite, and apatite. The name, given by O'Neill in 1914, is from Rouville County, Quebec. Obsolete.

ROV *Remotely Operated Vehicle.*

roweite (rowe'-ite) A light brown orthorhombic mineral: $Ca_2(Mn^{2+},Mg)_2B_4O_7(OH)_6$.

rowlandite-(Y) A grayish-white to dull green triclinic mineral: $Y_4Fe^{2+}F_2[Si_2O_7]_2$.

row stack nacreous structure A *nacreous structure* in which mutually parallel, elongate tablets show vertical stacking in vertical sections taken perpendicular to the tablet lengths but stair-step and/or brick-wall stacking in vertical sections taken parallel to the tablet lengths (Wise, 1970).

roxbyite (rox'-by-ite) A metallic blue-black monoclinic mineral: Cu_9S_5.

royal agate (roy'-al) A mottled obsidian.

royalty (roy'-al-ty) The landowner's share of the value of minerals produced on a property. It is commonly a fractional share of the current market value (oil and gas) or a fixed amount per ton (mining).

rozenite (ro'-zen-ite) A greenish-white monoclinic mineral of the *starkeyite* group: $Fe^{2+}SO_4 \cdot 4H_2O$.

rozhkovite (rozh'-kov-ite) A metallic orthorhombic mineral: $(Cu,Pd)_3Au_2$. Not approved by the International Mineralogical Association.

RQD *rock quality designation.*

RSA *ripe-snow area.*

r selection The evolutionary process favoring development of r strategists or *opportunistic species* (Wilson and Bossert, 1971, p.110-111).

RSI *ripple symmetry index.*

RSP *regular simple prismatic structure.*

r strategist *opportunistic species.*

RT *rotary table.*

RTE *reduction to the equator.*

RTP *reduction to the pole.*

ruarsite (ru-ar'-site) A metallic gray-white monoclinic mineral: RuAsS. Not approved by the International Mineralogical Association.

rubasse (ru-basse') (a) A crystalline variety of quartz, stained a ruby red by numerous small scales or flecks of hematite distributed within it. (b) An imitation produced by artificially staining crackled quartz red. Also spelled: rubase.

rubber sheeting (a) Any process in which a raster is stretched differentially to match a new set of geometric constraints. (b) A map transformation bringing a map into coincidence or near-coincidence with another reference map (Krzanowski et al., 1993). (c) A deformation algorithm forcing a registration of control points, distributed over a map, with their counterparts on the stable, better quality, map (Krzanowski et al., 1993). Also called warping.

rubbish A non-hazardous mixture of combustible solid waste, including paper, cardboard cartons, and wood scraps; contains up to 20% animal or food processing wastes (garbage) but little to no plastics or treated papers.

rubble (rub'-ble) (a) A loose mass of angular rock fragments, commonly overlying outcropping rock; the unconsolidated equivalent of a breccia. Cf: *talus [geol]; volcanic rubble.* (b) Loose, irregular pieces of artificially broken stone as it comes from the quarry. (c) Fragments of floating or grounded *sea ice,* in hard, roughly spherical blocks measuring 0.5 to 1.5 m in diameter, resulting from the breakup of larger ice formations. When afloat, it is commonly called *brash ice.*

rubble beach A beach composed of angular rock fragments or rubble.

rubble breccia (a) A breccia in which no matching fragments are parted by initial planes of rupture; the fragments are close-set and in touch (Norton, 1917, p.161). (b) A tectonic breccia characterized by prominent relative displacement of fragments and by some rounding (Bateman, 1959, p.133). Cf: *shatter breccia.*

rubble drift (a) An English term for a rubbly deposit or *congeliturbate* formed by solifluction under periglacial conditions; e.g. *head [mass move]* and *coombe rock.* (b) A coarse mass of angular debris and large blocks set in an earthy matrix of glacial origin.

rubble island *debris island.*

rubble ore (a) A term used in Brazil for iron ore found on the surface of *itabirite* and derived from it by "the breaking up of the thinner intercalated layers, more or less completely freed from the associated siliceous elements by rain and wind action" (Derby, 1910, p.818). Cf: *sandy ore; canga.* (b) Deposits of reworked banded residual enriched iron formation, similar to *canga* in tropical environments, found in the Knob Lake iron range of Quebec and Labrador, Canada. Cf: *blue ore; red ore; yellow to brown ore; treat rock.*

rubblerock (rub'-ble-rock) *breccia [geol].*

rubblestone (rub'-ble-stone) (a) A graywacke (Humble, 1843, p.224). (b) *rubble.*

rubble tract The part of a reef flat immediately behind and lagoonward of the *reef front,* paved with cobbles, pebbles, blocks, and other coarse reef-rock fragments; when consolidated it forms a *reef breccia.*

rubellite (ru-bel'-lite, ru'-bel-lite) A pale rose-red to deep ruby-red transparent lithian variety of tourmaline, used as a gemstone. Syn: *red schorl.*

rubicelle (ru-bi-celle') A yellow or orange-red gem variety of spinel. See also: *ruby spinel.*

rubicline A colorless triclinic mineral of the *feldspar* group: $(Rb,K)AlSi_3O_8$.

rubidium-strontium age method (ru-bid'-i-um-stron'-ti-um) Determination of an age for a mineral or rock in years based on the ratio of radiogenic strontium-87 to rubidium-87 and the known radioactive decay rate of rubidium-87. If ratios are measured for more than one phase of a single rock, or for a number of related rocks that differ in rubidium content, an isochron may be drawn. Syn: *rubidium-strontium dating; Rb-Sr age method.*

rubidium-strontium dating *rubidium-strontium age method.*

rubidium-vapor magnetometer (ru-bid'-i-um-va'-por) A type of *optically pumped magnetometer* that uses magnetic atoms of rubidium. Cf: *cesium-vapor magnetometer.*

rubinblende (ru'-bin-blende) A name applied to the red silver-sulfide minerals pyrargyrite, proustite, and miargyrite. Syn: *ruby blende.*

rubisco Enzyme responsible for fixing CO_2 in photosynthesis; most abundant protein in the world.

ruby (ru'-by) The red variety of corundum, containing small amounts of chromium, used as a gemstone, and found esp. in the Orient (Myanmar, Sri Lanka, Thailand). Cf: *sapphire.*

ruby blende (a) A brownish-red or reddish-brown transparent variety of sphalerite. Syn: *ruby zinc.* (b) *rubinblende.*

ruby copper Cuprous oxide; specif. *cuprite.*

ruby sand A red-colored beach sand containing garnets, as at Nome, Alaska.

ruby silver A red silver-sulfide mineral; specif. "dark ruby silver" (pyrargyrite) and "light ruby silver" (proustite).

ruby spinel A clear-red gem variety of magnesian spinel, Al_2MgO_4 containing small amounts of chromium and having the color but none of the other attributes of true ruby. See also: *spinel ruby; balas ruby; rubicelle.*

ruby zinc A deep-red transparent zinc mineral; specif. *ruby blende* and *zincite.*

rucklidgeite (ruck'-lidge-ite) A metallic white rhombohedral mineral: $(Bi,Pb)_3Te_4$. It is related to tellurobismuthite.

rudaceous (ru-da'-ceous) Said of a sedimentary rock composed of a significant amount of fragments coarser than sand grains; pertaining to a *rudite.* The term implies no special size, shape, or roundness of fragments throughout the gravel range, and is broader than "pebbly", "cobbly", and "bouldery". Term introduced by Grabau (1904, p.242). Cf: *psephitic.*

ruddle (rud'-dle) *red ocher.*

rudemark (rude'-mark) *rute-mark.*

rudimentary (ru-di-men'-ta-ry) Vestigial; said of structures that are not as highly developed in the descendants as they were in the ancestors; very imperfectly developed.

rudist (ru'-dist) Any bivalve mollusk belonging to the superfamily Hippuritacea, characterized by an inequivalve shell, usually attached to a substrate, and either solitary or gregarious in reeflike masses. "Although the first rudists were only slightly inequivalve, their descendants very early became strongly so, with the two valves of individuals usually differing greatly from each other in size, shape, and shell wall structure" (TIP, 1969, pt.N, P.751). They are frequently found in association with corals. Range, Upper Jurassic to Upper Cretaceous, possibly Paleocene.

rudite (rud'-ite) A general name used for consolidated sedimentary rocks composed of rounded or angular fragments coarser than sand (granules, pebbles, cobbles, boulders, or gravel or rubble); e.g. conglomerate, breccia, and calcirudite. The term is equivalent to the Greek-derived term, *psephite,* and was introduced as *rudyte* by Grabau (1904, p.242) who used it with appropriate prefixes in classifying coarse-grained rocks (e.g. "autorudyte", "autosilicirudyte", "hydrorudyte", and "hydrocalcirudyte"). Etymol: Latin *rudus,* "crushed stone, rubbish, debris, rubble". See also: *lutite; arenite.*

rudstone (rud'-stone) A carbonate rock composed of bioclasts or other fragments, over 2 mm in diameter, closely packed, in physical contact; the interstices may be open, or filled with fine carbonate sediment or crystalline cement (Embry and Klovan, 1971). Cf: *floatstone [sed]; grainstone; boundstone.*

rudyte (rud'-yte) Var. of *rudite.*

Rudzki anomaly (Rud'-zki) A gravity anomaly calculated by replacing the surface topography by its mirror image within the geoid.

ruffle (ruf'-fle) (a) A ripple produced by an eddy (Hobbs, 1917). Obsolete. (b) A roughness or disturbance of a surface, such as a ripple on a surface of water.

ruffled groove cast (ruf'-fled) A *groove cast* with a feather pattern, consisting of a groove with lateral wrinkles that join the main cast in the

downcurrent direction at an acute angle (Ten Haaf, 1959, p.32). See also: *vibration mark.*
ruga (ru'-ga) A visceral fold or wrinkle; e.g. a commarginal or oblique wrinkling of the external surface of a brachiopod shell, or a *growth ruga* on the shell of a bivalve mollusk. Pl: rugae.
ruggedness number A dimensionless number formed by the product of maximum basin relief and drainage density within a given drainage basin; it expresses the essential geometric characteristics of the drainage system, and implicitly suggests steepness of slope (Strahler, 1958, p.289).
rugose (ru'-gose) Coarsely wrinkled, uneven, rough (Swartz, 1971, p.409).
rugose coral Any zoantharian belonging to the order Rugosa, characterized by calcareous corallites that may be solitary and cone-shaped or cylindrical, curved or erect, compound and branching or massive. Range, Ordovician to Permian.
rugosity Roughness; irregularity of a borehole wall. The readings of logging tools which have to be held in contact with the hole wall (such as density and microresistivity logs) are affected by rugosity.
rugulate (ru'-gu-late) Said of sculpture of palynomorphs consisting of wrinklelike ridges that irregularly anastomose.
ruin agate (ru'-in) A brown variety of agate displaying on a polished surface markings that resemble the outlines of ruined buildings.
ruitenbergite A colorless to pale yellow monoclinic mineral: $Ca_9B_{26}O_{34}(OH)_{24}Cl_4 \cdot 13H_2O$. It is a dimorph of pringleite.
ruizite (ru'-iz-ite) An orange-brown monoclinic mineral: $Ca_2Mn^{3+}{}_2Si_4O_{11}(OH)_4 \cdot 2H_2O$.
rule of tautonymy *tautonymy rule.*
rule of V's The outcrop of a formation that crosses a valley forms an acute angle (a V) that points in the direction in which the formation lies underneath the stream. The V points upstream where the outcrops of horizontal beds parallel the topographic contours, where the beds dip upstream, or where the beds dip downstream at a smaller angle than the stream gradient; the V points downstream where the beds dip downstream at a larger angle than the stream gradient.
rumänite (ru-män'-ite) A brittle, yellow-brown to red or black variety of amber containing 1-3% sulfur and found in Romania. Also spelled: rumanite.
rumpffläche (rumpf'-flä-che) A plain extending across a region underlain by massive or undifferentiated rocks; a term used "purely to express relief, with no implication as to position in the cycle of erosion" (W. Penck, 1953, p.420). The term has often been used to indicate a *peneplain.* Etymol: German "Rumpffläche", "torso plain". Syn: *torso plain.*
run [ore dep] n. A flat irregular ribbonlike orebody following the stratification of the host rock.
run [streams] n. (a) A small, swiftly flowing watercourse; a brook or a small creek. (b) A *channel unit* in which downstream channel-bed gradient, flow depth, and grain size are intermediate between those characterizing *riffles* and *pools.*
rundkarren *Karren* having rounded channels and intervening rounded ridges, probably exhumed after formation under soil or peat. Etymol: German.
rundle (run'-dle) Var. of *runnel.*
runic (ru'-nic) *graphic [ign].*
runite (ru'-nite) A syn. of *graphic granite.* The term, first used by Pinkerton in 1811, was suggested by Johannsen (1939, p.273). Obsolete.
run-length coding A method of compressing data based on the reduction in the number of any type of repeating sequences of characters. Used frequently to compress image data (Krzanowski et al., 1993).
runlet (run'-let) *runnel [streams].*
runnel [beach] (run'-nel) (a) A trough formed landward of a ridge on the foreshore of an intertidal beach by the action of waves. It carries the water drainage off the beach as the tide retreats, and is flooded as the tide advances. Cf: *rill [beach].* (b) *swale.* (c) *trough [beach].*
runnel [snow] A furrow on a snow surface, which follow the fall line and forms during melt conditions.
runnel [streams] (a) A little brook; a rivulet or a streamlet. Syn: *runlet; rundle; rindle.* (b) The channel eroded by a runnel.
running ground (run'-ning) (a) In mining, incoherent surface material; earth, soil, or rock that will not stand, esp. when wet, and that tends to flow into mine workings. See also: *mudrush.* (b) A tunneler's term for soil that runs into a tunnel on removal of roof or side support; for example, dry cohesionless sand. See also: *firm ground; flowing ground; raveling ground; squeezing ground; swelling ground.*
running water Water that is flowing in a stream or that is not stagnant or brackish. Ant: *standing water.*
runoff [coal] (run'-off) The collapse of a pillar of coal in a steeply dipping coal seam; the pillar is said to have run off.
runoff [water] That part of precipitation appearing in surface streams. It is more restricted than *streamflow,* as it does not include stream channels affected by artificial diversions, storage, or other human works. With respect to promptness of appearance after precipitation, it is divided into *direct runoff* and *base runoff;* with respect to source, into *surface runoff, storm seepage,* and *groundwater runoff.* It is the same as *total runoff* used by some workers (Langbein and Iseri, 1960). May be expressed as average depth over watershed. Syn: *virgin flow.* Cf: *water yield.* See also: *runoff cycle.*
runoff coefficient The percentage of precipitation that appears as runoff. The value of the coefficient is determined on the basis of climatic conditions and physiographic characteristics of the drainage area and is expressed as a constant between zero and one (Chow, 1964, p.20-8, 21-37). Symbol: *C.*
runoff cycle That part of the hydrologic cycle involving water between the moment of its precipitation onto land and its subsequent evapotranspiration or discharge through stream channels. See also: *runoff [water].*
runoff desert An arid region in which local rain is insufficient to support any perennial vegetation except in drainage or runoff channels. Cf: *rain desert.*
runoff intensity Strictly, the volume of water derived from an area of land surface per hour; usually expressed in depth of rainfall per hour. Symbol: Q. Syn: *runoff rate.*
runoff rate Runoff per unit time. See: *runoff intensity.*
run-of-mine Said of ore in its natural, unprocessed state; pertaining to ore just as it is mined.
runout (run'-out) *water yield.*
runup (run'-up) The advance of water up the foreshore of a beach or structure, following the breaking of the wave. Cf: *backwash.* Syn: *uprush.*
runway (run'-way) The channel of a stream.
runzel mark (run'-zel) A wrinkled surface on intertidal-flat sediments, produced by the removal of sediment by windblown foam (Klein, 1985, p.206). Term first used by Reineck in 1963.
rupe (ru'-pe) A term established by the International Astronomical Union for a scarp of diverse form (USGS, 2003) and origin on planetary surfaces (Trask and Dzurisin, 1984). Pl: rupes.
Rupelian (Ru-pel'-i-an) A stage of the Standard Global Chronostratigraphic Scale: lower Oligocene (above Priabonian of the Eocene , below Chattian). The basal GSSP is near Ancona, Italy (Premoli Silva and Jenkins, 1993).
rupestral (ru-pes'-tral) Said of an organism living among rocks or in rocky areas. Syn: *rupestrine; rupicolous; lithophilous; saxicolous; saxigenous.*
rupestrine (ru-pes'-trine) *rupestral.*
rupicolous (ru-pic'-o-lous) *rupestral.*
rupture (rup'-ture) *fracture [exp struc geol].*
rupture envelope *Mohr envelope.*
rupture-front geometry The shape of the rupture front as indicated by the plumose morphology on a joint surface (Savalli and Engelder, 2004). The rupture front at any point during propagation is defined by a line drawn normal to plume lines or striae. Cf: *plumose surface morphology.*
rupture strength The differential stress that a material sustains at the instant of rupture. The term is normally applied when deformation occurs at atmospheric confining pressure and room temperature.
rupture velocity The speed with which rupture propagates along the fault surface during an earthquake.
rursiradiate (rur-si-ra'-di-ate) Said of an ammonoid rib inclined backward (adapically) from the umbilical area toward the venter. Cf: *prorsiradiate; rectiradiate.*
rusakovite (ru'-sa-kov-ite) A yellow-orange incompletely investigated mineral: $(Fe^{3+},Al)_5(VO_4,PO_4)_2(OH)_9 \cdot 3H_2O$.
Ruscinian (Rus-cin'-i-an) European stage: Lower Pliocene (above Turolian, below Villafranchian).
russellite (rus'-sel-lite) A pale yellow or greenish orthorhombic mineral: Bi_2WO_6.
rust An English term for a black shale discolored by ocher.
rustenburgite (rus'-ten-burg-ite') A metallic creamy cubic mineral of the *zvyagintsevite* group: $(Pt,Pd)_3Sn$.
rustumite (rus'-tum-ite) A colorless monoclinic mineral: $Ca_{10}(SiO_4)(Si_2O_7)_2Cl_2(OH)_2$.
rusty gold (rust'-y) Native gold that has a thin coat of iron oxide or silica that prevents it from amalgamating readily.
rute-mark (a) A type of polygonal ground found in arctic regions, consisting of a polygon enclosed by a row of stones. Etymol: Norwegian "rutemark", "route mark". Also spelled: rutmark; rutemark; rudemark. (b) A crack in soil or mud, similar in form to a rute-mark.
ruthenarsenite (ru-then-ar'-se-nite) A metallic orange-brown or brownish-gray orthorhombic mineral of the *westerveldite* group: (Ru,Ni)As.
rutheniridosmine (ru'-then-ir'-id-os'-mine) A metallic white hexagonal mineral: (Ir,Os,Ru).
ruthenium (ru-the'-ni-um) A rare metallic white hexagonal mineral: Ru. It is a metallic element of the *platinum* group.
ruthenosmiridium (ru'-then-os'-mi-rid'-i-um) A discredited name for ruthenium-bearing iridium.
Rutherford backscattering An analytical technique measuring the energies of backscattered particles from beam-induced nuclear reactions; the particles have collided elastically with atoms in the sample. Generally used to obtain information about the depths at which different elements occur in a sample. Abbrev: RBS.
rutherfordine (ruth'-er-ford-ine') A white, yellow, orange, or brown orthorhombic secondary mineral: $U^{6+}O_2(CO_3)$. Also spelled: rutherfordite.

rutilated quartz (ru'-ti-lat'-ed) Sagenitic quartz characterized by the presence of enclosed needle-like crystals of rutile. See also: *sagenite.* Syn: *Venus hairstone.*

rutile (ru-tile') A usually reddish-brown tetragonal mineral: TiO_2. It is trimorphous with anatase and brookite, and often contains a little iron. Rutile forms prismatic crystals in other minerals (esp. quartz); it occurs as a primary mineral in some acid igneous rocks (esp. those rich in hornblende), in metamorphic rocks, and as residual grains in sediments and beach sands. It is an ore of titanium. Syn: *red schorl.*

rutmark (rut'-mark) *rute-mark.*

rutterite (rut'-ter-ite) A medium-grained equigranular dark-pink nepheline-bearing *syenite,* composed chiefly of microperthite, microcline, and albite, with small amounts of nepheline, biotite, amphibole, graphite, and magnetite. The name, given by Quirke in 1936, is from Rutter, Ontario. Obsolete.

ruware (ru-ware') A term used in southern Africa for a low, flattish or gently domed, granitic pediment or outcrop of bare rock. It occurs where flat or gently dipping joint systems are prominent.

RV *right valve.*

R wave *Rayleigh wave.*

Ryazanian (Ry-a-zan'-i-an) European stage: lowermost Cretaceous (above Volgian, below Valanginian).

rynersonite (ry'-ner-son-ite) A creamy-white or reddish-pink orthorhombic mineral of the *aeschynite* group: $Ca(Ta,Nb)_2O_6$.

S

Saale (Saal'-e) The term applied in northern Europe to the third classical glacial stage of the Pleistocene Epoch, after the Elster glacial stage and before the Weichsel; equivalent to the *Riss* and *Illinoian* glacial stages.

Saalian (Saal'-i-an) North European climatostratigraphic and floral stage: Upper Pleistocene (above Holsteinian, below Eemian). Equivalent in time to Riss glaciation.

Saalic Orogeny (Saal'-ic) One of the 30 or more short-lived orogenies during Phanerozoic time recognized by Stille, in this case early in the Permian between the Autunian and Saxonian stages.

sabach (sa'-bach) A term used in Egypt for a calcareous accumulation; specif. caliche. Syn: sabath.

sabatierite (sa-ba-tier'-ite) A metallic gray-blue or yellow-brown orthorhombic mineral: Cu_6TlSe_4.

sabellariid reef (sab-el-lar'-i-id) A small *organic reef* composed of the closely packed, sand-walled or arenaceous tubes constructed by sabellarid worms (Polychaeta; Annelida), as off east-central peninsular Florida. Cf: *serpulid reef.*

sabelliite An emerald-green trigonal mineral: $(Cu,Zn)_2Zn(As,Sb)O_4(OH)_3$.

sabieite (sab'-ie-ite) A white trigonal mineral occurring in a cave deposit: $(NH_4)Fe^{3+}(SO_4)_2$. It is the Fe^{3+} analogue of godovikovite.

sabinaite (sa-bi'-na-ite) A silky-white monoclinic mineral: $Na_4Zr_2TiO_4(CO_3)_4$.

Sabinian (Sa-bin'-i-an) North American (Gulf Coast) stage: Paleocene-Eocene (above Midwayan, below Claibornian). It includes strata most commonly grouped as *Wilcoxian.*

sabkha (sab'-kha) (a) A supratidal environment of sedimentation, formed under arid to semiarid conditions on restricted coastal plains just above normal high-tide level (a saline marine marsh). It is the gradational zone between the land surface and the intertidal environment. Sabkhas are characterized by evaporite-saline minerals, tidal-flood, and eolian deposits, and are found on many modern coastlines, e.g. Persian Gulf, Gulf of California. (b) Any flat area, coastal or interior, where, through deflation and evaporation, saline minerals crystallize near or at the surface. (c) In the rock record, a sabkha facies may be indicated by evaporites, absence of fossils, thin flat-pebble conglomerates, stromatolitic laminae, desiccation features such as mud cracks, and diagenetic modifications, for example disrupted bedding, dissolution and replacement phenomena, and dolomitization. The sabkha environment may have been significant in the formation of certain petroleum and sulfide-mineral deposits (Kinsman, 1969; Renfro, 1974). Etymol: Arabic. Also spelled: sabkhah; sebkha.

sabkha process Shallow subsurface circulation, characteristic of brine of coastal sabkhas, suggested as a cause of transportation and deposition of metals in certain stratiform deposits (Renfro, 1974).

sabo dam *check dam.*

sabugalite (sa-bu'-ga-lite) A yellow monoclinic secondary mineral: $HAl(UO_2)_4(PO_4)_4{\cdot}16H_2O$.

sabulous (sab'-u-lous) Sandy or gritty; *arenaceous.*

sac A pouch within an animal or plant; e.g. *pollen sac* or *air sac.*

Saccardoan spore groups (Sac-car'-do-an) A classification of (dispersed) *conidia* and asexual spores produced by (extant) *Fungi imperfecti,* based on the number of cells, their disposition, and the organisation of *septa* in the spore. See also *amerospore; dictyospore; didymospore; helicospore; phragmospore; scolecospore; staurospore.*

saccate [paleont] (sac'-cate) Like or having the form of a sac or pouch; e.g. the "saccate mantle canal" of a brachiopod, without terminal branches and not extending to the anterior and lateral periphery of the mantle, or "saccate pollen" containing vesicles. See also: *vesiculate.*

saccate [palyn] Of a pollen grain, having a *saccus.*

saccharoidal (sac-cha-roi'-dal) (a) Said of a granular or crystalline texture resembling that of sugar; specif. said of the *xenomorphic-granular* texture typically developed in aplites, or said of the *crystalline granular texture* seen in some sandstones, evaporites, marbles, and dolomites. (b) Said of a white or nearly white equigranular rock having a saccharoidal texture. See also: *aplitic.* Syn: *sucrosic; sugary.* Etymol: Latin "saccharum", "sugar".

sacci Pl. of *saccus.*

saccus (sac'-cus) (a) A bladderlike expansion of the *ektexine* projecting beyond the main body of a pollen grain, usually with some form of internal structure; characteristically present in many coniferous genera. Syn: *air sac [palyn]; bladder; wing [palyn].* Less preferred syn: *vesicle [palyn].* (b) One of flattened, membrane-surrounded areas near the surface of a dinoflagellate cell (Williams et al., 2000, p.6-8, under "amphiesma"). (c) The main body of an acritarch (Tappan, 1980, p.976; Williams et al., 2000). Pl: sacci. Adj: saccate.

sackform chromite Pods or sacklike masses of chromite within the deformed peridotite component of ophiolite complexes.

sackung A ridge-parallel trough formed by surficial creep, believed in some cases to be activated by earthquake shaking. Etymol: German "versacken", "to sink" or "to give way". Pl: *sackungen* or sackungs.

sackungen (sack'-ung-en) Deep-seated rock creep which has produced a *ridge-top trench* by gradual settlement of a slablike mass into an adjacent valley. The top of the settled slab is usually parallel to the crest line of the ridge (Zischinsky, 1969).

sacrificial dune On the coast, an artificially constructed sand dune located so as to prevent property erosion by absorbing wave energy.

sacrofanite (sac'-ro-fan'-ite) A colorless hexagonal mineral of the *cancrinite* group: $(Na,Ca)_9(OH,SO_4)_4[(Si,Al)_{12}O_{24}]{\cdot}nH_2O$.

sadanagaite (sa-da-na'-ga-ite) A dark brown to black monoclinic member of the *amphibole* group: $NaCa_2(Fe^{2+}{}_3Fe^{3+}{}_2)(Si_5Al_3)O_{22}(OH)_2$.

saddle [coal] (sad'-dle) A less preferred syn. of *baum pot.*

saddle [geomorph] (a) A low point in the crest line of a ridge, commonly on a divide between the heads of streams flowing in opposite directions. (b) A broad, flat gap or pass, sloping gently on both sides, and resembling a saddle in shape; a *col.*

saddle [ore dep] *saddle reef.*

saddle [paleont] An element or inflexion of a *suture* line in a cephalopod shell that forms an angle or curve whose convexity is directed forward or toward the aperture (or away from the apex). Ant: *lobe [paleont].*

saddle [struc geol] A low point, sag, or depression along the surface axis or axial trend of an anticline.

saddleback (sad'-dle-back) A hill or ridge having a concave outline along its crest.

saddlebackite A metallic gray to black hexagonal mineral: $Pb_2Bi_2Te_2S_3$.

saddle dam A small dam constructed in a topographic saddle to increase the capacity of a reservoir compounded behind a main dam on a watercourse.

saddle dolomite Dolomite characterized by large crystal size (commonly 0.5 mm-16 mm), opaque white color (caused by abundant fluid inclusions and excessive cleavage), curving or saddle-shaped crystal faces, and undulose extinction. Common in sulfide ore deposits, veins, and evaporite-related sedimentary strata. Syn: *baroque dolomite* (Folk and Assereto, 1974).

saddle fold A type of fold that has an additional flexure near its crest, at right angles to that of the parent fold and much larger in radius.

saddle reef A mineral deposit associated with the crest of an anticlinal fold and following the bedding planes, usually found in vertical succession, esp. the gold-bearing quartz veins of Australia. Syn: *saddle [ore dep]; saddle vein.* Cf: *reverse saddle.*

saddle vein *saddle reef.*

Sado epithermal vein An *adularia-sericite type* mineral deposit hosted by felsic to intermediate volcanic rocks overlying older volcanic or intrusive rocks. Sado epithermal vein deposits, compared with Creede epithermal veins, have very low copper, lead, and zinc grades, higher gold and silver grades, and are smaller in size. Compared with Comstock epithermal veins, Sado epithermal veins are smaller in size and have lower gold and silver grades. Named for the Sado deposit in Japan. Cf: *Comstock epithermal vein, Creede epithermal vein.*

safety factor *factor of safety.*

safe yield A syn. of *economic yield* that is also applied to surface-water supplies. The volume of water that can be withdrawn annually from an aquifer (or groundwater basin or system) without (1) exceeding average annual recharge; (2) violating water rights; (3) creating uneconomic conditions for water use; or (4) creating undesirable side effects, such as subsidence or saline-water intrusion. Use of the term is discouraged because the feasible rate of withdrawal depends on the location of wells in relation to aquifer boundaries and rarely can be estimated in advance of development.

Saffir-Simpson hurricane intensity scale A five-category *hurricane* intensity scale developed by H. S. Saffir and R. H. Simpson that rates hurricanes from 1 to 5 on the basis of maximum sustained wind speed. Categories are 1 (weak), 2 (moderate), 3 (strong), 4 (very strong), 5 (devastating).

safflorite (saf'-flor-ite) A metallic tin-white orthorhombic mineral of the *löllingite* group: $(Co,Fe)As_2$. It is dimorphous with clinosafflorite and usually contains considerable iron.

sag [geomorph] (a) A saddlelike pass or gap in a ridge or mountain range. (b) A shallow depression in an otherwise flat or gently sloping land surface; a small valley between ranges of low hills or between swells and ridges in an undulating terrain.

sag [sed] (a) A depression in a coal seam. (b) A *sag structure.*

sag [struc geol] (a) A basin or downwarp of regional extent; a broad, shallow structural basin with gently sloping sides, such as the Michigan and Illinois basins. (b) A depression produced by downwarping of beds on the downthrown side of a fault such that they dip toward the fault. (c) *fault sag.*

sag-and-swell topography An undulating surface characteristic of till sheets, as in the landscape of midwestern U.S.; it may include moraines, kames, kettles, and drumlins. Cf: *swell-and-swale topography.*

sag correction A *tape correction* applied to the apparent length of a level base line to counteract the sag in the measuring tape. It is the difference between the effective length of the tape (or part of it) when supported continuously throughout its length and when supported at a limited number of independent points. Base tapes usually are used with three or five points of support and hang in catenaries between them.

sagenite (sag'-e-nite) (a) An acicular variety of rutile that occurs in reticulated twin groups of needlelike crystals crossing at 60 degrees, and is often enclosed in quartz or other minerals. See also: *Venus hair.* (b) A crystal of sagenite. Also, a similar crystal of tourmaline, goethite, actinolite, or other minerals penetrating quartz. (c) Sagenitic quartz; esp. *rutilated quartz.* Etymol: Latin sagena, "large fishing net".

sagenitic quartz (sag-e-nit'-ic) Transparent quartz, colorless to nearly colorless, containing needle-shaped crystals of rutile, tourmaline, goethite, actinolite, or other minerals. See also: *rutilated quartz.*

sagger (sag'-ger) A coarse *fireclay,* often forming the floor of a coal seam, so called because it is used for making saggers or protective boxes in which delicate ceramic pieces are placed while being baked. Etymol: corruption(?) of "safeguard". Also spelled: *seggar.*

sagittal (sag-it'-tal) Pertaining to or situated in the median anterior-posterior plane of a body having bilateral symmetry, or in any plane parallel thereto; e.g., a "sagittal plane" dividing an edrioasteroid or trilobite into two similar halves; a "sagittal axis" of the frustule of a pennate diatom; a "sagittal section" of a foraminiferal test in an equatorial plane; a "sagittal ring" in certain radiolarians; or a "sagittal triradiate" of a sponge, having two mirror-imaged rays and a coplanar third ray pointing away from them along the axis of fourfold symmetry.

sagittate (sag'-it-tate) Like an arrowhead in form; triangular, with the basal lobes pointing downward or concavely toward the stalk.

sag pond (a) A small body of water occupying an enclosed depression or sag formed where active or recent fault movement along a strike-slip fault has impounded drainage; specif. one of many ponds and small lakes along the San Andreas Fault in California. Cf: *swag.* Syn: *fault-trough lake; rift lake; rift-valley lake.*

sag structure A general term for load casts and related sedimentary structures.

sagvandite (sag-vand'-ite) A *carbonate rock* with a high content of enstatite and magnesite (Johannsen, 1939, p.278).

sahamalite-(Ce) (sa'-ha-ma-lite) A colorless monoclinic mineral: $(Mg,Fe)Ce_2(CO_3)_4$.

sahlinite (sah'-lin-ite) A dense sulfur-yellow monoclinic platy mineral: $Pb_{14}O_9(AsO_4)_2Cl_4$. It is the arsenate analogue of kombatite.

sahlite (sahl'-ite) *diopside.*

sai A term used in central Asia for a gravelly talus, a river bed filled with stones, and a dry wash (Stone, 1967, p.258); also for a piedmont plain covered with pebbles showing desert varnish (Termier and Termier, 1963, p.413).

saif *seif.*

sailaufite A dark red-brown to black monoclinic mineral: $(Ca,Na,[])_2Mn^{3+}{}_3O_2(AsO_4)_2CO_3{\cdot}3H_2O$.

sail line The line traversed by a ship acquiring geophysical data; the line of acquisition of marine data, esp. of a ship towing several seismic streamers.

saimaite A black monoclinic mineral of the *chevkinite* group: $(Sr,REE)_4Fe(Ti,Zr)_2Ti_2Si_4O_{22}$. It is probably *strontiochevkinite,* and has not been approved by the International Mineralogical Association.

sainfeldite (sain'-feld-ite) A colorless or light pink monoclinic mineral of the *huréaulite* group: $Ca_5(AsO_4)_2(AsO_3OH)_2{\cdot}4H_2O$.

Saint Venant substance (Saint Ven'-ant) A material that demonstrates *elasticoplastic* behavior: it behaves elastically below a *yield stress,* but deforms continuously under a constant stress equal to yield stress.

Sakaraulian (Sak-a-rul'-ian) Ponto-Caspian stage: lowermost Miocene (above Caucasian, below Takaraulian).

sakhaite (sak'-ha-ite) A vitreous gray, grayish-white, or colorless cubic mineral: $Ca_3Mg(BO_3)_2(CO_3){\cdot}nH_2O$.

sakharovaite (sak-ha-ro'-va-ite) An ill-defined metallic gray mineral: $(Pb,Fe)(Bi,Sb)_2S_4$.

Sakmarian (Sak-mar'-i-an) A stage of the Standard Global Chronostratigraphic Scale: middle Cisuralian Series, Lower Permian (above Asselian, below Artinskian).

sakuraiite (sa'-ku-rai-ite) A metallic gray cubic mineral of the *sphalerite* group: (Cu,Zn,Fe,In,Sn)S.

sal *sial.*

salada (sa-la'-da) A term used in SW U.S. for a salt-covered plain where a lake has evaporated. Etymol: Spanish, feminine of "salado", "salted, salty". See also: *playa [geomorph].*

sal ammoniac (sal am-mon'-i-ac) A cubic mineral: NH_4Cl. It is a white, gray, or yellow crystalline volatile salt that occurs esp. as an encrustation around volcanoes. Syn: *salmiac.* Also spelled: salammoniac.

salaquifer (sal-a'-qui-fer) *saline aquifer.*

salar (sa-lar') A term used in SW U.S. and in South America for a salt flat or for a salt-encrusted depression that may represent the basin of a salt lake. Etymol: Spanish, "to salt". Pl: salares; salars. See also: *playa [geomorph]; saline pan.*

salcrete (sal'-crete) A term suggested by Yasso (1966) for a thin, hard crust of salt-cemented sand grains as may occur on a marine beach that is periodically saturated by saline water.

saléeite (sa-lée'-ite) A lemon-yellow monoclinic mineral of the *autunite* group: $Mg(UO_2)_2(PO_4)_2{\cdot}10H_2O$.

salesite (sales'-ite) An orthorhombic blue-green mineral: $Cu(IO_3)(OH)$.

salic (sal'-ic) Said of certain light-colored silicon- or aluminum-rich minerals present in the norm of igneous rocks; e.g. quartz, feldspars, feldspathoids. Also, applied to rocks having one or more of these minerals as major components of the norm. Etymol: A mnemonic term derived from *s*ilicon + *al*uminum + *ic.* Cf: *femic; mafic; felsic.*

salic horizon A mineral soil horizon of enrichment with secondary salts more soluble in cold water than gypsum. A salic horizon is 15 cm or more in thickness, contains at least 20 g kg^{-1} salt, and the product of the thickness in centimeters and amount of salt by weight is > 600 g kg^{-1}.

Salids Aridisols which have a salic horizon that has its upper boundary within 100 cm of the soil surface. A suborder in the U.S. system of soil taxonomy.

salient [geomorph] adj. Projecting or jutting upward or outward; e.g. a salient point, or one formed by a conspicuous outward projection from the coast. n. A landform that projects or extends outward or upward from its surroundings; e.g. a cape along a shoreline, or a spur from the side of a mountain. Ant: *reentrant.*

salient [struc geol] (a) An area in which the axial traces of folds are convex toward the outer edge of the folded belt. (b) A broad, map-view curve in a mountain that is convex to the foreland. Ant: *recess [struc geol].*

Salientia (Sa-li-en'-ti-a) A superorder of lissamphibians that includes frogs and toads. Range, Lower Triassic to Holocene.

salina (sa-li'-na) (a) A place where crystalline salt deposits are formed or found, such as a salt flat or pan, a salada, or a salt lick; esp. a salt-encrusted playa or a *wet playa.* (b) A body of saline water, such as a salt pond, lake, well, or spring, or a playa lake having a high concentration of salts. (c) *saltworks.* (d) *salt marsh.* Etymol: Spanish, "salt pit, salt mine, saltworks". Anglicized equivalent: *saline.*

Salinan (Sa-li'-nan) North American provincial stage: Upper Silurian (above Lockportian, below Tonolowayan).

saline (sa'-line) n. (a) A natural deposit of halite or of any other soluble salt; e.g. an evaporite. Cf: *salines.* (b) An anglicized form of *salina.* In this usage, a "saline" may refer to various features such as a playa, a salt flat, a salt pan, a salt marsh, a salt lake, a salt pond, a salt well, or a saltworks. (c) *salt spring.* (d) A term used along the coast of Louisiana for a body of water behind a barrier island. adj. (a) Salty; containing dissolved sodium chloride, e.g., seawater. (b) Having a salinity appreciably greater than that of seawater, e.g. a brine. (c) Containing dissolved salts at concentrations great enough to allow the precipitation of sodium chloride; *hypersaline.* (d) Said of a taste resembling that of common salt, esp. in describing the properties of a mineral. (e) Used to describe a lake with an appreciable total of dissolved solids: 3,000 ppm (Geddes et al., 1981), 5,000 ppm (Beadle, 1974), or a concentration great enough to affect lake ecology.

saline-alkali soil (sa'-line-al'-ka-li) (a) A soil containing sufficient exchangeable sodium to interfere with the growth of most crop plants and containing appreciable quantities of soluble salts. The exchangeable-sodium percentage is >15, the conductivity of the saturation extract >4 dSm^{-1} (at 25°C), and the pH is usually 8.5 or less in the saturated soil. (b) A saline-alkali soil has a combination of harmful qualities of salts and either a high alkalinity or high content of exchangeable sodium, or both,

so distributed in the profile that the growth of most crop plants is reduced. See also: *saline-sodic soil.*
saline aquifer An aquifer containing salty water. Syn: *salaquifer.*
saline deposit *evaporite.*
saline giant A large evaporite deposit, the size of an ocean basin (tens of thousands of square kilometers or greater) for which there are no modern analogs (Hsu, 1972).
salinelle (sa-li-nelle') A *mud volcano* erupting saline mud.
saline minerals General term used to describe naturally occuring highly soluble minerals or salts. Syn: *evaporite mineral.*
saline mudflat A subaerial muddy plain in which saline minerals are precipitated within the sediment (Hardie et al., 1978). It is covered by an efflorescent salt crust. May occur in both marine and lacustrine settings. Syn: *sabkha.*
saline pan An ephemeral saline lake which precipitates a bottom crust that is subsequently modified during subaerial exposure (Lowenstein and Hardie, 1985).
salines (sa'-lines) (a) A general term for the naturally occurring soluble salts, such as common salt, sodium carbonate, sodium nitrate, potassium salts, and borax. (b) A general term for salt mines, salt springs, salt beds, salt rock, and salt lands.
saline seep Intermittent or continuous saline water discharge at or near the soil surface under dryland conditions which reduces or eliminates crop growth. It is differentiated from other saline soil conditions by recent and local origin, shallow water table, saturated root zone, and sensitivity to cropping systems and precipitation.
saline-sodic soil A soil containing sufficient exchangeable sodium to interfere with the growth of most crop plants and containing appreciable quantities of soluble salts. The exchangeable sodium ratio is greater than 0.15, conductivity of the soil solution, at saturated water content, of > 4dS m^{-1} (at 25°C), and the pH is usually 8.5 or less in the saturated soil. See also: *saline-alkali soil.*
saline soil (a) Soil in which saline minerals accumulate, commonly gypsum and halite. These minerals may form horizons of accumulation as isolated crystals, void fillings, and veins. (b) A nonsodic soil containing sufficient soluble salt to adversely affect the growth of most crop plants. The lower limit of saturation extract electrical conductivity of such soils is conventionally set at 4 dS m^{-1} (at 25°C). Actually, sensitive plants are affected at half this salinity and highly tolerant ones at about twice this salinity.
saline spring *salt spring.*
saline water Water that contains more than 1,000 milligrams per liter of dissolved solids (Solley et al., 1983). Cf: *fresh water; salt water.*
saliniferous (sal-i-nif'-er-ous) Said of a stratum that yields salt.
salinity (sa-lin'-i-ty) (a) An operational parameter used to make very accurate estimates on the total concentration of dissolved solids in seawater. The most accurate method for determining salinity is to measure the conductivity of seawater and, from its ratio to the conductivity of a standard seawater or to a standard KCl solution, calculate the salinity using an accepted equation or table, such as those published by UNESCO. In the strictest sense, a salinity determined this way is dimensionless although it is frequently given the unit "psu" (practical salinity units) in order to compare it with the older salinity unit "ppt" (parts per thousand). For most purposes, psu = ppt. Conductometric salinity measurements are usually well within 1% of the actual total concentration of dissolved solids. Salinity can also be determined by silver-nitrate titration of chlorinity which is then multiplied by 1.80655, the salinity-chlorinity ratio or seawater, or by refractometer or hydrometer. (b) The concentration of dissolved salts in seawater, measured by evaporation to dryness in parts per thousand by weight, with the following operational qualifications being applied in order to achieve reproducibility: all carbonates converted to oxides, all bromide and iodide replaced by chloride on a mole-for-mole basis, a known amount of NaF being added to precipitate magnesium ions, and all organic matter completely oxidized. This is the original definition of salinity and is also within 0.5% of the actual concentration of total dissolved solids in seawater. Cf: *salinometer; refractometer; salinity-chlorinity ratio; hydrometer; standard seawater.*
salinity-chlorinity ratio The slope of a linear regression of *salinity* against *chlorinity* for the oceans, with the regression being constrained to pass through the origin. Currently accepted value: 1.80655.
salinity log *chlorine log.*
salinity meter *salinometer.*
salinization (sal'-i-ni-za'-tion) In a soil of an arid, poorly drained region, the accumulation of soluble salts by evaporation of the waters that bore them to the soil zone.
salinometer (sal-i-nom'-e-ter) A device which measures the salinity of seawater from its electrochemical properties; usually measures conductance or conductivity. Cf: *salinity.* Syn: *salinity meter.*
saliotite A pearly white monoclinic mineral consisting of interlayered cookeite and paragonite: $(Li,Na)Al_3(Si_3Al)O_{10}(OH)_5$.
salite (sal'-ite) *diopside.*
salitral (sal-i-tral') A term used in Patagonia for a swampy place where salts (esp. potassium nitrate) become encrusted in the dry season. Etymol: Spanish, "saltpeter bed".
salitrite (sal'-i-trite) A *pyroxenite* composed chiefly of titanite and clinopyroxene, with accessory apatite, microcline, and occasionally anorthoclase and baddeleyite; a titanite-rich *jacupirangite.* Named by Tröger in 1928 for the Salitre Mountains, Minas Gerais, Brazil. Obsolete.
salmiac (sal'-mi-ac) *sal ammoniac.*
salmoite (sal'-mo-ite) *tarbuttite.*
salmonsite (salm'-ons-ite) A mixture of huréaulite and jahnsite, formerly thought to be a distinct mineral species.
Salopian (Sa-lo'-pi-an) European stage: Upper Silurian (above Llandoverian, below Devonian). Includes Wenlockian and Ludlovian.
salpausselkä A Finnish term for a steep recessional *moraine,* usually interpreted as a series of *end moraines,* like the one trending east-west across Finland.
salpingiform (sal-pin'-gi-form) Shaped like a trumpet; e.g. said of a cyrtolith coccolith with a trumpet-shaped central structure (as in *Discosphaera tubifer*).
salsima (sal-si'-ma) According to Van Bemmelen (1949), the theoretical layer of the Earth's crust beneath the sial and above the Mohorovicic discontinuity that is considered to be of basaltic composition. Also spelled: *sialsima.* Cf: *sifema.*
salsuginous (sal-su'-gi-nous) Said of a plant growing in soil or water with a high content of salts; i.e. a *halophyte.*
salt [sed] n. A general term for naturally occurring sodium chloride, NaCl. Syn: *halite; common salt; rock salt.* adj. Containing salt, as *salt water,* or containing salt water, as *salt marsh.*
salt-affected soil (salt-af-fect'-ed) A general term for a soil that is not suitable for the growth of crops because of an excess of salts, of exchangeable sodium, or both; e.g. saline sodic soil, saline soil, sodic soil.
salt-and-pepper Said of a sand or sandstone consisting of a mixture of light- and dark-colored particles, such as a strongly cherty graywacke (Krynine, 1948, p.152) or a lighter-colored and speckled subgraywacke (Pettijohn, 1957, p.319); e.g. the Bow Island Sandstone of Cretaceous age in Alberta.
salt anticline A diapiric or piercement structure, like a *salt dome* except that the salt core is linear rather than equidimensional, e.g. the salt anticlines in the Paradox basin of the central Colorado Plateau.
saltation [evol] (sal-ta'-tion) Sudden evolution of a new type of organism derived, in a single generation, from older ones without transitional intermediate forms. This process appears to be almost impossible genetically. See also: *saltatory evolution.*
saltation [sed] A mode of sediment transport in which the particles are moved progressively forward in a series of short intermittent leaps, jumps, hops, or bounces from a bottom surface; e.g. sand particles skipping downwind by impact and rebound along a desert surface, or bounding downstream under the influence of eddy currents that are not turbulent enough to retain the particles in suspension and thereby return them to the stream bed at some distance downstream. It is intermediate in character between suspension and the rolling or sliding of traction. Etymol: Latin saltare, "to jump, leap".
saltation load The part of the *bed load* that is bouncing along the stream bed or is moved, directly or indirectly, by the impact of bouncing particles.
saltation mark *skip mark.*
saltatory evolution (sal'-ta-to'-ry) The idea of evolution by saltation.
salt burst Rock destruction caused by soluble salts that enter pores and crystallize from nearly saturated solutions. In deserts, salt bursts may be due to crystallization pressure, to the volumetric expansion of salts in capillaries, and to hydration pressures of the entrapped salts (Winkler and Wilhelm, 1970).
salt cake Commercial term for sodium sulfate, Na_2SO_4.
salt canopy Composite diapiric structure formed by partial or complete coalescence of diapir *bulbs* or *salt sheets.* These bodies coalesce along salt sutures and may be connected to their source layer by feeder stems. If necessary, canopies can be further differentiated by their components (Jackson and Talbot, 1991).
salt corrie A cirquelike hollow, resembling a crater or caldera, produced by the solution of salt.
salt crust A salt deposit formed on an ice surface by ice-crystal growth which forces salt out of young sea ice and pushes it upward.
salt-crystal cast (salt'-crys-tal) A *crystal cast* formed by solution of a soluble salt crystal, followed by filling with mud or sand or by crystallization of a pseudomorph (such as calcite after halite). See also: *hopper crystal.*
salt-crystal growth The growth of salt crystals in openings in rock or soil, capable of exerting powerful stresses and producing granular disintegration in a dry climate. See also: *salt weathering.*
salt desert A desert with a *saliniferous* soil; e.g. a *kavir.*
salt diapir Mass of salt that has flowed ductilely and appears to have discordantly pierced or intruded the overburden. In its broadest sense, diapir includes lateral or vertical intrusion of any shape; upwelling of buoyant or nonbuoyant rock, migma, or magma; or emplacement by passive piercement or faulting of prekinematic overburden (Jackson and Talbot, 1991).
salt dome An imprecise, general term for a domal upwelling comprising a salt core and its envelope of deformed overburden. The salt may or may not be discordant (Jackson and Talbot, 1991).

salt-dome breccia A breccia found in deep shale sequences, occurring as a dome-shaped mass in a broad zone surrounding a salt plug. It is believed to be a result of differential pressure caused by diapiric intrusion of salt into shale (Kerr and Kopp, 1958).

saltern (sal′-tern) (a) A *saltworks* where salt is produced by boiling or evaporation of salt or brine. (b) A broad area of shallow water in which saline minerals precipitate (Warren, 1989); marine saline.

salt expulsion *salt withdrawal.*

salt fingering An instability of the interface of a warm salty layer above a cooler fresher layer above a cooler fresher layer, due to more rapid diffusion of heat than salt across the interface. The instability leads to small, finger-sized parcels of water dropping from the upper into the lower layer. Cf: *diffusive convection.*

salt flat (a) A level subaerial surface encrusted with salt. See also: *playa [geomorph]; alkali flat; saline mudflat; saline pan.* (b) A gently inclined segment of the stair-step basal contact of a *salt tongue.* Flats form where the ratio of aggradation to salt spreading is low (Jackson and Talbot, 1991).

salt flower An *ice flower* forming on surface sea ice around a salt-crystal nucleus.

salt garden A large, shallow basin or pond where seawater is evaporated by solar heat. Archaic. Syn: *saltern.*

salt glacier (a) A gravitational flow of salt down the slopes of an exposed salt plug or anticline, following the preexisting structure, commonly in a desert region. It can be compared with the coulee of a lava flow. (b) Sheetlike extrusion of salt flowing from an exposed diapir and spreading beneath air or water (Jackson and Talbot, 1991). Syn: *namakier.*

salt hill An abrupt hill of salt, with sinkholes and pinnacles at its summit (Thornbury, 1954, p.521).

saltierra (sal-ti-er′-ra) A deposit of salt left by evaporation of a shallow, inland lake. Archaic. Etymol: Spanish, "salt earth".

salting (salt′-ing) (a) A British term for the slightly higher part of a *salt marsh,* flooded only by spring tides, containing little bare mud, and supporting grassy vegetation. Syn: *high marsh.* (b) A term used in parts of Great Britain for land regularly covered by the tide, as distinguished from a salt marsh. The term is usually used in the plural.

salt lake An inland body of water situated in an arid or semiarid region, having no outlet to the sea, and containing a high concentration of dissolved salts (principally sodium chloride). Examples include the Great Salt Lake in Utah, and the Dead Sea in the Near East. See also: *alkali lake; bitter lake.* Syn: *brine lake.*

salt lick A place to which animals (e.g. deer, cattle, bison) go to lick up salt lying on the surface of the ground, as in an area surrounding a salt spring. The term has been used for the spring itself, but this usage is improper because a lick is dry. Syn: *lick.*

salt marsh Flat, poorly drained land that is subject to periodic or occasional overflow by salt water, containing water that is brackish to strongly saline, and usually covered with a thick mat of grassy halophytic plants; e.g. a coastal marsh periodically flooded by the sea, or an inland marsh (or *salina*) in an arid region and subject to intermittent overflow by water containing a high salt content. Cf: *tidal marsh; marine marsh.* Cf: *salting; open-coast marsh; tidal-delta marsh.*

salt meadow A grassy *tract* subject to overflow by salt water.

salt pan (a) A small, undrained, shallow natural depression in which water accumulates and evaporates, leaving a salt deposit. Also, a shallow lake of brackish water occupying a salt pan. See also: *playa [geomorph]; pan [geomorph]; marsh pan; saline pan.* (b) A large pan for recovering salt by evaporation. Also spelled: saltpan.

saltpeter (salt-pe′-ter) (a) Naturally occurring potassium nitrate; *niter.* Cf: *Chile saltpeter; Peru saltpeter; wall saltpeter.* (b) A speleologic term for earthy cave deposits of nitrate minerals. Also spelled: saltpetre.

saltpeter earth Cave sediment containing nitrocalcite or other nitrate minerals, either in a deliquesced or crystallized form; it can be mined.

salt pillow A subcircular upwelling of salt having concordant overburden (Jackson and Talbot, 1991).

salt pit (a) A pit in which saline lake water with sea water is received and evaporated and from which salt is obtained. (b) A body of salt water occupying such a pit. Also spelled: saltpit.

salt plug The salt core of a *salt dome.* It is nearly equidimensional, about one to two kilometers in diameter, and has risen through the enclosing sediments from a mother salt bed 5 to 10 kilometers below.

salt polygon A structure of salt on a *playa.* It has three to eight sides marked by ridges of material formed as a result of the expansive forces of crystallizing salt and ranges in width from several centimeters to 30 m (Stone, 1967, p.244; Tucker, 1981; Lugli and Schreiber, 1999).

salt pond (a) A body of salt water in a marsh or swamp along the seacoast. (b) An artificial pond used for evaporation in the production of salt from seawater.

salt proximity survey [geophys exploration] A survey to determine the location of the salt-sediment interface at a salt-dome flank. May involve recording offset and azimuthal VSP surveys in a borehole into the salt or near the salt flank from various surface locations, or a borehole gravimeter, long-spaced electrical log measurements, or other types of measurements.

salt ramp The steeply inclined segment of the stair-step basal contact of a *salt tongue.* Salt ramps cut up stratigraphic section in the direction of emplacement; they dip in a direction opposite to the spreading direction of a tongue during the stratigraphic time represented by strata truncated in the basal cutoff adjoining the salt ramp. Ramps form where the ratio of aggradation to salt spreading is high (Jackson and Talbot, 1991).

salt reduction Mass transfer of salt over time, resulting in obvious change in salt area in cross section by loss due to dissolution or extrusion, isochoric (constant volume) flow out of the plane of section, or isochoric flow within the plane of section but beyond the ends of the cross section. Reduction culminates in the formation of *salt welds* or *fault welds* (Jackson and Talbot, 1991). Cf: *salt withdrawal.*

salt sheet Allochthonous salt with breadth several times greater than its maximum thickness (O'Brien and Lerche, 1988).

salt spring A *mineral spring* whose water contains a large quantity of common salt; a spring of salt water. See also: *salt lick.* Syn: *saline; brine spring; saline spring.*

salt stock A pluglike salt diapir having subcircular planform (Jackson and Talbot, 1991).

salt suture Junction between individual salt structures that have coalesced laterally to form a canopy. Incomplete sutures have lensoid basins of overburden along them. Complete sutures are either overlain by a subsided, cuspate synform or the suture is the junction of a pair of appressed, raised salt antiforms.

salt table The flat upper surface of a salt stock, along which groundwater solution leads to the formation of cap rock by freeing anhydrite (Goldman, 1952).

salt tectonics *halokinesis.*

salt tongue A highly asymmetric variety of salt sheet fed by a single stem. The term is typically applied to wedgelike bodies having large taper angles (Jackson and Talbot, 1991).

salt wall An elongated upwelling of diapiric (discordant) salt, commonly forming sinuous, parallel rows (Jackson and Talbot, 1991).

salt-water encroachment (salt′-wa-ter) Displacement of fresh surface or groundwater by the advance of salt water due to its greater density, usually in coastal and estuarine areas but also by movement of brine from beneath a playa lake toward wells discharging fresh water. Encroachment occurs when the total head of the salt water exceeds that of adjacent fresh water. Syn: *sea-water intrusion; intrusion [grd wat]; salt-water intrusion; sea-water encroachment.* See also: *Ghyben-Herzberg ratio.*

salt-water front The interface between fresh and salty water in a coastal aquifer or in an estuary. Under certain conditions, a similar front may by found inland.

salt-water intrusion *salt-water encroachment.*

salt-water underrun A type of density current occurring in a tidal estuary, due to the greater salinity of the bottom water (ASCE, 1962).

salt-water wedge An intrusion of salty ocean water in the form of a wedge into an estuary or tidal river dominated by freshwater circulation. A salt-water wedge underlies the fresh water, slopes slightly downward in the upstream direction, and is characterized by a pronounced increase in salinity with depth.

salt weathering The granular disintegration or fragmentation of rock material effected by saline solutions or by *salt-crystal growth* (Wellman and Wilson, 1965). See also: *exsudation.*

salt wedge A wedge-shaped mass of salt water from an ocean or sea which intrudes the mouth and lower course of a river. The denser salt water underlies the fresher river water. Extent is regulated by river discharge and tides.

salt weld Surface or zone joining strata originally separated by salt. A weld results from the complete, or near-complete removal of intervening salt. The weld can consist of brecciated, insoluble residue with evaporite pseudomorphs or of salt too thin to be resolved in reflection seismic data. The weld is usually but not invariably marked by a structural discordance. Another distinctive feature of welds is a structural inversion above them (Jackson and Cramez, 1989).

salt well A drilled or dug well from which brine is obtained. See also: *brine pit.*

salt withdrawal Mass transfer of salt over time without obvious change in salt area in cross section. Examples are the migration of salt from the flanks of a *salt pillow* into its core as it evolves into a diapir or the flow of salt along a *salt wall* into local culminations that evolve into salt stocks (Jackson and Talbot, 1991). Syn: *salt expulsion.* Cf: *salt reduction.*

saltworks (salt′-works) Any installation where salt is produced commercially, as by extraction from seawater, saline lake water, from wells, or from the brine of salt springs. Syn: *salina; saltern.*

salty (salt′-y) Pertaining to, containing, or resembling salt; *saline.*

salvage (sal′-vage) In archaeology, the type of work on an archaeological excavation that is performed in an area soon to be built up or ruined, e.g. by a dam, with the aim to understand and protect cultural information that would otherwise be lost.

salzburgite A metallic orthorhombic mineral: $Cu_{1.6}Pb_{1.6}Bi_{6.4}S_{12}$. It is a member of the bismuthinite-aikinite series.

samara (sam'-a-ra) A dry, indehiscent, usually one-seeded winged fruit, such as that of elm or maple.

samarium-neodymium age method (sa-mar'-i-um-ne-o-dym'-i-um) A method of age determination based on the alpha decay of samarium-147 to neodymium-143 (λ=6.54 × $10^{-12}yr^{-1}$). The ratios $^{147}Sm/^{144}Nd$ and $^{143}Nd/^{144}Nd$ are measured and plotted on an *isochron* diagram.

samarskite (sa'-mar-skite) A group name for orthorhombic or monoclinic minerals with the general formula AB_2O_6, where A = Ca, Pb, Th, REE, Fe^{3+}, Fe^{2+}, Ti or U; and B = Nb, Ta, Ti or Zr.

samarskite-(Y) A velvet-black, brown, or yellowish-brown orthorhombic, but commonly metamict, mineral of the *samarskite* group: $(Y,Ce,U,Ca,Fe,Pb,Th)(Nb,Ta,Ti,Sn)_2O_6$. It has a splendent vitreous or resinous luster, and is found in granite pegmatites. Syn: *ampangabeite; uranotantalite.*

samfowlerite A vitreous colorless or white monoclinic mineral: $Ca_{14}Mn^{2+}{}_3Zn_2(Be,Zn)_2Be_6Si_{14}O_{52}(OH,F)_6$.

samiresite (sam-i-res'-ite) A lead-bearing *uranpyrochlore.*

sammelkristallization (sam'-mel-kris'-tal-li-za'-tion) The action, depending on surface tension or total free surface energy, by which smaller grains become unstable in relation to larger grains and will eventually be devoured by the larger grains (Barth, 1962, p.399). Etymol: German "Sammelkristallisation", "collecting, cumulative crystallization". There is no English equivalent for this term.

sample (sam'-ple) A subset of a population that, if properly selected, may be used to estimate the parameters of the population.

sampleite (sam'-ple-ite) A light blue to blue-green orthorhombic mineral of the *lavendulan* group: $NaCaCu_5(PO_4)_4Cl{\cdot}5H_2O$.

sample log A *log* depicting the sequence of lithologic characteristics of the rocks penetrated in drilling a well, compiled by a geologist from microscopic examination of *well cuttings* and cores. The information is referred to depth of origin and is plotted on a *strip log* form. See also: *interpretive log; percentage log.* Syn: *lithologic log; graphic log.*

sample splitter A device for separating dry incoherent material (such as sediment) into representative samples of workable size for laboratory study. Syn: *riffler.*

sampling (sam'-pling) In economic geology, the gathering of specimens of ore or wall rock for appraisal of the orebody. As the average of many samples may be used, representative sampling is crucial. The term is usually modified to indicate the mode or locality, e.g. hand sampling, mine sampling.

sampling error The error introduced into the definition of a variable due to the set of actual measurements not representing the population of the variable being determined. E.g., the determination of mean global sea-surface temperature requires measurements from all oceanic regions for many years, yet we have measurements from only a few areas and for limited periods.

samsonite (sam'-son-ite) A metallic steel-black monoclinic mineral: $Ag_4MnSb_2S_6$.

samuelsonite (sam'-u-el-son-ite') A sub-adamantine colorless or pale pink monoclinic mineral: $(Ca,Ba)Ca_8(Fe^{2+},Mn^{2+})_4Al_2(PO_4)_{10}(OH)_2$.

sanbornite (san'-born-ite) A peraly colorless or white platy orthorhombic mineral: $BaSi_2O_5$.

sancyite (san'-cy-ite) A light-colored *trachyte* containing tridymite (Sørensen, 1974, p.572). The name, given by Lacroix in 1923, is for Puy Sancy, Auvergne region, France. Obsolete.

sand [drill] A driller's term applied loosely to sandstone or any visibly granular sediment. Cf: *oil sand.*

sand [eng] Rock or mineral grains with diameters between 0.074 mm (retained on U.S. standard sieve no. 200) and 4.76 mm (passing through U.S. standard sieve no. 4). Cf: *coarse sand; fine sand; medium sand.*

sand [sed] (a) A detrital rock fragment or mineral particle smaller than a granule and larger than a coarse silt grain, having a diameter in the range of 1/16 to 2 mm (62-2,000 micrometers, or 0.0025-0.08 in., or 4 to -1 phi units, or a size between that at the lower limit of visibility of an individual particle with the unaided eye and that of the head of a small wooden match). In Great Britain, the range of 0.1-1 mm has been used. Cf: *very coarse sand; coarse sand; medium sand; fine sand; very fine sand.* (b) A loose aggregate of unlithified mineral or rock particles of sand size; an unconsolidated or moderately consolidated sedimentary deposit consisting essentially of medium-grained clastics. The material is most commonly composed of quartz resulting from rock disintegration, and when the term "sand" is used without qualification, a siliceous composition is implied; but the particles may be of any mineral composition or mixture of rock or mineral fragments, such as "coral sand" consisting of limestone fragments. Also, a mass of such material, esp. on a beach or a desert or in a stream bed. (c) *sandstone.*

sand [soil] (a) A term used in the U.S. for a rock or mineral particle in the soil, having a diameter in the range of 0.05-2 mm; prior to 1947, the range 1-2 mm was called "fine gravel". The diameter range recognized by the International Society of Soil Science is 0.02-2 mm. (b) A textural class of soil material containing 85% or more of sand, with the percentage of silt plus 1.5 times the percentage of clay not exceeding 15; specif. such material containing 25% or more of very coarse sand, coarse sand, and medium sand, and less than 50% of fine sand or very fine sand (SSSA, 1965, p.347). The term has also been used for a soil containing 90% or more of sand. See also: *very coarse sand; coarse sand; medium sand; fine sand; very fine sand.*

sand apron A deposit of sand, mostly carbonate, along the shore of the lagoon of a reef.

sandar (sand'-ar) Plural of *sandur.*

sandarac (san'-da-rac) A syn. of *realgar.* Also spelled: sandarach.

sand avalanche Movement of large masses of sand down a dune face when the angle of repose is exceeded, or when the dune is disturbed (Stone, 1967, p.245). Cf: *sand run.*

sandbag (sand'-bag) In the roof of a coal seam, a deposit of glacial debris formed by scour and fill subsequent to coal formation.

sandbank (sand'-bank) (a) A submerged ridge of sand in the sea, a lake, or a river, usually exposed during low tide; a *sandbar.* Syn: *bank [coast].* (b) A large deposit of sand, esp. in a shallow area near the shore.

sandbar (sand'-bar) A bar or low ridge of sand that borders the shore and is built to, or nearly to, the water surface by currents in a river or by wave action along the shore of a lake or sea. Syn: *sand reef.*

sandblast (sand'-blast) (a) A stream of windblown sand driven against an exposed rock surface. (b) A gust of wind, carrying sand.

sandblasting (sand'-blast-ing) A type of *blasting* in which the particles are hard mineral grains (usually quartz) of sand sizes.

sandblow (sand'-blow) A patch of coarse sandy soil denuded of vegetation by wind action.

sand blow A mixture of sand and water that has been vented through fractures to the surface. Sand-blow deposits are considered diagnostic evidence of extreme deep liquefaction and prehistoric seismicity (Li et al., 1996).

sand boil A spring that bubbles through a river levee, with an ejection of sand and water, as a result of water in the river being forced through permeable sands and silts below the levee during flood stage. Syn: *blowout [grd wat].*

sandbox model A less preferred term used to refer to a *scaled physical model.*

sand-calcite (sand-cal'-cite) A calcite crystal containing a large percentage of sand-grain inclusions; a *sand crystal* of calcite. See also: *Fontainebleau sandstone.*

sand cay A British syn. of *sand key.*

sand cone [geomorph] A conical deposit of sand, produced esp. in an alluvial cone.

sand cone [glaciol] A low *debris cone* whose protective veneer consists of sand.

sand crystal A large euhedral or subhedral crystal (as of barite, gypsum, and esp. calcite) loaded with detrital-sand inclusions (up to 60%), developed by growth in sand during cementation. Crystals or crystal clusters of gypsum and other saline minerals, e.g. glauberite may grow poikilitically encasing sandy matrix. See also: *sand-calcite; crystal sandstone.*

sand dike A sedimentary dike consisting of sand that has been squeezed or injected upward into a fissure. See also: *injection dike; neptunian dike.*

sand dome *dome [beach].*

sand drift (a) A general term for surface movement of wind-blown sand, occurring in deserts or along the shore. (b) An accumulation of sand formed in the lee of some fixed obstruction, such as a rock or a bush; usually smaller than a dune. See also: *sand shadow.*

sand dune An accumulation of loose sand heaped up by the wind, commonly found along low-lying seashores above high-tide level, more rarely on the border of large lakes or river valleys, as well as in various desert regions, where abundant dry surface sand is present during some part of the year. See also: *dune [geomorph]; sand hill.*

sanderite (san'-der-ite) An incompletely described mineral: $MgSO_4{\cdot}2H_2O$.

sandfall *slip face (a).*

sand fall An accumulation of sand swept over a cliff or escarpment (Stone, 1967, p.245). It may occur in a submarine canyon as well as on land.

sand field Extensive sand cover in which many dunelike irregularities characteristically are present on the surface. Sand fields lie behind some extensive sand beaches and also occur inland. In deserts they are called *ergs* or, by some, *sand seas.*

sand flag Fine-grained sandstone that can be readily split into flagstones.

sand flat A sandy *tidal flat* barren of vegetation. Cf: *mud flat.*

sand flood A vast body of sand moving over or borne along on a desert floor, as in the Arabian deserts.

sand flow [marine geol] In a submarine canyon, a discontinuous movement of sand down the axis, in a series of flows or slumps. See also: *sand fall.*

sand flow [mass move] A flow of wet sand, as along banks of noncohesive clean sand that is subject to scour and to repeated fluctuations in pore-water pressure due to rise and fall of the tide (Varnes, 1958, p.41).

sand flow [pyroclast] (a) An obsolete syn. of *ash flow.* (b) A term applied by Fenner in 1923 to an unsorted rhyolitic tuff in the vicinity of Mt. Katmai, Alaska.

sand gall *sand pipe.*

sand glacier (a) An accumulation of sand that is blown up the side of a hill or mountain and through a pass or saddle, and then spread out on the opposite side to form a wide fan-shaped plain. (b) A horizontal plateau of sand terminated by a steep slope.

sand hill A ridge of sand; esp. a *sand dune* in a desert region. See also: *chop hill.*

sandhills (sand'-hills) A region of sand hills, as in north-central Nebraska or the Midlands of the Carolinas.
sand hole A small pit, 7-8 mm in depth and a little less wide than deep, with a raised margin, formed on a beach by waves expelling air from a formerly saturated mass of sand; it resembles an impression made by a raindrop.
sand horn A pointed sand deposit extending from the shore into shallow water. Cf: *sand lobe.*
sandia (san-di'-a) An oblong, oval, or rounded mountain mass resembling a watermelon; e.g. the Sandia Mountains in New Mexico. Etymol: Spanish "sandía", "watermelon".
sanding (sand'-ing) The accumulation or building-up of sand, as by the action of currents in filling a harbor.
sanding up Filling-in or choking with sand, as in a well that produces sand mixed with oil and gas.
sand key A small sandy island close to the shore. Syn: *sand cay.*
sand levee A *whaleback* in the desert.
sand line [drill] A *wireline* used in cable-tool drilling to raise and lower such tools as a bailer, or in rotary drilling to operate a *swab.*
sand line [glac geol] An "easily overlooked" mark made by glacier ice, about 5-10 cm long, fine as a hair and similar to one of the marks made by the "finest sandpaper" (Campbell, 1865, p.4). Rarely used.
sand lobe A rounded sand deposit extending from the shore into shallow water. Cf: *sand horn.*
sand pavement A sandy surface derived from coarse-grained sand ripples, developed on the lower, windward slope of a dune or rolling sand area during a period of intermittent light, variable winds (Holm, 1957).
sand pipe A pipe formed in sedimentary rocks, filled with sand. Cf: *gravel pipe.* Syn: *sand gall.*
sand plain [geomorph] A sand-covered plain. The large sand plains in Western Australia have an uncertain origin: they may originate by deflation of sand dunes, the lower limit of erosion being governed by the groundwater level.
sand plain [glac geol] (a) A small *outwash plain* consisting chiefly of sand deposited by meltwater streams flowing from a glacier. (b) A term used in New England for an *esker delta* or *delta kame.*
sand plateau (a) *esker delta.* (b) *delta kame.*
sand plug A mass of sand that fills the upper end of a stream channel abandoned by the formation of a chute cutoff.
sandr *sandur.*
sand reef *sandbar.*
sand ridge (a) A generic name for any low ridge of sand formed at some distance from the shore, and either submerged or emergent. Examples include a *longshore bar* and a *barrier beach.* (b) One of a series of long, wide, extremely low, parallel ridges believed to represent the eroded stumps of former longitudinal sand dunes, as in western Rhodesia. (c) A seaward-pointing landform found on a sandy beach; e.g. a *beach cusp.* (d) *sand wave.*
sand ripple A ripple composed of sand. See also: *ripple.*
sand river A river that deposits much of its sand load along its middle course, to be subsequently removed by the wind; e.g. the Red River in Texas. Cf: *sand stream.*
sandrock (sand'-rock) (a) A field term for a sandstone that is not firmly cemented (Tieje, 1921, p.655). (b) A term used in southern England for a sandstone that crumbles between the fingers. (c) *sandstone.*
sand roll *pseudonodule.*
sand run (a) A fluidlike motion of dry sand. (b) A mass of dry sand in motion. Cf: *sand avalanche.*
sand sea [geog] An extensive assemblage of sand dunes of several types in an area where a great supply of sand is present, characterized by an absence of travel lines or other directional indicators, and by a wavelike appearance of dunes separated by troughs, much as though storm sea waves were frozen into place. See also: *erg.*
sand sea [pyroclast] The flat, rain-smoothed plain of volcanic ash and other pyroclastics on the floor of a caldera.
sand shadow A lee-side accumulation of sand, such as a small turret-shaped dune, formed in the shelter of, and immediately behind, a fixed obstruction, like a clump of vegetation. See also: *sand drift.*
sandshale (sand-shale') A sedimentary deposit consisting of thin alternating beds of sandstone and shale.
sand-shale ratio A term introduced by Sloss et al. (1949, p.100) for the ratio of the thickness or percentage of sandstone (and conglomerate) to that of shale in a stratigraphic section, disregarding the amount of nonclastic material; e.g. a ratio of 3.2 indicates that the section contains an average of 3.2 m of sandstone per meter of shale. Cf: *clastic ratio.*
sand sheet A large irregularly shaped plain of eolian sand, lacking the discernible slip faces that are common on dunes (Breed and Grow, 1979, p.281). The term is used in *remote sensing.*
sand sheet [sed] A sheetlike body of surficial sediment that veneers portions of marginal (pericontinental) and epeiric (epicontinental) sea floors and can range in thickness from a few centimeters to tens of meters, with a lateral persistence of a few meters to tens of kilometers.
sand size A term used in sedimentology for a particle size greater than that of a sphere with a diameter of $^{1}/_{16}$ mm (0.0025 in.) and less than that of a sphere with a diameter of 2 mm (0.08 in.).
sand snow Cohesionless dry snow that has fallen at such cold temperatures (usually below -25°C) that intergranular adhesion is inhibited and surface friction is high. Its surface has the consistency of dry sand. Cf: *powder snow.*
sand spit A *spit* consisting chiefly of sand.
sand splay A *floodplain splay* consisting of coarse sand particles.
sand stalagmite A stalagmite that is developed on sand and that is composed of sand cemented by calcite.
sandstone Originally defined by Lyell (1833, p.79) as "any stone which is composed of an agglutinization of grains of sand." (a) A medium-grained clastic sedimentary rock composed of abundant rounded or angular fragments of sand size with or without a fine-grained matrix (silt or clay) and more or less firmly united by a cementing material (commonly silica, iron oxide, or calcium carbonate); the consolidated equivalent of sand, intermediate in texture between conglomerate and shale. The sand particles are predominantly quartz, and the term "sandstone", when used without qualification, indicates a rock containing about 85-90% quartz (Krynine, 1940). The rock varies in color, may be deposited by water or wind, and may contain numerous primary features (sedimentary structures and fossils). Sandstone may be classified according to composition of particles, mineralogic or textural maturity, primary structures, and type of cement (Klein, 1963). (b) A field term for any clastic rock containing individual particles that are visible to the unaided eye or slightly larger. Syn: *sand [sed]; sandrock.*
sandstone-arenite (sand'-stone-ar'-e-nite) A term used by Folk (1968, p.124) for a *sedarenite* composed chiefly of sandstone fragments.
sandstone copper *sediment-hosted copper deposit.*
sandstone cylinder *sandstone pipe.*
sandstone dike (a) A *clastic dike* composed of sandstone or lithified sand; a lithified *sand dike.* (b) *stone intrusion.*
sandstone lead A *mineral-deposit type* characterized by disseminated galena in sandstone resting on sialic basement.
sandstone pipe A *clastic pipe* consisting of sandstone. It may originate in various ways: gravitational foundering of sand into underlying water-saturated mud; filling of a spring vent; filling of a cavity caused by solution of underlying limestone or by volcanic explosion; or penecontemporaneous sag due to removal of support by flowage. Syn: *sandstone cylinder.*
sandstone sill A tabular mass of sandstone that has been emplaced by sedimentary injection parallel to the structure or bedding of pre-existing rock in the manner of an igneous sill, such as one injected at the mud-water interface by the underflow of a dense slurry.
sandstone uranium A *mineral-deposit type* characterized by microcrystalline uranium oxide and silicate minerals concentrated in certain continental sandstone deposits. Cf: *roll-front orebody.*
sand streak A low, linear ridge formed at the interface of sand and air or water, oriented parallel to the direction of flow, and having a symmetric cross section.
sand stream A small sand delta spread out at the mouth of a gully, or a deposit of sand along the bed of a small creek, formed by a torrential rain (Stephenson and Veatch, 1915, p.112). Cf: *sand river.*
sand stretch A striation worn in a rock surface by windblown sand.
sand strip A long, narrow ridge of sand extending for a long distance downwind from each horn of a dune.
sand tank A *rock tank* filled with sand.
sand trap A device designed to remove sand and other particles from flowing water; e.g., for separating heavy, coarse particles from the cuttings-laden drilling fluid overflowing a drill collar.
sand tube A tubular *fulgurite* formed in sand.
sand tuff (a) A tuffaceous sandstone. (b) A tuff whose component fragments are in the size range of sand. This usage is obsolete.
sand twig A small, twiglike aggregate of sand that stands more or less upright on the surface of a sand dune undergoing wind scour, apparently forming around a root or stem of a plant exposed on the dune surface (Carroll, 1939, p.20-21).
sandur (sand'-ur) Icelandic term signifying "sand", but generally adopted for *outwash plain.* Pl: *sandar.* Also spelled: *sandr.*
sand volcano An accumulation of sand resembling a miniature volcano or low volcanic mound (maximum diameter 15 m), produced by expulsion of liquefied sand to the sediment surface. Examples are found on top of slump sheets or on the upper surface of highly contorted layers of laminated sediments.
sandwash (sand'-wash) A sandy or gravel stream bed, devoid of vegetation, containing water only during a sudden heavy rainstorm.
sand wave (a) A general term for a wavelike *bed form* in sand. (b) A generally large and asymmetrical bed form in sand, with a wavelike form but lacking the deep scour associated with dunes and megaripples (Boothroyd and Hubbard, 1974). Large sand waves generally have a high *ripple index,* and *lee* slopes inclined at less than the angle of repose of sand. (c) A general term to describe very large subaqueous sand ripples.
sand wedge A body of sand shaped like a vertical wedge with the apex downward. In areas of patterned ground, esp. Antarctica, it forms by infilling of debris in thermal contraction cracks.
sand-wedge polygon A *nonsorted polygon* formed by infilling of sand and gravel in intersecting fissures resulting from thermal contraction. Surface

diameter ranges from a few meters to tens of meters (Péwé 1959). Cf: *tessellation.*

sandy breccia A term used by Woodford (1925, p.183) for a breccia containing at least 80% rubble and 10% sand.

sandy chert Chert with oolithlike structures, formed when silica replaces cement or fills pore spaces in sandy beds and incorporates large, rounded sand grains in a cherty body or matrix. The perimeters of the sand grains are commonly resorbed, giving gradational contacts between them and the secondary silica.

sandy clay (a) An unconsolidated sediment containing 10-50% sand and having a ratio of silt to clay less than 1:2 (Folk, 1954, p.349). (b) An unconsolidated sediment containing 40-75% clay, 12.5-50% sand, and 0-20% silt (Shepard, 1954). (c) A soil containing 35-55% clay, 45-65% sand, and 0-20% silt (SSSA, 1965, p.347).

sandy clay loam A soil containing 20-35% clay, 45-80% sand, and less than 28% silt. Cf: *clay loam.*

sandy conglomerate (a) A conglomerate containing 30-80% sand and having a ratio of sand to mud (silt + clay) greater than 9:1 (Folk, 1954, p.347); a consolidated sandy gravel. (b) A conglomerate containing more than 20% sand (Krynine, 1948, p.141).

sandy desert An area of sand accumulation in an arid region, usually having an undulating surface of dunes; an *erg* or *koum.*

sandy gravel (a) An unconsolidated sediment containing 30-80% gravel and having a ratio of sand to mud (silt + clay) greater than 9:1 (Folk, 1954, p.346); if the ratio is between 1:1 and 9:1, the sandy gravel is "muddy". (b) An unconsolidated sediment containing more particles of gravel size than of sand size, more than 10% sand, and less than 10% of all other finer sizes (Wentworth, 1922, p.390). (c) An unconsolidated sediment containing 50-75% sand and 25-50% pebbles (Willman et al., 1942, p.343-344). Cf: *pebbly sand.*

sandy loam A soil containing 43-85% sand, 0-50% silt, and 0-20% clay, or containing at least 52% sand and no more than 20% clay and having the percentage of silt plus twice the percentage of clay exceeding 30, or containing 43-52% sand, less than 50% silt, and less than 7% clay (SSSA, 1965, p.347); specif. such a soil containing at least 30% very coarse sand, coarse sand, and medium sand, and less than 25% very coarse sand and less than 30% fine sand or very fine sand. It is subdivided into coarse sandy loam, fine sandy loam, and very fine sandy loam. Sandy loam contains sufficient silt or clay to make the soil somewhat coherent. Cf: *loamy sand.*

sandy mud An unconsolidated sediment containing 10-50% sand and the remainder having a ratio of silt to clay between 1:2 and 2:1 (Folk, 1954, p.349).

sandy ore A term used in Brazil for iron ore found along the bottoms of valleys and derived from *rubble ore* "by the natural sluicing of the streams" (Derby, 1910, p.818).

sandy silt (a) An unconsolidated sediment containing 10-50% sand and having a ratio of silt to clay greater than 2:1 (Folk, 1954, p.349). (b) An unconsolidated sediment containing 40-75% silt, 12.5-50% sand, and 0-20% clay (Shepard, 1954). (c) An unconsolidated sediment containing more particles of silt size than of sand size, more than 10% silt, and less than 10% of all other sizes (Wentworth, 1922).

sandy siltstone (a) A consolidated *sandy silt.* (b) A siltstone containing more than 20% sand (Krynine, 1948, p.141).

saneroite (sa-ne'-ro-ite) An orange triclinic mineral: $Na_2(Mn^{2+},Mn^{3+})_{10}V^{5+}Si_{11}O_{34}(OH)_4$.

Sangamon (San'-ga-mon) Pertaining to the third classical interglacial stage of the Pleistocene Epoch in North America, after the Illinoian glacial stage and before the Wisconsinan. Etymol: After Sangamon County, Ill. Cf: *Riss-Würm.*

sanidaster (san'-i-das'-ter) A spinose rodlike monaxonic sponge spicule (streptaster).

sanidine (san'-i-dine) A high-temperature mineral of the alkali *feldspar* group: $(K,Na)(Al,Si)_4O_8$. It is a highly disordered monoclinic form, occurring in clear, glassy, often tabular crystals embedded in unaltered acid volcanic rocks such as trachyte; it appears to be stable under equilibrium conditions above approximately 500°C. Sanidine forms a complete solid-solution series with high albite, and some sodium is always present. Syn: *glassy feldspar; ice spar; rhyacolite.*

sanidinite (san'-i-din-ite') An igneous rock (*syenite* or *trachyte*) composed almost entirely of sanidine. The term has also been applied to rocks composed of other alkali feldspars. The name was proposed by Nose in 1808. Obsolete.

sanidinite facies The facies (set of metamorphic mineral assemblages) in which are found tridymite, mullite, monticellite, larnite, and sanidine. The term was introduced by Eskola (1915) to cover occurrences of metamorphism at maximum temperature and minimum pressure, e.g. xenoliths in basic lavas, fragments in tuffs, or narrow contact zones bordering shallow basic pipes. Many rocks show evidence of partial to complete fusion; e.g., *buchite.*

sanitary landfill (san'-i-tar'-y) A land site where municipal solid waste is buried in a manner engineered to minimize environmental degradation. Commonly the waste is compacted and periodically covered with soil or other earth material.

sanjuanite (san-juan'-ite) A white triclinic mineral: $Al_2(PO_4)(SO_4)(OH){\cdot}9H_2O$.

sanmartinite (san-mar-tin'-ite) A dark brown or brownish-black monoclinic mineral of the *wolframite* group: $(Zn,Fe^{2+})WO_4$.

sannaite (san'-na-ite) (a) In the *IUGS classification,* a variety of *lamprophyre* in which alkali feldspar is more abundant than plagioclase, feldspathoid exists in the groundmass but is less abundant than feldspars; phenocrysts are any combination of olivine, clinopyroxene, amphibole, and biotite. (b) A mafic *phonolite* containing phenocrysts of amphibole, clinopyroxene, and biotite (in order of decreasing abundance) in a fine-grained groundmass of alkali feldspar, aegirine, chlorite, calcite, and pseudomorphs of mica after nepheline. The name, given by Brögger in 1921, is from Sannavand, in the Fen complex, Norway.

Sannoisian (San-nois'-i-an) Obs. European stage: Lower Oligocene (above Ludian, below Stampian). Syn: *Tongrian.*

sansicl (san'-sicl) An unconsolidated sediment consisting of a mixture of *san*d, *si*lt, and *cl*ay, in which no component forms 50% or more of the whole aggregate.

Sanson-Flamsteed projection (San'-son-Flam'-steed) A syn. of *sinusoidal projection.* Named after Nicolas Sanson (1600-1667), French geographer, and John Flamsteed (1646-1719), English astronomer, although first used in the 16th century.

Santa Ana wind A hot, dry *chinook* wind that blows from the desert plateaus of Utah and Nevada across the Sierra Nevada and downslope toward coastal California. This wind desiccates vegetation and contributes to outbreaks of forest and brush fires.

santabarbaraite A brown amorphous mineral: ${Fe^{3+}}_3(PO_4)_2(OH)_3{\cdot}5H_2O$.

santaclaraite (san-ta-cla'-ra-ite) A pale pink or reddish-orange triclinic mineral: $Ca{Mn^{2+}}_4Si_5O_{14}(OH)_2{\cdot}H_2O$.

santafeite (san-ta-fe'-ite) A subadamantine black bladed orthorhombic mineral: $(Na,Ca,Sr)_3(Mn^{2+},Fe^{3+})_2{Mn^{4+}}_2(VO_4)_4(OH,O)_5{\cdot}2\,H_2O$.

santanaite (sant-an'-a-ite) An adamantine yellow hexagonal mineral: $Pb_{11}CrO_{16}$.

Santernian (San-ter'-ni-an) An Italian substage: lower substage of regional Calabrian stage, lowermost Pleistocene (below Emilian substage).

santite (sant'-ite) A colorless orthorhombic mineral: $KB_5O_8(OH)_4{\cdot}2H_2O$.

Santonian (San-ton'-i-an) Stage of Standard Global Chronostratigraphic Scale: Upper Cretaceous (above Coniacian, below Campanian).

sanukite (sa-nu'-kite) An andesite characterized by sparse magnesian phenocrysts of bronzite, in a microcrystalline groundmass. Compositionally, sanukites have relatively high MgO, Ni, and Cr, and at Setouchi (former name: Sanuki) Shikoku, Japan, Tatsumi and Ishizaka (1981) call the rocks high-magnesian andesites (HMAs). Originally named by Weinschenk in 1890. Cf: adakite; boninite.

sanukitoid (sa-nu'-kit-oid) As defined by Koto (1916), at Setouchi (see: *sanukite*) sanukite and related relatively aphyric andesites and basalts with magnesian mafic phenocrysts that may include olivine, augite, and hornblende as well as bronzite. Occurrences in the Archean of Ontario, Canada, have the same high Mg, Ni, and Cr as those at Setouchi but differ in their higher K_2O, La/Yb, Sr, and Ba (Shirey and Hanson, 1984).

sanukitophone Sanukite stones assembled in a tuned series to produce a musical instrument played with mallets. It makes use of the stone's unusual property that it rings.

sapanthracite (sap-an'-thra-cite) Sapropelic coal of anthracitic rank; it is the highest stage in the sapropelic series. Cf: *humanthracite.*

sapanthracon (sap-an'-thra-con) Sapropelic coal of Carboniferous age; it is the fifth stage in the sapropelic series. Cf: *humanthracon.*

saphir d'eau (sa'-phir d'eau') A syn. of *water sapphire* (variety of cordierite). Etymol: French, "water sapphire".

saponite (sap'-o-nite) A trioctahedral member of the *montmorillonite* group that is somewhat different from the other members of the group in that it has a positive charge on the octahedral sheet that partially compensates for a large negative charge on the tetrahedral sheet. An ideal formula: $R_{0.33}(Mg_{2.67}Fe_{0.33})(Si_{3.34}Al_{0.66})O_{10}(OH)_2{\cdot}4H_2O$. Cf: *smectite.*

sappare (sap'-pare) (a) *sapphire.* (b) *kyanite.*

sapperite (sap'-per-ite) A natural pure white cellulose, $(C_6H_{10}O_5)_n$, which occurs in brown coal and fossil wood.

sapphire (sapph'-ire) (a) Any pure, gem-quality corundum other than *ruby;* esp. the fine blue transparent variety of crystalline corundum of great value, containing small amounts of oxides of cobalt, chromium, and titanium, used as a gemstone, and found esp. in the Orient (Kashmir, Myanmar, Thailand, Sri Lanka). Other colors, such as pink, purple, yellow, green, and orange, are included under *fancy sapphire.* Syn: *sappare.* (b) Any gem from a corundum crystal.

sapphire quartz (a) A rare, opaque, indigo-blue variety of quartz colored by included nonparallel fibers of silicified crocidolite. Cf: *hawk's-eye.* Syn: *azure quartz; blue quartz; siderite [mineral].* (b) A term used in the western U.S. for chalcedony having a light to pale sapphire-blue color.

sapphirine (a) A green or pale-blue dense, hard mineral crystallizing in monoclinic and triclinic forms: $Mg_7Al_{18}Si_3O_{40}$. It is a principal constituent of certain high-grade silica-deficient metamorphic rocks and occurs usually in granular form. (b) A name applied to certain blue minerals such as hauyne and blue chalcedony.

sapping [geomorph] (sap'-ping) (a) The natural process of erosion along the base of a cliff by the wearing-away of softer layers, often involving weakening by groundwater conducted along the contact between rock strata, and thus removing the support for the upper mass which breaks off into large blocks falling from the cliff face. Cf: *landslide sapping.* Syn: *cliff erosion; undermining.* (b) *spring sapping.*

sapping [glac geol] (a) *basal sapping.* (b) Sometimes used as a syn. of *plucking [glac geol].*

sapric material Organic soil material that contains less than 1/6 recognizable fibers (after rubbing) of undecomposed plant remains. Bulk density is usually very low, and water holding capacity very high.

Saprists Histosols that have a high content of plant materials so decomposed that original plant structures cannot be determined and a bulk density of about 0.2 Mg m^{-3} or more. Saprists are saturated with water for periods long enough to limit their use for most crops unless they are artificially drained. A suborder in the U.S. system of soil taxonomy.

saprobe (sap'-robe) *saprophyte.*

saprobic (sa-prob'-ic) *saprophytic.*

saprock Informal term for protolith of *hydrous-silicate nickel laterite* deposit.

saprocol (sap'-ro-col) Indurated sapropel; it is the second stage in the sapropelic series. Also spelled: *saprokol.* Cf: *humocoll.*

saprodil (sap'-ro-dil) A sapropelic coal of Tertiary age; it is the third stage in the sapropelic series. Cf: *humodil.*

saprodite (sap'-ro-dite) Sapropelic coal of brown-coal rank; it is the fourth stage in the sapropelic series. Cf: *humodite.*

saprogen (sap'-ro-gen) An organism that lives on dead organic matter and can cause its decay. Adj: *saprogenic.*

saprogenic (sap-ro-gen'-ic) Said of an organism that produces decay or putrefaction. Cf: *saprophytic.* n. *saprogen.*

saprohumolith series (sap-ro-hu'-mo-lith) Organic materials and coals intermediate between the *sapropelite series* and the *humolith series,* with sapropelic materials predominating. Cf: *humosapropelic series.*

saprokol (sap'-ro-kol) *saprocol.*

saprolite (sap'-ro-lite) A soft, earthy, typically clay-rich, thoroughly decomposed rock, formed in place by chemical weathering of igneous, sedimentary, and metamorphic rocks. It often forms a layer or cover as much as 100 m in thickness, esp. in humid and tropical or subtropical climates; the color is commonly some shade of red or brown, but it may be white or gray. Saprolite is characterized by preservation of structures that were present in the unweathered rock. The term was proposed by Becker (1895). Cf: *geest; laterite.* Also spelled: *sathrolith.*

sapromyxite (sap-ro-myx'-ite) *boghead coal.*

sapront (sap'-ront) *saprophyte.*

sapropel (sap'-ro-pel) An unconsolidated, jellylike ooze or sludge composed of plant remains, most often algae, macerating and putrefying in an anaerobic environment on the shallow bottoms of lakes and seas. It may be a source material for petroleum and natural gas. Cf: *dy; gyttja; humopel.*

sapropel-clay A sedimentary deposit in which the amount of clay exceeds that of sapropel.

sapropelic (sap-ro-pel'-ic) Pertaining to or derived from *sapropel;* indicating a high sulfate or reducing environment.

sapropelic coal Coal that is derived from organic residues (finely divided plant material, spores, algae) in stagnant or standing bodies of water. Putrefaction under anaerobic conditions, rather than peatification, is the formative process. The main types of sapropelic coal are *cannel coal, boghead coal,* and *torbanite.* Sapropelic coals are high in volatiles, generally dull, massive, and relatively uncommon. Cf: *humic coal; nonbanded coal.* See also: *sapropelite series.* Syn: *sapropelite.*

sapropelite (sap-ro-pel'-ite) *sapropelic coal.*

sapropelite series The organic materials of *sapropelic coal* in metamorphic rank: *sapropel, saprocol, saprodil, saprodite, sapanthracon,* and *sapanthracite* (Heim and Potonié, 1932, p.146). Cf: *humolith series; humosapropelic series; saprohumolith series.*

sapropel-peat *peat-sapropel.*

saprophilous (sa-proph'-i-lous) *saprophytic.*

saprophyte (sap'-ro-phyte) A plant that lives on decayed or decaying organic matter. Syn: *saprobe; sapront.* Adj: *saprophytic.*

saprophytic (sap-ro-phyt'-ic) Said of a plant that receives its nourishment from the products of decaying organic matter; i.e. a saprophyte. Syn: *saprophilous; saprobic; saprozoic.* Cf: *saprogenic.*

saprovitrinite (sap-ro-vit'-ri-nite) Vitrinite in vitrain of sapropelic coal. Cf: *humovitrinite.*

saprozoic (sap-ro-zo'-ic) *saprophytic.*

SAR *synthetic-aperture radar.*

sarabauite (sar-a-bau'-ite) A red monoclinic mineral: $CaSb_{10}O_{10}S_6$.

Saracen stone (Sar'-a-cen) A syn. of *sarsen.* The term originally signified a pagan stone or monument.

sarcodictyum (sar-co-dic'-ty-um) The outermost layer of cytoplasm in a radiolarian; a network of protoplasm on the surface of the *calymma* of a radiolarian.

Sarcodina (Sar-co-di'-na) A class of protozoans characterized mainly by their ability to form pseudopodia. Among the members of the class are rhizopods and actinopods. Range, Cambrian to Holocene.

sarcolite (sarc'-ol-ite) A colorless or pink tetragonal mineral: $Na_4Ca_{12}Al_8Si_{12}O_{46}(Si,P)O_4(OH,H_2O)_4(CO_3,Cl)$.

sarcopside (sarc'-op-side) A flesh-red, reddish brown, or dark brown monoclinic mineral: $(Fe^{2+},Mn^{2+},Mg)_3(PO_4)_2$. It is structurally related to *triphylite* and *heterosite.*

Sarcopterygii (Sar'-cop-te-ryg'-i-i) A subclass of bony fish, characterized by fins with an axial fleshy lobe supported by a skeletal projection from the body wall, and presumably by a paired swim bladder that functioned as a lung. Survivors are the living lungfish of Africa, Australia, and South America, and the coelacanth *Latimeria* of the Indian Ocean. Range, Devonian to Holocene. Syn: *lobefin.*

sarcotesta (sar-co-tes'-ta) The fleshy outer layer of a seed coat, as in *Trigonocarpus* (Scott, 1923, p.204). Cf: *sclerotesta.*

sard A translucent brown to deep orange-red variety of chalcedony, similar to *carnelian* but having less intense color; it is classed by some as a variety of carnelian. Syn: *sardius; sardine.*

Sardic Orogeny (Sar'-dic) One of the 30 or more short-lived orogenies during Phanerozoic time identified by Stille, in this case near the end of the Cambrian.

sardine *sard.*

sardius (sar'-di-us) Original name for *sard.*

sardonyx (sar-do'-nyx, sar'-do-nyx) A gem variety of chalcedony that is like *onyx* in structure but includes straight parallel brownish-red or reddish-brown bands of sard alternating with white, brown, black, or other colored bands of another mineral. The name is applied incorrectly to carnelian and sard.

Sargasso Sea (Sar-gas'-so) A warm region of the open North Atlantic Ocean to the east and south of the Gulf Stream, characterized by large masses of floating vegetation that is mainly sargasso (gulfweed), a seaweed (brown alga) of the genus *Sargassum.*

sarkinite (sark'-in-ite) A red, reddish-yellow, or yellow monoclinic mineral of the *triploidite* group: $Mn^{2+}_2(AsO_4)(OH)$.

Sarmatian (Sar-ma'-tian) (a) Central European stage: upper Miocene (above Karpatian, below Pontian). Equiv. Tortonian. (b) Ponto-Caspian stage: lower upper Miocene (above Konkian, below Maeotian).

sarmientite (sar-mi-en'-tite) A yellow-orange monoclinic mineral: $Fe^{3+}_2(AsO_4)(SO_4)(OH){\cdot}5H_2O$.

sarnaite (sar'-na-ite) A feldspathoid-bearing *syenite* containing cancrinite and clinopyroxene. Its name is derived from Särna, Sweden. Named by Brögger in 1890. Obsolete.

sarospatakite (sar-o-spat'-a-kite) A micaceous clay mineral composed of mixed layers of illite and montmorillonite, found in Sárospatak, Hungary. Cf: *bravaisite.*

sarsden stone (sars'-den) *sarsen.*

sarsen (sar'-sen) A large residual mass of stone (silcrete) left after the erosion of a once continuous bed of which it formed a part; specif. one of the large, rounded, gray blocks or fragments of silicified sandstone or conglomerate strewn over the surface of the English chalk downs (esp. in Wiltshire) and far from any similar beds, being the only remnants of the former Tertiary (Eocene?) cover. Specifically used by Megalithic people in U.K. and France, e.g. in Stonehenge. Similar stone blocks (but of calcrete beachrock) were used by Polynesian people, e.g. in Tonga, Palas. Etymol: alteration of "Saracen", a Moslem or "outlandish stranger". Syn: *Saracen stone; sarsden stone; greywether; druid stone.*

sartorite (sar'-to-rite) A gray monoclinic mineral: $PbAs_2S_4$. Syn: *scleroclase.*

sarule (sa'-rule) A *sceptrule* with a single spinose terminal outgrowth that resembles a brush.

saryarkite-(Y) A white tetragonal mineral: $Ca(Y,Th)A_{15}(SiO_4)_2(PO_4)_2(OH)_7{\cdot}6H_2O$.

sasaite (sa'-sa-ite) A chalky white orthorhombic mineral: $(Al,Fe^{3+})_6(PO_4,SO_4)_5(OH)_3{\cdot}36H_2O$.

sassolite (sas'-so-lite) A white or gray mineral consisting of native boric acid: H_3BO_3. It usually occurs in small pearly scales as an incrustation or as tabular triclinic crystals around fumaroles or vents of sulfurous emanations. Also called: sassoline.

sastrugi (sas'-tru-gi) Irregular ridges up to 50 cm high, formed in a level or nearly level snow surface by wind erosion, often aligned parallel to the wind direction, with steep, concave or overhanging ends facing the wind; or cut into snow dunes previously deposited by the wind. Cf: *wind ridge; skavl.* Also spelled: *zastrugi.* Usually used in this plural form; sing: "sastrugo".

sate To fill most of the pores between soil particles with liquid, the lack of complete filling being caused by the entrapment of air as water enters the soil. Syn: satiate.

satelite (sat'-e-lite) Fibrous serpentine with a slight chatoyant effect, being pseudomorphous after asbestiform tremolite that has been silicified. It occurs in Tulare County, Calif.

satellite (sat'-el-lite) A secondary celestial body, natural or man-made, that revolves about another, primary body, e.g., the Moon about the Earth.

satellite [volc] adj. Said of small monogenetic volcanic features (domes, cinder cones) distributed over the flanks of a large composite edifice (Davidson and De Silva, 2000). Syn: *parasitic [volc].*

satellite altimetry A remote method of measuring the height of sea level relative to the center of mass of earth. It is used to map sea-surface topography and its variability, from which surface *geostrophic currents,* their variability, and *tides* are calculated. See also: *altimeter satellite.*

satellite geodesy The part of geodesy dealing with the use of satellites to obtain geodetic data.

satellite gravity Measuring the Earth's surface relief over oceans and seas by radar from orbiting satellites, where the elevation of a satellite is nearly constant. These measurements reveal undulations of the geoid, which permit mapping of marine gravity anomalies. Capable of measuring the sea surface to an accuracy of about 10 cm, depending on the sea-surface roughness.

satellite navigation (SATNAV) Location determination by the *Global Positioning System.* Permits continuous determination of location based on observations of a navigation satellite.

satellitic crater (sat-el-lit'-ic) *secondary crater.*

sathrolith (sath'-ro-lith) *saprolite.*

satimolite (sa-tim'-o-lite) A white orthorhombic mineral: $KNa_2Al_4B_6O_{15}Cl_3{\cdot}13H_2O$.

satin ice (sat'-in) *acicular ice.*

satin spar (a) A white, translucent, fine fibrous variety of gypsum, characterized by chatoyancy or a silky luster. (b) A term used incorrectly for a fine fibrous or silky variety of calcite or aragonite. Also called: satin stone.

satpaevite (sat-pae'-vite) A yellow incompletely investigated mineral: $Al_{12}V^{4+}{}_2V^{5+}{}_6O_{37}{\cdot}30H_2O$ (?).

satterlyite (sat'-ter-ly-ite') A pale yellow to pale brown trigonal mineral: $(Fe^{2+},Mg,Fe^{3+})_{12}(PO_3OH)(PO_4)_5(OH,O)_6$.

saturated [geol] (sat'-u-rat'-ed) (a) Said of a rock having quartz in its norm. (b) Said of a mineral that can form in the presence of free silica, i.e. one that contains the maximum amount of combined silica. (c) Said of an igneous rock composed chiefly of such minerals, or of its magma. Cf: *oversaturated; undersaturated.*

saturated [water] Said of the condition in which the interstices of a material are filled with a liquid, usually water. It applies whether the liquid is under greater than or less than atmospheric pressure, so long as all connected interstices are full. See also: *saturated zone.*

saturated hydrocarbon Hydrocarbon with no double or triple bonds.

saturated permafrost Permafrost that contains no more ice than the ground could hold if the water were in the liquid state; permafrost in which all available pore spaces are filled with ice.

saturated pool An oil pool with an excess of gas, which forms a *gas cap* above the oil. Cf: *undersaturated pool.*

saturated zone A subsurface zone in which all the interstices are filled with water under pressure greater than that of the atmosphere. Although the zone may contain gas-filled interstices or interstices filled with fluids other than water, it is still considered saturated. This zone is separated from the *unsaturated zone* (above) by the *water table.* Syn: *zone of saturation; phreatic zone.*

saturation (sat-u-ra'-tion) A condition of dynamic equilibrium between evaporation and condensation rate; commonly assumed as the maximum possible content of water vapor in the Earth's atmosphere for a given temperature. See also: *dewpoint.*

saturation curve A curve showing the weight of solids per unit volume of a saturated earth mass as a function of its water content. Cf: *zero-air-voids curve.*

saturation line [glaciol] *wet-snow line.*

saturation line [petrology] The line on a variation diagram of an igneous-rock series that represents saturation with respect to silica. Rocks to the right of the line are oversaturated and those to the left are undersaturated. Obsolete.

saturation magnetization The maximum possible magnetization of a material, i.e., alignment of all magnetic cations.

saturation pressure *bubble-point pressure.*

saturation remanence The magnetization of an assemblage of permanently magnetized grains after the grains have obtained a saturation magnetization; measured in the absence of an applied DC field.

Saucesian (Sau-ces'-ian) Pacific Coast marine stage: lower Miocene (above Zemorrian, below Relizian).

sauconite (sau'-con-ite) The Zn analogue of the clay mineral saponite, with Zn in place of octahedral Mg: $Na_{0.3}Zn_3(Si,Al)_4O_{10}(OH)_2{\cdot}4H_2O$. It is monoclinic, and occurs in various pale colors.

saukovite (sauk'-o-vite) *metacinnabar.*

sault A waterfall or rapids in a stream. Etymol: French, old spelling of "saut" (from Latin "saltus"), "leap".

Saurischia (Saur-is'-chi-a) One of the two orders of archosaurian reptiles commonly treated as dinosaurs, characterized by a triradiate pelvis. It includes bipedal predators (Carnosauria, Coelurosauria), quadrupedal long-necked herbivores (Sauropods), and presumptive ancestors of the latter (Prosauropoda). Range, Middle Triassic to Upper Cretaceous. Cf: *Ornithischia.*

sausage structure (sau'-sage) *boudinage.*

saussurite (saus'-su-rite) A tough, compact, and white, greenish, or grayish mineral aggregate consisting of a mixture of albite (or oligoclase) and zoisite or epidote, together with variable amounts of calcite, sericite, prehnite, and calcium-aluminum silicates other than those of the *epidote* group. It is produced by alteration of calcic plagioclase. Saussurite was originally thought to be a specific mineral.

saussuritization (saus'-su-rit'-i-za'-tion) The replacement, esp. of plagioclase in basalts and gabbros, by a fine-grained aggregate of zoisite, epidote, albite, calcite, sericite, and zeolites. It is a metamorphic or deuteric process and is frequently accompanied by chloritization of the ferromagnesian minerals.

savanna (sa-van'-na) (a) An open, grassy plain with scattered trees, esp. as developed in tropical or subtropical regions. Usually it has distinct wet and dry seasons; its trees and shrubs are drought resistant. (b) Along the southeastern Atlantic Coast of the U.S., the term (often spelled "savannah") is used for marshy alluvial *flats* with occasional clumps of trees.

Savic Orogeny (Sav'-ic) One of the 30 or more short-lived orogenies during Phanerozoic time identified by Stille, in this case in the late Oligocene, between the Chattian and Aquitanian stages.

Savonius rotor current meter (Sa-vo'-ni-us) A sensor device for measuring the speed of a current. It utilizes an S-shaped rotor, or a pair of them with their axes perpendicular to each other. Now rarely used.

saw-toothed (a) *serrate [geomorph].* (b) Descriptive of *hacksaw structure.*

saxicavous (sax-i-ca'-vous) Said of an organism that bores into rock, i.e., of a *rock borer.*

saxicolous (sax-ic'-o-lous) *rupestral.*

saxifragous (sax-if'-ra-gous) Said of a plant that grows in crevices of rock and promotes its splitting; i.e., of a *chasmophyte.*

saxigenous (sax-ig'-e-nous) *rupestral.*

Saxonian (Sax-o'-ni-an) European stage: Middle Permian (above Autunian, below Thuringian).

Saxonian-type facies series Rocks produced in a type of dynamothermal regional metamorphism for which the classical locality is the granulite area in Saxony. The pressure range of such formation is extensive (200-800 MPa) as well as the temperature range (100°-700°C), probably involving polymetamorphism. The rocks differ little mineralogically from those formed in the *Barrovian-type facies series,* except at the highest pressure and temperature values, at which kyanite- and orthoclase-bearing granulites are formed (Hietanen, 1967, p.201). Rarely used.

saxonite (sax'-on-ite) A *peridotite* composed chiefly of olivine and orthopyroxene. It is considered by some petrologists as a syn. of *harzburgite* and by others as distinct from harzburgite owing to its lower percentage of opaque oxide minerals. The term was coined by Wadsworth in 1884, for Saxony, Germany. Obsolete.

sayrite (sayr'-ite) A yellow-orange to red-orange monoclinic mineral: $Pb_2(UO_2)_5O_6(OH)_2{\cdot}4H_2O$.

sazhinite-(Ce) (sa'-zhi-nite) A white, gray, or creamy orthorhombic mineral: $Na_2CeSi_6O_{14}(OH){\cdot}6H_2O$.

sazykinaite-(Y) A light green to yellow-green rhombohedral mineral of the *hilairite* group: $(Na,K)_5Y(Zr,Ti,Nb)Si_6O_{18}{\cdot}6H_2O$.

sborgite (sborg'-ite) An earthy white monoclinic mineral: $NaB_5O_6(OH)_4{\cdot}3H_2O$.

scabland (scab'-land) An elevated area, underlain by flat-lying basalt flows, with a thin soil cover and sparse vegetation, and usually with deep, dry channels scoured into the surface. An example is the Columbia lava plateau of eastern Washington, which was widely and deeply eroded by glacial meltwaters. See also: *channeled scabland; scabrock.*

scabrate (scab'-rate) A term for rough sculpture of palynomorphs, which consists of more or less isodiametric elements <1 µm in diameter.

scabrock (scab'-rock) (a) An outcropping of *scabland.* (b) Weathered material of a scabland surface.

scacchite (scacch'-ite) (a) A colorless, red, or brown rhombohedral mineral of the chloromagnesite group: $MnCl_2$. (b) A name applied to various minerals, including monticellite, a doubtful selenide of lead, and a brick-red powdery fluoride containing rare earths.

scaglia (scagl'-ia [skahl'-ya]) A dark, very fine-grained, more or less calcareous shale typically developed in the Upper Cretaceous and lower Tertiary of the northern Apennines. Etymol: Italian, "scale, chip".

scainiite A metallic bluish-black monoclinic mineral: $Pb_{14}Sb_{30}S_{54}O_5$.

scalariform (sca-lar'-i-form) Pertaining to the ladderlike thickenings with intervening pits on the walls of certain xylem cells.

Scaldisian (Scal-dis'-ian) North Sea basin stage: lower Pliocene, in some usage also middle Pliocene (above Hemmoorian, below Amstelian); based on warm-climate mollusk faunas.

scale [cart] The ratio between linear distance on a map, chart, globe, model, or photograph and the corresponding distance on the surface being mapped. It may be expressed in the form of a direct or verbal statement using different units (e.g. "1 inch to 1 mile" or "1 inch = 1 mile"); a *representative fraction* or numerical ratio (e.g. "1/24,000" or "1:24,000", indicating that one unit on the map represents 24,000 identical units on the ground); or a graphic measure (such as a bar or line marked off in feet, miles, or kilometers). The scale of a photograph is usually taken as the ratio of the principal distance of the camera to the altitude of the camera station above mean ground elevation. See also: *small-scale map; large-scale map.* Syn: *linear scale; map scale.*

scale [paleont] (a) A small platelike structure attached to septal grooves and interseptal ridges in some rugose corallites, as in *Tryplasma.* Also, a thin flat or nearly flat sclerite in an octocoral. (b) *scaphocerite.* (c) A flat to slightly curved projection from the lower lip of a pore, rising obliquely to the wall in archaeocyathids. (d) A small, more or less flattened, rigid to somewhat flexible, and definitely circumscribed plate forming part of the integumentary body covering (formed of keratinized epidermis, densely fibrous dermis, cartilage, or bone) of certain vertebrates.

scaled physical model In structural geology, an experimental model for which the size and time frame of the deformation process are reduced in size, allowing direct observation of the spatial and temporal development of geologic structures (e.g., faults, folds) for a variety of structural styles (extensional, compressional, strike-slip, salt tectonics, inversion tectonics). A model is properly scaled if it satisfies *geometric similarity, kinematic similarity,* and *dynamic similarity* (Hubbert, 1937). Consequently, modeling materials must have strengths that are scaled down from the prototype. Syn: *analogue model, sandbox model.*

scale factor (a) A multiplier for reducing a distance obtained from a map by computation or scaling to the actual distance on the datum of the map. (b) The ratio of the scale at a particular location and direction on a map to the stated scale of the map. At a standard parallel, or other standard line, the scale factor is 1.0.

scalenohedral (sca-le'-no-he'-dral) Having the form or symmetry of a *scalenohedron.*

scalenohedron (sca-le'-no-he'-dron) A closed crystal form whose faces are scalene triangles; the *hexagonal scalenohedron* has twelve faces, and the *tetragonal scalenohedron* has eight. Adj: *scalenohedral.*

scaler (scal'-er) An electronic instrument that counts the pulses from a *nuclear radiation* detector.

scaling (scal'-ing) A type of *exfoliation* that produces thin flakes, laminae, or scales.

scaling relations One of several algebraic equations that relate the depth and diameter of a *transient impact crater* to the dimensions, energy, density, and angle of impact of a meteoritic projectile (Melosh, 1989, chap.7).

scallop [paleont] (scal'-lop) (a) Any of many marine bivalve mollusks of the family Pectinidae having a shell that is characteristically rather flat, radially ribbed, and marginally undulated, that has a single large adductor muscle, and that is able to swim by opening and closing the valves. (b) One of the valves of the shell of a scallop.

scallop [sed] *scalloping.*

scallop [speleo] One of a mosaic of small shallow intersecting hollows formed on the surface of soluble rock by turbulent dissolution. They are steeper on the upstream side, and smaller sizes are formed by faster-flowing water. Syn: *facet [speleo]; flute [speleo]; solution ripple.*

scalloped upland (scal'-loped) "The region near or at the divide of an upland into which glacial cirques have cut from opposite sides" (Stokes and Varnes, 1955, p.129). Cf: *fretted upland.*

scalloping (scal'-lop-ing) A term used by Gruner et al. (1941, p.1621-1622) for a sedimentary structure of uncertain origin, superficially resembling oscillation ripple, and having a concave side that is always oriented toward the top of the bed. It may have formed by differential expansion or shrinkage of adjoining layers of mud before complete consolidation. Syn: *scallop [sed].*

scalped anticline *breached anticline.*

scaly (scal'-y) (a) Said of the texture of a mineral, esp. a mica, in which small plates break or flake off from the surface like scales. (b) Referring to the general appearance of a rock, as in *argille scagliose.*

scan n. A graphic or photographic depiction of the distribution of radioactivity of an object. See also: *autoradiograph.*

scan [geophys] (a) One complete sequence of events, such as sampling all record channels or beam-steering for all apparent dips. (b) A display of data calculated with different parameter values, used to select the optimum parameter value. (c) To examine in a systematic way, as to look through a data set to see the effects of a parameter or parameter change. (d) To traverse systematically as is done by an electron beam in a television tube.

scandent (scan'-dent) Said of a graptoloid with stipes that grew erect along or enclosing the *virgula.*

scandiobabingtonite A pale grayish-green triclinic mineral: $(Ca,Na)_2(Fe^{2+},Mn)(Sc,Fe^{3+})Si_5O_{14}(OH)$. It is the scandium analogue of babingtonite.

scan line (a) The narrow strip on the ground that is swept by the instantaneous field of view of a detector in a *scanner* system. One of the parallel tracks covered by the oscillating mirror of an airborne or satellite sensor. (b) The segment of an image produced as a result of a single sweep of a light source across a recording medium. The electronic signal of a single scan (Krzanowski et al., 1993). (c) A list of pixels forming a horizontal row of an image with a constant Y-coordinate, and ordered by increasing X-coordinate (Krzanowski et al., 1993).

scanner (scan'-ner) (a) An optical-mechanical imaging system in which a rotating or oscillating mirror sweeps the instantaneous field of view of a *detector* across the terrain. Scanners may be airborne or stationary. Nonrecommended syn: *line scanner.* (b) A digitizing input device that produces an image in raster format from flat input material such as photographs, maps, and drawings.

scanner distortion The geometric distortion that is characteristic of scanner images. The scale of the image is constant in the direction parallel with the flight direction of the aircraft or spacecraft. At right angles to this direction, however, the image scale becomes progressively smaller from the nadir line outward toward either margin of the image. Linear features, such as roads, that trend diagonally across a scanner image are distorted into S-shaped curves. Distortion becomes more pronounced with a larger angular field of view.

scanning electron microscope (scan'-ning) An *electron microscope* in which a finely focused beam of electrons is electrically or magnetically moved across the specimen to be examined, from point to point, again and again, and the reflected and emitted electron intensity measured and displayed, sequentially building up an image. The ultimate magnification and resolution is less than for the conventional electron microscope, but opaque objects can be examined, and great depth of field is obtained. Abbrev: SEM.

scanning force microscopy A type of *atomic resolution microscopy* whereby a fine metal-wire probe is forced down onto a sample and scanned across it. An optical laser system is used to detect up and down motions of the probe as it scans the surface of a sample. The optical laser data are then processed by a computer and transformed into an image showing the relative dimensions and distribution of atoms and the atomic-scale topography of the scanned surface. Scanning force microscopy can be used to analyze nonconductors of electricity, because it relies on physical motions of the probe rather than electron tunneling. Abbrev: SFM. See also: *scanning tunneling microscopy.*

scanning proton microprobe Instrumentation using a focused beam of charged particles (usually protons, but can include heavier particles such as alpha particles, oxygen ions, or sulfur ions) to excite samples. Generally used for *particle-induced X-ray emission* analysis, but may be used for *particle-induced gamma-ray emission* analysis as well. Abbrev: SPM.

scanning transmission electron microscope A *transmission electron microscope* that has the capability of forming the electron beam into a fine probe (<100Å in diameter) and scanning it across a thin specimen. The transmitted scanned beam is collected below the specimen by a solid-state detector and is reproduced electronically as an image on a cathode-ray tube. Image contrast is the same as in the conventional transmission electron microscope; the advantage of the scanning transmission electron microscope is the fine probe and the electronic manipulation of the detected transmitted beam. Electron diffraction patterns can be obtained from areas tens of angstroms in diameter, and, in conjunction with an X-ray detection system, compositional data can be obtained with spatial resolutions of 500-1,000Å. In some instruments, the electron-beam size can be reduced to several angstroms in diameter, resulting in high-resolution images of single large atoms, e.g. uranium. Abbrev: STEM.

scanning tunneling microscopy A type of *atomic resolution microscopy* whereby a fine metal-wire probe (usually composed of platinum or iridium) is used to scan over the surface of a sample. When the probe gets close to the electron cloud of an atom, electrons tunnel (travel in a fine stream) between the atom and the probe. When the probe moves away from the electron cloud of an atom, tunneling decreases or stops. A computer is used to translate the tunneling data from each line of scan into an image of atomic-scale topography showing the dimensions and distribution of atoms. This type of microscopy can be used only to analyze materials that are conductors (so their electrons will tunnel). Abbrev: STM. See also: *scanning force microscopy.*

scan skew Distortion of scanner images caused by forward motion of the aircraft or satellite during the time required to complete a scan.

scaphate element A laterally elaborate or platformed pectiniform *conodont element* with an *under side* marked by a capacious, commonly cuplike basal cavity (TIP, 1981, pt. W, supp.2).

scaphocerite (scaph-o-ce'-rite) A flattened plate on the second joint of the antennae of many crustaceans; e.g. a scalelike exopod of an antenna of a eumalacostracan. Syn: *scale [paleont].*

scapholith (scaph'-o-lith) An elongate, diamond- or boat-shaped *heterococcolith* with a central area of parallel laths. Syn: *rhombolith.*

scaphopod (scaph'-o-pod) Any benthic marine univalve mollusk belonging to the class Scaphopoda.

Scaphopoda A class of *diasome* mollusks with tapering tubular univalved shell closed dorsally and ventrally and open at both ends; radula and unique anterior feeding structures (captacula) present. Range, Middle Ordovician to Holocene; about 50 genera.

scapolite (scap'-o-lite) (a) A group of minerals of general formula: $(Na,Ca,K)_4Al_3(Al,Si)_3Si_6O_{24}(Cl,F,OH,CO_3,SO_4)$. It consists of generally white or gray-white minerals crystallizing in the dipyramidal class of the tetragonal system, and commonly forms an isomorphous series between marialite and meionite. Scapolite minerals characteristically occur in calcium-rich metamorphic rocks or in igneous rocks as the products of alteration of basic plagioclase feldspars. (b) A specific mineral of the scapolite group, intermediate in composition between marialite and meionite (Ma:Me from 2:1 to 1:3), containing 46-54% silica, and resembling feldspar when massive but having a fibrous appearance and a higher specific gravity. Syn: *wernerite.* (c) A member of the scapolite group, including scapolite, marialite, meionite, and mizzonite.

scar [geomorph] (a) A cliff, precipice, or other steep, rocky eminence or slope (as on the side of a mountain) where bare rock is well exposed to view; e.g,. a limestone face in northern England. Originally, the term referred to a crack or breach; later, an isolated or protruding rock. Etymol: Old Norse "sker", "skerry". (b) A rocky *shore platform.* (c) *landslide scar.* (d) *meander scar.*

scar [paleont] (a) *muscle scar.* (b) *cicatrix.*

scarbroite (scar'-bro-ite) A white triclinic or monoclinic mineral: $Al_5(CO_3)(OH)_{13}{\cdot}5H_2O$. Syn: *tucanite.*

scarp (a) A line of cliffs produced by faulting or by erosion. The term is an abbreviated form of escarpment, and the two terms commonly have the same meaning, although "scarp" is more often applied to cliffs formed by faulting. See also: *fault scarp; erosion scarp.* (b) A relatively straight, clifflike face or slope of considerable linear extent, breaking the general continuity of the land by separating surfaces lying at different levels, as along the margin of a *plateau* or *mesa.* A scarp may be of any height but does not describe properly a slope of highly irregular outline. Cf: *scarp slope.* (c) *beach scarp.* (d) The *headscarp* of a landslide.

scarped plain An area marked by a succession of faintly inclined or gently folded strata, e.g. the eastern part of the Great Plains of the U.S. The inclination of strata has perceptible influence on even the smaller elements of the topography.

scarped ridge *cuesta.*

scarp face *scarp slope.*

scarp-foot spring A spring that flows onto the land surface at or near the foot of an escarpment.

scarpland (scarp'-land) A region marked by a succession of nearly parallel cuestas separated by lowlands.

scarplet (scarp'-let) (a) A miniature scarp, ranging in height from several centimeters to 6 m or more; specif. a *piedmont scarp.* Also, a small scarp formed on a wave-cut platform by the outcropping of resistant rocks. (b) *earthquake scarplet.*

scarp retreat The *slope retreat* of a scarp.

scarp slope (a) The relatively steeper face of a *cuesta,* facing in a direction opposite to the dip of the strata. Cf: *dip slope; back slope.* Syn: *scarp face; inface; front slope.* (b) A *scarp* or an *escarpment.*

scarp stream An *obsequent stream* flowing down a scarp, such as down the scarp slope of a cuesta.

scarred pebble *pitted pebble.*

scatter diagram [stat] (scat'-ter) A graphic representation of paired measurements, usually along Cartesian axes, that aids in visualizing the relationships between two or more variables. Syn: *distribution scatter.*

scatter diagram [struc petrol] *point diagram.*

scattered ice (scat'-tered) An obsolete term for sea-ice concentration of $^1/_{10}$ to $^5/_{10}$; now replaced generally by *open pack ice* and *very open pack ice.*

scattering (a) The change in the direction of energy travel because of collisions or inhomogeneity or anisotropy of the medium. (b) The irregular and diffuse dispersion of energy caused by inhomogeneities in the medium through which the energy is traveling. A perturbation of a uniform field caused by irregularities in the medium.

scatterometer (scat-ter-om'-e-ter) A nonimaging radar device that records back scatter as a function of incidence angle.

scatterometer [oceanog] An instrument carried on a satellite that measures wind velocity at the sea surface by measuring the reflection of centimeter-wavelength radio waves from the sea surface from three or four angles, in a horizontal plane, relative to the direction the satellite is moving. The scatter is proportional to wind speed and the direction of the wind relative to the instrument.

scavenger (scav'-en-ger) An organism that feeds on dead animal matter, decaying organic matter, or refuse.

scavenger well A well located between a well or wells that yield usable water and a source of potential contamination; it is pumped (or allowed to flow) as waste to prevent the contaminated water from reaching the good wells. The most common application is in coastal areas, where scavenger wells are used to prevent salt water from reaching water-supply wells.

scawtite (scawt'-ite) A colorless, white, or grayish monoclinic mineral: $Ca_7Si_6O_{18}(CO_3){\cdot}2H_2O$.

scene The area on the ground that is covered by an image or photograph.

scenedesmacean algae (sce-ne-des-ma'-cean) Colonial, fresh- to brackish-water *green algae* (Chlorophyta) consisting of one, two, or a linear series of four to eight oval to elongate cells (e.g., *Scenedesmus*) (Batten, 1996).

sceptrule (scep'-trule) A hexactinellid-sponge spicule (microsclere) that consists of one long ray with one end containing an axial cross and usually bearing various anaxial outgrowths. See also: *sarule; scopule; clavule; lonchiole.*

S-C fabric A fabric consisting of two intersecting surfaces: a foliation (S surface, from French "schistosité") and a shear band (C surface, from French "cisaillement"). Common in foliated rocks that have undergone simple shear; e.g., mylonites.

schachnerite (schach'-ner-ite) A metallic gray hexagonal mineral: $Ag_{1.1}Hg_{0.9}$.

schafarzikite (scha-far'-zik-ite) A red to brown tetragonal mineral: $Fe^{2+}Sb^{3+}{}_2O_4$.

schäferite A vitreous red cubic mineral: $NaCa_2Mg_2(VO_4)_3$. It has the garnet-type structure.

schairerite (schai'-rer-ite) A colorless trigonal mineral: $Na_{21}(SO_4)_7F_6Cl$. Cf: *galeite.*

schallerite (schal'-ler-ite) A light brown or reddish-brown hexagonal mineral: $(Mn^{2+},Fe^{2+})_{16}As^{3+}{}_3Si_{12}O_{36}(OH)_{17}$. It is related to *friedelite.*

schalstein (schal'-stein) An altered tuff with shear structures; it is usually mafic or calcareous (Holmes, 1928). Rarely used. Etymol: German. Cf: *adinole; spotted slate.*

schapbachite (schap'-bach-ite) A metallic black cubic mineral, a dimorph of matildite: $AgBiS_2$. The term has also been applied to an intimate intergrowth of matildite and galena and to a high-temperature polymorph of matildite.

schaurteite (schaur'-te-ite) A silky white hexagonal mineral of the *fleischerite* group: $Ca_3Ge^{4+}(SO_4)_2(OH)_6{\cdot}3H_2O$.

scheelite (scheel'-ite) (a) A colorless or variously colored adamantine tetragonal mineral: $CaWO_4$. It is found in pneumatolytic veins associated with quartz, and fluoresces to show a blue color. Scheelite is isomorphous with powellite, and is an ore of tungsten. (b) A group name for tetragonal minerals with the general composition *AB*O4, where *A* = Ca, Pb or Y, and *B* = Mo, W, Nb, or Ta.

schefferite (schef'-fer-ite) A brown to black monoclinic mineral of the *pyroxene* group: $(Ca,Mn)_2(Mg,Mn,Fe)_2Si_4O_{12}$. It is a variety of diopside containing manganese and frequently much iron.

scheibeite (schei'-be-ite) *phoenicochroite.*

schematic mollusk (sche-mat'-ic) *hypothetical ancestral mollusk.*

S-chert Stratigraphically controlled chert, occurring in beds (bedded chert), or in groups of nodules (nodular chert) distributed parallel to bedding (Dunbar and Rodgers, 1957, p.248).

schertelite (scher'-tel-ite) A colorless or tan orthorhombic mineral: $(NH_4)_2Mg(PO_3OH)_2{\cdot}4H_2O$. It is found in cave deposits.

scheteligite (sche'-tel-i-gite) *betafite.*

schiavinatoite A grayish-pink tetragonal mineral: $(Nb,Ta)BO_4$. It is the Nb analogue of behierite.

schiefer (schief'-er) A general term referring to a rock's laminated or foliated structure, commonly used to describe rocks ranging from shale to schist: e.g. schieferton or argillaceous shale, tonschiefer or slate. Adj: *schiefrig*. Etymol: German.

schieffelinite (schief'-fe-lin-ite') An adamantine colorless orthorhombic mineral: $Pb(Te^{6+}O_4,SO_4){\cdot}H_2O$.

schiefrig (schief'-rig) Adj. of *schiefer.*

schiller (schil'-ler) A syn. of *play of color.* Etymol: German. See also: *schillerization.*

schillerization (schil'-ler-i-za'-tion) The development of *schiller* or play of color in a mineral, due to the arrangement of minute inclusions in the crystal.

schiller spar A syn. of *bastite.* Also spelled: schillerspar.

schirmerite (schir'-mer-ite) A metallic gray orthorhombic mineral, a member of the lillianite homologous series: $AgBi_3Pb_2S_7$.

schist A strongly foliated crystalline rock, formed by dynamic metamorphism, that can be readily split into thin flakes or slabs because of the well-developed parallelism of more than 50% of the minerals present, particularly those of lamellar or elongate prismatic habit (e.g., mica and hornblende). The mineral composition is not an essential factor in its definition unless specifically included in the rock name (e.g., quartz-muscovite schist). Varieties may also be based on general composition (e.g., calc-silicate schist, amphibole schist) or on texture (e.g., spotted schist).

schist-arenite (schist-ar'-e-nite) A light-colored sandstone containing more than 20% rock fragments derived from an area of regionally metamorphosed rocks (Krynine, 1940); specif. a lithic arenite having abundant fragments of schist. The term was attributed to Adolph Knopf and applied originally by Krynine (1937, p.427) to the medium-grained clastic rocks of the Siwalik Series in northern India, averaging about 40% quartz, 15% feldspar, 35-40% schist and phyllite fragments, and 5-10% accessory materials.

schistic (schist'-ic) *schistose.*

schistoid (schist'-oid) adj. Resembling *schist.*

schistose (schist'-ose) Said of a rock displaying *schistosity.* Cf: *gneissic.* Syn: *schistic.*

schistosity (schis-tos'-i-ty) The foliation in schist or other coarse-grained, crystalline rock due to the parallel alignment of platy mineral grains (mica) or inequant crystals of other minerals.

schizocarp (schiz'-o-carp) A dry fruit that, at maturity, splits apart into several one-seeded, indehiscent carpels.

schizochroal eye (schiz-o-chro'-al) A trilobite eye with large lenses, each with its own cornea, separated by scleral projections that ensure that each *ommatidium* receives light from only its own individual lens. Cf: *abathochroal eye; holochroal eye.*

schizodont (schiz'-o-dont) (a) Said of the dentition of a bivalve mollusk with the median tooth of the left valve broad and divided into two equal parts (bifid), and characterized by coarse, variable, and amorphous teeth diverging sharply from beneath the beak. (b) Said of a subclass of amphidont hinges in ostracodes, having anterior tooth and socket of one valve both bifid, and a reverse arrangement of elevations and depressions in the opposed valve (TIP, 1961, pt. Q, p.54).

schizohaline (schiz-o-hal'-ine) Said of an environment characterized by extreme variation from hypersaline to brackish or fresh-water conditions, as in coastal lagoons subjected to seasonal or sporadic drought or storms (Folk and Siedlecka, 1974). Etymol: Greek; "schizo", "split", + "haline", "salty".

schizolite (schiz'-o-lite) A light-red variety of pectolite containing manganese.

schizolophe (schiz'-o-lophe) A brachiopod lophophore indented anteriorly and medially to define a pair of brachia that bear a row of paired filamentary appendages, at least distally (TIP, 1965, pt. H, p.152).

schizomorph (schiz'-o-morph) In *acritarchs,* having a spherical, ellipsoidal, or discoidal shape and tending to divide into two symmetrical halves. Most of these forms are now considered the zygospores of *zygnematacean algae* (Grenfell, 1995). Also acritarchs of such type.

schizomycete (schiz'-o-my-cete') An organism of the class Schizomycetes, a group of unicellular or noncellular organisms that are variously classified with fungi, with cyanobacteria, or separately. See: Melchior and Werdermann, 1954, p.42. Cf: *myxomycete; eumycete.*

schizont The generation of foraminifers that divides sexually to form embryos; commonly with microspheric test (*B-form*). Cf: *gamont.*

schizoporellid (schiz'-o-po-rel'-lid) Said of bryozoans characterized by a median sinus at the proximal margin of the orifice, as in the cheilostome family Schizoporellidae (TIP, 1953, pt. G, p.14).

schizorhysis (schiz-o-rhy'-sis) A *skeletal canal* in dictyonine hexactinellid sponges passing completely through the dictyonal framework as well as connecting laterally. It is covered by exopinacoderm. Pl: schizorhyses.

schlanite (schlan'-ite) The soluble resin extracted from *anthracoxene* by ether. See also: *anthracoxenite.*

schlemaite A metallic black monoclinic mineral: $(Cu,Ag)_6(Pb,Bi)Se_4$.

schlieren (schlier'-en) Tabular or disclike concentrations of minerals (flow layers) within an igneous intrusion. They have the same general mineralogy as the plutonic rocks, but because of differences in mineral ratios they are darker or lighter; the boundaries with the rock tend to be transitional. Some schlieren are modified inclusions, others may be segregations of minerals. Etymol: German for a flaw in glass due to a zone of abnormal composition. Singular schliere. Also spelled: schliere. Adj: schlieric. Cf: *flow layer.*

schlieren arch A term introduced by Balk (1937, p.56) for an intrusive igneous body with flow layers along its borders but poorly developed or absent in its interior. Cf: *schlieren dome.* Obsolete.

schlieren dome A term introduced by Balk (1937, p.56) for an intrusive igneous body more or less completely outlined by flow layers which culminate in one central area. Cf: *schlieren arch.* Obsolete.

schlossmacherite (schloss'-mach-er-ite') A green or grayish-green rhombohedral mineral of the *beudantite* group: $(H_3O,Ca)Al_3(SO_4,AsO_4)_2(OH)_6$.

Schlumberger array An electrode array in which two closely spaced potential electrodes are placed midway between two current electrodes.

Schmidt hammer A portable, standardized tool for estimating compressive strength. The estimate is based on the height of rebound of a spring-loaded steel mass that impacts the surface of concrete or cemented soil or rock material.

Schmidt net A coordinate system used to plot a *Schmidt projection,* used in crystallography for statistical analysis of data obtained esp. from universal-stage measurements, and in structural geology for plotting azimuths as angles measured clockwise from north and about a point directly beneath the observer.

Schmidt projection A term used in crystallography and structural geology for a *Lambert azimuthal equal-area projection* of the lower hemisphere of a sphere onto the plane of a meridian. Named after Walter Schmidt (1885-1945), Austrian petrologist and mineralogist, who first used the projection in structural geology (Schmidt, 1925, p.395-399). See also: *Schmidt net.*

schmiederite (schmied'-er-ite) A light blue or greenish monoclinic mineral: $Pb_2Cu_2(Se^{4+}O_3)(Se^{6+}O_4)(OH)_4$. The name is sometimes incorrectly given as "schmeiderite".

schmitterite (schmit'-ter-ite) A colorless or pale yellow orthorhombic mineral: $(UO_2)Te^{4+}O_3$.

schneebergite An adamantine brown or beige mineral: $Bi(Co,Ni)_2(AsO4)_2(OH,H_2O)_2$.

schneiderhöhnite (schnei'-der-höhn-ite') A metallic dark brown, yellowish-brown, or yellowish-orange triclinic mineral: $Fe^{2+}Fe^{3+}{}_3As^{3+}{}_5O_{13}$.

schoderite (scho'-der-ite) A yellow or orange-yellow monoclinic mineral: $Al_2(PO_4)(VO_4){\cdot}8H_2O$.

schoenfliesite (schoen'-flies-ite) (a) A colorless cubic mineral: $MgSn^{4+}(OH)_6$. (b) A group name for cubic minerals with the general formula $ASn(OH)_6$, where A = Mg, Fe, Mn, Cu, Zn or Ca.

Schoenflies notation (Schoen'-flies) A system of describing crystal classes by means of symbols used esp. by physicists and chemists. Cf: *Hermann-Mauguin symbols.* Syn: *Schoenflies symbols; international symbols.*

Schoenflies symbols *Schoenflies notation.*

schoenite (schoen'-ite) A syn. of *picromerite.* Also spelled: schönite.

schoepite (schoep'-ite) An adamantine yellow to brown orthorhombic secondary mineral: $(UO_2)_4O(OH)_6{\cdot}6H_2O$. See also: *metaschoepite; paraschoepite.* Syn: *epiianthinite.*

schöllhornite (schöll'-horn-ite) A metallic gray rhombohedral meteorite mineral: $Na0.3CrS2{\cdot}H_2O$.

scholzite (scholz'-ite) A colorless to whitish-gray orthorhombic mineral: $CaZn_2(PO_4)_2{\cdot}2H_2O$.

schoonerite (schoo'-ner-ite) A pale tan, brown, or reddish-brown platy orthorhombic mineral: $Fe^{2+}{}_2ZnMn^{2+}Fe^{3+}(PO_4)_3(OH)_2{\cdot}9H_2O$.

schorenbergite (schor'-en-berg-ite') A hypabyssal nepheline *leucitite* containing phenocrysts of nosean, or sometimes leucite, in a groundmass of leucite, nepheline, and aegirine. Feldspar is absent. The name, given by Brauns in 1921, is from Schorenberg, Laacher See district, Germany. Obsolete.

schörl (a) A vitreous black rhombohedral mineral of the *tourmaline* group: $NaFe^{2+}{}_3Al_6(BO_3)_3Si_6O_{18}(OH)_4$. (b) An obsolete term for any of several dark minerals other than tourmaline; e.g. hornblende. Also spelled: *shorl;* schorl. Syn: *schorlite.*

schorlite *schörl.*

schorlomite (schor'-lo-mite) A black or brown-black cubic mineral of the *garnet* group: $Ca_3(Ti^{4+},Fe^{3+})_2(Si,Fe^{3+})_3O_{12}$. Cf: *melanite.*

schorl rock A term used in Cornwall, England, for a granular rock composed essentially of aggregates of needlelike crystals of black tourmaline (schorl) associated with quartz, and resulting from the complete tourmalinization of granite.

schorre (schor'-re) A Dutch term for that part of a sandy beach covered by the sea only during spring tides.

Schottky defect (Schott'-ky) In a crystal structure, the absence of an atom; it is a type of point defect. Cf: *Frenkel defect; interstitial defect; defect lattice.*

schreibersite (schrei'-ber-site) A silver-white to tin-white, highly magnetic, tetragonal meteorite mineral: $(Fe,Ni)_3P$. It contains small amounts of cobalt and traces of copper, and tarnishes to brass yellow or brown. Schreibersite occurs in tables or plates as oriented inclusions in iron meteorites. Syn: *rhabdite.*

Schreinemakers' analysis (Schrei'-ne-mak'ers') The method pioneered by H.A. Schreinemakers in the 1920s for determination of the topology of phase diagrams based on the number and composition of phases in the system.

schreyerite (schrey'-er-ite) A reddish-brown monoclinic mineral: $V^{3+}{}_2Ti_3O_9$.

schröckingerite (schröck'-ing-er-ite') A green-yellow triclinic secondary mineral: $NaCa_3(UO_2)(SO_4)(CO_3)_3F{\cdot}10H_2O$. It is an ore of uranium. Syn: *dakeite.*

schrötterite (schröt'-ter-ite) An opaline variety of allophane rich in aluminum. Material from the type locality has been shown to be a mixture of glassy halloysite and earthy variscite.

schrund *bergschrund.*

schrund line A term introduced by Gilbert (1904, p.582) for "the base of the bergschrund at a late stage in the excavation of the cirque basin". The line separates the steeper slope of the cirque wall from the gentler, usually scalable, slope below.

schubnelite (schub'-ne-lite) A brilliant black, yellowish-brown, or greensih-brown triclinic mineral: $Fe^{3+}VO_4{\cdot}H_2O$.

schuchardtite (schu'-chardt-ite) A discredited mineral consisting of irregularly interlayered vermiculite and chlorite.

schuetteite (schuet'-te-ite) A yellow, orange-yellow, or greenish trigonal mineral: $Hg_3O_2(SO_4)$.

schuilingite-(Nd) (schui'-ling-ite) A blue orthorhombic mineral: $Pb_2Cu_2(OH)_2(Nd,Gd,Y)_2(CO_3)_6{\cdot}3H_2O$.

schulenbergite (schu'-len-berg-ite') A green-blue trigonal mineral related to spangolite and namuwite: $(Cu,Zn)_7(SO_4)_2(OH)_{10}{\cdot}3H_2O$.

Schuler period (Schu'-ler) The time of 84 minutes that is a natural precession rate for gyrocompasses.

schultenite (schul'-ten-ite) An adamantine colorless to pale yellow monoclinic mineral: $Pb(AsO_3OH)$.

Schulze's reagent (Schul'-ze's) An oxidizing mixture, commonly used in maceration for paleopalynological studies. It consists of a saturated aqueous solution of $KClO_3$ and admixture of varying volumes of concentrated HNO_3 (Schulze, 1855). Named after Franz F. Schulze (1815-1873), German chemist. Also called: Schulze's mixture; Schulze's solution.

schumacherite (schu'-mach-er-ite') An adamantine yellow to yellowish-brown triclinic mineral: $Bi_3O(VO_4,AsO_4,PO_4)_2(OH)$.

schungite (schung'-ite) *shungite.*

schuppen structure (schup'-pen) A syn. of *imbricate structure [tect].* Etymol: German "Schuppenstruktur." The German word is sometimes used in English geologic literature as "schuppenstruktur."

schuppen zone An older term used for duplexes or imbricate fans, commonly with the implication that the faults comprising the structure are closely spaced and the rock involved in the structure has been pervasively sheared. Etymol: German. Syn: *schuppen structure.*

Schürmann series (Schür'-mann) A list of metals so arranged that the sulfide of any one is precipitated at the expense of the sulfide of any lower metal in the series.

schwagerinid (schwa-ger-in'-id) Any fusulinid belonging to the subfamily Schwagerininae.

schwartzembergite (schwartz'-em-berg-ite') An adamantine yellow or reddish-brown mineral: $H_2Pb_5I^{3+}O_6Cl_3$.

schwatzite (schwatz'-ite) A variety of tetrahedrite containing mercury.

schwertmanite An ocherous brownish-yellow tetragonal mineral: $Fe^{3+}_{16}O_{16}(OH)_{9.6}(SO_4)_{3.2}\cdot 10H_2O$.
sciaphilic (sci-a-phil'-ic) Shade-loving; referring to marine organisms encrusting hard substrates that are sheltered or protected from direct sunlight, such as the undersides of corals and the ceilings of reef caverns. Cf: *cryptic.*
scientific hydrology (sci-en-tif'-ic) Hydrologic study devoted principally to fundamental processes and relationships of the hydrologic cycle.
scientific method A general term for the lines of reasoning that scientists tend to follow in attempting to explain natural phenomena. It typically includes observation, analysis, synthesis, classification, and inductive inference, in order to arrive at a *hypothesis* that seems to explain the problem. Hypothesis becomes *theory* if it withstands repeated testing and application. Deductive use of the theory may then explain additional problems. Since the term actually covers several methods, it is often used in the plural. See also: *induction [philos]; deduction.*
scientific name The formal Latin name of a taxon. Cf: *vernacular name.*
scintillation [gem] (scin-til-la'-tion) The flashing, twinkling, or sparkling of light, or the alternating display of reflections, from the polished facets of a gemstone when the diamond, the observer, or the light source moves.
scintillation [radioactivity] A small flash of light produced by an ionizing agent (such as nuclear particles or photons of electromagnetic energy) in a phosphor or *scintillator.* See also: *scintilloscope.*
scintillation counter An instrument that measures ionizing radiation by counting individual scintillations of a substance. It consists of a phosphor and a photomultiplier tube that registers the phosphor's flashes. It may be smaller and more efficient than a Geiger-Müller counter. It is used in spectrometry as well as exploration. Syn: *scintillometer.* Cf: *lithium-drifted germanium detector.*
scintillation spectrometer An instrument for measuring a mass or energy spectrum, therein similar to a *gamma-ray spectrometer,* and determining its frequency distribution by the use of a *scintillation counter.*
scintillator (scin'-til-la'-tor) Any transparent material (crystalline, liquid, or organic) that emits small flashes of light when bombarded by an ionizing agent such as nuclear particles or radiation.
scintillite An helictitelike *speleothem* consisting of a microcrystalline silica core and a sparkling, euhedral quartz covering (Deal, 1964).
scintillometer (scin-til-lom'-e-ter) *scintillation counter.*
scintilloscope (scin'-til-lo-scope') An instrument that displays the *scintillation [radioactivity]* of a substance on a screen. Also spelled: scintilliscope.
sciophyte (sci'-o-phyte) A plant preferring growth in light of low intensity.
scissor fault (scis'-sor) A fault on which there is increasing offset or separation along the strike from an initial point of no offset, with reverse offset in the opposite direction. The separation may be due to a scissorlike or pivotal movement on the fault, or it may be the result of uniform strike-slip movement along a fault across a synclinal or anticlinal fold. The terminology is not rigorous, is often used in error, and should be avoided; *pivotal fault, hinge fault,* and *rotational fault* are similarly used. Obsolete syn: *differential fault.*
sclarite (sclar'-ite) A colorless monoclinic mineral: $(Zn,Mg,Mn^{2+})_4Zn_3(CO_3)_2(OH)_{10}$. It is the zinc analogue of loseyite.
scleracoma (scle-ra-co'-ma) A collective term for the hard skeletal parts of radiolarians.
scleractinian (scle-rac-tin'-i-an) Any zoantharian coelenterate belonging to the order Scleractinia, characterized by solitary and colonial forms with calcareous exoskeletons secreted by the ectoderm. This order includes most post-Paleozoic and living corals. Range, Middle Triassic to Holocene. Syn: *hexacoral.*
sclere A minute skeletal element; esp. a sponge *spicule.* See also: *megasclere; microsclere.*
sclerenchyma (scle-ren'-chy-ma, scle-ren-chy'-ma) Thick-walled strengthening tissue in a plant. It may consist of either elongate cells called fibers or shorter cells called stone cells.
sclerenchyme (scle'-ren-chyme) (a) The calcareous tissue of rugose corallites, esp. the notably thickened parts of the skeleton (TIP, 1956, pt.F, p.250). (b) The vesicular skeletal structure between corallites in colonial coralla, such as the stony substance secreted by the coenenchyme of a scleractinian coral (Shrock and Twenhofel, 1953, p.133). Cf: *mesenchyme; stereom; scleroderm.* Syn: *sclerenchyma.*
sclerine (scler'-ine) In palynomorphs, a term for the *sporoderm* which excludes the intine.
sclerite (scle'-rite) A hard chitinous or calcareous plate, piece, or *spicule* of an invertebrate, e.g. a hardened, chitinized cover forming part of the external skeleton of a merostome or arachnid; a calcareous *ossicle* (anchor, hook, rod, wheel, or disc) of a holothurian; a calcareous skeletal element of the *mesogloea* of an octocoral, irrespective of form; or a thickened line in the operculum, mandible, or frontal membrane of a bryozoan.
scleroblast (scle'-ro-blast) (a) One of the cells of a sponge by which a spicule is formed; a mother cell of one or more sclerocytes. Also, a *sclerocyte.* (b) One of the ectodermal cells of octocorallian *mesogloea* that produce calcareous spicules; e.g. *axoblast.*
scleroclase *sartorite.*
sclerocyte (scle'-ro-cyte) A cell that secretes all or part of a sponge spicule. Syn: *scleroblast; spiculoblast.*
scleroderm (scle'-ro-derm) The hard *sclerenchyme* of the skeleton of a scleractinian coral.
sclerodermite (scle-ro-der'-mite) (a) The center of calcification and surrounding cluster of calcareous (aragonitic) fibers making up a *septum* of a scleractinian coral. Sclerodermites are the apparent primary elements in septa and they are variously arranged in vertical series to make trabeculae. (b) A spine or plate of a holothurian. (c) The hard integument of an arthropod segment.
sclerometer (scle-rom'-e-ter) An instrument used in mineral analysis to determine hardness by measuring the pressure required to scratch a polished surface of the material with a diamond point.
sclerophyll (scle'-ro-phyll) An evergreen leaf that is tough and leathery due to the presence of *sclerenchyma,* typical of dry climates.
scleroseptum (scle-ro-sep'-tum) A calcareous radial *septum* of a coral.
sclerosome (scle'-ro-some) A continuous deposit in a calcareous sponge of nonspicular calcium carbonate that may form part or all of the skeleton.
sclerosponge (scle'-ro-sponge) A member of the sponge class Sclerospongiae, characterized by spherulitic tabular calcareous skeletons with embedded spicules.
sclerotesta (scle-ro-tes'-ta) The hard coat of a seed, e.g. of a cycad. Cf: *sarcotesta.*
sclerotia Pl. of *sclerotium.*
sclerotinite (scle-rot'-i-nite) A maceral of coal within the *inertinite* group, consisting of the sclerotia of fungi or of fungal spores characterized by a round or oval form and varying size. Term now replaced by *funginite* and *secretinite.*
sclerotium (scle-ro'-ti-um) In the eumycetes, a resting body composed of a hardened mass of hyphae, frequently rounded in shape; in the myxomycetes, a hard plasmodial resting stage. Pl: *sclerotia.*
sclerotized (scle'-ro-tized) Said of the covering of an invertebrate (esp. an arthropod) hardened by substances other than chitin.
sclit A Scottish term for slaty or fissile coal, or coaly *blaes.*
scolecite (scol'-e-cite) A silky colorless monoclinic zeolite mineral: $CaAl_2Si_3O_{10}\cdot 3H_2O$. It usually occurs in delicate radiating groups of white fibrous or acicular crystals, and in some forms it shows a worm-like motion when heated.
scolecodont (sco-le'-co-dont) The fossil jaw, with denticles, of an *annelid.* It is composed of silica and chitin, the chitin being carbonized to a jet black during fossilization.
scolecospore (sco-le'-co-spore) An elongate, needle- or worm-shaped, asexual *fungal spore.* Cf: *Saccardoan spore groups.*
scolite (sco'-lite) *scolithus.*
scolithus (sco'-li-thus, sco-li'-thus) Any of various tubular or vermiform trace-fossil structures found in Paleozoic (commonly Cambrian and Ordovician) quartz-rich sandstones (and also in upper Precambrian rocks), consisting of narrow, vertical and usually straight tubes or tube fillings, about 0.2-1 cm in diameter, commonly but not always closely crowded, and generally flaring out into cuplike depressions at their tops. They are thought to be fossil burrows of marine worms, possibly phoronids, and are assigned to the "genus" *Scolithus* (properly *Skolithos*). See also: *worm tube.* Syn: *scolite; pipe-rock burrow.*
scopule (sco'-pule) A *sceptrule* in which the terminal outgrowths are a pair or ring of spines whose ends may be clubbed and bear rings of recurved teeth.
scopulite (scop'-u-lite) A rodlike or stemlike crystallite that terminates in brushes or plumes. Obsolete.
scopulus A term established by the International Astronomical Union for a lobate or irregular scarp on planetary surfaces (USGS, 2003). Pl: scopuli.
score *scoring.*
scoria [coal] (sco'-ri-a) *clinker [coal].*
scoria [volc] A bomb-size pyroclast that is irregular in form and generally very vesicular (Macdonald, 1972, p.126). In less restricted usage, a vesicular cindery crust on the surface of andesitic or basaltic lava, the vesicular nature of which is due to the escape of volcanic gases before solidification; it is usually heavier, darker, and more crystalline than *pumice.* The adjective form, scoriaceous, is usually applied to pyroclastic ejecta. *Cinder* is sometimes used synonymously. Plural: scoriae. See also: *thread-lace scoria.*
scoriaceous [sed] Said of a sedimentary rock whose surface is pitted and irregular like that of volcanic scoria; e.g. "scoriaceous limestone" produced by partial dissolution of the nodules of a nodular limestone in the coastal spray zone.
scoriaceous [volc] Volcanic igneous vesicular texture involving relatively large gas holes such as in vesicular basalt. Coarser than pumiceous.
scoria tuff A deposit of fragmented scoria in a fine-grained ash matrix.
scoring (scor'-ing) (a) The formation of parallel scratches, lines, or grooves in a bedrock surface by the abrasive action of rock fragments transported by a moving glacier. (b) A scratch, line, or groove produced by scoring. Syn: *score.*
scorodite (scor'-o-dite) A variously colored orthorhombic mineral of the *variscite* group: $Fe^{3+}AsO_4\cdot 2H_2O$. It represents a lesser ore of arsenic.
scorzalite (scor'-za-lite) A blue, bluish-white, or bluish-green monoclinic mineral of the *lazulite* group: $(Fe^{2+},Mg)Al_2(PO_4)_2(OH)_2$.

Scotch pebble A rounded fragment of agate, carnelian, cairngorm, or other variety of quartz, found in the gravels of parts of Scotland, and used as a semiprecious stone.

Scotch topaz A yellow transparent variety of quartz resembling *topaz;* specif. *cairngorm.*

scotlandite (scot'-land-ite) An adamantine, pale yellow, grayish-white, or colorless monoclinic mineral, the first natural sulfite to be described: $PbS^{4+}O_3$.

scour [eng] (a) A current or flow of water that removes mud or other granular material from a streambed; also, the structure built to produce such a current. (b) The process of erosion that is controlled by the velocity of water flowing in a channel or being emitted under pressure from a nozzle.

scour [geomorph] (a) The powerful and concentrated clearing and digging action of flowing air, water, or ice, esp. the downward erosion by stream water in sweeping away mud and silt on the outside curve of a bend, or during time of flood. (b) A place in a streambed swept (scoured) by running water, forming a depression.

scour [tides] *tidal scour.*

scour and fill [geomorph] A process of alternate excavation and refilling of a channel, as by a stream or the tides; esp. such a process occurring in time of flood when the discharge and velocity of an aggrading stream are suddenly increased, causing the digging of new channels that become filled with sediment when the flood subsides. Cf: *cut and fill [geomorph].*

scour and fill [sed struc] A sedimentary structure consisting of a small erosional channel, generally ellipsoidal, that is subsequently filled; a small-scale washout.

scour cast A sole mark consisting of a cast of a scour mark; specif. a *flute cast.*

scour channel A large groovelike erosional feature produced in sediments by scour.

scour depression A crescentic hollow produced in a stream bed near the outside of a bend by water that scours below the grade of the stream (Bryan, 1920, p.191).

scouring (scour'-ing) The process of erosion by the action of flowing air, ice, or water, esp. *glacial scour.* Cf: *scour [geomorph].*

scouring rush *sphenopsid.*

scouring velocity The velocity of water that is necessary to dislodge stranded solids from the stream bed.

scour lineation A smooth, low ridge, 2-5 cm wide, formed on a sedimentary surface and believed to result from the scouring action of a current of water. It is characterized by symmetrical ends so that the line of current movement, but not its direction, can be ascertained.

scour mark A *current mark* produced by the cutting or scouring action of a current of water flowing over the bottom; e.g. a flute. See also: *transverse scour mark.*

scour pool Deepened area commonly located below a concave bank in the bed of a meandering stream, resulting from entrainment and removal by transport of sediment on the bed. Its depth depends on the scouring action of currents down to the scale of microturbulent. As river stage drops to a relatively low level the scour pool ordinarily accumulates sediment and decreases in depth. When stage rises and velocity increases, scour is initiated readily because the last sediment was deposited toward the end of a period of deposition and hence consists of small particles which are readily entrained. If bed materials are notably heterogenous, the windowing of finer sediment may leave a resistant pavement which stops or decreases the rate of pool deepening.

scour side The upstream, or *stoss,* side of a *roche moutonnée,* smoothed, striated, and rounded by glacial abrasion. Ant: *pluck side.*

scourway (scour'-way) A channel produced by a strong glacial stream near the margin of an ice sheet.

scrap mica Mica whose size, color, or quality is below specifications for sheet mica; e.g. flake mica, or the mica obtained as a by-product or waste from the preparation of sheet mica.

scratch *striation.*

scree A term commonly used in Great Britain as a loose equivalent of talus in each of its senses: broken rock fragments; a heap of such fragments; and the steep slope consisting of such fragments. Some authorities regard scree as the material that makes up the sloping land feature known as talus; others consider scree as a sheet of any loose, fragmental material lying on or mantling a slope (cf. *block field*), and talus as that material accumulating specif. at the base of, and obviously derived from, a cliff or other projecting mass (cf. *alluvial talus, avalanche talus, rockfall talus*).

scree creep The gradual and steady downhill movement of individual large blocks of rock on a slope that is often gentle; it is most noticeable where the rocks are massive or well-jointed. See also: *talus creep.*

screen [cart] To apply screen tints to copy for printing; a screen tint of the desired percent of full color is placed between the copy and the light-sensitive film or printing material, so that the copy will print in tints of the solid color. Cf: *screen tint.*

screen [eco geol] (a) An apparatus used to separate material according to size of its particles or to allow the passage of smaller particles while preventing that of larger (as in grading coal, ore, rock, or aggregate); it usually consists of a perforated plate or sheet, or of meshed wire or woven cloth, with regularly spaced holes of uniform size, mounted in a suitable frame. Cf: *sieve.* (b) *well screen.*

screen analysis Determination of the particle-size distribution of a soil, sediment, or ore by measuring the percentage of the particles that will pass through standard screens of various sizes.

screening The operation of passing loose materials (such as gravel or coal) through a screen so that constituent particles are separated into defined sizes.

screen tint A photomechanical impression on glass or plastic, consisting of evenly or randomly oriented squares or dots whose size increases in direct proportion to the intensity of tone to be printed. Screen tints may be negative (clear image and opaque background) or positive (opaque image and clear background). For printing, they are exposed on light-sensitized printing plates; screen tints may range from 4 to 95 percent of full color. See also: *screen [cart].*

screw axis A type of crystal symmetry element that is a combination of a rotation of $360°/n$ with a translation of a/mn where a is a lattice period (usually the a, b, or c crystal axis), n may be 1, 2, 3, 4, or 6, and m is an integer between 0 and n.

screw dislocation [cryst] A type of *line defect* in a crystal: a row of atoms along which a crystallographic plane seems to spiral. See also: *spiral growth.*

screw dislocation [struc petrol] A *dislocation [struc petrol]* in which the linear array of lattice imperfections is parallel to the glide, or slip, direction.

screw ice (a) Small ice fragments in heaps or ridges, produced by the crushing together of ice cakes. (b) A small formation of *pressure ice.*

screwing (screw'-ing) In glaciology, a general term used in the older literature to describe the processes giving rise to *pressure ridges.*

scribing (scrib'-ing) The process of preparing a map or other drawing for reproduction by cutting the detail to be shown into an opaque medium that coats a sheet of transparent plastic, using a scriber (an instrument holding one of a set of needles or blades of various diameters or cross-sectional shapes, sharpened to desired dimensions). The result of the process is a negative of the material to be reproduced.

scrobicule (scro'-bi-cule) One of the smooth, shallow, depressed rings or trenches surrounding the bases of echinoid tubercles and serving for attachment of muscles of spines. See also: *areole.*

scrobis septalis (scro'-bis sep-tal'-is) The inframarginal, asymmetrical, sometimes deep indentation or concave surface of the apertural face of a foraminiferal test (as in *Alabamina*) (TIP, 1964, pt. C, p.63). Etymol: Latin. See also: *infundibulum.* Syn: *inframarginal sulcus.*

scroll (a) One of a series of crescentic deposits built by a stream on the inner bank of a shifting channel; e.g. a *floodplain scroll.* (b) *meander scroll.*

scroll meander A *forced-cut meander* in which the building of meander scrolls on the inner bank is the cause of erosion on the outer bank of the meander (Melton, 1936, p.597). See also: *lacine meander.*

scroll topography A topography of alternating ridges and swales created by meander migration across a floodplain.

scrub Low-growing or stunted vegetation on poor soil or in semiarid regions, which sometimes forms inpenetrable masses.

scrutinyite (scru'-tin-y-ite) A submetallic reddish-brown or clove-brown orthorhombic mineral: αPbO_2.

SCS [sed] *swaley cross-stratification.*

SCS [seis] *single-channel seismic profiling.*

sculpture [geomorph] (sculp'-ture) (a) The carving-out of surficial features of the Earth's surface by erosive agents, such as rain, running water, waves, glaciers, and wind. The term has been loosely applied to include also the processes of deposition and earth movement. Syn: *land sculpture; glyptogenesis.* (b) A landform resulting from a modification or sculpturing of an existing form.

sculpture [paleont] Strongly developed ornamentation of preserved hard parts of an animal; e.g. the relief pattern on the surface of a gastropod shell or on superficial dermal bones of a vertebrate.

sculpture [palyn] The external textural modifications (such as spines, warts, granules, pila, pits, streaks, and reticulations) of the exine of pollen grains and spores. In pollen it is always a feature of the *ektexine.* The sculptural elements are predominantly less than a few micrometers in height. The term may also be used for similar features of the ectoderm or endoderm of an *acritarch* or for the surface of a *dinoflagellate theca* or *cyst* (Cramer and Diez Rodriguez, 1979; Williams et al., 2000). Cf: *structure [palyn].* Syn: *ornamentation [palyn].*

scum A film that floats on a liquid, such as a stagnant pool. The film may be composed of soap, of precipitated calcium carbonate, or of putrid matter.

scutulum (scu'-tu-lum) A very flat *shield volcano* complex, produced by highly fluid basaltic lavas. Term introduced by Noë-Nygaard in 1968.

scutum (scu'-tum) (a) One of a pair of opercular valves, adjacent to the rostrum in cirripede crustaceans, with adductor-muscle attachments. Cf: *tergum.* (b) A lateral *marginal spine,* generally broad and flat, overarching the frontal area in some anascan cheilostomes (bryozoans). Pl: scuta.

scyphozoan (scy-pho-zo'-an) Any marine coelenterate belonging to the class Scyphozoa, characterized by the predominance of medusoid forms. Range, Precambrian or Cambrian to Holocene.

Scythian (Scyth'-i-an) A series or stage of the Standard Global Chronostratigraphic Scale: equivalent to Lower Triassic Series (above Permian System, below Anisian Stage). Also spelled: *Skythian.*

se In structural petrology, a fabric defined by the preferred orientation of grains external to a porphyroblast. It may or may not be parallel to the preferred orientation of minerals within the porphyroblast. Cf: *si.*

sea [oceanog] (a) An inland body of salt water. (b) A geographic division of an *ocean.* (c) An ocean area of wave generation.

sea [planet] *mare.*

sea [waves] Irregular, short-crested waves within their generating area. Ant: *swell [waves].*

sea arch An opening through a headland, formed by wave erosion or solution (as by the enlargement of a *sea cave,* or by the meeting of two sea caves from opposite sides) and leaving a bridge of rock over the water. Syn: *natural arch; natural bridge; sea bridge.*

sea ball A spherical mass of somewhat fibrous living or dead vegetation (esp. algae), produced mechanically in shallow waters along a seashore by the compacting effect of wave movement. Cf: *lake ball.*

sea bank (a) *seashore.* (b) A *sandbank* adjacent to the sea. (c) seawall.

seaboard (sea'-board) (a) The strip of land bordering a seacoast. (b) *seacoast.*

sea breeze A relatively cool mesoscale surface wind directed from the ocean toward land in response to differential heating of land and sea; develops during daylight hours when the land surface warms more than the adjacent sea surface. Called a lake breeze when wind blows off a large lake.

sea bridge *sea arch.*

sea-captured stream (sea'-cap-tured) A stream, flowing parallel to the seashore, that is cut in two as a result of marine erosion and that may enter the sea by way of a waterfall (Cleland, 1925).

sea cave A cleft or cavity in the base of a sea cliff, excavated where wave action has enlarged natural lines of weakness in easily weathered rock; it is usually at sea level and affected by the tides. Syn: *cave [coast]; sea chasm.*

sea chasm A deep, narrow *sea cave.*

sea cliff A cliff or slope produced by wave erosion, situated at the seaward edge of the coast or the landward side of the wave-cut platform, and marking the inner limit of erosion. It may vary from an inconspicuous slope to a high, steep escarpment. Example: Gay Head on Marthas Vineyard, Mass. Cf: *wave-cut cliff; shore cliff.* Syn: *bluff; cliff [geomorph].*

sea coal An old British syn. of *bituminous coal,* named after coal washed ashore and used for fuel; the name was extended to mined coal, as well.

seacoast (sea'-coast) The coast adjacent to a sea or ocean. Syn: *seaboard.*

sea cucumber A *holothuroid* echinoderm having a body shape resembling a cucumber, a flexible body wall, and the ability to creep along the sea floor.

sea fan *submarine fan.*

sea-floor spreading A process whereby new oceanic crust is formed by upwelling of magma at the center of mid-ocean ridges and by a moving-away of the new material from the site of upwelling at rates of one to ten centimeters per year. This movement provides the source of seafloor within the theory of *plate tectonics,* which also contains a provision for destruction of sea-floor, by subduction. See also: *expanding Earth.* Syn: *ocean-floor spreading; spreading concept.*

sea-floor trench *trench [marine geol].*

sea-foam *sepiolite.*

seafront (sea'-front) The land, buildings, or section of a town along a seashore or bordering a sea.

sea gate (a) A restricted passage leading or giving access to the sea. (b) A gate that protects against the sea.

sea gully *slope gully.*

sea ice (a) Any form of ice originating from the freezing of seawater (thus excluding icebergs). Ant: *land ice.* See also: *field ice.* (b) A mariner's term for any ice that is floating in the sea or that has drifted to the sea.

sea-ice shelf Sea ice floating in the vicinity of its formation and separated from *fast ice* (of which it may have been a part) by a *tide crack* or a family of such cracks.

seaknoll (sea'-knoll) *knoll [marine geol].*

seal A rock whose petrophysical properties, mainly low permeability, make it capable or impeding the upward migration of oil or gas.

sea level (a) The height of the sea surface. The height can be relative to mean sea level, or the center of mass of earth. (b) More commonly, a syn. of *mean sea level.*

sea-level datum (sea'-lev-el) *Chart datum* referring to *mean sea level.*

sealing-wax structure (seal'-ing-wax) A term used by Fairbridge (1946, p.85, 87) for a primary sedimentary flow structure produced by slumping, characterized by a lack of a sharply defined slip plane at the base or a contemporaneous erosion plane at the top, and occupying a zone of highly fluid contortion in otherwise normal sedimentary succession.

sea loch A *fjord* along the western Highland coast of Scotland.

seam [ore dep] A particular bed or vein in a series of beds; it is usually said of coal but may also pertain to metallic minerals.

seam [stratig] (a) A thin layer or stratum of rock separating two distinctive layers of different composition or greater magnitude. (b) Strictly, the line of separation between two different strata, resembling the seam between two parts of a garment.

seamanite (sea'-man-ite) A pale-yellow to wine-yellow prismatic orthorhombic mineral: $Mn^{2+}_3B(PO_4)(OH)_6$.

sea mat A *bryozoan,* esp. an incrusting bryozoan.

sea meadows Upper layers of the open ocean that are rich in phytoplankton and provide abundant food for zooplankton. The term is usually used in the plural.

seamless data base A digital data base storing, as one continuous data structure, spatial information spanning two or more disjointed map sheets (Krzanowski et al., 1993).

sea moat *moat [marine geol].*

seamount (sea'-mount) An elevation of the sea floor, 1000 m or higher, either flat-topped (called a *guyot*) or peaked (called a *seapeak*). Seamounts may be either discrete, arranged in a linear or random grouping, or connected at their bases and aligned along a ridge or rise.

sea mud Mud from the sea; specif. a rich, slimy deposit in a salt marsh or along a seashore, sometimes used as a manure. Syn: *sea ooze.*

seam wave A seismic channel wave in a low-velocity member, such as a coal seam. See: *channel wave.*

sea ooze *sea mud.*

seapeak (sea'-peak) A type of *seamount* that has a pointed summit.

sea peat A rare type of peat, formed from seaweeds.

seapoose (sea'-poose) A term used along the shore of Long Island, N.Y., for a shallow inlet or tidal river. Etymol: Algonquian, akin to Delaware "sepus", "small brook".

seaquake (sea'-quake) An earthquake that occurs beneath the ocean and that can be felt on board a ship in the vicinity of the epicenter. Syn: *submarine earthquake.*

sea reach The *reach* of the lower course of a stream where it approaches the sea.

sea rim The apparent horizon as actually observed at sea; the sea-level horizon.

searlesite (searles'-ite) A white monoclinic mineral: $NaBSi_2O_5(OH)_2$.

Seasat (Sea'-sat) A NASA satellite for obtaining oceanographic data, launched in July 1978 and failed in October 1978. During its brief life it gave outstanding data on sea-surface elevation, which yielded much information on gravity over the ocean basins.

seascarp (sea'-scarp) A relatively long, high, rectilinear submarine cliff or wall.

seashore (sea'-shore) (a) The narrow strip of land adjacent to or bordering a sea or ocean. Syn: *seaside.* (b) A legal term for all the ground between the ordinary tide levels; the *foreshore.* Syn: *seastrand.*

seaside (sea'-side) *seashore.*

sea slick A smooth area on the surface of an ocean or body of fresh water, caused by organic material.

sea slope A slope of the land toward the sea.

sea-slug model A progradational or clinoform model used in sequence stratigraphy.

sea snow *marine snow.*

seasonally frozen ground (sea'-son-al-ly) Ground that is frozen by low seasonal temperatures and remains frozen only during the winter; it corresponds to the active layer in permafrost regions. Syn: *frost zone.*

seasonal recovery (sea'-son-al) Recharge to groundwater during and after a wet season, with a consequent rise in the level of the potentiometric surface.

seasonal stream An *intermittent stream* that flows only during a certain climatic season; e.g. a *winterbourne.*

sea stack *stack [coast].*

sea state A description of the roughness of the ocean surface, either numerical or in words.

seastrand (sea'-strand) *seashore.*

sea-surface topography The deviation of sea level from the geoid, due to tides, *storm surges, Kelvin waves, Rossby waves,* and surface geostrophic currents. Commonly called *dynamic topography.*

seat clay *underclay.*

seat earth A British term for a bed of rock underlying a coal seam, representing an old soil that supported the vegetation from which the coal was formed; specif. *underclay.* A highly siliceous seat earth is known as *ganister.* Syn: *seat rock; seat stone; coal seat.*

seat rock *seat earth.*

seat stone *seat earth.*

sea urchin An *echinoid* having a globular shape and a theca of calcareous plates, commonly with sharp movable spines.

seawall (a) A long, steep-faced embankment of shingle or boulders, built by powerful storm waves along a seacoast at the high-water mark. (b) A man-made wall or embankment of stone, reinforced concrete, or other material along a shore to prevent wave erosion.

seawater (sea'-wa-ter) The water of the oceans, characterized by its salinity and distinguished from the fresh water of lakes, streams, and rain. *Salt water* is sometimes used synonymously.

sea-water encroachment *salt-water encroachment.*

sea-water intrusion *salt-water encroachment.*

SeaWiFs The Sea-viewing Wide Field-of-view Sensor carried on the SeaStar spacecraft launched in 1997. It makes quantitative measurements of the bio-optical properties to the ocean including ocean color. The measurements are used to calculate the concentration of chlorophyll in the surface layer of the ocean, which is correlated with the abundance of phytoplankton in the layer.

seaworn (sea'-worn) Diminished or wasted away by the sea, as a *seaworn* shore.

sebastianite (se-bas'-ti-an-ite') A *gabbro* composed of euhedral anorthite, biotite, augite, and apatite, but without feldspathoids or quartz. This rock has been found as fragments in extrusive rocks of Monte Somma, Italy. The name was given by Lacroix in 1917, for San Sebastian, Monte Somma. Cf: *puglianite.* Obsolete.

sebkha (seb'-kha) *sabkha.*

Secchi disk (Sec'-chi) An instrument used to measure water transparency or clarity; a white disc of fixed diameter is lowered into the water, and an average is taken of the depth at which it disappears from view and at which it reappears when raised.

sechron (se'-chron) The maximum interval of geologic time occupied by a given depositional sequence, defined at the points where the boundaries of the sequence change laterally from unconformities to conformities along which there is no significant hiatus (Mitchum, 1977, p.210).

second (sec'-ond) (a) A unit of time equal to $^{1}/_{60}$ of a minute or $^{1}/_{3600}$ of an hour; specif. the cgs unit of time, originally equal to $^{1}/_{86400}$ part of the mean solar day but now defined as the duration of 9,192,631,770 cycles of frequency associated with the transition between two hyperfine levels of the fundamental state of the atom of cesium-133. Abbrev: sec; s (in physical tables). (b) A unit of angular measure equal to $^{1}/_{60}$ of a minute of arc or $^{1}/_{3600}$ of a degree. Symbol: ″. (c) An informal oceanographic term used to describe distance or depth, equal to about 1,463 m (4,800 ft) or the distance that sound will travel through seawater during one second of time.

Secondary (Sec'-ond-ar'-y) A term applied in the early 19th century as a syn. of *Floetz.* It was later applied to the extensive series of stratified rocks separating the older *Primary* and the younger *Tertiary* rocks, and ranging from the Silurian to the Cretaceous; still later, it was restricted to the whole of the Mesozoic Era. The term was abandoned in the late 19th century in favor of *Mesozoic.*

secondary [coast] (sec'-ond-ar'-y) Said of a mature coast or shoreline whose features are produced chiefly by present-day marine processes (Shepard, 1937, p.605); e.g. coasts shaped by wave erosion, marine deposition, or marine organisms. Cf: *primary.*

secondary [eco geol] (a) *supergene.* (b) Said of a fluid inclusion when it formed at a later time and in equilibrium with different fluids than the host mineral.

secondary [metal] Said of metal obtained from scrap rather than from ore. Ant: *primary [eco geol].*

secondary allochthony In coal formation, accumulation of plant remains in a region characterized by erosion, transport, and resedimentation of coal masses previously deposited elsewhere than in place. Cf: *primary allochthony.*

secondary aperture *deuteroforamen.*

secondary arc A mountain arc that lacks deep-seated connections. It is generally located behind the junction point of two primary arcs and is convex in the opposite direction. Obsolete.

secondary ash *extraneous ash.*

secondary axial septulum In a fusulinid foraminiferal test, an *axial septulum* located between primary axial septula. Cf: *secondary septulum.*

secondary carina Row of nodes or denticles on upper surface of palmate or scaphate pectiniform conodont elements marking position of a lateral process that is surrounded by platform extensions (TIP, 1981, pt. W, supp.2).

secondary clay A clay that has been transported from its place of formation and redeposited elsewhere. Cf: *residual clay; primary clay.*

secondary cleavage An obsolete term for cleavage, used by Leith (1905, p.11) to emphasize its development after consolidation of the rock, by deformation or metamorphism. See also: *metaclase.*

secondary consequent stream (a) A tributary of a *subsequent stream,* flowing parallel to or down the same slope as the original consequent stream; it is usually developed after the formation of a subsequent stream, but in a direction consistent with that of the original consequent stream. (b) A stream flowing down the flank of an anticline or syncline. Syn: *lateral consequent stream.*

secondary consolidation (a) Consolidation of sediments at essentially constant pressure, as it results from internal processes such as recrystallization. (b) In engineering, the plastic deformation of a soil mass under constant load. This contrasts with "primary consolidation", the reduction in volume of a soil mass, accompanied by outflow of water from pores between the soil particles.

secondary corner One of the minor convexities seen in the profile of a sedimentary particle. Fifteen to 30 may occur on angular particles, but secondary corners quickly disappear during abrasion and are absent at a roundness value of 0.60 (Pettijohn, 1957, p.58-59).

secondary crater An *impact crater* produced by the relatively low-velocity impact of fragments ejected from a large *primary crater.* Syn: *satellitic crater.*

secondary crater chain An aligned group, more or less straight, of *secondary craters;* often radial to the *primary crater* that ejected them.

secondary creep Deformation of a material under a constant differential stress, with the strain time relationship as a constant. Cf: *primary creep.* Syn: *steady-state creep.*

secondary enlargement Deposition, around a clastic mineral grain, of material of the same composition as that grain and in optical and crystallographic continuity with it, often resulting in crystal faces characteristic of the original mineral (Pettijohn, 1957, p.119); e.g. the addition of a quartz overgrowth around a silica grain in sandstone, or the growth of new material around detrital nuclei such as calcite, feldspar, and tourmaline. Cf: *rim cementation.* Syn: *secondary growth.*

secondary enrichment *supergene enrichment.*

secondary foliation A *foliation* that develops during deformation and/or metamorphism; includes slaty cleavage, crenulation cleavage, solution cleavage, schistosity, and gneissic banding. Cf: *foliation [struc geol].*

secondary fumarole *rootless fumarole.*

secondary geosyncline (a) A geosyncline appearing at the culmination of or after geosynclinal orogeny, such as an *exogeosyncline, epieugeosyncline,* or *intradeep* (Peyve and Sinitzyn, 1950). (b) Haug's term for a *sequent geosyncline* (1900, p.617-711). Obsolete.

secondary glacier A small *valley glacier* that joins a larger *trunk glacier* as a *tributary glacier.*

secondary growth *secondary enlargement.*

secondary inclusion A fluid inclusion formed by any process after crystallization of the host mineral is essentially complete. Most are formed during recrystallizational healing of fractures within a crystal (Roedder, 1967, p.522).

secondary interstice An *interstice* that formed after the formation of the enclosing rock. Cf: *original interstice.*

secondary ionization mass spectrometer *ion microprobe.* Abbrev: SIMS.

secondary layer The inner shell layer of a brachiopod deposited by the outer epithelium median of the outer mantle lobes. It may be fibrous, laminar, or prismatic. Cf: *primary layer.*

secondary limestone Limestone deposited from solution in cracks and cavities of other rocks; esp. the limestone accompanying the salt and gypsum of the Gulf Coast salt domes.

secondary maximum A term used by Udden (1914) for a particle size of a sediment or rock, having greater frequency than other size ranges surrounding it but not greater than the modal diameter. It can be obtained graphically by locating the second highest peak of the frequency curve. A given sample may have more than one secondary maximum. Syn: *secondary mode.*

secondary mineral A mineral formed later than the rock enclosing it, usually at the expense of an earlier-formed *primary mineral,* as a result of weathering, metamorphism, or solution.

secondary mode *secondary maximum.*

secondary optic axis One of two optic axes in a crystal along which all light rays travel with equal velocity. Secondary optic axes are close to but do not necessarily coincide with primary optic axes.

secondary oral A small to large oral plate in edrioasteroids, without intrathecal extensions and commonly with a fixed position relative to the primary orals; frequently only a surficial element (Bell, 1976).

secondary orifice The opening at the outer end of the *peristome* in some ascophoran cheilostomes (bryozoans). Syn: *peristomice.*

secondary permeability The permeability developed in a rock after its deposition or emplacement. Syn: *induced permeability.*

secondary pollutants In air pollution, those pollutants produced when *primary pollutants* react with normal atmospheric compounds; an example is ozone, which forms through reactions between primary pollutants, sunlight, and natural atmospheric gases (Keller, 2002).

secondary porosity The *porosity* developed in a rock after its deposition or emplacement, through such processes as solution or fracturing. Cf: *primary porosity.*

secondary production The organic matter that *zooplankton* herbivores produce within a given marine area or volume. Cf: *net primary production.*

secondary recovery Production of oil or gas as a result of artificially augmenting the *reservoir energy,* as by injection of water or other fluid. Secondary-recovery techniques are generally applied after substantial depletion of the reservoir. See also: *water flooding.*

secondary reflection *multiple reflection.*

secondary rocks (a) Rocks composed of particles derived from the erosion or weathering of pre-existing rocks, such as residual, chemical, or organic rocks formed of detrital, precipitated, or organically accumulated materials; specif. clastic sedimentary rocks. Cf: *primary rocks.* (b) A term applied by Lehmann (1756) to fossiliferous and stratified rocks, containing material eroded from the older *primitive rocks.*

secondary septulum A minor partition of a *chamberlet* in a foraminiferal test, reaching downward (adaxially) from the spirotheca (as in Neoschwagerininae); e.g. *secondary axial septulum* and *secondary transverse septulum.* Cf: *primary septulum.*

secondary spine An intermediate-sized echinoid spine, appearing later than the *primary spine.*

secondary stratification Stratification developed when sediments already deposited are resuspended and redeposited. Syn: *indirect stratification.*

secondary structure [geol] A structure that originated subsequent to the deposition or emplacement of the rock in which it is found, such as a fault, fold, or joint produced by tectonic movement; esp. an epigenetic *sedimentary structure*, such as a concretion or nodule produced by chemical action, or a sedimentary dike formed by infilling. Cf: *primary structure [geol].*

secondary structure [paleont] Coarse structure, commonly between distinct laminae, in the wall of a tintinnid *lorica.* Cf: *primary structure [paleont]; tertiary structure.*

secondary succession An association of plants that develops after the destruction of all or part of the original plant community.

secondary sulfide zone *sulfide zone.*

secondary tectogenesis Gravity-driven sliding off a *geotumor* and the resulting deformation (Haarmann, 1930). Obsolete.

secondary tectonite A tectonite whose fabric is *deformation fabric.* Most tectonites are of this type. Cf: *primary tectonite.*

secondary tissue Plant tissue developed from a lateral or secondary meristem, e.g., *vascular cambium* or *cork cambium* (Fuller and Tippo, 1954, p.970).

secondary transverse septulum In a foraminiferal test, a *transverse septulum* whose plane is approximately normal to the axis of coiling. Cf: *secondary septulum.*

secondary twinning *deformation twinning.*

secondary vein In mining law, a vein discovered subsequent to the one on which the mining claim was based; an incidental vein.

secondary wall A cell-wall layer deposited internally on the primary wall layer in a plant cell. It is generally of different composition from the primary wall (Cronquist, 1961, p.881).

secondary wastewater treatment The second step in most pubicly owned wastewater treatment systems is one in which bacteria consume the organic parts of the waste. It is accomplished by bringing together waste, bacteria, and oxygen in trickling filters or in the activated sludge process. This treatment removes floating and settleable solids and about 90 percent of the oxygen-demanding substances and suspended solids. Disinfection is the final stage of secondary wastewater treatment (USEPA, 1995). Cf: *advanced wastewater treatment; primary wastewater treatment.*

secondary wave *S wave.*

second boiling point The evolution of a gas or supercritical liquid phase from a liquid or melt upon cooling. During cooling, crystallization of large amounts of minerals low in or lacking volatile components (e.g., feldspars, quartz) results in a sufficient increase in concentration of the volatile component(s) (e.g., water) in the residual melt that the vapor pressure of the volatile component exceeds total pressure and a separate phase rich in this component forms, i.e., the liquid or melt boils. This is especially known to occur in hydrous felsic melts such as granite or syenite.

second bottom The first terrace above the normal flood plain (or *first bottom*) of a river.

second-class ore An ore that needs preliminary treatment before it is of a sufficiently high grade to be acceptable for shipment or market. Cf: *first-class ore.* Syn: *milling ore.*

second-cycle conglomerate A conglomerate made of clasts that themselves show evidence of having been derived from a pre-existing conglomerate.

second-derivative map (sec'-ond-de-riv'-a-tive) A contour or color-coded map of the second vertical derivative of a potential field, such as the Earth's gravity or magnetic field. Values represent a vertical rate of change of the vertical gradient of the potential field. This map enhances both short-wavelength anomalies and noise. See also: *first-derivative map.*

second law of thermodynamics The statement concerning entropy as a function of the state of the system, which says that for all reversible processes, the change in entropy is equal to the heat which the system exchanges with the outside world divided by the absolute temperature. In irreversible processes, the change in entropy is greater than the quotient of heat and temperature.

second-order leveling (sec'-ond-or'-der) Spirit leveling that has less stringent requirements than those of *first-order leveling,* in which lines between bench marks established by first-order leveling are run in only one direction using first-order instruments and methods (or other lines are divided into sections, over which forward and backward runnings are to be made) and in which the maximum allowable discrepancy is 8.4 mm times the square root of the length of the line (or section) in kilometers (or 0.035 ft times the square root of the distance in miles). Cf: *third-order leveling.*

second-order pinacoid In a monoclinic or triclinic crystal, any $\{h0l\}$ or $\{\bar{h}0l\}$ *pinacoid.* Cf: *first-order pinacoid; third-order pinacoid; fourth-order pinacoid.*

second-order prism A crystal form: in a tetragonal crystal, the $\{100\}$ prism; in a hexagonal crystal, the $\{11\bar{2}0\}$ prism; and in an orthorhombic crystal, any $\{h0l\}$ prism. Cf: *first-order prism; third-order prism; fourth-order prism.* See also: *macrodome.*

second-order transformation A *phase transformation* in which the second derivatives of Gibbs free energy with respect to temperature and pressure exhibit discontinuities. A polymorphic transformation in which the second derivatives of the free-energy function are discontinuous at the transformation temperature, reflecting a discontinuity in heat capacity. The first derivatives, entropy and volume, are continuous (Putnis and McConnell, 1980). Second-order transformations (e.g., that between alpha- and beta-quartz) are generally non-quenchable and transformation to the lower-temperature polymorph during cooling is spontaneous and complete. Cf: *first-order transformation; lambda transformation.*

second setting An orientation of a *crystal* or a *crystal structure* with monoclinic symmetry in which the *b* axis is the symmetrically unique axis (i.e., it is parallel to the *diad* or perpendicular to the mirror plane). This is the standard orientation for the *monoclinic system.* Cf: *first setting.*

second vertical derivative A rate of change of the *vertical gradient* of the potential field, such as the gravity or magnetic fields. A measure of the curvature of the potential field, where positive values indicate an increasing vertical gradient and negative values indicate a decreasing vertical gradient. It enhences both high-frequency components of the measured field and noise. See also: *first vertical derivative.*

second water An obsolete term for the quality or luster of a gemstone next below *first water,* such as that of a diamond that is almost perfect but contains slight flaws or turbid patches. Cf: *third water.*

second-year ice *Sea ice* that has survived only one summer's melt; it is thicker (2 meters or more) and less dense than first-year ice and therefore stands higher out of the water: any hummocks present show weathering. Syn: *two-year ice; young polar ice.*

secretinite An *inertinite*-group maceral showing round to oblong, noncellular forms with high reflectance. Previously grouped with *funginite* into *sclerotinite* (ICCP, 2001).

secretion [paleont] The act or process by which animals and plants transform mineral material from solution into skeletal forms.

secretion [sed struc] A secondary structure formed of material deposited from solution within a cavity in a rock, esp. a deposit formed on or parallel to the walls of the cavity; e.g. a mineral vein, an amygdule, or a geode. The space may be completely or only partly filled. Cf: *concretion.*

sectile (sec'-tile) Said of a mineral that can be cut with a knife; e.g. argentite.

section [geol] (sec'-tion) (a) An exposed surface or cut, either natural (such as a sea cliff or stream bank) or artificial (such as a quarry face or road cut), through a part of the Earth's crust; it may be vertical or inclined. (b) A description, or graphic representation drawn to scale, of the successive rock units or the geologic structure revealed by such an exposed surface, or as they would appear if cut through by any intersecting plane, such as a diagram of the geologic features or mine workings penetrated in a shaft or drilled well; esp. a *vertical section.* See also: *structure section.* Syn: *geologic section.*

section [stratig] (a) *columnar section.* (b) *geologic section.* (c) *type section.*

section [surv] (a) One of the 36 units of subdivision of a *township* of the U.S. Public Land Survey system, representing a piece of land normally one square mile in area (containing 640 acres as nearly as possible), with boundaries conforming to meridians and parallels within established limits. Sections within a normal township are numbered consecutively beginning with number one in the northeast section and progressing west and east alternately with progressive numbers in the tiers to number 36 in the southeast section. See also: *fractional section; half section; quarter section; quarter-quarter section.* (b) The part of a continuous series of measured differences of elevation that is recorded and abstracted as a unit. It always begins and ends on a bench mark.

section-gage log *caliper log.*

section line The boundary line of a section in surveying.

sector collapse Downslope movement of huge landslide blocks, with dimensions commonly measurable in kilometers, from the flanks of large volcanoes, e.g., Mauna Loa (Moore et al., 1995) or Mount Etna (Borgia et al., 1992).

sector graben (sec'-tor) A *volcanic graben* on the slope of a volcanic cone.

secular (sec'-u-lar) Said of a process, trend, or event lasting or persisting for an indefinitely long period of time, e.g. secular variation; progressive or cumulative rather than *cyclic.*

secular equilibrium A relationship between a long-lived radioactive *parent* and one or more short-lived radioactive daughters in which the activity of each daughter (or each set of daughters if decay is branching) is equal to the activity of the parent. Cf: *radioactive equilibrium.*

secular movements Systematic, persistent movements of the Earth's crust, either upward or downward, that take place slowly and imperceptibly over long periods of geologic time.

secular variation Very long-term changes in the Earth's magnetic field, measured in years, decades, and centuries. See also: *westward drift.* Syn: *geomagnetic secular variation.*

secule (sec'-ule) A syn. of *moment.* Term suggested by Jukes-Browne (1903, p.37) for the duration of a biostratigraphic zone. Etymol: Latin seculum, age.

secundine dike (sec'-un-dine) A dike commonly pegmatite or aplite, which has been intruded into hot country rock. Obsolete. See also: *welded dike.*

sedarenite (sed-ar'-e-nite) A term used by Folk (1968, p.124) for a *litharenite* composed chiefly of sedimentary rock fragments; e.g., *sandstone-arenite* and *shale-arenite.*

sedentary soil *residual soil.*

sederholmite (se'-der-holm-ite') A metallic hexagonal mineral of the *nickeline* group: beta-NiSe.

sedex-type ore deposit An informal term to refer to sedimentary exhalative mineral deposits.

sedge peat *carex peat.*

sedifluction (sed-i-fluc'-tion) The subaqueous or subaerial movement of material in unconsolidated sediments, occurring in the primary stages of diagenesis (Richter, 1952). Rarely used.

sediment (sed'-i-ment) (a) Solid fragmental material that originates from weathering of rocks and is transported or deposited by air, water, or ice, or that accumulates by other natural agents, such as chemical precipitation from solution or secretion by organisms, and that forms in layers on the Earth's surface at ordinary temperatures in a loose, unconsolidated form; e.g. sand, gravel, silt, mud, till, loess, alluvium. (b) Strictly, solid material that has settled down from a state of suspension in a liquid. In the singular, the term is usually applied to material held in suspension in water or recently deposited from suspension. In the plural, the term is applied to all kinds of deposits, and refers to essentially unconsolidated materials. Cf: *deposit.* Entymol: Latin "sedimentum," = what has settled, dregs.

sedimentary (sed-i-men'-ta-ry) adj. (a) Pertaining to or containing sediment; e.g. a "sedimentary deposit" or a "sedimentary complex". (b) Formed by the deposition of sediment (e.g. a "sedimentary clay"), or pertaining to the process of sedimentation (e.g. "sedimentary volcanism"). n. A sedimentary rock or deposit.

sedimentary ash *extraneous ash.*

sedimentary bentonite A *mineral-deposit type* consisting of smectite clay minerals formed by alteration of volcanic rocks in a marine environment. Sedimentary bentonite deposits occur in two varieties with different commercial uses: *calcium bentonite* and *sodium bentonite.*

sedimentary breccia A *terrigenous* rock formed by *lithification* of angular gravel (in which 30% of large particles are >2mm); e.g. a talus breccia. Syn: *sharpstone conglomerate.*

sedimentary-contact shoreline (sed-i-men'-ta-ry-con'-tact) A shoreline formed by the partial submergence of the slope left by the removal of weak beds from one side of a straight sedimentary contact (Johnson, 1925, p.19).

sedimentary cycle *cycle of sedimentation.*

sedimentary differentiation The progressive separation (by erosion and transportation) of a well-defined rock mass into physically and chemically unlike products that are resorted and deposited as sediments over more or less separate areas; e.g. the segregation and dispersal of the components of an igneous rock into sandstones, shales, limestones, etc.

sedimentary dike A tabular mass of sedimentary material that cuts across the structure or bedding of pre-existing rock in the manner of an igneous dike. It is formed by the filling of a crack or fissure from below, above, or laterally, by forcible injection or intrusion of sediments under abnormal pressure (as by gas pressure or by the weight of overlying rocks, or by earthquakes), or from above by simple infilling; esp. a *clastic dike.* See also: *sediment vein.*

sedimentary-exhalative mineral deposit Stratiform mineral deposit precipitated from brines exhaled by submarine faults. These brines accumulate in adjacent seafloor depressions.

sedimentary facies A term used by Moore (1949, p.32) for a stratigraphic *facies* representing any areally restricted part of a designated stratigraphic unit (or of any genetically related body of sedimentary deposits) which exhibits lithologic and paleontologic characters significantly different from those of another part or parts of the same unit. It comprises "one of any two or more different sorts of deposits which are partly or wholly equivalent in age and which occur side by side or in somewhat close neighborhood" (Moore, 1949, p.7).

sedimentary fault *growth fault.*

sedimentary injection *injection [sed].*

sedimentary insertion A term proposed by Challinor (1978, p.273) for the emplacement of sedimentary material among deposits or rocks already formed, such as by infilling, injection or intrusion, or localized solution subsidence.

sedimentary intrusion *intrusion [sed].*

sedimentary laccolith A term introduced by Raaf (1945) for an intrusion of plastic sedimentary material (such as clayey salt breccia) forced up under high pressure and penetrating parallel or nearly parallel to the bedding planes of the invaded formation, and characterized by a very irregular thickness.

sedimentary manganese A *mineral-deposit type* characterized by manganese oxides and carbonates deposited in shallow marine environments around the rims of basins during marine transgression.

sedimentary marble *crystalline limestone.*

sedimentary ore A sedimentary rock of ore grade; an ore deposit formed by sedimentary processes, e.g. saline residues, phosphatic deposits, or iron ore of the *Clinton ore* type.

sedimentary peat Peat formed under water, usually lacustrine, and consisting mainly of algae and related forms. Syn: *lake peat; pulpy peat; dredge peat.*

sedimentary petrography The description and classification of sedimentary rocks, esp. by means of microscopic study (Boggs, 1987, p.2).

sedimentary petrologic province An area underlain by sediments with a common provenance (Pettijohn, 1957, p.573). Cf: *dispersal shadow.* Syn: *petrologic province.*

sedimentary petrology The study of the composition, characteristics, occurrence, and origin of sediments and sedimentary rocks. Often miscalled "sedimentation".

sedimentary quartzite *orthoquartzite.*

sedimentary ripple *ripple [sed struc].*

sedimentary rock (a) A rock resulting from the consolidation of loose *sediment* that has accumulated in layers; e.g. a *clastic rock* (such as conglomerate or tillite) consisting of mechanically formed fragments of older rock transported from its source and deposited in water or from air or ice; or a *chemical rock* (such as rock salt or gypsum) formed by precipitation from solution; or an *organic rock* (such as certain limestones) consisting of the remains or secretions of plants and animals. The term is restricted by some authors to include only those rocks consisting of mechanically derived sediment; others extend it to embrace all rocks other than purely igneous and completely metamorphic rocks, thereby including pyroclastic rocks composed of fragments blown from volcanoes and deposited on land or in water. Syn: *stratified rock.* (b) Less restrictedly, a general term for any sedimentary material, unconsolidated or consolidated. This usage should be avoided.

sedimentary structure A structure in a sedimentary rock, formed either contemporaneously with deposition (a *primary sedimentary structure*) or by sedimentary processes subsequent to deposition (a secondary structure [geol]).

sedimentary tectonics The scientific study of the tectonic aspects of sedimentary strata (Friedman and Sanders, 1978, p.511) or sedimentary basins (Reading, 1986). Cf: *gravity orogenesis.*

sedimentary trap An area in which sedimentary material accumulates instead of being carried farther, as in an area between a high-energy and low-energy environment.

sedimentary tuff (a) A tuff containing a subordinate amount of nonvolcanic detrital material. (b) A deposit of tuff that was reworked by geomorphic agents, e.g. a river, after deposition from a volcanic eruption. Syn: *reworked tuff.*

sedimentary volcanism The expulsion or extrusion through overlying formations of a mixture of sediment (e.g. sand or clay), water, and gas, with gas under pressure furnishing the driving force. Also, the production of phenomena such as sand volcanoes and mud volcanoes. Sedimentary volcanism may result from diapiric intrusion, volcanism in a fumarolic stage, escaping hydrocarbons, oozing-out of material (during thaw) in a permafrost region, or orogenic pressure release as during an earthquake.

sedimentation (sed'-i-men-ta'-tion) (a) The act or process of forming or accumulating sediment in layers, including such processes as the separation of rock particles from the material from which the sediment is derived, the transportation of these particles to the site of deposition, the actual deposition or settling of the particles, the chemical and other (early diagenetic) changes occurring in the sediment, and the ultimate consolidation of the sediment into solid rock. (b) Less broadly, the process of deposition of sediment. (c) Strictly, the act or process of depositing sediment by mechanical means from a state of suspension in air or water or by ice and under the influence of gravity. (d) The accumulation of deposits of colluvium and alluvium derived from accelerated erosion of the soil. (e) *silting up.* (f) A term often used erroneously for "sedimentary petrology" and "sedimentology".

sedimentation analysis Determination of the particle-size distribution of a sediment by measuring the settling velocities of the various size fractions.

sedimentation balance An apparatus used to measure the settling rate of small particles dispersed in a liquid.

sedimentation curve An experimentally derived curve showing cumulatively the quantity of sediment deposited or removed from an originally uniform suspension in successive units of time (Krumbein and Pettijohn, 1938, p.112-115).

sedimentation diameter A measure of particle size, equal to the diameter of a hypothetical sphere of the same specific gravity and the same settling velocity as those of a given sedimentary particle in the same fluid; twice the *sedimentation radius.* Cf: *equivalent radius; nominal diameter.*

sedimentation radius One half of the *sedimentation diameter.*

sedimentation rate *rate of sedimentation.*

sedimentation trend The direction in which sediments were laid down.

sedimentation unit A layer or deposit resulting from one distinct act of sedimentation, defined by Otto (1938, p.574) as "that thickness of sediment which was deposited under essentially constant physical conditions"; the deposit made during a time period when the prevailing current has a mean velocity and deposits some mean size, such as a cross-bedded layer of sand formed under conditions of essentially constant flow and sediment discharge. It is distinguished from like units by changes in particle size and/or fabric indicating changes in velocity and/or direction of flow.

sediment binder A sessile-benthonic organism that encrusts several adjacent uncemented particles, producing a single larger mass less likely to be moved by water currents.

sediment budget The mass balance between inputs, outputs, and storage of sediment in a watershed; e.g., sediment yield = inputs - storage.

sediment charge The ratio of the weight or volume of sediment in a stream to the weight or volume of water passing a given cross section per unit of time (ASCE, 1962).

sediment concentration The ratio of the dry weight of the sediment in a water-sediment mixture (obtained from a stream or other body of water) to the total weight of the mixture. It is usually expressed in percent for high concentration values, or in parts per million for low concentration values.
sediment delivery The amount of material eroded from the land surface by runoff and delivered to a stream system. May be expressed in units of volume, mass, or volume or mass per unit drainage area of the watershed (depth).
sediment-delivery ratio (sed'-i-ment-de-liv'-er-y) (a) The ratio of sediment production in a basin to sediment yield in the basin (dimensionless). Historical measurements of sediment delivery ratios are typically less than 1.0, indicating net storage of historical sediment in modern watersheds on hillslopes, floodplains, and in channels. (b) The ratio of sediment yield of a drainage basin to the total amount of sediment moved by sheet erosion and channel erosion; expressed in percent (Chow, 1964, p.17-12).
sediment discharge The amount of sediment moved by a stream in a given time, measured by dry weight or by volume; the rate at which sediment passes a section of a stream. Syn: *sediment-transport rate.*
sediment-discharge ratio (sed'-i-ment-dis'-charge) The ratio between the total discharge of a stream and the discharge of its load.
sediment feeder *deposit feeder.*
sediment flood A stage in seismic *depth migration.*
sediment-hosted copper-cobalt deposit A *mineral-deposit type* characterized by lenses, veins, and disseminations of pyrite with minor copper- and cobalt-bearing minerals in sedimentary rocks, locally associated with chlorite alteration and tourmaline breccias (Höy, 1995).
sediment-hosted copper deposit A *mineral-deposit type* characterized by strata-bound, disseminated copper minerals in continental sedimentary rocks, typically interbedded with red-bed sandstone and evaporite rocks. Syn: *copper shale; kupferschiefer-type; red-bed copper; sandstone copper; sediment-hosted stratiform copper; shale-hosted copper.*
sediment-hosted stratiform copper *sediment-hosted copper deposit.*
sediment load The solid material (*load*) that is transported by a natural agent, esp. by a stream. The total sediment load of a stream is equal to *bed-material load* plus *wash load,* and is expressed as the dry weight of all sediment that passes a given point in a given period of time.
sedimento-eustasy (sed-i-men'-to-eu'-sta-sy) Worldwide changes in sea level produced by a change in the capacity of the ocean basins because of sediment accumulation. Term introduced by Fairbridge (1961, p.112). Cf: *glacio-eustasy; tectono-eustasy.*
sedimento-eustatism (sed-i-men'-to-eu'-stat-ism) *sedimento-eustasy.*
sedimentogenic A term proposed by Sangster (1976) to describe lead-zinc deposits formed by the transport of lead and zinc from a sedimentary source to a new host rock.
sedimentograph An automated device that gives directly the cumulative size curve of a sediment.
sedimentology (sed'-i-men-tol'-o-gy) The scientific study of sedimentary rocks and of the processes by which they were formed; the description, classification, origin, and interpretation of sediments. Sometimes miscalled "sedimentation".
sediment production The mass or volume of sediment produced in a basin by erosional processes including mass wasting, soil erosion by rain-drop splash and sheetflow processes, and by channel erosion.
sediment-production rate (sed'-i-ment-pro-duc'-tion) Sediment yield per unit of drainage area, derived by dividing the annual sediment yield by the area of the drainage basin; expressed as tons or acre-feet of sediment per square mile per year (Chow, 1964, p.17-11).
sediment stabilizer A sessile-benthonic organism that grows out on unconsolidated sediment and thereby protects it from being removed or eroded.
sediment station A vertical cross-sectional plane of a stream, usually normal to the mean direction of flow, where samples of suspended load are collected on a systematic basis for determining concentration, particle-size distribution, and other characteristics.
sediment transport The movement and carrying-away of sediment by mostly natural agents; e.g., the conveyance of a *stream load* by suspension, saltation, solution, or traction.
sediment-transport rate (sed'-i-ment-trans'-port) *sediment discharge.*
sediment trapper A sessile-benthonic organism that projects up into sediment-laden water, slows its flow, and thereby causes some of the suspended sediment to settle out around the organism and accumulate on the sea floor.
sediment vein A *sedimentary dike* formed by the filling of a fissure from above with sedimentary material.
sediment wave A concept of sediment transport through time in a form analogous to a storm hydrograph with a rising and falling limb and a peak instantaneous discharge rate. The concept was originated by G.K. Gilbert's study of hydraulic gold-mining sediment in California based on inferences from low-flow channel-bed elevations.
sediment yield The amount of material eroded from the land surface and passing the outlet of a watershed. May be expressed in units or volume, mass, or volume or mass per unit drainage area of the watershed (depth).
sedovite (se-dov'-ite) A brown to red-brown orthorhombic mineral: $U^{4+}(MoO_4)_2$.
seed [bot] The ripened ovule of a plant consisting of a seed coat, stored nutrients, and an embryo.
seed [cryst] *seed crystal.*
seed [geophys] (a) To select the *voxel* (or *pixel*) to be followed by an automatic picker. Usually the picked point is a maximum, minimum, or zero-crossing, preferably determined by fitting a curve to three or more successive data points spanning the maximum or minimum. (b) To initialize an iterative calculation.
seed crystal A small, suitably oriented piece of crystal used in *crystal seeding.* Syn: *seed [cryst].*
seed fern *pteridosperm.*
seed leaf *cotyledon.*
seed plant *spermatophyte.*
seeligerite (see'-li-ger-ite') A yellow orthorhombic mineral: $Pb_3O(IO_3)Cl_3$.
seelite A vitreous yellow monoclinic mineral: $Mg(UO_2)_2(AsO_3,AsO_4)_2 \cdot 7H_2O$.
seep n. An area, generally small, where water or oil percolates slowly to the land surface. For water, it may be considered as a syn. of *seepage spring,* but it is used by some for flows too small to be considered as springs. Cf: *oil seep.* Syn: *seepage [grd wat].* v. To move slowly through small openings of a porous material.
seepage [grd wat] (a) The act or process involving the slow movement of water or other fluid through a porous material such as soil. Cf: *influent seepage; effluent seepage.* (b) The amount of fluid that is involved in seepage. (c) *seep.* Cf: *percolation.*
seepage [petroleum] (seep'-age) *oil seep.*
seepage face (a) A belt along a slope, such as the bank of a stream, along which water emerges at atmospheric pressure and flows down the slope. See also: *seepage line.* (b) A boundary condition that is specific to flow systems in which saturated and unsaturated domains coexist. It is a constant potential boundary in the sense that the gage pressure of water is zero on it.
seepage force The force that is transferred from water flowing through a porous granular medium to the medium itself by means of viscous friction. Syn: *seepage stress.*
seepage lake (a) A *closed lake* that loses water mainly by seepage through the walls and floor of its basin. Cf: *drainage lake.* (b) A lake that receives its water mainly from seepage, as from irrigation waters in parts of western U.S.
seepage line The uppermost level at which flowing water emerges along a *seepage face;* an outcrop of the water table. Syn: *line of seepage; phreatic line.*
seepage loss Loss of water by *influent seepage* from a stream, canal, or other body of surface water.
seepage spring This term may be used as a syn. of *filtration spring,* but is often limited to springs of small discharge (Meinzer, 1923, p.50). See also: *seep.* Syn: *weeping spring.*
seepage stress *seepage force.*
seepage velocity The rate at which seepage water is discharged through a porous medium per unit area of pore space perpendicular to the direction of flow.
segelerite (se'-ge-ler-ite') A pale yellow-green or colorless orthorhombic mineral of the *overite* group: $CaMgFe^{3+}(PO_4)_2(OH) \cdot 4H_2O$.
Seger cone (Se'-ger) A small cone, made in the laboratory of a mixture of clay and salt, that softens at a definite, known temperature. It is used in the manufacture of refractories. It has also been used in volcanology to determine the approximate temperature of a molten lava. Syn: *pyrometric cone.*
seggar (seg'-gar) *sagger.*
segment (seg'-ment) (a) One of the constituent parts into which an invertebrate body is divided; esp. any of the succeeding or repeated body parts of an arthropod, many of which are likely to be similar in form and function (such as one of the components of the thorax of a trilobite, connected by articulation with adjoining segments), or the somite of a crustacean. See also: *podomere; article; joint [paleont].* (b) Any of the parts into which a heterococcolith naturally separates or divides. (c) In chambered cephalopod conchs, any part of the *siphuncle* between two successive septa.
segmentation (seg-men-ta'-tion) On the coast, the process by which a *lagoon* is subdivided into smaller patches of water by the accumulation of *transverse bars* (Price, 1947).
segminate element A pectiniform *conodont element* with only one primary process, which is anterior in all known forms (TIP, 1981, pt. W, supp.2).
segminiplanate element A segminate *conodont element* with lateral platform extensions and the attachment surface characteristic of planate pectiniform elements (TIP, 1981, pt. W, supp.2). See also: *segminate element; planate element.*
segminiscaphate element A segminate *conodont element* with lateral platform extensions and the attachment surface characteristic of scaphate pectiniform elements (TIP, 1981, pt. W, supp.2). See also: *segminate element; scaphate element.*
segnitite A greenish-brown, yellowish-brown, or brown rhombohedral mineral of the *crandallite* group: $PbFe^{3+}{}_3(AsO_3OH)_2(OH)_6$.
segregated ice (seg'-re-gat'-ed) Ice films, seams, lenses, pods, or layers, generally 1-150 mm thick, that grow in the ground by drawing-in water as the

ground freezes. Also known as segregation ice. Syn: *sirloin-type ice; Taber ice.*

segregated vein A fissure whose mineral filling is derived from the country rock by the action of percolating water. Syn: *exudation vein.* Cf: *infiltration vein.* See also: *lateral secretion.*

segregation [petrology] (seg-re-ga′-tion) (a) *magmatic segregation.* (b) A concentration of crystals of a particular mineral or minerals that accumulated during an early stage of consolidation as a result of magmatic segregation. (c) A "liquid segregation", e.g. from an early-formed immiscible liquid or a late-formed residual melt.

segregation [sed] A secondary feature formed as a result of chemical rearrangement of minor constituents within a sediment after its deposition; e.g. a nodule of iron sulfide, a concretion of calcium carbonate, or a geode.

segregation banding A compositional banding in gneisses that is not primary in origin, but rather is the result of segregation of material from an originally more nearly homogeneous rock (Billings, 1954). Cf: *cleavage banding.*

seiche (a) A free or standing-wave oscillation of the surface of water in an enclosed or semi-enclosed basin (as a lake, bay, or harbor) that varies in period, depending on the physical dimensions of the basin, from a few minutes to several hours, and in height from several centimeters to a few meters; that is initiated chiefly by local changes in atmospheric pressure, aided by winds, tidal currents, and small earthquakes; and that continues, pendulum fashion, for a time after cessation of the originating force. It usually occurs in the direction of longest diameter of the basin, but occasionally it is transverse. The term has also been applied to an oscillation superimposed on the tidal waves of the open ocean. (b) A standing wave at the metalimion in a lake, *internal seiche.* (c) A term used in the Great Lakes area for any sudden rise (whether oscillatory or not) in the water of a harbor or lake. Etymol: French, supposedly from Latin "siccus", "dry"; a term used locally to describe the occasional rise and fall of water at the narrow end of Lake Geneva, Switzerland, where the phenomenon was first observed. Pron: saysh. Cf: *internal seiche.*

seidite-(Ce) A yellow, pink-yellow, or cream monoclinic mineral: $Na_4(Ce,Sr)_2TiSi_8O_{18}(O,OH,F)_6{\cdot}5H_2O$.

seidozerite (sei-do′-zer-ite) A brownish-red, brownish-yellow, or orange monoclinic mineral: $(Na,Ca)_4Mn(Ti,Zr)_2(Si_2O_7)_2(O,F,OH)_4$.

seif [Pron. safe] A very large, sharp-crested, tapering *longitudinal dune* or chain of sand dunes, commonly found in the Sahara Desert; its crest in profile consists of a succession of *peaks* and *cols,* and it bears on one side a succession of curved slip faces produced by strong but infrequent cross winds that tend to increase its height and width. A seif dune may be as much as 200 m high, and from 400 m to >100 km long (300 km in Egypt). Etymol: Arabic "saif", "sword"; the term originated in North Africa but is applied elsewhere to similar dunes of appreciably smaller size. Syn: *seif dune; sif; saif; sword dune.*

seif dune *seif.*

seinäjokite (sei′-nä-jok-ite) A metallic gray orthorhombic mineral of the *löllingite* group: $(Fe,Ni)(Sb,As)_2$.

seis A colloquial syn. of *seismic detector.*

seism *earthquake.*

seismic (seis′-mic) Pertaining to an earthquake or Earth vibration, including those that are artificially induced.

seismic activity *seismicity.*

seismic area (a) An *earthquake zone.* (b) The region affected by a particular earthquake.

seismic array A network of seismographs arranged in a spatial pattern so as to enable the signal from earthquake ground motion to be enhanced with respect to the noise "network" of microseismic ground motion. Usually the outputs of the elements of the array are simply combined but sometimes they are recorded separately so that events arriving from different directions can be distinguished.

seismic belt An elongate *earthquake zone,* esp. a zone of subduction or sea-floor spreading.

seismic constant In building codes dealing with earthquake hazards, an amount of horizontal acceleration that a building must be designed to withstand.

seismic creep Relatively slow movement on a fault, as contrasted to the sudden movement associated with an earthquake.

seismic detector An instrument, e.g., a *seismometer* or *geophone* or *hydrophone,* that converts seismic wave energy into electrical voltage or otherwise makes them evident. Colloquial syn: *jug; pot [seis]; seis.*

seismic discontinuity *discontinuity [seis].*

seismic efficiency The percentage of earthquake-generated energy that goes into the production of elastic-wave energy.

seismic-electric effect (seis′-mic-e-lec′-tric) A voltage between two electrodes in the ground caused by passage of a seismic wave.

seismic event *event [seis].*

seismic exploration *applied seismology.*

seismic facies analysis The description and geologic interpretation of seismic reflection patterns, based on reflection configuration, continuity, amplitude, frequency, and interval velocity (Mitchum, 1977, p.210). See also: *seismic facies unit.*

seismic facies unit A mappable three-dimensional seismic unit composed of groups of reflections whose parameters, such as reflection configuration, continuity, amplitude, frequency, or interval velocity, differ from those of adjacent facies units. Once the interval reflection parameters, the external form, and the three-dimensional associations of the seismic facies are delineated, the unit can then be interpreted in terms of environmental setting, depositional processes, and estimates of lithology (Mitchum, 1977, p.210). See also: *seismic facies analysis.*

seismic gap A segment of an active fault zone that has not experienced a *major earthquake* during a time interval when most other segments of the zone have. Seismologists commonly consider seismic gaps to have high future-earthquake potential.

seismic intensity The average rate of flow of seismic wave energy through a unit cross section perpendicular to the direction of propagation.

seismicity (seis-mic′-i-ty) The phenomenon of movements in the Earth's crust. Syn: *seismic activity.*

seismicity map A map showing earthquake intensity.

seismic log *synthetic acoustic impedance log.*

seismic map A contour map constructed from seismic data. Values may be in either time or depth; data may be plotted with respect to the observing station (producing an "unmigrated map") or with respect to the subsurface reflecting or refracting point locations (a "migrated map"). Maps displaying seismic data themselves may also be made from 3-D volumes of data showing the distribution of amplitude along a picked horizon; this is called a *horizon slice* or *amplitude map.* Attributes may also be calculated from seismic data and displayed in map format. See also: *time slice.*

seismic moment A measure of the strength of an earthquake, particularly of the low-frequency wave motion. The seismic moment is equal to the product of the force and the moment arm of the double-couple system of forces that produces ground displacements equivalent to that produced by the actual earthquake dislocation. The seismic moment also is equal to the product of the rigidity modulus of the Earth material, the fault area, and the average dislocation along the fault surface (Aki and Richards, 1980).

seismic noise Any disturbance that is not signal. Sometimes refers to microseisms.

seismic profile *profile [seis].*

seismic prospecting *applied seismology.*

seismic pulse The signal generated by an impulsive seismic energy source (explosive, thumper, air gun, sparker, etc.); may include correlated vibroseis sweep signal. "Wavelet" is sometimes used as a synonym.

seismic pumping Fault-triggered fluid groundwater movement. Fluid moves into the fault zone, because of the increase in open space that accompanies fracturing of rock, and out of the fault zone when stresses cause fractures to close.

seismic record (a) In seismic prospecting, a recording, sometimes photographic or magnetic tape or other medium, of the energy received by a spread of geophone groups with time following the source energy release. (b) In earthquake seismology, a record of seismic activity during a period of time, including background noise, body waves, and surface waves, from both natural and artificial events. Syn: *record.*

seismic reflection *reflection.*

seismic reflection configuration The geometric patterns and relations of seismic reflections that are interpreted to represent configuration of strata generating the reflections (Mitchum, 1977, p.209).

seismic sea wave *tsunami.*

seismic section A plot of seismic data along a line. The vertical scale is usually arrival time but is sometimes depth. The data usually have been migrated.

seismic sequence analysis The seismic identification and interpretation of depositional sequences by subdividing the seismic section into packages of concordant reflections separated by surfaces of discontinuity, and interpreting them as depositional sequences (Mitchum, 1977, p.210).

seismic shooting A method of geophysical prospecting in which elastic waves are produced in the Earth by the firing of explosives or by other means. See also: *reflection shooting; refraction shooting.*

seismic source (a) A device for generating seismic waves. (b) The point at which seismic waves are generated. (c) The point of origin of an earthquake.

seismic spectrum A curve showing the amplitude of the ground motion as a function of frequency or period, obtained by a Fourier analysis of the ground motion.

seismic spread *spread [seis].*

seismic stratigraphy The study of stratigraphy and depositional facies as interpreted from seismic data (Mitchum, 1977, p.210).

seismic surveying The gathering of seismic data from an area. Cf: *reflection shooting; refraction shooting.*

seismic velocity The rate of propagation of an elastic wave, usually measured in km/s. The wave velocity depends on the type of wave, as well as the elastic properties and density of the Earth material through which it travels. Usually refers to *phase velocity* for a particular mode, generally P-wave unless otherwise stated. Cf: *interval velocity; normal-moveout velocity; average velocity [seis]; rms velocity; group velocity.*

seismic wave (a) A general term for elastic waves produced by earthquakes or generated artificially. It includes both *body waves* and surface waves and both P-waves and S-waves. (b) A seismic sea wave, or tsunami. Syn: *earthquake wave; elastic wave.*

seismite (seis'-mite) *Fault-graded beds* that are interpreted as an earthquake record, or "paleoseismogram" (Seilacher, 1969).

seismo-electric effect (seis'-mo-e-lec'-tric) The variation of resistivity because of elastic deformation of rocks.

seismogram (seis'-mo-gram) The record made by a *seismograph.* Syn: *earthquake record.*

seismograph (seis'-mo-graph) *seismometer.*

seismologist (seis-mol'-o-gist) One who applies the methods or principles of seismology, as in earthquake prediction or seismic exploration.

seismology (seis-mol'-o-gy) The study of earthquakes, and of the structure of the Earth, by both natural and artificially generated seismic waves.

seismometer (seis-mom'-e-ter) An instrument that detects, magnifies, and records vibrations of the Earth, either earthquakes or those generated for applied seismology purposes. The resulting record is a *seismogram..* Syn: *seismograph.* Cf: *seismic detector; geophone; hydrophone.*

seismometer plant *plant [seis].*

seismometer spread *spread [seis].*

seismometry (seis-mom'-e-try) The instrumental aspects of seismology.

seismoscope (seis'-mo-scope) An instrument that merely indicates the occurrence of an earthquake.

seismotectonics Study of the role of seismic activity in tectonics; includes examinations of the processes precursory to and accompanying earthquakes, the regionally significant geologic structures generated by earthquakes, and the temporal or spatial variations in processes or structures.

sekaninaite (se-ka-nin'-a-ite) A blue or violet-blue orthorhombic mineral of the cordierite group: $(Fe^{2+},Mg)_2Al_4Si_5O_{18}$.

Selachii (Se-la'-chi-i) An order, largely marine, of elasmobranch fishes characterized by a fusiform body and more or less predaceous habit; sharks. Range, Upper Devonian to Holocene.

Selandian (Se-lan'-di-an) A stage of the Standard Global Chronostratigraphic Scale: middle Paleocene (above Danian, below Thanetian). See also: *Montian.*

selbergite (sel'-berg-ite) A hypabyssal *phonolite* containing phenocrysts of leucite, nosean, sanidine, aegirine-augite, and biotite in a fine-grained groundmass of nepheline, alkali feldspar, and aegirine. The phenocrysts comprise a higher percentage of the volume of the rock than the groundmass. Its name, given by Brauns in 1922, is derived from Selberg, Laacher See district, Germany. The rock differs from *schorenbergite* in containing alkali feldspar. Obsolete.

selection (se-lec'-tion) *natural selection.*

selective association [eco geol] Particular combinations of minerals formed during replacement of earlier formed mineral assemblages.

selective fusion (se-lec'-tive) The fusion of only a portion of a mixture such as a rock. The liquid portion will generally contain a larger proportion of the more fusible components than the parent material did. Cf: *anatexis.*

selective preservation The process by which readily-labile molecules are broken down during diagenesis, leaving biologically-resistant ones.

selective weathering *differential weathering.*

selenate (sel'-e-nate) A mineral compound characterized by discrete $(SeO_4)^{-2}$ groups. An example is olsacherite, $Pb_2(SeO_4)(SO_4)$.

selenide (sel'-e-nide) A mineral compound that is a combination of selenium with a more positive element or radical. An example is eucairite, CuAgSe.

seleniferous plant (sel-e-nif'-er-ous) A plant that absorbs and retains large quantities of selenium from the soil. Syn: *selenophile.*

selenite (sel'-e-nite) (a) A mineral compound containing the oxyanion $(SeO_3)^{-2}$. An example is molybdenite, $PbSeO_3$. (b) The clear, colorless variety of gypsum, occurring (esp. in clays) in distinct, transparent monoclinic crystals or in large crystalline masses that easily cleave into broad folia. Syn: *spectacle stone.*

selenite butte A small tabular mound, rising 1-3 m above a playa, composed of lake sediments capped with a veneer of selenite formed by deflation of the playa or by the effects of rising groundwater (Stone, 1967, p.246).

selenite dome Mound-shaped accumulation of elongate gypsum crystals growing radially from a surface at the sediment-water interface. See also: *gypsum cavoli.*

selenite grass Individual crystals, small clusters of crystals, or regular layers of bladed selenitic gypsum growing upward from the sediment-water interface (Richter-Bernburg, 1973). Syn: grasslike gypsum.

selenite plate *gypsum plate.*

selenium (se-le'-ni-um) A metallic gray to purplish-gray trigonal mineral: Se.

selenizone (se-le'-ni-zone) A sharply defined spiral band of closely spaced crescentic growth lines or linear ridges generated by a narrow notch or slit in the outer lip of the aperture of a gastropod, generally at the periphery of a shell whorl. It typically marks the positions of the notch or *slit* during earlier stages of growth.

selenjoséite (sel'-en-jo-sé'-ite) *laitakarite.*

selenolite [mineral] (se-le'-no-lite) A name that had been used for minerals assumed to be SeO_2. That name has been rejected in favor of *downeyite,* naturally occurring SeO_2.

selenolite [sed] A sedimentary rock composed of gypsum or anhydrite.

selenology (sel-e-nol'-o-gy) A branch of astronomy that deals with the Moon; the science of the Moon, including *lunar geology.*

selenophile (se-le'-no-phile) *seleniferous plant.*

selenostephanite A metallic orthorhombic mineral: $Ag_5Sb(Se,S)_4$.

selen-tellurium (se'-len-tel-lu'-ri-um) A mixture of selenium and tellurium.

self-exciting dynamo A dynamo that produces a magnetic field around itself so that motions of an electrical conductor cutting the magnetic lines of force generate the current that keeps the dynamo turning. Such a mechanism is proposed as the source of the Earth's magnetic field in the Earth's core. The interactions of two or more such natural dynamos may explain the periodic reversals of the Earth's magnetic field.

self-mulching soil A soil in which the surface layer becomes so well aggregated that it does not crust and seal under the impact of rain but instead serves as a surface mulch upon drying.

self-organization In geomorphology, (a) a pattern formation that occurs independently of external forcings or controls (Hallet, 1990); (b) the development of patterns and structures due to the complex internal dynamics of Earth-surface systems. The term is used in this context in a variety of sometimes contradictory ways (Phillips, 1999).

self-organized criticality In geomorphology, a form of self-organization whereby systems evolve to critical (*threshold*) states where small events trigger reorganization of the entire system which nonetheless retains its macro-level properties. The term is often applied to studies of earthquakes, landslides, and drainage-basin and channel-network evolution (Hergarten, 2002).

self-potential curve (self-po-ten'-tial) *spontaneous-potential curve.*

self-potential method An electrical exploration method in which one determines the spontaneous electrical potentials (*spontaneous polarization*) that are caused by primary chemical, heat, or fluid flow. Syn: *spontaneous-potential method.*

self-reading leveling rod A *level rod* with graduations designed to be read by the observer at the leveling instrument. Syn: *speaking rod.*

self-reversal (self-re-ver'-sal) Acquisition by a rock of a natural remanent magnetization opposite to the ambient magnetic field direction at the time the rock was formed.

seligmannite (sel'-ig-mann-ite') A lead-gray orthorhombic mineral of the *bournonite* group: $PbCuAsS_3$.

Selinuntian (Sel-i-nun'-tian) An Italian stage: lowermost Pleistocene (above Gelasian, below Ionian). Syn: *Calabrian.* See also: *Santernian; Emilian; Sicilian.*

sellaite (sel'-la-ite) A colorless tetragonal mineral: MgF_2.

selvage [intrus rocks] A marginal zone of a rock mass, having some distinctive feature of fabric or composition; specif. the chilled border of an igneous mass (as of a dike or lava flow), usually characterized by a finer grain or sometimes a glassy texture, such as the glassy inner margins on the pillows in pillow lava.

selvage [ore dep] *gouge [ore dep].*

selvage [paleont] The principal ridge of the contact margin in an ostracode, serving to hermetically seal the carapace when it is closed.

selvage [struc geol] (sel'-vage) The altered, clayey material found along a fault zone; *fault gouge.*

selwynite A vitreous bluish-purple or pale lilac tetragonal mineral of the *gainesite* group: $NaK(Be,Al,Ca,Mn)Zr_2(PO_4)_4 \cdot 2H_2O$.

SEM *scanning electron microscope.*

semblance (sem'-blance) In seismic surveying, a measure of multichannel *coherence.* It is the energy of a summed trace divided by the average energy of the components of the sum.

semenovite-(Ce) (se-men'-o-vite) A colorless orthorhombic mineral: $(Na,Ca)_9Fe^{2+}(Ce,La)_2(Si,Be)_{20}(O,OH,F)_{48}$.

semianthracite (sem-i-an'-thra-cite) Coal having a fixed-carbon content of 86% to 92%. It is between bituminous coal and anthracite in metamorphic rank, although its physical properties more closely resemble those of anthracite.

semiarid (sem-i-ar'-id) Said of a type of climate with slightly more precipitation (25-50 cm) than in an arid climate, and in which sparse grasses are the characteristic vegetation. In Thornthwaite's classification, the moisture index is between -20 and -40.

semibituminous coal (sem'-i-bi-tu'-mi-nous) Coal that ranks between bituminous coal and semianthracite; it is harder and more brittle than bituminous coal. It has a high fuel ratio and burns without smoke. Syn: *smokeless coal.* Cf: *metabituminous coal.*

semibolson (sem-i-bol'-son) A wide desert basin or valley that is drained by an intermittent stream flowing through canyons at each end and reaching a surface outlet (such as another stream, a lower basin, or the sea); its central playa is absent or poorly developed. It may represent a *bolson* where the alluvial fill reached a level sufficient to permit occasional overflow across the lowest divide.

semibright coal (sem-i-bright') A type of banded coal defined microscopically as consisting of between 80% and 61% of bright ingredients such as vitrain, clarain, and fusain, with clarodurain and durain composing the remainder. Cf: *semidull coal; bright coal; dull coal; intermediate coal.*

semicannel coal (sem-i-can'-nel) *lean cannel coal.*

semiconductor radiation detector (sem'-i-con-duc'-tor) A solid-state detector of ionizing radiation, fabricated from material such as germanium or silicon. Lithium is commonly diffused ("drifted") into the semiconductor to compensate for impurities. Energy of the detected radiation is determined by collecting the electrical charge produced along the ionizing path. Semiconductor detectors usually yield one to two orders of magnitude greater energy resolution than scintillation-type detectors. See also: *silicon detector; lithium-drifted germanium detector.*

semicratonic (sem'-i-cra-ton'-ic) *quasicratonic.*

semicrystalline (sem-i-crys'-tal-line) *hyalocrystalline.*

semidesert (sem-i-des'-ert) An area intermediate in character between a desert and a grassland and often located between them.

semidiurnal (sem'-i-di-ur'-nal) *Tidal constituents* with a period near 12 hours.

semidiurnal tide A tide with two high waters and two low waters occurring in a day, such as at Bermuda.

semidull coal (sem-i-dull') A type of banded coal defined microscopically as consisting mainly of clarodurain and durain, with from 40% to 21% bright ingredients such as vitrain, clarain, and fusain. Cf: *semibright coal; bright coal; dull coal; intermediate coal.*

semifoliated structure An irregular *laminar structure* consisting of mutually parallel calcitic blades or slightly elongated, elliptical flakes dipping at the same angle and in the same general direction over a portion of the depositional surface but showing little aggregation into laterally extensive laminae. The blades sometimes appear longitudinally creased or keeled, e.g., locally in the transition between chalky and foliated structures in *Crassostrea virginica* (Carriker et al., 1980, Fig. 7). Calcitic semifoliated structure occurs in the cyclostome bryozoan *Entalophora,* where the blades are sometimes longitudinally keeled (Tavener-Smith and Williams, 1972, Figs. 116, 117, 122). Semifoliated structure also occurs in some articulate brachiopods in shell layers structurally intermediate between "standard fibrous" (i.e., fibrous prismatic) and crossed bladed, e.g., in the plectambonitacean *Aegiria grayi* (Brunton, 1972, Pl. 4, Fig. 3) and in the billingsellacean *Billingsella* (Carter, 1990, Fig. 42). It also occurs in rugosan and tabulate corals, where the basic structural units consist of variably scalloped blades or irregularly elliptical flakes which may be slightly cup shaped; in these corals, the three-dimensional configuration of the structure is yet incompletely known.

semifusain (sem-i-fu'-sain) A coal lithotype transitional between fusain and vitrain, but predominantly fusain. Cf: *fusovitrain.*

semifusinite (sem-i-fu'-si-nite) A maceral of coal within the *inertinite* group having a reflectance intermediate between that of fusinite and that of associated vitrinite. It shows plant-cell structure, with cavities generally oval or elongated in cross section, but in some specimens less well defined than in fusinite; and it has a particle size generally greater than about 50 micrometers except when isolated from other macerals. Semifusinite often occurs as a transitional material between fusinite and vitrinite.

semifusinoid (sem-i-fu'-si-noid) Fusinite, the optical properties of which are transitional between those of fusinoid and those of associated xylinoids, vitrinoids, and anthrinoids (ASTM, 1970, p.364).

semifusite (sem-i-fu'-site) A microlithotype of coal, consisting of 95% or more of *semifusinite.*

semi-infinite Extending so far that some of the boundaries have no effect. Thus a semi-infinite slab is a horizontal unit of finite thickness bounded by a vertical plane on one side but extending so far in other directions that the boundaries in those directions create no measurable effects. A semi-infinite prism is a vertical prism with a bottom so remote as to not affect measurements.

semilogarithmic (sem'-i-log'-a-rith'-mic) A graph in which equal intervals on one axis represent equal rates of change (i.e., a logarithmic scale) and equal intervals on the other axis represent equal amounts of change (i.e., an arithmetic scale). Syn: semilog.

semimajor axis The mean separation between two bodies in an elliptical orbit. Half the length of the major axis of an *ellipse.*

seminacreous structure A *laminar structure* in which imbricated sheets and/or horizontal tablets show a higher frequency of screw dislocations than in true nacre and their horizontal laminae are less continuous than in true nacre. Unlike true nacre, seminacre may be calcitic or aragonitic. Although a common calcitic layer in bryozoans and craniacean brachiopods, this structure is rare in molluscs. Aragonitic seminacreous structure occurs in the region of the dentition of trigoniacean bivalves where curvature of the laminae disrupts the table organization (Carter, 1990, Fig. 11).

semiopal (sem-i-o'-pal) A loosely used term for common opal, hydrophane, and any partly dehydrated or impure opal, as distinguished from precious opal and fire opal. Syn: *hemiopal.*

semiprecious stone (sem-i-pre'-cious) Generally regarded as any gemstone other than a *precious stone,* or any gemstone of lower commercial value than a precious stone; specif. a mineral that is less than 8 on the Mohs scale of hardness. A gemstone may also be regarded as semiprecious because of its comparative abundance, inferior brilliance, or unfamiliarity to the public, or owing to the whims of fashion. This arbitrary classification is misleading, as it does not recognize, for example, that a ruby of poor quality may be far less costly than a fine specimen of jadeite.

semisplint coal (sem-i-splint') A type of banded coal that is intermediate in composition and character between *bright-banded coal* and *splint coal;* it corresponds to *duroclarain.* It is defined quantitatively as having 20-30% opaque attritus and more than 5% *anthraxylon.* Cf: *clarodurain.*

semitropical (sem-i-trop'-i-cal) *subtropical.*

semivariance In *geostatistics,* a measure of spatial dependence calculated as one-half the variance of the differences between pairs of points separated by a constant distance or *lag.*

semivariogram In *geostatistics,* a plot of *semivariance* versus *lag* or distance. A measure of the continuity of spatial phenomena expressed as an average squared difference between measured quantities at different locations. Used in *kriging.* Syn: *variogram.*

semseyite (sem'-sey-ite) A metallic gray to black monoclinic mineral: $Pb_9Sb_8S_{21}$.

senaite (sen'-a-ite) A submetallic black rhombohedral mineral of the *crichtonite* group: $Pb(Ti,Fe,Mn)_{21}O_{38}$. It occurs in rounded crystals and grains in diamond-bearing sands.

senandorite (sen-an'-dor-ite) A metallic grayish-white orthorhombic mineral: $Ag_{12}Pb_{10}CuSb_{37}S_{72}$. Syn: andorite VI. A member of the ramdohrite-lillianite homologous series.

sénarmontite (sen-ar-mont'-ite) A colorless, white, or gray cubic mineral: Sb_2O_3. It is polymorphous with valentinite.

Senecan (Sen'-e-can) North American provincial series: lower Upper Devonian (above Erian, below Chautauquan).

senegalite (sen'-e-gal-ite') A colorless or pale yellow orthorhombic mineral similar to bolivarite but with less water: $Al_2(PO_4)(OH)_3{\cdot}H_2O$.

senescence [geomorph] (se-nes'-cence) The part of the developmental sequence of a landform, or the hypothetical *cycle of erosion* of a landscape, when the landform or region enters upon the stage of old age. Cf: *senility.*

senescence [paleont] Old age; esp. the later stages in the life cycle of a species or other group. See also: *racial senescence.*

senescent (se-nes'-cent) (a) Pertaining to the stage in the developmental sequence of a landform, or in the hypothetical *cycle of erosion,* when the processes of erosion become slow and ineffective in producing further topographic modification; esp. said of a landscape that is growing old or aging (such as one characterized by a pediplain). (b) Said of a lake that is nearing extinction, as from filling by the remains of aquatic vegetation.

senescent lake A lake that is changing from a body of standing water to a *mire.* Part of the terminology in an obsolete concept of a lake's life history, a *dying lake.* Cf: *dead lake; extinct lake.*

sengierite (seng'-i-er-ite') A yellow-green monoclinic mineral of the *carnotite* group: $Cu_2(UO_2)_2(V_2O_8)(OH)_2{\cdot}6H_2O$.

senile (se'-nile) Pertaining to the stage of *senility* of the cycle of erosion; esp. said of a landscape or topography that is approaching a base-level plain or the end of the erosion cycle, or of a stream approaching an ultimate stage that is seldom fully reached, characterized by a sluggish current and a tendency to meander through a peneplain of slight relief only a little above base level.

senility (se-nil'-i-ty) The stage of the hypothetical *cycle of erosion* in which erosion has reached a minimum and base level has been approached. Cf: *old age [geomorph]; senescence [geomorph].*

senior (sen'-ior) In zoological nomenclature, said of the earlier published of two synonyms or homonyms.

Senonian (Se-no'-ni-an) An informal European stage or series; spans most of Upper Cretaceous (Coniacian, Santonian, Campanian, and Maastrichtian stages).

sensible horizon (sen'-si-ble) (a) The plane tangent to the Earth's surface at the observer's position; the *apparent horizon.* (b) *astronomic horizon.*

sensitivity The effect of *remolding* on the shear strength and consolidation characteristics of a clay or cohesive soil. A "sensitive" clay is one whose shear strength is decreased to a fraction of its former value on remolding at constant moisture content. Cf: *sensitivity ratio.*

sensitivity ratio A measure of the degree of sensitivity of a clay or cohesive soil. It is the ratio between the unconfined compressive strength of an undisturbed specimen and that of the same specimen at the same moisture content but in a remolded state (Terzaghi and Peck, 1967, p.31). The sensitivity ratio for most clays ranges between 2 and 4, although extrasensitive clays may have values greater than 8. Symbol: S_t. Syn: *remolding sensitivity.*

sensu lato (sen'-su la'-to) Literally, "in a broad sense", referring to the wide application of the name of a *taxon.* Etymol: Latin. Abbrev: s.l. Cf: *sensu stricto.*

sensu stricto (sen'-su stric'-to) Literally, "in a narrow sense", referring to the narrow or restricted application of the name of a *taxon.* Etymol: Latin. Abbrev: s.s. Cf: *sensu lato.*

sepal (se'-pal) One of the separate parts of the calyx, the outermost part of the perianth of a flower; usually green and more or less leaflike in texture.

separate *soil separate.*

separate [cart] Traditionally, mylar scribecoats, peelcoats, or ink drawings that make up the original map manuscripts. The term is also used to refer to exact photographic copies and derived composites of these materials. It is also used for scanned versions of the original materials (USGS, 2002a, p.1A-3). Syn: *film separate, map separate.*

separation (sep-a-ra′-tion) The distance, measured in a specified direction, between two once-adjacent points on an index plane (e.g. a bed or vein) disrupted by a fault. See also: *horizontal separation; normal separation; reverse separation; vertical separation; stratigraphic separation.*
separation disc A breakage area in a filament of cyanobacteria, formed by the death of a cell.
separation layer *abscission layer.*
***Sepia* sp. laminar structure** A *plywood structure* similar to crossed bladed structure, except that the basic structural units within each sheet consist of straight or bifurcated rods and laths instead of blades. Similar, but not identical, to the lamello-fibrillar structure of *Spirula* and certain other coleoids (Carter, 1990, Fig. 14).
sepiolite (se′-pi-o-lite′) (a) An orthorhombic chain-structure clay mineral: $Mg_4Si_6O_{15}(OH)_2{\cdot}6H_2O$. It is a white to light gray or light yellow material, extremely lightweight, absorbent, and compact, that is found chiefly in Asia Minor and is used for making tobacco pipes, cigar and cigarette holders, and ornamental carvings. Sepiolite occurs in veins with calcite, and in alluvial deposits formed from weathering of serpentine masses. Syn: *meerschaum; sea-foam.* (b) A group name for minerals with the analogous composition, but with Mg replaced by Ni or Na_2.
septa (sep′-ta) Plural of *septum.*
septal angle (sep′-tal) The angle between tangents drawn from the apex of a planispiral nautiloid shell to two successive septa and measured on a section made along the plane of symmetry (TIP, 1964, pt. K, p.58).
septal cycle All septa belonging to a single stage in ontogeny of a scleractinian *corallite,* as determined by the order of appearance of septal groups (six protosepta comprising the first cycle, and later-formed exosepta and entosepta in constantly arranged succession being introduced in sextants) (TIP, 1956, pt. F, p.250).
septal flap The extension of each lamella in the tests of foraminifers of the superfamily Rotaliacea, formed on the inner side of the chamber over the distal face of the previous chamber, and resulting in secondarily doubled septa (TIP, 1964, pt. C, p.63).
septal fluting One of the folds, wrinkles, or corrugations of septa (and antetheca) in a fusulinid test transverse to the axis of coiling, generally strongest in the lower (adaxial) part of septa and toward the poles and decreasing in intensity toward the top.
septal foramen (a) An intercameral opening in the test of a foraminifer. It may be homologous with the aperture or be secondarily formed. Cf: *foramen.* (b) An opening in the *septum* at the *siphuncle* of a cephalopod, allowing passage of the *siphuncular cord.*
septal furrow (a) The narrow middorsal region of a nautiloid in which the mural part of the *septum* (attached to the wall of the conch) is lacking (TIP, 1964, pt. K, p.58). Syn: *dorsal furrow.* (b) An *external furrow* of a fusulinid.
septal groove A longitudinal groove on the outer surface of the wall of a corallite, corresponding in position to a *septum* on the inner surface of the wall. Cf: *interseptal ridge.*
septalial plate (sep-tal′-i-al) One of the crural plates forming the floor of the *septalium* of a brachiopod and united with the earlier-formed part of the *median septum.*
septalium (sep-tal′-i-um) A troughlike structure of the *brachial valve* of a brachiopod, situated between hinge plates (or homologues), consisting of septalial plates (or homologues) enveloping and buttressed by the *median septum.* It does not carry adductor muscles.
septal neck A funnel-like or tubelike forward or backward flexure or extension of a *septum* around the septal foramen of a cephalopod shell. Syn: *neck [paleont].*
septal pinnacle One of the upper tips of septa located in the zone between parent corallite and offset during the early stages of increase in Paleozoic corals. Cf: *pseudoseptal pinnacle* (Fedorowski and Jull, 1976, p.41).
septal pore A small perforation in a septum (and antetheca) of the test of a fusulinid. Cf: *mural pore.*
septal tooth A small projection along the upper margin of a *septum* of a scleractinian coral, formed by extension of a *trabecula* beyond calcareous tissue connecting it with others.
septaria (sep-tar′-i-a) Plural of *septarium.*
septarian (sep-tar′-i-an) Said of the irregular polygonal pattern of internal cracks developed in septaria, closely resembling the desiccation structure of mud cracks; also said of the epigenetic mineral deposits that may occur as fillings of these cracks.
septarian boulder *septarium.*
septarian nodule *septarium.*
septarium (sep-tar′-i-um) (a) A large roughly spheroidal concretion, 8 to 90 cm in diameter, usually of an impure argillaceous carbonate such as clay ironstone. It is characterized internally by irregular polyhedral blocks formed by a series of radiating cracks that widen toward the center and that intersect a series of cracks concentric with the margins, the cracks invariably filled or partly filled by crystalline minerals (most commonly calcite) that cement the blocks together. Its origin involves the formation of an aluminous gel, case hardening of the exterior, shrinkage cracking due to dehydration of the colloidal mass in the interior, and vein filling. The veins sometimes weather in relief, thus producing a septate pattern. Cf: *melikaria.* Syn: *septarian nodule; septarian boulder; beetle stone; turtle stone.* (b) A crystal-lined crack or fissure in a septarium. Pl: *septaria.*
septechlorite (sep-te-chlo′-rite) A name given to a group of minerals (amesite, cronstedtite, berthierine) having chlorite-like formulas; they have basal spacings of 7 Å and therefore belong to the kaolinite-serpentine group. Syn: *pseudochlorite.*
septic system A disposal method for domestic wastewater that commonly includes (1) a tank for anaerobic breakdown of waste and separation of solids and liquids and (2) a leach field for release of liquids into the soil.
septifer (sep′-ti-fer) Said of brachiopod crura having the form of septa that descend directly from the dorsal side of the hinge plates to the floor of the brachial valve.
septimembrate apparatus A conodont skeletal apparatus consisting of seven morphologically distinct element types, regardless of total number of elements in a complete apparatus (TIP, 1981, pt. W, supp.2).
septotheca (sep-to-the′-ca) A wall of a scleractinian or rugose *corallite,* formed by thickened outer parts of septa along the axis of trabecular divergence. Cf: *synapticulotheca; paratheca.*
septula (sep′-tu-la) A communication organ in cheilostome and ctenostome bryozoans, consisting of a *pore plate* and a *rosette* of cells. Pl: septulae. Syn: *septulum.*
septulum (sep′-tu-lum) (a) *septula.* (b) A small *septum;* e.g. a ridge or small partition extending adaxially downward from the lower surface of the spirotheca in the test of a fusulinid foraminifer so as to partially subdivide the chambers of the test. Pl: septula.
septum [bot] (sep′-tum) (a) In filamentous algae or fungi, a crosswall, generally perpendicular to the length of the trichome or hypha. (b) In thecae or cysts, a delicate linear projection on its wall.
septum [paleont] (a) One of the transverse internal calcareous partitions dividing the shell of a cephalopod, such as a partition that divides the *phragmocone* of a nautiloid into camerae and that is attached to the inside of the wall of the *conch.* (b) One of several radially disposed longitudinal calcareous plates or partitions of a *corallite,* occurring between or within mesenterial pairs. It presumably supported the basal disk and lower wall of the *polyp.* Also, a thin radial noncalcareous partition, composed of soft tissue, dividing the gastrovascular cavity of an octocorallian polyp. (c) A partition or wall dividing a foraminiferal test interiorly into chambers, commonly formed by previous outer wall or apertural face. (d) A relatively long, narrow, commonly bladelike elevation of the secondary shell of a brachiopod; a median ridge in either valve of a brachiopod. It is indicated in rhynchonelliforms (formerly articulates) (within the underlying floor of a valve) by persistent high narrow deflections of fibrous calcite originating near the primary layer, and in others by comparable deflections of shell lamellae. (e) The wall separating the two rows of thecae in a biserial graptoloid. (f) A radial, longitudinal, normally perforate plate connecting the inner and outer walls of an archaeocyathid. Formerly known as *pariety.* (g) A platelike structure in an echinoid spine, radiating from the axial zone toward the anterior of the spine and seen in a cross section of the spine. (h) A delicate linear projection on the wall of a *dinoflagellate cyst* or a partition that divides a generally elongate cavity into compartments in an acritarch (Cramer and Diez Rodriguez, 1979; Fensome et al., 1993; Williams et al., 2000). (i) Generally, any wall-like partition in multicellular organisms that separates two cavities (e.g., chambers of the heart) or tissue masses (e.g., lobules of the liver). Pl: *septa.* Adj: septal.
Sequanian (Seq-ua′-ni-an) Substage in Great Britain: Upper Jurassic (upper Lusitanian; above Rauracian Substage, below Kimmeridgian Stage).
sequence (se′-quence) (a) A succession of geologic events, processes, or rocks, arranged in chronologic order to show their relative position and age with respect to geologic history as a whole. (b) The unit immediately above a system in the chronostratigraphic hierarchy, now called *erathem* (Moore, 1933, p.54; Weller, 1960, p.418, 449). (c) A major informal lithostratigraphic unit of greater than group or supergroup rank, traceable over large areas of a continent, and bounded by unconformities of interregional scope, such as in the cratonic interior of North America (Sloss, 1963). (d) A geographically discrete succession of major rock units that were deposited under related environmental conditions (Silberling and Roberts, 1962). Cf: *sub-sequence.* Syn: *stratigraphic sequence; tectono-stratigraphic unit.* (e) In sequence stratigraphy, a relatively conformable succession of genetically related strata bounded by unconformities or their correlative conformities (Mitchum, 1977). Parasequences and parasequence sets are the stratal building blocks of the sequence (Van Wagoner et al., 1990). Cf: *parasequence; parasequence set.* (f) A faunal succession.
sequence set In sequence stratigraphy, a set of sequences arranged in a distinctive progradational, aggradational, or retrogradational stacking pattern (Mitchum and Van Wagoner, 1991).
sequence stratigraphy The study of rock relationships within a chronostratigraphic framework of repetitive, genetically related strata bounded by surfaces of erosion or nondeposition, or their correlative conformities (Van Wagoner et al., 1990).
sequent geosyncline (se′-quent) The constituent geosynclines of a *polygeosyncline,* separated from one another by the development of geanticlines (Schuchert, 1923). Obsolete. Syn: *secondary geosyncline (b).*

sequential infilling A process in which joints in a set initiate in stages, with later joints forming between preexisting joints.

sequential landform (se-quen'-tial) One of an orderly succession of smaller landforms that are developed by the erosion, weathering, and mass-wasting of larger initial landforms; it includes "erosional landforms" resulting from the progressive removal of earth materials, and "depositional landforms" resulting from the accumulation of the products of erosion. Cf: *ultimate landform; destructional.* Syn: sequential form.

sequum A B horizon together with any overlying eluvial horizons.

serac (se-rac' [Pron: seh-rahk']) A jagged pinnacle, sharp ridge, needlelike tower, or irregularly shaped block of ice on the surface of a glacier (commonly among intersecting crevasses, as on an icefall), formed where the glacier is periodically broken as it passes over a steep slope. Introduced by Saussure in 1787 on the basis of snow/ice features observed along the margins of snow cap; he soon extended the term to ice features at the surface of a glacier (Carozzi, 1996). Etymol: French "sérac", a solid curdy white cheese made in the Alps and sold in quadrangular blocks, which glacial seracs resemble in appearance and shape. Both the French form and the anglicized version are in regular use. Cf: *nieve penitente.*

seral (ser'-al) The adj. of *sere.*

seral succession *sere.*

sérandite (ser-and'-ite) A rose-red triclinic mineral of the *wollastonite* group: $Na(Mn^{2+},Ca)_2Si_3O_8(OH)$. Cf: *pectolite.*

sere A sequence of ecologic communities that succeed one another in development from the *pioneer* stage to *climax.* Adj: *seral.* See also: *paleosere; succession [ecol].* Syn: *seral succession.*

serendibite (ser-en'-di-bite) An indigo-blue to grayish blue-green triclinic mineral of the *aenigmatite* group: $(Ca,Na)_2(Mg,Fe)_3(Al,Fe^{3+})_{4.5}B_{1.5}Si_3O_{20}$.

sergeevite (ser-ge'-ev-ite) A white trigonal mineral: $Ca_2Mg_{11}(CO_3)_9(HCO_3)_4(OH)_4 \cdot 6H_2O$.

serial homology (se'-ri-al) The similarity that exists between members of a series of structures in an organism; e.g. the resemblance to each other of the vertebrae in the vertebral column.

serial sampling A method of gathering samples by a set pattern, such as a grid, to insure randomness.

seriate (se'-ri-ate) A variety of igneous porphyritic texture in which the sizes of the grains range gradually down to the size of ground mass grains. Cf: *hiatal [ign].*

sericite (ser'-i-cite) A white, fine-grained potassium mica occurring in small scales and flakes as an alteration product of various aluminosilicate minerals, having a silky luster, and found in various metamorphic rocks (esp. in schists and phyllites) or in the wall rocks, fault gouge, and vein fillings of many ore deposits. It is usually muscovite or very close to muscovite in composition, and may also include much illite.

sericitic alteration *phyllic alteration.*

sericitic sandstone (ser-i-cit'-ic) A sandstone in which sericite intermingles with finely divided quartz and fills the voids between quartz grains.

sericitization (ser'-i-cit'-i-za'-tion) A hydrothermal, deuteric, or metamorphic process involving the introduction of, alteration to, or replacement by sericitic muscovite.

series [cart] (se'-ries) *map series.*

series [geol] Any number of rocks, minerals, or fossils having characteristics, such as growth patterns, succession, composition, or occurrence, that make it possible to arrange them in a natural sequence. In order to avoid possible confusion with the stratigraphic term series, an alternative expression may be desirable, e.g. succession, sequence, collection.

series [radioactivity] *radioactive series.*

series [soil] *soil series.*

series [stratig] A conventional stratigraphic unit that ranks below a system and always is a division of a system. A series commonly constitutes a major unit of chronostratigraphic correlation within a province, between provinces, or between continents. Although many European series are being adopted increasingly for dividing systems on other continents, *provincial series* of regional scope continue to be useful (NACSN, 1983, Art. 73). The temporal equivalent of a series is an epoch. Most systems have been divided into series and most series into stages. The term is not restricted to stratified rocks but may be applied to intrusive rocks in the same time-stratigraphic sense. Formal series names are binomial, usually consisting of a geographic name (generally but not necessarily with the adjectival ending "-an" or "-ian") and the word "Series," the initial letter of both terms being capitalized. (b) An obsolete term for a lithostratigraphic unit more or less equivalent to a group or supergroup, esp. in the Precambrian. See also: *sub-sequence.* (c) A term used in England for a lithostratigraphic unit, generally large, throughout which the strata are conformable.

series circuit Elements in an electrical circuit so connected that there is a single continuous path through each element.

serir (se-rir') A *desert plain* strewn with rounded pebbles, older than the gravel-covered *reg;* a *stony desert* from which the sand has been blown away, as in the Sahara of Libya and Egypt. Etymol: Arabic, "dry". Pl: serir. Cf: *pebble armor; hammada.*

Serozem (Ser'-o-zem) *Sierozem.*

serpenticone (ser-pen'-ti-cone) An *evolute,* many-whorled cephalopod shell with the whorls hardly overlapping, resembling a coiled snake or rope.

serpentine (ser'-pen-tine) (a) A group of common rock-forming minerals having the general formula: $(Mg,Al,Fe,Mn,Ni,Zn)_{2-3}(Si,Al,Fe)_2O_5(OH)_4$. Serpentines have a greasy or silky luster, a slightly soapy feel, and a tough, conchoidal fracture; they are usually compact but may be granular or fibrous, and are commonly green, greenish-yellow, or greenish-gray and often veined or spotted with green, and white. Serpentines are always secondary minerals, derived by alteration of magnesium-rich silicate minerals (esp. olivines), and are found in both igneous and metamorphic rocks; they generally crystallize in the monoclinic system. Translucent varieties are used for ornamental and decorative purposes, often as a substitute for jade. (b) A mineral of the serpentine group, such as chrysotile, antigorite, lizardite, parachrysotile, and orthochrysotile. Etymol: Latin "serpentinus", "resembling a serpent", from the mottled shades of green.

serpentine asbestos *chrysotile.*

serpentine-hosted asbestos A *mineral-deposit type* characterized by stockworks of chrysotile asbestos in serpentinized ultramafic rocks. Syn: *ultramafic-hosted asbestos.*

serpentine-hosted magnesite A *mineral-deposit type* consisting of magnesite veins and massive magnesite, sometimes accompanied by talc, replacing serpentinite. Although typically small, these deposits have been an important source of magnesite in the past. Syn: *amorphous magnesite; bone magnesite; cryptocrystalline magnesite; ultramafic-hosted magnesite.*

serpentine-hosted talc A *mineral-deposit type* consisting of veins, massive bodies, and lenticular masses of talc, sometimes accompanied by magnesite, replacing serpentinite. Although the most abundant known resource of talc, these deposits have been of secondary importance for talc production. Syn: *ultramafic-hosted talc.*

serpentine jade A variety of the mineral serpentine resembling jade in appearance and used as an ornamental stone; specif. bowenite.

serpentine marble *verd antique.*

serpentine rock *serpentinite.*

serpentine spit A spit that is extended in more than one direction due to variable or periodically shifting currents.

serpentine-talc An unnecessary name for material with the composition: $Mg_6(OH)_6[Si_6O_{15}]$. It is intermediate between serpentine and talc in composition and physical characteristics.

serpentinite (ser-pen'-ti-nite) A rock consisting almost wholly of serpentine-group minerals, e.g. antigorite, chrysotile, or lizardite, derived from the hydration of ferromagnesian silicate minerals such as olivine and pyroxene. Accessory chlorite, talc, and magnetite may be present. Syn: *serpentine rock.*

serpentinization (ser-pen'-ti-ni-za'-tion) The process of hydrothermal alteration by which magnesium-rich silicate minerals (e.g., olivine, pyroxenes, or amphiboles) in dunites, peridotites, or other ultramafic rocks are converted to or replaced by serpentine minerals.

serpent kame (ser'-pent) A term introduced by Shaler (1889, p.549) for a sinuous *esker* and known in New England also as an *Indian ridge.*

serpent stone A stone, usually a highly absorbent aluminous gem, once believed to be formed by snakes and to be efficacious in drawing out poison; specif. adder stone.

serpierite (ser'-pier-ite) A bluish-green monoclinic mineral: $Ca(Cu,Zn)_4(SO_4)_2(OH)_6 \cdot 3H_2O$.

serpophite (ser'-po-phite) A metacolloidal variety of the mineral serpentine.

Serpukhovian A series or stage of the Standard Global Chronostratigraphic Scale: uppermost unit of Mississippian subsystem of the Carboniferous (above Visean, below Bashkirian).

serpulid (ser'-pu-lid) Any annelid that belongs to the family Serpulidae and that characteristically builds a contorted calcareous or leathery tube on a submerged surface. See also: *serpulid reef.*

serpulid reef A small reef patch, with a raised rim and cup-shaped central parts, constructed largely of calcareous tubes secreted by *serpulid* worms of the family Serpulidae (Polychaeta; Annelida). Cf: *sabellariid reef; vermetid reef; worm reef.*

serra (ser'-ra) (a) *sierra.* (b) A term used in Brazil for an elevated mountain zone supporting luxuriant vegetation. Etymol: Latin, "saw".

serrabrancaite A dark brown to dark greenish-black monoclinic mineral: $MnPO_4 \cdot H_2O$.

serrate [bot] (ser'-rate) With numerous small teeth frequently pointing toward the apex, as in some leaf margins.

serrate [geomorph] adj. Said of topographic features that are notched or toothed on the edge, or have a saw-edged profile; e.g. a serrate divide. Also called: serrated. Syn: *saw-toothed.* n. A rocky mountain summit having a serrate profile (Stone, 1967, p.246).

serrate [petrology] Said of saw-toothed contacts between minerals, usually resulting from replacement; e.g. the serrate texture of megacrysts in contact with plagioclase in igneous rocks.

Serravallian (Ser-ra-val'-lian) A stage of the Standard Global Chronostratigraphic Scale: upper middle Miocene (above Langhian, below Tortonian).

serule (ser'-ule) *microsere.*

sesquan (ses'-quan) A *cutan* consisting of sesquioxides or hydroxides (of aluminum or iron) (Brewer, 1964, p.214).

sessile (ses'-sile) Said of a plant or animal that is permanently attached to a substrate and is not free to move about. Cf: *vagile.*
sessile cruralium *Cruralium* united with the floor of the brachial valve of a brachiopod without intervention of the supporting *median septum.*
sessile spondylium *Spondylium* united with the floor of the pedicle valve of a brachiopod without intervention of the supporting *median septum.*
seston (ses'-ton) Suspended, nonsinking particulate material in ocean or lake water. It consists of living and dead algae, *particulate organic matter,* and mineral particles.
set [currents] The compass direction toward which a current is flowing; a direction of flow. Syn: *current direction.*
set [exp struc geol] *permanent set.*
set [stratig] (a) A term introduced by McKee and Weir (1953, p.382-383) for "a group of essentially conformable strata or cross-strata, separated from other sedimentary units by surfaces of erosion, nondeposition, or abrupt change in character"; it is composed of two or more consecutive beds of the same lithology, and is the smallest and most basic group unit. Cf: *coset.* (b) A unit composed of two or more consecutive beds of the same lithology (Bokman, 1956, p.126).
seta [bot] (se'-ta) (a) The sporophyte stalk in the mosses and liverworts. (b) A stiff, short hair on a plant.
seta [paleont] A slender, typically rigid or bristly and springy organ or part of an invertebrate, e.g. a movable whiplike part of a *vibraculum* in cheilostome bryozoans; a chitinous bristle arising from the invagination of the mantle groove of a brachiopod; or a hairlike process of the external membrane of a crustacean. Pl: setae.
setaceous (se-ta'-ceous) Bristlelike, as of a plant part covered with stiff, short hairs.
setal duct (se'-tal) A larger duct, up to 75 µm in diameter, through a trilobite exoskeleton. Cf: *pore canal.* Syn: *gland duct.*
seter (set'-er) A Norwegian term for a wave-cut rock terrace.
settled snow (set'-tled) A loose term indicating snow that has become more or less compacted under gravity to a stable density. Cf: *old snow.*
settlement (set'-tle-ment) (a) The subsidence of a structure, caused by compression or movement of the soil below the foundation. See also: *differential settlement.* (b) The lowering of the overlying strata in a mine, owing to extraction of the mined material.
settling [mass move] (set'-tling) The sag in outcrops of layered strata, caused by rock creep (Sharpe, 1938, p.33). Syn: *outcrop curvature.*
settling [sed] The deposition of sediment.
settling [snow] Time-dependent compaction of snow under its own weight.
settling basin (a) An artificial basin or trap designed to collect the suspended sediment of a stream before it flows into a reservoir and thereby prevent the rapid siltation of the reservoir; e.g. a *desilting basin.* It is usually provided with means to draw off the clear water. (b) A sedimentation structure designed to remove pollutant materials from mill effluents; a tailings pond.
settlingite (set'-tling-ite) A hard, brittle, pale-yellow to deep-red hydrocarbon (H:C about 1.53) found in resinous drops on the walls of a lead mine at Settling Stones in Northumberland, England. Also called: Settling Stones resin.
settling reservoir A reservoir consisting of a series of shallow basins arranged in steps and connected by long conduits, allowing the removal of only the clear upper layer of water in each basin.
settling velocity The rate at which suspended solids subside and are deposited. Syn: *fall velocity.*
setulf (set'-ulf) A *bed form* having positive relief and molded into a shape resembling the counterpart of a flute or groove, sculpted in mud and created by sheetflow of the incoming tide. Setulfs have high pointed ends on their upcurrent sides and flare out and become lower on their downcurrent sides. Etymol: reverse of flutes.
setup (set'-up) (a) The assembly and arrangement of the equipment and apparatus required for the performance of a surveying operation; specif. a surveying instrument (transit or level) placed in position and leveled, ready for taking measurements. (b) The actual physical placing of a leveling instrument over an instrument station. (c) *instrument station.* (d) The horizontal distance from the fiducial mark on the front end of a surveyor's tape (or the part of a tape which is in use at the time), measured in a forward direction to the point on the ground mark or monument to which the particular measure is being made. Also spelled: set-up.
Sevatian (Se-va'-tian) A European substage: upper substage of Norian Stage, Upper Triassic (above Alaunian, below Rhaetian Stage).
Sevier Orogeny (Se-vier') A name proposed by R.L. Armstrong (1958) for the well known deformations that occurred along the eastern edge of the Great Basin in Utah during times intermediate between the Nevadan Orogeny farther west and the Laramide Orogeny farther east, culminating early in the Late Cretaceous. The orogeny involved folding and eastward thrusting of the rocks. It is roughly coeval with the formation of the Sierra Nevada batholith. Thus, it is thought to represent the back-arc thrust belt of an Andean convergent margin.
sewardite A dark red to red-orange orthorhombic mineral: $CaFe^{3+}_2(AsO_4)_2(OH)_2$. It is the Ca analogue of carminite.
sexiform antapical plate (sex'-i-form) An antapical plate on a dinoflagellate theca or cyst that has five straight sides and one inwardly curved side, as in the modern genus *Gonyaulax.* The straight sides are in contact with the posterior intercalary plate and four postcingular plates; the curved side is in contact with the posterior sulcal plate (Williams et al., 2000, p.207-208).
seximembrate apparatus A conodont skeletal apparatus consisting of six morphologically distinct element types, regardless of the total number of elements in a complete apparatus (TIP, 1981, pt. W, supp.2).
sexine (sex'-ine) The outer, sculptured division of the exine of pollen, more or less equivalent to *ektexine* (but not including the *foot layer*). Cf: *nexine; perine.*
sexiradiate (sex-i-ra'-di-ate) A sponge spicule in the form of six equidistant coplanar rays arising from a common center.
sextant (sex'-tant) A double-reflecting, hand instrument used for measuring the angular distance between two objects in the plane defined by the two objects and the point of observation. It was originally characterized by having an arc of 60 degrees (and a range of 120 degrees), but the term is now applied to similar instruments regardless of range, esp. those used in navigation for measuring apparent altitudes of celestial bodies from a moving ship or airplane, or in hydrographic surveying for measuring horizontal angles at a point in a moving boat between shore objects. Cf: *quadrant [surv]; astrolabe.*
sexual dimorphism (sex'-u-al) A condition in which the two sexes of an organism are markedly dissimilar in appearance.
seybertite (sey'-bert-ite) *clintonite.*
seyrigite (sey'-rig-ite) A variety of scheelite containing molybdenum.
sferic (sfer'-ic) Natural "atmospheric" fluctuation of the Earth's electromagnetic field, generally at frequencies from 1 to 10^5 Hz, usually caused by lightning. Etymol: A short form of atmospheric. Also spelled: *spheric.*
SFM *scanning force microscopy.*
s-fold A fold whose profile has the form of the letter "s".
SF-tectonite A metamorphic rock with planar fabric in which the fabric elements were produced by fracture and/or shear along a pervasive set of subparallel surfaces. It is characterized by mesoscopic subparallel fractures, commonly slickensided, that are independent of the arrangement and/or orientation of mineral grains within the rock (Raymond, 1975, p.8).
shabaite-(Nd) (sha'-ba-ite) A yellowish-white monoclinic mineral: $Ca(Nd,Sm)_2(UO_2)(OH)_2(CO_3)_4 \cdot 6H_2O$.
shabka (shab'-ka) A desert landscape formed by wind erosion of alluvial basins. Etymol: Arabic, "network, fiber".
shabynite (sha'-by-nite) A white monoclinic mineral: $Mg_5(BO_3)(Cl,OH)_2(OH)_5 \cdot 4H_2O$.
shackanite (shack'-a-nite) An analcime *phonolite* containing rhombohedral phenocrysts of alkali feldspar in a groundmass of analcime, alkali feldspar, and glass. The name, given by Daly in 1912, is for Shackan railroad station, British Columbia. Obsolete.
shaded-relief map (shad'-ed-re-lief') A map of an area whose relief is made to appear three-dimensional by the method of *hill shading.*
shading (shad'-ing) *hill shading.*
shadlunite (shad'-lun-ite) A metallic grayish-yellow cubic mineral of the *pentlandite* group: $(Pb,Cd)(Fe,Cu)_8S_8$.
shadow weathering (shad'-ow) Mechanical weathering in which disintegration of rock occurs along the margin of sunlight and shade; a kind of insolation, or the effect of sunlight on weathering of rock.
shadow zone [geomorph] *wind shadow.*
shadow zone [seis] (a) An area in which there is little penetration of seismic waves. (b) A region 100°-140° from the epicenter of an earthquake in which, due to refraction from the *low-velocity zone* inside the core boundary, there is no direct penetration of seismic P-waves, although diffracted and multiple P-wave energy may be present. (c) Similar zone for S-waves extending from 100°-180°. Cf: *blind zone.*
shafranovskite (shaf-ra-nov'-skite) A dark green, olive green or yellowish green trigonal mineral: $(Na,K)_6(Mn^{2+},Fe^{2+})_3Si_9O_{24} \cdot 6H_2O$.
shaft [eng] A concrete foundation cast in a vertical drilled hole, usually of relatively large diameter. Some shaft foundations, enlarged at the bottom to increase the load-carrying capacity, are called belled shafts or belled *caissons.*
shaft [mining] A vertical or inclined excavation through which a mine is worked.
shaft [paleont] (a) The main part of the spine of an echinoid. (b) The ridgelike or stalklike proximal part of the cardinal process of a brachiopod, supporting the myophore. (c) The rodlike part of a *rhyncholite* (TIP, 1964, pt. K, p.22).
shaft [speleo] A passage in a cave that is vertical or nearly so.
shagreen (sha-green') *relief [optics].*
shagrinate *chagrenate.*
shakehole (shake'-hole) *sinkhole.*
shakhovite An adamantine bright green to olive-green monoclinic mineral: $Hg^{1+}_4Sb^{5+}O_3(OH)_3$.
shaking prairie (shak'-ing) A term used in Louisiana to describe delta land with a surface of matted vegetation resting on water, or on waterlogged peat or sand, which trembles when walked on. Syn: *trembling prairie.*

shale A laminated, indurated rock with >67% clay-sized minerals; a claystone with *fissility.* Cf: *clay [geol]; clay mineral; claystone [sed]; siltstone.*.
shale-arenite (shale-ar'-e-nite) A term used by Folk (1968, p.124) for a *sedarenite* composed chiefly of shale fragments.
shale-ball An iron meteorite largely converted to iron oxides by weathering. Cf: *oxidite.*
shale baseline The line drawn by a *well log* analyst through the spontaneous potential curve corresponding to shales or their electrochemical equivalents. Where freshwater drilling muds and saline formation waters prevail, this line forms the right-hand reference from which calculation of formation-water resistivity in beds invaded by mud filtrate may be attempted. Syn: *shale line.*
shale break In drilling, a thin layer or parting of shale between harder strata. Also a marked decrease in drilling rate thought to correspond to a change in the lithology being drilled. Cf: *break [drill].*
shale crescent A term used by Shrock (1948, fig.86) for a crescent formed by the filling of a ripple-mark trough by mud. It is a syn. of flaser structure.
shale-hosted copper *sediment-hosted copper deposit.*
shale ice A mass of thin, brittle plates of river or lake ice, formed when sheets of skim ice break up into small pieces which are gathered into bunches.
shale line *shale baseline.*
shale oil A crude oil obtained from *oil shale* by submitting it to destructive distillation.
shale-out A *stratigraphic trap* formed by lateral variation or facies change of a porous sandstone or limestone, in which the clay content increases until porosity and permeability disappear and the bed grades into claystone or shale. Cf: *pinch-out.*
shale shaker In *rotary drilling,* an inclined vibrating screen or screens in a steel game over which the *drilling mud* passes on its return to the surface and is screened to remove the *well cuttings* and be conditioned for recirculation.
shaley *shaly.*
shaliness (sha'-li-ness) The quality of being shaly; specif. the property of clay-rich rocks of splitting with concave or "shelly" surfaces roughly parallel to the bedding planes (Grabau, 1924, p.785).
shallower-pool test (shal'-low-er-pool') A well located within the known limits of an oil or gas pool and drilled with the object of searching for new producing zones above the producing zone of the pool (Lahee, 1962, p.134).
shallow-focus earthquake (shal'-low-fo'-cus) An earthquake with a focus at a depth of less than 70 km. Most earthquakes are of this type. Cf: *intermediate-focus earthquake; deep-focus earthquake.*
shallow percolation (shal'-low) Precipitation that moves downward and laterally toward streams. Syn: *storm seepage.* Cf: *deep percolation.*
shallow phreatic Said of cave formation near the top of the water-saturated zone. See also: *deep phreatic.*
shallows (shal'-lows) An indefinite term applied to a shallow place or area in a body of water, or to an expanse of shallow water; a *shoal.*
shallow scattering layer A stratified area of marine organisms over a continental shelf that scatter sound waves from an echo sounder. Cf: *deep scattering layer; surface scattering layer.*
shallow-water wave (shal'-low-wa'-ter) A wave on the surface of a body of water, the wave length of which is 25 or more times the water depth, and for which the water depth is an influence on the shape of the orbital and on the velocity. Cf: *deep-water wave; transitional-water wave.* Syn: *long wave [water].*
shallow well (a) A water well, generally dug by hand or by excavating machinery, or put down by driving or boring, that taps the shallowest aquifer in the vicinity. The water is generally unconfined groundwater. (b) A well whose water level is shallow enough to permit use of a shallow-well (suction) pump, the practical lift of which is taken as 7 m. Cf: *water-table well; deep well.*
shaly (a) Pertaining to, composed of, or having the character of shale; esp. readily split along closely spaced bedding planes, such as "shaly structure" or "shaly parting". Also, said of a fine-grained, thinly laminated sandstone having the characteristic fissility of shale owing to the presence of thin layers of shale; or said of a siltstone possessing bedding-plane fissility. (b) Said of bedding that consists of laminae ranging in thickness from 2 to 10 mm (Payne, 1942). Cf: *argillaceous.* Also spelled: *shaley.*
shandite (shand'-ite) A metallic white rhombohedral mineral: $Pb_2Ni_3S_2$.
Shand's classification A classification of igneous rocks based on crystallinity, degree of saturation with silica, degree of saturation with alumina, and color index. This system was developed in 1927 by S. J. Shand (Shand, 1947).
shank [paleont] (a) The connection between the flukes and the stock of an *anchor* of a holothurian. (b) The part connecting the eye and spear of a hook of a holothurian.
shank [struc geol] An obsolete syn. of *limb [struc geol].*
shannonite A white orthorhombic mineral: $Pb_2O(CO_3)$.
shantung (shan-tung') A monadnock in the process of burial by huangho deposits (Grabau, 1936, p.266). Type locality: Shantung rocky mass of northern China.
Shantung soil (Shan-tung') An early name for *Noncalcic Brown soil.* Var: Shantung Brown soil.
shape *particle shape.*
shape class The general group of shapes (oblate, prolate, or intermediate) to which a pollen grain belongs in terms of the ratio between its equatorial diameter and pole-to-pole dimension.
shape-preferred orientation (shape-pre-ferred') The preferred orientation of elongated or flattened axes of crystals, as a result of *crystal gliding, dynamic recrystallization* or magmatic settling or flow. Cf: *lattice-preferred orientation.*
shapometer (sha-pom'-e-ter) A device for measuring the shapes of sedimentary particles (Tester and Bay, 1931).
shard (a) A vitric fragment formed by volcanic processes; some have a characteristically curved surface of fracture. Shards generally consist of bubble-wall fragments produced by disintegration of pumice during or after the eruption. (b) A syn. of *sherd.*
shared character A character held in common by two or more organisms or taxa; it may be either a *homology* or an *analogy.*
shared derived character In *cladism,* a character shared by two or more organisms or taxa and inherited from an immediately preceding or recent common ancestor; an advanced character held in common by two or more taxa. Cf: *homology (c); derived character; shared character.* Syn: *synapomorphy.*
shared primitive character In *cladism,* a character shared by two or more organisms or taxa and inherited from a remote or much earlier common ancestor. Cf: *homology (c); primitive character; shared character.* Syn: *symplesiomorphy.*
shark A member of the order *Selachii.*
sharkskin pahoehoe (shark'-skin) A type of *pahoehoe* whose surface displays innumerable tiny spicules or spines produced by escaping gas bubbles.
shark-tooth projection A structure formed by the pulling or tearing apart of plastic lava into fine, sharp points several centimeters in length. Such projections may occur along the edge of a flow or along a slump scarp.
sharpite (sharp'-ite) A greenish-yellow, olive-green, or grayish-white orthorhombic mineral: $Ca(UO_2)_6(CO_3)_5(OH)_4{\cdot}6H_2O$.
sharp sand A sand composed of angular grains, nearly or wholly free from other particles (as of clay), and used in making mortar.
sharpstone (sharp'-stone) (a) A collective term proposed by Shrock (1948a) for any rock fragment larger than a sand grain (diameter greater than 2 mm) having angular edges and corners; a clastic constituent of detritus. Cf: *roundstone.* (b) A term used in Yorkshire, England, for a fine-grained, nonargillaceous sandstone that breaks into angular fragments. Also spelled: sharp stone.
sharpstone conglomerate *sedimentary breccia.*
Shasta (Shas'-ta) Provincial series in California: Lower Cretaceous.
shatter belt (shat'-ter) *fault zone.*
shatter breccia A tectonic breccia composed of angular fragments that show little rotation (Bateman, 1959, p.133). Cf: *rubble breccia.* Syn: *crackle breccia.*
shatter cone A distinctively striated conical structure in rocks, ranging in length from less than a centimeter to several meters, along which fracturing has occurred. It is generally found in nested or composite groups in the rocks of *impact structures* and formed by shock waves generated by *impact* (Dietz, 1959). Shatter cones superficially resemble *cone-in-cone structure* in sedimentary rocks. They are most common in fine-grained homogeneous rocks such as limestone and dolomite, but are also known in shale, sandstone, quartzite, and granite. The striated surfaces radiate outward from the apex in horsetail fashion; the apical angle varies but is close to 90° . Obsolete syn: *pressure cone.*
shatter-cone segment A part of an incompletely developed *shatter cone,* consisting of a single curved, striated surface, generally 10 to 45 degrees of cross section, of a cone whose apical angle may range from 90 to 120 degrees (Manton, 1965, p.1021). Most shatter-coned rocks display only segments.
shatter coning A mode of rock failure characterized by the development of shatter cones (Manton, 1965, p.1021).
shattering (shat'-ter-ing) The breaking-up into angular blocks of a hard rock that has been subjected to severe stresses; the fractures may cut across mineral grains and structures in the rock.
shatter zone An area, often a belt or linear zone, of fissured or cracked rock that may be filled by mineral deposits, forming a network pattern of veins.
shattuckite (shat'-tuck-ite) A blue orthorhombic mineral: $Cu_5(SiO_3)_4(OH)_2$. Cf: *planchéite.*
shcherbakovite (shcher'-bak-ov-ite) A dark brown orthorhombic mineral: $K_2NaTi^{4+}{}_2O(OH)Si_4O_{12}$.
shcherbinaite (shcher-bin'-a-ite) A yellow-green orthorhombic mineral: V_2O_5. It is classed as a "vanadium bronze".
sheaf structure A bundled arrangement of crystals that is characteristic of certain fibrous minerals, e.g. stilbite; the bundle is thicker at the ends than in the middle.
shear A deformation resulting from stresses that cause or tend to cause contiguous parts of a body to slide relatively to each other in a direction parallel to their plane of contact. It is the mode of failure of a body or mass whereby the portion of the mass on one side of a plane or surface slides past the portion on the opposite side. In geological literature the term refers almost invariably to strain rather than to stress. It is also used to refer to surfaces and zones of failure by shear, and to surfaces along which differential movement has taken place.

shear band Thin zone of very high shear strain occurring in rocks in a shear zone.

shear cleavage An obsolete term for cleavage whose formation had been interpreted to involve shear displacement along cleavage planes. The term should be avoided.

shear drag *shear resistance.*

shear fold A fold model of which the mechanism is shearing or slipping along closely spaced planes parallel to the fold's axial surface. The resultant structure is a *similar fold.* Rarely used syn: *slip fold; glide fold.* See also: *cleavage fold.*

shear fracture A brittle fracture that forms by the linkage and coalescence of microcracks along a plane of high shear stress. Cf: *tension fracture.*

shear-hosted lode gold *low-sulfide gold-quartz vein.*

shear modulus *modulus of rigidity.*

shear moraine A debris-laden surface or zone along the margin of an ice sheet or ice cap, dipping in toward the center of the ice sheet but becoming parallel to the bed at the base. The name originated because it was thought that such zones were formed by shearing at the boundary between stagnant ice and active ice. This has been disputed and it has been suggested that shear moraines form through the refreezing of meltwater at the base of a glacier. Cf: *shear plane [glaciol].* Syn: *Thule-Baffin moraine.*

shear plane [exp struc geol] *shear surface.*

shear plane [glaciol] A foliation band in a glacier, esp. one containing abundant rock debris. It is often believed (usually incorrectly) that shear stress rates are particularly high in shear planes, as on a fault. Such high stress rates in dirty rock, however, are likely only in ice near the melting point; in cold ice , the dirt stiffens the ice. Apparent displacements on some shear planes may simply be due to differential ablation. Cf: *shear moraine.*

shear resistance In fluid dynamics, a tangential stress caused by the fluid viscosity, taking place along a boundary of the flow in the tangential direction of local motion (Chow, 1957). Cf: *pressure resistance.* Syn: *shear drag.*

shear-sense indicator *kinematic indicator.*

shear slide A slide produced by shear failure, usually along a plane of weakness, as bedding or cleavage.

shear sorting Sorting of sediments in which the smaller grains tend to move toward the zone of greatest shear strain and the larger grains toward the zone of least shear.

shear strain A measure of the amount by which parallel lines have been sheared past one another by deformation, specif. the tangent of the change in angle between initially perpendicular lines (the ordinary or engineering shear strain); or half the tangent of the change in angle between initially perpendicular lines (the tensor measure that appears in *infinitesimal-strain theory*).

shear strength The internal resistance of a body to shear stress, typically including a frictional part and a part independent of friction called *cohesion.*

shear stress That component of *stress* which acts tangential to a plane through any given point in a body; any of the tangential components of the stress tensor. Symbol: σ_s or τ.

shear structure Any rock structure caused by shearing, e.g., foliated cataclasite, mylonite foliation, or crenulation cleavage.

shear surface A surface along which differential movement has taken place parallel to the surface. Syn: *shear plane; slip surface.*

shear velocity The square root of the product of the acceleration due to gravity, the hydraulic mean depth of flow, and the slope of the energy grade line (ASCE, 1962).

shear wave [seis] *S wave.*

shear-wave splitting *birefringence [seis].*

shear zone (a) A tabular zone of rock that has been brecciated by many parallel fractures; such an area is often mineralized by ore-forming solutions. Cf: *sheeted-zone deposit.* (b) A parallel-sided zone of localized shearing displacement, which may be recognized by sigmoidal mineral-filled veins, locally well-developed cleavage or foliation, wholesale grain-size reduction or mylonitization, or some combination of these features (Ramsay and Graham, 1970).

sheath [bot] (a) A tubular, enrolled part or organ of a plant, e.g. the lower part of a leaf in the grasses. (b) In cyanobacteria, a mucilaginous covering external to the cell wall.

sheath [paleont] An investing cover or case of an animal body or body part; e.g. a tentacle sheath of a bryozoan, a receptacle or container in radiolarians, an expanded basal part of a bar, blade, or limb of a conodont element, or a finely perforate external layer in archaeocyathids.

sheath fold A fold with an isoclinal profile whose hinge line also curves through more than 90° of arc. Sheath folds may result from superposed folding, but are often inferred to result from the distortion of folds with mildly curved hinges due to localized shearing (Cobbold and Quinquis, 1980).

shed A divide of land; e.g. a *watershed.*

sheen A subdued and often iridescent or metallic glitter that approaches but is just short of optical reflection and that modifies the surface luster of a mineral; e.g. the optical effect still visible in the body of a gem (such as tiger's-eye) after its silky surface appearance has been removed by polishing.

sheepback rock (sheep'-back) A term used as a syn. of *roche moutonnée* in reference to the fanciful comparison of the landform to a sheep's back. Syn: *sheep rock.*

sheep rock *roche moutonnée; sheepback rock.*

sheer *precipice.*

sheet [intrus rocks] A general term for a tabular igneous intrusion, esp. those that are concordant or only slightly discordant. In this general sense, the term dike is used for a discordant tabular body, and the term sill for a concordant one. Cf: *sole injection.*

sheet [ore dep] A term used in the Upper Mississippi lead-mining region of the U.S. for galena occurring in thin, continuous masses.

sheet [sed] *blanket.*

sheet [water] *sheetflood.*

sheet crack A planar crack attributed to shrinkage of sediment due to dewatering (Fischer, 1964, p.148). It is commonly parallel to the bedding and filled with sparry calcite or mud.

sheet deposit A mineral deposit that is generally stratiform, more or less horizontal, and areally extensive relative to its thickness.

sheet dip The dip of the median or enveloping surface of a train of folds. Rarely used.

sheet drift An evenly spread deposit of glacial drift that did not significantly alter the form of the underlying rock surface.

sheeted (sheet'-ed) Said of an igneous rock such as a granite that has undergone *pressure-release jointing* or *exfoliation,* sometimes giving it the appearance of being stratified.

sheeted dikes A *swarm* of parallel or subparallel igneous dikes so closely spaced that little or no intervening wall rock is preserved.

sheeted vein A set of closely spaced, distinct parallel fractures or veins filled with mineralization and separated from one another by thin screens of barren rock. Syn: *accretion vein.*

sheeted-zone deposit A mineral deposit consisting of veins or lodes filling a zone of shear faulting, or *shear zone.*

sheet erosion The removal of thin layers of surface material more or less evenly from an extensive area of gently sloping land, by broad continuous sheets of running water rather than by streams flowing in well-defined channels; e.g. erosion that occurs when rain washes away a thin layer of topsoil. Cf: *channel erosion; rill erosion; gully erosion.* Syn: *sheetflood erosion; sheetwash; unconcentrated wash; rainwash; slope wash; surface wash.*

sheet fabric One of three major types of materials recognized in electron-microscope study of graptolithine *periderm* as a fundamental structural element in the periderm. It is homogeneous and electron-dense; it commonly encloses or bounds *fusellar fabric* to form fusellar tissue (Urbanek and Towe, 1974, p.5). See also: *cortical fabric.*

sheetflood (sheet'-flood) A broad expanse of moving, storm-borne water that spreads as a thin, continuous, relatively uniform film over a large area in an arid region and that is not concentrated into well defined channels; its distance of flow is short and its duration is measured in minutes or hours. Sheetfloods usually occur before runoff is sufficient to promote channel flow, or after a period of sudden and heavy rainfall. Cf: *streamflood.* See also: *sheet flow [geomorph].* Also spelled: sheet flood. Syn: *sheetwash; sheet [water].*

sheetflood erosion A form of *sheet erosion* caused by a sheetflood. See also: *rock-floor robbing.*

sheet flow [geomorph] An *overland flow* or downslope movement of water taking the form of a thin, continuous film over relatively smooth soil or rock surfaces and not concentrated into channels larger than rills. Cf: *streamflow.* See also: *sheetflood.* Also spelled: sheetflow.

sheet flow [hydraul] *laminar flow [hydraul].*

sheet flow [volc] Submarine lava type with smooth, lineated, folded, or jumbled surface textures resembling subaerial pahoehoe (Batiza and White, 2000).

sheet fracture A type of joint produced by pressure release, or *exfoliation.* Sheet fracturing may separate large rock masses, e.g. of granite, into tabular bodies or lenses, roughly parallel with the rock surface, that become thicker, flatter, and more regular with depth. It is a useful characteristic of the rock in many quarries. Cf: *pressure-release jointing; lift joint; sheet structure [struc geol]; expansion joint; release joint; bedding [mining]; longitudinal splitting crack.*

sheet ground A term used in the Joplin district of Missouri for extensive disseminated low-grade zinc-lead deposits.

sheet ice Ice formed in a smooth, relatively thin layer by the rapid freezing of the surface layer of a body of water. Not to be confused with *ice sheet.* Cf: *ice table.*

sheeting pile *sheet pile.*

sheet mica Mica that is relatively flat and sufficiently free from structural defects so that it can be punched or stamped into shapes for use by the electronic and electrical industries.

sheet mineral *phyllosilicate.*

sheet nacreous structure A *nacreous structure* that shows irregular, stair-step, and/or brick-wall tablet stacking modes in all vertical sections (Carter, 1990, Fig. 10A).

sheet pile A *pile* with a generally flat cross section, which may be meshed or interlocked with adjacent similar members to form a diaphragm, wall, or bulkhead; it is designed to resist lateral earth pressure or to reduce groundwater seepage. Syn: *sheeting pile.*

sheet sand *blanket sand.*

sheet seminacreous structure A *seminacreous structure* in which the tablets show irregular stacking in all vertical sections (Carter, 1990, Fig. 11).

sheet silicate *phyllosilicate.*

sheet spar A sheet crack filled with spar (Fischer, 1964, p.148).

sheetstone Scleractinian reef growth framework dominated by sheetlike and lamellar colonies where calcification in the horizontal plane dominates over that of the vertical plane. Width-to-height ratio of dominant organisms is >30:1, with these growth forms constituting >60% of the skeletal volume (Insalaco, 1998).

sheet structure [sed struc] *vertical sheet structure.*

sheet structure [struc geol] A number of *sheet fractures.*

sheetwash (sheet'-wash) (a) A *sheetflood* occurring in a humid region. (b) The material transported and deposited by the water of a sheetwash. (c) A term used as a syn. of sheet flow (a movement) and *sheet erosion* (a process).

Sheinwoodian (Shein-wood'-i-an) A stage of the Standard Global Chronostratigraphic Scale: lowermost stage of Wenlockian Series, Middle Silurian (above Telychian, below Homerian). The basal GSSP is in southwest England (Holland, 1982).

Shelby tube sampler A standard steel tube that is 3 in. (76.2 mm) in diameter and 3 ft (914.4 mm) long and is used to collect samples of soft soil.

sheldrickite A colorless or white trigonal mineral: $NaCa_3(CO_3)_2F_3 \cdot H_2O$.

shelf [geomorph] (a) Bedrock or other solid rock beneath alluvial soil or deposits; a flat-surfaced layer or stratum. (b) A flat, projecting layer or ledge of rock, as on a slope.

shelf [marine geol] *continental shelf.*

shelf [paleont] The subhorizontal part of the whorl surface next to a *suture* in a gastropod shell, bordered on the side toward the periphery of the whorl by a sharp angulation or by a *carina.*

shelf [tect] A stable cratonic area that was periodically flooded by shallow marine waters and received a relatively thin, well-winnowed cover of sediments. Cf: *platform [tect].*

shelf atoll A ring shaped coral reef or reef complex rising from a continental shelf (e.g. Great Barrier Reef).

shelf break An abrupt change in slope, marking the boundary between the continental shelf and the continental slope. The physiographic province in a basin defined by a change in dip from the shelf (dipping less than 1:1000 landward of the shelf break) to the slope (dipping more than 1:40 seaward of the shelf break) (Heezen et al., 1959). Cf: *shelf edge.*

shelf channel A shallow, somewhat discontinuous valley along the continental shelf, e.g. the extension of the Hudson River Channel across the continental shelf.

shelf delta A delta-shaped feature standing above the surrounding shelf floor in the proximity of transverse shelf valleys, suggesting a deltaic origin during stillstands of a lower sea level (Swift et al., 1972, p.506).

shelf edge The demarcation between the continental shelf and the continental slope. Cf: *shelf break.*

shelf-edge delta A delta that has prograded to or formed at the edge of the continental shelf, characteristically during periods of falling to low sea level. Syn: *shelf-margin delta.*

shelf-edge reef A reef located along the break in slope between a shallow flat shelf and the adjacent deeper basin (Walker, 1973). Cf: *bank reef.*

shelf facies A sedimentary facies that contains sediments produced in the neritic environment of the shelf seas marginal to a low-lying, stable land surface. It is also known as *shelly facies* in recognition of the importance of its characteristic carbonate rocks and fossil shells. Cf: *geosynclinal facies.* Syn: *platform facies; foreland facies.*

shelf ice (a) The ice of an ice shelf; a term introduced by Nordenskjöld in 1909 to describe the type of floating freshwater ice formed in, or broken away from, the feature then known as an "ice barrier" but now referred to as an *ice shelf.* Syn: *barrier ice.* (b) Less preferred syn. of *ice shelf.*

shelf lagoon A broad, shallowly submerged marine shelf or platform, such as the interior of the modern Great Bahama Bank or the sea of Permian time across Kansas.

shelf-margin delta *shelf-edge delta.*

shelf sea A shallow *marginal sea* situated on the continental shelf, commonly not exceeding 200 m in depth.

shelfstone (shelf'-stone) A *speleothem* formed as a horizontally projecting ledge at the edge of a cave pool or on other speleothems in a pool. Cf: *lily pad; folia [speleo].*

shelf-valley complex (shelf-val'-ley) A group of morphologic elements that occur along the paths of retreat of estuary mouths on autochthonous shelves, composed of deltas, shelf valleys, and shoal-retreat massifs (Stanley and Swift, 1976, p.323).

shell [drill] A driller's term for a thin, hard layer of rock encountered in drilling a well. Cf: *shale break.* Syn: *shelly formation.*

shell [geol] (a) The crust of the Earth. Also, any of the continuous and distinctive concentric zones or layers composing the interior of the Earth (beneath the crust). The term was formerly used for what is now called the "mantle". Syn: *Earth shell.* (b) A thin and generally hard layer of rock; esp. such a stratum encountered in drilling a well.

shell [paleont] (a) The hard, rigid outer covering of an animal, commonly largely calcareous but in some cases chiefly or partly chitinous, siliceous, horny, or bony; e.g. the hard parts of an ammonoid (including the protoconch and the conch, but excluding the aptychus and the beaks or jaw structures), or of a cirripede crustacean (including compartmental plates, calcareous basis, and opercular valves). (b) A shell-bearing animal; esp. a shell-bearing mollusk.

shell bank A *bank [sed]* consisting largely of shells, esp. of corals and pelecypods.

shell ice Thin ice originally formed on a sheet of water and remaining as an unbroken shell after the underlying water has been withdrawn.

shell marl (a) A shell-rich deposit composed of a mixture of between 25% and 75% carbonate and 25-75% siliclastic, clay-rich sediment. (b) A sandy, clayey, or limy deposit, loose or weakly consolidated, containing abundant molluscan shells; a common term in the coastal plain of the southeastern U.S. (c) A light-colored calcareous deposit formed on the bottoms of small freshwater lakes, composed largely of uncemented mollusk shells and precipitated calcium carbonate, along with the hard parts of minute organisms.

shellstone (shell'-stone) A type of *reef rock, boundstone,* or *framestone* composed of shells, especially of brachiopods and pelecypods, cemented onto one another, with fine-grained sediment, coarse skeletal debris, or crystalline cement filling the interstices (Cuffey, 1985).

shelly (shel'-ly) (a) Pertaining to the shell of an animal; chitinous, siliceous, or testaceous. (b) Having a shell.

shelly facies A nongeosynclinal sedimentary facies that is commonly characterized by abundant calcareous fossil shells, dominant carbonate rocks (limestones and dolomites), mature orthoquartzitic sandstones, and paucity of shales. The term is frequently used in reference to lower Paleozoic strata, as in the upper Mississippi Valley and the Great Lakes area. The facies is also known as *shelf facies* in recognition of the presumed structural stability of the site of deposition.

shelly formation *shell [drill].*

shelly pahoehoe A type of *pahoehoe* whose surface contains large open tubes and blisters; its crust is 1-30 cm thick.

shelter cave (shel'-ter) *rockshelter.*

shelter porosity A type of primary interparticle porosity defined by Choquette and Pray (1970, p.249) as the porosity "created by the sheltering effect of relatively large sedimentary particles which prevent the infilling of pore space beneath them by finer clastic particles".

sherd A fragment or broken piece of pottery. Syn: *shard.*

shergottite (sher'-got-tite) A group of pigeonite-rich basaltic achondrites whose young age, petrologic, chemical, and isotopic composition suggest origin on the same parent body as the chassignites and nahklites, thought by many to be the planet Mars. See also: *SNC meteorite.*

sheridanite (sher'-i-dan-ite') *clinochlore.*

Shermanian North American stage: Upper Ordovician, above Kirkfeldian, below Edenian.

sherry topaz (sher'-ry) A variety of topaz resembling sherry wine in color. It is one of the more valuable and important varieties of topaz.

sherwoodite (sher'-wood-ite) A dark bluish black, bluish green or yellowish green tetragonal mineral: $Ca_9(AlV^{4+}{}_2V^{5+}{}_{12}O_{40})_2 \cdot 56H_2O$.

sheugh A Scottish term for a small ravine, esp. one containing a stream. Also spelled: sheuch.

shibkovite A colorless or white hexagonal mineral of the *milarite* group: $K(Ca,Mn,Na)_2(K,[\,])_2Zn_3Si_{12}O_{30}$.

Shidertinian (Shi-der-tin'-i-an) European stage: Upper Cambrian (above Tuorian, below Tremadocian).

shield [eng] (a) A framework or diaphragm of steel, iron, or wood, used in tunneling and mining in unconsolidated materials. It is moved forward at the end of the tunnel or adit in the process of excavation and is used to support the ground ahead of the *lining* and to aid in constructing the lining. (b) Also a braced steel device with open top, bottom and ends used in trenches primarily to protect workers in such excavations.

shield [paleont] (a) A protective cover or structure on an animal, likened to or resembling a shield; e.g. an ossicle of an ophiuroid, the carapace of a crustacean, or a large scale on the head of a lizard. (b) A flat or curved lateral outgrowth at one or more levels of a tangential rod or needle in the skeleton of an acantharian radiolarian, forming by fusion of the lattice shell. (c) One of the discoidal elements of the placolith coccolith.

shield [speleo] A *speleothem* composed of two parallel hemicircular plates separated by a thin, planar crack. Growth occurs radially along the rim, where water issues under pressure from the crack. Syn: *palette.*

shield [tect] A large area of exposed basement rocks in a craton, commonly with a very gently convex surface, surrounded by sediment-covered platforms; e.g. Canadian Shield, Baltic Shield. The rocks of virtually all shield areas are Precambrian. Syn: *continental shield; continental nucleus.*

shield basalt A basaltic lava flow that erupted from numerous small closely spaced shield-volcano vents, and coalesced to form a single unit. It is generally of smaller extent than a *flood basalt.*

shield field On Venus, an enhanced concentration of small volcanoes over areas >100 km across (Bougher et al., 1997).

shielding (shield'-ing) n. A grounded metallic enclosure intended to reduce noise, capacitively or inductively coupled into an electrical circuit.

shield volcano A volcano in the shape of a flattened dome, broad and low, built by flows of very fluid basaltic lava or by rhyolitic ash flows. Cf: *lava shield; ignimbrite shield; scutulum.*

shift The relative displacement of the units affected by a fault but outside the fault zone itself; a partial syn. of *slip [struc geol].* See also: *strike shift.* Little used.

shifting [coast] (shift'-ing) The fluctuation or oscillation of sea level; the change in position of a shoreline.

shifting [streams] The movement of the crest of a divide away from a more actively eroding stream (as on the steeper slope of an asymmetric ridge) toward a weaker stream on the gentler slope; the change in position of a divide (and of the stream channel) where one stream is captured by another. See also: *monoclinal shifting; leaping; creeping.* Syn: *migration [streams].*

shigaite (shi'-ga-ite) A light yellow rhombohedral mineral: $NaMn^{2+}{}_6Al_3(SO_4)_2(OH)_{18}{\cdot}[12]H_2O$. It is related to *woodwardite.*

shingle (shin'-gle) (a) Coarse loose well-rounded waterworn detritus or alluvial material of various sizes; esp. beach gravel, composed of smooth and spheroidal or flattened pebbles, cobbles, and sometimes small boulders, generally measuring 20-200 mm in diameter; it occurs typically on the higher parts of a beach. The term is more widely used in Great Britain than in the U.S. (b) A place strewn with *shingle;* e.g., a *shingle beach.* Etymol: probably Scandinavian, akin to "singel", "coarse gravel that sings or crunches when walked on".

shingle beach A narrow beach, usually the first to form on a coastline having resistant bedrock and cliffs, composed of *shingle,* and commonly having a steep slope on both its landward and seaward sides. Syn: *cobble beach; shingle.*

shingle-block structure An obsolete syn. of *imbricate structure [tect].*

shingle rampart A rampart or ridge, 1 or 2 m high, of *shingle* built up along the seaward edge of a reef.

shingle ridge A steeply sloping bank of *shingle* heaped up on and parallel with the shore.

shingle structure [eco geol] The arrangement of closely spaced veins overlapping in the manner of shingles on a roof.

shingle structure [sed] *imbricate structure [sed].*

shingling (shin'-gling) *imbrication [sed].*

Shipek bottom sampler (Shi'-pek) A commercial brand of *grab sampler.*

shipping channel (ship'-ping) The region of deeper water between the mainland and the principal offshore area of growth of a major barrier reef; as much as 100 feet deep and several miles wide, it is mostly free of patch reefs hazardous to navigation, so that large ocean-going ships can safely travel along it parallel to the coastline, as off northeastern Australia. Cf: *boat channel.*

shipping ore *first-class ore.*

shirokshinite A colorless or pale gray monoclinic mineral of the *mica* group: $K(Mg_2Na)Si_4O_{10}F_2$.

shirozulite A dark reddish-brown mineral of the *mica* group: $KMn^{2+}{}_3(Si_3Al)O_{10}(OH,F)_2$.

shiver An old English term for soft and crumbly shale, or slate clay approaching shale. Etymol: Middle English "scifre". Adj: shivery.

shiver spar A variety of calcite with slaty structure; specif. *argentine.* Syn: *slate spar.*

shkatulkalite A white, silvery, pink, or cream monoclinic mineral: $Na_{10}(Mn,Ca,Sr)Ti_3Nb_3(Si_2O_7)_6(OH)_2F{\cdot}12H_2O$.

shoal adj. Having little depth; shallow. n. (a) A relatively shallow place in a stream, lake, sea, or other body of water; a *shallows.* (b) A submerged ridge, bank, or bar consisting of or covered by sand or other unconsolidated material, rising from the bed of a body of water to near the surface so as to constitute a danger to navigation; specif. an elevation, or an area of such elevations, at a depth of 10 fathoms (formerly 6) or less, composed of material other than rock or coral. It may be exposed at low water. Cf: *reef.* (c) A rocky area on the sea floor within soundings. (d) A growth of vegetation on the bottom of a deep lake, occurring at any depth. v. To become shallow gradually; to cause to become shallow; to fill up or block off with a shoal; to proceed from a greater to a lesser depth of water.

shoal breccia A submarine breccia formed by the action of waves and tides on a shoal, commonly of limestone (Norton, 1917, p.177).

shoaling (shoal'-ing) A bottom effect that describes, in terms of wave heights, the alteration of a wave as it proceeds from deep water into shallow water; it is shown by an initial decrease in height of the incoming wave, followed by an increase in height as the wave arrives on the shore.

shoaling wave A wave that has progressed into water depths at which the interaction with the bottom affects the wave motion.

shoal-retreat massif A broad, shelf-transverse sand ridge that marks the retreat path of a zone of *littoral drift* convergence (Stanley and Swift, 1976, p.323).

shock n. *earthquake.*

shock breccia An obsolete term for fragmental rock formed by the action of shock waves; e.g. *suevite* formed by impact.

shock lamellae A less preferred synonym of *planer deformation features.*

shock lithification The conversion of originally loose fragmental materials into coherent aggregates, e.g. *instant rock,* by the action of shock waves, such as by those generated by explosions or impacts (Short, 1966). It involves such mechanisms as fracturing, compaction, and intergranular melting.

shock loading The process of subjecting material to the action of high-pressure shock waves generated by artificial explosions or by impact.

shock melting Melting of solid matter by shock-wave compression, resulting from high post-shock temperature after pressure release. Cf: *impact melt.*

shock-metamorphic facies (shock'-met-a-mor'-phic) A discredited term for an association of mineralogic features (such as planar deformation features and diaplectic glasses) formed by a particular degree of shock metamorphism. As it implies a near approach to equilibrium, the term is not recommended; the term "stages of shock metamorphism" is preferred.

shock metamorphism The totality of observed permanent physical, chemical, mineralogic, and morphologic changes produced in rocks and minerals by the passage of high-pressure shock waves acting over time intervals ranging from a few microseconds to a fraction of a minute (French, 1998). The only known natural mechanism for producing shock-metamorphic effects is *hypervelocity impact.* The term also includes identical effects produced by shock waves generated in small-scale laboratory experiments and in nuclear and chemical explosions. Cf: *impact metamorphism.*

shock remanent magnetization Magnetization caused by impulsive stress of a material in the presence of a magnetic field.

shock stage The degree of shock metamorphism of a rock within prograde (progressive) *shock metamorphism.*

shock wave A shock front, a steplike discontinuity in pressure, density, particle velocity, and internal energy, which propagates in gaseous, liquid, or solid matter with supersonic velocity. In rock, a shock wave is capable of vaporizing, melting, mineralogically transforming, or strongly deforming rock materials. Cf: *blast wave.*

shock wave velocity Velocity of a shock wave (shock front) propagating into material at rest.

shock zone A volume of rock in or around an impact crater or structure in which a distinctive shock-metamorphic deformation or transformation effect is present.

shoe stone A sharp-grained sandstone used as a whetstone by shoemakers and other leatherworkers. Also, the whetstone so used.

shoestring (shoe'-string) A long, relatively narrow, and straight or curving sedimentary body whose width/thickness ratio is less than 5 to 1, and is usually on the order of 1 to 1 or even smaller (Krynine, 1948, p.147); e.g. a channel fill, a bar, a dune, or a beach deposit. Cf: *prism [sed].*

shoestring rill One of several long narrow uniform channels, closely spaced and roughly parallel with one another, that merely score the homogeneous surface of a relatively steep slope of bare soil or weak clay-rich bedrock, and that develop wherever overland flow is intense.

shoestring sand A narrow linear deposit composed of sand or sandstone, usually buried in the midst of mud or shale; e.g. a buried sandbar or channel fill.

shomiokite-(Y) A colorless orthorhombic mineral: $Na_3(Y,Dy)(CO_3)_3{\cdot}3H_2O$.

shonkinite (shon'-kin-ite) (a) In the *IUGS classification,* a plutonic rock in which F is between 10 and 60, P/(A+P) is 10 or less, and M is between 60 and 90. (b) A dark-colored syenite composed chiefly of augite and alkali feldspar, and possibly containing olivine, hornblende, biotite, and nepheline. Its name, given by Weed and Pirsson in 1895, is derived from Shonkin, the Indian name for the Highwood Mountains of Montana.

shoot [ore dep] A syn. of *ore shoot.* Also spelled: *chute.*

shoot [seis] In seismic prospecting, to explore an area by employing seismic techniques; to generate seismic waves.

shoot [streams] (a) A place where a stream flows or descends swiftly. (b) A natural or artificial channel, passage, or trough through which water is moved to a lower level. (c) A rush of water down a steep place or a rapids. Etymol: Anglicized form of French "chute". See also: *chute [streams].*

shooting (shoot'-ing) *seismic shooting.*

shooting a well The procedure of measuring directly the traveltime from a source on the surface to a geophone positioned in a well. Cf: *vertical seismic profiling.*

shooting flow (a) *jet flow.* (b) *rapid flow.*

shooting-flow cast A term proposed by Wood and Smith (1958, p.169) for one of a series of "strong parallel ridges", representing the cast of a groove up to 10 cm deep, 30 cm wide, and 2 m long.

shooting star A visual *meteor* appearing as a thin, temporary streak or trace of light in the nighttime sky. Syn: *falling star.*

shor A salt lake in Turkestan. Etymol: Russian.

shoran (sho'-ran) A precise electronic measuring system for indicating distance from an airborne or shipborne station to each of two fixed ground stations simultaneously by recording (by means of cathode-ray screens) the time required for round-trip travel of radar signals or high-frequency radio waves and thereby determining the position of the mobile station. Its range is effectively limited to line-of-sight distances (about 40 nautical miles). Shoran is used in control of aerial photography, airborne geophysical prospecting, and offshore hydrographic surveys for measuring long distances. Cf: *loran.* Etymol: *sho*rt-*ra*nge *n*avigation.

shore [coast] (a) The narrow strip of land immediately bordering any body of water, esp. a sea or a large lake; specif. the zone over which the ground is alternately exposed and covered by tides or waves, or the zone between high water and low water. The shore is the most seaward part of the *coast;* its upper boundary is the landward limit of effective wave action at the base of the cliffs and its seaward limit is the low-water line. Subdivided into a *foreshore* and a *backshore.* See also: *beach; strand [coast].* (b) The term is commonly used in the sense of the *shoreline* and of the *foreshore.* (c) A nautical term for land as distinguished from the sea.

shore cliff A cliff at the edge of a body of water or extending along the shore. See also: *sea cliff.*

shore drift *littoral drift.*

shore dune A sand dune produced by wind action on beach sands along a shore.

shoreface (shore'-face) (a) The zone between the seaward limit of the shore and the more nearly horizontal surface of the *offshore* zone; typically extends seaward to storm wave depth or about 10 m. The term "shore face" was originally used by Barrell (1912, p.385-386), in his study of deltas, for the relatively narrow slope developed by breaking waves and separating the subaerial plain from the subaqueous one below. Not to be confused with *beach face.* See also: *inshore.* (b) A relatively steep but short concave inner portion of the continental shelf (Price, 1954, p.81).

shoreface ravinement surface In sequence stratigraphy, an abrupt, scoured surface that is interpreted to have been formed by the erosive action of shoreface currents and waves during landward movement of the shoreline. Marine strata, which in general display a deepening-upward trend, must directly overlie the surface (Embry and Johannessen, 1992).

shoreface terrace A wave-built terrace in the shoreface region, composed of gravel and coarse sand swept from the wave-cut bench into deeper water. See also: *offshore terrace.*

Shore hardness scale An empirical scale of hardness of rocks, metals, ceramics, or other materials as determined by a Shore scleroscope which utilizes the height of rebound of a small standard object (such as a diamond-tipped hammer) dropped from a fixed height onto the surface of a specimen. Named after Albert F. Shore, U.S. manufacturer, who proposed the technique in 1906.

shore ice (a) The basic form of *fast ice,* attached to the shore and, in shallow water, also grounded. (b) Sea ice that has been driven ashore and beached by wind, waves, currents, tides, or the pressure of adjacent ice. Cf: *stranded ice.* (c) Floating sea ice adjacent to the shore; it may or may not be attached to the shore.

shore-ice belt *icefoot [ice].*

shore lead A *lead [ice]* between pack ice and the shore, or between pack ice and an ice front. It formerly included what is now known as a *flaw lead.*

shoreline (shore'-line) (a) The intersection of a specified plane of water with the shore or beach; it migrates with changes of the tide or of the water level. The term is frequently used in the sense of "high-water shoreline" or the intersection of the plane of mean high water with the shore or beach, or the landward limit of the intermittently exposed shore. Syn: *waterline; shore [coast]; strandline.* Cf: *highstand shoreline; lowstand shoreline.* (b) The general configuration or outline of the shore. The terms "shoreline" and "coastline" are often used synonymously, but there is a tendency to regard "coastline" as a limit fixed in position for a relatively long time and "shoreline" as a limit constantly moving across the beach.

shoreline cycle The hypothetical succession of changes through which coastal features normally pass during the development of a shoreline, from the time when the water first assumed its level and rested against the new shore to the time when the water can do no more work (either erosion or deposition). Obsolete.

shoreline-development ratio A ratio indicating the degree of irregularity of a lake shoreline, given as the length of the shoreline to the circumference of a circle whose area is equal to that of the lake (Wetzel, 2001, p.34).

shoreline of depression A *shoreline of submergence* that implies an absolute subsidence of the land.

shoreline of elevation A *shoreline of emergence* that implies an absolute rise of the land. Not to be confused with *elevated shoreline.*

shoreline of emergence A hypothetical shoreline resulting from the dominant relative emergence of the floor of an ocean or lake; the water surface comes to rest against the partially emerged land which is marked by marine-produced forms and structures. The term carries no implication as to whether it is the land or the sea that has moved (Johnson, 1919, p.173). The concept is generally regarded now as obsolete, since Johnson's criteria have been shown as fundamentally incorrect. Cf: *shoreline of elevation.* Syn: *emerged shoreline; negative shoreline.*

shoreline of submergence A hypothetical shoreline resulting from the dominant relative *submergence* of a landmass; the water surface comes to rest against the partially submerged land which is marked by subaerially produced forms and structures. The term carries no implication as to whether it is the land or the sea that has moved (Johnson, 1919, p.173). The concept is generally regarded now as obsolete, since Johnson's criteria have been shown as fundamentally incorrect. Cf: *shoreline of depression; shoreline of emergence.* Syn: *submerged shoreline.*

shore platform A descriptive term for the horizontal or gently sloping surface produced along a shore by wave erosion; specif. a *wave-cut bench.* Also, sometimes used as a purely descriptive term for *wave-cut platform.* Cf: *scar [geomorph].*

shore polynya A *polynya* between pack ice and the coast, or between pack ice and an ice front, and formed by currents or by wind.

shore profile Intersection of the surface of land and nearby bottom with a vertical plane. Commonly extends inland to the top of a sea cliff or sand dune and seaward to some place on the lower (submerged) beach, but may be extended any distance across the continental shelf.

shore reef *fringing reef.*

shore terrace (a) A terrace produced by the action of waves and currents along the shore of a lake or sea; e.g. a *wave-built terrace.* (b) *marine terrace.*

shoreward (shore'-ward) Directed or moving toward the shore.

shorl *schörl.*

shortite (short'-ite) A colorless or pale yellow orthorhombic mineral: $Na_2Ca_2(CO_3)_3$.

short-path multiple A multiple reflection in which energy is reflected back and forth over only a small portion of the section so that the resulting wave blends with the primary pulse, changing its wave shape and adding a tail. Cf: *peg-leg multiple.*

short period A period of seismic oscillation that is less than six seconds in duration. Cf: *long period.*

short-range order The state of a crystal structure in which the probability of every atom having the correct nearest or second-nearest neighbors is high. Cf: *long-range order.*

short shot (a) *Weathering shot.* (b) A refraction profile where the source is at the end of the active line of geophones. The near geophone is often displaced a half interval away.

short wave *deep-water wave.*

shoshonite (sho-sho'-nite) (a) In the *IUGS classification,* a *basaltic trachyandesite* with Na_2O-2<K_2O in wt%. Cf: *mugearite.* (b) A *trachyandesite* composed of olivine and augite phenocrysts in a groundmass of labradorite with alkali feldspar rims, olivine, augite, a small amount of leucite, and some dark-colored glass. Shoshonite grades into *absarokite* with an increase in olivine and into *banakite* with more sanidine. Its name, given by Iddings in 1895, is derived from the Shoshone River, Wyoming.

shot [seis] The explosive charge or other energy source used in seismic prospecting.

shot [soil] Hard, rounded particles, generally of sand size, that occur in soils. They may be aggregates or concretions. Syn: *buckshot.*

shot break *time break.*

shot copper Small, rounded particles of native copper, molded by the shape of vesicles in basaltic host rock, and resembling shot in size and shape.

shotcrete (shot'-crete) (a) *Gunite* that commonly includes coarse aggregate (up to 2 cm). (b) Concrete specially designed to be applied under pressure through a nozzle. It may include fibers of steel or another strong material for reinforcement or may be applied onto a surface over which welded wire fabric has been placed for reinforcement.

shot depth In seismic work, the vertical distance from the surface to a seismic source.

shot elevation The elevation of the seismic source. Not to be confused with *shothole elevation.*

shothole (shot'-hole) In seismic prospecting, a borehole in which a source is placed for generating seismic waves.

shothole bridge An obstruction in a shothole that prevents a source from going deeper. It may be accidental or intentional.

shothole drill A drill (esp. a rotary drill) used for making a shothole.

shothole elevation The elevation of the top of a shothole. Not to be confused with *shot elevation.*

shot instant *time break.*

shotpoint (shot'-point) In seismic shooting, the location of the energy source.

shotpoint gap A greater distance between geophone groups on each side of the source in an otherwise uniform spread, so that the groups nearest the source will be far enough from it to minimize adverse effects of noise.

shott (a) A shallow and brackish or saline lake or marsh in southern Tunisia or on the plateaus of northern Algeria, usually dry during the summer; a playa lake, often many tens of kilometers in diameter. (b) A closed basin containing a shott; esp. the dried bed existing after the water has disappeared, characterized by salt deposits and frequently by absence of vegetation. Etymol: Arabic "shatt". See also: *sabkha.*

shoulder [geomorph] (shoul'-der) (a) A short, rounded spur projecting laterally from the side of a mountain or hill. (b) The sloping part of a mountain or hill below the summit. (c) *valley shoulder.*

shoulder [glac geol] A bench on the side of a glaciated valley, occurring at the marked change of slope where the steep side of the inner, glaciated valley meets the much gentler slope above the level of glaciation. Cf: *alp.* Syn: *trimline.*

shoulder [paleont] (a) The girdle of the gnathostome anterior limb, together with its musculature. (b) The salient angulation of a gastropod-shell whorl, parallel to the coiling and forming the abaxial edge of a subsutural ramp. (c) The ventral and lateral blunt angle of a whorl of an ammonoid shell (TIP, 1959, pt.L, p.5). See also: *umbilical shoulder.*

shoulder [struc geol] A structure formed on the face of a joint by the intersection of *plume structure* with fringe joints.

shoved moraine *push moraine.*

show (a) A trace of oil or gas detected in a core, cuttings, or circulated drilling fluid, or interpreted from the electrical or geophysical logs run in a

well. Partial syn: *oil show.* (b) A small particle of gold found in panning a gravel deposit.
show cave A cave that is open to the public for paid tours.
SHRIMP Acronym for super high resolution *ion microprobe.*
shrinkage (shrink'-age) The decrease in volume of clayey soil or sediment owing to reduction of void volume, principally by drying.
shrinkage crack A crack produced in fine-grained sediment or rock by the loss of contained water during drying, dehydration, or shrinkage of swelling-clay minerals; e.g., a *desiccation crack* and a *syneresis crack.*
shrinkage factor The factor that is applied to convert a barrel of oil in the reservoir into an equivalent amount of gas-free oil in a stock tank at the surface. Shrinkage factors generally range from 0.63 to 0.88. Cf: *formation-volume factor.*
shrinkage index The numerical difference between the plastic limit of a material and its shrinkage limit.
shrinkage limit That moisture content of a soil below which a decrease in moisture content will not cause a decrease in volume, but above which an increase in moisture will cause an increase in volume.
shrinkage polygon *desiccation polygon.*
shrinkage pore A term used by Fischer (1964, p.116) for an irregular pore formed in muddy sediment by shrinkage (desiccation). It may become a bird's-eye (in a limestone) when filled with sparry calcite. Syn: *fenestra [sed].*
shrinkage ratio The ratio of a volume change to the moisture-content change above the shrinkage limit.
shrub [sed] Shrub- or bushlike depositional form, commonly composed of calcite and ranging from microscopic to 5-10 cm in height. Shrubs display a highly irregular branching pattern away from their base, similar, morphologically, to the garden-variety woody plant. They are the products of microbially induced precipitation of mineral matter around clumps of bacteria (Chafetz and Guidry, 1984).
shrub-coppice dune (shrub-cop'-pice) A small, streamlined dune that forms to the lee of bush-and-clump vegetation on a smooth surface of very shallow sand. Syn: *coppice mound; nebkha.*
shuangfengite A metallic black rhombohedral member of the *melonite* group: $(Ir,Pt)Te_2$.
shubnikovite (shub'-ni-kov-ite) A light blue orthorhombic mineral: $Ca_2Cu_8(AsO_4)_6Cl(OH){\cdot}7H_2O$(?). It may be related to lavendulan.
shuga (shu'-ga) An accumulation of spongy lumps of white *sea ice,* measuring a few centimeters across, and formed from *grease ice* or *sludge,* and sometimes from *anchor ice* rising to the surface (U.S. Naval Oceanographic Office, 1968, p. B36).
shuiskite (shu-i'-skite) A dark brown monoclinic mineral of the *pumpellyite* group: $Ca_2(Mg,Al)(Cr,Al)_2(SiO_4)(Si_2O_7)(OH)_2{\cdot}H_2O$.
shungite (shung'-ite) A hard black amorphous material containing over 98% carbon, found interbedded among Precambrian schists. It is probably the metamorphic equivalent of bitumen, but it may represent merely impure graphite. Also spelled: *schungite.*
Shushwap-type deposit *Broken Hill-type deposit.*
shut-in A narrow, steep-sided gorge along the course of an otherwise wide and shallow stream valley.
shut-in pressure *Formation pressure* as recorded during a drill stem test at the wellhead when the valves are closed to the surface. Syn: *closed-in pressure.*
shutterridge (shut'-ter-ridge) A ridge formed by vertical, lateral, or oblique displacement on a fault traversing a ridge-and-valley topography, with the displaced part of a ridge "shutting in" the adjacent ravine or canyon on the opposite side of the fault. (Buwalda, 1937).
SH wave (a) An S wave which has only a horizontal component of motion. (b) The horizontal component of any S wave.
si In structural petrology, a fabric defined by the preferred orientation of grains within or internal to a porphyroblast. It may or may not be parallel to the preferred orientation of grains outside the porphyroblast. Cf: *se.*
SI *solidification index.*
sial (si'-al) A petrologic name for the upper layer of the Earth's crust, composed of rocks that are rich in silica and alumina; it may be the source of granitic magma. It is characteristic of the upper *continental crust.* Etymol: an acronym for *si*lica + *al*umina. Adj: *sialic.* Cf: *sialma.* Syn: *sal; granitic layer.*
sialic (si-al'-ic) Adj. of *sial.*
sialite (si'-a-lite) A syn. of *clay mineral.*
siallite (si'-al-lite) (a) A group name for the kaolin clay minerals and allophane. (b) A rock composed of siallite minerals.
sialma (si-al'-ma) A layer of the Earth's crust that is intermediate in both depth and composition between the *sial* and the *sima.* Archaic. Etymol: an acronym for *si*lica + *al*umina + *ma*gnesia.
sialsima (si-al-si'-ma) *salsima.*
SIB *super-solidus intrusion breccia.*
siberite (si-ber'-ite) A violet-red or purple lithian variety of tourmaline; rubellite from Siberia.
sibirskite (si-bir'-skite) A colorless or gray orthorhombic mineral: $CaH(BO_3)$. It is a dimorph of parasibirskite.
sibling species (sib'-ling) One of two or more species that are closely related, very similar morphologically, but reproductively isolated.
SIB-type PGE deposit *super-solidus intrusion breccia PGE deposit.*
sicherite A metallic gray orthorhombic mineral: $TlAg_2(As,Sb)_3S_6$.
Sicilian (Si-cil'-ian) An Italian substage: upper substage of regional Calabrian stage, lower Pleistocene (above Emilian substage, below Ionian stage).
sicklerite (sick'-ler-ite) A yellowish-brown or dark brown orthorhombic mineral of the *triphyllite* group: $Li(Mn^{2+},Fe^{3+})PO_4$. It is isomorphous with ferrisicklerite, a partial alkali-leached product of lithiophilite, and within the lithiophilite-purpurite series.
sickle trough (sick'-le) A flat-bottomed, crescent-shaped rock basin sculptured by a glacier. Syn: *skärtråg.*
sicula (sic'-u-la) The skeleton secreted by the initial *zooid* of a graptolite colony, divisible into a conical *prosicula* and a distal *metasicula.* Pl: siculae.
side (a) A slope of a mountain, hill, or bank; e.g. *hillside.* (b) A bank, shore, or other land bordering a body of water; e.g. *seaside.* (c) A geographic region; e.g. "countryside".
side canyon A ravine or other valley smaller than a canyon, through which a tributary flows into the main stream.
sidelap (side'-lap) The area common to two aerial or space photographs or images in adjacent parallel flight lines. Cf: *overlap [photo].*
side-looking airborne radar (side'-look-ing) An airborne radar system in which a long, narrow, stabilized antenna, aligned parallel to the motion of an aircraft or satellite, projects radiation at right angles to the flight path. It makes possible extremely fine-resolution photography and mapping of the ground surface. Abbrev: SLAR.
side moraine *lateral moraine.*
side muscle One of the lateral pedal and/or visceral muscles of primitive rostroconchs whose insertions form left and right linear connections between anterior and posterior median muscle insertions (Pojeta and Runnegar, 1976, p.47).
side pinacoid In an orthorhombic, monoclinic, or triclinic crystal, the *{010}* pinacoid. Cf: *front pinacoid; basal pinacoid.* Syn: *brachypinacoid.*
side plate A small ambulacral plate in a blastoid; esp. one of two (rarely three) serially repeated wedge-shaped plates lying between the central lancet plate and the adjacent radial or deltoid. Each pair of side plates bears a single erect *brachiole.*
siderazot (sid-er-az'-ot) A metallic white hexagonal mineral: Fe_5N_2. Also spelled: siderazote.
sidereal (si-de'-re-al) Pertaining to the stars. A sidereal period of rotation or orbital motion is the inertial angular motion with respect to the stars.
sidereal day The interval between two successive transits of a star over the meridian; the time required for the Earth to rotate once on its axis, or approx. 86,164 seconds. Cf: *solar day.*
sidereal month The mean time of the Moon's revolution in its orbit. It is equal to 27 days 7 hours 43 minutes 11.47 seconds of mean *solar time.*
sidereal time Time based on the *sidereal day.* Cf: *solar time.*
Siderian (Si-der'-i-an) A period of the Standard Global Geochronometric Scale: the oldest period of the *Paleoproterozoic* Era, characterized by an abundance of banded iron formations ("sideros", "iron"). It is defined as the time span between 2,500 Ma, the end of the *Archean* Eon, and 2,300 Ma, the beginning of the *Rhyacian* Period (Plumb, 1991).
siderite [meteorite] (sid'-er-ite) An obsolete general name for iron meteorites, composed almost wholly of iron alloyed with nickel.
siderite [mineral] (a) A rhombohedral mineral of the *calcite* group: $FeCO_3$. It is isomorphous with magnesite and rhodochrosite, and commonly contains magnesium and manganese. Siderite is usually yellow-brown, brown-red, or brown-black, but is sometimes white or gray; it is often found in impure form in beds and nodules (of *clay ironstone*) in clays and shales, and as a directly precipitated deposit partly altered into iron oxides. Siderite is a valuable ore of iron. Syn: *chalybite; spathic iron; sparry iron; rhombohedral iron ore; iron spar; siderose; white iron ore.* (b) An obsolete syn. of *sapphire quartz.* (c) An obsolete term formerly applied to various minerals, such as hornblende, pharmacosiderite, and lazulite.
siderodot (sid'-er-o-dot') A variety of siderite containing calcium.
sideroferrite (sid'-er-o-fer'-rite) A variety of native iron occurring as grains in petrified wood.
siderogel (sid'-er-o-gel') A mineraloid consisting of truly amorphous $FeO(OH)$ and occurring in some bog iron ores.
siderolite (sid'-er-o-lite') An obsolete syn. of *stony-iron meteorite.*
sideromelane (sid'-er-o-mel'-ane) *basaltic glass.*
sideronatrite (sid'-er-o-nat'-rite) An orange, straw-yellow, or yellowish-brown mineral: $Na_2Fe^{3+}(SO_4)_2(OH){\cdot}3H_2O$.
siderophile (sid'-er-o-phile') (a) Said of an element concentrated in the metallic rather than in the silicate and sulfide phases of meteorites, and probably concentrated in the Earth's core relative to the mantle and crust (in Goldschmidt's scheme of element partition in the solid Earth). Cf: *chalcophile; lithophile.* (b) Said of an element with a weak affinity for oxygen and sulfur, and readily soluble in molten iron (Rankama and Sahama, 1950, p.88). Examples are: Fe, Ni, Co, P, Pt, Au.
siderophyllite (sid-er-o-phyll'-ite) A black, dark brown, or dark green monoclinic or trigonal trioctahedral mica, of ideal composition $KFe^{2+}{}_2Al(Al_2Si_2)O_{10}(OH)_2$.
siderophyre (sid'-er-o-phyre') An obsolete term for a rare class *stony-iron meteorite* containing crystals of bronzite and tridymite in a network of

nickel-iron. The original siderophyre, Steinbach, is now known to belong to iron-meteorite group IVA.

siderose (sid′-er-ose) adj. Containing or resembling iron. The term was proposed to replace *ferruginous* when designating a form of iron other than iron oxide; e.g. "siderose cement" consisting of iron carbonate in a sandstone. n. A syn. of *siderite [mineral]*. Also spelled: sidérose.

siderosphere (sid′-er-o-sphere′) An archaic term for the *inner core* of the Earth.

siderotil (sid′-er-o-til′) A white, yellowish, or pale green triclinic mineral of the *pentahydrite* group: $(Fe^{2+},Cu)SO_4{\cdot}5H_2O$.

side-scanning sonar Active system for acquiring images of the seafloor using pulsed sound waves.

side-scan sonar An instrument, carried either on a ship or a towed vehicle, that creates sonic images of the sea floor and objects on the sea floor. The instrument transmits sound pulses in a beam that reflects from successive, narrow bands on the sea floor, that are perpendicular to the instruments track.

side-scan sonar mosaic Overlapping *side-scan sonar* images arranged to produce a continuous representation of an area.

side shot A reading or measurement from a survey station to locate a point that is off the traverse or that is not intended to be used as a base for the extension of the survey. It is usually made to determine the position of some object that is to be shown on a map.

side stream A *tributary* that receives its water from a drainage area separate from that of the main stream into which it flows.

sidetracked hole (side′-tracked) n. A well deviated at a previously drilled depth in order to bypass a fish. See also: *directional drilling; deviation [drill]*.

sidetracking (side′-track-ing) The deliberate process of deflecting and redrilling the lower part of a borehole away from a previous course; e.g. drilling to the side of and beyond a piece of drilling equipment that is permanently lost in the hole. Cf: *directional drilling*.

sidewall core (side′-wall) A small core extracted from the wall of a drill hole, either by shooting a retractable hollow projectile, or by mechanically drilling a sample.

sidewall crystallization Boundary layer crystallization along the lateral margins (or sidewall) or a magma body. Typically accompanied by *compositional convection* of the adjacent liquid magma.

sidewall sampling The process of obtaining *sidewall cores*, usually by percussion (shooting hollow retractable cylindrical bullets into the walls).

sidorenkite (sid-o-renk′-ite) A pale rose monoclinic mineral of the *bradleyite* group: $Na_3Mn^{2+}(PO_4)(CO_3)$.

sidpietersite A beige-cream triclinic mineral: $Pb^{2+}_4(S_2O_3)O_2(OH)_2$.

sidwillite (sid′-will-ite) A yellow monoclinic mineral: $MoO_3{\cdot}2H_2O$.

Siegenian (Sie-gen′-i-an) A European stage: Lower Devonian (above Gedinnian, below Emsian).

siegenite (sie′-gen-ite) A metallic gray cubic thiospinel mineral of the *linnaeite* group: $CoNi_2S_4$. It may contain copper or iron or both in appreciable amounts.

sieleckiite A deep blue triclinic mineral: $Cu_3Al_4(PO_4)_2(OH)_{12}{\cdot}2H_2O$.

sienna (si-en′-na) Any of various brownish-yellow earthy limonitic pigments for oil stains and paints. It becomes orange red to reddish brown when burnt, and is generally darker and more transparent in oils than ochers. Named after Siena, a town in Tuscany, Italy. Cf: *umber*.

Sierozem (Si-er′-o-zem) A zonal great soil group consisting of soils with pale grayish A horizons grading into calcareous material at a depth of 1 foot or less, and formed in temperate to cool, arid climates under a vegetation of desert plants, short grass, and scattered brush. Not used in current U.S. system of soil taxonomy.

sierra (si-er′-ra) (a) A high range of hills or mountains, esp. one having jagged or irregular peaks that when projected against the sky resemble the teeth of a saw; e.g. the Sierra Nevada in California. The term is often used in the plural, and is common in the SW U.S. and in Latin America. Syn: *serra*. (b) A mountainous region in a sierra. Etymol: Spanish, "saw", from Latin *serra*, "saw".

sierranite (si-er′-ra-nite) A rock, consisting of onyx and chert, found in the Sierra Nevada of California.

sieve An apparatus used to separate soil or sedimentary material according to the size of its particles; it is usually made of brass or stainless steel, with a wire-mesh cloth having regularly spaced square holes of uniform size. Cf: *screen [eco geol]*.

sieve analysis Determination of the particle-size distribution in a soil, sediment, or rock by measuring the percentage of the particles that will pass through standard sieves of various sizes.

sieve deposition A term proposed by Hooke (1967, p.454) for the formation of coarse-grained lobate masses on an alluvial fan whose material is sufficiently coarse and permeable to permit complete infiltration of water before it reaches the toe of the fan.

sieve diameter The size of a sieve opening (mesh) through which a given particle will just pass.

sieve element The basic food-conducting cell in the *phloem* of vascular plants. Sieve elements are alive at maturity and have specialized sieve areas on their walls through which movement of photosynthates takes place. Also called: sieve cells. Angiosperms have specialized sieve elements termed *sieve tube elements* which are linked end to end by *sieve plates* in series called *sieve tubes*. Gymnosperms have less specialized sieve elements.

sieve lobe A coarse-grained lobate mass produced by *sieve deposition* on an alluvial fan. Also called sieve deposits.

sieve membrane A sievelike, partly closing membrane in the areolae of a locular-walled diatom. It may occur at an outer or inner position of the wall.

sieve plate [bot] A part of vascular plants linking *sieve elements*.

sieve plate [paleont] (a) A perforated *diaphragm* extending across the oscular end of the *cloaca* of a sponge. (b) A unilaminar and circular, subcircular, or polygonal perforate plate of a holothurian (TIP, 1966, pt. U, p.653). (c) A minute discoidal plate with numerous circular, triangular, and polygonal micropores arranged in concentric rows, contained in a pore canal of certain foraminifers. Also, a *trematophore*. (d) A flat, circular, porous plate in spumellarian radiolarians.

sievert A dose of ionizing radiation equal to 100 rems or 1 J/kg. Abbrev: Sv.

sieve texture *poikiloblastic*.

sieve tube A phloem tube formed from several sieve elements set end to end (Cronquist, 1961, p.881).

sieve tube element A specialized *sieve element*.

sieving (siev′-ing) The shaking of loose materials in a sieve so that the finer particles pass through the mesh. It is the most common method of measuring particle sizes of sediments, esp. in the range 1/16 mm (very fine sand) to about 30 mm (coarse pebbles).

sif A syn. of *seif*. Pl: siuf.

sifema (si-fe′-ma) According to Van Bemmelen (1949), the theoretical ultrabasic layer underlying the sima; it is the equivalent of the sima of some authors, and of the ultrasima of others. Archaic. Cf: *salsima*.

siferna (si-fer′-na) A term that has been used for the sima, in a scheme in which the ultrasima is referred to as the sima (Schieferdecker, 1959, term 4547). Archaic.

sight (a) An observation (such as of the altitude of a celestial body) taken for determining direction or position. Also, the data obtained by such an observation; e.g. a bearing or angle taken with a compass or transit when making a survey. (b) A device with a small aperture through which objects are seen and by which their directions are determined; e.g. an "open sight" of an alidade.

sight rod *range rod*.

sigillarian (sig-il-lar′-i-an) n. An arborescent club moss of the genus *Sigillaria* that occurs in Carboniferous deposits. adj. Pertaining to *Sigillaria*. Cf: *lepidodendrid*.

sigismundite A greenish-gray monoclinic mineral of the *arrojadite* group: $(Ba,K,Pb)Na_3(Ca,Sr)(Fe^{2+},Mg)_{14}Al(OH)_2(PO_4)_{12}$.

sigloite (sig′-lo-ite) A straw-yellow to light brown triclinic mineral of the *laueite* group: $Fe^{3+}Al_2(PO_4)_2(OH)_3{\cdot}7H_2O$. It is an oxidation product of paravauxite.

sigma (sig′-ma) A C-shaped siliceous monaxonic sponge spicule (microsclere). Cf: *sigmaspire*. Pl: sigmata or sigmas.

sigma phi The verbalized expression for σ_ϕ or the standard deviation (sorting) of a particle-size distribution computed in terms of phi units of the sample. It is a measure of *degree of sorting*.

sigma porphyroclast An asymmetric porphyroclast with "wings" or "tails" of fine grained recrystallized material that extend tangentially away from the edge. The tails do not cross a line drawn through the center of the porphyroclast parallel to the external foliation. The asymmetry of the porphyroclast can be used to determine the shear sense. Cf: *delta porphyroclast*.

sigmaspire (sig′-ma-spire) An S-shaped siliceous monaxonic sponge spicule (microsclere), smooth or spinose; a *sigma* twisted in the form of a spiral of about one revolution. Cf: *spinispire*.

sigma-t Density of seawater divided by the maximum density of pure water minus one times 1,000 (1,000 (rho/rhomax 1)).

sigmoidal dune (sig-moi′-dal) An S-shaped, steep-sided, sharp-crested *sand dune* formed under the influence of alternating and opposing winds of roughly equal velocities (E. Holm, 1957); a dune transitional between a crescentic form and some of the dune complexes, being up to 50 m high, 1-2 km long, and 50-200 m wide.

sigmoidal fold A recumbent fold, the axial surface of which is so curved as to resemble the letter S.

signal [geophys] (sig′-nal) In geophysics, a desired indication on a reading. Cf: *noise*.

signal [surv] A natural or artificial object or structure located at or near a survey station and used as a sighting point or target for survey measurements; e.g. a flag on a pole, or a rigid structure erected over or close to a triangulation station.

signal correction In seismic analysis, a correction to eliminate the time differences between reflection times, resulting from changes in the outgoing signal from source to source.

signal-to-noise ratio The ratio of the amplitude of desired seismic energy (signal) to the amplitude of unwanted energy (noise). Abbrev: S/N.

signature (sig′-na-ture) (a) A characteristic or combination of characteristics by which a material or object or waveform may be identified, as on an image or photograph or seismic trace. (b) A wave form characteristic of an earthquake or other source of energy. (c) A graph of deflection versus time for points passed over by a wave.

significance In statistical testing, the probability that a stated statistical hypothesis, esp. the *null hypothesis*, will be rejected when in fact it is true. The level of significance is specified by the experimenter prior to running the test. Syn: *alpha level.* Also called: significance level.

significant wave height A statistical term for describing the height of ocean waves. It was originally defined to be the average of the highest one-third of waves passing a point. It corresponds to the height most commonly reported by an observer on a ship. Now, it is generally calculated from four times the standard deviation of waves measured at a point by a *wave gage.*

sign of elongation In hexagonal, tetragonal, and orthorhombic crystals of prismatic habit, the sign of the long crystallographic direction; a negative sign of elongation indicates that the trace of the vibration plane of the fast component is parallel to the long axis; a positive sign indicates that it is the slow component that is parallel to that axis. A crystal with a negative sign is said to be "length fast", and a crystal with a positive sign is termed "length slow". In crystals of lower symmetry, the crystal elongation may be only approximately parallel to vibration directions. See also: *negative elongation; positive elongation.* Syn: *elongation sign.*

Signor-Lipps effect Problems in sampling biostratigraphic ranges will mean that a sudden abrupt extinction event will have the appearance of a gradual extinction. Few organisms that were alive at a mass extinction will leave fossils right up to the final horizon.

sike (a) A British term for a small stream, esp. one that flows through flat or marshy ground and that is often dry in summer. (b) A British term for a gully, trench, drain, or hollow.

sikussak (si-kus'-sak) Very old, rough-surfaced sea ice trapped in a fjord, as along the north coast of Greenland; it resembles glacier ice because snow accumulation contributes to its formation and perpetuation. According to Koch (1926, p.100), to be called sikussak "the ice must be at least 25 years old". Etymol: Eskimo, "very old ice". Cf: *fjord ice.*

sil *yellow ocher.*

silcrete (sil'-crete) (a) A term suggested by Lamplugh (1902) for a conglomerate consisting of surficial sand and gravel cemented into a hard mass by silica. (b) A siliceous *duricrust.* Etymol: *sil*iceous + concrete. Cf: *calcrete; ferricrete.*

Silesian (Si-le'-sian) European stage: upper Lower Middle and Upper Carboniferous (above Dinantian, below Permian). Includes Namurian, Westphalian, and Stephanian.

silex (si'-lex) (a) The French term for flint. (b) Silica; esp. quartz, such as a pure or finely ground form for use as a filler. (c) An old term formerly applied to a hard, dense rock, such as basalt or compact limestone. Etymol: Latin, "hard stone, flint, quartz". The term was used by Pliny for quartz.

silexite [ign] (si'-lex-ite) An igneous rock composed essentially of primary quartz (60-100 percent). The term was first used by Miller (1919, p.30) to include a quartz dike, segregation mass, or inclusion inside or outside its parent rock. Syn: *peracidite; quartzolite.* Cf: *tarantulite. Quartzolite* is the preferred term, because silexite in French means *chert.*

silexite [sed] The French term for *chert;* specif. chert occurring in calcareous beds (Cayeux, 1929, p.554).

silhydrite (sil-hy'-drite) A white orthorhombic mineral: $3SiO_2 \cdot H_2O$.

silica (sil'-i-ca) The chemically resistant dioxide of silicon: SiO_2. It occurs naturally in several crystalline polymorphs (e.g. the minerals quartz, tridymite, cristobalite, coesite, and stishovite); in cryptocrystalline form (chalcedony); in amorphous and hydrated forms (opal); in less pure forms (e.g. sand, diatomite, tripoli, chert, flint); and combined in silicates as an essential constituent of many minerals.

silica coefficient In Osann's chemical classification of igneous rocks, the number that expresses the ratio of the total silica in a rock to the silica in the feldspars and metasilicates (Johannsen, 1939, p.61-82). Obsolete.

silica glass A glass or supercooled liquid consisting of pure or nearly pure silica, such as naturally occurring *lechatelierite* and artificially prepared vitreous silica. The term has been applied to impactites and to tektites.

silicalemma (sil'-i-ca-lem'-ma) The three-layered organic membrane of a diatom cell in which silica is deposited and which probably forms the basis of the organic skin of the mature diatom wall.

silicalite (sil'-i-ca-lite') A term used by Wadsworth (1893, p.92) for any rock composed of silica, such as quartz, jasper, or diatomaceous earth.

silica rock An industrial term for certain sandstones and quartzites that contain at least 95% silica (quartz). It is used as a raw material of glass and other products. Cf: *silica sand.*

silica sand An industrial term for a sand or an easily disaggregated sandstone that has a very high percentage of silica (quartz). It is a source of silicon and a raw material of glass and other industrial products. Cf: *silica rock.*

silicastone (sil'-i-ca-stone') A term suggested by Shrock (1948a, p.125) for any sedimentary rock composed of siliceous minerals.

silicate (sil'-i-cate) A compound whose crystal structure contains SiO_4 tetrahedra, either isolated or joined through one or more of the oxygen atoms to form groups, chains, sheets, or three-dimensional structures with metallic elements. Other elements, esp. Al, may substitute for some of the Si. Silicates are classified according to crystal structure (see *nesosilicate, sorosilicate, cyclosilicate, inosilicate, phyllosilicate, tectosilicate,* although alternative acceptable terms exist).

silicated (sil'-i-cat-ed) Said of a rock in which the process of *silication* has occurred.

silicate-facies iron formation (sil'-i-cate-fa'-cies) An *iron formation* in which the principal iron minerals are silicates, such as greenalite, stilpnomelane, minnesotaite, and iron-rich chlorite (James, 1954, p.263-272).

silicate tetrahedron A complex ion formed by four oxygen ions surrounding a silicon ion in a tetrahedral configuration, with a negative charge of 4 units. It is the basic unit of the silicates. It is commonly written as SiO_4.

silication (sil-i-ca'-tion) The replacement or breakdown of silicate minerals by reaction with free silica. Ant: desilication. Cf: *silicification [meta].* Adj: *silicated.*

siliceous [ecol] (si-li'-ceous) *silicicolous.*

siliceous [petrology] Said of a rock containing abundant silica, esp. free silica rather than as silicates.

siliceous cyst A resting stage common in various yellow-green algae that is endogenous, flasklike or bottle-shaped, and from six to ten micrometers or rarely as large as 20 micrometers in size. It is composed of cellulose or pectin that is highly impregnated with silica, and it is closed by an organic plug.

siliceous earth A friable, porous, fine-grained sediment, usually white, consisting chiefly of siliceous (opaline) material, having a dry earthy feel and appearance, and generally derived from the remains of organisms; e.g. *diatomite* and *radiolarian earth.* Cf: *tripoli.*

siliceous fireclay A fireclay composed mainly of fine white clay mixed with clean sharp sand.

siliceous limestone (a) A dense, dark, commonly thin-bedded limestone representing an intimate admixture of calcium carbonate and chemically precipitated silica that are believed to have accumulated simultaneously. (b) A silicified limestone, bearing evidence of replacement of calcite by silica.

siliceous ooze Any pelagic deep-sea sediment containing at least 30% siliceous skeletal remains, e.g. *radiolarian ooze, diatom ooze.*

siliceous residue An *insoluble residue* chiefly composed of siliceous material, such as quartz or chert.

siliceous sandstone A sandstone cemented with quartz or cryptocrystalline silica; e.g. an orthoquartzite.

siliceous sediment A sediment composed of siliceous materials that may be fragmental, concretionary, or precipitated, and of either organic or inorganic origin; e.g. chert, novaculite, geyserite, or diatomite. Siliceous sediments may be formed by primary deposition of silica or by secondary silicification and replacement.

siliceous shale A hard, fine-grained rock of shaly texture with an exceptional amount of silica (as much as 85%); it may have formed by silicification of normal shale (as by precipitation of silica derived from opal or devitrified volcanic ash) or by accumulation of organic material (such as diatom and radiolarian tests) at the same time the clay was deposited. Tarr (1938, p.20) prefers to describe such rock as *porcellanite* because it is not truly a shale. Syn: *phthanite.*

siliceous sinter The lightweight porous opaline variety of silica, white or nearly white, deposited as an incrustation by precipitation from the waters of geysers and hot springs. The term has been applied loosely to any deposit made by a geyser or hot spring. Syn: *sinter; geyserite; fiorite.*

siliceous sponge Any sponge having a skeleton composed of siliceous spicules.

silicic (si-lic'-ic) Said of a silica-rich igneous rock or magma. Although there is no firm agreement among petrologists, the amount of silica is usually said to constitute at least 65 percent or two-thirds of the rock. In addition to the combined silica in feldspars, silicic rocks generally contain free silica in the form of quartz. Granite and rhyolite are typical silicic rocks. The synonymous terms *acid* and *acidic* are used almost as frequently as "silicic". Syn: *oversaturated; persilicic.* Cf: *basic; intermediate; ultrabasic.*

siliciclastic (si-li'-ci-clas'-tic) Pertaining to clastic noncarbonate rocks "which are almost exclusively silicon-bearing, either as forms of quartz or as silicates" (Braunstein, 1961).

silicicolous (sil-i-cic'-o-lous) Said of an organism living in siliceous soil. Syn: *siliceous [ecol].*

silicide A mineral compound that is a combination of silicon with a metal. An example is gupeiite, Fe_3Si.

silicification [meta] (si-lic'-i-fi-ca'-tion) Hydrothermal alteration in which the quartz, opal, chalcedony, jasper, or other form of amorphous silica content of the rock increases. The term often refers to cases where there is a net addition of silica in the altered rock; but silicification may also occur by decomposition reactions where quartz is a byproduct (desilication) or by cation leaching in silica-stable systems. Often confused with *silication.* Adj: *silicified.* Var: *silification.*

silicification [paleont] A process of fossilization whereby the original organic components of an organism are replaced by silica, as quartz, chalcedony, or opal.

silicified (si-lic'-i-fied) Adj. of *silicification [meta].*

silicified wood A material formed by permineralization of wood by silica in such a manner that the original form and structure of the wood is preserved (Schopf, 1975, p.29). The silica is generally in the form of opal or chalcedony. Syn: *woodstone; agatized wood; opalized wood; wood agate.* Cf: *petrified wood.*

silicilith (si-lic′-i-lith) (a) A term suggested by Grabau (1924, p.298) for a quartz (sedimentary) rock. (b) A sedimentary rock composed principally of the siliceous remains of organisms (Pettijohn, 1957, p.429); e.g. a diatomite.

siliciophite (sil-ic′-i-o-phite′) A mixture consisting of serpentine penetrated by opal.

silicoflagellate (sil′-i-co-flag′-el-late) Any chrysomonad protozoan belonging to the family Silicoflagellidae and characterized by a skeleton composed of siliceous rings and spines. Range, Cretaceous to Holocene.

silicomagnesiofluorite (sil′-i-co-mag-ne′-si-o-flu′-o-rite) An ill-defined mineral: $Ca_4Mg_3Si_2O_5(OH)_2F_{10}$.

silicon-32 age method A method of age determination based on measurement of the activity of silicon-32 (half-life approximately 350 years), a nuclide formed in the upper atmosphere. The method has been applied to rapidly deposited siliceous oozes.

silicon detector (sil′-i-con) A *semiconductor radiation detector* composed of silicon rather than germanium.

silification (sil′-i-fi-ca′-tion) *silicification [meta].*

silinaite A colorless or white monoclinic mineral: $NaLiSi_2O_5{\cdot}2H_2O$.

silk Microscopic, needlelike inclusions of rutile crystals in a natural gem, such as ruby, sapphire, or garnet, from which subsurface reflections produce a whitish sheen resembling that of silk fabric.

silky luster (silk′-y) A type of mineral luster characteristic of certain fibrous minerals, e.g. chrysotile.

sill [intrus rocks] A tabular igneous intrusion that parallels the bedding or foliation of the sedimentary or metamorphic country rock, respectively. Cf: *dike [intrus rocks]; sheet [intrus rocks]; sole injection.*

sill [marine geol] (a) A submarine ridge or rise at a relatively shallow depth, separating a basin from another basin or from an adjacent sea and causing the basin to be partly closed, e.g. in the Straits of Gibraltar. (b) A ridge of bedrock or earth material at a shallow depth near the mouth of a fjord, separating the deep water of the fjord from the deep ocean water outside. Syn: *threshold [marine geol].*

sill [stat] In *geostatistics,* the upper limit of the *semivariance,* numerically equal to the global variance of a stationary variable. In an ideal *semivariogram,* the sill is the flat region beyond the range.

sillar (sil-lar′) (a) The deposit from a pyroclastic flow (*ignimbrite*) that became indurated by recrystallization due to escaping gases rather than by welding, as is the case with *welded tuff;* it is a type of ignimbrite. (b) A nonwelded ash-flow tuff. Etymol: Peruvian.

sill depth Depth to the top of a sill responsible for partly closing a basin.

silled basin *restricted basin.*

sillénite (sil-lén′-ite) A green or yellowish-green cubic mineral: Bi_2O_3. Some analyses suggest the presence of appreciable Si in its composition.

sillimanite (sil′-li-man-ite′) (a) A brown, gray, pale-green, or white orthorhombic mineral: Al_2SiO_5. It is trimorphous with kyanite and andalusite. Sillimanite occurs in long, slender, needlelike crystals often found in wisplike or fibrous aggregates in schists and gneisses; it forms at the highest temperatures and pressures of a regionally metamorphosed sequence and is characteristic of the innermost zone of contact-metamorphosed sediments. Syn: *fibrolite.* (b) A group of aluminum-silicate minerals including sillimanite, kyanite, andalusite, dumortierite, topaz, and mullite.

silt [eng] Nonplastic or slightly plastic material exhibiting little or no cohesive strength when air-dried, consisting mainly of particles having diameters less than 0.074 mm (passing U.S. standard sieve no. 200). Laboratory assessment of percentage silt in a sample is usually determined by *gradation* analysis in accordance with ASTM sieve analysis, test method D422. Cf: *clay [eng].*

silt [sed] (a) A rock fragment or detrital particle smaller than a very fine sand grain and larger than coarse clay, having a diameter in the range of 1/256 to 1/16 mm (4-62 micrometers, or 0.00016-0.0025 in., or 8 to 4 phi units; the upper size limit is approximately the smallest size that can be distinguished with the unaided eye). In Great Britain, the range of 0.01-0.1 mm has been used. Cf: *coarse silt; medium silt; fine silt; very fine silt.* (b) A loose aggregate of unlithified mineral or rock particles of silt size; an unconsolidated or moderately consolidated sedimentary deposit consisting essentially of fine-grained clastics. It varies considerably in composition but commonly has a high content of clay minerals in the finer fractions and a high content of quartz in the coarser fractions. The term is sometimes applied loosely to a sediment containing considerable sand- and clay-size particles in addition to silt-size particles. (c) Sedimentary material (esp. of silt-size particles) dispersed in running or standing water; mud or fine earth in suspension.

silt [soil] (a) A rock or mineral particle in the soil, having a diameter in the range of 0.002-0.05 mm; prior to 1937, the range was 0.005-0.05 mm. The diameter range recognized by the International Society of Soil Science is 0.002-0.02 mm. (b) A soil containing more than 80% silt-size particles, less than 12% clay, and less than 20% sand.

siltage (silt′-age) A mass of silt.

siltation (sil-ta′-tion) *silting.*

silting (silt′-ing) The deposition or accumulation of silt that is suspended throughout a body of standing water or in some considerable portion of it; esp. the choking, filling, or covering with stream-deposited silt behind a dam or other place of retarded flow, or in a reservoir. The term often includes sedimentary particles ranging in size from colloidal clay to sand. Syn: *siltation.*

silting up The filling, or partial filling, with silt, as of a reservoir that receives fine-grained sediment brought in by streams and surface runoff. The term has been used synonymously with *sedimentation* without regard to any specific grain size.

siltite (silt′-ite) A term used by Kay (1951) for a *siltstone.*

silt load A *suspended load* consisting essentially of silt.

silt loam A soil containing 50-88% silt, 0-27% clay, and 0-50% sand; e.g. one with at least 50% silt and 12-27% clay, or one with 50-88% silt and less than 12% clay (SSSA, 1965, p.347).

silt shale A consolidated sediment consisting of no more than 10% sand and having a silt/clay ratio greater than 2:1 (Folk, 1954, p.350); a fissile siltstone.

silt size A term used in sedimentology for a particle size greater than that of a sphere with a diameter of 1/256 mm (0.00016 in.) and less than that of a sphere with a diameter of 1/16 mm (0.0025 in.). See also: *dust size.*

siltstone An indurated silt having the texture and composition of shale but lacking its fine lamination or fissility; a massive *mudstone* in which the silt predominates over clay; a nonfissile silt shale. Pettijohn (1957, p.377) regards siltstone as a rock whose texture is intermediate between those of sandstone and shale and of which at least two-thirds is material of silt size; it tends to be flaggy, containing hard, durable, generally thin layers, and often showing various primary current structures. Syn: *siltite.*

silttil (silt′-til) A friable, brownish to buff, open-textured silt containing a few small siliceous pebbles, representing a chemically decomposed and eluviated till that may originally have been clayey, and developed in an undulatory, well drained area, as the drift sheets in Illinois (Leighton and MacClintock, 1930, p.41). Pronounced as if spelled "silt-till". Rarely used. Cf: *mesotil; gumbotil.*

silty breccia (silt′-y) A term used by Woodford (1925, p.183) for a breccia containing at least 80% rock fragments and 10% silt.

silty clay (a) An unconsolidated sediment containing 40-75% clay, 12.5-50% silt, and 0-20% sand (Shepard, 1954). (b) An unconsolidated sediment containing more particles of clay size than of silt size, more than 10% silt, and less than 10% of all other coarser sizes (Wentworth, 1922). (c) A soil containing 40-60% clay, 40-60% silt, and 0-20% sand (SSSA, 1965, p.347).

silty clay loam A soil containing 27-40% clay, 60-73% silt, and less than 20% sand. Cf: *clay loam.*

silty sand (a) An unconsolidated sediment containing 50-90% sand and having a ratio of silt to clay greater than 2:1 (Folk, 1954, p.349). (b) An unconsolidated sediment containing 40-75% sand, 12.5-50% silt, and 0-20% clay (Shepard, 1954). (c) An unconsolidated sediment containing more particles of sand size than of silt size, more than 10% silt, and less than 10% of all other sizes (Wentworth, 1922).

silty sandstone (a) A consolidated *silty sand.* (b) A sandstone containing more than 20% silt (Krynine, 1948, p.141).

Silurian (Si-lu′-ri-an) A system of the *Paleozoic* Erathem of the Standard Global Chronostratigraphic Scale, above the *Ordovician* and below the *Devonian.* Also the time during which these rocks were formed, the Silurian Period, covering the time span between 435 and 410 Ma. It is named after the Silures, a Celtic tribe.

silvanite (sil′-van-ite) *sylvanite.*

silver (sil′-ver) A soft white cubic or hexagonal mineral, the native metallic element Ag. It occurs in stringers and veins in volcanic and sedimentary rocks and in the upper parts of silver-sulfide lodes, and is often associated with small amounts of gold, mercury, copper, lead, tin, platinum, and other metals. Silver is ductile, malleable, and resistant to oxidation or corrosion, though it tarnishes brown; it has the highest thermal and electric conductivity of any substance. It is used for coinage, jewelry, and tableware, in photography, dentistry, and electroplating, and as a catalyst.

silver amalgam Naturally occurring *amalgam.*

silver Cape A *Cape diamond* having a very slight tint of yellow.

silver-copper glance (sil′-ver-cop′-per) *stromeyerite.*

silver glance *argentite.*

silver jamesonite *owyheeite.*

silver-lead ore Galena containing more than one percent silver; argentiferous galena.

silvialite A colorless or pale yellow tetragonal mineral of the *scapolite* group: $Ca_4Al_6Si_6O_{24}(SO_4)$.

silvicolous (sil-vic′-o-lous) Said of an organism that lives in wooded areas.

sima [interior Earth] (si′-ma) A petrologic name for the lower layer of the Earth's crust, composed of rocks that are rich in silica and magnesia. It is equivalent to the *oceanic crust* and to the lower portion of the *continental crust,* underlying the *sial.* Etymol: an acronym for *si*lica + *ma*gnesia. Adj: *simatic.* Cf: *sialma.* Syn: *intermediate layer; basaltic layer.*

sima [speleo] A Spanish word for a *natural well* that has vertical sides.

simatic (si-mat′-ic) Adj. of sima.

simetite (si′-me-tite) A deep-red to light orange-yellow or brown variety of amber, having a high content of sulfur and oxygen and a low content of succinic acid, and occurring in the waters off Sicily.

simferite A red orthorhombic mineral of the *triphylite* group: $Li(Mg,Fe^{3+},Mn^{3+})_2(PO_4)_2$.

similar fold (sim'-i-lar) A fold in which the orthogonal thickness of the folded strata is greater in the hinge than in the limbs, but the distance between any two folded surfaces is constant when measured parallel to the axial surface. Cf: *reverse similar fold; parallel fold.*

simmonsite A pale buff-cream monoclinic mineral: Na_2LiAlF_6.

simonellite (si-mo-nel'-lite) An orthorhombic hydrocarbon mineral: $C_{19}H_{24}$.

simonite (si-mon'-ite) A red-black monoclinic mineral: $TlHgAs_3S_6$.

simonkolleite (si'-mon-kolle'-ite) A colorless rhombohedral mineral: $Zn_5Cl_2(OH)_8{\cdot}H_2O$.

Simplan A method to simulate the response of the Earth to plane or cylindrical waves by summing of observations with spherical waves. Acronym for *sim*ulated *plane* waves (Sheriff and Geldart, 1995).

simple cone *coniform element.*

simple conic projection An atlas projection called "conic" is usually an equidistant or simple conic projection, on which parallels are equally space concentric circular arcs and meridians are equally spaced radii of these arcs. The poles are also circular arcs. This projection is one of the earliest types used for mid-latitude regions; it dates in prototype form to Claudius Ptolemy (150 A.D.) and especially to the 16th century. Syn: *equidistant conic projection.*

simple coral (sim'-ple) *solitary coral.*

simple crater An *impact crater* of relatively small diameter, characterized by a uniformly concave-upward shape and a maximum depth in the center, and lacking a central uplift and rim terraces (Dence, 1968, p.171). Simple craters differ from transient craters by the presence of a *breccia lens;* e.g., Barringer Crater (Meteor Crater) in Coconino County, Arizona. Cf: *complex impact crater.*

simple cross-bedding Cross-bedding in which the lower bounding surfaces are nonerosional (McKee and Weir, 1953, p.385); it is formed by deposition alone.

simple crossed lamellar structure A *crossed lamellar structure* in which the second-order lamellae are generally laminar and well defined. Simple crossed lamellar structures are generally linear and branching crossed lamellar structures. However, not all linear and branching crossed lamellar structures are simple crossed lamellar (some are rod-type crossed lamellar, especially in chiton shells) (Wise, 1968, p.325; Waller, 1978; Watabe, 1984).

simple cuspate foreland A foreland in which beach ridges, swales, and other symmetrical lines of growth are oriented parallel with both shores of the cusp (Johnson, 1919, p.325). Cf: *complex cuspate foreland.*

simple fold A single fold or flexure. Cf: *compound fold.*

simple impact crater A bowl-shaped *impact crater* with a relatively high depth/diameter ratio: close to 1/5 for fresh craters, less for degraded craters (Pike, 1974).

simple lamellar fibrous prismatic structure A *fibrous prismatic structure* in which the fibers are aggregated into mutually parallel sheets similar to the second-order lamellae of *crossed lamellar structure,* but the second-order lamellae and their contained fibers are all mutually parallel. This is a common structure in the hinge plate of arcacean bivalves (Carter, 1990, Fig. 2D).

simple lattice *primitive lattice.*

simple operculum A dinoflagellate operculum that consists of a single piece, regardless of how many plates are represented (Fensome et al., 1993, p.262). Cf: *compound operculum.*

simple ore An ore of a single metal. Cf: *complex ore.*

simple pit A pit [bot] in which there is no over-arching wall. Cf: *bordered pit.*

simple prismatic structure A *prismatic structure* in which each first-order prism has well-defined and generally noninterdigitating boundaries, the prisms show relatively low to moderate length/width ratios, the prism length axes are commonly oriented perpendicular or nearly perpendicular to the depositional surface, and the second-order structural units within each prism do not diverge toward the depositional surface. This category excludes the superficially similar regular spherulitic prismatic and vertical nondenticular composite prismatic structures, e.g., in the outer layer of *Neopilina galatheae* and in certain Unionacea, respectively (Carter and Tevesz, 1978; Carter, 1980b).

simple shear A particular type of constant-volume, plane-strain deformation characterized by rotational strain and fixed orientation of one of the circular sections of the *strain ellipsoid.* Simple shear can be closely approximated by shearing a deck of cards in one direction.

simple skeletal wall In stenolaemate bryozoans, a wall calcified by epidermis located on growing edges and one side (Boardman and Cheetham, 1969, p.211). Cf: *compound skeletal wall.*

simple spit A spit, either straight or recurved, without the development of minor spits at its end or along its inner side. Cf: *compound spit; complex spit.*

simple trabecula A *trabecula* of a scleractinian coral, composed of a series of single sclerodermites. Cf: *compound trabecula.*

simple twin A twinned crystal composed of only two individuals in twin relation.

simple valley A valley that maintains a constant relation to the general structure of the underlying strata; e.g. a longitudinal valley, or a transverse valley. Term introduced by Powell (1874, p.50). Cf: *complex valley; compound valley.*

simplicity (sim-plic'-i-ty) An equivalent term for the principle of *parsimony [philos].*

simplotite (sim'-plot-ite) A black, greenish-black or yellowish-green monoclinic mineral: $CaV^{4+}{}_4O_9{\cdot}5H_2O$.

simpsonite (simp'-son-ite) A colorless, cream, or yellowish-brown trigonal mineral: $Al_4(Ta,Nb)_3O_{13}(OH)$.

SIMS Abbrev. for *secondary ionization mass spectrometer.* See: *ion microprobe.*

simulant Any substance fashioned as a gemstone that imitates a known gemstone in appearance: an *imitation.*

simulated annealing Algorithms based on an analogy between optimization and the growth of long-range order, such as the growth of large crystals in a slowly cooling melt (Vasudevan et al., 1991).

simulation (sim-u-la'-tion) The representation of a physical system by a device such as a computer or model that imitates the behavior of the system; a simplified version of a situation in the real world. Cf: *model.*

simultaneous competitive growth A process of joint set development in which many joints are propagating at the same time and therefore mechanically interact during formation of the joint set.

simultaneous reconstruction technique A technique for solving problems in *tomography.* Abbrev: SIRT.

sincosite (sin'-cos-ite) A green, brownish-green, yellowish-green, or bluish-green tetragonal mineral: $Ca(VO)_2(PO_4)_2{\cdot}5H_2O$.

Sinemurian (Si-ne-mu'-ri-an) A stage of the Standard Global Chronostratigraphic Scale: Lower Jurassic (above Hettangian, below Pliensbachian). The basal GSSP is in southwest England (Bloos and Page, 2002).

singing (sing'-ing) A seismic resonance phenomenon that is produced by short-path multiples in a water layer. Syn: *reverberation; ringing.*

singing sand *sounding sand.*

single-chain silicate An *inosilicate* with Si:O ratio of 1:3.

single-channel seismic profiling A method of reflection seismic profiling in which the hydrophone receivers are grouped, in parallel, into a single section, which is recorded on a single channel. Abbrev: SCS.

single crystal diffractometer An instrument, often automated and computerized, that facilitates the orientation of a single crystal for diffraction of X-rays (or neutrons) by any set of lattice planes, and subsequent measurement of the intensity of the diffracted beam.

single crystal method The investigation of symmetry and *crystal structure* by any of several means that involve the diffraction of X-rays or neutrons by a single crystal of a substance. The term now usually implies the solution and/or refinement of a crystal structure using a *single crystal diffractometer.*

single cut (sin'-gle) A simplified *brilliant cut* consisting of 18 facets: a table, a culet, 8 bezel facets, and 8 pavilion facets. It is used mostly on small stones (*mêlée*) of low quality.

single-cycle mountain (sin'-gle-cy'-cle) A fold mountain that has been destroyed without re-elevation of any important part (Hinds, 1943).

single domains In a *ferromagnetic, ferroelastic,* or *ferroelectric* crystal, a region exhibiting a homogeneous *crystal structure* and uniform crystallographic orientation. Cf: *multidomains.*

single-ended spread (sin'-gle-end'-ed) A type of seismic spread in which the source is located at one end of the arrangement of geophones.

single-grain structure A type of structure of a noncoherent soil in which there is no aggregation or orderly arrangement. It is characteristic of coarse-grained soils.

single island silicate *nesosilicate.*

single-line stream A watercourse too narrow to depict (at the given scale on a map) by two lines representing the banks. Cf: *double-line stream; split stream.*

single refraction Refraction in an isotropic crystal, as opposed to the birefringence of an anisotropic crystal.

single tombolo A single, simple bar connecting an island with the mainland or with another island. Cf: *double tombolo.*

singular crystal form (sin'-gu-lar) *fixed form.*

singularity (sin-gu-lar'-i-ty) (a) A point where a function is not differentiable. (b) A value of a variable for which a function becomes infinite.

singular matrix A matrix that cannot be inverted because division by zero is encountered during the inversion process. A singular matrix has a determinant of zero.

sinhalite (sin'-ha-lite) A hard colorless, yellow, pink, greenish, or brown orthorhombic mineral: $MgAlBO_4$. It is structurally related to *monticellite.*

Sinian (Sin'-i-an) An obsolete geochronometric subdivision of the latest *Precambrian,* covering the time span from 800 to 570 Ma (Harland et al., 1990), therefore approximately equivalent to the *Cryogenian* and *Ediacaran* periods of the *Neoproterozoic* Era of the Standard Global Chronostratigraphic Scale.

sinistral (sin'-is-tral) Pertaining, inclined, or spiraled to the left; specif. pertaining to the reversed or counterclockwise direction of coiling of some gastropod shells. A sinistral gastropod shell in apical view (apex toward the observer) has the whorls apparently turning from the right toward the left; when the shell is held so that the axis of coiling is vertical and the apex or spire is up, the aperture is open toward the observer to the left of the axis. Actually, the definition depends on features of soft anatomy: with genitalia on the left side of the head-foot mass, the soft parts and shell are arranged as in a mirror image of a dextral shell (TIP, 1960, pt.I, p.133). Ant: *dextral.* Syn: *left-handed [paleont].*

sinistral element Component of a skeletal conodont apparatus on the left-hand side of the plane of bilateral symmetry (TIP, 1981, pt. W, supp.2).

sinistral fault *left-lateral fault.*

sinistral fold An asymmetric fold with the asymmetry of a Z as opposed to that of an S when seen in profile. The long limb appears to be offset to the left. Cf: *dextral fold.*

sinistral imbrication The condition in a *heterococcolith* in which each segment overlaps the one to the left when viewed from the center of the cycle. Ant: *dextral imbrication.*

sinjarite (sin'-jar-ite) A pink tetragonal hygroscopic mineral: $CaCl_2·2H_2O$.

sink [geog] A depression containing a central playa or saline lake with no outlet, as where a desert stream comes to an end or disappears by evaporation; e.g. Carson Sink in Nevada.

sink [glac geol] An obsolete term for a depression in a *terminal moraine.*

sink [karst] *sinkhole.*

sink [volc] A circular or ellipsoidal depression on the flank of or near to a volcano, formed by collapse. It has no lava flows or rim surrounding it. Cf: *collapse caldera.* Syn: *pit crater; volcanic sink; lava rise pit.*

sinkankasite (sin-kank'-as-ite) A colorless triclinic mineral: $Mn^{2+}Al(PO_3OH)_2(OH)·6H_2O$.

sinkhole (sink'-hole) A *closed depression* in a karst or pseudokarst area, commonly with a circular or ellipsoidal pattern. Its drainage is subterranean; its size is measured in meters or tens of meters; and it is commonly funnel shaped. Syn: *doline; sink [karst]; shakehole.* Cf: *collapse sinkhole; solution sinkhole.*

sinkhole karst The typical karst of temperate regions. Cf: *cockpit karst; tower karst.*

sinkhole plain Plain on which most of the local relief is due to closed depressions and nearly all drainage is subterranean.

sinkhole pond A small lake in a *closed depression* in limestone, resulting from an impervious clay floor or to the intersection of depression with the water table.

sinking (sink'-ing) (a) *subsidence.* (b) The downward movement of surface water, generally caused by converging currents or by a water mass that becomes denser than the surrounding water. Ant: upwelling. Syn: *downwelling.*

sinking stream A surface stream that disappears underground in a *karst* region. Cf: *disappearing stream; insurgence; lost stream; swallet; swallow hole.*

sinnerite (sin'-ner-ite) A metallic gray triclinic mineral: $Cu_6As_4S_9$. Cf: *luzonite.*

sinoite (sin'-o-ite) A colorless orthorhombic meteorite mineral: Si_2N_2O.

sinopite (sin'-o-pite) A brick-red earthy ferruginous clay mineral used by the ancients as a red paint.

sinople (sin'-o-ple) A red or brownish-red variety of quartz containing inclusions of hematite. Also spelled: sinopal; sinopel.

sinter (sin'-ter) A chemical sedimentary rock deposited as a hard incrustation on rocks or on the ground by precipitation from hot or cold mineral waters of springs, lakes, or streams; specif. *siliceous sinter* and *calcareous sinter* (travertine). The term is indefinite and should be modified by the proper compositional adjective, although when used alone it usually signifies "siliceous" sinter. Etymol: German *Sinter.* Cf: *tufa.*

sintering (sin'-ter-ing) The process by which bonds develop when grains of solid material are brought into contact. In principle, progressive growth of bonds, or "necks", between grains can be brought about by inelastic deformation under stress; molecular diffusion through the vapor phase; surface diffusion of molecules; or volume diffusion of molecules. In the sintering of ice grains, it appears that vapor diffusion is the dominant mechanism.

sinuosity (sin-u-os'-i-ty) Ratio of the length of the channel or thalweg to the down-valley distance (Leopold and Wolman, 1957, p.53). Channels with sinuosities of 1.5 or more are called "meandering".

sinuous rille A lava channel or tube in the lunar maria (Spudis, 2000).

sinupalliate (sin-u-pal'-li-ate) Said of a bivalve mollusk possessing a *pallial line* with a posterior embayment (*pallial sinus*). Cf: *integripalliate.*

sinus [bot] (si'-nus) The space or recess between two lobes or divisions of a leaf or other expanded organ (Lawrence, 1951, p.770).

sinus [paleont] (a) A curved, moderately deep groove, indentation, or re-entrant in the outer lip of the aperture of a gastropod shell. It is progressively filled in as the shell grows and forms a distinct band, and is distinguished from the *slit* by nonparallel sides. (b) Any part of a transverse feature (apertural margin, rib, growth line) of a cephalopod, concave toward the aperture. (c) A major undulation or rounded depression along the commissure of a brachiopod (generally found on the pedicle valve), with the crest directed ventrally and commonly but not invariably associated with the ventral fold and the dorsal sulcus. The term is also used, irrespective of commissure, as a syn. of *sulcus.* (d) A slit or a rounded or V-shaped notch in the proximal lip of the orifice in some ascophoran cheilostomes (bryozoans), serving as an inlet to the ascus; syn: *rimule.* Also, loosely used for a similar slit or notch in the *secondary orifice.* Also, in many stenolaemate bryozoans, a groove on either side of the keel in endozonal walls of zooids growing from encrusting or median colony walls. (e) A V-shaped indentation of blastoid ambulacrum along the margins of deltoid plates and radial plates. (f) A *pallial sinus* of a bivalve mollusk. (g) In vertebrate anatomy, the term applies to many cavities, depressions, or spaces. In the vascular system, these are areas within the body in which blood, lymph, or other fluids pool and may not be contained within well-defined walls of vessels. In mammalian skulls, cranial sinus are moist, membrane-lined, air-filled chambers that are connected by tubes to the buccal/pharyngeal/nasal parts of the head.

sinus [planet] A term established by the International Astronomical Union for a small plain on planetary surfaces (USGS, 2003).

sinusbed (si'-nus-bed) A term suggested by Engelund (1966) as a substitute for *antidune,* because, according to the definition adopted by hydraulic engineers, an antidune is a bed form in phase with surface water waves, and may move downstream, or be stationary, rather than move upstream.

sinusoidal climbing ripple A *climbing ripple* displaying internal *cross-lamination* that is composed of the deposits of both the upstream-dipping *stoss* and the downstream-dipping *lee* sides of migrating *ripples.* Laminae are continuous across many ripples and the stoss and lee deposits are of approximately equal thickness. The ripple may approach a symmetric form but show a slight lateral displacement of crests and troughs upward in a coset (Jopling and Walker, 1968). This structure requires a higher ratio of aggradation rate to migration rate than type-B or -C climbing ripples. Cf: *type-A climbing ripple; type-B climbing ripple; type-C climbing ripple.*

sinusoidal equal-area projection *sinusoidal projection.*

sinusoidal projection (si-nus-oi'-dal) An equal-area map projection representing the limiting form of the *Bonne projection,* using the equator as the standard parallel, and showing all parallels as equally spaced parallel straight lines drawn to scale. The meridians are sine curves, concaving toward the central meridian (a straight line, one half the length of, and at right angles to, the equator) along which the scale is true. The projection shows the entire globe but suffers from extreme distortion (shearing) in marginal zones at high latitudes; it is often used in atlases for the map of Africa. Syn: *Sanson-Flamsteed projection; Mercator equal-area projection; sinusoidal equal-area projection.*

siphon [hydraul] (si'-phon) A water conduit in the shape of an inverted U, in which the water is in hydrostatic equilibrium. Also spelled *syphon.* Cf: *inverted siphon.*

siphon [paleont] (a) Either of a pair of posterior tubelike extensions of the *mantle* in many bivalve mollusks, serving for the passage of water currents; e.g. an inhalant ventral tube that conducts water to the mouth and gills and confines the current flowing into the mantle cavity, and an exhalant dorsal tube that carries away waste water and confines the current flowing from the mantle cavity. (b) An anterior channel-shaped prolongation of the mantle in many gastropods, serving for the passage of water to the gills, and often being protected by a grooved extension of the margin of the shell. (c) The membranous *siphuncle* of cephalopods. (d) An internal tube extending inward from a foraminiferal aperture.

siphon [speleo] An obsolete term for *sump [speleo].*

siphonal canal (si'-phon-al) The tabular or troughlike extension of the anterior (abapical) part of the apertural margin of a gastropod shell, serving for the shielding of the inhalant *siphon.*

siphonal deposit Calcareous structures within the *siphuncle* of a cephalopod, attaining considerable thickness in some nautiloids.

siphonal fasciole A band of abruptly curved growth lines near the foot of the *columella* of a gastropod, marking successive positions of the *siphonal notch.*

siphonal notch The narrow sinus of the apertural margin near the foot of the columella of a gastropod, serving for the protrusion of the inhalant *siphon.* It virtually separates the inner lip and the outer lip.

siphonoglyph (si-pho'-no-glyph) A strongly ciliated groove extending down one side of the *pharynx* of a coral.

siphonostomatous (si'-pho-no-stom'-a-tous) Said of a gastropod shell with the apertural margin interrupted by a canal, spout, or notch for the protrusion of the *siphon;* e.g. said of various marine snails having the front edge of the aperture prolonged in the form of a channel for the protection of the siphon. Cf: *holostomatous.*

siphonozooid (si'-pho-no-zo'-oid) A degenerate octocorallian *polyp* with reduced or no tentacles and commonly with reduced septal filaments (thickened, convoluted edges of septa). It is usually much smaller than an *autozooid* and is believed to regulate the water supply of the colony.

siphuncle (si'-phun-cle) (a) A long membranous tube extending through all the camerae and septa from the protoconch to the base of the body chamber of a cephalopod shell, and consisting of soft and shelly parts, including septal necks, connecting rings, calcareous deposits, and siphuncular cord. (b) The tubular or funnel-shaped shelly septal structures that ensheathe and support the siphuncle.

siphuncular cord (si-phun'-cu-lar) The fleshy interior tissues of the *siphuncle* of a cephalopod.

sirloin-type ice (sir'-loin-type) A term used by Higashi (1958) and now regarded as a syn. of *segregated ice.*

SIRT *simultaneous reconstruction technique.*

siserskite (sis'-er-skite) *iridosmine.*

sismondite *magnesiochloritoid.*

sister group (sis'-ter) (a) Within a *cladogram,* any pair of taxa united at a single node. (b) In cladistic analysis, the group of organisms that is genealogically most closely related to the study group taxa, or ingroup (Brooks et al., 1984, p.81; Wiley et al., 1991, p.4). Cf: *outgroup, ingroup.*

sitaparite (si-ta-pa'-rite) *bixbyite.*

site The location of an atom in a *crystal structure,* or the location in a *unit cell* where, for the purposes of discussion, an atom might exist.

site characterization Analysis of the hydrogeological, geochemical, and/or geophysical properties of some volume of the earth, commonly for the purposes of subsurface environmental evaluation and/or remediation.

site investigation The collection of basic facts about, and the testing of, surface and subsurface materials (including their physical properties and distribution and their geologic relationships and structure) at a site to characterize the physical and chemical conditions affecting designs for an engineered structure or other use at this site.

site period In engineering geology, a characteristic property of a site, related to the thickness and stiffness of soil layers above bedrock. The stiffness typically is represented as the shear-wave velocity of the soil and rock layers above firm rock that is assumed to be effectively rigid.

site symmetry The local *point symmetry* about a *site* in a *unit cell;* that is, the point symmetry of the crystal when the site in question is taken as the origin and is required to remain invariant.

sitinakite A light brown, pale rose, or colorless tetragonal mineral: $KNa_2Ti_4Si_2O_{13}(OH)·4H_2O$.

sitting on a well (sit'-ting) Waiting at a well location for the well to be drilled into a producing formation. The geologist examines cuttings and cores, to ascertain what formations are penetrated and to look for signs of hydrocarbons.

six coordination The state of an atom when it is surrounded by six nearest neighbors. Cf: *octahedral coordination.*

sixfold axis An axis of rotational symmetry by which an object (crystal shape, *crystal structure,* etc.) is invariant to a rotation of 60°. Syn: *hexad.*

sixth-power law (sixth-pow'-er) A law asserting that the carrying power of a stream is proportional to the sixth power of its velocity; e.g. if the stream flows twice as rapidly, the size of the particles carried may be increased 64 times. The law postulates a complete transfer of kinetic energy from the water to the particle and makes no allowance for the effect of viscous drag.

size *particle size.*

size analysis *particle-size analysis.*

size distribution *particle-size distribution.*

size-frequency analysis *particle-size analysis.*

size-frequency distribution *particle-size distribution.*

sizing The arrangement or classification of particles according to size; e.g. the separation of mineral grains of a sediment into groups each of which has a certain range of size or maximum diameter, such as by sieving or screening.

sjögrenite (sjö'-gren-ite) A creamy white, yellowish, orange, or brownish platy hexagonal mineral of the *manasseite* group: $Mg_6Fe^{3+}{}_2(CO_3)(OH)_{16}·4H_2O$. It is dimorphous with pyroaurite.

S-joint *longitudinal joint.*

skarn As used by Fennoscandian geologists, an old Swedish mining term for silicate gangue (amphibole, pyroxene, garnet, etc.) of certain iron-ore and sulfide deposits of Archean age, particularly those that have replaced limestone and dolomite. Its meaning has been generally expanded to include calcium-bearing silicates of any geologic age, derived from nearly pure limestone and dolomite with the introduction of large amounts of Si, Al, Fe and Mg (Holmes, 1920, p.211). In American usage the term is more or less synonymous with *tactite.* Cf: *endoskarn; exoskarn.*

skärtråg (skär'-tråg) The Swedish term for *sickle trough.*

skavl A Norwegian term for a large wind-eroded ridge of snow on a glacier. Pl: skavler. Skavler are generally equivalent to *sastrugi.*

skeletal (skel'-e-tal) (a) Pertaining to material derived from organisms and consisting of the hard parts secreted by the organisms or of the hard material around or within organic tissue. (b) Synonymous with "bioclastic" (Leighton and Pendexter, 1962); but used by Nelson et al. (1962, p.234) to refer to a limestone that consists of, or owes its characteristics to, virtually in-place accumulation of skeletal matter (as distinguished from a fragmental limestone formed by mechanical transport); but regarded by Leighton and Pendexter (1962) as synonymous with "bioclastic", indicating faunal or floral fragments, or whole components of organisms, that are not in their place of origin.

skeletal apparatus *apparatus.*

skeletal canal A canal-like cavity in a coherent skeletal framework of a sponge. It may or may not correspond to a canal of the aquiferous system. Examples: *amararhysis; diarhysis; schizorhysis; surface groove.*

skeletal crystal growth Microscopic development of the outline or framework of a crystal, with incomplete filling in of the crystal faces. Crystals formed in this way are called skeleton crystals.

skeletal duplicature The outer exoskeletal layers or molted skin of a branchiopod crustacean, shed during ecdysis. See also: *duplicature.*

skeletal fiber Any fiberlike structure of the sponge skeleton, such as a spiculofiber, a spicule tract, a sclerosomal trabecula, or a spongin fiber.

skeletal framework A coherent meshwork in a sponge, built of sclerosomal trabeculae, fused spicules, interlocking spicules, spongin-cemented spicules or sand grains, or spongin alone.

skeletal pore An opening between spicules or between skeletal fibers of the regular skeletal framework in a sponge, as distinct from larger openings (such as ostia or oscula) that interrupt the regular net and as distinguished from the true pores of the soft parts, with which the skeletal pores may not correspond.

skeletal residue An insoluble residue whose constituent material comprises less than 25% of the volume, and containing rhombohedral (dolomoldic) or spheroidal (oomoldic) openings (Ireland et al., 1947, p.1482-1483). Cf: *lacy residue.*

skeletal stromatolite Stromatolite in which the organisms responsible for its formation are commonly preserved as calcified fossils (Riding, 1977).

skeletan (skel'-e-tan) A *cutan* consisting of skeleton grains adhering to the surface (Brewer, 1964, p.217); e.g. bleached sand and silt grains high in quartz and low in feldspar.

skeletogenesis (skel'-e-to-gen'-e-sis) The process of forming the hard or skeletal parts of an organism.

skeleton (skel'-e-ton) The hard or bony structure that constitutes the framework supporting the softer parts of an animal and protecting or covering its internal organs; e.g. the mesh of spicules of a sponge, the shell of a brachiopod or mollusk, the calcareous layers of the body wall of a bryozoan, the chitinous covering of an arthropod, or the bones of a vertebrate. See also: *endoskeleton; exoskeleton.*

skeleton grain A relatively stable and not readily translocated grain of a soil material, concentrated or reorganized by soil-forming processes (Brewer and Sleeman, 1960); e.g. a mineral grain, or a resistant siliceous or organic body larger than colloidal size. Cf: *plasma [soil].*

skeleton layer The structure formed at the bottom of sea ice while freezing, consisting of vertically oriented platelets of ice separated by layers of brine; this layer has almost no mechanical strength.

Skempton's coefficient B Similar to *tidal efficiency,* but it involves multidimensional deformation rather than one-dimensional deformation.

skerry (sker'-ry) A small rugged island or reef; an isolated rock detached from the mainland, rising above sea level from a shallow strandflat, and covered by the sea during high tides or stormy weather. Examples occur along the coasts of Scotland and Scandinavia. Etymol: Scandinavian, akin to Old Norse "sker". Cf: *stack [coast].*

skerry-guard A line, belt, or fringe of skerries, parallel to and extending along a coast for hundreds of kilometers, seemingly acting as a breakwater or "guard". The term is a common, but incorrect, translation of the Norwegian term *skjergaard,* "skerry enclosure", which properly refers to an area of calm water between a line of skerries and the mainland or enclosed by skerries.

sketch map A map drawn freehand from observation or from loose and uncontrolled surveys rather than from exact survey measurements, showing only the main features of an area. It preserves the essential space relationships, but does not truly preserve scale or orientation.

skewed projection Any standard projection, used in construction of maps or charts, that does not conform to a general north-south format with relation to the neat lines of the map or chart.

skewness (skew'-ness) (a) A measure of asymmetry in a frequency distribution, based on the third *moment* about the mean of the distribution or on the difference between the mean and the median. (b) A graphical measure of asymmetry in a grain-size distribution, calculated as the ratio $P_{25}P_{75}/(P_{50})^2$ where P_{25}, P_{50}, and P_{75} are the 25th, 50th and 75th percentiles of the distribution.

skiagite (ski'-a-gite) A hypothetical end-member of the *garnet* group: $Fe_3^{+2}Fe_2^{+3}(SiO_4)_3$.

skialith (ski'-a-lith) An indisdinct remnant, commonly rich in mafic minerals, in granitic rocks or leucosomes. Little used. Cf: *schlieren; xenolith.*

skid boulder *sliding stone.*

Skiddavian (Skid-dav'-i-an) *Arenigian.*

skim ice First formation of a thin layer of ice on the water surface. Syn: *skin.*

skimming (skim'-ming) (a) Diversion of water from a stream or conduit by shallow overflow in order to avoid diverting sand, silt, or other debris carried as bottom load (Langbein and Iseri, 1960). (b) Withdrawal of fresh groundwater from a thin body or lens floating on salt water by means of shallow wells or infiltration galleries.

skin *skim ice.*

skin depth The depth in a dissipative medium at which the electric field intensity associated with a plane electromagnetic wave is smaller than its value at the surface of the medium by 1/e (e is the base of the Napierian logarithm). See also: *skin effect [elect].*

skin effect [drill] (a) The frequency-dependent reduction of *resistivity log* measurements in conductive formations due to inductive interaction between the current paths; the induction logs now operating at about 20 KHz are most affected. (b) The alteration of permeability in the immediate vicinity of a well screen, due to the construction and operation of the well. By convention, positive skin effect implies permeability reduction and negative skin effect implies permeability enhancement.

skin effect [elect] The concentration of alternating current in a conductor near its exterior boundary. See also: *skin depth.*

skin friction [eng] (a) The frictional resistance developed between soil and an engineering structure, particularly applicable to design and construction of deep foundations. (b) The shearing resistance of the ground developed on the sides of a *pile,* concrete *shaft,* pipe, or probing rod. Cf: *negative skin friction.*

skin friction [hydraul] In hydraulics, the friction between a fluid and the surface of a solid moving through it (ASCE, 1962). Syn: *surface drag.*

skinnerite (skin'-ner-ite) A metallic monoclinic mineral: Cu_3SbS_3.

skiodrome (ski'-o-drome) A term used in optical mineralogy for an orthographic projection of curves of equal velocity as they would appear on a sphere, assuming that the light source is at the center of the sphere.

skiou (ski-ou') A facetious term used by Davis (1912, p.116) for a *morvan.*

skip cast The cast of a *skip mark.* Term originally proposed by Dzulynski et al. (1959, p.1117).

skip mark One of a linear series of regularly spaced, crescent-shaped tool marks produced by an object that intermittently impinged on or skipped along the bottom of a stream. Cf: *ring mark; roll mark.* Syn: *saltation mark.*

skippenite (skip'-pen-ite) A metallic gray rhombohedral mineral of the *tetradymite* group: Bi_2Se_2Te.

skirt Downward-facing synformal anticline fringing the *bulb* of a *mushroom diapir* and made up of deformed evaporites that envelop the stem of the diapir. A type of crescentic fold (Jackson and Talbot, 1991).

skjergaard (skjer'-gaard) *skerry-guard.*

sklodowskite (sklo-dow'-skite) A strongly radioactive, pale yellow to greenish-yellow, monoclinic secondary mineral: $Mg(UO_2)_2(SiO_3OH)_2{\cdot}6H_2O$.

skolite (sko'-lite) A scaly, dark-green variety of glauconite rich in aluminum and calcium and deficient in ferric iron.

skolochorate (skol-o-chor'-ate) Of dinoflagellate *chorate cysts,* having processes only or a combination of processes and shorter (lower) *septa* or ridges (Williams et al., 2000, p.211).

skomerite (sko'-mer-ite) An altered *andesite* containing microscopic grains and crystals of augite and olivine, and phenocrysts of decomposed plagioclase (probably albite), in a groundmass of plagioclase that is thought to be more calcic than the phenocrysts (Johannsen, 1939, p.280). The name, given by Thomas in 1911, is from Skomer Island, Pembrokeshire, Wales. Cf: *marloesite.* Obsolete.

skull A term generally applied to the bony skeleton of the vertebrate head, including cartilage-replacement and dermal components. It is sometimes not applied to the head skeleton of living Agnatha and Chondrichthyes because of the absence or perhaps secondary loss of bone.

skutterudite (skut'-te-rud-ite) A tin-white to silver-gray cubic mineral: $CoAs_{2-3}$. It may contain considerable iron, and it represents a minor ore of cobalt and nickel. See also: *smaltite.*

skylight (sky'-light) (a) The component of light that is scattered by the atmosphere and consists predominantly of shorter wavelengths. (b) A submariner's term for a *polynya* or *lead [ice]* during the winter; it is covered by relatively thin ice (usually less than 1 m thick) and has a normally flat undersurface. Cf: *lake [ice].*

Skythian *Scythian.*

sky wave Electromagnetic (radio) waves reflected from ionized layers in the ionosphere. Involved in sky-wave interference and in making radio waves receivable beyond the line-of-sight horizon.

s.l. *sensu lato.*

slab *downgoing plate.*

slab avalanche A snow avalanche that starts from a fracture line, in *wind-slab* snow possessing a certain amount of cohesion. Cf: *loose-snow avalanche; wind-slab avalanche.*

slab correction *Bouguer correction.*

slab joints Joints produced in rock by the formation of closely spaced parallel fissures dividing the rock into thin slabs.

slab pahoehoe A type of *pahoehoe* whose surface consists of a jumbled arrangement of plates or slabs of flow crust, presumably so arranged due to the draining-away of the underlying molten lava.

slab-pull force In plate tectonics, the force exerted on a plate by the downgoing slab, probably a key plate-driving force. It is a consequence of the negative buoyancy of the downgoing slab.

slabstone (slab'-stone) A rock that readily splits into slabs; *flagstone.*

slack [coast] A hollow or depression between lines of shore dunes or in a sandbank or mudbank on a shore.

slack [geomorph] (a) A British term for a hollow or dip in the ground; e.g. a pass between hills, a small shallow valley, or a depression in a hillside. (b) A British term for a soft, boggy piece of ground; a marsh or morass.

slack [oceanog] *slack water.*

slack [weath] *grus.*

slack ice Broken ice, esp. if floating on slowly moving water.

slacking index (slack'-ing) *weathering index.*

slack tide *slack water.*

slack water (a) The condition of a tidal current or horizontal motion of water when its velocity is very weak (less than 0.1 knot) or zero, esp. at the turn of the tide when there is a reversal between ebb current and flood current. Also, the interval of time during which slack water occurs. Syn: *slack tide.* (b) A quiet part of, or a still body of water in, a stream; e.g. on the inside of a bend, where the current is slight. Syn: *slack [oceanog].*

slag [materials] A byproduct of the fusion of ores, metals, flux, and fuel that contains noneconomic constituents of the furnace charge. Slag has many economic uses, including railroad ballast, sand-blasting media, and road-building materials.

slag [pyroclast] A scoriaceous or cindery pyroclastic rock.

slag [sed] A British term for a friable shale with many fossils.

slaking (slak'-ing) (a) The crumbling and disintegration of earth materials upon exposure to air or moisture; specif. the breaking-up of dried clay or soil when saturated with or immersed in water, or of clay-rich sedimentary rocks when exposed to air. (b) The disintegration of the walls of tunnels in swelling clay, owing to inward movement and circumferential compression (Stokes and Varnes, 1955, p.137). (c) The treating of lime with water to give hydrated (slaked) lime.

slant drilling To drill a well at an angle starting at the surface. Cf: *directional drilling.*

slant range In *SLAR,* the distance measured along a straight line between antenna and target.

slant-range image In *SLAR,* an image in which objects are located at positions corresponding to their *slant range* distances from the aircraft flight path. The scale in the range direction is compressed in the near-range region.

slant stack *tau-p mapping.*

SLAR *side-looking airborne radar.*

slash [beach] A term used in New Jersey for a wet or marshy *swale* between two parallel beach ridges.

slash [geog] (a) A local term in eastern U.S. for a *marsh* or a low swampy area overgrown with dense underbrush and often covered by water. (b) An open or *cutover* tract in a forest, strewn with debris (logs, bark, branches, etc.), as from logging or fire. Also, the debris in such a tract.

slate (a) A compact, fine-grained metamorphic rock that possesses *slaty cleavage* and hence can be split into slabs and thin plates. Most slate was formed from shale. (b) A coal miner's term for any shale accompanying coal; also, sometimes the equivalent of *bone coal.*

slate ground A term used in southern Wales for a dark fissile shale, resembling slate.

slate ribbon A relict ribbon structure on the cleavage surface of slate, consisting of varicolored and straight, wavy, or crumpled stripes. It is generally a trace of bedding.

slate spar *shiver spar.*

slatiness (slat'-i-ness) The quality of being slaty, such as a sedimentary rock splitting into thin layers or plates parallel to the bedding with essential regularity of surfaces similar to true slaty cleavage (Grabau, 1924, p.786).

slaty cleavage (slat'-y) A foliation defined by elongated domains of quartz or feldspar aggregates separated by anastomosing, mica-rich laminae. The width of the domains is typically less than 100 micrometers. Slaty cleavage is the characteristic foliation in fine-grained sedimentary and volcanic rock deformed under anchizone and lower greenschist metamorphic conditions.

slavikite (slav'-ik-ite) A greenish-yellow rhombohedral mineral: $NaMg_2Fe^{3+}_5(SO_4)_7(OH)_6{\cdot}33H_2O$.

slavyanskite (slav-yan'-skite) *tunisite.*

slawsonite (slaw'-son-ite) A colorless or gray monoclinic or triclinic mineral: $(Sr,Ca)Al_2Si_2O_8$. It is isomorphous to paracelsian.

slayanskite *tunisite.*

sleech An old English term for mud deposited by the sea or a river, or for a *mud flat.*

sleeve exploder A marine seismic-energy source in which propane or butane is exploded in a thick rubber bag (the sleeve), and from which the waste gases are vented to the air rather than into the water. Obsolete.

slice [stratig] An arbitrary informal division (either of uniform thickness or constituting some uniform vertical fraction) of an otherwise indivisible stratigraphic unit, distinguished for individual facies mapping or analysis.

slice ridge A narrow linear ridge, a meter to 100 m high, representing a slice of rock squeezed up within a fault zone (esp. along a strike-slip fault). Syn: *fault-slice ridge; pressure ridge [seis].*

slick n. A glassy smooth elongate patch or web-like net on an otherwise rippled water surface, occurring in coastal and inland waters. It is commonly caused by a monomolecular layer of organic material that has the effect of reducing surface tension.

slickenfibers Slickensides composed of fibers of secondary minerals deposited in the lee sides of asperities, steps, etc. on a fault surface, generally in the direction of the slip vector. Syn: *slip-fiber lineation.*

slickenline A lineation on a slickenside defined by grooves, ridges, and/or striations. Generally parallel to the direction of slip.

slickens (slick'-ens) Extremely fine-grained material, such as finely pulverized tailings discharged from hydraulic mines or a thin layer of extremely fine silt deposited by a stream during flood.

slickenside (slick'-en-side) Originally, a *polished fault surface* formed by frictional wear during sliding, but now used to denote any of several types of lineated fault surfaces. Slickensides commonly have groove lineations, or alternatively may have fibrous minerals; in both cases, the lineations indicate the direction of slip on the fault. Slickensides are also common below 50 cm in swelling clays subject to large changes in water content.

slickolite (slick'-o-lite) A field term proposed by Bretz (1940, p.338) for a vertically discontinuous striation produced by slippage and shearing and developed on strongly dipping bedding planes of limestone that forms the molding on the wall of a solution cavity. The structure resembles a slickenside but shows evidence of some dissolution and the development of incipient asymmetric stylolites oriented parallel to bedding.

slick spots Areas having a puddled or crusted, very smooth, nearly impervious surface. The underlying material is dense and massive.

slide [mass move] (a) A mass movement or descent resulting from failure of earth, snow, or rock under shear stress along one or several surfaces that are either visible or may reasonably be inferred; e.g. *landslide; snowslide; rockslide.* The moving mass may or may not be greatly deformed, and movement may be rotational or planar. A slide can result from lateral erosion, lateral pressure, weight of overlying material, accumulation of moisture, earthquakes, expansion owing to freeze-thaw of water in cracks, regional tilting, undermining, and human agencies. (b) The track of bare rock or furrowed earth left by a slide. See also: *landslide track.* (c) The mass of material moved in or deposited by a slide. (d) A shortened form of *landslide.*

slide [struc geol] An obsolete term proposed by Fleuty (1964) for a fault formed in close connection with folding, and conformable with the fold limb or axial surface. It is accompanied by thinning and/or disappearance of the folded beds.

slide cast The cast of a *slide mark,* commonly smoothly curved and less than a meter in length.

slide mark A scratch or groove left on a sedimentary surface by subaqueous sliding or slumping (Kuenen, 1957, p.251); it tends to be wider and shallower than a typical *drag mark,* and may be formed by sliding objects such as a mass of sediment, a plant mat, or a large soft-bodied animal. See also: *slide cast.*

sliding (slid'-ing) *gravitational sliding.*

sliding stone An isolated angular block of stone resting on the floor of a playa, derived from an outcrop near the playa margin, and associated with a trail or mark indicating that the stone recently slid across the playa surface, in some instances as much as 300 m (Sharp and Carey, 1976). Syn: *skid boulder.*

slieve An Irish term for a mountain.

slikke (slik'-ke) A Dutch term used also in France for a *tidal flat* or a *mud flat,* esp. one rich in decaying organic matter mixed with sand and crossed by tidal channels (Termier and Termier, 1963, p.414).

slim hole (a) A rotary borehole having a diameter of 5 in. or less. (b) A drill hole of the smallest practicable size, often drilled with a truck-mounted rig, used for mineral exploration, as a *stratigraphic test* or *structure test,* or monitor well in hard rock.

slingram A frequency-domain electromagnetic method in which coplanar (usually horizontal) transmitter and receiver coils separated by a constant distance are moved over a survey area. The primary field is nulled and in-phase and quadrature components are measured. Swedish for horizontal-loop method.

slip [clay] A suspension of fine clay in water, having the consistency of cream, and used in slip casting or the decoration of ceramic ware. See also: *slip clay.*

slip [coast] (a) An extension of navigable water into the space between adjacent structures (as piers) within which vessels can be berthed; a dock. (b) A sloping ramp extending into the water and serving as a landing place for vessels.

slip [geomorph] A narrow mountain pass; a defile.

slip [struc geol] (a) The relative displacement of formerly adjacent points on opposite sides of a fault, measured in the fault surface. Partial syn: *shift.* Syn: *total displacement.* (b) An individual increment of cumulative displacement. (c) Obsolete term for a small fracture along which there has been some displacement.

slipband (slip'-band) A planar region of microscopic size, within and parallel to which intracrystalline slip has occurred. Cf: *deformation lamella.*

slip bedding A term used by Ksiazkiewicz (1949) for *convolute bedding* supposedly produced by subaqueous sliding.

slip block A separate rock mass that has "slid away from its original position and come to rest some way down the slope without being much deformed" (Kuenen, 1948, p.371).

slip clay An easily fusible clay containing a high percentage of fluxing impurities, used to produce a natural glaze on the surface of clayware. See also: *slip [clay].*

slip cleavage An obsolete term for cleavage whose formation supposedly involved shear on cleavage planes. The term should be avoided. Syn: *strain-slip cleavage.* See also: *crenulation cleavage; spaced cleavage.*

slip face (a) The steeply sloping surface on the lee side of a dune, standing at or near the angle of repose of loose sand, and advancing downwind by a succession of slides wherever that angle is exceeded. Syn: *sandfall.* (b) The leeward surface of a sand wave, exhibiting foreset bedding. Syn: *slip slope.* Also spelled: slipface.

slip fiber Veins of fibrous minerals, esp. asbestos, in which the fibers are more or less parallel to slickensided vein walls. Cf: *cross fiber.*

slip-fiber lineation *slickenfibers.*

slip fold A rarely used syn. of *shear fold.*

slip-line field theory The concept, introduced by Tapponnier and Molnar, that collision of a rigid continent into the margin of another continent can generate regional-scale strike-slip faults that propagate away from the collision boundary into the overriding plate. The concept is based on application of well-established theories from soil mechanics to the pattern of faults in Asia whose activity may be related to the collision of India with Asia.

slip-line method A method for determining the shear sense on a fault or shear zone by measuring the relationship between vergence and plunge direction of folds formed within the zone (Hansen, 1967).

slip-mark A term proposed by Challinor (1978) for the markings on a rock surface made by the movement over it of another rock mass, but of a type other than *slickenside.* Cf: *groove [struc geol]; striation; chattermark [struc geol]; mullion structure.* Syn: *slip-scratch.*

slip-off slope The long, low, relatively gentle slope on the inside of a stream meander, produced on the downstream face of the meander spur by the gradual outward migration of the meander as a whole; located opposite to the *cutbank.*

slip-off slope terrace A local terrace on the *slip-off slope* of a meander spur, formed by a brief halt during the irregular incision by a meandering stream.

slip plane [mass move] A planar *slip surface.*

slip plane [struc petrol] *glide plane [struc petrol].*

slip-scratch *slip-mark.*

slip sheet A stratum or rock unit on the limb of an anticline that, having become fractured at its base, has slid down and away from the anticline. It is a gravity-collapse structure.

slip slope *slip face (b).*

slip surface A landslide displacement surface, often slickensided, striated, and subplanar. It is best exhibited in argillaceous materials and in those materials which are highly susceptible to clay alteration when granulated. See also: *slip plane [mass move].* Syn: *landslide shear surface; gliding surface; shear surface.*

slip-vector analysis (slip'-vec-tor) Determination of the movement on a fault from a study of the first motion from an earthquake. The first motion appears as a compression or a rarefaction, depending on the location of the observing station with respect to the epicenter and the direction of motion involved in the earthquake. See also: *fault-plane solution.*

slit A parallel-sided re-entrant in the outer lip of the aperture of a gastropod shell, ranging from a shallow incision to a deep fissure as much as half a whorl in extent (TIP, 1960, pt. I, p.133). Cf: *sinus [paleont].*

slither Loose rubble or talus; angular debris.

sliver *sliver polygon.*

sliver polygon A polygon created in an overlay process of two vector graphic files when boundaries are not in perfect registration. Such polygons are usually regarded as artifacts of the overlay process and not legitimate data. The generation of sliver polygons is often attributed to errors in the polygon boundaries of the original files (Krzanowski et al., 1993). Syn: *sliver.*

slob (a) A dense accumulation of heavy *sludge* of sea ice. Also called: slob ice. (b) A term used in Newfoundland for soft snow or mushy ice.

slob land A term used in Ireland for muddy ground or soil, or for a level tract of muddy ground; esp. alluvial land that has been reclaimed.

sloot *sluit.*

slope [geomorph] (a) gradient. (b) The inclined surface of any part of the Earth's surface, as a *hillslope;* also, a broad part of a continent descending toward an ocean, as the Pacific slope.

slope [streams] *stream gradient.*

slope correction *grade correction.*

slope curvature The rate of change of angle of slope with distance.

slope-discharge curve (slope'-dis-charge) A graphic presentation of the discharge at a given gaging station, taking into account the slope of the water surface as well as the gage height.

slope element A curved part of a slope profile; a smooth concave or convex area of a slope or portion of the slope profile. Cf: *slope segment.*

slope facet *slope segment.*

slope failure [eng] The gradual or rapid downslope movement of soil or rock in a constructed cut or fill slope under gravitational stress, contrary to the way it was designed. Cf: *slope movement.*

slope failure [mass move] Gradual or rapid downslope movement of soil or rock under gravitational stress, often as a result of man-caused factors, e.g. removal of material from the base of a slope.

slope fans In sequence stratigraphy, fans made up of turbidite-leveed channel and overbank deposits. They overlie the basin-floor fan and are downlapped by the overlying lowstand wedge (Posamentier and Vail, 1988).

slope gully A small, discontinuous submarine valley, usually formed by slumping along a fault scarp or the slope of a river delta. Syn: *sea gully.*

slope length (a) The ground-surface distance, measured along the line of steepest descent, from the top to the base of a hillslope or slope unit. (b) The linear distance from the divide to the stream channel at the base of a *valley-side slope.* (c) *length of overland flow.*

slope line An imaginary line on the ground surface indicating the direction of steepest gradient at a given point, and therefore intersecting the contour lines at right angles (Warntz, 1975, p.211).

slope map A map that shows the distribution of the degree of surface inclination; e.g. *isotangent map* and *isosinal map.*

slope movement In engineering geology, the gradual or rapid downslope movement of soil or rock in a natural slope under gravitational stress. Cf: *slope failure [eng].*

slope-onlap surface In sequence stratigraphy, a surface that is on a submarine slope on which there is *onlap.* There may or may not be a significant depositional hiatus associated with the surface (Embry and Johannessen, 1992).

slope retreat The progressive recession of a scarp or of the side of a hill or mountain; suggested causes include backwearing (*parallel retreat of slope*) and downwearing. Cf: *scarp retreat.*

slope sector The part of a slope element on which the slope curvature remains constant.

slope segment A rectilinear part of a slope or slope profile. Cf: *slope element.* Syn: *slope facet.*

slope sequence The part of a hillside surface that consists, in succession, of a waxing slope, a maximum slope segment, and a waning slope.

slope stability The tendency of soil or rock material comprising a natural or constructed slope to remain in place or to perform within acceptable limits. Cf: *bank stability.*

slope unit (a) The smallest feature of a slope profile, consisting either of a slope element or of a slope segment. (b) A system consisting of a base level of denudation and the correlated slope lying above it (Penck, 1953, p.129). Slope units are separated by breaks of gradient.

slope wash (a) Soil and rock material that is or has been transported down a slope by mass wasting assisted by running water not confined to channels. Cf: *colluvium.* (b) The process by which slope-wash material is moved; specif. *sheet erosion.*

slough [drill] (Pron: sluff) Fragmentary rock material that has crumbled and fallen away from the sides of a *borehole* or mine working. It may obstruct a borehole or be washed out during circulation of the drilling mud.

slough [geog] (Pron: sloo) (a) A small *marsh;* esp. a marshy tract lying in a swale or other local shallow undrained depression on a piece of dry land, as on the prairies of the Midwest U.S. Also, a dry depression that becomes marshy or filled with water. (b) A large *wetland,* as a swamp; e.g., in the Everglades of Florida. (c) A term used esp. in the Mississippi Valley for a creek or sluggish body of water in a tidal flat, bottomland, or coastal marshland. (d) A sluggish channel of water, such as a side channel of a river, in which water flows slowly through low, swampy ground, as along the Columbia River, or a section of an abandoned river channel, containing stagnant water and occurring in a flood plain or delta. Also, an indefinite term indicating a small lake, a marshy or reedy pool or inlet, a bayou, a pond, or a small and narrow backwater. (e) A small bay in eastern England. (f) A piece of soft, miry, muddy, or waterlogged ground; a place of deep mud, as a mudhole; or an obsolete syn. of *mire.*

slough [mass move] *sluff.*

slowness The reciprocal of the *phase velocity.* Slowness p is related to wave number κ by $\kappa = \omega p$, where ω is angular frequency.

slow ray In crystal optics, that component of light in any birefringent crystal section that travels with the lesser velocity and has the higher index of refraction. Cf: *fast ray.*

slud (a) The muddy material that has moved downslope by solifluction. (b) Ground that behaves as a viscous fluid, including material moved by solifluction as well as by mechanisms not limited to gravitational flow (Muller, 1947, p.221). Etymol: a provincial English word for a soft, wet, slippery mass, as mud or mire.

sludge [ice] A dense, soupy accumulation of new sea ice formed during an early stage of freezing, and consisting of incoherent floating *frazil crystals* that may or may not be slightly frozen together. It forms a thin gluey layer and gives the sea surface a steel-gray or leaden color. Cf: *ice gruel.* Syn: *sludge ice; slush [ice].*

sludge [sed] (a) A soft, soupy or muddy bottom deposit, such as found on *tideland* or in a stream bed; specif. black ooze on the bottom of a lake (Twenhofel, 1937, p.90). (b) A semifluid, slushy, murky mass of sediment and biomass resulting from treatment of water, sewage, or industrial and mining wastes, and often appearing as local bottom deposits in polluted bodies of water such as estuaries. (c) Used to describe the weathered material that commonly mantles an unconsolidated outcrop of, e.g. till or clay.

sludge cake An accumulation of hardened *sludge* (*sea ice*) strong enough to bear the weight of a person. Cf: *sludge floe.*

sludge cast *furrow flute cast.*

sludge floe A large *sludge cake.*

sludge ice *sludge [ice].*

sludge lump An irregular mass of *sludge* (*sea ice*) shaped by strong wind action.

sludging (sludg'-ing) *solifluction.*

sluff Small avalanches, commonly only referring to small, loose snow avalanches. Syn: *slough.*

slug flow Movement of an isolated body of water, such as *free water* moving downward in the unsaturated zone. The term is based on slang for a small amount of liquid, such as a slug of whiskey.

sluggish (slug'-gish) Said of a stream in which the peaks of a flood form slowly because of the decrease in the slope as the age of the stream system advances, or as the flow is reduced or retarded by withdrawal or storage in upstream reaches.

slug test An *aquifer test* that is conducted either by pouring a small volume, or slug, of water into a well or by withdrawing a small volume from a well and measuring the subsequent decline or rise of water level as a function of time. The introduction or withdrawal of water is considered to be instantaneous. Cf: *bail-down test.*

sluice (a) A conduit or passage for carrying off surplus water, often at high velocity. It may be fitted with a valve or gate for stopping or regulating the flow. (b) The abrupt release of water into a channel that normally is dry or carries little flow. (c) A body of water flowing through or stored behind a floodgate.

sluiceway (sluice'-way) *overflow channel.*

sluicing (sluic'-ing) Concentrating heavy minerals, e.g., gold or cassiterite, by washing unconsolidated material through boxes (sluices) equipped with riffles that trap the heavier minerals on the floor of the box.

sluit An African term for a narrow, usually dry ditch, gully, or gulch, produced by the washing of heavy rains in a large natural fissure; it is shallower than a ravine. Also, a similar watercourse produced artificially for irrigation or drainage. Etymol: Afrikaans, from Dutch "sloot", "ditch". Also spelled: *sloot.*

slump (a) A landslide characterized by a shearing and rotary movement of a generally independent mass of rock or earth along a curved slip surface (concave upward) and about an axis parallel to the slope from which it descends, and by backward tilting of the mass with respect to that slope so that the slump surface often exhibits a reversed slope facing uphill. Syn: *slumping.* (b) The sliding-down of a mass of sediment shortly after its deposition on an underwater slope; esp. the downslope flowage of soft, unconsolidated marine sediments, as at the head or along the side of a submarine canyon. This is the "commonest usage in geology in Britain", although "subaqueous slump would be more precise" (Challinor, 1978, p.283). Syn: *subaqueous gliding.* (c) The mass of material slipped down during, or produced by, a slump. See also: *slump block.*

slump ball A relatively flattened mass of sandstone resembling a large concretion, measuring 2 cm to 3 m across, commonly thinly laminated with internal contortions and a smooth or lumpy external form, and formed by subaqueous slumping (Kuenen, 1948, p.369). Cf: *crumpled ball; spiral ball.*

slump basin A shallow basin near the base of a canyon wall and on a shale hill or ridge, formed by small, irregular slumps, and usually containing a short-lived lake (Worcester, 1939).

slump bedding A term applied loosely to any disturbed bedding; specif. deformed bedding produced by subaqueous slumping or lateral movement of newly deposited sediment.

slump block The mass of material torn away as a coherent unit during *slumping.* It may be 2 km long and as thick as 300 m.

slump breccia A contorted sedimentary bed produced by slumping and exhibiting brecciation.

slump fold An intraformational fold produced by slumping of soft sediments.

slumping (slump'-ing) The downward movement, such as sliding or settling, of a slump. Syn: *slump.*

slump mark A mark made by sand (wet or dry) avalanching down the *lee* side of a *sand wave* or *dune.*

slump overfold A fold consisting of hook-shaped masses of sandstone produced during slumping (Crowell, 1957, p.998).

slump scarp A low cliff or rim of thin solidified lava occurring along the margins of a lava flow and against the valley walls, or around steptoes, after the central part of the lava crust collapsed due to outflow of still molten underlying layers; the inward-facing cliff may be several meters high. Term introduced by Finch (1933), but Sharpe (1938, p.70) prefers "lava subsidence scarp" or "lava slump scarp".

slump sheet A well-defined bed of limited thickness and wide horizontal extent, containing slump structures (Kuenen, 1948, p.373).

slump structure A generic term for any sedimentary structure produced by subaqueous slumping.

slurried bed (slur'-ried) A bed of muddy and sandy or pebbly material that is unsorted and ungraded except at the base, where there is commonly a thin graded layer (Wood and Smith, 1958, p.173). It is thought to form by slumping and fragmentation of partially consolidated sediments, and may move downslope as a subaqueous debris flow. Syn: *slurry bedding.*

slurry (slur'-ry) (a) A highly fluid mixture of water and finely divided material, e.g. a naturally occurring muddy lake-bottom deposit, or a manufactured mixture of pulverized coal and water, as for movement by pipeline. (b) *grout.*

slurry bedding *slurried bed.*

slurry slump A slump in which the incoherent sliding mass is mixed with water and disintegrates into a quasi-liquid slurry (Dzulynski and Slaczka, 1958, p.217).

slush [geog] A soft mud; *mire.*

slush [ice] *sludge [ice].*

slush [snow] Snow saturated with water, occurring on land or ice surfaces, or as a viscous floating mass in water after a heavy snowfall. Syn: *snow slush.*

slush avalanche *wet-snow avalanche.*
slush field *snow swamp.*
slushflow (slush'-flow) (a) A mudflowlike outburst of water-saturated snow along a stream course, commonly occurring in the Arctic after intense thawing produces more meltwater than can drain through the snow, and having a width generally several times greater than that of the stream channel (Washburn and Goldthwait, 1958). (b) A flow of clear slush on a glacier, as in Greenland.
slush pit *mud pit.*
slush pond A pool or lake, containing slush, on the ablation surface of a glacier. It is especially common during summer.
slush pump *mud pump.*
small boulder A *boulder* having a diameter in the range of 256-512 mm (10-20 in., or -8 to -9 phi units).
small circle A curve formed on the surface of a sphere by the intersection of any plane that does not pass through the center of the sphere; specif. a circle on the Earth's surface, the plane of which does not pass through the center of the Earth, such as any parallel of latitude other than the equator. Cf: *great circle.*
small-circle girdle (small-cir'-cle) *cleft girdle.*
small cobble A geologic term for a cobble having a diameter in the range of 64-128 mm (2.5-5 in., or -6 to -7 phi units).
smaller foraminifera (small'-er) An informal term generally used to designate those foraminifers that are studied without the aid of thin-sectioning. Cf: *larger foraminifera.*
small-scale map A map at a scale (in the U. S. Geological Survey, of 1:250,000 or smaller) such that a large area can be covered showing only generalized detail. The Defense Mapping Agency defines a small-scale map as one at 1:600,000 or smaller (Thompson, 1988, p.15). Cf: *intermediate-scale map; large-scale map.*
small spore An obsolete term for which *miospore* is a preferable near-synonym.
small watershed A drainage basin that is "so small that its sensitivities to high-intensity rainfalls of short durations and to land use are not suppressed by the channel-storage characteristics" and in which "the effect of overland flow rather than the effect of channel flow is a dominating factor affecting the peak runoff" (Chow, 1957, p.379); its size may be a few acres to 1000 acres, or even up to 130 sq km (50 sq mi).
smaltite (smalt'-ite) (a) *skutterudite.* Syn: *tin-white cobalt; gray cobalt; white cobalt; speisscobalt.* (b) A term applied to undetermined, apparently cubic arsenides of cobalt or to a mixture of cobalt minerals.
smaragd (sma'-ragd) *emerald.*
smaragdite (sma-rag'-dite) A fibrous or thinly foliated green amphibole (near actinolite) pseudomorphous after pyroxene (such as omphacite) in rocks such as eclogite.
smear slide A means of studying the mineral and microfossil constituents of silt with a polarizing microscope. A bit of sediment is placed on a glass slide, dispersed with water, dried on a hotplate, and covered with canada balsam and a glass cover slip.
smectite (smec'-tite) The accepted group name for 2:1 phyllosilicate clay minerals with layer charge between 0.2 and 0.6 per formula unit and which can take polar liquids into the interlayer space causing them to swell in a direction perpendicular to the 001 surface. They have a high cation exchange capacity, about 110 $cmol_c\ kg^{-1}$ for soil smectites, and variable interlayer spacing. The smectite minerals are derived from the alteration of volcanic glass and from the weathering of primary silicates. They are the chief constituents of bentonites and fuller's earth and are common in soils, sedimentary rocks, and some mineral deposits. Cf: *beidellite; montmorillonite; nontronite; saponite.*
smectite-to-illite transition The first carefully documented change of one clay mineral towards another. This transition can be used as an interpretive tool. The mechanism of transition is apparently transformation in some cases and neoformation in other cases. Hower, et al. (1976), who referred to the process as burial metamorphism, documented this change for U.S. Gulf Coast sediments. Inoue et al. (1987) documented it for hydrothermal conditions. Cf: *illite/smectite.*
smirnite (smir'-nite) A colorless, light gray, or light yellow orthorhombic mineral: $Bi_2Te^{4+}O_5$.
smirnovite (smir'-nov-ite) *thorutite.*
smirnovskite A dubious hexagonal mineral: $(Th,Ca)PO_4{\cdot}nH_2O$. It may be brockite, and has not been approved by the International Mineralogical Association.
Smithian (Smith'-i-an) Canadian substage and suggested lower substage of the Olenekian Stage: Lower Triassic (above Dienerian, below Spathian).
smithite (smith'-ite) An adamantine red monoclinic mineral: $AgAsS_2$.
smithsonite (smith'-son-ite) (a) A white to yellow, gray, brown, or greenish mineral of the *calcite* group: $ZnCO_3$. It is a secondary mineral associated with sphalerite and often found as a replacement in limestone; it is commonly reniform, botryoidal, stalactitic, or granular, and is distinguished from hemimorphite by its effervescence with acids. Smithsonite is an ore of zinc. Syn: *dry-bone ore; calamine; zinc spar; szaskaite.* (b) A term sometimes used as a syn. of *hemimorphite.*
smokeless coal (smoke'-less) (a) *semibituminous coal.* (b) Any coal that burns without smoke, from semibituminous to superanthracitic.
smokestone (smoke'-stone) *smoky quartz.*
smoking crest The crest of a dune, along which sand grains are being winnowed.
smoky quartz (smok'-y) A smoky-yellow to brown-gray and often transparent variety of crystalline quartz sometimes used as a semiprecious gemstone. It may contain inclusions of both liquid and gaseous carbon dioxide. The color is probably due to exposure to high-energy radiation, such as produced by radioactive minerals. Syn: *cairngorm; smokestone.*
smoky topaz A trade name for smoky quartz used for jewelry.
smolianinovite (smo'-li-a-ni'-nov-ite) A yellow orthorhombic mineral: $(Co,Ni,Mg,Ca)_3(Fe^{3+},Al)_2(AsO_4)_4{\cdot}11H_2O$. It is the Co analogue of fahleite.
smooth chert Hard, dense, homogeneous chert (as seen in insoluble residues), characterized by a conchoidal to even fracture surface that is devoid of roughness and by a lack of crystallinity, granularity, or other distinctive structure (Ireland et al., 1947, p.1484). See also: *chalcedonic chert; ordinary chert; porcelaneous chert.* Cf: *granular chert; chalky chert.*
smooth cleavage A *disjunctive cleavage* characterized by well-developed, continuous, planar, but spaced cleavage domains consisting of oriented platy minerals. Typical of the cleavage in fine-grained rocks such as shale or slate. Cf: *rough cleavage.*
smooth criterion In radar, the relationship between surface roughness, depression angle, and wavelength that determines whether a surface will scatter the incident radar pulse in smooth or intermediate fashion. Cf: *rough criterion; Rayleigh criterion.*
smoothing In *time series* analysis, the removal of minor fluctuations or erratic values from a series of observations, by replacing each value with a weighted *moving average* of surrounding values. An extension of this process is used in *computer contouring* to subdue erratic values in a grid.
smooth phase The part of stream traction wherein a mass of sediment travels as a sheet with gradually increasing density from the surface downward (Gilbert, 1914, p.30-34). Cf: *dune phase; antidune phase.*
smothered bottom (smoth'-ered) A term introduced by Shrock (1948, p.307-308) for a sedimentary surface on which complete, well-preserved, and commonly fragile and delicate fossils were saved by an influx of mud that buried them instantly. Such surfaces are common in sequences composed of alternating marine limestone and shale layers.
smrkovecite A white or yellow monoclinic mineral of the *atelestite* group: $Bi_2O(OH)PO_4$.
SMTP Simple mail transfer protocol.
SmX *synchrotron micro-XANES.*
smythite (smyth'-ite) A metallic bronze-yellow rhombohedral mineral: $(Fe,Ni)_{3+x}S_4$ (x = 0-0.3). It is related to pyrrhotite.
Sn A seismic phase that is a P wave, of velocities 7.6-8.4 km/s, associated with propagation in the uppermost mantle. Also, the velocity of that wave.
S/N *signal-to-noise ratio.*
snake hole A horizontal or nearly horizontal *borehole* used for blasting, drilled approximately on a level with the floor of a quarry or under a boulder to be broken up.
snaking stream *meandering stream.*
snapper (snap'-per) *grab sampler.*
SNC meteorite A group of igneous achondrites belonging to the shergottite, nahklite, and chassignite groups, thought by many to be from the planet Mars. Abbrev. for Shergottite-Nahklite-Chassignite.
Snell's law *law of refraction.*
snout [geog] A protruding mass of rock; a promontory.
snout [glaciol] The protruding lower extremity, leading edge, or front of a glacier. Syn: *terminal face; terminus; glacier snout; ice front; front [glaciol].*
snout [paleont] The enlarged anterior part of the shell of some rostroconchs, separated from the body of the shell by a *sulcus* and by differences in sculpture (Pojeta and Runnegar, 1976, p.74).
snow (a) A form of ice composed of small white or translucent hexagonal crystals of frozen water, formed directly by condensation of atmospheric water vapor around solid nuclei at a temperature below the freezing point. The crystals grow while floating or falling to the ground, and are often agglomerated into snowflakes. (b) A consolidated mass of fallen snow crystals. (c) *snowfall.* (d) A region covered with permanent snow, in which case the term is usually used in the plural, as "the high snows".
snow avalanche An *avalanche* consisting of relatively pure snow, although considerable earth and rock material may also be carried downward. Cf: *full-depth avalanche.* Syn: *snowslide.*
snowball garnet *rotated garnet.*
snowbank (snow'-bank) A mound of snow, often the remnant of a snowdrift.
snowbank glacier *nivation glacier.*
snow banner A stream of snow blowing off a mountain top, streaming out several miles from its source.
snow barchan A small crescentic or horseshoe-shaped *snow dune* with the ends pointing downwind.
snow blanket A surface accumulation of snow.
snowblink (snow'-blink) A bright, white glare in the sky near the horizon or on the underside of a cloud layer, produced by light reflected from a snow-covered surface (as a snowfield); brighter than *iceblink [meteorol].* Also spelled: snow blink. Syn: *snow sky; snow sheen.*

snow bridge An arch or layer of snow that has drifted across a *crevasse* in a glacier, or a connecting splinter of ice or snow which allows a person or vehicle to cross a crevasse.
snowcap (a) A covering of snow on a mountain peak or ridge when no snow exists at lower elevations. (b) An accumulation of snow on the surface of a frozen lake.
snow cornice *cornice.*
snow course A line or a series of connecting lines of regularly spaced observation stations (usually 10 or more) at which snow samples are taken for measuring depth and density; the *water equivalent* is then calculated for forecasting subsequent runoff. Cf: *snow survey.*
snow cover (a) All snow that has accumulated on the ground, including that derived from snowfall, drifting or blowing snow, avalanches, frozen or unfrozen rain stored in the snow, rime, and frost. (b) The areal extent of ground partly or wholly covered with snow in a particular area, usually expressed as a percent of the total area. (c) The average depth of accumulated deposited snow on the ground in a particular area, usually expressed in centimeters.
snowcreep (snow'-creep) The slow internal deformation of a *snowpack* resulting from the stress of its own weight and metamorphism of snow crystals; it usually involves shear parallel to the slope and compaction perpendicular to the slope.
snow crust A firm or hard surface of snow overlying a layer of softer snow; it may be formed by the melting and refreezing of surface snow, by wind action, or by freezing of water on the surface. Cf: *ice crust [glaciol]; wind crust; sun crust; rain crust.* Syn: *crust [snow].*
snow crystal (a) A single ice crystal nucleated and grown in the atmosphere. (b) A single crystal of deposited snow. Cf: *snowflake.*
snow cushion *snow pillow (b).*
snow density The mass of snow per unit volume, usually given as Mg/m^3. In these units it is numerically equal to specific gravity.
snowdrift (snow'-drift) Snow deposited by the wind in the lee of an obstacle or in other places where turbulent eddies are formed.
snowdrift glacier *drift glacier.*
snow dune Snow deposited by the wind as a gently rounded hummock, usually with its long axis parallel with the wind direction. A dune is much larger than ripples, sastrugi, or snow barchans.
snowfall (snow'-fall) (a) The deposition, on the ground or other surface, of snow precipitated out in the atmosphere. (b) The rate of deposition of snow.
snowfield (snow'-field) (a) A broad expanse of terrain covered with snow, relatively smooth and uniform in appearance, occurring usually at high latitudes or in mountainous regions above the snowline, and persisting throughout the year. (b) A region of permanent snow cover, as at the head of a glacier; the *accumulation area* of a glacier.
snowflake (snow'-flake) An aggregation of several single snow crystals that have collided and joined while falling through relatively still air.
snowflake obsidian An *obsidian* that contains white, gray, or reddish spherulites ranging in size from microscopic to a meter or more in diameter.
snow glide The slip of a sloping snow cover along the ground surface, or relative slip between snow layers. Cf: *snowcreep.* Syn: *glide.*
snow grain The smallest characteristic subunit of snow texture recognizable with a hand lens (e.g. 10 x); it can consist of one or more crystals of ice.
snow ice Ice that has been formed when snow slush, a mixture of snow and water, has frozen. It has a whitish appearance if air bubbles are included.
snow layer A stratum of snow that is different in at least one respect from the strata above and below.
snowline (snow'-line) (a) A temporary line delimiting or defining the altitude above which the area is snow covered; in a zone of patchy snow, the altitude above which the land is more than 50 percent covered with snow. Cf: *climatic snowline.* Syn: *transient snowline.* (b) The line separating areas in which deposited snow disappears in summer from areas in which it remains throughout the year. Cf: *regional snowline; equilibrium line.* (c) The ever-changing limit of the Earth's broad belt within which no snow falls. Its position depends on such physical conditions as altitude and nearness to the sea. The term is applied esp. to the winter snowline in the Northern Hemisphere.
snowmelt The water resulting from the melting of snow. Also spelled: snow melt. Syn: *snow water.*
snow moisture *liquid-water content.*
snow niche *nivation hollow.*
snow-on-roof texture Sphalerite occurring as coarse crystals preferentially coating the tops of fragments in ore-bearing collapse breccias or the bottoms of solution cavities in a *Mississippi Valley-type deposit.* This texture is diagnostic for the deposit type.
snowpack (snow'-pack) (a) Any snow cover. (b) The accumulation of deposited snow layers on the ground. (c) The amount of annual accumulation of snow at higher elevations in mountains, which is important for estimating the water available for human use (e.g., hydroelectric power and irrigation).
snow patch An isolated mass of perennial snow and firn not large enough to be called a glacier.
snow-patch erosion *nivation.*
snow penitente A *nieve penitente* consisting of compacted snow.
snow pillow (a) A device used to record the changing weight of the snow cover at a point. It consists of a fluid-filled bladder or metal container lying on the ground, the internal pressure of which measures the weight of overlying snow. (b) A snowdrift on a slope commonly within the starting zone of an avalanche. Syn: *snow cushion.*
snowquake (snow'-quake) The sudden collapse of one or more layers of surface or subsurface snow, often accompanied by a sound that may resemble that of a distant explosion (ADTIC, 1955, p.76). Syn: *snow tremor.*
snow roller A cylindrical mass of moist, cohesive snow, formed by rolling down a slope. It may be as large as 1 m in length and more than 2 m in circumference. Cf: *sun ball.*
snow sampler A snow-surveying instrument in the form of a tube, used to collect a core of deposited snow which is weighed to determine density.
snowshed (snow'-shed) A drainage basin primarily supplied by snowmelt.
snow sheen *snowblink.*
snow sky *snowblink.*
snowslide (snow'-slide) A *snow avalanche.* The term has also been used for a smaller mass of downward-moving snow.
snow slush *slush [snow].*
snow survey The process of determining the depth, density, and water equivalent of snow that has fallen on a particular area by sampling representative points along a *snow course.* Snow surveys made in the spring are used for forecasting subsequent runoff from snowmelt.
snow swamp An area of water-saturated snow having a soupy consistency, in which men, animals, and equipment readily sink. If saturated to the surface, the snow is bluish; if not, it is similar in color to the surrounding snow. Syn: *slush field.*
snow tremor *snowquake.*
snow water *snowmelt.*
snub-scar A term used by Wentworth (1936) for a "push-off" end or edge (or lee-end pressure spall) characteristic of a glacial cobble.
soak *water track.*
soakaway (soak'-a-way) A British term for a sink or depression in the Earth's surface, into which waters flow and drain away.
soaked zone *wet-snow zone.*
soap clay *bentonite.*
soap earth Massive talc; *steatite.*
soap hole A term used in Wyoming for a hole resulting from the wetting of the outcrop surface of bentonite.
soaprock (soap'-rock) (a) *soapstone [mineral].* (b) *soapstone [rock].*
soapstone [mineral] (soap'-stone) (a) A mineral name applied to *steatite,* or massive talc. Syn: *soaprock.* (b) *saponite.* (c) A term loosely applied to much agalmatolite.
soapstone [rock] (a) A metamorphic rock of massive, schistose, or interlaced fibrous or flaky texture and soft, unctuous feel, composed essentially of talc with varying amounts of micas, chlorite, amphibole, pyroxenes, etc. and derived from the alteration of ferromagnesian silicate minerals. It may be sawed into laboratory-bench tops, switchboards, and other types of special-purpose dimension stone. (b) A miner's and driller's term for any soft, unctuous rock such as micaceous shale or sericitic schist.
soapy (soap'-y) Said of a type of mineral texture that is slippery, smooth, and soft; e.g. that of talc. Syn:*unctuous.*
sobolevite (so'-bo-lev-ite) A brown triclinic mineral: $Na_{11}(Na,Ca)_4(Mg,Mn^{2+})Ti_4(Si_2O_7)_2(PO_4)_4O_3F_3$.
sobolevskite (so'-bo-lev'-skite) A metallic monoclinic mineral: PdBi. It was originally described as hexagonal.
sobotkite (so-bot'-kite) *saponite.*
socket (sock'-et) A recess or depression along the hinge line of a bivalve for the reception of a projecting hinge tooth from the opposite valve; esp. a *dental socket* of a brachiopod.
socket ridge A linear elevation of secondary shell extending laterally from the cardinal process of a brachiopod and bounding the margin of dental sockets.
soda (so'-da) Sodium carbonate, Na_2CO_3; especially the decahydrate, $Na_2CO_3{\cdot}10H_2O$. Loosely used for sodium oxide, sodium hydroxide, sodium bicarbonate, and even for sodium in informal expressions such as *soda spar.*
soda alum *sodium alum.*
soda ash Commercial term for sodium carbonate, Na_2CO_3. Prepared in processing plants from the mineral *trona* and from lake brines, or by synthesis from salt and limestone.
sodaclase (so'-da-clase) *albite.*
soda feldspar A misnomer for *sodium feldspar.*
soda hornblende *arfvedsonite.*
soda lake An *alkali lake* whose waters contain a high content of dissolved sodium salts, chiefly sodium carbonate accompanied by the chloride and the sulfate. Examples occur in Mexico and Nevada. Syn: *natron lake.*
soda leucite A hypothetical sodium-rich variety of leucite, postulated as the original material of some pseudoleucites.
soda-lime feldspar A misnomer for *sodium-calcium feldspar.*
sodalite (so'-da-lite) (a) A cubic mineral of the *feldspathoid* group: $Na_4Al_3Si_3O_{12}Cl$. It is usually blue or blue-violet, but may be white, greenish, gray, pink, or yellow, and it occurs in various sodium-rich igneous

rocks. (b) A group of bluish feldspathoid minerals containing sodium silicate, including sodalite, hauyne, nosean, and lazurite.
sodalite diorite In the *IUGS classification,* the variety of *foid diorite* in which sodalite is the dominant feldspathoid.
sodalite monzosyenite In the *IUGS classification,* the variety of *foid monzosyenite* in which sodalite is the dominant feldspathoid. Syn: *sodalite plagisyenite.*
sodalite plagisyenite In the *IUGS classification,* the variety of *foid plagisyenite* in which sodalite is the dominant feldspathoid. Syn: *sodalite monzosyenite.*
sodalithite (so-da-lith'-ite) An igneous rock in which sodalite is the only light-colored mineral present. The term was introduced by Ussing in 1911, for layered cumulus rocks of the Ilimaussaq complex, SW Greenland. Cf: *naujaite.* Obsolete.
soda mica *paragonite.*
soda microcline A variety of microcline in which sodium replaces potassium; specif. *anorthoclase.*
soda minette An alkalic *minette* named by Brögger in 1898, containing alkali feldspar (specif. cryptoperthite), dark-brown mica, aegirine, apatite, and titanite. This dike rock is more accurately called mafic *trachyte.* Obsolete.
soda niter *nitratine.*
soda orthoclase *loxoclase.*
soda sanidine A mineral of the alkali *feldspar* group, containing 40-60% albite in solid solution.
soda spar An informal term for sodic feldspar, i.e. *albite,* or for a feldspar mixture assaying at least 7% Na_2O (Rogers and Neal, 1975, p.638). Syn: *Na-spar.* Cf: *potash spar.*
soda straw A *tubular stalactite* that maintains the diameter of a drop of water and resembles a drinking straw in appearance. Syn: *straw stalactite.*
soda tremolite *richterite.*
soddyite (sod'-dy-ite) A greenish-yellow, yellow, or amber orthorhombic mineral: $(UO_2)_2SiO_4{\cdot}2H_2O$.
sodicanthophyllite A name for the theoretical Na-Mg anthophyllite end-member of the orthorhombic *amphibole* group:
$(Na,[])Mg_7(Si,Al)_8O_{22}(OH)_2$.
sodication The process whereby the exchangeable sodium content of a soil is increased.
sodic-ferri-clinoferroholmquistite A vitreous black monoclinic mineral of the *amphibole* group: $Na_{0.5}Li_2(Fe^{3+},Fe^{2+},Mg,Li,Al)_5Si_8O_{22}(OH,F)_2$. It represents clinoholmquistite with Na and Fe^{3+} dominant. Originally called ferri-clinoholmquistite.
sodic-ferri-ferropedrizite A vitreous black monoclinic mineral of the *amphibole* group: $(Na,[])Li_2(Fe^{3+},Fe^{2+},Mg,Li,Al)_5Si_8O_{22}(OH,F)_2$. It represents ferropedrizite with essential Na and Fe^{3+}.
sodic-ferripedrizite A vitreous green monoclinic mineral of the *amphibole* group: $NaLi_2Fe^{3+}{}_2(Mg_2Li)Si_8O_{22}(OH)_2$. It represents ferripedrizite with essential Na.
sodic-ferro-anthophyllite A name for the theoretical Na-Fe^{2+} anthophyllite end-member of the orthorhombic *amphibole* group:
$(Na,[])Fe^{2+}{}_7(Si,Al)_8O_{22}(OH)_2$.
sodic-ferrogedrite A name for the theoretical Na-Fe^{2+} gedrite end-member of the orthorhombic *amphibole* group: $Na(Fe^{2+},Mg)_5Al_2(Si_6Al_2)O_{22}(OH)_2$.
sodic-ferropedrizite A monoclinic mineral of the *amphibole* group: $NaLi_2(LiFe^{2+}{}_2Fe^{3+}Al)Si_8O_{22}(OH)_2$. It represents ferropedrizite with essential Na.
sodicgedrite A monoclinic member of the *amphibole* group:
$NaMg_5Al_2(Si_6Al_2)O_{22}(OH)_2$. It represents gedrite with essential Na.
sodicpedrizite A monoclinic member of the *amphibole* group: $NaLi_2(LiMg_2Fe^{3+}Al)Si_8O_{22}(OH)_2$. It represents pedrizite with essential Na.
sodic soil (so'-dic) A nonsaline soil containing sufficient exchangeable sodium to adversely affect crop production and soil structure under most conditions of soil and plant type. The sodium adsorption ratio of the saturation extract is at least 13.
sodium alum (so'-di-um) A colorless cubic mineral of the *alum* group: $Na(H_2O)_6Al(H_2O)_6(SO_4)_2$. Syn: *soda alum.* Cf: *mendozite.*
sodium autunite *natroautunite.*
sodium bentonite A *bentonite* with a high sodium to calcium ratio, that can absorb large quantities of water, swelling greatly.
sodium betpakdalite A dull yellow monoclinic mineral:
$(Na,Ca)_3Fe^{3+}{}_2(As_2O_4)(MoO_4)_6{\cdot}15H_2O$. It is the Na analogue of betpakdalite.
sodium boltwoodite A pale yellow orthorhombic mineral:
$(Na,K)(UO_2)(SiO_3OH){\cdot}2H_2O$. It is the Na analogue of boltwoodite.
sodium-calcium feldspar (so'-di-um-cal'-ci-um) A syn. of *plagioclase.* See also: *soda-lime feldspar; lime-soda feldspar.*
sodium feldspar An alkali feldspar containing the Ab molecule $(Na[AlSi_3O_8])$; specif. *albite.* See also: *soda feldspar.*
sodium illite *brammallite.*
sodium-pharmacosiderite A pale green cubic mineral of the *pharmacosiderite* group: $NaFe^{3+}{}_4(OH)_4(AsO_4)_3{\cdot}7H_2O$. It represents the Na analogue of pharmacosiderite.
sodium phlogopite *wonesite; aspidolite.*
sodium-uranospinite A yellow-green to lemon and straw-yellow tetragonal secondary mineral: $(Na_2,Ca)(UO_2)_2(AsO_4)_2{\cdot}5H_2O$. Also called: meta-natrium-uranospinite.
sodium-zippeite (so'-di-um-zip'-pe-ite) A yellow monoclinic mineral of the *zippeite* group: $Na_5(UO_2)_8(SO_4)_4O_5(OH)_3{\cdot}12H_2O$.
sofar (so'-far) A sound channel in the deep ocean that propagates acoustic waves (*channel waves*) for long distances with little attenuation. Etymol: an acronym for *so*und *f*ixing *a*nd *r*anging. Also spelled: SOFAR.
SOFAR (Sound Fixing and Ranging) A system of subsea microphones that record sounds in the *sound channel,* and that use the recording to determine the position of the sound source.
soffione (sof-fi-o'-ne) Steam-type fumaroles; the term was originally applied to boric-acid fumaroles of the Tuscany region of Italy. Etymol: Italian "soffio", "puff" or "blast".
sofiite *sophiite.*
soft coal (a) A syn. of *bituminous coal.* Cf: *hard coal.* (b) Outside the U.S., the term is sometimes used for brown coal and/or lignite, or for any coal with a calorific value of 23.86 MJ/kg (10,260 BTU/lb) or less on a moist, mineral-matter-free basis.
softening (sof'-ten-ing) Reduction of the hardness of water by removing hardness-forming ions (chiefly calcium and magnesium) by precipitation or ion exchange, or sequestering them as by combining them with substances such as certain phosphates, that form soluble but nonionized salts.
soft ground [eng] (a) Ground that is too moist or yielding to support weight and allows an object to sink in. (b) Soil that does not stand well and requires heavy timbering, such as that about an underground opening.
soft ground [mining] That part of an ore deposit that can be mined without drilling and blasting. It is usually the upper, weathered portion of the deposit.
soft hail *graupel.*
soft linkage The *geometric coherence* between faults achieved by ductile strain of the wall rocks, i.e., there is no linkage by faults visible at the scale of observation. A *relay ramp* is formed by soft linkage (Walsh and Watterson, 1991). Cf: *hard linkage.*
soft magnetization Magnetization that is easily destroyed; specifically, remanent magnetization dominated by grains with low coercivity. Cf: *hard magnetization.*
soft mineral A mineral that is softer than quartz, i.e. ranking less than seven on the Mohs scale. Cf: *hard mineral.*
soft ore A term used in the Lake Superior region for an earthy, incoherent iron ore mainly composed of hematite or limonite (goethite) and containing 45-60% iron. Cf: *hard ore.*
soft rock (a) A term used loosely for sedimentary rock, as distinguished from igneous or metamorphic rock. (b) A rock that is relatively susceptible to erosion. (c) Rock that can be removed by air-generated hammers, but cannot be handled economically by pick. (d) A term used loosely by drillers for a post-Cretaceous sedimentary rock that is drilled relatively rapidly and that produces samples difficult to classify as to exact depth. Cf: *hard rock.*
soft-rock geology A colloquial term for geology of sedimentary rocks, as opposed to *hard-rock geology.*
soft shore A shore composed of peat, muck, mud, or soft marl, or of marsh vegetation. Ant: *hard shore.*
soft slab avalanche An avalanche in which the snow pulverizes immediately after avalanche initiation.
soft water Water that lathers readily with ordinary soap; water containing not more than 60 mg/l of hardness-forming constituents expressed as $CaCO_3$ equivalent. Cf: *hard water; hardness [water].*
softwood (soft'-wood) The wood of a gymnospermous tree, lacking vessels and wood fibers. Actually, such wood may be either soft or relatively hard; it tends to be more uniform than the wood of angiosperms. Cf: *hardwood.*
sogdianite (sog'-di-an-ite') A violet hexagonal mineral of the *milarite* group: $(K,Na)_2(Li,Fe^{2+})_3(Zr,Fe^{3+},Ti)_2Si_{12}O_{30}$.
soggendalite (sog'-gen-dal-ite') A dark-colored dolerite containing abundant pyroxene. Named by Kolderup in 1896, for Soggendal, Norway. Obsolete.
sogrenite (so'-gre-nite) A black organic material containing uranium.
söhngeite (söhn'-ge-ite) A light brown cubic mineral: $Ga(OH)_3$. It is the Ga analogue of dzhalindite.
soil [eng] All unconsolidated materials above bedrock. This is the meaning of the term as used by early geologists and in some recent geologic reports, and has been vigorously advocated by Legget (1967, 1973). It is the common usage among engineering geologists (see, for example, *compaction [soil]; soil mechanics*). In recent years the approx. syn. *regolith* has come into wide geological use.
soil [planet] *regolith [planet]; lunar regolith.*
soil [soil] (a) The unconsolidated mineral or organic material on the immediate surface of the earth that serves as a natural medium for the growth of land plants. (b) The unconsolidated mineral or organic matter on the surface of the earth that has been subjected to and shows effect of genetic and environmental factors of: climate (including water and temperature effects), and macro- and microorganisms, conditioned by relief, acting on parent material over a period of time. A product-soil differs from the material from

which it is derived in many physical, chemical, biological, and morphological properties and characteristics.

soil aeration The condition, and sum of all processes affecting, soil pore-space gaseous composition, particularly with respect to the amount and availability of oxygen for use by soil biota and/or soil chemical oxidation reactions.

soil air The soil atmosphere; the gaseous phase of the soil, being that volume not occupied by solid or liquid.

soil amendment Any material such as lime, gypsum, sawdust, compost, animal manures, crop residue or synthetic soil conditioners that is worked into the soil or applied on the surface to enhance plant growth. Amendments may contain important fertilizer elements but the term commonly refers to added materials other than those used primarily as fertilizers.

soil association A kind of map unit used in soil surveys comprised of delineations, each of which shows the size, shape, and location of a landscape unit composed of two or more kinds of component soils or component soils and miscellaneous area, plus allowable inclusions in either case. The individual bodies of component soils an miscellaneous areas are large enough to be delineated at the scale of 1:24 000. Several to numerous bodies of each kind of component soil or miscellaneous area are apt to occur in each delineation and they occur in a fairly repetitive and describable pattern.

soil atmosphere The part of ground air that is in the soil and is similar to the air of the atmosphere but depleted or enriched in certain constituents, such as carbon dioxide. Cf: *subsurface air; ground air.*

soil blister *frost blister.*

soil caliche *caliche [soil].*

soil category A class of taxonomically related soils, defined at the same level of generalization that includes all soils. In the USDA soil taxonomy there are six categories: order, suborder, great group, subgroup, family, series.

soil circle A term used loosely for any circular form of patterned ground developed on a soil surface, either sorted or nonsorted, and with or without vegetation. Syn: *earth circle.*

soil classification The systematic arrangement of soils into groups or categories on the basis of their characteristics. Broad groupings are made on the basis of general characteristics and subdivisions on the basis of more detailed differences in specific properties. The U.S. Department of Agriculture soil classificatin system (Soil Taxonomy) was adapted for use in publications by the National Cooperative Soil Survey on 1 Jan. 1965. Abridged statements of diagnostic features, orders, and suborders are listed alphabetically. Great groups are named by adding a prefix to the suborder name.

soil climate The moisture and temperature of a soil.

soil colloids The inorganic and organic matter in soils having very small particle size and a correspondingly large surface area per unit of mass.

soil complex A kind of map unit used in soil surveys comprised of delineations, each of which shows the size, shape and location of a landscape unit composed of two or more kinds of component soils, or component soils and a miscellaneous area, plus allowable inclusions in either case. The individual bodies of component soils and miscellaneous areas are too small to be delineated at the scale of 1:24 000. Several to numerous bodies of each kind of component soil or the miscellaneous area are apt to occur in each delineation.

soil-cover complex (a) A group of similar areas in which the soils, slopes, plant litter, and vegetation cover have comparable physical characteristics (Chow, 1964, p.22-51). (b) The combination of a specific soil and a specific vegetation cover, used as a parameter in estimating the runoff in a drainage basin (Chow, 1964, p.21-11).

soil creep The gradual, steady downhill movement of soil and loose rock material on a slope that may be very gentle but is usually steep. Syn: *surficial creep.*

soil discharge The release of water from the soil by evaporation and transpiration. The water may have been derived from the soil or from the saturated zone by way of the *capillary fringe.* Syn: *soil evaporation.* See also: *vadose-water discharge.*

soil erosion Detachment and movement of topsoil, or soil material from the upper part of the profile, by the action of wind or running water, esp. as a result of changes brought about by human activity (such as unsuitable or mismanaged agricultural methods). It includes rill erosion, gully erosion, sheet erosion, and wind erosion.

soil evaporation *soil discharge.*

soil extract The solution separated from a soil suspension or from a soil by filtration, centrifugation, suction, or pressure. May or may not be heated prior to separation.

soilfall A *debris fall* involving soil material.

soil fertility The quality of a soil that enables it to provide nutrients in adequate amounts and in proper balance for the growth of specified plants or crops.

soil flow *solifluction.*

soil fluction *solifluction.*

soil formation *soil genesis.*

soil-formation factors (soil'-for-ma'-tion) The natural conditions and substances that interact to produce a soil: parent material, climate, plants and other organisms, topography, and time.

soil friction angle A property of soil strength that relates shear strength to normal or confining stress. Cf: *angle of repose; shear strength.* Syn: *angle of internal friction; friction angle.*

soil genesis (a) The mode of origin of the soil, with special reference to the processes of soil-forming factors responsible for the development of the solum from the parent material. (b) A division of soil science concerned with soil genesis. Syn: *soil formation; pedogenesis.*

soil horizon A layer of a soil that is distinguishable from adjacent layers by characteristic physical properties such as structure, color, or texture, or by chemical composition, including content of organic matter or degree of acidity or alkalinity. Soil horizons are generally designated by a capital letter, with or without a numerical annotation, e.g. A horizon, A2 horizon. Syn: *horizon [soil]; pedologic horizon.*

soil ice Any ice formed in situ during freezing of unconsolidated materials, where the moisture source was the contained moisture or that brought in by capillary action during the freezing process. *Segregated ice* and *pore ice* are the most common forms (Mackay, 1966, p.61).

soil interpretations Predictions of soil behavior in response to specific uses or management based on inferences from soil characteristics and qualities (e.g., trafficability, erodibility, productivity, etc.). They are either qualitative or quantitative estimates or ratings of soil productivities, potentials, or limitations.

soil loss tolerance (a) The maximum average annual soil loss that will allow continuous cropping and maintain soil productivity without requiring additional management inputs. (b) The maximum soil erosion loss that is offset by the theoretical maximum rate of soil development which will maintain an equilibrium between soil losses and gains.

soil map A map showing the distribution of kinds of soil in relation to prominent physical and cultural features of the Earth's surface. Kinds of soil are expressed in terms of soil taxonomic units, such as series, or as phases of series. Maps showing soil characteristics or qualities, such as slope, texture, depth, fertility, or erodibility are not soil maps.

soil mechanics The application of the principles of mechanics and hydraulics to engineering problems dealing with the behavior and nature of soils, sediments, and other unconsolidated accumulations of solid particles.

soil moisture *soil water.*

soil-moisture weathering Accelerated weathering of granite below an old soil line, often causing the steepening of margins of granitic *inselbergs* (Stone, 1967, p.249).

soil monolith A vertical section of a soil profile removed from the soil and mounted for display or study.

soil morphology (a) The physical constitution of a soil profile as exhibited by the kinds, thickness, and arrangement of the horizons in the profile, and by the texture, structure, consistence, and porosity of each horizon. (b) The visible characteristics of the soil or any of its parts.

soil order A group of soils in the broadest category. For example, in the 1938 classification system. The three soil orders were zonal soil, intrazonal soil, and azonal soil. In 1975 there were 10 orders, whereas in the current USDA classification scheme (Soil Survey Staff, 1994) there are 11 orders, differentiated by the presence or absence of diagnostic horizons: Alfisols, Andisols, Aridisols, Entisols, Histosols, Inceptisols, Mollisols, Oxisols, Spodosols, Ultisols, and Vertisols. Orders are divided into Suborders and the Suborders are further divided into Great Groups.

soil patterns Obsolete syn. of *patterned ground.*

soil permeability (a) The ease with which gases, liquids, or plant roots penetrate or pass through a bulk mass of soil or a layer of soil. Since different soil horizons vary in permeability, the particular horizon under question should be designated. (b) The property of a porous medium itself that expresses the ease with which gases, liquids, or other substances can flow through it, and is the same as intrinsic permeability.

soil physics The organized body of knowledge concerned with the physical characteristics of soils and with the methods and instruments used in determining these characteristics.

soil polygon A group term for forms of polygonal ground developed on a soil surface, frequently in permafrost areas but also in regions where contraction occurs (as in playa lakes and deserts), and occurring with or without a stone border. Diameter: a few millimeters to many tens of meters. The term is misleading because soil need not be present.

soil productivity The capacity of a soil, in situ, to produce a specified plant or sequence of plants under a specified system of management (SSSA, 1975, p.13). The term is not synonymous with *soil fertility.*

soil quality The capacity of a soil to function within ecosystem boundaries to sustain biological productivity, maintain environmental quality, and promote plant and animal health.

soil science That science dealing with soils as a natural resource on the surface of the earth including soil formation, classification and mapping; physical, chemical, biological, and fertility properties of soils per se; and these properties in relation to the use of management of soils.

soil separate A group of rock and mineral particles in the soil, obtained in separation (as in mechanical analysis), having equivalent diameters less than 2 mm, and ranging between specified size limits (from "very coarse sand" to "clay"). Cf: *coarse fragment.* Syn: *separate.*

soil series The lowest category of U.S. system of soil taxonomy; a conceptualized class of soil bodies (polypedons) that have limits and ranges more restrictive than all higher taxa. Soil series are commonly used to name dominant or codominant polypedons represented on detailed soil maps. The soil series serve as a major vehicle to transfer soil information and research knowledge from one soil area to another.

soil slip *debris slide.*

soil solution Soil water considered as a solution of various salts, organic compounds, gases, etc., that are of significance to plant growth or to the consequences of flushing the soil solution to a body of ground or surface water.

soil stabilization Chemical or mechanical treatment designed to increase or maintain the stability of a soil mass or otherwise to improve its engineering properties (ASCE, 1958, term 337); e.g., increasing its shear strength, reducing its compressibility, or decreasing its tendency to absorb water. Stabilization methods include physical compaction and treatment with cement, lime, bitumen, and other binding agents as needed.

soil-stratigraphic unit (soil'-strat-i-graph'-ic) A soil with physical features and stratigraphic relations that permit its consistent recognition and mapping as a stratigraphic unit. It is distinct from both rock-stratigraphic and pedologic units (ACSN, 1961, p.654).

soil strength *shear strength.*

soil strip *soil stripe.*

soil stripe A *sorted stripe* whose texture is considerably finer than that of a *stone stripe.* Syn: *soil strip; earth stripe.*

soil structure [pat grd] A term formerly used (usually in the plural) for *patterned ground* (Sharp 1942, p.275), but now discarded because humus and a soil profile may be absent in patterned ground. Syn: *structure soil.*

soil structure [soil] The combination or aggregation of primary soil particles into aggregates or clusters (peds), which are separated from adjoining peds by surfaces of weakness. Soil structure is classified on the basis of size, shape, and distinctness into classes, types, and grades.

soil survey (a) The systematic examination, description, classification, and mapping of soils in an area. Soil surveys are classified according to the kind and intensity of field examination. (b) The program of the National Cooperative Soil Survey that includes developing and implementing standards for describing, classifying, mapping, writing, and publishing information about soils of a specific area.

Soil Taxonomy U.S. Department of Agriculture soil classification system.

soil travertine A calcareous rock, a paleosol preciptate from lime-rich groundwater; may be part of a calcareous duricrust or caliche. Cf: *calcrete.*

soil type A phase or subdivision of a soil series based primarily on texture of the surface soil to a depth at least equal to plow depth (about 15 cm). In Europe, the term is roughly equivalent to the term *great soil group.*

soil vapor extraction A remediation technology designed to remove organic vapors from the *unsaturated zone* through extraction wells.

soil variant A soil whose properties are believed to be sufficiently different from other known soils to justify a new series name but comprising such a limited geographic area that creation of a new series is not justified. Use of this term was discontinued in 1988. See also: *taxadjunct.*

soil water Water in the *belt of soil water.* Syn: *soil moisture.*

soil-water belt (soil'-wa-ter) *belt of soil water.*

soil-water zone *belt of soil water.*

sol (a) A homogeneous suspension or dispersion of colloidal matter in a fluid (liquid or gas). (b) A completely mobile mud. A sol is in a more fluid form than a *gel.*

sol [planet] One solar day, whose duration varies among the planets. On Mars one sol equals 24.66 terrestrial hours (Kieffer et al., 1992).

sola (so'-la) Plural of *solum.*

solar altitude The angle of the sun above the horizon; influences the intensity of solar radiation that strikes Earth's surface. At the maximum solar altitude of 90 degrees, the solar radiation striking Earth's surface is most intense; the intensity declines with decreasing solar altitude and is zero at 0 degrees.

solar attachment (so'-lar) An auxiliary instrument which may be attached to a surveyor's transit or compass for determining the true meridian directly from the Sun. Cf: *solar compass.*

solar compass A surveying instrument that permits the establishment and surveying of the astronomic meridian or parallel directly by observation on the Sun. It has been replaced by the *solar attachment* in combination with a transit.

solar constant The rate at which solar radiant energy is received outside the atmosphere on a surface normal to the incident radiation, at the Earth's mean distance from the Sun. The currently accepted value of the mean solar constant is 1,370 watts per square meter (1.96 gram calories per minute per square centimeter). Cf: *insolation [meteorol].*

solar day The interval between successive passages of the sun over the meridian. It is slightly longer than the *sidereal day* because of the orbital motion of the Earth. A solar day is approx. 86,400 seconds. Cf: *sidereal day.*

solar lake A *meromictic lake* of high transparency, in which solar heating penetrates into the depths. The stability of the water column prevents efficient dissipation of heat to the surface where it would be lost by evaporation; instead, the heat accumulates, producing water temperatures at depth that may be some tens of degrees (C) higher than the surface temperature.

solar nebula The cloud of gases and dispersed solids from which the Sun, and other objects in the solar system, condensed and/or accreted. See: *dust-cloud hypothesis.*

solar prism *Roelof's prism.*

solar salt Crystalline salt obtained by evaporating seawater or other brine by the heat of the sun.

solar system All the objects gravitationally bound to the Sun and the volume of space they move in. Currently also used (often in plural, "solar systems") to refer to extra-solar planetary systems about other stars.

solar tide The part of the tide caused solely by the tide-producing force of the Sun. Cf: *lunar tide.*

solar time Time based on the *solar day.* Cf: *sidereal time.*

solar wind The motion of interplanetary plasma or ionized particles away from the Sun. When near the Earth, it interacts with the Earth's magnetic field.

sole [glac geol] The lower part or basal ice of a glacier, which often contains rock fragments, appears dirty, and is separated from clean ice by an abrupt boundary.

sole [mass move] The middle and lower parts of the shear surface of a landslide.

sole [stratig] The under surface of a rock body or vein, esp. the bottom of a sedimentary stratum.

sole [struc geol] The fault plane underlying a *thrust sheet.*

sole cast A *sole mark* preserved as a swelling or positive feature on the underside of a bed immediately overlying a finer-grained bed containing a primary sedimentary structure (depression).

soled boulder A stone with blunted corners and smoothed or flattened (and sometimes striated) sides, esp. one shaped by glacial grinding.

sole fault A low-angle thrust fault forming the base of a *thrust sheet;* also, the basal main fault of an imbrication. Syn: *detachment thrust; basal thrust plane.* Cf: *Décollement fault; detachment fault.*

sole injection An igneous intrusion that was emplaced along a thrust-fault plane. Obsolete.

sole mark A general descriptive term applied to a directional structure or to a small, wavelike, mainly convex irregularity or penetration found on the underside of a bed of sandstone or siltstone along its contact with a finer-grained layer such as shale. The term usually refers to a filling of a primary sedimentary structure, e.g. a crack, track, groove, or other depression, formed on the surface of the underlying mud by such agents as currents, organisms, and unequal loading, and preserved as a *sole cast* after the underlying material had consolidated and weathered away. Examples: *load cast; flute cast; groove cast.*

Solenhofen stone (So'-len-ho'-fen) A *lithographic limestone* of Late Jurassic age found at Solenhofen (Solnhofen), a village in Bavaria, West Germany. It is evenly and thinly stratified and contains little clay.

solfatara (sol-fa-ta'-ra) A type of *fumarole,* the gases of which are characteristically sulfurous. Cf: *solfataric stage.* Etymol: the Solfatara volcano, Italy.

solfatara field A group of solfataras (Rittmann, 1962, p.10). Cf: *fumarole field.*

solfataric stage (sol-fa-tar'-ic) A late or decadent type of volcanic activity, characterized by the emission of sulfurous gases from the vent. See also: *solfatara.* Cf: *fumarolic stage.*

solid (sol'-id) Sedimentary material that is in solution or suspension but when freed of solvent or suspending medium has the form and properties of a solid. The term is usually used in the plural; e.g. *dissolved solids.*

solid diffusion Diffusion of chemical species through a rock that remains essentially solid.

solid earth *lithosphere.*

solid flow Flow in a solid by rearrangement among or within the constituent particles. Cf: *liquid flow; viscous flow.*

solid geology A British term for bedrock geology.

solidification (so-lid'-i-fi-ca'-tion) The process of becoming solid or hard; esp. the change from the liquid to the solid state on the cooling of a magma. The term lithification is more generally applied in the case of sedimentary rocks. See also: *consolidation [geol].*

solidification front An inward growing region of high crystal volume density localized along the margins of a cooling magma body. Solidification fronts may occur along the floor, sidewalls, or roof of a magma body.

solidification index A chemical parameter of igneous rocks, equal to 100 X $MgO/(MgO+FeO+Fe_2O_3+Na_2O+K_2O)$. Its usual range is from about 40 (basalt) to 100 (rhyolite). It reflects the steady decrease (common to all trends) of MgO relative to total iron and to the alkalies throughout the greater part of fractionation. It appears to decrease at nearly the same rate as the amount of residual liquid in a crystallizing magma decreases; hence its name (Kuno, 1957). Abbrev: SI.

solid map A British term for a geological map showing the extent of *solid rock,* on the assumption that all surficial deposits, other than alluvium, are absent or removed (Nelson and Nelson, 1967, p.352). Cf: *drift map.*

solid precipitation The various kinds of solid water particles that develop in the atmosphere and fall earthward, for example, snow crystals or ice pellets, including freshly deposited particles that have not undergone perceptible transformation after being deposited on the ground; when clear morphological differences exist between falling and deposited particles, the term applies to precipitation while it remains air-borne.

solid rock A British term for bed rock.

solid solution A single crystalline phase that may be varied in composition within finite limits without the appearance of an additional phase. Syn: *mixed crystal.*

solid-solution series (sol'-id-so-lu'-tion) *isomorphous series.*

solidus (sol'-i-dus) On a temperature-composition diagram, the locus of points in a system at temperatures above which solid and liquid are in equilibrium and below which the system is completely solid. In binary systems without solid solutions, it is a straight line; in binary systems with solid solutions, it is a curved line or a combination of straight and curved lines.

soliflual (so-li-flu'-al) Said of debris resulting from solifluction (Baulig, 1957, p.927). Syn: *solifluidal.*

solifluction (so-li-fluc'-tion) The slow viscous downslope flow of waterlogged soil and other unsorted and saturated surficial material, normally at 0.5-5.0 cm/yr; esp. the flow occurring at high elevations in regions underlain by frozen ground (not necessarily permafrost) that acts as a downward barrier to water percolation, initiated by frost action and augmented by meltwater resulting from alternate freezing and thawing of snow and ground ice. The term was proposed by Andersson (1906, p.95-96) as "the slow flowing from higher to lower ground of masses of waste saturated with water", but as he did not state explicitly that it referred to flow over frozen ground, the term has been extended to include similar movement in temperate and tropical regions; also, it has been used as a syn. of *soil creep,* although solifluction is generally more rapid. It is preferable to restrict the term to slow soil movement in periglacial areas. Also spelled: *solifluxion.* Syn: *soil flow; soil fluction; sludging.*

solifluction lobe An isolated, tongue-shaped feature, up to 25 m wide and 150 m long, formed by more rapid solifluction on certain sections of a slope showing variations in gradient. It commonly has a steep front (15°-25°) and a relatively smooth upper surface. Syn: *solifluction tongue.*

solifluction mantle The unsorted, water-saturated, locally derived material moved downslope by solifluction. Syn: *flow earth.*

solifluction sediment A sediment that has resulted from solifluction.

solifluction sheet A broad deposit of *solifluction mantle,* occurring evenly across a wide slope.

solifluction slope A smooth curvilinear slope of 2° to 30°, produced by solifluction or along which solifluction occurs.

solifluction step The flattish area at the front of a small solifluction lobe; the tread of a small, turf-banked terrace usually restricted to immediately above timberline.

solifluction stream A narrow, laterally confined streamlike deposit of *solifluction mantle.*

solifluction stripe A form of *striped ground* associated with solifluction. The term was used by Washburn (1947, p.94) as a syn. of *nonsorted stripe,* but solifluction may also be associated with sorted stripes.

solifluction terrace A low terrace or bench formed by solifluction at the foot of a slope; it may have a lobate margin reflecting uneven movement.

solifluction tongue *solifluction lobe.*

solifluidal (so-li-flu'-id-al) *soliflual.*

solifluxion (so-li-flux'-ion) *solifluction.*

soligenous peat (so-lig'-e-nous) A *minerotrophic peat* similar to topogenous but with water moving through the *peat* (Gore, 1983).

solimixtion (so-li-mix'-tion) A term introduced by Rosauer (1957, p.65) for "a relatively homogeneous blending in the vertical plane of two different materials due to frost action with little or no macro-optical structures", as at the contact of two different layers in a *loess* profile.

soliqueous (so-liq'-ue-ous) A term proposed by Leet and Leet (1965, p.620) to describe a state of matter, e.g., that of the materials in the Earth's mantle, that is neither solid, liquid, nor gaseous, but a mixture of all three; it is maintained and controlled by pressure, and it has neither a crystalline structure nor the chilled-liquid molecular arrangements of a glass.

solitaire (sol'-i-taire) A single diamond or sometimes other gem, set alone, usually in a ring.

solitary coral (sol'-i-tar-y) A coral that does not form part of a colony; an individual *corallite* (of a polyp) that exists unattached to other corallites. Cf: *colonial.* Syn: *simple coral; cup coral; horn coral.*

solitary wave A nonlinear water wave or *internal wave* that propagates without change of shape.

Soller slits A series of closely spaced, parallel, and highly X-ray absorbing plates used to restrict the axial divergence of an X-ray beam. Some diffractometers have one at the tube and one at the detector; others have one only at the tube. This makes a difference in the X-ray diffraction tracings and must be taken into account when modeling X-ray diffraction tracings.

Solonchak A great soil group of the intrazonal order and halomorphic suborder, consisting of soils with gray, thin, salty crust on the surface, and with fine granular mulch immediately below being underlain with grayish, friable, salty soil; formed under subhumid to arid, hot or cool climate, under conditions of poor drainage, and under a sparse growth of halophytic grasses, shrubs, and some trees. Not used in current U.S. system of soil taxonomy.

Solonetz A great soil group of the intrazonal order and halomorphic suborder, consisting of soils with a very thin, friable, surface soil underlain by a dark, hard columnar layer usually highly alkaline; formed under subhumid to arid, hot to cool climates, under better drainage than Solonchaks, and under a native vegetation of halophytic plants. Not used in current U.S. system of soil taxonomy.

solongoite (so-lon'-go-ite) A colorless monoclinic mineral: $Ca_2B_3O_4Cl(OH)_4$.

solubility (sol-u-bil'-i-ty) The equilibrium concentration of a solute in a solution saturated with respect to that solute at a given temperature and pressure.

solubility product A syn. of *dissociation constant* that refers to a very slightly soluble compound.

solum (so'-lum) The upper part of a soil profile, including the A and B horizons, in which soil-forming processes occur. Plural: *sola.* Syn: *true soil.* Pl: sola.

solusphere (so'-lus-phere) That zone of the Earth in which water solutions affect geologic, chemical and life processes. Archaic.

solution (so-lu'-tion) A process of chemical weathering by which mineral and rock material passes into solution; e.g. removal of the calcium carbonate in limestone or chalk by carbonic acid derived from rainwater containing carbon dioxide acquired during its passage through the atmosphere. Cf : *dissolution.*

solution banding In roll-front uranium deposits, "concentric color bands smoothly cross-cutting a sand body in arcuate patterns" (Bailey and Childers, 1977, p.52). Color banding is in tints of red, yellow, and rusty brown; it probably reflects slight changes in the chemical composition of the groundwater during mineralization.

solution basin A shallow surface depression, either man-made or natural, produced by solution of surface material, or resulting from the settlement of a surface through the removal in solution of underlying material (such as salt or gypsum); specif. a "solution depression" in a karstic region.

solution bench A low bench produced on limestone coasts by the action of fresh water (Wentworth, 1939). Presence of such benches on many desert coasts requires broadening of the definition. Preferred term: *solution platform.*

solution bevel In speleology, a planar slope in a cave wall, caused by unequal rates of dissolution in adjacent strata of differing solubility; a geometric result of *solution* beneath slow-moving water (Lange, 1964).

solution breccia A *collapse breccia* formed where soluble material has been partly or wholly removed by dissolution, thereby allowing the overlying rock to fall or settle and become fragmented; e.g. a breccia consisting of chert fragments gathered from a limestone whose carbonate material has been dissolved away. See also: *evaporite-solution breccia.* Syn: *ablation breccia.*

solution cave A cave formed in a soluble rock. Cf: *lava tube [speleo]; sea cave; rockshelter.*

solution channel Tubular or planar channel formed by solution in carbonate-rock terranes, usually along joints and bedding planes. Cf: *solution opening.*

solution cleavage A secondary foliation defined by seams of insoluble residue, preferred orientation of dimensionally elongated grains, and/or mica beards. Generally believed to have formed by *diffusional mass transfer.* Syn: *pressure solution cleavage.*

solution collapse The abrupt collapse of nonsoluble strata due to the dissolution of soluble underlying rock. Cf: *solution subsidence.*

solution collapse breccia A rock composed of angular blocks formed by collapse of a cavern roof due to dissolution of an evaporite deposit, for instance halite or gypsum. The breccia may have a muddy or micritic matrix or may have sparry cement.

solution corridor In karst, a straight trench typically about 3 to 10 m wide. Syn: *bogaz; corridor [karst]; karst corridor; karst street; zanjón.* Cf: *karst valley; solution fissure.*

solution depression A general term for a solution basin that occurs in a *karst* region.

solution facet A nearly plane face, commonly bounded by a narrow rim or raised edge, developed on a soluble rock such as limestone from progressive dissolution by rainwater.

solution fissure In *karst,* one of a series of vertical open cracks, about 0.5 m wide, dissolved along joints and separating *limestone pavement* into *clints.* Syn: *cutter [karst]; grike; kluftkarren.* Cf: *solution corridor.*

solution gas drive Energy within an oil accumulation supplied by expansion of gas dissolved in the oil. Syn: *gas depletion; depletion drive.*

solution grooves *rillenkarren; rinnenkarren.*

solution lake (a) A syn. of *karst pond.* (b) A lake occupying a basin formed by surface dissolution of bedrock.

solution load *dissolved load.*

solution load structure A bulbous projection of sediment into an underlying layer that resembles a fluidization feature but was formed by sediment filling a depression as it was produced by dissolution of an efflorescent salt crust (Smoot and Castens-Seidell, 1994).

solution mining (a) The in-place dissolution of water-soluble mineral components of an ore deposit by permitting a leaching solution, usually aqueous, to trickle downward through the fractured ore to collection galleries at depth. It is a type of *chemical mining.* (b) The mining of soluble rock material, esp. salt, from underground deposits by pumping water down wells into contact with the deposit and removing the *artificial brine* thus created.

solution opening (a) An opening produced by direct solution by water penetrating pre-existing interstices. (b) An opening resulting from the decomposition of less soluble rocks by water penetrating pre-existing interstices, followed by solution and removal of the decomposition products. (c) *solution channel.*

solution pan A shallow solution basin formed on bare limestone and generally characterized by flat bottom and overhanging sides. Syn: *kamenitza; tinajita.* Cf: *opferkessel.*

solution pendant *pendant [speleo].*

solution pipe A vertical cylindrical hole, formed by solution and often without surface expression, that is filled with detrital matter.

solution pit An indentation up to about 1 mm in diameter formed on a rock surface by solution.

solution plane A plane containing lines of chemical weakness in a crystal, along which solution tends to occur under certain physical circumstances, e.g. great pressure.

solution platform An intertidal surface on carbonate rocks, nearly horizontal but not abraded, produced primarily by solution but with contributions by intertidal weathering and biological erosion and deposition (Bloom, 1978, p.448). See also: *solution bench.* Syn: *low-tide platform.*

solution ripple *scallop [speleo].*

solution scarp An escarpment formed on a higher slope by more active *solution* of a lower area or by dissolutional undercutting of the base of the escarpment.

solution sinkhole The most common type of *sinkhole,* which grows when water enlarges a fissure, or several of them, in the underlying soluble bedrock surface. Cf: *collapse sinkhole.*

solution subsidence The gradual subsidence of nonsoluble rock strata due to the solution of underlying rock. Cf: *solution collapse.*

solution transfer The process of *pressure solution* of detrital grains at points of contact, followed by chemical redeposition of the dissolved material on the less-strained parts of the grain surfaces. See also: *Riecke's principle.*

solution valley *karst valley.*

Solvan (Sol'-van) A British stage: Middle Cambrian (above Botomian, below Menevian) (Harland et al., 1990).

solvate (sol'-vate) A chemical compound consisting of a dissolved substance and its solvent, e.g. hydrated calcium sulfate.

solvation (sol-va'-tion) The chemical union of a dissolved substance and its dissolving liquid.

solvent extraction A hydrometallurgical separation of metals from ores and concentrates through the use of solvents. When followed by *electrowinning,* the procedure is referred to as SX-EW.

sölvsbergite (solvs'-berg-ite) A fine-grained, holocrystalline, rarely porphyritic hypabyssal *trachyte,* composed chiefly of sodic feldspar and a smaller amount of potassium feldspar, sodic pyroxene or amphibole, and little or no quartz. The name, given by Brögger in 1894, is for Sölvsberget, Norway. Obsolete. Cf: *lindoite.*

solvus (sol'-vus) Pl. solvi. A curved line in a binary system, or a surface in a ternary system, that separates a field of homogeneous solid solution from fields of two or more phases that form by exsolution of the homogeneous phase with decreases in temperature. Solvi typically widen with decreasing temperature, indicating decreased mutual solubilities of the two end-members. May also be referred to as a *miscibility gap.* Cf: *hypersolvus; subsolvus.* Cf: *hypersolvus; subsolvus.*

soma [palyn] (so'-ma) All of an organism except the germ cells. The assimilative body of a fungal organism, as distinguished from its reproductive organs or reproductive phase. Pl: *somata.* Adj: somatic.

somal unit (so'-mal) A stratigraphic unit that intertongues laterally with its neighbor; e.g. lithosome or biosome.

somata (so'-ma-ta) Pl. of *soma.*

sombric horizon A subsurface mineral horizon that is darker in color than the overlying horizon but that lacks the properties of a spodic horizon. Common in cool, moist soils of high altitude in tropical regions.

somite (so'-mite) One of the longitudinal series of body segments ("metameres") into which many animals (such as articulates and embryonic vertebrates) are more or less distinctly divided; esp. the basic embryologic unit of segmentation of the body of an arthropod, approximately equivalent to the part of the body covered by a single exoskeletal "body ring" and bearing no more than one pair of limbs. In vertebrates, mesoderm from somites forms important components of the skeleton, dermis, muscle, connective tissue, and vascular system.

somma (som'-ma) n. A circular or crescentic ridge that is steep on its inner side and represents the rim of an ancient volcanic crater or caldera. Etymol: Mt. Somma, the ancient crater rim that surrounds Vesuvius. Syn: *somma ring.* adj. Said of a volcanic crater with a central cone surrounded by a somma.

sommaite (som'-ma-ite) A leucite monzosyenite occurring as ejected blocks in lavas. Olivine and clinopyroxene form phenocrysts in a groundmass of leucite, alkali feldspar, and mafites. Its name, given by Lacroix in 1905, is derived from Monte Somma, Italy. Obsolete. Cf: *ottajanite.*

somma ring *somma.*

sonar (so'-nar) An acronym of *so*und *na*vigation and *r*anging, a method used in oceanography to study the ocean floor.

sondage (son'-dage) A deep trench, often of restricted area, dug to investigate the stratigraphy of an archaeological site. Syn: *test pit [arch].*

sondalite (son'-da-lite) A metamorphic rock consisting of cordierite, quartz, garnet, tourmaline, and kyanite (Holmes, 1928, p.213). Rarely used.

sonde The elongate cylindrical tool assembly used in a borehole to acquire a *well log.* It is 6 to 40 feet in length and 2 to 6 inches in diameter, and contains various energy-input devices and/or response sensors. The sonde is lowered into the borehole by a multi-conductor cable, or wire line. Syn: *logging tool.*

song of the desert The booming or roaring sound made by a *sounding sand* on a desert. Syn: *voices of the desert.*

sonic depth-finder (son'-ic) *echo sounder.*

sonic log An *acoustic log* showing the interval-transit time of compressional seismic waves in rocks near the well bore of a liquid-filled borehole. First used for seismic-velocity information, it is now used chiefly for estimating porosity and lithology by the empirical *Wyllie time-average equation.* Syn: *velocity log; continuous-velocity log.*

sonic wave *acoustic wave.*

sonobuoy (son'-o-buoy) A buoy, generally free-floating, that contains a hydrophone and a radio transmitter that broadcasts information picked up by the hydrophone. It is used in seismic refraction surveying.

sonolite (son'-o-lite) A brown, dark brown, reddish-brown, or reddish-orange monoclinic mineral of the *clinohumite* group: $Mn^{2+}_9(SiO_4)_4(OH,F)_2$.

sonoprobe (son'-o-probe) A type of echo sounder that generates sound waves and records their reflections. Obsolete.

sonoraite (so-no'-ra-ite) A yellowish-green monoclinic mineral: $Fe^{3+}Te^{4+}O_3(OH) \cdot H_2O$.

Sonstadt solution (Son'-stadt) A solution of mercuric iodide in potassium iodide that is used as a *heavy liquid;* its specific gravity is 3.2. Cf: *bromoform; Klein solution; Clerici solution.* Syn: *Thoulet solution.*

sooty chalcocite (soot'-y) *sooty ore.*

sooty ore A black, pulverulent variety of chalcocite, digenite, or djurleite, generally found coating pyrite and the supergene sulfides of porphyry copper deposits. Syn: *sooty chalcocite.*

sopcheite (sop'-che-ite) A metallic brownish-gray orthorhombic mineral: $Ag_4Pd_3Te_4$.

sophiite (so-phi'-ite) A colorless orthorhombic mineral: $Zn_2(Se^{4+}O_3)Cl_2$. Syn: *sofiite.*

sorbyite (sor'-by-ite) A gray metallic monoclinic mineral: $Pb_{19}(Sb,As)_{20}S_{49}$.

soredium (so-re'-di-um) In lichens, a mass of algal cells surrounded by fungus hyphae, extruded through the outer or upper cortex of a thallus.

sörensenite (sör'-en-sen-ite') A colorless or pinkish monoclinic mineral: $Na_4SnBe_2Si_6O_{18} \cdot 2H_2O$.

sörkedalite (sor'-ke-dal-ite') A feldspathoid-free *monzonite* or *monzodiorite* containing olivine, clinopyroxene, and antiperthitic andesine, resembling *kjelsasite* but with high titanium, iron, and phosphorus content as reflected by high percentages of opaque minerals and apatite. Brögger in 1933 derived the name from Sörkedal, Oslo district, Norway. Obsolete.

sorosilicate (so-ro-sil'-i-cate) A class or structural type of *silicate* characterized by the linkage of two SiO_4 tetrahedra by the sharing of one oxygen, with a Si:O ratio of 2:7. An example of a sorosilicate is hemimorphite, $Zn_4Si_2O_7(OH)_2 \cdot H_2O$. Cf: *nesosilicate; cyclosilicate; inosilicate; phyllosilicate; tectosilicate.* Syn: *double island silicate.*

sorosite A metallic hexagonal mineral of the bronze group: Cu(Sn,Sb).

sorption The general process by which solutes, ions, and colloids become attached to solid matter in a porous medium. Cf: *absorption; adsorption.*

sorption water (sorp'-tion) *pellicular water.*

sorted [part size] (sort'-ed) Said of an unconsolidated sediment or of a cemented detrital rock consisting of particles of essentially uniform size or of particles lying within the limits of a single grade; graded. See also: *well-sorted; moderately sorted; poorly sorted.*

sorted [pat grd] Said of a nongenetic group of *patterned ground* features displaying a border of stones (including boulders) commonly surrounding or alternating with fines (including sand, silt, and clay). Ant: *nonsorted.*

sorted circle A form of *patterned ground* "whose mesh is dominantly circular and has a sorted appearance commonly due to a border of stones surrounding finer material" (Washburn, 1956, p.827); developed singly or in groups. Diameter: a few centimeters to more than 10 m; the stone border may be 35 cm high and 8-12 cm wide. See also: *debris island; stone pit; plug [pat grd].* Syn: *stone circle; stone ring; stone wreath; rock wreath; frost circle.*

sorted crack A form of nearly horizontal *patterned ground* consisting of a concentration of boulders along a straight line, as in the Swedish Caledonides. Cf: *nonsorted crack.*

sorted net A form of *patterned ground* "whose mesh is intermediate between that of a sorted circle and a sorted polygon and has a sorted appearance commonly due to a border of stones surrounding finer material" (Washburn, 1956, p.830). Diameter: a few centimeters to 3 m.

sorted polygon A form of *patterned ground* "whose mesh is dominantly polygonal and has a sorted appearance commonly due to a border of stones surrounding finer material" (Washburn, 1956, p.831); never developed singly. Diameter: a few centimeters to 10 m. Syn: *stone polygon; stone ring.*

sorted step A form of *patterned ground* "with a steplike form and a sorted appearance due to a downslope border of stones embanking an area of finer material upslope" (Washburn, 1956, p.833); formed in groups, rarely if ever singly. Dimensions: 1-3 m wide; up to 8 m long in downslope direction. See also: *nonsorted step; stone garland; stone-banked terrace.*

sorted stripe One of the alternating bands of finer and coarser material comprising a form of *patterned ground* characterized by "a striped pattern and a sorted appearance due to parallel lines of stones and intervening strips of dominantly finer material oriented down the steepest available slope" (Washburn, 1956, p.836). It never forms singly, usually occurring as one of many evenly spaced, sometimes sinuous, stripes that often exceed 100 m in length on slopes as steep as 30°. An individual stripe may be a few centimeters to 2 m wide, with the intervening area two to five times wider. See also: *block stripe; soil stripe; stone stripe; striped ground.*

sorting (a) The dynamic process by which sedimentary particles having some particular characteristic (such as similarity of size, shape, or specific gravity) are naturally selected and separated from associated but dissimilar particles by the agents of transportation. (b) The result of sorting; the degree of similarity of sedimentary particles in a sediment. (c) A measure of sorting, or of the spread or range of the particle-size distribution on either side of an average. Cf: *gradation [part size]; degree of sorting.*

sorting coefficient A *sorting index* developed by Trask (1932), being a numerical expression of the geometric spread of the central half of the particle-size distribution of a sediment, and defined as the square root of the ratio of the larger quartile, Q_1 (the diameter having 25% of the cumulative size-frequency distribution larger than itself), to the smaller quartile, Q_3 (the diameter having 75% of the cumulative size-frequency distribution larger than itself). It is indicative of the range of conditions present in the transporting fluid (velocities, turbulence, etc.) and to some extent indicative of the distances of transportation. A perfectly sorted sediment is defined as having a coefficient of 1.0; less perfectly sorted sediments have higher coefficients. The Trask sorting coefficient is not considered a useful particle-size measure and is no longer used by sedimentologists, who now mainly use *phi standard deviation.*

sorting index A measure of the degree of sorting or of uniformity of particle size in a sediment, usually based on the statistical spread of the frequency distribution of particle sizes; e.g. *sorting coefficient* and *grading factor.*

sorus (so'-rus) A cluster of *sporangia,* as on a fern frond. Also used for the erupting fungal spore mass in the Uredinales.

sosedkoite (so-sed'-ko-ite) An adamantine colorless orthorhombic mineral: $(K,Na)_5Al_2(Ta,Nb)_{22}O_{60}$. It is isomorphous to rankamaite.

Sosie (so'-sie) A seismic method that employs a pseudo-random series of seismic impulses to generate seismic waves. The recorded data have to be correlated with this series to produce an interpretable result. (Barbier and Viallix, 1973, p.673-683)

sótano (so'-ta-no) In Mexico, a deep vertical shaft in a karst area that may or may not lead to a cave. Etymol: Spanish, "cellar". Syn: *vertical cave.*

soucekite (sou'-cek-ite) A metallic lead-gray orthorhombic mineral of the *bournonite* group: $PbCuBi(S,Se)_3$.

soufriere (sou-fri-ere') A common name for a volcanic crater or area of solfataric activity, used especially in the West Indies and other French-speaking regions.

sound [coast] (a) A relatively long, narrow waterway connecting two larger bodies of water (e.g., a sea or lake with the ocean or another sea); also two parts of the same body, or an arm of the sea forming a channel between a mainland and an island. It is generally wider and more extensive than a *strait [coast].* (b) A long, large, rather broad inlet of the ocean, generally extending parallel to the coast; e.g., Long Island Sound between New England and Long Island, N.Y. (c) A *lagoon* along the east coast of the U.S.; e.g., Pamlico Sound, N.C. (d) A long bay or arm of a lake; a stretch of water between the mainland and a long island in a lake.

sound [geophys] Acoustic waves in which the direction of particle motion is longitudinal, i.e., parallel with the direction of propagation. The term is sometimes restricted to such waves in gases, particularly air, and in liquids, particularly water, but it is also applied to wave motion in solids, where it is called a P-wave, the type of wave motion most often used in seismic exploration. Cf: *acoustics.*

sound channel A narrow depth zone in the water column centered on the depth at which sound speed is at a minumum. The minimum exists almost everywhere in the ocean except near the poles because sound speed increases above the channel due to warm water near the surface, and below the channel due to increasing pressure. Sounds generated in the channel tend to be trapped, and low-frequency sounds can travel great distances in the channel, up to halfway around the Earth.

sounding [elect] (sound'-ing) Mapping of (nominally) horizontal interfaces by resistivity, induced polarization, or electromagnetics. It involves variation of the electrode interval for resistivity and induced polarization soundings, but may involve variation of either frequency or coil separation in electromagnetics. See also: *parametric sounding; geometric sounding.* Cf: *profiling.*

sounding [eng] (a) Measuring the thickness of soil or depth to bedrock by driving a pointed steel rod into the ground or by using a *penetrometer* or other probe. (b) Measuring the depth of water using a weighted line or electronic device.

sounding [geophys] A determination of how some quantity varies with depth. Syn: depth probe.

sounding [meteorol] The measurement of one or more weather elements (temperature, humidity, pressure, and winds) with altitude, using either an instrument in situ (e.g. radiosonde) or a remote instrument platform such as a weather satellite or radar.

sounding [oceanog] The measurement of water depth taken from a ship by means of an *echo sounder* (or, in times long past, by a lead line).

sounding line *lead line* (of historical interest only).

sounding sand Sand, usually clean and dry, that emits a musical, humming, or crunching sound when disturbed, such as a desert sand when sliding down the slip face of a dune or a beach sand when it is stirred or walked over. Examples: *musical sand; booming sand; roaring sand; whistling sand.* Syn: *singing sand.*

sound ranging The location of a source of seismic energy by acoustic trilateration, i.e., by recording signals on receivers at known positions.

sound wave *acoustic wave.*

sour Said of crude oil or natural gas containing significant fractions of sulfur compounds. Cf: *sweet.*

source [seis] *seismic source.*

source [streams] (a) The point of origin of a stream of water; the point at which a river rises or begins to flow. Syn: *fountain.* (b) A *headwater,* or one of the headwaters, of a stream; e.g. a fountainhead.

source area *provenance.*

source-bed concept The theory of sulfide ore genesis that postulates an original syngenetic deposition of sulfides, and their later migration and concentration, due, for example, to a rise in temperature of the rock.

source bias An effect in which azimuthally dependent departures from standard traveltimes in the upper mantle beneath the hypocenter result in consistent errors in the estimated epicenter (Herrin and Taggart, 1968).

source-bordering lee dune (source-bor'-der-ing) *umbrafon dune.*

sourceland (source'-land) *provenance.*

source layer Layer supplying salt for the growth of salt structures; the source layer is a particular type of substratum (Jackson and Talbot, 1991).

sourceline (source'-line) In computer modelling of hydrothermal systems, a line formed by connecting the starting points of all fluid packets whose pathlines pass through a given fixed point or small volume in an area of interest in a hydrothermal system (Norton, 1978, p.23).

source link A magnitude-1 link associated with a first-order stream. Cf: *link [geomorph]; magnitude [geomorph].*

source point (a) The location where seismic energy is released. Refers to the location of a source of seismic energy, such as an explosive shot, an air gun release, a weight drop, a vibroseis excitation, etc. Where patterns of sources are used, usually refers to the center of the pattern. Abbrev: SP. (b) The area surrounding the source point.

source position The position of features as they exist on source materials, such as digital orthophotoquads (USGS, 2002b, p 1A-3). Cf: *cartographic position, geographic position, true position.*

source-receiver product (source-re-ceiv'-er) In seismic prospecting, the product of the number of detectors per trace and the number of sources used simultaneously or the records vertically stacked together.

sourceregion (source'-region) The region from which all fluids circulating through a pluton during some time interval were derived (Norton, 1978, p.25). Cf: *sourceline.*

source region That extensive area of the Earth's surface over which an *air mass* develops and acquires its distinctive characteristics.

source rock [petroleum] Sedimentary rock in which organic material under pressure, heat, and time was transformed to liquid or gaseous hydrocarbons. Source rock is usually shale or limestone.

source rock [sed] *parent rock [sed].*

source-rock index A term used by Dapples et al. (1953, p.297, 304) to indicate the extent of contribution to a sandstone of fragments from igneous and metamorphic rocks by measuring the degree of mixing of "arkose" and "graywacke" types. It is expressed as the ratio of sodic and potassic feldspar (arkose tendency) to the total of assorted rock fragments plus matrix of clay and micas (graywacke tendency). The index indicates source-rock types regardless of subsequent depositional history or diagenesis. Values greater than 3 indicate arkose; values less than 0.75 indicate graywacke. Rarely used.

South African ruby *Cape ruby.*

Southeast Missouri lead-zinc A variety of *Mississippi Valley-type deposit* characterized by strata-bound deposits of galena, sphalerite, and chalcopyrite filling primary and secondary porosity in dolomite and limestone reefs formed on paleotopographic highs.

southern oscillation Opposing swings of surface *air pressure* between the western and central tropical Pacific Ocean. When air pressure is low at Darwin (Australia) it is high at Tahiti (a south Pacific island) and when air pressure is high at Darwin it is low at Tahiti. Changes in the strength of the trade winds and occurrence of *El Niño* and *La Niña* are associated with the southern oscillation. See also: *El Niño Southern Oscillation.*

southing (south'-ing) A *latitude difference* measured toward the south from the last preceding point of reckoning; e.g. a linear distance southward from an east-west reference line.

south pole [astron] The south *celestial pole* representing the zenith of the heavens as viewed from the south geographic pole.

south pole [geog] The *geographic pole* in the southern hemisphere of the Earth at lat. 90°S, representing the southernmost point of the Earth or the southern extremity of its axis of rotation. Also spelled South Pole.

souzalite (sou'-za-lite) A dark green or blue-green triclinic mineral: $(Mg,Fe^{2+})_3(Al,Fe^{3+})_4(PO_4)_4(OH)_6{\cdot}2H_2O$. It is the Mg analogue of gormanite.

sövite (sö'-vite) A carbonatite that contains calcite as a dominant phase. The name, given by Brögger in 1921, is for Söve, Fen complex, Norway.

sowback (sow'-back) A long, low hill or ridge shaped like the back of a female pig; e.g. *hogback [geomorph]; horseback [glac geol]; drumlin.*

sowneck (sow'-neck) A narrow divide between two expanses of lowland, or a narrow boundary between two bodies of water, formed by a gentle rise of ground.

sp. *species [paleont].*

SP [geophys] *Spontaneous potential* or *self potential.*

SP [seis] *source point* or *shotpoint.*

spa (a) A *medicinal spring.* (b) A place where such springs occur, often a resort area or hotel. The name is derived from that of a town in eastern Belgium where medicinal springs occur.

spaced cleavage A cleavage with distinct cleavage domains or cleavage seams separated by regions lacking cleavage (microlithons). Cleavage domains often are defined by insoluble residues, aligned platy minerals, or a combination of the two. Cleavage domains may be closely spaced by non-penetrative diffuse zones, e.g. diffuse space cleavage or discretely bounded seams of finite thickness or crack-like surfaces that separate microlithons, e.g. disjunctive spaced cleavage. See also: *crenulation cleavage; slip cleavage.*

space-frequency domain The two-dimensional transform of a seismic section from location versus time to wave-number versus frequency.

space group A set of symmetry operations that map a three-dimensional periodic object (e.g., a crystal structure) into itself; there are 230 of them. Cf: *plane group.*

space lattice *crystal lattice.*

space-time unit A stratigraphic unit whose lateral limits are determined by geographic coordinates and whose vertical extent is measured in terms of geologic time (Wheeler, 1958).

spad An iron, brass, or tin nail, up to 5 cm long, with a hook or eye at the head for suspending a plumb line, used to mark an underground survey station (as in a mine or tunnel).

spadaite (spa'-da-ite) An ill-defined creamy white or pink mineral: $MgSiO_2(OH)_2{\cdot}H_2O$(?).

spaghetti (a) Data model for the storage of spatial information in vector data structures. In spaghetti models, points, lines and polygons are stored as simple lists of coordinates. (b) The approach of digitizing lines in an arbitrary sequence. (Krzanowski et al., 1993).

spall (a) A chip or fragment removed from a rock surface by weathering; esp. a small, relatively thin curved piece of rock produced by exfoliation. (b) A similar rock fragment produced by chipping with a hammer; e.g. a piece of ore broken by spalling.

spallation (spal-la'-tion) The ejection of atomic particles from a nucleus following the collision of an atom and a high-energy particle (e.g. a cosmic ray), which results in the formation of a different isotope that is not a *fission* product.

spalling (spall'-ing) (a) The chipping, fracturing, or fragmentation, and the upward and outward heaving, of rock caused by the interaction of a shock (compressional) wave at a free surface. (b) *exfoliation.*

spalmandite (spal'-man-dite) An unnecessary name for a variety of garnet intermediate in chemical composition between spessartine and almandine; a variety of spessartine rich in iron.

span [sed struc] A term introduced by Allen (1968, p.63) for the continuous length of the crest of a ripple, measured at right angles to the observed or inferred flow direction.

span [stratig] (a) The formal, hierarchical *diachronic unit* next lower in rank than phase and higher than *cline* (NACSN, 1983, Art. 93). (b) An informal term for an interval of geologic time.

spandite (span'-dite) An unnecessary name for a variety garnet intermediate in chemical composition between spessartine and andradite; a variety of spessartine rich in calcium and iron.

spangolite (span'-go-lite) A dark green trigonal mineral: $Cu_6Al(SO_4)Cl(OH)_{12}{\cdot}3H_2O$.

Spanish chalk (Span'-ish) A variety of steatite from the Aragon region of Spain.

Spanish topaz (a) Any orange, orange-brown, or orange-red variety of quartz resembling the color of topaz; e.g. heat-treated amethyst. (b) A wine-colored or brownish-red citrine occurring in Spain.

spar [mineral] A term loosely applied to any transparent or translucent light-colored crystalline mineral, usually readily cleavable and somewhat lustrous, esp. one occurring as a valuable nonmetallic mineral; e.g. Iceland spar (calcite), fluorspar (fluorite), heavy spar (barite), or feldspar. Obsolete syn: *spath.*

spar [mining] A miner's term for a small *clay vein* in a coal seam.

sparagmite (spa-rag'-mite) A collective term for the coarse, feldspar-rich, late Precambrian fragmental rocks of Scandinavia, esp. the conglomerates and sandstones of the Swedish Jotnian, consisting mainly of coarse arkoses and subarkoses (characterized by high proportions of microcline), together with polygenetic conglomerates and graywackes. Etymol: Greek "sparagma", "fragment, thing torn, piece".

sparite (spar'-ite) (a) A descriptive term for the crystalline transparent or translucent interstitial component of limestone, consisting of relatively coarse-grained calcite that was precipitated in situ. It is more coarsely crystalline than *micrite,* the grains having diameters that exceed 10 micrometers (Folk, 1959) or 20 micrometers (Chilingar et al., 1967, p.320). Syn: *sparry calcite; calcsparite.* (b) A limestone in which the sparite cement is more abundant than the micrite matrix. Syn: *sparry limestone.*

sparker (spark'-er) A marine seismic-energy source employing a high-voltage electrical discharge underwater.

spark spectrum The spectrum of light emitted by a substance, usually a gas or vapor, when an electric spark is passed through it. The spectrum is representative of the ionized atoms. Cf: *arc spectrum.*

Sparnacian (Spar-na'-cian) Obs. European stage: upper Upper Paleocene (above Thanetian, below Ypresian of Eocene or partially equivalent to the lower Ypresian).

sparry (spar'-ry) (a) Pertaining to, resembling, or consisting of spar; e.g. "sparry vein" or "sparry luster". (b) Pertaining to *sparite,* esp. in allusion to the relative clarity, both in thin section and hand specimen, of the calcite cement; abounding with sparite, such as a "sparry rock".

sparry calcite Clean, coarse-grained calcite crystal; *sparite.*

sparry iron *siderite [mineral].*

sparry limestone A limestone containing conspicuous coarse calcite crystals.

sparse biomicrite A *biomicrite* in which the skeletal grains make up 10-50% of the rock.

spartalite (spar'-ta-lite) *zincite.*

spasmodic turbidity current (spas-mod'-ic) A single, rapidly developed turbidity current, such as one initiated by a submarine earthquake. Cf: *steady turbidity current.*

spastolith (spas'-to-lith) A deformed oolith; e.g. a chamositic oolith that has been closely twisted or misshapen due to its soft condition at the time of burial (Pettijohn, 1957, p.97).

spate (a) A sudden flood on a river, caused by heavy rains or rapidly melting snow higher up the valley; a *freshet.* (b) A Scottish term for a flood.

spath An obsolete syn. of *spar [mineral].*

Spathian (Spath'-i-an) Canadian substage and suggested lower substage of the Olenekian Stage: Lower Triassic (above Smithian substage, below Anisian Stage).

spathic (spath'-ic) Resembling spar, esp. in regard to having good cleavage. Syn: *spathose.*

spathic iron A ferrous carbonate mineral with good rhombohedral cleavage; specif. siderite. Also called: spathose iron.

spathite (spath'-ite) A tubular stalactite, composed of aragonite, which consists of a vertical succession of small, petal-shaped, thin-walled cones.

spathose (spath'-ose) *spathic.*

spatial aliasing Misrepresentation of high-frequency data as low-frequency data due to undersampling; see *alias.* The aliasing frequency = velocity / (4 x bin size x sine of geologic dip).

spatial data base A data base containing information indexed by location (Krzanowski et al., 1993).

spatial dendrite (spa'-tial) A type of snow crystal somewhat like a *stellar crystal* except that branched arms form an irregular three-dimensional structure instead of building a pattern of hexagonal symmetry in a single plane.

spatial frequency The number of wave cycles per unit of distance in a given direction, often the direction of the seismic spread. Syn: *wave number.*

spatial information system An information system with the capability to manage spatially referenced information (Krzanowski et al., 1993). See also: *geographic information system.*

spatial resolution The ability to distinguish between closely spaced objects on an image. Commonly expressed as the most closely spaced line-pairs per unit distance that can be distinguished.

spatial sediment concentration The sediment contained in a unit volume of flow used to measure *transport concentration.*

spatial statistics Statistics applied to *multivariate* data characterized by two independent geographic coordinates. Spatial statistics are an extension of *time series* statistics, except the data are not ordered and *edge effects* often are significant. Cf: *geostatistics.*

spatio-temporal data Data defined with reference to both space and time (Krzanowski et al., 1993).

spatium (spa'-ti-um) A localized widening of an axial canal of a crinoid columnal opposite interarticular sutures. Pl: spatia.

spatter [meteorite] (spat'-ter) Droplets on the fusion-crust surface of meteorites.

spatter [pyroclast] An accumulation of initially very fluid pyroclasts, usually agglutinated, coating the surface around a vent. Syn: *driblet.*
spatter cone A low, steep-sided cone of spatter built up on a fissure or vent; it is usually of basaltic material. Syn: *volcanello; agglutinate cone.*
spatulate (spat'-u-late) Spoon-shaped; oblong with an attenuated base.
SPCS *State Plane Coordinate System.*
speaking rod (speak'-ing) *self-reading leveling rod.*
spear The recurved part of a hook of a holothurian.
Spearman's rank correlation A measure of the correlation between two variables when each variable has been replaced by the rank of the observation within the sample. Named for Charles Spearman (1863-1945), British psychologist.
spear pyrites A form of marcasite in twin crystals showing reentrant angles that resemble the head of a spear. Cf: *cockscomb pyrites.*
special creation (spe'-cial) The theory, strongly supported before the theory of evolution was generally accepted, that each species of organism inhabiting the Earth was created fully formed and perfect by some divine process.
special erosion Erosion effected by agents (such as wind, waves, and glaciers) that are important only within strictly limited areas or that work with help from the agents of *normal erosion* (Cotton, 1958, p.38). "The modern tendency ... is to regard any distinction between "normal" and "special" agencies as unreal" (Stamp, 1961, p.340).
specialist species (spe'-cial-ist) *equilibrium species.*
special position A location in a *unit cell* that falls on a symmetry element of the *space group*, and thus is mapped into itself by that symmetry element. Cf: *general equivalent position.*
special-purpose map (spe'-cial-pur'-pose) Any map designed primarily to meet specific requirements, and usually omitting or subordinating nonessential or less important information. Cf: *general-purpose geologic map.*
special-subject map *thematic map.*
speciation (spe-ci-a'-tion) (a) The production of new species of organisms from pre-existing ones during evolution. (b) The array of processes resulting in the origin of one or more new species (Wiley, 1981, p.8). Cf: *speciation event.* (c) The process of sorting a collection of many fossil or living specimens into groups, each of which represents one species.
speciation event The historical result of speciation; refers to a particular and historically unique speciation that has been preceded by one of the array of processes resulting in the origin of individual species. A speciation event may be instantaneous or protracted (Wiley, 1981, p.8). Cf: *speciation.*
species [mineral] (spe'-cies) A mineral distinguished from others by its unique combination of chemistry and crystal structure; it may have varieties.
species [paleont] A group of organisms, either plant or animal, that may interbreed and produce fertile offspring having similar structure, habits, and functions. As a fundamental unit in the hierarchy of classification, species ranks next below *genus.* The name of a species is a *binomen;* e.g. *Nuculana diversa.* Adj: *specific.* Abbrev: *sp.* Plural: species. Cf: *subspecies; epithet.*
species gap zone A zone designated by the gap between the end of the range of one zonal marker species and the commencement of the range of a second zonal species (McLean, 1968, p.10).
species index zone A zone designated by the total range of a designated species (McLean, 1968, p.10).
species overlap zone "A zone designated by the interval of range overlaps of two or more species whose ranges overlap in a restricted interval" (McLean, 1968, p.10).
species postlap zone "A zone designated by the part of the range of the marker species occurring after the cutoff of range of another marker species" (McLean, 1968, p.10).
species prelap zone "A zone designated by that portion of a marker species' range prior to the commencement of a marker which ends the zone by definition" (McLean, 1968, p.10).
species selection Natural selection operating at the level of the species, requiring emergent properties of the species as a whole. A process that generates differential proliferation of species within one clade relative to another clade.
species sorting Differential origination, extinction, or persistance of species. A pattern whereby species in one clade proliferate at a faster rate than species within another clade that can be driven by a variety of processes.
species zone The taxon-range-zone of a species (ISG, 1976, p.54).
specific (spe-cif'-ic) (a) The adj. of species. (b) Precise or limiting.
specific absorption The capacity of water-bearing material to absorb liquid, after removal of free water; the ratio of the volume of water absorbed to the volume of the saturated material. It is equal to *specific yield* except when the water-bearing material has been compacted due to the weight of overlying rocks.
specific activity The activity of a radioactive isotope, measured per unit mass of the element in the sample (or sample of radioactive material or pure radionuclide) and expressed in Bq/kg.
specific capacity The rate of discharge of a water well per unit of drawdown, commonly expressed in gallons per minute per foot. It varies slowly with duration of discharge. If the specific capacity is constant except for the time variation, it is proportional to the *hydraulic diffusivity* of the aquifer.
specific character A particular characteristic that serves to distinguish one species from others.
specific conductivity With reference to the movement of water in soil, a factor expressing the volume of transported water per unit of time in a given area.
specific discharge *Discharge [hydraul]* per unit area. It is often used to define the magnitude of a flood.
specific energy The energy of water in a stream; it equals the mean depth plus velocity head of the mean velocity (ASCE, 1962).
specific-gravity liquid (spe-cif'-ic-grav'-i-ty) *heavy liquid.*
specific head The height of the *energy line* above the bed of a conduit (ASCE, 1962).
specific heat capacity Heat capacity (c) divided by mass and measured in joules per kilogram kelvin (J $kg^{-1}K^{-1}$) at constant pressure (c_p), at constant volume (c_v), or at saturation (c_{sat}).
specific humidity The mass of water vapor per unit mass of moist air, usually expressed in g/kg. Cf: *absolute humidity; relative humidity.*
specific magnetization Magnetic moment per unit mass; magnetization divided by density.
specific mate recognition system The set of anatomical, behavorial, and chemical cues that allow males and females of a species to recognize one another for the purposes of reproduction.
specific name (a) The second term of a *binomen.* Cf: *epithet.* (b) A less preferred syn. of binomen.
specific permeability A factor expressing the permeability of a stream bed; it equals a constant times the square of representative pore diameter. Symbol: k. Syn: *intrinsic permeability.*
specific refractivity The *refractivity* of a substance divided by its density.
specific retention The ratio of the volume of water that a given body of rock or soil will hold against the pull of gravity to the volume of the body itself. It is usually expressed as a percentage. Cf: *field capacity.*
specific rotation The angle of rotation of plane-polarized light passing through a substance, measured in degrees per decimeter for liquids and solutions and in degrees per millimeter for solids. Syn: *specific rotatory power.* Cf: *optical rotation.*
specific rotatory power *specific rotation.*
specific storage The volume of water that a unit volume of aquifer releases from or takes into storage when the *pressure head* in the unit volume changes a unit amount (Domenico and Schwartz, 1990, p.115). Under saturated conditions, it has two components, pore-volume change and expansion of water. Under unsaturated conditions, a third component is dewatering of pores. Syn: *hydraulic capacitance.* Cf: *storage coefficient, storativity.*
specific susceptibility Susceptibility divided by density; the ratio of specific induced magnetization to the strength H of the magnetic field causing the magnetization. Syn: *mass susceptibility.*
specific tenacity The ratio of a material's tensile strength to its density.
specific yield The ratio of the volume of water that a given mass of saturated rock or soil will yield by gravity to the volume of that mass. This ratio is stated as a percentage. Cf: *effective porosity; storage coefficient; specific absorption.*
specimen (spec'-i-men) A sample, as of a fossil, rock, or ore; cf: *hand specimen.* Among miners, it is often restricted to selected or handsome samples, such as fine pieces of ore, crystals, or fragments of quartz showing visible gold.
specimen ore A particularly rich or well crystallized orebody.
speckstone (speck'-stone) An early name for talc or steatite. Etymol: German "Speckstein", "bacon stone", alluding to its greasy feel. Syn: *bacon stone.*
spectacle stone (spec'-ta-cle) *selenite.*
spectra (spec'-tra) Pl. of *spectrum.*
spectral (spec'-tral) Pertaining to a *spectrum [phys]*, e.g. *spectral line.*
spectral absorptance *Absorptance* measured at a specified wavelength. Cf: *total absorptance.*
spectral analysis *Fourier analysis.*
spectral band A well-defined, continuous wavelength range in the spectrum of reflected or radiated electromagnetic energy. The visible band contains the wavelengths for red, green, and blue light.
spectral emittance *Emittance* measured at a specific wavelength.
spectral gamma-ray log The *radioactivity log* curves of the intensity of natural gamma radiation within discrete energy bands characteristic of specific radioactive series (uranium-radium, thorium) or isotopes (potassium-40). It is used in correlation where other criteria fail; also in uranium exploration where thorium or potassium minerals contribute significantly to total gamma radiation. See also: *gamma-ray log; neutron-activation log.*
spectral line One component line in the array of intensity values of a spectrum emitted by a source. See also: *principal line [geochem].*
spectral log *spectral gamma-ray log.*
spectral radiance Radiance per unit wavelength interval at a particular wavelength. Symbol: N_λ.
spectral radiant flux *Radiant flux* measured at a specific wavelength.
spectral reflectance *Reflectance* measured at a specific wavelength.

spectral resolution Range of wavelengths recorded by a detector.
spectrochemical analysis (spec-tro-chem'-i-cal) Chemical analysis based on the spectral characteristics of substances. Syn: *spectrum analysis.*
spectrochemistry (spec-tro-chem'-is-try) The branch of chemistry concerned with *spectrochemical analysis.*
spectrocolorimeter (spec'-tro-col'-or-im'-e-ter) Essentially an absorption *spectrophotometer,* used to measure the absorptance of solutions over an entire spectrum, providing quantitative information about the composition of the solution. It is often used as a syn. of spectrophotometer. See also: *colorimeter.*
spectrocolorimetry (spec'-tro-col'-or-im'-e-try) The art or process of using the *spectrocolorimeter* for the quantitative study of color.
spectrogram (spec'-tro-gram) A map, photograph, or other picture of a spectrum, usually produced by a *spectrograph.*
spectrograph (spec'-tro-graph) A *spectroscope* designed to map or photograph a spectrum.
spectrography (spec-trog'-ra-phy) The art or process of using a *spectrograph* to photograph or map a spectrum.
spectrolite (spec'-tro-lite) A dark variety of the mineral labradorite that shows exceptional play of color and is used as a precious stone. It is found in Ylämaa, eastern Finland.
spectrometer (spec-trom'-e-ter) An instrument used for *spectroscopy* and the measuring, recording, and the comparing of the intensity of light in different parts of a spectrum as a function of wavelength; formerly called a spectrophotometer. See also: *spectrometry, flame photometer, spectrocolorimeter.*
spectrometry (spec-trom'-e-try) The process of using a spectrometer; also, the recording and quantitative components of *spectroscopy.*
spectrophotometer (spec'-tro-pho-tom'-e-ter) *spectrometer.*
spectrophotometry (spec'-tro-pho-tom'-e-try) *spectrometry.*
spectroreflectometer (spec'-tro-re'-flec-tom'-e-ter) An instrument for measuring and analyzing the *reflection spectrum* of a source.
spectroscope (spec'-tro-scope) An instrument for producing and visually observing a spectrum.
spectroscopy (spec-tros'-co-py) Analytical techniques used to produce, record, measure, analyze, and interpret spectra using *spectroscopes, spectrographs,* and spectrometers. See also: *absorption spectroscopy; atomic absorption spectroscopy; emission spectroscopy; flame spectroscopy; gamma-ray spectroscopy; infrared absorption spectroscopy; mass spectroscopy; Mössbauer spectroscopy; optical emission spectroscopy; Raman spectroscopy, visible region spectroscopy, X-ray emission spectroscopy, X-ray fluorescence spectroscopy.*
spectrum [phys] n. (a) An array of visible light ordered according to its constituent wavelengths (colors) by being sent through a prism or diffraction grating. (b) An array of intensity values ordered according to any physical parameter, e.g. energy spectrum, mass spectrum, velocity spectrum. (c) A portion of the electromagnetic spectrum observed relative to increasing or decreasing frequency or wavelength. Also any data or entity displayed relative to a continuum of increasing or decreasing magnitude. See also: *absorption spectrum; continuous spectrum; electromagnetic spectrum; emission spectrum.* (d) Amplitude and phase response as a function of frequency for the components of a wavetrain, such as given by Fourier analysis, or as used to specify filter-response characteristics. Pl: *spectra.* Adj: *spectral.*
spectrum analysis *spectrochemical analysis.*
specular coal (spec'-u-lar) *pitch coal.*
specular iron A syn. of *specularite.* Also called: specular iron ore.
specularite (spec'-u-lar-ite') A black or gray variety of hematite with a splendent metallic luster, often showing iridescence. It occurs in micaceous or foliated masses, or in tabular or disklike crystals. Syn: *specular iron; gray hematite; iron glance.*
specular schist Metamorphosed oxide facies iron formation characterized by a high percentage of strongly aligned flakes of specular hematite.
specular surface A surface that is smooth with respect to the wavelength incident upon it.
speculative resources (spec'-u-la'-tive) Undiscovered mineral resources that may occur either in known types of deposits in a favorable geologic setting where no discoveries have yet been made, or in as-yet-unknown types of deposits that remain to be recognized (Brobst and Pratt, 1973, p.2). Cf: *hypothetical resources; identified resources.*
speed In photography, the response or sensitivity of a photographic film, plate, or paper to light; also, the light-gathering power of a lens or optical system, expressed as the *relative aperture* of the lens. See also: *f-number.*
speed of light A geodetic constant with the value 299,792,458±1.2 m/s.
speisscobalt (speiss-co'-balt) *smaltite.*
spelean (spe'-le-an) Of, pertaining to, or related to caves.
speleochronology (spe'-le-o-chro-nol'-o-gy) The dating or chronology of a cave's formation, or of its mineral deposits or filling. The dating may be either relative or numerical.
speleofact A morphologic feature of a cave wall, including a *speleogen, speleothem,* or *petromorph.*
speleogen (spe'-le-o-gen') In a cave, any surface that is formed by solution, such as a *scallop, pendant,* or *pit.* Etymol: Greek,, "cave born".
speleogenesis (spe-le-o-gen'-e-sis) The process of cave formation.
speleologist (spe-le-ol'-o-gist) A scientist engaged in *speleology.* Cf: *caver.*
speleology (spe-le-ol'-o-gy) The exploration and scientific study of caves, both physical and biological, including geologic studies of their genesis, morphology, and mineralogy. The term was first published by Martel in 1896. Cf: *caving [speleo]; speleologist.*
speleothem (spe'-le-o-them) Any secondary mineral deposit that is formed in a cave (Moore, 1952). Obsolete syn: *cave formation; formation [speleo].* Cf: *cave onyx; dripstone.* Etymol: Greek, "cave deposit".
spelunker (spe-lunk'-er) A journalistic or an outsider's term for a *caver.*
spelunking (spe-lunk'-ing) *caving [speleo].*
spencerite (spen'-cer-ite) (a) A pearly-white monoclinic mineral: $Zn_4(OH)_2(PO_4)_2{\cdot}3H_2O$. (b) An artificial substance: $(Fe,Mn)_3(C,Si)$.
spencite (spen'-cite) *tritomite-(Y).*
spermatophyte (sper-mat'-o-phyte) A vascular plant that produces seeds, e.g. a gymnosperm or angiosperm; a *seed plant.* Such plants range from the Carboniferous. Cf: *pteridophyte; phanerogam.*
spermatozoid (sper'-mat-o-zo'-id) A motile male gamete characteristically produced in an antheridium by the gametophyte generation in pteridophytes. Fusion of the spermatozoid with an egg cell completes the fertilization process.
sperone (sper'-one) A vesicular *leucitite* that contains small andradite crystals. Obsolete.
sperrylite (sper'-ry-lite) A tin-white cubic mineral of the *pyrite* group: $PtAs_2$.
spertiniite (sper-ti'-ni-ite) A blue or blue-green orthorhombic mineral: $Cu(OH)_2$.
spessartine (spes'-sar-tine) The manganese-aluminum end-member of the *garnet* group: $Mn^{2+}{}_3Al_2(SiO_4)_3$. It has a brown-red to yellow-brown color, and usually contains some iron, magnesium, and other elements in minor amounts. Spessartine is rather rare; it occurs in pegmatites and granites. Syn: *spessartite [mineral].*
spessartite [mineral] (spes'-sar-tite) *spessartine.*
spessartite [rock] (a) In the *IUGS classification,* a variety of *lamprophyre* in which plagioclase is more abundant than alkali feldspar, no feldspathoid is present, and phenocrysts are mainly hornblende and clinopyroxene. (b) A *lamprophyre* composed of phenocrysts of green hornblende or clinopyroxene in a groundmass of sodic plagioclase, with accessory olivine, biotite, apatite, and opaque oxides. Rosenbusch derived the name in 1896 from Spessart, Germany.
spew frost *pipkrake.*
sphaeraster (sphae'-ras-ter) A many-rayed sponge spicule (euaster) in which the rays radiate from a prominent solid spherical centrum. Also spelled: *spheraster.*
sphaerite (sphae'-rite) A dubious light gray or bluish mineral consisting of hydrous aluminum phosphate in globular concretions. It is perhaps the same as *variscite.* Also spelled: spherite.
sphaerobertrandite A white, yellow, brownish, grayish, or beige monoclinic mineral: $Be_3SiO_4(OH)_2$.
sphaerobismoite An adamantine green or yellowish tetragonal mineral: Bi_2O_3. It is a dimorph of bismite.
sphaeroclone (sphae'-ro-clone) An *ennomoclone* in which six or more proximal arms radiate from one side of a frequently spherical centrum and terminate in cuplike zygomes.
sphaerocobaltite A peachblossom-red mineral of the *calcite* group: $CoCO_3$. It occurs in spherical masses. Also spelled: *spherocobaltite.* Syn: *cobaltocalcite.*
sphaerocone (sphae'-ro-cone) A coiled, depressed, *involute,* globular cephalopod shell that has a small nearly occluded *umbilicus* and a round venter, and that commonly opens out suddenly along the last whorl (as in *Sphaeroceras*).
sphaeromorph (sphae'-ro-morph) One of a heterogeneous group of *acritarchs* with a spherical to ellipsoidal *vesicle* that lacks *processes* but may show *sculpture.* Some sphaeromorphs may be related to *prasinophytes* (Downie et al., 1963; Mendelson, 1993).
sphaeroplast (sphae'-ro-plast) A lorica-forming granule representing part of the shieldlike mass formed during reproduction in tintinnids when a single cell divides into two theoretically equal parts.
sphaerosiderite (sphaer-o-sid'-er-ite) A variety of siderite occurring in globular concretionary aggregates of bladelike crystals radiating from a center, generally in a clayey matrix (such as those in or below underclays associated with coal measures). It appears to be the result of weathering of water-logged sediments in which iron, leached out of surface soil, is redeposited in a lower zone characterized by reducing conditions. Also spelled: *spherosiderite.*
sphagnum bog An acid freshwater bog containing abundant sphagnum, which may ultimately form a deposit of sphagnum peat (*highmoor peat*).
sphagnum moss A moss of the genus *Sphagnum,* often forming peat; *peat moss.*
sphagnum peat *highmoor peat.*
sphalerite (sphal'-er-ite) (a) A yellow, brown, or black cubic mineral: $(Zn,Fe)S$. It is dimorphous with wurtzite, and often contains manganese, arsenic, cadmium, and other elements. Sphalerite has a highly perfect dodecahedral cleavage and a resinous to adamantine luster. It is a widely distributed ore of zinc, commonly associated with galena in veins and other

deposits. Syn: *blende; zinc blende; jack [mineral]; blackjack [mineral]; steel jack; false galena; pseudogalena; mock ore; mock lead.* (b) A group name for cubic minerals with the formula *AX*, where *A* = Zn, Cd or Hg, and *X* = S, Se or Te.

Sphenacodontia (Sphe'-nac-o-don'-ti-a) A suborder of pelycosaurs characterized by adaptation to a predaceous habit. Range, Upper Pennsylvanian to ?Upper Permian.

sphene *titanite.*

spheniscidite (sphen-is'-cid-ite) A brown monoclinic mineral of the *leucophosphite* group: $(NH_4,K)(Fe^{3+},Al)_2(PO_4)_2(OH)\cdot 2H_2O$.

sphenitite (sphe'-ni-tite) A titanite-rich *jacupirangite.* The term was introduced by Allen in 1914. Obsolete.

sphenochasm (sphe'-no-chasm) A triangular gap of oceanic crust separating two continental blocks and converging to a point; it is interpreted by Carey (1958) as having originated by the rotation of one of the blocks with respect to the other. Cf: *rhombochasm.*

sphenoconformity (sphe'-no-con-form'-i-ty) A term used by Crosby (1912, p.297) for the relation between conformable strata that are thinner in one locality than in the other, though fully represented in both.

Sphenodontida (Sphen-o-don'-ti-da) An order of lepidosaurian reptiles that includes animals of small to moderate size in which the jaws form a beak and dentitions are modified for crushing. Range Upper Triassic to Holocene.

sphenoid (sphe'-noid) An open crystal form having two nonparallel faces that are symmetrical to an axis of twofold symmetry. It occurs in monoclinic crystals of the sphenoidal class. Cf: *dome [cryst]; disphenoid.*

sphenoidal class (sphe-noi'-dal) That crystal class in the monoclinic system having symmetry 2.

sphenolith [intrus rocks] (sphe'-no-lith) A wedgelike igneous intrusion, partly concordant and partly discordant. Obsolete.

sphenolith [paleont] A *coccolith* having a prismatic base formed by radial elements surmounted by a cone.

sphenophyte (sphen'-o-phyte) *sphenopsid.*

sphenopsid (sphe-nop'-sid) A *pteridophyte* of the class Sphenopsida, characterized by distinctly jointed stems, whorled appendages, and spores borne in cones (*strobili*). Sphenopsids range from the Middle Devonian to the present; the living members are species of *Equisetum* (horsetail, scouring rush). Syn: *equisetophytes;* informally *sphenophyte.*

spheraster (sphe'-ras-ter) *sphaeraster.*

sphere A standard shape taken as a reference form in the analysis of sedimentary-particle shapes; the limiting shape assumed by many rock and mineral fragments upon prolonged abrasion, being a solid figure bounded by a uniformly curved surface, any point on which is equidistant from the center. It has the least surface area for a given volume and the greatest settling velocity of any possible shape (under conditions of low velocity and of constant volume and density). See also: *spheroid [sed].*

sphere of reflection In X-ray crystallography, an imaginary sphere of radius $1/\lambda$, where λ is the wavelength of the X-rays, centered on the crystal. The Bragg law is satisfied for a set of lattice planes when the corresponding *reciprocal lattice point* is coincident with the surface of the sphere. Syn: *Ewald sphere.*

sphere ore *cockade ore.*

spheric (spher'-ic) *sferic.*

spherical bomb (spher'-i-cal) *spheroidal bomb.*

spherical cap In gravimetry, part of a spherical shell limited by a circular cone with the apex in the center of the sphere.

spherical coordinates (a) Three coordinates that represent a distance and two angles in space, consisting of two polar coordinates in a plane and the angle between this plane and a fixed plane containing the primary axis of direction (polar axis). The term includes coordinates on any spherical surface or on any surface approximating a sphere (such as the surface of the Earth). (b) A system of polar coordinates in which the origin is the center of a sphere and the points all lie on the surface of the sphere; a coordinate system used to define the position of a point with reference to two great circles that form a pair of axes (such as longitude and latitude) at right angles to each other or with reference to an origin and a great circle through the point.

spherical divergence The decrease in amplitude of seismic body waves with distance from the source. The amplitude varies inversely as the square of the distance. Cf: *cylindrical divergence.*

spherical harmonic analysis In spherical coordinates, decomposition into spherical harmonics analogous to Fourier analysis. Spherical harmonics are solutions to Laplace's equation.

spherical triangle A *triangle* on the surface of a sphere, having sides that are arcs of three great circles.

spherical wave A seismic wave propagated from a point source. Wavefronts are concentric spheres in a constant-velocity isotropic medium.

spherical weathering *spheroidal weathering.*

sphericity (sphe-ric'-i-ty) The relation to each other of the various diameters (length, width, thickness) of a particle; specif. the degree to which the shape of a sedimentary particle approaches that of a sphere. True sphericity, as originally defined by Wadell (1932), is the ratio of the surface area of a sphere of the same volume as the particle to the actual surface area of the particle. Due to the difficulty of determining the actual surface area and volume of irregular solids, Wadell (1934) developed an operational definition expressed as the ratio of the true nominal diameter of the particle to the diameter of a circumscribing sphere (generally the longest diameter of the particle). A perfect sphere has a sphericity of 1.0; all other objects have values less than 1.0. Many other measures of sphericity have been proposed. Not to be confused with *roundness* or *angularity.*

spherite [mineral] (sphe'-rite) *sphaerite.*

spherite [sed] (a) A sedimentary rock composed of gravel-sized aggregates of constructional (nonclastic) origin, simulating in texture a rudite of clastic origin; e.g. a rock formed of volcanic bombs. The term was introduced by Grabau (1911, p.1007). (b) An individual spherical grain in a sedimentary rock, such as a concentric oolith in an oolite or a radial spherulite in a limestone.

spheroclast (sphe'-ro-clast) A rounded *phenoclast,* such as a pebble or cobble of a conglomerate. Cf: *anguclast.*

spherocobaltite *sphaerocobaltite.*

spheroid [geodesy] (sphe'-roid) Any figure that generally differs little from a sphere, specif. in geodesy the *spherop* whose potential is identical to that of the *geoid.* It is sometimes used as a syn. of *ellipsoid of revolution.* Cf: *ellipsoid.*

spheroid [sed] A spherical, or equant or equiaxial, shape of a sedimentary particle, defined in *Zingg's classification* as having width/length and thickness/width ratios greater than $^2\!/_3$. See also: *sphere.*

spheroidal (sphe-roi'-dal) (a) Having the shape of a spheroid. (b) Composed of spherules.

spheroidal bomb A rotational volcanic bomb in the shape of an oblate spheroid. Syn: *spherical bomb.*

spheroidal joint *spheroidal parts.*

spheroidal parts A series of concentric and spheroidal or ellipsoidal cracks produced about compact nuclei in fine-grained, homogeneous rocks. Cf: *spheroidal weathering.* Syn: *spheroidal joint.*

spheroidal symmetry *axial symmetry.*

spheroidal weathering A form of chemical weathering in which concentric or spherical shells of decayed rock (ranging in diameter from 2 cm to 2 m) are successively loosened and separated from a block of rock by water penetrating the bounding joints or other fractures and attacking the block from all sides. It commonly forms a rounded *boulder of decomposition.* It is similar to the larger-scale *exfoliation* produced usually by mechanical weathering. See also: *spheroidal parts.* Syn: *onion-skin weathering; concentric weathering; spherical weathering.*

spherop (sphe'-rop) An equipotential surface in the normal gravity field of the Earth; a surface such that the spheropotential is constant and the normal gravity is perpendicular to it at every point. See also: *spheropotential number.* Syn: *spheropotential surface.*

spherophyre (sphe'-ro-phyre) An igneous rock in which the phenocrysts are aggregations of crystals in the form of spherulites. Obsolete.

spheropotential (sphe'-ro-po-ten'-tial) The potential function of the normal gravity defined as either the external gravity potential of a level ellipsoid approximating the Earth, or as a function consisting of the first few terms of some expansion of the *geopotential.* See also: *disturbing potential.*

spheropotential number The spheropotential difference between the marigraph *spherop* and the spherop through an observation point. It is expressed in geopotential units. Syn: *normal geopotential number.*

spheropotential surface *spherop.*

spherosiderite *sphaerosiderite.*

spherule (spher'-ule) A little sphere or spherical body; e.g., a "magnetic spherule" in a deep-sea sediment, or an object that appears to be an amygdule or a spherulite.

spherulite [petrology] (spher'-u-lite) A rounded or spherical mass of acicular crystals, commonly of feldspar, radiating from a central point. Spherulites may range in size from microscopic to several centimeters in diameter (Stokes and Varnes, 1955, p.140). Most commonly formed by the devitrication of volcanic glass. Cf: *variole; spheruloid; orbicule.* Also spelled: sphaerolite.

spherulite [sed] (a) Any more or less spherical body or coarsely crystalline aggregate with a radial internal structure arranged around one or more centers, varying in size from microscopic grains to objects many centimeters in diameter, formed in a sedimentary rock in the place where it is now found; e.g. a minute particle of chalcedony in certain limestones, or a large carbonate concretion or nodule in shale. Cf: *spherite [sed].* (b) A small (0.5-5 mm in diameter), spherical or spheroidal particle composed of a thin, dense calcareous outer layer with a sparry calcite core. It can originate by recrystallization or by biologic processes.

spherulitic (spher-u-lit'-ic) Volcanic igneous texture dominated by spherulites or spherical bodies of radiating mineral fibers. Cf: *variolitic.* Syn: *globular.*

spherulitic prismatic structure A *prismatic structure* in which each first-order prism consists of an aggregation of elongate second-order prisms radiating toward the depositional surface from a spherulite or from a single point of origin. This category includes regular, irregular, and planar spherulitic prismatic structures, but it excludes composite prismatic and spherulitic structures (Mutvei, 1964, p.257; Carter and Tevesz, 1978; Carter, 1980b; Watabe, 1984). Abbrev: SphP.

spheruloid (spher'-u-loid) n. A spherule or nodule that lacks radial structure, as in perlitic lava (Nelson and Nelson, 1967, p.355). Cf: *spherulite [petrology].*

sphinctozoan (sphinc′-to-zo′-an) Any calcisponge having a skeleton that consists of straight, curved, or branched series of hollow spheroidal bodies (TIP, 1955, pt.E, p.100).

SphP *spherulitic prismatic structure.*

spicularite (spic′-u-lar-ite′) A sediment or rock composed principally of the siliceous spicules of invertebrates; esp. a *spongolite* composed principally of sponge spicules. Syn: *spiculite [sed]; sponge-spicule rock.*

spiculation (spic-u-la′-tion) The formation, or the form and arrangement, of spicules. Also, a spicular component (as of a sponge).

spicule (spic′-ule) (a) One of the numerous minute calcareous or siliceous bodies, having highly varied and often characteristic forms, occurring in and serving to stiffen and support the tissues of various invertebrates, and frequently found in marine-sediment samples and in Paleozoic and Cretaceous cherts. Examples: a discrete skeletal element of a sponge, typically a needlelike rod or a fused cluster of such rods; a long sharp calcareous skeletal element of the mesogloea of an octocoral; a discrete elongate or needlelike skeletal element of many radiolarians; a scalelike calcareous object borne on the girdle of a primitive chiton; an irregular calcareous body secreted within the connective tissue of a brachiopod; and a minute cylindrical or radiate skeletal element of an asterozoan. (b) The empty siliceous shell of a diatom.

spicule tract A linear series or bundle of separate sponge spicules, usually held together by spongin.

spiculin (spic′-u-lin) The chemically undetermined protein substance that forms the axial filament of sponge spicules.

spiculite [sed] *spicularite.*

spiculoblast (spic′-u-lo-blast′) *sclerocyte.*

spiculofiber (spic′-u-lo-fi′-ber) A fiberlike structure built of a bundle of sponge spicules held together by mutual interlocking or fusion, or by *spongin* or *sclerosome.*

spiculoid (spic′-u-loid) A discrete autochthonous element of a sponge skeleton, resembling a *spicule* but made of organic material only. Syn: *pseudospicule.*

spider diagram A graph in which the vertical axis is the ratio of the concentration of a trace element in a sample to the concentration of that element in some reference material (average chondritic meteorite, estimated primitive mantle, etc.). On the horizontal axis the elements are arranged in order of their bulk partition coefficients. Also called: spidergram.

spike The known amount of an isotope added to a sample to determine the unknown amount present in analysis by *isotope dilution.* Syn: *tracer.*

spilite (spi′-lite) An altered basalt, characteristically amygdaloidal or vesicular, in which the feldspar has been albitized and is typically accompanied by chlorite, calcite, epidote, chalcedony, prehnite, or other low-temperature hydrous crystallization products characteristic of a greenstone. Spilite often occurs as submarine lava flows and exhibits pillow structure. Adj: spilitic. The name, given by Brongniart in 1827, is widely used, in spite of Johannsen's suggestion (1937, p.300) that it be dropped.

spilitic suite (spi-lit′-ic) A group of altered extrusive and minor intrusive basaltic rocks that characteristically have a high albite content. The group is named for its type member, *spilite.*

spilitization (spi′-lit-i-za′-tion) Albitization of a basalt to form a spilite.

spill bank A term used in Great Britain and India for a *natural levee.*

spilling breaker (spill′-ing) A breaker whose crest collapses gradually over a nearly flat bottom for a relatively long distance, forming a foamy patch at the crest, the water spilling down continuously over the advancing wave front. See also: *comber.* Cf: *plunging breaker; surging breaker.*

spillpoint (spill′-point) The point of maximum filling of a structural trap by oil or gas.

spill stream *overflow stream.*

spillway [eng] (spill′-way) A passage or outlet (usually constructed) through which surplus water flows from a pond, lake, or reservoir impounded by a dam or natural obstruction.

spillway [glac geol] *overflow channel.*

spilosite (spi′-lo-site) A rock representing an early stage in the formation of *adinole* or *spotted slate.* Rarely used.

spinach jade (spin′-ach) Dark-green nephrite.

spindle-shaped bomb (spin′-dle-shaped) A rotational volcanic bomb of fusiform shape, with earlike projections at its ends. See also: *fusiform bomb.* Syn: *almond-shaped bomb.*

spindle stage (spin′-dle) A single-axis stage of a microscope, consisting of a liquid-filled cell in which the crystal is immersed; the crystal is mounted on a horizontal needle (or spindle) and may be rotated 180°.

spindrift Spray blown from wave crests during gales at sea.

spine [bot] A thorn; a rigid, sharply pointed structure that may represent a modified leaf or leaf part, petiole, or stipule.

spine [paleont] (a) A projection of the shell surface found on various invertebrates, e.g. a movable calcareous shaft mounted on, and articulating with, a tubercle on the test of an echinoid or asteroid; a cylindrical or elongated triangular projection from the external shell surface of a brachiopod; or a hollow tubular skeletal projection with contained body cavity (*lumen*) at or near the margin of the orifice of a cheilostome bryozoan. (b) A sharply pointed process or an elongated, narrow ridge on bones. A bony or keratinous fin ray in fishes.

spine [palyn] A *sculpture* element in *echinate palynomorphs.* Cf: *spine [paleont].*

spine [volc] A pointed mass or monolith of solidified lava that sometimes occurs over the throat of a volcano. It may be formed by slow, forced extrusion of viscous lava, or it may represent magma in the pipe that was exposed by weathering. Syn: *peleean dome.*

spinel (spi-nel′) (a) A mineral: $MgAl_2O_4$. The magnesium may be replaced in part by ferrous iron, and the aluminum by ferric iron. Spinel has great hardness, usually forms octahedral crystals (cubic system), varies widely in color (from colorless to purple-red, green, and yellow to black), and is used as a gemstone. It occurs typically as a product of contact metamorphism of impure dolomitic limestone, and less commonly as an accessory mineral of basic igneous rocks; it also occurs in alluvial deposits. (b) A group of minerals of general formula: AB_2O_4 where A = Mg, Fe^{2+}, Zn, Mn^{2+}, Ni, Co, Cu and is in tetrahedral coordination by oxygen, and B = Al, Fe^{3+}, Cr, V^{3+}, Ti^{4+}, Ge, or Sb and is in octahedral coordination by oxygen. In "inverse" spinels, half of the B^{3+} atoms are in tetrahedral sites, and the remaining B^{3+} atoms and the B^{2+} atoms are in octahedral sites. (c) A member of the spinel group or spinel series. (d) A substance (such as a sulfide) that has a similar formula and the same crystal structure as a spinel. (e) An artificial substance, similar to the mineral spinel, that is used as a gemstone, a refractory, or instrument bearings; e.g. ferrospinel. Also spelled: spinelle; spinell.

spinel lherzolite A *lherzolite* containing spinel as the dominant aluminum-rich mineral; stable in the mantle at depths between *plagioclase lherzolite* and *garnet lherzolite.*

spinellid (spi-nel′-lid) A mineral of the *spinel* group.

spinellide (spi-nel′-lide) A name applied to the *spinel* group.

spinel ruby A deep-red gem variety of spinel. The term is sometimes used inappropriately as a syn. of *ruby spinel.*

spinel twin law A twin law in crystals of the hexoctahedral class, isometric system, e.g. spinel, having a twin axis of threefold symmetry with the twin plane parallel to one of the octahedron's faces.

spinifex texture (spin′-i-fex) Volcanic igneous texture consisting of needlelike olivine or pyroxene crystals forming a criss-crossing pattern, the result of rapid quenching from unusually high temperatures; typical of komatiites. Their disposition resembles the intermesh of an Australian grass for which the texture is named.

spinispire (spi′-ni-spire) A siliceous monaxonic sponge spicule (microsclere) in the form of a spiral of more than one revolution. Cf: *sigmaspire.* Syn: *spiraster.*

spinneret (spin-ner-et′) A spinning organ, or an organ for producing threads of silk from the secretion of the silk glands of an arachnid; specif. an abdominal appendage of spiders, with spinning tubes at the end, or a special spinning organ on the movable finger (galea) of a *chelicera* of pseudoscorpions.

spinner magnetometer (spin′-ner) An instrument used to measure the direction and magnitude of a moment in a material by continuously rotating the specimen, to produce an alternating voltage in a nearby coil by electromagnetic induction.

spinodal decomposition *Exsolution* that proceeds by small, pervasive compositional fluctuations that become increasingly divergent with time.

spinose (spi′-nose) Spinelike, or full of or armed with spines; e.g. said of a foraminiferal test having fine elongate solid spines on its surface (as in *Hastigerinella*) with each spine comprising a single calcite crystal elongated along the c axis.

S-P interval In earthquake seismology, the time interval between the first arrivals of longitudinal and transverse waves, which is a measure of the distance from the earthquake source.

spinulus (spi′-nu-lus) A closed-end process extending from the surface of a valve of a diatom frustule.

spionkopite (spi-on′-ko-pite) A metallic blue-gray trigonal mineral: $Cu_{39}S_{28}$. It and yarrowite were formerly known as "blaubleibend covellite".

spiracle [paleont] (spi′-ra-cle) (a) A large, generally rounded opening near the adoral tip of a deltoid plate of a blastoid and excavated within it. It opens into the space enclosed by a *hydrospire.* (b) A *stigma* of an arachnid.

spiracle [volc] A fumarolic vent in a lava flow, formed by a gaseous explosion in lava that is still fluid, probably due to generation of steam from underlying wet material. It is usually about one meter in diameter and up to five meters high, although in the NW U.S., where spiracles are common, they may be larger. Cf: *pipe vesicle.*

spiracular slit (spi-rac′-u-lar) An elongate spiracle at the side of a blastoid *ambulacrum,* excavated in adjoining radial and deltoid plates.

spiraculate (spi-rac′-u-late) Having spiracles.

spiral angle (spi′-ral) The angle formed between two straight lines tangent to the periphery of two or more whorls on opposite sides of a gastropod shell. It is commonly determined by drawing tangents to the lowermost whorls of the spire. Syn: spire angle.

spiral ball A term used by Hadding (1931, p.389) for a sandstone body having a rolled-up, spiral structure due to lateral mass flowage of thin interbedded sands and shales. Cf: *slump ball.*

spiral canal The part of the canal system in the umbilical region of a foraminiferal test (as in *Elphidium*) parallel to and inside the lateral-chamber margins.

spiral cone complex crossed foliated structure A *cone complex crossed foliated structure* in which each first-order lamella is an *Archimedes*-type conical spiral. (Wada, 1963a,b, 1968).

spiral cone complex crossed lamellar structure A *cone complex crossed lamellar structure* in which each first-order lamella is an *Archimedes*-type conical spiral. In radial and transverse sections, this structure is not readily distinguishable from nonspiral cone complex crossed lamellar structure (Carter, 1990, Fig. 38). Based on the "spiral crossed lamellar" structure described by Prezant and Tan Tiu (1986a,b).

spiral garnet *rotated garnet.*

spiral growth Growth of a crystal along a *screw dislocation.* It may result in a *growth island* on the surface of the crystal.

spiralium (spi-ral'-i-um) A spiral *brachidium;* one of a pair of spirally ribbonlike calcareous supports for the *deuterolophe* or the *spirolophe* of certain brachiopods, composed of secondary shell. Pl: spiralia. Syn: *spire.*

spiral lamina The coiled or winding part of the *lorica* of a tintinnid.

spiral side The part of a foraminiferal test where all whorls are visible (e.g., trochospiral forms), also commonly called dorsal side.

spiral suture A line of contact between two whorls in the coiled test of a foraminifer.

spiral tracheid *helical tracheid.*

spiramen (spi-ra'-men) A pore in the proximal wall of the *peristome* in some ascophoran cheilostome bryozoans, leading into the space enclosed by the peristome outward from the operculum-bearing *orifice.*

spiraster (spi'-ras-ter) A spiral sponge *spicule;* e.g. a *streptosclere* or a *spinispire.*

spire (a) The adapical visible upper part of a spiral gastropod shell, including the whole series of whorls except the last or body whorl. (b) A *spiralium* of a brachiopod. (c) *spirillum.*

spiriferoid (spi-rif'-er-oid) Any articulated brachiopod belonging to the order Spiriferida, characterized generally by a spiral brachidium and a biconvex, rarely planoconvex, shell. Range, Middle Ordovician to Jurassic. Var: spiriferid; spirifer.

spirilline (spi'-ril-line) Said of a foraminiferal test consisting of a planispiral nonseptate tube enrolled about a globular proloculus; specif. pertaining to the foraminiferal genus *Spirillina.*

spirillum (spi'-ril-lum) (a) A helical or coiled morphologic form of a bacterial cell. Syn: *spire.* (b) Any bacterium of the genus *Spirillum,* now restricted to elongated forms having tufts of flagella at one or both ends. Pl: spirilla.

spirit level (spir'-it) (a) A sensitive device for finding a horizontal line or plane, consisting of a small closed glass tube or vial of circular cross section, nearly filled with a liquid of low viscosity (ether or alcohol) with enough free space being left for the formation of a bubble of air or gas that will always assume a position at the top of the tube. See also: *circular level.* Syn: *level [surv].* (b) An instrument using a spirit level to establish a horizontal line of sight.

spirit leveling A type of *leveling* using a spirit level to establish a horizontal line of sight.

spiroffite (spi'-roff-ite) A deep pink to purplish-pink monoclinic mineral: $(Mn^{2+},Zn)_2Te^{4+}{}_3O_8$.

spirogyrate (spi-ro-gy'-rate) Said of the umbones of a bivalve mollusk, coiled outward from an anteriorly and posteriorly directed (sagittal) plane of symmetry.

spirolophe (spi'-ro-lophe) A brachiopod *lophophore* in which the brachia are spirally coiled and bear a single row of paired filamentary appendages. Cf: *deuterolophe; plectolophe.*

spirotheca (spi-ro-the'-ca) The outer or upper spiral wall of the test in fusulinids.

spiroumbilical aperture (spi'-ro-um-bil'-i-cal) An *interiomarginal aperture* in a foraminiferal test, extending from umbilicus to periphery and finally onto the spiral (dorsal) side where all whorls are visible.

spit (a) A small point or low tongue or narrow embankment of land, commonly consisting of sand or gravel deposited by longshore transport and having one end attached to the mainland and the other terminating in open water, usually the sea; a fingerlike extension of the beach. Syn: *barrier beach.* (b) A relatively long, narrow shoal or reef extending from the shore into a body of water.

spitskop (spits'-kop) A term used in South Africa for a hill with a sharply pointed top. Etymol: Afrikaans. Cf: *tafelkop.*

spitz Nonrecommended syn. of *tip [paleont].*

spitzkarren (spitz'-kar-ren) In *karst,* large deep grooves extending down from steep spires or pinnacles. Etymol: German, "peak tracks". Cf: *karren.*

s plane In structural geology, a nongenetic term for any planar fabric element; e.g., foliation or bedding. Syn: *s surface.*

splash cup In a cave, a depression in a *stalagmite,* mud, or other material caused by the impact of falling water.

splash erosion The dislodgment and movement of soil particles under the impact of falling raindrops.

splash zone The shoreline area that is affected by the splashing of seawater from breaking waves. Cf: *spray zone.*

splay [geomorph] *floodplain splay.*

splay [struc geol] A minor fault that branches off of a larger fault. Commonly major faults terminate in an array of splays. See also: *horsetail [struc geol].*

splaying crevasse (splay'-ing) A crevasse in a valley glacier that is parallel to the direction of flow in the center of the glacier but curves toward the margin downstream. Cf: *longitudinal crevasse; marginal crevasse; transverse crevasse.*

splaying out (a) The breakup of a fault into a number of minor faults. (b) The dying-out of a fault by its dispersal into a number of minor faults. See also: *splay [struc geol].*

splendent luster (splen'-dent) A mineral luster of the highest intensity.

splent coal *splint coal.*

spline In *residualizing,* a smooth curve passing through given data points.

splint *splint coal.*

splint coal A type of banded coal that is hard, dull, blocky, and grayish black, with rough, uneven fracture and granular texture. It is defined quantitatively as having more than 5% anthraxylon and more than 30% opaque attritus. Also spelled: *splent coal.* Cf: *semisplint coal; durain.* Syn: *splint.*

splintery fracture (splin'-ter-y) The property shown by certain minerals or rocks of breaking or fracturing into elongated fragments like splinters of wood.

split [coal] A coal seem that is separated from the main seam by a thick parting of other sedimentary rock. Syn: *coal split.*

split [eco geol] In exploration geology, typically one-half or one-quarter of a diamond drill core sectioned lengthwise. One split is typically crushed and sampled for assay, the remaining split is stored for archival purposes.

split [geochem] In sampling for geochemical analysis, a random fraction of crushed or ground rock taken for analysis. The remainder is stored for replicate or additional analysis.

split-barrel sampler A steel tool for sampling soil that is driven into the formation. The split barrel is held together by a bit at one end and a head at the other. The bit and head are removed to permit separation of the two sides of the split barrel to expose the sampled soil. Samplers commonly have core catchers made of plastic or steel that are located between the bit and the split barrel. Cf: *penetration test.* Syn: *split-spoon sampler.*

split-spoon sampler *split-barrel sampler.*

split spread A type of seismic spread in which the source is at the center of the arrangement of geophones. It is commonly used for continuous profiling and dip shooting. Syn: *straddle spread; symmetric spread.*

split stream (a) A stream shown on a map by a single line and containing an island that divides the stream into two channels. (b) A *single-line stream* that divides into branches in separate drainage areas. Cf: *braided stream.*

splitting [paleont] (split'-ting) In taxonomy, the practice of classifying species and genera on the basis of relatively minute differences. A taxonomist known for his preference for finely drawn distinctions is called a "splitter". Cf: *lumping.*

splitting [sed] (a) Abrasion of a rock fragment resulting in the production of two or three subequal parts or grains. (b) The property or tendency of a stratified rock of separating along a plane or surface of parting. (c) The sampling of a large mass of loose material (e.g. a sediment) by dividing it into two or more parts; e.g. quartering.

splotch Circular, radar dark or bright patches on the Venusian surface thought to result from impact shocks of disintegrated asteroids and/or comets without formation of an attendant impact crater (Bougher et al., 1997). See also: *halo [planet].*

SPM *scanning proton microprobe.*

spodic horizon (spod'-ic) A mineral soil horizon that is characterized by the illuvial accumulation of amorphous materials composed of aluminum and organic carbon with or without iron. The spodic horizon has a certain minimum thickness, and a minimum quantity of extractable carbon plus iron plus aluminum in relation to its content of clay.

spodiosite (spo'-di-o-site') An ill-defined ash-gray or brown orthorhombic mineral: $Ca_2(PO_4)F$.

Spodosols Mineral soils that have a spodic horizon or a placic horizon that overlies a fragipan. An order in the U.S. system of soil taxonomy.

spodumene (spod'-u-mene) A monoclinic mineral of the *pyroxene* group: $LiAlSi_2O_6$. It occurs mainly in white to green prismatic lath-like crystals, often of great size, esp. in granitic pegmatites. Spodumene is an ore of lithium. See also: *kunzite; hiddenite.* Syn: *triphane.*

spoil A nonspecific term typically referring to overburden, non-ore, or other waste material produced in mining, quarrying, dredging, excavating, construction, and other operations involving earth modification.

spoil bank A bank, mound, or other accumulation composed of *spoil;* e.g., a submerged embankment of waste earth material dredged from a channel and dumped along it. Cf: *spoil heap.*

spoil ground An area where dredged or excavated material is deposited or dumped.

spoil heap A pile of refuse material from an excavation or mining operation; e.g., a pile of dirt removed from, and stacked at the surface of, a mine in a conical heap or in layered deposits, such as a tip heap from a coal mine. Cf: *spoil bank.*

spoke A radial and typically flat component of a *wheel* of a holothurian, connecting the central part and the rim.

spondylium (spon-dyl'-i-um) A trough-shaped or spoonlike plate serving for muscle attachment and accommodating a ventral muscle field in the posterior part of the pedicle valve of a brachiopod, composed of dental plates in various stages of coalescence, usually with a median septum. Pl: spondylia. Cf: *pseudospondylium.*

spondylium duplex A *spondylium* formed by convergence of dental plates and supported by variably developed median septum arising from the floor of the pedicle valve.

spondylium simplex A *spondylium* formed by convergence of dental plates and supported ventrally by a variably developed simple median septum.

spong (Pron: spung) A term used in the Pine Barrens, N.J., for a *cripple* without a growth of cedar or with flowing water only after a rain. It is sometimes defined as any lowland area where highbush blueberries grow.

sponge A many-celled aquatic invertebrate belonging to the phylum *Porifera* and characterized by an internal skeleton composed most frequently of opaline silica and less commonly of calcium carbonate. Range, Precambrian to Holocene. Syn: *poriferan.*

sponge-spicule rock (sponge'-spic'-ule) A lithified *spicularite.*

spongework An entangled net of irregular interconnecting cavities of various sizes produced by solution in the walls of limestone caves and separated by intricate perforated partitions, as in pores of a sponge. Less preferred syn: *boneyard.* Cf: *anastomosis.*

spongework cave A cave consisting of irregular interconnected passages. See also: *anastomotic cave; network cave.*

spongin (spon'-gin) (a) A variety of collagen or scleroprotein (an insoluble fibrous protein) secreted by sponges and forming their skeletons. (b) A general term for any fibrous, organic skeletal material secreted by sponges. This usage is not recommended, although it is the sense in which the term has been used in all but the most recent literature.

spongioblast (spon'-gi-o-blast') A cell that secretes *spongin.*

spongocoel (spon'-go-coel) The *cloaca* of a sponge.

spongolite (spon'-go-lite) A sediment or rock composed principally of the remains of sponges; esp. a *spicularite.*

spontaneous fission (spon-ta'-ne-ous) A rare mode of *radioactive decay,* in which the nucleus of a heavy atom produces two fission products and several neutrons. Significant for ^{238}U and ^{244}Pu.

spontaneous fission dating A method of determining the age in years of uranium minerals based on the known rate of spontaneous fission of uranium-238 to two or more smaller nuclides.

spontaneous fission-track dating *fission-track dating.*

spontaneous generation An early concept in which living matter was thought to appear from dead material without the influence of outside or supernatural forces. Syn: *abiogenesis.*

spontaneous magnetization The magnetization within a domain of a magnetized grain in the absence of an applied magnetic field, due to spontaneous magnetic order caused by exchange forces.

spontaneous polarization Development of differences in static electrical potential between points in the Earth as a result of chemical reactions, differences in solution concentration, heat flow, or the movement of fluids through porous media. See also: *self-potential method.*

spontaneous-potential curve (spon-ta'-ne-ous-po-ten'-tial) The *electric log* curve that records changes in natural potential along an uncased borehole. Small voltages are developed between mud filtrate and formation water of an invaded bed, and also across the shale-to-mud interface. These electrochemical components are augmented by an electrokinetic potential (streaming potential) developed when mud filtrate moves toward a formation region of lower fluid pressure through the mud cake. Where formation waters are less resistive (more saline) than drilling-mud filtrate, the spontaneous-potential curve deflects to the left from the *shale baseline.* First used about 1932, the curve was added to the *resistivity log* to make up the basic electric log of well-logging practice. Syn: *self-potential curve.*

spontaneous-potential method *self-potential method.*

sporadic permafrost (spo-rad'-ic) A region of dominantly unfrozen ground containing scattered areas of permafrost (permafrost islands); it occurs along the southern limits of regions where summer frost conditions are usual, and in alpine areas. Cf: *discontinuous permafrost; continuous permafrost.*

sporae dispersae (spo'-rae dis-per'-sae) Dispersed fossil palynomorphs obtained by maceration of a rock sample, in contrast to those found in situ within fossil sporangia.

sporal (spor'-al) Pertaining to or having the special characteristics of a spore. Rarely used.

sporangia Pl. of *sporangium.*

sporangiospore (spo-ran'-gi-o-spore') A fungal spore produced in a sporangium. Rarely used.

sporangite (spo-ran'-gite) A fossilized spore case of a plant. Rarely used.

sporangium (spo-ran'-gi-um) An organ within which spores are produced and borne until their release; e.g., a pollen sac of a gymnosperm, an anther of an angiosperm, or the saclike structure in Fungi that produces spores endogenously. Pl: *sporangia.* Cf: *megasporangium; microsporangium.* Obsolete syn: *spore case.*

sporbo (spor'-bo) A term used in the San Joaquin Valley, Calif., for *oolite.* Pl: *sporbo.* Etymol: *s*mooth-*po*lished-*r*ound-*b*lack (*b*lue or *b*rown)-*o*bject (Galliher, 1931, p.257).

spore Any of a wide variety of minute, unicellular reproductive bodies or cells of plants, adapted to survive in unfavorable conditions and capable of independently developing new generations. Plant spores occur from the Ordovician to the present. In Fungi, spores may be sexual or asexual and single-celled or multicelled.

spore case *sporangium.*

spore coat *sporoderm.*

spore mother cell In *sporangia* of a spore-bearing plant, a diploid cell which, by reduction division, produces a tetrad of haploid spores.

sporinite (spo'-ri-nite) A maceral of coal within the *liptinite* group, consisting of spore exines that are generally compressed parallel to the stratification. Cf: *cutinite; alginite; resinite.*

sporite (spo'-rite) A coal microlithotype containing 95% or more of sporinite.

sporocyte (spo'-ro-cyte) *spore mother cell.*

sporoderm (spo'-ro-derm) The complete wall of a spore or pollen grain, generally consisting of a nonresistant inner layer (*endospore/intine*) and a resistant outer layer (*exospore/exine*). Sometimes, an extra third layer (*perispore/perine*) surrounds the exospore/exine.

sporogenous (spo-rog'-e-nous) Producing or adapted to the production of spores, or reproducing by spores; e.g. "sporogenous tissue" in sporangium from which spore mother cells originate.

sporomorph (spo'-ro-morph) A fossil pollen grain or spore. Cf: *palynomorph.*

sporophitic (spo-ro-phit'-ic) Said of the ophitic texture of an igneous rock, in which large pyroxene crystals enclose much smaller, widely separated plagioclase crystals (Walker, 1957, p.2); a form of poikilitic texture. Obsolete. Cf: *poikilophitic; hyalopilitic.*

sporophyll (spo'-ro-phyll) A spore-bearing leaf; commonly a much modified structure, as in seed plants.

sporophyte (spo'-ro-phyte) The asexual generation of a plant that produces spores; e.g. the diploid generation of an embryophytic plant, produced by fusion of egg and spermatozoid in lower vascular plants or by fusion of egg nucleus and the sperm nucleus produced by the pollen of seed plants. Cf: *gametophyte.*

sporopollenin (spo-ro-pol'-len-in) The very resistant and refractory organic substance of which the exine of palynomorphs is composed and which gives the sporomorph its extreme durability during geologic time, being readily destroyed only by oxidation. It is a high-molecular-weight polymer of C-H-O (primarily of monocarboxylic or dicarboxylic fatty acids), but its exact structural composition has not yet been resolved. Also spelled: sporopollenine.

sport An individual that exhibits a marked deviation well beyond the normal limits of individual variation and not anticipated or suggested by the characteristics of its parents; it is usually regarded as the result of a *mutation.*

S positions In locational notation for conodont skeletal apparatuses, *S* designates the positions occupied by ramiform elements that form a symmetry-transition series; *Sa* denotes the position of bilaterally symmetrical alate elements; *Sb* denotes the position of digyrate or tertiopedate elements; *Sc* denotes the position of bipennate or dolabrate elements; and *Sd* denotes the position of quadriramate elements (TIP, 1981, pt. W, supp.2).

SPOT The Systeme Probatoire d'Observation de la Terre. A French multispectral satellite system with pointable sensors. There are two sensors aboard the satellites. One has 10-m ground resolution in a single panchromatic spectral region; the other with 20-m ground resolution has three spectral bands for color-infrared mapping. The satellites may be pointed so as to produce stereo imagery for mapping uses. Cf: *Landsat.*

spot correlation In seismology, the correlation of reflections on isolated seismograms by noting similarities in character and interval.

spot elevation (a) An elevation shown on a topographic map at a critical point (such as a break in slope, a road intersection, or a point on a stream divide) to supplement the map information given by contour lines and bench marks from which contours can be correctly drawn. (b) A point on a map or chart whose elevation is noted; a *spot height.*

spot height A predominantly British term for a point, indicated on a map, whose elevation above a given datum has been correctly measured on the ground but, in contrast to a bench mark, is seldom indicated on the ground. See also: *spot elevation.*

spot medallion *frost scar.*

spotted (spot'-ted) *maculose.*

spotted schist *spotted slate.*

spotted slate A shaly, slaty, or schistose argillaceous rock the spotted appearance of which is the result of incipient growth of porphyroblasts in response to contact metamorphism of low to medium intensity. Cf: *adinole; desmosite; schalstein; spilosite; fleckschiefer; fruchtschiefer; garbenschiefer; knotenschiefer.* Syn: *knotted schist; knotted slate; spotted schist.*

spotty (spot'-ty) Said of a mineral deposit or mineralized zone in which the valuable constituent occurs in scattered masses of high-grade material.

spotty dendrite Pyrolusite or manganese wad deposited on joints or parting surfaces in the presence of low but geochemically significant concentrations of copper, resulting in a spotted fly-speck habit of the black oxides rather than a dendritic one.

spout (a) A discharge or jet of water ejected with some violence, either continuously (e.g. a spring) or periodically (e.g. a geyser). (b) A rush of water to a lower level; e.g. a waterfall. (c) waterspout.

spouter (spout'-er) *blowhole [coast].*

spouting horn (spout'-ing) A sea cave with a rearward or upward opening through which water spurts or sprays after waves enter the cave. Syn: *chimney; oven.*

spray ice Ice formed from ocean spray blown along the shore or upon floating ice.

spray print A small pit or depression similar to a *rain print* but formed by spray or windblown drops of water.

spray ridge An ice formation on an icefoot, formed by the freezing of wind-blown ocean spray.

spray zone The area along a coast affected by the spray of breaking waves. Cf: *splash zone.*

spread [gem] The surface or width at the girdle in proportion to the depth of a cut stone, e.g. a diamond.

spread [seis] The layout of geophone groups from which data from a single source are recorded simultaneously. Syn: *seismic spread.*

spread [streams] A marsh or shallow body of water resulting from the expansion in width of a stream, as where a natural obstruction, aquatic vegetation, or sediment infilling chokes or impedes streamflow. Syn: *widespread.*

spread correction A correction for *normal moveout.*

spreading center *mid-ocean ridge.*

spreading concept (spread'-ing) *sea-floor spreading.*

spread stone A diamond brilliant cut with a large table (greater than 60% of the stone's diameter) and a thin crown. Also called: fisheye.

spreite Nested concave-upward laminae. Formed by the upward- or downward movements of burrow-dwelling organisms that maintain a constant burrow depth below the water/sediment interface. A feature of some ichnofossils (Friedman et al., 1992, p.37).

spriggite A bright orange monoclinic mineral: $Pb_3(UO_2)_6O_8(OH)_2 \cdot 3H_2O$.

spring A place where groundwater flows naturally from a rock or the soil onto the land surface or into a body of surface water. Its occurrence depends on the nature and relationship of rocks, esp. permeable and impermeable strata, on the position of the water table, and on the topography.

spring alcove A term used along the Snake River Canyon in Oregon and Idaho for a short amphitheater or box-headed canyon formed by spring sapping along the edge of a basalt plateau.

springcreekite A black rhombohedral mineral of the *crandallite* group: $BaV^{3+}_3(PO_4)_2(OH,H_2O)_6$.

spring dome A descriptive term suggested by Williamson (1961) for a nondiastrophic limestone structure consisting of a circular or elliptical mound, usually with a central hollow or crater, believed to result from the expulsion of water from an underlying source (such as from a semirigid sediment).

Springeran (Spring'-er-an) North American stage: Lower Pennsylvanian (above Mississippian, below Morrowan).

spring-fed lake *spring lake.*

springhead *fountainhead.*

spring lake (a) A lake, usually of small size, that is created by the emergence of a spring or springs, esp. one having visibly flowing springs on its shore or springs rising from its bottom. (b) A lake that receives all or part of its waters directly from a spring. Syn: *spring-fed lake.*

spring line A line of springs marking the intersection of the water table with the land surface, as at the foot of an escarpment or along the base of a permeable bed at its contact with a *confining bed.*

spring mound A roughly circular mound of sand and silt, 5-6 m high and 10-12 m across, formed by a spring rising to the surface and depositing its load. See also: *mound spring.*

spring pit A small crater formed on a sand beach by ascending water, characterized by coarse sand in the center and finer sand around the edge, and measuring 30-60 cm across and about 15 cm deep (Quirke, 1930).

spring sapping The erosion of a hillside around the fountainhead of a strongly flowing stream, causing small landslides and resulting in the retreat of the valley head. Syn: *sapping [geomorph].*

spring snow A coarse, granular wet snow developed during the spring, generally by high temperatures and *melt metamorphism.* Syn: *corn snow; grain cluster.*

spring tide A tide occurring twice each month at or near the times of new moon (conjunction) and full moon (opposition) when the gravitational pull of the Sun reinforces that of the Moon, and having an unusually large or increased tide range. Cf: *neap tide.* Syn: *syzygy tide.*

springwood (spring'-wood) A syn. of *early wood.* Also spelled: spring wood. Cf: *summerwood.*

spruit A term used in southern and eastern Africa for a small stream, esp. one that is usually dry but nourished by sudden floods. Etymol: Afrikaans, from Middle Dutch "spruten", "to sprout".

SPT *standard penetration test.*

spud v. To start drilling a well with a drilling rig that is capable of drilling to total depth. Syn: *spud-in.*

spudder (spud'-der) (a) A small drilling rig, used primarily to start a new well. (b) A drill used for drilling seismic shotholes in hard rock or gravel. (c) The special drill bit used to begin a borehole.

spud-in *spud.*

spumellarian (spu-mel-lar'-i-an) Any radiolarian protozoan belonging to the suborder Spumellina, characterized by a thick-walled central capsule perforated by fine, evenly distributed pores.

spur [eng] An artificial obstruction, such as a pier dam, extending outward from the bank of a stream for the purpose of deflecting the current or of protecting the shore from erosion.

spur [geomorph] (a) A subordinate ridge or lesser elevation that projects sharply from the crest or side of a hill, mountain, or other high land surface; a small hill extending from a prominent range of hills or mountains. Syn: *prong [geomorph].* (b) *meander spur.*

spur [ice] *ram.*

spur [marine geol] (a) A ridge or other prolongation of a terrestrial mountain range, extending from the shore onto or across the continental shelf or insular shelf; e.g. the Bahama Spur in the Atlantic Ocean. (b) A subordinate ridge projecting outward from a larger submarine feature or elevation.

spur [ore dep] A small vein branching from a main one.

spur [paleont] A dependent projection of the basal margin of a *tergum* of a cirripede crustacean.

spur-and-groove structure A comb-tooth structure common to many reef fronts, best developed on the windward side, consisting of elongate channels or grooves a few meters wide and deep, separated by seaward-extending ridges or spurs (Maxwell, 1968, p.110). Grooves are commonly floored with coral rubble and skeletal sand and spurs covered with living corals and associated benthos (Shinn, 1963). Syn: *groove-and-spur structure.* Cf: *reef buttress.*

spur-end facet *triangular facet.*

spur furrow A groove on the outer surface of a *tergum* of a cirripede crustacean, extending along a spur to the apex.

spurrite (spurr'-ite) A white to light or lavender gray monoclinic mineral: $Ca_5(SiO_4)_2(CO_3)$. It is dimorphous to paraspurrite.

squall (a) A strong, sudden wind, often accompanied by precipitation, thunder, and lightning. There are many local names for various types of squalls. (b) A severe local storm of short duration.

squall line A narrow band of powerful cumulonimbus clouds traveling in advance of a *cold front.* Tornadoes are often associated with the squall line.

Squamata (Squa-ma'-ta) An order of lepidosaurian reptiles that includes the lizards and snakes and their marine relatives the mosasaurs, of Mesozoic age. Range, Upper Triassic to Holocene.

squamiform cast (squa'-mi-form) One of a group of "crowded, lobate casts overlapping downcurrent" (Ten Haaf, 1959, p.46), resembling sagged flute casts but supposedly having an opposite orientation with regard to current direction. It appears to be a variety of loaded *flute [sed], longitudinal furrow,* and *longitudinal ridge,* with a fleur-de-lis pattern.

squamose (squa'-mose) Scaly; said of a plant part that is covered with small scales. Cf: *lepidote.*

squamula (squam'-u-la) A small plate projecting subhorizontally in an eavelike manner from the wall of a tabulate corallite toward the axis, as in *Emmonsia.* Pl: squamulae.

square emerald cut An *emerald cut* with a square girdle outline modified by corner facets or with four equal sides modified by cut corners.

squawcreekite An adamantine yellow-brown tetragonal mineral of the *rutile* group: $(Fe^{3+},Sb^{5+},Sn^{4+},Ti)O_2$.

squeaking sand A term used by Humphries (1966, p.135) for *whistling sand.*

squeeze [drill] v. (a) To inject cement slurry under pressure into a cased well in order to recement a channeled area behind the casing or to close off perforations. Cf: *squeeze job.* (b) To inject fluid under high pressure, as in *hydraulic fracturing.* n. The plastic movement of a soft rock in the walls of a borehole. Cf: *heaving shale.*

squeeze [eng] (a) The rapid or gradual closing of a mine working by the displacement of weak floor strata from beneath supporting pillars into adjacent mine rooms. (b) A form of failure of bearing capacity, common in coal mines, caused by the mining-induced stress increase on supporting coal pillars. Bending of the roof strata is often reflected as subsidence at the ground surface. (c) An abnormal increase in load on tunnel or mine supports, often with some displacement of the floor around resisting roof-support members. (d) A mine area (such as a section in a coal seam) undergoing a squeeze.

squeeze job The forcing of cement slurry into a *borehole* or through perforations in a *cased hole.* Cf: *squeeze [drill].*

squeeze-up A small extrusion of viscous lava, or *toothpaste lava,* from a fracture or opening on the solidified surface of a flow, caused by pressure. It may take various forms, generally bulbous or linear, and may be from a few centimeters to almost a meter in height. It may be marked by vertical grooves. See also: *grooved lava.*

squeezeway (squeeze'-way) A small cave passage that is traversable only with difficulty.

squeezing ground (squeez'-ing) Soil or rock that tends to reduce the size of an underground opening, such as a tunnel, at an unspecified but noticeable rate. It is typical of clay, or rock containing much clay. Cf: *firm ground; flowing ground; raveling ground; running ground; swelling ground.*

squid magnetometer A sensitive magnetometer based on the use of a Josephson junction to detect changes in the magnetic field. Etymol: derived from the expression Superconducting Quantum Interference Device.

srebrodolskite (sreb-ro-dol′-skite) An adamantine black or brownish-red orthorhombic mineral: $Ca_2Fe^{3+}_2O_5$. It is a combustion product of coal.

srilankite (sri-lan′-kite) An adamantine black orthorhombic mineral of the *samarskite* group: $(Ti,Zr)O_2$.

s.s. *sensu stricto.*

SST Abbrev. for sea surface temperature.

s **surface** *s plane.*

stabile (sta′-bile) Resistant to chemical change, or decomposing with difficulty; e.g. "stabile protobitumen", a plant or animal product (such as wax, resin, spores, or leaf cuticle) that forms fossil carbonaceous deposits such as amber or cannel coal. Ant: *labile.*

stability [eng] (sta-bil′-i-ty) The tendency of a structure, slope, or embankment to perform within acceptable limits under external loading for which it was designed; e.g., *bank stability* and *slope stability.*

stability [geochem] In thermodynamics, an equilibrium state to which a system will tend to move from any other state under the same external conditions. Since it is never possible to examine all alternative states, assertions about the stability of real systems must always contain, at least implicitly, reference to the alternative states relative to which stability is claimed.

stability [oceanog] The resistance of a water column to mixing in the vertical. Numerically equal to the vertical gradient of density at a point in the water column divided by density at the point. The stability is greatest at the *pycnocline.*

stability field The range of conditions within which a mineral or mineral assemblage is stable.

stability frequency A measure of the importance of stability within the water column, especially important for theory of ocean dynamics. It is a function of the square root of *stability* and gravity. It has units of radians per second, which are commonly converted to the more useful units of cycles per hour. Typical values are a few cycles per hour. Syn: *Vaisala frequency; buoyancy frequency.*

stability series A grouping of minerals arranged according to their persistence in nature; i.e. to their resistance to alteration or destruction by weathering, abrasion during transportation, and postdepositional solution (Goldich, 1938); e.g. olivine (least stable), augite, hornblende, biotite (most stable). The most stable minerals are those that tend to be at equilibrium at the Earth's surface.

stability-time hypothesis The paleoecologic concept that long-term environmental stability, esp. of climate, tends to produce great faunal *diversity,* narrow ecologic niches, fast *speciation* rates, and ecosystems with complex tropic or feeding relationships. Syn: *time-stability hypothesis.*

stabilization [ecol] (sta′-bi-li-za′-tion) The characteristic of a *climax,* in which the greatest degree of adjustment between organisms and environment has been attained.

stabilization [eng] *soil stabilization.*

stabilized dune (sta′-bi-lized) *anchored dune.*

stabilizing force (sta′-bi-liz′-ing) The ordinary restoring force in an unstable gravimeter.

stabilizing selection *Natural selection* favoring a constant, near-the-mean phenotype.

stable [radioactivity] (sta′-ble) Said of a substance that is not spontaneously radioactive. Cf: *unstable [radioactivity].*

stable [sed] (a) Said of a constituent of a sedimentary rock that effectively resists further mineral change and that represents an end product of sedimentation (often resulting from more than one cycle of erosion and deposition); e.g., quartz, quartzite, chert, and accessory minerals such as zircon, rutile, muscovite, and tourmaline. (b) Said of a texturally and compositionally mature sedimentary rock (such as orthoquartzite) consisting of stable particles that are rounded or subrounded, well-sorted, and composed essentially of silica.

stable [tect] Said of an area or part of the Earth's crust that shows neither uplift nor subsidence or that is not readily deformed; e.g. a "stable shoreline" that is neither advancing nor receding.

stable equilibrium A state of equilibrium of a body, such as a pendulum, when it tends to return to its original position after being displaced. Cf: *unstable equilibrium; dynamic equilibrium.*

stable gravimeter A gravimeter with a stationary (i.e., stable) measuring system based on the use of a high-order optical and/or mechanical magnification so that an extremely small change in the position of a unit mass (weight) or associated property can be measured accurately.

stable isotope A nuclide that is not known to undergo *radioactive decay.* Cf: *unstable isotope.*

stable magnetization Remanent magnetization that does not change appreciably over geologic time, i.e., does not show magnetic viscosity and thus is carried by an assemblage of magnetized grains of typically high coercivity and/or unblocking temperature. Geologic stability must be proven through field-based tests. In practice it is nearly the same as *hard magnetization.*

stable relict A *relict [meta]* that was not only stable under the conditions of its formation but also under the newly imposed conditions of metamorphism. Cf: *unstable relict.*

stack [coast] An isolated pillar-like rocky island or mass near a cliffy shore, detached from a headland by wave erosion assisted by weathering; esp. one showing columnar structure and roughly horizontal stratification. Examples occur off the chalk cliffs of the Normandy coast. A stack may also form along the shore of a large lake. Cf: *skerry; chimney [coast].* Syn: *marine stack; rauk; rock stack; sea stack.*

stack [geomorph] An upstanding, steep-sided mass of rock rising above its surroundings on all sides from a slope or hill.

stack [seis] The sum of several seismic traces, usually traces that have been corrected for normal moveout and statics.

stack deposit A uranium deposit of irregular shape, associated with trend deposits and "frequently controlled in part by structure" (Bailey and Childers, 1977, p.27).

stacked profiles Display of magnetic or other profiles from a number of lines, to facilitate interpretation. Usually in a map form, sometimes in an isometric view.

stacking chart (stack′-ing) A diagram showing the interrelationships among the seismic traces from *common-midpoint shooting.* It is used to determine the proper traces for stacking, and for determining parameters for shifting traces.

stacking fault A type of *plane defect* in a crystal, caused by a mistake in the *stacking sequence.* Syn: *fault [cryst].*

stacking sequence The manner in which individual layers of a layered crystal structure are arranged in space.

stacking velocity The velocity determined from velocity analysis based upon normal-moveout measurements. Stacking velocity approaches the *rms velocity* as the offset approaches zero if velocity layering and reflectors are parallel and the layers are isotropic. Syn: *normal-moveout velocity.*

stade A substage of a glacial stage marked by a glacial readvance; "a climatic episode within a glaciation during which a secondary advance of glaciers took place" (ACSN, 1961, Art. 40). Adj: *stadial.*

stadia (sta′-di-a) (a) A surveying technique or method using a stadia rod in which distances from an instrument to the rod are measured by observing through a telescope the intercept on the rod subtending a small known angle at the point of observation, the distance to the rod being proportional to the rod intercept. The angle is usually defined by two fixed lines in the reticle of the telescope. (b) *stadia rod.* (c) An instrument used in a stadia survey; esp. an instrument with stadia hairs. Pl: stadias. The term is also used as an adjective in such expressions as "stadia surveying", "stadia distance", and "stadia station".

stadia constant (a) The ratio by which the intercept on the stadia rod must be multiplied to obtain the distance to the rod. On most surveying instruments, it is 100. (b) The ratio by which the sum of the stadia intervals of all sights of a continuous series of measured differences of elevation is converted to the length of the series in kilometers.

stadia hairs Horizontal cross hairs equidistant from the central horizontal cross hair; esp. two horizontal parallel lines or marks in the reticle of a transit telescope, arranged symmetrically above and below the line of sight, and used in the stadia method of surveying. Syn: *stadia wires.*

stadia interval The length of stadia rod subtended between the top and bottom cross hairs in the leveling instrument as these are projected against the face of the rod.

stadial (sta′-di-al) Pertaining to or formed during a *stade;* also used as a noun as an equivalent of *stade.*

stadial moraine *recessional moraine.*

stadia rod A graduated rod used with an instrument having stadia hairs to measure the distance from the observation point to the place where the rod is positioned. Syn: *stadia.*

stadia table A mathematical table from which may be found, with minimal computation, horizontal distance and the difference in elevation, knowing the stadia distance and the angle of sight.

stadia traverse A surveying traverse (such as a transit traverse or a traverse accomplished by planetable methods) in which distances are measured by the stadia method.

stadia wires *stadia hairs.*

stadimeter (sta-dim′-e-ter) A surveying instrument for determining the distance to an object of known height by measuring the angle subtended at the observer by the object.

staffelite (staf′-fel-ite) *carbonate-fluorapatite.*

staff gage A graduated scale or *gage* on a staff, wall, pier, or other vertical surface, used in gaging water-surface elevation. Cf: *chain gage.*

Staffordian (Staf-ford′-i-an) European stage: middle Upper Carboniferous (above Yorkian, below Radstockian). It is equivalent to part of upper Westphalian.

stage [geochron] An obsolete term for a *geochronologic unit* of lesser duration than *age [geochron],* during which the rocks of a *formation [stratig]* were formed. Cf: *substage [geochron].*

stage [geomorph] (a) A phase in the development of a *cycle of erosion* in which the features of the landscape have characteristic forms that distinguish them from similar features in other parts of the cycle, e.g., the hypothetical stages of youth, maturity, and old age in the development of a stream or region; also, the interval of time during which such a phase persists. The stage is one of the factors that determines the development of landforms in the Davisian cycle of erosion, although Davis (1899) originally referred to this factor as "time" (changed by later writers to "stage"). (b) A particular phase in the historical development of a geologic feature; e.g., the Calumet stage of Lake Chicago.

stage [glac geol] A time term for a major subdivision of a glacial epoch; specif. a major Pleistocene subdivision equated with a rock unit of formation rank. It includes glacial stage and interglacial stage. The ACSN (1961, art. 31b) rejects this usage because of conflict with the definition of "stage" (as a time-stratigraphic unit) and the requirement that stages be extended geographically on the basis of time-equivalent criteria.

stage [hydraul] The height of a water surface above an arbitrarily established datum plane. Syn: *gage height.*

stage [optics] In a microscope or similar optical apparatus, the small platform on which the object to be studied is placed. See also: *universal stage; mechanical stage; spindle stage.*

stage [stratig] (a) A chronostratigraphic unit of smaller scope and rank than a series. It is most commonly of greatest use in intra-continental classification and correlation, although it has the potential for worldwide recognition (NACSN, 1983, Art. 74). The temporal equivalent of stage is age. A stage may be, but need not be, divided into substages. Most stage names are based on lithostratigraphic units, although preferably a stage should have a geographic name not previously used in stratigraphic nomenclature; the adjectival ending for the geographic name is most commonly "-an" or "-ian", although it is permissible to use the geographic name without any special ending, such as "Claiborne Stage". (b) An informal term used to indicate "any sort" of chronostratigraphic unit of approximate stage rank "which is not a part of the standard hierarchy" of named chronostratigraphic units (ISST, 1961, p.24-25). (c) A *para-time-rock unit* consisting of two or more zones (Wheeler et al., 1950, p.2364). (d) A term used in England for a rock-stratigraphic unit.

stage-capacity curve (stage'-ca-pac'-i-ty) A graphic illustration of the relationship between the surface elevation of the water in a reservoir and the volume of water (Langbein and Iseri, 1960, p.17). Cf: *stage-discharge curve.*

stage-discharge curve (stage-dis'-charge) A graphic illustration of the relationship between gage height and volume of flowing water, expressed as volume per unit of time (Langbein and Iseri, 1960, p.17-18). Cf: *stage-capacity curve.* Syn: *rating curve; discharge-rating curve.*

staghorn coral (stag'-horn) Any coral (esp. a scleractinian belonging to the genus *Acropora*) characterized by large branching colonies which resemble antlers.

stagnant basin (stag'-nant) An isolated or barred basin containing essentially motionless water, rich in organic accumulations and noxious substances, but deficient in oxygen and capable of supporting only anaerobic organisms.

stagnant glacier *dead glacier.*

stagnant ice *dead ice.*

stagnation (stag-na'-tion) The condition or quality of water unstirred by a current or wave, or of a glacier that has ceased to flow.

stagnation point On the surface of a solid immersed in a flowing fluid, that point at which the stream lines separate.

stagnation-zone retreat A concept of glacier retreat set forth by Currier (1941). Ice in a marginal zone, 3 to 10 miles wide, ceases to move and is separated from the active ice mass by rapid thinning and burial by glacial sediments. The glacier continues to move forward behind this outer zone, and as successive outer zones progressively stagnate and become covered with debris, a complex system of ice-contact deposits forms, chiefly over and around immobile ice masses.

stagnicolous (stag-nic'-o-lous) Said of an organism that lives in or frequents stagnant water.

stagnum (stag'-num) A small lake or pool of water lacking an outlet.

stained stone A gemstone whose color has been altered by the use of a coloring agent, such as a dye, or by impregnation with a substance, such as sugar, followed by either chemical or heat treatment, which usually produces a permanent color; e.g. green chalcedony. Cf: *heated stone; burnt stone.*

stainierite (stai'-nier-ite) *heterogenite.*

staircase pond (stair'-case) One of a sequent group of a dozen to a hundred ponds following the approximate axis of a poorly developed watercourse on a sloping, thinly soil-mantled flat in a high-altitude valley, and resulting from the "armoring" and binding of naturally created bars by dense and rapidly growing grass (Ives, 1941, p.287-290); e.g. in Albion Valley, Colo.

stairway (stair'-way) (a) *glacial stairway.* (b) *cirque stairway.*

stalactite [speleo] (sta-lac'-tite) A conical or cylindrical *speleothem* that hangs from the ceiling or wall of a cave. It is deposited from drops of water and is usually composed of calcite but may be formed of other minerals. Cf: *stalagmite [speleo].* Etymol: Greek, "dripping". The term was first used by Worm in 1642.

stalactite [volc] A conical or tubular formation of lava hanging from the roof or walls of a lava tunnel or other cavity and developed by the dripping of fluid lava. It generally measures about 1-200 cm in length. Cf: *stalagmite [volc].*

stalagmite [speleo] (sta-lag'-mite) A conical or cylindrical *speleothem* that is developed upward from the floor of a cave by the action of dripping water. It is usually composed of calcite but may be formed of other minerals. Cf: *stalactite [speleo].* Etymol: Greek, "drop (something that has dripped down)". The term was first used by Worm in 1642.

stalagmite [volc] A conical or tubular formation of lava that is built up from the floor of a cavity in a lava flow, and formed as a corresponding feature to a stalactite of lava. It generally measures up to 30 cm in height and up to 200 cm in diameter.

stalderite A metallic black tetragonal mineral: $TlCu(Zn,Fe,Hg)_2As_2S_6$.

stalk That part of a plant by which a part is attached and supported, e.g. the petiole of a leaf, the stipe of an ovary, or the peduncle of a fruit.

stamen (sta'-men) That organ of a flower which produces pollen.

staminate (stam'-i-nate) Said of a flower that has stamens but no pistil. Cf: *pistillate.*

Stampian (Stamp'-i-an) *Rupelian.*

stamukha (sta-mu'-kha) A hummock or ridge of *grounded ice,* typically on an isolated shoal, formed by heaping-up of blocks. A stranded *floe* may act as a nucleus. On broad shelves in the Arctic Ocean, stamukhi occur along the 20-meter isobath and mark the seaward boundary of the *fast ice.* Pl: stamukhi. Etymol: Russian.

stand [drill] Two, three, or four lengths or connected joints of *drill pipe, casing,* or other tubulars handled as a unit in *rotary drilling.*

standard atmosphere (stand'-ard) A standard model of the atmosphere, which uses a temperature of 15°C, a pressure of 1,013.25 hectopascals (mb) measured at mean sea level, and a standard vertical gradient of temperature, pressure, and density. It is used as a representative model in various types of atmospheric analysis.

standard-cell method A method of studying the chemical relationships between rocks by calculating the number of various cations in the rock per 160 oxygen ions (Barth, 1948). Obsolete.

standard depth One of a series of depths (in meters) at which, by international agreement, physical measurements of seawater are to be taken.

standard deviation The square root of the average of the squares of deviations about the mean of a set of data. It is a statistical measure of *dispersion [stat].* Symbol: σ. Syn: *root-mean-square deviation.*

standard-deviation map (stand'-ard-de-vi-a'-tion) A *vertical-variability map,* or *moment map,* that shows the degree of statistical dispersion of one lithologic type (in a given stratigraphic unit) about its mean position in the unit. Cf: *center-of-gravity map.*

standard Earth An Earth model in which each surface of P or S wave velocity in the interior of the Earth is spherical, and encloses the same volume as the corresponding surface of equal velocity in the actual Earth (Runcorn et al., 1967, v. 2, p.1437).

standard error A measure of the accuracy of a sample mean as an estimator for the population mean; the *standard deviation* of the sampling distribution of a statistical parameter, or the standard deviation of a sample mean; the standard deviation divided by the square root of the number of observations of a sampled variable.

Standard Global Chronostratigraphic (Geochronologic) Scale The hierarchy of chronostratigraphic (and their equivalent geochronologic units) of worldwide scope which provides a standard scale of reference for the dating of rocks everywhere and for relating all rocks everywhere to world geologic history (ISSC, 1994, p.85-87). Equivalent to the General Stratigraphic Scale of the Stratigraphic Code of 1979 of the former USSR.

standardization Transformation of a value by subtracting the sample mean and dividing by the standard deviation. Converts a variable from its original units of measurement into a dimensionless form expressed in units of standard deviation.

standard meridian (a) The meridian used for determining standard time. (b) A meridian of a map projection, along which the scale is as stated.

standard normal form A *normal distribution* transformed to have a mean of 0 and variance of 1. This is the form of the z scores listed in tables of the standard normal distribution.

standard parallel (a) Any parallel of latitude that is selected as a standard axis on which to base a grid system; specif. one of a set of parallels of latitude (other than the base line) of the U.S. Public Land Survey system, passing through a selected township corner on a principal meridian, and on which standard township, section, and quarter-section corners are established. Standard parallels are usually at intervals of 24 miles north or south of the base line, and they are used to limit the convergence of range lines that intersect them from the south so that nominally square sections and townships can be laid out. Syn: *correction line.* (b) A parallel of latitude that is used as a control line in the computation of a map projection; e.g. the parallel of a normal-aspect conical projection along which the principal scale is preserved. (c) A parallel of latitude on a map or chart along which the scale is as stated for that map or chart.

standard penetration test *penetration test.* Abbrev: SPT.

standard port British term for *reference station.*

standard project flood The high-water stage expected from the most severe combination of meteorologic and hydrologic conditions that are considered reasonably characteristic of the geographic region.

standard rig A *cable-tool drilling* rig.

standard seawater A seawater adjusted so that its salinity is very close to the average salinity of the oceans (35) and its conductivity ratio relative to a standard KCl solution is very close to unity. Used to standardize conductometric salinometers. Prepared according to specifications of the International Association for the Physical Study of the Ocean. Cf: *salinity; salinometer*. Syn: *Copenhagen water; normal water*.

standard section A *reference section* showing as completely as possible a sequence of all the strata in a certain area, in their correct order, thus affording a standard for correlation. It supplements (and sometimes supplants) the type section, esp. for chronostratigraphic units.

standard state A condition in rocks in which the pressure is the same in all directions at any point, as a result of the weight of the overlying rocks.

standard zone A biostratigraphic zone "based on a type section in which specified beds yield characteristic fauna. The base of the lowest bed yielding this fauna is defined as the base of the zone. The top of the zone is not explicitly defined". It is named for "one member of the characteristic fauna, preferably one of the most common, or most characteristic" (Callomon, 1965, p.82).

standing crop (stand'-ing) Amount of living plant or animal material per unit area or per unit volume at any given time. Syn: *biomass*.

standing floe A separate floe standing vertically or inclined and enclosed by rather smooth ice.

standing level The water level in a well (or other excavation) penetrating the saturated zone, from which water is not being withdrawn, whether or not it is affected by withdrawals from nearby wells or other groundwater sources. Cf: *static level*. Syn: *standing water level*.

standing water Surface water that has no perceptible flow and that remains essentially in place, such as the water of some lakes and ponds; stagnant water, such as that enclosed in marshes and swamps. Ant: *running water*.

standing water level *standing level*.

standing wave A water wave, the wave form of which oscillates vertically between two points or nodes, without progressive movement.

standstill (stand'-still) *stillstand*.

stanekite A submetallic black monoclinic mineral of the *triploidite* group: $Fe^{3+}(Mn,Fe^{2+},Mg)O(PO_4)$.

stanfieldite (stan'-field-ite) An reddish or amber-colored monoclinic meteorite mineral: $Ca_7(Ca,Mg)_2(Mg,Fe^{2+})_9(PO_4)_{12}$.

stanleyite (stan'-ley-ite) A blue orthorhombic mineral: $V^{4+}O(SO_4)\cdot 6H_2O$.

stannic (stan'-nic) Relating to or containing tin in its tetravalent state. Cf: *stannous; stanniferous*.

stanniferous (stan-nif'-er-ous) Relating to or containing tin; e.g., stanniferous ore. Cf: *stannic; stannous*.

stannite (stan'-nite) (a) A steel-gray or iron-black tetragonal mineral: Cu_2FeSnS_4. Zinc often replaces iron. Stannite has a metallic luster and usually occurs in granular masses in tin-bearing veins, associated with cassiterite. Syn: *tin pyrites; bell-metal ore*. (b) Impure cassiterite. Also called: stannine.

stannoenargite An unnecessary name for a tin-bearing enargite. Not approved by the International Mineralogical Association.

stannoidite (stan-noid'-ite) A metallic brass-brown orthorhombic mineral: $Cu_8(Fe,Zn)_3Sn_2S_{12}$.

stannoluzonite An unnecessary name for a tin-bearing luzonite. Not approved by the International Mineralogical Association.

stannomicrolite A yellowish-brown cubic mineral of the *pyrochlore* group: $(Sn,Fe,Mn)_2(Ta,Nb,Sn)_2(O,OH,F)_7$. Syn: *sukulaite*.

stannopalladinite A metallic brown-rose hexagonal mineral: $(Pd,Cu)_3Sn_2$(?).

stannous (stan'-nous) Relating to or containing tin in its bivalent state. Cf: *stannic; stanniferous*.

stantienite (stan-tien'-ite) A black variety of retinite having a very high oxygen content (23%). Syn: *black amber*.

stapes (sta'-pes) (a) *columella [paleont]*. (b) The most lateral of three ossicles within the middle ear of mammals, or the only bone of the middle ear of other tetrapods. Known as columella in birds and certain groups of reptiles.

star n. (a) A rayed figure in a gemstone, consisting of two or more intersecting bands of light radiating from a bright center, and observed best under strong illumination; an optical phenomenon caused by reflected light from inclusions or channels, and brought to sharp lines in gem materials by cabochon cutting. Stars usually have four, six, or twelve rays, but three-, five-, seven-, or nine-rayed stars occur, or are possible due to the absence of inclusions in a portion of the stone. See also: *asterism [gem]*. (b) A gemstone showing such a figure. (c) *star facet*. (d) *star cut*. adj. Said of a mineral, crystal, or gemstone that exhibits asterism; e.g. "star sapphire". Syn: *asteriated*.

starburst gypsum A type of selenite wall crust consisting of facets of gypsum radiating from a common center.

star cluster A system of gravitationally interacting stars, numbering from hundreds to millions, sharing a common evolution. Two fairly distinct types are recognized. The globar clusters occupy a spherical, or halo, distribution with respect to the center of the galaxy; they are tightly gravitationally bound with a high central condensation, are very massive, and are relatively old (billions of years). The open, or galactic, clusters are distributed in the galactic plane, especially within the spiral arms; they are much less massive, show no central condensation, and are not tightly gravitationally bound. They are regarded as "stellar nurseries," often associated with clouds of gas and dust, where stars are currently being formed. They are sometimes as young as a few million years, and are relatively enriched in the heavy elements. Syn: *galactic cluster*.

star cut A general term that refers to any brilliant-cut stone whose pavilion facets present a star effect when viewed through the table. Syn: *star*.

star dune An isolated hill of sand, its base resembling in plan a star, and its sharp-crested ridges converging from the basal points to a central peak that may be as high as 100 m above the surrounding plain; it tends to remain fixed in place for centuries in an area where the wind blows from all directions. Syn: *pyramidal dune; heaped dune; stellate rose dune*.

star facet One of the eight triangular facets between the main bezel facets and bounding the table of a round brilliant-cut gem. Syn: *star*.

staringite (star'-ing-ite) A mixture of cassiterite and ferrotapiolite.

starkeyite (star'-key-ite) (a) A white or colorless monoclinic mineral: $MgSO_4\cdot 4H_2O$. Syn: *leonhardtite*. (b) A group name for minerals with the general formula $ASO_4\cdot 4H_2O$, where A = Mg, Fe^{2+}, Mn^{2+}, Co, or Zn.

star quartz A variety of quartz containing within the crystal whitish or colored starlike radiations along the diametral planes. The asterism is due to the inclusion of submicroscopic needles of some other mineral arranged in parallel orientation.

star ruby A semiopaque to semitransparent asteriated variety of ruby normally with six chatoyant rays.

star sapphire A semiopaque to semitransparent asteriated variety of sapphire normally with six rays resulting from the presence of microscopic crystals (such as rutile needles) in various orientations within the gemstone. Syn: *asteria*.

star stone (a) An *asteria*; esp. a star sapphire. (b) Any *asteriated* stone.

starting zone The area where unstable snow breaks loose from the snow cover and starts to slide. The starting zone of a particular slab avalanche can be outlined by the crown, flank, and stauchwall. The starting zone of an avalanche path includes the collective outline of all potential avalanche fracture surfaces.

starved basin A sedimentary basin in which the rate of subsidence is more rapid than the rate of sedimentation. Sediment thickness is greater at the margins than at the center (Adams et al., 1951).

starved ripple A ripple that preserves only the crest of the bed form due to an insufficient supply of sediment to allow construction of a complete ripple. Syn: *incomplete ripple*.

stasis (sta'-sis) Morphologic constancy of a species through a stratigraphically significant time interval.

stassfurtite (stass'-furt-ite) A massive variety of boracite from Stassfurt, Germany, sometimes having a subcolumnar structure and resembling a fine-grained white marble or granular limestone.

Stassfurt salt (Stass'-furt) Potassium salt from deposits in Stassfurt, Germany.

State Plane Coordinate System The plane-rectangular coordinate systems established by the National Geodetic Survey for each state for use in defining positions of geodetic stations in terms of plane-rectangular (x,y) coordinates. The zones for each state are divided on county lines and states may have more than one zone. The Transverse Mercator projection is used for zones with north-south extent and the Lambert Conformal Conic projection for zones with east-west extent. Abbrev: SPCS.

Statherian (Sta-ther'-i-an) A period of the Standard Global Geochronometric Scale: the youngest period of the *Paleoproterozoic* Era, characterized by craton expansion ("statheros", "stable, firm"). It is defined as the time span between 1,800 Ma, the end of the *Orosirian* Period, and 1,600 Ma, the beginning of the *Calymmian* Period of the *Mesoproterozoic* Era (Plumb, 1991).

static (a) Random noise which interferes with waves such as radio waves, seismic waves, etc. (b) In seismic exploration, a time correction which does not depend on time. See: *statics*.

static head (a) The total head without reduction for *velocity head* or losses; for example, the difference in elevation of headwater and tail water of a power plant. (b) The vertical distance between the free level of the source of supply and the point of free discharge or the level of the free surface (Rogers, 1981, p.366). (c) The sum of the *elevation head* and the *pressure head*. (d) *static level (c)*. See also: *hydrostatic head*.

static level (a) *hydrostatic level*. (b) *standing level*. (c) That water level of a well that is not being affected by withdrawal of groundwater.

static metamorphism A variety of *regional metamorphism* brought about by the action of heat and solvents at high lithostatic pressures with little or no deformation. Cf: *geothermal metamorphism; load metamorphism; thermal metamorphism*.

static modulus A *modulus of elasticity* that is produced by a very slow application of load.

static pressure Pressure that is "standing" or stabilized because it has reached the maximum possible from its source, and is not being diminished by loss.

static recrystallization Recrystallization in the absence of deformation.

static rejuvenation A kind of *rejuvenation* resulting from a decrease in stream load, an increase in runoff (owing to increased rainfall), or an increase in stream discharge through acquisition of new drainage; it involves neither uplift of the land nor eustatic lowering of sea level.

statics (stat'-ics) Time corrections applied to seismic traces to eliminate delays caused by variations in elevation, weathering-layer thickness, and/or near-surface velocity.
station [geophys] (sta'-tion) A location at which a geophysical observation is made.
station [surv] (a) A definite point on the Earth's surface whose position and location has been or will be determined by surveying methods; e.g. *triangulation station* and *instrument station*. It may or may not be marked on the ground. (b) A length of 100 ft, measured on a traverse along a given line, which may be straight, broken, or curved.
stationarity Having statistical properties which do not change with time and/or position. The statistics are the same if the time origin is changed.
stationary Said of a variable whose statistical characteristics remain unchanged with time or location.
stationary field (sta'-tion-ar'-y) A physical field that does not vary during the time period being considered, e.g., a magnetic field, either artificial or natural.
stationary flow *steady flow.*
stationary front A boundary between two air masses that does not move. Cf: *warm front; cold front.*
stationary mass The suspended or supported mass that tends to remain quiescent during passage of a seismic wave. Syn: *steady mass.*
stationary scanner A *scanner* designed for use from a fixed location. Coverage in the horizontal direction is provided by a faceted mirror that rotates about a vertical axis. Coverage in the vertical direction is provided by a planar mirror that tilts about a horizontal axis. Stationary-scanner images are individual frames, rather than continuous strips as acquired by airborne scanners.
stationary wave (a) A phenomenon produced by the interference of two continuous wavetrains showing alternating amplitude nodes and antinodes. (b) *standing wave.*
station error *deflection of the vertical.*
station pole A pole, rod, or staff for making stations in surveying; e.g. a *range rod* or a *level rod*. Also known as a "station rod" or "station staff".
statistical lithofacies (sta-tis'-ti-cal) A *lithofacies* that grades laterally into its neighbors and whose boundaries are vertical arbitrary-cutoff planes (Weller, 1958, p.633). Cf: *intertongued lithofacies.*
statistics (sta-tis'-tics) (a) The pure and applied science of devising, applying, and developing techniques such that the uncertainty of numerical inferences may be calculated. (b) The art of reducing numerical data and their interrelationships to comprehensible summaries or parameters. (c) Numbers describing samples taken from any population.
statoblast (stat'-o-blast) In the bryozoan class Phylactolaemata, a bud encapsulated in an envelope of chitin serving as an asexual reproductive body. It generally serves to preserve the species in winter, and develops into a new colony in the spring. See also: *hibernaculum.*
statospore (stat'-o-spore) A *resting spore*, e.g. a siliceous thick-walled resistant cyst formed within the frustules of various chiefly marine centric diatoms, and characterized by two overlapping convex valves and by absence of a girdle; or an intracellular cyst in various algae of the division Chrysophyta.
statute mile (stat'-ute) A measure of distance used on land, equal to 5,280 ft, 1,760 yd, 1,609.35 m, 1.61 km, 880 fathoms, 80 chains, 320 rods, or 8,000 links, and roughly equivalent to 0.87 international nautical miles. It is usually referred to as *mile.*
stauchwall The surface where the snow released in a slab avalanche rides up over the stable snow below.
stauractine (stau-rac'-tine) A spicule of hexactinellid sponges in the form of four coplanar rays at approximately right angles to one another.
staurodisc (stau'-ro-disc) A hexactinellid-sponge spicule (microsclere) in the form of two interpenetrating amphidiscs at right angles to one another. Cf: *hexadisc.*
staurolite (stau'-ro-lite) A dark reddish-brown, blackish-brown, yellowish-brown, or blue monoclinic (pseudo-orthorhombic) mineral: $(Fe,Mg)_4Al_{17}(Si,Al)_8O_{45}(OH)_3$. Twinned crystals often resemble a cross (six-sided prisms intersecting at 90° and 60°). It is a common constituent in rocks such as mica schists and gneisses that have undergone medium-grade metamorphism. Syn: *staurotide; cross-stone; grenatite; fairy stone.*
stauroscope (stau'-ro-scope) A type of *polariscope* that is used to determine the position of vibration-plane traces on a crystal for the accurate measurement of angles of extinction.
staurospore (stau'-ro-spore) A star-shaped, asexual fungal spore (conidiospore). Cf: *Saccardoan spore groups.*
staurotide (stau'-ro-tide) *staurolite.*
steacyite (stea'-cy-ite) A brown tetragonal mineral: $K_{0.3}(Na,Ca)_2ThSi_8O_{20}$.
stead A term used in south Wales for very thin bands of ironstone in coal measures.
steady flow (stead'-y) In hydraulics, flow that remains constant in magnitude or in direction of the velocity vector. Cf: *unsteady flow.* Syn: *stationary flow.*
steady mass *stationary mass.*
steady-state creep *secondary creep.*
steady-state stream *graded stream.*
steady-state theory A model of the Universe as a stationary, expanding world that does not change in appearance through space and time. It has no beginning or end and has a constant density of matter which is continuously and spontaneously created throughout space (Rogers, 1966, p.15). Syn: *continuous-creation hypothesis.*
steady turbidity current A persistent turbidity current, such as one produced where a stream heavily laden with sediment flows into a body of deep standing water. Cf: *spasmodic turbidity current.*
steam fog The general name for *fog* produced when extremely cold, dry air comes in contact with a relatively warm (unfrozen) water surface; has the appearance of rising streamers. Also known as Arctic sea smoke.
steam quality The quality of steam measured in terms of the weight of steam required to generate one kilowatt-hour of electrical energy. Measurement is in pounds or kilograms.
steam vent A type of hot spring from which superheated steam is rapidly and violently expelled.
steatite (ste'-a-tite) (a) A compact, massive, fine-grained, fairly homogeneous rock consisting chiefly of talc but usually containing much other material; an impure talc-rich rock. See also: *soapstone [rock].* (b) A term originally used as an alternative mineral name for *talc,* often restricted to gray-green or brown massive talc that can be easily carved into ornamental objects. Syn: *lardite; lard stone; soapstone [mineral]; soap earth.* (c) *steatite talc.*
steatite talc A relatively pure or high-grade variety of talc suitable for use in electronic insulators. It is the purest commercial form of talc. Syn: *steatite.*
steatitization (ste-at'-it-i-za'-tion) Introduction of, alteration to, or replacement by, talc (steatite); esp. the act or process of hydrothermal alteration of ultrabasic rocks that results in the formation of a talcose rock (such as steatite, soapstone, or relatively pure concentrations of talc).
Stebinger drum (Steb'-in-ger) A delicate vertical-angle adjustment for the *vernier* on the *explorer's alidade,* graduated in hundredths of a revolution. Cf: *tangent screw.*
S-tectonite A tectonite whose fabric is dominated by *planar elements* caused by deformation (no lineation is present), e.g., slate. Cf: *L-tectonite; B-tectonite; LS-tectonite.*
steel galena Galena having a fine-grained texture resulting from mechanical deformation or from incipient transformation to anglesite.
steel jack A colloquial syn. of *sphalerite.*
steel ore A name given to various iron ores (esp. to siderite) because they were readily used for making steel.
steenstrupine-(Ce) (steen-strup'-ine) A reddish-brown dark brown to black rhombohedral mineral: $Na_{14}Ce_6Mn^{2+}{}_2Fe^{3+}{}_2(Zr,Th)(PO_4)_6(PO_3OH)[Si_6O_{18}]_2 \cdot 3H_2O$. It is an occasional ore of thorium.
steephead (steep'-head) A nearly vertical, semicircular wall at the head of a *pocket valley,* at the base of which springs or a stream may emerge (Sellards and Gunter, 1918, p.27).
steep-to Said of a shore, coast, bank, shoal, or other coastal feature that has a precipitous, almost vertical, slope.
steer To introduce a time shift into an ensemble of seismic traces so that energy approaching from a given direction appears on all traces at the same time. It is used in studying earthquakes with large arrays and in other applications. See also: *beam steering.*
Stefan-Boltzmann law (Ste'-fan-Boltz'-mann) The statement that the radiant flux of a black body is equal to the temperature to the fourth power times the Stefan-Boltzmann constant of 5.68×10^{-12} watt cm^{-2} $°K^{-4}$.
stegidium (ste-gid'-i-um) The convex plate closing the gap between the delthyrial plate and the brachial valve of a spiriferoid brachiopod and consisting of a series of concentric layers deposited by the outer epithelium associated with the atrophying pedicle migrating dorsally (TIP, 1965, pt. H, p.153).
Stegocephalia (Steg'-o-ce-pha'-li-a) An obsolete term, which has been replaced by *Labyrinthodonta.* It formerly also included certain *Lepospondyli.*
steigerite (stei'-ger-ite) A yellow, greenish-yellow, or olive-green monoclinic mineral: $AlVO_4 \cdot 3H_2O$.
steilwand (steil'-wand) A syn. of *gravity slope.* Etymol: German "Steilwand", "steep wall".
steinkern (stein'-kern) Rock material consisting of consolidated mud or sediment that filled the hollow interior of a fossil shell (such as a bivalve shell) or other organic structure. Also, the fossil thus formed after dissolution of the mold. Etymol: German *Steinkern,* "stone kernel". Syn: *internal cast; endocast.* See also: *internal mold.*
Steinmann trinity (Stein'-mann) The association of *serpentines, pillow lavas,* and *radiolarian cherts,* which is characteristic of the axial belts of many geosynclines (Steinmann, 1926). The term was coined by Bailey in 1952. Now known to comprise elements of *ophiolites.*
stele The primary vascular structure of a plant stem or root, together with the tissues (such as the pith) which may be enclosed (Cronquist, 1961, p.873). Adj: stelar. Syn: *central cylinder.*
stell An English term for a brook.
stellar coal (stel'-lar) *stellarite.*
stellar crystal A common and beautiful type of snow crystal, having the shape of a flat hexagonal star, often with intricate branches. Cf: *spatial dendrite.*

stellarite (stel'-lar-ite) A variety of albertite from Stellarton, Nova Scotia. Syn: *stellar coal.*
stellate (stel'-late) Said of an aggregate of crystals in a starlike arrangement; e.g. wavellite.
stellate element A pectiniform *conodont element* with four primary processes, one of which is anterior and another posterior (TIP, 1981, pt. W, supp.2).
stellate rose dune *star dune.*
stellate structure In receptaculitids, a distal, calcified portion of the merome, generally consisting of four ribs.
stellerite (stel'-ler-ite) A colorless orthorhombic zeolite mineral: $CaAl_2Si_7O_{18}·7H_2O$. It is related to stilbite.
stelliplanate element A stellate *conodont element* with lateral platform extensions and the attachment surface, characteristic of planate pectiniform elements (TIP, 1981, pt. W, supp.2). See also: *stellate element; planate element.*
stelliscaphate element A stellate *conodont element* with lateral platform extensions and the attachment surface characteristic of scaphate pectiniform elements (TIP, 1981, pt. W, supp.2). See also: *stellate element; scaphate element.*
stem *scanning transmission electron microscope.*
stem [drill] *drill stem.*
stem [paleont] A narrow structure by which a sessile animal is made fast; e.g. the *column* of an echinoderm.
stem [tect] The comparatively slender part of a salt diapir below the *bulb* (Jackson and Talbot, 1991).
stemflow (stem'-flow) Water from precipitation that reaches the ground by running down the trunks of trees or plant stems. Cf: *throughfall.*
stem group A group of organisms that includes the common ancestor (stem species) of all taxa in a monophyletic group. Cf: *stem species.*
stemming The material (sand, clay, limestone) that fills a *shothole* after the explosive charge has been inserted. It is packed between the charge and the outer end of the shothole and is used to prevent the explosive from "blowing" out along the hole. Also the act of inserting the stemming into a shothole. Cf: *tamping.*
stem species In a monophyletic group, the common ancestor of all species in the clade (Wiley, 1981, p.222). Cf: *stem group, crown species.*
stem stream *main stream.*
stenecious (ste-ne'-cious) Said of an organism that can adjust or survive in only a limited range of environments.
stengel gneiss (sten'-gel) A syn. of *pencil gneiss.* Rarely used. Etymol: German "Stengel", "stalk".
stenhuggarite (sten-hug'-gar-ite) A bright orange tetragonal mineral: $CaFe^{3+}Sb^{3+}As^{3+}{}_2O_7$.
Stenian (Sten'-i-an) A period of the Standard Global Geochronometric Scale: the youngest period of the *Mesoproterozoic* Era, characterized by narrow belts of tectonic activity ("stenos", "narrow"). It is defined as the time span between 1,200 Ma, the end of the *Ectasian* Period, and 1,000 Ma, the beginning of the *Tonian* Period of the *Neoproterozoic* Era (Plumb, 1991).
steno- (sten'-o-) A prefix signifying "narrow", "close", "limited".
stenobathic (sten-o-bath'-ic) Referring to a marine organism that tolerates only a narrow range of depth. Cf: *eurybathic.*
stenobiontic (sten'-o-bi-on'-tic) Said of an organism that requires a stable and uniform environment.
stenogeographic (sten'-o-ge'-o-graph'-ic) Said of a plant or animal species with a limited geographic range. Cf: *eurygeographic.*
stenohaline (sten-o-hal'-ine) Referring to a marine organism that tolerates only a narrow range of salinity. Cf: *euryhaline.*
stenolaemate (sten-o-lae'-mate) Said of bryozoans belonging to the class Stenolaemata, characterized by colonies made up of long tubular members.
stenonite (sten'-o-nite) A colorless or white monoclinic mineral: $(Sr,Ba,Na)_2Al(CO_3)F_5$.
stenopalynous (sten'-o-pal-y-nous) Of plant taxa (families, etc.), having palynomorphs with only a slight variation in morphology. Ant: *eurypalynous.*
stenoplastic (sten-o-plas'-tic) Having limited capacity for modification and adaptation to new environmental conditions; incapable of major evolutionary differentiation. Cf: *euryplastic.*
stenopodium (sten-o-po'-di-um) A slender, elongate limb of a crustacean, composed of rodlike segments. Pl: stenopodia. Cf: *phyllopodium.*
stenosiphonate (sten-o-si'-phon-ate) Said of nautiloids with relatively narrow siphuncles. Cf: *eurysiphonate.*
Steno's principles Four principles relating to the deposition of sedimentary strata that were published by Nicolaus Steno in 1669. These include the superposition, original horizontality, and original lateral continuity of strata (Friedman et al., 1992, p.39).
stenothermal (sten-o-ther'-mal) Referring to a marine organism that tolerates only a narrow range of temperature. Cf: *eurythermal.*
stenotopic (sten-o-top'-ic) (a) Said of an organism that is restricted to one habitat or to relatively few habitats. (b) Said of an organism that has little adaptability to changes in the environment. Cf: *stenotropic.*
stenotropic (sten-o-trop'-ic) Said of an organism that has a narrow range of tolerance for variations in particular environmental factors. Cf: *stenotopic.* Ant: *eurytropic.*
stentorg (sten'-torg) A well-defined felsenmeer, usually on the crest and flanks of an esker, and often striped by former shorelines and beach ridges (Stamp, 1961, p.430). Etymol: Swedish, "stone marketplace". Rarely used.
step [coast] (a) The nearly horizontal section, not necessarily permanent, that roughly divides the beach from the shoreface, located just seaward of the low-tide shoreline, and in many cases marked by the presence of coarse sediment (such as gravel on a sand beach). Syn: *toe [coast].* (b) An abrupt downward inflection composed of coarse sand or gravel that marks the breakpoint of relatively small waves and swells prevailing during the summer. Cf: *plunge step.*
step [geomorph] (a) A *canyon bench* greatly broadened by erosion, such as those characteristic of the high plateaus of the western U.S.; a steplike landform on a hillside or valley slope that is otherwise smoothly rising. (b) *rock step.*
step [pat grd] A form of *patterned ground* characteristic of moderate slopes, and having a steplike form; it is transitional between a polygon (upslope) and a stripe (downslope). A step typically develops as a lobate solifluction terrace with the lower border convex downslope. See also: *sorted step; nonsorted step.* Obsolete syn: *terracette [pat grd].* Syn: *lobate soil.*
step-and-platform topography A landscape developed in a region in which the rocks dip very gently in one direction over a large area and are composed of alternating layers of hard and soft material; characterized by a sequence of lowlands and cuestas (Marbut, 1896, p.31-32).
stepanovite (step'-a-nov-ite) A greenish trigonal oxalate mineral: $NaMgFe^{3+}(C_2O_4)_3·8\text{-}9H_2O$. Cf: *zhemchuzhnikovite.*
step cline An irregular or broken *cline.*
step cut A style of cutting, widely used on colored gemstones, in which long, narrow four-sided facets are arranged in rows parallel to the girdle on both the crown and pavilion. They decrease in length as they recede above and below the girdle, giving the appearance of steps. The number of rows, or steps, may vary, although it is usually three on the crown and three on the pavilion. Different shapes of step cuts are described by their outline, such as rectangular step cut or square step cut. Cf: *emerald cut; brilliant cut; mixed cut.* See also: *baguette.* Syn: *trap cut.*
step delta One of a series of deltas built in a body of water whose level was alternately standing and falling, the upper delta being the oldest (Dryer, 1910, p.259); examples are numerous in the Finger Lakes region of New York. Syn: *hanging delta.*
step drawdown test A type of *aquifer test* in which the water level in a pumped well is measured while the well is pumped at a constant rate for a given period of time; the pumping rate is successively increased for each of several time intervals, the water level is measured during each interval, and drawdown is calculated for each pumping rate. The derived relationship between drawdown and pumping rate can be used to estimate aquifer properties, the practical yield of the pumping well, and the appropriate depth setting of the pump in a production well.
step fault (a) A little-used term for one of a set of parallel, closely spaced faults over which the total displacement is distributed. Cf: *distributive fault; fault zone.* (b) One of a series of low-angle thrust faults in which the fault planes step both down and laterally in the stratigraphic section to lower glide planes. Step faulting is due to variation in the competence of the beds in the stratigraphic section (Jones, 1971).
step fold An abrupt downward flexure of horizontal strata; it is a monoclinal structure.
Stephanian (Ste-pha'-ni-an) A European series or stage: uppermost Carboniferous (above Westphalian, below Asselian of the Permian).
stephanite (steph'-an-ite) A metallic black orthorhombic mineral: Ag_5SbS_4. It is an ore of silver. Syn: *brittle silver ore; black silver; goldschmidtine.*
stephanocolpate (steph'-a-no-col'-pate) Said of pollen grains having more than three colpi, meridionally arranged.
stephanocolporate (ste-phan'-o-col-por'-ate) Said of pollen grains having more than three colpi, meridionally arranged and provided with pores.
stephanolith (ste-phan'-o-lith) A crown- or star-shaped *coccolith.*
stephanoporate (steph'-a-no-por'-ate) In pollen grains, having more than three pores, centered on the equator. Syn: *zonoporate.*
step lake A lake occupying one of a series of rock basins on a glacial stairway; e.g. a *paternoster lake.*
stepout *moveout.*
step-out A well drilled at a distance from a producing oil or gas well in an effort to extend the productive limits of a field. Cf: *extension well.*
stepout correction *normal-moveout correction.*
steppe An extensive, treeless grassland area in the semiarid mid-latitudes of southeastern Europe and Asia. It is generally considered drier than the *prairie* which develops in the subhumid mid-latitudes of the U.S. Cf: *veld.*
stepped counter-regional system Major, listric landward-dipping growth "fault" or leaning salt diapir. The "fault" is actually a landward-dipping salt weld that passes downward like a series of steps into a flat salt weld and, even more deeply into another landward-dipping salt weld that eventually connects with the flat source layer. Sediment wedges in the system dip and thicken seaward (Schuster, 1995). Cf: *roho system.*

stepped crescent One of a succession of crescentic scarps (several centimeters to a meter high at the center) along the course of an arroyo, resembling together a stair with broad steps and shallow risers (Sharpe, 1938, p.25).

stepped plain A plain that has a succession of levels, "like the steps of a staircase belonging to one flight of stairs" (Schwarz, 1912, p.95). Syn: *klimakotopedion.*

stepping method A surveying method for determining the elevation of the *stadia rod* with respect to the *transit.* A point that is level with the transit is sighted through the telescope, which is then tilted so that the bottom cross-hair is aligned with the point. Another point in the line of the top cross-hair is sighted and the procedure is repeated until the stadia rod comes into the view of the telescope. The number of steps or times of retilting the telescope is counted and converted to Beamans, the full intercept between the stadia wires being one Beaman.

stepping stone An island used by a species that spread (or is spreading) from one region to another (MacArthur and Wilson, 1967, p.191).

step-pool sequence A downstream alternation between plunge pools and vertical down-steps along the streambed; common in steep channels.

step tablet *step wedge.*

step terrace (a) A man-made terrace having several steplike levels on which crops are cultivated. (b) A terrace with a vertical drop on the downhill side.

steptoe (step′-toe) An isolated hill or mountain of older rock surrounded by a lava flow. Also, a butte in the Columbia Plateau made by an older remnant surrounded by younger beds. Syn: *dagala.*

step vein A vein that alternately conforms with and cuts through the country-rock bedding. Similar in topology to ramp-flat faults, mineralized in structural geology.

step wedge An *optical wedge* whose transparency diminishes in discrete, graduated adjacent steps from one end to the other. Cf: *gray scale.* Syn: *step tablet.*

step zone A zone along a shoreline, located slightly below mean sea level and characterized by sediments that are coarser than those above (on the beach) or below (on the shoreface). Cf: *plunge step.*

sterane Hydrocarbon resulting from steroid molecules in the geological environment.

stercorite (ster′-co-rite) A white, yellowish, or brownish triclinic mineral: $(NH_4)Na(PO_3OH)\cdot4H_2O$. Syn: *microcosmic salt.*

stereocomparator (ster′-e-o-com-par′-a-tor) A stereoscope for accurately measuring the three space coordinates of the image of a point on a photograph; it is used in making topographic measurements by the comparison of stereoscopic photographs.

stereocompilation The production of a map from aerial photographs and geodetic control by means of photogrammetric instruments that use binocular vision and a pair of overlapping photographs to give the impression of depth. See also: *compilation.*

stereogenic (ster′-e-o-gen′-ic) Of the nature of or pertaining to a *stereosome.* Var: stereogenetic. Rarely used.

stereogram [geol] (ster′-e-o-gram′) A graphic diagram on a plane surface, giving a three-dimensional representation, such as projecting a set of angular relations; e.g. a block diagram of geologic structure, or a *stereographic projection* of a crystal.

stereogram [photo] A *stereoscopic pair* of photographs correctly oriented and mounted for viewing with a stereoscope. Syn: *stereograph.*

stereograph (ster′-e-o-graph′) (a) A stereometer with a pencil attachment, used to plot topographic detail from a stereogram. (b) *stereogram [photo].*

stereographic net (ster′-e-o-graph′-ic) *net [struc petrol].*

stereographic projection (a) A perspective, conformal, azimuthal map projection in which meridians and parallels are projected onto a tangent plane, with the point of projection on the surface of the sphere diametrically opposite to the point of tangency of the projecting plane. Any point of tangency may be selected (at a pole, on the equator, or a point in between). It is the only azimuthal projection that is conformal. Stereographic projections are much used for maps of a hemisphere and are useful in showing geophysical relations (such as patterns of island arcs, mountain arcs, and their associated earthquake epicenters). (b) A similar projection used in optical mineralogy and structural geology, made on an equatorial plane (passing through the center of the sphere) with the point of projection at the south pole. Syn: *stereogram [geol].* See also: *polar stereographic projection.*

stereom (ster′-e-om) (a) The calcareous tissue in the mesodermal endoskeleton of a living echinoderm. Cf: *stroma.* (b) The more or less dense calcareous skeletal deposit generally covering and thickening various parts of a scleractinian or rugose corallite. Cf: *sclerenchyme.* Also spelled: stereome.

stereometer (ster-e-om′-e-ter) A device used to measure heights of Earth features on a stereoscopic pair of aerial photographs, containing a micrometer movement by means of which the separation of two index marks can be changed to measure parallax dfference. Syn: *parallax bar.*

stereomodel (ster′-e-o-mod′-el) *stereoscopic image.*

stereonet (ster′-e-o-net′) A term used in crystallography for a *Wulff net.*

stereopair (ster′-e-o-pair′) *stereoscopic pair.*

stereo pair Two overlapping images or photographs that may be viewed stereoscopically.

stereophytic (ster′-e-o-phyt′-ic) Said of a sedimentary rock of organic origin (such as coral rocks and some algal limestones) that has been "built up directly, from the beginning, as a quite solid material" (Tyrrell, 1926, p.234).

stereoplasm (ster′-e-o-plasm′) Gelated protoplasm; specif. the relatively solid axis of granular-reticulose pseudopodia in foraminifers, surrounded by a granular fluid outer portion (rheoplasm). It is noted in foraminifers *Peneroplis* and *Elphidium,* but is not visible in most agglutinated types (TIP, 1964, pt.C, p.64).

stereoscope (ster′-e-o-scope′) A binocular optical instrument for assisting the observer to view two properly oriented photographs or diagrams to obtain the mental impression of a three-dimensional model (ASP, 1975, p.2105).

stereoscopic fusion (ster′-e-o-scop′-ic) The mental process by which two perspective views or images (one for each eye) are combined and brought to a focus on the retina of each eye to give the impression of a three-dimensional model. Syn: *fusion [photo].*

stereoscopic image The mental impression of a three-dimensional model that results from stereoscopic fusion. Syn: *stereoscopic model; stereomodel.*

stereoscopic model *stereoscopic image.*

stereoscopic pair An *overlapping pair* of photographs that, when properly oriented and used with a stereoscope, gives a three-dimensional view of the area of overlap. See also: *stereogram [photo].* Syn: *stereopair.*

stereoscopic principle The formation of a single, three-dimensional image by simultaneous vision with both eyes of two photographic images of the same terrain taken from different camera stations.

stereoscopic radius The limiting distance at which an object can be seen in stereoscopic relief; it is about 450 m with unaided human vision.

stereoscopic vision Simultaneous vision with both eyes in which the mental impression of depth and distance is obtained, usually by means of two different perspectives of an object (such as two photographs of the same area taken from different camera stations); the viewing of an object in three dimensions. Syn: *stereoscopy; stereovision.*

stereoscopy (ster-e-os′-co-py) (a) The science and art that deals with the use of simultaneous vision with both eyes for observation of a pair of overlapping photographs or other perspective views, and with the methods by which such viewing is performed and such effects are produced. (b) *stereoscopic vision.*

stereosome (ster′-e-o-some′) The part of a *chorismite* that remained solid at all times during the process of formation. Adj: *stereogenic.* Rarely used. Cf: *chymogenic; metaster; paleosome; restite.*

stereosphere (ster′-e-o-sphere′) A term that was originally proposed for the innermost shell of the Earth's mantle, but has also been used as equivalent to the lithosphere. Archaic. Cf: *chalcosphere.*

stereostatic pressure (ster′-e-o-stat′-ic) *geostatic pressure.*

stereotheca (ster′-e-o-the′-ca) The inner layer of a thecal plate of a cystoid. It is thicker than *epitheca.*

stereotriangulation (ster′-e-o-tri′-an-gu-la′-tion) A triangulation procedure that uses a stereoscopic plotting instrument to establish horizontal and/or vertical control by means of successive orientation of stereoscopic pairs of photographs in a continuous strip.

stereovision (ster′-e-o-vi′-sion) *stereoscopic vision.*

stereozone (ster′-e-o-zone′) An area of dense skeletal deposits in a *corallite,* generally peripheral or subperipheral in position.

steric (ster′-ic) Having to do with changes in density of sea water due to changes in temperature or salinity but not mass; e.g., sea level can change because the ocean warms and expands (a *steric* change) or because more water was added to the ocean (a mass change).

sterigma (ste-rig′-ma) A minute, spiculelike stalk at the apex of a basidium, on which the spore is developed in the basidiomycete fungi (Swartz, 1971, p.447).

sterile A term describing unmineralized material mined and stockpiled in the coarse of mining *ore.*

sterlinghillite A silky white to light pink monoclinic mineral: $Mn^{2+}_3(AsO_4)_2\cdot4H_2O$.

sternal pore (ster′-nal) A pore in radiolarian skeletons of the subfamily Trissocyclinae, directly below the vertical spine and partly framed by the tertiary-lateral bars. It is not present on specimens having a vertical pore.

sternbergite (stern′-berg-ite) A metallic dark-brown or black orthorhombic mineral: $AgFe_2S_3$. It occurs in tabular crystals or soft flexible laminae. It is a dimorph of argentopyrite. Syn: *flexible silver ore.*

Sternberg's law (Stern′-berg's) The decline in size of a clastic particle transported downstream is proportional to the weight of the particle in water and to the distance it has traveled, or to the work done against friction along the bed: $W = W_0 e^{-as}$, where W is the weight at any distance s, W_0 is the initial weight of the particle, and a is the coefficient of size reduction (Pettijohn, 1957, p.530). This relation was observed by H. Sternberg in 1875.

sternite (ster′-nite) The ventral part of a *somite* of an arthropod; e.g. the chitinous plate forming the ventral cover of an abdominal or thoracic segment of an insect, the sclerotized plate forming the ventral cover of a somite of an arachnid or merostome, or the sclerotized ventral surface of a single somite of a crustacean.

sternum (ster′-num) (a) In birds, mammals, and their closest extinct relatives, the ventral midsagittal bone with which the distal ends of most anterior ribs and proximal elements of the pectoral girdle articulate. (b) The ventral surface of the body, of a single tagma, or of a somite of an arthropod; the whole ventral wall of the thorax of an arthropod (as of an arachnid). Pl: sterna.

steroid (ster′-oid) Any one of several complex hydrocarbons occurring naturally in living organisms and having a polycyclic structure the same as found in the sterols. The fatty acids found in bile and vitamin D are examples of steroids.

sterraster (ster′-ras-ter) A sponge spicule (euaster) of globular or kidney shape with a granular surface, formed ontogenetically by expansion of a centrum to engulf all but the tips of the rays.

sterrettite (ster′-rett-ite) A syn. of *kolbeckite.* It was formerly described as a hydrous phosphate of aluminum: $Al_6(PO_4)_4(OH)_6{\cdot}5H_2O$.

sterryite (ster′-ry-ite) A metallic black orthorhombic mineral: $Ag_2Pb_{10}(Sb,As)_{12}S_{29}$.

stetefeldtite (stet′-e-feldt-ite′) A black, yellow, or brown cubic mineral of the *stibiconite* group: $Ag_2Sb_2(O,OH)_7$(?).

stevensite (ste′-vens-ite) A clay mineral, trioctahedral smectite characterized by no tetrahedral substitution of Al for Si: $(Ca_{0.5},Na)_{0.3}Mg_3Si_4O_{10}(OH)_2$. Its layer charge arises from octahedral vacancies. Syn: *aphrodite.*

stewartite (stew′-art-ite) (a) A brownish-yellow tabular triclinic mineral of the *laueite* group: $Mn^{2+}Fe^{3+}{}_2(PO_4)_2(OH)_2{\cdot}8H_2O$. It is a trimorph of *laueite* and *pseudolaueite.* (b) A steel-gray, ash-rich, fibrous variety of bort containing iron, having magnetic properties, and found in the diamond mines of Kimberley, South Africa.

stibarsen (stib-ar′-sen) A metallic white rhombohedral mineral: SbAs. Syn: *allemontite.*

stibianite (stib′-i-an-ite′) *stibiconite.*

stibiconite (stib′-i-con-ite′) (a) An earthy yellow, white, orange, gray, or brown cubic mineral with the pyrochlore structure: $Sb^{3+}Sb^{5+}{}_2O_6(OH)$. It is usually yellow to chalky white, and occurs as an alteration product of stibnite. Syn: *stibianite.* (b) A group name for minerals with the general formula $A_{2\text{-}3}Sb_2(O,OH)_7$, where A = Sb, Bi, Pb, Ca, Fe, or Ag.

stibiobetafite (stib′-i-o-bet′-a-fite) A subadamantine dark brown cubic mineral of the *pyrochlore* group: $(Ca,Sb^{3+})_2(Ti,Nb,Ta)_2(O,OH)_7$.

stibiocolumbite (stib′-i-o-co-lum′-bite) An adamantine dark brown, light yellowish-brown, reddish-yellow, reddish-brown or greenish-yellow orthorhombic mineral: $Sb(Nb,Ta)O_4$. It is isomorphous with stibiotantalite.

stibiocolusite A metallic dark gray cubic mineral: $Cu_{13}V(Sb,Sn,As)_3S_{16}$. It is the Sb analogue of colusite.

stibiomicrolite A white or greenish-white cubic mineral of the *pyrochlore* group: $(Sb,Ca,Na)_2(Ta,Nb)_2(O,OH,F)_7$.

stibiopalladinite (stib′-i-o-pal-la′-di-nite) A metallic silver-white to steel-gray hexagonal mineral: Pd_5Sb_2. Syn: *allopalladium.*

stibiotantalite (stib′-i-o-tan′-ta-lite) A dark brown, yellowish-brown, reddish-yellow, reddish-brown, or greenish-yellow orthorhombic mineral: $Sb(Ta,Nb)O_4$. It is isomorphous with stibiocolumbite.

stibium (stib′-i-um) An ancient name for *stibnite* used (as in Egypt) as a cosmetic for painting the eyes.

stibivanite (stib-i-van′-ite) An adamantine yellow-green monoclinic or orthorhombic mineral: $Sb^{3+}{}_2V^{4+}O_5$.

stibnite (stib′-nite) A metallic lead-gray mineral: Sb_2S_3. It has a brilliant metallic luster, differs from galena by ease of fusion, and often contains gold and silver. Stibnite occurs in massive forms and in prismatic orthorhombic crystals that show highly perfect cleavage and are striated vertically. It is the principal ore of antimony. Syn: *antimonite; antimony glance; gray antimony; stibium.*

stichtite (sticht′-ite) A lilac or rose-pink platy rhombohedral mineral of the *hydrotalcite* group: $Mg_6Cr_2(CO_3)(OH)_{16}{\cdot}4H_2O$. It is dimorphous with barbertonite and may contain some iron.

stick-slip A jerky, sliding motion associated with fault movement that may be a mechanism in shallow-focus earthquakes. A model of fault slip in which distinct slip episodes are separated by periods in which the fault is locked and stress increases.

sticky limit (stick′-y) The lowest water content at which a soil will adhere to a metal blade drawn across the surface of the soil mass (ASCE, 1958, term 353). Cf: *sticky point.*

sticky point A condition of consistency at which a soil material barely fails to adhere to a foreign object; specif. the moisture content of a well-mixed, kneaded soil material that barely fails to adhere to a polished nickel or stainless-steel surface when the shearing speed is 5 cm/sec (SSSA, 1965, p.349). Cf: *sticky limit.*

stictolith (stic′-to-lith) A *migmatite* with spotted appearance (Dietrich and Mehnert, 1961). Rarely used.

stiff clay Clay of low plasticity.

stigma [bot] (stig′-ma) That part of a pistil (usually at the tip of the style) which receives pollen grains and on which they germinate.

stigma [paleont] An opening of an arachnid; esp. an opening of an air-conveying tube forming the respiratory system or an opening into a saccular breathing organ (book lung). Pl: stigmata. Syn: *spiracle [paleont].*

Stigmaria (Stig-mar′-i-a) A genus of Permian and Carboniferous plants, identified by features of their rhizophores and roots. *Stigmaria* is regarded as a form genus; it is approximately correlated with *Sigillaria, Lepidodendron, Bothrodendron,* and other fossil lycopsids.

stilbite-Ca (stil′-bite) (a) A white, yellowish, gray, pink, reddish or brown, colorless to tan monoclinic zeolite mineral: $(Ca_{0.5},Na,K)_5(Si,Al)_{36}O_{72}{\cdot}30H_2O$. It represents the Ca-dominant form of stilbite, and occurs in sheaflike aggregates of crystals and also in radiated masses. Syn: *desmine; epidesmine.* (b) A term used by German mineralogists as a syn. of *heulandite.*

stilbite-Na The monoclinic Na-dominant form of stilbite: $(Na,Ca)_5(Si,Al)_{36}O_{72}{\cdot}30H_2O$.

stilleite (stil′-le-ite) A metallic gray cubic mineral of the *sphalerite* group: ZnSe.

stilling basin A pond at the downstream end of a *spillway* or *penstock* that allows flowing water to slow before it continues downstream in a natural, unlined channel.

stillstand (still′-stand) (a) Stability of an area of land, as a continent or island, with reference to the Earth's interior or mean sea level, as might be reflected, for example, by a relatively unvarying base level of erosion between periods of crustal movement. (b) A period of time during which there is a stillstand. Syn: *standstill.*

still water (a) A reach of a stream that is so nearly level as to have no visible current or other motion. Syn: *quiet reach.* (b) A sluggish stream, the water of which appears to be marked by little or no agitation.

stillwaterite (still′-wa-ter-ite′) A metallic trigonal mineral originally found in the Stillwater Complex: Pd_8As_3.

Stillwater-type nickel-copper deposit *stratiform nickel-copper deposit.*

Stillwater-type PGE deposit *stratiform PGE deposit.*

stillwellite-(Ce) (still′-well-ite) A reddish brown or pale pink trigonal mineral: $(Ce,La,Ca)BSiO_5$.

stilpnomelane (stilp-no′-mel-ane) A black, greenish-black, or bronze triclinic mineral, a trioctahedral Fe-rich phyllosilicate based on a modulation of a 2:1 T-O-T layer: $K(Fe^{2+},Mg,Fe^{3+})_8(Si,Al)_{12}(O,OH)_{36}{\cdot}nH_2O$. It occurs in micalike plates, fibrous forms, and velvety bronze-colored incrustations. Syn: *chalcodite.*

stinkdamp A mining term for hydrogen sulfide. The presence of this gas may indicate a gob fire in its early stages (Nelson, 1965).

stinkquartz (stink′-quartz) A variety of quartz that emits a fetid odor when struck.

stinkstein (stink′-stein) A syn. of *stinkstone.* Etymol: German *Stinkstein.*

stinkstone (stink′-stone) A stone that emits an odor on being struck or rubbed; specif. a *bituminous limestone* (or brown dolomite) that gives off a fetid smell (owing to presence of organic matter) when rubbed or broken. It may emit a "sweet-and-sour" smell if the carbonate rock is rich in organic-phosphatic material. See also: *anthraconite.* Syn: *stinkstein.*

stipe [bot] The stalk of a pistil or other small organ when axile in origin; also, the petiole of a fern leaf (Lawrence, 1951, p.771). Adj: *stipitate.*

stipe [paleont] One branch of a branched graptolite colony (rhabdosome), made up of a series of overlapping tubes (thecae). In an unbranched *rhabdosome,* a stipe is the entire graptolithine colony.

stipitate (stip′-i-tate) Said of a plant having a stipe or special stalk.

stipoverite (sti-pov′-er-ite) *stishovite.*

stipule (stip′-ule) A basal appendage of a petiole in some dicotyledenous plants. There are usually two stipules on a complete leaf.

stirrup-pore (stir′-rup-pore) In archaeocyathids, a pore excavated from the axial edges of a *septum* where it joins the inner wall (TIP, 1972, pt. E, p.19).

stishovite (sti′-shov-ite) A colorless tetragonal mineral: SiO_2. It is a high-pressure, extremely dense (4.35 g/cm^3) polymorph of quartz, produced under static conditions at pressures above about 100 kb and found naturally associated with coesite and only in shock-metamorphosed quartz-bearing rocks such as those from Barringer Crater (Meteor Crater), Ariz., and the Ries basin, Germany. Its occurrence provides a criterion for meteorite impact. Stishovite has a closely packed rutile type of structure in which the silicon has a coordination number of 6 (instead of 4 as in quartz and coesite); it forms at higher pressures than coesite and is apparently less stable at lower pressures after formation. Syn: *stipoverite.*

stistaite (stis′-ta-ite) A metallic gray rhombohedral mineral: SnSb.

STM *scanning tunneling microscopy.*

stochastic hydrology (sto-chas′-tic) That branch of hydrology involving the "manipulation of statistical characteristics of hydrologic variables with the aim of solving hydrologic problems, using the stochastic properites of the events" (Hofmann, 1965, p.120). Cf: *parametric hydrology; synthetic hydrology.*

stochastic process A process in which the *dependent variable* is random (so that prediction of its value depends on a set of underlying probabilities) and the outcome at any instant is not known with certainty. Ant: *deterministic process.* Syn: *random process.*

stock [intrus rocks] A relatively small, concordant and/or discordant plutonic body having an aerial extent less than 40 sq mi (100 sq km) and no known floor. Cf: *boss [ign].*

stock [ore dep] A rarely used term for a chimneylike orebody; a syn. of *pipe [ore dep].*

stock [paleont] The terminal bar of an *anchor* of a holothurian, perpendicular to the shank and of varying shape.

stockade (stock-ade′) Piling that serves as a breakwater.

stockwork (stock'-work) A mineral deposit consisting of a three-dimensional network of planar to irregular veinlets closely enough spaced that the whole mass can be mined. Cf: *chambered; reticulate [ore dep].* Syn: *network deposit; stringer lode.*

stockwork molybdenum deposit *porphyry molybdenum deposit.*

stoiberite (stoi'-ber-ite) A metallic black monoclinic mineral: $Cu_5V^{5+}{}_2O_{10}$.

stoichiometric (stoi'-chi-o-met'-ric) With reference to a compound or a phase, pertaining to the exact proportions of its constituents specified by its chemical formula. It is generally implied that a stoichiometric phase does not deviate measurably from its ideal composition.

stoichiometric coefficient One of the numerical coefficients in an equation that specify the combining proportions of the reactants and products of the reaction described.

Stokes' formula A formula first published in 1849 for computing the undulations of compensated geoids from gravity anomalies.

stokesite (stokes'-ite) A colorless or white orthorhombic mineral: $CaSnSi_3O_9{\cdot}2H_2O$.

Stokes' law A formula that expresses the rates of settling of spherical particles in a fluid: $V=Cr^2$, where V is velocity (in cm/sec), r is the particles' radius (in cm), and C is a constant relating relative densities of fluid and particle, acceleration due to gravity, and the viscosity of the fluid. It is named after Sir George Stokes, British mathematician and physicist. Cf: *impact law.*

stolidium (sto-lid'-i-um) The thin marginal extension of one or both valves of certain brachiopods (as in the superfamily Stenoscismatacea) forming a frill protruding at a distinct angle to the main contour of the shell (TIP, 1965, pt. H, p.153).

stolon (sto'-lon) (a) A slender internal threadlike tubule from which graptolithine thecae appear to originate, lying within the common canal. (b) A creeping and ribbonlike or membranous basal expansion from which certain octocorallian polyps arise (e.g. those of Stolonifera and Telestacea). (c) A slender tube of kenozooids bearing autozooids of ctenostomate bryozoans along its length. (d) A small calcareous tubelike projection serving as a connection between chambers in the test of an orbitoid foraminifer.

stolotheca (sto-lo-the'-ca) A type of graptolithine *theca* (tube) that encloses the main *stolon* and proximal parts of three new thecae (*autotheca* and *bitheca,* and daughter stolotheca). It is probably secreted by an immature autothecal zooid, constituting in effect the proximal part of the autotheca. Stolotheca is equivalent to *protheca* of graptoloids.

Stolt migration A migration method in the f-k domain assuming constant velocity. Vertical stretching before Fourier transformation is done to approximately accommodate vertical variations of velocity (Stolt, 1978).

stolzite (stolz'-ite) A sub-adamantine brown, yellow, red, green, etc. tetragonal mineral of the *scheelite* group: $PbWO_4$. It is isomorphous with wulfenite and dimorphous with raspite.

stoma (sto'-ma) A pore flanked by guard cells in the epidermis of a leaf, through which gases are exchanged. Pl: stomata. Adj: stomatal.

stomach stone (stom'-ach) *gastrolith.*

stomatal coccolith (sto'-ma-tal) One of the modified coccoliths surrounding the flagellar field in flagellate coccolithophores exhibiting dimorphism.

stomatal density In paleobotany, the number of stomata per a given area of a leaf cuticle. Cf: *stomatal index.*

stomatal index In paleobotany, the ratio of the number of epidermal cells to the number of stomata in a given area of a leaf, times 100. The index is used to compensate for the differences in stomatal density between sun and shade leaves, dry-climate and wet-climate leaves of the same species, and in the reconstruction of past levels of CO2. Cf: *stomatal density.*

stomodaeum (sto-mo-dae'-um) An esophaguslike tubular passageway or *pharynx* leading from the mouth of a coral polyp to the gastrovascular cavity (TIP, 1956, pt. F, p.251). Pl: stomodaea. Adj: stomodaeal.

stomostyle (sto'-mo-style) A thickened outer membrane invaginated in cytoplasm of the apertural region of some foraminifers, from which the mass of cytoplasm emerges giving rise to pseudopodia.

stone (a) A general term for rock that is used in construction, either crushed for use as *aggregate* or cut into shaped blocks as *dimension stone* or memorials. (b) One of the larger fragments in a variable matrix of a sedimentary rock; a *phenoclast* or *megaclast.* (c) A *stony meteorite.* (d) A cut and polished natural gemstone; a gem or precious stone. The term is used incorrectly for an artificial reproduction of, or a substitute for, a gem.

Stone Age In archaeology, a cultural level that was originally the first division of the *three-age system,* and was subsequently divided into the *Paleolithic* (*Old Stone Age*), *Mesolithic,* and *Neolithic* (*New Stone Age*). It is characterized by the use of materials other than metal, e.g. stone, wood, or bone, for technical purposes. Correlation of relative cultural levels with actual age (and, therefore, with the time-stratigraphic units of geology) varies from region to region; e.g., this oldest cultural level has been discovered to exist in recent times.

stone band *dirt band [coal].*

stone-banked terrace A *sorted step* whose steep slope is bordered by stones. The term should be reserved for a terracelike feature that lacks a regular pattern and is not a well-defined form of patterned ground (Washburn, 1956, p.833). Cf: *stone garland.*

stone bind An English term for interbedded layers of sandstone and shale, or for a rock (such as siltstone) intermediate between a sandstone and a mudstone (Arkell and Tomkeieff, 1953, p.115).

stone-bordered strip *stone stripe.*

stone canal A calcified, typically short tube or canal leading from the *madreporite* to the *ring canal* in the *water-vascular system* of an echinoderm.

stone cells Plant cells that have a strongly thickened wall, are usually lignified, commonly occur isolated in pulpy tissue, and are either isodiametric or highly irregular in form.

stone circle *sorted circle.*

stone coal *anthracite.*

stone eye *stone intrusion.*

stone field *block field.*

stone gall A clay concretion found in certain sandstones. Also spelled: stonegall.

stone garland A *sorted step* consisting of a tongue-shaped mass of fine material enclosed on the downslope side by a crescentic stony embankment; similar to but smaller than a *stone-banked terrace.* Syn: *stone semicircle; garland.*

stone guano A secondary deposit formed by leaching of guano and consequent enrichment in insoluble phosphates. Cf: *cave guano.*

stone intrusion An irregular, bulbous, sometimes much distorted *sandstone dike* occurring within a coal seam or penetrating it (frequently from top to bottom), and always connected with a similar sandstone in the roof or in higher strata. The British usage of the term intrusion for such a mass of sedimentary rock in a coal seam is not recommended (BSI, 1964, p.10). Cf: *drop [coal].* Syn: *stone eye.*

stone lace *stone lattice.*

stone land Legally, an area that is economically valuable for some variety of stone, such as granite or sandstone, that can be quarried. Cf: *mineral lands.*

stone lattice A *honeycomb structure [weath]* produced on a rock face in a desert by sandblast that "pecks away at the softer places and leaves the harder ones in relief" (Hobbs, 1912, p.205). Syn: *stone lace.*

Stoneley wave (Stone'-ley) A type of *guided wave* that is propagated along an internal surface of discontinuity; an interface wave. Cf: *channel wave.*

stone line The exposed edge of a broken line of angular and subangular rock fragments, paralleling a sloping topographic surface and lying just above the parent material of a soil at a depth of a few meters below that surface (Sharpe, 1938, p.24); it is exposed in natural and artificial cuts.

stone mesh *sorted net.*

stone net *sorted net.*

stone packing "A frost structure exclusively occurring on pebble beaches in arctic areas, consisting of a large, flat-lying boulder surrounded by a cluster of edgewise-lying flat stones, in an arrangement resembling the petals of a rose" (Schieferdecker, 1959, term 2403).

stone pavement An accumulation of rock fragments, esp. pebbles and boulders, in which the surface stones lie with a flat side up and are fitted together like a mosaic (Washburn, 1973, p.149). Cf: *subnival boulder pavement.*

stone peat The dark, compacted peat at the bottom of a bog.

stone pit A shallow *sorted circle,* less than a meter in diameter, consisting of a floor of isolated and dominantly circular stones (without finer material) surrounded by thick vegetation; term introduced by Lundqvist (1949, p.336).

stone polygon *sorted polygon.*

stone reef A longshore bar whose upper 3-4 m has been solidly cemented by calcium carbonate derived from organic material. Examples occur off the coast of Brazil near Recife.

stone ring A syn. of *sorted circle* and *sorted polygon;* the term refers to the circular or polygonal border of stones surrounding a central area of finer material.

stone river A term used in the Falkland Islands for a *rock stream* formed by solifluction. Cf: *block stream.*

stone semicircle *stone garland.*

stone stream *block stream.*

stone stripe A *sorted stripe* consisting of coarse rock debris, and occurring between wider stripes of finer material. Cf: *block stripe; soil stripe.* Syn: *stone-bordered strip; rock stripe.*

stone wall *hogback [geomorph].*

stoneware clay (stone'-ware) A clay suitable for manufacture of stoneware (ceramic ware fired to a hard, dense condition and with an absorption of less than 5%), used for items such as crocks, jugs, and jars. It is typically high in kaolinite and possesses good plasticity, fusible minerals, and a long firing range.

stonewort (stone'-wort) *charophyte.*

stone wreath *sorted circle.*

stony desert (ston'-y) A desert area whose surface has been deflated, leaving a concentration of coarse fragments after the removal of sand and dust particles, as a gravel-strewn *reg* or a pebble-strewn *serir;* a desert surface covered with *desert armor.* Cf: *rock desert.*

stony-iron meteorite (ston'-y-i'-ron) A general name for uncommon meteorites containing approximately equal amounts (by weight) of both nickel-iron and mafic silicates (such as pyroxene and olivine); e.g. the *pallasites* and *mesosiderites.* Syn: *iron-stony meteorite.* Obsolete syn: *siderolite; lithosiderite.*

stony meteorite A general name for meteorites consisting largely or entirely of silicate minerals (chiefly olivine, pyroxene, and plagioclase); e.g. chondrite and *achondrite.* Most stony meteorites resemble ultramafic rocks in composition, and they constitute more than 90% of all meteorites seen to fall. Syn: *meteoric stone.* Obsolete syn: *aerolite.*

stop A *dam* or *weir* that "stops", impounds, or otherwise impedes the flow of water.

stopbank (stop′-bank) An Australian term for a levee.

stope An underground excavation formed by the extraction of ore. Cf: *stoping [mining].*

stoping [intrus rocks] (stop′-ing) *magmatic stoping.*

stoping [mining] Extraction of ore in an underground mine by working laterally in a series of levels or steps in the plane of a vein. It is generally done from lower to upper levels, so that the whole vein is ultimately removed. The process is distinct from working in a shaft or tunnel or in a room in a horizontal drift, although the term is used in a general sense to mean the extraction of ore. See also: *stope.*

stoppaniite A vitreous light blue hexagonal mineral of the *beryl* group: $NaFe^{3+}_3(Mg,Fe^{2+})Be_6Si_{12}O_{36}{\cdot}2H_2O$.

storage (stor′-age) (a) Artificially impounded water, in surface or subsurface reservoirs, for future use. Also, the amount of water so impounded. (b) Water naturally detained in a drainage basin, e.g. groundwater in an aquifer, depression storage, and channel storage.

storage coefficient (a) For surface waters such as in a reservoir, a coefficient that expresses the relationship of the surface area to the mean annual flow that supplies it. (b) For groundwater *storativity.* Syn: *coefficient of storage.*

storage curve *capacity curve.*

storage ratio The net available storage of a body of water divided by the mean annual flow (Langbein and Iseri, 1960, p.18).

storage-required frequency curve (stor′-age-re-quired′) A graphic illustration of the frequency with which storage equal to or greater than selected amounts will be required to maintain selected rates of regulated flow (Langbein and Iseri, 1960, p.18).

storativity (stor-a-tiv′-i-ty) The volume of water an aquifer releases from or takes into storage per unit surface area of the aquifer per unit change in head. In a confined water body, the water derived from storage with decline in head comes from expansion of the water and compression of the aquifer; similarly, water added to storage with a rise in head is accommodated partly by compression of the water and partly by expansion of the aquifer. In an unconfined water body, the amount of water derived from or added to the aquifer by these processes generally is negligible compared to that involved in gravity drainage or filling of pores; hence, in an unconfined water body the storage coefficient is virtually equal to the specific yield (Lohman, 1972, p.13).

storied (sto′-ried) Said of a landform or landscape characterized by two or more adjacent levels; e.g. a *storied* peak plain with summit levels at different altitudes.

storis (sto′-ris [Pron: stur′-iis]) A floating mass of closely crowded icebergs and floes, esp. the remnants of the thickest polar ice drifting from the Arctic Ocean into the North Atlantic between Spitsbergen and Greenland. Etymol: Danish, "large ice".

storm In geophysics, a temporary, substantial disturbance of a geophysical field, e.g., a magnetic storm.

storm beach (a) A low, rounded ridge of coarse gravel, cobbles, and boulders, piled up by powerful storm waves behind or at the inner margin of a beach, above the level reached by normal high spring tides or by ordinary waves. Syn: *storm terrace.* (b) A beach as it appears immediately after an exceptionally violent storm, characterized either by removal or deposition of beach materials, commonly displaying concentrations of heavy minerals at the surface.

storm berm A low ridge along a beach, marking the limit of wave action during a storm. Cf: *winter berm.*

storm cusp A transient cusp, developed during a period of relatively heavy seas. Distance between storm cusps is 70 to 120 m. Term introduced by Evans (1938).

storm delta *washover.*

stormflow (storm′-flow) *direct runoff.*

storm icefoot An *icefoot* produced by the breaking of a heavy sea and consequent freezing of the wind-driven spray.

storm microseism A long-period microseism (25+ s) caused by ocean waves.

storm roller A term used by Chadwick (1931) for a wave-formed sedimentary structure "mismentioned" as a concretion. The feature is now regarded as a *ball-and-pillow structure* or *flow roll* formed by load deformation.

storm runoff *direct runoff.*

storm seepage The *runoff [water]* infiltrating the surface soil and moving toward streams as ephemeral shallow perched groundwater above the main groundwater level. It is usually considered part of direct runoff (Langbein and Iseri, 1960). Syn: *subsurface storm flow; subsurface runoff; subsurface flow; shallow percolation; interflow [water].* Cf: *surface runoff; groundwater runoff.*

storm surge An abnormal, sudden rise of sea level along an open coast during a storm, caused primarily by strong winds offshore or, less frequently, by a drop in atmospheric pressure, resulting in water piled up against the coast. It is most severe during high tide. Erroneous Syn: *tidal wave; storm tide.* Syn: *surge [waves]; hurricane surge; storm wave.*

storm terrace *storm beach.*

storm tide An erroneous syn. of *storm surge.*

storm water *direct runoff.*

storm wave *storm surge.*

story *level [speleo].*

stoss (Pron: shtohs) In glacial geology, describing the side of a hill or knob that faces the direction from which an advancing glacier or ice sheet moved; facing the upstream side of a glacier and most exposed to its abrasive action. Etymol: German "Stoss", "a push, thrust". Ant: *lee.*

stoss [sed struc] The term used for the relatively gentle, upstream-facing slope of an asymmetric bed form such as a *ripple* or *dune.* The stoss surface, or slope, extends from the deepest part of the *trough* to the next downstream *brinkpoint* of the bed form.

stoss-and-lee topography An arrangement, in a strongly glaciated area, of small hills or prominent rocks having gentle slopes on the stoss side and somewhat steeper, plucked slopes on the lee side; this arrangement is the reverse of *crag and tail.* Syn: *onset-and-lee topography.*

stottite (stott′-ite) (a) A brown, green, orange, or red tetragonal mineral: $Fe^{2+}Ge^{4+}(OH)_6$. (b) A group name for tetragonal minerals with the composition $AB(OH)_6$, where A = Fe, Mn, or Na, and B = Ge, Sn, or Sb.

straat A term used in the Kalahari region of southern Africa for the trough between dunes, often floored with clayey sand, and generally 100-150 m wide. Etymol: Afrikaans, "street". Pl: straate. Cf: *street.*

straczekite (stra′-czek-ite) A greenish-black monoclinic mineral: $(Ca,K,Ba)(V^{5+},V^{4+})_8O_{20}{\cdot}3H_2O$. It is classed as a "vanadium bronze".

straddle spread (strad′-dle) *split spread.*

straight hole A nearly vertical well drilled within the drilling contract specifications such as keeping with a 5° cone.

straight-line evolution *orthogenesis.*

straight suture An externally visible line of articular contact perpendicular to the longitudinal axis of adjoined crinoid ossicles.

strain In structural geology, the change in shape and/or size of a body. Strain is the nonrigid component of *deformation.*

strain axis *principal axis of strain.*

strain compatibility The geometric constraints required to keep a body of rock coherent and continuous during deformation.

strain ellipse An ellipse in the deformed state that is derived from a circle in the undeformed state.

strain ellipsoid An ellipsoid in the deformed state that is derived from a sphere in the undeformed state. The sphere is considered to have unit radius, and the ellipsoid accordingly has principal semi-axes equal in length to the principal strains expressed as stretches. Syn: *deformation ellipsoid.*

strain energy The recoverable energy associated with elastic deformation of a material. Elastic strain energy is related to the work required to stretch or compress a material.

strain energy density factor A parameter characterizing the intensity of the strain energy field surrounding a crack tip (Sih, 1973).

strain energy release rate The rate at which elastic strain energy is released during the growth of a crack (Griffith, 1920; Irwin, 1957). Rate is not measured in time but rather in unit length of crack growth.

strain factorization A way to break down the finite strain into component strains associated with different events, which when multiplied together yield the finite strain. The order of events is important because the multiplication is non-commutative.

strain field The array of states of strain at each point within a volume of material.

strain gage A general term for a device with which mechanical strain can be measured, commonly by an electrical signal, e.g. a wire strain gage.

strain hardening The behavior of a material whereby each additional increment of strain requires an additional increment of differential stress.

strain history A description of the continuous series of configurations that a deforming body of rock passes through during a finite period of time.

strainmeter A seisometer that is designed to detect deformation of the ground by measuring relative displacement of two points.

strain partitioning A way to partition the finite strain into component strains. Used in three ways: (1) to determine the contribution of different deformation mechanisms to the bulk strain, (2) to determine the contribution of different physical components in the rock (e.g. grains and matrix) to the bulk strain, and (3) to determine the contribution of different events to the finite strain (*strain factorization*).

strain ratio The ratio of the semi-axis lengths of the *strain ellipsoid.* Always expressed as long/short so that the ratio is ≥1.0. Used to describe the shape of the ellipsoid in any of its three principal planes.

strain recrystallization A *recrystallization* in which a deformed mineral changes to a mosaic of undeformed crystals of the same mineral; e.g., strained to unstrained calcite (Folk, 1965, p.15).

strain relaxation Gradual dissipation of strain over time by creep or viscous flow.

strain seismometer A seismometer that is designed to detect deformation of the ground by measuring relative displacement of two points.

strain shadow (a) *undulatory extinction.* (b) *pressure shadow.*

strain-slip cleavage *slip cleavage.*

strait [coast] A relatively narrow waterway connecting two larger bodies of water, as the Strait of Gibraltar linking the Atlantic Ocean with the Mediterranean Sea; a channel. The term is often used in the plural. Cf: *sound [coast].*

strait [geog] (a) A neck of land. (b) An obsolete term for a gorge.

strakhovite A black orthorhombic mineral: $NaBa_3(Mn^{2+},Mn^{3+})_4[Si_4O_{10}(OH)_2](Si_2O_7)O_2(F,OH)\cdot H_2O$.

strand [coast] (a) A syn. of shore and *beach;* the land bordering any large body of water, esp. the beach of a sea or an arm of the ocean, or the bank of a large river. (b) An Anglo-Saxon term for the narrow strip of land lying between high water and low water, being alternately exposed and covered by the tide. (c) A *barrier spit,* e.g. Silver Strand, San Diego, Calif.

strand [streams] A British term for a stream or current, and for a channel.

strand crack A fissure at the junction of a sheet of inland ice, an ice piedmont, or an ice rise with an ice shelf, the latter being subject to the rise and fall of the tide (Armstrong and Roberts, 1958, p.96).

stranded ice (strand'-ed) Floating ice that is deposited on the shore by retreating high water. Cf: *grounded ice; shore ice.*

strandflat (strand'-flat) (a) Any *wave-cut platform;* esp. a low, flat platform up to 65 km wide, extending for many hundreds of kilometers along the rocky coast of western Norway, either partly submerged or standing slightly above the present sea level as a result of isostasy, and supporting thousands of stacks, skerries, and other small islands. (b) A discontinuous shelf of land inside a fjord, reaching to about 30 m in height, having a rounded and dissected form.

strandline (strand'-line) (a) The ephemeral line or level at which a body of standing water, e.g. the sea, meets the land; the *shoreline,* esp. a former shoreline now elevated above the present water level. (b) A beach, esp. one raised above the present sea level. Cf: *raised beach.*

strand mark Any inorganic sole mark on a sedimentary surface along the shore (Clarke, 1918). Cf: *undertow mark.*

strand plain A prograded shore built seaward by waves and currents, and continuous for some distance along the coast (Cotton, 1958, p.431). It is characterized by subparallel beach ridges and swales, in places with associated dunes; some date back to about 6000 B.C. A glacio-isostatically uplifted or tilted strand plain may date back to the late Pleistocene (about 13,000 B.C.). Cf: *foreland [coast].*

stranskiite (stran'-ski-ite) A blue triclinic mineral: $Zn_2Cu(AsO_4)_2$. It is the arsenate analogue of mcbirneyite.

strashimirite (stra'-shi-mir-ite) A white or pale green monoclinic mineral: $Cu_8(AsO_4)_4(OH)_4\cdot 5H_2O$.

strata (stra'-ta) Plural of *stratum.*

strata-bound Said of a mineral deposit confined to a single stratigraphic unit. The term can refer to a stratiform deposit, to variously oriented orebodies contained within the unit, or to a deposit containing veinlets and alteration zones that may or may not be strictly conformable with bedding. See: Wolf, 1976. Cf: *bedded [ore dep].*

strata control *roof control.*

stratafer group Lithologic facies formed by chemical, biogenic, and hydrothermal effusive or exhalative processes, commonly composed of banded chert and quartz interbedded with oxide, sulfide, carbonate, and silicate minerals containing ferrous, nonferrous, and/or precious metals (Gross, 1990).

stratal (stra'-tal) Pertaining to a stratum or strata; e.g. "stratal dip" or "stratal unit".

stratal sequence In sequence stratigraphy, a stratigraphic unit that is defined exclusively with reference to bounding unconformities, without regard to their character. This is a descriptive sequence, in contrast to the interpretive *depositional sequence* (Berggren et al., 2001).

strata time *Geologic time* estimated from the thickness of strata and the rate of deposition (Kobayashi, 1944b, p.476).

strategic materials (stra-te'-gic) Materials that are vital to the security of a nation, but that must be procured entirely or in large part from foreign sources because the available domestic production will not meet the nation's requirements in time of war; e.g. *strategic minerals.*

strategic minerals Minerals that are considered to be *strategic materials;* e.g. chromium- and tin-bearing minerals, quartz crystal, and sheet mica were some of the "strategic minerals" during World War II.

strath [geomorph] (a) A terrace remnant of a broad, flat, former bedrock valley floor that now stands above the present drainage as a result of downcutting following uplift, change in base level, or climate change; e.g., a river terrace along a valley wall, interrupted in its development during the mature stage of a former erosion cycle. Bucher (1932) preferred the term "strath terrace" for this feature. Bascom (1931) proposed that "strath" be replaced by "berm". (b) A broad, flat valley bottom formed in bedrock and resulting from degradation, "first by lateral stream cutting and later by whatever additional processes of degradation may be involved" (Bucher, 1932, p.131); a level valley floor representing a local base level. It is usually covered by a veneer of alluvium, and is wider and flatter than a *glen.* Etymol: Scottish/Gaelic, "valley". Syn: *incipient peneplain.*

strath [marine geol] An elongate, broad, steep-sided depression on the continental shelf, usually glacial in origin. It is often deeper on its nearshore side.

strath stage The hypothetical stage in the *peneplanation* of a region when the main streams have carved broad, flat-floored valleys that are graded to the same regional base level. Cf: *maturity [geomorph]; old age [geomorph].*

strath terrace (a) A term used by Bucher (1932) for an extensive remnant of a strath (i.e., a flat valley bottom) that belonged to a former erosion cycle before rejuvenation of the stream following uplift, change in base level, or climate change. The term is synonymous with *strath* and *berm* as used by other writers. Cf: *fillstrath terrace.* (b) A strath (i.e., a remnant) of greater extension than that of a narrow ribbon along one valley (Engeln, 1942, p.222).

strath valley (a) A stream valley characterized by the development of a flat valley bottom (*strath*) resulting from degradation (Bucher, 1932, p.131). (b) A valley abandoned by a stream whose course was dislocated (Engeln, 1942, p.224).

stratic (strat'-ic) Pertaining to or designating the order or sequence of strata; stratigraphic, or pertaining to stratigraphy. Grabau (1924, p.821) referred to a disconformity as a "stratic unconformable relation".

straticulate (stra-tic'-u-late) Characterized by numerous very thin parallel layers, whether separable or not, either of sedimentary deposition (as a bed of clay) or of deposition from solution (as in a stalagmite or banded agate).

straticule (strat'-i-cule) A French term for a thin sedimentary layer, or lamina.

static unit *stratigraphic unit.*

stratification [lake] (strat'-i-fi-ca'-tion) The state of a body of water consisting of two or more horizontal layers of differing characteristics; esp. the arrangement of the waters of a lake in layers of differing densities (Wetzel, 2001, p.74). See also: *density stratification; thermal stratification.*

stratification [sed] (a) The formation, accumulation, or deposition of material in layers; specif. the arrangement or disposition of sedimentary rocks in strata. See also: *bedding [stratig].* (b) A structure produced by deposition of sediments in strata; a stratified formation, or stratum. It may be due to differences of texture, hardness, cohesion or cementation, color, internal structure, and mineralogic or lithologic composition. (c) The state of being stratified; a term describing a layered or bedded sequence, or signifying the existence of strata.

stratification [snow] Layering in a mass of snow, firn, or ice; it is caused by discontinuous deposition, by sedimentation of different kinds of snow, by accumulation of rock dust during summer periods, or by the development of layers of *depth hoar* at times of rapid changes in temperature.

stratification index A measure of the "beddedness" of a stratigraphic unit, expressed as the number of beds in the unit per 100 feet of section (Kelley, 1956, p.295). It is determined by multiplying the number of beds times 100, and dividing by the unit's thickness in feet. See also: *isostratification map; phyllofacies.* Syn: *beddedness index; bedding index.*

stratification plane *bedding plane.*

stratified (strat'-i-fied) Formed, arranged, or laid down in layers or strata; esp. said of any layered sedimentary rock or deposit. See also: *bedded [stratig].*

stratified cone A less preferred syn. of *stratovolcano.*

stratified drift Fluvioglacial, glaciolacustrine, or glaciomarine drift, consisting of sorted and layered material deposited by a meltwater stream or settled from suspension in a body of quiet water adjoining the glacier. Cf: *till.*

stratified estuary An estuary in which salinity increases with depth as well as along its length. An estuary is "highly stratified" if there is a density discontinuity separating surface river flow and bottom seawater; it is "partially stratified" if the amount of increase in salinity with depth is not significant. Ant: *vertically mixed estuary.*

stratified lake A lake exhibiting stratification of its waters.

stratified rock A rock displaying *stratification.* The term is virtually synonymous with *sedimentary rock,* although some sedimentary rocks (such as tillite) are without internal stratification. The term is sometimes, but not generally, applied to layered igneous rocks.

stratiform [ore dep] (strat'-i-form) Said of a special type of *strata-bound* deposit in which the desired rock or ore constitutes, or is strictly coextensive with, one or more sedimentary, metamorphic, or igneous layers; e.g. beds of salt or iron oxide, or layers rich in chromite or platinum in a layered igneous complex. See: Wolf, 1976. Cf: *bedded [ore dep].*

stratiform [sed struc] Having the form of a layer, bed, or stratum; consisting of roughly parallel bands or sheets, such as a "stratiform intrusion". Often found incorrect spellings: "strataform", "stratoform".

stratiform chromite deposit A *mineral-deposit type* characterized by stratiform chromite in the lower intermediate zone of layered mafic-ultramafic intrusive complexes. Syn: *Bushveld-type chromite deposit.*

stratiform intrusion *layered intrusion.*

stratiform iron-titanium-vanadium deposit A *mineral-deposit type* characterized by layers of titanium-vanadium-rich magnetite in the upper parts of layered mafic-ultramafic intrusive complexes. Syn: *Bushveld-type Fe-Ti-V deposit.*

stratiform nickel-copper deposit A *mineral-deposit type* characterized by stratiform nickel and copper sulfide minerals in the basal portion of layered mafic-ultramafic intrusive complexes. Syn: *Stillwater-type nickel-copper deposit.*

stratiform nonsulfide zinc deposit Zinc deposit composed of stratiform bodies or layers of zinc silicates, apparently formed by submarine volcanic exhalation (Hitzman et al., 2003).

stratiform PGE deposit A *mineral-deposit type* characterized by disseminated platinum group element sulfide minerals in olivine-rich cumulates within the anorthosite-gabbro zones of layered mafic-ultramafic complexes. Syn: *Merensky Reef-type deposit; Stillwater-type PGE deposit.*

stratify (strat'-i-fy) To lay down or arrange in strata.

stratignathy (stra-tig'-na-thy) The study of the time relationships of fish skeletal debris (ichthyoliths) (Doyle et al., 1974, p.825).

stratigrapher (stra-tig'-ra-pher) A geologist who studies or specializes in stratigraphy.

stratigraphic break (strat-i-graph'-ic) *break [stratig].*

stratigraphic classification The arbitrary but systematic arrangement, zonation, or partitioning of the sequence of rock strata of the Earth's crust into units with reference to any or all of the many different characters, properties, or attributes which the strata may possess (Hedberg, 1958, p.1881-1882).

stratigraphic code A usefully comprehensive, yet concisely stated, formulation of generally accepted views on stratigraphic principles, procedures, and practices, designed to obtain the greatest possible uniformity in applying such principles, etc.; specif. "a systematic collection of rules of formal stratigraphic classification and nomenclature" (ACSN, 1961, art.3). It is applicable to all kinds of rocks (sedimentary, igneous, metamorphic).

stratigraphic column *geologic column.*

stratigraphic condensation The mingling or intimate association, in a very thin stratigraphic interval or even a single bed, of fossils representative of different ages and environments (ISG, 1976, p.48).

stratigraphic control [ore dep] The influence of stratigraphic features on ore deposition, e.g. ore minerals selectively replacing calcareous beds. Cf: *structural control.*

stratigraphic control [stratig] The degree of understanding of the stratigraphy of an area; the body of knowledge that can be used to interpret its stratigraphy or geologic history.

stratigraphic correlation The process by which stratigraphic units in two or more separated areas are demonstrated or determined to be laterally similar in character or mutually correspondent in stratigraphic position, as based on geologic age (time of formation), lithologic characteristics, fossil content, or any other property; *correlation [geol]* in the usual or narrowest sense. See also: *lithostratigraphic correlation.*

stratigraphic cutoff *cutoff [stratig].*

stratigraphic facies Facies distinguished primarily on the basis of form, nature of boundaries, and mutual relations, to which appearance and composition are subordinated (Weller, 1958, p.627). These facies are all stratigraphic bodies of one kind or another; they may occur in vertical succession and have boundaries that are more or less horizontal stratigraphic planes (e.g. systems, formations, biostratigraphic zones, and lithostromes), or they may be laterally intergrading parts of some kind of a stratigraphic unit and separated at more or less arbitrary vertical cutoff boundaries (e.g., lithofacies), or they may bear both lateral and vertical relations to each other and have irregular boundaries (e.g. the magnafacies of Caster, 1934). See also: *facies.* Cf: *petrographic facies.*

stratigraphic geology *stratigraphy.*

stratigraphic guide In mineral exploration, a rock unit known to be associated with an ore. Cf: *lithologic guide.* See also: *ore guide.*

stratigraphic heave An obsolete syn. of *gap [struc geol]* and of overlap.

stratigraphic interval The body of strata between two stratigraphic markers (ISG, 1976, p.14). Syn: *interval [stratig].*

stratigraphic leak The deposition of sediments and/or fossils of a younger age within or under rocks of an older age; such a deposit may be said to be laid down in intraposition (Foster, 1966). It commonly involves microfossils, such as conodonts, which have descended through crevices or solution channels to lodge in a lower stratum where they become associated with fossils of greater age. Var: *stratigraphic leakage.* See also: *intrapositional deposit.*

stratigraphic leakage The process whereby sediments and/or fossils of a younger age are deposited within or under older rocks. The result or product of the process of stratigraphic leakage is stratigraphic leak (Foster, 1966, p.2604).

stratigraphic map A map that shows the areal distribution, configuration, or aspect of a stratigraphic unit or surface. It involves a span of geologic time. Examples include isopach map, structure-contour map, facies map, and vertical-variability map.

stratigraphic nomenclature The system of proper names given to specific stratigraphic units; for example, Trenton Formation, *Jurassic* System, Dibunophyllum Range-zone (ISG, 1976, p.13).

stratigraphic overlap An obsolete syn. of *overlap [stratig].*

stratigraphic paleontology The study of fossils and of their distribution in various geologic formations, emphasizing the stratigraphic relations (time and sequence) of the sedimentary rocks in which they are contained. Cf: *biostratigraphy.*

stratigraphic range The distribution or spread of any given species, genus, or other taxonomic group of organisms through geologic time, as indicated by its distribution in strata whose geologic age is known. Also, the persistence of a fossil organism through the stratigraphic sequence. Syn: *geologic range; time-rock span.*

stratigraphic record The *geologic record* based on or derived from a study of the stratigraphic sequence; the rocks arranged chronologically as in a geologic column. Syn: *record.*

stratigraphic reef A thick, laterally restricted mass of carbonate rock, without genetic connotations (Dunham, 1970, p.1931). Cf: *ecologic reef.*

stratigraphic section *geologic section.*

stratigraphic separation The thickness of the strata that originally separated two beds brought into contact at a fault. Syn: *stratigraphic throw.*

stratigraphic separation diagram A diagram showing the stratigraphic positions of the hanging-wall and footwall cutoffs of a fault. It is used to identify flats and ramps of thrusts and more generally to identify individual fault segments in an area of linked or interacting faults. It is also used to construct plots of displacement variation along strike.

stratigraphic sequence A chronologic succession of sedimentary rocks from older below to younger above, essentially without interruption; e.g. a *sequence* of bedded rocks of interregional scope, bounded by unconformities.

stratigraphic terminology The total of unit-terms used in stratigraphic classification, such as *formation, stage, biozone* (ISSC, 1994, p.14-15).

stratigraphic terrane A terrane characterized by coherent stratigraphic sequences in which depositional relations between successive units can be demonstrated; it may or may not include basement rocks (D.L. Jones et al., 1983).

stratigraphic test A hole drilled to obtain information on the thickness, lithology, sequence, porosity, and permeability of the rock penetrated, or to locate the position of a key bed. Cf: *structure test.* Colloquially shortened to: *strat test.*

stratigraphic throw *stratigraphic separation.*

stratigraphic trap A trap for oil or gas that is the result of lithologic changes rather than structural deformation. See also: *shale-out; pinch-out.* Cf: *structural trap; combination trap.* Syn: porosity trap.

stratigraphic unit A stratum or body of rock recognized as a unit in the classification of the Earth's rocks with respect to any of the many characters, properties, or attributes that rocks may possess (ISSC, 1994, p.14), for any purpose such as description, mapping, and correlation. Rocks may be classified stratigraphically on the basis of lithology (lithostratigraphic units), fossil content (biostratigraphic units), magnetic polarity (magnetostratigraphic polarity units), being bounded by unconformities (unconformity-bounded units), age (chronostratigraphic units), or properties (such as mineral content, radioactivity, seismic velocity, electric-log character, chemical composition) in categories for which formal nomenclature is lacking. Stratigraphic units based on one property will not necessarily coincide with those based on another. It is, therefore, essential that different terms be used for each so that their named units can be distinguished from each other. Clear definition of a stratigraphic unit is of paramount importance. A geologic-time unit is not a stratigraphic unit. Syn: *stratic unit.*

stratigraphy (stra-tig'-ra-phy) (a) The science of rock strata. It is concerned not only with the original succession and age relations of rock strata but also with their form, distribution, lithologic composition, fossil content, geophysical and geochemical properties indeed, with all characters and attributes of rocks as strata; and their interpretation in terms of environment or mode of origin, and geologic history. All classes of rocks, consolidated or unconsolidated, fall within the general scope of stratigraphy. Some nonstratiform rock bodies are considered because of their association with or close relation to rock strata (ISG, 1976, p.12). Syn: *stratigraphic geology.* (b) The science dealing with the description of all rock bodies forming the Earth's crust sedimentary, igneous, and metamorphic and their organization into distinctive, useful, mappable units based on their inherent properties or attributes. Stratigraphic procedures include the description, classification, naming, and correlation of these units for the purpose of establishing their relationship in space and their succession in time (ISSC, 1994, p.13). (c) The arrangement of strata, esp. as to geographic position and chronologic order of sequence. (d) The sum of the characteristics studied in stratigraphy; the part of the geology of an area or district pertaining to the character of its stratified rocks.

stratisphere (strat'-i-sphere) That part of the Earth's crust containing stratified rocks, both sedimentary and volcanic (Ronov, 1982). Not to be confused with *stratosphere.*

strätlingite (strät'-ling-ite) A olorless or light green rhombohedral mineral first described as occurring in a metamorphosed limestone inclusion: $Ca_2Al_2SiO_2(OH)_{10}·2.25H_2O$.

stratocladistics Method of evolutionary analysis in which relative stratigraphic position of a species or group is accorded the same status as morphologic or other characters.

stratofabric (strat-o-fab'-ric) The arrangement of strata in any body of stratified rock, "from the dimensions of a thin section to those of a sedimentary basin" (Fischer, 1964, p.148).

stratomere (strat'-o-mere) Any segment of a rock sequence, irrespective of magnitude (Geological Society of London, 1967, p.83).

stratomeric standard terms The terms of the formal chronostratigraphic hierarchy (Geological Society of London, 1967, p.83).

stratophenetics (strat′-o-phe-net′-ics) That method in *systematics* wherein ancestor-descendant (phylogenetic or evolutionary) relationships among taxa are inferred from comparison of morphologic, anatomic, and any other observable characteristics of fossils when the fossils are arranged according to their geochronologic ages or stratigraphic horizons. Cf: *phenetics.*

stratose (strat′-ose) Arranged in strata.

stratosphere (strat′-o-sphere) The layer of the atmosphere extending from 10 km to 50 km, overlying the *troposphere;* characterized by a region of constant temperature for the first several km. Between 20 km and 50 km the temperature increases with altitude, reaching a temperature maximum at the stratopause (approximately 50 km). The temperature of the stratosphere results from the absorption of ultraviolet radiation by the oxygen and ozone molecules between 30 km and 50 km.

stratotectonic (strat′-o-tec-ton′-ic) A term used, mainly in Australia, to describe tectonic evolution in relation to stratigraphy.

stratotype (strat′-o-type) The original or subsequently designated standard of reference of a named layered stratigraphic unit or of a stratigraphic boundary; a specific interval or point in a specific sequence of rock strata that constitutes the standard of the definition and characterization of the stratigraphic unit or boundary being defined (ISSC, 1994, p.26). Syn: *type section.*

stratovolcano (strat′-o-vol-ca′-no) A volcano that is constructed of alternating layers of lava and pyroclastic deposits, along with abundant dikes and sills. Viscous, acidic lava may flow from fissures radiating from a central vent, from which pyroclastics are ejected. Syn: *composite volcano; composite cone.* Less preferred syn: *bedded volcano; stratified cone.*

strat test *stratigraphic test.*

stratum (stra′-tum) A tabular or sheetlike body or *layer [stratig]* of sedimentary rock, visually separable from other layers above and below; a *bed [stratig].* It has been defined as a stratigraphic unit that may be composed of a number of beds (Dana, 1895, p.91), as a layer greater than 1 cm in thickness and constituting a part of a bed (Payne, 1942, p.1724), and as a general term that includes both "bed" and "lamination" (McKee and Weir, 1953, p.382). The term is more frequently used in its plural form, *strata.* Cf: *lamina [sed].*

stratum correlation Time correlation of beds or sequences of beds (Henningsmoen, 1961, p.64).

stratum plain A plain having a *stripped structural surface.* Examples in the Colorado Plateau region of the U.S. commonly form isolated buttes and mesas, or benchlike or terracelike areas along valley sides, but some have considerable areal extent. See also: *dip plain; cut plain.* Syn: *structural plain.*

Straumanis camera method (Strau-man′-is) In X-ray diffraction analysis, a method of mounting film in a cylindrical X-ray camera to allow for recording of both front and back reflections on both sides of the exit and entry ports, enabling the determination of film diameter from the measurements. Cf: *Wilson technique.*

straw mulching The use of straw to create a surface mulch on all or part of the soil surface for soil or water conservation, for soil temperature management or for weed supression.

straw stalactite *soda straw.*

stray (a) A lenticular or discontinuous rock unit encountered unexpectedly in *drilling.* (b) A thin, local rock unit separated by a short interval from a thicker, more persistent formation of similar lithology; e.g. *stray sand.*

stray current An electric current that is introduced into the Earth by leakage of currents from cultural installations.

stray sand A *stray* consisting of sandstone.

streak [mineral] The color of a mineral in its powdered form, usually obtained by rubbing the mineral on a *streak plate* and observing the mark it leaves. Streak is an important characteristic in mineral identification; it is sometimes different from the color of the sample, and is generally constant for the same mineral.

streak [sed] (a) A comparatively small and flattish or elongate sedimentary body, visibly differing from the adjacent rock, but without the sharp boundaries typical of a lens or layer (Stokes and Varnes, 1955, p.143). (b) A long, narrow body of sand, perhaps representing an old shoreline; a *shoestring.* (c) The outcropping edge of a coal bed.

streaking (streak′-ing) *mineral streaking.*

streak plate In mineral identification, a piece of unglazed porcelain used for rubbing a sample to obtain its powder color, or streak. It has a hardness of about seven.

stream [glaciol] (a) *ice stream.* (b) A stream of *meltwater.*

stream [streams] (a) Any body of running water that moves under gravity to progressively lower levels, in a relatively narrow but clearly defined channel on the surface of the ground, in a subterranean cavern, or beneath or in a glacier. It is a mixture of water and dissolved, suspended, or entrained matter. Cf: *river [streams].* (b) A term popularly applied to a *brook* (as in Maine) or to a small river. (c) The water flowing in a stream. (d) A term used in quantitative geomorphology interchangeably with *channel.*

stream action *fluviation.*

stream azimuth The orientation of the mean line of a stream from head to mouth, measured in degrees from some arbitrary direction, generally north (Strahler, 1954, p.346).

streambed The channel containing or formerly containing the water of a stream.

stream-built terrace *alluvial terrace.*

stream capture *capture [streams].*

stream channel The bed where a natural stream of water runs or may run; the long narrow depression shaped by the concentrated flow of a stream and covered continuously or periodically by water. Syn: *streamway.*

stream-cut terrace A *terrace* representing a former valley floor that has been abandoned by further erosion of a stream cutting either in rock or alluvium.

stream deposition The accumulation of fluvial sediment in channel deposits (e.g., channel bar, point bar, channel fill), bank deposits (e.g., levee, crevasse splay), or flood plain deposits (Reineck and Singh, 1963).

streamer (stream′-er) A marine seismic recording cable containing hydrophones and designed for recording seismic signals while under continuous tow in the water. Reflection surveys use streamers up to 6 km or more in length. Cf: *bottom cable.*

stream erosion The progressive removal, by a stream, of bedrock, overburden, soil, or other exposed matter from the surface of its channel, as by hydraulic action, corrasion, and corrosion.

streamflood (stream′-flood) A flood of water in an arid region, characterized by the "spasmodic and impetuous flow" of a *sheetflood* but confined to a definite, shallow channel that is normally dry (Davis, 1938).

streamflow (stream′-flow) A type of *channel flow,* applied to that part of surface runoff traveling in a stream whether or not it is affected by diversion or regulation. Also spelled: stream flow. Cf: *sheet flow [geomorph]; overland flow.*

streamflow depletion The amount of water that flows into a given land area minus the amount that flows out of that area.

streamflow wave A traveling wave caused by a sudden increase of water flow (ASCE, 1962).

stream frequency The ratio of the number of streams of all orders within a drainage basin to the area of that basin; a measure of topographic texture. Syn: *channel frequency.*

stream gaging Measurement of the velocity of a stream of water in a channel or open conduit and of the cross-sectional area of the water, in order to determine discharge. See also: *chemical gaging.* Syn: *gaging.*

stream gold Gold occurring in alluvial placers.

stream gradient The angle between the water surface or the channel floor and the horizontal, measured in the direction of flow; the "slope" of the stream. Cf: *law of stream gradients.* Syn: *stream slope; slope [streams].*

stream-gradient ratio The ratio of the gradient of a stream channel of a given order to that of a stream of the next higher order in the same drainage basin. Syn: *channel-gradient ratio.*

streamhead (stream′-head) The source or beginning of a stream.

streaming flow [glaciol] (stream′-ing) A *glacier flow* in which the ice moves without cracking or breaking into blocks, as where the walls and bottom are relatively smooth for a long distance. Obsolete.

streaming flow [hydraul] *tranquil flow.*

streaming lineation *parting lineation.*

streaming potential *electrofiltration potential.*

stream length The length of a stream segment of a given order *u.* Cf: *law of stream lengths.* Syn: *channel length.*

stream-length ratio The ratio of the mean stream length of a given order to the mean stream length of the next lower order within a specified drainage basin (Horton, 1945, p.296).

streamlet (stream′-let) A small stream.

stream line An imaginary line connecting a series of fluid particles in a moving fluid so that, at a given instant, the velocity vector of every particle on that line is tangent to it.

streamline flow (stream′-line) *laminar flow [hydraul].*

stream load (a) The solid material that is transported by a stream, either as visible sediment (carried in suspension, or moved along the stream bed by saltation and traction) or in chemical or colloidal solution. (b) The quantity or amount of such material at any given time or passing a point in a given period of time, and expressed as a weight or volume per unit time. Material in solution is sometimes excluded in the usage of the term. See also: *suspended load; bed load; dissolved load.*

stream mode The process of manual digitizing in which points are digitized automatically at preset time or distance intervals (Krzanowski et al., 1993). Cf: *point mode.*

stream morphology *river morphology.*

stream number A syn. of *number of streams;* see also: *law of stream numbers.*

stream order A classification of the relative position of streams in a channel network, assigning each link an integer order number determined by the pattern of confluences in the tributary network headward of the given link. The most widely employed ordering method was developed by Strahler (1952b): all exterior links are order 1; proceeding downstream the confluence of two links of order u generates a resultant stream of order u+1; the confluence of tributary orders u and v, where u>v, leads to a resultant stream of order u. The Horton (1945) ordering method modifies the Strahler stream orders by the headward extension of order numbers greater than 1: at each junction of equal-order tributary streams, the order of the resultant downstream link is extended headward along the tributary most nearly parallel to the downstream link; if both tributaries enter at the same angle, the order of the downstream link is assigned to the longer tributary. See also: *basin order; stream segment; link [geomorph].* Syn: *channel order.*

stream piracy *capture [streams].*
stream power A measure of the work done on the channel boundaries by flow. It may be expressed as total stream power, stream power per unit area, or specific stream power.
stream profile The *longitudinal profile* of a stream.
stream robbery *capture [streams].*
stream segment A link, or sequence of links, along a stream channel, extending from the fork where the stream achieves a given *stream order* to the downstream fork where it joins a stream of equal or higher order. Syn: *channel segment.*
stream slope *stream gradient.*
stream terrace A planar surface along the sides of a stream valley representing the remnants of an abandoned floodplain, stream bed, or valley floor produced during a former stage of erosion or deposition. Cf: *alluvial terrace; rock terrace; meander terrace; inset terrace.* Syn: *terrace [geomorph]; river terrace.*
stream tin Cassiterite occurring in the form of waterworn pebbles in alluvial or placer deposits or on bedrock along valleys or streams, such as that resulting from the wearing away of pneumatolytic veins associated with acid rocks. Cf: *lode tin.* Syn: *alluvial tin.*
stream transportation The movement by a stream of weathered or eroded rock material in chemical solution, in turbulent suspension, or by rolling, dragging, or bouncing along the stream bed.
stream underflow Percolating water in the permeable bed of a stream and flowing parallel to the stream (ASCE, 1962).
stream valley An elongate depression on the Earth's surface, carved by a stream during the course of its development.
stream velocity The rate of flow, measured by distance per time unit, e.g. ft/sec.
streamway (stream'-way) (a) The current of a stream. (b) *stream channel.*
street A part of a bare desert floor that forms a gap separating chains of sand dunes. See also: *straat.*
strelkinite (strel'-kin-ite) A yellow orthorhombic mineral of the *carnotite* group: $Na_2(UO_2)_2(V_2O_8)·6H_2O$.
strengite (streng'-ite) A colorless, red, or violet orthorhombic mineral of the *variscite* group: $Fe^{3+}PO_4·2H_2O$. It is dimorphous with phosphosiderite, and may contain some Mn^{3+} which imparts the red color.
strength A term used in experimental structural geology that is meaningful only when all the environmental conditions of the experiment are specified; in general, the ability to withstand differential stress, measured in units of stress.
streptaster (strep'-tas-ter) A sponge spicule (microsclere) having the form of a modified aster in which the rays do not arise from a common center but radiate from an axis; e.g. a streptosclere, a sanidaster, or a discorhabd. Cf: *euaster.*
streptosclere (strep'-to-sclere) A siliceous sponge spicule (streptaster) in which long raylike spines are given off in spiral succession about a central axis, and intergrade with simple euasters. Syn: *spiraster.*
streptospiral (strep-to-spi'-ral) Said of a foraminiferal test coiled like a ball of twine.
stress (a) In a solid, the force per unit area, acting on any surface within it, and variously expressed as pounds or tons per square inch, or dynes or kilograms per square centimenter; also, by extension, the external pressure which creates the internal force. The stress at any point is mathematically defined by nine values: three to specify the normal components and six to specify the shear components, relative to three mutually perpendicular reference axes. Cf: *strain.* See also: *normal stress; shear stress.* (b) A commonly used short form for *differential stress.*
stress axis *principal axis of stress.*
stress concentration The magnification of stress near the edge of a pore, crack, or other type of hole in an elastic material such as rock.
stress difference *differential stress.*
stress drop The decrease in differential stress that occurs in the vicinity of a fault following slip on the fault.
stress ellipsoid A geometric representation of the state of stress at a point that is defined by three mutually perpendicular principal stresses and their intensities.
stress field The state of stress, either homogeneous or varying from point to point, in a given domain.
stress intensity The stress concentration parameter at the tip of a joint that influences the rate at which a joint will propagate. Stress intensity accounts for joint driving stress, joint shape, and joint length.
stress mineral A term suggested by Harker (1918) for minerals such as chlorite, chloritoid, talc, albite, epidote, amphiboles, and kyanite, whose formation in metamorphosed rocks he believed was favored by shearing stress. The term has become obsolete. Cf: *antistress mineral.*
stress pillars *vertical sheet structure.*
stress reduction shadow (a) The zone next to a joint in which tensile stress is reduced as a consequence of a stress-free joint wall in the immediate vicinity. (b) The zone next to a fault characterized by a stress drop that inhibits other faults from propagating into it or nucleating within it (Ackermann and Schlische, 1997).
stress shadow *pressure shadow.*
stress-strain curve The plot of conventional strain in percent shortening or elongation (as the abscissa) and true longitudinal differential stress, i.e., the difference between the maximum and minimum principal stresses (as the ordinate). Syn: *stress-strain diagram.*
stress-strain diagram *stress-strain curve.*
stress tensor A description of the state of stress at a point, which involves nine components, referred to three orthogonal coordinate axes. Three components are normal stresses, acting perpendicular to the coordinate planes. The remaining six components are shear stresses, acting parallel to the coordinate planes.
stretch [exp struc geol] A measure of the change in length of a line, specifically the ratio of the final length to the initial length of the line (Malvern, 1969, p.164). Cf: *quadratic elongation; extension [exp struc geol].*
stretch [geog] A *reach* of water or land.
stretch [geophys] (a) n. The change in wavelet shape produced by applying a normal-moveout correction. (b) v. To change the scale to make comparisons easier, e.g, to display S- and P-wave sections at approximately double scale to roughly accommodate the differences between S- and P-wave velocities.
stretch [photo] A technique to enhance a digital image by changing the contrast by increasing or decreasing the range of values for a raster. Often the stretch is chosen based on a histogram of the brightness values in an image. Types of stretches include linear, contrast, histogram, or a custom designed stretch to enhance selected features in the image.
stretched Said of a structure or texture produced by *dynamic metamorphism,* in which the constituents are stretched and commonly broken in the same direction; e.g., stretch-pebble conglomerate. A stretched condition should not be confused with lineation.
stretch fault *stretch thrust.*
stretching lineation A lineation parallel to the maximum stretching direction, commonly defined by elongated mineral grains or mineral aggregates, or the long axes of deformed objects such as vesicles, ooids, or reduction spots.
stretch modulus *Young's modulus.*
stretch thrust A little-used term for a reverse fault formed by shear in the inverted limb of an overturned fold. Syn: *stretch fault.*
strewn field [impact] A dense field of small, separated *impact craters* produced by the atmospheric breakup of an incoming meteorite. Typically elliptical in plan, with dimensions on the Earth of a few km in the short axis and up to 10 km in the long axis. Similar but much larger fields are observed on Venus.
strewn field [meteorites] (a) A restricted geographic area within which a specific group of tektites is found. (b) An area over which a shower of meteorites is recovered.
stria [cryst] One of a series of parallel straight lines on the surface of a crystal, as in pyrite, indicative of an oscillation between two crystal forms; also, one of a series of such lines on the cleavage planes of a mineral, as of plagioclase, calcite, or corundum, indicative of *polysynthetic twinning.* Syn: *striation.* Adj: *striate.* Pl: *striae.*
stria [paleont] One of a series of fine grooved lines or threads on the surface of some shells, esp. on otherwise smooth shells, e.g. one of the parallel minute grooves on nautiloid and ammonoid conchs, separated by lirae and not easily discernible with the unaided eye.
striae (stri'-ae) Plural of *stria.*
striate (stri'-ate) (a) Adj. of *stria.* (b) Said of spores and pollen having a streaked sculpture characterized by multiple, usually parallel, grooves and ribs in the exine; specif. referring to the Striatiti. (c) *striated.*
striated (stri'-at-ed) (a) Adj. of *striation.* (b) *striate.*
striated ground *striped ground.*
striation (stri-a'-tion) (a) *stria.* (b) One of multiple scratches or minute lines, generally parallel, inscribed on a rock surface by a geologic agent, e.g., glaciers (*glacial striation*), streams (cf: *drag mark*), or faulting (Cf: *slickenside*). Syn: *scratch.* (c) The condition of being striated; the disposition of striations. Adj: *striated; striate.*
striation cast The cast of a striation produced on a sedimentary surface; it is usually found on the underside of a thin siltstone or fine sandstone bed interlayered with mudstone. Cf: *groove cast.* Syn: *microgroove cast.*
Striatiti (Stri-at'-i-ti) Mostly bisaccate (some mono- or trisaccate) pollen with characteristic striate sculpture that consists of raised *taeniae* crossing the proximal (and occasionally the distal) face of the central body. They are presumably produced by *glossopterids* and conifers and common in Permian and Lower Triassic strata.
stricture (stric'-ture) A contraction between successive shell joints of the skeleton of a nasselline radiolarian.
striding level (strid'-ing) (a) A spirit level so mounted that it can be placed above and parallel with the horizontal axis of a surveying instrument and so supported that it can be used for precise leveling of the horizontal axis of the instrument or for measuring any remaining inclination of the horizontal axis. (b) A demountable spirit level that can be attached to the telescope tube to level the line of sight.
strigovite (strig'-o-vite) An unnecessary name for a dark green variety of chlorite.

strike [eco geol] n. The discovery of a mineral deposit, esp. if sudden or unexpected. v. To discover or reach a mineral deposit suddenly or unexpectedly, e.g. to "strike" oil.
strike [struc geol] n. The direction or trend taken by a structural surface, e.g. a bedding or fault plane, as it intersects the horizontal. See also: *attitude [struc geol].* Cf: *trend [struc geol]; trace [struc geol].* Syn: *line of strike.* v. To be aligned or to trend in a direction at right angles to the *line of dip.*
strike fold *longitudinal fold.*
strike joint A joint that strikes parallel to the strike or lineation of the enclosing rock. Cf: *dip joint.*
strike-overlap (strike-o′-ver-lap) A term proposed by Melton (1947, p.1870) for truncation of sedimentary rocks below unconformities, esp. for a slow, extremely low-angle regional truncation of contrasting depositional strike below a regional unconformity. The term is essentially synonymous with *overstep* if it is assumed that "in most bodies of marine, or interfingering marine and nonmarine rock, angular unconformities eventually pass downdip into disconformities, which in turn disappear farther out in the basin" (Swain, 1949, p.634).
strike separation In a fault, the distance of *separation* of two formerly adjacent beds on either side of the fault surface, measured parallel to the strike of the fault. Cf: *dip separation; strike slip.*
strike-separation fault (strike′-sep-a-ra′-tion) An obsolete term for a *lateral fault.*
strike shift In a fault, the *shift* or relative displacement of the rock units parallel to the strike of the fault, but outside the fault zone itself; a partial syn. of *strike slip.* Little used.
strike-shift fault An obsolete term for a *strike-slip fault.*
strike slip In a fault, the component of the movement or slip that is parallel to the strike of the fault. Cf: *dip slip; strike separation; oblique slip.* Syn: *horizontal displacement; horizontal separation.* Partial syn: *strike shift.*
strike-slip fault A fault on which the movement is parallel to the fault's strike. Cf: *dip-slip fault; oblique-slip fault.* See also: *convergent fault; divergent fault; transcurrent fault.* Obsolete syn: *strike-shift fault.*
strike stream A *subsequent stream* that follows the strike of the underlying strata.
strike valley A subsequent valley eroded in, and developed parallel to the strike of, underlying weak strata; a valley containing a *strike stream.*
string (a) *drillstring.* (b) Numerous connected joints of tubulars such as *casing, tubing,* or *drill pipe,* of one size, used in a well.
string [geophys] *flyer.*
string bog A linear periglacial muskeg or moor with an undulating surface, occurring in the boreal needle-tree forest zone of the northern hemisphere (esp. western Siberia and the Hudson Bay area), and characterized by shallow water-filled depressions and festoons of lenticular ridges (up to 2 m high) consisting of floating fen or moss vegetation. Its origin is controversial, but it seems to be related to frost action and gravity movements in bog areas. See also: *ring moor.*
stringer [ore dep] (string′-er) A mineral veinlet or filament, usually one of a number, occurring in a discontinuous subparallel pattern in host rock. See also: *stringer lode.*
stringer [seis] A thin *high-speed layer,* usually with limited lateral continuity.
stringer [stratig] A thin sedimentary bed.
stringer lead *stringer zone.*
stringer lode A zone of shattered host rock containing a network of stringers; a *stockwork.*
stringer zone A shattered mineral zone cemented by many small, not persistent, veins. Cf: *stockwork.*
stringhamite (string′-ham-ite) A blue monoclinic mineral: $CaCuSiO_4 \cdot H_2O$.
strip [ice] A long narrow area of pack ice about 1 km or less in width, usually composed of small fragments detached from the main mass of ice, and run together under the influence of wind, swell, or current; it is more limited than a *belt.*
strip [photo] *flight strip.*
strip [meta] *ribbon [petrology].*
stripe [pat grd] One of the alternating bands of fine and coarse surficial material, or of rock or soil and vegetation-covered ground, constituting a form of patterned ground characteristic of slopes steeper than those of steps. It is usually straight, but may be sinuous or branching, and is probably the result of solifluction acting in conjunction with other processes, such as rillwork. See also: *sorted stripe; nonsorted stripe;* contraction stripe.
striped ground A form of *patterned ground* marked by alternating stripes produced on a sloping surface by frost action. See also: *sorted stripe.* Syn: *striped soil; striated ground.*
striped soil *striped ground.*
stripe hummock A *nonsorted stripe,* probably closely related to an earth hummock, but formed on sloping ground.
strip log A graphic record of a drilled well, plotted to scale on a log strip, e.g. *drilling-time log; sample log.*
strip mining (a) *opencut mining.* (b) Mining of coal by surface methods as distinguished from mining metal-bearing ores and industrial minerals, which is referred to as openpit mining.
stripped bedding plane The exposed top surface of a resistant stratum that forms a *stripped structural surface* when extended over a considerable area.
stripped illite *degraded illite.*
stripped peneplain *fossil peneplain.*
stripped structural surface A horizontal or gently inclined surface developed on rocks of unequal resistance, the overlying softer beds having been removed by erosion so as to expose the more or less smooth surface of a resistant stratum that has served as a local base level and thereby controlled the depth of erosion; specif. the surfaces produced on a *structural plateau, stratum plain,* and *structural terrace.* Syn: *stripped surface.*
stripped surface *stripped structural surface.*
stripping (strip′-ping) *layer stripping.*
stripping ratio A ratio of the volume or tonnage of overburden or waste to the underlying volume or tonnage of ore in an openpit mine. Syn: strip ratio.
strip ratio *stripping ratio.*
strip thrust An obsolete syn. of *décollement.*
strobili Pl. of *strobilus.*
strobilus (strob′-i-lus) An axis bearing a conelike aggregation of sporophylls, as in club mosses and many conifers. Pl: *strobili.* Cf: *cone [bot].*
stroma (stro′-ma) (a) The supporting framework of an animal organ, such as organic tissue in the mesodermal endoskeleton of a living echinoderm. Cf: *stereom.* (b) A compact mass of fungous cells, or of mixed host and fungous cells, in or on which spores or sporocarps are formed. Pl: stromata.
stromatactis (stro-ma-tac′-tis) A sedimentary structure characterized by a horizontal or nearly flat bottom, up to about 10 cm in diameter, and an irregular or convex-upward upper surface, consisting of sparry-calcite cement, usually in the central part of a *reef core* (Chilingar et al., 1967, p.321); sometimes called *reef tufa.* They have been interpreted as fillings of original cavities caused by the burial and decay of soft-bodied but rigid frame-building organisms (known as *Stromatactis* according to Lowenstam, 1950), although they may represent syngenetic voids in calcareous sediments; some examples represent recrystallized sheetlike bryozoan colonies.
stromatic *lit-par-lit.*
stromatite (stro′-ma-tite) A *migmatite* with two or more textural elements arranged in essentially parallel layers. Rarely used.
stromatolite (stro-mat′-o-lite) An organosedimentary structure produced by sediment trapping, binding, and/or precipitation as a result of the growth and metabolic activity of micro-organisms, principally cyanophytes (blue-green algae) (Walter, 1976, p.1). It has a variety of gross forms, from nearly horizontal to markedly columnar, domal, or subspherical. The term was introduced by Kalkowsky in 1908 as stromatolith. Cf: *oncolite.*
stromatolith [sed] *stromatolite.*
stromatology (stro-ma-tol′-o-gy) A term, now obsolete, proposed to embrace "the history of the formation of the stratified rocks" (Page, 1859, p.340).
stromatoporoid (stro-ma-top′-o-roid) A general name for any of a group of extinct sessile benthic marine organisms of uncertain biologic affinities (probably phylum Porifera, possibly Coelenterata or Cyanophyta). They secreted a calcareous skeleton, generally a few tens of centimeters across, of tabular, encrusting, dendroidal, domal, or bulbous form; internal structural elements were arranged in subhorizontal or concentric laminae or lines of dissepiments separated by small radial pillars, or in an irregular open network. Stromatoporoids were especially abundant in Ordovician-Devonian reefs. Range, Cambrian(?) to Cretaceous.
Strombolian-type bomb (Strom-bo′-li-an-type′) A general type of volcanic bomb produced from lava that is less fluid than that of a *Hawaiian-type bomb.*
Strombolian-type eruption A type of volcanic eruption characterized by jetting of clots or "fountains" of fluid, basaltic lava from a central crater. Etymol: Stromboli, Lipari Islands of Italy. Cf: *Hawaiian-type eruption; Peleean-type eruption; Vulcanian-type eruption.*
stromeyerite (stro′-mey-er-ite′) A dark steel-gray orthorhombic mineral with a blue tarnish: AgCuS. Syn: *silver-copper glance.*
stromoconolith (stro-mo-con′-o-lith) A layered igneous intrusion that is either conical or funnel-shaped (Tomkeieff, 1961). Obsolete.
stronalsite (stro′-nals-ite) A white orthorhombic mineral: $Na_2SrAl_4Si_4O_{16}$.
strong Said of large or important mineral veins or faults.
strong motion *Ground motion* that is sufficiently strong to be of interest in engineering seismology.
strongyle (stron′-gyle) A rodshaped sponge spicule (monaxon) with both ends blunt. Also spelled: strongyl.
strongylote (stron′-gy-lote) Said of a sponge spicule having one end rounded.
strontianite (stron′-ti-an-ite′) A pale green, white, gray, or yellowish orthorhombic mineral of the *aragonite* group: $SrCO_3$.
strontioborite (stron′-ti-o-bor′-ite) A colorless monoclinic mineral: $SrB_8O_{11}(OH)_4$.
strontiochevkinite (stron′-ti-o-chev′-kin-ite) A sub-metallic black monoclinic mineral of the *chevkinite* group: $(Sr,La,Ce,Ca)_4(Fe^{2+},Fe^{3+})(Ti,Zr)_4Si_4O_{22}$.
strontiodresserite A white orthorhombic mineral: $(Sr,Ca)Al_2(CO_3)_2(OH)_4 \cdot H_2O$. It is the Sr analogue of dresserite.
strontioginorite (stron′-ti-o-gi′-no-rite) A white monoclinic mineral of the *ginorite* group: $SrCaB_{14}O_{20}(OH)_6 \cdot 5H_2O$.

strontiohilgardite A mineral of the *hilgardite* group. Also called: hilgardite-1A. Cf: *tyretskite-1A.*
strontiojoaquinite (stron-ti-o-joa-quin′-ite) A green, yellow-green, or yellow-brown monoclinic mineral: $Sr_2Ba_2(Na,Fe^{2+})_2Ti_2Si_8O_{24}(O,OH)_2{\cdot}H_2O$. It is the Sr analogue of joaquinite-(Ce).
strontiomelane A sumetallic black monoclinic mineral: $(Sr,Ba,K)Mn_8O_{16}$. It is related to cryptomelane.
strontio-orthojoaquinite A yellow orthorhombic mineral: $Na(Ba,Sr)_4Fe^{3+}Ti_2Si_8O_{24}(OH)_4$. It is a dimorph of strontiojoaquinite.
strontiopiemontite A deep red monoclinic mineral of the *epidote* group: $CaSr(Al,Mn^{3+},Fe)_3(SiO4_)(Si_2O_7)(O,OH)_2$.
strontiopyrochlore A white to pale yellow cubic mineral of the *pyrochlore* group: $Sr_{0.6}Nb_2(O,OH)_7$. Previously called strontian hydropyrochlore. Not approved by the International Mineralogical Association.
strontiowhitlockite A white rhombohedral mineral of the *whitlockite* group: $Sr_9Mg(PO_3OH)(PO_4)_6$.
strontium-apatite (stron′-ti-um-ap′-a-tite) A pale-green to yellowish-green hexagonal mineral of the *apatite* group: $(Sr,Ca)_5(OH,F)(PO_4)_3$.
strophic (stroph′-ic) Said of a brachiopod shell in which the true *hinge line* is parallel to the *hinge axis.* Cf: *nonstrophic.*
strophomenid (stro-phom′-e-nid) Any articulate brachiopod belonging to the order Strophomenida, characterized chiefly by a plano- to concavo-convex shell that may be resupinate or geniculate. Range, Lower Ordovician to Lower Jurassic.
strophotaxis (stroph-o-tax′-is) *Taxis [ecol]* in which an organism tends to turn in response to some external stimulus. Cf: *phobotaxis; thigmotaxis.*
structural (struc′-tur-al) Of or pertaining to rock deformation or to features that result from it.
structural adjustment A term proposed by Salisbury (1904, p.710) for the rearrangement of the drainage of an area so as to conform to the geologic structure; esp. the flowing of streams along the strike of the strata. Cf: *topographic adjustment.*
structural analysis [stat] In *geostatistics,* inspection and modeling of the spatial dependence of a random variable, including determination of *stationarity* and modeling of the *semivariogram* or spatial *covariance.*
structural analysis [struc petrol] This term is meant to be an English translation of Sander's Gefügekunde. It includes the analysis of structural features on all scales from thin section to discontinuous outcrop. It also involves placing interpretations on the movements and stress fields that were responsible for deformation. Cf: *structural petrology; structural geology.*
structural basin *basin.*
structural bench A bench representing the resistant edge of a *structural terrace* that is being reduced by erosion (Cotton, 1958, p.94-95). Syn: *rock bench.*
structural closure *closure [struc geol].*
structural contour *structure contour.*
structural control The influence of structural features on ore deposition, e.g. ore minerals filling fractures. Cf: *stratigraphic control [ore dep].*
structural crystallography Study of the internal arrangement and spacing of atoms and molecules composing crystalline solids.
structural datum *datum horizon.*
structural depression A topographically low area resulting from structural deformation of the Earth's crust.
structural diagram A figure illustrating the spatial array of *fabric elements.*
structural dome *dome [struc geol].*
structural element A single class of structure in an outcrop or region that contains many classes of structures. For example, cleavage is a structural element in an outcrop that contains cleavage, joints, tilted beds, folds, and slip surfaces.
structural feature A feature produced by deformation or displacement of the rocks, such as a fold or fault. For such features the more colloquial term structure (used as a specific noun) is now generally accepted.
structural formula The chemical formula written to show the actual association of some of the atoms. For example schorl, $NaFe_3Al_6(BO_3)_3(Si_6O_{18})(OH)_3(OH)$. Cf: *chemical composition [mineral]; composition [mineral].*
structural geology The branch of geology that deals with the form, arrangement, and internal structure of the rocks, and especially with the description, representation, and analysis of structures, chiefly on a moderate to small scale. The subject is similar to *tectonics,* but the latter is generally used for the broader regional or historical phases.
structural high *high.*
structural lake *tectonic lake.*
structural landform A landform developed by erosion and controlled by the structure of the rocks. Cf: *tectonic landform.*
structural low *low [struc geol].*
structurally controlled nonsulfide zinc deposit Zinc deposits composed of primary zinc silicates which replace carbonate host rocks along and adjacent to faults (Hitzman et al., 2003).
structural map *structure-contour map.*
structural nose *nose [struc geol].*
structural petrology The analysis of fabric on the thin-section or micro scale. It includes the study of grain shapes and relationships (microstructure) and the study of crystallographic preferred orientations. Recently the transmission electron microscope has also been employed to examine the substructures of deformed crystals. Cf: *structural analysis [struc petrol].* Syn: *fabric analysis; petrofabric analysis; petrofabrics; microtectonics.*
structural plain *stratum plain.*
structural plateau A plateaulike landform with a *stripped structural surface.*
structural province A region whose geologic structure differs significantly from that of adjacent regions. It is generally coextensive with a *physiographic province.*
structural relief (a) The vertical distance between stratigraphically equivalent points at the crest of an anticline and in the trough of an adjacent syncline. (b) More generally, the difference in elevation between the highest and lowest points of a bed or stratigraphic horizon in a given region.
structural state The degree of *ordering* exhibited by a mineral that contains two or more symmetrically distinct sites that may be occupied by either of two species of atoms.
structural terms Structure is the concrete, real architectural framework of mineral material. For clay minerals, the assembly of one or more layers plus interlayer material is referred to as a unit structure. When using the terms plane, sheet and layer for clay minerals, the recommended designations are a single plane of atoms, a tetrahedral or octahedral sheet, and 1:1 or 2:1 layer. Plane, sheet, and layer refer to increasingly thicker arrangements. The fundamental building block is the structural component of greatest bond valence and it is designated in brackets. Lattice is not synonymous with structure. A lattice is a regular, uniform distribution of points in space invented to describe the patterns of the internal geometry of crystals.
structural terrace (a) A local shelf or steplike flattening in otherwise uniformly dipping strata, composed of a synclinal bend above and an anticlinal bend at a lower level. (b) A terracelike landform controlled by the structure of the underlying rocks; esp. a terrace produced by the more rapid erosion of weaker strata lying on more resistant rocks in a formation with horizontal bedding. Cf: *structural bench.*
structural trap A trap for oil or gas that is the result of folding, faulting, or other deformation. Cf: *stratigraphic trap; combination trap.*
structural unconformity *angular unconformity.*
structural valley A valley that owes its origin or form to the underlying geologic structure. Cf: *tectonic valley.*
structure [geomorph] (struc′-ture) A comprehensive term for the assemblage of rocks upon which erosive agents are, and have been, at work; the terrane underlying a landscape. The term indicates the product of all constructional agencies, and includes the arrangement and disposition of the rocks, their nature and mode of aggregation, and even their initial forms prior to erosion.
structure [mineral] (a) The form assumed by a mineral; e.g. bladed, columnar, tabular, or fibrous. (b) *crystal structure.*
structure [palyn] The internal architecture of the *exine* of palynomorphs, particularly of the *ektexine.* Some expressions of *sculpture* may intergrade with structural features.
structure [petrology] A megascopic feature of a rock mass or rock unit, generally seen best in the outcrop rather than in hand specimen or thin section, such as columnar structure, blocky fracture, platy parting, or foliation. The term is also applied to the appearance, or to a smaller-scale feature, of a rock in which the texture or composition is different in neighboring parts; e.g. banded structure, orbicular structure. The term texture is generally used for the smaller features of a rock or for the particles composing it; although the two terms are often used interchangeably, they should not be considered synonymous, even though some textures may parallel major structural features. See also: *sedimentary structure.*
structure [struc geol] (a) The general disposition, attitude, arrangement, or relative positions of the rock masses of a region or area; the sum total of the structural features of an area, consequent upon such deformational processes as faulting, folding, and igneous intrusion. (b) A term used in petroleum geology for any physical arrangement of rocks (such as an anticline or reef) that may hold an accumulation of oil or gas.
structure amplitude The magnitude of the *structure factor.*
structure contour A *contour* that portrays a structural surface such as a formation boundary or a fault. Syn: *subsurface contour.* See also: *structure-contour map.*
structure-contour map (struc′-ture-con′-tour) A map that portrays subsurface configuration by means of *structure contour* lines. See also: *contour map; tectonic map.* Syn: *structural map; structure map.*
structure factor A quantity that expresses the amplitude and phase angle of diffracted X-rays. It is symbolized *F(hkl).* The intensity of X-rays diffracted by planes with *diffraction indices hkl* is proportional to the square of the structure factor for that set of planes.
structure factor F The efficiency of a unit cell for scattering of X-rays, which is always less than the sum of the scattering of the individual atoms in the unit cell because of the beams that the individual atoms scatter are usually out of phase with one another. It is, then, a measure of the intensity of the observed diffracted beam. Because its value depends on the way atoms are arranged in the unit cell it is called the structure factor.
structure ground A term used by Antevs (1932, p.48) but now replaced by its syn. *patterned ground.*
structure map *structure-contour map.*

structure-process-stage (struc′-ture-proc′-ess-stage′) The name given to the trinitarian Davisian principle (Davis, 1899) that the development of all landforms in the cycle of erosion is a function of three basic factors: geologic structure, geomorphic process, and stage of development. Davis originally referred to "structure, process, and time", but later writers have changed this to "structure-process-stage".

structure refinement *crystal-structure refinement.*

structure section A *vertical section* that shows the observed geologic structure on a vertical or nearly vertical surface, or, more commonly, one that shows the inferred geologic structure as it would appear on a vertical plane cutting through a part of the Earth's crust. The vertical scale is often exaggerated.

structure soil *soil structure [pat grd].*

structure solution *crystal-structure solution.*

structure test A generally shallow hole drilled primarily to obtain information on geologic structure, although other types of information may be acquired during drilling. It is frequently drilled to a structural datum that is normally short of a known or expected oil-producing zone or zones. Cf: *stratigraphic test.*

structure type A group of crystals having the same atomic structure, i.e. having the constituent atoms arranged in a geometrically analogous way. An example is the NaCl structure type, in which equal numbers of cations and anions occur in six coordination; it includes sylvite, periclase, and galena.

struggle for existence (strug′-gle) The natural process by which members of a population compete for a limited supply of vital necessities, thus resulting in *natural selection.*

Strukturboden (Struk′-tur-bod′-en) A term formerly used for what is now known as *patterned ground.* Etymol: German, "structure ground" or "structure soil".

Strunian (Stru′-ni-an) European stage: uppermost Devonian, transitional into Carboniferous (below Tournaisian).

strunzite (strunz′-ite) (a) A yellow to brownish-yellow fibrous triclinic mineral: $Mn^{2+}Fe^{3+}_2(PO_4)_2(OH)_2 \cdot 6(H_2O,OH)$. (b) A group name for minerals with the analogous composition, but with Mn^{2+} replaced by Fe^{2+} or Fe^{3+}.

strut thrust An obsolete term for a fault initiated by the shearing of a strut, or competent bed.

strüverite (strü′-ver-ite) A black tetragonal mineral of the *rutile* group: $(Ti,Ta,Fe^{2+})O_2$. It is probably an unnecessary name for tantalum-bearing rutile.

struvite (stru′-vite) A colorless to yellow or pale brown orthorhombic mineral: $(NH_4)Mg(PO_4) \cdot 6H_2O$.

studenitsite A colorless monoclinic mineral: $NaCa_2B_9O_{14}(OH)_4 \cdot 2H_2O$.

Student's t test (Stu′-dent's) A statistical test used to determine whether two samples could have been drawn from the same *population.* Named after A. Student, pseudonym of William S. Gosset, 20th century Irish statistician. Syn: *t test.*

studtite (studt′-ite) A yellow monoclinic peroxide mineral: $(UO)_2O_2(H_2O)_2 \cdot 2H_2O$.

stuffed derivative A crystal structure derived from a "parent" structure by substitution of some atoms that have a lower oxidation number than those for which they substitute, and wherein charge balance is maintained by "stuffing" additional cations into available interstices in the structure; e.g., kalsilite, $KAlSiO_4$, is a stuffed derivative of tridymite, SiO_2 in which half of the Si is replaced by Al, and K is introduced into voids in the structure to maintain charge balance.

stumpflite (stumpf′-lite) A metallic hexagonal mineral of the *nickeline* group: Pt(Sb,Bi).

stunted fauna (stunt′-ed) *dwarf fauna.*

sturmanite (stur′-man-ite) A bright yellow trigonal mineral of the *ettringite* group: $Ca_6(Fe^{3+},Al,Mn^{2+})_2(SO_4)_2[B(OH)_4](OH)_{12} \cdot 25H_2O$.

Sturtian (Stur′-ti-an) An obsolete geochronometric subdivision of the late *Precambrian,* covering the time span from 800 to 610 Ma (Harland et al., 1982, 1990), therefore approximately equivalent to the *Cryogenian* Period of the Standard Global Chronostratigraphic Scale.

sturtite (sturt′-ite) An amorphous mineral, possibly related to *hisingerite*: $(Mn,Fe,Al,Ca)_3Si_4O_{10}(OH)_3 \cdot H_2O$.

sturzstrom (sturz′-strom) A huge mass of rapidly moving rock debris and dust, derived from the collapse of a rock cliff or mountainside, flowing down steep slopes and across gentle to flat ground, often for several kilometers. A sturzstrom may include compressed air, water, or mud between the rock fragments, but ordinarily it is considered to be a mass of cohesionless blocks of rock dispersed in a cloud of dust, flowing at speeds of more than 100 km/hr. Sturzstroms are the most catastrophic of all forms of mass movement. Several have been identified on the Moon. Term first used by A. Heim in 1881; (see Hsü, 1975). Cf: *debris flow; rockfall.* Syn: *rock avalanche; rockfall avalanche; rock-fragment flow.*

stützite (stütz′-ite) A metallic gray hexagonal mineral: $Ag_{5-x}Te_3$(x = 0.24 to 0.36). It was formerly regarded as identical with empressite. Also spelled: stuetzite.

stylaster (sty′-las-ter) Any one of a group of hydrozoan coelenterates belonging to the order Stylasterina, characterized by a calcareous skeleton and by sexual individuals that remain attached to the colony. Range, Cretaceous to Holocene. Cf: *hydroid; millepore.*

style [bot] The usually attenuated part of the pistil that connects the stigma to the ovary.

style [paleont] (a) A sponge spicule (monaxon) with one blunt end and one pointed end; e.g. *tylostyle.* (b) A tubule that arises from the galea in phaeodarian radiolarians. (c) A central calcareous process in certain pores of a stylaster coral. (d) The *telson* of a crustacean. (e) A spur of wrinkled enamel in mammalian teeth.

style [tect] *tectonic style.*

styliform columella (sty′-li-form) A solidly fused and longitudinally projecting coral columella. It is fused to scleractinian entosepta by secondary stereom.

styliform cyrtolith A cyrtolith *coccolith* with a long spinose central structure; e.g. a *pole coccolith.*

stylobreccia A nodular limestone which has undergone intense pressure dissolution to form many stylolites which give the rock the appearance of a breccia.

stylocerite (sty-lo-ce′-rite) A rounded or spiniform process on the outer part of the antennular *peduncle* in some decapod crustaceans.

stylocumulate (sty-lo-cu′-mu-late) A residue that has accumulated along stylolites (Logan and Semeniuk, 1976, p.68).

stylolinids Small, gradually tapering, narrow conical fossils with mostly smooth exteriors and interiors; generally assigned to the cricoconarids.

stylolite (sty′-lo-lite) (a) A surface or contact, usually occurring in homogeneous carbonate rocks and more rarely in sandstones and quartzites, that is marked by an irregular and interlocking penetration of the two sides: the columns, pits, and teeth-like projections on one side fit into their counterparts on the other. As usually seen in cross section, it resembles a suture or the tracing of a stylus. The seam is characterized by a concentration of insoluble constituents of the rock, e.g. clay, carbon, or iron oxides, and is commonly parallel to the bedding. Stylolites are supposedly formed diagenetically by differential vertical movement under pressure, accompanied by solution. See also: *microstylolite; suture joint.* (b) A straight, vertically grooved column, of the same material as the rock in which it occurs, commonly less than a centimeter in length, fitting into a corresponding socket in a stylolitic seam and being highly inclined or at right angles to the bedding plane. It often results from the slipping under vertical pressure of a part capped by a fossil shell through adjacent parts not so capped. Term introduced by Klöden (1828, p.28). Etymol: Greek "stylos", "pillar", + "lithos", "stone". Obsolete syn: *epsomite [sed]; crystallite; toenail; crowfoot.*

stylolitic (sty-lo-lit′-ic) Pertaining to a stylolite, as a "stylolitic seam" or "stylolitic column".

stylolitic fracture A fracture in which the opposite walls have moved toward each other, and interpenetrate due to localized *diffusional mass transfer.*

styloreactate (sty′-lo-re-ac′-tate) Dolomitization along stylolites (Logan and Semeniuk, 1976, p.67). Cf: *net fabric.*

stylotypite (sty-lo-typ′-ite) A syn. of *tetrahedrite.* The name has been applied esp. to tetrahedrite containing considerable silver.

S-type granitoid A granitic rock produced by magma derived from partial fusion of metasedimentary or supracrustal origin. The term, proposed by Chappell and White in 1974, is thoroughly discussed by Clarke (1992).

Styrian Orogeny (Styr′-i-an) One of the 30 or more short-lived orogenies during Phanerozoic time identified by Stille, in this case in the Miocene, between the Burdigalian and Aquitanian Stages.

suanite (su′-an-ite) A pearly-white triclinic mineral: $Mg_2B_2O_5$. Syn: *magnioborite.*

subaerial (sub-aer′-i-al) Said of conditions and processes, such as erosion, that exist or operate in the open air on or immediately adjacent to the land surface; or of features and materials, such as eolian deposits, that are formed or situated on the land surface. The term is sometimes considered to include *fluvial.* Cf: *subaqueous; subterranean.* See also: *surficial.*

subaerial bench A term used by Lawson (1915) for a nonalluviated, concave-upward pediment.

subaerialism (sub-aer′-i-al-ism′) The doctrine that the landscape and its landforms are formed chiefly by subaerial agents (esp. rainwash) and processes. Obsolete.

subage (sub-age′) A seldom-used term for a *geochronologic unit* shorter than an age, during which the rocks of the corresponding substage were formed. It is usually characterized by the occurrence of some specific phenomenon, such as the deposition of loess. Syn: *episode (b); time (d); phase [geochron] (a).*

subalkalic (sub-al′-ka-lic) (a) A group term applied to rocks of the tholeiitic and calc-alkaline series. (b) Said of an igneous rock that contains no alkali minerals other than feldspars. (c) Used to describe an igneous rock of the Pacific suite.

subalkaline basalt (sub-al′-ka-line) As proposed by Chayes (1964), a replacement for the terms *tholeiitic basalt* and *tholeiite.* Today it is the less preferred term. Basalts in which neither nepheline nor acmite appear in the CIPW norm fall in the subalkaline category.

suballuvial bench (sub-al-lu′-vi-al) A term used by Lawson (1915, p.34) for the outward or basinward extension of a pediment, covered by alluvium as the basin slowly filled (the thickness increasing basinward to several hundred meters), and exhibiting a convex-upward longitudinal profile. Cf: *concealed pediment.*

subalpine (sub-al'-pine) *montane.*

subalpine peat *hill peat.*

subaluminous (sub-a-lu'-mi-nous) Said of an igneous rock in which there is little or no excess of aluminum oxide over that required to form feldspars or feldspathoids; one of Shand's (1947) groups of igneous rocks, classified on the basis of the degree of aluminum-oxide saturation. Cf: *peralkaline; peraluminous; metaluminous.*

subangular (sub-an'-gu-lar) Somewhat angular, free from sharp angles but not smoothly rounded; specif. said of a sedimentary particle showing definite effects of slight abrasion, retaining its original general form, and having faces that are virtually untouched and edges and corners that are rounded off to some extent, such as a glacial boulder with numerous (10-20) secondary corners and a roundness value between 0.15 and 0.25 (midpoint at 0.200) (Pettijohn, 1957, p.59), or one with one-third of its edges smooth (Krynine, 1948, p.142). Also, said of the *roundness class* containing subangular particles.

subaquatic plant (sub-a-quat'-ic) A hydrophyte that is not a *submerged aquatic plant.*

subaqueous (sub-aq'-ue-ous) Said of conditions and processes, or of features and deposits, that exist or are situated in or under water, esp. fresh water, as in a lake or stream. Generally used to specify a process that occurs either on land or under water, e.g. slumping, gravitational slides. Cf: *subaerial.*

subaqueous gliding *slump.*

subaqueous levee Natural levee of prograding deltaic distributary that has not aggraded to the mean water level of the basin of deposition.

subaqueous sand dune *dune [streams].*

subaqueous till *Berg till;* also, till deposited from a glacier terminating in water.

subarctic (sub-arc'-tic) Pertaining or relating to the regions directly adjacent to the Arctic Circle, or to areas that have climate, vegetation, and animals similar to those of arctic regions, at temperate latitutes.

subarkose (sub-ar'-kose) A sandstone that does not have enough feldspar to be classed as an *arkose,* or a sandstone that is intermediate in composition between arkose and pure quartz sandstone. A quantitative definition: an arkosic sandstone containing 75-95% quartz and chert, less than 15% detrital clay matrix, and 5-25% unstable materials in which the feldspar grains exceed the rock fragments in abundance (Pettijohn, 1954, p.364). Other definitions have been given by Folk, (1954, p.354; 1968, p.124), and McBride (1963, p.667). Pettijohn (1957, p.322) later used 10-25% unstable fragments, so that a subarkose might have at least 5% feldspar. The rock is roughly equivalent to *feldspathic arenite* of Williams et al. (1954). Syn: *feldspathic quartzite; feldspathic sandstone.*

subarkosic wacke (sub-ar-ko'-sic) A wacke (over 15% matrix) with 5 to 25% feldspar; a species of feldspathic graywacke (Pettijohn et al., 1972, p.171). Syn: *feldspathic wacke; subfeldspathic lithic wacke.*

Subatlantic (Sub-at-lan'-tic) n. A term used primarily in Europe for a period of Holocene time (approximately the last 2,500 years, or from 500 B.C. to the present) following the Subboreal, during which the inferred climate became generally cooler and wetter; a unit of the *Blytt-Sernander sequence,* characterized by beech and linden vegetation. Also spelled: Sub-Atlantic. adj. Pertaining to the postglacial Subatlantic interval and to its climate, deposits, biota, and events.

subbase (sub'-base) A base or supporting material placed below that which ordinarily forms the base; specif. a layer of earth or rock placed between the *base course* and the *subgrade,* designed to give additional support, to distribute the load, or to form a pervious layer; e.g., the first layer of large-diameter crushed stone laid down in constructing a road, airstrip, or other graded surface. A subbase commonly contains material of lower quality, with less strict specifications, than a base course.

subbiozone (sub-bi'-o-zone) A subdivision of a *biozone* "to express finer biostratigraphic detail" (ISG, 1976, p.48-49).

subbituminous A coal (sub-bi-tu'-mi-nous) A type of *subbituminous coal,* characteristically nonagglomerating, having 10,500 to 13,000 BTU/lb (24.42 to 30.23 MJ/kg) (moist, mineral-matter-free). It is differentiated from high volatile C bituminous by agglomerating characteristics.

subbituminous B coal A type of *subbituminous coal* having 9,500 to 10,500 BTU/lb (22.10 to 24.42 MJ/kg) (moist, mineral-matter-free).

subbituminous C coal A type of *subbituminous coal* having 8,300 to 9,500 BTU/lb (19.30 to 22.10 MJ/kg) (moist, mineral-matter-free).

subbituminous coal A black coal, intermediate in rank between lignite and bituminous coals; or, in some classifications, the equivalent of *black lignite.* It is distinguished from lignite by higher carbon and lower moisture content. Further classification of subbituminous coal is made on the basis of calorific value. See also: *subbituminous A coal; subbituminous B coal; subbituminous C coal; lignite.* Cf: *gloss coal; metalignitous coal.*

Subboreal (Sub-bo'-re-al) n. A term used primarily in Europe for an interval of Holocene time (from about 5,000 to 2,500 calendar years ago) following the Atlantic and preceding the Subatlantic, during which the inferred climate became generally warmer and drier; a unit of the *Blytt-Sernander sequence,* characterized by oak, ash, and linden vegetation. Also spelled: Sub-Boreal. adj. Pertaining to the postglacial Subboreal interval and to its climate, deposits, biota, and events.

subboreal [climat] (a) Said of a climate that is very cold or approaching frigidity. (b) Pertaining to a biogeographic zone that approaches a *boreal* climatic condition. (c) Pertaining to the Subboreal postglacial period, and to the climate of such a period. Also spelled: sub-Boreal.

subbottom profiler An instrument that produces a high-resolution cross-section-like record of sediments below the sea floor. Syn: *pinger.*

subcannel coal (sub-can'-nel) Cannel coal of brown-coal to subbituminous rank. Cf: *metacannel coal; lean cannel coal.*

subcapillary interstice (sub-cap'-il-lar'-y) An *interstice* sufficiently smaller than a *capillary interstice* that the molecular attraction of its walls reaches across the entire opening. Water held in it by adhesive forces is immovable except by forces in excess of pressures commonly found in subsurface water. Cf: *supercapillary interstice.*

subchela (sub-che'-la) The distal prehensile or grasping part of a crustacean limb formed by folding dactylus against propodus or dactylus and propodus against carpus. Pl: subchelae.

subchondritic A term for metal-silicate meteorites or clasts in meteorites that have compositions differing from chondritic compositions in ways consistent with the gain or loss of a low melting fraction.

subchron A short (less than 10^5 years) time interval of constant polarity of the geomagnetic field.

subclass (sub-class') In the hierarchy of classification of plants and animals, a subcategory of class.

subcontinent (sub-con'-ti-nent) (a) A division or part of a continent having characteristics that distinguish it from the rest of the continent, e.g. the Indian subcontinent. This subdivision is typically based on geologic or geomorphic characteristics. (b) A large land mass, such as Greenland or Antarctica, that is smaller than any of the seven recognized continents.

subcortical crypt (sub-cor'-ti-cal) An inhalant aquiferous cavity lying beneath a cortex in a sponge and differentiated from a canal by virtue of its larger size and distinctive shape.

subcrevasse channel (sub-cre-vasse') A shallow channel eroded in subglacial material by a stream flowing along the bottom of a crevasse that completely penetrated a glacier (Leighton, 1959, p.340).

subcritical crack growth The slow growth of a crack as a consequence of stress corrosion at the crack tip.

subcritical flow (sub-crit'-i-cal) Water flow whose velocity is less than that of a long surface wave in still water; when the *Froude number* is <1. Syn: *tranquil flow.* Cf: *critical flow; supercritical flow.*

subcritically climbing translatent strata A term proposed by Hunter (1977) for *climbing ripple* with an angle of climb that is less than the angle of the *stoss* slope of the ripple bed form. This structure is comparable to *type-A climbing ripple* of Jopling and Walker (1968). Cf: *critically climbing translatent strata; supercritically climbing translatent strata.*

subcrop (sub'-crop) (a) An occurrence of strata in contact with the undersurface of an inclusive stratigraphic unit that succeeds an important unconformity on which *overstep* is conspicuous; a "subsurface outcrop" that describes the areal limits of a truncated rock unit at a buried surface of unconformity. (b) An area within which a formation occurs directly beneath an unconformity. The term, in common use in petroleum geology, appears to have been used first by Swesnik (1950, p.401) at the suggestion of Thom H. Green.

subcrop map A geologic map that shows the distribution of formations that have been preserved and remain covered beneath a given stratigraphic unit or immediately underlying an unconformity; properly, a map of an area where the overlapping formation is still present. The term "may be considered a generalization" of the term *paleogeologic map* (Krumbein and Sloss, 1963, p.448). Cf: *supercrop map.*

subcrustal (sub-crus'-tal) Said of a material or region beneath the Earth's crust.

subcutaneous zone *epikarst.*

subdelta (sub-del'-ta) A small delta, forming a part of a larger delta or complex of deltas.

subdeltoid (sub-del'-toid) An anal deltoid plate of a blastoid, typically horseshoe-shaped, located on the adoral and lateral margins of the anal opening and on the aboral border of the superdeltoid (TIP, 1967, pt. S, p.350). See also: *cryptodeltoid.*

subdermal space (sub-der'-mal) A *vestibule* of a sponge.

subdiabasic (sub'-di-a-bas'-ic) Said of an igneous-rock texture that is similar to ophitic texture except that the augite of the groundmass is not optically continuous but is divided into granular aggregates.

subdivide (sub-di-vide') A drainage divide between the tributaries of a main stream; a subordinate divide.

subdivision (sub-di-vi'-sion) A category in the hierarchy of botanical classification intermediate between division and class.

subdorsal carina (sub-dor'-sal) One of two longitudinal ridges on some decapod crustaceans, usually granulose or spinulose, extending backward from the submedian region of the rostrum and usually diverging.

subdrainage (sub-drain'-age) Drainage from beneath, either natural or artificial.

subdrift topography (sub-drift′) Topography of a bedrock surface underlying unconsolidated glacial drift.

subduction (sub-duc′-tion) The process of one lithospheric plate descending beneath another. The term was introduced by André Amstutz (1951). The idea of subduction had earlier roots in Otto Ampherer's concept of a crustal "swallowing zone" (Verschluckungs-zone). See also: *subduction zone.*

subduction [oceanog] The sinking of one water mass below another at the sea surface. *Potential vorticity* is conserved within the subducted water mass.

subduction erosion The tectonic removal of strata from near the leading edge of the upper lithospheric plate in a *subduction zone.*

subduction zone A long, narrow belt in which *subduction* takes place, e.g. along the Peru-Chile Trench or in the *volcanic arc* belts of the western Pacific Ocean.

subdued (sub-dued′) Said of a landform or landscape that is marked by a broadly rounded form and by moderate height, as if produced by long-continued weathering and erosion; esp. said of a mountain in the stage of senescence in a cycle of erosion, sufficiently worn down to have lost its peaks and cliffs, and having its moderately steep slopes covered with its own detritus. Cf: *feral.*

subepoch (sub-ep′-och) A term proposed by Sutton (1940, p.1402) for a geologic-time unit representing the first division of an epoch. It is applied only to a few portions of geologic time. Cf: *subseries.*

subera (sub-e′-ra) A little-used term referring to a "portion of an *era* comprised of two or more periods" (Sutton, 1940, p.1410).

suberain (su′-ber-ain) A kind of *provitrain* in which the cellular structure is derived from corky material. Cf: *periblain; xylain.*

suberin (su′-ber-in) An organic compound similar to *cutin* that occurs in corkified cell walls of bark and on roots, stems, and fruits, as a protection against desiccation. Cf: *suberinite.*

suberinite (su′-ber-in-ite′) (a) A variety of provitrinite characteristic of suberain and consisting of corky tissue. (b) A maceral of brown coal and lignite derived from the *suberin* layer in corkified cell walls of some Mesozoic and younger plants. (c) A maceral of the *liptinite* group. Cf: *periblinite; xylinite; telinite.*

subfabric (sub-fab′-ric) The array of only one kind of structural element in a rock. See also: *fabric [struc geol].*

subfacies (sub-fa′-cies) A subdivision of a facies, as of a broadly defined sedimentary facies, or of a metamorphic facies based on compositional differences rather than pressure-temperature relations.

subfamily (sub-fam′-i-ly) In the hierarchy of classification of plants and animals, a subcategory of family. In zoology, the name of a subfamily characteristically has the ending -inae; e.g. Cytheredeinae.

subfeldspathic (sub-feld-spath′-ic) (a) Said of a mature lithic wacke (or lithic graywacke) in which quartz grains and fragments of siliceous and argillaceous rocks predominate, and feldspars make up less than 10% of the rock and may be altogether lacking (Williams et al., 1954, p.302-303). Such rocks have also been called subgraywackes. (b) Said of a mature lithic arenite containing abundant quartz grains and fragments of the more stable rocks (such as cherts), and less than 10% feldspar grains (Williams et al., 1954, p.304, 307).

subfeldspathic lithic arenite An arenite with 10% or less of feldspar and a larger quantity of rock fragments (Williams et al., 1954, p.304).

subfeldspathic lithic wacke A lithic wacke containing less than 10% feldspar; a species of feldspathic graywacke (Williams et al., 1954, p.302). Syn: *subarkosic wacke.*

subfluvial (sub-flu′-vi-al) Situated or formed at the bottom of a river; e.g., a "subfluvial deposit".

subfossil (sub-fos′-sil) n. A fossil that is younger than what would be considered typical fossil age (i.e., preserved since about 6,000 years ago, by common convention) but not strictly recent or present-day. adj. Applied to an organism that would be considered a "subfossil". Cf: *fossil.*

subgelisol (sub-gel′-i-sol) The zone of unfrozen ground (*talik*) beneath permafrost.

subgenus (sub-ge′-nus) In the hierarchy of classification of plants and animals, a subcategory of *genus.* The name of a subgenus is placed in parentheses after the genus name and is followed by the name of the species; e.g., *Palaeoneilo (Koenenia) emarginata.*

subglacial (sub-gla′-cial) (a) Formed or accumulated in or by the bottom parts of a glacier or ice sheet; said of meltwater streams, till, moraine, etc. Syn: *infraglacial.* (b) Pertaining to the area immediately beneath a glacier, as *subglacial* eruption or *subglacial* drainage.

subgrade (sub′-grade) A layer, stratum, or surface immediately beneath some principal surface; specif. a layer of earth or rock that is graded to receive the foundation of an engineered structure. Often it is the soil or natural ground directly below a road, pavement, building, airfield, or railway that is prepared, *proof rolled,* and compacted to support the overlying structure. The "pavement section" includes the wearing course, the *base course,* and the *subbase.* Knowledge about the quality and conditions of the subgrade material is required for the proper design of the pavement.

subgrain A region within a mineral grain whose lattice is slightly misoriented (typically by <6°) with respect to the lattice of the surrounding mineral grain; a slightly misoriented portion of a crystal separated from the adjoining portion of the mineral grain by a *subgrain boundary.*

subgrain boundary A planar crystal defect bounding a subgrain consisting of a low angle boundary separating parts of a crystal whose lattices are tilted at low angles with respect to one another. The boundary is defined by a concentration of *dislocations*. Syn: *tilt boundary.*

subgrain rotation recrystallization Recrystallization resulting from the progressive rotation of subgrains, as dislocations are added to subgrain boundaries.

subgraphite (sub-graph′-ite) *meta-anthracite.*

subgraywacke (sub-gray′-wacke) (a) A term introduced by Pettijohn (1949, p.227, 255-256) for a sedimentary rock that has less feldspar and more and better-rounded quartz grains than *graywacke.* This rock, as originally defined, is equivalent to *quartz wacke* of Krumbein and Sloss (1963), to *low-rank graywacke* of Krynine (1948), to *lithic graywacke* of Pettijohn (1954), and to the *subfeldspathic* wackes of Williams et al. (1954). (b) A term redefined by Pettijohn (1957, p.316-320) as the most common type of sandstone, intermediate in composition between graywacke and orthoquartzite, containing less than 75% quartz and chert (commonly 30-65%), less than 15% detrital clay matrix, and an abundance (more than 25%) of unstable materials (feldspar grains and rock fragments) in which the rock fragments (at least 15%) exceed the feldspars; and having voids and/or mineral cement (esp. carbonates) exceeding the amount of clay matrix. The rock is lighter colored and better sorted and has less matrix than graywacke. Example: the Oswego Sandstone (Upper Ordovician) of central Pennsylvania. (c) A term used by Folk (1954, p.354) for a sedimentary rock that does not have enough rock fragments to be classed as a graywacke; specif. a sandstone with 5-25% micas and metamorphic-rock fragments and less than 10% feldspars and igneous-rock fragments, and having any clay content or sorting. This rock is equivalent to *quartzose graywacke* of Krynine (1951).

subgroup (sub-group′) A formally differentiated assemblage of formations within a *group* (ACSN, 1961, art.9d); a formally established subdivision of a group (ISSC, 1994, p.35). This term is not recognized by the North American Commission on Stratigraphic Nomenclature (1983).

subhedral (sub-he′-dral) A grain partly bounded by crystal faces; intermediate between *euhedral* and *anhedral.* Syn: *hypidiomorphic; hypautomorphic.*

subhepatic region (sub-he-pat′-ic) A part of the carapace of some decapod crustaceans extending below the edge of the hepatic region.

Subhercynian Orogeny (Sub-her-cyn′-i-an) One of the 30 or more short-lived orogenies during Phanerozoic time recognized by Stille, in this case in the Late Cretaceous, between the Turonian and Senonian stages.

subhumid (sub-hu′-mid) Said of a climate type that is transitional between humid and semiarid types according to quantity and distribution of precipitation. In Thornthwaite's classification, the moisture index is between zero and -20.

subhydrous (sub-hy′-drous) (a) Said of coal containing less than 5% hydrogen, analyzed on a dry, ash-free basis. (b) Said of a maceral of low hydrogen content, e.g. fusinite. Cf: *orthohydrous; perhydrous.*

subida (sub-i′-da) A rock-floored belt produced by wind scour and "potentially reaching to the base of a mountain range" (Stone, 1967, p.250). Etymol: Spanish, "ascent, acclivity".

subidioblast (sub-id′-i-o-blast′) *hypidioblast.*

subidiomorphic (sub-id′-i-o-mor′-phic) *hypidiomorphic; subhedral.*

subimposed (sub-im-posed′) Said of a subterranean stream that becomes a surface stream, as when the roof of a limestone cavern falls in. An obsolete term, originally proposed by Russell (1898b, p.246).

subirrigation (sub′-ir-ri-ga′-tion) Irrigation of plants with water delivered to the roots from underneath, either naturally or artificially.

subjacent [geomorph] (sub-ja′-cent) Being lower, but not necessarily lying directly below; e.g. "hills and subjacent valleys".

subjacent [intrus rocks] Said of an igneous intrusion, generally discordant and without a known floor, that presumably enlarges downward to an unknown depth. Obsolete.

subjacent [stratig] Said of a stratum situated immediately under a particular higher stratum or below an unconformity. Syn: *underlying.*

subjacent karst A *karst* landscape in non-carbonate rocks that is caused by the presence of karstified rocks beneath the surface formation. Cf: *interstratal karst.*

subjective synonym (sub-jec′-tive) In taxonomy, one of two or more competing names for the same taxon which are based on different types. Cf: *objective synonym.*

subkingdom (sub-king′-dom) In the hierarchy of classification of plants and animals, a subcategory of *kingdom.* It is sometimes considered synonymous with *phylum* and sometimes ranked above it.

sublacustrine (sub-la-cus′-trine) Existing or formed beneath the waters, or on the bottom, of a lake; e.g. a "sublacustrine channel" eroded in the lake bed by a surface stream before the lake was there or by a strong current within the lake.

sublevation (sub-le-va′-tion) Degradation of a sea floor composed of loose sediment (Dunbar and Rodgers, 1957). Etymol: Latin, "to lift up."

sublevel caving A type of underground mining in which a steeply inclined succession of blocks or panels of ore is undermined in sublevels and allowed to cave. See also: *block caving.*

sublimate (sub'-li-mate) A solid that has been deposited from a gas; in volcanology it refers to such a deposit made by a volcanic gas, e.g. metals around the mouth of a fumarole.

sublimation [chem] (sub-li-ma'-tion) The process by which a solid substance vaporizes without passing through a liquid stage. Cf: *evaporation.*

sublimation [meteorol] The phase-change process in which water molecules in ice change directly to atmospheric water vapor. Ant: *deposition [meteorol].*

sublimation [ore dep] The process of ore deposition, as of sulfur or mercury, by vapors; the volatilization and transportation of minerals followed by their deposition at reduced temperatures and pressures. Sublimation deposits are generally associated with fumarolic activity.

sublimation ice Hoarfrost crystals in open or closed cavities in permafrost, produced by condensation and freezing of water vapor.

sublimation loss Loss of water through the direct evaporation of ice and snow on lakes or from any body of ice or snow.

sublitharenite (sub-lith-ar'-e-nite) (a) A term introduced by McBride (1963, p.667) for a sandstone that does not have enough rock fragments to be classed as a *litharenite,* or a sandstone that is intermediate in composition between litharenite and pure quartz sandstone; specif. a sandstone with 5-25% fine-grained rock fragments, 65-95% quartz, quartzite, and chert, and less than 10% feldspar. (b) A term used by Folk (1968, p.124) for a sandstone, regardless of clay content or texture, with 75-95% quartz and metamorphic quartzite and a content (5-25%) of fine-grained volcanic, metamorphic, and sedimentary rock fragments (including chert) that exceeds that of feldspar and fragments of gneiss and granite.

sublithistid (sub-li-this'-tid) Said of a sponge containing desmoids.

sublithwacke (sub-lith'-wacke) A wacke with 5-25% detrital rock particles; a sublitharenite with over 15% matrix (Pettijohn et al., 1972).

sublittoral (sub-lit'-to-ral) *neritic.*

submarginal ambulacral suture (sub-mar'-gi-nal) The zone of contact between the ambulacral coverplates and floorplates in the Edrioasteroidea (Bell, 1976).

submarginal channel A channel formed by a meltwater stream flowing near the ice margin but also cutting across spurs or "behind small outlying hills" (Rich, 1908, p.528).

submarginal moraine *lodge moraine.*

submarginal resources Low-grade *resources* that are recoverable at prices more than 1.5 times those prevailing now, i.e. are of lower grade than *paramarginal resources* (Brobst and Pratt, 1973, p.3).

submarginal suture line In the Edrioasteroidea, the intersection of the *submarginal ambulacral suture* with the thecal surface. It extends along the line of contact of the adradial ends of the coverplates and the externally exposed parts of the floorplates (Bell, 1976).

submarine bar A *longshore bar* that is always submerged, even at low tide.

submarine barchan A large-scale lunate asymmetric bedform on the sea floor, ranging in length from 1 m to 100 m or more. Examples occur in the shallow-water areas of the Bahamas.

submarine canyon (a) A steep-sided, V-profile valley often heading in the continental shelf running down the continental slope, having tributaries and resembling an unglaciated, river-cut land canyon. (b) A general term for all valleys of the deep-sea floor. Syn: *submarine valley.*

submarine cone *submarine fan.*

submarine delta *submarine fan.*

submarine earthquake *seaquake.*

submarine fan A terrigenous, cone- or fan-shaped deposit located seaward of large rivers and submarine canyons. Syn: *submarine cone; abyssal cone; abyssal fan; subsea apron; deep-sea fan; submarine delta; sea fan; fan [marine geol]; cone [marine geol].*

submarine geology *marine geology.*

submarine geomorphology That aspect of marine geology which deals with the relief features of the ocean floor and with the forces that modify them.

submarine meadow An undersea grassland consisting of marine plants such as turtle grass.

submarine natural levee Deposit formed under water along sides of channel in areas where turbulence, and ordinarily the velocity of a current, diminishes in intensity, whether ahead of distributaries in a delta or elsewhere. Ordinarily there levees occur as ridges, shoal at first but increasing in size, in many cases becoming normal natural levees after their crests rise above water level. The patterns in which submarine ridges form commonly determine those of natural levees and stream courses between them in a delta or fill in an estuary.

submarine permafrost *subsea permafrost.*

submarine plain A syn. of *plain of marine erosion.* Term is not recommended because some of these plains have been uplifted. Syn: *submarine platform.*

submarine plateau *plateau [marine geol].*

submarine platform *submarine plain.*

submarine ridge *ridge [marine geol].*

submarine spring A large offshore emergence of fresh water, usually associated with a coastal karst area but sometimes with lava tubes.

submarine valley *submarine canyon.*

submarine volcano A volcano on the ocean floor, commonly of basalt. See also: *volcanic island.*

submarine weathering *halmyrolysis.*

submask geology (sub'-mask) The geology of the surface underlying a cover of alluvium, glacial drift, windblown sand, low-angle overthrust sheets, or water (as under shallow lakes and bays) (Kupsch, 1956). Rarely used.

submature [geomorph] (sub-ma-ture') Said of a topographic feature that has passed through the stage of youth but is not completely mature; e.g. a submature shoreline characterized by the cutting-back of headlands and by the near closing of baymouths by bars, thus simplifying an earlier intricately embayed shoreline (Cotton, 1958, p.456).

submature [sed] Pertaining to the second stage of textural maturity (Folk, 1951); said of a clastic sediment intermediate in character between an immature and a mature sediment, characterized by little or no clayey material and by poorly sorted and angular grains. Example: a clean "submature sandstone" containing less than 5% clay and commonly occurring in stream channels. Cf: *supermature.*

submeander (sub-me-an'-der) A small meander contained within the banks of the main channel (Langbein and Iseri, 1960, p.19); it is associated with relatively low discharges.

submerged aquatic plant (sub-merged') A *hydrophyte* the main part of which grows below the surface of the water. Cf: *subaquatic plant.*

submerged coastal plain The continental shelf representing the seaward continuation of a coastal plain on the land.

submerged contour A contour on the bed of a lake or reservoir, joining points of equal elevation where the elevation is related to a datum (usually mean sea level) used for mapping adjacent land (BNCG, 1966, p.14). Cf: *isobath [oceanog].* Syn: *underwater contour.*

submerged forest Forest remains, e.g. stumps still rooted in peaty soil, seen at low tide or found below sea level, indicating a rise in sea level or a subsidence of the coast.

submerged land A legal term for the land at the bottom of a lake, or the land covered by water when the lake is at its mean high-water level or at a level set by court decree (Veatch and Humphrys, 1966, p.324).

submerged rib A rib generated by the commissural denticles on the inside of the outer shell layers of some rostroconchs, and covered by the growth of inner shell layers (Pojeta and Runnegar, 1976, p.47).

submerged shoreline A former shoreline, identified by sandy beach deposits, lagoonal peats, or comparable littoral facies, that is today preserved in a submerged position. Such shorelines are commonly found at depths of 10 to 140 m and date from the late Pleistocene and early Holocene. Cf: *lowstand shoreline; shoreline of submergence.*

submerged unit weight The total unit weight of soil material minus the unit weight of water. Cf: *unit weight; wet unit weight.*

submerged valley A *drowned valley,* such as a ria.

submergence (sub-mer'-gence) A rise of the water level in relation to the land, so that areas formerly dry land become inundated; it results either from a sinking of the land or from a rise of the water level. Ant: *emergence [coast].*

submersible (sub-mers'-i-ble) A small self-propelled underwater vehicle for direct sea-floor observation and sampling.

submesothyridid (sub'-mes-o-thy'-ri-did) Said of a brachiopod *pedicle foramen* located mainly in the *delthyrium* and partly in the ventral *umbo* (TIP, 1965, pt. H, p.153). Cf: *permesothyridid.*

submetallic luster (sub-me-tal'-lic) A mineral luster between *metallic* and *nonmetallic.* Chromite, for example, has a metallic to submetallic luster.

subnival (sub-ni'-val) *periglacial.*

subnival boulder pavement A pavement of tightly packed blocks and boulders, with flat faces upward simulating a paved Roman highway, on a mountain valley floor in or near a stream channel and beneath a transient snowbank. Boulder pavements form downwind from sharp breaks in a valley floor and usually above treeline. They may be the result of the raising of blocks by *cryoturbation,* the rotation and flattening of the blocks by their own weight in fluid mud and by the weight of wind-packed snow and possible creep of the snowbank, and by meltwater saturation and later removal of fine sediment between blocks (White, 1972, p.195). Cf: *stone pavement.*

subnormal-pressure surface (sub-nor'-mal-pres'-sure) A potentiometric surface that is below the upper surface of the zone of saturation (Meinzer, 1923, p.39). The term is not in general use among hydrogeologists. Cf: *normal-pressure surface; artesian-pressure surface.*

subophitic (sub-o-phit'-ic) Igneous texture involving plagioclase laths only partly enclosed by pyroxene grains; grains of both minerals are roughly equal in size. Cf: *ophitic.*

suborder (sub-or'-der) In the hierarchy of classification of plants and animals, a subcategory of order. It is sometimes considered equivalent to *superfamily* and sometimes as the next higher rank.

suboutcrop (sub-out'-crop) *blind apex.*

subperiod (sub-pe'-ri-od) A geochronologic unit that is a portion of a *period [geochron],* but longer than an *epoch [geochron],* during which the rocks of the corresponding subsystem were formed.

subpermafrost water (sub-per'-ma-frost) Groundwater in the unfrozen ground beneath permafrost.

subphyllarenite (sub-phyl-lar'-e-nite) A *phyllarenite* containing 3-25% rock fragments.

subphylum (sub-phy'-lum) In the hierarchy of zoological classification, a category intermediate between *phylum* and class.

subplinian eruption A small-scale Plinian eruption intermediate between Strombolian and Plinian activity that is characterized by pumice and ash deposits covering less than 500 km^2.

subpolar glacier (sub-po'-lar) (a) *polythermal glacier.* (b) Originally, a glacier on which there is some surface melting during the summer but which is below the freezing temperature throughout most of its mass. Obsolete. Cf: *polar glacier; high-polar glacier.*

subpolar low pressure A semipermanent cell of low atmospheric pressure typically found over the oceans between 50° and 70° N and S.

subrosion (sub-ro'-sion) Subsurface erosion, apparently from solution of salts and subsequent adjustment/collapse. Obsolete.

subrounded (sub-round'-ed) Partially rounded; specif. said of a sedimentary particle showing considerable but incomplete abrasion and an original general form that is still discernible, and having many of its edges and corners noticeably rounded off to smooth curves, such as a cobble with a reduced number (5-10) of secondary corners, a considerably reduced area of the original faces, and a roundness value between 0.25 and 0.40 (midpoint at 0.315) (Pettijohn, 1957, p.59), or one with two-thirds of its edges smooth (Krynine, 1948, p.142). Also, said of the *roundness class* containing subrounded particles.

subsalt strata The sedimentary unit immediately underlying salt (Jackson and Talbot, 1991).

subsea apron (sub'-sea) *submarine fan.*

subsea permafrost Permafrost occurring beneath the sea bottom. Syn: *submarine permafrost; offshore permafrost.*

subseptate (sub-sep'-tate) Having imperfect or partial septa; e.g. having slight protuberances or incipient septa that form pseudochambers in a foraminiferal test (as in the family Tournayellidae).

sub-sequence (sub-se'-quence) A term applied by R.C. Moore (1958, p.80) to a Precambrian rock division (often called a series in Canada) that "cannot be correlated from one region to another". The term requires a hyphen in order to distinguish it from the noun "subsequence". See also: *sequence.*

subsequent [geomorph] (sub'-se-quent) Said of a post-consequent geologic or topographic feature whose development is controlled by differences in the erosional resistance of the underlying rocks; e.g., a subsequent ridge formed by differential erosion of a consequent ridge, a subsequent waterfall produced where a downcutting stream encounters rock of exceptional hardness, or a subsequent valley developed along the strike of a weakly resistant homoclinal bed. Cf: *consequent [geomorph].*

subsequent [streams] adj. Said of a stream, valley, or drainage system that is developed independently of, and subsequent to, the original relief of a land area, as by shifting of divides, stream capture, or adjustment to rock structure. The concept was originally discussed by Jukes (1862, p.393-395). n. *subsequent stream.*

subsequent fold *cross fold.*

subsequent stream A tributary that has developed its valley (mainly by headward erosion) along a belt of underlying weak rock and is therefore adjusted to the regional structure; esp. a stream that flows approximately in the direction of the strike of the underlying strata and that is subsequent to the formation of a consequent stream of which it is a tributary. Syn: *subsequent [streams]; strike stream; longitudinal stream.*

subsere (sub-sere') (a) A secondary ecologic *succession [ecol]* that arises on a denuded area following an ecologic climax. (b) A seral *community* that is prevented from reaching ecological climax by a temporary interference, e.g. by human activity.

subseries (sub-se'-ries) A term proposed by Sutton (1940, p.1402) for the first division of a series, representing the rocks formed during a *subepoch.*

subsidence (sub-sid'-ence, sub'-sid-ence) (a) The sudden sinking or gradual downward settling of the Earth's surface with little or no horizontal motion. The movement is not restricted in rate, magnitude, or area involved. Subsidence may be caused by natural geologic processes, such as solution (karst phenomena), thawing, compaction, slow crustal warping, or withdrawal of fluid lava from beneath a solid crust; or by man's activity, such as subsurface mining or the pumping of oil or groundwater. See also: *cauldron subsidence.* (b) A sinking or downwarping of a large part of the Earth's crust relative to its surrounding parts, such as the formation of a rift valley or the lowering of a coast due to tectonic movements. See also: *tectonic subsidence.* Syn: *sinking.*

subsidence caldera *collapse caldera.*

subsidence/head-decline ratio The ratio between land subsidence and the hydraulic head decline in the coarse-grained beds of the compacting aquifer system (Poland et al., 1972).

subsidence theory A theory of coral-atoll and barrier-reef formation according to which upward reef growth kept pace uninterruptedly over a long period with slow subsidence of a volcanic island, forming first a fringing reef that became a barrier reef and later an atoll when the island was completely submerged; it accounts satisfactorily for many Pacific reefs. Theory was proposed by Charles Darwin in 1842. Cf: *glacial-control theory; antecedent-platform theory.*

subsidiary fold (sub-sid'-i-ar'-y) *parasitic fold.*

subsilicic (sub-si-lic'-ic) A term proposed by Clarke (1908, p.357) to replace *basic.* Cf: *persilicic; mediosilicic.*

subsoil (sub'-soil) (a) A syn. of B horizon, in a soil profile having distinct horizons. (b) The soil below the *surface soil;* this is an older meaning. Cf: *topsoil.*

subsoil ice *ground ice.*

subsoil weathering A term used by Davis (1938) for the chemical decomposition that produces spheroidal boulders beneath the regolith in granitic areas of the desert by percolation of water along joints followed by exposure and exfoliation.

subsolidus (sub-sol'-i-dus) A chemical system that is below its melting point, and in which reactions may only occur in the solid state.

subsolution (sub-so-lu'-tion) Corrosion of a cemented carbonate sea floor by submarine dissolution (Bromley, 1975).

subsolvus (sub-sol'-vus) Refers to simultaneous crystallization of two separate feldspar minerals (typically potassic feldspar and sodic plagioclase) from granites or syenites. Such crystallization implies elevated water pressure (> about 2.5 kbar) during magma crystallization because it requires intersection of the plagioclase-potassic feldspar critical curve with the magma solidus curve. The general case is direct crystallization of a melt to form two phases related by a solvus. Cf: *solvus; hypersolvus.*

subspeciation (sub'-spe-ci-a'-tion) Division into or formation of subspecies.

subspecies (sub'-spe-cies) In the hierarchy of classification of plants and animals, a subcategory of *species.* Groups within a species that are geographically isolated from one another are geographic subspecies; groups separated in geologic time are chronologic subspecies. The name of a subspecies is a trinomen; e.g. *Bollia americana zygocornis.* Cf: *variety [taxon].*

substage [geochron] (sub'-stage) An obsolete term for a geologic-time unit of lesser duration than *stage [geochron],* during which the rocks of a *member* were formed.

substage [glac geol] A time term for a subdivision of a glacial stage during which there was a secondary fluctuation of glacial advance and retreat; specif. a Pleistocene subdivision equated with a rock unit of member rank, such as the "Woodfordian substage" of the Wisconsinan stage.

substage [optics] In a microscope, an attachment for holding polarizers or other attachments below the stage.

substage [stratig] A subdivision of a *stage [stratig];* the rocks formed during a *subage* of geologic time; the lowest unit of the conventional hierarchy of formed chronostratigraphic units (ISSC, 1994, p.80). The frequently used synonym zone is not recommended (ISST, 1961, p.13).

substitute (sub'-sti-tute) Any natural or synthetic material represented to be, or used to imitate, a gemstone; e.g. plastic, glass, doublet, synthetic ruby, or natural spinel all could be substitutes for natural ruby.

substitution (sub-sti-tu'-tion) *ionic substitution.*

substitutional transformation (sub-sti-tu'-tion-al) A type of *phase transformation* of a disordered phase (a substitutional solid solution) to an ordered phase. It is usually a slow transformation. Cf: *dilatational transformation; displacive transformation; reconstructive transformation; rotational transformation; order-disorder transformation.*

substitution solid solution A crystal in which a particular atomic site can be occupied by any of two or more elements. Cf: *omission solid solution.*

substrate [cryst] (sub'-strate) The surface on which *epitaxy* occurs.

substrate [ecol] The substance, base, or nutrient on which, or the medium in which, an organism lives and grows, or the surface to which a fixed organism is attached; e.g. soil, rocks, water, and leaf tissues, or perhaps a gel for the accumulation and preservation of prebiologic organic matter. Syn: *substratum [ecol].*

substratum [ecol] (sub-stra'-tum) *substrate [ecol].*

substratum [soil] Any layer lying beneath the solum. It is applied to both parent materials and to other layers unlike the parent material, below the B horizon or subsoil.

substratum [tect] In salt tectonics, the ductile layer below a brittle overburden and above the subsalt strata or basement. Substratum is a more general term than *source layer;* the substratum may or may not give rise to upwelling structures (Jackson and Talbot, 1991).

subsurface (sub-sur'-face) n. (a) The zone below the surface, whose geologic features, principally stratigraphic and structural, are interpreted on the basis of drill records and various kinds of geophysical evidence. (b) Rock and soil materials lying beneath the Earth's surface. adj. Formed or occurring beneath a surface, esp. beneath the Earth's surface. Cf: *surficial.* See also: *subterranean.*

subsurface air Gas in interstices in the unsaturated zone that open directly or indirectly to the surface and that is therefore at or near atmospheric pressure. Its composition is generally similar though not identical to that of the atmosphere (Meinzer, 1923, p.21). Cf: *soil atmosphere; included gas; ground air.*

subsurface contour A syn. of *structure contour,* used to distinguish it from a surface or topographic contour.

subsurface drainage The removal of surplus water from within the soil by natural or artificial means, such as by drains placed below the surface to lower the water table below the root zone.

subsurface flow *storm seepage.*

subsurface geology Geology and correlation of rock formations, structures, and other features beneath the land or sea-floor surface as revealed or inferred by exploratory drilling, underground workings, and geophysical methods. Ant: *surface geology.*

subsurface ice *ground ice.*

subsurface map A map depicting geologic data or features below the Earth's surface; esp. a plan of mine workings, or a structure-contour map of a petroleum reservoir or an underground ore deposit, coal seam, or key bed.

subsurface runoff *storm seepage.*

subsurface storm flow *storm seepage.*

subsurface water Water in the lithosphere in solid, liquid, or gaseous form. It includes all water beneath the land surface and beneath bodies of surface water. Syn: *subterranean water; underground water; groundwater.*

subsystem [chem] (sub-sys'-tem) Any part of a system that may be treated as an independent system.

subsystem [stratig] A formally established subdivision of a *system [stratig].* "Special circumstances have suggested the occasional need" for subsystems, e.g. the Mississippian Subsystem of the Carboniferous System (ISG, 1976, p.73-74).

subtalus buttress (sub-ta'-lus) The convex-upward rock surface developed under a rising talus slope as the cliff above it weathers back (Howard, 1942, p.27).

subterrane (sub-ter-rane') n. The bedrock beneath a surficial deposit or below a given geologic formation. adj. *subterranean.*

subterranean (sub-ter-ra'-ne-an) adj. Formed or occurring beneath the Earth's surface, or situated within the Earth. Cf: *subaerial.* See also: *subsurface.* Syn: *subterrestrial; subterrane.*

subterranean cutoff Diversion of a surface stream by the development of a shorter underground course across a meander neck.

subterranean ice *ground ice.*

subterranean river Underground stream of flowing water, not necessarily large.

subterranean stream A body of subsurface water flowing through a cave or a group of communicating caves, as in a *karstic* region. Cf: *underground stream.*

subterranean stream piracy The capture of a surface stream and its diversion through an underground channel to an adjacent entrenched stream. Cf: *tapoff passage.*

subterranean water (a) A syn. of *subsurface water.* (b) A syn. of *groundwater* in less preferred usage.

subterrestrial (sub-ter-res'-tri-al) *subterranean.*

subtractive primary colors The complementary colors (yellow, magenta, and cyan) used to produce other colors in the printing process. The three subtractive color primaries, cyan, magenta, and yellow, are produced by subtracting red, green or blue from white light. White light minus red gives cyan, white minus green gives magenta, and white minus blue gives yellow. Subtractive colors are used in lithographic printing among other applications.

subtropical (sub-trop'-i-cal) Said of the climate of the subtropics, which borders that of the tropics and is intermediate in character between tropical and temperate, though more like the former than the latter. Syn: *semitropical.*

subtropical high pressure A semipermanent cell of high atmospheric pressure typically found over the ocean basins at a latitude of approximately 30° N and S. Also called the horse latitudes; represents the base of the descending branches of the Hadley cell.

subturma (sub-tur'-ma) One of the subdivisions of *turma.* Pl: subturmae. Cf: *turma.*

subulate (su'-bu-late) Awl-shaped; said of a leaf that tapers from its base to a sharp point.

subvective system (sub-vec'-tive) A system for gathering and transporting food particles to the mouth of an echinoderm. It cannot be separated morphologically from the *ambulacral* system (TIP, 1966, pt. U, p.155).

subvolcanic (sub-vol-can'-ic) *hypabyssal.*

subweathering velocity (sub-weath'-er-ing) The seismic velocity of the layer immediately underlying the near-surface *low-velocity layer.* This velocity is usually 2-3 times larger than that in the weathered zone. Cf: *weathering velocity.*

subzone (sub'-zone) (a) A formal or informal subdivision of a *biozone,* (NACSN, 1983, Art. 53, 54). (b) A subdivision of stratigraphic zone of any kind (ISSC, 1994).

succession [ecol] (suc-ces'-sion) The progressive change in a biologic community as a result of the response of the member species to the environment. See also: *sere; faunal succession.* Syn: *ecologic succession.*

succession [stratig] (a) A number of rock units or a mass of strata that succeed one another in chronologic order; e.g., an inclusive stratigraphic sequence involving any number of stages, series, systems, or parts thereof, as shown graphically in a geologic column or seen in an exposed section. (b) The chronologic order of rock units.

successional speciation (suc-ces'-sion-al) The gradual evolution from one species to another, eventually leading to replacement by the latter.

succinic acid (suc-cin'-ic) A crystalline dicarboxylic acid, formula $HOOCCH_2CH_2COOH$, a constituent of wood bark and occurring in the amber group (but not in the retinite group) of fossil resins.

succinite (suc'-cin-ite) (a) An old name for *amber,* esp. that mined in East Prussia or recovered from the Baltic Sea. (b) A light-yellow, amber-colored variety of grossular garnet.

sucrosic (su-cros'-ic) A syn. of *saccharoidal.* The term is commonly applied to idiotopic dolomite rock. Syn: sucrose.

sudburite (sud'-bur-ite) An augite-bearing orthopyroxene basalt characterized by pillow structure and containing bytownite and magnetite. "It differs from normal basalts in containing neither glass nor olivine and in having an equigranular texture" (Johannsen, 1937, p.305). Its name, given by Coleman in 1914, is derived from the Sudbury district, Ontario. Obsolete.

sudburyite (sud'-bur-y-ite') A metallic hexagonal mineral of the *nickeline* group: (Pd,Ni)Sb.

Sudbury-type nickel-copper deposit *astrobleme-associated nickel-copper deposit.*

Sudetic Orogeny (Su-de'-tic) One of the 30 or more short-lived orogenies during Phanerozoic time identified by Stille, in this case between the Early and Late Carboniferous.

sudoite (su'-do-ite) A white or pale green monoclinic member of the *chlorite* group of minerals, in which the octahedral sheet within the 2:1 T-O-T layer is dioctahedral and the interlayer sheet is trioctahedral: $Mg_2(Al,Fe^{3+})_3Si_3AlO_{10}(OH)_8$. Sudoite is Li-poor relative to the species *cookeite.*

sudovikovite A metallic trigonal mineral of the *melonite* group: $PtSe_2$.

Suess effect A decrease in the concentration of carbon-14 in atmospheric carbon dioxide due to dilution by nonradioactive carbon from the burning of fossil fuels.

suessite (suess'-ite) A metallic cubic meteorite mineral: Fe_3Si. The crystal structure is related to that of kamacite.

suevite (sue'-vite) Polymict impact breccia with clastic matrix containing lithic and mineral clasts in various stages of *shock metamorphism,* including cogenetic impact-melt particles in a glassy or crystallized state but not supporting the clasts. Syn: *melt-fragment breccia, suevitic impact breccia; suevite breccia.*

suevite breccia *suevite.*

suevitic impact breccia *suevite.*

suffosion (suf-fo'-sion) The bursting-out on the surface in little eruptions of highly mobile or water-saturated material; esp. a destructive process operating under periglacial conditions, in which underground water, resulting from partial melting of ground ice, exerts upward pressure and bursts through a hard dried upper skin to deposit a mound of mud, clay, sand, and/or boulders. Suffosional forms due to corrasion by underground water include dimpling, pits, blind valleys, shafts, and cavern entrances. Other examples are associated with escape of artesian water. Cf: *mound spring; sand volcano; tundra crater.*

suffosion knob A *frost mound* (Muller, 1947, p.222).

sugarloaf (sug'-ar-loaf) A conical hill or mountain comparatively bare of timber, resembling the shape of a loaf of sugar; e.g. Sugarloaf Mountain, Maine.

sugar sand A sandstone that breaks up into granules that resemble sugar.

sugar stone Compact, white to pink datolite from the Michigan copper district.

sugary (sug'-ar-y) *saccharoidal.*

sugilite (su'-gi-lite) A brownish-yellow to magenta hexagonal mineral of the *milarite* group: $KNa_2(Fe^{3+},Mn^{3+},Al)_2Li_3Si_{12}O_{30}$.

suite [ign] (a) A set of apparently comagmatic igneous rocks. (b) A collection of rock specimens from a single area, generally representing related igneous rocks. (c) A collection of rock specimens of a single kind, e.g. granites from all over the world.

suite [stratig] (a) The lithodemic unit next higher in rank to *lithodeme.* It comprises two or more associated lithodemes of the same class (e.g. plutonic, metamorphic) (NACSN, 1983, Art. 35). For cartographic and hierarchical purposes, suite is comparable to *group.* The name of a suite combines a geographic term, the term "suite," and an adjective denoting the fundamental character of the suite, e.g. Idaho Springs Metamorphic Suite. (b) A local stratigraphic unit used in the former USSR and some eastern European countries, corresponding approximately or precisely to a formation. (c) A term used by Caster (1934, p.18) for a body of rocks intermediate between monothem (formation) and member, consisting of several intimately related members bracketed together; esp. a repeated sequence of such closely associated strata. (d) An association of apparently comagmatic intrusive igneous rock bodies of similar or related lithologies and close association in time, space, and origin.

sukulaite (su'-ku-la-ite) *stannomicrolite.*

sulcal notch (sul'-cal) A ventral indentation in the margin of some apical archeopyles in a dinoflagellate cyst, corresponding to the *sulcal tongue.*

sulcal plate One of the plates of the ventral furrow region in dinoflagellates possessing a theca. The plates are subdivided as to left or right, and anterior or posterior position.

sulcal tongue An extension of the operculum of a dinoflagellate cyst, in the position of the first apical plate of the theca. Cf: *sulcal notch.*

sulcate (sul′-cate) Pertaining to a sulcus, or scored with furrows, grooves, or channels, esp. lengthwise; e.g. said of a form of alternate folding in brachiopods, with the brachial valve bearing a median sulcus and an anterior commissure median sinus (TIP, 1965, pt. H, p.153). Ant: *uniplicate.*

sulcus [paleont] (sul′-cus) (a) A major longitudinal depression in the surface of either valve of a brachiopod, externally concave in transverse profile and radial from the umbo, and usually median in position. It is typically associated with the *fold [paleont].* See also: *sinus [paleont].* (b) An elongate shallow depression in the lateral surface of an ostracode valve, extending from the dorsal region toward the venter. (c) A radial depression of the surface of the shell of a bivalve mollusk. (d) A longitudinal groove on the venter of an ammonoid shell. (e) *main trough.* (f) In vertebrates, generally a deep groove (as on the surface of a bone) or the groove between convolutions (e.g., between convex gyri of the brain). Pl: sulci.

sulcus [palyn] (a) A relatively broad latitudinal furrow in the *exine* of pollen grains, situated at the distal pole. The term is nearly synonymous with *colpus* but is applied only to pollen grains with a single such aperture. (b) In dinoflagellates, the longitudinal posterior groove lying on the ventral surface of the theca, containing the trailing flagellum. It may also be used, with or without the prefix "para-" for the ventral groove of a dinoflagellate cyst (Williams et al., 2000). Adj: sulcal.

sulcus [planet] A term established by the International Astronomical Union for subparallel furrows and ridges on planetary surfaces (USGS, 2003). Pl: sulci.

sulfate (sul′-fate) A mineral compound characterized by the sulfate radical SO_4. Anhydrous sulfates, such as barite, $BaSO_4$, have divalent cations linked to the sulfate radical; hydrous and basic sulfates, such as gypsum, $CaSO_4 \cdot 2H_2O$, contain water molecules. Cf: *chromate.*

sulfate-reducing bacteria (sul′-fate-re-duc′-ing) *sulfur bacteria.*

sulfide (sul′-fide) A mineral compound characterized by the linkage of sulfur with a metal or semimetal, such as galena, PbS, or pyrite, FeS_2. See also: *sulfosalt.*

sulfide enrichment Enrichment of a deposit by replacement of one sulfide by another of higher value, as pyrite by chalcocite.

sulfide-facies iron formation (sul′-fide-fa′-cies) An *iron formation* consisting essentially of pyritic carbonaceous slate. It was formed in the deeper, reducing parts of seas or basins.

sulfide zone (a) That part of a sulfide deposit that has not been oxidized by near-surface waters. Cf: *oxidized zone; protore.* (b) A generally manto-shaped deposit in which secondary sulfide enrichment has occurred as a part of ore-deposit oxidation. Syn: *secondary sulfide zone.*

sulfidization (sul′-fid-i-za′-tion) *sulfurization.*

sulfite A mineral compound containing the oxyanion $(SO_3)^{-2}$. An example is scotlandite, $PbSO_3$.

sulfoborite (sul-fo-bor′-ite) A colorless orthorhombic mineral: $Mg_3B_2(SO_4)(OH)_9F$.

sulfohalite (sul-fo-hal′-ite) A colorless pale greenish, yellow, or gray cubic mineral: $Na_6(SO_4)_2FCl$. Also spelled: *sulphohalite.*

sulfosalt (sul′-fo-salt) A type of *sulfide* in which both a metal and a semimetal (Sb, As, or Bi) are present, forming a double sulfide, e.g. enargite, Cu_3AsS_4.

sulfur (sul′-fur) (a) An orthorhombic mineral, the native nonmetallic element S. It occurs in yellow crystals at hot springs and fumaroles, and in masses or layers associated with limestone, gypsum, and anhydrite, esp. in salt-dome caprock and bedded deposits. Sulfur exists in several allotropic forms, including the ordinary yellow orthorhombic alpha form stable below 95.5°C and the pale-yellow monoclinic crystalline beta form. See also: *rosickyite.* Syn: *brimstone.* (b) A mining term used for iron sulfide (pyrite) occurring in coal seams and with zinc ores in Wisconsin and Missouri. Also spelled: *sulphur.*

sulfur bacteria Anaerobic bacteria that obtain the oxygen needed in metabolism by reducing sulfate ions to hydrogen sulfide or elemental sulfur. Accumulations of sulfur formed in this way are *bacteriogenic* ore deposits. Cf: *iron bacteria.*

sulfur ball [coal] A pyritic impurity in coal, occurring as a spheroidal or irregular mass. Cf: *coal ball.*

sulfur ball [pyroclast] A sulfurous mud skin that forms on a bubble of hot volcanic gas and becomes firm on contact with the air.

sulfuric-acid aerosols Small droplets of sulfuric acid formed in the atmosphere from oxidation of sulfur-rich gases. These aerosols can form a haze in the upper atmosphere after volcanic eruptions, which can reduce incoming sunlight and cool the planet (Rampino and Self, 2000).

sulfuric-acid cave A cave formed by dissolution of carbonate rocks by rising deep-seated solutions carrying hydrogen sulfide which oxidizes to form sulfuric acid. Example: Carlsbad Caverns, N. Mex. (Hill, 1995).

sulfuric acid oil field karst A specific kind of porosity where carbonate reservoirs of cavernous size have been dissolved by a sulfuric acid mechanism (Hill, 1995).

sulfurization (sul′-fur-i-za′-tion) "Reaction between sulfur from an external source and cations such as iron, nickel, and copper in solid solution in common rock-forming minerals or in igneous magma," considered as an ore-forming process (Naldrett and Kullerud, 1965). Syn: *sulfidization.*

sulfur-mud pool *mud pot.*

sulfur spring A spring containing sulfur water.

sulfur water Generally, water containing enough hydrogen sulfide to smell and taste. Except for the hydrogen sulfide, it may not differ in mineral content from ordinary potable water, or it may qualify as *saline* water. In either case, it is usually considered a mineral water.

sulphatite (sul′-pha-tite) Free sulfuric acid (H_2SO_4) found in some waters.

sulphoborite (sul-pho-bo′-rite) *sulfoborite.*

sulphohalite *sulfohalite.*

sulphophile (sul′-pho-phile) Said of an element, e.g. fluorine or chlorine, that occurs preferentially in a mineral that is oxygen-free, i.e. as a sulfide, selenide, telluride, arsenide, or antimonide. The term incorporates the chalcophile elements and some of the siderophile elements of Goldschmidt's classification. The term is rarely used.

sulphotsumoite (sul-pho-tsu′-mo-ite) A metallic gray-white trigonal mineral: Bi_3Te_2S.

sulphur (sul′-phur) *sulfur.*

sulphur ore A mining term used for both pyrite and native sulfur.

sulrhodite *bowieite.*

sulvanite A metallic bronze-yellow cubic mineral: Cu_3VS_4. Not to be confused with *sylvanite.*

sumidero (a) A swallow hole. (b) In Latin America, any closed depression caused by solution. Etymol: Spanish.

summation method (sum-ma′-tion) In seismology, a method of correcting the arrival times of reflected waves for the time they spend in the low-velocity zone.

summer balance (sum′-mer) The change in mass of a glacier from the maximum value in a certain year to the following minimum value of that year; sometimes called *apparent ablation,* since it does not include losses during warm days in winter or loss of snow that falls during the melt season; or (erroneously) *net ablation.* Cf: *winter balance.*

summer beach Upper beach of maximum volume and steep front accumulated during summers in regions where lowest wave energies characterize that season, but the accumulation may occur at other seasons as well, and tendencies to develop summer beach characteristics occur during brief intervals whenever sea state is low and to a small degree with falling tide.

summer berm A berm built on the backshore by the uprush of waves during the summer. Cf: *winter berm.*

summer season *ablation season.*

summer surface An observable horizon (e.g. *dirt band*) in a glacier, marking the end of melting in one *balance year* and the beginning of snow accumulation in the next. Cf: *net balance.*

summerwood (sum′-mer-wood) A syn. of *late wood.* Also spelled: summer wood. Cf: *springwood.*

summit (sum′-mit) (a) The top, or the highest point or level, of an undulating land feature, as of a hill, mountain, volcano, or rolling plain; a peak. See also: *crest [geomorph].* (b) Loosely, a divide or pass; e.g. Donner Summit, Calif.

summit concordance Equal or nearly equal elevation of ridgetops or mountain summits over a region. The concordance is commonly thought to indicate the existence of an ancient erosion plain of which only scattered patches are preserved. See also: *accordant summit level; even-crested ridge.* Syn: *accordance of summit levels; concordance of summit levels.*

summit eruption An eruption at the top of a volcanic mountain. Cf:*flank eruption.*

summit graben A *volcanic graben* on the summit of a volcanic cone, more or less rectangular or triangular; also, a graben crossing the crest of a *resurgent cauldron* (Smith and Bailey, 1968). Cf: *sector graben.*

summit level [eng] The highest point of a road, railroad, or canal; the highest of a series of elevations over which a road or canal is carried.

summit level [geomorph] The elevation of a summit plane. See also: *accordant summit level.*

summit plain *peak plain.*

summit plane The plane passing through a series of accordant summits. See also: *gipfelflur.*

summitpoint (sum′-mit-point) The point of maximum elevation on the cross-section of a ripple. Term introduced by Allen (1968). Cf: *brinkpoint.*

sump [geog] (a) An excavation in which the drainage water of an area is collected for subsequent use in irrigation or wild-fowl conservation. (b) A dialect term for a swamp or morass, and for a stagnant pool or puddle of dirty water. (c) An English term for a cove or a muddy inlet.

sump [speleo] A part of a cave passage in which the ceiling is below the water level. Syn: *trap [speleo].*

sun ball A snowball formed by a lump of snow falling onto a slope of moist snow and rolling downhill. Cf: *snow roller.*

sun crack A crack in sediment or rock, formed by the drying action of the Sun's heat; esp. a *mud crack.*

sun crust A type of *snow crust* formed by refreezing of surface liquid that was melted by the sun; it is usually thin and has a smooth surface.

sun cup A cuspate hollow or depression in a snow surface, formed during sunny weather by complex ablation processes. In some environments, a sun cup may grow into a *nieve penitente.*

sundiusite (sun′-di-us-ite) An adamantine colorless monoclinic mineral: $Pb_{10}O_8(SO_4)Cl_2$.

sundtite (sundt'-ite) *andorite.*

sunken island A high-relief feature of a lake basin, such as a basin divide or the crest of a knob, covered by a shallow depth of water; it was "never originally above water level" and is therefore "not due to subsidence" (Veatch and Humphrys, 1966, p.317).

sun opal *fire opal.*

sun spike *nieve penitente.*

sunspot (sun'-spot) A relatively dark area on the Sun's surface, representing lower temperature and consisting of a dark central umbra surrounded by a penumbra which is intermediate in brightness between the umbra and the surrounding surface of the photosphere (NASA, 1966, p.47).

sunstone (sun'-stone) An *aventurine feldspar,* usually a brilliant, translucent variety of oligoclase that emits a reddish or golden billowy reflection from minute scales or flakes of hematite spangled throughout and arranged parallel to planes of repeated twinning. Cf: *moonstone.* Syn: *heliolite.*

sun-synchronous (sun-syn'-chro-nous) Said of an Earth-satellite orbit in which the orbit plane is near polar and the altitude such that the satellite passes over all places on Earth having the same latitude twice daily at the same local sun time.

suolunite (su'-o-lun-ite) A vitreous white or cream orthorhombic mineral: $Ca_2Si_2O_5(OH)_2 \cdot H_2O$.

superanthracite (su-per-an'-thra-cite) A syn. of *meta-anthracite.* Not listed by ASTM as a rank classification.

superbiozone (su-per-bi'-o-zone) A unit consisting of "several biozones with common biostratigraphic features" (ISG, 1976, p.48).

supercapillary interstice (su-per-cap'-il-lar'-y) An *interstice* sufficiently larger than a *capillary interstice* that surface tension will not hold water far above a free water surface. Water moving in it, as by *supercapillary percolation,* may develop currents and eddies (Meinzer, 1923, p.18). Cf: *subcapillary interstice.*

supercapillary percolation Percolation through supercapillary interstices.

supercell (su-per-cell') The unit cell of a *superlattice.*

supercell thunderstorm A relatively long-lived, large and intense *thunderstorm* cell characterized by an exceptionally strong updraft sometimes in excess of 240 km (150 mi) per hour; may produce a *tornado.*

superchron A long-duration (greater than about 10^7 years) time interval of constant polarity of the geomagnetic field.

superconducting magnetometer A high-sensitivity instrument used to measure the magnetic moment of permanently magnetized material, utilizing the principle that the resistance in a pick-up coil of superconducting material at a sufficiently low temperature is zero.

supercooling (su-per-cool'-ing) The process of lowering the temperature of a phase or assemblage below the point or range at which a phase change should occur at equilibrium, i.e. making the system metastable by lowering the temperature. It generally refers to a liquid taken below its liquidus temperature. Supercooling can also be achieved in hydrous magmas by decreasing pressure. Cf: *superheating.* Syn: *undercooling.*

supercritical (su-per-crit'-i-cal) Said of a system that is at a temperature higher than its critical temperature; also, said of the temperature itself.

supercritical flow Water flow whose velocity exceeds the velocity or propagation of a long surface wave in still water; when the *Froude number* is >1. Syn: *rapid flow; shooting flow.* Cf: *critical flow; subcritical flow.*

supercritical fluid Refers to a single-phase fluid that exists beyond the limits of a liquid-gas reaction curve which ends in a critical end point. Properties of such a fluid are continuously variable, from more gas-like at low pressure to more liquid-like at higher pressure. The classic geologic example is water above about 325°C.

supercritically climbing translatent strata A term proposed by Hunter (1977) for *climbing ripple* with an angle of climb that is greater than the angle of the *stoss* slope of the ripple bed form. Cf: *critically climbing translatent strata; subcritically climbing translatent strata.*

supercritical reflection A wide-angle reflection where the incident angle exceeds the *critical angle.*

supercrop map (su'-per-crop) A geologic map that shows the distribution of stratigraphic units lying immediately above a given rock body or a surface at a given time. Cf: *subcrop map.* Syn: *worm's-eye map.*

supercycle (su'-per-cy'-cle) (a) A group of cycles of eustatic or relative change of sea level in which a cumulative rise to a higher position is followed by a cumulative fall to a lower position (Mitchum, 1977, p.211). Commonly, one or two major falls rather than a succession of them are evident at the end of a supercycle. See also: *cycle of relative change of sea level; paracycle of relative change of sea level.* (b) A combination of several megacycles of the type present in the Vindhyan Supergroup of India (Mathur, 1981b). It may be considered equivalent to a supergroup, and may be several thousand meters thick.

superdeltoid (su-per-del'-toid) An anal deltoid plate on the border of the mouth opening of a blastoid, associated either with the subdeltoid or with a pair of cryptodeltoids abutting its aboral margin and in some genera bordering the anal opening (TIP, 1967, pt. S, p.350).

superfacies (su-per-fa'-cies) A large-scale stratigraphic facies, generally consisting of two or more subordinate facies; e.g. a laterally equivalent and contrasting part of a formation, within which smaller-scale, laterally equivalent, and contrasting parts are recognized.

superfamily (su-per-fam'-i-ly) In the hierarchy of classification of plants and animals, a category next above family. It may be considered as equivalent to *suborder,* or as intermediate between suborder and family.

superficial (su-per-fi'-cial) Pertaining to, or lying on or in, a surface or surface layer; e.g. "superficial weathering" of a rock, or a "superficial structure" formed in a sediment by surface creep. The term is used esp. in Great Britain; the syn. *surficial* is more generally applied in the U.S.

superficial deposit *surficial deposit.*

superficial fold A rarely used syn. of *detachment fold.*

superficial moraine *surficial moraine.*

superficial oolith An oolith with an incomplete or single layer; specif. one in which the thickness of the accretionary coating is less than the radius of the nucleus (Beales, 1958, p.1863).

superfluent lava flow (su-per-flu'-ent) A flow of lava issuing from a summit crater and streaming down the flanks of the volcano (Dana, 1890); an obsolete term. Cf: *effluent lava flow; interfluent lava flow.*

supergene (su'-per-gene) Said of a mineral deposit or *enrichment* formed near the surface, commonly by descending solutions; also, said of those solutions and of that environment. Cf: *hypogene; mesogene.* Syn: *hypergene; secondary [eco geol].*

supergene enrichment The *supergene* processes of mineral deposition. Near-surface oxidation produces acidic solutions that leach metals, carry them downward, and reprecipitate them, thus enriching sulfide minerals already present. Supergene enrichment has been important in upgrading porphyry copper deposits to the status of ore. Syn: *enrichment; secondary enrichment.* See also: *oxidized zone.*

superglacial (su-per-gla'-cial) Carried upon, deposited from, or pertaining to the top surface of a glacier or ice sheet; said of meltwater streams, till, drift, etc.

supergroup (su'-per-group) (a) A formal assemblage of related or superposed groups, or of groups and formations. The term should not be misused for series (NACSN, 1983, Art. 29). Cf: *megagroup.* (b) A lithostratigraphic unit composed of several associated groups or of associated formations and groups with significant lithologic features in common (ISSC, 1994, p.35).

superheating (su-per-heat'-ing) (a) The addition of more heat than necessary to complete a given phase change. (b) In a magma, the addition of more heat than is necessary to cause complete melting. The temperature increase above liquidus is called the superheat. (c) The process of increasing heat beyond that point at which a phase or assemblage changes at equilibrium, i.e. to a metastable state in the sense analogous to *supercooling.*

superimposed Said of a stream or drainage system let down from above by erosion, through the formations on which it was developed, onto rocks of different structure lying unconformably beneath. The term was first applied by Powell (1875, p.165-166) to the valley thus formed, although Maw (1866, p.443-444) earlier discussed the concept. Syn: *superposed; inherited.*

superimposed drainage Drainage by superimposed streams.

superimposed fan A term proposed by Blissenbach (1954, p.180-181) for a newly deposited alluvial fan that has a steeper gradient than the older fan upon which it is developed; it commonly results from tectonic movements that initiate a new stage of deposition.

superimposed fold *superposed fold.*

superimposed ice Ice formed when meltwater percolates down through a snowpack on a glacier and refreezes at the base of the snowpack, or as it is trapped on a lower horizon of reduced permeability such as a firn-ice boundary. This ice appears at the surface of a glacier in the *superimposed-ice zone.*

superimposed ice stream Ice from a tributary glacier that rises above the surface of a larger glacier but does not extend to its base, such as where a *tributary glacier* flows onto the surface of a *trunk glacier.* Cf: *inset ice stream.*

superimposed-ice zone The zone on a *glacier* or *ice sheet,* particularly a polar or polythermal one, where surface melting occurs, much of the meltwater is refrozen at the base of the snowpack as superimposed ice, and the *superimposed ice* is exposed at the surface when the snowpack is removed by melting. The zone is bordered at higher altitudes by the *firn line* and at lower altitudes by the *equilibrium line.*

superimposed metamorphism *polymetamorphism.*

superimposed profile A diagram on which a series of profiles drawn along several regularly spaced and parallel lines on a map are placed one on top of the other (Monkhouse and Wilkinson, 1952); it may emphasize such features as accordant summit levels and erosion platforms. Cf: *composite profile; projected profile.*

superimposed stream A stream that was established on a new surface and that maintained its course despite different lithologies and structures encountered as it eroded downward into the underlying rocks. Syn: *superposed* stream.

superimposition The establishment, originally on a cover of rocks now removed by erosion, of a stream or drainage system on existing underlying rocks independently of their structure. Gilbert (1877, p.144) recognized superimposition from an unconformable cover of sediments, from a surface of alluvium, and from a surface produced by planation.

superindividual (su'-per-in'-di-vid'-u-al) An aggregate of grains that behaves as a unit in the fabric of a deformed rock.

Superior iron *Lake Superior-type iron formation.*
superlattice (su-per-lat'-tice) The crystal lattice of a substance in which, because of chemical substitutions or atomic ordering, one or more translational periodicities are multiples of those in the related unsubstituted or disordered form. Syn: *superstructure [cryst].*
supermature (su'-per-ma-ture') Pertaining to the fourth and last stage of textural maturity (Folk, 1951); said of a mature clastic sediment whose well-sorted grains have become subrounded to well-rounded, such as a clay-free "supermature sandstone" whose sand-size quartz grains have an average roundness that exceeds 0.35 and that is presumed to form mainly as dune sands. Cf: *immature [sed]; submature [sed].*
superparamagnetism The *paramagnetic* behavior of an assembly of extremely small particles of ferromagnetic or ferrimagnetic minerals whereby the net magnetic moment is constantly fluctuating due to the small relaxation times of the magnetic moment of each grain.
superperiod (su-per-pe'-ri-od) A *geochronologic unit* intermediate in rank between period and *era* (NACSN, 1983, Table 2). It is the time represented by the equivalent *supersystem.* Cf: *subera.*
superperiodicity (su'-per-pe'-ri-o-dic'-i-ty) A lattice constant in an ordered superlattice which is a simple multiple of a corresponding direction in the disordered sublattice; the periodicity of a *superlattice.*
superplasticity (su'-per-plas-tic'-i-ty) The plastic behavior of very fine-grained material that yields a total strain larger than what can be expected from the internal deformation of the individual crystals; it implies grain-boundary sliding. This behavior is experimentally shown by ceramics and may explain the *fluidal texture* and inferred rheology of the deepest peridotite xenoliths in kimberlites.
superposed (su-per-posed') A term introduced by McGee (1888) as a shortened form of *superimposed.*
superposed fold A fold in which preexisting folds are affected by a younger folding event. Syn: *superimposed fold; complex fold.* Cf: *refolded fold.*
superposition (su'-per-po-si'-tion) (a) The order in which rocks are placed or accumulated in beds one above the other, the highest bed being the youngest. (b) The process by which successively younger sedimentary layers are deposited on lower and older layers; also, the state of being superposed. See also: *law of superposition.* Syn: *superimposition; supraposition.*
superprint (su'-per-print) *overprint [struc petrol].*
supersaturated permafrost (su-per-sat'-u-rat'-ed) Permafrost that contains more ice than the ground could possibly hold if the water were in the liquid state.
supersaturated solution A solution which contains more of the solute than is normally present when equilibrium is established between the saturated solution and undissolved solute.
superseries (su'-per-se'-ries) An infrequently used term for a group of two or more *series [stratig]* (ISG, 1976, p.72).
super-solidus intrusion breccia In mafic-ultramafic intrusive complexes, an *intrusive breccia* formed by the intrusion of volatile-rich ultramafic dikes into a mafic-ultramafic cumulate under super-solidus conditions resulting in partial melting of the cumulate rock.
super-solidus intrusion breccia PGE deposit A *mineral-deposit type* characterized by platinum-group element sulfide mineralization cutting mafic-ultramafic cumulates associated with an *intrusive breccia* formed by associated ultramafic dikes. Abbrev: *SIB-type PGE deposit.*
supersome An assemblage of contiguous and depositionally sequential *chronosomes* (Schultz, 1982, p.29-31).
superstage (su'-per-stage) A chronostratigraphic unit consisting of several adjacent stages (ISG, 1976, p.72).
superstructure [cryst] (su'-per-struc'-ture) *superlattice.*
superstructure [tect] The upper structural layer in an orogenic belt, subjected to relatively shallow or near-surface deformational processes, in contrast to an underlying and more complexly deformed and metamorphosed *infrastructure.* Also called: *suprastructure.*
supersuite (su-per-suite') The *lithodemic unit* next higher in rank to a suite. It comprises two or more suites or complexes having a degree of natural relationship to one another, either in the vertical or the lateral sense (NACSN, 1983, Art. 36). For cartographic and hierarchical purposes, supersuite is similar in rank to *supergroup.* Its name combines a geographic term with the term "supersuite."
supersystem (su'-per-sys'-tem) A chronostratigraphic unit next higher in rank than a *system [stratig].* "Special circumstances have suggested the occasional need" for supersystems, e.g. the Karroo Supersystem (ISG, 1976, p.73-74). See also: *erathem.*
superterranean (su'-per-ter-ra'-ne-an) Occurring on or above the Earth's surface. Rarely used.
supervised classification In digital mapping, a process for extracting information from an image by using a set of sites that represent known classes (e.g., lithologies in outcrops) to define a spectral signature that is used to search the image for regions with a similar signature. The goal is to produce a map of the distribution of the classes.
superzone (su'-per-zone) An assemblage of two or more stratigraphic zones of any kind. Syn: *megazone.*
supplementary contour (sup-ple-men'-ta-ry) A contour line, generally dashed or screened, drawn at less than the regular interval in order to increase the topographic expression of an area, such as in an area of extremely low relief. Syn: *auxiliary contour.*
supply-limited stream A stream that can carry more sediment than is supplied and therefore erodes its bed. Obsolete syn: *underloaded stream.* Cf: *transport-limited stream.*
support In *geostatistics,* a specification of the geometry of the observations being analyzed. Support includes the sample size (volume or area), shape, orientation, and spacing. Specification of the support is essential to the definition of the variable; i.e., permeability measured on horizontal core plugs is a different variable than permeability measured on vertical core plugs, or on whole cores.
suppressed (sup-pressed') Marked or affected by suppression, or being vestigial; e.g. said of aborted conodont denticles that could not develop into mature structures owing to crowded conditions along the growing edge of the structure.
supra-anal plate (su'-pra-a'-nal) The dorsal, posteriorly produced part of the *telson* of a branchiopod crustacean.
supracrustal (su-pra-crus'-tal) (a) Said of rocks that overlie the *basement.* (b) At the surface of the Earth. Cf: *infracrustal.*
supraembryonic area (su'-pra-em'-bry-on'-ic) Circular apical area over the megalospheric proloculus in some foraminifers of the family Orbitolinidae.
suprafan (su'-pra-fan) An upbulging zone at the downslope end of a *fan valley,* confined to fans composed of relatively coarse sediment (Kelling and Stanley, 1976, p.387).
supragelisol (su-pra-gel'-i-sol) *suprapermafrost layer.*
suprageneric name (su'-pra-ge-ner'-ic) The name of any taxon above the level of *genus.*
supraglacial (su-pra-gla'-cial) *superglacial.*
supralithion (su-pra-lith'-i-on) Animals that swim above rock but are dependent on it as their source of food.
supralittoral (su-pra-lit'-tor-al) Pertaining to the shore area marginal to the littoral zone, just above high-tide level. Syn: *supratidal.*
supraorbital carina (su-pra-or'-bit-al) The ridge extending backward from the supraorbital spine on some decapod crustaceans (Holthuis, 1974, p.735).
suprapelos (su-pra-pel'-os) Animals that swim above soft mud but are dependent on it as their source of food.
suprapermafrost layer (su-pra-per'-ma-frost) The layer of ground (or soil) above the permafrost, consisting of the active layer and, wherever present, taliks and the pereletok. Syn: *supragelisol.*
suprapermafrost water Groundwater existing above the permafrost table (Muller, 1947, p.222).
supraposition (su'-pra-po-si'-tion) *superposition.*
suprapsammon (su-pra-psam'-mon) Animals that swim above sand but are dependent on it as their source of food.
suprastructure (su'-pra-struc'-ture) *superstructure [tect].*
supratenuous fold (su-pra-ten'-u-ous) A pattern of fold in which there is thickening at the synclinal troughs and thinning at the anticlinal crests. It is formed by differential compaction on an uneven basement surface. See also: *compaction fold.*
supratidal (su-pra-tid'-al) *supralittoral.*
suranal plate (sur-a'-nal) One of the first-formed and largest plates of the periproctal system of an echinoid, often filling the central and anterior parts of the periproct; but not recognizable in many echinoids.
suredaite A metallic grayish-black orthorhombic mineral: $PbSnS_3$.
surf (a) The wave activity in the *surf zone.* (b) A collective term for breakers.
surface (sur'-face) (a) The exterior or outside part of the solid Earth or ocean; the top of the ground or the exposed part of a rock formation. (b) A two-dimensional boundary between geologic features such as formations or structures, including *bedding surface* and *fault surface,* or an imaginary surface such as the *axial surface* of a fold; usually an internal boundary, rather than one occurring on the outside of a feature (Challinor, 1978, p.302). It need not be flat. Cf: *plane.*
surface avalanche An avalanche involving only surface layers of the snowcover.
surface axis *axial trace.*
surface cauldron subsidence *Cauldron subsidence* in which the ring faults extend to the surface. Cf: *underground cauldron subsidence.*
surface conductivity Conduction along the surfaces of certain mineral grains, due to excess ions in the diffuse layer of adsorbed cations.
surface-consistent model A concept used for determining and applying seismic static corrections, amplitude correction, or deconvolution. Time delays, attenuations, or waveshape-changing factors are assigned to source and geophone (and other) locations based on a statistical study (Sheriff and Geldart, 1995, p.303-7, 315).
surface correction A correction of a geophysical measurement to remove the influence of varying surface elevation and of other irregularities at or near the Earth's surface.
surface creep The slow downwind advance of large sand grains along a surface by impact of smaller grains in saltation, as in the shifting or movement of a sand dune. See also: *surficial creep.* Syn: *reptation.*
surface curve *surface profile.*
surface density (a) The density of the surface material within the range of the elevation differences of the gravitational surface. Both the Bouguer cor-

rection and the terrain correction depend on the density of the surface material. (b) A quantity (as mass and electricity) per unit area distributed over a surface.

surface detention *detention.*

surface drag *skin friction [hydraul].*

surface drainage The removal of unwanted water from the surface of the ground, or the prevention of its entry into the soil, by natural or artificial means, such as by grading and smoothing the land to remove barriers and to fill in depressions, by terracing or digging ditches, or by diverting runoff from adjacent areas to natural waterways.

surface energy (a) The energy required to create new surface area in a material. Crack growth expends energy in the creation of new surface area. (b) A term that describes the energy at the surface (e.g., of a mineral). It is proportional to the surface area and is expressed in energy units/area. The size and shape of crystals depend in part on surface energy (a high surface energy tends to result in crystal faces and large crystals), but other factors are also important. Surface energy is equivalent to surface tension in liquids.

surface factor *fineness factor.*

surface forces Any of the forces acting over the surface of a body of material. Cf: *body force.*

surface geology (a) Geology and correlation of rock formations, structures, and other features as seen at the Earth's surface. Ant: *subsurface geology.* (b) *surficial geology.*

surface groove A *skeletal canal* in the form of a groove on the surface of the skeletal framework of a sponge and generally corresponding to an exhalant canal of the soft parts.

surface hoar A type of frost consisting of leaf- or plate-shaped ice crystals formed directly on a snow surface by condensation from vapor. Cf: *depth hoar; frost flower.*

surface moraine *surficial moraine.*

surface of concentric shearing In flexural folding, the bedding plane along which slip occurs. Var: *concentric shearing surface.*

surface of no strain A surface along which the original configuration of an array of points remains unchanged after deformation of the body in which it occurs. In two-dimensional models sometimes referred to as the *neutral axis.* Syn: *neutral surface.*

surface of rupture The projection of the *main scarp* surface under the displaced material of a landslide (Varnes, 1978).

surface of separation In a landslide, the surface separating displaced material from stable material but not known to have been a surface on which failure occurred; e.g. the surface beneath the foot of the slide (Varnes, 1978).

surface of unconformity The surface of contact between two groups of rocks displaying an unconformable relationship, such as a buried surface of erosion or of nondeposition separating younger strata from underlying older rocks. Syn: *unconformity.*

surface phase In metamorphism, a thin layer of material having properties that may differ from those of the phases on either side (Turner and Verhoogen, 1960, p.461).

surface pipe The uppermost length of *casing* set in a well, used to prevent unconsolidated sediments from entering the hole and to shut off and protect shallow freshwater sands from contamination by drilling mud and deeper saline waters. It also serves as a conductor to return the drilling mud through loose, near-surface formations during drilling deeper.

surface profile The longitudinal profile assumed by the surface of a stream of water in an open channel. See also: *backwater curve.* Syn: *surface curve.*

surface runoff The *runoff [water]* that travels over the soil surface to the nearest surface stream; runoff of a drainage basin that has not passed beneath the surface since precipitation. The term is misused when applied in the sense of *direct runoff* (Langbein and Iseri, 1960). Cf: *storm seepage; groundwater runoff.*

surface-scanning measurement Observation made during the movement of a radiation sensor over a surface.

surface scattering layer An area of marine organisms near the ocean surface that scatter sound waves from an echo sounder. Cf: *shallow scattering layer; deep scattering layer.*

surface-ship gravimeter An instrument designed to produce useful gravity observations aboard a surface ship that is underway. Accuracies attained range from about 0.2 to 1.0 milligals depending on quality of the navigation and state of the sea. Syn: *air-sea gravity meter.*

surface slope The inclination of the water surface expressed as change of elevation per unit of slope length; the sine of the angle which the water surface makes with the horizontal. The tangent of that angle is ordinarily used; no appreciable error results except for the steeper slopes (ASCE, 1962).

surface soil The uppermost part of the soil, ordinarily moved in tillage, or its equivalent in uncultivated soils and ranging in depth from 7 to 25 cm. Frequently designated as the "plow layer," the "surface layer," the "Ap layer," or the "Ap horizon."

surface spreading An artificial method of recharging groundwater by spreading water on the surface.

surface texture The aggregate of the surface features of sedimentary particles, independent of size, shape, or roundness; e.g. polish, frosting, or striations.

surface wash *sheet erosion.*

surface water (a) All waters on the surface of the Earth, including fresh and salt water, ice, and snow. (b) A *water mass* of varying salinity and temperature, occurring at the ocean surface and having a thickness of 300 meters or less. Cf: *intermediate water; deep water; bottom water [oceanog].*

surface wave [oceanog] (a) A progressive gravity wave in which the particle movement is confined to the upper limits of a body of water; strictly, a gravity wave whose celerity is a function of wavelength only. (b) *deep-water wave.*

surface wave [seis] A *seismic wave* that travels along the surface of the Earth, or along a subsurface interface. Surface waves include the *Rayleigh wave,* the *Love wave,* and the *coupled wave.* See also: *hydrodynamic wave.* Cf: *body wave; guided wave; P wave.* Obsolete syn: *circumferential wave; long wave [seis].* Syn: *L wave.*

surface-wave magnitude *Earthquake magnitude* determined at teleseismic distances, using the logarithm of the amplitude of 20-s-period surface waves.

surfactant A substance that affects the properties of the surface of a liquid or solid by concentrating on the surface layer. The use of surfactants can ensure that one substance is in thorough contact with the surface of another substance.

surf beat (a) The slow variability of the height of incoming *swell* due to interference among *swell* of nearly the same period. (b) The long-period (several minutes) change in water level along the coast due to the changing height of *breakers* in the *surf zone.*

surficial (sur-fi'-cial) Pertaining to, or occurring on, a surface, esp. the surface of the Earth. Cf: *subsurface.* See also: *subaerial.* Syn: *superficial.*

surficial creep A syn. of *soil creep.* See also: *surface creep.*

surficial deposit Unconsolidated and residual, alluvial, or glacial deposits lying on bedrock or occurring on or near the Earth's surface; it is generally unstratified and represents the most recent of geologic deposits. Syn: *superficial deposit.*

surficial geology Geology of surficial deposits, including soils; the term is sometimes applied to the study of bed rock at or near the Earth's surface. See also: *surface geology.*

surficial moraine A moraine, such as a lateral or medial moraine, in transit on the surface of a glacier. Syn: *superficial moraine; surface moraine.*

surf ripple A general term proposed by Kuenen (1950, p.292) for a ripple formed on a sandy beach by wave-generated currents in the surf zone.

surfusion (sur-fu'-sion) An obsolete term proposed by Fournet (1844) for a condition under which the fusing points of substances are lowered to temperatures much below the points at which they usually solidify (see Zittel, 1901, p.342).

surf zone The area bounded by the landward limit of wave uprush and the farthest seaward breaker. Syn: *surf; breaker zone.*

surge [glaciol] The period of very rapid flow of a *surging glacier;* also, the displacement or advance of ice resulting from the rapid flow. Syn: *glacier surge; catastrophic advance.*

surge [hydraul] In fluid flow, long-interval variations in velocity and pressure that are not necessarily periodic and may even be transient.

surge [waves] (a) *storm surge.* (b) Horizontal oscillation of water with a comparatively short period, accompanying a seiche. (c) Water transported up a beach by breaking waves.

surge channel Channel extending across the crest of a coral reef, in which water level fluctuates according to the height and arrival time of breakers. The in- and out-flow in surge channels is normally vigorous and commonly extremely dangerous. Surge channels and other irregularities grooves, buttresses, etc., as well as roughness of the reef face create violent turbulence which exhausts rapidly most of the energy of wave advance.

surgical mute A sharp division in offset-time space between elements that are retained unchanged in magnitude and those deleted entirely. As opposed to a tapered or ramped mute where the cutoff is gradual. Usually the data muted (deleted) are those preceding and immediately following the first breaks, including shallow head waves. Sometimes refers to a slice diagonally cutting across an x, t plot, where the values within the slice have been set equal to zero, as might be done to remove a ground-roll wavetrain.

surging breaker (surg'-ing) A type of breaker that peaks and then surges onto the beach face without spilling or plunging. It forms over a very steep bottom gradient. Cf: *plunging breaker; spilling breaker.*

surging glacier A glacier that alternates periodically between brief periods (usually one to four years) of rapid flow, called surges, and longer periods (usually 10 to 100 years) of near stagnation. During a surge, a large volume of ice from an *ice-reservoir area* is displaced downstream at speeds up to several meters per hour into an *ice-receiving area,* and the affected portion of the glacier is chaotically crevassed. Only in exceptional cases does the glacier advance beyond its former limit. In the interval between surges, the ice reservoir is slowly replenished by accumulation and normal ice flow, and the ice in the receiving area is greatly reduced by ablation.

surinamite (su'-ri-nam-ite') A dark blue to blue-green monoclinic mineral related to sapphirine: $(Mg,Fe^{2+})_3Al_4BeSi_3O_{16}$.

surite (sur'-ite) A white to plae green monoclinic mineral: $(Pb,Ca)_3Al_2(Si,Al)_4O_{10}(CO_3)_2(OH)_3{\cdot}0.3H_2O$.

surkhobite A brownish-red monoclinic mineral: $(Ca,Na)(Ba,K)(Fe^{2+},Mn)_4Ti_2O_2(Si_4O_{14})(F,OH,O)$.

sursassite (sur-sass′-ite) A copper-red monoclinic mineral: $Mn^{2+}_2Al_3(SiO_4)(Si_2O_7)(OH)_3$.

Surtseyan eruption An eruption characterized by steam explosions that eject new lava fragments as pyroclastic surges and fallout. The activity is typified by the eruptions of the volcano Surtsey in Iceland, where rising basaltic magma explosively vaporized near-surface water. Syn: *phreatomagmatic; hydrovolcanic.*

survey (sur′-vey) v. To determine and delineate the form, extent, position, boundaries, value, or nature of a tract of land, coast, harbor, or the like, esp. by means of linear and angular measurements and by the application of the principles of geometry and trigonometry. n. (a) The orderly and exacting process of examining, determining, finding, and delineating the physical or chemical characteristics of the Earth's surface, subsurface, or internal constitution by topographic, geologic, geophysical, or geochemical measurements; esp. the act or operation of making detailed measurements for determining the relative positions of points on, above, or beneath the Earth's surface. (b) The associated data or results obtained in a survey; a map or description of an area obtained by surveying. (c) An organization engaged in making surveys; e.g. a government agency such as the U.S. Geological Survey.

survey (2.5-D) *two-and-a-half-dimensional seismic.*

survey (2-D) *two-dimensional method.*

survey (3-D) *three-dimensional survey.*

survey (4-D) *four-dimensional survey.*

surveying (sur-vey′-ing) (a) The art of making a survey; specif. the applied science that teaches the art of making such measurements as are necessary to determine the area of any part of the Earth's surface, the lengths and directions of the boundary lines, and the contour of the surface, and of accurately delineating the whole on paper. (b) The act of making a survey; the occupation of one that surveys.

surveyor's compass (sur-vey′-or's) A surveying instrument used for measuring horizontal angles; specif. one designed for determining a magnetic bearing of a line of sight by the use of a sighting device, a graduated horizontal circle, and a pivoted magnetic needle. The surveyor's compass used on the early land surveys in U.S. had a pair of peep sights to define the line of sight and was usually mounted on a *Jacob's staff.* See also: *circumferentor.* Syn: *land compass.*

surveyor's level A *leveling instrument* consisting of a telescope (with cross hairs) and a spirit level mounted on a tripod, revolving on a vertical axis, and having leveling screws that are used to adjust the instrument to the horizontal.

surveyor's measure A system of measurement used in land surveying, having the surveyor's chain as a unit (one chain = 4 rods = 66 ft = 100 links = 1/80 mi). Cf: *Gunter's chain.*

surveyor's rod *level rod.*

survival of the fittest (sur-viv′-al) The tendency for the environmentally better-adapted members of a population to survive to reproductive age and thus to contribute more strongly to the genetic composition of the next generation.

susannite (su-san′-nite) A colorless, greenish, or yellowish trigonal mineral: $Pb_4(SO_4)(CO_3)_2(OH)_2$. It is polymorphous with leadhillite and macphersonite.

susceptibility [elect] (sus-cep′-ti-bil′-i-ty) The ratio of the electric polarization to the electric intensity in a polarized dielectric.

susceptibility [magnet] A measure of the degree to which a rock can be magnetized; the ratio of *induced magnetization* to the strength H of the magnetic field causing the magnetization. See also: *susceptibility anisotropy.* Syn: *magnetic susceptibility; volume susceptibility.*

susceptibility anisotropy In minerals of low crystal symmetry or in rocks with planar or linear fabric, magnetic susceptibility which is not perfectly parallel with the inducing magnetic field, because it depends on direction and induced magnetization. Syn: *magnetic anisotropy.*

suspect terrane (sus′-pect) A terrane whose spatial and genetic relations with respect to adjacent terranes during their time of formation is unknown or uncertain (D.L. Jones et al., 1983). The term was first used by Coney et al. (1980) for a terrane whose paleogeographic or paleotectonic setting with respect to the North American Craton is uncertain. Inasmuch as most terranes fall into this category, the term may be considered redundant.

suspended current (sus-pend′-ed) A *turbidity current,* in a body of standing water, that is not in contact with the bottom (as on a slope or where the current overrides denser underlying water) and that continues in suspension (Dzulynski and Radomski, 1955).

suspended load (a) The part of the total *sediment load* that is carried for a considerable period of time in suspension, free from contact with the bed; it consists mainly of clay, silt, and sand. (b) The material collected in, or computed from samples collected with, a suspended-load sampler. Syn: *suspension load; suspensate; silt load; wash load.*

suspended-load sampler A device that collects a sample of water with its sediment load without separating the sediment from the water.

suspended water *vadose water.*

suspensate (sus-pen′-sate) *suspended load.*

suspension (sus-pen′-sion) (a) A mode of sediment transport in which the upward currents in eddies of turbulent flow are capable of supporting the weight of sediment particles and keeping them indefinitely held in the surrounding fluid (such as silt in water or dust in air). Cf: *traction [sed].* (b) The state of a substance in such a mode of transport; also, the substance itself.

suspension current *turbidity current.*

suspension feeder An animal that obtains its food by removing particulate organic matter from water, but not necessarily through the use of a filter. Cf: *filter feeder.*

suspension flow Flow of a water-sediment mixture in which the sediment is maintained in suspension by the combination of water turbulence and relatively low settling velocities of particles. It commonly occurs in a turbidity current.

suspension load *suspended load.*

sussexite [mineral] (sus′-sex-ite) A white, buff, or straw-yellow monoclinic mineral: $Mn^{2+}BO_2(OH)$. It is isomorphous with szaibelyite.

sussexite [rock] A porphyritic tinguaitic rock composed chiefly of nepheline and aegirine, with subordinate feldspar; a feldspar-bearing *urtite.* Its name, given by Brögger in 1894, is derived from Sussex County, New Jersey. Obsolete. Cf: *muniongite.*

sustainability Managing soil and crop cultural practices so as not to degrade or impair environmental quality on or off site, and without eventually reducing yield potential as a result of the chose practice through exhaustion of either on-site resources or nonrenewable inputs.

sustained runoff (sus-tained′) *base runoff.*

sutural (su′-tur-al) Pertaining to a *suture,* or corresponding to sutures in position; e.g. "sutural supplementary apertures" in a foraminiferal test.

sutural element One of the major undulations of a *suture* of a cephalopod shell, such as a *lobe* or a *saddle.*

sutural pore (a) One of the pores bordered by sutures along the meeting branches of two or more adjacent radial spines in an acantharian radiolarian. Cf: *parmal pore.* (b) A pore lying along any suture between plates of an echinoderm theca, commonly restricted to nonambulacral areas.

suture [bot] (su′-ture) On a fruit, the line of dehiscence or splitting; a groove marking a natural division or union. Syn: *commissure.*

suture [paleont] (a) The line of junction of a *septum* of a cephalopod's shell with the inner surface of the shell wall. It is commonly more or less undulated or plicate, and is visible only when the shell wall is removed. (b) The line of contact between two whorls of a gastropod shell. It is typically a spiral on the outer surface, and also on the inner surface around the axis of coiling where the *umbilicus* is not closed. (c) A plane of junction between adjacent plates of an echinoderm. Also, the boundary line on the surface marking the area of contact between the adjacent plates. (d) A line of demarcation between two fused or partly fused limb segments or body somites of a crustacean; e.g. a line or seam at the juncture of two compartmental plates of a cirripede, or a weak or uncalcified narrow seam between parts of a trilobite exoskeleton that separate at time of molting. (e) A line of contact between two chambers or two whorls of a foraminiferal test. It may be reflected on the outer wall by a groovelike or ridgelike feature. (f) The boundary between segments of heterococcoliths. (g) In vertebrates, a line of junction between bones or an articulation in which bones of different embryonic centers of origin are united by fibrous connective tissue and function as a solid, nearly solid, or moderately movable unit in the adult. In many cases, the original suture in adulthood or old age may be completely fused by bone (ankylosis, synostosis).

suture [palyn] (a) The line along which the laesura of an embryophytic spore opens on germination; loosely, the *laesura.* See also: *commissure.* (b) The area of contact between two adjacent plates on a dinoflagellate theca or cyst.

suture [tect] A fault or complex *shear zone* that marks the boundary between two crustal blocks that were once widely separated, often on different plates. The presence of a suture implies that oceanic lithosphere once existed between the two blocks.

sutured (su′-tured) Said of the texture found in rocks in which mineral grains or irregularly shaped crystals interfere with their neighbors, producing interlocking, irregular contacts without interstitial spaces, resembling the sutural structures in bones. Also, said of the crystal contacts in rock with this texture. Syn: *consertal.*

suture joint A very small *stylolite.*

suture line Surficial line on an echinoderm theca formed by the intersection of sutures with the thecal surface (Bell, 1976).

suzukiite (su-zu′-ki-ite) A green orthorhombic mineral: $BaV^{4+}Si_2O_7$. It is the Ba analogue of haradaite.

sv *sverdrup.*

Sv [radiation] *sievert.*

svabite (svab′-ite) A colorless, yellowish-white, or gray hexagonal mineral of the *apatite* group: $Ca_5(AsO_4)_3(F,OH)$. It may contain phophorus, lead, magnesium, or manganese.

svanbergite (svan′-berg-ite) A colorless to yellow, rose, or reddish-brown rhombohedral mineral of the *beudantite* group: $SrAl_3(PO_4)(SO_4)(OH)_6$.

Svecokarelian Orogeny (Sve′-co-ka-re′-li-an) The Proterozoic orogeny that formed the Svecofennian and Karelian formations, consisting of metamorphosed and folded sedimentary and volcanic rocks, which cover most of Sweden and Finland.

sveite (sve'-ite) A white monoclinic mineral found as a cave efflorescence: $KAl_7(NO_3)_4Cl_2(OH)_{16}{\cdot}8H_2O$.
svenekite A colorless or white monoclinic mineral: $CaH_4(AsO_4)_2$.
sverdrup In oceanography, a unit of volume transport, equal to 10^6 m^3/s. Named for Harald U. Sverdrup. Abbrev: sv.
Sverdrup transport The southward (northward) transport of water in the upper one to two kilometers of the ocean due to the positive (negative) curl of the *wind stress* acting on large areas of the sea.
sverigeite (sver'-i-ge-ite) A yellow orthorhombic mineral: $NaMn^{2+}MgSn^{4+}Be_2Si_3O_{12}(OH)$.
svetlozarite (svet-lo'-za-rite) *dachiardite-Ca.*
sviatonossite (svi'-at-o-nos'-site) An andradite-bearing *syenite* in which the pyroxene is aegirine-augite and the plagioclase is oligoclase. The name, given by Eskola in 1922, is for Sviatoy Noss, Transbaikalia, Russia. Obsolete.
svitalskite (svi-tal'-skite) *celadonite.*
SV wave S-wave energy polarized so that motion is in the vertical plane which also contains the direction of wave propagation. Converted waves are (mostly) SV.
svyatoslavite (svy-a'-to-slav'-ite) A colorless orthorhombic mineral: $CaAl_2Si_2O_8$. It is a combustion product of a coal heap and a dimorph of anorthite.
svyazhinite (svy-a'-zhin-ite) A yellowish or colorless triclinic mineral of the *aubertite* group: $(Mg,Mn^{2+})(Al,Fe^{3+})(SO_4)_2F{\cdot}14H_2O$.
swab n. A pistonlike device equipped with an upward-opening check valve and provided with flexible circular suction cups, lowered into a cased well bore by means of a wireline for the purpose of cleaning out drilling mud or inducing flow of formation fluids. v. To clean out a well with a swab.
swag (a) A shallow pocket or closed depression in flat or gently rolling terrain, often filled with water (as in the bottomlands of the lower Mississippi Valley). (b) A shallow water-filled hollow produced by subsidence resulting from underground mining. Cf: *sag pond.*
swaknoite A vitreous white orthorhombic cave mineral: $(NH4)_2Ca(PO_3OH)_2{\cdot}H_2O$.
swale (a) A slight depression, sometimes swampy, in the midst of generally level land. (b) A shallow depression in an undulating ground moraine due to uneven glacial deposition. (c) A long, narrow, generally shallow, trough-like depression between two beach ridges, and aligned roughly parallel to the coastline. Syn: *furrow [geomorph]; runnel [beach]; slash [beach].*
swaley cross-stratification A type of cross-stratification resembling *hummocky cross-stratification* (HCS) but dominated by concave-up strata deposited in swales rather than convex-up strata deposited on hummocks as in HCS (Leckie and Walker, 1982). Abbrev: SCS.
swallet (swal'-let) The opening through which a sinking stream loses its water to the subsurface; or a place where such a stream may sink into alluvium in a streambed without the presence of a depression. Syn: *ponor; swallow hole.*
swallow hole (swal'-low) A *closed depression* or *doline* into which all or part of a stream disappears underground. Cf: *swallet; ponor.*
swallowtail gypsum Gypsum crystal habit of upward-opening chevron shaped termination due to twinning.
swamboite (swam'-bo-ite) A pale yellow monoclinic mineral related to the *uranophane* group: $U^{6+}(UO_2)_6(SiO_3OH)_6{\cdot}30H_2O$.
swamp An area of low, waterlogged ground having shrubs and trees, with or without the formation of peat. Cf: *bog; marsh; peat swamp.*
swamp ore *bog iron ore.*
swamp theory *in-situ theory.*
swarm [intrus rocks] A concentration of dikes of similar orientation (parallel, subparallel, or radiating) (Halls and Fahrig, 1987).
swarm [seis] *earthquake swarm.*
swartzite (swartz'-ite) A vitreous green monoclinic mineral: $CaMg(UO_2)(CO_3)_3{\cdot}12H_2O$. It turns dull yellow on dehydration.
swash bar (a) A small, transitory bar built above the stillwater level by wave uprush, and forming a tiny lagoon on its landward side (King and Williams, 1949, p.81). (b) A bar over which the sea washes, esp. on an *ebb-tidal delta.*
swash channel (a) A narrow sound or secondary channel of water lying within a sandbank or between a sandbank and the shore, or passing through or shoreward of an inlet or river bar. Syn: *swashway; swatch.* (b) A channel cut on an open shore by flowing water in its return to the parent body.
swash mark A thin, delicate, wavy or arcuate line or very small ridge on a beach, marking the farthest advance of wave uprush. It is convex landward and consists of fine sand, mica flakes, bits of seaweed, and other debris. Syn: *wave line; wavemark [coast]; debris line.* Cf: *trash line; wrackline.*
swash pool A shallow pool of water formed behind a swash bar.
swashway (swash'-way) *swash channel.*
swash zone The sloping part of the beach that is alternately covered and uncovered by the uprush of waves and where longshore movement of water occurs in a zigzag (upslope-downslope) manner.
swatch A British syn. of *swash channel.*
swathing kamacite Kamacite formed around a preexisting phase (e.g., around olivine in a pallasite).
S wave A seismic *body wave* propagated by a shearing motion that involves oscillation perpendicular to the direction of propagation. It does not travel through liquids, or through the outer core of the Earth. Its speed is often about half that of P-waves, about 3.0-4.0 km/s in the crust and 4.4-4.6 km/s in the upper mantle. The S stands for secondary; it is so named because it arrives later than the P wave (primary wave). There are two independent modes of S-waves: *SV waves* and *SH waves.* Syn: *shear wave [seis]; secondary wave; rotational wave; tangential wave; equivoluminal wave; distortional wave; transverse wave; converted wave.*
Swazian (Swa'-zi-an) An obsolete geochronometric subdivision of the early *Precambrian,* covering the time span from 3,500 to 2,800 Ma (Harland et al., 1990), therefore approximately equivalent to the combined *Paleoarchean* and *Mesoarchean* eras (early *Archean* Eon) of the Standard Global Chronostratigraphic Scale.
swedenborgite (swe'-den-borg-ite') A colorless or yellow hexagonal mineral: $NaBe_4SbO_7$.
Swedish mining compass (Swed'-ish) A compass in which a magnetic needle is suspended on a jewel and a stirrup so that it can rotate about both a horizontal and a vertical axis.
sweeping (sweep'-ing) The progressive down-valley movement or shift of a system of meanders. See also: *wandering.* Also called: sweep.
sweepstakes route (sweep'-stakes) A biogeographic dispersal path that constitutes a formidable obstacle to the migration of plants and animals but is occasionally conquered. The term implies that the odds against crossing the barrier are as great as those against someone winning a lottery, i.e., great but not impossible.
sweet Said of crude oil or natural gas that contains few or no sulfur compounds. Cf: *sour.*
sweetite (sweet'-ite) A white tetragonal mineral: $Zn(OH)_2$. It is polymorphous with wülfingite and ashoverite.
sweet water *fresh water.*
swell [eng] The increase in volume, or dilatancy, exhibited by soil or rock in response to removal of in-situ stress, as in the bottom of an excavation; or in the absorption of water; or in freezing (*frost heave*).
swell [marine geol] A smooth elevation of the sea floor, generally lower than a *rise [marine geol].*
swell [ore dep] An enlarged place in, or part of, an orebody, as opposed to a *pinch.*
swell [struc geol] A general, imprecise term and less-preferred syn. for *dome* and *arch.*
swell [waves] Series of regular, long-period (7-20 s), long-crested waves that have traveled out of their generating area. Cf: *ground swell.* Ant: *sea [waves].*
swell-and-swale topography A low-relief, undulating landscape characteristic of the ground moraine of a continental glacier, exhibiting gentle slopes and well-rounded hills interspersed with shallow depressions. Cf: *sag-and-swell topography.*
swelling chlorite (swell'-ing) A chlorite-like mineral, found in clays, that behaves like a chlorite on heating but has its basal spacing expand on glycerol treatment. It contains incomplete hydroxide (brucite or gibbsite) layers, and might be regarded as a special interlayering of chlorite with smectite or vermiculite (Martin Vivaldi and MacEwan, 1960). See also: *corrensite.* Syn: *pseudochlorite.*
swelling clay A generic term used to designate smectite or vermiculite minerals, both of which swell by the adsorption of H_2O or other polar molecules between the silicate layers (i.e., interlayer space) of the structure. The nature of the swelling is related to the relatively low excess negative charge of the layer, -0.25 to -0.60 for smectite and -0.60 to -0.90 for vermiculite. This term is used when the distinction between these two minerals cannot be made and when the particle size is sufficiently small to classify the material as clay. Cf: *clay mineral.*
swelling ground A term for rock or soil that increases in volume on excavation; it usually contains clay minerals with high swelling capacity. Cf: *firm ground; flowing ground; raveling ground; running ground; squeezing ground.*
swelling pressure The pressure exerted by a clay or shale when it absorbs water in a confined space. Also, the pressure produced by silt when it freezes; and that exhibited by other soils rich in complex salts (commonly sulfates) at crystallization temperatures under a road, slab, or foundation.
swept ripple A *ripple* that is sharply skewed relative to the general flow. The most common type, swept *catenary ripple,* appears to be generally formed in two stages: first by formation of a ripple transverse to the flow, and then by modification of the original crest to a catenary form as the flow changes direction. Term introduced by Allen (1968, p.66).
swimming leg (swim'-ming) The hindmost prosomal appendage of a merostome, serving as a swimming organ.
swimming stone *floatstone [mineral].*
swinefordite (swine'-ford-ite) A greenish-gray clay-like monoclinic mineral of the *montmorillonite* group: $Li_{0.5}(Al,Li,Mg)_2(Si,Al)_4O_{10}(OH,F)_2{\cdot}nH_2O$.
swinestone (swine'-stone) *anthraconite.*
swinging (swing'-ing) The steady lateral movement of a *meander belt* from one side of the valley floor to the other. See also: *wandering.* Also called: swing.
swing mark A circular or semicircular sedimentary structure formed by wind action on an anchored plant stem or root that is swept to and fro, or round and round, across a sandy surface.

swither (swith′-er) A colloquial term used in the Wisconsin lead-mining region for an offshoot or branch of a main lode.
switzerite (switz′-er-ite) A pale pink to light golden-brown monoclinic mineral: $(Mn^{2+},Fe^{2+})_3(PO_4)_2{\cdot}7H_2O$. It readily dehydrates to metaswitzerite.
sword dune *seif.*
sychnodymite (sych-no′-dy-mite) *carrollite.*
sycon (sy′-con) A sponge or sponge larva in which separate flagellated chambers open directly into a central *spongocoel* lined with *pinacoderm.* Cf: *ascon; leucon.* Adj: syconoid.
syenite (sy′-e-nite) (a) In the *IUGS classification,* a plutonic rock with Q between 0 and 5, and P/(A+P) between 10 and 35. (b) A group of plutonic rocks containing alkali feldspar (usually orthoclase, microcline, or perthite), a small amount of plagioclase (less than in *monzonite*), one or more mafic minerals (esp. amphibole), and quartz, if present, only as an accessory; also, any rock in that group; the intrusive equivalent of *trachyte.* With an increase in the quartz content, syenite grades into granite. Its name is derived from Syene, Egypt. A. G. Werner in 1788 applied the name in its present meaning to rock at Plauenscher Grund, Dresden, Germany; the Egyptian rock is a granite containing much more quartz.
syenodiorite (sy′-e-no-di′-o-rite) A group of plutonic rocks intermediate in composition between *syenite* and *diorite,* containing both alkali feldspar (usually orthoclase) and plagioclase feldspar, commonly more of the former; also, any rock in that group. Generally considered a syn. of *monzonite,* but may also include both monzonite and rocks intermediate between monzonite and diorite (Streckeisen, 1967, p.170).
syenogabbro (sy′-e-no-gab′-bro) A plutonic rock differing in composition from *gabbro* by the presence of alkali feldspar.
syenogranite In the *IUGS classification,* a plutonic rock with Q/(Q+A+P) between 20 and 60, and P/(A+P) between 65 and 90; in other words, a variety of granite in which alkali feldspar is more abundant than plagioclase. Cf: *monzogranite.*
Syltian (Sylt′-ian) [Silt′-ian]) North Sea stage: upper Miocene (above Langenfeldian, below Morsumian).
sylvanite (syl′-van-ite) A metallic grayish-white monoclinic mineral: $AuAgTe_4$. It often occurs in implanted crystals resembling written characters. Not to be confused with *sulvanite* or *sylvinite.* Also spelled: *silvanite.* Syn: *graphic tellurium; white tellurium; yellow tellurium; goldschmidtite.*
sylvine (syl′-vine) *sylvite.*
sylvinite (syl′-vin-ite) A mixture of halite and sylvite, mined as a potash ore; a rock that contains chiefly impure potassium chloride. Not to be confused with *sylvanite.*
sylvite (syl′-vite) A white, colorless, yellow, gray, or red cubic mineral of the *halite* group: KCl. It is the principal ore mineral of potassium. Sylvite occurs in beds as a saline residue with halite and other evaporites. It has a sharper taste than that of halite. Syn: *sylvine; leopoldite.*
symbiosis (sym-bi-o′-sis) The relationship that exists between two different organisms that live in close association, with at least one being helped without either being harmed. Cf: *mutualism; commensalism; parasitism.* Adj: *symbiotic.*
symbiotic (sym-bi-ot′-ic) The adj. of *symbiosis.*
symbol (sym′-bol) A diagram, design, letter, color hue, abbreviation, or other graphic device placed on maps, charts, and diagrams, which by convention, usage, or reference to a legend is understood to represent a specific characteristic, feature, or object, such as structural data, rock outcrops, or mine openings.
symesite A vitreous pink triclinic mineral: $Pb_{10}SO_4O_7Cl_4{\cdot}H_2O$.
symmetrical fold (sym-met′-ri-cal) (a) A fold whose axial surface is perpendicular to the *enveloping surface.* (b) A fold whose limbs have the same angle of dip relative to the axial surface. Cf: *asymmetric fold.*
symmetric bedding (sym-met′-ric) Bedding characterized by lithologic types or facies that follow each other in a "retracing" arrangement illustrated by the sequence 1-2-3-2-1-2-3-2-1. Cf: *asymmetric bedding.*
symmetric ripple A *ripple* having a symmetric profile in cross section, being similarly shaped on both sides of the crest, which in plan view is predominantly straight; specif. *oscillation ripple.* Ant: *asymmetric ripple.*
symmetric ripple mark *symmetric ripple.*
symmetric spread *split spread.*
symmetry [cryst] (sym′-me-try) The systematic repetition of objects (e.g., crystal faces or atoms of a *crystal structure*) by one or more geometrical operations. That aspect of a crystal that renders its appearance, structure, and directional properties invariant to certain operations, such as rotation, reflection, inversion, and their combinations.
symmetry [paleont] See *bilateral symmetry; radial symmetry.*
symmetry [struc petrol] The symmetry of a fabric is the combined symmetry of all the elements making up the fabric. There are five possible symmetries: (1) spherical, for fabrics having the symmetry of a sphere; (2) axial, for fabrics having the symmetry of a spheroid; (3) orthorhombic, for fabrics having the symmetry of a triaxial ellipsoid; (4) monoclinic, for fabrics having only one unique plane of symmetry; and (5) triclinic, for fabrics having no planes of symmetry (Turner and Weiss, 1963, p.43-44).
symmetry axis In a crystal, an imaginary line about which the crystal may be rotated, during which there may be two, three, four, or six repetitions of its appearance (lines, angles, or faces); it is one of the *symmetry elements.* Syn: *axis of symmetry; rotation axis; symmetry axis of rotation.*
symmetry axis of rotary inversion *rotoinversion axis.*
symmetry axis of rotation *symmetry axis.*
symmetry center *center of symmetry.*
symmetry elements The axes, planes, and center of symmetry, by which crystal symmetry can be described. There are 32 possible crystallographic arrangements of the elements of external symmetry; each arrangement is a crystal class. Syn: *elements of symmetry.*
symmetry operations Various movements of a crystal that leave the crystal and its directional properties invariant; these are rotation about an axis, reflection across a plane, inversion, and rotary inversion.
symmetry plane *plane of mirror symmetry.*
symmetry principle A statement of the way in which the symmetry of an observed fabric is related to the symmetry of the factors that produced it. "The symmetry of any physical system must include those symmetry elements that are common to all the independent factors (physical fields and physical properties of the medium) that contribute to the system, and it may include additional symmetry elements; however, any symmetry elements absent in the system must also be absent in at least one of the independent contributing factors" (Paterson and Weiss, 1961, p.859).
symmetry-transition series The array of three or four morphologically different but intergradational types of coniform or ramiform elements that occupy *S* positions in the skeletal apparatus of a conodont (TIP, 1981, pt. W, supp.2).
symmict (sym′-mict) Said of a structureless sedimentation unit, as in a varved clay or a graded series, composed of material in which coarse- and fine-grained particles are mixed to a greater extent in the lower part due to rapid flocculation. Also, said of the sedimentary structure so formed. Also spelled: *symminct.*
symmictite [sed] *diamictite.*
symmicton (sym-mic′-ton) *diamicton.*
symminct (sym′-minct) *symmict.*
symmixis (sym-mix′-is) Flocculation induced in sediments by certain electrolytes (esp. sodium chloride), resulting in the mixing of silt and clay particles and the formation of a nearly homogeneous or nonlaminated clay.
symon fault (sy′-mon) A syn. of *horseback [coal],* named after such a structure in the Coalbrookdale coalfield of England that was originally thought to be a large fault.
sympathetic release An avalanche which releases in direct response to the release of a neighboring avalanche.
sympatric (sym-pat′-ric) Said of populations occupying the same geographic area without losing their identity as a result of interbreeding. Noun: sympatry. Cf: *allopatric.*
sympatric species Related species whose geographic distributions overlap or coincide.
symphytium (sym-phy′-ti-um) A single plate formed in certain brachiopods by fusion of deltidial plates dorsally or anteriorly from the pedicle foramen and lacking a median line of junction.
symplectic (sym-plec′-tic) General rock texture involving wormy or irregular intergrowth of one mineral within another. Cf: *dactylitic; vermicular.*
symplectite (sym-plec′-tite) An intimate intergrowth of two different minerals, sometimes restricted to those of secondary origin; also, a rock (igneous or thermally metamorphosed) characterized by *symplectic* texture. Cf: *pegmatite.*
symplesiomorphy (sym-ple′-si-o-mor′-phy) (a) *Shared primitive character.* (b) In cladistic analysis, a character shared among a group of individuals that is present in their common ancestor and thought to have originated in that ancestor, not in an earlier ancestor (Brooks et al., 1984, p.81). Adj: symplesiomorphic. Cf: *plesiomorphy, synapomorphy.*
symplesite (sym′-ples-ite) A blue, blue-green, or pale indigo triclinic or monoclinic mineral: $Fe^{2+}_3(AsO_4)_2{\cdot}8H_2O$. Cf: *parasymplesite; ferrisymplesite; metaköttigite.*
sympod (sym′-pod) The *protopod* of a crustacean.
sympodium (sym-po′-di-um) A plant stem made up of a series of branches that grow on each other. It gives the effect of a simple stem.
syn- A prefix meaning "together", "with".
synadelphite (syn-a-del′-phite) A colorless, brown, deep red, or black orthorhombic mineral: $(Mn^{2+},Mg,Ca,Pb)_9(As^{3+}O_3)(As^{5+}O_4)_2(OH)_9{\cdot}2H_2O$.
synaeresis (syn-aer′-e-sis) *syneresis.*
synangium (syn-an′-gi-um) A compound *sporangium* composed of sporangia fused (*connate*) to one another as, e.g., marattialean ferns.
synantectic (syn-an-tec′-tic) Said of a primary mineral formed by the reaction of two other minerals, as in the formation of a reaction rim. Obsolete. See also: *deuteric.*
synantexis (syn-an-tex′-is) *Deuteric* alteration. Obsolete.
synapomorphy (syn-ap′-o-mor′-phy) (a) *Shared derived character.* (b) In cladistic analysis, a character shared among a group of individuals that is present in their common ancestor and thought to have originated in that ancestor, not an earlier ancestor (Brooks et al., 1984, p.80). Adj: synapomorphic. Cf: *symplesiomorphy, plesiomorphy.*

Synapsida (Syn-ap′-si-da) A subclass of reptiles characterized by a single lower temporal opening; mammal-like reptiles *sensu lato.* Range, Lower Pennsylvanian to Upper Triassic. Cf: *Therapsida.*

synapticula (syn-ap-tic′-u-la) (a) One of numerous transverse calcareous rods or bars that extend between and connect the opposed faces of adjacent septa of some corals and that perforate the mesenteries between them. A "compound synapticula" consists of a broad bar formed by fusion of opposed ridges on adjacent septa. (b) An anaxial bar of secondarily secreted silica connecting adjacent spicules in hexactinellid sponges. (c) A rodlike structure extending between septa of an archaeocyathid. (d) Plural of *synapticulum.* Pl: synapticulae.

synapticulotheca (syn-ap-tic′-u-lo-the′-ca) The porous outer wall of a scleractinian *corallite,* formed by union of one or more rings of synapticulae along the axis of trabecular divergence. Cf: *septotheca; paratheca.*

synapticulum (syn-ap-tic′-u-lum) A coral synapticula. Pl: *synapticula.*

syn-biotope The overlap area of the auto-biotopes of two or more systematic units (Van Morkhoven, 1966, p.306, 309).

syncarpous (syn-car′-pous) Having carpels united, as in a plant ovary; also, said of pistils partially united within a flower (Lawrence, 1951, p.772). Cf: *apocarpous.*

synchisite (syn′-chi-site) *synchysite.*

synchronal (syn′-chro-nal) *synchronous.*

synchrone (syn′-chrone) (a) A zone representing equal time. (b) A stratigraphic surface on which every point has the same geologic age; a *time plane.*

synchroneity (syn-chro-ne′-i-ty) The state of being *synchronous* or simultaneous; coincident existence, formation, or occurrence of geologic events or features in time, such as "glacial synchroneity". Syn: *synchronism.*

synchronism (syn′-chro-nism) *synchroneity.*

synchronogenic (syn′-chro-no-gen′-ic) Formed in the same part of geologic time (R.C. Moore, 1958, p.21); e.g. said of rocks possessing identical or nearly identical geologic ages. Cf: *syntopogenic.*

synchronous (syn′-chro-nous) Occurring, existing, or formed at the same time; contemporary or simultaneous. The term is applied to rock surfaces on which every point has the same geologic age, such as the boundary between two ideal time-stratigraphic units in continuous and unbroken succession. It is also applied to growth (or depositional) faults and to plutons emplaced contemporaneously with orogenies. Cf: *isochronous; diachronous.* Syn: *synchronal.*

synchrotron micro-XANES Microanalytical version of XANES spectrometer utilizing the energy of the pre-edge absorption of a given element to determine valence states of multivalent elements. Abbrev: SmX.

synchysite (syn′-chy-site) A group name for minerals with the chemical formula $Ca(REE)(CO_3)_2F$, where REE represents a rare-earth element. The members of this group crystallize in rhombohedral, trigonal, orthorhombic or monoclinic forms. Also spelled: *synchisite.*

synchysite-(Ce) (syn′-chy-site) A vitreous yellow to brown monoclinic or rhombohedral mineral of the *synchysite* group: $Ca(Ce,La,Nd,Y)F(CO_3)_2$.

synchysite-(Nd) A light gray-blue orthorhombic or trigonal mineral of the *synchysite* group: $Ca(Nd,La)(CO_3)_2F$.

synchysite-(Y) A yellow, brown or brownish-red monoclinic mineral of the *synchysite* group: $Ca(Y,Ce)(CO_3)_2F$. It is related to parisite. Also called: yttrosynchysite. Syn: *doverite.*

synclinal (syn′-cli-nal, syn-cli′-nal) n. An obsolete form of *syncline.* adj. Pertaining to a syncline.

synclinal axis Axis of a syncline.

syncline (syn′-cline) A fold of which the core contains the stratigraphically younger rocks; it is generally concave upward. Ant: *anticline.* See also: *synform; synclinal.*

synclinoria Pl. of *synclinorium.*

synclinorium (syn-cli-no′-ri-um) A composite synclinal structure of regional extent composed of lesser folds. Cf: *anticlinorium.* See also: *geosyncline.* Pl: *synclinoria.*

syncolpate (syn-col′-pate) Said of pollen grains in which the colpi join, at either one or both poles.

syndeposition (syn′-de-po-si′-tion) A term used by Chilingar et al. (1967, p.322) for that part of *syngenesis* comprising "processes responsible for the formation of the sedimentary framework". Cf: *prediagenesis.*

syndepositional fold (syn′-de-po-si′-tion-al) A fold structure that forms contemporaneously with sedimentation. Sedimentary layers typically thin across the crests of anticlines and thicken into the troughs of synclines.

syndiagenesis (syn′-di-a-gen′-e-sis) A term used by Bissell (1959) for the sedimentational, prediastrophic phase of diagenesis, including alterations occurring during transportation (halmyrolysis) and during deposition of sediments, and continuing through the early stages of compaction and cementation but ending before that of deep burial (less than 100 m). It is characterized by the presence of large amounts of interstitial or connate water that is expelled only very slowly and by extreme variations in pH and Eh. It is equivalent to *early diagenesis.* See also: *epidiagenesis; anadiagenesis.* Cf: *syngenesis.* Adj: syndiagenetic.

syndromous load cast (syn-drom′-ous) A term used by Ten Haaf (1959, p.48) for an elongate, shallow load cast having sharp creases that combine to form a dendritic pattern, the junctures always occurring downcurrent.

syneclise (syn′-e-clise) A negative or depressed structure of the continental platform; it is of broad, regional extent (tens to hundreds of thousands of square kilometers) and is produced by slow crustal downwarp during the course of several geologic periods. The term is used mainly in the Russian literature, e.g. the Caspian syneclise. Ant: *anteclise.*

synecology (syn-e-col′-o-gy) The study of the relationships between communities and their environments. Cf: *autecology.*

syneresis (syn-er′-e-sis) A process that produces subaqueous mud cracks (Jüngst, 1984) by shrinkage in a rapidly flocculating clay layer due to compaction (White, 1961) or an increase in salinity (Burst, 1965). It is common in *thixotropic clay.*

syneresis crack A subaqueous *shrinkage crack* formed by the loss of pore water from clays or shrinkage of swelling-clay mineral lattices due to changes in salinity of surrounding water (Burst, 1965).

syneresis joint A joint that grows as a consequence of the separation of liquid from a gel caused by contraction, usually in the mud of a seabed.

syneresis vug A vug formed by syneresis, esp. in a sedimentary carbonate rock.

synform (syn′-form) Any fold whose limbs close at the bottom. The term is usually used when the folded layers do not possess a stratigraphic order, when the stratigraphic order of the folded layers is not known, or when the fold core also contains the stratigraphically older rock.

synformal anticline (syn-form′-al) An *anticline* the limbs of which close downward as in a *synform* (Turner and Weiss, 1963, p.106).

syngenesis (syn-gen′-e-sis) (a) A term introduced by Fersman (1922) for the formation, or the stage of accumulation, of unconsolidated sediments in place, including the changes affecting detrital particles still in movement in the waters of the depositional basin. The term is in dispute by Russian geologists (Dunoyer de Segonzac, 1968, p.170): some would apply it to initial diagenesis (designating exchange phenomena between fresh sediment and the sedimentary environment), others would extend it to all the transformations undergone by a sediment before its compaction. The term is equivalent to *early diagenesis* as used in the U.S. Cf: *syndiagenesis.* (b) A term used by Chilingar et al. (1967, p.322) for the "processes by which sedimentary rock components are formed simultaneously and penecontemporaneously"; it includes *syndeposition* and *prediagenesis.*

syngenetic [ore dep] (syn-ge-net′-ic) Said of a mineral deposit formed contemporaneously with, and by essentially the same processes as, the enclosing rocks. Cf: *epigenetic [eco geol]; diplogenetic.*

syngenetic [sed] (a) Said of a primary sedimentary structure, such as a ripple, formed contemporaneously with the deposition of the sediment. (b) Pertaining to sedimentary syngenesis. Cf: *epigenetic [sed].*

syngenetic karst Karst that has developed simultaneously with the lithification of eolian calcarenite or reef limestone.

syngenite (syn′-ge-nite) A colorless, white, or pale yellow monoclinic mineral: $K_2Ca(SO_4)_2{\cdot}H_2O$. It is the K analogue of koktaite.

synkinematic (syn′-kin-e-mat′-ic) *syntectonic.*

synkinematic layer Strata interval, typically overlying the *prekinematic layer*, showing local stratigraphic thickening (above structures such as withdrawal basins that subside faster then their surroundings) or thinning (above relatively rising structures). Changes in thickness can also be recorded by onlap or truncation at all levels of the synkinematic layer. The synkinematic layer records sedimentation during salt flow or during any other type of deformation (Jackson and Talbot, 1991).

synmorph (syn′-morph) A smooth trough-shaped part of an undulatory thrust fault. It does not reflect warping or deformation of a once-planar fault surface. It is typically covered by erosional remnants of the upper plate (Kopf, 1982). Cf: *antimorph.*

synneusis (syn-neu′-sis) An igneous crystallization process in which adjacent, often touching, plagioclase crystals are enveloped by growing phenocrysts of larger feldspar. Also, said of the texture of a rock showing such crystals. Etymol: Greek, "to swim together". Cf: *gregaritic.*

synnyrite (syn′-ny-rite) A kalsilite syenite containing fine-grained intergrowths of alkali feldspar with kalsilite (Sørensen, 1974, p.573). The name, given by Zhidkov in 1962, is for Synnyr in the northern Baikal region, Russia. Obsolete.

synodic month (syn-od′-ic) The time between one new moon and the next; that is, the time necessary for the moon to complete one revolution about the earth. It is equal to 29 days 12 hours 44 minutes 2.78 seconds. It is longer than a *sidereal month,* because of the orbital motion of the earth.

synonym (syn′-o-nym) One of two or more names applied to the same *taxon* (Jeffrey, 1973, p.67). See also: *synonymy.*

synonymy (syn-on′-y-my) (a) The relationship between two or more different names that have been applied to the same *taxon.* (b) A list of *synonyms* that have been applied to a particular taxon.

synoptic (syn-op′-tic) Pertaining to simultaneously existing meteorologic conditions that together give a description of the weather; also, said of a weather map or chart that shows such conditions.

synoptic oceanography Continuous gathering and reporting of simultaneous oceanographic data. It has become more important and feasible with the development of satellites.

synorogenic (syn'-o-ro-gen'-ic) Said of a geologic process or event occurring during a period of orogenic activity; or said of a rock or feature so formed. Cf: *syntectonic.*

synorogenic pluton An igneous intrusion emplaced during an interval orogenic activity. Cf: *syntectonic pluton.*

synrhabdosome (syn-rhab'-do-some) An association of graptoloid rhabdosomes, commonly biserial, attached distally by their nemas around a common center (TIP, 1970, pt. V, p.11).

synsedimentary (syn'-sed-i-men'-ta-ry) Accompanying deposition; specif., said of a sedimentary ore deposit in which the ore minerals formed contemporaneously with the enclosing rock.

synsedimentary fault *growth fault.*

syntactic (syn-tac'-tic) Recommended adj. of *syntaxy.*

syntactic growth *syntaxy.*

syntaphral (syn-taph'-ral) Descriptive of a type of tectonics involving gravity sliding of unconsolidated sediments toward the axis of a geosyncline (Carey, 1963, p. A6). Obsolete. Cf: *diataphral; apotaphral.*

syntaxial (syn-tax'-i-al) Adj. of *syntaxy.*

syntaxial rim An optically oriented crystal overgrowth of a detrital grain, developed during diagenesis.

syntaxial vein A vein that fills by growth of minerals from vein walls towards the center. Vein minerals tend to be the same as those in the wall rock. The habit of the vein minerals is often fibrous. Cf: *antitaxial vein.*

syntaxic (syn-tax'-ic) Adj. of *syntaxy.*

syntaxis (syn-tax'-is) A sharp bend in an orogenic belt, accompanied by a fraying into several strands. The term is Sollas's translation of the term "Schaarung," first used by Suess in 1901. Staub in 1928 and Bucher in 1933 used the term as an ant. of *virgation.* Current usage follows Suess's original definition.

syntaxy (syn-tax'-y) Similar crystallographic orientation in a mineral grain and its overgrowth. Adj: *syntactic; syntaxic; syntaxial.* Cf: *topotaxy; epitaxy.* Syn: *syntactic growth.*

syntectite (syn-tec'-tite) A rock formed by syntexis. Obsolete. See also: *anatexite; protectite.*

syntectonic (syn-tec-ton'-ic) Said of a geologic process or event occurring during any kind of tectonic activity; or of a rock or feature so formed. Cf: *synorogenic.* Syn: *synkinematic.*

syntectonic joints Joints associated with intense deformation, associated with mountain building processes (Bahat, 1991). These are distinguished from regional joint sets which form on stable continental platforms.

syntectonic pluton An igneous intrusion emplaced during an interval of tectonic activity. Cf: *synorogenic pluton.*

syntexis (syn-tex'-is) Obsolete term. (a) The formation of magma by melting of two or more rock types and assimilation of country rock; *anatexis* of two or more rock types. (b) Modification of the composition of a magma by *assimilation.* (c) Any kind of reaction between a rising body of magma and the crustal rocks with which it comes into contact. Adj: syntectic.

syn-thanatotope The topographical area characterized by the co-presence of dead speciments of two or more taxa, formed by the overlap of their auto-thanatotopes (Van Morkhoven, 1966, p.309).

synthem (syn'-them) A body of rocks bounded above and below by specifically designated, significant and demonstrable discontinuities of regional or interregional magnitude in the stratigraphic succession; the basic unconformity-bounded unit (ISSC, 1994, p.46-48).

synthetic adj. Said of a substance produced artificially, such as a gemstone (ruby or sapphire made by the Verneuil process), or a diamond produced by subjecting a carbonaceous material to extremely high temperature and pressure. n. Something produced by synthesis; a *synthetic stone.*

synthetic acoustic impedance log A record of acoustic impedance, velocity, or reciprocal of velocity (*transit time* or *slowness*), derived by one-dimensional inversion of sonic-log data. With added assumptions as to density values, the output is called a *synthetic sonic log* or *seismic log* (Lindseth, 1979).

synthetic-aperture radar (syn-thet'-ic-ap'-er-ture) A *SLAR* system in which high-azimuth resolution is achieved by storing and processing the return information to give the effect of a very long antenna.

synthetic fault Term introduced by H. Cloos to describe faults that dip in the same direction as the rocks displaced and that rotate fault-bounded blocks so that the net slip on each fault is less than it would be without rotation (Dennis, 1967). Many authors now use the term to describe a fault that is subsidiary to a major fault, formed in the same stress regime, is oriented at a low angle to the major fault, and (for strike-slip faults) has the same sense of displacement as the major fault or (for normal faults) dips in the same direction (Biddle and Christie-Blick, 1985). Cf: *antithetic fault.*

synthetic group A lithostratigraphic unit consisting of two or more formations that are associated because of similarities or close relationships between their fossils or lithologic characters (Weller, 1960, p.434). Cf: *analytic group.*

synthetic hydrology A catchall term for new techniques in hydrologic analysis involving the generation of hydrologic information or sequences of hydrologic events by means other than direct measurement or observation. It has been suggested that the term be abandoned and replaced by the terms *parametric hydrology* and *stochastic hydrology* (Hofmann, 1965, p.119).

synthetic ore Material that is the equivalent of, or better than, natural ore, that can be put to the same uses, and is produced by means other than ordinary concentration, calcining, sintering, or nodulizing (Thrush, 1968).

synthetic seismogram An artificial seismic reflection record, usually manufactured from velocity-log data by convolving the reflectivity function with a waveform that includes the effects of filtering by the Earth and the recording system. It is compared with an actual seismogram to aid in identifying events or in predicting how stratigraphic variation might affect a seismic record.

synthetic sonic log *synthetic acoustic impedance log.*

synthetic stone A laboratory-grown gem material that has the same physical, optical, and chemical properties, and the same chemical composition, as the natural stone that it reproduces. Many gem materials have been made synthetically as a scientific experiment, but only diamond corundum, spinel, emerald, rutile, quartz, chrysoberyl (alexandrite), opal, and turquoise have been made commercially and cut as gemstones for the jewelry trade. Cf: *imitation; reconstructed stone.* Syn: *synthetic; reproduction.*

syntopogenic (syn'-to-po-gen'-ic) Formed in the same or similar place, or denoting similar conditions of origin (R.C. Moore, 1958, p.21); e.g. said of sedimentary rocks deposited in identical or nearly identical environments in marine waters or on land. Cf: *synchronogenic.*

syntype (syn'-type) Any of the specimens on which the description of a species or subspecies is based when no *holotype* has been designated. Less preferred syn: *cotype.*

synusia (syn-u'-sia) A subdivision of an ecologic community or habitat having a characteristic life pattern or uniform conditions. Pl: synusiae. Adj: synusial.

syphon (sy'-phon) *siphon [hydraul].*

syrinx (syr'-inx) A tube of secondary shell located medially on the ventral side of the delthyrial plate of certain brachiopods (as in *Syringothyris*) and split along its ventral and anterior surface. Pl: syringes or syrinxes.

sysertskite (sys-ert'-skite) A variety of iridosmine containing 50-80% osmium (or 20-50% iridium). Syn: *siserskite.*

system [chem] (sys'-tem) (a) Any portion of the material universe that can be isolated completely and arbitrarily from the rest for consideration of the changes that may occur within it under varied conditions. (b) A conceptual range of compositions defined by a set of components in terms of which all compositions in the system can be expressed, e.g. the system CaO - MgO - SiO_2. Ricci has proposed to distinguish this sense from the ordinary thermodynamic one by capitalizing the word.

system [cryst] *crystal system.*

system [geol] A group of related natural features, objects, or forces; e.g. a *drainage system* or a *mountain system.*

system [stratig] (a) The formal chronostratigraphic unit of rank next lower than *erathem* and above *series.* Rocks encompassed by a system represent a time span and an episode of Earth history sufficiently great to serve as a worldwide reference unit (NACSN, 1983, Art. 72). The temporal equivalent of a system is a period. The system is the fundamental unit of chronostratigraphic classification of Phanerozoic rocks.

systematic error (sys-tem-at'-ic) Any *error* that persists and cannot be considered as due entirely to chance, or an error that follows some definite mathematical or physical law or pattern and that can be compensated, at least partly, by the determination and application of a correction; e.g. an error whose magnitude changes in proportion to known changes in observational conditions, such as an error caused by the effects of temperature or pressure on a measuring instrument or on the object to be measured. Ant: *random error.* See also: *constant error; instrument error.*

systematic fractures Fractures belonging to a set that maintain consistent characteristics (e.g., orientation, morphology, mineralization, distribution) and can be traced from one outcrop to the next, on both a local and regional scale (Hodgson, 1961).

systematic joints Joints that are parallel and thus occur as one set. Their surfaces are planar or only broadly curved, they are oriented perpendicular to the boundaries of the enclosing rock unit, and the structures on their faces are oriented. Cf: *nonsystematic joints.*

systematic paleontology That branch of paleontology involving the classification and treatment of fossil forms according to their taxonomic position or order. See also: *systematics; taxonomy.*

systematics (sys-tem-at'-ics) The study of the types and diversity of organisms and their relationships. Cf: *taxonomy; classification [taxon].*

system-group A unit of the highest rank in the category of lithostratigraphic units (Wang, 1966).

systems tract A rock body that groups several intergradational and, in part, contemporaneous depositional systems; for example, fluvial, delta, shelf, and slope systems (Brown and Fisher, 1977, p.215). In sequence stratigraphy, linkage of contemporaneous depositional systems. Systems tracts are composed of parasequences and one or more parasequence sets forming the subdivisions of a sequence. Lowstand, transgressive, and highstand systems tracts are recognized. Systems tracts are interpreted based on parasequence stacking patterns, positions within the sequence, stratal geometry, and types of bounding surfaces. Systems tracts are not defined based on an inferred relationship with a cycle of sea level (Van Wagoner et al., 1987, 1990; Van Wagoner, 1995).

syzygial (sy-zyg'-i-al) (a) Pertaining to *syzygy;*e.g. a "syzygial pair" consisting of two crinoid ossicles joined by syzygy. (b) Pertaining to zygosis in sponges.

syzygy [paleont] (a) Ligamentary articulation of crinoid plates with opposed joint faces bearing numerous fine culmina that radiate from the axial canal, the culmina meeting one another instead of fitting into crenellae. It forms a single segment and allows very slight mobility of joined ossicles in all directions. (b) The segment formed by syzygy.

syzygy tide *spring tide.*

szaibelyite (szai'-bel-yite) A white to yellowish acicular monoclinic mineral occurring in nodular masses: $MgBO_2(OH)$. It is isomorphous with sussexite. Also spelled: szájbelyite; sjájbelyite. Syn: *ascharite.*

szaskaite (szas'-ka-ite) *smithsonite.*

szenicsite An adamantine dark green orthorhombic mineral: $Cu_3MoO_4(OH)_4$.

szmikite (szmik'-ite) A dirty white to rose-red monoclinic mineral of the *kieserite* group: $Mn^{2+}SO_4{\cdot}H_2O$.

szomolnokite (szo-mol'-nok-ite) A yellow, reddish-brown, blue, or colorless monoclinic mineral of the *kieserite* group: $Fe^{2+}SO_4{\cdot}H_2O$.

sztrokayite (sztro'-kay-ite) A metallic mineral: Bi_3TeS_2. It has not been approved by the International Mineralogical Association.

szymanskiite A blue-gray to blue-green hexagonal mineral: $Hg_{16}(Ni,Mg)_6(CO_3)_{12}(OH)_{12}(H_3O)_8{\cdot}3H_2O$.

T

taaffeite (taaffe′-ite) A group name for minerals with the formula $(Mg,Fe)_2BeAl_6O_{12}$. The minerals in this group crystallize in hexagonal or trigonal forms.
tabasheer (tab-a-sheer′) Translucent to opaque and white to bluish-white opaline silica of organic origin (deposited within the joints of the bamboo), valued in the East Indies as a medicine and used in native jewelry. Var: tabaschir; tabashir.
tabbyite (tab′-by-ite) A variety of solid *asphalt* found in veins in Tabby Canyon, Utah.
tabella (ta-bel′-la) One of several small subhorizontal plates in the central part of a *corallite,* forming part of an *incomplete tabula.* Pl: tabellae.
Taber ice (Ta′-ber) Ice films, seams, lenses, pods, or layers, generally 1-150 mm thick, that grow in the ground by drawing-in water as the ground freezes (Péwé, 1976, p.37). Named for Stephen Taber (1882-1963), American geologist. The preferred term is *segregated ice.* See also: *epigenetic ice.* Syn: *sirloin-type ice.*
tabetification (ta-bet′-i-fi-ca′-tion) The process of forming a *talik* (Bryan, 1946, p.640).
tabetisol (ta-bet′-i-sol) *talik.*
Tabianian (Ta-bi-an′-i-an) A European regional stage: lower Pliocene (above Messinian, below Plaisancian). Equivalent to *Zanclean* global stage.
table [gem] (ta′-ble) (a) The large facet that caps the crown of a faceted gemstone. In the standard round brilliant, it is octagonal in shape and is bounded by eight star facets. (b) *table diamond.*
table [geomorph] (a) The flat summit of a mountain, such as one capped with horizontal sheets of basalt. (b) A term used in the western U.S. for *tableland.*
table cut (a) An early style of fashioning diamonds, in which opposite points of an octahedron were ground down to squares to form a large culet and a larger table, and the remaining parts of the eight octahedral faces were polished. (b) A term sometimes used loosely to describe any one of the variations of the *bevel cut,* provided it has the usual large table of that cut.
table diamond A relatively flat, table-cut diamond. Syn: *table [gem].*
table iceberg *tabular iceberg.*
tableknoll (ta′-ble-knoll) *guyot.*
tableland (ta′-ble-land) (a) A general term for a broad, elevated region with a nearly level or undulating surface of considerable extent; e.g. South Africa. (b) A plateau bordered by abrupt clifflike edges rising sharply from the surrounding lowland; a mesa.
tablemount (ta′-ble-mount) *guyot.*
table mountain A mountain having a comparatively flat summit and one or more precipitous sides. See also: *mesa.*
table mountain [volc] A flat-topped and steep-sided intraglacial volcano composed typically of pillow lava overlain by hyaloclastite which, in turn, is overlain by subaerial sheet lava (Sigvaldason, 1968). Syn: *tuya.* Cf: *tindar.*
table reef A small, isolated, flat-topped organic reef, with or without islands, that does not enclose a lagoon. Cf: *patch reef; platform reef.*
tablet (tab′-let) (a) A tabular crystal. (b) A table-cut gem.
tabula (tab′-u-la) (a) One of the transverse and nearly plane or upwardly convex or concave partitions (septa) of a corallite, extending to the outer walls or occupying only the central part of the *corallite.* See also: *complete tabula; incomplete tabula.* (b) A subhorizontal perforate plate in the *intervallum* of an archaeocyathid, extending from one septum to another or in some genera supplanting the septa. (c) A six-sided crystalline element of a heterococcolith, with two dimensions equal and the third smaller. Pl: tabulae.
tabular (tab′-u-lar) (a) Said of a feature having two dimensions that are much larger or longer than the third, such as an igneous dike, or of a geomorphic feature having a flat surface, such as a plateau. (b) Said of the shape of a sedimentary body whose width/thickness ratio is greater than 50 to 1, but less than 1,000 to 1 (Krynine, 1948, p.146); e.g. a graywacke formation in a geosynclinal deposit. Cf: *blanket; prism [sed].* (c) Said of a sedimentary particle whose length is 1.5-3 times its thickness (Krynine, 1948, p.142). Cf: *prismatic.* (d) Said of a crystal form that shows one dimension markedly smaller than the other two. (e) Said of a metamorphic texture in which a large proportion of grains are tabular and have approximately parallel orientation (Harte, 1977).
tabula-rasa theory (ta′-bu-la-ra′-sa) A theory according to which during the Pleistocene Epoch the entire Scandinavian peninsula became covered with ice and its fauna and flora were completely destroyed, and subsequent immigration from central Europe, England, and Siberia produced an entirely new biota (Dahl, 1955, p.1500). Etymol: Latin, "a blank tablet".
tabular berg (tab′-u-lar) *tabular iceberg.*
tabular cross-bedding Cross-bedding in which the cross-bedded units, or sets, are bounded by planar, essentially parallel surfaces, forming a tabular body; e.g. *torrential cross-bedding.*
tabular dissepiment A nearly flat plate extending across an entire scleractinian *corallite* or confined to its axial part.
tabular iceberg A flat-topped iceberg that may be very large (up to 160 km long and more than 500 m thick), with clifflike sides. Tabular icebergs are usually detached from an ice shelf and common in the Antarctic. Cf: *ice island.* Syn: *tabular berg; table iceberg.*
tabularium (tab-u-lar′-i-um) The axial part of the interior of a *corallite* in which tabulae are developed. Cf: *marginarium.*
tabular spar *wollastonite.*
tabular structure The structure of a mineral or rock that makes it tend to separate into plates or laminae.
tabulate (tab′-u-late) adj. (a) Having tabulae; specif. said of a coral characterized by prominent tabulae. (b) Having plates; e.g. said of phaeodarian radiolarians having smooth plates, or of dinoflagellate thecae having armored plates. n. Any zoantharian belonging to the order Tabulata. Their stratigraphic range is Middle Ordovician to Permian, possibly also Triassic to Eocene.
tabulation [palyn] The pattern according to which the constituent plates of a dinoflagellate *theca* or *cyst* are arranged (Evitt, 1970; Williams et al., 2000, p.223).
tacharanite (ta-char′-a-nite) A white monoclinic mineral: $Ca_{12}Al_2Si_{18}O_{33}(OH)_{36}$.
tacheometer (tache-om′-e-ter) *tachymeter.*
tachygenesis (tach-y-gen′-e-sis) The extreme crowding and eventual loss of those primitive phylogenetic stages that are represented early in the life of an individual. Cf: *acceleration.*
tachygraphometer (tach′-y-graph-om′-e-ter) A tachymeter with alidade for surveying.
tachyhydrite (tach-y-hy′-drite) A colorless or yellow rhombohedral mineral: $CaMg_2Cl_6{\cdot}12H_2O$. Also spelled: tachydrite; tachhydrite.
tachylyte (tach′-y-lyte) *basaltic glass.*
tachymeter (ta-chym′-e-ter) A surveying instrument designed for use in the rapid determination from a single observation of the distance, direction, and elevation difference of a distant object; esp. a transit or theodolite with stadia hairs, or an instrument in which the base line for distance measurements is an integral part of the instrument. Tachymeters include range finders with self-contained bases, although range finders do not usually afford the means for the determination of elevation. Modern versions in the United States, usually with electronic readout, are often called total stations. Syn: *tacheometer.*
tachymetry (ta-chym′-e-try) A method of rapid surveying using the tachymeter; e.g. the stadia method of surveying used in U.S.
tachytely (tach′-y-tel′-y) A phylogenetic phenomenon characterized by a rapid temporary spurt in evolution that occurs as populations shift from one major zone of adaptation to another; episodic evolution. Cf: *bradytely; horotely; lipogenesis.*
Taconian Orogeny (Ta-co′-ni-an) *Taconic Orogeny.*
Taconic Orogeny (Ta-con′-ic) An orogeny in the latter part of the Ordovician period, named for the Taconic Range of eastern New York State, well developed through most of the northern Appalachians in the U.S. and Canada. In places it can be strictly defined as Late Ordovician by limiting fossiliferous strata, or it can be extended to include many pulsations that occurred from place to place from early in the Ordovician to early in the Silurian. Syn: *Taconian Orogeny.*

taconite (tac'-o-nite) (a) A local term used in the Lake Superior iron-bearing district of Minnesota for any bedded ferruginous chert or variously tinted jaspery rock, esp. one that enclosed the Mesabi iron ores (granular hematite); an unleached *iron formation* containing magnetite, hematite, siderite, and hydrous iron silicates (greenalite, minnesotaite, and stilpnomelane). The term is specifically applied to this rock when the iron content, either banded or disseminated, is at least 25%, so that natural leaching can convert it into a low-grade iron ore, with 50 to 60% iron. (b) Since World War II, a low-grade iron formation suitable for concentration of magnetite and hematite by fine grinding and magnetic treatment, from which pellets containing 62 to 65% iron can be manufactured.

tactical intelligence Mapping of surficial geologic environments for purposes of military analysis (Eastler, 1994).

tactite (tac'-tite) A rock of complex mineralogical composition, formed by contact metamorphism and metasomatism of carbonate rocks. It is typically coarse grained and rich in garnet, iron-rich pyroxene, epidote, wollastonite, and scapolite. Cf: *garnetite.* Approx. syn: *skarn.*

tactoid (tac'-toid) A spindle-shaped body, e.g. in vanadium pentoxide sol, visible under a polarizing microscope.

tactosol (tac'-to-sol) A sol that contains tactoids in a spontaneous, parallel arrangement.

tadpole nest (tad'-pole) A small, irregular *cross ripple* characterized by a polygonal or cell-like pattern, formed by the intersection of two sets of ripples approximately at right angles and once believed to have been made by tadpoles. The height of the ripple is considerably greater than in the equivalent form associated with *transverse ripple.* Term introduced by Hitchcock (1858, p.121-123).

tadzhikite-(Ce) (tad'-zhik-ite) A grayish brown, reddish brown, or dark brown monoclinic mineral: $Ca_3(Ce,Y)_2(Ti,Al,Fe)B_4Si_4O_{22}$.

taele (a) Older form of the Norwegian term *tele.* (b) Anglicized version of the Swedish term "tjäle"; see *tjaele.*

taenia [paleont] (tae'-ni-a) An irregularly bent small plate in the *intervallum* of an archaeocyathid (TIP, 1955, pt. E, p.7).

taenia [palyn] A straplike srip of inflated ektexine with intrareticulate sculpture like that of the sacci, characteristic for pollen of the *Striatiti.* Two or as many as a dozen taeniae may run, parallel to each other, between the proximal bases of the sacci. Pl: taeniae.

taeniolite (tae'-ni-o-lite) *tainiolite.*

taenite (tae'-nite) A metallic white cubic meteorite mineral consisting of the face-centered cubic gamma-phase of a *nickel-iron* alloy, with a nickel content ranging from about 27% up to 65%. It occurs in iron meteorites as lamellae or strips flanking bands of kamacite.

tafelberg (ta'-fel-berg) A term used in South Africa for a mesa or a table mountain; a large *tafelkop.* Etymol: Afrikaans.

tafelkop (ta'-fel-kop) A term used in South Africa for an isolated hill with a flat top; a butte. Etymol: Afrikaans. Cf: *spitskop.*

tafone (ta-fo'-ne) (a) A Corsican dialect term for one of the natural cavities in a honeycomb structure, formed by *cavernous weathering* on the face of a cliff in a dry region or along the seashore. The hole or recess may reach a depth of 10 cm, and is explained as due to solution of free salts in crystalline rock (granite, gneiss) following heating by insolation. (b) A granitic or gneissic block or boulder hollowed out by cavernous weathering. Pl: tafoni.

tafrogeny *taphrogeny.*

tagamite A melt-supported clastic breccia. First defined by Masaitis et al. (1975) at the Popigai structure in Siberia, this term is widely used in the Russian literature. Syn: *impact-melt breccia.*

Taghanican (Ta-ghan'-i-can) North American stage: uppermost Middle Devonian (above Tioughniogan, below Fingerlakesian).

tagilite (tag'-i-lite) *pseudomalachite.*

tagma (tag'-ma) A major division of the body of an arthropod, each composed of several somites; e.g. cephalon, thorax, and abdomen. Pl: tagmata.

tahitite (ta-hi'-tite) A feldspathoid-bearing *tephritic phonolite* that contains hauyne phenocrysts and generally more abundant sodic plagioclase than potassium feldspar. Its name, given by Lacroix in 1917, is derived from Tahiti. Obsolete.

tahoma (ta-ho'-ma) (a) A generic term indicating a high snowy or glacier-clad mountain in the NW U.S., like Mount Hood. (b) The V-shaped residual ridge between two cirques on a glacially carved volcanic cone in the NW U.S., e.g. one of several on Mount Rainier (Russell, 1898a, p.382-383). Etymol: An approximation of one of several Indian names for Mount Rainier; presumably the higher the mountain, the longer the second syllable was drawn out, as "ta-hoooom-ah".

TAI *thermal alteration index.*

taiga (tai'-ga) A swampy area of coniferous forest sometimes occurring between *tundra* and *steppe* regions. Etymol: Russian.

taikanite (tai'-kan-ite) A vitreous greenish-black monoclinic mineral: $BaSr_2Mn^{3+}_2O_2Si_4O_{12}$.

tail [coast] (a) A *bar* or *barrier* formed behind a small island or *skerry.* Syn: *trailing spit.* (b) The outermost part of a projecting bar.

tail [glac geol] A small ridge of rock, or an accumulation or streak of till, tapering down-glacier from the lee side of a knob of resistant rock. Cf: *crag and tail.*

tail [paleont] Any of various backwardly directed and usually posterior parts on the body of an invertebrate, esp. if attenuated.

tail [sed] The rear part of a turbidity current, which is less dense than the nose and moves more slowly.

tail [streams] (a) The downstream section of a pool or stream. (b) The comparatively calm water after a current or a reach of rough water.

tail coccolith A modified *coccolith* found at the end opposite the flagellar field in flagellate coccolithophores exhibiting dimorphism (such as *Calciopappus*). Cf: *pole coccolith.*

tail crack The tip of a joint that has kinked out of the plane of its parent joint. Cf: *kink [struc geol].*

tail dune A dune that accumulates on the leeward side of an obstacle and that tapers away gradually for a distance of up to 1 km. Cf: *head dune.*

tail fan *caudal fan.*

tailing (tail'-ing) adj: *leggy.*

tailings (tail'-ings) Those portions of washed or milled ore that are regarded as too poor to be treated further, as distinguished from the *concentrate,* or material of value.

tail water The water downstream from a structure, as below a dam.

taimyrite [mineral] (tai'-myr-ite) A metallic bronze-gray orthorhombic mineral: $(Pd,Cu,Pt)_3Sn$.

taimyrite [rock] (tai'-myr-ite) A nosean-bearing *phonolite.* It was named by Chrustschoff in 1894 after the Taimyr River, Siberia. Obsolete.

Taimyr polygon (Tai'-myr) An *ice-wedge polygon,* so-called from its occurrence in northern Siberia.

tainiolite (tai'-ni-o-lite') A greenish-brown monoclinic or trigonal member of the *mica* group: $KLiMg_2Si_4O_{10}F_2$. Syn: *taeniolite.*

takanelite (ta-ka-nel'-ite) A submetallic brownish-gray to black hexagonal mineral: $(Mn^{2+},Ca)_{0.2}Mn^{4+}O_2{\cdot}0.7H_2O$.

Takaraulian (Tak-a-rul'-i-an) Ponto-Caspian stage: upper lower Miocene (above Sakaraulian, below Chokrakian). Also Tarchanian; equiv. Kozachurian.

takedaite A grayish-white rhombohedral mineral: $Ca_3B_2O_6$.

takeout A connection point to a multiconductor cable where geophones or geophone flyers can be connected. Takeouts are usually polarized to reduce the likelihood of making the connection backwards.

takéuchiite (tak-é-u'-chi-ite) A metallic black orthorhombic mineral: $(Mn^{2+},Mg)_2Mn^{3+}BO_5$. Cf: *pinakiolite.*

takir A clay-silt playa (Cooke and Warren, 1973, p.215). Etymol: Russian. Also spelled: *takyr.*

takovite (ta'-kov-ite) A yellow-green rhombohedral mineral of the *hydrotalcite* group: $Ni_6Al_2(CO_3,OH)(OH)_{16}{\cdot}4H_2O$. Syn: *eardleyite.*

takyr *takir.*

tala (ta'-la) A term introduced by Berkey and Morris (1924, p.105) for a broad structural basin of internal drainage, formed in the Gobi Desert by subsidence or warping and bounded by inconspicuous divides or mountain ranges. Cf: *gobi.* Etymol: Mongolian, "open steppe-country".

talc (a) An extremely soft, light green or gray monoclinic mineral: $Mg_3Si_4O_{10}(OH)_2$. It has a characteristic soapy or greasy feel and a hardness of 1 on the Mohs scale, and it is easily cut with a knife. Talc is a common secondary mineral derived by alteration (hydration) of nonaluminous magnesium silicates (such as olivine, enstatite, and tremolite) in basic igneous rocks, or by metamorphism of dolomite rocks, and it usually occurs in foliated, granular, or fibrous masses. Talc is used as a filler, coating, and dusting agent, in ceramics, rubber, plastics, lubricants, and talcum powder. Originally spelled: talck. See also: *steatite.* (b) In commercial usage, a talcose rock; a rock consisting of talc, tremolite, chlorite, anthophyllite, and related minerals. (c) A thin sheet of muscovite mica.

talcite (talc'-ite) (a) A massive variety of talc. (b) *damourite; muscovite.*

talcoid (talc'-oid) (a) A discredited mineral reported as: $Mg_3Si_5O_{12}(OH)_2$. It is possibly a mixture of talc and quartz. (b) Resembling talc; e.g. "talcoid schist".

talcose (talc'-ose) (a) Pertaining to or containing talc; e.g. "talcose schist". (b) Resembling talc; e.g. a "talcose rock" that is soft and soapy to the touch.

talc schist A schist in which talc is the dominant schistose mineral. Common associates are mica and quartz.

talc slate An impure, hard, slaty variety of talc, a little harder than French chalk; "indurated talc".

taleola (ta-le-o'-la) A cylinder or rod of granular, nonfibrous calcite in the axial region of some pseudopunctae of brachiopods. Pl: taleolae.

talet A term used in the High Atlas of Morocco for a dried-out torrential gully (Termier and Termier, 1963, p.415). Etymol: Berber.

talik A Russian term for a layer or body of unfrozen ground above, within, or beneath permafrost; occurs in regions of *discontinuous permafrost.* It may be permanent or temporary. See also: *subgelisol; suprapermafrost layer.* Syn: *tabetisol.*

talmessite (tal'-mes-site) A colorless or pale green triclinic mineral of the *fairfieldite* group: $Ca_2Mg(AsO_4)_2{\cdot}2H_2O$. Syn: *belovite.*

talnakhite (tal-na'-khite) A metallic yellow cubic mineral: $Cu_9(Fe,Ni)_8S_{16}$.

talpatate (tal-pa-ta'-te) (a) A surficial rock formed by the cementing action of calcium carbonate on sand, soil, or volcanic ash, and partly equivalent to caliche. See also: *tepetate.* (b) A poor, thin soil consisting of partly decomposed volcanic ash that is more or less consolidated. Etymol: American Spanish, from Nahuatl (Aztec) "tepetatl", "stone matting". Also spelled: talpetate.

talus [geol] (ta'-lus) Rock fragments of any size or shape (usually coarse and angular) derived from and lying at the base of a cliff or very steep, rocky slope. Also, the outward sloping and accumulated heap or mass of such loose broken rock, considered as a unit, and formed chiefly by gravitational falling, rolling, or sliding. Cf: *alluvial talus; avalanche talus; rockfall talus.* See also: *scree.* Syn: *rubble.* Etymol: French "talu", later "talus", "a slope", originally in the military sense of fortification for the outside of a rampart or sloping wall whose thickness diminishes with height; from Latin "talutium", a gossan zone or slope indicative of gold (probably of Iberian origin). Pl: taluses.

talus [reef] *reef talus.*

talus apron A poorly sorted but well-bedded accumulation of coarse *reef detritus,* usually much larger in volume than the parent reef, with a surface inclination up to 40°.

talus breccia A *breccia* formed by the accumulation and consolidation of *talus.*

talus cave A cave consisting of the interconnected spaces between blocks of talus.

talus cone A small cone-shaped or apron-like landform at the base of a cliff, consisting of poorly sorted rock debris that has accumulated episodically by rockfall or alluvial wash. Also, a similar feature of fluvial origin, tapering up into a gully. Cf: *alluvial talus.*

talus creep The slow downslope movement of talus, either individual rock fragments or the mass as a whole. Cf: *scree creep.*

talus fan *alluvial fan.*

talus slope A steep, concave slope formed by an accumulation of loose rock fragments; esp. such a slope at the base of a cliff, formed by the coalescence of several rockfall taluses or alluvial taluses; the surface profile of an accumulation of talus. Etymologically, "talus slope" is a tautology, as the term "talus" originally signified a slope, although incorrectly used for the material composing the slope. See also: *scree.*

talus spring A spring occurring at the base of a *talus slope* and originating from water falling upon or seeping into the slope.

taluvium (ta-lu'-vi-um) A term introduced by Wentworth (1943) for a detrital cover consisting of talus and colluvium; the fragments range from large blocks to silt. Obsolete.

talweg (tal'-weg) Modern spelling of *thalweg.*

tamaite A colorless or pale yellowish-brown monoclinic mineral: $(Ca,K,Ba,Na)_{3.6}Mn_{24.2}(Si,Al)_{40}O_{95.3}(OH)_{16.6}\cdot21H_2O$.

Tamanian (Tam-an'-ian) Eastern European mammal age ("complex"): early Pleistocene (after Odessan, before Tiraspolian); equiv. Betfian.

tamaraite (ta-ma'-ra-ite) A dark-colored hypabyssal *lamprophyre* containing augite, amphibole, biotite, nepheline, plagioclase, alkali feldspar, minor accessories, and secondary cancrinite and analcime; a nepheline-rich *camptonite.* The name, given by Lacroix in 1918, is for Tamara Island, Los Archipelago, Guinea. Obsolete.

tamarugite (tam-a-ru'-gite) A colorless monoclinic mineral: $NaAl(SO_4)_2\cdot6H_2O$. It was originally named *lapparentite.*

tamping The act or an instance of filling up a drill hole above a blasting charge with moist, loose material (mud, clay, earth, sand) in order to confine the force of the explosion to the lower part of the hole. Also, the material used in tamping. Cf: *stemming.*

tancoite (tan'-co-ite) A colorless or pale pink orthorhombic mineral: $HNa_2LiAl(PO_4)_2(OH)$.

taneyamalite (ta-ne-yam'-a-lite) A vitreous greenish-yellow triclinic mineral, the Mn^{2+}-analogue of howieite: $Na(Mn^{2+},Mg,Fe^{2+})_{12}Si_{12}(O,OH)_{44}$.

tang A Scottish term for a low, narrow cape.

tangaite *redondite.*

tangeite (tange'-ite) A grayish green, yellowish green or olive green orthorhombic mineral of the adelite group: $CaCuVO_4(OH)$. Syn: *calciovolborthite.* Also spelled: tangueite; tangeïte.

tangent (tan'-gent) n. (a) A straight line that touches, but does not transect, a given curve or surface at one and only one point; a line that touches a circle and is perpendicular to its radius at the point of contact. (b) The part of a traverse included between the point of tangency (the point in a line survey where a circular curve ends and a tangent begins) of one curve and the point of curvature (the point in a line survey where a tangent ends and a circular curve begins) of the next curve. (c) A great-circle line that is tangent to a parallel of latitude at a township corner in the U.S. Public Land Surveys system. (d) A term sometimes applied to a long straight line of a traverse whether or not the termini of the line are points of curve. (e) The ratio of the length of the leg opposite an acute angle in a right-angled triangle to the length of the leg adjacent to the angle. adj. Said of a line or surface that meets a curve or surface at only one point.

tangential cross-bedding (tan-gen'-tial) Cross-bedding in which the foreset beds appear in section as smooth arcs meeting the underlying surface at low angles; large-scale tangential cross-bedding is commonly believed to imply deposition by wind. Cf: *angular cross-bedding.*

tangential longitudinal strain A pattern of intra-layer deformation in which the inner arc of a folded layer is shortened parallel to layering and the outer arc of the folded layer is elongated parallel to layering. A neutral surface or neutral line, where layering is neither shortened nor elongated parallel to layering, separates the shortened inner arc and extended outer arc.

tangential ray A sponge-spicule ray that lies approximately parallel to the surface of the sponge.

tangential section (a) A slice through part of a foraminiferal test parallel to the axis of coiling or growth but not through the proloculus. (b) A section at a right angle to the zooids in the exozone of stenolaemate bryozoans, generally parallel to and just under the surface of a colony. (c) A section of a cylindrical organ (such as a stem) cut lengthwise and at right angles to the radius of the organ.

tangential stress *shear stress.*

tangential wave *S wave.*

tangent law of refraction In steady-state groundwater flow systems, flow lines of water refract at the interface between materials of contrasting permeability. The refraction patterns obey a relation similar to Snell's law of light refraction, except that the relation involves tangents rather than sines.

tangent screw A very fine, slow-motion screw giving a tangential movement for making the final setting to a precision surveying instrument (such as for completing the alignment of sight on a theodolite or transit by gentle rotation of the reading circle about its axis).

tangi A term used in Baluchistan for a narrow, transverse *gorge* or *cleft* through which a stream penetrates a longitudinal ridge. Etymol: Persian "tang", "narrow".

tangiwai A term used by the Maoris of New Zealand for *bowenite.* Also spelled: tangiwaite; tangawaite.

tangle sheet (tan'-gle) Mica with intergrowths of crystals or laminae resulting in books that split well in some places but tear to produce a large proportion of partial films (Skow, 1962, p.170).

tangue (tong) Mud and sand occurring in the shallow bays along the coast of Brittany (NW France), consisting in part of a fluviolacustrine silt reworked by recent marine transgressions and in part of finely powdered molluscan shell material transported and deposited by the tides. It is coherent (even when dry), and has a permeability similar to that of fine sand.

tank (a) A term applied in SW U.S. to a natural depression or cavity in impervious rocks (usually crystalline) in which rainwater, floodwater, snowmelt, and seepage are collected and preserved during the greater part of the year. See also: *rock tank; charco.* (b) A natural or artificial pool, pond, or small lake occupying a tank; esp. an artificial reservoir for supplying water for livestock. (c) A term used in Ceylon and the drier parts of peninsular India for an artificial pond, pool, or lake formed by building a mud wall across the valley of a small stream to retain the monsoon rainwater.

tannbuschite (tann'-busch-ite) A rarely used term applied to dark-colored *olivine nephelinite* or *ankaratrite.* Johannsen (1938) coined the name after Tannbusch, Czech Republic. Obsolete.

tannin Reddish-brown organic compounds found in leaves.

tantalaeschynite-(Y) (tan-tal-aesch'-y-nite-y) A resinous brownish-black or black orthorhombic mineral of the *aeschynite* group: $(Ca,Y,Ce)(Ta,Ti,Nb)_2O_6$.

tantalcarbide (tan-tal-car'-bide) A metallic bronze cubic mineral: TaC.

tantalite (tan'-ta-lite) A sub-group name for minerals of the *columbite* group with Ta>Nb.

tantalum The element Ta. As a naturally occurring mineral it has been found in a placer.

tanteuxenite-(Y) (tant-eux'-e-nite) A brownish-black orthorhombic mineral of the *euxenite* group: $(Y,Ce,Ca)(Ta,Nb,Ti)_2(O,OH)_6$. It the Ta analogue of euxenite. Syn: *delorenzite; eschwegeite.*

tantite (tan'-tite) An adamantine colorless triclinic mineral: Ta_2O_5.

tanzanite (tan'-za-nite) A sapphire-blue gem variety of zoisite that exhibits strong pleochroism.

tape A continuous ribbon or strip of steel, invar, dimensionally stable alloys, specially made cloth, or other suitable material, having a constant cross section and marked with linear graduations, used by surveyors in place of a chain for the measurement of lengths or distances.

tape correction A quantity applied to a taped distance to eliminate or reduce errors due to the physical condition of the tape and the the manner in which it is used; e.g. a correction based on the length, temperature, tension, or alignment of the tape. See also: *sag correction; slope correction.*

tape person *chainperson.*

tapetum (ta-pe'-tum) Tissue of nutritive cells in the sporangium of embryophytic plants, digested during development of the spores. In angiosperms, it is the inner wall of the anther locules and provides nutritive substances for the developing pollen. Pl: tapeta.

taphocoenose (taph-o-coe'-nose) *thanatocoenosis.*

taphocoenosis (taph'-o-coe-no'-sis) *thanatocoenosis.*

taphoglyph (taph'-o-glyph) A *hieroglyph* representing the impression of the body of a dead animal (Vassoevich, 1953, p.72).

taphonomy (ta-phon'-o-my) The branch of paleoecology concerned with all processes occurring after the death of an organism until its discovery. It includes *biostratonomy* and *fossil diagenesis* (Behrensmeyer and Kidwell, 1985). Syn: *para-ecology.*

taphrogenesis (taph-ro-gen'-e-sis) *taphrogeny.*

taphrogenic (taph-ro-gen'-ic) Adj. of *taphrogeny.*

taphrogeny (taph-rog'-e-ny) A general term for the formation of rift phenomena, characterized by high-angle normal faulting and associated subsidence (Krenkel, 1922). It is often considered to be the first stage of continental rupture and plate separation. Etymol: Greek, "taphre", "trench". Also spelled: *tafrogeny.* Adj: *taphrogenic.* Syn: *taphrogenesis.*

taphrogeosyncline (taph'-ro-ge'-o-syn'-cline) A geosyncline developed as a rift or trough between faults (Kay, 1945). Obsolete. Cf: *aulacogen.*

taphrolith (taph'-ro-lith) A trough-shaped lava flow, poured out along the bounding faults into a trough or graben. Obsolete.

taping (tap'-ing) The operation of measuring distances on the ground by means of a surveyor's tape. Cf: *chaining.*

tapiolite (tap'-i-o-lite') A group name for tetragonal minerals with the general formula AB_2O_6, where $A = Fe^{2+}$, Mn^{2+}, Mg or Zn, and B = Nb, Ta or Sb.

tapoff passage A relatively short, high-gradient cave-stream passage, with dew or no tributaries, that pirates flow from a high-elevation stream to a lower stream along a locally steep and hydrologically more efficient gradient (Mylroie, 1975; Veni, 1997).

tapoon (ta-poon') A subsurface dam built in a dry wash, either to increase recharge to nearby wells or to impound water for direct use. In the latter case, a pipe is run from the dam to the point of use.

tar A thick brown to black viscous organic liquid, free of water, which is obtained by condensing the volatile products of destructive distillation of coal, wood, oil, etc. It has a variable composition, depending on the temperature and material used to obtain it.

taramellite (tar-a-mel'-lite) (a) A bronze-purple or reddish-brown platy orthorhombic mineral: $Ba_4(Fe^{3+},Ti,Fe^{2+},Mg)_4(B_2Si_8O_{27})O_2Cl_x$. (b) A group name for minerals with the analogous formula, but with Fe^{3+} replaced by Ti or V^{3+}.

taramite (tar'-a-mite) A black or blue-green monoclinic mineral of the *amphibole* group: $Na(NaCa)(Fe^{2+}{}_3AlFe^{3+})(Si_6Al_2)O_{22}(OH)_2$.

taranakite (tar-a-nak'-ite) A white, gray, or yellowish-white rhombohedral clayey mineral: $(K,Na)_3(Al,Fe^{3+})_5(PO_4)_2(PO_3OH)_6{\cdot}18H_2O$.

tarantulite (ta-ran'-tu-lite) A plutonic rock containing more than 50% quartz, less alkali feldspar, over half of which is potassium feldspar and the rest albite, and up to 5% dark minerals. The term was proposed by Johannsen (1920, p.54) as a substitute for *alaskite-quartz.* The rock is transitional between *alaskite* and *silexite [ign].* Its name is derived from Tarantula Spring, Nevada. Obsolete. Preferred usage is *quartz-rich granitoid.*

tarapacáite (tar'-a-pa'-ca-ite) A yellow orthorhombic mineral: K_2CrO_4.

tarasovite (ta'-ra-sov-ite) A mixed-layer clay mineral. An irregular interlayering of mica and smectite.

tarbuttite (tar'-butt-ite) A colorless, pale yellow, brownish, reddish, or greenish triclinic mineral: $Zn_2(PO_4)(OH)$. It is the phosphate analogue of paradamite. Syn: *salmoite.*

Tarchanian *Takaraulian.*

tar coal Resinous, bitumen-rich brown coal.

tare A discontinuity in data, indicating an error in measurement or computation rather than a sudden change in the quantity being measured. Sometimes spelled: *tear.*

target [photo] (tar'-get) (a) The distinctive marking or instrumentation of a ground point to aid in its identification on a photograph. It is a material marking so arranged and placed on the ground as to form a distinctive pattern over a geodetic or other control-point marker, on a property corner or line, or at the position of an identifying point above an underground facility or feature (ASP, 1975, p.2106). (b) The image pattern on an aerial photograph of the actual mark or target placed on the ground prior to photography.

target [surv] The vane or sliding sight on a surveyor's *level rod;* a device, object, or point upon which sights are made. (a) Any object at which a telescope is pointed for measurement of angle, direction, distance, height or elevation. The salient feature of a target is its marked visibility. Targets are selected or constructed to be clearly visible even when the lighting is poor. This definition makes signal and target almost synonymous. See, however, the next definition. (b) That point, on a signal, at which a telescope is pointed for measurement of angle, direction, distance, elevation and/or height. (c) A brightly painted, round disk which can be moved up and down a leveling rod until a mark on the disk is on the line of sight of the telescope in a leveling instrument. The rod person moves the disk up and down until the mark (e.g., a horizontal line) on it coincides, according to the instrument person, with the horizontal line in the telescope's reticle. The corresponding number on the leveling rod is then recorded. Targets are no longer used for precise leveling. In precise leveling, the instrument man himself reads the number at which the horizontal line on the reticle intersects the image of the rod (or interpolates to get that number). (d) photographic, target.

target rod A *level rod* with an adjustable target that is moved into position by the rodman in accordance with signals given by the man at the leveling instrument and that is read and recorded by the rodman when it is bisected by the line of collimation. Cf: *self-reading leveling rod.*

tarkianite A metallic black cubic mineral: $(Cu,Fe)(Re,Mo)_4S_8$.

tarn (a) A landlocked pool or small lake such as those that occur within tracts of swamp, marsh, bog, or muskeg in Michigan (Davis, 1907, p.116), and on moors in northern England. In unrestricted usage, "any small lake may be called a tarn" (Veatch and Humphrys, 1966, p.326). (b) A relatively small and deep, steep-banked lake or pool amid high mountains, esp. one occupying an ice-gouged rock basin amid glaciated mountains. (c) *cirque lake.* Etymol: Icelandic.

tarnish (tar'-nish) A thin film that forms on the surface of certain minerals, esp. those containing copper. Its color and luster are different from those of the fresh mineral.

tar pit An area in which an accumulation of natural bitumen is exposed at the land surface, forming a trap into which animals (esp. vertebrates) fall and sink, their hard parts being preserved in the bitumen.

tar sand A type of *oil sand* from which the lighter fractions of crude oil have escaped, leaving a residual asphalt to fill the interstices.

tarsus (tar'-sus) (a) The ankle or the ankle bones of a tetrapod. (b) The distal part of the limb of an arthropod; e.g. the last segment (sometimes subsegmented) of a leg of an arachnid, or a joint of the distal part of a prosomal appendage of a merostome. Pl: tarsi.

Tartarian (Tar-tar'-i-an) *Tatarian.*

TAS Abbrev. for total alkali-silica diagram. See: *IUGS classification.*

taseqite A clove-brown or yellowish-brown rhombohedral mineral of the *eudialyte* group: $Na_{12}Sr_3Ca_6Fe_3Zr_3NbSi_{25}O_{73}(O,OH,H_2O)_3Cl_2$.

tasmanite [coal] (tas'-ma-nite) An impure coal, transitional between cannel coal and oil shale. It resulted from massive blooms of *Tasmanites* and is mined commercially from Permian strata in Tasmania. Syn: *combustible shale; yellow coal; Mersey yellow coal; white coal.*

tasmanite [intrus rocks] An intrusive rock similar in composition to *ijolite* but containing melilite, and zeolites in place of nepheline. Among the zeolites are natrolite, thomsonite, and phillipsite. The rock is named after Tasmania (Johannsen, 1938). Obsolete.

Tasmanites (Tas'-ma-ni-tes) A *green algae* genus, belonging to the *prasinophytes.* Precambrian to present. *Tasmanites* cysts are preserved as large, spherical phycomata with a thick, perforate wall.

Tatarian (Ta-tar'-i-an) A Russian stage: middle Permian (above Kazanian, below Triassic).

tatarskite (ta-tar'-skite) A colorless or yellowish mineral: $Ca_6Mg_2(SO_4)_2(CO_3)_2Cl_4(OH)_4{\cdot}7H_2O$.

tatyanaite A metallic orthorhombic mineral: $(Pt,Pd,Cu)_9Cu_3Sn_4$. It is the Pt-dominant analogue of taimyrite.

tauactine (tau-ac'-tine) A spicule of hexactinellid sponges, having three coplanar rays of which two lie along the same axis.

tau-p mapping An unstacked seismic record or a common-midpoint gather can be described in terms of slope $dt/dx = p$ and intercept time τ, the arrival time obtained by projecting the slope back to $x = 0$, where x = source-geophone distance. The transform process is also called *slant stack,* the *Radon transform,* and *plane-wave decomposition.* Filtering can be done in the τ-p domain and the result transformed back into the time domain (Diebold and Stoffa, 1981).

tausonite (tau'-son-ite) An adamantine red, reddish-brown, or gray cubic mineral: $SrTiO_3$. It is the Sr analogue of loparite.

tautirite (tau'-ti-rite) A nepheline *trachyandesite* or hawaiite composed of potassium feldspar, andesine, nepheline, and amphibole, with abundant accessory titanite. Cf: *pollenite.* The name, given by Iddings in 1918, is for Tautira beach, Tahiti. Obsolete.

tautonym (tau'-to-nym) A *binomen* or *trinomen* in which the term designating the genus is the same as that for the species or subspecies; e.g. *Troglodytes troglodytes.* See also: *tautonymy rule.*

tautonymy rule (tau-ton'-y-my) A rule of binomial nomenclature by which a species having a name identical to the name of the genus to which it belongs automatically designates the type species of that genus. See also: *tautonym.*

tavistockite (tav'-is-tock-ite') *fluorapatite.*

tavolatite (ta-vo-la'-tite) A *phonolite,* occurring as blocks in volcanic breccia, containing large phenocrysts of leucite, hauyne, aegirine-augite, and garnet in a microphyric groundmass of the same minerals with interstitial leucite, alkali feldspar, hauyne, labradorite, aegirine-augite, biotite, garnet, and nepheline. The name, given by Washington in 1906, is for Osteria di Tavolato near Rome. Obsolete.

tavorite (ta-vor'-ite) A yellow, greenish-yellow, or green triclinic mineral of the *amblygonite* group: $LiFe^{3+}(PO_4)(OH)$.

tawite (taw'-ite) A plutonic rock, similar in composition to *ijolite,* containing 30 to 60 percent mafic minerals, but with sodalite in place of nepheline as the dominant sodium feldspathoid. The name, given by Ramsay in 1894, is from Tawojak, Kola Peninsula, Russia. Obsolete. A preferred name would be "sodalite ijolite".

tawmawite (taw'-maw-ite) A yellow or green to dark-green variety of epidote containing chromium and found in Tawmaw, upper Myanmar.

taxa (tax'-a) The plural of *taxon.*

taxadjunct A soil that is correlated as a recognized, existing soil series for the purpose of expediency. They are so like the soils of the defined series in morphology, composition, and behavior that little or nothing is gained by adding a new series.

taxis [ecol] (tax'-is) A movement or orientation of an organism with respect to a source of stimulation. Cf: *tropism.*

taxis [paleont] A definite linear series of plates in any part of the crown of a crinoid; e.g. anitaxis and brachitaxis. Pl: taxes.

taxite (tax'-ite) An obsolete term for a volcanic rock that appears to be clastic because of its mixture of materials of varying texture and structure from the same flow. See also: *ataxite [volc]; eutaxite.*

taxodont (tax'-o-dont) adj. Said of the dentition of a bivalve mollusk characterized by numerous short subequal hinge teeth forming a continuous row, with some or all of the teeth transverse to the margin of the hinge n. A taxodont mollusk; specif. a bivalve mollusk of the order Taxodonta, having numerous unspecialized hinge teeth and equally developed adductor muscles. Cf: *heterodont.*

taxon (tax'-on) A group of organisms of any rank, such as a particular species, family, or class; also, the name applied to that unit. A taxon is usually named, and may be designated, by a formal Latin name or by a letter, number, or other symbol. The term was proposed for general systematic use by H. J. Lam in 1948 (Lam, 1957, p.213; Morton, 1957, p.155; Rickett, 1958, p.37). Plural: *taxa.* See also: *parataxon.*

taxonomy (tax-on'-o-my) The theory and practice of classifying plants and animals. The terms taxonomy and *systematics* are usually distinguished, the latter having broader connotation, but they may also be used more or less synonymously. Cf: *classification [taxon].*

taxon-range zone The *interval zone* between the documented lowest and highest occurrences of a single taxon (NACSN, 1983, Art. 50). The body of strata representing the known range of occurrence (stratigraphic and geographic) of specimens of a certain taxon (species, genus, family, etc.) (ISSC, 1994, p.57-58). Cf: *concurrent-range zone; partial range zone; lineage zone.*

Tayloran (Tay'-lor-an) North American (Gulf Coast) stage: Upper Cretaceous (above Austinian, below Navarroan).

Taylor-Evitt tabulation system A way of describing the number and arrangement of plates in *dinoflagellates.* It was developed by Taylor (1979, 1980) and modified by Evitt (1985). Plates are labeled on the basis of a hypothetical, symmetrical model. Successive series are the polar series (A-C), preequatorial series (1-6), equatorial series (a-f), postequatorial series (I-VI), and antipolar series (X-Z). The system is used mostly for *gonyaulaceans.*

taylorite A white mineral variety: $(K,NH_4)_2SO_4$. It is a variety of arcanite containing ammonium ion, and occurs in compact lumps in the guano beds on certain offshore Peruvian islands.

Taylor tabulation system A way of describing the number and arrangement of plates in *dinoflagellates.* It was developed by Taylor (1979, 1980) as a "basic gonyaulacoid model" and formed the basis of the later *Taylor-Evitt tabulation system.*

tazheranite (tazh'er-an'-ite) An adamantine yellow-orange cubic mineral: $(Zr,Ti,Ca)O_2$.

TBM *temporary bench mark.*

tcheremkhite (tche-rem'-khite) An algal sapropelic deposit found in the vicinity of Cheremkhovo, USSR, which has been interpreted as an aggregation of peaty matter washed from other deposits (Twenhofel, 1939, p.434).

T-chert Tectonically controlled chert, occurring in irregular masses related to fractures and orebodies (Dunbar and Rodgers, 1957, p.248).

TCT Acronym for a group of plants in which it is difficult to differentiate the fossil pollen: Taxodiceae, Cupressaceae, and Taxaceae, which are therefore often grouped under this acronym.

TD *total depth.*

T.D. curve *time-distance curve;* sometimes *time-depth curve.*

T-deltaT (TΔT) analysis A method of measuring the velocity of seismic waves from *normal moveout* and *arrival time* measurements. Superseded by *velocity analysis.*

TDEM *Time-domain electromagnetic method,* a controlled-source method. Syn: *transient electromagnetic method.*

***t* direction** A term used in crystal plasticity to denote the direction of slip in a crystallographic slip plane. See also: *f axis; T plane.*

TE *transverse electric.*

teallite (teall'-ite) A black or black-gray orthorhombic mineral: $PbSnS_2$.

tear *tare.*

tear fault A steeply dipping fault that is linked to but generally strikes perpendicular or oblique to a thrust fault. Slip on tear faults commonly leads to apparent strike-slip offsets.

tear-shaped bomb A rotational volcanic bomb shaped like a teardrop and having an ear at its constricted end; it ranges in size from 1 mm to more than 1 cm in length. Cf: *Pele's tears; spindle-shaped bomb.*

technical scale (tech'-ni-cal) A standard of fifteen minerals by which the hardness of a mineral may be rated. The scale includes, from softest to hardest and numbered one to fifteen: talc; gypsum; calcite; fluorspar; apatite; orthoclase; pure silica glass; quartz; topaz; garnet; fused zircon; corundum; silicon carbide; boron carbide; and diamond. Cf: *Mohs scale.*

technogenic Man-made or industrially generated, in geology or in the landscape; any feature that is a physical reality, but is not of natural origin. For example, hills or hilly landscapes created by mine dumps or garbage disposal; basins or depressions created by open-cut (open-face) mining or quarrying, together with lakes and their sediments consequent upon the above. The term also covers anthropogenic lakes created by dam construction; also, all major railroad and highway routes. (The word, with classical Greek roots, has been widely used in Russian literature, and through translations has spread into English usage).

technological hazard A hazardous condition caused by anthropogenic development, such as chemical processing or power generation and transmission. Cf: *geologic hazard; risk.*

tecnomorph (tec'-no-morph) In dimorphic ostracodes, the adult form that has the general shape of the juvenile instars. It is generally presumed to be the male of the species. Cf: *heteromorph.*

tectate (tec'-tate) Said of a pollen grain whose ektexine has an outer surface supported by more or less complicated inner structure usually consisting of columellae that support the tectum.

tectine (tec'-tine) An albuminoid (proteinaceous) organic substance in the wall of some foraminifers, having the appearance of, but chemically distinct from, chitin (TIP, 1964, pt. C, p.64).

tectite (tec'-tite) Obsolete spelling of *tektite.*

tectocline (tec'-to-cline) *geotectocline.*

tectofacies (tec'-to-fa'-cies) A lithofacies that is interpreted tectonically. The term was introduced by Sloss et al. (1949, p.96) for "a group of strata of different tectonic aspect from laterally equivalent strata", and was defined by Krumbein and Sloss (1951, p.383) as "laterally varying tectonic aspects of a stratigraphic unit". The term appears to be of "very limited practical value" because generally the nature of a tectofacies is noted "only after the area of the tectofacies has been outlined on the basis of other considerations" (Weller, 1958, p.635). Not to be confused with *tectonic facies.* Cf: *facies.*

tectogene (tec'-to-gene) (a) A long, relatively narrow unit of downfolding of sialic crust considered to be related to mountain-building processes. The term was proposed by Haarmann (1926, p.107). Obsolete syn: *geotectogene.* (b) The downfolded portion of an orogene (Hess, 1938). Syn: *downbuckle.*

tectogenesis (tec-to-gen'-e-sis) *orogeny.*

tectonic (tec-ton'-ic) Said of or pertaining to the forces involved in, or the resulting structures or features of, *tectonics.* Partial syn: *geotectonic.*

tectonic accretion *accretion [struc geol].*

tectonic assemblage diagram A form of correlation diagram that characterizes the history of terrane accretion and *amalgamation* (Howell, 1995).

tectonic axis *fabric axis.*

tectonic block A mass of rock that has been transported with respect to adjacent rock masses through the operation of tectonic processes (Berkland et al., 1972, p.2296).

tectonic breccia A breccia formed as a result of crustal movements, usually developed from brittle rocks. Distinct from sedimentary breccias. Cf: *fault breccia; fold breccia; crush breccia.*

tectonic cave Any cave formed by mechanical movement of bedrock. Faulting and gravity sliding are common mechanisms involved.

tectonic conglomerate *crush conglomerate.*

tectonic creep Slow, apparently continuous movement on a fault, accomplished either aseismically or by numerous miroseismic events.

tectonic cycle *orogenic cycle.*

tectonic denudation The stripping of an underbody, such as basement or other competent rock, by the movement of an upper stratified layer over it. During movement of rootless masses of the upper rocks by gravity sliding, the surface of the underbody is laid bare in places. Cf: *décollement.*

tectonic discordance A term used in reconnaissance studies to indicate that an older region has been cross-cut by later tectonic events.

tectonic earthquake An earthquake due to faulting rather than to volcanic activity. The term is little used. Cf: *volcanic earthquake.*

tectonic enclave A body of rock that has become detached or isolated from its source by tectonic forces. Cf: *tectonic inclusion.*

tectonic erosion The removal of rock from the base of an *overriding plate* during subduction.

tectonic estuary An *estuary* formed by tectonic processes, e.g., San Fancisco Bay.

tectonic fabric *deformation fabric.*

tectonic facies A collective term for rocks that owe their present characteristics mainly to tectonic activity; e.g. mylonites and some phyllites. This concept was introduced by Sander (1912). Not to be confused with *tectofacies.*

tectonic flow *tectonic transport.*

tectonic framework The combination or relationship in space and time of subsiding, stable, and rising tectonic elements in sedimentary provenance and depositional areas. Var: *framework [tect].*

tectonic geomorphology The tectonic interpretation of the morphological or topographic features of the Earth's surface; it deals with their tectonic or structural relations and origins, rather than their origins by surficial processes of erosion and sedimentation. Cf: *orogeny.* Obsolescent and less preferred syn:*morphotectonics.*

tectonic inclusion A body of rock that has become detached or isolated from its source by tectonic disruption and that is enclosed or included in the surrounding rock, e.g. a boudin. The term was used by Rast (1956, p.401) to replace *boudin,* a term that is often misused for any such inclusion. Cf: *tectonic enclave.*

tectonic joints Joints which propagate when tectonic compaction, or a tectonically produced hydraulic head, is responsible for abnormal pressures leading to joint formation (Engelder, 1985).

tectonic lake A lake occupying a basin produced mainly by crustal movements; e.g. caused by the impoundment of a drainage system as a result of upwarping, or occupying a graben (such as Lake Baikal in Russia). Syn: *structural lake.*

tectonic land Linear fold ridges and volcanic islands that had an ephemeral existence in the internal parts of an orogenic belt during the early or geosynclinal phase. Kay (1951) compares them with modern island arcs, and proposes that their existence probably accounts for many of the features formerly ascribed to *borderlands.*

tectonic landform A landform produced by earth movements. Cf: *structural landform.*

tectonic lens A structurally isolated body of rock encased within a contrasting rock type.

tectonic line An older term for a major regional fault zone in an orogen, e.g. Cameron's line, in the New England Appalachians. Some tectonic lines are sutures.

tectonic map A map that portrays the architecture of the outer part of the Earth. It is similar to a structure-contour map, which primarily shows dipping strata, folds, faults, and the like, but the tectonic map also presents some indication of the ages and kinds of rocks from which the structures were made, as well as their historical development. Cf: *paleotectonic map.*

tectonic mélange A *mélange* produced by *tectonic* processes (Berkland et al., 1972, p.2296).

tectonic moraine An aggregation of boulders incorporated in the base or sole of an overthrust mass. It is often mistaken for a conglomerate as it has a local concordant relation to the associated strata (Pettijohn, 1957, p.281). Little used.

tectonic motif A *lithotectonic unit* consisting of an underlying precursory sedimentary mélange and an overlying allochthon. The precursory mélange contains olistoliths of rocks that comprise the allochthon (Drake, 1985). An example is the Annandale/Indian Run motif in Fairfax County, Virginia.

tectonic overpressure The pressure exceeding the load pressure, at times during metamorphism, by amounts that depend on the strength of the rocks (100-200 MPa).

tectonic profile *profile [struc geol].*

tectonics (tec-ton'-ics) A branch of geology dealing with the broad architecture of the outer part of the Earth, that is, the regional assembling of structural or deformational features, a study of their mutual relations, origin, and historical evolution. It is closely related to *structural geology,* with which the distinctions are blurred, but tectonics generally deals with larger features. Adj: *tectonic.* Partial syn: *geotectonics.*

tectonic stress Any component of a principal stress that makes that principal stress different from principal stress in a lithostatic state of stress (Engelder, 1993).

tectonic style The total character of a group of related structures that distinguishes them from other groups of structures, in the same way that the style of a building or an art object distinguishes it from others of different periods or influences. Syn: *style [tect].*

tectonic subsidence That part of the *subsidence* at a given point in a sedimentary basin caused by a tectonic driving mechanism. It is calculated by removing the component of subsidence produced by nontectonic processes such as sediment loading, sediment compaction, and changes in water depth (Biddle and Christie-Blick, 1985).

tectonic transport A kinematic term used strictly only for a *deformation plan* that possesses monoclinic symmetry. The direction of tectonic transport is the kinematic a-direction or direction of maximum displacement in the unique symmetry plane. Syn: *tectonic flow.*

tectonic valley A valley that is produced mainly by crustal movements, such as by faulting or folding. Cf: *structural valley.*

tectonism (tec'-ton-ism) *diastrophism.*

tectonite (tec'-ton-ite) (a) Any rock whose fabric reflects the history of its deformation; a rock whose fabric clearly displays coordinated geometric features that indicate continuous solid flow during formation (Turner and Weiss, 1963, p.39). (b) A rock whose fabric has been modified substantially by deformation processes. Also spelled: *tektonite.*

tectonization (tec'-ton-i-za'-tion) A term sometimes used as a generalized synonym of orogenesis, diastrophism, etc. Thus, deformed rocks in orogenic belts are said to have been "tectonized". This usage is not recommended.

tectono-eustasy (tec'-ton-o-eu'-sta-sy) Worldwide change of sea level produced by a change in the capacity of the ocean basins owing to plate tectonic motions, e.g. sea-floor spreading, subduction. Term introduced by Fairbridge (1961, p.111). Cf: *glacio-eustasy; sedimento-eustasy.* Syn: *diastrophic eustatism; tectono-eustatism.*

tectono-eustatism (tec'-ton-o-eu'-stat-ism) *tectono-eustasy.*

tectonophysics (tec'-ton-o-phys'-ics) A branch of geophysics that deals with the forces responsible for movements in, and deformation of, the Earth's crust.

tectonosphere (tec-ton'-o-sphere) The zone or layer of the Earth above the level of isostatic equilibrium, in which crustal or tectonic movements originate. Rarely used. Cf: *crust [interior Earth].*

tectono-stratigraphic unit (tec'-ton-o-strat'-i-graph'-ic) (a) A mixture of lithostratigraphic units resulting from tectonic deformation; e.g. a *mélange.* (b) *sequence (b).*

tectono-stratigraphy (tec-ton'-o-stra-tig'-ra-phy) Study of the relations of major lithostratigraphic units, e.g. sequences (b), with emphasis on tectonic effects on the stratigraphic record.

tectorium (tec-to'-ri-um) The internal lining of a foraminiferal chamber (as in fusulinids), composed of dense calcite formed at or near the same time as that in which the tunnel in the test was excavated. It may include the *lower tectorium* and the *upper tectorium.* Pl: tectoria.

tectosequent (tec-to-se'-quent) Said of a surface feature that reflects the underlying geologic structure. Ant: *morphosequent.*

tectosilicate (tec-to-sil'-i-cate) A class or structural type of *silicate* characterized by the sharing of all four oxygens of the SiO_4 tetrahedra with neighboring tetrahedra, and with a Si:O ratio of 1:2. Quartz, SiO_2, is an example. Cf: *nesosilicate; sorosilicate; cyclosilicate; inosilicate; phyllosilicate.* Syn: *framework silicate.* Also spelled: *tektosilicate.*

tectosome (tec'-to-some) A term used by Sloss (in Weller, 1958, p.625) for a "body of strata indicative of uniform tectonic conditions"; the sedimentary rock record of a uniform tectonic environment or of a tectotope. The term replaces *tectotope* as that term was originally defined.

tectosphere (tec'-to-sphere) A term introduced by T.H. Jordan (1975b) to refer to the layer of the Earth that moves during plate motion. In the case of oceanic plates, the tectosphere is equivalent to the *lithosphere.* Beneath continents, however, the tectosphere includes the lithosphere plus some underlying *asthenosphere* which is permanently affixed to the base of the lithosphere. The depth to the base of continental tectosphere is debatable, but may be at about 250 km below the surface.

tectostratigraphic (tec'-to-strat'-i-graph'-ic) Pertaining to facies aspects determined by tectonic conditions and influences; said of an interpretative (rather than an objective) stratigraphic facies characterized lithologically "in whatever way is considered to be most significant tectonically" (Weller, 1958, p.630).

tectotope (tec'-to-tope) An area of uniform tectonic environment. The term was originally defined by Sloss et al. (1949, p.96) as a "stratum or succession of strata with characteristics indicating accumulation in a common tectonic environment", but was used by Krumbein and Sloss (1951, p.381) to designate a tectonic environment. Sloss (in Weller, 1958, p.616) later regarded the term as referring to an area, rather than to a stratigraphic body or an environment, and notes that it "is an almost pure abstraction dependent upon interpretation" of a *tectosome.* Weller (1958, p.636) considers the term "superfluous" because tectonic areas are "generalized and extensive" and "not subject to the differentiation possible in the consideration or description of sedimentary environments".

tectum [paleont] (tec'-tum) (a) The thin, dense outermost layer of the spirotheca in fusulinids. Cf: *diaphanotheca.* (b) Marginal prolongation of a chamber in trochospirally coiled foraminiferal tests, making sutures of the dorsal (spiral) side more inclined than those of the ventral (umbilical) side. This usage is not recommended because of prior adoption of the term for fusulinids. Pl: tecta.

tectum [palyn] In the exine of pollen, a layer of sexine covering the columellae. A "tectum imperforatum" lacks perforations, whereas a "tectum perforatum" has perforations which are not >1 μm in diameter. Adj: *tectate.* Cf: *tegillum [palyn].*

tedhadleyite A dark red to black triclinic mineral: $Hg^{2+}Hg^{1+}_{10}O_4I_2(Cl,Br)_2$.

teepee butte (tee'-pee) *tepee butte.*

teepleite (tee'-ple-ite) A colorless tetragonal mineral: $Na_2B(OH)_4Cl$.

tegengrenite A dark red rhombohedral mineral: $(Mg,Mn^{2+})_2(Sb,Mn)O_4$.

tegillum [paleont] An umbilical covering in a planktonic foraminiferal test (as in *Globotruncana* and *Rugoglobigerina*), comprising an extension from a chamber comparable to a highly developed apertural lip but extending across the umbilicus, thus completely covering the primary aperture (main opening of the test) and attached at its farther margin or at the tegillum of an earlier chamber. It may have small openings along its margin or be pierced centrally. Pl: tegilla.

tegillum [palyn] (te-gil'-lum) A wall-structure term for outgrowths of the relatively homogeneous, innermost level of a *dinocyst* wall that form processes, crests, or columellae.

tegmen (teg'-men) The oral surface of an echinoderm body; strictly the calcareous adoral part of a crinoid calyx roofing the *dorsal cup,* situated at the origin of the free arms and occupying the space between them. It may include calcareous ambulacral and interambulacral plates or be composed entirely of soft tissue (Beerbower, 1968, p.400).

tegulation (teg-u-la'-tion) A term used for dinoflagellate thecae or cysts to describe the arrangement of *plates* into series on the basis of plate overlap.

teilchron (teil'-chron) A term proposed by Arkell (1933, p.33) for the locally recognizable time span of a taxonomic entity. It is synonymous with *teilzone* as defined by Pompeckj (1914).

teilzone (teil'-zone) (a) A time term introduced in the German literature as "Teilzone" by Pompeckj (1914) to designate the local duration of existence of a species. Syn: *teilchron.* (b) A syn. of *local-range zone.* Etymol: German *Teilzone,* "part zone". Also spelled: teil-zone.

teineite (tein'-e-ite) A blue orthorhombic mineral: $CuTe^{4+}O_3 \cdot 2H_2O$.

tejon (te-jon′ [tay-hone′]) A term used in the SW U.S. for a solitary, disk-shaped eminence separated by erosion from the mass of which it was originally a part. Etymol: Spanish "tejón", "round gold ingot". Cf: *huerfano.*

tektite (tek′-tite) A jet-black to greenish or yellowish body of silicate glass of nonvolcanic origin found in several widely separated areas of the Earth's surface and generally bearing no relation to the associated geologic formations. Most tektites are high in silica (68-82%) and very low in water content (average 0.005%): their elemental composition are similar to those of soils or shales. They exhibit two main structural categories: (1) the splash-form or spin-form tektites are relatively small (generally < 12 cm, <1 kg) and have shapes such as dumbbells, teardrops, oblate discoids, or spheroids that appear to reflect solidification of a spinning melt; (2) the layered (or Muong-Nong) tektites have irregular, fragmental shapes and large sizes (masses as high as 28 kg). In contrast to impact glasses found in or near craters, the tektites are essentially free of crystalline clasts. The tektites from four large strewn fields (ages): Australasian (0.8 Ma), Ivory Coast (1.1 Ma), Central European (Moldavite) (15 Ma) and North American (36 Ma). The Libyan Desert glass (ca. 98% SiO_2) appears to be a kind of layered tektite produced from a quartzite target, and should probably be included as a fifth field of tektites. Some, especially from Australia, have shapes strongly suggesting aerodynamic ablation during atmospheric entry. Tektites are generally believed to be the product of large hypervelocity meteorite impacts on sediments, although a few scientists have argued for a lunar origin. Term proposed by F. Suess (1900) who believed they were meteorites which at one time underwent thorough melting. Etymol: Greek tektos, "molten". Obs. spelling: *tectite.*

tektite field *strewn field [meteorites].*

tektonite (tek′-ton-ite) *tectonite.*

tektosilicate (tek-to-sil′-i-cate) Var. of *tectosilicate.*

telain (tel′-ain) A syn. of *provitrain.* It is used in the names of transitional coal lithotypes, e.g. clarotelain.

telalginite An *alginite* with distinct (microscopic) algal forms, usually *Tasmanites* or *Botryococcus.* A part of the *liptinite* maceral group.

telargpalite (tel-arg′-pa-lite) A metallic gray cubic mineral: $(Pd,Ag)_3Te$.

tele A Norwegian term for *frozen ground,* often used erroneously as a syn. of *permafrost.* Also spelled: *taele.*

telechemic (tel-e-chem′-ic) Said of the earliest minerals to crystallize during solidification of an extremely differentiated, high-silica magma; e.g. zircon, apatite, corundum. Obsolete.

teleconnection (tel′-e-con-nec′-tion) Identification and correlation of a series of varves, esp. over great distances or even worldwide, in order to construct a uniform time scale or hypothesize a causitive climatic link for a part of the Pleistocene Epoch.

teleconnection [clim] A linkage between changes in atmospheric circulation occurring in widely separated regions of the globe, often many thousands of kilometers apart.

telemeter (te-lem′-e-ter) n. A surveying instrument for measuring the distance of an object from an observer; e.g. a telescope with stadia hairs in which the angle subtended by a short base of known length is measured. See also: *range finder.* v. To transmit data by radio or microwave links.

teleoconch (tel′-e-o-conch′) The entire gastropod shell exclusive of the *protoconch.*

teleomorph (tel′-e-o-morph) The meiotic, sexual morph of a non-lichen-forming ascomycetous and basidiomycetous fungus (including Ustilaginales). Cf: *anamorph; holomorph.*

Teleostei (Tel-e-os′-te-i) An infraclass of ray-finned bony fish that includes the great bulk of living fishes, both marine and freshwater, in 9 superorders encompassing some 29 orders. Range, Upper Triassic to Holocene.

telescoped (tel′-e-scoped) Said of ore deposits in which the normal upward range of high- to low-temperature mineral assemblages is compressed, so that different zones overlap.

telescope structure (tel′-e-scope) A term proposed by Blissenbach (1954, p.180) for an alluvial-fan structure "characterized by younger fans with flatter gradients spreading out from between fan mesas of older fans with steeper gradients"; e.g. in the Santa Catalina Mountains, Ariz.

telescopic alidade (tel-e-scop′-ic) An *alidade* used with a planetable, consisting of a telescope mounted on a straightedge ruler, fitted with level bubble, scale, and vernier to measure angles, and calibrated to measure distances.

teleseism (tel′-e-seism) An earthquake whose epicenter is over 1,000 km away. Earthquakes originating nearer are "local" earthquakes.

telethermal (tel-e-ther′-mal) Said of a hydrothermal mineral deposit formed at shallow depth and relatively low temperatures, with little or no wall-rock alteration, presumably far from the source of hydrothermal solutions. Also, said of that environment. Cf: *hypothermal; mesothermal [eco geol]; epithermal; xenothermal; leptothermal.*

teleutospore (te-leu′-to-spore) Obsolete syn. of *teliospore.*

telinite (tel′-i-nite) (a) A maceral of coal within the *vitrinite* group, characteristic of vitrain and consisting of cell-wall material with visible cell structure. Cf: *suberinite; xylinite.* (b) A preferred syn. of *provitrinite.*

teliospore (tel′-i-o-spore′) A thick-walled *fungal spore* developed in the final stage of the life cycle of the rust fungi Uredinales and Ustilaginales. The wall may be composed of chitin. Cf: *urediospore.* Obsolete syn: *teleutospore.*

telite (tel′-ite) A microlithotype of coal, a variety of *vitrite,* consisting of 95% or more of telinite.

tell An Arabic term for *mound [arch].*

tellurantimony (tel-lur-an′-ti-mo′-ny) A metallic pink or cream rhombohedral mineral of the *tetradymite* group: Sb_2Te_3.

tellurate (tel′-lu-rate) A compound containing the radical $(TeO_4)^{-2}$. An example is montanite, $Bi_2TeO_6{\cdot}2H_2O$.

tellurbismuth (tel-lur-bis′-muth) *tellurobismuthite.*

telluric (tel-lu′-ric) Pertaining to the Earth, esp. the depths of the Earth, e.g. as applied to natural electric fields or currents.

telluric bismuth *tetradymite.*

telluric current Natural electric current that flows in large sheets on or near the Earth's surface. Syn: *earth current; ground current.*

telluric method An electrical exploration method in which the Earth's natural electric field is measured at two or more stations simultaneously to yield a quantitative estimate of the electric properties of the section.

telluric ocher *tellurite.*

telluric water Water formed by the combination of hydrogen with the oxygen of the atmosphere at high temperature and pressure (Swayne, 1956, p.137). Cf: *juvenile [water].*

telluride (tel′-lu-ride) A mineral compound that is a combination of tellurium with a metal. An example is hessite, Ag_2Te.

tellurite (tel-lur′-ite) A subadamantine white or yellowish orthorhombic mineral: TeO_2. It is dimorphous with paratellurite. Syn: *telluric ocher.*

tellurium (tel-lu′-ri-um) A silvery-white to brownish-black trigonal mineral, the native semi-metallic element Te. It is occasionally found in pyrites and sulfur, or in the fine dust of gold-telluride ores.

tellurium glance *nagyágite.*

tellurobismuthite (tel-lu′-ro-bis′-muth-ite) A metallic gray rhombohedral mineral of the *tetradymite* group: Bi_2Te_3. It is often intergrown with tetradymite. Syn: *tellurbismuth.*

tellurohauchecornite (tel-lu′-ro-hau′-che-cor′-nite) A metallic bronze tetragonal mineral of the *hauchecornite* group: Ni_9BiTeS_8.

telluroid (tel′-lu-roid) A surface near the terrain that is the locus of points in which the spheropotential coincides with the geopotential of corresponding points on the terrain.

telluromayingite *mayingite.*

Tellurometer (Tel-lu-rom′-e-ter) Trade name of a rugged portable electronic device that measures ground distances precisely by determining the velocity of a phase-modulated, continuous, microwave radio signal transmitted between two instruments operating alternately as master station and remote station. It has a range up to 65 km (35-40 miles). Cf: *geodimeter.*

telluronevskite A metallic steel-gray trigonal mineral: Bi_3TeSe_2.

telluropalladinite (tel-lu′-ro-pal-la′-di-nite) A metallic monoclinic mineral: Pd_9Te_4.

telmaro (tel-ma′-ro) A term proposed by Veatch and Humphrys (1966, p.326) for a river traversing a peat marsh or peat swamp. Etymol: Greek telma, "marsh".

telmatic peat (tel-mat′-ic) Peat formation of the water table due to plants growing under conditions of shallow water or periodic submergence. Cf: *limnic peat; terrestric peat.*

telmatology (tel-ma-tol′-o-gy) The study of wetlands, marshy areas and swamps.

teloclarain (tel-o-clar′-ain) A transitional lithotype of coal characterized by the presence of telinite, though in lesser amounts than other macerals. Cf: *clarotelain.* Syn: *teloclarite.*

teloclarite (tel-o-clar′-ite) *teloclarain.*

telodurain (tel-o-du′-rain) A coal lithotype transitional between durain and telain, but predominantly durain. Cf: *durotelain.*

telofusain (tel-o-fu′-sain) A coal lithotype transitional between fusain and telain, but predominantly fusain. Cf: *fusotelain.*

telogenetic (tel′-o-ge-net′-ic) A term proposed by Choquette and Pray (1970, p.220-221) for the period during which long-buried carbonate rocks are affected significantly by processes related to weathering and subaerial and subaqueous erosion and dissolution. Also applied to the porosity that develops during the telogenetic stage. Cf: *eogenetic; mesogenetic.*

telohuminite A subgroup of *huminite* showing cell structure (Pickel, 2004).

telome (tel′-ome) As an idealized evolutionary concept, a single terminal segment of a primitive plant branching axis. Cf: *telome theory.*

telome theory A theory holding that plants evolved through modifications of branches and terminal segments of the plant body, vascular tissue, and terminal sporangia, all derived from an originally undifferentiated axis.

telovitrinite A subgroup of *vitrinite* showing cell structure (ICCP, 1998).

telson (tel′-son) (a) The last *somite* of the body of a crustacean, bearing the anus and commonly the caudal furca. Syn: *postabdomen; style [paleont].* (b) A dorsal, postanal extension of the body of an arachnid, articulated to the last abdominal segment; a postanal spine or plate of a merostome. (c) A terminal or anal segment of a trilobite. The term is sometimes incorrectly used for a spine mounted on the terminal or one of the near-terminal segments of a trilobite and directed posteriorly along the midline, such as the first macrospine on the posterior part of the thorax in certain Olenellidae (TIP, 1959, pt. O, p.126).

telum (tel'-um) A guardlike structure in aulacocerid cephalopods, consisting of concentric layers of mostly organic material (conchiolin?) with a few interspersed calcitic layers. See: *rostrum.*

Telychian (Te-lych'-i-an) A stage of the Standard Global Chronostratigraphic Scale: uppermost stage of the Llandoverian Series: Lower Silurian (above Aeronian, below Sheinwoodian). The basal GSSP is in Wales (Holland, 1985).

telyushenkoite A vitreous white trigonal mineral: $CsNa_6Be_2Al_3Si_{15}O_{39}F_2$.

TEM *transmission electron microscope.*

temagamite (te-mag'-a-mite) A metallic orthorhombic mineral: Pd_3HgTe_3.

temblor (tem'-blor) A syn. of *earthquake.* Etymol: Spanish, a "trembling".

temiskamite (te-mis'-ka-mite) *maucherite.*

Temnospondyli (Tem-no-spon'-dy-li) An order of labyrinthodont amphibians characterized by reduction of the centrum and emphasis on the intercentrum in each vertebra. Range, Upper Mississippian to Middle Jurassic.

temper A material other than clay (usually sand, small rocks, or crushed pottery) added to clay in order to give it extra strength during the firing process and as finished pottery. Temper mineralogy can be used in archaeology to identify ceramic sources and trading.

temperate (tem'-per-ate) Said of a temperature that is moderate or mild. The term is also used to describe temperatures of the middle latitudes, whether moderate or not.

temperate glacier Originally, a glacier characteristic of the temperate zone, in which at the end of the ablation season the firn and ice of which it consists are near the melting point of ice (Ahlmann, 1933). Presently, a glacier that is at the melting point throughout, except for a surface layer, several meters thick, in which the temperature varies seasonally. Examples: almost all glaciers in Scandinavia and the Alps, and in the U.S. outside of northern Alaska. Cf: *polar glacier.* Syn: *warm glacier.*

temperate rain forest *mid-latitude rain forest.*

temperature (tem'-per-a-ture') A basic property of a thermodynamic system that defines the potential for heat transfer between two systems. Systems with the same temperature are in thermal equilibrium. Temperature is measured by various scales based on changes in volume, electrical resistance, thermal electromotive force, or length of the measuring instrument. The preferred units for temperature are degrees kelvin (K) or Celsius (C).

temperature factor An exponential expression that reflects the diminution of diffraction intensity with increasing thermal vibrations of an atom in a *crystal structure.* Alternatively, a temperature factor may represent diminished intensity owing to positional disorder of an atom over many unit cells. An *isotropic temperature factor* is an expression that models the vibrations or displacements of an atom as spherically symmetrical (i.e., random). An *anisotropic temperature factor* models the vibrational volume as a triaxial ellipsoid. Temperature factors are determined during *crystal-structure refinement.* Cf: *displacement parameters.*

temperature factor coefficients *displacement parameters.*

temperature gradient The rate of change of temperature with distance. When applied to the Earth, the term *geothermal gradient* is used.

temperature log A *well log,* usually run in a cased hole, that depicts temperature variation with depth. It is one of the prerequisites for determining *heat-flow density.* Useful for identifying the top of cement that is curing behind casing, or points of gas or fluid entry. Syn: *thermal log.*

temperature logging Measurement of temperature in drillholes either as a single reading near the bottom of the hole using nonelectric instruments such as mercury maximum-reading thermometers or as a temperature log using electrical-line systems or "slick-line" ("memory tool") computer systems connected to the surface by a solid cable. Those systems generally have a resolution of ~0.0001°C, precision of ~0.1°C, and a sample interval of ~0.1 m. For long-term monitoring of transient temperature conditions Distributed optical fibre Temperature Sensing (DTS) technique is suitable. Syn: *temperature measurement.*

temperature measurement *temperature logging.*

temperature-salinity diagram (tem'-per-a-ture'-sa-lin'-i-ty) A plot of temperature as a function of salinity within the water column (comprising one or many stations), useful in identifying water masses within it and the stability. Syn: *T-S diagram.*

temperature survey *thermal survey.*

temperature zone A general term for a region characterized by a relatively uniform temperature or temperature range. It may refer to a region of a particular climatic classification (a latitudinal division) or to a particular temperature belt on a mountainside (a vertical or altitudinal division). Cf: *climatic zone.*

tempestite (tem'-pes-tite) A storm deposit consisting of sandy layers, showing evidence of violent disturbance of pre-existing sediments followed by their rapid redeposition, all in a shallow-water environment, but deeper than fair-weather wave base. (Ager, 1974, p.86). Tempestites are vertically size graded. For the relation of tempestites to turbidites and inundites, see Einsele and Seilacher (1982, p.334).

templet (tem'-plet) In photogrammetry, a transparent celluloid overlay made on an aerial photograph and showing the center of the photograph and all radial lines through images of control points, as well as azimuth lines connecting the center with images of points that show on the photograph and are themselves the centers of other photographs. The term is a var. of "template".

temporal frequency Ordinary frequency (q.v.), i.e., cycles per unit time, as opposed to spatial frequency or wavenumber, cycles per unit distance.

temporal transgression (tem'-po-ral) A name given by Wheeler and Beesley (1948, p.75) to the principle that rock units and unconformities vary in age from place to place.

temporary base level (tem'-po-rar'-y) Any *base level,* other than sea level, below which a land area cannot be reduced, for the time being, by erosion; e.g. a level locally controlled by a resistant stratum in a stream bed, the surface of a lake, or the level of the main stream into which a tributary flows. Cf: *ultimate base level.* Syn: *local base level.*

temporary bench mark A supplementary *bench mark* of less enduring character than a *permanent bench mark,* intended to serve for only a comparatively short period of time, such as a few years; e.g. an intermediate bench mark established at a junction of sections of a continuous series of measured differences of elevation for the purpose of holding the end of a completed section and serving as an initial point from which the next section is run. Its elevation determination may not be precise. Examples include a chiseled square or cross on masonry, a nail and washer in the root of a tree, or a bolt on a bridge. Abbrev: TBM.

temporary hardness *carbonate hardness.*

temporary plankton *meroplankton.*

temporary stream *intermittent stream.*

temporary wilting A degree of wilting from which a plant can recover by decreasing its rate of transpiration and without adding water to the soil. Cf: *permanent wilting; wilting point.*

tenacity (te-nac'-i-ty) The property of the particles or molecules of a substance to resist separation; *tensile strength.*

tenebrescence (ten-e-bres'-cence) In optics, the absorption of light by a crystal under the influence of radiation.

tengchongite (teng'-chong-ite) A yellow orthorhombic mineral: $CaO_5(UO_2)_6(MoO_4)_2{\cdot}12H_2O$.

tengerite-(Y) (ten'-ger-ite) A chalky-white orthorhombic mineral: $Y_2(CO_3)_3{\cdot}2\text{-}3H_2O$. The name was applied originally to a supposed beryllium-yttrium carbonate.

tennantite (ten'-nant-ite) (a) A metallic black or gray cubic mineral: $(Cu,Ag,Zn,Fe)_{12}As_4S_{13}$. It sometimes contains zinc, silver, or cobalt replacing part of the copper. It is an important ore of copper. Syn: *fahlore; gray copper ore.* (b) A group name for cubic minerals with the analogous formula, but with Cu replaced by Ag or Hg; As replaced by Sb; and S replaced by Se.

tenor (ten'-or) *grade [ore dep].*

tenorite (ten'-o-rite) A metallic gray or black monoclinic mineral: CuO. It occurs in minute shining steel-gray or iron-gray scales, in black powder, or in black earthy masses, generally in the oxidized (weathered) zones or gossans of copper deposits. Tenorite is an ore of copper. Syn: *melaconite; black copper.*

tensile strength (ten'-sile) The maximum applied tensile stress that a body can withstand before failure occurs. Syn: *tenacity.*

tensile stress A *normal stress* that tends to cause separation across the plane on which it acts. Cf: *compressive stress.*

tensiometer (ten-si-om'-e-ter) A device to measure moisture tension in an unsaturated soil.

tension (ten'-sion) A state of stress in which tensile stresses predominate; stress that tends to pull a body apart.

tension crack A fracture caused by tensile stress.

tension fault A genetic term for any fault created in tension. Often used erroneously for faulting that results in extensional strain. Cf: *extension fault.*

tension fracture A fracture that is the result of tensional stress in a rock. Cf: *shear fracture.* See also: *extension fracture; joint [struc geol].*

tension gash A short extensional fracture along which the walls have been pulled apart. Gashes may be open or filled, and commonly have an en echelon pattern. They may occur in or adjacent to fault zones.

tension joint A joint that is a *tension fracture.* An example of redundancy in the geological literature because a joint only forms in tension.

tension zone *ecotone.*

tensor A set of quantities that relate different vector fields, usually expressed as a matrix. The state of stress at a point in an elastic material is a tensor.

tensor resistivity (ten'-sor) Representation of the resistivity of a material by a matrix of values, three of which represent resistance to current flow in the direction of the applied electric field and six of which represent resistance to current flow at right angles to the applied electric field. It is also used in two dimensions, with two principal values and two cross-coupling values.

tentacle (ten'-ta-cle) Any of various elongate and flexible processes of differing functions that are borne by invertebrates, e.g. a movable tubular extension of soft integument rising from the *oral disk* of a coral polyp, closed at the tip, and serving primarily for food getting; a short, slender armlike sensory appendage extending in front of the head of a cephalopod or gastropod;

or one of the numerous small ciliated processes borne on the arms of a brachiopod or the *lophophore* of a bryozoan and used primarily for food getting.

tentacle pore A term commonly used for *podial pore* of ophiuroids (TIP, 1966, pt. U, p.30).

tentacle sheath A thin, delicate membranous part of the body wall of a bryozoan, which introverts to enclose the tentacles when the *polypide* is retracted.

tentaculite (ten-tac'-u-lite) *tentaculitid.*

tentaculite zoomorph Organic remains of *tentaculites,* sometimes with robust walls, possibly chitinous. Cf: *zoomorph.*

tentaculitid (ten-tac'-u-lit'-id) A marine invertebrate animal characterized by radial symmetry, a small conical shell with transverse rings of variable size and spacing, longitudinal striae, an embryonic chamber with a bluntly pointed apex, and small pores in the shell wall. The tentaculitids are referred to the order Tentaculitida and questionably assigned to the mollusks. Some of the fossil forms are microscopic to near-microscopic and occur as palynomorphs with robust walls that possibly consist of chitin or a chitinlike substance. Range, Lower Ordovician to Upper Devonian, the oldest representatives belonging to the genus *Tentaculites.* Syn: *tentaculite; tentaculoid.*

tentaculoid *tentaculitid.*

tented ice (tent'-ed) Sea ice deformed by *tenting.*

tent hill A term used in Australia for a butte or flat-topped hill resembling a canvas tent; it often is capped by resistant rock from a former plateau surface. Cf: *tepee butte.*

tenting (tent'-ing) The vertical displacement upward of sea ice under lateral pressure to form a flat-sided arch over a cavity between the raised ice and the water beneath; a type of *ridging.*

tenuitas (te-nu'-i-tas) A thin area in the exine of palynomorphs, usually germinal in function, as the annular tenuitas of *Classopollis.* A tenuitas is a less distinct feature than a colpus or pore.

tepee butte (te'-pee) A conical hill or knoll resembling an American Indian tepee; esp. an isolated, residual hill formed by a capping of resistant rock that protects the underlying softer material from erosion, e.g. one of the partly exhumed bioherms in the Pierre Shale of Colorado, or one of the sandstone-capped hills in the Painted Desert, Ariz. Cf: *tent hill; klint [reef].* Also spelled: *teepee butte.*

tepee structure A disharmonic sedimentary structure consisting of a fold which in cross section resembles a chevron, an "inverted depressed V", or the profile of a peaked dwelling of the North American Indians. It is believed to be an early diagenetic structure formed at the margins of large polygons produced by expansion of surface sediments (Assereto and Kendall, 1977). The term was introduced by Adams and Frenzel (1950). Cf: *enterolithic.*

tepetate (te-pe-ta'-te) (a) An evaporite consisting of a calcareous crust coating solid rocks on or just beneath the surface of an arid or semiarid region; a deposit of *caliche.* (b) A term used in Mexico for a volcanic tuff, or a secondary volcanic or chemical nonmarine deposit, very commonly calcareous (Brown and Runner, 1939, p.338). Etymol: Mexican Spanish, from Nahuatl (Aztec) "tepetatl", "stone matting". Cf: *talpatate.*

tephra (teph'-ra) A collective term used for all pyroclastic material, regardless of size, shape or origin, ejected during an explosive volcanic eruption. Originally applied exclusively to fallout ejecta but now also applies to pyroclastic flow deposits. Introduced by Thorarinsson in 1944 (Thorarinsson, 1981). Etymol: from Greek term for ashes.

tephrite (teph'-rite) (a) In the *IUGS classification,* a volcanic rock with <10% modal olivine, defined in the QAPF diagram by F/(F+A+P) between 10 and 60, P/(P+A)>90; also defined in the TAS diagram as rock plotting in the irregular field bounded by points with the SiO_2 and total alkali coordinates: 41, 3; 41, 7; 45, 9.4; 49.4, 7.3; 45, 5; and 45, 3. (b) A group of extrusive rocks, of basaltic character, primarily composed of calcic plagioclase, augite, and nepheline or leucite as the main feldspathoids, with accessory alkali feldspar; also, any member of that group; the extrusive equivalent of *theralite.* With the addition of olivine, the rock would be called a basanite.

tephritic foidite In the *IUGS classification,* a collective term for volcanic rocks consisting of feldspathoids with some plagioclase and less than 10% modal olivine. Cf: *basanitic foidite.*

tephritic phonolite In the *IUGS classification,* a volcanic rock with F between 10 and 60 and P/(P+A)<50. Also defined in the TAS diagram as the field bounded by points with the SiO_2 and (Na_2O+K_2O) coordinates: 48.4, 11.5; 52.5, 14; 53, 9.3; and 57.6, 11.7.

tephrochronology (teph'-ro-chro-nol'-o-gy) The collection, preparation, petrographic description, and approximate dating of *tephra.* Chronological and correlation studies using tephra layers (Thorarinsson, 1944).

tephroite (teph'-ro-ite) An olive-green, bluish-green, gray, red, or reddish-brown orthorhombic mineral of the *olivine* group: $Mn^{2+}{}_2SiO_4$. It occurs with zinc and other manganese minerals.

teppe A Persian term for *mound [arch].* Variant form of the word "tepe"; equivalent to Arabic *tell.*

terebrataliiform (ter'-e-bra-ta'-lii-form) Said of the *loop,* or of the growth stage in the development of the loop, of a dallinid brachiopod (as in *Terebratalia*), consisting of long descending branches with connecting bands to the median septum, then recurving into ascending branches that meet in transverse band (TIP, 1965, pt. H, p.154). The terebrataliiform loop is morphologically similar to the *terebratelliform* loop.

terebratellacean (ter'-e-bra'-tel-la'-ce-an) n. Any terebratuloid belonging to the superfamily Terebratellacea, characterized by a long brachial loop. Range, Upper Cretaceous to Holocene. adj. Said of a terebratuloid having a long brachial loop, or of its shell. Not to be confused with *Terebratulacean.*

terebratellid (ter'-e-bra-tel'-lid) Any terebratuloid belonging to the suborder Terebratellidina, characterized by a loop that develops in connection with both cardinalia and median septum. Range, Lower Devonian to Holocene.

terebratelliform (ter'-e-bra-tel'-li-form) Said of the *loop,* or of the growth stage in the development of the loop, of a terebratellid brachiopod (as in the subfamily Terebratellinae), consisting of long descending branches with connecting bands to the median septum, then recurving into ascending branches that meet in transverse band (TIP, 1965, pt. H, p.154). The terebratelliform loop is morphologically similar to the *terebrataliiform* loop.

terebratulacean (ter'-e-brat'-u-la'-ce-an) Any terebratuloid belonging to the superfamily Terebratulacea, characterized by the development of the cardinal process and outer hinge plates and by the absence of inner hinge plates. Range, Upper Triassic to Holocene. Not to be confused with *terebratellacean.*

terebratuliform (ter-e-brat'-u-li-form') Said of a short, typically U- or W-shaped *loop* found in most terebratulacean brachiopods.

terebratuliniform (ter'-e-brat'-u-lin'-i-form) Said of a short brachiopod *loop* in which crural processes are fused medially to complete a ringlike or boxlike apparatus (TIP, 1965, pt. H, p.154).

terebratuloid (ter-e-brat'-u-loid) Any articulate brachiopod belonging to the order Terebratulida, characterized chiefly by a punctate shell with a teardrop-shaped outline, pointed at the posterior end. Range, Lower Devonian to present. Var: terebratulid.

tergal fold (ter'-gal) *epimere.*

tergite (ter'-gite) The dorsal plate or dorsal part of the covering of a segment of an articulate animal; e.g., the sclerotized dorsal surface of a single crustacean somite, a hardened chitinous plate on the dorsal surface of an arachnid body segment, or a plate forming the dorsal cover of a merostome somite.

tergopore (ter'-go-pore) A polymorph on the reverse side of some stenolaemate bryozoan colonies, with a cross section comparable in size to that of associated feeding zooids.

tergum (ter'-gum) (a) One of a pair of opercular valves, adjacent to the *carina* in cirripede crustaceans, lacking adductor-muscle attachments. Cf: *scutum.* (b) The back or dorsal surface of the body of an animal. Pl: terga.

terlinguaite (ter-lin'-gua-ite) An adamantine yellow, greenish-yellow, or brown monoclinic mineral: Hg_2OCl.

terminal curvature (ter'-mi-nal) A sharp, local change in the dip of strata or cleavage near a fault. Not commonly used in the U.S.

terminal diaphragm A membranous or skeletal partition in stenolaemate bryozoans, extending across the aperture of a *zooid* to seal off the living chamber from the environment.

terminal face The lower extremity, or *snout,* of a glacier.

terminal moraine (a) The end moraine, extending across a glacial valley as an arcuate or crescentic ridge, that marks the farthest advance or maximum extent of a glacier; the outermost end moraine of a glacier or ice sheet. It is formed at or near a more-or-less stationary edge, or at a place marking the cessation of an important glacial advance. Obsolete syn: *marginal moraine.* (b) A term sometimes used as a syn. of *end moraine.*

terminal plate A single plate at the end of an arm of an asteroid, appearing very early in ontogeny.

Terminal Proterozoic Former temporary name for the latest period of the *Neoproterozoic* Era prior to its formal definition (in 2004) as the *Ediacaran* Period. Also called: Neoproterozoic III.

terminal taxon *crown group.*

terminal tentacle A terminal *podium* of a radial vessel of the water-vascular system of an echinoid, extending through an *ocular pore.*

terminal velocity The limiting velocity reached asymptotically by a particle falling under the action of gravity in a still fluid (ASCE, 1962).

terminator (ter'-mi-na'-tor) The line separating the illuminated and the unilluminated parts of a celestial body (a planet, the Moon, etc.); the dividing line between day and night as observed from a distance.

terminus (ter'-mi-nus) The lower margin or extremity of a glacier; the snout. Syn: *glacier snout.*

termitarium (ter-mi-tar'-i-um) A mound of mud built by termites, ranging up to 4 m in height, and commonly found in the lateritic soil belts of the tropics and subtropics. Pl: termitaria. Syn: *anthill.*

ternary diagram (ter'-na-ry) A triangular diagram that graphically depicts the composition of a three-component mixture or ternary system.

ternary feldspar Any feldspar containing more than 5% of a third component; e.g. anorthoclase, soda sanidine, and potassian oligoclase or potassian andesine.

ternary system A system having three components, e.g. $CaO-Al_2O_3-SiO_2$.

ternesite A bright blue orthorhombic mineral: $Ca_5(SiO_4)_2SO_4$.

ternovite A silky white monoclinic mineral of the *franconite* group: $(Mg,Ca)Nb_4O_{11}\cdot 8\text{-}12H_2O$.

ternovskite (ter-nov'-skite) *magnesioriebeckite.*

terra (ter'-ra) A term established by the International Astronomical Union for an extensive land mass on planetary surfaces (USGS, 2003). Etymol: Latin, "earth". Pron: ter-uh. Pl: terrae. Syn: *continent.*

terra cariosa A syn. of *rottenstone.* Etymol: Latin, "rotten earth".

terrace [arch] (ter'-race) In archaeology, a field or habitation area artificially made level through the use of walls and fill. Often used for agriculture in steep terrain.

terrace [coast] (a) A narrow, gently sloping coastal *platform* veneered by a sedimentary deposit. Cf: *marine terrace.* (b) Loosely, a stripped *wave-cut platform* that has been exposed by uplift or by lowering of the water level; an elevated *wave-cut* or *wave-built* bench.

terrace [geomorph] (a) Any long, narrow, relatively level or gently inclined surface, generally less broad than a plain, bounded along one edge by a steeper descending slope and along the other by a steeper ascending slope; a large *bench* or steplike ledge that breaks the continuity of a slope. The term is usually applied to both the lower or front slope (the riser) and the flattish surface (the tread). A terrace commonly occurs along the margin and above the level of a body of water, marking a former water level; e.g., a *stream terrace.* (b) A term commonly but incorrectly applied to the deposit underlying the tread and riser of a terrace, esp. the alluvium of a stream terrace; "this deposit ... should more properly be referred to as a fill, alluvial fill, or alluvial deposit, in order to differentiate it from the topographic form" (Leopold et al., 1964, p.460). (c) *structural terrace.*

terrace [marine geol] A bench-like structure on the ocean floor.

terrace [soil] A horizontal or gently sloping ridge or embankment of earth built along the contours of a hillside for the purpose of conserving moisture, reducing erosion, or controlling runoff.

terrace cusp *meander cusp.*

terraced flowstone (ter'-raced) A series of rimstone dams or step-like flowstone levels.

terraced flute cast A flute cast with external sculpturing resembling differentially weathered laminae, but "in reality, a cast of differentially eroded laminations in the underlying shale and unrelated to internal structure of the cast" (Pettijohn and Potter, 1964, p.347).

terraced pool One of the shallow, rimmed pools on a reef surface, circular around the point where the water reaches the surface, produced by the growth of lime- and silica-secreting algae, and "arranged in successively lower terraces as sections of a circle around the overflow from the higher level" (Kuenen, 1950, p.431-432).

Terrace epoch (Ter'-race) An obsolete term formerly applied informally to the earlier part of the Holocene Epoch characterized by the formation of stream terraces in drift-filled valleys of the regions glaciated during the preceding Pleistocene Epoch. Syn: *Terracian.*

terrace flight A series of terraces resembling stairs steps, formed by the swinging meanders of a degrading stream that continuously excavates its valley. Syn: *flight.*

terrace line A fine raised ridge on the surface of a trilobite exoskeleton, particularly on the ventral surface of the *doublure.*

terrace meander A meander formed by the incision of a free meander associated with a former valley floor whose remnants now form a terrace (Schieferdecker, 1959, term 1492).

terrace placer *bench placer.*

terrace plain A well-developed stream terrace that represents a narrow but "true" plain (Tarr, 1902, p.88).

terrace slope The scarp or bluff below the outer edge of a terrace; the front or face of a terrace.

terracette [mass move] (ter-rac-ette') A small ledge, bench, or steplike form, or a series of such forms, produced on the surface of a slumped soil mass along a steep grassy slope or hillside, ranging from several centimeters to 1.5 m in height and averaging a meter in width, and developed as a result of small landslides and subsequent backward tilting of the soil surface. See also: *catstep.*

terracette [pat grd] An obsolete syn. of *step [pat grd].*

Terracian (Ter-ra'-ci-an) *Terrace epoch.*

terracing (ter'-rac-ing) (a) The formation of terraces, as by the shrinkage of a glacier. (b) A terraced structure, feature, or contour.

terra-cotta (ter'-ra-cot'-ta) A fired or kiln-burnt clay of a peculiar brownish-red or yellowish-red color, used for statuettes, figurines, and vases, and for ornamental work on the exterior of buildings. Also, an object made of terra-cotta. Etymol: Italian, "baked earth".

terra-cotta clay A term applied loosely to any fine-textured, fairly plastic clay that acquires a natural vitreous skin in burning and that is used in the manufacture of terra-cotta. It is characterized by low shrinkage, freedom from warping, strong bonding, and absence of soluble salts.

terrain (ter-rain') A tract or region of the Earth's surface considered as a physical feature, an ecologic environment, or a site of some planned activity of man, e.g. an engineering location; or in terms of military science, as in *terrain analysis* (Fairbridge, 1968, p.1145).

terrain analysis A study of the natural and man-made features of an area as they may be expected to affect military operations. Syn: *terrain assessment.*

terrain assessment *terrain analysis.*

terrain correction A correction applied to the simple *Bouguer correction* to compensate for the gravitational effects of topography around the point of gravity measurements. Calculated from a topographic map using the terrain-correction templates or zone charts (such as *Hammer charts*). Recent developments include computer algorithms based on the use of a digital topographic model called *Digital Elevation Model.*

terrain profile recorder *airborne profile recorder.*

terra Lemnia A clay, perhaps cimolite (Dana, 1892, p.689). Rarely used.

terra miraculosa *bole.*

terrane (ter-rane') A fault-bounded body of rock of regional extent, characterized by a geologic history different from that of contiguous terranes or bounding continents. A terrane is generally considered to be a discrete allochthonous fragment of oceanic or continental material added to a craton at an active margin by *accretion [struc geol].* See: D.L. Jones et al., 1983; Schermer et al., 1984. The term was first used in this sense by W.P. Irwin (1972). See also: *composite terrane; disrupted terrane; exotic terrane; metamorphic terrane; stratigraphic terrane; suspect terrane.* Syn: *accretionary terrane.* (b) Informally, a region where a particular rock or group of rocks predominates.

terranovaite A colorless or bluish orthorhombic zeolite mineral: $NaCa(Si,Al)_{20}O_{40}\cdot{\sim}8H_2O$.

terraqueous zone (ter-ra'-que-ous) That part of the lithosphere that is penetrated by water.

terra rosa *terra rossa.*

terra rossa A reddish-brown residual soil found as a mantle over limestone bedrock, typically in the karst areas around the Adriatic Sea, under conditions of Mediterranean-type climate. Also spelled: *terra rosa.* Etymol: Italian, "red earth".

terra verde A syn. of *green earth.* Etymol: Italian.

terreplein (ter'-re-plein) An embankment of earth with a broad level top. The French term is used primarily in military geology for an embankment behind which to emplace guns.

terrestrial (ter-res'-tri-al) (a) Pertaining to the Earth. Cf: *planetary; extraterrestrial.* (b) Pertaining to the Earth's dry land. Cf: *continental.*

terrestrial coordinates *geographic coordinates.*

terrestrial deposit (a) A sedimentary deposit laid down on land above tidal reach, as opposed to a marine deposit, and including sediments resulting from the activity of glaciers, wind, rainwash, or streams; e.g. a lake deposit, or a *continental deposit.* (b) Strictly, a sedimentary deposit laid down on land, as opposed to one resulting from the action of water; e.g. a glacial or eolian deposit. (c) A sedimentary deposit formed by springs or by underground water in rock cavities. Cf: *terrigenous deposit.*

terrestrial equator The equator on the Earth's surface. Cf: *astronomic equator; geodetic equator.* Syn: *equator.*

terrestrial geochron *geochron.*

terrestrial heat flow Flow of heat to the Earth's surface. Formerly used as synonym for terrestrial *heat-flow density.*

terrestrialization Peat formation by filling of a water body with organic remains, usually by gradual extension of peat-forming plant communities outwards from the shore of a lake. Cf: *paludification.*

terrestrial latitude Latitude on the Earth's surface; angular distance from the equator.

terrestrial longitude Longitude on the Earth's surface.

terrestrial magnetism *geomagnetism.*

terrestrial meridian A *meridian* on the Earth's surface; specif. an *astronomic meridian.*

terrestrial peat Peat that is developed above the water table. Syn: *terrestric peat.*

terrestrial planet One of the four inner planets of the solar system: Mercury, Venus, Earth, and Mars. Sometimes used to include all rocky (as opposed to gaseous) planetary bodies, and in that context the Moon and Pluto are sometimes regarded as terrestrial planets.

terrestrial pole *geographic pole.*

terrestrial radiation The total infrared radiation emitted from the Earth's surface. Cf: *counterradiation.* See also: *effective terrestrial radiation.* Syn: *Earth radiation.*

terrestric peat Peat formation above the general water table (Gore, 1983). Cf: *limnic peat; telmatic peat.* Syn: *terrestrial peat.*

terre verte A syn. of *green earth.* Etymol: French.

terrigenous (ter-rig'-e-nous) Derived from the land or continent.

terrigenous deposit Marine sediment consisting of material eroded from the land surface. Cf: *hemipelagic deposit; pelagic deposit.*

territorial sea (ter-ri-to'-ri-al) The coastal waters (and the accompanying seabed) under the jurisdiction of a maritime nation or state, usually measured from the mean low-water mark or from the seaward limit of a bay or river mouth and, as originally defined under international law, extending 3 nautical miles (about 5.6 km) outward to the high seas, although the U.S. officially accepts a 6-mile limit and attempts have been made to extend it to as much as 15 miles (28 km). See also: *marginal sea.* Syn: *territorial waters.*

territorial waters (a) The surface waters under the jurisdiction of a nation or state, including *inland waters* and *marginal sea.* (b) *territorial sea.*

tersia (ter'-si-a) A syn. of *tersioid outgrowth.* Pl: tersiae.

tersioid outgrowth (ter′-si-oid) In archaeocyathids, a fingerlike or roughly cylindrical holdfast, composed of close subparallel to curved plates connected by dissepiments, generally an expansion of the cup from the *intervallum* (TIP, 1972, pt. E, p.42). Syn: *tersia.*

terskite (ter′-skite) A pale lilac orthorhombic mineral: $Na_4ZrSi_6O_{15}(OH)_2{\cdot}H_2O$.

Tertiary (Ter′-tiar-y) The lower system of the *Cenozoic* Erathem of the Standard Global Chronostratigraphic Scale, above the *Cretaceous* System and below the *Quaternary.* Also the time during which these rocks were formed, the Tertiary Period, covering the time span between 65 and 1.75 Ma.

tertiary septum (ter′-ti-ar′-y) A third order of *septum* in scleractinian and rugose corals, normally shorter than a *major septum* or a *minor septum* (Fedorowski and Jull, 1976, p.62).

tertiary structure Very coarsely irregular shell material in the *lorica* of a tintinnid. Cf: *primary structure [paleont]; secondary structure [paleont].*

tertiopedate element Asymmetric ramiform *conodont element* with a posterior process and a lateral process on each side of the cusp; posterior process commonly long and denticulate (TIP, 1981, pt. W, supp.2).

tertschite (tertsch′-ite) A silky-white monoclinic mineral: $Ca_4B_{10}O_{19}{\cdot}20H_2O$.

teruggite (te-rugg′-ite) A colorless monoclinic mineral: $Ca_4MgAs_2B_{12}O_{22}(OH)_{12}{\cdot}14H_2O$.

teschemacherite (tesch′-e-mach′-er-ite) A yellowish to white orthorhombic mineral: $(NH_4)HCO_3$.

teschenite (tesch′-e-nite) A granular hypabyssal rock containing calcic plagioclase, augite, sometimes hornblende, and a small amount of biotite, with interstitial analcime. It is of darker color than *bogusite* and is distinguished from *theralite* by the presence of analcime in place of nepheline. Named by Hohenegger in 1861, for Teschen, Czech Republic. Syn: *analcime gabbro.* Obsolete.

tesla (tes′-la) A unit of *magnetic-field intensity,* equal to 10^4 gauss or 10^9 gammas. Cf: *nanotesla.*

tessellate (tes′-sel-late) Said of a plating arrangement in echinoderms, in which sutures between contiguous elements are vertical or nearly so and adjacent plates abut one another to form a mosaic, like tiles in a floor (Bell, 1976).

tessellation (tes-sel-la′-tion) A geomorphic feature resembling a mosaic pattern, for example a surface of sand-wedge polygons or of the salty crust of certain dry lakes in Australian deserts. Cf: *mud crack.* Etymol: Latin "tessella," small cube, mosaic piece.

tessera (tes′-sera) A term established by the International Astronomical Union for tile-like polygonal terrain on Venus having moderate to high relief and high radar reflectivity. Tessera surfaces are marked by a variety of patterns of linear scarps and troughs and linear to convoluted ridges interpreted as normal and strike-slip faults and folds of local to regional significance. Pl: tesserae.

tesserae (tes′-se-rae) A syn. of *felder.* Its singular form is tessera.

test [paleont] (a) The external shell, secreted exoskeleton, mesodermal endoskeleton, or other hard or firm covering or supporting structure of many invertebrates, such as the plates of the coronal, apical, periproctal, and peristomial systems of an echinoid; esp. a gelatinous, calcareous, or siliceous foraminiferal shell composed of secreted platelets or solid wall, agglutinated foreign particles, or a combination of these. A test may be enclosed within an outer layer of living tissue, such as a protozoan shell enclosed in cytoplasm. (b) The *theca* of a dinoflagellate.

test [petroleum] (a) Any procedure for sampling the content of an oil or gas reservoir; e.g. a drill-stem test or a wire-line test. (b) An informal syn. of test well.

testa (tes′-ta) The seed coat of a gymnosperm or a flowering plant developed from the *integument* of the ovule.

testaceous (tes-ta′-ceous) Having or consisting of a shell; specif. pertaining to the test of an invertebrate.

tester (test′-er) A service-company representative who supervises *drill-stem test* operations.

testibiopalladite (te-stib′-i-o-pal-lad′-ite) A metallic cubic mineral of the *gersdorffite* group: Pd(Sb,Bi)Te.

test pit [arch] *sondage.*

test pit [eng] A shallow excavation made with hand-operated shovel, backhoe, excavator, or bulldozer to expose soil and rock formations for observation and sampling, usually as part of a site investigation.

test reach A *reach [hydraul],* esp. one between two gaging stations, that is long enough to be used in the determination of slope.

test well [grd wat] (a) A well dug or drilled in search of water of adequate quality and quantity to meet the intended use. (b) A well dug or drilled for the purpose of making measurements of water levels and/or pumping rates during an aquifer test, obtaining geologic or water samples, running borehole geophysical logs, or conducting other types of tests.

test well [petroleum] An *exploratory well* drilled to determine the presence and commercial value of oil or gas in an unproven area. Informal syn: *test.*

tetartohedral (te-tar′-to-he′-dral) Said of that crystal class in a system, the general form of which has only one fourth the number of equivalent faces of the corresponding form in the *holohedral* class of the same system. Cf: *merohedral; hemihedral; ogdohedral.*

tetartohedron (te-tar′-to-he′-dron) Any crystal form in the tetartohedral class of a crystal system.

tetartoid (te-tar′-toid) An isometric, closed crystal form having 12 faces that correspond to one fourth of the faces of a hexoctahedron. A tetartoid may be right-handed or left-handed.

tetartoidal class (te-tar-toi′-dal) That crystal class of the isometric system having symmetry 23.

Tethys (Te′-thys) An ocean that occupied the general position of the Alpine-Himalayan orogenic belt between the Hercynian and Alpine orogenies. It was largely obliterated by the Alpine-Himalayan continental collision.

tetraauricupride (tet′-ra-au′-ri-cu′-pride) A metallic yellow tetragonal mineral: CuAu.

tetraclone (tet′-ra-clone) A four-armed desma of a sponge, built about a tetraxial crepis; e.g. a *trider.*

tetracoral (tet′-ra-cor′-al) (a) A coral with fourfold symmetry. (b) Any coral of the order Rugosa, a group restricted to the Paleozoic.

tetractine (te-trac′-tine) A sponge spicule having four rays.

tetrad [cryst] Noun or adj. for a symmetry axis that requires a rotation of 90° to repeat an object's appearance. Cf: *diad; triad; hexad.* Syn: *fourfold axis.*

tetrad [palyn] (tet′-rad) A cluster of four embryophytic spores or pollen grains resulting from meiotic division of one *spore mother cell.* Some pollen or spore types remain united in tetrads when dispersed, e.g., in the fossil *Classopollis* or the extant *Rhododendron.* Cf: *dyad; polyad.*

tetrad mark *laesura.*

tetrad scar *laesura.*

tetradymite (te-tra-dym′-ite) (a) A pale steel-gray rhombohedral mineral: Bi_2Te_2S. It occurs usually in foliated masses in auriferous veins, often with tellurobismuthite. Syn: *telluric bismuth.* (b) A group name for minerals with the general formula A_2B_2X, where A = Sb or Bi, B = Sb, Te, or Se, and X = S, Te, or Se.

tetraene (tet′-ra-ene) A sponge spicule with one long ray and four short rays at one end.

tetra-ferri-annite A reddish-brown to brownish-red monoclinic mineral of the *mica* group: $KFe^{2+}_3(Si_3Fe^{3+})O_{10}(OH)_2$. This represents annite with dominant Fe^{2+} in the octahedral structure site and essential Fe^{3+} in the tetrahedral structure site. Syn: *ferri-annite, monrepite.*

tetra-ferriphlogopite A monoclinic member of the *mica* group: $KMg_3Fe^{3+}Si_3O_{10}(OH)_2$. It represents phlogopite with essential Fe^{3+} in the tetrahedral structure site.

tetraferroplatinum (tet′-ra-fer′-ro-plat′-i-num) A metallic gray tetragonal mineral: PtFe.

tetragonal dipyramid (te-trag′-o-nal) A crystal form of eight faces consisting of two tetragonal pyramids repeated across a mirror plane of symmetry. A cross section perpendicular to the unique fourfold axis is ideally square. Its indices are $\{h0l\}$ or $\{hhl\}$ in most tetragonal crystals, also $\{hkl\}$ or $\{khl\}$ in class $4/m$.

tetragonal dipyramidal class That crystal class in the tetragonal system having symmetry $4/m$.

tetragonal disphenoid A crystal form consisting of four faces, ideally isosceles triangles, in which the unique $\bar{4}$-axis joins those two edges that are at right angles. Its indices are $\{hhl\}$, $\{h\bar{h}l\}$ in $\bar{4}2m$, or $\{h0l\}$, $\{hhl\}$, or $\{hkl\}$ in $\bar{4}$. Cf: *tetrahedron.*

tetragonal-disphenoidal class (te-trag′-o-nal-di-sphe-noi′-dal) That crystal class in the tetragonal system having symmetry $\bar{4}$.

tetragonal prism A crystal form of four equivalent faces parallel to the symmetry axis that are, ideally, square in cross section. Its indices are $\{100\}$ or $\{110\}$ with symmetry $4/m\ 2/m\ 2/m$, or $\{hk0\}$ in $4/m$.

tetragonal pyramid A crystal form consisting of four equivalent faces, ideally isosceles triangles, in a pyramid that is square in cross section. Its indices are $\{h0l\}$ and $\{hhl\}$ in symmetry 4mm, also $\{hkl\}$ in symmetry 4.

tetragonal-pyramidal class (te-trag′-o-nal-py-ram′-i-dal) That crystal class in the tetragonal system having symmetry 4.

tetragonal-scalenohedral class (te-trag′-o-nal-sca-le′-no-he′-dral) That crystal class in the tetragonal system having symmetry $\bar{4}2m$.

tetragonal scalenohedron A *scalenohedron* of eight faces, with symmetry $\bar{4}2m$ and indices $\{hkl\}$. It resembles a disphenoid. Cf: *hexagonal scalenohedron.*

tetragonal system One of the six *crystal systems,* characterized by three mutually perpendicular axes, the vertical one of which is a fourfold rotation or symmetry axis; it is longer or shorter than the two horizontal axes, which are of equal length. Cf: *isometric system; hexagonal system; orthorhombic system; monoclinic system; triclinic system.* Syn: *pyramidal system.*

tetragonal tetrad (te-trag′-o-nal) A uniplanar *tetrad* of palynomorphs, with all four members in one plane, as if fitting in a shallow rectangular box. This pattern is characteristic of monocotyledonous pollen and of monolete spores.

tetragonal-trapezohedral class (te-trag′-o-nal-tra-pe′-zo-he′-dral) That crystal class in the tetragonal system having symmetry 422.

tetragonal trapezohedron A crystal form of eight faces, each of which is a trapezium. Its indices are $\{hkl\}$ in symmetry 422, and it may be right- or left-handed.

tetragonal trisoctahedron *trapezohedron.*

tetragonal tristetrahedron *deltohedron.*

tetrahedral (tet-ra-he′-dral) Having the symmetry or form of a *tetrahedron.*
tetrahedral coordination An atomic arrangement in which an ion is surrounded by four ions of opposite sign, whose centers form the vertices of a tetrahedron around it. It is typified by SiO_4.
tetrahedral radius The radius of a cation when in tetrahedral coordination.
tetrahedral sheet A structural component of phyllosilicates that consists of a sheet of two anion (O or OH) planes in which the four-coordinated (i.e., tetrahedral) sites are occupied by cations (commonly Si, Al, or less commonly Fe^{3+}) in such a manner that individual tetrahedra share three corners with adjacent tetrahedra, and the fourth corner points in a direction normal to the anion planes. The resulting sheet forms a trigonal or hexagonal pattern involving six-fold rings of tetrahedra and has a siloxane surface (Sposito, 1989, p.129). Cf: *octahedral sheet.*
tetrahedral tetrad (tet-ra-he′-dral) A multiplanar *tetrad* of palynomorphs, with the proximal face of each member contacting those of the other members. The centers of the grains define a tetrahedron, e.g., in members of the Ericaceae.
tetrahedrite (tet-ra-he′-drite) A steel-gray to iron-black cubic mineral of the *tennantite* group: $(Cu,Fe,Ag,Zn)_{12}Sb_4S_{13}$. It often contains zinc, lead, mercury, cobalt, nickel, or silver replacing part of the copper. Tetrahedrite commonly occurs in characteristic tetrahedral crystals associated with copper ores. It is an important ore of copper and sometimes a valuable ore of silver. Syn: *fahlore; gray copper ore; panabase; stylotypite.*
tetrahedron (tet-ra-he′-dron) A crystal form in cubic crystals having symmetry $\bar{4}3m$ or 23. It is a four-faced polyhedron, each face of which is a triangle. Adj: *tetrahedral.*
tetrahedron hypothesis An obsolete hypothesis that the Earth, because of shrinking, tends to assume the form of a tetrahedron. For discussion see Holmes, 1965, p.32. Cf: *contracting Earth.*
tetrahexahedron (tet′-ra-hex′-a-he′-dron) An isometric crystal form having 24 faces that are isosceles triangles and that are arranged four to each side of a cube. Its indices are $\{hk0\}$ and its symmetry is $4/m\ \bar{3}\ 2/m$. Syn: *tetrakishexadron.*
tetrakalsilite (tet-ra-kal′-si-lite) *panunzite.*
tetrakishexadron (tet-ra′-kis-hex′-a-dron′) *tetrahexahedron.*
tetramorph (tet′-ra-morph) One of four crystal forms displaying *tetramorphism.*
tetramorphism (tet-ra-mor′-phism) That type of *polymorphism [cryst]* in which there are four crystal forms, known as *tetramorphs.* Adj: *tetramorphous.* Cf: *dimorphism [cryst]; trimorphism.*
tetramorphous (tet-ra-mor′-phous) Adj. of *tetramorphism.*
tetranatrolite (tet-ra-nat′-ro-lite) *gonnardite.*
tetrapod (tet′-ra-pod) n. An animal with four limbs; an informal term to distinguish amphibians, reptiles, and mammals from aquatic classes in which paired limbs are absent or are fins instead of legs. adj. Four-legged.
tetrarooseveltite An earthy white to yellowish-white tetragonal mineral: $BiAsO_4$.
tetrataenite A metallic creamy-white monoclinic meteorite mineral: FeNi.
tetrawickmanite (tet-ra-wick′-man-ite) A honey-yellow, brownish-orange, or bright yellow tetragonal mineral of the *stottite* group: $Mn^{2+}Sn^{4+}(OH)_6$. It is dimorphous with wickmanite.
tetraxon (te-trax′-on) A sponge spicule in which the rays grow along four axes arranged like the diagonals of a tetrahedron.
Texas bentonite A *sodium bentonite.*
texas tower (tex′-as) A radar-equipped offshore platform, mounted on the continental shelf or a shoal. These have been used in oil production in the Gulf of Mexico and to provide oceanographic and meteorological data. Etymol: texas, a structure on the deck of a steamboat.
textulariid (tex-tu-lar′-i-id) Any agglutinated foraminifer belonging to the family Textulariidae. Range, Carboniferous to Holocene.
textural lamina A form of horizontal lamination formed by deposition of sand on an upper plane bed that is characterized by systematic variation in grain size upward through the lamina; grain size may coarsen or fine upwards (Cheel and Middleton, 1986).
textural maturity (tex′-tur-al) A type of sedimentary maturity in which a sand approaches the textural end product to which it is driven by the formative processes that operate upon it. It is defined in terms of uniformity of particle size and perfection of rounding and depends upon the stability of the depositional site and the input of modifying wave and current energy; it is independent of mineral composition (Folk, 1951). Cf: *compositional maturity; mineralogic maturity.*
textural zoning Zonation of rocks and minerals by grain or crystal size.
texture [geomorph] (tex′-ture) *topographic texture.*
texture [petrology] The general physical appearance or character of a rock, including the geometric aspects of, and the mutual relations among, its component particles or crystals; e.g. the size, shape, and arrangement of the constituent elements of a sedimentary rock, or the crystallinity, granularity, and fabric of the constituent elements of an igneous rock. The term is applied to the smaller (megascopic or microscopic) features as seen on a smooth surface of a homogeneous rock or mineral aggregate. The term structure is generally used for the larger features of a rock. The two terms should not be used synonymously, although certain textural features may parallel major structural features. Confusion may arise because in some languages, e.g. French, the usage of texture and structure are the reverse of the English usage.
texture [remote sensing] The frequency of change and arrangement of tones on an image.
texture [sed] The group of properties of a *clastic deposit* that includes *particle shape, roundness,* surface features, grain size (and its distribution) and *fabric [sed]* (e.g., Pettijohn et al., 1972, p.68).
texture [soil] The physical nature of the soil according to the relative proportions of sand, clay, and silt. Cf: *loam; silt loam; clay [soil].*
texture [struc geol] Originally a metallurgical term synonymous with crystallographic preferred orientation. Geologists sometimes use the term, however, to denote the microscopic fabric of a deformed rock (see Wenk, 1985).
texture ratio The ratio of the greatest number of streams crossed by a contour line within a drainage basin to the length of the upper basin perimeter intercept (Smith, 1950, p.656); a measure of *topographic texture.*
TGA *thermogravimetric analysis.*
TGS *transcontinental geophysical survey.*
thadeuite (tha′-deu-ite) A yellow-orange orthorhombic mineral: $(Ca,Mn^{2+})(Mg,Fe^{2+},Mn^{2+})_3(PO_4)_2(OH,F)_2$.
thalassic (tha-las′-sic) (a) Pertaining to the deep ocean. (b) Pertaining to seas and gulfs. Not commonly used. Etymol: from the Greek word for ocean.
thalassocratic (tha-las′-so-crat′-ic) (a) Adj. of *thalassocraton.* (b) Said of a period of high sea level in the geologic past. Cf: *epeirocratic.*
thalassocraton (tha-las′-so-cra′-ton) Former continental craton that has subsided to submarine depths, interpreted as the result of crustal stretching and attenuation, e.g. the Rockall Plateau. It is the English version of "Tiefkraton" (Stille, 1940). Cf: *hedreocraton.* Adj: *thalassocratic.*
thalassogenesis (tha-las′-so-gen′-e-sis) A Russian term synonymous with *basification.*
thalassoid (tha-las′-soid) A lunar *mare basin* not filled, or only partly filled, with mare material; e.g. the Nectaris Basin.
thalcusite (thal′-cu-site) A metallic gray tetragonal mineral: $Tl_2(Cu,Fe)_4S_4$.
thalénite-(Y) (tha-len′-ite) A black, brown, greenish, or pink monoclinic mineral: $Y_3(OH)[Si_3O_{10}]$. Cf: *thortveitite; yttrialite-(Y), högtveitite.*
thalfenisite (thal-fe′-ni-site) A metallic brown cubic mineral, the Tl analogue of djerfisherite: $Tl_6(Fe,Ni,Cu)_{25}S_{26}Cl$.
thallite (thal′-lite) A yellowish-green *epidote.*
thallogen (thal′-lo-gen) *thallophyte.*
thallophyte (thal′-lo-phyte) A nonvascular plant without true roots, stems, or leaves. Algae are thallophytes. Cf: *bryophyte; pteridophyte.*
thallospore (thal′-lo-spore) A *fungal spore* formed on a *thallus (mycelium),* either singly or in chains within a *hypha,* and liberated by disintegration of the latter. The term is partly synonymous with *conidium* and *aleurospore.*
thallus [paleont] A plant body without differentiation into root, stem, and leaf, as in thallophytes.
thallus [palyn] In Fungi, the entire individual in its assimilative ("vegetative") phase; the *mycelium.*
thalweg [geomorph] The line of continuous maximum descent from any point on a land surface; e.g. the line of greatest slope along a valley floor, or the line crossing all contour lines at right angles, or the line connecting the lowest points along the bed of a stream. Etymol: German *Talweg,* "valley way". Also spelled: *talweg.*
thalweg [grd wat] A subsurface groundwater stream percolating beneath and generally in the same direction as the course of a surface stream or valley.
thalweg [streams] (a) The line connecting the lowest or deepest points along a stream bed or valley, whether under water or not; the *longitudinal profile* of a stream or valley; the line of maximum depth. Syn: *valley line.* (b) The median line of a stream; the *valley axis.* (c) *channel line.* Etymol: German, old spelling "Thalweg" (now "Talweg"), "valley way". Pron: [taal-vehk].
thamnasterioid (tham′-na-ste′-ri-oid) Said of a massive corallum characterized by absence of *corallite* walls and by confluent septa that join neighboring corallites, with a pattern of septa resembling lines of force in a magnetic field.
thanatocenosis (than′-a-to-ce-no′-sis) *thanatocoenosis.*
thanatocoenosis (than′-a-to-coe-no′-sis) (a) A set of fossils brought together after death by sedimentary processes, rather than by virtue of having originally lived there collectively. Cf: *biocoenosis.* Syn: *death assemblage.* (b) A group of fossils that may represent the *biocoenosis* of an area or the biocoenosis plus the thanatocoenosis of another environment; all the fossils present at a particular place in a sediment. The term was introduced by the German hydrobiologist Wasmund in 1926. Var: thanatocenosis; thanatocoenose; thanatocenose. Plural: thanatocoenoses. Etymol: Greek "thanatos", "death" + "koinos", "general, common". Syn: *taphocoenose; taphocoenosis.*
thanatotope (than′-a-to-tope′) The total area in which the dead specimens of a taxon or taxa are deposited.
Thanetian (Than-e′-tian) A stage of the Standard Global Chronostratigraphic Scale: upper Paleocene (above Selandian, below Ypresian Stage of the Eocene).

thaumasite (thau′-ma-site) A white to pale yellow acicular to compact hexagonal mineral: $Ca_3Si(OH)_6(SO_4)(CO_3)\cdot 12H_2O$. At present, it is the only known mineral with silicon octahedrally coordinated by hydroxyls.

thaw v. To go from a frozen state, as ice, to a liquid state; to melt. n. The end of a frost, when the temperature rises above the freezing point, and ice or snow undergo melting. Also, the transformation of ice or snow to water.

thaw depression A hollow in the ground resulting from subsidence following the local melting of ground ice in a permafrost region. See also: *cave-in lake.* Syn: *thermokarst depression.*

thaw hole A vertical hole in sea ice, formed where a surface puddle melts through to the underlying water.

thaw lake [glaciol] A pool formed on the surface of a large glacier by accumulation of meltwater.

thaw lake [permafrost] *thermokarst lake.*

thaw sink A closed *thaw depression* with subterranean drainage, believed to have originated as a thermokarst lake (Hopkins, 1949).

THDM *translucent humic degradation matter.*

theater-headed valley In glacial geology, a valley with an upstream end in a half-bowl shape, like a Greek hillside theater. Cf: *amphitheater.*

theca (the′-ca) (a) An echinoderm skeleton consisting of calcareous plates and enclosing the body and internal organs; e.g. the *dorsal cup* of the calyx of a crinoid. The term is generally applied to all fossilized parts, and includes ambulacra but excludes the column or stem and appendages such as free arms and brachioles. (b) An individual tube or cup (other than the sicula) that housed a zooid of a graptolite colony; the term is commonly used for the autotheca. (c) The external skeletal deposit of a coelenterate, such as the calcareous wall enclosing a corallite and presumably the sides of a coral polyp. (d) The sometimes resistant-walled coat or external covering, formed of numerous plates, of the nonencysted or actively swimming stage of the life cycle of some dinoflagellates. Syn: *test [paleont].* Pl: thecae.

thecal plate (the′-cal) Any of numerous calcareous plates that form an element in the *theca* of an echinoderm. It is usually distinguished from a plate of an ambulacrum or arm.

thecamoebian (the-ca-moe′-bi-an) Any one of a group of usually freshwater testaceous protozoans, the fossil members of which belong to the orders Arcellinida and Gromida and to part of the suborder Allogromiina of the foraminifers. Range, Eocene to Holocene.

thecodont (the′-co-dont) adj. Pertaining to vertebrate teeth set in sockets, or to the implantation of such teeth.

Thecodontia (The-co-don′-ti-a) An order of primitive archosaurian reptiles of varied body forms, some foreshadowing dinosaurs and some crocodilians. Range, Upper Permian to Upper Triassic.

Theis curve A log-log plot of drawdown or recovery of head against time, used in an aquifer test based on the *Theis equation.* Used in conjunction with the non-equilibrium type curve.

Theis equation A nonequilibrium differential equation relating drawdown or recovery of groundwater head to rate of withdrawal or addition of water to time since pumping began, and to the hydraulic characteristics of the aquifer, e.g., *transmissivity* and *storage coefficient* (Theis, 1935, p.520). Named after C.V. Theis, a groundwater engineer with the U.S. Geological Survey. See also: *Theis curve.*

theisite (theis′-ite) A pale blue-green orthorhombic mineral: $Cu_5Zn_5(AsO_4,SbO_4)_2(OH)_{14}$.

Theis method A graphical method of matching the *Theis type curve* with field data obtained during an *aquifer test,* used to determine values of *transmissivity* and *storage coefficient.* Cf: *Theis equation; Thiem equation.*

Theis type curve A log-log plot of the well function, W(u), versus the argument u, in which u describes for a *confined aquifer* the theoretical relationships among observation-well distance, time since pumping began, and the aquifer storativity coefficient and *transmissivity.* It is used to determine aquifer properties, drawdown, or pumping rate, based on the results of an *aquifer test* (Todd, 1959, p.90-94). Cf: *Theis equation; Thiem equation.*

thematic map A map designed to convey information about a special subject, such as geology, rainfall, or population. Syn: *special-subject map; topical map.*

thematic mapper (a) A remote-sensing device which measures the radiation in frequency bands which are sensitive to variation in vegetation (for example). (b) The sensor on Landsat D satellites that scans and stores seven individual images in spectral bands ranging from blue through thermal infrared. The ground resolution of TM pixels is 30 meters on a side. Aeroservice tradename.

thenardite (the-nard′-ite) A white or brownish orthorhombic mineral: Na_2SO_4. It occurs in masses or crusts often in connection with salt lakes.

The National Map An interactive online map service provided by the U. S. Geological Survey that offers public access to seamless, high-quality, nationally consistent geospatial, geographic names information, and software tools. Website at http://nationalmap.usgs.gov/index.html. Syn: *National Map.*

theodolite (the-od′-o-lite) A precision surveying instrument that can be rotated on a horizontal base so as to be sighted first upon one point and then upon another and that is used for measuring angular distances in both vertical and horizontal planes. It consists of a telescope so mounted on a tripod as to swivel vertically in supports secured to a revolvable table carrying a vernier for reading horizontal angles (azimuths), and usually includes a compass, a spirit level, and an accurately graduated circle for determining vertical angles (altitudes). See also: *transit.*

theoparacelsite A dark pistachio-green orthorhombic mineral: $Cu_3(OH)_2As_2O_7$.

theophrastite (the-o-phrast′-ite) A green trigonal mineral of the *brucite* group: $Ni(OH)_2$.

theory (the′-o-ry) An explanatory system of propositions, general principles or laws, inferred from the phenomena and linking known facts and observations; held to be true until contradicted or amended by new facts or observations. Examples: Quantum Theory, Theory of Evolution, Theory of Continental Drift.

theralite (ther′-a-lite) (a) In the *IUGS classification,* a plutonic rock with F between 10 and 60, and P/(A+P) greater than 90. Cf: *foid diorite.* Syn: *foid gabbro.* (b) A group of mafic plutonic rocks composed of calcic plagioclase, feldspathoids, and augite, with lesser amounts of sodic sanidine and sodic amphiboles and accessory olivine; also, any rock in that group; the intrusive equivalent of *tephrite.* Theralite grades into nepheline monzonite with an increase in the alkali feldspar content, into gabbro as the feldspathoid content diminishes, and into diorite with both fewer feldspathoids and increasingly sodic plagioclase. The term, defined by Rosenbusch in 1887, is derived from the Greek word for "eagerly looked for", and not from the island of Thera (Santorini).

Therapsida (The-rap′-si-da) An order of synapsid reptiles of nearly mammalian grade in limb structure, tail length, and general pattern of skull; mammal-like reptiles *sensu stricto.* Range, Upper Permian to Lower Jurassic. Cf: *Synapsida.*

thérèsemagnanite A pearly pink hexagonal mineral: $(Co,Zn,Ni)_6SO_4(OH,Cl)_{10}\cdot 8H_2O$.

Theria (The′-ri-a) A subclass that includes all living mammals except the egg-laying Monotremata. Range, Upper Jurassic to Holocene.

Theriodontia (The′-ri-o-don′-ti-a) Suborder of therapsid reptiles characterized by a tendency toward predaceous habit and a mammalian morphological and functional grade. Range, Upper Permian to Lower Jurassic.

thermal [glac geol] (ther′-mal) In glacial geology, used as a noun synonymous with *interglacial stage.*

thermal [meteorol] n. A vertically moving convective air current of air that is caused by differential heating of the ground below it.

thermal [phys] Pertaining to or caused by heat. Syn: *thermic.*

thermal alteration The chemical alteration of organic matter in sedimentary rock during diagenesis, resulting from pressure and temperature increase with depth and length of burial. Through study of the constituents of a palynological assemblage, such as palynomorphs and vitrinite particles, an assessment of the degree of thermal alteration of the organic matter in the rock and its relationship to hydrocarbon generation can be obtained.

thermal alteration index A scale based on color by transmitted light or on reflectance of particles of organic matter in sedimentary rocks. The data are used in palynological estimates of hydrocarbon maturity and source potential. Abbrev: TAI.

thermal analysis The study of chemical and/or physical changes in materials as a function of temperature, i.e. the heat evolved or absorbed during such changes. See also: *differential thermal analysis.* Syn: *thermoanalysis.*

thermal aureole *aureole.*

thermal band *thermal infrared.*

thermal bar A boundary separating regions of a large lake having considerable differences in surface temperature; commonly occurring in the spring and autumn in large lakes in temperate climates. In the spring, water shoreward of the thermal bar is commonly warmer than the main body of the lake, and in the autumn the shoreward water is colder.

thermal conduction *heat conduction.*

thermal conductivity (a) The time rate of heat flow by conduction, through unit thickness across unit area for unit difference of temperature; measured in watts per meter kelvin ($W\ m^{-1}K^{-1}$). (b) A measure of the ability of a material to conduct heat. A basic property of rocks that changes dependent on composition and geometry of rock matrix, porosity, and the pore medium. Usually measured on drill cores in the laboratory. Symbol: λ, K.

thermal contraction crack A nearly vertical fracture developed by thermal contraction in rock or in frozen ground with appreciable ice content. Thermal contraction cracks commonly intersect to form polygonal patterns in plan view. Syn: *ice crack; contraction crack; frost crack.*

thermal convection Mode of heat transfer by fluid flow as a consequence of pressure gradients (forced convection) or by buoyancy effects of density gradients caused by temperature gradients (free convection) (Jessop, 1990). Syn: *heat convection* (rarely used).

thermal convection [ign] Density driven convection within a magma chamber wherein density contrasts are generated by local differences in magma temperature.

thermal demagnetization A technique of partial *demagnetization* by progressively heating the specimen to a proscribed temperature, then cooling to room temperature in a nonmagnetic space, to destroy (unblock) a *partial thermoremanent magnetization* for that temperature interval, thus leaving unaffected a partial thermoremanent magnetization of higher unblocking temperatures. Cf: *alternating-field demagnetization; chemical demagnetization.*

thermal diffusivity The measure of the rate of change of temperature in a unit volume due to a specified heat flow. Thermal diffusivity is given by the *thermal conductivity* of a substance divided by the product of its density and specific heat capacity. In rock, the common range of values is 0.5-2.5 x 10^{-6} m^2s^{-1}.

thermal energy yield In volcanology, the thermal energy equivalent of a volcanic eruption, based on the thermal characteristics and volume of the volcanic products (Yokoyama, 1956-57, p.75-97).

thermal exfoliation A hypothetical type of *exfoliation* caused by the heating of rock during the day and its rapid cooling at night.

thermal exploration *geothermal exploration.*

thermal fracture The disintegration of, or the formation of a fracture or crack in, a rock as a result of sudden temperature changes. It is thought to occur where rock-forming minerals have varying coefficients of expansion, where a cliff face receives and loses radiation rapidly, or where there is a quick drop of temperature after sundown. Also, the result of such a process.

thermal inertia Measure of the response of a material to temperature changes, expressed in calories per square centimeter per square root of second. Symbol: P.

thermal infrared Referring to that portion of the infrared spectral region that ranges in wavelength from 3.0 to 15 μm. The thermal-infrared band includes the *radiant-power peak* of the Earth. Abbrev: *thermal IR.* Syn: *thermal band.*

thermal IR *thermal infrared.*

thermal log *temperature log.*

thermal maturation A rock or petroleum that has been exposed to high temperatures resulting in a different distribution of compounds.

thermal maximum A term proposed by Flint and Deevey (1951) as a substitute for *Climatic Optimum* and *Altithermal.*

thermal metamorphism A type of metamorphism resulting in chemical reconstitution controlled by a temperature increase and influenced to a lesser extent by the confining pressure; there is no requirement of simultaneous deformation. Cf: *geothermal metamorphism; load metamorphism; pyrometamorphism; static metamorphism.* Syn: *thermometamorphism.*

thermal-neutron log (ther'-mal-neu'-tron) A *radioactivity log* of the neutron-neutron type, in which the detector discriminates for slowed neutrons of thermal-energy level (about 0.025 ev). Cf: *neutron-gamma log; epithermal-neutron log.* Syn: *n-tn log.*

thermal niche A niche produced by rapid thaw and removal of permafrost at waterlines.

thermal parameters *displacement parameters.*

thermal pressure Pressure that is produced by the thermal expansion of fluids in a medium that is less thermally expansive. Syn: *aquathermal pressure.*

thermal resistance Temperature difference divided by heat-flow rate measured in kelvin per watt (KW^{-1}). Symbol: R.

thermal resistivity The reciprocal of thermal conductivity, measured in meter kelvin per watt (m K W^{-1}). Symbol: r.

thermal shock Failure of a material, esp. a brittle material, due to the thermal stress of rapidly rising temperature.

thermal spring A spring whose water temperature is appreciably higher than the local mean annual atmospheric temperature. A thermal spring may be a *hot spring* or a *warm spring* (Meinzer, 1923, p.54).

thermal stratification The stratification of a lake produced by changes in temperature at different depths and resulting in horizontal layers of differing densities. See also: *density stratification.*

thermal structure (a) Indicates the pattern of change of temperature conditions in the subsurface, e.g. by the arrangement of isotherms. (b) An arrangement of zones of increasing metamorphic grade in some distinct structural pattern, for example, a thermal anticline or a thermal dome. Such features are associated with and are produced by a localized heat source, possibly accompanied by anatexis or orogenesis (Winkler, 1967).

thermal survey Investigation of temperature conditions and their display as profiles, maps, or 3D graphs. Geothermal data are commonly shown on maps, usually as the final step of the data acquisition or data compilation, displaying local, regional, or global patterns of *temperature, temperature gradient, heat-flow density* as well as geothermal resources. Syn: *temperature survey.*

thermal water Water, generally of a spring or geyser, whose temperature is appreciably above the local mean annual air temperature.

thermic (ther'-mic) *thermal [phys].*

thermic temperature regime A soil temperature regime in which the mean annual temperature at 50-cm depth is at least 15°C but less than 22°C, with a variation of more than 5°C between summer and winter measurements (USDA, 1975). Cf: *isothermic temperature regime.*

thermistor chain (therm'-is-tor) A series of temperature-measuring instruments spaced along a cable or line and used to measure the vertical gradient of seawater temperature.

thermite (ther'-mite) An old name for any fossil substance that is combustible.

thermoanalysis (ther'-mo-a-nal'-y-sis) *thermal analysis.*

thermobarometer (ther'-mo-ba-rom'-e-ter) *hypsometer.*

thermochronology The determination of temperature histories in metamorphic rocks. It is based on *geochronology,* the determination of the age of a sample based on the rate of decay of a radioactive nuclide.

thermocline [lake] (ther'-mo-cline) (a) The horizontal plane in a thermally stratified lake located at the depth where temperature decreases most rapidly with depth. (b) The horizontal layer of water characterized by a rapid decrease of temperature and increase of density with depth; sometimes arbitrarily defined as the layer in which the rate of temperature decrease with depth is equal to at least 1°C/m. This definition describes an older and less preferred synonym of *metalimnion* but one that is often used in engineering literature (Wetzel, 2001, p.74).

thermocline [oceanog] (a) The depth of the maximum rate of decrease of temperature with depth, the depth of the maximum of the vertical temperature gradient. (b) More generally, the region of large vertical temperature gradients below the *mixed layer.*

thermodynamic equilibrium constant (ther'-mo-dy-nam'-ic) *equilibrium constant.*

thermodynamic potential Any thermodynamic function of state, an extremum of which is a necessary and sufficient criterion of equilibrium for a system under specified conditions. Thus, for example, the *Gibbs free energy* is the thermodynamic potential for a system at constant pressure and temperature, the *Helmholtz free energy* is the thermodynamic potential for a system at constant temperature and volume. See also: *free energy.*

thermodynamics (ther'-mo-dy-nam'-ics) The mathematical treatment of the relation of heat to mechanical and other forms of energy.

thermoelastic effect (ther'-mo-e-las'-tic) A fall in temperature under tension or a rise in temperature under compression during elastic deformation.

thermoelastic stress Stress in a body caused by a local temperature gradient within the body.

thermoerosional niche (ther'-mo-e-ro'-sion-al) A niche resulting from undercutting produced by bank erosion in north-flowing Arctic rivers during the limited summer flow, esp. in Alaska and Siberia where niches as wide as 8 m have been formed in one year (Hamelin and Cook, 1967, p.123).

thermogene (ther'-mo-gene) Pertaining to the formation of minerals primarily under the influence of temperature (Kostov, 1961).

thermogenesis (ther-mo-gen'-e-sis) A rise in temperature in a body from reactions within it, as by oxidation or the decay of radioactive elements.

thermogram (ther'-mo-gram) An image acquired by a stationary scanner operating in the thermal infrared spectral region. Thermogram and thermography refers to medical applications. The term is sometimes used erroneously as a syn. of a *thermal infrared* image.

thermograph [meteorol] A recording instrument that provides a continuous trace of *temperature* variations with time.

thermography (ther-mog'-ra-phy) A term suggested (R. Williams, 1972) as a replacement for the phrase "thermal infrared imagery", analogous to the term "photography". No confusion would then exist between "photograph" (a record of reflected solar energy) and "thermograph" (a record of emitted thermal energy), such as now exists between infrared photography and infrared imagery.

thermogravimetric analysis (ther'-mo-grav'-i-met'-ric) A syn. of *thermogravimetry.* Abbrev: TGA.

thermogravimetry (ther'-mo-gra-vim'-e-try) A method of analysis that measures the loss or gain of weight by a substance as the temperature of the substance is raised or lowered at a constant rate. Syn: *thermogravimetric analysis.*

thermohaline circulation (ther-mo-hal'-ine) Meridional circulation in the ocean fed by sinking of dense water at high latitudes and driven by deep mixing which reduces density and allows deep water to rise. "Thermohaline" refers to control of density by temperature and salt content (Wunsch, 2003).

thermokarst (ther'-mo-karst') (a) Karstlike topographic features produced in a permafrost region by the local melting of ground ice and the subsequent settling of the ground. Cf: *glaciokarst.* (b) A region marked by *thermokarst topography.* (c) The process of formation of a thermokarst topography. Syn: *cryokarst.*

thermokarst depression *thaw depression.*

thermokarst lake A shallow body of water whose basin is produced by collapse of the ground following thawing of ground ice in regions underlain by permafrost; a lake occupying a *thaw depression.* Cf: *kettle lake.* Syn: *thaw lake [permafrost]; cryogenic lake; cave-in lake.*

thermokarst mound A residual polygonal hummock, bordered by depressions that were formed by the melting of ground ice in a permafrost region. Cf: *cemetery mound.*

thermokarst topography An irregular land surface containing thermokarst lakes, bogs, caverns, pits, and other small depressions, formed in a permafrost region by the melting of ground ice; in exterior appearance, it resembles the uneven *karst topography* formed by the solution of limestone.

thermolabile (ther-mo-la'-bile) Said of a material that is decomposable by heat.

thermoluminescence (ther'-mo-lu'-mi-nes'-cence) The property possessed by many substances of emitting light when heated to a temperature below that of incandescence. It results from release of energy stored as electron displacements in the crystal lattice.

thermoluminescence dating A method of dating applicable to materials that have once been heated (e.g. pottery, lava flows) or which have been exposed to ultraviolet light from the sun. A fraction of the energy released by decay of long-lived radioactive nuclides is stored as trapped electrons, and this energy is released as light upon heating. The age of a sample can be determined if the natural thermoluminescence is measured, the thermoluminescence induced by a known radiation dose is measured, and the radiation dose received by the sample per unit time in the past is measured.

thermomer (ther'-mo-mer) A relatively warm period within the Pleistocene Epoch, such as an *interstade* (Lüttig, 1965, p.582). Ant: *kryomer.* Rarely used.

thermometamorphism *thermal metamorphism.*

thermometric depth (ther-mo-met'-ric) The depth in meters calculated from the difference in temperature between paired protected and unprotected reversing thermometers. The unprotected thermometer will read higher by about 1° per every hundred meters of depth as it is affected by hydrostatic pressure. The thermometric depth is used as a fast check to see if the water bottles tripped at the desired depth and did not pre- or post-trip at some unknown depth as they were being lowered or raised respectively. Cf: *Nansen bottle.*

thermometric leveling A type of indirect leveling in which elevations above sea level are determined from observed values of the boiling point of water. Cf: *barometric leveling.*

thermonatrite (ther-mo-na'-trite) A white, yelllowish or grayish orthorhombic mineral: $Na_2CO_3{\cdot}H_2O$. It is found in some lakes and alkali soils, and as a saline residue.

thermo-osmosis (ther'-mo-os-mo'-sis) Osmosis occurring under the influence of a temperature difference between fluids on either side of a semipermeable membrane, with movement from the warmer to the cooler side, as when water flows in small openings from the warmer to the cooler parts of a soil mass.

thermophilic (ther-mo-phil'-ic) Said of an organism that prefers high temperatures, esp. bacteria that thrive in temperatures between 45°C and 80°C. Noun: thermophile.

thermophyte (ther'-mo-phyte) A plant preferring high temperatures for growth.

thermoremanence (ther-mo-rem'-a-nence) *thermoremanent magnetization.*

thermoremanent magnetization (ther-mo-rem'-a-nent) Remanent magnetization acquired as a rock cools in a magnetic field from above the *Curie point* down to ambient temperature. It is very stable and is parallel to the ambient field at the time of cooling. Abbrev: TRM. Syn: *thermoremanence.*

thermosphere (ther'-mo-sphere) That part of the marine realm characterized by warm or seasonably variable hydroclimates that are concentrated in low and middle latitudes. Cf: *psychrosphere.*

thermotectonic Obsolete. adj. Said of the relationships which exist between thermal events and tectonic activity in an orogenic area, e.g. during an orogenic event a collisional plate boundary may involve subduction, thrusting, folding, doming, and magma intrusion. These processes would result in perturbations of the pre-orogenic thermal structure of the crust, which would eventually decay back towards steady state values in the post-orogenic phase.

thermuticle (ther-mu'-ti-cle) *porcellanite.*

theta porphyroclast Rigid *porphyroclast* devoid of recrystallized rim or tails in a high-strain rock such as mylonite.

Thetis hairstone (The'-tis) A variety of *hairstone* containing or penetrated by tangled or ball-like inclusions of green fibrous crystals of hornblende, asbestos, and esp. actinolite.

thetomorphic glass An obsolete syn. of *diaplectic glass.*

thick bands In banded coal, vitrain bands from 5.0 to 50.0 mm thick (Schopf, 1960, p.39). Cf: *thin bands; medium bands; very thick bands.*

thick-bedded A relative term applied to sedimentary beds variously defined as more than 6.4 cm (2.5 in.) to more than 100 cm (40 in.) in thickness; specif. said of a bed whose thickness is in the range of 60-120 cm (2-4 ft), a bed greater than 120 cm being "very thick-bedded" (McKee and Weir, 1953, p.383). Cf: *medium-bedded; stratification index; thin-bedded.*

thicket reef A reef consisting of closely interfingered delicate branching corals. Cf: *pillar reef; coral thicket.*

thickness [geol] (thick'-ness) The extent of a tabular unit from its bottom boundary surface to its top surface, usually measured normal to these surfaces; e.g. the distance at right angles between the hanging wall and the footwall of a lode, or the dimension of a stratigraphic unit measured at right angles to the bedding surface. See also: *true thickness; apparent thickness.*

thickness [paleont] (a) The greatest distance between the two valves of a brachiopod shell at right angles to the length and width. It is equal to the height in biconvex, plano-convex, and convexo-plane shells. (b) The distance between the inner and outer surfaces of the wall of a bivalve-mollusk shell. Also, a term sometimes used to denote the shell measurement of a bivalve mollusk commonly called inflation.

thickness contour *isopach.*

thickness line *isopach.*

thickness map *isopach map.*

thick-skinned deformation A style of regional deformation that involves crystalline basement, often by slip on basement-penetrating faults.

thick-skinned structure A term used for an interpretation of folds on faults in which faults are basement penetrating (perhaps dying out at mid-crustal levels). Folds are monoclines formed as cover strata are forced up by movement of the fault block.

thick/thin bed Classification of a reflecting bed depending on its thickness with respect to a quarter wavelength; thick is $> \lambda/4$, thin is $< \lambda/4$, where λ is the wavelength. Seismic interpretation methods are different above and below this thickness.

thief formation A very permeable rock unit penetrated during drilling operations where pressure is lower, and fluid, formation water, or hydrocarbons can be lost. Syn: *thief zone.*

thief zone *thief formation.*

Thiem equation An equilibrium equation that relates drawdowns in multiple observation wells to the discharge of a pumped well during an *aquifer test* and to the *hydraulic conductivity* of the aquifer (Todd, 1959, p.83). In order to apply the equation, pumping must continue at a steady rate for a sufficient time to approach steady-state conditions. Named after Günther Thiem. See also: *Theis equation.*

thigmation (thig-ma'-tion) A term describing the configuration of the boundaries between plates of adjacent series in dinoflagellate thecae or cysts.

thigmotaxis (thig-mo-tax'-is) *Taxis [ecol]* in response to mechanical or tactile stimuli. Cf: *strophotaxis; phobotaxis.*

thill A British term for the floor of a coal mine or coal seam; hence, *underclay.* The term is also used for a thin stratum of fireclay.

thin bands In banded coal, vitrain bands from 0.5 to 2.0 mm thick (Schopf, 1960, p.39). Cf: *medium bands; thick bands; very thick bands.*

thin-bedded A relative term applied to sedimentary beds variously defined as less than 30 cm (1 ft) to less than 1 cm (0.4 in.) in thickness; specif. said of a bed whose thickness is in the range of 5-60 cm (2 in. to 2 ft.), a bed less than 5 cm but more than 1 cm thick being "very thin-bedded" (McKee and Weir, 1953, p.383). Cf: *medium-bedded; stratification index; thick-bedded.*

thinic (thi'-nic) Pertaining to a sand dune (Klugh, 1923, p.374).

thin-layer chromatography An essentially adsorptive chromatographic technique for separating components of a sample by moving it in a mixture or solution through a uniformly thin deposit of an adsorbent on rigid supporting plates in such a way that the different components have different mobilities and thus become separated (May and Cuttitta, 1967, p.116). Abbrev: TLC. Cf: *chromatography.*

thinning The process of reducing the number of grid coordinates defining a line, most often used during the conversion of lines from grid to vector format. A line is "thinned" if each grid cell defining a line is adjoined to only two other cells that are defining the same feature (Krzanowski et al., 1993). See also: *smoothing; weeding.*

thinolite (thin'-o-lite) (a) A pale-yellow to light-brown variety of calcite, often terminated at both ends by pyramids. It may be pseudomorphous after gaylussite. (b) *thinolitic tufa.*

thinolitic tufa (thin-o-lit'-ic) A tufa deposit consisting in part of layers of delicate prismatic skeletal crystals of *thinolite,* up to 20 cm long and 1 cm thick; it occurs in the desert basins of NW Nevada, as in domelike masses along the shore of extinct Lake Lahontan, where it overlies *lithoid tufa* and underlies *dendroid tufa.*

thin out To grow progressively thinner in one direction until extinction. The term is applied to a stratum, vein, or other body of rock that decreases gradually in thickness so that its upper and lower surfaces eventually meet and the layer of rock disappears. The thinning may be original or due to truncation beneath an unconformity. Syn: *pinch out; wedge out.*

thin section A fragment of rock or mineral mechanically ground to a thickness of approximately 0.03 mm, and mounted between glasses as a microscope slide. This reduction renders most rocks and minerals transparent or translucent, thus making it possible to study their optical properties.

thin-skinned deformation A style of regional deformation in which faulting and folding occurs above a subhorizontal detachment and does not involve basement below the detachment.

thin-skinned structure A term used by Rodgers (1963) for an interpretation that folds and faults of miogeosynclinal and foreland rocks in an orogenic belt involve only the strata that occur above a *décollement* beneath which the structure differs. Proposed examples are in the Valley and Ridge and Plateau provinces of the Appalachian belt and in the Jura Mountains. Some authors use the term in a broader sense in reference to a thrust system occurring above a subhorizontal detachment, even in cases where thrust slices include basement rocks.

thiospinel (thi'-o-spi-nel') A general term for minerals with the spinel structure having the general formula: AB_2S_4.

third-law entropy The difference in entropy between a substance at some finite temperature and at absolute zero, as defined by the *third law of thermodynamics.*

third law of thermodynamics The statement that the entropy of any perfect crystalline substance becomes zero at the absolute zero of temperature. See also: *third-law entropy.*

third-order leveling Spirit leveling that does not attain the quality of *second-order leveling,* in which lines are not extended more than 30 miles (48.3 km) from lines of first- or second-order leveling and must close upon lines of equal or higher order of accuracy and in which the maximum allowable discrepancy is 12 mm times the square root of the length of the line in kilometers (or 0.05 ft times the square root of the distance in miles). It is used for subdividing loops of first- and second-order leveling and in providing local vertical control for detailed surveys. Cf: *first-order leveling.*

third-order pinacoid In a triclinic crystal, any $\{hk0\}$ or $\{\bar{h}k0\}$ pinacoid, with symmetry $\bar{1}$. Cf: *first-order pinacoid; second-order pinacoid; fourth-order pinacoid.*

third-order prism A crystal form: in a tetragonal crystal, a $\{hk0\}$ prism, with symmetry $4/m$, 4, or $\bar{4}$; in a hexagonal crystal, a $\{hki0\}$ prism, with symmetry $6/m$, 6, or $\bar{3}$; and in orthorhombic and monoclinic crystals, any $\{hk0\}$ prism. Cf: *first-order prism; second-order prism; fourth-order prism.*

third-order sequence In sequence stratigraphy, a sequence deposited in about one to three million years. This sequence has no internal unconformities and consists of lowstand, transgressive, and highstand systems tracts (Vail et al., 1977). These systems tracts are composed of parasequences. There are no high-frequency sequences or unconformities within a third-order sequence. However, composite sequences, deposited in 1-3 million years are common in the rock record.

third water An obsolete term for the quality or luster of a gemstone next below *second water,* such as that of a diamond that is clearly imperfect and contains flaws. Cf: *first water.*

thixotropic clay (thix-o-trop'-ic) A clay that displays *thixotropy,* i.e. weakens when disturbed and strengthens when left undisturbed. Syn: *false body.*

thixotropy (thix-ot'-ro-py) The property of certain colloidal substances, e.g. a *thixotropic clay,* to weaken or change from a gel to a sol when shaken but to increase in strength upon standing.

tholeiite (tho'-lei-ite) A *basalt* characterized by the presence of orthopyroxene in the norm and, in some cases, low-calcium pyroxenes (orthopyroxene and/or pigeonite) in the mode. Olivine or quartz may be present in the norm but not nepheline. The term, first used in 1840 by Steininger, was derived from Tholey, Saarland, Germany, and was applied to a sill of altered *andesite.* The term was given its present meaning by Kennedy in 1933. Chayes (1966) reviewed the history of the term and recommended its replacement by *subalkaline basalt.* Cf: *basalt [petrology].* Syn: *tholeiitic basalt; subalkaline basalt.*

tholeiitic basalt (tho-lei-it'-ic) The preferred name of the common *subalkaline basalt* (e.g., MORB). Not restricted to a particular tectonic environment. Also, any of the basaltic members of a tholeiitic magma series. Syn: *tholeiite.*

tholeiitic magma series (a) Subalkaline basalts, andesites, dacites, and rhyolites that show a trend of iron enrichment with increasing silica content, as commonly discriminated from the calc-alkaline magma series (also subalkalic) by Miyashiro's (1974) discriminant line on a plot of FeO/MgO vs. SiO_2. In specific occurrences, as in volcanoes or plutons, individual magmas in the series are often taken to be genetically related (e.g., by crystal fractionation, mixing, etc.). The series occurs in all tectonic environments; basaltic members from mid-ocean ridges are the commonest magma type on Earth. At convergent margins, usage is contentious at present, as some authors use as diagnostic (a) modal mineralogy (absence of hornblende) or (b) low K_2O or LREE contents. Others equate the pigeonite magma series (vs. hypersthene series) with the tholeiitic series. See also: *calc-alkaline magma series.*

tholoid (tho'-loid) *volcanic dome.*

tholus (tho'-lus) A term established by the International Astronomical Union for an isolated domical mountain or hill on planetary surfaces (USGS, 2003). Most are considered volcanic. Generally used as part of a formal name for a landform, such as Ceraunius Tholus on Mars. Etymol: Latin "tholus", "dome". See also: *dome [planet].* Pl: tholi.

tholus [volc] (tho'-lus) An isolated, domical, small mountain or hill, usually with slopes much steeper than those of a *patera* (Zimbelman, 2000). Pl: tholi.

thomasclarkite-(Y) A white or yellow monoclinic mineral: $Na(Y,La,Nd)(HCO_3)(OH)_3{\cdot}4H_2O$.

thometzekite (thom-et-zek'-ite) A bluish-green or green monoclinic mineral of the *tsumcorite* group: $Pb(Cu,Zn)_2(AsO_4)_2(OH,H_2O)_2$.

Thompson diagram (Thomp'-son) *AFM projection.*

Thomsen anisotropic parameters Combinations of anisotropic elastic parameters that relate fairly directly to observational differences (Thomsen, 2002).

thomsenolite (thom-sen'-o-lite) A white monoclinic mineral: $NaCaAlF_6{\cdot}H_2O$.

thomsonite (thom'-son-ite) A colorless to pink, yellowish, brown, or greenish orthorhombic zeolite mineral: $NaCa_2Al_5Si_5O_{20}{\cdot}6H_2O$. It has considerable replacement of CaAl by NaSi, and sometimes contains no sodium. It usually occurs in masses of radiating crystals. Syn: *ozarkite.*

thomsonite-Sr A colorless orthorhombic mineral: $Na(Sr,Ca)_2Al_5Si_5O_{20}{\cdot}6\text{-}7H_2O$. It represents thomsonite with Sr>Ca.

thoraceton (tho-rac'-e-ton) The *opisthosoma* or abdomen of a merostome.

thoracic (tho-rac'-ic) Pertaining to, located within, or involving the *thorax;* e.g. a "thoracic limb" attached to any somite of the thorax of a crustacean.

thoracomere (tho-ra'-co-mere) A *somite* of the thorax of a crustacean.

thoracopod (tho-ra'-co-pod) A limb of any thoracic *somite* of a crustacean; a thoracic limb, such as a *maxilliped* or a *pereiopod.*

thorax (tho'-rax) (a) The skeleton of the anterior trunk of a mammal, including ribs, sternum, and usually about 12 distinctive vertebrae to which movable free ribs are ligamentally attached. That part of the body of a mammal posterior to the neck and anterior to the muscular diaphragm (which separates the thorax from the abdomen). The thorax is not definable in other tetrapod groups. (b) The central *tagma* of the body of an arthropod, consisting of several generally movable segments, e.g. the nearly always limb-bearing tagma between cephalon and abdomen of a crustacean; the middle part of an exoskeleton of a trilobite, extending between cephalon and pygidium and consisting of several freely articulated segments; or the middle of the three chief divisions of the body of an insect. See also: *cephalothorax.* Syn: *trunk [paleont].* (c) The second joint of the shell of a nasselline radiolarian. Pl: thoraxes or horaces.

thorbastnäsite (thor-bast'-naes-ite) A brown hexagonal mineral of the *bastnäsite* group: $Th(Ca,Ce)(CO_3)_2F_2{\cdot}3H_2O$. It occurs as an accessory mineral in iron-rich albitites and in selvages of veinlets and stockworks. Also spelled: thorbastnaesite.

thoreaulite (thor-eau'-lite) A brown monoclinic mineral: $Sn^{2+}(Ta,Nb)_2O_6$. It is the Ta analogue of foordite.

thorianite (thor'-i-a-nite') A submetallic gray, brown, or black mineral of the uraninite group: ThO_2. It often contains rare-earth metals and uranium and is strongly radioactive.

thorikosite (thor'-ik-os-ite) A light yellow tetragonal mineral: $Pb_3(Sb^{3+},As^{3+})O_3Cl_2(OH)$.

thorite (thor'-ite) A black, dark brown, yellow, orange, or green tetragonal mineral of the *zircon* group: $(Th,U)SiO_4$. It is isostructural with thorogummite, strongly radioactive, and usually metamict, and may contain as much as 10% uranium. Thorite resembles zircon and occurs as a minor accessory mineral of granites, syenites, and pegmatites. It is dimorphous with huttonite.

thorium-230-deficiency method The calculation of an age in years for fossil coral or shell from 10,000 to 250,000 years old, based on the departure of ionium (thorium-230) from equilibrium with uranium-238 and uranium-234 as a result of carbonate recrystallization shortly after its formation or burial. The age depends on the growth toward the equilibrium ratio, which is directly related to the passage of time. Cf: *uranium-series age method; thorium-230/protactinium-231 deficiency method.* Syn: *ionium-deficiency method.*

thorium-230/protactinium-231 deficiency method Calculation of an age in years for fossil coral, shell, or bone 10,000 to 250,000 years old, based on the growth of uranium daughter products from uranium isotopes that enter the carbonate or phosphate material shortly after its formation or burial. The age depends on measurement of the thorium-230/uranium-234 and protactinium-231/uranium-235 activity ratios, which vary with time. See also: *uranium-series age method; ionium-deficiency method.*

thorium-230/protactinium-231 excess method *protactinium-ionium age method.*

thorium-230/thorium-232 age method *ionium-thorium age method.*

thorium-lead age method (tho'-ri-um-lead') Calculation of an age in years for geologic material based on the known radioactive rate of thorium-232 to lead-208. It is part of the more inclusive *uranium-thorium-lead age method* in which the parent-daughter pairs are considered simultaneously, esp. those of uranium.

thorium series (tho'-ri-um) The radioactive series beginning with thorium-232.

thorn A short, sharply pointed triangular or conical surface extension in the skeleton of a spumellarian radiolarian.

thornasite (thor'-nas-ite) A vitreous colorless or pale green rhombohedral mineral: $Na_{12}Th_3(Si_{18}O_{19})_4{\cdot}18H_2O$.

Thornthwaite's classification (Thorn'-thwaite's) A *climate classification,* formulated by the U.S. agricultural climatologist Warren Thornthwaite, that is based on ratios of precipitation to evaporation. Five humidity provinces are distinguished: perhumid, humid, subhumid, semiarid, and arid. Cf: *Köppen's classification.*

thorogummite (thor-o-gum'-mite) An earthy yellow tetragonal secondary mineral of the zircon group: $(Th,U^{6+})(SiO_4)_{1-x}(OH)_{4x}$. It is isostructural with thorite and may contain as much as 31.4% uranium. Syn: *mackintoshite; maitlandite.*

thorosteenstrupine (thor-o-steen'-strup-ine) A dark-brown to nearly black metamict mineral: $(Ca,Mn,Th)_3Si_4O_{11}(OH)F{\cdot}6H_2O$. Not to be confused with *steenstrupine.*

thorotungstite (tho-ro-tung'-stite) *yttrotungstite-(Y).*

thoroughfare (thor'-ough-fare) (a) A tidal channel or creek providing an entrance to a bay or lagoon behind a barrier or spit. (b) A navigable waterway, as a river or strait; esp. one connecting two bodies of water.

thortveitite (thort-veit'-ite) (a) A gray-green to black monoclinic mineral: $(Sc,Y)_2Si_2O_7$. It is a source of scandium. Cf: *thalénite; gittinsite.* (b) A group name for monoclinic minerals with the formula $A_2Si_2O_7$, where A = Sc, Y, Yb, Zr, or Th.

thorutite (thor'-u-tite) A black monoclinic mineral: $(Th,U,Ca)Ti_2(O,OH)_6$. Syn: *smirnovite.*

Thoulet solution (Thou-let') *Sonstadt solution.*

thread (a) A thin stream of water. (b) The middle of a stream. (c) The line along the surface of a stream connecting points of maximum current velocity. Cf: *channel line.*

thread flow Water flowing on an unchanneled surface at depths so shallow that the average sized grains on the surface are not submerged.

threadgoldite (thread'-gold-ite) A micaceous greenish-yellow monoclinic mineral: $Al(UO_2)_2(PO_4)_2(OH){\cdot}8H_2O$.

thread-lace scoria A scoria in which the vesicle walls have burst and are represented only by an extremely delicate three-dimensional network of glass threads. Syn: *reticulite.*

three-age system In Old World archaeology, the original classification scheme of relative prehistoric time, comprising *Stone Age, Bronze Age,* and *Iron age.*

three array An electrode array used in profiling, in which one current electrode is placed at infinity while one current electrode and two potential electrodes are in close proximity and are moved across the structure to be investigated.

three-component geophone A sensor that records the three mutually orthogonal components of particle velocity. Provides the information for determining the type of wave responsible for an event and its direction of approach.

three-component seismograph A sensor that records the three mutually orthogonal components of particle velocity. Provides information for determining the type of wave responsible for an event and its direction of approach.

three-dimensional survey A survey involving collecting data over an area with the objective of determining spatial relations in three dimensions, as opposed to determining components along separated survey lines. The data from such a survey constitute a volume which can be displayed in different ways. Abbrev: 3-D. See also: *time slice; horizon slice; fault slice; arbitrary line; attribute [seis].*

threefold axis An axis of rotational symmetry by which an object (crystal shape, *crystal structure,* etc.) is invariant to a rotation of 120°. Syn: *triad.*

threefold coordination (three'-fold) *triangular coordination.*

threeling (three'-ling) *trilling.*

three-mile limit The seaward limit of the *territorial sea* or *marginal sea* of 3 nautical miles (about 5.6 km), the one-time range of a cannon shot.

three-phase inclusion An inclusion in a gemstone consisting of a void containing solid, gas and liquid. Cf: *two-phase inclusion.*

three-point method A geometric method of calculating the dip and strike of a structural surface from three points of varying elevation along the surface.

three-point problem (a) The problem of locating the horizontal geographic position of a point of observation from data comprising two observed horizontal angles subtended by three known sides of a triangle (or situated between three points of known position). It is solved analytically by trigonometric calculation in triangulation, mechanically by means of a three-arm protractor, or graphically by trial-and-error change in the orientation of the board in planetable surveying. Cf: *two-point problem.* (b) A name applied to the method of solving the three-point problem in planetable surveying, commonly by taking backsights on three previously located stations. See also: *resection.*

three-swing cusp A *meander cusp* formed by three successive swings of a *meander belt,* the scar produced by the third swing meeting the point of the cusp formed by the first two swings (Lobeck, 1939, p.240-241); it may be seen on the edge of a rock-defended terrace.

three-taxon statement Basic unit of comparison in a cladistic analysis; two taxa connected to an outgroup (Brooks et al., 1984, p.81). Cf: *Wagner neighborhood.*

threshold [geochem] (thresh'-old) The entrance, boundary, or beginning of a new domain; the lowest detectable value; the point at which a process or effect commences.

threshold [geomorph] The point at which response to some forcing or input changes. It is generally based on a critical force, such as resistance relationships for sediment erosion, transport, and deposition, or on the relative rates of linked processes, such as accumulation and ablation in glaciers.

threshold [glac geol] *riegel.*

threshold [marine geol] *sill [marine geol].*

threshold [speleo] That part of a cave system to which light penetrates in some degree.

threshold pressure *yield stress.*

threshold velocity The minimum velocity at which wind or water, in a given place and under specified conditions, will begin to move particles of soil, sand, or other material. Also referred to as *critical velocity.*

throat plane A plane passed through the centers of the spheres in a layer of rhombohedral packing.

thrombolite (throm'-bo-lite) A texture of nonlaminated carbonate bioherms characterized by vague clotting, thought to be caused by microbial activity (Aitken, 1967). Cf: *rhodolith.*

through cave Cave through which a stream runs from entrance to exit or formerly did so.

throughfall (through'-fall) Water from precipitation that falls through the plant cover directly onto the ground or that drips onto the ground from branches and leaves. Cf: *interception; stemflow.*

throughflow Shallow subsurface downslope flow. See also: *storm seepage.*

through glacier A double-ended glacier, consisting of two valley glaciers situated in a single depression, from which they flow in opposite directions. A "through-glacier system" is a body of glacier ice consisting of interconnected through glaciers that may lie in two or more drainage systems. Cf: *transection glacier.*

through valley A flat-floored depression or channel eroded across a divide by glacier ice or by meltwater streams; a valley excavated by a through glacier.

throw The vertical component of displacement on a dip-slip fault. Cf: *heave [struc geol].* See also: *upthrow; downthrow.*

throwing clay (throw'-ing) Clay plastic enough to be shaped on a potter's wheel.

throwout (throw'-out) Fragmental material ejected from a crater during formation and redeposited on or outside the crater lip. Cf: *fallout [impact]; fallback.*

thrust (a) An overriding movement of one crustal unit over another, as in thrust faulting. (b) *thrust fault.*

thrust block *thrust sheet.*

thrust fault A fault with a dip of 45° or less over much of its extent, on which the hanging wall has moved upward relative to the footwall. Syn: *thrust.* Partial syn: *reverse fault.* Cf: *normal fault.*

thrust moraine (a) A *moraine* produced by the overriding and pushing forward, by a regenerated glacier, of dead ice and its deposits (Gravenor and Kupsch, 1959). (b) *push moraine.*

thrust nappe *thrust sheet.*

thrust outlier *klippe.*

thrust plate *thrust sheet.*

thrust pond A small, shallow, roughly circular pond on the floor of a slightly inclined mountain valley, and bordered by a raised rim of thick soil resulting from *ice push* and supporting a very dense growth of coarse alpine grass (Ives, 1941, p.290); e.g. in the high parts of the Rocky Mountains in Colorado.

thrust scarp The landform developed along the forward edge of a *thrust fault* or *nappe.*

thrust sheet The body of rock above a large-scale thrust fault whose surface is horizontal or very gently dipping. Syn: *thrust block; thrust nappe; thrust plate.*

thrust slice A relatively thin body of rock bounded above and below by thrust faults within a zone of thrusting.

thrust system An array of geometrically similar and kinematically related thrust faults.

thucholite (thu'-cho-lite) A brittle, jet-black mixture or complex of organic matter (hydrocarbons) and uraninite, with some sulfides, occurring esp. in gold conglomerates (as in the Witwatersrand of South Africa) or in pegmatites (as in Canada). It may contain up to 48% thorium in the ash.

thufa (thu'-fa) An Icelandic term for *earth hummock.* Pl: thufur.

Thulean province (Thu'-le-an) A region of Tertiary volcanic activity (basalt flood) including Iceland and most of Britain and Greenland. Etymol: Greek "Thule", "the northernmost part of the habitable world."

Thule-Baffin moraine (Thu'-le-Baf'-fin) *shear moraine.*

thulite (thu'-lite) A pink, rose-red, or purplish-red variety of zoisite containing manganese and used as an ornamental stone.

Thumper (Thump'-er) Trade name for a device for generating seismic waves by dropping a heavy weight (often a 3-megagram weight dropped from a 3-meter elevation).

thunder Sound accompanying *lightning.* A lightning discharge heats the air along the conducting path to tremendously high temperatures. Intense heating causes a tremendous rise in air pressure locally that generates a shock wave that propagates outward and produces sound waves.

thunder egg (thun'-der) A popular term for a small, geodelike body of chalcedony, opal, or agate that has weathered out of the welded tuffs of central Oregon.

thunderstorm A mesoscale weather system produced by strong *convection* currents surging to great altitudes within the *troposphere.* Consists of cumulonimbus clouds accompanied by *lightning* and *thunder* and often, locally heavy rainfall (or snowfall) and gusty surface winds. A thunderstorm consists of one or more convective cells, each of which progresses through the life cycle of cumulus stage, mature stage and dissipating stage usually in less than one hour.

Thuringian (Thu-rin'-gi-an) European stage: Upper Permian (above Saxonian, below Triassic).

thuringite (thur'-ing-ite) An olive-green or pistachio-green mineral of the *chlorite* group, a ferrian variety of chamosite. It is isomorphous with pennantite.

thurm A term used in Nova Scotia for a ragged and rocky headland swept by the sea. Also called: thurm cap.

thyriothecium (thy-ri-o-the'-ci-um) A shield-shaped fungal *fructification* (in Hemispherales and Microthyriales) that is oriented not by the mycelium but by the host, with the generative tissue hanging downward.

tibia (tib'-i-a) (a) The more medial of the two shank elements of the tetrapod leg (between the knee and tarsus). (b) The fifth segment of a typical leg or pedipalpus of an arachnid, following upon the patella which may be completely fused with it (TIP, 1955, pt. P, p.63). (c) A joint of the distal part of a prosomal appendage of a merostome. Pl: tibiae.

tibiscumibe (ti-bis'-cum-ite) A monoclinic mineral of doubtful validity: $(Na,Ca)(Al,Mg,Fe)_2(OH)_2(Si,Al)_4O_{10}$. It may be rectorite or allophone, and has not been approved by the International Mineralogical Association.

tickle (tick'-le) (a) Any narrow passage connecting larger bodies of water. (b) A term used in the Canadian maritime provinces for an inlet of the sea into a lagoon.

tidal basin (tid'-al) A dock or basin in a tidal region, in which water is maintained at a desired level by means of a gate; it is filled at high tide by water that is retained and then released at low tide.

tidal bedding Sedimentary bedding caused by tides in a tidal channel, a tidal flat, or a marsh; esp. bedding produced where currents of high tides are stronger than those of low tides flowing in the opposite direction, as where a layer of coarse sediments deposited by a high tide is not destroyed by the low tide.

tidal bench mark A durable bench mark fixed rigidly in stable ground and set to reference a tide gage at a tide station.

tidal bore *bore [tides].*

tidal bulge The theoretical increase in water level closest to and opposite the sun and moon.

tidal bundle The portion of cross strata produced by the stronger set of currents in one half of an asymmetric tidal cycle (currents of one half of the cycle much stronger than those in the other half). The cross laminae within a tidal bundle may display changes related to the range of changing currents from beginning to end of the tidal cycle. Within a tidal bundle, *reactivation surfaces* may separate the sets of cross laminae (Friedman et al., 1992, p.40).

tidal channel (a) A major channel followed by the tidal currents, extending from offshore into a tidal marsh or a tidal flat. (b) *tidal inlet.*

tidal compartment The part of a stream that "intervenes between the area of unimpeded tidal action and that in which there is a complete cessation or absence of tidal action" (Carey and Oliver, 1918, p.8).

tidal constant Set of parameters used to describe the tide at a point on the implicit assumption that the response of the ocean to tidal forcing does not change with time. See also: *tidal constituent.*

tidal constituent A tide wave of fixed frequency, amplitude, and phase used to describe the tide at a point. It is calculated from a harmonic analysis of the tide. The tide is the sum of all the tidal constituents.

tidal correction A correction to gravity measurements, applied to compensate for the effects of the Sun and Moon (i.e., for Earth tides). It is sometimes included in the *drift correction* and may be determined by a series of observations at a fixed base station.

tidal creek A relatively small tidal inlet or estuary. Syn: *creek.*

tidal current The periodic horizontal movement of ocean water associated with the vertical rise and fall of the tides and resulting from the gravitational attraction of the Moon and Sun upon the Earth. In the open ocean, its direction rotates 360° on a diurnal or semidiurnal basis; in coastal areas, however, topography influences its direction. Incorrect syn: *tide.* British syn: *tidal stream.*

tidal datum A *chart datum* based on a property of the local tide.

tidal delta A delta formed at the mouth of a tidal inlet on either the seaward or the lagoon side of a barrier island or baymouth bar by changing tidal currents that sweep sand in and out of the inlet. Cf: *ebb-tidal delta; flood-tidal delta.*

tidal-delta marsh (tid'-al-del'-ta) A *salt marsh* found around distributary patterns of tidal rivers inside a *tidal inlet.*

tidal divide A divide between two adjacent tidal channels.

tidal efficiency The ratio of the fluctuation of water level in a well to the tidal fluctuation causing it, expressed in the same units such as feet of water. Symbol: C. Cf: *barometric efficiency.*

tidal flat An extensive, nearly horizontal, barren tract of land that is alternately covered and uncovered by the tide, and consisting of unconsolidated sediment (mostly mud and sand). It may form the top surface of a deltaic deposit. Cf: *tidal marsh; mud flat.* Syn: *tide flat.*

tidal flushing The removal of sediment, as from an estuary, by the ebb and flow of tidal currents that are stronger, more constant, or more prevalent than the incoming river flow.

tidal friction The frictional effect of the tides, resulting in dissipation of energy. See: *tidal mixing.*

tidal glacier *tidewater glacier.*

tidal inlet Any *inlet* through which water alternately floods landward with the rising tide and ebbs seaward with the falling tide; specif. a natural inlet maintained by tidal currents. Syn: *tidal outlet; tidal channel.*

tidalite (tid'-a-lite) A sediment deposited by tidal tractive currents, by an alternation of tidal tractive currents and tidal suspension deposition, or by tidal slack-water suspension sedimentation. Tidalites occur both in the intertidal zone and in shallow, subtidal, tide-dominated environments. See also: *intertidalite.*

tidal marsh A marsh bordering a coast (as in a shallow lagoon or sheltered bay), formed of mud and of the resistant mat of roots of salt-tolerant plants, and regularly inundated during high tides; a marshy *tidal flat.* Cf: *salt marsh.*

tidal mixing Turbulent mixing driven by tidal currents. It dissipates tidal energy, slowing the Earth's rotation, which increases the length of day, and the rotation of the Moon about the Earth, which lengthens the lunar month. It helps drive the deep mass circulation. Mixing occurs in shallow seas, over mid-ocean ridges, at ocean edges, and near seamounts.

tidal outlet *tidal inlet.*

tidal pool *tide pool.*

tidal prism The volume of water that flows in or out of a harbor or estuary with the movement of the tide, and excluding any freshwater flow; it is computed as the product of the tide range and the area of the basin at midtide level, or as the difference in volume at mean high water and at mean low water.

tidal range *tide range.*

tidal resonance theory George Darwin's postulation of the cause of continental drift; the Moon separated from the Earth, leaving the Pacific Ocean as a scar, and the consequent shortage of continental crust caused a global tension, resulting in drifting (Fairbridge, 1966, p.83).

tidal river A river whose lower part for a considerable distance is influenced by the tide of the body of water into which it flows (but not necessarily by the salinity); also, the movement of water in and out of an estuary or other inlet as a result of the alternating rise and fall of the tide. Syn: *tidal stream.*

tidal sand ridge A low ridge formed by tidal action some distance from shore on a shelf, bank, or estuary. Commonly parallel or subparallel to current direction.

tidal scour The downward and sideward erosion of the sea floor by powerful tidal currents resulting in the removal of inshore sediments and the formation of deep channels and holes. Syn: *scour [tides].*

tidal stream (a) *tidal river.* (b) A British syn. of *tidal current.*

tidal swamp A swamp partly covered during high tide by the backing-up of a river (Stephenson and Veatch, 1915, p.37). Cf: *upland swamp.*

tidal water *tidewater.*

tidal wave An erroneous syn. of both *storm surge* and *tsunami.*

tidal wedge A tidal channel that is narrower and shallower at the downstream end.

tide (a) The periodic rise and fall of the surface of the ocean and bodies of water connected to the ocean that result from the gravitational attraction of sun and moon. Adj: tidal. (b) An incorrect syn. of *tidal current.* (c) *Earth tide.* (d) *atmospheric tide.*

tide amplitude A *tidal constant* representing one-half of the *tide range;* the elevation of tidal high water above mean sea level.

tide crack A *crack [ice],* usually parallel to the shore, at the junction line between an immovable *icefoot* or ice wall and *fast ice,* and caused by the rise and fall of the tide which moves the fast ice upward and downward.

tide curve A graphic record of the height (rise and fall) of the tide, with time as abscissa and tide height as ordinate.

tide-dominated delta A delta built in an environmental setting where fluvial sedimentation is significantly modified by tidal processes; characterized by bell-shaped estuarine river mouths and elongate tidal sand deposits. Cf: *fluvially-dominated delta; wave-dominated delta.*

tide flat A syn. of *tidal flat.*

tide gage A device for measuring the height (rise and fall) of the tide; esp. an instrument automatically making a continuous graphic record of tide height versus time.

tideland (tide'-land) (a) The coastal area that is alternately covered and uncovered by the ordinary daily tides; land that is covered by *tidewater* during a flood tide. (b) Land that underlies the ocean beyond the low-water mark, but within the territorial waters of a nation. The term is often used in the plural.

tide pool A pool of water, as in rock basin, left on a beach or reef by an ebbing tide. Syn: *tidal pool.*

tide race A type of race caused by a greater tide range at one end of the channel than at the other end.

tide range A *tidal constant* representing the difference in height between consecutive high water and low water at a given place; it is twice the *tide amplitude.* Cf: *mean range.* Syn: *tidal range.*

tide rip A rip caused by convergence of tidal current or by a tidal current suddenly entering shallow water.

tide station A place where tide observations are obtained. Cf: *reference station.*

tidewater (tide'-wa-ter) (a) Water that overflows the land during a flood tide; water that covers the *tideland.* Also, stream water that is affected by the rise and fall of the tide. Syn: *tidal water.* (b) A broad term for a seacoast low-lying coastal region traversed by tidewater streams; e.g., Tidewater region, Md./Va.

tidewater glacier A glacier that terminates in the sea, where it usually ends in an ice cliff from which icebergs are discharged. Syn: *tidal glacier.*

tidewell *ebb and flow spring.*

tie [geophys] In geophysics, the relating of a new station or value to ones already established.

tie [surv] (a) A survey connection from a point of known position to a point whose position is desired. (b) A survey connection to close a survey on a previously determined point.

tie bar *tombolo.*

tied island An island connected with the mainland or with another island by a tombolo. Syn: *land-tied island; tombolo island.*

tie line [chem] A line in a phase diagram that connects any two or more phases that are in equilibrium. By definition, the values of all intensive parameters must be the same in all phases. Syn: *conode.*

tie line [surv] A line measured on the ground to connect some object to a survey; e.g. a line joining opposite corners of a four-sided figure, thereby enabling its area to be checked by triangulation.

tiemannite (tie'-mann-ite) A dark-gray or nearly black cubic mineral of the *sphalerite* group: HgSe.

tienshanite (tien-shan'-ite) A green hexagonal mineral: $K(Na,K,[])_9Ca_2Ba_6Mn^{2+}_6Ti_6B_{12}Si_{36}O_{114}(O,OH,F)_{11}$.

tie point (a) A point to which a tie is made; esp. a point of closure of a survey either on itself or on another survey. (b) An image point identified on oblique aerial photographs in the overlap area between two or more adjacent strips of photography. They tie individual sets of photographs into a single flight unit and adjacent flights into a common network.

tier [speleo] *level [speleo].*

tier [surv] Any series of contiguous townships (of the U.S. Public Land Survey system) aligned east and west and numbered consecutively north and south from a base line. Also, any series of contiguous sections similarly situated within a township. Cf: *range [surv].*

tiering (tier'-ing) Spatial distribution of benthic organisms above the sea floor (epifaunal tiering) or below it (infaunal tiering). Tiering may allow for ecologic differentiation among benthic organisms (Ausich and Bottjer, 1982).

tietaiyangite A dark brown hexagonal mineral: $Fe^{3+}_4FeTiO_9$. Not approved by the International Mineralogical Association.

tiettaite A colorless, white, or gray orthorhombic mineral: $(Na,K)_{17}Fe^{3+}TiSi_{16}O_{29}(OH)_{30}\cdot 2H_2O$.

tiff A *sparry* mineral. The term is applied to calcite in SW Missouri and to barite in SE Missouri.

Tiffanian (Tif-fan'-i-an) North American land-mammal age: Upper Paleocene (after Torrejonian, before Clarkforkian).

TIGER Topologically Integrated Geographic Encoding and Referencing files. TIGER files are vector data files compiled and distributed by the U.S. Census Bureau.

tiger's-eye (ti'-ger's-eye) A chatoyant, translucent to semitranslucent, yellowish-brown or brownish-yellow gem and ornamental variety of quartz, pseudomorphous after crocidolite, whose fibers (penetrating the quartz) are changed to iron oxide (limonite); silicified crocidolite stained yellow or brown by iron oxide. Upon heating, the limonite turns to hematite and produces a red to brownish-red sheen. Cf: *hawk's-eye; cat's-eye [mineral].* Also spelled: tiger-eye; tigereye. Also called: tigerite.

tight fold A fold with an inter-limb angle between 0° and 30° (Fleuty, 1964, p.470).

tight hole A drilling or completed well about which information is kept secret by the operator.

tight sand A sand whose interstices are filled with fine grains or with matrix material, thus effectively destroying porosity and permeability. The term is used in petroleum geology.

Tiglian (Tig'-li-an) European warm-climate paleofloral stage: upper Pliocene (above Pretiglian, below Eburonian). Syn: Tegelian.

tikhonenkovite (tik-ho-nen'-kov-ite) A colorless or rosy monoclinic mineral: $SrAlF_4(OH)\cdot H_2O$.

tilasite (ti'-las-ite) (a) A gray, violet-gray or olive-green monoclinic mineral: $CaMg(AsO_4)F$. (b) A group name for monoclinic minerals with the analogous formula, but with CaMg replaced by NaAl or $NaFe^{3+}$; AsO_4 replaced by PO_4; and F replaced by OH.

tile ore A red or brownish earthy variety of cuprite often mixed with red iron oxide.

tilestone An English term for a flagstone (flaggy sandstone) used for roofing.

tiling Assembling large images from smaller segments that have common angular orientations, cell sizes, and map projections. Cf: *mosaic [photo].*

till Dominantly unsorted and unstratified *drift,* generally unconsolidated, deposited directly by and underneath a glacier without subsequent reworking by meltwater, and consisting of a heterogeneous mixture of clay, silt, sand, gravel, and boulders ranging widely in size and shape. Cf: *stratified drift; moraine [glac geol].* Syn: *boulder clay; glacial till; ice-laid drift.*

till ball An *armored mud ball* whose core is made of till, occurring in certain Pleistocene glacial deposits.

till billow An undulating or swelling accumulation of glacial drift irregularly disposed with regard to the direction of movement of the ice (Chamberlin, 1894b, p.523).

till crevasse filling A ridge of unstratified morainal material deposited in a crack of a wasting glacier and left standing after the ice melted; term introduced by Gravenor (1956, p.10). Cf: *crevasse filling.*

tilleyite (til'-ley-ite) A white monoclinic mineral: $Ca_5Si_2O_7(CO_3)_2$.

tillite (till'-ite) A consolidated or indurated sedimentary rock formed by lithification of glacial till, esp. pre-Pleistocene till (such as the Late Carboniferous tillites in South Africa and India).

tillmannsite An adamantine red to brownish-red tetragonal mineral: $HgAg_3(V,As)O_4$.

tilloid A term introduced by Blackwelder (1931a, p.903) for a till-like deposit of "doubtful origin", but redefined by Pettijohn (1957, p.265) as a nonglacial *conglomeratic mudstone* (also known as *geröllton*) varying from "a chaotic unassorted assemblage of coarse materials set in a mudstone matrix to a mudstone with sparsely distributed cobbles". Harland et al. (1966, p.251) urge the use of "tilloid" as a nongenetic term for a rock resembling tillite in appearance but whose origin is in doubt or unknown. Cf: *pebbly mudstone; pseudotillite.*

till plain An extensive area, with a flat to undulating surface, underlain by till with subordinate end moraines; such plains occupy parts of Indiana, Illinois, and Iowa. Cf: *ground moraine.*

till-shadow hill (till'-shad-ow) A glacial hill, without a core of resistant rock, that has a gentle south slope on which till thickens but does not form a well-developed tail (Coates, 1966); examples occur in central New York State.

till sheet A sheet, layer, or bed of *till,* without reference to its topographic expression. It may form a *ground moraine.*

tillstone (till'-stone) A boulder or other stone in a deposit of glacial till.

till tumulus In glacial geology, a low, stony mound representing the immature nucleus of a *drumlin* (Chamberlin, 1894b, p.522-523).

till wall A ridge consisting of morainal material squeezed upward into a crevasse by the pressure of overlying ice (Gravenor and Kupsch, 1959, p.58).

tilly (till'-y) Composed or having the nature of glacial till; e.g. *tilly* land.

tilt [photo] (a) The angle at the perspective center between the plumb line and the perpendicular from the interior perspective center to the plane of the photograph. (b) The lack of parallelism (or the angle) between the plane of the photograph from a downward-pointing aerial camera and the horizontal plane (normal to the plumb line) of the ground.

tilt [struc geol] The geometry of a fringe crack that propagates from that portion of a parent joint subject to mode II crack tip displacement (Pollard et al., 1982). Cf: *twist.*

tilt boundary *subgrain boundary.*

tilted iceberg A *tabular* iceberg that has become unbalanced due to melting or calving, so that its flat top is inclined.

tilted photograph An aerial photograph taken with a camera whose plane of film is not parallel (horizontal) with the plane of the ground at the time of exposure.

tilth The physical condition of a soil relative to its fitness for the growth of a specified plant or sequence of plants.

tiltmeter (tilt'-me-ter) An instrument that measures slight changes in the tilt of the Earth's surface, usually in relation to a liquid-level surface or to the rest position of a pendulum. It is used in volcanology and in earthquake seismology.

timberline (tim'-ber-line) The elevation (as on a mountain) or the latitudinal limits (on a regional basis) at which tree growth stops. Syn: *tree line.*

time (a) Measured or measurable duration; a nonmaterial dimension of the universe, representing a period or interval during which an action, process or condition exists or continues. Cf: *geologic time.* (b) A reference point from which duration is measured; e.g., the instant at which a seismic event occurs relative to a chosen reference time such as a shot instant. (c) A reckoning of time, or a system of reckoning duration. (d) An informal term proposed by the ISST (1961, p.13, 25) for the geologic-time unit next in order of magnitude below *age [geochron]* (a), during which the rocks of a substage (or of any time-stratigraphic unit of lesser rank than a stage) were formed. It is a syn. of *subage; episode (b); phase [geochron] (a).* (e) Any division of geologic chronology, such as "Paleozoic time" or "Miocene time".

time break The time of initiation of seismic waves.

time-correlation (time'-cor-re-la'-tion) *chronocorrelation.*

time-depth curve A tabular or graphical expression of the relation between velocity and arrival time of vertically travelling seismic reflections. It permits reflection time measurements to be converted to the corresponding depths.

time-distance curve (time-dis'-tance) A syn. of *traveltime curve.* Abbrev: T. D. curve.

time domain (time'-do-main) Measurements as a function of time, or operations in which time is the variable, in contrast to the *frequency domain.*

time-domain electromagnetic method (time'-do-main) Transmission of a single or repetitive pulse of electromagnetic energy, frequently of square wave form, and reception of electromagnetic energy, as a function of time, in the time interval after the transmitted wave form has been turned off. It is used with induced electrical polarization and electromagnetic methods. Abbrev: TDEM. Syn: *transient electromagnetic method.*

time lag A delay in the arrival time of seismic energy from the time expected. Time lags may be produced by an abnormal low-velocity layer, phase shifts in filtering, or other factors.

time-lapse seismology Repeating a survey to determine the changes that have occurred during the interval, such as may be caused by hydrocarbon production. Results are sometimes displayed as difference sections or *difference maps.* When using multiple 3-D surveys run at different times, this is sometimes called a 4-D survey, the fourth dimension being time.

time lead The arrival of seismic energy earlier than expected, indicating that part of the travel path involved high velocity. It is an indication of a salt dome in *fan shooting.* Syn: *lead [seis].*

time-level correlation Correlation of time-levels (Henningsmoen, 1961, p.64).
time line (a) A line indicating equal age in a geologic cross section or correlation diagram; e.g. a line separating two time-stratigraphic units. (b) A rock unit represented by a time line; e.g. an intraformational conglomerate formed by subaqueous slump and turbidity flows of but a few hours' duration.
time migration Migration that allows for velocity changes in the vertical direction only, as contrasted with *depth migration,* where horizontal velocity changes are also taken into consideration.
time of concentration *concentration time.*
time-parallel (time-par'-al-lel) (a) Said of a surface that is parallel to or that closely approximates a synchronous surface and that involves a geologically insignificant amount of time, such as the surface of a rapidly transgressed unconformity. (b) Said of a stratum bounded by time-parallel surfaces.
time plane A stratigraphic horizon identifying an *instant* in geologic time. Syn: *synchrone.*
time-rock span *stratigraphic range.*
time-rock unit "An undesirable term for chronostratigraphic unit" (ISSC, Rept. 6, 1971, p.6, footnote).
time sag A depression on a seismic section resulting from overlying lateral velocity changes, such as a localized area of low velocity. Also called: *gas sag.*
time scale *geologic time scale.*
time section A graphical representation of reflection seismic data along a line, in which the vertical scale is two-way seismic-traveltime and the horizontal scale is surface distance. Time sections may be either unmigrated or migrated and performed on prestacked/or poststacked data by either time or depth migration. See: *migration [seis].* Cf: *depth section.*
time series In *geostatistics,* a series of data collected at regular intervals of time; a frequency distribution in which the *independent variable* is time.
time-set unit A stratigraphic unit arranged in time-continuous sets ("time-sets") representing a continuous span of time (Henningsmoen, 1961, p.66).
time slice A display of the seismic measurements (usually amplitude) corresponding to a single arrival time (or single depth) through a volume of 3-D data. Cf: *depth slice; horizon slice.*
time-stability hypothesis (time-sta-bil'-i-ty) *stability-time hypothesis.*
time standard Any category of physical or biologic phenomena or processes by which segments of time can be measured or subdivided; e.g., radioactive decay of elements, orderly evolution of forms of life, rotation of the Earth on its axis, revolution of the Earth around the Sun, and human artifacts. All such time standards are "partial" with respect to time in the abstract (Jeletzky, 1956, p.681).
time-stratigraphic facies (time'-strat-i-graph'-ic) A stratigraphic facies that is recognized on the basis of the amounts of geologic time during which sedimentary deposition and nondeposition occurred; a facies that is a laterally segregated, statistical variant of a stratigraphic interval and whose boundaries (vertical surfaces or arbitrary cutoffs) extend from the bottom to the top of the interval (Wheeler, 1958, p.1060).
time-stratigraphic unit *chronostratigraphic unit.*
time-stratigraphy (time-stra-tig'-ra-phy) *chronostratigraphy.*
time tie The identification of seismic events on different records by their arrival times, when they possess common raypaths.
time-transgressive (time'-trans-gres'-sive) *diachronous.*
time-transitional (time-tran-si'-tion-al) Said of a rock unit including within itself an important geologic time plane and thus consisting of strata that belong to two adjacent chronostratigraphic units (such as systems).
time unit *geochronologic unit.*
time value The interval of geologic time represented by or involved in producing a stratigraphic unit, an unconformity, the range of a fossil, or any geologic feature or event. See also: *hiatus (b).*
timing line (tim'-ing) One of a series of marks or lines placed on seismic records at precisely determined intervals of time (usually 0.01 or 0.005 s) for the purpose of measuring the arrival time of recorded events.
tin (a) A metallic bluish-white tetragonal mineral, the native metallic element Sn. (b) A term used loosely to designate cassiterite and concentrates containing cassiterite with minor amounts of other minerals.
tinaja (ti-na'-ja [te-na'-ha]) (a) A term used in SW U.S. for a *water pocket* developed below a waterfall, esp. when partly filled with water. (b) A term used loosely in New Mexico for a temporary pool, and for a spring too feeble to form a stream. Etymol: Spanish, "large earthen jar".
tinajita (ti-na-ji'-ta [te-na-he'-ta]) *solution pan.* Etymol: Spanish, "little water jar".
tinaksite (ti-nak'-site) A vitreous colorless, pale yellow, or pale pink triclinic mineral: $K_2Na(Ca,Mn^{2+})_2TiSi_7O_{19}(OH)$.
tincal (tin'-cal) An old name for crude *borax* formerly obtained from Tibetan lake shores and deposits and once the chief source of boric compounds.
tincalconite (tin-cal'-co-nite) A colorless to dull-white rhombohedral mineral: $Na_2B_4O_5(OH)_4{\cdot}3H_2O$. Syn: *mohavite; octahedral borax.*
tind A Norwegian term for a glacial horn that is detached from the main mountain range by the lateral recession of cirques cutting through an upland spur between two glacial troughs (Thornbury, 1954, p.373). Syn: *monument [geomorph].*
tindar A type of *intraglacial volcano* that occurs as a steep-sided linear ridge composed typically of pillow lava and overlying hyaloclastite. It is one of the major forms of intraglacial volcano in Iceland (Jones, 1969). Etymol: Icelandic, "peak" or "pinnacle." Cf: *table mountain; tuya.*
tinder ore (tin'-der) An impure variety of jamesonite. Syn: *pilite.*
tin granite Intermediate to felsic intrusive rocks related to tin deposits.
tinguaite (tin'-gua-ite) A textural variety of *phonolite,* typically found in dikes, and characterized by conspicuous acicular crystals of aegirine arranged in radial or criss-cross patterns in the groundmass. The phenocrysts are of equigranular alkali feldspar and nepheline. The name, given by Rosenbusch in 1887, is derived from the Tingua Mountains (Serra de Tingua) near Rio de Janeiro, Brazil. Adj: tinguaitic. Cf: *muniongite.* Obsolete.
tinnunculite A yellowish-white orthorhombic organic mineral: $C_{10}H_{12}N_8O_8$. It is a combustion product of a coal heap, and has not been approved by the International Mineralogical Association.
tin ore *cassiterite.*
tin pyrites *stannite.*
tinsel (tin'-sel) A *flagellum* having a central axis from which extend many fine short hairs (mastigonemes) in one or two rows along its length.
tinsleyite (tins'-ley-ite) A magenta-red monoclinic mineral of the *leucophosphite* group: $KAl_2(PO_4)_2(OH){\cdot}2H_2O$.
tinstone (tin'-stone) *cassiterite.*
tint *hypsometric tint.*
tinticite (tin'-tic-ite) A white, pale yellow-green, or brown triclinic mineral: $Fe^{3+}{}_{5.3}(PO_4VO_4)_4(OH)_4{\cdot}6.7H_2O$.
tintinaite (tin-tin'-a-ite) A metallic gray orthorhombic mineral: $Pb_{22}Cu_4(Sb,Bi)_{30}S_{69}$.
tintinnid (tin-tin'-nid) A ciliate protozoan belonging to the family Tintinnidae and characterized by a lorica that is almost always inflated in the oral region. Range, Jurassic to Holocene.
tin-white cobalt *smaltite.*
tinzenite (tin'-zen-ite) A yellow to orange-red triclinic mineral of the *axinite* group: $(Ca,Mn^{2+},Fe^{2+})_3Al_2BSi_4O_{15}(OH)$. Cf: *manganaxinite.*
Tioughniogan (Ti-ough'-ni-o'-gan) North American provincial stage: Middle Devonian (above Cazenovian, below Taghanican).
tip [mass move] The point on the *toe [mass move]* farthest from the top of a slide (Varnes, 1978, pl. 1).
tip [paleont] The inversely conical initial part of the cup in archaeocyathids (TIP, 1972, pt. E, p.42). Nonrecommended syn: *spitz.*
tiphon (ti'-phon) *diapirism.*
tip line The line marking the tip or termination of a fault surface; a tip line separates faulted rock from unfaulted rock.
tip-line fold A fold found along the tip line of a fault; inferred to form as a result of slip on the fault.
tipper (tip'-per) Correlation matrix between the vertical and horizontal magnetic fields measured in a survey by the *magnetotelluric method.*
tiptopite (tip'-top-ite) A colorless hexagonal mineral: $K_2(Li,Na,Ca)_6Be_6(PO_4)_6(OH)_2{\cdot}H_2O$. It is structurally related to the *cancrinite* group.
tiragalloite (ti'-ra-gal'-lo-ite) A sub-adamantine orange or brownish monoclinic mineral: $Mn^{2+}{}_4As^{5+}Si_3O_{12}(OH)$.
Tiraspolian (Tir-as-pol'-ian) Eastern European mammal age ("complex"): early medial Pleistocene (after Tamanian); equiv. Cromerian.
tirodite (tir'-o-dite) *parvowinchite.*
tischendorfite A metallic orthorhombic mineral: $Pd_8Hg_3Se_9$.
tisinalite (ti-si'-na-lite) A yellow-orange trigonal mineral of the *lovozerite* group: $Na_2(Mn,Ca)_{1-x}(Ti,Zr,Nb,Fe^{3+})Si_6O_8(O,OH)_{10}$.
Tissot indicatrix *indicatrix.*
tissue (tis'-sue) An aggregate of similar cells into a structural unit that performs a particular function.
tissue preservation index A parameter derived from maceral percentages, defined as (telinite+telocollinite+semifusinite+fusinite)/(desmocollinite+macrinite+inertodetrinite), describing the ratio of material with remnant cellular structure to material lacking cellular structure. Developed by Diessel (1986) for Permian coals in the Sydney Basin (New South Wales, Australia) but widely applied to coals of other ages and locations.
titanaugite (ti-tan-au'-gite) A variety of augite rich in titanium and occurring in basaltic rocks: $Ca(Mg,Fe,Ti)(Si,Al)_2O_6$.
titanclinohumite An unnecessary name for Ti-bearing clinohumite.
titanic iron ore (ti-tan'-ic) A syn. of *ilmenite.* Also called: titaniferous iron ore.
titanite (ti'-ta-nite) (a) A usually yellow or brown monoclinic mineral: $CaTiSiO_5$. It often contains other elements such as niobium, chromium, fluorine, sodium, iron, manganese, and yttrium. Titanite occurs in wedge-shaped or lozenge-shaped monoclinic crystals as an accessory mineral in granitic rocks and in calcium-rich metamorphic rocks. Syn: *sphene; grothite.* (b) A group name for monoclinic minerals with an analogous composition, but with Ti replaced by Sn or V.

titanium The element with formula Ti, reported as a naturally occurring metallic silver-gray mineral, but not approved by the International Mineralogical Association.
titanochromite (ti'-tan-o-chro'-mite) An unnecessary name for a titanium-bearing chromite found in Apollo XI lunar rocks.
titanomaghemite (ti'-tan-o-mag'-hem-ite) A general term applied to an abnormal titanium-bearing magnetite with varying cation vacancies in an oxygen framework of spinel structure.
titanomagnetite (ti'-tan-o-mag'-net-ite) (a) A titaniferous variety of magnetite: $(Fe^{3+},Fe^{2+},Ti^{4+})_2[Fe^{2+}O_4]$. It is strictly a homogeneous cubic solid solution of ilmenite in magnetite. (b) A term loosely used for mixtures of magnetite, ilmenite, and ulvöspinel.
titanowodginite A vitreous dark brown to black monoclinic mineral of the *wodginite* group: $(Mn^{2+},Fe^{2+})(Ti,Sn,Ta)(Ta,Nb)_2O_8$.
titantaramellite A dark reddish-brown orthorhombic mineral of the *taramellite* group: $Ba_4(Ti,Fe^{3+},Mg)_4(O,OH)_2[B_2Si_8O_{27}]Clx$.
Tithonian (Ti-tho'-ni-an) A stage of the Standard Global Chronostratigraphic Scale: uppermost Jurassic (above Kimmeridgian, below Berriasian Stage of the Cretaceous).
tivanite (ti'-van-ite) A submetallic black monoclinic mineral: $V^{3+}TiO_3(OH)$.
tjaele (tjae'-le) A syn. of *frozen ground.* The term has been used erroneously as a syn. of *permafrost* (Bryan, 1951). Etymol: Swedish "tjäle", "frozen ground". Also spelled: tjäle; *taele.*
tjosite (tjos'-ite) A dark-colored porphyritic nepheline-bearing *phonolite,* tending toward the composition of *jacupirangite* or *lamprophyre,* forming dikes in *larvikite.* Its name, given by Brögger in 1906, is derived from Tjose, Larvik district, Norway. Obsolete.
tlalocite (tla'-loc-ite) A blue orthorhombic mineral: $(Cu,Zn)_{16}(Te^{4+}O_3)(Te^{6+}O_4)_2Cl(OH)_{25}{\cdot}27H_2O$.
tlapallite (tla'-pal-lite) A green monoclinic mineral: $H_6(Ca,Pb)_2Cu_3(SO_4)(Te^{4+}O_3)_4(Te^{6+}O_6)$.
TLC *thin-layer chromatography.*
TM *transverse magnetic.*
TMI *total magnetic intensity.*
toadback marl (toad'-back) A term used in Lancashire, England, for unlaminated marl with lumpy fracture. Rarely used.
toad's-eye tin A reddish or brownish variety of cassiterite occurring in botryoidal or reniform shapes that display an internal concentric and fibrous structure.
toadstone (toad'-stone) A fossilized object, such as a fish tooth or palatal bone, that was thought to have formed within a toad and was formerly worn as a charm or an antidote to poison.
toadstool rock (toad'-stool) *mushroom rock.*
Toarcian (To-ar'-ci-an) A stage of the Standard Global Chronostratigraphic Scale: uppermost Lower Jurassic (above Pliensbachian, below Aalenian).
tobacco jack (to-bac'-co) A miner's term for *wolframite.*
tobacco rock A term used in SW U.S. for a favorable host rock for uranium, characterized by light yellow or gray color and by brown limonite stains.
tobelite (to'-bel-ite) The white or yellowish-green monoclinic NH_4-analogue of muscovite, with NH_4 in place of K in the interlayer position: $(NH_4,K)Al_2(Si_3Al)O_{10}(OH)_2$.
tobermorite (to-ber-mor'-ite) A pale pinkish-white, white, or brown fibrous orthorhombic mineral: $Ca_5Si_6(O,OH)_{18}{\cdot}5H_2O$.
tochilinite (to-chil'-i-nite) A metallic bronze-black triclinic mineral: $6Fe_{0.9}S{\cdot}5(Mg,Fe^{2+})(OH)_2$.
tocornalite (to-cor'-na-lite) An incompletely investigated yellow mineral consisting of silver mercury and iodine.
toddite (todd'-ite) A mixture of columbite and samarskite.
todorokite (to'-do'-rok-ite) A metallic black monoclinic mineral: $(Mn^{2+},Ca,Na,K)(Mn^{4+},Mn^{2+},Mg)_6O_{12}{\cdot}3H_2O$. It may contain some barium and zinc. Syn: *delatorreite.*
toe [coast] *step [coast].*
toe [drill] The bottom of a *drill hole* (esp. one used for blasting), as distinguished from its open end.
toe [mass move] (a) The lower, usually curved, margin of the disturbed material of a landslide pushed over onto the undisturbed slope; it is most distant from the place of origin. Cf: *foot [geol]; tip [mass move].* (b) The intersection (sometimes buried) between the lower part of the *surface of rupture* of a landslide and the original ground surface (Varnes, 1978).
toe [slopes] The lowest part of a slope or cliff; the downslope end of an alluvial fan.
toe [struc geol] The leading edge of a thrust sheet.
toenail (toe'-nail) (a) A curved joint intersecting a sheet structure, usually along the strike or sometimes differing from it by 45° or more. (b) Obsolete syn. of *stylolite.*
toernebohmite Var. spelling of törnebohmite.
toeset (toe'-set) The forward part of a tangential foreset bed.
toe-tap flood plain The outer end of a meander lobe, built by a stream as it meanders down-valley.
toft A British term for an isolated hill, knoll, or other eminence in a flat region, esp. one suitable for a homesite.
toise An old French unit of length used in early geodetic surveys and equal to 6 French feet, 6.396 U.S. feet, or 1.949 meters.
tokeite (to-ke'-ite) A dark-colored picritic *basalt* similar in composition to schönfelsite, in which the plagioclase is labradorite. Cf: *oceanite.* The name, given by Duparc and Molly in 1928, is for Arête de Toké, Gouder Valley, Ethiopia. Obsolete.
tokkoite (tok'-ko-ite) A colorless, light brown, or pale yellow triclinic mineral: $K_2Ca_4Si_7O_{18}(OH,F)_2$. It is isomorphous with tinaksite.
tolbachite (tol'-bach-ite) A brown monoclinic mineral occurring in fumaroles: $CuCl_2$.
tolerance A numerical limit set on the value that a variable can assume.
Tolkowsky theoretical brilliant cut (Tol-kow'-sky) A style of diamond cutting having those proportions and facet angles calculated by Marcel Tolkowsky in 1919 to produce the maximum brilliancy with a high degree of fire in a round brilliant cut diamond. The "ideal" proportions are a table diameter of 53% of the girdle diameter; thickness above the girdle, 16.2%; thickness below the girdle 43.1%, girdle thickness 0.7-1.7%. Syn: *American brilliant cut; ideal cut.* Cf: *brilliant cut.*
tolovkite (to-lov'-kite) A metallic gray orthorhombic (pseudo-cubic) mineral: IrSbS. Its crystal structure is related to that of ullmannite.
tolt A term used in Newfoundland for an isolated peak rising abruptly from a plain.
tombarthite-(Y) (tom-barth'-ite) A brownish-black monoclinic mineral of the *huttonite* group: $Y_4(Si,H_4)_4O_{12}(OH)_4$.
tombolo (tom-bo'-lo) A sand or gravel bar or barrier that connects an island with the mainland or with another island. Etymol: Italian, "sand dune"; from Latin "tumulus", "mound". Pl: tombolos. Syn: *connecting bar; tie bar; tying bar.* Cf: *intertidal tombolo; island-tying.*
tombolo cluster *complex tombolo.*
tombolo island *tied island.*
tomichite (to'-mich-ite) An opaque black monoclinic mineral closely related to derbylite: $(V^{3+},Fe^{3+})_4Ti_3As^{3+}O_{13}(OH)$.
tomite (tom'-ite) *boghead coal.*
Tommotian (Tom-mo'-tian) A Russian stage: Lower Cambrian (above Nemakit, Daldynian, below Atdabanian).
tomography (to-mog'-ra-phy) A method for finding the velocity distribution from a multitude of observations using combinations of source and receiver locations. Space is divided into cells, and the data are expressed as line integrals along raypaths through the cells. Application is usually iterative beginning with a starting velocity model. Often involves borehole-to-borehole or surface-to-borehole observations. *Reflection tomography* involves observations of reflected events. The result is to determine a velocity for each cell to best match observed traveltimes. Sometimes amplitudes are measured to determine an attenuation factor for each cell. *Layer tomography* is a variation in which the locations of the cells are constrained by specifying layering. Tomographic methods include the algebraic reconstruction technique (ART), the *simultaneous reconstruction technique* (SIRT), and Gauss Seidel methods (Ivansson, 1986). Etymol: Greek "section drawing".
tonal balancing The act of adjusting the color, brightness, and contrast of an image in order to make the appearance of two or more images consistent.
tonalite (to'-nal-ite) In the *IUGS classification,* a plutonic rock with Q between 20 and 60, and P/(A+P) greater than 90. The name, given by Rath in 1864, is derived from Tonale Pass, northern Italy. See also: *adamellite.*
Tonawandan (Ton-a-wan'-dan) Stage in New York State: middle Middle Silurian.
tongbaite (tong'-ba-ite) A metallic pale brownish-yellow orthorhombic mineral: Cr_3C_2.
Tongrian (Ton'-gri-an) European stage: Lower Oligocene (above Ludian of Eocene, below Rupelian). Syn: *Sannoisian.*
tongue [coast] (a) A *point,* or long low strip of land projecting from the mainland into the sea or other body of water. (b) *inlet.*
tongue [glaciol] A long narrow extension of the lower part of a glacier, either on land or afloat. Cf: *glacial lobe.* Syn: *glacier tongue.* Less preferred syn: *ice tongue.*
tongue [ice] A projection of the *ice edge* up to several kilometers in length, caused by wind and current.
tongue [intrus rocks] An offshoot of a larger intrusive body. See also: *epiphysis [intrus rocks].* Syn: *apophysis [intrus rocks].*
tongue [oceanog] An extension of one type of water into water of differing salinity or temperature, e.g., salt water into the mouth of a river.
tongue [stratig] n. A projecting part of a lithostratigraphic unit extending beyond its main body (Stephenson, 1917). Cf: *lentil.* v. To thin laterally to disappearance.
tongue [streams] *meander lobe.*
tongue [volc] (a) A lava flow that is an offshoot from a larger flow; it may be as much as several kilometers in length. (b) A syn. of *coulee [volc].*
tongxinite A metallic gold-colored cubic mineral: Cu_2Zn of the brass group. Not approved by the International Mineralogical Association.
Tonian (To'-ni-an) A period of the Standard Global Geochronometric Scale: the oldest period of the *Neoproterozoic* Era, characterized by expansion of platform cover ("tonas", "stretch"). It is defined as the time span between 1,000 Ma, the end of the *Stenian* Period of the *Mesoproterozoic* Era, and 850 Ma, the beginning of the *Cryogenian* Period (Plumb, 1991).
tonnage-grade model *grade-tonnage model.*

tonnage-volume factor (ton′-nage-vol′-ume) In economic geology, the number of cubic feet in a ton of ore.
Tonolowayan (To-nol′-o-way′-an) North American provincial stage: Upper Silurian (above Salinan, below Keyseran).
tonsbergite (tons′-berg-ite) An altered red igneous rock that is sometimes porphyritic and resembles *larvikite,* the feldspar being represented by orthoclase and andesine. Brögger in 1898 derived the name from Tönsberg, Norway. Obsolete.
tonstein (ton′-stein [tohn-shtine]) A compact rock usually composed mainly of kaolinite together with occasional detrital and carbonaceous material. Tonsteins commonly occur as thin bands in coal seams and are used as an aid in correlation. They represent the alteration product of volcanic ash that has fallen into an acid environment, as it would be in a coal swamp in which organic acids have been generated by the breakdown of plant material (Bohor and Triplehorn, 1993). Etymol: German, "ton" = clay + "stein" = stone.
tooeleite A greasy orange orthorhombic mineral: $Fe^{3+}_{7.6}[(As,S)O_4]_6(OH)_6 \cdot 5H_2O$.
tool mark A *current mark* produced by the impact against a muddy bottom of a solid object swept along by the current, and generally preserved as a cast on the underside of the overlying bed. The mark may be produced by an object in continuous contact with the bottom (e.g. a *groove* or a *striation*), in intermittent contact with the bottom (e.g. a *skip mark* or a *prod mark*), or rolling along the bottom (e.g. a *roll mark*). The engraving "tools" include shell fragments, sand grains, pebbles, fish bones, seaweed, and wood chips. Originally defined by Dzulynski and Sanders (1962, p.72).
toolpusher (tool′-push-er) The supervisor of operations on a *drilling rig.*
tooth (a) In gnathostomes, any very hard element of mineral composition similar to bone but of divergent histologic structure and derived from ectodermal as well as mesodermal embryonic tissue, borne primarily on bones of the jaws and palate and specialized for seizing and/or processing food, social displays, grooming, or sound production. (b) Any of various horny, chitinous, or calcareous projections of an invertebrate that function like or resemble the vertebrate tooth, as on the radula of a gastropod. (c) A toothlike process on the margin of a bivalve shell; specif. *hinge tooth.* (d) A calcareous rod located in the pyramid of Aristotle's lantern of an echinoid. Its upper end is uncalcified. (e) A projection in the aperture of a foraminiferal test. It may be simple or complex, and single or multiple.
toothpaste lava (tooth′-paste) Viscous lava that is extruded as a *squeeze-up.*
tooth plate An internal, apertural modification of a foraminiferal test, commonly consisting of a contorted plate that extends from the aperture through the chamber to the previous septal foramen. One side may be attached to the chamber wall or to the proximal border of the foramen, the opposite side being free and folded (TIP, 1964, pt. C, p.64).
top [gem] *crown [gem].*
top [mass move] The highest point of contact between the disturbed material of a landslide and the scarp face along which it moved. Cf: *tip [mass move].*
top [ore dep] A quarrymen's syn. of *overburden.*
top [stratig] A term used in petroleum geology for the uppermost surface of a formation where it is encountered during drilling, usually characterized by the first appearance of a distinctive feature (such as a marked change in lithology or the occurrence of a guide fossil). It is often determined by a distinctive configuration on an electric log, and it is widely used in correlation and structure-contour mapping.
topaz (to′-paz) (a) A white or lightly colored orthorhombic mineral: $Al_2SiO_4(F,OH)_2$. It occurs as a minor constituent in highly siliceous igneous rocks and tin-bearing veins as translucent or transparent prismatic crystals and masses, and as rounded waterworn pebbles. Topaz has a hardness of 8 on the Mohs scale. (b) A transparent topaz used as a gemstone. (c) A yellow quartz that resembles topaz in appearance, such as smoky quartz turned yellow by heating; specif. *false topaz* and *Scotch topaz.* See also: *Spanish topaz.* (d) A term used for a green-yellow to orange-yellow mineral resembling topaz in appearance, such as "oriental topaz" (a yellow corundum).
topazolite (to-paz′-o-lite) A green-yellow to yellow-brown variety of andradite garnet, having the color and transparency of topaz.
topaz quartz Topaz-colored quartz; specif. *citrine.* Cf: *quartz topaz.*
topaz rhyolite A *peraluminous* rhyolite (Burt et al., 1982).
top conglomerate A *conglomerate* lying conformably at the top of a stratum (Twenhofel, 1939, p.203-204).
top discordance (top-dis-cord′-ance) A term used in seismic stratigraphy to refer to a lack of parallelism between a sequence of strata and its upper boundary, owing to either truncation or *toplap* (Mitchum, 1977, p.206). Cf: *base discordance.*
topical map *thematic map.*
toplap (top′-lap) Termination of strata against an overlying surface, mainly as a result of nondeposition (sedimentary bypassing) with perhaps minor erosion. Each unit of strata laps out in a landward direction at the top of the unit, but the successive terminations lie progressively seaward (Mitchum, 1977, p.211).
topocentric horizon (top-o-cen′-tric) *apparent horizon.*
topochemistry The pattern of locations of different atomic species in a crystal structure. Structures are said to be topochemically identical if they have (1) the same topology and (2) the same relative arrangement of atomic species, even if the structures are geometrically different.
topocline (top′-o-cline) A *cline* related to a geographic zone and usually unrelated to any ecologic condition.
topogenous peat (to-pog′-e-nous) Peat whose moisture content is largely dependent on surface water. It is relatively rich in plant nutrients, nitrogen, and mineral matter, is mildly acidic to nearly neutral, and contains little or no cellulose; forms in topographic depressions with an essentially stagnant or non-moving minerotrophic water supply. Topogenous peat is generally poorly preserved. Cf: *ombrogenous peat; soligenous peat; lowmoor peat; planar peat.*
topographic (top-o-graph′-ic) (a) Pertaining to *topography.* (b) Surveying or representing the topography of a region; e.g. a "topographic survey" or a "topographic map".
topographic adjustment The condition existing where the gradient of a tributary is harmonious with that of the main stream. Cf: *structural adjustment.*
topographic adolescence *adolescence.*
topographic-bathymetric map (top-o-graph′-ic-bath-y-met′-ric) A map with both the relief of the land and the bottom of the offshore areas shown by contours. Contour intervals on land and on sea bottom may not be the same (USGS, 2002b, p.2-50 to 2-51).
topographic contour *contour [cart].*
topographic correction [cart] A positive correction used with other corrections in making gravity reductions taking into account actual deviations from level terrain in the area surrounding a station.
topographic correction [geophys] *terrain correction.*
topographic datum *geoid [planet].*
topographic deflection of the vertical An expression used to indicate that the deflection of the vertical has been computed from the topography. This method may be used when it is not possible to compare astronomic and geodetic positions directly.
topographic depression *closed depression.*
topographic desert A desert of low rainfall because of its location in the middle of a continent, far from the ocean, or on the lee side of high mountains, cut off from prevailing winds. See also: *rain-shadow desert.*
topographic divide A drainage divide.
topographic expression The effect achieved by shaping and spacing contour lines on a map so that topographic features can be interpreted with the greatest ease and fidelity.
topographic feature A prominent or conspicuous *topographic form* or noticeable part thereof (Mitchell, 1948, p.80). Cf: *physiographic feature.*
topographic form A *landform* considered without regard to its origin, cause, or history (Mitchell, 1948, p.80). Cf: *physiographic form.*
topographic grain The alignment and direction of the topographic-relief features of a region.
topographic inversion An optical illusion that may occur on images with extensive shadows. Ridges appear to be valleys, and valleys appear to be ridges. The illusion is corrected by orienting the image so that shadows trend from the top margin of the image to the bottom.
topographic license The freedom to adjust, add, or omit contour lines, within allowable limits, in order to attain the best topographic expression; it does not permit the adjustment of contours by amounts that significantly impair their accuracy.
topographic map A map showing horizontal and vertical positions of the features being represented, with the relief being shown in measurable form, commonly by means of contour lines. It is generally on a sufficiently large scale to show in detail selected man-made and natural features, including relief and such physical and cultural features as vegetation, roads, and drainage. Cf: *contour map; planimetric map.*
topographic reversal A geomorphic phenomenon in which topographic lows coincide with structural highs and vice versa. Valleys are eroded on crests of anticlines to cause topographic lows, and synclines form ridges, or topographic high.
topographic survey A survey that determines the configuration (relief) of the Earth's ground surface and the location of natural and artificial features thereon. Also, an organization making such a survey.
topographic texture Disposition, grouping, or average size of the topographic units composing a given topography; usually restricted to a description of the relative spacing of drainage lines in stream-dissected regions. See also: *coarse topography; fine topography; texture ratio.* Syn: *texture [geomorph].*
topographic unconformity (a) The relationship between two parts of a landscape or two kinds of topography that are out of adjustment with one another, due to an interruption in the ordinary course of the erosion cycle of a region; e.g. a lack of harmony between the topographic forms of the upper and lower parts of a valley, due to rejuvenation. (b) A land surface exhibiting topographic unconformity.
topography (to-pog′-ra-phy) (a) The general configuration of a land surface or any part of the Earth's surface, including its relief and the position of its natural and man-made features. See also: *geomorphy.* Cf: *relief [geomorph].* Syn: *lay of the land.* (b) The natural or physical surface features of a region, considered collectively as to form; the features revealed by the contour lines

of a map. In nongeologic usage, the term includes man-made features (such as are shown on a topographic map). (c) The art or practice of accurately and graphically delineating in detail, as on a map or chart or by a model, selected natural and man-made surface features of a region. Also, the description, study, or representation of such features. Cf: *chorography.* (d) Originally, the term referred to the detailed description of a particular place or locality (such as a city, parish, or tract of land) as distinguished from the general geography of a country or other large part of the world, and also to the science or practice of such a description; this usage is practically obsolete. Etymol: Greek "topos", "place", + "graphein", "to write".

topological error Term used in spatial information systems to denote inconsistencies in the geometry of a map file as defined by the requirements of a particular software system. The following topological errors are commonly distinguished, dangle, overshoot, undershoot, leaking polygon, multiple centroid (Krzanowski et al., 1993).

topologically distinct (top-o-log'-i-cal-ly) Said of channel nets of a given magnitude "whose schematic map projections cannot be continuously deformed and rotated in the plane of projection so as to become congruent" (Shreve, 1966, p.27).

topologic path length (top-o-log'-ic) *link distance.*

topology [cart] (to-pol'-o-gy) Study of the properties of geometric configurations. A description of the relationship between nodes, lines (arcs), and polygons (areas) in a vector data file. In digital cartography, topology is concerned with the characteristics of data that describe the spatial relationships, such as dimensionality, adjacency, and connectivity, of map elements to each other.

topology [cryst] The connectivity of a *crystal structure,* manifested as chemical bonds between atoms, regardless of the chemical species of the atoms. Structures that have the same topology, even though not *isomorphous,* may be made geometrically identical by distortions that do not sever atomic bonds.

topology [geochem] The spatial relationships among invariant points, univariant lines, and divariant fields in a phase diagram (Zen, 1966). See also: *Schreinemakers' analysis.*

topometry (to-pom'-e-try) The art, process, or science of making large-scale, high-precision maps (1:20,000 or larger) upon which geomorphic features are "measured in with mathematical accuracy, practically nothing being 'sketched in' by eye" (Matthes, 1912, p.338). Cf: *topology [cart].*

toposequence (top-o-se'-quence) A sequence of kinds of soil in relation to position on a slope. See also: *catena [soil].*

topostratic unit The local part of a stratic unit, a "local unit" (Henningsmoen, 1961, p.67).

topo-stratigraphic unit (top'-o-strat'-i-graph'-ic) A term proposed by Jaanusson (1960, p.218) for a "convenient regional stratigraphic unit" consisting of a combined rock unit and biostratigraphic unit. Equiv. to the mixed topostratic unit of Henningsmoen (1961).

topostratigraphy (top'-o-stra-tig'-ra-phy) Preliminary or introductory stratigraphy, including lithostratigraphy and biostratigraphy; *prostratigraphy.*

topotactic (top-o-tac'-tic) Adj. of *topotaxy.*

topotaxial (top-o-tax'-i-al) Adj. of *topotaxy.*

topotaxy (top'-o-tax-y) The phenomenon of mutual orientation of two crystals of different species resulting from a solid-state transformation. It occurs in displacive transformations with no breaking of primary interatomic bonds, as in low- to high-quartz transformation. The degree of topotaxy is low in reconstructive transformations. Topotaxy is a more general term than epitaxy. Adj: *topotactic; topotaxial.* Cf: *epitaxy; syntaxy.*

topotype (top'-o-type) A specimen of a particular species that comes from the same locality as the *type specimen* of that species (Frizzell, 1933, p.665).

topozone (top'-o-zone) (a) A syn. of *local-range zone* and *teilzone.* The term was proposed by Moore (1957, p.1888) for a paleontologically defined horizon or zone recognizable in a single locality. (b) A biochronostratic unit (topobiozone, topohontozone) representing the local range of a taxon (Henningsmoen, 1961, p.69).

topozonite A biostratic unit representing the local development of a range zone; equivalent to local-range zone (Henningsmoen, 1961, p.68).

topple (top'-ple) A *mass movement* that consists of the forward rotation of a unit or units about some pivot point, below or low in the unit, under the action of gravity and forces exerted by adjacent units or by fluids in cracks. It is tilting without collapse (Varnes, 1978).

topset (top'-set) A *topset bed.*

topset bed One of the nearly horizontal layers of sediments deposited by a river on the top surface of an advancing delta and continuous with the landward alluvial plain; it truncates or covers the edges of the seaward-lying foreset beds. See also: *bottomset bed.* Also spelled: top-set bed. Syn: *topset.*

topsoil (top'-soil) (a) A presumably fertile soil used to cover areas of special planting. (b) A partial syn. of *surface soil.* (c) A syn. of A horizon. (d) The dark-colored upper portion of a soil, varying in depth according to soil type. Cf: *subsoil; loam.*

tor A high, isolated crag, pinnacle, or rocky peak; or a pile of rocks, much-jointed and usually granitic, exposed to intense weathering, and often assuming peculiar or fantastic shapes, e.g. the granite rocks standing as prominent masses on the moors of Devon and Cornwall, England. Linton (1955) suggested that a tor is a residual mass of bedrock resulting from subsurface rotting through the action of acidic groundwater penetrating along joint systems, followed by mechanical stripping of loose material. Periglacial processes may also be important in the formation of tors. Etymol: Celtic(?). Cf: *core-stone; elephant rock.*

torbanite (tor'-ban-ite) Essentially synonymous with *boghead coal,* but often considered as a highly carbonaceous oil shale. It is named from its type locality, Torbane Hill, in Scotland. Cf: *cannel coal; wollongongite.* Syn: *kerosine shale; bitumenite.*

torbernite (tor'-bern-ite) A green radioactive tetragonal mineral of the *autunite* group: $Cu(UO_2)_2(PO_4)_2{\cdot}8\text{-}12H_2O$. Torbernite is commonly a secondary mineral and occurs in tabular crystals or in foliated form. Syn: *chalcolite; copper uranite; cuprouranite; uran-mica.*

torca Large closed depression, more or less circular; a *doline.* Etymol: Spanish.

torch peat A waxy, resinous peat derived mainly from pollen.

Toreva block (To-re'-va) A slump block consisting essentially of a single large mass of unjostled material which, during descent, has undergone a backward rotation toward the parent cliff about a horizontal axis that roughly parallels it (Reiche, 1937). See also: *rotational landslide.*

tornado (tor-na'-do) A small-scale *cyclone,* generally less than 500 m in diameter and with very strong winds. Tornadoes commonly occur as dark funnel-like features suspended from low-lying cumulonimbus clouds. See also: *squall line.*

törnebohmite-(Ce) (tor'-ne-bohm-ite') An olive-green monoclinic mineral: $Ce_2Al(SiO_4)_2(OH)$.

törnebohmite-(La) A dark green or greenish-gray monoclinic mineral: $(La,Ce)_2Al(SiO_4)_2(OH)$.

toroid (to'-roid) A cast, commonly consisting of sand, of a circular scour pit made in firm, shallow-water sediments (such as hard mud) by an eddy or whirlpool in flowing water. It has a characteristic swirled shape like a folded bun, but with a homogeneous internal structure and texture.

torose load cast (to'-rose) One of a group of elongate load casts that pinch and swell along their trends and that may terminate downcurrent in bulbous, tear-drop, or spiral forms (Crowell, 1955, p.1360).

torque The effectiveness of a force that tends to rotate a body; the product of the force and the perpendicular distance from its line of action to its axis.

Torrands Andisols that have an aridic soil moisture regime. A suborder in the U.S. system of soil taxonomy.

Torrejonian (Tor-re-jo'-ni-an) North American land mammal age: Early Eocene (after Dragonian, before Tiffanian).

torrent (tor'-rent) (a) A violent and rushing stream of water; e.g. a flooded river, or a rapidly flowing stream in a mountain ravine, or a stream suddenly raised by heavy rainfall or rapid snowmelt and descending a steep slope. Also, any similar stream, as of lava. (b) A mountain channel that is intermittently filled with rushing water at certain times or seasons. Adj: torrential.

torrential cross-bedding (tor-ren'-tial) A variety of *angular cross-bedding* in which the beds make a nearly uniform but relatively large angle with the layers that enclose them (Hobbs, 1906, p.291). It is essentially planar *tabular cross-bedding.*

torrential plain An early term used by McGee (1897) for a feature now known as a *pediment.*

Torrerts Vertisols of arid regions that if not irrigated during the year have cracks in 6 or more out of 10 years that remain closed for less than 60 consecutive days during a period when the soil temperature at a depth of 50 cm from the surface is higher than 8°C. A suborder in the U.S. system of soil taxonomy.

torreyite (tor'-rey-ite) A colorless or bluish-white orthorhombic mineral: $(Mg,Mn^{2+})_9Zn_4(SO_4)_2(OH)_{22}{\cdot}8H_2O$. Cf: *mooreite.* Syn: *delta-mooreite.*

torricellian chamber (tor-ri-cel'-li-an) An air-filled cave room that is sealed by water and has a pressure less than atmospheric pressure and an air-water surface higher than adjacent free air-water surfaces. Cf: *periodic spring; siphon [speleo].*

Torrox (Tor'-rox) Oxisols that have a torric soil moisture regime. A suborder in the U.S. system of soil taxonomy.

torsion (tor'-sion) (a) The state of stress produced by two force couples of opposite moment acting in different but parallel planes about a common axis. (b) An early ontogenetic process in many gastropods, by which the posterior mantle cavity rotates laterally and anteriorly in a counterclockwise direction, so that it comes to lie above the head.

torsion balance A design concept for measuring the horizontal and vertical derivatives (i.e., gradients) of the *gravitational field.* In a non-uniform field, the forces on equal masses at opposite ends of a horizontal beam, suspended by a very fine torsion wire, differ and produce a torque (i.e., rotation of the suspended system about the fiber) that can be measured by counter balancing with a known torque. The measured gradients can be integrated to make a gravity map.

torsion balance gradiometer A gradiometer with a measuring system design based on the *torsion balance* concept. It consists of a pair of equal masses suspended by a sensitive torsion fiber and supported such that they are displaced both horizontally and vertically from each other. A measurement is made of the rotation of a suspended system about the fiber; the rotation is caused by slight differences in the gravitational attraction exerted upon the two masses. The torsion balance gradiometer is the oldest of all successful geophysical instruments. Syn: *Eötvös torsion balance.*

torsion coefficient The resistance of a material to torsional stress, measured as the work necessary to overcome it, in cgs units.

torsion fault An obsolete term for a *wrench fault.*

torsion magnetometer An instrument that is both a *horizontal-field balance* and a *vertical-field balance* and consists of a suspended permanent magnet.

torsion modulus *modulus of rigidity.*

torsion period The natural period of oscillation of the suspended system in a torsion balance.

torso mountain *monadnock.*

torso plain *rumpffläche.*

torta (tor'-ta) *low dome* (Francis, 1993). Etymol: Spanish, "cake."

torticone (tor'-ti-cone) A cephalopod shell coiled in a three-dimensional spiral with progressive twisting of the conch, like most gastropods, as distinguished from one coiled in a plane spiral. See also: *helicoid.* Syn: *trochoceroid.*

tortoise (tor'-toise) *camel back.*

Tortonian (Tor-to'-ni-an) (a) A stage of the Standard Global Chronostratigraphic Scale: lower upper Miocene (above Serravallian, below Messinian). The basal GSSP is near Ancona, Italy. (b) An obsolete middle European marine and continental stage: middle Miocene (above Helvetian, below Sarmatian). Marine equivalent: *Badenian.*

tortuosity [elect] (tor-tu-os'-i-ty) The inverse ratio of the length of a rock specimen to the length of the equivalent path of electrolyte within it.

tortuosity [hydraul] (a) In surface water, the ratio of the actual length of a river channel, measured along the middle of the main channel, to the axial length of the river (ASCE, 1962). (b) In groundwater, the radio of (1) the actual length of a groundwater flow path between two points, and (2) the direct-line distance between those two points.

tortuous flow (tor'-tu-ous) *turbulent flow.*

torus (to'-rus) (a) An invagination or protuberance of exine more or less paralleling the laesura of a spore. Cf: *kyrtome.* (b) The thickening of the closing membrane in a bordered pit.

torvane (tor'-vane) A hand-held vane shear device used in the field or laboratory as an aid in estimating shear strength of cohesive soils.

toryhillite (to-ry-hill'-ite) A plutonic rock containing albite, nepheline, clinopyroxene, garnet, iron oxides, apatite, and calcite. Sodalite may be present; potassium feldspar is absent. Johannsen in 1920 derived the name from Toryhill, Monmouth Township, Ontario. Obsolete.

tosca (tos'-ca) (a) A term used in Patagonia for a white deposit of calcium carbonate occurring in the loess of the pampas. (b) A term used in Mexico for various rocks, such as clayey vein matter, a talc seam, and soft, decomposed porphyry. (c) A soft coral limestone, used in Puerto Rico for masonry, road surfacing, and as fertilizer. Etymol: Spanish feminine of "tosco", "rough, coarse, unpolished".

toscanite (tos'-ca-nite) A volcanic rock named by Washington in 1897, and later considered a variety of *rhyodacite* with a glassy groundmass. Obsolete.

tosudite (to-sud'-ite) A regular 1:1 interstratification of chlorite and smectite that is dioctahedral on average.

total absorptance (to'-tal) *Absorptance* measured over all wavelengths of incident energy. Cf: *spectral absorptance.*

total depth The greatest depth reached by a wellbore, measured along its axis; not necessarily a vertical depth. Cf: *true vertical depth.* Abbrev: TD.

total displacement (a) Cumulative geological slip. (b) A syn. of *net slip.*

total dissolved solids A syn. of *dissolved solids.* Abbrev: TDS.

total field The vector sum of all components of a field such as a magnetic field. Syn: *total intensity [geophys].*

total field resistivity Using the bipole-dipole array, computation of *apparent resistivity,* ignoring information about the direction of the received electric field and using only its magnitude.

total-fusion age (to'-tal-fu'-sion) The $^{40}Ar/^{39}Ar$ age determination method in which the sample is fused and all of the argon analyzed in a single step. It is comparable to a *conventional potassium-argon age.*

total hardness *hardness [water].*

total head The sum of the *elevation head, pressure head,* and *velocity head* of a liquid. For groundwater, the velocity-head component is generally negligible. Cf: *static head.*

total intensity [geophys] *total field.*

total intensity [magnet] The total magnetic intensity, as opposed to components in the vertical or horizontal directions; the magnitude of the magnetic field, symbolized by F; it is one of the *magnetic elements.* Syn: *total magnetic intensity.*

total magnetic field *total intensity [magnet].*

total magnetic intensity *total intensity [magnet].*

total passing Transportation of all sediment across an area without any being deposited. Cf: *bypassing.*

total porosity *porosity.*

total-range zone The body of strata that comprises the complete horizontal and vertical range of occurrence of a specific fossil group (Geological Society of London, 1967, p.85).

total reflection *Reflection* in which all of the incident wave is returned.

total-rock *whole-rock.*

total runoff *runoff [water].*

total slip *net slip.*

total station *tachymeter.*

total strain The strain relating the configuration of a body at some point in its strain history to its initial, unstrained configuration. Also referred to as "finite strain" or "total finite strain".

total time correction The sum of all corrections applied to reflection traveltimes in seismic prospecting, to express times as those that would have been obtained if source and detectors were located on a selected datum plane in the absence of a low-velocity layer or variations in elevation.

total vacuity A missing interval or hiatus in a stratigraphic sequence caused by both erosion and nondeposition (Wheeler, 1958, p.1058).

T-O-T structure In crystallography, an acronym for a sheet-silicate structure consisting of units containing a sandwich of *t*etrahedral-*o*ctrahedral-*t*etrahedral layers.

touchstone (touch'-stone) A black, flinty stone, such as a silicified shale or slate, or a variety of quartz allied to chert or jasper, whose smoothed surface was formerly used to test the purity or fineness of alloys of gold and silver by comparing the streak left on the stone when rubbed by the metal with that made by an alloy of predetermined composition. Syn: *Lydian stone; basanite [sed]; flinty slate.*

tounkite A green hexagonal mineral of the *cancrinite* group: $(Na,Ca,K)_8(Si_6Al_6)O_{24}(SO_4)_2Cl{\cdot}0.5H_2O$.

tour (Pron: tower) The work shift of a *rotary drilling* crew.

tourelle A little tower; applied to small flat-topped buttes of limestone in karst areas. Contrasted with pitons, which have pointed tops, and with coupoles, which have rounded tops. Etymol: French.

tourmaline (tour'-ma-line) (a) A group of minerals of general formula: $(Na,Ca)(Mg,Fe^{2+},Fe^{3+},Al,Li)_3Al_6(OH)_4(BO_3)_3[Si_6O_{18}]$. It sometimes contains fluorine in small amounts. (b) Any of the minerals of the *tourmaline* group, such as buergerite, elbaite, and dravite. Tourmaline occurs in 3-, 6-, or 9-sided prisms, usually vertically striated, or in compact or columnar masses; it is commonly found as an accessory mineral in granitic pegmatites, and is widely distributed in acid igneous rocks and in metamorphic rocks. Its color varies greatly and gives a basis for naming the varieties; when transparent and flawless, it may be cut into gems. See also: *schörl; elbaite.* Also spelled: *turmaline.*

tourmalite (tour'-ma-lite) A rock composed almost entirely of tourmaline and quartz, with a mottled appearance and a texture ranging from dense to granular to schistose. It is of secondary origin, resulting from metasomatic and pneumatolytic effects along the margins of igneous intrusions (Johannsen, 1939, p.22). Cf: *luxullianite; schorl rock.*

Tournaisian (Tour-nai'-sian) A series or stage of the Standard Global Chronostratigraphic Scale: lowermost Carboniferous (above Famennian of the Devonian, below Visean). The basal GSSP is at Montagne Noir, southern France (Paproth et al., 1991).

tour report A record, filled out on a tabulated form by the driller on an oil or gas well drilling rig, showing drill progress (such as number of feet drilled each day), bit descriptions, drilling tools used, size of the hole, rock encountered (including its character, color, and description), personnel present during the *tour* of duty, and other pertinent facts having to do with the drilling (including any unusual event).

towan (tow'-an) A coastal sand dune in Cornwall, England.

tower (tow'-er) As a topographic feature, a very high rock formation or peak marked by precipitous sides; e.g. Devils Tower, Wyo.

tower coral A variety of subaqueous *coralloid* which forms in shallow pools and has a vertically oriented, towerlike appearance.

tower karst (a) A type of tropical *karst* that is characterized by isolated, steep-sided limestone hills surrounded by a flat alluvial plain. Syn: *fenglin.* (b) A cluster of peaks where towers rise from a common base and are separated by large *sinkholes.* Syn: *fengcong; turmkarst.* Cf: *cockpit karst; pinnacle karst; polygonal karst; sinkhole karst.*

towhead (tow'-head) A low alluvial island or shoal in a river; esp. a sandbar covered with a growth of cottonwoods or young willows. The term was first used by pilots on the Mississippi River to designate lenticular alluvial islands, which commonly are named.

township (town'-ship) The unit of survey of the U.S. Public Land Survey system, representing a piece of land that is bounded on the east and west by meridians approximately six miles apart (exactly six miles at its south border) and on the north and south by parallels six miles apart, and that is normally subdivided into 36 sections. Townships are located with reference to the initial point of a principal meridian and base line, and are normally numbered consecutively north and south from a base line (e.g. "township 14 north" indicates a township in the 14th *tier* north of a base line). The term "township" is used in conjunction with the appropriate range to indicate the coordinates of a particular township in reference to the initial point (e.g. "township 3 south, range 4 west" indicates the particular township which is the 3rd township south of the base line and the 4th township west of the principal meridian controlling the surveys in that area). Abbrev (when citing specific location): T. See also: *fractional township.*

township line One of the imaginary boundary lines running east and west at six-mile intervals and marking the relative north and south locations of townships in a U.S. public-land survey. Cf: *range line.*

toyohaite A metallic tetragonal mineral: $Ag_2FeSn_3S_8$. It is the Ag analogue of rhodostannite.

Toyonian Russian stage: Lower Cambrian, above Botomian, below Amgaian (Cowie and Bassett, 1989).

TP *tree pollen.*

TP [surv] *turning point.*

T phase A seismic phase with a period of 1 s or less, which travels through the ocean with the speed of sound in water. It is occasionally identified on the records of earthquakes in which a large part of the path from epicenter to station is across the deep ocean.

T plane A term used in crystal plasticity to denote the crystallographic slip plane. See also: *f axis; t direction.* Syn: *glide plane [cryst]; gliding plane; slip plane [cryst]; translation plane.*

trab A complex beam or rod in anthaspidellid sponges, formed by union of the ray tips of dendroclones. The trabs appear like siderails in ladders, with individual spicules for the rungs (Rigby and Bayer, 1971, p.609).

trabecula [bot] (tra-bec'-u-la) (a) In the wood of a gymnosperm, a small bar extending across the lumina of the ordinary tracheid from one tangential wall to another. It has no known function. (b) In a moss, a transverse bar of the teeth of the peristome. (c) In some dinoflagellate cysts, a narrow, solid rod connecting distal processes. (d) In the lycopod order Isoetales, a plate forming a partial septum in the microsporangium. (e) In the lycopod *Selaginella,* the lacunar tissue between the cortex and the central vascular strand. Pl: trabeculae.

trabecula [paleont] (a) A rod or pillar of radiating calcareous fibers forming a skeletal element in the structure of the *septum* and related components of a coral. See also: *simple trabecula; compound trabecula.* (b) A branch separating the fenestrae in reteporiform cheilostome bryozoans. (c) One of the individual anastomosing filaments of a hexactinellid sponge, which form a web in which the flagellated chambers are suspended and constitute the *pinacoderm.* Also, any rodlike or beamlike skeletal element of a sponge other than a ray or branch of a single spicule; esp. such a structure of *sclerosome.* (d) A tiny rodlike structure, smaller and less regular than a pillar, connecting layers of *sclerite* in holothurians. Pl: trabeculae.

trabecula [palyn] An interconnecting rod, ribbon, or tube between processes on a dinoflagellate cyst or acritarch (Williams et al., 2000, p.235). Pl: trabeculae.

trabecular columella (tra-bec'-u-lar) A spongy *columella* in scleractinian corals, formed of trabeculae loosely joined with synapticulae or paliform lobes.

trabecular linkage The connection between corallite centers in scleractinian corals, reflecting in the hard parts the *indirect linkage* of stomodaea.

trabecular network (a) A network of sponge trabeculae made of sclerosome. (b) The cellular web of a hexactinellid sponge.

trabeculate chorate cyst (tra-bec'-u-late) A dinoflagellate *chorate cyst* possessing trabeculae (e.g. *Cannosphaeropsis*).

trabzonite A vitreous colorless monoclinic mineral: $Ca_4Si_3O_{10}{\cdot}2H_2O$.

trace [geochem] A concentration of a substance that is detectable but too minute for accurate quantitative determination.

trace [meteorol] A quantity of precipitation that is insufficient to be measured by a gauge.

trace [paleont] A sign, evidence, or indication of a former presence; specif. a mark left behind by an extinct animal, such as a *trace fossil.*

trace [seis] (a) The record of the output of one geophone group with time after the source, displayed on paper, film, or magnetic tape. (b) The record of the processing output for one location or bin.

trace [struc geol] The intersection of a geological surface with another surface, e.g. the trace of bedding on a fault surface, or the trace of a fault or outcrop on the ground. Cf: *trend [struc geol]; strike [struc geol].*

trace element (a) An element that is not essential in a mineral but that is found in small quantities in its structure or adsorbed on its surfaces. Although not quantitatively defined, it is conventionally assumed to constitute significantly less than 1.0% of the mineral. Syn: *accessory element; guest element.* (b) An element that occurs in minute quantities in plant or animal tissue and that is essential physiologically. Syn: *minor element; microelement.*

trace fossil A sedimentary structure consisting of a fossilized track, trail, burrow, tube, boring, or tunnel resulting from the life activities (other than growth) of an animal, such as a mark made by an invertebrate moving, creeping, feeding, hiding, browsing, running, or resting on or in soft sediment. It is often preserved as a raised or depressed form in sedimentary rock. Many trace fossils were formerly assumed to be bodily preserved plants or animals. Syn: *ichnofossil; trace [paleont]; vestigiofossil; lebensspur; bioglyph.*

tracer (trac'-er) Any substance that is used in a process to trace its course, specif. radioactive material introduced into a chemical, biological, or physical reaction.

tracer test A technique that is commonly used in *karst* areas to define groundwater flow paths, travel times, and other aquifer parameters: A nontoxic, traceable substance, often a fluorescent dye, is injected into a groundwater system and recovered at a downgradient location (often a spring).

trace slip In a fault, that component of the net slip which is parallel to the trace of an index plane, such as bedding, on the fault plane. See also: *trace-slip fault.* Little used.

trace-slip fault A fault on which the net slip is *trace slip,* or slip parallel to the trace of the bedding or other index plane. Little used.

tracheid (tra'-che-id) A pitted, lignified, elongate, cylindrical cell that is the basic water-conducting and support cell in *xylem* tissue. Tracheids often have tapered end walls and are nonliving at maturity. *Cryptogams* and *gymnosperms* generally have only tracheids for conduction; most *angiosperms* also have *vessels.*

tracheophyte (tra'-che-o-phyte') A *vascular plant;* includes all plants that have a vascular system, i.e. xylem and phloem (Swartz, 1971, p.477).

trachographic map (trach-o-graph'-ic) A map using perspective symbols to show local relief and average slope of the Earth's surface, after the style of Erwin Raisz (1959). Syn: *physiographic pictorial map.*

trachyandesite (tra-chy-an'-de-site) In the *IUGS classification,* a volcanic rock defined in the TAS diagram by a field bounded by points with the SiO_2 and total alkali coordinates: 53, 9.3; 57, 5.9; 57.6, 11.7; 63, 7. Cf: *benmoreite; latite.*

trachybasalt (tra'-chy-ba-salt') In the *IUGS classification,* a volcanic rock defined in the TAS diagram by a triangular field bounded by points with the SiO_2 and total alkali coordinates: 45, 5; 49.4, 7.3; 52, 5. Cf: *hawaiite [petrology]; potassic trachybasalt.*

trachydacite In the *IUGS classification,* a volcanic rock defined in the TAS diagram by a field partly bounded by point with SiO_2 and total alkali coordinates: 57.6, 11.7; 61, 13.5; 63, 7; and 69, 8. The field is bounded at high silica contents by a vertical line with its lowest end at 69, 8. In addition, normative quartz is >20%. Cf: *trachyte.*

trachydiscontinuity (tra'-chy-dis'-con-ti-nu'-i-ty) A term proposed by Sanders (1957, p.293) for an unconformity characterized by an irregular surface. Cf: *leurodiscontinuity.* Etymol: Greek "trachys", "rough", + discontinuity.

trachyophitic (tra-chy-o-phit'-ic) Said of the ophitic texture of an igneous rock in which the feldspar grains enclosed by pyroxene have a parallel or subparallel alignment; it also applies to nesophitic textures with such a microlitic fabric (Walker, 1957, p.2).

trachyostracous (tra-chy-os'-tra-cous) Thick-shelled; esp. said of a thick-shelled gastropod.

trachyte (tra'-chyte) (a) In the *IUGS classification,* a volcanic rock defined in the QAPF diagram by Q/(Q+A+P) between 0 and 5 and P/(P+A) between 10 and 35, and in the TAS diagram by a field partly bounded by points with SiO_2 and total alkali coordinates: 57.6, 11.7; 61, 13.5; 63, 7; and 69, 8. The field is bounded at high silica contents by a vertical line with its lowest end at 69, 8. In addition, normative quartz is <20%. Cf: *trachydacite.* (b) A group of fine-grained, generally porphyritic, extrusive rocks having alkali feldspar and minor mafic minerals (biotite, amphibole, or pyroxene) as the main components, and possibly a small amount of sodic plagioclase; also, any member of that group; the extrusive equivalent of *syenite.* Trachyte grades into *latite* as the alkali feldspar content decreases, and into *rhyolite* with an increase in quartz. Etymol: Greek "trachys", "rough", in reference to the fact that rocks of this group are commonly rough to the touch.

trachytic (tra-chyt'-ic) Volcanic ground mass texture in which tightly-packed, strongly-aligned feldspar microlites occur with interstitial micro- or cryptocrystalline material. Cf: *trachytoid; pilotaxitic; orthophyric.*

trachytoid (trach'-y-toid) In the *IUGS classification,* a general term in the provisional "field" classification for rocks tentatively identified as *trachyte.* Specifying a texture in which feldspar crystals are parallel or subparallel.

track [paleont] (a) A fossil structure consisting of a mark left in soft material by the foot of a bird, reptile, mammal, or other animal. Cf: *trail [paleont].* Syn: *footprint.* (b) *muscle track.*

track [photo] n. A pathway, e.g. that of a ship, airplane, or satellite; or the data that can be read by a single magnetic head on such a path. v. To follow a pathway.

tracking Picking an event on a seismic section at a work station. The tracking may be manual or automatic by cross-correlating adjacent traces, by finding a local maximum, or in some other way. The interpreter needs to verify the results of autotracking and occassionally intervene because autotracking algorithms encounter difficulties at discontinuities (as at faults) and where waveshape changes (perhaps because of interference).

trackway (track'-way) A continuous series of tracks left by a single organism.

tract [geog] A region or area of land that may be precisely or indefinitely defined.

tract [streams] A part of a stream, such as a *plain tract* or a *valley tract.*

traction [exp struc geol] (trac'-tion) The stress vector acting across a particular plane in a body.

traction [sed] A mode of *bed load* transport in which the particles are swept along (on, near, or immediately above) and parallel to a bottom surface by rolling, sliding, dragging, pushing, or saltation; e.g. boulders tumbling along a stream bed, or sand carried by the wind over a desert surface or moved by waves and currents on a beach. The term was introduced into geology by Gilbert (1914, p.15) for the entire complex process of carrying material along a stream bed. Cf: *suspension.*

traction current A current, in standing water, that transports sediment along and in contact with the bottom, as in a stream. Cf: *turbidity current.* Syn: *tractive current.*
tractionite (trac′-tion-ite) A deposit of well-bedded, winnowed clastic sediments of sand size or larger, made by moving water or wind (Natland, 1976).
traction load *bed load.*
tractive current (trac′-tive) *traction current.*
tractive force In hydraulics, drag or shear developed on the wetted area of the stream bed, acting in the direction of flow. As measured per unit wetted area, unit tractive force equals the specific weight of water times hydraulic radius times slope of the channel bed (Chow, 1957). See also: *critical tractive force.*
trade-wind desert *tropical desert.*
trade winds A major system of tropical winds moving from the subtropical highs to the equatorial low-pressure belt. It is northeasterly in the Northern Hemisphere and southeasterly in the Southern Hemisphere. Cf: *antitrades.*
trafficability (traf′-fic-a-bil′-i-ty) A term used in the military to denote the feasibility of a terrain to support traffic.
traffic pan *pressure pan.*
trail [glac geol] "A line or belt of rock fragments picked up by glacial ice at some localized outcrop and left scattered along a more or less well defined tract during the movement and melting of a glacier" (Stokes and Varnes, 1955, p.154). Cf: *train.*
trail [mass move] *cryoturbate.*
trail [paleont] (a) A fossil structure consisting of a trace or sign of the passing of one or many animals; esp. a more or less continuous marking left by an organism moving over the bottom, such as a *worm trail.* Cf: *track [paleont].* (b) The extension of either valve of a brachiopod shell anterior to the geniculation (or anterior to the "visceral disc" or the part of the shell posterior to the geniculation) (TIP, 1965, pt.H, p.154, 155).
trailing edge (trail′-ing) The rear edge of a thrust sheet with respect to a given stratigraphic unit where it terminates against another thrust fault (Dahlstrom, 1970). In thrust faulting, *footwall cutoff.* Cf: *leading edge.*
trailing margin *passive margin.*
trailing spit *tail [coast].*
trail of a fault Crushed material along the fault surface that is used, often erroneously, as an indication of the direction of displacement. Such material can be a source of mineral deposits (*drag ore*). Little used.
train A narrow glacial deposit extending for a long distance, such as a *valley train* or a *boulder train.* See also: *trail [glac geol]; rock train.*
training data In *digital image processing,* a set of pixels already assigned to a distinct land-use or land-cover category used to "train" an *unsupervised classification* algorithm for remotely sensed data so that the pixels in the data set will be assigned an appropriate category.
trajectory [exp struc geol] (tra-jec′-to-ry) The curve that a moving body in a field, or a characteristic of a field, describes in space.
trajectory [seis] The path of a seismic wave. Syn: *raypath.*
tranquil flow (tran′-quil) *subcritical flow.*
tranquillityite (tran-quil′-li-ty-ite′) An opaque gray hexagonal lunar mineral: $Fe^{2+}_8(Zr,Y)_2Ti_3Si_3O_{24}$. It may be related to *långbanite.*
transapical (trans-ap′-i-cal) Of a suture or slit on the apex of a dinoflagellate cyst, running from side to side over the apex (Williams et al., 2000, p.236).
transceiver (trans-ceiv′-er) An instrument that converts one form of information into another, such as a sonar transmitter/receiver that changes electrical signals into acoustic waves or vice-versa, or a geophone that changes earth motion into electrical information or vice-versa.
transcontinental geophysical survey (trans′-con-ti-nen′-tal) A comprehensive geological and geophysical study from coast to coast across a continent; specif. the study of a band 4° wide (about 440 km) centered on lat. 37°N, extending across the U.S. and offshore into the Atlantic and Pacific oceans. Abbrev: TGS.
transcurrent fault (trans-cur′-rent) A term used for a continental strike-slip fault that does not terminate at plate boundaries. Cf: *transform fault.*
transducer A device which converts one form of energy into another. Many types of transducers are reversible, for example converting electrical energy into acoustical energy and vice versa.
transect (tran-sect′) n. In ecology, a sample area (usually elongate or linear) chosen as the basis for studying a particular assemblage of organisms. Cf: *quadrat.*
transecting cleavage A cleavage that cuts across a fold such that the fold axis does not lie in the plane of cleavage. Transecting cleavages may form prior to, during, or after the folding (Powell, 1974).
transection glacier (tran-sec′-tion) A glacier that fills an entire valley system, concealing the divides between the valleys. Cf: *through glacier.*
transfer (trans′-fer) A single process occurring continuously in space-time in which erosion is followed by transportation and deposition of sediment (Wilson, 1959).
transfer fault A strike-slip fault that links two segments of a rift that are offset relative to each other and/or have opposite dip directions for faults. Often erroneously interpreted; in many cases the two fault segments overlap and displacement is transferred kinematically by a *relay ramp.*
transfer function [geophys] The ratio of output to input as a function of frequency. The frequency-domain characteristics of a system (e.g., a filter). The transfer function usually is represented by amplitude-versus-frequency and phase-versus-frequency curves that contain the same information as the impulse response in the time domain.
transfer impedance The complex ratio of voltage at one pair of terminals to the current at another pair in a four-terminal network.
transfer percentage For any element, the ratio of the amount present in sea water to the amount supplied to sea water during geologic time by weathering and erosion, multiplied by 100.
transfer zone An area of deformation between two overstepping faults in which slip is transferred from one fault to the other. Cf: *accommodation zone; relay ramp; soft linkage.*
transfluence (trans-flu′-ence) The flowing of glacier ice through a breach made by the headward growth of cirques on both sides of a mountain ridge.
transformation [chem] *transmutation [chem].*
transformation [cryst] (trans-for-ma′-tion) Transformation is remodeling of an existing mineral structure in which major parts of at least the geometry, and usually structural elements of the parent mineral are retained. The change from one crystal polymorph to another, by one of several processes. See: *dilatational transformation; displacive transformation; reconstructive transformation; rotational transformation; substitutional transformation; order-disorder transformation.* Syn: *inversion [cryst]; phase transformation.*
transformation [petrology] *granitization.*
transformation [photo] The process of projecting a photograph mathematically, graphically, or photographically from its plane onto another plane by translation, rotation, and/or scale change. See also: *rectification [photo].*
transformation [stat] A mathematical or algorithmic procedure for changing values of a variable or statistic into another form. Examples include the logarithmic transformation, $y=\log x$, Fisher's z-transform of the *correlation coefficient,* $z=\tanh^{-1}(r)$, and *standardization.* The procedure or its result is often briefly called "transform".
transformational breccia (trans-for-ma′-tion-al) A term used by Landes (1945) for a breccia occurring in a stratigraphically vertical body and cutting through a stratigraphic section, and believed to have been produced by collapse, such as above a dissolved salt bed.
transformational faulting Refers to shear failure that occurs as a result of a rapid mineralogical phase transformation, e.g. the rapid transformation of olivine to a spinel structure along roughly planar anticracks has been postulated as a mechanism for producing deep-focus earthquakes (>350 km depth).
transformation series (a) *Morphocline.* (b) A group of homologous characters (Wiley et al., 1991, p.9). Cf: *polarized transformation series, unpolarized transformation series.*
transformation twin A crystal *twin* developed by a transformation from a higher to a lower symmetry; e.g. Dauphiné twinning in the transformation from high quartz to low.
transform fault (trans′-form) (a) A strike-slip fault that links two other faults or two other plate boundaries (e.g. two segments of a mid-ocean ridge). Transform faults often exhibit characteristics that distinguish them from transcurrent faults: (1) For transform faults formed at the same time as the faults they link, slip on the transform fault has equal magnitude at all points along the transform; slip magnitude on the transform fault can exceed the length of the transform fault, and slip does not decrease to zero at the fault termini. (2) For transform faults linking two similar features, e.g. if two mid-ocean ridge segments linked by a transform have equal spreading rates, then the length of the transform does not change as slip accrues on it. (b) A type of plate boundary at which lithosphere is neither created or destroyed, and plates slide past each other on a strike-slip fault. Cf: *transcurrent fault.*
transformism (trans-form′-ism) An obsolete theory that explains the origin of granite as a result of *granitization;* opposed to *magmatism.*
transgression (trans-gres′-sion) (a) The spread or extension of the sea over land areas, and the consequent evidence of such advance (such as strata deposited unconformably on older rocks, esp. where the new marine deposits are spread far and wide over the former land surface). Also, any change (such as rise of sea level or subsidence of land) that brings offshore, typically deep-water environments to areas formerly occupied by nearshore, typically shallow-water conditions, or that shifts the boundary between marine and nonmarine deposition (or between deposition and erosion) outward from the center of a marine basin. Ant: *regression [stratig].* Cf: *continental transgression; onlap.* Syn: *invasion [stratig]; marine transgression.* (b) A term used mostly in Europe for discrepancy in the boundary lines of continuous strata; i.e. *unconformity.*
transgressive (trans-gres′-sive) Said of a minor igneous intrusion, typically tabular, that cuts across the bedding or foliation of the country rock rather than confining itself to a single horizon.
transgressive lag A sedimentary deposit, commonly less than 2 ft. (0.61 m) thick, of relatively coarse-grained material composed of shells, shell fragments, clay rip-up clasts, calcareous nodules, siliciclastic gravel, or pebbles. This material derives from underlying strata by shoreface erosion during a marine transgression.
transgressive overlap *onlap.*
transgressive reef One of a series of nearshore reefs or bioherms superimposed on back-reef deposits of older reefs during the sinking of a landmass

or a rise of the sea level, and developed more or less parallel to the shore (Link, 1950). Cf: *regressive reef.*

transgressive sediments Sediments deposited during the advance or encroachment of water over a land area or during the subsidence of the land, and characterized by an *onlap* arrangement.

transgressive shoreline *retrograding shoreline.*

transgressive surface In sequence stratigraphy, the first major flooding surface across the shelf within a sequence, the surface that coincides with the change from regression to transgression on a regional scale. In most cases, it has a recognizable lithologic expression. The surface is usually a parasequence set boundary, separating an overlying retrogradational parasequence set from an underlying progradational or aggradational parasequence set. Note that there may be parasequence boundaries on the shelf within the lowstand systems tract; these relatively minor boundaries do not qualify as the transgressive surface. Nor does their presence require that the lowstand deposits on the shelf be placed in the transgressive systems tract (Van Wagoner et al., 1990; Embry and Johannessen, 1992).

transgressive systems tract In sequence stratigraphy, a *systems tract* bounded below by the transgressive surface and above by the downlap surface or maximum-flooding surface. Parasequences within the transgressive systems tract backstep in a retrogradational parasequence set. The systems tract progressively deepens upward as successively younger parasequences step farther landward. The downlap surface is the surface onto which the clinoform toes of the overlying highstand systems tract may merge and become very thin (Van Wagoner et al., 1987, 1988, 1990; Posamentier and Vail, 1988).

transient [elect] (tran′-sient) n. A pulse that is of short time duration.

transient [evol] n. A term occasionally used to denote a subdivision of a species whose members varied with time; comparable to a subspecies for a subdivision of a species in space; a chronologic subspecies.

transient beach A beach whose sand is removed by storm waves but is quickly restored by longshore currents.

transient-crater diameter The diameter of the *transient impact crater* measured at the pre-impact surface.

transient creep *primary creep.*

transient electromagnetic method An electromagnetic method in which the waveform of the transmitted signal is a train of pulses, step-functions, ramps, or other waveforms, and measurements are made in the off-times between pulses, usually after the primary field has stopped changing. Principal advantages of transient methods over continuous-wave methods are that the primary field is not present during the measurement of the secondary field and that measurements of the secondary field as a function of time are equivalent to continuous-wave measurements over a wide frequency range. Transient methods are used for both depth sounding and continuous profiling. Abbrev: TEM. Also called: time-domain electromagnetic method (TDEM).

transient impact crater The cavity formed immediately after the end of crater excavation, before it undergoes collapse and modification. Its outer extent is defined by the boundary between excavated and nonexcavated (but possibly displaced) target material, and its depth is the maximum achieved by material that will eventually form the floor of the *true crater* after cratering motions cease. Although it is by nature temporary, laboratory-derived scaling relations apply only to the diameter of the transient crater, not the final collapsed crater form (Melosh, 1989). Collapse of transient craters produces both *simple* and *complex impact craters.*

transient method Any electrical method of geophysical exploration that depends on measurement and intrepretation of a signal as a function of time.

transient snowline *snowline.*

transient strain A less precise, loosely used syn. of *creep recovery.*

transit (tran′-sit) n. (a) A *theodolite* in which the telescope can be reversed (turned end for end) in its supports without being lifted from them, by rotating it 180 degrees or more about its horizontal transverse axis. Syn: *transit theodolite.* (b) The act of reversing the direction of a telescope (of a transit) by rotation about its horizontal axis. v. To reverse the direction of a telescope (of a transit) by rotating it 180 degrees about its horizontal axis. Syn: *plunge [surv].*

Transition (Tran-si′-tion) A name, now obsolete, applied by Jameson (1808) from the teachings of A.G. Werner in the 1790's to the group or series of rocks occurring between the older and more crystalline *Primitive* rocks and the younger and better stratified *Floetz,* and roughly corresponding to the upper Precambrian and to the lower Paleozoic strata now assigned to the Cambrian, Ordovician, and Silurian. The rocks, consisting of dikes and sills, thick graywackes, and thoroughly indurated limestones, were considered to have been the first orderly deposits formed from the ocean during the passage (transition) of the Earth from its chaotic state to its habitable state; they were laid down with original steep dip and contained the first traces of organic remains, and were believed to extend uninterruptedly around the world.

transitional basalt Obsolete syn. of *olivine tholeiite.*

transitional crust Crust formed in the transition zone between continental and oceanic crust, during the history of continental rifting that culminates in the formation of a new ocean.

transitional fossils (tran-si′-tion-al) Fossils that collectively form sequences (*morphologic series*) showing gradual and continuous changes in morphologic characters from geologically earlier to later forms.

transitional-water wave A wave that is moving from deep water to shallow water, i.e. is transitional between a *deep-water wave* and a *shallow-water wave;* its wavelength is more than twice but less than 25 times the depth of the water, and the wave orbitals are beginning to be influenced by the bottom. Syn: *intermediate wave.*

transition metal Any of the elements scandium, titanium, vanadium, chromium, manganese, iron, cobalt, nickel, and copper, which have a partially filled d subshell of electrons in their atomic structure. They commonly give color to minerals, exhibit a variable oxidation state, and may give rise to magnetism.

transition point *inversion point.*

transition temperature *inversion point.*

transition zone (a) A region within the *upper mantle* bordering the lower mantle, at a depth of 410-660 km, characterized by a rapid increase in density of about 20% and an increase in seismic-wave velocities; it is equivalent to the *C layer.* (b) A region within the *outer core,* transitional to the inner core; the *F layer.* Archaic. (c) The region near the land-sea boundary where neither land nor marine operations can be carried out without significant modifications. May include marsh, shallow lagoons, surf zone, and marine waters that are too shallow or too obstructed for normal towed-streamer marine operations.

transit line Any line of a traverse that is projected, either with or without measurement, by the use of a transit or other device; an imaginary straight line between two transit stations.

transitory frozen ground (tran′-si-to′-ry) Ground that is frozen by a sudden drop of temperature and that remains frozen for a short period, usually hours or days (Muller, 1947, p.223).

transitory pygidium The posterior fused portion of a meraspis trilobite; it will become the true *pygidium* in the adult form, together with a varying number of segments released into the thorax during the meraspid stage of development.

transit theodolite *transit.*

transit time The travel time of a compressional acoustic wave over a unit distance recorded on a *sonic log.* The reciprocal of the compressional wave velocity.

transit traverse A surveying traverse in which the angles are measured with a transit or theodolite and the lengths with a metal tape. It is usually executed for the control of local surveys.

translation (trans-la′-tion) A shift in position without rotation, one of the components of *deformation..* When applied to *crystal plasticity,* it refers to the movement of one block of atoms past another.

translational (trans-la′-tion-al) Pertaining to or said of a uniform movement in one direction, without rotation.

translational fault An obsolete term for a fault in which there has been translational movement and no rotational component of movement; dip in the two walls remains the same. It can be strictly applied only to segments of faults (Dennis, 1967). Obsolete syn: *translatory fault.*

translational margin A continental margin formed by transform faults.

translational movement Apparent fault-block displacement in which the blocks have not rotated relative to one another, so that features that were parallel before movement remain so afterwards. Cf: *rotational movement.* See also: *translational fault.* Less-preferred syn: *translatory movement.*

translational slide A major group in the classification of landslides, involving the downslope displacement of soil-rock material on a surface that is roughly parallel to the general ground surface, in contrast to rockfalls and rotational landslides. The term includes such diverse landslide types as rock slides, block glides, slab or flake slides, and debris slides.

translation gliding *crystal gliding.*

translation lattice *crystal lattice.*

translation plane *glide plane [cryst].*

translation vector In tectonics, a term used by Bhattacharji (1958, p.626) for the vector representing the direction and the net displacement of material from a reference point; the sum of the vectors for compression and for flow.

translatory fault (trans′-la-to′-ry) An obsolete term for a *translational fault.*

translatory movement A less-preferred syn. of *translational movement.*

trans link A link in a trunk stream channel bounded by tributaries that enter from opposite sides (James and Krumbein, 1969). The trunk channel is traced upstream by following the link of greater magnitude at each fork. Cf: *cis link; cis-trans link.*

translucent (trans-lu′-cent) Said of a mineral that is capable of transmitting light, but is not transparent. Cf: *opaque; transopaque.*

translucent attritus Attritus consisting mostly of translucent humic degradation matter, mainly liptinite, with minor quantities of opaque materials. Cf: *opaque attritus.* Syn: *humodurite.*

translucent humic degradation matter *Humic degradation matter* that is translucent and of the same deep red color as anthraxylon; humic degradation matter that is less than 14 micrometers in width, measured perpendicular to the bedding. Abbrev: THDM.

translunar (trans-lu′-nar) Pertaining to phenomena, or to the space, beyond the Moon's orbit about the Earth. Cf: *cislunar.*

transmedian muscle (trans-me'-di-an) One of a pair of muscles in some lingulid brachiopods, anterior to the umbonal muscle. One muscle originates on the left side of the *pedicle valve* and rises dorsally to be inserted on the right side of the *brachial valve,* whereas the other muscle originates on the right side of the pedicle valve and is inserted on the left side of the brachial valve (TIP, 1965, pt. H, p.154).

transmissibility coefficient (trans'-mis-si-bil'-i-ty) *transmissivity.*

transmission capacity (trans-mis'-sion) In a column of soil of unit cross section, the volume of water that flows per unit of time, with a hydraulic gradient unity or with a hydraulic head equal to the length of the soil column (Horton, 1945, p.308).

transmission coefficient (a) The complex ratio of electric field intensity transmitted beyond to that incident upon an interface. (b) The ratio of the amplitude of a seismic wave transmitted through an interface to that of the wave incident upon it. This ratio can be greater than one. (c) A measure of the amplitude of a seismic wave passing through an interface restricted to the case of normal incidence. (d) With the seismic reflection method, users are usually interested in energy which has passed through an interface twice (once going down and once returning upward), and a two-way transmission coefficient T for normal incidence is commonly used. (e) A ratio of the energy densities (which involves the square of transmission coefficients as defined above).

transmission constant An expression of the ability of a permeable medium to transmit a fluid under pressure. As applied to groundwater, the discharge in cubic feet per minute through each square foot of cross-sectional area under a 100-percent hydraulic gradient (Tolman, 1937, p.564).

transmission electron microscope An electron-optical microscope that utilizes an assembly of magnetic lenses and a beam of high-energy electrons (80 keV to 3 MeV) that are transmitted through a thin specimen. The main advantage is high resolution, ~2Å, which results from the very small wavelengths of electrons (0.037Å at 100kV). Information is obtained on a fluorescent screen or photographic plate from images formed by diffraction-contract mechanisms using *bright field* or *dark field* modes, and from electron-diffraction patterns of selected areas. Abbrev: TEM.

transmission window *atmospheric window.*

transmissivity (trans-mis-siv'-i-ty) The rate at which water of the prevailing kinematic viscosity is transmitted through a unit width of the aquifer under a unit hydraulic gradient. It replaces *coefficient of transmissibility* because by convention it is considered a property of the aquifer, which is transmissive, whereas the contained liquid is transmissible. However, though spoken of as a property of the aquifer, it embodies also the saturated thickness and the properties of the contained liquid (Lohman et al., 1972, p.13). Transmissivity is equal to the product of *hydraulic conductivity* and aquifer thickness.

transmissometer (trans-mis-som'-e-ter) An instrument that measures the capability of a fluid to transmit light; esp. one that measures the turbidity of water by determining the percent transmission of a light beam. See also: *turbidimeter.*

transmutation [chem] The transformation of one element into another. Radioactive decay is an example. Transmutation can also be accomplished by bombardment of atoms with high-speed particles. Syn: *transformation [chem].*

transmutation [evol] (trans-mu-ta'-tion) The change from one species to another.

transopaque (trans-o-paque') Said of a mineral that is transparent in one part of the visible spectrum and *opaque* in another; e.g. goethite, hematite (Salisbury and Hunt, 1968).

transparency [oceanog] (trans-par'-en-cy) The ability of seawater to transmit light; the depth to which water is transparent may be measured by the use of a *Secchi disk.*

transparency [photo] A positive image, either black and white or in color, on a clear base (glass or film), intended to be viewed by transmitted light; a *diapositive.*

transparent (trans-par'-ent) Said of a mineral that is capable of transmitting light, and through which an object may be seen. Cf: *translucent; opaque; transopaque.*

transparent zone A zone of subsurface rocks from which no seismic reflections are obtained, but below which reflections are seen. Cf: *opaque zone.*

transpiration (tran-spi-ra'-tion) The process by which water absorbed by plants, usually through the roots, is evaporated into the atmosphere from the plant surface. Cf: *guttation.*

transponder A device which transmits a signal upon receiving another signal. When the receiver in the device detects the "interrogating signal", it triggers the transmitter which replies with a coded pulse or sequence of pulses. Transponders are used with both electromagnetic and sonar waves. A radar transponder is also called a radar beacon; a sonar transponder is also called a *pinger.*

transport [grd wat] The movement of solute, suspended matter, or heat by groundwater in a porous medium.

transport [sed] (trans'-port) A syn. of *transportation.* The term often occurs in combined terms such as *sediment transport* and *mass transport [sed].*

transport [struc petrol] *tectonic transport.*

transportation (trans-port-a'-tion) A phase of *sedimentation* that includes the movement by natural agents (such as flowing water, ice, wind, or gravity) of sediment or of any loose material, either as solid particles or in solution, from one place to another on or near the Earth's surface; e.g. the rolling of sand along a seashore under the influence of currents, the creeping movement of rocks on a glacier, or the conveyance of silt, clay, and dissolved salts by a stream. Syn: *transport [sed].*

transportation velocity *nonsilting velocity.*

transport concentration In a stream, the rate of flow of sediment passing through a given cross-sectional area perpendicular to the flow, compared with the rate of flow of the suspension of water and sediment passing through the same area (ASCE, 1962). See also: *spatial sediment concentration.*

transported (trans-port'-ed) Said of material that has been carried by natural agents from its former site to another place on or near the Earth's surface.

transported assemblage An *assemblage [ecol]* in which the specimens have been transported significant distances, thereby intermingling forms which did not originally live together. Cf: *fossil community; winnowed community; disturbed-neighborhood assemblage; mixed assemblage.* See also: *thanatocoenosis.*

transported kaolin Kaolin originally deposited in beds or formed as a residual product of weathering of granitic rocks and subsequently eroded, transported, and redeposited.

transported soil material *Parent material* that has been moved and redeposited from the site of its parent rock. The adjective "transported" is also applied to the soil formed from such a parent material. Cf: *residual material.*

transporting erosive velocity (trans-port'-ing) The velocity of water in a channel that both maintains silt in movement and scours the bed. Cf: *noneroding velocity.*

transport-limited stream A stream with more sediment available than it can carry, so that it is forced to deposit part of its load; e.g., the Platte River in Nebraska. Syn: *overloaded stream.* Cf: *supply-limited stream.*

transposed hinge (trans-posed') A *hinge* in bivalve mollusks in which certain hinge teeth present in one valve occupy positions of teeth usually found in the other valve.

transposition In structural petrology, disruption of preexisting fabric so that all fabric elements are rotated to a new orientation. Results in a transposition or transposed foliation.

transposition structure (trans-po-si'-tion) A primary sedimentary structure resulting from hydroplastic or fluid flow of sediment after deposition and sometimes after partial consolidation of the sediment (Hills, 1963, p.30).

transpression (trans-pres'-sion) (a) A system of stresses that tends to cause oblique shortening, i.e., combined shortening and strike slip. (b) In crustal deformation, an inetrmediate stage between compression and strike-slip motion; it occurs in zones with oblique compression and combines strike-slip or wrench movement with a superposed compression perpendicular to the wrench or stike-slip zone. The tectonic style in Caledonian Spitzbergen provides evidence for a transpression regime (Harland, 1971). Cf: *Transtension.*

transpressional fault (trans-pres'-sion-al) A strike-slip fault across which there is a component of shortening. The shortening may lead to formation of a *positive flower structure*.

transtension (trans-ten'-sion) (a) A system of stresses that tends to cause oblique extension, i.e., combined extension and strike slip (Peacock et al., 2000). (b) Crustal deformation that combines the two elements of superposed extension normal to the trend of the wrench or strike-slip zone with strike-slip or wrench motion (Harland, 1971). Cf: *transpression.*

transtensional fault (trans-ten'-sion-al) A strike-slip fault across which there is a component of extension. The extension may lead to the development of *negative flower structure*.

transuranium element Element whose atomic number is greater than 92, including neptunium, plutonium, americium, and curium, which can form in the fuel rods of reactors.

trans-vacant Said of a layered mineral with a vacant (M)1 site. Most of the dioctahedral phyllosilicates are monoclinic. The mirror present in the C2/m, is perpendicular to the platy or 001 surface of these minerals. This mirror bisects one of the three octahedral sites in the octahedral sheet. This is the (M)1 site and the other two are (M)2 sites. When the (M)1 site is vacant, the mineral is referred to as trans-vacant. Cf: *cis-vacant.*

transverse [geomorph] (trans-verse') Said of an entity that is extended in a crosswise direction; esp. of a topographic feature that is oriented at right angles to the grain or general strike of a region. Ant: *longitudinal.*

transverse [ore dep] Said of a vein or lode that is oriented across the bedding of the host rock or across any important planar feature.

transverse archeopyle An *archeopyle* in a dinoflagellate cyst that is formed by the opening of sutures along the anterior margins of the anterior intercalary plates, while no distinct *opercular* piece is formed (Williams et al., 2000, p.235).

transverse band The connecting *lamella* joining the posterior ends of the ascending branches of a brachiopod loop.

transverse bar (a) A slightly submerged sand ridge that extends more or less at right angles to the shoreline. It has been described as a "giant sand wave" and a "plateau-like sandbar". (b) In shallow rivers, a flat-topped sand or gravel body that grows by downcurrent additions to the slip-face margins.

transverse basin *exogeosyncline.*
transverse coastline *discordant coastline.*
transverse crevasse A crevasse developed roughly perpendicular to the direction of ice movement, and commonly convex on the downstream side. Cf: *marginal crevasse; splaying crevasse.*
transverse diameter The thecal diameter in edrioasteroids perpendicular to the *axial diameter* (Bell, 1976).
transverse dispersion Dispersion that is normal or transverse to the direction of groundwater flow. Cf: *longitudinal dispersion.*
transverse dune A strongly asymmetrical sand dune elongated perpendicularly to the direction of the prevailing winds, having a gentle windward slope and a steep leeward slope standing at or near the angle of repose of sand; it generally forms in areas of sparse vegetation.
transverse electric A mode of EM wave propagation in which the electric field is perpendicular to the direction of propagation. In 2-D modeling and processing of magnetotelluric (q.v.) data, the mode in which the electric field is parallel to strike. Abbrev: TE. Also called E-polarization. See also: *transverse magnetic.*
transverse fault A fault that strikes obliquely or perpendicular to the general structural trend of the region.
transverse flagellum The ribbon-like *flagellum* encircling the body of a dinoflagellate in a nearly transverse plane. It is usually lodged in a deep encircling groove (girdle), and is used mainly for propulsion.
transverse fold (a) *cross fold.* (b) A *fault-related fold* whose axis is perpendicular to the associated fault (Schlische, 1995).
transverse furrow An equatorial thinning in the exine of a dicotyledonous pollen grain, usually occurring at the equator, and always running perpendicular to a meridional colpus. Syn: *colpus transversalis.*
transverse isotropy Having the same property value when measured within a plane but a different value when measured perpendicular to the plane. Transverse isotropy can be produced by a sequence of isotropic layers (such as sedimentary bedding) for wavelengths that are appreciably (3 to 10 times) larger than the individual layer thicknesses. It can also be produced by parallel fracturing (Sheriff and Geldart, 1995, p.38, 55-6). Also called *polar anisotropy.* Abbreviated TIV or TIH when the axis of symmetry is vertical or horizontal.
transverse joint *cross joint.*
transverse lamination Lamination transverse to bedding. Syn: *oblique lamination.*
transverse magnetic A mode of EM wave propagation in which the magnetic field is perpendicular to the direction of propagation. In 2-D modeling and processing of magnetotelluric (q.v.) data, the mode in which the magnetic field is parallel to strike. Abbrev: TM. Also called: H-polarization. See also: *transverse electric.*
transverse Mercator projection A cylindrical conformal map projection, equivalent to the regular *Mercator projection* turned (transversed) 90° in azimuth, so that the cylinder is tangent along a given meridian (or any pair of opposing meridians) rather than along the equator. The central meridian is represented by a straight line and is divided truly; all other meridians and all parallels (except the 90th meridians from center and the equator, if shown) are curved lines intersecting at right angles. Lines of constant direction (rhumb lines) are also curved lines. The projection is designed to minimize scale error or variation along a narrow zone by using a great circle, centrally located to the area to be mapped, as the "theoretical equator". It is used for maps of areas extending short distances from the central meridian, for charts of polar regions, and as a worldwide standard for plotting military maps. A special case of the projection is used as the basis of the *universal transverse Mercator projection.* See also: *Gauss projection.*
transverse oral midline A perradial line in edrioasteroids that extends across the exterior of the oral region between opposing anterior and posterior orals, perpendicular to the anterior primary ambulacral radius. It extends from the proximal tip of one lateral ambulacral bifurcation plate to the other (Bell, 1976).
transverse profile *cross profile.*
transverse projection A projection in which the projection axis is rotated 90° in azimuth; e.g. "transverse Mercator projection" or "transverse polyconic projection". Syn: *inverse projection.*
transverse resistance The product of average resistivity and thickness of a rock layer. It is used in interpretation of direct-current resistivity soundings. Units are ohm-meters^2.
transverse resistivity Resistivity of rock measured across the direction of bedding. Cf: *longitudinal resistivity.*
transverse rib One of a series of ridges of pebbles, cobbles, or boulders oriented transverse to stream flow (McDonald and Banerjee, 1971, p.1290). Rib spacing is regular and generally ranges from about 0.2 to 1.5 m. Transverse ribs are commonly formed by upper-regime, shallow flows, for example on braided outwash plains, either by a hydraulic jump that moves progressively upstream, or by antidunes and clast stripes.
transverse ridge (a) A generally denticulate fulcral elevation on an articular face of an *ossicle* of a crinoid ray, disposed perpendicularly or slightly oblique to the axis extending from the dorsal toward the ventral side. (b) Ridge on upper surface of platformed pectiniform conodont element that trends more or less normal to longitudinal axis of element (TIP, 1981, pt. W, supp.2).
transverse ripple A *ripple* formed approximately at right angles to the direction of the current. Its profile may be asymmetric or symmetric.
transverse ripple mark *transverse ripple.*
transverse scour mark A *scour mark* whose long axis is transverse to the main direction of the current. The regular spacing of such marks may lead to confusion with ordinary transverse ripples. The term was apparently first used by Dzulynski and Sanders (1962, p.68). See also: *current-ripple cast.*
transverse section *cross section [geol].*
transverse septulum A minor partition of a chamber in a foraminiferal test, oriented transverse to the axis of coiling and observable in sagittal and parallel sections. See also: *primary transverse septulum; secondary transverse septulum.*
transverse septum One of a series of septa dividing the parietal tubes of a cirripede crustacean into a series of cells, oriented normal to a longitudinal septum and parallel to the basis (TIP, 1969, pt. R, p.103).
transverse valley (a) A valley having a direction at right angles to the general strike of the underlying strata; a *dip valley.* (b) A valley that cuts across a ridge, range, or chain of mountains or hills at right angles (Conybeare and Phillips, 1822, p.xxiv). Cf: *longitudinal valley.* Syn: *cross valley.*
transverse wall One of a pair of oppositely placed vertical walls bounding the ends of a cheilostome bryozoan zooid and commonly developed largely as an *interior wall.*
transverse wave *S wave.*
trap [cryst] An imperfection in a crystal structure that may trap a mobile electron, usually temporarily.
trap [eng] A device for separating suspended sediment from flowing water; e.g. a *sand trap.*
trap [ign] Any dark-colored fine-grained nongranitic hypabyssal or extrusive rock, such as a *basalt, peridotite, diabase,* or fine-grained *gabbro;* also, applied to any such rock used as crushed stone. Etymol: Swedish trappa, "stair, step", in reference to the stairstep appearance created by the abrupt termination of successive flows. Also spelled: trapp. Syn: *trap rock.* Cf: *whinstone.*
trap [petroleum] Any barrier to the upward movement of oil or gas, allowing either or both to accumulate. A trap includes a reservoir rock and an overlying or updip impermeable *roof rock;* the contact between these is concave as viewed from below. See also: *stratigraphic trap; structural trap; combination trap.*
trap [speleo] *sump [speleo].*
trap cut *step cut.*
trap-door fault A term for a curved fault bounding a block that is hinged along one edge; in the Little Rocky Mountains of Montana. Little used.
trapdoor subsidence caldera A *caldera* bounded by an incomplete arcuate fault and by a hinged segment (Lipman, 2000).
trap efficiency The ability of a storage reservoir to trap and retain sediment, expressed as the percent of incoming sediment (sediment yield) retained in the basin.
trapezohedral (tra-pe'-zo-he'-dral) Said of those crystal classes in the tetragonal and hexagonal systems in which the general form is a trapezohedron.
trapezohedron (tra-pe'-zo-he'-dron) (a) An isometric crystal form of 24 faces, each face of which is ideally a four-sided figure having no two sides parallel, or a trapezium. Syn: *tetragonal trisoctahedron; leucitohedron; icositetrahedron.* (b) A crystal form consisting of six, eight, or twelve faces, half of which above are offset from the other half below. Each face is, ideally, a trapezium. The tetragonal and hexagonal forms may be right- or left-handed.
trapezoidal projection (trap-e-zoi'-dal) A map projection in which equally spaced straight parallels and straight converging meridians divide the area into trapezoids.
trapped wave *channel wave.*
trapping temperature In fluid inclusion geothermometry, the actual temperature at which a fluid inclusion was trapped in a mineral. Rarely determined directly; most commonly inferred by a pressure correction on the homogenization temperature using salinity corrected isochores and assuming constant volume of the inclusion.
trap rock *trap [ign].*
trapshotten gneiss (trap'-shot-ten) A gneiss injected with flinty pseudotachylyte. The term was originated by King and Foote in 1864.
trash Highly combustible non-hazardous solid wastes, such as paper, wooden boxes, and cardboard cartons; contains up to 10% plastic and other combustible solid wastes, 10% moisture and 5% non-combustible solid wastes.
trash ice Broken or crumbled ice mixed with water.
trash line A line on a beach, marking the farthest advance of high tide, and consisting of debris (Pettijohn and Potter, 1964, p.350). Cf: *debris line; swash mark; wrackline.*
trash zone Accumulation of insoluble material at the bottom of ore-bearing collapse breccias in a *Mississippi Valley-type deposit.*

traskite (trask'-ite) A brownish-red hexagonal mineral: $Ba_{21}Ca_2(Fe^{2+},Mn,Ti)_4(Ti,Fe,Mg)_{12}(Si_{12}O_{36})_{12}(Si_2O_7)(O,OH)_{30}Cl_6 \cdot 14H_2O$.
trass A common name in the older literature for nonwelded massive ash and pumice-flow deposits (Fisher and Schmincke, 1984, p.189). Trass has been used in the production of pozzolan cement. Cf: *ignimbrite.*
trattnerite A dark blue hexagonal mineral of the *milarite* group: $(Fe,Mg)_2(Mg,Fe)_3Si_{12}O_{30}$.
traveled (trav'-eled) Removed from the place of origin, as by streams or wind, and esp. by glacier ice, as a *traveled* stone; a syn. of *erratic.*
traveling beach (trav'-el-ing) A beach that is continually moving in one general direction under the influence of floods.
traveling dune *wandering dune.*
traveling wave *kinematic wave.*
traveltime (trav'-el-time) The time required for a seismic wave to travel from its source to a point of observation. Cf: *traveltime curve.*
traveltime curve In seismology, a plot of wave *traveltime* against corresponding distance along the Earth's surface from the source to the point of observation. Syn: *time-distance curve.*
traverse [geol] (trav'-erse) (a) A line across a thin section or other sample along which grains of various minerals are counted or measured. (b) A vein or fissure in a rock, running obliquely and in a transverse direction.
traverse [surv] n. (a) A sequence or system of measured lengths and directions of straight lines connecting a series of surveyed points (or stations) on the Earth's surface, obtained by or from field measurements, and used in determining the relative positions of the points (or stations). (b) *traverse survey.* (c) A line surveyed across a plot of ground. v. To make a traverse; to carry out a traverse survey.
traverse map A map made from a traverse survey.
traverse survey A survey in which a series of lines joined end to end are completely determined as to length and direction, these lines being often used as a basis for triangulation. It is used esp. for long narrow strips of land (such as for railroads) and for underground surveys. Syn: *traverse [surv].*
traverse table A mathematical table listing the lengths of the two sides opposite the oblique angles for each of a series of right-angled plane triangles as functions of every degree of angle (azimuth or bearing) and of all lengths of the hypotenuse from 1 to 100. Traverse tables are used in computing latitudes and departures in surveying and courses in navigation.
travertine (trav'-er-tine) (a) Biotically and/or abiotically precipitated calcium carbonate (predominantly calcite and aragonite) from spring-fed, heated and/or ambient-temperature waters (Chafetz and Folk, 1984). The spongy or less compact variety is *tufa.* See also: *onyx marble.* Syn: *calcareous sinter; calc-sinter.* (b) A term sometimes applied to any cave deposit of calcium carbonate. (c) A term used inappropriately as a syn. of *kankar* (Viles and Goudice, 1990). Etymol: Italian "travertino" from Latin "(lapis) tiburtinus" = (stone) of Tibur (Tivoli).
travertine dam *rimstone dam.*
travertine terrace A series of *rimstone dams.*
tread (a) The flat or gently sloping surface of one of a series of natural steplike landforms, such as those of a glacial stairway or of successive stream terraces; a bench level. Cf: *riser.* (b) *flat [struc geol].*
treanorite (trea'-nor-ite) *allanite-(Ce).*
treasurite (treas'-u-rite) A metallic monoclinic mineral: $Ag_7Pb_6Bi_{15}S_{32}$. It is a member of the lillianite homologous series.
treated stone (treat'-ed) A gemstone that has been heated, stained, oiled, coated, or in-filled with a foreign substance such as a glass or polymer, or one that has been treated by various types of irradiation, in order to improve or otherwise alter its color; or laser-drilled to make flaws inconspicuous. Also, a stone that has been preserved from dehydration, such as an opal whose cracks have been filled with oil or other liquid; or one in which special effects have been produced, e.g. amber with "spangles" (tension cracks).
treat rock Partially oxidized and leached residual-enriched iron formation found in the Knob Lake iron range of Quebec and Labrador, Canada. Cf: *blue ore; red ore; rubble ore; yellow to brown ore.*
trechmannite (trech'-mann-ite) An adamantine red rhombohedral mineral: $AgAsS_2$. It is a dimorph of *smithite.*
tree [phylogen] *phylogenetic tree.*
tree agate A moss agate whose dendritic markings resemble trees.
tree line *timberline.* See: *krummholz.*
Tree of Life A figurative reference to the branching pattern of evolution, often represented by a fanciful picture of a tree.
tree optimization A means of deriving parsimonious inferences about phylogenetic sequences of character-state changes from a phylogenetic tree (Brooks et al., 1984, p.81).
tree ore A high-grade uranium ore consisting of buried *carbon trash* that has been replaced or enriched with uranium-bearing solutions.
tree pollen A syn. of *arboreal pollen.* Abbrev: TP.
tree ring *growth ring [geochron].*
tree-ring chronology *dendrochronology.*
trellis drainage pattern (trel'-lis) A drainage pattern characterized by parallel main streams intersected at or nearly at right angles by their tributaries, which in turn are fed by elongated secondary tributaries parallel to the main streams, resembling in plan the stems of a vine on a trellis. It is commonly developed where the beveled edges of alternating hard and soft rocks outcrop in parallel belts, as in a rejuvenated folded-mountain region or in a maturely dissected belted coastal plain of tilted strata; it is indicative of marked structural control emphasized by subsequent and secondary consequent streams. Examples are well displayed in the Appalachian Mountains region. Cf: *fault-trellis drainage pattern; rectangular drainage pattern.* Syn: *grapevine drainage pattern; espalier drainage pattern.*
trema (tre'-ma) An *orifice,* occurring singly or in series, in the outer wall of some gastropod shells for excretory functions. Pl: tremata.
Tremadocian (Tre-ma-do'-cian) A series or stage of the Standard Global Chronostratigraphic Scale: lowermost Ordovician (above Cambrian). The basal GSSP is in Newfoundland (Cooper et al., 2001). Also spelled: Tremadoc.
tremalith (tre'-ma-lith) A centrally and minutely perforate coccolith, such as a placolith or a rhabdolith. The term is sometimes restricted to *placolith* only. Cf: *discolith.* Syn: *trematolith.*
trematolith (tre-mat'-o-lith) *tremalith.*
trematophore (tre-mat'-o-phore) A perforated plate over the aperture of the test in some miliolid foraminifers. Syn: *sieve plate [paleont].*
trembathite A colorless or light blue rhombohedral mineral: $(Mg,Fe)_3B_7O_{13}Cl$. It is a dimorph of boracite.
trembling prairie (trem'-bling) *shaking prairie.*
tremocyst (trem'-o-cyst) A term used in the older literature of bryozoans for a calcareous layer with pseudopores, thought to form a secondary or tertiary deposit over an *olocyst* or a *pleurocyst* in the frontal shield of certain ascophoran cheilostomes.
tremolite (trem'-o-lite) A white to dark-gray monoclinic mineral of the amphibole group: $Ca_2Mg_5Si_8O_{22}(OH)_2$. It has varying amounts of iron, and may contain manganese and chromium. Tremolite occurs in long blade-shaped or short stout prismatic crystals and also in columnar, fibrous, or granular masses or compact aggregates, generally in metamorphic rocks such as crystalline dolomitic limestones and talc schists. It is a constituent of much commercial talc. Cf: *actinolite.*
tremopore (trem'-o-pore) A pseudopore in the *tremocyst* of an ascophoran cheilostome (bryozoan).
tremor (trem'-or) A minor earthquake, esp. a fore-shock or an aftershock. Syn: *earth tremor; earthquake tremor.*
tremor tract In coal mining, an area of complex folding, faulting, and gliding of coal seams and associated rocks. It may be formed by seismic shocks during the deposit's semicompacted state (Nelson and Nelson, 1967, p.387).
Trempealeauan (Trem'-pea-leau'-an) North American stage: Uppermost Cambrian, above Franconian, below Ordovician.
trench [arch] A regular area of archaeological excavation, normally carefully documented by occupation layer or stratigraphy and position. Cf: *sondage.*
trench [geomorph] (a) A long, straight, commonly U-shaped valley or depression between two mountain ranges, often occupied by two streams that drain in opposite directions. Syn: *trough [geomorph].* (b) A narrow, steep-sided canyon, gully, or other depression eroded by a stream. (c) Any long, narrow cut or excavation produced naturally in the Earth's surface by erosion or tectonic movements. Also, a similar feature produced artificially, such as a ditch.
trench [marine geol] A narrow, elongate depression of the deep-sea floor associated with a subduction zone. Oriented parallel to a *volcanic arc* and commonly to the edge of the adjacent continent, between the continental margin and the abyssal hills. Trenches are commonly two or more km deeper than the surrounding ocean floor, and may be thousands of kilometers long. Cf: *foredeep; trough [marine geol].* Syn: *oceanic trench; marginal trench; sea-floor trench.*
trench-slope basin A sedimentary basin between a *trench [marine geol]* and a *fore-arc basin.*
trend [geophys] That component in a geophysical anomaly map which is relatively smooth, generally produced by regional geologic features.
trend [paleont] In evolutionary paleontology, the evolution of a specific structure or morphologic characteristic within a group, esp. in taking an overall view of a large group, such as an order or a class; e.g., the evolution of the form of the septal suture, from simple to complex, in tracing the ammonoids as a group from the Devonian to the Triassic.
trend [stat] (a) The direction or rate of increase or decrease in magnitude of the individual members of a time series of data when random fluctuations of individual members are disregarded; the general movement through a sufficiently long period of time of some statistical progressive change. (b) *trend line.*
trend [struc geol] (a) n. The compass direction of a linear feature as projected onto a horizontal plane, i.e. the strike of a vertical plane containing the linear feature. Syn: *bearing.* (b) n. An alignment of exposures or occurrences of a specified geologic feature. (c) v. To have a specified trend, as in "The fold hinge trends north".
trend deposit A uranium orebody in sandstone along a mineralized belt or trend. Trend deposits are generally tabular and are subparallel to the gross stratification (Bailey and Childers, 1977, p.27).

trend line A straight line or curve expressing an empirical relationship between two variables, showing the tendency of one variable to increase or decrease with respect to the other.
trend map A stratigraphic map that displays the relatively systematic, large-scale features of a given stratigraphic unit, such as those indicating broad postdepositional structural and erosional changes or those controlled by regional deposition (Krumbein and Sloss, 1963, p.485-488). Cf: *residual map.*
trend-surface analysis A form of regression in which a spatially distributed dependent variable is modeled as a polynomial expansion of the geographic coordinates of the sample locations. A trend surface has the form of a dipping plane or smoothly curving surface and is fitted by least-squares. The equation of a first-order trend is $z=a+b_1x+b_2y$, a second-order trend is $z=a+b_1x+b_2y+b_3x^2+b_4y^2+b_5xy$, and so on. The difference between the fitted trend and the observed value at a location is the residual.
Trentonian (Tren-to'-ni-an) A North American provincial stage: Upper Ordovician (above Blackriverian, below Edenian). It has also been regarded as a substage (upper Mohawkian Stage).
trepostome (trep'-o-stome) Any ectoproct bryozoan belonging to the order Trepostomata, characterized by tubular zooecia with distinct endozone and exozone and a terminal aperture. Adj: trepostomatous. Range, Ordovician to Permian.
treppen (trep'-pen) Nacreous tablets in the mollusks arranged in a steplike pattern (Wise, 1970).
treppen concept The concept that, on a surface reduced to old age by streams and then uplifted, the rejuvenated streams develop second-cycle valleys first near their mouths and that these young valleys are extended headward to form piedmont steps. Etymol: German "Treppen" (sing: "Treppe"), "stair steps".
treptomorphic (trep-to-mor'-phic) Adj. of *treptomorphism.*
treptomorphism *isochemical metamorphism.*
trevor (tre'-vor) A dark-brown granitic rock from Wales, used to make curling stones. Local usage only.
trevorite (tre'-vor-ite) A black or brownish-black cubic mineral of the *spinel* group: $NiFe^{3+}{}_2O_4$.
triactine (tri-ac'-tine) A sponge *spicule* having three rays.
triad (tri'-ad) Noun or adj. for a symmetry axis that requires a rotation of 120° to repeat an object's appearance. Cf: *diad; hexad; tetrad [cryst].* Syn: *threefold axis.*
triaene (tri'-aene) (a) A tetraxon in which three similar rays differ from the fourth; specif. an elongated sponge spicule with one long ray (the rhabdome) and three similar short rays (the cladi), which are sometimes branched or otherwise modified, divergent at one end. See also: *dichotriaene; phyllotriaene; discotriaene.* (b) The initial four-rayed or tridentlike spicule of the ebridian skeleton.
triakisoctahedron (tri-a'-kis-oc'-ta-he'-dron) *trisoctahedron.*
triakistetrahedron (tri-a'-kis-tet'-ra-he'-dron) *trigonal tristetrahedron.*
triangle (tri'-an-gle) An ordinarily plane figure bounded by three straight-line sides and having three internal angles. Cf: *spherical triangle.*
triangle closure The amount by which the sum of the three measured angles of a triangle fails to equal exactly 180 degrees plus the spherical excess (the amount by which the sum of the three angles of a triangle on a sphere exceeds 180 degrees); the *error of closure* of triangle. Also known as "closure of triangle".
triangle zone A zone, often found at the external margin of a fold-thrust belt, where the hinterland-dipping thrusts of the thrust zone encounter foreland-dipping thrusts off the same detachment horizon (Price, 1981).
triangular coordination (tri-an'-gu-lar) An atomic arrangement in which an ion is surrounded by three ions of opposite sign. Syn: *threefold coordination.*
triangular facet As a physiographic feature, a triangular *face* having a broad base and an apex pointing upward; specif. the face on the end of a *faceted spur,* usually a remnant of a fault plane at the base of a block mountain. A triangular facet may also form by wave erosion of a mountain front or by glacial truncation of a spur. Syn: *spur-end facet.*
triangular organelle A small sensory structure near the peristome of a tintinnid.
triangular texture In mineral deposits, texture produced when exsolved or replacement mineral crystals are arranged in a triangular pattern, following the crystallographic directions of the host mineral.
triangulate (tri-an'-gu-late) To divide into triangles; esp. to use, survey, map, or determine by, triangulation. Etymol: back-formation from *triangulation.*
triangulation (tri-an'-gu-la'-tion) Creation of a spatial network by connecting points on a map to form a pattern of nonintersecting triangles. Used by some *contouring* algorithms to produce a contoured map of a surface.
triangulation net A net or series of adjoining triangles covering an area in such a manner that the lengths and relative directions of all lines forming the triangles can be computed successively from a single base line; arcs of triangulation connected together to form a system of loops or circuits extending over an area. See also: *base net.*
triangulation station A surveying station whose position is determined by triangulation. It is usually a permanently marked point that has been occupied (such as one identified by a bench mark), as distinguished from secondary points such as church spires, chimneys, water tanks, and prominent summits located by intersection. See also: *trigonometric point.*
triangulation tower An engineering structure used to elevate the line of sight above intervening obstacles (such as trees and topographic features), usually consisting of two separate towers built one within the other, the central one supporting the theodolite and the outer one supporting the observing platform; e.g. the Bilby steel tower consisting of two steel tripods that are demountable and easily erected.
triangulite (tri-an'-gu-lite) A bright yellow triclinic mineral belonging to the *ranunculite* group: $Al_3(UO_2)_4(PO_4)_4(OH)_5{\cdot}5H_2O$.
triangulum (tri-an'-gu-lum) In receptaculitids, a *merom* plate situated directly adapically of an interpositum.
Trias (Tri'-as) *Triassic.*
Triassic (Tri-as'-sic) The lower system of the *Mesozoic* Erathem of the Standard Global Chronostratigraphic Scale, above the *Permian* System of the *Paleozoic* Erathem and below the Jurassic System. Also the time during which these rocks were formed, the Triassic Period, covering the time span between 250 and 203 Ma. It was named because of its threefold lithostratigraphic division in Germany. Syn: *Trias.*
triaxial compression test (tri-ax'-i-al) A test in which a cylindrical specimen of rock encased in an impervious membrane is subjected to a confining pressure and then loaded axially to failure. See also: *unconfined compression test.* Cf: *extension test.* Syn: *compression test.*
triaxial extension test A test in which a cylindrical specimen of rock encased in an impervious membrane is subjected to a confining pressure and the axial load is decreased to failure. Cf: *triaxial compression test.* Syn: *extension test.*
triaxial state of stress A stress system in which none of the principal stresses is zero.
triaxial test (tri-ax'-i-al) In engineering geology, a laboratory test to measure shear strength of soil or rock in which confining stress can be applied to the sample. Cf: *uniaxial test.*
triaxon (tri-ax'-on) A siliceous sponge *spicule* in which six rays or their rudiments grow along three mutually perpendicular axes.
triboluminescence The emission of light by a substance when it is scratched, rubbed, or crushed.
tributary (trib'-u-tar'-y) n. (a) A stream feeding, joining, or flowing into a larger stream or into a lake. Ant: *distributary [streams].* Syn: *affluent; feeder [streams]; side stream.* (b) A valley containing a tributary stream. adj. Serving as a tributary.
tributary bifurcating link A link of magnitude μ that is formed at its upstream fork by the confluence of two links of unequal magnitude and that flows at its downstream fork into a link of a magnitude greater than or equal to 2μ (Mock, 1971, p.1559). Cf: *link [geomorph]; magnitude [geomorph].*
tributary glacier A glacier that flows into a larger glacier. See also: *secondary glacier.*
tributary link A link of magnitude μ that is formed at its upstream fork by the confluence of two links of unequal magnitude and that flows at its downstream fork into a link of a magnitude greater than or equal to 2μ (Mock, 1971, p.1559). Cf: *link [geomorph]; magnitude [geomorph].*
tributary source link A magnitude-1 link that joins a link of magnitude greater than 1 at its downstream fork (Mock, 1971, p.1558). Cf: *magnitude [geomorph]; link [geomorph].*
tricentric (tri-cen'-tric) Said of a corallite formed by a polyp retaining tristomodaeal condition permanently.
trichalcite (tri-chal'-cite) *tyrolite.*
trichite A straight or curved, hair-like crystallite in volcanic glass. Obsolete.
trichite [petrology] A straight or curved hairlike crystallite, usually black. Trichites occur singly or radially arranged in clusters and are found in glassy igneous rocks. Obsolete.
trichobothrium (trich-o-both'-ri-um) A sensory hair arising from the center of a disklike membrane on the legs or pedipalpi of an arachnid and serving for perception of air currents (TIP, 1955, pt. P, p.63). Also, a sensory organ consisting of one or more such hairs together with supporting structures. Pl: trichobothria.
trichocyst (tri'-cho-cyst) A rod-shaped or fusiform ejectile organelle found in many dinoflagellates (Williams et al., 2000, p.239).
trichome (tri'-chome) A bristle or hairlike outgrowth of the epidermis of plants.
trichotomocolpate (tri-chot'-o-mo-col'-pate) In certain pollen grains, having a triangular *colpus* that may simulate a trilete *laesura.* Syn: *trichotomosulcate.*
trichotomosulcate (tri-chot'-o-mo-sul'-cate) *trichotomocolpate.*
trichroic (tri-chro'-ic) Said of a mineral that displays *trichroism.*
trichroism (tri-chro'-ism) *Pleochroism* of a crystal that is indicated by three colors. A mineral showing trichroism is said to be *trichroic.* Cf: *dichroism.*
triclinic system (tri-clin'-ic) One of the six crystal systems, characterized by a onefold axis of symmetry, and having three axes whose lengths and angles of intersection are unconstrained.
tricolpate (tri-col'-pate) Said of pollen grains having three meridionally arranged colpi that are not provided with pores. Tricolpate pollen are typical

of dicotyledonous plants, and they first appeared in the fossil record in the Aptian-Albian interval (later part of Early Cretaceous).

tricolporate (tri-col-por'-ate) Said of pollen grains having three colpi that are provided with pores or with other, usually equatorial, modifications.

Triconodonta (Tri'-co-no-don'-ta) One of two orders of mammals (the other being Docodonta) of primitive structure and uncertain subclass assignment, mostly of Triassic and Jurassic age.

tricranoclone (tri-cran'-o-clone) An *ennomoclone* with three proximal rays.

trider (tri'-der) A *tetraclone* consisting of three similar arms differing from the fourth.

tridymite (trid'-y-mite) A vitreous colorless mineral: SiO_2. It is a high-temperature polymorph of quartz, and usually occurs as minute tabular white or colorless crystals or scales, in cavities in acidic volcanic rocks such as trachyte and rhyolite. Tridymite is stable between 870° and 1,470°C; it has an orthorhombic structure (alpha-tridymite) at low temperatures and a hexagonal structure (beta-tridymite) at higher temperatures. Cf: *cristobalite.* Syn: *christensenite.*

trigonal dipyramid (trig'-o-nal) A crystal form of six faces, ideally isosceles triangles, consisting of two trigonal pyramids repeated across a mirror plane of symmetry. It is trigonal in cross section, and its indices are {*h0l*} in symmetry $\bar{6}m2$, also {*hhl*} in $\bar{6}$ and 32, and {*hkl*} in $\bar{6}$.

trigonal dipyramidal class That crystal class in the hexagonal system having symmetry $\bar{6}$.

trigonal prism A crystal form of three faces that are parallel to a threefold symmetry axis. Its indices are {*100*} in symmetry $\bar{6}m2$, $\bar{6}$, 3*m*, and 3; or {*110*} in symmetry $\bar{6}$, 32, and 3; or {*hk*0} in $\bar{6}$ and 3.

trigonal pyramid A crystal form consisting of three faces, ideally isosceles triangles, in a pyramid with a triangular cross section. Its indices are <Φ255η0λ in symmetry 3*m*, and <Φ255η0λ, <Φ255ηηλ, and <Φ255ηκλ in symmetry 3.

trigonal-pyramidal class (trig'-o-nal-py-ram'-i-dal) That class in the rhombohedral division of the hexagonal system having symmetry 3.

trigonal-scalenohedral class (trig'-o-nal-sca-le'-no-he'-dral) *hexagonal-scalenohedral class.*

trigonal system A crystal system of threefold symmetry that is often considered as part of the *hexagonal system* since the lattice may be either hexagonal or rhombohedral. See also: *rhombohedral system.*

trigonal-trapezohedral class (trig'-o-nal-tra-pe'-zo-he'-dral) That crystal class in the rhombohedral division of the hexagonal system having symmetry 32.

trigonal trapezohedron A crystal form of six faces, having a threefold axis and three twofold axes, but neither mirror planes nor a center of symmetry. The top and bottom trigonal pyramids are rotated less than 30° about c with respect to each other. It may be either right-handed or left-handed. Its indices are <Φ255ηκλ.

trigonal trisoctahedron *trisoctahedron.*

trigonal tristetrahedron A *tristetrahedron* whose faces are triangular rather than quadrilateral, as in the *deltohedron.* Syn: *triakistetrahedron.*

trigoniid (tri-go'-ni-id) Any bivalve mollusk belonging to the family Trigoniidae, characterized by a variably shaped and ornamented shell, generally with opisthogyrate umbones and with the ornamentation of the posterior portion of the valves differing from that of the flank.

trigonite (trig'-o-nite) A yellow to brown monoclinic mineral: $Pb_3Mn^{2+}(As^{3+}O_3)_2(As^{3+}O_2OH)$. It occurs in triangular wedge-shaped crystals.

trigonododecahedron (trig'-o-no-do'-dec-a-he'-dron) An obsolete syn. of *deltohedron.*

trigonometric leveling (trig'-o-no-met'-ric) A type of indirect leveling in which differences of elevation are determined by means of observed vertical angles combined with measured or computed horizontal or inclined distances. Syn: *vertical angulation.*

trigonometric point A fixed point determined with great accuracy in the triangulation method of surveying; a *triangulation station* being the vertex of the triangle. Shortened form: trig point.

trigonometric survey A survey accomplished by triangulation and by trigonometric calculation of the elevations of points of observation. It is generally preliminary to a topographic survey, and is performed after careful measurement of a base line and of the angles made with this line by the lines toward points of observation.

trihedron (tri-he'-dron) A geometric form composed of three planes that meet at a central point, e.g. the trigonal pyramid crystal form.

trikalsilite (tri-kal'-si-lite) An hexagonal mineral: $(K,Na)AlSiO_4$. It is a polymorph of kalsilite. Cf: *tetrakalsilite.*

trilateration (tri'-lat-er-a'-tion) A method of surveying in which the lengths of the three sides of a series of touching or overlapping triangles are measured (usually by electronic methods) and the angles are computed from the measured lengths. Cf: *triangulation.*

trilete (tri'-lete) adj. Said of an embryophytic spore and some pollen having a laesura consisting of a three-pronged mark somewhat resembling an upper-case "Y". Cf: *monolete.* n. A trilete spore. The usage of this term as a noun is improper.

trilithionite (tri-lith'-i-o-nite') A colorless or lightly colored trioctahedral mineral of the *mica* group: $KLi_{1.5}Al_{1.5}(AlSi_3)O_{10}(OH)_2$.

trill *trilling.*

trilling (tril'-ling) A cyclic crystal twin consisting of three individuals. Cf: *twoling; fourling; fiveling; eightling.* Syn: *threeling; trill.*

trilobite (tri'-lo-bite) Any marine arthropod belonging to the class Trilobita, characterized by a three-lobed, ovoid to subelliptical exoskeleton divisible longitudinally into axial and side regions and transversely into cephalon (anterior), thorax (middle), and pygidium (posterior). Range, Lower Cambrian to Permian.

triloculine (tri-loc'-u-line) Having three chambers; specif. said of a foraminiferal *test* having three externally visible chambers, resembling *Triloculina* in form and plan.

trimaceral (tri-mac'-er-al) Said of a coal microlithotype consisting of three macerals. Cf: *monomaceral; bimaceral.*

trimacerite (tri-mac'-er-ite) A coal microlithotype consisting of macerals from the three groups, vitrinite, liptinite and inertinite, each more than 5% in abundance (ICCP, 1971). Cf: *duroclarite; clarodurite; vitrinertoliptite.*

trimerite (tri'-mer-ite) A salmon-colored, pale yellowish-red, or colorless monoclinic mineral: $CaMn^{2+}{}_2Be_3(SiO_4)_3$. It is structurally related to *beryllonite*.

trimerophyte (tri-mer'-o-phyte) A member of the Trimerophyta, one of the three major groups of early vascular plants as defined by Banks (1975), originally called Trimerophytina. Trimerophytes are more complex than *rhyniophytes* or *zosterophylls* and are believed to have given rise to the *sphenopsids, ferns,* and *progymnosperms.*

trimerous (trim'-er-ous) Said of the pentameral symmetry of certain echinoderms, such as edrioasteroids, characterized by three primary rays extending from the mouth, each of two lateral rays giving off two branches.

trimetric projection (tri-met'-ric) A projection based on representation of a spherical triangle by a plane triangle whose sides are lines at true scale, and in which the three spatial axes are represented as unequally inclined to the plane of projection (equal distances along the axes are drawn unequal).

trimetrogon (tri-met'-ro-gon) A system of three cameras arranged systematically at fixed angles and simultaneously taking photographs (one vertical photograph, and two oblique right and left photographs along the flight line at 60 degrees from the vertical) at regular intervals over the area being mapped. Etymol: originally equipped with wide-angle Metrogon lenses.

trimline (trim'-line) A sharp boundary line delimiting the maximum upper level of the margins of a glacier that has receded from an area. It usually coincides with a break in slope and the upper limit of unweathered rock on a valley wall or a nunatak; but the trimline of a long-extinct glacier may be marked by a sharp change in the age, constitution, or density of vegetation. Syn: *shoulder [glac geol].*

trimming (trim'-ming) The elimination of spurs that jut across a widening stream valley, effected by lateral erosion where a stream flows against and undercuts the sides of the spurs in going around meanders.

trimorph (tri'-morph) One of three crystal forms displaying *trimorphism.*

trimorphism (tri-mor'-phism) That type of *polymorphism [cryst]* in which there are three crystal forms, known as trimorphs. Adj: *trimorphous.* Cf: *dimorphism [cryst]; tetramorphism.*

trimorphous (tri-mor'-phous) Adj. of *trimorphism.*

trimounsite-(Y) An adamantine light brown monoclinic mineral: $(Y,Dy,Er,Yb)_2Ti_2O_5SiO_4$.

Trinitian (Tri-nit'-i-an) North American (Gulf Coast) stage: Lower Cretaceous (above Nuevoleonian, below Fredericksburgian).

trinomen (tri-no'-men) A name of an animal that consists of three words, the first designating the genus, the second the species, and the third the subspecies; e.g. *Odontochile micrurus clarkei.* See: ICZN, 1964, p.153. Syn: *trinomial.*

trinomial (tri-no'-mi-al) n. A syn. of *trinomen.*

trioctahedral (tri'-oc-ta-he'-dral) Pertaining to a layered-mineral structure in which all possible octahedral positions are occupied. Cf: *dioctahedral.*

triode (tri'-ode) (a) The initial triradial spicule (in which one ray is atrophied) of the ebridian skeleton. (b) A *triradiate* (sponge spicule). Also spelled: triod.

trip The operation in *rotary drilling* of pulling out ("trip out") the *drillstring,* as required to replace a worn bit, extract a core, or recover a *fish.* Syn: round trip.

tripartite method (tri-par'-tite) A method of determining the apparent surface velocity and direction of propagation of microseisms or earthquake waves by determining the times at which a given wave passes three non-colinear separated points.

tripestone (tripe'-stone) (a) A concretionary variety of anhydrite composed of contorted plates suggesting pieces of tripe. (b) A stalactite resembling intestines. (c) A variety of barite. Also spelled: tripe stone.

triphane (tri'-phane) *spodumene.*

triphylite (tri-phyl'-ite) (a) A gray-green or blue-gray orthorhombic mineral: $LiFe^{2+}PO_4$. It is isostructural with *monticellite.* Cf: *ferrisicklerite; heterosite.* (b) A group name for orthorhombic minerals of composition $AB(PO_4)$, where A = Li, Na or a vacant structure site, and $B = Fe^{2+}$, Mn^{2+} or Fe^{3+}.

triple junction (trip'-le) A point where three lithospheric plates meet.

triple point An invariant point at which three phases coexist in a unary (one component) system. When not otherwise specified, triple point usually refers to the coexistence of solid, liquid, and vapor of a pure substance, e.g. water. In metamorphic rocks, commonly refers to coexistence in the Al_2SiO_5 system of andalusite, sillimanite, and kyanite.

triple stomodaeal budding A type of budding in scleractinian corals similar to *tristomodaeal budding* in which the three stomodaea invariably form a triangle and only one interstomodaeal couple of mesenteries occurs between each pair of neighboring stomodaea.

triplet (trip′-let) An *assembled stone* composed of three bonded parts of natural or manufactured gem materials designed to imitate gems, or two main pieces and a colored cement layer (the third part of the triplet), which gives the stone color. Cf: *doublet.*

triple-tube core barrel A steel tool for coring rock that has a split inner tube to facilitate removal of recovered rock material.

triplite (trip′-lite) (a) A dark-brown, reddish-brown, or pink monoclinic mineral: $(Mn^{2+},Fe^{2+},Mg,Ca)_2(PO_4)(F,OH)$. Syn: *pitchy iron ore.* (b) A group name for monoclinic minerals with the general formula $A_2(PO_4)F$, where A = Mg, Fe^{2+} or Mn^{2+}.

triploblastic (trip-lo-blas′-tic) Said of the structure of embyonic animals having three primary germ layers (*ectoderm, mesoderm,* and *endoderm*).

triploidite (trip′-loid-ite) (a) A vitreous yellowish to reddish-brown monoclinic mineral: $(Mn^{2+},Fe^{2+})_2(PO_4)(OH)$. It is isomorphous with wolfeite. (b) A group name for monoclinic minerals with the general formula $A(XO_4)Z$, where A=Mg, Mn^{2+}, or Fe^{2+}, X=P or As, and Z=F, OH, or O.

tripod (tri′-pod) (a) A sponge *spicule* having three equal rays that radiate as if from the apex of a pyramid. (b) A stool-shaped shell formed from divergent rods united at a common center in nasselline radiolarians.

tripoli (trip′-o-li) (a) A light-colored porous friable siliceous (largely chalcedonic) sedimentary rock, which occurs in powdery or earthy masses that result from the weathering of chert or of siliceous limestone (Tarr, 1938, p.27). It has a harsh, rough feel, and is used for the polishing of metals and stones. (b) An incompletely silicified limestone from which the carbonate has been leached; *rottenstone.* (c) A term that was originally, but is now incorrectly, applied to a *siliceous earth* that closely resembles tripoli; specif. *diatomite,* such as the type deposit of northern Africa. See also: *tripolite.*

tripoli-powder (trip′-o-li-pow′-der) An obsolete syn. of *diatomite.*

tripolite (trip′-o-lite) A term that has been applied as a syn. of *diatomite,* in reference to the material from the north African location of Tripoli. It has also been used, less correctly, as a syn. of *tripoli.*

triporate (tri-por′-ate) Said of pollen grains having three pores, usually at 120° from each other and on or near the equator.

trippkeite (tripp′-ke-ite) A blue-green tetragonal mineral: $CuAs^{3+}{}_2O_4$. It has excellent prismatic cleavage that permits crystals to be broken into flexible fibers. It is the Cu-As analogue of schafarzikite.

Triprojectatites (Tri-pro-ject′-a-tites) A group of Late Cretacous to Early Tertiary pollen, characterized by three *colpi* occurring on the ends of flat extrusions from the cylindrical body of the grain.

tripuhyite (trip′-u-hy-ite′) A greenish-yellow to dark brown or black tetragonal mineral of the *rutile* group: $Fe^{2+}Sb^{5+}{}_2O_6$. Syn: *flajolotite.*

triradiate (tri-ra′-di-ate) n. A sponge spicule in the form of three coplanar rays radiating from a common center. Syn: *triode.* adj. Descriptive of trilete spores.

triradiate crest In a trilete spore, the three-rayed raised figure on the proximal surface caused by the intersection of the contact areas.

triserial (tri-se′-ri-al) Arranged in, characterized by, or consisting of three rows or series; specif. said of the chambers of a foraminiferal test arranged in three columns or in a series of three parallel or alternating rows, such as a trochospiral test with three chambers in each whorl.

trisoctahedron (tris′-oc-ta-he′-dron) An isometric crystal form of 24 faces, each of which is an isosceles triangle. Its indices are {*hhk*}. Syn: *triakisoctahedron; trigonal trisoctahedron.*

tristanite (tris′-tan-ite) Silica-saturated to undersaturated igneous rocks intermediate between *trachyandesite* and *trachyte,* with a differentiation index between 65 and 75 and K_2O:Na_2O greater than 1:2. The name, given by Tilley and Muir in 1964, is for the island of Tristan da Cunha. Obsolete.

tristetrahedron (tris′-tet-ra-he′-dron) An isometric crystal form having 12 faces that are either triangular (*trigonal tristetrahedron*) or quadrilateral (*deltohedron*). Its indices are {*hkk*} and its symmetry is $\bar{4}3m$ or 23.

tristomodaeal budding (tri-sto-mo′-dae-al) A type of budding in scleractinian corals in which three stomodaea are developed within a common tentacular ring and either occur in series or form a triangle, and two interstomodaeal couples of mesenteries are located between the original and each new stomodaeum. See also: *triple stomodaeal budding.*

tristramite (tris′-tram-ite) A yellow or greenish-yellow hexagonal mineral of the *rhabdophane* group: $(Ca,U^{4+},Fe^{3+})(PO_4,SO_4)\cdot 2H_2O$.

tritium (trit′-i-um) A radioactive isotope of hydrogen having two neutrons and one proton in the nucleus.

tritium dating Calculation of an age in years by measuring the concentration of radioactive hydrogen-3 (tritium) in a substance, usually water. Maximum possible age limit is about 30 years. The method also provides a means of tracing subsurface movements of water and of determining its velocities.

tritomite-(Ce) (tri′-tom-ite-Ce) A dark brown hexagonal mineral: $Ca_2(Ce,La)_3(SiO_4,BO_4)_3(OH,O)$.

tritomite-(Y) A greenish-black hexagonal mineral: $(Y,Ca,Ce)_5(SiO_4,BO_4)_3(OH,O)$.

tritonymph (tri′-to-nymph) The third developmental stage in the arachnid order Acarida.

trittkarren (tritt′-kar-ren) In *karst,* crescentic solution pockets about 30 cm in diameter on limestone surfaces. Etymol: German, "step tracks". Cf: *karren.*

trituration (trit-u-ra′-tion) *comminution.*

trivariant (tri-var′-i-ant) Pertaining to a chemical system having three degrees of freedom, i.e. having a variance of three.

trivium (triv′-i-um) (a) The three anterior ambulacra of an echinoid. (b) The part of an asterozoan containing three rays, excluding the bivium. The term is not recommended as applied to asterozoans (TIP, 1966, pt. U, p.30). Pl: trivia. Cf: *bivium.*

TRM *thermoremanent magnetization.*

trochanter (tro′-chan-ter) (a) The second segment of a *pedipalpus* or leg of an arachnid, so articulated to coxa and femur as to permit motion of the entire leg in any direction (TIP, 1955, pt. P, p.63). It corresponds physiologically to vertebrate hip articulation. (b) A joint of the proximal part of a prosomal appendage of a merostome. (c) One of up to five processes near the proximal end of the tetrapod femur.

trochiform (tro′-chi-form) Shaped like a top; e.g. said of a gastropod shell (e.g. *Trochus*) with a flat base, evenly conical sides, and a not highly acute spire.

trochoceroid (tro-choc′-e-roid) A syn. (esp. in the older literature) of *torticone.*

trochoid (tro′-choid) adj. (a) Said of a horn-shaped solitary *corallite* with sides regularly expanding from the apex at an angle of about 40 degrees. Cf: *turbinate; patellate.* (b) Said of a foraminiferal test with spirally or helically coiled chambers, *evolute* on one side of the test and *involute* on the opposite side. Syn: *trochospiral.* n. A trochoid corallite or foraminiferal test.

trocholophe (tro′-cho-lophe) A brachiopod *lophophore* disposed as a ring surrounding the mouth, bearing a single row of unpaired (or more rarely, a double row of paired) filamentary appendages (TIP, 1965, pt. H, p.154).

trochospiral (troch-o-spi′-ral) *trochoid.*

trochospiral suture (tro′-cho-spi-ral) An *excystment* structure present in some *netromorph* acritarchs (e.g., *Leiofusa, Eupoikilofusa*) with a spiral trace on the long axis of the *vesicle* (Miller and Wood, 1982).

troctolite (troc′-to-lite) (a) In the *IUGS classification,* a plutonic rock satisfying the definition of *gabbro,* in which pl/(pl+px+ol) is between 10 and 90 and px/(pl+px+ol) is less than 5. (b) A *gabbro* that is composed chiefly of calcic plagioclase (e.g. labradorite) and olivine with little or no pyroxene. Syn: *forellenstein; troutstone.* Such rocks commonly are speckled like trout, hence the three synonyms, derived from Greek, German, and English.

trögerite (trö-ger-ite) A micaceous lemon-yellow tetragonal mineral of the *natroautunite* group: $(H_3O)(UO_2)(AsO_4)\cdot 3H_2O$.

Tröger's classification (Trö′-ger's) A quantitative mineralogic classification of igneous rocks proposed by E. Tröger in 1935. Obsolete.

troglobiont (trog-lo-bi′-ont) A *troglodyte,* esp. one living in the lightless waters of caves.

troglobite (trog′-lo-bite) An organism that must live its entire life underground. See also: *troglodyte; troglophile; trogloxene.*

troglodyte (trog′-lo-dyte) Any organism that lives in a cave or rock shelter. Adj: troglodytic. Cf: *troglobite; troglophile; trogloxene.* Syn: *troglobiont.*

troglophile (trog′-lo-phile) Any organism that completes its life cycle in a cave but that also occurs in certain environments outside the cave. Cf: *troglodyte; trogloxene; troglobite.*

trogloxene (trog′-lo-xene) Any organism that regularly or accidentally enters a cave but that must return to the surface to maintain its existence. Cf: *troglophile; troglodyte; troglobite.*

trogschluss (trog′-schluss) The German equivalent of *trough end.*

trogtalite (trog′-tal-ite) A metallic cubic mineral of the *pyrite* group: $CoSe_2$. It is dimorphous with hastite.

troilite (tro′-ĭ-lite) A metallic bronze hexagonal mineral of the *pyrrhotite* group: FeS. It is present in small amounts in almost all meteorites.

trolleite (trol′-le-ite) A pale green monoclinic mineral: $Al_4(PO_4)_3(OH)_3$.

trommel (trom′-mel) A revolving cylindrical screen used in size classification of coal, gravel, and crushed stone.

trona (tro′-na) A white, gray, or yellow-white monoclinic mineral: $Na_3(CO_3)(HCO_3)\cdot 2H_2O$. It occurs in fibrous or columnar layers and thick beds in saline residues. Trona is a major source of sodium compounds.

trondhjemite (trond′-hjem-ite) (a) In the *IUGS classification,* a syn. of *plagiogranite* and leucocratic *tonalite.* (b) A light-colored plutonic rock primarily composed of sodic plagioclase (esp. oligoclase), quartz, sparse biotite, and little or no alkali feldspar. Its name, given by Goldschmidt in 1916, is derived from Trondhjem, Norway.

troostite (troost′-ite) A mineral variety: $(Zn,Mn)_2SiO_4$. It is a reddish variety of willemite containing manganese and occurring in large crystals.

Tropepts Inceptisols that have a mean annual soil temperature of 8°C or more, and <5°C difference between mean summer and mean winter temperatures at a depth of 50 cm below the surface. Tropepts may have an ochric epipedon and a cambic horizon, or an umbric epipedon, or a mollic epipedon under certain conditions but no plaggen epipedon, and are not

saturated with water for periods long enough to limit their use for most crops. A suborder in the U.S. system of soil taxonomy.
trophic (troph'-ic) Of or pertaining to nutrition.
trophic level A stage of nourishment representing one of the segments of the food chain, and characterized by organisms that all obtain food and energy in a basically similar fashion.
trophism (troph'-ism) Nutrition involving metabolic exchanges in the tissues.
trophocyte (troph'-o-cyte) An amoebocyte of a sponge that serves as a nourishing cell of an oocyte, embryo, or gemmule.
trophogenic (troph-o-gen'-ic) Said of the upper or illuminated zone of a lake in which photosynthesis converts inorganic matter to organic matter (Wetzel, 2001, p.164). Cf: *tropholytic.* See also: *euphotic zone [lake].*
tropholytic (troph-o-lyt'-ic) Said of the deeper part of a lake in which organic matter tends to be dissimilated. Cf: *trophogenic.* See also: *profundal.*
tropic (trop'-ic) n. The area of the Earth falling between the Ttropics of Cancer and Capricorn. Generally used in the plural. adj. Pertaining to features, climate, vegetation and animals characteristic of the tropics. Var: tropical.
tropical (trop'-i-cal) (a) Said of a climate characterized by high temperature and humidity and by abundant rainfall. An area of tropical climate borders that of equatorial climate. (b) Pertaining to the tropic regions.
tropical air mass An air mass originating over oceans and continents between 30° north and 30° south of the Equator; characterized by warm air. Cf: *polar air mass.*
tropical cyclone An intense *cyclone* that forms over the tropical oceans and ranges from 100 to 1,000 km in diameter. It moves first westward, then northeastward in the Northern Hemisphere and southeastward in the Southern Hemisphere; its wind speed is moderate at its fringe area but may increase to as much as 200 knots (230 mph) near its center. The center itself, or eye, is often calm and clear. Cf: *hurricane; typhoon; extratropical cyclone.*
tropical desert A hot, dry desert lying between latitude 15° and 30° north or south of the Equator, more specifically near the Tropics of Cancer or Capricorn, where subtropical high-pressure air masses prevail, producing conditions of very low, sporadic rainfall. Cf: *west-coast desert.* Syn: *low-latitude desert; trade-wind desert.*
tropical lake (a) A lake in the *tropics.* (b) A lake whose surface temperature never falls below the temperature of maximum-density water, about 3.96°C. The appropriateness of this sense of the term has been questioned, since it has identified lakes in Switzerland and Scotland. Cf: *polar lake.*
tropical rain forest A low latitude *biome* characterized by a warm, wet climate and high species diversity and biomass.
tropical spine A radial spine disposed according to the *Müllerian law* and marking a zone in an acantharian radiolarian comparable to the tropical zone of the terrestrial globe.
Tropic of Cancer The parallel of latitude approx. 23 ½° north of the Equator, which indicates the northernmost latitude reached by the Sun's vertical rays.
Tropic of Capricorn The parallel of latitude approx. 23 ½° south of the Equator, which indicates the southernmost latitude reach by the Sun's vertical rays.
tropism (tro'-pism) An involuntary orientational movement or growth in which an organism or one of its parts turns or curves as a positive or negative response to a stimulus. It may be indistinguishable from *taxis [ecol]* in motile organisms.
tropophilous (tro-poph'-i-lous) Said of an organism that has adapted physiologically to periodic changes in the environment, e.g. seasonal changes.
tropophyte (trop'-o-phyte) A plant adapted to seasonal changes in moisture and temperature; e.g. the deciduous trees of temperate and tropical regions.
troposphere (tro'-po-sphere) That portion of the atmosphere next to the Earth's surface, in which temperature generally decreases rapidly with altitude, clouds form, and convection is active. In middle latitudes the troposphere generally includes the first 10 to 12 km above the Earth's surface. Cf: *stratosphere.*
tropotaxis (trop-o-tax'-is) *Taxis [ecol]* resulting from the simultaneous comparison of stimuli of different intensity that are acting on separate ends of the organism.
trottoir (trot-toir') A narrow, organic, intertidal reef construction, composed of either a solid mass or a simple crust covering a rocky substratum, separating the shoreline from the sea in the same manner that a sidewalk separates the street from the adjoining houses. Etymol: French, "sidewalk". See also: *algal ridge.*
trough [beach] A small linear depression formed just offshore on the bottom of a sea or lake and on the landward side of a *longshore bar.* It is generally parallel to the shoreline, and is always under water. A trough may be excavated by the extreme turbulence of wave and current action in the zone where breakers collapse. Inappropriate syn: *runnel [beach]; low [beach].*
trough [geomorph] (a) Any long, narrow depression in the Earth's surface, such as one between hills; esp. a broad, elongate U-shaped valley, such as a *glacial trough* or a trench. (b) The channel in which a stream flows.
trough [marine geol] An elongate depression of the sea floor that is shallower than a trench, and with less steeply sloping sides.
trough [meteorol] An elongated region of relatively low atmospheric pressure at a given altitude, with cyclonic curvature of the winds and isobars. If a closed circulation occurs, the trough is called a *cyclone.* Ant: *ridge [meteorol].*
trough [paleont] (a) The furrow on the posterior part of the pedicle valve of some brachiopods beneath the apex which provides space for the *pedicle* (Moore et al., 1952, p.207). (b) *hinge trough.* (c) *main trough.*
trough [sed struc] The part of a *ripple* with an elevation of less than half its total height (Blatt et al., 1980).
trough [struc geol] (a) An obsolete term for a *graben.* (b) The lowest point of a given stratum in any profile through a fold. Cf: *trough line.*
trough [tect] A linear depression or basin that subsides as it receives clastic material, located not far from the source supplying the sediment. Obsolete syn: *geosynclinal trough.*
trough banding Rhythmic layering or alignment of minerals confined to troughlike depressions within an intrusion.
trough cross-bedding Cross-bedding in which the lower bounding surfaces are curved surfaces of erosion (McKee and Weir, 1953, p.385); it results from local scour and subsequent deposition. Cf: *festoon cross-bedding.* Syn: *crescent-type cross-bedding.*
trough end The steep, semicircular rock wall forming the abrupt head or end of a glacial trough. See also: *oversteepened wall.* Syn: *trough wall; trogschluss.*
trough fault An obsolete term for a fault, generally a normal fault, which bounds a graben or other structural depression.
trough-in-trough Said of a cross profile depicting two or more glaciations, each of which shaped its own troughlike valley, esp. where a steep-sided inner trough lies within a wider trough with a flatter bottom.
trough line The line joining the trough points of a given stratum. Cf: *trough [struc geol].*
trough reef *reverse saddle.*
trough surface A surface that connects the trough lines of the beds of a syncline.
trough valley *U-shaped valley.*
trough wall *trough end.*
troutstone (trout'-stone) *troctolite.*
T-R sequence A *sequence* bound by subaerial unconformities or *shoreface ravinement surfaces* on the basin flanks and by conformable *transgressive surfaces* farther basinward (Embry and Johannessen, 1992).
trudellite (tru-dell'-ite) A mixture of chloraluminite and natroalunite.
true azimuth The *azimuth [surv]* measured clockwise from true north through 360 degrees.
true bearing The *bearing* expressed as a horizontal angle between a geographic meridian and a line on the Earth; esp. a horizontal angle measured clockwise from true north. Cf: *magnetic bearing.*
true cleavage A quarrying term for the dominant cleavage in a rock, e.g. slaty cleavage, to distinguish it from minor or *false cleavage.* Geologically, the term is misleading and should be avoided.
true crater A depression, formed by impact or explosion, before modification by slumping or by deposition of material set in motion. "The true crater is defined as the boundary between the loose, allochthonous material and the underlying parautochthonous material that has been crushed and fractured but has not experienced significant displacement" (Nordyke, 1961, p.3447). Cf: *apparent crater.*
true dip A syn. of *dip,* used in comparison with *apparent dip.* Syn: *full dip.*
true folding An imprecise and therefore less desirable syn. of *buckle folding.*
true homology *homology.*
true horizon (a) A *celestial horizon.* Also, the horizon at sea. (b) A horizontal plane passing through a point of vision or a perspective center. The *apparent horizon* approximates the true horizon only when the point of vision is very close to sea level.
true north The direction from any point on the Earth's surface toward the geographic north pole; the northerly direction of any geographic meridian or of the meridian through the point of observation. It is the universal zero-degree (or 360-degree) mapping reference. True north differs from *magnetic north* by the amount of magnetic declination at the given point. Syn: *geographic north.*
true polar wandering The contribution to a time-sequence distribution of paleomagnetic poles that is common to all tectonic plates and thus records the motion of the Earth's outer core and mantle with respect to its axis of rotation. See also: *apparent polar wandering.*
true position The "exact" position of features as they exist on the ground (USGS, 2002b, p.1A-3). Cf: *cartographic position, geographic position, true position, source position.* See also: *positional accuracy.*
true resistivity The resistivity of a locally homogeneous medium. Cf: *apparent resistivity.*
true soil *solum.*
true thickness The thickness of a stratigraphic unit or other tabular body, measured at right angles to the direction of extension of the unit or body. Cf: *apparent thickness.*
true vertical depth The depth of a well, measured straight down. Cf: *total depth.* Abbrev: TVD.

truffite (truff'-ite) Nodular masses of woody lignite occurring within Cretaceous lignite of France. It is so named because of its trufflelike odor.
trumpet log (trum'-pet) *Microlaterolog.*
trumpet valley A narrow valley or gorge that cuts through the central part of a lobe (in the morainal landscape of a former piedmont glacier) and opens out into a broad funnel as it reaches the glaciofluvial sand and gravel cone or fan of the lower piedmont. Examples are numerous along the northern Alpine foothills in Bavaria.
truncate (trun'-cate) v. In crystal morphology, to replace the corner of a crystal form with a face of a different form. Such a crystal form is said to be truncated.
truncated [geomorph] (trun'-cat-ed) Said of a landform (such as a headland or mountain), or of a geologic structure, that has been abbreviated by tectonic or erosional processes; esp. said of a conical eminence (such as a volcano) whose apical part has been replaced by a plane section parallel to the land or ocean surface.
truncated [soil] Having lost all or part of the upper soil horizon or horizons by soil removal (erosion, excavation, etc.).
truncated spur A spur that projected into a preglacial valley and was partially worn away or beveled by a moving glacier as it widened and straightened the valley. Cf: *faceted spur.*
truncated wave-ripple lamina One of a set of parallel laminae that were deposited conformably with the original wave-generated *bed form* surface but were partly eroded before deposition of the next set of *bed form* laminae. Ripple wavelengths range from 10 cm to 10 m (Campbell, 1966, p.825). Cf: *hummocky cross-stratification.*
truncation [geomorph] (trun-ca'-tion) An act or instance of cutting or breaking off the top or end of a geologic structure or landform, as by erosion. Cf: *beveling.*
truncation [paleont] The natural loss, in life, of the apical portion of a nautiloid shell.
trunk [paleont] (a) The *thorax* of an arthropod; esp. the postcephalic part of the body of a crustacean. (b) The axial skeleton of a vertebrate, esp. a tetrapod, less neck and tail and with or without associated soft parts.
trunk [streams] The principal channel of a system of tributaries; the channel of a *main stream.*
trunk glacier A central or main valley glacier formed by the union of several tributary glaciers. Cf: *dendritic glacier.*
trunk stream *main stream.*
truscottite (trus'-cott-ite) A pearly white trigonal mineral related to reyerite: $(Ca,Mn^{2+})_{14}Si_{24}O_{58}(OH)_8 \cdot 2H_2O$.
trüstedtite (trü'-stedt-ite) A metallic cubic mineral of the *linnaeite* group: $(Ni,Co)_3Se_4$.
tsaregorodtsevite A colorless or pale yellow orthorhombic mineral: $N(CH_3)_4Si_4(Si,Al)_2O_{12}$.
tschermakite (tscher'-mak-ite) (a) A monoclinic mineral of the *amphibole* group with end-member composition $Ca_2(Mg_3AlFe^{3+})(Si_6Al_2)O_{22}(OH)_2$. Not to be confused with *Ca-Tschermak molecule.* (b) A gray-white feldspar (albite?) containing some magnesium but no calcium, from Bamble, Norway. (c) A plagioclase feldspar (oligoclase or albite) with composition ranging from $Ab_{95}An_5$ to $Ab_{80}An_{20}$.
Tschermak molecule (Tscher'-mak) *Ca-Tschermak molecule.*
tschermigite (tscher'-mig-ite) A colorless cubic mineral of the *alum* group: $(NH_4)Al(SO_4)_2 \cdot 12H_2O$. Syn: *ammonia alum; ammonium alum.*
tschernichite A colorless or white tetragonal mineral: $CaAl_2Si_6O_{16} \cdot 8H_2O$.
Tschokrakian *Chokrakian.*
tschörtnerite A light blue cubic mineral: $Ca_4(K,Ca,Sr,Ba)_3Cu_3Al_{12}Si_{12}O_{48}(OH)_8 \cdot 20H_2O$. $(Ca,K,Na)_{2-x}(Ti,Nb)_2Si_4O_{12}(OH,O)_2 \cdot 4H_2O$.
T-S diagram *temperature-salinity diagram.*
tsepinite-Ca A white or pale brownish monoclinic mineral of the *vuoriyarvite* group: $(Ca,K,Na)_{2-x}(Ti,Nb)_2Si_4O_{12}(OH,O)_2 \cdot 4H_2O$.
tsepinite-K A colorless monoclinic mineral of the *vuoriyarvite* group: $(K,Ba,Na)_2(Ti,Nb)_2Si_4O_{12}(OH,O)_2 \cdot 3H_2O$.
tsepinite-Na A colorless, white or light brown monoclinic mineral of the *vuoriyarvite* group: $(Na,H_3O,K,Sr,Ba,[])_{12}Ti_8(Si_4O_{12})_4(OH,O)_8 \cdot 12\text{-}16H_2O$.
tsilaisite (tsi'-lai-site) A yellow, yellowish-brown or greenish-yellow rhombohedral member of the *tourmaline* group: $Na(Mn,Al,Li)_3Al_6(BO_3)_3Si_6O_{18}(O,OH,F)$. It may be a Mn-bearing elbaite.
tsnigriite A metallic monoclinic mineral: $Ag_9SbTe_3(S,Se)_3$.
tsugaruite A metallic silvery lead-gray orthorhombic mineral: $Pb_4As_2S_7$.
tsumcorite (tsum'-cor-ite) (a) A reddish-brown monoclinic mineral: $Pb(Zn,Fe^{3+})_2(AsO_4)_2(OH,H_2O)_2$. (b) A group name for minerals with an analogous composition, but with Pb replaced by Ca; Zn and Fe^{3+} replaced by Fe^{2+}, Cu^{2+}, Co or Mn^{3+}; and AsO_4 replaced by VO_4.
tsumebite (tsu'-meb-ite) An emerald-green monoclinic mineral: $Pb_2Cu(PO_4)(SO_4)(OH)$.
tsumgallite A pale greenish-yellow or beige orthorhombic mineral of the *diaspore* group: GaOOH.
tsumoite (tsu'-mo-ite) A metallic creamy-white trigonal mineral: BiTe.
tsunami (tsu-na'-mi) A gravitational sea wave produced by any large-scale, short-duration disturbance of the ocean floor, principally by a shallow submarine earthquake, but also by submarine slumps, subsidence, or volcanic eruption. It is characterized by great speed of propagation (up to 950 km/hr), long wavelength (up to 200 km), long period (varying from 5 min to a few hours, generally 10-60 min), and low observable amplitude on the open sea, although it may pile up to heights of 30 m or more and cause much damage on entering shallow water along an exposed coast (often thousands of kilometers from the source). Etymol: Japanese, "harbor wave". Pl: tsunamis; tsunami. Adj: tsunamic. Erroneous syn: *tidal wave.* Syn: *seismic sea wave.*
t test *Student's t test.*
tube [paleont] (a) The central cylinder connecting the two shields of a placolith coccolith. (b) One of the siphons of a bivalve mollusk.
tube [speleo] A smooth-sided cave passage that is elliptical to circular in cross section. See also: *half-tube.*
tube foot One of numerous small tentacular flexible organs of echinoderms, being the end of a branch of the *water-vascular system* and serving for grasping, adhesion, locomotion, respiration, feeding, or combination of these.
tubercle (tu'-ber-cle) (a) One of the small rounded knoblike structures on the outer surface of the test plates of an echinoid, bearing a movable spine. (b) A low rounded prominence of intermediate size on the surface of an ostracode valve, commonly along the free margin (TIP, 1961, pt.Q, p.55). See also: *eye tubercle.* (c) Any fine low rounded protuberance on either surface of a brachiopod valve, irrespective of origin (TIP, 1965, pt.H, p.154). (d) A small, rounded, moderately prominent elevation on the surface of a gastropod shell (TIP, 1960, pt.I, p.133). (e) Any small process of a bone.
tubercle texture In mineral deposits, a texture in which gangue is replaced by automorphic minerals. Cf: *atoll texture.*
tuberose (tu'-ber-ose) Said of a mineral whose form is that of irregular, rootlike shapes or branches.
tube-wave [seis] A seismic surface wave traveling along the walls of a borehole; it is involved in vertical seismic profiling and seismic borehole logging.
tube well (a) *driven well.* (b) *tubular well.* Also spelled: tubewell.
tubing (tub'-ing) A small-diameter removable pipe, suspended and immobilized in a well inside a large-diameter *casing* through which fluids are produced (brought to the surface).
tubular spring (tu'-bu-lar) A gravity or artesian spring whose water issues from rounded openings, such as lava tubes or solution channels.
tubular stalactite *soda straw.*
tubular well General term for a drilled, driven, or bored well of circular cross section, and of a depth that is relatively great compared to diameter. Syn: *tube well.*
tubule (tu'-bule) A hollow twiglike calcareous concretion, characteristic of loess deposits.
tubules Canals that extend into the dentine, mainly at an angle toward the cusp of the tooth. They are filled with fluid and occupied by odontoblastic processes, i.e., the branched protoplasmic extensions of odontoblasts. The odontoblastic processes are thought to be sensitive to compression or shearing forces on the tooth. In some mammals, the tubules extend into the overlying enamel. The main tubular canals are nearly straight; resin casts of the tubules clearly show that the apparent sinuous nature of the canals is an artifact of sectioning. The main canals are branched, as are the odontoblastic processes. Current evidence indicates that only a fraction of any tubule is filled with an odontoblastic process; that fraction decreases with the growth (age) of the dentine. The older literature suggests that nerve fibers are present in dentinal tubules; however, current evidence indicates that this was an artifact induced by implosive forces generated when the tooth was extracted. Unmylinated fibers penetrate the odontoblastic layer to line the border between the odontoblastic cell body and the predentine (M. Crenshaw *in* Carter, 1990).
tubulospine (tu'-bu-lo-spine') A foraminiferal chamber produced radially into a long hollow extension (as in *Schackoina*).
tubulus (tu'-bu-lus) In the archaeocyathids, a radiating porous-walled prismatic tubule in the *intervallum* of the Syringocnemidida (TIP, 1972, pt. E, p.17). Pl: tubuli.
tucanite (tu'-ca-nite) A dehydrated *scarbroite.*
tucekite (tu'-cek-ite) A metallic pale-yellow tetragonal mineral of the *hauchecornite* group: $Ni_9Sb_2S_8$.
tufa (tu'-fa) A variety of *travertine* that is commonly spongy or porous due to precipitation around a variety of floral strictures, such as reeds, plant roots, leaves, etc. Tufa mounds or travertine mounds (as at Mono Lake) are associated with lake springs (Tucker and Wright, 1990, p.169). Analogous mounds in marine or freshwater settings may be organic, thus *bioherms.* The term is rarely applied to a similar deposit consisting of silica. It is not to be confused with *tuff.* Etymol: Italian, "tufo." Syn: *calcareous tufa; calc-tufa, tuft; petrified moss.*
tufaceous (tu-fa'-ceous) Pertaining to or like tufa. Not to be confused with *tuffaceous.*
tuff Consolidated or cemented volcanic ash and lapilli. Not to be confused with *tufa.* Adj: *tuffaceous.*
tuffaceous (tuff-a'-ceous) Said of nonvolcanic clastic sediments that contain mixtures of ash-size pyroclasts, e.g. crystal, vitric or lithic.
tuff ball An individual unit in a mass of *accretionary lapilli.*

tuff breccia A pyroclastic breccia consisting of more or less equal amounts of ash, lapilli, and larger volcanic fragments.

tuffeau (tuff-eau') A term used in France for tufa, micaceous chalk, and soft, very porous, extremely coarse limestone made up of bryozoan fragments.

tuffisite (tuff'-i-site) A term proposed by Cloos (1941) for fragmented country rock in pipes located in Swabia, southwest Germany. Syn: *intrusive tuff.*

tuffite (tuff'-ite) A tuff containing both pyroclastic and detrital material, but predominantly pyroclasts.

tufflava (tuff-la'-va) An extrusive rock containing both pyroclastic and lava-flow characteristics, so that it is interpreted to be an intermediate form between a lava flow and a welded-tuff type of *ignimbrite*. Whether or not it is actually a genetically distinct type of rock is a matter of debate. Also spelled: tuffolava; tuff lava; tuflava. Cf: *ignispumite.*

tuffolava (tuf'-fo-la'-va) *tufflava.*

tuff ring A broad, low volcano having low-angle outer slopes (2°-10°) and a wide crater that may be a maar. The pyroclastic deposits form aprons surrounding the broad crater. Syn: *maar.*

tuffstone (tuff'-stone) A sandstone that contains pyroclasts of sand-grain size.

tuft A term used in England for any porous or soft stone, such as the sandstone in the Alston district of Cumberland, and the rock now known as *tufa* (Arkell and Tomkeieff, 1953, p.121).

tugarinovite (tu-ga-ri'-nov-ite') A metallic lilac-brown monoclinic mineral: MoO_2.

tugtupite (tug'-tup-ite) A vitreous white, pink, blue, or green tetragonal mineral: $Na_4AlBeSi_4O_{12}Cl$. It possesses an ordered sodalite framework structure. Syn: *beryllosodalite.*

tuhualite (tu-hu'-al-ite) A violet orthorhombic mineral: $(Na,K)Fe^{2+}Fe^{3+}Si_6O_{15}$.

tuite A colorless rhombohedral meteorite mineral: $Ca_3(PO_4)_2$. It is a high-pressure polymorph of whitlockite.

tula (tu'-la) A relatively small but sacculike sexinous inflation occurring at the end of the sulcal axis of certain elliptical pollen, e.g. in *Ovalipollis.* Pl: tulae.

tulameenite (tu-la-meen'-ite) A metallic stell-gray tetragonal mineral: Pt_2FeCu. Syn: cuproplatinum.

tulare (tu-la'-re) A syn. of *tule land.* Etymol: Spanish, "tule field".

tule (tu'-le) *tule land.*

tule land A local term given in the Sacramento River valley, California, to a large tract of overflowed land (or *flood basin*) on which the tule, a variety of bulrush, is the dominant or characteristic native plant. The land is often referred to as *tule* or "the tules". Syn: *tulare.*

tuliokite A light to dark gray rhombohedral mineral: $Na_6BaTh(CO_3)_6{\cdot}6H_2O$.

tumchaite A vitreous colorless or white monoclinic mineral: $Na_2(Zr,Sn)Si_4O_{11}{\cdot}2H_2O$.

tumescence (tu-mes'-cence) The swelling of a volcanic edifice due to accumulation of magma in the reservoir. It may or may not be followed by eruption. Syn: *bulge; inflation [volc].*

tump (a) A mound, hummock, hillock, or other small rise of ground. (b) A clump of vegetation, such as trees, shrubs, or grass; esp. one forming a small island in a marsh or swamp.

tumuli (tu'-mu-li) Plural of *tumulus.*

tumulus [paleont] (tu'-mu-lus) (a) A secondary deposit on the chamber floor of a foraminiferal test, appearing in cross section as more or less a symmetric node with a rounded summit. (b) A minor bulge in the wall of the cup in archaeocyathids; it may be perforated by one or many pores. Pl: *tumuli.*

tumulus [volc] A doming or small mound on the crust of a lava flow, caused by pressure due to the difference in rate of flow between the cooler crust and the more fluid lava below. Unlike a *blister [volc]*, it is a solid structure. Pl: *tumuli.* Syn: *pressure dome.* Cf: *pressure ridge [volc].*

tundra (tun'-dra) A treeless, level or gently undulating plain characteristic of arctic and subarctic regions. It usually has a marshy surface, which supports a growth of mosses, lichens, and numerous low shrubs and is underlain by a dark, mucky soil and permafrost.

tundra climate A type of *polar climate* having a mean temperature in the warmest month of between 0° and 10°C. It represents a climate that does not have permanent ice cover but is too cold for tree growth. Cf: *perpetual frost climate.*

tundra crater A circular or shapeless "island" of silt (without vegetation) found in tundra and formed during the period of thawing by the forced rise of silt to the surface and its pouring-out like lava onto the surface. Cf: *suffosion; sand volcano.*

tundra ostiole A *mud circle* in the tundra soil of northern Quebec. Syn: *ostiole [geomorph].*

tundra peat Peat occurring in subarctic areas and derived from mosses, heaths, and birch and willow trees.

tundra polygon *ice-wedge polygon.*

Tundra soils (a) Soils characteristic of tundra regions. (b) A zonal great soil group consisting of soils with dark-brown peaty layers over grayish horizons mottled with rust and having continually frozen substrata; formed under frigid, humid climates, with poor drainage, and native vegetation of lichens, moss, flowering plants, and shrubs. Not used in current U.S. system of soil taxonomy.

tundrite-(Ce) (tun'-drite-Ce) A brownish-yellow, greenish-yellow, or light green triclinic mineral: $Na_2Ce_2TiO_2SiO_4(CO_3)_2$.

tundrite-(Nd) A brownish to yellowish triclinic mineral: $Na_2Nd_2TiO_2(SiO_4)(CO_3)_2$.

tunellite (tu-nell'-ite) A colorless or white platy monoclinic mineral: $SrB_6O_9(OH)_2{\cdot}3H_2O$.

tungomelane (tung-o'-mel-ane) A variety of psilomelane containing tungsten.

tungstate (tung'-state) A mineral compound characterized by the radical WO_4. An example of a tungstate is wolframite, $(Fe,Mn)WO_4$. Cf: *molybdate.*

tungsten (tung'-sten) (a) An obsolete term formerly applied to tungsten minerals such as scheelite and wolframite. Etymol: tung sten (Swedish), "heavy stone". (b) A metallic element with atomic number 74. Syn: *wolfram.* (c) A naturally-occurring mineral reported from the Ural Mountains, but not approved by the International Mineralogical Association.

tungstenite-2H (tung'-sten-ite-2H) An metallic gray hexagonal mineral of the *molybdenite* group: WS_2. It occurs in small lead-gray folia or scales. The term is "sometimes erroneously given as a translation from German or Russian for wolframite" (Fleischer, 1966, p.1317).

tungstenite-3R A metallic gray rhombohedral mineral of the *molybdenite* group: WS_2.

tungstibite A green orthorhombic mineral: Sb_2WO_6.

tungstic ocher (tung'-stic) (a) *tungstite.* (b) *ferritungstite.* Syn: *wolfram ocher.*

tungstite (tung'-stite) An orthorhombic earthy mineral: $WO_3{\cdot}H_2O$. It occurs in yellow, yellow-orange, or yellowish-green pulverulent masses. Syn: *tungstic ocher; wolframine.*

tungusite (tun-gus'-ite) A pearly-green or greenish-white triclinic mineral: $Ca_{14}Fe^{2+}_9(Si_8O_{20})_3(OH)_{22}$.

tuning effect A consequence of the interference from two or more closely spaced reflectors. The composite wavelet exhibits amplitude and phase effects that depend on the time differences between the successive reflection events, the magnitude and polarity of their associated reflection coefficients, and on the shape of the embedded wavelet.

tuning-fork spicule (tun'-ing-fork) A triradiate sponge spicule in which two of the rays are subparallel and at approximately 180 degrees to the third ray. They are found in the class Calcarea.

tuning thickness A bed that is a quarter-wavelength in thickness, for which reflections from its upper and lower surfaces interfere constructively when the overlying and underlying lithologies have similar properties, often resulting in an exceptionally strong reflection.

tunisite (tu'-nis-ite) A colorless tetragonal mineral: $NaCa_2Al_4(CO_3)_4Cl(OH)_8$. Syn: *slavyanskite.*

tunnel [mining] (tun'-nel) Strictly speaking, a passage in a mine that is open to the surface at both ends. It is often used loosely as a synonym for *adit* or drift.

tunnel [paleont] A low slitlike opening representing a resorbed area at the base of septa in the central part of the test in many fusulinids, serving to facilitate communication between adjacent chambers.

tunnel [speleo] *natural tunnel.*

tunnel cave *natural tunnel.*

tunneldal (tun'-nel-dal) A syn. of *tunnel valley.* Etymol: Danish, "tunnel valley".

tunnel erosion *piping.*

tunneling (tun'-nel-ing) (a) The operation of excavating, driving, and lining tunnels. (b) A form of failure occurring in earth dams and embankments, in which a tunnel is created when cracks, developed in the structure under dry conditions, collapse internally when brought into contact with water. It starts at the wet face or upstream side and proceeds downstream. Cf: *piping.*

tunnel lake A glacial lake occupying a *tunnel valley.*

tunnel valley A shallow trench cut in drift and other loose material, or in bedrock, by a subglacial stream not loaded with coarse sediment. Cf: *ice-walled channel.* Syn: *tunneldal; Rinnental.*

tunnerite (tun'-ner-ite) A mineral that may be identical with *woodruffite.*

Tuorian (Tu-or'-i-an) European stage: Upper Cambrian (above Mayan, below Shidertinian).

tuperssuatsiaite (tu-pers-suat'-si-a-ite) A reddish-brown monoclinic mineral of the *palygorskite* group: $Na(Fe^{3+},Mn^{2+})_3Si_8O_{20}(OH)_2{\cdot}4H_2O$.

turanite (tu-ran'-ite) An olive-green mineral: $Cu_5(VO_4)_2(OH)_4$(?).

turbation (tur-ba'-tion) Churning, stirring, mixing, or other modifications of a sediment or soil by agents not determined. The term is generally preceded by a prefix denoting agent, if known; e.g. congeli- (frost) or bio- (organisms).

turbid (tur'-bid) Stirred up or disturbed, such as by sediment; not clear or translucent, being opaque with suspended matter, such as of a sediment-laden stream flowing into a lake; cloudy or muddy in physical appearance, such as of a feldspar containing minute inclusions. Cf: *roily.*

turbidimeter (tur-bi-dim'-e-ter) An instrument for measuring or comparing the turbidity of liquids in terms of the reduction in intensity of a light beam passing through the medium. See also: *transmissometer.*

turbidimetry (tur-bi-dim'-e-try) The measurement of the amount of suspended or slow-settling matter in a liquid; the measurement of the decrease in intensity of a light beam passed through a medium. Cf: *nephelometry.*

turbidite (tur'-bid-ite) A sediment or rock deposited from, or inferred to have been deposited from, a *turbidity current.* It is characterized by graded bedding, moderate sorting, and well-developed primary structures in the sequence noted in the *Bouma cycle.* For the relation of turbidites to tempestites and inundites, see Einsele and Seilacher (1982, p.334).

turbidite-hosted gold vein A *mineral-deposit type* characterized by gold-quartz veins, segregations, and sheeted zones in fractures, faults, and openings in anticlines, synclines, and along bedding planes in *turbidites* and associated poorly-sorted clastic sedimentary rocks (McMillan, 1996). Syn: *Bendigo-type vein.*

turbidity (tur-bid'-i-ty) (a) The state, condition, or quality of opaqueness or reduced clarity of a fluid, due to the presence of suspended matter. (b) A measure of the ability of suspended material to disturb or diminish the penetration of light through a fluid.

turbidity current A *density current* in water, air, or other fluid, caused by different amounts of matter in suspension, such as a dry-snow avalanche or a descending cloud of volcanic dust; specif. a bottom-flowing current laden with suspended sediment, moving swiftly (under the influence of gravity) down a subaqueous slope and spreading horizontally on the floor of the body of water, having been set and/or maintained in motion by locally churned- or stirred-up sediment that gives the water a density greater than that of the surrounding or overlying clear water. Such currents are known to occur in lakes, and are believed to have produced the submarine canyons notching the continental slope. They appear to originate in various ways, such as by storm waves, tsunamis, earthquake-induced sliding, tectonic movement, over-supply of sediment, and heavily charged rivers in spate with densities exceeding that of sea water. The term was introduced by Johnson (1939, p.27), and is applied to a current due to turbidity, not to one showing that property. See also: *turbidity flow; suspended current.* Cf: *tractive current.* Syn: *suspension current.*

turbidity fan A local fan-shaped area of turbid water at the mouth of a stream flowing into a lake or adjacent to an eroding bank of a lake (Veatch and Humphrys, 1966, p.334).

turbidity flow A tongue-like flow of dense, muddy water moving down a slope; the flow of a *turbidity current.*

turbidity size analysis A kind of particle-size analysis based upon the amount of material in turbulent suspension, the turbidity decreasing as the particles settle.

turbinate (tur'-bi-nate) (a) Said of a horn-shaped solitary *corallite* with sides expanding from apex at an angle of about 70 degrees. Cf: *trochoid; patellate.* (b) Shaped like a top; e.g. said of a spiral gastropod shell with a generally rounded base, a broadly conical spire, and whorls decreasing rapidly from base to apex, or said of a protist shaped like a cone with the point down. (c) Pertaining to trabeculate, fragile, coiled sheets of bone that develop as scrolls from the inner walls of the nasal cavity in mammals. These support a large surface of mucous membrane in the air passage, serve to condition the temperature and moisture content of breathed air, and provide the substrate for olfactory chemoreceptive cells.

turbine motor In rotary drilling, a motor located at the bottom of a *drillstring* that turns the bit. Its turbine is driven by the circulating drilling mud pumped down the drillstring. It is commonly used in deviation drilling. Syn: *mud motor.*

turbostratic Refers to the stacking of layers of smectite in which the rotation of one sheet in relation to others has no constraints, rotations are an unpredictable n°. The degree to which this disorder exists is directly related to the smectite content of mixed-layered, smectite containing minerals. Cf: *rotational stacking disorder.*

turbulence (tur'-bu-lence) *turbulent flow.*

turbulence spectrum *eddy spectrum.*

turbulent diffusion (tur'-bu-lent) *eddy diffusion.*

turbulent flow (a) Water flow in which the flow lines are confused and heterogeneously mixed. It is typical of flow in surface-water bodies. (b) *cavern flow.* Cf: *laminar flow [hydraul]; mixed flow.* Syn: *turbulence; tortuous flow.*

turbulent flux *eddy flux.*

turbulent suspension A suspension of particles in turbulent water. Turbulence can be maintained up to the point where the density of the water/sediment mixture becomes ca. 2.0 g/cm$_3$. Addition of further sediment inhibits turbulence, and additional movement, if any, is as a thick paste or debris flow (Friedman et al., 1992, p.42).

turbulent velocity That velocity of water in a stream above which the flow is turbulent, and below which it may be either laminar or turbulent. Cf: *laminar velocity.*

turf Peat that has been dried for use as a fuel.

turf-banked terrace A *nonsorted step* whose riser is covered by vegetation and whose tread is composed of fine soil. The term should be reserved for an irregular, terracelike feature that is not a clearly defined form of patterned ground (Washburn, 1956, p.835). Syn: *turf garland.*

turf garland *turf-banked terrace.*

turf hummock A hummock of turf rather than soil. Cf: *earth hummock.*

turgite (tur'-gite) A red fibrous mineral mixture: $Fe_2O_3{\cdot}nH_2O$. It is equivalent to hematite with adsorbed water or to an iron ore intermediate between hematite and goethite (hematite being in excess). It occurs as a ferruginous cement in sandstones. Syn: *hydrohematite.*

turjaite (tur'-ja-ite) A dark-colored plutonic melilitolite containing nepheline and 60 to 90 percent mafic minerals. The presence of melilite distinguishes turjaite from *melteigite.* The name, given by Ramsay in 1921, is derived from Turja, on the Kola Peninsula, Russia. Obsolete.

turjite (tur'-jite) An *alnoite* containing calcite, andradite, and analcime. The name, given by Belijankin in 1924, is for Turij Mis (Turja), Kola Peninsula, Russia. Obsolete.

turkestanite A green to brown tetragonal mineral of the *steacyite* group: $Th(Ca,Na)_2KSi_8O_{20}{\cdot}nH_2O$.

turkey-fat ore (tur'-key-fat) A popular name used in Arkanasas and Missouri for smithsonite colored yellow by greenockite. Also called: turkey ore.

turkey shoot A comparison of the results of recording with two or more sets of seismic instruments simultaneously under the same field conditions.

Turkey stone (Tur'-key) (a) A very fine-grained siliceous rock, containing up to 25% calcite, quarried in central Turkey and used as a whetstone; *novaculite.* (b) *turquoise.*

turlough (tur'-lough) An Irish term for a winter lake that is dry or marshy in summer. Also, the ground or hollow periodically flooded to form a turlough. Etymol: Gaelic "turloch", "dry lake".

turma (tur'-ma) An artificial suprageneric grouping of form genera of fossil palynomorphs (mostly pre-Cenozoic) based on morphology. It is subdivided into other groups such as "subturmae" and "infraturmae". The system is not governed by the International Code of Botanical Nomenclature. Pl: turmae. See also: *anteturma.*

turmaline (tur'-ma-line) Var. of *tourmaline.* The mineral name was originally spelled: "turmalin".

turmkarst (turm'-karst) The German equivalent of *tower karst.*

turneaureite (turn-eaure'-ite) A vitreous colorless hexagonal mineral of the *apatite* group: $Ca_5(AsO_4,PO_4)_3Cl$.

turning point (turn'-ing) (a) A surveying point on which a level rod is held, after a foresight has been made on it, and before the differential-leveling instrument is moved to another station so that a backsight may be made on it to determine the height of instrument after the resetting; a point of intersection between survey lines, such as the intervening point between two bench marks upon which rod readings are taken. It is established for the purpose of allowing the leveling instrument to be moved forward (alternately leapfrogging with the rod) along the line of survey without a break in the series of measured differences of elevation. Abbrev: TP. (b) A physical object representing a turning point, such as a steel pin or stake driven into the ground.

turnover [ecol] (turn'-o-ver) (a) The process by which some species become extinct in a region and are replaced by other species (MacArthur and Wilson, 1967, p.191). (b) The number of animal generations that replace each other during a given length of time (Thorson, 1957, p.491).

turnover [lake] The complete *circulation* of the water column in a lake, such as occurs in spring and fall in a *dimictic* lake, as the surface temperature attains its maximum-density level (Wetzel, 2001, p.74). Cf: *overturn.*

turnover pulse hypothesis The hypothesis advanced by Elisabeth Vrba of Yale University that most evolutionary turnover (speciation and extinction) has occurred during discrete events associated with changes in the physical environment.

Turolian (Tu-ro'-li-an) European land-mammal age: Late Miocene (above Vallesian, below Ruscinian). See also: *Pontian; Pannonian.*

Turonian (Tu-ro'-ni-an) A stage of the Standard Global Chronostratigraphic Scale: Upper Cretaceous (above Cenomanian, below Coniacian). The basal GSSP is in Colorado, USA.

turquoise (tur'-quoise) (a) A triclinic mineral: $CuAl_6(PO_4)_4(OH)_8{\cdot}4H_2O$. It is blue, blue-green, or yellowish green; when sky blue it is valued as the most important of the nontransparent gem materials. It usually occurs as reniform masses with a botryoidal surface, in the zone of alteration of aluminum-rich igneous rocks (such as trachytes). Syn: *Turkey stone; calaite.* Also called: turquois; henwoodite. (b) A group name for triclinic minerals with the analogous formula, but with Cu replaced by Zn, Ca, Fe^{2+} or a vacant structure site; and Al replaced by Fe^{3+}.

turrelite (tur'-re-lite) An asphaltic shale found in Texas.

turriculate (tur-ric'-u-late) Turreted, or furnished with or as if with turrets; esp. said of a gastropod shell (e.g. of *Turritella*) with an acutely or highly conical spire composed of numerous flat-sided whorls.

turtleback (tur'-tle-back) An extensive smooth curved topographic surface, apparently unique to the Death Valley (Calif.) region, that resembles the carapace of a turtle or a large, elongate dome with an amplitude up to a few thousand meters. Turtlebacks were first mapped, described, and named by Curry (1938).

turtle stone (tur'-tle) A large, flattened oval *septarium* released from its matrix and so weathered and eroded that the interior vein-filled system of cracks may be seen. Its form has a rough resemblance to that of a turtle and its polished surface bears a fancied resemblance to a turtle's back. Such concretions are abundant in the Devonian shales of eastern North America. Also spelled: turtlestone. Formerly called: *beetle stone.*

turtle-structure anticline Mounded strata between salt diapirs having a flat base and rounded crest over a local primary increase in sedimentary thickness; the anticline may or may not be cored by a low salt pillow. The turtle structure forms between diapirs whose flanks subside because of regional extension (see: *diapir fall*) or between salt structures evolving from pillows to diapirs (Trusheim, 1960).

turtmannite A bright yellow or orange rhombohedral mineral: $(Mn,Mg)_{25.5}O_5[(V,As)O_4]_3(SiO_4)_3(OH)_{20}$.

tuscanite (tus'-can-ite) A colorless monoclinic mineral: $K(Ca,Na)_6(Si,Al)_{10}O_{22}(SO_4,CO_3)_2(OH)\cdot H_2O$.

tusionite (tu'-sion-ite) A yellow-brown rhombohedral mineral occurring in lamellar intergrowths with tetrawickmanite: $Mn^{2+}Sn^{4+}(BO_3)_2$. It is the manganese analogue of nordenskiöldine.

tussock (tus'-sock) A dense tuft of grass or grasslike plants usually forming one of many firm hummocks in the midst of a marshy or boggy area.

tussock-birch-heath polygon A *vegetation polygon* characterized by the assemblage indicated (Hopkins and Sigafoos, 1951, p.52-53, 87-92); permafrost appears essential for its formation. Diameter: 2-5 m.

tussock ring A *nonsorted circle* consisting of tussocks surrounding a patch of bare soil.

Tuttle lamellae (Tut'-tle) Planes of inclusions in quartz, oriented randomly rather than with reference to the enclosing crystal. Cf: *Boehm lamellae.*

tutvetite (tut'-vet-ite) A light-reddish trachytoid rock composed chiefly of alkali feldspars, a decomposed mafic mineral (possibly aegirine), and accessory pyrite and possibly anatase and tremolite; an altered *trachyte.* Its name (Johannsen, 1938) is derived from Tutvet, Norway. Obsolete.

Tuvalian (Tu-val'-i-an) A European substage: upper substage of Carnian Stage, Upper Triassic (above Julian, below Lacian Substage of Norian).

tuxtlite (tuxt'-lite) *omphacite.*

tuya (tu'-ya) A synonym of *table mountain [volc].* Examples of tuyas occur in northern British Columbia and Iceland.

tuyère (tu-yere' [Pron: twee-air]) In metallurgy, the tip of a bellows nozzle, designed to withstand high furnace temperatures; used in smelting.

tuzlaite A white or colorless monoclinic mineral: $NaCaB_5O_8(OH)_2\cdot 3H_2O$.

tvalchrelidzeite (tval-chre-lid'-ze-ite) An adamantine reddish-black triclinic mineral: $Hg_3(Sb,As)_2S_3$.

tvedalite A white, pale gray, or beige monoclinic mineral: $(Ca,Mn)_4Be_3Si_6O_{17}(OH)_4\cdot 3H_2O$.

tveitåsite (tvei-tås'-ite) A dark-colored, medium- to fine-grained contact igneous rock composed chiefly of a clinopyroxene and alkali feldspar (orthoclase, cryptoperthite, microperthite, albite), with accessory titanite, apatite, pyroxene, and possibly nepheline and calcite. The rock is probably a hybrid (Johannsen, 1939, p.285). The name, given by Brögger in 1921, is from Tveitåsen, Fen complex, Norway. Cf: *fenite.* Obsolete.

tveitite-(Y) (tveit'-ite) A white or pale yellow rhombohedral mineral: $Ca_{14}Y_5F_{43}$.

T wave A short-period (0.5 s) acoustic wave in the sea.

tweddillite A vitreous deep red monoclinic mineral of the *epidote* group: $CaSr(Mn^{3+},Fe^{3+})_2Al(Si_2O_7)(SiO_4)O(OH)$.

twenty-degree discontinuity (twen'-ty-de-gree') The break in the traveltime curve of seismic *P waves*, originally defined as occurring at an angular distance of about 20° and at a depth of 413 km. Later studies determined the angular distance to be 15°. The term is also written as "20° discontinuity".

twig *divining rod.*

twilight zone (twi'-light) *disphotic zone.*

twin A rational intergrowth of two or more single crystals of the same mineral in a mathematically describable manner, so that some lattices are parallel whereas others are in reversed position. The symmetry of the two parts may be related by a plane, axis, or center not present in the class of the crystal. See also: *twinning.* Syn: *twin crystal; twinned crystal.*

twin axis The crystal axis about which one individual of a twin crystal may be rotated (usually 180°) to bring it into coincidence with the other individual. It cannot be coincident with the axes of twofold, fourfold, or sixfold symmetry. Cf: *twin plane; twin center.* Syn: *twinning axis.*

twin center The crystal point about which the individuals of a twin may be symmetrically arranged. Cf: *twin plane; twin axis.*

twin crystal *twin.*

twin gliding *Crystal gliding* that results in the formation of crystal twins.

twin law A definition of a twin relationship in a given mineral or mineral group, specifying the twin axis, center, or plane, defining the composition surface or plane if possible, and giving the type of twin.

twinned crystal *twin.*

twinning (twin'-ning) The development of a twin crystal by growth, transformation, or gliding.

twinning axis *twin axis.*

twinning by merohedry A *twin* that has one or more symmetry elements that are not present in the class of the crystal, but that are present in the symmetry of the *crystal lattice.* This results in the exact superposition of the reciprocal lattices of the twinned individuals, and can lead to the incorrect interpretation of X-ray data. Syn: *merohedral twinning.*

twinning displacement Displacement in a crystal due to twin gliding.

twinning plane *twin plane.*

twinnite (twin'-nite) A metallic black orthorhombic mineral: $Pb(Sb,As)_2S_4$.

twin plane The common plane across which the individual components of a crystal twin are symmetrically arranged or reflected. It is parallel to a possible crystal face but not to a plane of symmetry of a single crystal. It is usually identical with the *composition surface.* Cf: *twin axis; twin center.* Syn: *twinning plane.*

twin shell Shell of a spumellarian radiolarian with a median transverse constriction.

twist The geometry of en echelon fringe cracks that propagate from that portion of a parent joint subject to mode III crack tip displacement (Pollard et al., 1982). Cf: *tilt [struc geol].*

twisted plywood structure *Plywood structure* in which fibrillar directions rotate from one horizontal sheet to the next by more or less uniform acute angles (Giraud et al., 1978a,b; Giraud-Guille, 1988). Also called: twisted plywoodlike structure.

twist hackle The en echelon fringe cracks at the edge of a joint that form with the twist of a crack rupture as the crack enters a zone in which the stress field has rotated. Cf: *abrupt twist hackle; gradual twist hackle.*

two-and-a-half-dimensional modeling Potential field modeling where end corrections are applied to the field of a model that extends to infinity perpendicular to the line. Also written: 2 1/2-D, 2.5-D modeling.

two-and-a-half-dimensional seismic A pseudo-3-D survey made by interpolation between data from 2-D surveys. Also written: 2 1/2-D, 2.5-D seismic.

two-circle goniometer (two-cir'-cle) A *goniometer* that measures the azimuth and polar angles for the pole of each face on a crystal by reflection of a parallel beam of light from a cross slit. The two circle angles are readily plotted in stereographic or gnomonic projection for indexing in any crystal system. Cf: *contact goniometer; reflection goniometer.*

two-cycle valley A valley produced by rejuvenation, as by headward erosion, and characterized by a *valley-in-valley* cross profile. Syn: *two-story valley.*

two-dimensional method (two-di-men'-sion-al) A simplified method for calculating the effect on gravity of geological structures in section, in which these structures are assumed to be infinitely long at right angles to the section. Abbrev: 2-D. See also: *survey (2-D).* Syn: *profile method.*

twofold axis An axis of rotational symmetry by which an object (crystal shape, *crystal structure,* etc.) is invariant to a rotation of 180°. Syn: *diad.*

twoling (two'-ling) A crystal twin consisting of two individuals. Cf: *trilling; fourling; fiveling; eightling.*

two-M (2M) polytype A 2M polytype is a crystal structure modification with overall monoclinic symmetry and a repeat of two layers along the stacking direction. There are two possible forms: 2M1 and 2M2 (Bailey, 1980a). The use of the 2M designation should be avoided if the precise form can be determined. The 2M1 polytype is a common form in the dioctahedral mica minerals but rare in the serpentine minerals. The 2M2 polytype is rare. Cf: *one-M (1M) polytype.*

two-mica granite A granite containing both dark mica (biotite) and light mica (muscovite). Cf: *aplogranite.*

two-phase inclusion A cavity (usually angular) in a gemstone consisting of any two states of matter (usually a gas bubble in a liquid); the cavity may or may not coincide with a possible crystal form of the host mineral. Examples occur in corundum and in many other gem minerals. Cf: *three-phase inclusion.*

two-point problem A problem in planetable surveying of determining the position of a point with the known factor being the length of one line that does not include the point to be located. Cf: *three-point problem.*

two-story valley *two-cycle valley.*

two-sweep cusp A *meander cusp* formed by the sweep of two successive meanders migrating downstream while the stream remains on the same side of the flood plain (Lobeck, 1939, p.241).

two-swing cusp A *meander cusp* formed by two successive swings of a meander belt, the scar produced by the first swing intersecting the scar of the second swing in such a way that a Y-shaped feature results, with the handle of the Y pointing either upstream or downstream (Lobeck, 1939, p.240-241).

two-to-one (2:1) layer A structural component of hydrous silicates composed of two continuous tetrahedral sheets of composition T_2O_5 (T commonly Si, Al, or Fe^{3+}) and one octahedral sheet where the octahedral sheet is composed of (a) medium-sized cations (e.g., Mg, Al, Fe^{2+}, Fe^{3+}, Li) or vacancies and (b) anions of O, and OH, and/or F (Bailey, 1980a). The ratio of tetrahedral to octahedral sheets is 2:1; the tetrahedral sheets oppose each other, and the octahedral sheet is sandwiched between them. Common mineral with 2:1 layers include micas, chlorites, smectites, pyrophyllite, and talc. In the older literature, the 2:1 layer was referred to as "three-layer clay," and this usage is no longer acceptable. Cf: *octahedral sheet; tetrahedral sheet; two-to-one (2:1) phyllosilicate.*

two-to-one (2:1) phyllosilicate A hydrous silicate composed of 2:1 layers. Different mineral structures are formed by the stacking of these layers, either without additional structural or chemical components (as in talc, pyrophyllite) or with cations (as in the mica minerals), cationic groups or polar molecules (as in the smectite minerals), and interlayer octahedral sheets (as in the chlorite minerals), etc., in the space between the 2:1 layers (the interlayer space) (Bailey, 1980a). Cf: *two-to-one (2:1) layer.*

two-year ice *second-year ice.*
tychite (tych'-ite) A white cubic mineral of the *northupite* group: $Na_6Mg_2(SO_4)(CO_3)_4$.
tychopotamic (ty'-cho-po-tam'-ic) Said of an aquatic organism adapted to living chiefly in still, fresh water. Cf: *autopotamic; eupotamic.*
tying bar (ty'-ing) *tombolo.*
tylaster (tyl'-as-ter) A small tylote aster (sponge spicule).
Tyler standard grade scale (Ty'-ler) A *grade scale* for the particle-size classification of sediments and soils, devised by the W.S. Tyler Company of Cleveland, Ohio; it is based on the square root of 2, with the midpoint values of each size class being simple whole numbers or common fractions. It is used as specifications for sieve mesh.
tylosis (ty-lo'-sis) A proliferation of a wood parenchyma cell through a pit-pair into the lumen of an adjacent inactive tracheary element, where it may or may not divide (Record, 1934, p.68) Cf: *tylosoid.*
tylosoid (ty'-lo-soid) A tylosislike intrusion of a parenchyma cell into an intercellular space. It differs from a *tylosis* in that it does not pass through the cavity of a pit (Record, 1934, p.70).
tylostyle (ty'-lo-style) A style (sponge spicule) in which the blunt end is swollen or knobbed.
tylote (ty'-lote) n. A slender, elongate sponge spicule (monaxon) with a knob at each end. adj. Said of a sponge spicule with the ends of the rays knobbed or swollen.
tylotoxea (ty-lot-ox'-e-a) A rodlike sponge spicule tapering to a sharp point at one end and to a knob at the other.
tympanoid (tym-pan'-oid) Said of a squat, drum-shaped scleractinian *corallite.*
Tyndall figure (Tyn'-dall) A water- and vapor-filled cavity in the shape of a snowflake, oriented parallel to the basal plane within an ice crystal, and formed through melting by radiation absorbed at points of defect in the ice lattice. It is called a "vapor figure" if the water in the cavity is refrozen. Also called: Tyndall flower; Tyndall star. Named after John Tyndall (1820-1893), British physicist. Syn: *negative snowflake.*
type [coal] A *coal classification* based on the constituent plant materials. Cf: *rank [coal]; grade [coal].*
type [taxon] The standard reference for determining the application of a scientific name. Unless otherwise qualified, a type usually refers to the *holotype* of a species. The type of a species is a single specimen and the species includes all organisms regarded as conspecific with the *type specimen.* The nomenclatural type of a genus is a species and the genus includes all species regarded as congeneric with its species type.
type-1 sequence In sequence stratigraphy, a sequence composed of lowstand, transgressive, and highstand systems tracts bounded beneath by a type-1 sequence boundary. Depending upon where the sequence is observed in the basin, not all of the systems tracts may be present due to truncation or nondeposition. A type-1 sequence boundary is interpreted to form when the rate of eustatic fall exceeds the rate of subsidence at the depositional-shoreline break, producing a relative fall in sea level at the position (Van Wagoner et al., 1987, 1988, 1990; Posamentier and Vail, 1988). This commonly results in major fluvial incision producing incised valleys in the updip end of the lowstand wedge and a basinward shift in facies associated with a type-1 sequence boundary.
type-2 sequence A sequence composed of shelf-margin, transgressive, and highstand systems tracts bounded by a type-2 unconformity and the correlative conformity. A type-2 sequence boundary is interpreted to form when the rate of eustatic fall is slightly less than or equal to the rate of basin subsidence at the existing depositional-shoreline break at the time of the eustatic fall (Jervey, 1988; Posamentier et al., 1992). This means that there is no relative fall in sea level at the depositional-shoreline break for the type-2 sequence boundary (Van Wagoner et al., 1987, 1988, 1990; Posamentier and Vail, 1988). Fluvial systems do not incise and there is no basinward shift in facies associated with a type-2 sequence boundary, although there is a significant decrease in accommodation in the coastal plain.
type-A climbing ripple A *climbing ripple* displaying internal *cross-lamination* and composed of laminae deposited on the downstream-dipping *lee* side of a migrating *ripple* but lacking preservation of *stoss*-side laminae (Jopling and Walker 1968). Cf: *type-B climbing ripple; type-C climbing ripple; sinusoidal climbing ripple.*
type area The geographic area or region that encompass the stratotype or type locality of a stratigraphic unit or stratigraphic boundary (ISSC, 1994, p.27). Syn: *type region.*
type-B climbing ripple A *climbing ripple* displaying internal *cross-lamination* and composed of laterally continuous laminae deposited both on the upstream-dipping *stoss* and downstream-dipping *lee* sides of a migrating *ripple* (Jopling and Walker 1968). This structure requires a higher ratio of aggradation rate to migration rate than type-A climbing ripple. Cf: *type-A climbing ripple; type-C climbing ripple; sinusoidal climbing ripple.*
type-C climbing ripple A *climbing ripple* displaying internal *cross-lamination* and composed of laterally continuous laminae deposited both on the upstream-dipping *stoss* and downstream-dipping *lee* sides of a migrating *ripple.* These are distinguished from type-B climbing ripples by the marked grading within *cosets* from fine sand and silt at the base to mud at the top and a gradual reduction in ripple amplitude upward, sometimes passing into *horizontal lamination* (Jopling and Walker 1968). This structure requires a higher ratio of aggradation rate to migration rate than type-A climbing ripple. Cf: *type-A climbing ripple; type-B climbing ripple; sinusoidal climbing ripple.*
type concept A principle for stabilizing the application of scientific nomenclature by recognizing a permanent association of a *taxon* with one of its constituent elements, designated as its nomenclatural type, which serves as a point of reference. The nomenclatural type is not necessarily the most typical or representative of a taxon; it is that element with which the name is permanently associated (ICBN, 1972, p.18).
type curve *master curve.*
type fossil A term occasionally used as a syn. of *index fossil.*
type genus The *genus* that serves as a permanent nomenclatural reference for application of the name of a family, and the ranks of super- and sub-taxa that have the same common point of nomenclatural reference.
type I error In statistical testing, the probability of rejecting a true *null hypothesis.* The probability is the *significance* of the test and is designated α.
type II error In statistical testing, the probability of accepting, or failing to reject, a false *null hypothesis.* This probability is designated β. Its complement $(1-\beta)$ is the probability of not committing a type II error, and is the *power* of the test.
type locality (a) The specific geographic locality in which the unit-stratotype of a layered stratigraphic unit or the boundary-stratotype of a boundary between layered units is situated; or lacking a designated stratotype, the locality where the unit or boundary were originally defined or named. In the case of stratigraphic units composed of nonlayered igneous or metamorphic rock bodies, the type locality is the specific geographic locality where the unit was originally defined and named (ISSC, 1994, p.26-27). (b) The place where a geologic feature (such as an ore occurrence, a particular kind of igneous rock, or the type specimen of a fossil species or subspecies) was first recognized and described.
type material All the specimens upon which the description of a new species is based. Cf: *hypodigm.*
type region *type area.*
type section (a) The originally described sequence of strata that constitute a stratigraphic unit. It serves as an objective standard with which spatially separated parts of the unit may be compared, and it is preferably in an area where the unit shows maximum thickness and is completely exposed (or at least shows top and bottom). Type sections for lithostratigraphic units can never be changed (ACSN, 1961, art.13h): there is only one type section, although there may be more than one typical section. Cf: *reference section.* (b) According to the International Stratigraphic Guide (ISSC, 1994), a syn. of *stratotype,* thus constituting not only the type representative of a stratigraphic unit but also that of a stratigraphic boundary or horizon.
type species That species on which the original description of a genus or subgenus is largely or entirely based; the type of a genus or subgenus. Syn: *genotype.*
type specimen The single specimen on which the original description of a particular species is based, which serves as a permanent point of nomenclatural reference for application of the name of that species. The type specimen may be a *holotype,* a *neotype,* or a *lectotype.*
typhoon (ty-phoon') A *tropical cyclone,* esp. in the western Pacific. Etymol: Chinese, "great wind".
typomorphic mineral (ty-po-mor'-phic) A mineral that is typically developed in only a narrow range of temperature and pressure. The term was originated by Becke. Cf: *critical mineral [meta]; index mineral.*
tyretskite-1A (ty-ret'-skite) A colorless, white, yellow, or pale brown triclinic mineral: $Ca_2B_5O_9(OH,Cl){\cdot}H_2O$. Cf: *strontiohilgardite.*
tyrolite (tyr-o'-lite) A green or blue orthorhombic mineral: $Ca_2Cu_9(AsO_4)_4(OH)_{10}{\cdot}10H_2O$. Syn: *trichalcite.* It is a dimorph of clinotyrolite.
tyrrellite (tyr-rell'-ite) A metallic bronze cubic selenospinel mineral of the *linnaeite* group: $(Cu,Co,Ni)_3Se_4$.
Tyrrhenian (Tyr-rhe'-ni-an) A European stage (Mediterranean): upper Pleistocene (above Ionian, below Versilian); base at oxygen stage 5e.
tysonite (ty'-son-ite) *fluocerite-(Ce).*
tyuyamunite (tyu-ya-mun'-ite) A yellow or greenish-yellow orthorhombic mineral of the *carnotite* group: $Ca(UO_2)_2(V_2O_8){\cdot}5\text{-}8H_2O$. It is an ore of uranium and occurs in yellow incrustations as a secondary mineral. Syn: *calciocarnotite.*

U

ubac A mountain slope so oriented as to receive the minimum available amount of light and warmth from the Sun; esp. a northward-facing slope of the Alps. Etymol: French dialect, "shady side". Cf: *adret.*
ubish bodies (u'-bish) *orbicule [palyn].*
uchucchacuaite (uch-uc-cha'-cua-ite) A metallic gray orthorhombic mineral of the *andorite* series: $AgPb_3MnSb_5S_{12}$.
Udalfs Alfisols that have a udic soil moisture regime and mesic or warmer soil temperature regimes. Udalfs generally have brownish colors throughout, and are not saturated with water for periods long enough to limit their use for most crops. A suborder in the U.S. system of soil taxonomy.
Udands Andisols that have a udic soil moisture regime. A suborder in the U.S. system of soil taxonomy.
Udden grade scale (Ud'-den) A logarithmic *grade scale* devised by Johan A. Udden (1859-1932), U.S. geologist; it uses 1 mm as the reference point and progresses by the fixed ratio of 1/2 in the direction of decreasing size and of 2 in the direction of increasing size, such as 0.25, 0.5, 1, 2, 4 (Udden, 1898). See also: *Wentworth grade scale.*
Uderts Vertisols of relatively humid regions that have wide, deep cracks that usually remain open continuously for < 60 days or intermittently for periods that total < 90 days. A suborder in the U.S. system of soil taxonomy.
udic A soil moisture regime that is neither dry for as long as 90 cumulative days nor for as long as 60 consecutive days in the 90 days following the summer solstice at periods when the soil temperature at 50 cm below the surface is above 5°C.
Udolls Mollisols that have a udic soil moisture regime with mean annual soil temperatures of 8°C or more. Udolls have no calcic or gypsic horizon, and are not saturated with water for periods long enough to limit their use for most crops. A suborder in the U.S. system of soil taxonomy.
Udox Oxisols that have a udic soil moisture regime. A suborder in the U.S. system of soil taxonomy.
Udults Ultisols that have low or moderate amounts of organic carbon, reddish or yellowish argillic horizons, and a udic soil moisture regime. Udults are not saturated with water for periods long enough to limit their use for most crops. A suborder in the U.S. system of soil taxonomy.
ufertite (u'-fer-tite) *davidite-(La).*
Ufimian (U-fim'-i-an) A Russian stage, lower Middle Permian.
ugandite (u-gan'-dite) An extrusive rock containing leucite, clinopyroxene, and abundant olivine, in a sodium-rich glassy groundmass; *olivine leucitite.* The name, given by Holmes and Harwood in 1937, is derived from Uganda. Obsolete.
ugrandite (u-gran'-dite) A group name for the calcium-bearing garnet minerals uvarovite, grossular, goldmanite, and andradite.
uhligite (uh'-lig-ite) (a) An ill-defined metallic black cubic mineral: $Ca_3(Ti,Al,Zr)_9O_{20}$(?). (b) An amorphous variscite or fischerite.
UHP metamorphism *ultra-high-pressure metamorphism.*
uintahite (u-in'-tah-ite) A black, shiny asphaltite, with a brown streak and conchoidal fracture, which is soluble in turpentine; it occurs primarily in veins in the Uinta Basin, Utah. See also: *wurtzilite.* Also spelled: *uintaite.* Syn: *gilsonite.*
uintaite (u-in'-ta-ite) *uintahite.*
Uintan (U-in'-tan) North American land mammal age: Late Eocene (after Bridgerian, before Duchesnean).
uklonskovite (u-klon'-skov-ite) A colorless monoclinic mineral: $NaMg(SO_4)F{\cdot}2H_2O$.
Ulatisian (U-la-tis'-ian) North American stage: Middle Eocene (above Penutian, below Narizian).
ulcerate (ul'-cer-ate) Having the characteristics of an *ulcus.*
ulcus (ul'-cus) A rounded, more or less porelike *ectoaperture* situated at a pole of a pollen grain. Adj: ulcerate.
ulexite (u-lex'-ite) A white triclinic mineral: $NaCaB_5O_6(OH)_6{\cdot}5H_2O$. It forms rounded reniform masses of extremely fine acicular crystals and is usually associated with borax in saline crusts on alkali flats in arid regions. Syn: *boronatrocalcite; natroborocalcite; cotton ball.*
uliginous (u-lin'-gi-nous) Said of an organism living in wet or swampy ground.
ullmannite (ull'-mann-ite) A metallic steel-gray to black cubic mineral of the *gersdorffite* group: NiSbS. It usually contains a little arsenic. Syn: *nickel-antimony glance.*
ulmain (ul'-main) A kind of *euvitrain* that consists completely of ulmin but that is not precipitated from solution. Cf: *collain.*
ulmic acid (ul'-mic) *ulmin.*
ulmification (ul'-mi-fi-ca'-tion) The process of peat formation. See also: *ulmin.*
ulmin (ul'-min) Vegetable-degradation material occurring in coal as an amorphous, brown to black substance or gel and that is insoluble in alkaline solution. It is abundant in peat and lignite, and forms vitrinite. Syn: *ulmic acid; humin; carbohumin; humogelite; fundamental jelly; jelly; gélose; fundamental substance; vegetable jelly.*
ulminite (ul'-min-ite) (a) A maceral of brown coal within the *huminite* group, consisting of gelified plant-cell walls (ICCP, 1971). (b) A variety of euvitrinite characteristic of ulmain and consisting of gelified but not precipitated plant material. Cf: *collinite.*
ulrichite (ul'-rich-ite) An apple-green monoclinic mineral: $CaCu(UO_2)(PO_4)_2{\cdot}4H_2O$.
ulrichite [mineral] (ul'-rich-ite) A syn. of *uraninite;* specif. the original unoxidized UO_2.
ulrichite [rock] A dark-colored hypabyssal rock composed of large phenocrysts of nepheline, alkali feldspar, sodic pyroxene, and amphibole, and smaller accessory olivine phenocrysts in a groundmass of feldspar, analcime, pyroxene, and amphibole; an olivine-bearing *phonolite.* The name, given by Marshall in 1906, is in honor of G. H. F. Ulrich, discoverer of alkaline rocks at Dunedin, New Zealand. Obsolete.
ULSEL *ultra-long-spaced electric log.*
Ulsterian (Ul-ste'-ri-an) North American provincial series: Lowermost Devonian (above Cayugan of Silurian, below Erian).
ultimate analysis (ul'-ti-mate) The determination of the elements in a compound; for coal, the determination of carbon, hydrogen, sulfur, nitrogen, ash, and oxygen. Cf: *proximate analysis.*
ultimate base level The lowest possible *base level;* for a stream, it is sea level, projected inland as an imaginary surface beneath the stream. Cf: *temporary base level.* Syn: *general base level.*
ultimate bearing capacity The average load per unit of area required to produce shear failure by rupture of a supporting soil mass. See also: *bearing capacity.*
ultimate landform The theoretical landform produced near the end of a cycle of erosion. Cf: *initial landform; sequential landform.*
ultimate recovery The quantity of oil or gas that a well, pool, field, or property will produce. It is the total obtained or to be obtained from the beginning of production to final abandonment.
ultimate shear strength The maximum shearing stress (i.e., half the differential stress) corresponding with the *ultimate strength.*
ultimate strength The maximum differential stress that a material can sustain under the conditions of deformation. Beyond this point, rock *failure* occurs. See also: *ultimate shear strength.*
Ultisols Mineral soils that have an argillic horizon with a base saturation of < 35% when measured at pH 8.2. Ultisols have a mean annual soil temperature of 8°C or higher. An order in the U.S. system of soil taxonomy.
ultrabasic (ul-tra-ba'-sic) Said of an igneous rock having a silica content lower than that of a basic rock. Percentage limitations are arbitrary and vary with different petrologists, although the upper limit was originally set at 44%. The term is frequently used interchangeably with *ultramafic.* Although most ultrabasic rocks are also ultramafic, there are some exceptions; e.g. monomineralic rocks composed of pyroxenes are ultramafic but are not ultrabasic because of their high SiO_2 content. A monomineralic rock composed of anorthite would be considered ultrabasic (SiO_2 = 43.2 percent) but not ultramafic. "Ultrabasic" is one subdivision of a widely used system for classifying igneous rocks on the basis of silica content; the other subdivisions are *acidic, basic,* and *intermediate.* Cf: *silicic.*
ultrabasic rock In the *IUGS classification,* any igneous rock with wt % SiO_2 less than 45.
ultrabasite (ul-tra-ba'-site) *diaphorite.*

ultra-Darwinism The proposition that processes at the level of the gene are the only important evolutionary processes.
ultra-high-pressure metamorphism The process that produces metamorphic rocks that contain relic grains of coesite or microdiamond, indicating formation at depths as great as 100 km. Such rocks are commonly found in mountains formed in continent-continent collision zones.
ultra-long-spaced electric log A *resistivity log* curve derived from 2-electrode spacing of as much as 1000 feet. See also: *normal log*. Abbrev: ULSEL.
ultramafic (ul-tra-maf'-ic) Said of an igneous rock composed chiefly of mafic minerals, e.g. monomineralic rocks composed of hypersthene, augite, or olivine. Cf: *ultrabasic.*
ultramafic-hosted asbestos *serpentine-hosted asbestos.*
ultramafic-hosted magnesite *serpentine-hosted magnesite.*
ultramafic-hosted talc *serpentine-hosted talc.*
ultramafic rock In the *IUGS classification,* a general name for plutonic rock with color index M greater than or equal to 90, including, among others, dunite, peridotite, and pyroxenite.
ultramafitite In the *IUGS classification,* a general term for volcanic rocks containing more than 90% mafic minerals. This name should be qualified by the names of the most abundant minerals. Obsolete.
ultramafitolite In the *IUGS classification,* a general term for plutonic rocks containing more than 90% mafic minerals. Includes *dunite, pyroxenite, peridotite,* etc. Obsolete.
ultramarine (ul'-tra-ma-rine') A syn. of *lazurite.* The term is also applied to artificial lazurite and to compounds allied to it; e.g. the brilliant blue pigment formerly made by powdering lapis lazuli and characterized by the durability of its color.
ultrametamorphism (ul'-tra-met'-a-mor'-phism) Metamorphic processes at the extreme upper range of temperatures and pressures, at which partial to complete fusion of the affected rocks (*anatexis*) takes place and magma is produced. The term was originated by Holmquist in 1909.
ultramicroearthquake (ul'-tra-mi'-cro-earth'-quake) An earthquake having a negative body-wave magnitude on the Richter scale. Such a limit is arbitrary, and may vary according to the user. Cf: *microearthquake; major earthquake.*
ultramylonite (ul-tra-my'-lo-nite) An ultra-crushed variety of *mylonite,* in which primary structures and porphyroclasts have been obliterated so that the rock becomes homogeneous and dense, with little if any parallel structure (Quensel, 1916). Cf: *protomylonite; pseudotachylyte.* Syn: *flinty crush rock.*
ultraplankton (ul-tra-plank'-ton) The smallest plankton organisms, less than a few µm in diameter. Cf: *nannoplankton; microplankton; macroplankton; megaplankton.* Syn: *picoplankton.*
ultraplinian eruption A highly energetic eruption in which the area of pumice and ash fall exceeds 50,000 km^2.
ultrapotassic rock An igneous rock in which the wt% of K_2O and MgO each exceed 3, and the wt% of K_2O is at least twice that of Na_2O (Foley et al., 1987). Cf: *perpotassic rock; potassic rock.*
ultrasima (ul-tra-si'-ma) The supposedly ultrabasic layer of the Earth below the *sima,* immediately below the *Mohorovicic discontinuity.* Archaic.
ultrasonic methods Methods involving sonic energy above 20 kHz.
ultrastructure (ul'-tra-struc'-ture) The internal structure and character of plant and animal tissues, esp. skeletal tissues, as revealed by the electron microscope. Cf: *microstructure.*
ultraviolet absorption spectroscopy (ul-tra-vi'-o-let) The observation of an *absorption spectrum* in the ultraviolet frequency region and all processes of recording and measuring that go with it.
ultravulcanian eruption (ul'-tra-vul-ca'-ni-an) A type of volcanic eruption characterized by violent, gaseous explosions of lithic dust and blocks, with little if any incandescent scoria. It is commonly observed during the opening or reopening of a volcanic vent. Its type occurrence is the explosion of Krakatoa in 1883.
ulvite (ul'-vite) *ulvöspinel.*
ulvöspinel (ul'-vö-spi-nel') A metallic cubic mineral of the *spinel* group: $TiFe^{2+}{}_2O_4$. It usually occurs as fine exsolution lamellae, intergrown with magnetite. Syn: *ulvite.*
umangite (u-mang'-ite) A metallic dark red or violet-blue tetragonal mineral: Cu_3Se_2.
umbel (um'-bel) (a) An umbrellalike structure consisting of multiple recurved teeth attached to the tip of a ray or the pseudoactine of a sponge *spicule* (such as of an amphidisc). (b) A sponge spicule consisting of a single shaft with an umbel at one end; e.g. a paraclavule or one type of a clavule.
umbelliferous (um-bel-lif'-er-ous) (a) Said of a tabulate corallum having corallites arranged like ribs of an umbrella, growing outward in whorls. (b) Producing umbels.
umber (um'-ber) A naturally occurring brown earth that is darker than ocher and sienna and that consists of manganese oxides as well as hydrated ferric oxide, silica, alumina, and lime. It is highly valued as a permanent paint pigment, and is used either in the greenish-brown natural state ("raw umber") or in the dark-brown or reddish-brown calcined state ("burnt umber"). Important in "Cyprus-type" massive sulfide deposits as a volcanic-associated marine sediment found with radiolarian cherts and clastic sediments filling irregularities on the surface of submarine lava flows. Often confused with *ocher* or *ochre.*
umbilical area (um-bil'-i-cal) The inner part or surface of a whorl of a cephalopod conch, between the umbilical shoulder and the umbilical seam. It is called an "umbilical wall" if it rises steeply from the spiral plane and "umbilical slope" if it rises gently (TIP, 1957, pt. L, p.6).
umbilical depression A closed, shallow, axial depressed area formed by the curvature of overlapping chamber walls in a foraminiferal involute spire; may be filled by thickenings or knobs.
umbilical lobe The large primary lobe of a *suture* of an ammonoid, centered on or near the *umbilical seam,* and forming part of both external and internal sutures.
umbilical perforation The vacant space or opening around the axis of coiling of a cephalopod shell, connecting the umbilici on opposite sides.
umbilical plug (a) The deposit of secondary skeletal or shell material in the axis or umbilical region of certain coiled foraminiferal tests (e.g. in *Rotalia*). (b) The calcareous deposit filling the *umbilicus* of a cephalopod. Syn: *plug [paleont].*
umbilical seam The helical line of junction or overlap of adjacent whorls of a coiled cephalopod conch. Syn: *umbilical suture.*
umbilical shoulder (a) The part of a cephalopod shell bordering the *umbilicus* and forming its outer margin; e.g. the strongly bent part of a whorl of a nautiloid shell between the flank and the inner part of the umbilical area. (b) The angulation of whorls at the margin of and within the umbilicus of a gastropod shell (Moore et al., 1952, p.289). (c) The part of a foraminiferal test bordering the umbilicus. See also: *shoulder [paleont].*
umbilical side The involute side in foraminiferal trochospiral forms, with only the chambers of the final whorl visible around the umbilicus. Also called ventral side; commonly with aperture.
umbilical suture (a) A continuous line separating successive whorls as seen in the *umbilicus* of *phaneromphalous* gastropod shells. (b) An *umbilical seam* of a cephalopod.
umbilical tooth One of the projections forming a triangular modification of the apertural lip of a foraminiferal test, with those of successive chambers in forms (e.g. *Globoquadrina*) with umbilical aperture giving a characteristic serrate border to the umbilicus.
umbilicus (um-bil'-i-cus) (a) A cavity or depression in the center of the base of a spiral shell of a univalve mollusk; e.g. the conical opening formed around the central axis of a spiral gastropod shell between faces of adaxial walls of whorls where these do not coalesce, or an external depression centered around the axis of coiling of a cephalopod shell and formed by the diminishing width of whorls toward the axis. (b) A circular depression or pit in the axis of a coiled foraminiferal test; e.g. the closed shallow depressed area formed by curvature of overlapping chamber walls in involute forms, or the space formed between inner margins of the walls of chambers belonging to the same whorl of the test. (c) In mammals, the navel, the only adult remnant of the umbilical cord. Pl: umbilici.
umbite (umb'-ite) A vitreous colorless or yellowish orthorhombic mineral: $K_2ZrSi_3O_9{\cdot}H_2O$. It is an inosilicate with wollastonite-like tetrahedral chains. Cf: *kostylevite; paraumbite.*
umbo (um'-bo) (a) The "humped" part of the shell of a bivalve mollusk, or the elevated and relatively convex part of a valve surrounding the point of maximum curvature of the longitudinal dorsal profile and extending to the beak when not coinciding with it. The term is sometimes used incorrectly as a synonym for *beak,* but with most shells two distinct terms are needed. (b) The relatively convex, apical part of either valve of a brachiopod, just anterior to or containing the beak. (c) The apical part of either valve of the bivalved carapace of a crustacean; e.g. the point on the plate from which successive growth increments extend in a cirripede. (d) A prominence on the *frontal shield* in some cheilostome bryozoans, proximal to the orifice or on the ovicell. (e) A central round elevated structure in discoidal foraminiferal tests. It is commonly due to lamellar thickening and may occur on one or both sides of the test. (f) A central projection on the thecal plate of an echinoderm, representing part of its ornamentation. Pl: umbones or umbos. Syn: *umbone.*
umbonal angle (um'-bo-nal) (a) The approximate angle of divergence of the posterior/dorsal and anterior/dorsal parts of the longitudinal profile of bivalve-mollusk shells; specif. the angle of divergence of umbonal folds in pectinoid shells. (b) The angle subtended at the *umbo* of a brachiopod by the region of the shell surface adjacent to the umbo.
umbonal chamber One of a pair of posteriorly and laterally located cavities in either valve of a brachiopod, bounded in the *pedicle valve* by dental plates and shell walls and limited medially in the *brachial valve* by crural plates (or homologues) and shell walls.
umbonal fold The ridge originating at the *umbo* of a pectinoid pelecypod shell and setting off the body of the shell from the auricle.
umbonal muscle A single muscle occurring in some lingulid brachiopods, thought to be homologous with the posterior adductor muscles, and consisting of two bundles of fibers, posteriorly and slightly asymmetrically placed (TIP, 1965, pt. H, p.154).
umbonate (um'-bo-nate) Having or forming an *umbo;* e.g. having an umbo on one or both sides of an enrolled foraminiferal test. Also, said of a foraminifer bearing a convex elevation in the center.
umbone (um'-bone) A syn. of *umbo* [paleont]. The term "umbones" is the usual plural for "umbo".

umbonuloid (um-bon′-u-loid) adj. Pertaining to an ascophoran cheilostome (bryozoan) characterized by a *frontal shield* formed by calcification of the inner wall of a double-walled fold overarching the frontal membrane; esp. said of a frontal shield formed in that way. n. An ascophoran cheilostome having such a structure.

umbozerite (um-bo′-zer-ite) An amorphous (metamict?) green or greenish-brown mineral: $Na_3Sr_4ThSi_8(O,OH)_{24}$.

umbra (um′-bra) (a) The completely shadowed region of an eclipse. (b) The inner, darker region of a sunspot. Cf: *penumbra.*

umbracer dune (um′-bra-cer′) A *lee dune* tapering to a point downwind, formed under constant wind direction commonly behind a clump of bushes or a prominent bedrock obstacle (Melton, 1940, p.120). See also: *wind-shadow dune.*

umbrafon dune (um′-bra-fon′) A *lee dune* developed to the leeward of a source or area of loose sand where the sand supply is constantly replenished (Melton, 1940, p.122); e.g. a dune on the lee side of a stream flood plain or landward from a sandy beach. Syn: *lee-source dune; source-bordering lee dune.*

Umbrepts Inceptisols formed in cold or temperate climates that commonly have an umbric epipedon, but they may have a mollic or an anthropic epipedon 25 cm or more thick under certain conditions. These soils are not dominated by amorphous materials and are not saturated with water for periods long enough to limit their use for most crops. A suborder in the U.S. system of soil taxonomy.

umbric epipedon (um′-bric) A diagnostic surface horizon that is similar to a *mollic epipedon* except for having a base saturation of less than 50%, measured at a pH of 7 (USDA, 1975). Cf: *ochric epipedon.*

umohoite (u-mo′-ho-ite) A bluish-black to dark green triclinic or orthorhombic mineral: $(UO_2)Mo^{6+}O_4{\cdot}2H_2O$.

umptekite (ump′-tek-ite) A *syenite* composed chiefly of microperthite and sodic amphibole, with accessory titanite, apatite, and opaque oxides, and occasionally small amounts of interstitial nepheline; a sodic syenite resembling *pulaskite.* Its name, given by Ramsay in 1884, is derived from Umptek (now Khibine), Kola Peninsula, Russia. Obsolete; "the name umptekite seems quite unnecessary since the composition of pulaskite is very similar" (Johannsen, 1938, p.9).

unaka (u-na′-ka) (a) A term proposed by Hayes (1899, p.22) for a large residual mass rising above a peneplain that is less advanced than one having a *monadnock,* and sometimes displaying on its surface the remnants of a peneplain older than the one above which it rises; an erosion remnant of greater size and height than a monadnock. (b) A group or sprawling mass of monadnocks, often occurring near the headwaters of stream systems where erosion has not yet reduced the area to the level of a peneplain (Lobeck, 1939, p.633). Type locality: Unaka Mountains of eastern Tennessee and western North Carolina.

unakite (u′-na-kite) An epidote-rich granite, which, besides epidote, contains pink orthoclase, quartz, and minor opaque oxides, apatite, and zircon. The name is derived from the type locality, the Unaka Range, Great Smoky Mountains, eastern Tennessee. Obsolete.

unalaskite (un-a-las′-kite) A melanocratic, undeformed, cumulate rock, found as xenolithic fragments in basaltic andesites of the Aleutian Islands. Adak occurrence has magnesian olivine and clinopyroxene chadacrysts in magnesian hornblende oikocrysts. Named after Unalaska, Aleutian Islands, Alaska. Obsolete.

unarmored (un-ar′-mored) Pertaining to *dinoflagellates* such as those of the order Gymnodiniales, lacking a prominent tabulate layer below the outer membrane of the theca (Edwards, 1993). Also called: naked; athecate. Ant: *armored.*

unary system (u′-na-ry) A chemical system that has only one component.

unavailable moisture (un-a-vail′-a-ble) *unavailable water.*

unavailable water Water that cannot be utilized by plants because it is held in the soil by *adsorption* or other forces; water in the soil in an amount below the *wilting point.* Syn: *unavailable moisture.*

unbalanced force (un-bal′-anced) A force that is not opposed by another force acting along the same line in the opposite sense of direction; an unbalanced force causes translation of a body.

unbiased estimate A statistical estimate whose *expected value* is equal to the population parameter it is intended to estimate. An unbiased estimate is neither consistently high nor consistently low.

uncertainty The precision with which a measurement or value is known. Often implies a 50-50 chance that any one of a series of measurements would fall within a given range. Uncertainty does not necessarily imply anything about accuracy, which is comparison with the true value. Cf: *risk; standard deviation.*

uncinate (un′-ci-nate) A hexactinellid-sponge spicule (diactinal monaxon) covered on all sides with short thornlike spines directed toward one end.

uncompahgrite (un-com-pah′-grite) (a) In the *IUGS classification,* a syn. of *pyroxene melilitolite.* (b) A plutonic rock composed chiefly of melilite, along with pyroxene, opaque oxides, perovskite, apatite, calcite, anatase, andradite, and occasionally phlogopite. The name, given by Larsen and Hunter in 1914, is for Mount Uncompahgre, Colorado.

unconcentrated flow (un-con′-cen-trat′-ed) *overland flow.*

unconcentrated wash *sheet erosion.*

unconditional probability The probability that an event will occur, without consideration of the occurrence or nonoccurrence of any other events. Cf: *conditional probability; joint probability.*

unconfined aquifer (un-con-fined′) An aquifer having a *water table;* an aquifer containing *unconfined groundwater.* Syn: *water-table aquifer.*

unconfined compression test A special condition of a *triaxial compression test* in which no confining pressure is applied. Syn: *crushing test.*

unconfined groundwater Groundwater that has a water table, i.e. water not confined under pressure beneath a confining bed. Ant: *confined groundwater.* Syn: *phreatic water; nonartesian groundwater; free groundwater; unconfined water.*

unconfined water *unconfined groundwater.*

unconformability (un′-con-form′-a-bil′-i-ty) The quality, state, or condition of being unconformable, such as the relationship of unconformable strata; *unconformity.*

unconformable (un-con-form′-a-ble) Said of strata or stratification exhibiting the relation of unconformity to the older underlying rocks; not succeeding the underlying rocks in immediate order of age or not fitting together with them as parts of a continuous whole. In the strict sense, the term is applied to younger strata that do not "conform" in position or that do not have the same dip and strike as those of the immediately underlying rocks. Also, said of the contact between unconformable rocks. Cf: *conformable [stratig].* Syn: *discordant [stratig].*

unconformity (un-con-form′-i-ty) (a) A substantial break or gap in the geologic record where a rock unit is overlain by another that is not next in stratigraphic succession, such as an interruption in the continuity of a depositional sequence of sedimentary rocks or a break between eroded igneous rocks and younger sedimentary strata. It results from a change that caused deposition to cease for a considerable span of time, and it normally implies uplift and erosion with loss of the previously formed record. An unconformity is of longer duration than a *diastem.* (b) The structural relationship between rock strata in contact, characterized by a lack of continuity in deposition, and corresponding to a period of nondeposition, weathering, or esp. erosion (either subaerial or subaqueous) prior to the deposition of the younger beds, and often (but not always) marked by absence of parallelism between the strata; strictly, the relationship where the younger overlying stratum does not "conform" to the dip and strike of the older underlying rocks, as shown specif. by an angular unconformity. Cf: *conformity.* Syn: *unconformability; transgression.* (c) *surface of unconformity.* Common types of unconformities recognized in U.S.: *nonconformity; angular unconformity; disconformity; paraconformity.* Since the essential feature of an unconformity, as understood in Great Britain, is structural discordance rather than a time gap, the British do not recognize disconformity and paraconformity as unconformities. For an historical study of unconformities, see Tomkeieff (1962). (d) In sequence stratigraphy, a surface separating younger from older strata along which there is evidence of subaerial-erosional truncation and, in some areas, correlative submarine erosion, a basinward shift in facies, onlap, truncation, or abnormal subaerial exposure, with a significant hiatus indicated. Local, contemporaneous erosion and deposition associated with geological processes such as point-bar development or aeolian-dune migration are excluded from the definition of unconformity (Mitchum et al., 1977; Van Wagoner et al., 1990).

unconformity-associated uranium-gold A *mineral-deposit type* characterized by uranium or gold minerals filling fractures and breccias in clastic metasedimentary rocks along an unconformity between Paleoproterozoic and Mesoproterozoic rock.

unconformity-bounded unit A body of rocks bounded above and below by specifically designated, significant and demonstrable discontinuities in the stratigraphic record, preferably of regional or interregional extent (ISSC, 1994, Chapter 6).

unconformity trap A trap for oil or gas associated with an unconformity.

unconformity-type uranium-gold *unconformity-associated uranium-gold.*

unconsolidated material (un-con-sol′-i-dat′-ed) (a) A sediment that is loosely arranged or unstratifid, or whose particles are not cemented together, occurring either at the surface or at depth. (b) Soil material that is in a loosely aggregated form.

uncontrolled mosaic (un-con-trolled′) An aerial *mosaic [photo]* formed solely by matching detail of overlapping photographs without spatial or directional adjustments to control points.

uncovers (un-cov′-ers) *dries.*

unctuous (unc′-tu-ous) *soapy.*

unda (un′-da) A term applied by Rich (1951, p.2) to the environment of sedimentation that lies in the zone of wave action. Etymol: Latin *unda,* "wave".

undae (un′-dae) A term established by the International Astronomical Union for dunes on planetary surfaces (USGS, 2003). Sing: unda.

undaform (un′-da-form) The subaqueous land form produced by the erosive and constructive action of the waves during the development of the subaqueous profile of equilibrium (Rich, 1951, p.2).

undathem (un′-da-them) Rock units formed in the *unda* environment of deposition (Rich, 1951, p.2).

undation theory (un-da′-tion) A theory proposed by Van Bemmelen (1933) that explains the structural and tectonic features of the Earth's crust by vertical upward and downward movements caused by waves that are generated

by deep-seated magma. The concept is no longer accepted. Cf: *blister hypothesis.*
undaturbidite (un-da-tur'-bid-ite) A term proposed by Rizzini and Passega (1964, p.71) for a sediment formed from a suspension produced by violent storms; a deposit intermediate between an ordinary wave deposit and a turbidite. Cf: *fluxoturbidite.*
underbalanced drilling Drilling with *mud weight* slightly less than *formation pressure.* This produces less formation damage but is dangerous because of the possibility of a *blowout.*
underclay (un'-der-clay) A stratum of soft, dispersible clay or claystone that typically underlies coal. Hughes et al. (1987) distinguished three facies: (1) shale-type or illitic; (2) soil-type or fireclay-type (kaolinitic); and (3) gley-type (rich in mixed-layered illite/smectite). They regarded some claystones without overlying coal to be underclays if they show alteration patterns typical of underclays. Seat rock is the equivalent term used by the British.
underclay limestone A thin, dense, nodular, relatively unfossiliferous freshwater limestone underlying coal deposits, so named because it is closely related to underclay.
undercliff [geomorph] (un'-der-cliff) A terrace or subordinate cliff along a coast, formed of material fallen from the cliff above; the lower part of a cliff whose upper part underwent landsliding.
undercliff [sed] A term used in southern Wales for a shale forming the floor of a coal seam.
underconsolidation (un'-der-con-sol'-i-da'-tion) Consolidation (of sedimentary material) less than that normal for the existing overburden; e.g. consolidation resulting from deposition that is too rapid to give time for complete settling. Ant: *overconsolidation.*
undercooling (un-der-cool'-ing) *supercooling.*
undercurrent (un'-der-cur'-rent) A current of water flowing beneath a surface current at a different speed or in a different direction; e.g. the Mediterranean Undercurrent off Gibraltar. See also: *equatorial undercurrent.*
undercut (un'-der-cut) A reentrant in the face of a cliff, produced by undercutting.
undercutting (un-der-cut'-ting) The removal of material at the base of a steep slope or cliff or other exposed rock by the erosive action of falling or running water (such as a meandering stream), of sand-laden wind in the desert, or of waves along the coast.
underearth (un'-der-earth) (a) A hard fireclay forming the floor of a coal seam; *underclay.* (b) The soil beneath the Earth's surface. (c) The depths of the Earth.
underfit stream (un'-der-fit) A *misfit stream* that appears to be too small to have eroded the valley in which it flows; a stream whose volume is greatly reduced or whose meanders show a pronounced shrinkage in radius. It is a common result of drainage changes effected by capture, by glaciers, or by climatic variations. Cf: *meandering valley.*
underflow (un'-der-flow) (a) The movement of groundwater in an *underflow conduit;* the flow of water through the soil or a subsurface stratum, or under a structure. (b) The rate of discharge of groundwater through an underflow conduit. (c) The water flowing beneath the bed or alluvial plain of a surface stream, generally in the same direction as, but at a much slower rate than, the surface drainage; esp. the water flowing under a dry stream channel in an arid region.
underflow conduit A permeable deposit that underlies a surface stream channel, that is more or less definitely limited at its bottom and sides by rocks of relatively low permeability, and that contains groundwater moving in the same general direction as the stream above it (Meinzer, 1923, p.43). See also *underflow.*
underground cauldron subsidence (un'-der-ground) *Cauldron subsidence* in which the ring faults do not extend to the surface (Billings, 1972, p.359). Cf: *surface cauldron subsidence.*
underground ice *ground ice.*
underground stream A body of water flowing as a definite current in a distinct channel below the surface of the ground, usually in an area characterized by joints or fissures; legally, such a stream discoverable by men without scientific instruments. Application of the term to ordinary aquifers is incorrect. Cf: *subterranean stream; percolating water.*
underground water (a) A syn. of *groundwater.* (b) A syn. of *subsurface water* in less preferred usage.
underhand stoping (un'-der-hand) The working of a block of ore from an upper to a lower level.
underlapping faults Two subparallel , noncollinear faults with traces that do not extend past each other (Peacock et al., 2000). Cf: *overlapping faults.*
underlay (un'-der-lay) In mining, the extension of a vein or ore deposit beneath the surface; also, the inclination of a vein or ore deposit from the vertical, that is, *hade.* Syn: *underlie.*
underlie [mining] (un'-der-lie) *underlay.*
underlie [stratig] v. To lie or be situated under, to occupy a lower position than, or to pass beneath. The term is usually applied to certain rocks over which younger rocks (usually sedimentary or volcanic) are spread out. Ant: overlie.
underloaded stream *supply-limited stream.*
underlying (un-der-ly'-ing) *subjacent [stratig].*
undermelting (un-der-melt'-ing) The melting from below of any *floating ice* (Huschke, 1959, p.601). Rarely used.
undermining (un-der-min'-ing) The action of wearing away supporting material, as the *undermining* of a cliff by stream erosion; *sapping.*
underplating (a) Addition of material to the base of an accretionary prism by development of a *duplex* at the base of the prism. (b) Addition of igneous rock to the base of the crust.
underplight (un'-der-plight) A substratum, once consisting of soft mud, that preserves the form of an overlying thin layer of sand or gravel that has been contorted by alternate freezing and thawing (Spurrell, 1887).
underpressured system A hydrogeologic system in which the fluid pressure is less than hydrostatic pressure. Cf: *overpressured system.*
undersaturated (un-der-sat'-u-rat'-ed) (a) Said of an igneous rock consisting of silica minerals, e.g. feldspathoids and olivine. (b) Said of a rock whose norm contains feldspathoids and olivine, or olivine and hypersthene. Cf: *critically undersaturated; oversaturated; saturated [geol].*
undersaturated permafrost Permafrost that contains less ice than the ground could hold if the water were in the liquid state.
undersaturated pool A pool in which all the gas present is dissolved in the oil. Cf: *saturated pool.*
undershooting Seismic surveying using a source on one side of a property with a spread on the opposite side in order to obtain subsurface coverage under the property itself. Used when the surface of the property is inaccessible (such as on a line crossing a river).
under side Side of a conodont element toward which the basal cavity or basal pit opens (TIP, 1981, pt. W, supp.2). Syn: lower side.
underthrust fault (un'-der-thrust) A type of thrust fault in which it is the lower rock mass that has been actively moved under the upper, passive rock mass. An underthrust may be difficult to distinguish from an *overthrust.*
undertow (un'-der-tow) The seaward return flow, near the bottom of a sloping beach, of water that was carried onto the shore by waves. Not to be confused with *rip current.*
undertow mark A channeled structure on a sedimentary surface, believed to have been made by currents dragging heavy objects in very shallow water adjacent to a beach (Clarke, 1918). Cf: *strand mark.*
underwater contour *submerged contour.*
underwater gravimeter (un-der-wa'-ter) An instrument capable of measuring gravity when lowered to the sea bottom from a stationary surface vessel; it is leveled and read in a few minutes by remote control and has an accuracy of about 0.1 milligal.
underwater ice Ice formed below the surface of a body of water; e.g. *anchor ice.*
undiscovered resources Any resource whose existence is only postulated. Subdivided into speculative and hypothetical resources.
undisturbed (un-dis-turbed') Said of a soil sample that has not been subjected to disturbance by boring, sampling, or handling, and that thus closely represents the in-situ characteristics of the material. Most sampling procedures produce, at best, only "relatively undisturbed samples".
undivided (un-di-vid'-ed) Referring to streams, said of a surface, landscape, or area that has no noticeable feature separating the drainage of neighboring streams.
undrained loading The deformation of a water-saturated geologic material by applying external stresses in such a way that water cannot exit the material. The ratio of the resulting change in fluid pressure to the applied external stresses is related to *tidal efficiency* and Skempton's coefficient B. It provides the logic for interpreting responses of aquifers to barometric changes, earth tides, and ocean tides.
undulate (un'-du-late) Wavy, as of the margin of a leaf or petal.
undulating fold (un'-du-lat'-ing) A minor fold with rounded apexes; a fold whose beds are bent into alternate elevations and depressions.
undulation [geodesy] (un-du-la'-tion) The separation or height of the geoid above or below the *reference ellipsoid.*
undulation [geomorph] (a) A landform having a wavy outline or form; e.g. a desert sand deposit similar to a *whaleback* but shorter and lacking the definite form of the whaleback (Stone, 1967, p.252). (b) A rippling or scalloped land surface, having a wavy outline or appearance, or resembling waves in form; e.g. a rolling prairie.
undulation [struc geol] An irregularity on a mirror surface of a joint that forms normal to the plume lines or striae. These subtle ridges are equaled to a hesitation or momentary arrest in joint propagation (Bahat et al., 2003).
undulatory extinction (un'-du-la-to'-ry) A type of extinction that occurs successively in adjacent areas of the same crystal, as the microscope's stage is turned. Cf: *parallel extinction; inclined extinction.* Syn: *strain shadow; oscillatory extinction; wavy extinction.*
undulose extinction The irregular distribution of *dislocations* in a mineral grain, resulting in small reorientations of the crystallographic lattice that is observable by means of a petrographic microscope with crossed polars.
unequilibrated chondrite A chondrite whose major mineral constituents show compositions out of equilibrium suggesting little or no exposure to metamorphic heating and recrystallization. These chondrites typically are of petrographic (petrologic) grade 1, 2, 3.
uneven fracture (un-e'-ven) A general type of mineral fracture that is rough and irregular.

ungarettiite A cherry-red to very dark red monoclinic mineral of the *amphibole* group: $NaNa_2(Mn^{2+},Mn^{3+},Mg)_5Si_8O_{22}O_2$.

ungemachite (un'-ge-mach'-ite) A colorless to yellowish rhombohedral mineral: $K_3Na_8Fe^{3+}(SO_4)_6(NO_3)_2 \cdot 6H_2O$. It is the Fe^{3+} analogue of humberstonite.

unglaciated (un-gla'-ci-at'-ed) Said of a land surface that has not been modified by the action of a glacier or an ice sheet; "never-glaciated". Cf: *deglaciation.*

ungrouped chondrite A chondrite whose mineralogy, bulk chemistry, or texture falls outside the ranges of currently described groups.

ungrouped iron An iron meteorite whose mineralogy, bulk chemistry, or texture falls outside the ranges of currently described groups.

ungursaite *calciotantite.*

uniaxial (u-ni-ax'-i-al) Said of a crystal having only one optic axis, e.g. a tetragonal or hexagonal crystal. Cf: *biaxial.*

uniaxial test (uni-ax'-i-al) In engineering geology, a laboratory device for measuring shear strength of soil or rock in which confining stress cannot be applied to the sample. Cf: *triaxial test.*

uniclinal shifting (u'-ni-cli'-nal) *monoclinal shifting.*

unicline (u'-ni-cline) An obsolete syn. of *monocline.*

unidirectional current A current flowing in one constant direction such as the current produced by water flowing in a stream. Cf: *bidirectional current; combined flow.*

uniform channel (u'-ni-form) In hydraulics, a channel having a uniform cross section and a constant roughness and slope (ASCE, 1962).

uniform development The production of a landscape where the rate of uplift is equal to the rate of downward erosion, characterized by constant relief and straight slopes. Cf: *accelerated development; declining development.*

uniform distribution In *geostatistics,* a frequency or probability distribution in which every value has an equal probability of occurrence; Also called: rectangular distribution.

uniform flow Flow of a current of water in which there is neither convergence nor divergence.

uniformitarian (u'-ni-form'-i-tar'-i-an) n. An adherent of the doctrine of *uniformitarianism.* adj. Pertaining to the doctrine of uniformitarianism.

uniformitarianism (u'-ni-form'-i-tar'-i-an-ism) The assumption that "the geological forces of the past differ neither in kind nor in energy from those now in operation" (Hooykaas, 1963), which was the basis advocated by Charles Lyell for interpreting past phenomena by analogy with modern ones-thus the cliché, "the present is the key to the past". The constraint of rate and intensity as well as of kinds of processes advocated by Lyell implied a vast extension of geologic time and earned the nickname *gradualism.* The terms "uniformitarianism" and its antonym *"catastrophism"* were coined by William Whewell in 1831 to express the divergent geologic systems of Lyell on the one hand and Cuvier and Élie de Beaumont on the other. Today, geology assumes only the uniformity of kinds of processes, so uniformitarianism should be used only in its historic context. Cf: *actualism; parsimony [philos]; uniformity of nature.*

uniformity (uni-for'-mi-ty) The principle of uniformity, which is the assumption of uniformity of causes or processes throughout time and space. "Amid all the revolutions of the globe the economy of Nature has been uniform, and her laws are the only thing that has resisted the general movement" (John Playfair, 1802). Uniformity is a special geological case of the principle of *parsimony* or *simplicity.* Not synonymous with the *uniformitarianism* of Charles Lyell, who constrained throughout geologic time both the intensity and frequency and the kinds of processes seen today. Cf: *actualism.*

uniformity coefficient (u-ni-form'-i-ty) A numerical expression of the variety in particle sizes in mixed natural soils, defined as the ratio of the sieve size through which 60% (by weight) of the material passes to the sieve size that allows 10% of the material to pass. It is unity for a material whose particles are all of the same size, and it increases with variety in size (as high as 30 for heterogeneous sand).

uniformity of nature The doctrine or principle of the invariability or regularity of nature, which holds that identical states or causes are followed by identical effects. Cf: *uniformity.*

uniform strain (u'-ni-form) *homogeneous strain.*

uniform suspension A turbulent suspension within which particle size is uniform.

unilaminate (u-ni-lam'-i-nate) Said of an encrusting or erect bryozoan colony consisting of a single layer of zooids with or without extrazooidal parts.

unilateral (u-ni-lat'-er-al) Said of a stream or drainage system in which all tributaries come in from one side, while the other walls of the main valleys are practically unbroken (Rich, 1915, p.145).

unilobite (u-ni-lo'-bite) A descriptive term for a *trace fossil* consisting of a one-lobed (unilobate) trail. About 80 percent of all invertebrate tracks are unilobites. The term is seldom used.

unilocular (u-ni-loc'-u-lar) Containing a single chamber or cavity; e.g. said of a single-chambered foraminifer. Syn: *monothalamous.*

unimembrate apparatus A conodont skeletal apparatus consisting of only one morphologically distinct element type, regardless of total number of elements in a complete apparatus (TIP, 1981, pt. W, supp.2).

unimodal sediment (u-ni-mod'-al) A sediment whose particle-size distribution shows no secondary maxima (modes).

uniplicate (u-ni-pli'-cate) Said of a form of alternate folding in brachiopods with the pedicle valve bearing a median *sulcus* and an anterior commissure median *plica* (TIP, 1965, pt. H, p.155). Ant: *sulcate.*

uniserial (u-ni-se'-ri-al) Arranged in, characterized by, or consisting of a single row or series, e.g. a "uniserial arm" of a crinoid composed of brachial plates arranged in a single row; a "uniserial test" of a foraminifer whose chambers are arranged in a single linear or curved series; or a "uniserial rhabdosome" of a graptoloid consisting of a single row of thecae. Cf: *biserial.*

unit bar (u'-nit) In a braided stream, a sand or gravel bar that is relatively unmodified, with a morphology "determined by mainly depositional processes" (Smith, 1974, p.210).

unit cell The smallest volume or parallelepiped within the three-dimensional repetitive pattern of a crystal that contains a complete sample of the atomic or molecular groups that compose this pattern, and that displays the symmetry present in the crystal structure. Crystal structure can be described in terms of the translatory repetition of this unit in space in accordance with one of the space lattices. Cf: *primitive unit cell; conventional unit cell; reduced unit cell.*

unit character A natural characteristic that is dependent on the presence or absence of a single gene.

unit circle [cart] In a gnomonic projection, the circle that is the projection of the equatorial plane of the sphere of projection. Its radius gives the scale used in plotting the projection.

unit circle [stat] In directional statistics, lines and other vector observations are standardized to have unit length and a common origin so they can be plotted on a graph as directional vectors or lines radiating from the center of a circle whose radius is defined as equal to 1.

unit coal Pure coal, free of moisture and noncoal mineral matter, calculated from analysis. Unit coal is expressed by the equation: unit coal=1.00-(W+1.08A+0.55S), in which W=water, A=ash, and S=sulfur.

unit dry weight *dry unit weight.*

unit form A crystal form in a system other than the cubic, having intercepts on the chosen crystal axes that define the axial ratio. Unit forms have Miller indices {111}, {110}, {011}, {101}.

unit hydrograph A simple, static runoff model that produces a storm hydrograph for a precipitation event of a given duration.

unit sphere A three-dimensional extension of the *unit circle,* usually represented as a projection of a hemisphere whose radius is defined as equal to 1. The tips of three-dimensional vectors form points where they intersect the unit sphere and the pattern of points is projected onto a flat diagram using an equal-angle or equal-area projection.

unit-stratotype The type section of a layered stratigraphic unit that serves as the standard of reference for the definition and characterization of a unit. The upper and lower limits of a unit stratotype are its boundary stratotypes (ISSC, 1994, p.26).

unit structure The assembly of one or more layers plus interlayer material to model a clay mineral is referred to as a unit structure (see structural terms). For using the terms for clay minerals of plane, sheet, layer, and unit structure, a single plane of atoms, a tetrahedral or octahedral sheet, and 1:1 or 2:1 layer are the recommended designations. Plane, sheet, and layer refer to increasingly thicker arrangements.

unit value The monetary value of a mineral or rock product per ton or other unit of measurement.

unit weight A term applied esp. in soil mechanics to the weight per unit of volume, such as grams per cubic centimeter; the density of a material. Cf: *dry unit weight; effective unit weight; wet unit weight.*

univalve (u'-ni-valve) adj. Having or consisting of one valve only. Cf: *bivalve.* n. (a) A univalve animal; specif. a mollusk with a univalve shell, such as a gastropod, a cephalopod, or a scaphopod. (b) A shell of a univalve animal; specif. a mollusk shell consisting of one piece.

univariant (u-ni-var'-i-ant) Said of a chemical system having one degree of freedom; said of an equilibrium system in which the arbitrary variation of more than one physical condition will result in the disappearance of one or more of the phases.

universal kriging In *geostatistics,* a form of spatial estimation in which the residuals from the drift are estimated by *kriging.* The drift is estimated simultaneously and effectively removes any local nonstationarity in the variable being estimated.

universal soil-loss equation An equation that includes the major factors affecting accelerated erosion, A=RKLSCP, where R = climatic erosivity (rainfall and runoff), K = soil erodibility, L = slope length, S = slope gradient or steepness, C = cover and management, and P = erosion control practice. These factors determine how much water enters the soil, how much runs off, and the manner and rate of its removal (Brady, 1990). Abbrev: USLE.

universal stage (u-ni-ver'-sal) A *stage [optics]* of three, four, or five axes, attached to the rotating stage of a *polarizing microscope,* that enables the thin section under study to be tilted about two horizontal axes at right angles. It is used for optical study of low-symmetry minerals or for determining the orientation of any mineral relative to the section surface and edge directions. Syn: *U-stage; Fedorov stage.*

Universal time Time defined by the rotation of the Earth and determined from its apparent diurnal motions. Because of variations in the rate of rotation, Universal time is not rigorously uniform. Twelve hours plus the Greenwich hour angle of a point on the Equator whose right ascension, measured from the mean equinox of date is $18^H38^M45.^S836 + 8640184.^S542\ T + 0.^S0929\ T^2$. Syn: *Greenwich mean time.* Abbrev: UT.

universal transverse Mercator projection A special case of the *transverse Mercator projection* used as a basis of the UTM grid. It consists of 60 north-south zones, each 6° wide in longitude, with the longitudes of the boundary edges integral multiples of 6°. The longitudes of the central meridians are therefore odd multiples of 3°. An overlap of about 25 miles is usually permitted beyond the edge of each zone. The zones are numbered consecutively, starting with zone 1 between 180° and 174° west longitude and increasing eastward to zone 60 between 174° and 180° east longitude. To reduce scale distortions near the edges of the zones, the scale at the central meridian is made 0.9996 of the nominal scale of the projection (USGS, 1965, p.86).

universe *population [stat].*

unloading (un-load′-ing) The removal by denudation of overlying material.

unloading joint A joint forming upon exhumation whose orientation is controlled by either a residual or contemporary tectonic stress, with propagation normal to least principal stress (Engelder, 1985). Syn: *neotectonic joint.*

unmatched terrace (un-matched′) *unpaired terrace.*

unmixing [chem] (un-mix′-ing) A syn. of *exsolution* that is also applied to the separation of immiscible liquids.

unmixing [sed] Segregation and concentration of sedimentary material during diagenesis.

unoriented [geol] (un-o′-ri-ent′-ed) Said of a rock or other geologic specimen whose original position in space, when collected, is unknown or not definitely ascertained.

unoriented [surv] Said of a map or surveying instrument whose internal coordinates are not coincident with corresponding directions in space.

unpaired terrace (un-paired′) A *stream terrace* with no corresponding terrace on the opposite side of the valley, usually produced by a meandering stream swinging back and forth across a valley. See also: *meander terrace.* Cf: *paired terrace.* Syn: *unmatched terrace.*

unpolarized transformation series A transformation series in which the direction of character evolution has not been specified (Wiley et al., 1991, p.9). Ant: *polarized transformation series.*

unprotected thermometer (un-pro-tect′-ed) A *reversing thermometer* that is not protected against hydrostatic pressure. Cf: *protected thermometer.*

unradiogenic lead (un′-ra-di-o-gen′-ic) Lead which has $^{207}Pb/^{204}Pb$ and $^{206}Pb/^{204}Pb$ ratios less than they would be in a single-stage development because of an integrated $^{238}U/^{204}Pb$ less than about 9.

unripe (un-ripe′) Said of peat that is in an early stage of decay, and in which original plant structures are visible. Cf: *ripe.*

unroofed anticline (un-roofed′) *breached anticline.*

unsaturated zone A subsurface zone containing water under pressure less than that of the atmosphere, including water held by capillarity; and containing air or gases generally under atmospheric pressure. This zone is limited above by the land surface and below by the surface of the *saturated zone,* i.e., the *water table.* The zone is subdivided into the *belt of soil water,* the *intermediate belt,* and the *capillary fringe.* Syn: *vadose zone; zone of aeration; zone of suspended water.*

unsaturated-zone flow The flow of water in the unsaturated zone by capillary action and gravity.

unsorted (un-sort′-ed) *poorly sorted.*

unstable [radioactivity] (un-sta′-ble) Said of a radioactive substance. Cf: *stable [radioactivity].*

unstable [sed] (a) Said of a constituent of a sedimentary rock that does not effectively resist further mineralogic change and that represents a product of intense weathering and rapid erosion and deposition (as in a region of tectonic activity and high relief); e.g. feldspar, pyroxene, hornblende, and various fine-grained rock fragments. (b) Said of an immature sedimentary rock (such as graywacke) consisting of easily weathered particles that are angular to subrounded, poorly to moderately sorted, and composed of feldspar grains or rock fragments. Cf: *labile.*

unstable equilibrium A state of equilibrium from which a chemical system, or a body (such as a pendulum), will depart in response to the slightest perturbation. Cf: *stable equilibrium.*

unstable gravimeter *astatic gravimeter.*

unstable isotope A syn. of *radioisotope.* Cf: *stable isotope.*

unstable relict A *relict [meta]* that is believed to be unstable under the newly imposed conditions of metamorphism but persists in a perhaps altered but still recognizable form owing to the low rate of reaction. Cf: *armored relict; stable relict.*

unstable remanent magnetization *viscous magnetization.*

unsteady flow (un-stead′-y) In hydraulics, flow that changes in magnitude or direction with time. Cf: *steady flow.* Syn: *nonsteady flow.*

unstratified (un-strat′-i-fied) Not formed or deposited in strata; specif. said of massive rocks or sediments with an absence of layering, such as granite or glacial till.

unsupervised classification In digital mapping, a process for extracting information from an image by letting a variety of algorithms attempt to define geologically meaningful classes based on their spectral signatures. Human intervention is usually required to produce meaningful results. As with *supervised classification,* the goal is to produce a map of the distribution of the classes.

unweathered (un-weath′-ered) *fresh [weath].*

Unwin's critical velocity (Un′-win's) *critical velocity (e).*

upalite (u′-pal-ite) An amber-yellow monoclinic mineral: $Al(UO_2)_3O(PO_4)_2(OH)\cdot 7H_2O$.

upbank thaw (up′-bank) A thaw or marked rise of temperature occurring at hill or mountain level while the frost is unbroken in the valley below.

upbuilding (up′-build-ing) (a) The building-up of a sedimentary deposit, as by a stream or in the ocean. Cf: *aggradation [geomorph].* (b) *active diapirism.*

upconcavity (up′-con-cav′-i-ty) The persistent downstream decrease in gradient as seen on the channel profiles of most streams.

upconing (up-con′-ing) *coning.*

updip (up-dip′) A direction that is upwards and parallel to the dip of a structure or surface. Cf: *downdip.*

updip block The rocks on the *upthrown* side of a fault. Cf: *downdip block.*

updrift (up-drift′) The direction opposite that of the predominant movement of littoral materials.

upfaulted (up-fault′-ed) Said of the rocks on the *upthrown* side of a fault, or the *updip block.* Cf: *downfaulted.*

upgrading (up-grad′-ing) *aggradation [geomorph].*

upheaved plug A type of *lava dome* in which the rock mass pushes bodily upward like a piston, breaks through the surface, but is too viscous to spread laterally once above the surface (Blake, 1989; Francis, 1993), e.g., Showa Sin-Zan, Hokkaido, Japan.

uphole (up-hole′) Said of any location in a *borehole* that is above a given depth. Cf: *downhole.*

uphole survey The setting-off of successive sources at varying depths in a shothole in order to determine velocities.

uphole time In seismic exploration, the time required for the seismic impulse to travel from a given depth in a shothole to the surface.

upland (up′-land) (a) A general term for high land or an extensive region of high land, esp. far from the coast or in the interior of a country. Sometimes used synonymously with *fastland.* (b) The higher ground of a region, in contrast with a valley, plain, or other low-lying land; a plateau. (c) The elevated land above the low areas along a stream or between hills; any elevated region from which rivers gather drainage. Also, an area of land above flood level, or not reached by storm tides. Ant: *lowland.*

upland plain A relatively level area of land lying at a considerable altitude; esp. a high-lying erosion surface.

upland swamp A swamp that "probably" occupies the site of a former shallow sound or coastal lagoon which has become land "through uplift and retreat of the sea" (Stephenson and Veatch, 1915, p.37). Cf: *tidal swamp.*

uplift [eng] (up′-lift) Any force that tends to raise an engineered structure and its foundation relative to its surroundings. It may stem from pressure of subjacent ground, surface water, or expansive soil under the base of the structure; or from lateral forces such as wind or earthquake shaking that tend to cause overturning.

uplift [tect] A structurally high area in the crust, produced by positive movements that raise or upthrust the rocks, as in a dome or arch. Cf: *depression [tect].*

uplift joint A joint forming upon exhumation and cutting through multilayers with wide and irregular spacing (Bahat, 1991).

uplimb thrust fault (up-limb′) A contractional fault developed on the limb of an anticline in which the direction of tectonic transport is uplimb and the fault dips initially in the same direction as the limb but at a steeper angle. Continued rotation of the limb toward the vertical results in associated limb-contraction faults that dip toward the fold axis (Perry and DeWitt, 1977). Syn: *flexural-slip thrust fault.*

upper (up′-per) Pertaining to rocks or strata that are normally above those of earlier formations of the same subdivision of rocks. The adjective is applied to the name of a chronostratigraphic unit (system, series, stage) to indicate position in the geologic column and corresponds to *late* as applied to the name of the equivalent geologic-time unit; e.g. rocks of the Upper Jurassic System were formed during the Late Jurassic Period. The initial letter of the term is capitalized to indicate a formal subdivision (e.g. "Upper Devonian") and is lowercased to indicate an informal subdivision (e.g. "upper Miocene"). The informal term may be used where there is no formal subdivision of a system or series. Cf: *lower; middle [stratig].*

upper break *head [struc geol].*

upper edge The denticulated edge of the process of a ramiform or pectiniform *conodont element* (TIP, 1981, pt. W, supp.2). Syn: upper margin.

upper flow regime The *flow regime* of a *unidirectional current* that is characterized by relatively high sediment transport rates (with considerable *suspended load*), *bed forms* that are in phase with the water surface, including *upper plane bed, in-phase waves,* and *antidunes;* flow *Froude number* is greater than unity (Simons and Richardson, 1961).

upper flow regime plane bed *upper plane bed.*

upper keriotheca The abaxial (upper) part of *keriotheca* in the wall of a fusulinid, characterized by fine alveolar structure (as in *Schwagerina*). Cf: *lower keriotheca.*

upper mantle That part of the mantle which lies above a depth of about 660 km and has a density of 3.4 g/cm^3 to 4.0 g/cm^3 with increasing depth. Similarly, P-wave velocity increases from about 8 to 11 km/sec with depth and *S wave* velocity increases from about 4.5 to 6 km/sec with depth. It is presumed to be peridotitic in composition. It includes the subcrustal *lithosphere* the *asthenosphere* and the *transition zone;* it is equivalent to the B and C layers. Syn: *outer mantle; peridotite shell.*

upper Paleolithic The third and most recent division of the *Paleolithic,* characterized by *Homo sapiens.* Cf: *lower Paleolithic; middle Paleolithic.*

upper plane bed A bed configuration under a *unidirectional current* that is characterized by a flat, almost featureless surface over which sediment transport is intense, both in suspension and on the bed. Some authors report very low, downstream migrating bed forms as being characteristic of this bed configuration (e.g., Best and Bridge, 1992). *Current lineations* are common on upper plane bed surfaces. Syn: *upper flow regime plane bed; upper stage plane bed.*

upper plate The hanging wall of a fault. Cf: *lower plate.*

upper side Side or surface of a *conodont element* opposite that toward which the basal cavity or pit opens; almost always the side or surface bearing denticles (TIP, 1981, pt. W, supp.2).

upper stage plane bed *upper plane bed.*

upper tectorium The abaxial secondary layer of spirotheca in the wall of a fusulinid, next above the tectum (as in *Profusulinella*). Cf: *tectorium; lower tectorium.*

upright fold (up'-right) A fold having an essentially vertical axial surface; a *vertical fold.*

uprush (up'-rush) *runup.*

upsetted moraine (up-set'-ted) *push moraine.*

upsiloidal dune (up-si-loi'-dal) A general term for a U-shaped or V-shaped dune whose form is concave toward the wind; e.g. a *parabolic dune.*

upslope (up-slope') n. A slope that lies upward; uphill. adj. In an upward or uphill direction, or ascending; e.g. an *upslope* ripple that climbed a sloping surface.

upslope fog *Fog* formed as a consequence of the expansional cooling of humid air that is forced to ascend a mountain slope.

upthrow (up'-throw) n. (a) The upthrown side of a fault. (b) The amount of upward vertical displacement of a fault. Cf: *downthrow; heave [struc geol].*

upthrown (up'-thrown) Said of that side of a fault that appears to have moved upward, compared with the other side. Cf: *downthrown.*

upthrown block *upthrow.*

upthrusting *active diapirism.*

up-to-basin fault (up-to-ba'-sin) A term used in petroleum geology for a fault whose upthrown side is toward the adjacent basin.

upward coarsening A succession of beds that increase in grain-size upward through a vertical column of sediment such as a beach sequence.

upward continuation (up'-ward) Calculation of the *potential field* at an elevation higher than that at which the field is known. The *continuation* involves the application of Green's theorem and is rigorous if the field is completely known over the lower surface and if no sources are present between the surfaces (as is usually true for gravity and magnetic fields). Upward continuation is used to smooth out near-surface effects and to tie aeromagnetic and gravity surveys obtained at different heights. Cf: *downward continuation.*

upward fining A succession of beds that decrease in grain size upward through a vertical column of sediment such as in a river or tidal bar.

upwarping (up'-warp-ing) Upward flexing of the Earth's crust, commonly as the result of release of isostatic pressure, e.g. the melting of an ice sheet. Cf: *downwarping.* Syn: *rebound.*

upwelling [currents] (up'-well-ing) The rising of subsurface water toward the surface, esp. along the western coasts of continents (as along the coast of southern California); it replaces surface water that is transported away from the coast by Ekman mass transport driven by equatorward winds. Upwelling also occurs in the open ocean along the equator and in association with cyclonic circulation. Open ocean upwelling is strong in the Eastern Pacific, where southern trade winds cross the equator. Areas of upwelling are commonly marked by anomalously cold temperature and high productivity. Ant: *sinking.*

upwelling [volc] The relatively quiet eruption of lava and volcanic gases, without much force.

upwelling-type phosphate deposit Phosphorite sediments deposited in the upwelling areas of marine basins.

Uquian (U-qui'-an [U-key'-an)] Obsolete South American mammal age; equiv. Marplatan.

uraconite (u-rac'-o-nite) A name that has been used for various uranium sulfates, but that "lacks specific meaning and should be abandoned" (Frondel et al., 1967, p.44).

uralborite (ur'-al-bor'-ite) A colorless monoclinic mineral: $CaB_2O_2(OH)_4$.

Uralian (U-ral'-i-an) Stage in Russia: uppermost Carboniferous (above Gzhelian, below Sakmarian of Permian). Equivalent to Orenburgian-Gzhelian.

Uralian emerald (a) Emerald from near Sverdlovsk in the Ural Mountains, Russia. (b) *demantoid.*

Uralian-type PGE deposit *zoned ultramafic PGE deposit.*

uralite (ur'-al-ite) A green, generally fibrous or acicular variety of secondary amphibole (hornblende or actinolite) occurring in altered rocks and pseudomorphous after pyroxene (such as augite).

uralitization (u-ral'-i-ti-za'-tion) The development of amphibole from pyroxene; specif. a late-magmatic or metamorphic process of replacement whereby uralitic amphibole results from alteration of primary pyroxene. Also, the alteration of an igneous rock in which pyroxene is changed to amphibole; e.g. the alteration of gabbro to greenstone by pressure metamorphism.

uralolite (ur'-al-o-lite) A silky colorless or white monoclinic mineral: $Ca_2Be_4(PO_4)_3(OH)_3 \cdot 5H_2O$.

Ural-type glacier *drift glacier.*

uramphite (ur-am'-phite) A bottle-green to pale green tetragonal mineral: $(NH_4)(UO_2)(PO_4) \cdot 3H_2O$.

urancalcarite (u-ran-cal'-car-ite) A yellow orthorhombic mineral: $Ca(UO_2)_3(CO_3)(OH)_6 \cdot 3H_2O$.

uraniferous pyritic quartz-pebble conglomerate A *mineral-deposit type* characterized by uraninite and pyrite in Precambrian quartz-pebble conglomerate. A variety of *paleoplacer* deposit.

uraninite (u-ra'-nin-ite) (a) A black, brown, or steel-gray cubic mineral, essentially UO_2, but usually partly oxidized. It is strongly radioactive, and is the chief ore of uranium. Uraninite often contains impurities such as thorium, radium, the cerium and yttrium metals, and lead; when heated, it yields a gas consisting chiefly of helium. It occurs in veins of lead, tin, and copper minerals and in sandstone deposits, and is a primary constituent of granites and pegmatites. See also: *pitchblende.* Syn: *ulrichite; coracite.* (b) A group name for cubic minerals with the general formula AO_2, where A = U, Th, Ce, or Zr.

uranite (u-ran'-ite) A general term for a mineral group consisting of uranyl phosphates and arsenates of the autunite, meta-autunite, and torbernite type.

uranium-234 age method The calculation of an age in years for fossil coral or shell (limited to those formed during the last million years), based on the assumption that the initial uranium-234/uranium-238 ratio is known for the fossil. The change in this ratio is directly related to passage of time as the two isotopes have very different half-lives. Cf: *uranium-series age method.* Syn: uranium-234 excess method; uranium-234/uranium-238 age method; uranium-238/uranium-234 disequilibrium method.

uranium-lead age method Calculation of an age in years for geologic material based on the known radioactive decay rate of uranium-238 to lead-206 and uranium-235 to lead-207. It is part of the more inclusive *uranium-thorium-lead age method* in which the parent-daughter pairs are considered simultaneously. Syn: *lead-uranium age method.*

uranium ocher *gummite.*

uranium series The radioactive series beginning with uranium-238.

uranium-series age method (u-ra'-ni-um-se'-ries) Calculation of an age in years for Quaternary materials based on the general finding that the decay products uranium-234, thorium-230, and protactinium-231 in natural materials form commonly in disequilibrium with their long-lived parent isotopes, uranium-238 and uranium-235. The age is determined from the measured activity ratios of these isotopes. Cf: *ionium-thorium age method; ionium-excess method; ionium-deficiency method; thorium-230/protactinium-231 deficiency method; protactinium-ionium age method; uranium-234 age method.*

uranium-thorium-lead age method (u-ra'-ni-um-tho'-ri-um-lead) Calculation of an age in years for geologic material, often zircon, based on the known radioactive decay rate of uranium-238 to lead-206, uranium-235 to lead-207, and thorium-232 to lead-208, whose ratios give three independent ages for the same sample. The determined lead-207/lead-206 ratio can be used to compute a useful but not independent fourth age (*lead-lead age*). Whether all four possible dates are concordant or discordant is useful for evaluating the results of this method, used alone or in comparison with other methods, and in determining whether the initially closed system has been disturbed. Most commonly, only the uranium-lead ages are measured. Partial syn: *thorium-lead age method; uranium-lead age method.*

uran-mica (u'-ran-mi'-ca) A uranite, esp. torbernite.

uranmicrolite A yellowish-brown, greenish-brown, or brownish-black cubic mineral of the pyrochlore group: $(U,Ca,Ce,[\,])_2(Ta,Nb)_2(O,OH,F)_7$. Syn: *djalmaite.*

uranocher (u'-ran-o'-cher) A general name used chiefly for uranium sulfates (such as uranopilite) and for some uranium oxides. Also spelled: uranochre.

uranocircite (u-ran'-o-cir'-cite) A yellow or yellowish-green tetragonal mineral of the *autunite* group: $Ba(UO_2)_2(PO_4)_2 \cdot 8\text{-}12H_2O$.

uranophane (u-ran'-o-phane) A strongly radioactive yellow monoclinic secondary mineral: $Ca(UO_2)_2(SiO_3OH)_2 \cdot 5H_2O$. It is isostructural with sklodowskite and cuprosklodowskite, and dimorphous with beta-uranophane. Syn: *uranotile.*

uranopilite (u-ran'-o-pil'-ite) A yellow triclinic secondary mineral: $(UO_2)_6SO_4O_2(OH)_6 \cdot 14H_2O$.
uranopolycrase A brownish-red orthorhombic mineral of the *euxenite* group: $(U,Y)(Ti,Nb,Ta)_2(O,OH)_6$.
uranosilite (u-ran'-o-sil'-ite) A yellow-white orthorhombic mineral: $U^{6+}Si_7O_{17}$.
uranospathite (u-ran'-o-spath'-ite) A yellow, pale green, or bluish-green tetragonal mineral: $HAl(UO_2)_4(PO_4)_4 \cdot 40H_2O$.
uranosphaerite (u-ran'-o-sphaer'-ite) An orange-yellow or brick-red monoclinic secondary mineral: $Bi(UO_2)O_2(OH)$. Also spelled: uranospherite.
uranospinite (u-ran'-o-spin'-ite) A green to yellow tetragonal secondary mineral of the autunite group: $Ca(UO_2)_2(AsO_4)_2 \cdot 10H_2O$. It is isomorphous with zeunerite.
uranotantalite (u'-ra-no-tan'-ta-lite) *samarskite-(Y).*
uranothallite (u'-ra-no-thal'-lite) *liebigite.*
uranothorianite (u'-ra-no-tho'-ri-a-nite') A variety of thorianite containing uranium; an intermediate member in the uraninite-thorianite isomorphous series.
uranothorite (u'-ra-no-tho'-rite) A variety of thorite containing uranium.
uranotile (u-ran'-o-tile) A syn. of *uranophane.* Also spelled: uranotil.
uranotungstite (u-ran'-o-tung'-stite) A yellow, orange, or brownish orthorhombic mineral: $(Fe^{2+},Ba,Pb)(UO_2)_2(WO_4)(OH)_4 \cdot 12H_2O$.
uranpyrochlore A red or red-brown mineral of the *pyrochlore* group: $(U,Ca,Ce,[])_2(Nb,Ta)_2(O,OH,F)_7$. Syn: *hatchettolite.*
urban geology (ur'-ban) The application of geologic knowledge and principles to the planning and management of cities and their surroundings. It includes geologic studies for physical planning, waste disposal, land use, water-resources management, and extraction of usable raw materials. Cf: *environmental geology.*
urban heat island An area of higher air temperature in a city setting, compared to the air temperatures of the suburban and rural surroundings; shows up as an island in the pattern of isotherms on a surface map.
urea (u-re'-a) A pale yellow or brown tetragonal mineral: $CO(NH_2)_2$.
urediospore (u-re'-di-o-spore') A thin-walled rust spore produced by the summer stage of rust fungi (Uredinales) and germinating without a rest period. Cf: *teliospore.*
ureilite (u-re'-i-lite) A group of *achondrites* rich in olivine and pyroxene and containing significant Ni-free metal, graphite, and occasionally diamond; thought to be the residue of incomplete melting on a chondritic parent body.
ureyite (u'-rey-ite) *kosmochlor.*
uricite (u'-ri-cite) A white monoclinic mineral: $C_5H_4N_4O_3$.
Urodela (U-ro-de'-la) An order of caudate lissamphibians that includes salamanders and newts. Range, Upper Jurassic to Holocene.
uropod (u'-ro-pod) Either of the flattened leaflike appendages of the last abdominal segment of various crustaceans, which with the telson form the *caudal fan;* e.g. limb of the sixth abdominal somite of a eumalacostracan, or one of the last three abdominal appendages of an amphipod. The term is sometimes applied to any abdominal appendage of a crustacean.
urosome (u'-ro-some) The part of the body of a copepod crustacean behind the major articulation that marks the posterior boundary of a *prosome.*
ursilite (ur'-sil-ite) A lemon-yellow orthorhombic mineral: $(Mg,Ca)_4(UO_2)_4(Si_2O_5)_{5.5}(OH)_5 \cdot 13H_2O$.
urstromtal (ur'-strom-tal) A wide, shallow, trenchlike valley or depression excavated by a temporary meltwater stream flowing parallel to the front margin of a continental ice sheet, esp. one of the east-west depressions across northern Germany; a large-scale overflow channel. Etymol: German "Urstromtal", "ancient river valley". Pl: urstromtäler. Syn: *pradolina.*
urtite (urt'-ite) (a) In the *IUGS classification,* a plutonic rock in which F is between 60 and 100, M is 30 or less, and sodium exceeds potassium. Cf: *italite.* (b) A light-colored member of the *ijolite* series that is composed chiefly of nepheline and 0-30% mafic minerals, esp. aegirine and apatite. Cf: *melteigite.* The name, given by Ramsay in 1896, is for Lujavr-Urt (now Lovozero), Kola Peninsula, Russia. Obsolete.
urusovite A vitreous light green monoclinic mineral: $CuAlO(AsO_4)$.
urvantsevite (ur-vant'-sev-ite) A metallic hexagonal mineral: $Pd(Bi,Pb)_2$.
usamerite (u-sa-mer'-ite) A term proposed by Boswell (1960, p.157) for a rock comparable to the type *graywacke* and characterized by size grades ranging from gravel to sand, by poor sorting with a "substantial" quantity of matrix, and by variable rock and mineral fragments that are predominantly angular to subangular. Etymol: United States of America + ite.
user line *arbitrary line.*
U-shaped dune A dune having the form of the letter "U", its open end facing upwind.
U-shaped valley A valley having a pronounced parabolic cross profile suggesting the form of a broad letter "U", with steep walls and a broad, nearly flat floor; specif. a valley carved by glacial erosion, such as a *glacial trough.* Cf: *V-shaped valley.* Syn: *U-valley; trough valley.*
ushkovite (ush'-kov-ite) An pale yellow, orange-yellow, or brown triclinic mineral of the *laueite* group: $MgFe^{3+}_2(PO_4)_2(OH)_2 \cdot 8H_2O$.
usovite (u-sov'-ite) A brown monoclinic mineral: $Ba_2CaMgAl_2F_{14}$.
Ussherian (Us-sher'-i-an) Pertaining to the biblical chronology compiled by James Ussher (d.1656), Irish archbishop, who calculated from studies of the Scriptures that the Earth was formed in 4004 B.C.
ussingite (us'-sing-ite) A pale pink, lilac blue, or dark violet-red triclinic mineral: $Na_2AlSi_3O_8(OH)$.
U-stage *universal stage.*
Ustalfs Alfisols that have an ustic soil moisture regime and mesic or warmer soil temperature regimes. Ustalfs are brownish or reddish throughout and are not saturated with water for periods long enough to limit their use for most crops. A suborder in the U.S. system of soil taxonomy.
Ustands Andisols that have an ustic soil moisture regime. A suborder in the U.S. system of soil taxonomy.
ustarasite (us-ta'-ra-site) An ill-defined metallic gray mineral: $Pb(Bi,Sb)_6S_{10}$.
Usterts Vertisols of temperate or tropical regions that have wide, deep cracks that usually remain open for periods that total > 90 days but do not remain open continuously throughout the year, and have either a mean annual soil temperature of 22°C or more or a mean summer and mean winter soil temperature at 50 cm below the surface that differ by < 5°C or have cracks that open and close more than once during the year. A suborder in the U.S. system of soil taxonomy.
ustic moisture regime (us'-tic) A soil moisture regime that is intermediate between the acidic and udic moisture regimes. It is one of limited moisture, but the moisture is present when conditions are suitable for plant growth (USDA, 1975).
Ustolls Mollisols that have an ustic soil moisture regime and mesic or warmer soil temperature regimes. Ustolls may have a calcic, petrocalcic, or gypsic horizon, and are not saturated with water for periods long enough to limit their use for most crops. A suborder in the U.S. system of soil taxonomy.
Ustox (Us'-tox) Oxisols that have an ustic moisture regime and either hyperthermic or isohyperthermic soil temperature regimes or have < 20 kg organic carbon in the surface cubic meter. A suborder in the U.S. system of soil taxonomy.
Ustults Ultisols that have low or moderate amounts of organic carbon, are brownish or reddish throughout, and have an ustic soil moisture regime. A suborder in the U.S. system of soil taxonomy.
UT *Universal time.*
utahite (u'-tah-ite) (a) A blue or blue-green triclinic mineral: $Cu_5Zn_3(TeO_4)_4(OH)_8 \cdot 7H_2O$. (b) *jarosite.* (c) *natrojarosite.*
utahlite (u'-tah-lite) A syn. of *variscite,* esp. that found in compact, nodular masses in Utah.
UTC *coordinated universal time.*
UTM projection *universal transverse Mercator projection.*
utricle (u'-tri-cle) Any bladder-shaped plant appendage.
uvala (u-va'-la) A Serbo-Croatian term for *karst valley.* Also spelled: *ouvala.*
U-valley *U-shaped valley.*
uvanite (u'-van-ite) A brownish-yellow orthorhombic mineral: $U^{6+}_2V^{5+}_6O_{21} \cdot 15H_2O(?)$.
uvarovite (u-va'-rov-ite) The calcium-chromium end-member of the *garnet* group, characterized by an emerald-green color: $Ca_3Cr_2(SiO_4)_3$. It may have considerable alumina. Also spelled: *uwarowite; ouvarovite.*
uvite (uv'-ite) A black, brown, green, or colorless mineral of the *tourmaline* group: $(Ca,Na)(Mg,Fe^{2+})_3Al_5Mg(BO_3)_3Si_6O_{18}(OH,F)_4$.
uwarowite (u-wa'-row-ite) *uvarovite.*
uytenbogaardtite (uy'-ten-bo-gaardt'-ite) A metallic grayish-white tetragonal mineral: Ag_3AuS_2. Syn: *liujinyinite.*
uzbekite (uz'-bek-ite) *volborthite.*
uzonite (uz'-on-ite) A yellow monoclinic mineral: As_4S_5.

V

VA Display in *variable-area* form.

vacancy (va'-can-cy) A vacant site in a crystal structure, due to the absence of an atom or ion from its normal structural position. Syn: *hole [cryst].*

vacuity (va-cu'-i-ty) *degradation vacuity; total vacuity.*

vacuole [paleont] (vac'-u-ole) A cavity in the cytoplasm of a cell of a plant or protozoan, often containing a watery solution enclosed by a membrane, and performing various functions such as digestion (food vacuole) and hydrostatic relation (contractile vacuole); e.g. one of the irregularly shaped alveoles in a foraminiferal-test wall. Also, the globular fluid inclusion or droplet enclosed in a vacuole.

vadose canyon A cave *passage* of high, narrow, commnoly twisting shape formed by a free-surface stream traversing the vadose zone of a karst aquifer.

vadose cement Cement precipitated in the vadose zone, typically showing a meniscus or dripstone fabric.

vadose solution (va-dose') Solution action by vadose water above the level of the water table. Cf: *phreatic solution.*

vadose water Water of the *unsaturated zone.* Syn: *suspended water.*

vadose-water discharge (va-dose'-wa'-ter) The release, by evaporation, of water not originating in the *saturated zone.* It may be in the form of *vegetal discharge* or *soil discharge.*

vadose zone *unsaturated zone.*

vaesite (vaes'-ite) A cubic metallic gray mineral of the *pyrite* group: NiS_2.

vagile (vag'-ile) Said of a plant or animal that is free to move about. Cf: *sessile.*

Vaisala frequency *stability frequency.*

vajdakite A green or greenish-gray monoclinic mineral: $(Mo^{6+}O_2)_2(As^{3+})_2O_5{\cdot}3H_2O$.

vake (vak'-e) The French term for *wacke* or soft, compact, mixed claylike material with a flat, even fracture, commonly associated with basaltic rocks.

val A longitudinal, synclinal valley in the folded Jura Mountains of the European Alps. Etymol: French, "narrow valley". Pl: *vaux.* Cf: *cluse.* See also: *combe.* Syn: *vallon.*

Valanginian (Va-lan-gin'-i-an) Stage of the Standard Global Chronostratigraphic Scale: Lower Cretaceous (above Berriasian, below Hauterivian).

vale (a) A lowland, usually containing a stream; e.g. the depression between two parallel *cuestas.* It often forms the wider and flatter part of a valley. (b) A rift valley or tectonic valley; e.g. the Vale of Arabia. (c) *valley.*

valencianite (va-len'-cia-nite) A variety of adularia from Guanajuato, Mexico.

valency The number of arcs that begin or end at a node (Kennedy et al., 2001, p.105).

valency table A table that lists the nodes in a *data layer* along with their valencies (Kennedy et al., 2001, p.105).

Valentian (Va-len'-tian) *Llandoverian.*

valentinite (val'-en-tin'-ite) An adamantine white, yellowish, or reddish orthorhombic mineral: Sb_2O_3. It is polymorphous with senarmontite. Syn: *antimony bloom; white antimony;* exitèle; exitèlite.

valid (val'-id) (a) Said of a name that must, under the rules of zoological nomenclature, be adopted for a *taxon* with a particular rank, position, and description. (b) Said of publication in accordance with articles 32-45 of the International Code of Botanical Nomenclature (ICBN, 1972), which cover effective publication, form of name, and descriptive and illustrative requirements.

valleriite (val-ler'-i-ite) (a) A metallic bronze rhombohedral mineral: $4(Fe,Cu)S{\cdot}3(Mg,Al)(OH)_2$. (b) A group name for similar minerals with structures composed of alternating sulfide and hydroxide layers.

Vallesian (Val-le'-sian) European stage: Upper Miocene (above Maremmian, below Turolian).

valleuse (val-leuse') A French term for a *hanging valley,* as on the chalk cliffs of France.

valley [geomorph] (val'-ley) (a) Any low-lying land bordered by higher ground; esp. an elongate, relatively large, gently sloping depression of the Earth's surface, commonly situated between two mountains or between ranges of hills or mountains, and often containing a stream with an outlet. It is usually developed by stream erosion, but may be formed by faulting. (b) A broad area of generally flat land extending inland for a considerable distance, drained or watered by a large river and its tributaries; a river basin. Example: the Mississippi Valley. Etymol: Latin *vallis.* Syn: *vale; dale.*

valley [marine geol] A wide, low-relief depression of the ocean floor with gently sloping sides, as opposed to a submarine canyon.

valley axis A term used by Woodford (1951, p.803a) to replace *thalweg,* signifying "the surface profile along the center line of the valley".

valley bottom *valley floor.*

valley braid *anabranch.*

valley drift Outwash material constituting a *valley train.*

valley fill The unconsolidated sediment deposited by any agent so as to fill or partly fill a valley.

valley flat (a) The low or nearly level ground lying between valley walls and bordering a stream channel; esp. the small plain at the bottom of a narrow, steep-sided valley. Howard (1959, p.239) recommended that the term be applied noncommittally to a flat surface that cannot be identified with certainty as a floodplain or terrace. Syn: *flat [geomorph].* (b) A bedrock surface produced by lateral erosion, commonly veneered with the alluvium of a *floodplain* (Thornbury, 1954, p.130).

valley floor The comparatively broad, flat bottom of a valley; it may be excavated and represent the level of a former erosion cycle, or it may be buried under a thin cover of alluvium. Syn: *bottom [geomorph]; valley bottom.*

valley-floor basement The gently sloping, degraded bedrock underlying the *valley-floor side strip* and the valley floor proper, developed in a humid climate by lateral extension of the valley floor at the expense of the enclosing slopes, and covered with slowly creeping soil and floodplain deposits (Davis, 1930).

valley-floor divide A divide in a valley; a dividing height located between two parts of the same valley, each part draining to a different river basin.

valley-floor increment The loose material coming to and lying upon the valley floor (Malott, 1928b, p.12).

valley-floor side strip The narrow, level to slightly concave surface between the wash slope and the valley floor proper (the floodplain), produced by degradation and recession of the valley-side slope. Cf: *valley-floor basement.*

valley glacier *alpine glacier.*

valley head The upper part of a valley.

valley-head cirque A cirque formed at the head of a valley. Cf: *hanging cirque.*

valley iceberg An iceberg eroded in such a manner that a large U-shaped slot, which may be awash, extends through the ice, separating pinnacles or slabs of ice. Syn: *drydock iceberg.*

valley-in-valley (a) Said of the condition, structure, or cross profile of a valley form whose side is marked by a *valley shoulder* separating a steep-sided, youthful valley below from a more widely opened, older valley above. (b) Pertaining to a *two-cycle valley.*

valley line *thalweg [streams].*

valley-loop moraine *loop moraine.*

valley meander One of a series of curves of a *meandering valley.*

valley-moraine lake (val'-ley-mo-raine') A glacial lake formed in a valley by the damming action of a recessional moraine produced by a mountain glacier. Cf: *drift-barrier lake.*

valley network On Mars, a system of interconnected drainages which may have carried fluids but lacks recognized bedform features that are direct indicators of fluid flow (Kieffer et al., 1992).

valley of elevation A syn. of *anticlinal valley.* The term was introduced by Buckland (1829, p.123).

valley of subsidence A syn. of *synclinal valley.* The term was used by Hitchcock (1841, p.178).

valley plain (a) A continuous flood plain (Cotton, 1940). (b) *valley floor.*

valley-plain terrace A term used by Cotton (1940, p.28-29) for the remnant of a formerly continuous flood plain or valley floor; it would include the features now known as a *strath terrace* and a *fillstrath terrace.*

valley plug A local constriction in a stream channel, which may be formed by any of several types of channel obstructions and may cause rapid deposition. See also: *plug [sed].*
valley profile The *longitudinal profile* of a valley.
valley shoulder A bedrock surface made in a *valley-in-valley* form, representing the sharp angle or break in slope between the side or floor of the upper, older valley and the side of the lower, newer valley. It is a remnant of the valley floor formed during a previous erosion cycle, marking the former base level of erosion, and extending across rocks of varying lithology. Syn: *shoulder [geomorph].*
valley-side moraine *lateral moraine.*
valley-side slope (a) A measure, generally expressed in degrees, of the steepest inclination of the side of a valley in stream-eroded topography. Maximum slope is measured at intervals along the valley walls on the steepest parts of the contour orthogonals running from divides to adjacent stream channels. (b) The surface between a drainage divide and the valley floor. Syn: *valley wall.*
valley spring A type of *depression spring* issuing from the side of a valley at the outcrop of the water table.
valley storage (a) The volume of water in a body of water below the water-surface profile. (b) The natural storage capacity or volume of water of a stream in flood that has overflowed its banks; it includes both the water within the channel and the water that has overflowed. (ASCE, 1962).
valley system A valley and all of its tributary valleys.
valley tract The middle part of a stream course, characterized by a moderate gradient and a fairly wide valley. Cf: *plain tract.*
valley train A long, narrow body of *outwash,* deposited by meltwater streams far beyond the terminal moraine or the margin of an active glacier and confined within the walls of a valley below the glacier; it may or may not emerge from the mouth of the valley to join an *outwash plain.* Cf: *gravel train; valley drift.* Syn: *outwash train.*
valley wall *valley-side slope.*
valley wind A daytime *anabatic wind* moving up a valley or mountain slope; alternates with a nighttime *mountain wind.*
vallis (val'-lis) A term established by the International Astronomical Union for a valley on planetary surfaces (USGS, 2003); generally used as a proper name, e.g. Ares Vallis on Mars. A system of valleys is labeled by the plural form, e.g. Valles Marineris on Mars. No particular origin is implied by the term (Mutch, 1970, p.224). Etymol: Latin. Pl: valles.
vallon (val'-lon) A syn. of *val.* Etymol: French, "small valley".
Valmeyeran (Val-mey'-er-an) Provincial series in Illinois: Lower and Upper Mississippian (equivalent to Osagian and Meramecian elsewhere).
valuable mineral In U.S. mining law, any mineral which adds sufficient value to a piece of land as to render it more valuable for mineral exploitation than for other, typically agricultural, purposes.
value (val'-ue) In economic geology, (a) the valuable constituents of an ore; (b) their percentage in an orebody, or *assay grade;* (c) their quantity in an orebody, or *assay value.* See also: *unit value.*
valva (val'-va) A noticeable thickening beyond the ends of the *laesura* branches of triangular *trilete* spores, generally applied in the plural: valvae. Cf: *auricula.* "Auriculate" is the extreme condition of "valvate."
valve (a) One of the distinct and usually articulated pieces that make up the shell of certain invertebrates, e.g. one of the two curved calcareous plates that constitute the shell of a bivalve mollusk; one of the two halves of the carapace of a crustacean, divided by articulation along the mid-dorsal line; or one of the two curved plates that form the shell of a brachiopod. (b) One of the two silicified pieces or encasing membranes forming the top or bottom surface of a diatom frustule; e.g. epivalve and hypovalve. (c) Generally, any of the diverse structures of the body that permit only unidirectional flow (as for blood through the heart).
valvular (val'-vu-lar) Resembling or having the function of a valve in the body of an invertebrate; e.g. "valvular pyramid" of a cystoid or edrioasteroid, composed of several more or less triangular plates covering the anus or a gonopore.
van A term used in the French Alps for *cirque [glac geol]* (Schieferdecker, 1959, term 1667).
vanadate (van'-a-date) A mineral compound characterized by pentavalent vanadium coordinated by four oxygen atoms. An example is vanadinite, $Pb_5(VO_4)_3Cl$. Cf: *arsenate; phosphate.*
vanadinite (va-nad'-in-ite) A red, yellow, or brown mineral of the *apatite* group: $Pb_5(VO_4)_3Cl$. It is isomorphous with pyromorphite, and commonly contains arsenic or phosphorus. Vanadinite often forms globular masses encrusting other minerals in lead mines, and is an ore of vanadium and lead.
vanadiumdravite A dark green to black rhombohedral mineral of the *tourmaline* group: $NaMg_3V_6Si_6O_{18}(BO_3)_3(OH)_4$. It represents dravite with essential V.
vanado-magnetite (va-nad'-o-mag'-net-ite) *coulsonite.*
vanadomalayaite An adamantine deep red monoclinic mineral of the *titanite* group: $Ca(V,Ti)O(SiO_4)$. It represents the V analogue of malayaite.
vanalite (van'-al-ite) A bright yellow-orange monoclinic mineral: $NaAl_8V_{10}O_{38}{\cdot}30H_2O$.
Van Allen belt (Van Al'-len) A zone of charged particles (protons and electrons) surrounding the Earth, beginning at about 1,000 km altitude; produced largely by geomagnetic trapping of solar and cosmic particulate radiation. Radiation belts are also known around Jupiter. Named for geophysicist James A. Van Allen.
vandenbrandeite (van-den-brande'-ite) A blackish-green triclinic mineral: $Cu(UO_2)(OH)_4$. Also spelled vandenbrandite.
vandendriesscheite (van-den-dries'-sche-ite) A yellow or amber-orange orthorhombic mineral: $Pb_{1.6}(UO_2)_{10}O_6(OH)_{11}{\cdot}11H_2O$.
Van der Grinten projection A compromise projection for world maps on which the world is shown enclosed in a circle and all the meridians and parallels, except for the central meridian and the equator, are shown as circular arcs. It resembles the *Mercator projection* but with a curved graticule of meridians and parallels, and the far polar regions are normally omitted. Presented by Alphons van der Grinten of Chicago in 1904 as his first projection of two.
van der Kolk method A test used in refractometry to determine the index of refraction of a mineral relative to that of the liquid medium in which it is immersed. When an obstacle blocks the light rays used for illumination, its shadow appears on the same side as itself when the mineral grain has a relatively higher refractive index, and on the opposite side when the mineral's refractive index is relatively lower than that of the medium. Syn: *oblique illumination method.*
Vandyke brown (Van-dyke') n. black earth. Etymol: its use by the 17th-Century Flemish painter Van Dyck.
vane (a) The target on a *level rod.* (b) One of the sights of a compass or quadrant.
vane test A test performed in the field (field vane test) or in the laboratory (laboratory vane test) to measure the shear strength of fine-grained cohesive soils and other soft deposits. A rod with four flat radial blades, or vanes, projecting at 90° intervals, is forced into the soil and rotated; the torque required to rotate the rod is a measure of the material's shear strength. A handheld device is available for rapid, approximate testing (*torvane* test).
Van Krevelen diagram A plot of elemental ratios (C/N, O/C) used to define the compositional space of organic matter.
vanmeersscheite (van-meers'-sche-ite) A yellow orthorhombic mineral: $U(UO_2)_3(PO_4)_2(OH)_6{\cdot}4H_2O$. Cf: *metavanmeersscheite.*
vanoxite (van-ox'-ite) An ill-defined black mineral: $V^{4+}{}_4V^{5+}{}_2O_{13}{\cdot}8H_2O$(?).
vantasselite (van-tas'-sel-ite) A pearly-white orthorhombic mineral: $Al_4(PO_4)_3(OH)_3{\cdot}9H_2O$.
van't Hoff equation An equation giving the temperature dependence of the *equilibrium constant* of a reaction: $d \ln K/dT=\Delta H°/RT^2$, where K=equilibrium constant, T=absolute temperature, ΔH°=enthalpy change for the hypothetical reaction with all substances in their standard states, and R=gas constant.
vanthoffite (vant-hoff'-ite) A colorless monoclinic mineral: $Na_6Mg(SO_4)_4$.
van't Hoff law The statement in phase studies that, when a system is in equilibrium, of the two opposed interactions, the endothermic one is promoted by raising the temperature, and the exothermic one by lowering it.
vanuralite (van-ur-al'-ite) A citron-yellow monoclinic mineral: $Al(UO_2)_2(V_2O_8)(OH){\cdot}11H_2O$.
vanuxemite (van-ux'-em-ite) A mixture of sauconite and hemimorphite.
Van Veen grab sampler A type of *grab sampler* that encloses ocean-bottom material in two hemicylindrical buckets that rotate shut on a hinge when the sampler strikes the bottom. Syn: *Peterson grab.*
vapor-dominated reservoir A geothermal reservoir in which the fractures and larger pores at depth are filled with steam to the extent that steam is the hydraulically dominant phase (Truesdell and White, 1973; White, Muffler, and Truesdell, 1971). Syn: *dry-steam reservoir.*
vaporization (va'-por-i-za'-tion) *evaporation.*
vara (va'-ra) Any of various old Spanish units of length used in Latin America and SW U.S., equal in different localities to between 31 and 34 inches; e.g. a unit equal to 33.3333 inches in Texas, to 33.372 inches in California, to 33.00 inches in Arizona and New Mexico, and to 32.9931 inches and 32.9682 inches (among others) in Mexico. For other values, see ASCE (1954, p.169-170).
Varangian (Va-ran'-gi-an) A near-global glacial interval at the end of the *Cryogenian* Period (middle *Neoproterozoic* Era). It is equivalent to the Marinoan glacial episode. Also called: Varanger.
varennesite A brownish-yellow or orange orthorhombic mineral: $Na_8(Mn,Fe^{3+},Ti)_2Si_{10}O_{25}(OH,Cl)_2{\cdot}12H_2O$.
variability [grd wat] (var'-i-a-bil'-i-ty) The ratio of the difference between maximum and minimum discharge of a spring to its average discharge, expressed as a percentage.
variability [paleont] The quality or attribute of an organism that causes it to exhibit variation.
variable (var'-i-a-ble) (a) Any measurable or changeable statistical quality or quantity; e.g., *independent variable* and *dependent variable.* Cf: *attribute [geol]; variate; parameter [stat].* (b) A quantity that can assume any of a given set of values at different stages in a computer program.
variable-amplitude trace (var'-i-a-ble-am'-pli-tude) *wiggle trace.*
variable-area (var'-i-a-ble-ar'-e-a) A display of seismic data in which the height of a blackened area on film or paper is proportional to the signal amplitude. Abbrev: VA.

variable-density (var'-i-a-ble-den'-si-ty) A display of seismic data in which the darkness of the image is proportional to the signal amplitude. Abbrev: VD.

variable density log A log of the acoustic wave train of compressional, shear and other wave-forms recorded in a variable density photographic display of relative amplitude.

variance [chem] (var'-i-ance) *degrees of freedom.*

variance [exp petrol] (Symbol: V) *degree of freedom.* See also: *phase rule.*

variance [stat] A measure of dispersion about the *mean.* In a sample, the average squared difference of each observation from the sample mean. Symbols: σ^2, s^2.

variance-covariance matrix In *multivariate* statistics, a square matrix whose diagonal elements are variances of the variables and the off diagonal elements are covariances of pairs of variables.

variant (var'-i-ant) An individual exhibiting *variation.*

variate (var'-i-ate) *random variable.*

variation Divergence in the structural or functional characteristics of an organism from those that are considered typical of the group to which it belongs. See also: *variant.*

variation diagram (var-i-a'-tion) A binary or ternary diagram that shows the relations among various chemical parameters (e.g., oxide percentages, Niggli numbers, differentiation indexes) of the igneous rocks in a suite.

variegated (var'-i-e-gat'-ed) Said of a sediment or sedimentary rock, such as red beds or sandstone, showing variations of color in irregular spots, streaks, blotches, stripes, or reticulate patterns. Cf: *mottled [sed].*

variegated copper ore *bornite.*

Variegerminantes (Var-i-e-ger-mi-nan'-tes) The name of an *anteturma* in the turmal system for fossil palynomorphs, including all *pollen* and *prepollen* produced by seed plants. Cf: *turma; Proximegerminantes.*

varietal mineral (va-ri'-e-tal) A mineral that is either present in considerable amounts in a rock or characteristic of the rock; a mineral which distinguishes one variety of rock from another. Syn: *distinctive mineral.*

variety [mineral] (va-ri'-e-ty) In gemology, a mineral that is a type of the mineral species, distinguished by color or other optical phenomenon or characteristic: e.g. emerald and aquamarine are varieties of beryl.

variety [taxon] A category in the hierarchy of botanical classification subordinate in rank to *subspecies.* Such infraspecific taxa are excluded from the current Code of Zoological Nomenclature.

varigradation (var'-i-gra-da'-tion) A term used by McGee (1891, p.261-267) for the process by which all streams of progressively increasing discharge tend constantly, in a degree varying inversely with the discharge, to depart slightly from the normal gradients.

varimax rotation In *factor analysis,* rotation of remaining factors to new positions in variable space after nonessential factors have been deleted. Cf: *rotated factor.*

variogram (var'-i-o-gram) A measure of spatial continuity that contains the information necessary for geostatistical estimation. A variogram is analogous to an *autocorrelation* function in *time series* analysis. Variograms may be anisotropic, i.e., may change in different directions.

variole (var'-i-ole) A pea-size spherule, usually composed of radiating crystals of plagioclase or pyroxene. This term is generally applied only to such spherical bodies in basic igneous rock, e.g. variolite. Cf: *spherulite [petrology].*

variolite (var'-i-o-lite') A general term, coined by Aldrovande in 1648 (Johannsen, 1938, p.300), for aphanitic and fine-grained igneous rocks containing *varioles.* The term is derived from the Latin for "smallpox", in allusion to the spotted appearance of weathered surfaces. Obsolete.

variolitic (var'-i-o-lit'-ic) Spherulitic texture confined to mafic volcanic rocks. Cf: *spherulitic.*

variometer (var-i-om'-e-ter) An obsolete instrument used for measuring small magnetic variations by slight rotation of a permanent magnet suspended on torsion fiber.

Variscan Orogeny (Var-is'-can) The late Paleozoic orogenic era of Europe, extending through the Carboniferous and Permian. By current usage, it is synonymous with the *Hercynian Orogeny.* Cf: *Armorican Orogeny; Altaides.*

Variscides (Var-is'-ci-des) Hercynian mountain chains in Europe. Approx. syn: *Hercynides.*

variscite (var'-is-cite) (a) A pale green, emerald green, bluish-green, or colorless orthorhombic mineral: $AlPO_4{\cdot}2H_2O$. It is isomorphous with strengite and dimorphous with metavariscite. Variscite is a popular material for cabochons and various kinds of carved objects, and is often used as a substitute for turquoise. See also: *sphaerite.* Syn: *lucinite; utahlite.* (b) A group name for orthorhombic minerals with the general formula $AXO_4{\cdot}2H_2O$, where A = Al, Fe^{3+}, or In and X = P or As.

varix (var'-ix) (a) One of the transverse elevations of the surface of a gastropod shell that is more prominent than a costa and that represents a halt in growth during which a thickened outer lip was developed (TIP, 1960, pt. I, p.134). (b) A thickening of an ammonoid shell marked on an internal mold by a transverse groove (Moore et al., 1952, p.366). Pl: varices.

varlamoffite (var'-lam-off-ite) An ill-defined orange-red tetragonal mineral: $(Sn,Fe)(O,OH)_2$. It is perhaps fine-grained cassiterite.

varnish (var'-nish) *rock varnish.*

varulite (va'-ru-lite) A dull olive-green monoclinic mineral of the *alluaudite* group: $CaMn^{2+}_3(PO_4)_3$.

varve (a) A sedimentary bed or lamina or sequence of laminae deposited in a body of still water within one year's time; specif. a thin pair of graded glaciolacustrine layers seasonally deposited, usually by meltwater streams, in a glacial lake or other body of still water in front of a glacier. A glacial varve normally includes a lower "summer" layer consisting of relatively coarse-grained, light-colored sediment (usually sand or silt) produced by rapid melting of ice in the warmer months, which grades upward into a thinner "winter" layer, consisting of very fine-grained (clayey), often organic, dark sediment slowly deposited from suspension in quiet water while the streams were ice-bound. Counting and correlation of varves have been used to measure the ages of Pleistocene glacial deposits. (b) Any cyclic sedimentary *couplet,* as in certain shales and evaporites. Cf: *rhythmite.* Etymol: Swedish "varv", "layer" or "periodical iteration of layers" (De Geer, 1912, p.242).

varved clay A distinctly laminated lacustrine sediment consisting of clay-rich varves; also the upper, fine-grained, "winter" layer of a glacial varve.

varvite (var'-vite) An indurated rock consisting of ancient varves.

varzea (var'-ze-a) A term used in Brazil and Portugal for an alluvial flood plain or the bank of a river; also, a field or a level tract of land, esp. one that is sowed and cultivated. Etymol: Portuguese "várzea".

varzea lake A lake on the alluvial floodplain of a river such as the Amazon, where the annual changes in discharge add great amounts of water to the varzea lakes during the season of high water (Wetzel, 2001, p.31).

vascular bundle (vas'-cu-lar) In a vascular plant, conductive tissue composed of a strand of xylem and a strand of phloem, commonly separated by cambium and containing sclerenchymatous supporting tissue (Swartz, 1971, p.492).

vascular cambium In *woody plants,* a *cambium* or lateral *meristem* beneath the bark that produces secondary *xylem* (wood) inwardly and secondary *phloem* (bark) outwardly. Cf: *cambium; cork cambium.*

vascular plant A plant with a well-developed conductive system and structural differentiation; a *tracheophyte.* The majority of visible terrestrial plants are vascular.

vascular ray A ribbonlike aggregate of cells extending radially in stems through xylem and, often, phloem for lateral transport of fluids (Fuller and Tippo, 1954, p.974). Syn: *ray [bot].*

vascular tissue The conducting tissue of vascular plants, composed of xylem and phloem.

vase (Pron: vahz) Freshwater silt deposited in estuaries along the Atlantic coast of Europe and Africa, consisting of a mixture of sandy and pulverulent grains of quartz, calcite, clay minerals, and diatom shells, with a binder of *algon* (Bourcart, 1941). Etymol: French, "slime, mud".

vashegyite (vas-heg'-y-ite) A white, pale greenish-yellow, or rust-brown orthorhombic mineral: $Al_{11}(PO_4)_9(OH)_6{\cdot}38H_2O$.

vasilite A metallic steel-gray cubic mineral: $(Pd,Cu)_{16}(S,Te)_7$.

vasilyevite An damantine or metallic silvery gray, red-black, or black triclinic mineral: $Hg^{2+}_{10}O_6I_3Br_2Cl(CO_3)$.

vastitas (vas'-ti-tas) A term established by the International Astronomical Union for an extensive plain on Mars (USGS, 2003). Used as part of a formal name for a Martian landform, such as Vastitas Borealis (Mutch, et al., 1976, p.57). Etymol: Latin "vastus", empty, immense. Pl: vastitates.

vaterite (va'-ter-ite) A relatively rare colorless hexagonal mineral: $CaCO_3$. It is trimorphous with calcite and aragonite, and is a relatively unstable form of calcium carbonate.

vaterite-A Artificial calcite.

vaterite-B Artificial vaterite.

vauclusian spring (vau-clu'-sian) A type of *karst spring* in which water is forced out under artesian pressure and rises along a deep, steeply inclined, water-filled passage. It has generally an exceptionally large discharge. Etymol: After Fontaine de Vaucluse in southern France.

vaughanite (vaugh'-an-ite) A metallic triclinic mineral: $TlHgSb_4S_7$.

vaugnerite (vaug'-ner-ite) A dark-colored, coarse-grained hypabyssal *quartz diorite* containing abundant biotite, along with green amphibole, white plagioclase, and quartz, with accessory alkali feldspar, apatite, magnetite, pyrite, and titanite. The name was given by Fournet in 1836 for Vaugneray, France. Obsolete.

vault [geomorph] A structure in the Earth's crust, resembling or suggesting a vault or an arched room; e.g. a cavern or a volcanic crater.

vault [paleont] (a) The part of a blastoid *theca* above the dorsal region (from aboral tips of ambulacra to dorsal pole). (b) An arched covering of calcareous plates between crinoid arms.

vauquelinite (vau-quel'-in-ite') A green to brownish-black monoclinic mineral: $Pb_2Cu(PO_4)(CrO_4)(OH)$. It is isomorphous with fornacite.

vaux Pl. of *val.*

vauxite (vaux'-ite) A blue triclinic mineral: $Fe^{2+}Al_2(PO_4)_2(OH)_2{\cdot}6H_2O$. It has less water than metavauxite and paravauxite.

väyrynenite (väy-ry'-nen-ite) A rose-red monoclinic mineral: $Mn^{2+}Be(PO_4)(OH)$.

V-bar A *cuspate bar* whose seaward angle is fairly sharp, as where a secondary spit trails abruptly back toward the shore from the point of a primary spit.
V-coal Microscopic coal particles that are predominantly vitrain and clarain, as found in miners' lungs. Cf: *F-coal; D-coal.*
VD Display in *variable-density* form.
VE *vertical exaggeration [cart].*
veatchite (veatch'-ite) A white monoclinic mineral: $Sr_2[B_5O_8(OH)]_2B(OH)_3{\cdot}H_2O$. It is dimorphous with p-veatchite. A triclinic type has been named veatchite-A.
vector (a) A matrix (or array of numbers) consisting of only one row or one column. (b) A format for processing and displaying graphic data. Vector data are represented by strings of coordinate pairs defining the true position of features represented by points, lines, and areas. (c) In digital terminology, the line connecting two nodes. (d) A quantity having both a magnitude and a direction.
vector structure (vec'-tor) *directional structure.*
veenite (veen'-ite) A metallic gray orthorhombic mineral: $Pb_2(Sb,As)_2S_5$.
veering A clockwise change in wind direction, e.g. north to northeast to east. Cf: *backing.*
vegasite (ve'-ga-site) A mineral that may be identical with plumbojarosite.
vegetable jelly (veg'-e-ta-ble) *ulmin.*
vegetal discharge (veg'-e-tal) The release, through the transpiration of plants, of water derived either from the zone of areation or from the zone of saturation by way of the capillary fringe. See also: *vadose-water discharge.*
vegetation anomaly (veg-e-ta'-tion) As seen on aerial or space photographs or images, a deviation from the normal distribution or properties of vegetation. It may be caused by conditions along faults, trace elements in the soil, or other factors.
vegetation arabesque *vegetation polygon.*
vegetation coast A coast that is being extended seaward by the growth of vegetation, such as the mangrove trees in southwest Florida and the salt marsh coast of the Big Bend in Florida.
vegetation polygon A small *nonsorted polygon* whose fissured borders are emphasized by thick vegetation (usually moss, lichen, or willow) and whose center consists of fine-textured material or a mixture of fines and stones. Diameter: about 1 m. See also: *lichen polygon; tussock-birch-heath polygon.* Syn: *vegetation arabesque.*
vegetation stripe (a) A form of *nonsorted stripe.* (b) A *sorted stripe* emphasized by vegetation (Sigafoos, 1951, p.289).
vegetative cyst An *algal* vegetative cell that also functions as a *cyst.*
vegetative reproduction (veg'-e-ta'-tive) Plant reproduction by vegetable parts such as buds and gemmae; nonsexual reproduction (Swartz, 1971, p.494). Cf: *budding.*
vei *volcanic explosivity index.*
veil [cryst] An aggregate of minute bubbles creating a whitish or cloudlike appearance in quartz.
veil [paleont] A variously formed weblike or netlike film in a radiolarian; e.g. *patagium.*
vein [bot] One of the vascular bundles of a leaf. See also: *venation.*
vein [ice] (a) A narrow water channel within *land ice;* also, the stream of water flowing through such a channel. (b) A narrow *lead [ice]* or *lane* in *pack ice.*
vein [intrus rocks] A thin, sheetlike igneous intrusion into a fissure. Not recommended usage as an igneous feature.
vein [ore dep] An epigenetic mineral filling of a fault or other fracture in a host rock, in tabular or sheetlike form, often with associated replacement of the host rock; a mineral deposit of this form and origin. Cf: *lode [eco geol].*
vein [streams] (a) A narrow waterway or channel in rock or earth. Also, a stream of water flowing in such a channel. (b) An archaic term for the flow or current of a stream.
vein [struc geol] Any fracture that contains mineralized material. Veins can display either crack-normal extension or shear displacement coupled with crack-normal extension.
vein bitumen Any one of the black or dark-brown bitumens that give off a pitchy odor, burn readily with a smoky flame, and occupy fissures in rocks or less frequently form basin-shaped deposits on the surface (Nelson and Nelson, 1967).
veined gneiss A composite gneiss with irregular layering. The term is generally used in the field and has no genetic implications (Dietrich, 1960, p.50). Cf: *venite; arterite; phlebite; composite gneiss.*
vein quartz A rock composed chiefly of sutured quartz crystals of pegmatitic or hydrothermal origin and commonly of variable size.
vein system An assemblage of veins of a particular area, age, or fracture system, usually inclusive of more than one lode.
velar (ve'-lar) Having the form of a veil, frill, or curtain.
velardeñite (ve-lar-deñ'-ite) *gehlenite.*
velar dimorphism In certain ostracodes, a kind of dimorphism in which the velar structures of the two sexes differ in size and/or curvature. Less preferred syn: velate dimorphism.
velar ridge A *velar structure* developed as a low, usually rounded ridge in an ostracode valve. Less-preferred syn: velate ridge.
velar structure Any elongate elevated structure parallel to the free edge of the valve in straight-backed ostracodes, commonly developed as a *frill* (velum) or a *velar ridge;* if a carina is present, the velar structure lies ventral to it.
Vela Uniform (Ve'-la U'-ni-form) A research program, sponsored by the Advanced Research Projects Agency of the U.S. Dept. of Defense, which had the objective of improving the capability of detecting underground nuclear explosions and of distinguishing them from earthquakes.
veld An open grassland area of South Africa with a diversity of elevation, vegetation, soil, etc. Also spelled: veldt.
veliger (vel-i-ger') Said of the swimming larval stage of mollusks that use a ciliated swimming organ, a velum.
velikite (vel'-ik-ite) A metallic dark gray tetragonal mineral: Cu_2HgSnS_4.
velocimeter (ve-lo-cim'-e-ter) An instrument for measuring the velocity of sound in water, sometimes used to correct Doppler-sonar data for salinity and temperature variations.
velocity analysis (ve-loc'-i-ty) Calculation of seismic velocity distribution using normal-moveout times. Involves finding the hyperbola which best fits arrival times.
velocity coefficient A numerical factor, always less than unity, that expresses the ratio between the actual velocity of water issuing from an orifice or other hydraulic structure or device and the theoretical velocity which would exist if there were no friction losses due to the orifice. The square of the velocity coefficient is a measure of the efficiency of a structure as a waterway. It is a dimensionless number (ASCE, 1962).
velocity contrast A change in velocity, such as produces a reflection or a change in wave direction.
velocity discontinuity *discontinuity [seis].*
velocity distribution The relationship between seismic velocity and horizontal and vertical location.
velocity filter Discrimination on the basis of *apparent velocity* (or *dip moveout*).
velocity function Seismic velocity approximated by a mathematical funtion of depth or time.
velocity gradient The rate of change of velocity with respect to distance normal to the direction of flow (ASCE, 1962).
velocity head Groundwater head resulting from kinetic energy of the water. It is negligible in most geological contexts. See also: *total head.*
velocity-head coefficient A correction factor applied to the velocity head of the mean velocity to correct for nonuniformity of velocity in a cross section. The factor is 1.0 where velocities are identical across a section and greater than 1.0 where velocities vary across a section (ASCE, 1962).
velocity inversion A decrease in velocity with depth. Such a situation can result in erroneous refractor depth calculation.
velocity log *sonic log.*
velocity model A specification of the value of P-wave velocity for each point in a volume, usually specified in depth, sometimes in two-way vertical traveltime.
velocity panels A display of the coherency when various normal moveouts (implying various velocities) are assumed.
velocity profile A seismic arrangement used to record reflections over a large range of source-to-geophone distances, which is used to determine seismic velocity from the time-distance relationship.
velu A term used in the Maldive Islands of the Indian Ocean for the lagoon of a *faro.*
velum (ve'-lum) (a) A sail- or frill-like structure along the distal part of the valve of an ostracode, typically developed as the double-walled outfold of the carapace. (b) A thin membrane in larval mollusks. Pl: vela.
velvet copper ore (vel'-vet) *cyanotrichite.*
venanzite (ve-nan'-zite) A holocrystalline porphyritic extrusive *leucitite* or melilitite composed of olivine and phlogopite phenocrysts in a fine-grained groundmass of these minerals plus melilite, leucite, and magnetite. The name, given by Sabatini in 1898, is for San Venanzo, Umbria, Italy. Woolley et al. (1996) replace this name with kalsilite-phlogopite-olivine-leucite melilitite. Obsolete.
venation (ve-na'-tion) The arrangement of the vascular bundles (veins) in a leaf. See also: *parallel venation; net venation.*
Vendian (Ven'-di-an) *Ediacaran.*
veneer (ve-neer') (a) A weathered or otherwise altered coating on a rock surface; e.g. desert varnish. (b) Said of a thin layer of sediment overlying thicker strata or beds.
Vening Meinesz zone (Ven'-ing Mei'-nesz) A belt of negative gravity anomalies, generally related to island arcs and/or oceanic deeps.
venite (ven'-ite) A *migmatite* of which the mobile portion(s) were formed by exudation (secretion) from the rock itself (Dietrich and Mehnert, 1961). Proposed, along with the term *arterite* (Sederholm), to replace *veined gneiss* with terms of genetic connotation (Mehnert, 1968, p.17). Cf: *composite gneiss; diadysite; phlebite.* Rarely used.
vent The opening at the Earth's surface through which volcanic materials are extruded; also, the channel or conduit through which they pass. Cf: *neck [volc]; pipe [volc].* See also: *feeder [volc]; chimney [volc].*
venter [bot] (ven'-ter) In the female gametophyte of certain plants, the enlarged base of an archegonium that contains the egg.

venter [paleont] (a) The outer and convex part of the shell of a curved or coiled cephalopod or gastropod, or the peripheral wall of a cephalopod whorl comprising the part of the shell radially farthest from the *protoconch;* the underside of a nautiloid and of its conch, distinguished generally by the hyponomic sinus and often by a conchal furrow (TIP, 1964, pt. K, p.59). (b) The median region of the shell of a productid brachiopod, situated between the valve surfaces on either side of the median sector of the shell (TIP, 1965, pt. H, p.155). (c) The belly region or the lower part of the carapace of an ostracode. Cf: *dorsum [paleont].*

ventifact (ven'-ti-fact) A general term introduced by Evans (1911) for any stone or pebble shaped, worn, faceted, cut, or polished by the abrasive or sandblast action of windblown sand, generally under desert conditions; e.g. a *dreikanter.* See also: *windkanter.* Syn: *glyptolith; wind-worn stone; wind-cut stone; wind-polished stone; wind-grooved stone; wind-scoured stone; wind-shaped stone.*

venting The escape through the Earth to the atmosphere of gases or radioactive products from an underground high-explosive or nuclear detonation.

ventral (ven'-tral) (a) Pertaining or belonging to the abdominal or lower surface of an animal or of one of its parts that is opposite the back, e.g. in the direction toward the pedicle valve from the brachial valve of a brachiopod; pertaining to the region of the shell of a bivalve mollusk opposite the hinge, where the valves open most widely; pertaining to the side of the stipe on which the thecal apertures of a graptoloid are situated or to the inferior (commonly umbilical or apertural) side of a foraminiferal test; in a vertebrate, pertaining to the underside of the body (opposite from the spinal cord). (b) Referring to the direction or side of an echinoderm toward or containing the mouth, normally upward; adoral or oral. Ant: *dorsal.*

ventral lobe The main adapical lobe or inflection of a *suture* on the venter of a cephalopod shell. See also: *external lobe.* Cf: *dorsal lobe.*

ventral process A medially located excessive thickening of secondary shell of a brachiopod underlying the pseudodeltidium and projecting dorsally to fit between lobes of the cardinal process.

ventral shield An *ossicle* of secondary origin on the oral side of an arm in an ophiuroid. Cf: *dorsal shield.*

ventral valve The *pedicle valve* of a brachiopod.

ventromyarian (ven'-tro-my-ar'-i-an) Said of a nautiloid in which the retractor muscles of the head-foot mass are attached to the shell along the interior areas of the body chamber adjacent to, or coincident with, its ventral midline (TIP, 1964, pt. K, p.59). Cf: *dorsomyarian; pleuromyarian.*

Venturian (Ven-tu'-ri-an) Western North American stage: Upper Pliocene (above Repettian, below Wheelerian).

Venus hair (Ve'-nus) Needlelike crystals of reddish-brown or yellow rutile, forming tangled swarms of inclusions in quartz. See also: *sagenite.*

Venus hairstone *rutilated quartz.*

verbeekite A metallic black monoclinic mineral: Cu_2HgSnS_4.

verd antique A dark green massive *serpentine,* commonly with veinlets of calcium carbonate and magnesium carbonate. It is capable of being polished and is commercially considered a marble. Var. spelling: verde antique. Cf: *ophicalcite.* Syn: *serpentine marble.*

verdelite (ver'-de-lite) A green variety of tourmaline.

verdite (verd'-ite) A deep green, relatively soft metamorphic rock, long known from the Barberton district of the Transvaal and more recently found in Zimbabwe (Harding and Jobbins, 1984). It is used by local artists and craftsmen for carvings. Verdite consists chiefly of finely intergrown chromian muscovite (*fuchsite*), with scattered grains of rutile.

vergasovaite An olive-green orthorhombic mineral: $Cu_3OMoO_4(SO_4)$.

vergence (ver'-gence) The direction of overturning or of inclination of a fold. The term is a translation of the German "Vergenz", "overturn", coined by Stille (1930, p.379) for the direction in which a geologic structure or family of structures is facing. Cf: *facing.*

verglas (ver'-glas) A thin film or layer of clear, hard, smooth ice on a rock surface, formed as a result of a frost following rain or snowmelt or by rime. Etymol: French. Cf: *black ice.*

verite (ver'-ite) A black extrusive *lamproite* containing phlogopite or biotite, clinopyroxene, and olivine crystals in a glassy groundmass; the glassy periphery of *fortunite.* Its name, given by Osann in 1889, is derived from the town of Vera, Cabo de Gata, Spain. Mitchell and Bergman (1991) suggest that this name should be replaced by glassy olivine-diopside-phlogopite *lamproite.* Obsolete.

vermeil (ver-meil') (a) An orange-red garnet. Syn: *vermilion.* (b) A reddish-brown to orange-red gem corundum; ruby. (c) An orange-red spinel. Syn: vermeille.

vermetid reef (ver'-me-tid) A small *organic reef* composed of the irregularly entwined tubelike calcareous shells of vermetid (wormlike) gastropods, as in Bermuda. Cf: *serpulid reef; worm reef.*

vermicular Microscopic igneous texture involving a wormy intergrowth of optically continuous quartz bodies with the outer fringes of K-feldspar crystals in contact with grains of plagioclase.

vermicular quartz (ver-mic'-u-lar) Quartz occurring in wormlike forms intergrown with or penetrating feldspar, as in myrmekite.

vermiculate (ver-mic'-u-late) Of a palynomorph *sculpture,* consisting of irregular raised areas and/or elongate, irregularly placed depressions (Williams et al., 2000, p.247).

vermiculated (ver-mic'-u-lat'-ed) Said of a stone, carbonate sediment, or any corroded geologic feature that has the appearance of having been eaten into by worms.

vermiculite (ver-mic'-u-lite) (a) A 2:1 layer monoclinic clay mineral distinguished from smectite on the basis of a higher layer charge, generally set at greater than 0.6 per formula unit: $Mg_{0.7}(Mg,Fe,Al)_6(Si,Al)_8O_{20}(OH)_4 \cdot 8H_2O$. (b) A group name for sheet silicates with exchangeable cations.

vermiform (ver'-mi-form) Wormlike or having the form of a worm; e.g. "vermiform problematica", consisting of long thin more or less cylindrical tubes.

vermiglyph (ver'-mi-glyph) A collective term used by Fuchs (1895) for a trace fossil consisting of a presumed worm trail appearing on the undersurface of flysch beds (mostly sandstones) as a threadlike, unbranched, and irregular relief form a few millimeters wide with a straight or variously winding course. Cf: *graphoglypt; rhabdoglyph.*

vermilion (ver-mil'-ion) (a) *cinnabar.* (b) An orange-red garnet; *vermeil.* Syn: vermillion.

vernacular name (ver-nac'-u-lar) In biologic nomenclature, the common name of a plant or animal as opposed to its formal Latin name; e.g. sugar maple is the vernacular name of *Acer saccharum.* Syn: *popular name.* Cf: *scientific name.*

vernadite (ver-nad'-ite) A sooty black hexagonal mineral: $(Mn^{4+},Fe^{3+},Ca,Na)(O,OH)_2 \cdot nH_20$. It may be a random-stacked birnessite.

vernadskite (ver-nad'-skite) *antlerite.*

Verneuil process (Ver-neuil') A method developed by Auguste V.L. Verneuil (1856-1913), French mineralogist and chemist, for the manufacture of large synthetic crystals of corundum and spinel, in which an alumina powder of the desired composition is melted in an oxyhydrogen flame, producing a series of drops that build up the boules of the synthetic gems.

vernier (ver'-ni-er) A short, uniformly divided, auxiliary scale that slides along the primary scale of a measuring device and that is used to measure accurately fractional parts of the smallest divisions of the primary scale or to obtain one more significant figure of a particular measurement. It is graduated such that the total length of a given number of divisions on a vernier is equal to the total length of one more or one less than the same number of divisions on the primary scale; parts of a division are determined by observing what line on the vernier coincides with a line on the measuring device. Named after Pierre Vernier (1580-1637), French mathematician.

vernier compass A surveyor's compass with a vernier, used for measuring angles without the use of the magnetic needle by means of a compensating adjustment made for magnetic variation.

verplanckite (ver-planck'-ite) A brownish-orange or light brownish-yellow hexagonal mineral: $Ba_4Mn^{2+}_2Si_4O_{12}(OH,H_2O)_3Cl_3$.

verrou A syn. of *riegel.* Etymol: French, "bolt".

verrucate (ver'-ru-cate) Warty, or covered with wartlike knobs or elevations; e.g. said of palynomorphs having sculpture consisting of such projections. Syn: *verrucose.*

verrucose (ver'-ru-cose) *verrucate.*

versant (ver'-sant) (a) The slope or side of a mountain or mountain chain. (b) The general slope of a region; e.g. the Pacific versant of the U.S. An obsolete term.

versiliaite (ver-sil'-i-a-ite) A metallic black orthorhombic mineral: $(Fe^{2+},Fe^{3+},Zn)_8(Sb^{3+},Fe^{3+},As)_{16}O_{32}S_{1.3}$.

Versilian (Ver-sil'-i-an) European stage: *Holocene* (above *Tyrrhenian*). Cf: *Flandrian.*

verst A Russian unit of distance equal to 0.6629 mile or 1.067 km.

vertebra (ver'-te-bra) (a) One of the bony or cartilaginous elements that together make up the spinal column of a vertebrate. (b) One of a fused pair of opposite ambulacral plates of an asterozoan, articulating with neighboring vertebrae by ball-and-socket joints. Pl: vertebrae.

Vertebrata (Ver-te-bra'-ta) A subphylum of the Chordata characterized by an internal skeleton of cartilage or bone, and by specialized organization of the anterior end of the animal; the front of the body is a head that bears organs of sight, smell, taste, and hearing, and the front of the central nervous system is a brain.

vertebrate paleontology (ver'-te-brate) The branch of paleontology dealing with fossil vertebrates. Syn: *vertebrate paleozoology.*

vertebrate paleozoology *vertebrate paleontology.*

vertex (ver'-tex) The culmination or high point of a feature. The term was used by Sollas as the English version of Suess's term "Scheitel", the apex or nucleus of a continental or other large structure; example, the *Angara Shield,* supposed to have been the vertex of Asia.

vertical (ver'-ti-cal) adj. Said of a direction that is perpendicular to a *horizontal* plane and parallel to which the force of gravity acts.

vertical accretion Upward growth of a sedimentary deposit; e.g. settling of sediment from suspension in a stream subject to overflow. Cf: *lateral accretion.*

vertical-accretion deposit (ver'-ti-cal-ac-cre'-tion) *flood-plain deposit.*

vertical accuracy Measure of the positional accuracy of a dataset with respect to a specified vertical datum (Geo-One-Stop, 2003, p.5).

vertical angle An angle in a vertical plane; the angle between the horizontal and an inclined line of sight, measured on a vertical circle either upward or downward from the horizon. One of the directions that form a vertical angle in surveying is usually either the direction of the vertical (the angle being termed the zenith distance) or the line of intersection of the vertical plane in which the angle lies with the plane of the horizon (the angle being termed the altitude). Cf: *horizontal angle.*

vertical angulation *trigonometric leveling.*

vertical axis The line through the center of a theodolite or transit about which the alidade rotates. Cf: *horizontal axis.*

vertical cable A cable used in deep water that is suspended vertically and contains several hydrophones. The result simulates a vertical seismic profile and permits imaging a large area around the vertical cable.

vertical cave A cave containing vertical or nearly vertical shafts, commonly but not necessarily at the entrance. Syn: *pothole [speleo]; pit [speleo]; pot [speleo]; sótano; gouffre.* Partial syn: *cenote; aven.*

vertical circle Any great circle of the celestial sphere passing through the zenith.

vertical collimator A telescope so mounted that its collimation axis may be made to coincide with the vertical (or direction of the plumb line). It may be used for centering a theodolite on a high tower exactly over a station mark on the ground. See also: *collimator.*

vertical control A series of measurements taken by surveying methods for the determination of elevation with respect to an imaginary level surface (usually mean sea level) and used as fixed references in positioning and correlating map features.

vertical coordinate The vertical distance (i.e., distance measured along a vertical; elevation) of a point above or below a surface of reference (datum).

vertical corrasion Corrasion of the stream bed, causing a deepening of the channel.

vertical dip slip *vertical slip.*

vertical erosion *downcutting.*

vertical exaggeration [cart] (a) A deliberate increase in the vertical scale of a relief model, plastic relief map, block diagram, or cross section, while retaining the horizontal scale, in order to make the model, map, diagram, or section more clearly perceptible. (b) The ratio expressing vertical exaggeration; e.g., if the horizontal scale is one inch to one mile (1:63,360) and the vertical scale is one inch to 2,000 ft (1:24,000), the vertical exaggeration is 2.64. Abbrev: VE.

vertical exaggeration [photo] The apparent increase in the relief as seen in a stereoscopic image.

vertical-field balance An instrument that measures the vertical component of the magnetic field by means of the torque that the field component exerts on a horizontal permanent magnet. Cf: *horizontal-field balance; torsion magnetometer.*

vertical fold A fold having a vertical axis.

vertical form index A term used by Bucher (1919, p.154) for a ratio also known as *ripple index.* Cf: *horizontal form index.*

vertical gradient A rate of change of the potential field, such as the Earth's gravity or magnetic field, in the vertical direction. It can be measured by a *gradiometer* or calculated from the horizontal gradient. The calculated vertical gradient is called *first vertical derivative.*

vertical gradiometer An instrument for measuring the vertical gradient of a *potential field.*

vertical intensity The vertical component of the *magnetic-field intensity;* it is one of the *magnetic elements,* and is symbolized by Z. It is usually considered positive if downward, negative if upward. Cf: *horizontal intensity.*

vertical interval The difference in vertical height between two points on a land surface; specif. *contour interval.* Abbrev: VI. Cf: *horizontal equivalent.* Syn: vertical distance.

vertical limb A graduated arc attached to a surveying instrument and used to measure vertical angles.

vertical-loop method A classical inductive electromagnetic method of prospecting in which the transmitting coil is in the vertical plane.

vertically mixed estuary (ver'-ti-cal-ly) An estuary in which there is no measurable variation in salinity with depth, although salinity may increase laterally from the head to the mouth; occurs where tidal currents are very strong relative to river flow. Ant: *stratified estuary.*

vertical nondenticular composite prismatic structure A *nondenticular composite prismatic structure* in which the first-order prisms are vertical or nearly vertical (Carter, 1990, Fig. 3D).

vertical photograph An aerial photograph made with the camera axis vertical (camera pointing straight down) or as nearly vertical as possible in an aircraft.

vertical pore A pore surrounding the vertical spine in the back of the lattice shell in the radiolarian skeletons of the subfamily Trissocyclinae.

vertical section (a) A natural or artificial vertical exposure of rocks or soil, as in a sea cliff or canyon wall. (b) A section or diagram representing a vertical segment of the Earth's crust either actually exposed or as it would appear if cut through by any intersecting vertical plane; e.g. a *columnar section* or a *structure section.*

vertical seismic profiling Measurements of the response of a geophone at various depths in a borehole to sources on the surface (Sheriff and Geldart, 1995, p.487-492). Where the source point is an appreciable horizontal distance from the wellhead, the result is an offset VSP. For an azimuthal VSP the sources are offset in different directions. A walkaway VSP involves moving the source away from the wellhead between energy injections. A reversed VSP has the source in the borehole and geophones on the surface. Abbrev: VSP.

vertical separation In a fault, the distance measured vertically between two parts of a displaced marker such as a bed. Cf: *horizontal separation.*

vertical shaft In a cave, a rounded vertical passage or high chamber, characterized by vertical solution grooves on its walls and usually by showering water. Cf: *aven; blind shaft; chimney [speleo]; pit [speleo].*

vertical sheet structure Slightly sinuous, subparallel, mainly vertical streaks in a sedimentary rock, distinguished from the surrounding rock by their lighter color and greater resistance to weathering. In plan view they show a straight or slightly wavy pattern of subparallel lines...and thus have a sheetlike three-dimensional form (Laird, 1970, p.428). The structure is thought to be formed after deposition of a sand bed by escape of fluid, which washes out some of the finer particles. Lowe (1975) called the same type of structures *stress pillars.* Cf: *pillar structure.*

vertical shift In a fault, the vertical component of the shift. Little used.

vertical slip In a fault, the vertical component of the net slip; it equals the vertical component of the dip slip. Cf: *horizontal slip.* Syn: *vertical dip slip.*

vertical stack A composite or mix of the records from several sources made in nearly the same location without correcting for offset differences. It is used especially with surface sources, in which the records from several successive weight drops or vibrations are added together.

vertical tectonics The tectonics of Pratt-type isostasy, in which vertical movements of crustal units of varying specific gravity cause topographic relief.

vertical-variability map (ver'-ti-cal-var'-i-a-bil'-i-ty) A stratigraphic map that depicts the relative vertical positions, thicknesses, and number of occurrences of specific rock types in a sequence of strata or within a designated stratigraphic unit; e.g. a "number-of-sandstones map" indicating the number of discrete sandstone units in a given stratigraphic body, or a "limestone mean-thickness map" indicating the average thickness of limestone units in a given stratigraphic body. The map gives information about the internal geometry of the stratigraphic unit in terms of a designated rock component or property, or it may show the degree of differentiation of the unit into subunits of different lithologic types. Cf: *facies map.* See also: *center-of-gravity map; standard-deviation map; multipartite map; interval-entropy map.*

vertical-velocity curve (ver'-ti-cal-ve-loc'-i-ty) A graphic presentation of the relationship between depth and velocity, at a given point along a vertical line, of water flowing in an open channel or conduit (ASCE, 1962). Syn: *mean velocity curve; depth-velocity curve.*

vertical wall One of the walls of a bryozoan *zooid* which are entirely or in part at high angles to the basal wall and the wall bearing the orifice.

verticil (ver'-ti-cil) A *whorl* or circular arrangement of similar parts around an axis, as leaves of arthrophytes. Adj: *verticillate.*

verticillate (ver-tic'-il-late) Arranged in or having verticils; e.g. having successive whorls of branches arranged like the spokes of a wheel. Syn: *whorled.*

vertikalschichtung (ver'-ti-kal-schich'-tung) Nacreous tablets in a mollusk shell, arranged in vertical columns (Wise, 1970). Etymol: German, vertical layering.

Vertisols Mineral soils that have 30% or more clay, deep wide cracks when dry, and either gilgai microrelief, intersecting slickensides, or wedgeshaped structural aggregates tilted at an angle from the horizon. An order in the U.S. system of soil taxonomy.

vertumnite (ver-tum'-nite) A colorless monoclinic mineral related to strätlingite and hexagonal hydrated gehlenite, but with a Si:Al ratio double that of those phases: $Ca_4Al_4Si_4O_6(OH)_{24}·3H_2O$.

very angular (ver'-y) A term used by Powers (1953) to describe a sedimentary particle with a roundness value between 0.12 and 0.17 (midpoint at 0.14). Also, said of the *roundness class* containing very angular particles. Cf: *angular.*

very close pack ice *Pack ice* in which the *concentration* approaches $^{10}/_{10}$; the floes are tightly packed with very little, if any, seawater visible. Cf: *close pack ice.*

very coarsely crystalline Descriptive of an interlocking texture of a carbonate sedimentary rock having crystals whose diameters are in the range of 1-4 mm (Folk, 1959).

very coarse pebble A term for a pebble having a diameter in the range of 32-64 mm (1.3-2.5 in., or -5 to -6 phi units).

very coarse sand (a) A term for a sand particle having a diameter in the range of 1-2 mm (1,000-2,000 micrometers, or zero to -1 phi units). Also, a loose aggregate of sand consisting of very coarse sand particles. (b) A soil term used in the United States for a sand particle having a diameter in the range of 1-2 mm. Obsolete syn: *fine gravel.*

very common In the description of coal constituents, 10-30% of a particular constituent occurring in the coal (ICCP, 1963). Cf: *rare; common; abundant; dominant [coal].*

very fine clay A term for a clay particle having a diameter in the range of $^{1}/_{4096}$ to $^{1}/_{2048}$ mm (0.24-0.5 micrometers, or 12 to 11 phi units). Also, a loose aggregate of clay consisting of very fine clay particles.

very finely crystalline Descriptive of an interlocking texture of a carbonate sedimentary rock having crystals whose diameters are in the range of 0.004-0.016 mm (Folk, 1959).

very fine pebble *granule [sed].*

very fine sand (a) A term for a sand particle having a diameter in the range of 0.062-0.125 mm (62-125 micrometers, or 4 to 3 phi units). Also, a loose aggregate of sand consisting of very fine sand particles. Syn: *flour sand.* (b) A soil term used in the United States for a sand particle having a diameter in the range of 0.05-0.10 mm. (c) Soil material containing 85% or more of sand-size particles (percentage of silt plus 1.5 times the percentage of clay not exceeding 15) and 50% or more of very fine sand (SSSA, 1965, p.347).

very fine silt A geologic term for a silt particle having a diameter in the range of $1/256$ to $1/128$ mm (4-8 micrometers, or 8 to 7 phi units). Also, a loose aggregate of silt consisting of very fine silt particles.

very large boulder A *boulder* having a diameter in the range of 2,048-4,096 mm (80-160 in., or -11 to -12 phi units).

very long baseline interferometry A pair of radio telescopes separated by a great distance, usually 1,000 km or more, that act as radio interferometers on signals of common origin. The radio signals are recorded and later compared to determine the difference in phase of radiation received at the two telescopes. From this information, the direction and structure of the source, and the distance and direction between telescopes can be determined. The ability to determine the location of one telescope relative to the other to within a few centimeters makes the device geodetically important. Abbrev: VLBI.

very open pack ice *Pack ice* in which the *concentration* is $1/10$ through $3/10$ with water predominating over ice; the floes are loose and widely spaced. Cf: *scattered ice.*

very thick bands In banded coal, vitrain bands exceeding 50.0 mm in thickness (Schopf, 1960, p.39). Cf: *thin bands; medium bands; thick bands.*

vesbite (ves'-bite) A melilite- and leucite-rich leucititolite containing clinopyroxene and accessory apatite and opaque oxides. Washington in 1920 derived the name from Vesbius (Latin for Vesuvius), Italy, where the rock occurs as ejected blocks. Obsolete.

vesecite (ve'-sec-ite) A monticellite-bearing melilitite that contains olivine, melilite, phlogopite, and nepheline, but no clinopyroxene. Cf: *polzenite.* The name, given by Scheumann in 1922, is for Vesec Svetla, Czech Republic. Obsolete.

vesicle [paleont] (a) A plant or animal structure having the general form of a membranous cavity; e.g. the space enclosed in the interior of a corallite or any of the fluid-filled cavities of the vertebrate brain. (b) A term incorrectly applied to a *dissepiment* in a corallite.

vesicle [palyn] Less preferred syn. of *saccus.*

vesicle [petrology] (ves'-i-cle) A cavity of variable shape in a lava, formed by the entrapment of a gas bubble during solidification of the lava.

vesicle cylinder A cylindrical zone in a lava, in which there are abundant vesicles, probably formed by the generation of steam from underlying wet material. This feature occurs in the lavas of the northwestern U.S.

vesicular [paleont] (ve-sic'-u-lar) (a) Pertaining to or containing vesicles in a plant or animal. (b) Having dissepiments in corals. This usage is not recommended.

vesicular [petrology] Said of the texture of a rock, esp. a lava, characterized by abundant vesicles formed as a result of the expansion of gases during the fluid stage of the lava. Cf: *cellular; scoriaceous [volc].*

vesiculate (ve-sic'-u-late) *saccate [paleont].*

vesiculated tuff Tuff containing open, fluid-form spaces among the fragments, reflecting entrapment of air in wet, fine-grained ash.

vesiculation (ve-sic'-u-la'-tion) (a) The process of forming vesicles. (b) The arrangement of vesicles in a rock.

vésigniéite (vé-sign-i-é'-ite) A vitreous yellow-green or dark olive-green monoclinic mineral: $BaCu_3(VO_4)_2(OH)_2$.

vessel (ves'-sel) A *xylem* tube, formed from several elements-modified *tracheid* cells set end to end-with end walls imperfect or nonexisting (Cronquist, 1961, p.883). It is the principal water-conducting cell structure in most *angiosperms,* also found in some *cryptogams* and *gymnosperms.*

vestibula Pl. of *vestibulum.*

vestibulate (ves-tib'-u-late) Having or resembling a vestibule or vestibulum.

vestibule (ves'-ti-bule) (a) An inhalant cavity, other than a canal, of the aquiferous system of a sponge, located close to the surface and receiving water from one or more ostia. Syn: *subdermal space.* (b) The distal (near-surface) part of the *zooecium* in a cryptostome bryozoan, often delimited basally by hemisepta. Syn: *vestibulum.* (c) In the bryozoans, a space surrounded by membranous wall below the *orifice* through which the lophophore and tentacles pass in protruding. (d) The space between the duplicature and the calcareous wall (outer lamella) composing the externally visible shell of an ostracode. (e) A subcylindrical prolongation of the *pedicle valve* dorsal of the *brachial valve* of a brachiopod (TIP, 1965, pt. H, p.155). (f) Generally, a cavity that leads into another cavity or passage. (g) In vertebrates, the term refers to the apparatus of the inner ear that always is involved with equilibrium and in many forms also functions in hearing.

vestibulum [paleont] The *vestibule* in a cryptostome bryozoan. Pl: *vestibula.*

vestibulum [palyn] The space between the external opening (*exopore*) in the ektexine and the internal opening (*endopore*) in the endexine of a pollen grain with a complex porate structure. The openings are about the same size. Cf: *atrium [palyn].* Pl: *vestibula.*

vestige (ves'-tige) A small and imperfectly developed or degenerate bodily part or organ that is a remnant of one more fully developed in an earlier stage in the life cycle of the individual, in a past generation, or in closely related forms. Adj: *vestigial.* Syn: *vestigial structure.*

vestigial (ves-tig'-ial) Of, pertaining to, or being a *vestige.*

vestigial structure *vestige.*

vestigiofossil (ves-tig'-i-o-fos'-sil) *trace fossil.*

vestured pit (ves'-tured) In certain plants, a *bordered pit* with a cavity wholly or partly lined with projections from a secondary wall (Swartz, 1971, p.496).

Vesulian (Ve-su'-li-an) Stage in Great Britain: Middle Jurassic (above Bajocian, below Bathonian).

vesuvian (ve-su'-vi-an) (a) *vesuvianite.* (b) A mixture of calcite and hydromagnesite. (c) *leucite.*

Vesuvian garnet (Ve-su'-vi-an) An early name for leucite whose crystal form resembles that of garnet.

vesuvianite (ve-su'-vi-an-ite') A dense, brittle tetragonal mineral: $(Ca,Na)_{19}(Al,Mg,Fe)_{13}(SiO_4)_{10}(Si_2O_7)_4(OH,F,O)_{10}$. It is usually brown, yellow, or green, sometimes contains iron and fluorine, and is commonly found in contact-metamorphosed limestones. Syn: *idocrase; vesuvian.*

Vesuvian-type eruption Many authors regard this term as a synonym for *Vulcanian-type eruption.* Others (e.g., Francis, 1993) describe it as a step up from Vulcanian, with longer sustained blasts and a higher (up to 30-km) eruption column. Syn: *subplinian eruption.*

vesuvite (ve-su'-vite) A *tephrite* containing abundant leucite. The name, given by Lacroix in 1917, is for Vesuvius, Italy. Obsolete.

veszelyite (ve-szel'-y-ite) A greenish-blue to dark blue monoclinic mineral: $(Cu,Zn)_3(PO_4)(OH)_3 \cdot 2H_2O$. Syn: *arakawaite.*

VGP *virtual geomagnetic pole.*

VHA basalt Acronym for a lunar rock type, very high alumina basalt. The term is based on chemical composition, and the material may be a mixture of other rock types (Hubbard, 1973, p.339).

VI *vertical interval.*

viaeneite A metallic monoclinic mineral: $(Fe,Pb)_4S_8O$.

vibertite (vi'-bert-ite) *bassanite.*

vibetoite (vi-be'-to-ite) A coarse-grained biotite-hornblende *pyroxenite* containing much calcite and apatite. It is either an altered alkalic ultramafic rock or a hybrid product of carbonatite with ultramafic silicate-rock affinities. The name, given by Brögger in 1921, is for Vibeto farm, Fen complex, Norway. Obsolete.

vibex (vi'-bex) A line, presumably marking a cuticular insertion, in the skeletal material on the basal side of a branch in the generally reteporiform colonies of certain ascophoran cheilostomes (bryozoans) (Lagaaij, 1952, p.16). Pl: vibices.

vibracore (vi'-bra-core) The descriptive term for a sampling technique in which high-frequency, low-amplitude vibration is used to liquefy saturated sediments, allowing a core barrel or sampling tube to penetrate into the liquefied sediments. A sample catcher is used to retain the sediment in the sampling tube. The technique is typically used for sampling sediments at the bottom of a body of water.

Vibracorer (Vi'-bra-cor'-er) Coring device used to obtain a continuous sample from coarse-grained or semilithified sediments. The core barrel is supported by a tripod which rests on the sediment surface, and the barrel is driven into the sediment by a vibrator.

vibraculum (vi-brac'-u-lum) A *heterozooid* in cheilostome bryozoans, having the equivalent of the operculum or the mandible setae in the form of a bristle or whip. Pl: vibracula.

vibrating-string gravimeter An exploration-type gravimeter that affords a measurement of relative changes of the gravity field by observation and recording the periods of transverse vibrations of a thin wire tensioned by the weight of a known mass. Used in some marine and airborne gravity systems.

vibrational parameters *displacement parameters.*

vibration direction *privileged direction.*

vibration magnetometer A type of magnetometer for individual rock specimens that uses the alternating voltage generated by relative vibration of the specimen and a coil. Syn: *Foner magnetometer.*

vibration mark A term used by Dzulynski and Slaczka (1958, p.234) for a sedimentary structure representing a modified groove consisting of crescentic depressions (concave upcurrent) presumed to result from the unsteady inscribing action of a solid object moved by the current. Cf: *chevron mark; chattermark [struc geol]; ruffled groove cast.*

vibration plane In optics, a plane of polarized light, including the directions of propagation and vibration. Syn: *plane of polarization; plane of vibration.*

vibrator An instrument which produces mechanical oscillations used as a seismic source for *vibroseis.*

vibroseis (vi'-bro-seis') A seismic method originally developed and patented by Continental Oil Company in which a vibrator is used as an energy source to generate a wavetrain of controlled frequencies.

vicanite-(Ce) A vitreous yellowish-green rhombohedral mineral: $(Ca,Ce,La,Th)_{15}As^{5+}(As^{3+},Na)_{0.5}Fe^{3+}{}_{0.7}Si_6B_4(O,F)_{47}$.

vicariance The process of allopatric speciation occurring when different populations of a species become progressively isolated by barriers.

vicariance biogeography A method of explaining common patterns of the geographic distribution of organisms that require general explanations, such as the disjunction of the distribution of a single ancestral species into two or more fragments (Wiley, 1981, p.291). Cf: *vicariance event, area cladogram.*

vicariance event A geographic separation of a continuous biota such that the biota becomes two or more geographic subunits (Wiley, 1981, p.8).

vicarious (vi-car'-i-ous) (a) Pertaining to closely related kinds of organisms that occur in similar environments or as fossils in corresponding strata but in distinct and often widely separated areas. Also, characterized by the prescence of, or consisting of, such organisms. (b) Said of an *avicularium* of a cheilostome bryozoan, which is intercalated in a linear series in a space as large as or larger than those occupied by autozooids.

vicinal face (vic'-i-nal) A crystal face that modifies a normal crystal face, which it closely approximates in orientation.

Vickers hardness test (Vick'-ers) A test of resistance to deformation of metals or minerals in which a pyramid-shaped diamond is forced into a polished surface of the specimen to be tested under various static loads. The result is a function of the average length of the diagonals of the resulting indentation. Cf: *Brinell hardness test; Knoop hardness test.*

Vicksburgian (Vicks-burg'-i-an) North American (Gulf Coast) stage: Oligocene (above Jacksonian, below Frio).

vicoite (vi'-co-ite) An extrusive rock composed of leucite, sodic sanidine, calcic plagioclase, and augite. The name, applied by Washington in 1906, is for Vico volcano, Italy. Obsolete.

vidicon (vid'-i-con) A storage-type electronically scanned photoconductive television camera tube, which often has a response to radiations beyond the limits of the visible region.

viewshed map (a) A map outlining those areas visible from a particular vista point (or set of vista points). Syn: *line-of-sight map; visibility map.* (b) A map outlining areas obscured from a particular vista point for a particular comminication device and medium, e.g. microwave visibility or radar visibility (Krzanowski et al., 1993).

vigezzite (vi-gezz'-ite) An orange-yellow orthorhombic mineral of the *aeschynite* group: $(Ca,Ce)(Nb,Ta,Ti)_2O_6$.

viitaniemiite A vitreous colorless, gray, or white monoclinic mineral: $Na(Ca,Mn^{2+})Al(F,OH)_3PO_4$.

vikingite (vi'-king-ite) A metallic monoclinic mineral: $Pb_5Ag_2Bi_6S_{15}$. It is a member of the lillianite homologous series.

Villafranchian (Vil-la-fran'-chi-an) European continental stage and land mammal age: upper Pliocene (Etouaires and Saint-Vallier units) and lower Pleistocene (Olivola and Tasso-Farneta units) (above Ruscinian, below Betfian). See also: *Villanyian.*

villamaninite (vil-la-man'-i-nite) A metallic gray triclinic (pseudo-cubic) mineral of the *pyrite* group: $(Cu,Ni,Co,Fe)S_2$.

Villanyian European land mammal age: late Pliocene-early Pleistocene (after Csarnotian, before Betfian). Equiv. medial and late Villafranchian.

villiaumite (vill'-iaum-ite) A vitreous red, pink, orange, or colorless cubic mineral of the *halite* group: NaF.

villyaellenite (vill-ya-el'-len-ite) A colorless or pale pink monoclinic mineral: $(Mn^{2+},Ca,Zn)_5(AsO_4)_2(AsO_3OH)_2{\cdot}4H_2O$.

vimsite (vim'-site) A colorless monoclinic mineral: $CaB_2O_2(OH)_4$.

vincentite (vin'-cen-tite) A metallic monoclinic or tetragonal mineral: $(Pd,Pt)_3(As,Sb,Te)$.

vinciennite (vin-cien'-nite) A metallic orange tetragonal mineral: $Cu_{10}Fe_4Sn(As,Sb)S_{16}$.

vincularian (vin-cu-lar'-i-an) *vinculariiform.*

vinculariiform (vin-cu-lar'-ii-form) Said of a rigid, ramose, erect colony in cheilostome bryozoans, firmly attached by a calcified base and having subcylindrical branches (Lagaaij and Gautier, 1965, p.51).

vinculum (vin'-cu-lum) Calcareous portion of connecting ring of siphuncle segment attached to adapical surface of septum in discosorid and, more rarely, in oncocerid cephalopod conchs in which it also may be present on the adoral side of septum (TIP, 1964, pt. K, p.59).

Vindobonian (Vin-do-bo'-ni-an) Obs. European stage: lower middle Miocene (above Burdigalian, below Sarmatian); equiv. Helvetian and Tortonian.

vinogradovite (vi-no-gra'-dov-ite) A white, colorless, mauve, or pink monoclinic mineral: $(Na,K)_4Ti_4(Si,Al)_8O_{26}{\cdot}(H_2O,K_3)$.

violaite (vi-o'-la-ite) *augite.*

violan (vi'-o-lan) A translucent massive blue or blue-violet Mg-rich variety of diopside or augite. Syn: violane.

violarite (vi-o-la'-rite) A metallic violet-gray cubic mineral of the *linnaeite* group: $FeNi_2S_4$.

virga Streak of water and ice particles falling from a *cloud* suggesting precipitation but vaporizing before reaching Earth's surface. Etymol: Latin, "rod," "streak."

virgal (vir'-gal) One of an articulated series of more or less rod-shaped or cylindrical ossicles forming a structure that extends outward from an ambulacral plate of certain asterozoans. Pl: virgals or virgalia.

virgation [geomorph] A sheaflike pattern, as shown on a map, of mountain ranges diverging from a common center. Ant: *syntaxis.*

virgation [struc geol] (vir-ga'-tion) (a) A divergent, branchlike pattern of fault distribution. (b) A fold pattern in which the axial surfaces diverge or fan out from a central bundle. Rarely used. The term is used in the Russian literature.

virgella (vir-gel'-la) A spine developed during formation of the *metasicula* of a graptolithine. It is embedded in the wall of the metasicula and projects freely from the apertural margin.

Virgilian (Vir-gil'-i-an) North American provincial series: uppermost Pennsylvanian (above Missourian, below Wolfcampian of Permian).

virgilite (vir'-gil-ite) A colorless hexagonal mineral: $LiAlSi_2O_6$.

virgin (vir'-gin) *primary [metal].*

virgin clay Fresh clay, as distinguished from that which has been fired.

virgin flow Streamflow unaffected by artificial obstructions, storage, or other works of man in the stream channels or drainage basin; a syn. of *runoff [water].*

Virglorian (Vir-glor'-i-an) *Anisian.*

virgula (vir'-gu-la) A tubular prolongation from the apex of the *prosicula* of scandent graptolites. It is homologous with *nema,* and the term is used where the prolongation is enclosed within a scandent rhabdosome (biserial rhabdosomes) or incorporated in the dorsal wall (as in monograptids).

viridine (vir'-i-dine) A grass-green variety of andalusite containing manganese. Syn: *manganandalusite.*

viridite (vir'-i-dite) (a) A general term formerly applied to the indeterminable or obscure green alteration products (such as chlorite and serpentine) occurring in scales and threads in the groundmasses of porphyritic rocks. Cf: *opacite; ferrite [ign].* (b) An iron-rich chlorite containing considerable ferric iron.

virtual geomagnetic pole (vir'-tu-al) A conventional form of expressing a measured instantaneous remanent magnetization; the pole location of the dipole magnetic field representing the instantaneous field direction at the rock's location is parallel to its measured remanent magnetization. A virtual geomagnetic pole is derived, for example, from the mean direction from a single, quickly cooled lava flow. Abbrev: VGP. See also: *paleomagnetic pole.*

visceral (vis'-cer-al) Pertaining to or located on or among the internal organs of a body; e.g. the "visceral mass" consisting of the part of the body of a mollusk behind the head and above the foot, in which the digestive and reproductive organs are concentrated.

visceral skeleton In the vertebrates, a system of serially arranged bars (visceral arches generally; branchial arches or "gill bars" in gill-bearing vertebrates) located in the wall of tissues lateral to the adult *pharynx* in fishes or in the adult derivatives of the pharynx in tetrapods. They are composed of cartilage or *endochondral bone* and usually jointed. The visceral skeleton forms from mesenchymal accumulations that derive in large part from ectoderm of lateral margins of the neural plate (as neural crest cells). It forms the *hyoid apparatus,* ear ossicles, and endochondral bones of the jaws in gnathostomes.

viscin thread (vis'-cin) An acetolysis- and acid-resistant thread arising from the *sexine* of a pollen grain, usually from the proximal surface, causing pollen grains to be released in clusters, e.g., in *Oenethera.*

viscoelastic behavior The response of a material to stress such that, after the application of a stress, strain builds up gradually and, after the removal of the stress, gradually returns to zero.

viscometer (vis-com'-e-ter) An instrument used to measure viscosity. Syn: *viscosimeter.*

viscometry (vis-com'-e-try) The measurement of viscosity. Syn: *viscosimetry.*

viscosimeter (vis-co-sim'-e-ter) *viscometer.*

viscosimetry (vis-co-sim'-e-try) *viscometry.*

viscosity (vis-cos'-i-ty) The property of a substance to offer internal resistance to flow. Specifically, the ratio of the shear stress to the rate of shear strain. This ratio is known as the *coefficient of viscosity.* See also: *Newtonian liquid.*

viscosity coefficient A numerical factor that measures the internal resistance of a fluid to flow; it equals the shearing force in dynes/sq cm transmitted from one fluid plane to another that is 1 cm away, and generated by the difference in fluid velocities of 1 cm/sec in the two planes. The greater the resistance to flow, the larger the coefficient. Syn: *absolute viscosity; dynamic viscosity.*

viscous behavior The response of a material to stress such that application of stress results in a permanent strain. In Newtonian (or ideal) viscous behavior, the relationship between stress and rate of strain accumulation is linear.

viscous creep (vis'-cous) Inelastic, time-dependent strain in which the rate of strain is constant at constant differential stress.

viscous flow A syn. of *Newtonian flow.* Cf: *liquid flow; solid flow.*

viscous magnetization A component of magnetization that behaves as *remanent magnetization* during the time needed for laboratory measurement yet comparable to an *induced magnetization* during geologic time, thus showing *magnetic viscosity.* Syn: *viscous remanent magnetization; unstable remanent magnetization.*

viscous remanent magnetization *viscous magnetization.* Abbrev: VRM.

viscous stress The resistive force of water. It is proportional to the speed of the current, but acts opposite to its direction of flow (U.S. Naval Oceanographic Office, 1966, p.175).

Viséan (Vi-se'-an) A series or stage of the Standard Global Chronostratigraphic Scale: Lower Carboniferous (above Tournaisian, below Serpukhovian). Also spelled: Visean.

viséite (vi-sé'-ite) A discredited name for a mixture of disordered crandallite and other phases.

vishnevite (vish'-nev-ite) A pale lilac, light blue, or dark gray mineral of the *cancrinite* group: $(Na,Ca,K)_8(AlSiO_4)_6O_{24}(SO_4){\cdot}2H_2O$.

visibility map *viewshed map.*

visible gold Gold that is visible to the naked eye or with the assistance of a low power (≤10x) magnifying lens. Used in exploration geology to indicate the presence of gold sufficiently coarse to be observed in hand specimen or while logging DDH core. Abbrev: vg.

visible horizon (vis'-i-ble) *apparent horizon.*

visible region spectroscopy The observation, recording, measurement, and analysis of an absorption spectrum in the visible frequencies. It is commonly used to determine the valence and coordination environments of transition metals in a crystal. Also called: *optical absorption spectroscopy.*

vismirnovite (vis-mir'-nov-ite) A pale-yellow cubic mineral of the *schoenfliesite* group: $ZnSn^{4+}(OH)_6$.

visor (vi'-sor) On the coast, the more or less inclined, overhanging surface directly above the wave-cut notch in a sea cliff, most commonly found in limestones in a tropical region.

vistepite A vitreous orange-yellow triclinic mineral: $Mn_4SnB_2O_2(Si_2O_7)_2(OH)_2$.

vitalism (vi'-tal-ism) The theory that some internal force or driving energy of organisms exerts a directional effect that more or less determines how variation and evolution will proceed. Cf: *holism.*

viterbite [mineral] (vi-ter'-bite) A mixture of allophane and wavellite(?).

viterbite [rock] An extrusive rock composed chiefly of sodic sanidine and large phenocrysts of leucite, with smaller quantities of calcic plagioclase, augite, biotite, apatite, and opaque oxides. Its name, given by Washington in 1906, is derived from Viterbo, Italy. Obsolete.

vitimite A vitreous white monoclinic mineral: $Ca_6B_{14}O_{19}(SO_4)(OH)_{14}{\cdot}5H_2O$.

vitr- A prefix meaning "glass". Cf: *hyalo-.*

vitrain (vit'-rain) A coal *lithotype* characterized macroscopically by brilliant, vitreous luster, black color, and cubic cleavage with conchoidal fracture. Vitrain bands or lenticles are amorphous, usually 3-5 mm thick, and their characteristic microlithotype is *vitrite.* Cf: *clarain; durain; fusain.* See also: *euvitrain; provitrain.* Syn: *pure coal.*

Vitrands Andisols that have 1,500-kPa water retention of less than 15 percent on air dry and less than 30 percent on undried samples throughout 60 percent of the thickness either; (1) within 60 cm of the soil surface or top of an organic layer with andic properties, whichever is shallower if there is no lithic, paralithic contact, duripan, or petrocalcic horizon within that depth, or (2) between the mineral soil surface or top of an organic layer with andic properties, whichever is shallower and a lithic, paralithic contact, duripan, or petrocalcic horizon. A suborder in the U.S. system of soil taxonomy.

vitreous [ign] *glassy.*

vitreous [paleont] Said of a hyaline foraminifer having the appearance and luster of glass.

vitreous copper *chalcocite.*

vitreous luster A type of luster resembling that of glass. Quartz, for example, has a vitreous or *glassy luster.*

vitreous silver *argentite.*

vitric (vit'-ric) Said of pyroclastic material that is characteristically glassy, i.e. contains more than 75% glass.

vitric tuff A tuff that consists predominantly of volcanic glass fragments. Cf: *crystal-vitric tuff.*

vitrification (vit'-ri-fi-ca'-tion) Formation of a glass.

vitrinertite (vit-ri-ner'-tite) A coal microlithotype that contains a combination of vitrinite and inertinite totalling at least 95%, and containing more of each than of liptinite. It generally occurs in high-ranking bituminous coals.

vitrinertoliptite (vit'-ri-ner'-to-lip'-tite) A microlithotype of coal, a variety of *trimacerite* consisting of liptinite in greater abundance than inertinite and vitrinite (ICCP, 1971). Cf: *duroclarite; clarodurite.*

vitrinite (vit'-ri-nite) (a) A coal maceral group that is characteristic of vitrain and is composed of humic material. It includes *provitrinite* and *euvitrinite* and their varieties. Cf: *inertinite; liptinite.* (b) A coal maceral group distinguished by a middle level of reflectance, higher than liptinite and lower than inertinite in the same coal. It includes *telinite, collinite,* and *vitrodetrinite* macerals (ICCP, 1971). Some coal petrologists distinguish "vitrinite A" and "vitrinite B", and others *pseudovitrinite.*

vitrinization (vit'-ri-ni-za'-tion) A process of *coalification* in which vitrain is formed. Cf: *incorporation; fusinization.*

vitrinoid (vit'-ri-noid) Vitrinite that occurs in bituminous caking coals and that has a reflectance of 0.5-2.0% (ASTM, 1970, p.466). Cf: *xylinoid; anthrinoid.*

vitriol peat (vit'-ri-ol) Peat that contains abundant iron sulfate.

vitrite (vit'-rite) A coal microlithotype group that contains vitrinite macerals totalling at least 95%. Cf: *inertite; liptite; vitrain.*

vitroclarain (vit-ro-clar'-ain) A transitional lithotype of coal characterized by the presence of vitrinite, but in lesser amounts than other macerals. Cf: *clarovitrain.* Syn: *vitroclarite.*

vitroclarite (vit-ro-clar'-ite) *vitroclarain.*

vitroclastic (vit-ro-clas'-tic) Pertaining to a pyroclastic rock structure characterized by fragmented bits of glass; also, said of a rock having such a structure.

vitrodetrinite (vit-ro-det'-ri-nite) A maceral of coal within the vitrinite group, consisting of small fragments with more or less angular and variable outlines due to extensive comminution during transport and deposition (ICCP, 1971). Particles are usually less than 10 micrometers across.

vitrodurain (vit-ro-du'-rain) A coal lithotype transitional between durain and vitrain, but predominantly durain. Cf: *durovitrain.*

vitrophyre (vit'-ro-phyre) Any porphyritic igneous rock having a glassy groundmass. Adj: *vitrophyric.* Cf: *granophyre.*

vitrophyric (vit-ro-phyr'-ic) A porphyritic igneous texture characterized by large phenocrysts in a glassy groundmass.

vitta (vit'-ta) One of a pair of oppositely placed, membrane-covered linear furrows near the lateral margins of the *frontal shield* in certain ascophoran cheilostomes (bryozoans). A series of pores in each furrow communicates with the zooidal body cavity. Pl: vittae.

vitusite-(Ce) (vi'-tus-ite) A pale pink, white, or pale green orthorhombic mineral: $Na_3(Ce,La)(PO_4)_2$.

vivianite (viv'-i-an-ite') A monoclinic mineral: $Fe^{2+}_3(PO_4)_2{\cdot}8H_2O$. It is colorless, blue, or green when unaltered, but grows darker on exposure; it occurs as monoclinic crystals, fibrous masses, or earthy forms in copper, tin, and iron ores and in clays, peat, and bog iron ore. Syn: *blue iron earth; blue ocher.*

vladimirite (vlad-i-mir'-ite) A pale rose, colorless, or white monoclinic mineral: $Ca_5(AsO_4)_2(AsO_3OH)_2{\cdot}5H_2O$.

vlasovite (vla'-sov-ite) A colorless, pink, pale brown, or silvery monoclinic or triclinic mineral: $Na_2ZrSi_4O_{11}$.

VLBI *very long baseline interferometry.*

vlei A Dutch word used in the Middle Atlantic states and in southern Africa for a shallow lake or marshy area, esp. one developed in the poorly drained valley of an intermittent stream. Also spelled: *vley; vly.*

VLEM Vertical-loop EM configuration where the transmitting loop is in a vertical plane. Called VCP when the receiving loop is coplanar or VCA when coaxial.

vley *vlei.*

VLF Very low frequency. (a) Radio transmission at frequencies of 3 to 30 kHz, used for communication with submerged submarines and for long-range radio positioning. Most VLF transmitters work in the range 15 to 24 kHz. (b) An electromagnetic prospecting method which uses such transmissions as plane-wave sources. The VLF receiver measures the tilt of the total field by nulling one of two small orthogonal coils in the plane of the primary field. (c) A radio-location system such as Omega which uses very low-frequency energy.

vlodavetsite A colorless tetragonal mineral: $Ca_2Al(SO_4)_2F_2Cl{\cdot}4H_2O$.

vloer (vlo'-er) A term used in South Africa for a flat surface of caked mud with a high salt content and generally destitute of vegetation; a *playa.* It has a more irregular shape, greater area, and shallower depth than a *salt pan* and usually has an outlet. Etymol: Afrikaans, "floor."

vltavite *moldavite [astron].*

vly *vlei.*

VOC *volatile organic compound.*

vochtenite (voch'-ten-ite) A bronzy-brown monoclinic mineral: $(Fe^{2+},Mg)Fe^{3+}(UO_2)_4(PO_4)_4(OH){\cdot}12\text{-}13H_2O$.

vogesite (vo'-ges-ite) (a) In the *IUGS classification,* a variety of *lamprophyre* in which alkali feldspar is more abundant than plagioclase, there is no feldspathoid, and the phenocrysts are mainly hornblende and clinopyroxene. Cf: *spessartite [rock].* (b) A *lamprophyre* composed of hornblende phenocrysts in a groundmass of alkali feldspar and hornblende. Clinopyroxene, olivine, and plagioclase feldspar also may be present. Vogesite contains less biotite than *minette.* The name, given by Rosenbusch in 1887, is for the Vosges Mountains, France. Obsolete.

voggite A vitreous colorless or white monoclinic mineral: $Ca_2Al(SO_4)_2F_2Cl{\cdot}4H_2O$.

voglite (vog'-lite) An emerald-green to bright grass-green monoclinic mineral: $Ca_2Cu(UO_2)(CO_3)_4{\cdot}6H_2O$.

voices of the desert (voic'-es) *song of the desert.*

void *interstice.*

voidal concretion (void'-al) A large tubelike iron-oxide concretion with a central hollow or cavity and a hard, dense limonitic rim, commonly found in sands and sandstones and in some clays. It appears to be a product of weathering (oxidation) of a sideritic concretion.

void ratio The ratio of the volume of void space to the volume of solid substance in any material consisting of voids and solid material, such as a soil sample, sediment, or sedimentary rock; a parameter used in soil mechanics. Note the contrast to *porosity,* which is the ratio of the volume of void space to the total volume of solid, liquid, and gas in a soil, sediment, or rock.

volatile (vol′-a-tile) (a) adj. Readily vaporizable. (b) n. A syn. of *volatile component.*
volatile combustible *volatile matter.*
volatile component A material in a magma, such as water or carbon dioxide, whose vapor pressures are sufficiently high for them to be concentrated in any gaseous phase. Syn: *fugitive constituent; volatile.*
volatile matter In coal, those substances, other than moisture, that are given off as gas and vapor when heated under reducing (oxygen-deficient) conditions. Standardized laboratory methods are used in analysis. Syn: *volatiles; volatile combustible.*
volatile organic compound A potentially toxic compound with high vapor pressure commonly detected in soils and groundwater and derived from spills or leaks of organic-liquid fuels or wastes. Abbrev: VOC.
volatiles (vol′-a-tiles) *volatile matter.*
volborthite (vol′-borth-ite) A dark olive-green, green, or yellowish-green monoclinic mineral: $Cu_3V_2O_7(OH)_2{\cdot}2H_2O$. It may contain some calcium and barium, and it represents an occasional ore of vanadium. Syn: *uzbekite.*
volcan (vol′-can) (a) The component of the Earth's crust made up of volcanoes and various hypabyssal rocks (Makiyama, 1954, p.146). (b) *volcano.* A little-used term.
volcanello (vol-ca-nel′-lo) (a) A small, active cone within the central crater of a volcano, e.g. Mount Nuevo in Vesuvius. (b) A *spatter cone.* Etymol: Italian, "small volcano".
volcanic (vol-can′-ic) (a) Pertaining to the activities, structures, or rock types of a volcano. (b) A syn. of *extrusive.*
volcanic accident A departure from the *normal cycle* of erosion, caused by the outbreak of volcanic activity. See also: *accident; climatic accident.*
volcanic aerosol A gaseous suspension of fine-grained solid or liquid particles, e.g. a stratospheric layer composed of tiny sulfuric acid droplets of about 0.1 micrometer in diameter occupying a zone between about 15 to 25 kilometers up. This aerosol is produced by a combination of volcanic sulfur dioxide from explosive eruptions and atmospheric water vapor. Sometimes called a "dust veil."
volcanic arc A generally curved linear belt of volcanoes above a *subduction zone.* Cf: *island arc; magmatic arc.*
volcanic arenite A term used by Williams et al. (1954, p.308) for a lithic arenite composed chiefly of volcanic detritus and having a low quartz content. Common among Tertiary and Mesozoic strata around the Pacific Basin. Folk (1968, p.124) used "volcanic-arenite" for a litharenite composed chiefly of volcanic rock fragments, and having any clay content, sorting, or rounding; for more detailed specification, terms such as "basalt-arenite" and "andesite-arenite" may be used.
volcanic ash *ash [volc].*
volcanic ball *lava ball.*
volcanic blowpiping *gas fluxing.*
volcanic bomb *bomb [pyroclast].*
volcanic breccia A *volcaniclastic* rock composed mostly of angular volcanic fragments greater than 2 mm in size. The name volcanic breccia is not synonomous with pyroclastic breccia. Volcanic breccia forms in different ways, e.g. *pyroclastic breccia; flow breccia.* (Fisher, 1958, p.1071).
volcanic butte An isolated hill or mountain resulting from the differential weathering or erosion and consequent exposure of a volcanic neck, or of a narrow, vertical igneous intrusion into overlying weaker rock; e.g. Ship Rock, N. Mex. Cf: *mesa-butte; neck [volc]; pipe [volc].*
volcanic center A site at which volcanic activity localized at one or several vents is occurring or has occurred in the past.
volcanic chain Linear arrangement of a number of volcanoes, apparently associated with a controlling geologic feature such as a fault or a hot spot.
volcanic clay *bentonite.*
volcanic cloud *eruption cloud.*
volcanic cluster A group of volcanic vents in apparent random arrangement.
volcanic conduit The channelway that brings volcanic material up from depth. Cf: *vent.*
volcanic cone A conical hill of lava and/or pyroclastics that is built up around a volcanic vent. Syn: *cone [volc].*
volcanic conglomerate A conglomerate composed of rounded volcanic fragments.
volcanic cycle A regular sequence of changes in the behavior of a volcano.
volcanic dome *lava dome.*
volcanic earthquake An earthquake associated with volcanic rather than tectonic forces. Cf: *tectonic earthquake.*
volcanic explosivity index The size of an explosive eruption, measured pricipally by the volume of erupted products and the height of the eruption column (Newhall and Self, 1982). Abbrev: VEI. Cf: *magnitude [volc].*
volcanic field A more or less well-defined area that is covered with volcanic rocks, e.g. the San Francisco volcanic field, Arizona. Cf: *ash field; lava field.*
volcanic gas Volatile matter, released during a volcanic eruption, that was previously dissolved in the magma. Water vapor forms about 30-90 mol% of the gas; other constituents include, among others, carbon gases, esp. carbon dioxide; sulfur gases, esp. sulfur dioxide at high temperatures and hydrogen sulfide at low temperatures; hydrogen chloride; nitrogen as a free element; and hydrogen fluoride (Krauskopf, 1967; Williams-Jones and Rymer, 2000).
volcanic glass A natural glass produced by the cooling of molten lava, or a liquid fraction of it, too rapidly to permit crystallization. Examples are obsidian, pitchstone, tachylyte, and the glassy *mesostasis* of many extrusive rocks.
volcanic graben A straight-walled collapse structure on the summit or flanks of a volcanic cone. See also: *summit graben; sector graben.* Cf: *gjá [volc].*
volcanic gravel Gravel composed mostly of volcanic particles. Particles may be of pyroclastic, detrital or any other origin.
volcanic graywacke *volcanic wacke.*
volcanic harbor A natural harbor formed by the sea breaking through a gap in the rim of a volcanic crater or caldera; e.g., that of St. Paul Rocks in the southern Atlantic Ocean or Thera (Santorini), Greece.
volcanic-hosted copper *basaltic copper deposit.*
volcanic island A *submarine volcano* that has been sufficiently built up to be exposed above sea level.
volcaniclastic (vol′-ca-ni-clas′-tic) Pertaining to all clastic volcanic materials formed by any process of fragmentation, dispersed by any kind of transporting agent, deposited in any environment, or mixed in any significant portion with nonvolcanic fragments (Fisher, 1961, p.1409).
volcaniclastic apron A wedge-shaped accumulation of volcaniclastic material in the back-arc area of an island arc (Carey, 2000).
volcanic mud A mixture of water and volcanic ash.
volcanic pisolite *pisolite [volc].*
volcanic plain Surface formed by extensive lava or ash flows that cover topographic irregularities.
volcanic province A petrographic province in which the visible igneous rocks are largely volcanic.
volcanic rain *eruption rain.*
volcanic redbed copper *basaltic copper deposit.* Abbrev: VRC.
volcanic rent A great volcanic depression, bordered by fissures that are usually concentric in plan, caused by the pressure of magmatic injection or by the overloading of cone material on a weak substratum.
volcanic rise A broad topographic rise on Venus or Mars characterized by shield volcanism and extensional tectonics. Volcanic rises are interpreted as hotspots associated with mantle plumes (Bougher et al., 1997).
volcanic rock (a) A generally finely crystalline or glassy igneous rock resulting from volcanic action at or near the Earth's surface, either ejected explosively or extruded as lava; e.g. basalt. The term includes near-surface intrusions that form a part of the volcanic structure. See also: *volcanics.* Cf: *plutonic rock.* (b) A general term proposed by Read (1944) to include the effusive rocks and associated high-level intrusive rocks; they are dominantly basic. Cf: *neptunic rock; plutonic rock.*
volcanic rubble The unconsolidated equivalent of volcanic breccia. Syn: volcanic debris.
volcanics (vol-can′-ics) Those igneous rocks that have reached or nearly reached the Earth's surface before solidifying. The common use of the term for volcanic rocks should be avoided (USGS, 1958, p.86).
volcanic sand An unconsolidated deposit of sandsize volcanic particles.
volcanic sandstone A rock composed mostly of sandsize volcanic particles.
volcanic shield *shield volcano.*
volcanic sink *sink [volc].*
volcanic tremor Continuous seismic signal with regular or irregular sine-wave appearance and low frequencies (0.5-5 Hz). Harmonic tremor has a uniform appearance, whereas spasmodic tremor is pulsating and consists of higher frequencies with a more irregular appearance (McNutt, 2000).
volcanic wacke A term used by Williams et al. (1954, p.303) for a lithic wacke composed chiefly of detritus derived from intermediate (andesitic) and basic (basaltic) volcanic rocks and having a low quartz content. Common among Tertiary and Mesozoic strata bordering the Pacific Ocean. Syn: *volcanic graywacke.*
volcanic winter A term used for the severe global and regional cooling that can be caused by large amounts of volcanic aerosols in the atmosphere after very large eruptions (Rampino and Self, 2000).
volcanism [planet] Quiescent or explosive emission of molten planetary materials from the solid planetary body, including admixtures of liquids with gases and solids. See also: *cryovolcanism.*
volcanism [volc] The processes by which magma and its associated gases rise into the crust and are extruded onto the Earth's surface and into the atmosphere. Cf: *vulcanism.*
volcanite [mineral] (vol′-can-ite) An old mineral name suggested as a synonym of pyroxene and of a variety of sulfur containing less than one percent selenium.
volcanite [rock] (a) An extrusive rock composed chiefly of anorthoclase, andesine, and augite phenocrysts in a glassy groundmass containing feldspar and augite microlites. The name, given by Hobbs in 1893, is derived from Vulcano in the Lipari Islands. Obsolete. (b) A syn. of *volcanic rock.* Also spelled: vulcanite.
volcano (vol-ca′-no) (a) A vent in the surface of the Earth through which magma and associated gases and ash erupt; also, the form or structure, usually conical, that is produced by the ejected material. (b) Any eruption of

material, e.g. mud, that resembles a magmatic volcano. Obsolete var: vulcano. Pl: volcanoes. Etymol: the Roman deity of fire, Vulcan.

volcanogenic (vol'-ca-no-gen'-ic) Formed by processes directly connected with volcanism; specif., said of mineral deposits (massive sulfides, exhalites, banded iron formations) considered to have been produced through volcanic agencies and demonstrably associated with volcanic phenomena. Also spelled: volcanigenic.

volcanogenic gold *arsenical sulfide-silicate iron formation.*

volcanologist (vol-ca-nol'-o-gist) One who studies *volcanology.* Obsolete syn: volcanist.

volcanology (vol-ca-nol'-o-gy) The branch of geology that deals with volcanism, its causes and phenomena. Also spelled: vulcanology. See also: *volcanologist.* Less preferred syn: *pyrogeology.*

volcano shoreline A roughly circular, steeply-sloping shoreline formed where fragmental volcanic materials or flows of lava occur in a coastal location, or where an active volcano projects above the water surface, building a cone or dome upward and outward by continued addition of ejected materials.

volcano-tectonic depression (vol-ca'-no-tec-ton'-ic) A large-scale depression, usually linear, that is controlled by both tectonic and volcanic processes (van Bemmelen, 1933). An example is the Toba trough in northern Sumatra.

volchonskoite *volkonskoite.*

volfsonite An ill-defined metallic tetragonal mineral: $Cu_{11}Fe_3Sn_3S_{16}$. It has not been approved by the International Mineralogical Association.

Volgian European stage: Uppermost Jurassic, equivalent to Tithonian.

Volhynian (Vol-hyn'-ian) Ponto-Caspian substage; lowermost upper Miocene, lower Sarmatian (above Konkian stage, below Bessarabian substage).

volkonskoite (vol-kon'-sko-ite) A green or bluish-green monoclinic clay mineral: a dioctahedral smectite in which Cr is the dominant trivalent octahedral cation present: $Ca_{0.3}(Cr^{3+},Mg,Fe^{3+})_2(Si,Al)_4O_{10}(OH)_2{\cdot}4H_2O$. Also spelled: *volchonskoite; wolchonskoite.*

volkovite A colorless monoclinic mineral of doubtful validity: $Sr_2B_{14}O_{17}(OH)_{12}{\cdot}2H_2O$. It is probably equivalent to strontioginorite.

volkovskite (vol-kov'-skite) A colorless or pink triclinic monoclinic mineral: $KCa_4[B_5O_8(OH)]_4[B(OH)_3]_4Cl{\cdot}4H_2O$..

voltage gradient (volt'-age) *electric-field intensity.*

voltaite (vol'-ta-ite) A dull oil-green to brown or black cubic mineral: $K_2Fe^{2+}_5(Fe^{3+},Al)_4(SO_4)_{12}{\cdot}18H_2O$.

volt-second *weber.*

voltzite (voltz'-ite) A yellowish or reddish material consisting of wurtzite plus an organometallic compound of zinc.

volume control (vol'-ume) *gain control.*

volume diffusion A *diffusional mass transfer* involving the movement of atoms or *vacancies* through the volume of a mineral grain.

volume-diffusion creep Grain-scale, ductile deformation accomplished by the diffusion of atoms through the crystal lattice of minerals. Syn: *Nabarro-Herring creep.*

volume elasticity *bulk modulus.*

volume law *Lindgren's volume law.*

volume magnetization *magnetization.*

volume phase In metamorphism, a phase with thermochemical properties determined by volumetric free energy.

volume susceptibility *susceptibility [magnet].*

volume transport The volume of water flowing through a vertical plane in the ocean perpendicular to the current, usually in units of millions of cubic meters per second. Cf: *sverdrup.*

volumetric analysis (vol-u-met'-ric) Quantitative chemical analysis where the amount of a substance in a solution is determined by adding a fixed volume of a standard solution to the prepared sample until a reaction occurs. The amount of standard solution needed to produce the desired reaction indicates the amount of substance in the original sample.

volumetric moisture content The volume of water divided by the total volume of the porous medium. Cf: *moisture content [soil]; gravimetric moisture content.*

volumetric shrinkage The decrease in volume, expressed as a percentage of the soil mass when dried, of a soil mass when the water content is reduced from a given percentage to the shrinkage limit (ASCE, 1958, term 411).

volution (vo-lu'-tion) A *whorl* of a spiral shell.

volynskite (vo-lyn'-skite) A metallic trigonal mineral of the *matildite* group: $AgBiTe_2$.

von Baer's law The principle stated by Karl E. von Baer (1792-1876), Estonian embryologist and geologist, according to which the rotation of the Earth causes an asymmetrical, lateral erosion of stream beds. Also called: *Baer's law.* Obsolete.

vonbezingite A vitreous dark blue monoclinic mineral: $Ca_6Cu_3(SO_4)_3(OH)_{12}{\cdot}2H_2O$.

von Kármán constant (von Kar'-man) A dimensionless number which relates the mixing length to the flow condition in turbulent flow (Middleton, 1965, p.252). It is symbolized by k in formulas for velocity and sediment distribution in turbulent flow.

von Mises criterion A shear failure criterion for ductile deformation where no deformation occurs until the second invariant of the stress tensor reaches a critical value. Once stresses attain that level, deformation will be continuous.

von Mises distribution *circular normal distribution.* Named for R.M.E. von Mises (1883-1953), Austrian mathematician.

von Schmidt wave *head wave.*

vonsenite (von'-sen-ite) A coal-black or greenish-black orthorhombic mineral of the *ludwigite* group: $Fe^{2+}_2Fe^{3+}BO_5$. Syn: *paigeite.*

von Wolff's classification A quantitative chemical-mineralogic classification of igneous rocks proposed in 1922 by F. von Wolff. Obsolete.

vorobievite (vo-ro'-biev-ite) A rose-red, purplish-red, or pinkish gem variety of beryl containing cesium. Appreciable amounts of sodium and other alkalies may be present. Also spelled: *worobieffite;* vorobyevite. Syn: *morganite; rosterite.*

vortex [hydraul] (vor'-tex) A fluid flow that has a revolving motion in which the stream lines are concentric circles, and in which the total head for each stream line is the same.

vortex [struc geol] A vertical, cylindrical fold formed in incompetent rock by late-stage deformation during deep-zone orogeny (Wynne-Edwards, 1957, p.643). Rarely used.

vortex cast *flute cast.*

vorticity The non-coaxial component of general shear. Can also be described as the rate of rotation of the incremental strain axes during *progressive deformation.*

vorticity [oceanog] (vor-tic'-i-ty) A measure of the rotation of a fluid. Cf: *absolute vorticity, planetary vorticity, relative vorticity, potential vorticity.*

voxel A volume element, the 3-D version of the 2-D *pixel.*

vozhminite (vozh'-mi-nite) A metallic brownish-yellow hexagonal mineral: $(Ni,Co)_4(As,Sb)S_2$.

Vraconian (Vra-co'-ni-an) European stage: uppermost Lower Cretaceous or lowermost Upper Cretaceous.

vrbaite (vr'-ba-ite) A gray-black to dark-red orthorhombic mineral: $Tl_4Hg_3Sb_2As_8S_{20}$.

VRC *volcanic redbed copper.*

vredenburgite (vred'-en-burg-ite') (a) *beta-vredenburgite.* (b) *alpha-vredenburgite.*

VRM *viscous remanent magnetization.*

V-shaped valley A valley having a pronounced cross profile suggesting the form of the letter "V", characterized by steep sides and short tributaries; specif. a young, narrow valley resulting from downcutting by a stream. The "V" becomes broader as the amount of mass wasting increases. Cf: *U-shaped valley.*

VSP *vertical seismic profiling.*

V's, rule of *rule of V's.*

vuagnatite (vua'-gna-tite) A vitreous white orthorhombic mineral: $CaAlSiO_4(OH)$.

vug A small cavity in a vein or in rock, usually lined with crystals of a different mineral composition from the enclosing rock. Etymol: Cornish "vooga", "underground chamber, cavern, cavity". Adj: *vuggy.* Cf: *druse; geode.* Syn: *bug hole.*

vuggy (vug'-gy) Pertaining to a *vug* or having numerous vugs.

vuggy porosity Pores in sedimentary rocks that result from dissolution of the rock. Range in size from pin point to cavernous.

vugh (a) A relatively large and usually irregular void in a soil material, not normally interconnected with other voids of comparable size (Brewer, 1964, p.189). (b) Var. of *vug.*

Vulcanian-type eruption (Vul-ca'-ni-an-type') A type of volcanic eruption characterized by the explosive ejection of fragments of new lava, commonly incandescent when they leave the vent but either solid or too viscous to assume any appreciable degree of rounding during their flight through the air. With these there are often breadcrust bombs or blocks, and generally large proportions of ash (Macdonald, 1972, p.223). Also spelled: *Vulcano-type eruption.* Syn: *Vesuvian-type eruption; paroxysmal eruption.*

vulcanism (vul'-can-ism) A term used in the early nineteenth century by *vulcanists,* those maintaining the volcanic origin of all basalt and related rocks, but not necessarily disputing the rest of A.G. Werner's *geognosy* including the *neptunian* origin of the crystalline rocks (Dean, 1998). Cf: *volcanism.*

vulcanist An adherent of *vulcanism.*

vulcanite [mineral] (vul'-can-ite) A metallic yellow orthorhombic mineral: CuTe.

vulcanorium (vul-ca-nor'-i-um) A structure transitional between a mid-ocean ridge and an island arc, e.g. the Arctic vulcanorium between the Nansen and Amundsen deeps in the Arctic Ocean (Runcorn et al., 1967, p.161).

vulcanospeleology The study of caves in volcanic rocks. Also spelled: volcanospeleology.

Vulcano-type eruption (Vul-ca'-no-type) *Vulcanian-type eruption.*

vulpinite (vul'-pi-nite) A scaly, granular, grayish-white variety of anhydrite. It sometimes has an admixture of silica.

vulsinite (vul'-si-nite) An extrusive *trachyte,* composed chiefly of alkali feldspar and also containing calcic plagioclase and augite with or without feldspathoids. The name, given by Washington in 1896, is from the Vulsinii, an Etruscan tribe in the Bolsena region of Italy. Obsolete.

vuonnemite (vu-on′-nem-ite) A vitreous light yellow triclinic mineral: $Na_{11}Ti^{4+}Nb_2(Si_2O_7)_2(PO_4)_2O_3(F,OH)$.

vuorelainenite (vu-o-re-lai′-ne-nite) A submetallic cubic mineral of the *spinel* group: $(Mn^{2+},Fe^{2+})(V^{3+},Cr^{3+})_2O_4$. It is the V-analogue of manganchromite and the Mn-analogue of coulsonite.

vuoriyarvite-K A vitreous white monoclinic mineral: $(K,Na,[])_{12}Nb_8(Si_4O_{12})_4O_8 \cdot 12\text{-}16H_2O$.

vyacheslavite (vya′-ches-lav-ite′) A green to dark green orthorhombic mineral: $U^{4+}(PO_4)(OH) \cdot 2.5H_2O$.

vyalsovite A metallic red orthorhombic mineral: $CaFeAlS(OH)_5$.

vysotskite (vy-sot′-skite) A metallic grayish-white tetragonal mineral of the braggite group: (Pd,Ni)S.

vyuntspakhite-(Y) (vy-unt′-spak-hite) An adamantine colorless monoclinic mineral: $Y(Al,Si)(SiO_4)(OH,O)_2$.

W

Waalian (Waal'-i-an) European warm-climate paleofloral stage: lower Pleistocene (above Eburonian, below Menapian).

wacke (wacke, wack'-e) (a) A "dirty" sandstone that consists of a mixed variety of unsorted or poorly sorted mineral and rock fragments and of an abundant matrix of clay and fine silt; specif. an impure sandstone containing more than 10% argillaceous matrix (Williams et al., 1954, p.290). The term is used for a major category of sandstone, as distinguished from *arenite.* (b) A term used by Fischer (1934) for a clastic sedimentary rock in which the grains are almost evenly distributed among the several size grades; or a mixed sediment of sand, silt, and clay in which no component forms more than 50% of the whole aggregate. (c) A term commonly used as a shortened form of *graywacke.* This usage is not recommended. (d) Originally, a term applied to a soft and earthy variety of basalt, or to the grayish-green to brownish-black clay-like residue resulting from the partial chemical decomposition in place of basalts, basaltic tuffs, and related igneous rocks. Syn: *vake.* Etymol: German *Wacke,* an old provincial mining term signifying a large stone or "stoniness" in general.

wackestone (wacke'-stone) A term for a mud-supported carbonate sedimentary rock containing more than 10% grains (particles with diameters greater than 20 micrometers); e.g. a calcarenite (Dunham, 1962). Cf: *mudstone; packstone.*

wad [coast] A Dutch term for *tidal flat.* Pl: *wadden.* Occasionally misspelled "wadd". Cf: German equivalent "Watt"/"Watten"; *wattenschlick.*

wad [mineral] (a) An earthy, dark-brown or black mineral substance consisting chiefly of an impure mixture of manganese oxides and other oxides, with varying amounts of other elements (such as copper, cobalt, and nickel) and 10-20% water. It is commonly very soft, soiling the hand, but is sometimes hard and compact, and it has a low apparent specific gravity. Wad generally occurs in damp, marshy areas as a result of decomposition of manganese minerals. Cf: *psilomelane.* Syn: *bog manganese; black ocher; earthy manganese.* (b) A general term applied to hydrated oxides of manganese (or of manganese and other metals) whose true nature is unknown or which have variable and uncertain compositions and at least some of which may be amorphous. (c) An English dialectal term for graphite.

wadalite A vitreous dark gray to black cubic mineral of the *garnet* group: $(Ca,Mg)_6(Al,Fe)_5Si_2O_{16}Cl_3$.

Wadati-Benioff zone A plane of seismicity associated with deep-sea trenches, dipping toward and beneath volcanic arcs, along which earthquake foci cluster. The deepest such earthquakes reach nearly 700 km depth. High seismic velocities are also associated with these zones, when plates of the lithosphere sink into the upper mantle causing earthquakes along the upper boundary of the plate. First described by Wadati (1935). Syn: *Benioff zone; Wadati zone.*

Wadati zone *Wadati-Benioff zone.*

wadden (wad'-den) Plural of *wad.* Occasionally misinterpreted as a singular form. The Wadden Sea on the North Sea coast of northern Europe takes its name from this term.

waddy (wad'-dy) *wadi.*

wadeite (wade'-ite) A colorless, pale pink, or lilac hexagonal mineral: $K_2ZrSi_3O_9$.

wadi (wa'-di) (a) A term used in the desert regions of SW Asia and northern Africa for a stream bed or channel, or a steep-sided and bouldery ravine, gully, or valley, or a dry wash, that is usually dry except during the rainy season, and that often forms an oasis. (b) The intermittent and torrential stream that flows through a wadi and ends in a closed basin. (c) A shallow, usually sharply defined, closed basin in which a wadi terminates. Variant plurals: wadis; wadies; wadian; widan. Also spelled: *waddy;* wady. Etymol: Arabic. Syn: *oued; widiyan.* Cf: *arroyo; nullah.*

wadsleyite (wads'-ley-ite) A pale fawn orthorhombic meteorite mineral: $(Mg,Fe^{2+})_2SiO_4$. It is a dimorph of forsterite.

wagnerite (wag'-ner-ite) A yellow, red, or greenish monoclinic mineral of the *triploidite* group: $(Mg,Fe^{2+})_2(PO_4)F$.

Wagner neighborhood A *three-taxon statement* used in quantitative phylogenetic analysis; e.g., any two taxa connected to an outgroup (Brooks et al., 1984, p.81). Cf: *three-taxon statement.*

Waiauan [Wy-ow'-an] South Pacific stage: upper middle Miocene (above Lillburnian, below Tongaporutuan).

Waipipian (Wai-pi-pi'-an) South Pacific stage: lower middle Pliocene (above Opoitian, below Mangapanian).

wairakite (wai-rak'-ite) A colorless monoclinic zeolite mineral: $Ca(Si_4Al_2)O_{12}{\cdot}2H_2O$. It is structurally related to analcime.

wairauite (wai-rau'-ite) A metallic white cubic mineral: CoFe.

Waitakian (Wai-tak'-ian) South Pacific stage: lowermost Miocene (above Duntroonian, below Otaian).

wakabayashilite (wa'-ka-bay-ash'-i-lite) A fibrous yellow monoclinic (pseudo-hexagonal) mineral: $SbAs_{10}S_{18}$.

wake dune A sand dune occurring on the leeward side of a larger dune, and trailing away in the direction of the wind.

wakefieldite-(Ce) A black tetragonal mineral of the *xenotime* group: $(Ce,La,Pb)VO_4$. Syn: *kusuite.*

wakefieldite-(Y) A yellow or pale tan tetragonal mineral of the *xenotime* group: YVO_4.

walchowite (wal'-chow-ite) A honey-yellow variety of retinite containing a little nitrogen, found in brown coal at Walchow in Moravia, Czechoslovakia.

walentaite (wa-len'-ta-ite) A bright yellow or light brownish orthorhombic mineral: $H_2Ca_2Fe^{3+}{}_6(AsO_4)_5(PO_4)_3{\cdot}14H_2O$.

walfordite An adamantine orange cubic mineral: $(Fe^{3+},Te^{6+},Ti^{4+},Mg)(Te^{4+})_3O_8$. It is the Fe^{3+} analogue of winstanleyite.

walkaway Noise analysis performed by moving source points to progressively larger offsets while keeping geophones fixed, or the source point can remain fixed while the geophones are moved to progressively larger offsets. Geophone spacing is usually quite small (5-20 m).

walkaway VSP A vertical seismic profile that involves moving the source away from the wellhead between energy injections.

walkerite A vitreous colorless or white orthorhombic mineral: $Ca_{16}(Mg,Li)_2[B_{13}O_{17}(OH)_{12}]_4Cl_6{\cdot}28H_2O$.

walker's earth (walk'-er's) A syn. of *fuller's earth.* Etymol: German "Walkerde", ("walk" = fulling, treating fibers for clothmaking) + "Erde" = earth.

walking beam (walk'-ing) A rigid, oscillating lever balanced on a fulcrum called a Samson post, used to activate the cable in cable-tool drilling, or to pump oil from a well, by alternating up-and-down motion.

walking leg (a) A prosomal appendage of a merostome, serving for walking. (b) A *pereiopod* of a malacostracan crustacean. (c) The inner branch of the biramous appendage of a trilobite.

walking out A simple method of correlation by which stratigraphic units are traced from place to place along continuous outcrops.

wall [mining] The side of a lode, or of mine workings. Cf: *footwall; hanging wall.*

wall [paleont] (a) An external layer surrounding internal parts of an invertebrate, e.g. a skeletal deposit, formed in various corallites, that encloses the column of a scleractinian polyp and unites the outer edges of the septa, such as septotheca, paratheca, and synapticulotheca; or the part of a cephalopod conch comprising the external shell. (b) The raised margin in a caneolith. (c) The limiting skeletal element of the cup in archaeocyathids.

wall [speleo] (a) The side of a cave passage. (b) A series of columns along a joint crack that have fused together into a solid mass. See also: *partition [speleo].*

wall [struc geol] The rock mass on one side of a fault, e.g. hanging wall, footwall. See also: *wall rock [struc geol].* Syn: *fault wall.*

Wallace's line (Wal'-lace's) The hypothetical boundary that separates the distinctly different floras and faunas of Asia and Australia. It is usually drawn between the islands of Bali and Lombok, through the Strait of Macassar, between Borneo and Celebes, and south of the Philippines. Named after Alfred Russel Wallace (1822-1913), English naturalist. Cf: *Weber's line.*

Wallachian Orogeny (Wal-la'-chi-an) One of the 30 or more short-lived orogenies during Phanerozoic time identified by Stille, in this case at the end of the Pliocene.

wall block A roughly cubical joint-controlled large block of limestone or dolomite, which has rotated outward from a cave wall. See also: *cave breakdown; wall slab.*

wall-contact log *contact log.*
walled lake A lake bordered along its shore by *lake ramparts.*
wallisite (wal'-lis-ite) A metallic gray triclinic mineral: $PbTl(Cu,Ag)As_2S_5$. It is the Cu analogue of hatchite.
wallkilldellite (wall-kill'-dell-ite) A deep red hexagonal mineral: $Ca_4Mn^{2+}_6(AsO_4)_4(OH)_8{\cdot}18H_2O$.
wallkilldellite-(Fe) A brownish-yellow hexagonal mineral: $(Ca,Cu)_4Fe_6(AsO_4,SiO_4)_4(OH)_8{\cdot}18H_2O$.
wall niche *meander niche.*
wallongite (wal-long'-ite) *wollongongite.*
wallow (wal'-low) A depression or area, often filled with water or mud, that suggests a place where animals have wallowed.
wall reef A narrow long reef away from shore, characterized by *reef walls* for sides.
wall rock [eco geol] The rock adjacent to, enclosing, or including a vein, layer, or dissemination of ore minerals. It is commonly altered by epigenetic processes. The term implies more specific adjacency than *host rock* or country rock.
wall rock [intrus rocks] *country rock [intrus rocks].*
wall rock [struc geol] The rock mass comprising the wall of a fault.
wall-rock alteration Alteration of country rock adjacent to hydrothermal ore deposits by the fluids responsible for or derived during the formation of the deposits; also, the alteration products themselves.
wall-rock-replacement nonsulfide zinc deposit A zinc deposit formed by migration of zinc derived from weathering of primary zinc sulfide into adjacent carbonate rocks and replacing calcite and dolomite with *smithsonite* (Hitzman et al., 2003).
wall saltpeter Naturally occurring calcium nitrate; nitrocalcite occurring on walls of limestone caves. Cf: *saltpeter.*
wall slab A thin but large block of rock which has fallen outward from the wall of a cave in limestone in which the dip is nearly vertical. See also: *cave breakdown.*
walpurgite (wal-pur'-gite) An adamantine colorless or yellow to yellow-orange triclinic mineral: $Bi_4(UO_2)O_4(AsO_4)_2{\cdot}2H_2O$. Syn: *waltherite.*
walstromite (wal'-strom-ite) A colorless or white triclinic mineral: $BaCa_2Si_3O_9$. It is the Ba analogue of *margarosanite.*
waltherite (wal'-ther-ite) A mineral formerly believed to be a poorly defined carbonate of bismuth, but now known to be identical to *walpurgite.*
Walther's law (Wal'-ther's) Only those facies and facies-areas can be superimposed which can be observed beside each other at the present time. This law has been much ignored or misunderstood. See Middleton (1973).
walthierite A white or pale yellow rhombohedral mineral of the *alunite* group: $Ba_{0.5}Al_3(SO_4)_2(OH)_6$.
Waltonian (Wal-to'-ni-an) North Sea Stage: middle Pliocene (above Icenian, below Ludhamian).
wandering (wan'-der-ing) The slow, winding compound movement of a stream, consisting of the *sweeping* of meanders and the *swinging* of a meander belt.
wandering dune A sand dune, such as a barchan, that is slowly shifted more or less as a unit in the leeward direction of the prevailing winds, and that is characterized by insufficient vegetation to anchor it. Cf: *anchored dune.* Syn: *migrating dune; traveling dune.*
waning development (wan'-ing) *declining development.*
waning slope The lower part of a hillside surface, tending to become concave below the *constant slope,* having an angle that decreases continuously downslope as the hillside stretches to the valley floor or other local base level (Wood, 1942). Cf: *wash slope.* Ant: *waxing slope.* Syn: *concave slope.*
want A zone in which the coal of a coal seam is missing, owing to a low-angle normal fault or a washout, squeeze, or roll. Cf: *nip [coal]; pinch.* Syn: *cutout.*
Warburg impedance (War'-burg) The ion-diffusion impedance that arises at the interface between an electronic conductor, such as a sulfide grain, and an electrolytic conductor (Sumner, 1976, p.255).
warden A term used in south Wales for a strong massive sandstone associated with coal.
wardite (ward'-ite) A vitreous blue-green, pale green, or colorless tetragonal mineral: $NaAl_3(PO_4)_2(OH)_4{\cdot}2H_2O$.
wardsmithite (ward-smith'-ite) A white hexagonal mineral: $Ca_5MgB_{24}O_{42}{\cdot}30H_2O$.
warikahnite (wa-ri-kahn'-ite) A pale yellow or colorless triclinic mineral: $Zn_3(AsO_4)_2{\cdot}2H_2O$.
warm-current-type phosphate deposit Phosphatic limestone and sandstone deposited from warm currents along the eastern coasts of continents.
warm front The gently sloping front between an advancing warm *air mass* and a colder air mass over which it moves. Its passage is usually preceded by considerable precipitation, whereas there is little or no precipitation afterwards. Cf: *cold front.*
warm glacier *temperate glacier.*
warm ice Ice at the *pressure-melting temperature.*
warm loess Continental loess composed of desert dust, such as that formed in the inland basins and steppes encircling the modern deserts of central Asia between lat. 52°N and 56°N. Cf: *cold loess.*
warm sector The warm tongue of air in a developing mid-latitude *cyclone* lying between the warm front and the trailing cold front.
warm spring A *thermal spring* whose temperature is appreciably above the local mean annual atmospheric temperature, but below that of the human body (Meinzer, 1923, p.54). Cf: *hot spring.*
warp [mass move] *cryoturbate.*
warp [sed] (a) An English provincial term for the fine mud and silt held in suspension in waters artificially introduced over low-lying land. (b) A general term for a bed or layer of sediment deposited by water; e.g. an estuarine clay, or the alluvium laid down by a tidal river. See also: *warping [sed].*
warp [tect] n. A slight flexure or bend of the Earth's crust, either upward or downward, and usually on a broad or regional scale. See also: *warping [tect].*
warped fault A fault, usually a thrust fault, that has been slightly folded.
warping [sed] (warp'-ing) (a) The intentional flooding at high tides of low-lying land near an estuary or tidal river by water loaded with fine mud and silt (warp), the water remaining until the warp is deposited and then being allowed to run off clear during low tides; a means of fertilizing or of raising the general level of large, low tracts, e.g. the conversion of a lagoon or tidal flat into a marsh. Cf: *colmatage.* (b) The filling up of hollows or the choking of a channel with warp.
warping [tect] The slight flexing or bending of the Earth's crust on a broad or regional scale, either upwards (*upwarping*) or downwards (*downwarping*); the formation of a warp [tect].
warpland (warp'-land) Low-lying land that has been built up or fertilized by the process of warping.
warrenite (war'-ren-ite) (a) A general term for gaseous and liquid bitumens consisting mainly of a mixture of paraffins and isoparaffins: a variety of petroleum rich in paraffins. (b) A pink variety of smithsonite containing cobalt. (c) *jamesonite.*
warwickite (war'-wick-ite) A dark brown to black prismatic orthorhombic mineral: $(Mg,Ti,Al)_2O(BO_3)$. It is isomorphous with yuanfuliite.
Wasatchian (Wa-satch'-i-an) North American continental stage: Lower Eocene (above Clarkforkian, below Bridgerian).
wash [coast] (a) A piece of land that is washed, or alternately covered and uncovered, by a sea or river; e.g. a sandbank or mudbank, or an area of such banks, alternately submerged and exposed by the tide. (b) The shallowest part of a river, estuary, or arm of the sea. (c) A bog, fen, or marsh.
wash [eco geol] An alluvial placer.
wash [geomorph] (a) Erosion effected by wave action. (b) The wearing away of soil by runoff water, as in gullying or sheet erosion; *rainwash.*
wash [sed] (a) Loose or eroded surface material (such as gravel, sand, silt) collected, transported, and deposited by running water, as on the lower slopes of a mountain range; esp. coarse alluvium. (b) A fan-shaped deposit, as an *alluvial fan* or an *alluvial cone,* or a mound of detritus below a cliff opening. (c) *downwash.*
wash [streams] (a) A term applied in the western U.S. (esp. in the arid and semiarid regions of the SW) to the broad, gravelly, normally dry bed of an intermittent stream, often situated at the bottom of a canyon; it is occasionally filled by a torrent of water. Syn: *dry wash; washout [streams].* (b) A shallow body of water; esp. a shallow creek.
washboard moraine (wash'-board) (a) One of several small, parallel, regularly spaced ridges that are oriented transverse to the ice movement in a general sense and that collectively resemble a washboard. They are abundant in north-central U.S. and in the western plains of Canada. (b) A subglacial disintegration feature formed at the base of a thrust plane in glacier ice by the periodic recession and readvance of a glacier which pushes previously deposited ground moraine into a ridge (Gravenor and Kupsch, 1959, p.54). Examples are common on the swell-and-swale topography of Alberta and Saskatchewan, Canada.
wash cone *outwash cone.*
washing (wash'-ing) (a) Erosion or wearing-away by the action of waves or running water. (b) The selective sorting, and removal, of fine-grained sediment by water currents. Cf: *winnowing.*
washings (wash'-ings) Material abraded or transported by the action of water.
Washitan (Wash'-i-tan) North American (Gulf Coast) stage: Lower and Upper Cretaceous (above Fredericksburgian, below Woodbinian).
Washita stone (Wash'-i-ta) A porous, uniformly textured *novaculite* found in the Ouachita ("Washita") Mountains, used esp. for sharpening woodworking tools. Syn: *Ouachita stone.*
washland (wash'-land) An embanked, low-lying land bordering a river or estuary, usually part of the natural floodplain, over which floodwaters are allowed to flow periodically in order to control high water levels in the river. Cf: *rond.*
wash load That part of a stream's sediment load that consists of grain sizes finer than those of the bed (Einstein, 1950, p.4). It is that part of the *suspended load* that is not derived from the bed but is supplied to the stream by bank erosion, sheetwash, and mass wasting. Cf: *bed-material load.*
wash ore *treat rock.*
washout [geomorph] (wash'-out) (a) The removal of surface materials by concentrated surface flow of water from *flash floods,* often causing extensive scouring and bank caving. (b) A place where part of a road or railway has been washed away by the waters of a freshet or local flood.

washout [mining] In a coal seam, a mass of shale, siltstone, or sandstone filling a channel that was cut into the coal swamp during the time of deposition. Cf: *cutout; horseback [coal].*

washout [sed struc] A channel or channel-like feature produced in a sedimentary deposit by the scouring action of flowing water and later filled with the sediment of a younger deposit. Cf: *channel cast.* Syn: *scour and fill [sed struc]; cut and fill [sed struc].*

washout [streams] A narrow channel or gully cut by a swiftly flowing stream during and after a heavy rainfall; a wash.

washover (wash'-o-ver) (a) Material deposited by the action of overwash; specif. a small fan built on the landward side of a bar or barrier, separating a lagoon from the open sea, produced by storm waves breaking over low parts of the bar or barrier and depositing sediment in the lagoon. Cf: *blowover.* Syn: *wave delta; storm delta.* (b) The process by which a washover is formed.

washover fan A fan-like deposit consisting of sand washed over a barrier island during a storm. Cf: *hurricane delta.* Syn: *overwash fan.*

wash plain (a) *outwash plain.* (b) An *alluvial plain* composed of coarse alluvium.

wash slope The lower, gentle slope of a hillside, lying at the foot of an escarpment or steep rock face and usually covered by an accumulation of talus; it is less steep than the *gravity slope* above and often consists of alluvial fans or pediments. Term introduced by Meyerhoff (1940). Cf: *waning slope.* See also: *foot slope.* Syn: *haldenhang; basal slope.*

wastage [geomorph] (wast'-age) A general term for the denudation of the Earth's surface. See also: *mass wasting.*

wastage [glaciol] *ablation [glaciol].*

waste [env geol] Any solid or liquid generated by human activity that has little or no economic value, usually the result of the manufacture, mining, or processing of a material to produce an economic product.

waste [geol] Loose material resulting from weathering by mechanical and chemical means, and moved down sloping surfaces or carried short distances by streams to the sea; esp. *rock waste.*

waste bank A bank or other accumulation composed of waste material; e.g., a bank where excess earth excavated during the digging of a ditch is dumped parallel to it. Cf: *spoil bank.*

waste-disposal well (waste'-dis-pos'-al) A less-preferred syn. of *waste-injection well.*

waste-injection well (waste'-in-jec'-tion) A well used for the injection of waste water or other fluids into the subsurface. Because the wastes rarely degrade to an innocuous condition but are simply stored, they are not truly disposed of; hence, this term is preferred to *waste-disposal well.*

waste mantle Disintegrated and decomposed rock material that overlies bedrock.

waste plain (a) *alluvial plain* (b) *bajada.*

waste rock In mining, rock that must be broken and disposed of in order to gain access to and excavate the ore; valueless rock that must be removed or set aside in mining. Syn: *mullock.*

waste stabilization lagoon A lined or unlined impoundment designed to treat municipal wastewater by a combination of aerobic and anaerobic processes. Also called: sewage lagoon.

waste stream The loose rock debris in transit to the sea or to rock-rimmed desert basins, consisting wholly of debris or of debris and water in varying proportions (Grabau, 1924, p.541).

wastewater (a) *return flow.* (b) Seepage of water from a ditch or reservoir. (c) Water that carries wastes from homes, businesses, and industries; a mixture of water and dissolved or suspended solids.

wasting (wast'-ing) The gradual destruction or wearing away of a landform or surface by natural processes, including removal by wind, gravity, and rill wash, but excluding stream erosion; e.g. of a glacier by melting, or of rocks by weathering. See also: *mass wasting; backwasting.* Cf: *wearing.*

watanabeite A metallic silvery lead-gray orthorhombic mineral: $Cu_4(As,Sb)_2S_5$.

watatsumiite A vitreous yellow-green monoclinic mineral of the *neptunite* group: $Na_2KLiMn_2V_2Si_8O_{24}$.

water [gem] (wa'-ter) An obsolete term referring to the color and clarity of a precious stone or pearl, and esp. of a diamond. Cf: *river [gem].*

water [geog] (a) A British term for lake, pond, pool, or other body of standing fresh water. (b) A Scottish term for a stream and for a stream bank or the land abutting a stream.

water agate *enhydros.*

water balance *hydrologic budget.*

water bar A low earth *berm,* or a berm-and-shallow-ditch combination, that is oriented nearly along a topographic contour to direct surface water away from a reclaimed construction area, usually on a relatively steep slope.

water-bearing formation *aquifer.*

water bed A term used in the upper Mississippi Valley for a bed of coarse gravel or pebbles occurring in the lower part of the upper till.

water biscuit *microbial biscuit.*

water bloom *plankton bloom.*

waterbreak The arrival of energy which travels in the water directly from the source.

water-break (a) A place in a stream where the surface of the water is broken by bottom irregularities. (b) *breakwater.*

water budget *hydrologic budget.*

water capacity The maximum amount of water that a rock or soil can hold.

water color The apparent color of surface waters of the ocean, due to dissolved and suspended material in the water. In the open ocean, it is due mostly to chlorophyll pigments. Often observed from space using specialized instruments on satellites. Cf: *SeaWiFS.*

water column *depth.*

water content [sed] Amount of water contained in a porous sediment or sedimentary rock, generally expressed as the ratio of the weight of the water in the sediment to that of the sediment when dried, multiplied by 100. See also: *moisture content [soil].*

water content [snow] This term is not recommended because it has been used for both *water equivalent* and *liquid-water content,* two different concepts.

watercourse (wa'-ter-course) (a) A natural, well-defined channel produced wholly or in part by a definite flow of water, continuous or intermittent. Also, a ditch, canal, aqueduct, or other artificial channel for the conveyance of water, as for the draining of a swamp. (b) Legally, a natural stream arising in a given drainage basin but not wholly dependent for its flow on surface drainage in its immediate area, flowing in a channel with a well-defined bed between visible banks or through a definite depression in the land, having a definite and permanent or periodic supply of water, and usually, but not necessarily, having a perceptible current in a particular direction and discharging at a fixed point into another body of water. (c) A legal right permitting the use of the flow of a stream, esp. of one flowing through one's land, or the receipt of water discharged upon land belonging to another.

water creep The movement of water under or around a structure, such as a dam, built on a semipermeable foundation. See also: *piping.*

water crop *water yield.*

water cupola A vaulted uprising of the surface of the ocean above a submarine volcanic explosion; the initial effect of a submarine explosion on the water surface. It immediately precedes the eruption of volcanic gases and ejectamenta. Syn: *water fountain.*

water cushion Water pumped into the drill pipe during a *drill-stem test* to retard fill-up and prevent collapse of the pipe under sudden pressure changes.

water cycle *hydrologic cycle.*

water-depth deepening In sequence stratigraphy, an increase in water depth relative to a depositional surface. The deepening may be caused by tectonic subsidence, eustasy, or compaction of strata below the depositional surface. This term is used when the process of sea-level rise is unknown, in doubt, or thought to be autocyclic.

water-depth shallowing In sequence stratigraphy, a decrease in water depth relative to a depositional surface. The shallowing may be caused by tectonic uplift, eustasy, or sedimentation onto the depositional surface.

water drive Energy within an oil or gas pool, resulting from hydrostatic or hydrodynamic pressure transmitted from the surrounding aquifer. Cf: *dissolved-gas drive; gas-cap drive.*

water equivalent The amount (or depth) of water that would result from the complete melting of a sample of deposited snow. Not to be confused with *liquid-water content.* See also: *water content [snow].*

water escape structure *fluid escape structure.*

water eye A small, shallow depression formed as a result of chemical weathering in crystalline rock (Russell, 1968, p.94).

water-faceted stone (wa'-ter-fac'-et-ed) *aquafact.*

waterfall (wa'-ter-fall) (a) A perpendicular or steep descent of the water of a stream, as where it crosses an outcrop of resistant rock overhanging softer rock that has been eroded, or flows over the edge of a plateau or cliffed coast. See also: *cascade [streams]; cataract.* Syn: *fall [streams].* (b) An obsolete term for a riffle or rapids in a swift stream. (c) A falling away of the ground such that water may be drained off.

waterfall lake *plunge pool.*

waterfinder (wa'-ter-find-er) (a) One who seeks sources of water supply, esp. a *dowser.* (b) Any instrument purported to indicate the presence of water (e.g. *divining rod*).

water-fit A Scottish term for the mouth of a river.

waterflood (wa'-ter-flood) Informal term for a *water flooding* operation.

water flooding A *secondary recovery* operation in which water is injected into a petroleum reservoir to force additional oil out of the reservoir rock and into producing wells. Syn: *waterflood.*

water fountain *water cupola.*

water gap A deep pass in a mountain ridge, through which a stream flows; esp. a narrow gorge or ravine cut through resistant rocks by an antecedent or superposed stream. Example: Delaware Water Gap, Pennsylvania. Cf: *wind gap.*

water gate A Scottish term for a natural watercourse.

water gun A seismic source which propels a slug of water into the water mass, producing an implosive effect.

water hammer An abnormally high pressure rise in a pipe when sudden changes in flow occur.

waterhead *headwater.*

water hemisphere That half of the Earth containing the bulk (about six-sevenths) of the ocean surface; it is mostly south of the equator, with New Zealand at its approx. center. Cf: *land hemisphere.*

water-holding capacity (wa'-ter-hold'-ing) The smallest value to which the water content of a soil can be reduced by gravity drainage (ASCE, 1958, term 414).

waterhole (wa'-ter-hole) (a) A natural hole, hollow, or small depression that contains water; esp. in an arid or semiarid region. (b) A spring in the desert. (c) A natural or artificial pool, pond, or small lake.

water-laid Deposited in or by water; sedimentary.

water level [grd wat] *water table.*

water level [petroleum] The surface below which the rock pores are virtually saturated with water and above which there is an exploitable concentration of hydrocarbons. Syn: *oil-water contact; gas-water contact.*

water level [surv] An instrument that shows the level by means of the surface of water in a trough or in the legs of a U-tube.

water leveling A type of leveling in which relative elevations are obtained by observing heights with respect to the surface of a body of still water (as of a lake).

water-level mark (wa'-ter-lev'-el) (a) A small horizontal wave-cut "terrace" on an inclined surface of unconsolidated sediment, marking a former water level. (b) watermark.

water-level weathering In coastal areas, a lateral widening of a pool of water, due to the alternate wetting and drying in and around it that causes the banks to retreat. By this process, beaches are created that are unrelated to the stands of the sea, esp. in porous or readily eroded rock (Russell, 1968, p.94).

waterline (a) The migrating line of contact between land and sea; the *shoreline.* (b) The actual line of contact, at a given time, between the standing water of a lake or sea and the bordering land. (c) The limit of backrush where waves are present on a beach.

waterlogged (wa'-ter-logged) Said of an area in which water stands near, at, or above the land surface, so that the roots of all plants except hydrophytes are drowned and the plants die.

waterlogging The process of accumulating water on top of soil where the water table and ground surface coincide.

water mass (a) A body of seawater having a characteristic temperature and salinity range defined by a temperature-salinity curve. (b) A mixture of two or more water types. See also: *intermediate water; deep water; bottom water [oceanog]; surface water.*

water of capillarity *capillary water.*

water of compaction *Rejuvenated water* originating from the destruction of interstices by compaction of sediments.

water of crystallization Water in a crystal structure that is chemically combined but may be driven off by heat; molecular water, e.g. in gypsum: $CaSO_4{\cdot}2H_2O$.

water of dehydration Water that has been set free from its chemically combined state. Cf: *water of crystallization.*

water of dilation *water of supersaturation.*

water of hydration Water that is chemically combined in a crystalline substance to form a hydrate, but that may be driven off by heat.

water of imbibition (a) The amount of water a rock can contain above the water table. (b) *water of saturation.*

water of retention That part of the interstitial water in a sedimentary rock that remains in the interstices under *capillary pressure* and under conditions of unhindered flow; usually (though incorrectly) called connate water.

water of saturation The amount of water that can be absorbed by water-bearing material without dilation of that material. Syn: *water of imbibition.*

water of supersaturation Water in excess of that required for saturation; water in sedimentary materials that are inflated or dilated, such as plastic clay or flowing mud whose particles are not in contact and are separated by water. Syn: *water of dilation.*

water opal (a) *hyalite.* (b) Any transparent precious opal.

water opening Any break in *sea ice* which reveals the water; e.g. a *lead [ice].*

water-plasticity ratio (wa'-ter-plas-tic'-ity) *liquidity index.*

water pocket A small, bowl-shaped depression on a bedrock surface, where water may gather; esp. a water hole in the bed of an intermittent stream, formed at the foot of a cliff by the action of falling water when the stream is in the flood stage. Syn: *tinaja.*

waterpower (wa'-ter-pow-er) The power of moving or falling water, once used to drive machinery directly, as by a water wheel, but now more commonly used to generate electricity by means of a power generator coupled to a turbine through which the water passes. Cf: *hydroelectric power; hydropower; white coal.*

water quality The fitness of water for use, being affected by physical, chemical, and biological factors.

water race A *race [water]* or *watercourse.*

water regimen *regimen [water].*

water reserve (a) An area of land set aside for feeding streams that are used for water supply. (b) A general term for a quantity or source of water regarded as a supplemental or reserve supply.

water resources A general term referring to the occurrence, replenishment, movement, discharge, quantity, quality, and availability of water.

water-resources region A natural drainage basin or hydrologic administrative region that contains either the drainage area of a major river or the combined drainage areas of a series of rivers. There are 18 water-resources regions in the conterminous United States and one each in Alaska, Hawaii, and the Caribbean (Solley et al., 1983).

water right The legal right to use a specific quantity of water, on a specific time schedule, at a specific place, and for a specific purpose (NAS-NRC, 1992).

water-rolled Said of round, smooth sedimentary particles that have been rolled about by water.

waters (wa'-ters) The marine *territorial waters* of a nation or state.

water sand A porous sand with high or total water content. Cf: *oil sand.*

water sapphire (a) A light-colored blue sapphire. (b) An intense-blue variety of cordierite occurring in waterworn masses in certain river gravels (as in Ceylon) and sometimes used as a gemstone. Syn: *saphir d'eau.* (c) A term applied to waterworn pebbles of topaz, quartz, and other minerals from Ceylon.

water saturation *liquid-water content.*

watershed (wa'-ter-shed) (a) A term used in Great Britain for a drainage *divide.* (b) A *drainage basin.* Etymol: probably German "Wasserscheide", "water parting". The original and "correct" meaning of the term "watershed" signifies a "water parting" or the line, ridge, or summit of high ground separating two drainage basins. However, the usage of the term, esp. in the U.S. and by several international agencies, has been changed to signify the region drained by, or contributing water to, a stream, lake, or other body of water. The term, when used alone, is ambiguous, and unless the context happens to suffice without aid from the word itself, "the uncertainty of meaning entailed by this double usage makes the term undesirable" (Meinzer, 1923, p.16).

watershed area The total area of the watershed above the discharge-measuring points. Cf: *basin area.*

watershed leakage Seepage or flowage of water underground from one drainage basin to an outlet in a neighboring drainage basin or directly to the sea.

watershed line A drainage divide.

watershed management Administration and regulation of the aggregate resources of a drainage basin for the production of water and the control of erosion, streamflow, and floods. Also includes the operational functions.

water sky Dark or gray streaks or patches in the sky near the horizon or on the underside of low clouds, indicating the small amount of light reflected from water features in the vicinity of sea ice; darker than *land sky.* See also: *blink.*

watersmeet (wa'-ters-meet) A meeting place of two streams.

waterspace (wa'-ter-space) The ecologic and social interplay among land, water, and social institutions (Padfield and Smith, 1968).

watersplash (wa'-ter-splash) A shallow ford in a stream.

water spreading Artificial recharge of groundwater by spreading water over an absorptive area. Generally broadened to include all methods of artificial recharge involving surficial structures or shallow furrows, pits, or basins, as opposed to injection of water through wells or deep pits or shafts.

water-stable aggregate A soil aggregate which is stable to the action of water such as falling drops, or agitation as in wet-sieving analysis.

waterstead (wa'-ter-stead) An English term for a stream bed.

water stone A mineral name that has been applied to moonstone, hyalite, enhydros, and jade.

water supply A source or volume of water available for use; also, the system of reservoirs, wells, conduits, and treatment facilities required to make the water available and usable. Syn: *water system.*

water system (a) *river system.* (b) *water supply.*

water table The surface between the *saturated zone* and the *unsaturated zone;* that surface of a body of *unconfined groundwater* at which the pressure is equal to that of the atmosphere. Syn: *water level [grd wat]; groundwater table; groundwater surface; level of saturation; phreatic surface; groundwater level; free-water elevation; free-water surface.*

water-table aquifer (wa'-ter-ta'-ble) *unconfined aquifer.*

water-table cement *groundwater cement.*

water-table decline The downward movement of the *water table* (Rogers, 1981, p.429). Syn: *phreatic decline.*

water-table divide *divide [grd wat].*

water-table map A map that shows the configuration of the upper surface of the saturated zone by means of contour lines.

water-table mound *groundwater mound.*

water-table rock Rock cemented at or near the level of the water table; a specific type of *hardpan* (Russell, 1968).

water-table well A well tapping *unconfined groundwater.* Its water level may, but does not necessarily, lie at the level of the water table. Cf: *artesian well; nonflowing well; shallow well.*

water tagging The introduction of foreign substances (tracers) into water to detect its movement by measurement of the subsequent location and distribution of the tracers.

water track Strip of fen with seepage of moving water, crossing bogs or separating bog areas from each other (Gore, 1983). Syn: *soak.*
water type A body of seawater having a characteristic combination of temperature and salinity of a very narrow range. Essentially a mass of water described by a single point on a temperature versus salinity diagram. Water type and *water mass* are often used synonymously.
water-vascular system (wa'-ter-vas'-cu-lar) A fluid-filled system of tubular vessels or canals peculiar to echinoderms, used to control the movement of tube feet and perhaps also functioning in excretion and respiration. Primary components include a stone canal, a ring canal, radial canals, and tube feet.
waterway (wa'-ter-way) (a) A way or channel, either natural (as a river) or artificial (as a canal), for conducting the flow of water. (b) A navigable body or stretch of water available for passage; a watercourse.
water well (a) A well that extracts water from the saturated zone or that yields useful supplies of water. (b) A well that obtains groundwater information or that replenishes groundwater. (c) A well drilled for oil but yielding only water.
water witch (a) A device for determining the presence of water, usually electrically. Cf: *divining rod.* (b) *dowser.* Nonpreferred syn: *witch.*
water witching *dowsing.*
waterwork (wa'-ter-work) A tank, dock, canal lock, levee, seawall, or other engineering structure built in, for, or as a protection against, water. Also applied to some water treatment facilities.
water yield The runoff from a drainage basin; precipitation minus evapotranspiration (Langbein and Iseri, 1960). Syn: *water crop; runout.*
wath A dialectal term for a *ford* in a stream.
watkinsonite (wat'-kin-son-ite) A metallic black monoclinic mineral: $PbCu_2Bi_4(Se,S)_8$.
watt A syn. of *tidal flat.* Pl: watten. Etymol: German "Watt/Watten". Cf: *wad; wadden.*
wattenschlick (wat'-ten-schlick) Tidal or intertidal mud. Etymol: German *Wattenschlick,* "tidal-flats mud".
wattersite A submetallic dark reddish-brown or black monoclinic mineral: $Hg^{1+}{}_4Hg^{2+}O_2(CrO_4)$.
wattevillite (watte'-vill-ite) An ill-defined silky white mineral: $Na_2Ca(SO_4)_2{\cdot}4H_2O(?)$. It occurs in hairlike crystals. Also spelled: wattevilleite.
Waucoban (Wau-co'-ban) North American series or stage: Lower Cambrian (above Precambrian, below Albertan). Also spelled: Waucobian. Syn: *Georgian.*
Waulsortian mound A particular type of carbonate mud mound, based on those in the Lower Carboniferous of Waulsort in Belgium. These mounds are gray with *stromatactis* structures and commonly contain zones rich in fossils such as crinoids.
wave [seis] A *seismic wave.*
wave [water] An oscillatory movement of water manifested by an alternate rise and fall of a surface in or on the water.
wave age The state of development of a wind-generated, water-surface wave, expressed as the ratio of wave velocity to wind velocity (measured at about 10 m above stillwater level).
wave attenuation A decrease in amplitude with distance from the source.
wave base The depth at which wave action no longer stirs the sediments; it is usually about 10 to 20 meters.
wave-built Constructed or built up by the action of lake or sea waves, assisted by their currents. Cf: *wave-cut.*
wave-built platform A syn. of *wave-built terrace.* The term is inconsistent because a platform is usually regarded as an erosional feature.
wave-built terrace A gently sloping coastal feature at the seaward or lakeward edge of a wave-cut platform, constructed by sediment brought by rivers or drifted along the shore or across the platform and deposited in the deeper water beyond. Cf: *marine terrace; beach plain.* Syn: *wave-built platform.*
wave cross ripple *oscillation cross ripple.*
wave-current ripple mark A longitudinal *compound ripple* in which the material forming the crest is believed to have accumulated by the oscillation produced by wave action on a preexisting transverse (current) ripple (Van Straaten, 1953a; and Kelling, 1958, p.124).
wave-cut Carved or cut away by the action of lake or sea waves, assisted by their currents. Cf: *wave-built.*
wave-cut bench A level to gently sloping narrow surface or platform produced by wave erosion, extending outward from above the base of the wave-cut cliff and occupying all of the shore zone and part or all of the shoreface (Johnson, 1919, p.162); it is developed mainly above water level by the spray and splash of storm waves aided by subaerial weathering and rainwash. The bench may be bare, freshly worn rock or it may be temporarily covered by a beach; it may end abruptly or grade into the *abrasion platform.* See also: *wave-cut platform.* Syn: *shore platform; beach platform; high-water platform.*
wave-cut cliff A cliff, esp. a *sea cliff,* produced by the breaking-away of rock fragments after horizontal and landward undercutting by waves.
wave-cut notch A notch produced along the base of a sea cliff by wave erosion.
wave-cut plain *wave-cut platform.*
wave-cut platform (a) A gently sloping surface produced by wave erosion, extending far into the sea or lake from the base of the wave-cut cliff. It represents both the wave-cut bench and the abrasion platform. Syn: *wave-cut terrace; erosion platform; shore platform; wave-cut plain; strandflat.* (b) A term sometimes used more restrictedly as a syn. of *abrasion platform.*
wave-cut terrace A syn. of *wave-cut platform.* The term is inconsistent because a terrace is usually regarded as a constructional feature.
wave delta *washover.*
wave dispersion *dispersion [seis].*
wave-dominated delta A delta built in an environmental setting where fluvial sedimentation is significantly modified by wave processes. Deltas of this category are characterized by simple shoreline configurations and are sand-rich. Cf: *arcuate delta; cuspate delta; fluvially-dominated delta; tide-dominated delta.*
wave drift The net translation of water in the direction of wave movement, caused by the open orbital motion of water particles with the passage of each surface wave.
wave energy The capacity of waves to do work. It is proportional to the square of the wave height. A high-energy coast is characterized by breaker heights greater than 50 cm and a low-energy coast is characterized by breaker heights less than 10 cm.
wave erosion *marine abrasion.*
wave-etched shoreline A relatively straight shoreline made irregular by differential wave erosion acting on coastal materials of varying resistance.
wavefield (a) The spatial perturbations at a given time that result from the passage of a wave, such as the pressure changes caused by a seismic wave. (b) The temporal perturbations over a given surface that result from seismic waves. Migration is sometimes referred to as "downward propagation of the wavefield."
wave forecasting Determination of future wave height, period, and direction, using surface winds calculated by numerical weather models plus measurements of present waves, often made from space using *satellite altimetry.*
waveform A plot (usually as a function of time) of a quantity involved in wave motion, such as voltage, current, seismic displacement, particle velocity, etc. The waveform involved in seismic work is called the *embedded wavelet.*
wavefront A surface representing the position of a traveling seismic disturbance at a particular time.
wave front In optics, the locus of all the points reached by light that is sent outward in all directions from a point. In an isotropic medium, the wave front is a sphere; if the light is constrained to a beam, the wave front will be a plane surface. Cf: *wave normal.* Syn: *wave surface.*
wavefront chart A diagram showing the position of a seismic disturbance at successive times. It usually shows raypaths also.
wave gage Instrument for measuring wave properties, usually wave height as a function of time at a point. Syn: *wave meter.*
wave generation The creation and growth of waves by natural or mechanical means, as by a wind blowing over a water surface for a certain period of time.
wave guide A region, usually a layer, in the atmosphere, ocean, or solid Earth that tends to channel acoustic or seismic energy. See *channel wave.*
wave interference ripple *interference ripple.*
wavelength (wave'-length) (a) The distance between successive wave crests, or other equivalent points. (b) In symmetrical, periodic fold systems, the distance between adjacent antiformal or synformal hinges.
wavelet processing (wave'-let) Processing of seismic data with the objective of shortening and simplifying the effective waveshape, making it constant along the entire line, or achieving a specific equivalent waveshape, often a zero-phase one.
wave line *swash mark.*
wavellite (wa'-vell-ite) A white to yellow, green, or brown orthorhombic mineral: $Al_3(PO_4)_2(OH,F)_3{\cdot}5H_2O$. It occurs usually in small hemispherical aggregates exhibiting a strongly developed internal radiating structure. See also: *fischerite.* Syn: *bialite.*
wavemark [coast] (wave'-mark) *swash mark.*
wavemark [sed] A *ripple* produced by wave action during the period of deposition.
wave meter *wave gage.*
wave normal In optics, the line at a given point perpendicular to a plane that is tangent to the surface of a light wave at that point. Cf: *wave front.*
wave number The number of wave cycles per unit of distance; the reciprocal of wavelength. Sometimes wave number (κ) is taken as 2π divided by the wavelength (λ). Syn: *spatial frequency.*
wave ogive *ogive.*
wave polarization The sense of displacement for a wave (such as an S-wave). Motion may be predominantly confined to one direction because of source and/or receiver orientation, as with the preferred generation of SH-waves by a vibrator vibrating perpendicular to the profile direction and horizontal geophones likewise oriented. A preferred orientation may also be imposed by anisoptropy, for example, S-waves involving displacement parallel (or perpendicular) to vertical fractures in rocks. See: *anisotropy.*
wave ray *orthogonal.*

wave refraction (a) The changing orientation of wave crests as a wave moves at an angle into shallower water. For example, crests in shallow water move more slowly than crests in deeper water, and the wave crests tend to align parallel to the shore. (b) The change in orientation of wave crests by currents.

wave ripple *oscillation ripple.*

wave spectrum The distribution of wave energy as a function of frequency.

wave splitting *birefringence [seis].*

wave steepness The ratio of the height of a water wave to its length. A wave with a ratio of 1/25 to 1/7 has "great" steepness; one with a ratio of less than 1/100 has "low" steepness.

wave surface *wave front.*

wave wash The erosion of shores or embankments by the lapping or breaking of waves; esp. the erosion of levees during floods.

wave-wave interaction The slow transfer, over hundreds of wave periods, of energy from one component to another of the *spectrum* of a *sea* due to the nonlinear nature of wave dynamics. Syn: *nonlinear interaction.*

wavy bedding (wav'-y) (a) Bedding characterized by undulatory bounding surfaces. Syn: *rolling strata.* (b) A form of interbedded mud and ripple-cross-laminated sand, in which "the mud layers overlie ripple crests and more or less fill the ripple troughs, so that the surface of the mud layer only slightly follows the concave or convex curvature of the underlying ripples" (Reineck and Wunderlich, 1968, p.101). Cf: *flaser structure [sed]; lenticular bedding.*

wavy extinction *undulatory extinction.*

wawayandaite (wa-wa-yan'-da-ite) A pearly colorless monoclinic mineral: $Ca_6Mn^{2+}_2BBe_9Si_6O_{23}(OH,Cl)_{15}$.

wax A solid, noncrystalline hydrocarbon of mineral origin such as ozocerite and paraffin wax, composed of the fatty acid esters of the higher hydrocarbons.

waxing development (wax'-ing) *accelerated development.*

waxing slope The upper part of a hillside surface, tending to become convex by being rounded off just above an escarpment, having an angle that increases continuously downslope as the hillside is worn back (Wood, 1942). Ant: *waning slope.* Syn: *convex slope.*

wax opal Yellow opal with a waxy luster.

waxy luster (wax'-y) A type of mineral luster that resembles the luster of wax, e.g. in chalcedony.

wayboard A British term for potassium bentonite.

waylandite (way'-land-ite) A white rhombohedral mineral of the *crandallite* group: $(Bi,Ca)Al_3(PO_4,SiO_4)_2(OH)_6$.

W-chert Chert nodules formed by weathering (Dunbar and Rodgers, 1957, p.249).

weak ferromagnetism *Antiferromagnetism* in which the opposing atomic magnetic moments do not cancel perfectly, because of spin canting or defects, so that there is a weak, spontaneous macroscopic magnetization. An example of a mineral displaying weak ferromagnetism is α Fe_2O_3, hematite. Syn: *parasitic ferromagnetism.*

weal A descriptive field term for one of the crisscrossing raised bands, 5-7.5 cm wide, occurring on a more or less evenly patterned sedimentary surface (Donaldson and Simpson, 1962, p.74). The bands in cross section are almost semicircular.

Wealden (Weald'-en) A European (Great Britain) lithostratigraphic unit: Lower Cretaceous (above Purbeckian, below Gault).

wear The reduction in size or the change in shape of clastic fragments by one or more of the mechanical processes of abrasion, impact, or grinding (Wentworth, 1931, p.24-25). See also: *wearing.*

wearing (wear'-ing) The gradual destruction of a landform or surface by friction or attrition. Cf: *wasting.* See also: *backwearing; downwearing; wear.*

weather [geol] (weath'-er) v. To undergo changes, such as discoloration, softening, crumbling, or pitting of rock surfaces, brought about by exposure to the atmosphere and its agents. See also: *weathering.*

weather [meteorol] n. The condition or state of the Earth's atmosphere, specif. its temperature, barometric pressure, wind velocity, humidity, clouds, and precipitation at a specific place and time. Cf: *climate.* adj. A syn. of *windward.*

weather chart *weather map.*

weather coal Brown coal that has been weathered and displays bright colors.

weathered ice (weath'-er-ed) *Sea ice* that has undergone a gradual elimination of surface irregularities by thermal and mechanical processes of removal and addition of material; ice whose *hummocks* and *pressure ridges* are smoothed and rounded.

weathered iceberg An iceberg that has undergone prolonged ablation, which generally gives it a very irregular but rounded shape.

weathered layer [seis] *low-velocity layer.*

weathered zone *low-velocity zone.*

weathering (weath'-er-ing) The destructive process or group of processes by which earthy and rocky materials on exposure to atmospheric agents at or near the Earth's surface are changed in color, texture, composition, firmness, or form, with little or no transport of the loosened or altered material; specif. the physical disintegration and chemical decomposition of rock that produce an in-situ mantle of waste and prepare sediments for transportation. Most weathering occurs at the surface, but it may take place at considerable depths, as in well-jointed rocks that permit easy penetration of atmospheric oxygen and circulating surface waters. Some authors restrict weathering to the destructive processes of surface waters occurring below 100°C and 1 kb; others broaden the term to include biologic changes and the corrasive action of wind, water, and ice. Obsolete syn: *demorphism; clastation.*

weathering [seis] A near-surface low-velocity layer (LVL), usually the portion where some air rather than water fills the pore spaces of rocks and unconsolidated earth (Sheriff, 2002). Syn: *low-velocity layer; weathered layer [seis].*

weathering boulder *boulder of weathering.*

weathering correction In seismic exploration, a correction applied to reflection and refraction data for variations in traveltime produced by irregularities in a low-velocity or weathered layer near the surface. Correction for elevation variations are usually also incorporated. Syn: *low-velocity-layer correction.*

weathering crust A term in use by European geologists to designate a regionally widespread and usually deep zone of weathered materials, formed over a geologically long interval by relatively uniform chemical weathering. The weathered materials typically consist of clay minerals, laterite (including bauxite), or both (Keller, 1977).

weathering escarpment An escarpment developed where gently dipping sedimentary rocks of varying resistance are subjected to degradation; the term is not appropriate because mass-wasting, sheetwash, and stream erosion are fully as important as weathering (Thornbury, 1954, p.71-72).

weathering front The interface of fresh and weathered rock; a term proposed by Mabbutt (1961) to replace *basal surface.*

weathering index A measure of the weathering characteristics of coal, according to a standard laboratory procedure. Syn: *slacking index.*

weathering out The exposing of relatively resistant rock as the surrounding softer rock is reduced by weathering.

weathering pit *weather pit.*

weathering-potential index (weath'-er-ing-po-ten'-tial) A measure of the degree of susceptibility to weathering of a rock or mineral, computed from a chemical analysis, and expressed as the mol-percentage ratio of the sum of the alkalies and alkaline earths (less combined water) to the total mols present exclusive of water (Reiche, 1943, p.66).

weathering rind An outer crust or layer on a pebble, boulder, or other rock fragment, formed by weathering.

weathering shot In seismic exploration, the use of a small seismic source to determine velocity characteristics and thickness of the weathering or low-velocity layer. Syn: *short shot; poop shot.*

weathering velocity That velocity with which a seismic *P wave* travels through the near-surface *low-velocity layer.* Weathering velocity is often 500-700 m/s. Cf: *subweathering velocity.*

weather map A chart that is used to show temperature, pressure, precipitation, wind direction and speed, air masses, and fronts of a given area for a given reference level. Syn: *weather chart.*

weather pit A small to tank-size depression in rock. A shallow depression on the flat or gently sloping summit of large exposures of granite or granitic rocks (as in the Sierra Nevada, Calif.), attributed to strongly localized solvent action of impounded water (Matthes, 1930, p.64); diameter is 30-45 cm, and depth ranges up to 15 cm. Syn: *weathering pit.* Cf: *rock doughnut; oven.*

weather shore A shore lying to the windward or in the direction from which the wind is blowing, and thereby exposed to strong wave action. Ant: *lee shore.*

weber (we'-ber) The SI unit of magnetic flux: 10^8 *maxwell.* Syn: *volt-second.*

weberite (we'-ber-ite) A pale gray orthorhombic mineral: Na_2MgAlF_7.

Weber number (We'-ber) The relationship of the forces of inertia to those of surface energy, expressed as the product of density times velocity of flow squared times length divided by surface energy. It is important in the movement of water in porous media and capillaries (Chow, 1964, p.7-5).

Weber's line (We'-ber's) A hypothetical boundary between the Asian and Australasian biogeographic regions. It generally coincides with the Australian-Papuan shelf and is sometimes used in preference to *Wallace's line.* Named after Max Weber (d. 1937), German zoologist.

websterite [mineral] (web'-ster-ite) *aluminite.*

websterite [rock] (a) In the *IUGS classification,* a plutonic rock with M equal to or greater than 90, ol/(ol+opx+cpx) less than 5, and both opx/(ol+opx+cpx) and cpx/(ol+opx+cpx) less than 90. (b) A pyroxenite composed chiefly of ortho- and clinopyroxene. The name, given by G. H. Williams in 1890, is from Webster, North Carolina.

weddellite (wed'-dell-ite) A colorless, white, or yellowish tetragonal oxalate mineral: $Ca(C_2O_4)\cdot 2H_2O$. It is found as small isolated crystals in urinary calculi and in mud at the bottom of Weddell Sea, Antarctica. Cf: *whewellite.*

wedge [optics] (a) *optical wedge.* (b) *quartz wedge.*

wedge [paleont] A five-sided crystalline element of a *heterococcolith,* having two dimensions subequal and the third dimension small at one edge and approaching zero at the other.

wedge [stratig] (a) The shape of a stratum, vein, or intrusive body that thins out; specif. a wedge-shaped sedimentary body, or *prism [sed].* (b) *sand wedge.*

wedge fault A *contraction fault* that cuts across a single bed; slip on the fault thickens the bed.
wedge ice *foliated ground ice.*
wedge-out n. The edge or line of *pinch-out* of a lensing or truncated rock formation.
wedge out v. To become progressively thinner or narrower to the point of disappearance; to *thin out.*
wedge theory A corollary of the obsolete *contracting Earth* theory; it supposes that shrinking of the crust breaks it into wedge-shaped blocks, which are uplifted and laterally compressed along their margins, resulting in two-sided orogenic structures.
wedgework (wedge'-work) The action of rock disintegration by the wedge-like insertion of agents such as roots and esp. ice. See also: *frost wedging.* Also, the results of wedgework action.
wedging (wedg'-ing) The splitting, breaking, or forcing apart of a rock as if by a wedge, such as by the growth of salt or mineral crystals in interstices; specif. *frost wedging.*
weedia (weed'-i-a) A type of stromatolite consisting of algal crusts that are nearly flat or essentially parallel to the bedding and appear in cross section as a branching network of bedding planes that join in an irregular manner (Pettijohn, 1957, p.222, 399).
weeding Reducing the number of points defining a line while preserving the essential line shape (Krzanowski et al., 1993). See also: *smoothing; thinning.*
weeksite (weeks'-ite) A yellow orthorhombic mineral: $(K,Ba)_{1-2}(UO_2)_2(Si_5O_{13})\cdot H_2O$.
weeping rock (weep'-ing) A porous rock from which water oozes.
weeping spring A spring of small yield. Syn: *seepage spring.*
Weertman creep A type of grain-scale, ductile deformation accomplished by the motion of dislocations where the rate of dislocation glide is limited by the rate at which dislocations can climb.
Wegener hypothesis (We'-ge-ner) The theory of *continental drift.*
wegscheiderite (weg'-schei-der-ite') A colorless triclinic mineral: $Na_5(HCO_3)_3(CO_3)$.
wehrlite [mineral] (wehr'-lite) A mineral that was earlier formulated Bi_2Te_3, but later shown to be a mixture of hessite and pilsenite.
wehrlite [rock] (a) In the *IUGS classification,* a plutonic rock with M equal to or greater than 90, ol/(ol+opx+cpx) between 40 and 90, and opx/(ol+opx+cpx) less than 5. (b) A peridotite composed chiefly of olivine and clinopyroxene with common accessory opaque oxides. Kobell in 1834 named the rock, which he thought to be a mineral, after Wehrle, who had analyzed it.
weibullite (wei'-bull-ite) A metallic gray orthorhombic mineral: $Ag_{0.3}Pb_{5.3}Bi_{8.3}Se_6S_{12}$.
Weichsel (Weich'-sel) The term applied in northern Europe to the classical fourth and last glacial stage of the Pleistocene Epoch, after the Saale glacial stage; equivalent to the *Würm* and *Wisconsinan.* Adj: *Weichselian.*
Weichselian (Weich-sel'-ian [Pron: Vi-xel'-ian]) European cold-climate paleofloral stage: uppermost *Pleistocene* (above *Eemian,* below *Flandrian*). Equiv. Würm-II glacial system.
weight dropping A method used in seismic prospecting, in which a heavy weight is dropped to create seismic waves. See also: *Thumper.*
weighted reciprocal lattice The *reciprocal lattice* of a crystal with each *reciprocal lattice point* weighted according to the intensity of X-ray diffraction from the corresponding planes in the *direct lattice.*
weighting (weight'-ing) The use of multipliers in statistical calculations to adjust the influence of certain observations. The multipliers may be constrained to sum to 1.00 to avoid bias.
weilerite (wei'-ler-ite) An earthy white mineral of the *beudantite* group: $BaAl_3(OH)_6(AsO_4)(SO_4)$.
weilite (weil'-ite) A white triclinic mineral: $CaHAsO_4$. It is the arsenate analogue of monetite.
weinebeneite A vitreous colorless monoclinic mineral: $CaBe_3(PO_4)_2(OH)_2\cdot 4H_2O$.
weinschenkite (wein'-schenk-ite) (a) *churchite-(Y).* (b) A dark-brown variety of hornblende low in ferrous iron and high in ferric iron, aluminum, and water.
weir (a) A small dam in a stream, designed to raise the water level or to divert its flow through a desired channel; e.g., a "leaping weir". (b) A notch in a levee, dam, embankment, or other barrier across or bordering a stream, through which the flow of water is regulated; e.g., a "wasteweir". Well-designed weirs usually have known dimensions and cross-section areas as a function of height above the base. Graduations of height allow rapid determination of hydraulic parameters, such as wetted perimeter, needed to calculate discharge.
weisbachite (weis'-bach-ite) A variety of anglesite containing barium.
weishanite (wei'-shan-ite) A metallic hexagonal mineral: $(Au,Ag)_3Hg_2$.
weissbergite (weiss-berg-ite) A metallic gray triclinic mineral: $TlSbS_2$.
Weissenberg pattern (Weis'-sen-berg) The pattern of X-ray diffraction spots obtained from a single crystal using a Weissenberg camera and monochromatic radiation, by a moving-film method which enables unambiguous indexing of all diffractions for any properly oriented single crystal.
weissite (weiss'-ite) A blue-black hexagonal or trigonal mineral: Cu_5Te_3.
weld A contact between two layers, or between what had been two sides of a salt intrusion, that are now in contact because the salt has flowed away.
welded dike A *secundine dike* whose boundaries have become obscured by continued mineral growth of the granitic country rock into the intrusion.
welded flow breccia The lower part of the fragmented crust of aa and block-lava flows, where the fragments are thoroughly welded together rather than being loose as in the upper part of the crust.
welded texture A texture of pyroclastic rocks, especially those derived from ash flows and nuées ardentes, that is formed by the heat and pressure of still-plastic particles as they are deposited.
welded tuff A glass-rich pyroclastic rock that has been indurated by the welding together of its glass shards under the combined action of the heat retained by particles, the weight of overlying material, and hot gases. It is generally composed of silicic pyroclasts and appears banded or streaky. Cf: *sillar.*
welding (weld'-ing) (a) Consolidation of sediments (esp. of clays) by pressure resulting from the weight of superincumbent material or from earth movement, characterized by cohering particles brought within the limits of mutual molecular attraction as water is squeezed out of the sediments (Tyrrell, 1926, p.196). (b) The diagenetic process whereby discrete crystals and/or grains become attached to each other during compaction, often involving pressure solution and solution transfer (Chilingar et al., 1967, p.322).
welinite (we-lin'-ite) A red-brown or reddish-black trigonal mineral: $(Mn^{4+},W)(Mn^{2+},Mg)(SiO_4)(O,OH)_3$.
well [eng] A hollow cylinder of reinforced concrete, steel, timber, or masonry constructed in a pit or hole in the ground that reaches to hardpan or bedrock and used as a support for a bridge or building. Also, the pit or hole in which the well is built.
well [gem] A dark, nonreflecting area in the center of a fashioned diamond or other gemstone cut with a deep pavillion. Also called: nailhead.
well [petroleum] A borehole or shaft sunk into the ground for the purpose of obtaining oil and/or gas from an underground source, or of introducing water or gas under pressure into an underground formation. See also: *oil well; gas well.*
well [water] (a) An artificial excavation (pit, hole, tunnel), generally cylindrical in form and often walled in, sunk (drilled, dug, driven, bored, or jetted) into the ground to such a depth as to penetrate water-yielding rock or soil and to allow the water to flow or to be pumped to the surface; a *water well.* (b) A term originally applied to a natural spring or to a pool formed by or fed from a spring; esp. a mineral spring. (c) A term used chiefly in the plural form for the name of a place where mineral springs are located or of a health resort featuring marine or freshwater activities; a spa.
well-bedded (well-bed'-ded) Said of a bedded rock whose beds are numerous and clearly defined.
wellbore *borehole.*
well casing *casing.*
well cuttings Rock chips cut by a bit in the process of well drilling, and removed from the hole in the *drilling mud* in rotary drilling or by the *bailer* in cable-tool drilling. Well cuttings collected at closely spaced intervals provide a record of the strata penetrated. Syn: *cuttings; drill cuttings; well samples.*
well-graded (a) A geologic term for *well-sorted.* (b) An engineering term pertaining to a graded soil or unconsolidated sediment with a continuous distribution of particle sizes from the coarsest to the finest, in such proportions that the successively smaller particles almost completely fill the spaces between the larger particles. Ant: *poorly graded.*
wellhead [drill] A steel casting located on the surface of an oil or gas well. It consists of a lower, larger *casinghead* and an upper, smaller *tubing head.*
wellhead [grd wat] (well'-head) The source from which a stream flows; the place in the ground where a spring emerges.
well log A graphic record of the measured or computed physical characteristics of the rock section encountered in a well, plotted as a continuous function of depth. Measurements are made by a *sonde* as it is withdrawn from the borehole by a wire line. Several measurements are usually made simultaneously, and the resulting curves are displayed side by side on the common depth scale. Both the full display and the individual curves are called logs. Well logs are commonly referred to by generic type, e.g. *resistivity log, radioactivity log,* or by specific curve type, e.g. *sonic log, gamma-ray log.* See also: *sample log; driller's log; mud log.* Syn: *borehole survey; geophysical log; wire-line log.*
well loss The decline in *hydraulic head* in a well bore below that in the adjacent aquifer or reservoir that is created by turbulent flow and by frictional head losses in the *well screen* and in the *gravel pack.*
well point A slotted or screened tube or pipe terminating in a pointed shoe that is driven into shallow, water-bearing strata for water-supply or excavation-dewatering purposes.
well record A concise statement of the available data regarding a well; a full history or day-by-day account, from the day the location was surveyed to the day production ceased.
well-rounded Said of a sedimentary particle whose original faces, edges, and corners have been destroyed by abrasion and whose entire surface con-

sists of broad curves without any flat areas; specif. said of a particle with no secondary corners and a roundness value between 0.60 and 1.00 (Pettijohn, 1957, p.59). The original shape may be suggested by the present form of the particle. Also, said of the *roundness class* containing well-rounded particles. Cf: *roundness.*

well samples *well cuttings.*

well screen A section of well casing perforated or slotted to allow inflow of water.

wellsite (well'-site) Barium-bearing phillipsite and calcium-bearing harmotome.

well site *location.*

well-sorted Said of a sorted sediment that consists of particles all having approximately the same size and that has a sorting coefficient less than 2.5. Based on the phi values associated with the 84 and 16 percent lines, Folk (1954, p.349) suggests sigma phi limits of 0.35-0.50 for well-sorted material. Ant: *poorly sorted.* Syn: *well-graded.*

wellspring *fountainhead.*

wellstrand (well'-strand) A Scottish term for a stream flowing from a spring.

well survey Determining seismic velocity as a function of depth by lowering a geophone into a borehole and recording energy from sources on or near the surface. See also: *check shot.* Cf: *vertical seismic profiling.*

well ties The comparison of seismic measurements with geologic datum points at well locations.

well water Water obtained from a well; water from the saturated zone or from a perched aquifer; *phreatic water.*

well yield The discharge of a well at nearly steady flow.

weloganite (we-lo'-ga-nite) A vitreous yellow to amber triclinic mineral of the *donnayite* group: $Na_2Sr_3Zr(CO_3)_6{\cdot}3H_2O$.

welshite (welsh'-ite) A reddish-brown to redish-black triclinic mineral of the *aenigmatite* group: $Ca_2Mg_4Fe^{3+}Sb^{5+}Si_4Be_2O_{20}$.

welt [speleo] In speleology, a small, disk-shaped or elongated deposit along a joint in bedrock or crack in a column; the initial phase of *shield* formation.

welt [tect] A nongenetic term used by Bucher (1933) for a raised part of the Earth's crust of any size with a distinct linear development. Cf: *furrow [struc geol].*

Wemmelian (Wem-mel'-i-an) European stage: Upper Middle Eocene (above Ledian, below Auversian).

wendwilsonite (wend-wil'-son-ite) A pink to red monoclinic mineral of the *roselite* group: $Ca_2(Mg,Co)(AsO_4)_2{\cdot}2H_2O$.

wenkite (wenk'-ite) A gray hexagonal zeolite mineral: $Ba_4Ca_6(Si,Al)_{20}O_{39}(OH)_2(SO_4)_3{\cdot}nH_2O$.

Wenlockian (Wen-lock'-i-an) A series of the Standard Global Chronostratigraphic Scale: Middle Silurian (above Llandoverian, below Ludlovian). The basal GSSP is in southwest England (Holland, 1982). Also called: Wenlock.

Wenner array (Wen'-ner) An electrode array in which the four electrodes are in line and equally spaced, and in which the outer pair is used to inject current into the ground while the inner pair is used to measure differences in potential.

Wentworth grade scale (Went'-worth) An extended version of the *Udden grade scale,* adopted by Chester K. Wentworth (1891-1969), U.S. geologist, who modified the size limits for the common grade terms but retained the geometric interval or constant ratio of 1:2 (Wentworth, 1922). The scale ranges from clay particles (diameter less than 1/256 mm) to boulders (diameter greater than 256 mm). It is the grade scale generally used by North American sedimentologists. Cf: *phi grade scale.*

werdingite (wer'-ding-ite) A vitreous brownish-yellow triclinic mineral: $(Mg,Fe^{2+})_2Al_{14}B_4Si_4O_{37}$. It is related to *sillimanite.*

Werfenian (Wer-fe'-ni-an) European stage: lowermost Triassic (above Thuringian, below Virglorian). See also: *Scythian.*

wermlandite (werm'-land-ite) A pale greenish-gray soft platy trigonal mineral: $(Ca,Mg)Mg_7(Al,Fe^{3+})_2(SO_4)_2(OH)_{18}{\cdot}12H_2O$. It is related to woodwardite.

Werner deconvolution (Wur'-nor) A method of inverting regularly spaced profiles of magnetic data, such as obtained from aeromagnetic surveys, for location, depth, and, sometimes, dip and *susceptibility* contrast of magnetic sources. Each calculation is run for two source model approximations: (1) thin dike and (2) magnetic contact, i.e., boundary between blocks of differing magnetic properties. Different window lengths provide solutions for different source depths. The realtion between an anomaly and its magnetic source is expressed as a convolution, hence, calling the operation a deconvolution or, sometimes, filtering. (Hartman et al., 1971; Ku and Sharp, 1983). See also: *Euler deconvolution.*

Wernerian (Wer-ne'-ri-an) Of or pertaining to *Wernerianism.*

Wernerianism The system of *geognosy* of Abraham G. Werner (1749-1817), German geologist and mineralogist who conceived of a sequential ordering of four (later five) classes of rocks originating in successive periods of Earth's history: (1) primitive rocks (granite, gneiss etc.), crystallized from a primeval global sea; overlain by (2) the flötz or layered rocks, deposited from the receding global sea; overlain by (3) volcanic rocks; and (4) alluvial deposits. Later, the class "Transition" was added between (1) and (2). These divisions were blended with other schemes to form the archaic time scale: Primitive or Primary, Transition, Secondary, and Tertiary. Werner's linear history was contradicted by the cyclic system of James Hutton, who claimed "no vestige of a beginning, no prospect of an end".

wernerite (wer'-ner-ite) A syn. of common *scapolite,* a specific mineral of the scapolite group intermediate between meionite and marialite.

wesselite (wes'-sel-ite) A hypabyssal *nephelinite* containing biotite, amphibole, titanian augite, hauyne, and nepheline. The name, given by Scheumann in 1922, is for Wesseln, Czech Republic. Obsolete.

wesselsite A vitreous blue tetragonal mineral of the *gillespite* group: $SrCuSi_4O_{10}$.

west-coast desert A *coastal desert* found on the western edge of continents and occurring in the tropical-desert latitude, i.e. near the tropics of Cancer or Capricorn. The temperature fluctuation, both annually and daily, is much lower than for inland tropical deserts (Strahler, 1963, p.335).

Western bentonite A *sodium bentonite.*

westerveldite (wes'-ter-veld-ite') (a) A metallic brownish-white or gray orthorhombic mineral: FeAs. (b) A group name for minerals with the general formula MAs, where M = Fe, Co, Ru, or Rh.

westgrenite (west'-gren-ite) *bismutomicrolite.*

westing (west'-ing) A *departure* (difference in longitude) measured to the west from the last preceding point of reckoning; e.g. a linear distance westward from the north-south (vertical) grid line that passes through the origin of the grid.

Westphal balance (West'-phal) In mineral analysis, a balance used to determine specific gravity of a *heavy liquid.* Syn: *beam balance.*

Westphalian (West-pha'-li-an) A European series or stage: Upper Carboniferous (above Namurian, below Stephanian).

westward drift (west'-ward) A component of the *secular variation* of the Earth's magnetic field; the movement is about 0.2° per year of the broad-scale departures of the actual geomagnetic field from an ideal dipole field.

wet analysis A method of estimating the effective diameters of particles smaller than 0.06 mm by mixing the sample in a measured volume of water and checking its density at intervals with a sensitive hydrometer (Nelson, 1965, p.512).

wet assay Any type of assay procedure that involves liquid, generally aqueous, as a means of separation. Cf: *dry assay.*

wet avalanche *wet-snow avalanche.*

wet blasting Abrasion or attrition effected by the impact of water against an exposed surface; e.g. the formation of an *aquafact* by wave action.

wet-bulb temperature The lowest temperature to which air can be cooled at constant pressure by the evaporation of water into it. During the process, the heat required for the evaporation is supplied by the air itself, which is thereby cooled. Measured by a psychometer. Cf: *dewpoint.*

wet chemical analysis Any of the methods for chemical determinations using water or other liquids as part of the process. Typically it refers to *gravimetric analysis* and *volumetric analysis.*

wet gas A natural gas containing liquid hydrocarbons. Cf: *dry gas; condensate.*

wetland An area that is regularly wet or flooded and has a water table that stands at or above the land surface for at least part of the year (NOAA, 2004). The term has supplanted more colloquial terms such as swamp and marsh.

wet loose avalanche A loose snow avalanche which involves wet snow.

wet playa A playa that is soft under foot, having a thin and puffy surface that is coated with white efflorescent salts indicating the active discharge of near-surface groundwater by evaporation (Thompson, 1929); a *salina.* It is underlain by loose granular silt, salt crystals, and moist clay. Cf: *dry playa.* Syn: *moist playa.*

wet slab avalanche A slab avalanche which involves wet snow.

wet snow Deposited snow that contains liquid water. Cf: *dry snow.*

wet-snow avalanche Any avalanche involving wet snow. Due to friction, it is the slowest-moving of the snow avalanches. Syn: *slush avalanche; wet avalanche.*

wet-snow line The lower boundary of the *percolation zone;* the line on a glacier below which the temperature of all of the previous winter's snow has been raised to the melting point by the end of the summer. Cf: *wet-snow zone.* Less preferred syn: *saturation line [glaciol].*

wet-snow zone The part of the *accumulation area* of a glacier where the temperature of all of the previous winter's snow has been raised to the melting point by the end of the summer.

wet surge A pyroclastic surge containing saturated steam. This designation reflects the physical state of the tephra after deposition. Wet surge deposits show textural and diagenetic evidence of having been wet when emplaced.

wettability (wet'-ta-bil'-i-ty) The ability of a liquid to form a coherent film on a surface, owing to the dominance of molecular attraction between the liquid and the surface over the cohesive force of the liquid itself.

wetted perimeter (wet'-ted) (a) The length of the wetted contact between a stream of flowing water and its containing conduit or channel, measured in a plane at right angles to the direction of flow. (b) The length of the perimeter of a conduit below the water surface. (c) The entire perimeter of a conduit flowing full. The wetted perimeter is used when computing the *hydraulic radius* (ASCE, 1962).

wetting front (wet'-ting) *pellicular front.*

wet unit weight The *unit weight* of soil solids plus water per unit of total volume of soil mass, irrespective of the degree of saturation.

WGS *world geodetic system.*

whaleback (whale'-back) (a) A large mound or hill having the general shape of a whale's back, esp. a smooth elongated ridge of desert sand having a rounded crest and ranging widely in size (about 300 km long, 1-3 km wide, and perhaps 50 m high). It forms a coarse-grained platform or pedestal built up and left behind by a succession of longitudinal (seif) dunes along the same path. Syn: *sand levee.* (b) A rounded, elongated rock mass, commonly granite, found in tropical areas associated with tors. (c) A *roche moutonnée,* often of granitic composition, such as those in Canada and Finland.

wheatleyite (wheat'-ley-ite) A bright blue triclinic oxalate mineral: $Na_2Cu(C_2O_4)_2{\cdot}2H_2O$.

wheel A holothurian *sclerite* in the form of a wheel, consisting of a *hub,* a rim, and spokes.

Wheelerian (Wheel-er'-i-an) Western North America stage: upper Pliocene and Pleistocene (above Venturian).

wheelerite (wheel'-er-ite) A yellowish variety of retinite that is soluble in ether and that fills fissures in, or is thinly interbedded with, lignite beds in northern New Mexico.

wheel ore The mineral bournonite, esp. when occurring in wheel-shaped twin crystals.

wherryite (wher'-ry-ite) A pale yellow to pale green monoclinic mineral: $Pb_7Cu_2(SO_4)_4(SiO_4)_2(OH)_2$.

whetstone (whet'-stone) Any hard fine-grained naturally occurring rock, usually siliceous, that is suitable for sharpening implements such as razors, knives, and mechanics' tools; e.g. *novaculite.*

whewellite (whew'-ell-ite) A white, colorless, or brown monoclinic oxalate mineral: $CaC_2O_4{\cdot}H_2O$. It occurs as a warty and somewhat opaline incrustation on marble, and is also a common phase in urinary calculi. Cf: *weddellite.*

whinstone (whin'-stone) A colloquial British term for dolerite, basalt, and other dark fine-grained igneous rocks. The term is derived from the Whin Sill in northern England. Cf: *trap [ign].*

whiplash (whip'-lash) A smooth-surfaced *flagellum* (without mastigonemes) of some algae and protozoans, having a long rigid basal part and a short thinner distal region.

whipstock (whip'-stock) n. A long wedge-shaped steel device with a concave groove along its inclined face, placed in an oil or gas well and used during drilling to deflect and guide the drill bit toward the direction in which the inclined grooved surface is facing. v. To use a whipstock to drill a *directional well.*

whirl ball A spindle-shaped, tubular, ellipsoidal, or spherical mass of fine sandstone embedded in silt, its long axis being vertical or steeply inclined. It is attributed to vortices in mudflows (Dzulynski et al., 1957).

whirlpool (whirl'-pool) A body of water moving rapidly in a circular path of relatively limited radius. It may be produced by a current's passage through an irregular channel or by the meeting of two opposing currents. Cf: *eddy; maelstrom.*

whirl zone A zone of transition between a slump sheet and the overlying strata.

whistling sand A *sounding sand,* often found on a beach, that gives rise to a high-pitched note when stepped on or struck with the hand, the sound apparently resulting from the translation of grain over grain. Syn: *squeaking sand; musical sand.*

Whitbian (Whit'-bi-an) Stage in Great Britain: upper Lower Jurassic (above Domerian, below Yeovilian).

Whitcliffian (Whit-cliff'-i-an) European stage: Upper Silurian (above Leintwardian, below Devonian).

white agate A term sometimes applied to white or whitish chalcedony.

white antimony *valentinite.*

white band A layer in a glacier consisting of ice that is white and opaque because it contains numerous air bubbles. Cf: *blue band [glaciol].*

white-bedded phosphate (white'-bed-ded) A term used in Tennessee for a phosphatic limestone characterized by partial replacement of calcite by calcium phosphate, and by a matrix consisting of cryptocrystalline quartz. It occurs in regular bands alternating with thinner beds of chert. Cf: *brown rock; hard-rock phosphate.*

whitecap (white'-cap) (a) White foam at the crests of waves in the open ocean due to intermittent breaking of steep waves when the wind speed exceeds 7 m/s. (b) The white foam produced when the wind blows the crest of the wave downwind.

white chert A light-colored *chert,* or chert proper, as distinguished from the dark variety or *black chert.*

white clay *kaolin [eco geol].*

white coal (a) A fanciful term for *waterpower.* (b) *tasmanite [coal].*

white cobalt (a) *cobaltite.* (b) *smaltite.*

white copperas (a) *goslarite.* (b) *coquimbite.*

whitedamp (white'-damp) A term for carbon monoxide in coal mines. Cf: *blackdamp; afterdamp; firedamp.*

white earth A siliceous earthy material that is used as a pigment in paint.

white feldspar *albite.*

white garnet (a) A translucent variety of grossular, sometimes resembling white jade in appearance. (b) *leucite.*

white gold A pale alloy of gold that resembles silver or platinum; esp. gold alloyed with a high proportion of nickel or palladium to give it a white color, with or without other alloying metals (such as tin, zinc, or copper).

white ice (a) *Sea ice* of not more than one winter's growth, and a thickness of 30-70 cm, also known as "thin" *first-year ice.* (b) Glacier ice containing abundant air bubbles that was formed by compaction of snow and appears white. Cf: *black ice; blue ice.* (c) Snow ice that forms on top of river and lake ice. Whiteness results from scattering of light by entrapped air bubbles.

white iron ore *siderite [mineral].*

white iron pyrites *marcasite [mineral].*

whiteite (white'-ite) A group name for monoclinic minerals with the general formula $A^{2+}B^{2+}C^{2+}D^{3+}(PO_4)_4(OH)_2{\cdot}8H_2O$. The subgroup with D^{3+} = Al is classified as whiteite, and that with $D^{3+} = Fe^{3+}$ is classified as jahnsite. The individual members of these subgroups are given the root name with chemical-element suffixes that indicate the dominant *A, B* and *C* structure-site occupancies.

whiteite-(CaFeMg) (white'-ite) A tan monoclinic mineral of the *whiteite* group: $CaFe^{2+}Mg_2Al_2(PO_4)_4(OH)_2{\cdot}8H_2O$.

whiteite-(CaMnMg) A yellow, greenish-yellow, pink or pale lavender monoclinic mineral of the *whiteite* group: $CaMn^{2+}Mg_2Al_2(PO_4)_4(OH)_2{\cdot}8H_2O$.

whiteite-(MnFeMg) A brown monoclinic mineral of the *whiteite* group: $Mn^{2+}Fe^{2+}Mg_2Al_2(PO_4)_4(OH)_2{\cdot}8H_2O$.

white lead ore *cerussite.*

white matter Term applied to portions of a conodont element that are more or less opaque in incident light and thus appear white or gray in contrast to adjacent or surrounding more or less clear and translucent areas. Areas of white matter correspond to internal parts of elements characterized by voids, which may be interlamellar spaces or concentrations of small, densely packed, irregularly shaped cellules (TIP, 1981, pt. W, supp.2).

white mica A light-colored mica; specif. *muscovite.*

white mundic *arsenopyrite.*

white nickel (a) *nickel-skutterudite.* (b) *rammelsbergite.*

white olivine *forsterite.*

white opal A form of precious opal of any light color, as distinguished from *black opal;* e.g. a pale blue-white gem variety of opal.

whiteout (white'-out) The diffusion of daylight by multiple reflection between falling snow and overcast clouds, so that the horizon and surface features are impossible to discern. See also: *blizzard.*

white pyrite *marcasite [mineral].*

white pyrites (a) *arsenopyrite.* (b) *marcasite [mineral].*

Whiterockian (White-rock'-ian) A North American stage: lowermost Middle Ordovician (above Canadian, below Chazyan).

white sand (a) Quartzitic sand, pure enough to resist heat, used in steel furnaces. (b) Gypsum sand, as at White Sands National Monument, New Mexico.

white sapphire The colorless or clear pure variety of crystallized corundum.

white schorl *albite.*

white smoker A plume of hydrothermal fluid issuing from the crest of an oceanic ridge; e.g., the East Pacific Rise at the mouth of the Gulf of California. The fluid is clouded by white precipitates, mostly barite and silica. It issues at rates of tens of cm/s and at temperatures of 100° to 350°C (Macdonald et al., 1980). Cf: *black smoker.*

white stone A clear, colorless imitation gem, resembling the diamond.

white tellurium (a) *sylvanite.* (b) *krennerite.*

white top In coal mining, light gray shale that occurs above a coal seam, between the coal and the overlying dark-shale roof. It may be arenaceous and is usually unlaminated.

white vitriol *goslarite.*

whiting (whit'-ing) (a) A mass of muddy water in which abundant carbonate material is suspended, producing a white color. Whitings typically occur over shallow carbonate platforms and are elongated by wind or tidal currents. Most of them consist of stirred-up bottom sediment (Blatt et al., 1972, p.426). (b) A coccolith bloom in open waters, as observed from satellites.

whitlockite (whit'-lock-ite) (a) A vitreous colorless, gray, pink, or yellowish rhombohedral mineral: $Ca_9(Mg,Fe^{2+})(PO_4)_6(PO_3OH)$. It is a constituent of dental enamel. Syn: *merrillite.* (b) A group name for rhombohedral minerals with the analogous composition, but with Ca replaced by Sr, and partially by Na.

whitmoreite (whit'-more-ite) (a) A pale tan, dark brown, or greenish-brown prismatic monoclinic mineral: $Fe^{2+}Fe^{3+}_2(PO_4)_2(OH)_2{\cdot}4H_2O$. It is isostructural with *arthurite.* (b) A group name for monoclinic minerals with the analogous composition, but with Fe^{2+} replaced by Mn^{2+}, Cu^{2+} or Zn, and PO_4 replaced by AsO_4.

Whitneyan (Whit'-ney-an) North American land mammal age: Late Oligocene (after Orellan, before Arikareean).

whittakerite A bluish-black monoclinic mineral of the *amphibole* group: $Na(NaLi)(Fe^{3+},Mg,Li)_5Si_8O_{22}(OH)_2$.

whole-rock Adj. Used to indicate that a portion of rock rather than individual minerals was analyzed. In the rubidium-strontium age method the rock may have remained a closed system for rubidium and strontium isotopes whereas the constituent minerals did not. Thus, a calculated age for the whole rock would give the apparent age of formation whereas the individual minerals might give discordant ages. This whole-rock, closed-system feature does not hold true for all isotopic systems. Syn: *total-rock.*

whorl (a) One of the turns of a spiral or coiled shell; specif. a single complete turn through 360 degrees of a gastropod shell, a cephalopod conch, or a foraminiferal test. See also: *body whorl.* Syn: *volution.* (b) An arrangement of two or more anatomical parts or organs of one kind in a circle around the same point on an axis; e.g. a circle of equally spaced branches around the stem of a plant, arranged like the spokes of a wheel. Syn: *verticil.*

whorl coccolith One of the modified coccoliths forming a whorl about the naked pole in nonmotile coccolithophores exhibiting dimorphism (such as *Ophiaster*).

whorled Having a circular arrangement of appendages, e.g. leaves, at nodes on an axis. Cf: *alternate; acyclic.* Syn: *verticillate.*

whorl height The height of a nautiloid or ammonoid whorl measured at right angles to the maximum *width,* comprising the distance from the middle of the venter to the middle of the dorsum plus the depth of the impressed area (TIP, 1959, pt. L, p.6). In practice, the "oblique whorl height" is commonly used, consisting of the distance from umbilical seam to the middle of the venter.

whorl section A transverse section of a cephalopod whorl.

whorl side The flank of a cephalopod conch; esp. the lateral wall of an ammonoid whorl between umbilical seam and ventral area.

wiborgite (wi'-borg-ite) *rapakivi.*

Wichita Orogeny (Wich'-i-ta) A name used by Van der Gracht and van Waterschoot (1931) for the first major phase of deformation in the Wichita and Ouachita orogenic belts of southern Oklahoma. In the Wichita belt it is dated by adjacent strata as early Pennsylvanian (Morrow); in the Ouachita Belt, it includes a Mississippian phase that produced the great flysch body of the Stanley and Jackfork formations.

wickenburgite (wick'-en-burg-ite') A colorless, white, or pink trigonal mineral: $Pb_3CaAl_2Si_{10}O_{27}{\cdot}4H_2O$.

wickmanite (wick'-man-ite) A brownish-yellow cubic mineral of the *schoenfliesite* group: $Mn^{2+}Sn^{4+}(OH)_6$.

wicksite (wicks'-ite) A dark blue, almost black orthorhombic mineral: $NaCa_2(Fe^{2+},Mn^{2+})_4MgFe^{3+}(PO_4)_6{\cdot}2H_2O$. It is isomorphous with grischunite.

wide-angle reflection Reflection where the angle of incidence is near or greater than the critical angle. Reflection coefficients may have large values near the critical angle so that reflection energy may be exceptionally strong.

widenmannite (wi'-den-mann-ite') A pale yellow, greenish-yellow, or colorless orthorhombic mineral: $Pb_2(UO_2)(CO_3)_3$.

widespread (wide'-spread) *spread [streams].*

widgiemoolthalite A bluish-green monoclinic mineral of the *hydromagnesite* group: $Ni_5(CO_3)_4(OH)_2{\cdot}4\text{-}5H_2O$.

widiyan (wi'-di-yan) Var. of *wadi* used in the north African deserts (Stone, 1967, p.264).

Widmanstätten structure (Wid'-man-stät'-ten) A geometric, generally triangular or trapezoidal, pattern observed on polished and etched surfaces of iron meteorites (octahedrites), composed of parallel bands or plates of kamacite bordered by thin layers of taenite that intersect one another in two, three, or four directions. The kamacite bands, arranged parallel to the octahedral planes in the host taenite, are produced by exsolution along 111 planes from an originally homogeneous face-centered cubic taenite crystal. Named after Aloys B. Widmanstätten (1753?-1849), Austrian mineralogist who discovered the structure in 1808. Also spelled: Widmanstatten structure. Syn: *octahedral structure;* Widmanstätten figure; Widmanstätten pattern.

width (a) On a brachiopod, the maximum dimension measured perpendicular to the plane of symmetry (at right angles to the length and thickness or height). (b) In cephalopod conchs, the horizontal distance between points located between ribs or spines on opposite whorl sides; the maximum dimension perpendicular to the height.

Wiechert-Gutenberg discontinuity (Wie'-chert-Gu'-ten-berg) *Gutenberg discontinuity.*

Wiener filter A causal filter that will transform an input into a desired output as nearly as possible, subject to certain constraints. "As nearly as possible" (in a least-squares sense) implies that the sum of the squares of differences between the filter output and the desired result is minimized. The filter optimizes standout of a signal S (which is a function of frequency, f) in the presence of random noise N (also a function of frequency). Also called a least-squares filter (Sheriff and Geldart, 1995). Named for Norbert Wiener (1894-1964), American mathematician.

Wien's displacement law The statement that when the temperature of a radiating blackbody increases, the wavelength corresponding to maximum radiance decreases in such a way that the product of the absolute temperature and wavelength is constant (Wien, 1894, p.132-165). See also: *color temperature.*

wiggle stick (wig'-gle) *divining rod.*

wiggle trace A graph of amplitude against time. Also called "squiggle recording". Syn: *variable-amplitude trace.*

wightmanite (wight'-man-ite) A vitreous colorless monoclinic mineral: $Mg_5O(BO_3)(OH)_5{\cdot}2H_2O$.

wiikite (wi'-ik-ite) A variable mixture of euxenite, yttropyrochlore and other minerals.

Wilcoxian (Wil-cox'-i-an) *Sabinian.*

wilcoxite (wil'-cox-ite) A colorless triclinic mineral that efflorescences on mine walls and in a breccia zone: $MgAl(SO_4)_2F{\cdot}18H_2O$.

wild Having very large and often unpredictable amplitudes, e.g., a noisy seismic channel at high gain.

wildcat (wild'-cat) *wildcat well.*

wildcat well An *exploratory well* drilled for oil or gas on a geologic feature not yet proven to be productive, or in an unproven territory, or to a zone that has never produced or is not known to be productive in the general area. Cf: *outpost well; deeper-pool test; shallower-pool test.* Syn: *wildcat.*

wilderness (wil'-der-ness) An area or tract of land that is uncultivated and uninhabited by man.

wilderness area An area set aside by government for preservation of natural conditions for scientific or recreational purposes. See also: *natural area.*

wildflysch (wild'-flysch) A type of *flysch* facies representing a mappable stratigraphic unit displaying large and irregularly sorted blocks and boulders resulting from tectonic fragmentation, and twisted, contorted, and confused beds resulting from slumping or sliding under the influence of gravity. The term was first applied by Kaufmann (1886) in the Alps.

wild land Uncultivated land, or land that is unfit for cultivation; e.g. a wasteland or a desert.

wild river (a) A river whose shores and waters remain essentially in a virgin condition, unmodified by human activity. (b) A torrential river. It may connote a legal definition in the U.S.

wilhelmkleinite An adamantine green monoclinic mineral of the *lazulite* group: $ZnFe^{3+}{}_2(AsO_4)_2(OH)_2$.

wilhelmvierlingite (wil'-helm-vier'-ling-ite) A yellow or pale brownish orthorhombic mineral of the *overite* group: $CaMn^{2+}Fe^{3+}(PO_4)_2(OH){\cdot}2H_2O$.

wilkeite (wil'-ke-ite) A rose-red or yellow variety of *apatite* group, containing hydroxyl, in which the phosphate is partly replaced by carbonate, sulfate, or silicate.

wilkinsonite (wil'-kin-son-ite) A lustrous black triclinic mineral of the *aenigmatite* group: $NaFe^{2+}{}_2Fe^{3+}Si_3O_{10}$.

wilkmanite (wilk'-man-ite) A metallic monoclinic mineral of the *brezinaite* group: Ni_3Se_4.

willemite (wil'-lem-ite) A rhombohedral mineral: Zn_2SiO_4. It was formerly a major ore of zinc at Franklin, Sussex County, NJ, and commonly contained manganese. Willemite varies in color from white or greenish yellow to green, reddish, and brown; it exhibits an intense green fluorescence in ultraviolet light.

willemseite (wil-lem'-se-ite) A light green monoclinic mineral of the *talc* group: $(Ni,Mg)_3Si_4O_{10}(OH)_2$.

willhendersonite (will-hen'-der-son-ite) A vitreous colorless triclinic zeolite mineral: $KCaAl_3Si_3O_{12}{\cdot}5H_2O$.

williamsite (wil'-liams-ite) A massive yellow to green, impure variety of antigorite resembling jade in appearance and used for decorative purposes. It usually contains specks of chromite.

Williston's law (Wil'-lis-ton's) The theory that evolution tends to reduce the number of similar parts in organisms and render them progressively more different from each other.

Willow Lake layering (Wil'-low) A type of *rhythmic layering* in which the layers are composed of elongate mineral grains oriented with their long axes roughly perpendicular to the layering (Taubeneck and Poldervaart, 1960, p.1295). Syn: *comb layering.*

willyamite (wil-lyam'-ite) A metallic steel-gray orthorhombic (pseudo-cubic) mineral of the *gersdorffite* group: (Co,Ni)SbS.

Wilson cycle (Wil'-son) A successive recurrence of plate-tectonic spreading and convergence with a period generally in the 100-million-year range. Named after J. Tuzo Wilson, Canadian geophysicist.

wilsonite (wil'-son-ite) A purplish-red material consisting of an aluminosilicate of magnesium and potassium, and representing an altered scapolite.

Wilson technique In X-ray diffraction analysis, a method of mounting film in a cylindrical X-ray powder camera which enlarges the area for recording back-reflection diffractions on both sides of the entry port. Cf: *Straumanis camera method.*

wilting coefficient (wilt'-ing) *wilting point.*

wilting percentage *wilting point.*

wilting point The point at which the water content of the soil becomes too low to prevent the *permanent wilting* of plants. As originally introduced, and to a certain extent today, the point at which a soil-water deficiency produces any degree of wilting is the wilting point. Syn: *wilting coefficient; wilting percentage.* Cf: *temporary wilting.*

wiluite (wi-lu'-ite) (a) A dark green tetragonal mineral of the *vesuvianite* group: $Ca_{19}(Al,Mg)_{13}(B,[\,],Al)_5(SiO_4)(Si_2O_7)_4(O,OH)_{10}$. (b) A green variety of grossular garnet.

winchite (winch'-ite) A blue or gray monoclinic member of the *amphibole* group: $NaCa(Mg_4Al)Si_8O_{22}(OH)_2$.

wind (a) Naturally moving air, of any direction or speed. (b) More specifically, a meteorologic term for that component of air that moves parallel to the Earth's surface. Its direction and velocity can be measured.
wind abrasion A process of erosion in which windblown particles of rock material or snow and ice scour and wear away exposed surfaces of any kind. Syn: *wind corrasion.*
wind corrasion *wind abrasion.*
wind crust A type of *snow crust* formed by the packing of previously fallen snow into a hard layer by wind action. Cf: *wind slab.*
wind current *drift [oceanog].*
wind-cut stone *ventifact.*
wind-deposition coast (wind'-de-po-si'-tion) A coast built out into the sea by sand dunes advancing in the direction of the prevailing winds; generally found on the lee side of a sandy neck of land.
wind direction That point of the compass from which the wind blows, e.g. a "westerly" wind is blowing from the west. It may also be stated in degrees, measured clockwise from true north, e.g. an east wind has a direction of 90°. Syn: *direction of the wind.*
wind drift (a) drift [oceanog]. (b) "That portion of the total vector drift of sea ice from which the effects of the current have been subtracted" (Baker et al., 1966, p.183).
wind-driven current (wind'-driv'-en) *drift [oceanog].*
wind erosion Detachment, transportation, and deposition of loose topsoil by wind action, esp. in duststorms in arid or semiarid regions or where a protective mat of vegetation is inadequate or has been removed. See also: *deflation.*
wind-faceted stone (wind'-fac'-et-ed) *windkanter.*
wind gap (a) A shallow notch in the crest or upper part of a mountain ridge, usually at a higher level than a *water gap.* (b) A former *water gap,* now abandoned (as by *piracy*) by the stream that formed it; a *pass* that is not occupied by a stream. Syn: *dry gap; air gap; wind valley.*
wind-grooved stone *ventifact.*
windkanter (wind'-kan-ter) A *ventifact,* usually highly polished, bounded by one or more smooth faces or facets, curved or nearly flat, that intersect in one or more sharp edges or angles. The faces may be cut at different times, as when the wind changes seasonally or the pebble is undermined and turned over on its flattened face. Etymol: German "Windkanter", "one having wind edges". Pl: windkanter; windkanters. See also: *einkanter; zweikanter; dreikanter; parallelkanter.* Syn: *faceted pebble; wind-faceted stone.*
wind noise (a) Random noise attributed principally to ground unrest caused by the wind moving plants and trees and shaking their roots. (b) Seismic background noise in the absence of a source.
window [gem] (a) A facet polished on a coated or rough-surfaced diamond to permit the designer to see into the interior and determine the color, clarity, and structure of the rough crystal before deciding how it will be fashioned. (b) The see-through effect which is the result of too-shallow pavilion angles; usually occurs in step cuts.
window [geomorph] (win'-dow) (a) The opening under a *natural bridge* (Gregory, 1917, p.134). (b) The opening in the top of a cave or tunnel by collapse of the ceiling.
window [geophys] (a) A portion of a seismic record free from certain disturbances, i.e., where noise trains are absent. (b) The portion of a data set chosen for consideration, such as for designing operators to be used for autocorrelation or frequency analysis. Syn: *gate [geophys].*
window [paleont] An opening in the skeleton of an ebridian, such as a "lower window" between the opisthoclades, a "middle window" between the mesoclades and the actines, and an "upper window" between the proclades.
window [photo] A rectangular portion of a larger mapped area selected for display in a geographic information system.
window [streams] A part of a river surrounded by river ice, remaining unfrozen during all or part of the winter, and caused by local inflow of warm water or by turbulence or a strong current.
window [tect] An eroded area of a thrust sheet that displays the rocks beneath the thrust sheet. Syn: *fenêtre; fenster.*
wind packing The compaction of snow by wind action.
wind polish *desert polish.*
wind-polished stone (wind'-pol-ished) A *ventifact* having a *desert polish.*
wind ridge A ridge of snow formed by the deposition of blowing snow at right angles to the direction of the prevailing wind. Its lee side is the steeper side. Cf: *sastrugi.*
wind-rift dune A sand dune produced in a shrub-covered area by a strong wind of constant direction, typically marked by a gap or "rift" at the very tip or downwind end of a hairpin-shaped sand rim (the hairpin or elongated chevron is opened toward the wind), and extending up to 2 km in length and about 100 m in width (Melton, 1940, p.129-130); it is usually found along a seashore, rarely on a desert. The term is also applied to the "doublet" of parallel sand ridges resulting from the formation of the "rift". The spelling "windrift" is not recommended owing to possible confusion with "wind-drift".
wind ripple [sed] One of many wavelike, asymmetric undulations produced on a sand surface by the saltatory movement of particles by wind and occasionally found in eolian rocks; it is generally longer and of smaller height than an *aqueous ripple,* but is similar in having a steep *lee* side (facing downcurrent) and a gentle windward side (facing upcurrent). See also: *sand ripple; granule ripple.* Cf: *antiripple.*
wind ripple [snow] One of a series of centimeter-scale, wavelike formations on a snow surface, lying at right angles to the wind direction, and formed as snow grains are moved along by the wind.
windrow (wind'-row) (a) A low bank, heap, or other accumulation of material, formed naturally by the wind (as a snowdrift) or the tide (as a pile of beach shells), or artificially (as a ridge of construction material along a road). (b) Part of a *slick* that has broken up into a narrower and shorter band or bands at wind speeds greater than about 7 knots, its long axis always oriented along the wind direction.
wind scoop A saucerlike depression in the snow near an obstruction (such as a tree or rock), caused by the eddying action of the deflected wind.
wind-scoured basin *deflation basin.*
wind-scoured stone *ventifact.*
wind set-up The vertical rise of sea level at the shore due to the force of the wind along the shore or just offshore. See also: *storm surge.*
wind shadow The area in the lee of an obstacle, where air motion is not capable of moving material (such as sand in saltation) and thus traps it when it falls; the zone that is gradually filled with sand drift during the formation of a dune, and determines the shape of the dune. Syn: *shadow zone [geomorph].*
wind-shadow dune (wind'-shad-ow) A longitudinal *umbracer dune.*
wind-shaped stone *ventifact.*
wind slab A layer of snow that is packed tightly by the wind while being deposited. Cf: *wind crust.*
wind-slab avalanche An avalanche started by the dislodging or slipping of a rigid *wind slab* from the underlying snow. Cf: *slab avalanche; loose-snow avalanche.*
windsorite (wind'-sor-ite) A light-colored quartz-monzonite *aplite* with a minor amount of biotite. It was named by Daly in 1903 after Windsor, Vermont. Obsolete.
wind streak Light or dark patches on planetary surfaces emanating from topographic obstacles and pointing in the direction that was downwind at the time of their formation. They can be either erosional or depositional features (Kieffer et al., 1992).
wind stress The force per unit area due to the wind's acting on the water surface. It produces waves and currents. It is proportional to the square of wind speed, air density, and surface roughness.
wind sweep The trough-shaped part of the windward slope of an advancing dune, up which the main wind currents pass.
wind valley *wind gap.*
wind vane An instrument used to measure *wind direction.*
windward (wind'-ward) adj. (a) Said of the side (as of a shore or reef) located toward the direction from which the wind is blowing; facing the wind, such as the "windward slope" of a dune, up which sand moves by saltation. Syn: *weather [meteorol].* (b) Said of a tide moving toward the direction from which the wind is blowing. n. The part or side (as of a hill or shore) from which the wind is blowing; the side facing the wind. Also, the direction from which the wind is blowing, or the direction opposite to that toward which the wind is blowing. Ant: *leeward.*
windward reef In a *reef complex* or *reef tract,* a reef on the side from which the prevailing winds blow. See also: *outer reef; inner reef.*
wind wave A *gravity wave* on the sea generated by the wind.
wind-worn stone *ventifact.*
winebergite (wine'-berg-ite) An ill-defined white mineral: $Al_4(OH)_{10}(SO_4)\cdot7H_2O$ (?).
wineglass valley (wine'-glass) A valley resembling in plan view a goblet or champagne glass. It flares broadly open at its upper end, where it has a cup-shaped or funnel-shaped head; narrows sharply to form a constricted lower section; and flares open again on a spreading alluvial fan. The valley commonly forms at right angles to a fault scarp in an arid region. Syn: *goblet valley; hourglass valley.*
wing [geomorph] The forward extending, outer end of a *barchan dune;* a *horn.*
wing [paleont] A solid or fenestrated extension from the side wall of the shell of a nasselline radiolarian.
wing [palyn] *saccus (a).*
wing bar A sandbar that partly crosses the entrance to a bay or the mouth of a river.
wing crack Forms as a consequence of out-of-plane propagation from a crack subject to mode II crack tip displacement.
wing dam *pier dam.*
winged headland A headland having spits extending from both sides in opposite directions. It may be produced by waves that are unable to move material to the bayhead.
wing wall A wall extending upstream or downstream from a culvert to aid in diverting water into the culvert or in protecting the adjacent embankment from erosion.
Winkler method (Wink'-ler) In oceanography, a chemical method that uses manganous and thiosulfate ions to determine the amount of dissolved oxygen in seawater.

wink technique Rapid alternation of two displays which are nearly alike, so that the places of difference appear to jitter whereas places which are alike remain stationary.

winnowed community (win'-nowed) An *assemblage [ecol]* in which the organisms lived in the same place where their fossils are found, but from which many small-sized forms (both immatures of large species and adults of small species) have been removed by gentle water movements. Cf: *fossil community; disturbed-neighborhood assemblage; transported assemblage; mixed assemblage.* See also: *biocoenosis.* Syn: *residual community.*

winnowing (win'-now-ing) The selective sorting, or removal, of fine particles by wind action, leaving the coarser grains behind. The term is often applied to removal by or sorting in water, but the term *washing* is more appropriate for such a process.

winonaite A group of chondritic meteorites consisting of silicate and metal phases in approximately chondritic proportions. Mineral and O-isotopic compositions of winonaites are similar to those IAB iron meteorites and some argue that they should be grouped together with the IAB irons. See: *primitive achondrite.*

winstanleyite (win-stan'-ley-ite) A yellow, tan, or cream cubic mineral: $TiTe^{4+}{}_3O_8$.

winter balance The change in mass of a glacier from the minimum value at the beginning of a *balance year* to the following maximum value at the beginning of the next melt season; sometimes called *apparent accumulation,* because it does not include snow fall during the summer or losses by melt during warm periods in winter; or (erroneously) *net accumulation.* Cf: *summer balance.*

winter beach Depleted upper beach resulting from erosion associated with high wave energy. Storms are particularly effective in removing the upper beach, lowering its level and flattening its profile. Storms may occur at any season, as may other periods of high sea state, but in many regions are most characteristic of winters. Tendencies toward winter beach development occur during short periods of high wave energy and also during rising tides.

winter berm A berm built on the backshore by the uprush of large storm waves during the winter; it is landward of, and somewhat higher than, the *summer berm.* See also: *storm berm.*

winterbourne (win'-ter-bourne) A regular *bourne* that breaks out every year at the same spot in the floor of a dry valley; specif. one that flows only or chiefly in winter, when the water table rises above the valley floor, as in the chalk regions of southern England.

winter moraine A minor end moraine formed during glacier readvance in a single winter, either on land or under water (Gravenor and Kupsch, 1959, p.54).

winter protalus ridge A wall-like *protalus rampart* formed of blocks and boulders derived from cliffs above a snowbank-occupied cirque before the summer heat melts the snow across which the blocks roll.

winter season *accumulation season.*

winze A steeply inclined opening, like a shaft, driven to connect one mine level with a lower level.

wire adj. A syn. of *capillary [mineral];* said of native metals, e.g. wire silver.

wireline A general term for any flexible steel line or cable connecting a surface winch to a tool assembly lowered in a wellbore. Cf: *sand line [drill].*

wireline coring Cutting and removing of a *core sample* (of soft sandstone or shale) with the drill bit still in place and without withdrawing and dismantling the drill pipes by raising the core in a retractable core barrel and lowering the same or an alternate barrel into place inside the drill pipe. Obsolescent, but still used in drilling small-diameter boreholes.

wire-line log *well log.*

wireline test A procedure for measuring the potential productivity of an oil reservoir by means of a tool lowered into a borehole by a *wireline,* in which a sample of fluid and the formation pressure are obtained. It is faster than a *drill-stem test.*

wireline well log A record made by remotely sensing rocks and fluid properties and the shape of a *wellbore* by lowering an instrument on a wireline into it.

wire strain gage An instrument consisting of a fine wire used to indicate minute changes in strain by detecting corresponding changes in electrical resistance via elongation of the wire.

Wisconsin (Wis-con'-sin) *Wisconsinan.*

Wisconsinan (Wis-con'-sin-an) Pertaining to the classical fourth glacial stage (and the last definitely ascertained, although there appear to be others) of the Pleistocene Epoch in North America, following the Sangamonian interglacial stage and preceding the Holocene. Cf: *Würm.* Syn: *Wisconsin.*

wiserite (wi'-ser-ite) A white, pinkish-brown, or reddish mineral: $(Mn^{2+},Mg)_{14}B_8(Si,Mg)O_{20}(OH)_8$.

witch A less preferred syn. of *water witch.*

witching stick (witch'-ing) *divining rod.*

withamite (with'-am-ite) A red to yellow variety of epidote containing a little manganese and occurring in andesites in Glencoe, Scotland. Cf: *piemontite.*

withdrawal (with-draw'-al) The act of removing water from a source for use; also, the amount removed.

witherite (with'-er-ite) A variously colored orthorhombic mineral of the *aragonite* group: $BaCO_3$.

witness butte (wit'-ness) *butte témoin.*

witness corner A monumented survey point near a *corner* and usually on a line of the survey, established as a reference mark where the true corner is inaccessible or cannot be monumented or occupied; e.g. a post set near the corner of a mining claim, with the distance and direction of the true corner indicated thereon.

witness mark A physical structure (such as a post, rock, stake, or tree) placed at a known distance and direction from a property corner, instrument, or other survey station, to aid in its recovery and identification; e.g. a blazed tree on the bank of a river, indicating the corner which is at the intersection of some survey line with the center line of the river and therefore cannot be marked directly.

witness point A monumented station on a line of survey, used to perpetuate an important location more or less remote from, and without special relation to, any regular corner.

witness rock *zeuge.*

witness tree *bearing tree.*

wittichenite (wit'-ti-chen-ite') A metallic steel-gray to tin-white orthorhombic mineral: Cu_3BiS_3.

wittite (witt'-ite) A metallic lead-gray monoclinic mineral: $Pb_3Bi_4(S,Se)_9$.

wobbling of the pole (wob'-bling) An expression sometimes used to describe the period polar motion identified by a Chandler term (approximately a 14-month period) and an annual term. A beat frequency between these two terms (avg. 6.2 yr) generates a cycle in climate and sedimentology. A longer term wobble, of 18.6 yr, is the *nutation* due to the nodal cycle of the Moon; it has an important effect on sea level, climate and sedimentation (Currie, 1995). See also: *Chandler wobble.*

wodanite (wo'-da-nite) A variety of biotite containing titanium.

wodginite (wodg'-i-nite) (a) A reddish-brown or dark brown monoclinic mineral: $Mn^{2+}(Sn^{4+},Ta)Ta_2O_8$. (b) A group name for minerals with the analogous composition, but with Mn^{2+} replaced by Fe^{2+} or Li, Sn replaced by Ti, and Ta replaced by Nb.

woebourne (woe'-bourne) A *bourne* that is regarded in some English localities as appearing only when some disaster is about to happen.

woehlerite (woeh'-ler-ite) *wöhlerite.*

wöhlerite (wöh'-ler-ite) (a) A vitreous white, yellow, grayish, or brown monoclinic mineral: $Na_2Ca_4ZrNb(Si_2O_7)_2(O,F)_4$. Cf: *låvenite.* (b) A name for organic matter in carbonaceous chondrites. Also spelled: *woehlerite.*

wolchonskoite (wol-chon'-sko-ite) *volkonskoite.*

wold A range of hills produced by differential erosion from inclined sedimentary rocks; a *cuesta.*

Wolfcampian (Wolf-camp'-i-an) North American series: lowermost Permian (above Virgilian of Pennsylvanian, below Leonardian).

wolfeite (wolfe'-ite) A vitreous orange, reddish-brown, dark brown, or green monoclinic mineral of the *triploidite* group: $(Fe^{2+},Mn^{2+})_2(PO_4)(OH)$.

wolfram (wolf'-ram) (a) *wolframite.* (b) The metallic element *tungsten.*

wolframine (wolf'-ram-ine) (a) *tungstite.* (b) *wolframite.*

wolframite (wolf'-ram-ite) A group name for monoclinic minerals with the general formula *M*WO4, where $M = Fe^{2+}$, Mn^{2+}, Co, or Zn; W may be replaced by Nb or Ta. Wolframite occurs in monoclinic crystals (commonly twinned so as to imitate orthorhombic tabular forms) and in granular masses or columnar aggregates (as in pneumatolytic veins near granite masses and associated with tin ores). Wolframite is the principal ore of tungsten. Syn: *wolfram; wolframine; tobacco jack.*

wolfram ocher *tungstic ocher.*

wolfsbergite (wolfs'-berg-ite) *chalcostibite.*

wolgidite (wol'-gid-ite) A *lamproite* that contains leucite, magnophorite, diopside, and minor amounts of olivine and phlogopite. The name, given by Wade and Prider in 1940, is from the Wolgidee Hills, Western Australia. Mitchell and Bergman (1991) suggest that this name should be replaced by diopside-leucite-richterite *madupitic lamproite.* Obsolete.

wollastonite (wol'-las-ton-ite') A triclinic or monoclinic chain silicate mineral of the pyroxenoid type: $CaSiO_3$. It dimorphous with parawollastonite. Wollastonite is found in contact-metamorphosed limestones, and occurs usually in cleavable masses or sometimes in tabular twinned crystals; it may be white, gray, brown, red, or yellow. It is not a pyroxene. Several polytypes have been characterized. Symbol: Wo. Syn: *tabular spar.*

Wollaston prism (Wol'-las-ton) In an optical system, a double-image prism consisting of two right-angled calcite prisms that produce two perpendicular beams of plane-polarized light.

wollongite (wol-lon'-gite) *wollongongite.*

wollongongite (wol-lon-gon'-gite) A coal-like shale similar to *torbanite.* It is named from its type locality, Wollongong, New South Wales, Australia. Also spelled: *wollongite; wallongite.*

wölsendorfite (wöl'-sen-dorf-ite') A red or orange-red orthorhombic mineral: $(Pb,Ba)_7(UO_2)_{14}O_{19}(OH)_4{\cdot}12H_2O$.

wonesite (wones'-ite) A greenish-brown monoclinic mineral of the *mica* group with the ideal formula $Na_{0.5}(Mg_{2.5}Al_{0.5})(Si_3Al)_{10}(OH)_2$. It is regarded as an interlayer-deficient trioctahedral mica. Syn: *sodium phlogopite; aspidolite.*

wood In *woody plants,* the tissue produced to the inside of a *vascular cambium,* i.e., the secondary xylem. Also syn. of *woody plant.*

wood agate *silicified wood.*
woodallite A magenta or purple rhombohedral mineral of the *meixnerite* group: $Mg_6Cr_2(OH)_{16}Cl_2{\cdot}4H_2O$.
Woodbinian (Wood-bin'-i-an) North American (Gulf Coast) stage: Upper Cretaceous (above Washitan, below Eaglefordian).
wood coal *woody lignite.*
wood copper A fibrous variety of olivenite.
woodfordite (wood'-ford-ite) *ettringite.*
wood hematite A finely radiated variety of hematite exhibiting alternate bands of brown or yellow or varied tints.
woodhouseite (wood'-house-ite) A colorless or pale orange rhombohedral mineral of the *beudantite* group: $CaAl_3(PO_4)(SO_4)(OH)_6$.
wood iron ore A fibrous variety of limonite from Cornwall, England.
wood opal A variety of common opal that has filled the cavities in, and replaced the organic matter of, wood and that often preserves the original features of the wood. See also: *opalized wood.* Syn: *xylopal; lithoxyl.*
woodruffite (wood'-ruff-ite) A sooty-black tetragonal mineral: $(Zn,Mn^{2+})_2Mn^{4+}{}_5O_{12}{\cdot}4H_2O$. It is related to *todorokite.*
woodstone (wood'-stone) *silicified wood.*
wood tin A nodular or reniform brownish variety of cassiterite, having a concentric structure of radiating fibers resembling dry wood in appearance. Syn: *dneprovskite.*
woodwardite (wood'-ward-ite) A greenish-blue to turquoise-blue rhombohedral mineral: $(Cu,Al)_9(SO_4)_2(OH)_{18}{\cdot}nH_2O$. It may be related to the *pyroaurite* group.
woodyard (wood'-yard) *forest bed.*
woody lignite (wood'-y) Lignite that shows the fibrous structures of wood. Cf: *earthy lignite.* Syn: *xyloid lignite; xyloid coal; wood coal; board coal; bituminous wood.*
woody peat *Fibrous peat* that contains abundant, partially degraded, recognizable wood fragments.
woody plant A usually perennial plant with significant development of secondary xylem.
wool An English term for a sandy shale or shaly flagstone with irregular curly bands or bedding.
wooldridgeite A vitreous blue-green orthorhombic mineral: $Na_2CaCu^{2+}{}_2(P_2O_7)_2{\cdot}10H_2O$.
Worden gravimeter (Wor'-den) A compact temperature-compensated gravity meter, in which a system is held in unstable equilibrium about an axis, so that an increase in the gravitational pull on a mass at the end of a weight arm causes a rotation opposed by a sensitive spring. The meter weighs 5 pounds and has a sensitivity of less than 0.1 mgal.
Wordian (Word'-ian) A stage of the Standard Global Chronostratigraphic Scale: middle stage of the Guadalupian Series, middle Permian (above Roadian, below Capitanian).
workings (work'-ings) The system of openings or excavations made in mining or quarrying; esp. the area where the ore is actually mined.
World Data Centers Centers under the auspices of the International Council of Scientific Unions (ICSU) for the collection, exchange, and general availability of data from various geophysical disciplines, e.g. solid-earth geophysics, solar-terrestrial geophysics, oceanography, glaciology, meteorology, tsunamis.
world geodetic system Any system that connects the major continental geodetic datums and land masses into a unified Earth-centered network. Abbrev: WGS.
world point A term proposed by Kobayashi (1944a, p.745) for a single restricted outcrop regarded as representative of a geologic province or part of the world. Rarely used.
world rift system A major tectonic element of the Earth, consisting of midoceanic ridges and their associated rift valleys, such as those along the Mid-Atlantic Ridge. It is believed to be the locus of extensional splitting and upwelling of magma that has resulted in *sea-floor spreading.* Cf: *rift [tect].*
world time *Geologic time* as indicated by the life range of a single cosmopolitan fossil species (Kobayashi, 1944a, p.745).
worm boring *worm tube.*
worm cast (a) A sinuous fossil trail of a worm, preserved as a sand cast on the bedding plane of an arenaceous rock. (b) *worm casting.* Also spelled: wormcast.
worm casting A cylindrical mass of earth or mud excreted by an earthworm. Syn: *erpoglyph; worm cast.*
worm reef A small *organic reef* built by wormlike organisms. Not recommended usage. Cf: *sabellariid reef; serpulid reef; vermetid reef.*
worm rock Hardened deposits of tubular structures originating at about sea level as a result of activities of various segmented worms. Typically calcareous secretions form or bind together the tubes by cementation of inorganic materials, such as coarse silt or fine sand. Most worm rock is gray, but some is amber or other colors. When first formed the masses may be crushed when stepped upon, but later acquire bearing strength, and in some cases become hard and durable.
worm's-eye map (a) A term applied to what is more formally known as a *supercrop map* or a *lap-out map,* in reference to the pattern of formations that would be visible to an observer looking upward at the bottom of the rocks overlying a given surface. (b) A map showing overlap of sediments, or of progressive transgressions of a sea over a given surface.
worm trail A marking in a fossiliferous rock, formed by the passage of an ancient worm or worms.
worm tube (a) A fossilized tubular structure built by a marine worm and preserved in the top of a bed that was exposed for some time as the sea floor; e.g. a *scolithus.* Syn: *worm boring.* (b) A membranous tube, usually of calcium carbonate or particles of mud or sand, built on a submerged surface by a marine worm.
worobieffite (wo-ro'-bieff-ite) *vorobievite.*
wrack A linear deposit including sand, gravel, seaweed, and other detritus that marks a former *shoreline* position.
wrackline The location of *wrack* deposits on a beach.
wrench fault A regional-scale strike-slip fault. Typically, use of the term carries with it the implication that the strike-slip movement resulted in formation of a complex band of subsidiary faults and *en echelon* folds. Obsolete and little-used syn: *basculating fault; torsion fault.*
Wright biquartz wedge *biquartz plate.*
wrinkle One of a set of complementary, but minor, ridges and furrows, with axes parallel to the longitudinal axis of a conodont element; commonly form a zone along the basal margin in certain panderodont elements, e.g. those of *Panderodus* (TIP, 1981, pt. W, supp.2).
wrinkle layer (wrin'-kle) Pattern of irregularly transverse, very fine ridges or wrinkles on the inside of the shell, especially the body chamber, of many cephalopod conchs.
wrinkle mark An irregular small ripplelike feature, composed of ridges 0.5 to 1 mm thick and a few millimeters long, parallel to each other or in a curved or honeycomb pattern. Described (as "Runzelmarken") by Häntzschel and Reineck in 1968 and named wrinkle mark by Teichert (1970).
wrinkle ridge A sinuous, irregular, segmented, crenelated ridge on planetary surfaces. Wrinkle ridges are tens of kilometers wide and tens to more than 100 m high, and may extend for hundreds of kilometers. They probably form by tectonic folding and thrust faulting of surface materials. Syn: *mare ridge.*
wrist *carpus.*
wroewolfeite (wroe-wolfe'-ite) A deep greenish-blue monoclinic mineral: $Cu_4(SO_4)(OH)_6{\cdot}2H_2O$.
wrong reading image An image that is the mirror image of the original.
W-shaped valley A valley having an inverted and faintly pan-shaped cross profile suggesting the form of the letter "W", such as the valley of a river having the highest parts of its flood plain immediately near both banks (Lane, 1923).
Wuchiapingian (Wuch-i-a-pin'-gi-an) A stage of the Standard Global Chronostratigraphic Scale: lower stage of Lopingian Series of Upper Permian (above Capitanian, below Changhsingian). The basal GSSP is in China.
wulfenite (wul'-fen-ite) A yellow, orange, or red (sometimes grayish or green) tetragonal mineral of the *scheelite* group: $PbMoO_4$. Wulfenite occurs in tabular crystals and in granular masses, and is an ore mineral of molybdenum. Syn: *yellow lead ore.*
Wulff net (a) A coordinate system used in crystallography to plot a *polar stereographic projection* with conservation of equal angles, such as for plotting angular relations obtained from universal-stage measurements. (b) *stereonet.* Named after Georgi Viktorovich Vulf (1862-1925), Russian crystallographer who introduced the net and whose name was transliterated into German.
wülfingite (wülf'-ing-ite) A colorless orthorhombic mineral: $Zn(OH)_2$.
wupatkiite A fibrous silky rose-colored monoclinic mineral of the *halotrichite* group: $(Co,Mg,Ni)Al_2(SO_4)_4{\cdot}22H_2O$.
Würm (a) A European stage: uppermost Pleistocene (above Riss, below Holocene). (b) The classical fourth glacial stage of the Pleistocene Epoch in the Alps, after the Riss-Würm interglacial stage. Cf: *Wisconsinan; Weichsel.* Etymol: From the river Würm which drains Lake Starnberg near Munich, Germany. Adj: Würmian.
wurtzilite (wurtz'-i-lite) A black, massive asphaltic pyrobitumen, sectile and infusible, that is closely related to *uintahite,* but insoluble in turpentine, and derived from the metamorphism of petroleum. It is found in veins in Uinta County in Utah.
wurtzite (wurtz'-ite) A reddish-brown to dark brown hexagonal mineral: $(Zn,Fe)S$. It is dimorphous with sphalerite. Wurtzite occurs in hemimorphic pyramidal crystals, or in radiating needles and bundles within lamellar sphalerite. Many hexagonal and rhombohedral polymorphs with slight variants on the wurtzite structure are known, and separate names proposed for some of these are "superfluous and not generally accepted" (Hey, 1962, 3.4.3).
wüstite (wüst'-ite) A metallic gray cubic mineral of the *periclase* group: $Fe^{2+}O$. Artificially prepared specimens are characteristically deficient in iron. Also spelled: wustite. Syn: *iozite.*
WWV Radio station of the U.S. Bureau of Standards, which broadcasts time and frequency standards. WWV (Ft. Collins, Colorado) and WWVH (Maui, Hawaii) broadcast continuously on 2.5, 5, 10, 15, 20, and 25 MHz (the last 2 only by WWV). WWV is off the air for 4 min commencing at 45 min 15 s after each hour and WWVH for 4 minutes commencing at 15 min 15 s after each hour. Each second is marked by a signal or tick. A voice announce-

ment is given every minute, e.g., "National Bureau of Standards, WWV, Fort Collins, Colorado. At the tone, 17 hr, 16 min, Coordinated Universal time."

wyartite (wy'-art-ite) A black or violet-black orthorhombic secondary mineral: $CaU^{5+}(UO_2)_2O_4(CO_3)(OH)·7H_2O$. It was erroneously called *ianthinite.*

wycheproofite A pale pinkish-orange or pale brownish-orange triclinic mineral: $NaAlZr(PO_4)_2(OH)_2·H_2O$.

wye level A leveling instrument having a removable telescope, with attached spirit level, supported in Y-shaped rests, in which it may be rotated about its longitudinal (or collimation) axis, and from which it may be lifted and reversed, end for end, for testing and adjustment. Cf: *dumpy level.* Syn: *Y level.*

wyllieite (wyl'-lie-ite) (a) A vitreous bluish-green to greenish-black monoclinic mineral with the ideal end-mamber formula: $Na_2Mn^{2+}{}_2Al(PO_4)_3$. Cf: *ferrowyllieite*. (b) A group name for minerals with the analogous composition, but with Mn^{2+} replaced by Fe^{2+} or Na, and Al replaced by Fe^{3+}.

Wyllie time-average equation (Wyl'-lie) The empirical formula proposed by Wyllie (1957) that linearly relates interval transit time (reciprocal velocity) from a *sonic log* (Δt) to porosity (ϕ) of granular rocks in terms of the travel time of the solid rock (Δt_m) and of the interstitial liquid (Δt_f): $\Delta t = \phi\ \Delta t_f + \Delta t_m (1-\phi)$.

Wyoming bentonite A *sodium bentonite.*

wyomingite (wy-o'-ming-ite) A hypabyssal *lamproite* containing phlogopite phenocrysts in a fine-grained groundmass of leucite and diopside. Its name, given by Cross in 1897, is derived from the state of Wyoming, U.S.A. Mitchell and Bergman (1991) suggest that this name should be replaced by diopside-leucite-phlogopite *lamproite*. Obsolete.

X

x In seismic prospecting, the distance from the source to a geophone.

X^2 - T^2 analysis A method of determining seismic velocity from *normal moveout* measurements.

x-acline B twin law A complex twin law in feldspars with twin axis parallel to (010) and composition plane (100).

xalostocite (xa-los'-to-cite) *landerite.*

XANES X-ray Absorption Near Edge Structure spectroscopy; a method using the detailed structure of an X-ray K, L, or M absorption edge for determination of the oxidations state, coordination, and other "local-structure" details of the environment of a cation in a mineral or glass.

xanthiosite (xan'-thi-o-site') A golden-yellow monoclinic mineral: $Ni_3(AsO_4)_2$.

xanthite (xan'-thite) A yellowish to yellowish-brown variety of vesuvianite.

xanthochroite (xan-tho-chro'-ite) *greenockite.*

xanthoconite (xanth-o-con'-ite) A brilliant red, orange-yellow, or brown monoclinic mineral: Ag_3AsS_3. Cf: *proustite.*

xanthophyllite (xan-tho-phyl'-lite) *clintonite.*

xanthosiderite (xan-tho-sid'-er-ite) *goethite.*

xanthoxenite (xanth-ox-en'-ite) (a) A pale yellow to brownish-yellow triclinic mineral of the *whiteite* group, original samples of which are probably in part stewartite. (b) Redefined as having composition $Ca_4Fe^{+3}{}_2(PO_4)_4(OH)_2{\cdot}3H_2O$.

XBT *expendable bathythermograph.*

x-Carlsbad twin law A complex twin law in feldspar, having a twin axis at right angles to [001] and a composition plane (100). It is supposedly a true interpretation of the *acline-B twin law.*

XCB *Extended Core Barrel.*

XCL *crisscross crossed lamellar structure.*

xenoblast (xen'-o-blast) A mineral of low *form energy* that has grown during metamorphism without development of its characteristic crystal faces. It is a type of *crystalloblast.* The term was originated by Becke (1903). Cf: *idioblast; hypidioblast.* Syn: *allotrioblast.*

xenocryst (xen'-o-cryst) A crystal that resembles a phenocryst in igneous rock but is foreign to the body of rock in which it occurs. Cf: *disomatic.*

xenogenous (xe-nog'-e-nous) A little-used syn. of *epigenetic.*

xenolith (xen'-o-lith) A fragment of country rock within a plutonic or volcanic rock. Cf: *autolith.* Syn: *inclusion; accidental inclusion.*

xenology (xe-nol'-o-gy) The study of the isotopic composition of xenon in terrestrial and extraterrestrial materials.

xenomorphic (xen-o-mor'-phic) A grain totally lacking crystal faces; poorly formed. Syn: *anhedral; allotriomorphic; anidiomorphic.* Cf: *automorphic; euhedral; hypautomorphic; hypidiomorphic.*

xenomorphic-granular (xen-o-mor'-phic-gran'-u-lar) A phaneritic igneous texture dominated by anhedral, equant grains. Syn: *allotriomorphic-granular.* Cf: *hypidiomorphic-granular; idiomorphic-granular.*

xenomorphism (xen-o-mor'-phism) The state or condition of special sculpture at the umbonal region of the unattached valve of a bivalve mollusk, resembling the configuration of the substratum onto which the attached valve is or was originally fixed. It is known on the left valves in the Anomiidae and on the right valves in the Gryphaeidae and Ostreidae. Erroneous syn: *allomorphism.*

xenon-xenon age method (xe'-non-xe'-non) A method of dating uranium-bearing minerals by means of mass spectrometric determination of xenon isotopic ratios. It is based on the compositional differences between xenon produced by spontaneous fission of uranium-238 in nature and xenon produced from uranium-235 fission induced by thermal neutrons in a nuclear reactor. Xenon is extracted incrementally, and an age spectrum produced. The method appears to be useful for rocks older than about 100 million years.

xenothermal (xen-o-ther'-mal) Said of a hydrothermal mineral deposit formed at high temperature but shallow depth; also, said of that environment. Cf: *telethermal; epithermal; mesothermal [eco geol]; hypothermal; leptothermal.*

xenotime (xen'-o-time) A group name for tetragonal minerals of the zircon structure type and with the general formula AXO_4, where A = Sc, Y, REE, or Bi and X = P, As, or V. Xenotime occurs as an accessory mineral in granites and pegmatites.

xenotime-(Y) A brown, yellow, or reddish tetragonal mineral of the *xenotime* group: YPO_4.

xenotime-(Yb) A colorless, yellowish, or brownish tetragonal mineral of the *xenotime* group: $YbPO_4$.

xenotopic (xen-o-top'-ic) Said of the fabric of a crystalline sedimentary rock in which the majority of the constituent crystals are anhedral. Also, said of an evaporite, a chemically deposited cement, or a recrystallized limestone or dolomite with such a fabric. The term was proposed by Friedman (1965, p.648). Cf: *idiotopic; hypidiotopic.*

Xeralfs Alfisols that have a xeric soil moisture regime. Xeralfs are brownish or reddish throughout. A suborder in the U.S. system of soil taxonomy.

xerarch (xe'-rarch) adj. Said of an ecologic succession (i.e., a *sere*) that develops under *xeric* conditions. Cf: *mesarch; hydrarch; xerosere.*

Xererts Vertisols that have a thermic, mesic, or frigid soil temperature regime and if not irrigated, cracks that remain both 5 cm or more wide through a thickness of 25 cm or more within 50 cm of the mineral soil surface for 60 or more consecutive days during 90 days following the summer solstice and closed 60 or more consecutive days during the 90 days following the winter solstice. A suborder in the U.S. system of soil taxonomy.

xeric (xe'-ric) Said of a habitat characterized by a low or inadequate supply of moisture; also, said of an organism or group of organisms existing in such a habitat. Cf: *mesic; hydric; xerarch.*

xeric moisture regime A soil moisture regime that is characteristic of the cool, moist winter and warm, dry summer of a Mediterranean climate. The soil is dry for at least 45 consecutive days in the summer and is moist for at least 45 consecutive days in the winter. Winter rainfall often penetrates deeply into the soil and is particularly effective for leaching (USDA, 1975).

xerochore (xe'-ro-chore) A climatic term for the part of the Earth's surface represented by waterless deserts.

xerocole (xe'-ro-cole) *xerophilous.*

Xerolls Mollisols that have a xeric soil moisture regime. Xerolls may have a calcic, petrocalcic, or gypsic horizon, or a duripan. A suborder in the U.S. system of soil taxonomy.

xeromorphic (xe-ro-mor'-phic) (a) Said of a plant characterized by the morphology of a *xerophyte.* (b) Said of conditions favorable for the growth of *xerophilous* organisms.

xerophile (xe'-ro-phile) n. *xerophyte.* adj. *xerophilous.*

xerophilous (xe-roph'-i-lous) Said of an organism adapted to dry conditions. Syn: *xerophile; xerocole.* Cf: *xerophobous.*

xerophobous (xe-roph'-o-bous) Said of a plant that cannot tolerate dry conditions. Cf: *xerophilous.*

xerophyte (xe'-ro-phyte) A plant adapted to dry conditions; a desert plant. Cf: *xerotherm; hydrophyte; mesophyte.* Syn: *eremophyte; xerophile.*

xerophytization (xe-roph'-y-ti-za'-tion) Adaptation, esp. in the development of a species, to conditions of low moisture supply, i.e. to xeric conditions.

xerosere (xe'-ro-sere) A *sere* that develops under extremely dry (i.e. xeric) conditions; a *xerarch* sere. Cf: *hydrosere; mesosere.*

xerotherm (xe'-ro-therm) A plant adapted to hot dry conditions. Cf: *xerophyte.*

Xerothermic (Xe-ro-ther'-mic) n. A term used to designate a postglacial interval of warmer and drier climate. Cf: *Altithermal; Hypsithermal.*

xerothermic [climat] adj. Said of a hot, dry climate; also, pertaining to the climate of the Xerothermic postglacial interval.

Xerults Ultisols that have low or moderate amounts of organic carbon, are brownish or reddish throughout, and have a xeric soil moisture regime. A suborder in the U.S. system of soil taxonomy.

xiangjiangite (xiang'-jiang-ite) An earthy yellow tetragonal mineral: $(Fe^{3+},Al)(UO_2)_4(PO_4)_2(SO_4)_2(OH){\cdot}22H_2O$.

xifengite (xi-feng'-ite) A metallic steel-gray hexagonal mineral, found in cosmic dust: Fe_5Si_3.

xilingolite (xi-ling′-o-lite) A metallic gray monoclinic mineral: $Pb_3Bi_2S_6$. It is a member of the lillianite homologous series.

ximengite A colorless trigonal mineral: $BiPO_4$.

xingsaoite A dark blue-violet rhombohedral mineral of the *phenakite* group: $(Zn,Co)_2SiO_4$. It is probably a cobalt-bearing willemite, and has not been approved by the International Mineralogical Association.

xingzhongite (xing-zhong′-ite) A metallic gray cubic mineral of the *linnaeite* group: $(Pb,Cu,Fe)(Ir,Pt,Rh)_2S_4$. It has not been approved by the International Mineralogical Association.

xiphosuran (xiph-o-su′-ran) Any *merostome* belonging to the subclass Xiphosura, characterized by a trilobate dorsal shield. Horseshoe crabs are included in this group. Cf: *eurypterid.* Range, Cambrian to Holocene.

xitieshanite (xit′-ie-shan-ite) A yellowish-green monoclinic mineral: $Fe^{3+}(SO_4)Cl{\cdot}6H_2O$.

xocomecatlite (xo′-co-me-ca′-tlite) A green orthorhombic mineral: $Cu_3(Te^{6+}O_4)(OH)_4$.

xonotlite (xon-ot′-lite) A pale pink, white, or gray prismatic monoclinic mineral: $Ca_6Si_6O_{17}(OH)_2$.

x-pericline twin law (x-per′-i-cline) A complex twin law in feldspar, now considered to be equivalent to the *Carlsbad B twin law.*

X-ray Non-nuclear electromagnetic radiation of very short wavelength, in the interval of 0.1-100 angstroms (10^{-11}-10^{-8}m), i.e. between that of gamma rays and ultraviolet radiation. Also spelled: x-ray.

X-ray diffraction The diffraction of a beam of X-rays, usually by the three-dimensional periodic array of atoms in a crystal that has periodic repeat distances (lattice dimensions) of the same order of magnitude as the wavelength of the X-rays. Cf: *electron diffraction.*

X-ray diffraction analysis A method of crystal analysis that relies on *X-ray diffraction* to obtain an *X-ray diffraction pattern*. Abbrev: XRD.

X-ray diffraction pattern The characteristic interference pattern obtained when X-rays are diffracted by a crystalline substance. The geometry of the pattern is a function of the repeat distances (lattice dimensions) of the periodic array of atoms in the crystal; the intensities of the diffracted beams give information about the atomic arrangement and unit-cell dimensions. Cf: *electron diffraction pattern.*

X-ray dispersion *anomalous dispersion.*

X-ray emission spectroscopy The qualitative study of a substance by exciting its characteristic X-ray spectrum and measuring the wavelengths present. Cf: *X-ray fluorescence spectroscopy.*

X-ray fluorescence spectroscopy A type of *X-ray emission spectroscopy* in which the characteristic X-ray spectrum of a substance is produced by using X-rays of short wavelength to induce the substance to fluoresce and emit secondary X-rays of a longer wavelength. A series of diffraction gratings collects the photons generated by the fluorescence of each element in the sample, and the concentration of each element is determined by comparison and standards. Abbrev: XRF.

X-ray photo-electron spectrometry The study of the photo-electron spectra produced by characteristic X-ray excitation. The energy carried by the emitted electrons is the difference between the energy of the exciting photon and the binding energy of the bound electron.

X-ray powder diffraction Diffraction of a beam of X-rays by planes of atoms in a powdered crystalline sample. The powders are prepared so they represent all possible crystal orientations to the X-ray beam and only the beam must move. For more precise analyses both the sample and the X-ray beam move.

X-ray powder pattern The pattern obtained in *X-ray powder diffraction.* Cf: *X-ray diffraction pattern.*

X-ray scattering The phenomenon of changes in direction of X-ray transmission by interaction of the waves with objects or with the transmitting medium due to reflection, refraction, or diffraction.

X-ray spectrograph An instrument for producing, recording, and analyzing an *X-ray spectrum* by reflecting X-rays from a given sample, measuring the angle of diffraction, and thence determining the wavelengths of the X-rays. Sometimes called an *X-ray spectrometer.*

X-ray spectrometer *X-ray spectrograph.*

X-ray spectroscopy The observation of an *X-ray spectrum* and all processes of recording and measuring that go with it.

X-ray spectrum The spectrum of X-rays emitted when a substance is bombarded with energetic particles or radiation. It consists of a "characteristic spectrum", resulting from specific electronic transitions within the atoms of the substance, superimposed on a "continuous spectrum" resulting from inelastic collisions of particles or incoherent scattering of the exciting radiation. The X-ray spectrum may be excited by electrons (cathode rays) as in standard X-ray tubes and electron microprobes, by other particles such as protons as in particle-induced X-ray emission, or by higher energy X-rays or gamma rays as in X-ray fluorescence.

XRD *X-ray diffraction analysis.*

XRF *X-ray fluorescence spectroscopy.*

x twin law A normal twin law in feldspar.

xylain (xy′-lain) A kind of *provitrain* in which the cellular structure is derived from woody material. Cf: *periblain; suberain.*

xylem (xy′-lem) In *vascular plants,* a complex tissue that is involved in water conduction, food storage, and support. Types of cells commonly found in xylem include *tracheids* (for conduction and support), vessels (for conduction, characteristic of most angiosperms, less commonly found in some other groups of vascular plants), *fibers* (for support), and *parenchyma* (for storage). Syn: *wood.*

xylinite (xy′-lin-ite) A variety of *provitrinite* characteristic of xylain and consisting of xylem or lignified tissue. Cf: *suberinite; periblinite; telinite.*

xylinoid (xy′-lin-oid) Vitrinite that occurs in noncaking subbituminous coals and lignite and that has a reflectance of less than 0.5% (ASTM, 1970, p.466). Cf: *vitrinoid; anthrinoid.*

xylith (xy′-lith) A type of lignite that is composed almost entirely of anthraxylon (Parks, 1951, p.30).

xyloid coal (xy′-loid) *woody lignite.*

xyloid lignite *woody lignite.*

xylopal (xy-lo′-pal) *wood opal.*

xylotile (xy′-lo-tile) A delicately fibrous mineral, approximately: $(Mg,Fe^{+2})_3Fe_2^{+3}Si_7O_{20}{\cdot}10H_2O$. It is an ill-defined serpentine mineral derived from alteration of asbestos or chrysolite.

xylovitrain (xy-lo-vit′-rain) *euvitrain.*

yafsoanite (yaf-so′-a-nite) A vitreous brown cubic mineral: $Ca_3Zn_3(Te^{6+}O_6)_2$.

yagiite (ya′-gi-ite) A colorless hexagonal meteorite mineral of the *milarite* group: $(Na,K)_{1.5}Mg_2(Al,Mg)_3(Si,Al)_{12}O_{30}$.

yaila A term used in central Kurdistan (of eastern Turkey) for a small, grassy upland plain.

yakatagite (ya-ka-tag′-ite) A name proposed by Miller (1953, p.26) for a "conglomeratic sandy mudstone" from Yakataga, SE Alaska. It is a poorly indurated tillitelike glaciomarine sedimentary rock containing angular gravel-sized fragments.

yakhontovite (yak′-hon-tov-ite) A dull green monoclinic mineral of the *montmorillonite* group: $(Ca,K)_{0.5}(Cu,Fe^{3+},Mg)_2Si_4O_{10}(OH)_2{\cdot}3H_2O$.

yamaskite (ya-mas′-kite) A medium- to fine-grained *pyroxenite* containing hornblende, titanian augite, a small amount of anorthite, and accessory biotite and iron oxides; an amphibole-bearing *jacupirangite.* The name, given by Young in 1906, is from Mount Yamaska, Quebec. Obsolete.

yamatoite (ya-ma′-to-ite) Hypothetical end-member of the *garnet* group: $Mn_3V_2(SiO_4)_3$.

yanomamite A pale green to yellowish-green orthorhombic mineral of the *variscite* group: $InAsO_4{\cdot}2H_2O$.

yardang (yar′-dang) (a) A long, irregular, sharp-crested, undercut ridge between two round-bottomed troughs, carved on a plateau or unsheltered plain in a desert region by wind erosion, and consisting of soft but coherent deposits (such as clayey sand); it lies in the direction of the dominant wind, and may be up to 6 m high and 40 m wide. Syn: *yarding; jardang.* (b) A landscape form produced in a region of limestone or sandstone by infrequent rains combined with wind action, and characterized by "a surface bristling with a fine and compact lacework of sharp ridges pitted by corrosion" (Stone, 1967, p.254). Etymol: A form of Turkic "yar", "steep bank".

yardang trough A long, shallow, round-bottomed groove, furrow, trough, or corridor excavated in the desert floor by wind abrasion, and separating two yardangs.

yardarm carina (yard′-arm) One of the oppositely placed *carinae* of a rugose coral that give cross sections of a septum the appearance of yardarms along a mast. Cf: *zigzag carina.*

yarding (yar′-ding) *yardang.*

Yarkovsky effect A radiation pressure effect on small rotating asteroids or meteoroids, in which absorbed solar energy is thermally re-radiated predominately in the direction in which such an object is turning, which differs from the direction to the sun and leads to a nonradial thrust on the body that can cause orbital migration inward toward or outward from the sun over long periods of time. The effect was first proposed by a Russian engineer named Yarkovsky in a now-lost paper in the early 1900s, but only recently has the effect been conclusively observed in the orbital and spin distributions of small asteroids. Cf: *Poynting-Robertson effect.*

Yarmouth (Yar′-mouth) Pertaining to the classical second interglacial stage of the Pleistocene Epoch in North America, after the Kansan glacial stage and before the Illinoian. Etymol: From Yarmouth, a town in Iowa. Cf: *Mindel-Riss.*

yaroslavite (ya-ro-slav′-ite) A vitreous white orthorhombic mineral: $Ca_3Al_2F_{10}(OH)_2{\cdot}H_2O$.

yarrowite (yar′-row-ite) A metallic bluish-gray trigonal mineral: Cu_9S_8. It and spionkopite were formerly known as "blaubleibend covellite".

yatalite (yat′-a-lite) A pegmatitic rock composed chiefly of amphibole (replacing clinopyroxene), albite, magnetite, titanite, and some quartz. Benson in 1909 proposed the name, derived from Yatala, South Australia. Obsolete.

yavapaiite (ya′-va-pai-ite) A pale pink monoclinic mineral: $KFe^{3+}(SO_4)_2$.

yaw [photo] The variation, or amount of variation, of the horizontal axis of a craft from the direction in which the craft is moving, i.e., the difference between the direction in which a craft is pointed and the direction in which it is moving (Maune, 2001, p.493). Cf: *pitch [photo], roll [photo].*

yazoo (yaz′-oo) (a) *yazoo stream.* (b) *deferred junction.*

yazoo stream A tributary that flows parallel to the main stream for a considerable distance before joining it at a *deferred junction;* esp. such a stream forced to flow along the outer base of a natural levee formed by the main stream. Type example: Yazoo River in western Mississippi, joining the Mississippi River at Vicksburg. Syn: *yazoo; deferred tributary.*

yeatmanite (yeat′-man-ite) A platy brown triclinic mineral closely related to katoptrite: $Mn^{2+}_9Zn_6Sb^{5+}_2Si_4O_{28}$.

yecoraite (ye-co-ra′-ite) A resinous yellow mineral: $Bi_5Fe^{3+}_3O_9(Te^{4+}O_3)(Te^{6+}O_4)_2{\cdot}9H_2O$.

yedlinite (yed′-lin-ite) A red-violet rhombohedral mineral: $Pb_6CrCl_6(O,OH)_8$.

ye'elimite (ye′-el′-im-ite) A colorless cubic mineral: $Ca_4Al_6O_{12}(SO_4)$.

yellow arsenic (yel′-low) *orpiment.*

yellow coal *tasmanite [coal].*

yellow copperas *copiapite.*

yellow copper ore *chalcopyrite.*

yellow earth [mineral] Impure *yellow ocher.*

yellow earth [sed] Loess of northern China.

yellow-green algae A small group of mostly freshwater algae of the class Xanthophyceae. Their color is due to the presence of β-carotene and xanthophylls and the lack of the brown pigment fucoxanthin. Cf: *brown algae; cyanobacteria; green algae; red algae.*

yellow ground Oxidized *kimberlite* of yellowish color found at the surface of diamond pipes (e.g. South Africa), above the zone of *blue ground.*

yellow lead ore *wulfenite.*

yellow ocher (a) A mixture of limonite usually with clay and silica, used as a pigment. See also: *yellow earth [mineral].* Syn: *sil.* (b) A soft, earthy, yellow variety of limonite or of goethite.

yellow ore A yellow-colored ore mineral; specif. carnotite and chalcopyrite.

yellow pyrites *chalcopyrite.*

yellow quartz *citrine.*

yellow snow Snow tinted yellow by algae such as *Raphidonema* and *Scotiella.* Cf: *green snow; red snow.*

yellow substance A portion of the dissolved organic matter in seawater; commonly carbohydrate-humic acids. Syn: *gelbstoff.*

yellow tellurium *sylvanite.*

yellow to brown ore Residually enriched silicate-carbonate iron formation composed mainly of goethite or hydrous iron oxide minerals and martite found in the Knob Lake iron range of Quebec and Labrador, Canada. Cf: *blue ore; red ore; rubble ore; treat rock.*

yenite (yen′-ite) *ilvaite.*

Yeovilian (Yeo-vil′-i-an) Stage in Great Britain: uppermost Lower Jurassic (above Whitbian, below Aalenian).

yftisite (yf′-ti-site) An ill-defined yellow orthorhombic mineral: $(Y,RE)_4(F,OH)_6(Ti,Sn)O(SiO_4)_2$ (?). Not approved by the International Mineralogical Association.

yield [exp struc geol] v. To undergo permanent deformation as a result of applied stress.

yield [lake] n. (a) The amount of water that can be taken continuously from a lake for any economic purpose. (b) The amount of organic matter (plant and animal) produced by a lake, either naturally or under management. See also: *production [lake].*

yield point *yield stress.*

yield scaling *Lampson scaling.*

yield strength A syn. of *yield stress;* the stress at which a material begins to undergo permanent deformation.

yield stress The differential stress at which permanent deformation first occurs in a material. Syn: *yield point; yield strength; threshold pressure.*

yimengite (yi′-meng-ite) A metallic black hexagonal mineral of the *magnetoplumbite* group: $K(Cr^{3+},Ti,Fe^{3+},Mg)_{12}O_{19}$.

yingjiangite A yellow orthorhombic mineral of the *phosphuranylite* group: $K_2Ca(UO_2)_7(PO_4)_4(OH)_6{\cdot}6H_2O$.

yixunite (yix′-u-nite) A metallic cubic mineral: Pt_3In.

Y level *wye level.*

Y-mark A trilete *laesura* on embryophytic spores and some pollen, consisting of a three-pronged mark somewhat resembling an upper-case "Y". It is commonly also a commissure or suture along which the spore germinates. The term is also applied to analogous marks, which are not laesurae, on pollen grains.

Ynezian (Y-nez'-i-an) North American provincial stage: Lower Paleocene (above Danian, below Bulitian).

yoderite (yo'-der-ite) A purple or green monoclinic mineral: $(Al,Mg)_4(Mg,Al,Fe^{3+})_3O_2(SiO_4)_4(OH)_2$.

yofortierite (yo-for'-tier-ite) A maroon monoclinic mineral of of the *palygorskite* group: $(Mn^{2+},Mg)_5Si_8O_{20}(OH)_2{\cdot}8\text{-}9H_2O$.

yoke-pass *joch.*

yosemite (yo-sem'-i-te) A portion of a glacial valley, esp. in the Sierra Nevada of California, that is deeply U-shaped, with sheer walls, hanging troughs, and a wide almost level floor, and hence resembles the Yosemite Valley, Calif.

yoshimuraite (yo-shi-mu'-ra-ite) An orange, red, or dark brown triclinic mineral: $(Ba,Sr)_2Mn^{2+}{}_2Ti(SiO_4)_2(PO_4,SO_4)(OH,Cl)$.

yoshiokaite (yosh-i-o'-ka-ite) A colorless trigonal Moon mineral: $Ca(Al,Si)_2O_4$.

young [geomorph] Pertaining to the hypothetical stage of youth of the cycle of erosion; esp. said of a stream that has not developed a profile of equilibrium, and of its valley. Syn: *youthful.*

young [struc geol] v. To face, in the sense "to present the younger aspect" of one formation toward another; e.g., if formation A "youngs" toward formation B, then B is younger than A unless some fold, fault, unconformity, or intrusion intervenes. Term coined by Bailey (1934, p.469) and used "as a record of observation and not of stratigraphic deduction".

young coastal ice *Sea ice* in the initial stage of *fast ice* formation, consisting of *nilas* or young ice of local origin and having a width ranging from a few meters to 100 or even 200 m from the shoreline.

Younger Dryas (Young'-er) n. An interval of late-glacial time, from about 13,000 to 11,600 calendar years ago (10,000 to 11,000 radiocarbon years), following the Allerød and preceding the Preboreal, during which the climate, as inferred from stratigraphic and pollen data (Iversen, 1954), deteriorated, favoring either expansion or retarded retreat of the waning continental glaciers and alpine glaciers. adj. Pertaining to the late-glacial Younger Dryas interval and to its climate, deposits, biota, and events.

younging (young'-ing) A colloquial syn. of *facing.*

youngland (young'-land) The land surface, with its plateaus and valleys, of the youthful stage of the cycle of erosion (Maxson and Anderson, 1935, p.90).

young mountain A mountain that was formed during the Tertiary or Quaternary periods, esp. a fold mountain produced during the last great period of folding (i.e., the Alpine orogeny). Ant: *old mountain.*

young polar ice *second-year ice.*

Young's modulus A *modulus of elasticity* of a body in tension or compression, involving a change of length. Cf: *elastic compliance.* Syn: *stretch modulus.*

youth [coast] A hypothetical stage in the development of a shore, shoreline, or coast characterized by an ungraded profile of equilibrium (Johnson, 1919). The concept is generally regarded now as obsolete, since Johnson's criteria have been shown as fundamentally incorrect. Cf: *primary [coast].*

youth [geomorph] The first stage of the hypothetical *cycle of erosion* in the topographic development of a landscape or region, in which the original surface or structure is still the dominant feature of the relief and the landforms are being accentuated or are tending toward complexity. It is characterized by a few small widely spaced young streams; broad, flat-topped interstream divides and upland surfaces, little modified by erosion; poorly integrated drainage systems, with numerous swamps and shallow lakes; and rapid and progressive increase of local relief, with sharp landforms, steep and irregular slopes, and a surface well above sea level. Cf: *infancy.*

youth [streams] The first stage in the hypothetical development of a stream, at which it has just entered upon its work of erosion and is increasing in vigor and efficiency, being able everywhere to erode its channel and having not reached a graded condition. It is characterized by (1) an ability to carry a load greater than the load it is actually carrying; (2) active and rapid downcutting, forming a deep, narrow, V-shaped valley (gorge or canyon) with a steep and irregular gradient and rocky outcrops; (3) numerous waterfalls, rapids, and lakes; (4) a swift current and clear water; (5) a few short, straight tributaries; (6) an absence of flood plains as the stream occupies all or nearly all of the valley floor; and (7) an ungraded bed.

youthful (youth'-ful) Pertaining to the stage of youth of the cycle of erosion; esp. said of a topography or region, and of its landforms (such as a plain or plateau), having undergone little erosion or being in an early stage of development. Cf: *infantile.* Syn: *young [geomorph]; juvenile [geomorph].*

Ypresian (Ypres'-ian [Pron: ee-pres'-ian]) A stage of the Standard Global Chronostratigraphic Scale: lowermost Eocene (above Thanetian Stage of the Paleocene, below Lutetian). The basal GSSP is near Luxor, Egypt.

Y-shaped valley A valley having a cross profile suggesting the form of the letter "Y", such as a rejuvenated valley in which the grade of the river has recently been increased by uplift of the headwaters (Lane, 1923).

Y-tombolo (Y-tom-bo'-lo) A tombolo consisting of two embankments that extend shoreward from an island or seaward from the mainland and that unite "to form a single ridge before the connection is completed" (Johnson, 1919, p.315); there is a body of water between the prongs of the "Y".

yttrialite-(Y) (yt'-tri-a-lite') An olive-green or orange-yellow monoclinic mineral: $(Y,Th)_2[Si_2O_7]$. Cf: *thalénite.*

yttrobetafite-(Y) A greenish metamict mineral of the *pyrochlore* group: $(Y,U,Ce,[])_2(Ti,Nb,Ta)_2(O,OH)_7$.

yttroceberysite-(Y) (yt'-tro-ce-ber'-y-site-Y) A white, pale-yellow, or light-green monoclinic mineral of the *datolite* group: $(Y,Ce)Be(OH)(SiO_4)$. It may be gadolinite-(Y) or hingganite-(Y), and has not been approved by the International Mineralogical Association.

yttrocerite (yt-tro-ce'-rite) A violet-blue variety of yttrofluorite containing cerium.

yttrocolumbite (yt'-tro-co-lum'-bite) An ill-defined vitreous black orthorhombic mineral: $(Fe,Mn,Y)(Nb,Ta)_2O_6$. It may be samarskite.

yttrocrasite-(Y) (yt-tro-cra'-site) A black or brown mineral of the *euxenite* group: $(Y,Th,U,Ca)Ti_2(O,OH)_6$.

yttrofluorite (yt-tro-flu'-o-rite) An ill-defined cubic mineral: $(Ca,Y)F_{2\text{-}3}$. It may be yttrium-bearing fluorite.

yttropyrochlore-(Y) A brown cubic mineral of the *pyrochlore* group: $(Y,Na,Ca,[])_2(Nb,Ta)_2(O,OH)_7$. Syn. *obruchevite.*

yttrotantalite-(Y) (yt-tro-tan'-ta-lite) A black or brown orthorhombic mineral: $(Y,U,Ca)(Ta,Fe)_2(O,OH)_6$. Cf: *yttrocolumbite.*

yttrotungstite-(Ce) An orange-yellow monoclinic mineral: $CeW_2O_6(OH)_3$. Syn: *cerotungstite-(Ce).*

yttrotungstite-(Y) (yt-tro-tung'-stite) A yellow monoclinic mineral: $(Y,La,Ca)(W,Fe,Al,Ti)_2(O,OH)_6$. It may contain a little thorium. Syn: *thorotungstite.*

yuanfuliite A submetallic black orthorhombic mineral: $Mg(Fe^{3+},Al)O(BO_3)$. It is the Fe^{3+} analogue of warwickite.

yuanjiangite A metallic silver-white hexagonal mineral: AuSn.

yugawaralite (yu-ga-war'-a-lite) A colorless monoclinic zeolite mineral: $CaAl_2Si_6O_{16}{\cdot}4H_2O$.

yukonite (yu'-kon-ite) An ill-defined reddish-brown or dark brown noncrystalline mineral: $Ca_7Fe^{3+}{}_{15}(AsO_4)_9OH_{16}{\cdot}25H_2O$(?).

yuksporite (yuk-spor'-ite) A yellowish-rose triclinic mineral: $(K,Na)_6(Ca,Na,Sr,Ba)_{20}(Ti,Fe)_6Si_{24}O_{72}(OH,F,Cl)$. It may be related to bafertisite.

yushkinite (yush'-kin-ite) A metallic pinkish-violet trigonal mineral: $V_{1-x}S{\cdot}n(Mg,Al)(OH)_2$. It is a layered structure allied to *valleriite.*

yvonite A vitreous turquoise-blue triclinic mineral: $Cu(AsO_3OH){\cdot}2H_2O$.

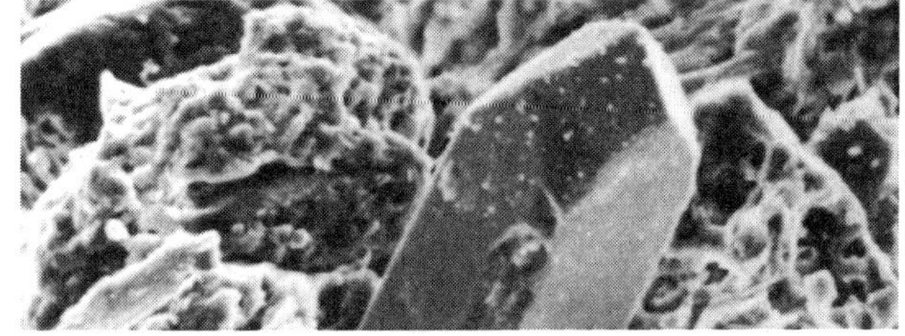

Z

Z *atomic number.*

zabuyelite A colorless monoclinic mineral: Li_2CO_3.

zaccagnaite A white hexagonal mineral of the *manasseite* group: $Zn_4Al_2(OH)_{12}(CO_3)\cdot3H_2O$.

zaherite (za'-her-ite) An earthy-white triclinic mineral: $Al_{12}(SO_4)_5(OH)_{26}\cdot20H_2O$.

zaïrite (za-ïr'-ite) A greenish rhombohedral mineral of the *crandallite* group: $Bi(Fe^{3+},Al)_3(PO_4)_2(OH)_6$.

zajacite-(Ce) A pale pink or colorless triclinic mineral: $Na(Ca,Ce,La,Nd)_2F_6$.

zakharovite (zak'-ha-rov-ite) A waxy-yellow trigonal mineral: $Na_4Mn^{2+}{}_5Si_{10}O_{24}(OH)_6\cdot6H_2O$.

zálesíite A pale green hexagonal mineral of the *mixite* group: $CaCu_6(AsO_4)_2(AsO_3OH)(OH)_6\cdot3H_2O$.

zanazziite A pale to dark olive-colored monoclinic mineral: $Ca_2Be_4(Mg,Fe^{2+})_5(PO_4)_6(OH)_4\cdot6H_2O$. It is the Mn^{2+}-Fe^{3+} analogue of roscherite.

Zanclean (Zan-cle'-an) A stage of the Standard Global Chronostratigraphic Scale: lowermost Pliocene (above Messinian Stage of the Miocene, below Piacenzian). The basal GSSP is in Sicily, Italy (Van Couvering et al., 2000).

zanjón (zan-jon') A Puerto Rican term for a *solution corridor.* Etymol: Spanish, "deep ditch".

zapatalite (za-pa'-ta-lite) A pale blue tetragonal mineral: $Cu_3Al_4(PO_4)_3(OH)_9\cdot4H_2O$.

zap crater An informal syn. of *microcrater.*

zaphrentid (za-phren'-tid) A solitary coral with marked pinnate septal arrangement, lacking dissepiments or axial structures.

zaratite (zar'-a-tite) An ill-defined emerald-green mineral: $Ni_3(CO_3)(OH)_4\cdot4H_2O$ (?). It occurs in secondary incrustations or compact masses. Syn: *emerald nickel.*

zastrugi (zas'-tru-gi) *sastrugi.*

zavaritskite (za-va-rit'-skite) A semi-metallic gray tetragonal mineral of the *matlockite* group: BiOF.

zdenekite A vitreous blue monoclinic mineral of the *lavendulan* group: $NaPbCu_5(AsO_4)_4Cl\cdot5H_2O$.

zeasite (ze'-a-site) An opal, formerly an old name for "fire opal" but now applied to "wood opal".

zebra dolomite (ze'-bra) An altered dolomite rock that shows conspicuous banding (generally parallel to bedding) consisting of light-gray coarsely textured layers alternating with darker finely textured layers. See also: *zebra rock.*

zebraic chalcedony (ze-bra'-ic) Chalcedony that under crossed nicols shows alternating black and white bands irregularly athwart the fibers, giving a zebralike appearance. The effect is produced by helical rotation of component quartz crystallites (McBride and Folk, 1977).

zebra layering *Rhythmic layering* in which dark and light bands alternate, generally reflecting differing amounts of pyroxene and plagioclase. Obsolete.

zebra limestone A limestone banded by parallel sheet cracks filled with calcite (Fischer, 1964, p.135).

zebra rock A dolomite that shows narrow banding consisting of black layers (indicative of organic matter) alternating with white, slightly coarse-grained, and somewhat vuggy layers. See also: *zebra dolomite.*

zebroid structure (ze'-broid) The banding typically shown in *zebra dolomite.*

Zechstein (Zech'-stein) European series (esp. in Germany): Upper Permian (above Rotliegende). It contains the Thuringian Stage.

zeilleriid (zeil-ler'-i-id) Said of a long brachiopod *loop* (as in the superfamily Zeilleriacea) not attached to the dorsal septum in adults.

zektzerite (zekt'-zer-ite) A colorless or pink orthorhombic mineral of the tuhualite group: $NaLiZrSi_6O_{15}$.

zellerite (zel'-ler-ite) A lemon-yellow orthorhombic secondary mineral: $Ca(UO_2)(CO_3)_2\cdot5H_2O$.

zemannite (ze'-mann-ite) (a) An adamantine brown hexagonal mineral: $Mg_{0.5}ZnFe^{3+}(Te^{4+}O_3)_3\cdot4.5H_2O$. It has zeolitic properties. (b) a group name for hexagonal minerals with the analogous composition, but with Zn replaced by Mn^{2+}, Fe^{2+}, or Ni, and a variable water content.

zemkorite A colorless hexagonal mineral: $(Na,K)_2Ca(CO_3)_2$.

Zemorrian (Ze-mor'-ian) Pacific Coast marine stage: lowermost Miocene (below Saucesian).

zenith (ze'-nith) The point on the celestial sphere that is directly above the observer and directly opposite the nadir. In a more general sense, the term denotes the stretch of sky overhead.

zenithal projection (ze'-nith-al) *azimuthal projection.*

zenzénite A black hexagonal mineral: $Pb_3(Fe^{3+},Mn^{3+})_4Mn^{4+}{}_3O_{15}$.

zeolite (ze'-o-lite) (a) A generic term for a large group of white or colorless (sometimes tinted red or yellow by impurities) hydrous aluminosilicate minerals that have an open framework structure of interconnected $(Si,Al)O_4$ tetrahedra with exchangeable cations and H_2O molecules in structural cavities. They have a ratio of (Al + Si) to nonhydrous oxygen of 1:2; and are characterized by their easy and reversible loss of water of hydration and by their ready fusion and swelling when strongly heated under the blowpipe. Zeolites have long been known to occur as well-formed crystals in cavities in basalt. Of more significance is their occurrence as authigenic minerals in the sediments of saline lakes and the deep sea and esp. in beds of tuff. They form "during and after burial, generally by reaction of pore waters with solid aluminosilicate materials (e.g., volcanic glass, feldspar, biogenic silica, and clay minerals)" (Hay, 1978, p.135). (b) Any of various silicates that are processed natural materials (such as glauconite) or artificial granular sodium aluminosilicates used in the base-exchange method of water softening and as gas adsorbents or drying agents. The term now includes such diverse groups of compounds as sulfonated organics or basic resins, which act in a similar manner to effect either cation or anion exchange. Etymol: Greek "zein", "to boil".

zeolite facies The facies (set of metamorphic mineral assemblages) that includes the zeolites analcime, heulandite, stilbite, laumontite, and wairakite (Coombs, 1960). Developed best in metagraywackes and metabasalts, it is the lowest grade of metamorphism, transitional between diagenesis (or unmetamorphosed rock) and the prehnite-pumpellyite facies or the *greenschist facies.* Various zeolite assemblages can be correlated with depth of burial (Miyashiro and Shido, 1970).

zeolitization (ze'-o-lit'-i-za'-tion) Introduction of, alteration to, or replacement by, a mineral or minerals of the zeolite group. Zeolitization is common in igneous protoliths. Cf: *zeolite facies.*

zeophyllite (ze-o-phyll'-ite) A pearly white rhombohedral mineral: $Ca_{13}Si_{10}O_{28}(OH)_2F_{10}\cdot6H_2O$. It sometimes contains iron.

zeravshanite A vitreous colorless monoclinic mineral: $Na_2Cs_4Zr_3Si_{18}O_{45}\cdot2H_2O$.

zerdeb (zer'-deb) A term used in Algeria for an interdune hollow floored with calcareous tufa (Capot-Rey, 1945, p.397).

zero-air-voids curve In engineering geology, the locus of points of *maximum dry density* of a compacted soil-water mixture plotted against water content for a single specific gravity of solids in which all pore space is filled with liquid (i.e., where the sediment is saturated).

zero curtain (ze'-ro) The zone immediately above the permafrost table, where zero temperature (0°C) lasts a considerable period of time during freezing and thawing of the overlying ground (Brown and Kupsch, 1974, p.42).

zero-energy coast (ze'-ro-en'-er-gy) A coast characterized by average breaker heights of 3 cm or less. Cf: *low-energy coast.*

zero-length spring A type of gravimeter spring whose effective length, as measured from its fixed point of support, is zero when the external forces acting upon it are zero. Gravimeters, using a spring of this type in their measuring systems, are linear and have extreme sensitivity combined with stability and reduced sensitivity to leveling errors. Cf: *LaCoste-Romberg gravimeter.*

zero-length-spring gravimeter An exploration-type gravimeter with a measuring system design based on the use of a quartz or metal *zero-length spring.* Cf: *relative-gravity instrument.*

zero meridian *prime meridian.*

zero-offset section A seismic section where each trace is the result of a coincident source and geophone, the section that a common-midpoint section simulates. Also called: ZSR (zero source-receiver distance) section.

zero-order basin An unchannelized upland valley that serves as a sediment storage site and may be important for debris-flow initiation. Cf: *hollow; stream order.*

zero-phase (a) Describing a filter for which the phase shift is zero for all frequencies. Zero-phase filters are anticipatory and hence are not physically realizable, i.e., half of the energy arrives before the time reference, so that one gets output before the input arrives. (b) Describing a wavelet symmetric about zero time. It is the shortest possible wavelet for any given spectrum.

zero time The reference time for a seismic trace, with respect to which arrival times are measured.

zetaform coast A coast of alternating headlands and beaches and a dominant swell that approaches obliquely causing beaches to assume a "half-heart" or zeta form curvature. Each zetaform beach segment seems to be offset relative to its neighbors with the straight segment of beach rotated from the mean trend of the coast toward the dominant swell direction (Bloom, 1978, p.457).

zeta potential (ze'-ta) Drop in potential across the diffuse layer in an electrolyte.

zeuge (zeu'-ge) A tabular mass of resistant rock left standing on a pedestal of softer rocks, resulting from differential erosion by the scouring effect of windblown sand in a desert region; it may be 2-50 m high. Etymol: German "Zeuge", "witness". Pl: zeugen. Cf: *mushroom rock.* Syn: *witness rock.*

zeugenberg (zeu'-gen-berg) A syn. of *butte témoin.* Etymol: German "Zeugenberg", "witness hill".

zeunerite (zeu'-ner-ite) A tetragonal green secondary mineral of the *autunite* group: $Cu(UO_2)_2(AsO_4)_2{\cdot}10\text{-}16H_2O$. It is isomorphous with uranospinite.

zeylanite (zey'-lan-ite) *ceylonite.*

zhamanshinite Glass associated with Zhamanshin crater, Russia, thought to have been produced by melting of local target rocks during an impact event. Cf: *impactite.*

zhanghengite (zhang'-heng-ite) A metallic yellow cubic meteorite mineral: CuZn.

zharchikhite (zhar'-chik'-hite) A colorless monoclinic mineral: $AlF(OH)_2$.

zhemchuzhnikovite (zhem-chuzh'-ni-kov-ite) A smoky-green or violet trigonal oxalate mineral: $NaMg(Al,Fe^{3+})(C_2O_4)_3{\cdot}8H_2O$. Cf: *stepanovite.*

zhonghuacerite-(Ce) (zhong-hua-cer'-ite-Ce) A vitreous yellow monoclinic mineral: $Ba_2CeF(CO_3)_3$. It may be the same as kukharenkoite-(Ce).

zibar Term applied to low-relief eolian sand dunes that lack discernible slip faces (Cooke and Warren, 1973, p.309). Etymol: Arabic.

ziesite (zies'-ite) A metallic black monoclinic mineral: $\beta Cu_2V^{5+}{}_2O_7$. It is a dimorph of blossite.

zietrisikite (zie-tri-si'-kite) Incorrect spelling of *pietricikite,* a variety of ozocerite.

zigzag carina (zig'-zag) One of the not quite oppositely placed *carinae* of a rugose coral on the two sides of a septum. Cf: *yardarm carina.*

zigzag commissure The line of articulation between brachiopod valves that is plicate rather than horizontal or semi-horizontal.

zigzag cross-bedding *chevron cross-bedding.*

zigzag fold A *kink fold,* the limbs of which are of unequal length. Cf: *chevron fold; knee fold.*

zigzag ridge A continuous ridge that trends first in one direction, then in another. Zigzag ridges are produced in folded mountains (e.g., the Appalachians) by truncation of *plunging folds.*

zigzag watershed A drainage divide through which rivers have broken by headward erosion, the divide retaining its original position between the drainage basins.

zimbabweite An adamantine yellow-brown orthorhombic mineral: $Na(Pb,Na,K)_2(Ta,Nb,Ti)_4As_4O_{18}$.

zinalsite (zin-al'-site) An ill-defined white to reddish-brown monoclinic clay mineral: $Zn_7Al_4(SiO_4)_6(OH)_2{\cdot}9H_2O$(?).

zinc A blue-white hexagonal mineral, the native metallic element Zn.

zincaluminite (zinc-a-lu'-mi-nite) An ill-defined white to pale blue mineral: $(Zn,Al)_9(SO_4)_2(OH)_{18}{\cdot}nH_2O$ (?).

zinc blende *sphalerite.*

zinc bloom *hydrozincite.*

zincgartrellite A vitreous green-yellow triclinic mineral: $Pb(Zn,Cu,Fe)_2(AsO_4)_2(H_2O,OH)_2$. It is the Zn analogue of gartrellite.

zincite (zinc'-ite) A red to yellow brittle hexagonal polar mineral: $(Zn,Mn^{2+})O$. It was an ore of zinc, as in New Jersey where it was associated with franklinite and willemite. Syn: *red zinc ore; red oxide of zinc; ruby zinc; spartalite.* It is isomorphous with bromellite.

zinckenite (zinc'-ken-ite) *zinkenite.*

zinclavendulan A blue orthorhombic mineral: $(Ca,Na)_2(Zn,Cu)_5(AsO_4)_4Cl{\cdot}4\text{-}5H_2O$. It is probably a zinc-bearing variety of lavendulan.

zinc-melanterite (zinc-me-lan'-ter-ite) A pale greenish-blue monoclinic mineral of the *melanterite* group: $(Zn,Cu,Fe)(H_2O)_6(SO_4){\cdot}H_2O$.

zincobotryogen (zinc-o-bot'-ry-o-gen') A bright orange-red monoclinic mineral of the *botryogen* group: $(Zn,Mg,Mn^{2+})Fe^{3+}(SO_4)_2(OH){\cdot}7H_2O$. It has not been approved by the International Mineralogical Association.

zincochromite (zinc-o-chrom'-ite) A submetallic brown cubic mineral of the *spinel* group: $ZnCr_2O_4$.

zincocopiapite (zinc'-o-co-pi-a'-pite') A yellowish-green triclinic mineral of the *copiapite* group: $ZnFe^{3+}{}_4(SO_4)_6(OH)_2{\cdot}20H_2O$.

zincohögbomite-2N2S A dark brown to black hexagonal mineral of the *högbomite* group: $(Al,Zn,Fe,Ti)_{22}(O,OH)_{32}$. Its structure is interpreted as being composed of two nolanite and two spinel modules.

zincohögbomite-2N6S An adamantine orange hexagonal mineral of the *högbomite* group: $Zn_{14}(Al,Fe^{3+},Ti,Mg)_8Al_{24}O_{62}(OH)_2$. Its structure is interpreted as being composed of two nolanite and six spinel modules.

zincostaurolite A colorless monoclinic mineral of the *staurolite* group: $Zn_4Al_{18}Si_8O_{40}(OH)_6$.

zincovoltaite (zinc-o-vol'-ta-ite) A green-black cubic mineral of the *voltaite* group: $K_2Zn_5Fe^{3+}{}_3Al(SO_4)_{12}{\cdot}18H_2O$.

zincowoodwardite A pale blue trigonal mineral of the *woodwardite* group: $Zn_{1-x}Al_x(OH)_2(SO_4)_{x/2}{\cdot}nH_2O$; x = 0.32-0.50. It occurs as a trigonal and a rhombohedral polymorph.

zincrosasite (zinc-ro'-sa-site) An ill-defined bluish monoclinic mineral, presumably the Zn analogue of rosasite: $(Zn,Cu)_2(CO_3)(OH)_2$ (?).

zincroselite (zinc-ro'-sel-ite) A colorless monoclinic mineral of the *roselite* group: $Ca_2Zn(AsO_4)_2{\cdot}2H_2O$.

zincsilite (zinc-sil'-ite) A white or bluish monoclinic mineral: $Zn_3Si_4O_{10}(OH)_2{\cdot}4H_2O$(?). It is the aluminum-free end-member of the montmorillonite-sauconite series.

zinc spar *smithsonite.*

zinc spinel *gahnite.*

zinc vitriol *goslarite.*

zinc-zippeite (zinc-zip'-pe-ite) An yellow orthorhombic mineral of the *zippeite* group: $Zn(UO_2)_2(SO_4)O_2{\cdot}3.5H_2O$.

Zingg's classification A classification of pebble shapes, devised by Theodor Zingg, a 20th century Swiss meteorologist and engineer, based on the graphical representation of the diameter ratio of intermediate (width) to maximum (length) plotted against the diameter ratio of minimum (thickness) to intermediate (Zingg, 1935). The classification distinguishes four shape classes: *spheroid [sed]; disk [sed]; blade [sed]; rod [sed].*

zinkenite (zin'-ken-ite) A steel-gray hexagonal mineral: $Pb_6Sb_{14}S_{27}$. Also spelled: *zinckenite.* Syn: *keeleyite.*

zinkosite An orthorhombic mineral of the *chalcocyanite* group: $ZnSO_4$. Also spelled: zincosite.

zinnwaldite (zinn'-wald-ite) A name for a series of trioctahedral mica minerals that includes siderophyllite and polylithionite: $KLiFe^{2+}Al(AlSi_3)O_{10}(F,OH)_2$. It is pale-violet, yellowish, brown, or dark gray, and is the characteristic mica of greisens.

zippeite (zip'-pe-ite) (a) A yellow, orange, red, or reddish-brown monoclinic mineral: $K_3(UO_2)_4(SO_4)_2O_3(OH){\cdot}3H_2O$. (b) A group name for monoclinic or orthorhombic minerals with the analogous composition, but with K replace by Na, Mg, Co, Ni, or Zn.

zipper termination The termination of a *chain-width error* in a *biopyribole* by adjustment of chain widths in the *c* crystallographic direction.

zircon (zir'-con) (a) A mineral: $ZrSiO_4$. It occurs in tetragonal prisms, has various colors and is a common accessory mineral in siliceous igneous rocks, crystalline limestones, schists, and gneisses, in sedimentary rocks derived therefrom, and in beach and river placer deposits. It is the chief ore of zirconium, and is used as a refractory; when cut and polished, the colorless varieties provide exceptionally brilliant gemstones. Syn: *zirconite; hyacinth;* jacinth. (b) A group name for tetragonal minerals with the general formula $ASiO_4$, where *A* = Zr, Hg, Th, or U.

zirconite (zir'-con-ite) A variety of gray or brownish zircon.

zirconolite (zir-con'-o-lite) A submetallic brown to black mineral: $CaZrTi_2O_7$. Monoclinic, orthorhombic, and tetragonal ploymorphs exist.

zircophyllite (zir-co-phyl'-lite) A brown platy triclinic mineral of the *astrophyllite* group: $(K,Na)_3(Fe^{2+},Mn^{2+})_7(Zr,Nb)_2Si_8O_{26}$. Also spelled: *zirkophyllite.*

zircosulfate (zir-co-sul'-fate) A colorless orthorhombic mineral: $Zr(SO_4)_2{\cdot}4H_2O$.

zirkelite (zir'-kel-ite) A black cubic mineral: $(Ti,Ca,Zr)O_{2-x}$.

zirklerite (zirk'-ler-ite) An ill-defined light gray rhombohedral mineral: $(Fe^{2+},Mg,Ca)_9Al_4Cl_{18}(OH)_{12}{\cdot}14H_2O$(?).

zirkophyllite (zir-ko-phyl'-lite) *zircophyllite.*

zirsilite-(Ce) A vitreous cream-colored rhombohedral mineral of the *eudialyte* group: $(Na,[])_{12}(Ce,Na)_3Ca_6Mn_3Zr_3NbSi_{25}O_{73}(OH)_3(CO3){\cdot}H_2O$.

zirsinalite (zir-sin'-a-lite) A vitreous colorless or yellowish-gray rhombohedral mineral: $Na_6(Ca,Mn^{2+},Fe^{2+})ZrSi_6O_{18}$.

zittavite (zit'-tav-ite) A type of lustrous black lignite. It is harder and more brittle than *dopplerite.*

zlatogorite A metallic silver-white trigonal mineral: $CuNiSb_2$.

znucalite A white, light yellow, or grayish-yellow triclinic mineral: $CaZn_{12}(UO_2)(CO_3)_3(OH)_{22}{\cdot}4H_2O$.

zoantharian (zo-an-thar'-i-an) Any anthozoan belonging to the subclass Zoantharia, characterized by paired mesenteries. They may or may not have a calcareous exoskeleton. Range, Ordovician to Holocene.

zoarium (zo-ar'-i-um) (a) The collective skeletal parts (zooecia and extrazooidal skeleton) of an entire bryozoan colony. (b) A colony of bryozoans. Pl: zoaria. Also spelled: *zooarium.*

ZoBell bottle A sterilized bottle used for collection of sea water for bacteriological analysis.

zobtenite (zob'-ten-ite) A gabbro-gneiss characterized by *augen* of diallage surrounded by uralite and embedded in granular epidote and plagioclase (saussurite) (Holmes, 1928, p.242). Rarely used. Cf: *flaser gabbro.*

zodacite (zo'-dac-ite) A yellow or colorless platy monoclinic of the *montgomeryite* group: $Ca_4Mn^{2+}Fe^{3+}_4(PO_4)_6(OH)_4 \cdot 12H_2O$.

zodiacal dust (zo-di'-a-cal) (a) *cosmic dust.* (b) The particulate matter in the ecliptic plane that reflects sunlight to create the zodiacal light, which forms a wedge-shaped bright region on the western horizon following sunset and on the eastern horizon just before sunrise.

zodiacal dust [astron] (zo-di'-a-cal) *interplanetary dust*, so-called because it lies predominantly in the ecliptic, or zodiacal, plane of the solar system.

zodiacal light A faint glow in the night sky near the ecliptic plane, visible immediately after sunset or prior to sunrise, caused by solar illumination of interplanetary dust inside Earth's orbit.

zoecium (zo-e'-ci-um) *zooecium.*

Zoeppritz's equations Equations that express the partition of energy when a plane wave impinges on an acoustic-impedance contrast. In the general case for an interface between two solids when the incident angle is not zero, four waves are generated, reflected P wave and S wave and transmitted P wave and S wave. The partition of energy among these is found from four boundary conditions that require continuity of normal and tangential displacement and stress.

zoisite (zo'-is-ite) A variously colored orthorhombic mineral related to epidote: $Ca_2Al_3(SiO_4)(Si_2O_7)O(OH)$. It often contains appreciable ferric iron, and is white, gray, brown, green, or rose red in color. Zoisite occurs in metamorphic rocks (esp. schists formed from calcium-rich igneous rocks), and in altered igneous rocks, and is an essential constituent of saussurite. Cf: *clinozoisite.*

zombie effect Apparent range extension of the fossils of an extinct organism due to reworking of those fossils into younger strata by sedimentary processes.

zona (zo'-na) An annular, equatorial extension of the *sexine* of a spore, having varying equatorial width and being relatively thin (a *cingulum* is much thicker). Pl: zonae. Cf: *auricula; corona [palyn]; crassitude.*

zonal In an east-west direction, along lines of constant latitude, e.g., zonal mean temperature.

zonal equation The statement that if a given crystal face (*hkl*) belongs to a zone with the axis [*uvw*], then $hu+kv+lw=0$.

zonal guide fossil A *guide fossil* that makes possible the identification of a specific biostratigraphic zone and that gives its name to the zone. It need not necessarily be either restricted to the zone or found throughout every part of it. Syn: *zone fossil.*

zonal profile *composite profile.*

zonal soil (a) A soil characteristic of a large area or zone. (b) One of the three primary subdivisions (orders) in soil classification as used in the United States. Not used in current U.S. system of soil taxonomy.

zonal structure *zoning [cryst].*

zonal theory A theory of hypogene mineral-deposit formation, and the spatial distribution patterns of mineral sequences to be expected from change in a mineral-bearing fluid as it migrates away from a magmatic source (Park and MacDiarmid, 1970, p.165). It also deals with thermal-chemical gradients associated with the genesis of ore deposits, whether of direct magmatic origin or not, and with metallogenic zoning on a regional scale. Cf: *zoning of ore deposits.*

zonate (zo'-nate) Said of spores possessing a zone.

zonation (zo-na'-tion) The condition of being arranged or formed in zones; e.g. the distribution of distinctive fossils, more or less parallel to the bedding, in biostratigraphic zones.

zone [cryst] *crystal zone.*

zone [ecol] (a) Part of a biogeographic region characterized by uniform climatic conditions, fauna, and flora; zones tend to occur in bands that are typically parallel to one another (e.g., parallel to a shoreline) or arrayed along (perpendicular to) ecological gradients; e.g., elevation or depth. (b) An area characterized by the dominance of a particular organism, such as the *Acropora palmata* zone on Caribbean reefs.

zone [geog] A term used generally, even vaguely, for an area or region of latitudinal character, more or less set off from surrounding areas by some special or distinctive characteristics; e.g., any of the five great belts or encircling regions into which the Earth's surface is divided with respect to latitude and temperature, viz. the torrid zone, the two temperate zones, and the two frigid zones.

zone [geol] A belt, band, or strip of earth materials, however disposed, characterized as distinct from surrounding parts by some particular property or content; e.g., the *zone of saturation,* the zone of fracture or a *fault zone.*

zone [meta] *metamorphic zone.*

zone [palyn] *zona.*

zone [stratig] A minor interval in any category of stratigraphic classification. There are many kinds of zones, depending on the stratigraphic characters under consideration biozones, lithozones, chronozones, magnetic-polarity zones, mineral zones, metamorphic zones, and so on (ISG, 1976, p.14; ISSC, 1994, p.15). The term should always be preceded by a modifier indicating the kind of zone to which reference is made.

zone axis That line or crystallographic direction through the center of a crystal which is parallel to the intersection edges of the crystal faces defining the *crystal zone.* Syn: zonal axis.

zone-breaking species A fossil species that is confined to a biostratigraphic zone in certain areas but transgresses the boundaries of that zone at other places (Arkell, 1933, p.32).

zoned ultramafic PGE deposit A *mineral-deposit type* characterized by platinum group element minerals associated with cumulate layers of chromite, magnetite, or clinopyroxenite in zoned ultramafic intrusive complexes. Syn: *Alaskan-type PGE deposit; Kachkanar-type PGE deposit; Uralian-type PGE deposit.*

zone fossil A fossil characteristic of a zone; a *zonal guide fossil.*

zone of ablation *ablation area.*

zone of accumulation [mass move] The area of a landslide within which the displaced material lies above the original ground surface (Varnes, 1978). Cf: *zone of depletion.*

zone of accumulation [snow] (a) *accumulation area.* (b) In respect to an avalanche, a syn. of *accumulation zone.*

zone of aeration *unsaturated zone.*

zone of astogenetic change A part of a bryozoan colony in which zooids show morphologic differences from generation to generation in a more or less uniform progression distally, ending with a pattern capable of endless repetition of one or more kinds of zooids.

zone of astogenetic repetition A part of a bryozoan colony in which zooids show one or more repeated morphologies from generation to generation in a pattern capable of endless repetition.

zone of capillarity *capillary fringe.*

zone of cementation The layer of the Earth's crust below the *zone of weathering,* in which percolating waters cement unconsolidated deposits by the deposition of dissolved minerals from above.

zone of depletion The area of a landslide within which the displaced material lies below the original ground surface (Varnes, 1978). Cf: *zone of accumulation [mass move].*

zone of deposition "The area in which continental glaciers deposit materials derived from the *zone of erosion.* It is usually covered with drift and has the general aspect of a plain" (Stokes and Varnes, 1955, p.164).

zone of erosion "The area from which continental glaciers have removed material by erosion. It is mostly a bare rock surface" (Stokes and Varnes, 1955, p.164). Ant: *zone of deposition.*

zone of flow *zone of plastic flow.*

zone of flowage *zone of plastic flow.*

zone of fracture The upper, brittle part of the Earth's crust, in which deformation is by fracture rather than by plastic flow; that region of the crust in which fissures can exist. Cf: *zone of plastic flow; zone of fracture and plastic flow.* Syn: *zone of rock fracture.*

zone of fracture and plastic flow That region of the Earth's crust intermediate in depth and pressure between the zone of fracture and the *zone of plastic flow,* in which deformation of the weaker rocks is by plastic flow, and of the stronger rocks by fracture.

zone of influence A circular or elliptical area surrounding a data point on a map. The values at all locations within the zone of influence are partially dependent upon the value at the central point. Beyond the zone of influence, points are statistically independent of the central point.

zone of intermittent saturation A term applied by Monkhouse (1965) to the temporary zone of saturation formed in the soil by infiltration from rainfall or snowmelt at a rate in excess of that at which the water can move downward to the main water table.

zone of mobility *asthenosphere.*

zone of plastic flow That part of the Earth's crust that is under sufficient pressure to prevent fracturing, i.e. is ductile, so that deformation is by flow. Cf: *zone of fracture; zone of fracture and plastic flow.* Syn: *zone of flow; zone of rock flowage; zone of flowage.*

zone of recessive basal margin In conodont morphology, faintly striated portion of an attachment surface flanking some part of, or surrounding, a basal pit, and formed as the basal margins of successively younger lamellae grew no farther downward than those of older lamellae, or actually retreated upward as the elements grew larger (TIP, 1981, pt. W, supp.2). Obsolete syn: *inverted basal cavity.*

zone of rock flowage *zone of plastic flow.*

zone of rock fracture *zone of fracture.*

zone of saturation *saturated zone.*

zone of soil water *belt of soil water.*

zone of suspended water *unsaturated zone.*

zone of weathering The superficial layer of the Earth's crust above the water table that is subjected to the destructive agents of the atmosphere, and in which soils develop. Cf: *zone of cementation.*

zone symbol The symbol of the zone axis of a crystal in terms of the crystal lattice, e.g. the symbols for the zone axis of a series of (*hk*0) faces would be [001]. Cf: *indices of lattice row.*

zone time A syn. of *moment.* The term was suggested by Kobayashi (1944a, p.742) for the average duration in years of a biostratigraphic zone in any given geologic system, ranging between about 300,000 and 5 million years. Cf: instant.

zoning [cryst] (zon'-ing) A variation in the composition of a crystal from core to margin, due to a separation of the crystal phases during its growth by loss of equilibrium in a continuous reaction series. The high-temperature phases of the isomorphic series form the core, with the low-temperature phases toward the margin. Cf: *armoring; normal zoning.* Syn: *zonal structure.*

zoning [eco geol] The spatial distribution of ore minerals and secondary alteration minerals or a variety of scales in a mineral deposit, e.g., vein zoning, ore deposit zoning, mineral district zoning.

zoning [meta] (a) A term commonly used to describe compositional variation within a single mineral grain. (b) The development of areas of metamorphosed rocks that may exhibit zones in which a particular mineral or suite of minerals is predominant or characteristic, reflecting the original rock composition, the pressure and temperature of formation, the duration of metamorphism, and whether or not material was added.

zoning of ore deposits Spatial distribution patterns of elements, minerals, or mineral assemblages; paragenetic sequences, either syngenetic or epigenetic (Park and MacDiarmid, 1970, p.165). Zoning is especially well developed in the mineralization-alteration assemblages about subvolcanic occurrences such as porphyry base-metal deposits. Cf: *zonal theory.* Syn: *mineral zoning.*

zonite (zon'-ite) A term proposed by Henningsmoen (1961, p.68); equivalent to *range zone.*

zonochlorite (zo-no-chlo'-rite) A syn. of *pumpellyite* occurring in green pebbles of banded structure (as in the Lake Superior region). It was previously thought to be an impure prehnite.

zonoporate (zo-no-por'-ate) *stephanoporate.*

zonotrilete (zo-no-tri'-lete) Said of a trilete spore characterized by an equatorial zone or other thickening.

zonule (zon'-ule) A small subdivision of a biostratigraphic *subzone.* The term was defined by Fenton and Fenton (1928, p.20-22) as the strata or stratum containing a *faunule* or *florule,* its thickness and area being limited by the vertical and horizontal range of that faunule or florule. Now generally used as as subdivision of a biozone or of a subbiozone. Use of the term is discouraged (ISSC, 1994, p.56).

zooarium (zo-o-ar'-i-um) *zoarium.*

zoobenthos (zo-o-ben'-thos) Animal forms of the benthos.

zoochore (zo'-o-chore) A plant whose seeds or spores are distributed by living animals.

zooclast (zo'-o-clast) Fragmentary acid-resistant material of definite animal origin (e.g., graptolite *periderm,* arthropod exoskeletal *cuticle*) that may be present in palynological preparations (Tyson, 1995; Miller, 1996).

zooeciule (zo-oe'-ci-ule) A minute polymorph in some cheilostome bryozoans, having a *polypide* and operculum-bearing *orifice* but apparently unable to feed or produce gametes.

zooecium (zo-oe'-ci-um) The skeleton of a bryozoan *zooid.* Pl: zooecia. Adj: zooecial. Also spelled: *zoecium.*

zooecology (zo'-o-e-col'-o-gy) The branch of ecology concerned with the relationships between animals and their environment. Cf: *phytoecology.*

zoogenic rock (zo-o-gen'-ic) A *biogenic rock* produced by animals or directly attributable to the presence or activities of animals; e.g., shell limestone, coral reefs, guano, and lithified calcareous ooze. Cf: *zoolith [paleont].* Syn: zoogenous rock.

zoogeography (zo'-o-ge-og'-ra-phy) The branch of *biogeography* dealing with the geographic distribution of animals. Cf: *phytogeography.*

zooid (zo'-oid) (a) A more or less independent animal produced by other than direct sexual methods and therefore having an equivocal individuality; any individual of a colony, irrespective of its morphologic specifications, such as an octocorallian polyp or a soft-bodied graptolite individual inhabiting a theca. (b) One of the physically connected, asexually replicated morphologic units which, together with any extrazooidal parts present, comprise a bryozoan colony. It performs major functions with a system of organs comparable to that of a solitary animal.

zoolite (zo'-o-lite) An animal fossil. Syn: *zoolith [paleont].*

zoolith [paleont] (zo'-o-lith) *zoolite.*

zoolith [sed] A *biolith* formed by animal activity or composed of animal remains; specif. *zoogenic rock.*

zoomorph (zo'-o-morph) The acid-resistant material that represents an essentially entire organism (e.g., organic-walled microforaminifera lining, tentaculite organic remains, chitinozoa) referable to an animal group (Tyson, 1995).

zoophyte (zo'-o-phyte) (a) Any member of the phylum Bryozoa. (b) In a broad sense, any nonmobile plantlike animal; e.g., sponges, sea anemones, hydroids, or bryozoans. Syn: *phytozoan.*

zooplankton (zo-o-plank'-ton) The animal forms of *plankton;* e.g., jellyfish. They consume *phytoplankton* or other zooplankton.

zootrophic (zo-o-troph'-ic) *heterotrophic.*

zooxanthella (zo'-o-xan-thel'-la) An algal cell living symbiotically in the cells of certain invertebrates and *protists,* e.g., in the endoderm of *hermatypic coral* polyps. Most of these organisms are dinoflagellates. Pl: zooxanthellae.

zorite (zor'-ite) A rosy orthorhombic mineral: $Na_6(Ti,Nb)_5(Si_6O_{17})_2(O,OH)_5{\cdot}11H_2O$.

zosterophyllophytes (zos-te-ro-phyl'-lo-phytes) *zosterophylls.*

zosterophylls A group of early vascular land plants from the Devonian, originally the Zosterophyllophytina of Banks (1975). Zosterophylls are leafless and have *enations* and lateral *sporangia.* They are considered to be the ancestors of the *lycopsids.* Cf: *rhyniophytes; trimerophytes.*

zoubekite (zou'-bek-ite) A metallic gray orthorhombic mineral: $AgPb_4Sb_4S_{10}$.

z scores Values of the standard normal distribution, with a mean of 0 and variance of 1. Test statistics calculated by standardization.

z-transform In digital seismic processing, a method of expressing time-dependent quantized data as a polynomial and calculating the effect of various operations. A transform useful for representing time series and calculating the effects of various operations. The z-transform technique is an easy way of converting (by inspection) from the time domain into a form which can be treated (in some ways) as in the frequency domain. The convolution operation can be accomplished by merely multiplying the z-transforms of the waveforms being convolved. The inverse of a filter can be found by finding the reciprocal of the filter's z-transform. If the sample values of a wavelet at successive times are: $X_t = (X_0, X_1, X_2, X_3, ..., X_n)$, then the z-transform of the wavelet is $x(z) = x_0 + x_1z + x_2z^2 + x_3z^3 + ... + x_nz^n$.

zugshunstite-(Ce) A pale pink to pale blue monoclinic oxalate mineral: $(Ce,Nd,La)Al(SO_4)_2(C_2O_4){\cdot}12H_2O$.

Zuloagan (Zu-lo-a'-gan) North American (Gulf Coast) stage: Upper Jurassic (above older Jurassic, below LaCasitan; it is equivalent to European Oxfordian) (Murray, 1961).

zunyite (zun'-y-ite) A colorless, gray, or pink cubic mineral: $Al_{13}Si_5O_{20}Cl(OH,F)_{18}$. It occurs in minute transparent tetrahedral crystals.

zussmanite (zuss'-man-ite) A pale green rhombohedral mineral: $K(Fe^{2+},Mg,Mn^{2+})_{13}(Si,Al)_{18}O_{42}(OH)_{14}$.

Z-value Commonly, a value for a spatially distributed phenomenon (surface elevation, depth to groundwater) at a particular location (Krzanowski et al., 1993).

zvyagintsevite (zvy'-a-gint'-sev-ite) A metallic white cubic mineral: $(Pd,Pt,Au)_3(Pb,Sn)$.

zweikanter (zwei'-kan-ter) A *windkanter* or stone having two faces intersecting in two sharp edges. Etymol: German "Zweikanter", "one having two edges". Pl: zweikanters; zweikanter.

zwieselite (zwie'-sel-ite) A resinous dark brown to black monoclinic mineral of the *triplite* group: $(Fe^{2+},Mn^{2+})_2(PO_4)F$.

Zwischengebirge (Zwisch'-en-ge-birg'-e) A term proposed by Kober in 1921 to designate an undeformed or little deformed area between the two marginal chains of a symmetrical *orogenic belt.* In Kober's original sense an orogen consists of two marginal chains ("Randketten") and an intervening Zwischengebirge. The term has been translated into English as "intramontane space" (Longwell, 1923), "betwixt mountains" (Collet, 1927), and "median mass" (Bucher, 1933). Staub (1928) has used the term for any undeformed area between deformed belts of similar age, e.g., the Swiss plain.

zygal ridge (zy'-gal) The ridge uniting the median lobe (L2) and the proximal posterior lobe (L3) in many Paleozoic ostracodes.

zygnematacean algae (zyg-ne-ma-ta'-cean) *Green algae* exclusively of fresh-water lakes, ponds, and rivers, producing thick-walled, resistant *zygospores* which are divided by an equatorial line of rupture into two more or less equal parts (Batten, 1996; Wood and Miller, 1997).

zygolith (zy'-go-lith) A coccolith in the form of an elliptic ring with a crossbar arching slightly or strongly upward (e.g., distally) and bearing a knob or short spine.

zygolophe (zy'-go-lophe) A brachiopod lophophore in which each *brachium* consists of a straight or crescentic side arm bearing two rows of paired filamentary appendages (TIP, 1965, pt. H, p.155).

zygome (zy'-gome) An articulatory structure of a desma in a sponge.

zygomorphic (zy-go-mor'-phic) Said of an organism or organ that is bilaterally symmetrical or capable of division into essentially symmetric halves by only one longitudinal plane passing through the axis. Cf: *actinomorphic.*

zygosis (zy-go'-sis) The interlocking of sponge desmas, without fusion, by means of zygomes.

zygospore (zy'-go-spore) A *resting spore* of various algae (such as desmids and Zygnemataceae), produced by sexual fusion of two protoplasts. It often has a thick, resistant wall and can therefore occur as a *palynomorph.* Zygnemataceous zygospores characteristically split into two symmetrical halves. Cf: *schizomorph; zygnematacean algae.*

zygote (zy'-gote) The fertilized *ovum,* resulting from the union of two *gametes.* In Fungi, a *resting spore* that results from the fusion of two sexually dissimilar gametes (*antheridium* and *oogonium*). Cf: *egg.*

zygous basal plate (zy'-gous) One of the two large plates of the basal circlet of a blastoid or the infrabasal or basal circlet of a crinoid, located in the right posterior (BD) or left anterior (DA) position and formed by fusion of a pair of antecedent small basal plates comparable to *azygous basal plate* in the AB interray (TIP, 1967, pt. S, p.350).

zykaite (zy'-ka-ite) A grayish-white orthorhombic mineral: $Fe^{3+}_4(AsO_4)_3(SO_4)(OH){\cdot}15H_2O$.

REFERENCES CITED

References Cited

A

Abbate, E., Bortolotti, V., and Passerini, P. (1970) *Olistostromes and olistoliths.* In: Sestini, G., ed., *Development of the Northern Appenines geosyncline.* Sedimentary Geology, v.4, no.3-4, p.521-557.

Abraham, Herbert (1960) *Asphalts and allied substances v.1*, 6th ed. Princeton, N.J.: Van Nostrand. 370 p.

Abraham, K., Schreyer, W., Medenbach, O., and Gebert, W. (1980) *Kulkeit, ein geordnetes 1:1 Mixed-Layer-Mineral zwischen Klinochlor und Talk.* (abstr.) Fortschritte der Mineralogie 58, 4-5.

Ackermann, Ernst (1951) *Geröllton!* Geologische Rundschau, v. 39, p. 237-239.

Ackermann, Ernst (1962) *Büssersteine — Zeugen vorzeitlicher Grundwasserschwankungen.* Zeitschrift für Geomorphologie, n. F., Bd. 6, p. 148-182.

ACSN: American Commission on Stratigraphic Nomenclature (1957) *Nature, usage, and nomenclature of biostratigraphic units.* American Association of Petroleum Geologists. Bulletin, v. 41, p. 1877-1889. (Its Report 5).

ACSN: American Commission on Stratigraphic Nomenclature (1959) *Application of stratigraphic classification and nomenclature to the Quaternary.* American Association of Petroleum Geologists. Bulletin, v. 43, p. 663-675. (Its Report 6).

ACSN: American Commission on Stratigraphic Nomenclature (1961) *Code of stratigraphic nomenclature.* American Association of Petroleum Geologists. Bulletin, v. 45, p. 645-660.

ACSN: American Commission on Stratigraphic Nomenclature (1965) *Records of the Stratigraphic Commission for 1963-1964.* American Association of Petroleum Geologists. Bulletin, v. 49, p. 296-300. (Its Note 31).

ACSN: American Commission on Stratigraphic Nomenclature (1967) *Records of the Stratigraphic Commission for 1964-1966.* American Association of Petroleum Geologists. Bulletin, v. 51, p. 1862-1868. (Its note 34).

Adams, J. A. S., and Weaver, C. E. (1958) *Thorium-to-uranium ratios as indicators of sedimentary processes—example of concept of geochemical facies.* American Association of Petroleum Geologists. Bulletin, vol. 42, p. 387-430.

Adams, J. E., and Frenzel, H. N. (1950) *Capitan barrier reef, Texas and New Mexico.* Journal of Geology, v. 58, p. 289-312.

Adams, J. E., and Rhodes, M. L. (1960) *Dolomitization by seepage refluxion.* American Association of Petroleum Geologists. Bulletin, v. 44, p. 1912-1920.

Adams, J. E., et al. (1951) *Starved Pennsylvanian Midland Basin.* American Association of Petroleum Geologists. Bulletin, v. 35, p. 2600-2607.

Adamson, A.W., and Gast, A.P. (1997) *Physical chemistry of surfaces.* New York, Wiley-Interscience, 784p.

ADTIC: U. S. Arctic, Desert, Tropic Information Center (1955) *Glossary of arctic and subarctic terms.* Maxwell Air Force Base, Ala.: Air University, Research Studies Institute. 90 p. (ADTIC Publication A-105).

Agassiz, Louis (1866) *Geological sketches.* Boston: Ticknor & Fields. 311 p.

Ager, D. V. (1963) *Principles of paleoecology: an introduction to the study of how and where animals and plants lived in the past.* New York: McGraw-Hill, Inc.

Ager, D. V. (1974) *Storm deposits in the Jurassic of the Moroccan High.* Palaeogeography, palaeoclimatology, palaeoecology, v. 15, p. 83-93.

AGI: American Geological Institute (1989) *AGI Data Sheets. 3rd ed.* Alexandria, VA: AGI.

Ahlmann, H. W. (1933) *Scientific results of the Swedish-Norwegian Arctic Expedition in the summer of 1931, Pt. 8.* Geografiska Annaler, v. 15, p. 161-216, 261-295.

AIME: American Institute of Mining, Metallurgical, and Petroleum Engineers (1960) *Industrial minerals and rocks (non-metallics other than fuels). 3rd ed.* New York: American Institute of Mining, Metallurgical, and Petroleum Engineers. 934 p.

Aitken, J. D. (1967) *Classification and environmental significance of cryptalgal limestones and dolomites, with illustrations from the Cambrian and Ordovician of southwestern Alberta.* Journal of Sedimentary Petrology, v. 37, p. 1163-1178.

Aki, K., and Richards, P.G., (1980) *Quantitative seismology: Theory and methods.* W.H. Freeman and Co.

Alden, W. C. (1928) *Landslide and flood at Gros Ventre, Wyoming.* American Institute of Mining and Metallurgical Engineers Transactions, v. 76, p. 347-361. (Technical publication no. 140).

Allan, R. S. (1948) *Geological correlation and paleoecology.* Geological Society of America Bulletin, v. 59, p. 1-10.

Allen, E. T., and Day, A. L. (1935) *Hot springs of the Yellowstone National Park. Microscopic examinations* by H. E. Merwin. Carnegie Institution of Washington, Publication 466. 525 p.

Allen, G.P., and Posamentier, H.W. (1993) *Sequence stratigraphy and facies model of an incised valley fill; the Gironde Estuary, France.* Journal of Sedimen-Tary Petrology, v.63, no.3, p.378-391.

Allen, J. R. L. (1960) *Cornstone.* Geological Magazine, v. 97, p. 43-48.

Allen, J. R. L. (1963) *Asymmetrical ripple marks and the origin of water-laid cosets of cross-strata.* Liverpool and Manchester Geological Journal, v. 3, p. 187-236.

Allen, J.R.L. (1963) *The classification of cross-stratified units with notes on their origin.* Sedimentology, v.2, p.93-114.

Allen, J. R. L. (1968) *Current ripples: their relation to patterns of water and sediment motion.* Amsterdam: North Holland Publishing Co. 433 p.

Allen, V. T. (1936) *Terminology of medium-grained sediments.* With notes by P. G. H. Boswell. National Research Council. Division of Geology and Geography. Annual report for 1935-1936, appendix I, exhibit B. 23 p. (Its Committee on Sedimentation. Report, exhibit B).

Allen, V. T., and Nichols, R. L. (1945) *Clay-pellet conglomerates at Hobart Butte, Lane County, Oregon.* Journal of Sedimentary Petrology, v. 15, p. 25-33.

Alley, R.B., Saltzman, E.S., Cuffey, K.M., and Fitzpatrick, J.J. (1990) *Summertime formation of depth hoar in central Greenland.* Geophysical Research Letters, v.17, p.2393-2396.

Alling, H. L. (1943) *A metric grade scale for sedimentary rocks.* Journal of Geology, v. 51, p. 259-269.

Alling, H. L. (1945) *Use of microlithologies as illustrated by some New York sedimentary rocks.* Geological Society of America Bulletin, v. 56, p. 737-755.

Allum, J. A. E. (1966) *Photogeology and regional mapping.* New York: Pergamon. 107 p.

Alvarez, Walter, et al. (1978) *Classification of solution cleavage in pelagic limestones.* Geology, v. 6, p. 263-266.

Amin, B. S., Lal, D., and Somayajulu, B. L. K. (1975) *Chronology of marine sediments using the ^{10}Be method: intercomparison with other methods.* Geochimica et Cosmochimica Acta, v. 39, p. 1187-1191.

Amiran, D. H. K. (1950-1951) *Geomorphology of the central Negev highlands.* Israel Exploration Journal, v. 1, p. 107-120.

Amstutz, A. (1951) *Sur l'évolution des structures alpines.* Archives des Sciences, v. 4, p. 323-329.

Ander, M.E., Summers, T., and Gruchalla, M.R. (1999) *LaCoste & Romberg gravity meter: System analysis and instrumental errors.* Geophysics, v.64, p.1708-1709.

Anderson, D. M., et al. (1969) *Bentonite debris flows in northern Alaska.* Science, v. 164, p. 173-174.

Anderson, E.M. (1951) *The dynamics of faulting and dyke formation, with applications to Britain*, 2nd ed. Edinburgh, Oliver and Boyd, 206p.

Anderson, J. A. R. (1964) The structure and development of the peat swamps of Sarawak and Brunei. Journal of Tropical Geography, v. 18, p. 7-16.

Andersson, J. G. (1906) *Solifluction, a component of subaerial denudation.* Journal of Geology, v. 14, p. 91-112.

Andreev, P. F., et al. (1968) *Transformation of petroleum in nature.* Translated from the Russian edition by R. B. Gaul and B. C. Metzner. Translation editors: E. Barghoorn and S. Silverman. New York: Pergamon. 468 p. (International series of monographs in earth sciences, v. 29.)

Andresen, M. J. (1962) *Paleodrainage patterns: their mapping from subsurface data, and their paleogeographic value.* American Association of Petroleum Geologists. Bulletin, v. 46, p. 398-405.

Anonymous (1984) *Shaded relief and surface markings of the Pelorus linea region of Europa.* U. S. Geological Survey Map I-1493 (Je-3).

Anonymous (1985) *Shaded relief and surface markings of the Philus Sulcus quadrangle of Ganymede.* U. S. Geological Survey Map I-1565 (Jg-4).

Anonymous (1995) *Dictionary of Computer Words.* Revised ed.: Houghton Mifflin, Boston, MA, 332p.

Antevs, E. V. (1932) *Alpine zone of Mt. Washington Range.* Auburn, Me.: Merrill & Weber. 118 p.

Antevs, E. V. (1948) *Climatic changes and pre-white man.* University of Utah. Bulletin, v. 38, no. 20, p. 168-191. (The Great Basin, with emphasis on glacial and postglacial times, 3.).

Antevs, E. V. (1953) *Geochronology of the Deglacial and Neothermal ages.* Journal of Geology, v. 61, p. 195-230.

Apfel, E. T. (1938) *Phase sampling of sediments.* Journal of Sedimentary Petrology, v. 8, p. 67-68.

Archie, G. E. (1942) *The electrical resistivity log as an aid in determining some reservoir characteristics.* American Institute of Mining, Metallurgical, and Petroleum Engineers. Transactions, v. 146, p. 54-60.

Aref, A.M.A. (2003) *Classification and depositional environments of Quaternary pedogenic gypsum crusts (gypcrete) from east of the Fayum Depression, Egypt.* Sedimentary Geology, v.155, p.87-108.

Arkell, W. J. (1933) *The Jurassic system in Great Britain.* Oxford: Clarendon Press. 681 p.

Arkell, W. J. (1956) *Jurassic geology of the world.* New York: Hafner. 806 p.

Arkell, W. J. and Tomkeieff, S. I. (1953) *English rock terms, chiefly as used by miners and quarrymen.* London: Oxford University Press. 139 p.

Arkley, R. J. and Brown, H. C. (1954) *The origin of Mima mound (hogwallow) microrelief in the far western states.* Soil Science Society of America. Proceedings, v. 18, p. 195-199.

Armstrong, F. C. (1974) *Uranium resources of the future—"porphyry" uranium deposits.* In: Formation of uranium ore deposits. Proceedings of a symposium, Athens, 6-10 May, 1974, p. 625-635.

Armstrong, J. (1969) *The crossed bladed fabrics of the shells of* Terrakea solida *(Etheridge and Dunn) and* Streptorhynchus pelicanensis *Fletcher.* Palaeontology, v.12, p.310-320.

Armstrong, R. L. (1958) *Sevier orogenic belt in Nevada and Utah.* Geological Society of America. Bulletin, v. 79, p. 429-458.

Armstrong, Terence, and Roberts, Brian (1958) *Illustrated ice glossary. Pt. 1.* Polar Record, v. 8, no. 52, p. 4-12.

Armstrong, Terence, et al. (1966) *Illustrated glossary of snow and ice.* Cambridge, England: Scott Polar Research Institute. 60 p.

Armstrong, Terence, et al. (1977) *Proposed new terms and definitions for snow and ice features.* Polar Record, v. 18, p. 501-502.

Arndt, N. T., et al. (1977) *Komatiitic and iron-rich tholeiitic lavas of Munro Township, northeast Ontario.* Journal of Petrology, v. 18, p. 319-369.

Arnold, C. A. (1947) *An introduction to paleobotany.* New York: McGraw-Hill. 433 p.

ASCE: American Society of Civil Engineers. Hydraulics Division. Committee on Hydraulic Structures (1962) *Nomenclature for hydraulics.* Fred W. Blaisdell and others, eds. New York: American Society of Civil Engineers. 501 p. (Manuals and reports on engineering practice, no. 43.)

ASCE: American Society of Civil Engineers. Soil Mechanics and Foundations Division. Committee on Glossary of Terms and Definitions in Soil Mechanics (1958) *Glossary of terms and definitions in soil mechanics; report of the committee.* R. E. Fadum, chairman. Its Journal, 1958, no. SM4, pt. 1. 43 p. (American Society of Civil Engineers. Proceedings. v. 84, paper 1826).

ASCE: American Society of Civil Engineers. Surveying and Mapping Division. Committee on Definitions of Surveying Terms (1954) *Definitions of surveying, mapping, and related terms.* Charles B. Breed, chairman. New York: American Society of Civil Engineers. 202 p. (Manuals of engineering practice, no. 34).

Ashley, G.M. (1990) *Classification of large-scale subaqueous bed forms: A new look at an old problem.* Journal of Sedimentary Petrology, v.60, p.160-172.

Ashworth, J. R., ed. (1985) *Introduction.* In: Migmatites. Blackie: Glasgow, p. 1-35.

ASP: American Society of Photogrammetry (1975) *Manual of remote sensing.* R. G. Reeves, ed. Falls Church, Va.: American Society of Photogrammetry. Vol. 1, 867 p.; vol. 2, 2144 p.

ASP: American Society of Photogrammetry. Committee on Nomenclature (1966) *Definitions of terms and symbols used in photogrammetry.* Editor: Robert D. Turpin. In: American Society of Photogrammetry. Manual of photogrammetry, v. 2, p. 1125-1161. 3rd ed. Falls Church, Va.: American Society of Photogrammetry. 1199 p.

Assereto, Riccardo, and Kendall, C. G. (1977) *Nature, origin, and classification or peritidal tepee structures and related breccias.* Sedimentology, v. 24, p. 153-210.

Association of Engineering Geologists (1969) *Definition of engineering geology.* Its Newsletter, v. 12, no. 4, p. 3.

ASTM: American Society for Testing and Materials (1970) *Annual book of ASTM standards, part 33: glossary of ASTM definitions and index to ASTM standards.* Philadelphia: American Society for Testing and Materials. 706 p.

Atterberg, Albert (1905) *Die rationelle Klassifikation der Sande und Kiese.* Chemiker-Zeitung, Jahrg. 29, p. 195-198.

Aubouin, Jean (1965) *Geosynclines.* Amsterdam: Elsevier. 335 p.

Aufrère, L. (1931) *Le cycle morphologique des dunes.* Annales de Géographie, v. 40, no. 226, p. 362-385.

Ausich, W. I., and Bottjer, D. J. (1982) *Tiering in suspension-feeding communities on soft substrata throughout the Phanerozoic.* Science, v. 216, p. 173-174.

Aydin, A., and Johnson, A.R. (1978) *Development of faults as zones of deformation bands and as slip surfaces in sandstone.* Pure and Applied Geophysics, v.116, p.931-942.

Aydin, A., and Nuir, A. (1982) *Evolution of pull-apart basins and their scale independence.* Tectonics, vol. 1, p. 91-105.

B

Båth, Markus (1966) *Earthquake seismology.* Earth-Science Reviews, v. 1, p. 69-86.

Back, William (1966) *Hydrochemical facies and ground-water flow patterns in northern part of Atlantic Coastal Plain.* U. S. Geological Survey. Professional Paper 498-A. 42 p.

Bagnold, R. A. (1941) *The physics of blown sand and desert dunes.* London: Methuen. 265 p.

Bagnold, R. A. (1956) *The flow of cohesionless grains in fluids.* Royal Society of London. Philosophical Transactions, ser. A, v. 249, p. 235-297.

Bahat, D. (1991) *Tectonofractography.* Elsevier Press, 250 p.

Bailey, E. B. (1934) *West Highland tectonics: Loch Leven to Glen Roy.* Geological Society of London. Quarterly Journal, v. 90, p. 462-525.

Bailey, E.B. (1967) *James Hutton—the founder of modern geology.* New York, Elsevier, 161p.

Bailey, R. V., and Childers, M. O. (1977) *Applied mineral exploration with special reference to uranium.* Boulder, Colo.: Westview Press. 542 p.

Bailey, S. W. (1980) *Summary of recommendations of the AIPEA Nomenclature Committee.* Clay and Clay Minerals, 28, p. 73-78.

Bailey, S.W. (1980a) *Structure in layer silicates.* In: Brindley, G.W., and Brown, G., eds., *Crystal structures of clay minerals and their X-ray identification.* London, Mineralogical Society Monograph 5, p.1-123.

Bailey, S.W. (1980b) *Summary of recommendations of AIPEA Nomenclature Committee.* Clay Minerals, v.15, p.85-93.

Bailey, S.W. (1982) *Nomenclature for regular interstratification.* American Mineralogist, v.67, p.394-398.

Bailey, S. W. (1988) *Odinite, a new dioctahedral-trioctahedral Fe^{3+} rich clay mineral.* Clay Minerals, v. 23, p. 237-247.

Bailey, S. W. (1989) *Halloysite - a critical assessment.* Farmer, V. C. and Tardy, T., editors. Sciences Géologiques Mém. 86, Vol. II, Proceedings 9th International Clay Conference, Strasbourg, 89-98.

Bailey, S. W., Banfield, J. F., Barker, W. W., and Katchan, G. (1995) *Dozyite, a 1:1 regular interstratification of serpentine and chlorite.* Amer. Minerl. 80, p. 65-77.

Bailey, S. W., Brindley, G. W., Kodama, H., and Martin, R. T. (1979) *Report of the Clay Minerals Society nomenclature committee for 1977 and 1978.* Clays and Clay Minerals, v. 27, p. 238-239.

Bailey, S. W., Brindley, G. W., Kodama, H., and Martin, R. T. (1982) *Report of the Clay Minerals Society Nomenclature Committee for 1980-81: nomenclature for regular interstratifications.* Clay and Clay Minerals, v. 30, p. 76-78.

Baker, A. A. (1959) *Imprisoned rocks: a process of rock abrasion.* Victorian Naturalist, v. 76, p. 206-207.

Baker, B. B., et al. (1966) *Glossary of oceanographic terms. 2nd ed.* Washington, D.C.: U. S. Naval Oceanographic Office. 204 p. (U. S. Naval Oceanographic Office. Special Publication 35).

Baker, H. A. (1920) *On the investigation of the mechanical constitution of loose arenaceous sediments by the method of elutriation, with special reference to the Thanet beds of the southern side of the London Basin.* Geological Magazine, v. 57, p. 321-332, 363-370, 411-420, 463-467.

Baker, P. E., et al. (1974) *Petrology and geochemistry of Easter Island.* Contributions to Mineralogy and Petrology, v. 44, p. 85-100.

Baker

Baker, Victor (1973) *Paleohydrology and Sedimentology of Lake Missoula Flooding in Eastern Washington.* Special Paper—Geological Society of America, 144; Geological Society of America, Boulder, CO, 79 p.

Balk, Robert (1937) *Structural behavior of igneous rocks (with special reference to interpretations by H. Cloos and collaborators).* Geological Society of America. Memoir 5. 177p.

Ballard, T. J., and Conklin, Q. E. (1955) *The uranium prospector's guide.* New York: Harper. 251 p.

Bally, A. W. (1981) *Thoughts on the tectonics of folded belts.* In: McClay, K. R. and Price, N. J., eds., Thrust and nappe tectonics : Geological Society of London Special Publication 9, p. 13-32.

Bambach, R.K. (1983) *Ecospace utilization and guilds in marine communities through the Phanerozoic.* In: Tevesz, M.J.S., and McCall, P.L., eds., *Biotic interactions in Recent and fossil benthic communities.* New York, Plenum Press, p.719-146.

Bandel, K. (1977a) *Übergänge von der Perlmutter-Schicht zu prismatischen Schichttypen bei Mollusken.* Biomineralization Research Reports, v.9, p.28-47.

Bandel, K. (1977b) *Die Herausbildung der Schraubenschicht der Pteropoden.* Biomineralization Research Reports, v.9, p.73-85.

Bandel, K. (1979a) *The nacreous layer in the shells of the gastropod-family Seguenziidae and its taxonomic significance.* Biomineralization Research Reports, v.10, p.49-61.

Bandel, K. (1979b) *Transitions from simple structural types to the crossed-lamellar structure in gastropod shells.* Biomineralization Research Reports, v.10, p.9-38.

Bandel, K. (1981a) *The structure and formation of the siphuncular tube of* Quenstedtoceras *compared with that of* Nautilus *(Cephalopoda).* Neues Jahrbuch für Geologie und Paläontologie, Abhandlungen, v.161, no.2, p.153-171.

Bandel, K. (1981b) *Struktur der Molluskenschale im Hinblick auf ihre Funktion.* Paläontologische Kursbücher, München, v.1, p.25-48.

Bandel, K., and Keupp, H. (1985) *Analogous mineralizations in molluscs and calcareous dinoflagellate cysts.* Neues Jahrbuch für Geologie und Paläontologie, Monatshefte, no.2, p.65-86.

Banks, C. J. and Warburton, J. (1986) *"Passive-roof" duplex geometry in the frontal structures of the Kirthar and Sulaiman mountain belts.* Journal of Structural Geology, vol. 8, p. 229-238.

Banks, H.P. (1975) *Reclassification of the Psilophyta.* Taxon, v.24, p.401-413.

Banta, W. C. (1970) *The body wall of cheilostome Bryozoa: III, the frontal wall of Watersipora arcuata Banta, with a revision of the Cryptocystidea.* Journal of Morphology, v. 131, p. 37-56.

Barbier, M. G., and Viallix, J. R. (1973) *Sosie: a new tool for marine seismology.* Geophysics, v. 38, p. 673-683.

Barenblatt, G. J. (1962) *The mathematical theory of equilibrium cracks in brittle fracture.* Advanced Applied Mechanics, v. 7, p. 55-129.

Barker, D. S. (1970) *Compositions of granophyre, myrmekite, and graphic granite.* Geological Society of America. Bulletin, v. 81, p. 3339-3350.

Barnes, H. L. (1979) *Geochemistry of hydrothermal ore deposits. 2nd ed.* New York: Wiley. 798 p.

Barnett, J.A.M., Mortimer, J., Rippon, J.H., Walsh, J.J., and Watterson, J. (1987) *Displacement geometry in the volume containing a single normal fault.* AAPG Bulletin, v.71, no.8, p.925-937.

Baronnet, A. (1982) *Ostwald ripening in solution; the case for calcite and mica.* Estudios geol. 38, p. 185-198.

Barrell, Joseph (1907) *Geology of the Marysville mining district, Montana: a study of igneous intrusion and contact metamorphism.* U. S. Geological Survey. Professional Paper 57. 178 p.

Barrell, Joseph (1912) *Criteria for the recognition of ancient delta deposits.* Geological Society of America. Bulletin, v. 23, p. 377-446.

Barrell, Joseph (1913) *The Upper Devonian delta of the Appalachian geosyncline.* American Journal of Science, ser. 4, v. 36, p. 429-472.

Barrell, Joseph (1917) *Rhythms and the measurements of geologic time.* Geological Society of America. Bulletin, v. 28, p. 745-904.

Barth, T. F. W. (1948) *Oxygen in rocks: a basis for petrographic calculations.* Journal of Geology, v. 56, p. 50-60.

Barth, T. F. W. (1959) *Principles of classification and norm calculations of metamorphic rocks.* Journal of Geology, v. 67, p. 135-152.

Barth, T. F. W. (1962) *Theoretical petrology. 2nd ed.* New York: Wiley. 416 p.

Barton, D. C. (1916) *The geological significance and classification of arkose deposits.* Journal of Geology, v. 24, p. 417-449.

Barton, D. C. (1929) *The Eötvös torsion balance method of mapping geologic structure.* American Institute of Mining, Metallurgical and Petroleum Engineers. [Transactions, v.18], Geophysical prospecting, p. 416-479.

Barton, D.C. (1933) *Mechanics of formation of salt domes, with special reference to Gulf Coast salt domes of Texas and Louisiana.* AAPG Bulletin, v.17, no.9, p.1025-1083.

Barton, P. B., Jr., and Toulmin, Priestley, III (1964a) *Experimental data from the system Cu-Fe-S and their bearing on exsolution textures in ores.* Economic Geology, v. 59, p. 1241-1269.

Barton, P. B., Jr., and Toulmin, Priestley, III (1964b) *The electron-tarnish method for the determination of the fugacity of sulfur in laboratory sulfide systems.* Geochimica et Cosmochimica Acta. v. 28, p. 619-640.

Basaltic Volcanism Study Project (1981) *Basaltic volcanism on the terrestrial planets.* Pergamon Press: New York, 1286 p.

Bascom, Florence (1931) *Geomorphic nomenclature.* Science. v. 74, p. 172-173.

Bateman, A. M. (1959) *Economic mineral deposits. 2nd ed.* New York: Wiley. 916 p.

Bates, R. L. (1938) *Occurrence and origin of certain limonite concretions.* Journal of Sedimentary Petrology, v. 8, p. 91-99.

Bathurst, R. G. C. (1958) *Diagenetic fabrics in some British Dinantian limestones.* Liverpool and Manchester Geological Journal, v. 2, pt. 1, p. 11-36.

Bathurst, R. G. C. (1959) *The cavernous structure of some Mississippian (Stromatactis) reefs in Lancashire, England.* Journal of Geology, v. 67, p. 506-521.

Bathurst, R. G. C. (1966) *Boring algae, micrite envelopes and lithification of molluscan biosparites.* Geological Journal, v. 5, p. 15-32.

Bathurst, R. G. C. (1971) *Carbonate sediments and their diagenesis.* New York: Elsevier. 620 p. (Developments in sedimentology, 12.)

Batiza, R., and White, J.D.L. (2000) *Submarine lavas and hyaloclastite.* In: Sigurdsson, H., ed., *Encyclopedia of volcanoes.* New York, Academic Press, p.361-381.

Batten, D.J. (1996) *Colonial Chlorococcales.* In: Jansonius, J., and McGregor, D.C., eds., *Palynology. Principles and applications 1.* Dallas, Tex., American Association of Stratigraphic Palynologists Foundation, p.191-203.

Batten, D.J., and Grenfell, H.R. (1996) *Botryococcus.* In: Jansonius, J., and McGregor, D.C., eds., *Palynology. Principles and applications 1.* Dallas, Tex., American Association of Stratigraphic Palynologists Foundation, p.205-214.

Batten, R.L., and Dumont, M.P. (1976) *Shell ultrastructure of the Atlantidae (Heteropoda, Mesogastropoda),* Oxygyrus *and* Protatlanta, *with comments on* Atlanta inclinata. American Museum of Natural History Bulletin, v.157, art.4, p.263-310.

Batten, Roger (1966) *The lower Carboniferous gastropod fauna from the Hotwells limestone of Compton Martin, Somerset. Part I.* Palaeontographical Society. Monographs, v. 119, p. 1-52.

Baulig, Henri (1956) *Vocabulaire franco-anglo-allemand de géomorphologie.* Paris: Soc. Ed. Belles Lettres. 229 p.

Baulig, Henri (1957) *Peneplains and pediplains.* Translated from the French by C. A. Cotton. Geological Society of America. Bulletin, v. 68, p. 913-929.

Baumhoff, M. A., and Heizer, R. F. (1965) *Postglacial climate and archaeology in the Desert West.* In: Wright, H. E., Jr., and Frey, D. G., eds. The Quaternary of the United States, p. 697-707. Princeton, N.J.: Princeton Univ. Press. 922 p.

Bayly, Brian (1968) *Introduction to petrology.* Englewood Cliffs, N. J.: Prentice-Hall. 371 p.

Bé, A.W.H., MacClintock, C., and Currie, D.C. (1972) *Helical shell and growth of the pteropod* Cuvierina columnella *(Rang) (Mollusca, Gastropoda).* Biomineralization Research Reports, v.4, p.47-79.

Beadle, L.C., (1974) *The Inland Waters of Tropical Africa.* Longman, London, 365p.

Beales, F. W. (1958) *Ancient sediments of Bahaman type.* American Association of Petroleum Geologists. Bulletin, v. 42, p. 1845-1880.

Beasley, H. C. (1914) *Some fossils from the Keuper Sandstone of Alton, Staffordshire.* Liverpool Geological Society. Proceedings, v. 12, p. 35-39.

Beattie, P., et al. (1993) *Terminology for trace-element partitioning.* Geochimica et Cosmochimica Acta, v. 57, p. 1605-1606.

Becke, F. (1903) *Über Mineralbestand und Struktur der Kristallinischen Schiefer.* Comptes Rendus, Congrès Géologique International, 9th, Vienna.

Becke, F. (1913) *Über Minerallesband und Struktur der krystallinischen Schiefer.* Akademie der Wissenschaften in Wien. Denkschriften, v. 75, p. 1-53.

Becker, G. F. (1895) *Reconnaissance of the gold fields of the southern Appalachians.* U. S. Geological Survey. Annual Report, 16th, pt. 3, p. 251-331.

Becker, Hans (1932) *Report on some work on sediments done in Germany in 1931.* National Research Council. Committee on Sedimentation. Report, 1930-1932, p. 82-89. (National Research Council. Bulletin, no. 89).

Beerbower, J. R. (1964) *Cyclothems and cyclic depositional mechanisms in alluvial plain sedimentation.* Kansas. State Geological Survey. Bulletin 169, v. 1, p. 31-42.

Beerbower, J. R. (1968) *Search for the past; an introduction to paleontology. 2nd ed.* Englewood Cliffs, N.J.: Prentice-Hall. 512 p.

Begon, M., Harper, J.L., and Townsend, C.R. (1996) *Ecology: Individuals, populations, and communities.* Oxford, Blackwell Science.

Behrensmeyer, A. K., and Kidwell, S. M. (1985) *Taphonomy's contributions to paleobiology.* Paleobiology, v. 11, p. 105-119.

Bell, B. M. (1976) *A study of North American Edrioasteroidea.* New York State Museum and Science Service. Memoir 21, 447 p.

Bell, F. G. (1992) *Engineering in rock masses.* Butterworth-Heinemann Ltd., Oxford, p. 54.

Bell, Keith, ed. (1989) *Carbonatites, genesis and evolution.* London: Unwin Hyman Ltd., 618 p.

Bell, Robert (1894) *Pre-Paleozoic decay of crystalline rocks north of Lake Huron.* Geological Society of America. Bulletin, v. 5, p. 357-366.

Bell, T. H. and Etheridge, M. A. (1973) *Microstructure of mylonites and their terminology.* Lithos, vol. 6, p. 337-348.

Below, R. (1987) *Evolution und Systematik von Dinoflagellaten-Zysten aus der Ordnung Peridinales, 1, Allgemeine Grundfragen und Subfamilie Rhaetogonyaulacoidea (Familie Peridiniaceae).* Palaeontographica, Abteilung B: Palaeophytologie, v.205, no.1-6, p.1-164.

Benedict, L. G., et al. (1968) *Pseudovitrinite in Appalachian coking coals.* Fuel, v. 47, p. 135-143.

Beneo, E. (1955) *Les résultats des études pour la recherche pétrolifère en Sicile.* World Petroleum Congress. 4th, Rome, 1955. Proceedings, sec. 1, p. 109-124.

Berg, E.L. (1938) *Notes on catlinite and the Sioux Quartzite.* American Mineralogist, v.23, p.258-268.

Berger, A. (1995) *Modeling the astronomical theory of paleoclimates.* Journal of Coastal Research, Special Issue 17, p. 355-362.

Berger, A., and Loutre, M.F. (1991) *Insolation values for the climate of the last 10 million years.* Quaternary Science Reviews, v.10, p.297-318.

Berger, W. H. (1974) *Deep-sea sedimentation.* In: Burk, C. A., and others, eds. The geology of continental margins, p. 213-241. New York: Springer-Verlag.

Berggren, W.A., Christie-Blick, N., Aubry, M.-P., Carter, R.M., Hallam, A., Miller, K.G., Owen, D.E., Van Couvering, J.A., and Watkins, J.S. (2001) *A proposed bipartite sequence stratigraphic nomenclature.* In: Mancini, E.A., ed., *Sequence stratigraphic and allostratigraphic principles and concepts.* AAPG Hedberg Research Conference, p.16.

Berkey, C. P., and Morris, F. K. (1924) *Basin structures in Mongolia.* American Museum of Natural History. Bulletin, v. 51, p. 103-127.

Berkey, C.P., and Morris, F.K., (1927) *Geology of Mongolia: Natural history of central Asia. v.2,* Am. Museum Nat. Hist., New York, 475 pp.

Berkland, J. O., et al. (1972) *What is Franciscan?* American Association of Petroleum Geologists. Bulletin, v. 56, p. 2295-2302.

Bernard, B. (1970) *ABC's of infrared.* Indianapolis: Howard W. Sames & Co.

Berry, L. G., and Mason, Brian (1959) *Mineralogy: concepts, descriptions, determinations.* San Francisco: Freeman. 630 p.

Berry, W. B. N. (1966) *Zones and zones; with exemplification from the Ordovician.* American Association of Petroleum Geologists. Bulletin, v. 50, p. 1487-1500.

Berthelsen, Asger (1970) *Globulith: a new type of intrusive structure, exemplified by metabasic bodies in the Moss area, SE Norway.* Norges Geologiske Undersokelse, [Publikasjoner]. No. 266 (Årbok 1969).

Bertrand, Marcel (1892) *Les récents progrés de nos connaissances orogéniques.* Revue Générale des Sciences. Tome 3, no. 1, p. 5-12.

Best, J.L., and Bridge, J.S. (1992) *The morphology and dynamics of low-amplitude bed waves upon upper stage plane beds and the preservation of planar laminae.* Sedimentology, v.39, p.737-752.

Betz, F., Jr. (1984) *Military geology.* In: Finkl, C. W., Jr., ed., *The encyclopedia of applied geology* , New York: Van Nostrand Reinhold Co., p. 355-358.

Beus, A. A. (1978) *Lithogeochemistry.* Journal of Geochemical Exploration, v. 9, p. 110-111.

Beus, A. A., et al. (1962) *Albitized and greisenized granite (apogranite).* (In Russian.) Moscow: Izd. Akad. Nauk SSSR. 195 p.

Bhattacharji, Somdev (1958) *Theoretical and experimental investigations on crossfolding.* Journal of Geology, v. 66, p. 625-667.

Biddle, K. T., and Christie-Blick, N. (1985) *Glossary—strike-slip deformation, basin formation, and sedimentation.* In: Biddle, K. T., and Christie-Blick, N., Strike-slip deformation, basin formation, and sedimentation. Society of Economic Paleontologists and Mineralogists. Special Publication 37, p. 375-386.

Bieniawski, Z.T. (1993) *Classification of rock masses for engineering. The RMR system and future trends.* In: Hudson, J.A., ed., *Comprehensive rock engineering.* Pergamon Press, v. 3, p.553-573.

Bilibin, S.I. (1998) *Application of high-precision magnetic susceptibility logging for well-log analysis in oil deposits.* CSEG, CSPG, CWLS Joint Convention Abstracts, p.438-439.

Billings, B. H. (1963) *Optics.* In: American Institute of Physics Handbook. 2nd ed.

Billings, M. P. (1954) *Structural geology. 2nd ed.* Englewood Cliffs, N.J.: Prentice-Hall. 514 p.

Billings, M. P. (1972) *Structural geology. 3rd ed.* Englewood Cliffs, N.J.: Prentice-Hall. 606 p.

Birch, A. F. (1952) *Elasticity and constitution of the Earth's interior.* Journal of Geophysical Research, v. 57, p. 227-286.

Bird, J. B. (1957) *The physiography of Arctic Canada, with special reference to the area south of Parry Channel.* Baltimore: Johns Hopkins Press. 336 p.

Birkenmajer, Krzysztof (1958) *Oriented flowage casts and marks in the Carpathian flysch and their relation to flute and groove casts.* Acta geologica Polonica, v. 8, p. 139-148.

Birkenmajer, Krzyysztof (1959) *Classification of bedding in flysch and similar graded deposits.* Studia geologica Polonica, v. 3, p. 81-133.

Bissell, H. J. (1959) *Silica in sediments of the Upper Paleozoic of the Cordilleran area.* In: Ireland, H. A., ed. Silica in sediments—a symposium. Tulsa: Society of Economic Paleontologists and Mineralogists. 185 p. (Its Special Publication, no. 7, p. 150-185).

Bissell, H. J. (1964) *Ely, Arcturus, and Park City groups (Pennsylvanian-Permian) in eastern Nevada and western Utah.* American Association of Petroleum Geologists. Bulletin, v. 48, p. 565-636.

Bissell, H. J. (1964a) *Patterns of sedimentation in Pennsylvanian and Permian strata of part of the eastern Great Basin.* Kansas State Geological Survey Bulletin 169, v. 1, p. 43-56.

Bissell, H. J., and Chilingar, G. V. (1967) *Classification of sedimentary carbonate rocks.* In: Chilingar, G. V., and others, eds. Carbonate rocks; origin, occurrence and classification, p. 87-168. Amsterdam: Elsevier. (Developments in Sedimentology 9A, ch. 4).

Black, R. F. (1954) *Permafrost—a review.* Geological Society of America. Bulletin, v. 65, p. 839-856.

Black, R. F. (1966) *Comments on periglacial terminology.* Biuletyn peryglacjalny, no. 15, p. 329-333.

Blackwelder, Eliot (1931) *Desert plains.* Journal of Geology, v. 39, p. 133-140.

Blackwelder, Eliot (1931a) *Pleistocene glaciation in the Sierra Nevada and Basin Ranges.* Geological Society of America. Bulletin, v. 42, p. 865-922.

Blackwelder, R. E. (1967) *Taxonomy.* New York: Wiley. 698 p.

Blake, S. (1989) *Viscoplastic models of lava domes.* In: *IAVCEI Proceedings in Volcanology, v.2, Lava flows and domes.* Heidelberg, Springer Verlag, p.88-126.

Blakely, R.J. (1995) *Potential theory in gravity and magnetic applications.* Cambridge University Press, 441p.

Blank, H. R. (1951) *"Rock doughnuts," a product of granite weathering.* American Journal of Science, v. 249, p. 822-829.

Blatt, H., Middleton, G., and Murray, R. (1980) *Origin of sedimentary rocks, 2nd ed.* Englewood Cliffs, N.J., Prentice-Hall, 766p.

Blatt, Harvey, et al. (1972) *Origin of sedimentary rocks.* Englewood Cliffs, N.J.: Prentice-Hall. 634 p.

Blench, Thomas (1957) *Regime behaviour of canals and rivers.* London: Butterworths. 138 p.

Blissenbach, Erich (1954) *Geology of alluvial fans in semiarid regions.* Geological Society of America. Bulletin, v. 65, p. 175-189.

Bloom, A.L. (1978) *Geomorphology: A system analysis of late Cenozoic landforms.* Englewood Cliffs, N.J., Prentice-Hall, 510p.

Bloom, Arthur L. (1967) *Pleistocene Shorelines: A New Test of Isostasy.* Geological Society of America Bulletin, v.78, p.1477-1494, December 1967.

Bloos, G., and Page, K.N. (2002) *Global Stratotype Section and Point for base of the Sinemurian Stage (Lower Jurassic).* Episodes, v.25, no.1, p.22-28.

BNCG: British National Committee for Geography (1966) *Glossary of technical terms in cartography.* London: The Royal Society. 84 p.

Bøggild, O.B. (1930) *The shell structures of the Mollusks.* Det Kongelige Danske Videnskabernes Selskab Skrifter. Naturvidenskabelig og Mathematisk Afdeling, ser.9, v.2, p.231-326.

Boardman, R. S. (1971) *Mode of growth and functional morphology of autozooids in some recent and Paleozoic tubular bryozoa.* Smithsonian Contributions to Paleobiology, no. 8, 51 p.

Boardman, R. S., and Cheetham, A. H. (1969) *Skeletal growth, intracolony variation, and evolution in bryozoa—a review.* Journal of Paleontology, v. 43, no. 2, p. 205-233.

Boardman, R. S., and Cheetham, A. H. (1973) *Degrees of colony dominance in stenolaemate and gymnolaemate Bryozoa.* In: Boardman, R. S., and others, eds. Animal colonies, p. 121-200. Stroudsburg, Pa.: Dowden, Hutchinson & Ross.

Boggs, S. Jr. (1987) *Principles of sedimentology and stratigraphy.* Columbus, Ohio, Merrill, 784p.

Bogli, A. (1964) *Mischungskorrosion. ein Beitrag zum Verkarstunsproblem:* Erdkunde, v. 18, p. 83-92.

Borgia, A., Ferrari, L., and Pasquare, G. (1992) *Importance of gravitational spreading in the tectonic and volcanic evolution of Mount Etna.* Nature. v.357, p.231-235.

Bohor, B. F., and Triplehorn, D. M. (1993) *Tonsteins; altered volcanic-ash layers in coal-bearing sequences.* Geological Society of America, Special Paper 285, Geological Society of America, Inc., Boulder, Colorado, 44 p.

Bokman, J. (1956) *Terminology for stratification in sedimentary rocks.* Geological Society of America Bulletin, v. 67, no. 1, p. 125-126.

Bolli, H. M. (1969) *Report of working group for a biostratigraphic zonation of the Cretaceous and Cenozoic as a basis for correlation in marine geology.* International Union of Geological Sciences. Geology Newsletter, v. 1969, no. 3, p. 199-207.

Bonney, T. G. (1886) *The anniversary address of the President. "Metamorphic" rocks.* Geological Society of London. Quarterly Journal, v. 42 (Proc.), p. 38-115.

Boothroyd, J. C., and Hubbard, D. K. (1974) *Bedform development and distribution pattern, Parker and Essex estuaries, Massachusetts.* U. S. Army Corps of Engineers. Coastal Engineering Research Center. Miscellaneous Paper 1-74.

Bosellini, A. (1966) *Protointraclasts: texture of some Werfenian (Lower Triassic) limestones of the Dolomites (northeastern Italy).* Sedimentology, v. 6, p. 333-337.

Bosellini, Alfonso, and Ginsburg, R. N. (1971) *Form and internal structure of recent algal nodules (rhodolites) from Bermuda.* Journal of Geology, v. 79, p. 669-682.

Bostick, N. H. (1970) *Measured alteration of organic particles (phytoclasts) as an indicator of contact and burial metamorphism in sedimentary rocks.* Geological Society of America. Abstracts with Programs, v. 2, no. 2, p. 74.

Boswell, P. G. H. (1960) *The term graywacke.* Journal of Sedimentary Petrology, v. 30, p. 154-157.

Boucot, A.J. (1983) *Does evolution take place in an ecological vacuum? II.* Journal of Paleontology, v.57, no.1, pt.1, p.1-30.

Boucot, A. J. (1984) *Ecostratigraphy.* In: Stratigraphy quo vadis? Studies in Geology No. 16, p. 55-60. American Association of Petroleum Geologists.

Boulton, G. S. (1970) *On the deposition of subglacial and melt-out tills at the margins of certain Svalbard glaciers.* Journal of Glaciology, v. 9, p. 231-245.

Bouma, A. H. (1962) *Sedimentology of some flysch deposits; a graphic approach to facies interpretation.* Amsterdam: Elsevier. 168 p.

Bourcart, Jacques (1939) *Essai d'une définition de la vase des estuaires.* Académie des Sciences, Paris. Comptes rendus hebdomadaires des séances, v. 209, no. 14, p. 542-544.

Bourcart, Jacques (1941) *Essai de définition des vases des eaux douces.* Académie des Sciences, Paris. Comptes rendus hebdomadaires des séances, v. 212, no. 11, p. 448-450.

Bowen, N. L. (1941) *Certain singular points on crystallization curves of solid solution.* National Academy of Science. Proceedings, v. 27, p. 301-309.

Bowen, R. (1966) *Oxygen isotopes as climatic indicators.* Earth Science Reviews, v. 2, p. 199-224.

Boydell, H. C. (1926) *A discussion of metasomatism and the linear "force of growing crystals."* Economic Geology, v. 21, p. 1-55.

Boyer, S. E. (1976) *Formation of the Grandfather Mountain window duplex thrusting (abstract).* Geological Society of America, Annual Meeting, Denver. Abstracts with Programs, p. 788-789.

Boyer, S. E., and Elliott, D. (1982) *Thrust systems.* American Association of Petroleum Geologists. Bulletin, v. 66, p. 1196-1230.

Bradley, R.S. (1999) *Paleoclimatology: Reconstructing climates of the Quaternary.* San Diego, Calif., Academic Press, 610p.

Bradley, R.S., and Jones, P.D., eds. (1992) *Climate since A.D. 1500.* New York, Routledge, 679p.

Bradley, R.S., and Jones, P.D., eds. (1993) *"Little Ice Age" summer temperature variations: Their nature and relevance to recent global warming trends.* The Holocene, v.3, p.367-376.

Bradley, W. C. (1958) *Submarine abrasion and wave-cut platforms.* Geological Society of America. Bulletin, v. 69, p. 967-974.

Bradley, W. C. (1963) *Large-scale exfoliation in massive sandstones of the Colorado Plateau.* Geological Society of America. Bulletin, v. 74, p. 519-527.

Bradley, W. H. (1930) *The behavior of certain mud-crack casts during compaction.* American Journal of Science, ser. 5, v. 20, p. 136-144.

Bradley, W. H. (1931) *Origin and microfossils of the oil shale of the Green River Formation of Colorado and Utah.* U. S. Geological Survey. Professional Paper 168. 58 p.

Brady, J. B. (1975) *Chemical components and diffusion.* American Journal of Science, v. 275, p. 1073-1088.

Brady, N. C. (1990) *Nature and properties of soils.* MacMillan Publishing Co., p. 437-438.

Braga, J.C., Martin, J.M., and Riding, R. (1995) *Controls on microbial dome fabric development along a carbonate-siliclastic shelf-basin transect, Miocene, SE Spain.* Palaios, v.10, p.347-361.

Bramlette, M. N. (1946) *The Monterey Formation of California and the origin of its siliceous rocks.* U. S. Geological Survey. Professional Paper 212. 57 p.

Branco, W., and Fraas, E. (1905) *Das kryptovulcanische Becken von Steinheim.* Akademie der Wissenschaften, Berlin. Physikalische Abhandlungen, Jahrg. 1905, Abh. 1. 64 p.

Braunstein, Jules (1961) *Calciclastic and siliciclastic.* American Association of Petroleum Geologists. Bulletin, v. 45, p. 2017.

Bray, Warwick, and Trump, David (1970) *The American Heritage guide to archaeology.* New York: American Heritage Press. 269 p.

Breddin, H. (1956) *Die tektonishe Deformatin der Fossilie im Rheinischen Schiefergebirge.* Z. dt. geol. Ges., 106, 227-305.

Breed, C. S., and Grow, T. (1979) *Morphology and distribution of dunes in sand seas observed by remote sensing.* In: E. D. McKee, ed. A study of global sand seas. U. S. Geological Survey Professional Paper 1052, p. 253-302.

Breithaupt, August (1847) *Handbuch der mineralogie. Vol. 3.* Dresden & Leipzig: Arnoldische Buchhandlung. 496 p.

Brett, C.E., Ivany, L.C., and Schopf, K.M. (1996) *Coordinated stasis: An overview.* Palaeogeography, Palaeoclimatology, Palaeoecology, v.127, p.1-20.

Bretz, J H. (1929) *Valley deposits immediately east of the channeled scabland of Washington.* Journal of Geology, v. 37, p. 393-427, 505-541.

Bretz, J H. (1940) *Solution cavities in the Joliet Limestone of northeastern Illinois.* Journal of Geology, v. 50, p. 675-811.

Bretz, J H. (1942) *Vadose and phreatic features of limestone caverns.* Journal of Geology, v. 50, p. 675-811.

Brewer, R. and Pawluk, S. (1975) *Investigations of some soils developed in hummocks of the Canadian sub-Arctic and southern Arctic regions: 1. Morphology and micromorphology.* Canadian Journal of Soil Science, Vol.55, No.3, pp.301-319.

Brewer, Roy (1964) *Fabric and mineral analysis of soils.* New York: Wiley. 470 p.

Brewer, Roy, and Sleeman, J. R. (1960) *Soil structure and fabric: their definition and description.* Journal of Soil Science, v. 11, p. 172-185.

Briggs, R. P. (1974) *Map of overdip slopes that can affect landsliding in Allegheny County, Pennsylvania.* U. S. Geological Survey Map MF-543.

Brigham, A. P. (1901) *A text-book of geology.* New York: Appleton. 477 p.

Brindley, G. W., and Pedro, G. (1976) *Meeting of the nomenclature committee of AIPEA; Mexico City, July 12, 1975.* AIPEA Newsletter No. 12, p. 5-6.

Brindley, G. W., Zalba, P. E., and Bethke, C. M. (1983) *Hydrobiotite, a regular 1:1 interstratification of biotite and vermiculite layers.* Amer. Minerl. 68, p. 420-425.

Brinker, R.C., and Wolf, P.R. (1977) *Elementary surveying,* 6th ed. Thomas Y. Cromwell Company, 568p.

Brobst, D. A., and Pratt, W. P. (1973) *United States mineral resources.* U. S. Geological Survey. Professional Paper 820. 722 p.

Broecker, W., Bond, G., Klas, M., Clark, E., and McManus, J. (1992) *Origin of the northern Atlantic's Heinrich events.* Climate Dynamics, v.6, p.265-273.

Bromley, R. G. (1975) *Trace fossils at omission surfaces.* In: R. W. Frey, ed. The study of trace fossils, p. 399-428. New York: Springer-Verlag.

Brongniart, Alexandre (1823) *Macigno. Dictionaire des sciences naturelles, v. 27,* p. 297-504.

Brook, G. A., and Ford, D. C. (1978) *The origin of labyrinth and tower karst and the climatic conditions necessary for their development.* Nature, v. 275, p. 493-496.

Brookfield, M. E. (1977) *The origin of bounding surfaces in ancient aeolian sandstones.* Sedimentology, v. 24, p. 303-332.

Brooks, D.R., Caira, J.N., Platt, T.R., and Pritchard, M.R. (1984) *Principles and methods of phylogenetic systematics: a cladistics workbook.* University of Kansas Museum of Natural History Special Publication 12, 92p.

Brooks, H. K. (1954) *The rock and stone terms limestone and marble.* American Journal of Science, v. 252, p. 755-760.

Brown, A.R., Edwards, G.S., and Howard, R.E., (1987) *Fault slicing—a new approach to the interpretation of fault detail.* Geophysics, 52, 1319-1327.

Brown, G., and Weir, A. H. (1963) *The identity of rectorite and allevardite.* International Clay Conf., Stockholm, Proceedings, New York: Macmillan, p. 27-35.

Brown, J. S. (1943) *Suggested use of the word microfacies.* Economic Geology, v. 38, p. 325.

Brown, L. F. and Fisher, W. L. (1977) *Seismic-stratigraphic interpretation of depositional systems; examples from Brazilian rift and pull-apart basins.* In: C. E. Payton, ed., Seismic stratigraphy applications to hydrocarbon exploration, American Association of Petroleum Geologists Memoir 26, p. 213-248.

Brown, R. J. E., and Kupsch, W. O. (1974) *Permafrost terminology.* National Research Council of Canada. Technical memorandum no. 111. 62 p. (NRCC 14274.)

Brown, R. W. (1946) *Salt ribbons and ice ribbons.* Washington Academy of Sciences. Journal, v. 36, p. 14-16.

Brown, V. J., and Runner, D. G. (1939) *Engineering terminology; definitions of technical words and phrases.* 2nd ed. Chicago: Gillette. 439 p.

Brunton, C.H.C. (1972) *The shell structure of chonetacean brachiopods and their ancestors.* British Museum (Natural History) Geology Bulletin, v.21, no.1, p.1-26.

Bruun, P. (1962) *Sea-level rise as a cause of shore erosion.* Journal of the Waterways and Harbors Division, American Society of Civil Engineers, 88 (WW1), p.117-130.

Bryan, A. B. (1937) *Gravimeter design and operation.* Geophysics, v. 2, no. 4, p. 301-308.

Bryan, Kirk (1920) *Origin of rock tanks and charcos.* American Journal of Science, 4th ser., v. 50, p. 186-206.

Bryan, Kirk (1923a) *Erosion and sedimentation in the Papago country, Arizona, with a sketch of the geology.* U. S. Geological Survey. Bulletin 730-B, p. 19-90.

Bryan, Kirk (1923b) *Geology and ground-water resources of Sacramento Valley, California.* U. S. Geological Survey. Water-supply Paper 495. 285 p.

Bryan, Kirk (1934) *Geomorphic processes at high altitude.* Geographical Review, v. 24, p. 655-656.

Bryan, Kirk (1940) *Gully gravure, a method of slope retreat.* Journal of Geomorphology, v. 3, p. 89-107.

Bryan, Kirk (1946) *Cryopedology, the study of frozen ground and intensive frost-action, with suggestions on nomenclature.* American Journal of Science, v. 244, p. 622-642.

Bryan, Kirk (1951) *The erroneous use of tjaele as the equivalent of perennially frozen ground.* Journal of Geology, v. 59, p. 69-71.

Bryant, D. G. (1968) *Intrusive breccias associated with ore, Warren (Bisbee) mining district, Arizona.* Economic Geology, v. 63, p. 1-12.

BSI: British Standards Institution (1964) *Glossary of mining terms; section 5:* Geology. London: British Standards Institution. 18 p. (B.S. 3618).

Bucher, W. H. (1919) *On ripples and related sedimentary surface forms and their paleogeographic interpretation.* American Journal of Science, ser. 4, v. 47. p. 149-210, 241-269.

Bucher, W. H. (1932) *"Strath" as a geomorphic term.* Science, v. 75, p. 130-131.

Bucher, W. H. (1933) *The deformation of the Earth's crust.* Princeton: Princeton University Press. 518 p.

Bucher, W. H. (1952) *Geologic structure and orogenic history of Venezuela; text to accompany the author's geologic tectonic map of Venezuela.* Geological Society of America. Memoir 49. 113 p.

Bucher, W. H. (1955) *Deformation in orogenic belts.* Geological Society of America. Special Paper 62, p. 343-368.

Bucher, W. H. (1963) *Cryptoexplosion structures caused from without or from within the Earth? ("astroblemes" or "geoblemes"?).* American Journal of Science, v. 261, p. 597-649.

Bucher, W. H. (1965) *Role of gravity in orogenesis.* Geological Society of America. Bulletin, v. 67, p. 1295-1318.

Buckland, William (1817) *Description of the Paramoudra, a singular fossil body that is found in the Chalk of the North of Ireland; with some general observations upon flints in chalk, tending to illustrate the history of their formation.* Geological Society of London. Transactions, v. 4, p. 413-423.

Buckland, William (1829) *On the formation of the Valley of Kingsclere and other valleys by the elevation of the strata that enclose them; and on the evidences of the original continuity of the basins of London and Hampshire.* Geological Society of London. Transactions, ser. 2, v. 2, p. 119-130.

Buckland, William (1829a) *On the discovery of coprolites, or fossil faeces, in the Lias at Lyme Regis, and in other formations.* Geological Society of London. Transactions, ser. 2, v. 3, p. 223-236.

Buckman, S. S. (1893) *The Bajocian of the Sherborne district; its relation to subjacent and superjacent strata.* Geological Society of London. Quarterly Journal, v. 49, p. 479-522.

Buckman, S. S. (1902) *The term "hemera".* Geological Magazine, dec. 4, v. 9, p. 554-557.

Bujak, J.P., and Davies, E.H. (1983) *Modern and fossiil Peridiniianeae.* American Association of Styratigraphic Palynologists, Contribution Series no.13, 203p.

Bujak, J.P., Downie, C., Eaton, G.L., and Williams, G.L. (1980) *Taxonomy of some Eocene dinoflagellate cysts and acritarchs from the Eocene of southern England.* Special Papers in Palaeontology, no.24, p.26-36.

Bull, W. B. (1975) *Allometric change of landforms.* Geological Society of America. Bulletin, v. 86, p. 1489-1498.

Bullard, E.C., (1947) *The time necessary for a borehole to attain temperature equilibrium.* Mon. Not. Roy. Astr. Soc., v.5, p.127-120.

Burollet, Pierre (1975) *Tectonique en radeaux en Angola [Raft tectonics in Angola].* Bulletin de la Sociéte Géologique de France, v. 17, p. 503-504.

Burst, J. F. (1965) *Subaqueously formed shrinkage cracks in clay.* Journal of Sedimentary Petrology, v. 35, p. 348-353.

Burst, J. F. and Hughes, R. E. (1994) *Clay-based ceramic raw materials.* In: D. D. Carr, editor, 6th edition, Industrial minerals and rocks : New York: American Institute of Mining, Metallurgical, and Petroleum Engineers, p. 317-324.

Burt, D. M. (1974) *Concepts of acidity and basicity in petrology; the exchange operator approach (abstract).* Geological Society of America. Abstracts, v. 6, p. 674-676.

Burt, F. A. (1928) *Melikaria: vein complexes resembling septaria veins in form.* Journal of Geology, v. 36, p. 539-544.

Bush, D.M., Pilkey, O.H., and Neal, W.J. (1996) *Living by the rules of the sea.* Durham, N.C., Duke University Press, 179p.

Butler, J.C., ed., (1995) *Internet resources for the geosciences.* Computers & Geosciences, v.21, no. 6, p.727-824.

Buwalda, J. P. (1937) *Shutterridges, characteristic physiographic features of active faults.* Geological Society of America. Proceedings, 1936, p. 307.

C

Cady, G. H. (1921) *Coal resources of District IV.* Illinois State Geological Survey. Cooperative Mining Series, Bulletin 26. 247 p.

Cairnes, D. D. (1912) *Some suggested new physiographic terms.* American Journal of Science, v. 34, p. 75-87.

Calkins, F. C. (1941) *"Band," "layer," and some kindred terms*. Economic Geology, v. 36, no. 3, p. 345-349.

Callomon, J. H. (1963) *Sexual dimorphism in Jurassic ammonoides*. Leicester Literary and Philosophical Society. Transactions, v. 57, p. 21-56.

Callomon, J. H. (1965) *Notes on Jurassic stratigraphical nomenclature*. Carpatho-Balkan Geological Association. 7th Congress, Sofia, Sept. 1965. Reports, pt. 2, v. 1, p. 81-85.

Cameron, W. S. (1972) *Comparative analyses of observations of lunar transient phenomena*. Icarus, v. 16, p. 339-387.

Campbell, C. V. (1966) *Truncated wave-ripple laminae*. Journal of Sedimentary Petrology, v. 36, p. 825-828.

Campbell, C. V. (1967) *Lamina, laminaset, bed and bedset*. Sedimentology, v. 8, p. 7-26.

Campbell, J. F. (1865) *Frost and fire; natural engines, tool-mark & chips with sketches taken at home and abroad by a traveller*. Vol. 2. Philadelphia: Lippincott. 519 p.

Campbell, K. M., et al. (1977) *Langmuir circulation as a factor in the formation of depositional beach cusps*. In: Tanner, W. F., ed. Coastal sedimentology, p. 245-262. Tallahassee: Department of Geology, Florida State University.

Campbell, M. R. (1896) *Drainage modifications and their interpretation*. Journal of Geology, v. 4, p. 567-581, 657-678.

Canada Soil Survey Committee, (1978) *The Canadian System of Soil Classification*. Agriculture Canada, Ottawa, 164p.

Capot-Rey, R. (1945) *Dry and humid morphology in the Western Erg*. Geographical Review, v. 35, p. 391-407.

Carey, A. E., and Oliver, F. W. (1918) *Tidal lands; a study of shore problems*. London: Blackie. 284 p.

Carey S. (2000) *Volcaniclastic sedimentation around island arcs*. In: Sigurdsson, H., ed., *Encyclopedia of volcanoes*. New York, Academic Press, p.627-642.

Carey, S. W. (1958) *A tectonic approach to continental drift*. In: Carey, S. W., convener. Continental drift: a symposium. Hobart: University of Tasmania, Geology Dept. p. 177-355.

Carey, S. W., convener (1963) *Syntaphral tectonics and diagenesis; a symposium*. Hobart: University of Tasmania, Geology Dept. 190 p.

Carmichael, I. S. E., et al. (1974) *Igneous petrology*. New York: McGraw-Hill. 739 p.

Carozzi, A. V. (1957) *Micro-mechanisms of sedimentation in epicontinental environment*. Geological Society of America. Bulletin, v. 68, p. 1706-1707.

Carozzi, A.V. (1960) *Microscopic sedimentary petrography*. New York, John Wiley and Sons, 458p.

Carozzi, A.V. (1996) *New historical data on H.B. de Saussure's introduction of the terms* moraine *(1764),* roches moutonnées *(1776), and* sérac *(1788) to geology*. Journal of Geoscience Education, v.44, no.2, p.169-174.

Carozzi, A. V., and Textoris, D. A. (1967) *Paleozoic carbonate microfacies of the eastern stable interior (U.S.A.)*. Leiden: Brill. 41 p. (International sedimentary petrographical series, v. 11.)

Carozzi, Albert Victor (1996) *New historical data on H.-B. de Saussure's introduction of the terms moraine (1764), roches moutonnées (1776), and serac (1788) to geology*. Journal of Geoscience Education, v. 44, p. 169.

Carr, M. H., et al. (1984) *The geology of the terrestrial planets*. National Aeronautics and Space Administration. NASA SP-469. 317 p.

Carriker, M.R., Palmer, R.E., and Prezant, R.S. (1980) *Functional ultramorphology of the dissoconch valves of the oyster* Crassostrea virginica. National Shellfisheries Association, Proceedings, v.70, no.2, p.139-183.

Carroll, Dorothy (1939) *Movement of sand by wind*. Geological Magazine, v. 76, p. 6-23.

Carter, J.G. (1976) *The structural evolution of the bivalve shell, with notes on the phylogenetic significance of crossed lamellar structures*. New Haven, Conn., Yale University doctoral dissertation, v.1, 255p.

Carter, J.G. (1980a) *Environmental and biological controls of bivalve shell mineralogy and microstructure*. In: Rhoads, D.C., and Lutz, R.A., eds., *Skeletal growth of aquatic organisms*, chap.2. New York, Plenum, p.69-113.

Carter, J.G. (1980b) *Guide to bivalve shell microstructures*. In: Rhoads, D.C., and Lutz, R.A., eds., *Skeletal growth of aquatic organisms*, Appendix 2, Part B. New York, Plenum, p.645-673.

Carter, J.G., and Tevesz, M.J.S. (1978) *Shell microstructure of a Middle Devonian (Hamilton Group) bivalve fauna from central New York*. Journal of Paleontology, v.52, p.859-880.

Carter, J.G., ed. (1990) *Skeletal biomineralization: Patterns, processes, and evolutionary trends*, v.1. New York, Van Nostrand Reinhold, 832p.

Cas, R.A.F., and Wright, J.V. (1987) *Volcanic successions, modern and ancient*. Winchester, Mass., Allen & Unwin, 528p.

Cashman, P.B. (1996) *Melanosclerites*. In: Jansonius, J., and McGregor, D.C., eds., *Palynology. Principles and applications 1*. Dallas, Tex., American Association of Stratigraphic Palynologists Foundation, p.365-371.

Cassidy, W. A. (1968) *Meteorite impact structures at Campo del Cielo, Argentina*. In: French, B. M., and Short, N. M., eds. Shock metamorphism of natural materials, p. 117-128. Baltimore: Mono Book Corp. 644 p.

Castens-Seidell, B., (1984) *The Anatomy of a Modern Marine Siliciclastic Sabkha in a Rift Valley Setting: North-west Gulf of California Tidal Flats*. Baja, California, Mexico, unpublished Ph.D. Dissertation, Johns Hopkins University, Baltimore, Maryland, 386p.

Caster, K. E. (1934) *The stratigraphy and paleontology of northwestern Pennsylvania. Pt. 1*. Bulletins of American Paleontology, v. 21, no. 71. 185 p.

Castradori, D., Rio, D., Hilgen, F.J., and Lourens, L.J. (1998) The *Global Standard Stratotype Section and Point (GSSP) of the Piacenzian Stage (middle Pliocene)*. Episodes, v.21, no.2, p.88-93.

Cayeux, Lucien (1929) *Les roches sedimentaires de France; roches siliceuses*. Paris: Imprimerie Nationale. 774 p.

Cecil, C. B., et al. (1985) *Paleoclimate controls on late Paleozoic sedimentation and peat formation in the central Appalachian Basin (U.S.A.)*. International Journal of Coal Geology, v. 5, p. 195-230.

CERC: U. S. Army. Coastal Engineering Research Center (1966) *Shore protection, planning and design. 3rd ed.* Washington, D.C.: Government Printing Office. 580 p. (Its Technical report no. 4.)

Cermák, V., and Haenel, R. (1988) *Geothermal maps*. In: Haenel, R., Rybach, L., and Stegena, L., eds. Handbook of terrestrial heat-flow density determinations. Kluwer Acad. Publ., Dordrecht, p.261-300.

Chadwick, G. H. (1931) *Storm rollers*. Geological Society of America. Bulletin, v. 42, p. 242.

Chadwick, G. H. (1939) *Geology of Mount Desert Island, Maine*. American Journal of Science, v. 237, p. 355-363.

Chadwick, G. H. (1948) *Ordovician "dinosaur-leather" markings (exhibit)*. Geological Society of America. Bulletin, v. 59, p. 1315.

Chafetz, H.S., and Folk, R.L. (1984) *Travertine: Depositional morphology and the bacterially constructed constituents*. Journal of Sedimentary Petrology, v.54, p.289-316.

Chafetz, H.S., and Guidry, S.A. (1999) *Bacterial shrubs, crystal shrubs, and ray-crystal shrubs: Bacterial vs. abiotic precipitation*. Sedimentary Geology, v.126, p.57-74.

Challinor, J., and Williams, K. E. (1926) *On some curious marks on a rock surface*. Geological Magazine, v. 63, no. 746, p. 341-343.

Challinor, John (1962) *A dictionary of geology. 1st ed.* New York: Oxford University Press. 235 p.

Challinor, John (1967) *A dictionary of geology. 3rd ed.* New York: Oxford University Press. 298p.

Challinor, John (1978) *A dictionary of geology. 5th ed.* New York: Oxford University Press. 365p.

Chamberlin, T. C. (1879) *Annual report of the Wisconsin Geological Survey for the year 1878*. Madison: Wisconsin Geological Survey. 52 p.

Chamberlin, T. C. (1883) *Terminal moraine of the second glacial epoch*. U. S. Geological Survey. Annual Report, 3rd, p. 291-402.

Chamberlin, T. C. (1888) *The rock-scorings of the great ice invasions*. U. S. Geological Survey. Annual Report, 7th, v. 3, p. 147-248.

Chamberlin, T. C. (1893) *The horizon of drumlin, osar and kame formation*. Journal of Geology, v. 1, p. 255-267.

Chamberlin, T. C. (1894a) *Pseudo-cols*. Journal of Geology, v. 2, p. 205-206.

Chamberlin, T. C. (1894b) *Proposed genetic classification of Pleistocene glacial formations*. Journal of Geology, v. 2, p. 517-538.

Chamberlin, T. C. (1897) *The method of multiple working hypotheses*. Journal of Geology, v. 5, p. 837-848.

Chang, K. H. (1975) *Unconformity-bounded stratigraphic units*. Geological Society of America Bulletin, v. 86, no. 11, p. 1544-1552.

Chao, E. C.-T. (1967) *Shock effects in certain rock-forming minerals*. Science, v. 156, p. 192-202.

Chao, E. C.-T. (1967a) *Impact metamorphism*. In: Abelson, P. H., ed. Researches in geochemistry, v. 2, p. 204-233. New York: Wiley. 663 p.

Chapin, D.A. (1996) *The theory of the Boguer gravity anomaly. A tutorial*. The Leading Edge, v.15, p.361-363.

Chapin, D.A. (1998) *Gravity instruments: past, present, future*. The Leading Edge, v.17, p.100-112.

Charlesworth, J. K. (1957) *The Quaternary era, with special reference to its glaciation*. London: Arnold. 2 vols., 1700 p.

Chayes, F. (1956) *Petrographic modal analysis*. New York: Wiley. 163 p.

Chayes, Felix (1964) *A petrographic distinction between Cenozoic volcanics in and around the open oceans*. Journal of Geophysical Research, v. 69, p. 1573-1588.

Chayes, Felix (1966) *Alkaline and subalkaline basalts*. American Journal of Science, v. 264, p. 128-145.

Chebotarev, I. I. (1955) *Metamorphism of natural waters in the crust of weathering. [Pt.] 1-3*. Geochimica et cosmochimica acta, v. 8, p. 22-48, 137-170, 198-212.

Cheel, R.J. (1984) *Heavy mineral shadows, a new sedimentary structure formed under upper flow regime conditions: its directional and hydraulic significance*. Journal of Sedimentary Petrology, v.54, p.1173-1180.

Cheel, R.J., and Middleton, G.V. (1986) *Horizontal lamination formed under upper flow regime plane bed conditions*. Journal of Geology, v.94, p.489-504.

Cheeney, R.F., (1983) *Statistical Methods in Geology*. George Allen & Unwin, London, 169p.

Chester, F. M., Friedman, M. and Logan, J. (1985) *Foliated cataclasites*. Tectonophysics, v. 111, p. 139-146.

Chilingar, G. V. (1957) *Classification of limestones and dolomites on basis of Ca/Mg ratio*. Journal of Sedimentary Petrology, v. 27, p. 187-189.

Chilingar, G. V., et al. (1967) *Diagenesis in carbonate rocks*. In: Larsen, Gunnar, and Chilingar, G. V., eds. Diagenesis in sediments, p. 179-322. Amsterdam: Elsevier. 551. (Developments in sedimentology 8.)

Chilingarian, G. V., and Wolf, K. H., eds. (1975, 1976) *Compaction of coarse-grained sediments.* Amsterdam: Elsevier. Vol. 1, 548 p.; vol. 2, 808 p. (Developments in sedimentology, vols. 18A and 18B.)
Chinner, G. A. (1966) *The significance of the aluminum silicates in metamorphism.* Earth Science Reviews, v. 2.
Chitwood, Lawrence (1994) *Inflated basaltic lava—examples of processes and landforms from central and southeast Oregon.* Oregon Geology, v. 56, no. 1, p. 11-21.
Chlupáè, I., and Oliver, W.A., Jr. (1989) *Decision on the Lochkovian-Pragian boundary stratotype (Lower Devonian).* Episodes, v.12, no.2, p.109-113.
Choquette, P. W. (1955) *A petrographic study of the "State College" siliceous oölite.* Journal of Geology, v. 63, p. 337-347.
Choquette, P. W., and Pray, L. C. (1970) *Geologic nomenclature and classification of porosity in sedimentary carbonates.* American Association of Petroleum Geologists. Bulletin, v. 54, p. 207-250.
Chow, Van Te, chairman (1957) *Report of the Committee on Runoff, 1955-1956.* American Geophysical Union. Transactions, v. 38, p. 379-384.
Chow, Van Te, ed. (1964) *Handbook of applied hydrology; a compendium of water-resources technology.* New York: McGraw-Hill. 1418 p.
Christner, B.C., Kvitko, B.H.,II, and Reeve, J.N. (2003) *Molecular identification of bacteria and eukarya inhabiting an Antarctic cryoconite hole.* Extremophiles, v.7, p.177-183.
Church, W. R. (1968) *Eclogites.* In: Hess, H. H., and Poldervaart, Arie, eds. Basalts—the Poldervaart treatise on rocks of basaltic composition, v. 2, p. 755-798. New York: Interscience.
Cisne, J.L., and Rabe, B.D. (1978) *Coenocorrelation: Gradient analysis of fossil communities and its applications in stratigraphy.* Lethaia, v.11, no.4, p.341-364.
Clapp, C. H. (1913) *Contraposed shorelines.* Journal of Geology, v. 21, p. 537-540.
Clark, K. F. (1972) *Stockwork molybdenum deposits in the western Cordillera of North America.* Economic Geology, v. 67, p. 731-758.
Clarke, D.B. (1992) *Granitoid rocks* (Topics in Earth Sciences, 7). London, Chapman and Hall, 283p.
Clarke, F. W. (1908) *The data of geochemistry.* U. S. Geological Survey Bulletin 330. 716 p.
Clarke, J. I. (1966) *Morphometry from maps.* In: Dury, G. H., ed. Essays in geomorphology, p. 235-274. New York: American Elsevier. 404 p.
Clarke, J. M. (1911) *Observations on the Magdalen Islands.* New York State Museum. Bulletin 149, p. 134-155.
Clarke, J. M. (1918) *Strand and undertow markings of Upper Devonian time as indications of the prevailing climate.* New York State Museum. Bulletin 196, p. 199-238.
Clarke, J. W. (1958) *The bedrock geology of the Danbury quadrangle.* Connecticut State Geological and Natural History Survey. Quadrangle report no. 7., 47 p.
Claus, George, and Nagy, Bartholomew (1961) *A microbiological examination of some carbonaceous chondrites.* Nature, v. 192, p. 594-596.
Cleland, H. F. (1910) *North American natural bridges, with a discussion of their origin.* Geological Society of America. Bulletin, v. 21, p. 313-338.
Cleland, H. F. (1925) *Geology, physical and historical.* New York: American Book Co. 718 p.
Cleland, H. F. (1929) *Geology; physical and historical.* New York: American Book Co. 718 p.
Clifton, H. E. (1985) *The sedimentologic relevance of convulsive geologic events.* Geological Society of America. Abstracts with Programs, v. 17, p. 549.
Cloos, Ernst (1946) *Lineation, a critical review and annotated bibliography.* Geological Society of America. Memoir 18. 122 p.
Cloos, Hans (1939) *Hebung, Spaltung, Vulkanismus; Elemente einer geometrischen Analyse indischer Grossformen.* Geologische Rundschau, v. 30, no. 4a, p. 405-527.
Cloos, Hans (1941) *Bau und Tätigkeit von Tuffschloten: Untersuchungen an dem schwäbischen Vulkan.* Geologische Rundschau, v. 32, p. 709-800.
Close, U., and McCormick, E. (1922) *Where the mountains walked.* National Geographic Magazine, v. 41, p. 445-464.
Cloud, P. E., Jr. (1957) *Nature and origin of atolls; a progress report. Pacific Science Congress. 8th, Philippines, 1953.* Proceedings, v. 3A, p. 1009-1036.
Cloud, P. E., Jr. (1974) *Rubey conference on crustal evolution.* Science, v. 183, p. 878-881.
Cloud, P. E., Jr., and Barnes, V. E. (1957) *Early Ordovician sea in central Texas.* Geological Society of America. Memoir 67, v. 2, p. 163-214.
Cloud, P. E., Jr., et al. (1943) *Stratigraphy of the Ellenburger Group in central Texas; a progress report.* University of Texas. Publication 4301, p. 133-161.
Clough, C. T., et al. (1909) *The cauldron-subsidence of Glen Coe, and the associated igneous phenomena.* Geological Society of London. Quarterly Journal, v. 65, p. 611-678.
Coates, D. R. (1966) *Glaciated Appalachian Plateau: till shadows on hills.* Science, v. 152, p. 1617-1619.
Coats, R. R. (1968) *The Circle Creek Rhyolite, a volcanic complex in northern Elko County, Nevada.* Geological Society of America. Memoir 116, p. 69-106.
Cobbold, P. R., and Quinquis, H. (1980) *Development of sheath folds in shear regions.* J. Struct. Geol., v. 2, p. 119-126.
Cocks, L.R.M. (1985) *The Ordovician-Silurian boundary.* Episodes, v.8, no.2, p.98-100.
Cody, R.C., (1979) *Lenticular gypsum: occurrences in nature and experimental determination of the effects of soluble green plant material on its formation.* Journal of Sedimentary Petrology, v.49, p.1015-1028.
Coffey, G. N. (1909) *Clay dunes.* Journal of Geology, v. 17, p. 754-755.
Coffin, M.F., and Eldholm, O. (1994) *Large igneous provinces: Crustal structure, dimensions, and external consequences.* Reviews of Geophysics, v.32, p.1-36.
Cohee, G. V. (1962) *Stratigraphic nomenclature in reports of the U. S. Geological Survey.* Washington: U. S. Geological Survey. 35 p.
Cohen, A.S. (2003) *Paleolimnology: The history and evolution of lake systems.* Oxford, U.K., Oxford University Press, 500p.
Cohen, Robert M., and Mercer, James W. (1993) *DNAPL site evaluation.* Boca Raton, FL: C. K. Smoley. 314 p.
Colbath, G.K., and Grenfell, H.R. (1995) *Review of biological affinities of Paleozoic acid-resistant, organic-walled eukaryotic algal microfossils (including "acritarchs").* Review of Palaeobotany and Palynology, v.86, p.287-314.
Colbeck, S. C. (1986) *Classification of seasonal snow cover crystals.* Water Resources Research, v. 22, p. 595-705.
Cole, J. P., and King, C. A. M. (1968) *Quantitative geography.* London: Wiley. 692 p.
Coleman, R. G. (1971) *Plate tectonic emplacement of upper mantle peridotites along continental edges.* Journal of Geophysical Research, v. 76, p. 1212-1222.
Coleman, R. G. (1977) *Ophiolites.* New York: Springer-Verlag. 229 p.
Coleman, R. G., et al. (1965) *Eclogites and eclogites—their differences and similarities.* Geological Society of America. Bulletin, v. 76, p. 483-508.
Collet, L. W. (1927) *The structure of the Alps.* London: Edward Arnold. 289 p.
Collie, G. L. (1901) *Wisconsin shore of Lake Superior.* Geological Society of America. Bulletin, v. 12, p. 197-216.
Collins, P. L. F. (1979) *Gas hydrates in CO_2 -bearing fluid inclusions and the use of freezing data for estimation of salinity.* Economic Geology, v. 74, p. 1435-1444.
Collinson, J. C. (1970) *Bedforms of the Tana River, Norway.* Geografiska Annaler, v. 52, p. 31-56.
Collinson, J. D. (1986) *Deserts.* In: H. G. Reading, ed. Sedimentary environments and facies, p. 95-112. Oxford: Blackwell.
Colp, J. L. (1967) *Terradynamics: a study of projectile penetration of natural earth materials.* Geological Society of America. Special Paper 115, p. 38.
Colton, R. B. (1978) *Geologic map of the Boulder-Fort Collins-Greeley area, Colorado.* U. S. Geological Survey Map I-855-G.
Compton, R. R. (1962) *Manual of field geology.* New York: Wiley. 378 p.
Comstock, T. B. (1878) *An outline of general geology, with copious references.* Ithaca: University Press. 82 p.
Conant, L. C., et al. (1976) *Sediment-filled pots in upland gravel of Maryland and Virginia.* U. S. Geological Survey. Journal of Research, v. 4, p. 353-358.
Coney, P. J. (1970) *The geotectonic cycle and the new global tectonics.* Geological Society of America. Bulletin, v. 81, p. 739-747.
Coney, P. J. (1980) *Cordilleran metamorphic complexes: an overview.* In: M. L. Crittenden et al., eds. Cordilleran metamorphic core complexes. Geological Society of America. Memoir 153.
Coney, P. J., et al. (1980) *Cordilleran suspect terranes.* Nature, v. 288, p. 329-333.
Conkin, J. E., and Conkin, B. M. (1973) *The paracontinuity and the determination of the Devonian-Mississippian boundary in the type Lower Mississippian area of North America.* University of Louisville Studies in Paleontology and Stratigraphy, no. 1, 36 p.
Conkin, J. E., and Conkin, B. M. (1975) *The Devonian-Mississippian and Kinderhookian-Osagean boundaries in the east-central United States are paracontinuities.* University of Louisville. Studies in Paleontology and Stratigraphy, no. 4. 52 p.
Conybeare, C. E. B., and Crook, K. A. W. (1968) *Manual of sedimentary structures.* Australia. Bureau of Mineral Resources, Geology and Geophysics. Bulletin 102. 327 p.
Conybeare, W. D., and Phillips, William (1822) *Outlines of the geology of England and Wales, with an introductory compendium of the general principles of that science, and comparative views of the structure of foreign countries. Part 1.* London: William Phillips. 470 p.
Cook, J. H. (1946) *Kame complexes and perforation deposits.* American Journal of Science, v. 244, p. 573-583.
Cook, P. L. (1965) *Notes on Cupuladriidae (Polyzoa, Anasca).* British Museum (Natural History). Bulletin, Zoology, v. 13, p. 159-160.
Cooke, R. U., and Warren, Andrew (1973) *Geomorphology in deserts.* London: Batsford. 374 p.
Coombs, D. S. (1960) *Lower grade mineral facies in New Zealand.* International Geological Congress. 21st, Copenhagen, 1960. Report, part 13, p. 339-351.
Coombs, D. S. (1961) *Some recent work on the lower grades of metamorphism.* The Australian Journal of Science, v. 24, no. 5, p. 203-215.
Coombs, D. S., et al. (1959) *The zeolite facies.* Geochimica et Cosmochimica Acta, v. 17, p. 53-107.
Cooper, B. N. (1945) *Industrial limestones and dolomites in Virginia; Clinch Valley district.* Virginia Geological Survey. Bulletin 66. 259 p.
Cooper, B. N., and Cooper, G. A. (1946) *Lower Middle Ordovician stratigraphy of the Shenandoah Valley, Virginia.* Geological Society of America. Bulletin, v. 57, p. 35-113.
Cooper, J. R. (1943) *Flow structure in the Berea Sandstone and Bedford Shale of central Ohio.* Journal of Geology, v. 51, p. 190-203.

Cooper, R.A., Nowlan, G.S., and Williams, S.H. (2001) *Global Stratotype Section and Point for base of the Ordovician System.* Episodes, v.24, no.1, p.19-28.

Cooper, W. S. (1958) *Terminology of post-Valders time.* Geological Society of America. Bulletin, v. 69, p. 941-945.

Corey, A.T. (1949) *Influence of shape on the fall velocity of sand grains.* Fort Collins, Colo., Colorado A&M College master's thesis, 102 p.

Cornish, Vaughan (1898) *On sea-beaches and sandbanks.* Geographical Journal, v. 11, p. 628-651.

Cornish, Vaughan (1899) *On kumatology.* Geographical Journal, v. 13, p. 624-628.

Correns, C. W. (1950) *Zur Geochemie der Diagenese; I: Das Verhalten von $CaCO_3$ und SiO_2.* Geochimica et Cosmochimia Acta, v. 1, p. 49-54.

Cottingham, Kenneth (1951) *The geologist's vocabulary.* Scientific Monthly, v. 72, p. 154-163.

Cotton, C. A. (1916) *Fault coasts in New Zealand.* Geographical Review, v. 1, p. 20-47.

Cotton, C. A. (1922) *Geomorphology of New Zealand. Pt. 1.* Wellington: Dominion Museum. 462 p. (New Zealand Board of Science and Art. Manual no. 3).

Cotton, C. A. (1940) *Classification and correlation of river terraces.* Journal of Geomorphology, v. 3, p. 27-37.

Cotton, C. A. (1942) *Climatic accidents in landscape-making.* Christchurch N.Z.: Whitcombe & Tombs. 354 p.

Cotton, C. A. (1948) *Landscape as developed by the processes of normal erosion. 2nd ed.* New York: Wiley. 509 p.

Cotton, C. A. (1950) *Tectonic scarps and fault valleys.* Geological Society of America. Bulletin, v. 61, p. 717-757.

Cotton, C.A. (1958) *Geomorphology.* Christchurch, New Zealand, Whitcombe and Tombs, 503p.

Covington, M. (1985) *New morphologic information on Cretaceous nannofossils from the Niobrara Formation (Upper Cretaceous) of Kansas.* Geology, v. 13, p. 683-686.

Covington, M.A., (1992) *Dictionary of Computer Terms. 3rd. ed.,* Barron's, Hauppauge, NY, 364p.

Cowan, S. T. (1968) *A dictionary of microbial taxonomic usage.* Edinburgh, Oliver & Boyd. 118 p.

Cowie, J. W. (1986) *Guidelines for boundary stratotypes.* Episodes, v. 9, no. 2, p. 78-82.

Cowie, J. W. and Bassett, M. G. (1989) *IUGS 1989 global stratigraphic chart.* Supplement to Episodes, v. 12, n. 2.

Cowie, J. W., et al. (1986) *Guidelines and statutes of the International Commission on Stratigraphy (ICS)* Cour. Forsch. —Inst. Senckenberg, no. 83. p. 1-14.

Cox, L.R., et al. (1969) *Bivalvia.* In: Moore, R.C., ed., *Treatise on invertebrate paleontology,* Part N, Mollusca 6, vols.1 and 2, p.1-952. Lawrence, Kan., University of Kansas Press and Geological Society of America.

Cram, D. J., and Hammond, G. S. (1959) *Organic chemistry.* New York: McGraw-Hill. 712 p.

Cramer, F.H., and Diez Rodriguez, M.V. (1979) *Términos morfográficos de acritarcos—definición y equivalencias en español e inglés.* Palinología, v.1, p.227-291.

Crampin, S., (1961) *A review of wave motion in anisotropic and cracked elastic media.* Wave Motion, 3, 363-390.

Cresta, S., Goy, A., Ureta, S., Arias, C., Barrón, E., Bernad, J., Canales, M.L., García-Joral, F., García-Romero, E., Gialanella, P.R., Gómez, J.J., Gonzáles, J.A., Herrero, C., Martínez, G., Osete, M.L., Perilli, N., and Villalaín, J.J. (2001) The *Global Boundary Stratotype Section and Point (GSSP) of the Toarcian-Aalenian boundary (Lower-Middle Jurassic).* Episodes, v.24, no.3, p.166-175.

Crickmay, C. H. (1933) *The later stages of the cycle of erosion; some weaknesses in the theory of the cycle of erosion.* Geological Magazine, v. 70, p. 337-347.

Crickmay, C. H. (1967) *A note on the term bocanne.* American Journal of Science, v. 265, p. 626-627.

Criswell, D. R., et al. (1975) *Seismic and acoustic emissions of a booming dune.* Journal of Geophysical Research, v. 80, p. 4963-4974.

Cronquist, Arthur (1961) *Introductory botany.* New York: Harper & Row. 892 p.

Crosby, W. O. (1912) *Dynamic relations and terminology of stratigraphic conformity and unconformity.* Journal of Geology, v. 20, p. 289-299.

Cross, Whitman, and Howe, Ernest (1905) *Description of the Silverton quadrangle.* U. S. Geological Survey. Geologic Atlas, folio no. 120. 34 p.

Cross, Whitman, et al. (1902) *A quantitative chemico-mineralogical classification and nomenclature of igneous rocks.* Journal of Geology, v. 10, p. 555-690.

Cross, Whitman, et al. (1906) *The texture of igneous rocks.* Journal of Geology, v. 14, p. 692-707.

Crowell, J. C. (1955) *Directional-current structures from the Prealpine Flysch, Switzerland.* Geological Society of America. Bulletin, v. 66, p. 1351-1384.

Crowell, J. C. (1957) *Origin of pebbly mudstones.* Geological Society of America Bulletin, v. 68, p. 993-1009.

Crowell, J. C. (1964) *Climatic significance of sedimentary deposits containing dispersed megaclasts.* In: Nairn, A. E. M., ed. Problems in palaeoclimatology, p. 86-99. New York: Interscience. 705 p.

Crumbler, L.S., and Aubele, J.C. (2000) *Volcanism on Venus.* In: Sigurdsson, H., ed., *Encyclopedia of volcanoes.* New York, Academic Press, p.727-769.

Cuffey, R. J. (1985) *Expanded reef-rock textural classification and the geologic history of bryozoan reefs.* Geology, v. 13, p. 307-310.

Cuif, J.-P., Dauphin, Y., Denis, A., Gaspard, D., and Keller, J.-P. (1980) *Continuité et périodicité du réseau organique intraprismatique dans le test de* Pinna muricata *Linné (Lamellibranches).* Comptes Rendus des Séances de l'Académie de Sciences, ser.D, v.290, p.759-762.

Cullison, J. S. (1938) *Origin of composite and incomplete internal moulds and their possible use as criteria of structure.* Geological Society of America. Bulletin, v. 49, p. 981-988.

Cumings, E. R. (1930) *List of species from the New Corydon, Kokomo, and Kenneth formations of Indiana, and from reefs in the Mississinewa and Liston Creek formations.* Indiana Academy of Science. Proceedings, v. 39, p. 204-211.

Cumings, E. R. (1932) *Reefs or bioherms?* Geological Society of America. Bulletin, v. 43, p. 331-352.

Cumings, E. R., and Shrock, R. R. (1928) *Niagaran coral reefs of Indiana and adjacent states and their stratigraphic relations.* Geological Society of America. Bulletin, v. 39, p. 579-619.

Curl, R. L. (1964) *On the definition of a cave.* National Speleological Society. Bulletin, v. 26, p. 1-6.

Curl, R.L. (1966) *Caves as a measure of karst.* Journal of Geology, v.74, p.798-830.

Curl, R.L. (1986) *Fractal dimensions and geometries of caves.* Mathematical Geology, v.18, p.765-783.

Curray, J. R. (1966) *Continental terrace.* In: Oceanography, p. 207-214. New York: Reinhold. 1021 p.

Currie, Robert G. (1995) *Variance contribution of M_n and S_c signals to Nile River data over a 30-8 year bandwidth.* In: Holocene cycles: climate, sea levels, and sedimentation. Journal of Coastal Research, Special Issue 17, p. 29-38.

Currier, L. W. (1941) *Disappearance of the last ice sheet in Massachusetts by stagnation zone retreat (abstract).* Geological Society of America. Bulletin, v. 52, p. 1895.

Curry, H. D. (1938) *"Turtleback" fault surfaces in Death Valley, California.* Geological Society of America. Bulletin, v. 49, p. 1875.

Curry, H. D. (1954) *Turtlebacks in the central Black Mountains, Death Valley, California.* California Division of Mines. Bulletin 170, ch. 4, [Pt.] 7, p. 53-59.

Cuvillier, Jean (1951) *Corrélations stratigraphiques par microfaciès en Aquitaine occidentale.* Avec la collaboration de V. Sacal. Leiden: E. J. Brill. 23 p. 90 plates.

Cvijic, Jovan (1924) *The evolution of lapiés, a study in karst topography.* Geographical Review, v. 14, p. 26-49.

Cys, J. M. (1963) *A new definition of the penesaline environment.* Compass, v. 40, p. 161-163.

D

Dahl, Eilif (1955) *Biogeographic and geologic indications of unglaciated areas in Scandinavia during the glacial ages.* Geological Society of America. Bulletin, v. 66, p. 1499-1519.

Dahlstrom, C. D. A. (1970) *Structural geology in the eastern margin of the Canadian Rocky Mountains.* Bulletin of Canadian Petroleum Geology, v. 18, p. 332-406. Reprinted, 1977, in Joint Wyoming-Montana-Utah Geological Associations Guidebook, p. 407-439.

Dale, T. N. (1894) *Mount Greylock, its areal and structural geology.* U. S. Geological Survey. Monograph 23, p. 119-203.

Dale, T. N. (1923) *The commercial granites of New England.* U. S. Geological Survey. Bulletin 738. 488 p.

Dalrymple, G. B., et al. (1974) *Origin of the Hawaiian Islands.* In: Greeley, Ronald, ed., Guidebook to the Hawaiian Planetology Conference, ch. 3, p. 23-36. Moffett Field, Calif.: NASA/Ames Research Center.

Dalrymple, G. B. (2004) *Ancient earth, ancient skies: the age of the Earth and its cosmic surroundings.* Stanford, Calif., Stanford University Press, 474p.

Daly, R. A. (1902) *The geology of the northeast coast of Labrador.* Harvard College. Museum of Comparative Zoology. Bulletin 38, Geological Series 5, p. 205-270.

Daly, R. A. (1917) *Metamorphism and its phases.* Geological Society of America. Bulletin, v. 28, p. 375-418.

Damon, P. E. (1968) *Potassium-argon dating of igneous and metamorphic rocks with applications to the Basin Ranges of Arizona and Sonora.* In: Hamilton, E. I., and Farquhar, R. M., eds. Radiometric dating for geologists, p. 1-71. New York: Interscience. 506 p.

Dana, E. S. (1892) *The system of mineralogy of J. D. Dana, 1837-1868; descriptive mineralogy. 6th ed.* New York: Wiley. 1134 p.

Dana, J. D. (1873) *On some results of the Earth's contraction from cooling including a discussion of the origin of mountains and the nature of the Earth 's interior.* American Journal of Science, v. 5, p. 423-443.

Dana, J. D. (1874) *A text-book of geology. 2nd ed.* New York: Ivison, Blakeman, Taylor. 358 p.

Dana, J. D. (1890) *Characteristics of volcanoes.* New York: Dodd, Mead. 399 p.

Dana, J. D. (1895) *Manual of geology; treating of the principles of the science with special reference to American geological history. 4th ed.* New York: American Book Co. 1088 p.

Dapples, E. C. (1962) *Stages of diagenesis in the development of sandstones.* Geological Society of America. Bulletin, v. 73, p. 913-933.

Dapples, E. C., et al. (1953) *Petrographic and lithologic attributes of sandstones.* Journal of Geology, v. 61, p. 291-317.

Darby, D. G., and Ojakangas, R. W. (1980) *Gastroliths from an Upper Cretaceous plesiosaur*. Journal of Paleontology, v. 54, p. 548-556.
Darrah, W. C. (1939) *Principles of paleobotany*. Leiden, Holland: Chronica Botanica Co. 239 p.
Daubrée, G. A. (1879) *Études synthétiques de géologie expérimentale*. Paris: Dunod, 828 p.
Davidson, J., and De Silva, S. (2000) *Composite volcanoes*. In: Sigurdsson, H., ed., *Encyclopedia of volcanoes*. New York, Academic Press, p.627-642.
Davies, H. G. (1965) *Convolute lamination and other structures from the Lower Coal Measures of Yorkshire*. Sedimentology, v. 5, p. 305-325.
Davies, J. L. (1963) *A morphogenic approach to world shorelines*. Zeit fur Geomorph. Bd. 8, p. 27-42.
Davis, C. A. (1907) *Peat, essays on its origin, uses, and distribution in Michigan*. Michigan Geological Survey. Annual Report 1906, p. 93-395.
Davis, G. H. (1980) *Structural characteristics of metamorphic core complexes, southern Arizona*. In: M. L. Crittenden et al., eds. Cordilleran metamorphic core complexes. Geological Society of America. Memoir 153.
Davis, G. H. (1984) *Structural geology of rocks and regions*. New York: Wiley. 492 p.
Davis, J. C. (1973) *Statistics and data analysis in geology*. New York: Wiley. 550 p.
Davis, J.C., (1986) *Statistics and Data Analysis in Geology. 2nd. ed.*: John Wiley & Sons, Inc., New York, 646p.
Davis, R. A., Jr. (1983) *Depositional systems: a genetic approach to sedimentary geology*. New York: Prentice-Hall. 669 p.
Davis, R.A.,Jr., and FitzGerald, D.M. (2004) *Beaches and coasts*. Oxford, Blackwell, 419p.
Davis, W. M. (1885) *Geographic classification, illustrated by a study of plains, plateaus, and their derivatives*. American Association for the Advancement of Science. Proceedings, v. 33, p. 428-432.
Davis, W. M. (1889a) *Topographic development of the Triassic formation of the Connecticut Valley*. American Journal of Science, 3rd ser., v. 37, p. 423-434.
Davis, W. M. (1889b) *The rivers and valleys of Pennsylvania*. National Geographic Magazine, v. 1, p. 183-253.
Davis, W. M. (1890) *The rivers of northern New Jersey, with notes on the classification of rivers in general*. National Geographic Magazine, v. 2, p. 81-110.
Davis, W. M. (1894) *Physical geography in the university*. Journal of Geology, v. 2, p. 66-100.
Davis, W. M. (1895) *The development of certain English rivers*. Geographical Journal, v. 5, p. 127-146.
Davis, W. M. (1897) *Current notes on physiography*. Science, n.s., v. 6, p. 22-24.
Davis, W. M. (1899) *The geographical cycle*. Geographical Journal, v. 14, p. 481-504.
Davis, W. M. (1902) *Base level, grade, and peneplain*. Journal of Geology, v. 10, p. 77-111.
Davis, W. M. (1909) *Geographical essays*. Boston: Ginn. 777 p.
Davis, W. M. (1911) *The Colorado Front Range; a study of physiographic presentation*. Association of American Geographers. Annals, v. 1, p. 21-83.
Davis, W. M. (1912) *Relation of geography to geology*. Geological Society of America. Bulletin, v. 23, p. 93-124.
Davis, W. M. (1918) *A handbook of northern France*. Cambridge: Harvard Univ. Press. 174 p.
Davis, W. M. (1922) *Peneplains and the geogeographical cycle*. Geological Society of America. Bulletin, v. 33, p. 587-598.
Davis, W. M. (1925a) *The undertow myth*. Science, v. 61, p. 206-208.
Davis, W. M. (1925b) *A Roxen lake in Canada*. Scottish Geographical Magazine, v. 41, p. 65-74.
Davis, W. M. (1927) *The rifts of southern California*. American Journal of Science, 5th ser., v. 13, p. 57-72.
Davis, W. M. (1930) *Rock floors in arid and in humid climates*. Journal of Geology, v. 38, p. 1-27, 136-158.
Davis, W. M. (1932) *Piedmont benchlands and Primärrümpfe*. Geological Society of America. Bulletin, v. 43, p. 399-440.
Davis, W. M. (1933) *Granite domes of the Mojave Desert, California*. San Diego Society of Natural History. Transactions, v. 7, no. 20, p. 211-258.
Davis, W. M. (1938) *Sheetfloods and streamfloods*. Geological Society of America. Bulletin, v. 49, p. 1337-1416.
Davydov, V.I., Glenister, B.F., Spinosa, C., Ritter, S.M., Chernykh, V.V., Wardlaw, B.R., and Snyder, W.S. (1998) Proposal of Aidaralash as *Global Stratotype Section and Point (GSSP) for base of the Permian System*. Episodes, v.21, no.1, p.11-18.
De Geer, G. (1912) *A geochronology of the last 12,000 years*. International Geological Congress. 11th, Stockholm, 1910. Report, v. 1, p. 241-253.
De la Beche, H. T. (1832) *Geological manual*. Philadelphia: Carey & Lea. 535 p.
De Paolo, D.J. (1981) *Trace-element and isotopic effects of combined wallrock assimilation and fractional crystallization*. Earth and Planetary Science Letters, v.53, p.189-202.
Deal, D.E., (1964) *Scintillites: A variety of quartz speleothem*. National Speleological Society Bulletin, v.26, p.75.
Dean, D.R. (1992) *James Hutton and the history of geology*. Ithaca, N.Y., Cornell University Press, 303p.
Dean, D.R. (1998) *Plutonists, Neptunists, Vulcanists*. In: Good, G., ed., *Sciences of the Earth: An encyclopedia of events, people, and phenomena*, v.2. New York, Garland Publishing, p.691-695.
Dean, W. E., et al. (1975) *Sedimentological significance of nodular and laminated anhydrite*. Geology, v. 3, p. 367-372.
Deane, R. E. (1950) *Pleistocene geology of the Lake Simcoe district, Ontario*. Canada. Geological Survey. Memoir 256. 108 p.
Deere, D.U., and Deere, D.W. (1988) *The rock quality designation (RQD) index in practice*. In: Kirkaldie, L., ed., *Rock classification systems for engineering purposes*. Philadelphia, Pa., American Society for Testing and Materials (ASTM) Special Publication 984, p.91-101.
Deevey, E.S., Jr. (1942) *Studies on Connecticut lake sediments. 3, The biostratonomy of Linsley Pond*. American Journal of Science, v.240, no.4, p.233-264.
Deevey, E. S., Jr., and Flint, R. F. (1957) *Postglaical hypsithermal interval*. Science, v. 125, p. 182-184.
Defant, M.J., and Drummond, M.S. (1990) *Derivation of some modern arc magmas by melting of young subducted lithosphere*. Nature (London), v.347 (no.6294), p.662-665.
DeFord, R. K. (1946) *Grain size in carbonate rock*. American Association of Petroleum Geologists. Bulletin, v. 30, p. 1921-1928.
Dellinger, J., and Noble, B. (1997) *Crossed dipole reciprocity paradox*. Leading Edge, v.16, p.1465-1471.
Dellwig, L. F. (1955) *Origin of the Salina salt of Michigan*. Journal of Sedimentary Petrology, v. 25, p. 83-110.
Delmelle, P., and Bernard, A. (2000) *Volcanic lakes*. In: Sigurdsson, H., ed., *Encyclopedia of volcanoes*. New York, Academic Press, p.877-895.
Deming, D., (1989) *Application of bottom-hole temperature corrections in geothermal studies*. Geothermic, v.18, no.5/6, p.775-786.
Dence, M. R. (1968) *Shock zoning at Canadian craters: petrography and structural implications*. In: French, B. M., and Short, N. M., eds. Shock metamorphism of natural materials, p. 169-184. Baltimore: Mono Book Corp. 644 p.
Dennis, J. G., and Atwater, T. M. (1974) *Terminology of geodynamics*. American Association of Petroleum Geologists. Bulletin, v. 58, p. 1030-1036.
Dennis, J. G., ed. (1967) *International tectonic dictionary; English terminology*. American Association of Petroleum Geologists. Memoir 7. 196 p.
Derby, O. A. (1910) *The iron ores of Brazil*. In: The iron ore resources of the world, v. 2, p. 813-822. International Geological Congress. 11th, Stockholm, 1910. Stockholm Generalstabens Litografiska Anstalt. 2 v. 1068 p.
Derr, J. S. (1977) *Earthquake lights*. U. S. Geological Survey. Earthquake Information Bulletin, v. 9, no. 3, p. 18-21.
DeSitter, L. U. (1954) *Gravitational sliding tectonics—an essay on comparative structural geology*. American Journal of Science, v. 252, p. 321-344.
DeVoe, J. R., and Spijkerman, J. J. (1966) *Mössbauer spectrometry*. Analytical Chemistry, v. 38, no. 5, p. 382R-393R.
Dewey, J. F. (1969) *Continental margins—a model for conversion of Atlantic type to Andean type*. Earth and Planetary Science Letters, v. 6, p. 189-197.
DeWiest, R. J. M. (1965) *Geohydrology*. New York: Wiley. 366 p.
Dewolf, Y. (1970) Les argiles à siles: paléosols ou pédolithes. Association francaise pour l'étude du Quaternaire. Bulletin, no. 2-3, p. 117-119.
Dick, H.J.B. (1989) *Abyssal peridotites, very slow spreading ridges, and ocean-ridge magmatism*. In: Saunders, A.D., ed., *Magmatism in the ocean basins*. Geological Society of America Special Publication 42, p.71-105.
Dickin, A.P. (1995) *Radiogenic isotope geology*. Cambridge, UK, Cambridge University Press, 452p
Dickinson, W. R. (1965) *Geology of the Suplee-Izee area, Crook, Grant, and Harney counties, Oregon*. Oregon Department of Geology and Mineral Industries. Bulletin 58. 109 p.
Dickinson, W. R. (1966) *Structural relationships of San Andreas fault system, Cholame Valley and Castle Mountain Range, California*. Geological Society of America. Bulletin, v. 66, p. 707-725.
Dickinson, W. R. (1970) *Interpreting detrital modes of graywacke and arkose*. Journal of Sedimentary Petrology, v. 40, p. 695-707.
Dickinson, W. R., and Vigrass, L. W. (1965) *Geology of the Suplee-Izee area, Crook, Grant, and Harney counties, Oregon*. Oregon Department of Geology and Mineral Resources. Bulletin, no. 58, 110 p.
Diebold, J.B. and Stoffa, P.L., (1981) *The traveltime equation, tau-p mapping, and inversion of complex midpoint data*. Geophysics, 46, 238-254.
Diegel, F. A. (1986) *Topological constraints on imbricate thrust networks, examples from the Mountain City window, Tennessee, U.S.A.* Journal of Structural Geology, v. 8, p. 269-280.
Diessel, C.F.K. (1986) *On the correlation of coal facies and depositional environments*. In: Symposium on Advances in the Study of Sydney Basin, 12th, Proceedings, p.19-22.
Dietrich, R. V. (1959) *Development of ptygmatic features with a passive host during partial anatexis*. Beiträge zur Mineralogie und Petrologie, v. 6, no. 6, p. 357-365.
Dietrich, R. V. (1960) *Nomenclature of migmatitic and associated rocks*. Geotimes, v. 4, no. 5, p. 36-37, 50-51.
Dietrich, R. V. (1960a) *Genesis of ptygmatic features*. International Geological Congress, 21st, Copenhagen, 1960. Report, Norden. Proceedings, sec. 14, p. 138-148.

Dietrich, R. V. (1963) *Banded gneisses of eight localities*. Norsk Geologisk Tidsskrift, v. 43, no. 1, p. 89-119.

Dietrich, R. V. (1969) *Hybrid rocks*. Science, v. 163, p. 557.

Dietrich, R. V., and Mehnert, K. R. (1961) *Proposal for the nomenclature of migmatites and associated rocks*. International Geological Congress, 21st, Copenhagen, 1960. Report, pt. 26, sec. 14, p. 56-67.

Dietz, R. S. (1959) *Shatter cones in cryptoexplosion structures (meteorite impact?)*. Journal of Geology, v. 67, p. 496-505.

Dietz, R. S. (1960) *Meteorite impact suggested by shatter cones in rock*. Science, v. 131, p. 1781-1784.

Dietz, R. S. (1961) *Astroblemes*. Scientific American, v. 205, p. 50-58.

Dietz, R. S. (1963) *Wave-base, marine profile of equilibrium, and wave-built terraces: a critical appraisal*. Geological Society of America. Bulletin, v. 74, p. 971-990.

Dietz, R. S., and Holden, J. C. (1966) *Miogeoclines (miogeosynclines) in space and time*. Journal of Geology, v. 74, no. 5, pt. 1, p. 566-583.

Dionne, J.-C. (1973) *Monroes: a type of so-called mud volcanoes in tidal flats*. Journal of Sedimentary Petrology, v. 43, p. 848-856.

Diver, W.L., and Peat, C.J. (1979) *On the interpretation and classification of Precambrian organic-walled microfossils*. Geology, v.7, p.401-404.

Dixon, E. E. L., and Vaughan, Arthur (1911) *The Carboniferous succession in Gower (Glamorganshire), with notes on its fauna and conditions of deposition*. Geological Society of London. Quarterly Journal, v. 67, p. 477-571.

Dixon, J. M., and Farrar, E. (1980) *Ridge subduction, eduction, and the Neogene tectonics of southwestern North America*. Tectonophysics, v. 67, p. 81-99.

Dobkins, J. E., Jr., and Folk, R. L. (1970) *Shape development on Tahiti-Nui*. Journal of Sedimentary Petrology, v. 40, p. 1167-1203.

Dobrin, M. B. (1952) *Introduction to geophysical prospecting*. New York: McGraw-Hill. 435 p.

DOD: U.S. Department of Defense. Army Topographic Command (1969) *Glossary of mapping, charting, and geodetic terms. 2nd ed.* Washington, D.C.: U. S. Department of Defense. 281 p.

DOD: U.S. Department of Defense. Army Topographic Command (1973) *Glossary of mapping, charting and geodetic terms. 3rd ed.* Washington: U. S. Department of Defense. 272 p.

Dodge, J.D. (1982) *Marine dinoflagellates of the British Isles*. London, Her Majesty's Stationery Office, 303p.

Doe, B. R. (1970) *Lead isotopes*. New York: Springer-Verlag. 137 p.

Domenico, Patrick A., Schwartz, Franklin W. (1990) *Physical and chemical hydrogeology*. John Wiley & Sons. New York, NY, 824 p.

Donaldson, Douglas, and Simpson, Scott (1962) *Chomatichnus, a new ichnogenus, and other trace-fossils of Wegber Quarry*. Liverpool and Manchester Geological Journal, v. 3, p. 73-81.

Donnay, Gabrielle, and Donnay, J. D. H. (1953) *The crystallography of bastnaesite, parisite, roentgenite, and synchisite*. American Mineralogist, v. 38, nos. 11 & 12, p. 932-963.

Dorr, J. V. N., II, and Barbosa, A. L. de Miranda (1963) *Geology and ore deposits of the Itabira district, Minas Gerais, Brazil*. U. S. Geological Survey. Professional Paper 341-C. 110 p.

Dott, R. H., Jr. (1964) *Wacke, graywacke and matrix—what approach to immature sandstone classification?* Journal of Sedimentary Petrology, v. 34, p. 625-632.

Douglas, R. J. W. (1950) *Callum Creek, Langford Creek, and Gap map areas, Alberta*. Geological Survey of Canada. Memoir 255. 124 p.

Downie, C., Evitt, W.R., and Sarjeant, W.A.S. (1963) *Dinoflagellates, hystrichospheres, and the classification of the acritarchs*. Stanford University Publications, Geological Sciences, no.7, 16p.

Downie, C., and Sarjeant, W.A.S. (1966) *The morphology, terminology, and classification of fossil dinoflagellate cysts*. In: Davey, R.J., Downie, C., Sarjeant, W.A.S., and Williams, G.L., *Studies on Mesozoic and Cainozoic dinoflagellate cysts*. British Museum (Natural History) Geology, Bulletin, Supplement 3, p.10-17.

Doyle, P. S., et al. (1974) *Stratigraphy. Deep Sea Drilling Project*. Initial Reports, v. 26, p. 825-905.

Drake, A. A. (1985) *Tectonic implications of the Indian Run Formation—a newly recognized sedimentary melange in the northern Virginia Piedmont*. U. S. Geological Survey. Professional Paper 1324, 12 p.

Draper, J. W. (1847) *On the production of light by heat*. Philosophical Magazine, ser. 3, v. 30, p. 345-360.

Drew, G. H. (1911) *The action of some denitrifying bacteria in tropical and temperate seas, and the bacterial precipitation of calcium carbonate in the sea*. Marine Biological Association of the United Kingdom. Journal, v. 9, p. 142-155.

Drewes, Harold (1959) *Turtleback faults of Death Valley, California; a reinterpretation*. Geological Society of America. Bulletin, v. 70, p. 1497-1508.

Dryer, C. R. (1901) *Certain peculiar eskers and esker lakes of northeastern Indiana*. Journal of Geology, v. 9, p. 123-129.

Dryer, C. R. (1910) *Some features of delta formation*. Indiana Academy of Science. Proceedings, 1909, p. 255-261.

Duff, P. M. D., and Walton, E. K. (1962) *Statistical basis for cyclothems: a quantitative study of the sedimentary succession in the East Pennine Coalfield*. Sedimentology, v. 1, p. 235-255.

Dulhunty, J. A. (1939) *The torbanites of New South Wales, Pt. 1*. Royal Society of New South Wales. Journal and Proceedings, 1938, v. 72, p. 179-198.

Dunbar, C. O., and Rodgers, John (1957) *Principles of stratigraphy*. New York: Wiley. 356 p.

Dunham, R. J. (1962) *Classification of carbonate rocks according to depositional texture*. American Association of Petroleum Geologists. Memoir 1, p. 108-121.

Dunham, R. J. (1970) *Stratigraphic reefs versus ecologic reefs*. American Association of Petroleum Geologists. Bulletin, v. 54, p. 1931-1932.

Dunn, J. A. (1942) *Granite and magmation and metamorphism*. Economic Geology, v. 37, p. 231-238.

Dunnet, D. (1969) *A technique of finite strain analysis using elliptical markers*. Tectonophysics, v. 7, p. 117-136.

Dunoyer de Segonzac, G. (1968) *The birth and development of the concepts of diagenesis (1866-1966)*. Earth-Science Reviews, v. 4, p. 153-201.

DuToit, A. (1920) *The Karroo dolerites of South Africa*. Geological Society of South Africa. Transactions, v. 23, p. 1-42.

DuToit, A. (1937) *Our wandering continents*. New York: Hafner. 379 p.

Duval, Bernard, Cramez, Carlos, and Jackson, M. P. A. (1990) *Raft tectonics in the Kwanza basin, Angola (abs.)*. Geological Society of America, Abstracts with Programs, v. 22, no. 7, p. A48.

Dziewonski, A. M. and Anderson, D. L. (1981) *Preliminary reference Earth model*. Physics of the Earth and Planetary Interiors, 25; 4, p. 297-356.

Dzulynski, Stanislaw, and Kotlarczyk, J. (1962) *On load-casted ripples*. Société Géologique Pologne. Annales, v. 32, p. 148-159.

Dzulynski, Stanislaw, and Radomski, Andrzej (1955) *Pochodzenie sladów wleczenia na tle teorii pradów zawiesinowych*. Acta geologica Polonica, v. 5, p. 47-66.

Dzulynski, Stanislaw, and Sanders, J. E. (1962) *Current marks on firm mud bottoms*. Connecticut Academy of Arts and Sciences. Transactions, v. 42, p. 57-96.

Dzulynski, Stanislaw, and Slaczka, Andrzej (1958) *Directional structures and sedimentation of the Krosno beds (Carpathian flysch)*. Polskie Towarzystwo Geologiczne. Rocznik, t. 28, p. 205-260.

Dzulynski, Stanislaw, and Walton, E. K. (1963) *Experimental production of sole markings*. Edinburgh Geological Society. Transactions, v. 19, p. 279-305.

Dzulynski, Stanislaw, and Walton, E. K. (1965) *Sedimentary features of flysch and greywackes*. New York: Elsevier. 274 p.

Dzulynski, Stanislaw, et al. (1957) *Sandstone whirl-balls in the silts of the Carpathian flysch*. Société Géologique Pologne. Annales, v. 26, p. 107-126.

Dzulynski, Stanislaw, et al. (1959) *Turbidites in flysch of the Polish Carpathian Mountains*. Geological Society of America. Bulletin, v. 70, p. 1089-1118.

E

Eakin, H. M. (1916) *The Yukon-Koyukuk region, Alaska*. U. S. Geological Survey. Bulletin 631. 88 p.

Eardley, A. J. (1962) *Structural geology of North America. 2nd ed.* New York: Harper & Row. 743 p.

Eastler, T. E. (1994) *Geologic assessment of military underground facilities as effective barriers to conventional weapons attack (abst.)*. Geological Society of America, Abstracts with Programs (Seattle, WA), v. 26, no. 7, p. 346.

Eaton, G.L. (1980) *Nomenclature and homology in peridinialean dinoflagellate plate patterns*. Palaeontology, v.23, Pt.3, p.667-688.

Eaton, J. E. (1929) *The by-passing and discontinuous deposition of sedimentary materials*. American Association of Petroleum Geologists. Bulletin, v. 13, p. 713-761.

Eaton, J. E. (1951) *Inadvisability of restricting terms that designate the relative*. Geological Society of America. Bulletin, v. 62, p. 77-80.

Echelmeyer, K., and Wang, Z. (1987) *Direct observation of basal sliding and deformation of basal drift at subfreezing temperatures*. Journal of Glaciology, v.33, no. 113, p.83-98.

Eckis, Rollin (1928) *Alluvial fans of the Cucamonga district, southern California*. Journal of Geology, v. 36, p. 225-247.

Eckstrand, O.R., Sinclair, W.D., and Thorpe, R.I. (1995) *Introduction*. In: Eckstrand, O.R., Sinclair, W.D., and Thorpe, R.I., eds., *Geology of Canadian mineral-deposit types*. Ottawa, Canada, Geological Survey of Canada, Geology of Canada, no.8, p.1-17.

Edwards, L.E. (1990) *Peridinialean dinoflagellate plate patterns, labels and homologies*. International Palynological Congress, 7th, Proceedings, Part II, Review of Palaeobotany and Palynology, v.65 (1-4), p.293-303.

Edwards, L.E. (1993) *Dinoflagellates*. In: Lipps, J.H., ed., *Fossil Prokaryotes and Protists*. New York, Blackwell Scientific Publications, p.105-127.

Edwards, M. (1986) *Glacial environments*. Chap. 13 in: Reading, H.G., ed., *Sedimentary environments and facies, 2nd. ed.* Oxford, U.K., Blackwell Scientific Publications, 615p.

Eggleton, R.A., ed. (2001) *The regolith glossary CRC for landscape evolution and mineral exploration*. Floreat Park, Western Australia, 144p.

Ehrenberg, C. G. (1854) *Mikrogeologie; das Erde und Felsen schaffende wirken des unsichtbar kleinen selbstständigen Lebens auf der Erde*. Leipzig: L. Voss. 347 p.

Ehrlich, H.L. (1990) *Geomicrobiology, 2nd ed.* New York, Marcel Dekker, 646p.

Einsele, G., and Seilacher, A., eds. (1982) *Cyclic and event stratification*. New York: Springer-Verlag. 536 p.

Einstein, H. A. (1950) *The bed-load function for sediment transportation in open channel flows*. U. S. Department of Agriculture. Technical Bulletin 1026. 70 p.

Eisenack, A. (1962) *Neue problematische Mikrofossilien*. Neues Jahrbuch für Geologie und Paläontologie, Abhandlungen, v.11, no.2, p.135141.

Eisenack, A., Cramer, F.H., and Diez, M. del C.R. (1973) *Katalog der fossilen Dinoflagellaten, Hystrichosphären, und verwandten Mikrofossilien, III, Acritarchs 1, E.* Stuttgart, Germany, Schweizerbartsche Verlagsbuchhandlung, 1104 p.

Eisenack, Alfred (1931) *Neue Mikrofossilien des baltischen Silurs. Pt. 1.* Palaeontologische Zeitschrift, Bd. 13, p. 74-118.

Ekdale, A.A., and Bromley, R.G. (1983) *Trace fossils and ichnofabric in the Kjølby Gaard Marl, uppermost Cretaceous, Denmark.* Meddelelser fra Dansk Geologisk Forening (Bulletin of the Geological Society of Denmark), v.31, no.3-4, p.107-119.

El-Etr, H. A. (1976) *Proposed terminology for natural linear features.* Utah Geological Association. Publication no. 5, p. 480-489.

Elias, M. K. (1965) *Cycleology—a discipline concomitant to paleoecology (abstract).* American Association of Petroleum Geologists. Bulletin, v. 49, p. 339.

Elliot, T. (1986) *Siliclastic shorelines.* In: Reading, H.G., ed., *Sedimentary environments and facies, 2nd ed.* Oxford, Blackwell Scientific Publications, 615p.

Elliott, D. W. (1976) *The energy balance and deformation mechanisms of thrust sheets.* Philosophical Transactions of the Royal Society of London, v. A283, p. 289-312.

Elliott, R. E. (1965) *A classification of subaqueous sedimentary structures based on rheological and kinematical parameters.* Sedimentology, v. 5, p. 193-209.

Elton, C. S. (1927) *The nature and origin of soil-polygons in Spitsbergen.* Geological Society of London. Quarterly Journal, v. 83, p. 163-194.

Embry, A.F., and Johannessen, E.P. (1992) *T-R sequence stratigraphy, facies analysis, and reservoir distribution in the uppermost Triassic-Lower Jurassic succession, western Sverdrup Basin, Arctic Canada.* In: Vorren, T.O., et al., eds. *Arctic geology and petroleum potential.* Norwegian Petroleum Society (NPF) Special Publication 2, p.121-146.

Embry, A. F., and Klovan, J. E. (1971) *A Late Devonian reef tract on northeastern Banks Island, N.W.T.* Bulletin of Canadian Petroleum Geology, v. 19, p. 730-781.

Emery, K. O. (1948) *Submarine geology of Bikini atoll.* Geological Society of America. Bulletin, v. 59, p. 855-859.

Emery, K.O. (1961) *A simple method of measuring beach profiles.* Limnology and Oceanography, v.6, p.90-93.

Emiliani, Cesare (1955) *Pleistocene temperatures.* Journal of Geology, v. 63, p. 538-578.

Emiliani, Cesare (1987) *Dictionary of the Physical Sciences.* Oxford University Press, p. 86.

Emiliani, Cesare (1992) *Planet Earth.* Cambridge University Press.

Emmons, W.H. (1936) *Hydrothermal zoning in metalliferous lodes.* International Geological Congress, 16th, Report Part 1, p.417-432.

Engel, A. E. J., et al. (1965) *Chemical characteristics of oceanic basalts and the upper mantle.* Geological Society of America. Bulletin, v. 76, p. 719-734.

Engelder, T. (1985) *Loading paths to joint propagation during a tectonic cycle: An example from the Appalachian Plateau, U.S.A.* Journal of Structural Geology, v. 7, p. 459-476.

Engelder, T. (1993) *Stress regimes in the lithosphere.* Princeton Press, Princeton, NJ, 451 p.

Engelder, T., and Fischer, M. P. (1996) *Loading configurations and driving mechanisms for joints based on the Griffith energy-balance concept.* Tectonophysics. 256; 1-4, p. 253-277.

Engelder, T., and Lacazette, A. (1990) *Natural hydraulic fracturing.* In Barton, N., and Stephansson, O., eds., Rock Joints : A.A. Balkema, Rotterdam, p. 35-44.

Engelhardt, Wolf von, and Stoffler, D. (1968) *Stages of shock metamorphism in crystalline rocks of the Ries basin, Germany.* In: French, B. M., and Short, N. M., eds. Shock metamorphism of natural materials, p. 159-168. Baltimore: Mono Book Corp. 644 p.

Engelund, Frank (1966) *On the possibility of formulating a universal spectrum function for dunes.* Technical University of Denmark. Hydraulics Laboratory. Basic Research Progress Rept. 18, p. 1-4.

Enos, Paul (1974) *Reefs, platforms, and basins of Middle Cretaceous in northeast Mexico.* American Association of Petroleum Geologists. Bulletin, v. 58, p. 800-809.

Erben, H.K. (1971) *Anorganische und organische Schalenkomponenten bei* Cittarium pica *L. (Archaeogastropoda).* Biomineralization Research Reports, v.3, p.51-64.

Erben, H.K. (1972) *On the formation and growth of nacre.* Biomineralization Research Reports, v.4, p.15-46.

Erben, H.K., and Krampitz, G. (1972) *Ultrastruktur und Aminosäuren-Verhältnisse in den Schalen der rezenten Pleurotomariidae (Gastropoda).* Biomineralization Research Reports, v.6, p.12-31.

Erhart, Henri (1955) *"Biostasie" et "rhexistasie," esquisse d'une théorie sur le rôle de la pédogenèse en tant que phénomène géologique.* Académie des Sciences, Paris. Comptes rendus hebdomadaires des séances, t. 241, no. 18, p. 1218-1220.

Erhart, Henri (1956) *La genèse des sols; esquisse d'une théorie géologique et géochimique: biostasie et rhexistasis.* Paris: Masson. 90 p.

Esau, Katherine (1965) *Plant anatomy. 2nd ed.* New York: Wiley.

Eskola, Pentti (1915) *Om sambandet mellan kemisk och mineralogisk sammansättning hos Orijävitraktens metamorfa bergarter.* With an English summary of the contents. Finland, Commission Géologique. Bulletin, v. 8, no. 44, 145 p.

Eskola, Pentti (1938) *On the esboitic crystallization of orbicular rocks.* Journal of Geology, v. 46, p. 448-485.

Eskola, Pentti (1939) *Die metamorphen Gesteine.* In: Barth, T. F. W., Correns, Carl W., and Eskola, Pentti. Die Entstehung der Gesteine, ein Lehrbuch der Petrogenese. Berlin: Julius Springer. 422 p.

Eskola, Pentti (1948) *The problem of mantled gneiss domes.* Geological Society of London. Quarterly Journal, v. 104, p. 461.

Eskola, Pentti (1961) *Granitentschung bei Orogenese und Epeirogenese.* Geologische Rundschau, v. 50, p. 105-113.

ESSA: U. S. Environmental Science Services Administration (1968) *ESSA.* Washington: U. S. Government Printing Office. [8] p.

Eugster, H.P. (1967) *Hydrous sodium silicates from Lake Magadi, Kenya: Presursors of bedded chert.* Science, v.157, p.1177-1180.

Evans, B. W. (1977) *Metamorphism of alpine peridotite and serpentinite.* Annual Review of Earth and Planetary Science, v. 5, p. 397-447.

Evans, J. W. (1911) *Dreikanter.* Geological Magazine, v. 8, p. 334-335.

Evans, O. F. (1938) *The classification and origin of beach cusps.* Journal of Geology, v. 46, p. 615-627.

Evitt, W. R. (1963) *A discussion and proposals concerning fossil dinoflagellates, hystrichospheres, and acritarchs. Pt. 2.* National Academy of Sciences, U.S.A. Proceedings, v. 49, p. 298-302.

Evitt, W.R. (1985) *Sporopollenin dinoflagellate cysts: their morphology and interpretation.* American Association of Stratigraphic Palynologists, Monograph Series, no.1, 333p.

Evitt, W.R., Lentin, J.K., Millioud, M.E., Stover, L.E., and Williams, G.L. (1977) *Dinoflagellate cyst terminology.* Geological Society of Canada Special Paper 76-24, 11p.

Eythórsson, Jón (1951) *Jökla-mys.* Journal of Glaciology, v. 1, p. 503.

F

Faegri, K., and Iversen, J. (1975) *Textbook of pollen analysis, 3rd rev. ed.* New York, Hafner Press, 295p.

Fairbairn, H. W. (1943) *Packing in ionic minerals.* Geological Society of America. Bulletin, v. 54, p. 1305-1374.

Fairbanks, H. W. (1906) *Practical physiography.* Boston: Allyn & Bacon. 542 p.

Fairbridge, R. W. (1946) *Submarine slumping and location of oil bodies.* American Association of Petroleum Geologists. Bulletin, v. 30, p. 84-92.

Fairbridge, R. W. (1951) *The Aroe Islands and the continental shelf north of Australia.* Scope, v. 1, no. 6, p. 24-29.

Fairbridge, R. W. (1954) *Stratigraphic correlation by microfacies.* American Journal of Science, v. 252, p. 683-694.

Fairbridge, R. W. (1958) *What is a consanguineous association?* Journal of Geology, v. 66, p. 319-324.

Fairbridge, R. W. (1961) *Eustatic changes in sea level.* In: L. H. Ahrens et al., eds. Physics and chemistry of the earth, v. 4, p. 99-185. New York: Pergamon Press.

Fairbridge, R. W. (1967) *Phases of diagenesis and authigenesis.* In: Larsen, Gunnar, and Chilingar, G. V., eds. Diagenesis in sediments. Amsterdam: Elsevier. 551 p. (Developments in sedimentology 8, p. 19-89).

Fairbridge, R. W. (1972) *Climatology of a glacial cycle.* Quaternary Research (Washington University), v. 2, p. 283-302.

Fairbridge, R. W. (1995) *Eolianites and eustasy: early concepts on Darwin's voyage of HMS Beagle.* Carbonates and Evaporites, v. 10(1), p. 92-101.

Fairbridge, R. W., ed. (1966) *The encyclopedia of oceanography.* New York: Reinhold. 1021 p. (Encyclopedia of earth sciences series, v. 1).

Fairbridge, R. W., ed. (1968) *The encyclopedia of geomorphology.* New York: Reinhold. 1295 p. (Encyclopedia of earth sciences series, v. 3).

Fairchild, H. L. (1904) *Geology under the planetesimal hypothesis of Earth origin.* Geological Society of America. Bulletin, v. 15, p. 243-266.

Fairchild, H. L. (1913) *Pleistocene geology of New York State.* Geological Society of America. Bulletin, v. 24, p. 133-162.

Fairchild, H. L. (1932) *New York moraines.* Geological Society of America. Bulletin, v. 43, p. 627-662.

Fairhead, J.D., and Odegard, M.E. (2002) *Advances in gravity survey resolution.* The Leading Edge, v.21, p.26-27.

Fannin, N. G. T. (1981) *Pockmarks in the North Sea.* Open Earth, no. 13, p. 28-29.

Farmin, Rollin (1934) *"Pebble dikes" and associated mineralization at Tintic, Utah.* Economic Geology, v. 29, p. 356-370.

Fay, A. H. (1918) *A glossary of the mining and mineral industry.* Washington: Government Printing Office. 754 p. (U. S. Bureau of Mines. Bulletin 95.)

Fay, A.H. (1920) *A glossary of the mining and mineral industry.* U.S. Bureau of Mines Bulletin 95, 754p.

Fedorowski, J., and Jull, R. K. (1976) *Review of blastogeny in Paleozoic corals and description of lateral increase in some Upper Ordovician rugose corals.* Acta Palaeontologica Polonica, v. 21, no. 1, p. 37-78.

Fenneman, N. M. (1909) *Physiography of the St. Louis area.* Illinois State Geological Survey. Bulletin, v. 12, 83 p.

Fenneman, N. M. (1914) *Physiographic boundaries within the United States.* Association of American Geographers. Annals, v. 4, p. 84-134.

Fenner, C. N. (1948) *Incandescent tuff flows in southern Peru.* Geological Society of America. Bulletin, v. 59, p. 879-893.

Fensome, R.A., Taylor, F.J.R., Norris, G., Sarjeant, W.A.S., Wharton, D.I., and Williams, G.L. (1993) *A classification of fossil and living dinoflagellates.* American Museum of Natural History, Micropaleontology, Special Paper 7, 351p.

Fenton, C. L., and Fenton, M. A. (1928) *Faunule and zonule.* American Midland Naturalist, v. 11, p. 1-23. (Ecologic interpretations of some biostratigraphic terms, 1).

Fenton, C. L., and Fenton, M. A. (1930) *Zone, subzone, facies, phase.* American Midland Naturalist, v. 12, p. 145-153. (Ecologic interpretations of some biostratigraphic terms, 2).

Ferguson, L. (1970) *"Armored snowballs" and the introduction of coarse terrigenous material into sea-ice.* Journal of Sedimentary Petrology, v. 40, p. 1057-1060.

Fernald, F. A. (1929) *Roundstone, a new geologic term.* Science, n.s., v. 70, p. 240.

Fernald, M. L. (1950) *Gray's manual of botany. 8th ed.* New York: American Book Company.

Fersman, A. E. (1922) *Geokhimiia Rossii.* St. Petersburg: Nauchnoe Khimichesko-Tekhnicheskoe izdatel'stvo. Approx. 210 p.

Fetter, Charles Willard (1994) *Applied hydrogeology.* MacMillan College Publishing Co., 691 p.

FGDC: Federal Geographic Data Committee (1998a) *Content standard for digital geospatial metadata* (version 2.0). Report FGDC-STD-001-1998, 78p. http://www.fgdc.gov/standards/documents/standards/metadata/v20698.pdf

FGDC: Federal Geographic Data Committee (1998b) Geospatial positioning accuracy standards, Part 3: National standard for spatial data accuracy. Report FGDC-STD-007.3-1998, p.3-1. http://www.fgdc.gov/standards/documents/standards/accuracy/chaper3.pdf

Field, M.S. (2002) *A lexicon of cave and karst terminology with special reference to environmental karst hydrology.* Washington, D.C., National Center for Environmental Assessment–Washington Office, Report EPA/600/R-02-003, 214p.

Fielder, Gilbert (1965) *Lunar geology.* Chester Springs, Penna.: Dufour. 184 p.

Finch, R. H. (1933) *Slump scarps.* Journal of Geology, v. 41, p. 647-649.

Fink, J.H., and Anderson, S.W. (2000) *Lava domes and coulees.* In: Sigurdsson, H., ed., *Encyclopedia of volcanoes.* New York, Academic Press, p.307-319.

Finsterwalder, Richard (1950) *Some comments on glacier flow.* Journal of Glaciology, v. 1, no. 7, p. 383-388.

Fischer, A. G. (1964) *The Lofer cyclothem of the Alpine Triassic.* Kansas State Geological Survey. Bulletin 169, v. 1, p. 107-149.

Fischer, A. G. (1969) *Geological time-distance rates: the Bubnoff unit.* Geological Society of America. Bulletin, v. 80, p. 549-551.

Fischer, A. G., and Teichert, Curt (1969) *Cameral deposits in cephalopod shells.* University of Kansas. Paleontological Contributions, Paper 37. 30 p.

Fischer, A. G., et al. (1954) *Arbitrary cut-off in stratigraphy.* American Association of Petroleum Geologists. Bulletin, v. 38, p. 926-931.

Fischer, Georg (1934) *Die Petrographie der Grauwacken.* Preussische Geologische Landesanstalt. Jahrbuch, 1933, Bd. 54, p. 320-343.

Fisher, O. (1866) *On the warp (of Mr. Trimmer)—Its age and probable connexion with the last geological events.* Geological Society of London. Quarterly Journal, v. 22. p. 553-565.

Fisher, R. V. (1958) *Definition of volcanic breccia.* Geologic Society of America Bulletin, vol. 69, p. 1071-1073.

Fisher, R. V. (1961) *Proposed classification of volcaniclastic sediments and rocks.* Geologic Society of America Bulletin, vol. 72, p. 1409-1414.

Fisher, R.V., and Schmincke, H.-U. (1984) *Pyroclastic rocks.* New York, Springer Verlag, 472p.

Fisher, W. B. (1950) *The Middle East; a physical, social, and regional geography.* London: Methuen. 514 p.

Fisher, W. L. and McGowan, J. H. (1967) *Depositional systems in the Wilcox Group of Texas and their relationship to occurrence of oil and gas.* Gulf Coast Association Geological Societies Transactions, v. 17, p. 105-125.

Fisk, H. N. (1959) *Padre Island and the Laguna Madre Flats, coastal south Texas.* Coastal Geography Conference, 2nd, Baton Rouge, La., 1959. [Proceedings], p. 103-151.

FitzGerald, D.M., Kraus, N.C., and Hands, E.B. (2001) *Natural mechanisms of sediment bypassing at tidal inlets.* Vicksburg, Miss., U.S. Army Engineer Research and Development Center, ERDC/CHL CHETN-IV-30 (*http:chl.wes.army.mil/library/publications/chetn*).

Flajs, G. (1972) *On the ultrastructure of the hinge of the Bivalvia,* 1. Biomineralization Research Reports, v.6, p.49-65.

Flanders, P. L., and Sauer, F. M. (1960) *A glossary of geoplosics: the systematic study of explosion effects in the Earth. Kirtland Air Force Base, N.Mex.:* U.S. Air Force Special Weapons Center. 34 p. (Technical note 60-20).

Flawn, P. T. (1953) *Petrographic classification of argillaceous sedimentary and low-grade metamorphic rocks in subsurface.* American Association of Petroleum Geologists. Bulletin, v. 37, p. 560-565.

Fleischer, Michael (1966) *Index of new mineral names, discredited minerals, and changes of mineralogical nomenclature in volumes 1-50* The American Mineralogist. American Mineralogist, v. 51, p. 1248-1357.

Fleischer, Michael (1975) *1975 Glossary of mineral species. Bowie, Md.: Mineralogical Record. 145 p. Additions and corrections*: Mineralogical Record, v. 7, no. 2, March-April 1976; v. 8, no. 5, Sept.-Oct. 1977.

Fleuty, M. J. (1964) *Tectonic slides.* Geological Magazine, v. 101, p. 452-456.

Fleuty, M. J. (1964a) *The description of folds.* Geologists' Association. Proceedings, v. 75, p. 461-492.

Flinn, D. (1962) *On folding during three-dimensional progressive deformation.* Geological Society of London. Quarterly Journal, v. 118, p. 385-433.

Flint, D. E., et al. (1953) *Limestone walls of Okinawa.* Geological Society of America. Bulletin, v. 64, p. 1247-1260.

Flint, R. F. (1928) *Eskers and crevasse fillings.* American Journal of Science, v. 15, p. 410-416.

Flint, R. F. (1955) *Pleistocene geology of eastern South Dakota.* U.S. Geological Survey. Professional Paper 262. 173 p.

Flint, R. F. (1957) *Glacial and Pleistocene geology.* New York: Wiley. 553 p.

Flint, R. F. (1971) *Glacial and Quaternary geology.* New York: Wiley. 892 p.

Flint, R. F., and Deevey, E. S., Jr. (1951) *Radiocarbon dating of late-Pleistocene events.* American Journal of Science, v. 249, p. 257-300.

Flint, R. F., et al. (1960a) *Symmictite: a name for nonsorted terrigenous sedimentary rocks that contain a wide range of particle sizes.* Geological Society of America. Bulletin, v. 71, p. 507-509.

Flint, R. F., et al. (1960b) *Diamictite, a substitute term for symmictite.* Geological Society of America. Bulletin, v. 71, p. 1809.

Fogg, G. E., et al. (1973) *The blue-green algae.* New York: Academic Press. 459 p.

Foley, S. F., Venturelli, G., Green, D. H., and Toscani, L. (1987) *The ultrapotassic rocks: characteristics, classification, and constraints for petrogenetic models.* Earth-Science Reviews, v. 24, p. 81-134.

Folk, R. L. (1951) *Stages of textural maturity in sedimentary rocks.* Journal of Sedimentary Petrology, v. 21, p. 127-130.

Folk, R. L. (1954) *The distinction between grain size and mineral composition in sedimentary-rock nomenclature.* Journal of Geology, v. 62, p. 344-359.

Folk, R. L. (1959) *Practical petrographic classification of limestones.* American Association of Petroleum Geologists. Bulletin, v. 43, p. 1-38.

Folk, R. L. (1962) *Spectral subdivision of limestone types.* American Association of Petroleum Geologists. Memoir 1, p. 62-84.

Folk, R. L. (1965) *Some aspects of recrystallization in ancient limestones.* Society of Economic Paleontologists and Mineralogists. Special Publication no. 13, p. 14-48.

Folk, R. L. (1968) *Petrology of sedimentary rocks.* Austin, Tex.: Hemphill's Book Store. 170 p.

Folk, R. L. (1973) *Evidence for peritidal deposition of Devonian Caballos Novaculite, Marathon Basin, Texas.* American Association of Petroleum Geologists. Bulletin, v. 57, p. 702-725.

Folk, R. L. (1974) *The natural history of crystalline calcium carbonate: effects of magnesium content and salinity.* Journal of Sedimentary Petrology, v. 44, p. 40-53.

Folk, R. L., and Assereto, Riccardo (1974) *Giant aragonite rays and baroque white dolomite in tepee-fillings, Triassic of Lombardy, Italy (abstract).* American Association of Petroleum Geologists. Annual meeting. Abstracts, v. 1, p. 34-35.

Folk, R. L., and Assereto, Riccardo (1976) *Comparative fabrics of length-slow and length-fast calcite and calcite-aragonite in a Holocene speleothem, Carlsbad Caverns, New Mexico.* Journal of Sedimentary Petrology, v. 46, p. 486-496.

Folk, R. L., and Siedlecka, Anna (1974) *The schizohaline environment: its sedimentary and diagenetic fabric as exemplified by late Paleozoic rocks of Bear Island, Svalbard.* Sedimentary Geology, v. 11, p. 1-15.

Folk, R. L., et al. (1973) *Black phytokarst from Hell, Cayman Islands, British West Indies.* Geological Society of America. Bulletin, v. 84, p. 2351-2360.

Folk, Robert L. (1993) *SEM imaging of bacteria and nannobacteria in carbonate sediments and rocks.* Journal of Sedimentary Petrology, 63;5, p. 990-999.

Ford, D.C., and Williams, P.W., (1989) *Karst geomorphology and karst hydrology.* London, England, Unwin Hyman, 601p.

Forgotson, J. M., Jr. (1954) *Regional stratigraphic analysis of the Cotton Valley Group of Upper Gulf Coastal Plain.* American Association of Petroleum Geologists. Bulletin, v. 38, p. 2476-2499.

Forgotson, J. M., Jr. (1957) *Nature, usage, and definition of marker-defined vertically segregated rock units.* American Association of Petroleum Geologists. Bulletin, v. 41, p. 2108-2113.

Forgotson, J. M., Jr. (1960) *Review and classification of quantitative mapping techniques.* American Association of Petroleum Geologists. Bulletin, v. 44, p. 83-100.

Forgotson, J. M., Jr., and Forgotson, J. M. (1975) *"Porosity pod": concept for successful stratigraphic exploration of fine-grained sandstones.* American Association of Petroleum Geologists. Bulletin, v. 59, p. 1113-1125.

Fortescue, J. (1965) *Exploration architecture.* Geological Survey of Canada. Paper 65-6, p. 4-14.

Foster, A. S., and Gifford, E. M., Jr. (1974) *Comparative morphology of vascular plants.* San Francisco: Freeman. 751 p.

Foster, N. H. (1966) *Stratigraphic leak.* American Association of Petroleum Geologists. Bulletin, v. 50, p. 2604-2611.

Fournet, J. (1844) *Sur l'état de surfusion du quartz dans les roches éruptives et dans les filons métallifères*. Académie des Sciences, Paris. Comptes rendus, v. 18, p. 1050-1057.

Fournet, J. (1847) *Résultats d'une exploration des Vosges. Société Géologique de France*. Bulletin, series 2, v. 4, p. 220-254.

Fowkes, W. W., and Frost, C. M. (1960) *Leonardite, a lignite by-product*. U. S. Bureau of Mines. Report of Investigations 5611. 12 p.

Fowler, G. M., and Lyden, J. P. (1932) *The ore deposits of the Tri-State district (Missouri-Kansas-Oklahoma)*. American Institute of Mining and Metallurgical Engineers. Transactions, v. 102, p. 206-251.

Fowler, H. W. (1937) *A dictionary of modern English usage. 1st ed, reprinted with corrections*. London: Oxford University Press. 742 p.

Fox, C.A. (1983) *Micromorphology of an orthic turbic cryosol—a permafrost soil*. In: Soil Micromorphology (P. Bullock and G.P. Murphy, Editors), A.B. Academic Publishers, Oxford, pp.699-705.

Fox, D. L. (1957) *Particulate organic detritus*. Geological Society of America. Memoir 67, v. 1, p. 383-389.

Francis, P. (1993) *Volcanoes: A planetary perspective*. New York, Oxford, 443p.

Frank, R., Sigogneau-Russell, D., and Vogel, J.C. (1984) *Tooth ultrastructure of Late Triassic Haramiyidae*. Journal of Dental Research, v.63, no.5, p.661-664.

Frank-Kamenetskii, V. A., Logvinenko, N. V., and Drits, V. A. (1965) *Tosudite—a new mineral, forming the mixed-layer phase in alushite*. Proceedings of the International Clay Conference, Stockholm 2, p. 181-186.

Freas, D. H., and Riggs, S. R. (1968) *Environments of phosphorite deposition in the central Florida phosphate district*. Texas Bureau of Economic Geology. Fourth Forum on Geology of Industrial Minerals. Proceedings, p. 117-128.

Freeburg, J. H. (1966) *Terrestrial impact structures—a bibliography*. U. S. Geological Survey. Bulletin 1220. 91 p.

Freeman, O. W. (1925) *The origin of Swimming Woman Canyon, Big Snowy Mountains, Montana, an example of a pseudo-cirque formed by landslide sapping*. Journal of Geology, v. 33, p. 75-79.

Freeze, R. A. and Cherry, J. A. (1979) *Groundwater*. Prentice-Hall, Englewood Cliffs, N.J., 604 p.

Freiberger, W.F., ed., (1960) *The International Dictionary of Applied Mathematics*. D. Van Nostrand Co., Inc., Princeton, NJ, 1173p.

French, B.M. (1998) *Traces of catastrophe: A handbook of shock-metamorphic effects in terrestrial meteorite impact structures*. Lunar and Planetary Institute, Houston, Tex., 120p.

Freund, J. E. (1960) *Modern elementary statistics, 2nd ed*. Englewood Cliffs, N.J.: Prentice-Hall.

Freund, J.E., and Williams, F.J., (1991) *Dictionary/Outline of Basic Statistics*. New York, Dover Publ., 195p.

Frey, M. (1987) *The reaction-isograd kaolinite + quartz = pyrophyllite + H_2O, Helvetic Alps, Switzerland*. Schweizerische Mineralogische und Petrographische Mitteilungen / Bulletin Suisse de Mineralogie et Petrographie, 67, 1-2, p. 1-11.

Friedman, G. M. (1965) *Terminology of crystallization textures and fabrics in sedimentary rocks*. Journal of Sedimentary Petrology, v. 35, p. 643-655.

Friedman, G. M., and Sanders, J. E. (1974) *Positive-relief bedforms on modern tidal flat that resemble molds of flutes and grooves; implications for geopetal criteria for origin and classification of bedforms*. Journal of Sedimentary Petrology, v. 44, p. 181-189.

Friedman, G. M., and Sanders, J. E. (1978) *Principles of sedimentology*. New York: Wiley. 792 p.

Friedman, G. M., Sanders, J. E., and Kopaska-Merkee, D. C., (1992) *Principles of sedimentary deposits: stratigraphy and sedimentology: glossary*. New York: MacMillan Publishing Co., 43 p.

Friedman, J. D. (1970) *The airborne infrared scanner as a geophysical research tool*. Optical Spectra, June, p. 35-44.

Friedman, M., et al. (1976) *Experimental folding of rocks under confining pressure: Part III. Faulted drape folds in multilithologic layered specimens*. Geological Society of America. Bulletin, v. 87, p. 1049-1066.

Fritsch, F. E. (1961) *The structure and reproduction of the algae. v. 1*. Cambridge, England: Cambridge University Press. 791 p.

Frizzell, D. L. (1933) *Terminology of types*. American Midland Naturalist, v. 14, p. 637-668.

Frondel, J. W., et al. (1967) *Glossary of uranium-and thorium-bearing minerals. 4th ed*. U. S. Geological Survey. Bulletin 1250. 69 p.

Fry, N. (1979) *Random point distributions and strain measurement in rocks*. Tectonophysics, 60, 69-105.

Frye, J. C., and Willman, H. B. (1960) *Classification of the Wisconsinian Stage in the Lake Michigan glacial lobe*. Illinois State Geological Survey. Circular 285. 16 p.

Fuchs, Theodor (1895) *Studien über Fucoiden und Hieroglyphen*. Akademie der Wissenschaften, Vienna. Mathematisch-Naturwissen-schaftliche Classe. Denkschriften, v. 62, p. 369-448.

Fuller, H.-J., and Tippo, Oswald (1954) *College botany*. Revised edition. New York: Holt, Rinehart & Winston.

Fuller, M. L. (1914) *The geology of Long Island, N.Y*. U. S. Geological Survey. Professional Paper 82. 231 p.

Fyfe, W. S., and Turner, F. J. (1966) *Reappraisal of the metamorphic facies concept*. Contributions to Mineralogy and Petrology, v. 12, p. 354-364.

Fyfe, W. S., et al. (1958) *Metamorphic reactions and metamorphic facies*. Geological Society of America. Memoir 73. 259 p.

G

Gajadhar, A.A., Marquardt, W.C., Hall, R., Gunderson, J., Carmona, E.V.A., and Sogin, M.L., (1991) *Ribosomal RNA sequences of* Sarcocystis muris, Theileria annulata, *and* Crypthecodinium cohnii *reveal evolutionary relationships among apicomplexans, dinoflagellates, and ciliates*. Molecular and Biochemical Parasitology, v.45, p.147-154.

Galkiewicz, T. (1968) *Geocosmology*. International Geological Congress. 23rd, Prague, 1968. Report, sec. 13, p. 145-149.

Galliher, E. W. (1931) *Collophane from Miocene brown shales of California*. American Association of Petroleum Geologists. Bulletin, v. 15, p. 257-269.

Galloway, J. J. (1922) *Value of the physical characters of sand grains in the interpretation of the origin of sandstones*. Geological Society of America. Bulletin, v. 33, p. 104-105.

Galton, Francis (1889) *On the principle and methods of assigning marks for bodily efficiency*. Nature, v. 40, p. 649-653.

Gammon, J. B. (1966) *Fahlbands in the Precambrian of southern Norway*. Economic Geology, v. 61, p. 174-188.

Gardiner, J. S. (1906) *The fauna and geography of the Maldive and Laccadive Islands*. Cambridge: University Press, 2 vols., 1079 p.

Gardner, Rita, and Scoging, Helen (1983) *Megageomorphology*. Oxford: Clarendon Press.

Garland, G. D. (1971) *Introduction to geophysics: mantle, core, and crust*. Philadelphia: W. B. Saunders. 420 p.

Gasche, H., and Tunca, Ö (1983) *Guide to Archaeostratigraphic classification and terminology: definitions and principles*. Journal of Field Archaeology, v. 12, p. 325-335.

Gault, D. E., et al. (1968) *Impact cratering mechanics and structures*. In: French, B. M., and Short, N. M., eds. Shock metamorphism of natural materials, p. 87-100. Baltimore: Mono Book Corp. 644 p.

Geddes, M.C., DeDeckker, P., Williams, W.D., Morton, D.W., and Toppings, M., (1981) *On the chemistry and biota of some saline lakes in Western Australia*. Hydrobiologia, v.81, p.201-222.

Geikie, J. (1885) *The ice age in Europe and North America*. Edinburgh, U.K., Transactions of the Edinburgh Geological Society, p.144-168.

Geikie, James (1898) *Earth sculpture, or the origin of land-forms*. New York: Putnam. 397 p.

Geiser, P. A. (1988) *The role of kinematics in the construction and analysis of geological cross sections in deformed terranes*. In: Geometries and mechanisms of thrusting, with special reference to the Appalachians , edited by Mitra, G. and Wojtal, S. Spec. Pap. Geol. Soc. Am., v. 222, p. 47-77.

Geissler, P. (2000) *Cryovolcanism in the outer solar system*. In: Sigurdsson, H., ed., *Encyclopedia of volcanoes*. New York, Academic Press, p.785-800.

Geo-One-Stop: E-Government Geospatial One-Stop (May 9, 2003) *Data content standards for digital elevation*. Draft, 33p. http://www.geo-one-stop.gov/Standards/Elevation/index.html

Geological Society of London (1967) *Report of the Stratigraphical Code Sub-Committee*. Geological Society of London. Proceedings, no. 1638, p. 75-87.

George, T. N., et al. (1969) *Recommendations on stratigraphical usage*. Geological Society of London. Proceedings no. 1656, p. 139-166.

Gignoux, M. (1950) *Géologie Stratigraphique*. Masson: Paris, 735 p.

Gilbert, G. K. (1875) *Report on the geology of portions of Nevada, Utah, California, and Arizona, examined in the years 1871 and 1872*. U. S. Geographical Surveys West of the One Hundredth Meridian. Report, v. 3, pt. 1, p. 17-187.

Gilbert, G. K. (1876) *The Colorado Plateau province as a field for geological study*. American Journal of Science, ser. 3, v. 12, p. 16-24, 85-103.

Gilbert, G. K. (1877) *Report on the geology of the Henry Mountains*. Washington: U. S. Geographical and Geological Survey of the Rocky Mountain Region. 160 p.

Gilbert, G. K. (1885) *The topographic features of lake shores*. U. S. Geological Survey. Annual Report, 5th, p. 69-123.

Gilbert, G. K. (1890) *Lake Bonneville*. U. S. Geological Survey. Monograph 1. 438 p.

Gilbert, G. K. (1898) *A proposed addition to physiographic nomenclature*. Science, n.s., v. 7, p. 94-95.

Gilbert, G. K. (1899) *Ripple-marks and cross-bedding*. Geological Society of America. Bulletin, v. 10, p. 135-140.

Gilbert, G. K. (1904) *Systematic asymmetry of crest lines in the high Sierra of California*. Journal of Geology, v. 12, p. 579-588.

Gilbert, G. K. (1914) *The transportation of debris by running water*. U. S. Geological Survey. Professional Paper 86. 263 p.

Gilbert, G. K. (1928) *Studies of Basin Range structure*. U. S. Geological Survey. Professional Paper 153. 92 p.

Gilbert, G. K., and Brigham, A. P. (1902) *An introduction to physical geography*. New York: Appleton. 380 p.

Giles, A. W. (1918) *Eskers in the vicinity of Rochester, N.Y*. Rochester Academy of Science. Proceedings, v. 5, p. 161-240.

Gill, J.B. (1981) *Orogenic andesites and plate tectonics*. Berlin, Springer Verlag, 401p.

Glaessner, M. F. (1945) *Principles of micropaleontology*. Melbourne: Melbourne Univ. Press. 296 p.
Glaessner, M. F., and Teichert, C. (1947) *Geosynclines: a fundamental concept in geology*. American Journal of Science, v. 245, p. 465-482, 571-591.
Gleeson, S.A., Butt, C.R.M., and Elias, M. (2003) *Nickel laterites, a review*. SEG Newsletter, no.54, p.1&12-18.
Glen, J. W. (1955) *The creep of polycrystalline ice*. Royal Society of London. Proceedings, ser. A, v. 228, p. 519-538.
Glenn, M. (1979) *Glossary*. In: A study of global sand seas, U.S. Geological Survey Professional Paper 1052. p. 399-407.
Glock, W. S. (1928) *An analysis of erosional forms*. American Journal of Science, ser. 5, v. 15, p. 471-483.
Glock, W. S. (1932) *Available relief as a factor of control in the profile of a landform*. Journal of Geology, v. 40, p. 74-83.
Glynn, P. W. (1974) *Rolling stones among the Scleractinia: mobile coralliths in the Gulf of Panama. Second International Coral Reef Symposium*. Proceedings, v. 2, p. 183-198. Brisbane: The Great Barrier Reef Committee. 753 p.
Godwin, Harry (1962) *Radiocarbon dating; fifth international conference*. Nature, v. 195, p. 943-945.
Goggin, D.J. (1993) *Probe permeametry: Is it worth the effort?* Marine and Petroleum Geology, v.10, p.299-308.
Goguel, J. (1962) *Tectonics*. San Francisco, Freeman, 384p. (Translation by H.E. Thalman of French original, *Traité de tectonique*, Paris, Masson, 1952).
Goldich, S. S. (1938) *A study in rock weathering*. Journal of Geology, v. 46, p. 17-58.
Goldman, M. I. (1921) *Lithologic subsurface correlation in the "Bend" series of north-central Texas*. U. S. Geological Survey. Professional Paper 129. p. 1-22.
Goldman, M. I. (1933) *Origin of the anhydrite cap rock of American salt domes*. U. S. Geological Survey. Professional Paper 175-D, p. 83-114.
Goldman, M. I. (1952) *Deformation, metamorphism, and mineralization in gypsum-anhydrite cap rock, Sulphur salt dome, Louisiana*. Geological Society of America. Memoir 50. 169 p.
Goldman, M. I. (1961) *Regenerated anhydrite redefined*. Journal of Sedimentary Petrology, v. 31, p. 611.
Goldring, Winifred (1943) *Geology of the Coxsackie quadrangle, New York*. New York State Museum. Bulletin, no. 332. 374 p.
Goldschmidt, V. M. (1937) *The principles of distribution of chemical elements in minerals and rocks*. Chemical Society (of London). Journal, v. 1937, p. 655-673.
Goldschmidt, V. M. (1954) *Geochemistry*. Oxford: Clarendon Press.
Goldsmith, Richard (1959) *Granofels: a new metamorphic rock name*. Journal of Geology, v. 67, p. 109-110.
Goldsmith, V. (1985) *Coastal dunes*. In: R. A. Davis, Jr., ed. Coastal sedimentary environments, p. 303-378. New York: Springer-Verlag.
Goldthwait, J. W. (1908) *Intercision, a peculiar kind of modification of drainage*. School Science and Mathematics, v. 8, p. 129-139.
Goode, H. D. (1969) *Geoevolutionism: a step beyond catastrophism and uniformitarianism*. Geological Society of America. Abstracts with programs for 1969, pt. 5, p. 29.
Goodspeed, G. E. (1947) *Xenoiths and skialiths*. American Journal of Science, v. 246, p. 515-525.
Goodspeed, G. E. (1948) *Origin of granites*. In: Gilluly, J. Origin of granite. Geological Society of America. Memoir 28. p. 55-78.
Gore, A. J. P. (1983) *Introduction*. in A. J. P. Gore (editor), Ecosystems of the world 4A General Studies; mires, swamp, bog, fen, and moor. Elsevier, chap. 1, p. 1-34.
Gornitz, V.M., and Schreiber, B.C., (1981) *Displacive halite hoppers from the Dead Sea: Some implications for ancient evaporite deposits*. Journal Sedimentary Petrology, v.51, p.787-794.
Govett, G. J. S. (1978) *Lithogeochemistry*. Journal of Geochemical Exploration, v. 9, p. 109-110.
Grabau, A. W. (1903) *Paleozoic coral reefs*. Geological Society of America. Bulletin, v. 14, p. 337-352.
Grabau, A. W. (1904) *On the classification of sedimentary rocks*. American Geologist, v. 33, p. 228-247.
Grabau, A. W. (1905) *Physical characters and history of some New York formations*. Science, v. 22, p. 528-535.
Grabau, A. W. (1906) *Types of sedimentary overlap*. Geological Society of America. Bulletin, v. 17, p. 567-636.
Grabau, A. W. (1911) *On the classification of sand grains*. Science, n.s., v. 33, p. 1005-1007.
Grabau, A. W. (1913) *Principles of stratigraphy*. Reprinted, 1960, by Dover Publishing Co., New York.
Grabau, A. W. (1920) *Principles of salt deposition. 1st ed*. New York: McGraw-Hill. 435 p. (Geology of the non-metallic mineral deposits other than silicates, v. 1.)
Grabau, A. W. (1920a) *General geology*. Boston: Heath. 864 p. (A textbook of geology. Pt. 1.)
Grabau, A. W. (1924) *Principles of stratigraphy. 2nd ed*. New York: A. G. Seiler. 1185 p.
Grabau, A. W. (1932) *Principles of stratigraphy. 3d ed*. New York: A. G. Seiler. 1185 p.
Grabau, A. W. (1936) *The Great Huangho Plain of China*. Association of Chinese and American Engineers. Journal, v. 17, p. 247-266.
Grabau, A. W. (1936a) *Oscillation or pulsation*. International Geological Congress. 16th, Washington, 1933. Report, v. 1, p. 539-553.
Grabau, A. W. (1940) *The rhythm of the ages; Earth history in the light of the pulsation and polar control theory*. Peking: Henri Vetch. 561 p.
Graham, A., Wood, G.D., Elsik, W.C., and Speed, R.C. (2000) *Petrofilaments in palynological preparations*. American Journal of Botany, v.87, no.5, p.752-753.
Graton, L. C., and Fraser, H. J. (1935) *Systematic packing of spheres, with particular relation to porosity and permeability*. Journal of Geology, v. 43, p. 785-909.
Gravenor, C. P. (1956) *Air photographs of the plains region of Alberta*. Research Council of Alberta. Preliminary Report 56-5. 35 p.
Gravenor, C. P., and Kupsch, W. O. (1959) *Ice-disintegration features in western Canada*. Journal of Geology, v. 67, p. 48-64.
Gray, H. H. (1955) *Stratigraphic nomenclature in coal-bearing rocks*. Geological Society of America. Bulletin, v. 66, p. 1567-1568.
Green, D. H., and Ringwood, A. E. (1963) *Mineral assemblages in a model mantle composition*. Journal of Geophysical Research, v. 68, p. 937-945.
Greene-Kelly, R. (1953) *Irreversible dehydration in montmorillonite; Part II*. Clay Mineral Bull. 2, p. 52-56.
Greenly, E. (1919) *The geology of Anglesey*. Geological Survey of Great Britain. Memoir, v. 1.
Grégoire C. (1972) *Structure of the molluscan shell*. In: Florkin, M., and Scheer, B.T., eds., *Chemical zoology, v.7, Mollusca*. New York, Academic Press, chap.2, p.45-102.
Grégoire C. (1974) *On the organic and mineral components of the shells of Aetheriidae(Mollusca, Bivalvia, Unionacea)*. Revue de Zoologie Africaine, v.88, pt.4, p.847-896.
Gregory, H. E. (1917) *Geology of the Navajo country; a reconnaissance of parts of Arizona, New Mexico, and Utah*. U. S. Geological Survey. Professional Paper 93. 161 p.
Gregory, J. W. (1912) *The relations of kames and eskers*. Geographical Journal, v. 40, p. 169-175.
Grenfell, H.R. (1995) *Probable fossil zygnematacean algal spore genera*. Review of Palaeobotany and Palynology, v.84, p.201-220
Gressly, Amanz (1838) *Observations géologiques sur le Jura Soleurois. [Pt. 1]*. Allgemeine Schweizerische Gesellschaft für die gesammten Naturwissenschaften. Neue Denkschriften, v. 2, [pt. 6]. 112 p.
Greuter, W., et al. (2000) *International Code of Botanical Nomenclature (Saint Louis Code)*. Regnum Vegetabile v.138. Königstein, Germany, Koeltz Scientific Books, 474p.
Greuter, Werner, ed. (1994) *International Code of Botanical Nomenclature (Tokyo Code)*. Regnum Vegetabile, vol. 131, 389 pp.; Adopted by the Fifteenth International Botanical Congress, Yokohama, Aug.-Sept. 1993.
Griffith, A. A. (1920) *The phenomena of rupture and flow in solids*. Philosophical Transactions of the Royal Society of London, Series A, v. 221, p. 163-198.
Griffith, A. A. (1924) *Theory of rupture. Proceedings of the First International Congress on Applied Mechanics*, Delft, the Netherlands, p. 55-63.
Griffiths, J. C., and Rosenfeld, M. A. (1954) *Operator variation in experimental research*. Journal of Geology, v. 62, p. 74-91.
Griggs, D. T. (1936) *Deformation of rocks under high confining pressures*. Journal of Geology, v. 44, p. 541-577.
Grim, R.E. (1953) *Clay mineralogy*. New York, McGraw-Hill, 596p.
Grim, R. E. (1953) *Clay mineralogy*. New York: McGraw-Hill. 384 p.
Grim, R.E. (1962) Applied *clay mineralogy*. New York, McGraw-Hill, 422p.
Grim, R. E. (1968) *Clay mineralogy. 2nd ed*. New York: McGraw-Hill. 596 p.
Grim, R. E., and Güven, Necip (1978) *Bentonites*. Amsterdam: Elsevier. 256 p.
Grim, R. E., et al. (1937) *The mica in argillaceous sediments*. American Mineralogist, v. 22, p. 813-829.
Grimshaw, R.W. (1971) *The chemistry and physics of clays and allied ceramic materials, 4th ed*. New York, Wiley-Interscience, 1,024p.
Grohskopf, J. G., and McCracken, Earl (1949) *Insoluble residues of some Paleozoic formations of Missouri, their preparation, characteristics and application*. Missouri Division of Geological Survey and Water Resources. Report of Investigations, no. 10. 39 p.
Gross, G. A. (1980) *A classification of iron formations based on depositional environments*. Canadian Mineralogist, v. 18, p. 215-222.
Gross, G.A. (1990) *Geochemistry of iron formation in Canada*. In: Chauvel, J.J., and others, eds., *Ancient banded iron formations*. Athens, Greece, Theophrastus Publications, p.3-26.
Gross, M. R. (1993) *The origin and spacing of cross joints: examples from the Monterey Formation, Santa Barbara Coastline, California*. Journal of Structural Geology, v. 15, p. 737-751.
Grossman, W. L. (1944) *Stratigraphy of the Genesee Group of New York*. Geological Society of America. Bulletin, v. 55, p. 41-75.
Grove, J.M. (1988) *The Little Ice Age*. London, Methuen, 498p.
Grove, J.M. (2001) *The onset of the Little Ice Age*. In: Jones, P.D., Ogilvie, A.E.J., Davies, T.D., and Brilla, K.R., eds., *History and climate: Memories of the future?* New York, Kluwer Academic, p.153-186.
Grubenmann, Ulrich (1904) *Die Kristallinen Schiefer*. Berlin: Gebrüder Borntraeger. 2 vol.
Grubenmann, Ulrich, and Niggli, P. (1924) *Die Gesteinsmetamorphose: I.* Allgemeiner Teil. Berlin: Gebrüder Borntraeger. 539 p.

Gruner, J. W. (1950) *An attempt to arrange silicates in the order of reaction energies at relatively low temperatures.* American Mineralogist, v. 35, p. 137-148.

Gruner, J. W., et al. (1941) *Structural geology of the Knife Lake area of northeastern Minnesota.* Geological Society of America. Bulletin, v. 52, p. 1577-1642.

Gümbel, C. W. von (1868) *Geognostische Beschreibung des ostbayerischen Grenzgebirges oder des bayerischen und oberpfèlzer Waldgebirges. Gotha:* Justus Perthes. 968 p. (Bavaria. Bayerisches Oberbergamt Geognostische Abteilung. Geognostische Beschreibung des Koenigreichs Bayern, v. 2.)

Gümbel, C. W. von (1874) *Die palèolithischen Eruptivgesteine des Fichtelgebirges* (als vorlèufige Mittheilung). Munich. 50 p.

Günthert, A., et al. (1976) *Isochemische granitgneisbildung im Maggia-Lappen (Lepontin der zentralalpen).* Schweizerische Mineralogische-petrographische Mitteilungen, v. 56, p. 105-143.

Güven, N. (1971) *Structural factors controlling stacking sequences in dioctahedral micas.* Clays and Clay Minerals 19, p. 159-165.

Guggenheim, S., and Martin, R.T. (1995) *Definition of clay and clay mineral. Joint report of AIPEA and CMS nomenclature committees.* Clays and Minerals, v.43, p.255-256.

Guilbert, J. M., and Park, C. F., Jr. (1986) *The geology of ore deposits.* New York: W. H. Freeman. 985 p.

Guilcher, A. (1988) *Coral reef geomorphology.* Chichester, New York: J. Wiley & Sons, 228 p.

Guilcher, André (1950) *Définition d'un type de volcan "écossais".* Association de Géographes Francais. Bulletin, no. 206-207, p. 2-11.

Guinier, A., et al. (1984) *Nomenclature of polytype structures; report of the International Union of Crystallography Ad-Hoc Committee on the nomenclature of disordered, modulated and polytype structures.* Acta Cryst. A40, p. 399-404.

Gulliver, F. P. (1896) *Cuspate forelands.* Geological Society of America. Bulletin, v. 7, p. 399-422.

Gulliver, F. P. (1899) *Shoreline topography.* American Academy of Arts and Sciences. Proceedings, v. 34, p. 149-258.

Guptill, S.C., Boyko, K.J., Domaratz, M.A., Feagas, R.G., Rossmeissl, H.J., and Usery, E.L. (1990) *An enhanced digital line graph design.* U.S. Geological Survey Circular 1048, 134p.

Gussow, W. C. (1954) *Differential entrapment of oil and gas: a fundamental principle.* American Association of Petroleum Geologists. Bulletin, v. 38, p. 816-853.

Gussow, W. C. (1958) *Metastasy or crustal shift.* Alberta Society of Petroleum Geologists. Journal, v. 6, p. 253-257.

Gussow, W. C. (1962) *Energy source of intrusive masses.* Royal Society of Canada. Transactions, v. 56, series 3, sec. 3, p. 1-19.

Gussow, William Carruthers (1972) *The rate of mountain building.* In: Tectonics—Tectonique, Section 3. Report of the ... Session - International Geological Congress, 24, 3, p. 355-362.

H

Haarmann, E. (1926) *Tektogenese oder Gefugebildung statt Orogenese oder Gebirgsbildung.* Deutsche Geologische Gesellschaft. Zeitschrift, v. 78, p. 105-107.

Haarmann, E. (1930) *Die Oszillationstheorie; eine erklarung der krustenbewegungen von erder und mond.* Stuttgart: F. Enke, 260 p.

Haas, W. H. (1962) *Conodonts.* In: Moore, R. C., ed. Treatise on Invertebrate Paleontology. Part W, p. 3-69.

Hack, J. T. (1960) *Interpretation of erosional topography in humid temperate regions.* American Journal of Science, v. 258a, p. 80-97.

Hadding, Assar (1931) *On subaqueous slides.* Geologiska Föreningen, Stockholm. Förhandlingar, Bd. 53, p. 377-393.

Hadding, Assar (1933) *On the organic remains of the limestones; a short review of the limestone forming organisms.* Acta universitatis Lundensis, n.s., Bd. 29, pt. 2, no. 4. 93 p. (The pre-Quaternary sedimentary rocks of Sweden, 5) .

Haeckel, Ernst (1862) *Die Radiolarien (Rhizopoda radiaria)*; eine Monographie. Berlin: Georg Reimer. 572 p.

Haefeli, R. (1962) *The ablation gradient and the retreat of a glacier tongue.* International Association of Scientific Hydrology Publication 58 (Symposium of Obergurgl), p.49-59.

Haenel, R., Rybach, L., and Stegena, L, eds., (1988) *Handbook of terrestrial heat-flow density determination.* Kluwer Acad. Publ., Dordrecht, 486p.

Hageman, B. P. (1969) *Development of western part of the Netherlands during the Holocene.* Geologie en Mijnbouw, v. 48, p. 373-388.

Hall, James (1843) *Geology of New York; part IV, comprising the survey of the fourth geological district.* Albany: Carroll and Cook. 683 p.

Hall, James (1859) *Description and figures of the organic remains of the lower Helderberg Group and the Oriskany Sandstone.* New York Geological Survey. Paleontology, v. 3. 532 p.

Hallet, B. (1990) *Spatial self-organization in geomorphology: from periodic bedforms and patterned ground to scale-invariant topography.* Earth Science Reviews, v.29, p.57-75.

Halliday, W.R. (2003) *Volcanic caves.* In: Gunn, J., ed., *Encyclopedia of caves and karst science.* New York, Routledge, 960p.

Halls, H.C., and Fahrig, W.F., eds. (1987) *Mafic dyke swarms: A collection of papers based on the proceedings of an international conference.* Geological Association of Canada Special Paper 34, 503p.

Hamblin, W. K. (1958) *The Cambrian sandstones of northern Michigan.* Michigan Geological Survey. Publication 51. 149 p.

Hamelin, Louis-Edmond (1961) *Périglaciaire du Canada: idées nouvelles et perspectives globales.* Cahiers de géographie de Québec, v. 5, no. 10, p. 141-203.

Hamelin, Louis-Edmond, and Clibbon, Peter (1962) *Vocabulaire périglaciaire bilingue (francais et anglais).* Cahiers de géographie de Québec, v. 6, no. 12, p. 201-226.

Hamelin, Louis-Edmond, and Cook F. A. (1967) *Le périglaciaire par l'image; illustrated glossary of periglacial phenomena.* Quebec: Presses de l'Université Laval. 237 p. (Centre d'Etudes Nordiques. Travaux et documents 4.)

Hamilton, E. J. (1965) *Applied geochronology.* New York: Academic Press. 267 p.

Hammer, W. (1914) *Das Gebiet der Bündnerschiefer im tirolischen Oberinntal.* Kaiserliche Köningliche Geologische Reichsanstalt (Austria). Jahrbuch, v. 64, p. 443-567.

Hammond, A. L. (1974) *Bright spot: better seismological indicators of oil and gas.* Science, v. 185, p. 515-517.

Hancock, P. L. (1972) *The analysis of en-echelon veins.* Geological Magazine, v. 109, p. 269-276.

Hancock, P. L. (1985) *Brittle microtectonics: principles and practice.* Journal of Structural Geology, v. 7, p. 437-457.

Hancock, P. L., and Engelder, T. (1989) *Neotectonic joints.* Geological Society of America. Bulletin, v. 101, p. 1197-1208.

Hansen, A. M. (1894) *The glacial succession in Norway.* Journal of Geology, v. 2, p. 123-143.

Hansen, E. (1967) *A method of deducing slip-line orientations from the geometry of folds.* Carnegie Inst. Yearbook, v. 65, p. 387-405.

Hansen, Wallace R. (1991) *Suggestions to authors of the reports of the United States Geological Survey. 7th ed.* Washington, DC: U.S. Government Printing Office, 289 p.

Hapke, B. W. (1966) *Optical properties of the Moon's surface.* In: Hess, W. N., and others, eds. The nature of the lunar surface, p. 141-154. Baltimore: Johns Hopkins Press. 320 p.

Haranczyk, C. (1969) *Noncolloidal origin of colloform textures.* Economic Geology, v.64, p.466-468.

Harbaugh, J. W., and Bonham-Carter, G. F. (1970) *Computer simulation in geology.* New York: Wiley. 575 p.

Harbaugh, J. W., and Merriam, D. F. (1968) *Computer applications in stratigraphic analysis.* New York: Wiley. 282 p.

Hardie, L.A., Smoot, J.P., and Eugster, H.P., (1978) *Saline lakes and their deposits: a sedimentological approach.* In: Matter, A., and Tucker, M.E., eds., Modern and Ancient Lake Sediments , International Association of Sedimentologist Special Paper 2, p.7-41.

Harding, R. R., and Jobbins, E. A. (1984) *Verdite and ruby-verdite from Zimbabwe.* Journal of Gemmology, v. 19, p. 150-159.

Hardy, C. T., and Williams, J. S. (1959) *Columnar contemporaneous deformation.* Journal of Sedimentary Petrology, v. 29, p. 281-283.

Hare, J.L., Ferguson, J.F., Aiken, C.L.V., and Brady, J.L. (1997) *4-D microgravity modeling and inversion for waterflood surveillance. A model study for the Prudhoe Bay Reservoir, Alaska.* Society of Exploration Geophysicists, 67th Annual International Meeting, Expanded Abstracts, p.513-516.

Harker, Alfred (1904) *Tertiary igneous rocks of Skye.* Geological Survey of the United Kingdom. Memoir. Glasgow: H. M. Stationery Office. 481 p.

Harker, Alfred (1909) *The natural history of igneous rocks.* New York: Macmillan. 377 p.

Harker, Alfred (1918) *The anniversary address of the President.* Geological Society of London. Quarterly Journal, v. 74, p. i-lxxx.

Harker, Alfred (1939) *Metamorphism. 2nd ed.* London: Methuen. 362 p.

Harland, W. B. (1971) *Tectonic transpression in Caledonian Spitzbergen.* Geological Magazine, v. 108, p. 27-42.

Harland, W. B., et al. (1966) *The definition and identification of tills and tillites.* Earth Science Reviews, v. 2, p. 255-256.

Harland, W. B., et al. (1982) *A geologic time scale.* Cambridge University Press. 131 p.

Harland, W. B., et al. (1990) *A geologic time scale 1989.* Cambridge University Press, 163 p.

Harland, W. B., et al., eds. (1964) *The Phanerozoic time-scale: a symposium.* Geological Society of London. Quarterly Journal, v. 120, Supplement. 458 p.

Harms, J. C., et al. (1975) *Depositional environments as interpreted from primary sedimentary structures and stratification sequences.* Society of Economic Paleontologists and Mineralogists. Short course, lecture notes 2. 161 p.

Harper, W. G. (1957) *Morphology and genesis of calcisols.* Soil Science Society of America. Proceedings, v. 21, p. 420-424.

Harrassowitz, H. (1927) *Anchimetamorphose, das Gebiet zwischen Oberflächen- und Tiefenumwandlung der Erdrinde.* Oberhessischen Gesellschaft für Natur- und Heilkunde zu Giessen. Naturwissenschaftliche Abteilung. Bericht, 12, p. 9-15.

Harris, D.P. (1984) *Mineral resources appraisal: Mineral endowment, resources, and potential supply; concepts, methods, and cases.* Oxford, UK, Clarendon Press, 455p.

Harris, J.G., Mylroie, J.E., and Carew, J.L., (1995) *Banana holes: Unique karst features of the Bahamas.* Carbonates and Evaporites, v.10, no.2, p.215-224.

Harris, S. E., Jr. (1943) *Friction cracks and the direction of glacial movement.* Journal of Geology, v. 51, p. 244-258.

Harrison, T. S. (1927) *Colorado-Utah salt domes.* American Association of Petroleum Geologists Bulletin, v. 11, p. 111-133.

Hart, S.R. (1988) *Heterogeneous mantle domains: Signatures, genesis, and mixing chronologies.* Earth and Planetary Science Letters, v.90, p.273-296.

Harte, Ben (1977) *Rock nomenclature with particular relation to deformation and recrystallization textures in olivine-bearing xenoliths.* Journal of Geology, v. 85, p. 279-288.

Hartman, R.R., Teskey, D.J. and Friedberg, J.J., (1971) *A system for rapid digital aeromagnetic interpretation.* Geophysics, 36, 891-918.

Hartshorn, J. H. (1958) *Flowtill in southeastern Massachusetts.* Geological Society of America. Bulletin, v. 69, p. 477-482.

Harvey, R. D. (1931) *Glacial chutes from the Peruvian Cordillera.* American Journal of Science, 5th ser., v. 31, p. 220-231.

Harvie-Brown, J. A. (1910) *"Caledonia rediviva"* Scottish Geographical Magazine, v. 26, p. 93-94.

Hass, W. H. (1962) *Conodonts.* In: Moore, R. C., ed. Treatise on Invertebrate Paleontology. Part W, p. 3-69.

Hatch, F. H., and Rastall, R. H. (1913) *The petrology of the sedimentary rocks; a description of the sediments and their metamorphic derivatives.* London: George Allen. 425 p.

Hatch, F. H., and Rastall, R. H. (1965) *Petrology of the sedimentary rocks. 4th ed.* revised by J. Trevor Greensmith. London: Thomas Murby. 408 p. (Textbook of petrology, v. 2.)

Hatch, F. H., Wells, A. K., and Wells, M. K. (1961) *Petrology of the igneous rocks, 12th ed.* London: Thomas Murby. 515 p.

Haüy, R. J. (1801) *Traité de minéralogie. T. 2.* Paris: Chez Louis. 444 p.

Haug, Emile (1900) *Les géosynclinaux et les aires continentales; contribution à l'étude des transgressions et des regressions marines.* Société Géologique de France. Bulletin, ser. 3, v. 28, p. 617-711.

Haug, Emile (1907) *Les phénomènes géologiques.* Paris: Librairie Armand Colin. 538 p. (His Traité de géologie, 1).

Haupt, A. W. (1953) *Plant morphology.* New York: McGraw-Hill.

Haupt, L. M. (1883) *The topographer, his instruments and his methods.* Philadelphia: H. C. Baird. 247 p.

Haupt, L. M. (1906) *Changes along the New Jersey coast.* New Jersey Geological Survey. Annual report of the State Geologist, 1905, p. 27-95.

Hawkes, H. E. (1958) *Principles of geochemical prospecting.* U. S. Geological Survey. Bulletin 1000, p. 225-353.

Hawkes, H. E., and Webb, J. S. (1962) *Geochemistry in mineral exploration.* New York: Harper & Row. 415 p.

Hay, R. L. (1978) *Geologic occurrence of zeolites.* In: Sand, L. B., and Mumpton, F. A., eds. Natural zeolites, p. 135-143. New York: Pergamon. 546 p.

Hayek, L.C., and Buzas, M.A. (1997) *Surveying natural populations.* New York, Columbia University Press

Hayes, C. W. (1899) *Physiography of the Chattanooga district, in Tennessee, Georgia, and Alabama.* U. S. Geological Survey. Annual Report, 19th, pt. 2, p. 1-58.

Hayes, M.O. (1975) *Morphology of sand accumulations in estuaries.* In: Cronin, L.E., ed., *Estuarine Research, 2.* New York, Academic Press, p.3-22.

Hayes, M. O. (1979) *Barrier island morphology as a function of tidal and wave regime.* In: S. P. Leatherman, ed. Barrier islands, p. 1-27. New York: Academic Press.

Hayes, M. O. (1980) *General morphology and sediment patterns in tidal inlets.* Sedimentary Geology, v. 26, p. 139-156.

Haynes, Vance (1973) *The Calico site: artifacts or geofacts?* Science, v. 181, p. 305.

Head, M.J. (1998) *New goniodomacean dinoflagellates with a compound hypotractal archeopyle from the late Cenozoic:* Capisocysta *Warny and Wrenn, emend.* Paleontology, v.72, p.795-807.

Heald, M. T. (1956) *Cementation of Simpson and St. Peter sandstones in parts of Oklahoma, Arkansas, and Missouri.* Journal of Geology, v. 64, p. 16-30.

Heath, R. C. (1983) *Basic ground-water hydrology.* U. S. Geological Survey Water-Supply Paper 2220, 84 p.

Heath, Ralph E. (1989) *The Piedmont ground-water system.* In: Proceedings of a conference on ground water in the Piedmont of the Eastern United States. Clemson University, p. 1-13.

Hedberg, H. D. (1958) *Stratigraphic classification and terminology.* American Association of Petroleum Geologists. Bulletin, v. 42, p. 1881-1896.

Hedberg, H. D. (1961) *The stratigraphic panorama (an inquiry into the bases for age determination and age classification of the Earth's rock strata).* Geological Society of America. Bulletin, v. 72, p. 499-517.

Heezen, B. C., and Menard, H. W. (1963) *Topography of the deep-sea floor.* In: The Sea, v. 3, ch. 12, p. 233-280. New York: Interscience.

Heezen, B. C., Tharpm, M., and Ewing, M. (1959) *The floors of the oceans, I: the North Atlantic.* Geological Society of America Special Paper 62, 122 p.

Heiken, G., Krier, D., McCormick, T., and Snow, M.G. (2000) *Intracaldera volcanism and sedimentation: Creede Caldera, Colorado.* Geological Society of America Special Paper 346, p.127-157.

Heiken, G., and Wohletz, K. (1985) *Volcanic ash.* Berkeley, University of California Press, 246p.

Heiken, G. H. (1971) *Tuff rings: examples from the Fort Rock—Christmas Lake valley basin, south-central Oregon.* Journal of Geophysical Research, v. 76, p. 5615-5626.

Heim, Arnold (1908) *Über rezente und fossile subaquatische Rutschungen und deren lithologische Bedeutung.* Neues Jahrbuch für Mineralogie, Geologie und Palèontologie, 1908, Bd. 2, p. 136-157.

Heim, Arnold (1934) *Stratigraphische Kondensation.* Eclogae Geologicae Helvetiae, v. 27, p. 372-383.

Heim, Arnold, and Potonié, Robert (1932) *Beobachtungen über die Entstehung der tertièren Kohlen (Humolithe und Saprohumolith) in Zentralsumatra.* Geologische Rundschau, v. 23, no. 3, 4, p. 145-172.

Heinrich, E. W. (1956) *Microscopic petrography.* New York: McGraw-Hill. 296 p.

Heinrich, E. W. (1966) *The geology of carbonatites.* Chicago: Rand McNally. 555 p.

Heinrich, Hartmut (1988) *Origin and consequences of cyclic ice rafting in the northeast Atlantic Ocean during the past 130,000 years.* Quaternary Research, v.29, p.142-152.

Hemley, J. J., and Jones, W. R. (1964) *Chemical aspects of hydrothermal alteration with emphasis on hydrogen metasomatism.* Economic Geology, v. 59, p. 538-569.

Henbest, L. G. (1952) *Significance of evolutionary explosions for diastrophic division of Earth history.* Journal of Paleontology, v. 26, p. 299-318.

Henbest, L. G. (1968) *Diagenesis in oolitic limestones of Morrow (Early Pennsylvanian) age in northwestern Arkansas and adjacent Oklahoma.* U. S. Geological Survey. Professional Paper 594-H. 22 p.

Henderson, E.P. (1949) *Some unusual formations in Skyline Caverns, Va.* National Speleological Society Bulletin, no.11, p.31-34.

Hendricks, C. L. (1952) *Correlations between surface and subsurface sections of the Ellenburger group of Texas.* Texas Bureau of Economic Geology. Report of Investigations, no. 11. 44 p.

Henkes, L. and Johannes, W. (1981) *The petrology of a migmatite (Arvika, Vaermland, western Sweden).* Neues Jarhbuch fur Mineralogie. Abhandlungen, 141, 2, p. 113-133.

Henningsmoen, Gunnar (1961) *Remarks on stratigraphical classification.* Norges Geologiske Undersokelse, no. 213, p. 62-92.

Hensen, B.J. (1971) *Theoretical phase relations involving cordierite and garnet in the system $MgO–FeO–Al_2O_3–SiO_2$.* Contributions to Mineralogy and Petrology, v.33, p.191-214.

Henson, F. R. S. (1950) *Cretaceous and Tertiary reef formations and associated sediments in Middle East.* American Association of Petroleum Geologists. Bulletin, v. 34, p. 215-238.

Hergarten, S. (2002) *Self-organized criticality in Earth systems.* Berlin, Springer, 272p.

Herrick, C. L. (1904) *The clinoplains of the Rio Grande.* American Geologist, v. 33, p. 376-381.

Herrin, Eugene, and Taggart, James (1968) *Source bias in epicenter determinations.* Seismological Society of America. Bulletin, v. 58, p. 1791-1796.

Hess, H. H. (1938) *Gravity anomalies and island arc structures, with particular reference to the West Indies.* American Philosophical Society. Proceedings, v. 79, p. 71-96.

Hess, H. H. (1960) *Stillwater igneous complex, Montana.* Geological Society of America. Memoir 80. 230 p.

Hess, W. N., et al., eds. (1966) *The nature of the lunar surface.* Baltimore: Johns Hopkins Press. 320 p.

Hesse, Richard (1924) *Tiergeographie auf ökologischer Grundlage.* Jena: G. Fischer. 613 p.

Hesse, Richard, et al. (1937) *Ecological animal geography.* New York: Wiley. 597 p.

Hey, M. H. (1962) *An index of mineral species & varieties arranged chemically, with an alphabetical index of accepted mineral names and synonyms. 2nd rev. ed., reprinted with corrections.* London: British Museum (Natural History). 728 p.

Hey, M. H. (1963) *Appendix to the second edition of An index of mineral species and varieties arranged chemically.* London: British Museum (Natural History). 135 p.

Hietanen, Anna (1967) *On the facies series in various types of metamorphism.* Journal of Geology, v. 75, p. 187-214.

Higashi, Akira (1958) *Experimental study of frost heaving.* U. S. Army. Snow, Ice & Permafrost Research Establishment. Research Report 45. 20 p.

Higashi, S. (1982) *Tobelite, a new ammonium dioctahedral mica.* Mineralogical Journal [Japan], v.11, no.3, p.138-146.

Hilgen, F.J., Iaccarino, S.M., Krijgsman, W., Villa, G., Langereis, C.G., and Zachariasse, W.J. (2000) *The Global Boundary Stratotype Section and Point (GSSP) of the Messinian Stage (uppermost Miocene).* Episodes, v.23, no.3, p.172-178.

Hill, C., and Forti, P., eds. (1997) *Cave minerals of the world,* 2nd ed.. Huntsville, Ala., National Speleological Society, 463p.

Hill, C.A., (1992) *Cave Pearls and Pisoliths: a Sedimentological Comparison.* West Texas Geological Society Bulletin, v.31, no.8. p.4-10.

Hill, C.A., (1995) *H_2S-Related Porosity and Sulfuric Acid Oil-Field Karst.* in Budd, D.A., and Harris, P.M., Unconformities in Carbonate Strata-Their Recognition and the Significance of Associated Porosity: American Association of Petroleum Geologists. Memoir 61, p.301-306.

Hill, C.A., and Fort, P., (1986) *Cave minerals of the world. Huntsville, Alabama.* National Speleological Society, 238p.

Hill, Carol A. (1987) *Geology of Carlsbad Cavern and other caves in the Guadalupe Mountains, New Mexico and Texas.* Bulletin - New Mexico Bureau of Mines and Mineral Resources, 117, 150 p.

Hill, R. T. (1891) *The Comanche series of the Texas-Arkansas region.* Geological Society of America. Bulletin, v. 2, p. 503-528.

Hill, R. T. (1900) *Physical geography of the Texas region.* U. S. Geological Survey. Topographic Atlas, folio 3. 12 p.

Hills, E. S. (1940) *The lunette, a new land form of aeolian origin.* Australian Geographer, v. 3, no. 7, p. 15-21.

Hills, E. S. (1963) *Elements of structural geology.* New York: Wiley. 483 p.

Himus, G. W. (1954) *A dictionary of geology.* Baltimore: Penguin Books. 153 p.

Hind, Wheelton, and Howe, J. A. (1901) *The geological succession and palaeontology of the beds between the Millstone Grit and the limestone-massif at Pendle Hill and their equivalents in certain other parts of Britain.* Geological Society of London. Quarterly Journal, v. 57, p. 347-404.

Hinds, N. E. A. (1943) *Geomorphology, the evolution of landscape.* New York: Prentice-Hall. 894 p.

Hitchcock, Edward (1841) *Elementary geology. 2nd ed.* New York: Dayton & Saxton. 346 p.

Hitchcock, Edward (1843) *The phenomena of drift, or glacioaqueous action in North America between the Tertiary and alluvial periods.* Association of American Geologists and Naturalists. Reports, 1st-3rd mtngs, p. 164-221.

Hitchcock, Edward (1844) *Report on ichnolithology or fossil footmarks, with description of several new species and the coprolites of birds, and of a supposed footmark from the valley of Hudson River.* American Journal of Science, v. 47, p. 292-322.

Hitchcock, Edward (1858) *Ichnology of New England; a report on the sandstone of the Connecticut Valley, especially its footmarks.* Boston: William White. 220 p.

Hitzman, M.W., Reynolds, N.A., Sangster, D.F., Allen, C.R., and Carman, C.E. (2003) *Classification, genesis, and exploration guides for nonsulfide zinc deposits.* Economic Geology, v.98, p.685-714.

Hobbs, B. E., et al. (1976) *An outline*

Hobbs, W. H. (1901) *The Newark system of the Pomperaug Valley, Connecticut.* U. S. Geological Survey. Annual Report, 21st, pt. 3, p. 7-160.

Hobbs, W. H. (1906) *Guadix formation of Granada, Spain.* Geological Society of America. Bulletin, v. 17, p. 285-294.

Hobbs, W. H. (1907) *On some principles of seismic geology.* Beiträge zur Geophysik, v. 7, p. 219-292.

Hobbs, W. H. (1911a) *Characteristics of existing glaciers.* New York: Macmillan. 301 p.

Hobbs, W. H. (1911b) *Repeating patterns in the relief and in the structure of the land.* Geological Society of America. Bulletin, v. 22, p. 123-176.

Hobbs, W. H. (1912) *Earth features and their meaning; an introduction to geology for the student and the general reader.* New York: Macmillan. 506 p.

Hobbs, W. H. (1917) *The erosional and degradational processes of deserts, with especial reference to the origin of desert depressions.* Association of American Geographers. Annals, v. 7, p. 25-60.

Hobbs, W. H. (1921) *Studies of the cycle of glaciation.* Journal of Geology, v. 29, p. 370-386.

Hodgson, R. A. (1961) *Classification of structures on joint surfaces.* American Journal of Science, v. 259, p. 493-502.

Hoffmeister, J. E., and Ladd. H. S. (1944) *The antecedent-platform theory.* Journal of Geology, v. 52, p. 388-402.

Hofmann, H. J. (1972) *Precambrian remains in Canada: fossils, dubiofossils, and pseudofossils.* International Geological Congress, 24th, Montreal. Section 1, p. 20-30.

Hofmann, Walter, chairman (1965) *Parametric hydrology and stochastic hydrology; report of the Committee on Surface Water Hydrology.* American Society of Civil Engineers. Proceedings, Hydraulics Division. Journal, v. 91, no. HY6, p. 119-122.

Holland, C.H. (1982) *The state of Silurian stratigraphy.* Episodes, [v.5], no.3, p.21-23.

Holland, C.H. (1985) *Series and stages of the Silurian system.* Episodes, v.8, no.2, p.101-103.

Holland, C. H., et al. (1978) *A guide to stratigraphical procedure.* Special Report - Geological Society, v. 11, 18 p.

Hollingsworth, S.E., Taylor, J.H., and Kellaway, G.A. (1945) *Large-scale superficial structures in the Northampton ironstone field.* Geological Society (London) Quarterly Journal, v.100, p.1-44.

Hollingsworth, S.E., Allison, J., and Godwin, H. (1950) *Interglacial deposits from the Histon road, Cambridge.* Geological Society (London) Quarterly Journal, v.105, p.495-509.

Hollister, C. D., and Heezen, B. C. (1972) *Geologic effects of ocean bottom currents: western North Atlantic.* In: Gordon, A. L., ed. Studies in physical oceanography , v. 2, p. 37-66. New York: Gordon and Breach.

Holm, D. A. (1957) *Sigmoidal dunes; a transitional form.* Geological Society of America. Bulletin, v. 68, p. 1746.

Holm, E. A. (1957) *Sand pavements in the Rub'al Khali.* Geological Society of America. Bulletin, v. 68, p. 1746.

Holmes, Arthur (1920) *The nomenclature of petrology. 1st ed.* London: Thomas Murby. 284 p.

Holmes, Arthur (1928) *The nomenclature of petrology. 2nd ed.* London: Thomas Murby. 284 p.

Holmes, Arthur (1959) *A revised geological timescale.* Edinburgh Geological Society. Transactions, v. 17, p. 183-216.

Holmes, Arthur (1965) *Principles of physical geology. 2nd ed.* New York: Ronald Press. 1288 p.

Holthuis, L. B. (1974) *The lobsters of the superfamily Nephropidea of the Atlantic Ocean.* Bulletin of Marine Science, v. 24, no. 4.

Hon, K., Kauahikaua, J., Deninger, R., and McKay, K. (1994) *Emplacement and inflation of pahoehoe sheet flow: Observations and measurements of active lava flows on Kilauea Volcano, Hawaii.* Geological Society of America Bulletin, v.106, p.351-370.

Honnorez, J., and Kirst, P. (1975) *Submarine basaltic volcanism: Morphometric parameters for discriminating hyaloclastites from hyalotuffs.* Bulletin Volcanologique, v.32, p.441-465.

Hooke, R. L. (1967) *Processes on arid-region alluvial fans.* Journal of Geology, v. 75, p. 438-460.

Hooke, R.L. (2005) *Principles of glacier mechanics, 2nd ed.* Cambridge University Press, 429p.

Hoover, G.M., (1972) *Acoustical holography using digital processing.* Geophysics, 37, 1-19.

Hooykaas, Reijer (1963) *Natural law and divine miracle: the principle of uniformity in geology, biology, and theology.* E. J. Brill. Leiden. 237 p.

Hopkins, D. M. (1949) *Thaw lakes and thaw sinks in the Imuruk Lake Area, Seward Peninsula, Alaska.* Journal of Geology, v. 57, p. 119-131.

Hopkins, D. M., and Sigafoos, R. S. (1951) *Frost action and vegetation patterns on Seward Peninsula, Alaska.* U. S. Geological Survey. Bulletin 974-C, p. 51-100.

Hopkins, D. M., et al. (1955) *Permafrost and ground water in Alaska.* U. S. Geological Survey. Professional Paper 264-F, p. 113-146.

Hoppe, Gunnar (1952) *Hummocky moraine regions, with special reference to the interior of Norbotton.* Geografiska Annaler, v. 34, p. 1-71.

Horberg, Leland (1954) *Rocky Mountain and continental Pleistocene deposits in the Waterton region, Alberta, Canada.* Geological Society of America. Bulletin, v. 65, p. 1093-1150.

Horner, D. R. (1951) *Pressure build-up in wells.* Proceedings of the Third World Petroleum Congress, The Hague, 2, p. 503-521.

Hornibrook, N. de B. (1965) *A viewpoint on stages and zones.* New Zealand Journal of Geology and Geophysics, v. 8, p. 1195-1212.

Horton, R. E. (1932) *Drainage basin characteristics.* American Geophysical Union. Transactions, v. 13, p. 350-361.

Horton, R. E. (1945) *Erosional development of streams and their drainage basins; hydrophysical approach to quantitative morphology.* Geological Society of America. Bulletin, v. 56, p. 275-370.

Hossack, J. (1968) *Pebble deformation and thrusting in the Bygdin area (S. Norway).* Tectonophysics, 5, 315-329.

Howard, A. D. (1942) *Pediment passes and pediment problem.* Journal of Geomorphology, v. 5, p. 1-31, 95-136.

Howard, A. D. (1959) *Numerical systems of terrace nomenclature; a critique.* Journal of Geology, v. 67, p. 239-243.

Howard, A.D. (1967) *Drainage analysis in geologic interpretation: A summation.* AAPG Bulletin, v.51, p.2246-2259.

Howard, W. V., and David, M. W. (1936) *Development of porosity in limestones.* American Association of Petroleum Geologists. Bulletin, v. 20, p. 1389-1412.

Howell, D.G. (1995) *Principles of terrance analysis: New applications for global tectonics.* London: Chapman & Hall. 245p.

Howell, J. V. (1922) *Notes on the pre-Permian Paleozoics of the Wichita Mountain area.* American Association of Petroleum Geologists. Bulletin, v. 6, p. 413-425.

Hower, John, Eslinger, E. V., Hower, M. E., and Perry, E. A. (1976) *Mechanism of burial metamorphism of argillaceous sediments.* Geol. Soc. Amer. Bull. 87, p. 725-737.

Höy, Trygve (1995) *Blackbird sediment-hosted Cu-Co.* In: Lefebure, D.V., and Ray, G.E., eds, *Selected British Columbia mineral deposit profiles. Volume 1, metallics and coal.* British Columbia Geological Survey Open-File Report 1995-20, p.41-43.

Hsü, K. J. (1955) *Monometamorphism, polymetamorphism and retrograde metamorphism.* American Journal of Science, v. 253, p. 237-239.

Hsü, K. J. (1968) *Principles of mélanges and their bearing on the Franciscan-Knoxville paradox.* Geological Society of America. Bulletin, v. 79, p. 1063-1074.

Hsü, K. J. (1975) *Catastrophic debris streams (Sturzstroms) generated by rockfalls.* Geological Society of America. Bulletin, v. 86, p. 129-140.

Hsü, K.J., (1972) *Origin of saline giants: A critical review after the discovery of the Mediterranean evaporite.* Earth Science Reviews, v.8, p.371-396.

Hsü, K.J., and Siegenthaler, C., (1969) *Preliminary experiments on hydrodynamic movement induced by evaporation and their bearing on the dolomite problem.* Sedimentology, v.12, p.11-25.

Hsu, T. C. (1966) *The characteristics of coaxial and non-coaxial strain paths.* Journal of Strain Analysis, v. 1, p. 216-222.

Huang, W. H., and Walker, R. M. (1967) *Fossil alpha-particle recoil tracks: a new method of age determination.* Science, v. 155, p. 1103-1106.

Huang, W. T. (1962) *Petrology.* New York: McGraw-Hill. 480 p.

Hubbard, N. J. (1973) *Chemistry of lunar basalts with very high alumina contents.* Science, v. 181, p. 339-342.

Hubbell, S.P. (2001) *The unified neutral theory of biodiversity and biogeography.* Princeton, N.J., Princeton University Press, 375p.

Hubbert, M.K. (1937) *Theory of scale models as applied to the study of geologic structures.* Geological Society of America Bulletin, v.48, p.1459-1519.

Hubbert, M. K., and Willis, D. G. (1957) *Mechanics of hydraulic fracturing.* Journal of Petroleum Technology, v. 9, p. 153-168.

Hubert, J. F. (1960) *Petrology of the Fountain and Lyons formations, Front Range, Colorado.* Colorado School of Mines. Quarterly, v. 55, no. 1, p. 1-242.

Hubral, O. and Krey, T., (1980) *Interval velocity from seismic reflection time measurements.* Soc. Expt. Geoph.

Hudleston, P. J. (1973) *Fold morphology and some geometrical implications of theories of fold development.* Tectonophysics, v. 16, p. 1-46.

Hudson, G. H. (1909) *Some items concerning a new and an old coast line of Lake Champlain.* New York State Museum. Bulletin 133, p. 159-163.

Hudson, G. H. (1910) *Joint caves of Valcour Island—their age and their origin.* New York State Museum. Bulletin 140, p. 161-196.

Hudson, R. G. (1924) *On the rhythmic succession of the Yoredale Series in Wensleydale.* Yorkshire Geological Society. Proceedings, n.s., v. 20, p. 125-135.

Hughes, N.F. (1994) *The enigma of angiosperm origins.* Cambridge, UK, Cambridge University Press, 303p.

Hughes, R.E., Berres, T.E., Moore, D.M., and Fransworth, K.B. (1998) *Revision of Hopewellian trading patterns in Midwestern North America based on mineralogical sourcing.* Geoarchaeology, v.13, p.709-729.

Hughes, R.E., DeMaris, P.J., White, W.A., and Cowin, D.K. (1987) *Origin of clay minerals in Pennsylvanian strata of the Illinois Basin.* In: Schultz, L.G., van Olphen, H., and Mumpton, F.A., eds., *Proceedings of the International Clay Conference, 1985.* Association Internationale pour l'Étude des Argiles, p.97-104.

Hughes, R. E., Moore, D. M., and Reynolds, R. C., Jr. (1993) *The nature, detection, and occurrence, and origin of kaolinite/smectite.* In: Murray, H. H., Bundy, W. M., and Harvey, C. C., editors, Kaolin genesis and utilization , Spec. Publication No. 1, Clay Minerals Society, Boulder, CO, p. 291-323.

Hulings, N. C., and Gray, J. S. (1971) *A manual for the study of meiofauna.* Smithsonian Contributions to Zoology, no. 78. Washington: Smithsonian Institution. 84 p.

Humble, William (1843) *Dictionary of geology and mineralogy, comprising such terms in botany, chemistry, comparative anatomy, conchology, entomology, palaeontology, zoology, and other branches of natural history, as are connected with the study of geology. 2nd ed., with additions.* London: Henry Washbourne. 294 p.

Humphries, D. W. (1966) *The booming sand of Korizo, Sahara, and the squeaking sand of Gower, S. Wales: a comparison of the fundamental characteristics of two musical sands.* Sedimentology, v. 6, p. 135-152.

Hunt, C. B. (1982) *Overthrust Belt—undefined.* Geotimes, v. 27, no. 1, p. 13.

Hunt, C. B., et al. (1953) *Geology and geography of the Henry Mountains region, Utah.* U. S. Geological Survey. Professional Paper 228. 234 p.

Hunter, R.E. (1977) *Terminology of cross-stratified sedimentary layers and climbing ripple structures.* Journal of Sedimentary Petrology, v.47, p.697-706.

Huschke, R. E., ed. (1959) *Glossary of meteorology.* Boston: American Meteorological Society. 638 p.

Hutchinson, G. E. (1957) *A treatise on limnology. Vol. 1.* New York: Wiley. 1015 p.

Hutchinson, G.E. (1967) *A treatise on limnology, v. II. Introduction to lake biology and the limnoplankton.* New York, John wiley and Sons, 1,115p.

Hutchinson, J. N. & Bhandari, R. K. (1971) *Undrained loading, a fundamental mechanism of mudflows and other mass movements.* Géotechnique, v. 21, no. 4, p. 353-358.

Hutchinson, J. N. (1967) *The free degradation of London Clay cliffs.* Geotechnical conference on shear strength properties of natural soils and rocks, Oslo. Proceedings, 1, p. 113-118.

Hutchinson, J. N. (1968) *Mass movement.* In: Fairbridge, R. W., ed. Encyclopedia of geomorphology, p. 688-696. New York: Reinhold. 1295 p.

Hutton, James (1788) *Theory of the Earth; or an investigation of the laws observable in the composition, dissolution, and restoration of land upon the globe.* Royal Society of Edinburgh. Transactions, v. 1, p. 209-304.

Huxley, T. H. (1862) *The anniversary address.* Geological Society of London. Quarterly Journal, v. 18, p. xl-liv.

Huxley, T. H. (1877) *Physiography: an introduction to the study of nature.* London: Macmillan. 384 p.

Hyde, H. A., and Williams, D. A. (1944) *The right word.* Pollen Analysis Circular, no. 8, p. 6.

I

IAEA: International Atomic Energy Agency (2003) *Guidelines for radioelement mapping using gamma ray spectrometry data.* Vienna, IAEA-TECDOC-1363, 173p.

ICBN: Greuter, W., ed. (1994) *International code of botanical nomenclature (Toyko code)* International Botanical congress, 15th, Yokohama, 1993. Regnum Vegetabile, vol. 131, 389 p.

ICBN: International Code of Botanical Nomenclature (2000) See: Greuter et al., 2000.

ICBN: Stafleu, F. A., ed. (1972) *International code of botanical nomenclature.* International Botanical Congress. 11th, Seattle, 1969. Utrecht, The Netherlands: Oosthoek. 426 p.

ICCP: International Committee for Coal Petrology (1963) *Handbook of coal petrology. 2nd ed.* Paris: Centre National de la Recherche Scientifique.

ICCP: International Committee for Coal Petrology (1971) *Handbook of coal petrology. 2nd ed., supplement.* Paris: Centre National de la Recherche Scientifique.

ICCP: International Committee for Coal and Organic Petrology (1998) *The new vitrinite classification (ICCP System 1994).* Fuel, v.80, p.349-358.

ICCP: International Committee for Coal and Organic Petrology (2001) *The new inertinite classification (ICCP System 1994).* Fuel, v.77, p.459-471.

ICZN: International Commission on Zoological Nomenclature (1964) *International code of zoological nomenclature, adopted by the XV International Congress of Zoology.* N. R. Stoll, Editorial Committee, chairman. 2nd ed. London: International Trust for Zoological Nomenclature. 176 p.

Illing, L. V. (1954) *Bahaman calcareous sands.* American Association of Petroleum Geologists. Bulletin, v. 38, p. 1-95.

Imbrie, J., and Imbrie, K.P. (1979) *Ice ages: Solving the mystery.* London, Macmillan.

Imbrie, John, and Buchanan, Hugh (1965) *Sedimentary structures in modern carbonate sands of the Bahamas.* Society of Economic Paleontologists and Mineralogists. Special Publication, no. 12, p. 149-172.

Imbt, W. C., and Ellison, S. P., Jr. (1947) *Porosity in limestone and dolomite petroleum reservoirs.* Drilling and Production Practice, 1946, p. 364-372.

Immirzi, C. P., Maltby, E., and Clymo, R. S. (1992) *The global status of peatlands and their role in carbon cycling.* Friends of the Earth Trust Limited, London, 145 p.

Ingle, J. C. (1966) *The movement of beach sand; an analysis using fluorescent grains.* New York: Elsevier. 221 p. (Developments in sedimentology, 5) .

Ingram, H. A. P. (1978) *Soil layers in mires; function and terminology.* Journal of Soil Science, v. 29, p. 224-227.

Inoue, A., Kohyama, N., Kitagawa, R., and Watanabe, T. (1987) *Chemical and morphological evidence for the conversion of smectite to illite.* Clays and Clay Minerals 35, p. 111-120.

Insalaco, E. (1998) *The descriptive nomenclature and classification of growth fabrics in fossil scleractinian reefs.* Sedimentary Geology, v.118, p.159-186.

Ireland, H. A., et al. (1947) *Terminology for insoluble residues.* American Association of Petroleum Geologist. Bulletin, v. 31, p. 1479-1490.

Ireland, H. O. (1969) *Foundations for heavy structures.* Reviews in Engineering Geology, v. 2, p. 1-15.

Irvine, T. R., and Baragar, W. R. A. (1971) *A guide to the chemical classification of the common volcanic rocks.* Canadian Journal of Earth Science, v. 8, p. 523-548.

Irwin, G. W. (1957) *Analysis of stresses and strains near the end of a crack traversing a plate.* Journal of Applied Mechanics, v. 24, p. 361-364.

Irwin, J. S. (1926) *Faulting in the Rocky Mountain region.* American Association of Petroleum Geologists. Bulletin, v. 10, p. 105-129.

Irwin, W.P. (1972) *Terranes of the western Paleozoic and Triassic belt in the southern Klamath Mountains, California.* U.S. Geological Survey Professional Paper 800-C, p.C103-C111.

Isacks, Bryan, et al. (1968) *Seismology and the new global tectonics.* Journal of Geophysical Research, v. 73, p. 5855-5899.

ISG: Hedberg, H. D., ed. (1976) *International stratigraphic guide. A guide to stratigraphic classification, terminology, and procedure.* International Union of Geological Sciences, Commission on Stratigraphy, International Subcommission on Stratigraphic Classification. New York: Wiley. 200 p.

ISSC: International Subcommission on Stratigraphic Classification (1970) *Preliminary report on lithostratigraphic units.* H. D. Hedberg, ed. International Union of Geological Sciences, ISSC, Rept. 3. 30 p.

ISSC: International Subcommission on Stratigraphic Classification (1971) *Preliminary report on biostratigraphic units.* H. D. Hedberg, ed. International Geological Congress. 24th, Montreal, 1972. Report 5. 50 p.

ISSC: International Subcommission on Stratigraphic Classification (1971) *Preliminary report on chronostratigraphic units.* H. D. Hedberg, ed. International Geological Congress. 24th, Montreal, 1972. Report 6. 39 p.

ISSC: International Subcommission on Stratigraphic Classification (1971a) *Preliminary report on chronostratigraphic units.* H. D. Hedberg, ed. International Geological Congress. 24th, Montreal, 1972. Report 6. 39 p.

ISSC: International Subcommission on Stratigraphic Classification (1987) *Unconformity-bounded stratigraphic units.* Geological Society of America Bulletin, v. 98, no. 2, p. 232-237.

ISSC: International Subcommission on Stratigraphic Classification of IUGS International Commission on Stratigraphy (1994) *International stratigraphic guide; a guide to stratigraphic classification, terminology, and procedure.* Amos Salvador, ed. 2nd ed; The International Union of Geological Sciences and the Geological Society of America, 214 p.

ISST: International Subcommission on Stratigraphic Terminology (1961) *Stratigraphic classification and terminology.* Edited by H. D. Hedberg. International Geological Congress. 21st, Copenhagen, 1960. Report, pt. 25. 38 p.

ISST: International Subcommission on Stratigraphic Terminology (1965) *Definition of geologic systems.* American Association of Petroleum Geologists. Bulletin, v. 49, p. 1694-1703. (American Commission on Stratigraphic Nomenclature. Note 32)

Ivansson, S., (1986) *Seismic borehole tomography—Theory and computational methods.* Proceedings of the IEEE, 74, 328-338.

Iversen, Johs (1954) *The late-glacial flora of Denmark and its relation to climate and soil.* Denmark. Geologiske Undersogelse. Danmarks geologiske undersogelse, ser. 2, no. 80, p. 87-119.

Ives, R. L. (1941) *Tundra ponds.* Journal of Geomorphology, v. 4, p. 285-296.

Iwata, K. (1981) *Ultrastructure and mineralization of the shell of* Lingula unguis *Linné (inarticulate brachiopod).* Journal of the Faculty of Science, Hokkaido University, Ser.4, v.20, no.1, p.35-65.

J

Jaanusson, Valdar (1960) *The Viruan (Middle Ordovician) of Öland.* Uppsala. University. Geological Institutions. Bulletin, v. 38, p. 207-288.

Jablonski, D. (1986) *Causes and consequences of mass extinction: a comparative approach.* In: D.K. Elliot, ed. Dynamics of extinction. New York: John Wiley & Sons, p. 183-229.

Jacks, G. V., et al., eds. (1960) *Multilingual vocabulary of soil science. 2nd ed., revised.* New York: United Nations, Food and Agriculture Organization, Land & Water Development Division. 430 p.

Jackson, B. D. (1928) *A glossary of botanic terms with their derivation and accent. 4th ed.* Reprinted 1949, 1953. New York: Hafner. 481 p.

Jackson, J. R. (1834) *Hints on the subject of geographical arrangement and nomenclature.* Royal Geographical Society. Journal, v. 4, p. 72-88.

Jackson, M. P. A., and Cramez, Carlos (1989) *Seismic recognition of salt welds in salt tectonics regimes (extended abs.).* Houston, Gulf Coast Section of the Society of Economic Paleontologists and Mineralogists Foundation, Tenth Annual Research Conference Program and Extended Abstracts, p. 66-71.

Jackson, M. P. A., Talbot, C. J., and Cornelius, R. R. (1988) *Centrifuge modeling of the effects of aggradation and progradation on syndepositional salt structures.* The University of Texas at Austin, Bureau of Economic Geology Report of Investigations No. 173, 93 p.

Jackson, M.P.A. and Talbot, C.J. (1991) *A glossary of salt tectonics.* Texas Bureau of Economic Geology, Austin: Geological Circulary 91-4, 44p.

Jacobsen, B.H. (1987) *A case for upward continuation as a standard separation filter for potential-field maps.* Geophysics, v.52, p. 1138-1148.

Jaeger, J. C. (1965) *Application of the theory of heat conduction to geothermal measurements.* In: Terrestrial heat flow, ch. 2. American Geophysical Union. Geophysical Monograph No. 8.

Jagodzinski, H. (1949) *Eindimensionale Fehlordnung in Kristallen und ihr Einuss auf die Rontgeninterferenzen.* I. Berechnung des Fehlordnungsgrades aus der Röntgenintensitaten. Acta Crystallographica, v. 2, p. 201-207.

Jahns, R. H. (1941) *Outwash chronology in northeastern Massachusetts (abstract).* Geological Society of America. Bulletin, v. 52, n. 12, pt. 2, p. 1910.

Jahns, R. H. (1953) *Surficial geology of the Ayer quadrangle, Massachusetts.* U. S. Geological Survey. Map GQ-21.

Jahns, R. H. (1967) *Serpentinites of the Roxbury district, Vermont.* In: Wyllie, P. J., ed. Ultramafic and related rocks, p. 137-160. New York: Wiley. 464 p.

James, H. L. (1954) *Sedimentary facies of iron-formation.* Economic Geology, v. 49, p. 235-293.

James, N. P. (1977) *The deep and the past. Third International Coral Reef Symposium.* Proceedings, v. 2, p. xxv-xxvii. Miami: University of Miami, Rosenstiel School of Marine and Atmospheric Science. 628 p.

James, W. R., and Krumbein, W. C. (1969) *Frequency distributions of stream link lengths.* Journal of Geology, v. 77, p. 544-565.

Jameson, Robert (1808) *Elements of geognosy.* Edinburgh: William Blackwood. 368 p. (His System of mineralogy, v. 3.)

Jamieson, J. A., et al. (1963) *Infrared physics and engineering.* New York: McGraw-Hill.

Jarvis, R. S. (1972) *New measure of the topologic structure of dendritic drainage networks.* Water Resources Research, v. 8, p. 1265-1271.

Jeffrey, Charles (1973) *Biological nomenclature.* New York: Crane, Russak. 69 p.

Jeletzky, J. A. (1956) *Paleontology, basis of practical geochronology.* American Association of Petroleum Geologists. Bulletin, v. 40, p. 679-706.

Jeletzky, J. A. (1966) *Comparative morphology, phylogeny, and classification of fossil Coleoidea.* University of Kansas. Paleontological Contributions, art. 7. 162 p.

Jennings, J. D. (1968) *Prehistory of North America.* New York: McGraw-Hill. 391 p.

Jervey, M. T. (1988) *Quantitative geological modeling of siliciclastic rock sequences and their seismic expressions.* In: C. K. Wilgus et al., eds, Sea level changes: an integrated approach. Society of Economic Paleontologists and Mineralogists Special Publication 42, p. 47-69.

Jessop, A.M., (1990) *Thermal geophysics.* Developments-in-Solid-Earth-Geophysics, v. 17, Calgary, 306p.

Johannsen, Albert (1917) *Suggestions for a quantitative mineralogical classification of igneous rocks.* Journal of Geology, v. 25, p. 63-97.

Johannsen, Albert (1920) *A quantitative mineralogical classification of igneous rocks—revised.* Journal of Geology, v. 28, p. 38-60, 158-177, 210-232.

Johannsen, Albert (1931) *Introduction, textures, classifications and glossary. 1st ed.* Chicago: University of Chicago Press. 267 p. (A descriptive petrography of the igneous rocks, v. 1.)

Johannsen, Albert (1931a) *The quartz-bearing rocks. 1st ed.* Chicago: University of Chicago Press. (A descriptive petrography of the igneous rocks, v. 2.)

Johannsen, Albert (1932) *A descriptive petrography of the igneous rocks, v. 2.* Chicago: University of Chicago Press. 428 p.

Johannsen, Albert (1937) *A descriptive petrography of the igneous rocks, v. 3.* Chicago: University of Chicago Press, 360 p.

Johannsen, Albert (1938) *The feldspathoid rocks; the peridotites and perknites.* Chicago: University of Chicago Press. 523 p. (A descriptive petrography of the igneous rocks, v. 4.)

Johannsen, Albert (1939) *Introduction, textures, classifications and glossary. 2nd ed.* Chicago: University of Chicago Press. 318 p. (A descriptive petrography of the igneous rocks, v. 1.)

Johnson, D.L. (1993) *Dynamic denudation evolution of tropical, subtropical, and temperate landscapes with three tiered soils: Toward a general theory of landscape evolution.* Quaternary International, v.17, p.67-78.

Johnson, D. W. (1916) *Plains, planes, and peneplanes.* Geographic Review, v. 1, p. 443-447.

Johnson, D. W. (1919) *Shore processes and shoreline development.* New York: Wiley. 584 p.

Johnson, D. W. (1925) *The New England-Acadian shore line.* New York: Wiley. 608 p.

Johnson, D. W. (1932) *Rock fans of arid regions.* American Journal of Science, 5th ser., v. 23, p. 389-416.

Johnson, D. W. (1939) *The origin of submarine canyons; a critical review of hypotheses.* New York: Columbia University Press. 126 p.

Johnson, P. H., and Bhappu, R. B. (1969) *Chemical mining—a study of leaching agents.* New Mexico Bureau of Mineral and Mineral Resources. Circular 99. 10 p.

Jones, C. R. (1973) *Lower Paleozoic.* In: D. J. Gobbett and C. S. Hutchison, eds. Geology of the Malay Peninsula, p. 25-60. New York: Wiley.

Jones, D. L., et al. (1983) *Recognition, character, and analysis of tectonostratigraphic terranes in western North America.* In: M. Hashimoto and S. Uyeda, eds. Accretion tectonics in the Circum-Pacific regions. Tokyo: Terra Scientific Publishing Co.

Jones, J.G. (1969) *Intraglacial volcanoes of the Laugarvatn region, southwest Iceland–I.* Quarterly Journal of the Geological Society of London, v.124, p.197-211.

Jones, J.G., and Nelson, P.H.H. (1970) *The flow of basalt lava from air into water—its structural expression and stratigraphic significance.* Geological Magazine, v.107, p.13-19.

Jones, O. T. (1937) *On the sliding or slumping of submarine sediments in Denbighshire, North Wales, during the Ludlow period.* Geological Society of London. Quarterly Journal, v. 93, p. 241-283.

Jones, P. B. (1971) *Folded faults and sequence of thrusting in Alberta foothills.* American Association of Petroleum Geologists. Bulletin, v. 55, p. 292-306.

Jones, P. H. (1969) *Hydrology of Neogene deposits in the northern Gulf of Mexico basin.* Louisiana Water Resources Research Institute. Bulletin GT-2, 105 p.

Jongman, R.H.G., Ter Braak, C.J.F., and Van Tongeren, O.F.R., eds. (1995) *Data analysis in community and landscape ecology.* Cambridge, UK, Cambridge University Press, 299p.

Joplin, G. A. (1968) *A petrography of Australian metamorphic rocks.* New York: American Elsevier. 262 p.

Jopling, A. V. (1961) *Origin of regressive ripples explained in terms of fluid-mechanic processes.* U. S. Geological Survey. Professional Paper 424-D, art. 299, p. 15-17.

Jopling, A.V., and Walker, R.G. (1968) *Morphology and origin of ripple-drift cross-laminations, with examples from the Pleistocene of Massachusetts.* Journal of Sedimentary Petrology, v.38, p.971-984.

JORC: Joint Ore Reserves Committee of the Australasian Institute of Mining and Metallurgy, the Australian Institute of Geoscientists, and the Minerals Council of Australia (1999) *Australasian code for reporting of mineral resources and ore reserves (The JORC Code),* 16p.

Jordan, T. H. (1975a) *The present-day motions of the Caribbean Plate.* Journal of Geophysical Research, v. 80, 32, p. 4433-4439.

Jordan, T. H. (1975b) *The continental tectosphere.* Review of Geophysics and Space Physics, v. 13, no. 3, Special Issue, National Report to IUGS, p. 1-12.

Joreskog, K. G., et al. (1976) *Geological factor analysis.* New York: Elsevier. 178 p. (Methods in geomathematics, 1.)

Journel, A.G. and Huijbregts, C.J., (1978) *Mining geostatistcs.* Academic Press.

Judson, S. S., Jr. (1953) *Geology of the San Jon site, eastern New Mexico.* Smithsonian Miscellaneous Collections, v. 121, no. 1. 70 p.

Juengst, H. (1934) *Zur geologischen Bedeutung der Synaerese. Ein Beitrag zur Entwaesserung der Kolloide im werdenden Gestein.* Geologische Rundschau, v.25, no.5, p.312-325.

Jukes, J. B. (1862) *On the mode of formation of some of the river-valleys in the south of Ireland.* Geological Society of London. Quarterly Journal, v. 18, p. 378-403.

Jukes-Browne, A. J. (1903) *The term "hemera."* Geological Magazine, Dec. 4, v. 10, p. 36-38.

Jull, R. K. (1967) *The hystero-ontogeny of Lonsdaleia McCoy and Thysanophyllum orientale Thomson.* Paleontology, v. 10, p. 617-628.

Jung, J., and Roques, M. (1936) *Les zones d'isometamorphisme dans le terrain cristallophyllien du Massif Central francais.* Rev. Sci. Nat. Auvergne, 1 (47).

Jung, J., and Roques, M. (1952) *Introduction a l'etude zoneographique des formations cristallogphylliennes*. Bull. Serv. Carte Geol. France, 235, p. 1-62.

Jurine, Louis (1806) *Réflexions sur la nécessité d'une nouvelle nomenclature en géologie, et l'exposé de celle qu'il propose*. Journal des Mines, v. 19, p. 367-378.

K

Kahn, J. S. (1956) *The analysis and distribution of the properties of packing in sand-size sediments*. Journal of Geology, v. 64, p. 385-395, 578-606.

Kalff, J. (2002) *Limnology*. Upper Saddle River, N.J., Prentice Hall, 592p.

Kalkowsky, Ernst (1880) *Über die Erforschung der archèischen Formationen*. Neues Jahrbuch für Mineralogie, Geologie und Palaeontologie, Jahrg. 1880, Bd. 1, p. 1-28.

Kalkowsky, Ernst (1908) *Oolith und Stromatolith im norddeutschen Buntsandstein*. Deutsche Geologische Gesellschaft. Zeitschrift, Bd. 60, p. 68-125.

Karcz, Iaakov (1967) *Harrow marks, current-aligned sedimentary structures*. Journal of Geology, v. 75, p. 113-121.

Karlstrom, T. N. V. (1956) *The problem of the Cochrane in late Pleistocene chronology*. U. S. Geological Survey. Bulletin 1021-J, p. 303-331.

Karlstrom, T. N. V. (1961) *The glacial history of Alaska: its bearing on paleoclimatic theory*. New York Academy of Sciences. Annals, v. 95, art. 1, p. 290-340.

Karlstrom, T. N. V. (1966) *Quaternary glacial record of the North Pacific region and worldwide climatic changes*. In: Blumenstock, D. I., ed. Pleistocene and post-Pleistocene climatic variations in the Pacific area, p. 153-182. Honolulu: Bishop Museum Press. 182 p.

Kauffman, E. G. (1969) *Cretaceous marine cycles of the Western Interior*. Rocky Mountain Association of Geologists. Mountain Geologist, v. 6, p. 227-245.

Kaufmann, F. J. (1886) *Emmen- und Schlierengegenden nebst Umgebungen bis zur Brünigstrasse und Linie Lungern-Grafenort geologisch aufgenommen und dargestellt*. Beitrège zur geologischen Karte der Schweiz, v. 24, pt. 1. 608 p. with atlas of 30 plates.

Kay, G. F. (1916) *Gumbotil, a new term in Pleistocene geology*. Science, v. 44, p. 637-638.

Kay, G. M. (1937) *Stratigraphy of the Trenton Group*. Geological Society of America. Bulletin, v. 48, p. 233-302.

Kay, G. M. (1945) *North American geosynclines: their classification [abstract]*. Geological Society of America. Bulletin, v. 56, p. 1172.

Kay, G. M. (1945a) *Paleogeographic and palinspastic maps*. American Association of Petroleum Geologists. Bulletin, v. 29, p. 426-450.

Kay, G. M. (1947) *Geosynclinal nomenclature and the craton*. American Association of Petroleum Geologists. Bulletin, v. 31, no. 7, p. 1287-1293.

Kay, G. M. (1951) *North American geosynclines*. Geological Society of America. Memoir 48. 143 p.

Kay, R. W. (1978) *Aleutian mangesian andesites: melts from the subducted Pacific ocean crust*. Jour. Volcan. Geotherm. Res., 4, p. 117-132.

Kay, R.W., and Kay, S.M.(2002) *Andean adakites: Three ways to make them*. Acta Petrologica Sinica, v.18, p.303-311.

Keeton, W. T. (1967) *Biological science*. New York: Norton.

Keith, Arthur (1895) *Description of the Knoxville sheet*. U. S. Geological Survey. Geologic Atlas, folio 16. 6 p.

Keith, M. L., and Degens, E. T. (1959) *Geochemical indicators of marine and freshwater sediments*. In: Abelson, P. H., ed. Researches in geochemistry, p. 38-61. New York: Wiley. 511 p.

Keller, B. M., et al. (1968) *The main features of the Late Proterozoic paleogeography of the USSR*. International Geological Congress. 23rd, Prague. Proceedings, sec. 4, p. 189-202.

Keller, E.A. (2002) *Introduction to environmental geology, 2nd. ed.* Saddle River, N.J., Prentice Hall.

Keller, G.V., and Frischknecht, F.C. (1966) *Electrical methods in geophysical prospecting*. Pergamon Press, 517p.

Keller, J.P., and Dauphin, Y. (1983) *Methodological aspects of the ultrastructural analysis of the organic and mineral components in mollusc shells*. In: Westbroek, P., and deJong, E.W., eds., *Biomineralization and biological metal accumulation*. Boston, D. Reidel Co. (533p.), p.255-260.

Keller, W. D. (1958) *Argillation and direct bauxitization in terms of concentrations of hydrolyzing aluminum silicates*. American Association of Petroleum Geologists. Bulletin, v. 42, p. 233-245.

Keller, W. D. (1963) *Diagenesis in clay minerals; a review*. National Conference on Clays and Clay Minerals. 11th, Ottawa, 1962. Proceedings, p. 136-157. (International series of monographs on earth sciences, v. 13.)

Keller, W. D. (1977) *Scan electron micrographs of kaolins collected from diverse environments of origin—IV, Georgia kaolin and kaolinizing source rocks*. Clays and Clay Minerals, v. 25, p. 311-345.

Keller, W. D., et al. (1954) *The origin of Missouri fireclays*. National Academy of Sciences—National Research Council. Publication 327, p. 7-46.

Keller, W. D., et al. (1977) *Texture of Arkansas novaculite indicates thermally induced metamorphism*. Journal of Sedimentary Petrology, v. 47, p. 834-843.

Kelley, J.T., Dickson, S.M., Belknap, D.F., Barnhardt, W.A., and Barber, D.C. (2003) *Distribution and volume of sand bodies on the rocky, glaciated inner continental shelf of the northwestern Gulf of Maine*. Journal of Coastal Research, v.19, p.41-56.

Kelley, V. C. (1956) *Thickness of strata*. Journal of Sedimentary Petrology, v. 26, p. 289-300.

Kelling, Gilbert (1958) *Ripple-mark in the Rhinns of Galloway*. Edinburgh Geological Society. Transactions, v. 17, p. 117-132.

Kelling, Gilbert, and Stanley, D. J. (1976) *Sedimentation in canyon, slope, and base of slope environments*. In: Stanley, D. J., and Swift, D. J. P., eds. Marine sediment transport and environmental management. Ch. 17.

Kellogg, C. E. (1950) *Tropical soils*. 4th International Congress on Soil Science. Transactions, v. 1, p. 266-276.

Kemp, J. F. (1896) *A handbook of rocks, for use without the microscope*. New York: J. F. Kemp. 176 p.

Kemp, J. F. (1900) *Handbook of rocks for use without the microscope*. New York: Van Nostrand. 185 p.

Kemp, J. F. (1934) *A handbook of rocks for use without the microscope, with a glossary of the names of rocks and other lithological terms. 5th ed.* New York: Van Nostrand. 300 p.

Kemp, J. F. (1940) *A handbook of rocks (for use without the petrographic microscope). 6th ed.*, completely revised & edited by F. F. Grout. New York: Van Nostrand.

Kendall, P. F. (1902) *A system of glacier-lakes in the Cleveland Hills*. Geological Society of London. Quarterly Journal, v. 58, p. 471-571.

Kendall, P. F., and Wroot, H. E. (1924) *Geology of Yorkshire; an illustration of the evolution of northern England. Vol. 1.* Vienna: Hollinek Brothers. 660 p.

Kennedy, J. F. (1963) *The mechanics of dunes and antidunes in erodible-bed channels*. Journal of Fluid Mechanics, v. 16, p. 521-544.

Kennedy, K.H., et al. (2001) *Dictionary of GIS terminology*. ESRI Press, 116p.

Kennedy, W. Q. (1933) *Trends of differentiation in basaltic magmas*. American Journal of Science, 5th series, v. 25, p. 239-256.

Keroher, G. C., et al. (1966) *Lexicon of geologic names of the United States for 1936-1960. Parts 1-3.* U. S. Geological Survey. Bulletin 1200. 4341 p.

Kerr, P. F., and Hamilton, P. K. (1949) *Glossary of clay mineral names*. New York: Columbia University. 66 p. (American Petroleum Institute. Project 49: Clay Mineral Standards. Preliminary Report no. 1.)

Kerr, P. F., and Kopp, O. C. (1958) *Salt-dome breccia*. American Association of Petroleum Geologists. Bulletin, v. 42, p. 548-560.

Kerr, R. C., and Nigra, J. O. (1952) *Eolian sand control*. American Association of Petroleum Geologists. Bulletin, v. 36, p. 1541-1573.

Kerr, W. C. (1881) *On the action of frost in the arrangement of superficial earthy material*. American Journal of Science, v. 21, p. 345-358.

Kesling, R. V. (1951) *Terminology of ostracod carapaces*. Ann Arbor: University of Michigan. Contributions from the Museum of Paleontology, v. 9, no. 4, p. 93-171.

Kesseli, J. E. (1941) *The concept of the graded river*. Journal of Geology, v. 49, p. 561-588.

Kessler, Paul (1922) *Über Lochverwitterung und ihre Beziehungen zur Metharmose (Umbildung) der Gesteine*. Geologische Rundschau, v. 12, p. 237-270.

Keyes, C. R. (1910) *Deflation and the relative efficiencies of erosional processes under conditions of aridity*. Geological Society of America. Bulletin, v. 21, p. 565-598.

Keyes, C. R. (1913) *Antigravitational gradation*. Science, n.s., v. 38, p. 206.

Kidwell, S.M. (1991) *The stratigraphy of shell concentrations*. In: Allison, P.A., and Briggs, D.E.G., eds., *Taphonomy: Releasing the data locked in the fossil record*. (Topics in Geobiology 9). New York, Plenum Press, p.211-290.

Kidwell, S.M., Fürsich, F.T., and Aigner, T. (1986) *Conceptual framework for the analysis and classification of fossil concentrations*. Palaios, v.1, no.3, p.228-238.

Kieffer, S. W. (1971) *Shock metamorphism of the Coconino sandstone at Meteor Crater, Arizona*. Journal of Geophysical Research, v. 76, p. 5449-5473.

Kilburn, C.R.J. (2000) *Lava flows and flow fields*. In: Sigurdsson, H., ed., *Encyclopedia of volcanoes*. New York, Academic Press, p.291-305.

Kim, S. J. (1975) *Rhodochrostone—a new sedimentary rock from the Janggun mine, Korea*. Journal of the Korean Institute of Mining Geology, v. 8, p. 63-71.

Kinahan, G. H. (1878) *Manual of the geology of Ireland*. London: C. K. Paul. 444 p.

Kindle, E. M. (1916) *Small pit and mound structures developed during sedimentation*. Geological Magazine, Dec. 6, v. 3, p. 542-547.

Kindle, E. M. (1917) *Recent and fossil ripple-mark*. Geological Survey of Canada. Museum Bulletin, no. 25. 121 p. (Canada Department of Mines. Geological Series, no. 34.)

Kindle, E. M. (1923) *Range and distribution of certain types of Canadian Pleistocene concretions*. Geological Society of America. Bulletin, v. 34, p. 609-648.

Kindle, E. M. (1923a) *Nomenclature and genetic relations of certain calcareous rocks*. Pan-American Geologist, v. 39, p. 365-372.

Kindle, E. M. (1926) *Contrasted types of mud cracks*. Royal Society of Canada. Proceedings and Transactions, ser. 3, v. 20, sec. 4, p. 71-75.

King, C. A. M., and Williams, W. W. (1949) *The formation and movement of sand bars by wave action*. Geographical Journal, v. 113, p. 70-85.

King, Clarence (1878) *Systematic geology*. Washington: U. S. Government Printing Office. 803 p. (U. S. Geological Exploration of the Fortieth Parallel. Report, v. 1.)

King, F.H. (1899) *Principles and conditions of the movements of ground water*. U.S. Geological Survey Annual Report 19, Part 2, p.59-294.

King, L. C. (1948) *A theory of bornhardts.* Geographical Journal, v. 112, p. 83-87.
King, L. C. (1959) *Denudational and tectonic relief in south-eastern Australia.* Geological Society of South Africa. Transactions, v. 62, p. 113-138.
King, L. C. (1962) *The morphology of the Earth; a study and synthesis of world scenery.* New York: Hafner. 699 p.
King, L. H., and MacLean, B. (1970) *Pockmarks on the Scotian Shelf.* Geological Society of America. Bulletin, v. 81, p. 3141-3148.
King, Philip (1959) *The evolution of North America.* Princeton: Princeton University Press. 190 p.
Kinsman, D. J. J. (1969) *Modes of formation, sedimentary associations, and diagnostic features of shallow-water and supratidal evaporites.* American Association of Petroleum Geologists. Bulletin, v. 53, p. 830-840.
Kirkham, R. V. (1971) *Intermineral intrusions and their bearing on the origin of porphyry copper and molybdenum deposits.* Economic Geology, v. 66, p. 1244-1249.
Kirkland, D.W., and Anderson, R.Y. (1970) *Microfolding in the Castile and Todilto evaporites, Texas and New Mexico.* Geological Society of America Bulletin, v.81, p.3259-3282.
Kirkland, D.W., and Evans, R., (1980) *Origin of castiles on the Gypsum Plain of Texas and New Mexico.* New Mexico Geological Society Guidebook, v.31, p.173-178.
Kirwan, Richard (1794) *Elements of mineralogy. Vol. 1. 2nd ed.* London: J. Nichols. 510 p.
Kjellström, G. (1971) *Ordovician microplankton (baltisphaerids) from the Grötlingbo Borehole Nr.1 in Gotland, Sweden, Sveriges Geologiska* Undersökning, *Series C, 65 (publication no.655), 75p.*
Klapper, G., Feist, R., and House, M.R. (1987) *Decision on the boundary stratotype for the Middle/Upper Devonian Series boundary.* Episodes, v.10, no.2, p.97-101.
Klapper, G., Feist, R., Becker, R.T., and House, M.R. (1993) *Definition of the Frasnian/Famennian Stage boundary.* Episodes, v.16, no.4, p.433-441.
Klein, E.M., and Langmuir, C.H. (1987) *Global correlations of ocean ridge basalt chemistry with axial depth and crustal thickness.* Journal of Geophysical Research, B, Solid Earth and Planets, v.92 (B8), p.8089-8115.
Klein, G. de V. (1963) *Analysis and review of sandstone classifications in the North American geological literature, 1940-1960.* Geological Society of America. Bulletin, v. 74, p. 555-575.
Klein, G. de V. (1985) *Intertidal flats and intertidal sand bodies.* In: R. A. Davis, Jr., ed. Coastal sedimentary environments , p. 187-224. New York: Springer-Verlag.
Klöden, K. F. von (1828) *Beitrège zur mineralogischen und geognostischen Kenntniss der Mark Brandenburg. [Pt. 1].* Berlin: W. Dieterici. 108 p.
Klotz, I. M. (1972) *Chemical thermodynamics, 3rd ed.* Menlo Park, Calif.: W. A. Benjamin. 444 p.
Klug, H.D., and Alexander, L.E. (1974) *X-ray diffraction procedures, 2nd. ed.* New York, John Wiley & Sons, 966p.
Klugh, A. B. (1923) *A common system of classification in plant and animal ecology.* Ecology, v. 4, p. 366-377.
Knight, C. L. (1957) *Ore genesis—the source bed concept.* Economic Geology, v. 53, p. 808-817.
Knight, J. B. (1941) *Paleozoic gastropod genotypes.* Geological Society of America. Special Paper 32. 510 p.
Knight, S. H. (1929) *The Fountain and the Casper formations of the Laramie Basin.* University of Wyoming. Publications in Science. Geology, v. 1, p. 1-82.
Knight, W. C. (1897) *"Mineral soap"* Engineering and Mining Journal, v. 63, p. 600-601.
Knight, W. C. (1898) *Bentonite.* Engineering and Mining Journal, v. 66, p. 491.
Knopf, E. B., and Ingerson, Earl (1938) *Structural petrology.* Geological Society of America. Memoir 6. 270 p.
Knox, Alexander (1904) *Glossary of geographical and topographical terms.* London: Edward Stanford. 432 p. (Stanford's compendium of geography and travel; suppl. vol.)
Knutson, R. M. (1958) *Structural sections and the third dimension.* Economic Geology, v. 53, p. 270-286.
Kobayashi, I. (1964) *Introduction to the shell structure of bivalvian molluscs.* Earth Science, v.73, p.1-12 (in Japanese with English summary).
Kobayashi, I. (1969) *Internal microstructure of the shell of bivalve molluscs.* American Zoologist, v.9, p.663-672.
Kobayashi, I. (1971) *Internal shell microstructure of Recent bivalvian molluscs.* Science Reports of Niigata University, ser.E, Geology and Mineralogy, no.2, p.27-50.
Kobayashi, Teiichi (1944) *An instant in the Phanerozoic Eon and its bearings on geology and biology.* Imperial Academy. Tokyo. Proceedings, v. 20, p. 742-750. (Concept of time in geology, 3.)
Kobayashi, Teiichi (1944a) *On the major classification of the geological age.* Imperial Academy, Tokyo. Proceedings, v. 20, p. 475-478. (Concept of time in geology, 1.)
Koch, Lauge (1926) *Ice cap and sea ice in north Greenland.* Geographical Review, v. 16, p. 98-107.
Kofoid, C.A. (1907) *The plates of* Ceratium *with a note on the unity of the genus.*Zoologischer Anzeiger, v.32, no.7, p.177-183.
Kofoid, C.A. (1910) *A revision of the genus* Ceratocorys, *based on skeletal morphology.* University of California Publications in Zoology, v.6, no.8, p.177-187.
Kofoid, C.A. (1911) *Dinoflagellata of the San Diego region, IV. The genus Gonyaulax, with notes on its skeletal morphology and a discussion of its generic and specific characters.* University of California Publications in Zoology, v.8, no.4, p.187-286.
Kopf, R. W. (1982) *Hydrotectonics: principles and relevance.* U. S. Geological Survey. Open-File Report 82-307.
Koppejan, A. W., et al. (1948) *Coastal flow slides in the Dutch province of Zeeland.* International Conference on Soil Mechanics and Foundation Engineering. 2nd, Rotterdam, 1948? Proceedings, 5, p. 89-96.
Kornicker, L. S., and Boyd, D. W. (1962) *Shallow-water geology and environments of Alacran reef complex, Campeche Bank, Mexico.* American Association of Petroleum Geologists. Bulletin, v. 46, p. 640-673.
Kostov, Ivan (1961) *Genesis of kyanite in quartz veins.* International Geology Review, v. 3, p. 645-651.
Koteff, Carl (1974) *The morphologic sequence concept and deglaciation of southern New England.* In: D. R. Coates, ed. Glacial geomorphology, p. 121-144. London: George Allen & Unwin.
Koteff, Carl, and Pessl, Fred, Jr. (1981) *Systematic ice retreat in New England.* U. S. Geological Survey Professional Paper 1179. 20 p.
Koto, B. (1916) *On the volcanoes of Japan, Part V.* Journal of the Geological Society of Japan, v.23, p.98-127.
Kotz, S., Johnson, N.L., and Read, C.B., eds., (1982-89) *Encyclopedia of Statistical Sciences. V. 1-10:* John Wiley & Sons, Inc., New York, 6672p.
Kovalenko, V. I., et al. (1971) *Topaz-bearing quartz keratophyre (ongonite), a new variety of subvolcanic igneous vein rock.* Doklady Akad. Nauk SSSR, v. 199, p. 132-135.
Kramm, U. (1976) *The coti cule rocks (spessartine quartzites) of the Venn-Stavelot Massif, Ardennes, a volcaniclastic metasediment?* Contributions to Mineralogy and Petrology, v. 56, p. 135-155.
Kranz, R. L. (1983) *Microcracks in rocks: a review.* Tectonophysics, v. 100, p. 449-480.
Krasheninnikov, G. F. (1964) *Facies, genetic types, and formations.* International Geology Review, v. 6, p. 1242-1248.
Krassilov, V. A. (1974) *Causal Biostratigraphy.* Lethaia, v. 7, no. 3, p. 173-179.
Kraus, Ivan (1979) *Deposits of clays, zeolites, limnoquartzites and perlites in Slovakia.* 10th International Kaolin Symposium, Bratislava. Excursion Guide, p. 40.
Krauskopf, K. B. (1967) *Introduction to geochemistry.* New York: McGraw-Hill. 721 p.
Kremp, G. O. W. (1965) *Morphologic encyclopedia of palynology.* University of Arizona, Program in Geochronology. Contribution 100. Tucson: University of Arizona Press. 185 p.
Krenkel, Erich (1922) *Die Bruchzonen Ostafrikas.* Berlin: Gebrüder Borntraeger.
Krishtofovich, A. N. (1945) *The mode of preservation of plant fossils and their bearing upon the problem of coal formation.* Akademiya Nauk SSSP. Izvestiya, Seriya Geologicheskaya, no. 2, p. 136-150.
Krivenko, V., and Lapchik, T. (1934) *On the petrography of the crystalline rocks of the rapids of the Dnipro.* Ukrainskii Naukovo-Doslidchnii Geologichnii Institute. Trudy, v. 5, no. 2, p. 67-77.
Kruger, F. C. (1946) *Structure and metamorphism of the Bellows Falls quadrangle of New Hampshire and Vermont.* Geological Society of America. Bulletin, v. 57, p. 161-205.
Krumbein, W. C. (1934) *Size frequency distributions of sediments.* Journal of Sedimentary Petrology, v. 4, p. 65-77.
Krumbein, W. C. (1955) *Composite end members in facies mapping.* Journal of Sedimentary Petrology, v. 25, p. 115-122.
Krumbein, W. C., and Graybill, F. A. (1965) *An introduction to statistical models in geology.* New York: McGraw-Hill. 475 p.
Krumbein, W. C., and Libby, W. G. (1957) *Application of moments to vertical variability maps of stratigraphic units.* American Association of Petroleum Geologists. Bulletin, v. 41, p. 197-211.
Krumbein, W. C., and Pettijohn, F. J. (1938) *Manual of sedimentary petrography.* New York: Appleton. 549 p.
Krumbein, W. C., and Sloss, L. L. (1951) *Stratigraphy and sedimentation.* San Francisco: Freeman. 497 p.
Krumbein, W. C., and Sloss, L. L. (1963) *Stratigraphy and sedimentation. 2nd ed.* San Francisco: Freeman. 660 p.
Krynine, P. D. (1937) *Petrography and genesis of the Siwalik Series.* American Journal of Science, ser. 5, v. 34, no. 204, p. 422-446.
Krynine, P. D. (1940) *Petrology and genesis of the Third Bradford Sand. Pennsylvania State College.* Mineral Industries Experiment Station. Bulletin 29. 134 p.
Krynine, P. D. (1941) *Differentiation of sediments during the life history of a landmass.* Geological Society of America. Bulletin, v. 52, p. 1915.
Krynine, P. D. (1945) *Sediments and the search for oil.* Producers Monthly, v. 9, p. 12-22.
Krynine, P. D. (1948) *The megascopic study and field classification of sedimentary rocks.* Journal of Geology, v. 56, p. 130-165.

Krynine, P. D. (1951) *Reservoir petrography of sandstones.* In: Payne, T. G., and others. Geology of the Arctic Slope of Alaska , sheet 2. U. S. Geological Survey. Oil and Gas Investigations, map OM 126. 3 sheets.

Krzanowski, Roman M., Palylyk, Claudia L., and Crown, Peter H. (1993) *Lexicon of terms for users of geographic information systems.* In: 1994 international GIS sourcebook; geographic information system technology , p. 595-605.

Ksiazkiewicz, Marian (1949) *Slip-bedding in the Carpathian flysch.* Société Géologique Polone. Annales, v. 19, p. 493-501.

Ksiazkiewicz, Marian (1958) *Submarine slumping in the Carpathian Flysch.* Polskie Towarzystwo Geologicne. Rocznik, t. 28, p. 123-151.

Kümmel, H. B. (1893) *Some rivers of Connecticut.* Journal of Geology, v. 1, p. 371-393.

Ku, C.C. and J.A. Sharp, (1983) *Werner devonvolution for automated magnetic interpretation and its refinement using Marquardt's inverse modeling.* Geophysics, 48, 754-774.

Kubiëna, W. L. (1953) *The soils of Europe.* London: Thomas Murby. 317 p.

Kudryavtsev, V. A., ed. (1978) *Obshcheye merzolotovendeniia (General Permafrost Science). In: Russian. Izd. 2.* (Edu 2) Moskva (Moskow), Izdatel'stvo Moskovskogo Universiteta (Moscow University Editions), 404p.

Kuenen, P. H. (1943) *Pitted pebbles.* Leidsche geologische mededeelingen, v. 13, p. 189-201.

Kuenen, P. H. (1947) *Water-faceted boulders.* American Journal of Science, v. 245, p. 779-783.

Kuenen, P. H. (1948) *Slumping in the Carboniferous rocks of Pembrokeshire.* Geological Society of London. Quarterly Journal, v. 104, p. 365-385.

Kuenen, P. H. (1950) *Marine geology.* New York: Wiley. 568 p.

Kuenen, P. H. (1953) *Significant features of graded bedding.* American Association of Petroleum Geologists. Bulletin, v. 37, p. 1044-1066.

Kuenen, P. H. (1957) *Sole markings of graded graywacke beds.* Journal of Geology, v. 65, p. 231-258.

Kuenen, P. H. (1958) *Experiments in geology.* Geological Society of Glasgow. Transactions, v. 23, p. 1-28.

Kulp, J. L. (1961) *Geologic time scale.* Science, v. 133, p. 1105-1114.

Kuno, Hisachi (1957) *Differentiation of Hawaiian magmas.* Japanese Journal of Geology and Geography, v. 28, p. 193-194.

Kuno, Hisachi (1960) *High-alumina basalt.* Journal of Petrology, v. 1, p. 121-145.

Kupsch, W. O. (1956) *Submask geology in Saskatchewan.* International Williston Basin Symposium. 1st, Bismarck (N.D.), 1956. Williston Basin Symposium, p. 66-75.

Kyle, J.R., and Posey, H.H., (1991) *Halokinesis, cap rock development, and salt dome mineral resources.* In: Melvin, J.L., ed., Evaporites, Petroleum and Mineral Resources, Elsevier, Amsterdam, p.413-474.

L

Labandeira, C.C. (2000) *The paleobiology of pollination and its precursors.* In: Gastaldo, R.A., and DiMichele, W.A., eds., *Phanerozoic terrestrial ecosystems.* Paleontological Society Paper 6, p.233-269.

Lachenbruch, A.H., and Brewer, M.C. (1959) *Dissipation of the temperature effect of drilling a well in Arctic Alaska.* U.S. Geol. Survey Bull. 1083-C, p.73-109.

Lacroix, A. (1922) *Minéralogie de Madagascar. v. 2: minéralogie appliquée; lithologie.* Paris: Augustin Challamel. 694 p.

Lacroix, A. V. (1980) *A short note on cryoseisms.* Earthquake Notes, v. 51, p. 15-16.

Lagaaij, R. (1952) *The Pliocene Bryozoa of the Low Countries and their bearing on the marine stratigraphy of the North Sea region.* Netherlands, Geologische Stichtung, Mededeelingen, serie C-V, no. 5, 233 p.

Lagaaij, R. and Gautier, Y. V. (1965) *Bryozoan assemblages from marine sediments of the Rhône delta,* France. Micropaleontology, v. 11, p. 39-58.

Lahee, F. H. (1923) *Field geology. 2nd ed.* New York: McGraw-Hill. 649 p.

Lahee, F. H. (1961) *Field geology. 6th ed.* New York: McGraw-Hill. 926 p.

Lahee, F. H. (1962) *Statistics of exploratory drilling in the United States 1945-1960.* Tulsa, American Association of Petroleum Geologists. 135 p.

Laird, M. G. (1970) *Vertical sheet structure—a new indicator of sedimentary fabric.* Journal of Sedimentary Petrology, v. 40, p. 428-434.

Lam, H. J. (1957) *What is a taxon?* Taxonomy, v. 6, p. 213-215.

Lamar, J. E. (1928) *Geology and economic resources of the St. Peter Sandstone of Illinois.* Illinois State Geological Survey. Bulletin no. 53. 175 p.

Lamb, H. H. (1970) *Volcanic dust in the atmosphere.* Royal Society of London, Philosophical Transactions, Series A, Mathematics and Physical Sciences, v. 266, p. 425-533.

Lambe, T. W. (1953) *The structure of inorganic soil.* American Society of Civil Engineers. Proceedings, v. 79, separate no. 315. 49 p.

Lamont, Archie (1957) *Slow anti-dunes and flow marks.* Geological Magazine, v. 94, p. 472-480.

Lamplugh, G. W. (1895) *The crush-conglomerates of the Isle of Man.* With a petrographical appendix by W. W. Watts. Geological Society of London. Quarterly Journal, v. 51, p. 563-599.

Lamplugh, G. W. (1902) *"Calcrete."* Geological Magazine, n.s., Dec. 4, v. 9, no. 462, p. 575.

Landes, K. K. (1945) *The Mackinac Breccia.* Michigan Geological Survey. Publication 44, Geological Series 37, ch. 3, p. 123-153.

Landes, K. K. (1957) *Chemical unconformities (abstract).* Geological Society of America. Bulletin, v. 68, p. 1759.

Landes, K. K. (1967) *Eometamorphism, and oil and gas in time and space.* American Association of Petroleum Geologists. Bulletin, v. 51, p. 828-841.

Lane, A. C. (1923) *Communication.* Journal of Geology, v. 31, p. 348.

Lane, A. C. (1928) *Isontic?* Science, v. 68, p. 37.

Lane, A.C. (1903) *Porphyritic appearance of rocks.* Geological Society of America. Bulletin, v. 14, p. 369-379.

Lang, W. T. B. (1937) *The Permian formations of the Pecos Valley of New Mexico and Texas.* American Association of Petroleum Geologists. Bulletin, v. 21, p. 833-898.

Langbein, W. B., and Iseri, K. T. (1960) *General introduction and hydrologic definitions.* U. S. Geological Survey. Water-supply Paper 1541-A. 29 p. (Manual of hydrology: pt. 1, p. 1-29).

Lange, A.L. (1955) *The role of caves in dating Grand Canyon.* Plateau, v.27, p.1-7.

Lange, A.L. (1964) *Solution bevels in limestone caves.* Cave Notes, v.6, p.34-38.

Laporte, L. F. (1975) *Encounter with the Earth.* San Francisco: Canfield Press. 538 p.

Larsen, E. S. (1938) *Some new variation diagrams for groups of igneous rocks.* Journal of Geology, v. 46, p. 505-520.

Larsen, Gunnar, and Chilingar, G. V., eds. (1967) *Diagenesis in sediments.* Amsterdam: Elsevier. 551 p. (Developments in sedimentology, 8.)

Larsen, Ole, and Sorensen, H. (1960) *Principles of classification and norm calculations of metamorphic rocks: a discussion.* Journal of Geology, v. 68, p. 681-683.

Larson, C. (1990) *A glossary of lava tube features.* In: Johnson, V., ed., *National Speleological Society convention guide book.* Huntsville, Ala., National Speleological Society, p.21-55.

Lasius, G. S. O. (1789) *Beobachtungen ueber die Harzgebirge, nebst einem Profilnisse, als ein Beytrag zur mineralogischen Naturkunde.* Hannover: Helwingischen Hofbuchhandlung. 559 p. in 2 v.

Lasky, S.G. (1950) *How tonnage and grade relations help predict ore reserves.* Engineering and Mining Journal, v.151, no.4, p.81-85.

Last, W.W., and Smol, J.P., eds. (2001) *Tracking environmental change using lake sediments, v. 2: Physical and geochemical methods.* Dordrecht, Kluwer, 504p.

Laudelout, H. (1987) *Cation exchange equilibria in clays.* in Newman, A. C. D., editor, Chemistry of clays and clay minerals : Monograph No. 6, Mineralogical Society, London, p. 225-236.

Lawn, B. R., and Wilshaw, T. R. (1975) *Fracture of brittle solids.* Cambridge University Press, London, 204 p.

Lawrence, G. H. M. (1951) *Taxonomy of vascular plants.* New York: Macmillan. 823 p.

Lawson, A. C. (1885) *Report on the geology of the Lake of the Woods region, with special reference to the Keewatin (Huronian?) belt of the Archean rocks.* Geological Survey of Canada. Annual Report, 1:CC. 151 p.

Lawson, A. C. (1894) *The geomorphogeny of the coast of northern California.* University of California. Department of Geology. Bulletin, v. 1, no. 8, p. 241-271.

Lawson, A. C. (1904) *The geomorphogeny of the upper Kern Basin.* University of California. Department of Geology. Bulletin, v. 3, p. 291-376.

Lawson, A. C. (1913) *The petrographic designation of alluvial-fan formations.* University of California. Department of Geology. Bulletin, v. 7, p. 325-334.

Lawson, A. C. (1915) *The epigene profiles of the desert.* University of California. Department of Geology. Bulletin, v. 9, p. 23-48.

Lawson, A. C., et al. (1908) *The California earthquake of April 18, 1906; report of the State Earthquake Commission. Vol. 1.* Carnegie Institution of Washington. Publication no. 87. 451 p.

Le Maitre, R.W., ed. (2002) *Igneous rocks: A classification and glossary of terms. Recommendations of the International Union of Geological Sciences Subcommission on the Systematics of Igneous Rocks.* Cambridge University Press, 252p.

Le Maitre, R.W. et al. (1989) *A classification of igneous rocks and glossary of terms: recommendations of the International Union of Geological Sciences Subcommission on the Systematics of Igneous Rocks.* Oxford, U.K.: Blackwell Scientific Publications, 193.

Leake, B. E. (1978) *Nomenclature of amphiboles.* American Mineralogist, v. 63, no. 11-12, p. 1023-1052.

Leckie, D. (1988) *Wave-formed, coarse-grained ripples and their relationship to hummocky cross-stratification.* Journal of Sedimentary Petrology, v.58, p.607-622.

Lee, C. A. (1840) *The elements of geology, for popular use; containing a description of the geological formations and mineral resources of the United States.* New York: Harper. 375 p.

Lee, W. T. (1900) *The origin of the débris-covered mesas of Boulder, Colorado.* Journal of Geology, v. 8, p. 504-511.

Lee, W. T. (1903) *The canyons of northeastern New Mexico.* Journal of Geography, v. 2, p. 63-82.

Lees, Alan (1975) *Possible influence of salinity and temperature on modern shelf carbonate sedimentation.* Marine Geology, v. 19, p. 159-198.

Lees, G. M. (1952) *Foreland folding.* Geological Society of London. Quarterly Journal, v. 108, p. 1-34.

Leet, L. D., and Leet, F. J. (1965) *The Earth's mantle.* Seismological Society of America. Bulletin, v. 55, p. 619-625.

Lefebure, D.V. (1995) *Iron oxide breccias and veins P-Cu-Au-Ag-U.* In: Lefebure, D.V., and Ray, G.E., eds, *Selected British Columbia mineral deposit profiles. Volume 1, metallics and coal.* British Columbia Geological Survey Open-File Report 1995-20, p.33-36.

Legget, R. F. (1962) *Geology and engineering. 2nd ed.* New York: McGraw-Hill. 884 p.

Legget, R. F. (1967) *Soil: its geology and use.* Geological Society of America. Bulletin, v. 78, p. 1433-1460.

Legget, R. F. (1973) *Soil.* Geotimes, v. 18, no. 9, p. 38-39.

Legrand, H. E. (1965) *Patterns of contaminated zones of water in the ground.* Water Resources Research, v. 1, no. 1, p. 83-95.

Lehmann, J. G. (1756) *Versuch einer Geschichte von Flötz-Gebürgen, betreffend deren Entstehung, Lage, darinne befindliche Metallen, Mineralien und Fossilien, gröstentheils aus eigenen Wahrnehmungen, chymischen und physicalischen Versuchen, und aus denen Grundsètzen der Natur-Lehre hergeleitet, und mit nöthigen Kupfern versehen.* Berlin: F. A. Lange. 329 p.

Leighton, M. M. (1959) *Stagnancy of the Illinoian glacial lake east of the Illinois and Mississippi rivers.* Journal of Geology, v. 67, p. 337-344.

Leighton, M. M., and MacClintock, Paul (1930) *Weathered zones of the drift-sheet of Illinois.* Journal of Geology, v. 38, p. 28-53.

Leighton, M. W., and Pendexter, C. (1962) *Carbonate rock types.* American Association of Petroleum Geologists. Memoir 1, p. 33-61.

Leighton, R. B., et al. (1969) *Mariner 6 and 7 television pictures: preliminary analysis.* Science, v. 166, no. 3901, p. 49-67.

Leith, C. K. (1905) *Rock cleavage.* U. S. Geological Survey. Bulletin 239. 216 p.

Leith, C. K. (1923) *Structural geology. Revised ed.* New York: Holt. 390 p.

Leith, C. K., and Mead, W. J. (1915) *Metamorphic geology; a textbook.* New York: Holt. 337 p.

Lentin, J.K., and Williams, G.L. (1976) *A monograph of fossil peridinioid dinoflagellate cysts.* Bedford Institute of Oceanography Report Series BI-R-75-16, 237p.

Leopold, L. B., and Langbein, W. B. (1962) *The concept of entropy in landscape evolution.* U. S. Geological Survey. Professional Paper 500-A, p. 1-20.

Leopold, L. B., and Maddock, Thomas, Jr. (1953) *The hydraulic geometry of stream channels and some physiographic implications.* U. S. Geological Survey. Professional Paper 252. 57 p.

Leopold, L. B., and Wolman, M. G. (1957) *River channel patterns—braided, meandering, and straight.* U. S. Geological Survey. Professional Paper 282-B, p. 39-85.

Leopold, L. B., et al. (1964) *Fluvial processes in geomorphology.* San Francisco: Freeman. 522 p.

Leopold, L. B., et al. (1966) *Channel and hillslope processes in a semiarid area, New Mexico.* U. S. Geological Survey. Professional Paper 352-G, p. 193-253.

Lesevich, V. V. (1877) *Opyt kriticheskogo izsledovaniia osnovonachal pozitivnoi filosofii.* St. Petersburg: M. Stasiulevich. 295 p.—Also in: Lesevich, V. V. (1915) Sobranie sochinenii, v. 1, p. 189-454. Moscow: IU. V. Leontovich.

Leverett, Frank (1903) *Glacial features of lower Michigan.* Journal of Geology, v. 11, p. 117-118.

Levinson, A. A. (1974) *Introduction to exploration geochemistry.* Calgary: Applied Publishing Co. 612 p.

Levorsen, A. I. (1933) *Studies in paleogeology.* American Association of Petroleum Geologists. Bulletin, v. 17, p. 1107-1132.

Levorsen, A. I. (1943) *Discovery thinking.* American Association of Petroleum Geologists. Bulletin, v. 27, p. 887-928.

Levorsen, A. I. (1960) *Paleogeologic maps.* San Francisco: Freeman. 174 p.

Levorsen, A. I. (1967) *Geology of petroleum, 2nd ed.* San Francisco: Freeman. 724 p.

Lewan, M. D. (1978) *Laboratory classification of very fine-grained sedimentary rocks.* Geology, v. 6, p. 745-748.

Lewis, D. W. (1964) *"Perigenic"; a new term.* Journal of Sedimentary Petrology, v. 34, p. 875-876.

Li, Yong, Craven, John, Schweig, Eugene S., and Obermeier, Stephen F. (1996) *Sand boils induced by the 1993 Mississippi River flood; could they one day be misinterpreted as earthquake-induced liquefaction?* Geology; v. 24, no. 2, p. 171-174.

Liddicoat, Richard T., Hummel, John H., and Aushalomov, D. M. A., eds. (1993) *The GIA diamond dictionary. 3rd ed.* Santa Monica: Gemological Institute of America.

Lilly, H. D. (1966) *Late Precambrian and Appalachian tectonics in the light of submarine exploration on the Great Bank of Newfoundland and in the Gulf of St. Lawrence; preliminary views.* American Journal of Science, v. 264, p. 569-574.

Lindgren, Waldemar (1912) *The nature of replacement.* Economic Geology, v. 7, p. 521-535.

Lindgren, Waldemar (1928) *Mineral deposits. 3rd ed.* New York: McGraw-Hill. 1049 p.

Lindgren, Waldemar (1933) *Mineral deposits. 4th ed.* New York: McGraw-Hill. 930 p.

Lindseth, R.O., (1979) *Synthetic sonic logs-a process for stratigraphic interpretation.* Geophysics, 44, 3-26.

Lindsey, A. A., et al. (1969) *Natural areas in Indiana and their preservation.* Lafayette: Indiana Natural Areas Survey, Department of Biological Sciences, Purdue University. 594 p.

Lines, L.R., and Treitel, S. (1984) *Tutorial: A review of least-squares inversion and its application to geophysical problems.* Geophysical Prospecting, v.32, p.159-186.

Link, T. A. (1950) *Theory of transgressive and regressive reef (bioherm) development and origin of oil.* American Association of Petroleum Geologists. Bulletin, v. 34, p. 263-294.

Linton, D. L. (1955) *The problem of tors.* Geographical Journal, v. 121, p. 470-487.

Lipman, P.M. (2000) *Calderas.* In: Sigurdsson, H., ed., *Encyclopedia of volcanoes.* New York, Academic Press, p.643-662.

Lipman, P. W. (1984) *The roots of ash flow calderas in western North America: windows into the tops of granitic batholiths.* Journal of Geophysical Research, v. 89, n. B-10, p. 8801-8841.

Lister, T.R. (1970) *The acritarchs and chitinozoa from the Wentlack and Ludlow series of the Ludlow and Millichope areas, Shropshire.* Palaeontographical Society Monograph 124, no.1(Publication 528), 100p.

Lliboutry, Louis (1958) *Studies of the shrinkage after a sudden advance, blue bands and wave ogives on Glaciar Universidad (central Chilean Andes).* Journal of Glaciology, v. 3, p. 261-272.

Lobeck, A. K. (1921) *A physiographic diagram of the United States.* Chicago: Nystrom. Map. 1:3,000,000.

Lobeck, A. K. (1924) *Block diagrams and other graphic methods used in geology and geography.* New York: Wiley. 206 p.

Lobeck, A. K. (1939) *Geomorphology; an introduction to the study of landscapes.* New York: McGraw-Hill. 731 p.

Lockwood, J. P. (1971) *Sedimentary and gravity-slide emplacement of serpentinite.* Geological Society of America. Bulletin, v. 82, p. 919-936.

Loeblich, A.R., Jr., and Tappan, H. (1969) *Acritarch excystment and surface ultrastructure with descriptions of some Ordovician taxa.* Revista española de micropaleontolgía, v.1, p.45-57.

Loewinson-Lessing, F. J. (1899) *Note sur la classification et la nomenclature des roches éruptives.* International Geological Congress. 7th, St. Petersburg, 1897. Compte rendu, p. 53-71.

Logan, B. W., and Semeniuk, V. (1976) *Dynamic metamorphism; processes and products in Devonian carbonate rocks, Canning Basin, western Australia.* Geological Society of Australia. Special Publication 6, 138 p.

Logan, W. E. (1863) *Report on the geology of Canada.* Geological Survey of Canada. Report of Progress to 1863.

Lohest, M. J. M., et al. (1909) *Compte rendu de la session extraordinaire de la Société Géologique de Belgique tenue à Eupen et à Bastogne les 29, 30 et 31 août et 1, 2 et 3 septembre 1908.* Société Géologique de Belgique. Annales, t. 35, p. 351-414.

Lohman, S. W., et al. (1970) *Definitions of selected ground-water terms.* U. S. Geological Survey. Open-File Report, Aug. 1970.

Lohman, S. W., et al. (1972) *Definitions of selected ground-water terms—revisions and conceptual refinements.* U. S. Geological Survey Water-Supply Paper 1988.

Lombard, Augustin (1963) *Laminites: a structure of flysch-type sediments.* Journal of Sedimentary Petrology, v. 33, p. 14-22.

Long, A. E. (1960) *A glossary of the diamond-drilling industry.* U. S. Bureau of Mines. Bulletin 583. 98 p.

Longwell, C. R. (1923) *Kober's theory of orogeny.* Geological Society of America. Bulletin, v. 34, p. 231-241.

Longwell, C. R. (1933) *Meaning of the term "roches moutonnées."* American Journal of Science, v. 225, p. 503-504.

Longwell, C. R. (1951) *Megabreccia developed downslope from large faults.* American Journal of Science, v. 249, p. 343-355.

Longwell, C. R., Knopf, et al. (1969) *Physical geology.* New York: Wiley. 685 p.

Loutit, T. S., Hardenbol, J., and Vail, P. R. (1988) *Condensed sections; the key to age determination and correlation of continental margin sequences.* In: C. K. Wilgus, et al., eds., Sea level changes; an integrated approach. Society of Economic Paleontologists and Mineralogists Special Publication 42, p. 183-213.

Lovén, S. L. (1874) *Etudes sur les Echinoïdées.* K. Svenska Venskaps-Akademien, Stockholm. Handlingar, ser. 4, Bd. 11, no. 7, p. 1-91. 53 plates.

Lovely, H. R. (1948) *Onlap and strike-overlap.* American Association of Petroleum Geologists. Bulletin, v. 32, p. 2295-2297.

Lovering, T. S. (1963) *Epigenetic, diplogenetic, syngenetic, and lithogene deposits.* Economic Geology, v. 58, p. 315-331.

Lowe, C. H., Jr. (1961) *Biotic communities in the Sub-Mogollon region of the inland Southwest. Arizona* Academy of Science. Journal, v. 2, p. 40-49.

Lowe, D., and Waltham, T. (1995) *A dictionary of karst and caves: A brief guide to the terminology and concepts of cave and karst science.* London, British Cave Research Association, Cave Studies Series No.6, 41p.

Lowe, D.J. (2000) *Role of stratigraphic elements in speleogenesis: the speleoinception concept.* In: Klimchouck, A.B., Ford, D.C., Palmer, A.N., and Dreybrodt, W., eds., *Speleogenesis evolution of karst aquifers.* Huntsville, Ala., National Speleological Society, p.65-76.

Lowe, D. R. (1975) *Water escape structures in coarse-grained sediments.* Sedimentology, v. 22, p. 157-204.

Lowe, D. R., and LoPiccolo, R. D. (1974) *The characteristics and origins of dish and pillar structures.* Journal of Sedimentary Petrology, v. 44, p. 484-501.

Lowell, J. D., and Guilbert, J. M. (1970) *Lateral and vertical alteration-mineralization zoning in porphyry ore deposits.* Economic Geology, v. 65, p. 373-408.

Lowenstam, H. A. (1950) *Niagaran reefs of the Great Lakes area.* Journal of Geology, v. 58, p. 430-487.

Lowenstein, T.K., and Hardie, L.A., (1985) *Criteria for the recognition of salt-pan evaporites.* Sedimentology, v.32, p.627-644.

Lowman, P. D., Jr. (1976) *Crustal evolution in silicate planets: implications for the origin of continents.* Journal of Geology, v. 84, p. 1-26.

Lozinski, W. V. (1909) *Das Sandomierz-Opalower Lössplateau.* Globus, v. 96, p. 330-334.

Lüttig, Gerd (1965) *Interglacial and interstadial periods.* Journal of Geology, v. 73, p. 579-591.

Lucia, F. J. (1962) *Diagenesis of a crinoidal sediment.* Journal of Sedimentary Petrology, v. 32, p. 848-865.

Luigi, S., Schreiber, B.C., and Triberti, B. (1999) *Giant polygons in the Messinian salt of the Realmonte Mine (Agrigento, Sicily): Implication for modeling the "Salinity Crisis" in the Mediterranean.* Journal of Sedimentary Research, v.69, p.764-771.

Lundqvist, G. (1949) *The orientation of the block material in certain species of flow earth.* Geografiska Annaler, v. 31, p. 335-347.

Lyell, Charles (1830) *Principles of geology. Vol. 1.* London: John Murray. 511 p.

Lyell, Charles (1833) *Principles of geology. Vol. 3.* London: John Murray. 398 p. plus 109 p. of appendix.

Lyell, Charles (1838) *Elements of geology.* London: John Murray. 543 p.

Lyell, Charles (1839) *Elements of geology. 1st American ed.*, from the 1st London ed. Philadelphia: J. Kay. 316 p.

Lyell, Charles (1840) *Principles of geology. 6th ed.* London: John Murray. 3 v.

Lyell, Charles (1854) *Principles of geology; or, the modern changes of the Earth and its inhabitants considered as illustrative of geology. 9th rev. ed.* New York: Appleton. 834 p.

Lynch, E. J. (1962) *Formation evaluation.* New York: Harper & Row. 422 p.

M

Mabbutt, J. A. (1961) *"Basal surface" or "weathering front".* Geologists' Association. Proceedings, v. 72, p. 357-358.

Macar, P. (1948) *Les pseudo-nodules du Famennien et leur origine.* Société Géologique de Belgique. Annales, v. 72, p. 47-74.

MacArthur, R. H. (1957) *On the relative abundance of bird species.* National Academy of Science. Proceedings, v. 43, p. 293-295.

MacArthur, R. H., and Wilson, E. O. (1967) *The theory of island biogeography.* Princeton, N. J.: Princeton University Press. 203 p. (Monographs in population biology.)

MacBride, T. H. (1910) *Geology of Hamilton and Wright counties.* Iowa Geological Survey. Volume 20, p. 97-149.

MacClintock, C. (1967) *Shell structure of patelloid and bellerophontoid gastropods (Mollusca).* Peabody Museum of Natural History, Yale University, Bulletin 22. p.1-140.

Macdonald, G. A. (1953) *Pahoehoe, aa, and block lava.* American Journal of Science. v. 251, p. 169-191.

Macdonald, G. A. (1960) *Dissimilarity of continental and oceanic rock types.* Journal of Petrology, v. 1, p. 172-177.

Macdonald, G. A. (1972) *Volcanoes.* Englewood Cliffs, N.J.: Prentice-Hall. 510 p.

Macdonald, K. C., et al. (1980) *Hydrothermal heat flux of the "black smoker" vents on the East Pacific Rise.* Earth and Planetary Science Letters, v. 48, p. 1-7.

Macdonald, Ray, and Bailey, D. K. (1973) *The chemistry of the peralkaline oversaturated obsidians.* U. S. Geological Survey. Professional Paper 440 (The data of geochemistry, 6th edition), ch. N, part 1. 37 p.

MacEwan, D. M. C. (1947) *The nomenclature of the halloysite minerals.* Mineralogical Magazine, v. 28, p. 36-44.

Mackay, J. R. (1966) *Segregated epigenetic ice and slumps in permafrost, Mackenzie Delta area, NWT.* Canada Dept. Mines & Technical Surveys, Geographical Branch. Geographical Bulletin, v. 8, p. 59-80.

Mackay, J. R. (1972) *The world of underground ice.* Association of American Geographers. Annals, v. 62, p. 1-22.

Mackie, William (1897) *On the laws that govern the rounding of particles of sand.* Edinburgh Geological Society. Transactions, v. 7, p. 298-311.

Mackin, J. H. (1937) *Erosional history of the Big Horn Basin, Wyoming.* Geological Society of America. Bulletin, v. 48, p. 813-893.

Mackin, J. H. (1948) *Concept of the graded river.* Geological Society of America. Bulletin, v. 59, p. 463-511.

Mackin, J. H. (1950) *The down-structure method of viewing geologic maps.* Journal of Geology, v. 58, p. 55-72.

Mackinder, H. J. (1919) *Democratic ideals and reality; a study in the politics of reconstruction.* London: Constable. 272 p.

MacNeil, F. S. (1954) *Organic reefs and banks and associated detrital sediments.* American Journal of Science, v. 252, p. 385-401.

MacRae, A. [1994] *Web reference* http://www.geo.ucalgary.ca/~macrae/palynology/.

Maiklem, W.R., (1971) *Evaporative drawdown-a mechanism for water-level lowering and diagenesis in the Elk Point Basin.* Bulletin of Canadian Petroleum Geology, v.19, p.487-503.

Makiyama, Jiro (1954) *Syntectonic construction of geosynclinal neptons.* Kyoto University. College of Science. Memoirs, ser. B, v. 21, p. 115-149.

Mallet, Robert (1838) *On the mechanism of glaciers, being an attempt to ascertain the causes and effects of their peculiar, and in part, unobserved motions.* Geological Society of Dublin. Journal, v. 1, p. 317-335.

Malmgren, B. A., et al. (1983) *Evidence for punctuated gradualism in the Late Neogene Globorotalia tumida lineage of planktonic foraminifera.* Paleobiology, v. 9, p. 377-389.

Malott, C. A. (1928a) *An analysis of erosion.* Indiana Academy of Science. Proceedings, v. 37, p. 153-163.

Malott, C. A. (1928b) *The valley form and its development.* Indiana University Studies, v. 15, no. 81, p. 3-34.

Malvern, L. E. (1969) *Introduction to the mechanics of a continuous medium.* Englewood Cliffs, N.J.: Prentice-Hall. 713 p.

Mandelbrot, B.B., (1982) *Fractal geometry.* W.H. Freeman & Co.

Mann, C. J. (1970) *Isochronous, synchronous, and coetaneous.* Journal of Geology, v. 78, p. 749-750.

Mann, K.O., and Lane, H.R. (1995) *Graphic correlation: A powerful stratigraphic technique comes of age.* In: Mann, K.O., Lane, H.R., and Scholle, P.A., eds., *Graphic correlation.* SEPM Special Publication 53, p.3-13.

Manton, W. I. (1965) *The orientation and origin of shatter cones in the Vredefort Ring.* New York Academy of Sciences. Annals, v. 123, p. 1017-1049.

Manutchehr-Danai, Mohsen (2000) *Dictionary of Gems and Gemology.* Berlin: Springer-Verlag.

Marbut, C. F. (1896) *Physical features of Missouri.* Missouri Geological Survey. Volume 10, p. 11-109.

March, A. (1932) *Mathematische Theorie der Regelung nach der Korngestalt.* Z. Krist., v. 81, 285-297.

Mariott, F.H.C., (1990) *A Dictionary of Statistical Terms. 5th ed.*: John Wiley & Sons, Inc., New York, 223p.

Marks, R. W. (1969) *The new dictionary and handbook of aerospace.* New York: Praeger. 531 p.

Marr, J. E. (1900) *The scientific study of scenery.* London: Methuen. 368 p.

Marr, J. E. (1901) *The origin of moels, and their subsequent dissection.* Geographical Journal, v. 17, p. 63-69.

Marr, J. E. (1905) *The anniversary address of the President.* Geological Society of London. Quarterly Journal, v. 61, p. xlvii-lxxxvi.

Marschner, Hannelore (1969) *Hydrocalcite ($CaCO_3 \cdot H_2O$) and nesquehonite ($MgCO_3 \cdot 3H_2O$) in carbonate scales.* Science, v. 165, p. 1119-1121.

Marshak, S., and Mitra, G. (1988) *Basic methods of structural geology.* Englewood Cliffs, N.J., Prentice-Hall, 2 pts., 466p.

Martel, E. A. (1896) *Speleology.* International Geographical Congress. 6th, London. Report, p. 717-722.

Martin Vivaldi, J. L., and MacEwan, D. M. C. (1960) *Corrensite and swelling chlorite.* Clay Minerals Bulletin, v. 4, no. 24, p. 173-181.

Martin, H. G. (1931) *Insoluble residue studies of Mississippian limestones in Indiana.* Indiana Department of Conservation. Publication no. 101. 37 p.

Martin, R., and de Sitter-Koomans, C. (1956) *Pseudotectites from Colombia and Peru.* Leidse Geologische Mededelingen, p. 151-164.

Martinsson, A., ed. (1977) *The Silurian-Devonian boundary. Final report of the Commission on the Silurian-Devonian boundary within IUGS Commission on Stratigraphy and a state-of-the-art report for Project Ecostratigraphy.* International Union of Geological Sciences, Series A (5), p.349.

Masaitis, V.L., Mikhaylov, M.V., and Selivanovskaya, T.V. (1975) *The Popigai meteorite crater (Popigaisjiy Meteoritniy Krater).* Moscow, Nauka Press (in Russian).

Mason, Brian (1958) *Principles of geochemistry.* New York: Wiley.

Mathur, S. M. (1981a) *A new and unusual sedimentary structure.* National Academy Science Letters, India, v. 4, p. 441-442.

Mathur, S. M. (1981b) *Cyclic sedimentation in the Vindhyan Supergroup.* Geological Survey of India. Records, v. 112, Part VIII, p. 1-6.

Matsukuma, Toshinori, and Horikoshi, Ei (1970) *Kuroko deposits in Japan, a review.* In: Tatsumi, T., ed. Volcanism and ore genesis, p. 153-179. Tokyo: University of Tokyo Press.

Matthes, F. É. (1912) *Topology, topography and topometry.* American Geographical Society. Bulletin, v. 44, p. 334-339.

Matthes, F. É. (1930) *Geologic history of the Yosemite Valley.* U. S. Geological Survey. Professional Paper 160. 137 p.

Matthes, F. É. (1939) *Report of Committee on Glaciers, April 1939.* American Geophysical Union. 20th annual meeting (1939). Transactions, pt. 4, p. 518-523.

Maune, D., ed. (2001) *Digital elevation model technologies and applications: The DEM users manual.* Bethesda, Md., American Society for Photogrammetry and Remote Sensing, 539p.

Maw, George (1866) *On subaerial and marine denudation.* Geological Magazine, v. 3, p. 439-451.

Mawe, John (1818) *A new descriptive catalogue of minerals, consisting of more varieties than heretofore published, and intended for the use of students, with which they may arrange the specimens they collect. 3rd ed.* London: Longman, Hurst, Rees, Orme, and Brown. 96 p.

Maxey, G. B. (1964) *Hydrostratigraphic units.* Journal of Hydrology, v. 2, p. 124-129.
Maxson, J. H. (1940) *Fluting and faceting of rock fragments.* Journal of Geology, v. 48, p. 717-751.
Maxson, J. H. (1940a) *Gas pits in non-marine sediments.* Journal of Sedimentary Petrology, v. 10, p. 142-145.
Maxson, J. H. (1950) *Physiographic features of the Panamint Range, California.* Geological Society of America. Bulletin, v. 61, p. 99-114.
Maxson, J. H., and Anderson, G. H. (1935) *Terminology of surface forms of the erosion cycle.* Journal of Geology, v. 43, p. 88-96.
Maxson, J. H., and Campbell, Ian (1935) *Stream fluting and stream erosion.* Journal of Geology, v. 43, p. 729-744.
Maxwell, W. G. H. (1968) *Atlas of the Great Barrier Reef.* New York: Elsevier. 258 p.
May, Irving, and Cuttitta, Frank (1967) *New instrumental techniques in geochemical analysis.* In: Abelson, P. H., ed. Researches in geochemistry, p. 112-142. New York: Wiley. 663 p.
McBirney, A. R., and Williams, H. (1969) *Geology and petrology of the Galapagos Islands.* Geological Society of America. Memoir 118. 197 p.
McBirney, A. R., Baker, B. H. and Nilson, R. H. (1985) *Liquid fractionation, Part I: Basic principles and experimental simulations.* Jour. Volc. Geotherm. Res., 24, p. 1-24.
McBride, E. F. (1962) *Flysch and associated beds of the Martinsburg Formation (Ordovician), central Appalachians.* Journal of Sedimentary Petrology, v. 32, p. 39-91.
McBride, E. F. (1962a) *The term graywacke.* Journal of Sedimentary Petrology, v. 32, p. 614-615.
McBride, E. F. (1963) *A classification of common sandstones.* Journal of Sedimentary Petrology, v. 33, p. 664-669.
McBride, E. F., and Folk, R. L. (1977) *Caballos Novaculite revisited, part II: Chert and shale members.* Journal of Sedimentary Petrology, v. 47, p. 1261-1286.
McBride, E. F., and Yeakel, L. S. (1963) *Relationship between parting lineation and rock fabric.* Journal of Sedimentary Petrology, v. 33, p. 779-782.
McCammon, R. B. (1968) *The dendrograph: a new tool for correlation.* Geological Society of America. Bulletin, v. 79, p. 1663-1670.
McConnell, Duncan (1973) *Biomineralogy of phosphates and physiological mineralization.* In: Griffith, E. J., and others, eds. Environmental phosphorus handbook. Ch. 25, p. 425-442. New York: Wiley.
McConnell, R. G., and Brock, R. W. (1904) *Report on the great landslide at Frank, Alberta.* Canada Department of the Interior. Annual report 1902-1903, pt. 8, appendix. 17 p.
McCullough, L. A. (1977) *Early diagenetic calcareous coal balls and roof shale concretions from the Pennsylvanian (Allegheny Series).* Ohio Journal of Science, v. 77, p. 125-134.
McDonald, B. C., and Banerjee, Indranil (1971) *Sediments and bed forms on a braided outwash plain.* Canadian Journal of Earth Sciences, v. 8, p. 1282-1301.
McElroy, C. T. (1954) *The use of the term "greywacke" in rock nomenclature in N.S.W.* Australian Journal of Science, v. 16, p. 150-151.
McGee, W J (1888) *The geology of the head of Chesapeake Bay.* U. S. Geological Survey. Annual Report, 7th, p. 537-646.
McGee, WJ (1888a) *The classification of geographic form by genesis.* National Geographic Magazine, v.1, p.27-36.
McGee, W J (1891) *The Pleistocene history of northeastern Iowa.* U. S. Geological Survey. Annual Report, 11th, pt. 1, p. 189-577.
McGee, W J (1897) *Sheetflood erosion.* Geological Society of America. Bulletin, v. 8, p. 87-112.
McGee, W J (1908) *Outlines of hydrology.* Geological Society of America. Bulletin, v. 19, p. 193-200.
McGraw-Hill (1984) *McGraw-Hill dictionary of scientific and technical terms,* 3rd. ed. McGraw-Hill Book Company, 1781p.
McIntosh, D. H., ed. (1963) *Meteorological glossary.* London: Her Majesty's Stationery Office. 288 p.
McIver, N. L. (1961) *Upper Devonian marine sedimentation in the central Appalachians.* PhD thesis, Johns Hopkins University. 347 p.
McKee, E. D. (1938) *The environment and history of the Toroweap and Kaibab formations of northern Arizona and southern Utah.* Carnegie Institution of Washington. Publication 492. 268 p.
McKee, E. D. (1939) *Some types of bedding in the Colorado River delta.* Journal of Geology, v. 47, p. 64-81.
McKee, E. D. (1954) *Stratigraphy and history of the Moenkopi Formation of Triassic age.* Geological Society of America. Memoir 61. 133 p.
McKee, E. D. (1965) *Experiments on ripple lamination.* Society of Economic Paleontologists and Mineralogists. Special Publication no. 12, p. 66-83.
McKee, E. D. (1966) *Structures of dunes at White Sands National Monument, New Mexico (and a comparison with structures of dunes from other selected areas.)* Sedimentology, v. 7, p. 1-69.
McKee, E. D., and Weir, G. W. (1953) *Terminology for stratification and cross-stratification in sedimentary rocks.* Geological Society of America. Bulletin, v. 64, p. 381-389.
McKelvey, V.E. (1960) *Relations of reserves of the elements to their crustal abundance.* American Journal of Science, v.258A, p.234-241.
McKelvey, V.E. (1973) *Mineral resource estimates and public policy.* In: Brobst, D.A., and Pratt, W.P., eds., *United States mineral resources.* U.S. Geological Survey Professional Paper 820, p.9-19.
McKinstry, H. E. (1948) *Mining geology; including a glossary of mining and geological terms.* New York: Prentice-Hall. 680 p.
McLean, J. D. (1968) *Foraminiferal zones and zone charts—an analysis and a compilation.* Manual of Micropaleontological Techniques, v. 7. Alexandria, Va.: McLean Paleontological Laboratory.
McManus, D. A. (1962) *A criticism of certain usage of the phinotation.* Journal of Sedimentary Petrology, v. 13, p. 670-674.
McMillan, R.H. (1996) *Turbidite-hosted Au veins.* In: Lefebure, D.V., and Höy, T., eds, *Selected British Columbia mineral deposit profiles. Volume 2, metallic deposits.* British Columbia Geological Survey Open-File Report 1996-13, p.59-61.
McNitt, J. R. (1965) *Review of geothermal resources, ch. 9* in: Terrestrial heat flow. American Geophysical Union. Geophysical Monograph No. 8, p. 240-266.
McVaugh, R., et al. (1968) *An annotated glossary of botanical nomenclature.* Utrecht: International Bureau for Plant Taxonomy and Nomenclature. 31 p. (Regnum Vegetabile, v. 56.)
McNutt, S.R. (2000) *Volcanic seismicity.* In: Sigurdsson, H., ed., *Encyclopedia of volcanoes.* New York, Academic Press, p.1015-1033.
McWilliams, M. O., and Howell, D. G. (1982) *Exotic terranes of western California.* Nature, v. 297, p. 215-217.
McWolfinvan, C., ed. (1971) *Orismologic progeny.* Bethesda (Md.)/Washington, D.C.: Rebecca Elizabeth Wolf and Matthew Joseph Sullivan.
Mead, D. W. (1919) *Hydrology, the fundamental basis of hydraulic engineering. 1st ed.* New York: McGraw-Hill. 647 p.
Means, W. D. (1976) *Stress and strain.* New York: Springer-Verlag. 339 p.
Means, W. D. (1987) *A newly recognized type of slickenslide striation.* J. Struct. Geol., v. 9, p. 585-590.
Mechtly, E. A. (1964) *International system of units: physical constants and conversion factors.* National Aeronautics and Space Administration: George C. Marshall Space Flight Center. 19 p.
Medlicott, H. B., and Blanford, W. T. (1879) *A manual of the geology of India. Pt. 1-2. 2 v.* Calcutta: Geological Survey of India. 817 p.
Mehnert, K. R. (1968) *Migmatites and the origin of granitic rocks.* Amsterdam: Elsevier. 393 p.
Meier, M. F. (1961) *Mass budget of South Cascade Glacier, 1957-1960.* U. S. Geological Survey. Professional Paper 424-B, p. 206-211.
Meinzer, O. E. (1923) *Outline of ground-water hydrology, with definitions.* U. S. Geological Survey. Water-supply Paper 494. 71 p.
Meinzer, O. E. (1939) *Discussion of question No. 2 of the International Commission on Subterranean Water: definitions of the different kinds of subterranean water.* American Geophysical Union. Transactions, pt. 4, p. 674-677.
Meinzer, O. E. (1942) *Hydrology.* New York: Dover. 712 p.
Meischner, K. D. (1964) *Allodapische Kalke, Turbidite in Riff-nahen Sedimentations-Becken.* In: Bouma, A. H., and Brouwer, A., eds. Turbidites, p. 156-191. New York: Elsevier. (Developments in Sedimentation, v. 3.)
Melchior, Hans, and Werdermann, Erich (1954) *A. Engler's syllabus der Pflanzenfamilien, v. 1.* Berlin: Gebrüder Borntraeger. 367 p.
Melchior, Hans, ed. (1954) *A. Engler's Syllabus der Pflanzenfamilien, v. 2.* Berlin: Gebrüder Borntraeger. 666 p.
Mellett, J. S. (1974) *Scatological origin of microvertebrate fossil accumulations.* Science, v. 185, p. 349-350.
Mellor, J. W. (1908) *A note on the nomenclature of clays.* English Ceramic Society. Transactions, v. 8, p. 23-30.
Melosh, H.J. (1989) *Impact cratering: A geologic process* (Oxford Monographs on Geology and Geophysics, v. 11). Oxford, UK, Oxford University Press, 245p.
Melton, F. A. (1936) *An empirical classification of flood-plain streams.* Geographical Review, v. 26, p. 593-609.
Melton, F. A. (1940) *A tentative classification of sand dunes; its application to dune history in the southern High Plains.* Journal of Geology, v. 48, p. 113-173.
Melton, F. A. (1947) *Onlap and strike-overlap.* American Association of Petroleum Geologists. Bulletin, v. 31, p. 1868-1878.
Mendelson, C.V. (1993) *Acritarchs and prasinophytes.* In: Lipps, J.H., ed., *Fossil prokaryotes and protists.* Oxford, UK, Blackwell Scientific, p.77-104.
Mercier, J.-C., and Nicolas, A. (1975) *Textures and fabrics of upper-mantle peridotites as illustrated by xenoliths from basalts.* Journal of Petrology, v. 16, p. 454-487.
Merriam, C. H. (1890) *Results of a biological survey of the San Francisco mountain region and desert of the Little Colorado in Arizona.* U. S. Department of Agriculture. Division of Ornithology and Mammalogy. North American fauna, no. 3. 136 p.
Merriam, D. F. (1962) *Late Paleozoic limestone "buildups" in Kansas.* Kansas Geological Society. Guidebook; field conference, 27th, p. 73-81.
Merriam, D. F. (1963) *The geologic history of Kansas.* Kansas State Geological Survey. Bulletin 162. 317 p.
Merriam, D. F. (1976) *CAI in geology.* Computers and Geoscience, v. 2, p. 3-7.
Merrill, G. P. (1897) *A treatise on rocks, rock-weathering and soils.* New York: Macmillan. 411 p.

Meunier, F.J. (1988) *Nouvelles données sur l'organisation spatiale des fibres de collagène de la plaque basale des écailles des Téléostéens*. Annales des Sciences Naturelles, Zoologie, ser.13, v.9, p.113-121.

Meunier, F.J., and Géraudie, J. (1980) *Les structures en contre-plaqué du derme et des écailles des Vertebres inférieurs*. Année Biologique, v.19, p.1-18.

Meyerhoff, H. A. (1940) *Migration of erosional surfaces*. Association of American Geographers. Annals, v. 30, p. 247-254.

Middleton, G. V. (1967) *Experiments on density and turbidity currents. III. Deposition of sediment*. Canadian Journal of Earth Science, v. 4, p. 475-505.

Middleton, G. V. (1973) *Johannes Walther's law of the correlation of facies*. Geological Society of America. Bulletin, v. 84, p. 979-987.

Middleton, G.V., ed. (1965) *Primary sedimentary structures and their hydrodynamic interpretation*. Society of Economic Paleontologists and Mineralogists. Special Publication no. 12, 265 p.

Mielenz, R. C., and King, M. E. (1955) *Physical-chemical properties and engineering performance of clays*. California Division of Mines. Bulletin 169, p. 196-254. (National Conference on Clays and Clay Technology. 1st, Berkeley, Calif., 1952. Proceedings.)

Milankovitch, Milutin (1920) *Théorie mathématique des phénomènes thermiques produits par la radiation solaire*. Paris: Gauthier-Villars. 338 p.

Milankovitch, Milutin (1941) *Kanon der Erdbestrahlung und seine Anwendung auf das Eiszeitenproblem*. Belgrade: Académie Royale serbe. 633 p.

Miller, D. J. (1953) *Late Cenozoic marine glacial sediments and marine terraces of Middleton Island, Alaska*. Journal of Geology, v. 61, p. 17-40.

Miller, Hugh (1841) *The Old Red Sandstone; or, new walks in an old field*. Edinburgh: J. Johnstone. 275 p.

Miller, Hugh (1883) *River-terracing: its methods and their results*. Royal Physical Society of Edinburgh. Proceedings, v. 7, p. 263-306.

Miller, J. A. (1972) *Dating Pliocene and Pleistocene strata using the potassium-argon and argon-40/argon-39 methods*. In: Bishop, W. W., and Miller, J. A., eds. Calibration of hominoid evolution, p. 63-76. Edinburgh: Scottish Academic Press.

Miller, K.G., Mountain, G.S., Browning, J.V., Kominz, M., Sugarman, P.J., Christie-Blick, N., Katz, M.E., and Wright, J.D. (1998) *Cenozoic global sea level, sequences, and the New Jersey transect: Results from coastal-plain and continental-slope drilling*. Review of Geophysics, v.36, p.569-601.

Miller, M.A. (1987) *A diagnostic encystment suture in the Silurian acritarch* Circinatisphaera aenigma *gen. et sp. nov.*Palynology, v.12, p.121-127.

Miller, M.A. (1996) *Invertebrate cuticular fragments*. In: Jansonius, J., and McGregor, D.C., eds., *Palynology; principles and applications 1*. Dallas, Tex., American Association of Stratigraphic Palynologists Foundation, p.381-382.

Miller, M.A., and Jansonius, J. (1996) *"Linotolypidae" and cenospheres*. In: Jansonius, J., and McGregor, D.C., eds., *Palynology. Principles and applications 1*. Dallas, Tex., American Association of Stratigraphic Palynologists Foundation, p.357-359.

Miller, M.A., and Wood, G.D. (1982) *Trochospiral suture: A new encystment structure in the lower Paleozoic Acritarcha* Leiofusa tumida *Downie, 1959, and* Eupoikilofusa cantabrica *(Cramer) Cramer, 1971*. Neues Jahrbuch für Geologie und Paläontologie, Monatshefte, v.9, p.547-552.

Miller, W. J. (1919) *Pegmatite, silexite, and aplite of northern New York*. Journal of Geology, v. 27, p. 28-54.

Milner, H. B. (1922) *The nature and origin of the Pliocene deposits of the county of Cornwall and their bearing on the Pliocene geography of the south-west of England*. Geological Society of London. Quarterly Journal, v. 78, p. 348-377.

Milner, H. B. (1940) *Sedimentary petrography, with special reference to petrographic methods of correlation of strata, petroleum technology and other economic applications of geology. 3rd ed.* London: Thomas Murby. 666 p.

Milton, D. J. (1969) *Astrogeology in the 19th century*. Geotimes, v. 14, no. 6, p. 22.

Mitchell, C.E., Chen Xu, Bergström, S.M., Zhang Yuan-dong, Wang Zhi-hao, Webby, B.D., and Finney, S.C. (1997) *Definition of a global boundary stratotype for the Darriwilian Stage of the Ordovician System*. Episodes, v.20, no.3, p.158-166.

Mitchell, H. C. (1948) *Definitions of terms used in geodetic and other surveys*. Washington: U. S. Government Printing Office. 87 p. (U. S. Coast and Geodetic Survey. Special Publication no. 242).

Mitchell, R.H., and Bergman, S.C. (1991) *Petrology of lamproites*. New York, Plenum Press, 447p.

Mitchum, R. M., Jr. (1977) *Glossary of terms used in seismic stratigraphy*. In: Payton, C. E., ed. Seismic stratigraphy—applications to hydrocarbon exploration, p. 205-212. American Association of Petroleum Geologists. Memoir 26.

Mitchum, R. M., Jr. and Van Wagoner, J. C. (1991) *High-frequency sequences and their stacking patters; sequence-stratigraphic evidence of high-frequency cycles*. Sedimentary Geology, v. 70, p. 131-160.

Miyashiro, A. (1974) *Volcanic rock series in island arcs and active continental margins*. American Journal of Science, v.274, p.321-355.

Miyashiro, Akiho (1961) *Evolution of metamorphic belts. Journal of Petrology*, v. 2, p. 227-311.

Miyashiro, Akiho (1968) *Metamorphism of mafic rocks*. In: Hess, H. H., and Poldervaart, Arie, eds. Basalts; the Poldervaart treatise on rocks of basaltic composition, v. 2, p. 799-834. New York: Interscience.

Miyashiro, Akiho (1973) *Metamorphism and metamorphic belts*. New York: Wiley. 492 p.

Miyashiro, Akiho, and Shido, F. (1970) *Progressive metamorphism in zeolite assemblages*. Lithos, v. 3, p. 251-260.

Mock, S. J. (1971) *A classification of channel links in stream networks*. Water Resources Research, v. 7, p. 1558-1566.

Molnia, B. F. (1983) *Subarctic glacial-marine sedimentation: a model*. In: Glacial-marine sedimentation. New York: Plenum Press. 844 p.

Monkhouse, F. J. (1965) *A dictionary of geography*. Chicago: Aldine. 344 p.

Monkhouse, F. J., and Wilkinson, H. R. (1952) *Maps and diagrams; their compilation and construction*. London: Methuen. 330 p.

Monroe, W.H., ed. (1970) *A glossary of karst terminology*. U.S. Geological Survey Water-Supply Paper 1899-K, 26p.

Montazer, Parviz, and Wilson, William E. (1984) *Conceptual hydrologic model of flow in the unsaturated zone, Yucca Mountain, Nevada*. Water-Resources Investigations 84-4345, U. S. Geological Survey, 55 p.

Moore, D. G., and Scruton, P. C. (1957) *Minor internal structures of some recent unconsolidated sediments*. American Association of Petroleum Geologists. Bulletin, v. 41, p. 2723-2751.

Moore, D.M., and Reynolds, R.C. (1997) *X-ray diffraction and the identification and analysis of clay minerals, 2nd. ed.* New York, Oxford University Press, 378p.

Moore, Derek (1966) *Deltaic sedimentation*. Earth-Science Reviews, v. 1, p. 87-104.

Moore, G.W. (1952) *Speleothem—a newcave term*. National Speleological Society News, v.10, p.2.

Moore, G. W. (1985) *Explanatory notes for the geodynamic map of the circum-Pacific region*. Tulsa: American Association of Petroleum Geologists. 12 p.

Moore, G. W., and Kennedy, M. P. (1975) *Quaternary faults at San Diego Bay, California*. U. S. Geological Survey. Journal of Research, v. 3, p. 589-595.

Moore, G. W., and Sullivan, G. N. (1997) *Speleology—Caves and the cave environment. 2nd ed.* St. Louis, MO: Cave Books. 176 p.

Moore, J. G., and Albee, W. C. (1981) *Topographic and structural changes, March-July, 1980—photogrammetric data*. In: P. W. Lipman and D. R. Mullineaux, eds. The 1980 Eruptions at Mount St. Helens, Washington, U. S. Geological Survey. Professional Paper 1250.

Moore, J.G., Bryan, W.B., Beeson, M.H., and Normark, W.A. (1995) *Giant blocks in the South Kona landslide, Hawaii*. Geology, v.23, p.125-128.

Moore, P. F. (1958) *Nature, usage, and definition of marker-defined vertically segregated rock units*. American Association of Petroleum Geologists. Bulletin, v. 42, p. 447-450.

Moore, R. C. (1926) *Origin of enclosed meanders on streams of the Colorado Plateau*. Journal of Geology, v. 34, p. 29-57.

Moore, R. C. (1933) *Historical geology*. New York: McGraw-Hill. 673 p.

Moore, R. C. (1936) *Stratigraphic classification of the Pennsylvanian rocks of Kansas*. Kansas State Geological Survey. Bulletin 22. 256 p.

Moore, R. C. (1949) *Meaning of facies*. Geological Society of America. Memoir 39, p. 1-34.

Moore, R. C. (1957) *Minority report*. In: American Commission on Stratigraphic Nomenclature. Nature, usage, and nomenclature of biostratigraphic units, p. 1888. American Association of Petroleum Geologists. Bulletin, v. 41, p. 1877-1889.

Moore, R. C. (1957a) *Modern methods of paleoecology*. American Association of Petroleum Geologists. Bulletin, v. 41, p. 1775-1801.

Moore, R. C. (1958) *Introduction to historical geology. 2nd ed.* New York: McGraw-Hill. 656 p.

Moore, R. C., et al. (1952) *Invertebrate fossils. 1st ed.* New York: McGraw-Hill. 246 p.

Moore, W. G. (1967) *A dictionary of geography; definitions and explanations of terms used in physical geography. 3rd ed.* New York: Praeger. 246 p.

Morgan, W.J. (1971) *Convection plumes in the lower mantle*. Nature, v.230, p.42-???.

Morlot, Adolphe von (1847) *Über Dolomit und seine künstliche Darstellung aus Kalkstein*. Naturwissenschaftliche Abhandlungen, gesammelt und durch Subscription hrsg. von Wilhelm Haidinger, v. 1, p. 305-315.

Morse, James (1977) *Deformation in ramp regions of overthrust faults: experiments with small-scale rock models*. Joint Wyoming-Montana-Utah Geological Associations Guidebook, Rocky Mountain thrust belt geology and resources, p. 457-470.

Morse, P.M., and Feshbach, H. (1967) *Methods of theoretical physics, Part I*. McGraw-Hill.

Morton, C. V. (1957) *The misuse of the term taxon*. Taxonomy, v. 6, p. 155. Also in: Rhodora, 1957, p. 43-44.

Moss, J. H. (1951) *Early man in the Eden valley*. Philadelphia: University of Pennsylvania. University Museum Monograph. 92 p.

Müller, G. (1966) *Die Verteilung von Eisenmonosulfid ($FeS{\cdot}nH_2O$) und organischer Substanz in den Bodensedimenten des Bodensees—ein Beitrag zur Frage der Eutrophierung des Bodensees*.

Müller, G., and Gees, R.A. (1970) *Distribution and thickness of quaternary sediments in the Lake Constance basin*. Sedimentary Geology, v.4, p.81-87.

Müller, German (1967) *Diagenesis in argillaceous sediments*. In: Larsen, Gunnar, and Chilingar, G. V., eds. Diagenesis in sediments, p. 127-177. Amsterdam: Elsevier. 551 p. (Developments in sedimentology 8.)

Müller, Johannes (1858) *Über die Thalassicollen, Polycystinen und Acanthometren des Mittelmeeres.* Akademie der Wissenschaften, Berlin. Abhandlungen (physikalische), 1858, p. 1-62.
Mueller, I. I., and Rockie, J. D. (1966) *Gravimetric and celestial geodesy; a glossary of terms.* New York: Ungar. 129 p.
Mueller, R. F., and Saxena, S. K. (1976) *Chemical petrology.* New York: Springer-Verlag. 394 p.
Muir, I. D., and Tilley, C. E. (1961) *Mugearites and their place in alkali igneous rock series.* Journal of Geology, v. 69, p. 186-203.
Muir-Wood, H. M., and Cooper, G. A. (1960) *Morphology, classification and life habits of the Productoidea (Brachiopoda).* Geological Society of America. Memoir 81. 447 p.
Muller, P. M., and Sjogren, W. L. (1968) *Mascons: lunar mass concentrations.* Science, v. 161, p. 680-684.
Muller, S. W. (1965) *Superposition of strata: part 1. Physical criteria for determining top and bottom of beds.* American Geological Institute: Data Sheet no. 10.
Muller, S. W., compiler (1947) *Permafrost or permanently frozen ground and related engineering problems.* Ann Arbor, Mich.: Edwards. 231 p.
Murchison, R. I. (1839) *The Silurian System. Pt. 1.* London: John Murray. 576 p.
Murray, A. N. (1930) *Limestone oil reservoirs of the northeastern United States and Ontario, Canada.* Economic Geology, v. 25, p. 452-469.
Murray, G. E. (1961) *Geology of the Atlantic and Gulf Coast Province of North America.* New York: Harper. 692 p.
Murray, G. E. (1965) *Indigenous Precambrian petroleum?* American Association of Petroleum Geologists. Bulletin, v. 49, p. 3-21.
Murray, R. (2004) *Evidence from the Earth: Forensic geology.* Missoula, Mont., Mountain Press.
Murray, R. C. (1960) *Origin of porosity in carbonate rocks.* Journal of Sedimentary Petrology, v. 30, p. 59-84.
Murzaevs, E., and Murzaevs, V. (1959) *Slovar' mestnykh geograficheskikh terminov.* Moscow: Geografizdat. 303 p.
Mutch, T. A. (1970) *Geology of the moon.* Princeton, N. J.: Princeton University Press. 324 p.
Mutch, T. A., et al. (1976) *The geology of Mars.* Princeton, N. J.: Princeton University Press. 400 p.
Mutvei, H. (1964) *On the shells of Nautilus and Spirula with notes on the secretions in noncephalopod molluscs.* Arkiv för Zoologi, v.16, no.4, p.221-278.
Mutvei, H. (1974) *SEM studies on arthropod exoskeletons. Part 1. Decapod crustaceans,* Homarus gammarus *L. and* Carcinus maenas *(L.).* Bulletin of the Geological Institutions of the University of Uppsala, new ser., v.4, p.73-80.
Mutvei, H. (1981) *Exoskeletal structure in the Ordovician trilobite* Flexicalymene. Lethaia, v.14, p.225-234.
Mylroie, J.E. (1975) *Hydrologic classification of caves and karst.* In: LaFleur, R.G., ed., *Groundwater as a geomorphic agent.* Boston, Mass., Allen and Unwin, p.157-172.
Mylroie, J.E., and Carew, J.L., (1990) *The Flank Margin Model for Dissolution Cave Development in Carbonate Platforms.* Earth Surface Processes and Landforms, v.15, p.413-424.
Mylroie, J.E., Carew, J.L., and Moore, A.I., (1995) *Blue holes. Definition and genesis.* Carbonates and Evaporites, v.10, no.2, p.225-233.

N

Nabholz, W.K. (1951) *Beziehungen zwischen Fazies und Zeit.* Eclogae geologicae Helvetiae, v. 44, p.131-158.
NACSN: North American Commission on Stratigraphic Nomenclature (1983) *North American Stratigraphic Code.* American Association of Petroleum Geologists. Bulletin, v. 67, p. 841-875.
Nadeau, P. H., Tait, J. M., McHardy, W. J., and Wilson, M. J. (1984) *Interstratified XRD characteristics of physical mixtures of elementary clay particles.* Clay Minerals, v. 19, p. 67-76.
Naldrett, A.J. (1973) *Nickel sulfide deposits—their classification and genesis, with special emphasis on deposits of volcanic association.* Canadian Institute of Mining and Metallurgy. Transactions, v.76, p.183-201.
Naldrett, A.J., and Kullerud, G. (1965) *Two examples of sulfurization in nature (abstract).* Economic Geology, v.60, p.1563.
Narr, W. and Suppe, J. (1991) *Joint spacing in sedimentary rocks.* Journal of Structural Geology, v. 13, p. 1037-1048.
NAS-NRC:National Academy of Science, National Research Council (1992) *Water transfers in the West.* Washington, DC:National Academy of Sciences Press.
NAS-NRC:National Academy of Sciences-National Research Council. Committee on Rock Mechanics (1966) *Rock-mechanics research; a survey of United States research to 1965, with a partial survey of Canadian universities.* Washington, D.C. 82p. (Its Publication 1466).
NASA: National Aeronautics and Space Administration, Goddard Space Flight Center (1971) NASA Earth resources data users' handbook. (Document 71SD4249.)
NASA:National Aeronautics and Space Administration (1966) *Short glossary of space terms.* Publication: SP-1.
National Research Council Panel on Land Burial (1976) *The shallow land burial of low-level radioactively contaminated solid waste.* Washington: National Academy of Sciences. 150p.
Natland, M.L. (1976) *Classification of clastic sediments (abstract).* American Association of Petroleum Geologists. Bulletin, v.60, p.702.
Natland, M.L., and Kuenen, P.H. (1951) *Sedimentary history of the Ventura Basin, California, and the action of turbidity currents.* Society of Economic Paleontologists and Mineralogists. Special Publication no.2, p.76-107.
Naum, T., et al. (1962) *Vulcanokarstul din Masivul Calimanului (Carpatii Orientali).* Bucharest. Universitatea. Analele; seria stiintele naturii: geologie-geografie, anul 11, no.32, p.143-179.
Naumann, C.F. (1850) *Lehrbuch der Geognosie. Bd. 1.* Leipzig: Wilhelm Engelmann. 1000p.
Naumann, C.F. (1858) *Lehrbuch der Geognosie. Bd. 1. 2nd ed.* Leipzig: Wilhelm Engelmann. 960p.
Nayak, V.K. (1970) *Geoplanetology: a new term for geology of the planets including the Moon.* Geological Society of America. Bulletin, v.81, p.1279.
Nelson, A. (1965) *Dictionary of mining.* New York, Philosophical Library, Inc., 523p.
Nelson, A., and Nelson, K.D. (1967) *Dictionary of applied geology, mining and civil engineering.* New York: Philosophical Library. 421p.
Nelson, H.F.,et al. (1962) *Skeletal limestone classification.* American Association of Petroleum Geologists. Memoir 1, p.224-252.
Nelson, T. H. (1989) *Style of salt diapirs as a function of the stage of evolution and the nature of the encasing sediments (extended abs.).* Houston, Gulf Coast Section of the Society of Economic Paleontologists and Mineralogists Foundation, Tenth Annual Research Conference Program and Extended Abstracts, p. 109-110.
Nelson, W.J. (1981) *Faults and their effect on coal mining in Illinois.* Illinois Geological Survey Circular 523, 38p.
Nelson, W.J. (1983) *Geologic disturbances in Illinois coal seams.* Illinois Geological Survey Circular 530, 47p.
Nettleton, L. L. (1934) *Fluid mechanics of salt domes.* American Association of Petroleum Geologists Bulletin, v. 18, p. 1175-1204.
Neumann, A.C. (1966) *Observations on coastal erosion in Bermuda and measurements of the boring rate of the sponge, Cliona lampa.* Limnology and Oceanography, v.11, p.92-108.
Nevin, C.M. (1949) *Principles of structural geology. 4th ed.* New York: Wiley. 410p.
Newell, N.D., et al. (1953) *The Permian reef complex of the Guadalupe Mountains region, Texas and New Mexico - a study in paleoecology.* San Francisco: Freeman. 236p.
Newhall, C.G., and Self, S. (1982) *The volcanic explosivity index (VEI): An estimate of explosive magnitude for historical volcanism.* Journal of Geophysical Research, v.87, p.1231-1238.
Nichols, R.A.H. (1967) *The "sparite" complex: eosparite vs. neosparite.* Journal of Sedimentary Petrology, v.37, p.1247-1248.
Nichols, R.L. (1936) *Flow-units in basalt.* Journal of Geology, v. 44, p.617-630.
Nichols, R.L. (1938) *Grooved lava.* Journal of Geology, v.46, p.601-614.
Nichols, R.L. (1946) *McCartys basalt flow, Valencia County, New Mexico.* Geological Society of America. Bulletin, v.57, p.1049-1086.
Nickel, E. H. (1995) *Definition of a mineral.* Mineralogical Magazine, v. 59, p. 767-768.
Nickel, Ernest H. and Nichols, Monte C. (1991) Mineral reference manual. New York, NY: Van Nostrand Reinhold, 250 p.
Nicodemus, F.E. (1971) *A proposed military standard on infrared terms and definitions.* Part I, Appendix C of the Philco-Ford Corp. final report, Contract No. DAAH01-71-C-0433. (Available as AD-758341 from the National Technical Information Service of the U.S. Department of Commerce, Springfield, Va.) 256p.
Nielsen, R. L. and Delong, S. E. (1992) *A numerical approach to boundary layer fractionation: application to differentiation in natural magma systems.* Contrib. Mineral. Petrol., 110, p. 335-369.
Niggli, Paul (1954) *Rocks and mineral deposits.* English translation by Robert L. Parker. San Francisco: Freeman. 559p.
Nikolaev, V.A. (1959) *Some structural characteristics of mobile tectonic belts.* International Geology Review, v.1, p.50-64.
Niles, H. A., and Goudie, A. S. (1990) *Tufas, travertines and allied carbonate deposits.* Progress in Physical Geography. 14; 1, p. 19-41.
NOAA: National Oceanic and Atmospheric Administration (2004) *Hydrology glossary of terms.* http//www.crh.noaa.gov/fgf/hydro/glossary/glossary.html, accessed Nov. 2004.
Noble, L.F. (1941) *Structural features of the Virgin Spring area, Death Valley, California.* Geological Society of America. Bulletin, v.52, p.941-999.
Nockolds, S.R. (1934) *The production of normal rock types by contamination and their bearing on petrogenesis.* Geological Magazine, v.71, p.31-39.
Nordenskjöld, Otto (1909) *Einige Beobachtungen über Eisformen und Vergletscherung der antarktischen Gebiete.* Zeitschrift für Gletscherkunde, Bd.3, p.321-334.
Nordyke, M.D. (1961) *Nuclear craters and preliminary theory of the mechancis of explosive crater formation.* Journal of Geophysical Research, v.66, p.3439-3459.
Norris, D.K. (1958) *Structural conditions in Canadian coal mines.* Geological Survey of Canada. Bulletin 44. 54p.
Norris, D.K. (1964) *Microtectonics of the Kootenay Formation near Fernie, British Columbia.* Bulletin of Canadian Petroleum Geology, v.12, p.383-398.

Northrop, J.I. (1890) *Notes on the geology of the Bahamas.* New York Academy of Sciences. Transactions, v.10, p.4-22.

Norton, Denis (1978) *Sourcelines, sourceregions, and pathlines for fluids in hydrothermal systems related to cooling plutons.* Economic Geology, v.73, p.21-28.

Norton, W.H. (1917) *A classification of breccias.* Journal of Geology, v. 25, p.160-194.

Noszticzius, Zoltan (1979) *A possible explanation for the origin of the earthquake lights.* Seismological Society of America. Bulletin, v. 69, p. 1627.

Nye, J.F. (1952) *The mechanics of glacier flow.* Journal of Glaciology, v.2, p.82-93.

Nye, J.F. (1958) *Surges in glaciers.* Nature, v.181, p.1450-1451.

Nye, J.F. (1958a) *A theory of wave formation on glaciers.* International Association of Scientific Hydrology Publication 47, p. 139-154.

O

O'Brien, J. J., and Lerche, Ian (1988) *Heat flow through and around salt sheets (abs.).* American Association of Petroleum Geologists Bulletin, v. 72, p. 230.

O'Leary, D.W., et al. (1976) *Lineament, linear, lineation: some proposed new standards for old terms.* Geological Society of America. Bulletin, v.87, p.1463-1469.

Oakes, H., and Thorp, J. (1951) *Dark clay soils of warm regions variously called rendzina, black cotton soils, regur, and tirs.* Soil Science Society of America. Proceedings, v.15, p.347-354.

Oberbeck, V.R. (1975) *The role of ballistic erosion and sedimentation in lunar stratigraphy.* Reviews of Geophsics and Space Physics, v.13, p.337-362.

Oberbeck, V.R., and Morrison, R.H. (1974) *Laboratory simulation of the herringbone pattern associated with lunar secondary crater chains.* The Moon, v.9, no.3-4, p.415-455.

Odin, G.S. (1982) *The Phanerozoic time scale revisited.* Episodes, v. 1982, no. 3, p. 3-9.

Odin, G.S., and Lamaurelle, M.A. (2001) The *global Campanian-Maastrichtian stage boundary.* Episodes, v.24, no.4, p.229-238.

Odin, G.S., and Letolle, R. (1980) *Glauconitization and phosphatization environments: a tentative comparison.* Society of Economic Paleontologists and Mineralogists. Special publications 29, p.227-237.

Odin, G.S., and Matter, A. (1981) *De glauconiarum origine.* Sedimentology, v. 28, p. 611-641.

Odom, I.E. (1984) *Glauconite and celadonite minerals.* In: S.W. Bailey, ed. Micas. Mineralogical Society of America. Reviews in Mineralogy, v. 13, 545-571.

Odum, E.P. (1959) *Fundamentals of ecology.* Philadelphia: W.B. Saunders.

Ogilvie, I.H. (1905) *The high-altitude conoplain; a topographic form illustrated in the Ortiz Mountains.* American Geologist, v.36, p.27-34.

Ohio, Legislative Service Commission (1969) *Preservation of natural areas, and Report of Committee to Study Natural Areas.* Project officer:John P. Bay. Columbus:Ohio Legislative Service Commission. 20p. (Its Staff Research Report no.89).

Ojakangas, R.W. (1985) *Evidence for Early Proterozoic glaciation: the dropstone unit—diamictite association.* Geological Survey of Finland. Bulletin 331, p. 51-72.

Oldham, Thomas (1879) *Geological glossary for the use of students.* Edited by R.D. Oldham. London: Stanford. 62p.

Olea, R.A., ed., (1991) *Geostatistical Glossary and Multilingual Dictionary.* Oxford Univ. Press, New York, 177p.

Oliver, W.A., jr. (1968) *Some aspects of colony development in corals.* Journal of Paleontology, v.42, no.5, pt.2, p.16-34.

Oppel, Albert (1856-1858) *Die Juraformation Englands, Frankreichs und des südwestlichen Deutschlands; nach ihren einzelnen Gliedern einigetheilt und verglichen.* Stuggart: Ebner & Seubert. 857p. (Württembergische Naturwissenschaftliche Jahreshefte. Jahrg.12, p.121-132, 313-556; 13, p.141-396; 14, p.128-291).

Oriel, S.S. (1949) *Definitions of arkose.* American Journal of Science, v.247, p.824-829.

Oriel, S.S., et al. (1976) *Stratigraphic Commission Note 44—Application for addition to code concerning magnetostratigraphic units.* American Association of Petroleum Geologists. Bulletin, v.60, p.273-277.

Orr, R.T. (1961) *Vertebrate biology.* Philadelphia, W.B. Saunders, 400p.

Osborn, J.W., and Hillman, J. (1979) *Enamel structure in some therapsids and Mesozoic mammals.* Calcified Tissue International, v.29, p.47-61.

Osborne, R.H. (1969) *Undergraduate instruction in geomathematics.* Journal of Geological Education, v.17, p.120-124.

Otterman, Joseph, and Bronner, F.E. (1966) *Martian wave of darkening: a frost phenomenon?* Science, v.153, p.56-60.

Otto, G.H. (1938) *The sedimentation unit and its use in field sampling.* Journal of Geology, v.46, p.569-582.

Otvos, E.G., Jr. (1965) *Types of rhomboid beach surface patterns.* American Journal of Science, v. 263, p. 271-276.

Otvos, E.G., jr. (1976) *"Pseudokarst" and "pseudokarst terrains": problems of terminology.* Geological Society of America. Bulletin, v.87, p.1021-1027.

Otvos, E.G., jr., and Price, W.A. (1979) *Problems of chenier genesis and terminology—an overview.* Marine Geology, v.31, p.251-263.

Oxburgh, E.R. (1972) *Flake tectonics and continental collision.* Nature, v.239, p. 202-204.

P

Packer, R.W. (1965) *Lag mound features on a dolostone pavement.* Canadian Geographer, v.9, p.138-143.

Packham, G.H. (1954) *Sedimentary structures as an important factor in the classification of sandstones.* American Journal of Science, v.252, p.466-476.

Packham, G.H., and Crook, K.A.W. (1960) *The principle of diagenetic facies and some of its implications.* Journal of Geology, v.68, p.392-407.

Padfield, H.I., and Smith, C.L. (1968) *Water and culture: new decision rules for old institutions.* Rocky Mountain Social Science Journal, v.5, no.2, p.23-32.

Page, David (1859) *Handbook of geological terms and geology.* Edinburgh: William Blackwood. 416p.

Palmer, A.N. (1972) *Dynamics of a sinking stream system: Onesquethaw Cave, New York.* National Speleological Society Bulletin, v.34, no. 3, p.89-110.

Palmer, A.N., (1991) *Origin and morphology of limestone caves.* Geological Society of America Bulletin, v.103, p.1-21.

Palmer, A.N., and Palmer, M.V., (1995) *The Kaskaskia Paleokarst of the northern Rocky Mountain and Black Hills, Northwestern U.S.A.* Carbonates and Evaporites, v. 10, no.2, p.148-160.

Palmer, A.R. (1965) *Biomere-a new kind of biostratigraphic unit.* Journal of Paleontology, v.39, p.149-153.

Palmer, A.R. (1983) *The Decade of North American Geology 1983 geologic time scale.* Geology, v. 11, p.503-504.

Palmer, L.A. (1920) *Desert prospecting.* Engineering and Mining Journal, v.110, p. 850-853.

Paproth, E., Feist, R., and Flajs, G. (1991) *Decision on the Devonian-Carboniferous boundary stratotype.* Episodes, v.14, no.4, p.331-336.

Park, C.F. jr., and MacDiarmid, R.A. (1970) *Ore deposits. 2d ed.* San Francisco: Freeman. 522p.

Park, C.F., jr. (1959) *The origin of hard hematite in itabirite.* Economic Geology, v.54, p.573-587.

Park, James (1914) *A text-book of geology; for use in mining schools, colleges, and secondary schools.* London: Charles Griffin. 598p.

Parks, B.C. (1951) *Petrography of American lignites.* Economic Geology, v.46, no.1, p.23-50.

Parsons, W. H. (1969) *Criteria for the recognition of volcanic breccias: a review.* Geological Society of America. Memoir 115, p. 263-304.

Passega, R. (1957) *Texture as characteristic of clastic properties.* American Association of Petroleum Geologists. Bulletin, v.41, p.1952-1984.

Paterson, M.S., and Weiss, L.E. (1961) *Symmetry concepts in the structural analysis of deformed rocks.* Geological Society of America. Bulletin, v.72, p.841-882.

Paterson, W.S.B. (1969) *The physics of glaciers.* New York: Pergamon. 250p.

Paterson, W.S.B. (1994) *Physics of glaciers, 3rd ed.* New York, Pergamon Press.

Paul, C.R.C. (1968) *Morphology and function of dichoporite porestructures in cystoids.* Palaeontology, v.11, p.697-730.

Paul, C.R.C. (1972) *Morphology and function of exothecal porestructures in cystoids.* Palaeontology, v.15, p.1-28.

Pavia, G., and Enay, R. (1997) *Definition of the Aalenian-Bajocian Stage boundary.* Episodes, v.20, no.1, p.16-22.

Pavlov, A.P. (1889) *Geneticheskie tipy materikovykh obrazovanii lednikovoi i posle-lednikovoi epokhi.* Geologicheskogo Komiteta (St. Petersburg). Izvestiia, v.7, p.243-261.

Pawluk, S. and Brewer, R. (1975) *Micromorphological and analytical characteristics of some soils from Devon and King Christian Islands, N.W.T.* Canadian Journal of Soil Science, Vol.55, No.3, pp.349-361.

Payne, T.G. (1942) *Stratigraphical analysis and environmental reconstruction.* American Association of Petroleum Geologists. Bulletin, v.26, p.1697-1770.

Péwé, T.L. (1959) *Sand-wedge polygons (tesselations) in the McMurdo Sound region, Antarctica—a progress report.* American Journal of Science, v.257, p.545-552.

Péwé, T.L. (1976) *Permafrost.* McGraw-Hill Yearbook of Science and Technology, p.30-42. New York: McGraw-Hill.

Peabody, F.E. (1947) *Current crescents in the Triassic Moenkopi Formation.* Journal of Sedimentary Petrology, v.17, p.73-76.

Peacock, D.C.P., Knipe, R.J., and Sanderson, D.J. (2000) *Glossary of normal faults.* Journal of Structural Geology, v.22, no.3, p.291-305.

Peacock, M.A. (1931) *Classification of igneous rock series.* Journal of Geology, v.39, p.54-67.

Pearce, T. H. (1968) *A contribution to the theory of variation diagrams.* Contrib. Mineral. Petrol., 19, p. 142-157.

Pearn, W.C. (1964) *Finding the ideal cyclothem.* Kansas State Geological Survey. Bulletin 169, v.2, p.399-413.

Peltier, L.C. (1950) *The geographic cycle in periglacial regions as it is related to climatic geomorphology.* Association of American Geographers. Annals, v.40, p.214-236.

Pelto, C.R. (1954) *Mapping of multicomponent systems.* Journal of Geology, v.62, p.501-511.

Penck, Albrecht (1900) *Geomorphologische Studien aus der Herzegowina.* Deutscher und Oesterreichischer Alpenverein. Zeitschrift, Bd. 31, p.25-41.

Penck, Albrecht (1919) *Die Gipfelflur der Alpen.* Akademie der Wissenschaften, Berlin. Sitzungsberichte, Jahrg. 1919, p.256-268.

Penck

Penck, Walther (1924) *Die morphologische Analyse; ein Kapitel der physikalischen Geologie.* Stuttgart:J. Engelhorns. 283p. (Geographische Abhandlungen, 2 Reihe, Hft.2).

Penck, Walther (1953) *Morphological analysis of land forms; a contribution to physical geology.* Translated by Hella Czech and K.C. Boswell. New York: St. Martin's Press. 429p.

Pennak, R.W. (1964) *Collegiate dictionary of zoology.* New York: Ronald. 583p.

Pennington, W. (1979) *The origin of pollen in lake sediments. An enclosed lake compared with one receiving inflow streams.* New Phytologist, v.83, no.1, p.189-211.

Pepper, J.F., et al. (1954) *Geology of the Bedford Shale and Berea Sandstone in the Appalachian basin.* U.S. Geological Survey. Professional Paper 259. 111p.

Perry, W.J., Jr., and DeWitt, Wallace (1977) *A field guide to thin-skinned tectonics in the central Appalachians.* American Association of Petroleum Geologists. Annual Convention, Washington, D.C. Field Trip 4, Guidebook. 54p.

Peters, L.J. (1949) *The direct approach to magnetic interpretation and its practical application.* Geophysics, v.14, p.290-320.

Peterson, D.W., and Tilling, R.I. (2000) *Lava flow hazards.* In: Sigurdsson, H., ed., *Encyclopedia of volcanoes.* New York, Academic Press, p.957-971.

Peterson, Donald; Holcomb, R. T.; Tilling, R. I.; and Christiansen, R. L. (1994) *Development of lava tubes in the light of observations at Mauna Ulu, Kilauea volcano.* Hawaii Bulletin of Volcanology, v. 56, p. 343-360.

Pettijohn, F.J. (1949) *Sedimentary rocks.* New York: Harper. 526p.

Pettijohn, F.J. (1954) *Classification of sandstones.* Journal of Geology, v.62, p.360-365.

Pettijohn, F.J. (1957) *Sedimentary rocks. 2nd ed.* New York: Harper. 718p.

Pettijohn, F.J. (1975) *Sedimentary rocks, 3rd ed.* New York, Harper & Row, 628p.

Pettijohn, F.J., and Potter, P.E. (1964) *Atlas and glossary of primary sedimentary structures.* New York: Springer-Verlag. 370p.

Pettijohn, F.J., Potter, P.E., and Siever, Raymond (1972) *Sand and sandstone.* New York: Springer-Verlag. 600p.

Pettijohn, F.J., Potter, P.E., and Siever, R. (1987) *Sand and sandstone, 2nd ed.* New York, Springer, 553p.

Peyer, B. (1968) *Comparative odontology.* Chicago, University of Chicago Press, 347p.

Peyve, A.V., and Sinitzyn, V.M. (1950) *Certains problèmes fondamentaux de la doctrine des géosynclinaux.* Akademiya Nauk SSSR. Izvestiya, Seriya Geologicheskaya, v.4, p.28-52.

Phemister, James (1956) *Petrography.* In: Great Britain. Geological Survey. The limestones of Scotland; chemical analyses and petrography. Edinburgh: H.M.S.O. 150p. (Its Memoirs; special reports on the mineral resources of Great Britain, v.37, p.66-74).

Philipsborn, H. von (1930) *Zur chemisch-analytischen Erfassung der isomorphen Variation gesteinbildender Minerale.* Die Mineralkomponenten des Pyroxengranulits von Hartmannsdorf (Sa.). Chemie der Erde, v.5, p.233-253.

Phillips, F.C. (1954) *The use of stereographic projection in structural geology.* London: Edward Arnold. 86p.

Phillips, J.D. (1999) *Divergence, convergence, and self-organization in landscapes.* Association of American Geographers, Annals, v.89, p.466-488.

Phillips, J.D. (2003) *Sources of nonlinear complexity in geomorphic systems.* Progress in Physical Geography, v.26, p.339-361.

Phillips, William (1818) *A selection of facts from the best authorities, arranged so as to form an outline of the geology of England and Wales.* London: William Phillips. 250p.

Phipps, M. (1989) *Dynamical behavior of cellular automata under the constraint of neighborhood coherence.* Geographical Analysis, v.21, p. 197-215.

Picard, M.D. (1953) *Marlstone-a misnomer as used in Uinta Basin, Utah.* American Association of Petroleum Geologists. Bulletin, v.37, p.1075-1077.

Pickel, W. (2004) *The new huminite classification (ICCP System 1994).* International Journal of Coal Geology.

Piel, K.M., and Evitt, W.R. (1980) *Paratabulation in the Jurassic dinoflagellate genus* Nannoceratopsis *and a comparison with modern taxa.* Palynology, v.4, p.79-104.

Pike, R.J. (1974) *Depth/diameter relations of fresh lunar craters: Revision from spacecraft data.* Geophysical Research Letters, v.1, p.291-294.

Pike, R.J., and Wilson, S.J. (1971) *Elevation-relief ratio, hypsometric integral, and geomorphic area-altitude analysis.* Geological Society of America. Bulletin, v.82, p.1079-1084.

Pinkerton, John (1811) *Petralogy; a treatise on rocks. 2 v.* London: White, Cochrane. 599p. and 656p.

Pirsson, L.V. (1896) *A needed term in petrography.* Geological Society of America. Bulletin, v.7, p.492-493.

Pirsson, L.V. (1915) *Physical geology.* New York: Wiley. 404p. (A text-book of geology, pt.1).

Pittock, A.B., et al., eds. (1978) *Climatic change and variability: a southern perspective.* New York, Cambridge University Press. 455p.

Playfair, John (1802) *Illustrations of the Huttonian theory of the Earth.* Edinburgh: Wm. Creech. 528p.

Playford, G. (1977) *Lower to Middle Devonian acritarchs of the Moose River Basin, Ontario.* Geological Survey of Canada Bulletin 279, 87p.

Playford, G., Ribecai, C., and Tongiorgi, M. (1995) *Ordovician acritarch genera* Peteinosphaeridium, Liliosphaeridium, and Cycloposphaeridium: *morphology, taxonomy, biostratigraphy, and palaeogeographic significance.* Bolletino della Societá Paleontologica Italiana, v.34, p.3-54.

Plint, A.G., and Nummedal, D. (2000) *The falling stage systems tract: Recognition and importance in sequence stratigraphic analysis.* In: Hunt, D., and Gawthorpe, R., eds., *Sedimentary responses to forced regressions.* Geological Society (London) Special Publication 172, p.1-17.

Plumb, K.A. (1991) *New Precambrian time scale.* Episodes, v.14, no.2, p.139-140.

Plumb, R. A., and Hickman, S. J. (1985) *Stress-induced borehole elongation: a comparison between the four-arm dipmeter and the borehole televiewer in the Auburn geothermal well.* Journal of Geophysical Research, v. 90, p. 5513-5522.

Plumley, W.J., et al. (1962) *Energy index for limestone interpretation and classification.* American Association of Petroleum Geologists. Memoir 1, p.85-107.

Poag, C. W., and Valentine, P. C. (1976) *Biostratigraphy and ecostratigraphy of the Pleistocene basin, Texas-Louisiana continental shelf.* Gulf Coast Assoc. Geol. Soc., Transactions, v. 26, p. 185-256.

Pojeta, J. (1986a) *Phylum Hyolitha.* In: Boardman, R.S., et al., Fossil invertebrates. Oxford: Blackwell.

Pojeta, J. (1986b) *Phylum Mollusca, Part 1.* In: Boardman, R.S., et al., Fossil invertebrates. Oxford: Blackwell

Pojeta, John, Jr., and Runnegar, Bruce (1976) *The paleontology of the rostroconch mollusks and the early history of the phylum Mollusca.* U.S. Geological Survey. Professional Paper 968. 88p.

Poland, J.F., et al. (1972) *Glossary of selected terms useful in studies of the mechanics of aquifer systems and land subsidence due to fluid withdrawal.* U.S. Geological Survey. Water-supply Paper 2025. 9p.

Poldervaart, Arie, and Parker, A.B. (1964) *The crystallization index as a parameter of igneous differentiation in binary variation diagrams.* American Journal of Science, v.262, p.281-289.

Pollard, D. D., and Segall, P. (1987) *Theoretical displacements and stresses near fractures in rock: with applications to faults, joints, veins, dikes, and solution surfaces.* In: Atkinson, B. K., ed., Fracture Mechanics of Rock : Academic Press, Orlando, p. 277-350.

Pollard, David D., and Aydin, Atilla (1988) *Progress in understanding jointing over the past century.* Geological Society of America Bulletin, v. 100, no. 8, p. 1181-1204.

Pompeckj, J.F. (1914) *Die Bedeutung des schwäbischen Jura für die Erdgeschichte.* Stuttgart: Schweizerbartsche. 64p.

Poole, D.F.G. (1956) *The structure of the teeth in some mammal-like reptiles.* Quarterly Journal of Microscopical Science, v.97, p.303-312.

Poole, D.F.G. (1967a) *Enamel structure in primitive mammals.* Journal of Dental Research, v.46, p.124.

Poole, D.F.G. (1967b) *Phylogeny of tooth tissues: Enameloid and enamel in Recent vertebrates, with a note on the history of cementum.* In: Miles, A.E.W., ed., *Structural and chemical organization of teeth,* v.I. New York, Academic Press, p.111-149.

Porsild, A.E. (1938) *Earth mounds in unglaciated Arctic northwestern America.* Geographical Review, v.28, p.46-58.

Porter, S.C., and Denton, G.H. (1967) *Chronology of neoglaciation in the North American Cordillera.* American Journal of Science, v.256, p.177-210.

Posamentier, H. W. and Vail, P. R. (1988) *Eustatic controls on clastic deposition II-sequence and systems tract models.* In: C. K. Wilgus et al., eds., Sea-level changes; an integrated approach. Society of Economic Paleontologists and Mineralogists Special Publication 42, p. 125-154.

Posamentier, H. W., Allen, G. P., James, D. P., and Tesson, M. (1992) *Forced regressions in a sequence stratigraphic framework; concepts, examples, and exploration significance.* American Association of Petroleum Geologists Bulletin, v. 76, p. 1687-1709.

Post, Lennart von (1924) *Ur de sydsvenska skogarnas regionala historia under postarkist tid.* Geologiska Föreningen, Stockholm. Förhandlingar, Bd. 46, p.83-128.

Postma, H. (1961) *Transport and acculmulation of suspended matter in the Dutch Wadden Sea.* Netherlands Journal of Sea Research, v.1, p.148-190.

Potter, P.E., and Glass, H.D. (1958) *Petrology and sedimentation of the Pennsylvanian sediments in southern Illinois-a vertical profile.* Illinois State Geological Survey. Report of Investigations 204. 60p.

Pough, F.H. (1967) *The story of gems and semiprecious stones.* Irvington-on-Hudson: Harvey House. 142p.

Powell, C. McA. (1974) *Timing of slaty cleavage during folding of Precambrian rocks, Northwest Tasmania.* Bull. Geol. Soc. Am., v. 85, 1043-1060.

Powell, C. McA. (1979) *A morphological classification of rock cleavage.* Tectonophysics, v. 58, p. 21-34.

Powell, J.W. (1873) *Some remarks on the geological structure of a district of country lying to the north of the Grand Cañon of the Colorado.* American Journal of Science, 3rd ser., v.5, p.456-465.

Powell, J.W. (1874) *Remarks on the structural geology of the Valley of the Colorado of the West.* Philosophical Society of Washington. Bulletin, v.1, p.48-51.

Powell, J.W. (1875) *Exploration of the Colorado River of the West and its tributaries.* Washington: Government Printing Office. 291p.

Powell, J.W. (1895) *Physiographic features.* National Geographic Monographs, v.1, no.2, p.33-64.

Powell, R.L., (1963) *Alluviated cave streams in south-central Indiana*. Indiana Academy of Science Proceedings, v.72, p.182-189.

Powell, W.G., Chapman, D.S., Balling, N., and Beck, A.E., (1988) *Continental heat-flow density*. In Haenel, R., Rybach, L., and Stegene, L., eds., Handbook of terrestrial heat-flow density determination. Kluwer Acad. Publ, Dordrecht, p. 167-222.

Power, F.D. (1895) *A glossary of terms used in mining geology*. Adelaide: Australasian Institute of Mining Engineers. 69p.

Power, W.R., Jr. (1961) *Backset beds in the Coso Formation, Inyo County, California*. Journal of Sedimentary Petrology, v.31, p.603-607.

Powers, M.C. (1953) *A new roundness scale for sedimentary particles*. Journal of Sedimentary Petrology, v.23, p.117-119.

Powers, R.W. (1962) *Arabian Upper Jurassic carbonate reservoir rocks*. American Association of Petroleum Geologists. Memoir 1, p.122-192.

Pratt, W.P., and Brobst, D.A. (1974) *Mineral resources: potentials and problems*. U.S. Geological Survey. Circular 698. 20p.

Premoli Silva, I., and Jenkins, D. (1993) *Decision on the Eocene-Oligocene boundary stratotype*. Episodes, v.16, no.3, p.379-382.

Prentice, J.E. (1956) *The interpretation of flow-markings and load-casts*. Geological Magazine, v.93, p.393-400.

Presnall, D.C. (1969) *The geometrical analysis of partial fusion*. American Journal of Science, v.267, p.1178-1194.

Preuschoft, H., Reif, W.-E., and Müller, W.H. (1974) *Funktionsanpassungen in Form und Struktur an Haifischzähnen*. Zeitschrift für Anatomie und Entwicklungsgeschichte, v.143, p.315-344.

Prezant, R.S., and Tan Tiu, A. (1986a) *Spiral crossed-lamellar shell growth in* Corbicula *(Mollusca, Bivalvia)*. American Microscopical Society Transactions, v.105, no.4, p.338-347.

Prezant, R.S., and Tan Tiu, A. (1986b) *Unique shell microstructure of* Corbicula fluminea. American Malacological Bulletin, v.4, p.116-117.

Price, R.A. (1965) *Flathead map-area, British Columbia and Alberta*. Geological Survey of Canada. Memoir 336. 221p.

Price, R.A. (1981) *The Cordilleran foreland thrust and fold belt in the southern Canadian Rocky Mountains*. In: Thrust and nappe tectonics , edited by McClay, K. R., and Price, N. J. Spec. Publ. Geol. Soc. Lond., v. 9, p. 293-310.

Price, W.A. (1947) *Equilibrium of form and forces in tidal basins of coast of Texas and Louisiana*. American Association of Petroleum Geologists. Bulletin, v.31, p.1619-1663.

Price, W.A. (1954) *Dynamic environments-reconnaissance mapping, geologic and geomorphic, of continental shelf of Gulf of Mexico*. Gulf Coast Association of Geological Societies. Transactions, v.4, p.75-107.

Prior, G.T. (1920) *The classification of meteorites*. Mineralogical Magazine, v.19, no.90, p.51-63.

Pritchard, D. (1952) *Estuarine hydrology*. Advances in Geophysics, v.1, p.243-280.

Proot, R.J., Visscher, H., and Dilcher, D.L. (1996) *Zoidogamy in fossil gymnosperms: The centenary of a concept. With special reference to prepollen of late Paleozoic conifers*. National Academy of Science, USA, v.93, p.11713-11717.

Pryor, E.J. (1963) *Dictionary of mineral technology*. London: Mining Publications. 437p.

Pumpelly, Raphael (1894) *Geology of the Green Mountains in Massachusetts*. U.S. Geological Survey. Monograph 23, p.5-34.

Pumpelly, Raphael, et al. (1894) *Geology of the Green Mountains in Massachusetts; part III, Mount Greylock: its areal and structural geology*. by T. Nelson Dale. U.S. Geological Survey. Monograph 23.

Purdy, R.C. (1908) *Qualities of clays suitable for making paving brick*. Illinois State Geological Survey. Bulletin no.9, p.133-278.

Purser, B.H., and Loreau, J.-P. (1973) *Aragonitic supratidal encrustations on the Trucial Coast, Persian Gulf*. In: Purser, B.H., ed. The Persian Gulf. New York: Springer-Verlag. 471p.

Pustovalov, L.V. (1933) *Geokhimicheskie fatsii, ikh znachenie v obshchei i prikladnoi geologii*. Problemy Sovetskoi geologii, 1933, [ser. 1], t.1, no.1, p.57-80.

Putnis, A., and McConnell, J.D.C. (1980) *Principles of mineral behavior*. New York: Elsevier. 257p.

Pyle, D.M. (2000) *Sizes of volcanic eruptions*. In: Sigurdsson, H., ed., *Encyclopedia of volcanoes*. New York, Academic Press, p.263-269.

Q

Quensel, P. (1916) *Zur Kenntnis der Mylonitbildung, erläutert an Material aus dem Kebnekaisegebiet*. Uppsala. Universitet. Geologiska Institut. Bulletin, v.15, p.91-116.

Quirke, T.T. (1930) *Spring pits, sedimentation phenomena*. Journal of Geology, v.38, p.88-91.

R

Raaf, J.F.M. de (1945) *Notes on the geology of the southern Rumanian oil district with special reference to the occurrence of a sedimentary laccolith*. Geological Society of London. Quarterly Journal, v.101, p.111-134.

Raistrick, Arthur, and Marshall, C.E. (1939) *The nature and origin of coal and coal seams*. London: English Universities Press Ltd. 282p.

Raisz, E.J. (1959) *Landform maps-a method of preparation*. U.S. Office of Naval Research, Geography Branch. Part one of Final Report, Contract Nonr 2339(00). 23p.

Ramage, A., and Craddock, P. (2000) *King Croesus' gold: Excavations and the history of gold refining*. Cambridge, Sardis, 272p.

Ramberg, Hans (1955) *Natural and experimental boudinage and pinch-and-swell structures*. Journal of Geology, v.63, p.512-526.

Rampino, M.R., and Self, S. (2000) *Volcanism and biotic extinctions*. In: Sigurdsson, H., ed., *Encyclopedia of volcanoes*. New York, Academic Press, p.1083-1091.

Ramsay, A.C. (1846) *On the denudation of South Wales and the adjacent counties of England. Great Britain*. Geological Survey. Memoirs, v.1, ch.2, p.297-335.

Ramsay, J. G. (1980) *The crack-seal mechanism of rock deformation*. Nature, v. 284, p. 135-139.

Ramsay, J. G. and Huber, M. I. (1983) *The techniques of modern structural geology; Volume 1, strain analysis*. London: Academic Press, 307 p.

Ramsay, J. G., and Graham, R. H. (1970) *Strain variation in shear belts*. Canadian Journal of Earth Sciences / Journal Canadien des Sciences de la Terre, v. 7, n. 3, p. 786-813.

Ramsay, J.G. (1967) *Folding and fracturing of rocks*. New York: McGraw-Hill. 568p.

Ramsbottom, W. H. C. (1977) *Major cycles of transgression and regression in the Namurian*. Yorkshire Geological Society Proceedings, v. 41, p. 261-291.

Rankama, Kalervo (1962) *Planetology and geology*. Geological Society of America. Bulletin, v.73, p.519-520.

Rankama, Kalervo (1967) *Megayear and gigayear: two units of geological time*. Nature, v.214, p.634.

Rankama, Kalervo, and Sahama, Th.G. (1950) *Geochemistry*. Chicago: University of Chicago Press. 912p.

Ransome, F.L., and Calkins, F.C. (1908) *The geology and ore deposits of the Coeur d'Alene district, Idaho*. U.S. Geological Survey. Professional Paper 62. 203p.

Rapp, Anders (1959) *Avalanche boulder tongues in Lappland*. Geografiska Annaler, v.41, p.34-48.

Rapp, G. (2000) *Archaeomineralogy*. New York, Springer, 326p.

Rast, Nicholas (1956) *The origin and significance of boudinage*. Geological Magazine, v.93, p.401-408.

Raunkiaer, C. (1934) *The life form of plants and statistical plant geography*. Oxford, Clarendon Press, 632p.

Raup, D.M. (1975) *Taxonomic diversity estimation using rarefaction*. Paleobiology, v.1, no.4, p.333-342.

Rausell-Colom, J.A., and Serratosa, J.M. (1987) *Reactions of clays with organic substances*. In: Newman, A.C.D., ed., *Chemistry of clay and clay mineral*. London, Mineralogical Society Monograph 6, p.371-422.

Raymond, L.A. (1975) *Tectonite and mélange—a distinction*. Geology, v. 3, p.7-9.

Raymond, L.A. (1978) *A classification of mélanges and broken formations (abstract)*. Geological Society of America. Abstracts with Programs, v.10, p.143.

Raymond, L.A. (1984) *Classification of melanges*. Geological Society of America. Special Paper 198, p. 7-20.

Read, H.H. (1931) *The geology of central Sutherland (east-central Sutherland and south-western Caithness). (Explanation of sheets 108 and 109)*. Geological Survey of Scotland. Memoir. 238p.

Read, H.H. (1944) *Meditation on granite, pt.2*. Geologists' Association. Proceedings, v.55, p.45-93.

Read, H.H. (1958) *A centenary lecture; Stratigraphy in metamorphism*. Geologists' Association. Proceedings, v.69, pt.2, p.83-102.

Read, J.F. (1981) *Carbonate ramps—types, evolution, and tectonic settings*. Geological Society of America. Abstracts with Programs, v.13, p.536.

Read, P.G. (1988) *Dictionary of Gemmology, 2nd ed*. Oxford: Butterworth-Heinemann Lts.

Reading, H.G., ed. (1986) *Sedimentary environments and facies, 2nd. ed*. Oxford, U.K., Blackwell Scientific Publications, 615p.

Rechard, P.A., and McQuisten, Richard, compilers (1968) *Glossary of selected hydrologic terms*. University of Wyoming. Water Resources Research Institute. Water Resources Series, no.1, 54p.

Reclus, Elisée (1872) *The Earth: a descriptive history of the phenomena of the life of the globe*. Translated by B.B. Woodward. Edited by Henry Woodward. New York: Harper. 573p.

Record, S.J. (1934) *Identification of the timbers of temperate North America, including anatomy and certain physical properties of wood*. New York: Wiley.

Reiche, Parry (1937) *The Toreva-block, a distinctive landslide type*. Journal of Geology, v.45, p. 538-548.

Reiche, Parry (1943) *Graphic representation of chemical weathering*. Journal of Sedimentary Petrology, v.13, p.58-68.

Reiche, Parry (1945) *A survey of weathering processes and products*. Albuquerque: University of New Mexico Press. 87p. (University of New Mexico. Publications in Geology, no.1).

Reid, A.B. (2003) *Euler magnetic structural index of a thin-bed fault*. Geophysics, v.68, p.1255-1256.

Reid, H.F. (1924) *Antarctic glaciers*. Geographical Review, v.14, p.603-614.

Reimnitz, Erk, et al. (1973) *Bottom features and processes related to drifting ice on the arctic shelf, Alaska*. U. S. Geological Survey. Map MF-532.

Reimnitz, Erk, et al. (1977) *Ice gouge recurrence and rates of sediment reworking, Beaufort Sea, Alaska*. Geology, v.5, p.405-408.

Reineck, H.-E., and Singh, I.B. (1973) *Depositional sedimentary environments.* New York: Springer-Verlag. 439p.

Reineck, H.-E., and Wunderlich, F. (1968) *Classification and origin of flaser and lenticular bedding.* Sedimentology, v.11, p.99-104.

Reinson, G.E. (1984) *Barrier Island and associated strand-plain systems.* In: Walker, R.D., ed., *Facies models, 2nd ed.* Geoscience Canada, Reprint Series 1, 317p.

Renevier, E., et al. (1882) *Rapport du comité suisse sur l'unification de la nomenclature.* International Geological Congress. 2nd, Bologna, 1881. Compte rendu, p.535-548.

Renfrew, J.M. (1973) *Palaeoethnobotany.* New York: Columbia University Press. 248p.

Renfro, A.R. (1974) *Genesis of evaporite-associated stratiform metalliferous deposits-a sabkha process.* Economic Geology, v.69, p.33-45.

Renwick, W.H. (1992) *Equilibrium, disequilibrium, and nonequilibrium landforms in the landscape.* Geomorphology, v.5, p.265-276.

Reusch, Hans (1894) *The Norwegian Coast Plain.* Journal of Geology, v.2, p.347-349.

Revelle, Roger, and Fairbridge, R.W. (1957) *Carbonates and carbon dioxide.* Geological Society of America. Memoir 67, v.1, p.239-295.

Reynolds, R. C., Jr. (1976) *The Lorentz factor for basal reflections from micaceous minerals in oriented powder aggregates.* American Mineralogist, v. 61, p. 484-491.

Reynolds, R. C., Jr. (1985) *NEWMOD*: A computer program for the calculation of one-dimensional diffraction patterns of mixed-layer clays.

Reynolds, R. C., Jr., DiStefano, M. P., and Lahann, R. W. (1992) *Randomly interstratified serpentine/chlorite: its detection and quantification by powder x-ray diffraction methods.* Clays and Clay Minerals 40, p. 262-267.

Rice, C.M. (1945) *Dictionary of geological terms. 1st ed.* Ann Arbor (Mich.): Edwards Brothers. 461p.

Rice, C.M. (1954) *Dictionary of geological terms. 2nd ed.* Ann Arbor (Mich.): Edwards Brothers.

Rich, J.L. (1908) *Marginal glacial drainage features in the Finger Lake region.* Journal of Geology, v.16, p.527-548.

Rich, J.L. (1914) *Certain types of stream valleys and their meaning.* Journal of Geology, v.22, p.469-497.

Rich, J.L. (1915) *Notes on the physiography and glacial geology of the northern Catskill Mountains.* American Journal of Science, 4th ser., v.39, p.137-166.

Rich, J.L. (1938) *A mechanism for the initiation of geosynclines and geo-basins.* Geological Society of America. Proceedings, p.106-107.

Rich, J.L. (1950) *Flow markings, groovings, and intra-stratal crumplings as criteria for recognition of slope deposits, with illustrations from Silurian rocks of Wales.* American Association of Petroleum Geologists. Bulletin, v.34, p.717-741.

Rich, J.L. (1951) *Geomorphology as a tool for the intrepretation of geology and Earth history.* New York Academy of Sciences. Transactions, ser. 2, v.13, p.188-192.

Rich, J.L. (1951) *Three critical environments of deposition; and criteria for recognition of rocks deposited in each of them.* Geological Society of America Bulletin, v.62, no.1, p.1-19.

Richter, C.F. (1958) *Elementary seismology.* San Francisco: Freeman. 768p.

Richter, Rudolf (1952) *Fluidal-Textur in Sediment-Gesteinen und über Sedifluktion überhaupt. Hessisches Landesamt für Bodenforschung, Wiesbaden.* Notizblatt, F.6, H.3, p.67-81.

Richter-Bernburg, G. (1973) *Facies and paleogeography of the Messinian evaporites in Sicily.* In: Drooger, C.W., ed., *Messinian events in the Mediterranean.* Amsterdam, Holland, Koninklijke Nederlandse Akademie Van Wetenschappen, p.124-141.

Richtofen, Ferdinand von (1868) *The natural system of volcanic rocks.* California Academy of Sciences. Memoirs, v.1, pt.2, 94p.

Rickett, H.W. (1958) *So what is a taxon?* Taxonomy, v.7, p.37-38.

Riding, R. (1977) *Skeletal stromatolites.* In: Flugel, E., ed. *Fossil algae, recent results and developments.* Berlin, Springer, p.57-60.

Riding, R. (1983) *Cyanoliths (cyanoids): oncoids formed by calcified cyanophytes.* In: Peryt-Tadeusz, M., ed. Coated grains. Berlin: Springer-Verlag, p. 276-283.

Riding, R. (2002) *Structure and composition of organic reefs and carbonate mud mounds: Concepts and categories.* Earth Science Reviews, v.58, p.163-231.

Riecke, Eduard (1894) *Über das Gleichgewicht zwischen einem festen, homogen deformirten Körper und einer flüssigen Phase, insbesondere über die Depression des Schmelzpunctes durch einseitige Spannung.* Gesellschaft der Wissenschaften zu Göttingen. Mathematisch-Physikalische Klasse. Nachrichten, 1894, no.4, p.278-284.

Riedel, W. (1929) *Zur Mechanik geologischer Brucherscheinungen.* Zentralblatt für Mineralogie, Geologie, und Palaeontologie B, p. 354-368.

Rieke, H. H., III, and Chilingarian, G. V. (1974) *Compaction of argillaceous sediments.* Amsterdam: Elsevier, 424 p. (Developments in sedimentology, 16).

Ries, Heinrich (1937) *Economic geology. 7th ed.* New York: Wiley. 720p.

Rigby, J.K. (1959) *Possible eddy markings in the Shinarump Conglomerate of northeastern Utah.* Journal of Sedimentary Petrology, v.29, p.283-284.

Rigby, J.K., and Bayer, T.N. (1971) *Sponges of the Ordovician Maquoketa Formation in Minnesota and Iowa.* Journal of Paleontology, v.45, p.608-627.

Rigon, R., Rinaldo, R., Rodriguez-Iturbe, I., Bras, R.L., and Ijjasz-Vasquez, E. (1993) *Optimal channel networks: A framework for the study of river basin morphology.* Water Resources Research, v.29, p.1635-1646.

Rio, D., Sprovieri, R., Castradori, D., and Di Stefano, E. (1998) *The Gelasian Stage (upper Pliocene): A new unit of the global standard chronostratigraphic scale.* Episodes, v.21, no.2, p.82-87.

Rittenhouse, Gordon (1943) *The transportation and deposition of heavy minerals.* Geological Society of America. Bulletin, v.54, p.1725-1780.

Rittmann, Alfred (1962) *Volcanoes and their activity.* New York: Interscience. 305p.

Rittmann, Alfred (1973) *Stable mineral assemblages of igneous rocks—a method of calculation.* Minerals, Rocks and Inorganic Materials, v.7. New York: Springer-Verlag.

Rivière, André (1952) *Expression analytique générale de la granulométrie des sédiments meubles.* Société Géologique de France. Bulletin, ser.6, v.2,p.155-167.

Rizzini, A., and Passega, R. (1964) *Évolution de la sédimentation et orogenése, vallé du Santerno, Apennin septentrional.* In: Bouma, A.H., and Brouwer, A., eds. Turbidites, p. 65-74. Amsterdam: Elsevier. 264p. (Developments in sedimentology 3).

Roberts, George (1839) *An etymological and explanatory dictionary of the terms and language of geology; designed for the early student, and those who have not made great progress in that science.* London: Longman, Orme, Brown, Green, & Longmans, 183p.

Roberts, R.J. (1951) *Geology of the Antler Peak quadrangle, Nevada.* U.S. Geological Survey. Map GQ10, scale 1:62,500, with text.

Robertson, R.H.S. (1986) *Fuller's earth: A history of calcium montmorillonite.* Hythe, Kent, U.K., Volturna Press, Mineralogical Society Occasional Publication, 421p.

Robinove, C.J. (1963) *Photography and imagery-a clarification of terms.* Photogrammetric Engineering, v.29, p.880-881.

Robinson, E.A., and Treitel, S., (1964) *Principles of digital filtering.* Geophysics, 29, 395-404.

Robson, D.A. (1971) *The structure of the Gulf of Suez (Clysmic) rift, with special reference to the eastern side.* Journal of the Geological Association, v. 127, p.247-276.

Robson, J., and Nance, R.M. (1959) *Geological terms used in S.W. England.* Royal Geological Society of Cornwall. Transactions, 1955-1956, v.19, pt.1, p.33-41.

Rodgers, D.A. (1980) *Analysis of pull-apart basin development produced by en echelon strike-slip faults.* In: P.F. Ballance and H.G. Reading, eds. Sedimentation in oblique-slip mobile zones. International Association of Sedimentologists. Special Publication 4, p.27-41. London: Blackwell.

Rodgers, John (1963) *Mechanics of Appalachian foreland folding in Pennsylvania and West Virginia.* American Association of Petroleum Geologists. Bulletin, v.47, p.1527-1536.

Rodgers, John (1967) *Chronology of tectonic movements in the Appalachian region of North America.* American Journal of Science, v.265, p.408-427.

Roedder, E.W. (1979) *Fluid inclusions as samples of ore fluids.* In: H.L. Barnes, ed. Geochemistry of hydrothermal ore deposits , 2nd ed., p.688-692. New York: Wiley.

Roedder, Edwin (1967) *Fluid inclusions as samples of ore fluids.* In: Barnes, H.L., ed. Geochemistry of hydrothermal ore deposits , p.515-574. New York: Holt, Rinehart, and Winston.

Rogers, B. G., et al. (1981) *Glossary — Water and wastewater control engineering.* American Public Health Association, American Society of Civil Engineers, American Water Works Association, and Water Pollution Control Federation, 441 p.

Rogers, C.P., Jr., and Neal, J.P. (1975) *Feldspar and aplite.* In: Lefond, S.J., ed. Industrial Minerals and Rocks, 4th ed., p.637-651. New York: American Institute of Mining, Metallurgical, and Petroleum Engineers.

Rogers, H.D. (1860) *On the distribution and probable origin of the petroleum, or rock oil, of western Pennsylvania, New York, and Ohio.* Royal Philosophical Society of Glasgow. Proceedings, v.4, p.355-359.

Rogers, H.H., compiler (1966) *Glossary of terms frequently used in cosmology.* New York: American Institute of Physics. 16p.

Rohrbach, C.E.M. (1885) *Über die Eruptivgesteine im Gebiete der schlesich-mèhrischen Kreideformation.* Tschermak's mineralogische und petrographische Mitteilungen, N.F., Bd.7, p.1-63.

Ronov, A.B. (1982) *The earth's sedimentary shell (Part 1).* International Geology Review, v.24, p.1313-1363.

Roques, M. (1961) *Nomenclature de J. Jung et M. Roques pour certains types de migmatites.* In: Sorensen, H., ed., Symposium on migmatite nomenclature , p.68. International Geological Congress, 21st, Copenhagen, 1960. Report, pt.26, sec. 14.

Rosalsky, M.B. (1949) *A study of minor beach features at Fire Island, Long Island, New York.* New York Academy of Sciences. Transactions, ser.2, v.12, p.9-16.

Rosauer, E.A. (1957) *Climatic conditions involved in glacial loess formation.* Doctoral dissertation, University of Bonn. 105p.

Rose, E. R. F. (1994) *British applications of military geology for 'Operation Overlord' and the Battle of Normandy (abst.)* Geological Society of America, Abstracts with Programs (Seattle, WA), v. 26, no. 7, p. 275.

Rosenbaum, M. S. (1994) *Recent applications of military geology for the British Armed Forces (abst.)* Geological Society of America, Abstracts with Programs (Seattle, WA), v. 26, no. 7, p. 277.

Rosenbusch, Harry (1887) *Mikroskopische Physiographie der massigen Gesteine. 2nd ed.* Stuttgart: Schweizerbart'sche Verlagshandlung (E. Koch). 877p. (His Mikroskopische Physiographie der Mineralien und Gesteine , Bd.2).

Rosenbusch, Harry (1888) *Microscopial physiography of the rock-making minerals: an aid to the microscopial study of rocks.* Translated and abridged by Joseph P. Iddlings. New York: Wiley. 333p.

Rosenbusch, Harry (1898) *Elemente der Gesteinlehre.* Stuttgart: Schweizerbart'sche. 546p.

Rosendahl, B.R., Reynolds, D.J., Lorber, P.M., Burgess, C.F., McGill, J., Scott, D., Lambiase, J.J., and Derksen, S.J. (1986) *Structural expressions of rifting: Lessons from Lake Tanganyika, Africa.* In: Frostick, L.E., Renaut, R.W., Reid, I., and Tiercelin, J.J., eds., *Sedimentation in the African rifts.* Geological Society [of London] Special Publication 25, p.29-43.

Ross, C.S., and Shannon, E.V. (1926) *The minerals of bentonite and related clays and their physical properties.* American Ceramic Society. Journal, v.9, p.77-96.

Ross, C.S., and Smith, R.L. (1961) *Ash-flow tuffs: their origin, geologic relations, and identification.* U.S. Geological Survey. Professional Paper 366. 81p.

Rousseau, Jacques (1949) *Modifications de la surface de la toundra sous l'action d'agents climatiques.* Revue Canadienne Géographie, v.3, p.43-51.

Roy, R.F., Decker, E.R., Blackwell, D.D., and Birch, F., (1968) *Heat flow in the United States.* Geophys. Res., v.73, no.16, p.5207-5221.

Royse, Frank, Jr., et al. (1975) *Thrust belt structural geometry and related stratigraphic problems, Wyoming-Idaho-northern Utah.* In: Bolyard, D.W., ed. Symposium on deep drilling in the central Rocky Mountains. Denver: Rocky Mountain Association of Geologists.

Rubey, W.W., and Hubbert, M.K. (1959) *Role of fluid pressure in mechanics of overthrust faulting.* Geological Society of America. Bulletin, V.70, p.167-206.

Ruhe, R.V. (1969) *Quaternary landscapes in Iowa.* Ames: Iowa State University Press. 255p.

Rumble, D., III (1976) *The use of mineral solid solutions to measure chemical potential gradients in rocks.* American Mineralogist, v.61, p.1167-1174.

Runcorn, S.K., et al., eds. (1967) *International dictionary of geophysics.* New York: Pergamon. 2 volumes. 1728p.

Runnegar, B., et al. (1975) *Biology of the Hyolitha.* Lethaia, v.8, p.181-191.

Russell, I.C. (1885) *Existing glaciers of the U.S.* U.S. Geological Survey. Annual Report, 5th p.303-355.

Russell, I.C. (1898a) *Glaciers of Mount Rainier.* U.S. Geological Survey. Annual Report, 18th, pt.2, p.349-415.

Russell, I.C. (1898b) *Rivers of North America.* New York: Putnam. 327p.

Russell, R.J. (1942) *Flotant.* Geographical Review, v.32, p.74-98.

Russell, R.J., ed. (1968) *Glossary of terms used in fluvial, deltaic, and coastal morphology and processes.* Coastal Studies Institute, Louisiana State University. Technical Report No. 63.

Rust, B.R. (1972) *Structure and process in a braided river.* Sedimentology, v.18, p.221-245.

Rutherford, S.R., and R.H. Williams, (1989) *Amplitude-versus offset variations in gas sands.* Geophysics, 54: 680-8.

Rutter, E.H. (1986) *On the nomenclature of mode of failure transitions in rocks.* Tectonophysics, v.122, p.381-387.

Ruxton, B.P., and Berry, Leonard (1959) *The basal rock surface on weathered granitic rocks.* Geologists' Association. Proceedings, v.70, p.285-290.

Rybach, L., (1988) *Determination of heat production rate.* In Haenel, R. Rybach, L., and Stegena, L., eds., Handbook of terrestrial heat-flow density determination. Kluwer Acad. Publ., Dordrecht, p.125-142.

Ryland, J.S. (1970) *Bryozoans.* London: Hutchinson University Library. 175p.

S

Srodon, J., Elass, F., McHardy, W. J. and Morgan, D. J. (1992) *Chemistry of illite-smectite inferred from TEM measurements of fundamental particles.* Clay Minerals 27, p. 137-158.

Sabins, F.F. (1978) *Remote sensing—principles and interpretation.* San Francisco: Freeman. 426p.

Sackin, M.J., and Merriam, D.F. (1969) *Autoassociation, a new geological tool.* International Association for Mathematical Geology. Journal, v.1, p.7-16.

Salisbury, J.W., and Hunt, G.R. (1968) *Martian surface materials: effect of particle size on spectral behavior.* Science, v.161, p.365-366.

Salisbury, R.D. (1904) *Three new physiographic terms.* Journal of Geology, v.12, p.707-715.

Salisbury, R.D., et al. (1902) *The glacial geology of New Jersey.* New Jersey Geological Survey. Final Report 5. 802p.

Salomon, W. (1915) *Die Definitionen von Grauwacke, Arkose und Ton.* Geologische Rundschau, Leipzig; VI, p.398-404.

Salvador, A. (1985) *Chronostratigraphic and geochronometric scales in COSUNA stratigraphic correlation charts of the United States.* American Association of Petroleum Geologists. Bulletin, v.69, p.181-189.

Sandberg, P.A. (1977) *Ultrastructure, mineralogy, and development of bryozoan skeletons.* In: Woollacott, R.M., and Zimmer, R.L., eds., *Biology of Bryozoans.* New York, Academic Press, chap.5, p.143-179.

Sandberg, P.A. (1983) *Ultrastructure and skeletal development in cheilostomate Bryozoa.* In: Boardman, R.S., Cheetham, A.H., Blake, D.B., et al., *Treatise on invertebrate paleontology,* Part G, Bryozoa, revised, v.1. Lawrence, Kans., Geological Society of America and University of Kansas Press, p.238-286.

Sander, Bruno (1911) *Über Zusammenhänge zwischen Teilbewegung und Gefüge in Gesteinen.* Tschermaks Mineralogische und Petrographische Mitteilungen, v.30, p.281-314.

Sander, Bruno (1912) *Über tektonische Gesteinsfazies.* Austria. Geologische Reichsanstalt. Verhandlungen, 1912, no.10, p.249-257.

Sander, Bruno (1930) *Gefügekunde der Gesteine mit besonderer Berücksichtigung der Tektonite.* Vienna: Julius Springer. 352p.

Sander, Bruno (1936) *Beiträge zur Kenntnis der Anlagerungsgefüge (rhythmische Kalke und Dolomite aus der Trias). [Pt.] 1-2.* Mineralogische und petrographische Mitteilungen, Bd. 48, p.27-139, 141-209.

Sander, Bruno (1951) *Contributions to the study of depositional fabric.* Translated by E.B. Knopf. Tulsa: American Association of Petroleum Geologists. 207p.

Sander, Bruno (1970) *An introduction to the study of fabrics of geological bodies.* Translated by F.C. Phillips and G. Windsor. New York: Pergamon. 641p.

Sander, N.J. (1967) *Classification of carbonate rocks of marine origin.* American Association of Petroleum Geologist. Bulletin, v. 51, p.325-336.

Sanders, J.E. (1957) *Discontinuities in the stratigraphic record.* New York Academy of Sciences. Transactions, ser.2, v.19, p.287-297.

Sanders, J.E. (1963) *Concepts of fluid mechanics provided by primary sedimentary structures.* Journal of Sedimentary Petrology, v.33, p. 173-179.

Sangster, D.F. (1972) *Precambrian volcanogenic massive sulphide deposits in Canada: a review.* Geological Survey of Canada. Paper 72-22. 44p.

Sangster, D.F. (1976) *Carbonate-hosted lead-zinc deposits.* In: Wolf, K.H., ed., *Handbook of strata-bound and stratiform deposits.* New York, Elsevier, v.6, p.447-456.

Sarjeant, W.A.S. (1969) *Microfossils other than pollen and spores in palynological preparations. In: Erdtman, G., Handbook of palynology.* New York, Hafner, p.165-208.

Sarjeant, W.A.S. (1970) *Xanthidia, palinosphere, and "Hystrix." A review of the study of fossil unicellular microplankton with organic cell walls.* Microscopy, v.31, p.221-223.

Sarjeant, W.A.S. (1982) *Dinoflagellate cyst terminology: A discussion and proposals.* Canadian Journal of Botany, v.60, no.6, p.922-245.

Sarjeant, W.A.S., and Stanclife, R.P.W. (1994) *The* Microhystridium *and* Veryhachium *complexes (Acritarcha: Acanthomorphitae and Polygonomorphitae): A taxonomic reconsideration.* Micropaleontology, v.40, p.1-77.

Sauer, C.O. (1930) *Basin and Range forms in the Chiricahua area.* University of California, Berkeley. Publications in Geography, v.3, p.339-414.

Saussure, H.-B. de (1786) *Voyages dans les Alpes, précédés d'un essai sur l'histoire naturelle des environs de Genève.* Vol.2. Genève: Barde, Manget. 641p.

Savrda, C.E., and Bottjer, D.J. (1987) *The exaerobic zone, a new oxygen-deficient marine biofacies.* Nature (London), v.327, p.54-56.

Saxov, S. (1982) *Marine slides and other mass movements.* S. Saxov and J.K. Nieuwenhuis, eds., p.1. New York: Plenum. 353p.

Scagel, R.F., et al. (1965) *An evolutionary survey of the plant kingdom.* Belmont, Calif: Wadsworth.

Schanz, John J., Jr., and Ellis, John G. (1983) *Assessing the mineral potential of the Federal public lands.* Congressional Research Service.

Schardt, Hans (1893) *Sur l'origine des Alpes du Chablais et du Stockhorn, en Savoie et en Suisse.* Comptes Rendus des Séances de l'Academie des Sciences, Tome 117, no. 21, p. 707-709.

Schearz, E.H.L. (1912) *South African geology.* Glasgow: Blackie. 200p.

Scheidegger, A.E. (1961) *Theoretical geomorphology.* Berlin: Springer-Verlag. 333p.

Scheidegger, A.E. (1963) *Principles of geodynamics. 2nd ed.* New York: Academic Press. 362p.

Schein-Fatton, E. (1988) *Un Pectinacea (Bivalvia) très primitif:* Bathypecten vulcani, *du site hydrothermal de 13° N (Pacific oriental).* Oceanologica Acta, special no.8, p.83-98.

Schenck, H.G., and Muller, S.W. (1941) *Stratigraphic terminology.* Geological Society of America. Bulletin, v.52, p.1419-1426.

Schenk, E. T. and McMasters, J. H. (1956) *Procedures in Taxonomy. 3rd ed.,* Stanford, CA: Stanford University Press, 119 pp.

Schenk, J.E.T., and McMasters, J.H. (1956) *Procedures in Taxonomy. 3rd ed.,* 119p. Stanford, Ca. Stanford University Press.

Schermer, E.R., et al. (1984) *The origin of allochthonous terranes: perspectives on the growth and shaping of continents.* Annual Review of Earth and Planetary Sciences, v.12, p. 107-131.

Schermerhorn, L.J.G. (1966) *Terminology of mixed coarse-fine sediments.* Journal of Sedimentary Petrology, v.36, p.831-835.

Schieferdecker, A.A.G., e.d. (1959) *Geological nomenclature.* Gorinchem: Royal Geological and Mining Society of the Netherlands. 523p.

Schiffman, A. (1965) *Energy measurements in the swash-surf zone.* Limnology and Oceanography, v.10, p.255-260.

Schindewolf, O.H. (1950) *Grundlagen und Methoden der paläontologischen Chronologie. 3rd ed.* Berlin: Gebrüder Borntraeger. 152p.

Schindewolf, O.H. (1954) *Über einige stratigraphische Grundbegriffe.* Roemeriana, H.1, p.23-38.

Schindewolf, O.H. (1957) *Comments on some stratigraphic terms.* American Journal of Science, v.255, p.394-399.

Schindewolf, O.H. (1959) *On certain stratigraphic fundamentals.* International Geology Review, v.1, no. 7, p.62-70.

Schipman, Henry, jr. (1968) *Buffalo rings.* National Parks Magazine, v.42, no.254, p.14-15.

Schlanger, S.O. (1957) *Dolomite growth in coralline algae.* Journal of Sedimentary Petrology, v.27, p.181-186.

Schlische, R.W. (1995) *Geometry and origin of fault-related folds in extensional settings.* AAPG Bulletin, v.79, no.11, p.1661-1678.

Schmidt, Volkmar (1965) *Facies, diagenesis, and related reservoir properties in the Gigas Beds (Upper Jurassic), northwestern Germany.* Society of Economic Paleontologists and Mineralogists. Special Publication no. 13, p.124-168.

Schmidt, Walter (1925) *Gefügestatistik.* Tschermak's mineralogische und petrographische Mitteilungen, N.F., Bd. 38, p.392-423.

Schmoll, H.R., and Bennett, R.H. (1961) *Axiometer-a mechanical device for locating and measuring pebble and cobble axes for macrofabric studies.* Journal of Sedimentary Petrology, v.31, p.617-622.

Schofield, W. (1920) *Dumb-bell islands and peninsulas on the coast of South China.* Liverpool Geological Society. Proceedings, v.13, p.45-51.

Schopf, J.M. (1960) *Field description and sampling of coal beds.* U.S. Geological Survey. Bulletin 1111-B, p.25-70.

Schopf, J.M. (1975) *Modes of fossil preservation.* Review of Palaeobotany and Palynology, v.20, p.27-53.

Schreiber, B.C. (1986) *Arid shorelines and evaporites.* In: H.G. Reading, ed. Sedimentary environments and facies, 2nd ed., p. 189-228. Oxford: Blackwell.

Schroeder, R.A., and Bada, J.L. (1976) *A review of the geochemical applications of the amino acid racemization reaction.* Earth-Science Reviews, v.12, p.347-391.

Schuchert, Charles (1910) *Paleogeography of North America.* Geological Society of America. Bulletin, v.20, p.427-606.

Schuchert, Charles (1923) *Sites and natures of the North American geosynclines.* Geological Society of America. Bulletin, v.34, p.151-260.

Schuchert, Charles (1924) *Historical geology. 2nd ed.* New York: Wiley. 724p. (A textbook of geology, pt.2.)

Schultz, E. H. (1982) *The chronosome and supersome; terms proposed for low-rank chronostratigraphic units.* Bulletin of Canadian Petroleum Geology, v. 30, no. 1, p. 29-33.

Schulze, Franz (1855) *Bemerkungen über das Vorkommen wohlerhaltener Cellulose in Braunkohle und Steinkohle.* Akademie der Wissenschaften, Berlin. Bericht über die zur Bekanntmachung geeigneten Verhandlungen, 5 Nov 1855, p.676-678.

Schumm, S.A. (1956) *Evolution of drainage systems and slopes in badlands at Perth Amboy, New Jersey.* Geological Society of America. Bulletin, v.67, p.597-646.

Schumm, Stanley A. (1977) *The fluvial system.* New York: John Wiley & Sons.

Schuster, D. C. (1995) *Deformation of allochthonous salt and evolution of related salt-structural systems, eastern Louisiana Gulf Coast.* In: M. P. A. Jackson, D. G. Roberts, S. Snelson, eds. Salt tectonics: a global perspective. AAPG Memoir 65, p. 177-198.

Schwarz, E. H. L. (1912) *South Africa geology.* Glasgow: Blackie. 200 p.

Schwarzbach, Martin (1961) *Das Klima der Vorzeit; eine Einführung in die Paläoklimatologie. 2nd ed.* Stuttgart: Ferdinand Enke. 275p.

Schytt, V. (1967) *A study of the ablation gradient.* Geographische Annalen, v.49A, nos.2-4, p.327-332.

Schytt, Valter (1963) *Fluted moraine surfaces.* Journal of Glaciology, v.4, p.825-827.

Sclater, J.G., Jaupart, C., and Galson, D., (1980) *The heat flow through oceanic and continental crust and the heat loss from the Earth.* Rev. Geophys. Space Phys., v.18, no.1, p.269-311.

Scott, D.H. (1923) *Studies in fossil botany, 3rd ed, v.2.* London: A. & C. Black. 446p.

Scott, H.W. (1947) *Solution sculpturing in limestone pebbles.* Geological Society of America. Bulletin, v.58, p.141-152.

Scott, W.B. (1922) *Physiography, the science of the abode of man.* New York: Collier. 384p.

Scrivenor, J.B. (1921) *The physical geography of the southern part of the Malay peninsula.* Geographical Review, v.11, p.351-371.

SCSA: Soil Conservation Society of America (1976) *Resource conservation glossary.* Journal of Soil and Water Conservation, v.31, no.4.

Seaber, P. R. (1996) *Background on hydrostratigraphic units.* Note 62: North American Commission on Stratigraphic Nomenclature. American Association of Petroleum Geologists Bulletin (in press).

Seaber, Paul R. (1988) *Hydrostratigraphic units.* In: Hydrogeology. Boulder, CO: Geological Society of America, p. 9-14.

Searle, A.B. (1923) *Sands and crushed rocks. Vol.1.* London: Henry Frowde and Hodder & Stoughton. 475p. (Oxford technical publications).

Sears, F.W. (1958) *Mechanics, wave motion and heat.* Reading, Mass.: Addison-Wesley. 664p.

SEC: Securities and Exchange Commission (2003) *Description of property by issuers engaged or to be engaged in significant mining operations.* Securities and Exchange Commission Industry Guide 7, 4p.

Sederholm, J.J. (1891) *Studien über Archäis die Eruptivgesteine aus dem sudwestlichen Finnland.* Tschermaks Mineralogische und Petrographische Mitteilungen, v.12, p.97-142.

Sederholm, J.J. (1907) *Om granit och gneis deras uppkomst, uppträdande och utbredning inom urberget i Fennoskandia.* Finland Commission Géologique. Bulletin, no.23, 110p.

Sederholm, J.J. (1924) *Granit-gneisproblemen belysta genom iakttagelser i Abo-Alands skärgard.* Geologiska Föreningen, Stockholm. Förhandlingar, v.46, p.129-153.

Seglund, J.A. (1974) *Collapse-fault systems of Louisiana Gulf Coast.* American Association of Petroleum Geologists. Bulletin, v.58, p.2389-2397.

Segovia, A. (1975) *Field classification and evaluation of seismicity of photogeologic lineaments and fracture traces (regmites).* Geological Society of America. Abstracts with Programs, v.7, p.1264.

Seilacher, A. (1981) *Towards an evolutionary stratigraphy.* In: J. Martinell, ed. Concept and method in paleontology. Acta Geologica Hispanica, v.16, p.39-44.

Seilacher, A. (1984) *Storm beds: their significance in event stratigraphy.* American Association of Petroleum Geologists. Studies in Geology No. 16, p. 49-54.

Seilacher, Adolf (1969) *Fault-graded beds interpreted as seismites.* Sedimentology, v.13, p.155-159.

Sellards, E.H., and Gunter, Herman (1918) *Geology between the Apalachicola and Ocklocknee rivers in Florida.* Florida Geological Survey. Annual Report, 10th-11th, p.9-56.

Sepkoski, J.J., Jr. (1978) *A kinetic model of Phanerozoic taxonomic diversity. I: Analysis of marine orders.* Paleobiology, v.4, no.3, p.223-251.

Sepkoski, J.J., Jr. (1988) *Alpha, beta, or gamma: Where does all the diversity go?* Paleobiology, v.14, no.3, p.221-234.

Sepkoski, J. John, Jr. (1981) *A factor analytic description of the Phanerozoic marine fossil record.* Paleobiology, v. 7, n. 1, p. 36-53.

Serra, Sandro (1977) *Styles of deformation in the ramp regions of overthrust faults.* Joint Wyoming-Montana-Utah Geological Associations Guidebook, Rocky Mountain thrust belt geology and resources , p.487-498.

Servais, T., and Eiserhardt, K.H. (1995) *A discussion and proposals concerning the lower Paleozoic "galeate" acritarch plexus.* Palynology, v.19, p.191-210.

Shackleton, R.M. (1958) *Downward-facing structures of the Highland Border.* Geological Society of London. Quarterly Journal, v.113, p.361-392.

Shaler, N.S. (1889) *The geology of Cape Ann, Mass.* U.S. Geological Survey. Annual Report, 9th, p.529-611.

Shaler, N.S. (1890) *General account of the fresh-water morasses of the United States, with a description of the Dismal Swamp district of Virginia and North Carolina.* U.S. Geological Survey. Annual Report, 10th, pt.1, p.255-339.

Shaler, N.S. (1895) *Beaches and tidal marshes of the Atlantic coast.* National Geographic Society. National Geographic Monographs, v.1, no.5, p.137-168.

Shand, R.D., and Baily, D.G. (1999) *A review of new offshore bar migration with photographic illustrations from Wanganui, New Zealand.* Journal of Coastal Research, v.15, p.365-378.

Shand, S.J. (1916) *The pseudotachylyte of Parijs (Orange Free State), and its relation to "trapshotten-gneiss" and "flint-crushrock".* Geological Society of London. Quarterly Journal, v.72, p.198-220.

Shand, S.J. (1947) *Eruptive rocks, their genesis, composition, classification, and their relation to ore deposits, with a chapter on meteorites. 3rd ed.* London: Thomas Murby. 488p.

Sharp, J.V.A., and Nobles, L.H. (1953) *Mudflow of 1941 at Wrightwood, Southern California.* Geological Society of America. Bulletin, v.64, no.5, p.547-560.

Sharp, R.P. (1942) *Soil structures in the St. Elias Range, Yukon Territory.* Journal of Geomorphology, v.5, p.274-301.

Sharp, R.P. (1954) *Physiographic features of faulting in southern California.* California Division of Mines. Bulletin 170, ch.5, [pt.] 3, p.21-28.

Sharp, R.P. (1973) *Mars: fretted and chaotic terrains.* Journal of Geophysical Research, v.78, p.4073-4083.

Sharp, R.P., and Carey, D.L. (1976) *Sliding stones, Racetrack Playa, California.* Geological Society of America. Bulletin, v.87, p.1704-1717.

Sharpe, C.F.S. (1938) *Landslides and related phenomena; a study of mass-movements of soil and rock.* New York: Columbia University Press. 136p.

Shatski, N.S. (1946a) *Basic features of the structure and development of the East European platform.* Comparative tectonics of ancient platforms. Izvestiya Akademii Nauk SSSR. Ser. Geol. no. 1, p.5-62.

Shatski, N.S. (1946b) *The Great Donets Basin and the Wichita System. Comparative tectonics of ancient platforms.* Izvestiya Akademii Nauk SSSR, Ser. Geol. no. 6, p.57-90.

Shatsky, N.S. (1945) *Outlines of the tectonics of Volga-Urals petroleum region and adjacent parts of the west slope of the southern Urals.* In: Mather, K.F., ed. Source book in geology 1900-1950 , p.263-267. Cambridge, Mass.: Harvard University Press.

Shatsky, N.S. (1955) *O proiskhozdenii Pachelmskogo progiba.* Moskovskoe Obshchestvo Ispytatelei Prirody. Biulleten'; otdel geologicheskii, n.s., v.30, no.5, p.3-26.

Shatsky, N.S., and Bogdanoff, A.A. (1957) *Explanatory note on the tectonic map of the U.S.S.R. and adjoining countries.* Moscow: State Scientific and Technical Publishing House. (English translation in International Geology Review, 1959, v.1, p.1-49).

Shatsky, N.S., and Bogdanoff, A.A. (1960) *La carte tectonique internationale de l'Europe au 2 500 000.* Akademiya Nauk SSSR, Izvestiya, Seriya Geologicheskaya, no.4.

Shaw, A.B. (1964) *Time in stratigraphy.* New York, McGraw-Hill, 365p.

Shaw, E.W. (1911) *Preliminary statement concerning a new system of Quaternary lakes in the Mississippi basin.* Journal of Geology, v.19, p.481-491.

Shawe, D.R. and Granger, H.C. (1965) *Uranium ore rolls—an analysis.* Economic Geology, v.60, p.240-250.

Shea-Albin, V.R. (1993) *Geological features that contribute to ground control problems in underground coal mines.* U.S. Bureau of Mines Information Circular 9370, 34p.

Shepard, F.P. (1937) *Revised classification of marine shorelines.* Journal of Geology, v.45, p.602-624.

Shepard, F.P. (1948) *Submarine geology.* New York: Harper. 348p.

Shepard, F.P. (1952) *Revised nomenclature for depositional coastal features.* American Association of Petroleum Geologists. Bulletin, v.36, p.1902-1912.

Shepard, F.P. (1954) *Nomenclature based on sand-silt-clay ratios.* Journal of Sedimentary Petrology, v.24, p.151-158.

Shepard, F.P. (1967) *The earth beneath the sea.* Revised ed. Baltimore: Johns Hopkins Press. 242p.

Shepard, F.P., and Cohee, G.V. (1936) *Continental shelf sediments off the mid-Atlantic States.* Geological Society of America. Bulletin, v.47, p.441-457.

Shepard, F.P., and Dill, R.F. (1966) *Submarine canyons and other sea valleys.* Chicago: Rand McNally. 381p.

Shepard, F.P., and Young, Ruth (1961) *Distinguishing between beach and dune sands.* Journal of Sedimentary Petrology, v.31, p.196-214.

Sheridan, R.E. (1983) *Phenomena of pulsation tectonics related to the breakup of the eastern North American continental margin.* In: Initial reports of the Deep Sea Drilling Project, v.76, p.897-909. Washington: U.S. Government Printing Office.

Sheriff, R.E., (1991) *Encyclopedic Dictionary of Exploration Geophysics.* Tulsa, Society of Exploration Geophysicists, 376 p.

Sheriff, R.E. (2002) *Encyclopedic dictionary of applied geophysics, 4th ed.* Tulsa, Okla., Society of Exploration Geophysicists, 429p.

Sheriff, R.E., and Geldart, L.P., (1995) *Exploration seismology. 2nd edition.* Cambridge University Press, 592 p.

Sherzer, W.H. (1910) *Criteria for the recognition of the various types of sand grains.* Geological Society of America. Bulletin, v.21, p.625-662, 775-776.

Shinn, E. (1963) *Spur-and-groove formation on the Florida Reef Tract.* Journal of Sedimentary Petrology, v.33, no.2, p.291-303.

Shipley, R.M. (1951) *Dictionary of gems and gemology including ornamental, decorative and curio stones. 5th ed.* Los Angeles: Gemological Institute of America. 261p.

Shipley, R.M. (1974) *Dictionary of gems and gemology. 6th ed.* Los Angeles: Gemological Institute of America.

Shirey, S.B., and Hanson, G.N. (1984) *Mantle-derived Archean monzodiorites and trachyandesites.* Nature, v.310, p.222-224.

Shirley, J. H., and Fairbridge, R. W. (eds.) *The encyclopedia of planetary science.* London & New York:Chapman Hall.

Shoemaker, E.M. (1963) *Impact mechanics at Meteor Crater, Arizona.* In: Middlehurst, B.M., and Kuiper, G.P., eds., *The Moon, meteorites and comets (The Solar System,* v.4). Chicago, University of Chicago Press, p.301-336

Short, N.M. (1966) *Shock-lithification of unconsolidated rock materials.* Science, v.154, p.382-384.

Short, Nicholas M., and Blair, Robert W., Jr. (1986) *Geomorphology from space; a global overview of regional landforms.* Washington, DC: National Aeronautics and Space Administration, Scientific and Technical Information Branch.

Shreve, R.L. (1966) *Statistical law of stream numbers.* Journal of Geology, v.74, p.17-37.

Shreve, R.L. (1967) *Infinite topologically random channel networks.* Journal of Geology, v.75, p.178-186.

Shreve, R.L. (1969) *Stream lengths and basin areas in topologically random channel networks.* Journal of Geology, v.77, p.397-414.

Shreve, R.L. (1984) *Glacier sliding at subfreezing temperatures.* Journal of Glaciology, v.30, no.106, p.341-347.

Shrock, R.R. (1947) *Loiponic deposits.* Geological Society of America. Bulletin, v.58, p.1228.

Shrock, R.R. (1948) *Sequence in layered rocks, a study of features and structures useful for determining top and bottom or order of succession in bedded and tabular rock bodies.* New York: McGraw-Hill. 507p.

Shrock, R.R. (1948a) *A classification of sedimentary rocks.* Journal of Geology, v.56, p.118-129.

Shrock, R.R., and Twenhofel, W.H. (1953) *Principles of invertebrate paleontology. 2nd ed.* New York: McGraw-Hill. 816 p.

Shvetsov, M.S. (1960) *K voprosu o diageneze.* In: Natsional'nyi komitet geologov Sovetskogo Soiuza. Voprosy sedimentologii; doklady sovetskikh geologv k VI Mezhdunarodnomu kongressu po sedimentologii , p.153-161. Moscow: Gos. nauchno-tekh. izd'vo lit'ry po geologii i okhrane nedr. 215p.

Sibson, R. H. (1977) *Fault rocks and fault mechanisms.* J. Geol. Soc. Lond., v. 133, p. 191-213.

Siegal, B.S., and Gillespie, A.R. (1980) *Remote sensing in geology.* New York: Wiley. 702p.

Siegel, R., and Howell, J.R. (1968) *Thermal radiation heat transfer.* National Aeronautics and Space Administration. SP-164. Vol. 1.

Sigafoos, R.S. (1951) *Soil instability in tundra vegetation.* Ohio Journal of Science, v.51, p.281-298.

Sigal, Jacques (1964) *Une thérapeutique homéopathique en chronostratigraphie: les parastratotypes (ou prétendus tels). France.* Bureau de Recherches Géologiques at Minières. Dépt. d'Information Géologique. Bulletin trimestriel, année 16, no.64, p.1-8.

Sigurdsson, H., ed. (2000) *Encyclopedia of volcanoes.* New York, Academic Press, 1417p.

Sih, G. C. (1973) *Handbook of stress intensity factors.* Institute of Fracture and Solid Mechanics, Lehigh University, Pennsylvania.

Silberling, N.J., and Roberts, R.J. (1962) *Pre-Tertiary stratigraphy and structure of northwestern Nevada.* Geological Society of America. Special Paper 72, 58p.

Sillitoe, R.H., et al. (1975) *Porphyry tin deposits in Bolivia.* Economic Geology, v.70, p.913-927.

Simons, D.B., and Richardson, E.V. (1961) *Forms of bed roughness in alluvial channels.* American Society of Civil Engineers. Proceedings, v.87, no.HY3, p.87-105.

Simons, D.B., et al. (1961) *Flume studies using medium sand (0.45mm).* U.S. Geological Survey. Professional Paper 1498-A. 76p.

Simpson, E. S. W. (1954) *On the graphical representation of differentiation trends in igneous rocks.* Geological Magazine, v.91, p.238-244.

Simpson, G.G. (1940) *Types in modern taxonomy.* American Journal of Science, v.238, p.413-431.

Sinclair, W.D. (1995) *Porphyry W.* In: Lefebure, D.V., and Ray, G.E., eds, *Selected British Columbia mineral deposit profiles. Volume 1, metallics and coal.* British Columbia Geological Survey Open-File Report 1995-20, p.101-104.

Sivenas, P., and Beales, F.W. (1982) *Natural geobatteries associated with sulfide ore deposits, I. Theoretical studies.* Journal of Geochemical Exploration, v.17, p.123-143. II. Field studies at the Viburnum Trend, southeast Missouri, U.S.A. Idem, p.145-160.

Sjogren, W.L. (1974) *Lunar gravity via the Apollo 15 and 16 subsatellites.* The Moon, v.9, p.115-128.

Skinner, C. E. (1993) *Open vertical volcanic conduits: a preliminary investigation of an unusual volcanic caveform with examples from Newberry volcano and the central High Cascades of Oregon.* Proceedings of the Third International Symposium of Vulcanospeleology, Bend Oregon, p. 7-17.

Skolnick, Herbert (1965) *The quartzite problem.* Journal of Sedimentary Petrology, v.35, p.12-21.

Skow, M.L. (1962) *Mica; a materials survey.* U.S. Bureau of Mines. Information Circular 8125. 241p.

Sloane, R.L., and Kell, T.R. (1966) *The fabric of mechanically compacted kaolin.* National Conference on Clays and Clay Minerals. 14th, Berkeley, Calif., 1965. Proceedings, p.289-296. New York: Pergamon Press. 443p. (International series of monographs on earth sciences, v.26).

Sloss, L.L. (1953) *The significance of evaporites.* Journal of Sedimentary Petrology, v.23, p.143-161.

Sloss, L.L. (1963) *Sequences in the cratonic interior of North America.* Geological Society of America. Bulletin, v.74, p.93-113.

Sloss, L.L. and Laird, W.M. (1947) *Devonian System in central and northwestern Montana.* American Association of Petroleum Geologists. Bulletin, v.31, p.1404-1430.

Sloss, L.L., et al. (1949) *Integrated facies analysis.* Geological Society of America. Memoir 39, p.91-123.

Smart, J.S. (1969) *Topological properties of channel networks.* Geological Society of America. Bulletin, v.80, p.1757-1774.

Smirnov, V.I. (1968) *The sources of the ore-forming fluid.* Economic Geology, v.63, p.380-389.

Smit, D.E., and Swett, Keene (1969) *Devaluation of "dedolomitization".* Journal of Sedimentary Petrology, v.39, p.379-380.

Smith, B. (1916) *Ball- or pillow-form structures in sandstones.* Geological Magazine, v.3, p.146-156.

Smith, J.T., Jr. (1968) *Glossary of aerial photographic terms.* In: Smith, J.T., Jr., and Anson, Abraham, eds. Manual or color aerial photography , p.489-509. Falls Church, Va: American Society of Photogrammetry. 550p.

Smith, K.G. (1950) *Standards for grading texture of erosional topography.* American Journal of Science, v.248, p.655-668.

Smith, M.L., Scales, J.A., and Fischer, T.L., (1992) *Global search and genetic algorithms.* Leading Edge, Jan 1992, 22-26.

Smith, N.D. (1974) *Sedimentology and bar formation in the upper Kicking Horse River, a braided outwash stream.* Journal of Geology, v.82, p.205-223.

Smith, R. (1991) *The application of cellular automata to the erosion of landforms.* Earth Surface Processes and Landforms, v.16, p.273-281.

Smith, R.A., et al. (1968) *The detection and measurement of infrared radiation. 2nd ed.* Oxford: Clarendon Press.

Smith, R.E. (1968) *Redistribution of major elements in the alteration of some basic lavas during burial metamorphism.* Journal of Petrology, v.9, p.191-219.

Smith, R.L. (1960) *Zones and zonal variations in welded ash flows.* U.S. Geological Survey. Professional Paper 354-F.

Smith, R.L., and Bailey, R.A. (1968) *Resurgent cauldrons.* Geological Society of America. Memoir 116, p.613-662.

Smith, W.O. (1961) *Mechanism of gravity drainage and its relation to specific yield of uniform sands.* U.S. Geological Survey. Professional Paper 402-A. 12p.

Smith, W.S.T. (1898) *A geological sketch of San Clemente Island.* U.S. Geological Survey. Annual Report, 18th, pt.2, p.459-496.

Smoot, J.P., and Castens-Seidell, B. (1994) *Sedimentary features produced by efflorescent salt crusts, Saline Valley and Death Valley, California.* In: Renaut, R., and Last, W.M., Sedimentology and Geochemistry of Modern and Ancient Saline Lakes : SEPM Special Publication 50, p.73-90.

Smyth, H.L. (1891) *Structural geology of Steep Rock Lake, Ont.* American Journal of Science, ser.3, v.42, p.317-331.

Sneed, E.D., and Folk, R.L. (1958) *Pebbles in the lower Colorado River, Texas.* A study in particle morphogenesis. Journal of Geology, v.66, p.114-140.

Snoke, A.W., and Calk, L.C. (1978) *Jackstraw-textured talc-olivine rocks, Preston Peak area, Klamath Mountains, California.* Geological Society of America. Bulletin, v.89, p.223-230.

Snyder, J.P. (1987) *Map projections—a working manual.* U.S. Geological Survey Professional Paper 1395, 383p.

Snyder, J.P., and Voxland, P.M. (1989) *An album of map projections.* U.S. Geological Survey Professional Paper 1453, 249p.

Sohl, N.F. (1977a) *Application for amendment of articles 8 and 10 of code, concerning smallest formal rock-stratigraphic unit.* American Association of Petroleum Geologists. Bulletin, v.61, p.252.

Sohl, N.F. (1977b) *Stratigraphic Commission Note 45—Application for amendment concerning terminology for igneous and high-grade metamorphic rocks.* American Association of Petroleum Geologists. Bulletin, v.61, p.248-251.

Soil Survey Staff (1994) *Soil taxonomy: A basic system of soil classification for making and interpreting soil surveys.* Soil Conservation Service, USDA. U.S. Government Printing Office, Washington, D.C.

Solley, W.B., et al. (1983) *Estimated use of water in the United States in 1980.* U.S. Geological Survey Circular 1001. 56p.

Sonder, R.A. (1956) *Mechanik der Erde; Elemente und Studien zur tektonischen Erdgeschichte.* Stuttgart: Schweizerbartsche. 291p.

Sorby, H.C. (1852) *On the oscillation of the currents drifting the sandstone beds of the south-east of Northumberland, and on their general direction in the coalfield in the neighborhood of Edinburgh.* West Yorkshire Geological Society. Proceedings, v.3, p.232-240.

Sorby, H.C. (1857) *On the physical geography of the Tertiary estuary of the Isle of Wight.* Edinburgh New Philosophical Journal, n.s., v.5, p.275-298.

Sorby, H.C. (1863) *Über Kalkstein-Geschiebe mit Eindrücken.* Neues Jahrbuch für Mineralogie, Geologie, und Palaeontologie, Jahrg. 1863, p.801-807.

Sorby, H.C. (1879) *The anniversary address of the President.* Geological Society of London. Proceedings, session 1878-1879, p.39-95. (Geological Society of London. Quarterly Journal, 1879, v.35).

Sorensen, H., ed. (1974) *The alkaline rocks.* New York: Wiley. 622p.

Southard, J.B. (1971) *Representation of bed configurations in depth-velocity size diagram.* Journal of Sedimentary Petrology, v.41, p.903-915.

Southgate, P.N., (1982) *Cambrian skeletal halite crystals and experimental analogues.* Sedimentology, v.29, p.391-407.

Spaeth, C., et al. (1971) *Some aspects of isotopic composition of belemnites and related paleotemperatures.* Geological Society of America. Bulletin, v.82, p.3139-3150.

Sparks, R. S. J. and Huppert, H. E. (1984) *Density changes during the fractional crystallization of basaltic magma: fluid dynamic implications.* Contrib. Mineral. Petrol., 85, p. 300-309.

Sparks, R.S.J., and Walker, G.P.L. (1973) *The ground surge deposit: a third type of pyroclastic rock.* Nature, v.241, p.62-64.

Spate, O.H.K. (1954) *India and Pakistan; a general and regional geography.* With a chapter on Ceylon by B.H. Farmer. London: Methuen. 827p.

Spector, A., and Grant, F.S. (1971) *Statistical models for interpreting aeromagnetic data.* Geophysics, v.35, p.293-302.

Speer, B.R. [1998] *Web reference* http://www.ucmp.berkeley.edu/protista/alveolates.html.

Speers, E.C. (1957) *The age relation and origin of common Sudbury breccia.* Journal of Geology, v.65, p.497-514.

Spencer, A.C. (1917) *The geology and ore deposits of Ely, Nevada.* U.S. Geological Survey. Professional Paper 96. 189p.

Spencer, E.W. (1969) *Introduction to the structure of the Earth.* New York: McGraw-Hill. 597p.

Spieker, E.M. (1956) *Mountain-building chronology and nature of geologic time scale.* American Association of Petroleum Geologists. Bulletin, v.40, no.8, p.1769-1815.

Spindler, W.M. (1977) *Structure and stratigraphy of a small Plio-Pleistocene depocenter, Louisiana continental shelf.* Gulf Coast Association of Geological Societies. Transactions, v.27, p.180-196.

Sposito, G. (1989) *The chemistry of soils.* New York, Oxford University Press, 277p.

Spotts, J.H., and Weser, O.E. (1964) *Directional properties of a Miocene turbidite, California.* In: Bouma, A.H., and Brouwer, A., eds. Turbidites , p.199-221. Amsterdam: Elsevier. 264p. (Developments in sedimentology 3.)

Sprinkle, James (1973) *Morphology and evolution of blastozoan echinoderms.* Harvard University Museum of Comparative Zoology. Special Publication. 283p.

Sproule, J.C. (1939) *The Pleistocene geology of the Cree Lake region, Saskatchewan.* Royal Society of Canada. Transactions, ser.3, v.33, sec.4, p. 101-109.

Spudis, P.D. (2000) *Volcanism on the Moon.* In: Sigurdsson, H., ed., *Encyclopedia of volcanoes.* New York, Academic Press, p.697-708.

Spurr, J.E. (1923) *The filling of fissure veins.* Engineering & Mining Journal, v.116, Aug.25, p.329-330.

Spurrell, F.C.J. (1887) *A sketch of the history of the rivers and denudation of West Kent, etc.* Geological Magazine, dec.3, v.4, no.273, p.121-122.

SSSA: Soil Science Society of America (1965) *Glossary of soil science terms.* Its Proceedings, v.29, p.330-351. Revised (1970), 27p.; (1975), 34p. Madison, Wis.: Its publication.

Stach, E. (1968) *Basic principles of coal petrology; macerals, microlithotypes and some effects on coalification.* In: Murchison, Duncan, and Westoll, T.S., eds. Coal and coal-bearing strata, p.3-17. New York: Elsevier. 418p.

Stamp, L.D. (1921) *On cycles of sedimentation in the Eocene strata of the Anglo-Franco-Belgian basin.* Geological Magazine, v.58, p.108-114, 146-157, 194-200.

Stamp, L.D., ed. (1961) *A glossary of geographical terms.* New York: Wiley. 539p.

Stamp, L.D., ed. (1966) *A glossary of geographical terms. 2nd ed.* London: Longmans, Green. 539p.

Stanek, W., and Worley, I. A. (1983) *A terminology of virgin peat and peatlands.* in C. H. Fuchsman and S. A. Spigarelli (eds.), International Symposium of Peat Utilization, Oct. 10-13, 1983, Bemidji State University Center for Environmental Studies, p. 75-102.

Stanger, Gordon (1994) *Dictionary of hydrology and water resources.* Adelaide, South Australia: Lochan, 208 p.

Stanley, D.J., and Swift, D.J.P., eds. (1976) *Marine Sediment transport and environmental management.* New York: Wiley. 602p.

Staub, Rudolf (1928) *Der bewegungsmechanismus der erde dargelegt am bau der irdischen gebirgssysteme.* Berlin: Gebrüder Borntraeger. 270p.

Stauffer, M.R., et al. (1976) *Rhomboidal lattice structure: a common feature on sandy beaches.* Canadian Journal of Earth Science, v.13, p.1667-1677.

Stearns, D.W. (1978) *Faulting and forced folding in the Rocky Mountains foreland.* Geological Society of America. Memoir 151, p.1-37.

Stearns, H. T., and Clark, W. O. (1930) *Geology and water resources of the Kau district, Hawaii.* U. S. Geological Survey Water Supply Paper 616.

Stearns, H.T., and Macdonald, G.A. (1942) *Geology and groundwater resources of the island of Maui, Hawaii.* Hawaii Division of Hydrography. Bulletin, no.7. 344p.

Stearns, N.D. (1927) *Report on the geology and ground-water hydrology of the experimental area of the United States Public Health Service at Fort Caswell, North Carolina.* U.S. Hygienic Laboratory Bulletin 147, p. 137-168.

Steen-McIntyre, Virginia (1977) *A manual for tephrochronology—collection, preparation, petrographic description, and approximate dating of tephra (volcanic ash).* Idaho Springs, Colo.: published by the author. 167p.

Stefan, Joseph (1879) *Über die Beziehung zwischen der Wärmestrahlung und der Temperatur. Akademie der Wissenschaften in Wien.* Sitzungsberichte, v.79, pt.2, p.391-428.

Steininger, F.F., Aubry, M.P., Berggren, W.A., Biolzi, M., Borsetti, A.M., Cartlidge, J.E., Cati, F., Corfield, R., Gelati, R., Iaccarino, S., Napoleone, C., Ottner, F., Rögl, F., Roetzel, R., Spezzaferri, S., Tateo, F., Villa, G., and Zevenboom, D. (1997) *The Global Stratotype Section and Point (GSSP) for the base of the Neogene.* Episodes, v.20, no.1, p.23-28.

Steinker, P. J. and Steinker, D. C. (1972) *The meaning of stratigraphic facies.* The Compass of Sigma Gamma Epsilon, v. 49, p. 45-53.

Steinmann, G. (1926) *Die ophiolitischen Zonen in den mediterranean Ketterigebirgen.* International Geological Congress. 14th, Madrid, 1926. Session 14, Report, p.637.

Steno, Nicolaus (1669) *Nicolai Stenonis de solido intra solidum naturaliter contento dissertationis prodromus.* Florence. 79 p.- See also: Steno, Nicolaus (1916) The prodromus of Nicolaus Steno's dissertation concerning a solid body enclosed by process of nature within a solid. English version with introduction & explanatory notes by J.G. Winter. Foreword by W.H. Hobbs. New York: Macmillan. 119p. (University of Michigan Studies; Humanistic Series, v.11, pt.2, p.165-283).

Stephens, N., and Synge, F.M. (1966) *Pleistocene shorelines.* In: Dury, G.H., ed. Essays in geomorphology. p.1-51. New York: Elsevier. 404p.

Stephenson, L. W. (1917) *Tongue, a new stratigraphic term, with illustrations from the Mississippi Cretaceous.* Washington Academy of Sciences Journal, v. 7, no. 9, p. 243-250.

Stephenson, L.W., and Veatch, J.O. (1915) *Underground waters of the Coastal Plain of Georgia.* U.S. Geological Survey. Water-supply Paper 341. 539p.

Sternberg, H. (1875) *Untersuchungen über Längen-und Querprofil geschiebeführender Flüsse.* Zeitschrift für Bauwesen, v.25, p.483-506.

Stevens, R.E., and Carron, M.K. (1948) *Simple field test for distinguishing minerals by abrasion pH.* American Mineralogist, v.33, p.31-50.

Stewart, D.B. (1975) *Apollonian metamorphic rocks—the products of prolonged subsolidus equilibration.* Abstracts of papers submitted to the Sixth Lunar Science Conference, March 17-21, 1975. Lunar Science VI, Part II, p.774-776. Houston: Lunar Science Institute.

Stewart, H.B., jr. (1956) *Contorted sediments in modern coastal lagoon explained by laboratory experiments.* American Association of Petroleum Geologists. Bulletin, v.40, p.153-161.

Stewart, J.H., and Poole, F.G. (1974) *Lower Paleozoic and uppermost Precambrian Cordilleran miogeocline, Great Basin, western United States.* Society of Economic Paleontologists and Mineralogists. Special Publication 22, p.28-57.

Stille, H.W. (1940) *Einführung in den Bau Amerikas.* Berlin: Gebrüder Borntraeger. 717p.

Stille, Hans (1930) *Über Einseitigkeiten in der germanotypen Tektonik Nordspaniens und Deutschlands.* Gesellschaft der Wissenschaften, Göttingen. Mathematisch-Physikalische Klasse. Nachrichten, 1930, H.3, p.379-397.

Stille, Hans (1935) *Der derzeitige tektonische Erdzustand.* Preussische Academie der Wissenschaften, Physikalisch-mathematische Klasse. Sitzungsberichte, v.13, p.179-219.

Stille, Hans (1936) *Present tectonic state of the Earth.* American Association of Petroleum Geologists. Bulletin, v.20, p.849-880.

Stille,Hans (1936) *Wege und Ergebnisse der geologisch-tektonisch-en Forschung.* Berlin: 25 Jahre Kaiser Wilhelm-Gesellschaft zur Förderung der Wissenschaft, v.2, p.77-97.

Stillwell, F.L. (1918) *The metamorphic rocks of Adelie Land.* Australasian Antarctic Expedition, 1911-1914, under the leadership of Sir Douglas Mawson. Scientific reports, Series A, v.3, pt.1, p.7-230.

Stolt, R.H. (1978) *Migration by Fourier transform.* Geophysics, v.43, p.23-48.

Stormer, Leif (1966) *Concepts of stratigraphical classification and terminology.* Earth-Science Reviews, v.1, p.5-28.

Stoces, Bohuslav, and White, C.H. (1935) *Structural geology, with special reference to economic deposits.* London: Macmillan. 460p.

Stockdale, P.B. (1939) *Lower Mississippian rocks of the east-central interior.* Geological Society of America. Special Paper 22. 248p.

Stockwell, C.H. (1964) *Fourth report on structural provinces, orogenics, and time-classification of rocks of the Canadian Precambrian Shield.* In: Age determinations and geological studies, part 2. Geological Survey of Canada. Geological studies. Paper 64-17.

Stokes, W. L. (1983) *Diastrophy—a word whose time has come.* Journal of Geological Education, v. 31, p. 35.

Stokes, W.L. (1947) *Primary lineation in fluvial sandstones, a criterion of current direction.* Journal of Geology, v.55, p.52-54.

Stokes, W.L. (1953) *Primary sedimentary trend indicators as applied to ore finding in the Carrizo Mountains, Arizona and New Mexico. Pt.1.* Salt Lake City: University of Utah. 48p. (U.S. Atomic Energy Commission. Report RME-3043, pt.1).

Stokes, W.L., and Judson, S.S. (1968) *Introduction to geology, physical and historical.* Englewood Cliffs, N.J.: Prentice-Hall. 530p.

Stokes, W.L., and Varnes, D.J. (1955) *Glossary of selected geologic terms, with special reference to their use in engineering.* Denver: Colorado Scientific Society. 165p. (Colorado Scientific Society. Proceedings, v.16).

Stone, R.O. (1967) *A desert glossary.* Earth-science Reviews, v.3, p.211-268.

Storey, T.P., and Patterson, J.R. (1959) *Stratigraphy—traditional and modern concepts.* American Journal of Science, v.257, p.707-721.

Stormer, Leif (1966) *Concepts of stratigraphical classification and terminology.* Earth-Science Reviews, v.1, p.5-28.

Strahan, Aubrey (1907) *The country around Swansea; being an account of the region comprised in sheet 247 of the map.* London: H.M.S.O. 170p. (Great Britain. Geological Survey. Memoirs; England and Wales: the geology of the South Wales coal-field, pt.8)

Strahler, A.N. (1946) *Geomorphic terminology and classification of land masses.* Journal of Geology, v.54, no.1, p.32-42.

Strahler, A.N. (1952a) *Dynamic basis of geomorphology.* Geological Society of America. Bulletin, v.63, p.923-938.

Strahler, A.N. (1952b) *Hypsometric (area-altitude) analysis of erosional topography.* Geological Society of America. Bulletin, v.63, p.1117-1142.

Strahler, A.N. (1954) *Quantitative geomorphology of erosional landscapes.* International Geological Congress. 19th, Algiers, 1952. Comptes rendus, sec. 13, pt.3, fasc.15, p.341-354.

Strahler, A.N. (1956) *The nature of induced erosion and aggradation.* In: Thomas, W.L., Jr., ed. Man's role in changing the face of the Earth , p.621-638. Chicago: University of Chicago Press. 1193p.

Strahler, A.N. (1958) *Dimensional analysis applied to fluvially eroded landforms.* Geological Society of America. Bulletin, v.69, p.279-299.

Strahler, A.N. (1963) *The earth sciences.* New York: Harper & Row. 681p.

Strahler, A.N. (1964) *Quantitative geomorphology of drainage basins and channel networks.* In: Chow, Ven Te, ed. Handbook of applied hydrology; a compendium of water-resources technology , sec.4, p.39-76. New York: McGraw-Hill. 29 sections.

Streckeisen, A. L. (1976) *To each plutonic rock its proper name.* Earth Science Review, v.12, p.1-33.

Streckeisen, A.L. (1967) *Classification and nomenclature of igneous rocks (final report of an inquiry).* Neues Jahrbuch für Mineralogie. Abhandlungen, v.107, no.2, p.144-214.

Strickland, Cyril (1940) *Deltaic formation with special reference to the hydrographic processes of the Ganges and the Brahmaputra.* New York: Longmans, Green. 157p.

Stringfield, V.T. (1966) *Hydrogeology-definition and application.* Ground Water, v.4, no.4, p.2-4.

Strom, K. (1966) *Geophysiography.* Atlas, v. 2, no. 1, p. 8-9.

Strother, P.K. (1996) *Acritarchs.* In: Jansonius, J., and McGregor, D.C., eds., *Palynology; principles and applications 1.* Dallas, Tex., American Association of Stratigraphic Palynologists Foundation, p.81-196.

Studdard, G. J. (compiler) (1993) *Common environmental terms—A glossary.* U.S. Environmental Protection Agency, EPA Region IV, Atlanta, GA, 23 p.

Stutzer, Otto, and Noé, A.C. (1940) *Geology of coal.* Chicago: University of Chicago Press. 461p.

Suess, F.E. (1900) *Die Herkunft der Moldavite und verwandter Gläser.* Austria. Geologische Reichsanstalt. Jahrbuch, Bd.50, H.2, p.193-382.

Suess, H.E., and Urey, H.C. (1956) *Abundances of the elements.* Reviews of Modern Physics, v.28, p.53-74.

Suggate, R.P. (1965) *The definition of "interglacial".* Journal of Geology, v.73, p.619-626.

Sullwold, H.H., Jr. (1959) *Nomenclature of load deformation in turbidites.* Geological Society of America. Bulletin, v. 71, p. 635-636.

Sullwold, H.H.,jr. (1960) *Load-cast terminology and origin of convolute bedding: further comments.* Geological Society of America. Bulletin, v.71, p.635-636.

Sumner, J.S. (1976) *Principles of induced polarization for geophysical prospecting.* New York: Elsevier. 277p.

Sutton, A.H. (1940) *Time and stratigraphic terminology.* Geological Society of America. Bulletin, v.51, p.1397-1412.

Suzuki, S., and Uozumi, S. (1981) *Organic components of prismatic layers in molluscan shells.* Journal of the Faculty of Science, Hokkaido University, ser.4, v.20, no.1, p.7-20.

Svensson, Nils-Bertil (1968) *Lake Lappajärvi, central Finland: a possible meteorite impact structure.* Nature, v.217, p.438.

Sverdrup, H.U., et al. (1942) *The oceans, their physics, chemistry, and general biology.* Englewood Cliffs, N.J.: Prentice Hall. 1087p.

Swain, F.M. (1949) *Onlap, offlap, overstep, and overlap.* American Association of Petroleum Geologists. Bulletin, v.33, p.634-636.

Swain, F.M. (1958) *Organic materials of early Middle Devonian, Mt. Union area, Pennsylvania.* American Association of Petroleum Geologists. Bulletin, v.42, p.2858-2891.

Swain, F.M. (1963) *Geochemistry of humus.* In: Breger, I.A. ed. Organic geochemistry, p.87-147. New York: Macmillan. 658p. (Earth science series 16.)

Swain, F.M., and Prokopovich, N. (1954) *Stratigraphic distribution of lipoid substances in Cedar Creek Bog, Minnesota.* Geological Society of America. Bulletin, v.65, p.1183-1198.

Swann, D.H., and Willman, H.B. (1961) *Megagroups in Illinois.* American Association of Petroleum Geologists, Bulletin, v.45, p.471-483.

Swartz, Delbert (1971) *Collegiate dictionary of botany.* New York: Ronald. 520p.

Swayne, J.C. (1956) *A concise glossary of geographical terms.* London: George Philip & Son. 164p.

Swenson, H.N., and Woods, J.E. (1971) *Physical science for liberal arts students. 2d ed.* New York: Wiley.

Swesnik, R.M. (1950) *Goldren Trend of south-central Oklahoma.* American Association of Petroleum Geologists. Bulletin, v.34, p.386-422.

Swift, D.J.P., et al. (1972) *Holocene evolution of the shelf surface, central and southern Atlantic shelf of North America.* In: Swift, D.J.P., Duane, D.B., and Pilkey, O.H., eds. Shelf sediment transport: process and pattern , ch.23, p.499-574. Stroudsburg, Pa.: Dowden, Hutchinson, & Ross.

T

Taber, Stephen (1943) *Perennially frozen ground in Alaska: its origin and history.* Geological Society of America. Bulletin, v.54, p.1433-1548.

Talbot, M., and Livingstone, D.A. (1989) *Hydrogen index and carbon isotopes of lacustrine organic matter as lake level indicators.* Palaeogeography, Palaeoclimatology, Palaeoecology, v.70, p.121-137.

Tan Sin Hok (1931) *On Cycloclypeus – its phylogeny and signification for the biostratigraphy in general and for the stratigraphy of the Tertiary of the Indo-Pacific region.* Overgedrukt vit de "Handelingen" van het Zesde Nederlandsch-Indish Natuurwetenschappelijk Congress, p. 641-644.

Tanaka, Kenneth L. (1986) *The stratigraphy of Mars.* In: Proceedings of the Seventeenth lunar and planetary science conference: Part 1. Washington, DC: American Geophysical Union.

Taner, M.T., Koehler, F., and Sheriff, R.E., (1979) *Complex trace analysis.* Geophysics, 44, 1041-1063.

Tanner, P. W. G. (1989) *The flexural-slip mechanism.* J. Struct. Geol., v. 11, p. 635-655.

Tanner, W.F. (1960) *Shallow water ripple mark varieties.* Journal of Sedimentary Petrology, v.30, p.481-485.

Tanner, W.F. (1975) *Beach processes, Berrien County, Michigan.* International Association for Great Lakes Research. Journal of Great Lakes Research, v.1, p.171-178.

Tanner, W.F. (1976) *Tectonically significant pebble types: sheared, pocked, and second-cycle examples.* Sedimentary Geology, v.16, p.69-83.

Tanton, T.L. (1944) *Conchilites.* Royal Society of Canada. Transactions, ser.3, v.38, sec.4, p.97-104.

Tappan, H. (1980) *The paleobiology of plant protists.* San Francisco, W.H. Freeman, 977 p.

Tappan, H., and Loeblich, A.R., Jr. (1971) *Surface sculpture on the wall in lower Paleozoic acritarchs.* Micropaleontology, v.17, no.4, p.385-410.

Tarr, R.S. (1902) *The physical geography of New York.* New York: Macmillan. 397p.

Tarr, R.S. (1914) *College physiography.* New York: Macmillan. 837p.

Tarr, R.S., and Martin, Lawrence (1914) *Alaskan glacier studies of the National Geographic Society in the Yakutat Bay, Prince William Sound and lower Copper River regions.* Washington: The National Geographic Society. 498p.

Tarr, R.S., and Von Engeln, O.D. (1926) *New physical geography*. New York: Macmillan. 689p.
Tarr, W.A. (1938) *Terminology of the chemical siliceous sediments. National Research Council. Division of Geology and Geography*. Annual report for 1937-1938, appendix A, exhibit A, p.8-27. (Its Committee on Sedimentation. Report, exhibit A).
Tator, B.A. (1949) *Valley widening processes in the Colorado Rockies*. Geological Society of America. Bulletin, V.60, p.1771-1783.
Tator, B.A. (1953) *Pediment characteristics and terminology; part II*. Association of American Geographers. Annals, v.43, p.47-53.
Tatsumi, Y., and Ishizaka, K. (1982) *Origin of high-magnesian andesites in the Setouchi volcanic belt, southwest Japan. I. Petrographical and chemical characteristics*. Earth and Planetary Science Letters, v.60, p.293-304.
Taubeneck, W.H., and Poldervaart, A. (1960) *Geology of the Elkhorn Mountains, northern Oregon: Part 2*. Willow Lake intrusion. Geological Society of America. Bulletin, v.71, p.1295-1322.
Tavener-Smith, R., and Williams, A. (1972) *The secretion and structure of the skeleton of living and fossil Bryozoa*. Royal Society of London Philosophical Transactions, ser.B, v.264, p.97-159.
Taylor, F.J.R. (1979) *The toxigenic gonyaulacoid dinoflagellates*. In: Taylor, D.L., and Seliger, H.H., eds., *Toxic dinoflagellate blooms*. New York, Elsevier/North Holland, p.47-56.
Taylor, F.J.R. (1980) *Dinoflagellate evolution*. BioSystems, v.13, p.65-108.
Taylor, G.L., and Reno, D.H. (1948) *Magnetic properties of "granite" wash and unweathered "granite"*. Geophysics, v.13, p.163-181.
Taylor, J.D., Kennedy, W.J., and Hall, A. (1969) *The shell structure and mineralogy of the Bivalvia. Introduction. Nuculacea-Trigonacea*. Bulletin of the British Museum (Natural History), Zoology, Supplement, no.3, p.1-125.
Taylor, J.D., Kennedy, W.J., and Hall, A. (1973) *The shell structure and mineralogy of the Bivalvia. II, Lucinacea-Clavagellacea, conclusions*. Bulletin of the British Museum (Natural History), Zoology, v.22, no.9, p.253-294.
Taylor, S.R. (1975) *Lunar science: a post-Apollo view*. New York: Pergamon. 372p.
Taylor, S.R. (1982) *Planetary science: a lunar perspective*. Houston: Lunar and Planetary Institute. 481p.
Taylor, T.G., ed. (1951) *Geography in the twentieth century*. New York: Philosophical Library. 630p.
Taylor, W.C. (1897) *Mineral soap*. Engineering and Mining Journal, v.63, p.600-601.
Teall, J.J.H. (1887) *On the origin of certain banded gneisses*. Geological Magazine, v.4, no.11, p.484-493.
Teall, J.J.H. (1903) *On dedolomitisation*. Geological Magazine, n.s., Dec.4, v.10, p.513-514.
Tebbutt, G.E., et al. (1965) *Lithogenesis of a distinctive carbonate rock fabric*. University of Wyoming. Contributions to Geology, v.4, no.1, p.1-13.
Teichert, Curt (1958) *Concept of facies*. American Association of Petroleum Geologists. Bulletin, v.42, p.2718-2744.
Teichert, Curt (1958a) *Some biostratigraphic concepts*. Geological Society of America. Bulletin, v.69, p.99-119.
Teichert, Curt (1970) *Runzelmarken (wrinkle marks)*. Journal of Sedimentary Petrology, v.40, p.1056-1057.
Telford, W.M., Geldart, L.P., and Sheriff, R.E. (1990) *Applied geophysics. Second edition*. Cambridge University Press, 770p.
Telford, W.M., Geldart, L.P., Sheriff, R.E., and Keys, D.A. (1984) *Applied geophysics*. Cambridge University Press, ???p.
Ten Haaf, Ernst (1956) *Significance of convolution lamination*. Geologie en Mijnbouw, v.18, p.188-194.
Ten Haaf, Ernst (1959) *Graded beds of the northern Apennines*. PhD thesis, Ricks University of Groningen. 102p.
Termier, Henri, and Termier, Geneviéve (1963) *Erosion and sedimentation*. Translated by D.W. Humphries and E.E. Humphries. New York: Van Nostrand. 433p.
Termier, Henri, and Termier, Geneviève (1956) *La notion de migration en paléontologie*. Geologische Rundschau, v.45, p.26-42.
Terzaghi, K.C. von (1943) *Theoretical soil mechanics*. New York: Wiley. 510p.
Terzaghi, Karl, and Peck, R.B. (1967) *Soil mechanics in engineering practice, 2nd ed.* New York: Wiley. 729p.
Tester, A.C., and Bay, H.X. (1931) *The shapometer; a device for measuring the shapes of pebbles*. Science, n.s., v.73, p.565-566.
Theis, C.V. (1935) *The relation between the lowering of the piezometric surface and the rate and duration of discharge of a well using ground-water storage*. American Geophysical Union. Transactions, 1935, p.519-524.
Theis, C.V. (1938) *The significance and nature of the cone of depression in ground-water bodies*. Economic Geology, v.33, p.889-902.
Thiessen, R.L. (1986) *Two-dimensional refold interference patterns*. Journal of Structural Geology, v.8, p.563-573.
Thomas, G.E. (1962) *Grouping of carbonate rocks into textural and porosity units for mapping purposes*. American Association of Petroleum Geologists. Memoir 1, p.193-223.
Thomas, H.D. (1960) *Misuse of "bioclastic limestone"*. American Association of Petroleum Geologists. Bulletin, v.44, p.1833-1834.
Thomas, H.D., and Larwood, G.P. (1956) *Some "uniserial" membraniporine polyzoan genera and a new American Albian species*. Geological Magazine, v.93, p.369-376.
Thompson, A.B., and Perkins, E.H. (1981) *Lambda transitions in minerals*. In: R.C. Newton et al., eds. Thermodynamics of minerals and melts. p. 35-62. New York: Springer-Verlag. 304p.
Thompson, D.G. (1929) *The Mohave Desert region, California; a geographic, geologic, and hydrologic reconnaissance*. U.S. Geological Survey. Water-supply Paper 578. 759p.
Thompson, J.B., jr. (1957) *The graphical analysis of mineral assemblages in pelitic schists*. American Mineralogist, v.42, p.842-858.
Thompson, J.B., jr., et al. (1982) *Reactions in amphibolite, greenschist and blueschist*. Journal of Petrology, v.23, p.1-27.
Thompson, J.B., Jr., Mullins, H.T., Newton, C.R., and Vercoutere, T.L. (1985) *Alternative biofacies model for dysaerobic communities*. Lethaia, v.18, p.167-179.
Thompson, M.M. (1988) *Maps for America*, 3rd ed. U.S. Geological Survey Special Publication, 265p.
Thomsen, L. (2002) *Understanding seismic anisotropy in exploration and exploitation*. Society of Explorational Geophysicists, ?????
Thorarinsson, S. (1981) *Tephra studies and tephrochronology; a historical review with special reference to Iceland*. In Self, S. and Sparks, R. S. J., eds., Tephra studies. D. Reidel Publ. Co., Dordrecht, Holland, p. 1-12.
Thorarinsson, Sigurdur (1944) *Tefrokronologiska studier pa Island: Thjorsardalur och dess foeroedelse; Tephrochronological studies in Iceland*. Geografiska Annaler, 1-2, p. 1-217.
Thorn, C.E. (1976) *Quantitative evaluation of nivation in the Colorado Front Range*. Geological Society of America. Bulletin, v.87, p.1169-1178.
Thornbury, W.D. (1954) *Principles of geomorphology*. New York: Wiley. 618p.
Thornton, C.P., and Tuttle, O.F. (1960) *Chemistry of igneous rocks—pt.1, differentiation index*. American Journal of Science, v.258, p.664-668.
Thorp, J., and Smith, G.D. (1949) *Higher categories of soil classification*. Soil Science, v.67, p.117-126.
Thorson, Gunnar (1957) *Bottom communities (sublittoral or shallow shelf)*. Geological Society of America. Memoir 67, v.1, p.461-534.
Thrush, P.W., compiler (1968) *A dictionary of mining, mineral, and related terms*. Compiled and edited by P.W. Thrush and the Staff of the Bureau of Mines, Washington: U.S. Bureau of Mines. 1269p.
Thurston, J.B., and Smith, R.S. (1997) *Automatic conversion of magnetic data to depth, dip, and susceptibility contrast using the SPI™ method*. Geophysics, v.62, p.807-813.
Thwaites, F.T. (1926) *The origin and significance of pitted outwash*. Journal of Geology, v.34, p.308-319.
Tiddeman, R.H. (1890) *On concurrent faulting and deposit in Carboniferous times in Craven, Yorkshire, with a note on Carboniferous reefs*. British Association for the Advancement of Science. Report, 1889, 59th, p.600-603.
Tieje, A.J. (1921) *Suggestions as to the description and naming of sedimentary rocks*. Journal of Geology, v.29, p.650-666.
Tilley, C.E. (1925) *Metamorphic zones in the southern highlands of Scotland*. Geological Society of London. Quarterly Journal, v.81, p.100-112.
Tilley, C.E., and Muir, I.D. (1964) *Intermediate members of the oceanic basalt-trachyte association*. Geologiska Föreningen i Stockholm. Förhandlingar, v.85, p.436-444.
Tilsley, J.E. (1977) *Placosols: another problem in exploration geochemistry*. Journal of Geochemical Exploration, v.7, p.21-30.
Tinkler, K.J. (1985) *A short history of geomorphology*. Totwa, N.J., Barnes and Noble Books, 317p.
TIP: Moore, R.C., editor 1953-1974 (subsequent editors, Curt Teichert and R.A. Robison) *Treatise on invertebrate paleontology*. Prepared under the guidance of the Joint Committee on Invertebrate Paleontology. Lawrence, Kan.: University of Kansas Press and the Geological Society of America. Issued in separate parts identified by letter.
Tippo, Oswald (1942) *A modern classification of the plant kingdom*. Chronica Botanica, v.7, p.203-206.
Titkov, Nikolai, et al. (1965) *Mineral formation and structure in the electrochemical induration of weak rocks*. New York: Consultants Bureau. 74p.
Törnebohm, A.E. (1880-1881) *Nagra ord om granit och gneis*. Geologiska Föreningen, Stockholm. Förhandlingar, v.5, p.233-248.
Tobi, A.C. (1971) *The nomenclature of the charnockitic rock suite*. Neues Jahrbuch für Mineralogie, Monatshefte, no.5, p.193-205.
Todd, David Keith (1959) *Ground water hydrology, xii*. New York: John Wiley & Sons.
Todd, J.E. (1902) *Hydrographic history of South Dakota*. Geological Society of America. Bulletin, v.13, p.27-40.
Todd, J.E. (1903) *Concretions and their geological effects*. Geological Society of America. Bulletin, v.14, p.353-368.
Tolman, C.F. (1937) *Ground water*. New York: McGraw-Hill. 593p.
Tomkeieff, S.I. (1943) *Megatectonics and microtectonics*. Nature, v.152, p.347-349.
Tomkeieff, S.I. (1946) *James Hutton's "Theory of the Earth", 1795*. Geologist's Association. Proceedings, v.57, p.322-328.
Tomkeieff, S.I. (1954) *Coals and bitumens and related fossil carbonaceius substances; nomenclature and classification*. London: Pergamon Press. 122p.

Tomkeieff, S.I. (1961) *Alkalic ultrabasic rocks and carbonatites of the U.S.S.R.* International Geology Review, v.3, p.739-758.

Tomkeieff, S.I. (1962) *Unconformity—an historical study*. Geologists' Association. Proceedings, v.73, p.383-417.

Tourtelot, H.A. (1960) *Origin and use of the word "shale"*. American Journal of Science, v.258-A (Bradley volume), p.335-343.

Towe, K.M., (1978) Tentaculites*: Evidence for a brachiopod affinity?* Science, v.201, p.626-628.

Tower, W.S. (1904) *The development of cutoff meanders*. American Geographical Society. Bulletin, v.36, p.589-599.

Trask, N.J., and Dzurisin, D. (1984) *Geologic map of the Discovery quadrangle of Mercury*. U.S. Geological Survey Map I-1658 (H-11).

Trask, P.D. (1932) *Origin and environment of source sediments of petroleum*. Houston: American Petroleum Institute. 323p.

Traverse, A. (1988) *Paleopalynology*. Boston, Unwin Hyman, 600p.

Treiber, Erich (1957) *Die Chemie der Pflanzenzellwand*. Berlin: Springer-Verlag. 511p.

Trenhaile, A.S. (1987) *The geomorphology of rocky coasts*. Oxford, Clarendon Press, 384p.

Trevisan, Livio (1950) *Genèse des terrasses fluviatiles en relation avec les cycles climatiques*. International Geographical Congress. 16th, Lisbon. Comptes Rendus, tome 2, p.511-528.

Tröger, W.E. (1935) *Spezielle Petrographie der Eruptivgesteine: Ein nomenklatur-Kompendium*. Berlin: Verlag der Deutschen Mineralogischen Gesellschaft e.V. 360p.

Troll, Carl (1944) *Strukturböden, Solifluktion und Frostklimate der Erde*. Geologische Rundschau, Bd.34, p.545-694.

Trowbridge, A.C. (1921) *The erosional history of the Driftless Area*. State University of Iowa. Studies in Natural History, v.9, no.3. 127p. (State University of Iowa. Studies, 1st ser., no.40).

Trümpy, Rudolf (1955) *Wechselbeziehungen zwischen Palaogeographie und Deckenbau*. Naturforschende Gesellschaft in Zürich, Vierteljahrschrift, pt.C, p.217-231.

Trueman, A.E. (1923) *Some theoretical aspects of correlation*. Geologists' Association. Proceedings, v.34, p.193-206.

Trusheim, Ferdinand (1957) *Über Halokinese und ihre Bedeutung für die strukturelle Entwicklung Norddeutschlands [On halokinesis and its significance for the structural development of North Germany]*. Zeitschrift für Deutsche Geologische Gesellschaft, v. 109, p. 111-151.

Tschudy, R.H., and Scott, R.D. (1969) *Aspects of palynology*. New York: Interscience. 510p.

Tsien, H.H. (1981) *Ancient reefs and reed carbonates*. Proceedings, 4th International Coral Reef Symposium (Manila, 1981), v.1, p.601-609.

Tucker, M. E., and Wright, V. P. (1990) *Carbonate Sedimentology*. Oxford: Blackwell Sci. Publ., 482 p.

Tucker, R.M. (1981) *Giant polygons in the Triassic salt of Cheshire, England: A thermal contraction model for their origin*. Journal of Sedimentary Petrology, v.51, p.779-786.

Turner, F.J. (1948) *Mineralogical and structural evolution of the metamorphic rocks*. Geological Society of America. Memoir 30. 342p.

Turner, F.J. (1968) *Metamorphic petrology*. New York: McGraw-Hill. 403p.

Turner, F.J., and Verhoogen, Jean (1951) *Igneous and metamorphic petrology. 1st ed.* New York: McGraw-Hill. 602p.

Turner, F.J., and Verhoogen, Jean (1960) *Igneous and metamorphic petrology. 2d ed.* New York: McGraw-Hill. 694p.

Turner, F.J., and Weiss, L.E. (1963) *Structural analysis of metamorphic tectonites*. New York: McGraw-Hill. 545p.

Tuttle, O.F. and Gittins, J., eds. (1966) *Carbonatites*. New York: Interscience. 591p.

Tuttle, O.F., and Bowen, N.L. (1958) *Origin of granite in the light of experimental studies in the system $NaAlSi_3O_8$ $KAlSi_3O_8$ SiO_3—H_2O*. Geological Society of America. Memoir 74. 153p.

Twenhofel, W.H. (1937) *Terminology of the fine-grained mechanical sediments*. National Research Council. Division of Geology and Geography. Annual report for 1936-1937, appendix I, exhibit F), p. 81-104 (Its Committee on Sedimentation Report, exhibit F).

Twenhofel, W.H. (1939) *Principles of sedimentation*. New York, McGraw-Hill, 610p.

Twenhofel, William Henry (1950) *Principles of sedimentation. 2nd ed.* New York: McGraw-Hill. 673 p.

Twiss, R. (1988) *Description and classification of folds in single surfaces*. J. Struct. Geol., v. 10, p. 607-623.

Tyrrell, G.W. (1921) *Some points in petrographic nomenclature*. Geological Magazine, v.58, p.494-502.

Tyrrell, G.W. (1926) *The principles of petrology; an introduction to the science of rocks*. London: Methuen. 349p.

Tyrrell, G.W. (1950) *The principles of petrology; an introduction to the science of rocks*. London: Methuen; New York: E.P. Dutton.

Tyrrell, J.B. (1904) *Crystosphenes or buried sheets of ice in the tundra of northern America*. Journal of Geology, v.12, p.232-236.

Tyrrell, J.B. (1910) *"Rock glaciers" or chrystocrenes*. Journal of Geology, v.18, p.549-553.

Tyrrell, J.B., and Dowling, D.B. (1896) *Report on the country between Athabasca Lake and Churchill River with notes on two routes travelled between the Churchill and Saskatchewan rivers*. Geological Survey of Canada. Annual report 1895, n.s., v.8, report D. 120p.

Tyson, R. (1995) *Sedimentary organic matter*. London, Chapman and Hall, 615p

U

U.S. Army Corps of Engineers (1991) *Glossary of hydrologic engineering terms*. Davis, CA: Hydrologic Engineering Center, 155 p.

U.S. Federal Committee on Research Natural Areas (1968) *A directory of research natural areas on Federal lands of the United State of America*. Washington: U.S. Government Printing Office. 129p.

U.S. Naval Oceanographic Office (1966) *Glossary of oceanographic terms. 2nd ed.* U.S. Naval Oceanographic Office, Special Publication 35. 204p.

U.S. Naval Oceanographic Office (1968) *Ice observations. 2nd ed.* Washington: U.S. Naval Oceanographic Office. 42p. (H.O. Publication no.606-d).

Udden, J.A. (1898) *The mechanical composition of wind deposits*. Rock Island, Ill: Augustana Library Publications, no.1. 69p.

Udden, J.A. (1914) *Mechanical composition of clastic sediments*. Geological Society of America. Bulletin, v.25, p.655-744.

Ulrich, E.O. (1911) *Revision of the Paleozoic systems*. Geological Society of America. Bulletin, v.22, p.281-680.

Umbgrove, J.H.F. (1933) *Verschillende Typen van tertiaire Geosynclinalen in den Indischen Archipel*. Leidsche Geologische Mededeelingen, v.5, no.1, p.33-43.

Underwood, L.B. (1957) *Rebound problem in the Pierre Shale at Oahe Dam, Pierre, South Dakota, Pt.1 (abstract)*. Geological Society of America. Bulletin, v.68, no.12, pt.2, p.1807-1808.

Uozumi, S., and Suzuki, S. (1981) *The evolution of shell structure in the Bivalvia*. In: Habe, T., and Omori, M., eds., *Studies of molluscan paleobiology*, Professor Omori Memorial Volume. Niigata, Japan, Niigata University Publication Committee, p.63-77.

Urbanek, Adam, and Towe, K.M. (1974) *Ultrastructural studies on graptolites, 1: The periderm and its derivatives in the Dendroidea and in Mastigograptus*. Smithsonian Contributions to Paleobiology, no.20. 48p.

USBLM: U.S. Bureau of Land Management (1973) *Manual of surveying instructions*. http://www.cadastral.com/73manl-1.htm

USDA: U.S. Department of Agriculture (1938) *Soils and men*. Washington: Government Printing Office. 1232p. (Its Yearbook of agriculture, 1938.)

USDA: U.S. Department of Agriculture (1957) *Soil*. Washington: Government Printing Office. 784p. (Its Yearbook of Agriculture, 1957).

USDA: U.S. Department of Agriculture (1975) *Soil taxonomy: a basic system of soil classification for making and interpreting soil surveys*. Agriculture Handbook 436. 754p.

USGS: U.S. Geological Survey (1947) *United States National Map accuracy standards*, rev. 6/47. http://rockyweb.cr.usgs.gov/nmpstds/nmas.html

USGS: U.S. Geological Survey (1958) *Suggestions to authors of the reports of the United States Geological Survey. 5th ed.* Washington, D.C.: U.S. Government Printing Office. 255p.

USGS: U.S. Geological Survey (1965) *Topographic instructions. 5C3 Definitions*. 166p.

USGS: U.S. Geological Survey (1970) *Geologic time—The age of the Earth*. Washington, D.C.: U.S. Government Printing Office. 20p.

USGS: U.S. Geological Survey (1980) *Principles of a resource/reserve classification for minerals*. USGS Circular C 831, 5p. (Replaces USGS Bulletin 1450-A).

USGS: U.S. Geological Survey (1991) *Standards for 1:12,000-scale orthophoto quarter quadrangles*, 8/91. USGS National Mapping Program (NMP) Standard. http://rockyweb.cr.usgs.gov/nmpstds/omastds.html

USGS: U.S. Geological Survey (1993) *Standards for 1:24,000-scale orthophoto quadrangles*, 2/93. USGS National Mapping Program (NMP) Standard. http://rockyweb.cr.usgs.gov/nmpstds/omastds.html

USGS: U.S. Geological Survey (1996) *Standards for 1:24,000-scale digital line graphs and quadrangle maps, Part 7: Hypsography, Appendix 7A: Background material for contour*. USGS National Mapping Program (NMP) Standard, 4/96, 34p. http://rockyweb.cr.usgs.gov/nmpstds/acrodocs/dlggmap/7dgm0401.pdf

USGS: U.S. Geological Survey (2000) *U.S. geodata digital elevation models*. USGS Fact Sheet FS-040-00 (04/2000). http://erg.usgs.gov/isb/pubs/factsheets/fs04000.pdf

USGS: U.S. Geological Survey (2001) *Standards for digital raster graphics*. USGS National Mapping Program (NMP) Standard, 4/01. http://rockyweb.cr.usgs.gov/nmpstds/drgstds.html

USGS: U.S. Geological Survey (2002a) *Standards for raster feature separates*. USGS National Mapping Program (NMP) Standard, 11/02. http://rockyweb.cr.usgs.gov/nmpstds/rfsstds.html

USGS: U.S. Geological Survey (2002b) *Standards for revised primary series quadrangle maps*. USGS National Mapping Program (NMP) Standard, 12/02. http://rockyweb.cr.usgs.gov/nmpstds/rfsstds.html

V

Vacher, H.L., and Mylroie, J.E. (2002) *Eogenetic karst from the perspective of an equivalent porous medium.* Carbonates and Evaporites, v.17, p.182-196.

Vail, P. R., Mitchum, R. M., and Thompson, S., III (1977) *Seismic stratigraphy and global changes of sea level, part 3; relative changes of sea level from coastal onlap.* In: C. E. Payton, ed., Seismic stratigraphy applications to hydrocarbon exploration , American Association of Petroleum Geologists Memoir 26, p. 63-82.

Valentine, J.W. (1969) *Patterns of taxonomic and ecological structure of the shelf benthos during Phanerozoic time.* Palaeontology, v.12, pt.4, p.684-709.

Vallance, J.W. (2000) *Lahars.* In: Sigurdsson, H., ed., *Encyclopedia of volcanoes.* New York, Academic Press, p.601-616.

Van Bemmelen, R.W. (1932) *De undatie theorie. Natuurkundig* Tijdschrift voor Nederlandsch-Indië, v.92, p.85-242, 373-402.

Van Bemmelen, R.W. (1933) *The undation theory of the development of the Earth's crust.* International Geological Congress. 16th, Washington, D.C. Proceedings, v.2, p.965-982, [1935].

Van Bemmelen, R.W. (1949) *The geology of Indonesia, vol. 1A.* The Hague: Government Printing Office. 732p.

Van Couvering, J.A., Castradori, D., Cita, M.B., Hilgen, F.J., and Rio, D. (2000) *The base of the Zanclean Stage and of the Pliocene Series.* Episodes, v.23, no.3, p.179-187.

Van de Poll, H.W., and Patel, I.M. (1981) *Flute casts and related structures on moulded silt injection surfaces in continental sandstone of the Boss Point Formation; southeastern New Brunswick, Canada.* Maritime Sediments and Atlantic Geology, v.17, p.1-22.

Van der Gracht, W.A.J.M. van Waterschoot (1931) *The Permo-Carboniferous orogeny in the south-central United States.* Koninklijke Akademie van Wetenschappen, Amsterdam. Afdeeling natuurkunde Verhandlingen, deel 27, no.3.

Van der Hammen, T., et al. (1971) *The floral record of the Late Cenozoic of Europe.* In: Turekian, K.K., ed. Late Cenozoic glacial ages, p.329-424. New Haven: Yale University Press.

Van Hinte, J. E. (1969) *The nature of biostratigraphic zones.* 1st International Conference on Planktonic Microfossils, Proceedings, v. 2, p. 267-272.

Van Hise, C.R. (1896) *Principles of North American pre-Cambrian geology.* U.S. Geological Survey. Annual Report, 16th, pt.1, p.571-843.

Van Hise, C.R. (1904) *A treatise on metamorphism.* U.S. Geological Survey. Monograph 47. 1286p.

Van Hise, C.R., and Leith, C.K. (1911) *The geology of the Lake Superior region.* U.S. Geological Survey. Monograph 52. 641p.

van Morkhoven, F. P. C. M. (1966) *The concept of paleoecology and its practical application.* Gulf Coast Assoc. of Geol. Soc. Trans., v. 16, p. 305-313.

Van Riper, J.E. (1962) *Man's physical world.* New York: McGraw-Hill. 637p.

Van Straaten, L.M.J.U. (1951) *Longitudinal ripple marks in mud and sand.* Journal of Sedimentary Petrology, v.21, p.47-54.

Van Straaten, L.M.J.U. (1953) *Megaripples in the Dutch Wadden Sea and in the basin of Arcachon (France).* Geologie en Mijnbouw, n.s., v.15, p.1-11.

Van Straaten, L.M.J.U. (1953a) *Rhythmic patterns on Dutch North Sea beaches.* Geologie en Mijnbouw, n.s., v.15, p.31-43.

van Straaten, L.M.J.U., and Kuenen, P.H. (1958) *Tidal action as a cause of clay accumulation.* Journal of Sedimentary Petrology, v.28, p.406-413.

Van Tuyl, F.M. (1916) *The origin of dolomite.* Iowa Geological Survey. Volume 25, p.251-421.

Van Wagoner, J. C. (1985) *Reservoir facies distribution as controlled by sea-level change.* Society of Economic Paleontologists and Mineralogists Mid-Year Meeting, Golden, CO, August 11-14, p. 91-92.

Van Wagoner, J. C. (1991) *Sequence stratigraphy and facies architecture of the Desert Member of the Blackhawk Formation and the Castlegate Formation in the Book Cliffs in eastern Utah and western Colorado.* In: J. C. Van Wagoner, C. R. Jones, D. R. Taylor, D. Nummedal, D. C. Jennette, G. W. Riley, eds., Sequence stratigraphy applications to shelf sandstone reservoirs; outcrop to subsurface examples. American Association of Petroleum Geologists Field Conference, September 21-25, p. 1-7 in section Day Four.

Van Wagoner, J. C., Mitchum, R. M., Campion, K. M., Rahmanian, V. D. (1990) *Siliciclastic sequence stratigraphy in well logs, cores and outcrops; concepts for high-resolution correlation of time and facies.* American Association of Petroleum Geologists Methods in Exploration Series, no. 7, 55 p.

Van Wagoner, J. C., Mitchum, R. M., Posamentier, H. W., and Vail, P. R. (1987) *An overview of sequence stratigraphy and key definitions, 1987.* In: A. W. Bailey, eds., Atlas of seismic stratigraphy, volume 1. American Association of Petroleum Geologists Studies in Geology, no. 27, p. 11-14.

Van Wagoner, J. C., Posamentier, H. W., Mitchum, R. M., Vail, P. R, Sarg, J. F., Loutit, T. S., and Hardenbol, J. (1988) *An overview of sequence stratigraphy and key definitions.* In: C. W. Wilgus et al., eds. Sea level changes; an integrated approach. Society of Economic Paleontologists and Mineralogists Special Publication, 42, p. 39-45.

Van Wagoner, John C. (1995) *Sequence stratigraphy and marine to nonmarine facies architecture of foreland basin strata, Book Cliffs, Utah., U.S.A.* In: Sequence stratigraphy of foreland basin deposits; outcrop and subsurface examples from the Cretaceous of North America. AAPG Memoir, 64; p. 137-223.

Van Wagoner, John C. (1995b) *Sequence stratigraphy of foreland basin deposits; outcrop and subsurface examples from the Cretaceous of North America.* AAPG Memoir, 64; 487 p.

Vanserg, Nicholas (1952) *How to write geologese.* Economic Geology, v.52, p.220-223.

Varnes, D.J. (1958) *Landslide types and processes.* National Research Council. Highway Research Board. Special Report 29, p.20-47.

Varnes, D.J. (1978) *Slope movement types and processes.* In: Landslides: analysis and control. National Academy of Sciences. Transportation and Research Board Special Report 176, Ch.2.

Vartiainen, H., and Paarma, H. (1979) *Geological characteristics of the Sokli carbonatite complex, Finland.* Economic Geology, v.74, p.1296-1306.

Vassoevich, N.B. (1948) *Evolyutsiya predstavlenii o geologicheskikh fatsiyakh.* Leningrad: Gostoptekhizdat. (Vsesouiznyi Neftianoi Nauchno-Issledovatel'skiy Geologo-Rezvedochnyi Institut. Litologicheskii sbornik, no.1).

Vassoevich, N.B. (1953) *O nekotorykh flishevykh teksturakh (znakakh). L'vovskoe Geologicheskoe Obshchestvo.* Trudy; geologicheskaia seriia, no.3, p.17-85.

Vassoevich, N.B. (1965) *Vernadskiy's views on the origin of oil.* International Geology Review, v.7, p.507-517.

Vasudevan, Wilson, and Laidlow (1991) *Simulated annealing statics computation using an order-based energy function.* Geophysics 56, 1831-39.

Veatch, J.O., and Humphrys, C.R. (1966) *Water and water use terminology.* Kaukauna, Wisc.: Thomas Printing & Publishing Co. 381p.

Veblen, D. R. (1991) *Polysomatism and polysomatic series: a review and applications.* American Mineralogist 76, p. 801-826.

Veblen, D.R., Buseck, P.R., and Burnham, C.W. (1977) *Asbestiform chain silicates: new minerals and structural groups.* Science, v.198, p.359-365.

Vella, P. (1964) *Biostratigraphic units.* New Zealand Journal of Geology and Geophysics, v.7, p.615-625.

Vendeville, B. C., and Jackson M. P. A. (1991) *Deposition, extension, and the shape of downbuilding salt diapirs (abs.)* American Association of Petroleum Geologists Bulletin, v. 75, p. 687-688.

Veni, G. (1997) *Geomorphology, hydrogeology, geochemistry, and evolution of the karstic Lower Glen Rose Aquifer, south-central Texas* (Pennsylavania State University doctoral dissertation). Austin, Tex., Texas Speleological Survey Monograph 1, 409p.

Veniale, F., and van der Marel, H. W. (1969) *Identification of some 1:1 regular interstratified trioctahedral clay minerals.* Proceedings of the International Clay Conference, Tokyo 1, p. 233-244.

Vening Meinesz, F.A. (1955) *Plastic buckling of the Earth's crust: the origin of geosynclines.* Geological Society of America. Special Paper 62, p.319-330.

Vergniolle, S., and Mangan, M. (2000) *Hawaiian and Strombolian eruptions.* In: Sigurdsson, H., ed., *Encyclopedia of volcanoes.* New York, Academic Press, p.447-461.

Vermeij, G.J. (1991) *When biotas meet: Understanding biotic interchange.* Science, v.253, no.5024, p.1099-1104.

Vernon, R.H. (1974) *Controls of mylonitic compositional layering during noncataclastic ductile deformation.* Geological Magazine, v.111, p.121-123.

Vernon, R.H. (1975) *Metamorphic processes, reactions and microstructure development.* New York: Wiley. 247p.

Vespermann, D., and Schmincke, H.-U. (2000) *Scoria cones and tuff rings.* In: Sigurdsson, H., ed., *Encyclopedia of volcanoes.* New York, Academic Press, p.683-694.

Viles, H.A., and Goudie, A.s. (1990) *Tufas, travertines, and allied carbonate deposits.* Progress in Physical Geography, v.14, no.1, p.19-41.

Viljoen, M.J., and Viljoen, R.P. (1969) *The geology and geochemistry of the lower ultramafic unit of the Onverwacht Group and a proposed new class of igneous rock.* Geological Society of South Africa. Special publication, Upper Mantle Project , v.2, p.221-244.

Vistelius, A.B. (1967) *Studies in mathematical geology.* New York: Consultants Bureau. 294p.

Vitaliano, D.B. (1968) *Geomythology: the impact of geologic events on history and legend, with special reference to Atlantis.* Indiana University. Journal of the Folklore Institute, v.5, p.5-30.

Vitaliano, D.B. (1973) *Legends of the earth: their geologic origins.* Bloomington: Indiana University Press. 305p.

Vogt, J.H.L. (1905) *Über anchi-eutektische und anchimonomineralische Eruptivgesteine.* Norsk geologisk tidsskrift, v.1, no.1, paper 2. 33p.

Von Bernewitz, M.W. (1931) *Handbook for prospectors. 2d ed.* New York: McGraw-Hill.

Von Engeln, O.D. (1942) *Geomorphology; systematic and regional.* New York: Macmillan. 655p.

W

Wada, K. (1963a) *On the spiral growth of the inner surface of the calcitic shell* Anomia lischkei – I. Japanese Society of Scientific Fisheries Bulletin, v.29, p.320-324.

Wada, K. (1963b) *On the spiral growth of the inner surface of the calcitic shell* Ostrea gigas – II. Japanese Society of Scientific Fisheries Bulletin, v.29, p.447-451.

Wada, K. (1968) *Spiral growth of calcitostracum.* Nature, London, v.219, p. 62.

Wada, K. (1989) *Allophane and imogolite.* in Dixon, J. B., and Weed, S. B., editors, Minerals in the soil environment , 2nd edition, Soil Science Society of America, p. 1051-1087.

Wadell, H.A. (1932) *Volume, shape, and roundness of rock particles.* Journal of Geology, v.40, p.443-451.

Wadell, H.A. (1934) *Shape determinations of large sedimental rock fragments.* Pan-American Geologist, v.61, p.187-220.

Wadsworth, M.E. (1893) *A sketch of the geology of the iron, gold, and copper districts of Michigan.* Michigan Geological Survey. Report of the State Geologist for 1891-1892, p.75-174.

Wager, L.R. (1968) *Rhythmic and cryptic layering in mafic and ultramafic plutons.* In: Hess, H.H., and Poldervaart, Arie, eds. Basalts: the Poldervaart treatise on rocks of basaltic composition , v.2, p.483-862. New York: Interscience.

Wager, L.R., and Brown, G.M. (1967) *Layered igneous rocks.* San Francisco: Freeman. 588p.

Wager, L.R., and Deer, W.A. (1939) *The petrology of the Skaergaard intrusion, Kangerdlugssuaq, east Greenland.* Meddelelser om Grönland, v.105, no.4. 352p. (Geological investigations in east Greenland, pt.3)

Wahlstrom, E. E. (1948) *Optical crystallography.* New York: Wiley. 206 p.

Wahlstrom, E.E. (1969) *Optical crystallography, 4th ed.* New York: Wiley. 489p.

Walcott, R.H. (1898) *The occurrence of so-called obsidian bombs in Australia.* Royal Society of Victoria. Proceedings, n.s., v.11, pt.1, p.23-53.

Walker, Frederick (1957) *Ophitic texture and basaltic crystallization.* Journal of Geology, v.65, no.1, p.1-14.

Walker, G., and Croasdale, R. (1972) *Characteristics of some basaltic pyroclastics.* Bulletin Volcanologique, v.35, p.303-317.

Walker, G.P.L. (1973) *Lengths of lava flows.* Philosophical Transactions, Royal Society of London, v. 274, p.107-118.

Walker, G.P.L. (1980) *The Taupo pumice—product of the most powerful known (ultraplinian) eruption.* Journal of Volcanology and Geothermal Research, v.8, p.69-94.

Walker, G. P. L. (1991) *Structure, and origin by injection of lava under surface crust, of tumuli, "lava rises", "lava-rise pits", and "lava inflation clefts" in Hawaii.* Bulletin of Volcanology, v. 53, p. 546-558.

Walker, G.P.L. (2000) *Basaltic volcanoes and volcanic systems.* In: Sigurdsson, H., ed., *Encyclopedia of volcanoes.* New York, Academic Press, p.283-289.

Walker, K.R. (1973) *Major Middle Ordovician reef tract in east Tennessee.* American Journal of Science, v.273-A (Cooper volume), p.294-325.

Walker, T.L. (1902) *The geology of Kalahandi state, central provinces.* India, Geological Survey. Memoirs, v.33, part 3.

Wall, D., and Dale, B. (1968) *Modern dinoflagellate cysts and evolution of the Peridiniales.* Micropaleontology, v.14, no.3, p.265-304.

Wallace, R.C. (1913) *Pseudobrecciation in Ordovician limestones in Manitoba.* Journal of Geology, v.21, p.402-421.

Waller, T.R. (1978) *Morphology, morphoclines, and a new classification of the Pteriomorphia.* Philosophical Transactions of the Royal Society of London, ser.B, v.284, p.345-365.

Walliser, O.H., Bultnyck, P., Weddige, K., Becker, R.T., and House, M.R. (1995) *Definition of the Eifelian-Givetian Stage boundary.* Episodes, v.18, no.3, p.107-115.

Walsh, J.J., and Watterson, J. (1991) *Geometric and kinematic coherence and scale effects in normal fault systems.* In: Roberts, A.M., Yielding, G, and Freeman, B., eds., *The geometry of normal faults.* Geological Society [of London] Special Publication 56, p.193-203.

Walter, M.R. (1976) *Stromatolites.* Amsterdam: Elsevier. 790p. (Developments in sedimentology, 20.)

Walther, Johannes (1893-1894) *Einleitung in die Geologie als historische Wissenschaft; Beobachtungen über die Bildung der Gesteine und ihrer organischen Einschlusse.* Jena G. Fischer. 1055p.

Walton, E.K. (1956) *Limitations of graded bedding and alternative criteria of upward sequence in the rocks of the Southern Uplands.* Edinburgh Geological Society. Transactions, v.16, p.262-271.

Walton, John (1940) *An introduction to the study of fossil plants.* London: Adam & Charles Black. 188p.

Wang Hong (1966) *On rock-stratigraphic units.* Acta Geologica Sinica, v. 46, no. 1, p. 1-13.

Wanless, H.R., and Weller, J.M. (1932) *Correlation and extent of Pennsylvanian cyclothems.* Geological Society of America. Bulletin, v.43, p.1003-1016.

Ward, F.K. (1923) *From the Yangtze to the Irrawaddy.* Geographical Journal, v.62, p.6-20.

Ward, W.H. (1953) *Glacier bands; conference on terminology.* Journal of Glaciology, v.2, no.13, p.229-232.

Warntz, William (1975) *Stream ordering and contour mapping.* Journal of Hydrology, v.25, p.209-227.

Warren, J.K., (1989) *Evaporite Sedimentology: Its importance in hydrocarbon accumulations.* Prentice-Hall Scientific Publications, Englewood Cliffs, New Jersey, 285p.

Washburn, A.L. (1947) *Reconnaissance geology of portions of Victoria Island and adjacent regions, arctic Canada.* Geological Society of America. Memoir 22. 142p.

Washburn, A.L. (1950) *Patterned ground.* Revue Canadienne Geographie, v.4, no.3-4, p.5-59.

Washburn, A.L. (1956) *Classification of patterned ground and review of suggested origins.* Geological Society of America. Bulletin, v.67, p.823-865.

Washburn, A.L. (1973) *Periglacial processes and environments.* London: Edward Arnold. 320p.

Washburn, A.L., and Goldthwait, R.P. (1958) *Slushflows (abstract).* Geological Society of America. Bulletin, v.69, p.1657-1658.

Watabe, N. (1984) *Shell.* In: Bereiter-Hahn, J., Matoltsy, A.G., and Richards, K.S., eds., *Biology of the integument.* Berlin, Springer, chap.25, p.448-485.

Waters, A.C., and Campbell, C.D. (1935) *Mylonites from the San Andreas fault zone.* American Journal of Science, 5th ser., v.29, no.174, p.473-503.

Waterston, C.D. (1975) *Gill structures in the Lower Devonian eurypterid Tarsopterella scotica.* Fossils and Strata, no.4, p.241-254.

Watson, R.A., and Wright, H.E., Jr. (1980) *The end of the Pleistocene: a general critique of chronostratigraphic classification.* Boreas, v.9, p.153-163.

Wayland, E.J. (1920) *Some facts and theories relating to the geology of Uganda.* Uganda Geological Dept. Pamphlet no.1. 52p.

Wayland, E.J. (1934) *Peneplains and some other erosional platforms.* Uganda Geological Survey Dept. Annual Report and Bulletin, 1933, Bull. no.1, p.77-79.

Weast, R.C., ed. (1970) *Handbook of chemistry and physics. 51st ed.* Cleveland: The Chemical Rubber Co.

Weatherley, A.H., ed. (1967) *Australian inland waters and their fauna; eleven studies.* Canberra: Australian National University Press. 287p.

Weaver, C. E. (1953) *Mineralogy and petrology of some Ordovician K-bentonites and related limestones.* Geological Society of America Bulletin, v. 64, p. 921-943.

Weaver, C. E. (1989) *Clays, muds and shales.* In Developments in Sedimentology. Elsevier Pub. Co., v. 44, 819 p.

Weaver, C. E. and Broekstra, B. R. (1984) *Illite-mica.* In: Weaver, C. E., et al., Shale-slate metamorphism in Southern Appalachians. Developments in Petrology 10 : Amsterdam: Elsevier, p. 67-97.

Weaver, C.E. (1980) *Fine-grained rocks: shales or physilites.* Sedimentary Geology, v.27, p.301-313.

Weaver, J.E., and Clements, F.E. (1938) *Plant ecology. 2nd ed.* New York: McGraw-Hill. 601p.

Webb, L.J. (1959) *A physiognomic classification of Australian rain forests.* Journal of Ecology, v.47, p.551-570.

Webb, R.W. (1936) *Kern Canyon Fault, southern Sierra Nevada.* Journal of Geology, v.44, p.631-638.

Webster's (1984) *Webster's II New Riverside University dictionary.* The Riverside Publishing Company.

Webster's third new international dictionary of the English language; unabridged (1967) P.B. Gove, ed. Springfield, Mass.; G. & C. Merriam. 2662p.

Weeks, L.G. (1952) *Factors of sedimentary basin development that control oil occurrence.* American Association of Petroleum Geologists. Bulletin, v.36, p.2071-2124.

Wegener, Alfred (1912) *The origin of the continents and oceans.* English translation, 1924. New York: Dutton.

Weik, M.H., (1977) *Standard Dictionary of Computers and Information Processing. 2nd. ed.:* Hayden Book Co., Inc., Rochelle Park, NJ, 390p.

Weiss, P., and Ferris, W.J. (1954) *Electronmicrograms of larval amphibian epidermis.* Experimental Cell Research, v.6, p.546-549.

Welch, P.S. (1952) *Limnology. 2nd ed.* New York: McGraw-Hill. 538p.

Weld, L.D., ed. (1937) *Glossary of physics. 1st ed.* New York: McGraw-Hill. 255p.

Weller, J.M. (1930) *Cyclical sedimentation of the Pennsylvanian Period and its significance.* Journal of Geology, v.38, p.97-135.

Weller, J.M. (1956) *Argument for diastrophic control of late Paleozoic cyclothems.* American Association of Petroleum Geologists. Bulletin, v.40, p.17-50.

Weller, J.M. (1958) *Stratigraphic facies differentiation and nomenclature.* American Association of Petroleum Geologists. Bulletin, v.42, p.609-639.

Weller, J.M. (1958a) *Cyclothems and larger sedimentary cycles of the Pennsylvanian.* Journal of Geology, v.66, p.195-207.

Weller, J.M. (1960) *Stratigraphic principles and practice.* New York: Harper. 725p.

Weller, J.M., et al. (1942) *Stratigraphy of the fusuline-bearing beds of Illinois.* Illinois State Geological Survey. Bulletin 67, p.9-34.

Wellman, H.W., and Wilson, A.T. (1965) *Salt weathering, a neglected geological erosive agent in coastal and arid environments.* Nature, v.205, p.1097-1098.

Wells, F.G. (1949) *Ensimatic and ensialic geosynclines (abstract).* Geological Society of America. Bulletin, v.60, p.1927.

Wells, J.W. (1944) *Middle Devonian bone beds of Ohio.* Geological Society of America. Bulletin, v.55, p.273-302.

Wells, J.W. (1947) *Provisional paleoecological analysis of the Devonian rocks of the Columbus region.* Ohio Journal of Science, v.47, p.119-126.

Wenk, H.-R., ed. (1985) *Preferred orientation in deformed metals and rocks: An introduction to modern texture analysis.* Orlando: Academic Press.

Wentworth, C.K. (1922) *A scale of grade and class terms for clastic sediments.* Journal of Geology, v.30, p.377-392.

Wentworth, C.K. (1922a) *A method of measuring and plotting the shapes of pebbles.* U.S. Geological Survey. Bulletin 730-C, p.91-114.

Wentworth, C.K. (1922b) *The shapes of beach pebbles.* U.S. Geological Survey. Professional Paper 131-C, p.75-83.

Wentworth, C.K. (1925) *Chink-faceting; a new process of pebbleshaping.* Journal of Geology, v.33, p.260-267.
Wentworth, C.K. (1931) *Pebble wear on the Jarvis Island beach.* Washington University Studies, n.s., Science and Technology, no.5, p.11-37.
Wentworth, C.K. (1935) *The terminology of coarse sediments.* With notes by P.G.H. Boswell. National Research Council. Division of Geology and Geography. Committee on Sedimentation. Report for 1932-1934, p.225-246. (National Research Council. Bulletin, no.98).
Wentworth, C.K. (1936) *An analysis of the shapes of glacial cobbles.* Journal of Sedimentary Petrology, v.6, p.85-96.
Wentworth, C.K. (1939) *Marine bench-forming processes: II, Solution benching.* Journal of Geomorphology, v.2, p.3-25.
Wentworth, C.K. (1943) *Soil avalanches on Oahu, Hawaii.* Geological Society of America. Bulletin, v.54, p.53-64.
Wentworth, C.K., and Williams, Howel (1932) *The classification and terminology of the pyroclastic rocks.* National Research Council. Division of Geology and Geography. Committee on Sedimentation. Report for 1930-1932, p.19-53. (National Research Council. Bulletin, no.89)
Wentworth, C.M. (1967) *Dish structure, a primary sedimentary structure in coarse turbidites.* American Association of Petroleum Geologists. Bulletin, v.51, p.485.
Wetzel, R.G. (2001) *Limnology: Lake and river ecosystems.* San Diego, Academic Press, 1006p.
Wheeler, H.E. (1958) *Time-stratigraphy.* American Association of Petroleum Geologists. Bulletin, v.42, p.1047-1063.
Wheeler, H.E. (1958a) *Primary factors in biostratigraphy.* American Association of Petroleum Geologists. Bulletin, v.42, p.640-655.
Wheeler, H.E. (1964) *Baselevel, lithosphere surface, and time-stratigraphy.* Geological Society of America. Bulletin, v.75, p.599-609.
Wheeler, H.E., and Beesley, E.M. (1948) *Critique of the time-stratigraphic concept.* Geological Society of America. Bulletin, v.59, p.75-85.
Wheeler, H.E., and Mallory, V.S. (1953) *Designation of stratigraphic units.* American Association of Petroleum Geologists. Bulletin, v.37, p.2407-2421.
Wheeler, H.E., and Mallory, V.S. (1956) *Factors in lithostratigraphy.* American Association of Petroleum Geologists. Bulletin, v.40, p.2711-2723.
Wheeler, H.E., et al. (1950) *Stratigraphic classification.* American Association of Petroleum Geologists. Bulletin, v.34, p.2361-2365.
Whitaker, J.H.McD. (1973) *"Gutter casts", a new name for scour-and-fill structures: with examples from Llandoverian and Rjngerike and Malmöya, southern Norway.* Norsk Geologisk Tidsskrift, v.53, p.403-417.
White, C.A. (1870) *Report on the geological survey of the State of Iowa. Vol.1.* Des Moines: Mills & Co. 391p.
White, D.E. (1957) *Thermal waters of volcanic origin.* Geological Society of America. Bulletin, v.68, p.1637-1657.
White, D.E., Thompson, G.A., and Sandberg, C.H. (1964) *Rocks, structure, and geologic history of Steamboat Springs thermal area, Washoe County, Nevada.* U.S. Geological Survey Professional Paper 458-B, 62p.
White, David (1915) *Some relations in origin between coal and petroleum.* Washington Academy of Sciences. Journal, v.5, no.6, p.189-212.
White, S.E. (1967) *Rockfall, alluvial, and avalanche talus in the Colorado Front Range (abstract).* Geological Society of America. Special Paper 115, p.237.
White, S.E. (1972) *Alpine subnival boulder pavements in Colorado Front Range.* Geological Society of America. Bulletin, v.83, p.195-200.
White, S.E. (1976) *Rock glaciers and block fields, review and new data.* Quaternary Research, v.6, p.77-97.
White, W.A. (1961) *Colloid phenomena in sedimentation of argillaceous rocks.* Journal of Sedimentary Petrology, v.31, no.4, p.560-570.
White, W.H. (1959) *Cordilleran tectonics in British Columbia.* American Association of Petroleum Geologists. Bulletin, v.43, p.60-100.
Whitney, J.D. (1888) *Names and places; studies in geographical and topographical nomenclature.* Cambridge, England: Cambridge University Press. 239p.
Whittaker, R.H. (1972) *Evolution and measurement of species diversity.* Taxon, v.21, 2.213-251.
Whitten, D.G.A., with Brooks, J.R.V. (1972) *The Penguin dictionary of geology.* Harmondsworth, England: Penguin Books. 495p.
Whitten, E.H.T. (1959) *A study of two directions of folding: the structural geology of the Monadhliath and mid-Strathspey.* Journal of Geology, v.67, p.14-47.
Whitten, E.H.T. (1966) *Structural geology of folded rocks.* Chicago: Rand McNally. 663p.
Whittington, H.B., and Rickards, R.B. (1968) *New tuboid graptolite from the Ordovician of Ontario.* Journal of Paleontology, v.42, p.61-69.
Wicander, R., Foster, C.B., and Reed, J.D. (1996) *Gloeocapsamorpha.* In: Jansonius, J., and McGregor, D.C., eds., *Palynology; principles and applications 1.* Dallas, Tex., American Association of Stratigraphic Palynologists Foundation, p.215-225.
Wickman, F.E. (1966) *Repose period patterns of volcanoes. Part 1: Volcanic eruptions regarded as random phenomena.* Arkiv for Mineralogi och Geologi, v.4, no.7, p.291-301.
Wiegel, R.L. (1953) *Waves, tides, currents and beaches: glossary of terms and list of standard symbols.* Berkeley, Calif.: Council on Wave Research, The Engineering Foundation. 113p.
Wien, Willy (1894) *Temperatur und Entropie der Strahlung.* Annalen der Physik, ser.2, v.52, p.132-165.
Wilcox, R.E. et al. (1973) *Basic wrench tectonics.* American Association of Petroleum Geologists. Bulletin, v.57, p.74-96.
Wiley, E.O. (1981) *Phylogenetics: The theory and practice of phylogenetic systematics.* New York, Wiley-Liss, 439p.
Wiley, E.O., Siegel-Causey, D., Brooks, D.R., and Funk, V.A. (1991) *The compleat cladist: A primer of phylogenetic procedures.* University of Kansas Museum of Natural History Special Publication 19, 158p.
Wilkinson, J.F.G. (1968) *The petrography of basaltic rocks.* In: Hess, H.H., and Poldervaart, A., eds. Basalts: the Poldervaart treatise on rocks of basaltic composition , v.1, p.163-214. New York: Interscience.
Willard, Bradford (1930) *Conglomerite, a new rock term.* Science, v.71, p.438.
Williams, Alwyn (1970) *Origin of laminar-shelled articulate brachiopods.* Lethaia, v.3, p.329-342.
Williams, B.J., and Prentice, J.E. (1957) *Slump-structures in the Ludlovian rocks of North Herefordshire.* Geologists' Association. Proceedings, v.68, p.286-293.
Williams, G.L. (1978) *Dinoflagellates, acritarchs, and tasmantids.* In: Haq, B.L., and Boersma, A., eds., *Introduction to marine micropaleontology.* Amsterdam, Netherlands, Elsevier, p.293-326; 358-366.
Williams, G.L., Fensome, R.A., Miller, M.A., and Sarjeant, W.A.S. (2000) *A glossary of the terminology applied to dinoflagellates, acritarchs, and prasinophytes, with emphasis on fossils.* 3rd ed. American Association of Stratigraphic Palynologists Contribution Series, no.37, 365p.
Williams, G.L., Sarjeant, W.A.S., and Kidson, E.J. (1973) *A glossary of the terminology applied to dinoflagellate amphiesmae and cysts and acritarchs.* American Association of Stratigraphic Palynologists Contribution Series, no.2, p.1-222.
Williams, G.L., Sarjeant, W.A.S., and Kidson, E.J. (1978) *A glossary of the terminology applied to dinoflagellate amphiesmae and cysts and acritarchs.* 1978 ed. American Association of Stratigraphic Palynologists Contribution Series, no.2A, p.1-121.
Williams, H.R., and Meyers, C.J. (1964) *Oil and gas terms; annotated manual of legal, engineering, tax words and phrases.* San Francisco: Matthew Bender. 449p.
Williams, H.S. (1893) *The elements of the geological time-scale.* Journal of Geology, v.1, p.283-295.
Williams, H.S. (1895) *Geological biology; an introduction to the geological history of organisms.* New York: Henry Holt. 395p.
Williams, H.S. (1901) *The discrimination of time-values in geology.* Journal of Geology, v.9, p.570-585.
Williams, Howel (1932) *The history and character of volcanic domes.* University of California, Department of Geological Sciences. Bulletin, v.21, no.5, p.51-146.
Williams, Howel (1941) *Calderas and their origin.* University of California, Department of Geological Sciences. Bulletin, v.25, no.6, p.239-346.
Williams, Howel, et al. (1954) *Petrography—an introduction to the study of rocks in thin sections.* San Francisco: Freeman. 406p.
Williams, M.Y. (1936) *Frost circles.* Royal Society of Canada. Transactions, ser.3, v.30, p.129-132.
Williams, P.W., (1983) *The role of the subcutaneous zone in karst hydrology.* Journal of Hydrology, v.61, p.45-67.
Williams, R.S., jr. (1972) *Thermography.* Photogrammetric Engineering, v.38, p.881-883.
Williams-Jones, G., and Rymer, H. (2000) *Hazards of volcanic gases.* In: Sigurdsson, H., ed., *Encyclopedia of volcanoes.* New York, Academic Press, p.997-1004.
Williamson, I.A. (1961) *Spring domes developed in limestone.* Journal of Sedimentary Petrology, v.31, p.288-291.
Willis, Bailey (1893) *The mechanics of Appalachian structure.* U.S. Geological Survey. Annual Report, 13th, pt.2, p.211-281.
Willis, Bailey (1903) *Physiography and deformation of the Wenatchee-Chelan district, Cascade Range.* U.S. Geological Survey. Professional Paper 19, p.41-97.
Willis, Bailey (1928) *Dead Sea problem: rift valley or ramp valley?* Geological Society of America. Bulletin, v.39, p.490-542.
Willis, Bailey (1938) *Asthenolith (melting spot) theory.* Geological Society of America. Bulletin, v.49, p.603-614.
Willman, H.B., et al. (1942) *Geology and mineral resources of the Marseilles, Ottawa, and Streator quadrangles.* Illinois State Geological Survey. Bulletin 66. 388p.
Wills, L.J. (1956) *Concealed coalfields; a palaeogeographical study of the stratigraphy and tectonics of mid-England in relation to coal reserves.* London: Blackie. 208p.
Wilmarth, M.G. (1938) *Lexicon of geologic names of the United States (including Alaska). Parts 1-2.* U.S. Geological Survey. Bulletin 896. 2396p.
Wilson, E.D., and Bossert, W.H. (1971) *A primer of population biology.* Stamford, Conn.: Sinauer Associates. 192p.
Wilson, Gilbert (1953) *Mullion and rodding structures in the Moine series of Scotland.* Geologists' Association. Proceedings, v.64, p.118-151.
Wilson, Gilbert (1961) *The tectonic significance of small scale structures and their importance to the geologist in the field.* Société Géologique de Belge. Annales, tome 84, p.423-548.
Wilson, I.G. (1972) *Aeolian bedforms—their development and origins.* Sedimentology, v.19, p.173-210.

Wilson, J.A. (1959) *Transfer, a synthesis of stratigraphic processes.* American Association of Petroleum Geologists. Bulletin, v.43, p.2861-2862.

Wilson, J.A. (1971) *Note 39—Records of the Stratigraphic Commission for 1968-1970.* American Association of Petroleum Geologists. Bulletin, v.55, p.1866-1872.

Wilson, J.T. (1950) *An analysis of the pattern and possible cause of young mountain ranges and island arcs.* Geological Association of Canada. Proceedings, v.3, p.141-166.

Wilson, J.T. (1965) *A new class of faults and their bearing on continental drift.* Nature, v.207, p.343-347.

Wilson, J.T. (1968) *Static or mobile earth. In: Gondwanaland revisited: New evidence for continental drift.* American Philosophical Society. Proceedings, v.112, p.309-320.

Wilson, L. (1976) *Explosive volcanic eruptions; III. Plinian eruption columns.* Royal Astronomical Society. Geophysical Journal, v.45, p.543-556.

Winchell, N.H., and Winchell, H.V. (1891) *The iron ores of Minnesota, their geology, discovery, development, qualities and origin, and comparison with those of other iron districts.* Minnesota Geological and Natural History Survey. Bulletin no.6 430p.

Winchester, S. (2002) *The map that changed the world: William Smith and the birth of modern geology.* New York, Perennial, 329 p.

Wind, F.H., and Wise, S.W., Jr. (1976) *Organic vs. inorganic processes in archeogastropod shell mineralization.* In: Watabe, N., and Wilbur, K.M., eds., *The mechanisms of mineralization in the invertebrates and plants.* Columbia, S.C., University of South Carolina Press, p.369-387.

Winkler, E.M., and Wilhelm, E.J. (1970) *Salt burst by hydration pressures in architectural stone in urban atmosphere.* Geological Society of America. Bulletin, v.81, p.567-572.

Winkler, H.G.F. (1967) *Petrogenesis of metamorphic rocks. 2d ed.* New York: Springer-Verlag. 237p.

Winsauer, W.O., et al. (1952) *Resistivity of brine saturated sands in relation to pore geometry.* American Association of Petroleum Geologists. Bulletin, v.36, p.253-277.

Wise, D. U., Dunn, D. E., Engelder, J. T., Geiser, P. A., Hatcher, R. D., Kish, S. A., Odon, A. L. and Schamel, S. (1984) *Fault-related rocks: suggestions for terminology.* Geology, v. 12, p. 391-394.

Wise, S.W., Jr. (1968) *Scanning electron microscopy and the ultrastructure of the molluscan crossed lamellar layer. Geological Society of America Programs with Abstracts, Annual Meeting, Mexico City, November 11-13, 1968, p.325-326.*

Wise, S.W., Jr. (1970) *Microarchitecture and mode of formation of nacre (mother of pearl) in pelecypods, gastropods, and cephalopods.* Eclogae Geologicae Helvetiae, v.63, p.775-797.

Wise, Sherwood, Jr. (1970) *Microarchitecture and mode of formation of nacre (mother-of-pearl) in pelecypods, gastropods, and cephalopods.* Eclogae Geologicae Helvetiae, v.63, p.775-797.

Withjack, M.O., and Callaway, J.S. (2000) *Active normal faulting beneath a salt layer: An experimental study of deformation in the cover sequence.* AAPG Bulletin, v.84, p.627-652.

Wohletz, K. and Heiken, G. (1992) *Volcanology and geothermal energy.* Berkeley, University of California Press, 432 pp.

Woldenberg, M.J. (1966) *Horton's laws justified in terms of allometric growth and steady state in open systems.* Geological Society of America. Bulletin, v.77, p.431-434.

Wolf, K.H. (1960) *Simplified limestone classification.* American Association of Petroleum Geologists. Bulletin, v.44, p.1414-1416.

Wolf, K.H. (1965) *Littoral environment indicated by open-space structures in algal limestones.* Palaeogeography, Palaeoclimatology, Palaeoecology, v.1, p.183-223.

Wolf, K.H., ed. (1976) *Handbook of strata-bound and stratiform ore deposits. Vol. 1-7.* Amsterdam: Elsevier. 2863p.

Wolfe, C.W., et al. (1966) *Earth and space science.* Boston: Heath. 630p.

Wolfe, J.A. (1993) *A method of obtaining climatic parameters from leaf assemblages.* U.S. Geological Survey Bulletin 2040, p.1-71.

Wolfe, J.A., and Spicer, R.A. (1999) *Fossil leaf character states: multivariate analyses.* In: Jones, T.P., and Rowe, N.P., eds., *Fossil plants and spores:modern techniques.* London, UK, Geological Society of London, p.233-239.

Wolff, J.A., and Sumner, J.M. (2000) *Lava fountains and their products.* In: Sigurdsson, H., ed., *Encyclopedia of volcanoes.* New York, Academic Press, p.321-329.

Wood, Alan (1935) *The origin of the structure known as guilielmites.* Geological Magazine, v.72, p.241-245.

Wood, Alan (1941) *"Algal dust": and the finer-grained varieties of Carboniferous limestone.* Geological Magazine, v.78, p.192-200.

Wood, Alan (1942) *The development of hillside slopes.* Geologists' Association. Proceedings, v.53, p.128-138.

Wood, Alan, and Smith, A.J. (1958) *The sedimentation and sedimentary history of the Aberystwyth Grits (upper Llandoverian).* Geological Society of London. Quarterly Journal, v.114, p.163-195.

Wood, G.D. (1996) *Biostratigraphic, paleoecologic, and biologic significance of the Silurian (Landovery) acritarch* Beromia redroadii *gen. emend. et sp. nov., mid-continent and eastern United States.* Palynology, v.20, p.177-189.2

Wood, G.D., and Miller, M.A. (1997) *Pre-Carboniferous Chlorophyta: New reports of Hydrodictyaceae, ?Scenedesmaceae, and ?Zygnemataceae.* In: Fatka, O., and Servais, T., eds., *Acritarcha in Praha.* Acta Universitatis Carolinae, v.40, p.703-717.

Wood, G.D., Palmer-Koleman, S.E., Alemán, A.M., and Padilla, H (2002) *Palynofacies and biomarker analysis of the lowermost Permo-Carboniferous Pular Formation, a volcanic-arc sequence in the Sierra de Almeida, northern chile.* Review of Palaeobotany and Palynology, v.118, p.323-333.

Wood, W.F., and Snell, J.B. (1960) *A quantitative system for classifying land forms.* Natick, Mass.: U.S. Army Natick Laboratory. Technical Report EP-124. 20p.

Woodford, A.O (1925) *The San Onofre Breccia; its nature and origin.* University of California. Department of Geological Sciences: Bulletin, v.15, no.7, p.159-280.

Woodford, A.O. (1925) *The San Onofre Breccia; its nature and origin.* University of California. Department of Geological Sciences: Bulletin, v.62, p.799-851.

Woodford, A.O. (1951) *Stream gradients and Monterey sea valley.* Geological Society of America. Bulletin, v.62, p.799-851.

Woodford, A.O. (1965) *Historical geology.* San Francisco, Freeman and Co., 512p.

Woodroffe, C.D. (2002) *Coasts.* Cambridge University Press, 623p.

Woodward, H.B. (1887) *Geology of England and Wales: with notes on the physical features of the country. 2d ed.* London: G. Philip. 670p.

Woodward, H.B. (1894) *The Jurassic rocks of Britain. Vol. 4.* London: Her Majesty's Stationery Office. 628p. (Great Britain. Geological Survey. Memoir).

Woodward, H.P. (1957) *Structural elements of northeastern Appalachians.* American Association of Petroleum Geologists. Bulletin, v.41, p.1429-1440.

Woodward, H.P. (1958) *Alleghany orogeny.* American Association of Petroleum Geologists. Bulletin, v.42, p.193.

Woodward, N.B., and Beets, J. W. (1988) *Critical evidence for southern Appalachian Valley and Ridge thrust sequence.* In: Geometries and mechanisms of thrusting, with special reference to the Appalachians , edited by Mitra, G. and Wojtal, S. Spec. Pap. Geol. Soc. Am., v. 222, p. 165-178.

Woodward, N.B., Boyer, S. E., and Supper, J. (1989) *Balanced geological cross sections.* American Geophysical Union, Short Course in Geology, v. 6, 132 p.

Woodward, N.B., et al. (1985) *An outline of balanced cross-sections.* University of Tennessee, Dept. of Geological Sciences. Studies in Geology 11, 2nd ed., p.3-4.

Woodworth, J.B. (1894a) *Postglacial eolian action in southern New England.* American Journal of Science, ser.3, v.47, p.63-71.

Woodworth, J.B. (1894b) *Some typical eskers of southern New England.* Boston Society of Natural History. Proceedings, v.26, p.197-220.

Woodworth, J.B. (1901) *Pleistocene geology of portions of Nassau County and Borough of Queens.* New York State Museum. Bulletin, v.48, p.618-670.

Woodworth, J.B. (1912) *Geological expedition to Brazil and Chile, 1908-1909.* Harvard College. Museum of Comparative Zoology. Bulletin, v.56, no.1. 137p.

Woollacott, R.M., and Zimmer, R.L. (1972) *Origin and structure of the brood chamber in Bugula meritina (Bryozoa).* Marine Biology, v.16, p.165-170.

Woolley, A.R., et al. (1996) *Classification of lamprophyres, lamproites, kimberlites, and the kalsilitic, melilitic, and leucitic rocks.* In: Eby, G.N., and Martin, R.F., eds., *Alkaline rocks: Petrology and mineralogy.* The Canadian Mineralogist, v.34, no.2, p.175-186.

Woolley, A.R. et al. (1996) *Classification of lamprophyres, lamproites, kimberlites, and the kalsilitic, melilitic, and leucitic rocks.* Canadian Mineralogist, 35, p. 175-186.

Woolley, A.R., and Kempe, D.R.C. (1989) *Carbonatites: Nomenclature, average chemical compositions, and element distribution.* In: Bell, K., ed., *Carbonatites: Genesis and evolution.* London, Unwin Hyman, p.1-14.

Woolnough, W.G. (1910) *Stone rolls in the Bulli coal seam of New South Wales.* Royal Society of New South Wales. Proceedings, v.44, p.334-340.

Woolnough, W.G. (1927) *The duricrust of Australia.* Royal Society of New South Wales. Journal and Proceedings, v.61, p.24-53.

Worcester, P.G. (1939) *A textbook of geomorphology.* New York: Van Nostrand. 565p.

Workman, W.H. (1914) *Nieve penitente and allied formations in Himalaya, or surface-forms of névé and ice created or modelled by melting.* Zeitschrift für Gletscherkunde, Bd. 8, p.289-330.

Wright, J.K. (1944) *The terminology of certain map symbols.* Geographical Review, v.34, p.653-654.

Wright, W.B. (1914) *The Quaternary ice age.* London: Macmillan. 464p.

Wright, W.B. (1926) *Stratigraphical diachronism in the Millstone Grit of Lancashire.* British Association for the Advancement of Science. Report, 94th, p.354-355.

Wulff, G.V. von (1902) *Untersuchungen im Gebiete der Optischen Eigenschaften isomorpher Krystalle.* Zeitschrift für Krystallographie und Mineralogie, Bd.36, p.1-28.

Wyllie, M.R.J. (1957) *The fundamentals of well log interpretation, 2nd ed.* New York: Academic Press. 176p.

Wyllie, M.R.J., et al. (1956) *Elastic wave velocities in heterogeneous and porous media.* Geophysics, v.21, p.41-70.

Wyllie, P.J. (1966) *Experimental petrology: an indoor approach to an outdoor subject.* Journal of Geological Education, v.14, p.93-97.

Wynne-Edwards, H.R. (1957) *Structure of the Westport concordant pluton in the Grenville, Ontario.* Journal of Geology, v.65, p.639-649.

Y

Yaalon, D.H. (1965) *Microminerals and micromineralogy*. Clay minerals, v.6, p.71.

Yanyi, G. (1987) *Raw materials for making porcelain and the characteristics of porcelain wares in North and South China in ancient times.* Archaeometry, v.29, p.3-19.

Yasso, W.E. (1966) *Heavy minerals concentration and sastrugi-like deflation furrows in a beach salcrete at Rockaway Point, New York.* Journal of Sedimentary Petrology, v.36, p.836-838.

Yeats, R.S. (1981) *Quaternary flake tectonics of the California Transverse Ranges.* Geology, v.9, p.16-20.

Yoder, H.S., jr., and Tilley, C.E. (1962) *Origin of basalt magmas: an experimental study of natural and synthetic rock systems.* Journal of Petrology, v.3, p.342-532.

Yokoyama, Izumi (1956-57) *Energetics and active volcanoes.* Tokyo University. Earthquake Research Institute. Bulletin, v.35, pt.1, p.75-97.

Yolkin, E.A., Kim, A.I., Weddige, K., Talent, J.A., and House, M.R. (1997) *Definition of the Pragian/Emsian Stage boundary.* Episodes, v.20, no.4, p.235-240.

Youell, R.F. (1960) *An electrolytic method for producing chlorite-like substances from montmorillonite.* Clay Minerals Bulletin, v.4, p.191-195.

Young, A.P. (1910) *On the glaciation of the Navis Valley in North Tirol.* Geological Magazine, v.7, p.244-258.

Z

Zankl, H., and Multer, H.G. (1977) *Origin of some internal fabrics in Holocene reef rocks, St. Croix, U.S. Virgin Islands.* Third International Coral Reef Symposium. Proceedings, v.2, p.127-133. Miami: University of Miami, Rosenstiel School of Marine and Atmospheric Science. 628p.

Zemansky, M.W. (1957) *Heat and thermodynamics.* New York: McGraw-Hill. 484p.

Zen, E-an (1966) *Construction of pressure-temperature diagrams for multicomponent systems after the method of Schreinemakers—a geometric approach.* U.S. Geological Survey. Bulletin 1225. 56p.

Zenger, D.H. (1983) *Burial dolomitization in the Lost Burro Formation (Devonian), east-central California, and the significance of late diagenetic dolomitization.* Geology, v.11, p.519-522.

Zernitz, E.R. (1932) *Drainage patterns and their significance.* Journal of Geology, v.40, p.498-521.

Ziegler, W., and Klapper, G. (1985) *Stages of the Devonian system.* Episodes, v.8, no.2, p.104-109.

Zimbelman, J.R. (2000) *Volcanism on Mars.* In: Sigurdsson, H., ed., *Encyclopedia of volcanoes*. New York, Academic Press, p.771-783.

Zingg, Theodor (1935) *Beitrag zur Schotteranalyse; die Schotteranalyse und ihre Anwendung auf die Glattalschotter.* Schweizerische mineralogische und petrographische Mitteilungen, Bd.15, p.39-140.

Zirkel, Ferdinand (1866) *Lehrbuch der Petrographie. Bd.1.* Bonn: Adolph Marcus. 607p.

Zirkel, Ferdinand (1876) *Microscopial petrography.* U.S. Geological Exploration of the Fortieth Parallel. Report, v.6, 297p. (U.S. Army. Engineer Dept. Professional Paper, no.18).

Zirkel, Ferdinand (1893) *Lehrbuch der Petrographie. Bd.1 2nd ed.* Leipzig: Wilhelm Engelmann. 845p.

Zischinsky, Ulf (1969) *Über Sackungen.* Rock Mechanics, v.1, p.30-52.

Zittel, K.A. von (1901) *History of geology and palaeontology to the end of the nineteenth century.* Translated by M.M. Ogilvie-Gordon. London: Walter Scott. 562p.